# Principles and Methods of
# Toxicology

FIFTH EDITION

# Principles and Methods of
# Toxicology

FIFTH EDITION

**A. Wallace Hayes**
Harvard School of Public Health
Andover, Massachusetts, U.S.A.

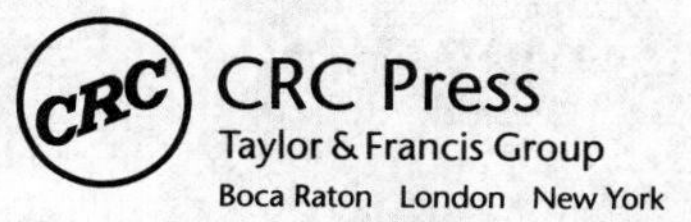


CRC Press is an imprint of the
Taylor & Francis Group, an **informa** business

CRC Press
Taylor & Francis Group
6000 Broken Sound Parkway NW, Suite 300
Boca Raton, FL 33487-2742


Printed in Canada on acid-free paper
10 9 8 7 6 5 4 3 2 1

International Standard Book Number-10: 0-8493-3778-X (Hardcover)
International Standard Book Number-13: 978-0-8493-3778-9 (Hardcover)



**Library of Congress Cataloging-in-Publication Data**

Principles and methods of toxicology / editor, A. Wallace Hayes. -- 5th ed.
p. ; cm.
Includes bibliographical references and index.
ISBN-13: 978-0-8493-3778-9 (alk. paper)
ISBN-10: 0-8493-3778-X (alk. paper)
1. Toxicology. I. Hayes, A. Wallace (Andrew Wallace), 1939-
[DNLM: 1. Toxicology--methods. 2. Poisoning. 3. Poisons. QV 600 P957 2007]

RA1211.P74 2007
615.9--dc22 2006100039

**Visit the Taylor & Francis Web site at**
**http://www.taylorandfrancis.com**

**and the CRC Press Web site at**
**http://www.crcpress.com**

# Contents

## PART 2 Agents



## PART 3 Methods

# Contributors

**Melvin E. Andersen**
Computational Biology Division
The Hamner Institutes for Health Sciences
Research Triangle Park, North Carolina

**Cheryl J. Bartleson**
Department of Biochemistry
Center in Molecular Toxicology
Vanderbilt University School of Medicine
Nashville, Tennessee

**Barbara D. Beck**
Gradient Corporation
Cambridge, Massachusetts

**Sven Beushausen**
Pfizer Global Research and Development
Chesterfield, Missouri

**Mark E. Blazka**
Drug Safety Evaluation
U.S. Science and Medical Affairs
sanofi aventis
Bridgewater, New Jersey

**Matthew S. Bogdanffy**
Toxicology and Safety Assessment
Boehringer Ingelheim Pharmaceuticals, Inc.
Ridgefield, Connecticut

**Joseph F. Borzelleca**
Department of Pharmacology and Toxicology
Medical College of Virginia
Richmond, Virginia

**Charles B. Breckenridge**
Syngenta Crop Protection, Inc.
Greenboro, North Carolina

**David J. Brusick**
Covance Laboratories, Inc.
Vienna, Virginia

**Margaret Buckalew**
ENVIRON International Corporation
Atlanta, Georgia

**Gary D. Byrd**
Targacept, Inc.
Winston-Salem, North Carolina

**Edward J. Calabrese**
Department of Public Health—Environmental Health Sciences Concentration
School of Public Health and Health Sciences
University of Massachusetts
Amherst, Massachusetts

**William S. Caldwell**
Targacept, Inc.
Winston-Salem, North Carolina

**Michel Charbonneau**
INRS–Institut Armand-Frappier
Laval, Québec, Canada

**Mildred S. Christian**
Argus International, Inc.
Horsham, Pennsylvania

**Eric Clegg**
Reprolific, LLC
Millboro, Virginia

**Lorris G. Cockerham**
Phenix Consulting & Services, Ltd.
Little Rock, Arkansas

**Ralph R. Cook**
RRC Consulting, LLC
Midland, Michigan

**Deborah A. Cory-Slechta**
Department of Environmental and Occupational Health
University of Medicine and Dentistry of New Jersey
Piscataway, New Jersey

**Peter A. Crooks**
College of Pharmacy
University of Kentucky
Lexington, Kentucky

**David L. Dahlstrom**
ENVIRON International Corp.
Atlanta, Georgia

**Cham E. Dallas**
Center for Mass Destruction Defense
Department of Pharmaceutical and Biomedical Sciences
University of Georgia
Athens, Georgia

**Jack H. Dean**
Department of Pharmacology and Toxicology
College of Pharmacy
University of Arizona
Tucson, Arizona

**J. Donald deBethizy**
Targacept, Inc.
Winston-Salem, North Carolina

**David J. Doolittle**
R.J. Reynolds
Winston-Salem, North Carolina

**Michael A. Dorato**
Lilly Research Laboratories
A Division of Eli Lilly and Company
Greenfield, Indiana

**Harald G. Enzmann**
German Federal Institute for Drugs
and Medical Devices (BfArM)
Bonn, Germany

**Nancy E. Everds**
Amgen, Inc.
Seattle, Washington

**Anne Fairbrother**
Western Ecology Division
National Health and Environmental Effects
Research Laboratory
U.S. Environmental Protection Agency
Corvallis, Oregon

**Wanda R. Fields**
R.J. Reynolds
Winston-Salem, North Carolina

**Bruce A. Fowler**
Division of Toxicology
Agency for Toxic Substances and Disease Registry
U.S. Department of Health and Human Services
Atlanta, Georgia

**Steven R. Frame**
Haskell Laboratories
E.I. DuPont de Nemours & Co.
Newark, Delaware

**Shayne C. Gad**
Gad Consulting Services
Cary, North Carolina

**David W. Gaylor**
Gaylor and Associates, LLC
Eureka Springs, Arkansas

**F. Peter Guengerich**
Department of Biochemistry
Center in Molecular Toxicology
Vanderbilt University School of Medicine
Nashville, Tennessee

**Mary L. Haasch**
National Research Council
Mid-Continent Ecology Division
Office of Research and Development
U.S. Environmental Protection Agency
Duluth, Minnesota

**Pertti J. (Bert) Hakkinen**
Gradient Corporation
Cambridge, Massachusetts

**Robert L. Hall**
Covance Laboratorites, Inc.
Madison, Wisconsin

**Jerry F. Hardisty**
Experimental Pathology Laboratories, Inc.
Research Triangle Park, North Carolina

**C. Terrance Hawk**
Department of Laboratory Animal Science
GlaxoSmithKline
King of Prussia, Pennsylvania

**A. Wallace Hayes**
Harvard School of Public Health
Boston, Massachusetts

**Benjamin B. Hayes**
Department of Dermatology
Vanderbilt University School of Medicine
Nashville, Tennessee

**Johnnie R. Hayes**
Consulting Toxicologist
Kernersville, North Carolina

**Robert V. House**
DynPort Vaccine Company, LLC
Frederick, Maryland

**Michael J. Iatropoulos**
Department of Pathology
New York Medical College
Valhalla, New York

**Y. James Kang**
School of Medicine, University of Louisville
Louisville, Kentucky

**Robert W. Kapp, Jr.**
BioTox
Lutherville, Maryland

**Raymond A. Kemper**
Toxicology and Safety Assessment
Boehringer Ingelheim Pharmaceuticals, Inc.
Ridgefield, Connecticut

**Gerald L. Kennedy**
Haskell Laboratory
E.I. DuPont de Nemours & Co.
Newark, Delaware

**Daniel E. Keyler**
Department of Medicine
Hennepin County Medical Center
Minneapolis, Minnesota

**Gary R. Klinefelter**
National Health and Environmental Effects
Research Laboratory
Office of Research and Development
U.S. Environmental Protection Agency
Research Triangle Park, North Carolina

**Kannan Krishnan**
Département de santé environnementale et santé au travail
Groupe de recherche interdisciplinaire en santé (GRIS)
Université de Montréal
Montréal, Québec, Canada

**Michael A. Landauer**
Department of Pathophysiology and Toxicology
Armed Forces Radiobiology Research Institute
Bethesda, Maryland

**Richard W. Lane**
Unilever Bestfoods NA
Englewood Cliffs, New Jersey

**Lawrence H. Lash**
Department of Pharmacology
School of Medicine, Wayne State University
Detroit, Michigan

**Michael A. Lewis**
Gulf Ecology Division
National Health and Environmental Effects
Research Laboratory
U.S. Environmental Protection Agency
Gulf Breeze, Florida

**Michael I. Luster**
National Institute of Occupational Safety and Health
Morgantown, West Virginia

**Amy K. Madl**
ChemRisk, Inc.
San Francisco, California

**Howard J. Maibach**
Department of Dermatology
School of Medicine
University of California, San Francisco
San Francisco, California

**Peter C. Mann**
Experimental Pathology Laboratories, Inc.
Seattle, Washington

**Carl L. McMillian**
Lilly Research Laboratories
A Division of Eli Lilly and Company
Greenfield, Indiana

**Harihara M. Mehendale**
Department of Toxicology
University of Louisiana at Monroe
Monroe, Louisiana

**Robert E. Menzer**
National Center for Environmental Research
U.S. Environmental Protection Agency
Washington, D.C.

**Jill C. Merrill**
Center for Drug Evaluation and Research
U.S. Food and Drug Administration
Rockville, Maryland

**G. Andrew Mickley, Jr.**
Department of Psychology
Baldwin-Wallace College
Berea, Ohio

**Joseph J.P. Morton**
Morton Associates, Inc.
Silver Spring, Maryland

**Brian C. Myhr**
Genotox Consulting
Bethesda, Maryland

**Patricia Nance**
Toxicology Excellence for Risk Assessment (TERA)
Cincinnati, Ohio

**Frederick W. Oehme**
Comparative Toxicology Laboratories
Kansas State University
Manhattan, Kansas

**Ann Parker**
Toxicology Excellence for Risk Assessment (TERA)
Cincinnati, Ohio

**Esther Patrick**
Amway Corporation
Ada, Michigan

**Jacqueline Patterson**
Toxicology Excellence for Risk Assessment (TERA)
Cincinnati, Ohio

**Dennis J. Paustenbach**
ChemRisk, Inc.
San Francisco, California

**William Pennie**
Drug Safety Research and Development
Pfizer Global Research and Development
Groton, Connecticut

**Sally D. Perreault**
National Health and Environmental Effects Research Laboratory
Office of Research and Development
U. S. Environmental Protection Agency
Research Triangle Park, North Carolina

**Gabriel L. Plaa**
Département de pharmacologie
Faculté de médecine
Université de Montréal
Montréal, Québec, Canada

**Chada S. Reddy**
Department of Veterinary Biomedical Sciences
College of Veterinary Medicine
University of Missouri
Columbia, Missouri

**Andrew Gordon Renwick**
School of Medicine
University of Southampton
Southampton, United Kingdom

**Don Robertson**
Investigative Toxicology
Pfizer Global Research and Development
Ann Arbor, Michigan

**Joseph V. Rodricks**
Environ International Corporation
Arlington, Virginia

**Ruthann Rudel**
Silent Spring Institute
Newton, Massachusetts

**Leonard M. Schechtman**
National Center for Toxicological Research
U.S. Food and Drug Administration
Rockville, Maryland

**Tracey M. Slayton**
Gradient Corporation
Cambridge, Massachusetts

**Stephen D. Soileau**
Environmental, Health, and Safety
The Gillette Company
Needham, Massachusetts

**Katherine S. Squibb**
Program in Toxicology
University of Maryland School of Medicine
Baltimore, Maryland

**James T. Stevens**
Department of Physiology and Pharmacology
School of Medicine
Wake Forest University
Winston-Salem, North Carolina

**William S. Stokes**
National Toxicology Program Interagency Center for the Evaluation of Alternative Toxicological Methods
National Institute of Environmental Health Sciences
National Institutes of Health
Research Triangle Park, North Carolina

**John A. Thomas**
School of Medicine
Indiana University
Indianapolis, Indiana

**Duncan Turnbull**
Environ International Corporation
Arlington, Virginia

**Rudolph Valentine**
Haskell Laboratory
E.I. DuPont de Nemours & Co.
Newark, Delaware

**Mary Ann Vasbinder**
Department of Laboratory Animal Science
GlaxoSmithKline
Research Triangle Park, North Carolina

**Mary Jo Vodicnik**
Lilly Research Laboratories
A Division of Eli Lilly and Company
Indianapolis, Indiana

**Thomas L. Walden, Jr.**
Cape Fear Valley Medical Center
Fayetteville, North Carolina

**Bernard Weiss**
Department of Environmental Medicine
University of Rochester
Rochester, New York

**Philip Wexler**
Toxicology and Environmental Health Information Program
National Library of Medicine
National Institutes of Health
Bethesda, Maryland

**William J. White**
Charles River Laboratories
Wilmington, Massachusetts

**Gary M. Williams**
Department of Pathology
New York Medical College
Valhalla, New York

**Nelson H. Wilson**
Experimental Pathology Laboratories, Inc.
Sterling, Virginia

**David R. Worthen**
College of Pharmacy
University of Kentucky
Lexington, Kentucky

# Acknowledgments

I would like to express my warm appreciation to the many people who contributed knowingly and otherwise to the fifth edition of this book. The editor most heartily thanks each of the contributors, who either revised chapters or prepared new chapters, for keeping in mind that thoughtfully worded information is greatly appreciated, especially by the student but also by the more advanced reader. I also am indebted to the contributors for their combined expertise that made a volume of this breadth possible. I thank Sandra Smith for her skillful editing of the manuscript. Appreciation also is expressed to the staff of Taylor & Francis.

# Foreword to the Third Edition

Until 1982 when the first edition of this book was published, there was no specific source to which a student or an investigator could turn for a comprehensive presentation of the methods used in modern toxicology. For anyone who was trying to teach the subject, the book filled a great void for both the teacher and the student. The book appeared at a time when technical achievements in the field related to toxicology were undergoing tremendous refinements. Techniques and the tools of experimental biology, pathology, mathematics, engineering, physics, and analytical/biological chemistry, which had been barely conceived 20 years earlier, were in common use. The rapid growth of toxicology at that same time created a need for scientists from all of the above fields to apply their expertise to the science of toxicology. Toxicology borrowed freely from these related sciences so a developing, modern scientifically acceptable body of procedures became identified as the methods of toxicology. Prior to the span of a single human life, the methods of toxicology consisted of some general, short-term test for the determination of the overall aspects of this difficult area of toxicology. The exponential rate of growth of toxicology continues, and the third edition of the book continues to be an authoritative and comprehensive source of the methods that are currently used in this science.

If toxicology can be appropriately defined as the study of the harmful effects of chemicals on biologic systems, it must then embody a systemized knowledge of the effects of chemicals which are introduced into the simplest, as well as the most complex, of all biologic systems, and methods must be available to accomplish these experiments. The availability of methods to detect the harmful effects of chemicals allows for the creation of data, but those data become useful in toxicology only after they are suitably interpreted. An additional link toward understanding the subject of toxicology is the placing of results obtained from the available methods in their proper relation and perspective to the whole picture of the role that toxicology can play for the improvement of mankind. In order to accomplish this function, the toxicologist must not only develop an understanding of the methods used but also determine the significance or insignificance of their data in the complete picture of the toxicity of each compound. My graduate school mentor, Dr. Roger Hubbard, once told me that no scientifically valid experiment creates erroneous results, but inappropriate application of those results can create erroneous conclusions. An understanding of the principles together with the methods involved in the science of toxicology prepare the critical scientist for developing an insight with regard to the proper application of experimental results. Results that are properly obtained by acceptable methodology and that are suitably weighted for the conditions under which they were obtained certainly contribute to the development of proper conclusions. In this book, very highly qualified toxicologists present the procedures in detail that are currently used and accepted in the science of toxicology. Discussions of each procedure or category of procedures enable the conduct of acceptable toxicologic tests that create the body of systemized knowledge essential to the science of toxicology. Properly applied, that knowledge serves to protect mankind and the biologic realm in general from sudden, as well as delayed, insidious chemically induced harm.

Ted Loomis, M.D., Ph.D.
Professor Emeritus
University of Washington
Seattle, Washington

# Foreword to the Fourth Edition

Publishing a toxicology book on the entrance into the third millennium is an appropriate time to reflect on the progress that has been made in that discipline since its inception. Almost five hundred years ago, Paracelsus published a treatise based on his intuitive observations; he merely argued that these observations should convince one that the dose makes the poison. Now science has advanced enough so that we can prove that Paracelsus was correct, because the law of mass action tells us that the degree of perturbation of a system is proportional to the chemical potential of a substance in that system. We still, however, must rely heavily on observational epidemiology to determine the details of exactly what dose effects what change in humans.

The fourth edition of this book is a magisterial, state of the art compilation of the principles and methods that toxicologists must use to identify whether a causal relationship exists between specific doses of a chemical and an alleged adverse effect, observed primarily in humans. Proper integration of principles and methods of toxicology is extremely important since the primary purpose of toxicology is to predict human toxicity. Previous editions of this book have delineated in a very useful detail the methods of toxicology and how these methods have been perfected steadily and rapidly in the last few decades. The necessarily heavy reliance on animal experimentation for determining causality in humans is obvious and certainly warranted.

This book was the first to chronicle the overall aspects of the use of animal experiments in toxicology. The exponential rate of growth of toxicology continued to be reflected in further editions of the book which served as the authoritative and comprehensive source of methods used in this science. Proper and critical conduct of acceptable toxicological tests still continue to create the body of systemized knowledge essential to the science of toxicology.

The current edition continues this tradition but adds some very significant new chapters. These chapters are on epidemiology and exposure assessment, and a chapter on repeat dosing combines previous chapters that subdivided multiple dosing into arbitrary intervals. It is remarkable that we have returned, almost full circle, to an emphasis on direct exposure and effects in human populations after finally, firmly establishing the basic scientific foundations of toxicology. This thorough, complete compendium is a necessary addition to the library of everyone interested in this subject.

William J. Waddell, M.D.
Professor and Chair, Emeritus
Department of Pharmacology and Toxicology
University of Louisville, Kentucky

# Foreword to the Fifth Edition

Toxicology is an evolving science with ongoing development of methods, concepts, and understanding. It has been only some five years since the fourth edition of this book was published, but a wealth of novel information has been reported in the scientific literature since that time. This is especially true regarding *toxicogenomics*, a term that was coined in 1999 to describe the marriage of toxicology and genomics. Since that time, gene expression analysis has been used as mechanistic toxicology screens, for more sensitive and earlier toxicity discovery, in drug discovery, and in drug and chemical safety assessments. Proteomic technologies have also recently been much applied in toxicology, allowing for examination of the entire complement of proteins in an organism, tissue, or cell type. Using large-scale, high-throughput methods, protein expression, posttranslational modifications, and protein interactions may be studied. Also coming to the fore is metabonomics, where high-resolution $^1$H-NMR spectroscopy is used in conjunction with pattern recognition to provide a fingerprint of the small molecules contained in a given body fluid that may be applied to define the dynamic phenotype of a cell, organ, or organism. These promising technologies are described in this edition in a new chapter on toxicopanomics.

During the last few decades, a wealth of toxicological information has become available, making it impossible for any individual toxicologist to keep abreast of all new information; for example, the TOXLINE file contains over 3 million bibliographic citations, thus the need for comprehensive and readily accessible information resources has become apparent. Many Web-based, searchable databases are now available that make it possible to retrieve information related to specific toxicological questions. This edition now contains a new chapter on information resources for toxicologists.

Already in antiquity it was well known that nature was not always benevolent but that human poisoning was possible via animal venoms and plant extracts. Such toxins were in earlier times used for hunting, waging war, and assassinations; in modern times, accidental poisoning with natural toxins is rather prevalent in many areas of the globe. This fifth edition includes a new chapter on plant and animal poisons.

Because of the considerable societal pressure with respect to reducing the use of animals in toxicology, non-animal methods are being developed in an attempt to predict what happens when whole animals, including humans, are exposed to toxic levels of drugs and other chemicals. Such methods include a number of cellular and subcellular systems, as well as mathematical models based on correlating a compound's chemical or structural variation with measured toxicological responses. One new chapter in this edition covers the use of non-animal methods in toxicology.

Erik Dybing, M.D., Ph.D.
Division Director and Professor
Division of Environmental Medicine
Norwegian Institute of Public Health
Oslo, Norway

# Preface

This edition of *Principles and Methods of Toxicology* was revised and updated while continuing to deal with evaluation of toxicological data. As was the case for past editions of the book, new chapters have been added to enhance the coverage of the book; these new chapters deal with genomics, natural toxins, pathology for toxicologists, toxicological information, and alternatives. A number of new authors have added considerably to the updating of the fifth edition. In addition, the glossary has been expanded significantly. As for each previous edition, every effort has been made to reflect the needs and issues in toxicology and to keep the book suitable for use as a textbook in graduate education.

Toxicology, the science of poisons or the study of the untoward effects of chemicals or physical agents on biological systems, has moved over the centuries from the art of food gathering and murder to the highly sophisticated science of mechanisms; yet, toxicology remains a paradox: "All substances are poisons; there is none which is not a poison. The right dose differentiates a poison and a remedy," so says Paracelsus, the sixteenth-century German–Swiss physician and alchemist.

Based on this paradox, toxicology must look beyond such a simple definition and focus its attention on determining safe doses from harmful or detrimental doses. To determine the safe use of a chemical or physical agent, it is necessary to have a sound understanding of biologic mechanisms and the methods employed to define these mechanisms. The vastness of the field of toxicology and the rapid accumulation of data preclude the possibility of absorbing and retaining more than a fraction of such techniques and information; however, an understanding of the principles underlying these methods is not only manageable but also essential, and it is to this end that this book was written.

This tome was designed to be used both in courses for general and advanced toxicology. Throughout the book, the basic cornerstones underlying toxicology—people differ, dose matters, and people change—are emphasized. The framework of the fifth edition follows that of the previous editions. The place of toxicology in history opens the book and is followed by sections on basic toxicological principles, agents that cause toxicity, target organ toxicity, and toxicological testing methods, including many of the test protocols required to meet regulatory needs worldwide. Each method or procedure is discussed from the standpoint of technique and interpretation of data. A state-of-the-art approach is emphasized, and discussions address the problems and pitfalls that may be encountered in performing a protocol. The organization of the book should facilitate its use by both new disciples of toxicology and the more advanced practitioners of our science. As with past editions, the addition of new authors has allowed a broader and more comprehensive coverage of the ever-changing and expanding field of toxicology.

The fifth edition of *Principles and Methods of Toxicology* continues to be a resource for research scientists who have used earlier editions as a reference source for updated materials in areas of their special or peripheral interests.

A. Wallace Hayes, Ph.D., DABT, FATS, FIBiol, FACFE, ERT
Harvard School of Public Health
Boston, Massachusetts

# Notes

# Part 1

## Principles of Toxicology

# 1 Harming and Helping through Time: The History of Toxicology

*Richard W. Lane and Joseph F. Borzelleca*

## CONTENTS



Continuity with the past is a necessity, not a duty.

**Oliver Wendell Holmes, Jr.**

## OVERVIEW

Poisonings—accidental, intentional, and unintentional—form the basis of toxicology, the science of poisons. It is more difficult to capture the breadth of the history of toxicology than any other single discipline because it undoubtedly began before recorded history, is found throughout recorded history, and is intertwined with so many important aspects of human life such as eating, healing and medicine, occupation, religion, folklore, murder, and suicide.

Toxicology passed through a number of phases during its maturation into a recognized scientific discipline. The various phases of toxicology include observation and phenomenology (lists of poisons and antidotes), detection of poisons, experimentation to deduce the mechanisms of toxicity, the development of rational therapeutic measures, and, finally, quantification (including dose–response, risk assessment) to establish safe exposure conditions and thereby protect the public from the adverse effects of chemical exposure at home, in the environment, and in the workplace. In this chapter, we will demonstrate that toxicology evolved from considering chemicals as agents of harm to understanding them in ways that can benefit humankind. Toxicology is now a critical and respected member of the scientific biomedical community.

These phases began at different times and have wound their way through history at their own pace. One phase does not stop when another develops but rather proceeds in its own fashion as others progress. Toxicology has evolved from a study of chemicals that induce harm (the art of poisoning) to a study of chemicals to prevent harm (the science of poisoning). The use, abuse, and misuse of chemical, physical, and biological agents by individuals, groups, and societies over time form the basis of toxicology and how it developed. We use examples of poisonings from accidental exposures (e.g., food, animals, plants), inadvertent exposures (e.g., occupational, environmental), and deliberate exposures (e.g., murders, suicides, capital punishment, chemical warfare, therapeutics) to explore the development of toxicology.

## INTRODUCTION

Toxicology! An exciting word with fascinating connotations. It cannot help but evoke thoughts of poisons, poisoners, intrigue, cloak-and-dagger, villains, victims and perpetrators, and plants and chemicals used as instruments of ill. It can evoke feelings of fear, tragedy, and fascination. It is a part of survival, religion, folklore. Toxicology is intellectually and emotionally seductive because it deals with agents that can harm or help depending on conditions of exposure. The toxicologist determines these conditions and thereby influences policy that impacts public health. But what is toxicology? A history of a discipline should begin with its definition. Toxicology is "the study of poisons," but what is "study" and what are "poisons"? *Study* may be defined as a branch of knowledge or the pursuit of knowledge by observation, experimentation, or reading. The definition of *poison* has generally meant a chemical that causes harm, but the term has evolved. Because what constitutes a poison has changed over time, one needs to understand and appreciate these changes to understand the history of toxicology.

The history of any discipline is evaluated in the context of that period and not by current standards. In the past, toxicology involved listing poisons, noting their onset of action and the nature of the harm produced (e.g., painful or painless, terminating in death or not). Today, the study of poisons generally includes an understanding of the mechanism or mode of action at the organ, cellular, or molecular level. Toxicology involves a knowledge of the physiological, biochemical, and morphological effects of a chemical. This approach—the physiological and biochemical understanding of toxicity—was first used by the famous French physiologist, Claude Bernard, more than 150 years ago. He also used poisons to investigate basic physiological phenomena. This marked a major turning point in toxicology because poisons were now used as experimental agents to dissect the basis of physiology and cell biology.

Another part of the study of toxicology is the "proper" use of chemicals and mixtures. In some cases, this involved how to kill or harm. Murders, suicides, chemical warfare, capital punishment, and the application of pesticides are all a part of toxicology. The proper use of chemicals also involves establishing safe limits of exposure so harm does not occur. In both instances, toxicology provides the basis for obtaining the desired outcome.

When investigating poisonings (accidental, inadvertent, or deliberate), it is essential to establish a cause-and-effect relationship, to demonstrate that a particular substance caused the adverse effects observed. When this relationship has been verified, then studies to elucidate the mechanism of action can be undertaken. This could lead to the development of appropriate treatment modalities (e.g., antidotes) or to identifying safe exposure limits. Discovering the cause of an illness has often been very difficult because cause and effect may not be evident. We have had to study hard to determine exactly what has caused an illness based on underlying assumptions about the nature of humans and the world at the time. Many early writings, based on keen observation, describe toxic plants and animals. Observations in humans following exposure to these toxic plants or animals and recommendations for treatment (sometimes based on limited evidence) are the hallmarks of the early phases of toxicology. Epidemiology is the modern extension of this approach, and it remains a difficult aspect of toxicology as society looks for the causes of more and more subtle effects.

So what is a "poison"? A generally accepted definition is a substance that causes harm. The word is usually thought of in terms of illness, injury, or death by animals and plants, but it also includes harm to the environment through contamination. Another form of harm is psychological. The mere presence of a chemical in food or in the workplace can sometimes be stressful even when no physical harm can be demonstrated. This may evolve to societal problems. Also, poisons have been used as a tool for social change by activist groups and the media. It can be argued that some chemicals (e.g., MTBE and Alar, or daminozide) have caused more societal problems than actual physiological or environmental harm due to incorrect and reckless application of toxicology.

The psychological aspects of poisons are difficult to evaluate. Poisons have generated fear throughout time and continue to do so. People still are suspicious of any chemical found in any amount. Thorough documentation and experimentation have become unnecessary in some quarters. Few data are required for hypotheses to become facts for some groups. Entire books are now written with little information and much conjecture. The significant advances in analytical chemistry and broader communication have led to a greater dissemination of information, so poisonings are better reported. Simply the presence of a minute amount of a chemical can be used by the media to inflame the public. The profusion of misinformation unnecessarily creates problems for the public and health authorities. Unfortunately, the misinformation frequently persists because proper communication of the correct interpretation of the data by scientists, including toxicologists, is rarely forthcoming. Discovering the causes of toxicity, understanding the mechanism of toxicity, and defining safe limits of exposure fall within the domain of toxicology. The fact that some of the most toxic chemicals make ideal agents for cancer therapy and others have been used to kill pests adds to the confusion experienced by the public.

Toxicology reflects the development of society: a progression from simplicity to sophistication, from crude to cultured, from elemental to elegant, from superstitious to scientific, and from taking lives to saving lives. Toxicology covers all aspects of chemical exposure, from accidental poisonings to occupational exposure to murders and suicide. It involves people, animals, the environment, and society. By the very nature of its breadth, it impacts medicine, ethics, law, and societal issues. To fully appreciate the evolution of toxicology, we need to address why we poison ourselves (e.g., all the many forms of addiction and suicide), each other (e.g., murders, mass killings, improper working conditions), and the environment (e.g., environmental illnesses). We also need to understand the events that shaped the course of toxicology. This chapter is an attempt to assess the influence of these many issues on the growth of toxicology. It looks at what was studied over the course of history and how it was studied. It also examines events that affected how toxicology developed as a science.

## HISTORY OF SOME WORDS

It helps to understand how certain words commonly used in toxicology today have evolved through history. A few examples are presented below.

*Toxic* was originally *tekw*, a word meaning to run or flee, later becoming *toxsa* in Persian and *toxon* in Greek, meaning bow (as in bow and arrow). The toxin meaning may have come from the poison used to tip the arrows, or, as Robert Graves suggested, from the yew tree (*taxus*), from which arrows were best made and whose berries were long known to be poisonous.

*Toxicology* was recently defined by the Society of Toxicology as the "study of the adverse effects of chemical, physical, and biological agents on living organisms and the ecosystem, including the prevention and amelioration of such effects." It is the science of poisons.

For clarity, the study of toxins is *toxinology*. Toxins are poisonous substances, consisting mainly of protein, that are products of certain microorganisms and some higher plant and animal species and which are highly toxic for other living organisms. Such substances are differentiated from simpler chemical poisons by their high molecular weight and antigenicity.

The word *poison* appropriately comes to us by a devious route, like a long-delayed afterthought. It derives from *poi* (to drink), becoming *potare* in Latin, from which came "potion," as well as "symposium" from *sym* (together) plus *posis* (to drink). Today's meaning did not come about until the notion of love potions evolved, and the idea of poison came to consciousness. The meaning of "poisoning" has evolved from the intentional use of chemicals by individuals to kill one or several to simply the exposure to any chemical by large groups of people.

There is also a strange history behind the word *venom*. It began as the simple word *wen*, meaning to wish or will, leading more or less directly to "win." Along the way, the word also developed into "venus," "venery," and "venerate," all indicating varieties of love. The love potion was called *venin*, which somehow gradually acquired today's sense of venom. Nobody can explain why the terms "poison" and "venom" come from love potions. Perhaps it was because the pharmacology of the day was primitive and chancy, a very fine line away from toxicity. Or, maybe there was a consensus that any sort of chemical additive intended to induce false love is, by nature, a fundamental poison. It tells something important about the common sense of earlier human beings that "venom" and "poison" were taken, with some resentment, out of the hands of artificial lovers and transferred to the stings of insects and the fangs of serpents. Of course, this was before the development of highly selective phosphodiesterase-5 inhibitors such as sildenafil and vardenafil!

*Noxious*, came from *nek*, meaning death, by way of *necare* and *nocere* in Latin, providing *necropsy* and cognate words. *Nectar* was the drink of the gods because it prevented death (*tar*, meaning to overcome).

*Chance* has its origin in "cadence," which comes from *kad*, meaning to fall, to die. *Kad* led to *cadere* in Latin and *cad* in Sanskrit, also meaning to fall (an allusion to the falling of dice), sometimes to die. The word "cadaver," a dead human body, especially one intended for dissection, comes from this base. Incidentally, *hazard* came from *dice*, by way of the Old French *hasard* and Spanish *azar*, an unforeseen disaster or accident or an unfortunate card or throw at dice, probably from the Arabic word *zahr* (a die).

*Antidote*, a remedy that stops or controls the effects of a poison, is from the Latin *antidotum*, from the Greek *antidoton*, from *antididonai*, meaning to give as a remedy against (*anti-* + *didonai*, to give).

*Alchemy* was the predecessor of chemistry, an earlier and unscientific form. It sought to transform base metals into gold and to discover a life-prolonging elixir, a universal cure for disease, and a universal solvent (*alkahest*). The word comes to us from the Old French *alquemie*, the medieval Latin *alchimia*, and from the Arabic *al-kīmiyā* (chemistry), which ultimately came from the Greek *khēmeia*.

*Remedy* is a treatment for a disease, a medication or treatment that cures a disease or disorder or relieves its symptoms. It comes from the Anglo-Norman *remedie* from, ultimately, the Latin *remedium* (medicine).

*Treacle* is an interesting word. Its current primary meaning is something cloying or excessively sentimental. This use came from a very sweet food, like molasses, which is its second meaning, and that use came from the preparation of a general antidote made of herbs in a thick, sweet base of honey to disguise the taste. So, treacle was an early antidote to poison, especially an ancient composition esteemed efficacious against the effects of (any) poison. It has specific reference to a certain compound of 64 drugs, prepared, pulverized, and reduced by means of honey to an electuary. Its derivation is unclear, but it probably comes from the Middle English word *triacle* (antidote for poison), from Old French, from the Latin *theriaca*, and from the Greek *theriake* (*antidotos*, an antidote against a poisonous bite from a wild animal), feminine of *theriakos* (of wild animals), from *therion* (diminutive of *ther*, beast). It is sometimes referred to as *theriaca Andromachi* and *Venice treacle*.

## HARMING AND HELPING THROUGH TIME

Initially, poisoning was accidental. As we learned about toxic agents through experience we could avoid them (a proper use of knowledge), but poisoning also could be deliberate (an abuse of knowledge). After the value of poisons was recognized, they became very attractive; they were appealing solutions to difficult problems. Initially, killing involved the use of physical agents (clubs, spears), and this required strength and skill, thus favoring large humans. How could smaller individuals level the field? Poisons were an answer. Knowledge of the poison as well as skill and cunning were required for their successful use.

As laws developed and people became more aware of poisons, subtler, more sophisticated (i.e., undetectable) poisons were needed (further abuse of knowledge). Once a need for poisons had been established, an industry of suppliers and practitioners developed them for implementation. The do-it-yourself poisoners were replaced by professional poisoners (early applied toxicologists), who offered advice, provided materials, and performed the required services. New agents developed by (al)chemists could make the poisoning fast or slow, painful or painless. As poisoning developed into an art, its practitioners became famous/infamous. The popularity of poisoning grew until it reached epidemic proportions in some countries. The resulting fear of poisoning was enhanced by the inability to detect poisons and prove that poisoning had occurred. Identification of the perpetrator was extremely difficult because determination of the cause of death (proof of poisoning) required analytical techniques that had not yet been developed.

The prevention of poisoning was accomplished by using bioassays (e.g., official tasters of prepared food and drink), taking precautions (only eating food of known origin, not eating foods that contained lumps or were highly seasoned), and developing tolerance or adaptation through the repeated ingestion of small doses of toxins.

Forensic toxicology began when advances in analytical techniques were applied to the detection of poisons. The Marsh test, developed in 1836 by British chemist James Marsh, allowed arsenic to be identified unambiguously (arsenic was the most popular poison at that time). More tests followed, and more poisons could be identified. This had a chilling effect on their use because poisoning could be proved. Practitioners became ever more sophisticated in their attempts to avoid detection, but they were no match for the chemists who continued to develop more sensitive and specific analytical methods. Perpetrators could now be identified and appropriate action taken. When the chemists turned their skills from developing poisons to detecting them, the popularity of such agents of death declined rapidly and poisoning became less common. Although subtle and ingenious means of poisoning are available today, forensic methods have made undetected poisoning quite challenging.

The development of sensitive analytical methods has continued into the present. Low levels of contaminants in soil, air, and water are now easily identified. These analyses challenge toxicologists as society calls for an understanding of the health significance of chemicals present at parts per billion or parts per trillion in the environment. Through these analytical techniques, the source of the exposure can now often be identified, the responsible party (or parties) determined, and appropriate action taken. These techniques are now being applied to body fluids and tissues (biomonitoring). Often, data are woefully inadequate—if any even exist—for a proper assessment of what these low levels might mean.

As the use of poisons to dispatch people declined, their use as tools to understand physiological and pathological processes increased. Their redeeming value emerged. Poisons have uses beyond harming! Agents commonly referred to as "poisons" have contributed to the health and safety of humankind and to the advancement of biological sciences (including medicine) in many ways. Claude Bernard, an outstanding early physiologist and probably the first and foremost mechanistic toxicologist who used curare to study the neuromuscular junction, wrote in 1878:

> Poisons can be used as agents for the destruction of life or as means to cure disease; but in addition to these uses—there is a third which particularly interests the physiologist. For him the poison becomes an instrument which

dissociates and analyses the most delicate phenomena of the living machine, and by careful study of the mechanism of death in different poisonings he can gain knowledge, indirectly, of the physiological mechanism of life [i.e., poisons can be used to explain physiological events].

**Translation of P.N. Magee (1965)**

Other poisons useful in our understanding of biology include radiation, a physical poison that has been an invaluable tool in elaborating some of the basic events in mutagenesis and carcinogenesis. Identifying the role of mixed-function oxidases and cytochrome P450 occurred through the use of carbon monoxide. Hepatotoxins helped us understand the workings of the liver at the molecular level. Botulinum toxin, the most acutely toxic material known, helped further our understanding of synaptic transmission. The list could go on and on.

Food production has evolved from subsistence farming that fed only a few people to a sophisticated process that feeds multitudes. The increased yields have caused increased competition with pests (weeds, insects, rodents) which reduces effective availability. The use of pesticides to help feed a growing population by controlling unwanted plants and animals has resulted in increased food production and subsequently better nutrition and health and a longer life expectancy. Chemicals (e.g., salt, smoke) have been used to preserve food. Combined with drying, chemical preservation has allowed a food to be maintained for long periods of time in a safe condition suitable for human consumption. Antimicrobials are used to control pathogens and preserve food. Again, humans are using toxic compounds in a useful fashion to control their environment to their benefit.

Chlorine gas, first used as a weapon by the German army against the French in April 1915 at Ypres, is a pulmonary irritant that causes acute damage in the upper and lower respiratory tract. It destroyed the respiratory systems of the French soldiers and led to slow death by asphyxiation. Today, chlorine is used as a disinfectant to treat public drinking water to prevent illness by dramatically reducing the threat of waterborne diseases. Its proper use has saved millions of lives.

People have been harmed when toxicological information has been ignored or misused. When it became known that nondeliberate exposure to chemicals could produce adverse health effects (e.g., in the workplace and environment), efforts were directed at the prevention of the effects by defining safe conditions of exposure to protect humans and other life forms from injury. Dose–response relationships were established as correlations were made between the level of the chemical in blood or tissues and biological activity. This was followed by the identification and quantification of the risk of adversity following exposure (risk identification, assessment, and management). Quantifying a risk (assigning a number to it) tends to decrease the uncertainty of extrapolation, lessen anxiety, and provide a degree of comfort. Quantification of the responses to toxic agents and the relationship of structure to biological activity is the basis for a great deal of scientific activity.

Moving away from practicing the art of poisoning and supplying and using poisons, the toxicologist now studies their mechanisms of action, develops analytical methods to identify and quantify poisons in body fluids and tissues, develops rational antidotes, establishes safe limits of exposure from carefully designed and executed studies, and quantifies and predicts adverse effects. The toxicologist now plays a critical role in the advancement of humankind.

## PREHISTORY

Poisonings predate recorded history and make toxicology arguably one of the oldest biological sciences, but no monuments, such as Stonehenge or the Pyramids, are dedicated to the work of the early people trying to understand the adverse effects of the substances that surrounded them. The best we can do to summarize the early aspects of toxicology is make educated guesses and extrapolate from sometimes meager writings.

It is not unreasonable to assume that harmful plants, moldy grains, and venomous animals were accidentally encountered with dire consequences. As with many aspects of everyday life, interpretations of the effects were frequently mixed with religion and mysticism. The earliest view of poisons, based on everyday life and needs, began when humans had only a rudimentary view of nature. Cause and effect were generally unknown. Early humans may have thought they were surrounded by poisons. Finding food was a matter of chance, and if you picked the wrong plant you could be poisoned. After some trial and error, the distinction between poisonous and nutritious plants became known, but even food that was wholesome at one point could become contaminated by mold and be rendered injurious; thus, a great deal was still unknown and uncontrollable. Life must have seemed capricious. It is easy to imagine how poisoning could be seen as magic. Mysticism and superstition took over to compensate for the lack of knowledge.

Later, toxic substances were intentionally used as tools to catch prey and dispose of unwanted persons. Killing for sustenance and survival also predates recorded history. It was, and still is, necessary to kill animals to survive. The initial instruments were physical weapons that required strength for effectiveness, but this put the small and less powerful at a great disadvantage. Later developments, such as the bow and arrow, required more skill and less physical strength, and these became something of an equalizer for the less physically endowed, but something more was needed to feed growing families and control one's surroundings. Might poisons be what was needed to solve some problems? As noted previously, early humans probably learned through experience which plants were beneficial and which were poisonous. The poisonous ones were used as aids in hunting (e.g., arrow poisons such as curare from the resinous extract obtained from several tropical American woody plants). Poisonous animals were also discovered. People in South America used the secretions from amphibia to kill animals for food. The adverse effects of venomous insects and animal were probably also noted, but the practical utility of these venoms was limited.

Poisons proved to be very useful in killing animals. This was a very important event! The age of poisoning, of practical toxicology, the seduction by toxicology, had begun. Poisons moved from being random problems to being tools. Poisons were initially used for survival, but humans would make use of poisons for their advancement throughout history.

Killing people was not sanctioned ("Thou shalt not kill"); nonetheless, it probably did not require too great a leap of reasoning to extrapolate from the effects seen in animals to humans. Could humans be dispatched as readily as animals with the use of poisons? It is unknown when the first human intentionally used a substance to kill another human, but humans have an instinct to control their own destinies and to satisfy their lust for power, wealth, and pleasure. The seduction of toxicology continued.

## OBSERVATION/RECORDING OF PHENOMENA

Poisonings in earlier times were not well reported for several reasons, including the inability to identify whether the cause of death was due to a poison, the inability to identify the poisoner, and limited means of communication. With the development of civilizations and writing, the known causes of toxicity could be recorded and others could learn from these writings. Many early cultures had lists (catalogs) of poisons and their effects in humans. Interest in plants that are harmful to health and as tools for vindication evolved, as did the beneficial use of plants, predominantly herbs, for medicinal purposes (there seemed to be less interest in animals, with the exception of avoiding venoms). The cures for the problems of humankind (healing or killing) could be found in nature. With time, the lists began to include detailed descriptions of the preparation, use, and effects of biologically active plant materials. Understandably, concerns about the prevention and treatment of poisoning began to emerge, but only when mechanisms of toxicity were understood (and toxicology became a science) could a rational basis for treatment develop.

### EGYPTIAN

Egyptian medicine was based on the work of the gods and the presence of evil spirits in the sick person. Physical medicines such as herbs were mostly expected to lessen pain, while magic effected the cure; however, a portion of Egyptian medicine was based on experimentation and observation, including the effects of poisons. Egyptian medicine was reputed to be the most advanced of the ancient world, and not unexpectedly the first known list of poisons and antidotes appears in Egyptian writings. Menes, the first Pharaoh of unified Egypt and the founder of Memphis, the capital, was reported in Egyptian papyri to have had an interest in poisons. He cultivated and studied the effects of poisonous and medicinal plants somewhere between 3500 and 3000 B.C. Unfortunately, no detailed written history of these activities exists. His son Athothis, a physician, wrote a textbook on medicine in which sanitation was stressed.

The Ebers papyrus (c. 1550 B.C.) is one of the oldest known writings pertaining to medicine. It contains 110 columns of hieratic (priestly) script (equivalent to about 110 pages). It reveals many customs, practices, and traditions of Egyptian doctors and describes over 800 recipes, many containing recognizable poisons such as hemlock, aconite, opium, and some of the heavy metals. The formulas also contain over 700 drugs (medicinal substances), specific indications, and dosages, together with appropriate spells and incantations. Forty-seven case histories are presented. Modes of administration include snuffs, inhalations, gargles, pills, troches, suppositories, enemas, fumigations, lotions, ointments, and plasters. Vehicles included beer, wine, milk, and honey. Drugs were identified on the basis of origin as plant (e.g., acacia, castor bean, wormwood, fennel, garlic), animal (e.g., honey, grease, milk, excrement), or mineral (e.g., alum, iron oxide, limestone, sodium bicarbonate, salt, sulfur). Insect and animal venoms were described.

Egyptian physicians were much sought after in the ancient world despite the fact that little new knowledge was added after about 2000 B.C. The Egyptians had some correct general principles of toxicology, but the concept of cause and effect was missing which led to poor treatments for disease. Their methods were based on examination followed by diagnosis. Treatment was generally conservative; if no remedy was known, then no steps were taken that would endanger the patient. Herbs played a major role in Egyptian medicine as antidotes, as were some minerals.

The Egyptians used chemicals in the administration of justice. The "penalty of the peach" involved having the accused ingest the distillate from crushed pits of peaches (high in hydrocyanic acid). If the accused died, it was a presumption of guilt. If the accused lived, it was a presumption of innocence. The practice of using chemicals in the administration of justice continued into other cultures (e.g., Greek, Roman) and persists to the present day; for example, lethal injections of chemicals are used for some state executions in the United States.

Thousands of years later, Cleopatra (c. 69–31 B.C.) poisoned her second brother after Caesar was murdered so she could jointly rule Egypt with her infant son. She committed suicide by injecting asp venom directly from the asp.

### The Bible

Pagan medical practice can be considered valid insofar as it is derived from generations of experience, but medical science can only flourish when it is divorced from superstition. When the conditions of illness and health are no longer ascribed to the whims of deities or to the malevolence of demons or to the magical ministrations of shamans, then science can come forth. The Judaic concept that the universe was created by a single, undefinable force (monotheism) was a profound leap in human thinking and may have started moving medicine away from superstition. With toxicology, this was probably more important than with any other biological science because the confusion between the divine and the secular made it difficult to distinguish between an adverse effect that was natural from one that was assisted. Although humans would go through repeated periods when adverse effects were seen as "God's will" and religion would be invoked to provide some explanation and structure to the adverse effects that were occurring, in general the direction was always to understand the physical basis for such problems.

The Bible introduces three basic premises for healthful living: rest, cleanliness, and prophylaxis. Food regulations are also part of the Bible. In the Book of Genesis, God is portrayed as the Supreme Regulator and protector as He proscribes certain foods. The regulated were Adam and Eve. Had they accepted the regulations imposed upon them, had they more faith in their Regulator, there would have been no dire consequences.

Most of the specific references to poisons in the Bible are limited to venoms, but the general concept of poisons and poisoning was clearly understood. The Bible contains 16 references to "poison" and 8 to "venom," depending on which version one consults. Bitter water to test the fidelity of a wife suspected of unfaithfulness is described in Numbers 5:11–31. Venoms and plant poisons appear in Deuteronomy 32:24 ("with the venom of reptiles gliding in the dust") and 32:32 ("poisonous are their grapes and bitter their clusters"). Arrow poisons are mentioned in Job 6:4 ("for the arrows of the Almighty pierce me, and my spirit drinks in their poison") and again later in Job 20:16 ("The poison of asps he shall drink in; the viper's fangs shall slay him"). "The venom of asps lies behind their lips" appears in Romans 3:13. In Psalms 58:5, "theirs is poison like a serpent's, like that of a stubborn snake that stops its ears," and in Psalms 140:4, "they make their tongues sharp as those of serpents; the venom of asps is under their lips." Jeremiah mentions chemical and biological poisons: "He has given us poison to drink" (8:14) and "I will send against you poisonous snakes against which no charm will work when they bite you" (8:17). In James 3:8, we find "the tongue no man can tame. It is restless evil, full of deadly poison."

An awareness of the natural occurrence of poisons and the effects produced is evident, but no mention is made of the deliberate use of them. Antidotes are not mentioned, only charms. The Bible includes no list of poisons, although it does contain proscriptions about foods and food practices (e.g., Deuteronomy 14; Leviticus 11, 17, and 19), apparently based on potential adverse health effects. The Bible reveals observation, but no listing or experimentation.

### Chinese

Legend has it that the second of China's mythical emperors, Shen Nung, is the father of Chinese medicine and agriculture. He is credited with writing a 40-volume work entitled *Pen Ts'ao* or *Pun Tsao* (the Great Herbal, a Chinese *materia medica*) around 2735 B.C. It contained lists of poisonous plants, plants with medicinal value (365), and drugs (265, of which 240 are vegetable in origin). The effects of plants and drugs and the appropriate antidotes were described. Drugs and poisons were presented together, presaging the concept that the dose differentiates a poison from a remedy. Included among the drugs were iodine, aconite (also used as an arrow poison), opium, cannabis, rhubarb, alum, camphor, iron, sulfur, and mercury. Shen Nung was also reputed to have discovered a number of drugs and experimented upon himself.

Another emperor, Huang Ti (2650 B.C.), reportedly wrote *Huang Ti Nei Ching* (*The Yellow Emperor's Medicine Classic*), the oldest extant classic of traditional Chinese medicine. Although authorship of *Huang Ti Nei Ching* is attributed to the Yellow Emperor, it was more likely written by several authors over a long period of time, compiled roughly 2000 years ago. The book is divided into two sections, the second being *Lingshu* (*The Vital Axis*), and was written sometime in the second century B.C. with revisions taking place up to the Han Dynasty (206 B.C.–25 A.D.). This great work forms the theoretical basis of traditional Chinese medicine. As traditional Chinese medicine's history developed over the millennia, nearly all significant medical works benefited from the enlightenment of this unparalleled book. The *Yellow Emperor's Medicine Classic* demonstrates that, even in ancient times, people accomplished scientific achievements that are still applicable, relevant, and innovative in modern times. Beyond medicine, this book also presents ethical, philosophical, and religious considerations. The three themes that run through the book are the theory of Taoism, yin and yang, and the five elements. It still remains one of the most respected and studied texts on Chinese medicine.

Another medical text, found during an excavation of the Mawangdui tombs and dating back to 168 B.C., is the *Wushier Bingfang* (*The Fifty-Two Prescriptions*). It details 52 ailments and 52 prescriptions and is an early written reference of Chinese pharmacology.

The Chinese may have been the originators of chemical warfare. Chinese writings contain hundreds of recipes for the production of poisonous or irritating smokes for use in war, as well as accounts of their use. There are reports from the fourth century B.C. of the Chinese using bellows to pump smoke from mustard and other noxious vegetable matter into tunnels being dug by a besieging army to discourage the diggers. The use of cacodyl ($As_2(CH_3)_4$, a colorless liquid possessing an intensely disagreeable garlic-like odor) smoke, is also mentioned in early Chinese manuscripts.

## Hindu

The *Rig Veda*, a Sanskrit document written between 1500 and 1200 B.C., is the earliest account of Hindu medicine. It contains many references to alchemy, science, and magic in the treatment of disease. Included are discussions of many diseases, including cough, fever, diarrhea, seizures, tumors, and skin ailments, as well as treatments for specific diseases. These treatments include spells and incantations—again, the combination of rational and mystical therapies. Medicinal and poisonous plants and antidotes (e.g., for snake bites) are listed. The influence of gold as a therapeutic agent and for longevity is discussed. A later work, the *Ayur Veda*, the Veda of long life, is a Sanskrit document written about 700 B.C. It discusses medicine and all its branches in eight parts; drugs and poisons are also mentioned.

> He who knows only one branch of his art is like a bird with one wing.
>
> **Susruta**

Susruta (c. 380–450 A.D.), a Hindu surgeon, authored a medical/surgical text called *Susruta Samhita*. The text was divided into six sections, in which he identified 1120 diseases and gave fever great importance. He stressed the importance of hygiene and presented many surgical procedures in great detail. The section on drugs listed 760 indigenous medicinal plants, of which many were used externally as ointments, baths, sneezing powders, and inhalations. Also listed are animal and mineral remedies. The fifth section, the *Kalpa Sthana*, addressed toxicology, as it dealt with the nature of poisons and their management.

## Greek

The Greeks borrowed heavily from the medicine of Egypt, and they took it forward by developing a system of philosophy (*philos*, friend; *sophia*, wisdom) that defined the place of humans in nature. Philosophy for the Greeks encompassed a wide range of intellectual activities, including what we would now call science. All major philosophers had an interest in medicine and the healing arts, as they were thought to be directly related to the soul, which governed vital functions and thought. Over the centuries this formed the theoretical basis of their attempts at a causal foundation for explaining disease. It was another step away from the whims of the gods. The classical Greek definition of medicine was to "prolong life and prevent disease." In other words, to keep people healthy. Physicians gave advice on how to live that would now be considered holistic, as the advice included exercise, diet, and rest.

The Greeks made many significant contributions to the advancement of medicine that led to contributions to toxicology. They had lists of poisons and antidotes, lists that were consulted by citizens and the government. They had a great deal of knowledge about poisons (especially plant poisons) and metals (especially arsenic, antimony, mercury, gold, copper, and lead). Other and more significant contributions to the advancement of toxicology include detailed descriptions of the effects of various agents in humans, antidotes, and principles for the management of poisonings (e.g., hot oil and vomiting).

Suicide and murder by poisoning were not uncommon because poisons were readily available. The Greeks also used chemicals in warfare. Around 590 B.C., Solon of Athens poisoned the water of an aqueduct during the siege of Cirrha with hellebore root, which contains two glucosides: helleborin, which is narcotic, and helleborcin, which is a highly active cardiac poison, similar in its effects to digitalis and a purgative.

> There is only one good, knowledge; there is only one evil, ignorance.
>
> **Socrates**

The Greeks executed criminals with poisons, Socrates (469–399 B.C.) being the most famous victim of state poisoning in history. Socrates had a strong contempt for conventional ideas and lifestyles. His iconoclastic attitude did not sit well with everyone, and at age 70 he was charged with heresy and corruption of local youth. Convicted, he carried out the death sentence by drinking hemlock (the state poison), becoming one of history's earliest martyrs of conscience. His idealistic philosophy was passed on through the writings of Plato, his most famous student (Socrates left no writings of his own). Plato spent his entire life restoring Socrates' good name and banishing Lycon for initiating the process that had condemned his former master.

Pythagoras of Samos (c. 570–480 B.C.) may be the first Greek to have an influence on toxicology. Although best known as the mathematician who developed the theory of numbers and considered to be the founder of arithmetic, he was also a physician and scientist who was especially interested in procreation and animal physiology. His theory of harmony may be the basis for the theory that health is the result of a perfect balance among the various elements and humors, and disease is a disruption of this harmony—the beginning of the humoral theory of medicine. His contributions to toxicology include his studies of the effects of metals (e.g., tin, iron, mercury, silver, lead, gold, copper) in the body. Because he left few, if any, writings, all of his teachings have come through his pupils.

Empedocles from Agrient (Akragas, a Greek city in Sicily) was the son of Meton and grandson of an Empedocles who was a victor at Olympia. Empedocles (c. 492–432 B.C.) was famous for his medical skills and healing powers and is the best known representative of the Pythagorean philosophical school. In his works, he presents himself as a wandering healer offering promises to thousands of eager followers, such as "You will learn remedies [*pharmaka*] for ills and help against old age." To what degree this represents the real Empedocles is not known, but a tradition grew up of him as a renowned physician, as a practitioner of magical cures, and as a charlatan. His teachings became the basis for the four body humors—blood, phlegm, yellow bile (choler), and black bile (melancholy)—and for the theory that health was the result of harmony among these humors and among the four elements. Empedocles was a strong advocate of hygiene, personal and social, and of public health measures to prevent epidemics (e.g., draining of swamps). His work shows considerable interest in biology and especially in embryology. The stories of his wonder-working such as curing entire plagues, reviving the dead, and controlling the elements are clearly exaggerated, but at that time no unambiguous separation existed between nature and the mystical, theological aspects of life. So, it may well have seemed that no great difference existed between healing ills through empirical understanding of human physiology or by means of sacred incantations and ritual purifications. He was eminent enough as a writer on medicine to be attacked by Hippocrates in *On Ancient Medicine*, in which Hippocrates attempts to separate medicine from philosophy and rejects Empedocles' work as irrelevant.

> Life is short, and Art long; opportunity fleeting; experiment dangerous, and judgment difficult.
>
> **Hippocrates**

Hippocrates (460–377 B.C.), regarded as the father of medicine, was born on the island of Cos, the son of Heraclides, a physician, and Phenarete. His contributions to the advancement of medicine are legendary, due in great measure to his belief that the causes of diseases were natural and not supernatural. Writings attributed to him rejected the superstition and magic of primitive medicine and laid the foundations of medicine as a branch of science. In addition, he stressed the importance of nutrition and diet and believed that too little or too much food was equally harmful. What he lacked in instrumentation was more than compensated for by his use of sound observation and logical reasoning. He is the presumed author of a number of significant texts and treatises that are characterized by advanced scientific and practical thinking and skillful clinical observation. Like other Greek physicians, he believed that health was the result of an equilibrium or balance in the body among the humors (blood, black bile, yellow bile, and mucus) and that disequilibrium resulted in ill health.

Hippocrates taught that the body is maintained by air and nutriments, that it is nature that heals, and that the role of the physician is to assist nature in the healing process by increasing nature's healing forces ("help nature to help herself"). This could best be done by following a proper diet and modifying one's lifestyle to get sufficient exercise and adequate rest (which makes a great deal of sense even today!). Drugs could be used to assist the dietetic cure. He established a code of medical ethics, the Hippocratic oath, which has survived to the present.

Hippocrates apparently was the first physician of record who believed that environmental factors should be considered as probable causes of disease. For example, in his book *Airs, Waters, and Places*, he argued that environmental factors (overall weather, local weather conditions, and drinking water) can influence health: "Every disease has its own nature and arises from external causes, from cold, from the sun, or from changing winds."

According to Hippocrates, the first step in treating disease, including poisonings, should be to purify the body of disease-producing humors by purgation (catharsis, purification), diet, or drugs. Cleansing was the first step in restoring equilibrium (including the management of poisoning) and was a first step in recognizing the role of pharmacokinetics in treating poisoning. Hippocrates probably foreshadowed the current practice of the clinical ecologists, who espouse thorough cleansing of xenobiotics from the body as the first step in restoring health.

Hippocrates identified about 400 drugs, mostly of plant origin, that included narcotics (e.g., poppy, henbane, mandragora [mandrake]), purgatives, and sudorifics (inducing perspiration, diaphoretics). He also advocated the use of emetics and enemas as part of the cleansing process. His contributions to toxicology include the use of sound observation, logical reasoning, and basic approaches to the management of intoxication (decrease absorption; if ingested, induce vomiting) and the use of proper antidotes.

Diocles of Carystus (375–300 B.C.), also known as the "younger Hippocrates," was one of the most prominent medical authorities in antiquity. A pupil of Aristotle, he wrote extensively on a wide range of areas such as anatomy (including the first systematic textbook on animal anatomy), physiology, pathology, therapeutics, embryology, gynecology, dietetics, foods, and poisons. Diocles wrote the *Rhizotomikon*, one of the earliest *materia medica* ("materials of medicine," a reference that lists the curative indications and therapeutic actions of medicines). It is considered to be the first work on botany to include the names of the plants, their habitats, means of collection, and medical uses. His second book on plants described those used for food, and his third dealt with poisonous plants. His works indicate that serious attention to the pharmacology and toxicology of plants had begun.

Hippocrates and Diocles extended toxicology beyond merely listing poisons and antidotes. Rational methods for the study of the effects of poisons and the treatment of poisoning were proposed. Experimental studies to assess the biological effects of plants had begun.

> Nature does nothing without a purpose.
>
> **Aristotle**

Aristotle (384–322 B.C.) was the son of the court physician to Amyntas II (King of Macedonia in 394 and 392–370 B.C.), a student of Plato, a teacher of Alexander the Great, and a philosopher. A scientific genius, he made important contributions to biology by establishing the foundations for comparative anatomy and embryology. He influenced medicine by stressing the importance of biology as a science. Like other Greek scientists and physicians of his day, he believed that the human body possessed four qualities—hot, cold, dry, and moist—and that it was composed of four humors—blood, phlegm, yellow bile, and black bile. He thought that disease resulted from an imbalance of these.

Heraclides Ponticus of Tarentum (387–312 B.C.), a philosopher and student of Plato, was reported to have spent a great deal of time studying poisons and antidotes. Heraclides belonged to the "Empiric" school, which rejected anatomy as useless and which relied entirely on the use of drugs. He may have been the first physician to suggest the value of opium in the treatment of certain painful diseases.

Theophrastus (372–287 B.C.) studied in Athens under Plato and afterwards under Aristotle. He became the favorite pupil of Aristotle and one of his chief collaborators in the attempt to achieve a complete study of all the known fields of wisdom. Aristotle named Theophrastus his successor and bequeathed to him his library and manuscripts of his writings. Theophrastus was probably the most famous Greek botanist or herbalist. He wrote *De Causis Plantarum* (*About the Reasons of Vegetable Growth*) and *De Historia Plantarum* (*A History of Plants*) in 300 B.C. These works may be considered the beginning of modern botany and served as excellent texts of medicinal and poisonous plants, as well as indications for the use of medicinal plants. His contributions to toxicology include a list of poisonous plants and the recognition of adulterated food.

An important figure in toxicology was Nicander of Colophon (185–135 B.C.), a Greek physician, poet, and grammarian who wrote, among other things, two didactic poems about poisons. They are the most ancient works devoted exclusively to poisons. The longest, *Theriaca*, is a hexameter poem (958 lines) on the nature of venomous animals and the wounds they inflict. The other, *Alexipharmaca*, consists of 630 hexameters on the properties of poisonous plants, including opium, henbane, poisonous fungi, colchicum, aconite, and conium (poison hemlock), and their antidotes. Although the work included some fanciful parts, much of the text was accurate and reflected upon his powers of observation and his experiences. Nicander divided poisons into those that killed quickly and those that killed slowly. He recommended emetics in the treatment of poisoning. So important were these works that *theriac* has come to mean antidote against all poisons, a concept that survived into the eighteenth century and was considered by some to be a tonic and a means of maintaining good health.

An interesting figure of the times, although not a Greek *per se*, was Mithridates VI, or Mithridates Eupator (132–63 B.C.), king of Pontus (now the northeastern part of Turkey). In 120 B.C., while still a child, Mithridates became king. His mother, said to have assassinated her husband Mithridates V, ruled in her young son's stead. Afraid his mother would try to kill him, Mithridates went into hiding, at which time he started ingesting small doses of various poisons in order to develop an immunity. When Mithridates returned (c. 115–111 B.C.), he took command, had his mother imprisoned (and possibly executed), and set about extending his dominion. In him, Rome found its most formidable enemy, save only Hannibal.

Mithridates was obsessively possessed with a fear of poisons. As protection, he took poisons daily, beginning with very small doses and increasing the amounts ingested to develop a polyvalent tolerance. He drank the blood of ducks fed toxic chemicals and took mixtures of antidotes. It has been reported that he may have ingested all the known poisons and their antidotes every day of the year, starting early in his life. Mithridatum, his universal antidote, was to be taken each morning before breakfast to effectively prevent poisoning. The term *mithridate*, meaning an antidote or preventive for poisoning containing many ingredients, immortalizes his contribution to toxicology.

He was a student of toxicology and one of the first to systematically study poisons in humans. He tested the effects of poisons and their potential antidotes on slaves, criminals, and prisoners. Mithridates used his knowledge of poisons against his enemies. In 67 B.C., the Roman general Pompey led a large army against Mithridates. Mithridates slowly retreated over the course of a year, until he reached the southern shores of the Black Sea.

Along the way, near the outskirts of the city of Trabzon, he left a large supply of locally produced honey in clay pots knowing that it would be found and eaten by the advancing Roman army. Three squadrons of Pompey's army found and ate the honey, then became violently ill. The next morning the sick Roman troops were easily annihilated by King Mithridates' army. That honey, locally called "mad honey," was purposely left for the Romans because it was produced from the nectar of local rhododendron plants. King Mithridates knew about the mad honey in that region because his adviser, the Greek physician Kateuas, had read about Xenophone's experiences in the same area in 401 B.C. when his entire army became sick after eating the local honey. (The "mad honey" contained grayanotoxin; symptoms of grayanotoxin poisoning include vomiting, loss of coordination, muscular weakness, low blood pressure, and hallucinations. Although grayanotoxin poisoning is rarely fatal, the physical effects often last for 24 hours or longer.) Part of Pompey's army was rendered helpless to repel their attackers and many were killed.

After his eventual victory over the king, Pompey found the prescription for the mithridatum with the king's body. The prescription was sent to Rome, where efforts to improve upon it were made and these are described below. It is said that, when Mithridates saw that people supported his son over him, he attempted to take his own life but failed because of the resistance to poison he had built up. He had to ask one of his mercenary soldiers to kill him with a sword.

Dioscorides (Pedanius Dioscorides; 40–90 A.D.) was born in Anazarbia in Cilicia (today's Turkey). He was a Greek physician, pharmacologist, and botanist who practiced in Rome at the time of Nero. He was a surgeon with the army of the emperor, so he had the opportunity to travel extensively, seeking medicinal substances (plants and minerals) from all over the Roman and Greek world. Dioscorides is famous for writing a text on botany and pharmacology free from superstition; *De Materia Medica* (*On Medical Matters*) was a precursor to all modern pharmacopeias. This five-volume set (*On Plant Materials*, *On All Manner of Animal*, *On All Manner of Oils*, *On Materials Derived from Trees*, and *On Wines and Minerals and Other Similar Substances*) became the leading text in pharmacology for 16 centuries. It covered 4740 medical uses of the materials and included descriptions of about 600 plants and 1000 simple drugs. Also discussed are the dietetic and therapeutic value of animal products (e.g., milk, honey) and mineral drugs (e.g., mercury, arsenic, lead acetate, calcium hydrate, copper oxide). He also described a surgical anesthetic made from opium and mandragora (mandrake). He was the first to recognize the toxicity of mercury. His contributions to toxicology include classifying poisons into three major classes (animal, plant, mineral), identifying antidotes, and recommending decreasing absorption to control intoxication (e.g., by inducing vomiting or purgation; *cf.* Hippocrates of Cos).

Galen of Pergamum (129–c. 216 A.D.) may be second only to Hippocrates of Cos in his importance to the development of medicine. If the work of Hippocrates can be taken as representing the foundation of Greek medicine, then the work of Galen, who lived six centuries later, is the apex of that tradition. He knew all of the medical knowledge of his day, gathered it together, and wrote voluminously (and well) about it. Galen summed up the medicine of antiquity. It is essentially in the form of Galenism that Greek medicine was transmitted to the Renaissance scholars. It was Galen who first introduced the notion of experimentation to medicine. His fame as an outstanding healer in Rome led to his being given a dissecting room in which to study comparative anatomy using the bodies of slain gladiators. He was the first to prove that the arteries carried blood and not air and conducted other experiments involving the nervous system, heart, and liver.

Galen argued that, although apothecaries knew drugs, only the physician understood both the drug and the patient and, further, that drugs were tools only for physicians (hence, Greece had few experimental nonphysician pharmacologists). He introduced rationality into drug therapy. He recommended mixtures of drugs for treating disease (precursor of polypill?), which is the basis for the term *galenicals* (medicinal preparations or remedies composed mainly of herbal or vegetable matter). He further developed the theriaca, the universal antidote, to include 100 substances; it was to be administered in honey and wine. Galen warned against the adulteration of herbs and spices.

Galen's physiological theories proved extremely seductive, and few possessed the skills needed to challenge them in succeeding centuries. Galenic physiology continued Hippocratic concepts and was a powerful influence in medicine for the next 1400 years. Galen and his work *On the Natural Faculties* remained the authority on medicine until Vesalius in the sixteenth century, even though many of his views about human anatomy were incorrect, as he had performed his dissections on pigs, Barbary apes, and dogs. Galen mistakenly maintained, for example, that humans have a five-lobed liver (which dogs do) and that the heart had only two chambers. His writings were a blessing to the ancient world, but they became a curse when, for more than a millennium, they were held to be an unassailable authority and paralyzed the progress of medicine, something Galen would have greatly deplored.

Paul of Aegina (625–690 A.D.) was a celebrated Greek physician during the Byzantine period and was probably the last Greek compiler. He was the quintessential student of the best medical authorities, Hippocrates and Galen, and authored a seven-volume medical encyclopedia, *Epitome*. His only extant work, it is based on the 70 books by

Oribasius and others. Book 5 deals with toxicology, specifically bites and wounds of venomous animals. He also displayed a peculiar genius in the field of surgery, and the sixth book, a treatise on surgery, influenced European and Arabic surgery in the Middle Ages. Paul of Aegina's medical handbook or pragmateia was transmitted and transformed through Syriac and Arabic translations, to become one of the cornerstones of the Islamic medical tradition. Paul's influence on the development of medical theory in the Islamic world and beyond makes him an important contributor to both Greek and Arabic medicine.

### Roman

The Romans had an intense interest in poisons. Records back to the fourth century B.C. indicate that poisoning was common as a means of suicide and murder. Cicero's court speeches confirm the high incidence of murder by poison in the first century B.C. Poisoning during the first century A.D. reached a peak during the reign of the Julio–Claudian emperors. The emperors poisoned members of their families and others who displeased them. Horace tells of the professional poisoner Canidia, who with Martina and Locusta became the infamous trio of women poisoners in Roman times. Locusta in particular gained infamy as a poisoner in Rome. Convicted of multiple crimes under Claudius, she was sentenced to death, but the sentence had not been carried out when Claudius died. The new emperor, Nero, made use of Locusta to eliminate many of his subjects, including his half-brother, Britannicus. Once Britannicus was dead, Nero suspended Locusta's death sentence and made her his advisor on poisons. Locusta organized a school of poisoning where she could tutor others and conduct experiments to determine how to poison and defend the Emperor against poison. Locusta became one of the first to systematically investigate the use of poisons with state sponsorship.

Reports of poisoning continued during the reign of subsequent emperors during the first century A.D., and poisoning almost became a status symbol with the moral decay of Rome. Suicide by poisoning was not uncommon, but Pliny the Elder defended euthanasia by poison in the elderly when so desired. Mass poisonings were recorded as well. However, during the second century A.D., when tensions and fear of the previous two centuries gave way to peace and prosperity, very few deaths by poisoning were recorded.

Our understanding of poisons available during Roman times is derived from the writings of Dioscorides, Scribonius Largus, Nicander, Pliny the Elder, and Galen (note the influence of the Greeks). Poisoners preferred plant poisons rather than animal or mineral poisons. Favorites included belladonna, aconite (Wolfbane, monkshood), hemlock, hellebore, colchicum, yew extract, and opium. The specific poisons used in poisoning incidents are infrequently mentioned, but it is known that hemlock in honey was the poison favored by Canidia and that Seneca drank hemlock. Ovid referred to aconite as "mother-in-law's poison."

The first effort at improving the mithridatum was made by Damocrates, one of Nero's body physicians, and it is known as *mithridatum damocratis*. Andromachus the Elder (c. 60 A.D.) was another archiater (the chief physician of some cities and first body physician of princes), in this case the royal physician to Nero. He too was ordered by Nero to improve on the existing antidote. Andromachus removed some ingredients from the mithridatum and added others: squill, opium, and the most important: viper's flesh. It was administered in honey to Nero. This became known as Theriaca Andromachi, or Venice treacle. (The name *Theriaca* or *Tiriaca* could have come from the work of Nicander of Colophon. It is also reportedly derived from the snake called Tyrus, the flesh of which was added to the mixture by Andromachus.) The theriaca contained 70 substances and was used until the eighteenth century.

Mercurialism as an occupational disease was recognized by the Romans. Mining in the Spanish cinnabar mines of Almadén, 225 km southwest of Madrid, was regarded as being akin to a death sentence due to the shortened life expectancy of the miners, who were slaves or convicts. We now know that this was due to the exposure of the miners to mercury. (Cinnabar, mercuric sulfide, is the principal ore of mercury and was used as a red pigment. Later, it was a source of mercury metal, which was used for centuries as the best way to extract gold and silver from their ores. When Spanish prospectors discovered rich cinnabar deposits in central California, they named the site after the mines of Almadén, Spain. The ready availability of New Almaden mercury was a crucial ingredient in the California gold rush.)

Chemical warfare of a type not common today (or perhaps it was early biological warfare?) was used by and against the Romans. The Romans catapulted bees and hornets at their enemies. Against the Romans in the Punic Wars, Hannibal of Carthage hurled pots of snakes on the decks of Roman ships in a sea battle in 184 B.C. When the pots broke, the Romans were forced to fight both the snakes and Hannibal's forces. The Romans were not above poisoning wells when it suited them.

Cato the Elder (234–149 B.C.) was a Roman statesman and moralist renowned for his devotion to the old Roman ideals: simplicity of life, honesty, and unflinching loyalty. Even though the Romans derived a great deal of their knowledge of medicine from the Greeks (and other Mediterranean societies), Cato nonetheless considered everything Greek to be suspect and even mistrusted Greek doctors, claiming they only wanted to poison Romans. Cato was interested in food preservation (salting) and in detecting adulterations, especially of wine, and recommended that a method be developed to determine whether wine had been watered down.

The first law against poisoning, *Lex Cornelia de sicariis et veneficis* (concerning assassins and sorcerers), was passed in the time of Sulla (82 B.C.). The law not only provided for cases of poisoning but also contained provisions against those who made, sold, bought, possessed, or gave poison for the purpose of poisoning.

Aurelius Cornelius Celsus (30 B.C.–50 A.D.) was the author of the first systematic Roman treatise on medicine. It is the most important historical source of knowledge of Alexandrian and Roman medicine. Little is known of Celsus. It appears that he was not esteemed as a scientist in his time, and whether or not he was even a physician is disputed. His fame rests entirely upon the eight volumes of his *De Medicina*. *De Medicina* was among the first medical books to be printed early in the Renaissance (in Florence in 1478), and more than 50 editions appeared. It became very influential largely because of its splendid Latin style. It was required reading in most medical schools into the 1800s. The surgical section, which even Joseph Lister studied in the nineteenth century, is perhaps the best part of the treatise, and his description of the use of skin from other parts of the body for facial plastic surgery makes him the father of plastic surgery. His four classical signs of inflammation—calor, dolor, rubor, and tumor (heat, pain, redness, and swelling)—are still used today. Book 5, *Toxicology and Rabies*, includes the works of Nicander and Dioscorides and covers poisons and antidotes. He cited others who believed that poisons and animal venoms depressed the vital factor resulting in the loss of innate heat. His contributions to toxicology include his list of poisons and antidotes and the management of poisoning. Consistent with Hippocratic teaching, Celsus advocated eliminating the poison as quickly as possible (acrid materials applied to wounds, cupping severe wounds, suction with palms of the hand, and the use of hypertonic salt solutions). In addition, he recommended the use of appropriate antidotes, including the antidote of Mithridates: 37 ingredients in honey.

Gaius Plinius Secundus, better known as Pliny the Elder (23–79 A.D.), was a famous Roman naturalist, historian, military tactician, philosopher, and one of the most learned men of his time (he wrote 160 books). His most famous and one surviving work, *Historia Naturalis* (*Natural History*), was published in 77 A.D. *Historia Naturalis* consists of 37 books and includes all that the Romans knew about the natural world in the fields of astronomy, geography, zoology, botany, mineralogy, medicine, metallurgy, and agriculture. Despite its flaws, *Historia Naturalis* remains a key resource on Roman life. The unifying thread of this work was anthropocentrism. A new and important feature of the *Historia Naturalis* was the care taken by Pliny in naming his sources. Book 1 consists of an index of topics and authorities for each of the succeeding 36 books. In it, he cited nearly 4000 authors. Book 2 deals with cosmology, meteorology, and terrestrial phenomena. It is followed by books 3 to 6 (geography), 7 (man), 8 to 11 (animals), 12 to 17 (botany), 18 to 19 (agriculture), 20 to 27 (*materia medica* from botanical sources), 28 to 32 (*materia medica* from animal sources), and 33 to 37 (metals, stones, and their uses in medicine and architecture).

Pliny's contributions to toxicology include lists of poisons and their biological effects and his questioning of the value of nonspecific antidotes such as mithridatics. He was also interested in adulteration of foods and developed methods for the detection of adulteration (e.g., chalk in flour, herbs, and spices). His last assignment was that of commander of the fleet in the Bay of Naples. Learning of the eruption of Mt. Vesuvius (responsible for the burying of Pompeii), Pliny went ashore to ascertain the cause and to reassure the terrified citizens. He was overcome by the fumes from the volcano and died, indicating that inhalation toxicology was not fully appreciated in Rome at this time.

Galen (see above) became the chief physician of Rome in 164 A.D. and is credited with systematizing Roman medical practice. With that, the totality of Greek medicine became part of the Roman world, and it was the Roman medicine that was passed down to posterity as "Western" medicine.

After the collapse of the western Roman Empire in the fifth century, Europe lost touch with much of its medical heritage. The center of Europe's view became the Church, which exerted a profound influence on medicine. The Church viewed care for the soul as far more important than care of the body, so much so that medical treatments and even physical cleanliness were little valued. In time, illness became seen as a condition caused by supernatural forces, and cures could only be effected by holy men. This pre-Hippocratic belief that disease was punishment by God and treatable only by prayer and penance meant that licensed medicine as an occupation vanished. It would be centuries before its return to Europe, and many people interested in medicine moved to other locales.

### Arab

At roughly the same time that Europe was moving away from medicine, a new civilization was rising to the east. Pre-Islamic medicine in the Arab region had been negligible due to the unsettled, nomadic life. As Islam spread and conditions changed, the Arabs attempted to collect all knowledge that was available. Greek medicine was one of the first sciences studied by Islamic scholars. Translators rendered the entire body of Greek medical texts into Arabic by the end of the ninth century. These translations established the foundations of Arab medicine. Based on the Greek teachings, Muslim physicians came to look upon medicine as the science that helps recognize the dispositions of the human body, with the goal of preserving health and, if health was lost, assisting in recovering it. The Arabs also learned from the Indo-Persian practices farther east. They built on these traditions and made significant contributions to all of the health professions.

The Arabs excelled in chemistry and are credited with inventing distillation, sublimation, and crystallization. Jabir ibn Hayyan (Latinized to "Gerber," c. 705–769) may be the father of Arab alchemy. He was an expert in chemical procedures and was the first to discover mercury. He produced arsenic trioxide (arsenious oxide, $As_4O_6$) from realgar (arsenic sulfide, $As_4S_4$), a naturally occurring, red-colored ore found in lead and iron mining, and thus made available to mankind one of the most widely administered poisons for homicide. He wrote one of the first pharmacological treatises in Arabic.

The Arab pharmacopoeia of the time was extensive and gave descriptions of the geographical origin, physical properties, and methods of application of everything found useful in curing disease. Arab pharmacists introduced a large number of new drugs to clinical practice, including senna, camphor, sandalwood, musk, myrrh, cassia, tamarind, nutmeg, cloves, benzoin, saffron, laudanum, naphtha, and mercury. They were familiar with the anesthetic effects of cannabis and henbane, when taken either as a liquid or inhaled.

The practice of pharmacy was extended by Arab physicians and eventually became a separate profession run by highly skilled specialists who were licensed. Arab pharmacies are considered to be the forerunners of modern pharmacies. To keep patients happy, make the physician's job easier, and promote more effective healing, Arab pharmacists are credited with developing or perfecting syrups (from the Arabic word for "to drink") and juleps (from the Persian word for "rose water"), tinctures, confections, pomades, plasters, and ointments as means of administering drugs. They were the first to wrap medicines (pellets) in silver foil.

Among the famous physicians of Umayyad times (661–750) were Ibn Uthal and Abu al-Hakam al-Dimashqi. Ibn Uthal was a Christian and physician to the first Umayyad caliph. He was skilled in the science of poisons and during the reign of the first caliph many prominent men and princes died mysteriously. Ibn Uthal was later killed out of revenge. Abu al-Hakam al-Dimashqi was a Christian physician skilled in therapeutics and served the second Umayyad caliph.

Abu Bakr Muhammad ibn Zakariya Al-Razi (c. 841–926) was born in Persia and took up the study of medicine at the age of 40. He is regarded as Islamic medicine's greatest clinician and most original thinker. He was a strong proponent of experimental medicine and the beneficial use of previously tested medicinal plants and other drugs. He was instrumental in the introduction of mercurial ointments to treat scabies. His first major work was the ten-part *Al-Kitab al-Mansuri*, which discussed, among other things, diet and drugs, the effect of environment on health, and epidemiology and toxicology.

Abu Ali Husain ibn Abdullah ibn Sina (Latinized to "Avicenna," 980–1037), the "Prince of Physicians," was born in Bokhara, Persia (today Bukhara, Uzbekistan). He was a child prodigy who knew the Koran by heart at the age of 10. He studied philosophy, jurisprudence, and mathematics. At the age of 16, he turned to the study of medicine. By 18, his fame as a physician was so great that he was appointed physician to the prince and became physician-in-chief to the hospital in Baghdad. Avicenna was given access to the library of the prince as a token of gratitude for healing the prince of a serious ailment. He was also the personal physician to other caliphs.

He was a logical thinker and an astute observer. Some have referred to him as a second Aristotle. Avicenna wanted to develop a system of medicine, to make medicine "a quasi-mathematical discipline." This would remove uncertainty from medical decisions (*cf.* Hippocrates, Galen). By the age of 21, he had written a 20-volume encyclopedia. His most significant medical works were the *Book of Healing*, a medical and philosophical encyclopedia, and *The Canon of Medicine*, a codification of all existing medical knowledge. It included *The Theory of Medicine*, *Simpler Drugs*, *Special Pathology and Therapeutics*, *General Diseases*, and *Pharmacopoeia*. The *Canon* included descriptions of some 760 medicinal plants and drugs that could be derived from them. The *Canon* rapidly became the standard medical reference of the Islamic world.

Avicenna laid out the basic rules of drug trials that are still followed today. He discussed oral and parenteral poisons, as well as bites and stings and their treatment, and he classified and discussed poisons as plant, animal, or mineral. He developed his own psychiatry and believed that psychic alterations were the result of changes in the brain. He espoused the importance of keeping the patient happy.

His contributions to toxicology include mechanisms of action of poisons including neurotoxicity and metabolic effects. He also recommended the bezoar (from the Arabic *bazahr*, from the Persian *padzahr* "counter-poison," from *pad*, "protecting, guardian, master," and *zahr*, "poison") stone as an antidote for venoms and preventive of disease. Originally the term meant "antidote," but later it referred specifically to a solid mass found in the stomachs and intestines of ruminants, which was held to have antidotal qualities. His work was the authoritative text on poisons and antidotes for 500 years.

Avicenna mentions a girl, the "poison maiden," who was so poisonous that insects that bit her died. This concept is also found in a manuscript at the Khalidi Library in Jerusalem, copied from the library of the Zengid ruler Nur al-Din Arslan Shah (1193–1211). Written by the Indian physician Canakya and entitled *Canakya's Book on Poisons and Antidotes*, it was presented to a ruler as a warning and as a "recipe" book intended to protect him from assassination. One anecdote in the manuscript tells of a beautiful young maiden who was fed slowly increasing amounts of a poison so her system could resist its effects but would retain it, until she became so saturated with the deadly substance that any contact with her would be fatal. She was then presented to the king as a gift.

Jewish medical erudition was in the background of Greek, Roman, and Arabic works and was exemplified by Isaac Israeli of Kairouan (c. 855–955), an Egyptian Jew who emigrated to West Africa and brought science with him. Israeli has been classed among the great physicians of the Middle Ages. One of his surviving works is on pharmacology, *De Gradibus Simplicum*. It was a standard reference and served as the foundation for most of the medieval works on the subject. Also, Rabbi Moses ben Maimon (Moses Maimonides, 1135–1204) was a famous Jewish philosopher and physician, court physician to Saladin, and rabbi of Cairo. His book on poisons, *Poisons and Antidotes: Upon Poisoning and Its Treatment*, was translated into Latin by Armend and Blasii in 1305, into German in 1813 by Steinschneider, and into French in 1865. He taught that the simplest method to poison someone was to add a single or compound poison to a highly spiced and/or chopped dish or in a victim's glass of wine, under the reasoning that the strong flavors and uneven texture would mask the bitter taste or consistency. He described poisonous insects and animals and noted that the most dangerous bite was that of a fasting human. His treatment of poisons included ligature of the bite, sucking out the poison by means of cupping glasses or with oiled lips (another extension of Hippocratic teaching to decrease absorption), and the use of external (e.g., salt, onions, asafetida) and internal (e.g., emetics) remedies. It was a much-cited text. His books on health were very advanced and resemble modern medical texts. He believed in the importance of preventive medicine and stressed the importance of hygiene. He wrote a four-volume treatise upon hygiene and diet (*Sepher Rephuoth*). His aphorisms are as pertinent today as when they were written:

> Man should believe nothing that is not attested to (1) by rational proof as in mathematical science, (2) by evidence of the senses, or (3) by authority of prophets and saints.
>
> **Maimonides**

The medical works of Hippocrates and Galen were returned to the West by way of the Middle East and North Africa, recovered through translation of Arab medical references in Sicily, southern France, and Spain. Avicenna's *The Canon of Medicine* had a great influence on Europe during the Middle Ages and was a standard European medical text for centuries. Its *materia medica* was the pharmacopoeia of Europe. So great was the reputation of Arab physicians that Chaucer names four in *The Canterbury Tales* and Dante, in *The Inferno*, placed Avicenna next to two other great physicians from ancient times, Hippocrates and Galen. Much of what we take for granted in medicine today came through Al-Razi and Avicenna.

## Middle Ages (c. 500–1450) and the Renaissance (c. 1450–1600)

As little medical work was performed in Europe during the Middle Ages, information related to drugs and poisons is meager. European works on poisons were largely based on the remnants of classical works available and on the works of the Arabs. Dispensing drugs during the monastic period was to a great extent under the control of religious orders, particularly the Benedictines, who preserved Greco–Roman herbalism by copying ancient texts and learned pharmacy, pharmacology, and therapeutics from contact with Arab physicians. (Large towns under Arab rule in the south of Europe had pharmacies; these were placed under severe legal restriction when they fell back to European control.) They adopted the Arab practice of making tinctures. They flavored wines with digestion-promoting herbs and produced the forerunners of today's liqueurs, one of which is still known as Benedictine. Charlemagne (742–814) was so impressed by the Benedictines' herb gardens and practice of herbal medicine that he ordered all of the monasteries to plant such gardens to ensure an adequate supply of medicines throughout his realm.

Academic texts on poisoning were often written by monks because monasteries were the main seats of learning in a largely illiterate population. One example is *The Book of Venoms*, written by Magister Santes de Ardoynis in 1424. This was a reasonably comprehensive account of the poisons known at the time (e.g., arsenic, aconite, hellebore, laurel, opium, mandrake, cantharides), their effects, and treatment.

Petri (or Pietro) d'Abano (c. 1250–1316) was a teacher of science and medicine at the University of Padua and one of the most famous teachers and skillful physicians of his time. His famous work, *Conciliator differentiarium*, attempted to reconcile Arabic medicine and Greek natural philosophy. In another book, *De remedis venenorum* (*De venenis eorumque remedies*), he classified poisons as mineral, vegetable, and animal, and, although many innocuous substances were often in the lists of ingredients thought to be poisonous, these were side-by-side with many truly deadly plants and minerals. He further noted that poisons can be absorbed from air and through the skin ("poisoning by touch"). The book was very popular and went through 14 editions. The power of the Church was great, and he was tried twice by the Inquisition on charges of heresy and practicing magic. Acquitted at the first trial, he was found guilty at the second, after his death.

Humoral theories dating back to the Greeks had led physicians to regard each patient's disease as unique, requiring unique therapy. During the Renaissance, a shift in the notion of illness occurred. Disease was no longer viewed as a unique experience, but as a process essentially similar in all patients. Science and medicine began to look more like how we think of them today. The period of observation (recording phenomena) and categorizing and listing poisons started to give way to the period of challenge and active investigation—experimental toxicology. God was still used to describe conditions and cures, but the grip of the Church was weakening and interpretation and experimentation were becoming more important.

The first known work on industrial hygiene and toxicology is Ulrich Ellenbog's *Treatise on Industrial Hygiene*, written in 1473 (but not published until 1524). It dealt with occupational diseases and injuries among gold miners. Ellenbog also wrote about the toxicity of carbon monoxide, mercury, lead, and nitric acid.

This work was followed by that of two influential men in toxicology, Agricola and one of the most controversial yet influential figures in medicine, Paracelsus. George Bauer (Latinized to Georgius Agricola, 1494–1555) was born in Germany during the early years of the European Renaissance, an exciting period of exploration and rediscovery of learning. Agricola was highly educated. In 1522, he began to study medicine, first at Leipzig and then at Bologna and Padua, Italian universities being the centers for science, medicine, and philosophy. He took his degree in 1526 and became a practicing physician; however, he never seems to have been terribly enthusiastic about his profession and devoted most of his energy to studies of mining and geology. He initially hoped to discover new medical drugs from mine ores. Agricola, who is considered the father of mineralogy, observed first hand the ill effects of the mining operations on miners. The publication of his book, *De Re Metallica*, advanced the science of industrial hygiene as he described the diseases of miners and prescribed preventive measures. In the book, Agricola recounts how the miners suffered from diseases as a consequence of their work. He attributed the diseases to the dusts, stagnant air, and gases in the mines. He also describes other effects that seem to be consistent with hypoxia or asphyxiation. The book included suggestions for mine ventilation and worker protection, discussed mining accidents, and described diseases associated with mining occupations such as silicosis. He died one year before the publication of his great work.

Agricola was among the first to found a natural science-based observation and field experience, as opposed to dogma or conjecture. His readiness to discard received authority, even that of classical authors, is impressive. Agricola's scholarly contemporaries regarded him highly. Later, a mining engineer who became a U.S. president, Herbert Hoover, translated *De Re Metallica* into English in 1912 and regarded Agricola as the originator of the experimental approach to science, "the first to found any of the natural sciences upon research and observation, as opposed to previous fruitless speculation."

The universities do not teach all things.

**Paracelsus**

Philippus Theophrastus Aureolus Bombastus von Hohenheim (1492–1541) was born outside the village of Einsiedeln (near Zurich), Switzerland, the son of Wilhelm Bombast von Hohenheim, a German physician/alchemist. Following the death of his mother when he was still very young, Paracelsus moved with his father to Villach in southern Austria, where his father taught chemistry, practiced medicine, and became interested in the health problems of the local miners, eventually becoming an expert in occupational medicine. Paracelsus attended the universities of Basel, Tubingen, Wittenberg, Leipzig, Heidelberg, Cologne, and Vienna, from which he received a baccalaureate in medicine in 1510, at the age of 17. He received his doctorate from the University of Ferrara in 1516. It was the custom of that time to Latinize one's name after receiving a degree, and Philippus von Hohenheim chose the name Paracelsus (*para*, "above," and *Celsus*), as he considered himself greater than Celsus. Deichmann et al. [20] claimed that *Paracelsus* "is a Greco–Roman translation of Hoheriheim and says 'next to heaven.'" He traveled throughout Europe, England, Scotland, Egypt, the Holy Land, and Constantinople, attempting to learn the most effective means of medical treatment and the latest findings in alchemy. He wanted to discover "the latent forces of nature" and wrote: "He who is born in imagination discovers the latent forces of Nature … besides the stars that are established, there is yet another—Imagination—that begets a new star and a new heaven." He returned to Villach in 1524 and became town physician and lecturer in medicine at the University of Basel in 1527.

Paracelsus was a peripatetic physician and was always trying to learn more medicine. He was a keen student of human behavior (reflected in his writings on psychiatry) who believed that practicing physicians needed to use common sense, gain experience, travel, and practice humility (good advice even today!). As his fame spread, his lectures became very popular and students thronged to them.

His approach to medicine and the body was chemical; for example, he taught that it was more important to learn about the chemical composition of the body than about the muscles (i.e., more chemistry/biochemistry and less emphasis on anatomy and physiology). Paracelsus wrote that "nature hints at cures." Because God created (caused) diseases, God (nature) also provided cures. This is the basis for the doctrine of signatures (*cf.* Pliny the Elder: "that an agent of nature shows by its external forms its unique qualities"). For example, because turmeric is yellow, it should be used to treat jaundice. It was the role of the alchemist to find these and convert them to effective remedies, and Paracelsus, the physician/alchemist, began to do so with simple materials, the metals. He tried to bring chemistry into therapeutics by encouraging the use of mineral salts, acids, and chemical therapeutic agents. He believed that the body was a chemical laboratory. When the "conservative physicians" of his day warned that metals would poison patients, Paracelsus replied, "This poison, as you call it, has a far better effect than the wagon grease … with which you are so fond of smearing your patients."

An iconoclast, he believed all physicians who preceded him were incompetent, liars, or fakers. "I am to be the monarch, and the monarchy will belong to me." His disdain for established authorities, for everything that had been said by his predecessors, reached its climax on June 24, 1527, when he publicly burned the books of Avicenna and Galen in front of the university. He discarded the old ways, including humoral pathology, but upheld Hippocrates. He attacked medical principles of his time, trusted only his own observations, ideas, and works.

Paracelsus was a deeply religious man. He was intensely concerned with the eternity, or soul, in man and felt that a doctor was neither "pillmaker" nor businessman but a legate of God, the supreme physician. Medicine was therefore a divine mission, and the doctor must raise his eyes from "excrements and salvepots to the stars." The perfect physician, he felt, was a philosopher, an astrologer, an alchemist, and, above all, a virtuous man. The character of such a doctor, Paracelsus proclaimed, was far more effective than mere mechanical skill [8].

Paracelsus was a free thinker. He developed his own system of medicine and boasted about his contempt for science. One of his methods of learning was theosophical intuition: All knowledge is the result of mystical insight, all wisdom comes directly from God, and one should be in intimate contact with God and God's creation. Submit to the will of God and all knowledge will flow. This intuitive process of learning was also part of Gnosticism (the doctrine of salvation by knowledge). He appeared to be contemptuous of the established way of thinking and believed that man was a little world (microcosm) that contained all knowledge. This was based on man's direct decadency from Adam, who had within himself all sciences as he contained the germs of all creatures. In his book *Paramirum*, Paracelsus described his system of medicine. Health, disease, and human destiny depend upon five entities (*cf.* Shen Nung); *ens astrale* (influence of the stars), *ens venini* (influence of nutrition/poisons in food), *ens naturale* (nature and functions of the body), *ens spirituale* (spirits, demons), and *ens Dei* (acts of God directly upon us to restore order and health). His theory of humors included three elementary principles: salt (representing stability), sulfur (representing combustibility), and mercury (representing liquidity). Disease is a separation of one principle from the other two. He also believed that there are five phlegms, five hydropathies, five jaundices, five fevers, five cancers, and so forth. Diseases tend to be localized in a particular "target" organ. Although he tried to bring more chemistry into medicine (e.g., by the use of inorganic salts), he also believed that God provided cures because God is benevolent.

Cope [17] described Paracelsus as arrogant and conceited "almost to the point of insanity … extremely effective in [his] criticisms of the then accepted doctrines … reveled in the wildest speculations and taught [his] mad conjectures as unassailable truths … bitter and unscrupulous controversialist … mystic … his writings … so confused and obscure as to be often quite unintelligible … braggart, scorner of authority … that Paracelsus scarcely ever lectured except when he was half drunk or attended a patient until he was wholly drunk." He seems somewhat like Empodecles of Agrient: part medic, part mystic. Paracelsus defended himself in his *Seven Arguments, Answering to Several of the Detractions of His Envious Critics*, written in 1537.

Paracelsus' positive contributions to medicine and toxicology outweigh their incongruity with his mystical approach. He is considered by some to be the father (founder) of chemistry and/or medicinal chemistry and the reformer of *materia medica*. He forever destroyed the doctrine of the four humors as the basis for disease but believed that diseases were specific/discrete conditions and are cured by specific/discrete treatments. He taught that observation and experience are essential for success in medicine. He is credited with the introduction of the following into the practice of medicine: mineral baths, laudanum, mercury, lead, arsenic, copper sulfate, and iron. Although others made observations in humans (often after deliberately administering a poison), Paracelsus encouraged the use of animals. He also developed and promulgated certain basic principles of the action of chemicals (e.g., dose–response) that still form the scientific underpinnings of modern experimental toxicology.

His principal works include *Chirurgia magna* (1536), *De gradibus* (1568), and *A Treatise on Diseases of Miners* (1567). In one of his books, *Paragranum*, he presents the four pillars upon which medicine should be based: philosophy (knowledge of nature; disease and healing are part of nature); astronomy (heaven deals with us paternalistically); chemistry (provide drugs and insight into biological events; "nature is the ideal chemist"); and virtue (love is the foundation of medicine).

In his *Third Defense*, he wrote, "What is there that is not poison? All things are poison and nothing [is] without poison. Solely, the dose determines that a thing is not a poison" [20]. This is often misquoted as "The dose makes the poison," when it should really be "The dose unmakes the poison." This concept has been expanded to include no-effect level, threshold, extrapolation, and dose–response relationship. His other contributions include target-organ toxicity, animal experimentation to study the effects of chemicals, and the use of inorganic salts in medicine. He was thoroughly seduced by the complexity of chemical–biological interactions and spent his lifetime trying to solve the mysteries of these interactions.

Paracelsus' teachings in psychiatry are often overlooked and may be as significant as his contributions in other medical areas; for example, he believed that two antagonistic forces dwelled inside man—animal and godly—and that man has to suppress the animal spirit if he is to be successful. He also believed that psychoses are not demonic in origin, the mind (will or spirit) can influence the state of the body (psychosomatic disease) and not the existence of a subconscious, and that women are different from men and must be treated differently (cutting-edge insight).

Paracelsus died at age 49, some say in a brawl at the White Horse Tavern in Salzburg on December 24, 1541, presumably exhausted. Despite his early death, Paracelsus drastically and permanently changed the course of medicine and toxicology.

Attempts to explain the action of toxins attracted the attention of other giants in the biomedical sciences of that time (intellectual seduction), such as Ambrose Paré (1510–1590). Paré practiced surgery in France and is considered a founding father of modern surgical practice, the greatest surgeon of the sixteenth century, and one of the most famous anatomists of all time. He introduced the use of ligatures of blood vessels as opposed to cauterization and authored books on many medical topics. Paré is remembered mainly for innovations in treating war wounds and for the treatment for skin ulcers, but he also investigated carbon monoxide poisoning and published a report in 1575. Another French physician, Jacques Grevin (1538–1570), considered the father of modern biotoxicology, published his classical work, *Deux Livres des Venins*, in 1568 and further developed the concept of chemical–biological interactions.

In the Middle Ages and even much later, eastern France saw a string of epidemics of ergotism, then called "holy fire" or "hell's fire." This was such a great affliction that the monastic Order of St. Anthony was founded to care for the sufferers of a malady which came to be named St. Anthony's Fire. In a Europe that could not differentiate religion from magic from medicine, this was another instance where God was invoked to help explain what was occurring. In the 1500s and 1600s, ergotism involving the central nervous system was likely to be the cause of behavior that was sometimes associated with witches.

The beginning of the Renaissance also began a period of a great number of notorious poisonings. Life was not valued as it is today, and poisons became a leading weapon, due to their relative inconspicuousness, to remove rivals or partners. Poisons and their effects were being studied by alchemists mainly to create the most potent concoction. Perhaps the most notorious of the Italian poisoners were Cesare (1476–1507) and Lucrezia (1480–1519) Borgia, the illegitimate children of Rodrigo Lenzuoli Borgia (1431–1503), who become Pope Alexander VI in 1492. They dispatched several of their rivals with a secret poison, "La Canterella." The exact composition of "La Canterella" is not known but it may have included copper, arsenic, and phosphorous, which reflected the trend in alchemy at that time to make the most potent mixtures from known toxic substances. The poor quality of historical records makes it difficult to know who actually committed the crimes, but Cesare is the primary suspect. The death of Pope Alexander VI, by the way, was likely due to poisoning, but it is not known whether he drank poisoned wine intended for a cardinal, or the cardinal had him poisoned, or there was a mix-up in the kitchen!

In Venice during the sixteenth century, a body of alchemists known as the "Council of Ten" met regularly to arrange poisonings for the State. The Council's written records have been preserved, showing they planned, voted on, and carried out the eradication of any chosen person for a sum of money. Victims were named, prices agreed upon, and contracts with poisoners recorded. Payment was made after the deed was accomplished. The Council seems to have had a number of poisonous ingredients available: corrosive sublimate (mercuric chloride), white arsenic (arsenic trioxide), arsenic trisulfide, and arsenic trichloride. In 1543, John of Ragusa, mercenary poisoner, confronted the Council, declaring that with his collection of poisons he could remove any person from society. He also added, "The farther the journey, the more eminent the man, the more it is necessary to reward the toil and hardship undertaken, and the heavier must be the payment." The ultimate consultant! His estimate was carefully considered by the Doge and Council of Venice.

Poisoning had become such an art (and so rampant) that schools for poisoners were established in Venice and Rome. A publication on the art of poisoning appeared in 1589. Written by Giovanni Battista Porta (1535–1615), *Neopoliani Magioe Naturalis* describes various methods of poisoning, particularly drugging wine, as this was perhaps the most popular method at that time. Porta gives a formula for *Veninum Lupinum*, which he described as a "very strong poison." This was a concoction of aconite, yew, caustic lime, arsenic, bitter almonds, and powdered glass. Mixed with honey, it was made into pills the size of walnuts.

Even Leonardo da Vinci (1452–1519) experimented with poisons. He tried to make them more potent by passage through animals and did some musing about their use for chemical warfare in the form of throwing powdered chalk and arsenic trisulfide on enemy ships.

Royalty used poisons on their rivals and on the poor, just to study their effects on humans. Catherine de Medici (1519–1589) of Florence, and later queen of France, tested and carefully studied the effects of various toxic concoctions on the poor and the sick, noting the onset of action, potency, site of action, and signs and symptoms.

Marquise de Brinvilliers (1630–1676) was a French poisoner who worked with her lover, Jean-Batiste de Godin de St. Croix. Brinvilliers poisoned her father, two brothers, and a sister for their inheritance. She attempted to poison her children's tutor, Briancourt, with whom she had shared romantic relations, but his quick wits saved him. His intelligence also saved the lives of Brinvilliers' sister-in-law and sister, cloistered in a convent, whom she also tried to poison. Brinvilliers even went so far as to poison her own daughter, merely because she thought her stupid! She regretted it immediately afterward, however, and made her drink a great quantity of milk as an antidote. St. Croix finally betrayed her upon his death when incriminating documents were found among his belongings. After several years on the run in England and The Netherlands, Madame de Brinvilliers was tried and convicted on all charges of poisoning. She was forced to do public penance, was put to torture, both ordinary and extraordinary, and was beheaded.

The pinnacle of this period in France occurred when Catherine Deshayes (1638–1680), popularly known as *La Voisin*, developed a flourishing trade selling poisons to wives who wished to rid themselves of their husbands. It is said that she was responsible for the deaths of many thousands. La Voisin was burned at the stake.

Throughout time woman have taken a particular interest in poison for criminal purposes, and the extent of this can be seen in an account from Rome in 1659. A society of women was formed in secret, meeting regularly at the house of a reputed witch, Hieronyma Spara. Usually married, the members of this society were issued the poison they required with instructions for its use. Spara was eventually arrested by the Papal police and tortured on the rack. She refused to confess; nevertheless, she was hanged along with a dozen other women suspected of having been her aides.

Perhaps the most notorious poisoner of seventeenth-century Italy was an woman named Giulia Toffana (c. 1635–1719). In 1690, she invented a poisonous mixture, "Aqua Toffana." The solution was sold in vials that bore the representation of a saint, usually Saint Nicholas of Bari. She managed to sell it under the pretense that it helped a woman's complexion. This was not a complete misrepresentation, as the active ingredient was arsenic, which was used to treat various skin disorders; however, the real purpose of her vials was made known to those with whom Toffana had a "rapport." It has been estimated that Toffana aided the murder of over 600 people, usually husbands, making her one of the most prolific homicidal poisoners of all time. She was executed in Naples in 1719.

By the end of the Renaissance, despite the poisonings, science was flourishing; it was the beginning of the period of enlightenment. Competition among ideas allowed old thinking to be discarded. The new ideas provided a path forward to allow for the development of theories that could be tested (cause and effect) and the practical application of scientific information.

## SYTEMATIC TOXICOLOGY

Some of the concepts developed by Paracelsus were further developed by others. For example, Felice Fontana (1720–1805), an abbot, physician, physiologist, naturalist, and professor of philosophy at Pisa and director of the Natural History Museum at Florence, investigated the physiological action of poisons, particularly of snakes. He is the first modern scientist to study venoms (*Ricerche fisiche sopra il veleno della vipera*, 1767). After a series of impressive and ingenious experiments, Fontana believed the action of the bite of the viper to be an alteration in the irritability of the fibers, which he maintained was mediated by the blood; in other words, the poison directly alters the blood, coagulating it, and this in turn alters all parts of the organism—especially the nerve fibers—that the blood would normally nourish. Through this work he advanced the concept of target-organ toxicity and secondary toxicity; that is, the symptoms of poisoning may not be the direct result of poisons acting on a particular organ but may occur as a result of effects on other organs or tissues. Fontana extended his toxicological experiments to other substances, especially to the laurel berry and curare. Although he did not hold a chair in chemistry, Fontana was perhaps the greatest Italian chemist of the end of the eighteenth century.

Richard Mead (1673–1754) was a British physician (medical degree from Padua) who worked at St. Thomas' Hospital and was physician to many of the leading figures of the day, including George I, George II, Newton, and Robert Walpole (the first British prime minister). He attempted to explain the action of poisons (venoms) in his book *A Mechanical Account of Poisons* (1702). The book was well received and established Mead's reputation, although it has been said that the rules of treatment laid down are sounder than the arguments. Its publication excited so much attention that an abstract of it was printed in the *Philosophical Transactions* for 1703. Mead dissected vipers and gave an exact account of the mechanism that provides for the erection of the fang when the snake opens its mouth. He described snake poisoning and noted that the venom is only effective parenterally. He swallowed the poison and confirmed Galen's experiment on fowls, that puncture is necessary to produce the effect (a performance that almost certainly would not be allowed today). Mead also considered other poisonous animals, plants, including opium, and toxic natural gases. *A*

*Mechanical Account of Poisons* was later republished with many additions in 1743.

Ellenbog, Agricola, and Paracelsus drew attention to the plight of miners, but little attention was focused on the effects of nondeliberate exposure to chemicals in the workplace. It was the brilliant Italian physician Bernardino Ramazzini (1633–1714) who effectively and convincingly brought the workplace situation to the attention of the world, especially to the field of medicine. He was the first to describe in a comprehensive, systematic, and detailed fashion, industrial health problems in his *De Morbis Artificum Diatriba* (*Diseases of Workers*). It is not known when Ramazzini commenced work on his book, but it is known that he lectured on this topic as early as 1690. This is the first comprehensive work on occupational diseases, and it was published in 1700. It is considered a milestone in the history of occupational medicine. *De Morbis Artificum Diatriba* outlines the health hazards of irritating chemicals, dust, metals, and abrasive agents encountered by workers such as miners, potters, masons, farmers, nurses, soldiers, and many others. He noted the high incidence of breast cancer among nuns which he attributed to their celibate life (which is now known to be due to the unabated presence of estrogen). In discussing the etiology, treatment, and prevention of these diseases, Ramazzini often cites Hippocrates, Celsus, and Galen and, after summarizing their observations, relates his own experience with the various diseases. By recognizing the social significance of occupational diseases, he earned the title of father of industrial hygiene.

The observations of Ramazzini concerning the relationship between workplace exposure and disease were extended by the classical studies of Sir Percival Pott (1714–1788), British physician and surgeon to St. Bartholomew's Hospital (from 1749 to 1787) who achieved fame in two areas: surgery and occupational medicine/toxicology. He sustained an ankle fracture following a fall and described it so well that it became known as Pott's fracture. Pott's disease, a characteristic fracture of the spine that results from tuberculosis, was first described by him. His contributions to toxicology include describing the relationship between squamous cell carcinoma of the scrotum ("sooty warts") in London chimney sweeps—men (boys!) whose job it was to clean the soot and other residue from the chimneys and fireplaces in the houses in London—and soot from burning coal in 1775. This was the first identification of occupational chemical carcinogenesis and represents the beginning of the study of occupational cancer. He also noted the increased sensitivity of children to some chemicals. The sweeps were usually children who worked from age 8 to adulthood, although apprenticeships could begin at age 4. Pott noted that this was an occupational disease and postulated that the cancer was caused by an ingredient in the residue from the burning coal. Pott's pioneering work resulted in the Chimney-Sweepers Act of 1788, the year of his death.

With the Industrial Revolution, which occurred between 1760 and 1830, workers no longer owned the means of production. The demand for goods had grown to a point whereby the only means of meeting it was through mass production. This production was achieved through the invention of machines, such as James Hargreave's spinning jenny, which could do the work of several individual workers. With the machines came the textile mills and factories, which in turn generated a proportionate increase in the exposure to chemicals needed for processing textiles such as acids, alkalis, soaps, and mordants (substances that fix a dye in and on textiles and leather by combining with the dye to form a stable, insoluble compound). As more workers were used to increase production, exposures to chemicals and dusts were increased. The factory owners realized the benefits of increased production, but the risks were borne by the workers—not an acceptable or sustainable risk–benefit relationship. Charles Turner Thackrah (1795–1833) developed an interest in the diseases he came to see among the poorer classes of people living in the city of Leeds. His observations led him to develop some of the basic principles of occupational hygiene to improve the health of his patients. He advocated the elimination of lead as a glaze in the pottery industry (something we still have not learned), advised installing ventilation and respiratory protection for knife grinders, and suggested a change in the work practices of tailors and in the design of their work stations to eliminate their cramped postures, which he felt contributed to their high prevalence of tuberculosis. He published a book in 1831 entitled *The Effects of the Principal Arts, Trades and Professions, and of Civic States and Habits of Living, on Health and Longevity, with Suggestions for the Removal of Many of the Agents Which Produce Disease and Shorten the Duration of Life*. Although Ramazzini recognized the relationship between a worker's occupation and health, Thackrah's importance lies in the fact that the Industrial Revolution began in England and he was the first physician in the English-speaking world to establish the practice of industrial medicine. Also, his writing led to a raised public awareness of the plight of many of the new working class who toiled under conditions often so abominable that public outcry and the efforts of reformers such as Thackrah led to the passing of the Factory Act in 1833 and the Mines Act in 1842.

Forensic toxicology, the application of analytical techniques to the detection of poisons, had its beginning with the Austrian, Joseph Jacob Plenck (c. 1735–1807), who noted in his text, *Elementa Medicinae et Chirurgiae Forensis*, that the only proof of poisoning is the identification of the poison in the organs of the body. This remains a basic principle of forensic toxicology. Unfortunately, it was not accepted by the medical or scientific communities until the work of Orfila (see below). He also wrote a treatise, *Icones Plantarum Medicinalium secundum systema Lynnaei cum enumeratione virium et usus medici, chirurgici et diaetetici*, dealing with therapy based on plants, in which Plenck discusses the medical use of about 800 plants. It is centered on 111 plants with diuretic properties that still appear in many pharmacopoeias. Plenck is a forerunner of modern European dermatology and also worked in ophthalmology.

Fredrick Accum (1769–1838) is representative of a chemist who is largely forgotten these days but nevertheless contributed to important changes in society, in his case by raising awareness of food safety. The application of analytical chemistry to matters of food and drug safety formally began with Accum, although earlier attempts were made by Theophrastus, Cato, Pliny the Elder, Dioscorides, and Galen. Born in Buckebourg, Germany, Accum moved to London in 1797 as a pharmacist and in 1801 began working with Sir Humphrey Davy (of electrochemistry fame). In Accum's time it was common to add all sorts of materials to food to make it less expensive to produce and yet still pass it off as a quality product. He was the first to use analytical chemistry to detect adulterants in food, and he published *A Treatise on Adulterations of Food and Culinary Poisons* in 1820, a very successful book that was acclaimed worldwide. Accum and fellow campaigners fought against food fraud and paved the way for the 1875 Sale of Food and Drugs Act in Britain. He also published *An Attempt To Discover the Genuineness and Purity of Drugs and Medicinal Preparations*. Accum also had an entrepreneurial bent and was very successful in his business of selling laboratory chemicals and equipment. He equipped the first chemistry laboratories of both Harvard and Yale. He left England and returned to Germany because of unsubstantiated charges of embezzlement directed against him related to his position as a librarian.

## ANALYTICAL AND MECHANISTIC TOXICOLOGY

> In physical science, the first essential step in the direction of learning any subject is to find principles of numerical reckoning and practicable methods for measuring some quality connected with it. I often say that when you can measure what you are speaking about, and express it in numbers, you know something about it; but when you cannot measure it, when you cannot express it in numbers, your knowledge is of a meager and unsatisfactory kind; it may be the beginning of knowledge, but you have scarcely in your thoughts advanced to the state of Science, whatever the matter may be.
>
> **Sir William Thomson, Lord Kelvin (1883)**

Advances in chemistry, physiology, pathology, and clinical medicine in the eighteenth and nineteenth centuries resulted in significant advances in toxicology. The time had come for analytical techniques to be formally incorporated into toxicology. Up until this time, it had been difficult to establish poisons as the cause of death because they could not be identified in tissues, the only scientifically valid proof. Analytical (forensic) toxicology had its formal origins in the outstanding work of Orfila.

> Measure what can be measured; make measurable what is not so.
>
> **Galileo**

Mathieu Joseph Bonaventure Orfila (1787–1853) was born on the island of Minorca, Spain. He was educated in Valencia and Barcelona and studied chemistry and medicine in Paris, receiving his medical degree from the University of Paris in 1811. Upon graduation, he became a private lecturer on chemistry. In 1813, when he was only 26, he published his monumental two-volume work, *Traité de Toxicologie: Traité des poisons tires des regnes mineral, vegetal at animal ou toxicologie generale considerée sous les rapports de la physiologie, de la pathologie et de la médecine légale* (Crochard, Paris). This classic work, the first of its kind, effectively combined forensic and clinical toxicology with analytical chemistry. It is a vast mine of experimental observation on the symptoms of poisoning of all kinds, on the effects poisons have in the body, on their physiological action, and on the means of detecting them. It earned Orfila the title of father of forensic toxicology. It is the first book devoted entirely to toxicology and established toxicology as an experimental science separate from pharmacology. He summed up everything known about poisons at the time and classified poisons into six categories: corrosives, astringents, acrids, narcotics, narcotico-acids, and septica and putreficants. He presented the chemical, physical, physiological, and toxic properties of each chemical; methods of treatment; and chemical tests for their identification. An English translation of his work first appeared in 1816, and American editions were published in 1817 and 1826. In 1816, Orfila published *Eléments de chimie médecale* and in 1818 *Secours à donner aux personnes empoisonnées au asphyxiés*. He provided a rational basis for some antidotes and demonstrated the toxicity of strychnine in numerous experiments on dogs. At that time, strychnine was widely used in prescriptions and in tonics and was considered by practitioners of medicine to be a safe drug (Magendie later established the mechanism of action of strychnine). He later published *Leçons de médecine lègale* (1821), *Trait des exhumations juridiques* (1830), and *Recherches sur l'empoisonnement par l'acide arsenieux* (1841). Orfila's books were translated into many languages, and this helped internationalize toxicology.

Becoming professor of medical jurisprudence in 1819, Orfila helped develop tests for the presence of poisons in blood and used a microscope to assess blood and semen stains. Four years later, he was professor of chemistry in the faculty of medicine at Paris. In 1830, he was nominated dean of that faculty, a high medical honor in France. Orfila was an excellent analytical chemist and very capable physician. He had tried the various tests for poison detection and had found them to be highly unreliable. He was also an experimental toxicologist and administered known doses of poisons to animals, carefully observed the effects produced, examined organs for evidence of toxicity (target-organ toxicity), and chemically analyzed tissues and body fluids to establish relationships among dose, response, and tissue levels. He was able to demonstrate conclusively and quantitatively that poisons are absorbed from the gastrointestinal tract and accumulate in tissues. He refined Rose's method for arsenic detection (in 1806, Valentine Rose showed how arsenic could be detected in human organs) to achieve greater testing accuracy. It was Orfila who showed with tests on animals that, after ingestion, arsenic is distributed throughout the body. He consulted on many criminal cases due to his fame as an analyst and to his prominent university position. A French woman, Madame Lafarge, was accused of poisoning her husband with arsenic and put on trial. Chemical tests conducted shortly after his death were inconclusive. During the trial, Orfila had the body exhumed and found traces of arsenic in the man's organs (using the Marsh test, developed in 1836). Madame Lafarge was found guilty and sentenced to the penitentiary. This may be the first time forensics was successfully used in a court case.

Orfila's significant contributions to toxicology include the chemical detection of poisons in tissues and fluids, thereby permitting better diagnoses; furthering the concept of target-organ toxicity by evaluating tissues grossly and histologically; relating symptoms to specific tissue injury; and extending the concept of dose–response. His investigations were also the forerunner of modern toxicokinetics and dynamics. He retired in 1848 and died 5 years later. His influence on modern toxicology is equal to that of Paracelsus and Bernard. His books were published in many languages and used in many countries. Few branches of science can be said to have been created and raised at once to a state of high advancement by the labors of a single man as Orfila did for toxicology. He spawned other works that brought to toxicology the recognition it deserved and needed.

Isidore Geoffroy Saint-Hilaire (1805–1861), a French zoologist noted for his work studying anatomical abnormalities in humans and animals, coined the term "teratology" in 1832. His father had initiated studies on chicken eggs, but it was Isidore who first published an extensive work on teratology, *Histoire Générale et Particulière des Anomalies de l'Organisation chez l'Homme et les Animaux*, organizing all known human and animal malformations. Many of the principles governing abnormal development were described for the first time, and many hundreds of names for specific malformations are still in use.

An unresolved allegation from this period is that Napoleon Bonaparte was poisoned with arsenic or antimony while exiled on St. Helena in 1821. It is argued that the British governor of the island, Hudson Lowe, had conspired with Count Charles de Montholon to assassinate Napoleon for fear he would escape and return to France. Despite extensive sophisticated testing on locks of hair that supposedly belonged to him, the official cause of death is still stomach cancer.

In 1836, James M. Marsh, an English chemist, developed a method for the detection of arsenic so sensitive that it can be used to detect minute amounts of arsenic in foods or in stomach contents. The sample is placed in a flask with arsenic-free zinc and sulfuric acid. Arsine gas (also hydrogen) forms and is led through a drying tube to a hard glass tube in which it is heated. The arsenic is deposited as a "mirror" just beyond the heated area and on any cold surface held in the burning gas emanating from the jet. Antimony gives a similar test, but the deposit is insoluble in sodium hypochlorite, whereas arsenic will dissolve. Importantly, this test was rigorous enough that it held up in court.

The mid-1800s were an era of high-profile poisonings in Great Britain that left the public almost in a panic. The ready availability of poisons and the accessibility of scientific knowledge due to increased literacy and scientific education led the public to believe that poisoning was something new. The press fanned the panic by capturing and embellishing every detail. Additionally, poisoning seemed to pose a special problem to the new social order because of the rise of life insurance. People could now be murdered for money, not for being highly placed in society or upsetting a powerful ruler. Poisoning was simply a commercial transaction (remember the Council of Ten). This public furor stimulated the development of forensic toxicology in Great Britain. The Marsh test for arsenic was one specific result, but, on a larger scale, the result was the evolution of toxicology and medical and legal practices. Some of these changes are apparent today—for example, the registration of deaths by physicians, the banning of the sale of poisons by chemists (pharmacists), and rules of evidence in courts. A realistic appraisal of the situation would lead to the conclusion that there really was no poisoning epidemic in Britain at that time.

Not all deaths from arsenic were by homicide. A vast amount of circumstantial evidence from newspaper and medical articles suggests mass poisoning of the Victorian world (mainly Germany, France, and England) by arsenic. During the nineteenth century, a number of cases of arsenic poisoning occurred that were rather puzzling. Some people became sick and some even died. Arsenic was found, using the Marsh test, and foul play was sometimes suspected, but in many cases it just did not seem possible that the person had been deliberately poisoned. In 1893, an Italian physician and biochemist named Bartolomeo Gosio (1863–1944) worked out what was happening. Arsenic greens—copper arsenite (Scheele's green) and copper aceto arsenite (emerald green)—were first synthesized in 1778 by the renowned Swedish chemist Karl Scheele (1742–1786). Scheele's green was a coloring that was used in fabrics and wallpapers. The pigment was easy to make and had a bright color. By 1863, 500 to 700 tons of arsenic green were manufactured in Britain. Gosio discovered that under certain circumstances, when wallpaper containing Scheele's green became damp, mold grew and converted the copper arsenite to a gaseous form of arsenic. If a person breathed in enough of the vapor, arsenic poisoning occurred. A garlic odor was noticed in some rooms with these wallpapers, and it was linked to the deaths of people who slept in these rooms. This may be the first finding of metabolism by a microbe to a more toxic form of a material. Gosio isolated a gas and found that it had a garlic smell, but he could not identify it. Frederick Challenger (1887–1983) identified the gas in the 1940s as trimethylarsine ($As(CH_3)_3$). This gas had killed many, mainly children, who died in their green bedrooms. But, it was not just wallpapers that were dyed green with arsenic; clothes were, also. A campaign was run by *The Lancet* to banish arsenic greens, as many illnesses and deaths were attributed to rooms wallpapered with arsenic-dyed papers. It was not until the turn of the twentieth century that arsenic greens were finally phased out, only to be replaced by lead-based wall paint, which would become the next heavy-metal issue in the home environment.

Alfred Swaine Taylor (1806–1880), British physician and the founder of British forensic medicine, is also the founder of modern medical jurisprudence, a natural continuation to the development of forensic toxicology. He received a diploma from the Apothecaries Society in 1828 and his certificate to practice from the Royal College of Surgery in 1830, and he presented the first course in medical jurisprudence in England in 1831 at Guy's Hospital. He was probably the most famous expert witness of his time. He published his *Manual of Medical Jurisprudence* in 1842, which became very popular; the tenth edition was published in 1879. Taylor was the leading author on the subject of medical jurisprudence in the nineteenth century. In his books on medical jurisprudence and on poisons, the standard works throughout the world, he codified the legal precedents, judicial rulings, and anatomical and chemical data that bore on the subject.

Sir Robert Christison (1797–1882), a noted Scottish physician with a medical degree from Edinburgh, studied toxicology under Orfila. He was appointed to a Chair of Medical Jurisprudence at the University of Edinburgh in 1822. In 1832, he transferred to a Chair of Materia Medica at the same institution, which he held for the next 45 years. He published *A Treatise on Poisons* in 1829. The fourth edition, published in 1845, became the first American edition. He became a recognized authority on poisons, and in the course of his inquiries he did not hesitate to experiment on himself. He took large doses of calabar bean, the seed of a western Africa tropical woody vine, *Physostigma venenosum*, which is the source of the drug physostigmine. African natives used this as an ordeal: If the accused person ate the bean and vomited within half an hour, he was judged innocent; if he succumbed, he was found guilty. His attainments in medical jurisprudence and toxicology procured him the appointment of medical officer to the crown in Scotland, and from that time until 1866 he was called as a witness in many celebrated criminal cases (more high-priced consultations). His works helped develop the basis for expert witnessing. He strove to provide a further scientific basis for toxicology. He also was one of the three main pioneers of modern nephrology along with Bright and Rayer.

(An interesting aside about the use of poison by Christison: The Scottish firm of W. and G. Young wanted a harpoon that would kill a whale quickly to prevent it from diving under the ice. The Youngs approached Christison in 1831and asked him to invent a harpoon for them, one that would utilize poison as the killing agent. Christison finally accepted the challenge and chose pure prussic acid (hydrogen cyanide) because of its extreme potency. Christison's harpoon contained two glass cylinders for the liquid poison, each approximately 10 cm long and 2 cm in diameter, together containing almost 60 g. In 1833, the whaleship *Clarendon* was sent out with prussic acid harpoons. According to Christison, the harpoon gun was fired for the first and only time. The harpoon was buried deeply in the whale, which immediately "sounded," or dived perpendicularly downwards. But, in a very short time, the rope relaxed, and the whale rose to the surface quite dead. Apparently the men were so appalled by the terrific effect of the harpoon that they declined to use any more of them. Christison was required by the Youngs to keep his invention secret, and he remained silent until 1860 (although the crewmen must have discussed it because a U.S. patent was granted in 1835 for a prussic acid harpoon very similar to the Christison design), when the Youngs had died and their heirs were no longer involved in whaling and had released Christison from his promise of silence. In 1860, Christison explained his harpoon design in an article, "On the Capture of Whales by Means of Poison," in the *New Edinburgh Philosophical Journal*.)

Other forensic toxicologists of note include Henry Coley, a New York City forensic toxicologist, who in 1832 published *A Treatise on Medical Jurisprudence, Part I: Comprising the Consideration of Poisons and Asphyxia*. Included in this book were mineral acids, caustic alkalis, ammonia, nitrates, phosphorus, cyanide, metals, and alkaloids, as well as their chemistry and uses, signs and symptoms of poisoning, causes of death, postmortem findings, and treatments. In 1867, Theodore George Wormley

(1826–1897) published *Microchemistry of Poisons*, the first American toxicology textbook which became the standard on the subject worldwide. In 1884, A.W. Blyth published *Poisons: Their Effects and Detection*, an excellent analytical toxicological text. Rudolph A. Witthaus and Tracy C. Becker edited a four-volume text, *Medical Jurisprudence, Forensic Medicine, and Toxicology*, which became the standard reference text in the field (1894–1896). Walter S. Haines and Frederick Peterson wrote a toxicology text, *Textbook of Legal Medicine and Toxicology*, which was published in 1903. Alexander Gettler (1883–1968), who probably influenced the development of forensic toxicology in America more than anyone else, began working in the Office of the Chief Medical Examiner in New York City in 1918.

Toxicology was being recognized as a scientific discipline. The advances in forensic toxicology paralleled advances made in analytical techniques; however, true understanding of the basic mechanisms of action of chemicals and drugs lagged. Little had been done to answer the basic question, "How do poisons kill?" But, that changed as the era of mechanistic toxicology began with the classical studies of the two most famous physiologists in medical history: François Magendie and his pupil Claude Bernard. Although important contributions had been made by others, they were not as systematic, fundamental, and far-reaching as those of Magendie and Bernard.

François Magendie (1783–1855), French physician and experimental physiologist, contributed significantly to the advancement of physiology, medicine, and toxicology. He was the first to prove the functional difference of the spinal nerves (Magendie proved Charles Bell's theory on the motor function of anterior roots and the sensory function of dorsal roots of spinal nerves), and he studied blood flow, swallowing, and vomiting, as well as the dynamics of movement across body membranes. His interest in the functioning of the nervous system led him to establish the mechanisms of action of emetine and strychnine, leading to the scientific introduction of these compounds into medical practice. He also experimented on the effects and uses of morphine, quinine, and other alkaloids, for which he is sometimes called the founder of experimental pharmacology. He was also the first, or one of the first, to observe and describe anaphylactic shock.

Magendie's most famous pupil was Claude Bernard (1813–1878), the son of a Burgundian vinegrower who studied pharmacy and enjoyed science but wanted to be a playwright. His critics told him to study medicine and, fortunately for humankind, he accepted their advice. In 1834, Bernard enrolled in the Paris School of Medicine, and after a few years he obtained a position at a lab at the Collège de France, where he worked under Magendie. He enthusiastically endorsed Magendie's philosophy that physiologists must discover the laws of "vital manifestations," or physiological functions, and that observation and experimentation were the only methods of investigation. He received his degree in 1843. His thesis dealt with gastric juice and digestion. He and Magendie did much to advance physiology, especially of the autonomic nervous system and the gastrointestinal tract. In 1854, he accepted the newly created chair of physiology at the Sorbonne. When Magendie died in 1855, Bernard took over his post at the Collège de France. He held the positions at the Sorbonne and the Collège de France concurrently until 1868.

Bernard studied both normal and pathological physiology. He discovered vasomotor nerves, the action of curare and other poisons in human body, and the functions of pancreatic juice in digestion, and he elucidated the glycogenic function of the liver. Also, in trying to understand how the systems of an organism maintain a state of balance, he was the first to propose the concept that later became known as homeostasis.

In addition to his work in experimental physiology, Bernard insisted that an experiment should be designed to either prove or disprove a guiding hypothesis. He also maintained that an experiment should produce the same results again and again, so long as the starting conditions are the same. These two points are integral parts of the modern scientific method.

His specific contributions to toxicology include furthering the concept of target-organ toxicity, establishing approaches to defining the mechanism of action of drugs and other chemicals (e.g., curare, nicotine, carbon monoxide), demonstrating that the basic principles of pharmacology and toxicology are identical, and showing that drugs and other chemicals can modify the function and structure of tissues. He believed that "the physiological analysis of organic systems ... can be done with the aid of toxic agents" (a new use for poisons!). His works were published in 18 volumes. One of his most famous, *An Introduction to the Study of Experimental Medicine*, was published in 1865 and translated into English in 1949. It is a clear and penetrating presentation of the basic principles of scientific research. Bernard introduces his idea of homeostasis in this book, and he explains how and why it works and how humans, as well as animals, could not live without it. It is a classic in the field of experimental biology and must reading for all students of biology and medicine. Medicine eventually adopted the concepts of cellular physiology pioneered by Magendie and Bernard.

It can be argued that toxicology as we know it today began with Bernard. Paracelsus took some important first steps, but much of his work is instructional only in hindsight based on what we know today. Orfila made the crucial introduction of analytical chemistry and jurisprudence, but Bernard completely changed the focus of toxicology from a science of poisoning (inducing harm) to a biomedical science that could help to explain basic physiological processes. Substances that were used to induce harm were now being used to help mankind.

> One must break the bonds of philosophic and scientific systems as one would break the chains of scientific slavery. Systems tend to enslave the human spirit.
>
> **Claude Bernard, Introduction to *l'Etude de la Medicine Experimental* (1865)**

Magendie's and Bernard's work stimulated others to experimentally establish the mechanisms of action of toxic agents and to publish textbooks—for example, the Florentine physician and scientist Ranieri Bellini Pisano (1817–1878) who, in addition to promoting research in toxicology, was also a pioneer in experimental pharmacology. He authored the first experimental toxicology text, entitled *Lezioni Sperimentali de Tossicologia*. He was also the founder of the Istituto Tossicologico Fiorentino.

The discipline of toxicology was also advanced by the outstanding research efforts of such noted northern European pharmacologists as Rudolph Buchheim (1820–1879), the founder of modern pharmacology; Oswald Schmiedeberg (1838–1921), another of the founders of modern pharmacology and toxicology and a student of Buchheim; and Rudolf Kobert (1854–1918), a student of Schmiedeberg, who published a number of textbooks in the 1890s (e.g., *Practical Toxicology for Physicians and Students*). Kobert spoke of essential oils stimulating the creation of antitoxins which he felt would eventually prove to be one of their most important functions. A contemporary, Louis Lewin (1850–1929), regarded as the father of psychopharmacology, reported on the toxicology of alcohols, chloroform, opiates, and plant-derived hallucinogens, and also wrote a toxicology text published in 1929.

The mechanistic studies of Claude Bernard were furthered by the brilliant German chemist, microbiologist, and immunologist Paul Ehrlich (1854–1915), who significantly advanced mechanistic toxicology and pharmacology. His keen interest in chemistry and biological structure and function led him to propose the concept of a receptor as the sensitive site for chemical–biological interaction and that "chemical substances in organisms had specific points of attachment"; once these were known, specific remedies could be developed. His most famous remedy was the use of arsenic in the management of syphilis (compound 606, or arsphenamine). He subsequently identified several receptors. His successful bout with tuberculosis stimulated his interest in immunity (as did his association with Koch), and he subsequently formulated the concepts of active and passive immunity. In addition to his originating the concept of receptors, his contributions to toxicology and pharmacology include underscoring the importance of mechanistic studies and structure–activity relationships. He shared the Nobel Prize in Physiology or Medicine with E. Metchnikoff in 1908.

The epidemiological study of chemical carcinogenesis that began with Pott in 1775 continued in 1822 when Dr. John Ayrton Paris (1785–1856) surmised that arsenic fumes might be the cause of the frequent occurrence of scrotal cancer in copper and tin smelter workers in Cromwell, England. In 1875, Richard von Volkmann (1830–1889) observed occupational skin tumors among workers in the tar and paraffin industry at Halle, Germany. In 1876, Joseph Bell (1837–1911) of Edinburgh suggested that shale oil was responsible for certain skin cancers in Scotland. Bladder cancer among aniline dye workers was first described by Dr. Ludwig Rehn (1849–1930) of Germany in 1895.

The experimental study of chemical carcinogenesis began in 1915 when Katsusaburo Yamagiwa and Koichi Ichikawa produced malignant epithelial tumors by applying coal tar to the ears of rabbits. This was the first demonstration of chemical carcinogenesis and *was experimental proof of Pott's hypothesis of 150 years earlier*. In 1922, R.D. Passey produced malignant growths by painting the skin of mice with coal tar ether extracts. In 1925, Murphy and Sturm [50] reported that they had produced a high incidence of lung tumors in mice when coal tar was applied to the skin without local irritation. In 1932, Cook and his colleagues published their findings with regard to pure hydrocarbons causing cancer in mouse skin [16]. In 1935, Sasaki and Yoshida [61] showed that *o*-amidoazotoluene caused liver tumors in rats. In 1938, Hueper et al. [39] first reported the successful induction of bladder cancer in dogs by repeated injections of 2-naphthylamine.

Despite the rise of toxicology as a science to help, a certain element of mankind still wanted to use chemicals to harm. Advances in chemistry allowed people to use new agents for chemical warfare. These were generally gaseous compounds intended to indiscriminately debilitate or kill hundreds or thousands of people. It seems to be the worst scenario for toxicology: The knowledge gained by advances intended to help were deviously twisted to harm. Early chemical warfare agents were crude destroyers of tissues, but later agents became quite elegant in their destructiveness. In one sense, what occurred beginning in World War I was a continuation of the use of smokes and irritants against one's enemies. In another, it changed the way chemicals and toxicology would be viewed.

Chlorine was the first lethal chemical used in modern warfare. At 5 p.m. on April 22, 1915, German troops at Ypres discharged 180,000 kg of chlorine gas from 5730 cylinders, creating a gas cloud that blew with the wind. It either killed the French and Algerian troops in the opposing trenches or caused them to flee, opening a gap in the Allied line. On April 24, the Germans conducted a second chlorine gas attack at Ypres, this time against Canadian troops. On May 31, chlorine was employed on the eastern front by the Germans approximately 50 km southeast of Warsaw. This attack employed 12,000 cylinders, releasing 264 tons of chlorine along a 12-km line. In total, nearly 200 chemical attacks were carried out during World War I using gas released from cylinders. The largest attack occurred in October of 1915, when the Germans released 550 tons of chlorine from 25,000 cylinders at Rhiems. The effect of chlorine on the lungs has been described in many places. Chlorine is now considered obsolete as a chemical warfare agent.

Prof. Fritz Haber was chief of the German chemical warfare service during World War I and personally directed the first chlorine gas attack. Haber, a Nobel laureate and known for his discovery of a process for synthesizing ammonia from nitrogen and hydrogen, is often referred to as the father of chemical warfare. As the war continued, many other compounds were tested for utility as chemical warfare agents. Of those agents tried, chlorine, phosgene (on December 19, 1915, 88 tons of phosgene were released from 4000 cylinders at Nieltje in Flanders), diphosgene, chloropicrin, hydrogen cyanide, cyanogen chloride, and mustard were produced and used in large quantities. Mustard gas was first used in an artillery attack on July 12, 1917, by the Germans. This agent caused the most casualties of any agent used during World War I.

It is estimated that close to 1,300,000 casualties were produced by approximately 125,000 tons of chemical warfare agents used by the combatants, but it is known that in many cases the official figures underestimate the true number of casualties. Furthermore, it is unclear to what degree the official figures include individuals who were injured in gas attacks but who developed serious symptoms only after the war; however, considering the estimated 10,000,000 battle deaths due to the war, it is arguable as to whether chemical warfare was more or less horrific than the other methods used. Nonetheless, a line had been crossed, and the use of toxicology to deliberately and indiscriminately inflict harm on humans was now part of governmental programs. The next 50 years produced some of the most lethal chemicals and combinations of chemicals imaginable. They will not be discussed here, as they were not used and are now limited by international treaty, although not all countries abide by the treaty.

A number of nonlethal materials developed during World War I were chemical irritants (lachrimators) that irritate the mucous membranes of the eyes, causing a stinging sensation and tears. They may also irritate the upper respiratory tract, causing coughing, choking, and general debility. These effects are short lived and rarely disabling, and these agents are used today by law-enforcement personnel. Tear gas has gained widespread acceptance as a means of controlling civilian crowds and subduing barricaded criminals. The most widely used forms of tear gas have been *o*-chlorobenzylidenemalononitrile and *w*-chloroacetophenone. Proponents of their use claim that, when they are used correctly, the effects of exposure are transient and of no long-term consequence, but exposure is difficult to control and is indiscriminate. They may not always be used correctly, and lethal toxic injury has been documented. In 1969, 80 countries voted to include tear gas agents among chemical weapons banned under the Geneva Protocol.

During this era, the need for laws to protect the public from unscrupulous purveyors of foods and drugs became apparent, and the tools required to prosecute against those laws were being developed. In the United States, Harvey Washington Wiley (1844–1930), physician and chemist, served as head of the Bureau of Chemistry of the U.S. Department of Agriculture from 1883 to 1912. His main goal was to provide effective food and drug legislation to protect the unsuspecting public. His efforts culminated in the first U.S. Food and Drug Act (1906), which has been expanded and forms the basis for much of food safety legislation worldwide. He issued a number of bulletins summarizing his studies of the effects of food chemicals in human subjects, tested in his "Poison Squad." He wrote, "Injury to public health, in my opinion, is the least important question in the subject of food adulteration, and it is the one which should be considered last of all. The real evil of food adulteration is deception of the consumer." Wiley also served as Director of Foods, Health and Sanitation for *Good Housekeeping* magazine from 1912 to 1930. He wrote a number of books, including *Principles and Practices of Agricultural Analysis* (three volumes, 1894–1897), *Foods and Their Adulteration* (1907), and *History of a Crime Against the Food Law* (1929).

In the 1910s and 1920s, the U.S. Radium Corporation produced glow-in-the-dark watch dials painted with radioluminescent paint consisting of zinc sulfide and radium 226. The paint was applied with a small brush. Many of the young women employed in this work added points to the brushes by licking them between applications, and they ingested a small quantity of radium each time. Nobody knew it was harmful, except the owners of the U.S. Radium Corporation and scientists who were familiar with the effects of radium because the scientific and medical literature contained ample information about the hazards of radium. Most people, however, thought radium was a miracle elixir that could cure cancer and many other medical problems; instead, radium accumulated in the bone marrow of the women, eventually producing bone cancer. Alice Hamilton (see below) worked on the case to obtain just compensation for the workers. The Consumers League and the news media, as represented by Walter Lippmann, served the process well. This is one of the first instances of an occupational hazard from radioactivity. Radium watches were manufactured into the 1950s, but with strict controls.

Mechanistic studies led to a better understanding of the toxic action of many chemicals and to the development of specific antidotes. A classic example is the development of British antilewisite (BAL, 2,3-dimercaptopropanol, dimercaprol, and, for security reasons, OX 217) as an antidote for lewisite ($CHCl{=}CHAsCl_2$), an arsenic-based gas that was synthesized too late to be used in World War I. Expanding on the work of Carl Voegtlin (1879–1969), a developer of the "arsenic receptor" in chemotherapy and National Cancer Institute director from 1938 to 1943, Rudolph Peters headed the Oxford University laboratory that searched for antidotes to chemical warfare agents. They developed BAL in 1940.

Understanding the mechanism of poisoning by organophosphorus compounds, the basis for many insecticides and chemical warfare agents, also led to a rational antidote. Atropine, a drug that blocks muscarinic acetylcholine receptors, counteracts the vomiting and diarrhea, excessive salivation and bronchial secretions, sweating, and bronchospasm. It is administered intravenously, if possible, in high doses at frequent intervals until signs of intoxication diminish. Pralidoxime chloride (2-PAM), a drug that reactivates the nerve-agent-inhibited cholinesterase, is administered at the same time. Diazepam or another anticonvulsant is administered in severe cases to control seizures and thereby prevent seizure-induced brain damage.

A biochemical mechanism for cyanide antagonism was described by Chen et al. [12,13]. They suggested using a combination of amyl nitrite, sodium nitrite, and sodium thiosulfate. Nitrite converts hemoglobin to methemoglobin, which in turn competes effectively for cyanide with the mitochondrial cytochrome oxidase complex. Cyanide is then removed from cyanomethemoglobin by intravenous sodium thiosulfate, which serves as a sulfur donor for rhodanese (thiosulfate sulfur transferase). Rhodanese accelerates cyanide detoxification by forming the nontoxic metabolite thiocyanate. This represented the development of one of the first antidotes based on knowledge of toxicological mechanisms. This combination of antidotes has stood the test of time and still represents one of the most efficacious antidotal combinations for the treatment of cyanide intoxication.

Josef Warkany (1902–1992), of Cincinnati's Children's Hospital Research Foundation, was the first person to demonstrate that exposures to environmental chemicals and dietary deficiencies and excesses can be responsible for the production of congenital malformations. Until that time, it was widely believed that birth defects were due to chance or "God's will." Warkany was born and educated in Vienna, Austria, and by 1930 had published 23 scientific papers. Warkany accepted a one-year fellowship at Cincinnati's Children's Hospital Research Foundation in 1932 and ended up staying for 60 years. In the late 1930s, Warkany and Rose Cohen Nelson began attempts to produce endemic cretinism in rats. Although they failed, they obtained a syndrome of congenital skeletal malformations that was even more interesting. More than 3 years of painstaking research were needed to show that the skeletal malformations were not caused by a dietary iodine deficiency in the mother before birth, as in endemic cretinism, but instead were due to a riboflavin deficiency in the diet fed the pregnant animals. At that time, medical scientists believed that malformations were always genetic in origin, and most were reluctant to believe that environment could have such a dramatic effect on fetal development. For his work in this area, Dr. Warkany is known as the father of teratology. His 1300-page textbook *Congenital Malformations* is a medical classic.

In 1937, Rolla Harger (1890–1983) developed the Drunkometer for testing drivers presumed to be under the influence. This was important, as Prohibition had ended in the United States in 1933 and at the same time cars became more available and attained higher speeds.

Occupational medicine and industrial toxicology were identified by Agricola and Paracelsus, systematized and advanced by the pioneering efforts of Bernardo Ramazzini, and further advanced by one of America's foremost physicians, Alice Hamilton (1869–1970). Physician and pathologist, she researched occupational diseases, publicized the hazards of industrial chemicals to workers, and wrote several books on industrial toxicology. She was the foremost woman occupational physician and industrial hygienist, the first woman faculty member of the Harvard Medical School, and the only woman to serve on the Health Committee of the League of Nations. She graphically described the history of industrial toxicology/occupational medicine in the United States in her autobiography, *Exploring the Dangerous Trades* [37]. Others who contributed significantly to industrial toxicology include Cecil Drinker (1887–1919), also of Harvard, who believed that toxicological information was accumulating very rapidly, that mechanisms of toxicity were being elaborated, and that exposure to chemicals was increasing due to advances in manufacturing. Ethel Browning (1891–1979), of Great Britain, received her doctorate in medicine in 1927 and wrote the *Toxicity of Industrial Organic Solvents* in 1937. Interestingly, this was the first book on this subject and was written when Browning had no personal occupational medical experience. Her other publications included *Ionizing Radiations* (1959), *Toxicity of Industrial Metals* (1961), and her greatest work, *Toxicity and Metabolism of Industrial Solvents* (1965).

In 1933, more than a dozen women were blinded and one woman died from using a permanent mascara called "Lash Lure," which contained *p*-phenylenediamine, an untested chemical that caused blisters, abscesses, and ulcers on the face, eyelids, and eyes of Lash Lure users. It led to blindness for some, and in one case the ulcers were so severe that the woman developed a bacterial infection and died. Although the factual basis of the story was disputed by the industry and never confirmed, the so-called incident provided a "smoking gun," and Congress passed legislation for a new Food, Drug, and Cosmetic Act in 1938 (the death of children by sulfanilamide in ethylene glycol was another crucial incident). Lash

Lure was the first product seized under its authority. The tragedy showed that a better way to test for eye and skin irritation was needed.

In 1944, John M. Draize, a U.S. Food and Drug Administration (FDA) scientist, standardized the scoring system of a preexisting test for ocular irritation. Frequently referred to as the Draize test, a liquid or solid substance is placed in one eye of a rabbit, and changes in the cornea, conjunctiva, and iris are observed and scored compared to the untreated eye. Despite differences between the rabbit eye and the human eye, the Draize test, when performed by trained personnel, has proven quite accurate in predicting human eye irritants, particularly slightly to moderately irritating substances, which are difficult to identify using other methods. The Draize test performs its primary function of assessing both the damage and potential for recovery after exposure to irritants very well. Many toxicologists are reluctant to reject the Draize test because it is predictive, reliable, and verified, but they recognize its scientific and humane shortcomings. A battery of *in vitro* alternatives may eventually replace the Draize test, but much more data are needed. Draize and his colleagues also standardized the scoring of skin reactions as a method to evaluate skin irritation or corrosion using rabbits. Attempts to replace this test with a battery of *in vitro* alternatives are in progress.

By this time, the discipline of toxicology was now recognized by the scientific community and society as a distinct entity, separate from pharmacology and drawing upon the chemical, biological, and physical sciences. The stage was set for the application of toxicological principles and findings to the protection of the public and especially workers from the adverse effects of chemical exposure. Consumers also needed protection from the potentially adverse effects of chemicals found in foods and other consumer products.

## POST-WORLD WAR II

> Irrespective of its location and source of identity, toxicology as a field must studiously cross many traditional boundaries. ...The science of toxicology is located somewhere between medicine and the sciences, but drawn toward law by forensic uses and the need to regulate various human activities. Societal values and needs tug the field more into the center of the tetrahedron.
>
> **James Gillett (1987)**

As society developed following World War II, the demands on toxicology as a science grew. Much of this demand was a result of the enormous growth of the chemical industry. New synthetic chemicals were being produced and new uses were discovered for older chemicals. The amounts of chemicals produced were greater than ever. The production, use, and disposal of chemicals were not always conducted in the best interest of humans or the environment. This led to new laws and regulations and a new type of activism. Advances in analytical chemistry and the biomedical sciences also dramatically impacted toxicology. Significant contributions to the advancement of toxicology came from academia, government, and industry, but some of the issues raised outstripped the capacity to generate data and form coherent theories. Some of the questions being asked of toxicology helped move the field forward, while others were beyond the scope of any science. Some responses to those challenges have been to the detriment of the science of toxicology, and, as ever, chemicals were still intentionally used to harm others.

### Basic Science

Academic, industrial, and governmental research laboratories and private research foundations advanced the frontiers of toxicology by seeking the molecular basis for toxic action. Almost all the work in the field of xenobiotic biotransformation grew from the life-long research of a Welshman, Richard Tecwyn Williams (1909–1979). His early work on determining the ring structure of glucuronic acid by isolating bornyl glucuronide from the urine of dogs fed borneol allowed him to crystallize the conjugate and use it as a source material for his elucidation of its pyranoid structure in 1931. This work stimulated his interest in the metabolism of foreign compounds and led to a series of papers on the fate of phenols, terpenes, and sulfonamides. Williams became convinced that the biotransformation of foreign compounds was just as important as the metabolism of natural body constituents. During the late 1930s, Williams began writing a book on the detoxication of foreign compounds but, because of the war, it was not published until 1947. It was a slim volume summarizing much of the work that had been done to date. In the 1950s, studies on a broad range of compounds added considerably to the systemization of the biotransformation routes of xenobiotics, culminating in publication of an expanded *Detoxication Mechanisms* in 1959 which provided a systematic approach to biotransformation based on organic chemistry classification. Williams also expanded his concepts of the principal biochemical reactions whereby drugs and other foreign compounds are biotransformed in the body. He proposed that foreign compounds were biotransformed in two distinct phases: one of oxidation, reduction, and hydrolysis and the other of conjugation reactions.

A major class of oxidative transformations was characterized in 1955 by O. Hayaishi in Japan and H.S. Mason in the United States. This class of oxygenases had requirements for both an oxidant (molecular oxygen) and a reductant (reduced NADP) and hence was given the trivial name "mixed-function oxidases." The chemistry of these reactions grew out of studies on liver pigments by D. Garfinkel and M. Klingenberg who, in 1958, observed in liver microsomes an unusual carbon monoxide binding pigment with an absorbance maximum at 450 nm. This pigment was characterized as a cytochrome with typical absorption bands by T. Omura and R. Sato in Japan in 1964 through the use of detergent solubilization of microsomes and interaction with isocyanide ligands. O. Rosenthal, D.Y. Cooper, and R.W. Estabrook in 1965 studied the metabolism of codeine, monomethyl-4-aminopyrine, and acetanilide and found them to be inhibited by carbon monoxide, and the inhibition could be reversed by yielding the same action spectrum, demonstrating that cytochrome P450 is the oxygen-activating enzyme in xenobiotic metabolism as well as in steroid hydroxylation. From the isolation of membrane-bound P450 by A.Y. Lu and M.J. Coon in 1969 to the first crystallization of a mammalian P450 in 1999 by E. Johnson and coworkers, this area of research has established the important role of P450s in the disposition of drugs and other xenobiotics. In the years since the identification of cytochrome P450 as the terminal component of oxygenation reactions, the field has grown from one of narrow interest to toxicologists and pharmacologists to one that has captured the attention of molecular biologists, biochemists, and physicians.

A key figure in the xenobiotic biotransformation was James R. Gillette (1928–2001), who worked with Bert LaDu, Jr., at the Laboratory of Chemical Pharmacology/Heart and Lung Institute at the National Institutes of Health (NIH), where B.B. Brodie was chief. Gillette's studies on cytochrome P450 were influential, and he succeeded Brodie as chief of the laboratory in 1972. He had a prolific career and published around 300 papers and chapters and coedited seven books. His vision in the fields of drug biotransformation and pharmacokinetics earned him numerous honors and awards.

James A. and Elizabeth C. Miller (1915–2000; 1920–1987) made seminal discoveries related to the biotransformation of synthetic and naturally occurring chemicals to toxic or carcinogenic electrophilic metabolites and to the regulation of xenobiotic metabolism. They developed the important unifying concept that most carcinogenic and mutagenic chemicals are not carcinogenic or mutagenic *per se* but must undergo biotransformation to reactive electrophilic intermediates that exert their effects by covalently binding to critical sites on cellular macromolecules. Their work began in the late 1940s, when they demonstrated that a foreign chemical could be biotransformed to intermediates that covalently bind to macromolecules. The administration of hepatocarcinogenic aminoazo dyes to rats resulted in the covalent binding of metabolites to protein in the liver. Little or no covalent binding occurred in nontarget tissues that did not exhibit tumorigenesis. The Millers found that factors influencing the *in vivo* binding of aminoazo dyes to protein also influenced hepatocarcinogenicity, which led them to suggest that covalent binding of metabolites to liver macromolecules was required for carcinogenicity. This line of thinking was extended to carcinogenic polycyclic aromatic hydrocarbons when they found that metabolites covalently bound to protein only in the skin.

James Miller demonstrated the oxidation of a foreign compound in a cell-free system by enzymes that were later identified as cytochrome P450. He demonstrated that liver microsomes reduced the azo linkage of 4-dimethylaminoazobenzene and that NADPH was required for catalytic activity. He also reported that flavin-adenine dinucleotide was required for azo dye reductase activity, and these results provided a mechanistic explanation for the protective effect of riboflavin on the carcinogenicity of aminoazo dyes. These observations suggested that a dietary vitamin can inhibit the carcinogenicity of a chemical by influencing its biotransformation (an early example of cancer chemoprevention). In the 1950s, Miller continued to make fundamental discoveries on the properties of enzyme systems that biotransform foreign chemicals. He discovered that the *N*-demethylation of an aminoazo dye by liver homogenate was an oxidative process. This study provided the basis for subsequent investigations that demonstrated the NADPH-dependent oxidative metabolism of drugs and carcinogens by liver microsomes. In 1957, the Millers made the important observation that the demethylase system was inhibited by carbon monoxide. Although they did not pursue this line of research, these early observations and the other studies described above helped pave the way for the discovery of cytochrome P450.

The Millers discovered that foreign chemicals can induce the synthesis of liver enzymes that biotransform the compound administered and other foreign chemicals. Studies on microsomal enzyme induction provided a mechanistic understanding of the inhibitory effects of certain polycyclic aromatic hydrocarbons on aminoazo dye carcinogenesis. The induction of these enzymes is important toxicologically because it leads to an accelerated biotransformation of drugs and environmental chemicals *in vivo* and so alters their action and toxicity.

In the 1960s, the Millers elucidated the molecular events leading to the activation of 2-acetylaminofluorene, aminoazo dyes, aflatoxin $B_1$, safrole, estragole, and ethyl carbamate to chemically reactive products that react with macromolecules in cells. The later studies were the start of research on the biotransformation of naturally occurring carcinogens in our diet. Studies on the activation of several structurally diverse carcinogens led to the important unifying concept proposed in 1969 by the Millers that most carcinogenic and mutagenic chemicals are not carcinogenic or mutagenic *per se*, but these compounds must undergo biotransformation to reactive electrophilic intermediates that exert their toxic effects by covalently binding to critical sites on DNA, RNA, and protein. This discovery laid the foundation for subsequent research indicating a relationship between the mutagenicity of chemicals after activation and initiated a new era of carcinogenesis research that led to the development of rapid mutagenicity tests for the screening of potential human carcinogens. James and Elizabeth Miller contributed extensively to the characterization of DNA adducts that resulted from the biotransformation of several chemical. They also trained many students who made significant contributions to science.

Other early studies in the field of carcinogenesis took a different approach. In the 1940s, J.C. Mottram, Isaac Berenblum, and Philippe Shubik studied the development of tumors in mouse skin; their findings gave rise to the two-stage initiator–promotor model. Cancer models involving two or more stages helped researchers understand carcinogenesis at the level of the whole organ and serve as a basis for the classical model of carcinogenesis, which now includes initiation, promotion, and progression. The initiation phase has traditionally been described as involving the induction of mutations and escape from DNA repair. Cell proliferation in the promotion phase plus additional genetic events and angiogenesis in the progression phase are required for the process of carcinogenesis to result in cancer. Roswell Boutwell, Stuart Yuspa, Henry Hennings, Thomas Slaga, and others have contributed to this field.

In the mid-1950s, Peter Armitage and Richard Doll used the two-stage model to investigate the age distribution of human cancer incidences by using a simple power function of age. Later, Armitage and Doll and Suresh H. Moolgavkar found better equations to fit cancer incidence data. Richard Peto and Doll used epidemiologic data to estimate the fundamental causes of human cancer.

Ernst Wynder (1922–1999), born in Germany and raised in New Jersey when his family fled to escape Nazi persecution, attended medical school at Washington University, St. Louis. During a summer internship at New York University, his curiosity was piqued during the autopsy of a two-pack-a-day smoker who had died from lung cancer. Wynder began collecting case histories of lung cancer victims, first in New York City and then in St. Louis. His research brought him to thoracic surgeon Evarts Graham, who, despite initial skepticism about Wynder's premise, granted access to his extensive case records, and agreed to sponsor the medical student. In 1950, the *Journal of the American Medical Association* published Wynder and Graham's "Tobacco Smoking as a Possible Etiologic Factor in Bronchiogenic Carcinoma: A Study of 684 Proven Cases." Wynder and Graham's retrospective study was not the first to link smoking and cancer (in 1950, Doll also demonstrated that smoking causes cancer), but its sophisticated design, impressive population size, and unambiguous findings demanded attention. During the next decade, hundreds of reports were published linking cancer and smoking, including large prospective studies and animal investigations. The preponderance of evidence convinced many doctors that the health risks of smoking were serious and led to the publication of the first Surgeon General's Report on Smoking and Health in January 1964, the first official recognition in the United States that cigarette smoking causes cancer and other serious diseases. This seminal report prompted a series of public health actions reflecting changes in societal attitudes toward the health hazards of tobacco use. Among the actions were banning tobacco advertising on broadcast media, raising the age of legal purchase, developing effective treatments for tobacco dependence, and issuing 27 Surgeon General's reports on such topics as environmental (second-hand) tobacco smoke, which led to the creation of smoke-free public places, restaurants, and bars.

Another husband–wife team was John and Elizabeth Weisburger, who worked at the National Cancer Institute (NCI) of the NIH. John studied the mode of action of chemical carcinogens, in general, and arylamines, in particular. Some of these chemicals were important occupational carcinogens, and more recently it has been discovered that they occur in cooked meats, as the cooking process generates heterocyclic amines. These products undergo a two-step biochemical activation to DNA-reactive chemicals: *N*-oxidation, usually in the liver of animals and humans, followed by *N*,*O*-acylation in target organs such as the liver, intestinal tract, mammary gland, and pancreas. Between 1961 and 1972, as Director of the Bioassay Segment of the Carcinogenesis Contract Program Management Group of the National Cancer Institute, John Weisburger was also involved with testing methods for environmental and industrial compounds, including the role of dose levels and the species and strains of animals to be used in these tests. He introduced the F344 rat and the $B6C3F_1$ mouse as the standard animals. With the discovery that many carcinogens are mutagens, he organized national programs to develop rapid *in vitro* bioassay systems to test for carcinogenicity. When he left the NCI to become Director of Research at the American Health Foundation in 1972, the NCI Bioassay Program was transferred to the National Institute of Environmental Health Sciences (NIEHS) in North Carolina, where it continues today as the National Toxicology Program.

Elizabeth Weisburger identified biochemical pathways of malignant growth and mechanisms of carcinogenesis, and she synthesized reference metabolites and analogs of research carcinogens. In 1951, she was appointed to the NCI Laboratory of Biochemistry. Ten years later, along with John, she formed a research group to test for carcinogenic activity in environmental and industrial compounds. Elizabeth stayed at the NCI and became head of the Laboratory of Carcinogen Metabolism. She was appointed assistant director for chemical carcinogenesis in the NCI Division of Cancer Etiology in 1981, where she remained until her retirement in 1988.

During the late 1960s and early 1970s, many questions were being raised about rodent carcinogenicity testing: the use of the maximum tolerated dose (MTD), the appropriateness of the test species, the cost and amount of time required, and the significance, predictiveness, and reproducibility of results (these are still being asked!). A breakthrough of sorts occurred in 1975 when Bruce Ames and his colleagues at the University of California in Berkeley introduced a standard bacterial mutagenesis system utilizing a special type of *Salmonella*. The Ames test, as it is commonly known, is an easy, exquisitely sensitive biological method for measuring the mutagenic potency of chemical substances. The great interest in the Ames test was based on the proposition that any substance that is mutagenic to the bacteria may also be carcinogenic because the DNA bases are the same in all cells. The test is based on inducing growth in genetically altered strains of the bacterium *Salmonella typhimurium* that are unable to synthesize the amino acid histidine from the ingredients in its culture medium. When a test material is applied to the bacteria in the presence and absence of a mammalian microsomal enzyme system for metabolism, some undergo a back mutation ("revert") such that the bacteria can grow like the original "wild" (unaltered) strains without histidine, seen as visible colonies. By simply counting the colonies after a standard time (e.g., 48 hours) under standard growing conditions (37°C), the mutagenic potential of the parent compound and its metabolites can be estimated.

The test initially gave the impression that it correlated very highly with rodent carcinogenicity. This correlation diminished as more data became available; for example, many substances that caused cancer in laboratory animals did not elicit a positive response in the Ames test and *vice versa*. The Ames test alone does not demonstrate cancer risk, but the mutagenic potency does correlate with the carcinogenic potency for certain types of chemicals. Further, the ease and low cost of the test make it invaluable for screening substances in the environment and new substances in the laboratory. It also set a new paradigm by distinguishing between genotoxic carcinogens and nongenotoxic carcinogens.

James G. Wilson (1915–1987) is one of the fathers of modern teratology. He introduced a slicing technique (free-hand slicing) that standardized the examination of fetuses from teratology studies, especially the soft tissues. Fetuses used for the Wilson's soft-tissue sectioning technique are first fixed in Bouin's solution, which is a mixture of saturated picric acid, formaldehyde, and glacial acetic acid. This fixes the tissues, hardens the soft tissues, and softens the bones in order to preserve the specimens and make it possible to slice them cleanly into thin sections with a razor blade. The effect is much like taking a computed axial tomography (CAT) scan by hand except that the slices are examined under a dissecting microscope. All the internal organs are examined, and any abnormalities or developmental variations are documented. The sections can be saved for further examination, if desired. This technique has its limitations (e.g., the original coloration of the tissues is lost), but its simplicity made it very popular and it was accepted internationally. In 1973, Wilson outlined the four major manifestations of abnormal development: growth alterations, functional deficits, structural malformations, and death.

The growth in toxicology continued to attract more students and practitioners, and this resulted in more research. As with other recognized, independent scientific disciplines, toxicologists realized a need for a learned society to provide a forum for the exchange of the burst of new scientific information. The Society of Toxicology was founded in 1961. This was the first international society for and by toxicologists. Since its founding and as a result of the tremendous growth of toxicology, other societies of toxicology have been established, and almost every developed country has its own version—a testimony of the recognition of its importance and its growth. The need for appropriate journals in which to publish, and thus disseminate, the results of investigations was acute and noted by the Society of Toxicology. The journal *Toxicology and Applied Pharmacology* was founded by Fred Coulston; this was followed by *Fundamental and Applied Toxicology* (now *Toxicological Sciences*).Other journals included *Food and Cosmetic Toxicology* (now *Food and Chemical Toxicology*, founded by Leon Golberg), *Journal of Applied Toxicology*, and *Human and Experimental Toxicology*. Board certification began in the early 1980s to ensure a minimal level of competency among people calling themselves toxicologists.

### Regulatory

> The safety of the people shall be the highest law.
>
> **Cicero (106–43 B.C.)**

A prominent regulatory pharmacologist/toxicologist was Arnold J. Lehman (1900–1979), who earned his doctorate from the University of Washington in Seattle in 1930 and his medical degree from Stanford University in 1936. He taught at a number of universities and joined the FDA as Director of the Division of Pharmacology in 1946. He and his staff published *Appraisal of the Safety of Chemicals in Foods, Drugs and Cosmetics* in 1955, the first attempt by the agency to provide guidelines for toxicological studies (this presaged the Redbooks). He and his colleagues, most notably O. Garth Fitzhugh (1901–1994), laid the foundations of an acceptable intake of a material (the ADI) in 1952. They developed the concept of safety factors (a number applied to the highest dose that did not elicit an adverse effect in a properly designed and performed toxicological study, the no-observed-adverse-effect level, or NOAEL) for use in extrapolating animal data to humans. He is also renowned for the expression on his office wall, "You too can become a toxicologist in two easy lessons, each ten years long." He was succeeded by Leo Friedman (formerly of MIT), who advocated the use of *in utero* exposure in lifetime studies. Sandford Miller (also of MIT) was successor to Friedman. Other notable scientists at the FDA who contributed to the advancement of toxicology include Alan Rulis, Laura Tarentino, Antonia Mattia, Michael DiNovi, Herb Blumenthal, and Gary Flamm.

To assist toxicologists concerned with the safety of food and color additives, the FDA issued a series of guidelines, referred to as "Redbook I" and "Redbook II," to standardize the minimum requirements for the conduct of proper toxicological/safety studies. These guidelines are designed to encourage sound science and the conservation of resources while providing adequate data for determining safe exposure limits for consumers. Following the disclosure that a contract laboratory had falsified data used to support the safety of regulated materials, the FDA instituted good laboratory practices (GLPs) to be followed for the proper conduct of all toxicology studies to be submitted for regulatory review. The Redbooks and GLPs moved sound science into regulatory toxicological research.

The Organization for Economic Cooperation and Development (OECD) issued guidelines similar to those in Redbooks I and II (the two guidelines are generally consistent although there are some differences—for example, in the caging of test animals) that are now accepted internationally. Government/regulatory toxicologists have made significant contributions especially in the area of safety evaluation, including the quantification of risk. Regulatory agencies demand adequate data of high quality to serve as the basis for establishing safe exposure levels. The extent of testing was and is often determined by the nature of the chemical, its chemical and physical properties, and the extent of exposure; for example, if absorption, distribution, metabolism, and excretion (ADME) studies were conducted properly and it was determined that the test material was not absorbed from the gastrointestinal tract, then the need for long-term studies was not compelling. This allowed toxicologists who were conducting safety studies to design rational programs that addressed basic issues early, often resulting in considerable conservation of resources. The regulators were contributing to the advancement of toxicology.

At the international level, the World Health Organization (WHO), through the efforts of Frank Lu, Gaston Vettorazzi, and John Herrman, applied sound toxicological thinking to establishing safe exposure conditions for food chemicals including pesticides. They more thoroughly developed the concept of an acceptable daily intake (ADI) based on sound toxicological data and the proper use of appropriate safety factors. This concept, recognized and effectively used worldwide, has resulted in no significant problems with food chemicals, including pesticides, evaluated in this manner. The evaluations are conducted through the auspices of the International Programme on Chemical Safety (IPCS) and are implemented by the Joint Expert Committee on Food Additives (JECFA) and the Joint Meeting on Pesticide Residues (JMPR).

## Problems and Responses

Despite the successes of toxicology in understanding the mechanisms of toxicity and being incorporated into regulations that should have prevented or minimized health problems, a number of incidents occurred that influenced society and caused toxicology to be modified and take on new responsibilities. After issues arose about the proper use of food colors and other food additives in the 1950s, the U.S. Congress passed a significant revision to the Food, Drug, and Cosmetic Act requiring approval by the FDA of all new food additives based on safety before they could be added to foods. A noteworthy part of the law was the Delaney Clause, which stipulated that no additive could be deemed safe (i.e., given FDA approval) if it were found to cause cancer in humans or experimental animals. This section was initially opposed by the FDA and by scientists who agreed that an additive used at very low levels need not necessarily be banned because it may cause cancer at high levels. Proponents justified the clause on the basis that some cancer experts were not able to determine a safe level for any carcinogen.

The Delaney Clause has created problems for the food and chemical industries and for regulatory officials since the law was enacted. The trouble began on November 6, 1959, right before Thanksgiving, when the FDA announced that residues of a pesticide called amitrole, a rodent carcinogen, had been found in cranberries and recommended that the public stop buying them. The scare passed quickly (but an entire year's sales were wiped out because people eat little cranberry sauce except at Thanksgiving and Christmas), tainted supplies were withdrawn, and new inspection procedures were put in place. Notwithstanding publicity critical of the FDA, this action had beneficial results, particularly in convincing farmers that pesticides must be used with care. The use of amitrole was phased out during the 1960s; nonetheless, the cranberry scare of Thanksgiving 1959 left an indelible residue of suspicion and worry in the public mind.

A month after the cranberry scare, federal officials learned that diethylstilbestrol (DES; a nonsteroidal synthetic estrogen) had been shown to cause cancer in laboratory animals. At that time, DES was widely used as an additive in chicken feed, and DES residues were measurable in chickens sold in grocery stores. Officials banned DES from chickens, and the DES story faded from the newspapers (only to reappear later due to its delayed carcinogenic effects in humans when mothers had used it to prevent miscarriage). DES was still permitted as a feed additive for beef and sheep because residues had not been measured in those animals due to analytical difficulties. Some criticized the FDA, claiming that it had stopped sampling to avoid getting squeezed between the meat industry and the Delaney Clause. The politics of toxicology was heating up.

The environment presented other problems. Minamata disease was the most massive pollution problem to strike Japan in the post-World War II period. Minamata is a small factory town dominated by the Nippon Chisso Corporation. From 1932 to 1968, Chisso dumped an estimated 27 tons of mercury compounds into Minamata Bay. The town consists of mostly farmers and fisherman, so, as Chisso Corporation dumped this massive amount of mercury into the bay, thousands of people whose normal diet included fish from the bay unexpectedly developed unusual symptoms. In May 1956, four patients suffering from an unknown disease were brought to the city hospital. They all had in common severe convulsions, intermittent loss of consciousness, repeated lapses into crazed mental states, and then finally permanent coma. Death was usually preceded by a very high fever. It was discovered that the same type of patients had been seen in the fishing villages surrounding Minamata City and that 17 people died after showing the same signs and symptoms. The illness became known as "Minamata disease," and eventually it was determined to be caused by methyl mercury. The same syndrome was discovered again at Niigata City, Japan, in 1965. The probable cause of the disease in Niigata was methylmercury from effluent from the Showa Denko Company's Kase factory, located on the upper reaches of the Agano River. The second occurrence of Minamata disease was recognized at an earlier stage, so fewer cases were reported. Both incidents were attributed to the production of acetaldehyde using mercury as catalyst. Methylmercury had been produced by plankton and accumulated in fish and shellfish. Those who ate the contaminated seafood developed methylmercury poisoning.

The long-term effects in those who did not die were sensory disturbances and constriction of the visual field, incoordination and walking difficulties, dysarthria, hearing problems, and tremors, but patients had various combinations of symptoms and various degrees of symptoms that ranged from mild to serious, and the population of patients with atypical symptoms was greater than those with typical symptoms. The total picture in relation to the epidemiology of the problem has yet to unfold because of the persistence of methylmercury in the environment and because of the long-term effects in children conceived at that time. These cases of organic mercury poisoning were the first known to occur through the food chain transfer of environmental pollution.

In the early 1960s, the drug thalidomide was used by some pregnant women in Europe and Canada as a sedative/hypnotic to treat morning sickness. At that time, it was not approved in the United States because Dr. Frances Kelsey of the FDA insisted that there was insufficient proof of the safety of the drug in humans. Women who took the drug in early pregnancy delivered children with severe birth defects such as missing or shortened limbs. Thalidomide was soon banned worldwide. A widely varying but recognizable pattern of limb deformities emerged. It is estimated that more than 10,000 children around the world were born with major malformations. The most well-known pattern was the absence of most of the arm with the hands extending flipper-like from the shoulders (phocomelia). Another frequent arm malformation, known as radial aplasia, was the absence of the thumb and the adjoining bone in the lower arm. Similar limb malformations occurred in the lower extremities. The affected babies almost always had both sides affected and often had both the arms and the legs malformed. In addition to the limbs, the drug caused malformations of the eyes and ears, heart, genitals, kidneys, digestive tract (including the lips and mouth), and nervous system. The first published suggestion of teratogenicity in humans was W.G. McBride's letter in *The Lancet* in 1961. Thalidomide is now recognized as a powerful human teratogen. Taking even a single dose of thalidomide during early pregnancy could cause major birth defects. As a result of the thalidomide tragedy, teratology studies are now a requirement for new drugs (and may be recommended for other chemicals). Thalidomide has recently been approved to treat the painful, disfiguring skin sores associated with leprosy and to prevent and control their return.

*Silent Spring*, written by Rachel Carson and published in 1962, offered the first look at widespread ecological degradation. It touched off an environmental awareness that still exists. The book focused on chemicals used in agriculture that sometimes led to high levels in the environment. Carson argued that those chemicals were more dangerous than radiation and that for the first time in history humans were exposed to chemicals that remained in their bodies from birth to death. Well-written and presented with thorough documentation, the book alerted a large audience to the environmental and human dangers of the indiscriminate use of chemicals. *Silent Spring* became a bestseller with international impact; spurred revolutionary changes in the laws affecting our air, land, and water; and remains a landmark work. Carson defended her book in the face of an assault from the chemical industry from the its publication until her untimely death in 1964.

Following the publication of *Silent Spring*, rising concern about the environment swept the nation's university campuses. The intensity of the discontent compared to that raised by the war in Vietnam. Earth Day, a national day of observance of environmental problems, was held in the spring of 1970. This was a nationwide grassroots demonstration on behalf of the environment. The American people finally had a forum to express their concern about what was happening to the environment, and they did so with exuberance. Earth Day is now an annual observance. As a result of Carson's book and the ensuing activism and in recognition of the problems in the land, air, and water (e.g., lead, dioxin, DDT, PCBs), President Richard Nixon established the U.S. Environmental Protection Agency, which was inaugurated on December 2, 1970.

Occupational issues began to emerge during the 1960s. In the United States and elsewhere, asbestos was widely used for its heat-resistant characteristics in a wide range of building materials (roofing shingles, ceiling and floor tiles, paper products, and asbestos cement products), friction products (automobile clutch, brake, and transmission parts), heat-resistant fabrics, packaging, gaskets, and coatings. Its usage peaked during World War II and into the 1970s. During the late 1960s, evidence emerged indicating that asbestos fibers were a health risk. Breathing high levels of asbestos fibers for a long time could result in asbestosis, a serious disease that can eventually lead to disability and death. Breathing asbestos also increased the risk of lung cancer, mesothelioma, and possibly cancers in other parts of the body (stomach, intestines, esophagus, pancreas, and kidneys). During the 1980s, the concern regarding asbestos resulted in the spawning of a new industry, asbestos removal and abatement.

In the late 1960s, implications surfaced regarding a cover-up by the chemical industry about the adverse effects of vinyl chloride, a major commodity. Early research conducted by producers and users of vinyl chloride focused on its toxicological properties. Later studies investigated the potential for chronic toxicity and carcinogenicity. Carcinogenic responses were observed in long-term rodent inhalation studies at almost the same time that case reports were published on a finding of a rare cancer, hepatic angiosarcomas, in workers exposed to high levels of vinyl chloride. More stringent occupational exposure limits were instituted and further research on vinyl chloride was initiated, including epidemiological studies of workers, animal carcinogenicity bioassays, and mechanistic investigations. The studies firmly established an association between prolonged exposure to high levels of vinyl chloride and angiosarcomas of the liver. More detailed investigation showed that workers who inhaled high levels of vinyl chloride for several years had altered liver function, nerve damage, poor blood flow in the hands, and unusual immune reactions. Animal studies showed that long-term exposure to vinyl chloride might adversely affect male reproductive organs. Vinyl chloride is now classified as a known human carcinogen.

The carcinogenicity findings revealed marked differences in potency between humans and rodents. Research on the metabolic kinetics and molecular dosimetry of vinyl chloride and its biotransformation products provided a basis for reconciling the species differences in potency and provided a mechanistic basis for the very specific carcinogenic response, hepatic angiosarcomas. The research conducted on vinyl chloride may be viewed as a success story for mechanistic-based findings and their importance in establishing appropriate health protective standards. This work was seminal, and Perry Gehring and Phil Watanabe of Dow Chemical, among others, brought a new level of understanding and importance of the role of biotransformation and pharmacokinetics in toxicity including the dose–response relationship. More stringent exposure standards have been effective in protecting workers. Moreover, the research approach used with vinyl chloride has served as a template for evaluating the toxicity and carcinogenicity of other chemicals.

As a result of workplace problems involving asbestos, vinyl chloride, and other chemicals, Congress passed the Occupational Safety and Health Act of 1970, a new effort to protect workers from harm. The Act established for the first time a nationwide, federal program to protect almost the entire work force from job-related injury, illness, and death. The Occupational Safety and Health Administration (OSHA) was established within the Labor Department to administer the Act, effective April 28, 1971. Building on the Bureau of Labor Standards, the new agency took on the difficult task of creating a program that would meet the legislative intent of the Act.

Another potential food safety issue arose in 1969. Eleven years prior, cyclamate (a non-nutritive, high-intensity sweetener) was classified as generally recognized as safe (GRAS). Cyclamate was initially marketed as a tabletop sweetener for diabetics. Later, a mixture of cyclamate and saccharin, which had been found to have synergistic sweetening properties and improved taste, was marketed for use in special dietary foods. Diet soft drinks were introduced that used a cyclamate/saccharin blend. The market grew rapidly and soon accounted for about 30% of soft drink sales. In 1969, the results of a chronic toxicity study of a mixture of cyclamate and saccharin was interpreted (in the United States) as implicating cyclamate as a bladder carcinogen in rats. Cyclamate had its GRAS status removed, and in 1970 it was banned from use in foods, beverages, and drugs, as it remains today. Many other countries did not act on these incomplete data, and cyclamate continues to be used as a sweetener in those countries. (Saccharin remained an approved sweetener in the United States.)

Because of uncertainty of the safety of GRAS compounds in general, the FDA commissioned the Select

Committee on GRAS Substances (SCOGS) of the Life Sciences Research Office (LSRO) of the Federation of American Societies for Experimental Biology to conduct scientific reviews of hundreds of substances that were then used as food ingredients. Its charter was to review the available scientific literature and made recommendations to the FDA. Over a period of 10 years, SCOGS forwarded to the FDA detailed reports on 468 food substances (of which 422 were direct ingredients). This Select Committee, after creating an array of five standardized recommendations, concluded that 72% of the food substances under review should remain GRAS, and only 1% should immediately become subject to food additive requirements.

Dioxin (2,3,7,8-tetrachlorodibenzodioxin [TCDD] and its congeners) first achieved notoriety in the 1970s when it was discovered as a contaminant in some batches of Agent Orange, an herbicide to which armed forces personnel had been exposed when it was used to defoliate trees in large areas of Vietnam that would otherwise provided cover for the enemy. Dioxin is unintentionally produced by many processes, such as the manufacture of certain industrial chemicals and pesticides, as well as the chlorine bleaching process of pulp and paper mills. It is also produced by burning of wastes and forest fires. Dioxin is highly lipophilic, resists environmental degradation, and in some species of laboratory animals is very acutely toxic. It also causes cancers in animals and possibly humans. It continues to generate concern because of its widespread distribution as an environmental contaminant, its persistence within the food chain, and its toxicity. For these reasons, it has driven many aspects of food, environmental, and forensic toxicology, even though interpretation of its effects at low levels has been controversial. It is interesting to note that humans appear to be among the least sensitive species studied.

Most of our information about the effects of dioxin in humans comes from occupational accidents. Workers exposed to dioxin after a March 8, 1949, explosion at a Monsanto plant in Nitro, West Virginia, developed skin lesions (chloracne), eye irritations, headaches, dizziness, and breathing problems in the immediate aftermath of the incident. On July 10, 1976, an explosion at an Icmesa factory in Seveso, Italy, released 1.3 kg of dioxin into the air. The residents of the area were not evacuated immediately after the accident. Studies later confirmed that the residents exhibited the highest levels of dioxin ever found in human serum and that the soil in the zone was heavily contaminated. Epidemiological studies of the residents of Seveso over a quarter of a century indicate increases in certain cancers. Because of exposure to Agent Orange, the Veterans Administration set up the Agent Orange Registry, a health examination program for Vietnam veterans who were concerned about the possible long-term medical effects of exposure to Agent Orange. The National Academy of Sciences, in its 1994 report on Agent Orange, concluded that individual dioxin levels in Vietnam veterans are usually not meaningful because of background exposures to dioxin, poorly understood variations among individuals in dioxin metabolism, relatively large measurement errors, and exposure to herbicides that did not contain dioxin. Thus, the Veterans Administration will treat a number of diseases presumed to have resulted from exposure to herbicides like Agent Orange (with some limitations): chloracne or other acneform disease consistent with chloracne (occurring within one year of exposure to Agent Orange), Hodgkin's disease, multiple myeloma, non-Hodgkin's lymphoma, acute and subacute peripheral neuropathy (temporary peripheral neuropathy that appears within weeks or months of exposure to an herbicide agent and resolves within 2 years of the date of onset), porphyria cutanea tarda (within 1 year of exposure to Agent Orange), prostate cancer, respiratory cancers, and some soft-tissue sarcomas.

Environmental contamination by dioxin has been extremely controversial. Dioxin was implicated in vague illnesses in Love Canal, New York. Love Canal, near Niagara Falls, was built on and around a chemical waste site. Some epidemiologists claimed high rates of cancers and birth defects in the town, and the residents were evacuated in 1980. Some houses were torn down and the rest boarded up. Careful studies subsequently showed that diseases in Love Canal were exactly what would be expected in a community of that size. Some parts of the area are now repopulated. The entire town of Times Beach, Missouri, was purchased by the federal government in 1983 and bulldozed because of dioxin in the soil in unpaved roads (and nowhere else). The buyout caused the only biological effect ever identified at Times Beach: huge populations of wild turkey and deer in the area of the former town. In 1996, the EPA announced the purchase of 158 homes and 200 apartments in the Escambia section of Pensacola, Florida, and relocated their residents because dioxin-like chemicals are present in the soil of a former wood treatment plant. The EPA, worried about the chemicals possibly contaminating ground water, dug up the soil, and covered it with plastic. The residents demanded that their homes be purchased and that they be relocated, but they refused examinations by U.S. Public Health Service doctors.

So, even though little hard evidence exists regarding the adverse effects in humans from dioxin in the environment, toxicology has been challenged to deal with it by the public. The controversy over dioxin involves the extrapolation of exposure to high levels (such as occupational or accidental) to the low levels generally found in the environment. For most people, the major exposure to dioxin is from food, mainly dairy and meat. This knowledge has led to increased surveillance of the food supply with resulting decreases in levels. Levels in the environment have been lowered tenfold through new manufacturing methods and controlled incineration; nonetheless, the concern remains.

In the early morning hours of December 3, 1984, 200,000 people in Bhopal, India, were exposed to methyl isocyanate gas from a Union Carbide plant. The 90-minute exposure resulted in at least 2500 deaths and countless cases of severe eye and lung damage. Most of the deaths were caused by pulmonary edema (excess accumulation of fluid in the lungs) or its effects. Medical treatment of the victims was handicapped by a lack of information on the effects of methyl isocyanate in human beings. The long-term effects on the residents of Bhopal are still being tallied.

During 1989, environmental groups waged an aggressive campaign to ban the spraying of daminozide (trade name Alar™), a chemical used on apples to promote uniform ripening and prolong shelf life. They wanted it banned because a breakdown product, unsymmetrical dimethyl hydrazine (UDMH), had been shown to cause liver tumors in mice and was portrayed as posing a cancer risk to humans, especially children. The core of the dispute was in the risk figures and risk interpretations being used by each organization. Dimethyl hydrazine was a rodent carcinogen only at doses exceeding the maximum dose that can be tolerated by test animals during their lifetime, a dose equivalent to humans drinking 4000 gallons of apple juice per day for life. The Alar controversy once again heightened people's awareness—and anxiety—about cancer risks of manmade chemicals in our environment. After hearing charges from the Natural Resources Defense Council (NRDC) that eating Alar-containing apples significantly increased a child's risk of developing cancer, numbers of school districts dropped apples from their menus and parents poured apple juice down the drain. Apple sales plummeted. The NRDC's charges, which were disseminated by a well-planned and effective public relations campaign, brought counter-charges from the EPA, which accused the NRDC of basing its study on poor data, among other things. Nonetheless, the manufacturer, Uniroyal Chemical Co., Inc., gave in to public pressure and requested the EPA to voluntarily cancel all food-use registrations of the pesticide.

Not all environmental problems are the result of industrial misconduct. Wells drilled in Bangladesh to provide fresh, clean water to the local people have resulted in arsenic poisoning on a massive scale. Efforts are underway to reduce the arsenic concentration in the water, but this will take years.

## Safety and Risk Assessment

Predicting what will happen in the future is the hallmark of any science. Given the same starting conditions, what happened before should happen again. It works with chemistry and physics, why not in toxicology? Reproducibility may occur in the laboratory, but the public does not care about laboratories. The questions they pose are about different species, different routes of exposures, different exposure levels, and different exposure periods. How, then, can a toxicologist explain to the public what may or may not happen under conditions that have not been tested and without theories that are as solid as in other sciences?

The elements of health risk assessment can be traced back to when early humans had to make choices about what to eat: dangerous animals vs. less-than-edible plants vs. starving. There is no historical record that allows us to know all the thinking behind the decisions made, but one would assume that those who failed to successfully conduct these assessments fared poorly. During the Renaissance, the foundations for the current approaches to health risk assessment were being established with the development of modern science and the introduction of probability into mathematics. Blaise Pascal and Pierre de Fermat invented probability theory in the mid-1600s to answer questions about gambling. Probability allowed people to make decisions about the future based on mathematics and in so doing moved human beings further away from the whims of the gods.

During the early decades of the twentieth century (1900 to 1940), qualitative understanding of health risks improved as scientists learned of the hazards of occupational exposure to the chemicals then routinely used in the workplace. This knowledge was often successfully extrapolated to other forms of exposure, but almost all of it used data generated accidentally, intentionally, or unintentionally in humans. With more chemicals being synthesized and exposure escalating, a different approach to generating toxicity data was needed. Animals were the obvious choice.

The emergence of modern safety assessment can be traced to the early 1950s with the concept of the NOAEL and ADI developed by Lehman and Fitzhugh and others. The basic premises have not changed in 50 years. For example, to determine the ADI, take the NOAEL (the highest dose tested that causes no adverse effects in a test species in a properly designed and executed study) and divide it by an appropriate safety/uncertainty factor (generally 100, to account for inter- and intraspecies differences). The resulting value, expressed as mg of chemical per kg body weight per day, is an amount that can be safely consumed for a lifetime by all segments of society. This approach has withstood the test of time and has effectively protected the public.

Risk assessment for carcinogens started to take a different route in the 1960s. It tried to take information from lifetime studies in male and female rats and mice administered a material at the maximum tolerated dose (MTD) and extrapolate it to millions of people exposed at levels orders of magnitude lower. Since then, the considerable knowledge gained from scientists in many disciplines has been slowly transferred to regulators, who often take a lead role in formulating improvements in various risk assessment methods. Unfortunately, standardizing assessments resulted in inflexibility, numerous levels of conservatism, and ultimately the inability to properly characterize likely risks because the layers of default assumptions caused calculations to be unrealistic.

These efforts were often based on theories that were not always well tested. For example, the most basic tenet of toxicology, the dose–response relationship, with its origins dating back to Paracelsus, is apparently not as well understood and appreciated as toxicologists might think. What exactly is a dose, and is the nature of the response linear or curved? Is there a threshold for every effect? Further, is the response always in one direction; that is, as the dose is increased, is the severity of the response increased? Should hormesis be considered in risk assessment? Edward Calabrese, at the University of Massachusetts, has taken the ideas of Mithridates to another level and proposed that some "toxicants" in small amounts may actually confer benefits rather than harm. This concept has given rise to a great deal of debate, and much more data are needed. How risk assessments are eventually conducted, by science or dogma, will greatly influence public policy. Much more research into these basic concepts of toxicology and risk assessment is necessary.

## Intentional Poisonings: Politics, Mysticism, and Revenge

One might think that intentional poisonings would have disappeared as analytical and forensic procedures improved, but they continued; in fact, in some ways, they became worse than ever. We note here a few that made an impact either on world history or the science of toxicology. The ultimate chemical genocide was carried out by the Nazis during World War II, when they used cyanide, carbon monoxide, and engine exhaust to kill millions of Jews and other groups deemed to be undesirable. Between 1942 and 1945, the Nazis also produced 12,000 tons of tabun (an organophosphate, ethyl *N,N*-dimethylphosphoramidocyanidate) but fortunately never used it.

Suicide is usually an act of lonely desperation, carried out in isolation or near isolation by those who see death as an acceptable alternative to the burdens of continued existence (a permanent solution to a temporary problem). It can also be an act of self-preservation among those who prefer a dignified death to the ravages of illness or some perceived humiliation. It is even, occasionally, a political statement. But, rarely, if ever, is it a social event. On November 18, 1978, members of the People's Temple died at Jonestown, Guyana, after orders were issued by their leader, Rev. Jim Jones. Those who refused to voluntarily drink the cyanide-laced grape punch were forced to drink at gunpoint or be shot. Mothers administered the poison to their babies using eye-droppers. Of the 912 dead, 276 were children. Jones ordered the deaths after People's Temple members killed Congressman Leo Ryan, who had gone to Jonestown on a fact-finding mission.

On September 7, 1978, a Bulgarian dissident, Georgii Markov, was stabbed with an umbrella by an operative of the Bulgarian secret service on Waterloo Bridge in London. The umbrella fired a tiny pellet into his thigh. The pellet contained ricin and Markov died a few days later.

During the Iran–Iraq war, from July 1983 to January 1984, localized and limited Iraqi chemical attacks were carried out using mustard gas ($ClCH_2$–$CH_2$–S–$CH_2$–$CH_2Cl$). From February 1984 until the end of the war (1988), mustard and nerve agents were used on a large scale. The spring of 1988 saw the last major offensive, when Iranian troops broke through in Kurdistan. Iraq was on the verge of defeat and used gas to stop the Iranians. On March 16, 1988, the Iraqis killed from several hundred to 7000 people. A United Nations report later confirmed the use of poison gas on a civilian target by the Iraqis. This was the largest chemical weapons attack against a civilian population in modern times. The Halabja attack involved multiple chemical agents, including mustard gas, and the nerve agents sarin, tabun, and VX. The vast majority of the casualties resulted from the use of the mustard agents, and most of the medical and casualty treatment data obtained from the conflict pertain to mustard agents.

On March 20, 1995, a terrorist cult group released sarin, an organophosphate nerve gas developed in Germany in the 1930s, at several points in the Tokyo subway system, killing 12 and injuring more than 5500 people. The gas was concealed in lunch boxes and soft-drink containers and was released in commuter trains on three different Tokyo subway lines by Aum Shinrikyo ("Supreme Truth"). Sarin can be absorbed through any body surface, but vaporized sarin is mainly absorbed through the respiratory tract and conjunctiva. Because most victims encountered vaporized sarin, they showed ophthalmic and respiratory signs and symptoms. Many suffer from long-term effects. More deaths and injuries would have occurred had the sarin been more completely volatilized. This may be the first use of a weapon of mass destruction against civilians by a nongovernment entity.

In the United States, some states use chemicals as a means of execution. The use of cyanide in gas chambers has evolved to lethal injection. The three classes of drugs that are generally used in lethal injections are a general anesthetic to induce unconsciousness (e.g., sodium thiopental), a paralyzing agent to stop breathing (e.g., pancuronium bromide), and a cardiotoxic agent to stop the heart (e.g., potassium chloride).

Unfortunately, after hundreds of years, arsenic is still a common homicidal poison. More than a dozen members of Gustaf Adolph Lutheran Church, in New Sweden, Maine, became nauseated Sunday afternoon, April 27, 2003, shortly after drinking coffee and eating sandwiches and sweets at the church. The 78-year-old head usher died the next day from what Maine health officials identified as arsenic poisoning. Arsenic is available in rural areas of Maine, where it was once used as a pesticide. Maine State Police told reporters that arsenic was found only in a church coffee pot. The reason for the killing may never be known as the prime suspect committed suicide.

In late 2004, Viktor A. Yushchenko, a Ukrainian politician, appeared to be poisoned based on the appearance of a broad array of painful and disfiguring conditions that plagued him during the last several months of his presidential campaign. Yushchenko's blood dioxin level was reportedly more than 1000 times the upper limits of normal, and his initial severe abdominal pain suggested that he had eaten the poison. This poisoning must have been intended to serve as a message rather than to kill him because no cases of dioxin being lethal to humans have been recorded.

A case of potential poisoning in the Soviet Union that just recently came to light may have occurred in 1953 when Stalin was allegedly given warfarin by members of the Politburo. Supposedly they allowed him to bleed to death before calling for an ambulance.

A possible new age of toxicity is upon us with weapons of mass destruction being used by terrorists. Cyanide, ricin, and other chemicals are now seen in the light of mass killings. Twenty-five years after the ricin incident in London, six Arab men were arrested on January 5, 2003, during a raid by British police at two addresses in London based on a tip from the French Secret Services. Another man, 33 years old, was arrested days later. All were said to have attended an al-Qa'eda training camp in Afghanistan or received terrorist training in Chechnya and the Pankisi Gorge region of Georgia. In the apartment where the original six were arrested, several castor oil beans and equipment that could be used to process those beans were found, as were traces of ricin. The United States went to war against Iraq in March 2003 ostensibly to remove weapons of mass destruction (thought to be mostly nerve gases) from the hands of Saddam Hussein. He had used them against Iran, but little to none was found. Where this will take toxicology cannot be said.

## Training

> Even while they teach, men [they] learn.
>
> **Seneca**

As toxicology became a recognized scientific discipline, many training programs began. Although it was very difficult to develop programs that could address all of the many facets of toxicology, including chemistry and biochemistry, physiology and pharmacology, pathology, statistics, and epidemiology, excellent training programs were developed at some of the most prestigious universities. Examples included, in the United States, the University of California at Davis, the University of Chicago, Harvard University, Iowa State University, the University of Kansas Medical Center, the Medical College of Virginia, New York University, the University of Rochester, and Vanderbilt; in Germany, Freiburg, Hannover, Tubingen, and Wurzburg; in Sweden, Karolinska; in Denmark, Copenhagen; in Switzerland, Zurich; in Italy, Bologna, Milan, and Padua; in England, Guy's Hospital, London, and St. Mary's, Surrey; in Ireland, Dublin; and in Australia, Canberra. The continuing (and increasing) popularity of the study of toxicology is due to many factors, including its intellectual and emotional appeal. It is multidisciplinary in nature, challenging, and demanding, and its experimental findings are relevant and applicable to public health issues, including improving the quality of life and (it is hoped) extending it.

The need for a standard, modern textbook became evident. Although several texts were available, none appeared adequate. This issue was addressed by Louis J. Casarett (1927–1972) and John Doull. Casarett received his doctorate in 1958 from the University of Rochester, where he studied respiratory toxicodynamics and morphological changes following exposures to potentially toxic materials, especially polonium. In 1967, he moved to the University of Hawaii, where he developed a program in toxicology. His research involved drugs of abuse and pesticides. He was considered an excellent researcher and teacher. Doull received both his doctorate in pharmacology and his medical degree from the University of Chicago. He remained at Chicago for a number of years, then moved to the University of Kansas Medical Center, where he established one of the most outstanding programs in toxicology in the world (and attracted and trained a number of internationally famous toxicologists). Casarett and Doull published *Toxicology: The Basic Science of Poisons* in 1975. It is now in its sixth edition and is edited by Curtis Klaassen, a colleague of John Doull. The first comprehensive text to address the principles and practices of toxicology was edited by A. Wallace Hayes.

Many corporations developed centers of excellence in toxicology to study product and workplace safety and produced scientists of great renown. Some of these laboratories include DuPont's Haskell Labs (established in 1935), Dow Chemical (V.K. Rowe and Perry Gehring), and Union Carbide/Carnegie Mellon Bushy Run (Carroll Weil).

Contract toxicology laboratories also made a significant contribution to toxicology by providing unique opportunities for those interested in the pragmatic (applied) aspects of toxicology—namely, the conduct of appropriate tests to establish safe conditions of exposure. These studies must consider the latest developments and advances in toxicology and related disciplines, as well as the needs of regulators internationally. This is especially challenging in an era of increased international trade and harmonization. Laboratories such as Hazleton Laboratories (now Covance), founded by Lloyd Hazleton; Food and Drug Research Laboratories, founded by Ben Oser; Biodynamics, founded by Tom Russell (now known as Huntingdon Life Sciences, USA); and International Research and Development Laboratories, founded by Frank Wazeter also trained toxicologists that populated a number of influential positions in academia, industry, and government.

Toxicologists also worked with and within trade associations to assist industry (and ultimately regulatory authorities) in establishing safe limits of exposure by using the best science possible. These associations include the Flavor and Extract Manufacturers Association (FEMA), Cosmetic Toiletry and Fragrance Association (CTFA), and the International Association of Color Manufacturers (IACM). Toxicologists were also involved with organizations to promote the science, such as the International Life Sciences Institute, founded by Alex Malaspina and John Kirschman.

> Toxicology is the ultimate Renaissance science.
>
> **Gillett (1987)**

Toxicology continues to grow. Its critical position in society and the uniqueness of the issues it faces continues to attract (yes, to seduce by virtue of its attractiveness) some of the brightest minds. There is something for everyone: from the molecular to the macro, from the gene to the whole animal to the human, from SAR to QSAR. Toxicology's strengths derive from the integration of many chemical and biological sciences and supporting disciplines. Toxicology is also one of the few sciences in which academic, industrial, and regulatory scientists can and do effectively interact to protect the public. The importance of toxicology is recognized by governments worldwide. Toxicology has evolved from listing poisons to protecting the public, from simply identifying effects (qualitative toxicology) to identifying and quantifying human risks from exposure, and from observing phenomena to experimenting and determining mechanisms of action of toxic agents and rational management for intoxication. As Claude Bernard noted in 1896 [6]:

> Where then, you will ask, is the difference between observers and experimenters? It is here: we give the name observer to the man [human] who applies methods of investigation, whether simple or complex, to the study of phenomena which he [she] does not vary and which he [she] therefore gathers as nature offers them. We give the name experimenter to the man [human] who applies methods of investigation, whether simple or complex, so as to make natural phenomena vary, or so as to alter them with some purpose or other, and to make them present themselves in circumstances or conditions in which nature does not show them. In this sense, observation is investigation of a natural phenomenon, and experiment is investigation of a phenomenon altered by the investigator.

Toxicology has come a long way! As science continues to advance, toxicology will continue to draw from these advances in its constant quest to protect the public from harm.

## REFERENCES AND SELECTED READINGS

This is not meant to be an exhaustive list of citations for all the details in the chapter; rather, it is a list of some key books and articles that the reader may wish to consult to gain a fuller appreciation of the history and meaning of toxicology.


1. Accum, F. (1820): *A Treatise on Adulterations of Food and Culinary Poisons*. ABM Small, Philadelphia, PA.
2. Ackerknecht, E. H. (1982): *A Short History of Medicine*. The Johns Hopkins University Press, Baltimore, MD.
3. Albert, A. (1985): *Selective Toxicity*, 7th ed., Chapman & Hall, New York; 1st ed., Methuen, London, 1951.
4. Baas, J. H. (1889): *Outlines of the History of Medicine* (trans. by H. E. Handerson). J. H. Vail, New York.
5. Beeson, B. B. (1930): Orfila: pioneer toxicologist. *Ann. Med. Hist.*, 2:68–70.
6. Bernard, C. (1865/1957): *An Introduction to the Study of Experimental Medicine* (trans. by H. C. Greene). Dover, New York.
7. Bernstein, P. L. (1996): *Against the Gods: The Remarkable Story of Risk*. John Wiley & Sons, New York.
8. Bettmann, O. L. (1979): *A Pictorial History of Medicine*, Charles C Thomas, Springfield, IL.
9. Breathnach, C. S. (1987): Orfila. *Irish Med. J.*, 80:99.
10. Casarett, L. J. (1975): Origin and scope of toxicology. In: *Toxicology: The Basic Science of Poisons*, edited by L. J. Casarett and J. Doull. Macmillan, New York.
11. Castiglioni, A. (1941): *A History of Medicine* (trans. by E. B. Krumbhaar). Alfred A. Knopf, New York.
12. Chen K. K., Rose, C. L., and Clowes, G. H. A. (1933): Methylene blue, nitrites and sodium thiosulfate against cyanide poisoning. *Proc. Soc. Exp. Biol. Med.*, 31:250–252.
13. Chen K. K., Rose, C. L., and Clowes, G. H. A. (1934): Comparative values of several antidotes in cyanide poisoning. *Am. J. Med. Sci.*, 188:767.
14. Christison, R. A. (1845): *A Treatise on Poisons*. Barrington & Howell, Philadelphia, PA.
15. Clendening, L. (1942): *Source Book of Medical History*. Paul B. Hober, New York; Dover, New York, 1960.
16. Cook J. W., Hieger, I., Kennaway, E. L., and Mayneord, W. V. (1932): The production of cancer by pure hydrocarbons, Part I. *Proc. Royal Soc. Lond. (Biol.)*, 111:455–484.
17. Cope, Z. (1957): *Sidelights on the History of Medicine*. Butterworth, London.
18. Debus, A. G. (1999): *Paracelsus and the Medical Revolution of the Renaissance: A 500th Anniversary Celebration*. Paracelsus: Five Hundred Years, Three American Exhibits. National Library of Medicine, Washington, D.C.
19. Decker, W. J. (1987): Introduction and history. In: *Handbook of Toxicology*, edited by T. J. Haley and W. O. Berndt. Hemisphere, Washington, D.C.

20. Deichmann, W. B., Henschler, D., Holmstedt, B., and Keil, G. (1986): What is there that is not poison? A study of the Third Defense by Paracelsus. *Arch. Toxicol.*, 58: 207–213.
21. Doull, J. and Bruce, M. C. (1986): Origin and scope of toxicology. In: *Casarett & Doull's Toxicology: The Basic Science of Poisons*, 3rd ed., edited by C. D. Klaassen, M. O. Amdur, and J. Doull. Macmillan, New York.
22. DuBois, K. and Geiling, E. M. K. (1959): *Textbook of Toxicology.* Oxford University Press, New York.
23. Eckert, W. G. (1980): Historical aspects of poisoning and toxicology. *Am. J. Forens. Med. Pathol.*, 1:261–264.
24. Gallo, M. A. and Doull, J. (1991): History and scope of toxicology. In: *Casarett and Doull's Toxicology*, 4th ed., edited by C. D. Klaassen, M. O. Amdur, and J. Doull. Pergamon Press, New York.
25. Garrison, F. H. (1929): *An Introduction to the History of Medicine*, 4th ed. W. B. Saunders, Philadelphia, PA.
26. Gettler, A. O. (1953): The historical development of toxicology. *J. Forens. Sci.*, 1:1–25.
27. Gillett, J. (1987): *The ICET Newsletter.*
28. Glaister, J. (1954): *The Power of Poison*. William Morrow, New York.
29. Godon, B. L. (1959): *Medieval and Renaissance Medicine*. Philosophical Library, New York.
30. Goulding, R. (1978): Poisoning as a fine art. *SO Med. Leg. J.*, 46:6–17.
31. Goulding, R. (1987): Poisoning as a social phenomenon. *J. R. Coll. Phys. Lond.*, 21:282–286.
32. Gunther, R. T. (1959): *The Greek Herbal of Disocorides*. Hafner, New York.
33. Guthrie, D. A. (1946): *A History of Medicine*. J. B. Lippincott, Philadelphia, PA.
34. Haggard, H. W. (1933): *Mystery, Magic and Medicine*. Doubleday, Doran & Co., Garden City, NY.
35. Hamilton, A. (1925): *Industrial Poisons in the United States*. Macmillan, New York.
36. Hamilton, A. (1934): *Industrial Toxicology.* Harper & Brothers, New York.
37. Hamilton, A. (1943): *Exploring the Dangerous Trades: The Autobiography of Alice Hamilton, M.D.* Little, Brown, Boston.
38. Holmstedt, B. and Liljestrand, G. (1981): *Readings in Pharmacology.* Raven Press, New York.
39. Hueper, W. C. et al. (1938). Experimental production of bladder tumors in dogs by administration of beta-naphthylamine. *J. Ind. Hyg. Toxicol.*, 20. 46–84.
40. Hutt, P. B. and Hutt, P. B. II (1984): A history of governmental regulation of adulteration and misbranding of food. *Food Drug Cosmet. Law J.*, 39:2–73.
41. LaWall, C. H. (1924): *Four Thousand Years of Pharmacy.* J. B. Lippincott, Philadelphia, PA.
42. Lewin, L. (1920): *Die Gifte in der Weltgeschichte: Toxikologische, allgemeinverstandliche Untersuchungen der historischen Qhellen*. Springer, Berlin.
43. Lewin, L. (1929): *Gifte und Vergiftungen*. Stilke, Berlin.
44. Loomis, T. A. and Hayes, A. W. (1996): *Essentials of Toxicology.* Academic Press, San Diego, CA.
45. Macht, D. J. (1931): Louis Lewin: pharmacologist, toxicologist, medical historian. *Ann. Med. Hist.*, 3:179–194.
46. Massengill, S. E. (1943): *A Sketch of Medicine and Pharmacy.* S. E. Massengill, Bristol, TN.
47. McBride, W.G. (1961). Thalidomide and congenital abnormalities. *Lancet*, 278:1358.
48. Meek, W. J. (1954): *Medico-Historical Papers: The Gentle Art of Poisoning*. Department of Physiology, University of Wisconsin, Madison.
49. Mettler, C. C. and Mettler, F. A. (1947): *History of Medicine*. Blakiston, Philadelphia, PA.
50. Murphy, J. B. and Sturm, E. (1925): Primary lung tumors in mice following the cutaneous application of coal tar. *J. Exp. Med.*, 42:693–700.
51. Neuberger, A. and Smith, R. L. (1983): Richard Tecwyn Williams: the man, his work, his impact. *Drug Metab. Rev.*, 14:559–607.
52. Neuburger, M. (1910): *History of Medicine* (trans. by Ernest Playfair). Oxford University Press, London.
53. Olmsted, J. M. D. (1938): *Claude Bernard: Physiologist.* Harper & Brothers, New York.
54. Oser, B. L. (1987): Toxicology then and now. *Regul. Toxicol. Pharmacol.*, 7:427–443.
55. Osius, T. G. (1957): The historic art of poisoning. *Univ. Mich. Med. Bull.*, March:111–116.
56. Pagel, W. (1982): *An Introduction to Philosophical Medicine in the Era of the Renaissance*. Karger, Basel, Switzerland.
57. Peters, R. A., Stocken, L. A., and Thompson, R. H. S. (1945): British anti-lewisite (BAL). *Nature*, 156:616–619.
58. Ramazzini, B. (1713): *Diseases of Workers* (Latin text trans. by W. C. Wright).
59. Rhodes, P. (1985): *An Outline History of Medicine*. Buttersworth, London.
60. Rosenfield, L. (1985): Alfred Swaine Taylor (1806–1880), pioneer toxicologist—and a slight case of murder. *Clin. Chem.*, 31:1235–1236.
61. Sasaki T. and Yoshida, T. (1935): Experimentelle Erzeugung der Lebercarcinomas durch Fütterung mit o-Aminoazotoluol. *Virchows Arch. Pathol. Anat.*, 295:175–200.
62. Sigerist, H. E. (1958): *The Great Doctors: A Biographical History of Medicine*. Doubleday, New York.
63. Sonnedecker, G. (1976): *Kremers and Urdang's History of Pharmacy*, 4th ed. J. B. Lippincott, Philadelphia, PA.
64. Talbott, J. H. (1970): *A Biographical History of Medicine: Excerpts and Essays on the Men and Their Work*. Grone & Stratton, New York.
65. Temkin, C. L., Rosen, G., Zilboorg, G., and Sigerist, H. W. (1996): *Four Treatises of Theophrastus von Hohenheim Called Paracelsus*, edited by H. W. Sigerist. The Johns Hopkins University Press, Baltimore, MD.
66. Thomas, L. (1979): *The Medusa and the Snail*. Viking Press, New York.
67. Thompson, C. J. S. (1931): *Poisons and Poisoners, with Historical Accounts of Some Famous Mysteries in Ancient and Modern Times*. H. Shaylor, London.
68. Voegtlin, C., Dyer, H. A., and Leonard, C. S. (1923): On the mechanism of the action of arsenic upon protoplasm. *Public Health Rep.*, 38:1882–1912.
69. von Oettingen, W. F. (1952): *Poisoning: A Guide to Clinical Diagnosis and Treatment*. Paul B. Hoeber, Harper & Brothers, New York.

70. Williams, R. T. (1959): *Detoxication Mechanisms*. John Wiley, New York.
71. Wilson, J. G. (1973). *Environment and Birth Defects*. Academic Press, New York.
72. Wooton, A. C. (1910): *Chronicles of Pharmacy*. Macmillan, London.
73. Wynder, E.L. and Graham, E. (1950). Tobacco smoking as a possible etiologic factor in bronchiogenic carcinoma: a study of 684 proven cases. *JAMA*, 143:329–36.

# Notes

# 2 The Use of Toxicology in the Regulatory Process

*Barbara D. Beck, Edward J. Calabrese, Tracey M. Slayton, and Ruthann Rudel*

**CONTENTS**

## BACKGROUND

Regulatory toxicology is that area of toxicology directed at protecting public health by regulating exposure to potentially harmful materials. Historically, regulatory toxicology has developed in a manner that has reflected humankind's ability to relate exposure to certain agents with adverse health effects. Thus, early regulatory attention generally focused on preventing the acute effects of chemical agents, because these effects were observable and could be easily associated with exposure. Once the germ theory of disease was developed and the disease potential of human and animal waste was recognized, the disposal of these materials began to be regulated. Food and drugs were also the focus of early regulation due, no doubt, to the relative ease in associating acute health effects, such as food poisoning, with exposure to materials in the diet or in medications. Hutt [1] notes that adulteration of the food supply was a serious problem in the ancient world and quotes Pliny the Elder, who, writing in the first century A.D., said, "So many poisons are employed to force wine to suit our taste—and we are surprised that it is not wholesome!"

Occupational exposures were also an early focus of regulation, due again to the fact that the relationship between exposure and effect was often observable. Early industrial hygiene efforts were typically intended to prevent overt or frank effects of materials in the workplace. Some of the first observations of the effects of chronic human exposures to certain chemicals were also made in occupational settings. Hutt [1] notes that, during the sixteenth century, Paracelsus wrote about diseases characteristic of miners. The fact that certain chronic occupational hazards affected the exposed individual at the point of contact also made the connection between agent and effect easier to discern. The first epidemiological study linking human cancer to a specific cause is attributed to Sir Percival Pott, who identified occupational exposure to soot as being responsible for scrotal cancers in young British chimney sweeps [2].

The development of regulatory toxicology during the twentieth century and up through the present has continued to shadow the ability to detect both chemicals and effects; that is, as we have become able to detect chemicals at lower and lower levels and to detect smaller biochemical and physiological changes, we have turned regulatory attention to these "new problems." For example, small increases in airway resistance following exposure to certain air pollutants are currently used as one basis for regulating these air pollutants; historically, no one would have been aware of these subtle effects. Similarly, guidelines for occupational exposures to benzene have decreased by two orders of magnitude from 100 ppm in 1927 to the current Occupational Safety and Health Administration (OSHA) standard of 1 ppm. Ambient criteria for nonoccupational benzene exposures are currently as low as 0.03 ppm in some states [3].

Because of the dramatic increase in our ability to detect smaller effects and lower concentrations, programs to regulate chemicals in the environment have increased at an astronomical rate during the last 30 years. Factors contributing to the recent increase in regulatory activity include the following:

- The realization of the vast number of chemicals that humans have dispersed into the environment and to which humans have been exposed. Approximately 70,000 commercial chemicals in commerce have been identified and listed under the Toxic Substances Control Act (TSCA) [4]. Advances in analytical chemistry have allowed part-per-billion levels of chemicals to be detected in "pristine" areas, in wildlife, in food products, and in human body tissues. This message was delivered initially by Rachel Carson in 1962 with the publication of *Silent Spring* [5]. Recently, the U.S. Centers for Disease Control and Prevention (CDC) [6] began a series of biomonitoring studies (i.e., the National Report on Human Exposure to Environmental Chemicals) of human serum and urine, demonstrating the presence in biological media of multiple exogenous chemicals, including anthopogenic compounds such as perfluorinated chemicals, as well as chemicals such as lead or cadmium, which have both natural and anthropogenic sources.

- The realization that historical chemical management practices might today be associated with low-level risks, even though such practices were consistent with the state of knowledge at the time. For example, during the 1970s, residents of Love Canal realized that they had unknowingly been exposed to chemicals that had migrated from a nearby landfill into their basements. The Comprehensive Emergency Response, Compensation, and Liability Act (CERCLA), also known as Superfund, was enacted shortly after the Love Canal incident.
- The establishment over the past 50 years of causal relationships between certain diseases and chronic chemical exposures, such as leukemia and benzene, or mesothelioma and asbestos.
- The reduction in illness and mortality due to microbial diseases and the improved standard of living, which have focused increasing attention on other causes of ill health.

The rapid increase in the number and complexity of regulatory programs to address potential health effects from chemical exposures is also a result of the increased scientific uncertainty about toxicology and risk that has evolved with our increased understanding of these subjects (i.e., the more we learn, the more we realize how much more there is to learn). As we have come to better understand the complexities of toxicology, we have developed more complex procedures for characterizing toxic responses, such as the use of probabilistic risk assessment methods (e.g., see the Monte-Carlo example provided later in this chapter).

In 1958, when the Delaney Clause forbidding the addition to food of any substance found to induce cancer in animals or humans was passed [7], the public generally believed that the intent of the law to provide a "zero risk" food supply was achievable. No one foresaw that, 20 years later, scientists would have identified over 500 animal carcinogens, been able to detect chemical concentrations between two and five orders of magnitude lower than they could detect in the 1950s, and found that many naturally occurring chemicals in food could be considered animal carcinogens [8]. These developments have forced the recognition that absolute safety is impossible to achieve, even in products regulated to ensure safety. Thus, Scheuplein [9] noted that "the vast improvement in our methods of analytical detection have [*sic*] exposed carcinogens in the food supply in amounts too minuscule for our carcinogen bioassays and our risk assessment procedures to evaluate with comparable precision. We are capable now, more than before, of asking questions that we cannot answer."

## CURRENT REGULATORY FRAMEWORK

This chapter focuses on regulatory approaches in the United States; however, the authors recognize the importance of the globalization of risk assessment and risk management procedures. We therefore provide examples of regulatory frameworks outside the United States, although we by no means presume that this discussion of frameworks outside the United States is anywhere near complete.

At the federal level in the United States, four agencies bear most of the direct responsibility for the regulation of toxic chemicals: the Consumer Product Safety Commission (CPSC), the Environmental Protection Agency (EPA), the Food and Drug Administration (FDA), and OSHA. Table 2.1 describes the acts that empower these and several other federal agencies.

It is clear from Table 2.1 that federal regulatory authorities are concerned with a broad range of chemical exposures. Chemicals may be regulated on the basis of environmental medium (e.g., air, water), activity (e.g., food manufacture, chemical transport, ocean dumping), or type of exposure environment (e.g., workplace, residential). Although the statutes in Table 2.1 represent about 100 years of federal legislative history, 19 of the 25 have been written since 1970 (and some of the earlier ones have been updated since 1970), illustrating the recent increase in public concern about chemical exposures.

The language of each statute provides the implementing agency with the basis for issuing regulations under the law. Some statutes instruct the agency to limit chemical release or exposure by requiring the use of certain control technologies. Other statutes require the agency to develop and implement risk-based standards, and still others require the agency to balance risks with the costs of regulating or the benefits of not regulating. The latter two types of statutes are the most likely to involve regulatory toxicology in their implementation.

Section 307 of the Clean Water Act (CWA) is an example of a statute that requires technology-based standards for pollution control. Under this portion of the CWA, industries discharging to surface water must use the best available control technology to limit their pollutant discharges; installation of the appropriate control technology is required from the discharger to obtain a National Pollutant Discharge Elimination System (NPDES) permit.

Other statutes specify the standard for safety that regulations and standards issued under the law are supposed to provide. A commonly cited example of a law that required health-based, or risk-based, standards for pollution control is Section 112 of the 1970 Clean Air Act

**TABLE 2.1**
**Federal Laws Related to Exposures to Toxic Substances**

| Legislation | Agency | Area of Concern |
|---|---|---|
| Food, Drug and Cosmetics Act (1906, 1938, amended 1958, 1960, 1962, 1968, 1976, 1996 [also known as the Food Quality Protection Act], 1997) | FDA | Food, drugs, cosmetics, food additives, color additives, new drugs, animal and food additives, and medical devices |
| Federal Insecticide, Fungicide and Rodenticide Act (1948; amended 1972, 1975, 1978, 1996) | EPA | Pesticides |
| Dangerous Cargo Act (1952) | DOT, USCG | Water shipment of toxic materials |
| Atomic Energy Act (1954) | NRC | Radioactive substances |
| Federal Hazardous Substances Act (1960; amended 1981) | CPSC | Toxic household products |
| Federal Meat Inspection Act (1967), Poultry Products Inspection Act (1968), Egg Products Inspection Act (1970) | USDA | Food, feed, color additives, and pesticide residues |
| National Environmental Policy Act (1970; amended 1975, 1985, 1989, 1996, 1997) | EPA | Ecosystems and natural resources |
| Occupational Safety and Health Act (1970; amended 1974, 1978, 1982, 1983, 1984, 1986, 1987, 1990, 1992, 1995, 1996, 1997, 1998, 2002) | OSHA, NIOSH | Workplace toxic chemicals |
| Poison Prevention Packaging Act (1970; amended 1981) | CPSC | Packaging of hazardous household products |
| Clean Air Act (1970; amended 1974, 1977, 1990) | EPA | Air pollutants |
| Hazardous Materials Transportation Act (1972) | DOT | Transport of hazardous materials |
| Clean Water Act (formerly Federal Water Pollution Control Act, 1972; amended 1977, 1978, 1987) | EPA | Water pollutants |
| Marine Protection, Research and Sanctuaries Act (1972) | EPA | Ocean dumping |
| Consumer Product Safety Act (1972; amended 1981) | CPSC | Hazardous consumer products |
| Lead-Based Paint Poison Prevention Act (1973; amended 1976) | CPSC, HEW (HHS), HUD | Use of lead paint in federally assisted housing |
| Residential Lead-Based Paint Hazard Reduction Act (1992) | EPA | Use of lead paint in all housing |
| Safe Drinking Water Act (1974; amended 1977, 1986, 1996) | EPA | Drinking water, contaminants |
| Resource Conservation and Recovery Act (1976; amended 1984) | EPA | Solid waste, including hazardous wastes |
| Toxic Substances Control Act (1976), Asbestos Information Act (1988) | EPA | Hazardous chemicals not covered by other laws, includes premarket review |
| Federal Mine Safety and Health Act (1977) | DOL, NIOSH | Toxic substances in coal and other mines |
| Comprehensive Environmental Response, Compensation, and Liability Act (1981), Superfund Amendments and Reauthorization Act (1986), Emergency Planning and Community Right-to-Know Act (1986) | EPA | Hazardous substances, pollutants and contaminants |
| Radon Gas and Indoor Air Quality Research Act (1986) | EPA | Indoor air |
| Oil Pollution Act (1990) | DOT | Oil pollution |
| Pollution Prevention Act (1990) | EPA | Toxics use reduction |
| Bioterrorism Act (2002) | FDA, CDC, USDA, EPA | Biological agents and toxins used in acts of war |

*Note:* CDC, Centers for Disease Control and Prevention; CPSC, Consumer Product Safety Commission; DOL, Department of Labor; DOT, Department of Transportation; EPA, Environmental Protection Agency; FDA, Food and Drug Administration; HEW, Department of Health, Education, and Welfare; HHS, Health and Human Services; HUD, Housing and Urban Development; NIOSH, National Institute of Occupational Safety and Health; NRC, National Research Council; OSHA, Occupational Safety and Health Administration; USCG, United States Coast Guard; USDA, U.S. Department of Agriculture.

(CAA), which required the EPA to set emission standards for hazardous air pollutants (under the National Emissions Standards for Hazardous Air Pollutants [NESHAPS] program) that would "protect public health" with an "ample margin of safety." Implementation of this standard of safety for carcinogenic air pollutants proved to be so troublesome for the agency that, between 1970 and 1990, NESHAPS were set for only seven air pollutants. The difficulty in setting the risk-based standards was that the statute provided no indication of what an "ample margin of safety" was or how such a concept might be applied to carcinogens, given that the agency considered carcinogens to act by a no-threshold mechanism. The 1990 amendments to the CAA replaced the NESHAPS health-based standards with specified technology-based standards for controlling hazardous air pollutants. The statute states that, after installation of the control technology, health-based standards must be set to further control emissions where unacceptable risks remain.

The Federal Food, Drug, and Cosmetic Act (FFDCA) is another example of a law requiring health-based standards for limiting the public's exposure to chemicals. Section 409 of the FFDCA requires the sponsor of a food additive to show that no harm to consumers will result when the additive is put to its intended use, and it contains the Delaney Clause (discussed earlier), a special provision that forbids the use of any food additives that have been found to induce cancer in humans or animals. The Delaney Clause essentially specifies that the acceptable risk from carcinogens as food additives is zero. This bright line has proven to be very difficult for both the FDA and (until passage of the Food Quality Protection Act in 1996) the EPA to implement, because the law does not allow the implementing agencies to specify *de minimis*, or acceptable, levels of risk. The EPA regulates pesticides under both the Federal Insecticide, Fungicide, and Rodenticide Act (FIFRA) and the FFDCA. Under the FIFRA, the agency was required to balance the risk from the pesticide with the benefit associated with its use; however, under the FFDCA, the EPA was bound by the Delaney Clause with regard to pesticides that may concentrate in processed foods above the level allowed on the raw agricultural commodity. This dichotomous standard (known as the "Delaney Paradox") forced the EPA to regulate to zero risk for pesticides applied to foods that may concentrate during processing, while regulating using risk–benefit analysis for the same pesticides on raw agricultural commodities [10]. In 1992, a Circuit Court ruled that the Delaney Clause does not allow the EPA to permit use of carcinogenic pesticides under the FFDCA, even if their use is associated with negligible risk [11]. The Delaney Clause has also been difficult for the FDA to implement as more chemicals (including naturally occurring chemicals in foods) have been determined to be carcinogens, and the FDA has searched for ways to establish acceptable risk levels for food additives. In 1988, the U.S. Court of Appeals for the District of Columbia struck down an effort by the FDA to interpret the Delaney Clause as allowing the agency to set a *de minimis* risk level for two color additives for use in cosmetics and drugs [12].

Other laws require the implementing agency to balance the risks and benefits of alternative regulatory choices. Examples of balancing statutes include Section 408 of the FFDCA, which, until the Food Quality Protection Act (FQPA) of 1996, required tolerances for pesticide residues on raw agricultural commodities to be set at levels necessary to protect the public health, while considering the need for "an adequate, wholesome, and economical food supply" [10]. Section 6 of the TSCA requires the EPA to consider the potential benefits of using a chemical and the economic consequences of restricting its use when determining whether the manufacture, distribution, use, or disposal of a substance presents an unreasonable risk of injury to health or the environment.

Language in the Occupational Safety and Health Act (OSH Act) specifies that the Agency must "adequately assure(s) to the extent feasible ... that no employee will suffer material impairment of health or functional capacity" [13]. This statutory language also requires balancing of risks and costs but is particularly interesting because the dual requirements of "feasibility" and "no employee will suffer" may be impossible to reconcile in certain situations [14].

Although the narrative terms "unreasonable risk" and "ample margin of safety" have not been clearly or consistently defined across agencies or across statutes, over the past decade agencies have generally interpreted this language as requiring a qualitative, and frequently quantitative, estimate of the health risks associated with an exposure and the reduction in risks resulting from regulatory action. A major factor in the increased use of risk analysis by regulatory agencies was the 1980 Supreme Court decision in *Industrial Union Department v. American Petroleum Institute*. In this case, OSHA proposed lowering the occupational standard for benzene from 10 ppm to 1 ppm on the basis that benzene was a carcinogen, that any reduction in exposure would result in a reduction in risk, and that 1 ppm was technologically feasible. The Supreme Court did not find for the Union, stating that, "Before he can promulgate any permanent health or safety standard, the Secretary [of Labor] is required to make a threshold finding that a place of employment is unsafe—in the sense that significant risks are present and can be eliminated or lessened by a change in practices" [15]. The Court left the decision of what constitutes a "significant risk" to OSHA. This landmark decision has had a major impact on agencies in addition to OSHA, resulting in an increase in the development and use of tools to quantify risks from exposure to environmental chemicals.

Perhaps one of the most far-reaching (and likely precedent-setting) statutes in the past 10 years is the FQPA of 1996. This statute, which addresses risks from pesticides in food through the setting of tolerance limits, is primarily risk based, with limitations on the extent to which the EPA can consider benefits. This is in contrast to the risk–benefit balancing requirements of the FFDCA noted earlier. Only in certain narrow circumstances under the FQPA can the EPA set pesticide tolerance levels that do not meet health-based criteria. Specifically, the circumstances include those situations where the use of the pesticide prevents even greater risks from occurring to consumers (a risk–risk balancing) or where the lack of the pesticide would result in "a significant disruption in domestic production of an adequate, wholesome, and economical food supply." In addition, the FQPA eliminates certain aspects of the "Delaney Paradox" discussed earlier. Tolerance limits for pesticides in both raw agricultural products and in processed foods, for both carcinogens and noncarcinogens (the Delaney Clause considered carcinogens only), are now to be based on health only. Other important provisions of the FQPA include the requirement that the EPA specifically consider exposures and risks to infants and young children in setting pesticide tolerance limits, allowing an additional safety factor, up to tenfold; the need to consider all pathways of exposure (e.g., drinking water, soil/dust ingestion) to a pesticide in setting tolerance limits for that pesticide in food; the need to consider the cumulative risk for multiple pesticides that act via a common mechanism of action when setting a tolerance limit for any single pesticide of the "common mechanism" class; and the establishment of a very ambitious comprehensive screening program for pesticides that exert estrogenic and possibly other endocrine effects [16]. The statute represents a landmark piece of legislation not only in terms of the regulatory implications but also with respect to the advancement in scientific understanding required to implement the statute.

The combined effect of the increasing use of risk assessment to help make regulatory decisions and the significant uncertainty that accompanies most quantitative estimates of toxicological risk has resulted in considerable debate about the practice of risk assessment. On one side of the debate, the EPA, the FDA, and other agencies have been criticized by the Office of Management and Budget (OMB) and many representatives of the regulated community for being too conservative in their risk assessment procedures [17]. On the other side, environmental advocacy groups such as Greenpeace have claimed that, "In the real world, quantitative risk assessments are used almost exclusively to justify pollution" [18]. Others have noted that "current risk estimates are by no means routinely exaggerated, either for the entire populations they apply to or for highly exposed or highly susceptible individuals within those populations" [19]. Much of this difference in opinion is perhaps due to the fact that risk estimates are frequently inadequately defined and presented. Often, risk assessors produce single-value estimates of risk that may apply to some unknown percentage of the population. Because the variability in the exposure and dose–response characteristics of a population are so large, the risk estimates for a small, highly exposed, or sensitive subpopulation will generally be very different from the estimates of the most likely risks for the entire population. Although the risk assessment results are supposed to be qualified and uncertainty discussed, the risk number is often used without appropriate qualification. To address this problem, the EPA prepared guidance to risk assessors on the need to provide fuller, more explicit descriptions of risk when providing such information not only to risk managers but also to the general public [20]. For example, the guidance recommends multiple descriptions of risk (e.g., high-end and general population) as well as clear descriptions of the uncertainties in the risk assessment.

Governmental and nongovernmental agencies other than those already discussed can influence the regulatory process as well. The American Conference of Governmental Industrial Hygienists (ACGIH) sets exposure limits based solely on health protection for approximately 600 workplace chemicals. These exposure limits, known as Threshold Limit Values (TLVs®) [21], do not carry any regulatory weight, but it is not uncommon for workplaces to adhere to the TLVs for chemicals that OSHA does not regulate or which have an exposure limit that has not been revised since the inception of OSHA in 1970. The TLVs have also been used by several state environmental agencies to derive acceptable ambient levels for toxic air pollutants.

Agencies in the Department of Health and Human Services that influence the regulatory process include the National Cancer Institute; the National Institute of Environmental Health Sciences, in particular the National Toxicology Program (NTP); the National Institute for Occupational Safety and Health (NIOSH) and the Center for Environmental Health (part of the CDC); and the Agency for Toxic Substances and Disease Registry [22]. These agencies affect the regulatory process in several ways, ranging from decisions on which chemicals to test in long-term cancer bioassays to defining principles for evaluating carcinogens to conducting site-specific (as with a hazardous waste site) and chemical-specific risk assessments. International organizations such as the World Health Organization (WHO) and the International Agency for Research on Cancer (IARC) also have a significant role in the use of information by regulatory agencies.

The primary focus of this chapter is on the use of regulatory toxicology at the federal level; however, state governments have also been active in regulating exposure to toxicants in the environment. For example, in 1986, voters in California overwhelmingly adopted Proposition 65, the Safe Drinking Water and Toxic Enforcement Act of 1986. This act contains two major provisions: one prohibiting the "discharge or release [of] a chemical known to the state to cause cancer or reproductive toxicity into water," and the other, a labeling requirement, mandating that no person expose another individual to any carcinogen or reproductive toxicant without providing "clear and reasonable warning." Exemptions for the discharge requirements are provided for carcinogens at discharge levels that will pose lifetime cancer risk to a person drinking the water of less than $1 \times 10^{-5}$, or for reproductive toxicants for discharges resulting in exposure levels less than 1000

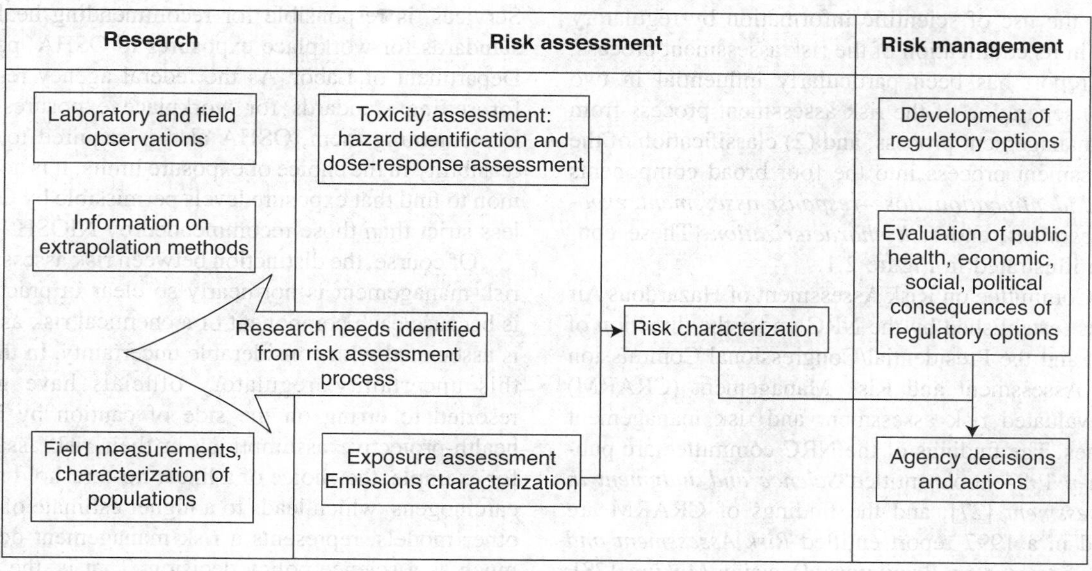


**FIGURE 2.1** NAS/NRC risk assessment/management paradigm. (Adapted from National Research Council, Science and Judgment in Risk Assessment, National Academy Press, Washington, D.C., 1994.)

times smaller than the *no-observable-effect level* (NOEL) for reproductive effects.

Some states have also developed their own risk assessment procedures or standards. In the absence of emissions standards for hazardous air pollutants at the federal level, many states developed their own emissions standards for hazardous air pollutants. Under the CWA, states can develop water quality criteria for certain water bodies in the state. Many states also have Superfund-type laws for the cleanup of abandoned hazardous waste sites and have consequently developed risk assessment procedures for these sites.

Of particular interest with respect to non-U.S. agencies are risk assessment and risk management approaches undertaken by the European Commission (EC). The precautionary principle was presented in principle 15 from the Declaration of the 1992 Rio Conference on the Environment and Development which states that: "In order to protect the environment, the precautionary approach shall be widely applied by States according to their capability. Where there are threats of serious or irreversible damage, lack of full scientific certainty shall not be used as a reason for postponing cost-effective measures to prevent environmental degradation" [23].

The precautionary principle has engendered much debate [24]; for example, under what conditions is action to mitigate risks appropriate? Under what conditions is development of a new technology considered so potentially "risky" that restrictions on development are warranted? The Registration, Evaluation and Authorization of Chemicals (REACH) program in the European Union (EU) may be said to embody elements of the precautionary principle. This program requires companies that manufacture or import more than one ton of a particular chemical substance per year to register the chemical in a central database. Specifically, the regulation "is based on the principle that it is up to manufacturers, importers, and downstream users of substances to ensure that they manufacture, place on the market, import, or use such substances that do not adversely affect human health or the environment" [25]. Thus, overall, the emphasis of the precautionary principle is on demonstrating that a chemical does not present a significant risk and risk management action is unnecessary. This is a subtle, but important, difference from current approaches that typically require demonstration that a chemical does present a significant risk and hence risk management action may be necessary.

## RISK ASSESSMENT PARADIGM

In response to a directive from the U.S. Congress, the FDA contracted with the National Research Council (NRC) of the National Academy of Sciences (NAS) to evaluate the risk assessment process in the federal government and to make recommendations on how the process could be improved. As a result of this effort, the Committee on the Institutional Means for Assessment of Risks to Public Health published a book in 1983 entitled *Risk Assessment in the Federal Government: Managing the Process* [26]. The book summarized past experiences, and, although it did not propose new ways to evaluate risks from environmental chemicals, it has nevertheless had an important effect on the use of scientific information by regulatory agencies in its codification of the risk assessment process.

The report has been particularly influential in two areas: (1) separation of the risk assessment process from the risk management process, and (2) classification of the risk assessment process into the four broad components of *hazard identification*, *dose–response assessment*, *exposure assessment*, and *risk characterization*. These concepts are illustrated in Figure 2.1.

The Committee on Risk Assessment of Hazardous Air Pollutants (established by the NRC under the direction of the EPA) and the Presidential/Congressional Commission on Risk Assessment and Risk Management (CRARM) have reevaluated risk assessment and risk management approaches. The findings of the NRC committee are published in a 1994 book entitled *Science and Judgment in Risk Assessment* [27], and the findings of CRARM are published in a 1997 report entitled *Risk Assessment and Risk Management in Regulatory Decision-Making* [28]. Although the separation of risk assessment and risk management and the four components making up the basic risk assessment paradigm remain key underlying principles, both committees recommended refinements in risk assessment and risk management approaches. For example, the NRC committee highlights the importance of an iterative approach to risk assessment to reduce uncertainties, with each iteration incorporating fewer default assumptions and more specific information, balancing the use of "better science" with the constraints of the available resources [27]. CRARM proposes a framework for risk management that encourages early and frequent involvement of all groups affected by the risk management problem and decision making based on consideration of the risk management problem in the context of the broader, real-world goals of risk reduction and improved health status [28].

Risk assessment is defined as the "systematic, scientific characterization of potential adverse effects of human or ecological exposures to hazardous agents or activities" and involves assessment of the strength of the evidence as well as evaluation of the uncertainties associated with risk estimates [28]. In contrast, risk management is the "process of identifying, evaluating, selecting, and implementing actions to reduce risk to human health and ecosystems" [28]. Risk managers choose actions that will mitigate risks, considering not only the information derived from risk assessment but also cultural, ethical, political, social, economic, and engineering information in the decision process.

The distinction between risk assessment and risk management is critical [26]. The influence of risk management issues, such as the economic significance of a product, on the risk assessment process can seriously undermine the credibility of the risk assessment. This concern is exemplified in the separation between NIOSH and OSHA. NIOSH, part of the Department of Health and Human Services, is responsible for recommending health-based standards for workplace exposures to OSHA, part of the Department of Labor. As the federal agency responsible for setting standards for workplace exposures and for implementing them, OSHA also is required to consider feasibility in the choice of exposure limits. It is not uncommon to find that exposure levels permissible by OSHA are less strict than those recommended by NIOSH [29].

Of course, the distinction between risk assessment and risk management is not nearly so clear in practice. This is because each component of a chemical risk assessment is associated with considerable uncertainty. In the face of this uncertainty, regulatory officials have generally resorted to erring on the side of caution by including health-protective assumptions in their risk assessments; for example, the choice of a linear no-threshold model for carcinogens, which leads to a higher estimate of risk than other models, represents a risk management decision as much as a science policy decision. That is, the approach is conservative and provides the regulator with a greater level of confidence that the true risk to the human population is likely to be less than that expressed through the model. This approach has historically been justified as consistent with prudent public health policy when uncertainty is so great that it is difficult to provide a precise estimate of risk (i.e., in the face of uncertainty, it is easier to say the risk is less than $x$ than to say the risk equals $y$); however, this practice can lead to inconsistent levels of protection for different chemicals and may direct resources away from the more significant risks [30]. For example, the potential cancer risks associated with chemical disinfectants should be compared to the risks of waterborne microbial diseases when making decisions about treating public drinking water supplies; yet, such risk–risk trade-offs cannot be accurately weighed if health-protective assumptions have been used to different extents in the underlying risk assessments [31]. The practice of using health-protective assumptions in conducting risk assessments has been described by some as an inappropriate application of risk management to the risk assessment process [17].

Although risk assessments are commonplace at many federal and state agencies, no uniform guidelines exist that specify how regulatory officials should calculate chemical risks, nor do any uniform criteria indicate how the findings of a risk assessment should influence regulatory decisions [32]. As a result, cancer potency estimates (i.e., the estimated upper bound on lifetime cancer risk associated with the lifetime daily dose of a chemical) developed by different regulatory agencies for the same chemical can vary substantially [33]. Differences in cancer potency estimates for chemicals can also vary among European agencies [34]. Furthermore, the level of risk sufficient to trigger regulatory action can vary considerably among agencies and even among different programs within a single agency

[35]. The EPA's post-regulatory "acceptable" risk levels for arsenic under one statute (the CAA) vary by four orders of magnitude [36].

As discussed above, risk assessment is commonly broken down into four components. The first component of risk assessment, *hazard identification*, involves an evaluation of whether a particular chemical can cause an adverse health effect in humans. The hazard identification process can be considered to be a qualitative risk assessment. It involves identifying the potential for exposure as well as the nature of the adverse effect expected. The types of information used in hazard identification include all categories described in the previous section. In hazard identification, the risk assessor must evaluate the quality of the studies (e.g., choice of appropriate control groups, sufficient numbers of animals), the severity of the effect described, the relevance of the toxic mechanisms in animals to those in humans, and many other factors.

The result is a scientific judgment that the chemical can, at some exposure concentrations, cause a particular adverse health effect in humans. The result is not a simple yes-or-no evaluation but a weight-of-evidence estimation of the likelihood that the particular chemical has the potential to cause the particular effect. For example, studies showing that ozone can suppress pulmonary defenses against microbial agents in several species of animals [37] and information on similarities in pulmonary defenses between humans and animals [38] would lead to the conclusion that ozone exposure in humans could, under certain conditions, result in an increased susceptibility to infection [39]. The hazard identification process has been codified mainly for carcinogens, as exemplified in the classification schemes from a variety of agencies including IARC [40], the EPA [41], and OSHA [42]. These schemes are discussed in more detail later in this chapter.

*Dose–response evaluation*, the second component of the risk assessment process, involves quantitative characterization of chemical potency—that is, the relationship between the dose of a chemical administered or received and the incidence or severity of an adverse health effect in the exposed population. Characterizing the dose–response relationship involves understanding the importance of the intensity of exposure, the concentration × time relationship, whether a chemical has a threshold, and the shape of the dose–response curve. The metabolism of a chemical at different doses, its persistence over time, and an estimate of the similarities in disposition of a chemical between humans and animals are also important aspects of a dose–response evaluation. Although the NAS report considers dose–response estimates mostly in terms of carcinogens, the evaluation of the dose–response relationships has long been a key component of pharmacology and toxicology for many chemicals [26]. Recent advances in dose–response assessment for non-cancer risk assessment are described in Chapter 9 of this book.

In *exposure assessment*, the third component of the risk assessment process, a determination is made of the amount of a chemical to which humans are exposed. Data can be very limited for exposure assessment. Measures of chemicals in environmental media, such as air or soil, or in food may be available; however, the extrapolation of those levels to a dose received by humans has many uncertainties. Models exist that can describe the movement of chemicals through a particular medium, and assumptions can be made regarding inhalation, ingestion, or dermal contact rates and the bioavailability of the chemical. This information can then be used to derive an estimate of the dose taken up by humans. Host factors, such as exercise, the use of certain consumer products, or the consumption of particular foodstuffs, will complicate the exposure assessment.

The use of biological monitoring (e.g., measurement of volatile organic chemicals in exhaled breath [43]), as well as personal sampling devices, such as respirable particulate monitors [44], represent ways in which the uncertainties of exposure assessment can be reduced. Blood lead testing is another example of biological monitoring that has the ability to reduce the uncertainty in quantifying exposure and in extrapolating from exposure to dose.

The last stage of the risk assessment process, *risk characterization*, involves a prediction of the frequency and severity of effects in the exposed population. The information from the dose–response evaluation (what dose is necessary to cause the effect?) is combined with the information from the exposure assessment (what dose is the population receiving?) to produce an estimate of the likelihood of observing the effect in the population being studied. Most risk assessments, particularly for cancer, performed in the regulatory arena produce a single-number estimate of risk (e.g., lung cancer risk of 1 in a million). These are often designed to represent the risk to the reasonable maximally exposed (RME) individual in a potentially exposed population.

Substantial *variability* exists within any potentially exposed population in exposure rates, intake and uptake rates, and sensitivity to the effect. This variability is such that the risk to the most highly exposed and sensitive portion of the population may be orders of magnitude higher than the risks to the majority of the population. For example, some individuals in a population may never eat locally caught fish, but other individuals may subsist on locally caught fish. The fish intakes of these respective individuals will consequently vary by orders of magnitude. Information should generally be provided on both the risk to individuals and the aggregate risk of the exposed population. Point estimates of risk to a single individual in the population can be misleading when no information is provided to indicate whether that individual's exposure is typical of 50% or 0.001% of the exposed population.

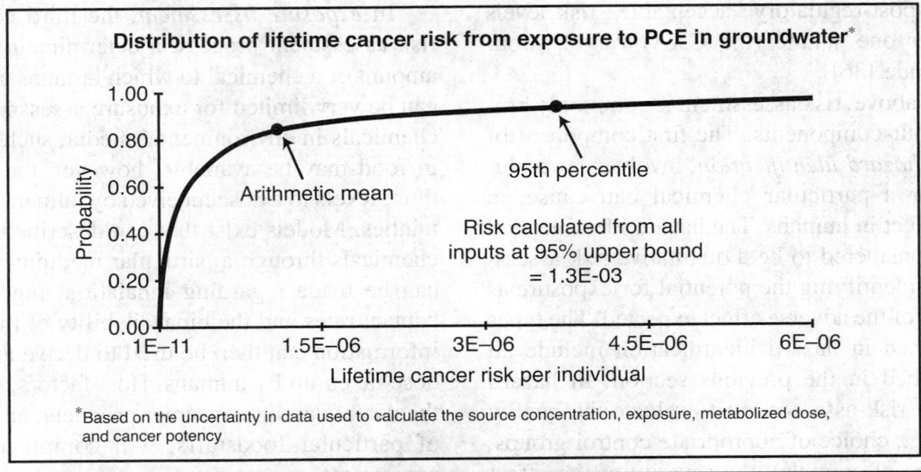


**Selected input parameters and distributions**

| Parameter | Units | Distribution type | Min | Max | Arithmetic mean | Arithmetic std. dev. | Geometric std. dev. |
|---|---|---|---|---|---|---|---|
| PCE concentration in water | mg/L | Empirical | | | 0.0003 | | 2.5 |
| Breathing rate per unit body weight | m³/kg-day | Lognormal | | | 0.4 | 0.00035 | 2.7 |
| PCE metabolized cancer potency | kg-day/mg | Empirical | | | 0.11 | 0.5 | 4.8 |
| Exposure time in house | h/day | Uniform | 8 | 20 | 14 | 0.14 | |
| Skin permeability | m/h | Uniform | 0.004 | 0.01 | 0.007 | | |

**FIGURE 2.2** Monte Carlo analysis of risk: tetrachloroethylene (PCE) in groundwater. (Adapted from McKone, T.E. and Bogen, K.T., *Environ. Sci. Technol.*, 25, 1674, 1991.)

In addition to population variability, a significant *uncertainty* is also present in risk estimates, due to uncertainty in many of the risk assessment components (e.g., model and measurement error). It is critical that the risk characterization step of the risk assessment process describes the biological and statistical uncertainties in the final estimation and identifies which component of the risk assessment process (hazard identification, dose–response, or exposure) involved the greatest degree of uncertainty. For example, the dose–response evaluation is generally highly uncertain; this is often due to model error in extrapolating from animals to humans or short-term to lifetime exposures. Information may not be available to characterize the active species, mechanism of effect, effective dose, or absorption, metabolism, and excretion rates. McCone and Bogen [45] determined that 65% of the variance in the risk assessment results for a case study of tetrachloroethylene in groundwater was due to variance in the estimate of the chemical potency. Because the degree of uncertainty varies greatly among risk assessments for different chemicals, lack of consideration of uncertainty can lead to inappropriate levels of concern for different chemicals.

Monte Carlo uncertainty analysis techniques have been applied to the risk assessment process as one method of attempting to more fully characterize the distribution of potential risks in a population. Rather than using single values to represent input parameters such as contaminant ingestion rates, body weights, and chemical potencies, Monte Carlo analysis uses probability density functions to characterize the input parameters and to produce a probability density function for risk [46–49]. Figure 2.2 shows a probability density function, produced by a Monte Carlo analysis, of risk as a function of input parameters characterized by probability density functions. Although these techniques provide more information on the distribution of potential risk than a single-number risk estimate, they are limited by the availability of information with which to characterize the input probability density functions. Particularly uncertain are estimates of chemical potency, which can vary by orders of magnitude depending on different interpretations regarding carcinogenic mechanisms.

The Committee on Risk Characterization, convened by the NRC, makes recommendations for improving the risk characterization process in their 1996 book entitled *Understanding Risk, Informing Decisions in a Democratic Society* [50]. Rather than just presenting numerical risk results and associated uncertainties, a risk characterization should also convey the information in a clear and easily understandable way that is useful to risk managers in making informed decisions and addresses the concerns of interested and affected parties. The rigorous scientific analyses involved in risk characterization, therefore, must be performed in conjunction with frequent deliberations with all interested and affected parties. As explained by the NRC Committee on Risk Characterization, developing an accurate, balanced, and informative synthesis involves "getting the science right," "getting the right science," "getting the right participation," and "getting the participation right" [50]. The EPA adopted similar values in their 1995 risk characterization guidance [51].

## TOXICOLOGY INFORMATION USED IN THE REGULATORY PROCESS

Three main categories of scientific information are employed by agencies in the evaluation and regulation of toxic chemicals in the environment: (1) epidemiology, (2) controlled clinical exposures, and (3) animal toxicology. *In vitro* studies and structure–activity relationships are typically used by regulatory agencies to support the interpretation of information from the three major categories and are only occasionally used as a primary source of information.

Epidemiology, studies of clinical exposures, and animal toxicology provide qualitatively different information, with unique advantages and limitations. Environmental epidemiology studies, which attempt to associate disease or other adverse health outcomes with an environmental exposure, have the advantage of measuring an effect in humans at exposure conditions that are by definition realistic. The first demonstration that benzene was a carcinogen came from epidemiological studies of rubber workers [52]. It was not until several years after these studies that benzene was shown to cause cancer in animal studies [53]. Studies of the London smog pollution episode in 1952 demonstrated that high levels of pollution from coal combustion could cause mortality, particularly in the very young, the elderly, and those individuals with preexisting cardiopulmonary disease [54]. Evaluation of similar effects in animal studies would be difficult, given the complexity of the exposure in London and the lack of good animal models for susceptible populations, such as asthmatics. In general, epidemiology has been particularly helpful in the evaluation of working environments or other environments where exposure concentrations are relatively high.

Several factors limit the use of epidemiological studies by regulatory agencies. One of the major limitations is the lack of well-defined exposure information, for both chemical species and for actual concentrations; for example, the lack of accurate total exposure information limits the ability to quantify the effects of ambient air pollution in the United States. The Harvard University Six City Study showed that outdoor NO-monitoring devices are inadequate in accurately assessing total exposure to NO, due to the importance of indoor exposure [55]. A recent study, the Total Exposure Assessment, a Columbia and Harvard study (TEACH study), has quantified indoor, outdoor, and personal exposures of inner-city residents to a number of pollutants, including formaldehyde, dichlorobenzene, and benzene [56]. This study has demonstrated that, depending on the contaminant, outdoor exposure levels may underestimate, overestimate, or relatively well predict personal exposures; for example, due to the importance of indoor sources, outdoor measures of formaldehyde typically underestimated personal exposures. As discussed earlier in this chapter regarding the CDC biomonitoring study, the use of biological markers of exposure, such as measurements of arsenic levels in urine or lead levels in blood, can provide more accurate information about exposure and help reduce uncertainties in the results of epidemiological studies.

It is also difficult to define the causal element in epidemiological investigations, particularly when complex exposures are involved. For example, several indicators of pollution were measured during the London smog episodes that occurred between 1958 and 1972. Initial evaluations focused on the role of total particulate and $SO_2$ as causative agents for the elevated mortality levels; however, subsequent analyses of the London studies, as well as studies from other cities, indicate the importance of acid sulfates on mortality [57]. Large-scale epidemiological studies have demonstrated associations between various indicators of air quality, in particular particulate matter of 2.5 μm or less (PM2.5), but debate continues regarding the nature and magnitude of causality due to the complexity of PM composition and difficulties in replicating the findings in animal studies (see, for example, Valberg [58]).

Another limitation is that epidemiological studies are frequently of worker populations, and such studies can be difficult to apply to prediction of health effects in the general population. Occupational studies, in general, focus on healthy adult male workers. The general population is more heterogeneous than the worker population and, for some pollutants, may exhibit a greater range in susceptibility. In general, only more recent epidemiology studies have considered adverse health effects of chemicals specific to women and children, such as developmental, reproductive, or hormonally mediated effects, including cancer. An example of the limitations of occupational epidemiology involves studies of peripheral nerve function in lead-exposed workers which would underestimate the risk of lead exposure in young children, for whom the primary concern is neurobehavioral effects resulting from relatively low-level exposures [59]. At the same time, it must be recognized that the young and elderly are not always more susceptible to the effects of chemicals. As noted by Calabrese [60], adults, rather than the young or the elderly, are more susceptible to the renal toxicity of fluorides and mercury. High-risk subgroups are discussed in more detail in a later section of this chapter.

Epidemiology studies are frequently limited by the need for a relatively large increase in disease incidence (twofold or more), given the sample sizes generally available for such investigations. Enterline [61] notes that it would require a large population (1000 deaths, using the

Peto model) to detect a 50% excess in deaths from lung cancer at an asbestos level of 2 fibers/cm$^3$ air.

Controlled clinical studies of humans exposed to pollutants address some of the difficulties of epidemiology studies. The exposures can be controlled and quantified, the effects are observed in humans, and the exposed population can be chosen to consist of susceptible individuals, such as asthmatics or exercising individuals. Thus, changes in airway resistance in asthmatics exposed to $SO_2$ during exercise [62,63], have been important in the EPA's evaluation of the National Ambient Air Quality Standards (NAAQS) for $SO_2$ [64], because these effects reflect the response of the susceptible population, using an appropriate exposure concentration and a relevant averaging time. Given the subtlety of these changes (nonsymptomatic bronchoconstrictions) and the fact that they occur only in a selected subset of the general population (asthmatics constitute about 4% of the total population), these effects would not have been detectable in the general population.

One of the advantages of controlled clinical exposure studies—that they are performed with humans—is also a major limitation. Because these studies must be limited to short-term effects that are readily reversible, they cannot be used to evaluate the potential of a chemical to cause chronic disease. There is debate within the scientific and regulatory community as to the circumstances under which data from controlled human studies, even considering that such studies present negligible risk to the participants, may be used by agencies in decision making. For example, intentional dosing studies in human subjects (e.g., cholinesterase inhibition tests in individuals exposed to organophospate pesticides) were considered by the NAS [65] to be acceptable under some circumstances (e.g., when there was "reasonable certainty" of no adverse effects to the participants and the studies addressed an important question that could not be answered with animal studies); however, others [66] have questioned the appropriateness of such testing, in particular because of concerns for testing on children. In addition, because of the nature of the changes observed in clinical exposure studies, the health significance of the indicators is open to discussion. As an example, perchlorate, a chemical used in the treatment of certain thyroid disease and also found as a groundwater contaminant, inhibits iodide uptake at the thyroid. Studies in humans [67] have demonstrated that doses of perchlorate greater than 0.007 mg/kg/day are required to inhibit iodide uptake; however, the inhibition of iodide uptake must be of sufficient magnitude and for sufficient duration (e.g., in adults up to 75% inhibition over several months may be required to inhibit thyroid hormone synthesis) before impacts on thyroid hormone synthesis (the physiologically relevant effect) may occur [68]. Thus, the clinical significance of low-level inhibition of iodide uptake for a modest period of time (e.g., days to weeks) is uncertain.

Another issue with the use of clinical studies is that, although some susceptible populations, such as mild asthmatics, can be tested, individuals with a greater degree of impairment, such as asthmatics who require continual medication, are usually not considered to be appropriate subjects for these studies because of the greater potential for harm during exposure. Later sections in this chapter address the questions of severity of effect in susceptible populations in greater detail.

Animal toxicology studies constitute the third major source of information for assessing the toxicity of chemicals. Animal toxicology studies allow the investigator the greatest degree of control over the exposure conditions, the population exposed, and the effects measured. One can readily evaluate subtle effects of acute and chronic exposure. For example, morphological and numerical changes in the pulmonary type I and type II cell populations, as well as interstitial fibrosis, have been observed in rats exposed to 0.06 to 0.25 ppm $O_3$ [69,70]. It would have been very difficult to describe this effect with other experimental approaches, and yet the effect is clearly of concern for humans who are exposed to comparable concentrations of ozone in certain urban environments.

In animal experiments, the ability to manipulate the experimental conditions permits the evaluation of many variables on the response to toxic chemicals. Thus, Elsayed and Mustafa [71] were able to demonstrate the protective effect of vitamin E on the acute toxicity of nitrogen dioxide ($NO_2$) in mice. In addition, the role of metabolism in susceptibility to polycyclic aromatic hydrocarbon (PAH)-induced carcinogenesis has been evaluated in studies of genetic variants in mice [72,73]. Such studies can be important in predicting modifiers of toxicity in humans and in identifying the susceptible human populations.

The use of toxicogenomics, in which multiple changes in gene transcription, protein synthesis, and metabolite profiles as a consequence of toxicant exposure in animals can be quantified and measured even at a cellular level, will allow for even more refined understanding of molecular responses to chemicals in animal models. Toxicogenomics allows for a better characterization of the linkage between chemical exposure and toxicological effects on a number of levels, such as identifying biomarkers of susceptibility, improving the understanding of a chemical's mode of action, and identifying changes prior to histopathological events. For example, transcriptional changes have been observed after low- and high-dose acetaminophen exposure, prior to the development of histopathological changes [74,75]. At present however, it is difficult to use such information directly in risk assessment, although future developments, including improved integration of toxicogenomics findings with toxic responses, will fuel its incorporation into risk assessment.

The limitations of animals studies fall into two broad categories: (1) those due to uncertainties in extrapolating from animals to humans, and (2) those due to uncertainties in extrapolating from the high exposures in animal studies to the lower exposures typically experienced by humans. Interspecies extrapolation is complicated by the greater homogeneity of laboratory animals than humans, the controlled conditions of housing and diet, innate genetic factors, and other variables. The relevance of di-(2-ethylhexyl) phthalate (DEHP)-induced hepatocarcinogenesis in rodents to humans has been questioned on the basis of differences in peroxisomal proliferation in the liver in the two species as a consequence of differences in peroxisome proliferator-activated receptor alpha (PPARα) binding [76,77]. Similarly, high exposure concentrations typically used in animal studies may result in saturation of detoxification pathways and thus may produce effects that are not relevant to effects produced at ambient exposure concentrations, where detoxification pathways are not saturated. Increased numbers of macrophages and impairment of alveolar clearance are observed in rats exposed to relatively high concentrations of diesel particulates [78]. The significance to humans who are exposed to ambient levels of diesel particulates much lower than those employed in the animal studies is uncertain [79].

Historically, *in vitro* studies and analysis of structure–activity relationships have been used to help set priorities for chemical testing; for example, structure–activity relationships have been used to predict mutagenicity, lethality, and carcinogenicity [80]. This type of information can be very useful, for example, in selecting compounds for longer term testing in animals or eliminating chemicals being considered for potential industrial or pharmaceutical applications due to toxicological concerns.

Short-term tests have typically been used indirectly in the regulatory process to support decision making rather than as a basis *per se* for decision making; for example, evidence that a chemical is a point mutagen in an *in vitro* test system might be used to support the classification of a chemical as a possible human carcinogen or the use of a linear dose–response model for carcinogenesis. Metabolism, pharmacokinetic, and mechanistic studies can also provide information to reduce uncertainties in the use of toxicology information. Metabolic studies showing that a critical reactive metabolite in rodents is also formed in humans could reduce uncertainties in extrapolating from animals to humans, while mechanistic studies could indicate whether a subtle effect observed in a clinical study is a precursor for later, more serious health endpoints and therefore of concern as a biomarker of effect.

In some circumstances, short-term tests and structure–activity relationships may be used directly to provide a basis for decision making. Beck and coworkers [81] estimated permissible levels for alkylphenols in water, based on the ability of different alkylphenols to inhibit cyclooxygenase and by comparison with toxic effects of aspirin, a well-known cyclooxygenase inhibitor. A critical element in this example is that adequate toxicological data were available for some members of the classes of chemicals being studied and that estimates of risk were applied within a class of chemicals of similar physical and chemical properties.

A summary comparing the differences between epidemiology, controlled clinical exposure, and animal toxicology studies is provided in Table 2.2, Table 2.3, and Table 2.4. We can conclude from the preceding discussion that there is no "best" source of information for regulatory agencies. The rational approach is to examine all available sources of information in the evaluation of toxic chemicals. Some kinds of information may be especially useful in *hazard identification*, the likelihood that a chemical will be toxic to humans, whereas other types of information will be more appropriately applied to estimation of the *dose–response relationship*.

## EVALUATION OF CARCINOGENS

### BACKGROUND

The frequent public demand for zero risk has made the regulation of carcinogens a formidable challenge. Within the scientific community, there is ongoing debate on how to define a potential human carcinogen, as well as on how to estimate cancer risks under practical conditions of chemical exposure. This uncertainty is due largely to the fact that mechanisms of carcinogenesis for many chemicals are still poorly understood, and different carcinogens act in different ways to induce cancer. The task of regulating carcinogens has been complicated, rather than simplified, by many of the mechanistic discoveries of recent years. The simple picture of the 1950s, when only a relatively small number of chemicals were thought to be carcinogens, has been replaced by the realization that chemical carcinogenesis takes place in multiple stages, some with reversible steps, which have different dose–response relationships. Essential nutrients and hormones can be carcinogenic in some circumstances. The same chemical can promote or inhibit carcinogenesis, depending on the circumstances of exposure [82,83]. The evidence for hormesis (see later discussion) has added further complexity to stimulatory and antagonistic carcinogenic responses. Public pressure to regulate carcinogens, even where very little toxicological information exists, has in many circumstances compelled regulatory agencies to treat carcinogens as though they all act by the same mechanisms, even as it has become apparent that they do not.

From a public health standpoint, regulatory agencies have generally regulated carcinogens at exposure levels that reflect a very low probability of tumor production (e.g., excess cancer risk of 1 per million exposed); however, for practical reasons, it is impossible to conduct

**TABLE 2.2**
**Advantages and Disadvantages of Epidemiological Studies**

| Advantages | Disadvantages |
|---|---|
| Exposure conditions realistic | Costly and time consuming |
| Occurrence of interactive effects among individual chemicals | *Post facto*, not protective of public health[a] |
| Effects measured in humans | Difficulty in defining exposure, problems with confounding exposure |
| Full range of human susceptibility frequently expressed | Difficult to see less than twofold increase in risk except in very large populations |
| | Effects measured can be relatively crude (morbidity, mortality) |

[a] Use of biomarkers in epidemiological studies, rather than disease endpoints, can allow such studies to be public health protective.

**TABLE 2.3**
**Advantages and Disadvantages of Controlled Clinical Studies**

| Advantages | Disadvantages |
|---|---|
| Well-defined, controlled exposure conditions | Costly |
| Responses measured in humans | Relatively low exposure concentrations and short-term exposures |
| Potential to study subpopulations (e.g., asthmatics) | Limited to relatively small groups (usually <50 individuals) |
| Ability to measure relatively subtle effects | Limited to short-term, minor, reversible effects |
| | Usually most susceptible group not appropriate for study |

**TABLE 2.4**
**Advantages and Disadvantages of Animal Toxicological Studies**

| Advantages | Disadvantages |
|---|---|
| Readily manipulated exposure conditions | Uncertainties in relevance of animal response to human exposure |
| Ability to measure many types of responses | Controlled housing, diet, etc., of questionable relevance to humans |
| Ability to assess effect of host characteristics (e.g., gender, age, genetics) and other modifiers (e.g., diet) of response | Exposure concentrations and time frames often very different from those experienced by humans |
| Potential to evaluate mechanisms | |

animal studies of a size that would allow observation of effects following treatment at such low doses. The practice has therefore been to conduct animal studies at relatively high dose levels and then extrapolate the results from high to low dose and from animals to humans; thus, the chronic animal bioassay results, the extrapolation from high to low doses, and the extrapolation across species are used to derive potency factors (i.e., indicators of carcinogenic potency) for carcinogens. These potency factors allow one to relate an estimated chemical dose in humans to an upper-bound probability of tumor occurring as a result of that dose. It should be noted, that even with established human carcinogens, extrapolation procedures are still used to extrapolate carcinogenic response from high to low dose (e.g., workplace to ambient) or from one type of exposure condition (e.g., intermittent, subchronic) to another exposure condition (e.g., continuous chronic).

This section on carcinogens first provides some basic information on mechanisms of carcinogenesis. We then describe some of the key issues that agencies address in the interpretation and application of scientific data on carcinogens. These issues fall into the categories of hazard identification and dose–response assessment [26]. Hazard identification for carcinogens addresses two questions: (1) What is the evidence that a particular chemical is an animal carcinogen? and (2) What is the likelihood that an animal carcinogen is a human carcinogen (and under what circumstances of exposure pathway and dose)? Dose–response assessment has traditionally attempted to determine the probability of tumor production, given a particular exposure or dose level (i.e., assuming a low-dose linear response).* The dose–response assessment section

* As discussed in later sections of this chapter, for carcinogens that act through a threshold mechanism such probabilistic low-dose models are not appropriate.

of this chapter discusses mathematical models used to extrapolate from high to low doses, physiologically based pharmacokinetic (PBPK) modeling to relate administered and effective doses in animals and humans, and issues concerning the relationship between effective dose and response.

## Mechanisms of Carcinogenesis

Carcinogenesis is currently understood to be a multistage process which has been described as involving the initiation, promotion, and progression of normal cells into neoplastic cells. Chemicals can act at one or more of these stages and can act directly (e.g., mutagen) or indirectly (e.g., immune suppression) and through both indirect and direct mechanisms. Initiation is generally understood to be a permanent and irreversible event involving DNA mutation and the first step in the process of carcinogenesis. Genotoxic agents are typically considered to be capable of initiating activity, thus having the potential to begin the transition from normal to cancer cells; therefore, genotoxic agents have been considered to act via a nonthreshold mechanism, and this belief has formed the basis for linear extrapolation of effects seen at high doses down to low doses. Inferences as to the absence of a threshold for initiating agents comes from the study of mutations that result from these agents. In addition, studies investigating the number of preneoplastic focal lesions induced by an initiating agent did not find a measurable threshold [84]. Ionizing radiation is an example of an initiating agent. In addition, certain chemicals (e.g., aflatoxin $B_1$, diethylnitrosamine, tobacco smoke) are considered to be complete carcinogens, capable of initiation, promotion, and progression. Potential factors modifying the efficiency of initiation include rates of cell division and DNA synthesis as well as the rate of metabolism of a chemical to its active form or rate of metabolic detoxification. (It should therefore be noted that, due to the existence of repair mechanisms and other factors that reduce or eliminate responses at low exposure levels, even the no-threshold concept may not be applicable to all genotoxic carcinogens.)

The promotion stage is characterized by clonal expansion of the initiated cells. Promoting agents can act by various mechanisms to increase rates of cell proliferation or decrease rates of cell death. For example, cell proliferation can be induced by cytoxic agents or mitotic agents. Interference with intercellular communication may also be involved in clonal expansion of initiated cells [85]. An important feature of this stage is its reversibility and, in some cases, the existence of a threshold for the effect. In the standard tumor initiation/promotion model, withdrawal of the promoting agent halts the development of tumors. The promotion stage can also be modulated by environmental factors, including frequency of dosing, age of test animal, and diet [84]. Promoting agents are generally thought to exhibit a threshold (or inflection point) in the dose–response curve. Examples of promoting agents include hormones, alcohol, and dietary fat.

More recent studies indicate that the above paradigm is likely to be simplistic in some cases. Some chemicals may act primarily through nongenotoxic mechanisms that nonetheless result in secondary genotoxicity; for example, the rat-specific bladder carcinogen dimethylarsinic acid (DMA) induces cytotoxicity to the urothelial cells of the bladder, followed by necrosis, cell regeneration, proliferation, and hyperplasia, leading to tumor production. In this case, genotoxicity occurs subsequent to cytotoxicity in both dose and in time and appears to have a limited, if any, relationship to carcinogenicity. Thus, with respect to DMA, genotoxicity would not be indicative of a linear dose–response relationship [86]. In the case of hormonal carcinogens, tumors may arise as a consequence of prolonged stimulation of cell division in which genetic damage occurs as a secondary event [87,88].

Progression is an irreversible stage characterized by the development of malignant neoplasms and is understood to require a second genetic mutation. Agents that act only during progression or advance a cell from promotion to progression have not yet been definitively characterized. It has been hypothesized that malignant neoplasms may all exhibit an abnormal expression of one or more proto- and cellular oncogenes [84]. In this scenario, initiation is defined by the first mutation event and progression as the second mutation, resulting in homozygosity at the anti-oncogene locus and total loss of growth control [89].

## Hazard Identification

The question of how to decide whether a particular chemical is a potential human carcinogen is currently the subject of considerable scientific debate. It is an important question because the act of labeling some chemicals "carcinogens" and not labeling others can have profound regulatory and societal implications [90]. The regulatory paradigm whereby chemicals are regulated either as carcinogens or as noncarcinogens requires that the question of whether a particular chemical is a carcinogen or not typically be answered with a "yes" or a "no." In the United States, most regulatory agencies have historically regulated all carcinogens as though they operate via the same no-threshold mechanism; however, different chemicals may act in different ways during the various stages of cancer formation to affect the development of tumors. These various mechanisms of tumor formation are not all consistent with the mechanistic assumptions that form the basis of the regulatory framework for carcinogens. Thus, the more we learn about carcinogenesis, the less the standard regulatory approach is able to accommodate the new information. A chemical may be carcinogenic via certain routes of exposure and not others, or only above certain dose levels. More flexible classification approaches are being developed [91] that allow the incorporation of more science into the classification process.

In the next section, we describe current classification approaches, but the reader is reminded that the current scientific debate on these schemes continues to fuel new approaches. Regulatory agencies generally classify potential carcinogens based on an evaluation of both human and animal studies, as well as on supporting information from short-term tests for mutagenicity and structure–activity relationships. Because human evidence exists for so few chemicals, animal studies generally provide most of the available information about the potential carcinogenicity of a chemical to humans.

## Animal Studies

The evidence that a chemical is an animal carcinogen frequently derives from long-term animal bioassays. Such studies usually consist of exposing groups of about 50 animals (typically rats or mice) to two concentrations of a chemical over the lifetime of the animals. Sex- and age-matched unexposed animals constitute the control group. At the termination of the bioassay, the animals are killed and the number of tumor-bearing animals and the number and type of tumors per animal are quantified. All tumors are recorded, including those that are present as a consequence of spontaneous processes. Interim examinations may be performed, particularly on animals that appear moribund. Alternatives to the standard bioassay are being developed. As an example, genetically engineered strains of mice in which tumor suppressor genes are inactivated (knockouts) or activated oncogenes are introduced (transgenics) [92] may allow detection of carcinogens in shorter periods of time than the standard bioassay; however, the use of transgenics and knockouts remains limited for a number of reasons, such as the limited histopathological analysis (EPA guidelines). The alternative models provide useful information on carcinogenic modes of action and can complement the standard bioassay but are unlikely at present to supplant the more typical approaches.

### *The Maximum Tolerated Dose*

Dose selection plays a key issue in the design and interpretation of the animal bioassay. Animals are typically exposed at two dose levels: the maximum tolerated dose (MTD), one half of the MTD, and, in recent years, one quarter of the MTD. The MTD is predicted from subchronic toxicity studies as the dose "that causes no more than a 10% weight decrement, as compared to the appropriate control groups, and does not produce mortality, clinical signs of toxicity or pathologic lesions (other than those related to a neoplastic response) that would be predicted [in the long-term bioassay] to shorten an animal's natural lifespan" [93]. The MTD is not a nontoxic dose and is expected to produce some level of acceptable toxicity to indicate that the animals were sufficiently challenged by the chemical. The MTD has been justified as a means of increasing the sensitivity of an animal bioassay involving limited numbers of animals so as to be able predict risks in large numbers of humans [94].

An objection to the use of MTDs has been that metabolic overloading may occur at high-dose levels, leading to an abnormal handling of the test compound [95]; for example, toxic metabolites could be produced as a consequence of saturation of detoxification pathways. Organ toxicity could occur that might not happen at lower concentrations of the chemical [96], particularly at those concentrations to which humans are typically exposed. Thus, it has been argued that nongenotoxic agents that are determined to be positive in rodent carcinogenicity bioassays may exert their carcinogenicity via target-organ toxicity and subsequent cell proliferation and should not be assumed to be carcinogenic at low doses [97].

Ames and coworkers [98,99], have suggested that target-organ toxicity and subsequent mitogenesis are responsible for the fact that over half of all chemicals tested in chronic bioassays at the MTD are determined to be carcinogens in rodents. They observed that both genotoxic and nongenotoxic agents tested at the MTD cause increased rates of mitogenesis, thus increasing the rate of mutation. For several chemicals, induction of tumors was more strongly correlated with cell division than with DNA adducts or mutagenic activity. Others have reported that cancer potency and MTD are inversely correlated and that, consequently, the potency estimate is simply an artifact of the experimental design [100]. Goodman and Wilson [101] found that cancer potency and the MTD were more strongly related for nonmutagens than for mutagens in rat bioassays, indicating that the carcinogenic effect and toxicity were more closely associated for nonmutagens than for mutagens; however, they noted that even for most mutagens, their findings suggested that at high doses carcinogenicity is induced via mechanisms associated with toxicity.

Gaylor [102] noted that, given sufficient animals (e.g., about 200 per group), it is estimated that about 92% of all chemicals tested would, if tested at the MTD, yield a positive response at one or more tumor sites in rats or mice. Gaylor observed that "this MTD bioassay screen is not distinguishing between true carcinogens and noncarcinogens." The author further suggests a common mechanistic explanation for this result; that is, for nongenotoxic carcinogens in particular, the mode of action involves cytotoxicity followed by regenerative hyperplasia. Thus, the relevant question is not so much whether a chemical causes cancer at the MTD (i.e., is a chemical a carcinogen?), but what is the dose at which the chemicals induces cancer?

The EPA91 cancer guidelines noted that bioassay results at doses that exceed the MTD can be rejected if toxic damage to target organs compromises study interpretation. The reason is that dosing above the MTD in a study may result in tumor production secondary to tissue damage rather than a direct carcinogenic influence of the agent tested. Thus, use of information from testing at fractional doses of the MTD may yield results that are more relevant to human risk. Importantly, the use of information on mode of action, metabolism, and other biological processes is being used as a more scientifically grounded approach to dose selection [103].

### *Other Issues in Hazard Identification*

Another key issue in the evaluation of animal bioassays is the analysis of the tumors themselves. Considerations include the categorization of benign tumors and whether tumor analysis should be site-specific or based on all sites. The position of the IARC is that "few, if any chemicals exist which produce only benign tumors and no malignant tumors in any species" and that chemicals that cause a marked increase in the number of benign tumors "are now viewed with almost as much suspicion as potential human hazards as they would have been if the induced tumors had been malignant." Thus, it has been the general policy of regulatory agencies to accord almost the same weight to benign tumors as to malignant tumors, especially if evidence suggests that the benign tumors could progress to malignancy [105].

It is sometimes stated that one should consider only the overall incidence of tumors, because, from a public health perspective, the concern is with total cancer risk for humans rather than risk at any one site. Although this position has an innate appeal, it is difficult to apply in practice for two main reasons:

- This approach greatly decreases the ability of the bioassay to detect a positive effect, given the high background incidence of some tumor types in rodents; for example, testicular tumors can be as high as 82% in rats and liver tumors can be as high as 25% in mice [106].
- The grouping of tumor types that do not share a common cellular origin is of questionable biological relevance, because the mechanisms involved in the production of the different tumor types could differ. Furthermore, the metastatic potential of different tumor types is highly variable and would have an important influence on the lethality of a particular type of cancer.

It should be noted that chemical carcinogens produce specific types of tumors that are characteristic of that chemical, exposure route, and dose. No convincing evidence has been found of a chemical agent that, in animal bioassays, increases overall tumor incidence rather than increasing the incidence at specific sites. Thus, the classification of 2,3,7,8-tetrachlorodibenzodioxin as a human carcinogen based on an increase in all cancers as observed in epidemiology studies [107–109] has been questioned on the basis of the biological plausibility of this finding [110].

Interestingly, reductions in tumor incidence are frequently observed in the same cancer bioassays in which tumor increases are observed. Linkov and coworkers [83] concluded that the anticarcinogenic effects observed in rodent bioassays are not explained by random effects. The basis for the reduction in tumors is not known, and it could be a consequent perturbation in the animal's physiology. These observations lend credence to the concept that animal bioassays must be interpreted with special attention as to whether biological phenomena are induced at high doses that may not occur (or occur with a relatively much lower frequency) at low doses. A similar observation is found in the evaluation of some human carcinogens, in particular those that act through hormonal processes; for example, oral contraceptives are associated with an increased risk of breast cancer but a decreased risk of ovarian and endometrial cancer (Table 2.5). Anticarcinogenic properties of carcinogens are typically not considered as part of the regulatory process for carcinogens.

In addition, the standard NTP-type 2-year cancer bioassay may not be sensitive to hormonally regulated cancers such as breast cancer. This is because the mouse and rat strains used to perform these bioassays are selected because they are known to be susceptible to liver, kidney, or lung tumors in particular. It has not been demonstrated that these strains are also susceptible to cancers at hormonally sensitive sites. Thus, it is possible that bioassays may be less sensitive for detecting certain cancer effects that are hormonally regulated [111].

## Carcinogen Classification Schemes

The IARC, the EU, the EPA, the NTP, the German Commission for Investigation of Health Hazards, Health Canada, and the ACGIH have developed classification schemes for carcinogens, based on a weight-of-evidence or strength-of-evidence evaluation of available human and animal studies. These seven classification systems are shown in Table 2.6. The IARC [40] and some other agencies typically conclude that a chemical demonstrating "sufficient evidence of carcinogenicity" from animal experiments is a potential human carcinogen. To some degree, this conclusion is supported by evaluation of known human carcinogens in animal bioassays. For the 73 chemicals or processes or environmental factors associated with cancer indication in humans by the IARC [112,113], about half of those that have been tested have also been positive in animal bioassays (Table 2.5). More current understanding of carcinogenesis indicates that this

**TABLE 2.5**
**Chemicals, Industrial Processes, and Environmental Factors Associated with Cancer Induction in Humans: Target Organs and Main Routes of Exposure in Humans and Degree of Supporting Evidence in Animals (IARC)**

| | Humans | | Animals |
|---|---|---|---|
| Chemical or Industrial Process | Main Type of Exposure[a] | Target Organ(s)[b] | Degree of Evidence for Carcinogenicity |
| 1-(2-Chloroethyl)-3-(4-methylcyclohexyl)-1-nitrosourea (methyl-CCNU; semustine) | Medicinal | Leukemia | Limited |
| 1,4-Butanediol dimethanesulphonate (busulphan; Myleran®) | Medicinal | Leukemia | Limited |
| 2-Naphthylamine | Occupational | Bladder (liver) | Sufficient |
| 4-Aminobiphenyl | Occupational | Bladder | Sufficient |
| 8-Methoxypsoralen (methoxsalen) plus UVA radiation | Medicinal | Skin | Sufficient |
| Aflatoxins [255] | Environmental | Liver (lung) | Sufficient |
| Alcoholic beverages | Cultural | Oral cavity, pharynx, larynx, esophagus, liver (breast) | Inadequate |
| Aluminum production | Occupational | Lung, bladder (lymphoma, esophagus, stomach) | No data |
| Arsenic compounds and gallium arsenide[c] [256,257] | Occupational, environmental, and medicinal | Urinary bladder, skin, lung, (liver, hematopoietic system, gastrointestinal tract, kidney) | Limited |
| Asbestos | Occupational | Lung, pleura, peritoneum, gastrointestinal tract, larynx | Sufficient |
| Auramine manufacture | Occupational | Bladder | No data |
| Azathioprine | Medicinal | Lymphoma, skin, mesenchymal tumors, hepatobiliary system | Limited |
| Benzene | Occupational | Leukemia | Sufficient |
| Benzidine | Occupational | Bladder | Sufficient |
| Beryllium and beryllium compounds [258] | Occupational | Lung | Sufficient |
| Betel quid (with and without tobacco), areca nut [259] | Cultural | Oral cavity (pharynx, larynx, esophagus) | Sufficient |
| *bis*(Chloromethyl) ether and chloromethyl methyl ether (technical grade) | Occupational | Lung | Sufficient |
| Boot and shoe manufacture and repair | Occupational | Leukemia, nasal sinus (bladder, digestive tract) | No data |
| Cadmium and cadmium compounds [260] | Occupational | Lung | Sufficient |
| Chlorambucil | Medicinal | Leukemia | Sufficient |
| Chromium compounds (hexavalent)[c] | Occupational | Lung, (gastrointestinal tract) | Sufficient |
| Ciclosporin [261] | Medicinal | Lymphoma, Kaposi's sarcoma | Limited |
| Coal tars and pitches, mineral and shale oils, and soots[d] | Occupational, environmental | Skin, lung, bladder (gastrointestinal tract, leukemia, colon) | Sufficient |
| Cyclophosphamide | Medicinal | Bladder, leukemia | Sufficient |
| Diethylstilboestrol | Medicinal | Cervix/vagina, breast, testis (endometrium) | Sufficient |
| Epstein–Barr virus [262] | Environmental | Lymphomas, nasopharyngeal carcinoma | Sufficient |
| Erionite | Environmental, cultural | Pleura, peritoneum | Sufficient |

| | | | |
|---|---|---|---|
| Estrogens (steroidal, nonsteroidal) [3] | Medicinal | Endometrium, breast, cervix/vagina, testis | Sufficient |
| Estrogen therapy, postmenopausal [263] | Medicinal | Endometrium, breast | No data |
| Ethylene oxide [264] | Occupational | Lymphatic and hematopoietic systems | Sufficient |
| Etoposide (with cisplatin and bleomycin) [265] | Medicinal | Leukemia | Sufficient |
| Formaldehyde [266] | Occupational, environmental | Nasopharyngeal (leukemia, sinonasal) | Sufficient |
| Furniture and cabinet making | Occupational | Nasal sinus | Inadequate |
| Helicobacter pylori | Environmental | Stomach | No data |
| Hematite mining (with radon exposure) | Occupational | Lung | Inadequate |
| Hepatitis B virus [268] | Environmental | Liver | Inadequate |
| Hepatitis C virus [269] | Environmental | Liver | Inadequate |
| Herbal remedies containing *Aristolochia* [270] | Medicinal | Renal pelvis/ ureter, bladder | No data |
| Human immunodeficiency virus type 1 [271] | Environmental | Kaposi's sarcoma, non-Hodgkin's lymphoma | Inadequate |
| Human papilloma virus types 16 and 18 [272] | Environmental | Cervix | None |
| Human T-cell lymphotropic virus type 1 [273] | Environmental | Adult T-cell leukemia/lymphoma | Inadequate |
| Internally deposited radionuclides[c] [274] | Environmental, occupational | Bone sarcoma, liver, leukemia, liver, thyroid, paranala sinuses and mastoid process | Sufficient |
| Involuntary smoking [275] | Cultural | Lung | Limited (mixture of main- and sidestream smoke); sufficient (sidestream condensates) |
| Ionizing radiation (neutrons, X- and gamma-radiation) [276] | Environmental | Leukemia, skin, various internal organs | Sufficient |
| Iron and steel founding | Occupational | Lung (digestive tract, genitourinary tract, leukemia) | No data |
| Isopropyl alcohol manufacture (strong acid process) | Occupational | Nasal sinus (larynx) | Inadequate |
| Magenta manufacture | Occupational | Bladder | Inadequate |
| Melphalan | Medicinal | Leukemia | Sufficient |
| MOPP and other combined chemotherapy, including alkylating agents | Medicinal | Leukemia | No data |
| Mustard gas (sulfur mustard) | Occupational | Lung, larynx, pharynx | Limited |
| Nickel and nickel compounds[c] | Occupational | Nasal sinus, lung (larynx) | Sufficient |
| *N,N-bis*(2-chloroethyl)-2-naphthylamine (chlornaphazine) | Medicinal | Bladder | Limited |
| *Opisthorchis viverrini* [277] | Environmental | Liver | Limited |
| Oral contraceptives, combined [278,279], | Medicinal | Liver (also protective effect against cancers of the ovary and endometrium) | Sufficient |
| Oral contraceptives, sequential | Medicinal | Endometrium | Sufficient |
| Painters (occupational exposures as) | Occupational | Lung (esophagus, stomach, bladder) | No data |
| Phenacetin (in analgesic mixtures) | Medicinal | Renal pelvis/ureter, bladder | Limited |
| Radon and its decay products | Environmental | Lung | Sufficient |

**(cont.)**

**TABLE 2.5 (cont.)**
**Chemicals, Industrial Processes, and Environmental Factors Associated with Cancer Induction in Humans: Target Organs and Main Routes of Exposure in Humans and Degree of Supporting Evidence in Animals (IARC)**

| | Humans | | Animals |
|---|---|---|---|
| Chemical or Industrial Process | Main Type of Exposure[a] | Target Organ(s)[b] | Degree of Evidence for Carcinogenicity |
| Rubber industry | Occupational | Bladder, leukemia (lymphoma, lung, renal tract, digestive tract, skin, liver, larynx, brain, stomach) | Inadequate |
| Salted fish (Chinese-style) [280] | Environmental | Nasopharynx | Limited |
| *Schistosoma haematobium* [281] | Environmental | Urinary bladder | Limited |
| Silica, crystalline [282] | Occupational | Lung | Sufficient |
| Solar radiation [283] | Environmental | Skin | Sufficient |
| Strong-inorganic-acid mists containing sulfuric acid [284] | Occupational | Nasal sinus, laryngeal, lung | No data |
| Talc containing asbestos fibers | Occupational | Lung (pleura) | Inadequate |
| Tamoxifen [285] | Medicinal | Endometrium (reduces risk for contralateral breast cancer in women with previous diagnosis of breast cancer) | Sufficient |
| 2,3,7,8-Tetrachlorodibenzo-*para*-dioxin [286] | Occupational | Multi-site with no site predominating | Sufficient |
| Thiotepa [287] | Medicinal | Leukemia | Sufficient |
| Tobacco products, smokeless | Environmental, cultural | Oral cavity (pharynx, esophagus) | Inadequate |
| Tobacco smoke [275] | Environmental, cultural | Lung, bladder, oral cavity, larynx, pharynx, esophagus, pancreas, kidney, nasal cavity, stomach, liver, cervix, ureter, leukemia | Sufficient |
| Treosulphan | Medicinal | Leukemia | No data |
| Vinyl chloride | Occupational | Liver, lung, brain, lymphatic and hematopoietic system (gastrointestinal tract) | Sufficient |
| Wood dust [288] | Occupational | Nasal cavity, paranasal sinus | Inadequate |

[a] The main types of exposure mentioned are those by which the association has been demonstrated; exposures other than those mentioned may also occur.
[b] Suspected target organs in parentheses.
[c] The evaluation of carcinogenicity to humans applies to the group of chemicals as a whole and not necessarily to all individual chemicals within the group.
[d] Not all chemicals in this group are associated with all cancers listed.

*Sources*: IARC [112,292,289], Tomatis et al. [113], and individual citations referenced above.

**TABLE 2.6**
**Summary of the Classification Schemes for Carcinogens**

| Agency | Classification | Meaning |
|---|---|---|
| Germany (DFG/MAK)[a] | 1 | Carcinogenic to humans |
| | 2 | Carcinogenic in animal studies |
| | 3 | Suspected carcinogenic potential |
| | 4 | Nongenotoxic carcinogens |
| | 5 | Weak potency genotoxic carcinogens |
| EU [290] | 1 | Carcinogenic to humans |
| | 2 | Should be regarded as if carcinogenic to humans |
| | 3 | Cause for concern in humans |
| | | 3A. Substances that are well investigated |
| | | 3B. Substances that are insufficiently investigated |
| IARC [40,291–293] | 1 | Carcinogenic to humans |
| | 2A | Probably carcinogenic in humans; limited human; sufficient animal evidence |
| | 2B | Possibly carcinogenic in humans; limited human evidence; less than sufficient animal evidence |
| | 3 | Not classifiable |
| | 4 | Probably not carcinogenic to humans |
| ACGIH [294] | A1 | Confirmed human carcinogen |
| | A2 | Suspected human carcinogen, limited human evidence and sufficient relevant animal evidence |
| | A3 | Confirmed animal carcinogen with unknown relevance to human, epidemiologic studies does not confirm risk to humans |
| | A4 | Not classifiable |
| | A5 | Not suspected as human carcinogen, based on properly conducted epidemiologic studies or evidence in animal studies |
| Health Canada [295] | Group 1 | Carcinogenic to humans |
| | Group 1I | Probably carcinogenic to humans. Inadequate epidemiologic evidence; sufficient evidence in animal species |
| | Group 1II | Possibly carcinogenic to humans. Inadequate or flawed epidemiologic studies; limited animal evidence, or adequate animal evidence, but involves epigenetic mechanisms |
| | Group 1V | Unlikely to be carcinogenic in humans. No evidence in adequate epidemiologic studies; positive animal studies of limited or unlikely relevance to humans |
| | Group V | Probably not carcinogenic in humans; no evidence in adequate epidemiologic studies; no evidence or inadequate evidence in animal studies |
| U.S. EPA [91] | | Carcinogenic to humans |
| | | Likely to be carcinogenic to humans |
| | | Suggestive evidence of carcinogenic potential |
| | | Inadequate information to assess carcinogenic potential |
| | | Not likely to be carcinogenic in humans |
| NTP [296] | 1 | Known to be a carcinogen |
| | 2 | Reasonably anticipated to be a carcinogen |
| | | A. Limited evidence in human studies indicating credible causal relationship evidence in human studies |
| | | B. Sufficient evidence in animal studies |

[a] DFG/MAK, Deutsche Forschungsgemeinschaft/maximale arbeitsplatz-Konzentration (German Commission for the Investigation of Health Hazards of Chemical compounds in the work area), as discussed in Seeley et al. [34].

assumption is not valid for all animal carcinogens. Species-specific responses or high-dose-only effects provide evidence that positive animal results are not always evidence of human carcinogenicity.

A more sophisticated understanding of carcinogenicity with an emphasis on understanding the biology of carcinogenesis on mode of action is embodied in the 2005 EPA guidelines [91]. The 2005 EPA guidelines are groundbreaking. In their flexibility and incorporation of new science, these guidelines represent a significant advance in carcinogen classification schemes. The guidelines have a number of important new features; in particular, the *mode*

**TABLE 2.7**
**Classifications for Specific Carcinogens in Various Countries and Scientific Organizations**

| Chemical | IARC | European Union | Germany | The Netherlands | Sweden | Norway |
|---|---|---|---|---|---|---|
| Acrylonitrile | 2B | n.c. | 2 | I | n.c. | K2 |
| Asbestos | 1 | 1 | 1 | I | 1 | K2 |
| Benzene | 1 | 1 | 1 | I | 1 | K2 |
| Benzo(*a*)pyrene | 2A | 2 | 2 | I | 2 | K1 |
| Cadmium | 1 | 2 | 2 | II | 2 | I1 |
| Chloroform | 2B | 3 | 3 | II | 3 | K3 |
| Chromium (VI) | 1 | 2 | 2 | I | 1 | K1 |
| 1,2-Dichloroethylene | 2B | n.c. | 2 | I | n.c. | K2 |
| Ethylene oxide | 1 | 2 | 2 | I | 2 | K1 |
| Methylene chloride | 2B | 3 | 3 | II | 3 | K3 |
| Propylene oxide | 2B | 2 | 2 | I | 2 | K2 |
| Tetrachloroethylene | 2A | 3 | 3 | II | 3 | K2 |
| Trichloroethylene | 2A | 3 | 3 | II | 3 | K3 |
| Vinyl chloride | 1 | 1 | 1 | I | 1 | K2 |

*Note:* n.c., not classified. Acetonitrile and 1,2-dichloroethylene are classified as dangerous, but not carcinogenic, by the EU and Sweden.

*Source*: Seeley, M. R. et al., *Regul. Toxicol. Pharmacol.*, 34, 153–169, 2001. With permission.

*of action** forms the underpinning of other elements wherever possible.

The guidelines use five standard narrative descriptors to assess the carcinogenic hazard to humans: (1) carcinogenic to humans, (2) likely to be carcinogenic to humans, (3) suggestive evidence of carcinogenic potential, (4) inadequate information to assess carcinogenic potential, and (5) not likely to be carcinogenic to humans. The guidelines take a weight-of-evidence approach that evaluates all human, animal, and other relevant toxicological information. As part of this evaluation, the quality of an individual study as well as the overall consistency across studies is considered. Confidence that a specific chemical is the cause of cancer in human studies is enhanced by positive findings at the same organ site in multiple studies, with well-characterized exposures. In contrast to the 1986 EPA guidelines [105], other evidence relevant to carcinogenicity, importantly the mode of action information in animals that attests to the relevance (or lack of relevance) of a particular tumor response, is explicitly considered. Data from human, animal, and other sources are combined to weigh the totality of evidence to classify the human carcinogenic potential of a particular chemical.

It is noteworthy that the guidelines allow for multiple descriptors of carcinogenicity; that is, a chemical may be classified as "not likely to be carcinogenic" by one route of exposure, but "likely to be carcinogenic" by another route of exposure. Dose may also be an element of the descriptor in which a chemical may be carcinogenic but only above a specified dose. (Note that this is equivalent to assuming a threshold dose–response relationship.)

In the face of a lack of information, the EPA has specified certain default assumptions (e.g., positive findings in animals are assumed to be relevant to humans); however, as noted earlier, well-founded mode of action information (such as findings that a kidney tumor response in rodents is a consequence of α-2-macroglobulin accumulation) can be used to avoid use of such a default assumption.

It is not uncommon for different agencies to apply different criteria to the evaluation of carcinogens. In an analysis of regulatory approaches used by different European scientific agencies and countries, Seeley and coworkers [34] noted that, although nearly all agencies and countries considered mechanistic data in interpreting the relevance of animal studies of humans, some (e.g., IARC) considered species-specific but not dose-dependent differences in mechanisms, whereas others (e.g., EU) applied a more inclusive approach to use of data. They also observed that, in general, agreement was greater for classifications based mostly on human data (e.g., asbestos); however, the organizations differed more with respect to classifications based more on animal data (e.g., 1,2-dichloroethylene). Table 2.7 presents a comparison of the classification scheme applied by European agencies and countries to different chemicals.

## Dose–Response Assessment

One of the most contentious aspects of the evaluation of animal carcinogens by regulatory agencies is characterizing the dose–response relationship at the exposure levels

* *Mode of action* is defined by the EPA as a sequence of biochemical and cellular event resulting in tumor formation. It may be contrasted with mechanism of action, which implies a detailed understanding of the carcinogenic process, often at the molecular level.

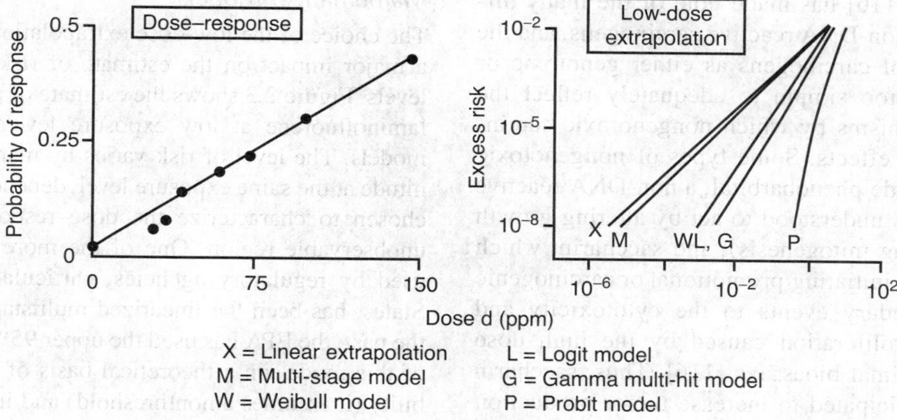


**FIGURE 2.3** Low-dose extrapolation for 2-acetylaminofluorene under several mathematical models. (From Bickis, M. and Krewski, D., in *Toxicological Risk Assessment*, Vol. 1, Clayson, D.B. et al., Eds., CRC Press, Boca Raton, FL, 1985. With permission.)

to which humans are likely to be exposed. Animals are typically exposed to carcinogens at levels that are orders of magnitude greater than those likely to be encountered in the environment by humans. It would be impossible to perform animal experiments with a large enough number of animals to directly estimate the level of risk at low exposure levels. Thus, to obtain a quantitative estimate of the risks humans are likely to encounter at ambient exposures requires the extrapolation of effects observed at high doses to low doses and from effects observed in animals to humans. Even the use of carcinogenicity data from human studies (mostly occupational studies) frequently requires the use of extrapolation models to estimate risks to humans exposed at lower ambient levels.

Mechanistic models are being developed to assist in dose–response assessment. Pharmacokinetic models attempt to describe the relationship between exposure and biologically relevant dose to the target tissue. These models characterize absorption, distribution, metabolism, and excretion of chemicals. Pharmacodynamic models attempt to describe the relationship between the dose to target tissue and response. Both of these types of models can assist in extrapolation from high to low doses and across species.

## Low-Dose Extrapolation

Extrapolation from high to low dose is done using mathematical models that are hypothesized to characterize the dose–response relationship of carcinogens at both the high-dose and response levels observed in animal or human occupational studies and the low-dose and response levels of interest for human exposures. The choice of mathematical model depends on two factors: (1) the hypothesis for the mechanism of carcinogenesis for a particular chemical, and (2) the science policy decision to choose, in the absence of data firmly supporting one model or another, the more conservative model (of several biologically plausible models) or to present results from a range of plausible models.

### *Threshold vs. Nonthreshold Mechanisms*

The determination of whether carcinogenesis is a threshold or nonthreshold phenomenon is a key consideration in the choice of model to characterize the dose–response relationship. It is considered plausible by some scientists that carcinogenesis could be a nonthreshold phenomenon for genotoxic agents, particularly those that act directly to cause mutations. For example, the human carcinogen vinyl chloride is an electrophilic agent and is understood to interact with DNA [114], although even in the case of vinyl chloride cell proliferation may play a significant role in carcinogenesis [115], possibly resulting in a nonlinear dose–response even for this agent. There is much debate over whether carcinogenesis is a threshold phenomenon, especially for the many chemicals that do not interact directly with DNA (not directly genotoxic) and which may induce cancer through epigenetic mechanisms.

In addition, for many carcinogens, it is likely that the dose–response relationship observed at high doses is not necessarily the same as the dose–response relationship that might occur at low doses. Because the measure of the carcinogenic potency of a chemical is typically determined by fitting a model to the observed data and then extrapolating to low doses, the implicit assumption is that the dose–response relationship is the same at high and low doses. For chemicals that cause cell damage at high doses or for chemicals for which detoxification pathways become saturated at high doses, it is likely that a different dose–response relationship will be observed at high and low doses, even for those chemicals where a non-zero slope is plausible at any dose. As discussed earlier, work by Gaylor (as well as others) highlights the predominant role of cell proliferation in the carcinogenicity of many chemicals that yield positive tumor responses at the MTD.

Butterworth [116] has made note of the many different classes of non-DNA reactive carcinogens, and the characterization of carcinogens as either genotoxic or nongenotoxic is too simple to adequately reflect the numerous mechanisms by which nongenotoxic carcinogens exert their effects. Some types of nongenotoxic carcinogens include phenobarbital, a non-DNA reactive carcinogen that is understood to act by altering growth control (increasing mitogenesis), and saccharin, which appears to exhibit initiating/promotional or carcinogenic activity as secondary events to the cytotoxicity and increased cell proliferation caused by the high dose levels used in animal bioassays [116]. Thus, saccharin would not be anticipated to increase tumor production at doses that do not cause cytotoxicity, and phenobarbital might not be expected to cause cancer at doses that do not affect growth control. Zeise and coworkers [117] reviewed the experimental evidence for various shapes of dose–response relationships for carcinogens. They concluded that "reliable high dose data from human studies contains examples of superlinearity (radium injections and bone cancer), linearity (various radiation exposures), and sublinearity (smoking)." Their analysis of animal studies indicated that the "variety of shapes of dose–response curves observed for humans was also seen for animals." Zeise and coworkers noted the lack of data to indicate the shape of dose–response relationships at doses corresponding to lifetime risks of 1 in a million; in humans, some data exists for incidence rates as low as 1%, and in animals two large studies have provided data at lifetime risks of a few tenths of a percent. Because carcinogens act by different mechanisms, it is very likely that different carcinogens, or classes of carcinogens, will exhibit different types of dose–response relationships. Moreover the same carcinogen may cause cancer at different tumor sites via different mechanisms and hence be associated with different dose–response relationships.

A striking example of different dose–response relationships for a single carcinogen is the example of 2-acetylaminofluorene (2-AAF) [118], which is a potent mutagenic carcinogen. The dose–response relationship for 2-AAF-induced liver cancer exhibits the expected (for a genotoxic carcinogen) linear dose–response relationship, whereas the dose–response relationship for bladder cancer is highly nonlinear, demonstrating an apparent threshold. The mechanistic basis for the different dose–response relationships appears to involve differences in the relative importance of genetic damage (the likely key event in liver cancer) vs. genetic damage *and* hyperplasia of the bladder urothelium (the likely key events in bladder cancer). Thus, selection of the appropriate shape of the dose–response relationship for *any* chemical requires an understanding of the mechanism by which tumors are induced.

### *Mathematical Models*

The choice of the low-dose extrapolation model can have a major impact on the estimate of risk at low exposure levels. Figure 2.3 shows the estimate of risk from 2-acetylaminofluorene at low exposure levels, using different models. The level of risk varies by many orders of magnitude at the same exposure level, depending on the model chosen to characterize the dose–response curve in the unobservable region. One of the more common models used by regulatory agencies, particularly in the United States, has been the linearized multistage model [33]. In the past, the EPA has used the upper 95% confidence limit of this model on a theoretical basis of biological plausibility (it assumes a nonthreshold) and its conservatism (it is unlikely to underestimate risk at low exposure levels) [105]. Cancer potency estimates for chemicals are highly dependent on model choice; for example, Anderson [33] analyzed cancer potency estimates for 2,3,7,8-tetrachlorodibenzo-*p*-dioxin (TCDD), derived by using different low-dose extrapolation models and different selection and treatment of bioassay data, and found that model choice alone (Weibull, multistage, log probit) would account for a difference in the calculated cancer potency of 13 orders of magnitude.

Because of the uncertainties in dose–response modeling for low-dose cancer risk and an increased emphasis on mode of action in cancer risk assessment along with a growing acceptance of a threshold dose–response relationship for some carcinogens, efforts are being made to incorporate a greater biological understanding of tumorigenesis into cancer dose–response assessment. Some of the most comprehensive efforts in this area have been conducted by scientists at the CIIT Centers for Health Research with respect to the development of a biologically motivated computational model for formaldehyde in the F344 rat. These investigators have developed a model that incorporates information on nasal dosimetry, cell replication, and DNA cross-links into a two-stage clonal growth model (see Figure 2.4) [119]. This modeling yields a J-shaped dose–response relationship that reflects the highly nonlinear dose–response relationship for tumorigenicity. The biologically based modeling for nasal tumors from formaldehyde results in maximum likelihood estimates (MLEs) for cancer risk at 0.1 ppm formaldehyde in air that are, in some cases, as much as 1000-fold lower than the values used in the current EPA Integrated Risk Information System (IRIS).

In the 2005 cancer risk assessment guidelines, the EPA has revised its approach to dose–response assessment [91]. If sufficient data are available, a biologically based dose–response model (such as the one for formaldehyde described earlier) is chosen as the most appropriate method for evaluating both the observed data and for extrapolating to exposures below the observed dose range. On the other hand, the more likely case is that data are

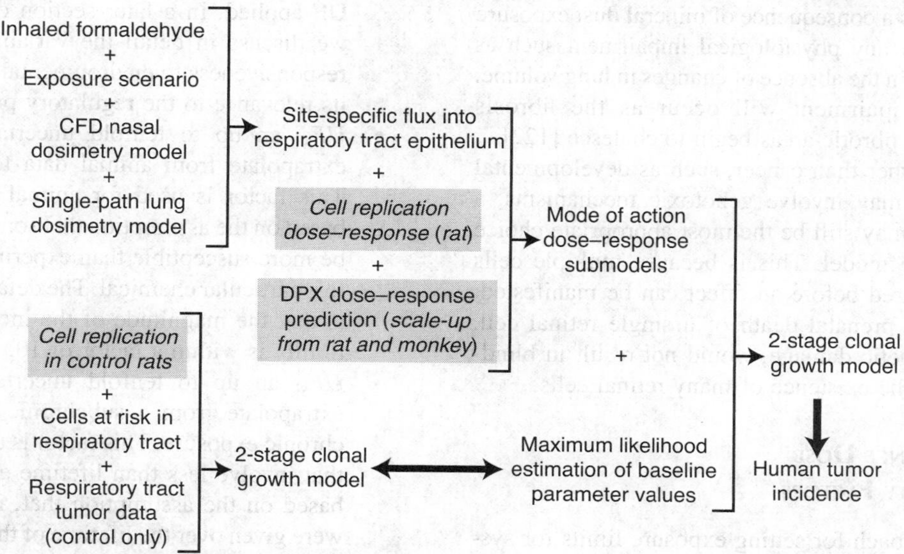


**FIGURE 2.4** Interrelationships of the major components of the human dose–response model. (From Conolly, R.B. et al., *Toxicol. Sci.*, 82, 279, 2004. With permission.)

not available for development of biologically based models. In this situation, a "point of departure" approach is recommended. The point of departure represents a dose, within the range of observed data, associated with a specified extra tumor risk. The point of departure is developed using mathematical models, such as the linearized multistage model (although other models can be used) and is typically expressed as the lower 95% confidence limit on the dose associated with specified extra risk (e.g., $LED_{10}$ would be the lower 95% confidence limit on the dose associated with 10% extra risk above background.). Risks below the $LED_{10}$ are characterized either through linear extrapolation, for chemicals believed to act via a linear dose–response relationship (e.g., genotoxic carcinogens), or through a margin-of-exposure analysis, for chemicals whose dose–response relationship is likely to be either threshold or nonlinear. For chemicals where data might support either linear extrapolation or a margin-of-exposure analysis, both analyses are to be presented.

A panel organized by the International Life Sciences Institute (ILSI) applied the earlier version of the 2005 EPA guidelines to an assessment of the dose–response relationship for chloroform [120]. The ILSI panel evaluated the large database of information relevant to the mode of action by which ^chloroform induces liver tumors in rodents. The group identified a number of key elements important to the likely mode of action of chloroform: lack of evidence of genotoxicity, tumor induction at doses associated with frank toxicity at the tumor sites, and the role of cytotoxicity and compensatory cell proliferation in tumor induction. The group concluded that the evidence did not support a linear dose–response relationship for chloroform. A margin-of-exposure analysis was considered to represent the most appropriate method for evaluating the potential hazards of chloroform at low doses. The significance of such an approach is potentially quite large with respect to regulatory decision making. In the case of chloroform, use of the margin-of-exposure analysis would yield a permissible level in drinking water in the United States of 300 μg/L, a 60-fold increase beyond the present permissible level of 5 μg/L.

## EVALUATION OF SYSTEMIC TOXICANTS

In its broadest sense, systemic toxicity refers to all adverse effects, but in general it is applied only to chemicals that are postulated to induce adverse effect through a threshold mechanism; that is, these chemicals have a level of exposure below which there is minimal, if any, chance for an adverse effect. The effects range from skin and eye irritation to subchronic or chronic damage to any organ system, such as pulmonary fibrosis.

The underlying hypothesis for the threshold model for systemic toxicants is that multiple cells must be injured before an adverse effect is experienced and that the injury must occur at a rate that exceeds the rate of repair. This is in contrast to the approach for carcinogenesis, in which a genotoxic insult involving direct DNA damage to a single cell is theoretically sufficient to allow that cell to grow to a malignant tumor [121]. (As discussed earlier in this chapter, it should be noted that this model for carcinogenesis is now viewed as too limiting.) An example of a threshold-type injury can be seen with pulmonary fibrosis due to mineral dust exposure. Fibrotic areas may be present and observed as radiographic or histopathologic changes in the lungs of miners as a consequence of mineral dust exposure in the absence of any physiological impairment such as reduced FEV1 or in the absence of changes in lung volume. Physiological impairment will occur as the fibrosis increases and the fibrotic areas begin to coalesce [122].

For effects other than cancer, such as developmental effects that still may involve genotoxic mechanisms, a threshold model may still be the most appropriate choice of dose–response model. This is because multiple cells must still be injured before an effect can be manifested; for example, the prenatal death of a single retinal cell, even through genetic damage, would not result in blindness because of the existence of many retinal cells.

## The Risk Reference Dose and Uncertainty Factors

The general approach for setting exposure limits for systemic toxicants is based on developing a permissible daily intake, the *acceptable daily intake* (ADI), a daily intake level of a chemical in humans that is associated with minimal or no risk of adverse effects. Basically, the ADI is developed through the use of appropriate uncertainty factors (UFs), also termed *safety factors*, which are applied to an experimental exposure. The ADI is expressed in terms of milligrams of chemical per kilogram of body weight per day [123]. The EPA refers to such an exposure level as the *risk reference dose* (RfD). The basis for the change in terminology is that the ADI does not represent a magic dividing line between safe and not safe but represents an estimated dose, derived through a consistent methodology, at which the chance of adverse effects is estimated to be negligible. The lack of precision is reflected in the EPA's description of the RfD as having an uncertainty perhaps spanning an order of magnitude [123].

An ADI or RfD is typically based on either a *no-observed-adverse-effect level* (NOAEL) or a *lowest-observed-adverse-effect level* (LOAEL) from an epidemiology or animal toxicology study. UFs are then applied to the NOAEL or LOAEL to account for uncertainties in the relationship between exposure to a chemical in an animal study and a particular effect, and the relationship between lifetime daily exposure to the same chemical in the general population of humans and the likelihood of a particular effect.

The history as well as the experimental support for UFs in developing a permissible dose for production against systemic toxicity in humans have been described by Dourson and Stara [124]. These authors describe four categories of UFs:

- $UF_H$, an up to tenfold uncertainty factor to account for variations in susceptibility in humans. If a NOAEL was defined from a long-term study in humans, this would be the only UF applied. In a later section of this chapter, we discuss in detail the variability in human responsiveness to environmental pollutants and its relevance to the regulatory process.
- $UF_A$, an up to tenfold uncertainty factor to extrapolate from animal data to human data. This factor is used for animal studies and is based on the assumption that some humans may be more susceptible than experimental animals to a particular chemical. The default assumption is that the magnitude of the increased susceptibility is within a factor of 10.
- $UF_S$, an up to tenfold uncertainty factor to extrapolate from a subchronic exposure to a chronic exposure. This factor is used for studies that involve less than lifetime exposure and is based on the assumption that, if the chemical were given over the lifetime of the animal rather than over a fraction of the lifetime, a smaller amount of chemical would result in the same NOAEL.
- $UF_L$, an up to tenfold uncertainty factor to extrapolate from a LOAEL to a NOAEL. This factor is used for studies in which a NOAEL was not identified. The default assumption is that a dose at 1/10 the LOAEL would result in a NOAEL.

The EPA has developed a fifth uncertainty factor, $UF_D$, which may be applied when the database is incomplete [125]. The assumption here is that, when the database for a chemical is limited, there is uncertainty as to whether the identified NOAEL might be significantly lower if other studies were performed or whether a different NOAEL might have been identified if additional health endpoints (such as reproductive toxicity) had been evaluated. A complete database is defined as having two chronic mammalian studies, one mammalian multigeneration study, and two mammalian developmental toxicity studies. If these five studies are available, then there is a high degree of confidence that one has approximated the lowest NOAEL.

Uncertainty factors are used multiplicatively. To derive the RfD, the EPA divides the exposure level from the toxicity study by the UFs. Mathematically, this is represented as:

$$\mathrm{RfD} = \frac{\mathrm{LOAEL\ or\ NOAEL}}{\mathrm{UF}_1 \times \mathrm{UF}_2 \times \mathrm{UF}_n}$$

The use of all five UFs ($\mathrm{UF_H}$, $\mathrm{UF_A}$, $\mathrm{UF_S}$, $\mathrm{UF_L}$, $\mathrm{UF_D}$), each representing an order of magnitude, could in theory lead to a total uncertainty factor of 100,000. This would occur if data were from a subchronic animal study that identified only a LOAEL and the database was limited. However, because the multiplication of 4 or 5 factors of 10 each is likely to yield unrealistically conservative RfDs, the EPA has restricted the total uncertainty factor calculation as follows: When uncertainty exists in four areas, the EPA uses a 3000-fold total uncertainty factor; when uncertainty exists in five areas, the EPA uses a 10,000-fold uncertainty factor [125].

Although 10 represents the default value for a UF, values less than 10 may also be used, depending on the nature of the available information; for example, in deriving an RfD for chromium (VI), a factor of 3 was applied for the UFs to account for the fact that the 1-year exposure duration in the principal study was less than lifetime but longer than typical subchronic studies in rodents [126]. No UFs were used in developing the RfD for fluoride because the NOAEL for the critical effect (dental fluorosis) was observed in the sensitive population (children) for a sufficiently long exposure duration [127].

In addition, a *modifying factor* (MF) that is greater than 0 and less than or equal to 10 may be used to address uncertainties not addressed in the other factors; for example, the use of a very large number of animals in a study may enhance certainty in the RfD, resulting in the use of a MF less than 1 but greater than 0. Alternatively, when an RfD is based on a very limited number of animals, an MF greater than 1 but less than or equal to 10 may be appropriate [125].

The RfD approach represents a generally accepted (NAS, FDA, and EPA among others) method for setting lifetime exposure limits for humans, and the use of default tenfold UFs has some experimental support, particularly as upper-bound estimates [124]. For example, the ratio between the subchronic and chronic NOAEL or LOAEL for 52 chemicals was less than 10 in 96% of the cases, as described in the analysis by Dourson and Stara [124]; thus, the uncertainty factor of 10 would be an underestimate for only 4% of these chemicals. A similar analysis regarding subchronic to chronic was performed by Lewis [128], who observed that for 18 chemicals the ratio of the subchronic to chronic NOAEL was 3.5 or less for 14 chemicals, and only one had a ratio of greater than 10. If the chemical with a ratio greater than 10 were excluded from the analysis, the mean subchronic to chronic NOAEL ratio was 3.3. Thus, the default UF of 10 for extrapolating from subchronic to chronic exposures would be very protective for most chemicals, and an UF of 3 may be more appropriate than the default value of 10 for many chemicals.

There are several limitations in the RfD approach, the net result of which is that exposures resulting in the same RfD do not imply the same level of risk for all chemicals and that exposures above the RfD do not represent the same increase in risk for all chemicals. First of all, the choice of a LOAEL or a NOAEL does not take into consideration the greater experimental confidence associated with, for example, studies using more experimental animals. An exposure dose defined as a NOAEL in one experiment could turn out to be a LOAEL if more experimental animals had been used (i.e., an effect may have been detected using more animals). As a result, poor experiments may yield anticonservative RfDs, as studies using fewer animals may result in a higher RfD than studies using larger numbers of animals [129].

In addition, the RfD approach does not make use of dose–response information, which is a key determinant in assessing the likelihood of effects. Thus, a chemical with a steep dose–response curve would be associated with a greater likelihood of effects as exposure increased above the RfD and a smaller likelihood of effects with exposures below the RfD than would a chemical with a more shallow dose–response curve [130].

There are also difficulties with the implications of specific UFs. The default value of 10 for the interspecies uncertainty factor ($UF_A$) may be reasonable in some cases, but in other cases may not be appropriate. For chemicals for which metabolism is a key determinant of toxicity, interspecies differences may be due mainly to physiological and metabolic differences across species [131]. Under this assumption, interspecies differences are believed to scale according to allometric principles; that is, when the dose is expressed on a dose-per-unit surface area, different species are presumed equally sensitive to a chemical [124]. According to this interpretation, then, a scaling factor based on surface area (body weight$^{2/3}$) should account for interspecies differences [124,131]. For rodent toxicity studies (most commonly used in toxicological risk assessments), such a scaling factor would be about 8 for rats and 13 for mice, similar to the default value of 10 for the $UF_A$; however, using this interpretation, the default value of 10 would not be adequately protective when using studies of much smaller animals and would be overly protective when using studies of much larger animals.

If, however, pharmacokinetic modeling has been used to estimate a biologically effective dose and extrapolate dose across species, the use of an interspecies $UF_A$ may be unnecessary; for example, this may apply to chloroform, for which a mechanistically based dose–response model for hepatotoxicity has been developed [132]. A similar conclusion has been reached by Jarabek and coworkers [133], who have developed a methodology for estimating the reference concentration (RfC). The RfC is an air concentration of a chemical that is expected to be associated with minimal risk, if any, for adverse effects in humans, including susceptible populations [133,134]. As such, the RfC is the functional equivalent of the RfD, except that it is based on an air concentration rather than an administered dose. Overton and Jarabek [135] noted that, when dosimetric adjustments (e.g., through physiologically based pharmacokinetic modeling) are made, use of the value of 10 for the $UF_A$ for cross-species extrapolation may be inappropriate. This is because the dosimetric adjustment has already addressed some of the basis for interspecies variability.

The process of identifying appropriate UFs has been undergoing refinements over the past years. The $UF_A$ and the $UF_H$ can be considered as having two components: a toxicokinetic (TK; exposure to dose) component and a toxicodynamic (TD; dose to response) component. In 1993, the values of $10^{0.6}$ (i.e., 4) and $10^{0.4}$ (i.e., 2.5), respectively, were proposed for species differences, and equal values of $10^{0.5}$ (i.e., 3.16) for human variability [136,137]. This approach allows for greater use of chemical-specific information and allows the use of data-derived uncertainty factors or chemical-specific uncertainty factors. The EPA [138] recently applied the data-derived UFs for developing an RfD for boron. The endpoint of concern was developmental effects in rodents; hence, the pregnant female was considered the sensitive population and the basis for the NOAEL. Data were available for differences in clearance rates of boron across species (the TK component of the $UF_A$), allowing a $UF_A$ for TK of 3.3, only slightly higher than the default value of 3.16. For the TD component of the $UF_H$, data on differences in glomerula filtration rate among women resulted in the use of a value of 2.0 (vs. the default value of 3.16). Default values of 3.16 were used for the TD component for the $UF_A$ and for the $UF_H$. The total UFs in this analysis were 66. Thus, the use of data-derived UFs resulted in a more scientifically founded RfD that was somewhat lower than would have resulted from the use of default values.

## Alternative Approaches to the Risk Reference Dose

Alternative approaches to the standard RfD approach exist. By employing dose–response modeling and statistics, these alternative approaches can address issues of experimental quality, the shape of the dose–response curve, and other limitations of the RfD approach. Examples include the benchmark dose method [139,140], probabilistic RfD approaches [141,142], and distributional population approaches. These methods are described below.

The EPA defines a benchmark dose (BMD) as "a statistical lower confidence limit on the dose producing a predetermined level of change in adverse response compared with the response in untreated animals" [140]; for example, a BMD could represent the 95% lower confidence limit on a dose that produces a 10% increase in a particular adverse health effect. The BMD is then used like a NOAEL or LOAEL, and appropriate UFs are applied to derive an RfD based on a BMD. Calculation of a BMD is illustrated in Figure 2.5. Typically, a value of 10% has been selected as the appropriate parameter, as 10% represents a value at or near the limit of sensitivity in most cancer and non-cancer bioassays [143].

The BMD approach overcomes many of the weaknesses of the RfD approach [139,140]. Because BMDs

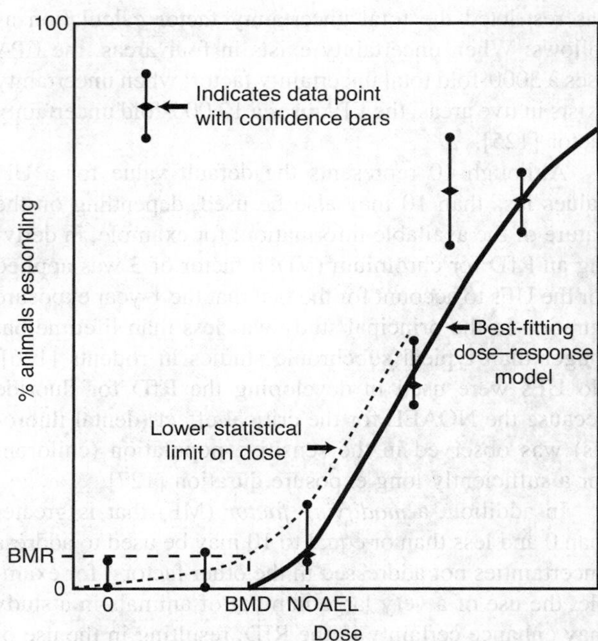


**FIGURE 2.5** Example of calculation of a BMD. (From EPA, *The Use of the Benchmark Dose Approach in Health Risk Assessment*, EPA/630/R-94/007, Risk Assessment Forum, Office of Research and Development, U.S. Environmental Protection Agency, Washington, D.C., 1995.)

are determined based on statistical modeling of dose–response data, the approach incorporates information on the sample size and the shape of the dose–response curve, information that is not taken into account using the standard RfD approach. Unlike NOAELs or LOAELs, BMD values are not constrained to be based on one of the experimental doses tested and are less dependent on the study design. Also, the BMD approach can be used for both threshold and nonthreshold adverse health effects, as well as for both quantal and continuous toxicity data. The BMD approach allows for greater consistency between values derived for different chemicals. A further benefit of the BMD approach is that it allows for possible future harmonization of cancer and noncancer risk assessment methods [143,144].

For any particular chemical, UFs applied to NOAELs or LOAELs using the standard RfD approach may not be appropriate to apply to a BMD. This is because additional dose–response information has already been incorporated into the BMD. When using a BMD, for example, a steeper dose–response curve or a less severe critical effect may warrant a smaller value for $UF_L$ than would a more shallow dose–response curve or a more severe effect [130].

The EPA has used the BMD approach in developing RfDs for a number of chemicals, including beryllium, methylmercury, and tributyltin oxide [145]. Similarly, benchmark concentrations (BMCs) have been used to

**TABLE 2.8**
**Heart Person-Days with at Least One Hourly COHb Estimate ≥ Value for Four Alternative Scenarios**

| | "As Is" Air Quality | | "Just Attain" Air Quality | |
|---|---|---|---|---|
| **Exposure Indicators** | **"Ambient Air" Plus Internal Sources** | **"Ambient Air" without Internal Sources** | **"Ambient Air" Plus Internal Sources** | **"Ambient Air" without Internal Sources** |
| COHb ≥ 2.1% | 488 | 72 | 457 | 0 |
| COHb ≥ 3.0% | 37 | 0 | 24 | 0 |

*Source:* Adapted from EPA, *Review of the National Ambient Air Quality Standards for Carbon Monoxide: 1992 Reassessment of Scientific and Technical Information*, OAQPS staff paper, U.S. Environmental Protection Agency, Washington, D.C., 1992.

develop RfCs for antimony trioxide, carbon disulfide, methylene diphenyl diisocyanate, methyl methacrylate, phosphoric acid, 1,1,1,2-tetrafluoroethane, and chromium (VI) particulates [145]. For all of the chemicals listed, a 10% relative change was chosen as the benchmark response. Although the EPA has performed a BMD analysis for naphthalene, the resulting RfD ($3 \times 10^{-2}$ mg/kg-day) was very similar to the value derived using a standard NOAEL/RfD approach ($2 \times 10^{-2}$ mg/kg-day), and the EPA decided to use the NOAEL/RfD approach in deriving the final EPA-recommended value [146].

A probabilistic approach for developing RfD or ADI values, combining distributions for each uncertainty factor using a probabilistic approach to generate an overall distribution for no adverse effect in sensitive human subpopulations, has been recommended by The Netherlands National Institute of Public Health and the Environment [141]. This approach allows consideration of all available data on various uncertainties and minimizes the conservatisms introduced by multiplying numerous UFs. A similar approach has also been recommended by Baird and coworkers [142].

For compounds that have been well studied, particularly in humans (both in terms of exposure and in terms of toxicity), distributional population approaches have been used to evaluate toxicity and to provide input into the decision-making process. Such approaches have been applied mainly to evaluation of the NAAQS, standards for moderately toxic air contaminants that are ubiquitous in the United States. Much of the basis for the selection of NAAQS is human toxicological data, although animal data are used in a supporting role.

An example of a distributional population approach can be seen in the evaluation of carbon monoxide (CO) toxicity and exposure by the EPA Ambient Standards Branch [147]. As part of this assessment, the EPA reviewed several studies that evaluated the relationship between exposure to CO, using carboxyhemoglobin (COHb) in blood as an indicator of exposure, and percent decrease in time to angina or pain in the chest, as an indicator of effect. Most of the studies showed an impact of low COHb levels on angina; however, no consistent dose–response relationship was observed when studies were analyzed in the aggregate. This may have been due to differences in study design, study populations, and other factors. Because of the lack of a clear dose relationship, the EPA evaluated the impact of different concentrations of CO in air on various cut-off points of COHb, from 2.1 to 3.0%. These cut-off points are conceptually similar to the LOAEL used in RfD development. Levels of CO that result in COHb of 2.9 to 3.0% or higher might constitute frank effect levels (FELs). This is because levels of COHb of 2.9 to 3.0% or higher in persons with heart disease are considered as possibly increasing the risk of myocardial ischemia and diminishing blood flow to the heart.

The risk of CO exposure to people with heart disease in Denver (36,345 individuals at the time) was estimated under different CO levels [147]. The number of person-days when individuals might have at least one hourly COHb level greater than or equal to a defined percent COHb was estimated. Table 2.8 presents some of the results of this analysis. Under conditions at the time (considering both indoor and outdoor sources of CO), approximately 488 person-days occurred in which the Denver population with preexisting heart disease would experience COHb levels greater than 2.1%. If only ambient air is considered, then the person-days drops to 72. If the NAAQS for CO is attained, then the person-days drops to 457 for all sources and 0 for ambient air only. This type of analysis is useful in showing the benefits of CO reduction, as well as identifying the significance of different sources.

Distributional population approaches to evaluating environmental chemicals provide a more comprehensive evaluation of risks than the RfD approach. Rather than focusing on point estimates (e.g., above or below the RfD), this method allows one to more fully characterize variability in responsiveness to chemicals and variability in exposure levels among defined populations; however, this approach is feasible only for a limited number of chemicals and is quite resource intensive.

## Endpoints: Developmental Toxicity

Refinements in the evaluation of systemic toxicants have focused on specific target organ systems, such as the nervous system or the developing fetus. In this section, we focus on developmental toxicity as illustrative of the use of a specific endpoint in the regulatory toxicology, including, for example, classification schemes and approaches to setting permissible levels. Evaluation of developmental toxicants presents some unique challenges in risk assessment. This is because standard chronic and subchronic toxicity tests cannot be used to provide information on developmental effects in the offspring of exposed mothers; rather, animal selection and exposure conditions must be designed specifically to assess developmental effects.

Developmental toxicity has been defined to include "any detrimental effect produced by exposures during embryonic stages of development" [123]. These effects may include structural abnormalities, functional abnormalities, growth retardation, or lethality [148]. The effects may be reversible, such as a temporary reduction in growth rate, or irreversible, such as an overt physical malformation.

Our discussion on the evaluation of developmental toxicants in the regulatory process focuses on four main issues:

- Selection of testing protocols
- Relevance of animal studies to hazard identification in humans, including weight-of-evidence classifications
- Use of the RfD approach to assess risk
- Use of other approaches to assess risk

### Selection of Testing Protocols

The use of animal studies is particularly critical in evaluating developmental toxicants, as compared to chemicals that induce other effects, for several reasons, including the emphasis on adult males in occupational epidemiology studies, the lack of a long-term national registry of birth defects, and the difficulty in identifying certain endpoints, such as resorptions. Types of studies used to evaluate developmental toxicity include the conventional segment 2 study, in which the dams are typically exposed during the period of fetal organogenesis and litters are evaluated for a number of endpoints, including number of viable offspring, types and incidence of skeletal or visceral malformations or variations, and body weight [148,149]. Segment 1 tests focus on fertility and reproductive performance of males and females. Segment 3 tests include perinatal and postnatal study after the treatment of females only and, as such, may be considered sequential to the segment 2 test. In addition, multigeneration studies are also performed to assess fertility, reproductive performance and, sometimes, teratology [149].

In the segment 2 tests, maternal toxicity endpoints, such as organ weights and clinical histopathology, are also evaluated. Of particular concern are those compounds that induce toxicity in the offspring in the absence of significant maternal toxicity. Significant differences can be found in the protocols for segment 2 tests, particularly between countries. These include differences in animal species, dosing regimen, and specific endpoints evaluated. For example, the Japanese protocol requires that some females be allowed to litter and the pups in the litters are examined for physical, reproductive, and functional development [149], which contrasts with typical U.S. protocols where littering does not occur and only *in utero* pups are examined. These differences may be significant, with potential implications for regulatory action; for example, functional deficits in offspring could be observed in the Japanese segment 2 protocol but not in the standard U.S. segment 2 protocol, where additional testing (segment 3) would be required to detect such effects.

Because of the implication of test differences between countries, an expert panel of scientists has proposed that efforts be made toward the harmonization of guidelines for reproductive and developmental toxicity testing [149]. Harmonization would result in international guidelines for reproductive and developmental toxicity testing to improve the comparability of data from studies. In addition, harmonization of testing schemes would allow for more efficient use of resources and possibly reduce the need for animal testing, as one country could more readily use the results of studies performed in other countries than is now possible.

In addition to the standard methods, specialized developmental toxicity methods involving, for example, developmental neurotoxicity testing are available. These can be either an addition to a segment 2 test or a distinct test. Developmental neurotoxicity testing may be especially critical in evaluating certain chemicals such as lead [59] or PCBs [150], which have been associated with subtle neurobehavioral changes in offspring from relatively low-level prenatal exposures.

Short-term screening tests and *in vitro* tests, such as the Chernoff/Kavlock [151], have also been used to evaluate developmental toxicants. In general, these tests are insufficient for performing quantitative risk assessments; however, they may be useful in selecting chemicals for further analysis and in helping to guide the nature of further analyses.

Short-term tests are of particular use when one considers the complexities of animal testing and the relatively high frequency of developmental abnormalities. For example, about 3% of infants are born with major congenital malformations that are recognized in the first year of life [152].

**TABLE 2.9**
**Classification of Chemicals Based on Teratogenic Potential**

| Criteria | Category A | Category B | Category C | Category D |
|---|---|---|---|---|
| 1. Ratio, minimum maternotoxic dose to minimum teratogenic dose | Much greater than 1 | Generally greater than 1; teratogenic range starts below the maternotoxic dose range[a] and overlaps it | ≤1 | No teratogenicity even at maternotoxic doses |
| 2. Incidence of malformations | Dose related and high | Dose related and high | Dose relatedness of each malformation less obvious; incidence low | — |
| 3. Type of malformation at lower doses | Organ systems involved are specific | Characteristics possibly specific, generally multiple | Nonspecific involving different organ systems | |
| 4. Target cell | Specific cells | Specific cells | Nonspecific and generalized | Not known |
| 5. Range of safety factor | 1–400 | 1–300 | 1–250 | 1–100 |

[a] The maternotoxic dose range extends between the dose initiating signs of toxicity and the dose causing 50% mortality ($LD_{50}$).

*Source:* Khera, K. S. et al., *Current Issues in Toxicology: Interpretation and Extrapolation of Reproductive Data To Establish Human Safety Standards*, Springer-Verlag, New York, 1989, p. 94. With permission.

## Relevance of Animal Studies to Hazard Identification in Humans, Including Weight-of-Evidence Classifications

Significant species differences have been observed with respect to the susceptibility of chemicals to induce developmental toxicity. Perhaps the classic example of species differences is thalidomide. Thalidomide exposure induces comparable target-organ specificity for limb defects in rabbits and various primates, but not rats. Had initial toxicity tests on thalidomide involved more appropriate animal species, the human tragedy of thalidomide might have been mitigated [152].

Attempts have been made to develop categorical classification schemes and to provide interpretative descriptions of developmental toxicity data. Overall, the aim of such evaluations is to assess the likelihood that a chemical can cause developmental effects in humans. Although not as codified as the cancer classification schemes described earlier, the use of such schemes for developmental toxicants could be further developed to form a more integral part of the regulatory process. For example, weight-of-evidence schemes could be used for developing regulatory priorities.

A classification scheme for developmental toxicants is shown in Table 2.9. This scheme categorizes chemicals based on teratogenic potential. Given equal exposure levels, chemicals in category A would present the greatest concern for teratogenic potential. This particular scheme includes several elements: relationship of maternally toxic dose to developmentally toxic dose, shape of the dose–response curve, and the nature of the malformations. The scheme differs from most carcinogen classification schemes in that it takes into consideration elements of the dose–response relationship; for example, the scheme considers the relationship between the maternally toxic and the developmentally toxic dose and proposes a range of safety factors for extrapolating to human risks, depending on category.

An alternative to classification schemes for developmental toxicants is the use of text descriptors of hazard and other risk elements. This approach was used by the Institute for Evaluating Health Risk (IEHR) in the report *An Evaluative Process for Determining Human Reproductive and Developmental Toxicity of Agents* [153]. This document contains the deliberations of an *ad hoc* group of industry and government scientists. Using a hypothetical chemical ("Terminator"), the relevance of animal toxicological data for human risk was evaluated, using expert opinion and consensus development and considering factors such as pharmacokinetic differences between humans and animals, absorption potential through different body interfaces, and biological monitoring data. As noted earlier, rather than yielding a rigid categorization scheme, the analysis resulted in text descriptors to assist in the interpretation of animal toxicological studies with respect to potential for hazard in humans.

The use of classification schemes or interpretive text descriptors to evaluate the relevance of animal developmental studies to humans, as described earlier, could significantly improve the use of developmental toxicity data in the regulatory process. In addition, similar approaches are warranted for other endpoints, such as immunotoxicity.

## Use of the Risk Reference Dose To Assess Risk

Risk reference doses may be derived for developmental effects, using essentially the same methodology as described earlier for systemic toxicants in which case the value is an RfDDT [148]. The RfDDT is derived from a LOAEL or NOAEL from a developmental toxicity study; however, because the relevant exposure period is not chronic, but the *in utero* and possibly earlier time period, the RfDDT applies not to a lifetime exposure but only to the study exposure period. At present, RfDDTs are available for only a limited number of chemicals. In general, the use of RfDs for specific endpoints and periods of exposure duration could allow greater comparability of RfDs and better use of toxicological information.

### Use of Other Methods To Assess Risk

Alternative methods to the RfDDT have been developed to assess risks or to develop protective levels for developmental toxicants. One method described earlier in Table 2.9 involves the use of different safety factors applied to the minimum teratogenic dose depending on the teratogenic potential category (A to D) to which the chemical belonged. The use of variable safety (now termed "uncertainty") factors for developing permissible levels from a NOAEL or a LOAEL allows better use of information about the specific developmental endpoint and the likelihood of hazard for humans. Nonetheless, this approach still suffers from the same basic flaws as the RfD approach, such as the difficulty in evaluating excursions above the RfD level. The benchmark dose approach has also been proposed to improve assessments of developmental toxicants [154]. Briefly, the BMD is developed through the use of dose–response modeling and reflects the dose (or the lower confidence limit on the dose) associated with a certain percent response in the population.

## Incorporating Information on Severity of Effect

A critical difference in evaluating risks for carcinogenicity vs. risks for systemic effects is that, from a regulatory perspective, almost all types of cancer are considered equally severe, but the severity of systemic effects can vary significantly. In general, little basis exists from a regulatory perspective for distinguishing among carcinogens on the basis of malignancy or tumor type. Despite advances in earlier diagnosis and treatment, the fatality rate for cancer is still relatively high. The relative 5-year survival rates for all cancers from 1995 to 2000, excluding nonmelanoma skin cancer, were 66% for whites and 55% for blacks [155], meaning that 34% of whites and 45% of blacks did not survive 5 years past diagnosis. Nonmelanoma skin cancer includes squamous and basal cell carcinoma, which can be induced by agents such as ultraviolet light and arsenic; it has relatively low (<10%) fatality rates, even when untreated [156]. In contrast, target organ effects vary greatly in severity; for example, using the same target organ and susceptible population (namely, airways in asthmatics), responses may range from imperceptible mild bronchoconstriction induced by low levels of $SO_2$ to a fatal asthmatic response, as may have been due to acid sulfate pollution in the London smog episode [57,64].

Consideration of severity then becomes important for regulatory decision making in several ways. For RfD development, is an effect such as a 2% decrease in weight a NOAEL or a LOAEL? Is an effect of sufficient severity to warrant protection of 95% of the population or 99%? Several agencies and organizations have developed approaches to incorporate information on severity of effect into the risk assessment or risk management process for environmental chemicals.

In 1985 and 2000, the American Thoracic Society (ATS) defined an adverse respiratory effect [157,158]. Rather than providing a clear demarcation between non-adverse and adverse, the ATS described a continuum of respiratory effects from mild effects of limited, if any, medical significance (e.g., occasional cough, runny nose) to effects of obvious adverseness and medical significance (e.g., an asthmatic attack). To the extent that the effect caused discomfort and impaired daily function and quality of life, it was viewed as more adverse. ATS also observed that changes in biomarkers could indicate a homeostatic response and would reflect injury only if the magnitude of change in biomarkers exceeded certain threshold limits.

Similarly, the EPA has considered severity in its evaluation on the effects of ozone as part of the NAAQS setting process. For example, based on lung function changes, duration of effect, symptoms, and impact on activity level, the EPA categorized ozone responses into four categories: mild, moderate, severe, and incapacitating. The mild category includes FEV1 declines of 5 to 10% and no impact on activity. The EPA recommended that the responses in the mild category *not* be considered an adverse respiratory effect in adults for purposes of defining the NAAQS for ozone [39].

As noted earlier, the concept of severity of effect is also incorporated into the process for developing the RfD. Specifically, the RfD is based on an effect that, by definition, considers adverseness. The critical effect is adverse, because it may result in functional or structural impairment or is a precursor state to irreversible toxicity [125]; for example, fatty infiltration of the liver or a greater than 10% reduction in weight gain vs. controls would be considered adverse effects, and the associated dose would be a LOAEL. As the dose increases, the fraction of that population experiencing such effects would increase. FELs are dose levels that result in overt, often clinically apparent toxicity and are considered "too adverse" to be used in the development of the RfD. Examples of frank effects include liver necrosis or cirrhosis, which are severe and may be irreversible. FELs are not considered appropriate for RfD development because the protection level would be inadequate.

**TABLE 2.10**
**Rating Values for NOAELS, LOAELS, and FELs Used To Rank Chronic Toxicity**

| Rating | Effects |
|---|---|
| 1 | Enzyme induction or other biochemical change with no pathologic changes and no change in organ weights |
| 2 | Enzyme induction and subcellular proliferation or other changes in organelles, but no other apparent effects |
| 3 | Hyperplasia, hypertrophy, or atrophy, but no change in organ weights |
| 4 | Hyperplasia, hypertrophy, or atrophy with changes in organ weights |
| 5 | Reversible cellular changes: cloudy swelling, hydropic change, or fatty changes |
| 6 | Necrosis, or metaplasia with no apparent decrement of organ function; any neuropathy without apparent behavioral, sensory, or physiologic changes |
| 7 | Necrosis, atrophy, hypertrophy, or metaplasia with a detectable decrement of organ functions; any neuropathy with a measurable change in behavioral, sensory or physiologic activity |
| 8 | Necrosis, atrophy, hypertrophy, or metaplasia with definitive organ dysfunction; any neuropathy with gross changes in behavior, sensory, or motor performance; any decrease in reproductive capacity; any evidence of fetotoxicity |
| 9 | Pronounced pathologic changes with severe organ dysfunction; any neuropathy with loss of behavioral or motor control or loss of sensory ability; reproductive dysfunction; any teratogenic effect with maternal toxicity |
| 10 | Death or pronounced life shortening; any teratogenic effect without signs of maternal toxicity |

*Source:* deRosa, C.T. et al., *Toxicol. Indust. Health*, 1, 177–192, 1985. With permission.

Information on severity has been incorporated into the *reportable quantity* (RQ) definition, under CERCLA. Under this statute, releases of chemicals in amounts greater than some predetermined level, defined as the RQ, require that the EPA be notified of the release [159]. The amount of release that triggers notification is based on an assessment of the potency of the chemical and the severity of the effect at the dose level where the potency was quantified. The ranking of severity is shown in Table 2.10, where it can be seen that effects range from slight biochemical changes through gross toxicity, including lethality [159]. Unlike the RfD process, this scoring is not restricted to datasets containing information on mildly adverse effects, from subchronic or chronic studies. The RQ process can result in the development of scoring indicators from lower quality datasets, involving shorter time periods of exposure and more severe toxicity. The RQ process demonstrates the use of severity information in both risk assessment (developing RQ indicators) and risk management (defining release levels requiring notification as associated with defined RQ values).

Efforts involving the use of categorical exposure response modeling demonstrate additional approaches to the consideration of severity. Guth and coworkers [128] analyzed acute effects resulting from methyl isocyanate exposures of less than 8 hours in duration (as seen in Figure 2.5). Effects were separated into three categories: NOAEL (circles), adverse effect level (triangles), and lethal (squares). Effect categories were then analyzed on the basis of concentration and time, using logistic regression. Figure 2.6 presents a line above which there is a 90% probability that the true NOAEL lies. This method allows the use of data from a range of severity endpoints and considers various combinations of exposure level and exposure duration. Conceivably, this type of approach could lead to the development of concentration time nomograms for definition of NOAELs for different exposure durations.

## PHYSIOLOGICALLY BASED PHARMACOKINETIC MODELS

One of the areas of regulatory attention is that of physiologically based pharmacokinetic models and their potential use in risk assessment. This issue is discussed in more detail in Chapter 5 in this book; we discuss the topic here in terms of regulatory implications. PBPK models are essentially mechanistic models that describe quantitatively the pharmacokinetic processes affecting the disposition of a chemical and its metabolism from the time it is absorbed to the interaction with different and various body tissues. When it has been determined whether the likely cause of a carcinogenic response is the parent compound or its metabolites, a PBPK model may be developed to quantify the magnitude and the time course of exposure to this agent at the critical target site in the animal model. After the estimates of target tissue dose in the animal model have been made and validated, the information can then be scaled to the human to obtain an estimate of target organ dose in humans. This estimate may then be used to predict human cancer risk under different exposure conditions or to develop more precise estimates of the reference dose. It should be emphasized that the PBPK model does not offer an explanation of the most appropriate dose–response relationship, once the dose to target is estimated [160,161]. Furthermore,

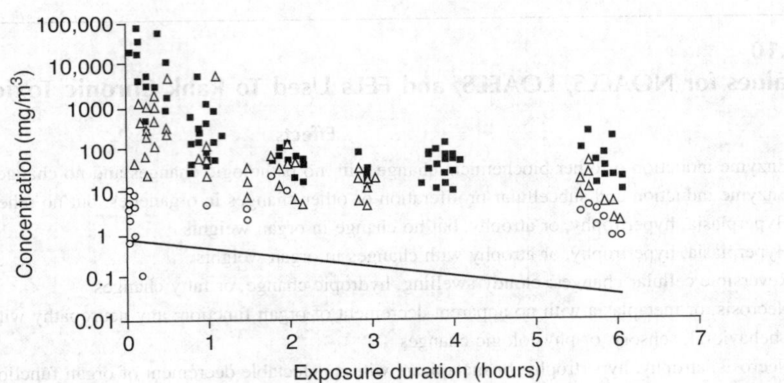


**FIGURE 2.6** Categorical data from published results on methyl isocyanate for exposures of less than 8 hr in duration and shown as NOAEL (circles), AEL (triangles), or lethality (squares). The maximum likelihood model fit is shown by the line representing the model prediction of $p = 0.1$ that severity is greater than the NOAEL category at the corresponding exposure concentration and duration. (From Beck, B.D. et al., *Fund. Appl. Toxicol.*, 20, 1, 1993. With permission.)

full validation of the model at the relatively low levels of environmental chemicals to which humans are exposed can be difficult.

Despite these limitations, PBPK modeling does offer an important tool for researchers and regulators alike. PB-PK models can be used to quantify target organ doses between species and to extrapolate from high to low doses. Of added significance is that new information about the pharmacokinetics of a chemical can be incorporated into the model without affecting the basic structure of the model, thus enhancing its predictive capability.

The use of PBPK models also provides important advantages over conventional pharmacokinetic analyses [160,161]. In typical pharmacokinetic modeling, time-course curves are determined for the concentration of the administered agent or its metabolites in blood or some other body compartment. The resulting curves are then described by curve-fitting biostatistical techniques. The approach of conventional pharmacokinetics may be criticized for being more dependent on the mathematical model than on the biological system it purports to represent; however, PBPK models are designed to predict kinetic behavior over a wide range of doses and exposure conditions and are based on basic physiologic and metabolic parameters. This modeling requires many data on anatomical and physiological parameters, the partitioning of test agents into selected tissues, and the biochemical constants for tissue binding and metabolism in various organs. From these data, a series of mass-balance differential equations can be written to describe the interactions between the chemical and the animal model.

Physiologically based pharmacokinetic modeling can improve dose–response assessment by accounting for sources of change in the proportions of applied to delivered dose in animals vs. humans and at high vs. low doses. Although this approach does not account for the fact that the sensitivity of the target tissue to the delivered dose may differ in humans and animals or between high and low doses, it still addresses some major areas of uncertainty in risk assessment. In fact, many sources of potential nonlinearity in applied dose–response involve saturation or induction of enzymatic processes at high doses or differences in toxification/detoxification pathways between humans and animals or across doses.

Physiologically based pharmacokinetic modeling has been applied to several agents, including methylene chloride and ethylene dichloride [160–162]. A look at the methylene chloride case illustrates the powerful implications of this approach. Andersen et al. [160] developed a PBPK model based on data indicating two routes of metabolism, one dependent on oxidation by mixed function oxidase (MFO) and the other dependent on glutathione *S*-transferase (GST) in four species (mouse, rat, hamster, and human). Models were designed to quantify the contributions of the two metabolic pathways in the lung and liver and to allow for extrapolation from rodents to humans. Kinetic constants for the model were obtained from experiments or the literature, with model validation involving a comparison of predicted blood concentration time-course data in rats, mice, and humans with experimental data from these species.

The capacity of methylene chloride to cause tumors in mice was associated with the target tissue dose and was closely related to the amount of methylene chloride metabolized by the GST but not the MFO pathway. Using the PBPK model, the target tissue doses in humans exposed to low concentrations of methylene chloride were between approximately 50- and 200-fold lower than would have been predicted by the linear extrapolation and body surface area factors used in conventional risk assessment methods. Thus, the PBPK analysis suggested that conventional risk analysis greatly overestimated the risk to humans exposed to low levels of methylene chloride. One of the major uncertainties, however, is the metabolic capacity of the body at low exposure levels where metabolism may not be saturated. Also, the dominant pathway for methylene chloride metabolism at other organ sites has not been determined. Still, the PBPK approach represents an attractive development, as it can increase the biological plausibility of predictive approaches while still incorporating biomathematical approaches for low-dose risk prediction. The EPA has incorporated this pharmacokinetic information into its cancer risk assessment for methylene chloride exposure via inhalation [163].

It is important to note the substantial uncertainties in PBPK modeling; for example, Hattis et al. [164] compared PBPK models for perchloroethylene developed by seven different authors and found appreciable differences among the model predictions. Given identical exposure levels in humans, the range of values for metabolized perchloroethylene span a 50-fold range, with 6 of the 7 models having predictions with a 14-fold range. With respect to methylene chloride, Clewell [162] noted the importance of the tissue distribution of GST enzyme activity across species, especially in humans, as a source of model uncertainty. Studies to refine estimates of GST enzyme activity across species and within the human population will serve to provide more refined estimates of dose across humans and hence of potential differences in susceptibility.

## ROLE OF HIGH-RISK GROUPS

In a previous section, we described the RfD concept as used by regulatory agencies to estimate acceptable levels for non-cancer effects. One of the factors in the derivation of this level was to account for variations in population susceptibility. The purpose of this section is to expand upon that issue, to describe the basis for variations in susceptibility and the magnitude of that variation, and to demonstrate the relationship of this issue to the regulatory process. (For more details, the reader is referred to Calabrese [60], Beck [165], and Neumann and Kimmel [166]).

A high degree of variability exists in the response of humans to different exposure levels of environmental pollutants [60,167]. In fact, the variability in the dose–response relationship in a heterogeneous population makes it difficult to estimate an acceptable level for chemical contaminants that would be protective of the entire population. Perhaps the most critical question is not what is a safe numerical standard but how many individuals are adversely affected at different levels of exposure [168]. Knowing which groups of individuals are at high risk with respect to pollutants is very important in answering this question, as these individuals will be the first to experience morbidity and mortality as pollutant levels increase. If the high-risk segments are protected, then the entire population is also protected. Information concerning both the identification and quantification of high-risk groups should play an integral role in the derivation of environmental health standards.

### Consideration of Specific High-Risk Groups

A better approach, when data are available to support it, is to consider specific groups at high risk on a chemical-by-chemical basis. Clear examples can be found of groups more susceptible to particular chemicals and cancer and non-cancer health effects [169–171]. These include:

- Individuals with genetic variations in metabolism; for example, a slow acetylator phenotype is associated with an increased risk of bladder cancer following exposure to aromatic amine dyes [169].
- Individuals with enzymatic genetic polymorphisms; for example, polymorphisms in cytochrome P450 enzymes can result in differential detoxification or bioactivation of environmental chemicals [171].
- Individuals with inherited genetic defects; for example, xeroderma pigmentosum, an autosomal recessive disease, results in altered DNA repair capacity and increases the risk of skin cancer by more than 1500-fold [171].
- Individuals with preexisting illness; for example, asthmatics may be more susceptible to ozone [170], and those with hepatitis B are more susceptible to liver cancer [169].

Other factors that can affect susceptibility to environmental chemicals include gender, age, and lifestyle (e.g., cigarettes, alcohol, diet). The role of diet and certain types of cancer is shown in studies demonstrating an inverse relationship between the amount of vitamin A in the diet and susceptibility to hydrocarbon-induced epithelial cancers [172]. Also, certain subgroups may be at greater risk, not because of an inherent difference in toxicological susceptibility but because they are more likely to be exposed. Young children, for example, are at greater risk from soil contaminants because they tend to accidentally ingest more soil and dust than older children and adults due to their significant hand-to-mouth activity. Thus, it is likely that, even given the same exposure, individuals are not equally susceptible to the induction of cancer and other adverse health effects, and in many cases the differential susceptibility may be very large.

Regulatory agencies have focused in particular on the potential for children to be more susceptible to environmental chemicals. In 1996, the EPA emphasized their focus of protecting infants and children in a report entitled *A National Agenda To Protect Children's Health from Environmental Threats* [173]. The 1996 FQPA requires the use of an additional tenfold UF for pesticides to account for potential prenatal and postnatal developmental toxicity [174] for agents. The UF is not used if the agent has not been demonstrated to exhibit developmental toxicity in a reliable testing program. As noted by Roberts [175], children may be more susceptible because many cells and organs are undergoing growth and development and have not yet matured. A child's diet and physical environment, and therefore his or her exposure potential, may vary significantly from that of an adult. For many routes of exposure (air, food, water, and dermal exposures), chemical intake (on a per-kilogram body weight basis) is generally greater for infants and children than adults [176].

A subgroup at high risk for one chemical is not necessarily at high risk for other chemical exposures; for example, although children are often assumed to be more sensitive than adults, this is not always the case. Reactions to pharmaceuticals, because they are more widely studied than responses to environmental exposures, can be considered as examples. Acute overdoses of acetaminophen result in less hepatotoxicity in children than adults with comparable plasma concentrations, possibly due to differences in metabolism [177].

Currently being debated is whether current risk cancer assessment methods adequately account for more highly susceptible groups. Only recently has evaluation of carcinogen exposure addressed the role of population variability in susceptibility to carcinogens; consequently, groups at high risk to environmental carcinogens, with the obvious exceptions of smoking as a risk factor for exposure to asbestos, uranium, and coke-oven emission-related cancer, have not generally been addressed. It should be noted, however, that the conservatism of the cancer risk assessment process has generally been believed to result in adequate protection of high-risk groups.

For non-cancer effects, a subject of debate is the appropriate UFs to account for high-risk subgroups. At a conference organized by the ILSI and the EPA, "it was suggested that, in many cases, genetic variation in human susceptibility may be greater than an order of magnitude when comparing differences between children and adults" [178]. A coalition of farm food, manufacturing, and pest management organizations concluded that the additional UF of ten required by the FQPA is not necessary to use across the board to protect infants and children [179]. They also concluded that the standard default UFs are adequate for a pesticide with a complete and reliable database and that an additional UF "should only be applied to an endpoint that is relevant to protection of fetuses, infants, and/or children" [179]. The EPA is looking into establishing criteria for appropriate use of the tenfold additional FQPA uncertainty factor [174,180]. Overall, the best approach is to consider susceptible subgroups on a case-by-case basis when data are available, for both carcinogens and noncarcinogens.

### Regulatory Implications

The role of population variability should be considered by regulatory agencies in risk assessments for both carcinogens and noncarcinogens. Identification and quantitative characterization of susceptible populations could provide decision makers with a theoretical framework on which to base regulatory action. For example, Tamplin and Gofman [181] have employed knowledge of susceptible populations in predicting the incidence of cancer from radiation pollution in drinking water to help define acceptable levels of exposure. They assumed that the latency period is shorter for *in utero* exposure than for all radiation exposure beyond birth (i.e., 5 years vs. 15 years). Consideration of the increased susceptibility of the fetus to radiation-induced cancer resulted in greater estimates of cancer risk, as compared with traditional methodological approaches which predict carcinogenic effects at low doses, based on high levels of exposure in adults [182]. The EPA has specifically evaluated the increased sensitivity of specific high-risk groups in setting NAAQSs for CO, lead, $NO_2$, ozone, particulates, and sulfur dioxide, and in establishing drinking water standards for some environmental chemicals. Examples of the high-risk groups considered are shown in Table 2.11. For example, the NAAQS for lead considers high-risk populations in a more quantitative way by estimating the fraction of the susceptible subpopulation (children) that would be protected at different air levels of lead [183]. Following is a detailed description of the EPA's consideration of high-risk groups in the derivation of drinking water standards for nitrates and cadmium.

#### Nitrates in Drinking Water

The drinking water standard of 10 mg nitrate ($NO_3^{-2}$) as mg nitrogen per liter is designed to prevent the formation of elevated levels of methemoglobin (MetHb) in infants. In the presence of nitrite ($NO_2^{-1}$), formed from $NO_3^{-2}$ in infants, hemoglobin is oxidized to MetHb, which is not able to reversibly combine with oxygen. Levels of 1 to 2% and 2 to 5% MetHb are typical in the blood of adults and infants, respectively. When concentrations are less than 5% MetHb, there are no obvious indications of toxicity; however, with levels of MetHb from 5 to 10%, clinical signs of toxicity (e.g., cyanosis) may appear [60].

Infants are at considerable risk for nitrate-related toxicity, as compared to adults. Factors that predispose infants to the development of MetHb formation include:

- The incompletely developed ability to secrete gastric acid; this permits the gastric pH to be high enough (5 to 7 pH) to permit the growth of nitrate-reducing bacteria in the gastrointestinal tract, thereby converting $NO_3^{-2}$ to $NO_2^{-1}$ before absorption into the circulation [184].

**TABLE 2.11A**
**High-Risk Groups in the Derivation of Standards by the EPA: Drinking Water Standards**

| Substance | High-Risk Condition Considered |
|---|---|
| Arsenic | None |
| Barium | No specific groups, but a safety factor of two is incorporated to account for variation (or increased susceptibility) within the human population |
| Cadmium | None |
| Fluoride | Children, to prevent mottling of teeth |
| Lead | Children, to prevent neurological disorders |
| Mercury | Based on humans who exhibited toxicity at the lowest level of exposure from a group of mercury-poisoned adults |
| Nitrate | Infants, to protect against methemoglobinemia |
| Selenium | None |
| Sodium (no standard) | Individuals with heart and kidney disease |
| Chlorinated hydrocarbon insecticides (noncarcinogenic) | None |
| Chlorophenoxy herbicides (noncarcinogenic) | None |

**TABLE 2.11B**
**High-Risk Groups in the Derivation of Standards by the EPA: National Ambient Air Quality Standard**

| Substance | Original Group | Primary Groups Currently Considered |
|---|---|---|
| Carbon monoxide | Individuals with neurological or visual impairment | Adults with heart disease (angina, coronary artery disease) |
| Lead | Children, to protect against neurological and hematological impairment | Same |
| Nitrogen dioxide | Children, to protect against respiratory infections, also concern for changes in lung structure | Same |
| Ozone | Asthmatics | Exercisers, individuals with preexisting disease |
| Particulates | Elderly, individuals with cardiopulmonary disease | Same |
| Sulfur dioxide | Elderly, individuals with cardiopulmonary disease | Asthmatics |

- The higher levels of fetal hemoglobin in infants; this form of hemoglobin is more susceptible than adult hemoglobin to oxidation to MetHb [185].
- The diminished enzymatic capability of infants to reduce MetHb to hemoglobin [186].

Research has revealed that levels of $NO_3^{-2}$ beyond 20 mg/L result in a marked upshift in the frequency of methemoglobinemia in infants but not in adults [60]; consequently, a standard of 10 mg/L is principally designed to prevent the occurrence of elevated levels of MetHb in infants. Concentrations twice as great would still protect adults.

## Cadmium

Studies with rats show that renal damage is initiated at a kidney concentration of 200 ppm Cd [187]. The EPA calculated that humans would need to ingest 50 g Cd per day for 50 years to reach a level of 200 ppm in their kidneys. In the derivation of the Cd drinking water standard, the EPA assumed a daily Cd exposure of 75 μg from the diet and 20 μg from water. This 20 μg Cd per day from drinking water would occur at a level of 0.01 mg/L. The total daily Cd exposure is therefore approximately 95 μg Cd per day, thus a safety factor of 4 was assumed. In proposing their drinking water standard for Cd, the EPA requested feedback from the public as to whether the standard should include additional protection for cigarette smokers, as smoking is a source of appreciable Cd exposure (e.g., approximately 1.5 μg Cd per cigarette) [60]. It is interesting to note that, of the 52 comments received by the EPA on this issue, only 3 suggested that this standard be modified to include protection for cigarette smokers. The EPA decided not to incorporate additional safety factors to protect smokers [105]; thus, this example describes a situation in which protection of a high-risk group was not taken into account in derivation of a standard.

### Susceptible Groups and Early-Life-Stage Exposure to Carcinogens

Recent analyses by Ginsberg [188] and others (e.g., see Preston [189]) have sought to determine whether children might represent a susceptible population with respect to carcinogen exposure. Answering this question is complicated by limitations in the available data. Relevant human data are derived primarily from epidemiological studies of radiation exposure and cancer development in atom bomb survivors. Such studies provide evidence for increased risk of certain cancers from early life exposure; however, it is difficult to extrapolate from studies of ionizing radiation, a direct-acting mutagen that can induce mutation at any stage of the life cycle, to chemical carcinogens, the majority of which are not direct-acting mutagens and require cell proliferation for indication of mutation (often via indirect mechanisms) [189].

Animal bioassays nonetheless provide some insights into early-life-stage exposures to carcinogenic agents; for example, acute exposure to 12 mutagenic agents yields a greater tumorigenic response from early-life-stage than from later-life-stage exposures [188,190]. Mechanistic information also suggests the possibility of a greater susceptibility for early-life-stage exposure from mutagenic agents. The high proliferation rate early in life may increase the likelihood that a cell containing damaged DNA could replicate before the DNA is repaired. Studies of carcinogens with other modes of action (e.g., hormonally mediated modes of action) or genotoxic agents that are not direct acting but lead to genetic damage through other means, such as generation of reactive oxygen species, provide a more complicated picture with evidence for and against early-life-stage susceptibility.

In response to concerns that children may be more susceptible to certain carcinogens than adults, the EPA has developed guidance for early-life-stage exposures to carcinogens. Figure 2.7 [190] provides a schematic for this approach. As with the 2005 EPA cancer guidelines (discussed earlier), an understanding of the mode of action is an important component of the approach. For agents with either a nonlinear mode of action or a linear, but nonmutagenic, mode of action, the dose–response approach is unchanged; however, for agents that the EPA concluded were likely to be linear due to a mutagenic mode of action, the first approach recommended is chemical-specific adjustment, where feasible. Where such an adjustment is not feasible, then age-dependent adjustment factors are proposed: a tenfold increase in the potency factor for ages up to 2 years and a threefold increase in potency for ages 2 to <16 years of age [190]. The underlying premise is that direct-acting mutagens are likely to exhibit a linear, no-threshold dose–response, an assumption that, as discussed earlier in this chapter, does not consider repair mechanisms nor the potential for hormetic responses and thus may be especially conservative. The implications of these recommendations for regulatory decision making remain to be seen.

## HORMESIS

In the previous edition of this chapter, the concept of hormesis was introduced for the first time into a major general toxicology textbook. Over the intervening years, considerable progress has occurred with respect to the documentation of hormetic dose–responses in not only toxicology but in essentially all other biological and biomedical subdisciplines relying on the dose–response. Furthermore, the most recent edition of *Casarett & Doull* included substantial discussion of hormesis for the first time [191]. Likewise, other toxicological monographs are including chapters entirely devoted to the topic of hormesis [192,193], as well as the forthcoming edition of the major text entitled *Fundamentals of Aquatic Toxicology* [194]. In parallel fashion, major professional societies, such as the U.S. Society of Toxicology, have had major sessions at their annual conferences addressing hormesis. The annual conference of the Canadian Society of Toxicology recently included a major presentation on hormesis during its Plenary Session [195]; similar activities have taken place in Europe. Based on website listings of academic courses, numerous universities are now incorporating hormesis into their toxicology and radiation health curricula. Such collective recognition of the hormetic concept has also extended to the general public, with prominent stories on hormesis appearing in *Discover* [196], *The American Spectator* [197], *Forbes* [198], *Fortune* [199], *The Wall Street Journal* [200], *The Boston Globe* [201], *The Baltimore Sun* [202], *U.S. News and World Report* [203], *The London Times* [204], and numerous other publications. The topic has also been featured on National Public Radio's environment program "Living on Earth" and on the Canadian Broadcasting Station's "Quirks & Quarks." Hormesis has been featured in articles for the general scientific community in *Scientific American* [205], *Environmental Sciences and Technology* [206], *Chemistry and Industry* [207], and the *European Molecular Biology Organization Journal* [208,209]. Perhaps most importantly for promoting the hormesis concept was its becoming the object of a commentary article in *Nature* [210] and an extensive four-page news story in the journal *Science* [211], as well as being highlighted in *Toxicological Sciences* [212,213]. In 2003, the journal *Nonlinearity in Biology, Toxicology, and Medicine* was created to provide a focused peer-reviewed vehicle for the publication of research relating to the hormesis concept, and in 2005 the International Hormesis Society (www.hormesissociety.org) was created.

The above brief summary of progress concerning knowledge about and acceptance and integration of hormesis within the scientific community over the past

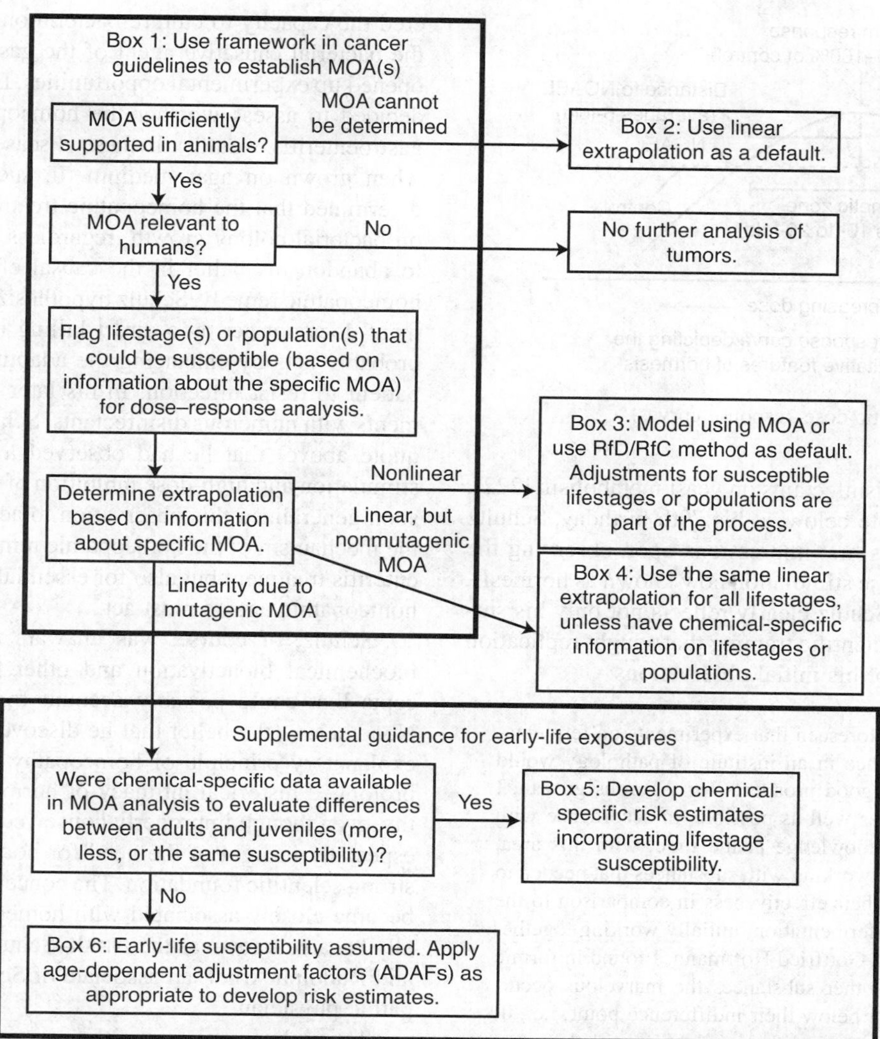

Box 1: Use framework in cancer guidelines to establish MOA(s)
MOA sufficiently supported in animals?
MOA cannot be determined
Box 2: Use linear extrapolation as a default.
Yes
MOA relevant to humans?
No
No further analysis of tumors.
Yes
Flag lifestage(s) or population(s) that could be susceptible (based on information about the specific MOA) for dose–response analysis.
Determine extrapolation based on information about specific MOA.
Nonlinear
Box 3: Model using MOA or use RfD/RfC method as default. Adjustments for susceptible lifestages or populations are part of the process.
Linear, but nonmutagenic MOA
Box 4: Use the same linear extrapolation for all lifestages, unless have chemical-specific information on lifestages or populations.
Linearity due to mutagenic MOA
Supplemental guidance for early-life exposures
Were chemical-specific data available in MOA analysis to evaluate differences between adults and juveniles (more, less, or the same susceptibility)?
Yes
Box 5: Develop chemical-specific risk estimates incorporating lifestage susceptibility.
No
Box 6: Early-life susceptibility assumed. Apply age-dependent adjustment factors (ADAFs) as appropriate to develop risk estimates.

**FIGURE 2.7** Flow chart for early-life risk assessment using mode of action framework. (From EPA, *Supplemental Guidance for Assessing Susceptibility from Early Life Stage Exposure to Carcinogens*, EPA/630/R-03/003F, Risk Assessment Forum, U.S. Environmental Protection Agency, Washington, D.C., 2005.)

decade illustrates the remarkable progress that has occurred with respect to this concept. It also suggests that a revolution is occurring with respect to the dose–response relationship, the central pillar within the field of toxicology. The argument over the nature of the dose–response relationship, especially in the low-dose zone, is basic to toxicology. Hormesis is challenging fundamental and long-held beliefs concerning the nature of the dose–response and, in so doing, has generated much interest, as the outcome of the debate has the potential to affect how toxicologists design experiments, model research findings, and perform risk assessments. It is the purpose of this section to assess briefly the concept of hormesis, its definition, how it may be assessed, documentation within the toxicological/pharmacological literature, quantitative features of the dose–response, possible mechanism explanations, and implications for risk assessment and regulatory decision making. Detailed scientific assessments of hormesis exist and should be referred to [214–220].

## What Is Hormesis?

*Hormesis* is a term created by Southam and Ehrlich in 1943 [221] to describe a low-dose stimulation and a high-dose inhibition (Figure 2.8) based on observations relating to the effects of extracts of the red cedar plant on fungal metabolism. The term *hormesis* was selected, because its etymological root, *horm-*, means "to excite" in Greek. However, the person who first published reproducible data on this concept seems to have been Hugo Schultz in the 1880s at the University of Griswald in northern Germany in his assessment of the effects of

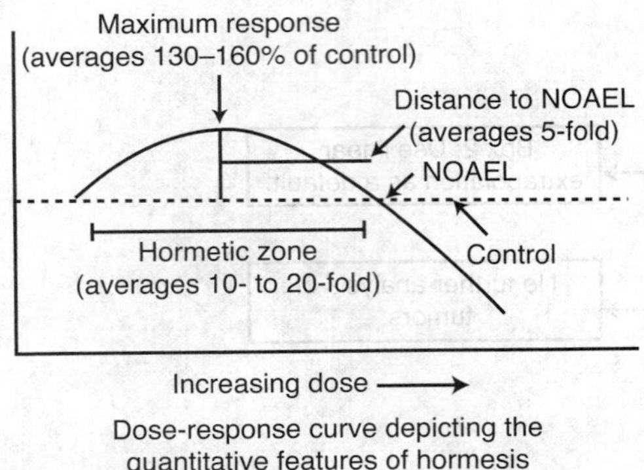


**FIGURE 2.8** Hormetic dose–response curves.

various chemical disinfectants on yeast metabolism [222]. As seen in the quote below, at his 70th birthday, Schultz [223] recounted his first impressions upon observing the unexpected low-dose stimulation now known as hormesis. The comment by Schulz clearly reflects not only his surprise but also his scientific training that sought replication and confirmation of his initial observations:

> Since it could be foreseen that experiments on fermentation and putrescence in an institute of pathology would offer particularly good prospects for vigorous growth, I occupied myself as well as possible, in accordance with the state of our knowledge at the time, with this area. Sometimes, when working with substances that needed to be examined for their effectiveness in comparison to the inducers of yeast fermentation, initially working together with my assistant, Gottfried Hoffmann, I found in formic acid and also in other substances the marvelous occurrence that as I got below their indifference point, i.e., if, for example, I worked with less formic acid than was required in order to halt the appearance of its anti-fermentive property, that all at once the carbon dioxide production became distinctly higher than in the controls processed without the formic acid addition. I first thought, as is obvious, that there had been some kind of experimental or observation error. But the appearance of the overproduction continually repeated itself under the same conditions. First I did not know how to deal with it, and in any event at that time still did not realize that I had experimentally proved the first theorem of Arndt's fundamental law of biology.

Although a university professor of pharmacology and trained by notable leaders in the German scientific community, such as Pflueger (physiology), under whose direction he published his first scientific paper, Schulz had an academic interest in the medical practice of homeopathy. He believed that these observations provided the underlying explanatory principle of the medical practice of homeopathy. In the early 1880s, Schulz came to accept that a successful homeopathic treatment existed for a form of gastroenteritis. Robert Koch's laboratory had both discovered the capacity to culture bacteria on agar and isolated the bacterial causative agent of the gastroenteritis which opened up experimental opportunities. In this case, Schulz decided to assess whether the homeopathic remedy for gastroenteritis would kill these disease-causing bacteria when grown on agar medium. In such testing, Schulz determined that the homeopathic treatment had no effect on bacterial colony growth, regardless of dose. Refusing to abandon his belief in the causal effectiveness of the homeopathic remedy, Schulz hypothesized that if the treatment does not act by a direct killing mechanism then it probably acts by enhancing the adaptive capacity of the patient to resist infection. In his later laboratory experiments with numerous disinfectants, Schulz concluded (see quote above) that he had observed a reliable low-dose stimulation and high-dose inhibition of colony growth. He soon generalized this observation to account not only for the mechanism of the homeopathic remedy for the gastroenteritis treatment but also for essentially how most other homeopathic agents must act.

Schulz, of course, was unaware of the concept of biochemical bioactivation and other toxicological concepts that could possibly account for his observations. Nonetheless, the belief that he discovered the underlying explanatory principle of homeopathy was born, and the prolonged historical intimacy of hormesis and homeopathy, even though improperly conceived, was the outcome, especially in light of the need for homeopathy to have a strong scientific foundation. The concept of hormesis soon became closely associated with homeopathy and eventually became known as the Arndt–Schulz law, after Schulz and Rudolph Arndt, an associate of Schulz and a homeopathic physician.

The early association of the concept (but not the name of hormesis) with homeopathy proved disastrous for hormesis because traditional medicine, from which pharmacology and toxicology would emerge, vehemently opposed homeopathy [215]. In the ensuing historical struggles, homeopathy became profoundly marginalized and discredited because of its lack of theoretical foundation, unproven mechanistic bases, and its focus on the high dilution school of homeopathic thought. This high dilutionist perspective, originated by Samuel Hahnemann, asserted that biological/medical responses occurred at concentrations so dilute as to lack the theoretical presence of molecules (i.e., below 10 to 23 *M*). In this conflict, the concept of hormesis received the equivalent of near-fatal collateral damage, leading to its failure to compete and thrive in the area of scientific ideas and practices. Several assessments of hormesis have detailed the historical foundations of this concept in the fields of chemical toxicology and radiation biology [224–228]. More recent papers have extended these historical assessments into an integrated and detailed evaluation of how the toxicological community rejected the concept of hormesis in the early decades

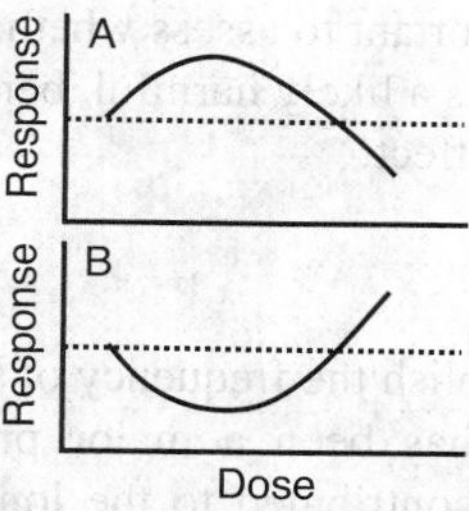


(A) The most common form of the hormetic dose–response curve depicting low-dose stimulatory and high-dose inhibitory responses, the β- or inverted U-shape curve.

(B) The hormetic dose–response curve depicting low-dose reduction and high-dose enhancement of adverse effects, the J- or U-shaped curve.

**FIGURE 2.9** Typical dose–response curve depicting the quantitative features of hormesis.

of the twentieth century and how this rejection was sustained for the remainder of the past century [192,193].

## The Hormetic Dose–Response

To document the occurrence of hormetic dose–response relationships, its quantitative features, and its potential for generalization, an hormetic database has been created based on *a priori* evaluative criteria [214,229]. Based on the nearly 6000 hormetic dose–response examples entered into the database, the stimulatory component of the dose–response begins immediately below the quasi-threshold, the toxicological NOAEL. The low-dose stimulation has often been observed to occur following a modest disruption in homeostasis—that is, a toxic effect. This consistent observation suggests that hormesis may be defined as a compensatory or adaptive response to most, if not all, types of induced injury. Such observations require study designs with adequate numbers of doses and temporal measures for proper evaluation.

The magnitude of the stimulation or overcompensation is generally quite modest, with the maximum response typically being in a range approximately 30 to 60% greater than the control (Figure 2.9). This modest stimulatory response and its relationship to the NOAEL are the most toxicologically distinguishable characteristics of the hormetic response regardless of the type of biological model, endpoint, or underlying agent used. The maximum stimulatory response frequently occurs within a factor of 5 to 10 of the NOAEL (i.e., 1/5 to 1/10 of the NOAEL dose). Because the hormetic zone is contiguous with the NOAEL and human variability may extend over a tenfold dose range, it is likely that hormetic responses may be occurring for some members of the population while others may still be in a zone of low-level toxicity. This may be particularly the case in a more heterogeneous study population.

The modest nature of the stimulatory response reflects a process that has been selected to allocate slightly more resources than are necessary to restore the altered biological system to its homeostatic equilibrium state. Such a response offers enhanced survival advantages by helping to ensure that induced biological lesions were repaired with a careful allocation of biological resources such that the extra amount, if not required to repair the damage, could be then employed for other biological advantage, such as enhanced resistance to subsequent exposures to similar toxicants.

The modest nature of the stimulatory response has made the hormetic phenomenon difficult to confidently establish, especially when observed within the context of weak study designs that have employed only two to three doses and aimed at defining LOAEL and NOAEL values and a concurrent control. Nonetheless, the modest stimulatory response is precisely what would be predicted for a stimulatory response that is compensatory to a toxic insult.

To properly assess hormesis in an experimental setting, it is therefore necessary to carefully consider animal model selection, the endpoint measured, and the study design, including the number and spacing of doses and the number of repeat measures made (i.e., temporal features). In the case of animal models, it is necessary to select one in which the endpoint has a background occurrence that permits the evaluation of the hormetic hypothesis. If the animal model endpoint displays little or no background disease incidence, it will not be possible to study hormesis because the response could not decrease significantly or at all below that of controls. This is a serious challenge for a number of commonly used animal models that have a very low incidence for a wide range of diseases or have background diseases that take a long time to develop. Given historical conditions under which animal models were selected by agencies such as the NCI, FDA, and EPA, a low disease background was considered an attractive characteristic because one could obtain higher statistical power with fewer animals. Thus, it is possible that the goals of an hormetic assessment can conflict with other laudatory aims, forcing the investigator to carefully clarify the goals of the study prior to selecting the animal model and study design. The use of such low disease incidence models reduces the likelihood of observing the hormetic response independent of the study design.

## Hormetic Stimulatory Range

The hormetic stimulatory range is generally within 20-fold of the dose between the apparent start of the stimulatory response to where the NOAEL occurs. We have estimated the stimulatory dose range by determining the dose that results in an estimated 10% increase on the ascending part of the hormetic curve and when the stimulatory response is estimated to return to 10% on the descent as it

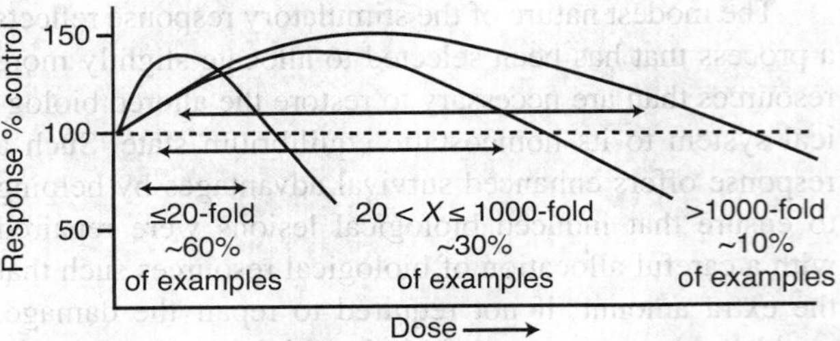


**FIGURE 2.10** Range of possible hormetic dose–responses.

approaches the NOAEL; however, it has not been uncommon for the width of the stimulatory range to exceed a factor of a 100-fold and occasionally exceed greater than 1000-fold (Figure 2.10) [214]. The causes underlying the occasional wide range in the stimulatory response have been generally unexplored; however, it is likely that such variation may be explained by the extent of the heterogenicity of the population studied. In a series of simulations, we observed that the shape of the dose–response curve was a function of the number of subgroups comprising the population, the nature of the specific dose–response relationship of each subgroup, and the relative proportion of each subgroup in the population. Depending on the complexity of the underlying subgroup characteristics (i.e., number of subgroups, range of susceptibilities, and relative proportion contributing to the entire population), it was possible to observe almost any type of dose–response relationship. Based on these findings, it may be concluded that the reason for most of the dose–responses in the hormetic database displaying a modest stimulatory response range is the likely widespread use of the highly homogenous biological test systems.

Given this situation, it is not unexpected that the use of highly heterogeneous models (e.g., highly outbred rodent strains) as well as epidemiological data of the highly heterogeneous human population could yield not only a wide hormetic stimulatory range but also highly complex dose–response curves such as multiphase curves, as observed in our more complicated simulations. This multiphase dose–response feature may result from the inclusion of multiple subgroups with some or all displaying an hormetic dose–response over different or overlapping dose–response ranges. When all such groups are combined into a single population dose–response, the outcomes can be highly complex and difficult to interpret initially.

The observation that the width of the hormetic stimulatory zone may be highly variable has important implications for study design, risk assessment, and clinical medicine; that is, if an anti-tumor agent enhances tumor development at low doses, then it would be important to consider this possibility when developing the patient treatment strategy [230]. In the case of environmental risk assessment, it is important to assess whether the hormetic stimulation represents a likely harmful, beneficial, or generally insignificant effect.

## FREQUENCY

The inability to establish the frequency of hormesis in the toxicological field has been a major problem for the hormetic model. It contributed to the long and strongly held opinion that hormetic dose–responses were simply manifestations of chance variation and at best an occasional paradoxical phenomenon. Despite this unstudied, but conclusionary judgment, much data exist supporting hormesis using a systematic evaluative assessment methodology [214,229,231]. In the toxicological literature, such reports have objectively supported the conclusion that hormetic dose–responses were not only reproducible but also quite common when the study design and biological model were appropriate for the evaluation of this phenomenon. These findings, however, did not adequately answer several major challenges that were necessary for hormesis to become a more established toxicological concept. Two areas of weakness were the need to establish the frequency of hormesis in the toxicological literature and its mechanistic basis.

To assess the frequency of hormesis in the toxicological literature it is necessary to establish *a priori* entry and evaluative criteria. In the case of hormesis, it was not possible to offer unequivocally clear and absolute entry as well as evaluative criteria; that is, it is not possible to conclude without some degree of uncertainty that dose–response relationships displaying the characteristics of hormesis are indisputably real cases of hormesis. This is because it is not possible to derive unequivocal criteria defining hormesis. In the cases of *a priori* entry and evaluative criteria, it is necessary to have rigorous but reasonable and objective criteria. With respect to entry criteria, it was reasoned that each dose–response would be required to display at a minimum a LOAEL, NOAEL, and at least two doses below the NOAEL along with a control group. Ideally, it would be better to have both upper and lower ends of the dose–response better defined (i.e., the use of more doses). This would require even more doses (not including controls) than the minimum number of four. In addition, it would have been ideal to require a temporal component to address properly the dynamic aspects of the overcompensation process; however, even with the minimum entry criteria only a very small proportion of dose–responses satisfy this standard [232]. If more rigorous entry criteria had been applied, the proportion satisfying the criteria would be even lower, possibly resulting in not enough dose–responses to evaluate the hormetic hypothesis. If the criteria are too rigorous for the entry criteria, then one is left with too few dose–responses to assess. In the case of the evaluative criteria, consideration was given to whether or not hypothesis testing was present. If hypothesis testing was present, the findings were given a higher value than responses lacking such evaluation (see Calabrese and Baldwin [232] for a detailed consideration of evaluative criteria). If the evaluation criteria are too rigorous, most dose–responses would fail to satisfy a definition of hormesis; if the criteria are too lax, the result would have little credibility. Despite such necessary compromises, it was possible to provide an optimization of choices to continue the assessment process in a credible manner.

The frequency assessment paper of Calabrese and Baldwin [232] found that only 2% of nearly 21,000 toxicology studies satisfied the minimum *a priori* entry requirement necessary for being able to be evaluated for the presence or absence of hormetic responses. Of the nearly 800 dose–responses satisfying these *a priori* entry criteria, about 40% satisfied the evaluative criteria, thereby establishing, for the first time, a frequency of hormetic responses in the toxicological literature. Follow-up research using these 800 dose–response relationships tested whether the hormetic or the threshold dose–response model better accounted for the distribution of responses of the doses below the NOAEL than the threshold model. According to the threshold model, the responses to doses below the NOAEL should be randomly distributed above and below control values; however, the response data were nonrandomly distributed in the direction strongly supporting the hormetic hypothesis with the ratio of above to below control values being 2.5 to 1. Similar findings using 13 yeast cell lines and 70 human tumor cell lines with over 3 million dose–response relationships have likewise confirmed both the inability of the threshold model to account for responses to doses below the threshold and the consistency of the findings with the hormetic model (Calabrese, unpublished data). These three separate studies dealing with different biological models, endpoints, and agents seriously challenge the predictive capacity of the threshold model and weaken the assumption that it is the most fundamental model in the biological sciences, including toxicity and pharmacology, while offering strong support for the hormetic model.

## Mechanisms

For hormesis to be taken seriously by the toxicological and pharmacological communities, the hormesis dose–response model needed to have a strong mechanistic foundation. Unfortunately, toxicological research has not historically been directed toward assessing underlying switching mechanisms that could have relevance in potentially explaining hormetic-like biphasic dose–response relationships. The discipline of pharmacology, however, has been successful in developing methods and procedures to account for hormetic-like biphasic dose–response relationships. Our evaluation indicates that mechanisms have been reported for hormetic-like biphasic dose–responses for several dozen receptor-based systems. These pharmacological dose–response relationships conform to the general quantitative features of the hormetic dose–response and are likely the means by which toxic substances induce hormetic responses. A partial listing of receptor-based biphasic dose–response relationships is given in Table 2.12.

**TABLE 2.12**
**Representative Receptor Systems Displaying Biphasic Dose–Response Relationships**

| | |
|---|---|
| Adenosine | Neuropeptides |
| Adrenergic | Nitric oxide |
| Bradykinin | *N*-methyl-D-aspartate |
| Cholecystokinin | Opioid |
| Corticosterone | Platelet-derived growth factor |
| Dopamine | Prolactin |
| Endothelin | Prostaglandin |
| Epidermal growth factor | Somatostatin |
| Estrogen | Spermine |
| 5-Hydroxytryptamine (serotonin) | Testosterone |
| Human chorionic gonadotrophin | Transforming growth factor β |
| Muscarinic | Tumor necrosis factor α |

A general explanation accounting for the receptor-based hormetic-like biphasic dose–response relationship involves the following. Biphasic dose–responses are commonly explained by the presence of two receptor subtypes which lead to opposing activity pathways (e.g., smooth muscle relaxation and contraction). These receptor subtypes display profoundly different affinities for the agonist; for example, at low concentrations of agonist, one receptor subtype that has a high affinity for the agonist becomes occupied, leading to activation and possible muscle contraction. As the concentration of agonist increases, a second receptor subtype that has lower affinity but greater receptor capacity than the first receptor subtype becomes more occupied and dominant, leading to relaxation of the smooth muscle. This would represent an hormetic-like biphasic dose–response relationship when studied over a broad dose–response framework. This basic pattern broadly exists within the pharmacological literature to account for biphasic dose–responses. Greater complexity could be added by including other receptors via receptor cross-talking mechanisms; nonetheless, the net result, despite different degrees of complexity, yields the hormetic dose–response.

Although numerous types of hormetic mechanisms exist, the question must be raised as to whether there is a single hormetic mechanism or underlying mechanistic foundation. The most distinctive feature of the hormetic dose–response is the modest nature of the stimulation. This "biological regularity" conforms to the expression of an allometric relationship that is quantitatively predictable across species based on body weight (Calabrese, unpublished data). This suggests that the hormetic dose–response is ultimately related to underlying biological processes that are basic to species survival, most likely related to energy regulatory processes. Although further work is necessary in this area, the recognition that hormetic responses are a manifestation of an allometric framework is an important step toward providing a basic understanding of its evolutionary origin and significance.

The striking consistency of the modest nature of the stimulatory response across all experimental settings strongly supports the hypothesis that hormetic responses reflect allometric biological patterns that are highly predictive within and across species. Such an allometric evaluation framework may reflect a unifying biological mechanism to account for the highly generalizable quantitative features of hormetic dose–responses.

## Hormesis and Risk Assessment

Hormesis challenges the way the toxicology community thinks about risk assessment. It indicates that the traditional NOAEL is truly a misnomer and that biological activity occurs below this apparent threshold. The key feature is that the hormetic dose–response is highly generalizable, occurring in all types of biological systems, for all types of endpoints, and across all chemical classes and physical agents. Furthermore, whenever the hormetic and threshold models have been directly compared, the threshold model has been strikingly outperformed. Given these circumstances, it is appropriate to consider the risk assessment implications of hormesis. This issue has been addressed in some detail by Calabrese [233], Calabrese and Baldwin [234], Gaylor [235], Sielken and Stevenson [236], and Calabrese and Cook [237]. The most significant implication will be for the carcinogen risk assessment affecting both the derivation process itself and estimates of possible risk.

Prior to the consideration of hormesis, it has been difficult, if not impossible, to convincingly support a threshold model for carcinogens. This is principally because too few doses are found in traditional animal bioassays such that either model (threshold vs. linear) tends to fit the data. Under such circumstances, a protectionist philosophy has typically defaulted to the linear low-dose model estimate; however, the hormetic dose–response, with its notable dip below control values, offers a much greater opportunity to establish the biological validity of a threshold for carcinogens than could be normally established via the traditional model. Application of the hormetic dose–response model to carcinogen risk assessment reveals a threshold below which the risk becomes less than the control before regressing back to control risk values at even lower concentrations. For example, the banned pesticide dichlorodiphenyltrichloroethane (DDT) has long been known to cause liver cancer in animal models, presumably acting linearly at low doses; however, in studies by Sukata et al. [238] that assessed the high- and low-dose effects of DDT on liver foci formation, the hormetic curve became quite clear (Figure 2.11).

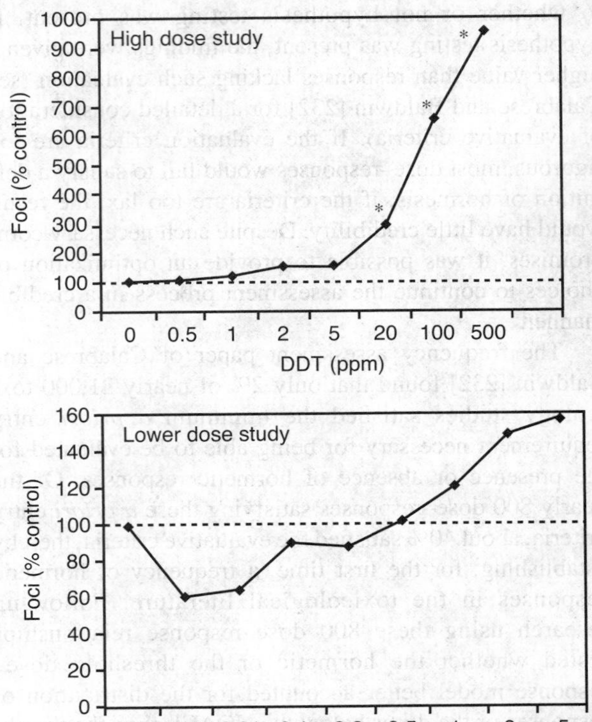


**FIGURE 2.11** Effects of DDT on hepatic foci formation in the F344 Rat. (From Sukata, T. et al., *Int. J. Cancer*, 99, 112, 2002. With permission.)

Follow-up experiments also offered possible mechanistic insights for the switch in biological activity as the dose was lowered into the hormetic zone. This type of mechanism-based prospective study offers a serious challenge to the current linear low-dose model for cancer risk assessment but also strongly supports the hormetic properties. Similar hormetic dose–response relationships were reported for other carcinogens agents [231], including strong evidence in the ED01 megamouse study concerning AAF-induced bladder cancer [239]. Although it is obvious that the hormetic dose–response model supports a threshold for carcinogens, the case with noncarcinogens, while important, is more subtle. In this case, the hormetic response occurs immediately below the NOAEL and extends for a dose range that is typically less than 20-fold from the NOAEL but which could far exceed 100-fold in specific circumstances. Such an hormetic dose–response would be expected to occur for all population subgroups, including those considered at high risk [240].

The current risk assessment approach for noncarcinogens is blind to the possibility that hormesis exists. Such a lack of consideration denies the risk assessor the opportunity to become informed on whether the hormetic response offers the affected population a response that enhances health, is harmful, or is uncertain. Such knowledge provides the risk assessor with a set of more flexible, data-based options that could be used to target UF size (i.e., interindividual variation) for the optimal population response, an option that is currently ignored by regulatory agencies [237].

## IMPLICATIONS OF CHEMICAL INTERACTIONS FOR THE REGULATORY PROCESS

One of the major difficulties in current environmental public health practice is that the focus is on a limited number of environmental contaminants, with limited consideration of interactive effects among pollutants. In fact, the number of environmental pollutants in different media is large, making it difficult to estimate the degree of public health protection afforded by our current regulatory apparatus. Still, it is clear that the scientific and regulatory communities must address the issue of multiple chemical exposures. In fact, animal models and human epidemiological studies show that interactions do occur among chemicals and that this can result, under certain circumstances, in greater than additive effects; for example, uranium miners who do not smoke have a fourfold greater risk of cancer than nonsmokers; however, uranium miners who smoke display a 40-fold greater cancer risk than the general population of nonsmokers [241].

Interactions have been studied for many years by the drug industry, insecticide manufacturers, and forensic/clinical toxicologists. Given the widespread use of multiple-drug therapy, the need to anticipate possible interactions has been essential; thus, much of the basis of our current understanding of toxicological interactions derived from the pharmaceutical industry. The uncertainty should be recognized when extrapolating from drug exposures where doses are relatively high (by definition, at pharmacologically active doses) to environmental exposures, which are typically much lower than doses with established effects.

Chemical interactions have been broadly classified by three general terms: *addition* (additivity), when the toxic effect produced by two or more chemicals in combination is equivalent to that expected by simple summation of their individual effects; *antagonism*, when the effect of a combination is less than the sum of the individual effects; and *synergism*, when the effect of the combination is greater than would be predicted by summation of the individual effects. Other terms have been used, such as

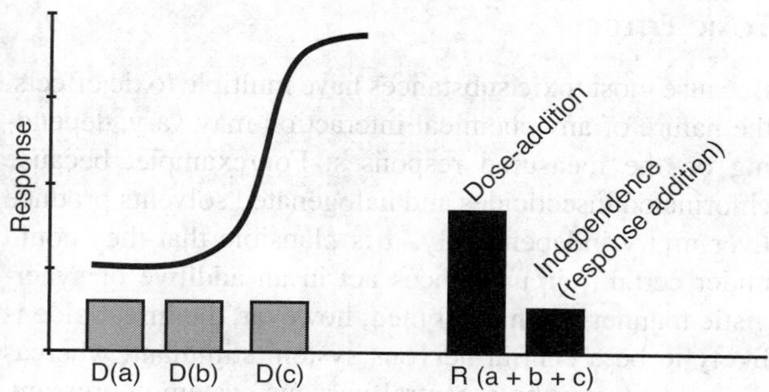


**FIGURE 2.12** A model dose–response curve for an effect with a threshold. Below the curve, bars indicate doses D(a), D(b), and D(c), each of which is one half the toxic threshold for chemicals a, b, and c, respectively. The calculated combined response *R* (a + b + c) would be greater or less than the threshold depending on whether a dose addition or independence model was used to predict the combined action of chemicals a, b, and c in a mixture. (From Borgert, C.J. et al., *Toxicol. Appl. Pharmacol.*, 201(2), 85–96, 2004. With permission.)

*indifference* and *potentiation*, which represent specialized aspects of antagonism and synergism, respectively.

It should also be noted that interactions in chemical mixtures can be characterized through a *dose-addition* or *response-addition* framework. The underlying assumption of the dose-addition framework is that the potency of the mixture is a function of the sum of the potencies of the individual chemicals. An implication of this approach, as reflected in Figure 2.12, is that even when chemicals result in no toxicological response individually, toxicity may result when multiple chemicals (all below their NOAELs) are present in a mixture. In contrast, the response-addition framework assumes that the toxicity of a mixture is a function of the toxic effects of the individual compounds; in the example in Figure 2.12, response additions would imply that no toxicity would ensue, as no individual chemical is associated with any toxicological responses [242]. In testing for possible interactions, several important considerations must be addressed. These include temporal (time) factors and response-endpoint considerations.

### TIME FACTOR

Although most screening tests for interactions employ simultaneous exposure, this type of exposure approach has the chance of reducing the likelihood of detecting some potential interactions; for example, two agents may affect the same cellular mechanism but may have markedly different times of onset to expression. If a critical threshold of reversible cellular injury is required for the adverse effect, tests of acute toxicity of combinations given simultaneously may show antagonism, whereas an additive action would be observed if the dosing and observation periods were spaced to cause the maximum effect.

### Toxic Effect

Because most toxic substances have multiple toxic effects, the nature of any chemical interaction may vary, depending on the measured responses. For example, because chlorinated insecticides and halogenated solvents produce liver injury independently, it is plausible that they could under certain circumstances act in an additive or synergistic manner when combined; however, the insecticide is likely to be a central nervous system stimulant, whereas the solvent may be a central nervous system depressant. As measured by neurological tests, these chemicals could interact in an antagonistic way.

### Predictive Models

Development of predictive models of chemical interaction must rely on an understanding of the basic toxicological principles concerning kinetics of reactions of chemicals with primary sites of action (tissue receptor sites) and with secondary tissue sites of reaction. Four factors have been identified as being of central importance:

- Relative affinities of the individual chemicals for sites of action (e.g., target enzymes, cellular membranes)
- Relative affinities for sites of loss of the chemical (e.g., detoxifying enzymes, non-vital tissue binding sites, pathways of excretion, and storage sites)
- Intrinsic activity of the agents at their sites of action
- Sites of bioactivation

Many of the examples are derived from the pharmacology literature, but similar interactions could occur among environmental chemicals.

### Pharmacokinetic Drug/Pollutant Interactions

The above four factors allow us to predict how toxicological interactions may occur. The biological damage caused by a toxic agent is proportional to the amount of the biologically active form of the agents able to react with critical cellular macromolecules. An interaction may occur when the availability of an active chemical is altered by the presence of another agent or when its reactivity with critical macromolecules is altered by the presence of another agent. The first case involves a site of loss of active chemical, whereas the second involves an interaction at a site of action; thus, considerable research activity has investigated the capacity of a chemical to affect the absorption, distribution, metabolism, and excretion of another chemical.

#### Absorption

Absorption of an agent may be affected by a second drug that alters pH or gut motility. Aspirin, for example, is absorbed more rapidly at low pH because more of the drug is present in the readily diffusible nonionized, lipid-soluble form. Agents that cause an increase in the pH will slow down the gastric absorption of aspirin when taken simultaneously. Similarly, the absorption of tetracycline is reduced by aluminum hydroxide gels and readily ionized salts of calcium and magnesium. In contrast, the gastrointestinal absorption of acetaminophen is enhanced in the presence of sorbitol.

#### Protein Binding

Drugs may compete for the same protein binding sites in plasma. When this occurs, the effective biological concentrations of the displaced drug can rise markedly. For example, usually 98% of the anticoagulant drug warfarin is bound to the plasma protein albumin, so only 2% of the total drug in the plasma is biologically active. If the effect of another drug competing for the same plasma albumin site is to reduce the binding of warfarin from 98 to 96%, the concentration of pharmacologically active warfarin would be doubled. This interaction would have approximately the same effect on clotting time as would doubling the dose of the anticoagulant. This type of interaction with an anticoagulant drug has resulted in a number of clinical incidents, with some resulting in fatal hemorrhagic complications.

#### Metabolism

Many chemicals, including drugs and environmental contaminants, enhance the metabolic capacity of the liver. Other chemicals may diminish the metabolic capacity of the liver. These interactions could have profound implications. In fact, it is now recognized that several of the insecticide synergists (i.e., agents that, when administered along with insecticides, markedly enhance the ability of the insecticide to kill insects) act by blocking the enzymes normally affecting insecticide detoxification [243]. Knowledge of synergy has been used to develop more effective insecticide formulations; for example, the toxicity of the insecticide carbaryl against susceptible female houseflies is enhanced by over 200-fold by certain chemical synergists.

Insecticides provide another example of how agencies use information on chemical interactions. In 1957, Frawley et al. [244] reported the first synergistic interaction of two organophosphate insecticides (i.e., malathion and ethyl-*p*-nitrophenyl phenyl phosphonothionate [EPN]) which led to the development of the FDA requirement that all newly registered organophosphate insecticides be evaluated for possible synergisms against all the already-registered organophosphate insecticides. As more organophosphate insecticides were developed, this regulatory requirement became an extreme testing burden; however, with elucidation of the biochemical mechanism* for this interaction, it became possible to assess possible interactions of organophosphate insecticides via biochemical means and thereby circumvent the time-consuming and costly toxicological testing of whole animals. In both examples, that of making a more efficient insecticidal preparation and that of predicting adverse public health effects from multiple agent exposures, predictions of chemical interactions were markedly enhanced by a clear understanding of the mechanisms of toxicity.

### Implications

The examples given above represent ideal situations, as the mechanisms of toxicity of the insecticides were very well characterized. More frequently, little information is available on toxic mechanisms. Regulatory agencies need to develop approaches in such situations when reasonable mechanistic predictions cannot be made. To this end, Finney [245] developed a theoretical mathematical approach for predicting the degree of toxicity derived from various types of chemical interactions. Pozzani and coworkers [246] indicated that only 2 of the 36 pairs of mixtures of industrial vapors tested for acute toxicity in rats deviated significantly from the calculations of Finney's theoretical approach for additive joint toxicity. According to Smyth et al. [247], the study by Pozzani and coworkers supported the hypothesis that the acute toxicity of chemical mixtures randomly chosen has a high likelihood of being accurately predicted by Finney's theoretical formula for additive joint toxicity. In an attempt "to evaluate the overall confidence that can be placed on the prediction of the joint toxicity of many chemical pairs," Smyth et al. [247] studied the toxicity of 27 industrial chemicals in all possible pairs to rats. Their results were consistent with the prediction of Finney [245] that most interactions should be considered as additive until proven otherwise. Smyth et al., in agreement with the general findings of Pozzani and coworkers [246], concluded that approximately 5% of the various combinations tested exhibited more or less than additive effects.

It should be noted, however, that recent studies of chemicals in mixtures at doses below the NOAEL provide evidence that dose additivity may overpredict the toxicity of a mixture (in which case, response additivity is a more appropriate characterization of the interaction). For example, Feron and coworkers [248], for example, observed that exposure to a mixture of nephrotoxic chemicals, each below the NOAEL, did not result in a toxic response, suggesting that dose additivity may not appropriately describe interaction potential at low doses. A similar conclusion was reached by Borgert et al. [298], who postulated that the lack of additivity may reflect a different mode of action at low doses. Thus, it is possible that synergistic interactions, the type of interactions of greatest regulatory concern, may be less likely to occur at environmentally relevant exposure levels than at higher (e.g., pharmaceutical or some occupational) exposure levels.

### Approaches Used by Regulatory Agencies To Assess Interactions

Despite the frequent lack of a clear mechanistic understanding of how chemicals may interact, regulatory agencies have developed interim approaches to facilitate the decision-making process. In this section, we highlight some of the typical approaches used by agencies with illustrative examples.

#### The Hazard Index Approach

Perhaps the simplest approach is the assumption of additivity of hazard. In this approach (applied to noncarcinogens), hazard indices (the ratio between the estimated dose and the reference dose) are summed across chemicals and across routes of exposure to obtain a total hazard index for a particular exposure setting [249]. If the decision criterion of a total hazard index of 1 is exceeded, then further review is performed to determine which chemicals act at the same target organ. A subsequent summation is performed for those chemicals only. Although this approach is useful as a screening approach, it has several limitations that must be considered when interpreting the results. These include (but are not limited to) the following:

- Reference doses (the denominator in the hazard index) for different chemicals contain different types and magnitudes of UFs; thus, differences in the magnitude of a hazard index for a particular chemical or a pathway of exposure may reflect intrinsic differences in hazard as well as differences in the uncertainty of a particular toxicity value.
- Different types of interactive effects are possible, even for chemicals that act at the same target organ; for example, organophosphates that act via the inhibition of acetylcholinesterase at nerve endings would generally be presumed to act in an additive manner [250]. In contrast, trichloroethylene (TCE) and alcohol, both of which affect the central nervous system, can, when consumed simultaneously,

* EPN inhibits the nonspecific enzyme carboxyesterase, which detoxifies malathion; thus, in the presence of EPN, malathion is more persistent and causes a greater effect as a cholinesterase inhibitor than would have occurred had enzymatic detoxification mechanisms not been affected.

act synergistically (e.g., producing "degreaser's flush"); however, chronic alcohol consumption can, by induction of metabolizing enzymes, diminish the response to TCE [77].

### The Toxicity Equivalency Factor Approach

The toxicity equivalency factor (TEF) approach has been applied to a mixture containing toxicologically and structurally similar chemicals. Perhaps one of the best known examples of the TEF approach is the approach developed by the EPA [251] for polychlorinated dibenzodioxins (PCDDs) and for polychlorinated dibenzofurans (PCDFs). This approach is based on the assumption that PCDDs and PCDFs exert their toxicity through binding to the Ah receptor with subsequent effects on transcription and translational events responsible for toxicity. The most potent PCDD, 2,3,7,8-tetrachlorodibenzodioxin (2,3,7,8-TCDD), has the greatest affinity for the Ah receptor and, hence, is the most potent member of this class. TEFs are developed for individual PCDDs and PCDFs, expressed as a fraction (typically in orders of magnitude) of that of TCDD, which is given a TEF of 1.0. Thus, RfDs and cancer slope factors (CSFs) are calculated as a ratio to the RfD and CSF for 2,3,7,8-TCDD. Experimental data, based on mixtures of PCDDs and PCDFs, support this approach, but few, if any, data are available from long-term studies [251]. Other uncertainties in this approach include the assumption of additivity, where competitive interactions may occur at sufficiently high doses, as well as the choice of a particular TEF value, which can be influenced by selection of endpoint, exposure duration and dose [252].

### The Complex Mixture Approach

For some classes of chemicals, toxicological data exist primarily for the complex mixture itself, with limited data for individual constituents. An example of this type of mixture is polychlorinated biphenyls (PCBs). PCBs were manufactured in the United States under the trade name of Aroclor® for use in electrical capacitors. Different Aroclor mixtures contained different percentages of chlorine; for example, Aroclor 1242 contained approximately 42% chlorine [253]. Much of the toxicity testing of PCBs consists of studies of various Aroclor mixtures [253,254]. As a result, toxicity criteria for PCBs are typically expressed as Aroclor-specific values. In the case of Aroclor 1254, for example, the ATSDR developed a chronic minimum risk level (MRL), a value conceptually comparable to the RfD, based on immunological effects in monkeys exposed to Aroclor 1254 in feed for 23 months [252]. Although this approach does not require assumptions on how individual constituents will interact, it does assume that the characteristics of the mixture in the environment are the same as in the laboratory studies. Unfortunately, this assumption is not always correct, as complex mixtures frequently undergo chemical transformations in the environment; moreover, the individual constituents may partition differently in the environment. In the case of PCBs, for example, the more chlorinated forms bioaccummulate in fish more readily than the less chlorinated forms [162].

Consideration of the interactive effects in the regulatory arena is an evolving process. Because data are limited in many cases, simplifying assumptions are often used (e.g., the assumption of hazard index additivity for chemicals that act via the same target organ). As scientists acquire a greater mechanistic understanding of the interactive effects of complex mixtures, approaches that better reflect molecular events can be developed, such as use of the TEF approach. It must be recognized, however, that uncertainty still exists regarding the extent to which such effects occur at environmentally relevant exposure levels and under exposure conditions that do not mimic those tested in the laboratory (e.g., intermittent vs. chronic exposures).

## CONCLUSIONS

In this chapter, we have demonstrated the multiple applications of toxicology to the regulatory process. Applications include developing and evaluating chemical testing protocols, such as for developmental toxicants; developing classification schemes (so far mainly for carcinogens) aimed at characterizing the types of toxic effects that might be observed in humans; and developing health-based criteria for chemicals in various media (food, water, air, soil) or notification levels for the release of chemicals under accidental circumstances.

In addition, toxicology is used in the regulatory process to help assess potential risk associated with defined exposure levels. The traditional paradigm for assessing such risks is as follows: For carcinogens, the potential risk is defined as an upper-bound estimate of excess cancer risk based on cancer incidence at high-dose levels; for noncarcinogens, the potential risk is defined as the ratio of the estimated exposure to an exposure level associated with negligible, if any, risk. Advances in the understanding of toxicological mechanisms indicate that these methodologies are not appropriate in all circumstances. Some carcinogens, such as those that operate by receptor-mediated or cytotoxic mechanisms, may exhibit a threshold or nonlinear dose–response relationship; thus, exposure levels associated with virtually zero risk might be defined. Examples of chemicals with these types of dose–response relationships are saccharin and phenobarbital. The EPA's recent cancer risk assessment guidelines, which consider different dose–response relationships for different carcinogens, represent an important development in this area. The concept of hormesis, in which simulatory effects may occur at doses below the quasi-threshold, has challenged current thinking on dose–response modeling and may have significant implications for the use of toxicology in risk assessment and risk management decision making. In particular, evidence for the existing hormetic dose–response relationships is casting doubt on the appropriateness of linear no-threshold modeling for carcinogens.

Our understanding of certain noncarcinogenic effects, such as angina associated with CO exposure, is reasonably advanced; in this example, risks from CO are more fully described in terms of the number of individuals with heart disease who might be expected to exceed defined COHb levels under certain exposure conditions. BMD and categorical exposure–response modeling represent additional examples of recent advances in noncarcinogenic risk assessment.

Toxicology is frequently applied in the regulatory context of developing permissible exposure levels in different exposure media, such as ambient air, drinking water, or food. As discussed in this chapter, defining the health-based permissible exposure level is only one part of developing a regulatory standard. Other important factors in the regulatory process include risk management issues such as the definition of acceptable risk, the weighing of the costs and technical feasibility of reducing risk, the availability of alternatives, and the new risks possibly created by reducing the original risk (e.g., use of a less well tested substitute chemical). Issues of equity and whether certain members of the population are unfairly burdened by chemical exposure represent other considerations.

A critical role for toxicologists participating in the regulatory process is to effectively communicate not only the results of a risk assessment but also the uncertainties associated with risk evaluations to provide risk managers with the full information needed for making sound decisions. In addition, despite pressure to employ older methods for the sake of consistency, toxicologists must work to develop and encourage the use of new methodologies reflecting the advances in our understanding of toxicological mechanisms. We hope this chapter is useful as a guide to the use of better science in the regulatory process.

## QUESTIONS

1. How do different approaches used for noncancer risk assessment (e.g., benchmark dose, reference dose, and the distributional population approach) address susceptible populations?
2. What is *mode of action* and how is it used in carcinogen classification and in selecting dose–response models for cancer risk assessment?
3. What types of challenges does the concept of hormesis present to traditional paradigms for cancer and non-cancer risk assessment? How robust is the support for hormesis?

## REFERENCES


1. Hutt, P. B. (1985): Use of quantitative risk assessment in regulatory decisionmaking under federal health and safety statutes. In: *Risk Quantitation and Regulatory Policy*, edited by D. G. Hoel, R. A., Merrill, and F. P. Perera, Banbury Report 19. Cold Spring Harbor Laboratory, Cold Spring Harbor, New York, p. 15.
2. Pott, P. (1779): Cancer scroti. In *The Chirurgical Works*, A New Edition in Three Volumes, Volume I. London.
3. NYSDEC (1986): *New York State Air Guide 1: Guidelines for the Control of Toxic Ambient Air Contaminants*. Division of Air Resources, New York State Department of Environmental Conservation, Albany, NY.
4. EPA (2004). *Chemical Information Collection and Data Development (Testing), Master Testing List: Introduction*. U.S. Environmental Protection Agency, Washington, D.C. (http:/www.epa.gov/opptintr/chemtest/mtlintro.htm).
5. Carson, R. L. (1962): *Silent Spring*. Houghton Mifflin, Boston, MA.
6. CDC (2005): *Third National Report on Human Exposure to Environmental Chemicals*. NCEH Pub. No. 05-0570, National Center for Environmental Health, Centers for Disease Control and Prevention, Atlanta, GA.
7. Public Law 85-929 (1958): Food Additives Amendment of 1958.
8. U. S. Food and Drug Administration (1986): Listing of D&C orange no. 17 for use in externally applied drugs and cosmetics, final rule. *Fed. Reg.*, 51:28331.
9. Scheuplein, R. J. (1987): Risk assessment and food safety: a scientist and regulator's view. *Food Drug Cosm. Law J.*, 42:237.
10. National Research Council (1987): *Regulating Pesticides in Food: The Delaney Paradox*, National Academy Press, Washington, D.C.
11. Abelson, P. H. (1993): Pesticides and food. *Science*, 259:1235.
12. U.S. Food and Drug Administration (1988): Color additives: denial of petition for listing of D&C red no. 19 for use in externally applied drugs and cosmetics. *Fed. Reg.*, 53:26831.
13. OSHA (1970): Occupational Safety and Health Act of 1970, 29 U.S.C. 655.
14. Rodricks, J. V. and Taylor, M. R. (1989): Comparison of risk management in U.S. regulatory agencies. *J. Haz. Mater.*, 21:239.
15. Industrial Union Department (1980): *AFL-CIO v. American Petroleum Institute*, 448 U.S. 60165 L. Ed. 2d 1010, 100 S. Ct. 2844.
16. EPA (1999): Summary of FQPA amendments to FIFRA and FFDCA, U.S. Environmental Protection Agency, Washington, D.C. (http://www.epa.gov/oppfead1/fqpa/fqpa-iss. htm).
17. OMB (1990): Current regulatory issues in risk assessment and management. In: *Regulatory Program of the United States Government*. Executive Office of the President, Office of Management and Budget, Washington, D.C.
18. Thorton, J. (1991): Written testimony of J. Thornton [Greenpeace U.S.A.] for the U.S. House of Representatives Committee on Science, Space, and Technology, Subcommittee on Environment, Hearing on Risk Assessment: Strengths and Limitations of Utilization for Policy Decisions, May 21.

19. Finkel, A. M. (1991): Testimony before the U.S. House of Representatives Committee on Science, Space, and Technology, Subcommittee on Environment, Hearing on Risk Assessment: Strengths and Limitations of Utilization for Policy Decisions, May 21.
20. EPA (1999): *Reference Dose (RfD): Description and Use in Health Risk Assessments*, (Background Document 1A). Integrated Risk Information System, U.S. Environmental Protection Agency, Washington, D.C. (http://www.epa.gov.ngispgm3/iris/rfd.htm).
21. ACGIH (1992): *Threshold Limit Values for Chemical Substances and Physical Agents*. American Conference of Governmental Industrial Hygienists, Cincinnati, OH.
22. DHHS (1985): *Risk Assessment and Risk Management of Toxic Substances*. Report to the Secretary of DHHS from the Executive Committee of the DHHS Committee to Coordinate Environmental and Related Programs. U.S. Department of Health and Human Services, Washington, D.C.
23. Commission on the European Communities (2000): *Communication from the Commission on the Precautionary Principle*. Brussels, Belgium (http://europa.eu.int/comm/dgs/health_consumer/library/pub/pub07_en.pdf).
24. European Union (2000): *EU's Communication on Precautionary Principle*. European Union, Brussels, Belgium (http://www.gdrc.org/u-gov/precaution-4.html).
25. Europa (2003): *The New EU Chemicals Legislation: REACH*. European Commission, Brussels, Belgium (http://europa.eu.int/comm/enterprise/reach/index_en.htm).
26. National Research Council (1983): *Risk Assessment in the Federal Government: Managing the Process*. National Academy Press, Washington, D.C.
27. National Research Council (1994): *Science and Judgment in Risk Assessment*, National Academy Press, Washington, D.C.
28. The Presidential/Congressional Commission on Risk Assessment and Risk Management (1997): *Risk Assessment and Risk Management in Regulatory Decision-Making*, Vol. 2, Final Report.
29. U.S. Department of Health and Human Services (1986): NIOSH recommendations for occupational safety and health standards. *Morbid. Mortal. Weekly Rep.*, 35:1S.
30. Nichols A. L. and Zeckhauser, R. J. (1986): The perils of prudence: how conservative risk assessments distort regulation, *Regulation*, 10:13.
31. Graham, J. D. and Wiener, J. B., Eds. (1995): *Risk vs. Risk: Tradeoffs in Protecting Health and the Environment*. Harvard University Press, Cambridge, MA.
32. Rosenthal, A., Graf, G. M., and Graham, J. D. (1992): Legislating acceptable cancer risk from exposure to toxic chemicals. *Ecol. Law Q.*, 19:269.
33. Anderson, P. D. (1988): Scientific origins of incompatibility in risk assessment. *Stat. Sci.*, 3:320.
34. Seeley, M. R. et al. (2001): Procedures for health risk assessment in Europe. *Regul. Toxicol. Pharmacol.*, 34:153.
35. Travis, C. C. et al. (1987): Cancer risk management. *Environ. Sci. Technol.*, 21:415.
36. Travis, C. C. and Hattemer-Fry, H. A. (1988): Determining an acceptable level of risk. *Environ. Sci. Technol.*, 22:873.
37. Miller, F. J., Illing, J. W., and Gardner, D. E. (1978): Effect of urban ozone levels on laboratory-induced respiratory infections, *Toxicol. Lett.*, 2:163.
38. Green, G. M. (1984): Similarities of host defense mechanisms against pulmonary disease in animals and man. *J. Toxicol. Environ. Health*, 13:471.
39. EPA (1989): *Review of the National Ambient Air Quality Standard for Ozone: Assessment of Scientific and Technical Information*. Office of Air Quality Planning and Standards, U.S. Environmental Protection Agency, Research Triangle Park, N.C.
40. IARC (1982): *Evaluation of Carcinogenic Risk of Chemicals to Humans*, IARC Monographs, Suppl. 4. International Agency for Research on Cancer, Lyons, France.
41. EPA (2005): *Guidelines for Carcinogen Risk Assessment*, EPA/630/P-03-001F. Risk Assessment Forum, U.S. Environmental Protection Agency, Washington, D.C.
42. Occupational Safety and Health Administration (1980): Identification, classification, and regulation of potential occupational carcinogens, *Fed. Reg.*, 45:5002.
43. Wallace, L. A. et al. (1983): Personal exposure of volatile organics and other compounds indoors and outdoors: the TEAM study. In: *Proc. of the 76th Annual Meeting of the Air Pollution Control Association*. Air Pollution Control Association, Pittsburgh, PA, 1983.
44. Tosteson, T., Spengler, J. D., and Weber, R. A. (1982): Aluminum, iron, and lead content of respirable particulate samples from a personal monitoring system. *Environ. Int.*, 2:265.
45. McKone, T. E. and Bogen, K. T. (1991): Predicting the uncertainties in risk assessment, *Environ. Sci. Technol.*, 25:1674.
46. Van Landingham, C. B., Lawrence, G. A., and Shipp, A. M. (2004): Estimates of lifetime-absorbed daily doses from the use of personal-care products containing polyacrylamide: a Monte Carlo analysis. *Risk Anal.*, 24:603.
47. Thompson, K. M., Burmaster, D. E., and Crouch, E. A. C. (1992): Monte-Carlo techniques for quantitative uncertainty analysis in public health risk assessments. *Risk Anal.*, 12:53.
48. EPA (1997): *Guiding Principles for Monte Carlo Analysis*, EPA/630/R-97/001. Risk Assessment Forum, U.S. Environmental Protection Agency, Washington, D.C.
49. Cullen, A. C. and Frey, H. C., Eds. (1999): *Probabilistic Techniques in Exposure Assessment: A Handbook for Dealing with Variability and Uncertainty in Models and Inputs*. Plenum Press, New York.
50. National Research Council (NRC), *Understanding Risk: Informing Decisions in a Democratic Society*, edited by P. C. Stern and H. V. Fineberg. National Academy Press, Washington, DC, 1996.
51. EPA (1995): *Guidance for Risk Characterization*. Science Policy Council, U.S. Environmental Protection Agency, Washington, D.C.
52. Infante, P. F. et al. (1977): Leukemia in benzene workers. *Lancet*, ii:76.
53. Maltoni, C., Conti, B., and Cotti, G. (1983): Benzene: a multi-potential carcinogen. Results of long-term bioassays performed at the Bologna Institute of Oncology. *Am. J. Ind. Med.*, 4:589.

54. Lipfert, F. W. (1980): Sulfur oxides, particulates and human mortality: synopsis of statistical correlations. *J. Air Pollut. Control Assoc.*, 30:366.
55. Quakenboss, J. J. et al. (1982): Personal monitoring for nitrogen dioxide exposure: methodological considerations for a community study. *Environ. Int.*, 8:249.
56. Sax, S. N. et al. (2004): Differences in source emission rates of volatile organic compounds in inner-city residences of New York City and Los Angeles. *J. Expo. Anal. Environ. Epidemiol.*, 14:S95.
57. Thurston, G. D. et al., Reexamination of London, England, mortality in relation to exposure to acidic aerosols during 1963–1972 winters. *Environ. Health Perspect.*, 79:73.
58. Valberg, P. A. (2004): Is PM more toxic than the sum of its parts? Risk-assessment toxicity factors vs. PM-mortality "effect functions." *Inhal. Toxicol.*, 16:19.
59. CDC (1991): *Preventing Lead Poisoning in Young Children: A Statement by the Centers for Disease Control*. U.S. Public Health Services, Washington, D.C.
60. Calabrese, E. J. (1978): *Pollutants and High Risk Groups*. John Wiley & Sons, New York.
61. Enterline, P. E. (1983): Epidemiologic basis for the asbestos standard. *Environ. Health Perspect.*, 52:93.
62. Bethel, R. A. et al. (1983): Sulfur dioxide-induced bronchoconstriction in freely breathing exercising, asthmatic subjects, *Am. Rev. Respir. Dis.*, 128:987.
63. Roger, L. J. et al. (1985): Bronchoconstriction in asthmatics exposed to sulfur dioxide during repeated exercise. *J. Appl. Physiol.*, 59:784.
64. EPA (1986): *Review of the National Ambient Air Quality Standards for Sulfur Oxides: Updated Assessment of Scientific and Technical Information*, EPA-450/5-86-013, Addendum to the 1982 OAQPS Staff Paper. Office of Air Quality, Planning, and Standards, U.S. Environmental Protection Agency, Washington, D.C.
65. National Academy of Sciences (2004): *Intentional Human Dosing Studies for EPA Regulatory Purposes: Scientific and Ethical Issues*. Committee on the Use of Third-Party Toxicity Research with Human Research Participants; Science, Technology, and Law Program; Policy and Global Affairs Division, National Research Council of the National Academies, Washington, D.C.
66. Wiles, R. [Environmental Working Group] (2005): Letter to Taylor, M. R. [National Academy of Sciences] re: using humans in laboratory tests, February 23, 2004.
67. Greer, M. A. et al. (2002): Health effects assessment for environmental perchlorate contamination: the dose–response for inhibition of thyroidal radioiodine uptake in humans. *Environ. Health Perspect.*, 110:927.
68. National Academy of Sciences (2004): *Committee to Assess the Health Implications of Perchlorate Ingestion*. National Academy of Sciences, Washington, D.C. (http://www4.nas.edu/webcr.nsf/ProjectScopeDisplay/BEST-K-03-05-A?OpenDocument).
69. Chang, L.-Y. et al. (1992): Epitheliel injury and interstitial fibrosis in the proximal alveolar regions of rats chronically exposed to a simulated pattern of urban ambient ozone. *Toxicol. Appl. Pharmacol.*, 115:241.
70. Germolec, D. R. et al. (1989): Toxicology studies of a chemical mixture of 25 groundwater contaminants. II. Immunosuppression in B5C3F1 mice. *Fundam. Appl. Toxicol.*, 13:377.
71. Elsayed, N. M. and Mustafa, M. G. (1982): Dietary antioxidants and the biochemical response to oxidant inhalation. I. Influence of dietary vitamin E on the biochemical effects of nitrogen dioxide exposure in rat lung. *Toxicol. Appl. Pharmacol.*, 66:319.
72. Kouri, R. E. and Nebert, D. W. (1977): Genetic regulation of susceptibility to polycyclic hydrocarbon induced tumors in the mouse. In: *Origins of Human Cancer*, edited by H. H. Hiatt, J. D. Watson, and J. A. Winstyen. Cold Spring Harbor Laboratory, Cold Spring Harbor, NY, p. 811.
73. Nebert, D. W. (1989): The Ah locus: genetic differences in toxicity, cancer, mutation. *Crit. Rev. Toxicol.*, 20:153.
74. Waters, M. D. and Fostel, J. M. (2004): Toxicogenomics and systems toxicology: aims and prospects. *Nature Rev. Gen.*, 5:936.
75. International Programme on Chemical Safety Workshop (2003): *Toxicogenomics and the Risk Assessment of Chemicals for the Protection of Human Health, Summary*. World Health Organization, International Labour Organization, United Nations Environment Programme, held at the Federal Institute for Risk Assessment, Berlin, Germany, November 17–19.
76. Cohen, S. M. et al. (2003): The human relevance of information on carcinogenic modes of action: overview. *Crit. Rev. Toxicol.*, 33:581.
77. Agency for Toxic Substances and Disease Registry (1997): *Toxicological Profile for Trichloroethylene: Update*. Sciences International, Inc., and U.S. Public Health Services, U.S. Environmental Protection Agency, Washington, D.C.
78. Heinrich, U. et al. (1986): Chronic effects on the respiratory tract of hamsters, mice and rats after long-term inhalation of high concentrations of filtered and unfiltered diesel engine emissions. *J. Appl. Toxicol.*, 6:383.
79. ILSI Risk Science Institute (2000): The relevance of the rat lung response to particle overload for human risk assessment: a workshop consensus report, ILSI Risk Science Workshop Participants. *Inhal. Toxicol.* 12:1.
80. Enslein, K. (1987): Computer-assisted prediction of toxicity. In *Toxic Substances and Human Risk: Principals of Data Interpretation*. edited by R. G. Tardiff and J. V. Rodricks. Plenum Press, New York, p. 317.
81. Beck, B. D. et al. (1991): Utilization of quantitative structure activity relationships (QSARs) in risk assessment. *Reg. Toxicol. Pharmacol.*, 14:273.
82. Hart, R. W. and Turturro, A. (1988): Introduction. In: *Banbury Report 31: Carcinogen Risk Assessment—New Directions in the Qualitative and Quantitative Aspects*, edited by R. W. Hart and F. D. Hoerger. Cold Spring Harbor Laboratory, Cold Spring Harbor, NY.
83. Linkov, I., Wilson, R., and Gray, G. M. (1998): Anticarcinogenic responses in rodent cancer bioassays are not explained by random effects. *Toxicol. Sci.*, 43:1.
84. Pitot, H. C. and Dragan, Y. P. (1991): Facts and theories concerning the mechanisms of carcinogenesis. *FASEB J.*, 5:2280.

85. Trosko, J. E. and Chang, C. C. (1988): Nongenotoxic mechanisms in carcinogenesis: role of inhibited intercellular communication. In: *Banbury Report 31: Carcinogen Risk Assessment—New Directions in the Qualitative and Quantitative Aspects*, edited by R. W. Hart and F. D. Hoerger. Cold Spring Harbor Laboratory, Cold Spring Harbor, NY.
86. Cohen, S. M. et al. (2005): Methylated arsenicals: the implications of metabolism and carcinogenicity studies in rodents to human risk assessment. *Crit. Rev. Toxicol.*, 36(2):99.
87. Cohen, S. M. et al. (2004): Evaluating the human relevance of chemically induced animal tumors. *Toxicol. Sci.*, 78:181.
88. Roe, F. J. (1989): Non-genotoxic carcinogenesis: implications for testing and extrapolation to man. *Mutagenesis*, 4:407.
89. Moolgavkar, S. H. (1986): Carcinogenesis modeling: from molecular biology to epidemiology. *Ann. Rev. Public Health*, 7:151.
90. Graham, J. D. (1992): *Recommendations for Improving Cancer Risk Assessment*. Center for Risk Analysis, Harvard School of Public Health, Boston, MA.
91. EPA (2005): *Guidelines for Carcinogen Risk Assessment (Final)*, EPA/630/P-03/001B. Risk Assessment Forum, U.S. Environmental Protection Agency, Washington, D.C. (http:// cfpub.epa.gov/ncea/cfm/recordisplay.cfm?deid=116283).
92. Tennant, R. W., French, J. E., and Spalding, J. W. (1995): Identifying chemical carcinogens and assessing potential risk in short-term bioassays using transgenic mouse models. *Environ. Health Perspect.*, 103:942.
93. Sontag, J. M., Page, N. P., and Sanotti, U. (1976): *Guidelines for Carcinogen Bioassays in Small Rodents*, DHHS Publ. (NIH) 76-801. National Cancer Institute, Bethesda, MD.
94. Haseman, J. K. (1985): Issues in carcinogenicity testing: dose selection. *Fund. Appl. Toxicol.*, 5:66.
95. Munro, I. C. (1977): Considerations in chronic toxicity testing: the chemical, the dose, the design. *J. Environ. Pathol. Toxicol.*, 1:183.
96. Melnick, R.L. et al. (1984): Toxicity and carcinogenicity of melamine in F344 rats and B5C3F1 mice. *Toxicol. Appl. Pharmacol.*, 72:292.
97. Clayson, D. B. and Clegg, D. J. (1991): Classification of carcinogens: polemics, pedantics, or progress? *Reg. Tox. Pharm.*, 14:147.
98. Ames, B. N. and Gold, L. S. (1990): Too many rodent carcinogens: mitogenesis increases mutagenesis. *Science*, 249:970.
99. Ames, B. N., Swirsky-Gold, L., and Shigenaga, M. K. (1996): Cancer prevention, rodent high-dose cancer tests, and risk assessment. *Risk Anal.*, 16:613.
100. Rieth, J. P. and Starr, T. B. (1989): Chronic bioassays: relevance to quantitative risk assessment of carcinogens. *Reg. Tox. Pharm.*, 10:160.
101. Goodman, G. and Wilson, R. (1992): Comparison of the dependence of the TD50 on maximum tolerated dose for mutagens and nonmutagens. *Risk Anal.*, 12:525.
102. Gaylor, D. W. (2005): Are tumor incidence rates from chronic bioassays telling us what we need to know about carcinogens? *Regul. Toxicol. Pharmacol.*, 41:128.
103. Foran, J. A. and the ILSI Risk Science Working Group on Dose Selection (1997): Principles for the selection of doses in chronic rodent bioassays. *Env. Health Perspect.*, 105.
104. IARC (1980): Long-term and Short-term Screening Assays for Carcinogens: A Critical Appraisal, IARC Monographs, Supplement 2, International Agency for Research on Cancer, Lyons, France.
105. U.S. Environmental Protection Agency (1986): Guidelines for carcinogen risk assessment, *Fed. Reg.*, 51:33992.
106. Hart, R. W. and Fishbein, L. (1985): Interspecies extrapolation of drug and genetic toxicity data. In: *Toxicological Risk Assessment*, Vol. I, edited by D. B. Clayson, D. Krewski, and I. Munro. CRC Press, Boca Raton, FL, p. 3.
107. IARC (1997): *Polychlorinated Dibenzo-Para-Dioxins and Polychlorinated Dibenzofurans*, Vol. 69, IARC Monographs on the Evaluation of Carcinogenic Risks to Humans. International Agency for Research on Cancer, World Health Organization, Geneva, Switzerland.
108. EPA (2003): *Exposure and Human Health Risk Assessment of 2,3,7,8-Tetrachlorodibenzo-p-Dioxin (TCDD) and Related Compounds*. Part II. *Health Assessment of 2,3,7,8-Tetrachlorodibenzo-p-Dioxin (TCDD) and Related Compounds*, NAS Review Draft (Dioxin Reassessment). National Center for Environmental Assessment, U.S. Environmental Protection Agency, Washington, D.C.
109. EPA (2003): *Exposure and Human Health Risk Assessment of 2,3,7,8-Tetrachlorodibenzo-p-Dioxin (TCDD) and Related Compounds*. Part III. *Integrated Summary and Risk Characterization for 2,3,7,8-Tetrachlorodibenzo-p-Dioxin (TCDD) and Related Compounds*, NAS Review Draft (Dioxin Reassessment). National Center for Environmental Assessment, U.S. Environmental Protection Agency, Washington, D.C.
110. Cole, P. (2003): Dioxin and cancer: a critical review. *Regul. Toxicol. Pharmacol.*, 38:378.
111. Strauss, H. S. (1993): Sex biases in the risk assessment of toxic chemicals, presented at the Annual Meeting of the American Association for the Advancement of Science, Boston, MA, February 12.
112. IARC (1987): *Overall Evaluations of Carcinogenicity: An Updating of IARC Monographs Volumes 1 to 42*, Suppl. 7, IARC Monographs. International Agency for Research on Cancer, Lyons, France.
113. Tomatis, L. et al. (1989): Human carcinogens so far identified. *Jpn. J. Cancer Res.*, 80:795.
114. Waring, M. J., DNA modification and cancer. *Annu. Rev. Biochem.*, 50:159.
115. Popper, H., Maltoni, C., and Selikoff, I. J. (1980): Vinyl chloride-induced hepatic lesions in man and rodents: a comparison, *Liver*, 1:7.
116. Butterworth, B. E. (1990): Consideration of both genotoxic and nongenotoxic mechanisms in predicting carcinogenic potential. *Mutat. Res.*, 239:117.
117. Zeise, L., Wilson, R., and Crouch, E. A. C. (1987): Dose–response relationships for carcinogens: a review, *Environ. Health Perspect.*, 73, 259, 1987.

118. Cohen, S. M. and Ellwein, L. B., Biological theory of carcinogenesis: implications for risk assessment. In: *Low-Dose Extrapolation of Cancer Risks*, edited by S. Olin et al. ILSI Press, Washington, D.C., p. 145.
119. Conolly, R. B. et al., Human respiratory tract cancer risks of inhaled formaldehyde: dose–response predictions derived from biologically motivated computational modeling of a combined rodent and human dataset. *Toxicol. Sci.*, 82, 279, 2004.
120. Golden, R. J. et al. (1997): Chloroform in mode of action: implications for cancer risk assessment. *Regul. Toxicol. Pharmacol.*, 26:142.
121. U. S. Interagency Staff Group of Carcinogens (1986): Chemical carcinogens: a review of the science and its associated principals. *Environ. Health Perspect.*, 67:201.
122. Ziskind, M., Jones, R. N., and Weil, H. (1976): Silicosis. *Am. Rev. Respir. Dis.*, 113:643.
123. National Research Council (1986): *Drinking Water and Health*, Vol. 6. National Academy Press, Washington, D.C.
124. Dourson, M. L. and Stara, J. F. (1986): Regulatory history and experimental support of uncertainty (safety) factors. *Regul. Toxicol. Pharmacol.*, 3:224.
125. EPA (1990): *General Quantitative Risk Assessment Guidelines for Noncancer Health Effects*, ECAO-CIN-538. Technical Panel for the Development of Risk Assessment Guidelines for Noncancer Health Effects, U.S. Environmental Protection Agency, Washington, D.C.
126. EPA (1999): *IRIS Substance File for Chromium (VI)*, CASRN 18540-29. U.S. Environmental Protection Agency, Washington, D.C. (http://www.epa.gov/ngispgm3/iris/subst/0144.htm).
127. EPA (1999): *IRIS Substance File for Fluorine (Soluble Fluoride)*, CASRN 7782-41-4. U.S. Environmental Protection Agency, Washington, D.C. (http://www.epa.gov/ngispgm3/iris/subst/0053.htm).
128. Beck, B. D. et al. (1993): Improvements in quantitative noncancer risk assessment. *Fund. Appl. Toxicol.*, 20:1.
129. Crump, K. S. (1984): A new method for determining allowable daily intake. *Fund. Appl. Toxicol.*, 4:854.
130. Dourson, M. L. (1986): New approaches in the derivation of acceptable daily intake (ADI), *Comments Toxicol.*, 1:35.
131. Calabrese, E. J., Beck, B. D., and Chappell, W. R. (1992): Does the animal-to-human uncertainty factor incorporate interspecies differences in surface area? *Reg. Toxicol. Pharmacol.*, 15:172.
132. Conolly, R. B. and Butterworth, B. E. (1995): Biologically based dose response model for hepatic toxicity: a mechanistically based replacement for traditional estimates of noncancer risk, *Toxicol. Lett.*, 82/83:901.
133. Jarabek, A. M. et al. (1990: The U.S. Environmental Protection Agency's inhalation RfD methodology: risk assessment for air toxics. *Toxicol. Ind. Health*, 6:279.
134. EPA (1990): *Interim Methods for Development of Inhalation Reference Concentrations*, Review Draft, EPA/600/8-90/066A. Environmental Criteria and Assessment Office, U.S. Environmental Protection Agency, Washington, D.C.
135. Overton, J. H. and Jarabek, A. M. (1989): Estimating equivalent human concentrations of no observed adviser effect levels: a comparison of several methods. *Exp. Pathol.*, 37:89.
136. WHO (1994): *International Programme on Chemical Safety: Assessing Human Health Risks of Chemicals—Derivation of Guidance Values for Health-Based Exposure Limits*, Environmental Health Criteria 170. International Programme on Chemical Safety, World Health Organization, Geneva, Switzerland.
137. WHO (1999): *International Programme on Chemical Safety: Assessing Human Health Risks of Chemicals—Principles for the Assessment of Risk to Human Health from Exposure to Chemicals*, Environmental Health Criteria 210. World Health Organization, Geneva, Switzerland.
138. EPA (2004): *Boron and Compounds*, CASRN 7440-42-8. Integrated Risk Information System, U.S. Environmental Protection Agency, Washington, D.C. (http://cfpub.epa.gov/iris/quickview.cfm?substance_nmbr=0410).
139. Faustman, E. M. (1996): *Review of Noncancer Risk Assessment: Application of Benchmark Dose Methods*, prepared for the Presidential/Congressional Commission on Risk Assessment and Risk Management, Washington, D.C.
140. EPA (1995): *The Use of the Benchmark Dose Approach in Health Risk Assessment*, EPA/630/R-94/007. Risk Assessment Forum, Office of Research and Development, U.S. Environmental Protection Agency, Washington, D.C.
141. Slob, W. and Pieters, M. N. (1997): *A Probabilistic Approach for Deriving Acceptable Human Intake Limits and Human Health Risks from Toxicological Studies: General Framework*, Report No. 620110-005. National Institute of Public Health and the Environment, Bilthoven, The Netherlands.
142. Baird, S. J. S. et al. (1996): Noncancer risk assessment: a probabilistic alternative to current practice. *Hum. Ecol. Risk Assess.*, 2:79.
143. EPA (2000): *Benchmark Dose Technical Guidance Document*, Preliminary Draft, EPA/630/R-00/001. Risk Assessment Forum, U.S. Environmental Protection Agency, Washington, D.C.
144. Crump, K. S., Clewell, H. J., and Andersen, M. E. (1997): Cancer and non-cancer risk assessment should be harmonized. *Hum. Ecol. Risk Assess.*, 3:495.
145. EPA (1999): *Integrated Risk Information System (IRIS)*, U.S. Environmental Protection Agency, Washington, D.C.
146. EPA (1999): *IRIS Substance File for Naphthalene*, CASRN 91-20-3. U.S. Environmental Protection Agency, Washington, D.C. (http://www.epa.gov/ngispgm3/iris/subst/0436.htm).
147. EPA (1992): *Review of the National Ambient Air Quality Standards for Carbon Monoxide: 1992 Reassessment of Scientific and Technical Information*, OAQPS staff paper. U.S. Environmental Protection Agency, Washington, D.C.
148. U.S. Environmental Protection Agency (1991): Guidelines for developmental toxicity, *Fed. Reg.*, 56:63798.
149. Khera, K. S., Grice, H. C., and Clegg, D. J. (1989): *Current Issues in Toxicology: Interpretation and Extrapolation of Reproductive Data To Establish Human Safety Standards*. Springer-Verlag, New York.
150. U.S. PHS (1989): *Toxicological Profile for Selected PCBs (Arochlor-1260, -1254, -1248, -1242, -1232, -1221, and -1016)*, Report No. ATSDR/TP-8821. Agency for Toxic Substances and Disease Registry, U.S. Public Health Service, Washington, D.C.

151. Chernoff, N. and Kavlock, R. J. (1982): An *in vivo* teratology screen utilizing pregnant mice. *J. Toxicol. Environ. Health*, 10:541.
152. Manson, J. M. and Wise, L. D. (1991): Teratogens. In: *Casarett and Doull's Toxicology*, edited by M. O. Amdur, J. Doull, and C. D. Klaassen. Pergamon Press, New York, p. 226.
153. IEHR (1992): *An Evaluative Process for Determining Human Reproductive and Developmental Toxicity of Agents*, Draft Version. Institute for Evaluating Health Risks, Washington, D.C.
154. Kavlock, R. J., Schmid, J. E., and Setzer, R. W. (1996): A simulation study of the influence of study design on the estimation of benchmark doses for developmental toxicity. *Risk Anal.*, 16:399.
155. American Cancer Society (2005): *Cancer Facts and Figures 2005*. American Cancer Society, Washington, D.C.
156. EPA (1988): *Special Report on Ingested Inorganic Arsenic: Skin Cancer; Nutritional Essentiality*, EPA-625/3-87-013F. Risk Assessment Forum, U.S. Environmental Protection Agency, Washington, D.C.
157. American Thoracic Society (1985): Guidelines as to what constitutes an adverse respiratory health effect with special reference to epidemiologic studies of air pollution. *Am. Rev. Respir. Dis.*, 131:666.
158. American Thoracic Society (2000): What constitutes an adverse respiratory health effect for air pollution? *Am. J. Respir. Crit. Care*, 161:665.
159. deRosa, C. T., Stara, J. F., and Durkin, P. R. (1985): Ranking chemicals based on chronic toxicity data. *Toxicol. Ind. Health*, 1:177.
160. Andersen, M. E. et al. (1987): Physiologically based pharmacokinetics and the risk assessment process for methylene chloride. *Toxicol. Appl. Pharmacol.*, 87:185.
161. Calabrese, E. J. (1987): Animal extrapolation: a look inside the toxicologist's black box. *Environ. Sci. Technol.*, 21:618.
162. Clewell, H. J. (1995): The use of physiologically based pharmacokinetic modeling in risk assessment: a case study with methylene chloride. In: *Low-Dose Extrapolation of Cancer Risks*, edited by S. Olin et al. ILSI Press, Washington, D.C., p. 199.
163. EPA (1999): *IRIS Substance File for Dichloromethane*, CASRN 75-09-2. U.S. Environmental Protection Agency, Washington, D.C. (http://www.epa.gov/ngispgm3/iris/subst/0070.htm).
164. Hattis, D. et al. (1990): Uncertainties in pharmacokinetic modeling for perchloroethylene. I. Comparison of model structure, parameters, and predictions for low-dose metabolism creates for models derived by different authors. *Risk Anal.*, 10:449.
165. Beck, B. D. (1997): The use of information on susceptibility in risk assessment: state of the science and potential for improvement. *Environ. Toxicol. Pharmacol.*, 4:229.
166. Neumann D. A. and Kimmel, C. A., Eds. (1998): *Human Variability in Response to Chemical Exposure*. ILSI Press, Washington, D.C.
167. Cooper, W. C. (1973): Indicators of susceptibility to industrial chemicals. *J. Occup. Med.*, 15, 355, 1973.
168. Carnow, B. W. (1976): Panel discussion on TLVs: lead. In: *Health Effects of Occupational Lead and Arsenic Exposure: A Symposium*, edited by B. Carnow. U.S. Public Health Service, National Institute for Occupational Safety and Health, Washington, D.C., p. 197.
169. Grassman, J. A., Kimmel, C. A., and Neumann, D. A. (1998): Accounting for variability in responsiveness in human health risk assessment. In: *Human Variability in Response to Chemical Exposure*, edited by D. A. Neumann and C. A. Kimmel. ILSI Press, Washington, D.C., p. 1.
170. Bromberg, P. A. (1998): Risk assessment of the effects of ozone exposure on respiratory health: dealing with variability in human responsiveness to controlled exposures. In: *Human Variability in Response to Chemical Exposure*, edited by D. A. Neumann and C. A. Kimmel. ILSI Press, Washington, D.C., p. 139.
171. Frame, L. T. et al. (1998): Host–environment interactions that affect variability in human cancer susceptibility. In: *Human Variability in Response to Chemical Exposure*, edited by D. A. Neumann and C. A. Kimmel. ILSI Press, Washington, D.C., p. 165.
172. Colditz, G. A., Stampfer, M. J., and Green, L. C. (1988): Diet. In: *Variations in Susceptibility to Inhaled Pollutants*, edited by B. D. Brain, A. J. Waven, and R. A. Shaiker. The Johns Hopkins University Press, Baltimore, MD, p. 314.
173. EPA (1996): *A National Agenda To Protect Children's Health from Environmental Threats*. U.S. Environmental Protection Agency, Washington, D.C. (http://occ-env-med.mc.duke.edu/oem/content/epa.htm).
174. U.S. Environmental Protection Agency (1998): Framework for addressing key scientific issues presented by the Food Quality Protection Act (FQPA) as developed by the Tolerance Reassessment Advisory Committee (TRAC). *Fed. Reg.*, 63:58038.
175. Roberts, R. J. (1992): Overview of similarities and differences between children and adults: implications for risk assessment. In: *Similarities and Differences Between Children and Adults: Implications for Risk Assessment*, edited by P. S. Guzelian, C. J. Henry, and S. S. Olin. ILSI Press, Washington, D.C., p. 11.
176. Plunkett, L. M., Turnbull, D., and Rodricks, J. V. (1992): Differences between adults and children affecting exposure assessment. In: *Similarities and Differences Between Children and Adults: Implications for Risk Assessment*, edited by P. S. Guzelian, C. J. Henry, and S. S. Olin. ILSI Press, Washington, D.C., p. 79.
177. Kauffman, R. E. (1992): Acute acetaminophen overdose: an example of reduced toxicity related to developmental differences in drug metabolism. In: *Similarities and Differences Between Children and Adults: Implications for Risk Assessment*, edited by P. S. Guzelian, C. J. Henry, and S. S. Olin. ILSI Press, Washington, D.C., p. 97.
178. Guzelian, P. S. and Henry, C. J. (1992): Conference summary; similarities and differences between children and adults: implications for risk assessment (November 5–7, 1990, Hunt Valley, MD). In: *Similarities and Differences Between Children and Adults: Implications for Risk Assessment*, edited by P. S. Guzelian, C. J. Henry, and S. S. Olin. ILSI Press, Washington, D.C., p. 1.

179. Implementation Working Group (1998): *A Science-Based, Workable Framework for Implementing the Food Quality Protection Act*, Implementation Working Group's "Road Map" Report, prepared by Jellinek, Schwartz & Connolly, Inc.; McDermott, Will & Emery; and Morgan, Lewis & Bockius.
180. EPA (1998): *Health Effects Division on FQPA Safety Factor for Infants and Children*, presentation for FIFRE Scientific Advisory Panel by Office of Pesticide Programs, U.S. Environmental Protection Agency, Washington, D.C. (http://www.epa.gov/pesticides/SAP/march/10x.htm).
181. Tamplin, A. R. and Gofman, J. W. (1970: *Population Control through Nuclear Pollution*. Nelson-Hill, Chicago, IL.
182. Riddiough, C. R., Musselmann, R., and Calabrese, E. J. (1977): Is EPA's radium-226 drinking water standard justified? *Med. Hypotheses*, 3:171.
183. EPA (1986): *Air Quality Criteria for Lead*, Vols. I–IV, EPA-600/8-83/028adf. U.S. Environmental Protection Agency, Washington, D.C.
184. U.S. HEW (1962): *Public Health Drinking Water Standards*. Public Health Service, U.S. Department of Health, Education, and Welfare, Rockville, MD.
185. Betke, J., Kleihaver, E., and Lipps, M. (1956): Vergleichende Untersucheg uber Sportanozydation von Nabelschnur and Erwachsenenhamoglobin. *Ztschr. Kinderh.*, 77:549.
186. Ross, J. D. and Des Forges, J. F. (1959): Reduction of methemoglobin by erythrocytes from cord blood: further evidence of deficient enzyme activity in newborn period. *Pediatrics*, 23:218.
187. U.S. Environmental Protection Agency (1975): Interim primary drinking water standards. *Fed. Reg.*, 40:11990.
188. Ginsberg, G. L. (2003): Assessing cancer risks from short-term exposures in children. *Risk Anal.*, 23:19.
189. Preston, R. J. (2004): Children as a sensitive subpopulation for the risk assessment process. *Toxicol. Appl. Pharmacol.*, 199:132.
190. EPA (2005): *Supplemental Guidance for Assessing Susceptibility from Early Life Stage Exposure to Carcinogens*, EPA/630/R-03/003F. Risk Assessment Forum, U.S. Environmental Protection Agency, Washington, D.C.
191. Eaton, D. L. and Klaassen, C. D. (2003): Principles of toxicology. In: *Casarett & Doull's Essentials of Toxicology*, edited by C. D. Klaassen and J. B. Watkins. McGraw-Hill, New York, chap. 2.
192. Calabrese, E. J. (2005): Hormesis: implications for risk assessment. In: *Inhalation Toxicology*, 2nd ed., edited by H. Salem and S. Katz. Marcel Dekker, New York.
193. Calabrese, E. J., Hormesis: a key concept in toxicology. In: *Biological Concepts and Techniques in Toxicology: An Integrated Approach*, edited by J. E. Riviere. Marcel Dekker, New York.
194. Rand, G., Ed. (2006): *Fundamentals of Aquatic Toxicology*,. Taylor & Francis, Boca Raton, FL, 2006.
195. Calabrese, E. J. (2003): Hormesis and Its Role in Toxicology, paper presented at the Society of Toxicology of Canada, Plenary Session, Montreal, Canada, December 8.
196. Hively, W. (2002): Is radiation good for you? *Discover*, 23(12):74.
197. Bethell, T. (2002): Underdosed: could toxins and radiation be good for you? *Am. Spect.*, July/Aug.:54.
198. Lambert, E. (2003): A pinch of poison. *Forbes Life Online* (www.forbes.com).
199. Stipp, D. (2003): A little poison can be good for you. *Fortune*, 147:54.
200. Begley, S. (2003): Scientists revisit idea that a little poison could be beneficial. *The Wall Street J.*, p. CCXLII, Dec. 19.
201. Cook, G. (2003): A scientist finds benefit in small doses of toxins. *The Boston Globe*, p. A16, Dec. 12.
202. Bell, J. (2004): Can a low dose of poison be a good thing? *Baltimore Sun*, March 15.
203. Boyce, N. (2004): Is there a tonic in the toxin? *U.S. News World Rep.*, Oct. 18:74.
204. Ahuja, A. (2003): Science: some toxins may be good for us. *Times Online (Lond.)*, Oct. 1.
205. Renner, R. (2003): Hormesis: Nietzsche's toxicology. *Sci. Am.*, 289(3):28.
206. Renner, R. (2004): Redrawing the dose–response curve. *Env. Sci. Technol.*, 38(5):90A.
207. Butler, R. (2004): When toxic turns to treatment. *Chem. Indust.*, 3:10.
208. Hadley, C. (2003): What doesn't kill you makes you stronger. *EMBO J.*, 4:924.
209. Calabrese, E. J. (2004): Hormesis: a revolution in toxicology, risk assessment and medicine. *EMBO J.*, 5:S37.
210. Calabrese, E. J. and Baldwin, L. A. (2003): Toxicology rethinks its central belief: hormesis demands a reappraisal of the way risks are assessed. *Nature*, 421:691.
211. Kaiser, J. (2003): Sipping from a poisoned chalice. *Science*, 302:376.
212. Calabrese, E. J. and Baldwin, L. A. (2003): The hormetic dose–response model is more common than the threshold model in toxicology. *Toxicol. Sci.*, 71:246.
213. Rodricks, J. V. (2003): Hormesis and toxicological risk assessment. *Toxicol. Sci.*, 71:134.
214. Calabrese, E. J. and Blain, R. (2005): The occurrence of hormetic dose responses in the toxicological literature: the hormesis database: an overview. *Toxicol. Appl. Pharmacol.*, 202:289.
215. Calabrese, E. J. (2005): Paradigm lost, paradigm found: the re-emergence of hormesis as a fundamental dose–response model in the toxicological sciences. *Environ. Pollut.*, 138:379.
216. Calabrese, E. J. and Baldwin, L. A. (2001): Special issue: scientific foundations of hormesis. *Crit. Rev. Toxicol.*, 31:351.
217. Calabrese, E. J. and Baldwin, L. A. (2003): Special issue: hormesis: environmental and biomedical perspectives. *Crit. Rev. Toxicol.*, 33:213.
218. Calabrese, E. J. (2005): Hormetic dose–response relationships in immunology: occurrence, quantitative features of the dose response, mechanistic foundations, and clinical implications. *Crit. Rev. Toxicol.*, 35:89.
219. Luckey, T. D. (1980): *Ionizing Radiation and Hormesis*. CRC Press, Boca Raton, FL.
220. Luckey, T. D. (1991): *Radiation Hormesis*. CRC Press, Boca Raton, FL.

221. Southam, C. M. and Ehrlich, J. (1943): Effects of extracts of western red-cedar heartwood on certain wood-decaying fungi in culture. *Phytopathology*, 33:517.
222. Schulz, H. (1887): Zur Lehre von der Arzneiwirdung. *Virchows Archiv. fur Pathol. Anatomie und Physiol. Fur Klin. Med.*, 108:423.
223. Schulz, H. (1923/2003): *Contemporary Medicine as Presented by Its Practitioners Themselves*. Leipzig, pp. 217–250; reprinted in *Nonlinear Biol. Toxicol. Med.*, 1(3):295 (NIH Library Translation NIH-98-134, trans. by T. Crump).
224. Calabrese, E. J. and Baldwin, L. A., Chemical hormesis: its historical foundations as a biological hypothesis, *Hum. Exper. Toxicol.*, 19, 2, 2000.
225. Calabrese, E. J. and Baldwin, L. A. (2000): The marginalization of hormesis. *Hum. Exp. Toxicol.*, 19:32.
226. Calabrese, E. J. and Baldwin, L. A. (2000): Radiation hormesis: its historical foundations as a biological hypothesis. *Hum. Exp. Toxicol.*, 19:41.
227. Calabrese, E. J. and Baldwin, L. A. (2000): Radiation hormesis: the demise of a legitimate hypothesis. *Hum. Exp. Toxicol.*, 19:76.
228. Calabrese, E. J. and Baldwin, L. A. (2000): Tales of two similar hypothesis: the rise and fall of chemical and radiation hormesis. *Hum. Exp. Toxicol.*, 19:85.
229. Calabrese, E. J. and Baldwin, L. (1997): The dose determines the stimulation (and poison): development of a chemical hormesis database. *Int. J. Toxicol.*, 16:545.
230. Paoletti, P. et al. (1990): Characteristics and biological role of steroid hormone receptors in neuroepithelial tumors. *J. Neurosurg.*, 73:736.
231. Calabrese, E. J., Baldwin, L. A., and Holland, C. D. (1999): Hormesis: a highly generalizable and reproducible phenomenon with important implications for risk assessment. *Risk Anal.*, 19:261.
232. Calabrese, E. J. and Baldwin, L. A. (2001): The frequency of U-shaped dose–responses in the toxicological literature. *Tox. Sci.*, 62:330.
233. Calabrese, E. J. (1996): Expanding the RfD concept to incorporate and optimize beneficial effects while preventing toxic responses from non-essential toxicants. *Ecotox. Environ. Safety*, 34:94.
234. Calabrese, E. J. and Baldwin, L. (1998): Hormesis as a default parameter in RfD derivation. *BELLE Newslett.*, 7:1.
235. Gaylor, D. (1998): Safety assessment with hormetic effects. *Hum. Exper. Toxicol.*, 17:251.
236. Sielken, R. L., Jr., and Stevenson, D. E. (1998): Some implications for quantitative risk assessment if hormesis exists. *Hum. Exp. Toxicol.*, 17:259.
237. Calabrese, E. J. and Cook, R. (2005): Hormesis: how it could affect the risk assessment process. *BELLE Newslett.*, 12:22.
238. Sukata, T. et al. (2002): Detailed low-dose study of 1,1-bis(*p*-chlorophenyl)-2,2,2-trichloroethane carcinogenesis suggests the possibility of a hormetic effect. *Int. J. Cancer*, 99:112.
239. Society of Toxicology Panel (1981): Re-examination of the ED01 study: adjusting for time on study. *Fund. Appl. Toxicol.*, 1:67.
240. Calabrese, E. J. and Baldwin, L. A. (2002): Hormesis and high risk groups. *Reg. Tox. Pharmacol.*, 35:414.
241. National Research Council (1980): *Principles of Toxicological Interactions Associated with Multiple Chemical Exposures*. National Academy Press, Washington, D.C.
242. Monosson, E. (2005): Chemical mixtures: considering the evolution of toxicology and chemical assessment. *Environ. Health Perspect.*, 113:383.
243. Wilkinson, C. F. (1971): Effects of synergists on the metabolism and toxicity of anticholinesterase. *Bull. WHO*, 40:171.
244. Frawley, J. P. et al. (1957): Marked potentiation in mammalian toxicity from simultaneous administration of two anti-cholinesterase compounds. *J. Pharmacol. Exp. Ther.*, 121:96.
245. Finney, D. J. (1952): *Probit Analysis*. Cambridge University Press, London.
246. Pozzani, U. S., Weil, C. S., and Carpenter, C. P. (1959): The toxicological basis of TLVs. 5. The experimental inhalation of vapor mixtures by rats, with notes upon the relationship between single dose inhalation and single dose oral data, *Am. Ind. Hyg. Assoc. J.*, 20:364.
247. Smyth, H. F., Jr. et al. (1969): An exploration of joint toxic action: 27 industrial chemicals in rats in all possible pairs. *Toxicol. Appl. Pharmacol.*, 14:340.
248. Feron, V. J. et al. (1995): Toxicology of chemical mixtures: challenges for today and the future. *Toxicology*, 105:415.
249. EPA (1988): *Technical Support Document on Risk Assessment of Chemical Mixtures*, EPA/600/8-90/064. National Technical Information Service, U.S. Environmental Protection Agency, Washington, D.C.
250. Mileson, B. E. et al. (1998): Common mechanism of toxicity: a case study of organophosphorus pesticides. *Toxicol. Sci.*, 41:8.
251. Bellin, J. S. et al. (1989): *Interim Procedures for Estimating Risks Associated with Exposures to Mixtures of Chlorinated Dibenzo-p-dioxins and Dibenzofurans (CDDs and CDFs): A 1989 Update*, EPA-625/3-89-016. Risk Assessment Forum, U.S. Environmental Protection Agency, Washington, D.C.
252. Pohl, H. R., Hansen, H., and Chou, C. H. (1997): Public health guidance values for chemical mixtures: current practice and future directions. *Reg. Toxicol. Pharmacol.*, 26:322.
253. Research Triangle Institute (1998): *Toxicological Profile for Polychlorinated Biphenyls (PCBs)*, Updated Draft for Public Comment. National Technical Information Service, Agency for Toxic Substances and Disease Registry (ATSDR), U.S. Public Health Service, Washington, D.C.
254. Cogliano, V. J. (1996): *PCBs: Cancer Dose–Response Assessment and Application to Environmental Mixtures*, NTIS PB96-140603, EPA/600/P-96/001A. National Center for Environmental Assessment, U.S. Environmental Protection Agency, Washington, D.C.
255. IARC (2002): *IARC Monograph Summary for Some Traditional Herbal Medicines, Some Mycotoxins, Naphthalene, and Styrene*, International Agency for Research on Cancer, World Health Organization, Geneva, Switzerland (http://www-cie.iarc.fr/monoeval/allmonos.html).

256. IARC (2003): *Cobalt in Hard Metals and Cobalt Sulfate, Gallium Arsenide, Indium Phosphide and Vanadium Pentoxide*, Vol. 86. International Agency for Research on Cancer, World Health Organization, Geneva, Switzerland (http://www-cie.iarc.fr/htdocs/announcements/vol86.htm).
257. IARC (2004): *IARC Monograph Summary for Some Drinking-Water Disinfectants and Contaminants, Including Arsenic*. International Agency for Research on Cancer, World Health Organization, Geneva, Switzerland (http://www-cie.iarc. fr/monoeval/allmonos.html).
258. IARC (1999): *IARC Monograph Summary for Beryllium and Beryllium Compounds (Group 1)*. International Agency for Research on Cancer, World Health Organization, Geneva, Switzerland (http://193.51.164.11/cgi/iltound/chem/ilt_chem_frames.html).
259. IARC (2004): *IARC Monograph Summary for Betel-Quid and Areca-Nut Chewing and Some Areca-Nut-Derived Nitrosamines*. International Agency for Research on Cancer, World Health Organization, Geneva, Switzerland (http://www-cie.iarc.fr/monoeval/allmonos.html).
260. IARC (1999): *IARC Monograph Summary for Cadmium and Cadmium Compounds (Group 1)*. International Agency for Research on Cancer, World Health Organization, Geneva, Switzerland (http://193.51.164.11/cgi/iltound/chem/ilt_chem_frames.html).
261. IARC (1999): *IARC Monograph Summary for Ciclosporin (Group 1)*. International Agency for Research on Cancer, World Health Organization, Geneva, Switzerland (http://193.51.164.11/cgi/iltound/chem/ilt_chem_frames.html).
262. IARC (1999): *IARC Monograph Summary for Epstein–Barr Virus (Group 1)*. International Agency for Research on Cancer, World Health Organization, Geneva, Switzerland (http://193.51.164.11/cgi/iltound/chem/ilt_chem_frames.html).
263. IARC (1999) *IARC Monograph Summary for Oestrogen Replacement Therapy (Group 1)*. International Agency for Research on Cancer, World Health Organization, Geneva, Switzerland (http://193.51.164.11/cgi/iltound/chem/ilt_chem_frames.html).
264. IARC (1999): *IARC Monograph Summary for Ethylene Oxide (Group 1)*. International Agency for Research on Cancer, World Health Organization, Geneva, Switzerland (http://193.51.164.11/cgi/iltound/chem/ilt_chem_frames.html).
265. IARC (2000): *IARC Monograph Summary for Some Antiviral and Antineoplastic Drugs, and Other Pharmaceutical Agents*. International Agency for Research on Cancer, World Health Organization, Geneva, Switzerland (http://www-cie.iarc.fr/monoeval/allmonos.html).
266. IARC (2004): *IARC Monograph Summary for Formaldehyde, 2-Butoxyethanol and 1-tert-Butoxy-2-propanol*. International Agency for Research on Cancer, World Health Organization, Geneva, Switzerland (http://www-cie.iarc.fr/monoeval/allmonos.html).
267. IARC (1999): *IARC Monograph Summary for Infection with Helicobacter pylori (Group 1)*. International Agency for Research on Cancer, World Health Organization, Geneva, Switzerland (http://193.51.164.11/cgi/iltound/chem/ilt_chem_frames.html).
268. IARC (1999): *IARC Monograph Summary for Hepatitis B Virus (Group 1)*. International Agency for Research on Cancer, World Health Organization, Geneva, Switzerland (http://193.51.164.11/cgi/iltound/chem/ilt_chem_frames.html).
269. IARC (1999): *IARC Monograph Summary for Hepatitis C Virus (Group 1)*. International Agency for Research on Cancer, World Health Organization, Geneva, Switzerland (http://193.51.164.11/cgi/iltound/chem/ilt_chem_frames.html).
270. IARC (2004): *IARC Monograph Summary for Some Traditional Herbal Medicines, Some Mycotoxins, Naphthalene, and Styrene*. International Agency for Research on Cancer, World Health Organization, Geneva, Switzerland (http://www-cie.iarc.fr/monoeval/allmonos.html).
271. IARC (1999): *IARC Monograph Summary for Human Immunodeficiency Viruses: HIV-1 (Group 1); HIV-2 (Group 2B)*. International Agency for Research on Cancer, World Health Organization, Geneva, Switzerland (http://193.51.164.11/cgi/iltound/chem/ilt_chem_frames.html).
272. IARC (1999): *IARC Monograph Summary for Human Papilloma Viruses (HPV): HPV Types 16 and 18 (Group 1); HPV Types 31 and 33 (Group 2A); Some HPV Types Other Than 16, 18, 31 and 33 (Group 2B)*. International Agency for Research on Cancer, World Health Organization, Geneva, Switzerland (http://193.51.164.11/cgi/iltound/chem/ilt_chem_frames.html).
273. IARC (1999): *IARC Monograph Summary for Human T-Cell Lymphotropic Viruses: HTLV-I (Group 1); HTLV-II (Group 3)*. International Agency for Research on Cancer, World Health Organization, Geneva, Switzerland (http://193.51.164.11/cgi/iltound/chem/ilt_chem_frames.html).
274. IARC (2001): *IARC Monograph Summary for Some Internally Deposited Radionuclides*. International Agency for Research on Cancer, World Health Organization, Geneva, Switzerland (http://www-cie.iarc.fr/monoeval/allmonos.html).
275. IARC (2002): *IARC Monograph Summary for Tobacco Smoke and Involuntary Smoking*. International Agency for Research on Cancer, World Health Organization, Geneva, Switzerland (http://www-cie.iarc.fr/monoeval/allmonos.html).
276. IARC (2000): *IARC Monograph Summary for Ionizing Radiation*. Part 1. *X- and gamma (γ)-Radiation, and Neutrons*. International Agency for Research on Cancer, World Health Organization, Geneva, Switzerland (http://www-cie.iarc.fr/monoeval/allmonos.html).
277. IARC (1999): *IARC Monograph Summary for Infection with Liver Flukes (Opisthorchis viverrini, Opisthorchis felineus, and Clonorchis sinensis): Opisthorchis viverrini (Group 1); Opisthorchis felineus (Group 3); Clonorchis sinensis (Group 2A)*. International Agency for Research on Cancer, World Health Organization, Geneva, Switzerland (http://193.51.164.11/cgi/iltound/chem/ilt_chem_frames.h tml).
278. IARC (1998): *IARC Monograph Summary for Hormonal Contraception and Postmenopausal Hormonal Therapy*. International Agency for Research on Cancer, World Health Organization, Geneva, Switzerland (http://193.51.164.11/cgi/iltound/chem/ilt_chem_frames.html).

279. IARC (1999): *IARC Monograph Summary for Oral Contraceptives, Combined (Group 1)*. International Agency for Research on Cancer, World Health Organization, Geneva, Switzerland (http://193.51.164.11/cgi/iltound/chem/ilt_chem_frames.html).
280. IARC (1999): *IARC Monograph Summary for Salted Fish: Chinese-Style Salted Fish (Group 1); Other Salted Fish (Group 3)*. International Agency for Research on Cancer, World Health Organization, Geneva, Switzerland (http://193.51.164.11/cgi/iltound/chem/ilt_chem_frames.html).
281. IARC (1999): *IARC Monograph Summary for Infection with Schistomsomes (Schistosoma haematobium, Schistosoma mansoni, and Schistosoma japonicum): Schistosoma haematobium (Group 1); Schistosoma mansoni (Group 3); Schistosoma japonicum (Group 2B)*. International Agency for Research on Cancer, World Health Organization, Geneva, Switzerland (http://193.51.164.11/cgi/iltound/chem/ilt_chem_frames.html).
282. IARC (1999): *IARC Monograph Summary for Silica— rystalline Silica: Inhaled in the Form of Quartz or Cristobalite from Occupational Sources (Group 1); Amorphous Silica (Group 3)*. International Agency for Research on Cancer, World Health Organization, Geneva, Switzerland (http://193.51.164.11/cgi/iltound/chem/ilt_chem_frames.html).
283. IARC (1999): *IARC Monograph Summary for Solar and Ultraviolet Radiation: Solar Radiation (Group 1), Ultraviolet A Radiation (Group 2A), Ultraviolet B Radiation (Group 2A), Ultraviolet C Radiation (Group 2A), Use of Sunlamps and Sunbeds (Group 2A), Exposure to Fluorescent Lighting (Group 3)*. International Agency for Research on Cancer, World Health Organization, Geneva, Switzerland (http://193.51.164.11/cgi/iltound/chem/ilt_chem_frames.html).
284. IARC (1997): *IARC Monograph Summary for Occupational Exposures to Mists and Vapours from Strong Inorganic Acids and Other Industrial Chemicals*. International Agency for Research on Cancer, World Health Organization, Geneva, Switzerland (http://www-cie.iarc.fr/monoeval/allmonos.html).
285. IARC (1999): *IARC Monograph Summary for Tamoxifen (Group 1)*. International Agency for Research on Cancer, World Health Organization, Geneva, Switzerland (http://193.51.164.11/cgi/iltound/chem/ilt_chem_frames.html).
286. IARC (1999): *IARC Monograph Summary for Polychlorinated Dibenzo-para-dioxins: 2,3,7,8-Tetrachlorodibenzo-para-dioxin (Group 1); Polychlorinated Dibenzo-para-dioxins (Other than 2,3,7,8-Tetrachlorodibenzo-para-dioxin): 2,7-DCDD, 1,2,3,6,7,8-/1,2,3,7,8,9-HxCDD, 1,2,3,4,6,7,8-HpCDD (Group 3), Dibenzo-para-dioxin (Group 3)*. International Agency for Research on Cancer, World Health Organization, Geneva, Switzerland (http://193.51.164.11/cgi/iltound/chem/ilt_chem_frames.html).
287. IARC (1999): *IARC Monograph Summary for Thiotepa (Group 1)*. International Agency for Research on Cancer, World Health Organization, Geneva, Switzerland (http://193.51.164.11/cgi/iltound/chem/ilt_chem_frames.html).
288. IARC (1999): *IARC Monograph Summary for Wood Dust (Group 1)*. International Agency for Research on Cancer, World Health Organization, Geneva, Switzerland (http://193.51.164.11/cgi/iltound/chem/ilt_chem_frames.html).
289. IARC (2004): *Overall Evaluations of Carcinogenicity to Humans*. International Agency for Research on Cancer, World Health Organization, Geneva, Switzerland (http://www-cie.iarc.fr/monoeval/crthall.html).
290. European Union (1983), as cited in Neumann, H.-G. et al. (1997): Proposed changes in the classification of carcinogenic chemicals in the work area. *Reg. Toxicol. Pharmacol.*, 26:288.
291. IARC (1992): Meeting report: working group on mechanisms of carcinogenesis and the evaluation of carcinogenic risks. *Cancer Res.*, 52:2357.
292. IARC (1999): *Overall Evaluations of Carcinogenicity to Humans*. International Agency for Research on Cancer, World Health Organization, Geneva, Switzerland (http://193.51.164.11/monoeval/crthal.html).
293. IARC (2004): Preamble. In: *IARC Monographs on the Evaluation of Carcinogenic Risks to Humans*. International Agency for Research on Cancer, World Health Organization, Geneva, Switzerland (http://www-cie.iarc.fr/monoeval/preamble.html).
294. ACGIH (1998): *Guide to Occupational Exposure Values: 1998*, American Conference of Governmental Industrial Hygienists Cincinnati, OH.
295. Health Canada (1994), as cited in TERA (1999): *Cancer Risk Assessment Methods*. Toxicology Excellence for Risk Assessment, Cincinnati, OH (http://www.tera.org/iter/methods/cancer.htm).
296. National Toxicology Program (1998), as cited in American Cancer Society (1998): *Cancer Facts & Figures*, Publ. No. 98-300M-No.5008.98. American Cancer Society, Atlanta, GA.
297. Bickis, M. and Krewski, D. (1985): Statistical design and analysis of the long-term carcinogenicity bioassay. In: *Toxicological Risk Assessment*, Vol. 1, edited by D. B. Clayson, D. Krewski, and I. Munro. CRC Press, Boca Raton, FL, p. 125.
298. Borgert, C. J. et al. (2004): Can mode of action predict mixture toxicity for risk assessment? *Toxicol. Appl. Pharmacol.*, 201:85.

# 3 Metabolism: A Determinant of Toxicity

*Raymond A. Kemper, Johnnie R. Hayes, and Matthew S. Bogdanffy*

## CONTENTS

## INTRODUCTION

Understanding the metabolism, or biotransformation, of xenobiotics has come to be regarded as fundamental to appreciating the toxic mechanisms of chemicals, be they drugs, industrial chemicals, pesticides, or other molecule foreign to the body. Although it is not clear whether xenobiotic metabolism is better regarded as the introduction of foreign chemicals into a metabolic machinery that was intended by nature for natural substrates or whether the vast array of xenobiotic metabolizing enzymes and their wide substrate specificity is a product of evolutionary adaptation to exposure chemicals in their environment, what is clear is that the complicated interaction of oxidation, reduction, hydrolysis, and conjugation reactions can have the net result of activation of chemicals to toxic species or their detoxification and enhanced elimination from the body.

Xenobiotic metabolism became recognized as an important aspect of pharmacology and toxicology in the mid-twentieth century when substrates such as amphetamine and steroids were found to bind to and undergo oxidative metabolism by a hemoprotein found in the particulate or microsomal fraction of liver and adrenal gland [72]. This protein was later isolated, characterized, and given the descriptive name *cytochrome P450*. In the late 1940s, the research team of Elizabeth and James Miller demonstrated that the aminoazo dye commonly known as butter yellow (*N,N*-dimethyl-4-aminoazobenzene) is metabolized to a reactive product that covalently binds to protein and that its metabolic activation is an important step in the carcinogenic activity of the substance. The work of the Millers and many other scientists of the time was the beginning of an awareness that xenobiotic metabolism could be a fundamental contributor to toxic mechanisms. In the half century that has since passed, the role of xenobiotic biotransformation as a determinant of toxicity has only become reinforced. With ever more eloquent research into the molecular genetics of these enzymes and the factors controlling their expression, this element of toxicology has evolved into one of greater understanding of the differences between species and individuals in biotransformation enzyme phenotypes that render each more or less susceptible to these toxic mechanisms. Fundamental knowledge of the chemistry of metabolic activation and detoxification pathways and the association of chemical structure with predisposition for metabolism via these pathways are converging to yield computer-based models aimed at simulating and predicting the major metabolic transformations.

This chapter discusses the major pathways of xenobiotic metabolism with an emphasis on the role that biotransformation plays as a determinant of toxicity. The material is organized in the traditional view of the functionalization of xenobiotics, that being biological oxidations, reductions, and hydrolytic conversions, frequently referred to as *phase I reactions*, and biochemical conjugation reactions, known as *phase II reactions*.

## GENERAL FEATURES AND BASIC CONCEPTS OF XENOBIOTIC METABOLISM

The majority of organisms studied have biotransformation enzymes, although there is diversity in the occurrence, function, and rates of specific enzymes. Certain bacteria contain more primitive or less highly developed systems and may lack certain pathways altogether. Even mammals demonstrate diversity in the activity or rates of specific systems, and, as would be expected of genetically controlled functions, there are species and individual differences. This diversity extends to the organ level in multicellular organisms. Specific organs show different levels of activity, and specific cell types within organs demonstrate variation in biotransformation capability. There is even subcellular diversity, in that certain of these enzymes are compartmentalized whereas others are free in the cytoplasm.

The variety of chemicals to which organisms may be exposed requires that the biotransformation enzymes have broad substrate specificity. This characteristic is not shared by the majority of enzymes involved in anabolic and catabolic metabolism. In addition, the types of reactions catalyzed are diverse, as shown in Table 3.1, including oxidation, reduction, epoxidation, deamination, hydroxylation, sulfoxidation, dehalogenation, and conjugation with endogenous compounds, to name a few. Although it is logical to initially focus on xenobiotic metabolizing systems one at a time, in many cases xenobiotic metabolism involves more than a single metabolic route. In addition, the eventual toxicity of a xenobiotic may be modified by a number of factors, including age, gender, physiological status, nutrition, diet, and the presence or absence of disease, among others.

**TABLE 3.1**
**Reaction Types and Enzymes That Participate in Xenobiotic Metabolism**

| Phase I Reactions | |
|---|---|
| *Oxidation* | *Ester Hydrolysis* |
| Cytochromes P450 | Carboxylesterases |
| Flavin-containing monooxygenases | Amidases |
| Xanthine oxidase | *Dehydrogenases* |
| Amine oxidase | Alcohol dehydrogenases |
| Monoamine oxidase | Aldehyde dehydrogenases |
| Semicarbazide-sensitive oxidases | *Hydration* |
| *Reduction* | Epoxide hydrolase |
| Cytochromes P450 | *Miscellaneous* |
| NADP-quinone oxidoreductase | Cysteine conjugate β-lyase |
| Carbonyl reductases | Superoxide dismutase |
| Glutathione peroxidases | Catalase |
| **Phase II Reactions** | |
| UDP-Glucuronosyltransferases | *Methylation* |
| Sulfotransferases | *O*-Methyltransferase |
| Glutathione *S*-transferases | *N*-Methyltransferase |
| Glucosyltransferase | *S*-Methyltransferase |
| Thiotransferase | *Acetylation* |
| Transacylases | *N*-acetyltransferases |
| | Acyltransferases |
| | *Miscellaneous* |
| | Rhodanase |

Exposure of an animal to certain xenobiotics can result in the induction of specific enzymes associated with xenobiotic metabolism. When induced, their activity can dramatically increase, compared to their basal level. Induction is sometimes coordinated with more than one enzyme induced. Induction results in an increase in the ability of animals to metabolize a xenobiotic, and in most cases this reduces their susceptibility to its toxicity. Induction generally lasts only a few days. When exposure ceases, the enzymes return to their basal levels.

Because xenobiotic metabolism does not always represent detoxification, the term *biotransformation* has come into general use to denote the actions of xenobiotic metabolizing enzymes, although it is still not semantically specific for xenobiotic metabolism. Biotransformation is divided into two distinct phases. Phase I reactions result in *functionalization*, the addition or the uncovering of specific functional groups that are required for subsequent metabolism by phase II enzymes. Phase II reactions are biosynthetic. These phase I and II reactions are often coordinated, with the product of phase I reactions becoming the substrate for phase II enzymes. A commonality of biotransformation reactions is the conversion of hydrophobic xenobiotics into more polar, more easily excreted compounds. Because the composition of cells is more lipophilic than their environment, nonpolar compounds tend to accumulate. This could lead to bioconcentration of chemicals within the cell to levels higher than that of the environment and increase the likelihood of a cytotoxic event; however, conversion of nonpolar chemicals to more polar metabolites allows them to be more easily excreted by the cell. Conjugation of a xenobiotic with an endogenous compound, a phase II reaction, increases water solubility, and, in some cases, the added chemical group is recognized by specific carrier proteins or proteins involved in facilitated diffusion or active transport. This increases the ability of the cell to remove the xenobiotic.

Many diverse examples exist of xenobiotics whose toxicity is directly dependent on the activity of the biotransformation enzymes. For most chemicals, increases in the activity of these enzymes result in decreases in toxicity, whereas decreases in activity result in increased toxicity; however, in some instances the product of xenobiotic metabolism is more toxic than the parent compound. Conversion of a foreign compound to a more toxic metabolite is termed *metabolic activation*; for example, the majority of genotoxic and carcinogenic chemicals require metabolic activation to highly reactive species capable of interacting with DNA. The enzymes that protect the animal from the toxicity of certain compounds may be responsible for the toxicity of others. The susceptibility of an organism to the toxicity of a particular chemical is dependent, in many cases, on the delicate balance between detoxification and metabolic activation that exists during exposure to the xenobiotic. Due to the sensitivity of the enzymes of xenobiotic biotransformation to both endogenous and exogenous factors, this balance may differ among individuals and at different points in time.

# BIOLOGICAL OXIDATION

## Cytochrome P450-Dependent Monooxygenase System

The P450-dependent monooxygenase system is central to the metabolism of most xenobiotics. Not only is it the primary enzymatic system for metabolism of many xenobiotics, but it is also involved as the initial functionalization step in the further metabolism of many others. Consequently, P450 plays essential roles in several areas of research, including biochemistry, pharmacology, toxicology, physiology, and medicine. Several names for the P450 system exist in the literature. The names most commonly encountered include:

- Mixed function oxidase
- P450 system
- P450-dependent monooxygenase system

Generally these names are related either to a specific function or are descriptive of a biochemical mechanism. Currently, it generally is referred to in terms of a monooxygenase system to denote its ability to incorporate one atom of molecule oxygen into its substrates.

## Components of the Cytochromes P450 System

The history of the discovery of P450 and elucidation of its functions and mechanisms of action is intriguing and has been reviewed by Estabrook [1]. P450 was first described independently in 1958 in microsomes isolated from rat and pig liver homogenates. The name P450 derived from the occurrence of a pigment that, when reduced and treated with carbon monoxide, yielded a spectrophotometric Soret band at 450 nm. The laboratory of Britton Chance at the Johnson Foundation was the first to observe the pigment. Six years later, Omura and Sato [195,196] published their pivotal papers describing P450 as a b-type hemocytochrome and demonstrated that the cytochrome was located in hepatic microsomes, which form from the endoplasmic reticulum upon cellular disruption. Upon isolation from the membrane, through the use of proteases, P450 is converted to an inactive form whose reduced carbon monoxide complex produces a spectrophotometric peak at 420 nm.

Before the discovery of P450, Julius Axelrod and his colleagues [12], in the laboratory of Chemical Pharmacology at the National Heart Institute, were involved with studies on the metabolic disposition of drugs [12]. They found that the oxidative metabolism of amphetamine required the cofactor NADPH (nicotinamide adenine dinucleotide phosphate, reduced form) and the presence of oxygen. Estabrook et al. [73] established that P450 was the terminal oxidase involved with the C-21 hydroxylation of steroids in adrenal cortical microsomes, giving P450 a role in endobiotic metabolism. Many individuals and laboratories have since played major roles in the development of the current knowledge concerning P450.

It soon became obvious that, although P450 played a major role in the activity of the monooxygenase, it did not act alone. In 1950, Horecher [106] isolated a flavoprotein from the liver, but no function was identified. This flavoprotein used reducing equivalents from NADPH and was termed *NADPH–cytochrome c reductase*. In 1955, La Du et al. [138] showed that cytochrome *c* could inhibit dealkylation of aminopyrine. This was followed by the studies of Gillette et al. [87] in 1957, which presented additional evidence that cytochrome *c* reductase was involved in xenobiotic metabolism. In the 1960s, it was reported that NADPH–cytochrome *c* reductase occurred in the endoplasmic reticulum of liver cells, that antibodies to the reductase inhibited xenobiotic metabolism, and that the reductase is required for monooxygenase activity when reconstituted from isolated components [137,194, 203,292].

Although the major components of the P450-dependent monooxygenase system appear to be P450 and P450 reductase, other components may also be involved with the metabolism of specific xenobiotics. Cytochrome $b_5$ reductase has been proposed to participate in monooxygenase activity through electron transport to cytochrome $b_5$ and, subsequently, to P450; however, several systems of electron transport in the endoplasmic reticulum and isolated microsomes, as well as other activities such as peroxidation, have greatly complicated the elucidation of the role of cytochrome $b_5$. In many cases, cytochrome $b_5$ has been found to enhance NADPH-dependent substrate oxidation by P450, although the effect of cytochrome $b_5$ is dependent on both the specific P450 isoform and the substrate under investigation [301]. Cytochrome $b_5$ may enhance electron transfer to P450 by at least four different mechanisms. These include faster provision of the second electron to P450, the rate-limiting step in catalysis; enhanced coupling between NADPH consumption and substrate oxidation; formation of a complex with P450 that is capable of accepting two electrons from NADPH–P450 reductase; and allosteric activation of P450 [225]. An elucidation of the role of cytochrome $b_5$ must await further understanding of the complex electron transfer pathways that exist in the endoplasmic reticulum.

Although the catalytic activity of the monooxygenase system appears to require only two proteins, NADPH–P450 reductase and P450, it is capable of carrying out a variety of different reactions on a large number of substrates. This ability is based on the occurrence of a variety of P450 isozymes, but it also is based on the basic reaction

$$R\text{–}H + O_2 + NADPH + H^+ \xrightarrow{CYP} R\text{–}OH + H_2O + NAD^+$$

**FIGURE 3.1** Global reaction and stoichiometry of cytochrome P450.

mechanism of the cytochromes and their overlapping substrate specificity. The nonspecificity of the monooxygenase provides important flexibility to xenobiotic metabolism, but this flexibility comes with a price. Generally, the enzymatic reactions of anabolism and catabolism are both extremely specific in substrate specificity and catalytically efficient, resulting in high activity and high substrate turnover number. The turnover number and efficiency of P450 are considerably lower than most enzymes. This is probably related to the inefficient electron transfer due to the presence of a water molecule at the active site. Some substrates are more efficiently oxidized because they exclude water from the active site [206]. Inefficiency of metabolism is more than made up for by the ability to metabolize a variety of chemical structures and the ability to catalyze a variety of reactions. An additional factor that compensates for the relatively low substrate turnover number is the high concentration of the system in organs important in detoxification.

Before discussing the various reactions catalyzed by P450, a discussion of the catalytic cycle is appropriate. Knowledge of the catalytic cycle will assist in understanding the various reactions catalyzed by the system and in predicting metabolic pathways for specific xenobiotics.

### Catalytic Cycle of the P450-Dependent Monooxygenase System

The reaction catalyzed by the cytochrome P450-dependent monooxygenase system and its stoichiometry is illustrated in Figure 3.1. One molecule of substrate reacts with one molecule of molecular oxygen and NADPH to yield oxidized substrate containing one atom from molecular oxygen, water (containing the other oxygen atom), and $NADP^+$. The incorporation of one oxygen atom from molecular oxygen into the substrate is the source of the term *monooxygenase*. Oxidation of substrate and concomitant reduction of one atom of oxygen to water is the source of the name *mixed function oxidase*. Although the reaction stoichiometry appears simple, obtaining this stoichiometry in the laboratory is difficult [91]. The main difficulty is the number of oxidation–reduction reactions that occur simultaneously in the endoplasmic reticulum. These reactions use oxygen and NADPH and may yield water and $NADP^+$. When these diverse reactions have been accounted for, the predicted stoichiometry has been obtained.

It is recommended that the reader carefully follow the reaction sequence illustrated in Figure 3.2 during this discussion of the catalytic cycle. The initial step of the cycle

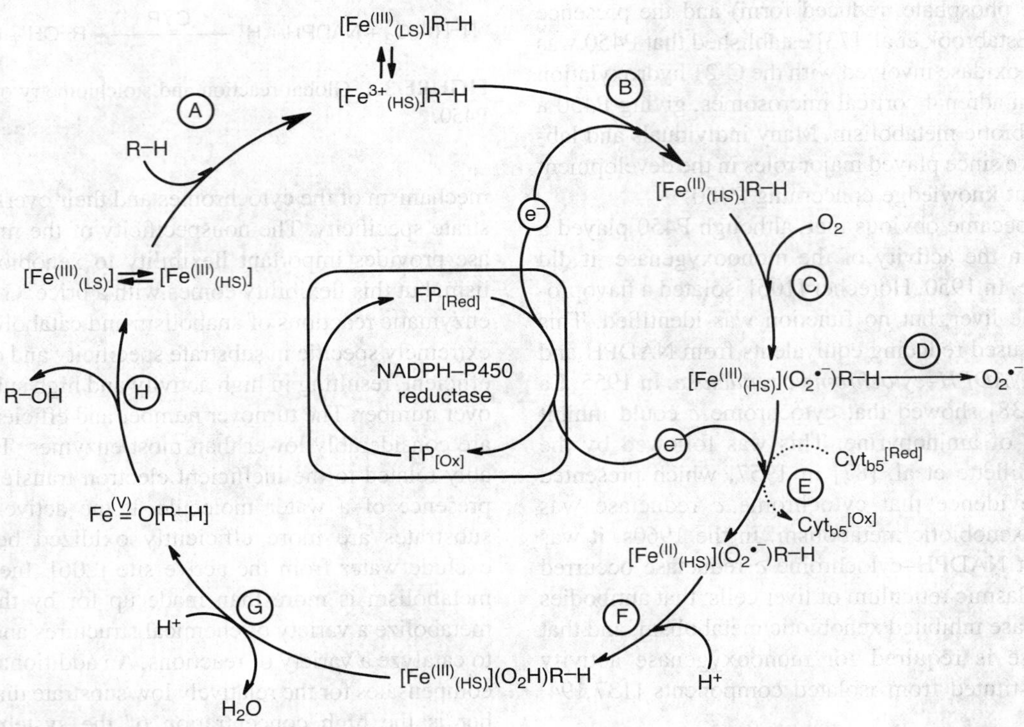


**FIGURE 3.2** Catalytic cycle of cytochrome P450-dependent substrate oxidation. HS, high spin; LS, low spin; Cytb5, cytochrome $b_5$. See text for details. (Adapted from Testa, B., *The Metabolism of Drugs and Other Xenobiotics: Biochemistry of Redox Reactions*, Academic Press, London, 1995.)

is binding of the substrate (R–H) to P450 (Figure 3.2A). As previously mentioned, P450 exists as a series of closely related isozymes, each of which demonstrates a degree of substrate specificity. This substrate specificity is not absolute, and overlapping is evident. At any one time, several isozymes of P450 exist in the endoplasmic reticulum. This is dependent on the specific genetic, environmental, and physiological conditions of the organism; therefore, binding of the substrate to the active site of P450 may represent binding to a single isozyme predominantly but not exclusively. The activity of the catalytic process (as well as the specific metabolites produced) is a function of the particular isozyme profile. Although our understanding of the structure of the active site of P450 is developing [148,149], more remains to be learned. From the nature of the hemoprotein and its substrates, the active site contains the heme and a hydrophobic region. The substrate must have a specific orientation within the active site.

As occurs with many other enzymes, the binding of the substrate to the hemoprotein appears to produce conformational alterations in the enzyme that assist its catalytic activity; for example, substrate binding facilitates the reduction of P450 by NADPH–P450 reductase, in part, by lowering its redox potential. Binding of the substrate to the active site changes the absorption spectrum of the cytochrome. Because the oxidized heme iron is paramagnetic, electron paramagnetic resonance (EPR) can be applied to probe the environment of the iron in the heme. These studies have revealed alterations in the EPR signal that correlate with the blue shift in the Soret band from about 419 nm to 390 nm, observed when substrates bind the cytochrome. EPR and visible spectra changes result from the substrate binding in close proximity to the heme iron with a concomitant displacement of a water molecule from the iron. Substrate binding is rapid, the heme is transformed from its low-spin form to the high-spin form, and the substrate is placed in close spatial proximity to the oxygen activation site on the heme. The relationships between the spin state of the cytochrome, interaction with the amino acids at the binding site, and substrate binding are more complex than described here. The reader is referred to discussions of changes in the spin state of P450 in Lewis [148], Poulas and Raag [206], Rein and Jung [210], and Sligar and Murray [242].

The next step in the catalytic cycle after substrate binding is the one-electron reduction of the substrate–P450 binary complex (Figure 3.2B). As mentioned, substrate binding and the concomitant alterations in P450 may facilitate this reduction step. The ferric (Fe3±) hemocytochrome P450–substrate complex is reduced by a single electron to the ferrous ($Fe^{2+}$) hemocytochrome P450–substrate complex. This electron is provided by NADPH through NADPH–P450 reductase. This flavoprotein contains two flavins: flavin adenine dinucleotide (FAD) and flavin mononucleotide (FMN). The flavoprotein appears to exist in its half-reduced (one-electron reduced) form and, upon reaction with NADPH, is fully reduced (two-electron reduced). The intramolecular electron flow appears to be from FAD to FMN. It is interesting that, whereas the flavoprotein is a one-electron donor, its substrate, NADPH, provides two electrons. The mechanism for the two-electron shuttle by the one-electron donor flavoprotein is incompletely understood [13].

NADPH–P450 reductase has at least two domains, one of which is imbedded in the endoplasmic reticulum membrane and the other above the plane of the membrane on the cytosolic side. The domain solubilized in the membrane consists mainly of hydrophobic amino acids. The actual interaction with NADPH and oxidation–reduction takes place outside the membrane. Another interesting aspect of P450 reductase is that the quantity of P450 is in large excess to the quantity of reductase (as much as 15- to 20-fold or more, depending on conditions). This means that each flavoprotein must reduce several P450 molecules, indicating that the interaction between the reductase and P450 is an important consideration, as discussed later.

Upon reduction of the ferric hemocytochrome P450–substrate binary complex to the ferrous state by the reductase, the enzyme binds oxygen (Figure 3.2C) to form a ternary complex. The oxygen binds at the free ligand of the heme iron and is believed to be oriented spatially with the substrate binding portion of the active site. Uncoupling (interrupting the flow of electrons) the catalytic cycle at this point can produce the oxidized ferric P450 and a reduced form of oxygen, the superoxide radical (Figure 3.2D). Other reactive oxygen species can be generated by P450, including hydrogen peroxide and the hydroxyl radical. Generation of active oxygen species by P450 has been reviewed in Bernhardt [17].

At this stage of the catalytic cycle, highly critical reactions take place that are still incompletely understood [206]. The major event is activation of the oxygen molecule. The ternary complex accepts a second electron required for reaction (Figure 3.2E). The source of this electron can be either NADH or NADPH, depending on the mediator of electron transport. Because the purified, reconstituted system consisting of isolated P450, P450 reductase, and phospholipid requires only the presence of NADPH, NADPH–P450 reductase can mediate this step; however, as previously mentioned, in some systems it appears that cytochrome $b_5$ can mediate the electron transfer employing reducing equivalents from NADH through NADH–cytochrome $b_5$ reductase. Whichever the source of the second electron, it results in the production of the peroxy P450–substrate complex, which has a net charge of −2 (Figure 3.2F). Of the variety of mechanisms proposed for oxygen activation and insertion into the substrate, two appear to be generally accepted. The first mechanism involves heterolytic cleavage of diatomic oxygen with the abstraction of hydrogen from the substrate and the insertion of oxygen into the substrate. The second mechanism involves homolytic cleavage, whereby two oxygen radicals are generated. Whatever the mechanism, one atom of this reactive oxygen is introduced into the substrate, whereas the other is reduced to water (Figure 3.2G). The oxidized substrate and water are released, regenerating the oxidized ferric P450, which can again initiate the catalytic cycle (Figure 3.2H).

It must be emphasized that other pathways of electron transport in the endoplasmic reticulum can have significant impact on the catalytic activity of the monooxygenase by altering the availability of reducing equivalents. The interested reader is encouraged to consult other sources for a more comprehensive discussion of these pathways [13,197].

This catalytic cycle is common to cytochrome P450-dependent monooxygenase activity associated with xenobiotic metabolism in a variety of organs and among different species; however, certain of these monooxygenases, especially the more specific forms associated with anabolic and catabolic metabolism, have different mediators of electron transport. For example, the adrenal cortex mitochondria systems use a non-heme iron protein in addition to the P450 reductase in the electron transport chain, as does the monooxygenase in certain microorganisms [206,221].

The P450 system is not totally independent, and its activity is affected by a number of factors. One of these factors is the availability of reducing equivalents. The monooxygenase is primarily dependent on NADPH, as previously discussed, and possibly, to a lesser extent, on NADH. NADPH is generated from the pentose–phosphate shunt, isocitrate dehydrogenase, and the malic enzyme. Under most conditions, these pathways provide saturating levels of NADPH; however, certain conditions can stress the ability of the cell to provide NADPH, and it may become rate limiting. Under conditions of high monooxygenase activity, starvation may reduce the activity toward certain substrates due to reduced levels of NADPH. It is generally believed that the decreased activity due to limiting NADH is an unlikely condition. A discussion of these and other factors that regulate monooxygenase activity can be found in Thurman [262].

An additional factor that influences monooxygenase activity is the endoplasmic reticulum membrane. The asymmetric nature of the protein components of the system with respect to the membrane surface, coupled with the disproportionality of the concentrations of the components (i.e., a ratio of 1 to 15–20 between the flavoprotein and P450), indicates an interesting topology and interaction between the components. The membrane topology of the P450 system has been a topic of research for a number of years. The interaction between the protein components of the system and the interaction of these components with the lipid matrix of the membrane are important in the overall reactions of this system. P450 appears to be anchored into the membrane of the endoplasmic reticulum by an anchor peptide at the $NH_2$-terminal end of the protein with the anchor peptide transversing the membrane. The active site, including the heme, is on the cytoplasmic side of the membrane. The active site portion is rich in alpha helix content, globular in nature, and not associated with membrane lipids. The area around the active site may be associated with the cytosolic surface of the membrane, providing a somewhat rigid character.

P450 appears to exist as multicomponent complexes of six P450 molecules clustered around a single P450 reductase. The $NH_2$-terminal regions on the opposite side of the membrane may interact to anchor this complex together. This allows for a catalytic advantage because of the close association of the components. This organization implies that each reductase would be capable of sequentially reducing several P450s. P450 may form a transient complex with the reductase that has an extremely short, non-rate-limiting half-life [21,97].

## Isozyme Heterogeneity and Substrate Specificity of Cytochrome P450

For many years, the apparent lack of substrate specificity of P450 intrigued investigators. It appeared that one of the major features of substrate specificity was lipid solubility. There appeared to be few other structural restraints for substrates. Intensive research on the nature of the hemoproteins has revealed that much of this apparent lack of substrate specificity results from the existence of multiple families and multiple subfamilies of P450 isozymes.

As the array of individual isozymes grew in number, nomenclature became a problem. It was sometimes difficult for investigators to know what exact P450 they were working with because of inconsistencies in nomenclature. This led to attempts to develop a systematic nomenclature for the isozymes. P450 nomenclature has evolved from identifications based on spectral peaks to species-dependent nomenclature based on isolated and semi-purified P450s to the current system, which is based on amino acid sequences that result from specific gene sequences [185]. P450 are now placed in families, which are further divided into subfamilies.

Names are based on the root CYP (derived from cytochrome P450). The CYP is followed by a number identifying the gene family to which it belongs, such as CYP1, CYP2, CYP3, etc. The number for the gene family is followed by a letter denoting the subfamily to which the P450 belongs, such as CYP1A and CYP2A. The subfamilies are further defined by the addition of a number identifying the gene, such as CYP1A1 and CYP1A2; thus, P450 nomenclature is based on genetic relationships defined by protein and gene sequences. All P450s within a single family must exhibit a protein sequence similarity greater than 40%. P450s within the same subfamilies have sequence similarities greater than 55% within the same species. Subfamilies have sequence similarity that may be somewhat less than 55% when comparing species that are more distantly related. Members of subfamilies within a species appear to be located on a specific chromosome, and different subfamilies within a gene family may be clustered on the same chromosome.

As is generally found in biology, the classification system has exceptions resulting from P450s that do not fit the usual patterns. Although this nomenclature provides information on genetic and evolutionary relationships, it provides little information about substrate specificities and the reactions catalyzed by the different P450s. In fact, more is known about the protein and gene sequences of many P450s than about their specific roles in metabolism. With the advent of methodologies such as polymerase chain reaction (PCR) techniques, our ability to sequence specific P450s has outgrown our ability to define their specific roles in the metabolism of xenobiotics and endogenous compounds.

Different species may contain a CYP gene or protein that appears to be highly related; these are termed *orthologous genes*, or *orthologs*. Orthologs are believed to have evolved from a single gene that existed before the two species diverged from a single species. Although these genes and their proteins may contain a high degree of sequence homology, it is not necessarily true that they share a catalytic similarity, or *vice versa*. A small change in an important amino acid sequence can result in a large change in the activity of a P450. Humans and rats have the CYP2 family and the CYP2D subfamily. The rat subfamily contains five genes, one of which is CYP2D1. This P450 has catalytic activity toward debrisoquine metabolism. The human P450 that has the highest catalytic activity toward debrisoquine is CYP2D6, which makes up less than 5% of the complement of P450 in the liver (Figure 3.3). Because these rat and human P450s have similar substrate specificity, it might be assumed that they have a high degree of sequence similarity, but this was found not to be true; therefore, even though these two isozymes have similar catalytic activity, they may have been derived from different ancestral genes. Sequence orthologs do not always predict similar catalytic activities. This is important for toxicologists who extrapolate toxicity from animal models to humans, as noted later.

The section that follows provides a brief description of the major P450 families involved in xenobiotic metabolism. Those families predominately involved in metabolism of endogenous substrates have been excluded. For a more complete discussion of the P450 families, the reader is referred to the excellent reviews in Ioannides [112], Lewis [150], and Smith et al. [247].

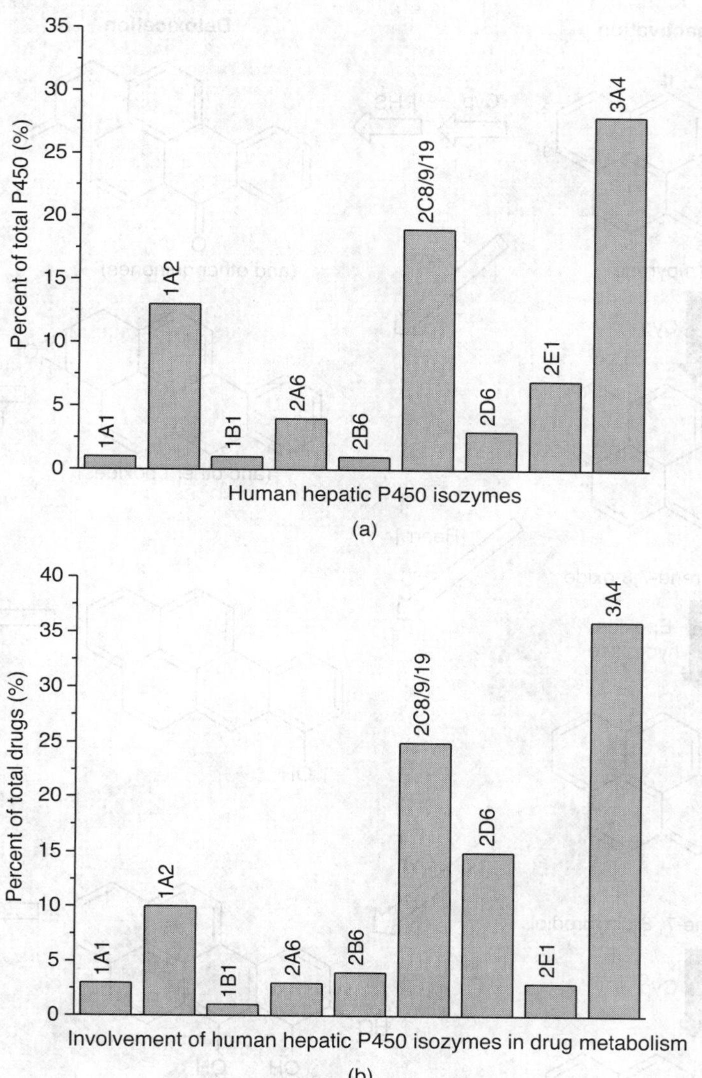


**FIGURE 3.3** Cytochrome P450 isoforms and xenobiotic metabolism: (A) Hepatic distribution of P450 isoforms as a percentage of total hepatic P450. (B) Distribution of isozymes involved in drug metabolism as a percentage of the total number of drugs investigated. (Data from Rendic [211].)

## CYP1 Family

The CYP1 family contains two subfamilies of P450s: CYP1A and CYP1B. CYP1A has been much more extensively characterized than the CYP1B subfamily.

### *CYP1A Subfamily*

The CYP1A subfamily contains CYP1A1 and CYP1A2, which appear to occur in all mammals. These two P450s may have been derived from a common ancestor approximately 120 million years ago. Both P450s are important in the metabolism of environmental xenobiotics. Historically, these two P450s were known collectively as cytochrome P448, due to their characteristic CO binding spectrum. Although these hemoproteins share a number of physiochemical characteristics, such as similar primary structures, they demonstrate different substrate specificities. For example, CY1A1 is highly active in the metabolism of planar polycyclic aromatic hydrocarbons such as benzo(*a*)pyrene (Figure 3.4), whereas CYP1A2 is active in the metabolism of acetanilide, caffeine, and other aromatic and heteroaromatic amines [128].

Both CYP1A1 and CYP1A2 also play a role in catabolism of estrogens [168,264], and CYP1A1 may be involved in heme catabolism [150]. Although low levels of CYP1A1 have been detected in liver, this isoform is constitutively expressed primarily in extrahepatic tissues, including lung, kidney, gastrointestinal tract, and breast. In contrast, expression of CYP1A2 is found primarily in

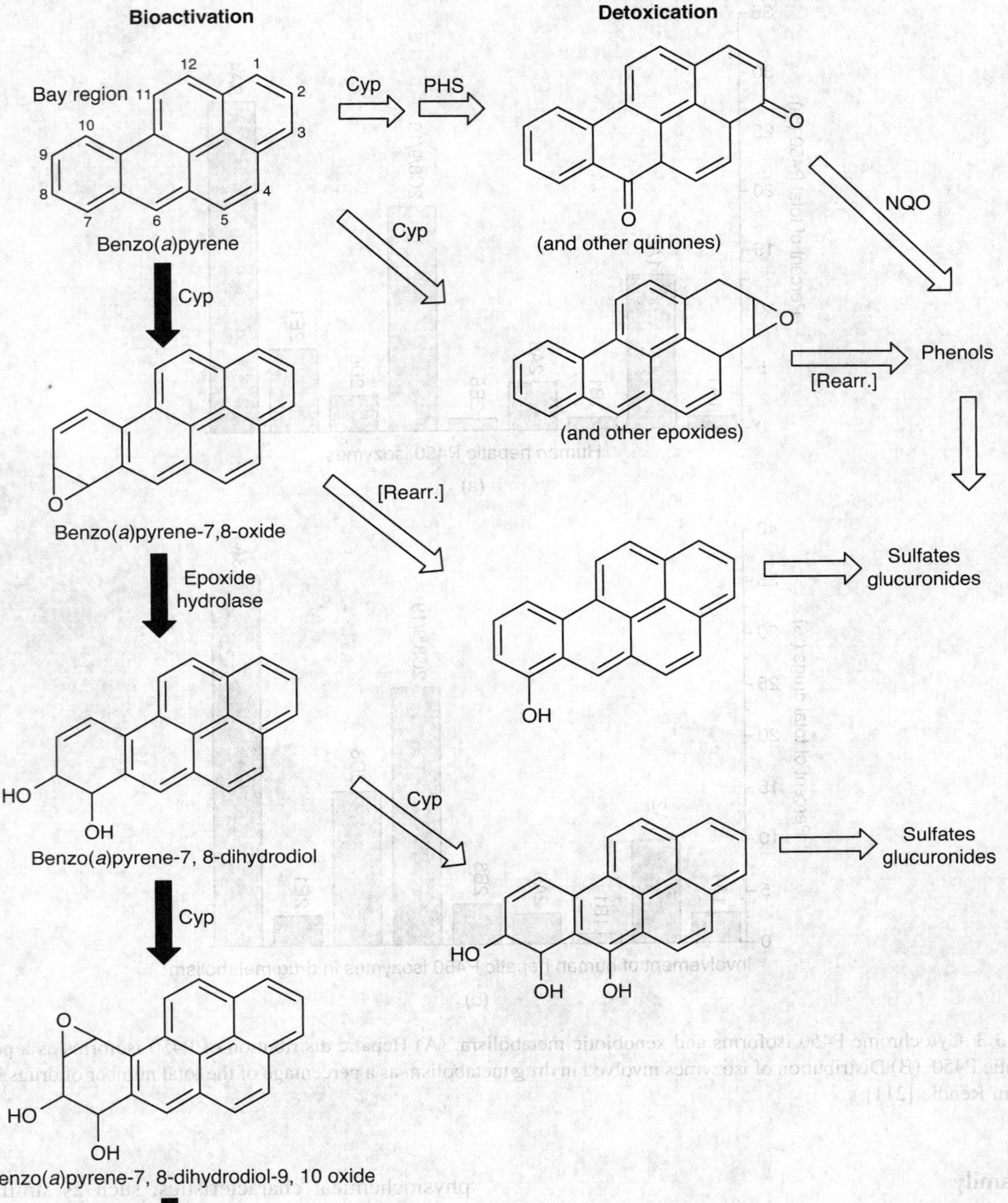


**FIGURE 3.4** Bioactivation and detoxification of the carcinogen benzo(*a*)pyrene. Bioactivation pathways are indicated by filled arrows; detoxification pathways are indicated by open arrows. PHS, prostaglandin H synthase; NQO, NAD(P)H:quinine oxidoreductase.

liver, where it accounts for approximately 15% of total hepatic CYP (Figure 3.3). Both CYP1A isoforms are highly inducible by planar polycyclic aromatic compounds such as 2,3,7,8-tetrachlorodibenzo-*p*-dioxin (TCDD) and 3-methylcholanthrene. In addition, isosafrole is a specific inducer of CYP1A2. The *O*-dealkylation of ethoxyresorufin has been used as a functional marker for CYP1A1 activity *in vitro*, while $N_3$-demethylation has been used as an *in vivo* marker for CYP1A2 activity. A variety of CYP1A1 inhibitors have been characterized, including furocoumarins [15], flavonoids [68], and acetylenic aromatic hydrocarbons [233]. However, the specificity of these inhibitors for CYP1A1 remains to be demonstrated. Furafylline has been used as a selective inhibitor of CYP1A2 [136]. For a number of years, there has been an interest in the role of CYP1A1 in the metabolic activation of polycyclic hydrocarbons, such as benzo(*a*)pyrene.

Although many of the investigations on the role of CYP1A1 in the activation of polycyclic hydrocarbons have been done in animals with induced CYP1A1, polycyclics may be metabolized by other P450s in uninduced animals. Because its concentration in the liver is low, CYP1A1 may be more important in the metabolic activation of polycyclics in extrahepatic tissues, such as the lung. In humans, CYP1A1 demonstrates genetic polymorphism (discussed later). CYP1A2 is active in the metabolism of a wide variety of drugs and environmental contaminants in humans [211]. Good substrates for CYP1A2 tend to be relatively planar molecules containing one or more hydrogen bond donor/acceptors, which play a significant role in the regioselectivity of substrate oxidation [150]. CYP1A2 has been shown to be associated with the mutagenic activation of aromatic and heterocyclic amines, such as 4-aminobiphenyl, 2-aminonaphthalene, and 2-amino-3,5-dimethylimidazo[4,5-*f*]quinoline (MeIQ). It also *O*-dealkylates phenacetin and 4-hydroxylates acetanilide. Although CYP1A2 does play a major role in the bioactivation of aromatic and heteroaromatic amines, other CYP isoforms, notably CYP1A1, CYP1B1, and CYP3A4, can also contribute to the mutagenic activity of this chemical class [128]. Humans can show large differences in the activity of this P450, suggesting it also may be polymorphic.

### *CYP1B Subfamily*

CYP1B1 is the only known member of the CYP1B subfamily, which was identified and characterized in the mid-1990s [222,256]. CYP1B1 has been identified in mouse rat and human tissues. Expression of CYP1B1 mRNA has been detected in a number of normal human tissues including liver, kidney, gastrointestinal tract, endometrium, breast epithelium, and brain [183]; however, expression of functional protein in normal human tissues has been much more difficult to demonstrate [183]. Interestingly, CYP1B1 has been shown to be highly expressed in a number of human tumors, suggesting potential roles as a tumor biomarker or a chemotherapeutic target [184]. CYP1B1 is regulated by both Ah-dependent and -independent mechanisms (addressed later), and thus is inducible by TCDD [297] and polycyclic aromatic hydrocarbons, as well as by estradiol [46]. Recombinantly expressed human CYP1B1 is capable of activating a variety of polycyclic aromatic hydrocarbons, aromatic amines, and heterocyclic amines to mutagenic metabolites, as demonstrated by the *Salmonella* (Ames) mutagenesis assay [54]. In addition, CYP1B1 is active in the metabolism of estrogens, particularly 4-hydroxylation of estradiol, which is the most specific functional assay for this isoform demonstrated to date. 4-Hydroxyestradiol is carcinogenic in male hamsters and may play a role in estrogen-induced tumorigenesis. The potential of CYP1B1 as a chemotherapeutic target has fueled interest in the discovery and development of specific inhibitors. Although a number of highly potent inhibitors have been identified, realization of sufficient selectivity for CYP1B1 remains a challenge [47].

## CYP2 Family

The CYP2 family contains a number of subfamilies important in xenobiotic metabolism, including CYP2A, CYP2B, CYP2C, CYP2D, and CYP2E.

### *CYP2A Subfamily*

The CYP2A subfamily contains at least 12 members that differ in their substrate specificity, tissue distribution, and response to inducers and inhibitors, and they demonstrate species differences. CYP2A1, CYP2A2, and CYP2A3 are rat P450s; Cyp2a4, Cyp2a5, and CYP2a12 are found in mice; and humans have CYP2A6 and CYP2A7. Rat CYP2A1, along with CYP2A2, hydroxylates testosterone, progesterone, and androsteredione. CYP2A1 and CYP2A2 can also metabolize aminopyrine, benzphetamine, ethylmorphine, aniline, acetanilide, and *N*-nitrosodimethylamine. CYP2A1 has low activity toward 3-hydroxylation of benzo(*a*)pyrene and 7-ethoxycoumarin *O*-deethylation, and it does not metabolize 7-ethoxyresorufin. It occurs in liver and testis, but not kidney and lung. In adult rats, it predominates in females and appears to be under endocrine control. CYP2A3 appears to be lung specific, and its substrate specificity has not been well characterized. Human CYP2A6 demonstrates coumarin 7-hydroxylase activity but no activity toward testosterone, in contrast to rat CYP2A1 and CYP2A2. It has no activity toward substrate probes such as 7-ethoxyresorufin, 7-benzyloxyresorufin, ethylmorphine, and testosterone.

Typical substrates for CYP2A6 include both planar and nonplanar molecules, and the preferred site of oxidation is generally within six topological steps of a hydrogen bond acceptor. In addition to coumarin, other substrates for

CYP2A6 include nicotine, cotinine, acetaminophen, and methoxyflurane [150]. Coumarin 7-hydroxylation is commonly used as a functional marker for CYP2A6 activity, and inhibitors include pilocarpine and 8-methoxypsoralin. In primary human hepatocytes, CYP2A6 can be induced by phenobarbital and rifampicin, and 7-hydroxylation of coumarin was found to be accelerated in patients taking anticonvulsant drugs [202]. CYP2A6 accounts for approximately 4% of hepatic P450 (Figure 3.3) and is also expressed in skin [66] and respiratory tract [63], where it may play a role in activation of tobacco-specific carcinogens. Initially thought to be monomorphic, recent data indicate that CYP2A6 is a polymorphic enzyme and that polymorphism of this enzyme may be a determinant of lung cancer risk in smokers [123], although other authors have not observed this association [284]. As mentioned previously, studies in human microsomes have indicated that this P450 may play a role in the metabolic activation of a number of nitrosamines and possibly in the activation of aflatoxin $B_1$ and 1,3-butadiene to carcinogenic epoxides.

### *CYP2B Subfamily*

This subfamily contains P450s, such as rat CYP2B1 and rat CYP2B2, that are highly induced by phenobarbital. Hydroxylation of testosterone at the 16β position is used as a specific substrate probe for this P450 in rats. Other substrates for this subfamily include benzyloxyresorufin, ethoxycoumarin, and pentoxyresorufin. In rats, CYP2B1 has been detected in the lung, adrenal gland, testis, and brain, whereas CYP2B2 occurs in liver and brain. Although its role in xenobiotic toxicity has not been thoroughly investigated, the CYP2B subfamily can metabolically activate xenobiotics, such as bromobenzene, carbon tetrachloride, benzo(*a*)pyrene, aflatoxin $B_1$, and some nitrosamines in the rat. Although the rat P450s in the CYP2B subfamily have been studied for a number of years, members of the CYP2B subfamily in humans have received far less attention.

Constitutive expression of CYP2B6 in human liver is low, accounting for only about 0.2% of total hepatic P450 [69]. In addition to in liver, CYP2B6 has been detected in a variety of extrahepatic tissues, including various regions of the respiratory tract [63], skin [66], kidney, brain [86], placenta, and endometrium [69]. CYP2B6 catalyzes oxidation of a variety of drugs and other xenobiotics, including several coumarin derivatives, methoxychlor, bupropion, benzyloxyresorufin, and benzphetamine [150]. Although debenzylation of 7-benzyloxyresorufin has been used as a marker for CYP2B6 activity, the selectivity of this appears to be questionable [69]; rather, *N*-demethylation of *S*-mephenytoin and ring hydroxylation of the phosphodiesterase inhibitor RP 73401 appear to be more selective probes for CYP2B6. Orphenadrine and 9-ethynylphenathrene have been used as inhibitors, although the former appears to lack selectivity for this isoform. CYP2B6 is inducible by phenobarbital, as well as classic CYP3A inducers such as dexamethasone and rifampicin, but a selective inducer of CYP2B6 has not been identified.

### *CYP2C Subfamily*

At least eight members of the CYP2C subfamily have been identified in rats, and P450s in this subfamily are expressed in both hepatic and extrahepatic tissues. Expression of rat CYP2C isoforms is gender specific, in contrast to humans. Four functional CYP2C subfamily members have been identified in humans: CYP2C8, CYP2C9, CYP2C18, and CYP2C19. Together, isoforms of the CYP2C subfamily are estimated to account for metabolism of approximately 20% of prescribed drugs in humans. All four genes are expressed in liver and small intestine, and CYP2C8, CYP2C9, and CYP2C18 mRNA has been detected in a variety of extrahepatic tissues. To date, CYP2C18 protein has not been detected in any tissue examined. Marker activities for CYP2C8 include Paclitaxel 6α-hydroxylation and *N*-dealkylation of amiodarone, and both of these compounds have been used as probe substrates for this isoforms [263]. Relatively selective competitive inhibitors of CYPC8 include gemfibrozil and montelukast.

Polymorphism of CYP2C8 was first reported in 2001 [55], and since this time a variety of variant alleles have been described. One of these variants, CYP2C8*5, has been associated with development of rhabdomyolosis following administration of cerivastatin. CYP2C9 is quantitatively the most important member of the human CYP2C subfamily in liver, second only to CYP3A4 [213]. CYP2C9 oxidizes a wide variety of clinically important drugs, many of which contain weakly acidic groups. Among these, warfarin and tolbutamide have been used as probe substrates for this P450. CYP2C9 also catalyzes *S*-oxidation of Tienellic acid, resulting in bioactivation of this compound to a reactive electrophile. Benzbromarone derivatives and sulfaphenazole are used as high-affinity selective inhibitors of CYP2C9.

In common with other members of this subfamily, CYP2C9 is polymorphic, and it has been estimated that up to 40% of Caucasians carry at least one variant CYP2C9 allele [254]. Because of its prominent role in the clearance of several low therapeutic index drugs such as warfarin and phenytoin, polymorphisms of CYP2C9 represent a significant challenge in the development of drugs metabolized by this isoform. CYP2C19 catalyzes oxidation of a number of different drug classes, including proton pump inhibitors, antidepressants, and benzodiazepines. *S*-Mephenytoin is a prototypical substrate for CYP2C19, and 4′-hydroxylation of this compound is frequently used as a selective marker activity. Both omeprazole and ticlopidine have been used as inhibitors of CYP2C19, although the selectivity of these compounds has been questioned. Recently, *N*-benzylnirvanol has been shown to be a potent and selective inhibitor of CYP2C19 suitable for diagnostic purposes [281]. Induction of CYP2C8, 2C9, and 2C19 has been observed following treatment with rifampicin, and the former two isoforms are also induced by phenobarbital. Like other members of the human CYP2C subfamily, CYP2C19 is polymorphic, and at least eight allylic variants have been described. Although adverse drug reactions have been observed following coadministration of CYP2C19 substrates and inhibitors, a recent extensive review failed to find convincing evidence of adverse drug reactions related to CYP2C19 polymorphisms [62].

*CYP2D Subfamily*

Rats have six members in the CYP2D subfamily, whereas three have been identified in humans: CYP2D6, CYP2D7, and CYP2D8. Mice have five CYP2D members, and this subfamily has been identified in other mammals. Human CYP2D6 is the most important member of the subfamily for xenobiotic metabolism in humans and was the first human P450 shown to be polymorphic, as discussed later. CYP2D6 is expressed in liver, lung, small intestine, and skin [63,66]. CYP2D6 accounts for less than 5% of total hepatic P450 in humans (Figure 3.3) but is estimated to participate in metabolism of approximately 20% of clinical drugs [110]. Substrates for CYP2D6 are relatively lipophilic and contain a basic amine group, and hydroxylation generally occurs within 5 to 7 Å of the basic nitrogen [150]. Typical substrates include debrisoquine, dextromethorphan, and tricyclic antidepressants. The *O*-demethylation of dextromethorphan has been used as a functional marker for CYP2D6 activity. Numerous alkaloids are potent ligands for CYP2D6, and quinidine has been used as a selective inhibitor of this activity. Reactions catalyzed by CYP2D6 range from aryl hydroxylation to *N*- and *O*-dealkylation.

*CYP2E Subfamily*

The CYP2E subfamily is one of particular interest to toxicologists, due to its involvement in the metabolism and bioactivation of a wide variety of industrial and environmental chemicals. Currently, CYP2E1 is the only member of this subfamily in rats, mice, and humans. CYP2E1 appears restricted to mammals and may have evolved more recently than certain other gene families. It is expressed in liver and kidney and occurs at low levels in a number of other tissues including lung, skin, esophagus, and small intestine [63,66]. Although it normally represents less than 10% of total P450 in human liver, it is induced by a broad array of its substrates. Its hepatic concentration can vary up to 50% between different humans. CYP2E1 is highly conserved across species, and rodent and human forms of CYP2E1 share many similarities, including similar substrate specificities. It is known to metabolize more than 70 different chemicals with diverse structures.

Structural requirements for CYP2E1 substrates appear limited to small molecules with hydrophobic character [150]. CYP2E1 does not appear to be active in the metabolism of many drugs but does metabolize a wide array of alcohols, aldehydes, alkanes, aromatic hydrocarbons, ethers, fatty acids, halogenated hydrocarbons (including anesthetics), heterocyclics, and ketones. Aniline hydroxylation and *p*-nitrophenol hydroxylation have been used as marker activities for CYP2E1, but both substrates are oxidized by other isoforms of P450. More recently, 6-hydroxylation of the muscle relaxant chlorzoxazone has been used as a diagnostic activity for CYP2E1, and, although the selectivity of this substrate is not absolute, this reaction has become the marker of choice for CYP2E1 activity [26]. Carbon-tetrachloride-dependent lipid peroxidation has also been used to follow CYP2E1 activity *in vitro*. CYP2E1 can be competitively inhibited by many of its substrates, and a number of sulfur-containing compounds, including disulfiram and diallyl sulfate, have been shown to be mechanism-based inhibitors (metabolism-dependent) [286]. As noted previously, CYP2E1 is induced by many of its substrates, including ethanol, and it has been suggested that chronic alcoholics may be more sensitive to chemicals that undergo CYP2E1-mediated bioactivation than nonalcoholics.

Interest in the role of CYP2E1 as a mediator of toxicity comes from two of its actions. First, it is known to be important in the metabolic activation/detoxification of a number of carcinogens and heptatoxins. Second, it may have an important role in free-radical production and oxidative stress [36]. For example, CYP2E1 is believed to be involved in the metabolic activation associated with the carcinogenicity of benzene, butadiene, nitrosamines, and azoxymethane, as well as the hepatotoxicity of nitrosamines, acetaminophen, halothane, and enflurane.

With respect to free radical production, CYP2E1 is involved in the formation of a reactive hydroxyethyl radical produced during its metabolism of ethanol to acetaldehyde. This hydroxy radical is believed to play a role in ethanol-related liver damage. It also appears to be involved in the production of a trichloroethyl radical produced by chlorine removal during the metabolism of carbon tetrachloride. This radical may initiate membrane lipid peroxidation associated with carbon-tetrachloride-induced hepatotoxicity. An additional mechanism by which CYP2E1 could produce reactive radicals is associated with its potential for futile cycling in the absence of substrate. CYP2E1 appears more loosely coupled than some of the other P450s. Oxygen activation during the catalytic cycle in the absence of substrate results in the production of highly reactive hydroxyl radicals, superoxide anions, and hydrogen peroxide. Indeed, cells that constitutively overexpress CYP2E1 exhibited a 40 to 50% increase in generation of reactive oxygen species compared to wild-type cells in the absence of substrate [36]. If these reach concentrations that overcome cellular protection mechanisms, they can initiate oxidative stress leading to tissue damage.

P450s are not evenly expressed in the liver but occur in specific zones; for example, oxygen tension varies significantly across the hepatic lobule, ranging from ~13% in the periportal region to ~4% in the centrilobular region, and this gradient is thought to be an important determinant of CYP expression [162]. The highest concentration of P450 is generally found in a layer surrounding the terminal hepatic venules. This is especially true for induced CYP2E1. Enhanced CYP2E1 activity in the centrilobular region appears related to the centrilobular necrosis produced by hepatotoxins, such as ethanol, carbon tetrachloride, benzene, nitrosamines, and acetaminophen. It may appear that CYP2E1 is predominately involved with metabolic activation; however, this is not necessarily true. As noted before, P450-mediated xenobiotic metabolism is generally associated with production of less toxic metabolites, but in some cases more toxic metabolites are produced. CYP2E1 is no exception to this rule and participates not only in metabolic activation but also in detoxification.

#### *CYP2F Subfamily*

Isozymes of the CYP2F subfamily have been identified in humans, nonhuman primates, rodents, and ruminants [43]. Unlike other members of the CYP2 family, only a single isoform of CYP2F has been identified in each of these species. CYP2F is expressed almost exclusively in lung with very little expression in liver. The human form, CYP2F1, has been shown to be involved in bioactivation of a number of environmental toxicants, including 3-methylindole, naphthylene, styrene, dichloroethylene, and benzene, and the cellular localization of CYP2F1 correlates with the site of injury of these toxicants [282]. Thus, this isoform may play a significant role in the pneumotoxicity of inhaled xenobiotics. To date, no selective substrates for CYP2F1 have been identified; however, the pneumotoxicant 3-methylindole has been found to be a selective mechanism-based inactivator of this P450. No data on the inducibility of CYP2F isoforms has been published.

### CYP3 Family

The CYP3 family of P450s encompasses CYP3A1 and CYP3A2 in rats, Cyp3a-11 and Cyp3a-13 in mice, and CYP3A3, CYP3A4, CYP3A5, and CYP3A7 in humans, along with others from rabbits, dogs, and other species. The CYP3 family contains P450s that are important in the metabolism of many xenobiotics, especially drugs.

#### *CYP3A Subfamily*

This subfamily contains at least four genes in humans, CYP3A4, CYP3A5, and CYP3A7. Together, these isoforms constitute approximately 30% of total hepatic P450 and are estimated to mediate metabolism of around 50% of prescribed drugs, as well as a variety of environmental chemicals and other xenobiotics. CYP3A4 is the major form of P450 expressed in human liver. It is also the major P450 expressed in the human gastrointestinal tract, and intestinal metabolism of CYP3A4 substrates can contribute significantly to first-pass elimination of orally ingested xenobiotics. Small amounts are found in several other organs, such as the kidney and skin. X-ray crystallography studies have demonstrated that CYP3A4 has a cavernous active site, allowing it to oxidize very large substrates such as erythromycin (MW = 734) and cyclosporin A (MW = 1203) [228]. In addition, the large active site allows for simultaneous binding of multiple ligands and is thought to account for homotropic (same ligand) and heterotropic (different ligands) cooperativity in substrate oxidation [258]. This cooperativity is though to be responsible for the non-Michaelis–Menten (sigmoid) enzyme kinetics observed for some CYP3A4 substrates. As such, CYP3A isozymes do not demonstrate a high degree of structural selectivity with respect to their substrates, and the substrate selectivity of CYP3A4 has been difficult to generalize.

Prototypical substrates include erythromycin and midazolam, both of which have been used as probes for CYP3A4 activity. Of significance to toxicologists, CYP3A isoforms are also capable of metabolically activating carcinogens, such as aflatoxin $B_1$ and benzo(*a*)pyrene. A number of selective mechanism-based inhibitors for CYP3A4 have been identified, and ketoconazole is frequently used for this purpose. CYP3A4 and other members of the CYP3A subfamily are induced by a number of drugs, including rifampicin, phenobarbital, and phenytoin [211]. Because of the large number of drugs metabolized by CYP3A4, it frequently plays a role in a number of drug–drug interactions that may result in adverse effects, and this has become an important factor in development of therapeutic agents.

An example of how dietary constituents can affect specific isozymes is provided by the interaction between the consumption of grapefruit juice and CYP3A4. Consumption of grapefruit juice can cause an increase in the oral availability of a number of drugs that are CYP3A4 substrates. Increased bioavailability is produced by inhibition of intestinal CYP3A4 activity by 6′,7′-dihydroxybergamottin, which is a component of grapefruit juice. This dietary compound is a mechanism-based inhibitor of CYP3A4 that results in the rapid partial loss of CYP3A4 activity [226]. Inhibition of metabolism of the CYP3A4 substrates during their intestinal absorption accounts for the higher than anticipated plasma concentrations of the drugs.

#### *Other CYP3A Isoforms*

CYP3A5 is the most extensively studied of the minor isoforms in this subfamily. This P450 may be polymorphically expressed in humans (discussed below). It does not appear to have the broad substrate specificity of CYP3A4 and has lower activity. CYP3A5 phenotype did have any significant effect on *in vivo* metabolism of a variety of CYP3A4 substrates, including midazolam, nifedipine, cyclosporin A, and docetaxel [56]. However, clearance of tacrolimus was more rapid in individuals expressing CYP3A5, suggesting that this isoform may be active in drug metabolism in humans. CYP3A7 is active in the metabolism of steroids and retinoids, but a role in xenobiotic metabolism has not been demonstrated.

#### *Other P450s*

A large number of P450 families and subfamilies have not been discussed here. Most of these are involved with the metabolism of endogenous substrates or occur in species that are beyond the scope of the topic of this chapter; see Ioannides [112] and Lewis [150] for a more complete discussion.

## Role of the Cytochrome P450-Dependent Monooxygenase in Toxicity

The toxicity of any agent is dependent on its concentration at its target site. This is a function of many factors, including the route of exposure, the pharmacokinetics of the xenobiotic, the excretion of both the parent compound and its metabolites, and the sensitivity of the target site. The ability of the organism to clear the xenobiotic through excretion will have a profound influence on the concentration at the target site. Directly associated with the ability to clear many xenobiotics is the ability to metabolize the xenobiotic to more water-soluble metabolites.

Without doubt, the P450-dependent monooxygenase plays a pivotal role in the metabolism of xenobiotics. It is the prime metabolic route for the majority of xenobiotics, acting either directly in detoxification or indirectly by priming the xenobiotic for further metabolism through functionalization, as illustrated in other sections of this chapter.

The original interest in the P450 system was associated with its ability to metabolize drugs and decrease both their toxicity and duration of action. It soon became evident that, in certain cases, this enzyme system converted certain drugs from pharmacologically inactive forms to active forms. Examples of the metabolic activation of toxicants, such as the *in vivo* conversion of the inactive insecticide parathion to its active form, paraoxon, were soon encountered. It was also discovered that this enzyme system could activate stable molecules such as benzo(*a*)pyrene to highly reactive metabolites capable of damaging cellular macromolecules, as shown in Figure 3.4. Further studies have indicated that metabolic activation plays an important role in the toxicity of a number of xenobiotics. Recently, an extensive review of bioactivation has been conducted, and a compendium of bioactivation reactions has been published [122].

Studies undertaken to understand the biochemistry of P450 played a large role in the development of the modern fields of biochemical and molecular toxicology. Currently, much effort is being placed on determination of the balance between metabolite activation and detoxification and the detoxification of activated metabolites. This is providing new insight for toxicologists seeking to understand the toxicity of xenobiotics. Studies on the active sites of P450 and other xenobiotic metabolism enzymes and the factors that influence their activity and their expression are bringing toxicologists closer to being able to predict potential toxicity with more accuracy. These efforts are also aiding toxicologists in the difficult task of predicting human toxicity from studies done with cellular and animal models.

## Reactions Catalyzed by the Cytochrome P450-Dependent Monooxygenase System

On first inspection, it appears that P450 can catalyze a bewildering number of reactions (Table 3.2); however, on closer inspection, a degree of commonality exists among these reactions. The first area of commonality is that most of the reactions represent oxidations. Second, the reactions convert lipophilic substrates to more hydrophilic products. Third, many of the reactions can be understood as hydroxylations, as pointed out by Mannering [165]. For a detailed review of P450 reactions, see Testa [261] and Guengerich [95]. Representative examples of the various reactions catalyzed by P450s are illustrated in Figure 3.5.

### Aliphatic Hydroxylation

Aliphatic hydroxylation may be thought of as a special case of the oxidation of an $sp^3$ hybridized carbon atom, and examination of aliphatic hydroxylation reactions is illustrative of several important aspects of monooxygenase activity. Hydroxylation of aliphatic carbon atoms represents one of the most common reactions in phase I metabolism of xenobiotics. The reaction mechanism, which may be common to several other types of monooxygenase metabolism, appears to occur by a hydrogen (or electron) abstraction mechanism. (Figure 3.6). Oxygen activation produces a $[FeO]^{3+}$ at the heme of P450. Hydrogen abstraction from the substrate results in production of the carbon-centered radical. This radical interacts with activated oxygen (through oxygen rebound) to yield hydroxylation. Other reactions, such as *O*-dealkylation of ethers and carboxylic acid esters, may proceed through this mechanism with decomposition of unstable hydroxylation products. Hydrogen abstraction is site selective, resulting in a nonrandom hydroxylation. The specific hydroxylation site is determined by structure and the specific spacial orientation of the substrate at the active site. Different isozymes of P450 show different degrees of site selectivity. For example, *n*-hexane hydroxylation can occur at $C_1$,

**TABLE 3.2**
**Distribution of Reaction Types Catalyzed by Major Human P450s Involved in Xenobiotic Metabolism**

| Reactions | 1A1 | 1A2 | 2C9 | 2C19 | 2D6 | 2E1 | 3A4 |
|---|---|---|---|---|---|---|---|
| *N*-Dealkylation | 19 | 24.4 | 19.8 | 23.6 | 23.8 | 15.2 | 26.5 |
| *O*-Dealkylation | 9.8 | 10 | 8.8 | 11.8 | 22.7 | 8.3 | 9.6 |
| *S*-Oxidation | 3.4 | 3.4 | 3.9 | 3.9 | 6.5 | 2.3 | 5.8 |
| Aromatic hydroxylation | 25.9 | 25.2 | 23.7 | 21.4 | 24.5 | 21.2 | 12.1 |
| Aliphatic hydroxylation | 13.7 | 13.2 | 24 | 27.9 | 11.9 | 24.9 | 24.6 |
| *N*-Oxidation | 6.8 | 7.2 | 4.9 | 2.6 | 5.4 | 6.0 | 6.5 |
| Nitro reduction | 0.5 | 1.1 | 0.4 | 0.9 | 0 | 0.5 | 1 |
| Peroxidation | 14.1 | 9.7 | 7.1 | 4.8 | 0.7 | 14.7 | 3.4 |
| Hydroxycarbonyl oxidation | 2.9 | 2.3 | 4.2 | 1.3 | 2.5 | 4.6 | 4.1 |
| Desaturation | 2.4 | 2.9 | 3.2 | 1.3 | 1.4 | 1.4 | 5.6 |
| Aldehyde oxidation | 1.5 | 0.6 | 0 | 0.4 | 0.4 | 0.9 | 0.9 |
| Total reactions (%) | 9.6 | 16.3 | 13.2 | 10.7 | 12.9 | 10.1 | 27.3 |

*Source:* Adapted from Lewis, D.F.V., *Pharmacogenomics*, 5(3):305–318, 2004.

$C_2$, $C_3$, or $C_4$. P450 isozymes induced by phenobarbital metabolized *n*-hexane to yield a four- to fivefold increase in the 2-, 3-, and 4-hydroxylated metabolites and only a slight increase at the 1 position. On the other hand, benzo(*a*)pyrene-induced isozymes result in decreased yields of the 1- and 2-hydroxylated products but increased yields of the 3- and 4-hydroxylated products [80]. Hydroxylation of aliphatic compounds is generally considered detoxification because of the greater water solubility of the products, but one must be cautioned against overgeneralization, as products that are more toxic could be produced by subsequent metabolism.

## Aliphatic Desaturation

Aliphatic desaturation is another special case of the oxidation of an $sp^3$ hybridized carbon atom. Mechanistically, the first step of the reaction is abstraction of a hydrogen atom, similar to aliphatic hydroxylation; however, instead of oxygen rebound leading to insertion of oxygen, the second step in the mechanism results in abstraction of a second hydrogen atom and formation of a double bond. Compounds of toxicological significance that undergo CYP-mediated desaturation include the carcinogen ethyl carbamate and valproic acid.

## Aromatic Oxidation

The mechanisms of aromatic hydroxylation are not completely understood but probably involve several alternative mechanisms. The exact mechanism for a given hydroxylation may be based on a number of factors, such as the steric features of the substrate and configuration of the active site of the specific P450. Potential mechanisms include direct oxygen insertion into the C–H bond to form an epoxide through radicaloid reactions or through intermediates bonded to $[FeO]^{3+}$. Evidence for intermediacy of an epoxide in the mechanism of aromatic hydroxylation comes from the so-called *NIH shift*, in which deuterium or halogen substituents are observed to migrate around the aromatic ring in a characteristic pattern during arene oxidation [103]; however, the necessity of an epoxide intermediate in the NIH shift has been challenged [302]. The production of arene oxides has been widely studied because of their importance in the formation of epoxide ultimate carcinogens.

## Oxidation of Alkenes and Alkynes

In addition to aromatic compounds, both alkenes (aliphatic double bond) and alkynes (aliphatic triple bond) are subject to epoxidation by P450. The mechanism of these reactions involves abstraction of an electron to form the first C–O bond and a carbon-centered radical, followed by oxygen rebound giving rise to the epoxide. Epoxidation generally, though not always, occurs with retention of stereochemistry. With some terminal alkenes and alkynes, the carbon-centered radical may also react with one of the porphyrin nitrogens in the P450 heme, resulting in covalent binding and inactivation of the enzyme. In the case of polyhalogenated alkenes, migration of a halogen atom may occur with rearrangement of the product to the corresponding aldehyde, as seen in the metabolism of 1,1,2-trichloroethylene to 1,1,1-trichloroacetaldehyde. Other important examples of toxicological significance include oxidation of the industrial monomer 1,3-butadiene and the mycotoxin aflatoxin $B_1$. It is generally agreed that these reactive epoxides are the ultimate carcinogenic metabolites of the parent molecules.

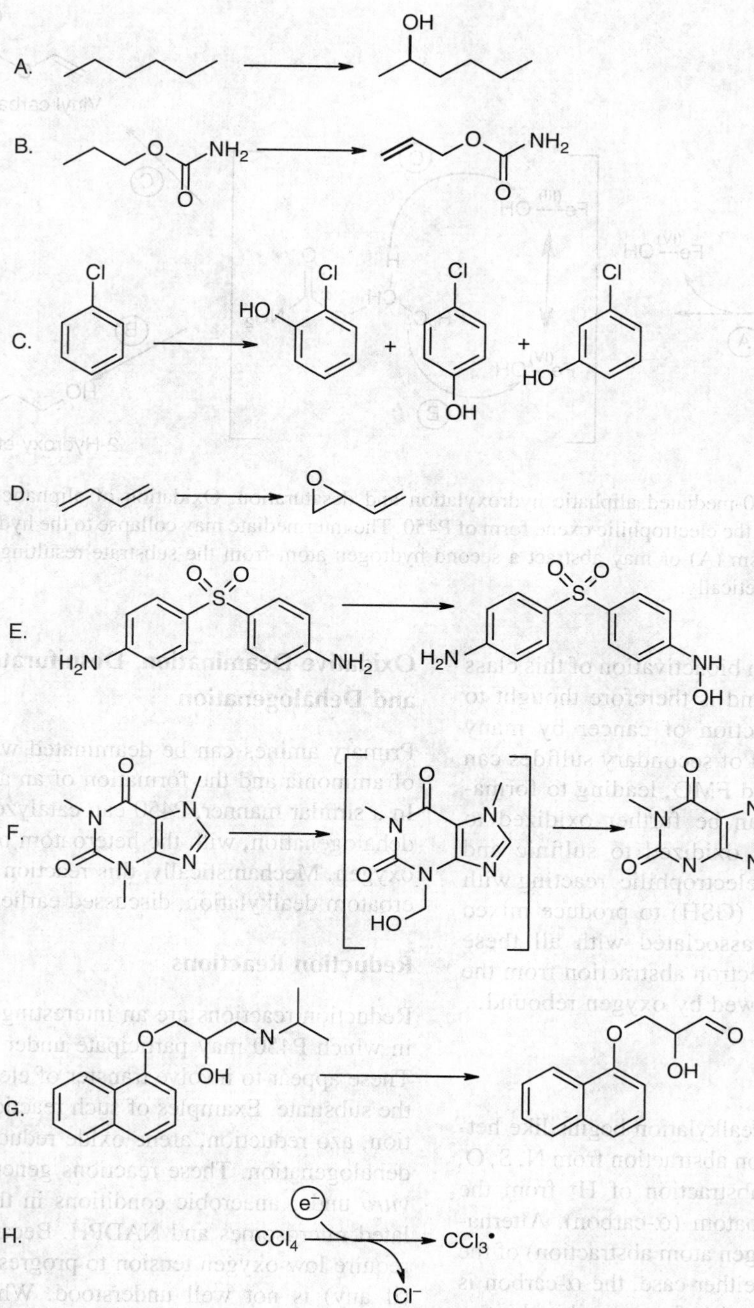

**FIGURE 3.5** Examples of reaction types catalyzed by P450: (A) aliphatic hydroxylation, (B) desaturation, (C) aromatic oxidation (hydroxylation), (D) epoxidation of alkenes, (E) *N*-hydroxylation, (F) heteroatom dealkylation (*N*-dealkylation), (G) oxidative deamination, (H) reductive dehalogenation.

## Heteroatom Oxidation

P450 not only oxidizes carbon atoms but also nitrogen, sulfur, phosphorus, and halogen atoms. A number of nitrogen-containing compounds can be oxidized to stable *N*-oxides. Another hepatic enzyme, flavin-containing monooxygenase (FMO), can also catalyze this reaction, though the substrate specificities of these two enzymes are different (discussed later). P450 and FMO may form *N*-oxides from the same xenobiotic; however, FMO generally prefers substrates with an electron-deficient nitrogen, whereas P450 prefers an electron-rich nitrogen. P450-mediated *N*-oxidation is possible with primary and secondary aromatic amines to produce hydroxylamines.

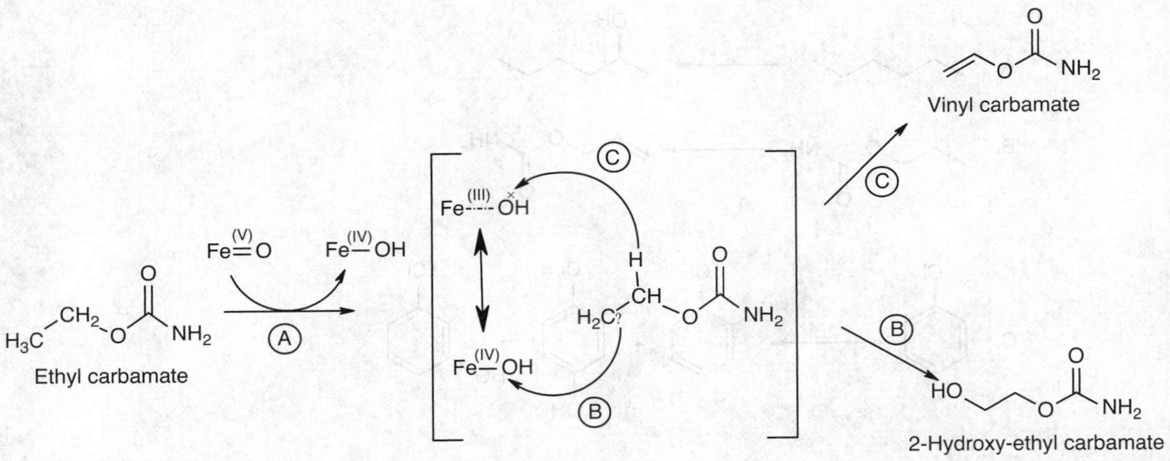

**FIGURE 3.6** Mechanism of P450-mediated aliphatic hydroxylation and desaturation. Oxidation of aliphatic carbon begins with abstraction of a hydrogen atom by the electrophilic oxene form of P450. The intermediate may collapse to the hydroxylated metabolite via the oxygen rebound mechanism (A) or may abstract a second hydrogen atom from the substrate resulting in desaturation (B). These two pathways compete kinetically.

This reaction is the first step in bioactivation of this class of compounds to mutagens and is therefore thought to play a central role the induction of cancer by many aromatic amines. *S*-Oxidation of secondary sulfides can also be catalyzed by P450 and FMO, leading to formation of sulfoxides, which can be further oxidized to sulfones. Free thiols can be oxidized to sulfinic and sulfenic acids, which may be electrophilic, reacting with protein thiol and glutathione (GSH) to produce mixed disulfides. The mechanism associated with all these reactions is believed to be electron abstraction from the heteroatom by $(FeO)^{3+}$, followed by oxygen rebound.

### Heteroatom Dealkylation

P450-dependent heteroatom dealkylation begins like heteroatom oxidation with electron abstraction from N, S, O, or Si. This is followed by abstraction of $H^+$ from the carbon attached to the heteroatom ($\alpha$-carbon). Alternatively, direct oxidation (hydrogen atom abstraction) of the $\alpha$-carbon may also occur. In either case, the $\alpha$-carbon is hydroxylated by oxygen rebound to form a carbinol intermediate. Carbinolamines and related intermediates are generally unstable and undergo carbon–heteroatom bond cleavage, followed by rearrangement to the corresponding aldehyde or ketone. Regardless of the identity of the heteroatom, the products of this reaction are the hydrogenated heteroatom compound and an aldehyde or ketone. Sulfur and silicon atoms generally are not as readily dealkylated as nitrogen and oxygen atoms. Dealkylations unmask more polar functional groups, facilitating conjugation and excretion; however, dealkylation can also facilitate bioactivation, as in the case of *N*-dealkylation of secondary aromatic amines.

### Oxidative Deamination, Desulfuration, and Dehalogenation

Primary amines can be deaminated with the elimination of ammonia and the formation of an aldehyde or ketone. In a similar manner, P450 can catalyze desulfuration and dehalogenation, with the heteroatom being replaced with oxygen. Mechanistically, this reaction is identical to heteroatom dealkylation, discussed earlier.

### Reduction Reactions

Reduction reactions are an interesting series of reactions in which P450 may participate under special conditions. These appear to involve transfer of electrons from $Fe^{2+}$ to the substrate. Examples of such reactions are nitro reduction, azo reduction, arene oxide reduction, and reductive dehalogenation. These reactions generally are studied *in vitro* under anaerobic conditions in the presence of isolated microsomes and NADPH. Because these reactions require low oxygen tension to progress, their *in vivo* role (if any) is not well understood. Whether or not these reactions represent simply a curious phenomenon associated with P450 or a viable metabolic pathway is not known. It may be possible that under certain cellular conditions of low oxygen tension these reactions could proceed *in vivo*.

## Induction and Inhibition of Cytochromes P450

### Induction

When animals are exposed to certain xenobiotics, their ability to metabolize a variety of xenobiotics is increased. This phenomenon is termed *induction*. Induction produces

**TABLE 3.3**
**Inducers of Cytochrome P450**

| Structural Class | Primary Example | Other Examples | Receptor | CYPs Induced |
|---|---|---|---|---|
| Polycyclic hydrocarbon type | 3-Methylcholanthrene | TCDD, Benzo(*a*)pyrene, β-naphthaflavone, chlorpromazine, isosafrole, ketoconazole | AhR | 1A1, 1A2, 1B1 |
| Phenobarbital type | Phenobarbital | Phenytoin, griseofulvin, chlorpromazine, ketoconazole, dieldrin, BHT | CAR–RXR | 2A1, 2B1, 2B2, 2B6, 2C6, 3A4 |
| Glucocorticoid type | Rifampicin[a] | Dexamethasone, pregnenalone 16a-carbonitrile, spironolactone, prednisolone, methylprednisolone | PXR–RXR (GR) | 3A |
| Ethanol type | Ethanol | Acetone, heptane, pyrazole | None | 2E1 |
| Clofibrate type | Clofibrate | Phorbol esters, WY-14,634 | PPAR–RXR | 4A |

[a] Significant species differences exist in inducer selectivity. Rifampicin is the prototypical glucocorticoid type inducer in humans; dexamethasone is a prototypical inducer in rodents.

a transitory resistance to the toxicity of many compounds; however, this may not be the case with compounds that require metabolic activation because their toxicity may increase. The exact toxicological outcome of this increased metabolism will be dependent on the specific xenobiotic and its metabolic pathway. Because the toxicological outcome of a xenobiotic exposure can depend on the balance between those reactions that represent detoxification and those that represent activation, increases in metabolic capacity may, at times, produce unpredictable results. Induction of P450 has been reviewed in Bresnick [31], Wang and LeCluyse [283], and Waxman [289].

One of the initial reports of increased metabolic capacity associated with xenobiotic exposure suggests how induction may provide a survival advantage. In 1954, Brown et al. [32] were studying the metabolism of methylated aminoazo dyes and found that xenobiotics in the animal diets enhanced the P450-dependent demethylation of these compounds. Free-living animals consume a variety of feeds that may contain toxic constituents. If the animal can respond rapidly to these toxic compounds by developing resistance, it can continue to use the feed source and obtain a survival advantage. One mechanism of rapidly developing such resistance is through increased detoxification resulting from stimulation of xenobiotic metabolizing enzyme activity. Conney [52] published a pivotal review in 1967 that indicated that more than 200 chemicals could induce P450-dependent metabolism, and most of these chemicals were monooxygenase substrates.

The classical definition of enzyme induction requires transcriptional activation at the level of DNA and increased production of mRNA, followed by an increase in the synthesis of the enzyme. The term has taken on a broader definition when used in respect to xenobiotic metabolism. This broader definition includes mechanisms such as mRNA and enzyme stabilization, all of which are associated with xenobiotic induction. The classes of P450 inducers are listed in Table 3.3; however, the concept of inducer classes has become less meaningful in light of recent advances in our understanding of CYP induction.

The polycyclic aromatic hydrocarbon class of inducers includes 3-methylcholanthrene, benzo(*a*)pyrene, and 2,3,7,8-tetrachlorodibenzo-*p*-dioxin, and their mechanism of induction in animals has been extensively investigated. These inducers induce CYP1A1, CYP1A2, and CYP2B1, which are expressed in the liver or extrahepatic tissues of rodents and humans. The low constitutive hepatic concentrations of CYP1A1 result from suppression of transcription by a nuclear repressor protein. Within the cytoplasm of the hepatocyte exists a receptor protein termed the *Ah receptor*, which is complexed with heat-shock protein (Hsp 90). When a polycyclic-hydrocarbon-type inducer enters the hepatocyte, it binds and activates the Ah receptor, resulting in release of hsp90. The Ah receptor is phosphorylated and subsequently binds to the Ahr nuclear translocator protein (Arnt), which is also activated by phosphorylation. This complex then moves to the nucleus of the hepatocyte. In the nucleus, this complex binds to a DNA regulatory sequence termed the xenobiotic response element (XRE). A DNA segment similar to rat XRE has been found in mouse and human cells. The XRE has also been found in genes of other xenobiotic metabolism enzymes, such as glutathione *S*-transferase, aldehyde dehydrogenase, and uridine diphosphate (UDP)–glucuronosyltransferase (UGT), where it may be involved in regulation of their expression. Binding of the ligand-bound Ah–Arnt complex to XRE enhances transcription of the CYP1A1 gene, resulting in increased quantities of CYP1A1 mRNA followed by an increase in the hepatic concentration of CYP1A1.

CYP2A isoforms are induced by a structurally diverse array of xenobiotics from several different inducer classes [255]; for example, human CYP2A6 is induced by phenobarbital, dexamethasone, and rifampicin in primary hepatocyte cultures, and CYP2A3 was induced in rat lung by 3-methylcholanthrene. In addition, several metal salts induce CYP2A5 in mice. Induction of CYP2A is poorly understood and appears to occur by transcriptional activation, RNA stabilization, and protein stabilization. Further, the mechanisms of induction vary with the isoforms, species, and tissues examined. Roles in CYP2A induction have been proposed for several of the orphan nuclear receptors (discussed below), as well as for a variety of RNA binding proteins, which appear to bind in the 3′-untranslated region of CYP2A mRNA in response to pyrazole treatment.

Phenobarbital (PB) and other compounds of diverse structure induce expression of CYP2B1 and CYP2B2, as well as, to a lesser extent, CYP2A1, CYP2C6, CYP3A1, and CYP3A2. PB also produces a general pleiotropic response in the liver, resulting in proliferation of the smooth endoplasmic reticulum in hepatocytes, increases in total microsomal protein, and increases in NADPH–cytochrome P450 reductase, as well as other xenobiotic metabolizing enzymes, such as UDP–glucuronosyltransferase and epoxide hydrolase. Induction of CYPs and other enzymes by PB occurs by transcriptional activation, but until relatively recently no cytoplasmic receptor had been identified for the phenobarbital-type inducers. Over the last 10 years, however, significant gains have been made with respect to mechanisms of CYP induction by the phenobarital-like inducers. Convincing evidence now suggests that PB-mediated CYP induction is associated with a cytoplasmic receptor known as the constitutively active receptor (CAR), so named because early studies in HepG2 cells indicated that this transcription factor was active even in the absence of ligand [289]. It has since been demonstrated that certain androgens, including androstanol and androstenol, bind to CAR and inhibit transcriptional activation of CYP2B and other genes [78]; hence, CAR is also known as the *constitutive androstane receptor*. CAR is one of the so-called orphan nuclear receptors, whose endogenous ligands are unknown. Other orphan receptors include the pregnane X receptor (PXR; discussed below), the retinoid X receptor (RXR), the peroxisome proliferator-activated receptor (PPAR; discussed below), and several others.

The details of CAR-mediated induction of CYP are unclear. As is the case with several other orphan receptors, CAR forms a heterodimer with RXR, and the heterodimer forms a DNA-binding complex with the steroid receptor coactivator SRC-1. CYP induction in physiological systems is strictly dependent on the presence of PB or other PB-like inducers, but actual binding of the inducer does not appear to be required. It is thought that PB somehow causes dissociation of the androstanes from CAR, resulting in depression of transcriptional activation [283]. Compounds in the phenobarbital inducer class, such as terpenes, organochlorine pesticides, and polychlorinated biphenyls, may act through a common pathway of induction [83].

CYP2E1 induction has been studied in detail in the rat and represents an interesting situation where induction is controlled at the transcription, mRNA stabilization, translation, and enzyme stabilization levels. CYP2E1 is induced by the ethanol-type inducers. Although not true for all P450s, the CYP2E1 inducers generally are substrates for the isozyme. In many cases, the regulation of expression of CYP2E1 is controlled by stabilization of CYP2E1 mRNA and stabilization of the enzyme apoprotein, along with possible increased efficiency of translation. Cycloheximide, which blocks translation, blocked the increase in CYP2E1 apoprotein when mRNA was unchanged, indicating that the increase in apoprotein was related to increased translation. Actinomycin D, which blocks transcription, did not block the apoprotein increase, indicating that it was not transcription related. Many of the CYP2E1 inducers act by posttranslational stabilization, including acetone, low ethanol doses, pyridine, and pyrazole. With these inducers, CYP2E1 concentration increases, whereas no change in mRNA occurs. This indicates that CYP2E1 degradation decreases while synthesis remains constant, with the net result being increased CYP2E1.

Nutritional factors and disease conditions can also result in increased activity of CYP2E1. High-fat diets and starvation produce an induction of CYP2E1, as does insulin-dependent diabetes and obesity. One common factor in all of these conditions is increased plasma ketone body concentrations. Whether or not this induction is produced by increased ketone bodies, including acetone, or by other factors is currently under investigation.

The glucocorticoid-type inducers, such as dexamethasone and pregnenolone-16α-carbonitrile, induce CYP3A1 and CYP3A2 in rodents but not in human hepatocytes. In contrast, rifampicin is an effective inducer of CYP3A isoforms in humans and rabbits but not in rodents. Regulation of CYP3 induction is now known to be under the control of PXR, another of the orphan nuclear receptors. In humans, PXR is found primarily in the liver and gastrointestinal tract and in lower amounts in kidney and lung; thus, the tissue distribution of PXR tracks with the tissue distribution of CYP3A4 in humans. As with CAR and other orphan nuclear receptors, PXR consists of a DNA-binding domain, which is highly conserved across species, and a ligand-binding domain, with lower sequence homology across species. Differences in the selectivity of CYP3A inducers are due to structural differences in the ligand-binding domain of PXR, and site-directed mutagenesis has been used to convert the rodent PXR ligand-binding spectrum to a human-like ligand selectivity that no longer responds to rodent CYP3A inducers [288]. Like other orphan receptors, PXR forms a functional heterodimer with RXR. Evidence suggests that the glucocorticoid receptor also mediates CYP3A induction, although the role played by this receptor is unclear. One hypothesis is that binding to the glucocorticoid receptor upregulates PXR, leading to induction of CYP3A and other PXR-dependent genes; however, some studies demonstrate glucocorticoid induction of CYP3A by a PXR-independent mechanism. Thus, the exact role of the glucocorticoid receptor in induction of CYP3A genes remains to be resolved. Because of the large number of drugs metabolized by CYP3A4 and the wide array of natural and synthetic PXR ligands, drug–drug interactions due to CYP3A4 induction are common, and PXR reporter assays have become popular in the pharmaceutical industry as a screen for CYP3A4 induction.

Clofibrate-type inducers induce the CYP4A subfamily that is, in the most part, associated with metabolism of endogenous compounds. The clofibrate-type inducers are structurally diverse but in general are highly lipophilic and possess a carboxylic acid functional group. They also cause hepatocyte peroxisome proliferation in rodents.

Induction of CYP4A and enzymes related to peroxisomal fatty acid β-oxidation is mediated by another member of the orphan nuclear receptor superfamily, PPAR. Three members of the PPAR family have been identified and designated as PPARα, PPARβ, and PPARγ. PPARα is responsible for induction of CYP4A in rodents and rabbits. Upon ligand binding, PPAR forms an activated heterodimer with RXR; the heterodimer binds to the peroxisome proliferator response element upstream from CYP4A and related genes, activating transcription. Many compounds that cause peroxisome proliferation in rodent liver are also hepatocarcinogens; however, no CYP4A induction has been observed in humans by PPARα ligands, and humans are resistant to peroxisome proliferation, suggesting that this phenomenon may have little relevance for humans. The low concentrations of CYP4A in human liver and its limited number of xenobiotic substrates reduce its role in drug–drug interactions in humans.

Although CAR and PXR were originally identified as mediators of CYP2B and CYP3A enzymes respectively, there is significant overlap in the selectivity profiles of inducers of these two subfamilies. As such, many phenobarbital-type inducers are capable of inducing CYP3A genes, and several glucocorticoid-type inducers are effective inducers of CYP2B genes. This overlap may be related in part to the ability of CAR and PXR to recognize and bind to each other's response elements, a phenomenon referred to as *cross talk*. Another mechanism for cross talk may be the overlapping ligand-binding selectivity of CAR and PXR; for example, phenobarbital is a ligand for both CAR and PXR in humans.

### Inhibition

Just as induction of xenobiotic metabolism can have important toxicological ramifications, inhibition of the ability to metabolize a xenobiotic can result in profound changes in its toxicity. Inhibition of the metabolism of a compound can result in a higher plasma concentration than predicted and unexpected toxicity. During treatment with multiple drugs, unexpected adverse effects can be produced through drug–drug interactions where one drug inhibits the metabolism of another, resulting in higher than expected plasma concentrations.

Four mechanisms are generally associated with inhibition of P450-mediated detoxification. First, two xenobiotics may be substrates for the same P450 isozyme and will compete for the active site of the enzyme, a phenomenon known as *competitive inhibition*. An example of this type of mechanism is the inhibition of bioactivation of the rodent carcinogen ethyl carbamate by ethanol, both of which are found in fermented beverages [294]. A second mechanism of competitive inhibition is the binding of a xenobiotic to the active site of a P450, although it is not a substrate for that P450. The presence of the nonsubstrate at the active site blocks the binding of the true substrate, inhibiting its metabolism. Examples of this mechanism of CYP inhibition are rare. A third mechanism of inhibition involves the metabolism of a xenobiotic to a product that has a higher affinity for the active site than the parent compound, forming a so-called metabolite–inhibitor (MI) complex. The active site is then occupied and additional substrate cannot bind. This essentially makes the enzyme inactive and is an example of noncompetitive inhibition. Compounds such as erythromycin, niacardipine, and diltiazem reversibly inhibit human CYP3A4 by this mechanism [64]. The fourth mechanism is another example of noncompetitive inhibition resulting from the production of a highly reactive metabolite that binds (often covalently) to the heme or apoprotein of P450, destroying its activity. This type of inhibitor is termed a *suicide substrate*. As mentioned previously, inhibition of CYPs by trichloroethylene is an example of suicide inhibition. Both MI complex formation and suicide inhibition are examples of mechanism-based inhibition.

Other, less common mechanisms can result in inhibition of P450-mediated xenobiotic metabolism, including compounds that may modify protein or heme synthesis or degradation, those that may uncouple electron transport to P450, those that may interfere with cofactor availability, and those that may directly inhibit NADPH–P450 reductase activity. Just as some substrates may demonstrate a higher affinity for specific P450s and others may not, inhibitors may show a narrow or broad range of affinity for a specific P450. Inhibitors have been useful tools in determining mechanisms associated with xenobiotic metabolism and in attempts to predict specific drug–drug interactions. Induction and inhibition of human cytochromes P450 have been recently reviewed in Pelkonen et al. [201].

## Pharmacogenetics, Human Polymorphism of P450 Isozymes, and Their Toxicological Significance

Pharmacogenetics is the study of the hereditary basis of the observed differences in response (both therapeutic and adverse) to drugs by individuals and populations. The term can be expanded to include not only drugs but also dietary and environmental chemicals. This field has seen a large expansion over the last decade, as the understanding of the genetics, genetic regulation, and interindividual variations in P450 and other xenobiotic metabolism enzymes has increased. Maturation of methodologies from molecular biology and refinement of other methodologies to study genetic differences in individuals and populations have spurred interest in the pharmacogenetics of xenobiotic metabolism.

Many studies of the adverse effects of chemicals in animals and humans have indicated that highly significant differences exist between animal strains, among individual animals, and especially in individual humans and different human population groups. These genetic polymorphisms can result in unexpected drug and environmental toxicities and complicate safety assessments and the extrapolation of data from animal studies to humans. It has been estimated that almost 50% of drugs for which adverse reactions have been reported are metabolized by significantly polymorphic P450s [109]. As discussed later, this has led to recommendations that the specific family and subfamily of P450 that metabolizes a specific drug candidate be determined during early drug discovery efforts. This could avoid unexpected interactions and suggest potential adverse effects before additional developmental efforts with a drug or other chemical product are undertaken.

Several alleles in the CYP2D6 family, for example, are known to contain specific nucleotide deletions that result in inactive genes and a lack of production of the CYP2D6 protein [57]. Individuals homozygous for these gene variations will be *poor metabolizers* (PMs) of CYP2D6 substrates. In contrast, some individuals have multiple copies of the CYP2D6 gene, possibly due to gene duplication [120]. These individuals have enhanced capability to metabolize CYP2D6 substrates and are *ultra-rapid metabolizers* (UMs) [18]. A chemical whose detoxification depends on CYP2D6 would be more toxic than expected in PM, but less toxic than expected in UM individuals. In contrast, a chemical that is metabolically activated would be less toxic in poor metabolizers and more toxic in ultra-rapid metabolizers. To predict toxicity, it is obvious that not only must the role of metabolism in the toxicity of a compound be known but also the potential genotype of exposed individuals.

Several human P450s have been shown to exhibit significant polymorphic expression, including CYP1A1, CYP2A6, CYP2C9, CYP2C19, CYP2D6, CYP3A5, and CYP2E1. Many other CYP polymorphisms are known, encompassing most of the known human CYP-mediating xenobiotic metabolism, but the functional significance of many of these polymorphisms is unclear. The most up-to-date information on CYP polymorphisms can be found on the homepage of the Human Cytochrome P450 (CYP) Allele Nomenclature Committee (http://www.imm.ki.se/CYPalleles). The chromosomal locations of these P450 genes have been identified and the genetic basis for the altered P450 activity is becoming understood. From a toxicological perspective, the most important P450 polymorphisms are those of the CYP2 and to a lesser degree the CYP3 subfamilies. These polymorphisms have been the subject of several recent reviews [37,109,249] and are summarized here.

### *CYP2A6*

CYP2A6 is the major isoforms responsible for C-oxidation of nicotine. Several significant allelic variants of this isoforms have been identified, including CYP2A6*2, which is inactive due to a Leu–His substitution at position 160, and CYP2A6*4A, which is a complete deletion of the CYP2A6 gene. The former allele is found at a frequency of 1 to 3% in Europeans, and the latter allele is more prevalent in Asians. Because of the role of CYP2A6 in nicotine metabolism, polymorphism of this isoform has been implicated in differences in smoking behavior and the incidence of lung cancer among smokers.

### *CYP2C9*

CYP2C9 plays a major role in metabolism of the anticoagulant drug warfarin, and defects in this isoform can lead to excessive plasma concentrations and increased risk of bleeding episodes in affected individuals. The major allelic variants of CYP2C9 are CYP2C9*2 and CYP2C9*3, both of which are due to missense mutations. In Caucasians, the frequency of these two alleles are approximately 11% and 7%, respectively, and the frequency of homozygotes, who express the poor metabolizer (PM) phenotype, is around 3 to 4%.

### *CYP2C19*

Major polymorphisms of CYP2C19 are associated with deficient 4-hydroxylation of *S*-mephenytoin and the sensitivity of affected individuals to excessive sedation by this anticonvulsant agent. Deficient mephenytoin metabolism is found in 2 to 5% of Caucasians but is much more prevalent in Asians, with a frequency of 18 to 23%. Two major allelic variants were identified as CYP2C19*2 and CYP2C19*3 [60]. CYP2C19*2 is a splice mutant that codes a truncated protein, and CYP2C19*3 contains a premature stop codon in exon 4. In addition to mephenytoin, CYP2C19 is involved in metabolism of a variety of other drugs, including proton pump inhibitor and tricyclic antidepressants, although the functional consequences of CYP2C19 polymorphism for these agents is unclear.

*CYP2D6*

High inter-individual variability in the metabolism of the antihypertensive agent debrisoquine led to the discovery of one the first CYP polymorphisms. Since its initial discovery, CYP2D6 polymorphism has been the subject of intensive research efforts and is probably the most thoroughly characterized CYP polymorphisms. CYP2D6 plays a major role in the metabolism of tricyclic antidepressants and antipsychotics, and PM status has been associated with a higher incidence of extrapyramidal effects, while UM status is associated with a lack of therapeutic effect of these agents. One of the major defective CYP2D6 variants is CYP2D6*4, a splice mutation that is found in Caucasians at a frequency of almost 21%. Another common variant is CYP2D6*5, in which the entire gene is deleted. The frequency of this allele is 4 to 6%. The UM phenotype is due to gene duplication, and frequencies for this phenotype range from 1 to 2% in Caucasians to almost 30% in Arabian and North African populations [110].

*CYP3A5*

Functional CYP3A5 is expressed in approximately 20% of Caucasians and about 67% of African-Americans. Although this isoform contributes to total CYP3A activity in these individuals, the clinical implications of CYP3A5 status are unclear in most cases. Recently, a specific role for CYP3A5 in the metabolism of tacrolimus has been described, suggesting that CYP3A5 expression status may play a role in the high inter-individual variability seen with many CYP3A substrates.

## Species, Strain, and Gender Differences in Monooxygenase Activity

The activities of cytochromes P450 play a central role in the expression of the toxicity of the majority of xenobiotics. One factor that complicates extrapolation of toxicity between species is the quantitative and qualitative differences in how species metabolize xenobiotics. Generally, the basic reactions and major metabolites of a xenobiotic are similar between species; however, subtle differences in metabolism can lead to major differences in susceptibility to the toxicity of a xenobiotic. Mechanisms that may account for species differences include: (1) lack of a metabolic pathway or a genetic defect in a particular metabolic pathway; (2) differences in the $K_m$ and $V_{max}$ (i.e., level of expression) of specific enzymes; (3) the existence of different isozymes and differences in the ratios of specific isozymes of important enzymes, such as P450; and (4) differences in the ratio of activities of separate enzyme systems that act together to metabolize a specific xenobiotic.

When one metabolite represents a metabolically activated form and another a detoxicated form, the ratio of these metabolites can dictate a species susceptibility to a xenobiotic. This type of species difference is most commonly encountered when the P450-dependent monooxygenase acts in coordination with another pathway. Species may differ in either the initial monooxygenase functionalization reaction or in the activity of the secondary pathway. This is illustrated by the metabolic activation of benzo(*a*)pyrene (BP) (see Figure 3.4) in rats and mice. The metabolic activation of BP requires initial epoxidation by the P450-dependent monooxygenase at the 7,8 position, followed by hydration of the epoxide by epoxide hydrolase to yield the 7,8-diol. This diol is then epoxidated by the monooxygenase to yield the ultimate carcinogen of BP, the 7,8-dihydrodiol 9,10-oxide. When mouse hepatic microsomes were used for metabolic activation in the Ames assay for mutagenicity, BP was highly mutagenic, indicating a high degree of metabolic activation; however, when rat hepatic microsomes were employed in the same assay, only slight mutagenicity was evident. This indicates a significantly lower ability for the rats to metabolically activate BP *in vitro* [192]. Although mice do metabolize BP to a greater extent than rats, rats have six- to sevenfold more epoxide hydrolase activity. Further studies [191,192] indicated that both species have adequate monooxygenase to metabolically activate BP and that higher epoxide hydrolase activity in the rat may have been responsible for the lower mutagenicity; therefore, the species differences in the secondary pathway, epoxide hydrolase, may have controlled the mutagenicity, as opposed to differences in the monooxygenase activity.

Different strains of the same species may demonstrate differences in metabolism; for example, if a different strain of mouse had been used in the studies described above, the conclusions may have been different. It is important to recognize these strain differences when designing toxicological studies. The mechanisms associated with strain differences may be diverse. Without an understanding of these species and strain differences, it will be difficult to extrapolate toxicological studies performed with animals to humans.

Studies of species differences in animals are difficult to design and interpret, and those involving humans are even more complex. This complexity results from the large differences in xenobiotic biotransformation found in humans. Many factors contribute to these individual differences in metabolism, including the following: (1) Humans are free-living and have few restraints to reproductive diversity, diminishing the development of small genetic pools that result in genetically less diverse, more homogeneous control of metabolism; (2) environmental factors, such as diet, nutrition, and xenobiotic exposure, are diverse among humans; and (3) humans generally have more control and probably more interest in consumption of varied non-nutritive materials, such as alcohol or drugs. These, as well as other factors, result in a large diversity in susceptibility to xenobiotic exposure. This is, in part, why such large safety factors are employed in risk or hazard assessments of xenobiotics to which humans may be exposed. These safety factors are used to attempt to protect the vast majority of individuals at risk. For further discussions of species differences, the reader is directed to several recent review articles [303–305].

Gender differences in xenobiotic metabolism may be an important factor in gender-dependent differences in toxicity. The best example of gender differences in xenobiotic metabolism, especially cytochrome P450-mediated metabolism, is the rat. Because the rat is commonly used in toxicological safety assessments, it is important to realize the gender differences in this species and understand how they relate to the extrapolation of rat data to humans.

Generally, male rats have a higher capacity to metabolize xenobiotics than females. This difference is primarily related to the cytochromes P450. Although females have 10 to 30% less total P450 than males, this difference is not high enough to explain the 2- to 20-fold differences seen in metabolism. Much of the differential seen between males and females can be explained by differences in P450 isozymes between the sexes; for example, males express CYP2C11 but females do not. Isozymes that predominate in males are CYP2A1, CYP2A2, and CYP3A2. Adult females also have predominant P450s, such as CYP2C12, which occurs in juvenile and older males but not in young adult males. These differences are under hormonal control and can be altered by procedures such as castration and administration of sex hormones. They also are developmentally controlled, and the stage of life at which these procedures are done can influence their outcome. Neonatal castration of male rats results in different expression of P450 when they become adults. The adult expression of P450s can actually be imprinted during the neonatal period. Although sex hormones play an important role in the expression of P450 in rats, other hormones, including growth hormone, thyroxine, insulin, and somatostatin, may play important roles.

These differences between male and female rats also show an age dependency. As male rats age, their P450 isozyme profiles begin to appear more like females. Toxicologists using rats as a model in safety assessments need to be cognizant of these gender- and age-dependent changes. The toxicity data from young rats, generally used in toxicity studies, may not reflect the toxicity seen in old rats. During chronic toxicity and carcinogenicity studies, the response of rats to the toxicity of a test material may change as the study progresses. This is especially true of carcinogenicity studies that begin with *in utero* exposure. Early developmental changes in P450 in animals and humans may be important in responses to teratogens and embryotoxic compounds [174].

If gender differences in xenobiotic metabolism in rats can complicate toxicity assessments, what about other species, including humans? Rats have been the most intensively investigated species in regard to gender differences; however, studies with other species suggest that they generally do not demonstrate such large gender differences. Mice, another species important in toxicology studies, generally do not show the exaggerated gender differences in xenobiotic metabolism seen in rats. Gender differences in mice seem to be dependent on the specific strain of mouse investigated. Where gender differences do exist in mice, it is generally the female that has the higher metabolic capacity, but the differences are not as great as that seen in rats.

Other species used in toxicological investigations, such as dogs, appear to demonstrate some gender differences in the expression of P450 isozymes, but, again, they are not as exaggerated as in rats. Although there are few reported studies, monkeys have not been reported to demonstrate significant gender differences in xenobiotic metabolism.

Humans have not been shown to demonstrate gender-dependent differences in the expression of P450 isozymes. Although there can be significant differences between human males and females in xenobiotic biodisposition, these appear to be more based on anatomical and physiological differences that affect absorption, distribution, and excretion. Individual humans can display large differences based on life-style factors and exposure to environmental chemicals, foods, and drugs; however, inherent gender differences in the expression of P450 are not apparent.

This raises the question as to how species, such as rats, can be used to predict toxicity in humans. With care and knowledge of the differences between rats and humans, the rat can serve as a model for human toxicity. This has been shown through decades of use; for example, rats and humans share similarities between the CYP isozyme subfamilies CYP1A1, CYP1A2, and CYP2E1, and these subfamilies are not expressed in a highly gender-dependent manner in rats. On the other hand, CYP2C is not found in humans but is a major subfamily in rats. Xenobiotics metabolized by CYP1A or CYP2E may reflect human metabolism because regulatory control over these isoforms has been highly conserved between rodents and humans; however gender-dependent differences are generally not reflected when extrapolating from rats to humans [181]. Gender-dependent differences in xenobiotic metabolism are but one of the reasons toxicologists must use both sexes in the safety assessment of chemicals. Gender differences in xenobiotic metabolism have recently been reviewed [181].

## Microsomal Flavin-Containing Monooxygenase

Since 1960, it has been apparent that a microsomal monooxygenase other than P450 could catalyze the oxygenation of nucleophilic nitrogen, sulfur, and phosphorus compounds. Purification to homogeneity indicated it was

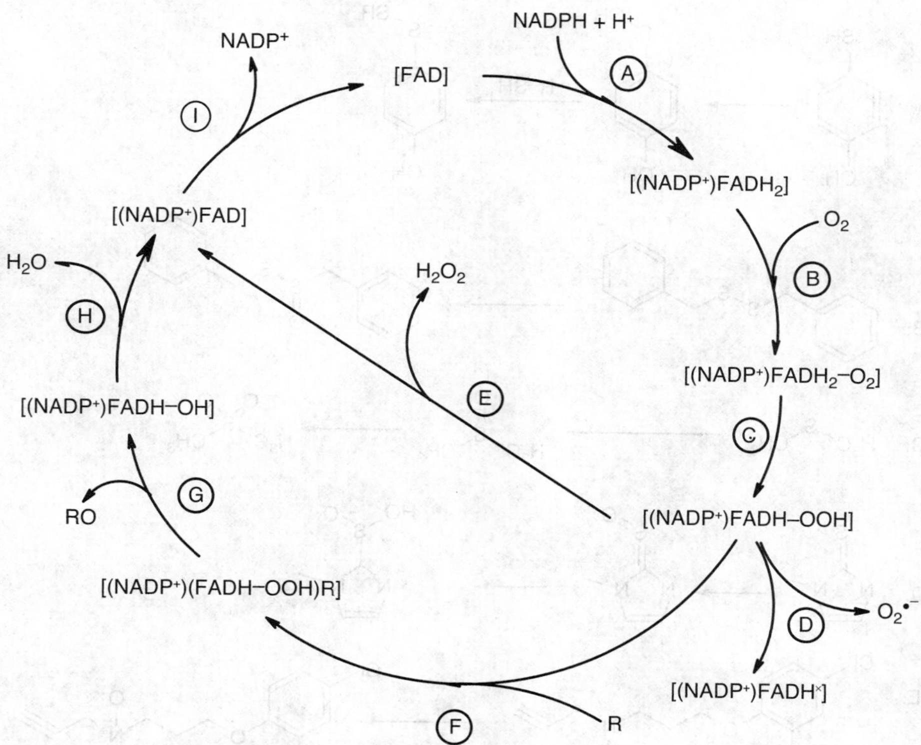


**FIGURE 3.7** Catalytic cycle of the flavin-containing monooxygenases (FMO). See text for details. (Adapted from Masnil, M. and Testa, B., in *Advances in Drug Research*, Vol. 13, Testa, B., Ed., Academic Press, London, 1984, pp. 95–207.)

an NADPH-dependent, flavin-containing monooxygenase distinct from P450. This monooxygenase has been referred to as amine oxidase, Ziegler enzyme, dimethylaniline monooxygenase, and flavin-containing monooxygenase. This enzyme may be a good example of proteins involved in normal anabolic and catabolic metabolism being recruited for xenobiotic metabolism. The flavin prosthetic group that is characteristic of these enzymes is especially versatile at carrying out redox functions.

The catalytic cycle for the flavin-containing monooxygenase is shown in Figure 3.7. NADPH reduces the FAD of the enzyme, and the oxidized $NADP^+$ remains bound (Figure 3.7A). Oxygen then binds to yield FAD peroxide (Figure 3.7B,C) followed by substrate binding (Figure 3.7F). Prior to substrate binding, the peroxide complex can decompose, releasing superoxide (Figure 3.7D). One oxygen atom from the peroxide is transferred to the substrate, leaving the hydroxyflavin (Figure 3.7G). The final and rate-limiting step of the cycle is the dehydration to regenerate FAD, yield water, and release the bound $NADP^+$ (Figure 3.7H,I). NADH can substitute for NADPH, but with lower affinity and activity.

The flavin-containing monooxygenase has at least five isoforms (designated as FMO1 to FMO5), whose genes are expressed across several species and tissues. These forms have different substrate specificities and are probably related to the species-dependent toxicity of certain substrates, such as the pyrrolizidine alkaloids. In addition to the five functional forms of FMO, at least six other FMO genes have been identified, all of which appear to be pseudogenes [38]. Humans, rats, and mice have relatively high activity of FMO1 in kidney. Humans and mice have low activity for this form in the liver, whereas the rat has high activity. This isoform is also expressed in human lung and brain. FMO2 is the primary isoform expressed in human lung, with lower amounts found in kidney, liver, small intestine, and brain. Humans and mice have high activity of FMO3 in the liver, whereas the rat has low activity. FMO3 activity has also been detected in human kidney, lung, and small intestine, and rat and mouse kidney both show high activity. In mice, only females express FMO3 and have two- to threefold higher activity of FMO1 compared to males. Human FMO4 expression levels are highest in liver and kidney, with lesser amounts in small intestine and lung and very low levels in brain. FMO5 is quantitatively the most important isoform in human liver and is also expressed in small intestine, kidney, and lung to a lesser, though significant, degree. FMO5 shows no gender differences. Male rats have higher total FMO than females and two- to threefold more FMO1 than females, but no differences are seen in FMO3 [44]. Humans can show

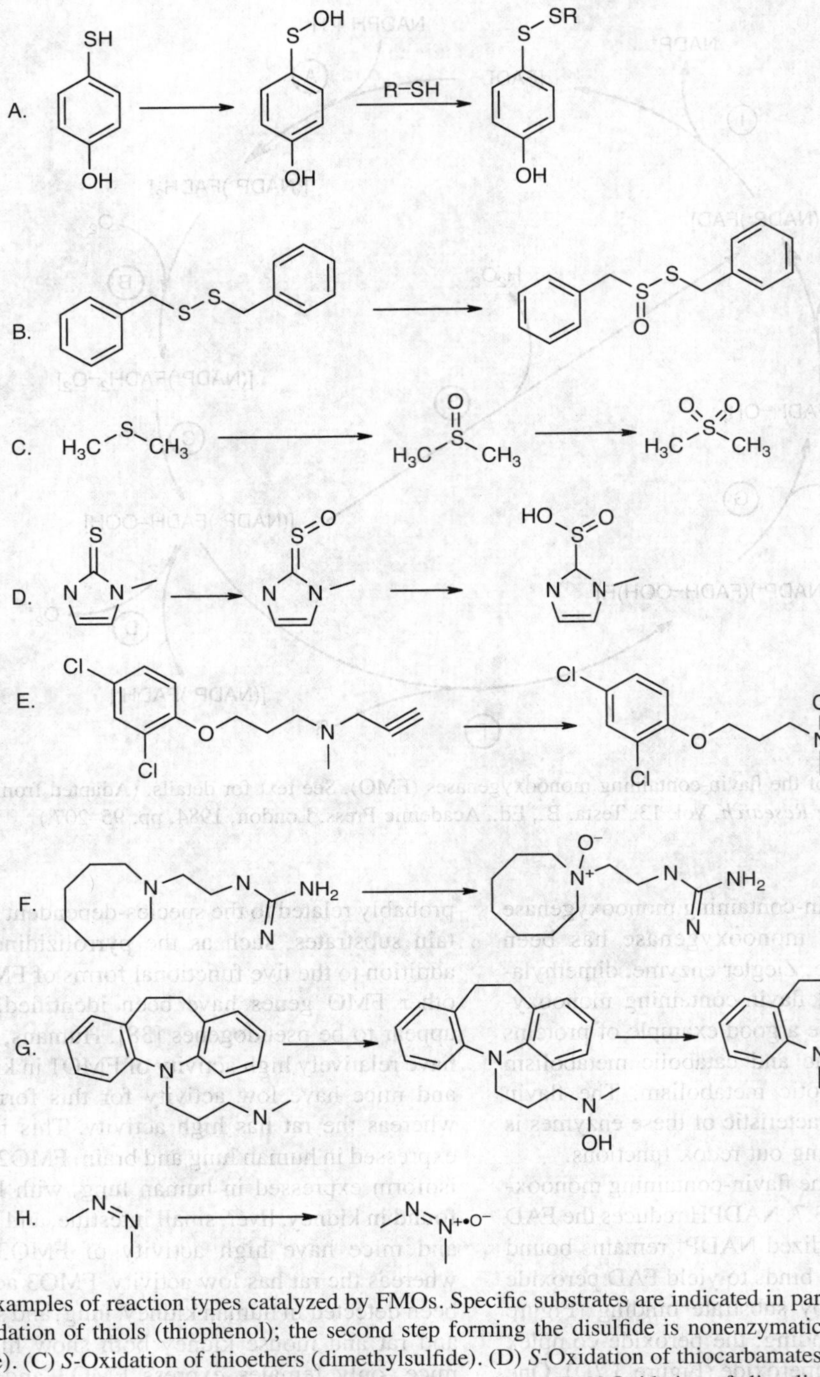

**FIGURE 3.8** Examples of reaction types catalyzed by FMOs. Specific substrates are indicated in parentheses following the reaction type: (A) *S*-Oxidation of thiols (thiophenol); the second step forming the disulfide is nonenzymatic. (B) *S*-Oxidation of disulfides (benzyl disulfide). (C) *S*-Oxidation of thioethers (dimethylsulfide). (D) *S*-Oxidation of thiocarbamates, thiocarbamides, and thioureas (methimazole). (E) *N*-Oxidation of aliphatic tertiary amines (clorgyline). (F) *N*-Oxidation of alicyclic tertiary amines (guanethidine). (G) *N*-Oxidation of aliphatic secondary amines (desipramine). (H) *N*-Oxidation of azo compounds (azomethane).

considerable individual differences, but no gender differences have been demonstrated. This information illustrates the species, gender, and tissue differences that can be encountered with this monooxygenase and emphasizes the importance of choosing an appropriate animal model for toxicological studies of compounds that are potential substrates for this enzyme.

Flavin-containing monooxygenases catalyze the oxidative attack on the nucleophilic nitrogen and sulfur heteroatom of a variety of xenobiotics (Figure 3.8) [300]. It was once believed that oxidations of basic aliphatic and tertiary aromatic amines were carried out by the FMO, primary aromatic amines and the acidic nitrogens of amides were catalyzed by P450, and secondary amines were oxidized by both enzyme systems. More recent studies with the purified enzymes have demonstrated no clear division between the types of substrates preferred by the two enzymes; therefore, the metabolism of each nitrogen-containing xenobiotic must be considered on an individual basis. The thermal instability of the flavin-containing monooxygenase in the absence of NADPH (above 35°C) has provided a tool to separate the activity of this enzyme *in vitro* from that of P450; however, defining the relative contribution of the flavin-containing monooxygenase and P450 to the metabolism of many xenobiotics is difficult because some inhibitors of P450, such as SKF-525A, are substrates for the flavin-containing monooxygenases. More selective inhibitors of P450, such as *N*-benzylimidazole and aminobentotriazole, are a better choice for distinguishing these two enzymes [300]. Antibodies to specific P450 isozymes can be used to inhibit P450 and determine the role of the flavin-containing monooxygenase.

Many nitrogen- and sulfur-containing xenobiotics are metabolized by this phase I enzyme, as seen in Figure 3.8. N-Oxidation of nucleophilic tertiary amines yields *N*-oxides, and primary and secondary amines are oxidized to hydroxylamines. In addition, primary amines can be oxidized to oximes and secondary amines to nitrones. Thiols, thioethers, and other xenobiotics containing sulfur can be oxidized to sulfur oxides. In addition to the functional groups shown in Figure 3.8, FMOs are also capable of oxidizing organic phosphines, boronides, selenides, and iodides. The flavin-containing monooxygenases have a relatively broad substrate specificity, but individual isozymes demonstrate some specificity. Broad substrate specificity and its occurrence in several tissues indicate that it can be a major determinant in oxidative xenobiotic metabolism.

Transcriptional regulation of FMO has received much less attention than the regulation of P450s, and the mechanisms governing the expression of FMOs are unclear. Basal expression of FMOs is under hormonal control. To date, little evidence exists to suggest xenobiotic-mediated induction of FMOs. Few isoform-selective FMO substrates are known, although stereoselective *N*-oxidation of nicotine and *S*-oxidation of cimetidine have been used as marker activities for FMO3. Further, *N*-oxidation of trimethylamine is catalyzed by FMO3, and a genetic defect in this isoforms results in trimethylaminuria (fish odor syndrome). Few selective inhibitors of FMOs are known, although indole-3-carbinol and *N,N*-dimethylaminostilbene carboxylates have been used for this purpose. It has been suggested that targeting drugs for FMO-mediated metabolism may result in fewer adverse drug reactions, due to the lack of induction and selective inhibition of FMOs [38]. Numerous single nucleotide polymorphisms for FMOs have been identified, but because of the limited role of FMO in drug metabolism the clinical significance of polymorphic FMO expression is uncertain.

## Cooxidation of Xenobiotics by Prostaglandin H Synthase and Other Peroxidases

Pathways other than the monooxygenases may be involved in xenobiotic oxidation. These include myeloperosidases, eosinophil peroxidase, uterine peroxidase, lactoperoxidase, thyroid peroxidase, and the prostaglandin synthases. Prostaglandin synthase, also known as cyclooxygenase, is the initial enzyme in arachidonate metabolism and the formation of prostanoids such as prostaglandins, prostacyclins, and thromboxanes. Marnett and Reed [166] demonstrated that prostaglandin H synthetase, an enzyme system responsible for prostaglandin biosynthesis, was capable of oxidizing benzo(*a*)pyrene to quinines (Figure 3.4). The following cycle of reactions is involved in the oxidation of xenobiotics [190]:

$$\text{Peroxidase} + \text{ROOH} \rightarrow \text{Compound I} + \text{ROH}$$

$$\text{Compound I} + \text{ROH} \rightarrow \text{Compound II} + \text{RO}^-$$

$$\text{Compound II} + \text{ROH} \rightarrow \text{Peroxidase} + \text{RO}^- + H_2O$$

The cyclooxygenase activity of prostaglandin synthase catalyzes the oxygenation of arachadonic acid to form the hydroperoxy endoperoxide prostaglandin $G_2$ ($PGG_2$). With a xenobiotic acting as an electron donor, $PGG_2$ is reduced to the hydroxyl endoperoxide $PGH_2$, with the coordinate oxidation of the hydroxyl group of the xenobiotic as illustrated in Figure 3.7, for acetaminophen cooxidation.

Two prostaglandin H synthase enzymes, PGHS-1 and PGHS-2, have been characterized, and both are homodimeric integral membrane proteins. PGHS-1 and PGHS-2 are localized to the luminal surface of the endoplasmic reticulum, and PGHS-2 is also found in the inner and outer membranes of the nuclear envelope. Both enzymes share about 60% primary sequence identity but are encoded by separate genes on separate chromosomes. They are under regulatory control of cytokines and growth factors [277]. Two catalytic activities copurify with the synthase: fatty acid cyclooxygenase and prostaglandin hydroperoxidase. The cyclooxygenase catalyzes arachidonic acid oxidation to prostaglandin $G_2$, and the hydroperoxidase reduces the hydroperoxide (–OOH) to the corresponding alcohol in prostaglandin $H_2$, as shown in Figure 3.9. Oxidation of xenobiotics results from a one-electron pathway involving an oxidizing agent produced during the hydroperoxidase-catalyzed reduction of prostaglandin $G_2$ to the hydroxyl endoperoxide $PGH_2$. Prostaglandin synthetase is a major source of alkyl hydroperoxides produced during normal metabolism [167]. Most tissues possess prostaglandin synthetase activity and are capable of oxidizing certain xenobiotics, even if the tissue is low in P450 content. In fact, acetaminophen, which is activated to a reactive intermediate by

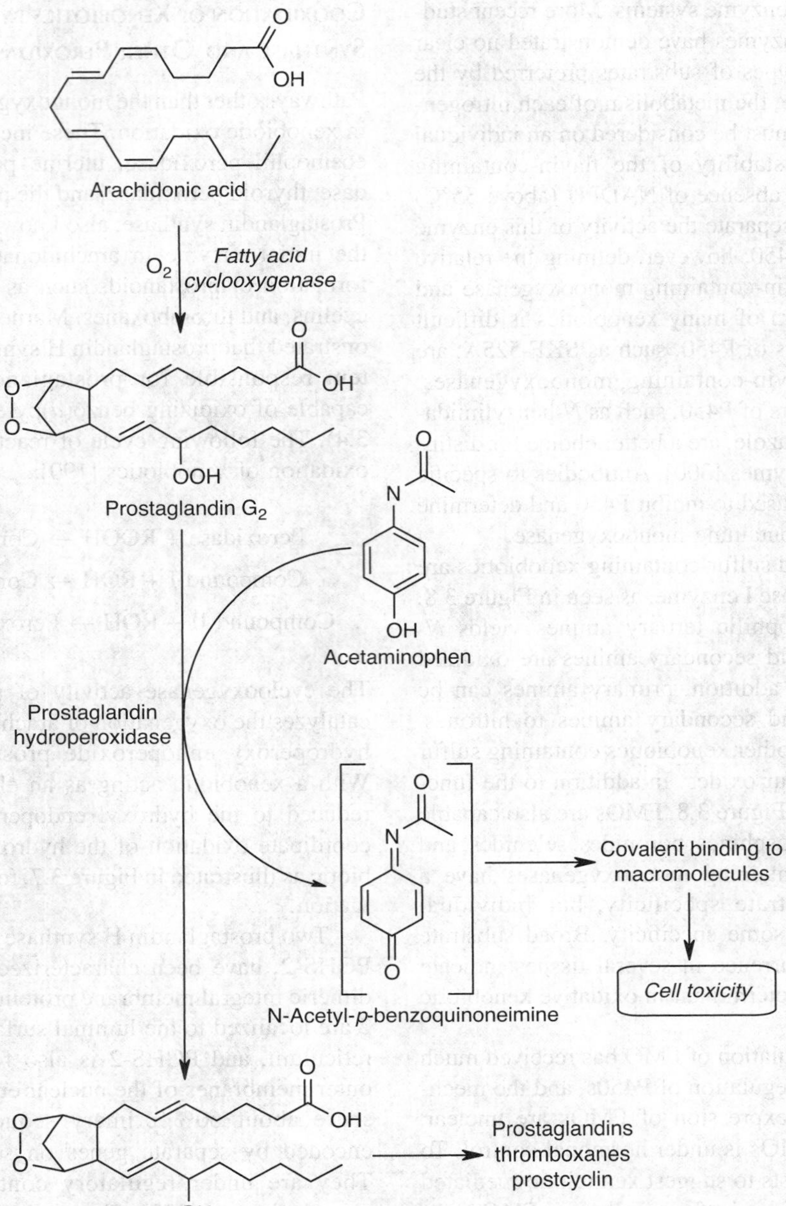

**FIGURE 3.9** Cooxidation of acetaminophen by prostaglandin endoperoxide synthase.

P450, can also be activated by prostaglandin synthetase in the medulla of the kidney. This tissue is low in P450 activity, but in the presence of arachidonic acid the medulla activates acetaminophen to a reactive intermediate that covalently binds to tissue macromolecules [29]. The localization of prostaglandin synthetase in the inner medulla and papilla may be a contributing factor to the toxicity produced by other chemicals in this region of the nephron [152]. Other compounds that undergo cooxidation include aminopyrine, benzphetamine, oxyphenbutazone, benzidine, and benzo(*a*)pyrene.

In addition to kidney, other extrahepatic tissues including bladder, intestinal mucosa, spleen, and blood cells, such as peripheral blood mononuclear cells, and macrophages possess prostaglandin synthase activity. The bladder also possesses high prostaglandin synthetase activity. Mattammal et al. [169] proposed that several structurally diverse renal and bladder carcinogens are metabolically activated by prostaglandin synthetase; for example, the bladder carcinogen 2-amino-4-(5-nitro-2-furyl)thiazole is believed to be activated by prostaglandin synthetase cooxidation in bladder transitional epithelium

**TABLE 3.4**
**Human Peroxidases**

| Peroxidases | Cells | Subcellular Location |
|---|---|---|
| MPO | Neutrophils, leukocytes | Lysosomes |
| EPO | Eosinophils | Lysosomes |
| LPO | Mammary ductal epithelial cells, secretory cells of exocrine glands | Extracellular milk, saliva, tears |
| TPO | Thyroid follicular cells | Rough endoplasmic reticulum, Golgi apical membrane, perinuclear membrane |
| PGHS-1 | Platelets, seminal vesicles | — |
| PGHS-2 | Inflammatory cells | — |

*Note:* MPO, myeloperoxidase; EPO, eosinophil peroxidase; LPO, lactoperoxidase; TPO, thyroid peroxidase; PGHS-1, prostaglandin H synthase 1; PGHS-2, prostaglandin H synthase 2.

Source: Adapted from O'Brien, P.J., *Chemico-Biol. Interact.*, 129, 113–139, 2000.

to metabolites capable of covalently modifying RNA and DNA. Feeding aspirin to rats can inhibit the bladder lesion induced by 5-nitrofuran, the ultimate carcinogen. This suggests that prostaglandin synthetase is involved in the metabolic activation, as aspirin is a specific inhibitor of prostaglandin synthetase.

Use of the analgesic *p*-phenetidine has declined because of reports of kidney damage in humans following prolonged use of the drug. Andersson et al. [7] proposed a mechanism by which phenetidine is activated by prostaglandin synthetase in the kidney. The primary amine nitrogen of phenetidine undergoes a one-electron oxidation similar to that shown in Figure 3.9 for acetaminophen. This leads to hydrogen abstraction, yielding a reactive nitrenium radical. A radical intermediate is postulated based on its rate of reaction phenacetin with reduced glutathione in the presence of arachidonic acid and microsomes from sheep seminal vesicles. Benzene can be hydroxylated to phenol in the liver by P450, and the phenol can be further oxidized to hydroquinone. The phenol and hydroquinone can enter the blood stream and be distributed to other tissues. In the bone marrow, the phenol stimulates prostaglandin synthetase peroxidative activation of hydroquinone to reactive metabolites that form adducts with nucleophiles, such as protein and DNA. This is believed to result in the bone marrow suppression seen with chronic exposure to benzene. Phenolic compounds may be converted to reactive phenoxyl radicals by the one-electron oxidative process.

Another example of a PGHS-2-mediated metabolic activation reaction is the biotransformation of procainamide. Drugs such as procainamide undergo an *N*-acetylation reaction as a primary means of elimination; however, as discussed in another section, individuals with the slow acetylator phenotype do not readily eliminate procainamide via the *N*-acetylation pathway, leaving more drug to reach extrahepatic tissues. It is in tissues such as monocytes and macrophages that PGHS-2 can oxidize procainamide to the hydroxylamine and nitroso derivatives. These reactive molecules are proposed to form haptens and subsequently sensitize T-cells. The enhanced neoantigen formation and T-cell sensitization seen in slow acetylators might be explained by the higher concentration of procainamide that is available for extrahepatic *N*-oxidation in antigen-presenting cells [89].

In the developing embryo and other conceptal tissues, levels of cytochrome P450 expression are very low, especially in the first trimester. Oxidative metabolism, however, can proceed through peroxidative mechanisms dependent on prostaglandin synthase as well as lipoxygenase, peroxidase, and lipid peroxidation-coupled cooxidation [134,190]. Metabolic activation of xenobiotics to toxic intermediates through these mechanisms may be responsible for certain terata. Phenytoin, an antiepileptic drug and known teratogen, was shown *in vitro* to be less effective when mouse embryos were cultured in the presence of inhibitors of the prostaglandin synthase and lipoxygenase pathways.

Myeloperoxidase (MPO), eosinophil peroxidase (EPO), and lactoperoxidase (LPO) are unique among the peroxidases in that they are primarily found in lysosomes of neutrophils, eosinophils, and secretory cells of the exocrine glands, respectively (Table 3.4) [190]. Neutrophil MPO catalyzes the oxidation of halides by hydrogen peroxide to produce hypohalous acid. Leukemias induced by benzene exposure have been attributed to DNA prooxidant phenoxyl radicals formed by the MPO/$H_2O_2$-mediated oxidation of the benzene CYP2E1 product phenol. Thyroid peroxidase (TPO) is a membrane-bound enzyme localized to the thyroid follicular cells and is under regulatory control of thyroid-stimulating hormone. TPO catalyzes the iodination of thyroglobin tyrosine residues.

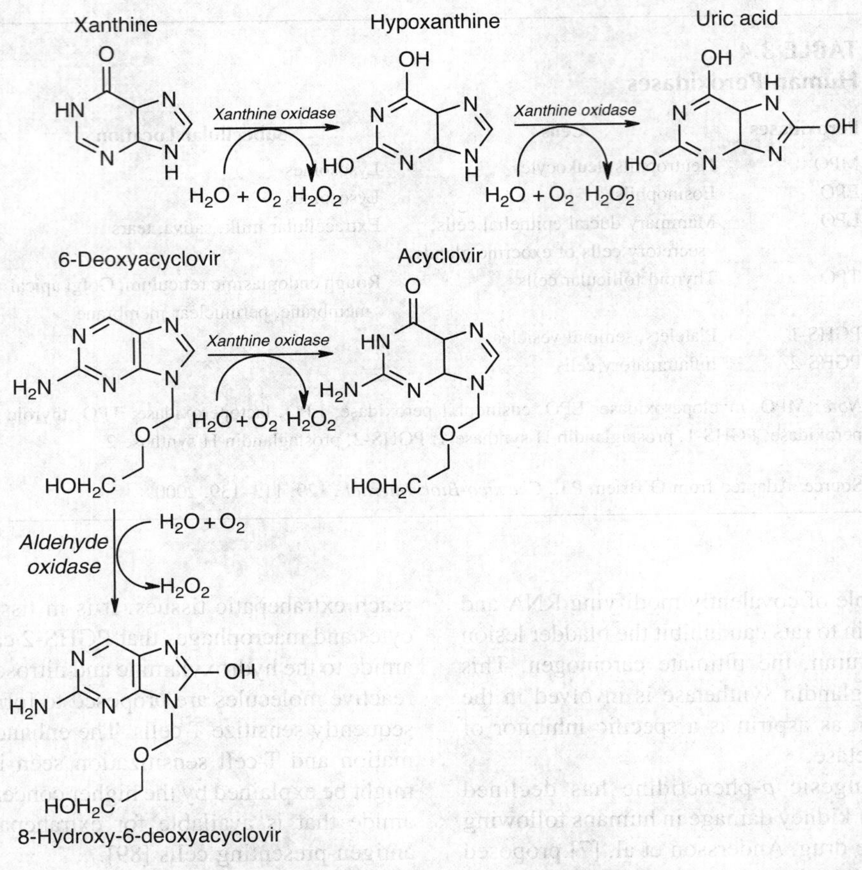

**FIGURE 3.10** Oxidation of xanthine, hypoxanthine, and 6-deoxyacyclovir by xanthine oxidase or aldehyde oxidase using molecular oxygen as the electron acceptor.

## Other Enzymes Associated with Oxidative Metabolism

### Xanthine Oxidoreductase

Xanthine oxidase and xanthine dehydrogenase are members of the molybdenum hydroxylase flavoprotein family commonly referred to as the *xanthine oxidoreductase* (XOR) family [207]. Oxidation reactions carried out by xanthine oxidase and aldehyde oxidase (discussed below) are different from that of cytochrome P450-catalyzed oxidations in that hydroxylation of the substrate is derived from water rather than molecular oxygen.

Xanthine oxidase and xanthine dehydrogenase are actually different enzymes derived from the same gene product. Conversion of the dehydrogenase to the oxidase involves oxidation of critical protein thiol groups followed by the cleavage of a 20-kDa fragment from each of two subunits. Although both forms of the enzyme have been recognized for years, comparatively little information exists for this enzyme system, especially the dehydrogenase. XOR has been identified in tissues from all species studied to date, and among mammals the highest activity is found in lactating mammary gland and cow's milk, liver, and intestine. The predominant form is the dehydrogenase, and both forms are localized in the cytoplasm. Human XOR is generally less active than that of other species.

The XOR enzyme system is the rate-limiting enzyme in purine catabolism but is also well known to metabolize xenobiotics. XOR carries out the oxidation of hypoxanthine to xanthine and xanthine to uric acid (Figure 3.10). Anticancer drugs including substituted and unsubstituted purines, pyrimidines, pteridines, azopurines, and heterocyclic compounds such as doxorubicin and menadione are well-known substrates of xanthine oxidase. More recently the generation of nitric oxide from *S*-nitrosothiols and nitrite has been demonstrated.

Both the oxidase and reductase forms are capable of metabolizing xenobiotics with the preference determined by the preference of each enzyme for a different electron acceptor [295]. The oxidase utilizes molecular oxygen as an electron acceptor with negligible reactivity toward $NAD^+$. Reoxidation of the oxidase enzyme takes place via two one-electron reductions of molecular oxygen to yield hydrogen peroxide. On the other hand, the dehydrogenase utilizes $NAD^+$ as an electron acceptor to produce NADH through a two-electron reduction. Thus, the xanthine dehydrogenase enzyme has been shown to participate in the redox cycling of doxorubicin and menadione, resulting in the formation of their hydroquinones, which are generally unstable and generate oxygen radicals. The efficient utilization of oxygen and the production of oxygen radicals have been proposed to contribute to the cytotoxic action of these drugs [295].

## Amine Oxidases

Amine oxidases can play a significant role in the metabolism of specific xenobiotics [253]. Monoamine oxidase (MAO) and related amine oxidases catalyze the oxidative deamination of endogenous amines. They can also be involved in the metabolism of primary, secondary, and tertiary xenobiotic amines. Two of the amine oxidases (MAO and semicarbazide-sensitive amine oxidase) will be used as examples of amine oxidases.

Most tissues express two forms of the mitochondrial enzyme MAO (termed MAO-A and MAO-B), each being expressed by a separate gene. Although MAOs are expressed in most tissues, expression in various tissues is isoform specific. The highest concentrations of both isoforms are found in the liver and gastrointestinal tract. Only MAO-B is expressed in human platelets. MAO is a flavoprotein capable of oxidative deamination of primary, secondary, and tertiary amines. Metabolism of primary amines yields an aldehyde and ammonia, whereas secondary amines yield an aldehyde and a primary amine. The aldehyde products may be further metabolized by other enzymes to carboxylic acids or alcohols. Unlike the monooxygenases, the oxygen used in the reaction is derived from water. During the oxidation, the FAD prosthetic group is reduced (FAD $\rightarrow$ $FADH_2$) then reoxidized by oxygen with the production of hydrogen peroxide.

Several amine drugs have been shown to be substrates for MAO. Some of these act as pro-drugs and require MAO metabolism to produce the active form; others have their activities limited by MAO metabolism. MAO-A and MAO-B have different substrate specificities, but overlaps in specificity can occur.

Induction of MAO by drugs or other xenobiotics has not been observed. Basal expression of MAOs is under hormonal control and can be perturbed by steroid analogs such as prednisone. Nonselective inhibitors of MAO include iproniazid and phenelzine, and selective inhibition of MAO-A and MAO-B can be achieved with clorgyline and pargyline, respectively. Although these inhibitors are useful for *in vitro* diagnostic studies of drug metabolism, they also inactivate some CYP isoforms, diminishing their usefulness as *in vivo* probes of MAO involvement. Polymorphisms of both MAO isoforms are known, although the consequences for drug metabolism and toxicity are not well understood. MAOs play a role in the metabolism of a variety of compounds, including β-adrenergic agonists/antagonists and phenyethylamine derivatives such as mescaline. The quantitative contribution of MAOs to xenobiotic clearance is unknown but is likely to be low compared to P450s.

A well-publicized example of an MAO-related toxicity was initiated by individuals attempting to synthesize a narcotic related to demerol. Instead of the intended product, 1-methyl-4-phenyl-1,2,3,6-tetrahydropyridine (MPTP) resulted from the synthesis. Individuals who self-administered MPTP demonstrated symptoms similar to those for Parkinson's disease. This was related to neurocytotoxicity in dopaminergic neurons produced by brain MAO-B metabolism of MPTP to 1-methyl-4-phenyl-2,3-dihydropyridine ($MPDP^+$), which oxidizes to the neurotoxic 1-methyl-4-phenylpyridine ($MPP^+$). The cytotoxicity of $MPP^+$ results from its inhibition of mitochondrial respiration.

Semicarbazide-sensitive amine oxidases (SSAOs), like monoamine oxidase, catalyze the oxidative deamination of endogenous amines but can also metabolize xenobiotic amines [159,253]. The SSAOs do not contain a flavin but do contain copper. They demonstrate a more limited activity than MAO by only catalyzing deamination of primary aliphatic and aromatic monoamines. They are sensitive to inhibition by semicarbazide but insensitive to the classic MAO inhibitors. The products of their reaction are an aldehyde, ammonia, and hydrogen peroxide. They occur in most species, including bacteria, fungi, plants, and animals. In animals, they occur in plasma and may be bound to tissues. Although they can metabolize several endogenous substrates [296], their exact physiological role is currently unknown.

Considerable species differences exist for SSAO; for example, rats have relatively low concentrations of plasma SSAO compared to humans. SSAO can metabolize certain xenobiotics to more toxic metabolites. 3-Aminopropene has been used in the manufacture of pharmaceuticals and in rubber vulcanization; chronic exposure to this compound can produce lesions similar to acute myocardial necrosis and atherosclerosis. SSAO appears to metabolize 3-aminopropene to 2-propenal (acrolein), which alkylates and inactivates glutathione *S*-transferase and allows excessive peroxidative damage [99]. Damage occurs in the heart and aortic tissue, which have high SSAO activity [51]. The tissue specificity of this effect is related to relatively high expression levels of SSAO in heart. This is in contrast to acrolein produced from allyl alcohol by alcohol dehydrogenase, which results in liver toxicity (discussed below).

## Aldehyde Oxidase

Aldehyde oxidase is similarly a cytosolic, sulfur-containing molybdenum hydroxylase, is closely related structurally and catalytically to xanthine oxidase, and is present in highest quantities in liver, particularly of the rabbit. Despite its ability to oxidize aldehydes *in vitro*, especially

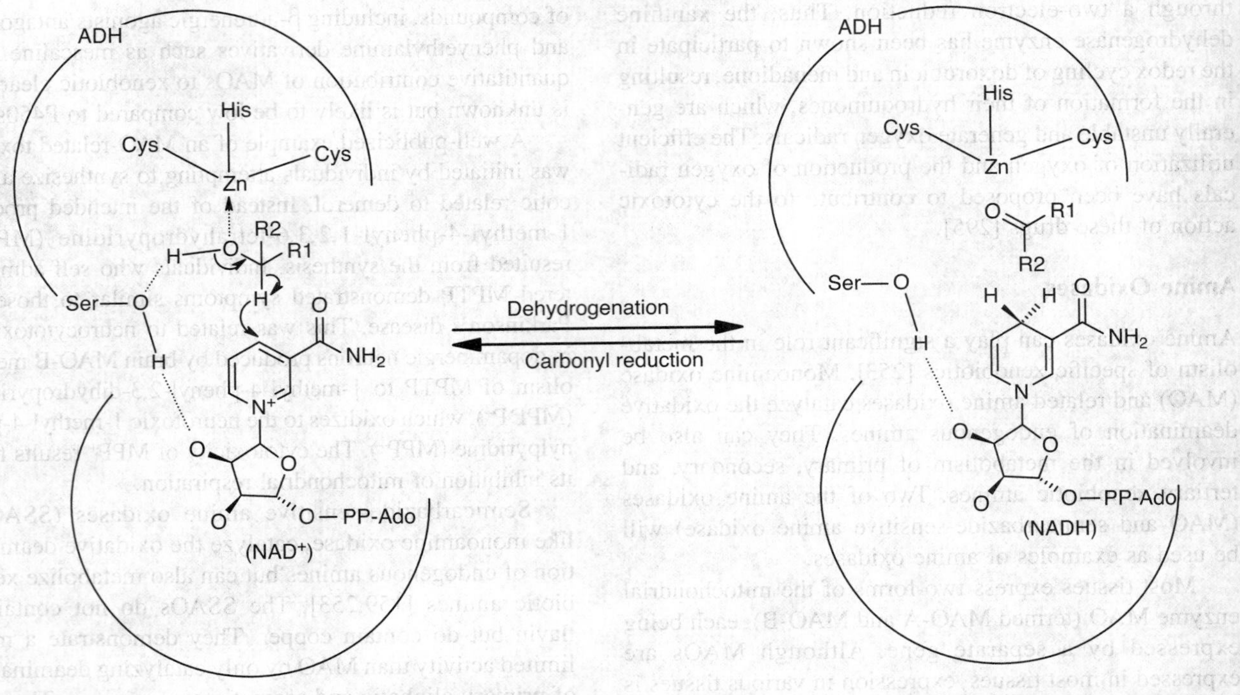

**FIGURE 3.11** Catalytic mechanism of cytosolic alcohol dehydrogenase. Initial deproteination of the alcohol group is accomplished by a proton shuttle involving an active site serine, the ribosyl group of the cofactor, and an active site histidine residue (not shown). As shown, the reaction is reversible so ADH can also function as a carbonyl reductase.

aromatic aldehydes (e.g., vanillin to vanillic acid), this misnamed enzyme preferentially catalyzes the oxidation of purines and other heterocyclic amines. Unlike xanthine oxidase, aldehyde oxidase is able to catalyze hydroxylations at the C-8 position of purines (Figure 3.10); thus, the pro-drug 6-deoxyacyclovir undergoes hydroxylation at the C-8 position by aldehyde oxidase to yield active acyclovir [133]. The most efficient substrates for aldehyde oxidase are aromatic heterocycles with two fused six-membered rings [199]. There is also evidence that iminium ions can be metabolized to lactams by aldehyde oxidase [275].

## Alcohol and Aldehyde Dehydrogenases

### Alcohol Dehydrogenase

Alcohol dehydrogenases (ADHs) catalyze the NAD$^+$-dependent oxidation of primary and secondary alcohols to aldehydes and ketones, respectively [261]. ADHs are dimeric cytosolic proteins with a molecular weight of approximately 40,000 and contain one structural and one catalytic zinc atom. Human ADHs are encoded by 6 different genes (ADH1 to ADH6), each of which codes for an individual subunit (designated $\alpha$, $\beta$, $\gamma$, $\pi$, $\chi$, and $\sigma$). The $\alpha$, $\beta$, and $\gamma$ subunits have >90% sequence homology and can thus form both homodimers and heterodimers. The $\pi$ and $\chi$ subunits have lower homology and can only form homodimers. ADHs are divided into four classes (I, II, III, and IV) based on their subunit composition. Class I ADHs, composed of $\alpha$, $\beta$, and $\gamma$ hetero- and homodimers, are the most important isoforms involved in ethanol metabolism. Class I ADHs are expressed at high levels in liver and adrenal gland and at lower levels in a variety of other tissues. Class IV ADH, composed of $\sigma$ homodimers, is expressed primarily in the gastrointestinal tract in adult humans and may be involved in the induction of gastrointestinal cancer following chronic alcohol abuse [229].

The catalytic mechanism involves initial deproteination of the hydroxyl group, followed by hydride transfer to the NAD$^+$ cofactor (Figure 3.11). Enzyme activity can be followed by monitoring formation of NADH spectrophotometrically, and this has been used as a convenient nonspecific assay for ADH activity. ADHs have wide substrate specificity and can catalyze the dehydrogenation of a variety of primary and secondary aliphatic alcohols and aromatic alcohols, as well as diols and aminoalcohols [27,126,204]. Primary alcohols are more readily dehydrogenated compared to secondary alcohols, and within a series catalytic efficiency appears to be correlated with lipophilicity. Pyrazole and 4-methylpyrazole have been used as selective inhibitors of ADH, although at higher concentrations these compounds can inhibit P450 activity as well.

Alcohol dehydrogenases are active in the metabolism of a variety of drugs and other xenobiotics. The most obvious example is conversion of ethanol to acetaldehyde which is detoxified by aldehyde dehydrogenases (dis-

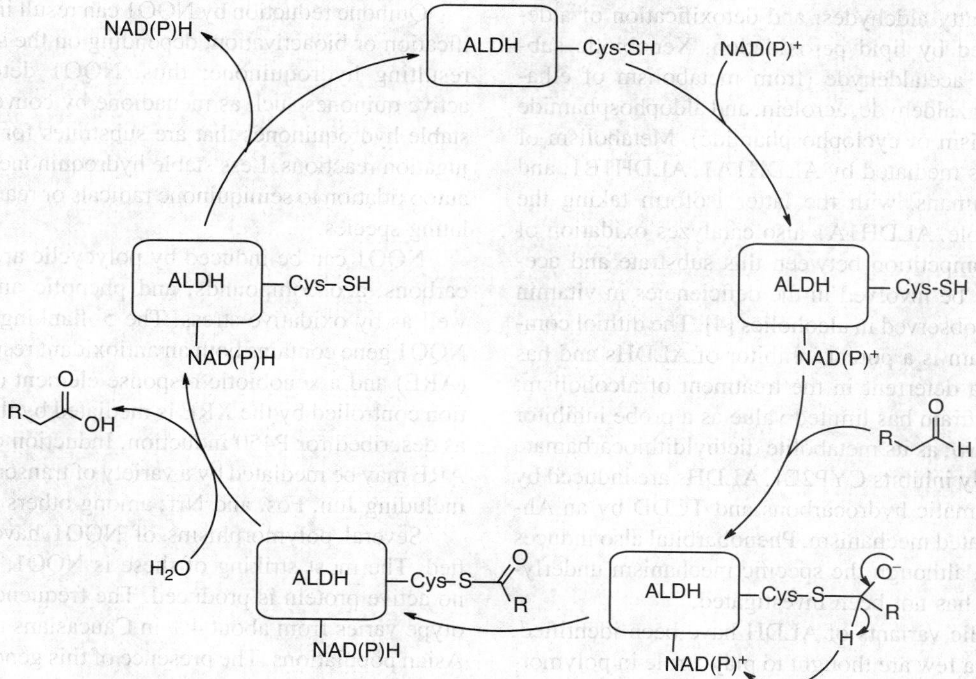


**FIGURE 3.12** Catalytic mechanism of cytosolic aldehyde dehydrogenase. The oxidation of aldehydes to carboxylic acids occurs in two steps: dehydrogenation and thioester hydrolysis. By virtue of the second activity, ALDH can also function as an esterase.

cussed below). ADH mediates toxicity of a number of alcohol-containing toxicants; for example, ADHs oxidize methanol to formaldehyde, which is converted to formic acid, resulting in metabolic acidosis. Similarly, oxidation of ethylene glycol to glyoxal, which is ultimately converted to oxalic acid, results in kidney toxicity [113]. The ADH inhibitor 4-methylpyrazole is used to treat accidental and intentional ingestion of methanol and ethylene glycol. A third example is the ADH-mediated dehydrogenation of allyl alcohol to the hepatotoxin acrolein [11].

Expression of ADH is polymorphic, with allelic variants occurring at the ADH2 and ADH3 loci (β and γ subunits). Only a single allele has been identified for ADH1. ADH2*2 isoforms, containing at least 1 $\beta_2$ allele and collectively known as *atypical ADH*, occur at a frequency of approximately 90% in Pacific rim Asian populations and are responsible for the rapid oxidation of ethanol to acetaldehyde in these individuals [1]. These isoforms are found at a much lower frequency in Caucasians and African-Americans. ADH3*1 and ADH3*2 variants also occur at a high frequency in Pacific rim Asians, but these variants have little impact on ethanol metabolism.

## Aldehyde Dehydrogenase

Aldehyde dehydrogenases (ALDHs) are the major enzymes responsible for oxidation of aldehydes to carboxylic acids. ALDHs are found in the cytosol, mitochondria, and endoplasmic reticulum, and both constitutive and inducible forms are expressed in numerous tissues [261]. ALDHs exist as tetramers with an approximate molecular weight of 200,000 to 250,000 Da. ALDHs can use either $NAD^+$ or $NADP^+$ as a cofactor, and cofactor preference is isoform specific. ALDH genes are classified into families and subfamilies based on sequence homology. Isoforms with >40% homology are assigned to the same family (designated by an Arabic number), while enzymes with >60% similarity are assigned to the same subfamily (designated by a letter). Individual subfamily members are designated by an Arabic number. At least 17 ALDH genes have been identified in humans. These are arranged into 10 families and 13 subfamilies [273].

The mechanism of ALDH catalysis is illustrated in Figure 3.12. In the initial step, the NAD(P) cofactor binds with the enzyme, followed by binding of the aldehyde substrate, which forms a covalent bond with an active site cysteine sulfhydryl group. The second step is transfer of a hydride ion from the substrate to the pyridine moiety of NAD(P), which effectively oxidizes the substrate to an acyl compound. In the final step, the acyl compound is hydrolyzed, giving rise to a carboxylic acid.

Similar to ADH, ALDH can oxidize a broad array of substrates, including aliphatic and aromatic aldehydes and dialdehydes. ALDHs are involved in a number of endogenous biosynthetic pathways, such as the synthesis of retinoids, amino acids, and neurotransmitters; metabolism of folate and fatty aldehydes; and detoxification of aldehydes generated by lipid peroxidation. Xenobiotic substrates include acetaldehyde (from metabolism of ethanol), *p*-nitrobenzaldehyde, acrolein, and aldophosphamide (from metabolism of cyclophosphamide). Metabolism of acetaldehyde is mediated by ALDH1A1, ALDH1B1, and ALDH2 in humans, with the latter isoform taking the predominant role. ALDH1A1 also catalyzes oxidation of retinal, and competition between this substrate and acetaldehyde may be involved in the deficiencies in vitamin A metabolism observed in alcoholics [4]. The dithiol compound disulfiram is a potent inhibitor of ALDHs and has been used as a deterrent in the treatment of alcoholism; however, disulfiram has limited value as a probe inhibitor of ALDH *in vivo*, as its metabolite diethyldithiocarbamate also irreversibly inhibits CYP2E1. ALDHs are induced by polycyclic aromatic hydrocarbons and TCDD by an Ah-receptor-mediated mechanism. Phenobarbital also induces some ALDHs, although the specific mechanism underlying this effect has not been investigated.

Many allelic variants of ALDH have been identified, although only a few are thought to play a role in polymorphic xenobiotic metabolism. The most thoroughly studied ALDH polymorphism involves a deficiency in ALDH2. The ALDH2*2 variant is caused by a point mutation in exon 12, resulting in synthesis of an inactive enzyme. This allele is common in Asian populations, and individuals carrying this variant have a compromised ability to detoxify acetaldehyde following consumption of ethanol. This deficiency is manifested as the so-called flushing syndrome common in Asians following ethanol consumption. ALDH1A1 polymorphism may also play a role in ethanol tolerance, although to a lesser degree.

## Quinone Oxidoreductases

### NAD(P)H:Quinone Oxidoreductase 1

NAD(P)H:quinone oxidoreductase 1 (NQO1; also known as DT diaphorase) catalyzes the two-electron reduction of quinones to the corresponding hydroquinones [214,215]. Functional NQO1 is a cytosolic homodimer containing two FAD prosthetic groups. High levels of NQO1 are found in liver of rodents and other common laboratory species. In contrast, human NQO1 was detected in lung, breast, and gastrointestinal epithelium; vascular endothelium; adipocytes; bone marrow; and several areas of the eye but not in liver. Thus, this enzyme does not appear to play a role in hepatic metabolism of xenobiotics. In addition to quinones, NQO1 can catalyze two-electron reductions of a wide variety of functional groups including quinone-imines, glutathione conjugates of naphthoquinones, azo compounds, and aromatic nitro compounds. NQO1 can also catalyze four-electron reductions of aromatic azo and nitro compounds. Dicumerol is a potent inhibitor of NQO1.

Quinone reduction by NQO1 can result in either detoxification or bioactivation, depending on the stability of the resulting hydroquinone; thus, NQO1 detoxifies redox active quinones such as menadione by converting them to stable hydroquinones that are substrates for phase II conjugation reactions. Less stable hydroquinones can undergo autooxidation to semiquinone radicals or rearrange to alkylating species.

NQO1 can be induced by polycyclic aromatic hydrocarbons, azo compounds, and phenolic antioxidants, as well as by oxidative stress. The 5′ flanking region of the NQO1 gene contains both an antioxidant response element (ARE) and a xenobiotic response element (XRE). Induction controlled by the XRE is mediated by the Ah receptor as described for P450 induction. Induction of NQ1 by the ARE may be mediated by a variety of transcription factors, including Jun, Fos, and Nrf, among others [287].

Several polymorphisms of NQO1 have been identified. The most striking of these is NQO1*2*2, in which no active protein is produced. The frequency of this genotype varies from about 4% in Caucasians to over 20% in Asian populations. The presence of this genotype has been shown to be a risk factor for the development of hematotoxicity of benzene.

### NAD(P)H:Quinone Oxidoreductase 2

NAD(P)H:quinone oxidoreductase 2 (NQO2) has approximately 82% sequence similarity to NQO1, although it is 43 amino acids shorter at the carboxy terminal of the protein. In humans, NQO2 is expressed in a variety of tissues, including liver, kidney, and lung, although the highest expression levels were found in skeletal muscle. NQO2 is also similar to NQO1 in many respects, such as substrate specificity and response to inducers; however, these enzymes are different in a number of important ways. For example, NQO2 uses dihydronicotinamide riboside as a cofactor instead of NAD(P)H. Inhibitor selectivity is also different; NQO2 is inhibited by quercetin and benzo(*a*)pyrene but not by dicoumerol. Despite gains in understanding the expression and functional activity of this enzyme, the significance of NQO2 for endobiotic and xenobiotic metabolism remains to be demonstrated [153,214].

# BIOCHEMICAL CONJUGATIONS

Mammals can synthesize xenobiotic conjugates that are more polar and readily excreted compared with the parent compound. Conjugate synthesis is finely controlled through various feedback pathways. Two major reactants are required for conjugate synthesis: a xenobiotic with the appropriate functional group and a cosubstrate that can be conjugated with the xenobiotic. If the xenobiotic does not have a functional group amenable to conjugation, such as

**TABLE 3.5**
**Selected Human and Rodent CYPs, UGTs, and Glucuronide Transporters Induced by the Ah, CAR, and PXR Receptors**

| Nuclear Receptor | Uptake Transporter | CYPs | UGTs | Export Transporter |
|---|---|---|---|---|
| AhR | Not determined | CYP1A1<br>CYP1A2 | UGT1A1<br>UGT1A6 (rUGT1A7) | Not determined |
| CAR | OATO2 (mOatp2) | CYP2B6 (rCYP2B6, mCyp2b10) | UGT1A1 (rUGT2B1) | MPR2 (mMrp2) |
| PXR | OATP2 (mOatp2) | CYP3A4 (rCYP3A23, mCyp3a11) | UGT1A1 (mUgt1a6, mUgt1a9) | MRP2<br>MRP3 |

*Note:* Inducible rodent enzymes are listed in parentheses.

*Source:* Bock, K.W. and Köhle, C., *Drug Metab. Rev.*, 36, 595–615, 2004. With permission.

a hydroxyl group, it may be oxidized (functionalized) by cytochromes P450. The oxidized product and the cosubstrate must be simultaneously available for conjugation. Both functions must be tightly integrated for rapid excretion of the xenobiotic. Although the forthcoming sections will discuss each conjugation system as a separate entity, it must be emphasized that *in vivo* metabolism is integrated. Examples showing the integration of the conjugation systems with related pathways will be presented.

## Glucuronidation: Uridine Diphospho-Glucuronosyltransferases

P450s are the principle phase I oxidative enzymes. Similarly, uridine diphospho-glucuronosyltransferases (also known as UDP-glycosyltransferases, or UGTs) are the principal phase II enzymes. Glucuronosyltransferases can use monooxygenase products to form glucuronides; however, it is not a necessity for substrates of the glucuronosyltransferase to be monooxygenase products. Significant numbers of xenobiotics and certain endobiotics possess the necessary functional groups for glucuronidation and do not require functionalization. This pathway has been estimated to account for 35% of all drugs metabolized by phase II drug metabolizing enzymes [74]. UDP-glucuronosyltransferases occur in several tissues, but their highest activity is found in the liver.

Whereas the multi-enzyme complex of the P450 monooxygenase is termed a system because the enzymes are closely linked, the multiple enzymes of glucuronidation are not linked but are interdependent. The general reaction mechanism of the conjugating enzymes involves the activation of an endogenous molecule. Subsequent reaction of this activated form of the endogenous molecule with the xenobiotic produces the conjugate. Activation may occur in a different cellular compartment than conjugation, as is the case with glucuronidation. Activation of glucose occurs in the cytosol, whereas conjugation occurs in the lumen of the endoplasmic reticulum.

Although the products of P450 are more water soluble than their parent compounds, some still possess considerable lipophilicity. Subsequent conjugation produces metabolites with higher water solubility. These metabolites can generally be readily excreted in the bile or urine. Transport proteins recognize the glucuronic acid moiety of the glucuronide and aid in excretion from the liver and kidney. An additional method by which glucuronidation produces less toxic metabolites is via the addition of a bulky moiety to the xenobiotic. This can result in both the shielding of reactive portions of the xenobiotic and in the blocking of reactions between the xenobiotic and the site responsible for the toxicological sequelae. In some cases, the product of glucuronidation has more toxicological activity than the parent compound, and conjugation can be considered metabolic activation, although examples are far fewer than with P450 oxidation. Similarly, conjugation with glucuronic acid results in significant structural change so pharmacologic activity is generally abolished, although in a few cases glucuronidation will result in a molecule with similar or even greater pharmacologic activity [50].

Recent findings on regulation of P450, UDP-glucuronosyltransferases, and transporters suggest that, although nuclear receptor signaling induces different cytochromes P450, regulation may converge on single UGTs and transporters [22]. The nuclear receptors CAR, PXR, and AhR coordinate the induction of several CYP, UGT, and drug transporters (Table 3.5) and thus lead to differential expression of various UGT forms. As an example, rifampicin induction of CYP3A4 is PXR mediated and is responsible for the conversion in the liver of lithocolic acid to the less toxic form hydrodeoxycholic acid. Hydrodeoxycholic acid is in turn conjugated by several UGTs, and transport out of the hepatocyte is mediated by the rifampicin inducible transporter multidrug resistance protein 2 (MRP2). Similarly, comedication with rifampicin leads to reduced effects of ezetimibe, an inhibitor of the cholesterol uptake transporter, by faster elimination via glucuronidation and subsequent intestinal or hepatic secretion via the efflux transporter P-glycoprotein (P-gp) and MRP2 [198].

Glucuronides are secreted either by the liver into the bile and consequently found in the feces or by the kidney into the urine. The excretion route is generally dependent on the molecular weight of the xenobiotic. In both cases, secretion is via specific organic anion transporters, members of the adenosine triphosphate (ATP)-binding cassette superfamily such as MRP2, at the apical plasma membrane, and MRP3, at the basolateral membrane of hepatocytes and enterocytes. The rat excretes glucuronides of xenobiotics with molecular weights greater than about 250 to 300 into the bile and those with lower molecular weights in the urine. Higher molecular weight xenobiotics, such as morphine, chloramphenicol, and endogenous steroids, are excreted in bile and enter the intestine. Biliary excretion can result in enterohepatic circulation, which can cause prolonged plasma half-lives for some compounds. Intestinal microflora express the enzyme β-glucuronidase, which catalyzes the hydrolysis of glucuronide conjugates. This releases the xenobiotic (referred to as the *aglycone*) in the intestine, where it can be absorbed into the blood. The xenobiotic can then be taken up by the liver, where it is reconjugated and excreted into the bile, where the cycle is again initiated. This can cause prolonged exposure to target organs, such as the liver, and result in unanticipated toxicity.

## Nomenclature for UDP-Glucuronosyltransferase Gene Superfamily

Nomenclature for the UDP-glucuronosyltransferases has progressed similarly to that for the P450 superfamily [160,265]. It has been proposed that each gene be identified by the root symbol UGT for UDP-glucuronosyltransferase. The gene family is identified by a number and a letter is added to designate the subfamily (e.g., UGT2B) followed by a number to identify the gene (e.g., UGT2B1). This system, as with the P450 nomenclature, is an attempt to provide isoforms with a name that is not only specific but also reflects the evolutionary divergence of the genes.

The two primary gene families are denoted UGT1 and UGT2. These are further divided into UGT1A, UGT2A, and UGT2B. The entire UGT1A family is derived from a single gene locus and encodes eight different proteins via alternative splicing of the UDPGA-binding domain with different substrate-binding domains. Most of the UGT1A isoforms have been isolated from human liver, whereas three isoforms have been isolated from extrahepatic sources such as bile ducts, tissues of the entire gastrointestinal tract, olfactory epithelium, brain, and fetal lung. UGT2 genes encode seven proteins and have also been isolated from liver, gastrointestinal tract, mammary gland, prostate, and adrenal tissues.

## Biochemistry of Glucuronidation

Glucuronidation (illustrated in Figure 3.13) requires the availability of three reactants:

- UDP-α-D-glucuronic acid (UDPGA), generated in the cytoplasm
- UDP-glucuronosyltransferase (UDPGT), bound to the endoplasmic reticulum
- Substrate with the requisite functional group and some hydrophobic character

Maximal enzyme activity is dependent on optimal concentrations of these reactants at the membrane site of catalysis.

As seen in Figure 3.13, D-glucose is the original precursor of UDPGA. During anabolic metabolism, D-glucose is converted to β-D-glucose-1-phosphate. This compound serves as substrate for UDP-glucose pyrophosphorylase, which catalyzes its reaction with uridine triphosphate to yield the high-energy phosphate-containing UDP-D-glucose and pyrophosphate. UDP-D-glucose then reacts with nicotinamide adenine dinucleotide (NAD) catalyzed by UDP-glucose dehydrogenase to yield UDP-D-glucuronic acid, which completes glucose activation. This compound is termed the *glycone*, indicating its source. The xenobiotic that is conjugated is termed the *aglycone*. Glucose activation occurs within the cytoplasm, whereas glucuronidation of the aglycone occurs at the endoplasmic reticulum. Because UDP-D-glucose is also used in glycogen synthesis, it generally is available in the cell. This is not true for all conjugation reactions and may be one of the reasons why glucuronidation is a major conjugation pathway.

The topology of UDP-glucuronosyltransferases is important for understanding substrate specificity and the need to disrupt microsomes with detergents or other means before assaying these enzymes *in vitro*. It is believed that the large interlaboratory variation of *in vitro* glucuronidation data comes in part from variations in detergent-released latency. UDP-glucuronosyltransferases are oriented in the endoplasmic reticulum in such a way that the majority of the protein protrudes into the lumen of the endoplasmic reticulum. The intraluminal portion of the protein possesses the UDP-glucuronic acid-binding domain as well as the xenobiotic- or endobiotic- (endogenous substrates) binding domain. This means that UDP-glucuronic acid must pass through the membrane, possibly by carrier mediation, and that the substrate must also pass through the membrane [96]. Latency is an *in vitro* phenomenon characterized by low glucuronidation activity due to the inability of the cofactor or substrate to diffuse into the microsomal lumen. Latency can be mitigated by disruption of the microsomal membrane with detergents or by inclusion of a pore-forming peptide such as alamethicin in the reaction mixture. Molecular biology studies indicate that the *C*-terminal half of the protein is highly

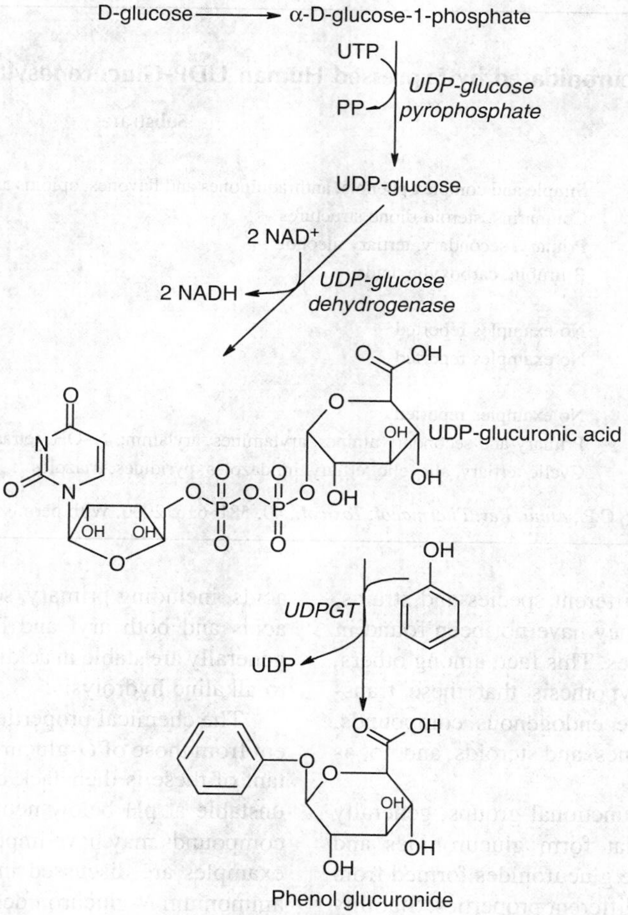

**FIGURE 3.13** Glucuronidation of phenol; an example of the pathway leading to production of glucuronic acid conjugates.

conserved among different UDP-glucuronosyltransferases, whereas the *N*-terminal region is highly variable. The *C*-terminal half of the protein contains the transmembrane sequences that anchor the enzyme within the membrane and the short portion of the *C*-terminus that protrudes from the outside surface of the endoplasmic reticulum into the cytoplasm. The *C*-terminal half of the enzyme may contain a UDP-glucuronic-acid-binding site. The broad substrate specificity is believed to come from variation in the primary sequence of the *N*-terminal region where the substrate-binding domain resides [171].

Uridine diphosphate glucuronic acid and the aglycone (xenobiotic or endobiotic) must be present for the conjugation reaction to be initiated. The number of xenobiotics that have been shown to be substrates for UDPGTs is large and continues to grow [175]. The major functional groups forming glucuronides are (1) hydroxyl, (2) carboxyl, (3) amino, and (4) sulfhydryl. The substituents to which these functional groups are attached can be quite variable (see Table 3.6). Similar to the substrate requirements for monooxygenases of the endoplasmic reticulum, the aglycone must be somewhat lipid soluble to be a substrate for the UDPGTs. This requirement reflects the need for the xenobiotic to penetrate the endoplasmic reticulum to gain access to the active site. All of the endobiotics associated with normal metabolism and homeostasis that are substrates for the UDPGTs are lipid soluble and include bilirubin, catechols such as 3-*O*-methyladrenaline, serotonin, and 17-hydroxy-containing steroids.

### Reactions Catalyzed by the UDP-Glucuronosyltransferases

As with many of the enzymes of detoxification, the glucuronosyltransferases have a low order of substrate specificity. This lack of substrate specificity makes them ideally suitable as detoxification enzymes. Whether or not they evolved as detoxification enzymes or represent enzymes of normal metabolism whose lack of specificity make them suitable for detoxification is open to debate. Of interest in this respect is that they occur only in higher organisms. Glucuronosyltransferases have been found in all mammals, birds, and reptiles that have been investigated, although their specific activities toward specific

**TABLE 3.6**
**Xenobiotic Substrates Glucuronidated by Expressed Human UDP-Glucuronosyltransferases (UGTs)**

| Human Glucuronides | Substrates |
|---|---|
| *Linkage through –O–* | |
| Aryl hydroxyl (ether) | Simple and complex phenols, anthraquinones and flavones, opioids and steroids, hydroxylated coumarins |
| Aryl or alkyl enolic | Coumarins, steroid-dione structures |
| Alkyl hydroxyl | Primary, secondary, tertiary alcohols |
| Acyl hydroxyl (carboxylic esters) | Bilirubin, carboxylic acids |
| *Linkage through –S–* | |
| Aryl and alkyl thiols | No examples reported |
| *Linkage through –C–* | No examples reported |
| *Linkage through –O–* | |
| Sulfonamides | No examples reported |
| Nonquaternary | Primary and secondary amines, arylamines, arylamine N–OH, tetrazoles |
| Quaternary | Cyclic tertiary, alicyclic tertiary, imidazoles, pyridines, triazoles |

*Source:* Tukey, R.H. and Strassburg, C.P., *Annu. Rev. Pharmacol. Toxicol.*, 40, 581–616, 2000. With permission.

substrates may vary among different species and strains. Unlike the monooxygenase, they have not been found in bacteria and other lower species. This fact, among others, lends support to Dutton's hypothesis that these transferases evolved to metabolize endogenous compounds, such as bilirubin, catecholamines, and steroids, and not as detoxification enzymes [67].

Table 3.6 illustrates the functional groups, generally nucleophilic heteroatoms, that form glucuronides and examples of the reactions. The glucuronides formed from these functional groups have different properties. Stability is among the most important with respect to detoxification. Breakdown of the glucuronide can lead to reformation of the parent compound and in certain cases the production of highly reactive electrophilic species. These reactive species may be responsible for the production of acute and chronic toxicity by covalent binding to nucleophilic sites on tissue macromolecules.

Among the most commonly encountered glucuronides are those involving linkage of glucuronic acid and the xenobiotic through an oxygen atom. These *O*-glucuronides may form with a number of chemical classes, including aryl, alkyl, and acyl compounds, as illustrated in Table 3.6.

The alkyl-*O*-glucuronides are ether-linked glucuronides that can form from a variety of primary, secondary, and tertiary alcohols. Although generally stable at physiological conditions, they can be hydrolyzed under acidic conditions.

The enolic glucuronides are formed from aglycones without a free hydroxyl group. Glucuronides are formed from the enolized keto group. These conjugates lack the stability of the ether glucuronides and are susceptible to both acid and alkaline hydrolysis. They are more stable at neutral and alkaline pH than in acid conditions. Ester glucuronides can be produced from a variety of carboxylic acids, including primary, secondary, and tertiary aliphatic acids and both aryl and heterocyclic compounds. They generally are stable in acidic conditions but are susceptible to alkaline hydrolysis.

The chemical properties of *N*-glucuronides are different from those of *O*-glucuronides. One of the most important of these is their lack of stability. They are especially unstable at pH below neutrality. The instability of these compounds may have important biological consequences; examples are discussed in more detail later. Quaternary ammonium *N*-glucuronides are formed by *N*-glucuronidation of cyclic and acyclic tertiary amines. These charged metabolites may be formed in higher primates while not being found in other animal models, such as the rat.

The *S*-glucuronides are not as commonly encountered as the *O*-glucuronides, but they represent important detoxification pathways for thiolic compounds. Their stability is similar to that of the *O*-glucuronides.

The *C*-glucuronides represent recently recognized conjugates, and only a few examples are known, such as phenylbutazone. Generally, they appear to be formed by the transferase, but other possible mechanisms of formation have been suggested.

### Role of UDP-Glucuronosyltransferases in Detoxification and Metabolic Activation

The foregoing discussion indicates that the UDP-glucuronosyltransferases play a critical role in the metabolism and detoxification of xenobiotics. Some substrates require functionalization by the monooxygenase before metabolism by the transferase, whereas others can be directly conjugated. The conjugates are more water soluble than the parent xenobiotic, and some readily form salts. Addition of the glycone may enable some of the conjugates to be more readily excreted through carrier-mediated mechanisms. Mechanisms other than increased excretion rates may also be important. The addition of the relatively bulky glycone may hide or hinder the biological reactivity of particular functional groups on the xenobiotic. In addition, binding of the toxicant to particular receptors responsible for toxicity may be blocked. Overall, these mechanisms represent an efficient system for detoxification. On the other hand, glucuronidation of certain compounds represents a metabolic activation where the product is more toxic than the parent compound.

Aromatic amines are among the most studied examples of the role glucuronidation plays in metabolic activation of carcinogens. These glucuronides transport the proximate carcinogen to the target site, where it decomposes to the species that react with cellular macromolecules producing the biochemical lesion responsible for generating the pathological lesion.

Several of the arylamines are potent bladder carcinogens, including 4-aminobiphenyl, 1-naphthylamine, and benzidine. Metabolic activation of these carcinogens to the ultimate carcinogen appears similar and requires the action of UDP-glucuronosyltransferase. Metabolic activation begins with P450-dependent activation of the arylamine to the proximate carcinogen, an *N*-hydroxyarylamine. Other specific ring hydroxylated forms may be produced and may represent more stable products. The unstable *N*-hydroxyarylamines are then converted to stable *N*-glucuronides. These *N*-glucuronides are transported to the bladder. In the bladder, the *N*-glucuronides are subject to β-glucuronidase activity, which splits off the aglycone. They are also subject to hydrolysis in acidic urine producing the *N*-hydroxyarylamine. The *N*-hydroxyarylamine spontaneously converts to the electrophilic arylnitrenium ion. A similar mechanism involving sulfonation-mediated formation of this reactive species is illustrated later in the chapter. The electrophilic arylnitrenium ion can then react with nucleophilic centers on macromolecules of the bladder epithelium, especially DNA, to initiate tumor formation. The concentration of the glucuronide in the bladder, in combination with the time the glucuronide remains in the bladder, can modify the potential for tumor formation. Glucuronides may function in this manner with a number of carcinogens and be important in explaining why certain target organs are susceptible to a specific carcinogen and others are not susceptible. In the above example, glucuronidation may protect the liver but makes the bladder, the target organ, susceptible.

Glucuronidation has also been implicated in adverse drug reactions of certain carboxylic drugs that have resulted in a toxic immunological response. It is believed that a reactive glucuronide covalently binds to cellular proteins that act as haptens, producing an anaphylactic reaction. Glucuronidation of the carboxylic acid moiety of drugs such as diclofenac, a nonsteroidal antiinflammatory drug, leads to an unstable and reactive acyl glucuronide metabolite. The conjugate then undergoes transacylation of protein nucleophiles by the 1-*O*-acyl-glucuronide or glycation of proteins via mechanisms that involves open-chain aldedhyde reactions with protein amino groups. These drug–protein adducts are believed to be recognized as foreign by the immune system, resulting in an immune response and thereby leading to the associated idiosyncratic hepatoxicity [14,94].

### Species, Gender, and Genetic Differences in UDP-Glucuronosyltransferase Activity

Studies of species, strain, and gender differences in glucuronidation are complicated by a number of factors. Activity may be affected by age, hormonal status, environmental exposure to xenobiotics in the diet and other sources, and by nutritional status. Factors associated with the methodology to determine differences in glucuronidation also play a role, including substrate, assay method, method of freeing latent activity, and the method of isolating the preparation employed to measure activity. This has led to a number of reports of differences in activity that could be artifactual; however, the large number of reports concerning differences in glucuronidation among species, strains, and the sexes indicate that certain of these differences are real and may have a genetic basis.

As mentioned previously, lower animals, including prokaryotes and invertebrates, do not produce glucuronides. Fish and reptiles do demonstrate glucuronidation of xenobiotics but vary dramatically in activity, which is generally at least tenfold lower than mammalian activity. Birds have glucuronidation ability similar to that of mammals.

Differences among mammalian species in their ability to glucuronidate a xenobiotic may be quite large, as is the case for the HIV drug zidovudine. This drug is eliminated in the rat and dog primarily unchanged, whereas the glucuronide represents the majority of metabolites in monkeys and humans [187]. The guinea pig generally has higher activity than most other laboratory species. This higher activity may be associated with less latent enzyme activity, as its UDP-glucuronosyltransferases can be activated by much gentler methods than other species. Cats are well known for their extremely low transferase activity. Although capable of forming glucuronides with endogenous compounds, they form only low levels of or no glucuronides with xenobiotics.

Glucuronidation of amines is divided into two groups: nonquaternary *N*-conjugates and quaternary *N*-conjugates. Major qualitative species differences do not appear to exist in the conjugation of the primary and secondary amines, sulfonamides, arylamines, and cyclic and heterocyclic amines to form nonquaternary *N*-conjugates, although quantitative differences do exist. Quaternary glucuronidation occurs in primates, including humans, but not in other species. In humans, quaternary-ammonium-linked glucuronides of aliphatic amines appear to be produced by UGT1A3 and UGT1A4 [45,92].

A well-known example of a strain difference is the almost complete lack of bilirubin glucuronidation in the Gunn rat. This rat strain also has low activity toward a number of xenobiotic substrates but normal activity toward others. There is a genetic component to this, with the low activity being autosomally recessive. The mutation in the Gunn rat responsible for its lack of bilirubin conjugation occurs in the UGT1 family and affects this entire group of isozymes. A frameshift mutation occurs because of a deleted guanine that results in a TGA stop codon occurring sooner than normal. This mutation results in a protein missing 115 amino acids that constitute a hydrophobic region associated with insertion of the protein into the membrane. Lack of insertion negates the activity of this enzyme form and results in degradation of the incomplete protein. The genes in the UGT2 family are normally expressed in the Gunn rat.

Similar defects occur in humans and result in unconjugated hyperbilirubinemias. Gilbert's syndrome is a milder form of the disease that occurs in 2 to 5% of the population. This large prevalence in the population makes it an important human genetic deficiency when considering inter-individual variation in xenobiotic metabolism. These patients are characterized by mild, chronic, unconjugated hyperbilirubinemia that produces jaundice and an impaired ability to metabolize menthol. Decreased clearance of several drugs, including tolbutamide, rifamycin, josamycin, and paracetamol, has been observed. Crigler–Najjar syndrome is a familial form of severe unconjugated hyperbilirubinemia. Infants often develop severe neurological damage from bilirubin encephalopathy (kernicterus). Patients are divided into two types. Type I is more severe (unconjugated bilirubin, >20 mg/dL) and not responsive to barbiturate or glutethimide therapy. Type II patients respond to induction by phenobarbital, which suggests a fundamental difference from type I in the molecular basis of the genetic defect. Type I results from mutations in the UGT1 family that produces a loss of bilirubin conjugation [48], whereas less severe mutations occur in type II that produce a decrease, but not a loss, of activity.

Gender differences appear hormonally related [252] and can be substrate dependent. Although it is sometimes stated that males have higher glucuronidation activity than females, this is substrate dependent, and no general classification should be made. Like monooxygenase activity, activity may be sensitive to imprinting or programming during the neonatal period. As with species and strain differences, care must be taken when extrapolating data obtained with one substrate to other substrates. Glucuronidation of estradiol and estrone is higher in female rats than male rats [299]. Paracetamol, oxazepam, and diflunisal are cleared 30 to 50% faster in males, due primarily to enhanced glucuronidation.

### Induction of the Glucuronosyltransferases

Uridine diphosphate glucuronosyltransferases are inducible enzymes, much like cytochrome P450, and are inducible by some of the same chemicals. Evidence of a true induction process involving *de novo* protein synthesis and increases in mRNA has been observed for induction of the UDP-glucuronosyltransferases by phenobarbital. Most inducers of CYP1A, CYP2B, CYP3A, and CYP4A can induce these transferases. Rat UGT1A6 and UGT1A7 and human UGT1A6 and UGT1A9 are polycyclic-hydrocarbon-inducible transferases. Induction appears mediated by the Ah receptor. Rat UGT1A7 and human UGT1A9 have high activity toward the phenolic and diphenolic metabolites of polycyclics, such as benzo(*a*)pyrene [23]. Few specific inducers of the transferases that do not also induce the monooxygenase are known. For example, *trans*-stilbene oxide and ethoxyquin appear to only induce the transferases, but more studies are needed to determine if this is a true induction. Induction of the transferases modifies the toxicity of xenobiotics in a manner similar to induction of P450, as previously discussed.

## Sulfonation: Sulfotransferases

Sulfonation of xenobiotics and endobiotics is catalyzed by a set of enzymes called *sulfotransferases*. These enzymes belong to a multigene family and occur in prokaryotes, plants, and animals. Some of the enzymes are membrane bound and others occur in the cytosol. The membrane-bound sulfotransferases are found in the Golgi membranes and are involved in the sulfonation of endogenous compounds, such as glycosaminoglycans, glycoproteins, and proteins, and peptides secreted by the Golgi apparatus; they are not involved in xenobiotic metabolism. The soluble or cytostolic sulfotransferases catalyze the sulfate conjugation of a variety of substrates, including steroid hormones such as 17β-estradiol and dehydroepiandrosterone; thyroid hormones; catecholamines and xenobiotics, such as *N*-hydroxy-2-acetylaminofluorene; isoflavones; and many drugs, including acetaminophen and minoxidil. For the most part, sulfonation of xenobiotics results in metabolites that are less toxic than the parent compound; however, the sulfotransferases, like many xenobiotic metabolism enzymes, can produce metabolically activated products that have mutagenic and carcinogenic potential.

Until recently, the sulfotransferases have not been as intensely investigated as some of the other xenobiotic metabolism enzymes. Lately, interest has been renewed in these enzymes, particularly their description at the gene level. Utilization of the tools of molecular biology has provided new insight into their roles in metabolism, has revealed the complexity of their gene family, and has enabled development of a nomenclature system [65,82,209]. The ability to sequence the sulfotransferase

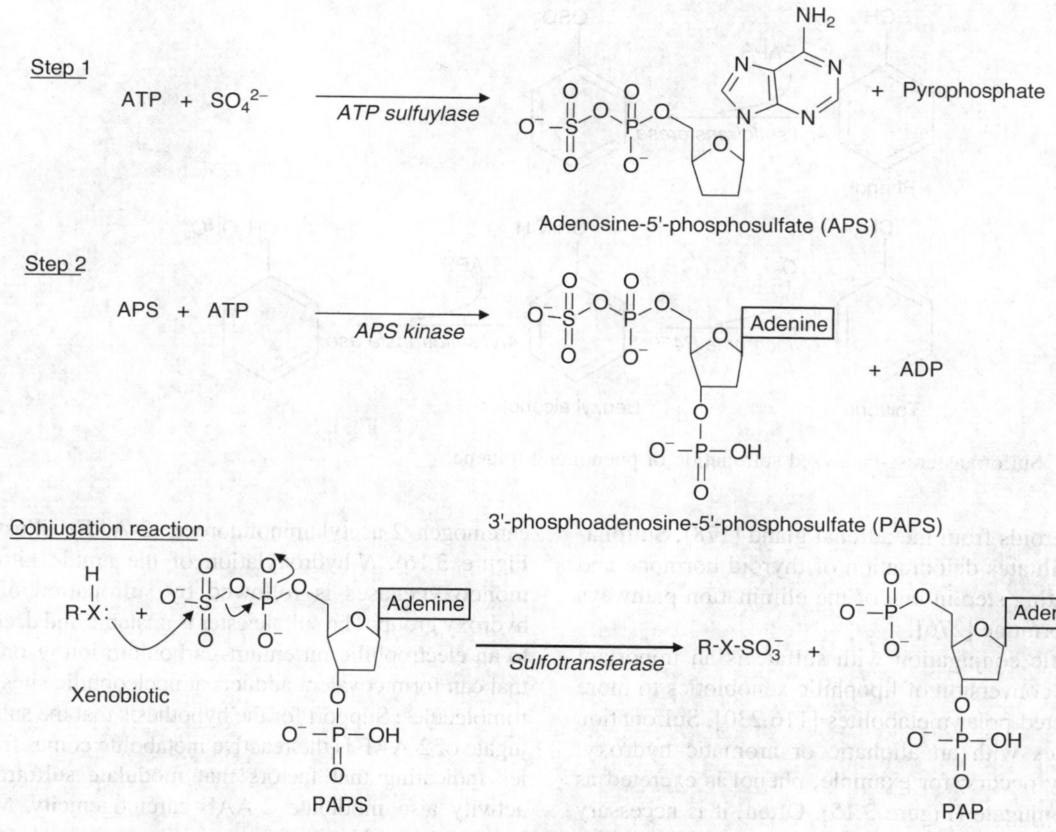

**FIGURE 3.14** Reactions catalyzing the formation of PAPS and the xenobiotic–sulfate conjugate. *X* in the conjugation reaction represents a nucleophilic atom in a functional group such as oxygen in a hydroxyl group or nitrogen in an amine group.

and identify new isoforms of these enzymes has progressed faster than our understanding of their individual roles in xenobiotic metabolism.

## Biochemistry of Sulfonation

A limiting factor in the sulfonation of xenobiotics by the sulfotransferases is the availability of 3-phosphoadenosine-5′-phosphosulfate (PAPS) (reviewed in Klaassen and Boles [129] and Schwartz [227]). As illustrated in Figure 3.14, PAPS is synthesized in a two-step process. The first step is formation of adenosine-5′-phosphosulfate (APS) catalyzed by ATP-sulfurylase. Although the synthesis of APS from sulfate and ATP is not energetically favored, the rapid hydrolysis of pyrophosphate and the rapid utilization of APS as a substrate for APS-kinase drives the reaction toward APS synthesis. APS-kinase catalyzes synthesis of PAPS from APS and ATP. This enzyme is tightly coupled with the ATP-sulfurylase, which results in the rapid utilization of APS.

Tissue concentrations of PAPS are relatively low compared to UDPGA, the active form of glucuronic acid used in glucuronidation. During active sulfonation, PAPS becomes rapidly depleted; for example, the sulfotransferase has a high affinity for acetaminophen, which forms a sulfate conjugate. At low doses, rats excrete the sulfated acetaminophen as a major urinary metabolite. As the dose of acetaminophen is increased, the sulfate metabolite does not increase, whereas the glucuronide of acetaminophen increases dramatically; this is believed to be due to the limited availability of PAPS. The limitation in the synthesis of PAPS is sulfate. The major sources of sulfate include diet and degradation of sulfur amino acids (methionine and cysteine). These sources are inadequate to maintain sulfate concentrations for PAPS synthesis during rapid sulfotransferase activity. In the mouse, sulfonation appears more limited by sulfotransferase activity than by PAPS and sulfate.

## Reactions Catalyzed by Sulfotransferases

As mentioned, sulfotransferases esterify a variety of endogenous substrates, including steroids, carbohydrates, and proteins. Sulfonation also plays a role in the disposition of hormones. Sulfonation directs lipophilic compounds, such as the steroidal hormones, to more polar environments, including the active sites of enzymes and to body fluids; for example, sulfonation enhances the elim-

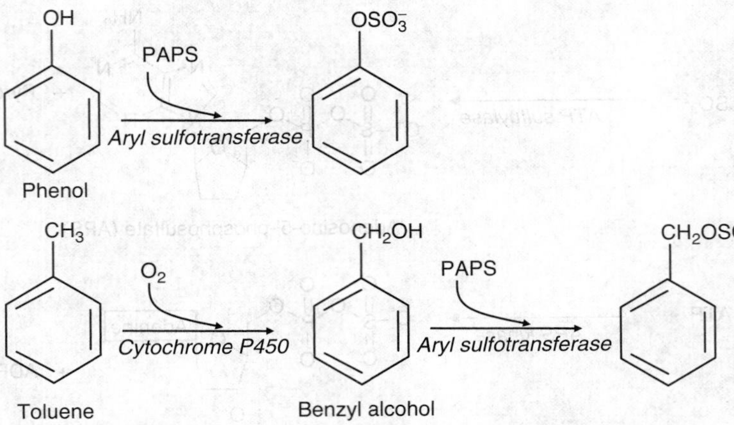

OH PAPS $OSO_3^-$ Aryl sulfotransferase Phenol $CH_3$ $O_2$ Cytochrome P450 $CH_2OH$ PAPS Aryl sulfotransferase $CH_2OSO_3^-$ Toluene Benzyl alcohol

**FIGURE 3.15** Sulfotransferase-catalyzed sulfonation of phenol and toluene.

ination of steroids from the adrenal gland [178]. Sulfonation also facilitates deiodination of thyroid hormone and is a rate-limiting step in one of the elimination pathways of thyroid hormone [276].

Xenobiotic conjugation with sulfate is an important route for the conversion of lipophilic xenobiotics to more readily excreted polar metabolites [116,230]. Sulfonation of xenobiotics with an aliphatic or aromatic hydroxyl group readily occurs; for example, phenol is excreted as its sulfate conjugate (Figure 3.15). Often, it is necessary for phase I metabolism to functionalize a xenobiotic with a hydroxyl group before it can be sulfated; for example, toluene is oxidized to benzyl alcohol before conjugation with sulfate (Figure 3.15).

## Role of Sulfotransferases in Detoxification and Metabolic Activation

Alcohols, phenols, aliphatic and aromatic amines, and aromatic hydroxylamines and hydroxylamides can be sulfated. These same groups can form glucuronides. At low doses, sulfonation may play an important role in detoxification of xenobiotics; however, as acetaminophen demonstrates, at high doses glucuronidation becomes more important because of sulfate limitations. Secondary effects may be produced by the sulfonation lowering sulfate availability for the sulfonation of endogenous substrates. Depletion of sulfate pools as a result of the metabolism of high doses of drugs has been proposed to interfere with the normal biosynthesis of glycosaminoglycans during development, resulting in teratogenic effects in animals [278]. Sulfotransferases can be involved in the conversion of pro-drugs to their active forms; for example, minoxidil is sulfoconjugated to its active form, which is more active as an antihypertensive and hair-growth stimulant than the parent drug.

Sulfotransferases can be involved in the metabolic activation of a number of mutagens and carcinogens. One of the best known examples is the metabolic activation of the carcinogen 2-acetylaminofluorene (2-AAF) (illustrated in Figure 3.16). *N*-hydroxylation of the amide nitrogen by monooxygenases is followed by sulfonation of the *N*-hydroxy group. The sulfate ester is unstable and decomposes to an electrophilic nitrenium–carbonium ion resonance ion that can form covalent adducts at nucleophilic sites on macromolecules. Support for the hypothesis that the sulfate conjugate of 2-AAF is the reactive metabolite comes from studies indicating that factors that modulate sulfotransferase activity also modulate 2-AAF carcinogenicity. Male rats have higher sulfotransferase activity and develop more 2-AAF-induced tumors than females. Reduction of sulfotransferase activity in male rats by castration, hypophysectomy, thyroidectomy, or steroid hormones reduces 2-AAF covalent adducts. These results are consistent with the hypothesis that sulfonation of 2-AAF is required for covalent modification of DNA. This mechanism is at least partially responsible for the activation of several other xenobiotics, including aromatic amines, mono- and dinitrotoluene, *N*-hydroxyphenacetin, 1′-hydroxysafrole, $N^3$-hydroxyxanthine, and other *N*-hydroxyarylamides [182]. Secondary nitroalkanes, such as 2-nitrobutane and 3-nitropentane, can be metabolically activated to mutagens by aryl sulfotransferase and hepatocarcinogens. Primary nitroalkanes, such as 1-butane and 1-nitropentane, are not activated by aryl sulfotransferase [76]. Mutagenicity testing is frequently hampered by the fact that phase II metabolic activation systems are typically not present in standard Ames bacterial mutagenesis assays which do simulate phase I activation. New systems are being developed in which sulfotransferase genes are expressed in *Salmonella* strains and Chinese hamster V79 cells and are yielding a diverse set of structures capable of being activated by sulfotransferases to mutagenic metabolites [88].

Sulfotransferases can metabolically activate certain products of CYP1A1 metabolism of polycyclic hydrocarbons; for example, 9-hydroxymethylbenzo(*a*)pyrene can be sulfated to yield a highly reactive sulfate ester that is heterolytically cleaved to produce an electrophilic cation

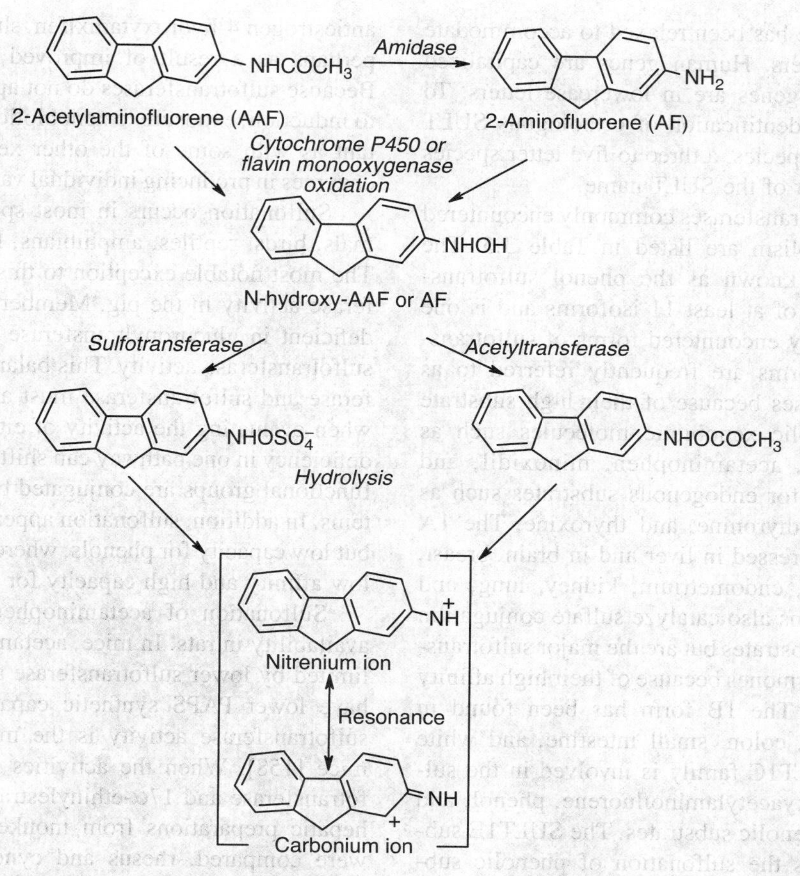

Amidase
NHCOCH3
NH2
2-Acetylaminofluorene (AAF)
2-Aminofluorene (AF)
Cytochrome P450 or
flavin monooxygenase
oxidation
NHOH
N-hydroxy-AAF or AF
Sulfotransferase
Acetyltransferase
NHOSO4−
NHOCOCH3
Hydrolysis
NH+
Nitrenium ion
Resonance
NH
Carbonium ion

**FIGURE 3.16** Metabolic activation of 2-acetylaminofluorene to a reactive metabolite capable of covalent modification of macromolecules.

that damages DNA, RNA, and protein. In addition, 6-hydroxymethylbenzo(*a*)pyrene can be activated to the carcinogenic 6-sulfooxymethylbenzo(*a*)pyrene by rat and mouse sulfotransferase [77]. Other examples include 5-hydroxymethylchrysene and 7,12-dihydroxymethyl benz(*a*)anthracene [306].

## Sulfotransferase Isoforms, Genetics, and Species Differences

Sulfotransferases belong to a multigene family that produces a number of distinct enzymes that have different, but overlapping, substrate specificities. Some of these enzymes demonstrate species and tissue specificity in their expression. The nomenclature used to describe these enzymes is still evolving, but recently a system for classifying the cytosolic sulfotransferase superfamily has been proposed [20]. Membrane-bound sulfotransferases that are localized to the Golgi apparatus exhibit a low degree of amino acid sequence identity with the cytosolic sulfotransferases, and, although they exhibit some structural similarity to cytosolic isoforms, they are generally considered a separate superfamily. The cytosolic sulfotransferases are typically involved in xenobiotic metabolism, thus their genetics and nomenclature is discussed here. The cytosolic form is identified by the abbreviation SULT.

More than 56 distinct eukaryotic sulfotransferase isoforms have been identified and functionally characterized [20]. SULT families are identified by the Arabic numeral immediately following the name and subfamilies identified by alphabetical categories (Figure 3.17). Unique subforms are further identified by an additional Arabic numeral; however, in some cases, such as SULT2A1, the

Isoform
Family
(HUMAN)SULT2B1*1A
Species
Superfamily
Subfamily
Suballele

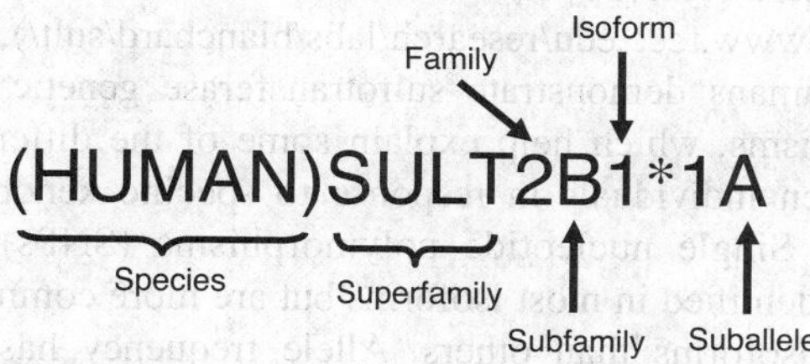

**FIGURE 3.17** Naming convention illustrated for a representative cytosolic sulfotransferase (SULT) allele name. A complete SULT allele name contains species, superfamily, family, sioform, allel, and suballele designations as shown. (Adapted from Blanchard, R. et al., *Pharmacogenetics*, 14, 199–211, 2004.)

standard nomenclature has been relaxed to accommodate more historic identifiers. Human genes are capitalized while rat and mouse genes are in lowercase letters. To further facilitate the identification of orthologous SULT isoforms in different species, a three to five letter species code is placed in front of the SULT name.

Many of the sulfotransferases commonly encountered in xenobiotic metabolism are listed in Table 3.6. The SULT1 family, also known as the phenol sulfotransferases, is comprised of at least 11 isoforms and is one of the most commonly encountered forms of sulfotransferase. The 1A isoforms are frequently referred to as phenol sulfotransferases because of their high substrate specificity for phenolic xenobiotic molecules such as 17α-ethinylestradiol, acetaminophen, minoxidil, and isoflavones, but also for endogenous substrates such as 17β-estradiol, triiodothyronine, and thyroxine. The 1A isoform is highly expressed in liver and in brain, breast, intestinal epithelium, endometrium, kidney, lung, and platelets. The 1B forms also catalyze sulfate conjugation of typical phenolic substrates but are the major sulfotransferases for thyroid hormones because of their high affinity for these substrates. The 1B form has been found in tissues such as liver, colon, small intestine, and white blood cells. The SULT1C family is involved in the sulfonation of *N*-hydroxyacetylaminofluorene, phenol, and other prototypical phenolic substrates. The SULT1E subfamily also catalyzes the sulfonation of phenolic substrates but has a much lower affinity than that of the 1A subfamily. The SULT1E isoforms are typically found in liver and small intestine.

The SULT2 family, also known as the hydroxysteroid sulfotransferases, consists of two subfamilies, 2A and 2B, and typically catalyzes sulfate conjugation of 3β-hydroxy groups of steroids. The 2A family catalyzes the sulfonation of a range of xenobiotics, including benzylic alcohols of polycyclic aromatic hydrocarbons, and is expressed in adrenal gland, liver, brain, and intestine. 2B isoforms have only been identified in human prostate, placenta, and trachea and in mouse intestine, epididymis, and uterus.

Additional information on the SULT gene family, their nomenclature and substrate specificities can be found in Blanchard et al. [20] or at the SULT nomenclature website (http://www.fccc.edu/research/labs/blanchard/sult/).

Humans demonstrate sulfotransferase genetic polymorphisms, which help explain some of the differences between individuals in response to specific xenobiotics [188]. Single nucleotide polymorphisms (SNPs) have been identified in most isoforms but are more common in some isoforms than others. Allele frequency has been associated with certain ethnic groups and may contribute to differential drug responses in these individuals. For example, evidence is emerging that women carrying the SULT1A1*2 allele, which is associated with diminished capacity to sulfate SULT1A1 substrates such as the active antiestrogen 4-hydroxytamxifen, show increased survival, perhaps as a result of improved drug exposure [290]. Because sulfotransferases do not appear to be as sensitive to induction, exposure to xenobiotics may not be as important as with some of the other xenobiotic metabolizing enzymes in producing individual variations in metabolism.

Sulfonation occurs in most species, including mammals, birds, reptiles, amphibians, fish, and invertebrates. The most notable exception to this is the low sulfotransferase activity in the pig. Members of the cat family are deficient in glucuronyltransferase activity but have high sulfotransferase activity. This balance of glucuronyltransferase and sulfotransferase must always be kept in mind when evaluating the activity of either enzyme system. A deficiency in one pathway can shift metabolism, as similar functional groups are conjugated by the two enzyme systems. In addition, sulfonation appears to have high affinity but low capacity for phenols, whereas glucuronidation has low affinity and high capacity for these substrates.

Sulfonation of acetaminophen is limited by PAPS availability in rats. In mice, acetaminophen sulfonation is limited by lower sulfotransferase activity. Although mice have lower PAPS synthetic capability than rats, lower sulfotransferase activity is the major limiting factor in mice [158]. When the activities of acetaminophen sulfotransferase and 17α-ethinylestradiol sulfotransferase in hepatic preparations from monkeys, dogs, and humans were compared, rhesus and cynomolgus monkeys and dogs had higher acetaminophen sulfotransferase activity than humans [231].

## Factors Modifying Metabolism

Sulfotransferases are not induced by the classical inducers, phenobarbital and 3-methylcholanthrene, and these compounds may actually suppress their expression [219]. Several inhibitors of sulfotransferase have been discovered and exploited experimentally to study these enzymes. Pentachlorophenol and 2,6-dichloro-4-nitrophenol are potent sulfotransferase inhibitors. Only 0.2-μ*M* pentachlorophenol is required for 50% inhibition of 2-dichloro-4-nitrophenol sulfonation by purified arylsulfotransferase [118]. Pentachlorophenol and 2,6-dichloro-4-nitrophenol are effective inhibitors because the *ortho*- and *para*-aromatic ring positions are substituted with electron-withdrawing groups. This effect is consistent with the mechanism whereby the sulfotransferases facilitate electrophilic attack of the hydroxyl oxygen by the sulfur.

## Gender Differences

Major gender differences have been observed in the sulfate conjugation of steroid hormones; for example, female rats have fivefold higher activity for cortisol metabolism than do male rats [237]. This gender difference in cortisol

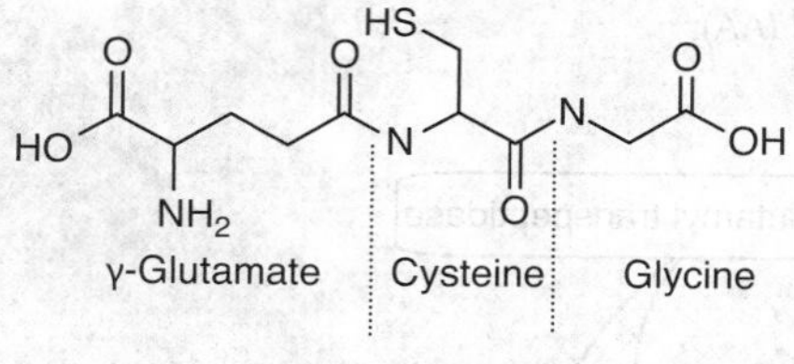

HS, O, O, O, HO, N, N, OH, NH2, O
γ-Glutamate | Cysteine | Glycine

**FIGURE 3.18** Structure of glutathione. Note the unusual γ-configuration of the linkage between glutamate and cysteine.

metabolism is apparently due more to suppression of sulfotransferase by male hormone levels than to stimulation by the ovaries [235,236]. Three steroid sulfotransferases have been isolated from rat liver, and it is the relative amounts of these isozymes that account for the large gender difference. Aryl sulfotransferase concentrations in the livers of male rats were higher than in females; in contrast, hydroxysteroid sulfotransferase concentration was higher in the liver of female rats compared to males [42].

Lower sulfotransferase activity observed in neonatal rats has been attributed to sexual immaturity because, as gonads develop, sulfotransferase activity increases. Newborn infants, who characteristically exhibit pronounced immaturity in glucuronidation, have a fully developed phenol sulfotransferase activity; for example, newborns excrete acetaminophen as a sulfate conjugate, whereas adults primarily excrete it as a glucuronide conjugate. Chloramphenicol is extremely toxic in neonates because it is a poor substrate for sulfotransferase and is primarily cleared by glucuronidation in adults.

## Glutathione *S*-Transferases

A family of cytosolic enzymes known as glutathione *S*-transferases (GST) is capable of conjugating relatively hydrophobic electrophilic molecules with the reduced form of the intracellular nucleophile glutathione (Figure 3.18) [40]. These enzymes are found in highest concentrations in the liver, kidney, intestines, and lung, but they occur in most tissues. Glutathione conjugates have higher molecular weights and are more water soluble and more likely to be excreted in urine and bile than are the parent compounds. Further, glutathione conjugates are substrates for transporters involved in biliary and renal excretion which facilitates their clearance from the body. In general, conjugation with glutathione decreases the likelihood that a xenobiotic will react with toxicological targets.

### Synthesis and Regulation of Glutathione

Glutathione is a tripeptide composed of glutamate, cysteine, and glycine. In contrast to the α-linkage normally found in most peptides, the glutamate and cysteine of glutathione are joined by a γ-linkage, which confers resistance to hydrolysis by peptidases (Figure 3.18). Synthesis of glutathione is carried out in two sequential steps (Figure 3.19). The first of these is catalysis by γ-glutamylcysteine ligase (GCL), which results in formation of the γ-linkage between glutamate and cysteine. The reaction requires ATP, and is the rate-limiting step in glutathione synthesis. GCL is a dimeric protein composed of a catalytic subunit that provides ligase activity and ATP hydrolysis and a regulatory subunit that lowers the $K_m$ for glutamate [248]. The second step of glutathione synthesis, addition of the glycine residue, is catalysis by glutathione synthetase, another ATP-dependent enzyme. Regulation of glutathione synthesis is mediated by GCL, which is sensitive to cellular redox status and glutathione concentration. GCL is also regulated at the transcriptional level in conjunction with glutathione *S*-transferases. Buthionine *S*-sulfoxime is a potent inhibitor of GCL that has been used extensively to study the biological and toxicological roles of glutathione. Cysteine, which is the rate-limiting substrate for glutathione synthesis, may come from several sources, including cystine, methionine, and the recycling of glutathione itself, as depicted in Figure 3.19 [157].

### Organization, Structure, and Localization of GSTs

The mammalian glutathione *S*-transferase superfamily is composed of three major families of proteins which are expressed in the cytosol, mitochondria, and endoplasmic reticulum. The cytosolic and mitochondrial proteins are dimeric, while the microsomal forms may exist as monomers, trimers, or higher order aggregates [100]. The current nomenclature is based on amino acid sequence similarity and subunit composition. Enzymes within a class share >40% sequence similarity (generally, ~70%), while different classes have <25% sequence similarity. In humans, 11 major classes of GSTs have been identified. The cytosolic GSTs make up the largest family and include the alpha (α), mu (μ), pi (π), sigma (σ), theta (θ), zeta (ζ), and omega (ω) classes. The mitochondrial isoform, only a single member of which has been identified, is designated as the kappa (κ) class. The microsomal GSTs are designated as the membrane-associated proteins in eicosanoid and glutathione metabolism (MAPEGs), to reflect their predominant role in eicosanoid synthesis.

#### *Cytosolic GSTs*

The cytosolic GSTs are the major isoforms involved in xenobiotic metabolism, comprising over 95% of total cellular GSTs. Cytosolic GSTs are dimeric proteins with subunit molecular weights of approximately 22 to 26 kDa. Seven classes of cytosolic GSTs have been identified in humans, with varying numbers of distinct subunits in each class, each of which is encoded by a separate gene. The α and μ classes each have five subunits, the θ and ω classes each have two subunits, and only a single subunit has been identified for each

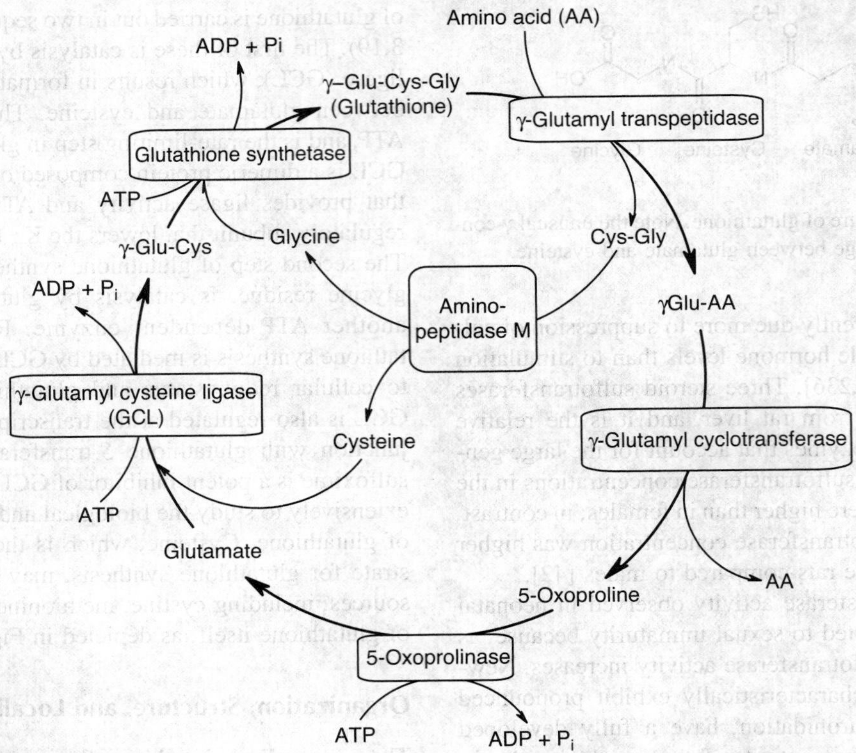


**FIGURE 3.19** The γ-glutamyl cycle: biosynthesis of glutathione. Cysteine for glutathione synthesis may come from other sources, including cystine and methionine.

of the π, σ, and ζ classes. Subunits from the α and μ classes can form heterodimers with other members of their class, but other classes of cytosolic GSTs exist only as homodimers. Cytosolic GSTs are named using an uppercase letter to designate class (A, M, P, S, T, Z, or O), followed by Arabic numerals to designate subunit composition; for example, GSTA1-1 is an α-class homodimer of subunit 1. Previous GST nomenclature systems were based either on apparent substrate selectivity or elution order from chromatography columns [114], and these designations are found in the older GST literature.

Like the cytochromes P450, the cytosolic GSTs exhibit broad and overlapping substrate specificities. The substrates for all isoforms are hydrophobic compounds possessing an electrophilic center. In addition to conjugation of xenobiotics, cytosolic GSTs play a role in metabolism of endogenous compounds, including amino acids, steroid hormones, and eicosanoids. Cytosolic GSTs are also capable of binding chemicals on the enzyme surface. This binding may or may not inhibit the catalytic activity of the enzyme, but it prevents the xenobiotic from interacting with other critical cellular sites, such as proteins and nucleic acids. Some GSTs had previously been termed *ligandins* because of this property [244,245].

### *Mitochondrial GSTs*

In rodents and humans, mitochondrial GST is a homodimeric κ-class enzyme (GSTK1-1). Structurally, mitochondrial GST is more similar to bacterial GSH-dependent isomerases and disulfide-forming oxidoreductases than it is to the mammalian cytosolic GSTs [100]. In rodents, GSTK is expressed at high concentrations in liver, kidney, stomach, and heart, but in humans the tissue expression of this enzyme is ubiquitous. GSTK has high activity for 1-chloro-2,4-dinitrobenzene (CDNB) and cumene hydroperoxide, as well as eicosanoid peroxides such as (*S*)-15-hydroperoxy-5,8,11,13-eicosatetraenoic acid. In addition to mitochondria, GSTK has also been found associated with peroxisomes, where it may play a role in fatty acid β-oxidation. Its role in the metabolism of environmental chemicals and drugs remains to be elucidated.

### *Microsomal GSTs*

Microsomal GSTs, currently referred to as MAPEGS, are membrane-bound enzymes involved primarily in eicosanoid synthesis. MAPEGs are divided into four subgroups, of which three (I, II, and IV) are expressed in mammals. MGST1, MGST2, and MGST3 are all capable of conjugating a variety of xenobiotic electrophiles as well as protecting against oxidative stress; for example,

MGST1 has been shown to catalyze regioselective and stereoselective conjugation of fluoroalkenes [121]. MGST2 and MGST3 also function as leukotriene $C_4$ synthetases. Although these enzymes are clearly capable of detoxification of xenobiotics, their quantitative contribution to xenobiotic metabolism is unclear.

## Biochemistry of Glutathione *S*-Transferases

Glutathione *S*-transferases involved in xenobiotic metabolism catalyze the conjugation of glutathione with substrates bearing an electrophilic atom [115]. The catalytic mechanism of GSTs has been reviewed in detail by Armstrong [10], and involves stabilization of the *S*-deprotein-ated form of glutathione (GS–). In the cytosolic classes of GST, this is accomplished by donation of a hydrogen bond to the glutathione cysteine sulfur, which effectively lowers the pKa of the thiol to between 6 and 7.5. Thus, in the enzyme-bound state, the thiolate ion is the predominant form of glutathione at physiological pH. The hydrogen bond may be provided by either tyrosine or serine, depending on the GST class. In the α-class enzymes, the thiolate receives additional stabilization from electrostatic interactions with an active-site arginine residue.

The substrate specificities of GSTs are broad and overlapping. The general structural characteristics of substrates for the cytosolic enzymes are that: (1) they are relatively hydrophobic, (2) they possess an electrophilic center, and (3) they will react nonenzymatically with glutathione at some measurable rate. The substrate selectivity of these enzymes is sensitive to small changes in the primary structure of the enzymes; for example, a glutathione *S*-transferase of the π-class contains a tyrosine at a site important for selectivity (Tyr108), whereas a transferase of the α-class contains a valine. When the tyrosine of the π-class enzyme is replaced with a valine, its substrate selectively is changed toward that of the α-class [189]. Similarly, by replacing a tryptophan residue in human GSTT1-1 with an arginine (found in the rodent homolog), the substrate selectivity can be changed to reflect the rodent enzyme [234].

Selective inhibitors of GST have been sought both as research tools and as potential therapeutic agents. Some commonly used inhibitors include ethacrynic acid, triphenyltinchloride, bromosulfophthalein, cibarcon blue, and hematin [269]. Considering the broad substrate specificity of GSTs, it should not be surprising that development of isoform-selective GST inhibitors has been challenging. A variety of chemotypes have been evaluated as selective GST inhibitors. These include GS–R conjugates (e.g., *S*-octylglutathione), glutathione peptide analogs (e.g., substitution of $SO_2NH$ for CONH in the γ-glu-cys linkage), nonpeptide GSH analogs, bivalent inhibitors that interact with both subunits of the dimeric protein simultaneously, and ligandin-type inhibitors [161]. These efforts have met with varying degrees of success, although truly isoform selective inhibitors remain elusive.

Although not highly efficient in its reactions (relatively high $K_m$ for the xenobiotic), glutathione *S*-transferases are capable of catalyzing or reacting with a number of reactive chemical functional groups. Any lack of efficiency is made up for by the high cellular concentration of glutathione and glutathione *S*-transferase. Liver glutathione concentrations are high (up to 10 m*M*) and glutathione *S*-transferases can represent as much as 10% of the total hepatocellular proteins [2]; however, it is possible for glutathione conjugation to become capacity limited at high doses of xenobiotics. Glutathione utilization can be faster than its synthesis, resulting in decreased conjugation and increased toxicity.

## Reactions of Glutathione *S*-Transferases

Some xenobiotics contain sufficiently electrophilic groups to react directly with glutathione, whereas others must first undergo phase I metabolism. Most xenobiotics react with glutathione through the catalytic activity of the glutathione *S*-transferases. Glutathione *S*-transferase reactions fall into four broad categories, as depicted in Figure 3.20: reaction with (1) electrophilic carbon, (2) nitrogen, (3) sulfur, or (4) oxygen [81,98,117]. For electrophilic heteroatoms, reaction with GSH results in formal reduction of the heteroatom with comcomitant oxidation of glutathione to its disulfide form (GSSG).

### *Reaction with Electrophilic Carbon*

The reactions of glutathione with electrophilic carbon can be divided into three types: (1) displacement reactions, (2) opening of strained rings, and (3) addition to activated double bonds:

- *Displacement of functional groups such as halides, sulfates, sulfonates, phosphates, and nitro groups from saturated or unsaturated carbon atoms*. When carbon is bonded to an electronegative atom or group of atoms, electron density is drawn away from the carbon nucleus, creating an electrophilic center. Many electron-withdrawing groups make good leaving groups for nucleophilic substitution reactions. Such reactions occur for alkyl ($sp^3$-hybridized), olefinic, and aromatic ($sp^2$-hybridized) carbons. Displacement of the leaving group is facilitated if the saturated carbon atom bearing the leaving group is allylic or benzylic. Displacement of halide or nitro groups on aromatic rings can occur via an addition–elimination mechanism if the aromatic ring contains additional elec-

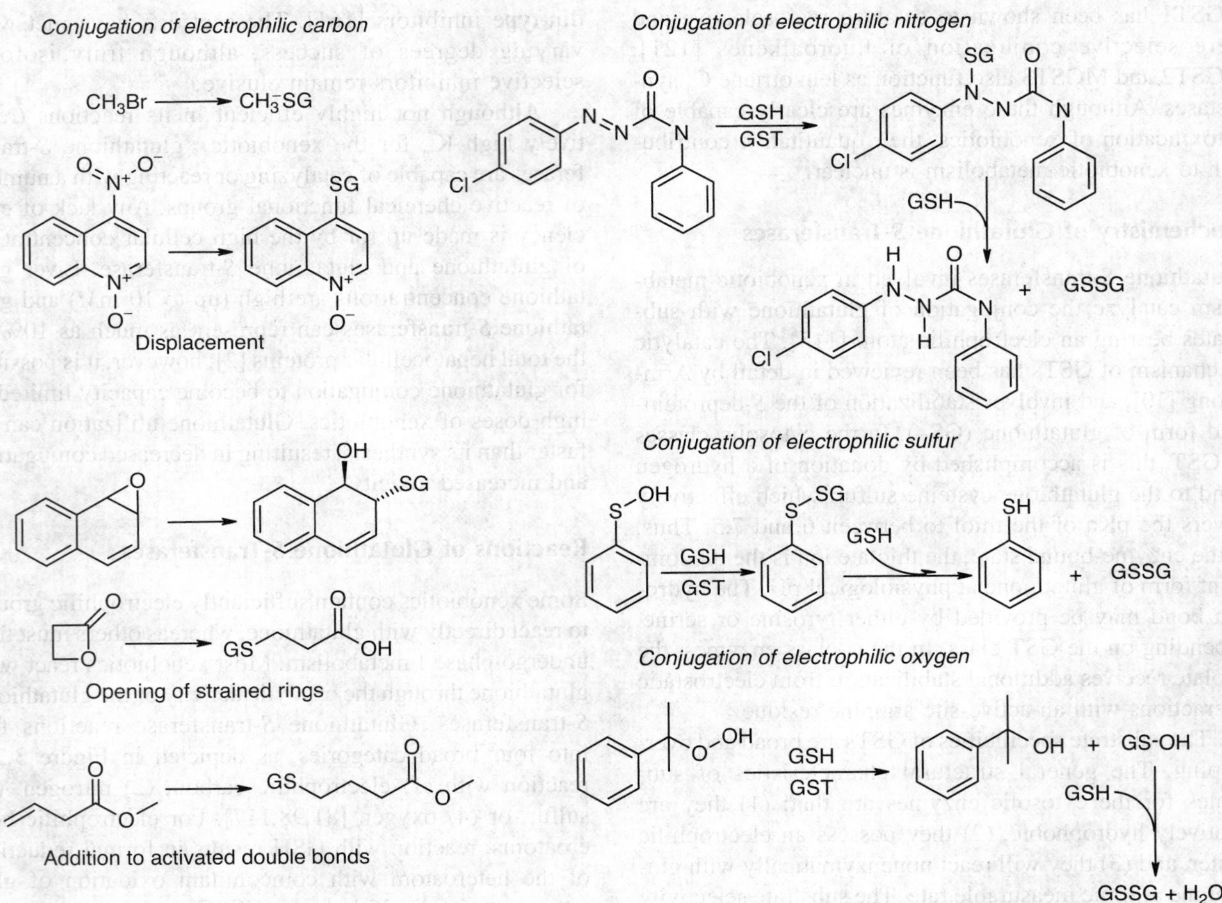


**FIGURE 3.20** Examples of the reaction types catalyzed by glutathione *S*-transferases.

tron-withdrawing groups of sufficient strength. (Figure 3.21). In this case, the rate of formation of a carbanion intermediate of the aromatic ring governs the overall rate of the reaction. Functional groups that withdraw electrons from the ring system stabilize the carbanion and are considered good leaving groups. On the other hand, electron donating substituents destabilize the anionic intermediate, deactivating the aromatic ring and making displacement of the leaving group by glutathione less likely.

- *Opening of strained rings, such as epoxides and four-membered lactones*. Ring systems with three or four members are highly strained, as the bond angles are forced from their minimum energy configuration (normally ~109° for $sp^3$-hybridized carbon). The presence of a heteroatom in such ring systems decreases electron density at the adjacent carbon and predisposes it to nucleophilic attack and ring opening. An example, shown in Figure 3.20, is conjugation of the 1,2-epoxide of naphthalene, resulting in a 1-naphthol conjugate of glutathione. These reactions can be stereoselective; for example, phenanthrene-9,10-epoxide is converted exclusively to the 9*S*,10*S*-diasteromeric conjugate by rat GSTM1-1, while GSTM2-2 produces approximately equal amounts of each diastereomer [49]. Epoxide products of P450 are detoxified by this reaction and are an example of a phase II conjugation of a phase I activated metabolite.
- *Addition to activated double bonds via Michael addition*. The presence of a carbonyl or cyano group adjacent to an olefinic group polarizes the electrons in the olefin double bond, creating an electron-deficient center on the β-carbon. The glutathione thiolate anion will attack β-unsaturated xenobiotics due to this partial positive charge on the β-carbon, leading to 1,2-addition of glutathione across the double bond, as shown in Figure 3.21.

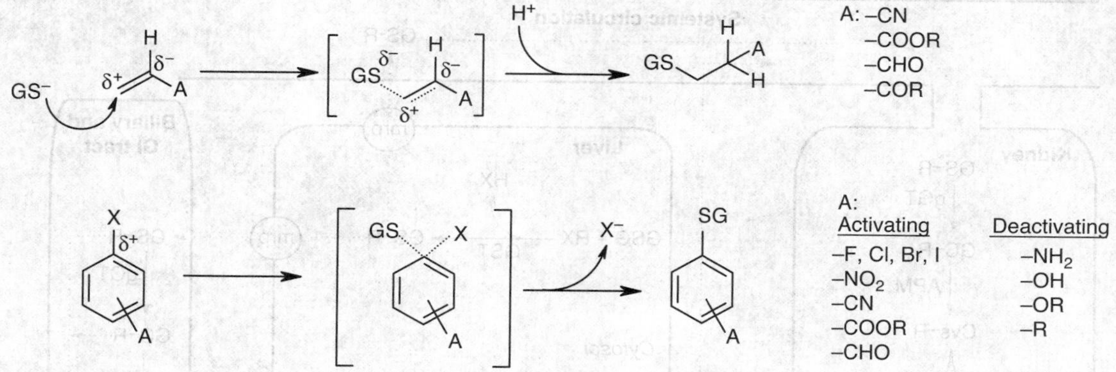

**FIGURE 3.21** Putative reaction mechanisms for the glutathione *S*-transferase-catalyzed Michael addition and aromatic substitution reactions. *X* represents a halogen atom, *A* represents the substituents listed to the right of each reaction, and *R* represents an alkyl group.

### *Reaction with Electrophilic Nitrogen*

Glutathione can react with electrophilic nitrogen atoms, such as diazenes [119]. The reaction is exemplified by the GSH-dependent reduction of the diazenecarboxamide JK-914 [180]. As seen in Figure 3.20, the first step of the reaction is analogous to the Michael addition of glutathione to polarized olefins. In the second step, the initial addition product reacts nonenzymatically with a second molecule of reduced glutathione, resulting in reduction of the diazenecarboxamide to a hydrazide with concomitant formation of oxidized glutathione.

### *Reaction with Electrophilic Sulfur*

Alkyl and aryl thiocyanates are substrates for glutathione *S*-transferase-catalyzed conjugations, as shown in Figure 3.20. Products of this nucleophilic attack of the thiolate ion on the sulfur of the xenobiotic result in a mixed disulfide and hydrogen cyanide. The mixed disulfide can react nonenzymatically with another molecule of glutathione to yield a thiol of the xenobiotic (RSH) and oxidized glutathione (GSSG).

### *Reaction with Electrophilic Oxygen*

Figure 3.20 illustrates how glutathione reacts with organic hydroperoxides in a two-step sequence. The first step is catalysis by glutathione *S*-transferase which forms an alcohol or phenol and a glutathione sulfenic acid intermediate (G-SOH). Another glutathione reacts nonenzymatically with the sulfenic acid to form oxidized glutathione and water. An example of this reaction with endogenous hydroperoxides is the conversion of hydroperoxy–prostaglandin $F_{2c}$ ($PGF_{2c}$) to prostaglandin $F_\alpha$ ($PGF_\alpha$). Cumene hydroperoxide is metabolized as rapidly by purified glutathione *S*-transferases as the classical transferase substrate probe 1-chloro-2,4-dinitrobenzene. Denitrosation of trinitroglycerol is another example of a reaction with electrophilic oxygen.

## The Metabolic Fate of Glutathione Conjugates: Mercapturic Acid Formation

Mercapturic acids are *N*-acetylated, *S*-substituted, cysteine conjugates that arise from conjugation of a xenobiotic with glutathione [30]. The glutathione conjugates formed in the liver and other tissues are polar and partition into the aqueous phase of cells and blood. Because 25% of the blood flow passes through the kidney, glutathione conjugates are transported to the kidney via systemic circulation. There the glutathione conjugate undergoes a series of reactions (shown in Figure 3.22) generally resulting in mercapturic acid formation; however, in some cases, bioactivation and nephrotoxicity are the outcomes.

The initial step in mercapturic acid synthesis is cleavage of glutamic acid from cysteine catalyzed by γ-glutamyltranspeptidase. This enzyme is located in the brush border of the proximal tubules in the kidney [291]. Evidence that this enzyme is involved in glutathione degradation comes from observations of pronounced glutathionemia and glutathionuria (high levels of glutathione in the blood and urine, respectively) in patients who lack detectable γ-glutamyltranspeptidase. This enzyme not only hydrolyzes the glutathione moiety, but also transfers the γ-glutamyl group to a variety of amino acids and dipeptides. These two reactions have been shown to proceed at equivalent rates under physiological conditions [293].

Next, the glycine group is cleaved from the resulting cysteinylglycine conjugate by aminopeptidase M, yielding the *S*-substituted cysteine conjugate of the xenobiotic. The cysteine conjugate is a substrate for *N*-acetyltransferases that acetylate the free amino group of cysteine to yield the mercapturic acid, which is excreted in the urine. Alternatively, the cysteine conjugate can be cleaved by renal cysteine conjugate β-lyase, possibly resulting in bioactivation and nephrotoxicity (discussed later). These two enzymes, γ-glutamyltransferase and aminopeptidase M,

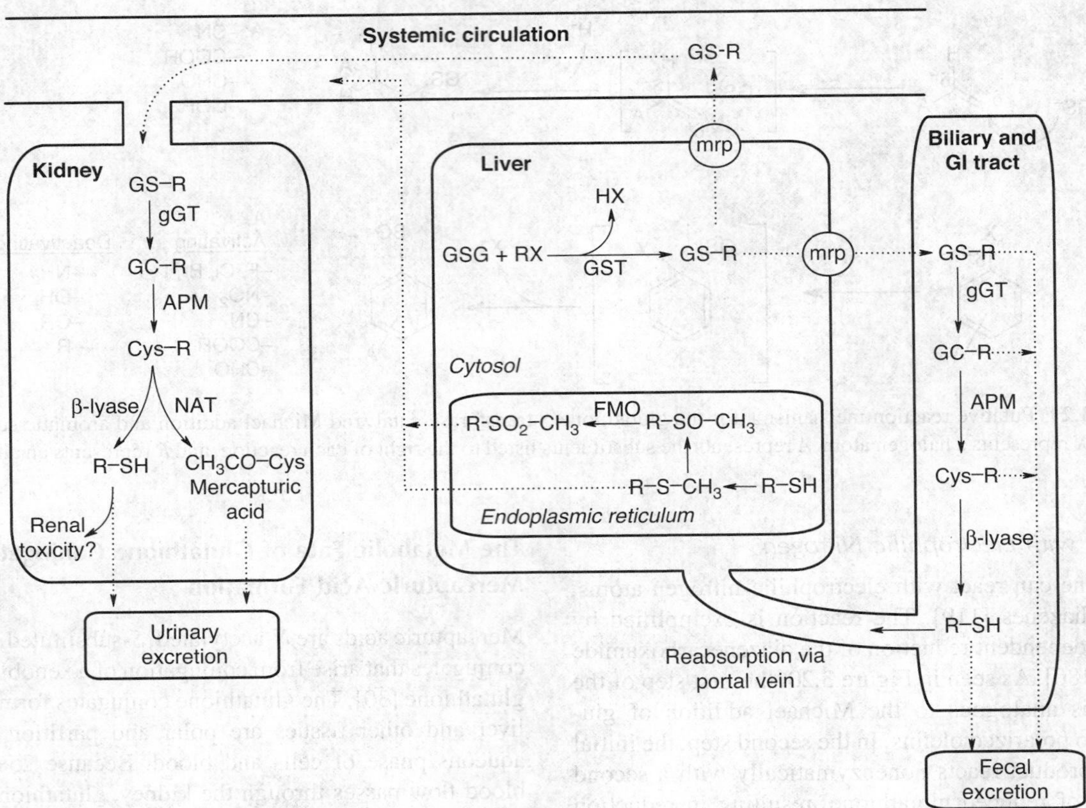


**FIGURE 3.22** Summary of the metabolic fate of glutathione conjugates. GST, glutathione *S*-transferase; gGT, γ-glutamyltransferase; APM, aminopeptidase M; FMO, flavin-containing monooxygenases; NAT, *N*-acetyltransferase. Refer to text for details.

are also responsible for the normal turnover of glutathione in mammalian cells previously shown in Figure 3.19.

## Role of Glutathione *S*-Transferase in Detoxification

Free reactive electrophilic intermediates of xenobiotics can produce damage to important cellular constituents. Reduced glutathione and the glutathione *S*-transferases protect cells from this damage by capturing the reactive electrophiles before they can react at nucleophilic sites critical to cell viability.

The metabolism of acetaminophen, an analgesic that at high doses can produce hepatic necrosis, serves as an example of this protective system. A large body of work has shown that one of the principal ways in which acetaminophen produces its hepatotoxicity is via the reactive intermediate, *N*-acetyl-*p*-benzoquinoneimine (illustrated later). This intermediate is an electrophile that reacts readily with glutathione and other tissue nucleophiles. As long as the amount of glutathione present at the site of activation of acetaminophen is sufficient to bind the reactive intermediate, no toxicity ensues; however, as demonstrated in the classic study by Mitchell [177], when glutathione is depleted by pretreatment with diethyl maleate, the benzoquinone imine covalently binds to tissue proteins, resulting in tissue necrosis. Understanding of the role of glutathione in protection against acetaminophen-induced hepatotoxicity led to the introduction of *N*-acetylcysteine (Mucomyst®) as a standard antidote for acetaminophen poisoning. Mitchell [177] was among the first to propose that glutathione plays a fundamental role in protecting tissues against electrophilic attack by xenobiotics.

Since these early studies demonstrating the protective role of glutathione, many compounds have been shown to form conjugates with glutathione. For a comprehensive review of these reactions, see Chasseaud [41] and Koob and Dekant [132].

## Factors Affecting Metabolism

Glutathione *S*-transferases have been found in most species, including reptiles, birds, insects, amphibians, and plants. Factors that influence the availability of reduced glutathione drastically alter the effectiveness of glutathione *S*-transferases. As was discussed previously, the toxicity of acetaminophen is modulated by the availability of reduced glutathione. Most xenobiotics that are highly reactive nonenzymatically with glutathione can deplete glutathione. Other mechanisms can also lower glutathione availability; for example, certain individuals have genetic defects in the γ-glutamyl cycle, resulting in low tissue concentrations of glutathione. These individuals generally are anemic due to the lack of glutathione and the resulting loss of protection from oxidative damage to erythrocytes [172].

As discussed earlier, cysteine is the limiting factor for synthesis of glutathione via the cycle shown in Figure 3.19. Nutritional factors that limit sulfur amino acid availability decrease glutathione *S*-transferase activity by reducing the availability of glutathione [260]. Methionine is an essential amino acid that can be used to synthesize cysteine and cystine via the transsulfuration pathway. Diets low in sulfur amino acids can decrease the availability of glutathione for conjugation with reactive intermediates of xenobiotics.

## Regulation of Glutathione *S*-Transferases

Glutathione *S*-transferases are inducible by a wide variety of xenobiotics including phenolic antioxidants, phenobarbital, and planar aromatic hydrocarbons. Dietary ingredients, such as cruciferous vegetables, specific components of coffee, butylated hydroxyanisole (BHA), and organosulfur compounds of allium vegetables, can also induce glutathione *S*-transferases. When cafestol and kahweol (diterpenes found in coffee) were administered to rats for up to 90 days, DNA adducts produced by aflatoxin $B_1$ were inhibited 50%. This appeared related to induction of glutathione *S*-transferase and a decrease in P450 isozymes involved in the metabolic activation of aflatoxin [35]. Coffee consumption has been shown to increase salivary concentrations of the glutathione *S*-transferases in humans [250]. Induction may be specific for one or more of the transferases and may be tissue specific.

Mechanisms of GST induction are complex and incompletely understood. Induction of GSTs by antioxidants such as BHA occurs at the level of transcription and is controlled in large part by the antioxidant response element (ARE), also known as the electrophile response element (EpRE) [186,205]. The major transcription factor that recognizes and binds to this response element is nuclear factor-E2-related factor 2 (Nrf2). Under basal conditions, this transcription factor is retained in the cytoplasm bound to the redox-sensitive protein Kelch-like ECH-associated protein 1 (Keap 1) and is targeted for proteosome degradation. Binding of Nrf2 to Keap 1 is controlled by one or more cysteine residues. Oxidation or binding of electrophiles to these cysteines destabilizes the Keap 1/Nrf2 complex, leading to release of Nrf2 and its translocation to the nucleus, allowing interaction with the ARE. Nrf2 can form functional heterodimers with several binding partners, including Maf and c-Jun. In addition to the ARE, other regulatory elements play a role in regulation of GSTs and other phase II enzymes that respond to oxidative stress and electrophiles, including C/EBPβ and the AP-1 family of transcription factors. Some GST genes also contain an XRE in their 5′-flanking region, and this response element may play a role in GST induction by 3-methylcholanthrene and other planar aromatic compounds [186]. Phenobarbital is thought to act through an AP-1 (*Fos*/*Jun*)-related mechanism [16], rather than through the constitutive androstane receptor (CAR), which mediates CYP induction by phenobarbital. As mentioned earlier, cysteine, which catalyzes the rate-limiting step in glutathione synthesis, is also regulated by the Nrf2 and other transcription factors involved in GST induction.

## Polymorphisms of Glutathione *S*-Transferases

Glutathione *S*-transferases exhibit polymorphic expression in humans. Among the most toxicologically relevant polymorphisms are those of the μ and θ class enzymes. Both of these isoforms exhibit null genotypes resulting from homozygous deletion of the corresponding gene. Because these enzymes play a critical role in protecting the cell from cytotoxic and mutagenic damage, a number of population studies have been done to determine relationships between genotype and disease. Several studies have attempted to correlate lung cancer risk and transferase expression, with mixed results. A better correlation has been found between transferase genotype and diseases associated with oxidative stress, especially for GSTM1 or GSTT1 polymorphism and colon cancer [85] and esophageal cancer [151]. No correlations were found between breast cancer and GSTM1 polymorphism [3]. The GSTM1*0 variant occurs at relatively high frequencies in Australians, Caucasians, and Africans [246] and has been associated with increased risk of cancers of the lung, colon, and bladder [170]. A rapid metabolizer phenotype for GSTM has also been described in Arabian populations, resulting from duplication of the GSTM1 gene. The GSTT1*0 genotype has been associated with an increased incidence of acute and chronic myelogenous leukemia. Polymorphic expression of other GST isoforms may also be of toxicological significance; for example, GSTO1-1 is the rate-limiting enzyme in the metabolism of inorganic arsenic, and variation of this enzyme may compromise an individual's ability to metabolize this toxic metal [257]. Similarly, GSTZ1-1 is involved in the detoxification of α-halocarboxylic acids such as dichloroacetic acid (DCA), and single nucleotide polymorphisms in the promoter region of this gene may impact DCA metabolism [75]. GSTP isoforms are overexpressed in a variety of human tumors, and this has led to their investigation as a chemotherapeutic target. GSTP1*A has been associated with the development of resistance to cisplatin and decreased response rate, while GSTP*B has correlated with decreased cisplatin metabolism and increased chemotherapeutic response. Overall, it appears that some of the highest correlations between genotype and cancer susceptibility are those where P450 genotype and transferase genotype are combined for analysis. This again emphasizes the close relationship between metabolic activation by phase I enzymes and detoxification of reactive metabolites by phase II enzymes.

## Species and Gender Differences

As mentioned earlier, glutathione *S*-transferases have been found in most species investigated. Species differences in the expression, substrate specificity, and activity of these transferases can have a significant role in the toxicity of xenobiotics. For example, rats are susceptible to the potent hepatocarcinogen aflatoxin $B_1$ ($AFB_1$), whereas mice are extremely resistant. This species difference results from the expression in mice of mGSTA3-3, which has a high activity toward the P450-generated activated metabolite of $AFB_1$ (the 8,9 epoxide). Although rats express a closely related transferase (rGSTA3-3), it has low activity toward the epoxide. These two transferases have equivalent activity toward a probe substrate (1-chloro-2,4-dinitrobeneze), but the rat form has 1000-fold less activity toward the $AFB_1$ epoxide compared to the mouse. This difference in activity between the transferases from the two species appears to be based on differences in as few as six critical amino acids [272].

Hepatic glutathione *S*-transferase activities are low in prepubertal male and female rats. As the rats reach sexual maturity between 30 and 50 days of age, glutathione-conjugating activity toward dichloronitrobenzene is two- to threefold higher in males than in females [140]. This difference in glutathione *S*-transferase activity was not related to sex steroids but was dependent on pituitary secretions. Growth hormone may play a role in establishing glutathione *S*-transferase activities [139], as it does with P450. Although growth hormone is important in regulating adult levels of glutathione *S*-transferase in the rat, it appears that other factors also play a role. The student of toxicology should be aware of the multifaceted way that xenobiotics can affect organisms; for example, monosodium glutamate, which produces lesions in the arcuate nucleus of the hypothalamus, can lower the glutathione *S*-transferase activity in male rats. This, in turn, could increase their sensitivity to electrophilic chemicals.

Gender differences in the expression of glutathione *S*-transferase have been suggested to be responsible for the higher susceptibility of female mice to the carcinogenicity of benzo(*a*)pyrene compared to males. Males express higher mGSTP1-1, mGSTA3-3, mGSTM1-1, and mGSTA4-4 compared to females. At higher doses of benzo(*a*)pyrene this gender difference is lost, possibly due to the higher doses overcoming the protective role of the higher transferase activity in males [238].

Some studies suggest that humans do not demonstrate gender differences in glutathione *S*-transferases. No gender or age differences were seen in GSTM and GSTP activity in human lymphocytes, but an age-dependent decrease in glutathione was detected [271].

## Role of Glutathione *S*-Transferases in Metabolic Activation

Glutathione conjugation does not always produce an innocuous and readily excreted metabolite; for example, Elfarra and Anders [70] compiled a list of 1,2-dihaloalkanes and halogenated alkenes whose glutathione or cysteine conjugates were nephrotoxic. Glutathione reacts with these 1,2-dihaloalkanes via a glutathione *S*-transferase-catalyzed reaction that yields sulfur mustards. An electrophilic episulfonium ion can be formed from the mustard when the second halogen atom is displaced by a cellular nucleophile (Figure 3.23). The episulfonium ion intermediate has been implicated in the toxicity of these chemicals. The major DNA adduct resulting from exposure to the carcinogen 1,2-dibromoethane was *S*-2-$N^7$-guanylethylglutathione [111]. A brief review of this bioactivation pathway is included in Anders et al. [6] and Vambakas and Anders [268].

As shown in Figure 3.22, glutathione and cysteine conjugates (GSR and CySR, respectively) formed in the liver can be excreted in the bile. Glutathione conjugates can be hydrolyzed to cysteine conjugates by γ-glutamyltranspeptidase/aminopeptidase M, present in the bile duct epithelia, or by pancreatic peptidases in the small intestine. Cysteine conjugates originating from the bile and those formed by hydrolysis of glutathione conjugates are good substrates for microfloral β-lyase. β-Lyase, an enzyme found in liver, kidney, and intestinal microflora, cleaves thioether linkages in cysteine conjugates of xenobiotics [259]. The resulting thiol compounds are more hydrophobic than the conjugates and can be readily absorbed in the small intestine. These thiol metabolites return to the liver via the portal circulation and act as substrates for thiol *S*-methyltransferase that methylates the thiol group. Enterohepatic circulation of glutathione conjugates accounts for some of the unusual sulfur-containing metabolites that have been found in the urine of animals treated with xenobiotics, such as propachlor [208] and acetaminophen [19]. A portion of the glutathione-derived sulfur-containing metabolites formed in the small intestine is excreted in the feces.

Reactions of glutathione and cysteine conjugates of compounds shown in Figure 3.24 are believed to play a role in the nephrotoxicity of several xenobiotics. Cysteine conjugates are actively transported into renal tubular epithelia, where they may be bioactivated by renal β-lyase [144]; for example, the cysteine conjugate of trichloroethylene, *S*-(1,2-dichlorovinyl)-L-cysteine (DCVC), is a potent nephrotoxin and a β-lyase substrate. Inhibition of renal β-lyase with aminooxyacetic acid, an inhibitor of pyridoxyl phos-

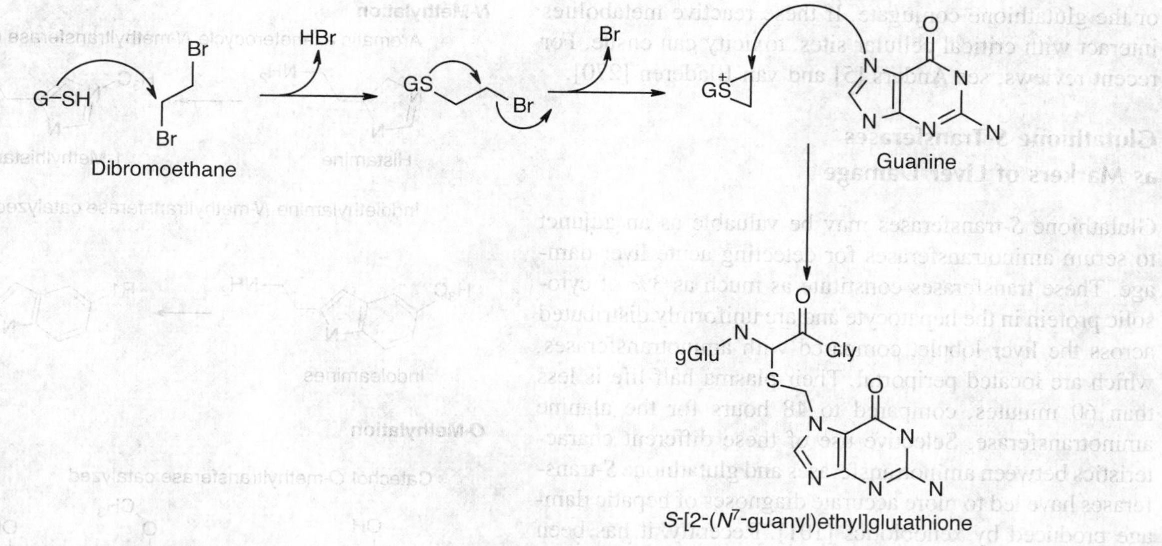

FIGURE 3.23 Bioactivation of 1,2-dibromoethane by conjugation with glutathione, resulting in formation of a DNA-reactive episulfonium ion.

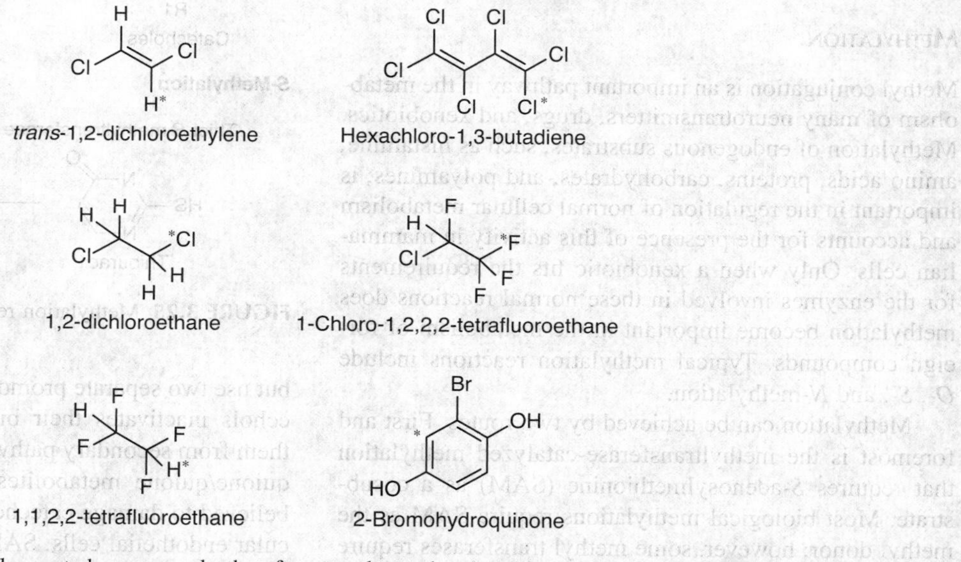

FIGURE 3.24 Representative halogenated compounds that form nephrotoxic glutathione conjugates. The site of reaction with glutathione is indicated by an asterisk (*).

phate-dependent enzymes, protected against DCVC-induced nephrotoxicity [70]. In contrast, the nephrotoxicity of hexachloro-1,3-butadiene was enhanced by the γ-glutamyltranspeptidase inhibitor acivicin, suggesting that the nephrotoxicity of this polyhaloalkene may not be mediated by its glutathione conjugate [5,59].

Another glutathione-dependent bioactivation mechanism involves the reversible conjugation of isocyanates such as methylisocyanate, the chemical responsible for the Bhopal disaster of 1984. Isocyanates are excellent substrates for glutathione conjugation, resulting in formation of an *S*-carbamoylated glutathione; however, the reaction is reversible, which can lead to regeneration of the free isocyanate. Furthermore, both the glutathione and cysteine participate in transcarbamoylation reactions with tissue free sulfhydryl groups and other nucleophiles [200]; thus, it has been suggested that glutathione conjugation may actually increase the toxicity of isocyanates by facilitating distribution and release of isocaynate within the body [241].

In general, glutathione conjugate synthesis results in readily excreted polar metabolites; however, in some cases, the residence time of a glutathione conjugate in the body is prolonged. This can result in formation of metabolites that are more reactive than the original parent xenobiotic or the glutathione conjugate. If these reactive metabolites interact with critical cellular sites, toxicity can ensue. For recent reviews, see Anders [5] and van Bladeren [270].

### Glutathione *S*-Transferases as Markers of Liver Damage

Glutathione *S*-transferases may be valuable as an adjunct to serum aminotransferases for detecting acute liver damage. These transferases constitute as much as 3% of cytosolic protein in the hepatocyte and are uniformly distributed across the liver lobule, compared with aminotransferases, which are located periportal. Their plasma half-life is less than 60 minutes, compared to 48 hours for the alanine aminotransferase. Selective use of these different characteristics between aminotransferases and glutathione *S*-transferases have led to more accurate diagnoses of hepatic damage produced by xenobiotics [101]. Recently, it has been suggested that determination of glutathione *S*-transferase should be included in toxicology studies, and validated methods for rats and dogs have been developed [127].

## Methylation

Methyl conjugation is an important pathway in the metabolism of many neurotransmitters, drugs, and xenobiotics. Methylation of endogenous substrates, such as histamine, amino acids, proteins, carbohydrates, and polyamines, is important in the regulation of normal cellular metabolism and accounts for the presence of this activity in mammalian cells. Only when a xenobiotic fits the requirements for the enzymes involved in these normal reactions does methylation become important in the metabolism of foreign compounds. Typical methylation reactions include *O*-, *S*-, and *N*-methylation.

Methylation can be achieved by two routes. First and foremost is the methyltransferase-catalyzed methylation that requires *S*-adenosylmethionine (SAM) as a cosubstrate. Most biological methylations require SAM as the methyl donor; however, some methyl transferases require SAM as a cosubstrate but vary in other requirements for optimal activity [164]. Reactions involving four of these SAM-dependent methyltransferases are shown in Figure 3.25. A secondary source of methylation is $N^5$-methyltetrahydrofolate (5-$CH_3$-THF)-catalyzed methylation. This methylation is important in the synthesis of nucleic acids; however, 5-$CH_3$-THF is 1000 times less reactive toward soft nucleophiles than SAM, suggesting that it plays a smaller role in xenobiotic metabolism.

The methylation of catechol oxygen atoms is catalyzed by catechol *O*-methyltransferase (COMT), which is most widely known as the enzyme catalyzing the methylation and deactivation of dopamine, other catecholamines, and catechol estrogens. Both cytosolic and membrane-bound forms exist and are encoded by a single gene

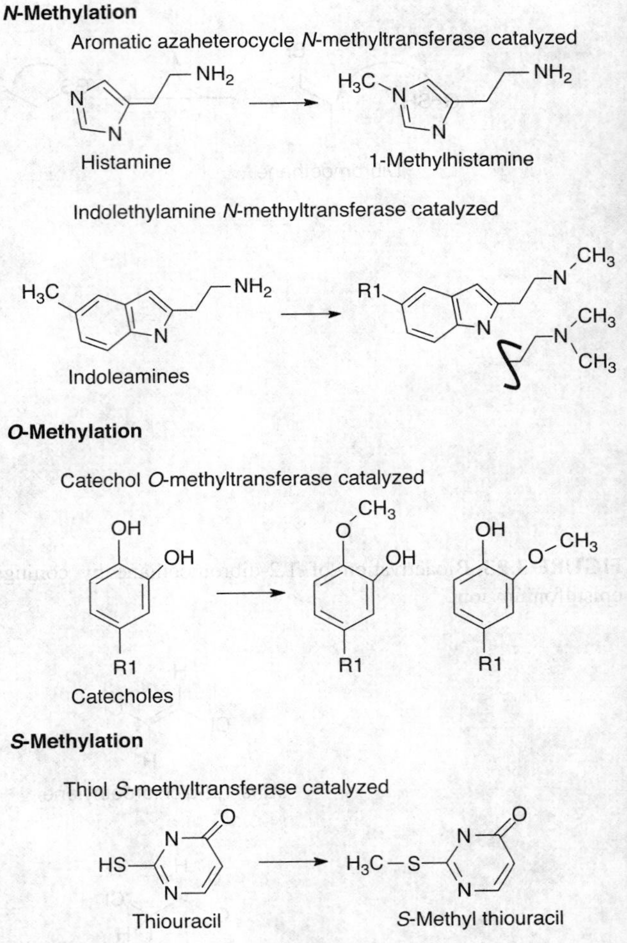

**N-Methylation**

Aromatic azaheterocycle *N*-methyltransferase catalyzed

Histamine → 1-Methylhistamine

Indolethylamine *N*-methyltransferase catalyzed

Indoleamines

**O-Methylation**

Catechol *O*-methyltransferase catalyzed

Catecholes

**S-Methylation**

Thiol *S*-methyltransferase catalyzed

Thiouracil → *S*-Methyl thiouracil

**FIGURE 3.25** Methylation reactions.

but use two separate promoters [298]. Methylation of catechols inactivates their biological function and diverts them from secondary pathways that lead to reactive semiquione/quione metabolites and oxyradicals, which are believed to damage catechol-containing neurons and vascular endothelial cells. SAM availability as a co-factor is critical to COMT activity. SAM provides the methyl group via COMT liberating *S*-adenosyl homocysteine (SAH). Because SAH can have an inhibitory effect on SAM, nutritional states that impair the removal of SAH and resynthesis of SAM can have negative consequences on SAM homeostasis and SAM-dependent methylation pathways. Thus, nutritional and biochemical deficiencies in $B_6$, $B_{12}$, and folate will disrupt SAM/SAH homeostasis.

Two enzymes catalyze nitrogen methylation: aromatic azaheterocycle *N*-methyltransferase and indolethylamine *N*-methyltransferase. *N*-methylation of azaheterocycles typically leads to quaternary azaheterocycles. Examples where biotransformation has led to toxic products are more well known and include the biosynthesis of paraquat the potent lung toxin, and 1-methyl-4-phenyl-1,2,3,6-

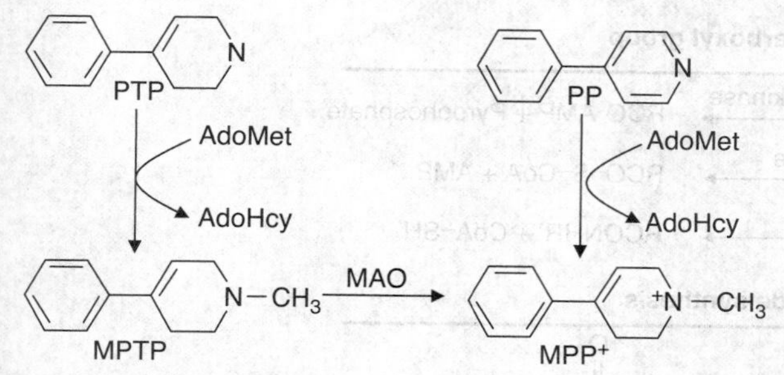

**FIGURE 3.26** Metabolic conversion of 4-phenyl-1,2,3,6-tetrahydropyridine (PTP) and 4-phenylpyridine (PP) to the Parkinsonian neurotoxin 1-methyl-4-phenylpyridinium ion ($MPP^+$): MPTP is oxidized by monoamine oxidase in the brain to $MPP^+$, which is selectively taken up into dopaminergic cells of the nigrostriatum via the dopaminergic reuptake transporter. Once in the pathway of the cells, $MPP^+$ becomes concentrated in mitochondria, where it causes cell death by inhibiting mitochondrial respiration. (From Hoffman, J.L., *Adv. Pharmacol.*, 27, 449–477, 1994. With permission.)

tetrahydropyridine (MPTP), which, following oxidation via monoamine oxidase (MAO), leads to the formation of 1-methyl-4-phenylpyridinium ions ($MPP^+$). $MPP^+$ is the active metabolite associated with Parkinsonian syndrome which has been observed in humans after self-administration of designer drugs contaminated with MPT (Figure 3.26) [104]. These *N*-methylation reactions form inhibitors of mitochondrial respiration.

The second pathway, catalyzed by indolethylamine *N*-methyltransferase, catalyzes the methylation of primary amines to secondary and tertiary amines. This enzyme *N*-methylates endogenous biogenic amines such as serotonin, tyrptamin, tyramine, norepinephrine, and dopamine, as well as drugs and xenobiotics such as amphetamine, normorphine, and aniline.

Sulfur methylation occurs via thiol *S*-methyltransferase and is also dependent on SAM as the methyl donor. Substrates for *S*-methyltransferase reactions range from hydrogen sulfide to thiopurine. Hydrogen sulfide, produced by the anaerobic metabolic activity of gut microflora, is initially methylated by gut mucosal *S*-methyltransferase to yield methanthiol. Although a poorer substrate, methanthiol can be further methylated to dimethylsulfide. *S*-Methyltransferase activity is a well-known detoxification pathway for the thiopurine drug 6-metcaptopurine and its pro-drug azathioprine, the activity being known as thiopurine *S*-methyltransferase (TPMT). These cytotoxic drugs are effective against acute lymphoblastic leukemia and autoimmune diseases but have narrow therapeutic indexes, and toxic overdoses can be induced in patients with genetic variants of TPMT [285]. Through relatively simple measurements of TPMT activity in red blood cells, patient dose can be effectively modulated to match patient genotype and phenotype.

## Amide Synthesis

Amide biosynthesis can take place via two principal routes:

- Conjugation of a carboxylic-acid-containing xenobiotic with the free amino group of an amino acid such as glycine
- Acetylation of a xenobiotic containing a primary amine ($-NH_2$)

### Amino Acid Conjugation

Xenobiotics that contain a carboxylic acid moiety are susceptible to conjugation with endogenous amino acids. Xenobiotic conjugation occurs in hepatic mitochondria. The free carboxylic acid is activated by reaction with ATP followed by reaction with coenzyme A (CoA), as shown in reactions 1 and 2 of Figure 3.27. For example, the carboxylic acid of benzoic acid is activated to a thioester CoA intermediate that reacts with the primary amine of glycine to form the amide hippuric acid.

Glycine has historical significance in xenobiotic conjugation because it is one of the earliest reactions attributed to xenobiotic metabolism. Keller [125], in 1842, administered benzoic acid to himself and then isolated and characterized the major metabolite, hippuric acid, a glycine conjugate. This reaction has been used as a liver function test in humans. The liver is the principal site of glycine conjugation. Other amino acids, such as taurine, can be used for conjugating aliphatic, aromatic, and heterocyclic carboxylic acids. Taurine conjugates of pioglitazone metabolites were identified in the bile of treated dogs [232], and arginine is used by arachnids, glutamine by chimpanzees, and ornithine by certain birds.

### Acetylation

Acetylation, catalyzed by *N*-acetyltransferases, is the principal pathway of amide formation for primary aromatic amines, endogenous primary aliphatic amines, anutrient amino acids, hydrazines, hydrazides, and sulfonamides. *N*-Acetyltransferase catalyzes the two-step transfer of an acetyl group from the donor (acetyl-CoA) to the aromatic amine. These enzymes are cytosolic, occur in many tissues, and are comprised of at least three families of isoenzymes. Acetyltransferases can also catalyze *O*-acetylation of xenobiotics, as has been shown especially for the acetylation of aryl hydroxylamines such as *N*-hydroxy-2-aminofluorene and 4-aminobiphenyl (Figure 3.28).

Some species, including humans, rabbits, and hamsters, express two independently regulated transferases, NAT1 and NAT2; other species, such as mouse, express three: NAT1, NAT2, and NAT3. NAT 1 and NAT 2 have been studied most extensively. They are structurally similar proteins with a cysteine residue at the active site, but they have different substrate specificities, although some

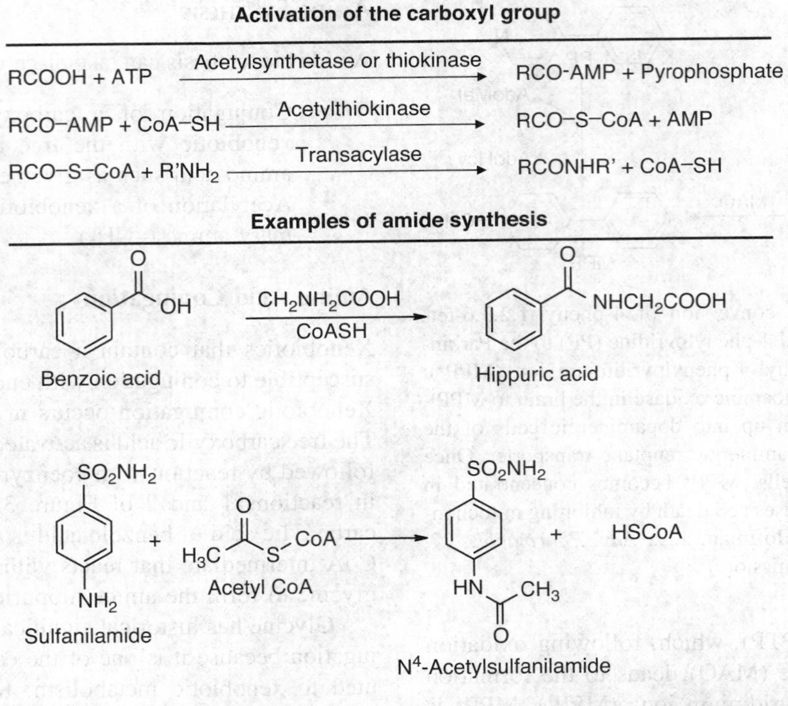

**Activation of the carboxyl group**

RCOOH + ATP —(Acetylsynthetase or thiokinase)→ RCO-AMP + Pyrophosphate

RCO–AMP + CoA–SH —(Acetylthiokinase)→ RCO–S–CoA + AMP

RCO–S–CoA + R′$NH_2$ —(Transacylase)→ RCONHR′ + CoA–SH

**Examples of amide synthesis**

Benzoic acid —($CH_2NH_2COOH$, CoASH)→ Hippuric acid ($NHCH_2COOH$)

Sulfanilamide ($SO_2NH_2$, $NH_2$) + Acetyl CoA ($H_3C$–CO–S–CoA) → $N^4$-Acetylsulfanilamide (HN–CO–$CH_3$) + HSCoA

FIGURE 3.27 Series of reactions leading to amide formation from either a xenobiotic containing a carboxylic functional group (RCOOH) or a primary amine group (R′$NH_2$).

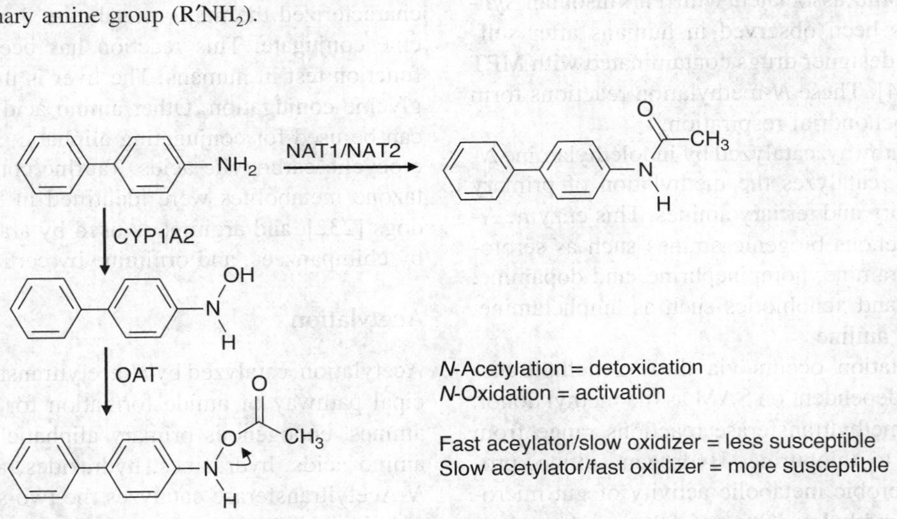

4-Aminobiphenyl ($NH_2$) —(NAT1/NAT2)→ N-acetyl-4-aminobiphenyl (NH–CO–$CH_3$)

4-Aminobiphenyl —(CYP1A2)→ N-hydroxy-4-aminobiphenyl (N(H)–OH) —(OAT)→ N-acetoxy-4-aminobiphenyl (N(H)–O–CO–$CH_3$) —(Heterolytic cleavage)→ nitrenium ion ($\overset{+}{N}$–H) + acetate ($CH_3COO^-$) → Reactions with DNA, protein

*N*-Acetylation = detoxication
*N*-Oxidation = activation

Fast aceylator/slow oxidizer = less susceptible
Slow acetylator/fast oxidizer = more susceptible

FIGURE 3.28 Metabolism of the bladder carcinogen 4-aminobiphenyl. *N*-Acetyltransferase (NAT) conjugation is a detoxification pathway that competes with CYP1A2 *N*-oxidation, which is the first step in metabolic activation that ultimately forms a highly reactive nitrenium ion. Some ethnic groups, such as those of Middle Eastern populations, have a high incidence of the slow acetylator phenotype, which, when coupled with a fast *N*-oxidation phenotype, results in a much higher risk of bladder cancer.

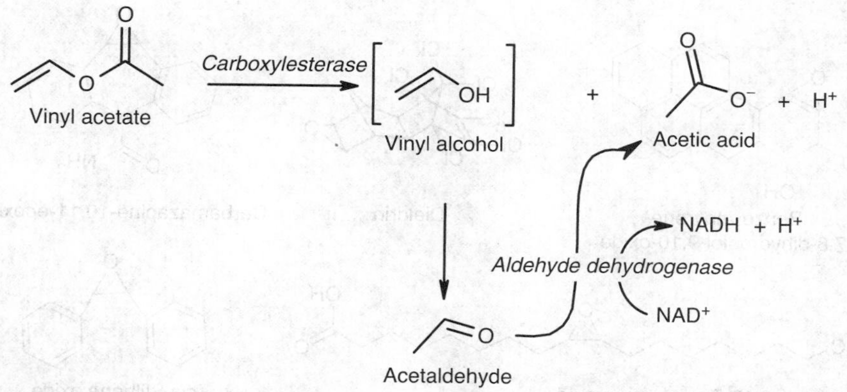

**FIGURE 3.29** Carboxylesterase-mediated hydrolysis of vinyl acetate. Esterase hydrolysis is so efficient that cytochrome P450-mediated epoxidation of the double bond does not occur. Vinyl alcohol is an unstable intermediate and readily rearranges to form acetaldehyde, which undergoes further oxidation to acetic acid. The potential toxicity of acidic metabolites is often overlooked, but, in the case of vinyl acetate, the acetic acid generated contributes substantially to the mechanism of toxicity. (From Bogdanffy, M.S. and Valentine, R., *Toxicol. Lett.*, 140/141, 83–98, 2003. With permission.)

overlap occurs. NAT1 is expressed in most tissues, whereas NAT2 is expressed only in the liver and gut. Polymorphisms in *N*-acetyltransferases are well described for the human population, and more than 26 different alleles have been reported [28]. NAT2 and more recently NAT1 have been shown to have several allelic variants resulting in fast and slow acetylator status. Especially well described for the NAT2 enzyme, acetylator status can affect susceptibility to drugs such as dapsone and isoniazid in which slow acetylation leads to peripheral neuropathy.

Acetylation of aromatic amines is generally a detoxification pathway because the added acetyl group blocks further oxidation of the amide nitrogen; however, once oxidized, the resulting hydroxylamine can undergo *O*-acetylation. The resulting acetoxy ester is unstable, and heterolytic loss results in the formation of highly reactive nitrinium ions that form adducts with cellular DNA and consequently are mutagenic (Figure 3.28). It is frequently the balance between *N*-oxidation and *N*-acetylation that determines susceptibility to aromatic amine carcinogenesis. Dogs, which lack NAT activity, are highly susceptible to arylamine-induced carcinogenesis. Human susceptibility has been demonstrated for 4-aminobiphenyl in which *N*-oxidation and *N*-acetylation status is tightly associated with risk of bladder cancer among smokers. High *N*-oxidation status coupled with slow acetylator phenotype yields the highest bladder cancer risk among smokers.

Mercapturic acid formation in the kidney is an example of acetylation that has been presented and represents one of the unusual circumstances in which aliphatic amines undergo *N*-acetylation. In this reaction, the primary amine group of the cysteine conjugate of the xenobiotic is acetylated to form the mercapturic acid. This is an exception to the rule that aliphatic primary amines generally are not good substrates for the *N*-acetyltransferases.

## HYDROLYSIS

Many xenobiotics and their phase I metabolites contain a carboxyl ester, an amide bond, or an epoxide that masks hydrophilic functional groups, such as alcohols, carboxylic acids, and amines. The rate at which an organism can hydrolyze these bonds and unmask these function groups can influence their toxicity. In fact, pesticides and therapeutic drugs have been synthesized with intent to modulate the bioavailability of the active species by affecting the rate of hydrolysis of the parent compound.

Hydrolysis normally competes with other detoxification reactions, but esterases are in very high content in many tissues, especially liver, and their affinity is low enough such that esterase-/amidase-mediated hydrolysis typically predominates. An example of competition is demonstrated by the metabolism of vinyl acetate (Figure 3.29). This molecule contains both a double bond and an ester group and therefore would be expected to be a substrate for both epoxidation and carboxylesterase-mediated hydrolysis; however, the metabolism of this compound via the carboxylesterase pathway is so efficient that no epoxide is formed. Hydrolysis products such as alcohols, amines, or thiols and carboxylic acids and are typically further metabolized.

### Epoxide Hydrolase

Organisms may be exposed to epoxides in the environment or they may be produced during the oxidative metabolism of specific xenobiotics from their environment. Epoxides generally are reactive electrophilic compounds due to the highly strained oxirane ring. Excess strain energy can be released by ring opening in the presence of nucleophiles. Ring opening may follow either a $SN_1$-type mechanism, with the formation of an intermediate with carbonium ion character, or an $SN_2$ mechanism, with bond formation with

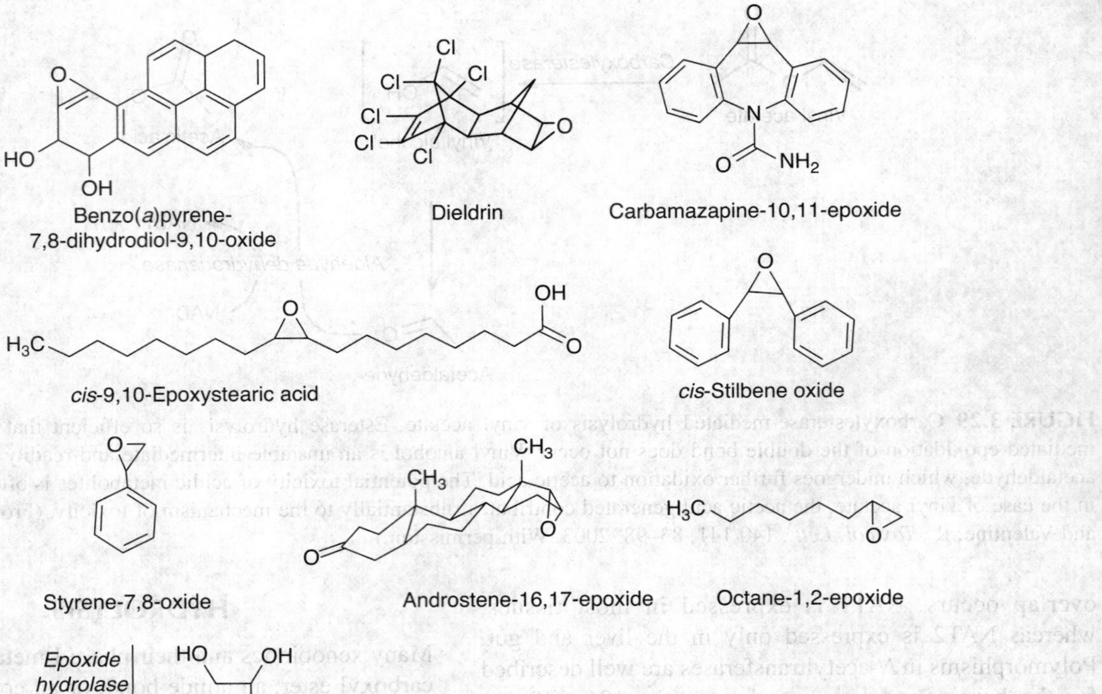

**FIGURE 3.30** Structural diversity of epoxides that are substrates for epoxide hydrolase. Benzo(*a*)pyrene-7,8-dihydrodiol-9,10-oxide illustrates a potent mutagen which is first oxidized to the 7,8-epoxide, followed by epoxide hydrolase hydrolysis of the epoxide and then secondary oxidation at the 9,10 position. The dihydrodiol imparts a steric hindrance toward the 9,10 epoxide, leaving the epoxide relatively, although not completely, resistant to hydrolysis. (Adapted from Arand, M. et al., *Drug Metab. Rev.*, 25, 365–383, 2003.)

the attacking nucleophile. The latter case has important toxicological consequences when the nucleophile is on a critical tissue macromolecule, such as DNA, because it results in covalent modification of the macromolecule. Modification of DNA results in a biochemical lesion that may be the precursor to a number of pathological lesions, including cancer. Reaction of the epoxide with cellular nucleophiles, such as proteins, could also lead to other mechanisms producing acute or subchronic toxicity.

The chemical reactivity and, consequently, the biological activity of epoxides are influenced by the constituents attached to the oxirane ring carbons. Asymmetric substitution and electron withdrawing substituents near the oxirane ring tend to destabilize the epoxide and enhance its reactivity. Epoxides that hydrolyze in water are among the most reactive and electrophilic but may not be toxic when hydrolyzed. Alternatively, if they are generated close to or have a long enough half-life to reach and react with critical cellular macromolecules, epoxides can have both cytotoxic and genotoxic consequences. An example of a relatively stable epoxide is styrene-7,8-oxide, which is the substrate typically used as a generic assay for epoxide hydrolase activity (Figure 3.30).

Although not always the case, the epoxides that are formed *in vivo* appear to be more toxicologically important than those that occur in the environment. Highly reactive epoxides would most likely interact with nucleophilic sites in the environment, such as proteins in food, and not be absorbed in their active form. Epoxides formed *in vivo* are generally produced close to their sites of action and require only diffusion or short transport to their target. Epoxides most frequently formed *in vivo* represent alkene and arene oxides produced by P450. Their efficient detoxification is important to cellular survival.

Detoxification of epoxides may follow several routes:

- Spontaneous decomposition
- Nonenzymatic reaction with glutathione
- Reaction with glutathione catalyzed by glutathione transferase
- Hydration by epoxide hydrolase
- Minor mechanisms such as a P450 hydrolysis

Nonenzymatic and enzymatic conjugations with glutathione have been previously discussed.

A major route for biodisposition of epoxides is hydration catalyzed by epoxide hydrolase to vicinal (from the Latin *vicinalis*, "neighboring") dihydrodiols. The membrane-bound and cytosolic forms were first characterized in 1973 and 1976, respectively. They are now known to comprise a large and heterogeneous group of enzymes. Their structure, function, and mechanism have been reviewed recently [9]. This microsomal enzyme catalyzes the biotransformation of a diverse group of arene oxides and aliphatic epoxides (Figure 3.30). In most cases, this enzymatic pathway results in less reactive diol metabolites that are more readily excreted from the organism, either as the diol or as a glucuronide or sulfate conjugate of the diol.

Epoxide hydrolases occur as membrane-bound proteins located in the endoplasmic reticulum and as a soluble enzyme in the cytosol of most mammalian cells. The membrane-bound microsomal form has broad substrate specificity, while the soluble, or cytosolic, form has a higher affinity for nonbulky *trans*-substituted oxiranes [107,220].

Membrane-bound epoxide hydrolase has a 20-amino-acid sequence at the *N*-terminal end that anchors it to the membrane. The active site of the enzyme occurs outside of the membrane. Unlike P450, if the anchor sequence of the protein is not present, the enzyme retains a portion of its catalytic activity [79]. Polymorphic forms of both the microsomal and soluble forms exist, but these variations appear to affect protein half-life rather than catalytic activity.

Epoxide hydrolase has been found in a variety of tissues, including liver, kidney, lung, skin, intestine, colon, testis, ovary, spleen, thymus, heart, and brain. The activity of liver microsomal enzyme is relatively low in newborn rats and increases during neonatal development until adult males have about twice the activity of females. This sexual dimorphism is remarkably similar to that seen in the rat for P450. In contrast, the renal epoxide hydrolase of male and female rats does not demonstrate age-dependent changes or gender differences. Human hepatic microsomal epoxide hydrolase activities increase during gestation, but no gender difference in humans has been observed [193].

Humans demonstrate considerable variation in epoxide hydrolase activity. This, in part, is associated with the inducibility of epoxide hydrolase and environmental exposures and life-style differences among individuals. In addition, evidence suggests that genetic polymorphisms with the enzymes result from amino acid sequence differences. These human polymorphisms may not result in significantly altered enzyme activity or posttranscriptional regulation [145], although more work is required in this area. Polymorphic expression of epoxide hydrolase has been related to specific human diseases [141,243].

The activity of this enzyme is induced by the classical inducers of cytochromes P450. Although *trans*-stibene oxide has been shown to be an inducer of epoxide hydrolase, no specific inducer of epoxide hydrolase has been reported. Two widely used inhibitors of epoxide hydrolase are trichloropropane oxide and cyclohexene oxide.

$$R\text{-}C(=O)\text{-}O\text{-}R' + H_2O \longrightarrow R\text{-}C(=O)\text{-}OH + HOR'$$

Carboxylesterase hydrolysis

$$R\text{-}C(=O)\text{-}N(R'')\text{-}R' + H_2O \longrightarrow R\text{-}C(=O)\text{-}OH + HNR'R''$$

Carboxyamide hydrolysis

$$R\text{-}C(=O)\text{-}S\text{-}R' + H_2O \longrightarrow R\text{-}C(=O)\text{-}OH + HSR'$$

Carboxythioester hydrolysis

**FIGURE 3.31** Reactions catalyzed by esterases and amidases.

Because of its localization in the endoplasmic reticulum, microsomal epoxide hydrolase is ideally situated to catalyze the detoxification of lipophilic epoxides formed by P450; however, it can also be involved in metabolic activation. As noted earlier, an example of metabolic activation is the biotransformation of benzo(*a*)pyrene to the ultimate mutagen benzo(*a*)pyrene *trans*-7,8-dihydrodiol-9,10-oxide (Figure 3.4). These diol epoxides are poor substrates for further metabolism by epoxide hydrolase, and, as shown in Figure 3.4, they react with critical cellular macromolecules.

An immunologically distinct epoxide hydrolase has also been identified in the cytosol of some species. This enzyme may play a role in the hydrolysis of more water-soluble epoxides that partition out of the endoplasmic reticulum. Epoxide hydrolases may compete with glutathione transferases for cytosolic epoxides; however, epoxide hydrolase has a higher affinity for many epoxide substrates than glutathione transferase and is therefore generally a more efficient detoxification mechanism.

### Esterases and Amidases

Hydrolysis of xenobiotics containing ester linkages and amide bonds is catalyzed by a group of enzymes with broad substrate specificity. In general, these enzymes perform endogenous functions and appear to metabolize xenobiotics that have structural similarities to endogenous substrates. The reactions carried out by this diverse group of enzymes are illustrated in Figure 3.31. The specificity of carboxylesterases depends on the nature of the R groups rather than on the atom (O, N, or S) adjacent to the carbonyl carbon [102]. The esterases have been broadly grouped into three categories based on their reactivity with organophosphorous compounds [280]. Those esterases preferring carboxylesters with aryl groups in the R position and that can use organophosphate esters as substrates are classified as A-esterases (Table 3.7). Those esterases

**TABLE 3.7**
**Classification of Esterases by How They Interact with Organophosphates and Substrate Specificity**

| Esterase | Interaction with Organophosphates | Substrates | Examples |
|---|---|---|---|
| A-Esterases (arylesterases) | Substrates | Aromatic esters (aryl ester: ring bearing R, with C(=O)–O–R1) | Organophosphate and carbamate insecticides |
| B-Esterases (carboxylesterases including cholinesterase) | Inhibitors | $R–CH_2–C(=O)–O–R'$<br>Aliphatic esters | Acetylcholine, acrylate esters, succinylcholine, propanidid |
| C-Esterases (acetylesterases) | No interaction | $CH_3–C(=O)–O–R'$<br>Acetate esters | *p*-Nitrophenyl acetate, *n*-propylchloroacetate |

preferring esters with alkyl groups in the R position and that are inhibited by organophosphate esters are classified as B-esterases. Another group of esterases that prefer acetate esters and do not interact with organophosphates are referred to as C-esterases. This classification has been devised to help organize this multifarious group of enzymes. It also has some practical value in toxicology. The mechanism of organophosphate and carbamate insecticide toxicity is inhibition of acetylcholinesterase, a B-type esterase. Organophosphate insecticides, such as malathion, target serine hydrolases (B-esterases) and are detoxified in mammals by A-esterase hydrolysis [39]. Many insects have lower levels of A-esterases than mammals. The selective toxicity of malathion in birds and insects can be explained by the low activity of A-esterases compared to mammals [279].

Carboxylesterases are widely distributed in body including tissues lining the major portals of entry (i.e., skin, gastric mucosa, and respiratory tract). Liver has the highest capacity for esterase hydrolysis, but on a tissue-weight basis other tissues, such as the olfactory mucosa, contains comparable levels [24]. Hydrolytic activity at these sites can be used to improve drug bioavailability by designing ester-containing pro-drugs that are more lipid soluble than their alcohol or carboxylic acid analogs and therefore are more readily absorbed; however, carboxylesterase activity at the portal of entry can also result in metabolic activation when the hydrolysis products are toxic. Such is the case for vinyl acetate, a volatile organic monomer that, when inhaled, is absorbed and metabolized by carboxylesterases within the nasal cavity mucosal lining. The protons produced by both the liberated acetic acid and the further oxidation of acetaldehyde lead to low toxic intracellular pH that ultimately causes tissue cytotoxicity (Figure 3.29) [143].

## MICROFLORAL METABOLISM

Xenobiotic metabolism by microorganisms can be divided into reactions occurring in the environment and reactions occurring inside the body [212]. Metabolism of chemicals by microorganisms in the environment has become familiar through the use of microorganisms to degrade chemical spills [251]. The *in vivo* metabolism of chemicals by microorganisms is not as familiar. Mammals are colonized by microorganisms (only those animals raised in a germ-free environment [*gnotobiotic*] are microbe free) [105]. The metabolic reactions carried out by these microorganisms are dependent on the substrate and environment in which they are growing. Microbes growing in an aerobic environment are capable of cleavage of aromatic nuclei and can use these xenobiotics as sole carbon sources for biosynthetic reactions and growth. Microbes growing in an anaerobic environment are more likely to carry out reductive metabolism. The hallmark of metabolism by organisms colonizing the intestinal tract of mammals is reduction (Table 3.8).

Because the majority of microbes that colonize various surfaces of the mammalian body reside in the intestinal tract, most of this discussion will center around intes-

**TABLE 3.8**
**Types of Metabolic Reactions Carried Out by Intestinal Bacteria**

| Reaction | Representative Substrate |
|---|---|
| | *Hydrolysis* |
| Glucuronides | Estradiol-3-glucuronide |
| Glycosides | Cycasin |
| Sulfamides | Cyclamate, amygdalin |
| Amides | Methotrexate |
| Esters | Acetydigoxin |
| Nitrates | Pentaerythitol trinitrate |
| | *Dehydroxylation* |
| *C*-Hydroxy groups | Bile acids |
| *N*-Hydroxy groups | *N*-Hydroxyfluorenylacetamide |
| Decarboxylation | Amino acids |
| *N*-Demethylation | Biochanin A |
| Deamination | Amino acids |
| Dehydrogenase | Cholesterol, bile acids |
| Dehalogenation | DDT |
| | *Reduction* |
| Nitro groups | *p*-Nitrobenzoic acid |
| Double bonds | Unsaturated fatty acids |
| Azo groups | Food dyes |
| Aldehydes | Benzaldehydes |
| Alcohols | Benzyl alcohols |
| *N*-Oxides | 4-Nitroquinoline-1-oxide |
| | *Other Reactions* |
| Nitrosamine formation | Dimethylnitrosamine |
| Aromatization | Quinic acid |
| Acetylation | Histamine |
| Esterification | Gallic acid |

tinal microflora metabolism. The intestinal microflora can alter xenobiotic bioavailability by metabolizing the parent compound to a metabolite that may be absorbed to a greater or lesser extent. Intestinal microflora can also metabolize products of xenobiotic biotransformation that are secreted into the intestine directly from the blood or via the bile, saliva, or swallowing respiratory tract mucus. Metabolism of secreted metabolites is a common mechanism by which microflora influence xenobiotic toxicity.

## Xenobiotic Biotransformation by Microbes Colonizing Mammals

The intestinal tract of mammals contains a variety of microorganisms. The location, total number, and species diversity of microflora vary among mammals. They can range from ruminants that have evolved to be dependent on microflora metabolism for energy needs to monogastric mammals, such as humans, that have great numbers of bacteria only in the large intestine. Because of this variation in location within the intestinal tract, the types of microorganisms present and hence the types of microflora metabolism vary with the mammalian species being studied. Another factor that relates to the location of the microflora is the disposition of the xenobiotic and its microflora metabolites. Chemicals metabolized by microflora located in the stomach will be distributed differently than chemicals metabolized in the large intestine.

The majority of mammals have a gradient of microflora that increases in numbers and species diversity along the intestinal tract from the foregut to the hindgut. Most research on microflora metabolism has focused on microorganisms that colonize the large intestine of humans, as most of the research in toxicology is directed toward understanding the toxicity of chemicals in humans. *In vivo* and *in vitro* models have been developed for studying human colonic flora [218].

## Role of Diet and Other Factors in Modulating Microflora Metabolism

The microflora colonizing the digestive tract of mammals play a major role in the digestion of plant cell wall constituents that are indigestible by mammalian enzymes. These dietary fibers provide energy substrates that support the large bacterial populations in the gut. These energy sources also influence the microflora metabolism of xenobiotics. Certain types of dietary fiber, such as the fermentable carbohydrate pectin, can influence the toxicity of xenobiotics that require microflora metabolic activation by increasing the number of anaerobic bacteria colonizing the large intestine [61]. This diet-induced elevation in the number of bacteria increases the total metabolic capacity of the large intestine for metabolizing xenobiotics. For reviews of this topic, see Rowland and Wise [217] and Rowland [216].

## Examples of Xenobiotics Whose Toxicity Is Dependent on Microflora Metabolism

### Nitroaromatics

The toxicity of many nitroaromatic compounds is dependent on microflora metabolism. One of the most studied nitroaromatics is 2,6-dinitrotoluene (DNT), which is hepatocarcinogenic in male rats [146]. DNT is metabolized to the 2,6-dinitrobenzylalcohol glucuronide conjugate that is preferentially excreted in the bile of male rats (Figure 3.32) [154]. The glucuronide conjugate is hydrolyzed by gut microflora β-glucuronidase, and one or both of the nitro groups are reduced by microflora nitroreductase to a reduced aglycone. The resulting aminobenzyl alcohol is relatively nonpolar and reabsorbed in the intestine, where it returns to the liver via the portal circulation. In the liver, the aglycone is activated to the putative proximate carcinogen by *N*-hydroxylation of the amine functional group followed by sulfation of the *N*-hydroxy group [124]. Evi-

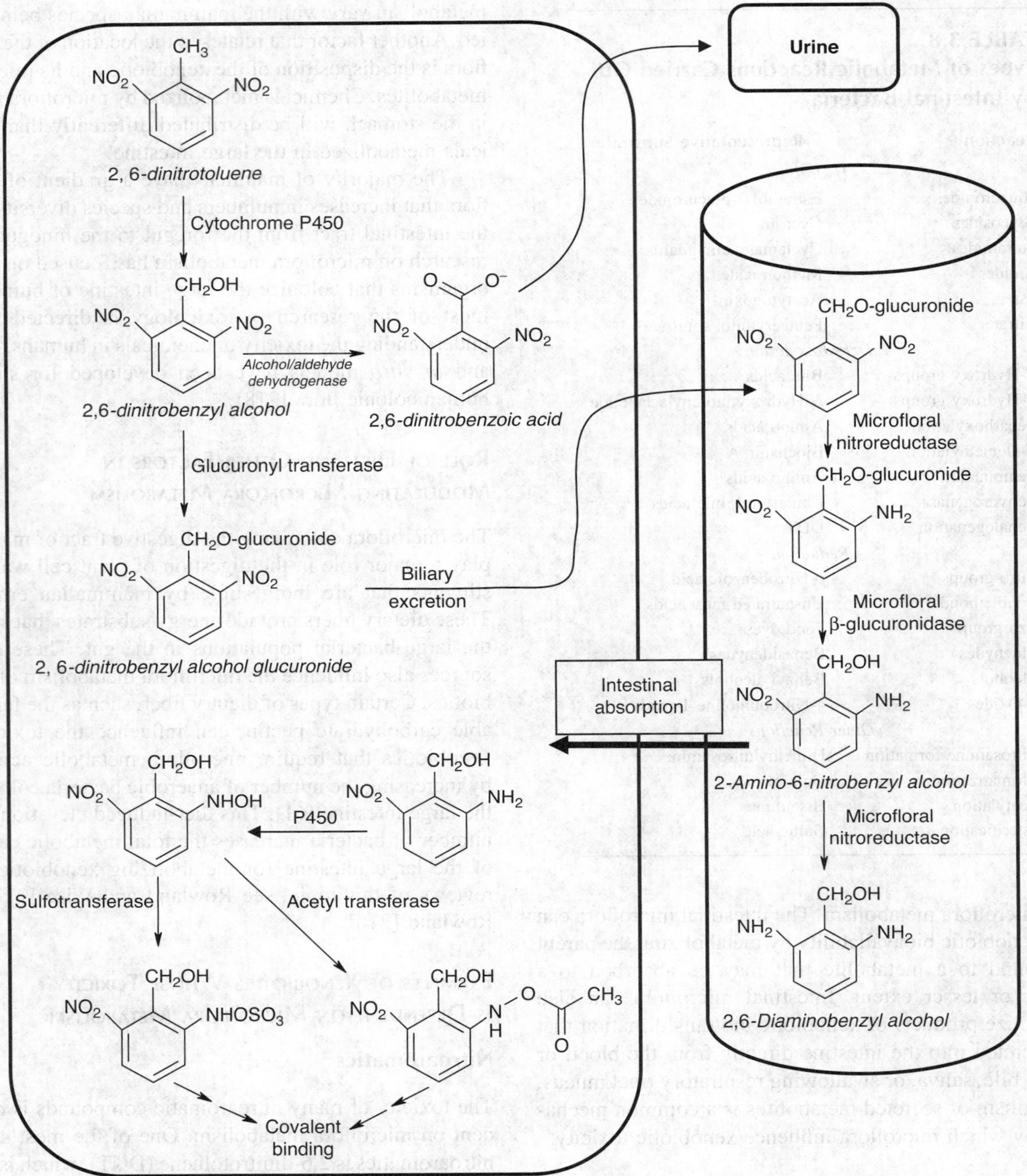


**FIGURE 3.32** Putative route of disposition of 2,6-dinitrotoluene.

dence that intestinal microflora were required for the activation of DNT was provided by studies indicating that the genotoxicity of DNT in hepatocytes isolated from rats treated with DNT was dependent on the presence of bacteria in the intestinal tract [176]. Rats raised in a germ-free environment showed minimal levels of genotoxicity. Additional evidence emphasizing the role of microflora in the metabolic activation of DNT was the observation that DNT was not genotoxic when tested *in vitro* in isolated hepatocytes [176]. These results indicated that liver metabolism was not sufficient to activate the molecule to the ultimate carcinogen. The genotoxicity of DNT to liver cells only occurred when the compound was administered to the animal and was allowed to undergo enterohepatic circulation involving intestinal microflora.

The level of DNT-derived radioactivity covalently bound to DNA, RNA, and protein isolated from the livers of rats treated with DNT was also dependent on the presence of intestinal microflora [154]. Dietary treatments that increased the microbial metabolic capacity of the rat's large intestine also increased the covalent binding of DNT-derived radioactivity to hepatic macromolecules [61].

The importance of complementary *in vitro* short-term toxicity tests with suitable *in vivo* tests for predicting the toxicity of a chemical is well illustrated by this example, where the toxicity of DNT was dependent on the disposition within the host rather than metabolic activation within a single organ.

### Cyclamate

The sodium and calcium salts of cyclamic acid (cyclohexylsulfamic acid) were used as an artificial sweetening agent until 1969 in the United States, when it was removed from the market because a metabolite (cyclohexylamine) was suspected of being a bladder carcinogen. Most of the hydrolysis of cyclamate to cyclohexylamine takes place in the gut by the microflora. Cyclohexylamine is more lipophilic than the parent acid and is readily absorbed from the intestine and excreted in the urine. Minor urinary metabolites include cyclohexanol and *trans*-cyclohexane-1,2-diol.

Although only trace amounts of the cyclohexylamine could be detected in humans administered cyclamate, chronic exposure to the acid increased the capacity to produce this metabolite [163]. It was found that certain individuals possessed a greater capacity to metabolize cyclamate to cyclohexylamine; these individuals were called *converters*. Thus, cyclamate is a good example of how prior exposure to a xenobiotic can alter the disposition of the xenobiotic. For additional reading on intestinal microflora xenobiotic metabolism, see Goldman [90] and Scheline [223].

## INTEGRATION OF METABOLIC PATHWAYS

From the sections presented above, it is clear that pathways of xenobiotic metabolism range from simple, such as carboxylesterases, to complex, such as the various pathways capable of introducing oxidations. Also clear is the fact that a xenobiotic is not necessarily subject to a single metabolic transformation; rather, xenobiotics are more typically metabolized and eliminated following multiple transformations. Thus, it is important that an integrated view of xenobiotic metabolism be considered when assessing the extent to which metabolism is a determinant of toxicity.

Consider for example the case of acetaminophen, a drug whose three primary metabolic transformations (oxidation, sulfonation, and glucuronidation) have been discussed in multiple sections above. It is the balance of the transformations, the predominance of one pathway over the other, and factors that determine the predominant pathway that determine toxicity (Figure 3.33). Acetaminophen is eliminated metabolically in the liver primarily by sulfate conjugation of the phenyl hydroxyl group; however, competing with this high-affinity, low-capacity pathway is elimination via glucuronidation also of the hydroxyl group. At high doses of acetaminophen, PAPS pools become depleted and rate limiting, and the secondary pathway of glucuronidation predominates. Glucuronidation thus provides a back-up route of elimination. Saturation of these two elimination pathways allows more acetaminophen to flow through the oxidation pathways.

In liver, acetaminophen is oxidized predominantly by CYP2E1 to yield the *N*-acetyl-*p*-benzoquinoneimine. This pathway of metabolic activation forms reactive species that bind to critical macromolecules in the liver and is believed responsible for hepatotoxicity. Thus, inducers of CYP2E1, such as ethanol, increase hepatic capacity for CYP2E1-mediated oxidation and may enhance hepatotoxicity.

Acetaminophen has been associated with analgesic nephropathy. Acetaminophen can also be oxidized by prostaglandin synthase to a reactive metabolite capable of binding to macromolecules. The realization that a pathway other than the well-characterized P450 route was involved occurred through observations that acetaminophen was covalently bound in the inner medulla of the rabbit kidney, a site nearly devoid of P450 activity [179]. Covalent binding has been attributed to prostaglandin synthase-mediated oxidation in the inner medulla; however, acetaminophen-induced renal toxicity is largely limited to the renal cortex, which contains P450 activity. This is because prostaglandin synthase activity of the inner medulla is inhibited at high acetaminophen doses, but the P450 pathway of the cortex is not inhibited and is thus responsible for the cortical damage. This gradient of oxidation pathways across the kidney (with the cortex possessing higher P450 activity than prostaglandin synthase, the outer medulla being intermediate, and the inner medulla possessing far greater prostaglandin synthetase activity) results in certain xenobiotics being more toxic to one region of the kidney than another [307]. The example of acetaminophen highlights that many factors, including species, organ of metabolism, and dose, can influence the expression of xenobiotic-metabolism-mediated toxicity.

## COMPUTATIONAL APPROACHES FOR PREDICTION OF BIOTRANSFORMATION

Currently, tens of thousands of chemical entities can be found in commerce and in the environment for which little or no toxicological or metabolism data are available, and chemical manufacturers and pharmaceutical companies continue to develop new compounds at a prodigious rate.

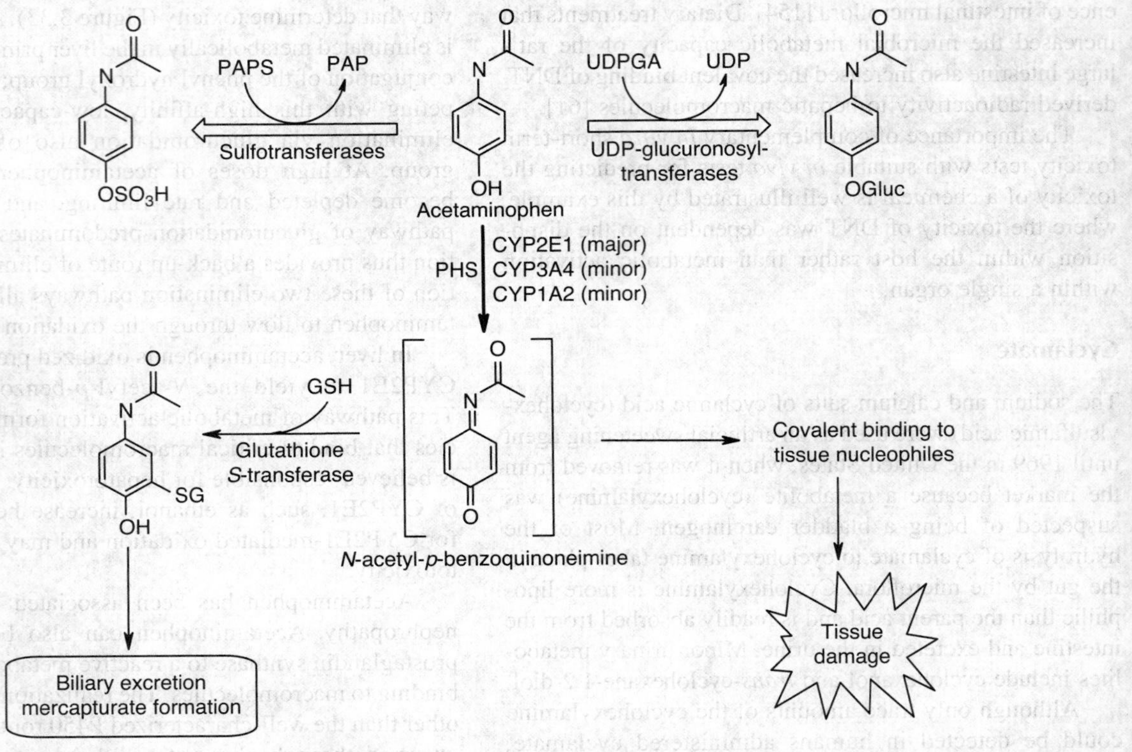

FIGURE 3.33 Integration of metabolic pathways: acetaminophen. Major detoxification pathways are indicated by open arrows, and include sulfonation and glucuronidation. Bioactivation of acetaminophen involves oxidation to *N*-acetyl-*p*-benzoquinoneimine (NAPQI), and it may be catalyzed by several P450 isozymes as well as prostaglandin H synthase (PHS). At therapeutic doses, conjugation pathways predominate and bioactivation is negligible. Following overdose, conjugation pathways become saturated and oxidation to the reactive intermediate NAPQI becomes significant. NAPQI may be detoxified by reaction with reduced glutathione (GSH), either by a spontaneous or GST-catalyzed reaction and excreted as the corresponding mercapturate. At high exposure levels, GSH may become depleted, allowing NAPQI to react with tissue nucleophiles, resulting in toxicity.

Individual testing of such large numbers of chemicals and drugs is time consuming, costly, resource intensive, and, in the case of whole animal testing, ethically unsupportable. For these reasons, a great deal of interest has developed in computational (*in silico*) methodologies to predict chemical toxicity and, more recently, metabolism based on structure and physicochemical properties. The two basic computational approaches to the prediction of chemical metabolism are quantitative structure–activity relationships (QSARs) and expert systems. QSAR, as the name implies, seeks to quantitatively correlate physicochemical, electronic, quantum mechanical, spatial, and other descriptors of a molecule with rates of metabolism, generally focusing on a particular reaction at a particular site within a series of structurally related compounds. These techniques can be very useful in understanding the molecular and electronic determinants of substrate–enzyme interactions and in optimizing (or minimizing) the interaction of a compound with a particular enzyme.

Perhaps of greater interest to the toxicologist is the so-called expert system approach, which seeks to predict global biotransformation pathways based on chemical structure [142]. Expert systems are based on libraries of generalized metabolism "rules" distilled from large sets of experimental metabolism data. Each rule describes an individual metabolic transformation on a particular substructure or molecular fragment. Several metabolism expert systems are currently commercially available, including METEOR® [93], TIMES® [173], META® [131], and MetabolExpert® [58]. These programs allow the user to draw structures using a graphical interface or to input structures in a variety of files formats. Most are capable of calculating some physicochemical parameters based on the query structure and use these as additional inputs for metabolite prediction. In general, expert system predictions are qualitative, but some programs, such as TIMES, also provide a quantitative estimate of metabolite production.

A major strength of rule-based expert systems is that they incorporate mechanistic insight into predictions and can be used to guide chemistry efforts in designing desirable biotransformation properties into new compounds. A downside to these systems is the potential to generate excessive numbers of unrealistic metabolites via multiple pathways, leading to a so-called combinatorial explosion

[135]. To address this problem, most programs have cut-off filters based on properties such as molecular weight and lipophilicity, or they allow the user to constrain the analysis as to types of transformations predicted (phase I vs. phase II) or the total number of metabolites predicted in any given pathway. In addition to predicting the products of biotransformation, many of these programs also allow assessment of potential toxicity of metabolites, either through an integrated toxicity prediction engine or through links to external companion programs. Newer approaches to combined prediction of metabolism and toxicity seek to integrate rule-based systems for metabolism prediction with biochemical pathway analysis and high-throughput biology techniques ("omics" technologies) in a systems biology approach [33]. Although still in its relative infancy, systems biology promises significant advances in our ability to understand and predict the interactions among biotransformation, biological activity, and toxicity of chemicals and drugs.

## REGULATORY AND PRODUCT DEVELOPMENT ASPECTS OF XENOBIOTIC METABOLISM

International regulatory agencies require information about the metabolism of drugs and other chemicals that fall under their jurisdiction. The amount of data required will depend on the use of the chemical, its potential exposure to humans, and the potential role of metabolism in its efficacy and toxicity. Early information on the metabolism of drugs is becoming essential for the selection of drug candidates for further development. With the development of combinatorial synthetic chemistry and high-throughput pharmacological screening, the number of potential drug candidates that must be rapidly screened for their potential metabolism has increased dramatically. This is leading to the development of rapid methods to predict metabolism and potential drug–drug interactions. Data are needed from *in vitro* and *in vivo* animal models that will be used in safety assessments to allow the toxicologist to design appropriate studies during the preclinical phase of a safety assessment. Data are also needed concerning the potential human metabolism of the drug candidate to allow the toxicologist to extrapolate animal safety data to humans. In the recent past, it was difficult to obtain *in vitro* data from human tissues. With the current knowledge of the human P450 isozymes and the commercial availability of human hepatic preparations and cells that express human P450s, it is possible to obtain data concerning human metabolism of drug candidates, pesticides, and other chemicals.

The U.S. Food and Drug Administration (FDA) has recently released a guidance document concerning *in vitro* drug metabolism and drug–drug interaction studies during the drug development process [266]. This document stresses the importance of obtaining information on the metabolism of a drug candidate during the early stages of development. This information is important in predicting potential individual differences based on polymorphic expression of xenobiotic metabolism enzymes and in predicting drug–drug interactions. The guidance document is based on the following general observations:

- The concentrations of a drug or its active metabolite circulating in the body determine the extent of its desirable and/or adverse effects.
- A major determinant of the concentration of a drug is clearance, and metabolism is a major determinant of clearance.
- Drugs that are not substantially metabolized may impact the metabolism of other drugs.
- Large differences in blood concentrations can occur because of polymorphic metabolism. Drug–drug interactions can also produce large changes in the blood concentration of a drug.
- Major advances have been made in availability of human tissue and recombinant enzymes for *in vitro* studies of drug metabolism.

The guidance document suggests that the goals of *in vitro* metabolism and interaction studies should be to (1) identify major metabolic pathways and the specific isozymes involved, and (2) explore and extrapolate the effects of the drug candidate on the metabolism of other drugs and the effect of other drugs on the metabolism of the candidate. To accomplish these goals, the FDA suggests starting from human hepatic microsomes, now commercially available, then moving to cell-based systems that express specific human P450s, which are also commercially available. They note that it is possible to move to hepatocytes and precision-cut liver slices but recognize the technical difficulties associated with these preparations.

Coincubation of the drug candidate with substrate probes having known metabolic pathways can be used to determine its effects on their metabolism. In addition, assessment of the metabolism of the drug candidate by CYP3A4 may provide information on intestinal metabolism that can affect drug bioavailability. Parallel studies using preparations from animal models can aid the toxicologist in choice of species for the toxicological studies. The document emphasizes the point that *in vitro* studies currently cannot replace *in vivo* studies but can give direction for the proper design of *in vivo* studies. For example, if the drug candidate is not metabolized by CYP2D6, then it will not be important to study the impact of the slow metabolizer phenotype on potential adverse effects. If the drug candidate is not a substrate for CYP3A4, then there is less concern that inhibition of CYP3A4 by drugs such as ketoconazole and erythromycin or induction by rifampin and anticonvulsants could cause problems.

A companion guidance document outlining recommendations for *in vivo* characterization of drug metabolism and potential drug–drug interactions was released shortly after the *in vitro* guidance document [267]. The conceptual basis for this document is similar to the *in vitro* guidance and stresses the importance of determining whether excretion or metabolism is the major mechanism of drug clearance, as well as the identification of major metabolic routes in the latter case. Further, the guidance lays out recommendations on study designs for the investigation of both inhibition and induction of metabolism of the drug of interest. These studies are useful in determining changes in dosage regimen for the target and interacting drugs during polypharmacy.

In addition to these published guidance documents, the FDA is developing guidance documents on the submission of pharmacogenomic data and on characterization and nonclinical safety testing of unique human drug metabolites. As of this writing, both guidance documents are still in draft form. The draft guidance documents may be viewed on the FDA website (http://www.fda.gov/cder/guidance/index.htm).

## QUESTIONS

1. During the development of a new drug, it was decided that the introduction of a hydroxyl group onto the molecule would make the compound more water soluble, which offered the advantages of increasing gastric solubility and improving the pharmaceutics properties. An initial study of the plasma concentrations of the nonhydroxylated analog in rats had been completed in anticipation of beginning a subchronic toxicity study. When the plasma concentrations of the less lipophilic, hydroxylated drug were determined, it was found that the plasma concentrations were maintained for a longer period of time than with the nonhydroxylated analog. How would you explain this finding?
2. As a toxicologist, you have been asked to design a program to assess the potential hazard of a chemical. What type of information concerning its metabolism would you want to have before you design the hazard assessment program? Based on the metabolism information you have requested, how would you choose the species to be used in the hazard assessment program?
3. A cancer chemotherapeutic drug has been shown effective in treating a specific type of cancer; however, the drug is also cytotoxic and produces severe side effects if it is not rapidly metabolized by cytochromes P450. It is important, therefore, not to treat a patient with doses of the drug that are too high for the patient's capacity to rapidly metabolize it to the less toxic product. What characteristics of the patient should be considered when attempting to choose a dose that will minimize the side effects?
4. A compound is functionalized by the P450 system and then forms sulfate and glucuronide conjugates and a mercapturic acid before being excreted. How may its metabolism be altered as the dose is increased from a no-observable-effect level (NOEL) to a dose that produces severe toxicity?
5. Many chemical carcinogens are metabolized by routes that represent detoxification and by other routes that represent metabolic activation. What are the various phenomena that may shift the balance between detoxification and metabolic activation?

## REFERENCES


1. Agarwal, D. P. and Goedde, H. W. (1992): Pharmacogenetics of alcohol metabolism. *Pharmacogenetics*, 2(2): 48–62.
2. Akerboom, T. P. M. and Sies, H. (1981): Assay of glutathione, glutathione disulfide, glutathione mixed disulfides in biological samples. In: *Methods in Enzymology*, Vol. 77, *Detoxication and Drug Metabolism: Conjugation and Related Systems*., edited by S. P. Colowick and N. O. Kaplan. Academic, Press, New York, p. 376.
3. Ambrosone, C. B., Coles, B. F., Freudenheim, J. L., and Shields, P. G. (1999): Glutathione *S*-transferase (GSTM1): genetic polymorphisms do not affect human breast cancer risk regardless of dietary antioxidants. *J. Nutr.*, 129:565–568.
4. Ambroziak, W., Izaguirre, G., and Pietruzko, R. (1999): Metabolism of retinaldehyde and other aldehydes in soluble extracts of human liver and kidney. *J. Biol. Chem.*, 274(47):33366–33373.
5. Anders, M. W. (2004): Glutathione-dependent bioactivation of haloalkanes and haloalkenes. *Drug Metab. Rev.*, 36(3–4):583–594.
6. Anders, M. W., Lash, L., Dekant, W., Wlfarra, A. A., and Dohn, D. R. (1988): Biosynthesis and biotransformation of glutathione *S*-conjugates to toxic metabolites. *CRC Crit. Rev. Toxicol.*, 18:311–341.
7. Andersson, B., Nordenskjold, M., Rahimtula, A., and Moldeus, P. (1982): Prostaglandin synthetase-catalyzed activation of phenacetin metabolites to genotoxic products. *Mol. Pharmacol.*, 22:479–485.
8. Arand, M., Cronin, A., Adamska, M., and Oesch, F. (2005): Epoxide hydrolases: structure, function, mechanism, and assay. *Meth. Enzymol.*, 400:569–588.
9. Arand, M., Cronin, A., Oesch, F., Mowbray, S., and Jones, T. A. (2003): The telltale structures of expoxide hydrolases. *Drug Metab. Rev.*, 25:365–383.
10. Armstrong, R. N. (1997): Structure, catalytic mechanism, and evolution of the glutathione transferases. *Chem. Res. Toxicol.*, 10:2–18.

11. Atzori, L., Dore, M., and Congiu, L. (1989): Aspects of allyl alcohol toxicity. *Drug Metab. Drug Interact.*, 7(4):295–319.
12. Axelrod, J. (1983): The discovery of the microsomal drug-metabolizing enzymes. In: *Drug Metabolism and Distribution: Current Reviews in Biomedicine* 3, edited by J. W. Lamble. Elsevier, New York, pp. 1–6.
13. Backes, W. L. (1993): NADPH-cytochrome P450 reductase: function. In: *Cytochrome P450*, edited by J. B. Schenkman and H. Greim. Springer-Verlag, Berlin, pp. 15–34.
14. Baily, M. J. and Dickinson, R. G. (2003): Acyl glucuronide reactivity in perspective: biological consequences. *Chem. Biol. Interact.*, 145:117–137.
15. Baumgart, A., Schmidt, M., Schmitz, H.-J., and Schrenk, D. (2005): Natural furocoumarins as inducers and inhibitors of cytochrome P450 1A1 in rat hepatocytes. *Biochem. Pharmacol.*, 69(4):657–667.
16. Bergelson, S., Pinkus, R., and Daniel, V. (1994): Induction of AP-1 (Fos/Jun) by chemical agents mediates activation of glutathione *S*-transferase and quinone reductase gene expression. *Oncogene*, 9(2):565–571.
17. Bernhardt, R. (1995): Cytochrome P450: structure, function, and generation of reactive oxygen species. *Rev. Physiol. Biochem. Pharmacol.*, 127:137–221.
18. Bertilsson, L., Dahl, M. L., Sjoqvist, F., Abergwisted, A., Humble, M., Johansson, I., Lundqvist, E., and Ingelman-Sundberg, M. (1993): Molecular basis for rational megaprescribing in ultrarapid hydroxylators of debrisoquine. *Lancet*, 341:63.
19. Bessems, J. G. M. and Vermeulen N. P. E. (2001): Paracetamol (acetaminophen)-induced toxicity: molecular and biochemical mechanisms, analogues and protective approaches. *Crit. Rev. Toxicol.*, 31(10):55–138.
20. Blanchard, R., Freimuth, R. R., Buck, J., Weinshilboum, R. M., and Coughtrie, W. H. (2004): A proposed nomenclature system for the cytosolic sulftransferase (SULT) superfamily. *Pharmacogenetics*, 14:199–211.
21. Blanck, J. and Ruckpaul, K. (1993): Lipid-protein interactions. In: *Cytochrome P450*, edited by J. B. Schenkman and H. Greim. Springer-Verlag, Berlin, pp. 581–597.
22. Bock, K. W. and Köhle, C. (2004): Coordinate regulation of drug metabolism by xenobiotic nuclear receptors: UGTs acting together with CYPS and glucuronide transporters. *Drug Metab. Rev.*, 36:595–615.
23. Bock, K. W., Gschaidmeier, H., Heel, H., Lehmkoster, T., Munzel, P. A., Raschko, F., and Bock-Hennig, B. (1998): Ah receptor-controlled transcriptional regulation and function of rat and human UDP-glucuronosyltransferase isoforms. *Adv. Enzyme Regul.*, 38:207–222.
24. Bogdanffy, M. S. and Keller, D. A. (1999): Metabolism of xenobiotics by the respiratory tract. In: *Toxicology of the Lung*, 3rd ed., edited by D. E. Gardiner et al. Raven Press, New York, pp. 85–123.
25. Bogdanffy, M. S. and Valentine, R. (2003): Differentiating between local cytotoxicity, mitogenesis, and genotoxicity in carcinogen risk assessments: the case of vinyl acetate. *Toxicol. Lett.*, 140/141:83–98.
26. Bolt, H. M., Roos, P. H., and Their, R. (2003): The cytochrome P-450 isozyme CYP2E1 in the biological processing of industrial chemicals: consequences for occupational and environmental medicine. *Int. Arch. Occup. Environ. Health* 76:174–185.
27. Bosron, W. F. and Li, T. K. (1980): Alcohol dehydrogenase. In: *Enzymatic Basis of Detoxication*, Vol. 1, edited by W. B. Jakoby. Academic Press, New York.
28. Boukoubala, S. and Fakis, G. (2005): Arylamine *N*-acetyltransferases: what we learn from genes and genomes. *Drug Metab. Rev.*, 37:511–564.
29. Boyd, J. A. and Eling, T. E. (1981): Prostaglandin endoperoxide synthetase-deficient cooxidation of acetaminophen to intermediates which covalently bind *in vitro* to rabbit renal medullary microsomes. *J. Pharmacol. Exp. Ther.*, 219:659–664.
30. Boyland, E. and Chasseaud, L. F. (1969): The role of glutathione and glutathione *S*-transferases in mercapturic acid biosynthesis. *Adv. Enzymol.*, 32:173–219.
31. Bresnick, G. (1993): Induction of cytochromes P450 1 and P450 2 by xenobiotics. In: *Cytochrome P450*, edited by J. B. Schenkman and H. Greim. Springer-Verlag, Berlin, pp. 503–524.
32. Brown, R. R., Miller, J. A., and Miller, E. C. (1954): The metabolism of methylated aminoazo dyes. IV. Dietary factors enhancing demethylation *in vitro*. *J. Biol. Chem.*, 209:211–217.
33. Bugrim, A., Nikolskaya, T. and Nikolsky, Y. (2004): Early prediction of drug metabolism and toxicity: systems biology approach and modeling. *Drug Disc. Today*, 9(3):127–135.
34. Caldwell, J. (1980): Comparative aspects of detoxication in mammals. In: *Enzymatic Basis of Detoxication*, edited by W. B. Jakoby. Academic Press, New York, pp. 85–114.
35. Calvin, C., Holzhauser, D., Constable, A., Huyggett, A. C., and Schilter, B. (1998): The coffee-specific diterpenes cafestol and kahweol protect against aflatoxin $B_1$ induced genotoxicity through a dual mechanism. *Carcinogenesis*, 19:1369–1375.
36. Caro, A. A. and Cederbaum, A. I. (2004): Oxidative stress, toxicology and pharmacology of CYP2E1. *Annu. Rev. Pharmacol. Toxicol.*, 44:27–42.
37. Cascorbi, I. (2006): Genetic basis of toxic reactions to drugs and chemicals. *Toxicol. Lett.*, 162:16–28.
38. Cashman, J. R. and Zhang, J. (2006): Human flavin-containing monooxygenases. *Annu. Rev. Pharmacol. Toxicol.*, 46:65–100.
39. Casida, J. E. and Quistad, G. B. (2005): Serine hydrolase targets of organophosphorus toxicants. *Chemico-Biol. Interac.*, 157/158:277–283.
40. Chasseaud, L. F. (1973): The nature and distribution of enzymes catalyzing the conjugation of glutathione with foreign compounds. *Drug Metab. Rev.*, 2:185–220.
41. Chasseaud, L. F. (1978): The role of glutathione and glutathione *S*-transferases in the metabolism of chemical carcinogens and other electrophilic agents. *Adv. Cancer Res.*, 29:175–274.

42. Chen, G., Baron, J., and Duffet, M. W. (1995): Enzyme and sex-specific differences in the interlobular localization and distributions of aryl sulfotransferase IV (tyrosine-ester sulfotransferase) and alcohol (hydroxysteroid) sulfotransferase a in rat liver. *Drug Metab. Dispos.*, 12:1346–1353.
43. Chen, N., Whitehead, S. E., Caillat, A.W., Gavit, K., Isphording, D. R., Kovacevic, D., McCreary, M. B., and Hoffman, S. M. G. (2002): Identification and cross-species comparisons of *CYP2F* subfamily genes in mammals. *Mut. Res.*, 499:155–161.
44. Cherrignton, J. J., Cao, Y., Cherrington, J. W., Rose, R. L., and Hodgson, E. (1998): Physiological factors affecting protein expression of flavin-containing monooxygenases 1, 3, and 5. *Xenobiotica*, 28:673–682.
45. Chiu, S. H. and Huskey, S. W. (1998): Species differences in *N*-glucuronidation. *Drug Metab. Dispos.*, 26:838–847.
46. Christou, M., Savas, U., Schroeder, S., Shen, X., Thompson, T., Gould, M. N., and Jefcote, C. R. (1995): Cytochromes CYP1A1 and CYP1B1 in the rat mammary gland: cell-specific expression and regulation by polycyclic aromatic hydrocarbons and hormones. *Mol. Cell. Endocrinol.*, 115(1):41–50.
47. Chun, Y.-J. and Kim, Sanghee (2003): Discovery of cytochrome P450 1B1 inhibitors as new promising anti-cancer agents. *Medicinal Res. Rev.*, 23(6):657–668.
48. Ciotti, M., Obaray, R., Martin, M. G., and Owens, I. S. (1997): Genetic defects at the UGT1 locus associated with Crigler–Naffar type I disease, including a prenatal diagnosis. *Am. J. Med. Genet.*, 68:173–178.
49. Cobb, D., Boehlert, C., Lewis, D., and Armstrong, R. N. (1983): Stereoselectivity of isozyme C of glutathione *S*-transferase toward arene and azaarene oxides. *Biochemistry*, 22:806–812.
50. Coffman, B. L., King, C. D., Rios, G. R., and Tephly, T. R. (1998): The glucuronidation of opioids, other xenobiotics and androgens by human UGT2B7Y(268) and UGT2B7H(268). *Drug Metab. Dispos.*, 26, 27–77.
51. Conklin, D. J., Langford, S. D., and Boor, P. J. (1998): Contribution of serum and cellular semicarbazide-sensitive amine oxidase to amine metabolism and cardiovascular toxicity. *Toxicol. Sci.*, 49(46):386–392.
52. Conney, A. H. (1967): Pharmacological implications of microsomal enzyme induction. *Pharmacol. Rev.*, 19:317–350.
53. Cooper, D. Y. (1964): Photochemical action spectrum of the terminal oxidase of mixed function oxidase systems. *Science*, 147:400–402.
54. Crespi, C. L., Penman B. W., Steimer, D. T., Smith, T., Yang, C. S., and Sutter, T. R. (1997): Development of a human lymphoblastoid cell line constitutively expressing human CYP1B1 cDNA: substrate specificity with model substrates and promutagens. *Carcinogenesis*, 12:83–89.
55. Dai, D., Zeldin, D. C., Blaisdell, J. A., Chanas, B., Coulter, S. J., Ghanayem, B. I., and Goldstein, J. A. (2001): Polymorphisms in human CYP2C8 decrease metabolism of the anticancer drug paclitaxel and arachidonic acid. *Pharmacogenetics*, 11(7):597–607.
56. Daly, A. K. (2006): Significance of the minor cytochrome P450 3A isoforms. *Clin. Pharmacokin.*, 45(1):13–31.
57. Daly, A. K., Brockmuller, J., Broly, F., Eichelbaum, M., Evans, W. E., Gonzalez, F. J., Huang, J. D., Idle, J. R., Ingelman-Sundberg, M., Ishizaki, T., Jacqzaigrain, E., Meyer, U. A., Nebert, D. W., Steen, V. M., Wolf, C. R., and Zanger, U. M. (1996) Nomenclature for human CYP2D6 alleles. *Pharmacogenetics*, 6:193–201.
58. Darvas, F. (1988): Predicting metabolic pathways by logic programming. *J. Mol. Graphics.*, 6:80–86.
59. Davis, M. E. (1988): Effects of AT-125 on the nephrotoxicity of hexachloro-1,3-butadiene in rats. *Toxciol. Appl. Pharmacol.*, 95:44–52.
60. De Morais, S. M. F., Wilkenson, G. R., Baisdell, J., Meyer U. A, Nakamura, K., and Goldstein, J. A. (1994): Identification of a new genetic defect responsible for the polymorphism of (*s*)-mephenytoin in Japan. *Mol. Pharmacol.* 46:594–598.
61. deBethizy, J. D., Sherrill, J. M., Rickert, D. E., and Hamm T. E., Jr. (1983): Effects of pectin containing diets on the hepatic macromolecular covalent binding of 2,6-dinitro[$^3$H]toluene in Fischer-344 rats. *Toxicol. Appl. Pharmacol.*, 69:369–376.
62. Desta, Z., Zhao, X., Shin, J.-G., and Flockhart, D. A. (2002) Clinical significance of cytochrome P450 2C19 genetic polymorphism. *Clin. Pharmacokinet.*, 41(12): 913–958.
63. Ding, X. and Kaminsky L. S. (2003): Human extrahepatic cytochromes P450: function in xenobiotic metabolism and tissue-selective chemical toxicity in the respiratory and gastrointestinal tracts. *Annu. Rev. Pharmacol. Toxicol.* 43:2149–173.
64. Donavon, J.M. II, Lin, Y.S., Allen, K., Kunze, K., and Thummel, K.E. (2004): Differences in the inhibition of cytochromes P450 3A4 and 3A5 by metabolite-inhibitor complex-forming drugs. *Drug Metab. Disp.*, 32:1083–1091
65. Dooley, T. P. and Huang, Z. (1996): Genomic organization and DNA sequences of two human phenol sulfotransferase genes (STP1 and STP2): on the short arm of chromosome 16. *Biochem. Biophys. Res. Commun.*, 228:134–140.
66. Du, L., Hoffman, S. M. G., and Keeney, D. S. (2004) Epidermal CYP2 family cytochromes P450. *Toxicol. Appl. Pharmacol.*, 195(3):278–287.
67. Dutton, G. J. (1980): *Glucuronidation of Drugs and Other Compounds*. CRC Press, Boca Raton, FL.
68. Dvorak, Z., Vrzal, R., and Ulrichova, J. (2006): Silybin and dehydrosilybin inhibit cytochrome P450 1A1 catalytic activity: a study in human keratinocytes and human hepatoma cells. *Cell Biol. Toxicol.*, 22(2):81–90.
69. Ekins, S. and Wrighton, S. (1999): The role of CYP2B in human xenobiotic metabolism. *Drug Metab. Rev.* 31(3):719–754.
70. Elfarra, A. A. and Anders, M. W. (1984): Renal processing of glutathione conjugates: role in nephrotoxicity [commentary]. *Biochem. Pharmacol.*, 33:3729–3732.
71. Estabrook, R. W. (1996): Cytochrome P450: from a single protein to a family of proteins, with some personal reflections. In: *Cytochromes P450: Metabolic and Toxicological Aspects*, edited by C. Ioannides. CRC Press, Boca Raton, FL, pp. 3–28.
72. Estabrook, R. W. (2003): A passion for P450s (remembrances of the early history of research on cytochrome P450). *Drug Metab. Dispos.*, 31:1461–1473.

73. Estabrook, R. W., Cooper, D. Y., and Rosenthal, O. (1963): The light reversible carbon monoxide inhibition of the steroid C-21-hydroxylase system of adrenal cortex. *Biochem. Z.*, 338:741–755.
74. Evans, W. E. and Relling, M. V. (1999): Pharmacogenomics: translating functional genomics into rational therapeutics. *Science*, 286:487–491.
75. Fang, Y.-Y., Kashkarov, U., Anders, M. W., and Board, P. G. (2006): Polymorphisms in the human glutathione transferase zeta promoter. *Pharmacogenet. Genom.*, 16:307–313.
76. Fiala, E. S., Sodum, R. S., Hussain, N. S., Rivenson, A., and Dolan, L. (1995): Secondary nitroalkanes: induction of DNA repair in rat hepatocytes, activation by aryl sulfotransferase and hepatocarcinogenicity of 2-nitrobutane and 3-nitropentane in male F344 rats. *Toxicology*, 99:89–97.
77. Flesher, J. W., Horn, J., and Lehner, A. F. (1997): 6-sulfooxymethylbenzo(*a*)pyrene is an ultimate electrophilic and carcinogenic form of the intermediary metabolite 6-hydroxymethylbenzo(*a*)pyrene. *Biochem. Biophys. Res. Commun.*, 234:554–558.
78. Forman, B. M., Tzameli I., Choi H. S., Chen, J., Simba, D., Seol, W., Evans, R. M., and Moore, D. D. (1998): Androstane metabolites bind to and deactivate the nuclear receptor CAR-beta. *Nature*, 395:612–615.
79. Friedberg, T., Lollmann, B., Becker, R., Holler, R., and Oesch, F. (1994): The microsomal epoxide hydrolase has a single membrane signal anchor sequence which is dispensable for the catalytic activity of this protein. *Biochem. J.*, 303:967–972.
80. Frommer, U., Ullrich, V., Staudinger, H., and Orrenius, S. (1972): The monooxygenation of *n*-heptane by rat liver microsomes. *Biochem Biophys. Acta*, 280:487–494.
81. Fukami, J. I. (1984): Metabolism of several insecticides by glutathione *S*-transferase. *Int. Encycl. Pharmacol. Ther.*, 113:223–264.
82. Gamage, N., Barnett, A., Hempel, N., Duggleby, R. G., Windmill, K. F., Martin, J. L., and McManus, M. E. (2006): Human sulfotransferases and their role in chemical metabolism. *Toxicol. Sci.*, 90:5–22.
83. Ganem, L. G., Trottier, E., Anderson, A., and Jefcoate, C. R. (1999): Phenobarbital induction of CYP2B1/2 in primary hepatocytes: endocrine regulation and evidence for a single pathway for multiple inducers. *Toxicol. Appl. Pharmacol.*, 155:32–42.
84. Garfinkel, D. (1958): Studies on pig liver microsomes. I. Enzymic and pigment composition of different microsomal fractions. *Arch. Biochem. Biophys.*, 77:493–509.
85. Gertig, D. M., Stampfer, M., Haiman, C., Hennekens, X. H., Kelsey, K., and Huner, D. J. (1998): Glutathione *S*-transferase GSTM1 polymorphisms and colorectal cancer risk: a prospective study. *Cancer Epidemiol. Biomarker Prev.*, 11:1001–1005.
86. Gervot, L., Rochat, B., Gauteir, J. C., Bohnenstengel, F., Kroemer, H., De Berardinis, V., Martin, H., Beaune, P., and De Waziers, I. (1999): Human CYP2B6: expression, inducibility and catalytic activities. *Pharmacogenetics*, 9(3):295–306.
87. Gillette, J. R., Brodie, B. B., and La Du, B. N. (1957): The oxidation of drugs by liver microsomes: on the role of TPNH and oxidase. *J. Pharmacol. Exp. Ther.*, 119–540.
88. Glatt, H., Boeing, H., Engelke, C. E. H., Ma, L., Kuhlow, A., Pabel, U., Pomplun, D., Teubner, W., and Meinl, W. (2001): Human cytosolic sulphotransferases: genetics, characteristics, toxicological aspects. *Mutat. Res.*, 482:27–40.
89. Goebel, C., Vogel, C., Wulferink, M., Mittmann, S., Sachs, B., Schraa, S., Abel, J., Degen, G., Uetrecht, J., and Gleichmann, E. (1999): Procainamide, a drug causing lupus, induces prostaglandin H synthase-2 and formation of T cell-sensitizing drug metabolites in mouse macrophages. *Chem. Res. Toxicol.*, 12:488–500.
90. Goldman, P. (1978): Biochemical pharmacology of the intestinal flora. *Annu. Rev. Pharmacol.*, 18:523–539.
91. Gorsky, L. D., Koop D. R., and Coon, M. J. (1984): On the stoichiometry of the oxidase and monooxidase reaction catalyzed by liver microsomal cytochrome P450. *J. Biol. Chem.*, 259:6812–6817.
92. Green, M. D. and Tephly, T. R. (1998): Glucuronidation of amine substrates by purified and expressed UDP-glucuronosyltransferase proteins. *Drug Metab. Dispos.*, 26:860–867.
93. Greene, N., Judson, P. N., Langowski, J. J., and Marchant, C. A. (1999): Knowledge based expert systems for toxicity and metabolism prediction: DEREK, StAR, and METEOR. *SAR QSAR Evviron. Res.*, 10:299–313.
94. Grillo, M. P., Hua, F., Knutson, C. G., Ware, J. A., and Li, C. (2003): Mechanistic studies on the bioactivation of dichlofenac: identification of dichlofenac-*S*-acyl-glutathione *in vitro* in incubations with rat and human hepatocytes. *Chem. Res. Toxicol.*, 16:1410–1417.
95. Guengerich, F. P. (1996): The chemistry of cytochrome P450 reactions. In: *Cytochrome P450: Metabolic and Toxicological Aspects*, edited by C. Ioannides. CRC Press, Boca Raton, FL, pp. 55–74.
96. Gueraud, F. and Paris, A. (1998): Glucuronidation: a dual control. *Gen. Pharmacol.*, 31:683–688.
97. Gut, J. (1982): Rotation of cytochrome P450. II. Specific interactions of cytochrome P450 with NADPH-cytochrome P450 reductase in phospholipid vesicles. *J. Biol. Chem.*, 257:7030–7036.
98. Habig, W. H. (1982): Glutathione *S*-transferases: versatile enzymes of detoxification. In: *Radioprotectors and Anticarcinogens*, edited by O. F. Nygaard. Academic Press, New York, pp. 169–190.
99. Haenen, G. R. M., Vermeulen, N. P. E., Tai Tin Tsoni, J. N. L., Regetti, H. M. N., Timmerman, H., and Bast, A. (1988): Activation of the microsomal glutathione *S*-transferase and reduction of the glutathione dependent protection against lipid peroxidation by acrolein. *Biochem. Pharmacol.*, 37:1933–1938.
100. Hayes, J. D., Flanagan, J. U., and Jowsey, I. R. (2005): Glutathione transferases. *Annu. Rev. Pharmacol. Toxicol.*, 45:51–88.
101. Hayes, P. C., Bouchier, I. A. D., and Becket, G. J. (1991): Glutathione *S*-transferases in human health and disease. *Gut*, 32:813–818.
102. Heyman, E. (1982): Hydrolysis of carboxylic esters and amides. In: *Metabolic Basis of Detoxication*, edited by W. B. Jakoby, J. R. Bend, and J. Caldwell. Academic Press, New York, pp. 229–245.

103. Hinson, J.A., Freeman, J.P., Potter, D.W., Mitchum, R.K. and Evans, F.E. (1985): Mechanism of microsomal metabolism of benzene to phenol. *Molec. Pharmacol.*, 27:574–577.
104. Hoffman, J. L. (1994): Bioactivation by *S*-adenosylation, *S*-methylation, or *N*-methylation. *Adv. Pharmacol.*, 27:449–477.
105. Hooper, L. V., Wong, M. H., Thelin, A., Hanson, L., Falk, P. G., and Gordon, G. I. (2001): Molecular analysis of commensal host–microbial relationships in the intestine. *Science*, 291:881–884.
106. Horecher, B. L. (1950): Triphosphopyridine nucleotide cytochrome c reductase in liver. *J. Biol. Chem.*, 183:593–605.
107. Hosagrahara, V. P., Rettie, A. E., Hassett, C., and Omiecinski, C. J. (2004): Functional analysis of human microsomal epoxide hydrolase genetic variants, *Chem. Biol. Interact.*, 150:149–159.
108. Huttner, W. B. (1982): Sulphation of tyrosine residues: a widespread modification of proteins. *Nature*, 299: 273–276.
109. Ingelman-Sundberg, M. (2004): Pharmacogenetics of cytochrome P450 and its applications in drug therapy: the past, present and future. *Trends Pharmacol. Sci.*, 25(4):193–200.
110. Ingelman-Sundberg, M. (2005): Genetic polymorphisms of cyotchrome P450 2D6 (CYP2D6): clinical consequences, evolutionary aspects and functional diversity. *Phamacogenom. J.*, 5:6–13.
111. Inskeep, P. B., Koga, N., Cmarik, J. L., and Guengerich, F. P. (1986): Covalent binding of 1,2-dihaloalkanes to DNA and stability of the major DNA adduct, S42-(N7-guanyl)ethyl glutathione. *Cancer Res.*, 46:2839–2844.
112. Ioannides, C., Ed. (1996): *Cytochromes P450: Metabolic and Toxicological Aspects*. CRC Press, Boca Raton, FL.
113. Jacobsen, D. and McMartin, K. E. (1986): Methanol and ethylene glycol poisonings: mechanisms of toxicity, clinical course, diagnosis and treatment. *Med. Toxicol.*, 1(5):309–334.
114. Jakoby, W. B. (1976): Glutathione *S*-transferases: catalytic aspects. In: *Glutathione Metabolism and Function*, edited by I. M. Arias and W. B. Jakoby. Raven Press, New York, pp. 189–211.
115. Jakoby, W. B. (1978): The glutathione *S*-transferases: a group of multifunctional detoxification proteins. *Adv. Enzymol. Relat. Areas Mol. Biol.*, 46:383–414.
116. Jakoby, W. B. (1980): Sulfotransferases. In: *Enzymatic Basis of Detoxification*, Vol. II, edited by W. B. Jakoby. Academic Press, New York, pp. 199–228.
117. Jakoby, W. B. and Habig, W. H. (1980): Glutathione transferases. In: *Enzymatic Basis of Detoxification*, Vol. II, edited by W. B. Jakoby. Academic Press, New York, pp. 63–94.
118. Jakoby, W. B., Duffel, M. W., Lyon, E. S., and Ramaswamy, S. (1984): Sulfotransferases active with xenobiotics: comments on mechanism. In: *Progress in Drug Metabolism*, Vol. 8, edited by J. W. Bridges and L. F. Chasseaud. Taylor & Francis, London, pp. 11–33.
119. Jocelyn, P. C. (1980): Glutathione and the mitochondrial reduction of some diazenes. *Biochem. Pharmacol.*, 29(3):331–333.
120. Johansson, I., Lundquist, E., Bertilsson, L., Dahl, M.-L., Sjoqvist, F., and Ingrelman-Sundberg, M. (1993): Inherited amplification of an active gene in the cytochrome P450 CYP2D locus as a cause of ultrarapid metabolism of debrisoquine. *Proc. Natl. Acad. Sci. USA*, 90:11825–11829.
121. Jolivette, L. J. and Anders, M. W. (2003): Computational and experimental studies on the distribution of addition and substitution products of the microsomal glutathione transferase 1-catalyzed conjugation of glutathione with fluoroalkenes. *Chem. Res. Toxicol.*, 16:137–144.
122. Kalgutkar, A. S., Gardner, I., Obach, R. S., Shaffer, C. L., Callegari, E., Henne, K. R., Mutlib, A. E., Dalvie, D. K., Le, J. S., Nakai, Y., O'Donnell, J. P., Boer, J., and Harriman, S. P. (2005): A comprehensive listing on bioactivation pathways of organic functional groups. *Curr. Drug Metab.*, 6:161–225.
123. Kamataki, T., Fujieda, M., Kiyotani, K., Iwano S., and Kunitoh, H. (2005): Genetic polymorphism of *CYP2A6* as one of the potential determinants of tobacco-related cancer risk. *Biochem. Biophys. Res. Comm.*, 338:306–310.
124. Kedderis, G. L., Dyroff, M. C., and Rickert, D. E. (1984): Hepatic macromolecular covalent binding of the hepatocarcinogen 2,6-dinitrotoluene and its 2,4-isomer *in vivo*: modulation by the sulfotransferase inhibitors pentachlorophenol and 2,6-dichloro-4-nitrophenol. *Carcinogenesis*, 5:1199–1204.
125. Keller, W. (1842): Ueber verwandlung der Benzoesaure in hippursaure. *Justus Liebig's Ann. Chem.*, 43:108–111.
126. Kemper, R. A. and Elfarra, A.A. (1996): Oxidation of 3-butene-1,2-diol by alcohol dehydrogenase. *Chem. Res. Toxicol.*, 9:1127–1134.
127. Kilty, C., Doyle, S., Hassett, B., and Manning, F. (1998): Glutathione *S*-transferases as biomarkers of organ damage: applications of rodent and canine GST enzyme immunoassays. *Chem. Biol. Interact.*, 112:123–135.
128. Kim, D. and Guengerich, F. P. (2005): Cytochrome P450 activation of arylamines and heterocyclic amines. *Annu. Rev. Pharmacol. Toxicol.*, 45:27–49.
129. Klaassen, C. D. and Boles, J. W. (1997): The importance of 3′-phosphoadenosine 5′-phosphosulfate (PAPS): in the regulation of sulfation. *FASEB J.*, 11:404–418.
130. Klingenberg, M. (1958): Pigments of rat liver microsomes. *Arch. Biochem. Biophys.*, 75:179–386.
131. Klopman, G., Dimayuga, M., and Talafous, J. (1994): META 1: a program for evaluation of metabolic transformation of chemicals. *J. Chem. Info. Comp. Sci.*, 34:1320–1325.
132. Koob, M. and Dekant, W. (1991): Bioactivation of xenobiotics by formation of toxic glutathione conjugates. *Chem. Biol. Interact.*, 77:107–136.
133. Krenitsky, T. A., Hall, W. W., de Miranda, P., Beauchamp, L. M., Schaeffer, H. J., and Whiteman, P. D. (1984): 6-Deoxyacyclovir: a xanthine oxidase-activated prodrug of acyclovir. *Proc. Natl. Acad. Sci.*, 81, 3209–3213.
134. Kulkarni, A. P. (2001): Role of biotransformation in conceptal toxicity of drugs and other chemicals. *Curr. Pharmaceut. Des.*, 7:833–857.
135. Kulkarni, S. A., Zhu, J., and Blechinger, S. (2005): *In silico* techniques for the study and prediction of xenobiotic metabolism: a review. *Xenobiotica*, 35(10–11):955–973.

136. Kunze, K. L. and Trager, W. F. (1993): Isoform-selective mechanism-based inhibition of human cytochrome P450 1A2 by furafylline. *Chem. Res. Toxicol.*, 6(5):649–656.
137. Kuriyama, Y., Omura, T., Siekevitz, P., and Palade, G. E. (1969): Effects of phenobarbital on the synthesis and degradation of protein components of rat liver microsomal membranes. *J. Biol. Chem.*, 244:2017–2026.
138. La Du, B. N., Gaudette, L., Trousof, N., and Brodie, B. B. (1955): Enzymatic dealkylation of aminopyrine (Pyramidon) and other alkylamines. *J. Biol. Chem.*, 214: 741–752.
139. Lamartiniere, C. A. (1981): The hypothalamic–hypophyseal–gonadal regulation of hepatic glutathione *S*-transferases in the rat. *Biochem. J.*, 198:211–217.
140. Lamartiniere, C. A. and Lucier, G. W. (1983): Endocrine regulation of xenobiotic conjugation enzymes. *Basic Life Sci.*, 24:295–312.
141. Lancaster, J. M., Brownlee, H. A., Bell, D. A., Futreahs, R. A., Marks, J. R., Berchuchk, A., Wiseman, R. W., and Taylor, J. A. (1996): Microsomal epoxide hydrolase polymorphisms as a risk factor for ovarian cancer. *Mol. Carcinog.*, 3:160–162.
142. Langowski J. and Long, A. (2002): Computer systems for the prediction of xenobiotic metabolism. *Adv. Drug Delivery Rev.*, 54:407–415.
143. Lantz, R. C., Orozco, J., and Bogdanffy, M. S. (2003): Vinyl acetate induces decreases in intracellular pH in rat nasal epithelial cells. *Toxicol. Sci.*, 75, 423–431.
144. Lash, L. H. and Anders, M. W. (1989): Uptake of nephrotoxic *S*-conjugates by isolated rat proximal tubular cells. *J. Pharmacol. Exp. Ther.*, 248:531–537.
145. Laurenzana, E. M., Hassett, C., and Omiecinski, C. J. (1998): Post-transcriptional regulation of human microsomal epoxide hydrolase. *Pharmacogenetics*, 8:157–167.
146. Lenoard, T. B. and Popp, J. A. (1981): Investigation of the carcinogenic initiation potential of dinitrotoluene: structure–activity study. *Proc. Am. Assoc. Cancer Res.*, 22:82.
147. Lew, D. F. V., Lake, B. G., and Dickins, M. (2004): Substrates of human cytochromes P450 from families CYP1 and CYP2: analysis of enzyme selectivity and metabolism. *Drug Metab. Drug Interact.*, 20(3):111–142.
148. Lewis, D. F. V., Ed. (1996): *Cytochromes P450: Structure, Function and Mechanism.* Taylor & Francis, London.
149. Lewis, D. F. V. (1998): The CYP2 family: models, mutants and interactions. *Xenobiotica*, 28:617–661.
150. Lewis, D. F. V. (2004): 57 varieties: the human cytochromes P450. *Pharmacogenomics*, 5(3):305–318.
151. Lin, D. X., Tang, Y. M., Peng, Q., Lu, S.X., Ambrosone, C. B., and Kadlubar, F. F. (1998): Susceptibility to esophageal cancer and genetic polymorphisms in glutathione *S*-transferases T1, P1, and M1 and cytochrome P450 2E1. *Cancer Epidemiol. Biomarkers Prev.*, 11:1013–1018.
152. Lock, E. A. and Reed, C. A. (1998): Xenobiotic metabolizing enzymes of the kidney. *Toxicol. Pathol.*, 26:18–25.
153. Long, D. J. and Jaiswal, A. K. (2000): NRH:quinone oxidoreductase 2 (NQO2). *Chem. Biol. Interact.*, 129:99–112.
154. Long, R. M. and Rickert, D. E. (1982): Metabolism and excretion of 2,6-dinitro-[$^{14}$C]toluene *in vivo* and in isolated perfused rat livers. *Drug Metab. Dispos.*, 10:455–458.
155. Lu, A. Y. H, Strobel, H. W., and Coon, M. J. (1969): Hydroxylation of benzphetamine and other drugs by a solubilized form of cytochrome P450 from liver microsomes: lipid requirement for drug demethylation. *Biochem. Biophys. Res. Commun.*, 36:545–551.
156. Lu, A. Y. H., Strobel, A. H. W., and Coon, M. J. (1970): Properties of a solubilized form of the cytochrome P450-containing mixed-function oxidase of liver microsomes. *Mol. Pharmacol.*, 6:213–220.
157. Lu, S. C. (1999): Regulation of hepatic glutathione synthesis: current concepts and controversies. *FASEB J.*, 13:1169–1183.
158. Lui, L. and Klaassen, C. D. (1996): Different mechanism of saturation of acetaminophen sulfate conjugation in mice and rats. *Toxicol. Appl. Pharmacol.*, 139: 128–134.
159. Lyles, G. A. (1996): Mammalian plasma and tissue-bound semicarbazide sensitive amine oxidase: biochemical, pharmacological and toxicological aspects. *Int. J. Biochem. Cell Biol.*, 28:259–274.
160. Mackenzie, P. I., Owns, I. S., Burchell, B., Bock, K. W., Bairoch, A., Belanger, A. et al. (1997): The UDP glycosyltransferase gene superfamily: recommended nomenclature update based on evolutionary divergence. *Pharmacogenetics*, 7:255–269.
161. Mahajan, S. and Atkins, W. M. (2005): The chemistry and biology of inhibitors and prodrugs targeted to glutathione *S*-transferases. *Cell. Mol. Life Sci.*, 62:1221–1233.
162. Maier, P., Saad, B., and Schawalder, H.P. (1994): Effect of periportal- and centrilobular-equivalent oxygen tension on liver specific functions in long-term rat hepatocyte cultures. *Toxicol. In Vitro*, 8(3):423–435.
163. Mallett, A. K. (1985): Metabolic adaptation of rat faecal microflora to cyclamate *in vitro*. *Food Chem. Toxicol.*, 23:1029–1034.
164. Mandell, H. G. (1981): Pathways of drug biotransformation: biochemical conjugations. In: *Fundamentals of Drug Metabolism and Drug Disposition*, edited by B. N. La Du, H. G. Mandel, and E. L. Way. Kreiger, Malabar, FL, pp. 169–171.
165. Mannering, G. T. (1971): Microsomal enzyme systems which catalyze drug metabolism. In: *Fundamentals of Drug Metabolism and Drug Disposition*, edited by B. N. La Du, H. G. Mandel, and E. L. Way. Kreiger, Malabar, FL, pp. 206–252.
166. Marnett, L. J. and Reed. G. A. (1979): Peroxidative oxidation of benzo(*a*)pyrene and prostaglandin biosynthesis. *Biochemistry*, 18:2923–2929.
167. Marnett, L. J., Reed, G. A., and Johnson, J. T. (1977): Prostaglandin synthetase dependent benzo(*a*)pyrene oxidation: products of the oxidation and inhibition of their formation by antioxidants. *Biochem. Biophys. Res. Commun.*, 79:569–576.
168. Mason, L. F., Sharp, L., Cotton, S. C., and Little, J. (2005): CYP1A1 gene polymorphism and risk of breast cancer: a HuGE review. *Am. J. Epidemiol.*, 161(10):901–915.
169. Mattammal, M. B., Zenser, T. V., and Davis, B. B. (1981): Prostaglandin hydroperoxidase-mediated 2-amino-4-(5–nitro-2-furye[$^{14}$C]thiazole metabolism and nucleic acid binding. *Cancer Res.*, 41:4961–4966.

170. McIlwain, C. C., Townsend, D. M., and Tew, K. D. (2006): Glutathione *S*-transferase polymorphisms: cancer incidence and therapy. *Oncogene*, 25:1639–1648.
171. Meech, R. and Mackenzie, P. I. (1997): Structure and function of uridine diphosphate glucuronosyltransferases. *Clin. Exp. Pharmacol. Physiol.*, 24:907–915.
172. Meister, A. and Tate, S. S. (1976): Glutathione and related gamma-glutamyl compounds: biosynthesis and utilization. *Annu. Rev. Biochem.*, 45:559–604.
173. Mekenyan, O. G., Dimitrov, S. D., Pavlov, T. S. and Veith, G. D. (2004): A systematic approach to simulating metabolism if computational toxicology. I. The TIMES heuristic modeling framework. *Curr. Pharmacol. Design*, 10:1273–1293.
174. Miller, M. S., Juchau, M. R., Guengerich, P., Nebert, D. W., and Raucy, J. L. (1996): Symposium overview: drug metabolism enzymes in developmental toxicology. *Fund. Appl. Toxicol.*, 34:165–175.
175. Miners, J. O. and Mackenzie, P. I. (1991): Drug glucuronidation in humans. *Pharmacol. Ther.*, 51:347–369.
176. Mirsalis, J. C. and Butterworth, B. E. (1982): Induction of unscheduled DNA synthesis in rat hepatocytes following *in vivo* treatment with dinitrotoluene. *Carcinogenesis*, 3:241–245.
177. Mitchell, J. R. (1973): Acetaminophen-induced hepatic necrosis. IV. Protective role of glutathione. *J. Pharmacol. Exp. Ther.*, 187:211–217.
178. Miyazaki, M., Yoshizawa, I. I., and Fishman, J. (1969): Direct methylation of estrogen catechol sulfates. *Biochemistry*, 8:1669–1672.
179. Mohandas, J., Duggin, G. G., Horvath, J. S., and Tiller, D. J. (1981): Metabolic oxidation of acetaminophen (paracetamol): mediated by cytochrome P450 mixed-function oxidase and prostaglandin endoperoxide synthetase in rabbit kidney. *Toxicol. Appl. Pharmacol.*, 61:252–259.
180. Moskatello, D., Polanc, S., Kosmrlj, J., Vukovic, L., and Osmak, M. (2002): Diazenecarboxamide UP-91, a potential anticancer agent, acts by reducing cellular glutathione content. *Pharmacol. Toxicol.*, 91(5):258–263.
181. Mugford, C. A. and Kedderis, G. L. (1998): Sex-dependent metabolism of xenobiotics. *Drug Metab. Rev.*, 30:441–498.
182. Mulder, G. J. (1981): Generation of reactive intermediates from xenobiotics by sulfate conjugation: their potential role in chemical carcinogenesis. In: *Sulfation of Drugs and Related Compounds*, edited by G. J. Mulder. CRC Press, Boca Raton, FL, pp. 213–226.
183. Murray, G. I., Melvin, W. T., Greenlee, W. F., and Burke, M. D. (2001): Regulation, function, and tissue-specific expression of cytochrome P450 CYP1B1. *Annu. Rev. Pharmacol. Toxicol.*, 41:297–316.
184. Murray, G. I., Taylor, M. C., McFayden, M. C., McKay, J. A., and Greenlee, W. F. (1997): Tumor-specific expression of cytochrome P450 CYP1B1. *Cancer Res.*, 57:3026–3031.
185. Nelson, D. R., Koymang, L., Kamataki, T., Stegeman, J. J., Feyerelsin, R., Waxman, D. J., Waterman, M. R., Gotoh, O., Coon, M. J., Estabrook, R. W., Gunsalus, I. C., and Nebert, D. W. (1996): P450 superfamily: update on new sequences, gene mapping, accession numbers, and nomenclature. *Pharmacogenetics*, 6:1–42.
186. Nguyen, T., Sherrat, P. J., and Pickett, C. B. (2003): Regulatory mechanisms controlling gene expression mediated by the antioxidant response element. *Ann. Rev. Pharmacol. Toxicol.*, 43:233–260.
187. Nicolas, F., De Sous, G., Thomas, P., Placidi, M., Lorenzon, G., and Rahmani, R. (1995): Comparative metabolism of 3′azido-3-deoxythymidine in cultured hepatocytes from rats, dogs, monkeys, and humans. *Drug Metab. Dispos.*, 23:308–313.
188. Nowell, S. and Falany, C. N. (2006): Pharmacogenetics of human cytosolic sulfotransferases. *Oncogene* 25:1673–1678.
189. Nuccetelli, M., Mazzetti, A. P., Rossjohn, J., Parker, M. W., Beard, P., Caccuri, A. M., Federiai, G., Ricci, G., and LoBello, M. (1998): Shifting substrate specificity of human glutathione transferase (from class pi to class alpha): by a single point mutation. *Biochem. Biophys. Res. Commun.*, 252:184–189.
190. O'Brien, P. J. (2000): Peroxidases. *Chemico-Biol. Interact.*, 129:113–139.
191. Oesch, F. (1980): Species differences in activating and inactivating enzymes related to *in vitro* mutagenicity mediated by tissue preparations from these species. *Arch. Toxicol., Suppl.*, 3:179–194.
192. Oesch, F., Bentley, P., and Glatt, H. R. (1970): Prevention of benzo(*a*)pyrene induced mutagenicity by homogenous epoxide hydratase. *Int. J. Cancer*, 18:448–452.
193. Omiecinski, C. J., Aicher, L., and Swenson, L. (1994): Developmental expression of human microsomal epoxide hydrolase. *J. Pharmacol. Exp. Ther.*, 269:417–423.
194. Omura, T. (1969): Discussion. In: *Microsomes and Drug Oxidations*, edited by J. R. Gillette. Academic Press, New York, pp. 160–161.
195. Omura, T. and Sato, R. (1964): The carbon monoxide-binding pigment of liver microsomes. I. Evidence for its hemoprotein nature. *J. Biol. Chem.*, 239:2370–2378.
196. Omura, T. and Sato, R. (1964): The carbon monoxide-binding pigment of liver microsomes. II. Solubilization purification and properties. *J. Biol. Chem.*, 2379–2385.
197. Ortiz de Montellano, P. R. (1986): Oxygen activation and transfer. In: *Cytochrome P450: Structure, Mechanism and Biochemistry*, edited by P. R. Ortiz de Montellano. Plenum Press, New York, pp. 217–271.
198. Oswald, S., Haenishch, S., Fricke, C., Sudhop, T., Remmler C., Giessmann, T. et al. (2006): Intestinal expression of p glycoprotein (ABCB1), multidrug resistance associated protein 2 (ABCC2), and uridine diphosphate-glucuronosyltransferase 1A1 predicts the disposition and modulates the effect of the cholesterol absorption inhibitor ezetimibe in humans *Clin. Pharmacol. Ther.*, 79:206–217.
199. Panoutsopoulos, G. I., Kouretas, D., and Beedham, C (2004): Contribution of aldehyde oxidase, xanthine oxidase, and aldehyde dehydrogenase to the oxidation of aromatic aldehydes. *Chem. Res. Toxicol.*, 17:1368–1376
200. Pearson, P. G., Slatter, J. G., Rashed, M. S., Han, D. H., and Baillie, T. A. (1991): Carbamoylation of peptides and proteins *in vitro* by *S*-(*N*-methylcarbamoyl)glutathione and *S*-(*N*-methylcarbamoyl)cysteine, two electrophilic *S*-linked conjugates of methyl isocyanate. *Chem. Res. Toxicol.*, 4(4): 436–444.

201. Pelkonen, O., Maenpan, J., Taavitsainen, P., Rautio, A., and Rautio, H. (1998): Inhibition and induction of human cytochrome P450 (CYP): enzymes. *Xenobiotica*, 28:1203–1253.
202. Pelkonen, O., Rautio, H., and Pasanen, M. (2000): CYP2A6: a human coumarin 7-hydroxylase. *Toxicology*, 144:139–147.
203. Phillips, A. H. and Langdon, R. G. (1962): Hepatic triphosphopyridine nucleotide-cytochrome reductase: isolation, characterization and kinetic studies. *J. Biol. Chem.*, 237A:2652–2660.
204. Plapp, B. V., Parsons, M. Leidal, K. G., Baggenstoss, B. A., Ferm, J. R. G., and Waer, S. S. (1987): Characterization of alcohol dehydrogenase from cultured rat hepatoma cells. In: *Enzymology and Molecular Biology of Carbonyl Metabolism*, edited by H. Weiner and T. G. Flynn. Liss, New York.
205. Pool-Zobel, B., Veeriah, S., and Böhmer, F.-D. (2005): Modulation of xenobiotic metabolizing enzymes by anti-carcinogens: focus on glutathione *S*-transferases and their role as targets of dietary chemoprevention in colorectal cancer. *Mut. Res.*, 591:74–92.
206. Poulas, T. L. and Raag, R. (1992): Cytochrome $P450_{cam}$: crystallography, oxygen activation, and electron transfer. *FASEB J.*, 6:674–679.
207. Pritsos, C. (2000): Cellular distribution, metabolism and regulation of the xanthine oxidoreductase enzyme system. *Chem. Biol. Interac.*, 129:195–208.
208. Rafter, J. J. (1983): Studies on the reestablishment of the intestinal microflora in germ-free rats with special reference to the metabolism of *N*-isopropyl-alpha-choloracetanilide (Propachlor), *Xenobiotica*, 13:171–178.
209. Raftogianis, R. B., Her, C., and Weinshiboum, R. M. (1996): Human phenol sulfotransferase pharmacogenetics: STP1 gene cloning and structural characterization. *Pharmacogenetics*, 6:473–478.
210. Rein, H. and Jung, C. (1993): Metabolic reactions: mechanism of substrate oxidation. In: *Cytochrome P450*, edited by J. Schenkman and H. Greim. Springer-Verlag, Berlin, pp. 106–122.
211. Rendic, S. (2002): Summary of information on human CYP enzymes: human P450 metabolism data. *Drug Metab. Rev.*, 34(1–2):83–448.
212. Renwick, A. G. (1977): Microbial metabolism of drugs. In: *Drug Metabolism: From Microbe to Man*, edited by D. V. Parke and R. L. Smith. Proceeding of the International Symposium, Guilford, England. Taylor & Francis, London, pp. 169–189.
213. Rettie, A. E. and Jones, P. (2005): Clinical and toxicological relevance of CYP2C9: drug–drug interactions and pharmacogenetics. *Annu. Rev. Pharmacol. Toxicol.*, 45:477–494.
214. Ross, D. (2004): Quinone reductases multitasking in the metabolic world. *Drug Metab. Rev.*, 36(3–4): 639–654.
215. Ross, D., Kepa, J. K., Winski, S. L., Beall, H. D., Anwar, A., and Siegel, D. (2000): NAD(P)H:quinone oxidoreductase 1 (NQO1): chemoprotection, bioactivation, gene regulation and genetic polymorphisms. *Chem. Biol. Interact.*, 129:77–97.
216. Rowland, I. R. (1988): Factors affecting metabolic activity of the intestinal microflora. *Drug Metab. Rev.*, 19:243–261.
217. Rowland, I. R. and Wise, A. (1985): The effect of diet on the mammalian gut flora and its metabolic activities. *CRC Crit. Rev. Toxicol.*, 16:31–103.
218. Rumney, C. J. and Rowland, I. R. (1992): *In vivo* and *in vitro* models of the human colonic flora. *CRC Food Sci. Nutr.*, 31:299–331.
219. Runge-Morris, M. A. (1997): Regulation of expression of the rodent cytosolic sulfotransferases. *FASEB J.*, 11:109–117.
220. Sandberg, M., Hassett, C., Adman, E. T., Meijer, J., and Omiecinski, C. J. (2000): Identification and functional characterization of human soluble epoxide hydrolase genetic polymorphisms, *J. Biol. Chem.*, 275:28873–28881.
221. Sanglard, D. and Kappeli, O. (1993): Cytochrome P450 in unicellular organisms. In: *Cytochrome P450*, edited by J. Schenkman and H. Greim. Springer-Verlag, Berlin, pp. 325–349.
222. Savas, Ü., Bhattacharyya, K. K., Christou, M., Alexander D. L., and Jefcote, C. R. (1994): Mouse cytochrome P450EF, representative of a new 1B subfamily of cytochrome P450s. Cloning, sequence determination and tissue expression. *J. Biol. Chem.*, 269:14905–14911.
223. Scheline, R. R. (1973): Metabolism of foreign compounds by gastrointestinal microorganisms. *Pharmacol. Rev.*, 25:451–523.
224. Schenkman, J. and Greim, H. (1993): *Cytochrome P450*. Springer-Verlag, Berlin.
225. Schenkman, J.B. and Jansson, I. (2003): The many roles of cytochrome $b_5$. *Pharmacol. Therap.*, 97(2):139–152.
226. Schmiedlin-Ren, P., Edwards, D. J., Fitzsimmons, M. E., He, K., Lown, K. S., Woster, P. M., Rahman, A., Thummel, K. E., Fisher, J. M., Hollenberg, P. F., and Watkins, P. B. (1997): Mechanisms of enhanced oral availability of CYP3A4 substrates by grapefruit constituents: decreased interocyte CYP3A4 concentration and mechanism-based inactivation by furanocoumarins. *Drug Metab. Disp.*, 25:1228–1233.
227. Schwartz, N. B. (2005): PAPS and sulfoconjugation. In *Human Cytosolic Sulfotransferases*, edited by G. M. Pacifici and M. W. Coughtrie. CRC Press, Boca Raton, FL, pp. 43–61.
228. Scott, E. E. and Halpert, J. R. (2005): Structures of cytochrome P450 3A4. *Trends Biochem. Sci.*, 30(1):5–7.
229. Seitz, H. K. and Oneta, C. M. (1998): Gastrointestinal alcohol dehydrogenase. *Nutr. Rev.*, 56:52–60.
230. Sekura, R. D., Marcus, C. J., Lyon, E. S., and Jakoby, W. B. (1979): Assay of sulfotransferases. *Anal. Biochem.*, 95:82–86.
231. Sharer, J. E., Shipley, L. A., Vanderbranden, R. R., Binkley, S. N., and Wrighton, S. A. (1995): Comparisons of phase I and phase II *in vitro* hepatic enzyme activities of human, dog, rhesus monkey and cynomolgus monkey. *Drug Metab. Dispos.*, 11:1231–1241.
232. Shen, Z., Reed, J. R., Creighton, M., Liu, D. Q., Tang, Y. S., Hora, D. F., Feeney, W., Szewczyk, J., Bakhtiar, R., Franklin, R. B. and Vincent, S. H. (2003): Identification of novel metabolites of pioglitazone in rat and dog. *Xenobiotica*, 33(5):499–509.

233. Shimada, T., Yamazaki, H., Foroozesh, M., Hopkins, N. E., Alworth, W. L., and Guengerich, F. P. (1998): Selectivity of polycyclic inhibitors for human cytochrome P450s 1A1, 1A2 and 1B1. *Chem. Res. Toxicol.*, 11:1048–1056.
234. Shokeer, A., Larsson, A.-K., and Manervik, B. (2005): Residue 234 in glutathione transferase T1-1 plays a pivotal role in the catalytic activity and selectivity against alternative substrates. *Biochem. J.*, 388(1):387–392.
235. Singer, S. S. and Brun, L. (1978): Enzymatic sulfation of steroids. VII. Hepatic cortisol sulfation and glucocorticoid sulfotransferases in old and young male rats. *Exp. Gerontol.*, 13:425–429.
236. Singer, S. S. and Sylvester, S. (1976): Enzymatic sulfation of steroids. II. The control of the hepatic cortisol sulfotransferase activity and of the individual hepatic steroid sulfotransferases of rats by gonads and gonadal hormones. *Endocrinology*, 99:1346–1352.
237. Singer, S. S., Giera, D., Johnson, J., and Sylvester, S. (1976): Enzymatic sulfation of steroids. I. The enzymatic basis for the sex difference in cortisol sulfation by rat liver preparations. *Endocrinology*, 98:963–974.
238. Singh, S. V., Benson, P. J., Hy, X., Pal, A., Srivastava, S. K., Awasthi, S., Zaren, H. A., and Orchard, J. L. (1998): Gender-related differences in susceptibility of A/J mouse to benzo(*a*)pyrene-induced pulmonary and forestomach tumorigenesis. *Cancer Lett.*, 128:197–204.
239. Skipper, P. L. and Tannenbaum, S. R. (1994): Molecular dosimetery of aromatic amines in human populations. *Environ. Health Perspect.*, 102 Suppl 6:17–21.
240. Skipper, P. L. and Tannenbaum, S. R. (1994): Molecular dosimetry of aromatic amines in human populations. *Environ. Health Persp.*, 102 (Suppl., 6), 17–21.
241. Slatter, J. G., Rashed, M. S., Pearson, P. G., Han, D. H., and Baillie, T. A. (1991): Biotransformation of methyl isocyanate in the rat: evidence for glutathione conjugation as a major pathway of metabolism and implications for isocyanate-mediated toxicities. *Chem. Res. Toxicol.*, 4(2):157–161.
242. Sligar, S. G. and Murray, R. I. (1986): Cytochrome $P450_{cam}$ and other bacterial P450 enzymes. In: *Cytochrome P450: Structure, Mechanism, and Biochemistry*, edited by P. R. Ortiz de Montellano. Plenum Press, New York, pp. 429–503.
243. Smith, C. A. and Harrison, D. J. (1997): Association between polymorphism in gene for microsomal epoxide hydrolase and susceptibility to emphysema. *Lancet*, 350:630–633.
244. Smith, G. J. and Litwack, G. (1980): Roles of ligandin and the glutathione *S*-transferases in binding steroid metabolites, carcinogens and other compounds. *Rev. Biochem. Toxicol.*, 2:1–47.
245. Smith, G. J., Ohl, V. S., and Litwack, G. (1977): Ligandin, the glutathione *S*-transferases, and chemically induced hepatocarcinogenesis: a review. *Cancer Res.*, 37:8–14.
246. Smith, G., Stanley, L. A., Sim, E., Strange, R. C., and Wolf, C. R. (1995): Metabolic polymorphisms and cancer susceptibility. *Cancer Surv.*, 25:27–65.
247. Smith, G., Stubbins, M. J., Harries, L. W., and Wolf, C. R. (1998): Molecular genetics of human cytochrome P450 monooxygenase superfamily. *Xenobiotica*, 28:1124–1165.
248. Soltaninassab, S. R., Sekhar, K. R., Meredith, M. J., and Freeman, M. L. (2000): Multi-faceted regulation of γ-glutamylcysteine synthetase. *J. Cell. Physiol.*, 182:163–170.
249. Solus, J. F., Arietta, B. J., Harris, J. R., Sexton, D. P., Steward, J. Q., McMunn, C., Ihrie, P., Mehal, J. M., Edwards, T. L., and Dawson, E. P. (2004): Genetic variation in eleven phase I drug metabolizing genes in an ethnically diverse population. *Pharmacogenomics*, 5(7):895–931.
250. Sreeravan, L., Hedge, M. W., and Sladek, N. E. (1995): Identification of a class 3 aldehyde dehydrogenase in human saliva and increased levels of this enzyme, glutathione *S*-transferases, and DT-diaphorase in the saliva of subjects who continually ingest large quantities of coffee or broccoli. *Clin. Cancer Res.*, 1:1153–1163.
251. Stirling, L. A. (1980): Microorganisms and environmental pollutants. In: *Introduction to Environmental Toxicology*, edited by F. E. Guthrie and J. J. Perry. Elsevier, New York, pp. 329–342.
252. Strasser, S. I., Smid, S. A., Mashford, M. L., Desmond, P. V. (1997): Sex hormones differentially regulate isoforms of UDP-glucuronosyltransferase. *Pharm. Res.*, 14:1115–1121.
253. Strolin Benedetti, M. and Dostert, P. (1994): Contribution of amine oxidases to the metabolism of xenobiotics. *Drug Metab. Rev.*, 26:507–535.
254. Stubbins, M. J., Harries, L. W., Smith, G., Tarbit, M. H and Wolf, C. R. (1996): Genetic analysis of human cytochrome P450 CYP2C9 locus. *Pharmacogenetics*, 6; 429–239.
255. Su, T. and Ding, X. (2004): Regulation of the cytochrome P450 2A genes. *Toxicol. Appl. Pharmacol.*, 199:285–294.
256. Sutter, T. R., Tang, Y. M., Hayes, C. L., Wo, Y.-Y. P., Jabs E. W., Li, X., Yin, H., Cody, C. W., and Greenlee, W. F (1994): Complete cDNA sequence of a human dioxin-inducible mRNA identifies a new gene subfamily of cytochrome P450 that maps to chromosome 2. *J. Biol. Chem.* 269:13092–13099.
257. Tanaka-Kagawa, T., Jinno, H., Hasegawa, T., Makino, Y. Seko, Y., Hanoika, N., and Ando, M. (2003): Functional characterization of two variant human GSTO 1-1s (Ala140Asp and Thr217Asn), *Biochem. Biophys. Res Comm.*, 301:516–520.
258. Tang, W. and Stearns, R. A. (2001): Heterotropic cooperativity if cytochrome P450 3A4 and potential drug-drug interactions. *Curr. Drug Metab.*, 2:185–198.
259. Tateishi, M., Suzuki, S., and Shimizu, H. (1978): Cysteine conjugate beta-lyase in rat liver: a novel enzyme catalyzing formation of thiol-containing metabolites of drugs. *J. Biol Chem.*, 253:8854–8859.
260. Tateishi, N. and Sakamoto, Y. (1983): Nutritional significance of glutathione in rat liver. In: *Glutathione: Storage Transport, and Turnover in Mammals*, edited by S. Y. Sakamoto, T. Higashi, and N. Tateishi. Japan Science Society Press, Tokyo/VNH Science Press, Utrecht, pp. 13–38.
261. Testa, B. (1995): *The Metabolism of Drugs and Other Xenobiotics: Biochemistry of Redox Reactions*. Academic Press, London.
262. Thurman, R. G. (1987): Regulation of monooxygenation in intact cells. In: *Mammalian Cytochromes P450*, Vol. I edited by F. P. Guengerich. CRC Press, Boca Raton, FL pp. 131–152.

263. Totah, R. A. and Rettie, A. E. (2005): Cytochrome P450 2C8: substrates, inhibitors, pharmacogenetics, and clinical relevance. *Clin. Pharmacol. Therapeut.*, 77(5):341–352.
264. Tsuchiya, Y., Nakajima, M., and Yokoi, T. (2005): Cytochrome P450-mediated metabolism of estrogens and its regulation in human. *Cancer Lett.*, 227(2):115–124.
265. Tukey, R. H. and Strassburg, C. P. (2000): Human UDP-glucuronosyltransferases: metabolism, expression and disease. *Annu. Rev. Pharmacol. Toxicol.*, 40:581–616.
266. U.S. FDA (1997): *Guidance for Industry: Drug Metabolism/Drug Interaction Studies in the Drug Development Process, Studies in Vitro*. The Drug Information Branch, Center for Drug Evaluation and Research, U.S. Food and Drug Administration, Rockville, MD.
267. U.S. FDA (1998): *Guidance for Industry: In Vivo Drug Metabolism/Drug Interaction Studies, Study Design, Data Analysis and Recommendations for Dosing and Labeling*. The Drug Information Branch, Center for Drug Evaluation and Research, U.S. Food and Drug Administration, Rockville, MD.
268. Vamvakas, S. and Anders, M. W. (1990): Formation of reactive intermediates by phase II enzymes: glutathione-dependent bioactivation reactions. In: *Biological Reactive Intermediates IV*, edited by C. M. Witmer et al. Plenum Press, New York, pp. 13–24.
269. van Bladeren, P. J. and van Ommen, B. (1991): The inhibition of glutathione *S*-transferases: mechanisms, toxic consequences and therapeutic effects. *Pharmacol. Ther.*, 51:35–46.
270. van Bladeren, P. J. (2000): Glutathione conjugation as a bioactivation reaction. *Chem. Bio. Inter.*, 129:61–76.
271. van Lieshout, E. M. and Peters, W. H. (1998): Age and gender dependent levels of glutathione and glutathione *S*-transferases in human lymphocytes. *Carcinogenesis*, 19:1873–1875.
272. Van Ness, K. P., McHugh, T. E., Bammier, T. K., and Eaton, D. L. (1998): Identification of amino acid residues essential for high aflatoxin B1-8,9-epoxide conjugation activity in alpha class glutathione *S*-transferases through site-directed mutagenesis. *Toxicol. Appl Pharmacol.*, 152:166–174.
273. Vasiliou, V., Pappa, A., and Estey, T. (2004): Role of human aldehyde dehydrogenases in endobiotic and xenobiotic metabolism. *Drug Metab. Rev.*, 36(2):279–299.
274. Vesell, E. S. and Penno, M. B. (1983): Intraindividual and interindividual variations. In: *Biological Basis of Detoxication*, edited by J. Caldwell and W. B. Jakoby. Academic Press, New York, pp. 369–410.
275. Vickers, S. and Polsky, S. L. (2000): The biotransformation of nitrogen containing xenobiotics to lactams. *Curr. Drug Metab.*, 1:357–389.
276. Visser, T. J., van Buuren, J. C. J., Rutger, M., Rooda, S. J. E., and deHerder, W. W. (1990): The role of sulfation in thyroid hormone metabolism. *Trends Endocrinol. Metab.*, 1:211–218
277. Vogel, C. (2000): Prostaglandin H synthetases and their importance in chemical toxicity. *Curr. Drug Metab.*, 1:391–404.
278. Waddell, W. J. and Marlow, C. (1981): Biochemical regulation of the accessibility of teratogens to the developing embryo. In: *The Biochemical Basis of Chemical Teratogenesis*, edited by M. R. Jachau. Elsevier, New York, pp. 1–62.
279. Walker, C. H. and Desch, F. (1983): Enzymes in selective toxicology. In: *Biological Basis of Detoxication*, edited by J. Caldwell and W. B. Jakoby. Academic Press, New York, pp. 349–368.
280. Walker, C. H. and Mackness, M. I. (1983): Esterases: problems of identification and classification. *Biochem. Pharmacol.*, 32:3265–3269.
281. Walsky, R. L. and Obach, R. S. (2003): Verification of the selectivity of (+)N-3-benzylnirvanol as a CYP2C19 inhibitor. *Drug Metab. Disp.*, 31:343.
282. Wan, J., Carr, B. A., Cutler, N. S., Lanza, L. D., Hines, R. N., and Yost, G. S. (2005): SP1 and SP3 regulate basal transcription of the human CYP2F1 gene. *Drug Metab. Disp.*, 33:1244–1253.
283. Wang, H. and LeCluyse, E. L. (2003): Role of orphan nuclear receptors in the regulation of drug-metabolizing enzymes. *Clin. Pharmacokinet.*, 42(15):1331–1357.
284. Wang, H., Tan, W., Hao, B., Miao, X., Zhou, G., He, F., and Lin, D. (2003): Substantial reduction of risk of lung adenocarcinoma associated with genetic polymorphism in CYP2A13, the most active cytochrome P450 for the metabolic activation of tobacco-specific carcinogen NNK. *Cancer Res.*, 63(22):8057–8061.
285. Wang, L. and Weinshilboum, R. (2006): Thiopurine *S*-methyltransferase pharmacogenetics: insights, challenges and future directions. *Oncogene*, 26:1629–1638.
286. Wargovich, M. J. (2006): Diallylsulfide and allylmethylsulfide are uniquely effective among organosulfur compounds in inhibiting CYP2E1 protein in animal models. *J. Nutr.*, 136 (3, Suppl.):832S–834S.
287. Wasserman, W. W., Fahl, W. E. (1997): Functional antioxidant responsive elements. *Proc. Natl. Acad. Sci. USA*, 94:5361–5366.
288. Watkins, R. E., Wisely, G. B., Moore, L. B., Collins, J. L., Lambert, M. H., Williams, S. P., Willson, T. M., Kliewer, S. A., and Redinbro, M. R. (2001): The human nuclear xenobiotic receptor PXR: structural determinants of directed promiscuity. *Science*, 292:2329–2333.
289. Waxman, D. J. (1999): P450 gene induction by structurally diverse xenochemicals: central role of nuclear receptors CAR, PXR, and PPAR. *Arch. Biochem. Biophys.*, 369(1):11–23.
290. Wegman, P., Vainikka, L., Stal, O., Nordenskjold, B., Skoog, L., Rutqvist, L. E., and Wingren, S. (2004): Genotype of metabolic enzymes and the benefit of tamoxifen in postmenopausal breast cancer patients. *Breast Cancer Res.*, 7:R284–R290.
291. Wendel, A., Heinle, H., and Silbernagl, S. (1977): The degradation of glutathione derivatives in the rat kidney. *Curr. Probl. Clin. Biochem.*, 8:73–84.
292. Williams, C. H., Jr., and Kamin, H. (1962): Microsomal triphosphopyridine nucleotide-cytochrome c reductase of liver. *J. Biol. Chem.*, 237:587–595.
293. Wood, J. I. (1970): Biochemistry of mercapturic acid formation. In: *Metabolic Conjugation and Metabolic Hydrolysis*, Vol. 2, edited by W. H. Fishman. Academic Press, New York, pp. 261–299.
294. Yamamato, T., Pierce, W. M., Jr., Hurst, H. E., Chen, D., and Waddell, W. J. (1988): Inhibition of the metabolism of urethane by ethanol. *Drug Metab. Disp.*, 16:355–358.

295. Yee, S. B., and Pritsos, C. A. (1997): Comparison of oxygen radical generation from the reductive activation of doxorubicin, streptonigrin, and menadione by xanthine oxidase and xanthine dehydrogenase. *Arch. Biochem. Biophys.*, 347:235–241.

296. Yu, P. H. and Zuo, D.-M. (1997): Formation of formaldehyde from adrenaline: a potential risk factor for stress-related antipathy. *Neurochem. Res.*, 22:615–620.

297. Zhang, L., Savas, Ü., Alexander, D. L., and Jefcote, C. R. (1998): Characterization of the mouse *Cyp1B1* gene. *J. Biol. Chem.*, 273:5174–5183.

298. Zhu, B. T. (2002): Catechol-*O*-methyltransferase (COMT)-mediated methylation metabolism of endogenous bioactive catechols and modulation by endobiotics and xenobiotics: importance in pathophysiology and pathogenesis. *Curr. Drug Metab.*, 3:321–349.

299. Zhu, B. T., Suchar, L. A., Huang, M. T., and Conney, A. H. (1996): Similarities and differences in the glucuronidation of estradiol and estrone by UDP-glucuronosyltransferase in liver microsomes from male and female rats. *Biochem. Pharmacol.*, 51:1195–1202.

300. Ziegler, D. M. (1993): Recent studies on the structure and function of multisubstrate flavin-containing monooxygenase. *Annu. Rev. Pharmacol. Toxicol.*, 33:179–199.

301. Schenkman, J. B. and Jansson, I. (2003): The many roles of cytochrome $b_5$. *Pharmacol. Therap.*, 97:139–152.

302. Meunier, B., de Visser, S. P., and Shaik, S. (2004): Mechanism of oxidation reactions catalyzed by cytochrome P450 enzymes. *Chem. Rev.*, 104:3947–3980.

303. Dorne, J. L., Walton, K., and Renwick, A. G. (2004): Human variability for metabolic pathways with limited data (CYP2A6, CYP2C9, CYP2E1, ADH, esterases, glycine and sulphate conjugation). *Food Chem. Toxicol.*, 42:397–421.

304. Dorne, J. L. C. M., Walton, K., and Renwick, A. G. (2005): Human variability in xenobiotic metabolism and pathway-related uncertainty factors for chemical risk assessment: a review. *Food Chem. Toxicol.*, 43: 203–216.

305. Lipscomb, J. C. and Kedderis, G. L. (2002): Incorporating human interindividual biotransformation variance in health risk assessment. *Sci. Total Environ.*, 288:13–21.

306. Watabe, T. (1985): Metabolic activation of 7,12-dimethylbenz(*a*)anthracene and 7-methylbenz(*a*)anthracene via hydroxymethyl sulfate esters by P450-sulfotransferase. *Gann Monogr.*, 30:125–139.

307. Zenser, T. V., Mattammal, M. B., and Davis, B. B. (1979): Demonstration of separate pathways for the metabolism of organic compounds in rabbit kidney. *J. Pharmacol. Exp. Ther.*, 208:418-421.

# 4 Toxicokinetics

*Andrew Gordon Renwick*

## CONTENTS

## INTRODUCTION

It has been recognized for many years [1,2] that the relationship between drug administration and therapeutic response can be subdivided into two aspects:

- *Pharmacokinetics*, which relates to the movement of the drug within the body
- *Pharmacodynamics*, which is concerned with the pharmacological effects once the drug is delivered to its site of action/receptor.

The processes concerned with the absorption, distribution, and elimination of therapeutic drugs are nonspecific and shared with other types of non-nutrients. The principles of pharmacokinetics apply to any environmental non-nutrient (xenobiotic), and many of the biological processes, such as the metabolizing enzymes, are shared by drugs and low-molecular-weight organic molecules, such as additives, pesticides, and contaminants. It is valid to apply the term pharmacokinetics to all foreign compounds, although the term *toxicokinetics* now has wide usage [3] and is useful for potentially toxic chemicals that do not have a therapeutic effect. Most of the basic principles of pharmaco-/toxicokinetics were established in relation to the absorption, distribution, and elimination of therapeutic drugs. The introductory chapters of many undergraduate clinical pharmacology textbooks contain much useful basic information. Toxicokinetics, in the current context, is the application of pharmacokinetic principles to animal toxicity studies and to human toxicity data to provide information on exposure to the parent compound and its metabolites and on other aspects, such as accumulation during chronic exposure. The incorporation of toxicokinetic data from animal studies into risk assessment requires data from related *in vivo* studies in humans at appropriate doses or from the results of in vitro data incorporated into a physiologically based pharmacokinetic (PBPK) model (see below and Chapter 5).

The understanding and interpretation of toxicological findings requires information on two key areas: (1) delivery of the compound to its site of action (toxicokinetics), and (2) the mechanism of action and potency of the chemical at the site of action (toxicodynamics) (Figure 4.1). Such information may assist in understanding the dose–response relationship in animal toxicity studies and its relevance to humans and may allow identification of potentially at risk subgroups of the exposed human population.

Risk assessment has traditionally used different approaches for cancer (and other nonthreshold effects) compared with toxicity believed to show a biological

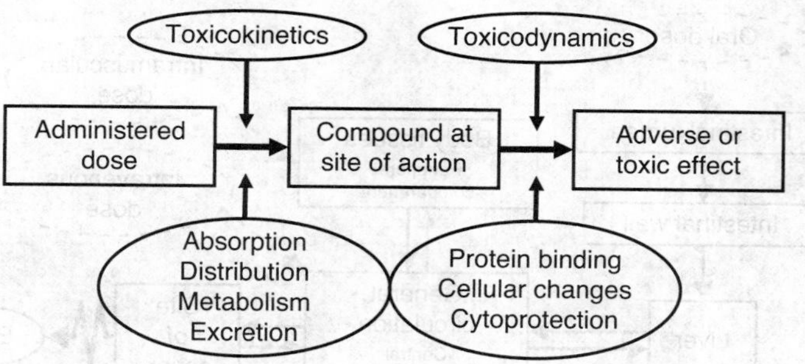


**FIGURE 4.1** The relationship between *in vivo* response and toxicokinetics and toxicodynamics.

threshold [4]. For cancer and similar effects, the dose–response relationship in animals is extrapolated down to a very low level of risk, or a virtually safe dose. Toxicokinetics are sometimes incorporated into the risk assessment process for genotoxic cacinogens by the use of a PBPK model that corrects the external administered doses in the animal study to internal doses and also allows for differences between animals and humans in the relationship between external dose and target organ dose.

For threshold toxicity, the risk assessment approach is to estimate a level of human intake without appreciable health effects (i.e., a health-based guidance value). Depending on the context and jurisdiction, health-based guidance values are termed the reference dose (RfD), the acceptable daily intake (ADI), or the tolerable daily intake (TDI); in each case, the guidance value for human exposure is usually derived by dividing the intake of animals treated at the no-observed-adverse-effect level (NOAEL) (on a mg/kg basis) by an appropriate uncertainty factor. An uncertainty factor of 100 is usually applied with a 10-fold factor to allow for extrapolation from animals to humans and a 10-fold factor to allow for inter-individual differences in the exposed human population [5,6]. A scheme has been developed [7–10] that allows appropriate toxicokinetic or mechanistic data to be incorporated into the derivation of a chemical-specific adjustment factor by the replacement of part of the relevant 10-fold default factor. This is illustrated in the examples discussed at the end of the chapter. The use of such a scheme would produce a more secure basis for the establishment of the ADI, TDI, or RfD while providing a direct return for the investment necessary to produce more than the minimum toxicity database required for regulatory purposes. Greater use of toxicokinetic data and PBPK models in regulatory decisions will provide a more scientific basis for risk assessment and will encourage the increased generation of such data.

Toxicokinetic studies are important in compound development, and such information is considered necessary before proceeding with long-term and carcinogenicity tests [11]. If the kinetic evidence indicates tissue accumulation on prolonged dosing, saturation of elimination at subtoxic doses, the formation of chemically reactive metabolites, or that the main route of metabolism is via an enzyme showing a major genetic polymorphism, then chemical analogs without these problems may be selected for development, because these properties mitigate against a high therapeutic index (for drugs) or a high ADI or RfD (for food additives and pesticides).

The term *toxicokinetics* means the movement of a toxicant around the body and relates to any information on the fate of the chemical in the body. Such information may be derived by various methods and approaches, including: (1) data from studies where a radiolabeled dose of the chemical (usually $^{3}$H-, $^{14}$C-, or $^{35}$S-labeled) is given to experimental animals and the fate of the radiolabel followed over time, and (2) chemical-specific data using a sensitive analytical method.

Radiolabeled studies are valuable for following the fate of the chemical skeleton as it moves from the site of administration into the blood, is distributed to the tissues, and finally is eliminated from the body in air, urine, or bile. Measurement of total radiolabel is nonspecific and reflects both the chemical and its metabolites; this is both an advantage and a disadvantage. The advantage is that it allows quantitative balance studies to be performed—for example, to determine how much of the dose is absorbed, which organs accumulate the compound or its metabolites, the pathways of metabolism, and the routes of excretion. The disadvantage is that radiolabeled absorption, distribution, metabolism, and excretion (ADME) studies do not allow an assessment of how much of the chemical is absorbed intact and how much is distributed around the body as the parent chemical. Initial radiolabeled ADME studies are useful to define the overall fate of the chemical in the body and identify the main chemical species (parent compound or metabolites) that are present in the circulation and delivered to the site of toxicity. In recent years, it has been recognized that chemical-specific measurements of the circulating concentrations of the chemical or its metabolites can provide useful information on both the magnitude and the duration of exposure of targets for toxicity.

The term *toxicokinetics* is sometimes restricted to studies based on chemical-specific measurements of the concentrations of the chemical or its metabolites in blood

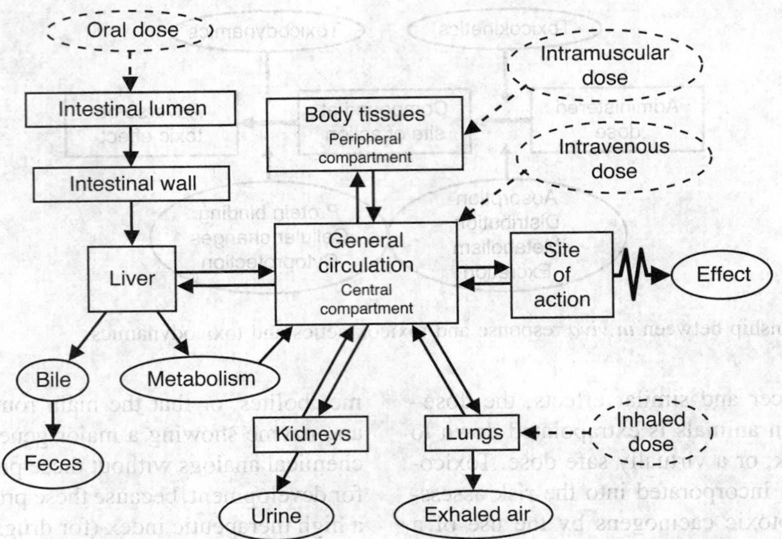


**FIGURE 4.2** Toxicity in relation to pharmacokinetics. The chemical may be given orally or by injection or inhalation. The concentration at the target organ is in equilibrium with that in the systemic circulation, which is itself in equilibrium with a large number of other physiological processes, which can increase or decrease that concentration. The transfer from one tissue to another usually involves transfer across a lipid membrane and frequently entails entering a tissue with high elimination capacity such as the liver or kidneys. The parent chemical is eliminated from the body when it is converted into a different structure by metabolism or is eliminated in the urine, bile, or expired air.

or plasma, as these provide a vital link between the dosing of experimental animals and the amounts of the chemical in the general circulation and delivered to the site of toxicity. This chapter concentrates on the derivation and interpretation of chemical-specific toxicokinetic data based on measurements of the concentrations in plasma and urine. Such information can be of great value in the interpretation of species differences in toxic response and in evaluating the possible risk to humans of hazards identified in animal experiments. Toxicokinetic data are also useful in extrapolating across different routes of exposure or administration, as well as from single doses to chronic administration. Chemical-specific toxicokinetic measurements are essential if the results of *in vitro* toxicity tests are to be interpreted logically, because they can define the biological upper limit of plausible *in vivo* concentrations.

Toxicokinetics is concerned with the relationship between the external dose, as usually measured in toxicity studies (e.g., mg/kg body weight per day), and a measure of the internal dose of the active compound delivered to the target for toxicity (Figure 4.1), such as the concentration in the general circulation or at the target for toxicity. The concentration in the general circulation is influenced by absorption, distribution, and elimination processes, as shown in Figure 4.2, and frequently the compound has to pass many lipid and metabolic barriers prior to reaching the site of toxicity. An understanding of the extent and nature of these processes may be derived from serial analysis of the concentrations of the chemical in plasma and urine. Knowledge of the concentrations of the parent compound and any metabolites in plasma and tissues, allied to the rate of change on further dosing or cessation of administration, allows logical selection of the animal species most appropriate for toxicity testing and extrapolation of any toxicity observed in animals to the likely risk for humans [5–10,12–15].

Toxicokinetic studies, therefore, are designed to produce information on the profile of exposure of the site of toxicity to the active moiety, under the conditions that produce the toxicity and which are the basis for determining the NOAEL. Toxicokinetic data have the potential to define

- The internal exposure (internal dose) in animals based on plasma or blood concentrations of the parent compound or its active metabolite in relation to the dose given to the animals
- The relationship between plasma or blood concentrations and those at the site of toxicity
- Information allowing quantitative interspecies comparisons derived from appropriate blood/plasma data after the administration of tracer doses to human volunteers

The aim of this chapter is to introduce the underlying principles in both biological and mathematical terms and subsequently to describe methods of obtaining suitable samples from certain animal species, primarily the rat. The final section of the chapter discusses examples where the generation of suitable toxicokinetic data has been important in the risk assessment of the chemical.

In the past, animal and human disposition studies during drug development utilized slightly different approaches, with radiolabeled ADME studies in animals and studies using chemical-specific data to construct the plasma concentration–time curves in humans. The key to interpreting plasma concentration–time curves was the development of suitable mathematical models that allowed the derivation of rates of absorption, metabolism, and excretion in humans. It was thus possible, using pharmacokinetic methods, to gain an insight into the rates and extents of kinetic processes in humans that had been shown to occur in animals using serial sacrifice and tissue analysis and which had been detected in metabolic studies.

The development of automated analytical techniques of high sensitivity and specificity (such as HPLC, LC-MS, and LC-MS-MS) and the expansion of laboratories undertaking plasma drug analyses allowed pharmacokinetics to reveal information on *in vivo* drug absorption, distribution, and elimination. Today these techniques are being applied increasingly to toxicology problems in laboratory animals. Problems of accumulation on repeated dosing and saturation of elimination are particularly pertinent to high-dose animal toxicity studies, and information on these areas can be obtained only from suitably designed *in vivo* toxicokinetic studies. It must be emphasized at the outset that the key to successful kinetic studies is the development of an assay of high specificity that measures the chemical without interference by its metabolites and that is of sufficient sensitivity to define the terminal slope accurately.

## BIOLOGICAL PRINCIPLES

Certain general principles governing the disposition of therapeutic drugs are applicable to nearly all low-molecular-weight organic compounds that are not substrates for normal intermediary metabolism. These general properties of absorption, distribution, and elimination are valuable concepts, but it should be emphasized that they are not universally applicable, and researchers should be alert for exceptions. Exceptions arise when the foreign compound is structurally similar to an endogenous body constituent, because it may then undergo a specific carrier-mediated uptake process or metabolism. Examples of compounds showing such characteristics include the drug levodopa, the amino acid metabolites of the intense sweetener aspartame, and the purine and pyrimidine base analogs used in cancer chemotherapy and antiviral agents, many of which undergo active uptake into cells and also may be metabolized to phosphorylated products, which accumulate within the cells of the body.

### Absorption

Absorption describes the processes involved in the transfer of a chemical from the site of administration into the systemic blood circulation. Because most toxicity studies are performed using oral administration, absorption from the gut is of greatest importance, although absorption from other sites is appropriate for certain toxicological studies.

#### Absorption from the Gut

For significant absorption to occur, the compound has to be present in the gut lumen as a molecular solution. Slow dissolution and release of compounds from oral sustained-release formulations gives slow absorption, which can maintain blood concentrations for prolonged periods compared with the administration of the same amount of compound in solution. Sustained-release formulations are particularly useful for therapeutic drugs that are eliminated very rapidly. A similar phenomenon can occur in toxicity studies when high doses are administered as a suspension, because dissolution of the chemical may be slow and the rate-limiting process. Under such circumstances, slow dissolution of the compound in the gut lumen can affect the rate of absorption, the peak concentrations, and even the magnitude of acute effects.

The rate and extent of absorption are determined largely by the pH of the gut lumen and the $pK_a$ and lipid solubility of the compound. Other biological variables, such as the presence of food, gastric emptying time, intestinal transit time, and the metabolic activity of the gut microflora, may also play important roles in limiting the rate of absorption and/or the amount of compound absorbed unchanged. The absorption of chemicals requires passage across lipoid membranes, which can involve (1) passive diffusion through the membrane, (2) passage through membrane pores, and (3) specialized carrier-mediated processes.

The rate of diffusion of a chemical across a membrane, given by Fick's law, is proportional to the concentration gradient, the membrane surface area, and the permeability coefficient of the compound. Most environmental anutrients are absorbed in the small intestine because of its large surface area. The permeability coefficient depends on the diffusivity of the molecule through the membrane, the membrane/aqueous medium partition coefficient, and the thickness of the membrane [16]; thus, it is a characteristic for that particular compound and corresponds to a rate constant. For weak acids and bases, the membrane/aqueous partition coefficient varies with the pH of the medium. For such compounds, the diffusivity of the ionized molecular species may be regarded as insignificant compared with the uncharged or un-ionized species.

Because only the uncharged species readily diffuses across membranes, absorption is faster under conditions in which ionization is suppressed (i.e., at low pH for acids and high pH for bases). When the two compartments separated by a lipoid membrane are maintained at different pH values, the total concentrations differ in each compart-

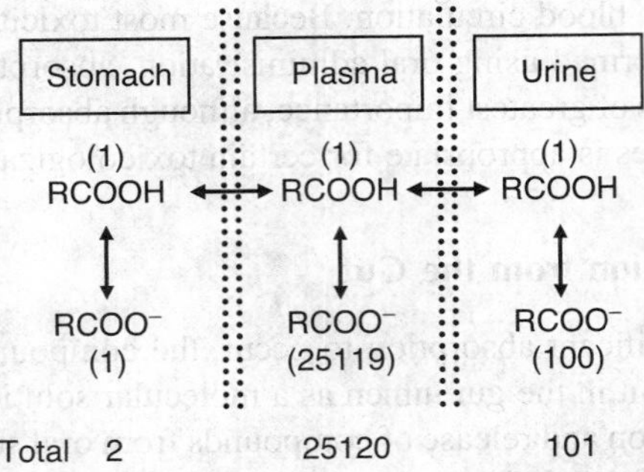


**FIGURE 4.3** pH partitioning. The numbers give the relative concentration of unionized and ionized species in each compartment, as determined by the Henderson–Hasselbalch equation for a weak acid ($pK_a$ 3.0) at the pH of stomach (3.0), plasma (7.4), and urine (5.0). The total concentration is the concentration of compound in each compartment at equilibrium assuming that the ionized form undergoes negligible diffusion.

ment at equilibrium. The extent of ionization of a weak acid may be related to the environmental pH and its $pK_a$ by the Henderson–Hasselbalch equation:

$$pH = pK_a + \log \frac{[\text{conjugate base}]}{[\text{conjugate acid}]}$$

At equilibrium, the concentrations of the diffusible form (unionized) on each side of the membrane are equal, and the concentration of the ionized form is given by the Henderson–Hasselbalch equation (Figure 4.3).

It is apparent from Figure 4.3 that weak acids should be absorbed rapidly and extensively in the low pH of the stomach, whereas weak bases should undergo absorption in the intestine and not the stomach. Although this is true for bases, absorption of acids from the stomach is limited, possibly because of the relatively small surface area of the gastric mucosa and the presence of a higher pH at the surface of the mucosal cells. Strong organic acids and bases frequently show incomplete absorption from the gut, because they are extensively ionized at all pH values of the gut.

The absorption of foreign compounds by passage through membrane pores, which are about 4 Å in diameter, is largely applicable to small water-soluble molecules (less than 200 Da) [16]. Bulk passage of water across the membrane may act as a driving force and carry small molecules with it [17], and this should be borne in mind when studying the absorption kinetics of large doses of sparingly water-soluble compounds. Under such circumstances, the oral administration of high doses in large volumes of hypotonic solution could result in enhanced absorption.

In contrast, reduced absorption may occur for compounds undergoing carrier-mediated absorption, because high concentrations may saturate the carrier, so the rate and extent of absorption may be reduced at high doses. Food can also affect carrier-mediated absorption by competition of the natural substrate for the carrier. Examples of foreign compounds undergoing active absorption are rare and usually apply when the chemical resembles a nutrient (e.g., levodopa).

Highly lipid soluble compounds, such as 2,3,7,8-tetrachlorodibenzo-*p*-dioxin (TCDD) and β-carotene, show incomplete absorption, and a significant fraction of an oral dose is eliminated unchanged in the feces. The main reason for this is that a molecular solution is not formed; therefore, the interaction between the chemical and the gut wall is less than might be expected based on its lipid solubility. For such compounds, the formation of mixed micelles and uptake via the lymphatic system may be important in the initial transfer from the gut lumen into the general circulation.

A number of factors may limit the amount of a compound that reaches the systemic circulation as the parent compound after oral administration (bioavailability; see later discussion):

- *Extremes of pH, which may affect the stability of the compound*. Species differences can then arise between rats (gastric pH 3.8 to 5.0), rabbits (gastric pH 3.9), and humans (gastric pH 1 to 2) [18].
- *Hydrolytic enzymes*. The gut is rich in nonspecific proteases and lipases, which may affect foreign compounds.
- *Gut microflora*. The gut flora can perform a wide range of largely degradative metabolic reactions on foreign compounds [19], which may reduce the amount of parent chemical available for absorption or result in the formation of potentially toxic metabolites [20]. The intestinal microflora show differences between host species in both the organisms present and their distribution along the gut [18], with larger numbers in the stomach and upper intestine of rats and mice, compared with rabbits, dogs, and humans, in which the upper intestinal tract is almost sterile because of the low gastric pH.
- *Metabolism by the gut wall*. The gut wall has the capacity to inactivate metabolically certain compounds prior to their reaching the hepatic portal vein (first-pass effect) [21]. The intestinal wall is rich in enzymes catalyzing general hydrolysis reactions, monoamine oxidase, conjugation reactions (such as glucuronidation and sulfation), and some oxidative enzymes, particularly CYP3A4/5.
- *Metabolism by the liver*. Many compounds are removed from the hepatic portal vein, to some extent, during the single passage through the

liver that occurs as part of the overall absorption process (first-pass effect) [22,23]. The liver is the main site of foreign compound metabolism in the body (see below) and represents the main metabolic barrier to the parent compound reaching the general circulation. First-pass metabolism is not always associated with a decrease in biological activity; for example, the antihistamine terfenadine undergoes essentially complete first-pass metabolism to its active metabolite fexofenadine.

- *Food present in the gut lumen*. Food in the gut lumen may affect the absorption rate, gastric pH, or gut motility.
- *Efflux transporters*. Efflux transporters such as P-glycoprotein are expressed on the luminal surface of the intestinal epithelium and can act as efflux pumps for compounds entering the enterocyte (this contrasts with the transporters for specific nutrients which carry substrates from the gut lumen into the enterocyte or from the enterocyte into the splanchnic blood vessels).

These barriers to the establishment of high blood concentrations of the compound may be associated with suppression of systemic pharmacological and toxic properties [22] and thus render dietary administration an inappropriate route for the toxicity testing of compounds for which exposure of humans is parenteral.

### Absorption from the Nasal Cavity

Although the nasal cavity has a relatively small surface area, the mucous membrane is highly permeable; after nasal administration, even quaternary ammonium compounds (which are poorly absorbed from the gut) show blood levels approaching those present following intravenous administration [24]. In addition, the nature of this site and its venous drainage directly into the systemic circulation allow increased absorption of compounds extensively metabolized in the gut lumen (e.g., proteins) or in the liver. Local toxicity may be a problem with this route at high doses [24], as is exemplified by the necrosis of the nasal septum in cocaine snuffers.

### Absorption from the Lung

The lung represents a poor barrier to a chemical entering the blood, as it has a large surface area of thin membrane, a limited capacity to metabolize foreign compounds, and an excellent blood supply. The epithelium acts as a limited permeability barrier, allowing only slow absorption of highly water-soluble compounds [25,26], although the rate may be greater than that from the gastrointestinal tract. The lung is a major site of inactivation of circulating local hormones such as peptides and prostaglandins, but such substances would not be likely to be given by this route for toxicity testing. Major problems exist with quantitative analysis of the extent and rate of absorption from the lung due to poor measurement of the dose delivered to the airways following inhalation. Inhaled particulate matter is largely trapped by the cilia in the upper airways and passed back to be swallowed and absorbed in the gut. Volatile compounds are absorbed only partially, and the unabsorbed fraction is eliminated in the expired air and not retained for subsequent absorption, such as would occur in the gut.

### Absorption Across the Skin

The extent of percutaneous absorption is highly dependent on the lipophilicity of the compound, because the stratum corneum of the epidermis acts as an effective barrier [27,28]. This route is important for the therapeutic administration of potent lipophilic drugs that undergo extensive first-pass metabolism (e.g., organic nitrates in angina) and for workers exposed to environmental aerosols and particulates. Studies in animals and humans suggest that the rate-limiting step is the initial penetration of the stratum corneum [28], which may result in very slow absorption and "flip-flop" kinetics (see following discussion).

## Distribution

Foreign compounds are distributed largely via the blood. The rate of uptake of a compound by the tissues may be limited by either the diffusion rate across the cell membrane or the perfusion rate of the tissue:

- *Diffusion rate*—If the diffusion of the chemical across membranes is slow, the rate of entry into tissues is limited by this property of the molecule.
- *Perfusion rate*—If the diffusion of the chemical across membranes is rapid, the rate of entry into tissues is limited by the rate of delivery to the tissue (i.e., the perfusion rate).

As a generalization, diffusion rate limitation applies to highly water-soluble compounds, whereas perfusion rate limitation applies to the entry of lipid-soluble compounds into slowly perfused tissues, such as adipose tissue. The perfusion rates of the major organ systems of humans (Table 4.1) can be readily divided into well and poorly perfused tissues [29,30].

The extent to which a chemical leaves the blood and enters a tissue depends on its relative affinities for blood and that tissue. Thus, compounds that are highly bound to plasma protein but not to tissue show a relatively high concentration in the plasma, whereas chemicals with a high affinity for tissue components such as proteins or fat have a low plasma concentration [30,31]. It should be

**TABLE 4.1**
**Relative Organ Perfusion Rates in Humans**[a]

| Organ | % Body Weight[b] | Blood Flow[c] (mL/min) | % Cardiac Output[b,c] | Blood Flow[b,c] (mL/min/100 g) |
|---|---|---|---|---|
| *Well perfused* | | | | |
| Lung | 1.2 | 5000 | 100 | 1000 |
| Adrenals | 0.02 | 25 | 1 | 550 |
| Kidneys | 0.4 | 1260 | 23 | 450 |
| Thyroid | 0.04 | 50 | 2 | 400 |
| Liver | | | | |
| Total | 2 | 1350 | 25 | 75 |
| Via portal vein | | 1050 | 20 | 60 |
| Heart | 0.4 | 252 | 5 | 70 |
| Intestines | 2 | 1050 | 20 | 60 |
| Brain | 2 | 750 | 15 | 55 |
| *Poorly perfused* | | | | |
| Skin | 7 | 462 | 9 | 5 |
| Skeletal muscle | 40 | 840 | 16 | 3 |
| Connective tissue | 7 | — | — | 1 |
| Fat | 15 | 95 | 2 | 1 |

[a] The results are for an adult male under resting conditions and are approximate values only.
[b] Data from Butler [30].
[c] Data from Bard [29].

remembered, however, that it is the *relative* affinity that determines the extent of distribution to tissues. The dye Evans Blue has a high affinity for plasma protein, and its distribution (Table 4.2) is restricted to the plasma volume (3 L in humans); the β-blocker drug propranolol is also highly bound to plasma protein (95%), but it also shows a higher affinity for the tissues due to its lipid solubility, so relatively low concentrations remain in the plasma after distribution.

The volumes of body fluids and chemicals that distribute in them are given in Table 4.2. Only rarely, however, do compounds distribute to a single physiologically recognizable volume, and usually some degree of tissue selectivity is observed. Thus, a compound may appear to have dissolved in total body water because the apparent volume of distribution (see below) corresponds to about 60% of body weight, but it may actually show a nonuniform tissue distribution. Highly lipid soluble chemicals show very high adipose-to-blood concentration ratios, such that after tissue distribution the blood contains only a very small fraction of the total body burden (see later TCCD discussion).

Many foreign compounds bind reversibly to plasma proteins, with albumin being of the greatest importance, although acid glycoproteins may be important for certain organic bases [32]. Foreign compounds bind at specific sites in a reversible, saturable fashion, and the bound material represents an inactive depot of the chemical. Extensive protein binding lowers the concentration of unbound chemical in the blood, which may increase the concentration gradient and thus the rate of diffusion into blood from the gut (during absorption) or reabsorption from the kidney tubules into blood (during elimination) [33,34]. The dissociation of the chemical–protein complex occurs within milliseconds and by comparison with tissue perfusion times may be regarded as instantaneous. Active uptake processes that rapidly lower the unbound or free plasma concentration can effectively strip a compound from plasma proteins during a single passage through an organ.

The presence of transporters in the vasculature of certain organs can influence the rate of uptake of chemicals by that tissue and the equilibrium between the tissue and blood. Transporters can either actively take up the chemical from the circulation (e.g., in the liver or kidneys; see below) or transport chemicals that have diffused into the tissues back into blood (e.g., from the brain) [35–37]. The induction of tissue efflux transporters, such as P-glycoprotein and multidrug-resistance-associated protein (MRP1), on chronic administration is associated with resistance to anticancer drugs [38]; because of the relatively nonspecific nature of these transporters, a similar effect might occur with other large and complex foreign compounds and result in a reduction in response on repeated administration. Membrane transporters can show species differences, sex differences and genetic polymorphisms [35,39,40], although the toxicological consequences of these have not been clearly established.

**TABLE 4.2**
**Volumes of Body Fluids with Chemicals Showing Restricted Distribution**

| Fluid | Volume (L) | % Body Weight | Compound[a] |
|---|---|---|---|
| Total body water | 41 | 58 | $D_2O$, antipyrine, ethanol, urea |
| Extracellular water | 12 | 17 | $Na^+$, $Br^-$, tubocurarine, sucrose |
| Plasma | 3 | 4 | Evans Blue, [$^{131}$I]albumin |

[a] Compounds for which the distribution is restricted to a particular body fluid.

*Source:* Adapted from Butler, T.C., in *Fundamentals of Drug Metabolism and Drug Disposition*, LaDu, B.N. et al., Eds., Williams & Wilkins, Baltimore, MD, 1971, pp. 44–62; Gehring, P.J. and Young, D.J., in *Proc. of the First Int. Congress on Toxicology: Toxicology as a Predictive Science*, Plaa, G.L. and Duncan, W.A.M., Academic Press, New York, 1978, pp. 119–141.

## Elimination

The two main mechanisms by which the circulating levels of a foreign compound may be reduced are metabolism and excretion. *Metabolism* is a major source of both species differences and human variability [41–44], and its toxicological consequences are discussed in an earlier chapter. Certain mathematical implications are discussed on the following pages. The principal routes of excretion are via the urine and feces and in the case of volatile compounds the expired air.

### Excretion via Urine

Three major processes affect elimination in the kidney.

- *Glomerular filtration*. The glomerular membrane has pores of 70 to 80 Å. Under the positive hydrostatic conditions that exist in the glomerulus, all molecules smaller than about 20,000 Da are filtered; thus, proteins and protein-bound compounds remain in the plasma, and about 20% of the nonbound chemical is carried together with 20% of the plasma water into the glomerular filtrate.
- *Reabsorption*. Because the glomerular filtrate contains many essential body constituents, such as glucose, specific active uptake processes transport them from urine back into blood. Foreign chemicals are not normally substrates for these specific reuptake transport processes, but lipid-soluble chemicals can diffuse back from the renal tubule into the blood, especially as the urine becomes more concentrated because of water reabsorption. The pH of the urine is generally lower than that of the plasma; therefore, weak acids are more ionized in plasma, so pH partitioning tends to increase the reabsorption of weak acids. The pH of the urine can be altered appreciably by treatment with ammonium chloride (decreases pH) or sodium carbonate or sodium hydrogen carbonate (increases pH). In contrast, the plasma shows little change because its pH is buffered by the high protein content. It is therefore relatively easy to affect the pH partitioning of foreign compounds between tubule contents and plasma and either increase or decrease the elimination rate. This possibility should be considered when preparing dose solutions, because the use of excess acid or alkali to dissolve the test compound could alter its renal elimination.
- *Tubular secretion*. Foreign compounds may be secreted actively into the renal tubule against a concentration gradient by anion and cation carrier processes. Because the dissociation rate for the chemical–albumin complex is rapid, it is possible for highly protein-bound compounds to be almost completely cleared at a single passage through the kidney. These processes are saturable and of relatively low specificity; many basic or acidic compounds and their metabolites (especially the phase 2 or conjugation products) are removed by them. Transporters on the apical and luminal membranes of the renal tubule are important in extracting chemicals from blood and transferring them into the tubule lumen [45,46]. It is now recognized that there are a number of different transporters for organic anions (OATs, transporters for acids), organic cations (OCTs, transporters for bases), peptide transporters, and nonspecific transporters (members of the MRP family). These may occur on either the basolateral or apical membranes of the renal tubule or on both, and they show species and sex differences [40].

### Excretion via the Gut

The bile is the most important route allowing foreign compounds to move from the general circulation into the gut. The biological aspects of this mechanism have been reviewed [47], and certain pertinent points have emerged. Organic cation transporters on the sinusoidal membrane transfer large polar cations into the hepatocyte and from the hepatocyte in the bile [40,45]. The bile may be regarded as a complementary pathway to the urine, with small molecules being eliminated by the kidney and large molecules in the bile; thus, the bile becomes the principal excretory route for many xenobiotic conjugates. Species differences exist in the molecular weight requirement for significant biliary excretion, which has been estimated as 325 ± 50 Da in the rat, 440 ± 50 Da in the guinea pig, 475 ± 50 Da in the rabbit, and about 500 Da in humans. In the rat, small molecules (less than 350 Da) are not eliminated in the bile, and large molecules (more than 450 Da) are not excreted in the urine, even if the principal excretory mechanism is blocked by ligation of the renal pedicles or bile duct, respectively. Compounds of intermediate molecular weight (350 to 450 Da) are excreted by both routes, and ligation of one pathway results in increased use of the other [48].

Foreign compounds may also enter the gut by direct diffusion or secretion across the gut wall, elimination in the saliva, pH partitioning of bases into the low pH of the stomach, and elimination in the pancreatic juice. In most cases these routes are quantitatively of minor importance, although diffusion into fecal fat is the main route of elimination in humans for polyhalogenated compounds, such as TCDD, which are resistant to metabolism. Transfer from the blood into the gut lumen may play an important role in toxicity by allowing a foreign compound to undergo metabolism by the gastrointestinal flora [49]. The toxicological implications of the gut microflora have been reviewed by Scheline [19].

## MATHEMATICAL PRINCIPLES

To describe adequately the changes in blood or plasma concentrations of foreign compounds, it is necessary to assign a suitable mathematical model that accurately describes the shape of the plasma concentration–time curve; however, certain aspects are model independent and are considered first, because they are usually constituent parts of the various mathematical models. In recent years, there has been a marked trend away from multicompartmental mathematical analysis, which offers little apart from mathematical predictability, toward physiologically more relevant model-independent concepts such as clearance [13,50]. Physiologically related parameters such as clearance and bioavailability represent an intermediate stage between mathematical multicompartment models and full physiologically based pharmacokinetic (PBPK) models.

### MODEL-INDEPENDENT CONSIDERATIONS

Biochemical and physiological processes are usually either zero-order or first-order reactions. In zero-order reactions, the rate of change in concentration with time occurs at a fixed amount per unit of time:

$$\frac{dC}{dt} = k$$

where $C$ is concentration, $t$ is time, and $k$ is a constant with units of amount per time (e.g., μg/min). In first-order reactions, the rate of change in concentration is proportional to the concentration of the chemical available for the reaction:

$$\frac{dC}{dt} = kC$$

where $k$ is a constant that represents a proportional change with time and has units of time$^{-1}$ (e.g., min$^{-1}$).

Most kinetic processes (e.g., diffusion, carrier-mediated uptake, metabolism, excretion) are first-order reactions at low concentrations. *Most of the equations given below make this assumption*. Zero-order reactions are particularly important at high concentrations, when enzymes are working at maximum rate and an increase in $C$ cannot result in an increase in rate. This situation produces nonlinear or saturation kinetics, which can assume considerable importance in toxicity studies, as is discussed below.

First-order reactions can be described by equations that include exponential functions. In many cases, the entry of a foreign compound into the body or into a tissue follows an exponential increase, which may be described mathematically by:

$$\text{Uptake} = 1 - e^{-kt} \tag{4.1}$$

where the uptake is the concentration present at time divided by the final concentration when all the compound has entered the body or tissue. This equation assumes that no elimination process is occurring. The elimination of a compound (by a single mechanism) once it has entered the body or tissue may be described by an exponential with a negative slope:

$$C = C_0 e^{-kt} \tag{4.2}$$

where $C$ is the concentration present at time $t$, and $C_0$ is the initial concentration. In Equation 4.1 and Equation 4.2 $k$ is the rate constant for that process.

Exponential equations of the type given in Equation 4.2 may be solved as:

$$\ln C = \ln C_0 - kt$$

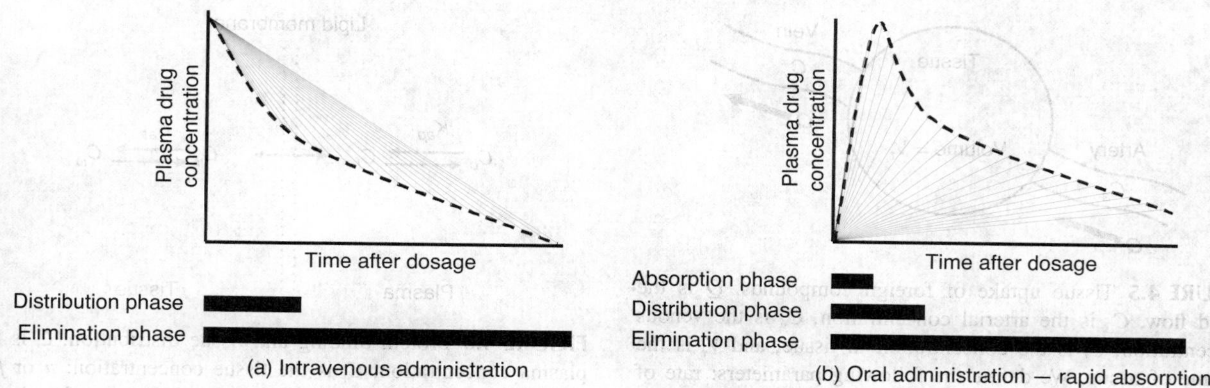


**FIGURE 4.4** The plasma concentration–time profile of a chemical following intravenous and oral dosage. The concentration is on a logarithmic scale. Rapid processes, such as absorption and distribution, do not significantly affect later time points which are determined largely by the slowest process (elimination in the diagrams shown).

or using $\log_{10}$:

$$\log C = \log C_0 - \frac{kt}{2.303}$$

These are equations of the generalized form:

$$y = C + mx$$

where $x$ and $y$ are variables, and $C$ and $m$ are constants. In such cases, a plot of $x$ against $y$ gives a straight-line graph with a slope of $m$ and an intercept of $C$; thus, for toxicokinetics, a graph of $\ln C$ against time gives a slope of $-k$ and an intercept of $\ln C_0$. If such a graph is drawn using log–linear graph paper (see Figure 4.9), then the slope must be calculated either by taking natural logarithms of the concentrations when the slope will be $-k$ or by taking the $\log_{10}$ of the concentration terms and dividing by the time when the slope will be $-k/2.303$.

The units of $k$ are time$^{-1}$, which is a difficult unit to visualize. The rate of a first-order process is usually described by its half-life; substituting values of 0.5 and 1.0 for $C$ and $C_0$ into the above equations give the relationship $t_{1/2} = 0.693/k$.

Usually, the equation necessary to describe the kinetics of a compound in the body requires the use of at least two exponential rate terms. This is illustrated in Figure 4.4, in which two phases are seen in the plasma concentration–time curve after an intravenous dose and three different phases after an oral dose. Absorption into and elimination from a single compartment and elimination from a two-compartment system (see below) require two different rate constants, and in such cases the early time points in the concentration–time curve are influenced by both rates. However, provided the rate constants are sufficiently dissimilar, eventually the influence of the component with the higher rate (the faster component) becomes negligible, while the slower component (with the smaller rate constant) still affects the concentration. Thus, the terminal phase of the concentration–time curve is determined by the slower process (i.e. the one with the smaller rate constant and the earlier phase by the sum of both processes). This process allows both rate constants to be determined by the procedure known as the method of residuals or stripping.

### Tissue Extraction

Removal of a compound from the blood by a tissue is schematized in Figure 4.5. On constant infusion, the rate of entry into the tissue may be regarded as equivalent to a first-order absorption rate [51]:

$$\text{Uptake} = 1 - e^{-kt}$$

where uptake is the fractional uptake = $C_t/C_{equilibrium}$.

In *perfusion-limited uptake*, the value $k$ is related to the flow rate ($Q$) as follows (Figure 4.5):

$$\text{Fractional uptake} = 1 - e^{-(Q/PV_t)t} \tag{4.3}$$

where $Q/V_t$ is the volume-adjusted flow rate, and $P$ is the partition ratio. The uptake half-time may be derived as described below for Equation 4.19:

$$t_{1/2}(\text{uptake}) = \frac{0.693}{k} = \frac{0.693PV_t}{Q} = \frac{0.693P}{Q/V_t} \tag{4.4}$$

For *diffusion-limited uptake*, the value $k$ is related to the diffusion rate constant and thus is not readily measurable.

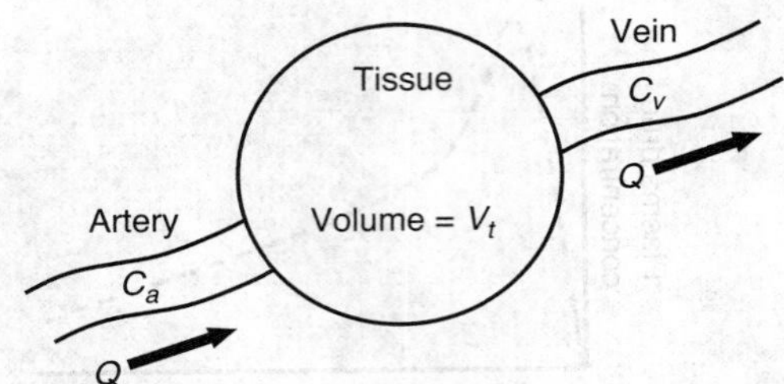


**FIGURE 4.5** Tissue uptake of foreign compounds. $Q$ is the blood flow, $C_a$ is the arterial concentration, $C_v$ is the venous concentration, $C_t$ is the concentration in tissue, and $V_t$ is the volume of tissue. We define the following parameters: rate of delivery of compound, $QC_a$; rate of outflow of compound, $QC_v$; rate of uptake, $QC_a - QC_v$. The units for these rates are mass time$^{-1}$ (e.g., μg min$^{-1}$). Thus,

$$\text{Extraction ratio } (E) = \frac{\text{Rate of uptake}}{\text{Rate of delivery}} = \frac{QC_a - QC_v}{QC_a} = \frac{C_a - C_v}{C_a}$$

$$\text{Partition ratio } (P) = \frac{\text{Concentration in tissue}}{\text{Concentration in blood supply}}$$

These relationships are an essential part of physiologically based toxicokinetic (PBTK) models.

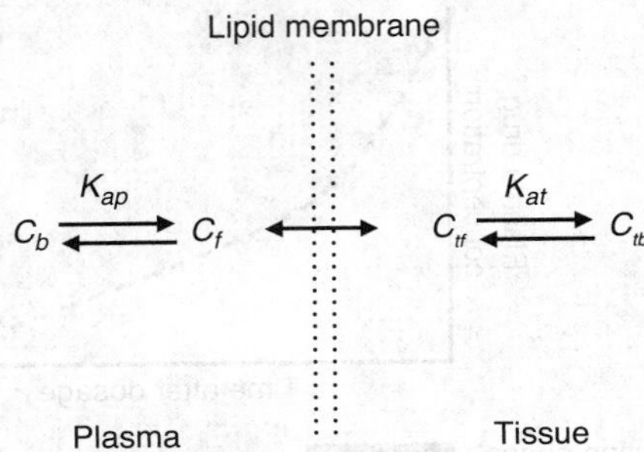


**FIGURE 4.6** Protein binding and tissue distribution. $C$ is the plasma concentration; $C_t$ is the tissue concentration; $u$ or $f$ is unbound; $b$ is bound; $K_{ap}$ is the association constant for plasma protein binding; $K_{at}$ is the association constant for tissue protein binding. At equilibrium, $C_f = C_{tf}$, and the relative distribution between plasma and tissue is determined by the values of $K_{ap}$ and $K_{at}$.

## Plasma Protein Binding

The extent of protein binding may be represented by an equilibrium reaction:

$$P^r + C_f \leftrightarrow C_b$$

where $P^r$ is the free protein, $C_f$ is the unbound or free chemical, and $C_b$ is the chemical–protein complex. The equilibrium constant $K$ is given by:

$$K = \frac{[C_b]}{[P^r][C_f]} = \frac{[\text{product}]}{[\text{reactants}]} \tag{4.5}$$

The fraction unbound is given by:

$$\alpha = \frac{C_f}{C_f + C_b} = \frac{C_f}{C_p} \tag{4.6}$$

where $C$ is total plasma concentration. Scatchard plots of $r/C_f$ against $r$ (where $r$ is the moles of chemical bound per mole of protein, and $C_f$ is the free concentration of the chemical in moles) can be used to derive the number of binding sites and their association or affinity constants [34]. Normally, for the binding of organic compounds to albumin, two or more binding sites are revealed. This approach is valuable for detailed studies on binding to and displacement from albumin, but it could yield multiple binding sites if applied to a complex protein mixture such as plasma, containing different proteins at different concentrations.

Because it is the unbound compound in plasma that undergoes equilibration with the unbound compound in tissues (Figure 4.6), adequate information may be obtained from knowledge of α, or of the percentage bound. Detailed knowledge is not required for pharmacokinetic analysis of the plasma data because the plasma concentrations used to calculate kinetic parameters are usually the total concentrations ($C_f + C_b$). (*Note:* When the nonbound concentration is used to calculate kinetic parameters, this has to be specified in the description of the parameters, such as "clearance [non-protein bound]"; in the absence of such a description, it is assumed that the total plasma concentration has been used.) Because plasma protein binding is a saturable process, *in vitro* binding studies should be performed over a range of concentrations.

## Clearance

Clearance ($CL$) is defined as the ratio:

$$CL = \frac{\text{Rate of elimination}}{\text{Plasma concentration}}$$

and may be regarded as the volume of plasma or blood that is cleared of compound in unit time by the route under consideration. The units are volume time$^{-1}$; for example, the unit is frequently mL min$^{-1}$, because if the rate is μg min$^{-1}$ and plasma concentration is μg/mL, then the plasma clearance will be mL min$^{-1}$.

## Renal Clearance

The renal clearance ($CL_R$) is given by:

$$CL_R = \frac{\text{Rate of elimination in urine}}{\text{Plasma concentration}} = \frac{C_u \times F_u}{C} \quad (4.7)$$

where $C_u$ is the urine concentration, $F_u$ is the urine flow (volume in unit time), and $C$ is the plasma concentration at the midpoint of the urine collection period. The concentration in urine is dependent on a number of variables, which are now described.

### *Glomerular Filtration*

The rate of filtration is given by:

$$\text{GFR} \times C_f = \text{GFR} \times C \times \alpha \quad (4.8)$$

where GFR is the glomerular filtration rate, $C_f$ is the unbound concentration in plasma, $C$ is the total plasma concentration, and $\alpha$ is the fraction unbound.

Thus, compounds that are extensively bound to plasma proteins show limited elimination by glomerular filtration. The protein binding equilibrium is not disturbed in the glomerulus after filtration of 20% of the free compound and 20% of the plasma water so $C_f$ is unaltered, whereas the concentrations of both free protein and compound–protein complex increase by 20%:

| Before filtration | After filtration |
|---|---|
| $K_{ap} = \frac{[C_b]}{[C_f][P^r]}$ | $K_{ap} = \frac{1.2[C_b]}{[C_f]1.2[P^r]}$ |

where $K_{ap}$ is the protein-binding association constant. Thus, the chemical–protein complex does not dissociate within the glomerular vasculature. The chemical–protein complex dissociates to release free compound when the plasma is diluted by water reabsorbed in the distal parts of the renal tubule. Under such circumstances, about 99% of the plasma water is reabsorbed, so the concentrations of the protein and complex almost return to their initial levels, whereas the concentration of unbound compound is diluted to about 80% of its former level; that is, after reabsorption of most of the water filtered in the glomerulus:

$$K_{ap} \neq \frac{[C_b]}{0.8[C_f][P^r]}$$

Therefore, the complex dissociates to restore the equilibrium. The glomerular filtration rate is about 130 mL min$^{-1}$ in men and 120 mL min$^{-1}$ in women, or approximately 2 mL min$^{-1}$ kg$^{-1}$, which is lower than that of the Wistar rat (3.4 mL min$^{-1}$ kg$^{-1}$) [52].

### *Reabsorption*

Reabsorption from the renal tubule back into the blood is variable and dependent on the lipid solubility of the compound, the pH of the urine, the p$K_a$ of the chemical, and the extent of concentration of the urine (i.e., water reabsorption). In rare instances, the administered foreign compound may be a substrate for carrier-mediated reuptake or reabsorption, in which case the renal elimination is dose dependent and is greater at high doses when this reuptake is saturated. Such a saturable reuptake process is obviously ideal for maintaining a constant low body load of an essential compound (e.g., glucose or riboflavin) that might show adverse effects at high concentrations in the body.

### *Tubular Secretion*

Saturable carrier-mediated processes are present in the proximal part of the tubule, and a number of different transporters have been identified. They show relatively low substrate specificities, and the extent of their involvement for a particular compound is dependent on the affinity between the compound and the carrier protein. The transporters are active saturable processes, and saturation of secretion causes a dose-dependent decrease in elimination at high plasma concentrations.

All three processes described above can alter, simultaneously and independently, the concentration in urine for any given value of $C$ delivered by the blood to the kidney (see Equation 4.7). The overall renal clearance may be regarded as a composite expression:

Renal excretion = glomerular filtration − reabsorption + tubular secretion

Rate of excretion = GFR $C\alpha$ − rate of reabsorption + rate of tubular secretion

The values of GFR, $C$, and $\alpha$ can be determined experimentally. Measurement of inulin clearance (or creatinine clearance in humans) determines the GFR, because this compound does not undergo significant reabsorption, tubular secretion, or protein binding, so:

$$\text{Rate of renal excretion of inulin} = \text{GFR} \times C$$

and because:

$$CL_R = \frac{\text{Rate of excretion in urine}}{C}$$

$$CL_R \text{ for inulin} = \text{GFT}$$

The extent of reabsorption and secretion of a compound may be inferred by comparison of its renal clearance with the value of GFR × $\alpha$:

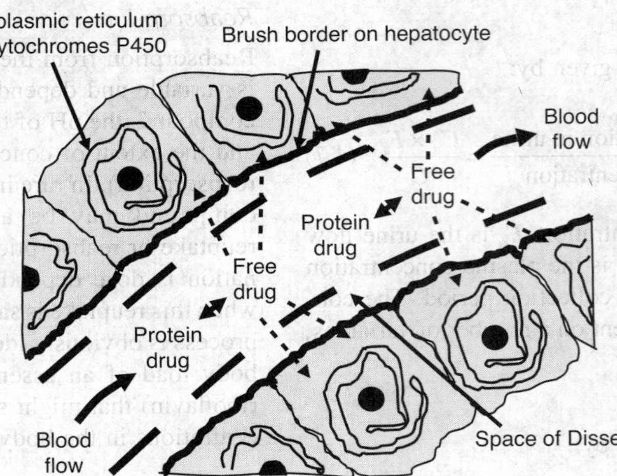


**FIGURE 4.7** The processes involved in the hepatic extraction and elimination of chemicals delivered via the hepatic portal vein. The equilibrium between protein-bound and free chemical in the blood is essentially instantaneous, so that the chemical can be stripped from plasma proteins in a single passage through the sinusoid.

If $CL_R < \text{GFR} \times \alpha$, reabsorption must be occurring and is greater than any secretion (which may or may not be present).

If $CL_R = \text{GFR} \times \alpha$, reabsorption, which may or may not be present, is negated by an equal rate of secretion.

If $CL_R > \text{GFR} \times \alpha$, tubular secretion must be occurring and is greater than any reabsorption (which may or may not be present).

The mathematical implications of the renal elimination process have been the subject of a number of reviews [53,54].

## Hepatic Clearance

The clearance of a compound by the liver may be regarded as dependent on the rate of delivery to the organ (blood flow) and the efficiency of removal from the blood (extraction ratio; see Figure 4.5); thus,

$$CL_H = QE \tag{4.9}$$

where $CL_H$ is the hepatic (metabolic) clearance, $Q$ is the hepatic blood flow, and $E$ is the extraction ratio.

The uptake and metabolism of nutrients, hormones, and absorbed chemicals is a primary function of the liver, and hepatocytes and hepatic sinusoids show a number of features that facilitate these processes (Figure 4.7). These include fenestrae in the endothelium that allow even large molecules to leave the circulation and enter the space of Disse, a fluid collagen-containing matrix in the space of Disse, a brush-border on the hepatocytes that greatly increases the surface area for absorption, active uptake transporters for some chemicals, and very high enzyme activity (intrinsic clearance) within the hepatocytes. Many of these properties are lost in chronic liver disease.

The simple relationship in Equation 4.9 has been verified experimentally for a number of compounds; however, it is complicated by the finding that the variables $Q$ and $E$ are not independent, because for certain compounds an increase in blood flow decreases the extraction efficiency. This finding led Rowland et al. [55] to propose the following relationship, known as the *perfusion-limited model*:

$$CL_H = Q\left(\frac{\alpha CL_{int}}{Q + \alpha CL_{int}}\right) \tag{4.10}$$

where $\alpha$ is the fraction unbound in plasma, $CL_{int}$ is the intrinsic metabolic clearance by the hepatocytes from the cell water, and $Q$ is the blood flow (as plasma).

If the metabolic clearance ($CL_{int}$) is high, then the value $Q + \alpha CL_{int}$ approximates to $\alpha CL_{int}$; therefore, the term in parentheses, which is equivalent to the term $E$ in Equation 4.9, approaches unity. Under these circumstances, the hepatic clearance approximates to the hepatic blood flow and becomes dependent on the hepatic blood flow. However, if the metabolic clearance is low, then $Q + \alpha CL_{int}$ approximates to $Q$; therefore, the extraction ratio ($E$) decreases with an increase in blood flow, and the hepatic clearance remains relatively constant. These equations adequately explain the effects of changes in perfusion rate on the extraction ratio and clearance of compounds that show a range of extraction ratios. In addition, comparison of the hepatic clearance (calculated by measurements of $Q$ and $E$) with "nonrenal" clearance (calculated as plasma clearance minus renal clearance; see later discussion) can indicate the role of extrahepatic tissues in the elimination of the compound.

Further analysis of this equation indicates that:

$$CL_{int} = \frac{V_{max}}{K_m + C_f} \quad (4.11)$$

where $V_{max}$ and $K_m$ are Michaelis–Menten constants for the enzyme metabolizing the chemical, and $C_f$ is the hepatic venous concentration of unbound chemical. If $C_f$ is low and is much less than the $K_m$ for the enzyme (i.e., well below saturation levels), this term may be ignored and

$$CL_{int} = \frac{V_{max}}{K_m} = \text{constant}$$

When the value of $C_f$ approaches or exceeds $K_m$, the substrate concentration is sufficient to saturate the enzyme, and the kinetics are grossly altered and become nonlinear. This situation is a distinct possibility in high-dose toxicity testing and is discussed later in more detail. The concepts given above are important for the use of $V_{max}$ and $K_m$ values in interspecies comparisons. $V_{max}/K_m$ is taken as a measure of enzyme activity; species differences in clearance will relate to differences in this measure of enzyme activity for low-clearance compounds but will relate to liver blood flow for high-clearance compounds (regardless of differences in $V_{max}/K_m$). This possible source of error in extrapolation across species is avoided when both the enzyme activity and organ blood flows are part of a PBPK model (see below).

## Biliary Clearance

The clearance via the bile $CL_B$ is given, by analogy with renal clearance, as:

$$CL_B = \frac{\text{Rate of elimination in bile}}{C} = \frac{C_B \times F_B}{C} \quad (4.12)$$

where $C_B$ is the concentration in bile, $F_B$ is the volume of bile in unit time (bile flow), and $C$ is the plasma concentration.

## Plasma Clearance

Plasma clearance ($CL$) may be defined as:

$$CL = \frac{\text{Rate of elimination from plasma}}{C}$$

The plasma clearance is the sum of the various contributory clearance processes:

$$CL = CL_R + CL_H + CL_B + \ldots \quad (4.13)$$

Plasma clearance, which is one of the most valuable toxicokinetic constants, is determined from the plasma concentration–time curve and is discussed in detail later. It may be used to derive other model-independent variables, such as the mean residence time, which are given later under the Statistical Moment Analysis section.

## Physiologically Based Pharmacokinetic Models

In recent years, models have been developed that are based on the principles discussed above (i.e., organ blood flow, tissue extraction, and rates of metabolism and excretion). These models are derived from the physiology of the test animal and are discussed in detail in Chapter 5. PBPK models have been applied successfully to a number of compounds and have been particularly successful for organic solvents (e.g., benzene) [56]. This approach represents a powerful method, capable of dealing with saturation of metabolism [57,58] and valuable for the extrapolation of animal data to humans [57–62]; however, its ability to predict concentrations is dependent on the precision of the parameter estimates used and the model chosen [63]. PBPK modeling should therefore be considered as one of three possible approaches to the analysis and interpretation of toxicokinetic data:

- Simple physiologically related concepts, such as bioavailability and clearance (this chapter)
- Compartmental analysis, which gives mathematical precision but is difficult to relate directly to metabolic or physiological processes (this chapter)
- PBPK modeling, which allows the prediction of target organ concentrations (Chapter 5).

## Compartmented Systems: Modeling

To describe plasma concentration–time curves mathematically, an appropriate predictive model has to be fitted to the data. The correlation between the actual data and the plasma or blood concentration–time curve generated using the model shows the suitability of the model in describing the experimental results. Thus, considering the data presented in Figure 4.8 and Table 4.3, it is apparent that the same model cannot describe the properties of both compounds, although in both cases the initial and final plasma measurements were the same. The differences in the plasma concentration–time profiles originate in the number of rates at which the compound may leave and enter the plasma. When the tissues show instantaneous equilibration with plasma, a simple exponential

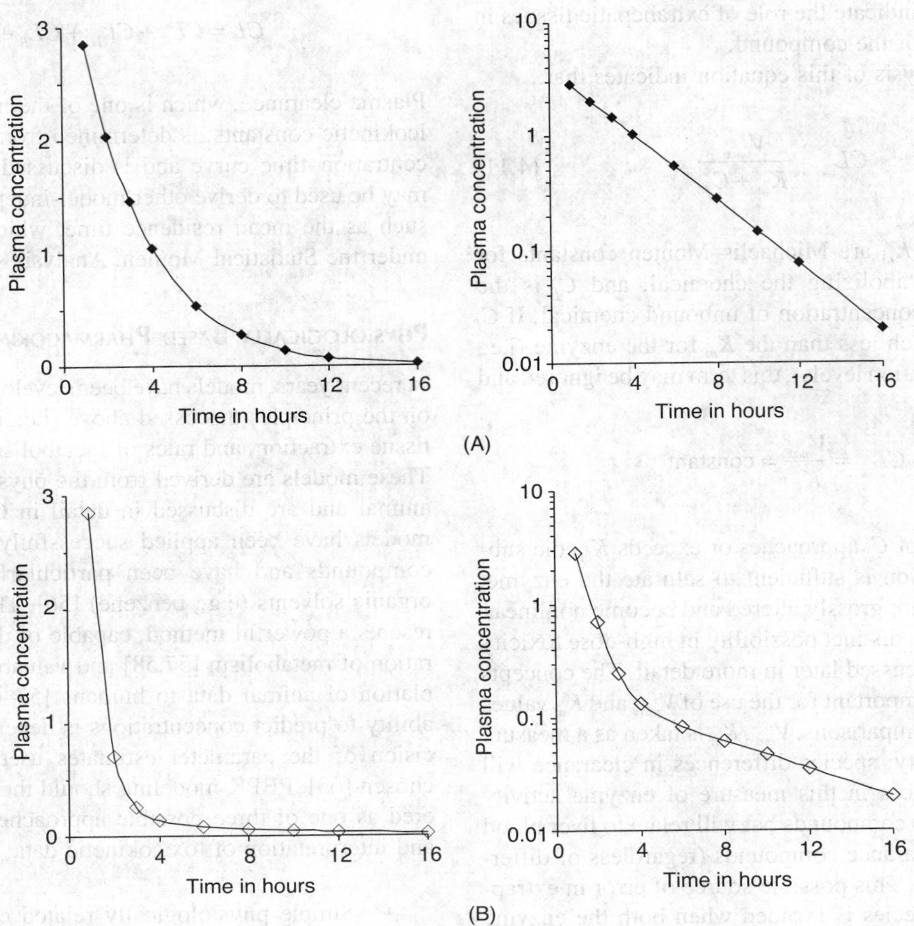


**FIGURE 4.8** Plasma concentration–time data for two compounds; results are plotted in linear and semilogarithmic forms. The data used to generate the curves are given in Table 4.3.

decrease in plasma concentrations results from the elimination processes (Scheme 4.1A). Alternatively, the compound may leave the plasma to enter "other tissues" at measurable rates, as well as undergoing elimination from

**TABLE 4.3**
**Data Used for Figure 4.8**

| Time After Intravenous Dosing (hr) | C (μg/mL) Compound A | Compound B |
|---|---|---|
| 1 | 2.850 | 2.850 |
| 2 | 2.040 | 0.705 |
| 3 | 1.470 | 0.250 |
| 4 | 1.050 | 0.136 |
| 6 | 0.540 | 0.082 |
| 8 | 0.280 | 0.062 |
| 10 | 0.145 | 0.047 |
| 12 | 0.075 | 0.035 |
| 16 | 0.020 | 0.020 |

the plasma. Under such circumstances, the "other tissues" may be adequately described mathematically by a second compartment in addition to the plasma (plus tissues that reach equilibrium before the first plasma measurement, known as the *central compartment*) (Scheme 4.1B). In some cases, two or more additional compartments are required. It is important to realize that these "other tissues" share only one criterion (i.e., their associated rates of uptake and transfer back into plasma), and biologically diverse tissues may be part of the same compartment. In addition, elimination may occur from compartments other than the central compartment (Scheme 4.1C, D, E, G). In most cases the processes of elimination and distribution of foreign compounds occur via first-order reactions; that is, the rate of the reaction is proportional to the amount of substrate available for the reaction.

Wagner [64] reviewed compartmental models and showed that 17 linear models existed to describe one-, two-, and three-compartment systems (i.e., models in which the plasma concentration–time curve could be resolved into a number of log–linear components). However, if the input

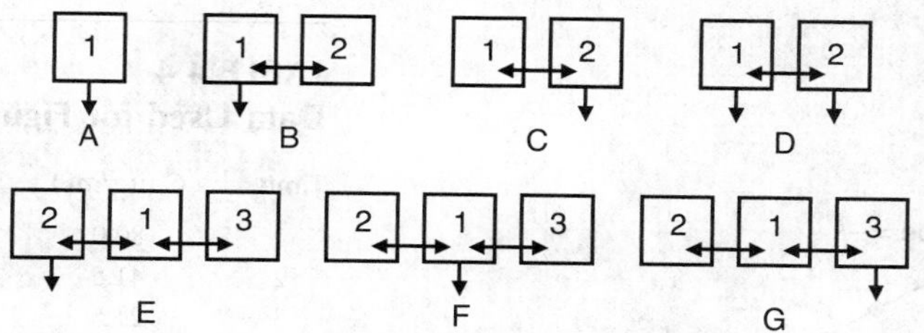


**SCHEME 4.1** Compartmental models. Linear disposition models showing one (A), two (B, C, D), or three (E, F, G) compartments. Only 3 of the 13 possible three-compartment models are shown; others are derived by variable elimination from any or all compartments and by compartment 3 equilibrating via compartment 2, not compartment 1. In all cases, compartment 1 is taken as the blood and tissues undergoing essentially instantaneous equilibration.

into the model is not instantaneous, then an additional rate constant to describe the input is necessary, and additional models would be generated. Wagner [64] concluded that 760 possible pharmacokinetic models comprised up to three distribution compartments and two input compartments, but that in many cases and for many calculations knowledge of the model that fitted the data best was not necessary. Because the aim of this chapter is to provide an introduction to toxicokinetics (i.e., samples needed, data handling, and the type of information that can be obtained), only simple models are discussed in detail. Readers are referred to standard texts on pharmacokinetics if the simple models discussed in this chapter do not fit the plasma data. The two models discussed show widespread applicability, and an understanding of the principles underlying these simple models is essential if the data generated by computer analysis using more complex models are to have any meaning. In addition, more complex models contain greater numbers of variables, and blood sampling must be increased to define the rate constants accurately.

Texts recommended for further reading include those by Rescigno and Segre [51], a mathematical approach with few drug illustrations; Gibaldi and Perrier [23], a classic text, which is a mathematical approach that is well explained and illustrated using actual experimental data; Rowland and Tozer [65], a well-written, readable text with many excellent illustrations and study problems at the end of each section; and Wagner [66], an approach similar to that of Gibaldi and Perrier [23] but with a useful biological introductory chapter and expanded sections on dosage regimen calculations, pharmacological response, and automated pharmacokinetic analysis. Additional useful sources for background reading include Benet et al. [67], a collection of papers from a symposium to honor S. Riegelman, and Rowland and Tucker [68], which has useful sections on interspecies scaling and time- and dose-dependent kinetics. A particularly valuable book is Gabrielsson and Weiner [69], which gives a clear account of different models in relation to the use of WinNonlin and provides excellent explanations of the basics of data fitting. All volumes provide references, either at the end of each chapter or for each illustration.

## ONE-COMPARTMENT OPEN MODEL

### Intravenous Bolus Dose

The compound is dissolved in and evenly distributed within a single compartment of volume *V*. Elimination of the compound, by both excretion and metabolism, is by first-order processes, and changes in plasma concentration are reflected in similar and simultaneous decreases in the tissue concentrations, because all tissues represent part of the single compartment (Scheme 4.1A and Scheme 4.2). In Scheme 4.2, *V* is the volume of distribution, $k_{ex}$ is the excretion rate constant, and $k_m$ is the metabolism rate constant. The plasma concentration–time curve for a one-compartment system is given in Figure 4.9, and the data are presented in Table 4.4. In mathematical terms, such a system may be described adequately by a simple first-order equation, where the rate of removal of a compound from the body (e.g., in milligrams per hour) is proportional to the body load (e.g., in milligrams):

$$\frac{dAb}{dt} = -kAb \tag{4.14}$$

where *Ab* is the amount of compound in the body, and *k* is the elimination rate constant ($k = k_{ex} + k_m$). A solution to this equation to give the amount remaining in the body at time *t* after injection is given by:

$$Ab = Ab_0 e^{-kt} \tag{4.15}$$

where *Ab* is the amount of compound at time *t*, and $Ab_0$ is the amount at time zero.

Assuming uniform distribution of the chemical within a single compartment, the concentration in the plasma (*C*)

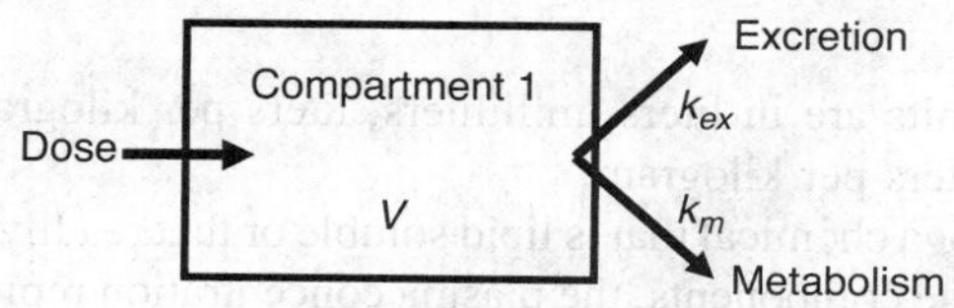


**SCHEME 4.2** One-compartment model.

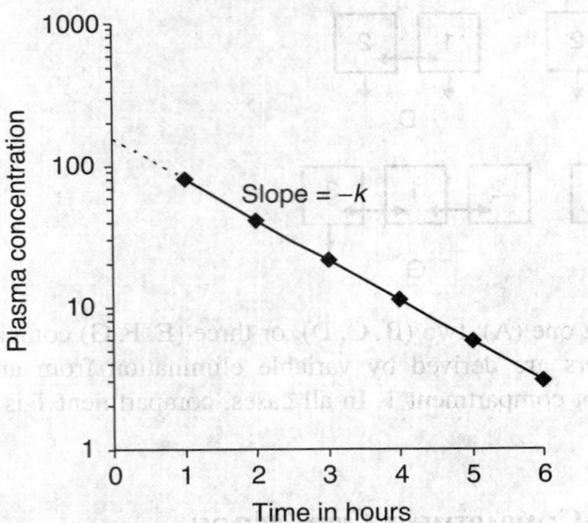


**FIGURE 4.9** Plasma concentration–time curve after a bolus intravenous dose for a one-compartment system. The data are given in Table 4.4 The concentrations have been shown on a logarithmic scale for illustrative purposes; the slope $-k$ would be calculated by plotting the natural logarithms of the concentrations against time, or $-k/2.303$ by plotting the logarithms to the base 10 of the concentrations against time.

**TABLE 4.4**
**Data Used for Figure 4.9**[a]

| Time | $C$ (μg/mL) | ln C | |
|---|---|---|---|
| 1 | 80.0 | 4.382 | |
| 2 | 41.5 | 3.726 | $k = 0.656$ hr$^{-1}$ |
| 3 | 21.5 | 3.068 | $t_{1/2} = 1.06$ hr |
| 4 | 11.2 | 2.416 | |
| 5 | 5.8 | 1.758 | $C_0 = 154.3$ μg/mL |
| 6 | 3.0 | 1.099 | $V = 324$ mL/kg |

[a] The plasma concentrations were obtained after an intravenous bolus dose of 50 mg/kg.

may be related to $Ab$ by the *apparent volume of distribution* ($V$). This volume may be regarded as the volume of plasma in which the body burden (body load) would have to be dissolved to give the plasma concentration measured:

$$C = \frac{Ab}{V} \tag{4.16}$$

where $C$ is the plasma concentration, and $V$ is the apparent volume of distribution. Thus, Equation 4.15 may be rewritten in its more usual form:

$$C = C_0 e^{-kt} \tag{4.17}$$

where $C$ is the plasma concentration at time $t$, and $C_0$ is the concentration at time zero. For such a system we can define the following parameters.

### *Apparent Volume of Distribution*

The apparent volume of distribution ($V$) is the apparent volume into which the dose would have been dissolved to give the initial plasma concentration, $C_0$:

$$V = \frac{Ab}{C} = \frac{\text{Dose}}{C_0} \tag{4.18}$$

The units are in liters, milliliters, liters per kilogram, or milliliters per kilogram.

For a chemical that is lipid soluble or that readily binds to tissue components, the plasma concentration represents a small fraction of the total amount in the body; thus, the compound appears to have been dissolved in a large volume of plasma that may greatly exceed the physiological volume of plasma. (Such compounds usually require a model with at least two compartments to describe the concentration–time curve; the concept of apparent volume of distribution applies to multicompartment models but requires a different method of calculation.)

### *Elimination Rate Constant*

The elimination rate constant ($k$) represents the fractional loss of compound from the body per unit time:

$$k = \frac{\text{Amount of chemical eliminated in unit time}}{\text{Amount of chemical in the body}}$$

$$= \frac{(dAb/dt)}{Ab}$$

Equation 4.17 may be rewritten as either:

$$\ln C = \ln C_0 - kt$$

or

$$\log C = \log C_0 - \frac{kt}{2.303}$$

Thus, a graph of ln $C$ against time has a slope of $-k$ and an intercept of ln $C_0$; a graph of log $C$ against time has a slope of $-k/2.303$ and an intercept of log $C_0$ (Figure 4.9). The units of $k$ are time$^{-1}$ (e.g., hr$^{-1}$ or min$^{-1}$). Thus, if the elimination rate constant is determined as 0.4 hr$^{-1}$, it means that 40% of the body load is removed each hour. The value of $k$ is the summation of component elimination rate constants (e.g., $k_{ex}$, $k_m$).

### *Elimination Half-Life*

The elimination half-life ($t_{1/2}$) is the time taken for the amount in the body ($Ab$) or the plasma concentration ($Ab/V$) to decrease to one half. Thus, after one half-life $C$ in Equation 4.17 equals $C_0/2$:

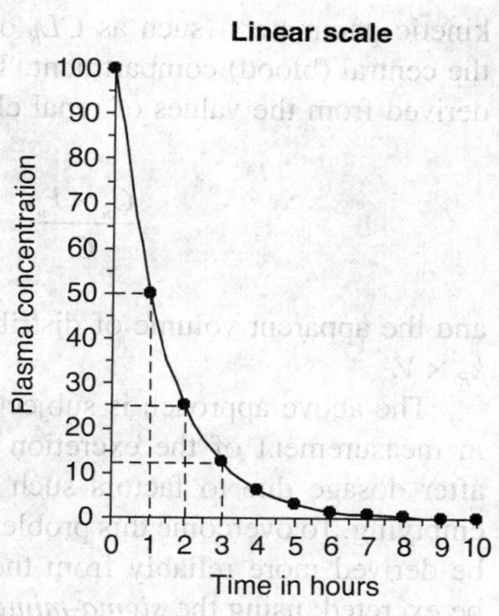


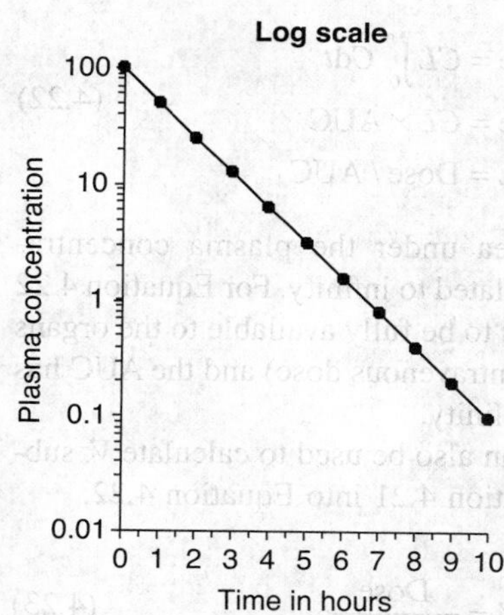


**FIGURE 4.10** The elimination of a chemical with a half-life of 1 hr from a one-compartment model, plotted on linear and semilogarithmic scales.

$$\frac{C_0}{2} = C_0 e^{-kt1/2} \quad \text{or} \quad \frac{1}{2} = e^{-kt1/2}$$

Therefore,

$$\ln 0.5 = -kt_{1/2} \quad \text{or} \quad -0.693 = -kt_{1/2}$$

$$t_{1/2} = \frac{0.693}{k} \tag{4.19}$$

where the units are time (e.g., hours or minutes). For first-order reactions, the half-life is independent of dose, body burden, and plasma concentration (Figure 4.10).

### *Plasma Clearance*

Plasma clearance (*CL*) is the amount of chemical eliminated in unit time related to the plasma concentration and may be regarded as the *volume of blood that is cleared of chemical in unit time*. *CL* is a constant for first-order reactions. In many respects, this measurement is a better reflection of the inherent capacity of the tissues to eliminate the compound than is the half-life or elimination rate constant:

$$CL = \frac{\text{Rate of elimination from the body}}{\text{Plasma concentration}} = \frac{(\mathrm{d}Ab/\mathrm{d}t)}{C} \tag{4.20}$$

Substituting from Equation 4.14,

$$CL = \frac{kAb}{C}$$

The amount in the body at any time (*Ab*) is given by Equation 4.16; therefore,

$$CL = \frac{kCV}{C} = kV \tag{4.21}$$

where the units are in liters $\text{hr}^{-1}$, liters $\text{min}^{-1}$, mL $\text{hr}^{-1}$, or mL $\text{min}^{-1}$. Rearranging Equation 4.21:

$$k = \frac{CL}{V}$$

This equation shows clearly that the elimination rate constant (*k*) is derived from two independent variables, each of which can be related to physiological processes: the *clearance*, which reflects the capacity of the organs of elimination to remove the compound from the plasma, and the *apparent volume of distribution*, which reflects the proportion of the total body burden that is circulated to the organs of elimination. Plasma clearance may depend on the rate of the active processes in the organs of elimination or on the plasma flow to the principal organs of elimination.

Clearance may be obtained without knowing the value of *V*. Rearranging Equation 4.20,

$$\frac{\mathrm{d}Ab}{\mathrm{d}t} = CL \times C$$

or in time d*t*, the amount lost d*Ab* = *CL* × *C* × d*t*. Integrating between time = 0 and infinity (∞), the total dose will have been eliminated, so d*Ab* = *dose*:

$$\text{Dose} = CL\int_0^{\infty} C\text{d}t$$
$$\text{Dose} = CL \times \text{AUC} \qquad (4.22)$$
$$CL = \text{Dose / AUC}$$

where AUC is the area under the plasma concentration–time curve extrapolated to infinity. For Equation 4.22 to be valid, the dose has to be fully available to the organs of elimination (i.e., an intravenous dose) and the AUC has to be extrapolated to infinity.

This relationship can also be used to calculate $V$; substituting $CL$ from Equation 4.21 into Equation 4.22:

$$V = \frac{\text{Dose}}{\text{AUC} \times k} \qquad (4.23)$$

The importance of Equation 4.22 and Equation 4.23 is that both the clearance and the apparent volume of distribution can be derived from infusion administration, where the determination of $V$ using Equation 4.18 is not possible. These equations may also be applied to oral administration, providing that allowance is made for incomplete absorption of the dose. This method of calculating $CL$ is also applicable to multicompartment linear systems with elimination from the central compartment.

*Information Obtainable from Urinary Data*

From Equation 4.7, we obtain:

$$\text{Rate of urinary excretion} = CL_R \times C$$

where $CL_R$ is the renal clearance. Thus, from Equation 4.21 and Equation 4.18:

$$CL_R \times C = k_R \times V \times C = k_R \times Ab$$

where $k_R$ is the renal excretion rate constant. However, from Equation 4.15, $Ab$ at any time = dose × $e^{-kt}$. Therefore,

$$\text{Rate of urinary excretion} = k_R \times \text{dose} \times e^{-kt} \qquad (4.24)$$

or

$$\ln(\text{rate of urinary excretion}) = (\ln k_R \times \text{dose}) - kt \qquad (4.25)$$

A plot of the natural logarithm of the rate of urinary excretion (amount excreted per time interval) against time gives a straight line with a slope of $-k$, and an intercept of ($\ln k_R$ × dose). It is important to note that the slope of this graph gives the overall elimination rate constant, not the specific urinary elimination rate constant. In other words, the decrease in the amount appearing in the urine mirrors the overall decrease in the plasma concentration. It is not possible to obtain information regarding other kinetic parameters (such as $CL_R$ or $V$) without sampling the central (blood) compartment. The value of $k_R$ may be derived from the values of renal clearance:

$$\frac{C_u \times F_u}{C}$$

and the apparent volume of distribution ($V$) when $CL_R$ = $k_R \times V$.

The above approach is subject to considerable errors in measurement of the excretion rate at different times after dosage due to factors such as incomplete bladder emptying. To overcome this problem, the rate constant can be derived more reliably from the amount remaining to be excreted, using the *sigma-minus* method. This method is based on the equation below, which is derived from integration of Equation 4.24:

$$A_{ex} = \frac{k_R \times \text{dose}}{k}\left[1 - e^{-kt}\right] \qquad (4.26)$$

where $A_{ex}$ is the total amount excreted up to time $t$. At infinite time, $(1 - e^{-kt})$ equals unity; therefore,

$$A_{ex}^{\infty} = \frac{k_R \times \text{dose}}{k}$$

where $A_{ex}^{\infty}$ is the cumulative total amount excreted in urine up to time infinity. Substituting back into Equation 4.26:

$$A_{ex} = A_{ex}^{\infty}\left[1 - e^{-kt}\right]$$

or

$$A_{ex}^{\infty} - A_{ex} = A_{ex}^{\infty} e^{-kt} \qquad (4.27)$$

The left-hand side of Equation 4.27 is equivalent to the amount finally excreted minus the amount excreted up to that time ($\Delta A_{ex}$). Taking natural logarithms,

$$\ln \Delta A_{ex} = \ln A_{ex}^{\infty} - kt \qquad (4.28)$$

A plot of $\ln \Delta A_{ex}$ against time gives a straight line of slope $-k$.

By analogy with Equation 4.22, $CL_R$ may be calculated from the total amount excreted and the plasma AUC:

$$CL_R = \frac{A_{ex}}{\text{AUC}}$$

where $A_{ex}$ and AUC refer to the same time interval. Combining this with Equation 4.22 gives the following relationships:

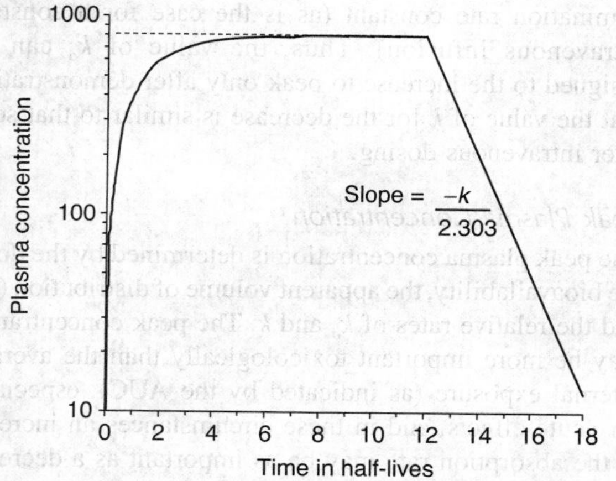


**FIGURE 4.11** Plasma concentration–time curve for constant intravenous infusion into a single-compartment system. The foreign compound was infused at a constant rate from time = 0 to time = 12 half-lives when the infusion was stopped. The concentrations have been shown on a logarithmic scale for illustrative purposes; the slope $-k$ would be calculated by plotting the natural logarithms of the concentrations against time, or $-k/2.303$ by plotting the logarithms to the base 10 of the concentrations against time.

$$\text{AUC}_0^\infty = \frac{\text{dose}}{CL} = \frac{A_{ex}^\infty}{CL_R}$$

or

$$CL_R = CL \times \frac{A_{ex}^\infty}{\text{dose}}$$

where $\text{AUC}_0^\infty$ is the AUC from zero to infinity after an intravenous dose. In other words, renal clearance ($CL_R$) equals plasma clearance ($CL$) multiplied by the fraction of the dose eliminated unchanged in urine.

## Constant Intravenous Infusion

During infusion the plasma concentration ($C$) increases to reach a plateau or steady-state concentration ($C_{ss}$), at which time the rate of infusion equals the rate of elimination (Figure 4.11). The extent of accumulation to steady-state is given by analogy with Equation 4.1:

$$\frac{C}{C_{SS}} = \left(1 - e^{-kt}\right)$$

or

$$C = C_{ss}\left(1 - e^{-kt}\right) \qquad (4.29)$$

The various kinetic parameters may be derived from the plasma concentration–time curve for infusion.

*Increase to Plateau*

Rearranging Equation 4.29 as used in Equation 4.27,

$$C_{ss} - C = C_{ss}e^{-kt}$$

Therefore, a plot of the natural logarithm of ($C_{ss} - C$) against time gives a straight line with a slope equal to $-k$. The time taken to reach the plateau is therefore similar to the time taken to eliminate the compound after infusion, or about 97% of the final steady-state level within five half-lives.

*Steady-State Plasma Concentration ($C_{ss}$)*

At steady state, the rate of infusion ($R$) equals the rate of elimination:

$$R = CL \times C_{ss}$$

or

$$CL = \frac{R}{C_{ss}} = V \times k$$

*Decrease at End of Infusion*

The slope of the decrease at the end of the infusion equals $-k$ because on cessation of entry into the single compartment, $C = C_0e^{-kt}$. The same slope would be obtained if the infusion was stopped at any stage during the infusion.

*Area Under the Curve*

Both $CL$ and $V$ may be derived using Equation 4.22 and Equation 4.23.

## Oral Administration

Absorption frequently obeys first-order kinetics [70] but may involve a lag time due to delayed gastric emptying. The plasma concentration–time profile may thus resemble Figure 4.12 (see also Table 4.5), and the various pharmacokinetic parameters are related by the equation:

$$C = \frac{F \times \text{dose} \times k_a\left(e^{-kt} - e^{-kat}\right)}{V\left(k_a - k\right)} \qquad (4.30)$$

where $F$ is the fraction of the dose absorbed and $k_a$ is the absorption rate constant.

*Increase to Peak*

The increase to peak is determined by the more rapid of the two processes, absorption and elimination; for lipid-soluble compounds, absorption is usually more rapid than elimination. Measurement of the absorption rate constant must make allowance for the excretion occurring throughout the post-dosing period; the method of residuals is used (see Gibaldi and Perrier [23] for the mathematical basis

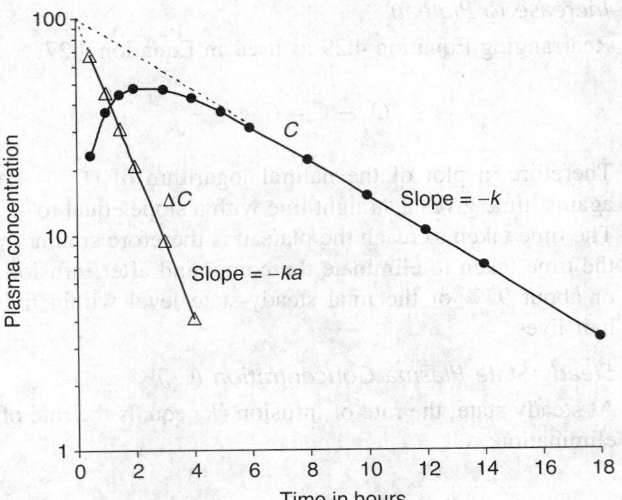


**FIGURE 4.12** Use of the method of residuals to calculate the absorption rate constant for a one-compartment system. The dose was given at time 0, and plasma levels (*C*) were measured at intervals. The linear terminal phase was extrapolated to yield the values corresponding to the measurement times. The difference values (*C* extrapolated – *C* measured) are plotted (Δ*C*) to yield slopes –*ka* or –*k*; see text. The concentrations have been shown on a logarithmic scale for illustrative purposes; the slopes –*k* and –*ka* would be calculated by plotting the natural logarithms of the concentrations against time, or –*k*/2.303 and –*ka*/2.303 by plotting the logarithms to the base 10 of the concentrations against time. The data are given in Table 4.5.

of this method). The method is illustrated and explained in Figure 4.12 and Table 4.5. In cases where absorption is slow, the rate of increase may be determined by the elimination rate constant (as is the case for a constant intravenous infusion). Thus, the value of $k_a$ can be assigned to the increase to peak only after demonstration that the value of $k$ for the decrease is similar to that seen after intravenous dosing.

### *Peak Plasma Concentration*

The peak plasma concentration is determined by the dose, the bioavailability, the apparent volume of distribution ($V$), and the relative rates of $k_a$ and $k$. The peak concentration may be more important toxicologically than the average internal exposure (as indicated by the AUC), especially for acute effects, and in these circumstances an increase in the absorption rate may be as important as a decrease in the elimination rate.

### *Decrease After the Peak*

The decrease after the peak concentration is determined by the slower of the two processes (absorption or elimination), but it is usually elimination, and the slope is equal to $-k$. (*Note:* For a polar compound showing slow absorption and rapid elimination, the decrease is equivalent to $-k_a$, a situation described by Gibaldi and Perrier [23] as "flip-flop" kinetics; see Figure 4.13).

### *Area Under the Curve*

Both $CL$ and $V$ may be derived using Equation 4.22 and Equation 4.23, providing that the dose used in the calculation is adjusted for the fraction absorbed ($F$):

$$CL = \frac{\text{Dose}_{oral} \times F}{\text{AUC}_{oral}}$$

**TABLE 4.5**
**Data Used for Figure 4.12**

| Time (hr) | C | ln C | ln $C_{ex}$ | $C_{ex}$ | ΔC | ln ΔC | |
|---|---|---|---|---|---|---|---|
| 0.5 | 23.0 | — | 4.501 | 90.1 | 67.1 | 4.206 | $k_a$ = 0.797 hr$^{-1}$ |
| 1 | 36.5 | — | 4.406 | 82.0 | 45.5 | 3.818 | |
| 1.5 | 43.9 | — | 4.312 | 74.6 | 30.7 | 3.424 | |
| 2 | 47.2 | — | 4.218 | 67.9 | 20.7 | 3.030 | |
| 3 | 46.8 | — | 4.029 | 56.2 | 9.4 | 2.241 | |
| 4 | 42.4 | — | 3.840 | 46.5 | 4.1 | 1.411 | |
| 5 | 36.7 | — | 3.652 | 38.5 | 1.8 | | |
| 6 | 31.1 | — | 3.463 | 31.9 | 0.8 | | |
| 8 | 21.8 | 3.082 | $k$ = 0.1887 hr$^{-1}$ | | | | |
| 10 | 15.0 | 2.708 | | | | | |
| 12 | 10.3 | 2.332 | | | | | |
| 14 | 7.1 | 1.960 | | | | | |
| 18 | 3.3 | 1.194 | | | | | |

*Note:* ln $C_{ex}$, data generated by linear regression analysis of the terminal phase of ln C against time; $C_{ex}$, antilogs; Δ*C*, the values ($C_{ex}$ – *C*) used to draw the residuals line. The 5- and 6-hr points are not included in the residuals analysis, as an error of 3% in the original value of *C* would translate into an error of 61 and 117%, respectively, for the Δ*C* value.

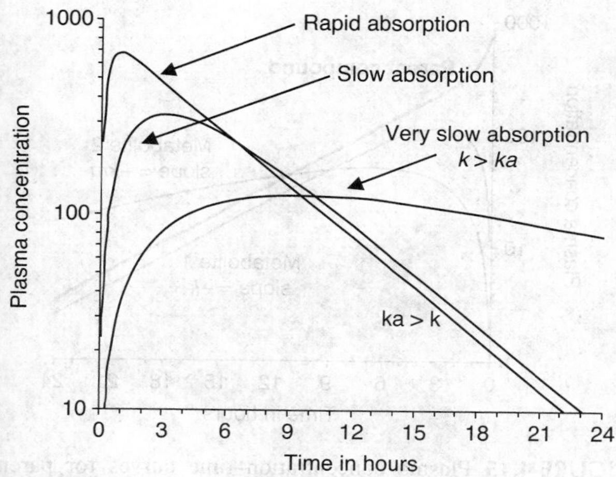


**FIGURE 4.13** The effect of absorption rate on the shape of the plasma concentration–time curve.

It is common to see $CL_{oral}$ calculated as (oral dose/$AUC_{oral}$) in the absence of any information on $F$. Such a term is meaningless because what is calculated is $CL/F$, which is dependent on two physiologically processes that are often unrelated. If the value of $F$ is unknown, then oral AUC data should be compared as such, without the calculation of a parameter that cannot be interpreted readily in physiological terms. Intravenous data are necessary to relate a nonlinear change in AUC at high oral doses to either altered $CL$ or $F$.

The value of $F$ is determined by comparison of the plasma concentration–time curves after oral and intravenous dosing. Because of the different shapes after oral and intravenous administration, comparisons at any single time point are not valid (although they have been used as a measure of absorption in some early studies). Instead, the overall systemic exposure, as indicated by the AUC extrapolated to infinity (Figure 4.14), arising from each dose is compared, with the assumption that systemic clearance ($CL$) is the same on both dosing occasions:

$$CL = \frac{\text{Dose}_{oral} \times F}{\text{AUC}_{oral}} = \frac{\text{Dose}_{iv}}{\text{AUC}_{iv}}$$
$$F = \frac{\text{Dose}_{iv} \times AUC_{oral}}{\text{AUC}_{iv} \times \text{dose}_{oral}} \tag{4.31}$$

These relationships are valid only if the AUC/dose ratio is constant; if this ratio is dose dependent, then the value of either $F$ or $CL$ must change with an increase in dose, suggesting saturation of absorption or elimination.

An alternative method to estimate the fraction of the dose absorbed as the parent compound ($F$) may be derived from the cumulative urinary excretion as the parent compound:

$$F = \frac{A^{\infty}_{ex\,oral}}{A_{ex\,iv}} \times \frac{\text{dose}_{iv}}{\text{dose}_{oral}} \tag{4.32}$$

## Metabolite Kinetics

As discussed elsewhere, the biotransformation of xenobiotics usually results in detoxification but is frequently associated with the formation of a toxic metabolite. Measurement of the rate of metabolism *in vivo* can provide much useful information on detoxification or bioactivation processes. In most cases, the rate of metabolite formation is governed by *in vivo* enzyme kinetics; enzyme reactions are first-order only over a limited substrate concentration range. Saturation of metabolism is discussed in more detail below, and the following analysis relates to metabolite formation under first-order reaction conditions and when $CL$ depends on enzyme activity rather than liver blood flow.

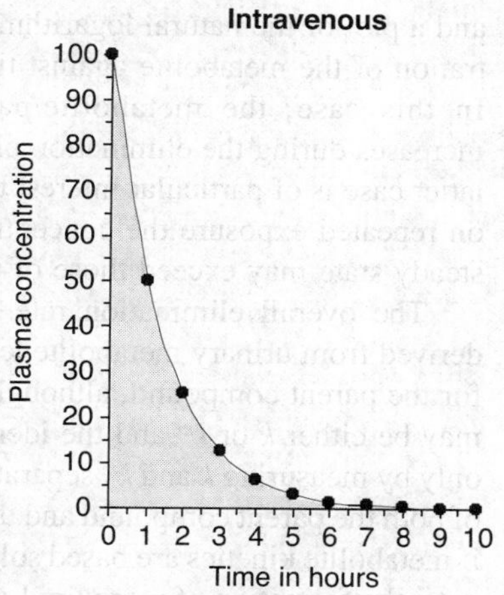


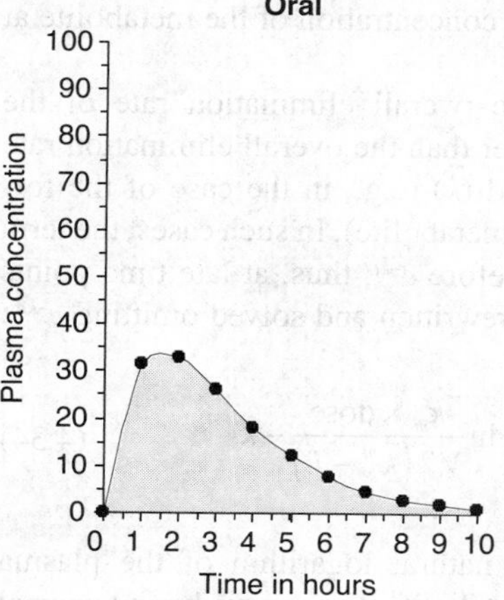


**FIGURE 4.14** The shape of the plasma concentration–time curve after intravenous and oral administration. The shaded area is the AUC, which is used to calculate the bioavailability ($F$).

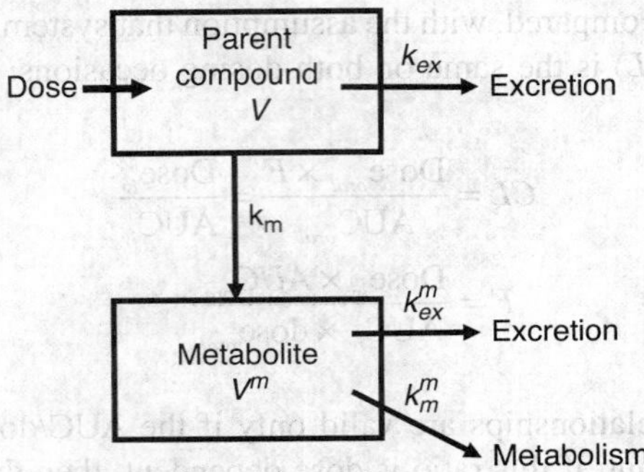


**SCHEME 4.3** One-compartment model with metabolite formation.

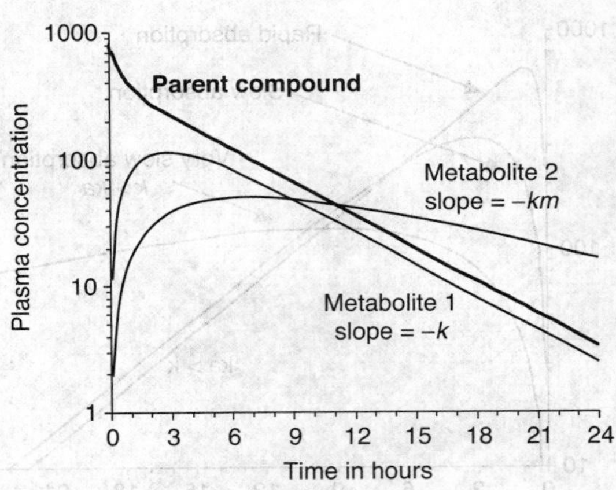


**FIGURE 4.15** Plasma concentration–time curves for parent compound and metabolites after intravenous dosing. The parent compound was given as an intravenous bolus dose at time = 0. The concentrations have been shown on a logarithmic scale for illustrative purposes; the slopes –*k* and –*km* would be calculated by plotting the natural logarithms of the concentrations against time, or –*k*/2.303 and –*km*/2.303 by plotting the logarithms to the base 10 of the concentrations against time.

Data that may be available for the analysis of metabolite kinetics include plasma levels of the parent compound ($C^p$) and of its metabolite ($C^m$). The simple system given earlier, for the parent compound (Scheme 4.2) can be extended into Scheme 4.3, where $V$, $k_{ex}$, and $k_m$ are, respectively, the apparent volume of distribution, excretion rate constant, and metabolism rate constants for the parent compound, and $V^m$, $k_{ex}^m$, and $k_m^m$ are the same parameters for the metabolite. The time course for the metabolite is given by:

$$\frac{dM}{dt} = k_m Ab - k^m M$$

where $Ab$ and $M$ are the amount of parent compound and metabolite in the body, respectively, and $k^m$ is the overall elimination rate constant for the metabolite; that is, $k^m = k_{ex}^m + k_m^m$. This equation may be solved to yield:

$$C^m = \frac{k_m \text{dose}\left(e^{-kmt} - e^{-kt}\right)}{V^m\left(k - k^m\right)} \tag{4.33}$$

where $C^m$ is the plasma concentration of the metabolite at time ($t$).

In many cases, the overall elimination rate of the metabolite ($k^m$) is greater than the overall elimination rate of the parent compound ($k$) (e.g., in the case of the formation of a more polar metabolite). In such cases, the term $e^{-kmt}$ approaches zero before $e^{-kt}$; thus, at late time points Equation 4.33 may be rewritten and solved omitting $e^{-km}$:

$$\ln C^m = \ln \frac{k_m \times \text{dose}}{V^m\left(k^m - k\right)} - kt \tag{4.34}$$

Thus, the plot of the natural logarithm of the plasma concentration of the metabolite–time curve has a terminal slope similar to that of the parent compound (i.e., –*k*) (Figure 4.15). In this case, the metabolite is present only as long as the parent compound remains in the body. The rate of elimination of the metabolite is limited by the elimination of the parent compound, and the metabolite/parent compound ratio remains constant during the elimination phase (Figure 4.15).

In those cases where the elimination rate of the metabolite ($k^m$) is less than that of the parent compound ($k$), the term $e^{-kt}$ approaches zero before $e^{-kmt}$; thus, Equation 4.33 may be written as:

$$\ln C^m = \ln \frac{k_m \times \text{dose}}{V^m\left(k - k^m\right)} - k^m t \tag{4.35}$$

and a plot of the natural logarithm of the plasma concentration of the metabolite against time has a slope of $-k^m$. In this case, the metabolite/parent compound ratio increases during the elimination phase (Figure 4.15). The latter case is of particular interest to toxicologists because on repeated exposure the concentrations of metabolite at steady state may exceed those of the parent compound.

The overall elimination rate constants may also be derived from urinary metabolite levels as described above for the parent compound, although again the derived rate may be either $k$ or $k^m$ and the identity can be determined only by measuring $k$ and $k^m$ separately after administration of both the parent compound and the metabolite. However, if metabolite kinetics are based solely on urinary excretion data, the formation of more lipid-soluble metabolites may be missed because negligible amounts of such metabolites would be excreted in the urine.

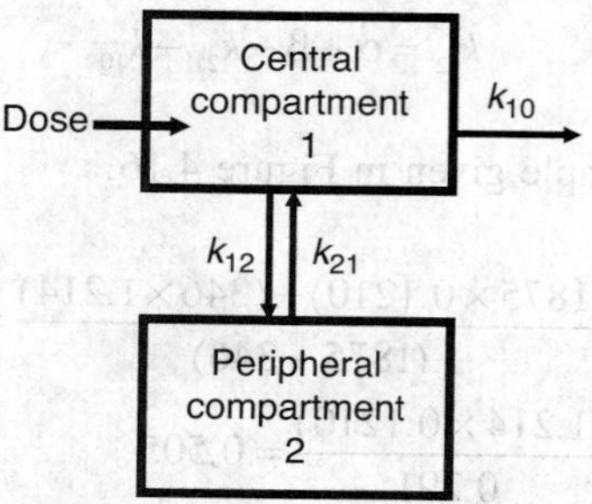


**SCHEME 4.4** Two-compartment model.

## Two-Compartment Open Model

Mathematically and physiologically, it is often more appropriate to regard the body as representing a simple two-compartment open system in which the distribution to certain peripheral tissues is not an instantaneous process. In such a system, the chemical initially enters a central compartment (the plasma and those tissues in which distribution is instantaneous) and is subsequently distributed to a second, peripheral compartment. Elimination occurs from the central compartment, so chemical in the peripheral compartment must transfer back to the central compartment to be eliminated (Scheme 4.lB or Scheme 4.4). In Scheme 4.4, $k_{12}$ and $k_{21}$ are the rate constants for transfer from compartment 1 to 2 and from 2 to 1, respectively, and $k_{10}$ is the elimination rate from the central compartment.

### Intravenous Bolus Dose

After a single intravenous bolus dose into a two-compartment system, the plasma concentration ($C$) at time $t$ may be described by:

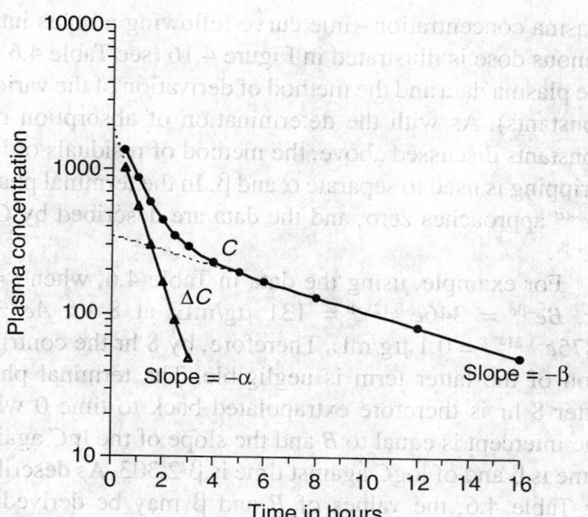


**FIGURE 4.16** Plasma concentration–time curve for two-compartment system. The concentrations have been shown on a logarithmic scale for illustrative purposes; the slopes α and β would be calculated by plotting the natural logarithms of the concentrations against time, or α/2.303 and β/2.303 by plotting the logarithms to the base 10 of the concentrations against time. The data are given in Table 4.6.

$$C = Ae^{-\alpha t} + Be^{-\beta t} \tag{4.36}$$

where $A$ and $B$ may be regarded as analogous to $C_0$ for each compartment, and $A + B = C_0$; α and β correspond to hybrid rate constants, each influenced by all the individual distribution, redistribution, and elimination rate constants (i.e., $k_{12}$, $k_{21}$, and $k_{10}$). The shape of a typical

**TABLE 4.6**
**Data Used for Figure 4.16**

| Time (hr) | C (μg mL⁻¹) | ln C | ln $C_{ex}$ | $C_{ex}$ | ΔC | ln ΔC | |
|---|---|---|---|---|---|---|---|
| 0.5 | 1345 | | 5.788 | 326 | 1019 | 6.927 | |
| 1 | 864 | — | 5.727 | 307 | 557 | 6.323 | By linear regression |
| 1.5 | 593 | — | 5.666 | 289 | 304 | 5.717 | α = 1.214 hr⁻¹ |
| 2 | 438 | — | 5.606 | 272 | 166 | 5.112 | ln $C_0$ = 7.537 |
| 2.5 | 346 | — | 5.545 | 256 | 90 | 4.500 | ∴ $A$ = 1875 μg/mL |
| 3 | 290 | — | 5.485 | 241 | 49 | 3.892 | |
| 4 | 228 | — | 5.364 | 214 | 15 | 2.708 | |
| 5 | 193 | — | 5.243 | 189 | 4 | 1.386 | |
| 6 | 168 | 5.122 | Terminal phase; by linear regression | | | — | |
| 8 | 131 | 4.879 | β = 0.1210 hr⁻¹ | | | — | |
| 12 | 81 | 4.395 | −ln $C_0$ = 5.848 | | | — | |
| 16 | 50 | 3.911 | −∴ $B$ = 346 μg/mL | | | — | |

*Note:* ln $C_{ex}$, data generated by linear regression analysis of the terminal phase data for ln C against time; $C_{ex}$, antilogs of these extrapolated points (similar values may be obtained from the extrapolated line on the graph); ΔC, the values ($C - C_{ex}$), which may be used to derive the $\log_{10}$ residuals line (slope – α/2.303) or may be converted to natural logarithms and analyzed by linear regression.

plasma concentration–time curve following a bolus intravenous dose is illustrated in Figure 4.16 (see Table 4.6 for the plasma data and the method of derivation of the various constants). As with the determination of absorption rate constants discussed above, the method of residuals or line stripping is used to separate α and β. In the terminal phase, $Ae^{-\alpha t}$ approaches zero, and the data are described by $C = Be^{-\beta t}$.

For example, using the data in Table 4.6, when $t$ = 8 hr, $Be^{-\beta t} = 346e^{-0.121\times 8} = 131$ μg/mL; at 8 hr, $Ae^{-\alpha t} = 1875e^{-1.214\times 8} = 0.1$ μg/mL. Therefore, by 8 hr the contribution of the latter term is negligible. The terminal phase after 8 hr is therefore extrapolated back to time 0 when the intercept is equal to $B$ and the slope of the ln$C$ against time is β and of log$C$ against time is β/2.303. As described in Table 4.6, the values of $B$ and β may be derived by least-squares linear regression analysis of the terminal phase, after graphical analysis to determine the point at which linearity commences.

At early time points, the difference between the actual $C$ values and the concentrations derived by back extrapolation of the $Be^{-\beta t}$ line are due to the contribution from $Ae^{-\alpha t}$. The values of $A$ and α may be similarly derived by calculated linear regression or graphical analysis of the residuals or Δ$C$ ($C$ actual – $C$ extrapolation). In the analysis of the residuals (Table 4.6), the Δ$C$ values for 4 and 5 hr were not included, as these values represent only about 5% or less of the original value of $C$ and are subject to large inaccuracies (up to +100%) due to the errors inherent in all methods of analysis of foreign compounds in biological fluids. (The pharmacokinetic computer programs outlined later will estimate *the best fit overall for both α and β phases simultaneously.)*

*The plasma concentration*–time curve in Figure 4.16 may be represented by the equation:

$$C = 1875e^{-1.214t} + 346e^{-0.1210t}$$

The rate constants α and β are composite rate constants, from which it is possible to derive $k_{12}$, $k_{21}$, and $k_{10}$ given in Scheme 4.4 using the following equations [23]:

$$\begin{aligned} C_0 &= A + B \\ \alpha + \beta &= k_{12} + k_{21} + k_{10} \\ V_1 &= \frac{\text{Dose}}{A + B} \end{aligned} \tag{4.37}$$

where $V_1$ is the volume of the central compartment, and

$$k_{21} = \frac{A\beta + B\alpha}{A + B} \tag{4.38}$$

$$k_{10} = \frac{\alpha\beta}{k_{21}} \tag{4.39}$$

$$k_{12} = \alpha + \beta - k_{21} - k_{10} \tag{4.40}$$

For the example given in Figure 4.16,

$$k_{21} = \frac{(1875 \times 0.1210) + (346 \times 1.214)}{(1875 + 346)} = 0.291$$

$$k_{10} = \frac{(1.214 \times 0.1210)}{0.291} = 0.505$$

$$k_{12} = 1.214 + 0.1210 - 0.291 - 0.505 = 0.539$$

It is important to note that $k_{10}$ (0.505) and β (0.121) do not relate to the same process, because $k_{10}$ refers to the elimination from the central compartment, whereas β refers to the overall elimination from the body (and is slower due to transfer out of tissues as well as elimination from the central compartment). The relation between β and $k_{10}$ is given by Equation 4.40, which may be rewritten as:

$$\beta = k_{10} + k_{21} + k_{12} - \alpha$$

which clearly shows that β is a hybrid rate constant; however, β is a very valuable constant and can be used to derive the terminal half-life (0.693/β).

As with the one-compartment system, an intravenous bolus allows derivation of most pertinent pharmacokinetic parameters:

1. $A$, $B$, α, and β may be derived from plasma data (see above).
2. $k_{10}$, $k_{12}$, $k_{21}$, and $V_1$ may be derived by manipulation of α, β, etc. (see above).
3. α, β, $k_{10}$, $k_{12}$, and $k_{21}$ may be derived from urine by plotting the excretion rate against time. In this case, the intercept values of excretion rate ($A' + B'$) do not equate to $A + B$, so $V_1$ cannot be deduced; however, $k_{10}$, $k_{12}$, and $k_{21}$ can be obtained from Equation 4.38, Equation 4.39, and Equation 4.40 by substitution of $A$ and $B$ by $A'$ and $B'$. The renal elimination rate constant ($k_R$) is given by:

$$k_R = \frac{A' + B'}{\text{Dose}}$$

4. α, β, $k_{10}$, $k_{12}$, and $k_{21}$ may be derived from urine by the sigma-minus method, where ln $(A_{ex}^{\infty} - A_{ex})$ (see Equation 4.27) is plotted against time. Again, α and β may be derived by the method of residuals; $k_{10}$, $k_{12}$, and $k_{21}$ can be calculated from α and β, and the intercepts ($A''$ and $B''$) by substitution in Equation 4.38, Equation 4.39, and Equation 4.40. The renal elimination rate constant ($k_R$) is given by:

$$k_R = \frac{A_{ex}^{\infty}}{\text{Dose}} \times k_{10}$$

5. The renal elimination constant, $k_R$, may be derived also from the renal clearance:

$$CL_R = \frac{C_u \times F_u}{C}$$

and the value of $V_1$ as $CL_R = k_R V_1$.

6. The amount in the peripheral compartment may be calculated from the following equation (which is similar to Equation 4.30 for absorption into a single compartment):

$$C_2 = \frac{\text{Dose} \times k_{12}\left(e^{-\beta t} - e^{\alpha t}\right)}{V_2(\alpha - \beta)} \quad (4.41)$$

where $C_2$ and $V_2$ are the concentrations in and volume of the peripheral or deep compartment, respectively.

During the terminal phase of the concentration–time curve, $e^{-\alpha t}$ approaches zero; therefore, Equation 4.41 may be simplified as:

$$C_2 = \frac{\text{Dose} \times k_{12} \times e^{-e\beta t}}{V_2(\alpha - \beta)}$$

Therefore, a graph of ln $C_2$ against time has a slope of $-\beta$. Thus, the terminal rate of decrease in the peripheral compartment of a two-compartment system is identical to the decrease in the central compartment.

In absolute terms, the calculation of $C_2$ is not particularly valuable, because the peripheral tissues comprising the deep compartment are not homogeneous, and the compound may not show a uniform concentration. Thus, $C_2$ should not be regarded as the effective compound concentration, even if the target organ lies within the deep compartment; rather, the concentration in the target organ should be measured, from which subsequent concentrations may be calculated using $\beta$ defined from the central compartment.

A further useful kinetic parameter ($V_\beta$), which relates the total amount of chemical in the body to the plasma concentration, is given by the equation:

$$V_\beta \times \beta = V_1 \times k_{10} = \frac{\text{dose}}{\text{AUC}} = CL$$

Just as $\beta$ is a hybrid term reflecting overall elimination from the body, so $V_\beta$ is a composite but valuable function:

$$V_\beta = \frac{\text{Dose}}{\text{AUC} \times \beta}$$

## Intravenous Infusion

The shape of the plasma concentration–time curve on intravenous infusion into a two-compartment open system is similar to that given in Figure 4.11, but with a biphasic increase at the start of the infusion and a biphasic decrease at the end. The kinetic parameters may be derived from the graph similarly to the one-compartment model, as follows.

### *Increase to Plateau*

The increase to plateau follows a complex exponential function with 90 and 99% of the steady-state concentration being reached after four and seven half-lives, respectively.

### *Plateau Level ($C_{ss}$)*

At steady state the rate of infusion, ($R$) equals the rate of elimination; therefore,

$$\frac{R}{C_{SS}} = CL = V_1 \times k_{10} = V_\beta \times \beta$$

### *Decrease After Plateau*

The decrease after plateau follows the equation:

$$C = A^* e^{-\alpha t^*} + B^* e^{-\beta t^*}$$

where $A^*$ and $B^*$ are the intercepts by back extrapolation to the end of the infusion of the $\alpha$ and $\beta$ slopes (determined as described for Figure 4.16), and $t^*$ is the time since cessation of infusion.

In many cases, two-compartment characteristics seen after a bolus dose are obscured in post-infusion data, because much of the distribution phase will have occurred during the infusion, so the duration of the $\alpha$ phase may be reduced.

### *Area Under the Curve*

The AUC can be used to derive the plasma clearance using Equation 4.22.

## Oral Administration

Assuming first-order absorption into compartment 1, the plasma concentration at time $t$ is given by:

$$C = A^{\ddagger} e^{-\alpha t} + B^{\ddagger} e^{-\beta t} + C^{\ddagger} e^{-kat}$$

Graphical analysis by a semilogarithmic plot of $C$ against time may reveal three separate phases from which $\alpha$, $\beta$, and $k_a$ should be measurable using the method of residuals applied to the ln concentration–time data. However, in practice the value of $k_a$ is frequently similar to α, and compounds that require a two-compartment model after intravenous administration appear to fit first-order absorption into a one-compartment model following oral dosing [71]. Thus, analysis is not possible without reference to intravenous data to determine which rate constant refers to the absorption rate. An example of linear regression analysis to obtain the three rate constants was given by Wagner [66]. An alternative method (deconvolution method) may be used that derives the absorption rate constant by a comparison of plasma concentrations for intravenous and oral administration and does not require fitting the data to a particular one-, two-, or three-compartment model. This method [23,66] requires analysis of the plasma concentrations at the same time points after both oral and intravenous dosing. Various methods of calculating the absorption rate are discussed in Gibaldi and Perrier [23].

The absorption rate is likely to be of greatest importance in acute toxicity studies, whereas the bioavailability ($F$) may be more significant in chronic studies; the latter may be measured using model-independent equations (Equation 4.31 or Equation 4.32). However, absorption from the gastrointestinal tract is complex, as it involves physiologically different membranes at differing luminal pH values. Thus, the process may involve more than one first-order rate or a zero-order component, or both; statistical moment methods provide an alternate approach to compartmental analysis and a valuable measure is the mean absorption time (see below).

### Metabolite Kinetics

Frequently metabolites of foreign compounds fit a two-compartment open model, in which case a second compartment for the metabolite is in equilibrium with the central metabolite compartment, as well as a second compartment for the parent compound (see Scheme 4.3). The equation describing these four compartments requires four exponential terms, but often the concentration–time curve for the metabolite appears as a bi-exponential decrease. The slow terminal phase of the metabolite is given by either β for the parent compound or the terminal rate for the metabolite (see earlier); the faster rate is a composite of the other three rate constants.

## Multiple Oral Dosing: Chronic Administration

On multiple dosing or continuous intake, the plasma levels increase over a period of four to five half-lives to establish a plateau concentration, similar to that seen with intravenous infusion (Figure 4.11). The average plateau (steady-state) level is subject to variations around a mean because material is eliminated between "doses." In oral toxicity studies, these "doses" may represent either repeated single gavage doses or the feeding habits of the animals if the test compound is incorporated into the diet and fed *ad libitum*. On cessation of chronic intake, the rate of decrease in blood levels is usually but not always similar to that seen after a single dose.

### One-Compartment Open Model

The time taken to reach plateau plasma levels is four to five times the half-time of the terminal phase of the plasma concentration–time curve. The average plateau level is given (by analogy with intravenous infusion) as:

$$C_{ss} = \frac{\text{Dose} \times F}{V \times k \times T} \tag{4.42}$$

where $F$ is the fraction absorbed, $T$ is the dose interval, and $k$ is the elimination rate constant. However, it is important to realize that this equation includes the term $k$, which may or may not be the terminal phase following oral administration. When the compound exhibits slow absorption and rapid elimination (flip-flop kinetics), the terminal decrease in plasma levels is determined by the slower absorption rate ($k_a$). An alternative form of this equation can be derived from the fact that at steady state the rate of input ($F \times \text{dose}/T$) is balanced by the rate of elimination ($C_{ss\ mean} \times CL$); therefore,

$$C_{ss\ mean} = \frac{\text{Dose} \times F}{T \times CL}$$

The fluctuations around the mean plateau level depend on the dosing interval in relation to the terminal elimination rate; thus, compounds with a short half-life show much larger fluctuations, as more of the chemical is eliminated between each dose. In the case of compounds with a short half-life (2 to 3 hr), single daily dosing gives plasma levels approaching zero prior to each dose. Inter-dose fluctuations may be reduced and blunted by slow absorption. The equations relating to these processes were detailed by Gibaldi and Perier [23] and Wagner [66]. In summary, at steady state after repeated intravenous doses, the minima and maxima are given by:

$$C_{ss\ minimum} = \frac{Dose}{V}\left(\frac{e^{-kT}}{1-e^{-kT}}\right) \tag{4.43}$$

$$C_{ss\ maximum} = \frac{Dose}{V}\left(\frac{1}{1-e^{-kT}}\right) \tag{4.44}$$

When absorption from the gut occurs as a first-order process, the fluctuating concentrations at steady-state concentration–time curve can be described by the following equation:

$$C = \frac{F \times \text{dose} \times k_a}{V(k_a - k)} \times \left[ \left( \frac{1}{1 - e^{-kT}} \right) e^{-kt} - \left( \frac{1}{1 - e^{-k_a T}} \right) e^{-kt} \right] \tag{4.45}$$

where $C$ is the concentration at time $t$ and $T$ is the dose interval. The similarity between this equation and Equation 4.30 for absorption of a single dose is apparent.

The value of the mean plasma concentration at steady state ($C_{ss\ mean}$) may be calculated without knowledge of $F$, $V$, $k_a$, or $k$ by measuring the area under the plasma concentration–time curve for a single oral dose:

$$AUC_{oral} = \frac{\text{dose} \times F}{V \times k} = \frac{\text{dose} \times F}{CL}$$

where $\text{AUC}_{oral}$ is the area under the plasma concentration–time curve between $t = 0$ and $t$ = infinity for a single oral dose. Substituting into Equation 4.42,

$$C_{ss\ mean} = \frac{\text{AUC}_{oral}}{T} \tag{4.46}$$

It is important to realize, however, that substitution of Equation 4.23 into Equation 4.42 assumes that the AUC is directly proportional to the dose; that is, dose-dependent kinetics are absent and $CL$ does not change during chronic administration of the compound. A large number of compounds can induce their own metabolism on chronic treatment. When this happens, $CL$ increases over the first few days of the study so the steady-state concentrations and body burden are lower than would have been predicted from the AUC, $CL$, or $CL/F$ measured after a single dose. Induction of cytochrome P450 (CYP) isoenzymes is a well-recognized phenomenon that can have toxicological implications in humans [72] and could also affect the outcome of animal toxicity studies [73]. The latter possibility may be assessed by comparison of the $\text{AUC}_{0-\infty}$ for a single dose, with the AUC for a dose interval at steady state (i.e., $\text{AUC}_{0-T}$ for chronic administration):

$$CL = \frac{\text{Dose(single)} \times F}{\text{AUC}_{0-\infty}} = \frac{\text{Dose(chronic)} \times F}{C_{ss\ mean} \times T} = \frac{\text{Dose(chronic)} \times F}{\text{AUC}_{0-T}}$$

$\text{AUC}_{0-T}$ (chronic) < $\text{AUC}_{0-\infty}$ (single) indicates either induction of metabolism or a decrease in bioavailability; conversely, if $\text{AUC}_{0-T}$ (chronic) > $\text{AUC}_{0-\infty}$ (single) then inhibition or saturation of metabolism or an increase in bioavailability is suggested.

The extent of accumulation on repeated intake may be measured by the average amount in the body at steady state ($Ab_{ss\ mean}$) divided by the amount in the body after a single dose ($Ab$):

$$\text{Extent of accumulation} = \frac{Ab_{ss\ mean}}{Ab} = \frac{Ab_{ss\ mean}}{\text{dose} \times F}$$

The amount in the body at the plateau is given by Equation 4.42:

$$Ab_{ss\ mean} = VC_{ss\ mean} = \frac{F \times \text{dose}}{k \times T}$$

Therefore,

$$\text{Extent of accumulation} = \frac{1}{k \times T} = \frac{1}{0.693/t_{1/2} \times T} = \frac{1.44 \times t_{1/2}}{T}$$

## Two-Compartment Open Model

The equations giving the plasma concentration at time $t$ at steady state into a two-compartment system with first-order absorption are considerably more complex than those for the one-compartment system; however, the simplified equation (Equation 4.42) applies in the form:

$$C_{ss\ mean} = \frac{\text{Dose} \times F}{V_1 \times k_{10} \times T} = \frac{\text{Dose} \times F}{V_B \times \beta \times T}$$

and the value of $C_{ss\ mean}$ may still be derived from Equation 4.46:

$$C_{ss\ mean} = \frac{\text{AUC}_{oral}}{T} = \frac{\text{Dose} \times F}{CL \times T} \tag{4.47}$$

In addition, the relationship between the AUC between $t = 0$ and $t = \infty$ for a single dose and the AUC for a dose interval at steady state applies on the condition that neither $CL$ or $F$ changes on chronic intake; a difference between these AUC estimates indicates changes in $CL$ or $F$ during chronic treatment.

## Statistical Moment Analysis

In recent years, both clinical pharmacokinetic and animal toxicokinetic studies have moved away from compartmental analyses because they involve multiple variables which require numerous properly timed blood samples to characterize them adequately. Also, curve fitting is dependent on the terminal slope, which is frequently measured using plasma concentrations that approach the limit of detection of the assay method (i.e., the weakest data). In contrast, terms such as clearance are measured from dose and AUC, the latter being determined largely from the highest and most accurately measured concentrations. Such "time-averaged" parameters may be extended to "time-related" parameters by the use of statistical moment theory, which allows assessment of additional useful kinetic parameters such as *mean residence time* (MRT) and *mean absorption time* (MAT). The plasma concentration–time curve may be regarded as a statistical distribution curve for which the zero and first moments are the AUC and MRT, respectively:

$$\text{AUC} = \int_0^\infty C\,dt$$
$$\text{MRT} = \frac{\text{AUMC}}{\text{AUC}} \tag{4.48}$$

where AUMC is the area under the first moment of concentration time curve; that is,

$$\int_0^\infty t \times C\,dt$$

The AUC and AUMC may be calculated using the trapezoid rule applied to the observed data.

The measured plasma concentrations and time points are used for AUC calculation. The AUC between two consecutive time points is given by $[(C_1 + C_2)/2] \times (t_2 - t_1)$ and the total AUC for the period of observation is the sum of the calculated AUC segments. The AUC from the last data point to infinity can be calculated as $C_{last}/\beta$.

For AUMC calculation, the measured plasma concentrations and time points are multiplied to give *Ct* values for each time point, and these are used with the time points. The AUMC for two consecutive time points is given by $[(C_1 \times t_1 + C_2 \times t_2)/2] \times (t_2 - t_1)$, and the total AUMC for the period of observation is the sum of the calculated AUMC segments. The AUMC from the last data point to infinity has to be calculated as:

$$\frac{t_{last} \times C_{last}}{\beta} + \frac{C_{last}}{\beta^2}$$

Clearly any inaccuracy in the value of β affects the extrapolation of AUMC to infinity more than the extrapolation

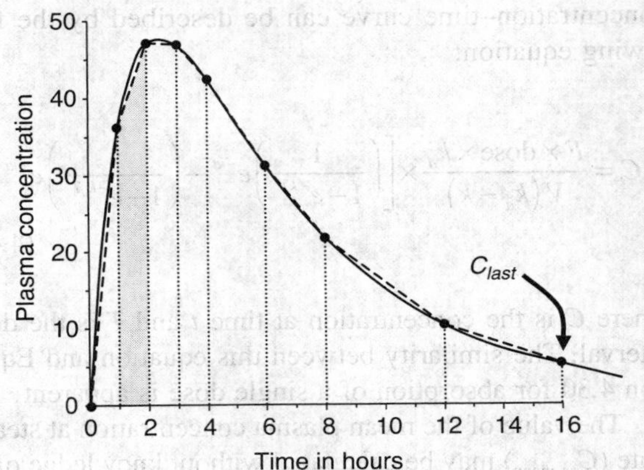


**FIGURE 4.17** The estimation of the AUC using the trapezoidal rule. The measured data points are shown as circles and the true concentration–time curve is shown as a continuous line. The AUC for each time interval is calculated separately and the values summed over the period of observation (0 to 16 hr); for example, the AUC for the highlighted second time interval is $(2 - 1) \times (36.5 + 47.2)/2 = 41.85$; if the units of concentration are μg/mL and time is in hr then the units of AUC are (μg/mL)·hr. The AUC from $C_{last}$ to infinity is calculated as described in the text.

of AUC. Another source of inaccuracy is that the method in effect draws a straight line between adjacent points, which will tend to underestimate the true AUC as concentrations increase and overestimate the true AUC as concentrations decrease (Figure 4.17). This error is reduced when the AUC and AUMC are estimated using log-transformed data.

### Intravenous Administration

Following an intravenous bolus dose, the MRT can be calculated by Equation 4.48. The *apparent volume of distribution at steady state* ($V_{ss}$) may be regarded as the volume of plasma in which the compound appears to be dissolved and that has to be "removed" from the body; that is, it is the product of clearance (mL min$^{-1}$) and MRT (min):

$$V_{ss} = CL \times \text{MRT} = \frac{\text{Dose}}{\text{AUC}} \times \frac{\text{AUMC}}{\text{AUC}} = \frac{\text{Dose} \times \text{AUMC}}{\text{AUC}^2} \tag{4.49}$$

If the compound is too toxic to be given as an instantaneous bolus, the MRT can be calculated from the AUMC following an intravenous infusion using the equation:

$$\text{MRT}_{infusion} = \text{MRT} + \frac{T}{2} \tag{4.50}$$

where $MRT_{infusion}$ is calculated from the AUMC and AUC by Equation 4.48 from the infusion data, and $T$ is the infusion time.

The $V_{ss}$ cannot be derived directly from the AUMC and AUC data from infusions, because the AUMC value contains a component due to the infusion time. The following equation, therefore, applies:

$$V_{ss} = \frac{\text{Infused dose} \times \text{AUMC}}{\text{AUC}^2} - \frac{\text{infused dose} \times T}{2 \times \text{AUC}} \qquad (4.51)$$

In the same way that $CL$ may be related to $V$ by the rate constant $k$ (Equation 4.21), so $CL$ may be related to $V_{ss}$ by the first-order rate constant $k_{ss}$ [23,74]:

$$CL = k_{ss} V_{ss} = \frac{V_{ss}}{\text{MRT}}$$

Therefore, $k_{ss}$ is equivalent to 1/MRT; for a two-compartment system, $k_{ss}$ is intermediate between α and β. The half-life derived from $k_{ss}$ (0.693/$k_{ss}$ or 0.693 × MRT) is therefore a composite half-life and may be regarded as the "effective" half-life; it represents a useful kinetic parameter, particularly for a single dose.

## Oral Administration

A major strength of the statistical moment approach is its ability to derive meaningful data following oral administration, because it is both more reliable and easier to use than most other methods [71] and does not rely on assumptions about the presence of a first-order or zero-order process. The most useful absorption parameter is the *mean absorption time* (MAT), which is the difference between the mean residence times following oral and intravenous dosing:

$$\text{MAT} = \text{MRT}_{oral} - \text{MRT}_{iv} \qquad (4.52)$$

The MAT may be used to derive apparent first-order rate constants and half-lives:

$$k_a = \frac{1}{\text{MAT}}$$

$$\text{Absorption } t_{1/2} = 0.693 \times \text{MAT}$$

Alternatively, if absorption appears to be zero order, by analogy with Equation 4.50:

$$\text{MAT} = \frac{T_{abs}}{2}$$

where $T_{abs}$ is the duration of the absorption process. The measurement of MAT is generally applied to absorption from a solution. If a sparingly soluble compound is given,

$$\text{MRT}_{oral} = \text{MRT}_{iv} + \text{MAT} + \text{MDT}$$

where MDT is the mean dissolution time. The statistical moment theory is therefore a valuable technique for comparisons on the influence of dosage formulations on absorption [71,75].

## Dose-Dependent or Nonlinear Kinetics

Whereas simple diffusion obeys first-order kinetics at all concentrations, many of the other processes fundamental to toxicokinetics involve interactions between the foreign chemical and a specific site on a protein (examples being active transport across the gut, plasma and tissue protein binding, metabolism, and renal tubular secretion). Because of the limited availability of the protein, these processes have a finite capacity, and all of the specific sites on the protein may be occupied at high concentrations of chemical. Addition of further chemical cannot result in further interaction between the chemical and protein, and the concentration of free compound increases rapidly.

Depending on the nature of the protein–chemical interaction, a number of consequences are possible, as summarized in Table 4.7. This table represents a considerable simplification because the effect of saturation at one site may affect another protein–chemical interaction; for example, saturation of renal tubular secretion gives increased AUC/dose and elevated plasma concentrations. However, the resultant high concentrations may saturate plasma protein binding, resulting in an increase in free compound and increased glomerular filtration or hepatic clearance. Thus, the decreased elimination in the renal tubule may be overcome to some extent by increased elimination elsewhere.

Almost all of the processes listed in Table 4.7 may be described by a Michaelis–Menten equation of the type introduced into Equation 4.11:

$$\frac{-\text{d}C}{\text{d}t} = \frac{V_{max} \times C}{K_m + C} \qquad (4.53)$$

where $V_{max}$ is the theoretical maximum rate of the reaction and $K_m$ is the Michaelis constant (which reflects the concentration giving 50% saturation of the protein). At low concentrations, $C << K_m$, and $K_m + C$ approximates to $K_m$ so:

$$\frac{-\text{d}C}{\text{d}t} = \frac{V_{max} \times C}{K_m}$$

and $V_{max}/K_m$ is equivalent to the first-order rate constant $k$.

At higher concentrations, $C >> K_m$, and $K_m + C$ approximates to $C$ so:

**TABLE 4.7**
**Consequences of Saturation of Chemical–Protein Interactions**

| Site | Interaction | Possible Consequences of Saturation at High Dose |
|---|---|---|
| Absorption | Active uptake | Reduced plasma levels and AUC after oral but not i.v. doses |
| | First-pass metabolism | Increased plasma levels and AUC after oral but not i.v. doses |
| Distribution | Plasma protein | Increased volume of distribution; increased glomerular filtration; increased hepatic clearance if extraction ratio is low |
| | Tissue protein | Decreased volume of distribution; a graph of $C_t/C$ against C will be nonlinear |
| Metabolism | Metabolizing enzyme (saturation by substrate, depletion of cofactors, product inhibition) | Decreased clearance; AUC/dose ratio increases for parent compound, whereas AUC of metabolite/dose ratio may decrease for both oral and i.v. doses; enzymes with high $K_m$ values may handle a larger proportion of the dose |
| Excretion | Renal tubular secretion | Decreased renal clearance; AUC/dose ratio increases for oral and i.v. doses; nonrenal routes of elimination become of more importance; total excretion in urine per dose may decrease depending on the availability of other routes of elimination |
| | Renal tubular reabsorption (rare) | Opposite of effects for saturation of renal tubular secretion |
| | Biliary excretion | Decreased biliary clearance; decreased enterohepatic recirculation; renal route may become more important; AUC/dose ratio increases for oral and i.v. doses |

$$\frac{-dC}{dt} = \frac{V_{max} \times C}{C} = V_{max}$$

Thus, the elimination is a zero-order reaction. The shape of the plasma concentration–time curve for a hypothetical compound showing saturation kinetics is given in Figure 4.18, which clearly shows that, although low doses are indistinguishable from first-order elimination, the decrease at high plasma concentrations shows zero-order and then first-order reaction components.

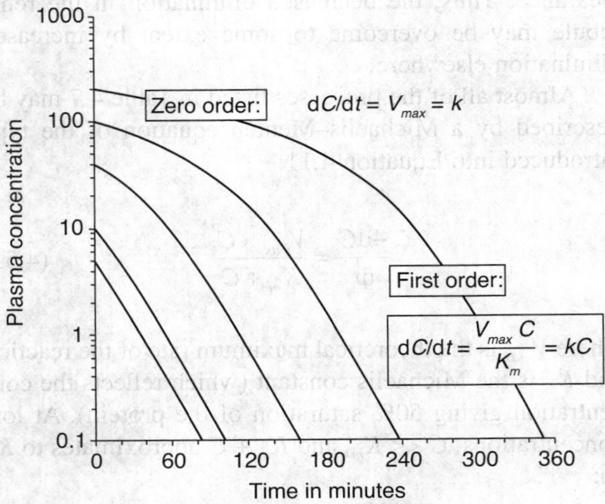


**FIGURE 4.18** Plasma concentration–time curve for a compound showing saturation kinetics. The data were generated using an apparent $V_{max}$ of 1 µg/min and a $K_m$ of 20 µg/mL for initial concentrations of 5, 10, 40, 100, and 200 µg/mL. Data points were obtained using a derivative of Equation 4.53; that is, $V_{max}(t - t_0) = C_0 - K_m \ln(C_0/C)$.

It is important to note that the terminal slope and terminal half-life are derived from low plasma concentrations and do not provide evidence of dose dependent kinetics. However, the plasma clearance, which is derived from AUC data and which reflects the capacity of the organs of elimination to remove the chemical from plasma, provides the best evidence of saturation. This is shown clearly by derivation of the appropriate rate constants, for the example given in Figure 4.18 (Table 4.8), which shows a fivefold dose-dependent change in *CL* over the doses calculated; it also illustrates the power of the statistical moment approach, which shows a fourfold increase in MRT. The value of *k* (0.0485) approximates to $(V_{max}/K_m)$ (0.050).

An increased understanding of saturation kinetics can be obtained by the determination of $K_m$ and $V_{max}$ from *in vivo* data. The value of $K_m$ that reflects the plasma concentration necessary to give 50% saturation of the active process is particularly useful for interpreting toxicity dose–response relationships. These constants can be determined following a single intravenous bolus dose using various equations, provided that the elimination is by a single saturable process. The simplest method, applicable to a one-compartment model, is by calculation directly from the plasma concentration–time curve using the equations:

$$\ln C = \ln C_{0e} - \frac{V_{max} \times t}{K_m}$$

and

$$K_m = \frac{C_{0a}}{\ln(C_{oe}/C_{0a})}$$

**TABLE 4.8**
**Pharmacokinetic Parameters Derived from Data Showing Saturation Kinetics (Figure 4.18)**

| Parameter | Curve 1 | Curve 2 | Curve 3 | Curve 4 | Curve 5 |
|---|---|---|---|---|---|
| Dose (mg/kg) | 5 | 10 | 40 | 100 | 200 |
| $C_0$ (μg/mL) | 5 | 10 | 40 | 100 | 200 |
| $k$ (min$^{-1}$)[a] | 0.0486 | 0.0486 | 0.0486 | 0.0485 | 0.0486 |
| Half-life (min)[a] | 14.3 | 14.3 | 14.3 | 14.3 | 14.3 |
| AUC (μg/mL min$^{-1}$)[b] | 115 | 254 | 1614 | 7020 | 24,027 |
| AUMC (μg/mL min$^{-1}$)[b] | 2437 | 5702 | 50,682 | 355,788 | 2,002,493 |
| $CL$ (mL/min/kg)[c] | 43.5 | 39.4 | 24.8 | 14.2 | 8.3 |
| MRT (min)[d] | 21.2 | 22.4 | 31.4 | 50.7 | 83.3 |

[a] Derived from data between 2.0 and 0.1 μg/mL for each dose.
[b] Calculated by the trapezoid rule with extrapolation to infinity.
[c] $CL$ = dose/AUC (Equation 4.22).
[d] MRT = AUMC/AUC (Equation 4.48).

*Note:* The parameters were calculated assuming a one-compartment model with a volume of distribution of 1 L/kg, which is not dose dependent.

where $C_{0e}$ is the value of $C$ at $t$ = 0 derived by back extrapolation of the terminal linear phase, and $C_{0a}$ is the actual concentration measured at $t$ = 0. A plot of ln $C$ against time has a terminal first-order slope of $V_{max}/K_m$.

Applying these equations to the data in Figure 4.18 for the highest dose gives values of 0.0486 for the slope, 200 for $C_{0a}$, and 2,701,271 for $C_{0e}$. Thus,

$$K_m = \frac{200}{\ln(2{,}701{,}271/200)} = 21\ \mu g/mL$$

and

$$V_{max} = 0.0486 \times 21 = 1.0\ \mu g\ min^{-1}$$

Alternative equations for the calculation of $K_m$ and $V_{max}$ require calculation of the rate of change of concentration from one sample to the next ($\Delta C/\Delta t$) as well as the plasma concentration at the midpoint ($C_m$).

**Lineweaver–Burk Plot**

$$\frac{1}{\Delta C/\Delta t} = \frac{K_m}{V_{max} \times C_m} + \frac{1}{V_{max}}$$

Therefore a plot of 1/($\Delta C/\Delta t$) against 1/$C_m$ has a slope of $K_m/V_{max}$ and an intercept of 1/$V_{max}$.

**Hanes–Woolf Plot**

$$\frac{C_m}{\Delta C/\Delta t} = \frac{K_m}{V_{max}} + \frac{C_m}{V_{max}}$$

Therefore, a plot of $C_m(\Delta C/\Delta t)$ against $C_m$ has a slope of 1/$V_{max}$ and an intercept of $K_m/V_{max}$.

**Woolf–Augustinsson–Hofstee Plot**

$$\frac{\Delta C}{\Delta t} = V_{max} - \frac{(\Delta C/\Delta t)K_m}{C_m}$$

Therefore, a plot of ($\Delta C/\Delta t$) against ($\Delta C/\Delta t$)/$C_m$ has a slope of $-K_m$ and an intercept of $V_{max}$.

When the data for the highest dose in Figure 4.18 are analyzed by these techniques (Table 4.9), the following values were obtained (Figure 4.19).

*Lineweaver–Burk Plot*

$$x\text{-intercept} = -\frac{1}{K_m} = -0.0537; \quad K_m = 18.6\ \mu g/mL$$

$$y\text{-intercept} = \frac{1}{V_{max}} = 1.04; \quad V_{max} = 0.96\ \mu g\ min^{-1}$$

$$\text{Slope} = \frac{V_{max}}{K_m} = 0.0517; \quad \frac{0.96}{18.6} = 0.516$$

*Hanes–Woolf Plot*

$$\text{Slope} = \frac{1}{V_{max}} = 0.999; \quad V_{max} = 1.001\ \mu g\ min^{-1}$$

$$\text{Intercept} = \frac{K_m}{V_{max}}; \quad K_m = 19.9\ \mu g/mL$$

**TABLE 4.9**
**Calculation of $K_m$ and $V_{max}$ from Plasma Concentration Time Data**

| Time[a] | C[a] | ΔC/Δt[b] | $C_m$[c] | 1/(ΔC/Δt) | $1/C_m$ | $C_m$/(ΔC/Δt) | (ΔC/Δt)/$C_m$ |
|---|---|---|---|---|---|---|---|
| 0 | 200 | 0.905 | 188 | 1.10 | 0.0053 | 207.7 | 0.0048 |
| 22.1 | 180 | 0.893 | 168 | 1.12 | 0.0060 | 188.1 | 0.0053 |
| 44.5 | 160 | 0.885 | 147 | 1.13 | 0.0068 | 166.1 | 0.0060 |
| 67.1 | 140 | 0.865 | 127 | 1.16 | 0.0079 | 146.8 | 0.0068 |
| 90.2 | 120 | 0.844 | 107 | 1.18 | 0.0093 | 126.8 | 0.0079 |
| 113.9 | 100 | 0.820 | 87 | 1.22 | 0.0115 | 106.1 | 0.0094 |
| 138.3 | 80 | 0.775 | 68 | 1.29 | 0.0147 | 87.7 | 0.0114 |
| 164.1 | 60 | 0.712 | 49 | 1.40 | 0.0204 | 68.8 | 0.0145 |
| 192.2 | 40 | 0.637 | 34.2 | 1.57 | 0.0292 | 53.7 | 0.0186 |
| 207.9 | 30 | 0.549 | 24.5 | 1.82 | 0.0408 | 44.6 | 0.0224 |
| 226.1 | 20 | 0.467 | 17.7 | 2.14 | 0.0565 | 37.9 | 0.0264 |
| 236.8 | 15 | 0.382 | 12.5 | 2.62 | 0.0800 | 32.7 | 0.0306 |
| 249.9 | 10 | 0.265 | 7.3 | 3.77 | 0.1370 | 27.5 | 0.0363 |
| 268.8 | 5 | 0.141 | 3.3 | 7.09 | 0.3030 | 23.4 | 0.0427 |
| 290.1 | 2 | 0.067 | 1.4 | 14.93 | 0.7140 | 20.9 | 0.0479 |
| 305.0 | 1 | 0.035 | 0.70 | 28.57 | 1.4286 | 20.0 | 0.0500 |
| 319.3 | 0.5 | 0.012 | 0.225 | 83.33 | 4.44 | 18.8 | 0.0530 |
| 351.9 | 0.1 | — | — | — | — | — | — |

[a] Raw data.
[b] Calculated as 200 – 180/22.1 – 0 = 0.905, etc.
[c] Read off the concentration time curve at midpoint of interval.

*Woolf-Augustinsson-Hofstee Plot*

$$\text{Slope} = -K_m = -20.3; \quad K_m = 20.3\ \mu\text{g/mL}$$

$$\text{Intercept} = V_{max} = 1.005\ \mu\text{g min}^{-1}$$

## Calculating $V_{max}$ and $K_m$ Using Data from an Intravenous Infusion

The values of $V_{max}$ and $K_m$ may be derived from plateau levels on intravenous infusion, providing that elimination is essentially by a saturable process only, because at steady state the rate of input = rate of elimination:

$$R = \frac{V_{max} \times C_{ss}}{K_m + C_{ss}}$$

where $R$ is the rate of infusion, or

$$R = V_{max} - K_m \times \frac{R}{C_{ss}}$$

Thus, a plot of $R$ for different rates of infusion against $R/C_{ss}$ gives a straight line with a slope of $K_m$ and an intercept of $V_m$ on the $R$ axis.

## Other Considerations Related to Nonlinear Kinetics

Frequently, the rate of elimination can be described by a combination of saturable and nonsaturable processes when:

$$\frac{-\mathrm{d}C}{\mathrm{d}t} = \frac{V_{max} \times C}{(K_m + C)} + k'C$$

where $k'$ is the rate constant for the nonsaturable process: $k'$ may be replaced by $CL'/V$, where $CL'$ is the clearance by the nonsaturable process; for example, for glomerular filtration, the value (GFR × $\alpha/V$) may be substituted for $k'$.

Of greatest importance for toxicology is the clear demonstration of saturation at high doses, an estimation of the plasma concentration above which first-order kinetics cease to apply, and the plasma concentrations present in animals showing overt toxicity. Wagner [66] proposed five tests for the establishment of saturation or nonlinear kinetics:

- Graphs of $C$/dose against time should be superimposable for linear kinetics at different doses. Although considerable scatter is seen, an overall trend to increased or decreased levels at higher doses should be apparent for nonlinear systems.
- Administer different intravenous doses and estimate $C_0$ by fitting only the first two or three

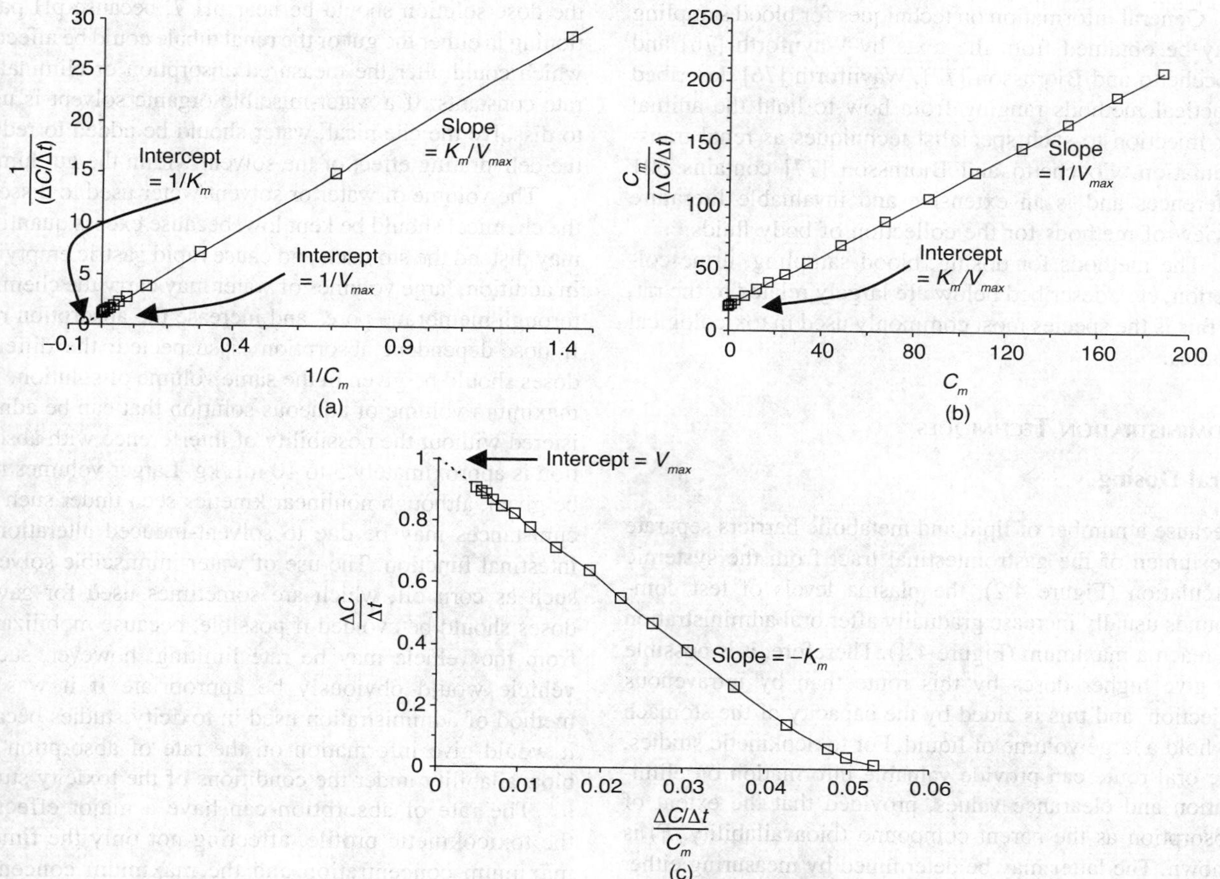


**FIGURE 4.19** Analysis of the maximum dose given in Figure 4.18 to derive $K_m$ and $V_{max}$ using the data in Table 4.9 and the methods of (a) Lineweaver–Burk, (b) Hanes–Woolf, and (c) Woolf–Augustinsson–Hofstee, which are outlined in the text.

early time points to the equation $\ln C = \ln C_0 - kt$. Graphs of $C/C_0$ against time should be superimposable if linear kinetics apply.

- Fit each set of concentration–time data to a linear model and derive the appropriate kinetic parameters ($CL$, $V$, $k$, $k_{12}$, $k_{21}$, $V_1$, …). A dose-dependent change in a parameter indicates nonlinearity or saturation kinetics (which will invalidate some of the derived parameters, such as $k$).
- If Michaelis–Menten kinetics apply, the percentage metabolized by that pathway decreases with an increase in dose (provided other elimination routes are available), the value of AUC/dose is not constant, and plots of log $C$ or ln $C$ against time curve downward, as shown in Figure 4.18.
- Measure the tissue and unbound plasma concentrations over a range of doses. A graph of tissue concentration against unbound concentration in plasma should be a straight line for a linear tissue extraction. Saturation of tissue binding is shown by the tissue concentration having a smaller increase at higher concentrations.

A consequence of nonlinear kinetics is that the time to reach steady state is also dose dependent. In simple terms, this is because the effective half-life, which can be calculated by 0.693 × MRT, increases with an increase in dose (e.g., Table 4.8) so the time to steady state (four to five half-lives) must also increase. This situation should be borne in mind when planning short-term studies.

## PRACTICAL METHODS

It is essential that any necessary legal and ethical approvals are obtained prior to any *in vivo* experiment in animals. This applies to noninvasive procedures, such as the incorporation of a chemical into an animal's feed, as well as invasive procedures, such as the collection of blood samples. Under U.K. legislation, separate Home Office licenses are required for the premises, the individual, and the procedure. These issues are discussed in Chapter 16.

General information on techniques for blood sampling may be obtained from the texts by Waynforth [76] and Cocchetto and Bjornsson [77]. Waynforth [76] described practical methods ranging from how to hold the animal for injection to such specialist techniques as renal transplantation. Cocchetto and Bjornsson [77] contains 501 references and is an extensive and invaluable literature review of methods for the collection of body fluids.

The methods for dosing, blood sampling, urine collection, etc., described below are largely related to the rat, as this is the species most commonly used in toxicological studies.

## Administration Techniques

### Oral Dosing

Because a number of lipid and metabolic barriers separate the lumen of the gastrointestinal tract from the systemic circulation (Figure 4.2), the plasma levels of test compounds usually increase gradually after oral administration to reach a maximum (Figure 4.4). Therefore, it is possible to give higher doses by this route than by intravenous injection, and this is aided by the capacity of the stomach to hold a large volume of liquid. For toxicokinetic studies, the oral route can provide valuable information on elimination and clearance values, provided that the extent of absorption as the parent compound (bioavailability, $F$) is known. The latter may be determined by measuring either the area under the plasma concentration–time curve or the total amount of the test substance excreted in the urine unchanged after both oral and intravenous administration at low doses. The fraction absorbed is given by Equation 4.31, and this figure can then be used to measure the clearance as described in the derivative of Equation 4.22. The possibility of saturation of absorption may be determined by plotting the value $C$/dose against time for doses up to and including those producing overt toxicity.

Rats, guinea pigs, and mice may be dosed by gavage using a syringe fitted with a suitable intubation needle; in rabbits, a polyethylene cannula is passed into the stomach while the jaws are held open by a gag. Certain precautions should be taken to prevent artifacts; for example, it is important that the test chemical is completely dissolved, because if the chemical is given as a suspension the apparent absorption rate may include a component due to dissolution of the chemical. If this factor is rate limiting, the measured absorption rate reflects the dissolution rate and is not related to the biological availability of the chemical. The ideal vehicle for dissolution is water or a small volume of a water-miscible solvent such as ethanol, propylene glycol (propane-1,2-diol), or dimethylsulfoxide, although for very lipid-soluble compounds it may be necessary to give the dose in corn oil or as an emulsion. Excess acids or bases should not be used to dissolve the test compound, and the pH of the dose solution should be near pH 7, because pH partitioning in either the gut or the renal tubule could be affected, which could alter the measured absorption or elimination rate constants. If a water-miscible organic solvent is used to dissolve the chemical, water should be added to reduce the dehydrating effect of the solvent within the gut lumen.

The volume of water or solvent/water used to dissolve the chemical should be kept low, because excess quantities may distend the stomach and cause rapid gastric emptying. In addition, large volumes of water may carry the chemical through membrane pores and increase the absorption rate. If dose-dependent absorption is suspected, the different doses should be given in the same volume of solution. The maximum volume of aqueous solution that can be administered without the possibility of interference with absorption is approximately 5 to 10 mL/kg. Larger volumes may be given, although nonlinear kinetics seen under such circumstances may be due to solvent-induced alteration of intestinal function. The use of water-immiscible solvents, such as corn oil, which are sometimes used for gavage doses should be avoided if possible, because mobilization from the vehicle may be rate limiting; however, such a vehicle would obviously be appropriate if it was the method of administration used in toxicity studies because it would give information on the rate of absorption and bioavailability under the conditions of the toxicity study.

The rate of absorption can have a major effect on the toxicokinetic profile, affecting not only the time to maximum concentration and the maximum concentration but also the total amount entering the systemic circulation by saturating hepatic uptake and first-pass metabolism. The hepatotoxicity of oral carbon tetrachloride is markedly higher after a bolus dose compared with gastric infusion [78].

When toxicity studies are performed by mixing the compound into the animals' diet, it is important to measure the concentration–time curve over a 24-hr period at steady state using dietary administration, because both the peak concentration and the AUC may be different from data obtained from bolus gavage studies.

### Nasal Administration

Methods have been described for assessing absorption from the nasal cavity based on plasma pharmacokinetics following intranasal and intravenous dosing and by *in situ* perfusion experiments [24,79]. A technique for inhalation with nose-only exposure has been described for studies in guinea pigs [80].

### Rectal Administration

Because a number of therapeutic compounds are given in the form of suppositories, an indication of the bioavailability after rectal administration is sometimes required

Normally, toxicity studies and initial drug formulations of such compounds are performed by the oral route, and the rectal formulation comes late in development and marketing. Animal bioavailability studies late in drug development are of limited value because of the differences between laboratory animals and humans in intestinal anatomy and microflora of the colon and rectum.

## Inhalation

A major problem associated with determining the kinetics of inhalation concerns the measurement of the extent to which the chemical is absorbed across the lung, rather than passed back into the mouth to be swallowed, exhaled in the expired air, or absorbed across the skin. Comparison of the plasma AUC or the total urinary excretion of unchanged compound after a period of inhalation with the same parameter after a known intravenous dose can be used to determine the total dose entering via the lungs plus gut.

A method used successfully by McKenna et al. [81] to obtain kinetic data involved a 6-hr exposure to the vapor of [$^{14}$C]vinylidene chloride in rats, after which the animals were transferred to a metabolism cage. The body load at the time of removal was determined by the total recovery of radioactivity in the expired air, excreta, cage washings, and carcass. This method is appropriate for determining the total dose because the nonspecific measurement of $^{14}$C includes the parent compound and all metabolites. If the parent compound alone is measured, the inhalation data must be compared to intravenous data to measure the extent of exposure after inhalation. The approach of McKenna et al. [81] was capable of revealing differences between fasted and fed animals in their capacity to metabolize vinylidene chloride, which correlated well with the toxicity of this compound.

The absorption of most compounds given by inhalation is rapid, although certain compounds (e.g., the highly polar antiasthmatic drug sodium cromoglycate) may show zero-order absorption across the lungs [26]. It should be remembered that the observed absorption rate after instillation of a micronized powder into the trachea may be that of the formulation, not of the chemical moiety itself.

The metabolism rate constants of inhaled 1,1-dichloroethylene were determined by measurement of the rate of removal of the compound from circulating air in a closed chamber system containing the experimental animal [82]. The air was recirculated, with oxygen added to maintain the concentration at 19 to 21%, and the air was sampled at regular intervals and analyzed for unabsorbed 1,l-dichloroethylene by gas–liquid chromatography. The rate of removal showed two phases: (1) a rapid uptake phase proportional to the mass of the animal and the concentration of chemical, and (2) a slow phase, which represented metabolism of the compound. The slow phase showed saturation (Michaelis–Menten) kinetics, and rate constants ($K_m$ and $V_{max}$) were derived in terms of the concentration of chemical in the chamber. This approach is interesting because the data are obtained by a noninvasive method, and the kinetic constants are derived in terms of vapor or gas concentrations, which are most appropriate when interpreting inhalation studies in relation to human exposure to volatile agents.

## Percutaneous Absorption

The percutaneous route is likely to be of increasing importance in drug formulation in the future. An added pharmacokinetic advantage to this route is that the fraction absorbed may be measured as described above for the oral route based on AUC data and also by analysis of the amount remaining at the site of administration. The dermal absorption of vapors can be assessed in rats using a body-only chamber [83], but there may be major species differences related to the presence of hair follicles and the barrier function of the stratum corneum. Shaving the hair from the backs of rats can provide a suitable site for *in vivo* absorption studies [84], but this can change the permeability characteristics of the stratum corneum. In reality, *in vitro* data can provide a suitable model for extrapolation to humans.

## Intravenous Injection

The bolus intravenous dose is the most important single technique for deriving information concerning the kinetics of the distribution and elimination of chemicals. Aqueous or aqueous miscible solvents should be used, although the maximum dosage volume is about 2 mL/kg for aqueous and 1 mL/kg for solvent–aqueous mixtures. Ideally, the solution is made isotonic by the addition of sodium chloride, although in practice dissolution of low doses in isotonic saline is adequate.

Two important parameters must be considered in such studies. First, the data processing assumes that the material was administered instantaneously at time zero. In practice, a rapid injection may produce acute toxicity, which a slower injection can prevent. Generally, a bolus dose, given over a finite period of up to a few minutes, is regarded as instantaneous provided that the total injection time does not represent more than about 5% of the half-life of the most rapid phase of the plasma concentration–time curve. The other parameter to be considered is the true location of the dose, because in kinetic studies it is important that 100% of the dose is intravenous and none ends up in a perivascular site.

The tail and hind paw veins of rats are convenient for dosing but neither is particularly easy to use or gives 100% intravascular dosing repeatedly and routinely without the necessary expertise. Cannulation of a vein, such as the external jugular or femoral vein, under anesthesia provides a more secure method of intravenous administration. The same cannula can be used for subsequent sampling, *providing that the compound is not adsorbed onto the cannula*. The external jugular vein may also be used for intravenous dosing of guinea pigs. The vein running around the periphery of the ear lobe is of sufficient size and visibility to give reliable intravenous dosing for rabbits.

### Intravenous Infusion

For intravenous infusion studies the dose must be given via an indwelling cannula. If the infusion period is prolonged, such that recovery from anesthesia is envisaged, the cannula can be run under the skin from the ventral surface of the neck and exteriorized on the dorsal surface behind the ears. If the cannula is then secured on the dorsal surface, the animal is prevented from damaging it during infusion while being permitted a degree of restricted movement. This method of exteriorization is also valuable as a method of long-term sampling. The delivery of compound during intravenous infusion must be at a constant but low rate such that the animal is not subjected to excessive hemodilution. To date, the standard method has been to use a high-quality infusion pump. The development of osmotically driven minipumps (Alzet®), which can be implanted into the animal and deliver a constant rate as low as 0.5 µL/hr for up to 2 weeks, opened up further possibilities. Minipumps allow investigations of steady-state plasma and tissue concentrations associated with toxicity and the pharmacokinetics under similar steady-state conditions [85].

## Sampling Techniques

### Blood (Plasma and Serum)

When considering the frequency, timing, and duration of blood sampling, it is important that an adequate number of samples are taken to define each section of the plasma concentration–time curve (Figure 4.8). It has been suggested that plasma samples be collected during the first four to five half-lives, during which time 93 to 97% of the compound will have been eliminated [86]; however, it is possible that such a restriction may mask a quantitatively minor distribution component (e.g., a third compartment) capable of significant accumulation on continuous ingestion as part of a toxicity test. If such a tissue was the site of toxicity on chronic administration, a single-dose kinetic investigation restricted to four half-lives would have failed to define the most relevant processes.

As a general guideline, the plasma concentrations should be measured until the limit of detection of the analytical method is reached. Obviously, if the limit of detection allows analysis over a large number of half-lives, less frequent sampling is required during the slow terminal phase, which can be defined adequately by about four samples. A three-compartment system can be accurately analyzed by about 12 samples if they are correctly timed. The corollary to this situation is that a relatively insensitive analytical method is incapable of yielding full pharmacokinetic data. Considering the data in Figure 4.8, if the limit of detection for compounds *A* and *B* were 0.1 µg/mL, both would appear to be represented adequately by a one-compartment model, with different values of *k*. Indeed, under these circumstances the plasma concentration of both *A* and *B* would fall from 2.85 to 0.10, a decrease of 97%, equivalent to about five half-lives.

Using the methods described below, it is possible to withdraw a significant fraction of the total blood volume (64 mL/kg in the rat), thereby modifying the perfusion of the organs of elimination and corrupting the derived pharmacokinetic data. This problem can be avoided by taking the smallest samples consistent with accurate analysis and the minimum number of samples necessary to define adequately the various phases (i.e., smaller samples at early time points). As a general rule, individual blood samples should be restricted to a maximum of about 0.5 mL/kg body weight if the total number of samples is small (i.e., less than 10). This general rule takes no account of the duration of the experiment, which may indicate either smaller or larger sample sizes.

#### *Rats*

Various methods have been used successfully to obtain small serial blood samples from anesthetized and conscious rats. Whole-blood samples, of approximately 100 to 200 µL can be obtained from the tail vein, either by snipping off the very end of the tail under local anesthesia or by making a small incision closer to the base of the tail. An advantage of this method is that usually only a single manipulation is necessary, as washing the tail vein with warm water will often remove the blood clot and reinstigate blood flow. A disadvantage is that the sampling site may be contaminated by urine and feces, although washing the tail may remove the polar metabolites excreted in urine and feces. Other methods that have been used to obtain blood samples include clipping the toe nail into the vascular bed, multiple cardiac sampling, and rupture of the sinus membrane at the back of the orbit, although these produce considerably more trauma and depending on national or local legislation may have to be performed in nonrecovery animals (i.e., under terminal anesthesia).

Alternative and more reliable methods require the insertion, under anesthesia, of a cannula into an exposed vein, such as the external jugular vein. Such tubing can then be used for sampling during and after recovery from anesthesia and may remain patent for periods up to 2 months. The use of silicon tubing is preferred to polyethylene for long-term studies because it is more flexible for exteriorization on the dorsal surface and less apt to cause thrombosis. For long-term studies, the cannula is exteriorized such that the animal cannot damage the tubing [77,87].

Cannulation of both the external jugular vein and carotid artery under general anesthesia in nonrecovery animals can be used for both intravenous dosage (venous) and blood sampling (arterial).

*Other Species*

For experiments performed under anesthesia, a major vein or artery (e.g., jugular, carotid, femoral) can be cannulated. The orbital sinus is a reliable site for collection of blood from the mouse while under terminal anesthesia. The marginal ear vein can be used for the rabbit without anesthesia.

### Urine

Knowledge of the urinary excretion rate is necessary for calculating the overall renal clearance of a compound. The bladder causes variable slowing of the output; for compounds with a short half-life, a method of overcoming sporadic urination is necessary. Calculating results by the sigma-minus method (Equation 4.27), rather than using excretion rate data, reduces the importance of incomplete bladder emptying and the resultant scatter in the data. For compounds with a half-life of many hours, sufficiently frequent samples may be obtained merely by placing the animals in a metabolism cage, which gives adequate separation of urine and feces, and by encouraging reflex urination [77].

In nonrecovery animals under anesthesia, the effect of the bladder may be overcome by cannulation of the urinary bladder via the urethra and allowing the urine to be expelled naturally or with the aid of gentle massage or by cannulation of both ureters directly. Renal clearance studies may be performed either after single doses or during infusion at steady state (when the clearance can be related to total clearance and plasma concentration). Insights into the extent of reabsorption and tubular secretion can be obtained by measuring the renal clearance of inulin given simultaneously (1 to 20 μCi of [$^{14}$C]inulin per kg or 50 to 100 μCi of [$^{3}$H]inulin per kg).

### Bile

In rats, bile may be collected from a cannula inserted into the common bile duct under general anesthesia such that the tip is located at the point of bifurcation near the hilar region of the liver. Bile flow, which is usually 0.5 to 1.0 mL/hr in the rat, may be collected either by exteriorizing the cannula or by passing the tubing into a suitable container (sealed plastic sachet) placed subcutaneously. A problem with such studies is that changes in bile composition occur if the bile salts are not allowed to recirculate. To avoid any influence of this on the pattern of biliary excretion, the test chemical is usually given soon after establishing the cannula. In a modified technique, the bile cannula can be exteriorized and joined to a second cannula, which passes back into the body cavity and enters the duodenum via the greater curvature of the stomach. The animals are then left for 4 days, after which feed intake and body weight are constant and biliary excretion can be studied under more physiologically normal conditions [88]. In animal species that possess a gallbladder (e.g., guinea pig and rabbit), it is necessary to prevent this organ from delaying elimination by ligation around its base so bile has to pass directly down the cannula.

## DATA HANDLING

Data on the concentrations of a chemical in plasma and urine can be analyzed graphically, as illustrated in this chapter or by linear regression applied to discreet phases (as illustrated in Table 4.6). The advent of powerful personal computers means that such approaches are largely redundant. Concentration–time curves, such as those in Figure 4.12 and Figure 4.16, can be analyzed by fitting multiexponential equations; these simultaneously fit all phases of the profile, such that a decision of where the initial rate is not significantly affecting the later phases is not necessary. Computerized analysis is the method of choice because data handling is optimized.

### Computation

Several suitable programs are available for nonlinear least-squares regression analysis, which is the most appropriate method (e.g., BLIN, NONLIN, SIPHAR); readers are referred to Gibaldi and Perrier [23], Wagner [66], and Gabrielsson and Weiner [69] for further details. Such programs automatically put the best fit line through the data. In the analysis of data by computer program, it is common to apply a suitable weight to each data point to ensure the most appropriate fit. The weights that can be applied to the concentration data include: (1) all weights equal, which is applicable if the errors in measurement are a constant amount (e.g., ± 2 μg/mL); (2) weighted by $1/y$, which is applicable if the errors of measurement are a constant proportion (e.g., ± 2%); or (3) weighted by $1/y^2$, which can be used to force the fit through the later time points at the expense of the early higher values. The second option ($1/y$) closely represents the accuracy of most assay procedures and is used most frequently.

It is important with computer fitting of data that some indication of appropriateness of fit is obtained by either a graphical representation or analysis of the deviation between observed and calculated concentrations (error analysis). With the latter approach, a consistent positive or negative deviation is more important than wider but randomly distributed deviations because it indicates an inadequate fit. Reasons for this situation could be the choice of an inappropriate model to fit to the data or incorrect weighting. Another factor to consider is that, although adoption of a more complex model may give a closer fit to the data, the sampling times may be inadequate to provide accurate parameter estimates.

It should be realized that, although kinetic constants derived from sophisticated computerized curve-fitting contain the minimum possible errors due to data handling, any errors in the raw data due to methodological problems will still be present. Indeed, the adage "rubbish in, rubbish out" is particularly pertinent to the use of sophisticated data handling to analyze inaccurate or badly designed animal toxicokinetic experiments.

## THE USE AND INTERPRETATION OF *IN VIVO* TOXICOKINETIC DATA

The three principle aims of *in vivo* toxicokinetic studies are:

- Toxicokinetics can provide an understanding of the physiological processes that are involved in the fate of the chemical in the body.
- The relation between dose and toxicokinetics may be the key to either the establishment of appropriate dose levels for chronic studies or the interpretation of such studies.
- Comparative toxicokinetics may be used to assess potential human risks with a more secure basis by reducing the number of unknown variables involved in the extrapolation from animal to humans.

The relationship between kinetics, dose, and toxicity is probably the single most important contribution that kinetics can make to the field of toxicity testing. Although a few therapeutic chemicals show nonlinear kinetics at the doses normally given to humans (notable examples being salicylates, phenytoin [diphenylhydantoin], and ethanol), the plasma levels of foreign chemicals in humans are usually well below those necessary to saturate any protein-mediated reactions. However, in toxicity testing, when the maximum dose tested is designed to show some degree of toxicity, nonlinear kinetics are a distinct possibility and should be fully and carefully investigated. At doses above saturation, the body load of free compound increases steeply with an increase in dose. Under such circumstances, effective tissue concentrations of the chemical will also be considerably higher than predicted by extrapolation from doses showing first-order kinetics. The presence of dose-dependent kinetics may result in an extremely steep dose–response curve for the toxic effect observed. In such circumstances, the nonlinearity in kinetics shown in animal toxicity testing must be taken into account when extrapolating the effects to humans.

A possible cause of toxicity associated with large saturating doses of chemicals is that normally minor pathways of metabolism may become of major significance. Thus, if a chemical undergoes metabolism by two routes, one with a low $K_m$ (high affinity) and the other with a high $K_m$ (low affinity), at low doses most chemical in the cell is eliminated by the former route. If the levels increase to saturate the high-affinity enzyme, any further input will exceed removal and levels will rise such that the low-affinity enzyme will metabolize the excess. (A useful analogy is that of water pouring into a bucket that has two holes in the side at different heights; little escapes through the upper hole until the rate of input exceeds the rate of removal by the lower hole.)

Finally, toxicokinetic data can be invaluable for the extrapolation of animal toxicity to possible human risk. Interspecies comparisons may be made in the absence or the presence of human kinetic data. In the absence of human data, extrapolation may be by either a PBPK model or compartmental modeling methods [89]. The physiological approach relies on the scale-up between animals and humans of such parameters as tissue volume and blood flow (Figure 4.5) and by their relationship to body weight. Assuming that the uptake from blood to tissue (extraction ratio) is a function of the chemical and not the animal species, it is possible to derive complex models involving all the major tissues of the body [90]. These models may then be scaled up from known animal data to humans based on the known physiological differences. Alternatively, the plasma kinetics in various species may be fitted by compartmental modeling and then scaled up empirically according to the body mass of the species studied and extrapolated to humans [91,92].

The most secure interspecies comparison is when kinetic data are available in animals and humans. In general, species differences in basic physiological processes such as cardiac output and relative tissue weights usually result in lower clearances and longer half-lives in humans than in animals [7]. Thus, comparisons of animals and humans on the basis of plasma levels or AUC values rather than intake or exposure data (expressed per kg body weight), removes important variables from interspecies comparisons and provides a more secure basis for the safety assessment [7,93].

Traditionally, 10-fold uncertainty factors have been used in risk assessment to allow for possible species differences and for human variability, and the NOAEL or intake in an animal study (in mg/kg body weight) would normally be divided by 100 to calculate the health-based guidance value, such as the RfD or ADI. The usual 10-fold factors have to allow for both toxicokinetics and toxicodynamics. Unlike toxicodynamics, toxicokinetics can be studied ethically in humans, and this allows the production of chemical-specific data on this aspect of interspecies differences and human variability. Subdivision of

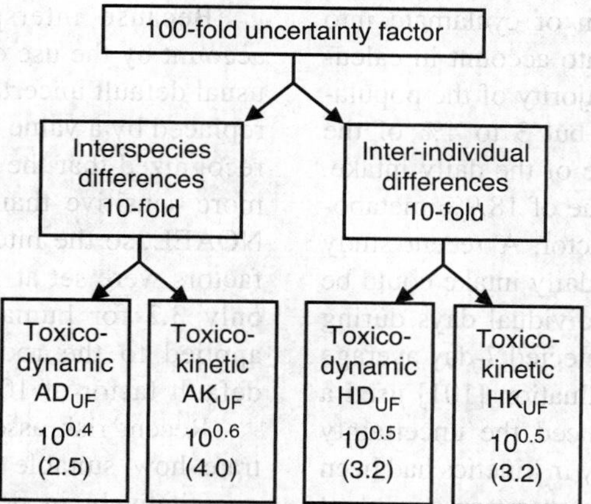


**FIGURE 4.20** Subdivision of the 10-fold uncertainty factors to allow for species differences and human variability in toxicokinetics or toxicodynamics. The total composite factor would be the product of any chemical-specific adjustment values and the remaining default uncertainty factors that had not been replaced: for example, if the interspecies toxicokinetic UF were replaced by chemical specific value of 20, then the total factor would be $2.5 \times 20 \times 3.2 \times 3.2 = 500$.

each 10-fold factor into toxicokinetic and toxicodynamic aspects [8,9] allows relevant chemical-specific data on toxicokinetics to replace the relevant default (Figure 4.20). Although this subdivision has the potential to replace uncertainty with real data, replacement of one of the factors for toxicokinetics requires an extensive database. This subdivision has been used in recent evaluations of the sweetener cyclamate (see cyclohexylamine, below) and of dioxins. The following examples illustrate the contribution of toxicokinetic studies to understanding the biological basis of toxicity and to the interpretation of toxicity data in terms of risk assessment.

## Cyclohexylamine

Cyclohexylamine is a metabolite of the intense sweetener cyclamate that is formed by the intestinal bacteria in the lower gut. Cyclohexylamine produces testicular toxicity when given chronically to rats, but not to mice [94]. Toxicokinetic studies indicated that the plasma clearance was higher in mice than in rats and that rats but not mice showed evidence of nonlinear kinetics at high doses [95]. The steady-state concentrations in the plasma and testes during chronic administration confirmed dose-dependent kinetics in the rat (Figure 4.21), which coincided with the dose–response for testicular atrophy in this species [95]. These toxicokinetic data thus provide a possible explanation for the steepness of the dose–response curve in the rat and the apparent species difference in sensitivity.

Cyclohexylamine is an indirectly acting sympathomimetic amine in rats [96], but it does not increase blood pressure in humans following its formation from cyclamate metabolism [97]. The apparent difference in response arises from the different concentration–time profiles when the metabolite is administered orally or when it is formed by the intestinal microflora from cyclamate [97], illustrating further the importance of kinetics in the interpretation of dose–effect relationships.

The ADI for cyclamate established by the Joint Expert Committee on Food Additives (JECFA) [98] and the Scientific Committee on Foods (SCF) in Europe [99] is based on the NOAEL for testicular toxicity of cyclohexylamine in a 90-day study in rats. The extremely wide person-to-

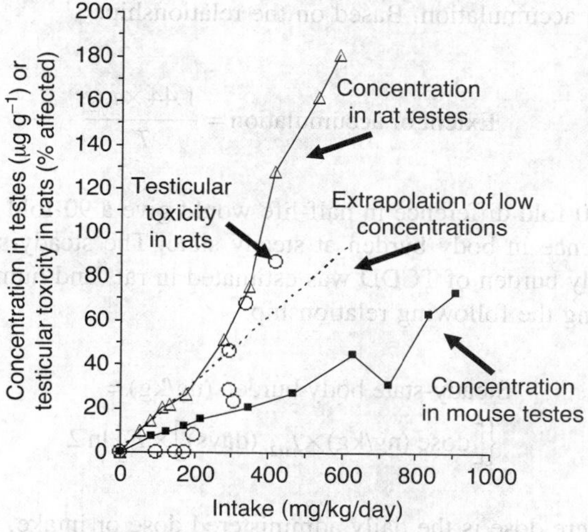


**FIGURE 4.21** Relation between dose, toxicity, and target organ concentrations of cyclohexylamine during chronic administration: (△) concentration in rat testes; (■) concentration in mouse testes; (○) testicular toxicity in rats. (Data from Roberts and Renwick [95].)

person differences in the conversion of cyclamate into cyclohexylamine have to be taken into account in calculating an ADI for cyclamate. The majority of the population cannot form cyclohexylamine, but 3 to 4% of the population metabolizes 20% or more of the daily intake. Early evaluations [98,99] used a value of 18.9% metabolism with a 100-fold uncertainty factor. A recent study [100] showed that up to 85% of the daily intake could be converted to cyclohexylamine on individual days during chronic intake, but that the highest detected 7-day average was 58%. The most recent SCF evaluation [101] used a value of 85% metabolism and reduced the uncertainty factor to 32 because human variability in kinetics had been taken into account by the use of a worst-case percent metabolism.

## Dioxins (TCDD)

A major problem with deriving a health-based guidance value for 2,3,7,8-tetrachlorodibenzo-*p*-dioxin (TCDD) from animal toxicity data is the considerable accumulation due to the very long half-life and the massive species differences in elimination half-life. The usual 10-fold default uncertainty used for extrapolating from rats to humans would be inadequate for TCDD, because of the approximately 90-fold difference in half-life between rats (about 30 days) and humans (about 7.5 years). Recent evaluations of dioxins by the JECFA [102] and the SCF [103,104] illustrate how toxicokinetic data can be used, and to some extent misused, in the risk assessment.

The recent risk assessments have been based on the body burden, rather than the daily intake, because this would take into account species differences in half-life and accumulation. Based on the relationship:

$$\text{Extent of accumulation} = \frac{1.44 \times t_{1/2}}{T}$$

a 90-fold difference in half-life would give a 90-fold difference in body burden at steady state. The steady-state body burden of TCDD was estimated in rats and humans using the following relationship:

$$\text{Steady-state body burden (ng/kg)} = \left\{\left[\text{dose (ng/kg)} \times t_{1/2}\ \text{(days)}\right] \times F\right\} \ln 2$$

where dose is the daily administered dose or intake, $F$ is the fraction absorbed, and $t_{1/2}$ is the species-specific half-life of TCDD. This relationship is based on Equation 4.42, using one day as the dose interval, with the body burden as given $C_{mean} \times V$ and $1/k$ converted to $t_{1/2}/\ln 2$ (see Equation 4.4).

Because interspecies differences are taken into account by the use of body burden as a dose metric, the usual default uncertainty factor of 4.0 (Figure 4.20) was replaced by a value of 1.0 [102–104]. In addition, it was recognized that the most sensitive human would not be more sensitive than the rat strains used to define the NOAEL, so the interspecies and human toxicodynamic factors were set at 1.0. As a consequence, the factor of only 3.2 for human variability in toxicokinetics was applied to the rodent NOAEL, instead of the usual default factor of 100-fold.

Recent risk assessments of dioxins [102–104] illustrate how suitable chemical-specific data can be used quantitatively in derivation of a health-based guidance value. Despite this welcome scientific development, a number of difficulties remained and were only partially resolved:

1. The majority of human exposure to dioxin-like compounds (polyhalogenated dibenzodioxins, polyhalogenated dibenzofurans, and coplanar PCBs) is to congeners other than TCDD. These are taken into account by the use of toxic equivalency factors [105], which relate the *in vivo* potency of the individual congener in animals to that of TCDD. The application of the relationship given above for TCDD to non-TCDD congeners assumes that the 90-fold species difference in the half-life of TCDD applies equally to the TEFs of other congeners.
2. The toxicity used by the JECFA and SCF to set the health-based guidance value for dioxins was based on postnatal effects in male rats following *in utero* exposure to a single oral dose given on day 15 of gestation (GD15) [106–110]. Initially, these studies could not be interpreted because of potential and undefined differences in the distribution of the body burden following a bolus dose and when the same body burden arose from slow accumulation over four to five half-lives. TCDD and related compounds are highly lipid soluble and show very high fat-to-blood concentration ratios, such that the majority of the body burden at steady state is in adipose tissue and not in potential sites of toxicity, such as the fetus. Publication of toxicokinetic studies on TCDD performed in pregnant rats specifically to address this issue [111,112] allowed the fetal concentrations on GD16 to be estimated following different methods of exposure. Risk assessments were based on conversion of the maternal body burden in rats following a bolus oral dose on GF15 [106–108,110] or following a loading dose and weekly doses [109] into a chronic maternal daily intake that would give the same fetal concentra-

tions at steady state. Extrapolation to humans was made using the relationship given above and the known species differences in half-life. This complex issue was addressed comprehensively by both the JECFA and the SCF.

3. The risk of cancer was determined from occupational cohort studies [102] in which the body burden was estimated some time after exposure had ceased and the relationship between exposure and cancer risk was determined by back-extrapolation of the body burden to the period of occupational exposure. The average half-life in humans was used to back extrapolate, but TCDD is a known enzyme inducer [73], and if enzyme induction had occurred at occupational exposures then a shorter half-life should have been used to back extrapolate. A shorter half-life would result in a higher body burden during occupational exposure, so any cancer risk could have been overestimated by the use of an inappropriately long half-life (because the association would be with a lower predicted body burden).
4. The best dose metric for a chemical with a very long half-life would be the steady-state concentration in plasma, which is given by:

$$C_{ss\ mean} = \frac{\text{Dose} \times F}{T \times CL}$$

However, $CL$ is not known for humans; therefore, the JECFA [102] and SCF [103,104] used the half-life instead of $CL$, which had the effect of changing $C_{ss\ mean}$ into $Ab_{ss\ mean}$. Half-life, though, is affected by both $CL$ and $V$; rearrangement of Equation 4.19 and Equation 4.21 gives the following relationship between half-life, $CL$, and $V$:

$$t_{1/2} = \frac{0.693 \times V}{CL}$$

Risk assessments based on estimated body burdens in animals and humans assume that the pattern of tissue distribution is similar, so the concentration at the site of action represents a similar proportion of the body burden in animals and humans (Figure 4.2). However, rats and humans have different body compositions, with fat representing about 10% of the body mass of rats but about 20 to 25% of humans. Consequently, the value of $V$ would be over twofold higher per kg body weight in humans, so the concentrations in all tissues, including sites of toxicity, would be twofold lower in humans for any given body burden. Body composition was not taken into account in the JECFA [102] and SCF [103,104] evaluations, which ignored the contribution of differences in $V$ to differences in half-life.

### Paracetamol (Acetaminophen)

Some notable examples of toxicity have occurred with doses of foreign chemicals that saturate metabolic or physiological processes. In a series of classic papers by Brodie and coworkers [113–116] it was shown that the hepatotoxicity of paracetamol (acetaminophen) was related to the oxidative metabolism of the compound and occurred only at high doses, which were associated with extensive covalent binding of the compound to tissue components. Little binding occurred at low doses or if the oxidative metabolism of toxic doses was inhibited by treatment with piperonyl butoxide. The binding at high doses arose from saturation of the capacity of the hepatocytes to protect themselves from the reactive metabolite produced. The protective mechanism was conjugation with glutathione, and saturation of this system at toxic doses was caused by depletion of the available glutathione.

### Retinol

Retinol illustrates how toxicokinetic data can help to define the contributions of a parent compound and its metabolites to toxicity. Retinol (vitamin A) is an animal and human teratogen that is metabolized to a number of biologically active metabolites, such as all-*trans*-retinoic acid, 13-*cis*-retinoic acid, and 9-*cis*-retinoic acid. The principle sources of exposure to retinol (rather than the precursor carotenoids) are from the consumption of animal liver and from vitamin supplements. The evidence for teratogenicity of retinol in humans is from case reports of excessive consumption of vitamin supplements and the teratogenicity of 13-*cis*-retinoic acid (isotretin), which is used as a therapeutic drug. Little evidence suggests a teratogenic risk from liver consumption, despite the fact that this route may give similar or even higher intakes of retinol than those causing malformations after excessive supplement consumption.

The areas under the plasma concentration–time curve of retinol and its metabolites in young women following oral doses of retinyl palmitate as an oral solution (fasting) and as cooked liver (as part of a meal) [117] are given in Table 4.10. The main difference was the 5- to 10-fold higher AUC of all-*trans*-retinoic acid following supplements, which is recognized to be the major teratogenic metabolite, although there was no major difference in the AUC of 13-*cis*-retinoic acid. Therefore, the key issues for the assessment of human risk were the plasma concentration–treatment relationship for the main metabolites and whether the teratogenicity of retinol and 13-*cis*-retinoic acid could be explained by their metabolism to all-*trans*-retinoic acid.

A subsequent study [118] gave single oral doses of retinol, all-*trans*-retinoic acid, and 13-*cis*-retinoic acid to pregnant rats and determined the dose–response relation-

**TABLE 4.10**
**AUC Values for Retinol and Its Metabolites in Young Women After Oral Doses of 50 mg and 150 mg of Retinol Given as Retinyl Palmitate in a Supplement or in Cooked Calf Liver**

| | AUC Values (ng/mL/hr) | | | | | |
|---|---|---|---|---|---|---|
| **Dose** | **Retinyl Palmitate** | **Retinol** | **All-*trans*-Retinoic Acid** | **13-*cis*-Retinoic Acid** | **All-*trans*-4-Oxo-Retinoic Acid** | **13-*cis*-4-Oxo-Retinoic Acid** |
| 50 mg S | 10,400 (5300) | 2070 (1900) | 86 (63) | 359 (104) | 30 (33) | 877 (207) |
| 50 mg L | 5900 (3700) | 1530 (2490) | 6 (5) | 243 (70) | 11 (15) | 492 (145) |
| 150 mg S | 18,900 (20,200) | 2300 (1180) | 170 (118) | 674 (156) | 133 (47) | 2385 (563) |
| 150 mg L | 14,500 (8800) | 2430 (2090) | 23 (11) | 596 (131) | 120 (84) | 1820 (294) |

*Note:* S, supplement given under fasting conditions; L, cooked liver as part of a meal. The results are the mean (with SD in parentheses) for 10 subjects dosed at 4-week intervals.

Source: Data from Buss, N.E. et al., *Human Exp. Toxicol.*, 13, 33–43, 1994.

ships (Figure 4.22) and plasma kinetics of the different metabolites for each compound. Both retinol and 13-*cis*-retinoic acid gave measurable levels of all-*trans*-retinoic acid in the plasma. A graph of teratogenic response for the three compounds against the AUC of all-*trans*-retinoic acid formed following each treatment (Figure 4.23) showed that, whereas the teratogenicity of 13-*cis*-retinoic acid could be explained on the basis of the circulating all-*trans*-retinoic acid, this was not the case for retinol. Retinol was considerably more potent than would be predicted from the plasma AUC of all-*trans*-retinoic acid after retinol administration, possibly due to local bioactivation within the fetus. Consequently, the 5- to 10-fold difference in the AUC of all-*trans*-retinoic acid between liver and supplements (Table 4.10) would not be expected to be translated into a similar difference in teratogenic risk. However, using the comparative potency ratios in rats (Figure 4.22) of about 25:3:1 for all-*trans*-retinoic acid, 13-*cis*-retinoic acid, and retinol and the AUC data in Table 4.10, the doses of 50 mg and 150 mg as supplements would be expected to be about 2.2 and 1.8 times more active than the equivalent doses given in cooked liver. This analysis, however, is heavily biased by the AUC and response data for retinol, as there could be major species differences in local activation of retinol in the fetus or vehicle-dependent differences (liver compared with supplements) in the proportions bound to retinol-binding protein (which could alter this conclusion).

## Saccharin

Saccharin is a non-nutritive sweetener that causes an increased incidence of tumors of the urinary bladder in male rats when fed as the sodium salt at high dietary

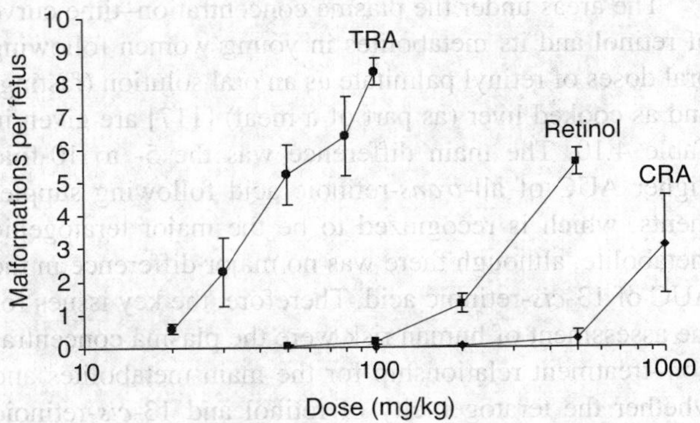


**FIGURE 4.22** Dose–response relationships for teratogenicity of all-*trans*-retinoic acid (TRA), retinol, and 13-*cis*-retinoic acid (CRA) given as a single oral dose to rats on day 10 of gestation. (Data from Tembe et al. [118].)

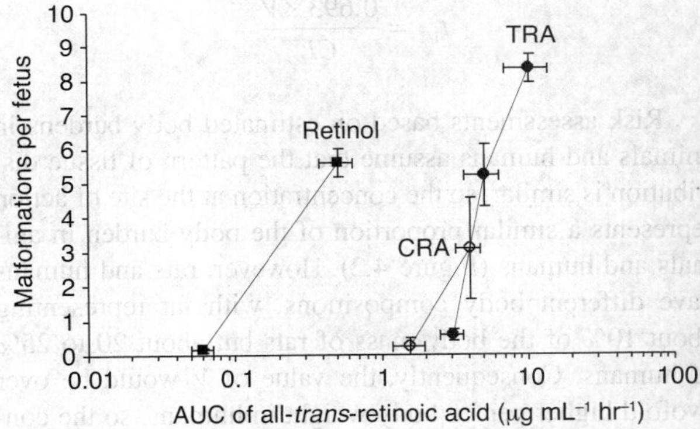


**FIGURE 4.23** Relationship between AUC of all-*trans*-retinoic acid in maternal plasma and teratogenicity in female rats given all-*trans*-retinoic acid (TRA), retinol, and 13-*cis*-retinoic acid (CRA) on day 10 of gestation. (Data from Tembe et al. [118].)

concentrations (more than 3%) for two generations or from birth [119]. The toxicokinetics [120] were investigated in Charles River CD-derived rat in order to investigate the cause of this phenomenon. Saccharin is a highly polar unmetabolized compound, and kinetic parameters were derived by measuring total $^3$H following administration of $^3$H–saccharin. These data showed that saccharin is absorbed slowly from the gut. The absorbed saccharin has a low volume of distribution so the concentrations in most tissues are similar to or lower than those in plasma. The urinary bladder tissue is part of the central compartment. The sweetener is cleared rapidly from plasma with an elimination half-life of about 30 minutes after an intravenous bolus dose. During chronic dietary administration, only slight diurnal fluctuations occur due to the absorption rate producing "flip-flop" kinetics. Intravenous infusion at different rates showed that clearance decreased at steady-state plasma concentrations above about 200 μg/mL, which corresponded to the levels in animals given about 5% in the diet.

Inhibition studies showed that renal tubular secretion was saturated at the high dietary intakes that are necessary to produce an increase in bladder tumors. One possible consequence of saturation of a major route of elimination—that a normally minor route of metabolism may become significant and have toxicological consequences (see 2,4,5-T, below)—was not found for saccharin. Renal tubular secretion is a general mechanism for the elimination of organic acids, and saturation of renal clearance of saccharin at high dietary concentrations is accompanied by decreased renal clearance of indican, a major urinary metabolite of tryptophan [121]. Subsequent studies demonstrated that the slow absorption of saccharin from the gut results in altered metabolism of essential nutrients within the gastrointestinal tract [121–123]. Thus, the dietary levels necessary throughout life (≥3%) to increase the incidence of bladder tumors in male rats produced profound perturbations of the physiology and biochemistry of the test animal, although these are not directly related to bladder tumor formation. Neither saturation of renal clearance [124] nor altered excretion of amino acid metabolites [125] has been found in humans following doses equivalent to the highest likely human intake.

### 2,4,5-Trichlorophenoxyacetic Acid (2,4,5-T)

Another example of saturation in toxicology is seen in the studies of Gehring and coworkers [31,126] on the herbicide 2,4,5-trichlorophenoxyacetic acid (2,4,5-T). This compound showed a higher toxicity in dogs than in rats because the former species exhibits a longer half-life, reduced renal and increased biliary elimination, and the presence of metabolites. In the rat, embryotoxicity is seen with doses of 100 mg kg$^{-1}$, and nonlinear kinetics were seen at similar doses due to saturation of elimination by renal tubular secretion. Metabolites for 2,4,5-T were detected in the urine of rats given saturating doses (i.e., 100 or 200 mg kg$^{-1}$).

## THE USE AND INTERPRETATION OF *IN VITRO* DATA

*In vitro* data can provide important qualitative insights into the metabolic fate of the compound, but care must be taken in their quantitative incorporation into risk assessment. A wide variety of *in vitro* systems of increasing cell integrity [127] can be used—for example, subcellular fractions (such as microsomes), cell homogenates, isolated cells and cell lines, and tissue slices. Each preparation has strengths and weaknesses that can be exploited to provide useful information [128,129].

### Strengths of *In Vitro* Systems

Microsomes comprise the smooth endoplasmic reticulum and its associated enzymes: cytochromes P450 and uridine diphosphate (UDP)–glucuronyltransferases. The rates of reaction *in vitro* are determined by the availability of appropriate cofactors; therefore, it is possible to study oxidation by the addition of NADPH in the absence of glucuronidation, which requires uridine diphosphate glucuronic acid (UDPGA), and *vice versa*. This allows the rates of P450-mediated oxidation to be studied directly, because part of the primary oxidation produced is not lost due to conjugation. An additional major value for such simple systems is that they can be used to generate metabolites for structural analysis.

More complex systems, such as isolated cells and tissue slices, provide a comprehensive picture of the metabolic fate of the compound. All enzyme systems are present, including cytoplasmic and mitochondrial enzymes, and the cell architecture can affect cell uptake and intracellular distribution of the chemical. Perhaps the most integrated *in vitro* system is the isolated perfused rat liver, which can give excellent correlations with *in vivo* clearance [130]; thus, these systems can provide information on the relative importance of alternative metabolic pathways.

A major advantage of *in vitro* systems is that they allow data to be generated on the potential metabolism in humans, without the need for *in vivo* exposure. This has been particularly valuable for carcinogens, where the generation of *in vivo* data would be unethical. Identification of the specific isoenzymes of cytochrome P450 is important in understanding the potential variability in metabolism within the human population because of the genetic polymorphism in some of the isoforms (e.g., CYP2D6). Such information can be generated by *in vitro* studies in three ways [131]:

- Comparisons of the rates of metabolism in stored (banked) liver preparations from individuals with characterized isoenzyme profiles
- Use of isoenzyme-specific inhibitors or inducers (in cell-intact preparations)
- Use of expression systems in which the DNA for specific isoenzymes is incorporated and expressed by a suitable host, such as a yeast or bacteria

The generation of *in vitro* data using human tissues allows characterization of species differences both qualitatively and quantitatively by the generation of the appropriate enzyme constants $V_{max}$ and $K_m$. Such data represent critical components of PBPK models [132] and for the prediction of *in vivo* clearance [133–136]. The discussion provided here should be sufficient to indicate the huge potential for *in vitro* studies and explain why these methods have been the basis for much of our understanding of pathways of xenobiotic metabolism.

### Precautions with *In Vitro* Systems

A number of limitations need to be remembered when considering *in vitro* data:

- Many studies give data on the extent of metabolism at a single high concentration *in vitro* and therefore represent $V_{max}$, which may be of limited relevance to *in vivo* concentrations. A full analysis of the enzyme kinetics is necessary to give both $V_{max}$ and $K_m$.
- The strengths of simple systems (such as the microsomes outlined above) will be weaknesses if the data are overinterpreted in relation to the fate *in vivo*.
- Changes in enzyme expression occur *in vitro*; for example, isolated cell lines show a different complement of cytochrome P450 activities than those in the same cells at isolation.
- Many human data are generated from stored liver samples obtained at postmortem. The *in vitro* enzyme activity could be affected by both *in vivo* aspects, such as drugs given in attempts at resuscitation or the presence of disease, and *ex vivo* aspects, such as the period between death and freezing and storage [137].
- *In vitro* data may still be misleading, even if all of these aspects are optimum. This occurs when the clearance of compound by an organ is blood flow limited rather than enzyme limited (see earlier). Under these conditions, both interspecies differences and inter-individual variability will reflect organ blood flow rather than $V_{max}$ and $K_m$. A good example of this is furan [138], for which the rate of oxidation *in vitro* would greatly exceed delivery via the liver blood flow. This problem can be avoided if the *in vitro* data are incorporated into a PBPK model that will allow for organ blood flow, partitioning between blood and tissue, and enzyme kinetics.

The increasing use of *in vitro* test systems facilitates a quantitative analysis of the dose–toxicity curve and may provide information on mechanisms of action [128,139]. The logical interpretation of such data with respect to human risk requires information on the steady-state concentrations of the active chemical species in the target organ and plasma of the test animal during chronic toxicity testing combined with knowledge of the toxicokinetics of the chemical in the test animal at toxic doses and in humans at the likely exposure level. It must be emphasized that large safety factors have been introduced to protect us from our own ignorance. The increased use of kinetic data, especially when combined with knowledge of the mechanism of toxicity, will allow the future use of potentially toxic chemicals to be based on scientific principles and understanding [7–10,93,140].

## QUESTIONS

1. A new chemical has been administered to rats and humans by both oral and intravenous routes. Basic toxicokinetic measurements (extrapolated to infinity) are given below:

| | Rat | Human |
|---|---|---|
| *Intravenous* | | |
| Dose (mg kg$^{-1}$) | 10 | 1 |
| AUC (µg mL$^{-1}$ min) | 2000 | 500 |
| Terminal slope (min$^{-1}$) | 0.0025 | 0.001 |
| % Dose excreted unchanged in urine | 1 | 15 |
| *Oral* | | |
| Dose (mg/kg) | 100 | 1 |
| AUC (µg mL$^{-1}$ min$^{-1}$) | 8000 | 490 |
| Terminal slope (min$^{-1}$) | 0.0025 | 0.001 |

   Calculate appropriate toxicokinetic parameters and suggest biochemical and physiological mechanisms that could explain the species difference.

2. The pharmaceutical company for which you work has synthesized a new anti-anxiety drug, which is a basic compound structurally related to the old drug debrisoquine. The parent drug, which is the active form, causes enzyme (cytochrome P450) induction and liver enlargement; the hydroxylated metabolite, which is formed on incubation of the drug with liver microsomes, is inactive. After an oral dose, 40% is excreted in the urine within 24 hr as the parent compound,

40% is in urine as a hydroxylated metabolite, and 20% is in feces as the parent drug. After an intravenous dose, 80% is in the urine as the parent drug, and 20% is in the urine as the metabolite. What advice would you give the company about the following issues:

a. Is the drug likely to be toxic after oral dosage?
b. Would the oral and intravenous doses associated with toxicity be the same?
c. What are the likely sources of variability in kinetics in young physically healthy adults (20 to 30 years old)?
d. Would the kinetics be different in the elderly (70 to 80 years old)?
e. How much would a 50% decrease in liver or kidney function affect the kinetics, and would the toxicity be increased or decreased?
f. How much would a 50% increase in liver or kidney function affect the kinetics, and would the toxicity be increased or decreased?
g. Should the pharmaceutical group develop a slow-release formulation, and would this be likely to affect the toxicity?

3. The company you work for has developed a novel opioid for the treatment of intractable pain. The drug is 20 times more potent than morphine in relation to both analgesia and respiratory depression when given to rats by intravenous injection, and binding studies show that it has a high and similar affinity for μ-receptors of rats and humans. Initial kinetic studies in humans after a single intravenous bolus dose of 10 mg gave the following data:

| Time After Dose (hr) | Plasma Concentration (ng/mL) |
|---|---|
| 0.5 | 367 |
| 1.0 | 336 |
| 2 | 283 |
| 4 | 200 |
| 6 | 141 |
| 10 | 71 |
| 24 | 6.3 |

The area under the plasma concentration–time curve (AUC) extrapolated to infinity was 2310 ng $mL^{-1}$ $hr^{-1}$. Urine was collected over the period of 2 to 4 hr after dosing and contained a total of 1.85 mg of the parent drug and 0.1 mg of a hydroxymetabolite; the plasma concentration of the parent drug at 3 hr was 238 ng/mL. After a single oral dose of 10 mg, the maximum plasma concentration occurred at 8 hr and was only 48 ng/mL; the blood concentration reached 6 ng/mL by 36 hr. The AUC to infinity was 1155 ng $mL^{-1}$ $hr^{-1}$. Plot the intravenous data on graph paper. Calculate the appropriate pharmacokinetic parameters to describe the elimination rate, clearance, distribution, and absorption of the drug. Describe the probable overall fate of the drug in the body (e.g., routes of elimination). Your research director needs the following advice:

a. What extra studies or data could support your description of the fate of the drug?
b. What route of administration should the company use for its first trials of clinical effect for pain relief?
c. How should the drug be administered to provide relief of chronic pain?

## Hints and Clues

*Question 1*

a. Calculate clearance (per kg body weight). Why is it different? (See next hint.)
b. Use urinary excretion data to think about pathways of elimination.
c. Use clearance and "terminal slope" ($k$ or β; we don't know) to calculate the apparent volume of distribution.
d. Use AUC data to calculate bioavailability.
e. Are terminal rates different after oral dosage? What would it mean if they were?
f. What are the likely causes of differences between species? Could clearance and bioavailability be interrelated? If so, how?
g. Would scaling to body surface area affect the calculations and conclusions? If so, how?

*Question 2*

a. Use urinary excretion data to interpret the potential for exposure (or not) of the liver to the parent compound. (Obviously, the dose will affect the response, but is toxicity possible?)
b. Use urinary excretion data to calculate bioavailability. What processes are giving rise to the low bioavailability?
c. Variability in adults—what are the routes of elimination? What is the relevance of debrisoquine?
e. Consider 50% changes in liver in relation to bioavailability and clearance, then consider changes in renal function similarly. Will kidney function affect bioavailability?
g. A slow-release formulation is necessary when a drug has a very short half-life (e.g., 3 to 4 hr or less). Information on the rate of elimination is provided in the question; what can you conclude about the half-life? (*Clue:* Could the half-life be 24 hr?)

*Question 3*

a. You can calculate clearance, apparent volume of distribution, and half-life from the intravenous data, but what route is important for elimination? (*Clue:* Use urine data to calculate renal clearance and compare with plasma clearance.)
b. The extra studies should relate to kinetics. (*Clue:* What studies would we normally have before giving the first dose to humans?)
c. What is happening with the oral data? What is the extent of absorption? Why are blood levels at 36 hr higher after oral dosage? (*Clue:* You can calculate the concentration at 36 hr after i.v. dosage using the exponential terms derived from the i.v. data.)
d. Phase I studies (initial human studies) are usually by the oral route. Is this likely to produce analgesia or side effects with this compound?
e. Chronic pain relief requires the maintenance of constant concentrations of the analgesic. Which route would be likely to give this profile? If oral dosage could not give effective plasma levels without unacceptable side-effects (such as constipation), how could you give the drug parenterally to provide similar constant concentrations?

## REFERENCES


1. Peck, C. C., Barr, W. H., Benet, L. Z., Collins, J., Desjardins, R. E., Furst, D. E., Harter, J. G., Levy, G., Ludden, T., Rodman, J. H., Sanathanan, L., Schentag, J. J., Shah, V. P., Sheiner, L. B., Skelly, J. P., Stanski, D. R., Temple, R. J., Viswanathan, C. T., Weissinger, J., and Yacobi, A. (1992): Opportunities for integration of pharmacokinetics, pharmacodynamics, and toxicokinetics in rational drug development. *Clin. Pharmacol. Ther.*, 51:465–473.
2. Abdel-Rahman, S. M., and Kauffman, R. E. (2004): The integration of pharmacokinetics and pharmacodynamics: understanding dose–response. *Ann. Rev. Pharmacol. Toxicol.*, 44:111–136.
3. WHO (1986): *Principles of Toxicokinetic Studies*, Vol. 57, Environmental Health Criteria. World Health Organization, Geneva, Switzerland.
4. Edler, L., Poirier, K., Dourson, M., Kleiner, J., Mileson, B., Nordmann, H., Renwick, A., Slob, W., Walton, K., and Würtzen. G. (2002): Mathematical modelling and quantitative methods. *Food Chem. Toxicol.*, 40:193–236.
5. WHO (1987): *Principles for the Safety Assessment of Food Additives and Contaminants in Food*, Vol. 70, Environmental Health Criteria. World Health Organization, Geneva, Switzerland.
6. Renwick, A. G. (1991): Safety factors and establishment of acceptable daily intakes. *Food Addit. Contamin.*, 8:135–150.
7. Renwick, A. G. (1993): Data derived safety factors for the evaluation of food additives and environmental contaminants. *Food Addit. Contamin.*, 10:275–305.
8. WHO (1994): *Assessing Human Health Risks of Chemicals: Derivation of Guidance Values for Health-Based Exposure Limits*, Vol. 170, Environmental Health Criteria. World Health Organization, Geneva, Switzerland.
9. WHO (1999): *Assessing Human Health Risks of Chemicals: Principles for the Assessment of Risk to Human Health from Exposure to Chemicals*, Vol. 210, Environmental Health Criteria. World Health Organization, Geneva, Switzerland.
10. WHO (2005): *Chemical-Specific Adjustment Factors for Interspecies Differences and Human Variability: Guidance Document for Use of Data in Dose/Concentration–Response Assessment*, Harmonization Document No. 2. World Health Organization, Geneva.
11. Spurling, N. W., and Carey, P. F. (1992): Dose selection for toxicity studies: a protocol for determining the maximum repeatable dose. *Human Exp. Toxicol.*, 11:449–457.
12. Anderson, M. W., Hoel, D. G., and Kaplan, N. L. (1980): A general scheme for the incorporation of pharmacokinetics in low-dose risk estimation for chemical carcinogens: example—vinyl chloride. *Toxicol. Appl. Pharmacol.*, 55: 154–161.
13. Campbell, D. B., and Ings, R. M. J. (1988): New approaches to the use of pharmacokinetics in toxicology and drug development. *Human Toxicol.*, 7:469–479.
14. Garattini, S. (1987): Toxic effects of chemicals: difficulties in extrapolating data from animals to man. *CRC Crit. Rev. Toxicol.*, 16:1–29.
15. Munro, A. M. (1990): Interspecies comparisons in toxicology: the utility and futility of plasma concentrations of the test compound. *Reg. Toxicol. Pharmacol.*, 12:137–160.
16. Cohn, V. H. (1971): Transmembrane movement of drug molecules. In: *Fundamentals of Drug Metabolism and Drug Disposition*, edited by B. N. LaDu, H. G. Mandel, and E. L. Way. Williams & Wilkins, Baltimore, MD, pp. 3–43.
17. Ochsenfahrt, H., and Winne, D. (1972): Solvent drag influence on the intestinal absorption of basic drugs. *Life Sci.* 11:1115–1122.
18. Smith, J. W. (1965): Observations on the flora of the alimentary tract of animals and factors affecting its composition. *J. Pathol. Bacteriol.*, 89:95–122.
19. Scheline, R. R. (1973): Metabolism of foreign compounds by gastrointestinal microorganisms. *Pharmacol. Rev.* 25:451–523.
20. Renwick, A. G. (1982): First pass metabolism within the lumen of the gastrointestinal tract. In *Presystemic Drug Elimination*, edited by C. F. George, D. G. Shand, and A. G. Renwick. Butterworth, Boston, pp. 3–28.
21. Caldwell, J., and Varwell Marsh, M. (1982): Metabolism of drugs by the gastrointestinal tract. In: *Presystemic Drug Elimination*,, edited by C. F. George, D. G. Shand, and A. G. Renwick. Butterworth, Boston, pp. 29–42.
22. Gibaldi, M., and Perrier, D. (1974): Route of administration and drug disposition. *Drug Metab. Rev.*, 3:185–199.
23. Gibaldi, M., and Perrier, D. (1982): *Pharmacokinetics*, 2nd ed. Marcel Dekker, New York.
24. Su, K. S. E., Campanale, K. M., and Gries, C. L. (1984): Nasal drug delivery system of a quaternary ammonium compound: clofilium tosylate. *J. Pharm. Sci.*, 73: 1251–1254.

25. Taylor, A. E., and Gaar, K. A. (1970): Estimation of equivalent pore radii of pulmonary capillary and alveolar membranes. *Am. J. Physiol.*, 218:1133.
26. Richards, R., Dickson, C. R., Renwick, A. G., Lewis, R. A., and Holgate, S. T. (1987): Absorption and disposition kinetics of cromolyn sodium and the influence of inhalation technique. *J. Pharmacol. Exp. Therap.*, 241:1028–1032.
27. McCarley, K. D., and Bunge, A. L. (2001): Pharmacokinetic models of dermal absorption. *J. Pharm. Sci.*, 90:1699–1719.
28. Guy, R. H., Hadgraft, J., and Maibach, H. I. (1984): Percutaneous absorption in man: a kinetic approach. *Toxicol. Appl. Pharmacol.*, 78:123–129.
29. Bard, P. (1956): In: *Medical Physiology*, 10th ed. Henry Kimpton, London, p. 221.
30. Butler, T. C. (1971): The distribution of drugs. In: *Fundamentals of Drug Metabolism and Drug Disposition*, edited by B. N. LaDu, H. G. Mandel, and E. L. Way. Williams & Wilkins, Baltimore, MD, pp. 44–62.
31. Gehring, P. J., and Young, D. J. (1978): Application of pharmacokinetic principles in practice. In: *Proc. of the First Int. Congress on Toxicology: Toxicology as a Predictive Science*, edited by G. L. Plaa and W. A. M. Duncan. Academic Press, New York, pp. 119–141.
32. Piafsky, K. M., Borga, O., Odar-Cedelof, I., Johansson, C., and Sjoqvist, F. (1978): Increased plasma protein binding of propranolol and chlorpromazine mediated by disease-induced elevations of plasma $\alpha_1$ acid glycoprotein. *N. Engl. J. Med.*, 299:1435–1439.
33. Anon. (1984): Clinical implications of drug protein binding. *Clin. Pharmacokinet.*, 9(Suppl. 1):1–104.
34. Davison, C. (1971): Protein binding. In: *Fundamentals of Drug Metabolism and Drug Disposition*, edited by B. N. LaDu, H. G. Mandel, and E. L. Way. Williams & Wilkins, Baltimore, MD, pp. 63–75.
35. Lee, G., Dallas, S., Hong, M., and Bendayan, R. (2001): Drug transporters in the central nervous system: brain barriers and brain parenchyma considerations. *Pharmacol. Rev.* 53:569–596.
36. de Boer, A. G., van der Sandt, I. C. J. and Gaillard, P. J. (2003): The role of drug transporters at the blood–brain barrier. *Annu. Rev. Pharmacol. Toxicol.*, 43:629–656.
37. Fromm, M. F. (2004): Importance of P-glycoprotein at blood–tissue barriers. *Trends Pharmacol. Sci.*, 25: 423–429.
38. Gottesman, M. M. (2002): Mechanisms of cancer drug resistance. *Annu. Rev. Med.*, 53:615–627.
39. Morris, M. M., Lee, H-J., and Predko, L. M. (2003): Gender differences in the membrane transport of endogenous and exogenous compounds. *Pharmacol. Rev.*, 55: 229–240.
40. Lee, W., and Kim, R. B. (2004): Transporters and renal drug elimination. *Annu. Rev. Pharmacol. Toxicol.*, 44:137–166.
41. Daly, A. K. (2003): Pharmacogenetics of the major polymorphic metabolizing enzymes. *Fundam. Clin. Pharmacol.*, 17:27–41.
42. Dorne, J. L. C. M., Walton, K., and Renwick, A. G. (2004): Human variability in xenobiotic metabolism and pathway-related uncertainty factors for chemical risk assessment: a review. *Food Chem. Toxicol.*, 43:203–216.
43. Pirmohamed, M., and Park, B. K. (2001): Genetic susceptibility to adverse drug reactions. *Trends Pharmacol. Sci.*, 22:298–305.
44. Park, B. K., Kitteringham, N. R., Powell, H., and Pirmohamed, M. (2000): Advances in molecular toxicology: towards understanding idiosyncratic drug toxicity. *Toxicology*, 153:39–60.
45. Zhang, L., Brett, C. M., and Giacomini, K. M. (1998): Role of organic cation transporters in drug absorption and elimination. *Annu. Rev. Pharmacol. Toxicol.*, 38:431–460.
46. Burckhardt, B. C., and Burckhardt, G. (2003): Transport of organic anions across the basolateral membrane of proximal tubule cells. *Rev. Physiol. Biochem. Pharmacol.*, 146:95–158.
47. Smith, R. L. (1973): *The Excretory Function of Bile: The Elimination of Drugs and Toxic Substances in Bile*. Chapman & Hall, London.
48. Hirom, P. C., Millburn, P., and Smith, R. L. (1976): Bile and urine as complementary pathways for the excretion of foreign organic compounds. *Xenobiotica*, 6:55–64.
49. Renwick, A. G. (1986): Gut bacteria and the enterohepatic circulation of foreign compounds. In: *Microbial Metabolism in the Digestive Tract*, edited by M. J. Hill. CRC Press, Boca Raton, FL, pp. 135–153.
50. Wilkinson, G. R. (1987): Clearance approaches in pharmacology. *Pharmacol. Rev.*, 39:1–47.
51. Rescigno, A., and Segre, B. (1966): *Drug and Tracer Kinetics*. Blaisdell, London.
52. Solomon, S. (1977): Developmental changes in nephron number. proximal tubular length and superficial glomerular filtration rate of rats. *J. Physiol. (Lond.)*, 272:573–589.
53. Weiner, I. M. (1967): Mechanisms of drug absorption and excretion: the renal excretion of drugs and related compounds. *Annu. Rev. Pharmacol.*, 7:39–56.
54. Garrett, E. R. (1978): Pharmacokinetics and clearance related to renal processes. *Int. J. Clin. Pharmacol.*, 16:155–172.
55. Rowland, M. et al. (1973): Clearance concepts in pharmacokinetics. *J. Pharmacokinet. Biopharm.*, 1:123–136.
56. Travis, C. C., Quillen, J. L., and Arms, A. D. (1990): Pharmacokinetics of benzene. *Toxicol. Appl. Pharmacol.*, 102:400–420.
57. Clewell, H. J., Gentry, P. R., Gearhart, J. M., Allen, B. C., and Andersen, M. E. (2001): Comparison of cancer risk estimates for vinyl chloride using animal and human data with a PBPK model. *Sci. Total Environ.*, 274:37–66.
58. Johanson, G., and Filser, J. G. (1996): PBPK model for butadiene metabolism to epoxides: quantitative species differences in metabolism, *Toxicology*, 113:40–47.
59. Jonsson, F., and Johanson, G. (2001): A Bayesian analysis of the influence of GSTT1 polymorphism on the cancer risk estimate for dichloromethane. *Toxicol. Appl. Pharmacol.*, 174:99–112.
60. Mann, S., Droz, P.-O., and Vahter, M. (1996): A physiologically based pharmacokinetic model for arsenic exposure. *Toxicol. Appl. Pharmacol.*, 140:471–486.
61. Pierce, C. H., Dills, R. L., Morgan, M. S., Nothstein, G. L., Shen, D. D., and Kalman, D. A. (1996): Inter-individual differences in $^2H_8$–toluene toxicokinetics assessed by a semi empirical physiologically based model. *Toxicol. Appl. Pharmacol.*, 139:49–61.

62. Wang, X, Santostefano, M. J., Evans, M. V., Richardson, V. M., Diliberto, J. J., and Birnbaum, L. S. (1997): Determination of parameters responsible for pharmacokinetic behavior of TCDD in female Sprague–Dawley rats. *Toxicol. Appl. Pharmacol.*, 147:151–168.
63. Bois, F. Y., Woodruff, T. J., and Spear, R. C. (1991): Comparison of three physiologically based pharmacokinetic models of benzene disposition. *Toxicol. Appl. Pharmacol.*, 110:79–88.
64. Wagner, J. G. (1975): Do you need a pharmacokinetic model and, if so, which one? *J. Pharmacokinet. Biopharm.*, 3:457–478.
65. Rowland, M., and Tozer, T. N. (1980): *Clinical Pharmacokinetics: Concepts and Applications*. Lea & Febiger, Philadelphia, PA.
66. Wagner, J. G. (1975): *Fundamentals of Clinical Pharmacokinetic*. Drug Intelligence Publications, Hamilton, IL.
67. Benet, L. Z., Levy, G., and Ferraiolo, B. L., Eds. (1984): *Pharmacokinetics: A Modern View*. Plenum Press, New York.
68. Rowland, M., and Tucker, G., Eds. (1986): *Pharmacokinetics: Theory and Methodology*. Pergamon Press, New York.
69. Gabrielsson, J., and Weiner, D. (1997): *Pharmacokinetic/Pharmacodynamic Data Analysis: Concepts and Applications*, 2nd ed. Swedish Pharmaceutical Society, Swedish Pharmaceutical Press, Sweden.
70. Riegelman, S., Loo, J. C. K., and Rowland, M. (1968): New method for calculating the intrinsic absorption rate of drugs. *J. Pharm. Sci.*, 57:918–928.
71. Chan, K. K. H., and Gibaldi, M. (1985): Assessment of drug absorption after oral administration. *J. Pharm. Sci.*, 74:388–393.
72. Park, B. K., Kitteringham, N. R., Pirmohamed, M., and Tucker, G. T. (1996): Relevance of induction of human drug-metabolising enzymes: pharmacological and toxicological implications. *Br. J. Clin. Pharmacol.*, 41:477–491.
73. Santostefano, M. J., Wang, X., Richardson, V. M., Ross, D. G., DeVito, M. J., and Birnbaum, L. S. (1998): A pharmacodynamic analysis of TCDD-induced cytochrome P450 gene expression in multiple tissues: dose- and time-dependent effects. *Toxicol. Appl. Pharmacol.*, 151:294–310.
74. Benet, L. Z., and Galeazzi, R. L. (1979): Noncompartmental determination of the steady state volume of distribution. *J. Pharm. Sci.*, 68:1071–1074.
75. Riegelman, S., and Collier, P. (1980): The application of statistical moment theory to the evaluation of *in vivo* dissolution time and absorption time. *J. Pharmacokinet. Biopharm.*, 8:509–534.
76. Waynforth, H. B. (1980): *Experimental and Surgical Technique in the Rat*. Academic Press, New York.
77. Cocchetto, D. M., and Bjornsson, T. D. (1983): Methods for vascular access and collection of body fluids from the laboratory rat. *J. Pharm. Sci.*, 72:465–492.
78. Sanzgiri, U. Y., Kim, H. J., Muralidhara, S., Dallas, C. E., and Bruckner, J. V. (1995): Effect of route and pattern of exposure on the pharmacokinetics and acute hepatotoxicity of carbon tetrachloride. *Toxicol. Appl. Pharmacol.*, 134:148–154.
79. Huang, C. H., Kimcera, R., Nassar, R. B., and Hussain, A. (1985): Mechanisms of nasal absorption of drugs. I. Physicochemical parameters influencing the rate of in situ nasal absorption of drugs in rats. *J. Pharm. Sci.*, 74: 608–611.
80. Langenberg, J. P., Spruit, H. E. T., van der Wiel, H. J., Trap, H. C., Helmich, R. B., Bergers, W. W. A., van Helden, H. P. M., and Benschop, H. P. (1998): Inhalation toxicokinetics of soman stereoisomers in the atropinized guinea pig with nose-only exposure to soman vapour. *Toxicol. Appl. Pharmacol.*, 151:79–87.
81. McKenna, M. J., Zempel, J. A., Madrid, E. O., and Gehring, P. J. (1978): The pharmacokinetics of [$^{14}$C]vinylidene chloride in rats following inhalation exposure. *Toxicol. Appl. Pharmaco1.*, 45:599–610.
82. Andersen, M. E., Gargas, M. L., Jones, R. A., and Jenkins, L. J. (1979): The use of inhalation techniques to assess the kinetic constants of 1,l-dichloroethylene metabolism. *Toxicol. Appl. Pharmacol.*, 47:395–409.
83. McDougal, J. N., Jepson, G. W., Clewell, H. J., and Andersen, M. E. (1985): Dermal absorption of dihalomethane vapours. *Toxicol. Appl. Pharmacol.*, 79:150–158.
84. Jepson, G. W., and McDougal, J. N. (1997): Physiologically based modeling of nonsteady state dermal absorption of halogenated methanes from an aqueous solution. *Toxicol. Appl. Pharmacol.*, 144:315–324.
85. Shen, S. K., Williams, S., Onkelinx, C., and Sunderman F. W. (1979): Use of implanted minipumps to study the effects of chelating drugs on renal $^{63}$Ni clearance in rats *Toxicol. Appl. Pharmacol.*, 51:209–217.
86. Withey, J. R. (1978): Pharmacokinetic principles, in *Proc of the First Int. Congress on Toxicology: Toxicology as a Predictive Science*, edited by G. L. Plaa and W. A. M Duncan. Academic Press, New York, pp. 97–117.
87. Bakar, S. K., and Niazi, S. (1983): Simple reliable method for chronic cannulation of the jugular vein for pharmacokinetic studies in rats. *J. Pharm. Sci.*, 72:1027–1029.
88. Light, H. G., Witmer, C., and Vars, H. M. (1959): Interruption of the enterohepatic circulation and its effects on rat bile, *Am. J. Physiol.*, 197:1330–1332.
89. Bachmann, K. (1989): Predicting toxicokinetic parameter in humans from kinetic data acquired in three small mammalian species. *J. Appl. Toxicol.*, 9:331–338.
90. Gerlowski, L. E., and Jain, R. K. (1983): Physiologically based pharmacokinetic modelling: principles and application. *Pharm. Sci.*, 72:1103–1127.
91. Grene-Lerouge, N. A. M., Bazin-Redureau, M. I., Debray M., and Scherrmann, J. M. G. (1996): Interspecies scaling of clearance and volume of distribution for digoxin-specific Fab. *Toxicol. Appl. Pharmacol.*, 138:84–89.
92. Mordenti, J. (1985): Pharmacokinetic scale up: accurate prediction of human pharmacokinetic profiles from animal data. *J. Pharm. Sci.*, 74:1097–1099.
93. Scheuplein, R. J., Shoaf, S. E., and Brown, R. N. (1990) Role of pharmacokinetics in safety evaluation and regulatory decisions. *Annu. Rev. Pharmacol. Toxicol.*, 30 197–218.
94. Bopp, B. A., Sonders, R. C., and Kesterson, J. W. (1986) Toxicological aspects of cyclamate and cyclohexylamine *CRC Crit. Rev. Toxicol.*, 16:213–306.

95. Roberts, A., and Renwick, A. G. (1989): The pharmacokinetics and tissue concentrations of cyclohexylamine in rats and mice. *Toxicol. Appl. Pharmacol.*, 98:230–242.
96. Buss, N. E., and Renwick, A. G. (1992): Blood pressure changes and sympathetic function in rats given cyclohexylamine by intravenous infusion. *Toxicol. Appl. Pharmacol.*, 115:211–215.
97. Buss, N. E., Renwick, A. G., Donaldson, K. M., and George, C. F. (1992): The metabolism of cyclamate to cyclohexylamate and its cardiovascular consequences in human volunteers. *Toxicol. Appl. Pharmacol.*, 115: 199–210.
98. JECFA (1982): *Toxicological Evaluation of Certain Food Additives*, WHO Food Additives Series No 17, WHO Technical Report Series No. 683, Twenty-Sixth Report of the Joint FAO/WHO Expert Committee on Food Additives, World Health Organization, Geneva, Switzerland (http://www.who.int/pcs/jecfa/JECFA_publications.htm).
99. SCF (1995): *Opinion on Cyclamic Acid and Its Salts*. Scientific Committee for Food of the European Commission (http://europa.eu.int/comm/food/fs/sc/scf/outcome_en.html).
100. Renwick, A. G., Thompson, J. P., O'Shaughnessy, M., and Walter, E. J. (2004): The metabolism of cyclamate to cyclohexylamine in humans during long-term administration. *Toxicol. Appl. Pharmacol.*, 196:367–380.
101. SCF (2000): *Opinion on Cyclamic Acid and Its Sodium and Calcium Salts*. Scientific Committee on Food of the European Commission (ehttp://europa.eu.int/comm/food/fs/sc/scf/outcome_en.html).
102. JECFA (2002): *Polychlorinated Dibenzodioxins, Polychlorinated Dibenzofurans, and Coplanar Polychlorinated Biphenyls*, WHO Food Additives Series No. 48. Joint FAO/WHO Expert Committee on Food Additives, pp. 451–658.
103. SCF (2000): *Opinion on the Risk Assessment of Dioxins and Dioxin-Like PCBs in Food*. Scientific Committee for Food of the European Commission.
104. SCF (2001): *Opinion on the Risk Assessment of Dioxins and Dioxin-Like PCBs in Food*. Scientific Committee for Food of the European Commission.
105. van den Berg, M., Birnbaum, L., Bosveld, B. T. C., Brunström, B., Cook, P., Feeley, M., Giesy, J. P., Hanberg, A. et al. (1998): Toxic equivalency factors (TEFs) for PCBs, PCDDs, PCDFs for humans and for wildlife. *Environ. Health Perspect.*, 106:775–792.
106. Gray, L. E., Ostby, J. S., and Kelce, W. R. (1997): A dose–response analysis of the reproductive effects of a single gestational dose of 2,3,7,8-tetrachlorodibenzo-*p*-dioxin in male Long Evans Hooded rat offspring. *Toxicol. Appl. Pharmacol.*, 146:11–20.
107. Gray, L. E., Jr., Wolf, C., Mann, P., and Ostby, J. S. (1997): *In utero* exposure to low doses of 2,3,7,8-tetrachlorodibenzo-*p*-dioxin alters reproductive development of female Long-Evans Hooded rat offspring. *Toxicol. Appl. Pharmacol.*, 146:237–244.
108. Mably, T. A., Bjerke, D. L., Moore, R. W. (1992): Gendron-Fitzpatrick, A., and Peterson, R. E., *In utero* and lactational exposure of male rats to 2,3,7,8-tetrachlorodibenzo-*p*-dioxin. 3. Effects on spermatogenesis and reproductive capability. *Toxicol. Appl. Pharmacol.*, 114:118–126.
109. Faqi, A. S., Dalsenter, P. R., Merker, H.-J., and Chahoud, I. (1998): Reproductive toxicity and tissue concentrations of low doses of 2,3,7,8-tetrachlorodibenzo-*p*-dioxin in male offspring rats exposed throughout pregnancy and lactation. *Toxicol. Appl. Pharmacol.*, 150: 383–392.
110. Ohsako, S., Miyabara, Y., Nishimura, N., Kurosawa, S., Sakaue, M., Ishimura, R., Sato, M., Takeda, K., Aoki, Y., Sone, H., Tohyama, C., and Yonemoto, J. (2001): Maternal exposure to a low dose of 2,3,7,8-tetrachlorodibenzo-*p*-dioxin (TCDD) suppressed the development of reproductive organs of male rats: dose-dependent increase of mRNA levels of 5α-reductase type 2 in contrast to decrease of androgen receptor in the pubertal ventral prostate. *Toxicol. Sci.*, 60:132–143.
111. Hurst, C. H., De Vito, M. J., Setzer, R. W., and Birnbaum, L. (2000): Acute administration of 2,3,7,8-tetrachlorodibenzo-*p*-dioxin (TCDD) in pregnant Long-Evans rats: association of measured tissue concentrations with developmental effects. *Toxicol. Sci.*, 53:411–420.
112. Hurst, C. H., DeVito, M. J., and Birnbaum, L. S. (2000): Tissue disposition of 2,3,7,8-tetrachlorodibenzo-*p*-dioxin (TCDD) in maternal and developing Long-Evans rats following subchronic exposure. *Toxicol. Sci.*, 57:275–283.
113. Mitchell, J. R., Jollow, D. J., Potter, W. Z., Davis, D. C., Gillette, J. R., and Brodie, B. B. (1973): Acetaminophen-induced hepatic necrosis. I. Role of drug metabolism. *J. Pharmacol. Exp. Ther.*, 187:185–194.
114. Jollow, D. J., Mitchell, J. R., Potter, W. Z., Davis, D. C., Gillette, J. R., and Brodie, B. B. (1973): Acetaminophen-induced necrosis II. Role of covalent binding *in vivo*. *J. Pharmacol. Exp. Ther.*, 187:195–202.
115. Potter, W. Z., Davis, D. C., Mitchell, J. R., Jollow, D. J., Gillette, J. R., and Brodie, B. B. (1973): Acetaminophen-induced hepatic necrosis. III. Cytochrome P-450–mediated covalent binding *in vitro*, *J. Pharmacol. Exp. Ther.*, 187:203–210
116. Mitchell, J. R., Jollow, D. J., Potter, W. Z., Gillette, J. R., and Brodie, B. B. (1973): Acetaminophen-induced hepatic necrosis. IV. Protective role of glutathione. *J. Pharmacol. Exp. Ther.*, 187:211–217.
117. Buss, N. E., Tembe, E. A., Prendergast, B. D., Renwick, A. G., and George, C. F. (1994): The teratogenic metabolites of vitamin A in women following supplements and liver. *Human Exp. Toxicol.*, 13:33–43.
118. Tembe, E. A., Honeywell, R., Buss, N. E., and Renwick, A. G. (1996): All-*trans*-retinoic acid in maternal plasma and teratogenicity in rats and rabbits. *Toxicol. Appl. Pharmacol.*, 141:456–472.
119. Schoenig, G. P., Goldenthal, E. I., Geil, R. G., Frith, C. H., Richter, W. R., and Carlborg, F. W. (1985): Evaluation of the dose response and *in utero* exposure to saccharin in the rat. *Food Chem. Toxicol.*, 23:475–490
120. Sweatman, T. W., and Renwick, A. G. (1980): The tissue distribution and pharmacokinetics of saccharin in the rat. *Toxicol. Appl. Pharmacol.*, 55:18–31.
121. Sims, J., and Renwick, A. G. (1983): The effects of saccharin on the metabolism of dietary tryptophan to indole, a known cocarcinogen for the urinary bladder of the rat. *Toxicol. Appl. Pharmacol.*, 67:132–151.

122. Sims, J., and Renwick, A. G. (1985): The microbial metabolism of tryptophan in rats fed a diet containing 7.5% saccharin in a two-generation protocol. *Food Chem. Toxicol.*, 23:437–444.
123. Lawrie, C. A., Renwick, A. G., and Sims, J. (1985): The urinary excretion of bacterial amino-acid metabolites by rats fed saccharin in the diet. *Food Chem. Toxicol.*, 23: 445–450.
124. Renwick, A. G. (1985): The disposition of saccharin in animals and man: a review. *Food Chem. Toxicol.*, 23: 429–435.
125. Roberts, A., and Renwick, A. G. (1985): The effect of saccharin on the microbial metabolism of tryptophan in man. *Food Chem. Toxicol.*, 23:451–455.
126. Sauerhoff, M. W., Braun, W. H., Blau, G. E., and Gehring, P. J. (1976): The dose-dependent pharmacokinetic profile of 2,4,5-trichlorophenoxy acetic acid following intravenous administration to rats. *Toxicol. Appl. Pharmacol.*, 36:491–501.
127. Eisenbrand, G., Pool-Zobel, B., Baker, V., Balls, M., Blaauboer, B. J., Boobis, A., Carere, A., Kevekordes, S., Lhuguenot, J.-C., Pieters, R., and Kleiner, J. (2002): Methods of *in vitro* toxicology. *Food Chem. Toxicol.*, 40:283–326.
128. Davila, J. C., Rodriguez, R. J., Melchert, R. B., and Acosta, D. (1998): Predictive value of *in vitro* model systems in toxicology. *Annu. Rev. Pharmacol. Toxicol.*, 38:63–96.
129. Lake, B. G. (1997): *In vitro* methods. In: *Comprehensive Toxicology*. Vol. 9. *Hepatic and Gastrointestinal Toxicology*, edited by R. S. McCuskey and D. L. Earnest. Pergamon Press, New York, pp. 233–246.
130. Damian, P., and Raabe, O. G. (1996): Toxicokinetic modeling of dose-dependent formate elimination in rats: *in vivo–in vitro* correlations using the perfused rat liver. *Toxicol. Appl. Pharmacol.*, 139:22–32.
131. Gonzalez, F. J., and Korzekwa, K. R. (1995): Cytochrome P450 expression systems. *Annu. Rev. Pharmacol. Toxicol.*, 35:369–390.
132. Ploemen, J.-P. H. T. M., Wormhoudt, L. W., Haenen, G. R. M. M., Oudshoorn, M. J., Commandeur, J. N. M., Vermeulen, N. P. E., de Waziers, I., Beaune, P. H., Watabe, T., and van Bladeren, P. J. (1997): The use of human *in vitro* metabolic parameters to explore the risk assessment of hazardous compounds: the case of ethylene dibromide. *Toxicol. Appl. Pharmacol.*, 143:56–69.
133. Houston, J. B., and Carlile, D. J. (1997): Prediction of hepatic clearance from microsomes, hepatocytes, and liver slices. *Drug Metab. Rev.*, 29:891–922.
134. Ito, K., Iwatsubo, T., Kanamitsu, S., Nakajima, Y., and Sugiyama, Y. (1998): Quantitative prediction of *in vivo* drug clearance and drug interactions from *in vitro* data on metabolism, together with binding and transport. *Annu. Rev. Pharmacol. Toxicol.*, 38:461–499.
135. Worboys, P. D., Bradbury, A., and Houston, J. B. (1994): Kinetics of drug metabolism in rat liver slices: rates of oxidation of ethoxycoumarin and tolbutamide, examples of high- and low-clearance compounds. *Drug Metab. Disp.*, 23:393–397.
136. Worboys, P. D., Bradbury, A., and Houston J. B. (1996): Kinetics of drug metabolism in rat liver slices. II. Comparison of clearance by liver slices and freshly isolated hepatocytes. *Drug Metab. Disp.*, 24:676–681.
137. Olinga, P, Merema, M., Hof, I. H., de Jong, K. P., Slooff, M. J. H., Meijer, D. K. F., and Groothuis, G. M. M. (1997): Effect of human liver source on the functionality of isolated hepatocytes and liver slices. *Drug Metab. Disp.*, 26: 5–11.
138. Kedderis, G. L., and Held, S. D. (1996): Prediction of furan pharmacokinetics from hepatocyte studies: comparison of bioactivation and hepatic dosimetry in rats, mice and humans, *Toxicol. Appl. Pharmacol.*, 140:124–130.
139. Flamm, W. G., and Lorentzen, R. J. (1987): The use of *in vitro* methods in safety evaluation, *In Vitro Toxicol.*, 1:1–3
140. Lipscomb, J. C. and Ohanian, E. V. (2007): *Toxicokinetics and Risk Assessment*. Informa Healthcare Press, New York

# 5 Physiologically Based Pharmacokinetic and Toxicokinetic Models

*Kannan Krishnan and Melvin E. Andersen*

## CONTENTS

## INTRODUCTION

The dose to target tissues is the net result of absorption, distribution, metabolism, and excretion (ADME) of substances in biota. Knowledge of target tissue dose or internal exposure provides a better basis than the administered dose for relating to toxic effects or potential risks associated with animal and human exposure. It is not always feasible or possible to measure the target tissue dose associated with various routes, doses, and exposure scenarios in each species of interest. In such cases, a scientifically sound alternative would involve the development and use of pharmacokinetic (PK) or toxicokinetic (TK) models—quantitative descriptions of the temporal change in the concentration of chemicals and/or their metabolites in biological matrices (e.g., blood, tissue, urine, alveolar air) of the exposed organism. The development of these models can be accomplished on the basis of either available *in vivo* kinetic data or knowledge of mechanistic determinants of ADME.

The data-based PK/TK modeling considers the organism as a single homogeneous compartment or as a multicompartmental system with elimination occurring in specific compartments of the model [128,281]. The number, behavior, and volumes of these hypothetical compartments are derived on the basis of the equation used to describe the data and not necessarily by the physiological characteristics of the organism in which the blood/tissue concentration data were collected. Whereas these data-based PK/TK models can be used for interpolation, the ability of these models to predict the behavior of chemicals outside the range of doses, exposure routes, and species used to generate data for constructing these models is limited. These various extrapolations, which are essential for conducting dose–response assessment of chemicals, can be performed more confidently with a physiologically based pharmacokinetic (PBPK), alternatively referred to as a physiologically based toxicokinetic (PBTK), modeling approach [5,31, 50,173]. This chapter presents the basic principles and methods of PBPK modeling as applied in toxicology and risk assessment.

Physiologically based pharmacokinetic modeling refers to the development of mathematical descriptions of the uptake and disposition of chemicals based on quantitative interrelations among the critical determinants of these processes. These determinants include partition coefficients, rates and affinities for biochemical reactions, and physiological characteristics of the species. The biological and mechanistic basis of the PBPK models enables them to be used, with limited animal experimentation, for extrapolation of the kinetic behavior of chemicals from high dose to low dose, from one exposure route to another, and from test animal species to people.

The development of PBPK models for volatile and gaseous anesthetics dates back to the research work of Haggard [136], who mathematically described the uptake of inhaled diethyl ether from a physiological perspective. Further developments in PBPK modeling with vapors were contributed by Kety [163] and Riggs [239], who provided mathematical descriptions of the kinetics of chemicals in the body based on parameters such as blood flow rates, tissue volumes, and chemical partitioning into tissues, as well as by Mapleson [185], who developed PBPK models for inert gases utilizing an electric analog. This electric analog approach was expanded by Fiserova-Bergerova [101,102] to describe the pharmacokinetic behavior of metabolized vapors and gases relying on numerical integration of mass balance equations. In the pharmaceutics area, PBPK modeling traces back to Teorell's pioneering work in the 1930s [262,263]. Beginning in the early 1960s, scientists trained in chemical engineering also developed PBPK models of various drugs, particularly antineoplastic agents such as methotrexate, 5-fluorouracil, and cisplatin [31,32,64,98]. Subsequently, the PBPK modeling approach has found extensive application in toxicology, particularly for conducting various extrapolations essential for the dose–response assessment of chemicals [182].

In this chapter, the process of PBPK model development will be described in terms of the following interconnected steps: representation, parameterization, simulation evaluation/validation, and refinement (Figure 5.1). *Model representation* involves the development of conceptual and mathematical descriptions of the relevant compartments of the animal as well as the exposure and metabolic pathways of the chemical. *Model parameterization* involves obtaining estimates or measures of the mechanistic determinants, such as physiological, physicochemical, and biochemical parameters, which are included in one or more of the PBPK model equations. *Model simulation* involves prediction of the uptake and disposition kinetics of a chemical for defined exposure scenarios using a numerical integration algorithm. The *model evaluation and validation* step involves comparison of the *a priori* predictions of the PBPK model with experimental data a

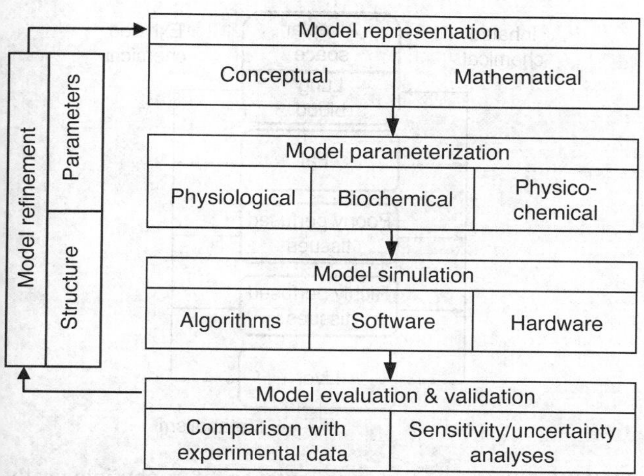


**FIGURE 5.1** Schematic of the steps involved in the development of physiologically based pharmacokinetic models.

well as conducting uncertainty, sensitivity, and variability analyses to refute, support, or refine the model description and parameters. After appropriate validation or refinement, PBPK models have been used successfully to conduct extrapolations of the PK behavior of chemicals from one exposure route or scenario to another, from high dose to low dose, and from one species to another [5,50].

The PBPK model development for a chemical is preceded by the definition of the problem, which in toxicology may often be related to the apparent complex nature of toxicity. Examples of such apparent complex toxic responses include nonlinearity in dose–response, sex and species differences in tissue response, differential response of tissues to chemical exposure, qualitatively or quantitatively different responses for the same cumulative dose administered by different routes or scenarios, etc. In these instances, PBPK models can be utilized to evaluate the pharmacokinetic basis of the apparent complex nature of toxicity induced by the chemical. One of the advantages of PBPK modeling, in fact, is that the accurate description of target tissue dose often resolves behavior that appears complex at the administered dose level [52,252]. The *a priori* definition of the objective of a modeling effort is critical because it will essentially guide the level of detail and complexity to be included in the model. Once the goals of the PBPK modeling effort are established, the model development process begins with the model representation step.

## MODEL REPRESENTATION

Model representation refers to the development of conceptual (i.e., diagrammatic) and mathematical descriptions of the relationships among system elements as they relate to the system response of interest (e.g., tissue dose).

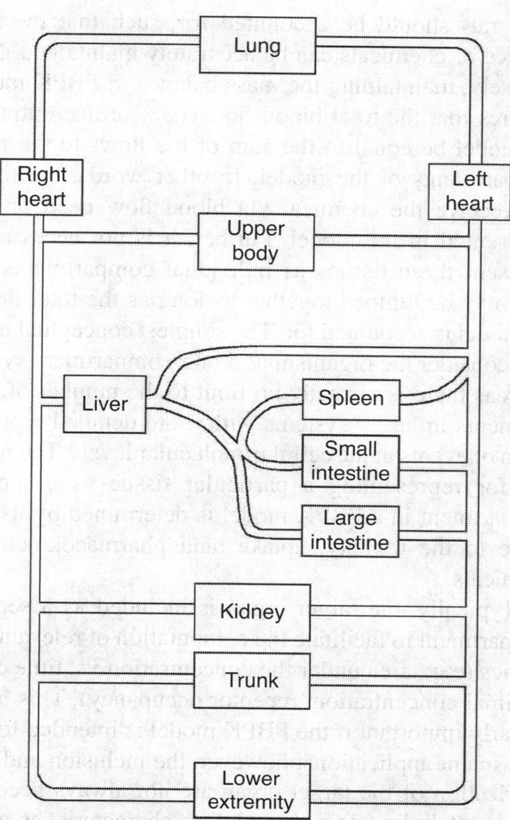


**FIGURE 5.2** Flow diagram for mammals. (Adapted from Bischoff, K.B., *Drinking Water Health*, 8, 36–64, 1987. With permission.)

### Conceptual Representation

This step involves the diagrammatic representation of the relevant anatomical and physiological features of the organism and the uptake and disposition pathways of the chemical. Whereas the organism is represented as a network of compartments, each of which is physically, physiologically, or biochemically distinct, the pathways of uptake and disposition are indicated by adding arrows to the appropriate compartments. The conceptual representation of the PBPK model for a chemical requires an understanding of the anatomical and physiological characteristics of the test animal species and the pathways of uptake and disposition of the chemical under study such that both the animal and the chemical can be represented adequately.

#### Representing the Animal

The diagrammatic representation of the organism (e.g., rat) should correspond to the real system; in other words, it should clearly show how the relevant individual compartments are placed and interconnected in the test organism (Figure 5.2). To represent the animal system, the organism as a whole in terms of its body weight (e.g., 250 g for a rat) should be accounted for, such that the mass balance of chemicals can be accurately maintained. More precisely, maintaining the mass balance in PBPK models requires that the total blood flow (i.e., cardiac output) in the model be equal to the sum of the flows to the tissue compartments of the model. In other words, the tissues that receive the chemical via blood flow need only be represented in the model. Further, it is not necessary to represent these tissues as individual compartments, and they may be lumped together as long as the total flow in the model is accounted for. The simplest conceptual model may consider the organism as a one-compartment system, whereas there is virtually no limit to the number of compartments in larger systems with more detailed representation of events at the cellular/molecular levels. The necessity for representing a particular tissue as a separate compartment in a PBPK model is determined by its relevance to the toxicity, uptake, and pharmacokinetics of chemicals.

Typically, the target organ is included as a separate compartment to facilitate the computation of relevant dose metrics (e.g., area under the concentration vs. time curve, maximal concentration, receptor occupancy). This is particularly important if the PBPK model is intended for risk assessment applications; however, the inclusion and characterization of the target organ are not always necessary if the metabolic and accumulation characteristics are no different from other organs (e.g., brain vs. other richly perfused tissues). When the target organ has not been identified or represented as a separate compartment, the investigator should have an idea of the compartments or variables that could be used as a surrogate of target organ exposure (e.g., blood concentration).

Next, the portals of entry or uptake of chemicals should be considered for representation as separate compartments. In this regard, the lungs, skin, and gastrointestinal tract are generally represented as separate compartments in PBPK models to describe the rate of chemical uptake in the body through specific exposure routes. In certain situations (e.g., oral absorption), the portal of entry is not represented as a separate compartment; instead, the rate of the amount absorbed (i.e., input to the system) is calculated.

The necessity for representing other tissues as separate compartments is usually evaluated on the basis of their unique (1) chemical, (2) biochemical, and (3) physiological characteristics that might contribute significantly to the uptake and disposition of the chemical being modeled. The *chemical composition* of tissues in the present context refers primarily to the water and nonpolar lipid contents. These tissue constituents are nonreactive but account for the differential solubility of a chemical among various tissues. In this respect, the adipose tissue is represented as a separate compartment in many PBPK models (Figure 5.3, fat) because of its ability to sequester lipophilic chemicals during exposure and release them after the cessation of exposure. In the case of a hydrophilic chemical, adipose tissue may be combined with the rest of the body because it may not show any particular kinetic behavior that is unique and different from the rest of the body. If other chemical components of the tissue (e.g., chloride levels or pH) are critical determinants of disposition, then individual or groups of tissue compartments defined with this kind of information should be included in the model as necessary [207].

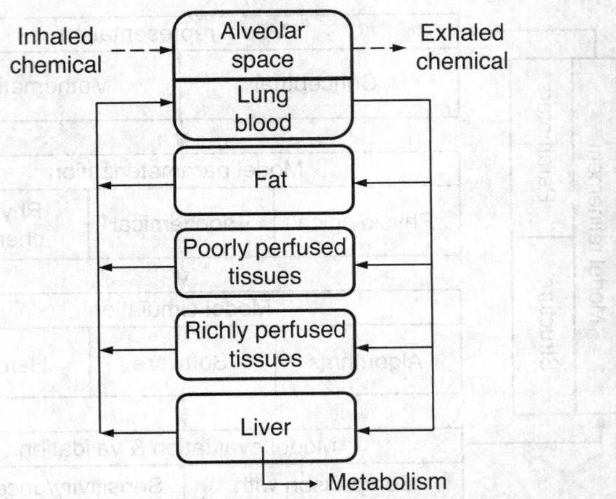


**FIGURE 5.3** Conceptual representation of a physiologically based pharmacokinetic model for styrene. (Adapted from Ramsey, J.C. and Andersen, M.E., *Toxicol. Appl. Pharmacol.*, 73, 159–175, 1984. With permission.)

The *biochemical characteristics*, in the current context, refer primarily to the binding and metabolizing capacities of the tissues. These properties account primarily for the removal of chemicals from the circulating blood by mechanisms other than chemical partitioning; for example, liver is often represented as a compartment in the PBPK models because of its central role in the metabolism of many organic chemicals. Representation of other tissues as separate compartments may be required according to the extent of macromolecular binding and the expression of specific enzyme activities of relevance to the metabolism of the chemical being modeled (e.g., P450 and glutathione *S*-transferase in lung or kidney, epoxide hydrolase in testis, or myeloperoxidase in bone marrow).

The *physiological characteristics* refer to breathing rate, cardiac output, glomerular filtration rate, tissue blood flow rates, etc. These characteristics essentially determine the biodisposition of chemicals. The tissue compartments possessing these properties (i.e., lung, heart, and kidney) or a quantitative description of these physiological processes should be included in the PBPK model. Respiration, urinary excretion, and blood circulation often are represented as quantitative descriptions of the processes themselves. Depending on the proposed use of the model, the tissues involved may be represented individually and characterized for particular aspects; for example, lung is represented as both an uptake and a metabolizing tissue in PBPK models

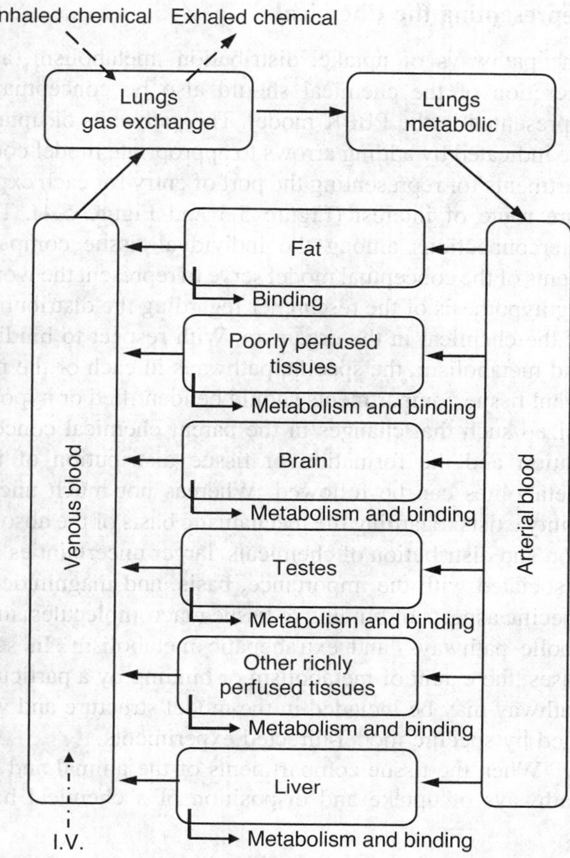


FIGURE 5.4 Conceptual representation of a physiologically based pharmacokinetic model for ethylene oxide. (Adapted from Krishnan, K. et al., *Toxicol. Indust. Health*, 8, 121–140, 1992. With permission.)

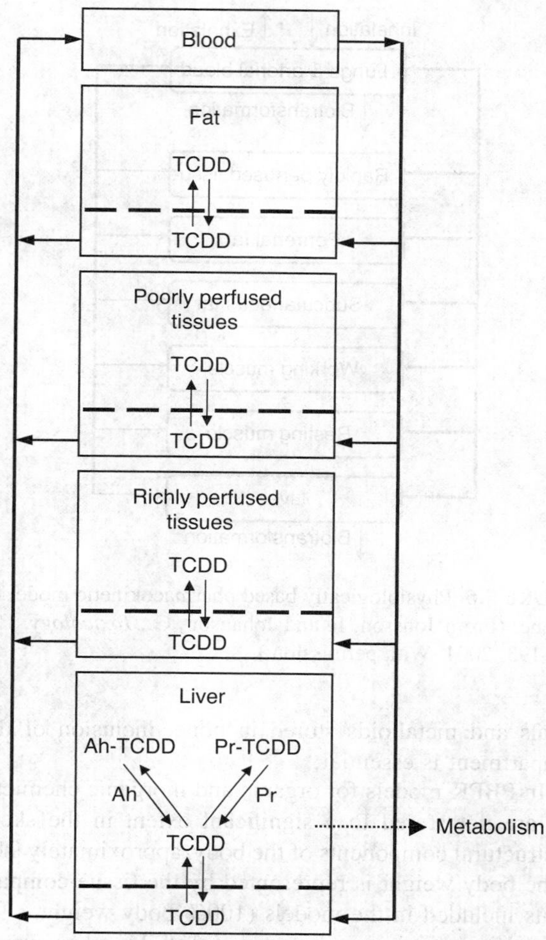


FIGURE 5.5 Conceptual representation of a physiologically based pharmacokinetic model for 2,3,7,8-tetrachlorodibenzo-*p*-dioxin (TCDD). Pr, protein; Ah, Ah receptor. (Adapted from Andersen, M.E. et al., *Risk Anal.*, 13, 25–36, 1993. With permission.)

for certain volatile organics (Figure 5.4), whereas it is not characterized separately in the case of nonvolatile organics that are neither eliminated by exhalation nor metabolized significantly by this tissue (Figure 5.5).

In principle, if the characteristic time constants of tissue disposition (i.e., the product of partition coefficient and volume divided by blood flow rate) are similar for various tissues, they can be lumped to form a single tissue group. In other words, when the critical determinants of pharmacokinetics do not vary quantitatively among several tissues, the time course of the chemical in these tissues will be similar. That is why fat depots such as perirenal, epididymal, and omental fat are frequently grouped and represented as a single "fat" compartment (Figure 5.3, Figure 5.4, and Figure 5.5). If necessary, a "fat" compartment may be subdivided into two or more groups according to the perfusion rates (e.g., perirenal and subcutaneous adipose tissues) (Figure 5.6). Another example of this kind involves tissues such as adrenal, kidney, thyroid, brain, lung, heart, testis, and hepatoportal system, which often are pooled into one compartment and referred to as "richly or rapidly perfused tissues" (Figure 5.3 and Figure 5.5). When the contents of relevant metabolizing enzymes quantitatively differ among the richly perfused tissues, the individual organs are represented as separate compartments even though the blood flow rate and solubility characteristics are somewhat similar (Figure 5.4). Tissues with poor blood perfusion characteristics (muscle, skin) are frequently grouped as "slowly or poorly perfused tissues." Muscle compartment may be split into two to provide better description of the time course of chemicals, particularly during physical exercise (Figure 5.6 and Figure 5.7). Various levels of splitting or lumping of tissues in PBPK models may be performed depending on the chemical, and a sensitivity analysis is likely to be useful in this regard [33].

Because the skeletal and structural components of the body have only a negligible perfusion and do not play a significant role in the disposition of many organic chemicals, they have not been included in the PBPK model descriptions for these chemicals. When describing certain

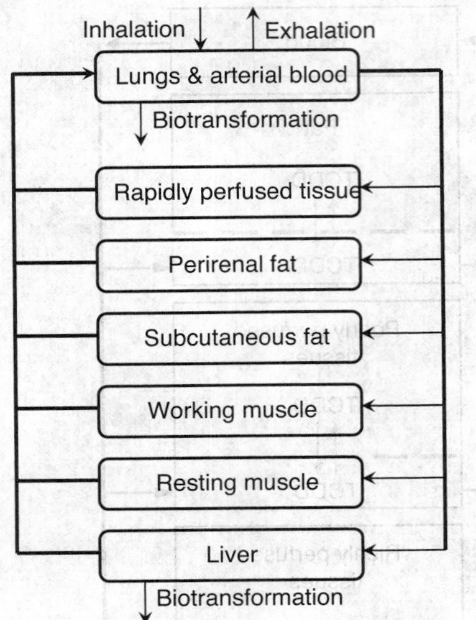


**FIGURE 5.6** Physiologically based pharmacokinetic model for toluene. (From Jonsson, F. and Johanson, G., *Toxicology*, 157, 177–193, 2001. With permission.)

metals and metalloids stored in bone, inclusion of this compartment is essential.

In PBPK models for organic and inorganic chemicals that are not stored to a significant extent in the skeletal/structural components of the body, approximately 91% of the body weight is represented by the tissue compartments included in the models (100% body weight – 9% skeletal/structural component weight). In other words, 91% of the body weight is fractionated into the four, five, or nine compartments included in the models presented in Figure 5.3, Figure 5.4, and Figure 5.5. The compartment volumes in some models may only correspond to those of neutral lipids, if they are the sole determinants of the tissue accumulation of chemicals (Figure 5.7). In PBPK models, blood is not routinely characterized as a separate compartment (Figure 5.3 vs. Figure 5.4) even though blood concentrations are calculated. When blood is not described as a separate compartment, its total volume is apportioned among the tissues implicitly or explicitly.

Physiologically based pharmacokinetic models are mechanistically based and more detailed than the classical data-based pharmacokinetic models. They still represent a significant simplification of the true complexities of the biological systems. Model complexity and the number of compartments should not be equated with accuracy and usefulness of the model description; often, model complexity yields a multitude of parameters to be estimated and greater uncertainty of the model description. Parsimony in PBPK modeling, on the other hand, refers to the choice of a model structure that has minimal but necessary elements that together adequately describe the pharmacokinetics of a chemical.

## Representing the Chemical

The pathways of uptake, distribution, metabolism, and excretion of the chemical should also be conceptually represented in the PBPK model. The pathways of uptake are indicated by adding arrows to appropriate model compartments for representing the port of entry for each exposure route of interest (Figure 5.3 and Figure 5.4). The interconnections among the individual tissue compartments of the conceptual model serve to represent the working hypothesis of the researcher regarding the distribution of the chemical in the organism. With respect to binding and metabolism, the specific pathways in each of the relevant tissue compartments should be identified or hypothesized such that changes in the parent chemical concentration and the formation or tissue distribution of the metabolites can be followed. Whereas not much uncertainty exists regarding the mechanistic basis of the absorption and distribution of chemicals, larger uncertainties are associated with the importance, basis, and magnitude of specific aspects of binding to tissue macromolecules, metabolic pathways, and extrahepatic metabolism. In such cases, the extent of metabolism or binding by a particular pathway may be included in the model structure and verified by specific model-directed experiments.

When the tissue compartments of the animal and the pathways of uptake and disposition of a chemical have

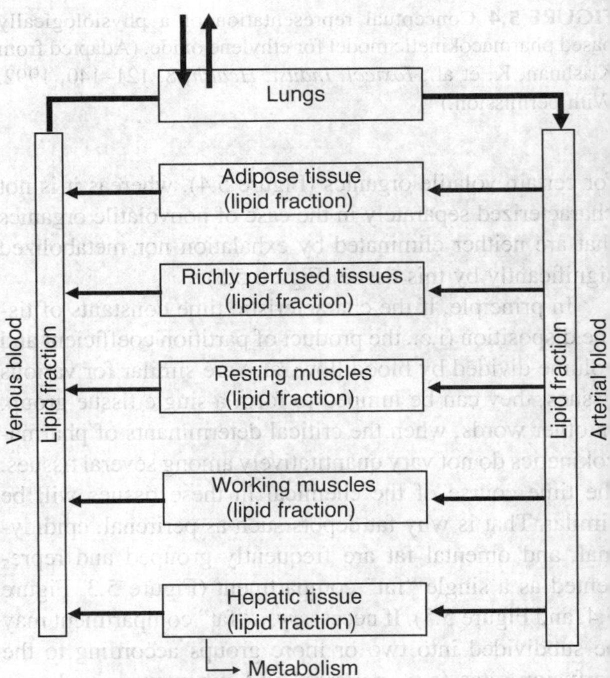


**FIGURE 5.7** Representation of a physiological pharmacokinetic model for simulating inhalation pharmacokinetics of highly lipophilic volatile organic chemicals. (From Emond, C. and Krishnan, K. *Toxicol. Mech. Methods*, 16(8), 395–403, 2006. With permission

been identified or hypothesized and conceptually represented, mathematical descriptions of pharmacokinetic processes are developed.

## Mathematical Representation

The mathematical representation of a PBPK model requires a rudimentary knowledge of calculus. A brief review of basic mathematics and differential calculus required for PK/TK modeling has been provided by O'Flaherty [206]. In PBPK modeling, each tissue compartment is described with a mass balance differential equation (MBDE) that consists of a series of clearance terms. Clearance concepts have been conventionally applied to describe the function of organs, such as kidney and liver. Clearance of compounds from the blood by these organs, primarily by metabolism or filtration, is expressed in units of volume per time. The interpretation here is that clearance is the volumetric flow of blood from which all chemical would have to be removed to account for the loss of chemical from the blood. In many ways, PBPK models might just as appropriately be referred to as physiological clearance models for the kinetics of compounds.These clearances may be related to flows, metabolism, excretion, or secretion. A typical model-based design and evaluation for a tissue in a PBPK model includes clearance terms for input from arterial blood, metabolism by saturable or linear pathways, excretion, and efflux of compound in venous drainage to the mixed venous pool as follows:

$$\frac{dA_t}{dt} = \text{Chemical in} - \text{chemical out} - \text{excretion} - \text{metabolism} \quad (5.1)$$

The above equation can be written on the basis of clearance terms as follows:

$$\frac{dA_t}{dt} = CL_uC_a - CL_eC_a - CL_mC_a - CL_fC_a \quad (5.2)$$

The symbols and abbreviations used in Equation 5.1 through Equation 5.50 are presented in Table 5.1.

In the above equation, the unit for each of the clearance terms is flow per time (i.e., L/hr or mL/min). In each one of these terms, clearance multiplied by the chemical concentration gives rise to a flux (i.e., mass per unit time) either into or out of the compartment. With each term, we have the same idea for clearance (i.e., the volume of the compartment cleared of chemical per unit time).

### Descriptions of Tissue Uptake

Of the various clearance processes described above, the basic process that applies to all PBPK model compartments is chemical uptake (i.e., inter- and intra-tissue transfers of chemicals). The tissue uptake of a chemical from the blood is described according to Fick's law of simple diffusion, which states that the flux of a chemical is proportional to its concentration gradient:

$$V_t\frac{dC_t}{dt} = k\Delta C \quad (5.3)$$

For high-molecular-weight compounds, diffusion is often the rate-limiting process; therefore, their uptake through the tissue subcompartments (Figure 5.8) must be considered. This requires that tissue blood and cellular matrix be described separately.

The rate of change in the amount of chemical in the cellular matrix is equal to the product of the diffusion rate constant and the net flux from tissue blood:

$$V_{cm}\frac{dC_{cm}}{dt} = PA_t\left(C_{vt} - \frac{C_t}{P_t}\right) \quad (5.4)$$

The rate of change in the tissue blood subcompartment equals the sum of the net retention from blood flow plus the net gain from cellular matrix:

$$V_{tb}\frac{dC_{tb}}{dt} = Q_t\left(C_a - C_{vt}\right) - PA_t\left(C_{vt} - \frac{C_t}{P_t}\right) \quad (5.5)$$

In diffusion-limited tissue uptake descriptions, each tissue compartment has a specified cellular matrix volume and tissue blood volume. The movement of a chemical from tissue blood into cellular matrix is described as being proportional to the permeation coefficient–surface area cross-product ($PA_t$) for the tissue ($t$). Tissue uptake is diffusion limited when $PA_t < Q_t$.

If the diffusion of a chemical from tissue blood to cellular matrix is slow with respect to total tissue blood flow, both equations are necessary. On the other hand, if tissue blood flow (i.e., perfusion) is slow with respect to diffusion, tissues are described as homogeneous, well-mixed compartments such that the rate of change in the amount of chemical in the tissue is described with a single equation for the whole tissue mass (cellular matrix plus tissue blood; that is, Equation 5.4 plus Equation 5.5) as follows:

$$V_t\frac{dC_t}{dt} = Q_t\left(C_a - C_{vt}\right) \quad (5.6)$$

In the perfusion-limited tissue descriptions, the transfer constant is the rate of blood flow to the compartment, and the effluent venous blood concentration ($C_{vt}$) is assumed to be in equilibrium with the tissue concentration ($C_t$) as specified by the tissue–blood partition coefficient ($P_t$) such that $C_{vt} = C_t/P_t$.

**TABLE 5.1**
**Symbols and Abbreviations Used in the Mathematical Representations of PBPK Models Described in This Chapter**

| Symbol or Abbreviation | | Description |
|---|---|---|
| *A* | | Amount (mg) |
| | *Subscripts* | |
| | *f* | Fat |
| | *r* | Richly perfused tissues |
| | *s* | Poorly perfused tissues |
| | *l* | Liver |
| | *met* | Metabolized |
| | *sk* | Skin |
| | *stom* | Remaining in the stomach |
| | *t* | Tissue "*t*" |
| | *v* | Mixed venous blood |
| *C* | | Concentration (e.g., mg/L, mmol/L, mg/mL) |
| | *Subscripts* | |
| | *a* | Arterial blood |
| | *a,ss* | Arterial blood at steady state |
| | *air* | Air contacting skin |
| | *alv* | End-alveolar air |
| | *cf* | Cofactor in tissue "*t*" |
| | *cm* | Cellular matrix |
| | *f* | Fat |
| | *f,ss* | Fat at steady state |
| | *inh* | Inhaled air |
| | *l* | Liver |
| | *l,ss* | Liver at steady state |
| | *r* | Richly perfused tissues |
| | *r,ss* | Richly perfused tissues at steady state |
| | *s* | Poorly perfused tissues |
| | *s,ss* | Poorly perfused tissues at steady state |
| | *sk* | Skin |
| | *t* | Tissue "*t*" |
| | *tb* | Tissue blood |
| | *v* | Mixed venous blood |
| | *vf* | Venous blood leaving fat |
| | *vl* | Venous blood leaving liver |
| | *vr* | Venous blood leaving richly perfused tissue |
| | *vs* | Venous blood leaving poorly perfused tissue |
| | *vt* | Venous blood leaving tissue "*t*" |
| *CL* | | Clearance ($L.hr^{-1}$) |
| | *Subscripts* | |
| | *e* | Efflux clearance |
| | *f* | Functional clearance |
| | *h* | Hepatic clearance |
| | *int* | Intrinsic clearance |
| | *m* | Metabolic clearance |
| | *u* | Uptake clearance |
| *E* | | Hepatic extraction ratio |
| *IV* | | Intravenous dose ($mg\ hr^{-1}$) |
| *k* | | Transfer constant ($L.hr^{-1}$) |

**TABLE 5.1 (cont.)**
**Symbols and Abbreviations Used in the Mathematical Representations of PBPK Models Described in This Chapter**

| Symbol or Abbreviation | | Description |
|---|---|---|
| $K$ | | Binding, metabolism or absorption rate constant |
| | *Subscripts* | |
| | $f$ | First-order metabolism ($hr^{-1}$) |
| | $m$ | Michaelis constant (mg/L) |
| | $o$ | Oral absorption ($hr^{-1}$) |
| | $p$ | Skin permeability coefficient (cm $hr^{-1}$) |
| | $s$ | Second-order metabolism (L $mg^{-1}$ $hr^{-1}$) |
| $P$ | | Partition coefficient |
| | *Subscripts* | |
| | $b$ | Blood–air |
| | $f$ | Fat–blood |
| | $l$ | Liver–blood |
| | $r$ | Richly perfused tissue–blood |
| | $s$ | Slowly or poorly perfused tissue–blood |
| | $s{:}a$ | Skin–air |
| | $s{:}b$ | Skin–blood |
| | $t$ | Tissue–blood |
| $PA1$ | | Permeation area cross product for tissue "$t$" ($L.hr^{-1}$) |
| $Q$ | | Flow rate ($L.hr^{-1}$) |
| | *Subscripts* | |
| | $c$ | Cardiac output |
| | $f$ | Blood flow to fat |
| | $fc$ | Blood flow to fat (fraction of cardiac output) |
| | $k{:}b$ | Tissue–blood |
| | $l$ | Blood flow to liver |
| | $lc$ | Blood flow to liver (fraction of cardiac output) |
| | $p$ | Alveolar ventilation |
| | $r$ | Blood flow to richly perfused tissue |
| | $rc$ | Blood flow to richly perfused tissue (fraction of cardiac output) |
| | $s$ | Blood flow to poorly perfused tissue |
| | $sc$ | Blood flow to poorly perfused tissue (fraction of cardiac output) |
| | $sk$ | Blood flow to skin |
| | $t$ | Blood flow to tissue "$t$" |
| $S$ | | Exposed skin surface area ($cm^2$) |
| $t$ | | Elapsed time (hr) |
| $V$ | | Volume (L) |
| | *Subscripts* | |
| | $cm$ | Cellular matrix in a tissue |
| | $t$ | Tissue "$t$" |
| | $tb$ | Tissue blood |
| $V_{max}$ | | Maximal velocity of enzymatic reaction (mg $hr^{-1}$) |

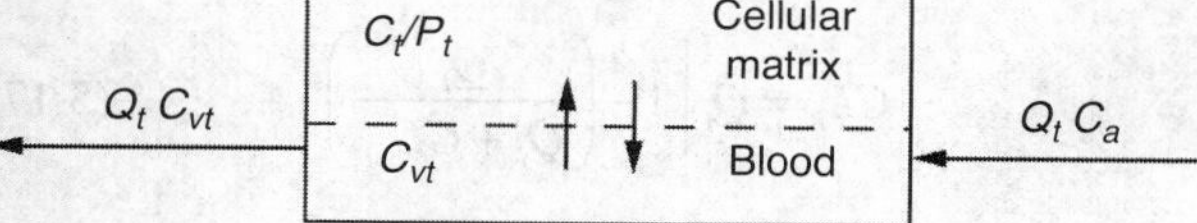


IGURE 5.8 Schematic of a tissue compartment. $Q_t$ is tissue lood flow rate, $C_a$ is arterial blood concentration, $C_{vt}$ is the oncentration of the chemical in the venous blood leaving tissue, $C_t$ is the tissue concentration, and $P_t$ is the tissue-blood partition oefficient.

## Description of Metabolism and Elimination

In the case of metabolizing and eliminating tissues, additional terms are included to represent chemical loss due to specific biochemical processes. The rate of the amount of chemical consumed by macromolecular binding has been calculated either as a second-order reaction or by using equations based on reversible equilibrium relationships. The rate of the amount metabolized can be described as a first-order, second-order, or saturable process as follows:

$$\frac{dA_{met}}{dt} = K_f C_{vt} V_t \quad (5.7)$$

$$\frac{dA_{met}}{dt} = K_s C_{vt} V_t C_{cf} \quad (5.8)$$

$$\frac{dA_{met}}{dt} = \frac{V_{max} C_{vt}}{K_m + C_{vt}} \quad (5.9)$$

Conjugation reactions have often been described as a second-order process (Equation 5.8) with respect to the concentration of the cofactor and the chemical [87,171]. Alternatively, descriptions based on a ping-pong mechanism have also been used successfully [69].

The saturable (Equation 5.9) or first-order (Equation 5.7) metabolism descriptions presented above use the venous blood concentration (i.e., free concentration). When these descriptions are integrated within Equation 5.6, the blood flow limitation (i.e., perfusion limitation) of metabolism is automatically accounted for. The role of blood flow may be explicitly accounted for by using an equation of the following type [154,227,283]:

$$\frac{dA_{met}}{dt} = CL_h \times C_a \quad (5.10)$$

where:

$$CL_h = \frac{Q_l CL_{int}}{Q_l + CL_{int}} \quad (5.11)$$

The derivation of the above relationship is presented in the following paragraphs.

Consider a typical liver compartment for which the rate of input equals $Q_t C_a$, the rate of output via venous blood equals $Q_t C_{vl}$, and the rate of removal by metabolism equals $Cl_{int} C_{vl}$ (Figure 5.9). In this case, hepatic clearance equals the volumetric flow of blood from which all chemical would have to be removed or extracted to account for the loss of chemical. Therefore,

$$Cl_h = \text{Blood flow to liver} \times \text{extraction ratio}$$

*Extraction* here refers to the difference between the input and output (i.e., $Q_l C_a - Q_l C_{vl}$). The extraction ratio ($E$) then is the fraction of the input removed by a tissue. The hepatic extraction ratio can be calculated as (input – output)/input, or $(Q_l C_a - Q_l C_{vl})/Q_l C_a$.

Simplifying the above relationship,

$$E = \frac{C_a - C_{vl}}{C_a} \quad (5.12)$$

For example, if the $C_a$ = 1 mg/L and $C_{vl}$ = 0.5 mg/L, then the $E$ value is 0.5. Theoretically, the $E$ value can only be between 0 and 1. When $C_{vl} = C_a$, the $E$ = 0 implies no uptake or removal by the tissue. On the other hand, if $C_{vl}$ is close to zero, then the $E$ value would be about 1, indicating flow-limited clearance in the tissue.

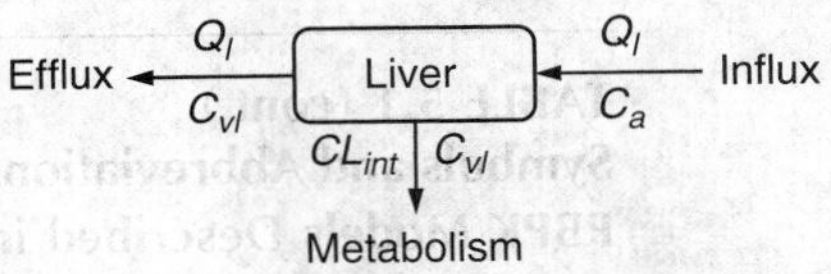

**FIGURE 5.9** Schematic of liver compartment. $Q_l$ is liver blood flow rate, $C_a$ is arterial blood concentration, $C_{vl}$ is the concentration of the chemical in the venous blood leaving liver (i.e., free concentration), and $CL_{int}$ denotes intrinsic clearance.

Combining Equation 5.11 and Equation 5.12, we get:

$$CL_h = \frac{Q_l(C_a - C_{vl})}{C_a} \quad (5.13)$$

To solve the above equation, $C_{vl}$ may be derived on the basis of the following mass balance equation:

$$\text{Input} = \text{Output} + \text{removal}$$

Mathematically,

$$Q_l C_a = Q_l C_{vl} + CL_{int} C_{vl} \quad (5.14)$$

where $Q_l C_{vl}$ represents output and $Cl_{int} C_{vl}$ represents the rate of removal by metabolism (Figure 5.9). Rearranging the above equation,

$$C_{vl} = \frac{Q_l C_a}{Q_l + CL_{int}} \quad (5.15)$$

Inserting Equation 5.(15) in Equation 5.(13), we obtain:

$$CL_h = Q_l \left[ \frac{C_a - \left( \frac{Q_l C_a}{Q_l + CL_{int}} \right)}{C_a} \right] \quad (5.16)$$

Simplifying,

$$CL_h = Q_l \left[ 1 - \left( \frac{Q_l}{Q_l + CL_{int}} \right) \right] \quad (5.17)$$

Rewriting,

$$CL_h = Q_l \left( \frac{Q_l + CL_{int} - Q_l}{Q_l + CL_{int}} \right) \quad (5.18)$$

which upon simplification becomes:

$$CL_h = \left(\frac{Q_l CL_{int}}{Q_l + CL_{int}}\right) \tag{5.19}$$

In other terms [175],

$$CL_h = Q_l\left(\frac{CL_{int}}{Q_l + CL_{int}}\right) \tag{5.20}$$

where:

$$E = \frac{CL_{int}}{Q_l + CL_{int}}$$

Accordingly, metabolism in PBPK models has also been described using the following equation:

$$\frac{dA_{met}}{dt} = Q_l \times E \times C_a \tag{5.21}$$

Because $Q_l \times E$ = hepatic clearance (L/hr), the above equation represents the classical way of calculating the amount of chemical metabolized from knowledge of hepatic clearance and arterial blood concentration (Equation 5.10). Because $C_{vl} = C_a(1 - E)$, and $V_{max}/K_m$ or $CL_{int} = Q_l \times E/(1 - E)$, Equation 5.9, Equation 5.10, and Equation 5.21 are mathematically equivalent [227]. Figure 5.10 represents mathematically equivalent forms of equations frequently employed in PBPK models to calculate the rate of metabolism.

The rate of metabolism calculated using Equation 5.9 is based on the venous blood concentrations of chemicals (i.e., $C_{vl}$). This is equivalent to the venous equilibration model for hepatic metabolism. Other types of physiological descriptions of liver metabolism include the parallel tube model and the distributed sinusoidal perfusion model [240,241]. The parallel tube model describes the flow of substrate through the sinusoids lined with enzymes by considering them to be functionally homogeneous. In contrast, the distributed sinusoidal model accounts for the functional heterogeneity among sinusoids by including statistical distributions of enzyme contents and sinusoidal blood flow [240]. Although the distributed sinusoidal perfusion model is physiologically more realistic than the venous equilibration model, the latter simpler model may often be sufficient. Whereas the predictions of tissue dose might vary in cases where the metabolizing organ alone is considered in isolation [241], the difference in predictions between these models might not be significant when considering the whole-body clearance of chemicals. Recent effort has produced a geometric multicompartmental description for liver, which can be used to simulate regional protein induction [9]. The decision to use a multicompartmental liver depends on the objective and intended use of the model.

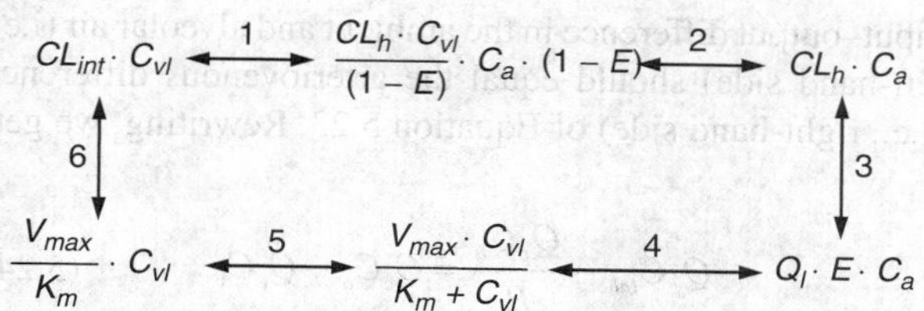


**FIGURE 5.10** Mathematically equivalent descriptions employed in PBPK models to calculate the rate of metabolism. (1) $CL_{int} = CL_h/(1-E)$ and $C_{vl} = C_a(1 - E)$; (2) simplification of the equation on the left-hand side; (3) $CL_h = Q_l \times E$; (4) for first-order conditions (i.e., when $C_{vl} << K_m$); and (5) $CL_{int} = V_{max}/K_m$ for first-order conditions. $CL_{int}$, intrinsic clearance; $CL_h$, hepatic clearance; $E$, hepatic extraction ratio; $C_{vl}$, free (venous) concentration in liver; $C_a$, arterial blood concentration; $V_{max}$, maximal velocity for metabolism; and $K_m$, Michaelis constant.

## Calculation of Arterial Blood Concentration

In PBPK models for volatile organic compounds (VOCs), arterial blood is frequently calculated on the basis of the steady-state solution for the MBDE for the combined lung tissue–alveolar air compartments (Figure 5.11) as follows [230]:

$$C_a = \frac{Q_p C_{inh} + Q_c C_v}{\left(Q_c + \frac{Q_p}{P_b}\right)} \tag{5.22}$$

The above algebraic expression is derived from the following mass conservation equation for lung, which specifies that the loss of chemical from the air is balanced by an identical gain of the chemical in the pulmonary blood:

$$Q_p(C_{inh} - C_{alv}) = Q_c(C_a - C_v) \tag{5.23}$$

Because the lung equilibrates vapor between alveolar air and blood, $C_{alv} = C_a/P_b$. This relationship assumes rapid equilibrium between alveolar air and arterial blood as defined by the blood–air partition coefficient, no significant metabolism by the lung tissue, and negligible storage capacity in the lungs. The above equation implies that

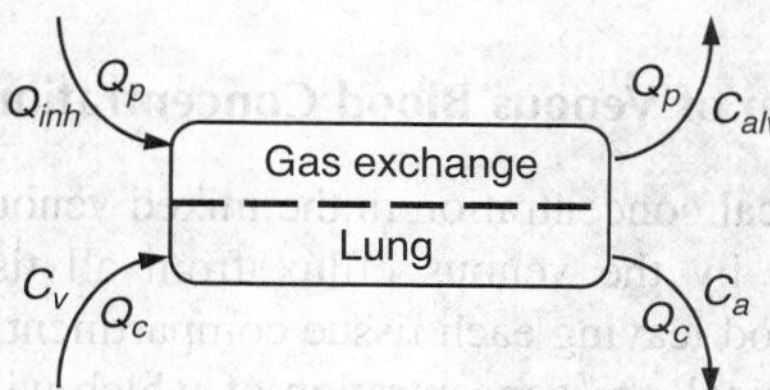


**FIGURE 5.11** Schematic of the combined lung tissue-alveolar air compartments. $Q_p$ is the alveolar ventilation rate, $Q_c$ is cardiac output, $C_a$ is arterial blood concentration, $C_v$ is the mixed venous concentration, and $C_{alv}$ is the alveolar air concentration.

input–output difference in the ambient and alveolar air (i.e., left-hand side) should equal the arteriovenous difference (i.e., right-hand side) of Equation 5.23. Rewriting, we get:

$$Q_p C_{inh} - \frac{Q_p C_a}{P_b} = Q_c C_a - Q_c C_v \quad (5.24)$$

Combining the two terms based on $C_a$ on the right-hand side gives:

$$Q_p C_{inh} + Q_c C_v = Q_c C_a + \frac{Q_p C_a}{P_b} \quad (5.25)$$

Isolating $C_a$, Equation 5.25 can be rewritten as follows:

$$C_a\left(Q_c + \frac{Q_p}{P_b}\right) = Q_p C_{inh} + Q_c C_v \quad (5.26)$$

Isolating $C_a$, we get Equation 5.22:

$$C_a = \frac{Q_p C_{inh} + Q_c C_v}{\left(Q_c + \frac{Q_p}{P_b}\right)}$$

This algebraic expression facilitating the calculation of $C_a$ (as dose divided by clearance) has been used frequently in PBPK models for lipophilic VOCs. In the case of nonvolatile chemicals, for which $P_b$ is very large and $C_{inh}$ is essentially zero, the above equation simplifies to $C_a = C_v$, unless significant uptake or metabolism occurs in the lung compartment. Pulmonary metabolism, in addition to uptake, can be included by describing both functions of the lung [171]. The concentration of the chemical appearing in the systemic arterial blood is then affected by the pulmonary first-pass effect associated with metabolic processes [8,171]. Sample mathematical descriptions used for calculating the rate of uptake by other routes (oral, intravenous, and dermal) are given in Table 5.2. Following uptake through these routes, the arterial blood concentration is calculated according to Equation 5.22 provided the venous blood concentration ($C_v$) is known.

## Calculation of Venous Blood Concentration

The chemical concentration in the mixed venous blood is determined by the venous efflux from all tissues. The venous blood leaving each tissue compartment brings out some chemical, the concentration of which will be equal to or lower than $C_a$. The $C_{vt}$ will be lower than $C_a$ when clearance processes are operative in a tissue. On the other hand, if the tissue does not consume any chemical, then $C_{vt}$ will be equal to $C_a$. The mixed venous concentration ($C_v$) is the weighted average of the chemical concentration exiting tissues:

$$C_v = \frac{Q_l}{Q_c} C_{vl} + \frac{Q_f}{Q_c} C_{vf} + \frac{Q_s}{Q_c} C_{vs} + \frac{Q_r}{Q_c} C_{vr} \quad (5.27)$$

Because $Q_l/Q_c = Q_{lc}$, $Q_f/Q_c = Q_{fc}$, $Q_s/Q_c = Q_{sc}$, and $Q_r/Q_c = Q_{rc}$, Equation 5.27 can be rewritten as:

$$C_v = Q_{lc} \cdot C_{vl} + Q_{fc} \cdot C_{vf} + Q_{sc} \cdot C_{vs} + Q_{rc} \cdot C_{vr} \quad (5.28)$$

If the numerical value representing the fraction of cardiac output flowing through each of the tissues is identical, then $C_v$ equals the average of the $C_{vt}$ of the four tissues. Because the $Q_{tc}$ is different from one tissue to another (e.g., $Q_{lc} = 0.25$, $Q_{fc} = 0.09$, $Q_{sc} = 0.15$, and $Q_{rc} = 0.51$ in the rat), we need to calculate the weighted average. In other words, in the adult rat,

$$C_v = (0.25 \times C_{vl}) + (0.09 \times C_{vf}) + (0.15 \times C_{vs}) + (0.51 \times C_{vr}) \quad (5.29)$$

where $C_{vt} = C_t/P_t$. This means that the chemical concentrations in the tissue matrix and the venous blood leaving the tissue are in equilibrium as specified by the tissue–blood partition coefficient. If the $C_t$ is 100 mg/L and the $P_t$ is 10, then $C_{vt}$ will be equal to 10 mg/L. In other words, at steady state, the tissue concentration of the chemical will be 10 times greater than the venous blood concentration, as specified by the $P_t$.

**TABLE 5.2**
**Examples of Equations Used To Describe Uptake by Intravenous, Oral, and Dermal Routes in PBPK Models**

| Exposure Route | Differential Equation |
|---|---|
| Intravenous | $\frac{dA_v}{dt} = \left(\sum_{i=1}^{n} Q_t C_{vt} + IV\right) - (Q_c C_v)$ |
| Oral | $\frac{dA_l}{dt} = Q_l (C_a - C_{vl}) - \left(\frac{dA_{met}}{dt}\right) + K_o A_{stom}$ |
| Dermal | $\frac{dA_{sk}}{dt} = K_p S\left(C_{air} - \frac{C_{sk}}{P_{s:a}}\right) + Q_{sk}\left(C_a - \frac{C_{sk}}{P_{s:b}}\right)$ |

*Note:* All abbreviations are defined in Table 5.1.

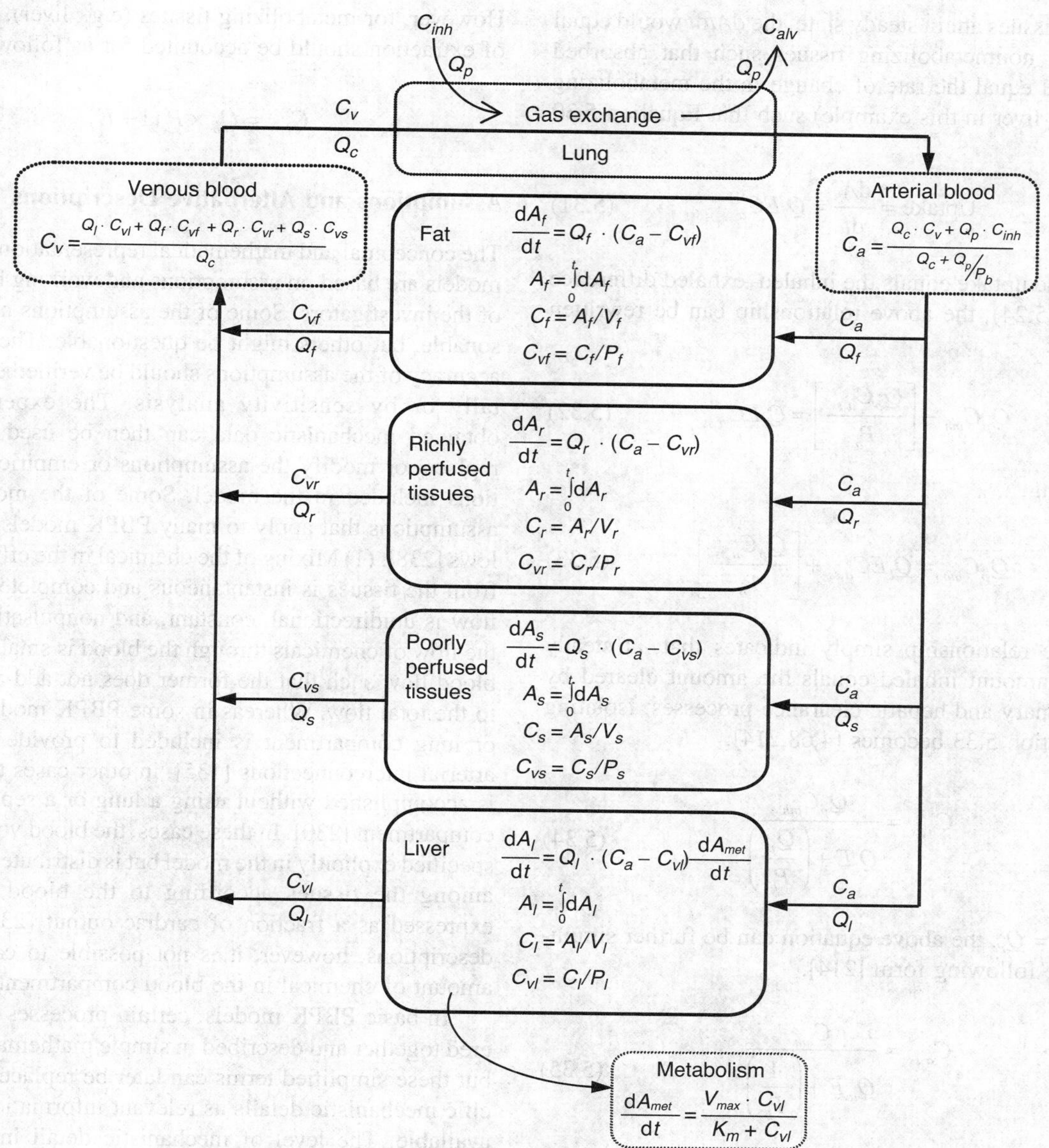


**FIGURE 5.12** A schematic of the PBPK model for styrene. In this model, the rat is represented as a four-compartmental system interconnected by systemic circulation. The input for the system is the product of the inhaled concentration of styrene times the alveolar ventilation rate. The resulting arterial blood concentration is in turn provided as input to the tissue compartments, the effluent venous blood concentrations of which are provided as input for the calculation of mixed venous concentration. All abbreviations are defined in Table 5.1.

Even though "full-blown" PBPK models are useful in simulating the time course of blood and tissue concentrations or amounts, the use of simple algebraic expressions may be sufficient for steady-state conditions. The algebraic expressions of steady-state concentrations may also be used to verify the proper functioning of PBPK models. Such steady-state relationships have been developed for inhaled VOCs [4,68,214] and more recently for chemicals exhibiting significant nasal clearance [16].

## Calculating Steady-State Concentrations

Regardless of whether the tissues are at steady state or not, the rate of chemical uptake would equal the rate of consumption or removal by the tissues. For the four-compartment PBPK model illustrated in Figure 5.12, the absorbed dose will be equal to:

$$\text{Uptake} = \frac{dA_f}{dt} + \frac{dA_r}{dt} + \frac{dA_s}{dt} + \frac{dA_l}{dt} \qquad (5.30)$$

When the tissues attain steady state, the $dA/dt$ would equal zero in all nonmetabolizing tissues such that absorbed dose would equal the rate of change in the metabolizing tissue (i.e., liver in this example) such that Equation 5.30 becomes:

$$\text{Uptake} = \frac{dA_l}{dt} = Q_l E C_a \tag{5.31}$$

Because the uptake equals the inhaled–exhaled difference (Equation 5.24), the above relationship can be rewritten as follows:

$$Q_p C_{inh} - \left[\frac{Q_p C_{a,ss}}{P_b}\right] = Q_l E C_{a,ss} \tag{5.32}$$

Rearranging,

$$Q_p C_{inh} = Q_l E C_{a,ss} + \left[\frac{Q_p C_{a,ss}}{P_b}\right] \tag{5.33}$$

The above relationship simply indicates that, at steady state, the amount inhaled equals the amount cleared by the pulmonary and hepatic clearance processes. Isolating $C_{ass}$, Equation 5.33 becomes [4,68,214]:

$$C_{a,ss} = \frac{Q_p C_{inh}}{Q_l E + \left(\frac{Q_p}{P_b}\right)} \tag{5.34}$$

When $Q_p = Q_c$, the above equation can be further simplified to the following form [214]:

$$C_{a,ss} = \frac{C_{inh}}{Q_{lc} E + \left(\frac{1}{P_b}\right)} \tag{5.35}$$

The preceding equations simply indicate that the steady-state arterial blood concentration equals the dose rate divided by the clearance (pulmonary + hepatic + renal, if applicable), and this relationship holds true regardless of the exposure route and species.

At steady state, the venous concentration exiting the nonmetabolizing tissues equals the arterial concentration entering the tissue (i.e., $C_{vt} = C_a$). Because tissue concentration $C_t = C_{vt} * P_t$, the chemical concentration in nonmetabolizing tissues at steady state (slowly or poorly perfused tissues [$C_{s,ss}$], richly perfused tissues [$C_{r,ss}$], and fat [$C_{f,ss}$]) can be computed as follows [46,214]:

$$C_{s,ss} = C_a \times P_s \tag{5.36}$$

$$C_{r,ss} = C_a \times P_r \tag{5.37}$$

$$C_{f,ss} = C_a \times P_f \tag{5.38}$$

However, for metabolizing tissues (e.g., liver), the extent of extraction should be accounted for as follows [214]:

$$C_{l,ss} = C_a \times P_l (1 - E) \tag{5.39}$$

### Assumptions and Alternative Descriptions

The conceptual and mathematical representations of PBPK models are based on assumptions and working hypotheses of the investigators. Some of the assumptions may be reasonable, but others might be questionable. The impact or accuracy of the assumptions should be verified experimentally or by sensitivity analysis. The experimentally obtained mechanistic data can then be used to accept, replace, or modify the assumptions or empirical descriptions included in the model. Some of the more general assumptions that apply to many PBPK models are as follows [238]: (1) Mixing of the chemical in the effluent blood from the tissues is instantaneous and complete; (2) blood flow is unidirectional, constant, and nonpulsatile; and (3) the flow of chemicals through the blood is smaller than the blood flow such that the former does not add appreciably to the total flow. Whereas in some PBPK models a blood or lung compartment is included to provide venous-to-arterial interconnections [135], in other cases this linkage is accomplished without using a lung or a separate blood compartment [230]. In these cases, the blood volume is not specified explicitly in the model but is distributed implicitly among the tissues according to the blood flow rates expressed as a fraction of cardiac output [230]. In such descriptions, however, it is not possible to calculate the amount of chemical in the blood compartment.

In basic PBPK models, certain processes are considered together and described in simple mathematical terms but these simplified terms can later be replaced with specific mechanistic details as relevant information becomes available. The level of mechanistic detail in the model description conforms to the intended use of the model. PBPK models, then, are of varying complexities according to the particular model's intended purpose. Lists of toxicologically important chemicals for which PBPK models have been developed in one or more species are provided in Table 5.3 and Table 5.4. The following section presents some prototypical representations of PBPK models for diverse groups of chemicals.

## Sample Model Representations

### Organic Chemicals

#### *Lipophilic, Volatile Organic Chemicals*

The development of PBPK models has been accomplished for more low-molecular-weight, lipophilic volatile organic chemicals than for other groups of chemicals. The tissue uptake of these VOCs is described as a perfusion-limited

**TABLE 5.3**
**List of Environmental Chemicals (Organic) for Which PBPK Models Have Been Developed in One or More Species**

- Acetone
- Acrylamide
- Acrylic acid
- Acrylonitrile and cyanoethylene oxide
- Allyl chloride
- Amyl methyl ether (*tert*) and *tert*-amyl alcohol
- Benzene
- Benzo(*a*)pyrene
- Benzoic acid
- Bisphenol A
- Bromochloromethane
- Bromodichloromethane
- Bromoform
- Bromotrifluoromethane
- Butadiene (1,3–) and metabolites
- Butanol (tertiary)
- Butoxyacetic acid (2–)
- Butoxyethanol (2-) and metabolites
- Butyl compounds (*n*-butyl acetate, *n*-butanol, *n*-butyraldehyde, *n*-butyric acid)
- Carbon tetrachloride
- Chlordecone
- Chlorfenvinphos
- Chloro (1-)-1,1-difluoroethane
- Chloro (2-)-1,1,1,2-tetrafluoroethane
- *p*-Chlorobenzotrifluoride
- Chlorofluorohydrocarbons
- Chloroform
- Chloromethane
- Chloroethane
- Chloropentafluorobenzene
- β-Chloroprene
- Chlorpyrifos
- Cyclohexane
- DDE
- Decane
- Diazinon
- Dibromochloromethane
- Dibromoethane (1,2-)
- Dibromomethane
- Dichloro (1,1-)-1-fluoroethane
- Dichlorodiphenylsulfone (*p*-*p*′-)
- Dichloroethane (1,2-)
- Dichloroethylene (1,1-, 1,2-)
- Dichloromethane
- Dichlorophenoxyacetic acid (2,4-)
- Dieldrin
- Diethylether
- Difluoromethane
- Diisopropylfluorophosphate
- Dimethyl sulfate
- Dioxane (1,4-)
- Dioxins
- Ethanol
- Ethyl acrylic acid
- Ethylbenzene
- Ethylene dibromide
- Ethylene dichloride
- Ethylene glycol and its metabolites
- Ethylene glycol mono butyl ether and intermediates
- Ethylene oxide
- Fluazifop-butyl
- Formaldehyde
- Furans
- Glycol ethyl ether acetate
- Glycol monomethyl ether
- Heptafluoropropane
- Hexachlorobutadiene
- Hexachloroethane
- *n*-Hexane
- Hexanedione (2,5-)
- Hexanemethyldisiloxane
- Hydroquinone
- Isofenphos
- Isoflurane
- Isopropanol
- Isopropene
- Lindane
- Malathion
- Methanol
- Methoxyacetic acid (2-)
- Methoxyethanol (2-)
- Methyl chloroform
- Methyl *tertiary*-butyl ether
- Methylene chloride
- Methylene dianiline (4,4′-)
- Methylethylketone
- Methylmetacrylate
- Naphthalene and naphthalene oxide
- Nicotine
- Nitropyrene
- Octamethylcyclotetrasiloxane
- Paraoxon
- Parathion
- PCBs
- Pentachloroethane
- Pentafluoroethane
- Phthalates (diethylhexyl-, dibutyl-, monobutyl-)
- Polychlorotrifluoroethylene
- Propylene glycol methyl and propylene glycol methyl ether acetate
- Pyrene
- Styrene
- Tetrachloroethane (1,1,1,2-)
- Tetrachloroethane (1,1,1,2-)
- Tetrachloroethylene
- Toluene
- Toluene diamine
- Trichloroethylene and its metabolites
- Trichlorofluoromethane
- Trichloropropane (1,2,3-)
- Trifluoroethane
- Trifluoroiodomethane
- Trifluralin
- Trimethyl-2-pentanol (2,4,4-)
- Trimethylbenzene (1,2,4-)
- Vinyl acetate
- Vinyl chloride
- Vinyl fluoride
- Xylene

*Source:* Based on compilations by Krishnan and Anderson [169] and Reddy et al. [232] from which original citations can be obtained with the exception of the following references: 11, 17, 18, 22, 40, 43, 56, 83, 91, 92, 104, 106, 119, 120, 149, 164, 165, 177, 186, 187, 216, 217, 246, 250, 260, 261, 267, 278, 279, and 286.

process, whereas the inhalational uptake may either be blood flow or ventilation limited [230]. The basic mathematical representation that is applicable to many members of this category is provided in Figure 5.12. In this example, the model consists of four tissue compartments—liver, fat, richly perfused tissue group, and poorly perfused tissue group—similar to the conceptual representation shown in Figure 5.3. Here, the chemical input to the model results from the inspiration of the chemical in the inhaled air at a flow rate equal to the alveolar ventilation rate. The chemical in alveolar air is assumed to equilibrate rapidly with arterial blood, so the concentration of chemical in arterial blood and in alveolar air leaving the lungs maintains a constant ratio specified by the blood–air partition

**TABLE 5.4**
**List of Metals, Metalloids, and Anions for Which PBPK Models Have Been Developed in One or More Species**

| | |
|---|---|
| Arsenic | Lead |
| Bromide | Manganese |
| Chloride | Methyl mercury |
| Chromium | Nickel |
| Fluoride | Perchlorate |
| Iodide | Zinc |
| Iron | |

*Source:* Based on compilations by Krishnan and Andersen [169] and Reddy et al. [232] from which original citations can be obtained, with the exception of references 61, 62, 193, and 194.

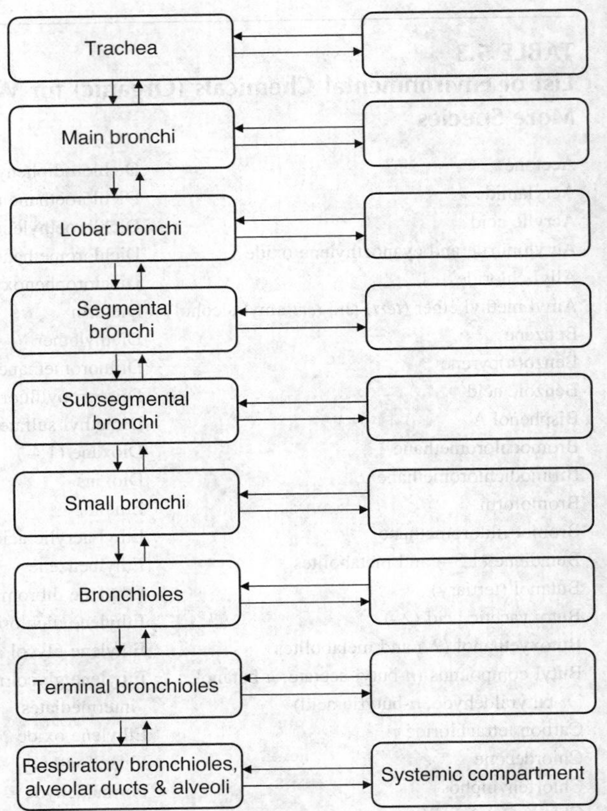


**FIGURE 5.13** Conceptual representation of a physiologically based model to describe the retention and excretion of solvent vapors in the lung. (Adapted from Johanson, G., *Ann. Occup. Hyg.*, 35, 323–339, 1991. With permission.)

coefficient. Arterial blood, flowing at a rate equal to the cardiac output, is apportioned among liver, fat, richly perfused tissues, and poorly perfused tissues. Venous blood leaving each tissue compartment mixes simultaneously to yield the chemical concentration in the mixed venous blood returning to the lungs at a flow rate equal to cardiac output. For the chemical depicted in Figure 5.12, the tissue uptake is assumed to be perfusion limited and metabolism is assumed to occur only in the liver by a single saturable process. It is entirely possible that extrahepatic metabolism is important [40,69,243] and that it is necessary to estimate the total amount of a chemical in the blood [171]. In such cases, the metabolic capacity of each tissue included in the model is characterized, and blood is described explicitly as a separate compartment [135,171].

### *Polar Solvents*

In the models presented above, pulmonary uptake is represented by assuming that all the chemical disappearing from the inspired air appears in the arterial blood and that the chemicals in alveolar air and arterial blood are in instantaneous equilibrium [230]. In these descriptions, the conducting airways (i.e., nasal passages, larynx, trachea, bronchi, and bronchioles) are considered as inert tubes that carry the chemical to the pulmonary region, where diffusion occurs. Mounting evidence suggests that this kind of a simple model is not predictive of either total respiratory uptake or regional uptake of highly soluble polar solvents [150]. These chemicals feature complex relationships between uptake and the blood–air partition coefficient. Further, several studies have shown that the total respiratory uptake is less than 100% as predicted by the continuous ventilation equilibrium model [150]. The reduced pulmonary uptake of polar solvents has been suggested to be due to their adsorption or dissolution in the surface of the respiratory epithelium during inhalation and their desorption during exhalation [150]. This adsorption–desorption mechanism is a consequence of both the aqueous solubility of the chemicals and the cyclic nature of respiratory exchange and should be taken into consideration in addition to the anatophysiological characteristics of the respiratory tract, blood-flow rates, and partition coefficients of the chemical [127,150,197]. The PBPK model for polar solvents developed by Johanson [150] consists of nine serially connected central compartments each one corresponding to an anatomical level of the respiratory tree (Figure 5.13). The first central compartment corresponds to the airway and the outermost layer of mucus lining the airway wall. Radial diffusion of solvent from the outermost layer and deeper portions of the airway wall is accounted for by linking a peripheral compartment with each of the first eight compartments. The central compartment of the ninth and final region corresponds to the pulmonary or gas-exchange region of the respiratory tract (respiratory bronchioles, alveolar ducts, and alveoli) the volume of which increases during inhalation and decreases during exhalation. The peripheral compartment of this ninth region represents the rest of the body, where immediate equilibrium between alveolar air and arterial blood is assumed. Either a single compartmental or a multicompartmental physiological description can be used to account for chemical disposition in the body.

*Nonvolatile Organic Chemicals: Uncharged Species*

Physiologically based pharmacokinetic models for nonvolatile organics describe chemical uptake as a diffusion-limited or as a perfusion-limited process and accommodate descriptions for chemical input via oral, dermal, intraperitoneal, and intravenous routes [15,135]. These models typically consist of the following compartments: liver, fat, poorly perfused tissues, richly perfused tissues, and blood. Evidently, lung tissue or a description for pulmonary uptake is not included if its contribution to the overall kinetic behavior of the chemical is negligible [15]. When there is evidence to the contrary and when there is evidence for pulmonary metabolism, lungs should be separated from the richly perfused tissues group and described as a separate compartment. Binding to tissue macromolecules and other clearance processes are described as appropriate [14,162,244]. For several persistent organic pollutants and highly lipophilic VOCs, partitioning into neutral lipid components of tissues and blood is the critical determinant whereas their solubility in water components of tissues and blood is negligible. The volumes and flows in PBPK models for such chemicals have been set equal to that of neutral lipid (Figure 5.7).

*Nonvolatile Organic Chemicals: Charged Species*

The distribution of water-soluble, charged species, such as weak acids and weak bases, is determined primarily by the $pK_a$ of the compound and pH of the body fluids. The tissue uptake of these chemicals can be defined by the conventional flow-limited exchange between plasma/blood and tissues, with partitioning being determined by the $pK_a$ of the chemical and the pH of the body fluids in accordance with the Henderson–Hasselbach equation [207]. The elimination of these chemicals is mainly via urine; therefore, time-course information regarding this process may be necessary to adequately describe the kinetics of excretion.

## Inorganic Chemicals

*Gases and Vapors*

For inorganic gases that do not interact with walls of the conducting airways, simple dosimetry descriptions similar to those discussed in the preceding sections have been employed [7,150]. For reactive inorganic gases, however, simulation models should incorporate critical elements of local absorption in the lower respiration tract and reactions with the biological constituents [196,208]. Typically, these dosimetry models use quantitative information on the physiology of the lower respiratory tract (i.e., ventilation parameters and varying airway dimensions during the breathing cycle). Lung dimensions are taken into account by making use of the airway models. In these models, airways of the lower respiratory tract are represented by a sequence of sets of right circular cylinders. All cylinders in series corresponding to a particular generation are of the same size. The upper respiratory tract (if used) consists of pretracheal generations or sequential segments. For each generation, the simulation model requires the specification of the number of airways or segments and their diameters and lengths. Additionally, for the pulmonary region, the alveolar volume and the surface area for each generation are needed. Similarly, the surface area of each upper respiratory segment is included. The processes of transport and chemical reactions are described with a series of partial differential equations. In the liquid lining, tissue, and blood compartments, where only the processes of molecular diffusion and chemical reactions are considered, the form of the equation is the same. In the lumen of the airway and in the alveolar air spaces, axial convection, axial dispersion, the loss of chemical to the liquid lining, and lung expansion and contraction are taken into account. With quantitative information on the boundary conditions, initial conditions, and the physical, chemical, and biological parameters, these equations are solved to simulate dose and dose patterns. A hybrid computational fluid dynamics and PBPK model has also been constructed to estimate the regional tissue dose of organic acids in rodent and human nasal cavity [18,108]. In this case, the computational requirements are greater than those associated with the simpler PBPK models discussed above.

*Metals*

The functional description used in PBPK models for metals is basically similar to that for organics [205]; however, the common assumptions of flow-limited tissue uptake and linear partitioning into tissues may not be applicable for some metals. Further, the systemic uptake of metals may be mediated by ion channels or carrier-mediated transport mechanisms, and metabolism may be limited to oxidation state transitions and alkylation/dealkylation reactions. In the PBPK models for metals, the uptake has been described as a diffusion-limited process in some tissues and as a flow-limited process in others to obtain adequate fitting of the model to experimental data [148]. Clearance associated with binding to subcellular proteins occurs to a greater extent in the case of metals. Another important phenomenon associated with certain metals is their storage in bone. The metals for which PBPK models have been developed include arsenic, nickel, lead, chromium, zinc, and mercury (Table 5.4).

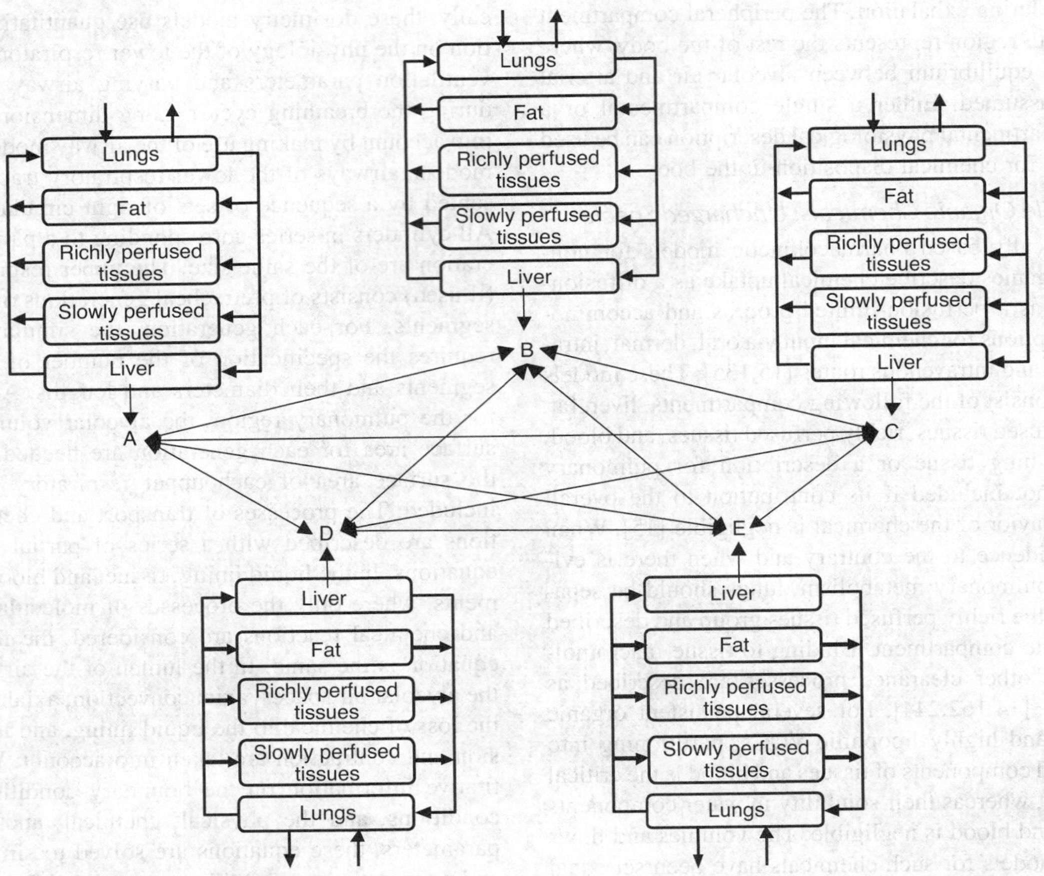


**FIGURE 5.14** Conceptual representation of a PBPK model for a hypothetical mixture of five chemicals (ABCDE). In this example, all mixture components are assumed to be metabolized in liver where they compete with each other for enzyme binding sites. (From Krishnan, K. and Johanson, G., *J. Environ. Sci. Health*, 23(C), 31–53, 2005. With permission.)

## Mixtures

Physiologically based pharmacokinetic models have been developed for several binary mixtures and are only beginning to be developed for more complex mixtures [23,81,84,134,258,259]. The PBPK model for a chemical mixture essential consists of individual chemical models interconnected by the interaction mechanisms (Figure 5.14). The mixture PBPK models developed to date have investigated metabolic inhibition (competitive, noncompetitive, or uncompetitive) as the mechanism of interaction. Accordingly, for constructing a binary mixture PBPK model, two individual chemical models have been interconnected by modifying the equation for computing the rate of hepatic metabolism. For modeling more complex mixtures, all plausible binary interactions should be characterized [172]. In a mixture of three chemicals, for example, three two-way interactions occur. Individual chemical models should first be constructed and then interconnected by modifying the equations for the rate of metabolism for each chemical (Table 5.5). It is important to note that all linkages are of binary nature only; for example, Figure 5.15A represents the interconnections among the components of a quaternary mixture (benzene [B], toluene [T], ethyl benzene [E], and *m*-xylene [X]). Here, when we have described the inhibitory effect of B on T, this would lead to a reduction in the rate of T metabolism and consequently an increase in the free concentration of T ($C_{vl}$). The $C_{vl}$ of toluene is the numerator of the term representing the inhibitory effect of T on E, C, and B (i.e., $C_{vl}/K_i$) (Figure 5.15B). Co-exposure to B increases the $C_{vl}$ of toluene, thus leading to a greater interactive effect between toluene and other chemicals. Similarly, B may affect the $C_{vl}$ of other chemicals which in turn will have an impact on their rate of metabolism and an interactive effect on each of the other mixture components. It is important to note that the $K_i$ for binary interaction is not modified during mixture exposure situations; rather, the $C_{vl}$ of each of the components is modified, leading to a greater impact according to their inhibition potency [172]. To solve the PBPK models of individual chemicals and mixtures, knowledge of the various input parameters is essential.

**TABLE 5.5**
**Mathematical Descriptions of the Rate of the Amount Metabolized (RAM) of a Chemical in the Presence of Other Chemicals That Compete for the Same Enzymic Catalytic Sites**

| Number of Chemicals in Mixture | Chemical | RAM Equation[a] |
|---|---|---|
| 2 | A | $\frac{V_{maxA} \cdot C_{vlA}}{K_{mA}\left(1+\frac{C_{vlB}}{K_{iBA}}\right)+C_{vlA}}$ |
| | B | $\frac{V_{maxB} \cdot C_{vlB}}{K_{mB}\left(1+\frac{C_{vlA}}{K_{iAB}}\right)+C_{vlB}}$ |
| 3 | A | $\frac{V_{maxA} \cdot C_{vlA}}{K_{mA}\left(1+\frac{C_{vlB}}{K_{iBA}}+\frac{C_{vlC}}{K_{iCA}}\right)+C_{vlA}}$ |
| | B | $\frac{V_{maxB} \cdot C_{vlB}}{K_{mB}\left(1+\frac{C_{vlA}}{K_{iAB}}+\frac{C_{vlC}}{K_{iCB}}\right)+C_{vlB}}$ |
| | C | $\frac{V_{maxC} \cdot C_{vlC}}{K_{mC}\left(1+\frac{C_{vlA}}{K_{iAC}}+\frac{C_{vlB}}{K_{iBC}}\right)+C_{vlC}}$ |
| 4 | A | $\frac{V_{maxA} \cdot C_{vlA}}{K_{mA}\left(1+\frac{C_{vlB}}{K_{iBA}}+\frac{C_{vlC}}{K_{iCA}}+\frac{C_{vlD}}{K_{iDA}}\right)+C_{vlA}}$ |
| | B | $\frac{V_{maxB} \cdot C_{vlB}}{K_{mB}\left(1+\frac{C_{vlB}}{K_{iAB}}+\frac{C_{vlC}}{K_{iCB}}+\frac{C_{vlD}}{K_{iDB}}\right)+C_{vlB}}$ |
| | C | $\frac{V_{maxC} \cdot C_{vlC}}{K_{mC}\left(1+\frac{C_{vlA}}{K_{iAC}}+\frac{C_{vlB}}{K_{iBC}}+\frac{C_{vlD}}{K_{iDC}}\right)+C_{vlC}}$ |
| | D | $\frac{V_{maxD} \cdot C_{vlD}}{K_{mD}\left(1+\frac{C_{vlA}}{K_{iAD}}+\frac{C_{vlB}}{K_{iBD}}+\frac{C_{vlD}}{K_{iCD}}\right)+C_{vlD}}$ |

[a] $V_{maxA}$ is the maximal velocity for metabolism of chemical A; $K_{mx}$ is the Michaelis constant of chemical $x$; $C_{vlx}$ is the free concentration of chemical $x$; and $K_{iXY}$ is the inhibition constant for the effect of chemical $x$ on chemical $y$.

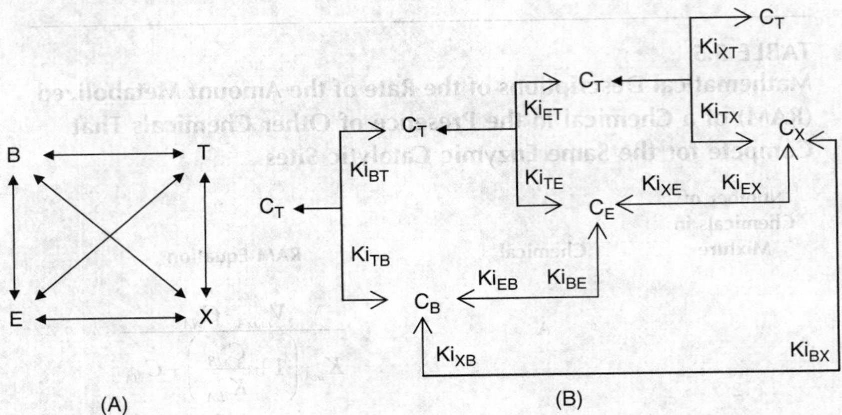


**FIGURE 5.15** (A) Illustration of the interconnections among components of a quaternary mixture of benzene (B), toluene (T), ethylbenzene (E), and m-xylene (X). Here, a change in concentration of one component will affect all other components because the chemicals are part of a network of binary interactions. (B) Illustration of the gradually increasing effective concentration of toluene (see $C_r$, the size of which increases from right to left due to the increasing number of mixture components that act as metabolic inhibitors). $Ki_{XY}$ is the inhibition constant for the effect of chemical $X$ on chemical $Y$. (From Haddad, S. et al., *Toxicol. Appl. Pharmacol.*, 167, 199–209, 2000. With permission.)

## MODEL PARAMETERIZATION

Model parameterization refers to obtaining estimates or measurements of the mechanistic determinants (namely, physiological parameters, physicochemical parameters, and biochemical rate constants) that are included in one or more of the PBPK model equations.

### Physiological Parameters

Physiological parameters included in most PBPK models include alveolar ventilation rate, cardiac output, tissue blood flow rates, and tissue volumes. Additional parameters, (e.g., tissue DNA levels, hematocrit) may be required in certain cases. The physiological parameters can generally be measured directly in the animal species of interest [41,45,85,144,242,249,269]. For example, breathing rates can be measured with the use of a spirometer, plethysmograph, pneumotachograph, hotwire anemometer, or nonbreathing valves [188]. Cardiac output has been determined from dye dilution curves using oximeters [80]. Compilations of numerical values of physiological parameters for laboratory animals and humans have appeared in the literature [19,41,72]. Table 5.6, Table 5.7, and Table 5.8 present the reference values for mice, rat, and humans, respectively.

**TABLE 5.6**
**Reference Physiological Parameters for Mice, Rats, and Humans**

| Physiological Parameters | Mouse | Rat | Human |
|---|---|---|---|
| Body weight (BW) (kg) | 0.025 | 0.25 | 70.0 |
| Tissue volume (fraction of BW) | | | |
| Liver | 0.055 | 0.04 | 0.026 |
| Fat | 0.10 | 0.07 | 0.190 |
| Organs | 0.05 | 0.05 | 0.05 |
| Muscle | 0.70 | 0.75 | 0.62 |
| Cardiac output ($Q_C$) (L/min) | 0.017 | 0.083 | 6.20 |
| Tissue perfusion (fraction of $Q_C$) | | | |
| Liver | 0.25 | 0.25 | 0.26 |
| Fat | 0.09 | 0.09 | 0.05 |
| Organs | 0.51 | 0.51 | 0.44 |
| Muscle | 0.15 | 0.15 | 0.25 |
| Minute volume (L/min) | 0.037 | 0.174 | 7.50 |
| Alveolar ventilation (L/min) | 0.025 | 0.117 | 5.00 |

*Source:* Travis, C.C. and Hattemer-Frey, H.A., in *Statistics in Toxicology*, Krewski, D. and Franklin, C., Eds., Gordon and Breach, New York, 1991, p. 170. With permission.

**TABLE 5.7**
**Range of Plausible Values of the Volume and Perfusion of Selected Tissues in the Mice**

| | Volume (% Body Weight) | | Regional Blood Flow (% Cardiac Output) | |
|---|---|---|---|---|
| **Tissue** | **Mean** | **Range** | **Mean** | **Range** |
| Adipose | 7 | 5–14[a] | — | — |
| Brain | 1.7 | 1.35–2.03 | 3.3 | 3.1–3.5 |
| Heart | 0.5 | 0.4–0.6 | 6.6 | 5.9–7.2 |
| Kidneys | 1.7 | 1.35–1.88 | 9.1 | 7.0–11.1 |
| Liver | 5.5 | 4.19–7.98 | 16.1 | — |
| Lungs | 0.7 | 0.66–0.86 | 0.5 | — |
| Muscle | 38.4 | 35.8–39.9 | 15.9 | 12.2–19.6 |
| Skin | 16.5 | 15.9–20.8 | 5.8 | 3.3–8.3 |

[a] Varies proportionately with body weight.

*Source:* Adapted from Brown, R.P. et al., *Toxicol. Indust. Health*, 13, 407–484, 1997.

**TABLE 5.8**
**Range of Plausible Values of the Volume and Perfusion of Selected Tissues in the Rat**

| | Volume (% Body Weight) | | Regional Blood Flow (% Cardiac Output) | |
|---|---|---|---|---|
| **Tissue** | **Mean** | **Range** | **Mean** | **Range** |
| Adipose | 7 | 4.6–12.0[a] | 7 | — |
| Brain | 0.6 | 0.38–0.83 | 2.0 | 1.5–2.6 |
| Heart | 0.3 | 0.27–0.40 | 5.1 | 4.5–5.1 |
| Kidneys | 0.7 | 0.49–0.91 | 14.1 | 9.5–19.0 |
| Liver | 3.4 | 2.14–5.16 | 18.3 | 13.1–22.1 |
| Lungs | 0.5 | 0.37–0.61 | 2.1 | 1.1–17.8 |
| Muscle | 40.4 | 35.4–45.5 | 27.8 | — |
| Skin | 19.0 | 15.8–23.6 | 5.8 | — |

[a] Varies proportionately with body weight.

*Source:* Adapted from Brown, R.P. et al., *Toxicol. Indust. Health*, 13, 407–484, 1997.

as well as the ranges of physiological parameter values for rats and mice. More recently, a flurry of activity has addressed compiling physiological data in children of various age groups [126,228,229,245] to facilitate data interpretations in the context of developmental risk assessment. Figure 5.16 presents examples of age-related change in the volumes of tissues included in PBPK models for children.

## Physicochemical Parameters

The physicochemical parameters required for PBPK models refer primarily to the partition coefficients, which represent the relative distribution of a chemical between two phases at equilibrium. Partitioning between two media (e.g., blood and air) as described by Henry's law for gases is a balance of the solubility of a chemical in the two media. Partition coefficients are represented as the ratio of the concentration of a chemical in the two media (e.g., blood to air, tissue to blood) at equilibrium. These physicochemical parameters are necessary to describe the tissue distribution of most uncharged xenobiotics as well as the pulmonary uptake of volatile organic chemicals. Several *in vitro*, *in vivo*, and animal replacement methods are available for estimating the partition coefficients of chemicals.

### *In Vitro* Methods

#### *Vial Equilibration*

This *in vitro* method involves comparison of the equilibrium concentration of a chemical in the headspace of test vials containing tissues with empty/reference vials [103,115,118,151,248]. The experimental procedure for

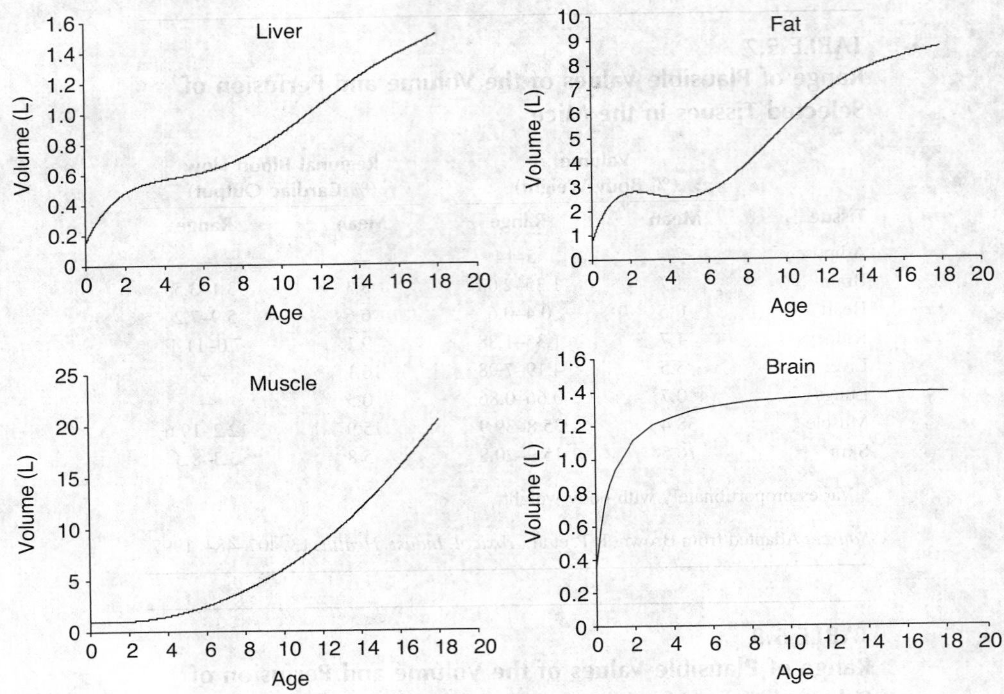


**FIGURE 5.16** Tissue volumes in children as a function of age in years. (From Krishnan, K. and Johanson, G., *J. Environ. Sci. Health*, 23(C), 31–53, 2005. With permission.)

determining the tissue–air and blood–air partition coefficients of VOCs by vial equilibration is given below:

- Prepare a batch of 16 glass vials (volume of ~25 mL) stoppered with Teflon® septa. Transfer a measured quantity of raw ground tissue (e.g., 200 mg) or blood (e.g., 200 μL) into even-numbered vials. The odd-numbered empty vials are used as reference vials. All vials are placed in a shaker–incubator at 37°C.
- Aerate the vials after a 5-minute initial equilibrium at 37°C, and cap them again.
- Remove a predetermined volume of air (e.g., 0.5 or 1.0 mL) from each vial individually with a gas-tight syringe inserted through the septa, and replace it with an equal volume of air containing a known quantity of the chemical.
- Draw a sample of headspace atmosphere (e.g., 1.0 mL) from one set of eight individual vials (four reference plus four sample) at either of the sampling time points (e.g., hr and 2 hr).

The tissue–air (or blood–air) partition coefficient is calculated as follows [115]:

$$P_{t:a} = \frac{\left(C_{ref}V_{ref}\right) - \left[C_{sam}\left(V_{ref} - V_{sam}\right)\right]}{C_{sam}V_{sam}} \quad (5.40)$$

where $C_{ref}$ is the chemical concentration in the headspace of the reference vial, $C_{sam}$ is the chemical concentration in the headspace of the sample vial, $V_{ref}$ is the volume of the reference vial, and $V_{sam}$ is the volume of the sample.

If the $P_{t:a}$ values obtained at both time points are not significantly different from each other, they can be averaged and standard deviations calculated. If the $P_{t:a}$ values obtained at the later time point are significantly different from those obtained at the earlier time point, they are not considered to represent a true measure of solubility. An increase in $P_{t:a}$ values with incubation time indicates chemical reactions in the aqueous phase or the non-attainment of equilibrium during the time points chosen for sampling. In such cases, additional experiments have been conducted to generate a time course of the distribution ratio to estimate $P_{t:a}$ by extrapolating to time 0 [171]. Metabolic inhibitors also may be added in excess before introducing the chemical. The latter approach has been shown to eliminate chemical uptake by tissues due to specific and known reactions (e.g., glutathione conjugation) [109,123]. On the other hand, if the calculated $P_{t:a}$ value decreases with incubation time, then deterioration of the tissue sample during incubation is likely; in such cases, headspace sampling for the determination of partition coefficients should be conducted at earlier time points.

The tissue and blood samples do not have to be used as raw preparations in these experiments. Especially for chemicals with low solubility, the tissue and blood sam ples have been prepared as 1:2 or 1:3 homogenates in saline (0.9% NaCl). The corresponding reference vials will contain saline alone. The $P_{t:a}$ value in this case is obtained using the following formula [115]:

$$P_{t:a} = \frac{\left[C_{ref}\left(V_{ref} - V_{liq}\right)\right] - \left[C_{sam}\left(V_{ref} - V_{liq} - V_{sam}\right)\right] + \left[\left(C_{ref} - C_{sam}\right)\left(V_{liq} P_{liq}\right)\right]}{C_{sam} V_{sam}} \quad (5.41)$$

where $V_{liq}$ is the volume of the diluent liquid, and $P_{liq}$ is the diluent liquid–air partition coefficient. Partition coefficients for nonvolatile chemicals have been determined *in vitro* using equilibrium dialysis or ultrafiltration techniques.

### *Equilibrium Dialysis*

In this technique, the dialysis cell cavities are separated with a membrane of desired molecular weight specifications. The experimental procedure consists of dialyzing the tissue homogenate or whole blood prepared in a buffer (e.g., Tris HCl, 0.1 *M*, pH 7.4) against the same buffer in a metabolic shaking bath [142,178,254,255]. These experiments are conducted for several different initial concentrations of the chemical (radiolabeled plus cold). Preliminary studies using different dialysis durations should be conducted to determine the time needed to attain equilibrium. At the end of the incubation, radioactivity in both the tissue and buffer is determined separately by liquid scintillation counting. The sum of the buffer and tissue radioactivity should account for all of the radioactivity initially added to the dialysis cells. The unbound fraction of radioactive chemical in the tissue is determined by dividing the chemical concentration in buffer by the concentration in the tissue homogenate [178]. The $P_t$ values obtained from these experiments will be accurate only if the ratio of the bound to the free form of the chemical is constant over a wide range of concentrations.

### *Ultrafiltration*

The ultrafiltration assembly consists of an ultrafiltration device and a semipermeable membrane. Here, the tissue homogenates are spiked with a known amount of a chemical (radiolabeled plus cold) and allowed to equilibrate for a predetermined time. Following equilibration, they are transferred into the reservoir portion of the ultracentrifugation device placed in an angle rotor and spun in a superspeed centrifuge. The concentrations of the chemical in the tissue homogenate and the buffer are determined and solved for the concentration of the chemical in the tissue [178]. The time and speed of ultracentrifugation should be determined on a case-by-case basis such that a desired volume of the ultrafiltrate is collected. Untreated tissues should be used to generate a protein-free ultrafiltrate for measuring the nonspecific binding onto the surface of the ultrafiltration device.

## *In Vivo* Methods

Methods for estimating tissue–blood partition coefficients ($P_t$) based on data on blood and tissue concentrations of the parent chemical after a single-bolus dose or at steady-state conditions have been published [46,110,111,176]. $P_t$ values have also been estimated from the slope of a best-fit straight line with a unit slope drawn through the log–log plot of tissue concentration vs. blood concentration of the parent chemical for each tissue [74]. The steady-state approach will only work if the chemical is not removed by active binding or metabolic processes in one or more tissues. In cases of active tissue metabolism, the estimated $P_t$ values tend to underestimate the true $P_t$ values. Estimates of $P_t$ for these tissues can be obtained if the amount of chemical consumed by the metabolic process is accounted for. Thus, for metabolizing tissues, the partition coefficients are determined after adjusting for clearance [46]. A potential problem associated with determination of the partition coefficients relates to the presence of residual blood in the tissues. The contamination of tissues with blood in the tissue vasculature might introduce errors in the estimated partition coefficients. The importance of this problem has been investigated and the means of correcting these errors proposed [166].

## Animal-Replacement Approaches

The *in silico* approaches for estimating tissue–air, blood–air, and tissue–blood partition coefficients may be one of the following types: empirical, mechanistic, or semiempirical [26,213]. The empirical methods relate the partition coefficients to structural features or physicochemical properties of chemicals through a mathematical function. These methods are essentially data based, implying that the model and parameter estimates are derived from a training dataset. Additional datasets may not have been used for evaluating the predictive ability of such models. Further, these algorithms are not comprised of parameters that vary between species or among individuals of a population; rather, they only attempt to explain the variation of partition coefficients among chemicals on the basis of variations in the structure and properties of the chemicals [1–3,24,25, 28,195,212]. As such, they may be applied to calculate the partition coefficients of chemicals that are within the "predictive domain" of the developed equations.

Mechanistic and semiempirical *in silico* approaches, on the other hand, are based on properties that are specific to chemicals as well as characteristics that are specific to an individual or a population [20,27,79,222–224]. Thus, these approaches account for the determinants of blood and tissue partitioning amenable for estimating the magnitude of the interspecies and inter-individual variation in $P_{b:a}$, $P_{t:a}$, and $P_{t:b}$. Mechanistic *in silico* approaches have been successfully applied to calculate the $P_{b:a}$, $P_{t:a}$, and $P_{t:b}$ of low-molecular-weight volatile organic substances (alkyl benzenes, chloroethanes, chloroethylenes, ketones, acetate esters, and alcohols), as well as some highly lipophilic organochlorine compounds in rats and humans (PCBs, PBBs, dieldrin, DDT, dioxins) [27,133,222–224]. These mechanistic algorithms for computing tissue–air, blood–air, and tissue–blood partition coefficients are described below.

### *Tissue–Air Partition Coefficients*

The partitioning of a chemical between two matrices can be predicted if its solubility and binding in each of the two matrices can be estimated with reasonable accuracy. Using this basic premise, mechanistic animal-replacement approaches for predicting tissue–air, blood–air, and tissue–blood partition coefficients have been developed. Accordingly, the tissue–air partition coefficients (PCs) of low-molecular-weight volatile organic chemicals, for which macromolecular binding may be negligible, have been calculated as follows [96,103,212]:

$$P_{t:a} = \left(P_{l:a} \times F_{lt}\right) + \left(P_{w:a} \times F_{wt}\right) \qquad (5.42)$$

where $P_{l:a}$ is the lipid–air partition coefficient, $P_{w:a}$ is the water–air partition coefficient, $F_{lt}$ is the volume fraction of lipid in tissue, and $F_{wt}$ is the volume fraction of water in tissue.

In the above equation, $P_{l:a} \times F_{lt}$ represents the partitioning of a chemical between the tissue lipids and air, and $P_{w:a} \times F_{wt}$ represents the partitioning between the tissue aqueous phase and air. The *n*-octanol–water or vegetable oil–air partition coefficient ($P_{o:a}$) has been used as a predictor of $P_{l:a}$, and $P_{w:a}$ as a surrogate of chemical partitioning between tissue water and air [96,212]. However, the use of "tissue lipids" is too generic to encompass the differential lipophilicity characteristics of neutral lipids (e.g., triglycerides) and polar lipids (e.g., phospholipids); therefore, the partitioning of a chemical into neutral lipids and polar lipids may have to be considered separately. The physiochemical properties of phospholipids are influenced by the presence of hydrophobic (e.g., glyceride) or hydrophilic (e.g., phosphomonoester) groups; thus, the use of either $P_{o:a}$ or $P_{w:a}$ alone cannot adequately predict tissue phospholipids–air partition coefficients. The partitioning of a chemical between tissue or blood polar lipids (i.e., phospholipids) and air can be calculated as a fractional additive function of its partitioning into neutral lipids ($0.3 \times P_{o:a}$) and water ($0.7 \times P_{w:a}$). This approximation of chemical partitioning into tissue phospholipids is based on the assumption that the lipophilicity/hydrophilicity characteristics of tissue phospholipids are similar to those of commercial lecithin [222]. Based on this working hypothesis, Poulin and Krishnan [222–225] proposed the following equation to predict the $P_{t:a}$ of VOCs, considering separately the partitioning of chemicals into neutral lipid and polar lipid portions:

$$\begin{aligned} P_{t:a} &= P_{o:a} \times F_{nt} + P_{o:a} \times 0.3F_{pt} \\ &\quad + P_{w:a} \times 0.7F_{pt} + P_{w:a} \times F_{wt} \end{aligned} \qquad (5.43)$$

where $F_{nt}$ is the volume fraction of neutral lipids in tissue, and $F_{pt}$ is the volume fraction of phospholipids in tissue. The above equation can be rewritten as:

$$P_{t:a} = P_{o:a}\left(F_{nt} + 0.3F_{pt}\right) + P_{w:a}\left(F_{wt} + 0.7F_{pt}\right) \qquad (5.44)$$

In Equation 5.43 and Equation 5.44, the partitioning of a chemical between tissue neutral lipids and air is assumed to correspond directly to $P_{o:a}$, whereas the partitioning between tissue water and air is assumed to correspond to $P_{w:a}$. Accordingly, $P_{t:a}$ can be calculated with knowledge of tissue composition data ($F_{nt}$, $F_{pt}$, $F_{wt}$), and physicochemical properties of chemicals ($P_{o:a}$ and $P_{w:a}$). Compilations of species-specific tissue composition data [90,222–225] as well as the $P_{o:a}$ and $P_{w:a}$ for several VOCs at 37°C are available in the literature [115,212]. To facilitate the use of *n*-octanol–water partition coefficient ($P_{o:w}$) values instead of $P_{o:a}$ values, which are not readily available in the literature, Equation 5.44 can be rewritten as follows:

$$P_{t:a} = P_{o:w}P_{w:a}\left(V_{nt} + 0.3V_{pt}\right) + P_{w:a}\left(V_{wt} + 0.7V_{pt}\right) \qquad (5.45)$$

Equation 5.45 has been used to predict the rat and human $P_{t:a}$ (liver, muscle, fat) of several alkanes, haloalkanes, and aromatic hydrocarbons [220,224–227]. In general, the predicted $P_{t:a}$ values were within a factor of two of the experimentally determined PCs. For chemicals such as alcohols, acetate esters, and ketones, the values of rat and human fat–air PCs calculated using Equation 5.45 differed substantially from the experimental data [221,224,225]. These results have been explained based on the choice of the surrogate of biotic lipid used in Equation 5.45. *n*-Octanol, being an alcohol, would appear to solubilize other alcohols to a greater extent than biotic neutral lipids. Based on it hydrophilicity/lipophilicity characteristics and its fatty acid composition, vegetable oil has been suggested to be an acceptable alternative to *n*-octanol as a surrogate of biotic neutral lipids, especially for hydrophilic organics [221]. Then, to predict the $P_{t:a}$ of hydrophilic VOCs, especially for fatty tissues, $P_{o:w}$ in Equation 5.45 should represent vegetable oil–water PCs. However, in the case of relatively lipophilic VOCs ($\log P_{o:w} > 1.25$) there is little difference between the *n*-octanol–water PCs and vegetable oil–water PCs; therefore, either one of these PCs can be used as the biotic lipid surrogate for solving Equation 5.45 to predict the tissue–air PCs of these chemicals [225].

### Blood–Air Partition Coefficients

Based on Equation 5.45, Poulin and Krishnan [223,225] proposed the following equation for predicting the $P_b$ of VOCs that do not bind significantly to blood proteins:

$$P_b = \left[P_{o:w} P_{w:a} \left(F_{nleb}\right)\right] + \left[P_{w:a}\left(F_{web}\right)\right] \tag{5.46}$$

where $F_{nleb}$ is the fractional volume of neutral lipid equivalents in blood, calculated as the sum of neutral lipids plus 0.3 × phospholipid content, and $F_{web}$ is the fractional volume of water equivalents in blood, calculated as the sum of water content plus 0.7 × phospholipid content.

Accordingly, the $P_b$ of VOCs can be calculated with the knowledge of the blood composition data, $P_{o:w}$, and $P_{w:a}$. The data on lipid and water levels in rat and human blood are available in the literature [222,223] and so are the numerical values of $P_{o:w}$ and $P_{w:a}$ at 37°C for several VOCs [115,151,212]. The predictions of rat $P_b$ obtained using Equation 5.46 are adequate for relatively hydrophilic VOCs (e.g., alcohols, ketones, acetate esters), but it is not the case for relatively lipophilic organic chemicals (e.g., alkanes, haloalkanes, aromatic hydrocarbons). The $P_b$ of a chemical is a composite number that potentially represents two processes occurring in the blood: solubility and binding. Whereas chemical solubility is likely to be determined by the neutral lipid, phospholipid, and water contents in blood, the binding would appear to be associated with plasma proteins and/or hemoglobin. For alcohols, acetate esters, and ketones, rat and human $P_b$ values appear to be adequately predicted by solubility-based algorithms (i.e., Equation 5.46) [223,225]. For more lipophilic VOCs (e.g., alkanes, haloalkanes, aromatic hydrocarbons), however, the rat $P_b$ values calculated using Equation 5.46 were lower (60 to 80%) than the experimental data [223,225]. The fact that the rat $P_b$ values of lipophilic VOCs are underpredicted could be explained by the potential binding of these substances to blood proteins [223], a phenomenon not considered in Equation 5.46. This situation can be rectified by adding another term representative of protein binding to Equation 5.46. This term can be based on the binding association constant or the binding capacity of proteins and free concentration of chemical [220,223], or it can be empirically calculated as ($P_{p:a} \times F_p$) where $P_{p:a}$ is the protein–air partition coefficient and $F_p$ is the fractional volume of binding protein in blood [227]. Poulin and Krishnan [223] derived the binding association constants for alkanes, haloalkanes, and aromatic hydrocarbons in rat blood (1930 ± 819 $M^{-1}$) that are applicable to VOCs possessing the following characteristics: $P_{o:w} > 1$, molecular volume < 300°A, and lacking an oxygen in their molecular formula [220]. This is consistent with the fact that, for several alcohols, acetate esters, and ketones, rat $P_{b:a}$ can be adequately predicted only by considering solubility in blood lipids and water; however, for lipophilic VOCs, binding in blood should be additionally considered. This process would appear to be more important in rat blood than in human blood [115]. At the present time, an *in silico* approach does not exist for predicting protein association constants of organic chemicals in rat or human blood and tissues.

### Tissue–Blood Partition Coefficients

Tissue–blood PCs of VOCs for which macromolecular binding in tissue and blood is negligible can be estimated from *n*-octanol–water PCs or vegetable–water PCs ($P_{o:w}$) as follows:

$$P_t = \frac{\left(P_{o:w} \times F_{nlet}\right) + F_{wet}}{\left(P_{ow} \times F_{nleb}\right) + F_{web}} \tag{5.47}$$

where $F_{nlet}$ is the fractional volume of neutral lipid equivalents in tissue, calculated as the sum of neutral lipid plus 0.3 × phospholipid content, and $F_{wet}$ is the fractional volume of water equivalents in tissue, calculated as the sum of tissue water content plus 0.7 × phospholipid content.

The numerator and denominator of Equation 5.47 essentially correspond to Equation 5.45 and Equation 5.46. In the case of VOCs, $P_t$ values can be obtained by dividing $P_{t:a}$ by $P_b$. The predictions of Equation 5.47 will therefore be identical to the ratio of the predictions obtained using Equation 5.45 and Equation 5.46 [221,222, 225]. The above equation implies that, for extremely hydrophilic chemicals, $P_t$ would be equal to the ratio of the water content of tissues to blood, whereas for highly lipophilic chemicals the $P_t$ values would equal the ratio of the neutral lipid content of tissues to blood [133].

In the above equation, the neutral lipid equivalents have, in some cases, been considered to be equivalent to total lipid content [96]; however, the sum of $F_{nlet} + F_{wet}$ is not always equal to 1 due to the presence of other tissue components such as proteins. This aspect should be appropriately considered when using the tissue composition data to calculate the PCs of chemicals.

When the tissue–air, blood–air, and tissue–blood PCs of un-ionized organic chemicals are not known, Equation 5.45, Equation 5.46, and Equation 5.47 can be used to provide first-cut estimates. Because the tissue and blood composition data can be estimated experimentally or obtained from literature (Table 5.9), only the numerical values of $P_{o:w}$ and $P_{w:a}$ are needed for each new chemical. The $P_{o:w}$ and $P_{w:a}$ in turn can be predicted from molecular structure information [225]. An example of the prediction of tissue–blood PCs from molecular structure information

**TABLE 5.9**
**Neutral Lipid and Water Equivalent Contents of Major Tissues and Blood of Adults, Humans, and Rats**

| | Water Equivalent | | Neutral Lipid Equivalent | |
|---|---|---|---|---|
| Tissues | Rat | Human | Rat | Human |
| Blood | 0.8423 | 0.8217 | 0.0020 | 0.0040 |
| Fat | 0.1215 | 0.1514 | 0.8536 | 0.7986 |
| Liver | 0.7176 | 0.7400 | 0.0425 | 0.0473 |
| Muscle | 0.7471 | 0.7573 | 0.0117 | 0.0378 |

*Source:* Adapted from Poulin, P. et al., in *Toxicity Assessment Alternatives: Methods, Issues, Opportunities*, Salem, H. and Katz, S.A., Eds., Humana Press, Totowa, NJ, 1999, pp. 115–139.

of 1,1,1-trichloroethane is presented in Figure 5.17. More recently, Beliveau et al. [27] have developed QSARs for $P_{w:a}$, $P_{o:a}$, and $P_{p:a}$ to facilitate molecular-structure-based computation of tissue–blood partition coefficients of VOCs containing one of more of the following fragments: $CH_3$, $CH_2$, CH, C, C=C, H, Cl, Br, F, benzene ring, and H in benzene ring. The application of this approach is limited to some lipophilic VOCs and can be extended to other groups of chemicals if the fragment constants can be estimated.

## Biochemical Parameters

Biochemical parameters such as the rates of absorption, biotransformation, macromolecular binding, and excretion can be determined by conducting time-course analysis *in vivo* or *in vitro*. One strategy for accurate estimation of specific biochemical parameters *in vivo* is to conduct experiments under conditions where the pharmacokinetic behavior of a chemical is related to one or two dominant factors and thereby derive estimates of these parameters.

The rate constant for dermal absorption of volatile organic chemicals has been determined by conducting body-only exposure of animals covered with a latex face mask. The total amount of the chemical absorbed through the skin during exposure is calculated by analyzing the blood time-course data collected during the exposure with a PBPK model that has all parameters defined except the skin permeability coefficient ($K_p$). The value of $K_p$ is estimated by fitting PBPK model simulations to the blood time-course concentration data obtained experimentally [189,190].

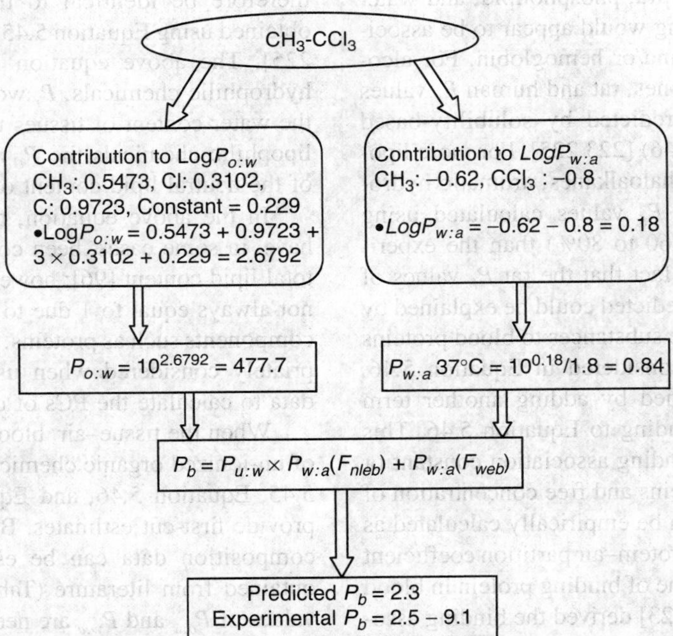


**FIGURE 5.17** Prediction of the human blood–air partition coefficient ($P_b$) of 1,1,1-trichloroethane from knowledge of its molecular structure, according to Poulin and Krishnan [225].

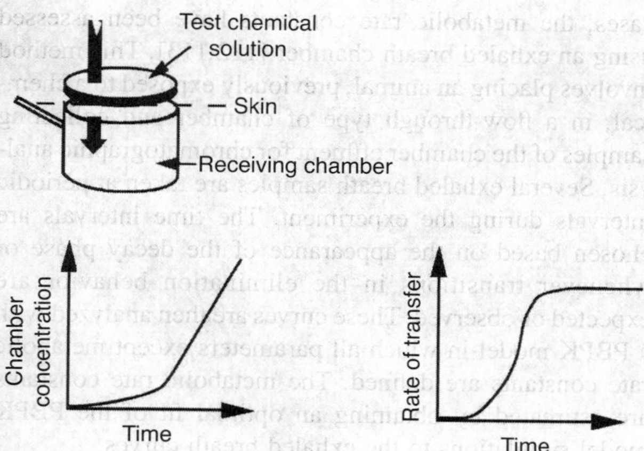


**FIGURE 5.18** Schematic representation of *in vitro* approaches to estimate skin permeability constants. (Adapted from Andersen, M.E. and Keller, W.C., in *Cutaneous Toxicity*, Drill, V.A. and Lazar, P., Eds., Raven Press, New York, 1984, pp. 9–27. With permission.)

The skin permeability constant can also be determined *in vitro* using excised skin tissue. In these experiments, the test material is placed on excised skin in a vehicle and its appearance in the bathing medium determined [12]. A plot of concentration in the bath vs. skin has a time lag, a period of increasing slope, and a final phase of constant slope (Figure 5.18). The plot of rate of uptake (the first derivative of this curve) gives the maximum uptake rate. The permeability coefficient $K_p$ (cm/hr) is calculated by dividing the uptake rate (mg/cm$^2$/hr) with the applied concentration (mg/cm). A number of predictive algorithms and QSARs have been developed for estimating the $K_p$ of organic chemicals (reviewed in Béliveau and Krishnan [26]). The adequacy of these approaches for incorporation within PBPK models to predict dermal absorption in intact animals and humans has not been evaluated systematically.

The rate constant for gastrointestinal absorption ($K_o$) of volatile organics has been estimated by analyzing the time course of exhalation of the parent chemical after oral administration, with a PBPK model that had all parameters except $K_o$ defined [112]. The estimation of the rate of gastrointestinal absorption for less volatile chemicals ($P_b$ > 90) has been performed by determining the blood concentration following oral dosing and analyzing the data with a PBPK model that had all parameters defined except $K_o$ [107,112].

The rate constants for metabolism can be determined *in vivo* or *in vitro*. Two innovative noninvasive methods have been devised for the estimation of the *in vivo* metabolic rate constants of volatile organic chemicals: (1) the closed chamber or gas uptake method, and (2) the exhaled breath chamber method.

The closed chamber or gas uptake method uses a desiccator-type chamber with a recirculating atmosphere for exposing groups of animals to volatile chemicals. Several gas uptake systems have been described in the literature [10,70,99,100]. The gas uptake approach involves periodical monitoring of the chamber concentration of chemicals during exposures and analysis of the time-course data to derive metabolic constants. The rate of change in the chamber concentration of the chemical both in the absence and in the presence of animals is determined for various starting concentrations. The net difference in the rates determined in these two sets of experiments represents the loss of chemical due to uptake and metabolism by the animals. When animals are placed in the gas uptake chamber, the rate of decline in chamber concentration of the chemical increases, with the magnitude being proportional to the rate of metabolism once the chemical has equilibrated within the organism. The rate of spontaneous loss of chemicals in the empty chamber, corresponding to degradation or adsorption to chamber surface, should not exceed ~2% per hour; otherwise, the decline in the chamber concentration may not be sensitive enough to enable determination of the metabolic rate constants.

The gas uptake data analysis is conducted with a PBPK model that has all parameters defined except the metabolic rate constants [114]. Initially, the PBPK model for closed chamber exposures is run for several starting concentrations with the metabolic rate constants set to 0 (i.e., $V_{max}$ = 0, $K_f$ = 0). The model simulations obtained at this stage reflect chemical uptake by the organism in the absence of tissue metabolism. Then, by setting $V_{max}$, $K_m$, and $K_f$ to some numerical values, it can be seen that the model simulations correspond to experimentally observed decline in chamber chemical concentrations (Figure 5.19). By optimization to get the best fit to the set of gas uptake curves, the values of metabolic rate constants are estimated. A sensitivity analysis may be performed in this context to identify key determinants of the estimation of metabolic constants [94]. These results are also likely to be useful in selecting the appropriate initial concentrations to be used in gas uptake studies. A single set of $K_m$ and $V_{max}$ values obtained in these *in vivo* experiments has been considered to represent the role of a single isoenzyme or the average of the sum of the activities of multiple enzymes. The gas uptake method has been used successfully to obtain metabolic rate constants of volatile organic chemicals that are biotransformed by a single first-order process, a saturable process, or a combination of both [8,16].

It is important to monitor the oxygen concentration, humidity level, and chamber pressure during gas uptake studies. Any change in the respiratory rate should be investigated such that modeling of gas uptake exposure provides reliable estimates of metabolic rates [70,152]. Control gas uptake runs, during which naive animals are placed in the chamber without any added chemical, are necessary to ensure the absence of interfering chromatographic peaks

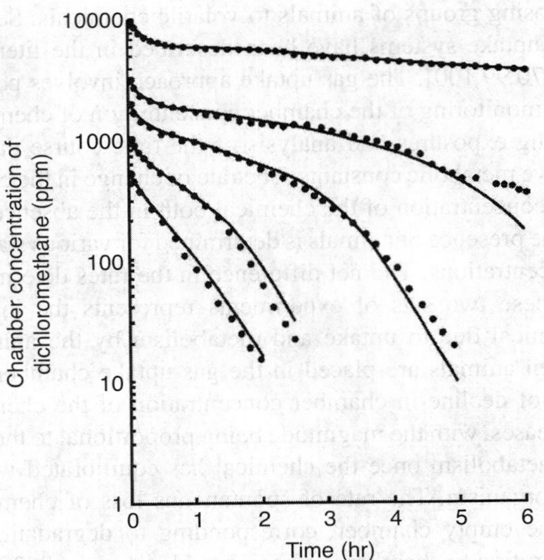


**FIGURE 5.19** Dichloromethane (DCM) gas uptake by groups of 14 mice in a 9.6-L chamber with starting concentrations of 490, 960,2000, 3200, and 10,000 ppm DCM. Experimental data are shown as symbols, and PBPK model simulation is presented as solid lines. Values of kinetic constants for DCM metabolism ($V_{max}$ = 1.054 mg/hr, $K_m$ = 0.396 mg/L) were obtained by computer optimization of these data with a PBPK model that had all parameters defined except metabolic rate constants. (From Andersen, M.E. et al., *Toxicol. Appl. Pharmacol.*, 87, 185–205, 1987. With permission.)

of exhaled endogeneous chemicals. Because the gas uptake studies involve whole-body exposures, adsorption to fur and dermal uptake are possible. If dermal absorption during inhalation exposures is significant or if the PBPK model cannot satisfactorily simulate the chamber decline data with only pulmonary uptake description, then rate constants for the dermal absorption process should be determined independently and included in the model. To determine whether or not adsorption to fur has contributed significantly to chemical uptake during whole-body exposures, the animals, after termination, should be placed in a clean chamber and the time course of the appearance of the chemical determined [113]. Recently, Dennison et al. [82] reported modifications to the design of the gas uptake system, which involved the use of a steel ring to improve connection to the autosampler and adsorbent for controlling the levels of $CO_2$ in the chamber and the simultaneous sampling of chamber gas and blood for analysis.

The gas uptake method is not particularly well suited for use with organic chemicals that (1) have low vapor pressure, (2) exhibit high chamber loss rate, or (3) are highly soluble in blood and tissue. For example, in the case of a chemical with $P_b$ greater than 60, the equilibration phase may become prolonged, thus occupying most part of the gas uptake curves. The behavior in these gas uptake curves is restricted largely to tissue uptake. In such cases, the metabolic rate constants have been assessed using an exhaled breath chamber [112,113]. This method involves placing an animal, previously exposed to a chemical, in a flow-through type of chamber and collecting samples of the chamber effluent for chromatographic analysis. Several exhaled breath samples are taken at periodic intervals during the experiment. The time intervals are chosen based on the appearance of the decay phase or whenever transitions in the elimination behavior are expected or observed. These curves are then analyzed with a PBPK model in which all parameters except metabolic rate constants are defined. The metabolic rate constants are estimated by obtaining an optimal fit of the PBPK model simulations to the exhaled breath curves.

Metabolic rate constants have also been determined by measuring the production of a stable metabolite resulting from the conversion of the parent compound *in vivo* [117]. For a particular metabolite to provide a quantitative measure of the metabolism of the parent chemical, it is ideal if it is produced in the first step of metabolism and is resistant to further biotransformation. Only very few metabolites exhibit these attributes. An example of this kind is the bromide ion resulting from the initial metabolism of dibromomethane. This metabolite is distributed almost exclusively in the extracellular fluid spaces and is excreted in the urine slowly. In this case, the metabolic rate constants were estimated by fitting the simulations of a PBPK model that accounted for the formation, distribution, and excretion of bromide to experimental data on plasma bromide levels [117].

The blood time-course data of the parent chemical obtained after intravenous administration have also been used to determine the rate constants of metabolism [235]. Accordingly, the blood or tissue time course of the parent chemical is obtained following its intravenous administration over a dose range. The experimental data are then analyzed with a PBPK model that has all parameters defined except metabolic rate constants. A single combination of metabolic rate constants (i.e., $V_{max}$, $K_m$, and/or $K_f$) that best describes the data is obtained by fitting the model simulations to the set of blood time-course curves (Figure 5.20). Similarly, pharmacokinetic data obtained following other modes of administration can be used to determine the metabolic rate constants, provided the rate constant of absorption for the particular exposure pathway is known.

In the above methods, we know or hypothesize that the decline in the blood/chamber concentration of parent chemical is determined by the magnitude of the metabolic rate constants and that metabolism occurs via a single first-order, second-order, or saturable process, or via a combination of any two processes in a particular tissue. In the case of chemicals that are metabolized by more than two competing metabolic pathways to varying extents in several tissues, the gas uptake, exhaled breath chamber, o

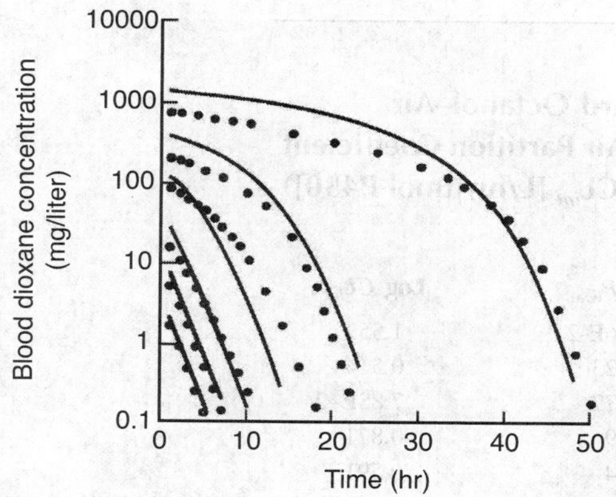


**FIGURE 5.20** Comparison of PBPK model simulations (solid lines) with the experimental data (symbols) of blood concentrations of 1,4-dioxane following intravenous dosing (3, 10, 30, 100, 300, and 1000 mg/kg). Model simulations were obtained using metabolic rate constants estimated by statistical optimization of the model fit to the experimental data. (From Reitz, R.H. et al., *Appl. Pharmacol*., 105, 37–54, 1990. With permission.)

intravenous pharmacokinetic studies alone cannot be used to determine the rate constants for each of the multiple pathways occurring in several tissues. The use of these methods in this case will yield one set of rate constants that represent the overall metabolic clearance of the chemical (i.e., the sum total of metabolism via all metabolic pathways). Alternatively, if the rate constants for all metabolic pathways except one have been determined independently, then an intravenous dosing or a gas uptake study can be conducted to obtain the rate constants for this pathway [171]. The rate constants for individual metabolic pathways can potentially be obtained from *in vitro* studies.

Subcellular fractions, postmitochondrial preparations, isolated cells, tissue slices, and isolated perfused organs are all potentially useful as *in vitro* systems for the estimation of metabolic rate constants [73,75,77,78,123,139,146,179, 180,198–200,234,236,247]. The relevance of rate constants determined *in vitro* to the intact animal is not clear in all cases. The *in vitro* and *in vivo* comparison of intrinsic clearance can be performed with knowledge of the free fraction of chemical in the medium, diffusion rate, volumes of the biological matrix, and the incubation medium, as well as the partition coefficient of the chemical [277]. Several studies using microsomes, postmitochondrial fractions or hepatocytes, have succeeded in determining metabolic rate constants for direct incorporation into PBPK models [13,34,170, 181,218]. The $K_m$ values obtained in *in vitro* studies have been used directly (or scaled to reflect the *in vitro/in vivo* ratios in a test species [234]), but the $V_{max}$ obtained *in vitro* has been scaled to the whole organism based on the mass recovery of the particular fraction. For example, the $V_{max}$ for the intact animal has been estimated from the $V_{max}$ obtained using rat liver microsomes as follows [170]:

$$V_{max(in\ vivo)} = V_{max(in\ vitro)} \times 60 \times C_{prot} \times V_t \tag{5.48}$$

where $V_{max(in\ vivo)}$ is expressed in mg/hr, $V_{max(in\ vitro)}$ is expressed in mg/min/mg protein, 60 is the factor for converting the per-minute rate to per-hour rate, $C_{prot}$ is the concentration of protein in the microsomal sample (mg protein per g tissue), and $V_t$ refers to the volume of the tissue (g tissue).

Care must be taken to check the validity of the various *in vitro* systems to adequately predict the kinetics of chemicals *in vivo* [170]. For example, the clearance of dichloromethane (DCM) by oxidative metabolism estimated *in vivo* using rat liver microsomes is lower than the actual clearance estimated by *in vivo* methods. The examination of the *in vitro*- and *in vivo*-derived rate constants for oxidative metabolism revealed that the $V_{max}$ agreed well, but differences of up to four orders of magnitude in the $K_m$ values were observed between the two approaches. Use of these *in vitro* $K_m$ values in a PBPK model would severely underestimate the amount of DCM metabolized via the oxidative pathway at low exposure concentrations [7,8]. Such a description would have the oxidative pathway competing less efficiently with the glutathione (GSH) conjugation pathway, thus overpredicting metabolite production via the GSH pathway. Products from the GSH pathway have been correlated with tumor outcome, and this incorrect parameter specification would lead to substantial errors in assessing the carcinogenic risk associated with low-level DCM exposures [7,8].

The identification of *in vitro* systems for determining metabolic rate constants that give values consistent with those operative *in vivo* is crucial for eventually predicting human dosimetry. Recent studies indicate that freshly isolated hepatocytes and postmitochondrial fractions can provide rate constants comparable to the *in vivo* estimates. Our ability to predict metabolic rates from one *in vitro* system to another is a significant step toward predicting the rate constants for the *in vivo* system [181].

Mechanistic animal-replacement approaches for predicting the numerical values of $E$, $V_{max}$, $K_m$, or $Cl_{int}$ of chemicals are not available yet; however, semiempirical approaches relating the molecular structure information to PBPK model parameters such as partition coefficients and metabolic rate constants have been developed [27,28,118, 184,210,211,282]. Table 5.10 presents the numerical values of the fragment constants for estimating the $Cl_{int}$ and partition coefficients for a series of VOCs. A major limitation of these approaches is that experimental data for a series of related chemicals should be collected before developing equations that consistently describe the relationship between molecular structure and PBPK model parameters. Truly predictive approaches can only be developed as our understanding of the biochemical processes improve. At the present time, however, the hepatic

**TABLE 5.10**
**Fragment-Specific Contributions Toward Octanol–Air Partition Coefficient ($P_{o:a}$) and Water–Air Partition Coefficient ($P_{w:a}$), As Well As Intrinsic Clearance ($CL_{int}$ [L/hr/μmol P450]) of Some Volatile Organic Chemicals**

| Fragments | Log $P_{o:a}$ | Log $P_{w:a}$ | Log $CL_{int}$ |
|---|---|---|---|
| $CH_3$ | 0.354 | –3.76E-2 | 1.552 |
| $CH_2$ | 0.441 | –0.223 | 0.514 |
| CH | 0.377 | –0.477 | 7.83E–2 |
| C | –0.354 | –1.49 | –0.871 |
| C=C | 0.197 | –1.94 | 0.591 |
| H (ON=) | 0.134 | 0.555 | 0.383 |
| Br | 0.174 | 0.622 | 1.00 |
| Cl | 0.776 | 0.468 | 0.522 |
| F | 0.136 | 0.229 | |
| AC | 3.729 | 0.650 | –7.646 |
| H-AC | –0.19.0 | –0.0624 | 1.535 |

*Source:* Adapted from Béliveau, M. et al., *Chem. Res. Toxicol.*, 18, 475–485, 2005.

extraction can be assumed to be complete or negligible in PBPK models to generate simulations. Accordingly, the numerical value of $E$ in Equation 5.21 may be set to 0 or 1 during the model simulations. The region encompassed by the simulated lines obtained with $E = 0$ and $E = 1$ will naturally contain the experimental data for that particular exposure scenario (e.g., Figure 5.21). This approach is particularly useful for simulating the range of internal dose of chemicals that are not rapidly cleared at the port of entry (e.g., first-pass metabolism following oral uptake). Poulin and Krishnan [227] used this approach successfully to simulate the range of blood concentration profiles of dichloromethane, ethyl benzene, trichloroethylene, tetrachloroethylene, *m*-xylene, and 1,1,1-trichloroethane in humans following inhalation exposure. Their results indicate that the ratio of blood AUC values for these chemicals would only vary by a factor of about 2 to 5 when simulations are conducted based on assumptions of $E = 0$ and $E = 1$ (i.e., when the actual rate of metabolism is not known). Such screening level approaches to parameter estimation might help determine the extent of improvement in model predictions that can be obtained while investing time and energy to refine or estimate specific input parameters for PBPK models.

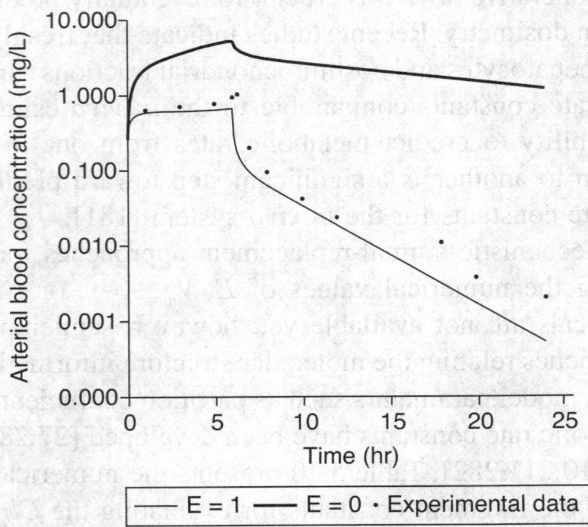


**FIGURE 5.21** Comparison of experimental data (closed circles) on venous blood concentration with the envelope simulated by rat PBPK model for a 6-hr exposure to 80 ppm styrene (solid lines). The upper line corresponds to the simulation obtained when the hepatic extraction was set to zero, whereas the lower line represents simulation obtained when the hepatic extraction ratio was set to 1. The experimental data were obtained from Ramsey and Andersen [230].

The rate constants of chemical reactions with hemoglobin, tissue proteins, etc., determined *in vitro* or *in vivo* have been incorporated into the PBPK model to make predictions of these phenomena *in vivo* [109,171]. Attempts have also been made to include receptor binding and DNA binding properties of chemicals into a PBPK modeling framework based on *in vitro*-derived data [98,264]. *In vitro* and *in vivo* assays are also useful for estimating the binary level inhibition constants for developing PBPK models of mixtures [23,258].

Some PBPK models in which perfusion-limited tissue descriptions are used may predict greater tissue uptake of chemicals than what actually is observed. In such cases, tissue uptake is described as a diffusion-limited process and the mass transfer coefficient for each tissue is required. In this regard, the permeation-area cross product for tissues ($PA_t$) is estimated by fitting model simulations to experimental data on tissue concentrations of the parent chemical

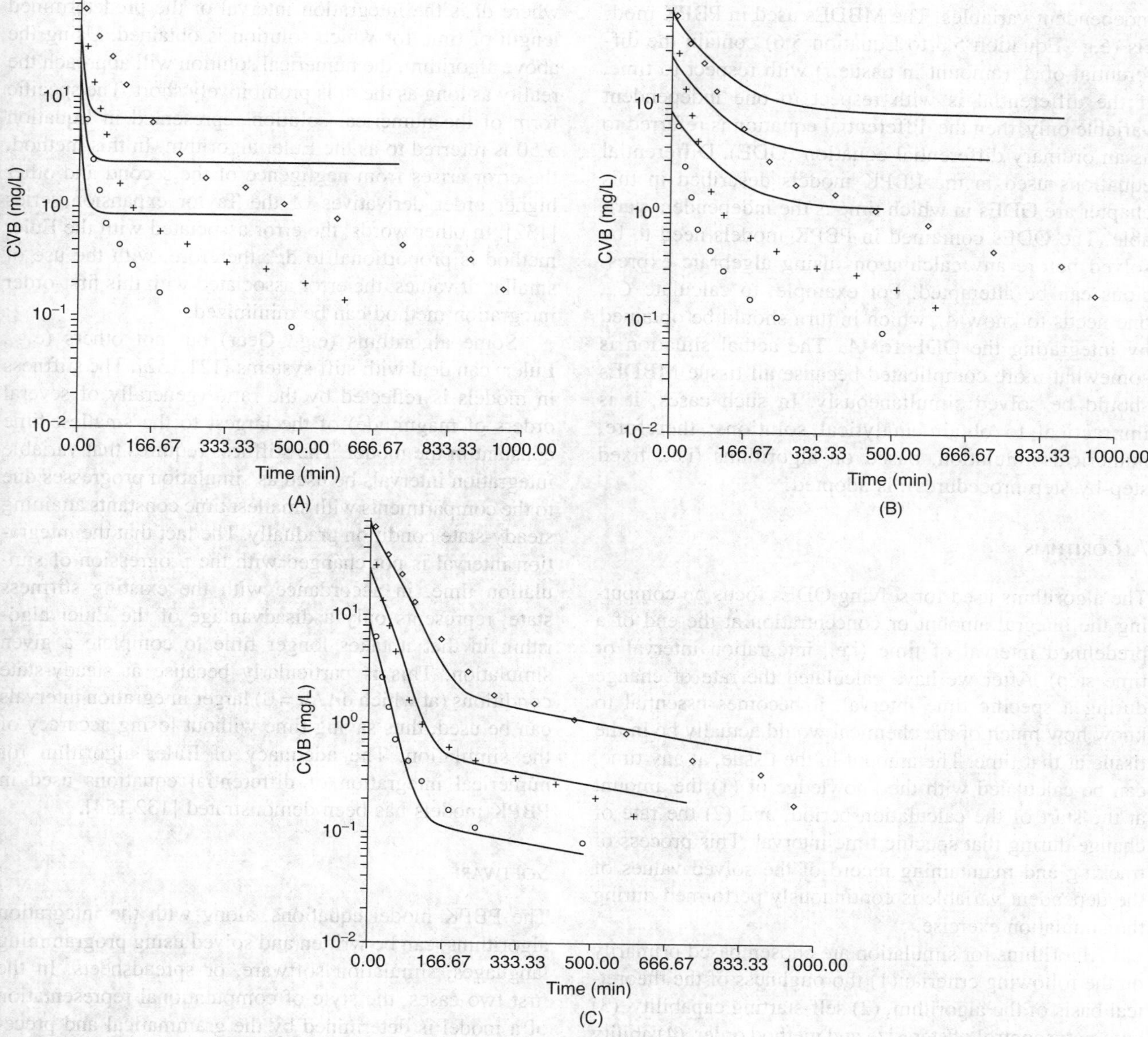


**FIGURE 5.22** Comparison of the experimental data (symbols) on the venous blood concentration of unchanged pyrene in rats with PBPK model simulations (solid lines). The simulations were obtained with a model that described uptake as a perfusion-limited process in all tissues (A) or as a diffusion-limited process in adipose tissues and slowly perfused tissues (B). With the additional description of metabolism in appropriate tissues, the PBPK model simulations corresponded to the experimental data (C). The dose levels were 2, 6, or 15 mg/kg (i.v.). (From Haddad, S. et al., *Environ. Toxicol. Pharmacol.*, 5, 245–255, 1998. With permission.)

and using a PBPK model in which all other parameters are defined (Figure 5.22) [15,74,135]. When the mathematical representation of a PBPK model is prepared and its parameters estimated, the model can be used to simulate the kinetic behavior of a chemical in the test species.

## MODEL SIMULATION

Simulation predicts system behavior by solving the differential equations representing the quantitative interrelations among the various model parameters. Simulation, in the context of PBPK modeling, refers to the prediction of the kinetic profiles of chemicals in blood and tissues by solving the set of MBDEs. Typically, simulation is an option when (1) the real system does not exist, (2) the real system exists but experimentation is expensive, (3) a forecasting model is required to analyze events occurring over long periods of time in a compressed format, and (4) the model does not have practical analytical solutions (e.g., nonlinear systems) [238]. PBPK models often contain differential equations that contain the differentials of dependent variables with respect to independent variables. The MBDEs used in PBPK models (e.g., Equation 5.3 to Equation 5.6) contain the differential of $A_t$ (amount in tissue $t$) with respect to time. If the differential is with respect to one independent variable only, then the differential equation is referred to as an ordinary differential equation (ODE). Differential equations used in the PBPK models described in this chapter are ODEs in which time is the independent variable. The ODEs contained in PBPK models need to be solved before any calculations using algebraic expressions can be attempted. For example, to calculate $C_{vt}$, one needs to know $A_t$, which in turn should be obtained by integrating the ODE for $A_t$. The actual situation is somewhat more complicated because all tissue MBDEs should be solved simultaneously. In such cases, it is impractical to obtain analytical solutions; therefore, numerical simulation, based on algorithms (i.e., fixed step-by-step procedures), is adopted.

## Algorithms

The algorithms used for solving ODEs focus on computing the integral amount or concentration at the end of a predefined interval of time (i.e., integration interval or time step). After we have calculated the rate of change during a specific time interval, it becomes essential to know how much of the chemical would actually be in the tissue at that time. The amount in the tissue, at any time, can be calculated with the knowledge of (1) the amount at the start of the calculation period, and (2) the rate of change during that specific time interval. This process of tracking and maintaining record of the solved values of the dependent variable is continuously performed during the simulation exercise.

Algorithms for simulation are chosen based primarily on the following criteria: (1) thoroughness of the theoretical basis of the algorithm, (2) self-starting capability, (3) automatic control of step size and method order, (4) ability to deal with both stiff and nonstiff problems and to detect stiffness automatically, and, finally, (5) proof that the algorithm works for test problems of the same kind as the ones under consideration [238]. Some of the commonly used algorithms include Euler, Gear, Runge-Kutta routines and predictor–corrector methods.

The general principle underlying these algorithms used for solving first-order ordinary differential equations can be represented in simple terms as follows:

$$\text{New value} = \text{Old value} + (\text{slope} \times \mathrm{d}t) \tag{5.49}$$

For a tissue compartment in PBPK model,

$$A_{t,1} = A_{t,0} + \left(dA_t/\mathrm{d}t \times \mathrm{d}t\right) \tag{5.50}$$

where d$t$ is the integration interval or the predetermined length of time for which solution is obtained. Using the above algorithm, the numerical solution will approach the reality as long as the d$t$ is prohibitively short. The specific form of the numerical solution represented in Equation 5.50 is referred to as the Euler algorithm. In this method, the error arises from negligence of the second and other higher order derivatives of the Taylor expansion series [132]. In other words, the error associated with the Euler method is proportional to $\mathrm{d}t^2$; therefore, with the use of smaller d$t$ values, the error associated with this first-order integration method can be minimized.

Some algorithms (e.g., Gear) but not others (e.g., Euler) can deal with stiff systems [121,132]. The stiffness in models is reflected by the ratio (generally of several orders of magnitude) of the largest to the smallest time constant in the model. The stiffness requires that variable integration intervals be used as simulation progresses due to the compartments with smallest time constants attaining steady-state condition gradually. The fact that the integration interval is not changed with the progression of simulation time, in accordance with the existing stiffness state, represents only a disadvantage of the Euler algorithm in that it takes longer time to complete a given simulation. This is particularly because at steady-state conditions (at which $\mathrm{d}A_t/\mathrm{d}t = 0$) larger integration intervals can be used, thus saving time without losing accuracy of the simulation. The adequacy of Euler algorithm for numerical integration of differential equations used in PBPK models has been demonstrated [132,154].

## Software

The PBPK model equations, along with the integration algorithms, can be written and solved using programming languages, simulation software, or spreadsheets. In the first two cases, the style of computational representation of a model is determined by the grammatical and precedence rules of the programming language (e.g., FORTRAN, BASIC) or simulation language to be used. Simulation languages are computer programming packages that are general in nature but may have special features for modeling certain types of systems. Examples of simulation languages that possess features particularly useful for PBPK modeling are listed in Table 5.11. When selecting a particular simulation language for modeling, it is important to ensure that it (1) provides a convenient means for initializing the status of the model (e.g., generating random numbers in case of stochastic models); (2) permits the introduction of changes in both the status and temporal structure of the model as simulation time evolves (i.e., scheduling the occurrence of events); (3) provides simple methods by which model results and statistical summaries can be obtained; (4) allows considerable flexibility in conducting sensitivity and other types of model analyses; and

**TABLE 5.11**
**Examples of Simulation Software Used for PBPK Modeling**

| Name | Refs. |
|---|---|
| ACSL, ACSL-Tox or acslXterme (Advance Continuous Simulation Language) | 86, 266 |
| AVS (Application Visualization System) | 203 |
| BASICA | 86 |
| Berkely Madonna | 231 |
| CMATRIX | 21 |
| Fortran compiler with IMSL library packages, C, Pascal, Basic | 141, 161 |
| Matlab | 88, 280 |
| MCSim | 160 |
| Microsoft Excel® | 132, 154 |
| ScoP (Simulation Control Program) | 167, 192 |
| SimuSolv | 237 |
| SONCHES (Simulation of Nonlinear Complex Hierarchical Ecological Systems) | 285 |
| Stella | 141 |

*Source:* EPA, *Approaches for the Application of Physiologically Based Pharmacokinetic (PBPK) Models and Supporting Data in Risk Assessment*, EPA 600/R-05/043F, U.S. Environmental Protection Agency, Washington, D.C., 2006.

(5) contains error detection facilities [238]. The choice of simulation software is up to the individual as long as the software package provides a framework for creating and solving the type of model equations under consideration.

Several commercially available simulation or programming software packages can be used for conducting PBPK model simulations [192]. These programs and packages are easily accessed and understood by individuals who are familiar with the techniques of programming or mechanics of simulation. The models constructed and solved using simulation packages appear like a "black box" to the beginner-level analyst, who gets to see the end result but not the temporal evolution of solutions to the complex mathematical formulations constituting the basis of the end results generated. The internal mechanics of computer simulation in such cases can be visualized if the user can reconstruct the way in which the simulation software (1) solves each equation in the model, and (2) takes the output of one equation and provides it as input to other equations of the model. This can be accomplished using spreadsheet programs such as Lotus 1-2-3®, QuattroPro®, and Microsoft Excel®. The limitations of the spreadsheet approach relate to (1) the number of cells required for conducting the analysis, and (2) the runtime required to solve complex PBPK models. This approach, therefore, is appropriate for individuals who do not have sufficient knowledge of numerical integration algorithms and simulation software. The beginner who develops an understanding of how the PBPK models work using the spreadsheet-based methodology can then move on to using advanced techniques and specialized simulation languages offering flexibility, speed, and additional features (e.g., sensitivity analysis, optimization routines).

### Hardware

The equations and algorithms constituting PBPK models can be solved using programming languages, simulation software, or spreadsheet programs on mainframes, workstations, desktops, or laptops. In general, for running large models, dedicated workstations and multi-user mainframe computers offer the lowest execution times, thus providing overall time savings. For a PBPK modeler working with small models and simple descriptions, Macintosh- or IBM PC-based modeling packages are sufficient. The processing speed, hard-disk space, and runtime memory of desktops and laptops marketed today are quite adequate for PBPK modeling. The sample simulations obtained using spreadsheets are presented in the following section.

### Sample Simulations

Let's consider a simple, two-compartment PBPK model for simulating the inhalation pharmacokinetics of styrene in the rat (Figure 5.23). First, the numerical values of each of the PBPK model parameters should be entered into a specific cell in the spreadsheet and identified appropriately; for example, the numerical value contained in cell C5 is referred to as $Q_p$ (Table 5.12). Because the alveolar ventilation is referred to as $Q_p$ in this example, whenever $Q_p$ is typed in any other part of the spreadsheet, the numerical value found in cell C5 will be imported automatically.

Table 5.13 lists the manner in which the model equations are written in the spreadsheets. In addition to the equations for computing blood concentrations, four equations per compartment are written. These equations correspond to the tracking of: (1) the rate of change in the amount of chemical in tissue, (2) the amount of chemical

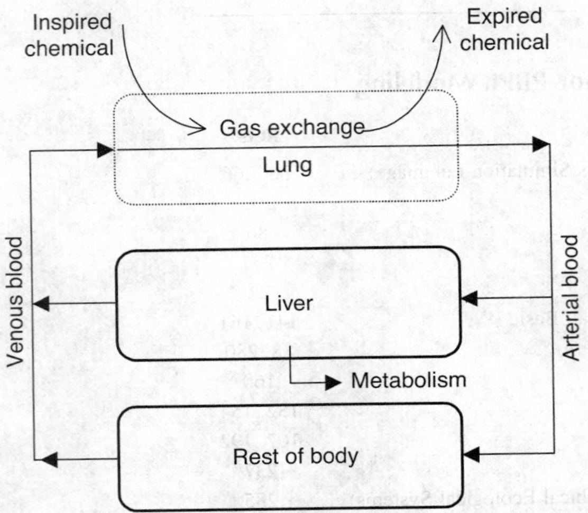


**FIGURE 5.23** Conceptual representation of a two-compartmental PBPK model for styrene in the rat.

in tissue, (3) the concentration of chemical in tissue, and (4) the concentration of chemical in venous blood leaving tissue. The calculation of the amount of chemical in tissue is based on Euler algorithm. If intended, other integration algorithms can be used. In the various equations (Table 5.13), the model parameters are referred to by using the appropriate abbreviations (e.g., $Q_p$, $Q_c$, $V_l$) whereas the variables are referred to with the use of relative reference expressions. The relative expression involves referring to a cell according to its location relative to another cell where the calculation is being carried out. This option is particularly useful when the output of an equation contained in one cell is to be provided as input for equation contained in another (e.g., adjacent) cell. Thus, the use of relative reference expressions in spreadsheets may be useful to facilitate loop-type calculations essential for advancing the state of a system during simulations.

Table 5.14 presents a part of the spreadsheet depicting how equations presented in Table 5.13 are actually entered into spreadsheets. Accordingly, in the spreadsheet, the descriptions of the two tissue compartments occupy eight columns (columns E to L, in Table 5.14 and Figure 5.24), and the calculation/representation of the simulation time, exposure concentration, arterial concentration, and venous blood concentration occupy one column each (columns B, C, D, and M, respectively). The mixed venous blood concentration resulting from that of the venous blood exiting the two tissue compartments described in a particular row is calculated in the same row (i.e., cell M35 in Table 5.14 and Figure 5.24). This $C_v$ then is used along with $C_{inh}$ (i.e., cell C36) to calculate $C_a$ in the subsequent row (e.g., cell D36). In this structure then, according to the schematics shown in Figure 5.23 and Figure 5.24, all model equations are interconnected by specifying the proven or hypothetical input–output connections among them.

The simulation begins when the numerical values of the model parameters have been provided, the equations in the first and subsequent rows of the spreadsheet are entered, the time interval for integration is specified, and the required number of cells is chosen. One has only to repeat the calculations shown in row 36 of Figure 5.24 for each time interval of integration until the end of the desired duration of simulation. In the present example, the time interval of integration was fixed at 0.005 hr. Each line in the Excel® spreadsheet then represents calculations

**TABLE 5.12**
**List of Parameters for the Two-Compartment PBPK Model, Their Numerical Values, and Location in Excel® Spreadsheet**

| Parameters | Abbreviation[a] | Numeric Values[b] | Place of Cell[c] |
|---|---|---|---|
| Cardiac output | Qc | 5.64 L/hr | C4 |
| Alveolar ventilation rate | Qp | 4.5 L/hr | C5 |
| Hepatic blood flow | Ql | 2.11 L/hr | C6 |
| Blood flow in rest of body | Qbo | 0.261 L/hr | C8 |
| Liver volume | Vl | 0.012 L | D6 |
| Volume of rest of body | Vbo | 0.012 L | D6 |
| Liver–blood partition coefficient | Pl | 2.7 | E6 |
| Rest of body–blood partition coefficient | Pf | 50 | E8 |
| Blood–air partition coefficient | Pb | 40 | F7 |
| Maximal velocity of metabolism | Vmax | 3.6 mg/hr | G6 |
| Michaelis–Menten affinity constant | Km | 0.36 mg/L | H6 |

[a] The various model parameters are referred to, with the use of these abbreviations in the spreadsheet.
[b] All parameters estimates were based on Ramsey and Andersen [230].
[c] The cell locations provided here correspond to the column and row coordinates respectively (i.e., the alphabetical letters denote the columns and the Arabic numerals correspond to the rows of the spreadsheet).

**TABLE 5.13**
**Equations Used in the Calculation of Tissue, Arterial, and Venous Blood Concentrations of Styrene, and Their Expression in EXCEL® Spreadsheet**

| Compartment | Equations[a] | Expression in EXCEL® [b] |
|---|---|---|
| Arterial blood | $C_{a,n} = \dfrac{Q_c \times C_{v,n-1} + Q_p \times C_{inh,n}}{Q_c + Q_p/P_b}$ | D36 = ((Qc*M35) + (Qp*C36))/(Qc + (Qp/Pb)) |
| Liver | $dA_l/dt_n = Q_l \times (C_{a,n} - C_{vl,n-1}) - \dfrac{V_{max} \times C_{vl,n-1}}{K_m + C_{vl,n-1}}$ | E36 = (Ql*(D36-H35)) - (Vmax*H35/(Km+H35)) |
| | $A_{l,n} = dA_l/dt_n \times t + A_{l,n-1}$ | F36 = E36 *t + F35 |
| | $C_{l,n} = \dfrac{A_{l,n}}{V_l}$ | G36 = F36 / Vl |
| | $C_{vl,n} = \dfrac{C_{l,n}}{P_l}$ | H36 = G36 / Pl |
| Rest of body | $dA_{bo}/dt_n = Q_{bo} \times (C_{a,n} - C_{vbo,n-1})$ | I36 = Qbo*(D36 - L35) |
| | $A_{bo,n} = dA_{bo}/dt_n \times t + A_{bo,n-1}$ | J36 = I36 *t + J35 |
| | $C_{bo,n} = \dfrac{A_{bo,n}}{V_{bo}}$ | K36 = J36 / Vbo |
| | $C_{vbo,n} = \dfrac{C_{bo,n}}{P_{bo}}$ | L36 = K36 / Pbo |
| Venous blood | $C_{v,n} = \dfrac{Q_l \times C_{vl,n} + Q_{bo} \times C_{vbo,n}}{Q_c}$ | M36 = ((Ql*H36) + (Qbo*L36))/Qc |

[a] All abbreviations and symbols used in the equations, except *n* and *n*–1, are defined in Table 5.1. Subscripts *n* and *n*–1 refer to the current and previous simulation times. The difference between *n* and *n*–1 in the styrene example was 0.005 hr.

[b] The components of these equations refer either to absolute references (in the case of constant input parameters, as defined in Table 5.12) or to relative references (in the case of state variables).

characterizing the state of the system at every 0.005 hr. In this case, simulations were conducted for 24 hr using 0.005 hr as the integration interval (Figure 5.24). In total, then, 4800 (=24/0.005) lines were used to conduct these PBPK simulations. The solution to the set of PBPK model equations is generated every time the numerical values in cells corresponding to input parameters are changed, as these cells are specified in one or more equations appearing in the spreadsheet (Figure 5.24). Figure 5.25 presents simulations of the pharmacokinetics of styrene obtained using the parameters and equations for a four-compartment PBPK model developed by Ramsey and Andersen [230].

The PBPK modeling framework uniquely captures and simulates interactions at multiple levels when the individual chemical descriptions are linked at the binary level. This approach has been applied to simulate the kinetics of benzene and dichloromethane in mixtures of varying complexities [134]. Figure 5.26 illustrates the change in the blood concentration profile of benzene in rats exposed to 50 ppm of this chemical alone or in combination with 50 ppm each of toluene, dichloromethane,

**TABLE 5.14**
**A Portion of the Spreadsheet Depicting the Entry of Model Equations**

| | B | C | D | E | F | G | H | I | J | K | L | M |
|---|---|---|---|---|---|---|---|---|---|---|---|---|
| 33 | | | | Liver | | | | Body | | | | |
| 34 | Time (hr) | Cinh (mg/L) | Ca | dAl/dt | Al | Cl | Cvl | dAo/dt | Abo | Cbo | Cvbo | Cv |
| 35 | =t | =Cinh | =((Qc*O)+ (Qp*C36))/ (Qc+(Qp/Pb)) | =Ql*(D36–0))– (Vmax*0/ (Km+0)) | =E36*t+0 | =F36/VI | =G36/Pl | =Qbo*(D36-0) | =136*t+0 | =J36/Vbo | =K36/Pbo | =((Ql*H36)+ (Qbo*L36))Qc |
| 36 | =t+B35 | =Cinh | =((Qc*M35)+ (Qp*C36))/ (Qc+(Qp/Pb)) | =(Ql*(D36–H35)) –(Vmax*H35/ (Km+H35)) | =E36*t+F35 | =F36/VI | =G36/Pl | =Qbo*(D36-L35) | =136*t+J35 | =J36/Vbo | =K36/Pbo | =((Ql*H36)+ (Qbo*L36))/Qc |

*Note:* The row and column coordinates are designated by an Arabic numeral and an alphabetical letter, respectively. The equations found in rows 35 and 36 correspond to calculations at the first and second integration intervals. For continuing the simulation, the set of equations in row 36 should be copied onto the desired number of subsequent rows. In this table, column B represents the state of the system, which advances during each time interval (*t*). Column C contains the exposure concentration at any given time during the simulation, whereas column D and U represent calculations of arterial and venous blood concentrations of the chemical. In between, four columns per compartment (e.g., liver, columns E to H; rest of the body, columns I to L) are devoted to calculation of: (1) the rate of change in the amount of chemical in tissue, (2) amount of chemical in tissue, (3) concentrations of chemical in tissue, and (4) concentrations of chemical in venous blood leaving tissue. All abbreviations are defined in Table 5.12.

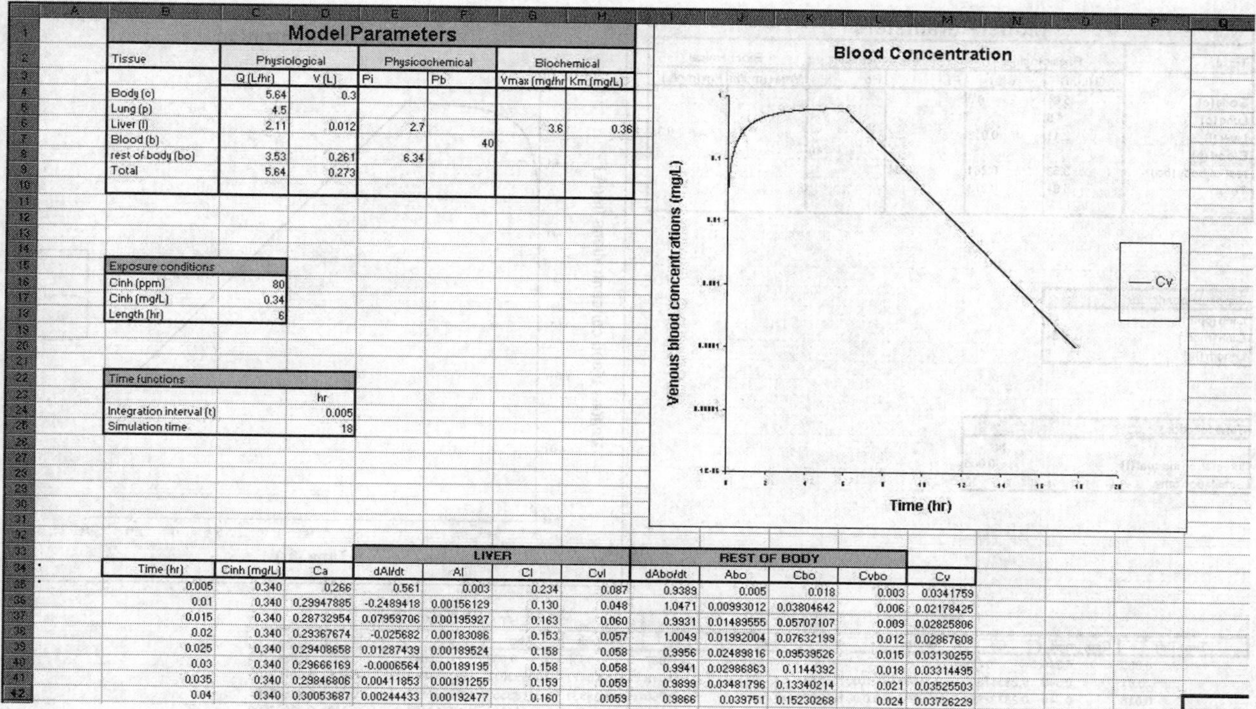

| Model Parameters | | | | | | |
|---|---|---|---|---|---|---|
| Tissue | Physiological | | Physicochemical | | Biochemical | |
| | Q (L/hr) | V (L) | Pi | Pb | Vmax (mg/hr | Km (mg/L) |
| Body (c) | 5.64 | 0.3 | | | | |
| Lung (p) | 4.5 | | | | | |
| Liver (l) | 2.11 | 0.012 | 2.7 | | 3.6 | 0.36 |
| Blood (b) | | | | 40 | | |
| rest of body (bo) | 3.53 | 0.261 | 6.34 | | | |
| Total | 5.64 | 0.273 | | | | |

| Exposure conditions | |
|---|---|
| Cinh (ppm) | 80 |
| Cinh (mg/L) | 0.34 |
| Length (hr) | 6 |

| Time functions | |
|---|---|
| | hr |
| Integration interval (t) | 0.005 |
| Simulation time | 18 |

| | | | LIVER | | | | REST OF BODY | | | | |
|---|---|---|---|---|---|---|---|---|---|---|---|
| Time (hr) | Cinh (mg/L) | Ca | dAl/dt | Al | Cl | Cvl | dAbo/dt | Abo | Cbo | Cvbo | Cv |
| 0.005 | 0.340 | 0.266 | 0.561 | 0.003 | 0.234 | 0.087 | 0.9389 | 0.005 | 0.018 | 0.003 | 0.0341759 |
| 0.01 | 0.340 | 0.29947885 | -0.2489418 | 0.00156129 | 0.130 | 0.048 | 1.0471 | 0.00993012 | 0.03804642 | 0.006 | 0.02178425 |
| 0.015 | 0.340 | 0.28732954 | 0.07959706 | 0.00195927 | 0.163 | 0.060 | 0.9931 | 0.01489555 | 0.05707107 | 0.009 | 0.02825806 |
| 0.02 | 0.340 | 0.29367674 | -0.025682 | 0.00183086 | 0.153 | 0.057 | 1.0049 | 0.01992004 | 0.07632199 | 0.012 | 0.02867608 |
| 0.025 | 0.340 | 0.29408658 | 0.01287439 | 0.00189524 | 0.158 | 0.058 | 0.9956 | 0.02489816 | 0.09539526 | 0.015 | 0.03130255 |
| 0.03 | 0.340 | 0.29666169 | -0.0006564 | 0.00189195 | 0.158 | 0.058 | 0.9941 | 0.02986863 | 0.1144392 | 0.018 | 0.03314495 |
| 0.035 | 0.340 | 0.29846806 | 0.00411853 | 0.00191255 | 0.159 | 0.059 | 0.9899 | 0.03481796 | 0.13340214 | 0.021 | 0.03525503 |
| 0.04 | 0.340 | 0.30053687 | 0.00244433 | 0.00192477 | 0.160 | 0.059 | 0.9866 | 0.039751 | 0.15230268 | 0.024 | 0.03726229 |



**FIGURE 5.24** Printout of a computer screen depicting an EXCEL® spreadsheet with (1) the plot of two-compartment PBPK model simulation of venous blood concentrations of styrene during and following a 6-hr exposure of rats to 80 ppm of this chemical; (2) the numerical values of the PBPK model parameters; and (3) a portion of the raw numbers corresponding to the rate of change in the amount of styrene in tissue, amount in tissue, concentration in tissue, and venous blood concentration leaving the tissues, generated during simulations between time 0.005 and 18 hr.

*m*-xylene, and ethylbenzene. This figure also depicts the PBPK model simulations of the change in blood concentration kinetics of dichloromethane during mixed exposures [134].

## MODEL EVALUATION AND VALIDATION

When the model structure, equations, and parameters have been developed and the code has been written in a simulation or programming language, the task becomes one of evaluating whether the model is useful. The primary objective of the model evaluation or validation process is to determine whether all major determinants and processes that are essential for describing the system behavior have been adequately identified and characterized. Some investigators use the term *validation*, but others prefer to refer to the process as *evaluation*. Even though no consensus or guideline regarding the terminology exists, the intent here is the same: to evaluate whether the model is a useful tool to simulate the behavior of a chemical under specific conditions. Such an evaluation is generally done by comparing the model simulations with diverse sets of experimental data that have not been previously used for estimating parameters and by conducting sensitivity/uncertainty/variability analyses. No single, accepted method of model evaluation or validation is available; the choice of methods for evaluating and validating models depends on the purposes for which the models are to be used [275].

### COMPARISON WITH EXPERIMENTAL DATA

Physiologically based pharmacokinetic models, as other mechanistic models, are simplified representations of the system under study; therefore, not all but only those system variables that the investigator hypothesizes to be critical determinants are accounted for by these models. A high degree of concordance between model predictions and diverse sets of experimental data leads to greater confidence in the predictive capability of the model. Meaningful comparisons of PBPK model simulations with experimental data can be performed by visual inspection, statistical tests, or discrepancy indices. It is important to ensure that such comparisons be made for the dose metrics of relevance to the intended use of the model. For example, if a model for predicting metabolite concentrations in tissues is validated or evaluated with data on mixed venous blood concentrations in test species and humans, it will be of only limited use [57]. Further, when multiple models

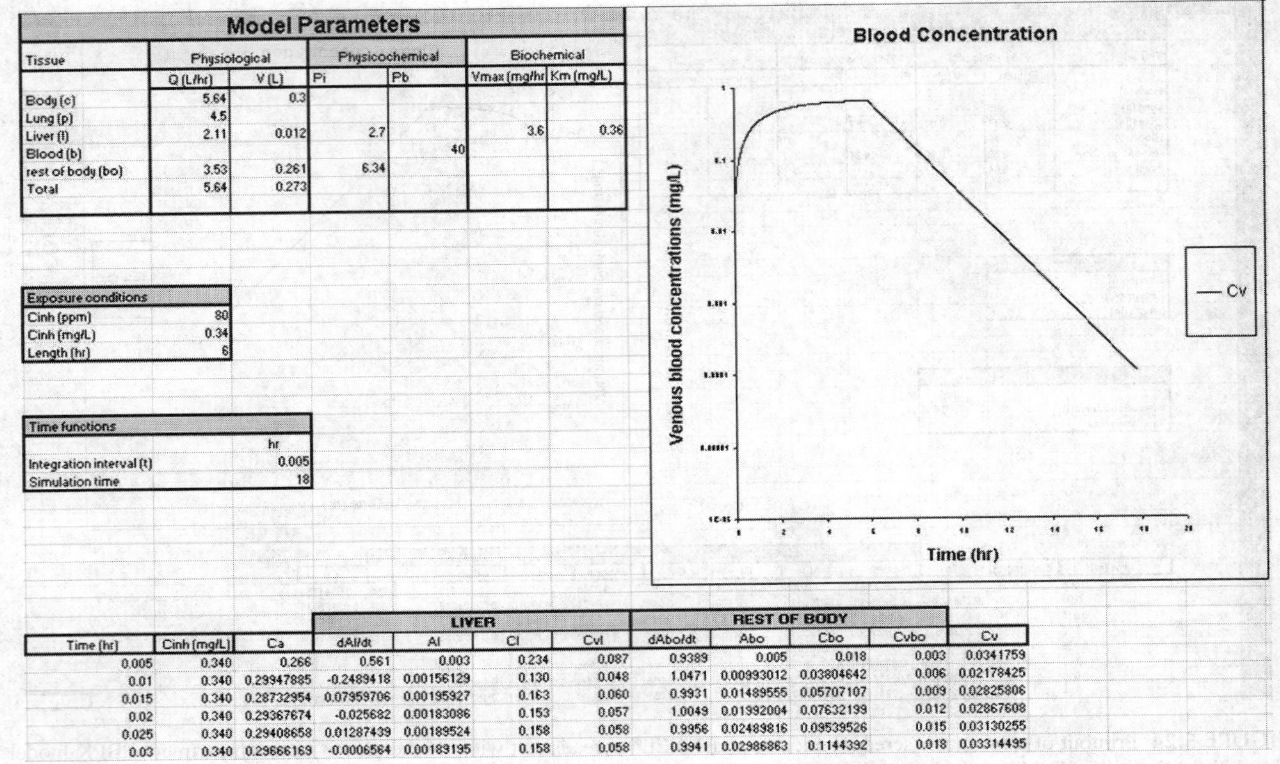

**Model Parameters**

| Tissue | Physiological | | Physicochemical | | Biochemical | |
|---|---|---|---|---|---|---|
| | Q (L/hr) | V (L) | Pi | Pb | Vmax (mg/hr | Km (mg/L) |
| Body (c) | 5.64 | 0.3 | | | | |
| Lung (p) | 4.5 | | | | | |
| Liver (l) | 2.11 | 0.012 | 2.7 | | 3.6 | 0.36 |
| Blood (b) | | | | 40 | | |
| rest of body (bo) | 3.53 | 0.261 | 6.34 | | | |
| Total | 5.64 | 0.273 | | | | |

| Exposure conditions | |
|---|---|
| Cinh (ppm) | 80 |
| Cinh (mg/L) | 0.34 |
| Length (hr) | 6 |

| Time functions | |
|---|---|
| | hr |
| Integration interval (t) | 0.005 |
| Simulation time | 18 |

| | | | LIVER | | | | REST OF BODY | | | | |
|---|---|---|---|---|---|---|---|---|---|---|---|
| Time (hr) | Cinh (mg/L) | Ca | dAl/dt | Al | Cl | Cvl | dAbo/dt | Abo | Cbo | Cvbo | Cv |
| 0.005 | 0.340 | 0.266 | 0.561 | 0.003 | 0.234 | 0.087 | 0.9389 | 0.005 | 0.018 | 0.003 | 0.0341759 |
| 0.01 | 0.340 | 0.29947885 | -0.2489418 | 0.00156129 | 0.130 | 0.048 | 1.0471 | 0.00993012 | 0.03804642 | 0.006 | 0.02178425 |
| 0.015 | 0.340 | 0.28732954 | 0.07959706 | 0.00195927 | 0.163 | 0.060 | 0.9931 | 0.01489555 | 0.05707107 | 0.009 | 0.02825806 |
| 0.02 | 0.340 | 0.29367674 | -0.025682 | 0.00183086 | 0.153 | 0.057 | 1.0049 | 0.01992004 | 0.07632139 | 0.012 | 0.02867608 |
| 0.025 | 0.340 | 0.29408658 | 0.01287439 | 0.00189524 | 0.158 | 0.058 | 0.9956 | 0.02489816 | 0.09539526 | 0.015 | 0.03130255 |
| 0.03 | 0.340 | 0.29666169 | -0.0006564 | 0.00189195 | 0.158 | 0.058 | 0.9941 | 0.02986863 | 0.1144392 | 0.018 | 0.03314495 |



**FIGURE 5.25** Printout of a computer screen depicting EXCEL® spreadsheet with (1) the plot of experimental data (symbols) and a four-compartment PBPK model simulation (solid line) of arterial blood concentrations of styrene during and following a 6-hr exposure of rats to 80 ppm of this chemical; (2) the numerical values of the PBPK model parameters; and (3) a portion of the raw numbers corresponding to the rate of change in the amount of styrene in tissue, amount in tissue, concentration in tissue, and venous blood concentration leaving the tissues, generated during simulations between time 0.005 and 24 hr. Experimental data were obtained from Ramsey and Andersen [173].

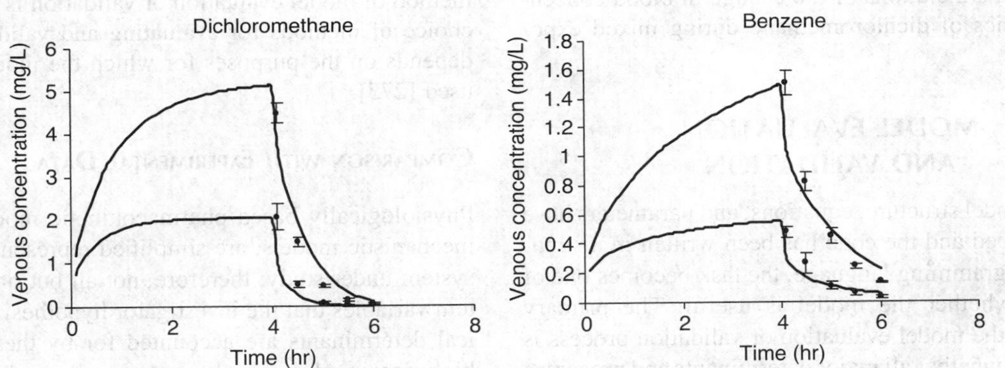



**FIGURE 5.26** PBPK model simulations of blood concentrations profiles of dichloromethane and benzene in rats exposed to 50 ppm of these chemicals alone (lower lines and data points) or in combination along with 50 ppm each of toluene, *m*-xylene, and ethylbenzene (upper lines and corresponding data points). (From Krishnan, K. and Johanson, G., *J. Environ. Sci. Health*, 23(C), 31–53, 2005. With permission.)

are available for a given chemical, their ability to adequately predict the risk-relevant dose surrogate is critical [38,57]. This aspect should be kept in mind while designing studies and strategies for conducting model evaluation or validation. In the past, a number of studies have used venous blood concentrations to access the predictive capability of PBPK models by visual inspection, even though there have also been instances involving the use of data on metabolites and tissue concentrations for this purpose [71].

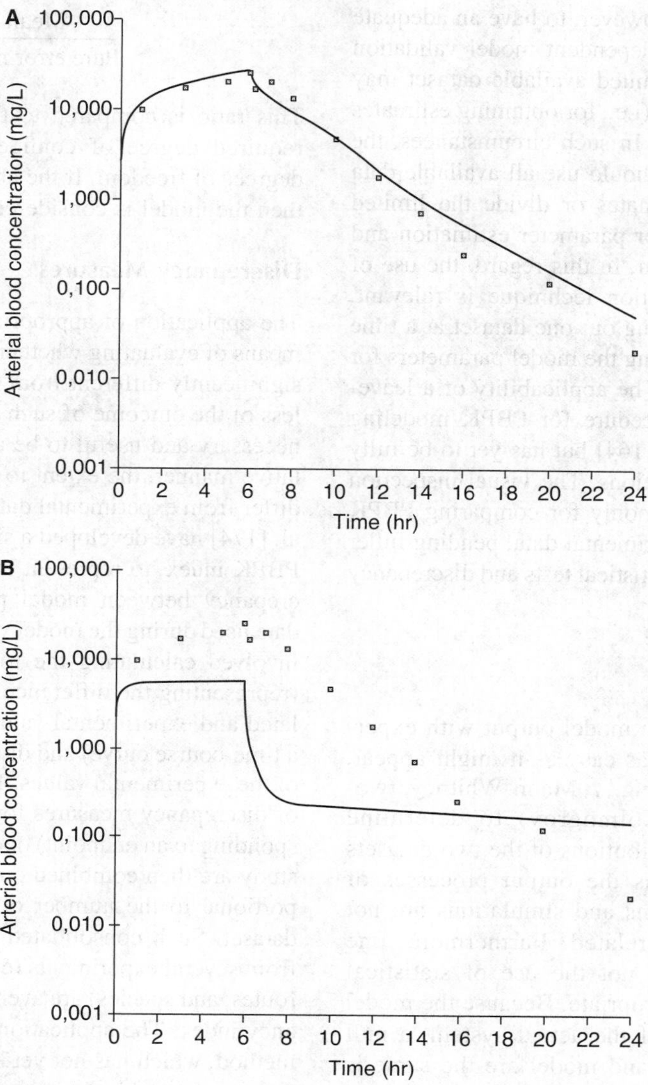


**FIGURE 5.27** Comparison of PBPK model simulations (solid lines) with experimental data (symbols) for a hypothetical chemical.

## Visual Inspection Approach

The visual inspection approach involves eyeballing or visual comparison of the plots of simulated data (usually continuous and represented by solid lines) with experimental values (usually discrete and represented by symbols) against a common independent variable (usually time). The rationale behind this approach is that the greater the commonality between the simulated and experimental data, the greater our confidence will be in the model. Figure 5.27 presents two hypothetical examples of comparisons of PBPK model simulations with experimental data. Based on visual inspection, the model simulation presented in Figure 5.27A is considered adequate because it captures the behavior in the data even though the simulation line does not touch upon each experimental data point; however, the model simulations presented in Figure 5.27B are considered inadequate and would require more work or refinement to make them adequate. Instead of simply visualizing the degree of concordance between experimental data and simulation results, the residuals (i.e., difference between experimental and simulated data) may be examined. The residual analysis can be applied either to the whole model as it is or, in some other cases (as in diagnostic checkup during model building), to some estimated parameters of a tentatively ascertained theoretical model so as to pinpoint inadequacies [147]. The residuals should be random if the model is adequate. Time plots of residuals as well as the plots of residuals with respect to various controllable variables can detect possible model inadequacies, which can shed light on how to improve the model [147].

It may not be possible, however, to have an adequate dataset for conducting an independent model validation exercise. In some cases, a limited available dataset may be used to develop the model (i.e., for obtaining estimates of parameters of the model). In such circumstances, the issue becomes whether one should use all available data to get better parameter estimates or divide the limited dataset into two parts, one for parameter estimation and the other for model validation. In this regard, the use of a leave-one-out cross-validation technique is relevant. This procedure involves pulling out one dataset at a time from the series and reevaluating the model parameters for their adequacy [27,28,284]. The applicability of a leave-one-out cross-validation procedure for PBPK modeling has recently been illustrated [164] but has yet to be fully evaluated for routine applications. The visual inspection approach is used more commonly for comparing PBPK model simulations with experimental data, pending fuller development of applicable statistical tests and discrepancy indices.

## Statistical Tests

Statistical comparison of the model output with experimental observations is not as easy as it might appear. None of the classical tests (e.g., *t*, Mann-Whitney, two-sample $\chi^2$, two-sample Kolmogrov) to determine whether the underlying distributions of the two datasets are similar is applicable, as the output processes of almost all real-world systems and simulations are not stationary and are autocorrelated. Furthermore, one might question whether or not the use of statistical hypothesis tests is even appropriate. Because the model is only an approximation of the actual system, a null hypothesis that the system and model are the same is clearly false. The more appropriate formulation is whether or not the differences between the system and the model are significant enough to affect conclusions derived from the model. In this regard, Haddad et al. [131] screened various statistical procedures (correlation, regression, confidence interval approach, lack-of-fit *F* test, univariate analysis of variance, and multivariate analysis of variance) for their potential usefulness in testing the degree of agreement of PBPK model simulations and experimental data obtained in intact animals. According to these authors, the multivariate analysis of variance represents the most appropriate test, depending on the simulation data variance. Alternatively, the lack-of-fit *F* test represents a useful way of evaluating the adequacy of simulation models. This simple procedure permits the consideration of multiple datasets (e.g., data for several endpoints collected at various time intervals) in conducting such an evaluation of model validity. The *F* statistic in model fitting may be defined as lack of fit:

$$\frac{\text{Mean square}}{\text{Pure error mean square}} \tag{5.51}$$

This ratio is compared to the critical value of *F* at the required degree of confidence and the corresponding degrees of freedom. If the above ratio is greater than $F_{crit}$, then the model is considered to be inadequate [147].

## Discrepancy Measures

The application of appropriate statistical tests provides a means of evaluating whether or not model simulations are significantly different from experimental values. Regardless of the outcome of such statistical analyses, it is often necessary and useful to be able to represent, in a quantitative manner, the extent to which the model simulations differ from experimental data. In this context, Krishnan et al. [174] have developed a simple index, referred to as the PBPK index, to represent the degree of closeness or discrepancy between model predictions and experimental data used during the model validation phase. The approach involved calculating the root mean square of the error (representing the difference between the individual simulated and experimental values for each sampling point in a time-course curve) and dividing by the root mean square of the experimental values. The resulting numerical values of discrepancy measures for several datasets (each corresponding to an endpoint) obtained in a single experimental study are then combined on the basis of a weighting proportional to the number of data points contained in the dataset. Such consolidated discrepancy indices obtained from several experiments (e.g., exposure scenarios, doses, routes, and species) are averaged to get an overall discrepancy index. The application of this kind of a quantitative method, which has not yet been routinely done, may help remove the ambiguity when communicating the degree of concordance or discrepancy between PBPK model simulations and experimental data. The routine calculation of this kind of index for PBPK models at the end of the model development phase should result in tagging each model with an index value, and such an open declaration of the face value of a particular PBPK model might be appreciated by the end users.

The use of statistical test or a discrepancy index to show that *a priori* predictions of a particular endpoint are in agreement with the experimental data is not sufficient proof of the validity of the assumptions and model building approaches used. These approaches are only useful in providing a quantitative measure of difference and do not provide any information on either the model robustness or the reliability of the model structure. In this context, it is important to ensure the adequacy of the model development process (see box) and to ascertain the influence of uncertainty, sensitivity, and variability associated with model parameters.

### Questions and Issues To Be Considered When Evaluating the Adequacy of a PBPK Model

Is an appropriate target organ or a surrogate tissue included as one of the model compartments?
Are the known, major sites of storage, transformation, and clearance included in the model structure?
Is the form of equation used to describe chemical uptake justified based on the hypothesis of tissue uptake of the chemical?
Have enzymatic processes been described appropriately?
Are the units throughout differential equations consistent?
Are the input parameters related to the characteristics of the host, chemical, or environment?
Is the sum total of compartment volumes within 100% of the body weight?
Do the tissue blood flow rates add up to the cardiac output?
Does the ventilation-to-perfusion ratio specified in the model fall within physiological limits?
Is the volume of each tissue compartment within known physiological limits?
Is the approach used to establish partition coefficients valid?
Is the method used for estimating biochemical parameters adequate?
Is the allometric scaling done appropriately?
Is the integration algorithm used in the study known for solving differential equations and appropriate?
Does the shape of the pharmacokinetic curve generated by the model match that obtained experimentally?
Is the model, with its set of parameters, able to consistently describe available pharmacokinetic data?

*Source:* Thompson, K.M. et al., in *Toxicokinetics and Risk Assessment*, Lipscomb, J. C. and Ohanian, E. V., Eds., Informa Health Care, New York, 2007, pp. 123–140. With permission.

## Uncertainty Analysis

Uncertainty analysis in the context of PBPK modeling refers to the evaluation of the impact of the lack of precise knowledge about the numerical value of a parameter or model structure on tissue dose simulations. The lack of precise knowledge above the parameter values may contribute to uncertain predictions of dose metrics. Such an analysis becomes extremely important when the models are developed using point estimates of parameters whose reliability and precision might be questionable or uncertain.

When there is a lack of confidence about the numerical value of one or more input parameters, a quantitative analysis of the uncertainty should be performed. Such an analysis permits an assessment of the direction and magnitude of the impact of parameter uncertainty on tissue dose simulations. The approaches to uncertainty analysis include: (1) Monte Carlo simulation, (2) the *p*-bounds approach, and (3) the fuzzy simulation approach. The approach to be chosen for the conduct of uncertainty analysis depends on the extent of available information on the input parameters of concern [275].

If the statistical characteristics and probability distributions of the uncertain parameter as well as other input parameters are known, then a quantitative analysis can be performed using a Monte Carlo simulation approach [97,143,168]. This approach involves repeated computations using inputs selected at random from statistical distribution for each parameter to provide a statistical distribution of the output (Figure 5.28). If the statistical distributions of parameters cannot be defined reliably, then a stochastic response surface method or a fuzzy simulation approach may be implemented [145,202]. When the minimal and maximal values of the parameters can be ascertained, a probability-bounds (*p*-bounds) approach may be applied to evaluate the impact of parameter uncertainty on the PBPK simulation approach. The *p*-bounds approach is more readily applied with steady-state models. For each one of these methods, however, the endpoint for the uncertainty analysis should be established *a priori*, and the impact of uncertainty associated with one or several parameters may have to be assessed depending on the PBPK model. If a model has been shown to simulate adequately the dose surrogate of interest in various exposure situations and species, then the benefits of uncertainty analysis might be limited [275]. Conversely, the uncertainty analysis is likely to be useful when a PBPK model does not adequately simulate the experimental data. Such a situation may arise due either to a lack of precise estimates of parameter values or the inadequacy of the model structure chosen for the study. In these cases, a quantitative uncertainty analysis or model-directed mechanistic studies should help improve the predictive ability and robustness of the PBPK model [51,135,253].

## Variability Analysis

Variability analysis in the context of PBPK modeling refers to evaluating the impact on tissue dose simulations of the range of parameter values expected among individuals of a population. PBPK models have frequently been constructed using typical or average values of physiological, physiological, and physicochemical parameters. It is well known that several of these parameters vary as a function of age, sex, disease state, physical exercise, etc. In such cases, the PBPK model developed for an average

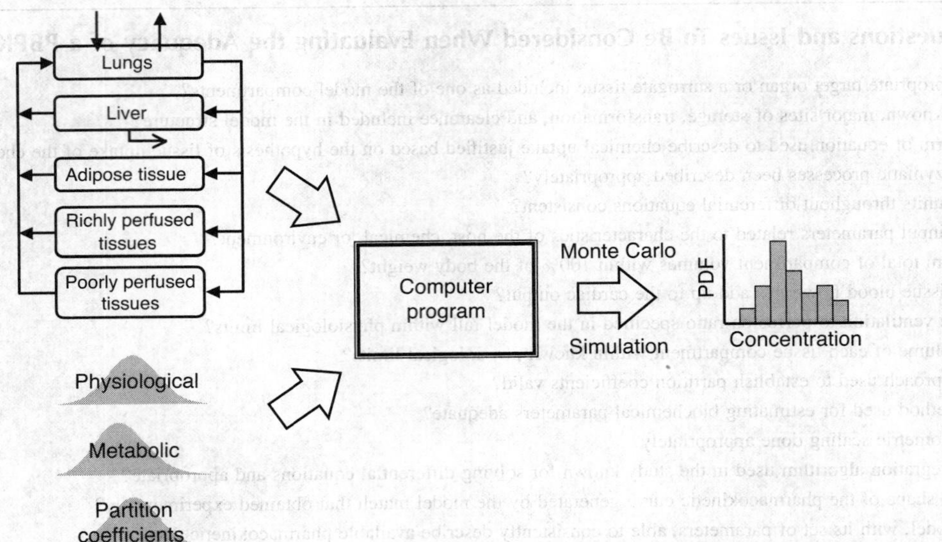


**FIGURE 5.28** Illustration of the use of Monte Carlo simulation along with PBPK modeling to compute probability density functions (PDF) of tissue dose.

individual may not adequately simulate tissue dose in a population. Depending on the availability of data on input parameters, one of the following methods may be used for conducting variability analysis in PBPK models: (1) individual-specific modeling, (2) Monte Carlo simulation, or (3) Markov chain Monte Carlo (MCMC) simulation. Individual-specific parameters (e.g., enzyme levels) may be used for constructing PBPK models to simulate the tissue dose in each individual [204]. PBPK models for an average individual representing specific subgroups of populations (e.g., adult women, pregnant women, lactating women, children) have been developed by accounting for the subgroup-specific physiological, biochemical, and physicochemical parameters [42,54,67,105,228,229,245]. Such analyses, however, would not provide the probability or likelihood of a particular output for a population. To attain this goal, a Monte Carlo simulation approach based on the probability distributions of the input parameters (physiological parameters or enzyme content, activity, and polymorphism) has been applied [89,129,183,219,265,268].

When subject-specific kinetic data are available, the information contained in such data can be used to improve upon the parameter estimates for the population. This is effectively facilitated by the Bayesian approach, in which the investigators' prior knowledge of input parameter distributions is refined by taking into account the behavior of new pharmacokinetic data. The resulting probability distributions ("posteriors") are consistent not only with the new data but also with the prior distributions because they are a function of the likelihood of the new data and prior probability of the input parameters [124,153,160]. The Markov chain Monte Carlo technique is being used to perform Bayesian analyses of PBPK models to obtain estimates of population distributions of parameters along with individual parameter vectors [30,35]. The MCMC technique involves the generation of parameter values from the prior distributions, assessment of the likelihood of the data given these particular parameter values, and generation of a posterior distribution after several thousand iterations (Figure 5.29). Table 5.15 presents a list of chemicals for which PBPK modeling has been performed on the basis of MCMC analysis.

## Sensitivity Analysis

Sensitivity analysis refers to evaluation of the effect of a change in the value of a particular input parameter on tissue dose estimates provided by a PBPK model. Sensitivity is expressed as the magnitude of change in the endpoint of interest (e.g., tissue dose) as a function of change in the value of a particular model parameter. Sensitivity analysis involves determination of the system response to defined changes in the parameter values to identify the most critical model parameters. The conventional approach to sensitivity analysis does not provide an indication of the likelihood of a particular output or range of outputs. Monte Carlo methods are designed to provide an estimate of this probability, the idea being to make repeated computations using inputs selected at random that have the same statistical properties expected of each input parameter. After performing a large number of such computations, a statistical distribution of the output can be generated.

Sensitivity analysis is informative for identifying the key parameters that are likely to affect the performance of the model. If the output is sensitive to certain aspects

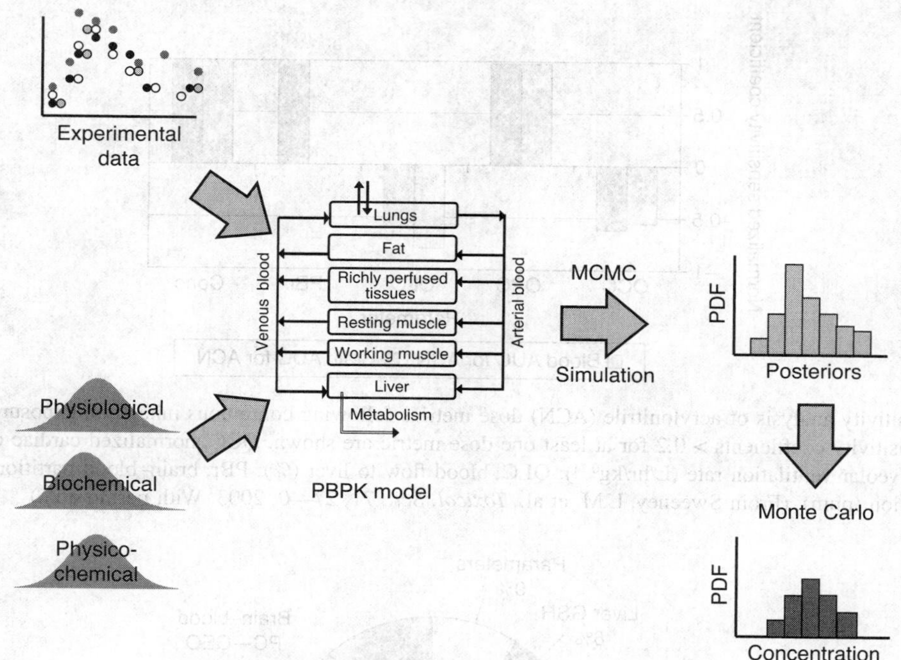


**FIGURE 5.29** Illustration of the Markov chain Monte Carlo (MCMC) simulation approach in PBPK modeling.

of the model, then those aspects must be modeled carefully. Sensitivity and uncertainty analyses with PBPK models have been conducted with respect to specific endpoints such as dose surrogate, pharmacokinetic behavior, or cancer risk estimate [38,39,59,63,88,97,137,138,143, 168]. Figure 5.30 depicts the sensitivity ratios associated with some input parameters of a PBPK model for acrylonitrile. The larger the numerical value of the sensitivity ratio, the greater the impact of the parameter on tissue dose simulations. In this example, even though the PBPK model consisted of more than 40 parameters, the dose surrogates (area under the blood concentration vs. time curve and area under the brain concentration vs. time curve for acrylonitrile) were sensitive to only a very few parameters. Similarly, only seven parameters accounted for about 98% of the variance in the simulations of the metabolite-related dose surrogate (i.e., area under the brain concentration vs. time curve) (Figure 5.31). The greatest potential for uncertainty or sensitivity analysis is perhaps in improving experimental design and resource allocation in risk-assessment-oriented research.

**TABLE 5.15**
**List of Chemicals for Which the Markov Chain Monte Carlo Approach to PBPK Modeling Has Been Applied**

| Chemicals | Species | Refs. |
|---|---|---|
| Dichloromethane | H | 155, 157 |
| Methylchloride | H | 156 |
| Styrene | R | 159 |
| Toluene | H | 158 |
| Trichloroethylene | H, M, R | 36, 37 |

## MODEL APPLICATIONS

The principal application of PBPK models is to predict the target tissue dose of the toxic parent chemical or its reactive metabolite. Using the tissue dose of the toxic moiety of a chemical in risk assessment calculations provides a better basis of relating to the observed toxic effects than the external or exposure concentrations of the parent chemical [29,52,182]. Because PBPK models facilitate the prediction of target doses for various exposure scenarios, routes, doses, and species [14,50,52,209], they can help reduce the uncertainty associated with the conventional extrapolation approaches employed in cancer and noncancer risk assessments. Even though no guidelines have been established regarding the acceptability of PBPK models, it is essential that the model intended for use in risk assessment (1) be developed or calibrated for the species and life stage of relevance to the risk assessment, (2) be peer-reviewed and evaluated for its structure and parameters, (3) contain parameters essential for simulating uptake via routes associated with human exposures as well as the critical study chosen for the assessment, and (4) be able to provide predictions of the time course of concentration of toxic moiety (parent chemical or metabolite) in the target organ or a surrogate compartment [275].

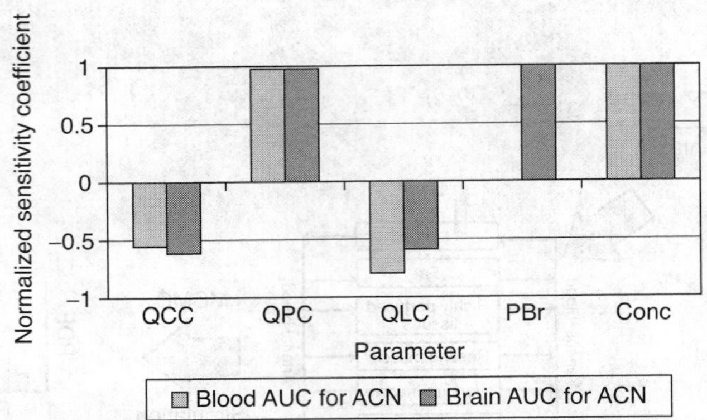


**FIGURE 5.30** Sensitivity analysis of acrylonitrile (ACN) dose metrics following continuous inhalation exposure to 0.4 ppm. Only parameters with sensitivity coefficients > 0.2 for at least one dose metric are shown. QCC, normalized cardiac output (L/hr/kg$^{0.74}$); QPC, normalized alveolar ventilation rate (L/hr/kg$^{0.74}$); QLC, blood flow to liver (%); PBr, brain–blood partition coefficient; Conc, exposure concentration (ppm). (From Sweeney, L.M. et al., *Toxicol. Sci.*, 71, 27–40, 2003. With permission.)

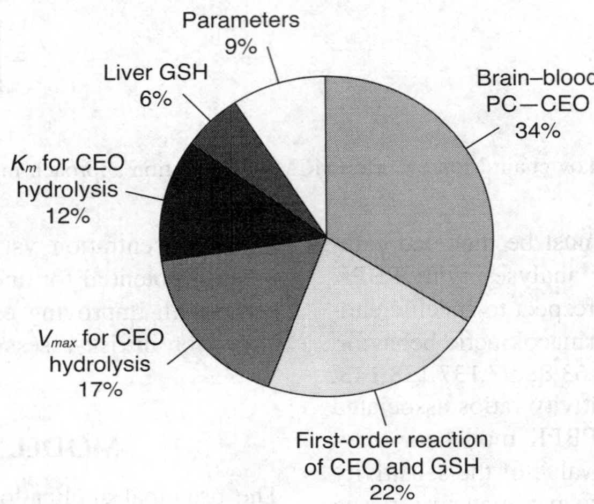


**FIGURE 5.31** Contributors to predicted variability of area under the cyanoethylene oxide (CEO) concentration vs. time curve following continuous inhalation exposure to 0.12 acrylonitrile. (From Sweeney, L.M. et al., *Toxicol. Sci.*, 71, 27–40, 2003. With permission.)

## Cancer Risk Assessment

The assessment of risk associated with human exposure to carcinogens is often based on data collected in bioassays in which the animals are administered high doses. In such cases, the consideration of dose-dependent transitions in mechanisms of toxicity is critical to the conduct of scientifically sound cancer risk assessment [252]. For example, in the case of vinyl chloride the dose–response is linear up to about 500 ppm, and above that level it is essentially flat (Figure 5.32). If linear extrapolation of the response incidence at 6000 ppm is performed, the low dose risk would have been underestimated. The nonlinearity in this case arises from the saturable metabolism of the chemical. Because PBPK models are based on mechanistic determinants of ADME, they are potentially useful in reducing the uncertainties associated with high dose to low dose, route to route, and interspecies extrapolations. This aspect was initially demonstrated with dichloromethane [7]. DCM caused liver and lung tumors in mice exposed to 2000 or 4000 ppm, 6 hr/day, for their lifetime [201]. DCM is metabolized by two processes: (1) oxidation leading to the production of highly reactive formyl chloride, as well as carbon monoxide and small amounts of carbon dioxide; and (2) GSH conjugation, yielding chloromethylglutathione (a reactive intermediate) and carbon dioxide. Either of the reactive metabolites resulting from the oxidation or GSH conjugation could be involved in the mutagenic changes leading to cancer. In the PBPK model for DCM [7], these metabolic pathways were described according to their different kinetic characteristics (i.e., a saturable term for oxidation and with a linear term for reaction with GSH).

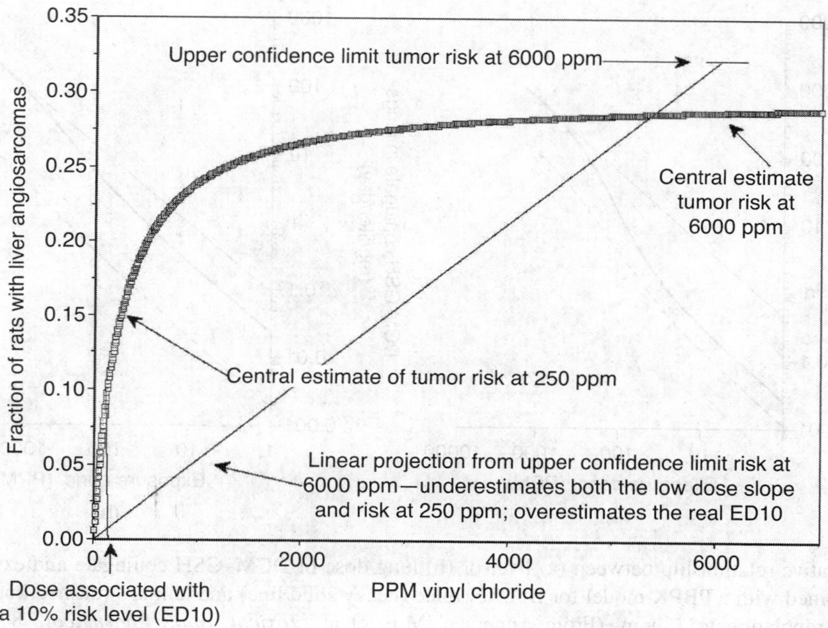


**FIGURE 5.32** Under- and overestimations of cancer risk for vinyl chloride exposure. (From Slikker, Jr., W. et al., *Toxicol. Appl. Pharmacol.*, 201, 203–225, 2004. With permission.)

The PBPK model for DCM developed for the mouse by integrating quantitative information on physiological characteristics, partition coefficients, and metabolic rate constants adequately described the disposition of DCM. The mouse DCM model then was used to calculate the tissue dose of metabolites and parent chemical arising from exposure scenarios comparable to those of the NTP bioassay studies (Table 5.16). Their relationship to the observed tumor incidence then was examined. Because the parent chemical is unreactive, it is unlikely to be directly involved in the tumorigenicity; hence, the relationship between the tissue exposure to its metabolites and tumor incidence was examined (Table 5.16). Whereas the dose surrogate based on the oxidative pathway did not vary between DCM exposure concentrations of 2000 and 4000 ppm, the flux through the GSH pathway did correspond well with the degree of DCM-induced cancer at these exposure concentrations. These observations are consistent with a role for the metabolites arising from the GSH conjugation pathway in DCM-induced lung and liver

**TABLE 5.16**
**Tumor Incidence and Calculated Tissue Dose of Dichloromethane (DCM) Metabolites Following Inhalation Exposures of 0, 2000, or 4000 ppm DCM in Female Mice**

| DCM Exposure Concentration | Tumor Incidence (%) | Tissue Dose (amount in mg per L tissue per day) GSH Pathway | Oxidative Pathway |
|---|---|---|---|
| *A. Liver* | | | |
| 0 | 6 | — | — |
| 2000 | 33 | 851 | 3537 |
| 4000 | 83 | 1800 | 3701 |
| *B. Lung* | | | |
| 0 | 6 | — | — |
| 2000 | 63 | 123 | 1531 |
| 4000 | 85 | 256 | 1583 |

*Source:* Andersen, M.E. et al., *Toxicol. Appl. Pharmacol.*, 87, 185–205, 1987. With permission.

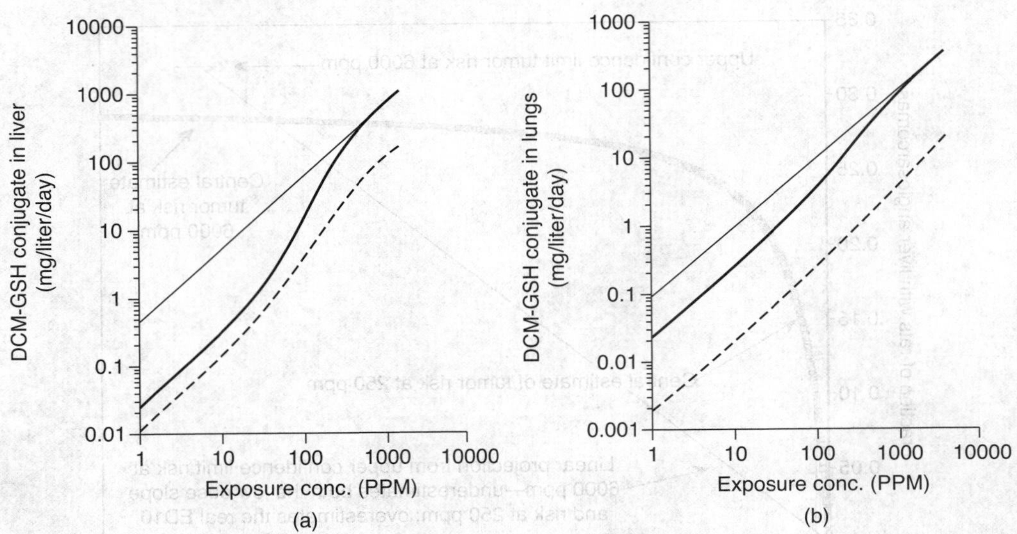


**FIGURE 5.33** Quantitative relationship between (a) liver or (b) lung dose of DCM–GSH conjugate and external exposure concentration of DCM determined with a PBPK model for B6C3F$_1$ mice (heavy solid line) and humans (dashed line). The lighter solid line depicts linear back-extrapolation to 1 ppm. (From Andersen, M.E. et al., *Toxicol. Appl. Pharmacol.*, 87, 185–205, 1987. With permission.)

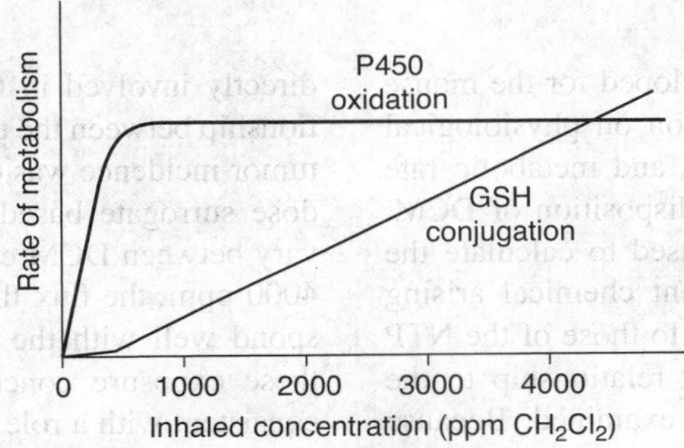


**FIGURE 5.34** Representation of the rate of metabolism of dichloromethane (DCM) at various inhaled concentrations. DCM is metabolized by P-450 mediated oxidation and by conjugation with GSH. These two metabolic pathways have different characteristics. At low inhaled concentrations, oxidation is the preferred manner of metabolism, but at higher concentrations GSH conjugation becomes more favored. (From Krishnan, K. and Andersen, M.E., in *New Trends in Pharmacokinetics*, Rescigno, A. and Thakur, A.K., Eds., Plenum Press, New York, 1991, pp. 335–354. With permission.)

cancer. The GSH conjugation of DCM yields formaldehyde as a metabolite. Casanova et al. [44] reported DNA–formaldehyde–protein cross-links resulting from DCM exposure, further strengthening the case for GSH conjugation as the pathway leading to potentially carcinogenic metabolites; therefore, high dose to low dose extrapolation and interspecies extrapolation of DCM-induced cancer risk were conducted with the tissue dose of the GSH pathway metabolite predicted by the PBPK model.

The model prediction of the target tissue dose of the DCM–GSH conjugate resulting from 6-hr inhalation exposures to 1 to 4000 ppm of DCM is presented in Figure 5.33. Estimation of the target tissue dose of the DCM–GSH conjugate by linear back-extrapolation gives rise to a 21-fold higher estimate than that obtained by the PBPK modeling approach. This discrepancy arises from the nonlinear behavior of DCM metabolism at high exposure concentrations (Figure 5.34). At exposure concentrations exceeding 33 ppm, the CYP-mediated oxidation pathway is saturated, giving rise to a corresponding disproportionate increase in the flux through GSH conjugation pathway.

The interspecies extrapolation of DCM kinetics was possible because the critical biological determinants were first identified in the test species: the mouse. Thus, the physiological parameters in the mouse PBPK model were scaled allometrically, the metabolic parameters were determined experimentally, and the tissue–blood partition

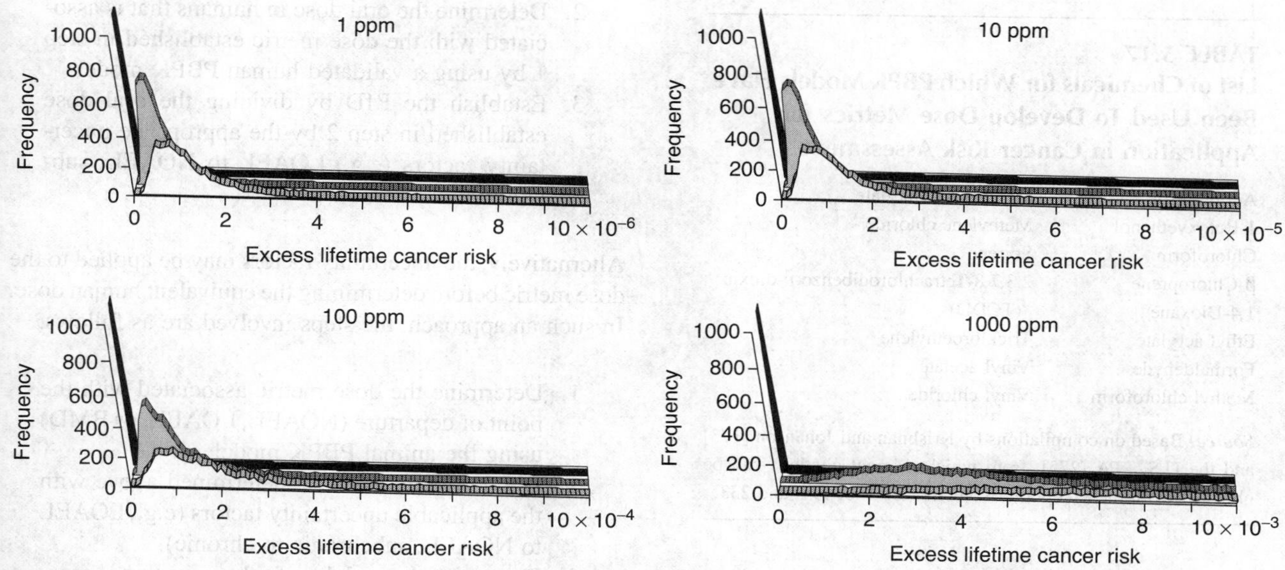


**FIGURE 5.35** PBPK model-based predictions of population distribution of cancer risk associated with lifetime exposure to 1 to 1000 ppm dichloromethane [157]. The predictions are grouped according to GSTT1 genotype (rear: ■ 0/0, middle: ■ +/0, front: □ +/+). (From Krishnan, K. and Johanson, G., *J. Environ. Sci. Health*, 23(C), 31–53, 2005. With permission.)

coefficients were assumed to be species invariant. The PBPK model adequately simulated the blood levels of DCM observed in humans after a 6-hr inhalation exposure to 100 or 350 ppm DCM. The target tissue dose for humans was estimated to be some 2.7 times lower than that for the mouse. Considered together, the human tissue dose of DCM–GSH conjugate for a 6-hr exposure to 1 ppm DCM is expected to be some 57 times lower than that expected by linear extrapolation of its behavior at high exposure concentrations, such as the ones used in the mouse cancer bioassay.

The cancer risk assessment for DCM then was conducted using the linearized multistage (LMS) model to relate the tissue dose of DCM–GSH metabolite (rather than DCM exposure concentration) to the tumor incidence rates observed at high exposure concentrations in the mouse. When assessing the tumorigenic risks associated with human exposure to this chemical, it was assumed that humans are as sensitive as the most sensitive test species; therefore, equal target tissue doses are expected to produce similar tumor incidence regardless of the species. The PBPK model-based DCM risk assessment predicted the human low-dose risk to be about 100- to 200-fold less than that predicted by the U.S. Environmental Protection Agency (EPA) using standard default assumptions [251]. After further refinement of the model with estimation of the metabolic rate constants for humans *in vitro* [236], the PBPK model-based approach using tissue dose of the DCM–GSH conjugate predicted a cancer risk of $3.7 \times 10^{-8}$ for a lifetime inhalation exposure of 1 μg/m$^3$. This risk estimate was still lower, by more than two orders of magnitude, than that calculated by the EPA using the standard default assumptions and exposure concentrations of DCM [270].

The initial efforts of DCM modeling did not treat or address variability and uncertainty associated with model parameters in an optimal manner. The issues of variability and uncertainty in parameter estimates, as well as their impact on tissue dose and cancer risk have been addressed using a Bayesian approach [30,35, 153,157]. Jonsson and Johanson [157] used the Markov chain Monte Carlo simulation approach to estimate the population distribution of tissue dose and excess cancer risk associated with DCM exposures, while taking into account the polymorphism associated with glutathione *S*-transferase theta 1 (GSTT1). The Monte Carlo simulations predicted median excess lifetime cancer risks of 0.61, 6.2, 85, and 4700 per million for 1, 10, 100, and 1000 ppm dichloromethane exposures, respectively, for the Swedish population (Figure 5.35). Using the MCMC approach it was also possible to obtain an estimate of the distribution of risk in the population. The simulations suggested that the distribution is considerably wider at 1000 ppm than at 1 ppm (Figure 5.35) and is explained by a shift from flow-limited metabolism at 1 ppm to capacity-limited metabolism at 1000 ppm, due to saturation of the CYP2E1 pathway [157].

Following the DCM example, several reports were published describing the use of PBPK models in the prediction of the dose metric to enhance the scientific basis of cancer risk assessment of environmental agents (Table 5.17) [42,52,140,173,191,233,275]. It is a misconception

**TABLE 5.17**
**List of Chemicals for Which PBPK Models Have Been Used To Develop Dose Metrics for Application in Cancer Risk Assessment**

| | |
|---|---|
| Acrylonitrile | Methyl methacrylate |
| 1-Butoxyethanol | Methylene chloride |
| Chloroform | Styrene |
| β-Chloroprene | 2,3,7,8-Tetrachlorodibenzo-*p*-dioxin (TCDD) |
| 1,4-Dioxane | |
| Ethyl acrylate | Trichloroethylene |
| Formaldehyde | Vinyl acetate |
| Methyl chloroform | Vinyl chloride |

*Source:* Based on compilations by Krishnan and Johanson [173] and the U.S. EPA [275], from which original citations can be obtained, with the exception of References 140, 191, and 233.

that the use of PBPK models in quantitative risk assessment will always lead to estimation of lower risk than the conventional approach adopted by regulatory agencies. If the test chemical is a direct-acting agent, the PBPK approach actually could predict greater risk to humans than conventional methods because enzyme-mediated metabolic clearance (detoxification) is expected to be lower in larger species. Similarly, if the toxicity of a chemical is mediated by reactive intermediates resulting from a saturable metabolic process, then the high dose to low dose extrapolation conducted with the PBPK modeling approach would actually predict a greater risk than that predicted by the linear extrapolation procedure.

## Non-Cancer Risk Assessment

One principal use of PBPK models in non-cancer risk assessment relates to the estimation of the magnitude of the pharmacokinetic component of the interspecies and intraspecies uncertainty factors [52,55,58,125,182,275]. When the pharmacokinetic-based dose adjustment is performed in a risk assessment using PBPK models, the uncertainty factor (UF) related to the pharmacodynamic component is set equal to 3 [271,275]. The current approaches relating to the use of PBPK models in non-cancer risk assessment (i.e., derivation of reference dose [RfD] and reference concentration [RfC]) are now presented.

The use of PBPK models in RfD derivation, with the objective of replacing the default $UF_{PK}$, involves the following steps [275]:

1. Determine the dose metric associated with the point of departure (no-observed-adverse-effect level [NOAEL], lowest-observed-adverse-effect level [LOAEL], or benchmark dose [BMD]) using a validated animal PBPK model.
2. Determine the oral dose in humans that is associated with the dose metric established in step 1 by using a validated human PBPK model.
3. Establish the RfD by dividing the oral dose established in step 2 by the appropriate uncertainty factors (e.g., LOAEL to NOAEL, subchronic to chronic).

Alternatively, the uncertainty factors may be applied to the dose metric before determining the equivalent human dose. In such an approach, the steps involved are as follows:

1. Determine the dose metric associated with the point of departure (NOAEL, LOAEL, or BMD) using the animal PBPK model.
2. Divide the dose metric determined above with the applicable uncertainty factors (e.g., LOAEL to NOAEL, subchronic to chronic).
3. Determine the oral dose for humans that is associated with the target dose metric established in step 2 by using a validated human PBPK model.

The U.S. EPA [272] used the above approach for establishing the RfD for ethylene glycol monobutyl ether. Based on Corley et al. [66], the maximal concentration of the metabolite (butoxy acetic acid in blood [$BAA_{max}$]) associated with the $LOAEL_{animal}$ (59 mg/kg/day), was determined using an animal PBPK model (103 μ*M*). Using the human PBPK model for ethylene glycol monobutyl ether, the oral dose associated with a maximal concentration of 103 μ*M* butoxy acetic acid in blood was determined (7.6 mg/kg/day). This dose was then divided by the appropriate uncertainty factors (10 for inter-individual differences and 3 for LOAEL to NOAEL extrapolation) to establish the RfD for ethylene glycol monobutyl ether (0.3 mg/kg/day). Note that, in this particular case, the interspecies pharmacodynamic factor was set to 1 because *in vitro* studies suggested that humans are less sensitive than rats to the hematologic effects of ethylene glycol monobutyl ether [272,275]. The above approach has also been used to establish the RfD from the benchmark dose [272]. The magnitude of the PK component of the inter-individual variability factor for RfD derivation can be computed if the probability or population distributions of the input parameters of the human PBPK model are known. PBPK model-based determination of the inter-individual variability factor has been applied in the derivation of RfD for methyl mercury [53].

The U.S. EPA's reference concentration documentation explicitly includes routine application of interspecies differences in dosimetry in assessing RfCs [271]. The use of PBPK models in RfC derivation is primarily related to the estimation of the human equivalent concentration of inhaled chemicals. This estimation corresponds to the $UF_{PK}$ used in RfD derivation. The approach for the use of

PBPK models in the RfC derivation for systemically acting gases involves:

1. Determination of the dose metric associated with the point of departure (unadjusted or duration-adjusted NOAEL, LOAEL, BMC) in the test animal species using a validated animal PBPK model
2. Determination of the human exposure concentration that corresponds to the dose metric established in step 1 by using a validated human PBPK model

The EPA established the RfC for *m*-xylene using this approach [274]. Accordingly, first, the dose metric (steady-state blood concentration of *m*-xylene = 0.144 mg/L) associated with the NOAEL of 39 mg/m$^3$ was determined using a validated rat PBPK model [258]. Subsequently, using this dose metric as the target, the human PBPK model was applied to determine the corresponding human equivalent concentration (HEC = 41 mg/m$^3$) [274]. The RfC for *m*-xylene (0.1. mg/m$^3$) was established by dividing HEC by the appropriate uncertainty factors (3 for interspecies difference in pharmacodynamics, 10 for interindividual variability, 3 for subchronic to chronic extrapolation, and 3 for database deficiency). The RfC derivations for acetone, isopropanol, vinyl chloride, and acrylic acid also exemplify the unique usefulness of PBPK models in this process [18,125,273]. An increasing number of reports demonstrate the use of PBPK models in deriving appropriate dose metrics and their application in setting occupational, acute, and chronic guideline values of systemic toxicants [48,49,122,257,260,276].

## Mixture Risk Assessment

The assessment of health risks associated with human exposure to chemical mixtures is performed frequently by considering the mixture as a single entity or by summing the response associated with the components. The commonly used component-based risk assessment approaches are response addition and dose addition for carcinogenic and noncarcinogenic chemical mixtures. During mixed exposures, the tissue dose or tissue response of certain or all components might be altered due to pharmacokinetic or pharmacodynamic interactions, the magnitude of which depend on the dose, route, and species. The data on PK interactions have not been accounted for in the health risk assessment of chemical mixtures, primarily because the consequences of binary interactions in mixtures of greater complexity could not be easily evaluated or predicted. Several researchers have studied and developed PBPK models of metabolic interactions among chemicals in mixtures [81,84,130–134,258,259]. Most of the initial studies related to modeling binary mixtures. Tardif et al. [258] developed and validated for the first time a PBPK model for a ternary mixture in rats and humans, only on the basis of knowledge of binary level interactions. This modeling approach enables prediction of the change in tissue dose of chemicals in mixtures by taking into account the pharmacokinetic interactions at various levels of complexity. Simulations of altered dose surrogates of chemicals in mixtures have been used to compute the risk associated with mixtures based on either response additivity or dose additivity.

Haddad et al. [130] developed a methodology for conducting cancer and non-cancer risk assessments of mixtures on the basis of dose metrics simulated by PBPK model. Accordingly, for systemic toxic effects, an interaction-based hazard index was calculated using the data on (1) tissue dose of mixture constituents associated with human exposure, and (2) tissue dose corresponding to the guideline value of individual chemicals. In this approach, the tissue dose of mixture constituents during mixed exposures was obtained using a PBPK model that accounted for the mechanism and magnitude of multichemical interactions. The interaction-based hazard index was computed for each toxic effect by summing the ratio of the tissue dose associated with mixed exposure to the guideline value of each relevant mixture constituent as follows:

$$HI_{interaction\text{-}based} = \sum_{i=1} \frac{TMi}{TRi} \quad (5.52)$$

where $TR_i$ is the tissue dose estimated by PBPK models for human exposure to guideline values of individual mixture constituents, and $TM_i$ is the tissue dose of each mixture constituent during human exposure to mixtures as simulated by PBPK models. Table 5.18 presents sample calculations of a PBPK-based hazard index for the central nervous system (CNS) effect associated with human exposure to various mixtures of benzene, toluene, dichloromethane, ethylbenzene, and *m*-xylene.

The PBPK model-based cancer risk assessment approach for mixtures proposed by Haddad et al. [130] is based on the response addition approach. It involves computation of the risk level, $P(d)$, using knowledge of the tissue dose-based slope factor ($q^*_{tissue}$) and tissue dose of each mixture component associated with human exposures ($TM_i$) as follows:

$$P(d) = \sum q^*_{tissue} TM_i \quad (5.53)$$

Here, the $TM_i$ for each mixture constituent computed with the PBPK models might vary from single exposures, depending on the extent of interaction. Table 5.19 presents sample calculations of the extent of change in risk level associated with dichloromethane in mixtures. For the mixture exposure scenarios presented in this table, the cancer

**TABLE 5.18**
**Comparison of Interaction-Based and Conventional Hazard Index (HI) for Central Nervous System Effect Calculated on the Basis of Area Under the Concentration in Richly Perfused Tissue vs. Time Curve ($AUC_{RTP}$) for Different Mixtures of Dichloromethane (D), Benzene (B), Toluene (T), Ethylbenzene (E), and *m*-Xylene (X)**

| Exposure Concentration | | | | | $AUC_{RPT}$ (mg/L × hr) | | | | HI | |
|---|---|---|---|---|---|---|---|---|---|---|
| | | | | | | | | | Interaction-Based[a] | Conventional[b] |
| D | B | T | E | X | D | T | E | X | | |
| 50 | 0.5 | 50 | 100 | 100 | 13.9 | 54.4 | 64.0 | 94.4 | 6.8 | 4.0 |
| 25 | 0.5 | 25 | 50 | 50 | 5.99 | 21.0 | 24.6 | 35.2 | 2.7 | 2.0 |
| 16 | 0.5 | 16 | 33 | 33 | 3.36 | 11.3 | 14.1 | 19.5 | 1.5 | 1.3 |
| 12.5 | 0.5 | 12.5 | 25 | 25 | 2.41 | 8.06 | 10.1 | 13.6 | 1.1 | 1.0 |
| 10 | 0.5 | 5 | 40 | 20 | 2.04 | 3.35 | 16.5 | 11.1 | 0.94 | 0.90 |
| 20 | 0.5 | 10 | 10 | 10 | 3.32 | 5.77 | 3.79 | 4.98 | 0.82 | 0.80 |
| 10 | 0.5 | 10 | 10 | 10 | 1.62 | 5.64 | 3.71 | 4.85 | 0.58 | 0.60 |

[a] Calculated as the sum of the ratio of the exposure concentration to the TLVs of D (50 ppm), T (50 ppm), E (100 ppm), and X (100 ppm).
[b] Calculated as the sum of the ratio of the $AUC_{RPT}$ determined during mixture exposures to that associated with single exposures to the TLVs of D, T, E, and X.

*Note:* Exposure scenario, 8 hours of inhalation per day; simulation period, 24 hours.

*Source:* Haddad, S. et al., *Toxicol. Sci.*, 63, 125–131, 2001. With permission.

risk attributed to dichloromethane could increase by up to a factor of 4 compared to single chemical exposures [130]. The PBPK-based approach, however, suggests that the cancer risk associated with benzene would decrease during mixed exposures compared to individual chemical exposures, because the rate of benzene metabolism is reduced during concurrent exposure to other P450 2E1 substrates [130].

## CONCLUDING REMARKS

Modeling, in general terms, involves a mathematical description of the interrelationships among critical parameters that determine the behavior of the system under study. Mathematical models can be constructed to fit experimental data by adjusting one or more model parameters or by deriving an equation that describes the data. The latter approach reflects the methodological basis of conventional pharmacokinetic models. This approach might well be sufficient to describe the kinetics of prescription drugs and other pharmaceutical products, as these substance are often tested in humans in prescription dose range. Such empirical models are not sufficient in the case of environmental contaminants for which human health risk assessments have to be performed based on data obtained in animal studies conducted by administering high doses of chemicals by routes often different from anticipated human exposures. In this respect, mechanically based PBPK models are useful for conducting the required extrapolations. So, the kind of modeling approach (physiological or nonphysiological) that is required or sufficient to describe the kinetics of a chemical depends on the intended use.

The potential use of PBPK/TK-type dosimetry models in risk-assessment-related evaluations can be viewed as being exploratory, interpretive, or mechanistic [5]. Exploratory applications involve evaluating possible modes of action, correlating risk projections from rodent and human studies, and investigating the appropriate dose surrogate for risk assessments. Interpretive evaluations occur when applying estimated dose metrics to derive RfCs, RfDs, or cancer risk estimates. Here, the dose surrogate associated with the point of departure in the critical study is evaluated using an animal PBPK model, and then the equivalent human dose (associated with the target dose surrogate) is estimated using a human PBPK model. The mechanistic evaluations are intended to characterize the relationship of tissue dose and response and to determine consistency among specific hypotheses regarding toxicity and biochemical responses. We are only beginning to see such applications of PBPK models (e.g., evaluation of Hill coefficients and the switching phenomenon with dioxins).

If any of these interpretations (interpretive, exploratory, or mechanistic) is the goal of a study, then the choice will be PBPK modeling; however, the ability to conduct extrapolations with PBPK models will be compromised if the methods employed for developing and validating the models are inappropriate. For example, PBPK models should not be constructed by assembling sets of equations in which the parameters are not interpretable in terms of physicochemical, biochemical, or physiological properties. That sort of an approach will compromise the very

**TABLE 5.19**
**Effect of Pharmacokinetic Interactions on the Cancer Risk Level Associated with Dichloromethane (D) When This Chemical Is Present Along with Benzene (B), Toluene (T), Ethylbenzene (E), and *m*-Xylene (X)**

| Exposure Concentration | | | | | $A_{GSH(24\text{-}hr)}$ (mg/L × hr)[a] | |
|---|---|---|---|---|---|---|
| D | B | T | E | X | Mixture | D Single |
| 50 | 0.5 | 50 | 100 | 100 | 110 | 26.2 |
| 25 | 0.5 | 25 | 50 | 50 | 42.9 | 10.9 |
| 16 | 0.5 | 16 | 33 | 33 | 21.4 | 6.53 |
| 12.5 | 0.5 | 12.5 | 25 | 25 | 13.9 | 4.98 |
| 10 | 0.5 | 5 | 40 | 20 | 12.6 | 3.92 |
| 20 | 0.5 | 10 | 10 | 10 | 15.2 | 8.39 |
| 10 | 0.5 | 10 | 10 | 10 | 7.06 | 3.92 |

*Note:* Exposure scenario, 8-hr inhalation per day; simulation period, 24 hr.

[a] Integrated amount of D conjugated with GSH per tissue volume over 24 hr.

*Source:* Haddad, S. et al., *Toxicol. Sci.*, 63, 125–131, 2001. With permission.

basis of PBPK modeling (i.e., the mechanistic basis). In other words, the mathematical equations employed in PBPK modeling should clearly show the interrelationships among the critical biological determinants. One should be able to dissect each equation into subsections, each of which describes a particular phenomenon (e.g., tissue uptake, metabolism). Further, the dissociation of each term should provide parameters that are biologically meaningful (e.g., breathing rate, tissue volume). If the mathematical descriptions employed in the model do not satisfy this basic criterion (i.e., use of biologically relevant parameters), then the model should not be considered as a true physiologically based model, and the appropriateness of the use of such a model to conduct extrapolations is questionable. The guiding principles of modeling in toxicology and the characteristics of a good modeling paper are summarized elsewhere [6,47,275]. Genuine PBPK modeling efforts should ensure that (1) the assumptions upon which the model is based are appropriate, (2) the coding of model equations is errorless, (3) the model parameter values are accurate, and (4) the model is adequately evaluated or validated.

Needless to say, the model is as good as the input parameters; therefore, accurate parameterization is fundamentally important for constructing useful PBPK models. Parameter identifiability and model overspecification are problems inherent in these PBPK models or in any other multiparameter model. Direct measurement of model parameters by experimental methods, independent of analysis of tissue time-course curves, is the preferred approach. Nonetheless, limited numbers of parameters often will still have to be estimated by analysis of time-course data using curve-fitting techniques under well-defined experimental conditions where the curves are particularly sensitive to the parameters of interest.

Methodological aspects of some of the important and widely used techniques for model parameterization were provided in the earlier sections of this chapter. Several prototypical descriptions have also been provided to serve as examples for developing PBPK models for other chemicals of interest. These prototypical structures and descriptions may not be directly applicable to a chemical of interest. The model structure and the phenomena to be represented in a particular model depend on the chemical for which the kinetics is being modeled. Each chemical may possess some unique properties, thus posing some very different problems and requiring the modification of existing model structures and functional representations. This might lead to the development of totally new and novel descriptions; thus, PBPK modeling is as much an art as a science. The creativity of the researcher is as much implied in the formulation of these models as the experimental techniques to obtain parameter estimates, such that novel model structures and descriptions will continually evolve in this field.

The motivation for the use of PBPK models in toxicology research is to uncover the biological determinants of tissue dosimetry. These models are part of a systematic approach to studying how chemicals gain entry into, distribute within, and are eliminated from the body. A major advantage of PBPK models is their use in designing critical mechanistic toxicological studies. With respect to the design of studies, PBPK and other biologically based models provide an opportunity to evaluate the various plausible hypotheses by computer simulation. We can ask questions of an "if–then" nature; for example, if the model structure is correct and the rate of a reaction or another process is varied, what is the expected impact on tissue dosimetry? The PBPK model can be used to generate quantitative predictions of the expected experimental outcome based on the most attractive hypothesis of the researchers, and this can then be verified experimentally. In this case, the model serves as a tool in designing experimental studies that allow efficient resource utilization and maintaining a focus on human health risk assessment endpoints.

Unlike the mandated mathematical models used in conventional risk assessment, the biologically based dosimetry and response models are versatile and often, but not always, difficult to validate. In contrast to the mandated models, which are useful only for generating a risk number, the biologically based models allow integration of various observations, identification of critical data gaps, and estimation of risk numbers, along with attendant appreciation of areas of significant biological uncertainty [65].

## QUESTIONS

1. Calculate the fat–blood partition coefficient for chemicals with $P_{o:w}$ values of 1, 100, or 1000. Interpret your results.
2. Develop a conceptual representation of a PBPK model for *n*-octane ($P_{o:w}$ = 151356, $P_{w:a}$ = 0.00762).
3. Calculate the alveolar ventilation rate ($Q_p$) for a human weighing 64 kg, knowing that the body-weight-normalized $Q_p$ for mammals is 15 L/hr/kg.
4. The $V_{max}$ and $K_m$ of pyrene determined *in vitro* using rat liver postmitochondrial fractions were $5.935 \times 10^{-4}$ μmol/min per mg protein and 27.73 μmol/L, respectively. Convert these potentially useful *in vitro* values for incorporation within a PBPK model for the rat (protein concentration = 88 mg protein per g liver; liver weight = 10g).
5. Using the rat PBPK model presented in Figure 5.25 of this chapter, determine the external exposure concentration of styrene corresponding to an area under the liver concentration vs. time curve (AUC) of 150 μg·hr/L (for the parent chemical). Set the exposure duration to 6 hr and the length of simulation to 24 hr.
6. Determine the human exposure concentration of styrene that yields the same AUC in animals (i.e., 150 μg·hr/L). Set the exposure duration and length of simulation to 24 hr.

## REFERENCES


1. Abraham, M. H., Ibrahim, A., and Acree, W. E., Jr. (2005): Air to blood distribution of volatile organic compounds: a linear free energy analysis. *Chem. Res. Toxicol.*, 18: 904–911.
2. Abraham, M. H., Kamlet, M. J., Taft, R. W., Doherty, R. M., and Weathersby, P. K. (1985): Solubility properties in polymers and biological media. 2. The correlation and prediction of the solubilities of nonelectrolytes in biological tissues and fluids. *J. Med. Chem.*, 28:865–870.
3. Abraham, M. H., and Weathersby, P. K. (1994): Hydrogen bonding. 30. Solubility of gases and vapors in biological liquids and tissues. *J. Pharm. Sci.*, 83:1450–1456.
4. Andersen, M. E. (1981): A physiologically based toxicokinetic description of the metabolism of inhaled gases and vapors, *Toxicol. Appl. Pharmacol.*, 60:509–526.
5. Andersen, M. E. (2003): Toxicokinetic modeling and its applications in chemical risk assessment. *Toxicol. Lett.*, 138:9–27.
6. Andersen, M. E., Clewell, H. J., III, and Gargas, M. L. (1995): Contemporary issues in toxicology. Applying simulation modeling in toxicology and risk assessment: a short perspective. *Toxicol. Appl. Pharmacol.*, 133:181–187.
7. Andersen, M. E., Clewell, H. J., III, and Gargas, M. L. (1991): Physiologically based pharmacokinetic modeling with dichloromethane, its metabolite carbon monoxide and blood carboxyhemoglobin in rats and humans. *Toxicol. Appl. Pharmacol.*, 108:14–27.
8. Anderson, M. E., Clewell, H. J., III, Gargas, M. L., Smith, F. A., and Reitz, R. H. (1987): Physiologically based pharmacokinetics and risk assessment process for methylene chloride. *Toxicol. Appl. Pharmacol.*, 87:185–205.
9. Andersen, M. E., Eklund, C. R., Mills, J. J., Barton, H. A., and Birnbaum, L. S. (1997): A multicompartment geometric model of the liver in relation to regional induction of cytochrome P-450s. *Toxicol. Appl. Pharmacol.*, 144: 135–144.
10. Andersen, M. E., Gargas, M. L., Jones, R. A., and Jenkins, L. J. (1980): Determination of the kinetic constants for metabolism of inhaled toxicants *in vivo* by gas uptake measurements. *Toxicol. Appl. Pharmacol.*, 54:116.
11. Andersen, M. E., Green, T., Frederick, C. B., and Bogdanffy, M. S. (2002): Physiologically based pharmacokinetic (PBPK) models for nasal tissue dosimetry of organic esters assessing the state-of-knowledge and risk assessment applications with methyl methacrylate and vinyl acetate. *Regul. Toxicol. Pharmacol.*, 36:234–245.
12. Andersen, M. E., and Keller, W. C. (1984): Toxicokinetic principles in relation to percutaneous absorption and cutaneous toxicity. In: *Cutaneous Toxicity*, edited by V. A. Drill and P. Lazar. Raven Press, New York, pp. 9–27.
13. Andersen, M. E., and Krishnan, K. (1995): Relating *in vitro* to *in vivo* exposures with physiologically based tissue dosimetry and tissue response models. In: *Animal Test Alternatives: Refinement, Reduction, Replacement*, edited by H. Salem. Marcel Dekker, New York, pp. 9–25.
14. Andersen, M. E., MacNaughton, M. G., Clewell, H J., III, and Paustenbach, D. J. (1987): Adjusting exposure limits for long and short exposure period using a physiological pharmacokinetic model. *Am. Indust. Hyg. Assoc. J.*, 48: 335–343.
15. Andersen, M. E., Mills, J. J., and Gargas, M. L. (1993): Modeling receptor-mediated processes with dioxin: implications for pharmacokinetics and risk assessment. *Risk Anal.*, 13:25–36.
16. Andersen, M. E., and Sarangapani, R. (2001): Physiologically based clearance/extraction models for compounds metabolized in the nose: an example with methyl methacrylate. *Inhal. Toxicol.*, 13:397–414.

17. Andersen, M. E., Sarangapani, R., Frederick, C. B., and Kimbell, J. S. (1999): Dosimetric adjustment factors for methyl methacrylate derived from a steady-state analysis of a physiologically based clearance-extraction model. *Inhal. Toxicol.*, 11:899–926.
18. Andersen, M., Sarangapani, R., Gentry, R., Clewell, H., Covington, T., and Frederick, C. B. (2000): Application of a hybrid CFD–PBPK nasal dosimetry model in an inhalation risk assessment: an example with acrylic acid. *Toxicol. Sci.*, 57:312–325.
19. Arms, A. D., and Travis, C. C. (1988): *Reference Physiological Parameters in Pharmacokinetic Modeling*, NTIS PB 88-196019. Office of Health and Environmental Assessment, U.S. Environmental Protection Agency, Washington, D.C.
20. Balaz, S., and Lukacova, V. (1999): A model-based dependence of the human tissue/blood partition coefficients of chemicals on lipophilicity and tissue composition. *Quant. Structure-Activity Relat.*, 18:361–368.
21. Ball, R., and Schwartz, S. L. (1994): Cmatrix: software for physiologically based pharmacokinetic modeling using a symbolic matrix representation system. *Comput. Biol. Med.*, 24:269–276.
22. Barton, H. A., Deisinger, P. J., English, J. C., Gearhart, J. M., Faber, W. D., Tyler, T. R., Banton, M. I., Teeguarden, J., and Andersen, M. E. (2000): Family approach for estimating reference concentrations/doses for series of related organic chemicals. *Toxicol. Sci.*, 54:251–261.
23. Barton, H. A., Creech, J. R., Godin, S., Randall, G. M., and Seckel, C. S. (1995): Chloroethylene mixtures: pharmacokinetic modeling and *in vitro* metabolism of vinyl chloride, trichloroethylene, and *trans*-1,2-dichloroethylene in rat. *Toxicol. Appl. Pharmacol.*, 130:237–247.
24. Basak, S. C., Mills, D., Hawkins, D. M., and El-Marsi, H. A. (2003): Prediction of human blood– air partition coefficient: a comparison of structure-based and property-based methods. *Risk Anal.*, 23:1173–1184.
25. Basak, S. C., Mills, D., Hawkins, D. M., and El-Marsi, H. A. (2002): Prediction of tissue-air partition coefficients: a comparison of structure-based and property-based methods. *SAR QSAR Environ. Res.*, 13:649–665.
26. Béliveau, M., and Krishnan, K. (2003): *In silico* approaches for developing physiologically based pharmacokinetic (PBPK) models. In: *Alternative Toxicology Methods*, edited by H. Salem and S. Katz. CRC Press, Boca Raton, FL, pp. 479–532.
27. Béliveau, M., Lipscomb, J., Tardif, R., and Krishnan, K. (2005): Quantitative structure-property relationships for interspecies extrapolation of the inhalation pharmacokinetics of organic chemicals. *Chem. Res. Toxicol.*, 18:475–485.
28. Béliveau, M., Tardif, R., and Krishnan, K. (2003): Quantitative structure–property relationships for physiologically based pharmacokinetic modeling of volatile organic chemicals in rats. *Toxicol. Appl. Pharmacol.*, 189:221–32.
29. Benignus, V. A., Boyes, W. K., and Bushnell, P. J. (1998): A dosimetric analysis of behavioral effects of acute toluene exposure in rats and humans. *Toxicol. Sci.*, 43:186–195.
30. Bernillon, P., and Bois, F. Y., (2000): Statistical issues in toxicokinetic modeling: a Bayesian perspective. *Environ. Health Perspect.*, 108(Suppl.):883–893.
31. Bischoff, K. B. (1987): Physiogically based pharmacokinetic modeling. *Drinking Water Health*, 8:36–64.
32. Bischoff, K. B., Dedrick, R. L., Zakharo, D. S., and Longstreth, J. A. (1971): Methotrexate pharmacokinetics. *J. Pharm. Sci.*, 60:1128–1133.
33. Björkman, S. (2003): Reduction and lumping of physiologically based pharmacokinetic models: prediction of the disposition of fentanyl and pethidine in humans by successively simplified models. *J. Pharmacokinet. Pharmacodyn.*, 30:285–307.
34. Bogaards, J. J., Freidig, A. P., and van Bladeren, P. J. (2001): Prediction of isopropene diepoxide levels *in vivo* in mouse, rat, and man using enzyme kinetic data *in vitro* and physiologically based pharmacokinetic modeling. *Chem. Biol. Interact.*, 138:247–265.
35. Bois, F. Y. (1999): Analysis of PBPK models for risk characterisation. *Ann. N.Y. Acad. Sci.*, 895:317–337.
36. Bois, F. Y. (2000): Statistical analysis of Clewell et al. PBPK model of trichloroethylene kinetics. *Environ. Health Perspect.*, 108(Suppl. 2):307–316.
37. Bois, F. Y. (2000): Statistical analysis of Fisher et al. PBPK model of trichloroethylene kinetics. *Environ. Health Perspect.*, 108(Suppl. 2):275–282.
38. Bois, F. Y., Woodruff, T. J., and Spear, R. C. (1991): Comparison of three physiologically based pharmacokinetic models for benzene disposition. *Toxicol. Appl. Pharmacol.*, 110:79–88.
39. Bois, F. Y., Zeise, L., and Tozer, T. N. (1990): Precision and sensitivity of pharmacokinetic models for cancer risk assessment: tetrachloroethylene in mice, rats and humans. *Toxicol. Appl. Pharmacol.*, 102:300–315.
40. Bond, J. A., Himmelstein, M. W., Seaton, M., Boogaard, P., and Medinsky, M. A. (1996): Metabolism of butadiene by mice, rats, and humans: a comparison of physiologically based toxicokinetic model predictions and experimental data. *Toxicology*, 113:48–54.
41. Brown, R. P., Delp, M. D., Lindstedt, S. L., Rhomberg, L. R., and Belisle, R. P. (2005): Physiological parameter values for physiologically based pharmacokinetic models. *Toxicol. Indust. Health*, 13:407–484.
42. Byczkowski, J. Z., and Fisher, J. W. (1995): A computer program linking physiologically based model with cancer risk assessment for breast-fed infants. *Comput. Methods Programs Biomed.*, 46:155–163.
43. Cahill, T. M., Cousins, I., and Mackay, D. (2003): Development and application of a generalized physiologically based pharmacokinetic model for multiple environmental contaminants. *Environ. Toxicol. Chem.*, 22:26–34.
44. Casanova, M., d'Heck, H., and Deyo, D. F. (1992): Dichloromethane (methylene chloride): metabolism to formaldehyde and formation of DNA–protein crosslinks in $B6C3F_1$ mice and Syrian golden hamsters. *Toxicol. Appl. Pharmacol.*, 114:162–165.
45. Caster, W. O., Poncelet, J., Simon, A. B., and Armstrong, W. B. (1956): Tissue weights of the rat. I. Normal values determined by dissection and chemical methods. *Proc. Soc. Exp. Biol. Med.*, 91:122–126.
46. Chen, H. S. G., and Gross, J. F. (1979): Estimation of tissue to plasma partition coefficients used in physiological pharmacokinetic models. *J. Pharmacokin. Biopharm.*, 7:117–125.

47. Clark, L. H., Setzer, R. W., and Barton, H. A. (2004): Framework for evaluation of physiological-based pharmacokinetic models for use in safety or risk assessment. *Risk Anal.*, 24:1697–1717.
48. Clarke, D. O. et al. (1992): 2-Methoxyacetic acid dosimetry-teratology relationships in CD-1 mice exposed to 2-methoxyethanol. *Toxicol. Appl. Pharmacol.*, 114:77–87.
49. Clarke, D. O. et al. (1993): Pharmacokinetics of 2-methoxyethanol and 2-methoxyacetic acid in the pregnant mouse: a physiologically based mathematical model. *Toxicol. Appl. Pharmacol.*, 121:239–252.
50. Clewell, H. J., III, and Andersen, M. E. (1987): Dose, species and route extrapolation using physiologically based pharmacokinetic models. *Drinking Water and Health*, 8:159–182.
51. Clewell, H. J., III, and Andersen, M. E. (1994): Physiologically based pharmacokinetic modeling and bioactivation of xenobiotics. *Toxicol. Indust. Health*, 10:1–24.
52. Clewell, H. J., III, Andersen, M. E., and Barton, H. A. (2002): A consistent approach for the application of pharmacokinetic modeling in cancer and noncancer risk assessment. *Environ. Health Perspect.*, 110:85–93.
53. Clewell, H. J., III, Gearhart, J. M., Gentry, P. R., Covington, T. R., Van Landingham, C. B., Crump, K. S., and Shipp, A. M. (1999): Evaluation of the uncertainty in an oral reference dose for methylmercury due to interindividual variability in pharmacokinetics. *Risk Anal.*, 19:547–558.
54. Clewell, H. J., III, Gentry, P. R., Covington, T. R., Sarangapani, R., and Teeguarden, J. G. (2004): Evaluation of the potential impact of age- and gender-specific pharmacokinetic differences on tissue dosimetry. *Toxicol. Sci.*, 79:381–393.
55. Clewell, H. J., III, Gentry, P. R., and Gearhart, J. M. (1997): Investigation of the potential impact of benchmark dose and pharmacokinetic modeling in noncancer risk assessment. *Toxicol. Environ. Health*, 52:475–515.
56. Clewell, H. J., III, Gentry, P. R., Gearhart, J. M., Covington, T. R., Banton, M. I., and Andersen, M. E. (2001): Development of a physiologically based pharmacokinetic model of isopropanol and its metabolite acetone. *Toxicol. Sci.*, 63:160–172.
57. Clewell, H. J., III, Gentry, P. R., Kester, J. E., and Andersen, M. E. (2005): Evaluation of physiologically based pharmacokinetic models in risk assessment: an example with perchloroethylene. *Crit. Rev. Toxicol.*, 35:413–433.
58. Clewell, H. J., III, and Jarnot, B. M. (1994): Incorporation of pharmacokinetics in noncancer risk assessment: example with chloropentafluorobenzene. *Risk Anal.*, 14:265–276.
59. Clewell, H. J., III, Lee, T. S., and Carpenter, R. L. (1994): Sensitivity of physiologically based pharmacokinetic models to variation in model parameters—methylene chloride. *Risk Anal.*, 14:521–531.
60. Clewell, R. A., and Gearhart, J. M. (2002): Pharmacokinetics of toxic chemicals in breast milk: use of PBPK models to predict infant exposure. *Environ. Health Perspect.*, 110:A333–A337.
61. Clewell, R. A., Merrill, E. A., Yu, K. O., Mahle, D. A., Sterner, T. R., Fisher, J. W., and Gearhart, J. M. (2003): Predicting neonatal perchlorate dose and inhibition of iodide uptake in the rat during lactation using physiologically based pharmacokinetic modeling. *Toxicol. Sci.*, 74:416–436.
62. Clewell, R. A., Merrill, E. A., Yu, K. O., Mahle, D. A., Sterner, T. R., Mattie, D. R., Robinson, P. J., Fisher, J. W., and Gearhart, J. M. (2003): Predicting fetal perchlorate dose and inhibition of iodide kinetics during gestation: a physiologically based pharmacokinetic analysis of perchlorate and iodide kinetics in the rat. *Toxicol. Sci.*, 73:235–255.
63. Cohn, M. S. (1987): Sensitivity analysis in pharmacokinetic modeling. *Drinking Water Health*, 8:265–272.
64. Collins, J. M., Dedrick, R. L., Flessner, M. F., and Guarino, A. M. (1982): Concentration dependent disappearance of fluorouracil from peritoneal fluid in the rat: experimental observations and distributing modeling. *J. Pharm. Sci.*, 71:735–738.
65. Conolly, R. B., and Andersen, M. E. (1991): Biologically based pharmacodynamic models: tools for toxicological research and risk assessment. *Annu. Rev. Toxicol. Pharmacol.*, 31:503–523.
66. Corley, R. A., Markham, D. A., Banks, C., Delorme, P., Masterman, A., and Houle, J. M. (1997): Physiologically based pharmacokinetics and the dermal absorption of 2-butoxyethanol vapor by humans. *Fundam. Appl. Toxicol.*, 39:120–130.
67. Corley, R. A., Mast, T. J., Carney, E. W., Rogers, J. M., and Daston, G. P. (2003): Evaluation of physiologically based models of pregnancy and lactation for their application in children's health risk assessments. *Crit. Rev. Toxicol.*, 33:137–211.
68. Csanady, G. A., and Filser, J. G. (2001): The relevance of physical activity for the kinetics of inhaled gaseous substances. *Arch. Toxicol.*, 74:663–672.
69. Csanady, G. A., Kreuzer, P. E., Baur, C., and Filser, J. G., (1996): A physiological toxicokinetic model for 1,3-butadiene in rodents and man: blood concentrations of 1,3-butadiene, its metabolically formed epoxides, and of haemoglobin adducts—relevance of glutathione depletion. *Toxicology*, 113:300–305.
70. Dallas, C. E., Bruckner, J. V., Megden, J. L., and Weir, F. W. (1986): A method for direct measurements of systemic uptake and elimination of volatile organics in small mammals. *J. Pharmacol. Meth.*, 16:239–250.
71. Dallas, C. E., Chen, X. M., O'Bass, K., Muralidhara, S., Varkonyi, P., and Bruckner, J. (1994): Development of a physiologically based pharmacokinetic model for perchloroethylene using tissue concentration–time data. *Toxicol. Appl. Pharmacol.*, 128:50–59.
72. Davies, B., and Morris, T. (1993): Physiological parameters in laboratory animals and humans. *Pharm. Res.*, 10:1093–1095.
73. Dedrick, R. L., Forrester, D. D., and Ho, D. H. W. (1972): *In vitro/in vivo* correlation of drug metabolism: deamination of 1-β-D-arabinosyl cytosine. *Biochem. Pharmacol.*, 21:1–16.
74. Dedrick, R. L., Zaharko, D. S., and Lutz, R. J. (1973): Transport and binding of methotrexate *in vivo*. *J. Pharm. Sci.*, 62:882–890.
75. DeJongh, J., and Blaauboer, B. J. (1996): *In vitro*-based and *in vivo*-based simulations of benzene uptake and metabolism in rats. *ATLA*, 24:179–190.
76. DeJongh, J., and Blaauboer, B. J. (1997): Simulation of lindane kinetics in rats. *Toxicology*, 122:1–9.

77. DeJongh, J., and Blaauboer, B. J. (1996): Simulations of toluene kinetics in the rat by a physiologically based pharmacokinetic model with application of biotransformation parameters derived independently *in vitro* and *in vivo*. *Fundam. Appl. Toxicol.*, 32:260–268.
78. DeJongh, J., and Blaauboer, B. J. (1997): Evaluation of *in vitro*-based simulations of toluene uptake and metabolism in rats. *Toxicol. In Vitro*, 11:485–489.
79. DeJongh, J., Verhaar, H. J. M., and Hermens, J. L. M. (1997): A quantitative property–property relationship (QPPR) approach to estimate *in vitro* tissue–blood partition coefficients of organic chemicals in rats and humans. *Arch. Toxicol.*, 72:17–25.
80. Delp, M. D., Manning, R. O., Bruckner, J. V., and Armstrong, R. B. (1991): Distribution of cardiac output during diurnal changes of activity in rats. *Am. J. Physiol.*, 261:H1487–1493.
81. Dennison, J. E., Andersen, M. E., and Yang, R. S. H. (2003): Characterization of the pharmacokinetics of gasoline using PBPK modeling with a complex mixtures chemical lumping approach. *Inhal. Toxicol.*, 15:961–986.
82. Dennison, J. E., Andersen, M. E., and Yang, R. S. H. (2005): Pitfalls and related improvements of *in vivo* gas uptake pharmacokinetic experimental systems. *Inhal. Toxicol.*, 17:539–548.
83. Dietz, K. F., Rodriguez-Giaxola, M., Traiger, G. J., Stella, V. J., and Himmelstein, K. J. (1981): Pharmacokinetics of 2-butanol and its metabolites in the rat. *J. Pharmacokin. Biopharm.*, 9:553–573.
84. Dobrev, I. D., Andersen, M. E., and Yang, R. S. (2001): Assessing interactions thresholds for trichloroethylene in combination with tetrachloroethylene and 1,1,1-trichloroethane using gas uptake studies and PBPK modeling. *Arch. Toxicol.*, 75:134–144.
85. Domenech, R. J., Hoffman, J. E., Noble, M. M., Saunder, K. B., Hensen, J. R., and Subijanto, S. (1996): Total and regional coronary blood flow measured by radioactive microsphere in conscious and anesthetized dogs. *Circul. Res.*, 25:581–596.
86. Dong, M. H. (1994): Microcomputer programs for physiologically based pharmacokinetic (PB-PK) modeling. *Comput. Methods Programs Biomed.*, 45:213–221.
87. D'Souza, R. W., Franci, W. R., and Andersen, M. W. (1988): Physiological model for tissue glutathione depletion and decreased resynthesis after ethylene dichloride exposures. *J. Pharmacol. Exp. Ther.*, 245:563–568.
88. Easterling, M. R., Evans, M. V., and Kenyon, E. M. (2000): Comparative analysis of software for physiologically based pharmacokinetic modeling: simulation, optimization, and sensitivity analysis. *Toxicol. Methods*, 10:203–229.
89. El-Masri, H. A., Bell, D. A., and Portier, C. J. (1999): Effects of glutathione transferase theta polymorphism on the risk estimates of dichloromethane to humans. *Toxicol. Appl. Pharmacol.*, 158:221–230.
90. El-Masri, H. A., and Portier, C. J. (1998): Physiologically based pharmacokinetics model of primidone and its metabolites phenobarbital and phenylethylmalonamide in humans, rats and mice. *Drug Metab. Dispos.*, 26:585–594.
91. Emond, C., Birnbaum, L. S., and DeVito, M. J. (2004): Physiologically based pharmacokinetic model for developmental exposures to TCDD in the rat. *Toxicol. Sci.*, 80:115–133.
92. Emond, C., Charbonneau, M., and Krishnan, K. (2005): Physiologically based modeling of the accumulation in plasma and tissue lipids of a mixture of PCB congeners in female Sprague–Dawley rats. *J. Toxicol. Environ. Health*, 68:1393–1412.
93. Emond, C., and Krishnan, K. (2006): A physiological pharmacokinetic model based on tissue lipid content for simulating inhalation pharmacokinetics of highly lipophilic volatile organic chemicals. *Toxicol. Mech. Methods*, 16(8):395–403.
94. Evans, M. V., and Andersen, M. E. (1995): Sensitivity analysis and the design of gas uptake inhalation studies. *Inhal. Toxicol.*, 7:1075–1094.
95. Evans, M. V., and Andersen, M. E. (2000): Sensitivity analysis of a physiological model for 2,3,7,8-tetrachlorodibenzo-*p*-dioxin (TCDD): assessing the impact of specific model parameters on sequestration in liver and fat in the rat. *Toxicol. Sci.*, 54:71–80.
96. Falk, A., Gullstrand, E., Löf, A., and Wigaeus-Hjelm, E. (1990): Liquid/air partition coefficients of four terpenes. *Br. J. Indust. Med.*, 47:62–64.
97. Farrar, D., Allen, B., Crump, K., and Shipp, A. (1989): Evaluation of uncertainty in input parameters to pharmacokinetic models and the resulting uncertainty in output. *Toxicol. Lett.*, 49:371–385.
98. Farris, F. F., Dedrick, R. L., and King, F. G. (1988): Cisplatin pharmacokinetics: applications of a physiological model. *Toxicol. Lett.*, 43:117–137.
99. Filser, J. G., and Bolt, H. M. (1979): Pharmacokinetics of halogenated ethylenes in rats. *Arch. Toxicol.*, 42:123–136.
100. Filser, J. G., Kessler, W., and Csanady, G. A. (2004): The "Tuebingen dessicator" system, a tool to study oxidative stress *in vivo* and inhalation toxicokinetics. *Drug Metab. Rev.*, 36:787–803.
101. Fiserova-Bergerova, V. (1975): Mathematical modeling of inhalation exposure. *J. Combust. Toxicol.*, 32:201–210.
102. Fiserova-Bergerova, V. (1985): Toxicokinetics of organic solvents. *Scand. J. Work Environ Health*, 11(Suppl. 1–7):7–12.
103. Fiserova-Bergerova, V., and Diaz, M. L. (1986): Determination and prediction of tissue–gas partition coefficients. *Int. Arch. Occup. Environ. Health*, 58:75–87.
104. Fisher, J. W., Dorman, D. C., Medinsky, M. A., Welsch, F., and Conolly, R. B. (2000): Analysis of respiratory exchange of methanol in the lung of the monkey using a physiological model. *Toxicol. Sci.*, 53:185–193.
105. Fisher, J., Mahle, D., Bankston, L., Greene, R., and Gearhart, J. (1997): Lactational transfer of volatile chemicals in breast milk. *Indust. Hyg. Assoc. J.*, 58:425–431.
106. Fisher, J. W., Mahle, D., and Abbas, R. (1998): A human physiologically based pharmacokinetic model for trichloroethylene and its metabolites, trichloroacetic acid and free trichloroethanol. *Toxicol. Appl. Pharmacol.*, 152:339–359.
107. Fisher, J. W., Whittaker, T. A., Taylor, D. H., Clewell, H. J., and Andersen, M. E. (1989): Physiologically based pharmacokinetic modeling of the pregnant rat: multiroute exposure model for trichloroethylene and trichloroacetic acid. *Toxicol. Appl Pharmacol.*, 99:395–414.

108. Frederick, C. B., Bush, M. L., Lomax, L. M., Black, K. A., Finch, L., Kimbell, J. S., Morgan, K. T., Subramaniam, R. P., Morris, J. B., and Ultman, J. S. (1998): Application of a hybrid computational fluid dynamics and physiologically based inhalation model for interspecies dosimetry extrapolation of acidic vapors in the upper airways. *Toxicol. Appl. Pharmacol.*, 152:211–231.
109. Frederick, C. B., Potter, D. W., Chang-Mateu, M. I., and Andersen, M. E. (1992): A physiologically based pharmacokinetic and pharmacodynamic model to describe the oral dosing of rats with ethyl acrylate and its implications for risk assessment. *Toxicol. Appl. Pharmacol.*, 114:246–260.
110. Gabrielsson, J. L., Paalkow, L. K., and Nordstrom, L. (1987): A physiologically based pharmacokinetic model for theophylline disposition in the pregnant and nonpregnant rat. *J. Pharmacokin. Biopharm.*, 12:149–165.
111. Gallo, J. M., Lam, F. C., and Perrier, D. G. (1987): Area method for the estimation of partition coefficients for physiological pharmacokinetic models. *J. Pharmacokin. Biopharm.*, 15:271–280.
112. Gargas, M. L. (1990): An exhaled breath chamber system for assessing rates of metabolism and rates of gastrointestinal absorption with volatile chemicals. *J. Am. Coll. Toxicol.*, 9:447–453.
113. Gargas, M. L., and Andersen, M. E. (1989): Determining the kinetic constants of chlorinated ethane metabolism in the rat from rates of exhalation. *Toxicol. Appl. Pharmacol.*, 99:344–353.
114. Gargas, M. L., Andersen, M. E., and Clewell, H. J. (1986): A physiologically based simulation approach for determining metabolic rate constants from gas uptake data. *Toxicol. Appl. Pharmacol.*, 86:341–352.
115. Gargas, M. L., Burgess, R. J., Voisard, D. E. Cason, G. H., and Andersen, M. E. (1989): Partition coefficients of low molecular weight volatile chemicals in various liquids and tissues. *Toxicol. Appl. Pharmacol.*, 98:87–99.
116. Gargas, M. L., Clewell, H. J., and Andersen, M. E. 1986): Gas uptake inhalation techniques and the rates of metabolism of chloromethanes, chloroethanes, and chloroethylenes in the rat. *Inhal. Toxicol.*, 2:319.
117. Gargas, M. L., Clewell, H. J., and Andersen, M. E. (1986): Metabolism of inhaled dihalomethanes *in vivo*: differentiation of kinetic constants for two independent pathways. *Toxicol. Appl. Pharmacol.*, 87:211–223.
118. Gargas, M. L., Seybold, P. G., and Andersen, M. E. (1988): Modeling the tissue solubilities and metabolic rate constants of halogenated methanes, ethanes and ethylenes. *Toxicol. Lett.*, 43:235–256.
119. Gargas, M. L., Tyler, T. R., Sweeney, L. M., Corley, R. A., Weitz, K. K., Mast, T. J., Paustenbach, D. J., and Hays, S. M. (2000): A toxicokinetic study of inhaled ethylene glycol ethyl ether acetate and validation of physiologically based pharmacokinetic model for rat and human. *Toxicol. Appl. Pharmacol.*, 165:63–73.
120. Gargas, M. L., Tyler, T. R., Sweeney, L. M., Corley, R. A., Weitz, K. K., Mast, T. J., Paustenbach, D. J., and Hays, S. M. (2000): A toxicokinetic study of inhaled ethylene glycol monomethyl ether (2-ME) and validation of a physiologically based pharmacokinetic model for the pregnant rat and human. *Toxicol. Appl. Pharmacol.*, 165:53–62.
121. Gear, C. W. (1971): *Numerical Initial Value Problems in Ordinary Differential Equations*. Prentice-Hall, Englewoods Cliffs, NJ.
122. Gearhart, J. M., Clewell, H. J. I., Crump, K. S., Shipp, A. M., and Silvers, A. (1995): Pharmacokinetic dose estimates of mercury in children and dose–response curves of performance tests in a large epidemiology study. *Water Air Soil Pollut.*, 80:49–58.
123. Gearhart, J. M., Jepson, G. W., Clewell, H. J., Andersen, M. E., and Conolly, R. B. (1990): A physiologically based model for the *in vivo* inhibition of acetylcholinesterase by diisopropylfluorophosphate. *Toxicol. Appl. Pharmacol.*, 16:295–310.
124. Gelman, A., Bois, F., and Jiang, J. (1996): Physiological pharmacokinetic analysis using population modeling and informative prior distributions. *J. Am. Stat. Assoc.*, 91:436.
125. Gentry, P. R., Covington, T. R., Andersen, M. E., and Clewell, H. J., III. (2002): Application of a physiologically based pharmacokinetic model for isopropanol in the derivation of a reference dose and reference concentration. *Regul. Toxicol. Pharmacol.*, 36:51–68.
126. Gentry, P. R., Haber, L. T., McDonald, T. B., Zhao, Q., Covington, T., Nance, P., Clewell, H. J., III, and Lipscomb, J. C. (2004): Data for physiologically based pharmacokinetic modeling in neonatal animals: physiological parameters in mice and Sprague–Dawley rats. *J. Child Health*, 2:363–411.
127. Gerde, P., and Dahl, A. R. (1991): A model for the uptake of inhaled vapors in the nose of the dog during cyclic breathing. *Toxicol. Appl. Pharmacol.*, 109:276–288.
128. Gibaldi, M., and Perrier, D. (1982): *Pharmacokinetics*. Marcel Dekker, New York.
129. Haber, L. T., Maier, A., Gentry, P. R., Clewell, H. J., and Dourson, M. L. (2002): Genetic polymorphisms in assessing interindividual variability in delivered dose. *Regul. Toxicol. Pharmacol.*, 35:177–197.
130. Haddad, S., Béliveau, M., Tardif, R., and Krishnan, K. (2001): A PBPK modeling approach to account for interactions in the health risk assessment of chemical mixtures. *Toxicol. Sci.*, 63:125–131.
131. Haddad, S., Gad, S. C., Tardif, R., and Krishnan, K. (1995): Statistical approaches for the validation of physiologically based pharmacokinetic (PBPK) models. *Toxicologist*, 15:258.
132. Haddad, S., Pelekis, M., and Krishnan, K. (1996): A methodology for solving physiologically based pharmacokinetic models without the use of simulation softwares. *Toxicol. Lett.*, 85:113–126.
133. Haddad, S., Poulin, P., and Krishnan, K. (2000): Ratio of lipid content in adipose tissues and blood as the sole determinant of the adipose tissue–blood partition coefficients of highly lipophilic organic chemicals. *Chemosphere*, 40:839–843.
134. Haddad, S., Tardif, R., Charest-Tardif, G., and Krishnan, K. (2000): Validation of a physiological modeling framework for simulating the toxicokinetics of chemicals in mixtures. *Toxicol. Appl. Pharmacol.*, 167:199–209.
135. Haddad, S., Withey, J., Laparé, S., Law, F. C. P., and Krishnan, K. (1998): Physiologically based pharmacokinetic modeling of pyrene in the rat. *Environ. Toxcol. Pharmacol.*, 5:245–255.

136. Haggard, H. W. (1924): The absorption, distribution and elimination of ethyl ether: analysis of the mechanism of the absorption and elimination of such a gas or vapor as ethyl ether. *J. Biol. Chem.*, 59:753–770.
137. Hattis, D., White, P., Marmorstein, L., and Koch, P. (1990): Uncertainties in pharmacokinetics modeling for perchloroethylene. I. Comparison of model structure, parameters, and predictions for low dose metabolic rates for models by different authors. *Risk Anal.*, 10:449–458.
138. Hetrick, D. M., Jarabek, A. M., and Travis, C. C. (1991): Sensitivity analysis for physiologically based pharmacokinetic models. *J. Pharmacokin, Biopharm.*, 19:1–20.
139. Hilderbrand, R. L., Andersen, M. E., and Jensen, L. J. (1981): Prediction of *in vivo* kinetic constants for metabolism of inhaled vapors from kinetic constants measured *in vitro*. *Fundam. Appl. Toxicol.*, 1:403–409.
140. Himmelstein, M. W., Carpenter, S. C., Evans, M. V., Hinderliter, P. M., and Kenyon, E. M. (2004): Kinetic modeling of beta-chloroprene metabolism. II. The applications of physiologically based modeling for cancer dose response analysis. *Toxicol. Sci.*, 79:28–37.
141. Hoang, K. C. T. (1995): Physiologically based pharmacokinetic models: mathematical fundamentals and simulation implementations. *Toxicol. Lett.*, 79:87–98.
142. Igari, Y., Sugiyama, Y., Sawada, Y., Iga, Y., and Hanano, M. (1983): Prediction of diazepam disposition in rat and man by a physiologically based pharmacokinetic model. *J. Pharmacokin. Biopharm.*, 11:577–593.
143. Iman, R., and Helton, J. (1998): An investigation of uncertainty and sensitivity analysis techniques for computer models. *Risk Anal.*, 8:71–90.
144. International Commission on Radiation Protection (1975): *Report of the Task Group on Reference Man*, ICRP Publication No. 23. Pergamon Press, New York.
145. Isukapalli, S. S., Roy, A., and Georgopoulos, P. G. (1998): Stochastic response surface methods (SRSMs) for uncertainty propagation: applications to environmental and biological systems. *Risk Anal.*, 18:351–363.
146. Iwatsubo, T., Suzuki, H., and Sugiyama, Y. (1997): Prediction of species differences (rats, dogs, humans) in the *in vivo* metabolic clearance of YM796 by the liver from *in vitro* data. *J. Pharmacol. Exp. Ther.*, 283:462–469.
147. Iyengar, S., and Rao, M. S. (1983): Statistical techniques in modeling of complex systems: single versus multiresponse models. *IEEE Trans. Syst. Man. Cybernet.*, 13:175–189.
148. Jain, R. K., Gerlowski, L. E., Weissbrod, J. M., Wang, J., and Pierson, R. N. (1982): Kinetics of uptake, distribution and excretion zinc in rats. *Ann. Biomed. Eng.*, 9:347–361.
149. Jang, J. Y., Droz, P. O., and Kim, S. (2001): Biological monitoring of workers exposed to ethylbenzene and co-exposed to xylene. *Int. Arch. Occup. Environ. Health*, 74:31–37.
150. Johanson, G. (1991): Modeling of respiratory exchange of polar solvents. *Ann. Occup. Hyg.*, 35:323–339.
151. Johanson, G., and Dynesius, B. (1988): Liquid: air partition coefficients for six commonly used glycol ethers. *Br. J. Indust. Med.*, 45:561–564.
152. Johanson, G., and Filser, J. G. (1992): Experimental data from closed chamber gas uptake studies in rodents suggest lower uptake rate of chemical than calculated from literature values on alveolar ventilation. *Arch. Toxicol.*, 66:291–295.
153. Johanson, G., Jonsson, F., and Bois, F. (1999): Development of new technique for risk assessment using physiologically based pharmacokinetic models. *Am. J. Indust. Med.*, 36(Suppl. 1):101–103.
154. Johanson, G., and Naslund, P. H. (1988): Spreadsheet programming: a new approach in physiologically based modeling of solvent toxicokinetics. *Toxicol. Lett.*, 41:115–127.
155. Jonsson, F., Bois, F., and Johanson, G. (2001): Physiologically based pharmacokinetic modeling of inhalation exposure of humans to dichloromethane during moderate to heavy exercise. *Toxicol. Sci.*, 59:209–218.
156. Jonsson, F., Bois, F. Y., and Johanson, G. (2001): Assessing the reliability of PBPK models using data from methyl chloride-exposed, non-conjugating human subjects. *Arch. Toxicol.*, 75:189–199.
157. Jonsson, F., and Johanson, G. (2001): A Bayesian analysis of the influence of GSTT1 polymorphism on the cancer risk estimate for dichloromethane. *Toxicol. Appl. Pharmacol.*, 174:99–112.
158. Jonsson, F., and Johanson, G. (2001): Bayesian estimation of variability in adipose tissue blood flow in man by physiologically based pharmacokinetic modeling of inhalation exposure to toluene. *Toxicology*, 157:177–193.
159. Jonsson, F., and Johanson, G. (2002): Physiologically based modeling of the inhalation kinetics of styrene in humans using a Bayesian population approach. *Toxcol. Appl. Pharmacol.*, 179:35–49.
160. Jonsson, F., and Johanson, G. (2003): The Bayesian population approach to physiological toxicokinetic–toxicodynamic models: an example using the MCSim software. *Toxicol. Lett.*, 138:143–150.
161. Karba, R., Zupancic, B., and Bremsak, F. (1990): Simulation tools in pharmacokinetic modeling. *Acta Pharm. Jugosl.*, 40:247–262.
162. Kedderis. L. B., Mills, J. J., Andersen, M. E., and Birnbaum, L. S. (1993): A physiologically based pharmacokinetic model of 2,3,7,8-tetrabromodibenzo-*p*-dioxin (TBDD) in the rat: tissue distribution and CYPIA induction. *Toxicol. Appl. Pharmacol.*, 121:87–98.
163. Kety, S. S. (1951): The theory and application of the exchange of inert gas at the lungs. *Pharmacol. Rev.*, 3:1–41.
164. Keys, D. A., Bruckner, J. V., Muralidhara, S., and Fisher, J. W. (2003): Tissue dosimetry expansion and cross-validation of rat and mouse physiologically based pharmacokinetic models for trichloroethylene. *Toxicol. Sci.*, 76:35–50.
165. Keys, D. A., Wallace, D. G., Kepler, T. B., and Conolly, R. B., (2000): Quantitative evaluation of alternative mechanisms of blood disposition of di(*n*-butyl) phthalate and mono(*n*-butyl) phthalate in rats. *Toxicol. Sci.*, 53:173–184.
166. Khor, S. P., and Mayersohn, M. (1991): Potential error in the measurement of tissue to blood distribution coefficients in physiological pharmacokinetic modeling: residual tissue blood. I. Theoretical considerations. *Drug Metab. Disp.*, 19:478–485.
167. Kootsey, J. M., Kohn, M. C., Feezor,M. D., Mitchell, G. R., and Fletcher, P. R. (1986): ScoP: an interactive simulation control program for micro- and minicomputers. *Bull. Math. Biol.*, 48:427–441.

168. Krewski, D., Wang, Y., Barlett, S., and Krishnan, K. (1995): Uncertainty, variability, and sensitivity analysis in physiological pharmacokinetic models. *J. Biopharm. Statist.*, 5:245–271.

169. Krishnan, K., and Andersen, M. E. (2001): Physiologically based pharmacokinetic modeling in toxicology. In: *Principles and Methods of Toxicology*, edited by A. W. Hayes, Taylor & Francis, New York, pp. 193–241.

170. Krishnan, K., Gargas, M. L., and Andersen, M. E. (1993): *In vitro* toxicology and risk assessment. *Altern. Meth. Toxicol.*, 9:185–203.

171. Krishnan, K., Gargas, M. L., Fennell, T. R., and Andersen, M. E. (1992): A physiologically based description of ethylene oxide dosimetry in the rat. *Toxicol. Indust. Health*, 8:121–140.

172. Krishnan, K., Haddad, S., Beliveau, M., and Tardif, R. (2002): Physiological modeling and extrapolation of pharmacokinetic interactions from binary to more complex chemical mixtures. *Environ. Health Perspect.*, 110(Suppl. 6):989–994.

173. Krishnan, K., and Johanson, G. (2005): Physiologically based pharmacokinetic and toxicokinetic models in cancer risk assessment. *J. Environ. Sci. Health*, 23(C):31–53.

174. Krishnan, K., Pelekis, M. L., and Haddad, S. (1995): A simple index for describing the discrepancy between PBPK model simulations and experimental data. *J. Toxicol. Indust. Health*, 11:413–421.

175. Labaune, J. P. (1988): *Pharmacocinétique: principes fondamentaux*. Masson, Paris.

176. Lam, G., Chen, M. L., and Chiou, W. L. (1982): Determination of tissue– blood partition coefficients in physiologically based pharmacokinetic models. *J. Pharm. Sci.*, 71:454–456.

177. Licata, A. C., Dekant, W., Smith, C. E., and Borghoff, S. J. (2001): A physiologically based pharmacokinetic model for methyl *tert*-butyl ether in humans: implementing sensitivity and variability analyses. *Toxicol. Sci.*, 62: 191–204.

178. Lin, J. H., Sugiyama, Y., Awazu, S., and Hanano, M. (1982): *In vitro* and *in vivo* evaluation of the tissue to blood partition coefficients for physiological pharmacokinetic models. *J. Pharmacokin. Biopharm.*, 10:637–647.

179. Lin, J. H., Sugiyama, Y., Awazu, S., and Hanano, M. (1982): Physiological pharmacokinetics of ethoxybenzamine based on biochemical data obtained *in vitro* as well as on physiological data. *J. Pharmacokin. Biopharm.*, 10:649–661.

180. Lipscomb, J. C., Barton, H. A., Tornero-Velez, R., Evans, M. V., Alcasey, S., Snawder, J. E., and Laskey, J. (2004): The metabolic rate constants and specific activity of human and rat hepatic cytochrome P-450 2E1 toward toluene and chloroform. *J. Toxicol. Environ. Health*, 67(A):537–553.

181. Lipscomb, J. C., Fisher, J. W., Confer, P. D., and Byczkowski, J. Z. (1998): *In vitro* to *in vivo* extrapolation for trichloroethylene metabolism in humans. *Toxicol. Appl. Pharmacol.*, 152:376–387.

182. Lipscomb, J. C., and Ohanian, E. V., Eds. (2007): *Toxicokinetics and Risk Assessment*. Informa Health Care, New York.

183. Lipscomb, J. C., Teuschler, L. K., Swartout, J., Popken, D., Cox, T., and Kedderis, G. L. (2003): The impact of cytochrome P450 2E1-dependent metabolic variance on a risk-relevant pharmacokinetic outcome in humans. *Risk. Anal.*, 6:1221–1238.

184. Loizou, G. D., Eldirdiri, N. I., and King, L. J. (1996): Physiologically based pharmacokinetics of uptake by inhalation of a series of 1,1,1-trihaleothanes: correlation with various physicochemical parameters. *Inhal. Toxicol.*, 8:1–19.

185. Mapleson, W. W. (1963): An electric analog for uptake and elimination in man. *J. Appl. Physiol.*, 18:197–204.

186. Maruyama, W., Yoshida, K., Tanaka, T., and Nakanishi, J. (2002): Determination of tissue–blood partition coefficients for a physiological model for humans, and estimation of dioxin concentration in tissues. *Chemosphere*, 46:975–985.

187. Maruyama, W., Yoshida, K., Tanaka, T., and Nakanishi, J. (2003): Simulation of dioxin accumulation in human tissues and analysis of reproductive risk. *Chemosphere*, 53:301–313.

188. Mauderly, J. L. (1990): Measurement of respiration and respiratory responses during inhalation exposures. *J. Am. Coll. Toxicol.*, 9:397–406.

189. McDougal, J. N., Jepson, G. W., Clewell, H. J., and Andersen, M. E. (1985): Dermal absorption of dihalomethane vapors. *Toxicol. Appl. Pharmacol.*, 79:150–158.

190. McDougal, J. N., Jepson, G. W., Clewell, H. J., MacNaughton, M. G., and Andersen, M. E. (1986): A physiological pharmacokinetic model for dermal absorption of vapors in the rat. *Toxicol. Appl. Pharmacol.*, 85:286–294.

191. Meek, M. E., Beauchamp, R., Long, G., Turner, L., and Walker, M. (2002): Chloroform: exposure estimation, hazard characterisation, and exposure–response analysis. *J. Toxicol. Environ. Health*, 5(B):283–334.

192. Menzel, D. B., Wolpert, R. L., Boger, J. R., and Kootsey, J. M. (1987): Resources available for simulation in toxicology: specialized computers, generalized software and communication networks. *Drinking Water and Health*, 8:229–254.

193. Merrill, E. A., Clewell, R. A., Gearhart, J. M., Robinson, P. J., Sterner, T. R., Yu, K. O., Mattie, D. R., and Fisher, J. W. (2003): PBPK predictions of perchlorate distribution and its effect of thyroid uptake of radioiodide in the male rat. *Toxicol. Sci.*, 73:256–269.

194. Merrill, E. A., Clewell, R. A., Robinson, P. J., Jarabek, A. M., Gearhart, J. M., Sterner, T. R., and Fisher, J. W. (2005): PBPK model for radioactive iodide and perchlorate kinetics and perchlorate-induced inhibition of iodide uptake in humans. *Toxicol. Sci.*, 83:25–43.

195. Meulenberg, C. J., and Vijverberg, H. P. (2000): Empirical relations predicting human and rat tissue–air partition coefficient of volatile organic compounds. *Toxicol. Appl. Pharmacol.*, 165:206–216.

196. Miller, F. J., Overton, J. H., Jaskot, R. H., and Menzel, D. B. (1985): A model for the regional uptake of gaseous pollutants in the lung. I. The sensitivity of the uptake of ozone in the human lung to lower respiratory tract secretions and exercise. *Toxicol. Appl. Pharmacol.* 79:11–27.

197. Mork, A. K., and Johanson, G. (2006): A human physiological model describing acetone kinetics in blood and breath during various levels of physical exercise. *Toxicol. Lett.*, 164:6–15.
198. Mortensen, B., Lokken, T., Zahlsen, K., and Nilsen, O. G. (1997): Comparisons and *in vivo* relevance of two different *in vitro* headspace metabolic systems: liver S9 and liver slices. *Pharmacol. Toxicol.*, 81:35–41.
199. Mortensen, B., and Nilsen, O. G. (1988): Allometric species comparison of toluene and *n*-hexane metabolism: prediction of hepatic clearance in man from experiments with rodent liver S9 in headspace vial equilibration system. *Pharmacol. Toxicol.*, 82:183–188.
200. Nakajima, T., and Sato, A. (1979): Enhanced activity of liver drug-metabolizing enzymes for aromatic and chlorinated hydrocarbons following food deprivation. *Toxicol. Appl. Pharmacol.*, 50:549–556.
201. National Toxicology Program (1985): *NTP Technical Report on the Toxicology and Carcinogenesis Studies of Dichloromethane in Fisher-344 Rats and B6C3F$_1$ Mice (Inhalation Studies)*, NTP Tech. Rep. No. 306. National Toxicology Program, Washington, D.C.
202. Nestorov, I. A. (2001): Modeling and simulation of variability and uncertainty in toxicokinetics and pharmacokinetics. *Toxicol. Lett.*, 120:411–420.
203. Nichols, J., Rheingans, P., Lothenbach, D., McGeachie, R., Skow, L., and McKim, J. (1994) Three-dimensional visualization of physiologically based kinetic model outputs. *Environ. Health Perspect.*, 102:952–956.
204. Nong, A., McCarver, D. G., Hines, R. N., and Krishnan, K. (2005): Modeling interchild differences in pharmacokinetics on the basis of subject-specific data on physiology and hepatic CYP2E1 levels: a case study with toluene. *Toxicol. Appl. Pharmacol.*, 214(1):78–87.
205. O'Flaherty, E. (1998): Physiologically based models of metal kinetics. *CRC Crit. Rev. Toxicol.*, 28(3):271–317.
206. O'Flaherty, E. J. (1981): *Toxicants and Drugs: Kinetics and Dynamics*. John Wiley & Sons, New York.
207. O'Flaherty, E. J., Scott, W., Schreiner, C., and Beliles, R. P. (1992): A physiologically based kinetic model of rat and mouse gestation: disposition of a weak acid. *Toxicol. Appl. Pharmacol.*, 112:245–256.
208. Overton, J. H., Graham, R. C., and Miller, F. J. (1987): Mathematical modeling of ozone absorption in the lower respiratory tract. *Drinking Water Health*, 8:302–311.
209. Page, N. P., Singh, D. V., Farland, W., Goodman, J. I., Conolly, R. B., Andersen, M. E., Clewell, H. J., Frederick, C. B., Yamasaki, H., and Lucier, G. (1997): Implementation of EPA revised cancer assessment guidelines: incorporation of mechanistic and pharmacokinetic data. *Fundam. Appl. Toxicol.*, 37:16–36.
210. Parham, F. M., Kohn, M. C., Matthews, H. B., DeRosa, C., and Portier, C. J. (1997): Using structural information to create physiologically based pharmacokinetic models for all polychlorinated biphenyls. I. Tissue:blood partition coefficients. *Toxicol. Appl. Pharmacol.*, 144:340–347.
211. Parham, F. M., and Portier, C. J. (1998): Using structural information to create physiologically based pharmacokinetic models for all polychlorinated biphenyls. II. Rates of metabolism. *Toxicol. Appl. Pharmacol.*, 151:110–116.
212. Paterson, S., and MacKay, D. (1989): Correlation of tissue, blood, and air partition coefficients of volatile organic chemicals. *Br. J. Indust. Med.*, 46:321–328.
213. Payne, M. P., and Kenny, L. C. (2002): Comparison of models for the estimate of biological partition coefficients. *J. Toxicol. Environ. Health*, 65(Part A):897–931.
214. Pelekis, M., Krewski, D., and Krishnan, K. (1997): Physiologically based algebraic expressions for predicting steady-state toxicokinetics of inhaled vapors. *Toxicol. Meth.*, 7:205–225.
215. Pelekis, M., Poulin, P., and Krishnan, K. (1995): An approach for incorporating tissue composition data into physiologically based pharmacokinetic models. *Toxicol. Indust. Health*, 11:511–522.
216. Poet, T. S., Kousba, A. A., Dennison, S. L., and Timchalk, C. (2004): Physiologically based pharmacokinetic/pharmacodynamic model for the organophosphorus pesticide diazinon. *Neurotoxicology*, 25:1013–1030.
217. Poet, T. S., Weitz, K. K., Gies, R. A., Edwards, J. A., Thrall, K. D., Corley, R. A., Tanojo, H., Hui, X., Maibach, H. I., and Wester, R. C. (2002): PBPK modeling of the percutaneous absorption of perchloroethylene from a soil matrix in rats and humans. *Toxicol. Sci.*, 67:17–31.
218. Poet, T. S., Wu, H., English, J. C., and Corley, R. A. (2004): Metabolic rate constants for hydroquinone in F344 rat and human liver isolated hepatocytes: application to a PBPK model. *Toxicol. Sci.*, 82:9–25.
219. Portier, C. J., and Kaplan, N. L. (1989): Variability of safe estimated when using complicated models of carcinogenic processes. A dose study: methylene chloride. *Fundam. Appl. Toxicol.*, 13:533–544.
220. Poulin, P., Beliveau, M., and Krishnan, K. (1999): Mechanistic animal replacement approaches for predicting pharmacokinetics of organic chemicals. In: *Toxicity Assessment Alternatives: Methods, Issues, Opportunities*, edited by H. Salem and S. A. Katz. Humana Press, Totowa, NJ, pp. 115–139.
221. Poulin, P., and Krishnan, K. (1995): An algorithm for predicting tissue–blood partition coefficients of organic chemicals from *n*-octanol:water partition coefficient data. *J. Toxicol. Environ. Health*, 46:117–129.
222. Poulin, P., and Krishnan, K. (1995): A biologically based algorithm for predicting human tissue–blood partition coefficients of organic chemicals. *Human Exp. Toxicol.*, 14:273–280.
223. Poulin, P., and Krishnan, K. (1996): A mechanistic algorithm for predicting blood–air partition coefficients of organic chemicals with the consideration of reversible binding in hemoglobin. *Toxicol. Appl. Pharmacol.*, 136:131–137.
224. Poulin, P., and Krishnan, K. (1996): A tissue composition-based algorithm for predicting tissue–air partition coefficients of organic chemicals. *Toxicol. Appl. Pharmacol.*, 136:136–130.
225. Poulin, P., and Krishnan, K. (1996): Molecular structure-based prediction of the partition coefficients of organic chemicals for physiological pharmacokinetic models. *Toxicol. Meth.*, 6:117–137.
226. Poulin, P., and Krishnan, K. (1998): A quantitative structure-toxicokinetic relationship model for highly metabolised chemicals. *ATLA*, 26:45–59.

227. Poulin, P., and Krishnan, K. (1999): Molecular structure-based prediction of the toxicokinetics of inhaled vapors in humans. *Int. J. Toxicol.*, 18:7–18.
228. Price, K., Haddad, S., and Krishnan, K. (2003): Physiological modeling of age-specific changes in the pharmacokinetics of organic chemicals in children. *J. Toxicol. Environ. Health*, 66(A):417–433.
229. Price, P. S., Conolly, R. B., Chaisson, K., Gross, E. A., Young, J. S., Mathis, E. T., and Tedder, D. R. (2003): Modeling interindividual variation in physiological factors used in PBPK models of humans. *Crit. Rev. Toxicol.*, 33:469–503.
230. Ramsey, J. C., and Andersen, M. E. (1984): A physiologically based description of the inhalation pharmacokinetics of styrene in rats and humans. *Toxicol. Appl. Pharmacol.*, 73:159–175.
231. Reddy, M. B., Andersen, M. E., Morrow, P. E., Dobrev, I. D., Varaprath, S., Plotzke, K. P., and Utell, M. J. (2003): Physiological modeling of inhalation kinetics of octamethylcyclotetrasiloxane in humans during rest and exercise. *Toxicol. Sci.*, 72:3–18.
232. Reddy, M. B., Yang, R. S. H., Clewell, H. J., III, and Andersen, M. E. (2005): *Physiologically Based Pharmacokinetic Modeling: Science and Application*. John Wiley & Sons, New York, p. 420.
233. Reitz, R. H., Gargas, M. L., Andersen, M. E., Provan, W. M., and Green, T. L. (1996): Predicting cancer risk from vinyl chloride exposure with a physiologically based pharmacokinetic model. *Toxicol. Appl. Pharmacol.*, 137:253–267.
234. Reitz, R. H., Mandrela, A. L., and Guengerich, F. P. (1989): *In vitro* metabolism of methylene chloride in human and animal tissues: use in physiologically based pharmacokinetic models. *Toxicol. Appl. Pharmacol.*, 97:230–246.
235. Reitz, R. H., McCroskey, P. S., Park, C. N., Andersen, M. E., and Gargas, M. L. (1990): Development of a physiologically based pharmacokinetic model for risk assessment with 1,4-dioxane. *Toxicol. Appl. Pharmacol.*, 105:37–54.
236. Reitz, R. H., Mendrala, A. L., Park, C. N., Andersen, M. E., and Guengerich, F. P. (1988): Incorporation of *in vitro* enzyme data into the physiologically based pharmacokinetic (PBPK) model for methylene chloride: implications for risk assessment. *Toxicol. Lett.*, 43:97–116.
237. Rey, T. D., and Havranek, W. A. (1996): Some aspects of using the SimuSolv program for environmental, pharmacokinetics and toxicological applications. *Ecolog. Model.*, 86:277–282.
238. Rideout, V. C. (1991): *Mathematical and Computer Modeling of Physiological Systems*. Prentice Hall, New York.
239. Riggs, D. S. (1970): *The Mathematical Approach to Physiological Problems: A Critical Treatise*. MIT Press, Cambridge, MA.
240. Robinson, P. J. (1991): Effect of microcirculatory heterogeneity in the determination of pharmacokinetic parameters: implications for risk assessment. *Drug Metab. Rev.*, 23:43–64.
241. Robinson, P. J. (1992): Physiologically based liver modeling and risk assessment. *Risk Anal.*, 12:139–148.
242. Ross, R., Leger, L., Guardo, R., de Guise, J., and Pike, B. G. (1991): Adipose tissue volumes measured by magnetic resonance imaging and computerized tomography in rats. *J. Appl. Physiol.*, 70:2164–2172.
243. Roth, R. A., and Vinegar, A. (1990): Action by the lungs on circulating xenobiotic agents with a case study of physiologically based pharmacokinetic modeling of benzo(*a*)pyrene disposition. *Pharmacol. Therap.*, 48: 143–155.
244. Santostefano, M. J., Wang, X., Richardson, V. M., Ross, D. G., DeVito, M. J., and Birnbaum, L. S. (1998): A pharmacodynamic analysis of TCDD-induced cytochrome P450 gene expression in multiple tissues: dose- and time-dependent effects. *Toxicol. Appl. Pharmacol.*, 151: 294–310.
245. Sarangapani, R., Gentry, P. R., Covington, T. R., Teeguarden, J. G., and Clewell, H. J., III (2003): Evaluation of the potential impact of age- and gender-specific lung morphology and ventilation rate on the dosimetry of the vapor. *Toxicol. Sci.*, 15:987–1016.
246. Sarangapani, R., Teeguarden, J. G., Gentry, P. R., Clewell, H. J., III, Barton, H. A., and Bogdanffy, M. S. (2004): Interspecies dose extrapolation for inhaled dimethyl sulfate: a PBPK model-based analysis using nasal cavity N7-methylguanine adducts. *Inhal. Toxicol.*, 16:593–605.
247. Sato, A., and Nakajima, T. (1997): A vial equilibration method to evaluate the drug metabolizing enzyme activity for volatile hydrocarbons. *Toxicol. Appl. Pharmacol.*, 47:41–46.
248. Sato, A., and Nakajima, T. (1979): Partition coefficients of some aromatic hydrocarbons and ketones in water, blood and oil. *Br. J. Indust. Med.*, 36:231–234.
249. Schoeffner, D. J., Warren, D. A., Muralidhara, S., Bruckner, J. V., and Simmons, J. E. (1999): Organ weights and fat volume in rats as a function of strain and age. *J. Toxicol. Environ. Health*, 56(Part A):449–462.
250. Shin, B. S., Kim, C. H., Jun, Y. S., Kim, D. H., Lee, B. M., Yoon, C. H., Park, E. H., Lee, K. C., Han, S. Y., Park, K. L., Kim, H. S., and Yoo, S. D. (2004): Physiologically based pharmacokinetics of bisphenol A. *J. Toxicol. Environ. Health A*, 67:1971–1985.
251. Singh, D. V., Spitzer, H. L., and White, P. D. (1987): *Addendum to the Health Risk Assessment for Dichloromethane: Updated Carcinogenicity Assessment for Dichloromethane*, EPA 600/8-82/004F. U.S. Environmental Protection Agency, Washington, D.C.
252. Slikker, Jr., W., Andersen, M. E., Bogdanffy, M. S., Bus, J. S., Cohen, S. D., Conolly, R. B., David, R. M., Doerrer, N. G., Dorman, D. C., Gaylor, D. W., Hattis, D., Rogers, J. M., Woodrow, Setzer, R., Swenberg, J. A., and Wallace, K. (2004): Dose-dependent transitions in mechanisms of toxicity. *Toxicol. Appl. Pharmacol.*, 201:203–225.
253. Slob, W., Janssen, P. H. M., and Van den Hof, J. M. (1997): Structural identifiability of PBPK models: practical consequences for modeling strategies and study designs. *Crit Rev. Toxicol.*, 27(2):261–272.
254. Sultatos, L. G. (1990): A physiologically based pharmacokinetic model for parathion based on chemical specific parameters determined *in vitro*. *J. Am. Coll. Toxicol.* 9:611–617.
255. Sultatos, L. G., Kim, B., and Woods, L. (1990): Evaluation of estimations *in vitro* of tissue–blood distribution coefficients for organothiophosphate insecticides. *Toxicol. Appl Pharmacol.*, 103:52–55.

256. Sweeney, L. M., Gargas, M. L., Strother, D. E., and Kedderis, G. L. (2003): Physiologically based pharmacokinetic model parameter estimation and sensitivity and variability analyses for acrylonitrile disposition in humans. *Toxicol. Sci.*, 71:27–40.
257. Sweeney, L. M., Tyler, T. R., Kirman, C. R., Corley, R. A., Reitz, R. H., Paustenbach, D. J., Holsen, J. F., Whorton, M. D., Thompson, K. M., and Gargas, M. L. (2001): Proposed occupational exposure limits for select ethylene glycol ethers using PBPK models and Monte Carlo simulations. *Toxicol. Sci.*, 62:124–139.
258. Tardiff, R., Charest-Tardif, G., Brodeur, J., and Krishnan, K. (1997): Physiologically based pharmacokinetic modeling of a ternary mixture of alkyl benzenes in rats and humans. *Toxicol. Appl. Pharmacol.*, 144:120–134.
259. Tardiff, R., Lapare, S., Charest-Tardif, G., Brodeur, J., and Krishnan, K. (1995): Physiologically based pharmacokinetic modeling of a mixture of toluene and xylene in humans. *Risk Anal.*, 15:335–342.
260. Teeguarden, J. G., Deisinger, P. J., Poet, T. S., English, J. C., Faber, W. D., Barton, H. A., Corley, R. A., and Clewell, H. J., III (2005): Derivation of a human equivalent concentration for *n*-butanol using a physiologically based pharmacokinetic model for *n*-butyl acetate and metabolites *n*-butanol and *n*-butyric acid. *Toxic. Sci.*, 85:429–446.
261. Teeguarden, J. G., Waechter, J. M., Jr., Clewell, H. J., III, Covington, T. R., and Barton, H. A. (2005): Evaluation of oral and intravenous route pharmacokinetics, plasma protein binding, and uterine tissue dose metrics of bisphenol A: a physiologically based pharmacokinetic approach. *Toxicol. Sci.*, 85:823–838.
262. Teorell, T. (1937): Kinetics of distribution of substances administered to the body. I. The extravascular modes of administration. *Arch. Int. Pharmacodyn.*, 57:205–225.
263. Teorell, T. (1937): Kinetics of distribution of substances administered to the body. II. The intravascular modes of administration. *Arch Int. Pharmacodyn.*, 57:226–240.
264. Terasaki, T., Iga, T., Sugiyama, Y., Sawada, Y., and Hanano, M. (1984): Nuclear binding as a determinant of tissue distribution of adriomycin, daunomycin, adramycinol, daunorubicinol and actinomycin D. *J. Pharmacodyn.*, 7:269–277.
265. Thomas, R. S., Bigelow, P. L., Keefe, T. J., and Yang, R. S. H. (1996): Variability in biological exposure indices using physiologically based pharmacokinetic modeling and Monte Carlo simulation. *Am. Indust. Hyg. Assoc. J.*, 57:23–32.
266. Thompson, C., Sonawane, B., Nong, A., and Krishnan, K. (2007): Considerations for applying physiologically based pharmacokinetic models in risk assessment. In: *Toxicokinetics and Risk Assessment*,edited by J. C. Lipscomb and E. V. Ohanian, pp. 123–140. Informa Health Care, New York.
267. Thrall, K. D., Soelberg, J. J., Weitz, K. K., and Woodstock, A. D. (2002): Development of a physiologically based pharmacokinetic model for methyl ethyl ketone in F344 rats. *J. Toxicol. Environ. Health A*, 65:881–896.
268. Timchalk, C., Kousba, A., and Poet, T. S. (2002): Monte Carlo analysis of the human chlorpyrifos-oxonase (PONI) polymorphism using a physiologically based pharmacokinetic and pharmacodynamic (PBPK/PD) model. *Toxicol. Lett.*, 135:51–59.
269. Travis, C. C., and Hattemer-Frey, H. A. (1991): Physiological pharmacokinetic models. In: *Statistics in Toxicology*, edited by D. Krewski and C. Franklin. Gordon and Breach, New York, p. 170.
270. U.S. EPA (1987): *Update to the Health Risk Assessment Document and Addendum for Dichloromethane: Pharmacokinetics, Mechanism of Action and Epidemiology*, EPA 600/8-87/030A. U.S. Environmental Protection Agency, Washington, D.C.
271. U.S. EPA (1994): *Methods for Derivation of Inhalation Reference Concentrations and Application of Inhalation Dosimetry*, EPA 600/8-90/066F. Office of Research and Development, U.S. Environmental Protection Agency, Washington, D.C.
272. U.S. EPA (1999): *Toxicological Review of Ethylene Glycol Monobutyl Ether: In Support of Summary Information on the IRIS*, National Center for Environmental Assessment, U.S. Environmental Protection Agency, Washington, D.C. (http://www.epa.gov/iris).
273. U.S. EPA (2000): *Toxicological Review of Vinyl Chloride: In Support of Summary Information on the IRIS*, EPA/635/R-00/004. National Center for Environmental Assessment, U.S. Environmental Protection Agency, Washington, D.C. (http://www.epa.gov/iris).
274. U.S. EPA (2003): *Toxicology Review of Xylenes: In Support of Summary Information on the IRIS*, EPA/635/R-03/001. U.S. Environmental Protection Agency, Washington, D.C. (http://www.epa.gov/iris).
275. U.S. EPA (2006): *Approaches for the Application of Physiologically Based Pharmacokinetic (PBPK) Models and Supporting Data in Risk Assessment*, EPA 600/R-05/043F. U.S. Environmental Protection Agency, Washington, D.C.
276. Van Asaperen, J., Rijcken, W. R. P., and Lammers, J. H. C. M. (2003): Application of physiologically based toxicokinetic modeling to study the impact of the exposure scenario on the toxicokinetics and the behavioural effects of toluene in rats. *Toxicol. Lett.*, 138:51–68.
277. Van Eijkeren, J. C. H. (2003): Estimation of metabolic rate constants in PBPK-models from liver slice experiments: what are the experimental needs? *Risk Anal.*, 22(1):159–173.
278. Vinegar, A. (2001): PBPK modeling of canine inhalation exposures to halogenated hydrocarbons. *Toxicol. Sci.*, 60:20–27.
279. Vinegar, A., and Jepson, G. W. (1996): Cardiac sensitization thresholds of halon replacement chemicals predicted in humans by physiologically based pharmacokinetic modeling. *Risk Anal.*, 16:571–579.
280. Wada, D. R., Stanski, D. R., and Ebling, W. F. (1995): A PC-based graphical simulator for physiological pharmacokinetic models. *Comput. Meth. Programs Biomed.*, 46:245–255.
281. Wagner, J. G. (1975): *Fundamentals of Clinical Pharmacokinetics*. Drug Intelligence, Hamilton, IL.
282. Waller, C. L., Evans, M. V., and McKinney, J. D. (1996): Modeling the cytochrome P450-mediated metabolism of chlorinated volatile organic compounds. *Drug Metab. Dispos.*, 24:203–210.
283. Wilkinson, G. R., and Shand, D. G. (1975): A physiological approach to hepatic drug clearance. *Clin. Pharmacol. Ther.*, 18:377–390.
284. Wold, S. (1991): Validation of QSARs, *QSAR*, 10: 191–193.

285. Wünscher, G., Kersting, H., Heberer, H., Westmeier, I., Wenzel, V., Flechsig, M., and Matthaeus, E. (1991): Simulation system SONCHES-based toxicokinetic model and data bank as a tool in biological monitoring and risk assessment. *Sci. Total Environ.*, 101:101–109.

286. You, L., Gazi, E. (1999): Archibeque-Engle, S., Casanova, M. Connolly, R. B., and Heck, H. A., Transplacental and lactational transfer of *p,p′*-DDE in Sprague–Dawley rats. *Toxicol. Appl. Pharmacol.*, 157:134–144.

# 6 Toxicopanomics: Applications of Genomics, Proteomics, and Metabonomics in Predictive and Mechanistic Toxicology

*Don Robertson, Sven Beushausen, and William Pennie*

## CONTENTS

## INTRODUCTION: THE GENOMIC ERA

The successes of genome sequencing efforts have been widely discussed in the popular and scientific press, including predictions that genetic characterizations of individual risk of disease (and likely responses to therapeutic agents) will soon become routine medical procedures. We have clearly arrived in a new "genomic" era, where our understanding of the mammalian genome and its products is expanding at an increasing rate, and we are challenged to put mountains of data in a true biological context. This impact is being felt in our discipline of toxicology where, over the last decade, large-scale analysis of genes, proteins, and metabolites has become a more integrated, widespread, and mainstream approach in predictive and mechanistic toxicology.

In this chapter, we briefly outline the major "omics" applications: technologies that characterize changes in gene transcripts (toxicogenomics), proteins (proteomics), and metabolites (metabonomics). We also attempt to illustrate the current and future utility of these tools. It is not an exaggeration to state that this suite of technologies, coupled with the data analysis and bioinformatics tools that support them, is key to the continuing evolution of the toxicology discipline to a science based more on mechanistic understanding than on empirical observations. Toxicogenomics applications and technologies have become increasingly common research tools over the last decade or so (e.g., reagents and RNA isolation methods available in kit form). Toxicogenomics is increasingly becoming a technique used in molecular toxicology laboratories, in contrast to the niche technical disciplines that proteomics and metabonomics continue to be. For this reason, our chapter outlines the practicality of toxicogenomics but also provides additional emphasis on the emerging science and technology that underpin proteomics and metabonomics.

What can these capabilities help the toxicologist accomplish? Current applications are largely focused on three broad areas:

- Better *in vivo* and *in vitro* tools for predicting adverse outcomes as a consequence of chemical exposure
- Assistance in investigating and confirming specific mechanisms of toxicity
- Identifying safety-related biomarkers with possible utility in both animal models and clinical settings to monitor, for example, potential adverse side-effects of a new drug

The complementary approaches of toxicogenomics, -proteomics, and -metabonomics bring a high level of context in which to better understand the biology of toxicity responses (see Figure 6.1).

## TOXICOGENOMICS: CHARACTERIZING CHANGES IN GENE EXPRESSION AS A RESULT OF TOXICANT EXPOSURE

The human genome is estimated to consist of somewhere between 15,000 and 30,000 genes, an estimate that continues to be refined as our understanding of genes and their variants improves based on mining of data from the Human Genome Project. The expression pattern of the transcribed messenger RNAs (mRNAs) of these genes varies from cell to cell and determines both cell identity and function. In response to stimuli, cells may induce or repress certain genes for functional needs to adapt to a change in local environmental or repair damage resulting from a toxic insult. By comparing the expression levels of genes at baseline and in response to stimuli, one can identify genes that change in response to the stimuli. With the development of technologies such as microarrays, the expression levels of tens of thousands of genes, effectively at a scale of the entire mammalian genome, can be measured accurately and efficiently. These genes may serve as biomarkers for exposure and also aid in understanding the mechanism of action of the stimuli and the cellular pathways involved in response. From such data we can hope to derive mechanistic understanding, predictive tools, and diagnostic biomarkers.

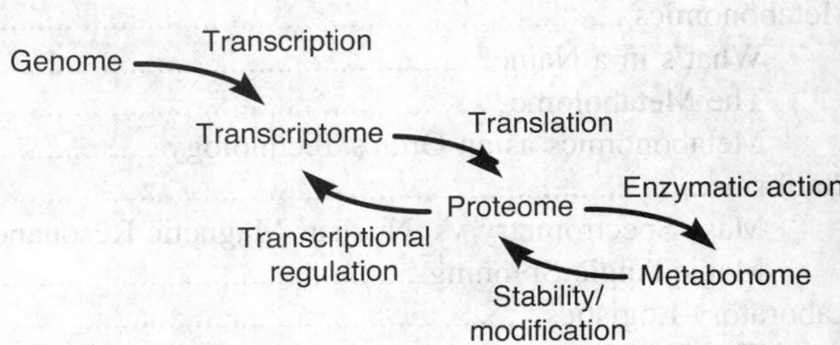


**FIGURE 6.1** Example processes by which information flow through different categories of molecules in a cell. Xenobioti exposure has the potential to cause alterations at different orga nizational levels of a cell or tissue. Genome, the chromosoma DNA information; transcriptome, the messenger RNA (mRNA from actively transcribed genes; proteome, the entire protei complement of a biological sample; metabonome, the constituen metabolites in a biological sample.

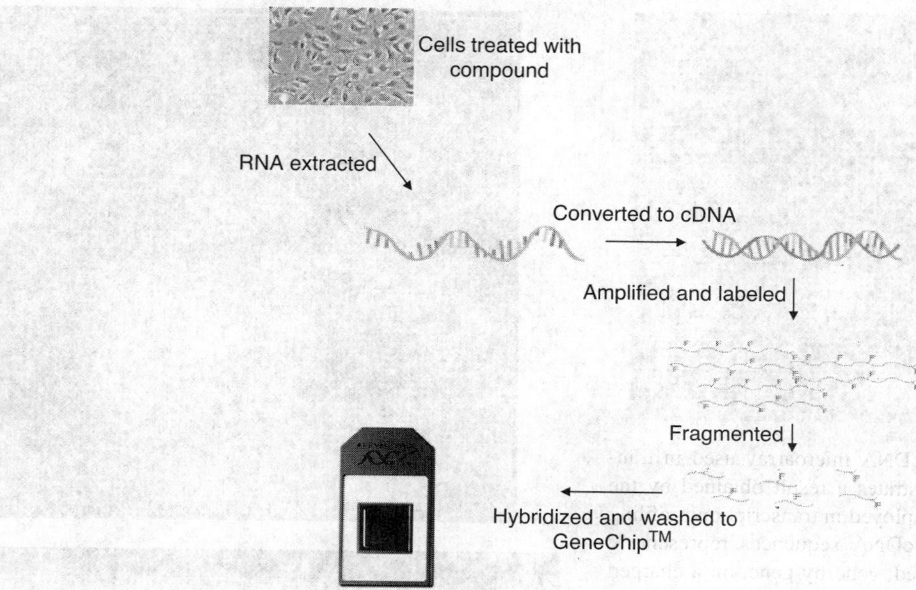


**FIGURE 6.2** GeneChip® hybridization workflow. The figure illustrates the procedure involved in generating gene expression data. Total RNA is isolated from treated cell lines or tissues and converted to double-stranded complementary DNA (cDNA). This then serves as a template for amplification via *in vitro* transcription, during which a biotinylated label is incorporated. The resulting complementary RNA (cRNA) is fragmented and hybridized to GeneChips overnight. When the hybridization is complete, the GeneChips are washed and scanned with a laser to generate a data file that is computationally processed to give expression values for all genes expressed in the originating sample.

## PRACTICAL CONSIDERATIONS IN TRANSCRIPT PROFILING

### Microarray Technology: Basic Methods and Platforms

#### What Is a Microarray?

DNA chips (or microarrays) are the basic tool for transcript profiling. Use of these chips allows quantitative comparison of the expression levels of potentially thousands of individual genes between different biological samples, thus facilitating comparisons, for example, of normal tissue compared with diseased and control with toxicant-treated cell lines. Physical construction of such arrays involves the immobilization of DNA sequences (representative of the coding sequence of genes of interest) on a solid medium such as a glass or quartz slide or a nylon membrane filter. The mRNA prepared from cells or tissues can be labeled (usually in the form of a reverse-transcribed cDNA copy) and then hybridized to the previously "spotted" chip. Software analysis then allows determination of the extent of hybridization of the labeled probes to the corresponding arrayed cDNA or oligonucleotide spots, and a comparison of control with treated samples allows quantitative measurement of treatment-associated changes in gene expression (see Figure 6.2) [1]. Microarrays designed to profile genes have been developed by commercial vendors, pharmaceutical companies, and academic institutions alike.

#### cDNA Microarrays

In general, two major differences exist between the complementary DNA (cDNA) microarray platform and the now more widely used oligonucleotide platforms. First, on a cDNA microarray, the length of the DNA fragment is generally 500 to 1500 bp, as opposed to the 25 to 60 nucleotides on an oligonucleotide array. Second, in cDNA microarray experiments, two RNA samples (control and experimental) are usually labeled with different fluorophores (Cy5 and Cy3) and competitively hybridized to the same microarray slide. This distinguishes the cDNA microarray platform from most oligonucleotide-based platforms, where one sample is hybridized to one chip. The cDNA microarray output data provide a ratio of the gene expression levels of a given RNA in the experimental vs. control sample, directly measured on a single slide (see Figure 6.3).

#### Oligonucleotide-Based Microarrays

Currently, perhaps the most widespread technical platform for gene expression work is the oligonucleotide-based gene array, or GeneChip®. Such microarrays, exemplified by those manufactured by Affymetrix, consist of small overlapping oligonucleotide fragments corresponding to the genes of interest which are assembled on a 5 × 5-inch quartz wafer through a photolithographic chemical process. In this process the chip is coated with a light-sensitive chemical that blocks the coupling of nucleotides to

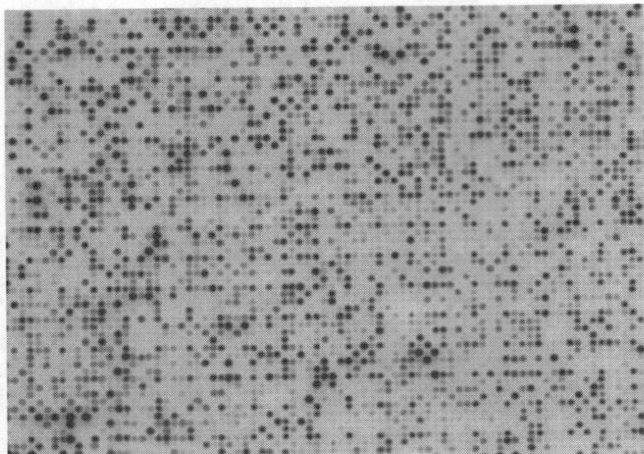

**FIGURE 6.3** Illustration of a cDNA microarray used in transcript profiling. The figure illustrates a result obtained by the DNA microarray technology employed in transcription profiling (transcriptome). In this case, cDNA sequences representing genes of interest are immobilized, gene by gene, on a charged nylon membrane. Hybridization with labeled cDNA preparations from treated and untreated tissue will show different intensities of the spots when gene activities have changed due to compound treatment. The relative intensities on the same array show the difference of messenger RNA levels within a given sample. RNAs of high copy number (e.g., for structural proteins such as albumin) yield highly intensive spots.

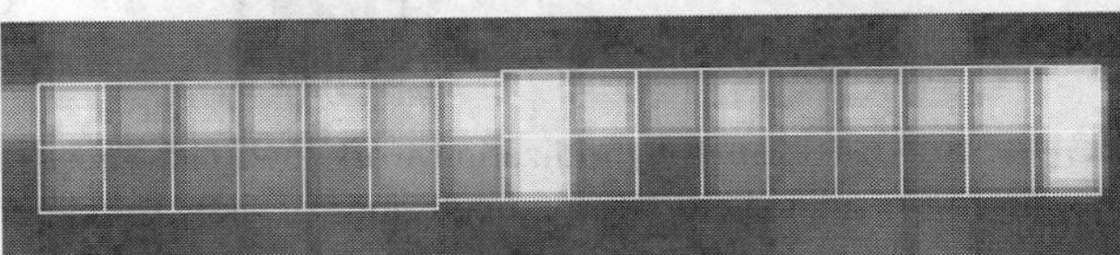

**FIGURE 6.4** Scanned Affymetrix GeneChip®. An image of a hybridized GeneChip scanned with a laser where each bright square represents the hybridization of fragmented cRNA to a specific oligomer. The intensity is proportional to the quantity of the RNA species in the sample. The magnified inset shows perfect-match oligomers on the top row and mismatch oligomers on the bottom row. The mismatch oligomers are used to control for nonspecific hybridizations.

the surface. The chip is then masked with a solid template to expose particular regions of the chip to light and to protect other regions from it. The light-sensitive chemical in the exposed areas is then deactivated by illumination and the chip is exposed sequentially to solutions containing adenine, thymine, cytosine, or guanine. These nucleotides themselves bear light-sensitive blocking groups so specific nucleotide chains can be built up across the chip surface following the sequence of masking, illumination (to deprotect), then nucleotide addition. In this way, the microarray is built as the probes are synthesized through repeated cycles of deprotection and coupling, resulting in a chip with literally millions of immobilized probes for specific gene sequences (see Figure 6.4).

## Experimental Design Considerations and Limitations

- Determine the study design, time, and dose; compound groupings; use of inhibitor molecules; etc.
- Determine whether the study will be *in vivo* or *in vitro*.
- Determine sample sources, surrogate analysis, use of microdissection.
- Determine analytical methods to be used.
- Recognize that compound-induced transcript changes may not accurately reflect the response of the corresponding organ *in vivo*.
- Realize that the availability of appropriate cell lines may be limited; however, when mechanistic information is not sought, however, generic cell lines may still be useful.
- Determine whether metabolism is required to produce the active species; often this requirement can be overcome by preincubation with metabolically active cell extracts.
- Importantly, where the toxicity is specific to species, strain, sex, or route of administration, be aware that *in vitro* modeling is unlikely to be diagnostic.

A major practical consideration of microarray applications is the use of appropriate statistical analysis and data evaluation tools. Principal component analysis (PCA) is a visualization method useful in the reduction of complex multivariate data to a three-dimensional plot. PCA helps the investigator visualize how related different datasets are to each other (see Figure 6.5). In terms of analyzing individual experiments, other visualization tools such as the volcano plot and ubiquitous heatmap/hierarchical cluster

**Sample Applications of Gene Expression Analysis in Toxicology Research**

*Transcriptional changes associated with exposure to toxicants/biomarkers of exposure and effects*

- Mouse hepatoma cells to chromium [3]
- Liver of transgenic mice with constitutively active dioxin-aryl hydrocarbon receptor [4]
- MCF-7 cells to estrogen [5]
- Human keratinocytes to inorganic arsenic [6]

*Dose–response assessment*

- Dose-dependent expression changes in kidney HEK293 cells to arsenite [7]

*Extent of individual variation in response to exposure*

- Susceptibility genes for resistance to DDT in *Drosophila* [8,9]

*Targets of toxicants and mode of action*

- Proposes that chromium inhibits transcription by blocking the release of histone deacetylase and preventing the binding of p300 to chromatin [3]

ng methods are useful in understanding the spread of genomics data and helping to visualize the contributions of individual genes or gene groups when clustering datasets together (see Figure 6.6 and Figure 6.7).

### Example: Applications of Toxicogenomics in Investigative and Mechanistic Toxicology

Several groups, including the International Life Sciences nstitute (ILSI) and the Health and Environmental Sciences Institute (HESI), have initiated genomic-scale risk ssessment projects [2]. Microarray-based experiments have been performed to address questions in toxicology (see box).

The ability of transcript profiling to distinguish between distinct classes of compounds has now been demonstrated by many laboratories [10], and the development and application of predictive toxicology models based on gene transcript changes have been effectively commercialized by a number of biotechnology companies. Such pattern recognition facilitates the discovery and subsequent validation of biomarkers useful for application in higher throughput approaches to help reduce risk for novel chemical series in the discovery process. One illustration of this pproach, by Burczynski and colleagues [11], involved the analysis of a prototypic single representative from two compound classes to look for consistent diagnostic expression changes while avoiding background noise. Computer-based prediction tools were employed to expand the list of consistent gene expression events to those genes that ould distinguish accurately between the chemical classes (DNA-damaging agents and the antiinflammatory drugs) n a 100-compound learning set. More recently, Thomas and colleagues [12] took a broadly comparable approach to distinguishing five hepatotoxicant classes, based on a learning set assembled from 24 reference compounds, profiled using a custom 1200-gene microarray. Perhaps surprisingly, these results suggest that the gene expression fingerprint that allows such classification can consist of merely dozens of genes, again raising the possibility that the approach could be modified to a high-throughput format. Correlating gene expression changes with clinical chemistry and pathology findings should put gene expression data in context with more established endpoints, as demonstrated by Waring and colleagues [13].

A more fundamental application of these technologies is the investigation of the regulation events that underpin the development of an adverse biological response, rather than those that allow prediction of the outcome. This approach should allow a more mechanism-based assessment of risk, particularly when applied to characterize a finding found in a regulatory study performed in a preclinical test species (e.g., a rodent or canine study). With appropriate experimental design, this approach can generate candidate gene lists that can be used to formulate hypotheses as to the mechanism by which a compound gives rise to a toxicity finding. Proving the causative involvement of any of these candidates requires detailed follow-up work, most probably employing more traditional functional genomics and biochemical approaches. Transcript profiling is also being employed to understand the relationship between *in vivo* and *in vitro* models. For example, the dedifferentiation of hepatocytes following explant has been characterized over time at the transcriptional level [14], demonstrating that the isolation of hepatocytes can have marked effects on pathways known to be involved in toxicant responses.

Concern has been voiced already that a potential problem is the misinterpretation, or over-interpretation, of genomic analyses, particularly in the context of determining product safety. It must be recognized that the interaction of chemicals with biological systems will in many instances result in some changes in gene expression, even under circumstances where such interactions are benign with respect to adverse effects. The challenge again is to ensure that sound judgment and the appropriate toxicological skills and experience are brought to bear on the data generated so toxicologically relevant changes in gene expression are distinguished from those that are of no concern.

## PROTEOMICS

In its simplest form, proteomics can be defined as a systematic approach to the separation, identification, and cataloging of proteins within a biological sample. This task, in and of itself, can be a monumental undertaking given that the individual protein complement of a cell lies some-

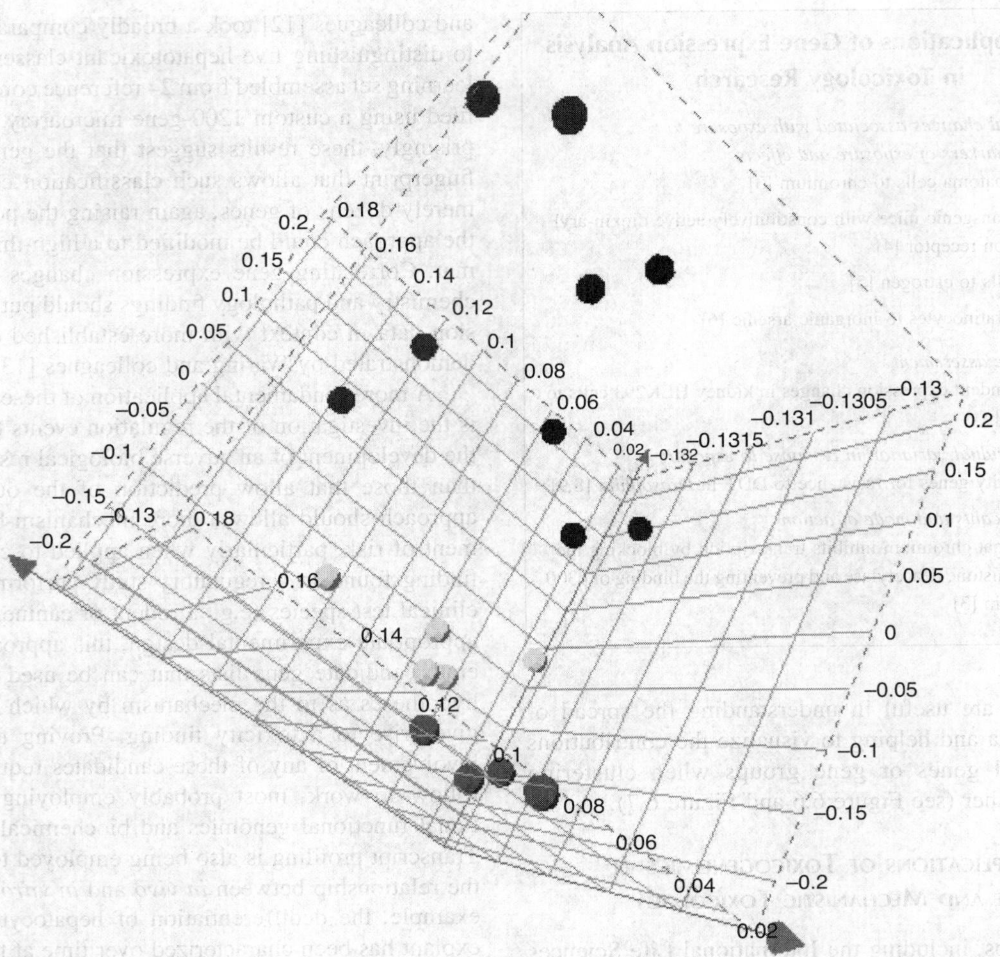

**FIGURE 6.5** Principal component analysis (PCA) is a statistical approach used to find patterns in high dimensional data, such as expression values for a large number of genes. The results of the computation are more often used to reduce data to three dimensions allowing for the visualization of similarities and differences between individual samples. The samples are shaded according to dose groups; dose groups clustered together indicates less individual sample variability, whereas more dispersion in the dose groups indicates greater sample variability.

where between 5000 and 10,000 proteins and that the difference in level of individual protein expression can span several orders of magnitude. Proteome analysis is complicated even further when dealing with a particular tissue or biological fluid where several different cell types exhibiting disparate proteome signatures contribute to the complexity of the proteome of the tissue or biofluid under investigation. In addition, proteome perturbation in response to a changing environment is often instantly manifested by changes in levels of individual protein expression, protein–protein interaction, and, most importantly, posttranslational modification, the latter having profound effects on protein function, intracellular compartmentalization, and residence time within a given cell or tissue. Toxicoproteomics aims to look at such changes in an effort to develop tools that will help predict or develop an understanding of the mechanism of toxicity o unanticipated adverse events during the course of drug development (see references 15 through 35).

Drug safety evaluation is one of the most critical components of drug development. Compounds are monitored for safety at every stage when a lead compound or serie for a particular indication has been identified. Although great care has been taken to document chemical liabilities that have contributed to past safety issues and screens are in place to test for toxicity preclinically, many compound induced adverse events often present first in human because the targets they are designed to hit are metabolically networked somewhat differently between species Consequently, the outcomes of these metabolic differences may be manifested as an adverse event in one species while appearing quite normal in another (e.g., theo

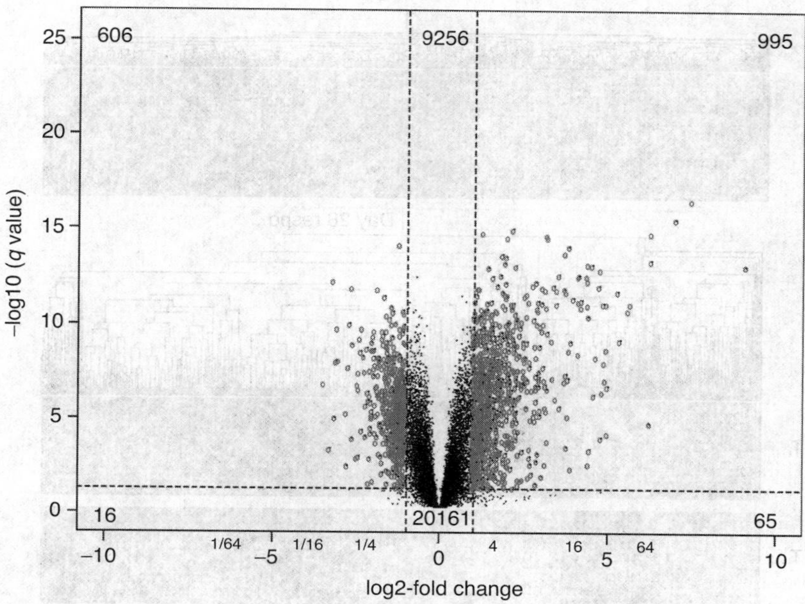


**FIGURE 6.6** Volcano plot. The results of a significance calculation such as an ANOVA analysis are plotted with the fold-change (log2) along the *x*-axis and statistical significance (*q*, log10) along the *y*-axis, giving rise to a volcano pattern. This type of plot allows a rapid assessment of genes that are differentially regulated with a high statistical confidence (denoted by circles). The horizontal dashed line marks the statistical significance threshold, and the vertical dashed lines mark the ± twofold change thresholds.

bromine in chocolate is toxic to dogs but not to humans). This becomes particularly confounding when a compound is placed on clinical hold because of an observed preclinical liability for which no apparent precedence exists in humans and a marker to monitor or predict the event is unavailable. These are often the most difficult compounds to move forward because the risk of causing an adverse event in humans often outweighs the benefit of further development.

Traditionally, drug testing begins preclinically and often well before testing in animals begins. Compounds are tested *in vitro* against panels of known proteins (e.g., receptors, kinases, key metabolic proteins) to determine if they are specific for the intended target or if crossover onto other proteins might result in triggering an undesirable adverse event. Following *in vitro* screening, compounds are dosed preclinically into rats and dogs. Blood and urine chemistry is routinely monitored to report changes in biofluid metabolites that flag potential compound-induced, organ-specific toxicities. Many of these metabolic markers can be considered as the original biomarkers. Examples include elevated blood urea nitrogen and creatinine to indicate renal injury. Most current biomarkers have been useful in diagnosing or reporting organ-specific injury. The need to develop more sophisticated biomarkers to predict compound-induced, organ-specific injury is driven by the increasing cost of drug development due to compound attrition and increasing pressure to develop safer and more selective compounds. Predictive biomarkers would be of tremendous value in the clinic insofar as alerting physicians to the potential of organ toxicity before irreversible damage occurs. The value to the pharmaceutical industry would be to limit attrition by culling toxic compounds from the drug development pipeline before they reach the clinic and to deliver better diagnostics to monitor organ-specific injury as a consequence of disease or drug therapy. The end result would be the development of better, less expensive, safer medicines with the potential to save many more lives.

Recently, many pharmaceutical companies have invested heavily in proteomics platform technologies with the goal of developing better tools to support predictive and mechanistic toxicology. While both are ambitious objectives, developing tools for predictive toxicology is much more complex and time consuming than developing tools for understanding mechanistic toxicology. The reason for this is that the former requires an unbiased whole proteome analysis approach that is analogous to looking for a needle in a haystack, whereas the latter benefits from a targeted approach that uses known proteins or compounds as bait to identify off-target proteins or metabolites with which each specifically interacts. The following discussion attempts to describe platform technologies currently used in toxicoproteomics laboratories and their advantages, limitations, and compatibility with other platform technologies.

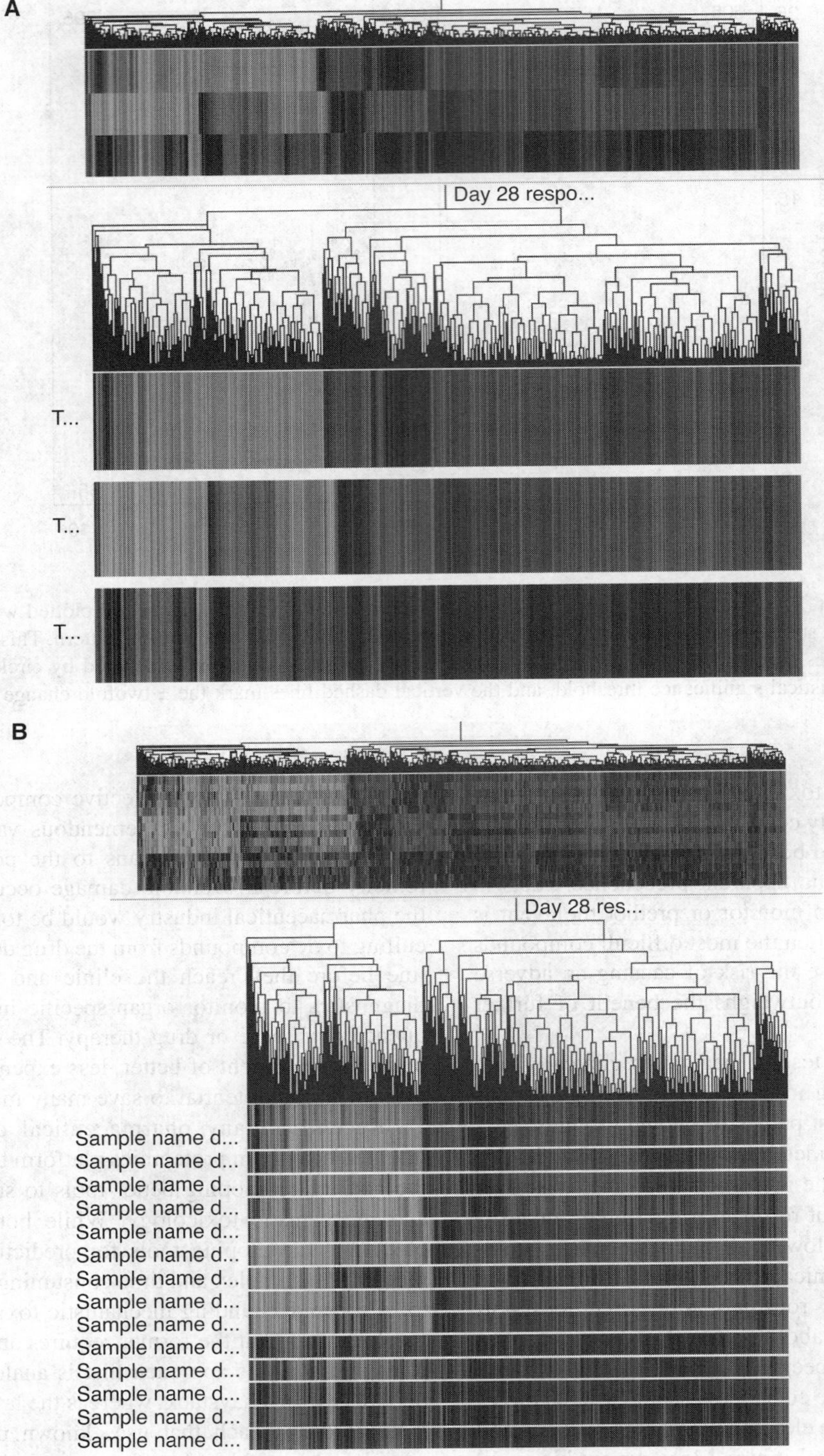


**FIGURE 6.7** Hierarchical clustering. With hierarchical clustering, genes are grouped according to expression values using the Pearson correlation algorithm; thus, genes having a similar expression pattern are clustered together. In (A), the mean fold-change value for each dose group is used, and in (B) the fold-change values for each individual samples are used. Each gene is typically represented by a colored rectangle (reproduced here in grayscale), usually red for upregulated genes, green for downregulated, and black for no change. Each column represents a sample or a cohort. This type of analysis allows the identification of groups of genes that are regulated in a similar manner across treatments.

## WHOLE PROTEOME ANALYSIS

### Two-Dimensional Polyacrylamide Gel Electrophoresis

Two-dimensional polyacrylamide gel electrophoresis (2DGE) is widely credited as the earliest proteomics technology platform developed for whole proteome analysis. 2DGE has the capacity to separate and visualize up to 3000 proteins at a time. Proteins are separated in the first dimension by isoelectric focusing based on the pH at which proteins exhibit a neutral charge and then further separated in a second dimension based on size. This technology provided investigators with a very powerful and convenient tool to ask how proteomes are affected as a consequence of external perturbation. Examples of questions addressed with 2DGE include protein changes due to bacterial or viral infection, physical or electrical shock, changes in nutrition, or, more relevant to this discussion, proteins affected in response to drug treatment. Many recent innovations have been developed to help increase sample throughput using 2DGE; for example, the development of immobilized pH gradient (IPG) strips made it much simpler to separate proteins in the first dimension. IPG strips are available in a number of different lengths to accommodate different second-dimension gel sizes and, even more critically, with variable pH gradients so the investigator has the option of looking at very broad pH ranges (3 to 11) or the ability to zoom in using narrow pH range strips to focus on areas of interest (4 to 7, 6 to 10). The development of image acquisition software to capture gel images has also been extremely useful in streamlining the 2DGE process. Images are digitized, and the subsequent output files are easily amenable for analysis using state-of-the-art image analysis software. Image analysis software allows the investigator to simultaneously compare samples between control and treated animals to determine which protein spots were affected and by how much.

Quantitative analysis of the gels is based on spot area and relative intensity. The final calculation is reported as a fold change to normalized controls. The three most common issues vendors faced when creating 2DGE analysis software included how to deal with proteins that are present in some gel sets but absent in others, how to overcome differences in spot registration due to slight variations in run time in each dimension, and how to deal with spot warping due to artifacts caused by imperfect gel polymerization. One of the most ingenious ways of dealing with the first two issues was by differentially staining two different sample sets with fluorescent dyes (Cy3 and Cy5), mixing the samples, and separating them simultaneously on the same gel. Differential staining of sample sets would be a very popular technique in the development of proteomics tools; however, the most challenging aspect has been developing algorithms that can deal with the warping phenomenon. Investigators have found a way to deal with gel warping by running several gels per sample and analyzing gels by PCA to determine which should be excluded from further analysis. Consequently, gel analysis technology has reduced analysis times from months to a matter of days depending on the size of the project.

Although 2DGE provided a tremendous technological advantage due to its capacity to resolve so many proteins at once, it became clear that it could not be used to look at all of the proteins present in any one tissue or biofluid simultaneously because the dynamic range of protein expression can span up to 12 orders of magnitude. Furthermore, 2DGE is limited to resolving proteins within a restricted pH and size range, eliminating the possibility of looking at very acidic or very basic proteins or very large or very small proteins. In addition to these shortcomings, 2DGE is best suited to looking at soluble proteins and is not well suited for integral membrane proteins, as separation in the first dimension relies on proteins in the native state. The technology is also limited by its sensitivity of detection. Initially, protein visualization following 2DGE was accomplished with Coomassie™ Blue, which could detect proteins in the range of tens to hundreds of nanograms. Greater sensitivity came with the introduction of silver staining, but this technology was incompatible with the subsequent protein sequencing technologies of Edman degradation or mass spectrometry. A later innovation in staining technology delivered the SYPRO® series of protein dyes, which demonstrated detection sensitivities equal to the nanogram sensitivity of silver stains. SYPRO Ruby, in particular, is the stain of choice, and this technology is compatible with downstream protein identification technologies. Another recent development in protein gel staining technology is Deep Purple™, which offers up to eight times greater sensitivity than SYPRO Ruby and has also demonstrated excellent compatibility with mass spectrometry. Although staining technology has improved the sensitivity of protein detection almost 1000-fold in the past three decades, it is still not enough to even begin to see even moderately abundant proteins in an environment that spans 12 orders of magnitude in protein levels.

### Protein Fractionation

Because of the shear number of proteins present in any given proteome coupled with the dynamic range of protein abundance therein, it soon became clear that investigators would need to develop technologies that would allow them to drill deeper into the proteome. Consequently, technologies were developed to enrich protein families with similar physiochemical properties, and others were designed to eliminate abundant proteins so less abundant proteins could be displayed and investigated for potential biomarker utility. To this end, many efforts were focused on applying traditional protein fractionation or biochemical chromatographic technologies and adapting innovations in this area to existing proteomics platforms. For example, in toxicoproteomics research, investigators often rely on biofluids such as plasma or urine to discover biomarkers that predict organ-specific damage. Each biofluid has its own complement of abundant proteins that in most instances are irrelevant as far as biomarker discovery is concerned. Plasma contains serum albumin, immunoglobulins, and transferrin, which together comprise almost 60% of the mass of total protein within the sample. Removal of these proteins by immunoaffinity chromatography would allow one to enrich the remaining protein complement by almost two orders of magnitude. There are commercially available immunoaffinity columns that will remove up to six of the most abundant proteins from rat serum, including albumin, transferrin, haptoglobin, $\alpha_1$-antitrypsin, immunoglobulin G (IgG), and IgA, as well as up to 12 from human serum, including the latter six in addition to IgM, $\alpha_2$-macroglobulin, $\alpha_1$-acid glycoprotein, high-density lipoprotein (HDL; Apo A1 and Apo AII), the latter comprising up to 96% of the total protein mass in human serum. Although immunoaffinity removal of proteins may be very useful for sample preparation, the technology is not without its limitations; for example, removing such a large amount of protein mass from a sample can make the sample protein limiting for further analysis. This is particularly true of 2DGE where submilligram quantities of protein are required per gel, multiple gels are run per sample, and quantities of starting material are limited to less than a milliliter or 30 to 60 mg/mL of protein. Also, commercially available columns are prepared with polyclonal antibodies that can vary from lot to lot; consequently, the 2DGE pattern or mass spectrometric profile generated from a sample prepared by immunoaffinity depletion using a column from lot A of a specific vendor is very likely to be remarkably different from the same sample passed through a column from lot B manufactured by that same vendor. Care must be taken to ensure that all samples compared within a given sample set are processed using the same column the first and the *n*th time if one wishes to compare the data. A third limitation rather specific to the toxicoproteomics laboratory is the quality and range of immunoaffinity depletion columns developed for preclinical species; for example, only one vendor currently manufactures a column for removal of rodent proteins, and that column incorporates antibodies made to mouse-specific antigens that cross react with the rat. Although not optimal for the removal of rat proteins, this is the best column available for the purpose. No such columns are currently available for the removal of canine proteins, the second most popular nonrodent animal model for preclinical testing.

A particularly novel innovation resulting from the removal of abundant or high-molecular-weight proteins from plasma or serum was the cataloguing of the low-molecular-weight (LMW) proteome and to determine if this subproteome could be used to help in the diagnosis of a diseased state. The LMW proteome is obtained through the removal of high-molecular-weight proteins by centrifugation of serum through LMW cut-off membranes in the presence of a buffer optimized for the minimization of protein–protein interaction, usually by inclusion of up to 20% acetonitrile into the buffer. Efficient removal of high-molecular-weight proteins from the resulting filtrate is subsequently analyzed by 2DGE and can be further processed for analysis by mass spectrometry. This process has been used to report greater than 340 different proteins in human serum; although not yet published, it may have potential for biomarker identification in the field of toxicoproteomics.

Two-dimensional liquid chromatography (2D-LC) is another technology adapted to protein enrichment for proteomics applications. This technology maximizes the potential for protein separation by sequentially exploiting differences in the physiochemical properties of proteins. The process works by identifying and collecting a prospective peak from an initial column and applying that peak onto a second column for further separation and enrichment. Care must be taken to retain compatibility between the mobile phases of the first and second columns. Typical separation scenarios include size-exclusion chromatography onto a reversed-phase column, reversed-phase chromatography onto an ion-exchange column, and size-exclusion chromatography followed by ion-exchange chromatography. Once again, this technology is a front-end preparative approach for protein enrichment in samples that will be analyzed using platform technology with far greater sensitivity, such as mass spectrometry.

Clearly, protein fractionation is a mainstay for any successful toxicoproteomics laboratory. The technologies developed in this discipline support all of the principal proteomics platforms, including 2DGE, SELDI, and mass-spectrometry-based platforms. The latter two are discussed below. Protein fractionation technologies provide the investigator with an opportunity to design a systematic and comprehensive approach to the mapping of any proteome. The only other elements required are resources time, and patience.

## Surface-Enhanced Laser Desorption Ionization

Surface-enhanced laser desorption ionization (SELDI was first introduced in the early 1990s and positioned as an instrument that would shorten sample analysis times by combining the selectivity of traditional ligand-based chromatography with the sensitivity of high-resolution mass spectrometry. The advantages of the system included that analysis required very little sample volume (microliters as opposed to milliliters), protein fractionation wa performed on a chip within minutes rather than hours or days, and several samples could be analyzed sequentially on a single chip. In addition, the technology was excellent at resolving peptides and proteins at the low end of the molecular weight spectrum (<10,000 Daltons) thereby complementing the 2DGE approach. SELDI chips were engineered such that small depressions or wells within each chip were coated with a specific ligand (ion-exchange, hydrophobic, metal chelator). Subsequent single-step fractionation of complex samples could be performed using different binding or elution buffers, and the remaining bound protein complement could be analyzed by mass spectrometry using the SELDI instrument. Data output consisted of a typical mass spectrum offering relative peak heights and mass over charge ratio (*m/z*) values for each observed peak.

The technology first demonstrated success in the area of disease biomarker discovery through serum profiling of patients with ovarian cancer. Evidence was provided that the instrument could distinguish serum profiles between unaffected and benign patients from those diagnosed in different stages of ovarian cancer with an extremely high degree of accuracy. SELDI was embraced by many toxicoproteomics labs because it offered entry into the field of proteomics with a relatively inexpensive and highly sophisticated technology. Serum and urine profiling was relatively straightforward using the SELDI instrument. Profiles generated from samples of control and compound-treated animals were first analyzed by PCA, and further computation would reveal peaks of interest that were either increased or diminished as a consequence of treatment. The *m/z* values of singly charged peaks would be entered into databases to aid in the identification of the isolated peptides. It is important to emphasize that samples were often completely digested with trypsin prior to profiling to help reduce the time spent searching the database for a match and to maximize the possibility for peak identification; however, profiling *per se* could not be routinely applied to preclinical samples to predict compound-induced, organ-specific toxicity as it had in human patients to diagnose disease. In the human, it was sufficient to compare profiles alone without having to identify constituent proteins because comparisons were restricted to the same species.

In the preclinical toxicoproteomics environment, particularly as it applies to biomarker discovery, one is looking for concordance between species that can be translated to the clinic; therefore, at the very least one is looking for biomarker concordance across three species, including humans. Given that serum profiles are completely disparate between species and that many of the peptides comprising those profiles would also be different, the technique could only provide value if all of the novel peaks in the treatment groups were identified for each species and compared and a subset or fingerprint of shared peaks corresponding to identical proteins were assembled to predict the injury. Although this might appear as a fairly systematic and comprehensive approach toward solving this issue, SELDI technology suffered from two principle liabilities that prevented practical implementation of the concept. The first was chip-to-chip variability and the second was poor mass resolution. The former made it difficult to reproduce profiles from one chip to the next unless chips were purchased in bulk and experiments were limited to chips within the same lot. Poor mass resolution often precluded proper protein identification. A SELDI protein chip interface was offered as a front-end bolt-on to higher end time-of-flight mass spectrometers to overcome the problem of poor mass resolution; however, the high cost of the interface coupled with the additional time investment in sample fractionation and peak enrichment required for peak identification made it impractical to pursue the technology. A third liability unrelated to the technology but relevant to the industry in general was the lack of robust canine databases for peptide/protein sequence information, making it difficult to adapt this approach to the second species.

## Mass Spectrometry

Mass spectrometry is widely considered to be the cornerstone platform technology central to toxicoproteomics research. Mass spectroscopy provides several key advantages over all other existing technologies that make it an attractive platform from a perspective of project implementation and platform development. Chief among them is the incredible sensitivity that can be achieved with the technology and the speed with which samples can be processed and raw data generated. Proteins or peptides can be detected and identified in subfemtogram quantities from complex biological samples including tissues and biofluids. Because of the dynamic range of proteins in a given sample, however, this technology also suffers from the limitation that more abundant proteins swamp out those expressed at much lower levels. Furthermore, mass spectrometry only provides relative ion or peak height information in a spectrum and is therefore a poor first choice as a quantitative proteomics platform. As a result, much of the development work around mass spectrometry has focused on protein separation technologies prior to mass spectrometric analysis and differential labeling technologies so it is possible to compare and quantify like peptides from disparate samples. This ideally positions the platform for toxicoproteomics analyses where simultaneous comparisons between control and compound-treated samples are key to increasing our understanding of how compounds affect the proteome and improving the identification of potential biomarkers. What follows is a brief discussion that attempts to outline some of the advantages and limitations of some of the major mass-spectrometric-based proteomics technologies that have recently emerged and been adapted for use in toxicoproteomics laboratories today. These include MudPIT, ICAT, GIST, ITRAQ, SISCAPA, and AQUA.

Multidimensional protein identification technology (MudPIT) was initially developed to utilize a liquid chromatography–tandem mass spectrometry (LC–MS/MS) approach for greater in-depth coverage of the yeast proteome. Protein samples are initially separated by 2D-LC (strong cation exchange chromatography followed by reversed-phase chromatography) prior to tandem mass spectrometric analysis, resulting in the production of thousands of mass spectrometry spectra. This platform has been used to identify many low abundance proteins that cannot be detected using the more conventional but less sensitive 2DGE–MS technology. Although the platform offers great potential in the area of discovery, it has not been developed to its full potential for binary sample analysis in the toxicoproteomics laboratory. Coupling MudPIT with a robust differential labeling platform would maximize the potential for toxicoproteomics applications, particularly toward safety biomarker discovery.

Isotope-coded affinity tag (ICAT™) technology was specifically developed for the quantitative comparison of two samples simultaneously. Identical linkers are used to label protein-containing sample sets on cysteine residues, with the exception that one set of linkers has eight deuterium atoms incorporated into the structure whereas the other retains hydrogen in identical positions. The samples are then mixed, digested with trypsin, and analyzed by LC–MS/MS. Quantitative differences in labeled ions can be calculated by comparing differences in peak heights of like ions that are separated by 8 atomic mass units (amu) afforded them by the incorporation of the eight deuterium ions in the linker. Algorithms developed to support the platform take the raw data and perform quantitative analysis on selected peaks. One of the shortcomings of the technology is that the software cannot discriminate between a peptide containing two labeled cysteines from a similar peptide containing an oxidized cysteine because the net increase in atomic mass units (+16) is indistinguishable from the former. Consequently, a more robust platform was developed where linkers were substituted at nine positions with heavy $^{13}C$, resulting in a net increase of 9 amu over standard linkers. An additional advantage to this improvement was that like peptides coelute from reversed-phase columns, making peptide quantification and identification much simpler. Although the ICAT technology seemed ideally suited to the analysis of binary samples, the high degree of technical difficulty associated with its implementation has precluded its adoption by many toxicoproteomics laboratories.

Global internal standard technology (GIST) is another differential peptide labeling technology that was developed at the turn of the century to facilitate proteomics research. GIST technology uses three deuterium atoms per linker to discriminate like peptides labeled with cold linkers. GIST differs from ICAT in that the sample sets to be analyzed are globally digested with trypsin prior to labeling with the heavy or light linkers and are subsequently mixed and submitted for LC–MS/MS analysis. The greatest distinction between the two technologies is that GIST employs a post-tryptic digest peptide labeling strategy that covalently links tags to primary amino groups whereas ICAT is a pretryptic peptide labeling technology that covalently links tags to the protein through alkylation of free cysteines. An advantage of GIST over ICAT is that not all peptides contain cysteine residues, thereby limiting peptide labeling. Also, post-tryptic peptide labeling offers greater peptide coverage within a sample by generating a new N-terminal primary amino group for each peptide that can be targeted for labeling. Labeling done prior to tryptic digestion precludes the labeling of many peptides that do not contain lysine residues as the only other targets with primary amines. GIST, like ICAT, was not readily adopted by the toxicoproteomics community because of the steep learning curve and the continued development of more robust and sophisticated technologies.

Isotope tags for relative and absolute quantization (iTRAQ) is a second-generation ICAT platform that provides for the labeling and quantitative analysis of four sample sets simultaneously. The technology utilizes four isobaric (same mass) reagents to label different sample sets that are cleaned up using strong cation ion-exchange chromatography prior to MS/MS analysis. iTRAQ is also a post-tryptic digest, primary amino, peptide labeling technology designed to maximize peptide coverage within any given sample. The tags are cleverly designed by differential incorporation of $^{13}C$, $^{15}N$, and $^{18}O$ into two-part linkers comprised of a reporter and balance piece that add up to yield a net mass of 145 amu. The reporter pieces of each of the four tags have mass values of 114, 115, 116, and 117 amu, respectively. The corresponding balance pieces have mass values of 31, 30, 29, and 28 amu, respectively. Following liquid chromatography of a mixture of four labeled samples, peaks of interest are sent into the collision cell of the mass spectrometer cleaving the labeled peptides between the balance and reporter pieces. The four different reporter tag ions can then be detected by the corresponding atomic mass units (114, 115, 116, and 117) and quantified using software specifically written to support the platform. Relative quantification of peptides can be calculated by comparing relative peak heights in different biological samples, and absolute quantification can be achieved by doping samples with a known amount of control peptide. The latter requires knowledge of the specific peptide of interest. Although a relative newcomer to the field, iTRAQ technology provides deep coverage into the proteome and promises to be of tremendous value in toxicoproteomics laboratories.

### Targeted Proteomics

In an effort to take some of the guesswork out of biomarker discovery and understanding mechanisms of toxicity, investigators have been developing new and innovative ways to focus their research efforts on particular targets. Targeted proteomics approaches fall into two categories. The first is identification of off-target proteins through direct interaction between specific compounds with proteins in an affected tissue or biological matrix. The second approach involves identification of a peptide biomarker or group of peptides that might provide a signature that predicts compound-induced, organ-specific injury or disease. Interestingly, most of these new, targeted technologies find their roots in well-established biochemical methods. The improvements come in the form of more sophisticated detection reagents offering greatly increased sensitivity.

### Affinity Chromatography

One of the simplest and most effective ways to determine what a compound interacts with in a complex biological matrix (other than the target for which it was intended) is to immobilize it on a solid support and capture targets from a solution of the target tissue or biofluid as it passes over the column. This is a widely practiced and very well characterized technology. For example, adenosine triphosphate (ATP) affinity matrices are commonly used to identify many different ATP-binding proteins from a variety of biological source materials. This concept has transitioned into the toxicoproteomics laboratory, where investigators are interested in identifying compound-specific off-target binding partners to determine the selectivity of a target or glean a greater understanding of the mechanism of toxicity should an unexplained toxicity accompany compound development. Success in using this platform requires a critical understanding of the compound to be immobilized and how it interacts with the target. Having a clear understanding of how the business end of the compound interacts with the binding pocket of the target is essential to developing an immobilization strategy that will capture off-target proteins with similar compound binding pockets. Linking to a single site, however, precludes the investigator from identifying additional off-target proteins that do not bind to targets in the predicted manner but might, in fact, bind to that portion of the molecule to which a linker has been attached.

### Reverse Pharmacology

Reverse pharmacology was developed to overcome the shortcoming of limiting off-target protein binding by immobilizing the compound of interest to a solid support and effectively eliminating one portion of the compound for interaction with other potential off-targets. The technology is an adaptation of photoaffinity labeling, which is another rather widely accepted and well characterized technology. Compounds selected for investigation are constructed to incorporate a photosensitive cross-linker, often an azide, and a radioisotope for sensitive detection. Two of the most common approaches include direct incorporation of the cross-linker and radioisotope into the scaffolding of the molecule or purchasing a commercially available photoaffinity reagent and cross-linking it to the compound in-house. The advantages of direct incorporation include maintenance of compound architecture such that the new compound retains the selectivity and toxicity profile of the parent molecule. The disadvantages lie in the costs associated with procurement or chemistry support. The advantages of purchasing a photoaffinity reagent off the shelf are reduced cost and time of synthesis; however, the clear disadvantage is that this approach suffers from the same shortcomings as compound immobilization to a solid substrate—namely, the linker precludes potential off-target binding to that portion of the molecule. Methods of protein separation, visualization, and identification include 2DGE or column chromatography followed by MS/MS.

### Stable Isotope Standards and Capture by Anti-Peptide Antibodies

Stable isotope standards and capture by anti-peptide antibodies (SISCAPA) is an immunoaffinity capture technology developed for use with mass spectrometry. This platform was developed for the isolation and absolute quantification of a known target from a complex biological mixture. The greatest advantage of this technology is that it is positioned to reveal information regarding the absolute concentration of diminishing amounts of protein (<10 pg/mL) within a sample. This is achieved by first subjecting proteins from tissues or biofluids to global tryptic digestion, as is done in the ICAT, GIST, and iTRAQ differential labeling methodologies. The digests, spiked with stable [$^{13}$C]-isotope-labeled peptide for use as an internal standard, are passed over a nanoliter nanoaffinity column and washed extensively prior to elution. Subsequently, the enriched target peptide and internal standard are coeluted, analyzed by MS/MS, and quantified using the appropriate software. The internal isotopes are synthesized by fully substituting carbons of an amino acid near the C-terminus of the target peptides with [$^{13}$C]. The corresponding differences in molecular weight between the internal standard and control are reflected by the number of carbons in that amino acid (e.g., Pro = 6+, Val = +6, Phe = +10).

The nanoaffinity columns have demonstrated excellent reproducibility and peptide enrichment in excess of 100-fold. This technology has great potential if one has a working hypothesis about which proteins are affected by compound-induced adverse events or disease. Recent improvements to the technology include a multiplexing capability by concatenation of disparate peptides such that each is tagged in sequence with a stable isotope and separated by a trypsin-specific cleavage site. A known quantity of the concatamer is introduced into the biological sample and processed simultaneously over a nanoaffinity column containing the requisite peptide specific antibodies to capture all of the peptides represented in the concatamer. Subsequent elution and analysis allow for the quantification and fingerprinting of several peptides in a single sample. The resulting signature may be useful as a predictor of compound-induced toxicity or a valuable diagnostic of disease.

### Protein-AQUA™

Protein-AQUA™ is another example of an absolute quantification technology using stable isotope labeling technology developed for the mass spectrometry platform. It was developed by the same team that introduced GIST. Like SISCAPA, absolute target quantification is achieved by inclusion of a stable-isotope-labeled peptide internal standard into the complex mixture of proteins to be analyzed. The stable isotopes incorporated into the peptides include $^{13}C$ and $^{15}N$. Prior optimization of separation and analysis schemes are recommended for best results. Unlike SISCAPA, Protein-AQUA does not employ an immuno-enrichment step prior to analysis by MS/MS. A second application of the Protein-AQUA technology is a methodology developed specifically for differential proteome profiling. In this application, binary analysis of complex protein mixtures is achieved by the differential incorporation of molecular oxygen ($^{16}O$ and $^{18}O$) into peptides of either control or treated samples during the process of tryptic digestion. The two labeled samples are then mixed in a 1:1 ratio and analyzed simultaneously by MS/MS. Protein-AQUA and SISCAPA are two relatively recent technologies that have received much interest and are currently under intense scrutiny for application in the toxicoproteomics laboratory. The platforms are limited to downstream quantification of known targets that are suspected of contributing to an adverse event or disease. Although they remain technically challenging, increased success in mass-spectrometry-based proteomics laboratories should prompt greater acceptance of the technologies and transition into toxicoproteomics laboratories.

### Antibody Arrays

A great deal of attention has been focused on the development of antibody or protein array technology and applications to the toxicoproteomics laboratory. Most new innovations are based on standard sandwich-based enzyme-linked immunosorbent assay (ELISA) technology and are aimed at focusing on particular families or groups of proteins. Currently, many different platforms are commercially available and many more are under development; for example, some antibody arrays are printed onto slides, others are linked to microspheres, and in other instances some are coated onto multiwell plates. It is beyond the scope of this chapter to discuss the advantages and limitations of each specific platform; however, a few common generalizations apply to all. The principle advantage of array technology is the ability to multiplex—that is, to be able to analyze many different analytes simultaneously. Array development was initially targeted toward the analysis of proteins in clinical samples, but vendors are making progress toward developing panels of antibodies that address specific therapeutic areas preclinically. As an example, panels for antibodies to chemokines and cytokines known to be affected as a consequence of stimulating an inflammatory response have been made to recognize rodent antigens that cross-react with the rat. Additional areas of development include diabetes, cardiovascular disease, and obesity.

The principle differences in the technologies lie in signal detection. Most are fluorescent based but others rely on epi-chemiluminescence technology. The more sophisticated platforms come with their own proprietary software to support data acquisition and analysis. The primary limitation shared by all platforms is the quality of the antibodies used to capture the analytes. Initial attempts at building arrays incorporated commercially available antibodies that were often poorly characterized or insufficiently validated. This led to lot-to-lot variability in kit performance. Increasing rigor in quality control, antibody, and kit validation has resulted in the manufacture of superior platforms. In addition, lot-to-lot variability is encountered when different batches of a particular antibody are incorporated into a defined product. This cannot be avoided when using polyclonal antibodies generated in different animals that can result in the production of antibodies with differing binding characteristics to the antigen, different epitope recognition sites and ratios, and differing immunoglobulin subtypes; consequently, the kits are only as good as the reagents that comprise them and the rigor that goes into validating them. Another consideration that must be taken into account when building arrays is the concentration of the analytes to be measured. Care must be taken to build arrays with a panel of antibodies that capture analytes found within the same dynamic range of abundance. An acceptable range spans approximately three orders of magnitude from least to most abundant analyte. This ensures linearity of quantification and will maximize the capacity for multiplexing. Optimal multiplexing capacity for many array platforms is currently at about 25 analytes per assay. Array technology is meant to support questions investigators have about known analytes and whether expression signatures or sets of analytes can be used to predict predisposition to injury or aid in the diagnosis of disease.

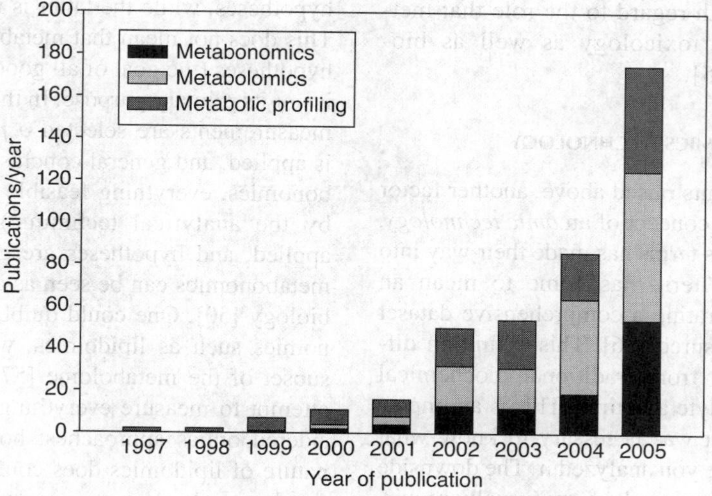


**FIGURE 6.8** Growth of metabonomics-related publications as indicated by using *metabonomics*, *metabolomics*, or *metabolic* for MEDLINE keyword searches.

## METABONOMICS

Metabonomics as a science has grown exponentially over the past several years (Figure 6.8). The applications of the technology are quite varied, ranging from environmental and botanical applications [36–42] all the way to clinical applications and epidemiological investigations [43–46]. Although all of these applications are interesting in their own right, it is beyond the scope of this section to cover the breadth of the science in any detail. Instead, the focus here is on more universal principles of the technology; however, because the expertise of the authors is in the area of preclinical toxicological applications of $^1$H nuclear magnetic resonance (NMR)-based metabonomics, examples and illustrations are somewhat biased toward that subfield of the science.

### What's in a Name?

*Metabonomics* has been defined as the "quantitative measurement of the time-related multi-parametric metabolic response of living systems to pathophysiological stimuli or genetic modification" [12]; however, a quick perusal of the literature quickly reveals other terms that seem to describe the same thing, such as *metabolomics* or *metabolic profiling*. What is the difference? Amplifying the confusion, metabolomics and metabonomics have been described as subsets of each other [36,48]. Although distinctions have been made in an attempt to differentiate the meanings of the terms [36,37,49,50], common usage is quickly homogenizing the three terms so they have come to be used interchangeably. For the sake of clarity, metabonomics will be used throughout this discussion with the understanding that metabolomics or metabolic profiling could just as appropriately be used.

### The Metabolome

Although the name of the technology may get all the attention, the concept of the *metabolome* is just as important. The metabolome has been defined as the "set of all low-molecular-weight compounds synthesized by an organism" [37,51]. Problems with this definition are immediately apparent; for example, what constitutes a metabolite (e.g., do you include peptides, DNA, or RNA?) and what constitutes an organism? If an animal consumes plants, do transient plant metabolites constitute part of the organism or not? What about metabolites generated by symbiotic gut flora? Do they count? What about xenobiotic metabolites (e.g., drug metabolites)? Where do they stand? While several have tried to parse out a more comprehensive definitions of the metabolome [52,53], for many this is an academic discussion.

On a more pragmatic note, the metabolome can be thought of in terms of what can be measured. In other words, the metabolome represents the non-artifact metabolites that can be measured in a sample derived from an organism, including metabolites from any bacteria, parasites, etc. that the organism may host and anything the animal may have consumed or had administered. The data then become defined by the analytical platforms and their attendant sensitivity and selectivity rather than by rote definition. If a peptide, plant-derived phenolic, or drug metabolite is measurable in the sample, it will be assessed as part of the data regardless of whether or not it is officially part of the metabolome. No attempt is made to eliminate or restrict metabolites, with the possible exception of artifacts (e.g., plasticizers from sample tubes). This does not mean that thoughtful discussion of what constitutes the metabolome is not warranted. Metabonomics opens up whole new areas of thinking with regard to the role that metabolic analysis plays in toxicology as well as biosciences in general [54,55].

### Metabonomics as an Omics Technology

Beyond the significant points raised above, another factor requiring discussion is the concept of an *omic technology*. An impressive list of omics terms has made their way into the literature. An omics term has come to mean an approach capable of generating a comprehensive dataset of whatever is being measured [56]. This definition differentiates metabonomics from traditional biochemical assays measuring one analyte at a time. This is an important point. Traditionally, it was necessary to know what you were measuring before you analyzed it. The downside to that approach is that it tends to be slow (usually requiring several separation steps), and it is usually costly on a per-sample basis. Metabonomics enables the rapid measurement of numerous metabolites. Downsides to metabonomics include the fact that researchers frequently do not know what they are measuring, and the assays in some cases are not quantitative. Investigators need to be aware that the output of a metabonomics analysis is often a handful of identified metabolites and a much longer list of unknowns.

One final distinction needs to be made with regard to what differentiates metabonomics from traditional analytical measurements of various metabolites. Biofluids have been sampled for various constituents since the days of the Egyptian dynasties; for example, urine tasting was used to assess diabetes in medieval times (gustomics perhaps?). With the advent of modern biochemistry, scientists have developed quite a repertoire of analytical assays to probe all kinds of fluids and tissues. Are those assays just metabonomics by another name? Clearly not, but the question still remains as to when analytical assays of various metabolites graduate to metabonomic status. Metabonomics is not an assay or analytical technique; metabonomics is an approach, not a platform. Nuclear magnetic resonance spectrometry and mass spectrometry are most often used for analytical approaches other than metabonomics. Metabonomics makes use of these tools (and others) but is not defined by them. Metabonomics endeavors to understand the response of the metabolome to "pathophysiological stimuli or genetic modification," to use Nicholson's definition. If one measures cholesterol, lipoproteins, and triglycerides, which are traditional clinical pathology measurements, after treatment with a lipid-modifying drug, do those measures constitute a metabonomic analyses? Most would conclude that they do not. If, however, one measures all biomolecular changes in response to the same drug as determined by MS or NMR, that would clearly constitute a metabonomic approach. The difference is that the former approach aims to (or at least is able to) refute hypotheses, while the latter is intended to generate them. This does not mean that metabonomic data cannot refute hypotheses (the goal of all good experimentation) but that is not its primary purpose. In the former approach, analyte measurements are selected *a priori*, inductive reasoning is applied, and general conclusions are made. With metabonomics, everything feasible is measured (limited only by the analytical technique), deductive reasoning is applied, and hypotheses are generated. For this reason, metabonomics can be seen as a tool for studying systems biology [50]. One could quibble that subsets of metabonomics such as lipidomics, which focuses on the lipid subset of the metabolome [57,58], by definition, do not attempt to measure everything feasible but still are considered omics approaches; however, the comprehensive nature of lipidomics does conform to the rest of the distinction made above.

## PLATFORMS

Theoretically, any technique capable of resolving metabolites can be used as an analytical platform for metabonomic analyses. Two analytical platform properties of critical importance to metabonomic analysis are the sensitivity and selectivity of the technique. The goal for most metabonomics applications is to get as unbiased a representation of the depth (sensitivity) and breadth (selectivity) of the metabolome as possible. Although a number of analytical techniques have been applied to metabonomics applications [59], techniques based on $^1$H (proton) NMR [60–63] and mass spectrometry [36,64–70] are the platforms most cited in the literature for metabonomics applications. Although it is beyond the scope of this chapter to go into a significant discussion of these two analytical techniques, a high-level comparison of the advantages and disadvantages of the two approaches is warranted.

### Mass Spectrometry vs. Nuclear Magnetic Resonance

Table 6.1 presents a summary of the advantages and disadvantages of NMR vs. MS as platforms for metabonomic analyses in the toxicology laboratory. The table is only a general guideline, because the MS and NMR configurations used by various laboratories make almost every line of the table debatable to some extent. In short, MS provides a huge sensitivity advantage while NMR offers several smaller advantages relative to quantitation, reproducibility, and lack of sample bias.

At first glance, it may seem obvious that MS would be the platform of choice for metabonomics studies because of the tremendous sensitivity advantage; however in reality, the MS sensitivity advantage is not all it is cracked up to be. The advantage of sensitivity depends

**TABLE 6.1**
**Comparison of Nuclear Magnetic Resonance vs. Mass Spectrometry for Metabonomics Applications**

| | Nuclear Magnetic Resonance | Mass Spectrometry |
|---|---|---|
| *Logistics* | | |
| Capital cost | No advantage | No advantage |
| Routine operating costs | No advantage | No advantage |
| Maintenance | Advantage | — |
| Per sample cost | No advantage | No advantage |
| Footprint | No advantage | No advantage |
| Required technical skill[a] | — | Advantage |
| Instrument uptime | Advantage | — |
| Instrument lifespan | No advantage | No advantage |
| *Analytical considerations* | | |
| Sensitivity | — | Big advantage |
| Reproducibility (within lab) | Advantage | — |
| Reproducibility (across labs) | Big advantage | — |
| Quantitation | Big advantage | — |
| Average run speed | No advantage | No advantage |
| Capacity (samples/day) | No advantage | No advantage |
| Sample preparation requirements | Advantage | — |
| Sample analysis automation | Advantage | — |
| Versatility[b] | Advantage | — |
| Selectivity[c] | — | Advantage |
| Nonselectivity[c] | Advantage | — |
| *Metabonomics* | | |
| Resolvable metabolites | — | Big advantage |
| Identification of unknowns | — | Advantage |
| Potential for sample bias[d] | Big advantage | — |
| Data analysis automation | — | Advantage |

[a] Pool of qualified analysts is much smaller for NMR than for MS.
[b] Generally, any NMR instrument can be configured for most applications, but different MS instrumentation may be required for specific applications.
[c] MS excels at selective identification of a molecular entity, and NMR excels at identification of all proton-containing species in a sample; therefore, selectivity can be an advantage or disadvantage, depending on the nature of the application.
[d] Potential exists for misleading, incomplete, or nonreproducible dataset due to bias inherent to the technology (e.g., ion suppression in MS).

*Source:* Robertson, D.G., *Toxicol. Sci.*, 85(2), 809–822, 2005. With permission.

argely on the ability to identify the thousands of resolv-ıble metabolites. Routine application of MS in metabo-ıomics applications will depend heavily on a viable spec-ral database, a task that has proven very elusive for MS ›ractitioners. The main problem is the bewildering variety ›f separation techniques and MS configurations available or metabonomic applications. This diversity makes stan-lardization for database applications a daunting task. Still, number of efforts are underway to generate databases 71,72] and several online resources are available; how-ver, the success (or not) of these efforts will only be vident as they gain wider acceptance within the metabo-ıomic user communities.

Nuclear magnetic resonance has the distinct advantage of database development ease, as the technique is quite reproducible across laboratories [73]. A primary reason for this is that typical biofluid $^1$H NMR requires no separation step (e.g., chromatography) prior to analysis; however, this advantage is tempered by the lack of sensitivity inherent to the technique. Another key advantage of NMR is that the integrated areas of spectral peaks are absolutely proportional to concentration, which is not the case for MS spectra. Quantitation for MS requires reference standards, which can be difficult to come by in some cases.

The bottom line is that serious practitioners of metabonomics are not limited by the instrumentation. Appli-

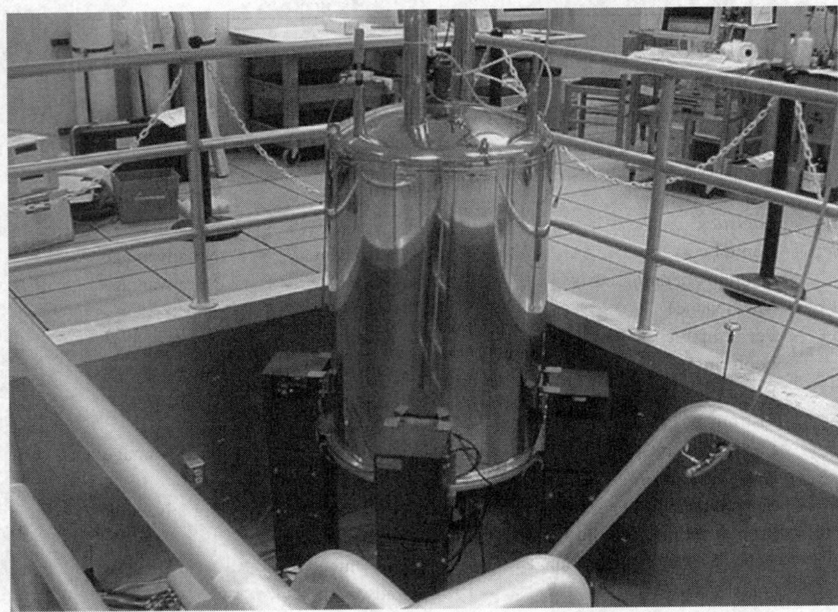

**FIGURE 6.9** A 600-MHz high-field magnet with active shielding, a type frequently used in metabonomics studies.

cations and endpoints are the best determinants of which platform is most suitable. Correspondingly, most metabonomics laboratories are now, or soon will be, employing both NMR and MS platforms in their research efforts. In recent publications comparing platforms for the metabonomic evaluation of serum from normal and Zucker (*fa/fa*) rats concluded that analysis of samples by both platforms led to little overlap in identified metabolites, indicating that the sum of the two technologies was far greater than either one independently [74,75].

As the authors of this chapter have greatest familiarity with NMR data, the focus of this section is biased toward NMR-derived spectral data. Figure 6.9 is a photograph of a 600-MHz NMR instrument used for metabonomics applications. The instruments themselves and their attendant magnetic field have a relatively large footprint; the higher the field, the larger the footprint required. Active shielding can mitigate the space requirements substantially, but there is no getting around the size and costs of the instrumentation. MS configurations are generally a bit smaller but, dependent on the configuration, are also quite expensive. In either case, an important requirement is trained personnel to operate the equipment and analyze the data. Although the pool of talented MS-capable scientists is significantly larger than NMR-trained scientists, it is not something that should be taken for granted with either technology.

### Magic Angle Spinning

One distinct advantage for NMR is that properly equipped instruments enable the nondestructive analysis of the biomolecular makeup of intact tissue using a technique called *magic angle spinning* (MAS). The name is derived from the fact that, when samples are spun rapidly at 54.7° relative to the applied magnetic field (the so-called magic angle), line-broadening effects that would ordinarily obfuscate a proton spectra of a solid sample can be reduced [76]. In practice, the technique is carried out by the placement of a few milligrams of intact tissue into a specially designed rotor that is spun at high speed within the bore of the magnet. This technique provides the ability to follow metabolic changes across tissues and the various biofluids that percolate through them, greatly enhancing the ability to gain mechanistic understanding [77–83].

## LABORATORY LOGISTICS

A detailed overview of laboratory logistics is available [84] so only the basic concepts are presented here. Figure 6.10 presents a typical laboratory configuration for a laboratory conducting rodent metabonomics studies. Key features include a capacity for refrigerated sample collection in metabolic cages, sample preparation, intermediate sample storage, the analytical platform (NMR, in this example), and, finally, some capacity for spectral data analysis and interpretation.

### The Sample

Any biofluid can potentially be used for metabonomic [62]. In practice, most analyses are conducted on urine, plasma, or serum. Urine is particularly useful for rodent studies where serial blood samples are either impractical or impossible (e.g., in the mouse); however, bacterial con

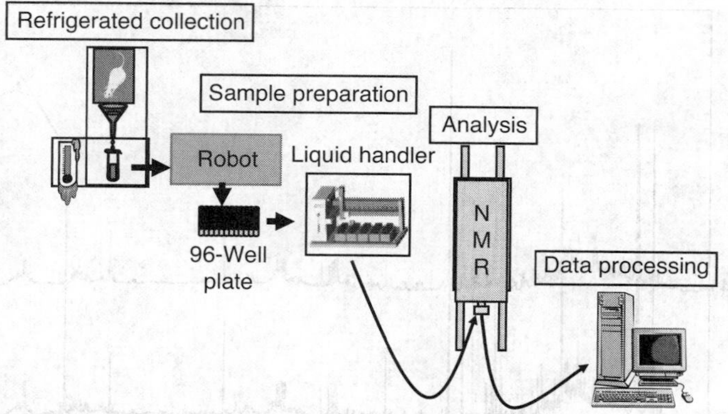


FIGURE 6.10 Schematic diagram of the metabonomics laboratory. (From Robertson, D.G., *Toxicol. Sci.*, 85(2), 809–822, 2005. With permission.)

amination is a significant problem when dealing with animal urine samples, and both refrigerated collection as well as the collection of samples in tubes containing an NMR-friendly bacteriostatic agent (typically $NaN_3$) are necessary. Other options may be considered for mass spectrometry analyses, but the caution against bacterial contamination still remains. A metabolism cage set-up for continuous collection of urine is shown in Figure 6.11. Such set-ups allow interval urine collection in rats over periods as short as 2 hours, although 4 hours more typically ensures adequate sample volumes. Urine volume can vary dramatically across rat strains bred for various purposes from as little as 0.5 mL/6 hr in rats with a gulonolactone oxidase deficiency to over 25 mL/6 hr in certain diabetic rat strains [85]. More realistically, typical laboratory rat strains such as Wistar and Sprague–Dawley produce approximately 15 mL of urine over a 24-hr period, and common mouse strains (A/J, C57BL/6, CD-1, and B6C3$F_1$) produce urine volumes in the range from 0.3 to 1.0 mL over a 24-hr period [86]. Importantly, approximately two thirds of the daily urinary output of the rat takes place during the dark cycle.

Biofluids other than urine require similar care in collection, although bacterial contamination tends to be less of an issue. In particular, the choice of anticoagulant is important when blood or blood-derived fluids (serum/plasma) are being assessed. Ethylenediaminetetraacetic acid (EDTA) is particularly problematic for NMR analyses, and citrate con-

FIGURE 6.11 Metabolic cage rack with refrigerated sample collection capability. Samples are collected in tubes in a refrigerated fraction collector positioned under the cages, allowing for collection over varying intervals from minutes to hours.

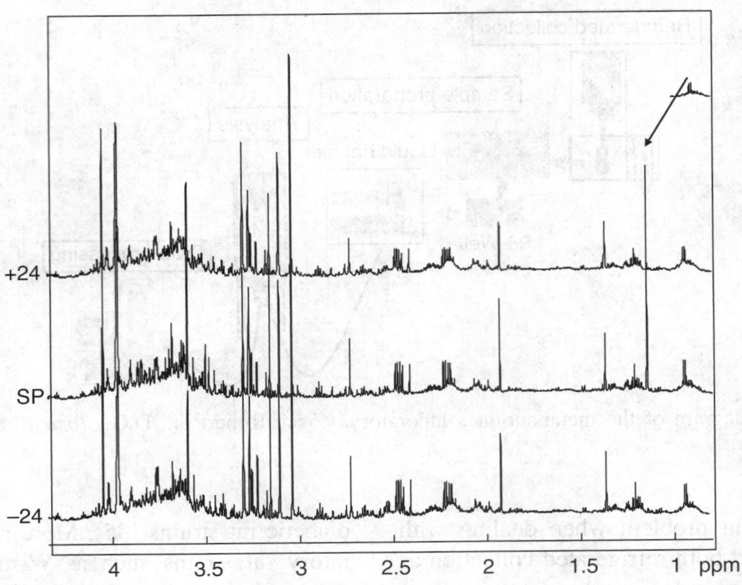


**FIGURE 6.12** NMR spectra taken from control guinea pig 24-hr urine samples collected 24 hours prior to (–24), immediately after (SP), and 24 to 48 hours after (+24) administration of a sweet potato treat. Note the prominent metabolite evident after the sweet potato administration (arrow).

founds the usually sensitive response of this tricarboxylic acid (TCA) cycle intermediate in toxicological analyses.

Regardless of the biofluid, metabonomics is particularly sensitive to many ancillary effects that are frequently ignored in toxicology studies; in other words, anything that goes into an animal for any reason frequently will find its way into the urine or other biofluids. The effect of vehicles used to deliver drugs or toxicants can have a dramatic effect on urinary spectra that goes beyond their simple appearance in the biofluid [87], suggesting that they may not be as innocuous as we would like to believe. Excipients in drug solutions or anesthetics, such as propylene glycol, an oft-found ingredient of barbiturate formulations, can markedly contaminate urine and serum spectra obtained after animal anesthesia either for interim procedures or euthanasia [56]. Xylazine used either alone or in combination with ketamine is another frequently used anesthetic that produces a significant glucosuria within the first few hours after administration. Although this effect is clearly documented, it is often forgotten. Another example of unintended consequences is presented in Figure 6.12. In some laboratories, it is standard operating procedure to occasionally give guinea pigs treats to supplement their diets. In this particular case, all animals in a guinea pig study (including controls) were noted to have a prominent peak in their urinary spectra that was absent the day before and the day after the finding. Further detective work revealed the peak to be due to the guinea pigs being given sweet potatoes on that particular day. This example reinforces the observation made earlier that anything going into an animal can affect the data. Also, it highlights the need for thorough documentation, as records of what treats were given on what days allowed for relatively easy elucidation of the cause of this effect.

## The Data

Perhaps the most unappreciated task in metabonomics analysis is data processing. Data processing typically proceeds in a stepwise fashion with varying paths dependent on what endpoint the investigator is seeking. Figure 6.13 presents a diagram describing data analyses for NMR-based metabonomic studies. MS-based approaches might vary in some details, particularly as to how raw spectra are derived, but they would be similar. When raw spectra are obtained, various chemometric approaches are applied to prepare for analysis, including phasing, baseline correction, and normalization. These processes are not dealt with here as they vary significantly across analytical platforms and the task can be accomplished in many ways [88–92]. When corrected spectra are obtained, they can be fed directly into multivariate statistical analyses. These approaches can be used directly to generate predictive metabonomic models based strictly on the unannotated spectra. Much of the early work in metabonomics was undertaken to generate such models [93–99], and there is no shortage of literature on the subject [38,88,100–106]. A detailed discussion of multivariate techniques applicable to metabonomic analyses is beyond the scope of this chapter. Many of these types of analyses are not unique to metabonomics and are covered in more detail elsewhere in this chapter.

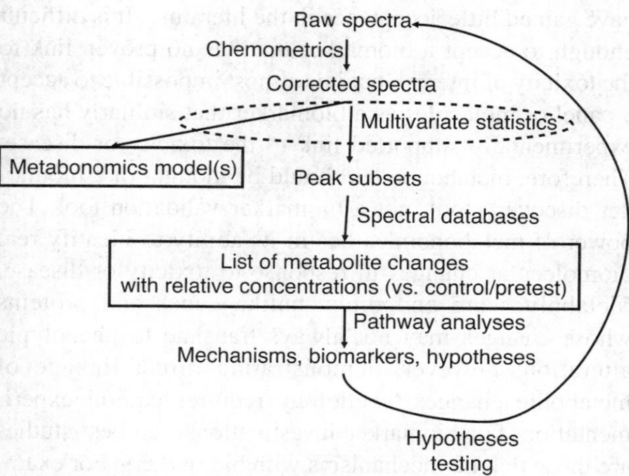


**FIGURE 6.13** Metabonomics data analysis. Schematic diagram of data analysis in a metabonomics setting showing various pathways the data can take and the process they undergo prior to delivery of the final analytical product (in boxes).

In recent years, with the expansive growth of MS-based metabonomics and the shortcomings of statistically based screening becoming evident, a concerted effort has been made to expand the capability to annotate spectral peaks with metabolite identifications. A variety of spectral databases for both NMR and MS applications are publicly available [107–111], not to mention commercially available software solutions. Metabolite identification is perhaps the slowest part of the analytical aspect of metabonomics technology. It frequently involves more than one analytical technology, and the absence of standards is problematic for confirming identification. The problem was bad enough using NMR-based metabonomics when frequently several dozen to several hundred unidentified peaks changed with treatment. With MS, that number increases at least an order of magnitude or more. In these cases, statistical approaches can be used to focus on those peaks whose changes are most significant (in a statistical sense) in the analyses. Although it is not a perfect solution because many minor changes might be quite biologically significant, it is at least a practical approach to the problem. Another approach is to combine both NMR and MS spectral data using statistical heterospectroscopy, a technique that compares variance across spectral datasets (from different platforms) and identifies those peaks that change either proportionally or inversely proportionally to each other between the two platforms [112]. In this fashion, the two techniques can be used to help focus analyses on only those metabolites that consistently change, removing a great deal of biologic noise from the data. Regardless of the method, the final product of the metabonomics analysis, from an analytical perspective, is a table of identified metabolites and unknowns and their relative concentrations relative to a pretest or control condition.

### Data Interpretation

Although an annotated metabolite list may be the end of the process from an analytical perspective, it is frequently far from satisfactory from a toxicologic perspective. Lightning sometimes strikes and the metabolite list reveals an easily understandable and accessible biomarker, but that is typically not the case. The metabolite list is more often just the jumping-off point for a great deal of basic biochemical research and hypothesis testing. Although the desire of most investigators is to find a unique mechanistically linked biomarker that is both sensitive and specific, such a finding is exceedingly rare in the toxicology setting. More often common metabolic intermediates change with both dose and time. They seldom are specific, although they may be quite sensitive. The trick is to integrate the biomolecular changes into a mechanistic story that makes sense. Online metabolite pathway resources are available to help in this effort [113–116], as well as an evolving array of commercial software solutions, but nothing is automatic. Concurrent transcriptomics or proteomic data can help to shape the story, as the data can suggest involvement of numerous pathways [117–119]. Regardless of the means by which a biomarker or hypothesis is generated, it is important to experimentally test the hypotheses. Although it would seem obvious, the design of any such experiment should be to disprove the hypothesis, not validate the biomarker. This is a basic tenet of scientific inquiry, but it is frequently ignored, leading to the devotion of substantial resources (both human and otherwise) to hypotheses that do not really merit it.

## APPLICATIONS

Metabonomics has broad applicability across the spectrum of biological sciences, including applications in botany [36,120–123] and environmental sciences [40,42, 124–126], such as environmental toxicology [41,127,128]. Various biomedical applications, including phenotyping [82,129–134], nutritional applications [117,135–139], and evaluation of parasitic infections [54,140–142], are all well represented in the literature. Any one of these fields could serve as the subject of a chapter of its own, but this section focuses on the three applications of metabonomics technology of interest to toxicologists: screening, mechanisms, and biomarkers.

### Screening

Toxicity screening means different things to different people. It can range from a comprehensive screen on a new chemical entity to a target-organ-focused screen of a structural analog to a compound known to produce a well-defined toxicity. Although a comprehensive screen for any compound type that would be indicative of any and all toxicity would be ideal, it will also never happen. More realistically, screens attempt to identify target organs and generate therapeutic indices.

Early metabonomics screening efforts focused on predictive models that supported the concept of rapid throughput safety screening [143–145]; however, experience gained by the COMET consortium [146,147] and others suggested that this was not as simple an exercise as originally conceived. Basic questions as to how toxicity is classified (e.g., chemical structure of toxin, pathophysiology, target organ, mechanism) greatly complicate the generation of predictive models. The development of comprehensive rapid-throughput screens has not been abandoned, but the reality is that this application has gained little acceptance by the toxicology community to date. More realistically, metabonomics has found application in screening for toxicities that are otherwise difficult to screen for; for example, drug-induced vasculitis is a problematic toxicity evident in the development of many drug classes [148]. Importantly, no definitive peripheral biomarkers for the lesion have been identified, meaning that the only sure way to assess the toxicity is to obtain tissues at necropsy for microscopic analysis. Metabonomics, however, readily identified animals with vasculitis when given a phosphodiesterase type IV inhibitor, distinguishing between animals with and without the lesion at a given dose [149]. The metabolic differentiation was not simply a reflection of the concurrent vascular/perivascular inflammation associated with the lesion [150]. Similarly metabonomics has been proposed as a tool for the identification of various other problematic toxicities, including peroxisome proliferation [151] and phospholipidosis [152].

### Mechanisms and Biomarkers

Mechanisms and biomarkers are naturally linked. If you identify a mechanism, you have identified a biomarker. Of course, that biomarker may be inaccessible or otherwise impossible to analyze. On the other hand, it might be possible to define a biomarker without necessarily defining the mechanism of toxicity or the etiology of a disease. For example, urinary D-β-hydroxybutyrate in the absence of other ketone bodies was proposed as a marker of proximal tubule renal damage, and trimethylamine and dimethylamine were suggested as markers of renal papillary damage [153,154]. Although the associations were convincing, the mechanistic link for these biomarkers was dubious; also, because they are quite commonly affected by other toxins, they have gained little acceptance as biomarkers of renal toxicity. The argument has been made that several biomarkers in concert may be more specific than any one in isolation. In a sense, the pattern of biomolecular changes that underlies the screening models discussed above is nothing more than a pattern biomarker. Pattern-based biomarkers may intuitively make sense, but they too have gained little acceptance in the literature. It is difficult enough to accept a biomarker that has no proven link to the toxicity of interest, but it is almost impossible to accept a panel of molecules as a biomarker that similarly has no experimentally supported link to the toxicity or disease. Therefore, metabonomics should be thought of a biomarker discovery tool, not a biomarker validation tool. The power of metabonomics lies in its ability to identify real biomolecular changes in response to toxicity or disease. Metabolites are endpoints, unlike genes and proteins whose changes may not always translate to phenotypic alteration; however, demonstrating direct linkage of metabolite changes to etiology requires careful experimentation. For biomarker investigations, the best studies are those that tie mechanisms with biomarkers. For example, hydrazine was shown to markedly increase levels of 2-aminoadipate, which is known to affect kynurenic acid levels in the brain. This provided a plausible hypothesis for the heretofore unexplained neurotoxic effects of the compound [155], providing a mechanism-linked biomarker for the neurotoxicity of the compound. Mortishire-Smith et al. linked urinary dicarboxylic aciduria to impaired fatty acid metabolism, which may be common to some hepatotoxic mechanisms [156]. Clayton et al. [157,158] mechanistically linked elevated urinary creatine to hepatotoxicity via effects on cysteine synthesis. Creatine levels in serum and urine were also shown to be associated with hepatotoxicity and nutritional effects. This work points to a pressing need in metabonomics science. Many biomolecules commonly seem to respond to toxins or physiological disruption in a similar fashion. These have come to be dubbed the "usual suspects" and include creatine, citrate, 2-oxoglutarate, and succinate, among others [56]. The commonality of the usual suspects has been attributed to such things as inappetence and weight loss [159], but these are certainly not the only reasons why these molecular changes are so common across toxicity studies, as the temporal displacement and indeed the direction of change can vary by compound. Clayton et al. [157,158] laid out a very plausible (and, importantly, testable) hypothesis that could underlie the role of creatine as a usual suspect in toxicity studies. This type of basic research—investigating these common metabolic intermediates—will be vitally important in understanding their potential role in the etiology of toxicity and disease as well as in realizing their potential as biomarkers.

## THE FUTURE

The role of metabonomics in toxicology has expanded exponentially over the past several years. The science is moving beyond its first tentative steps directly into the mainstream of toxicology research. Whether or not it can hold its own against the flood of new techniques and approaches remains to be seen; however, the science has moved far enough ahead so MIAME-like standardization [160] has been promulgated for the technology [161]. The National Institutes of Health (NIH) has recognized the significance of the technology, and it is well represented in the RoadMap initiative [162,163]. The Metabolomics Society that was recently established involves a wide array of metabonomics practitioners, including toxicologists [164]. With all of this interest, it is clear that metabonomics will not simply fade away like so many other technologies; however, the science will have to move from simply being "cool" to being practical with demonstrated impact both scientifically and from a business perspective.

## The Challenges of Data Analysis: Informatics and Systems Biology

As a discipline, bioinformatics is evolving from a genome information annotation, comparison, and analytical tool to playing a significant role in understanding the fundamental biology behind disease processes and mechanisms to identify and test new therapeutic strategies in the pharmaceutical industry [165]. The genomics sciences, by definition, generate large-volume datasets and therefore offer the opportunity to characterize biological processes in terms of patterns or of changes rather than the traditional biomarker approaches that have tended to concentrate on changes in a single or discrete number of molecules as the measured endpoint. This potential has challenged our existing definition of biomarkers. We now recognize that the patterns or fingerprints of individual changes (themselves composed of potentially dozens of individual markers) may be the biomarkers of the future [166].

Creating a more holistic view of biological processes, including an understanding of the regulation of and interactions between regulatory pathways is an emerging discipline often described under the broad term of *systems biology*. As a consequence of the availability of higher volume omics data, these approaches appear to be maturing at a rapid pace and are beginning to have demonstrable impact in the interpretation of large-volume datasets generated in the course of drug discovery and development [167]. Clearly, the sharing of nonproprietary genomics data among academia, industry, and regulators will be critical in the development of this field. Public software and databases are being developed at a number of institutions such as the National Centre for Toxicogenomics Research's ArrayTrack software for toxicogenomics data management and analysis [168] and the European Bioinformatics Institute's ArrayExpress database [169]. Consortia efforts among academia, industry, and regulatory scientists are also generating data to seed these public domain databases. A notable example is the International Life Sciences Institute (ILSI) Genomics Consortium, which has worked with some 30 member companies to develop a toxicogenomics dataset and release it into the public domain through collaboration with the European Bioinformatics Institute [170]. It is hoped that integrating genomics data with more traditional endpoints will help with developing a holistic understanding of the genotype–phenotype relationship, and this approach has been included in a number of efforts to build analytical tools, databases, and data exchange standards [168–170]. This issue is particularly important if genomics data are going to be compared or extrapolated across species [171].

Although the development of robust bioinformatics analysis tools continues apace, several examples of using pattern-matching tools from other disciplines have been used to characterize omics datasets. For example, voice–speech-pattern algorithms have been used in concert with transcript profiling experiments to classify and predict the therapeutic response of ovarian cancer patients [172].

One step in the maturation of systems biology may be more complete and better descriptions of functional units in biology (i.e., a characterization of the major pathways and physiological processes into a discrete number of units). This is an important extension of ongoing efforts to take biochemical pathway information (such as that found in the KEGG database) to a more highly annotated and linked relational database [173]. Biological processes can then be defined in terms of interactions between major pathways rather than individual genes or proteins [174,175]. Such initiatives are complemented by protein informatics, which aims to predict the putative function of uncharacterized (or hypothetical) proteins based on structural features and pathway mapping [176,177].

In addition to the multiple technical platforms and designs in microarray construction, a great many analytical methods are being employed to generate the final data on the number and magnitude of induced changes in gene expression changes [178]. The analysis of a standard microarray dataset by multiple methods has been proposed to facilitate discussions on standardizing approaches [179–181].

Broad application of the pattern-recognition approach to predictive toxicology is quite likely to require the application of complex computer algorithms and statistical approaches. Several resources exist in both the academic and commercial sectors to assist in managing reference compound gene expression datasets to facilitate pattern recognition. Scientists at the National Human Genome Research Institute were among the first to develop such a system: the software platform ArrayDB, which facilitates the storage, retrieval, and analysis of microarray data along with information linking tens of thousands of genes to public domain sequence and pathway databases. In addition, collaborative efforts involving pharmaceutical, agrochemical, and chemical industries, such as the International Life Sciences Institute committee on the Application of Genomics and Proteomics to Mechanism-Based Risk

Assessment, are generating publicly accessible datasets (see http://www.ilsi.org). The feasibility of such an ambition, including the time scales required before the realization of databases that are truly predictive of a broad range of toxicities, remains to be seen and will most likely involve contributions from a very large number of laboratories worldwide.

To facilitate data sharing between laboratories using different microarray designs, standard data format standards are evolving. For example, Gene Expression Markup Language (GEML), developed by a number of organizations, is designed to separate data reporting and collection from the methodology used to collect the data, to be independent of analysis software, and to serve as a general-purpose format for data exchange. The broad acceptance of a common data exchange format is perhaps the biggest bottleneck to be addressed before public domain repositories for microarray data are realized. Various laboratories are developing such data repositories for gene expression data. An example of such is the Stanford Microarray Database (SMD), which stores both raw and normalized data from multiple microarray experiments. This platform also allows access to analysis and visualization tools and electronic links to published microarray experiments (http://genome-www4.stanford.edu/MicroArray/SMD).

## Current and Future Issues

The careful application of transcript profiling technologies will assist research associated with safety assessment by providing:

- Enhanced ability to extrapolate accurately between experimental animals and humans in the context of risk assessment
- More detailed appreciation of molecular mechanisms of toxicity
- Facilitation of more rapid screens for compound toxicity
- Provision of new research leads

Clearly, technical issues that must be resolved before this promise is fully realized include reproducibility, variability, gene annotation, and platform differences, among others.

In the context of mechanism-based research, it is probably best to regard results obtained using transcript-profiling technologies as the springboard to more detailed and focused investigations that would confirm or refute the significance of the observed changes. Concern has been voiced regarding possible misinterpretation, or over-interpretation, of such high-volume data analyses, particularly in the context of safety assessment. It must be recognized that the interaction of xenobiotics with biological systems will without fail result in some changes in gene expression, even under circumstances where such gene expression events are benign. It is therefore important that emerging data from these technologies are followed through with traditional approaches and analyzed fully to establish if the measured changes are background noise, adaptive, beneficial, or potentially harmful. Such expression changes clearly do not compromise previously characterized no-adverse-effect levels (NAELs) of exposure under circumstances where there are no physiological or pathological indicators of harmful effect. These are points on which it is critically important to foster the development of consensus and common understanding among industry, academia, and regulatory bodies.

It is arguably the case that the most valuable aspect of these approaches is that they will encourage scientists to consider biological responses in a more holistic fashion. The approaches, being open, remove to a great extent the bias that takes place when a few distinct endpoints are chosen in advance, opening our minds to consideration of changes in the genes, proteins, and metabolites that were believed to be of no apparent relevance. Perhaps the most exciting advances in defining toxicological mechanisms and identifying new research leads will come from the open-minded and integrated toxicology that panomics approaches will undoubtedly encourage.

## QUESTIONS

1. Compared to transcriptomics and proteomics, metabonomics can be characterized by which of the following attributes:
   a. Sensitivity.
   b. Selectivity.
   c. It is a biomarker discovery tool.
   d. It is suitable only for biofluids.
2. Which of the following is/are true with regard to metabonomics:
   a. MS is much more useful than NMR as the base analytical platform.
   b. Identifying biomarkers of toxicity is more useful then identifying mechanisms of toxicity.
   c. The technology is not subject to the unintended effects of experimental design to the same extent as other technologies.
   d. All of the above.
   e. None of the above.
3. What must be considered in the experimental design for transcript profiling experiments? What precautions can be taken to determine if gene expression changes are cause or effect?
4. What methods are routinely used to visualize the multivariate datasets used in toxicopanomics approaches?
5. Compare and contrast the practical advantages of transcript profiling vs. proteomics vs. metabonomics for the discovery of biomarkers that may be useful for monitoring toxicity in humans.

## REFERENCES


1. Castle, A. L., Carver, M. P., and Mendrick, D. L. (2002): Toxicogenomics: a new revolution in drug safety. *Drug Discov. Today*, 7:728.
2. Pennie, W., Pettit, S. D., and Lord, P. G. (2004): Toxicogenomics in risk assessment: an overview of an HESI collaborative research program. *Environ. Health Perspect.*, 112:417–419.
3. Wei, Y. D., Tepperman, K., Huang, M. Y., Sartor, M. A., and Puga, A. (2004): Chromium inhibits transcription from polycyclic aromatic hydrocarbon-inducible promoters by blocking the release of histone deacetylase and preventing the binding of p300 to chromatin. *J. Biol. Chem.*, 279: 4110–4119.
4. Moennikes, O., Loeppen, S., Buchmann, A., Andersson, P., Ittrich, C., Poellinger, L., and Schwarz M. (2004): A constitutively active dioxin/aryl hydrocarbon receptor promotes hepatocarcinogenesis in mice. *Cancer Res.*, 64: 4707–4710.
5. Terasaka, S., Aita, Y., Inoue, A., Hayashi, S., Nishigaki, M., Aoyagi, K., Sasaki, H., Wada-Kiyama, Y., Sakuma, Y., Akaba, S., Tanaka, J., Sone, H., Yonemoto, J., Tanji, M., and Kiyama R. (2004): Using a customized DNA microarray for expression profiling of the estrogen-responsive genes to evaluate estrogen activity among natural estrogens and industrial chemicals. *Environ. Health Perspect.*, 112: 773–781.
6. Rea, M. A., Gregg, J. P., Qin, Q., Phillips, M. A., and Rice, R. H. (2003): Global alteration of gene expression in human keratinocytes by inorganic arsenic. *Carcinogenesis*, 24:747–756.
7. Zheng, X. H., Watts, G. S., Vaught, S., and Gandolfi, A. J. (2003): Low-level arsenite induced gene expression in HEK293 cells. *Toxicology*, 187:39–48.
8. Daborn, P. J., Yen, J. L., Bogwitz, M. R., Le Goff, G., Feil, E., Jeffers, S., Tijet, N., Perry, T., Heckel, D., Batterham, P. et al. (2002): A single p450 allele associated with insecticide resistance in *Drosophila*. *Science*, 297:2253–2256.
9. Pedra, J. H., McIntyre, L. M., Scharf, M. E., Pittendrigh, B. R. (2004): Genome-wide transcription profile of field- and laboratory-selected dichlorodiphenyltrichloroethane (DDT)–resistant *Drosophila*. *Proc Natl. Acad. Sci. USA*, 101:7034–7039.
10. Goodsaid, F. M. (2003): Genomic biomarkers of toxicity. *Curr. Opin. Drug Discov. Devel.*, 6(1):41.
11. Burczynski, M. E., McMillian, M., Ciervo, J., Li, L., Parker, J. B. et al. (2000): Toxicogenomics-based discrimination of toxic mechanism in HepG2 human hepatoma cells. *Toxicol. Sci.*, 58:399–415.
12. Thomas, R. S., Rank, D. R., Penn, S. G., Zastrow, G. M., Hayes, K. R., Pande, K., Glover, E., Silander, T., Craven, M. W. et al. (2001): Identification of toxicologically predictive gene sets using cDNA microarrays. *Mol. Pharmacol.*, 60(6):1189.
13. Waring, J. F., Jolly, R. A., Ciurlionis, R., Lum, P. Y., Praestgaard, J. T., Morfitt, D. C., Buratto, B., Roberts, C., Schadt, E., and Ulrich, R. G. (2001): Clustering of hepatotoxins based on mechanism of toxicity using gene expression profiles. *Toxicol. Appl. Pharmacol.*, 175(1):28.
14. Baker, T. K., Carfagna, M. A., Gao, H., Dow, E. R., Li, Q., Searfoss, G. H., and Ryan, T. P. (2001): Temporal gene expression analysis of monolayer cultured rat hepatocytes. *Chem. Res. Toxicol.*, 14:1218.
15. Bandara, L., and Kennedy, S. (2002): Toxicoproteomics: a new preclinical tool. *Drug Discov. Today*, 7:411.
16. Kennedy, S. (2002): The role of proteomics in toxicology: identification of biomarkers of toxicity by protein expression analysis. *Biomarkers*, 7:269.
17. Hale, J. E., Gelfanova, V., Ludwig, J. R., and Knierman, M. D. (2003): Application of proteomics for discovery of protein biomarkers. *Brief Funct. Genom. Proteomics*, 2:185.
18. Walgren, J. L., and Thompson, D. C. (2004): Application of proteomic technologies in the drug development process. *Toxicol. Lett.*, 149(1–3):377.
19. Kennedy, S. (2001): Proteomic profiling from human samples: the body fluid alternative. *Toxicol. Lett.*, 120(1–3):379.
20. Fountoulakis, M., and Suter, L. (2002): Proteomic analysis of the rat liver. *J. Chromatogr. B*, 782:197.
21. Chaurand, P., DaGue, B., Pearsall, R., Threadgill, D., and Caprioli, R. (2001): Profiling proteins from azoxymethane-induced colon tumors at the molecular level by matrix–assisted laser desorption/ionization mass spectrometry. *Proteomics*, 1:1320.
22. Petricoin, E., Rajapaske, V., Herman, E., Arekani, A., Ross, S., Johann, D., Knapton, A., Zhang, J., Hitt, B., Conrads, T., Veenstra, T., Liotta, L., and Sistare, F. (2004): Toxicoproteomics: serum proteomic pattern diagnostics for early detection of dug induced cardiac toxicities and cardioprotection. *Toxicol. Pathol.*, 32(Suppl. 1):122.
23. Meneses-Lorente, G., Guest, P., Lawrence, J., Muniappa, N., Knowles, M., Skynner, H., Salim, K., Cristea, I., Mortishire–Smith, R., Gaskell, S., and Watt, A. (2004): A proteomic investigation of drug-induced steatosis in rat liver. *Chem. Res. Toxicol.*, 17:605.
24. Gao, J., Garulacan, L., Storm, S., Hefta, S., Opiteck, G., Lin, J., Moulin. F., and Dambach D. (2004): Identification of *in vitro* protein biomarkers of idiosyncratic liver toxicity. *Toxicol. In Vitro*, 18:533.
25. Dare, T., Davies, H., Turton, J., Lomas, L., Williams, T., and York, M. Application of surface-enhanced laser desorption/ionization technology to the detection and identification of urinary parvalbumin: a biomarker of compound-induced skeletal muscle toxicity in the rat. *Electrophoresis*, 23:3241.
26. Jones, J., Kaphalia, L., Treinen-Moslen, M., and Leibler, D. (2003): Proteomic characterization of metabolites, protein adducts, and biliary proteins in rats exposed to 1,1-dichloroethylene or diclofenac. *Chem. Res. Toxicol.*, 16:1306.
27. Fountoulakis, M., Berndt, P., Boelsterli, U., Crameri, F., Winter, M., Albertini, S., and Suter, L. (2000): Two-dimensional database of mouse liver proteins: changes in hepatic protein levels following treatment with acetaminophen or its nontoxic regioisomer 3-acetamidophenol. *Electrophoresis*, 21:2148.
28. Da Cruz, S., Xenarios, I., Langridge, J., Vilbois, F., Parone, P., and Martinou, J. (2003): Proteomic analysis of the mouse liver mitochondrial inner membrane. *J. Biol. Chem.*, 278:41566.

29. Leonoudakis, D., Conti, L., Anderson, S., Radeke, C., McGuire, L., Adams, M., Froehner, S., Yates, J., and Vandenberg, C. (2004): Protein trafficking and anchoring complexes revealed by proteomic analysis of inward rectifier potassium channel (Kir2. x)-associated proteins. *J. Biol. Chem.*, 279:22331.
30. Nisar, S., Lane, C., Wilderspin, A., Welham, K., Griffiths, W., and Patterson, L. (2004): A proteomic approach to the identification of cytochrome P450 isoforms in male and female rat liver by nanoscale liquid chromatography–electrospray ionization–tandem mass spectrometry. *Drug Metab. Disp.*, 32:382.
31. Zhu, H., Bilgin, M., and Snyder, M. (2003): Proteomics. *Annu. Rev. Biochem.*, 72:783.
32. Baggerly, K., Morris, J., and Coombes, K. (2004): Reproducibility of SELDI–TOF protein patterns in serum: comparing datasets from different experiments. *Bioinformatics*, 20:777.
33. Heijne, W., Stierum, R., Slijper, M., van Bladeren, P., and van Ommen, B. (2003): Toxicogenomics of bromobenzene hepatotoxicity: a combined transcriptomics and proteomics approach. *Biochem. Pharmacol.*, 65:857.
34. Ruepp, S., Tonge, R., Shaw, J., Wallis, N., and Pognam, F. (2002): Genomics and proteomics analysis of acetaminophen toxicity in mouse liver. *Toxicol. Sci.*, 65:135.
35. Omenn, G. (2004): The Human Proteome Organization Plasma Proteome Project pilot phase: reference specimens, technology platform comparisons, and standardized data submissions and analyses. *Proteomics*, 4:1235.
36. Fiehn, O. (2001): Combining genomics, metabolome analysis, and biochemical modeling to understand metabolic networks. *Comp. Funct. Genom.*, 2:155–168.
37. Fiehn, O. (2002): Metabolomics: the link between genotypes and phenotypes. *Plant Mol. Biol.*, 48(1–2):155–171.
38. Weckwerth, W. et al. (2004): Differential metabolic networks unravel the effects of silent plant phenotypes. *Proc. Natl. Acad. Sci. USA*, 101(20):7809–7814.
39. Roberts, J. K. (2000): NMR adventures in the metabolic labyrinth within plants. *Trends Plant Sci.*, 5(1):30–34.
40. Viant, M. R., Rosenblum, E. S., and Tieerdema, R. S. (2003): NMR-based metabolomics: a powerful approach for characterizing the effects of environmental stressors on organism health. *Environ. Sci. Technol.*, 37(21):4982–4989.
41. Bundy, J. G. et al. (2004): Environmental metabonomics: applying combination biomarker analysis in earthworms at a metal contaminated site. *Ecotoxicology*, 13(8): 797–806.
42. Griffin, J. L. et al. (2000): NMR spectroscopy based metabonomic studies on the comparative biochemistry of the kidney and urine of the bank vole (*Clethrionomys glareolus*), wood mouse (*Apodemus sylvaticus*), white toothed shrew (*Crocidura suaveolens*) and the laboratory rat. *Comp. Biochem. Physiol. B Biochem. Mol. Biol.*, 127(3): 357–367.
43. Coen, M. et al. (2005): Proton nuclear magnetic resonance-based metabonomics for rapid diagnosis of meningitis and ventriculitis. *Clin. Infect. Dis.*, 41(11):1582–1590.
44. Brindle, J. T. et al. (2002): Rapid and noninvasive diagnosis of the presence and severity of coronary heart disease using $^{1}$H-NMR-based metabonomics. *Nat. Med.*, 8(12):1439–1444.
45. German, J. B., Roberts, M. A., and Watkins, S. M. (2003): Genomics and metabolomics as markers for the interaction of diet and health: lessons from lipids. *J. Nutr.*, 133(6, Suppl. 1):2078S–2083S.
46. German, J. B., Watkins, S. M., and Fay, L. B. (2005): Metabolomics in practice: emerging knowledge to guide future dietetic advice toward individualized health. *J. Am. Diet. Assoc.*, 105(9):1425–1432.
47. Nicholson, J. K., Lindon, J. C., and Holmes, E. (1999): 'Metabonomics': understanding the metabolic responses of living systems to pathophysiological stimuli via multivariate statistical analysis of biological NMR spectroscopic data. *Xenobiotica*, 29(11):1181–1189.
48. Lindon, J. C., Holmes, E., and Nicholson, J. K. (2003): So what's the deal with metabonomics? *Anal. Chem.*, 75(17):384A–391A.
49. Villas–Boas, S. G., Rasmussen, S., and Lane, G. A. (2005): Metabolomics or metabolite profiles? *Trends Biotechnol.*, 23(8):385–386.
50. Goodacre, R. et al. (2005): Metabolomics by numbers: acquiring and understanding global metabolite data. *Trends Biotechnol.*, 22(5):245–252.
51. Oliver, S. G. et al. (1998): Systematic functional analysis of the yeast genome. *Trends Biotechnol.*, 16(9):373–378.
52. Beecher., C. W. W. (2003): The human metabolome. In: *Metabolic Profiling: Its Role in Biomarker Discovery and Gene Function Analysis*, edited by G. G. H. A. R. Goodacre. Kluwer Academic, Boston, pp. 311–319.
53. Nicholson, J. K., and Wilson, I. D. (2003): Opinion: understanding 'global' systems biology: metabonomics and the continuum of metabolism. *Nat. Rev. Drug Discov.*, 2(8): 668–676.
54. Nicholson, J. K. et al. (2004): The challenges of modeling mammalian biocomplexity. *Nat. Biotechnol.*, 22(10): 1268–1274.
55. Nicholson, J. K., Holmes, E., and Wilson, I. D. (2005: Gut microorganisms, mammalian metabolism and personalized health care. *Nat. Rev. Microbiol.*, 3(5):431–438.
56. Robertson, D. G. (2005): Metabonomics in toxicology: a review. *Toxicol. Sci.*, 85(2):809–822.
57. Watkins, S. M. et al. (2002): Lipid metabolome-wide effects of the PPARgamma agonist rosiglitazone. *J. Lipid Res.*, 43(11):1809–1817.
58. Morris, M., and Watkins, S. M. (2005): Focused metabolomic profiling in the drug development process: advances from lipid profiling. *Curr. Opin. Chem. Biol.*, 9(4): 407–412.
59. Robertson, D. G., Reily, M. D., and Baker, J. D. (2005): Metabonomics in preclinical drug development. *Expert Opin. Drug Metab. Toxicol.*, 1(3):363–376.
60. Reily, M. D., and Lindon., J. C. (2005): NMR spectroscopy: principles and instrumentation. In: *Metabonomics in Toxicity Assessment*, edited by E. Holmes et al. Taylor & Francis, Boca Raton, FL, pp. 75–104.

61. Lindon, J. C. et al. (2004): Metabonomics technologies and their applications in physiological monitoring, drug safety assessment and disease diagnosis. *Biomarkers*, 9(1): 1–31.
62. Lindon, J. C., Nicholson, J. K., and Everett, J. R. (1999): NMR spectroscopy of biofluids. *Annu. Rep. NMR Spect.*, 38:1–88.
63. Reo, N. V. (2002): NMR-based metabolomics. *Drug Chem. Toxicol.*, 25(4):375–382.
64. Fernie, A. R. et al. (2004): Metabolite profiling: from diagnostics to systems biology. *Nat. Rev. Mol. Cell. Biol.*, 5(9):763–769.
65. Harrigan, G. (2002): Metabolic profiling: pathways in drug discovery. *Drug Discov. Today*, 7(6):351–352.
66. Plumb, R. S. et al. (2003): Use of liquid chromatography/time-of-flight mass spectrometry and multivariate statistical analysis shows promise for the detection of drug metabolites in biological fluids. *Rapid Commun. Mass Spectrom.*, 17(23):2632–2638.
67. Plumb, R. S. et al. (2005): A rapid screening approach to metabonomics using UPLC and oa-TOF mass spectrometry: application to age, gender and diurnal variation in normal/Zucker obese rats and black, white and nude mice. *Analyst*, 130(6):844–8449.
68. Plumb, R. S. et al. (2002): Metabonomics: the use of electrospray mass spectrometry coupled to reversed-phase liquid chromatography shows potential for the screening of rat urine in drug development. *Rapid Commun. Mass Spectrom.*, 16(20):1991–1996.
69. van der Greef, J. et al. (2004): The role of mass spectrometry in systems biology: data processing and identification strategies in metabolomics. *Adv. Mass Spectrom.*, 16:145–165.
70. Dunn, W. B., Bailey, N. J., and Johnson, H. E. (2005): Measuring the metabolome: current analytical technologies. *Analyst*, 130(5):606–625.
71. Wagner, C., Sefkow, M., and Kopka, J. (2003): Construction and application of a mass spectral and retention time index database generated from plant GC/EI–TOF–MS metabolite profiles. *Phytochemistry*, 62(6):887–900.
72. Smith, C. A. et al. (2005): METLIN: a metabolite mass spectral database. *Ther. Drug Monit.*, 27:747–751.
73. Keun, H. C. et al. (2002): Analytical reproducibility in (1)H NMR-based metabonomic urinalysis. *Chem. Res. Toxicol.*, 15(11):1380–1386.
74. Williams, R. et al. (2006): A multi-analytical platform approach to the metabonomic analysis of plasma from normal and Zucker (*fa*/*fa*) rats. *Mol. BioSyst.*, 2:174–183.
75. Williams, R. E. et al. (2005): A combined (1)H NMR and HPLC–MS-based metabonomic study of urine from obese (*fa*/*fa*) Zucker and normal Wistar-derived rats. *J. Pharm. Biomed. Anal.*, 38(3):465–471.
76. Shockcor, J. P., and Holmes, E. (2002): Metabonomic applications in toxicity screening and disease diagnosis. *Curr. Top. Med. Chem.*, 2(1):35–51.
77. Coen, M. et al. (2003): An integrated metabonomic investigation of acetaminophen toxicity in the mouse using NMR spectroscopy. *Chem. Res. Toxicol.*, 216(3):295–303.
78. Griffin, J. L. et al. (2002): Spectral profiles of cultured neuronal and glial cells derived from HRMAS (1)H-NMR spectroscopy. *NMR Biomed.*, 15(6):375–384.
79. Griffin, J. L. et al. (2000): The biochemical profile of rat testicular tissue as measured by magic angle spinning $^1$H NMR spectroscopy. *FEBS Lett.*, 486(3):225–229.
80. Griffin, J. L. et al. (2001): High-resolution magic angle spinning $^1$H-NMR spectroscopy studies on the renal biochemistry in the bank vole (*Clethrionomys glareolus*) and the effects of arsenic ($As^{3+}$) toxicity. *Xenobiotica*, 31(6):377–385.
81. Waters., N. J. et al. (2000): High-resolution magic angle spinning (1)H NMR spectroscopy of intact liver and kidney: optimization of sample preparation procedures and biochemical stability of tissue during spectral acquisition. *Anal. Biochem.*, 282(1):16–23.
82. Griffin, J. L., Cemal, C. K., and Pook, M. A. (2004): Defining a metabolic phenotype in the brain of a transgenic mouse model of spinocerebellar ataxia 3. *Physiol. Genom.*, 16(3):334–340.
83. Bollard, M. E. et al. (2003): A study of metabolic compartmentation in the rat heart and cardiac mitochondria using high-resolution magic angle spinning $^1$H-NMR spectroscopy. *FEBS Lett.*, 553(1–2):73–78.
84. Robertson, D. G. et al. (2002): Metabonomic technology as a tool for rapid throughput *in vivo* toxicity screening. In: *Comprehensive Toxicology*, edited by J. P. Vanden Heuvel, G. J. Perdew., W. B. Mattes, and W. F. Greenlee. Elsevier Science, Amsterdam, pp. 583–610.
85. Anon. (2005): *Kyoto Rat Phenotype Database (or Genotype Database)*, National Bio Resource Project for the Rat in Japan, Institute of Laboratory Animals, Graduate School of Medicine, Kyoto University, Japan (www. anim.med. kyoto-u.ac.jp/nbr).
86. Stevens, G. J., Deese, A. J., and Robertson, D. G. (2005): The application of metabonomics as an early *in vivo* toxicity screen. In: *Metabonomics in Toxicity Assessment*, edited by D. G. Robertson and J. Lindon. Taylor & Francis, Boca Raton, FL, pp. 195–224.
87. Beckwith-Hall, B. M. et al. (2002): NMR-based metabonomic studies on the biochemical effects of commonly used drug carrier vehicles in the rat. *Chem. Res. Toxicol.*, 15(9):1136–1141.
88. Purohit, P. V. et al. (2004): Discrimination models using variance-stabilizing transformation of metabolomic NMR data. *Omics*, 8(2):118–130.
89. Spraul, M. et al. (1994): Automatic reduction of NMR spectroscopic data for statistical and pattern recognition classification of samples. *J. Pharm. Biomed. Anal.*, 12(10): 1215–1225.
90. Jonsson, P. et al. (2005): High-throughput data analysis for detecting and identifying differences between samples in GC/MS-based metabolomic analyses. *Anal Chem.*, 77(17): 5635–5642.
91. Webb-Robertson, B. J. et al. (2005): A study of spectral integration and normalization in NMR-based metabonomic analyses. *J. Pharm. Biomed. Anal.*, 39(3–4): 830–836.

92. Sandusky, P., and Raftery, D. (2005): Use of selective TOCSY NMR experiments for quantifying minor components in complex mixtures: application to the metabonomics of amino acids in honey. *Anal. Chem.*, 77(8): 2455–2463.
93. Beckwith-Hall, B. M. et al. (1998): Nuclear magnetic resonance spectroscopic and principal components analysis investigations into biochemical effects of three model hepatotoxins. *Chem. Res. Toxicol.*, 11(4):260–272.
94. Gartland, K. P. et al. (1991): Application of pattern recognition methods to the analysis and classification of toxicological data derived from proton nuclear magnetic resonance spectroscopy of urine. *Mol. Pharmacol.*, 39(5): 629–642.
95. Jansen, J. J. et al. (2004): Analysis of longitudinal metabolomics data. *Bioinformatics*, 20(15):2438–2446.
96. Scholz, M. et al. (2004): Metabolite fingerprinting: detecting biological features by independent component analysis. *Bioinformatics*, 20(15):2447–2454.
97. Holmes, E. et al. (2000): Chemometric models for toxicity classification based on NMR spectra of biofluids. *Chem. Res. Toxicol.*, 13(6):471–478.
98. Hart, B. A. et al. (2003): $^1$H-NMR spectroscopy combined with pattern recognition analysis reveals characteristic chemical patterns in urines of MS patients and non-human primates with MS-like disease. *J. Neurol. Sci.*, 212(1–2): 21–30.
99. Baumgartner, C., Bohm, C., and Baumgartner, D. (2005): Modelling of classification rules on metabolic patterns including machine learning and expert knowledge. *J. Biomed. Inform.*, 38(2):89–98.
100. Viant, M. R. (2003): Improved methods for the acquisition and interpretation of NMR metabolomic data. *Biochem. Biophys. Res. Commun.*, 310(3):943–948.
101. Jonsson, P. et al. (2004): A strategy for identifying differences in large series of metabolomic samples analyzed by GC/MS. *Anal. Chem.*, 76(6):1738–1745.
102. Holmes, E., and Antti, H. (2002): Chemometric contributions to the evolution of metabonomics: mathematical solutions to characterising and interpreting complex biological NMR spectra. *Analyst*, 127(12):1549–1557.
103. Beckwith-Hall, B. M. et al. (2002): Application of orthogonal signal correction to minimise the effects of physical and biological variation in high resolution $^1$H NMR spectra of biofluids. *Analyst*, 127(10):1283–1288.
104. Holmes, E. et al. (1994): Automatic data reduction and pattern recognition methods for analysis of $^1$H nuclear magnetic resonance spectra of human urine from normal and pathological states. *Anal. Biochem.*, 220(2):284–296.
105. Holmes, E., and Antti, H. (2002): Chemometric contributions to the evolution of metabonomics: mathematical solutions to characterising and interpreting complex biological NMR spectra. *Analyst*, 127(12):1549–1557.
106. Smilde, A. K. et al. (2005): ANOVA simultaneous component analysis (ASCA): a new tool for analyzing designed metabolomics data. *Bioinformatics*, 21(13):3043–3048.
107. Comprehensive Systems-Biology Database. (2006): *Golm Metabolome Database* (http://csbdb.mpimp-golm.mpg.de/csbdb/gmd/gmd.html).
108. Human Metabolome Project. (2006): *Human Metabolite Database* (http://www.hmdb.ca/).
109. METLIN. (2006): *METLIN Metabolite Database* (http://metlin.scripps.edu/).
110. The National Center for Plant and Microbial Metabolomics. (2006): *NMR Spectral Database* (http://www.metabolomics.bbsrc.ac.uk/currentactivities.htm).
111. Metabolomics Database of Linkoping, Sweden. (2006): *NMR Metabolite Database* (http://mdl.imv.liu.se/main/).
112. Crockford, D. J. et al. (2006): Statistical heterospectroscopy, an approach to the integrated analysis of NMR and UPLC–MS data sets: application in metabonomic toxicology studies. *Anal. Chem.*, 78(2):363–371.
113. GenomeNet. (2006): *Kegg Pathway Database* (http://www.genome.jp/kegg/pathway.html).
114. ExPASy. (2006): *ExPASy Biochemical Pathways* (http://www.expasy.org/cgi-bin/show_thumbnails.pl).
115. MetaCyc. (2006): *MetaCyc Encyclopedia of Metabolic Pathways* (http://metacyc.org/).
116. International Union of Biochemistry and Molecular Biology. (2006): *IUBMB–Nicholson Animaps* (http://www.tcd.ie/Biochemistry/IUBMB-Nicholson/animaps.html).
117. Hirai, M. Y. et al. (2004): Integration of transcriptomics and metabolomics for understanding of global responses to nutritional stresses in Arabidopsis thaliana [see comment]. *Proc. Natl. Acad. Sci. USA*, 101(27): 10205–10210.
118. Verhoeckx, K. C. et al. (2004): A combination of proteomics, principal component analysis and transcriptomics is a powerful tool for the identification of biomarkers for macrophage maturation in the U937 cell line. *Proteomics*, 4(4):1014–1028.
119. Matsuzaki, K. et al. (2005): Transcriptomics and metabolomics of dietary leucine excess. *J. Nutr.*, 135(6, Suppl.): 1571S–1575S.
120. Fiehn, O., Kloska, S., and Altmann, T., Integrated studies on plant biology using multiparallel techniques. *Curr. Opin. Biotechnol.*, 12(1):82–86.
121. Fiehn, O., and Weckwerth, W., Deciphering metabolic networks. *Eur. J. Biochem.*, 270(4):579–588.
122. Roberts, J. K., and Xia, J. H. (1995): High-resolution NMR methods for study of higher plants. *Meth. Cell. Biol.*, 49:245–258.
123. Trethewey, R. N. (2004): Metabolite profiling as an aid to metabolic engineering in plants. *Curr. Opin. Plant Biol.*, 7(2):196–201.
124. Bundy, J. G., Ramlov, H., and Holmstrup, M. (2003): Multivariate metabolic profiling using $^1$H nuclear magnetic resonance spectroscopy of freeze-tolerant and freeze-intolerant earthworms exposed to frost. *Cryo-Letters*, 24(6): 347–358.
125. Bundy, J. G. et al. (2002): Earthworm species of the genus *Eisenia* can be phenotypically differentiated by metabolic profiling. *FEBS Lett.*, 521(1–3):115–120.
126. Viant, M. R. et al. (2002): Utilizing *in vivo* nuclear magnetic resonance spectroscopy to study sublethal stress in aquatic organisms. *Mar. Environ. Res.*, 54(3–5):553–557.
127. Bundy, J. G. et al. (2002): Metabonomic assessment of toxicity of 4-fluoroaniline, 3,5-difluoroaniline and 2-fluoro-4-methylaniline to the earthworm *Eisenia veneta* (Rosa): identification of new endogenous biomarkers. *Environ. Toxicol. Chem.*, 21(9):1966–1972.

128. Viant, M. R. et al. (2002): Sublethal actions of copper in abalone (*Haliotis rufescens*) as characterized by *in vivo* $^{31}$P NMR. *Aquat. Toxicol.*, 57(3):139–151.
129. Gavaghan, C. L. et al. (2000): An NMR-based metabonomic approach to investigate the biochemical consequences of genetic strain differences: application to the C57BL10J and Alpk:ApfCD mouse. *FEBS Lett.*, 484(3):169–174.
130. Plumb, R. et al. (2003): Metabonomic analysis of mouse urine by liquid chromatography–time of flight mass spectrometry (LC–TOFMS): detection of strain, diurnal and gender differences. *Analyst*, 128(7):819–823.
131. Robosky, L. C. et al. (2005): Metabonomic identification of two distinct phenotypes in Sprague–Dawley (Crl:CD(SD)) rats. *Toxicol. Sci.*, 87(1):277–284.
132. Griffin, J. L. (2004): Metabolic profiles to define the genome: can we hear the phenotypes? *Philos. Trans. R. Soc. Lond. B Biol. Sci.*, 359(1446):857–871.
133. Phelps, T. J., Palumbo, A. V., and Beliaev, A. S. (2002): Metabolomics and microarrays for improved understanding of phenotypic characteristics controlled by both genomics and environmental constraints. *Curr. Opin. Biotechnol.*, 13(1):20–24.
134. Zeisel, S. H. et al. (2005): The nutritional phenotype in the age of metabolomics. *J. Nutr.*, 135(7):1613–1616.
135. Davis, C. D., and Milner, J. (2004): Frontiers in nutrigenomics, proteomics, metabolomics and cancer prevention. *Mutat. Res.*, 551(1–2):51–64.
136. German, J. B., Roberts, M. A., and Watkins, S. M. (2003): Personal metabolomics as a next generation nutritional assessment. *J. Nutr.*, 133(12):4260–4266.
137. Go, V. L., Butrum, R. R., and Wong, D. A. (2003): Diet, nutrition, and cancer prevention: the postgenomic era. *J. Nutr.*, 133(11, Suppl. 1):3830S–3836S.
138. German, J. B. et al. (2002): Metabolomics and individual metabolic assessment: the next great challenge for nutrition. *J. Nutr.*, 132(9):2486–2487.
139. Whitfield, P. D., German., A. J., and Noble, P. J. (2004): Metabolomics: an emerging post-genomic tool for nutrition. *Br. J. Nutr.*, 92(4):549–555.
140. Raso, G. et al. (2004): Multiple parasite infections and their relationship to self-reported morbidity in a community of rural Cote d'Ivoire. *Int J. Epidemiol.*, 33(5): 1092–1102.
141. Cleary, M. D. et al. (2002): *Toxoplasma gondii* asexual development: identification of developmentally regulated genes and distinct patterns of gene expression. *Eukaryot. Cell*, 1(3):329–340.
142. Wang, Y. et al. (2004): Metabonomic investigations in mice infected with *Schistosoma mansoni*: an approach for biomarker identification. *Proc. Natl. Acad. Sci. USA*, 101(34): 12676–12681.
143. Anthony, M. L. et al. (1995): Classification of toxin-induced changes in $^1$H NMR spectra of urine using an artificial neural network. *J. Pharm. Biomed. Anal.*, 13(3):205–211.
144. Holmes, E. et al. (1998): Development of a model for classification of toxin-induced lesions using $^1$H NMR spectroscopy of urine combined with pattern recognition. *NMR Biomed.*, 11(4–5):235–244.
145. Robertson, D. G. et al. (2000): Metabonomics: evaluation of nuclear magnetic resonance (NMR) and pattern recognition technology for rapid *in vivo* screening of liver and kidney toxicants. *Toxicol. Sci.*, 57(2):326–337.
146. Lindon. J. C. et al. (2005): The Consortium for Metabonomic Toxicology (COMET) aims, activities, and achievements. *Pharmacogenomics*, 6(7):691–699.
147. Lindon, J. C. et al. (2003): Contemporary issues in toxicology the role of metabonomics in toxicology and its evaluation by the COMET project. *Toxicol. Appl. Pharmacol.*, 187(3):137–146.
148. Kerns., W. et al. (2005): Drug-induced vascular injury: a quest for biomarkers. *Toxicol. Appl. Pharmacol.*, 203(1): 62–87.
149. Robertson, D. G. et al. (2001): Metabonomic assessment of vasculitis in rats. *Cardiovasc. Toxicol.*, 1(1):7–19.
150. Slim, R. M. et al. (2002): Effect of dexamethasone on the metabonomics profile associated with phosphodiesterase inhibitor-induced vascular lesions in rats. *Toxicol. Appl. Pharmacol.*, 183(2):108–109.
151. Connor, S. C. et al. (2004): Development of a multivariate statistical model to predict peroxisome proliferation in the rat, based on urinary $^1$H-NMR spectral patterns. *Biomarkers*, 9(4–5):364–385.
152. Nicholls, A. W. et al. (2000): A metabonomic approach to the investigation of drug-induced phospholipidosis: an NMR spectroscopy and pattern recognition study. *Biomarkers*, 5:410–423.
153. Anthony, M. L. et al. (1994): Studies of the biochemical toxicology of uranyl nitrate in the rat. *Arch. Toxicol.*, 68(1):43–53.
154. Gartland, K. P., Bonner, F. W., and Nicholson, J. K. (1989): Investigations into the biochemical effects of region-specific nephrotoxins. *Mol. Pharmacol.*, 35(2):242–250.
155. Nicholls, A. W. et al. (2001): Metabonomic investigations into hydrazine toxicity in the rat. *Chem. Res. Toxicol.*, 14(8):975–987.
156. Mortishire-Smith, R. J. et al. (2004): Use of metabonomics to identify impaired fatty acid metabolism as the mechanism of a drug-induced toxicity. *Chem. Res. Toxicol.*, 17(2):165–173.
157. Clayton, T. A. et al. (2003): An hypothesis for a mechanism underlying hepatotoxin-induced hypercreatinuria. *Arch. Toxicol.*, 77(4):208–217.
158. Clayton, T. A. et al. (2004): Hepatotoxin-induced hypercreatinaemia and hypercreatinuria: their relationship to one another, to liver damage and to weakened nutritional status. *Arch Toxicol.*, 78(2):86–96.
159. Connor, S. C. et al. (2004): Effects of feeding and body weight loss on the $^1$H-NMR-based urine metabolic profiles of male Wistar Han rats: implications for biomarker discovery. *Biomarkers*, 9(2):156–179.
160. Brazma, A. et al. (2001): Minimum information about a microarray experiment (MIAME): toward standards for microarray data. *Nat. Genet.*, 29(4):365–371.
161. Lindon, J. C. et al. (2005): Summary recommendations for standardization and reporting of metabolic analyses. *Nat. Biotechnol.*, 23(7):833–838.
162. Zerhouni, E. (2003): Medicine: the NIH RoadMap. *Science*, 302(5642):63–72.

163. National Institutes of Health. (2006): *The NIH RoadMap Initiative* (http://nihroadmap.nih.gov/initiatives.asp).
164. The Metabolomics Society. (2006): http://www.metabolomicssociety.org/.
165. Whittaker, P. A. (2003): What is the relevance of bioinformatics to pharmacology? *Trends Pharmacol. Sci.*, 24(8):434.
166. Bailey, W. J., and Ulrich, R. (2004): Molecular profiling approaches for identifying novel biomarkers. *Expert Opin. Drug Safety*, 3(2):137.
167. Butcher, E. C., Berg, E. L., and Kunkel, E. J. (2004): Systems biology in drug discovery. *Nat. Biotechnol.*, 22(10):1253.
168. Tong, W., Harris, S., Cao, X., Fang, H., Shi, L., Sun, H., Fuscoe, J., Harris, A., Hong, H., Xie, Q., Perkins, R., and Casciano, D. (2004): Development of public toxicogenomics software for microarray data management and analysis. *Mutat. Res.*, 549(1–2):241.
169. Rocca-Serra, P., Brazma, A., Parkinson, H., Sarkans, U., Shojatalab, M., Contrino, S., Vilo, J., Abeygunawardena, N., Mukherjee, G., Holloway, E., Kapushesky, M., Kemmeren, P., Lara, G. G., Oezcimen, A., and Sansone, S. A. (2003): ArrayExpress: a public database of gene expression data at EBI. *Crit. Rev. Biol.*, 326(10–11):1075.
170. Pennie, W., Pettit, S. D., and Lord. P. G. (2004): Toxicogenomics in risk assessment: an overview of an HESI collaborative research program. *Environ. Health Perspect.*, 112(4):417.
171. Twigger, S. N., Nie, J., Ruotti, V., Yu, J., Chen, D., Li, D., Mathis, J., Narayanasamy, V., Gopinath, G. R., Pasko, D., Shimoyama, M., De La Cruz, N., Bromberg, S., Kwitek, A. E., Jacob, H. J., and Tonellato, P. J. (2004): Integrative genomics: *in silico* coupling of rat physiology and complex traits with mouse and human data. *Genome Res.*, 14(4):651.
172. Selvanayagam, Z. E., Cheung, T. H., Wei, N., Vittal, R., Kit Lo, K. W., Yeo, W., Kita, T., Ravatn, R., Hung Chung, T. K., Wong, Y. F., and Chin, K. V. (2004): Prediction of chemotherapeutic response in ovarian cancer with DNA microarray expression profiling. *Cancer Genet. Cytogenet.*, 154(1):63.
173. Kanehisa, M., Goto, S., Kawashima, S., Okuno, Y., and Hattori, M. (2004): The KEGG resource for deciphering the genome. *Nucleic Acids Res.*, 32:D277.
174. Ge, H., Walhout, A. J., and Vidal, M. (2003): Integrating 'omic' information: a bridge between genomics and systems biology. *Trends Genet.*, 19(10):551.
175. Ozsoyoglu, Z. M., Nadeau, J. H., and Ozsoyoglu, G. (2003): Pathways database system. *Omics*, 7(1):123.
176. Kinoshita, K., and Nakamura, H. (2003): Protein informatics: toward function identification. *Curr. Opin. Struct. Biol.*, 13(3):396–400.
177. Wu, C. H., Huang, H., Yeh, L. S., and Barker, W. C. (2003): Protein family classification and functional annotation. *Comput. Biol. Chem.*, 27(1):37.
178. Brazma, A., and Vilo, J. (2001): Gene expression data analysis. *Microbes Infect.*, 3:823–829.
179. Ermolaeva, O., Rastogi, M., Pruitt, K. D., Schuler, G. D., Bittner, M. L., Chen, Y., Simon, R., Meltzer, P., Trent, J. M., and Boguski, M. S. (1998): Data management and analysis for gene expression arrays. *Nat. Genet.*, 20(1):19–23.
180. Johnson, K. F., and Lin, S. M. (2001): Call to work together on microarray data analysis. *Nature*, 411(6840):885.
181. Johnson, K. F. and Lin, S. M. (2001): Critical assessment of microarray data analysis: the 2001 challenge. *Bioinformatics*, 17:857–885.


## Further Reading

### General: Individualized Medicine


Ardekani, A. M., Petricoin, E. F. 3rd, and Hackett, J. L. (2003): Molecular diagnostics: an FDA perspective. *Expert Rev. Mol. Diagn.*, 3(2):129.

Evans, W. E., and Relling, M. V. (2004): Moving towards individualized medicine with pharmacogenomics. *Nature*, 429(6990):464.

Halapi, E., Stefansson, K., and Hakonarson, H. (2004): Population genomics of drug response. *Am. J. Pharmacogenom.*, 4:73.

International Human Genome Sequencing Consortium. (2004): Finishing the euchromatic sequence of the human genome. *Nature*, 431(7011):931.

Lesko, L. J., and Woodcock, J. (2004): Translation of pharmacogenomics and pharmacogenetics: a regulatory perspective. *Nat. Rev. Drug Discov.*, 3(9):763.

Nebert, D. W., and Vesell, E. S. (2004): Advances in pharmacogenomics and individualized drug therapy: exciting challenges that lie ahead. *Eur. J. Pharmacol.*, 500(1–3):267.

Smith, L. L. (2001): Key challenges for toxicologists in the 21st century. *Trends Pharmacol. Sci.*, 22(6):281–285.

Weinshilboum, R., and Wang, L. (2004): Pharmacogenomics: bench to bedside. *Nat. Rev. Drug Discov.*, 3(9):739.


### Microarray Applications, Intergration, and Data Analyses


Baker, T. K., Carfagna, M. A., Gao, H., Dow, E. R., Li, Q., Searfoss, G. H., and Ryan, T. P. (2001). Temporal gene expression analysis of monolayer cultured rat hepatocytes. *Chem. Res. Toxicol.*, 14:1218–1231.

Burczynski, M. E., McMillian, M., Ciervo, J., Li, L., Parker, J B., Dunn, R. T. 2nd., Hicken, S., Farr, S., and Johnson, M. D. (2000): Toxicogenomics-based discrimination of toxic mechanism in HepG2 human hepatoma cells. *Toxicol. Sci.* 58:399–415.

Pennie, W. D., Woodyatt, N. J., Aldridge, T. C., and Orphanides G. (2001): Application of genomics to the definition of the molecular basis for toxicity. *Toxicol. Lett.*, 120:353.

Thomas, R. S., Rank, D. R., Penn, S. G., Zastrow, G. M., Hayes K. R., Pande, K., Glover, E., Silander, T., Craven, M. W. Reddy, J. K., Jovanovich, S. B., and Bradfield, C. A. (2001): Identification of toxicologically predictive gene sets using cDNA microarrays. *Mol. Pharmacol.*, 60(6):1189–1194.

Ulrich, R., and Friend, S. H. (2004): Toxicogenomics and drug discovery: will new technologies help us produce better drugs? *Nat. Rev. Drug Discov.*, 1(1):84.

Waring, J. F., Jolly, R. A., Ciurlionis, R., Lum, P. Y., Praestgaard J. T., Morfitt, D. C., Buratto, B., Roberts, C., Schadt, E. and Ulrich, R.G. (2001): Clustering of hepatotoxins based on mechanism of toxicity using gene expression profiles *Toxicol. Appl. Pharmacol.*, 175(1):28–42.

## Proteomics


Harris, R.A., Yang, A., Stein, R. C., Lucy, K., Brusten, L., Herath, A., Parekh, R., Waterfield, M. D., O'Hare, M. J., Neville, M. A., Page, M. J., and Zvelebil, M. J. (2002): Cluster analysis of an extensive human breast cancer cell line protein expression map database. *Proteomics*, 2(2):212.


## Metabonomics


Forster, J. et al. (2002): A functional genomics approach using metabolomics and in silico pathway analysis. *Biotechnol. Bioeng.*, 79:703.

Lindon., J. et al. (2004): Metabonomics technologies and their applications in physiological monitoring, drug safety assessment and disease diagnosis. *Biomarkers*, 9:1.

Nicholson, J. K. et al. (2002): Metabonomics: a platform for studying drug toxicity and gene function. *Nat. Rev. Drug Discov.*, 1(2):153.

Reo, N. V. (2002): NMR-based metabolomics. *Drug Chem., Toxicol.*, 25(4):375.

Robertson, D. G., Reily, M. D., Albassam, M., and Dethloff, L. A. (2001): Metabonomic assessment of vasculitis in rats. *Cardiovasc. Toxicol.*, 1(1):7.

Robosky, L. C. et al. (2002): *In vivo* toxicity screening programs using metabonomics. *Comb. Chem. High Throughput Screen*, 5:651.

Slim, R. M., Robertson, D. G., Albassam, M., Reily, M. D., Robosky, L., and Dethloff, L. A. (2002): Effect of dexamethasone on the metabonomics profile associated with phosphodiesterase inhibitor induced vascular lesions in rats. *Toxicol. Appl. Pharmacol.*, 183(2):108.


## Integrated Data Mining


Bassett, D. E., Jr., Eisen, M. B., and Boguski, M. S. (1999): Gene expression informatics: it's all in your mind. *Nat. Genet.*, 21(Suppl. 1):51–55.

Coen, M. S. et al. (2004): Integrated application of transcriptomics and metabonomics yields new insight into the toxicity due to paracetamol in the mouse. *J. Pharm. Biomed. Anal.*, 35(1):93.

Parsons, L., and Orban, J. (2004): Structural genomics and the metabolome: combining computational and NMR methods to identify target ligands. *Curr. Opin. Drug Discov. Devel.*, 7(1):62.

Yu, J. K., Chen, Y. D., and Zheng, S. (2004): An integrated approach to the detection of colorectal cancer utilizing proteomics and bioinformatics. *World J. Gastroenterol.*, 10(21):3127.


## Toxicopanomic Databases


Burgoon, L. D., Boutros, P. C., Dere, E., Zacharewski, T. R. (2006): dbZach: a MIAME-compliant toxicogenomic supportive relational database. *Toxicol. Sci.*, 90(2):558–568.

Fonger, G. C., Stroup, D., Thomas, P. L., Wexler, P. (2000): TOXNET: a computerized collection of toxicological and environmental health information. *Toxicol. Indust. Health*, 16(1):4–6.

Fostel, J. et al. (2005): Chemical effects in biological systems–data dictionary (CEBS-DD): a compendium of terms for the capture and integration of biological study design description., conventional phenotypes, and omics data. *Toxicol. Sci.*, 88(2):585–601.

Ganter, B. et al. (2005): Development of a large-scale chemogenomics database to improve drug candidate selection and to understand mechanisms of chemical toxicity and action. *J. Biotechnol.*, 119(3):219–244.

Hayes, K. R., Vollrath, A. L., Zastrow, G. M., McMillan, B. J., Craven, M., Jovanovich, S., Rank, D. R., Penn, S., Walisser, J. A., Reddy, J. K., Thomas, R. S., and Bradfield, C. A. (2005): EDGE: a centralized resource for the comparison, analysis, and distribution of toxicogenomic information. *Mol. Pharmacol.*, 67(4): 1360–1368.

Lindon, J. C., Keun, H. C., Ebbels, T. M., Pearce, J. M., Holmes, E., Nicholson, J. K. (2005): The Consortium for Metabonomic Toxicology (COMET): aims, activities, and achievements. *Pharmacogenomics*, 6(7):691–699.

Mattingly, C. J., Colby, G. T., Rosenstein, M. C., Forrest, J. N., Jr., and Boyer, J. L. (2004): Promoting comparative molecular studies in environmental health research: an overview of the comparative toxicogenomics database (CTD). *Pharmacogenom. J.*, 4(1):5–8.

Pennie, W., Pettit, S. D., and Lord, P. G. (2004): Toxicogenomics in risk assessment: an overview of an HESI collaborative research program. *Environ. Health Perspect.*, 112(4): 417–419.

Salter, A. H. (2005): Large-scale databases in toxicogenomics. *Pharmacogenomics*, 6(7):749–754.

Tomasulo, P. (2005): ITER, International Toxicity Estimates for Risk: new TOXNET database. *Med. Ref. Serv. Q.*, 24(1): 55–66.

Tong, W., Harris, S., Cao, X., Fang, H., Shi, L., Sun, H., Fuscoe, J., Harris, A., Hong, H., Xie, Q., Perkins, R., and Casciano, D. (2004): Development of public toxicogenomics software for microarray data management and analysis. *Mutat. Res.*, 549(1–2):241–253.

Waters, M., Boorman, G., Bushel, P., Cunningham, M., Irwin, R., Merrick, A., Olden, K., Paules, R., Selkirk, J., Stasiewicz, S., Weis, B., Van Houten, B., Walker, N., and Tennant, R. (2003): Systems toxicology and the Chemical Effects in Biological Systems (CEBS) knowledge base. *EHP Toxicogenom.*, 111(1T):15–28.

Wexler, P. (2004): The U.S. National Library of Medicine's Toxicology and Environmental Health Information Program. *Toxicology*, 198(1–3):161–168.

Wexler, P. (2001): TOXNET: an evolving web resource for toxicology and environmental health information. *Toxicology*, 157(1–2):3–10.


## Predictive Toxicopanomics


Fielden, M. R., and Kolaja, K.L. (2006): The state of the art in predictive toxicogenomics. *Curr. Opin. Drug Discov. Devel.*, 9(1):84–91.

Maggioli, J., Hoover, A., and Weng, L. (2005): Toxicogenomic analysis methods for predictive toxicology. *J. Pharmacol. Toxicol. Meth.*, 53(1):31–37.

Sawada, H., Takami, K., and Asahi, S. (2005): A toxicogenomic approach to drug-induced phospholipidosis: analysis of its induction mechanism and establishment of a novel *in vitro* screening system. *Toxicol. Sci.*, 83(2):282–92 (erratum in *Toxicol. Sci.*, 89(2), 554, 2006).

Yang, Y., Abel, S. J., Ciurlionis, R., and Waring, J. F. (2006): Development of a toxicogenomics *in vitro* assay for the efficient characterization of compounds. *Pharmacogenomics*, 7(2):177–186.

# 7 The Toxicologic Assessment of Pharmaceutical and Biotechnology Products

*Michael A. Dorato, Carl L. McMillian, and Mary Jo Vodicnik*

## CONTENTS

## GENERAL OVERVIEW OF DRUG DEVELOPMENT

The World Health Organization (WHO) Scientific Group has defined a drug as "any substance or product that is used or intended to be used to modify or explore physiological systems or pathological states for the benefit of the recipient" [154]. The drug discovery process covers a wide range of therapeutic areas and treatment regimens and is a risky, multifaceted, expensive undertaking. The goal is to develop a new product with therapeutic benefits (efficacy) and few side effects (toxicity) [4]. The drug development process for a new chemical entity (NCE) starts at the chemist's bench with its isolation, moves through efficacy pharmacology testing using various *in vivo* and *in vitro* models, then proceeds through an abbreviated toxicology profile, including pharmacologic profiling (the determination of pharmacologic effects other than the desired therapeutic effect), based on the proposed clinical plan for the first human dose (FHD).

The principal aim of nonclinical safety testing is to understand the toxicity of the candidate drug well enough to make a judgment that the risk/benefit profile is adequate to initiate clinical trials [111]. Provided the efficacy pharmacology and initial toxicology profiles are acceptable, clinical safety, pharmacokinetic, and pharmacodynamic studies (phase I studies) are initiated. As the human clinical trials progress through phase II (proof of concept and safety studies) and phase III (pivotal registration studies), the drug candidate moves through nonclinical subchronic studies, chronic and developmental toxicology studies, and oncogenic evaluations. Zbinden [241] has provided a summary of the biological parameters that should be evaluated for new drug candidates (Table 7.1).

Accelerating the development of safe and effective drugs is not a new topic to the pharmaceutical industry [38]. The technical risks in new drug development programs are enormous. Drug development is complicated by

**TABLE 7.1**
**Biological Properties of Chemicals That Should Be Considered in Safety Evaluations**

| |
|---|
| Acute toxicity |
| Cumulative toxicity |
| Absorption from various routes |
| Elimination $t_{1/2}$ and accumulation in deep compartments |
| Penetration of barriers |
| Milk excretion |
| Teratogenicity |
| Mutagenicity |
| Carcinogenicity |
| Sensitization |
| Local irritation |

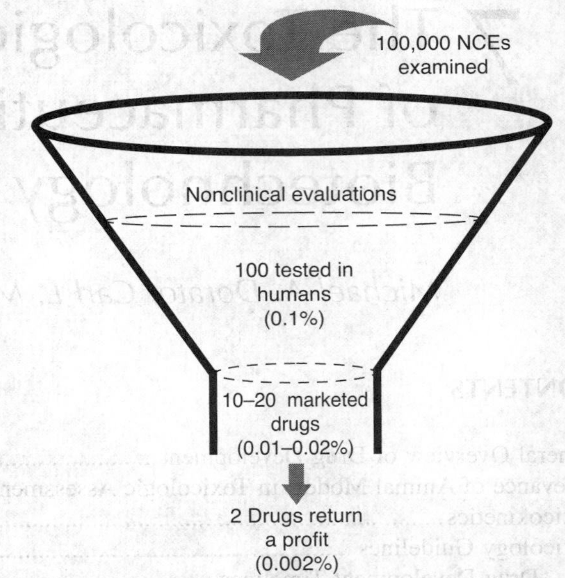


**FIGURE 7.1** Attrition rate of new drug candidates. (Data from Chien [35] and DiMasi et al. [67,68].)

the requirement to simultaneously address complex issues related to potency, selectivity, reversibility, solubility, duration, metabolic stability, permeability, toxicity, physical stability, patentability, and manufacturability for each drug candidate. The risk of failure related to one or more of these aspects has been reviewed by Chien [35], where it was reported that <0.02% of NCEs result in marketed drug products, and even fewer (i.e., 0.002%) return a profit to support continued drug research (Figure 7.1). Very little of what enters the drug development pipeline ever enters the marketplace [67]. It has been estimated that the cost of developing a NCE could exceed $1.7 billion [171].

By its nature, a drug must modify a biological process (i.e., alter or adjust a physiologic system in some way) [44]. Toxicology is a critical part of both early- and late-phase drug development, and the role of toxicology in that process has been extensively reviewed [65,70,102]. The initial purpose of toxicology testing programs is to identify the circumstances (e.g., dose, treatment duration, route) under which a NCE produces potentially harmful effects [45]. A general approach to developing a toxicity profile for a pharmaceutical agent is given in Figure 7.2 and will be discussed more extensively below.

During the discovery process, toxicologists employ rapid, quantitative screening methods, focusing on a limited spectrum of toxicity, to help identify, out of a myriad of molecules, potential drug candidates with the best safety profile. These early screening procedures, however, are only a prelude to the required comprehensive safety assessments expected from the toxicologist. Regulatory requirements, termed good laboratory practices (GLPs), dictate many aspects of the toxicology study protocol and

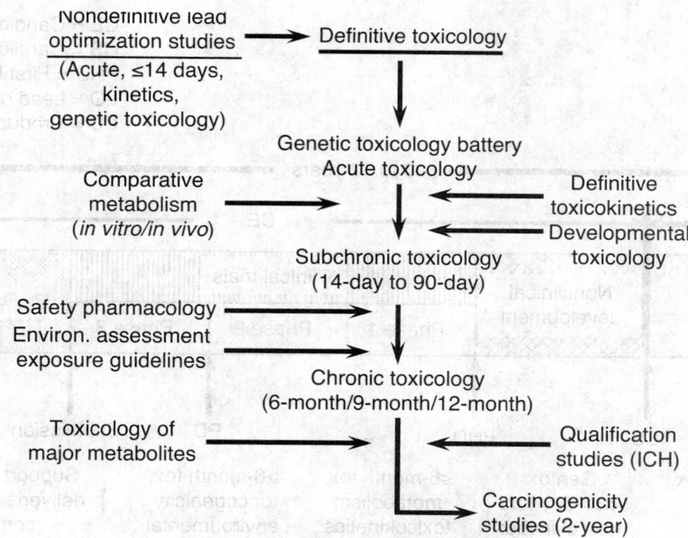


**FIGURE 7.2** General approach to developing a toxicity profile for pharmaceutical agents.

must be followed closely for all definitive (those that support human studies) toxicology studies [80]. The early toxicology studies, conducted in the discovery phase, are not required to be in full compliance with the GLPs, but good research practices ensuring validation, documentation, reproducibility, etc. should be applied.

Prior to initiating clinical trials, physicians need to understand the toxicologic profile produced by the drug candidate in relevant animal models. Prior to the FHD, determination of the relevance of animal models (e.g., metabolism relative to humans) is limited to *in vitro* evaluations using human and animal tissue preparations. On the basis of the clinicians' needs, the toxicology profile of a NCE is characterized by a number of questions [155]:

- What dose/exposure produces toxic effects in animals?
- What dose/exposure does not produce toxic effects in animals?
- Were the animals relevant models for predicting human toxicity?
- What were the signs and duration of toxic responses?
- Did effects differ following single or multiple dosing?
- Were the toxic responses reversible?
- What were the target organs or systems?
- Was the toxicity expected for this chemical class?
- Are toxic metabolites produced?
- Was accommodation to the toxic effects observed?

The answers to these questions form the basis of the toxicology profile supporting initial and continued clinical trials.

The major objectives of toxicologic evaluation change according to the stage of the development process [211]. The relationships of the studies used to develop a toxicology profile (Figure 7.2) to clinical trial phases are shown in Figure 7.3. The early stages of development focus on toxicologic screening (Table 7.2). Definitive toxicology studies are very time consuming and costly; thus, relatively inexpensive, short-term screening procedures are used to eliminate the most toxic compounds [240]. Inherent to these initial approaches to evaluate potential drug toxicity are a number of imperfections: The affected systems may not be routinely examined; the assay procedures are inadequate or improperly timed relative to the onset of the toxic response; target organ exposure is insufficient; identifying and measuring adverse effects (lack of functional evaluations) are not possible; predicting metabolic, anatomic, and physiologic differences between species is not possible; and the test animals may not express human-specific responses [244]. There is no simple answer to the often-asked question: "What toxicity profile would cause a company to stop development of a new drug candidate?" [124]. However, the demonstrated toxicity of other compounds in the class, if available, and the gravity of the disease state under study often provide guidance as to what might be an acceptable safety profile for a NCE. A comprehensive overview of the international pharmaceutical industry's nonclinical testing strategies, in relation to clinical trial phase, is provided in the *Pharmaceutical R&D Compendium* [95].

Varied opinions on the occurrence of drug toxicity in the human population have been reported [39,146]. Although the magnitude of adverse responses seems small (i.e., ~1 per 10,000 patients demonstrate untoward adverse reactions [146]), focus must be placed on the identification of all potentially adverse effects [32].

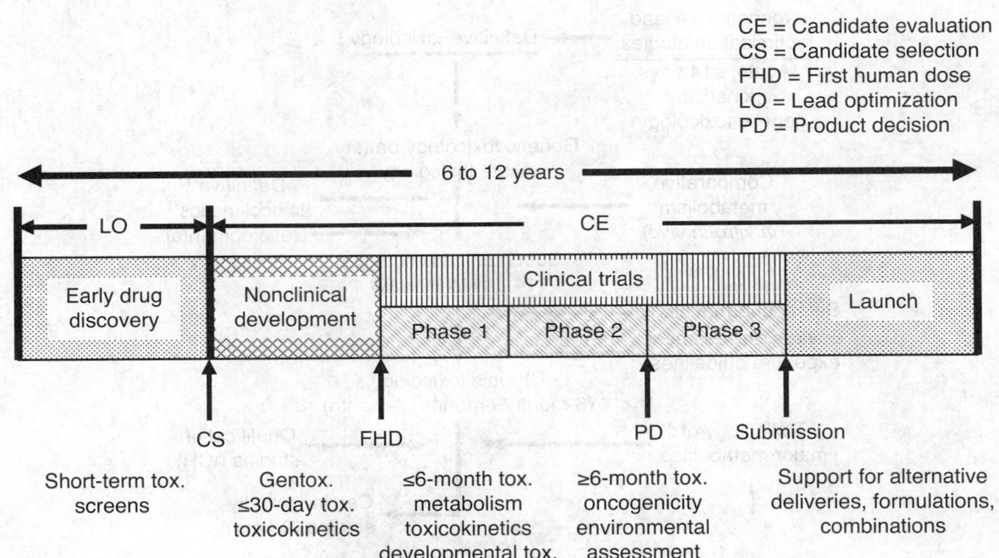


**FIGURE 7.3** Duration of toxicology involvement in drug discovery and development showing the major milestones and study types by phase. The goal is to reduce development time while maintaining a focus on nonclinical and clinical safety assessment.

In addition to the drug substance, the delivery system may also require nonclinical evaluation because it may alter pharmacodynamic (action of the drug on the body) and pharmacokinetic (action of the body on the drug) relationships. The regulatory requirements for the safety assessment of known and novel drug delivery systems have been reviewed by Weissinger [226].

Commercially advantageous forms of genetic manipulation date back to antiquity (e.g., inbreeding, cross fertilization). The introduction of modern recombinant DNA (rDNA) technology has had a major impact on life science research and has allowed for the large-scale production of protein pharmacologic agents that would have been very difficult to produce by normal chemical synthetic means.

A new biological entity (NBE) is defined as a complex, high-molecular-weight material that cannot be fully characterized by standard chemical analysis and which may require immunologic, biochemical, or bioassay techniques to measure the quantity present and to assess activity [47]. The development, utility, and relative safety of human insulin [105], human growth hormone [166], and interferon [173] have fueled the current interest in the production of biologically active peptides. As the interests of pharmacologists in biotechnology research expand, the difficulties of producing a comprehensive set of safety guidelines increase. Legal definitions of biotechnology products and regulatory guidance for nonclinical assessments have been reviewed by Tsang and Beers [218].

When NBEs were introduced, beginning with insulin, regulatory concepts were not in place to address the problems forthcoming from this new technology [242]. The regulatory issues relating to recombinant products are formidable, and the possibility that each biotechnology product might require customized safety testing has been given serious consideration [52,199,206]. Clearly, the immunologic response to foreign proteins may compromise the utility of using traditional animal models in the safety assessment of these agents. Other major regulatory issues include the assurance that recombinant production methods do not result in addition of contaminants and the demonstration to regulators that biosynthetic products are identical to natural substances [104]. In the biosynthetic

**TABLE 7.2**
**Purpose of Toxicology Evaluations of New Drugs**

| Phase | Principal Activity | Purpose |
|---|---|---|
| Discovery | Identification of candidates | Toxicologic screening |
| Before first human dose (FHD) | Safety and principal target organs | Regulatory prerequisites for human exposure |
| During clinical trial | Toxicologic spectrum | Cumulative effects and mechanisms |
| Premarketing | Complete routine test program | Regulatory requirements |
| Postmarketing | Identify special risks due to population or use circumstances | Improve utility and safety |

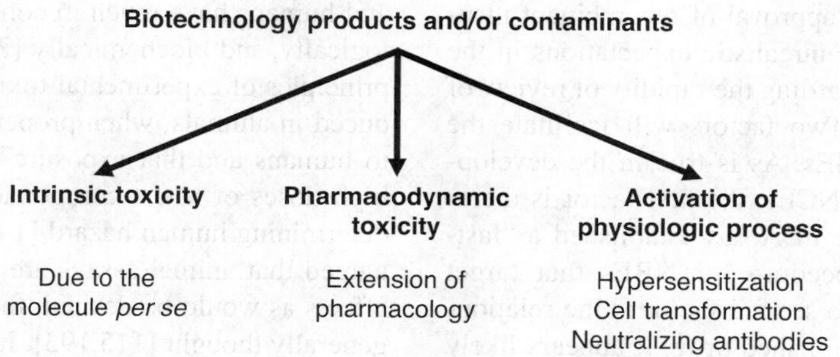


**FIGURE 7.4** Main areas of concern with response to biotechnology products. (Adapted from Zbinden, G., in *Nonclinical Safety of Biotechnology Products for Human Use*. Graham, C. E., Ed., Alan R. Liss, New York, 1987, pp. 143–159.)

human insulin approval process, meetings between regulatory agency and industry scientists to review the manufacturing process, molecular biology, and purification of the hormone, as well as clinical trial programs, were critical in facilitating eventual approvals. Industry and regulatory agency representatives agreed that the chemistry of a NBE should prove its identity [223]. The identity and purity of rDNA insulin, therefore, received much attention [33]. Anticipation of problems and the communication of concerns were key to the rapid New Drug Application (NDA) approval for biosynthetic human insulin (5.5 months). The U.S. Food and Drug Administration (FDA) has strongly recommended that it be involved early in the nonclinical and clinical development plan to facilitate the approval process for both NCEs and NBEs [109].

The U.S. biotechnology policy states that "the same physical and biological laws govern the response of organisms modified by modern molecular and cellular methods and those produced by classical methods … no conceptual distinction exists between genetic modification of plants and microorganisms by classical methods or by molecular techniques that modify DNA and transfer genes" [219]. Thus, it would not be expected that NBEs *per se* pose an unusual risk to human health and the environment [219]. The toxicologist should be aware, however, that compounds made via rDNA techniques are not necessarily identical to the natural material, as might be assumed [232]. Dayan [56] suggested that the toxicology profile for a NBE should be defined in terms of chemical identity of the material, extent of prior knowledge, and intended use. The U.S. Pharmaceutical Research and Manufacturers Association (PhRMA) has recommended that nonclinical toxicologic evaluations of NBEs should be decided on a case-by-case basis [206], and regulatory and industry representatives attending the first International Conference on Harmonisation of Technical Requirements for Registration of Pharmaceuticals for Human Use (ICH) have also supported this position [138]. The established toxicology information will guide the clinical trial and address possible hazards in the workplace, where humans are exposed to the compound, and its precursors and contaminants that are also contained in the bulk material to be tested during the chemical or biological synthetic process.

The role of the toxicologist is usually less routine and requires more innovation in study design when dealing with NBEs than when dealing with NCEs [57]. As is the case with NCEs, however, the principal goals of the toxicologic evaluation of recombinant products are to detect major toxicity, to identify lesser toxicity, to determine the dose relationship of toxic effects and their duration to guide the clinical dose schedule, and to investigate the mechanisms of action related to the toxic response. Cavagnaro [31] has provided a comprehensive overview of the challenges and approaches to nonclinical safety evaluations of NBEs.

The three main areas of concern relative to the toxicity of NBEs are toxicity *per se*, exaggerated pharmacodynamic effects (anticipated toxicity based on the pharmacologic mechanism of action), and allergic reactions (Figure 7.4) [242]. *Intrinsic toxicity* has been defined as undesirable effects having no obvious relationship to the molecule's pharmacodynamic properties. *Pharmacodynamic toxicity* is defined as an exaggerated pharmacologic response (i.e., hypoglycemic shock from insulin). *Immunotoxicity* has been related to hypersensitivity, cell transformation, and the production of neutralizing antibodies. The loss of the biological activity of a recombinant therapeutic agent through production of neutralizing antibodies and the development of immune complex disease in experimental animals are factors that must be given individual attention [206]. It has been suggested that animal models of immunotoxicity are of limited usefulness in that no animal model may be fully suitable for predicting the toxicity of highly species-specific proteins. Friedmann [100] has indicated, however, that the lack of hypersensitivity reactions in response to small peptides in animal experiments may be viewed as an indication of their acceptability in humans. Graham [110] emphasized the use of a case-by-case approach to toxicologic evaluations of NBEs based on their similarity to natural human proteins, immune response in animal models, and production of neutralizing antibodies in nonclinical and clinical studies.

The unique regulatory approval of recombinant insulin most likely resulted in unrealistic expectations in the biotechnology industry regarding the rapidity of review of NBE applications [153]. Two factors will facilitate the regulatory approval of NBEs. As is true in the development requirements for all NCEs, the first factor is therapeutic importance, and the FDA has established a "fast-track" rapid-approval procedure for NBEs that target unsatisfied indications. The second factor is the relationship of the NBE to an established drug. It appears likely that new therapeutic agents derived from biotechnology will have to satisfy all the traditional demands of regulatory agencies [153]. The possibility that subtle changes in chemical structure may exist and may thus influence pharmacokinetics, pharmacodynamics, or immunogenicity is used to support this regulatory position [103].

Questions of safety are not only properly asked about the NBE *per se* but also about contaminants or residues resulting from the manufacturing or purification processes, antigenic variation, or reversion to the wild type of a living organism [47]. Worker exposure in the production process may be of concern due to relatively high-level, long-term exposure to various end-products of the biotechnology process (i.e., live and dead microorganisms and mammalian cells and their derivatives) [232]. This leads to the area where traditional scientific approaches and techniques do not provide a satisfactory toxicologic profile (e.g., transfer of an immortalization factor from a mammalian cell, allergic reactions) [99].

As we have learned with new technologies, every new technology seems "better" the less we know about it. As we develop an understanding of the true potential contributions of new technology over the long term, we realize that the promises of generating meaningful data in the short-term were generally overstated. The prime example of this is the expectation that the Ames *in vitro* genotoxicity assay would replace the 2-year carcinogenicity assays. New focus also takes us in interesting directions; for example, biomarkers have become a battle cry for those seeking innovation in drug development who forget that biomarkers are an old concept. If we think broadly enough, the common quote, "The dose alone makes a thing *not* a poison" [210], would have no meaning if Paracelsus (1493–1541) were not thinking about biomarkers. Toxicology studies have always depended on the evaluation of biomarkers, which have traditionally been used in toxicology studies to confirm exposure, to monitor susceptibility to a toxic agent and to assess adverse effects [3].

## RELEVANCE OF ANIMAL MODELS IN TOXICOLOGIC ASSESSMENT

The suitability of experimental animal data for assessing risk to humans as well as animal welfare concerns [187] are important contemporary issues in toxicology. Animals and humans have much in common anatomically, physiologically, and biochemically [235]. The two main guiding principles of experimental toxicology are that effects produced in animals, when properly qualified, are applicable to humans and that exposure of experimental animals to high doses of a test compound is necessary and valid in determining human hazard [148]. Although it is generally agreed that animal assays are not as predictive of human effects as would be desired, they are more predictive than generally thought [115,193]. It has been reported that animal assays are predictive of human toxicity in all but 10% of comparisons [159]. It must be recognized, however, that major differences in response to chemical agents can exist both within and between species [127]. The most serious differences between laboratory animal studies and human clinical trials are related to biochemical and physiological species differences, such as metabolism and genetics (hypersensitivity responses), as well as differences in experimental design, including quantity, route, and duration of drug administration [154]. Humans can be as much as 50 times more sensitive on a milligram-per-kilogram basis than experimental animals [154].

Regulatory agencies and research-based pharmaceutical companies consider laboratory animal toxicology studies as a critical part of the assessment of new drug candidates [98,160]. Confidence in the validity of experimental toxicology is based on the large inventory of chemically induced lesions that occur both in animals and humans [247]. It may be incorrect to assume that what is demonstrated in animal toxicology studies will occur in human clinical trials, but until it is shown that the toxicity expressed is not relevant to humans, that assumption must be made [11]. Also, until our knowledge base expands, animal data must be extrapolated to the human situation using a conservative approach (e.g., use of relatively high doses, assuming that humans are more sensitive than the most sensitive species) [17].

The ultimate goals of the toxicology assessments are to characterize toxicity in animal models to identify potential problems in short- and long-term clinical studies, identify the circumstances under which toxicity occurs, evaluate the extent to which the data warrant extrapolation to humans, recommend safe levels of exposure, and contribute to the decision to test the new drug candidate in humans [9,55,161,188,214]. It has been recognized that qualitative extrapolation of drug toxicity from animals to humans is more reliable than estimation of the magnitude of dose producing a similar effect in animals and humans; that is, the pharmacodynamics of an agent are more predictable than its pharmacokinetics [181]. Complicating the ability to extrapolate data from animals to humans are the excessive doses sometimes used, and often required, in animal studies. As a result, adverse effects are described that may be the result of frank intoxication of the animal and are irrelevant in humans. Zbinden [236] has proposed

**TABLE 7.3**
**Common Untoward Reactions to Drugs**

| Clinical Side Effect | Predictable from Animal Studies? (Y/N) | Clinical Side Effect | Predictable from Animal Studies? (Y/N) | Clinical Side Effect | Predictable from Animal Studies? (Y/N) |
|---|---|---|---|---|---|
| Drowsiness | Y | Hypertension | Y | Anorexia | Y |
| Nausea | N | Insomnia | Y | Depression | Y |
| Dizziness | N | Fatigue | N | Increased appetite | Y |
| Sedation | Y | Constipation | Y | Tremor | Y |
| Dry mouth | Y | Tinnitus | N | Perspiration | Y |
| Nervousness | Y | Weight gain | Y | Dermatitis | Y |
| Epigastric distress | N | Hypotension | Y | Increased energy | Y |
| Headache | N | Dryness of nasopharynx | Y | Vertigo | N |
| Vomiting | Y | Heartburn | N | Palpitation | Y |
| Weakness | Y | Diarrhea | Y | Blurred vision | Y |
| Nasal stuffiness | Y | Skin rash | Y | Lethargy | Y |

*Source:* Data from Ronneberger and Hilfenhaus [182] and Zbinden [237,239].

that the ability of animal toxicity studies to predict potential human toxicity is related to the mechanism of drug action. Within limitations, animals and humans respond in ways similar enough, from a pharmacodynamic perspective, for animal toxicity evaluations to serve as useful predictors of human toxicity [53,119,157,161,174]. However, toxicologic evaluations in animals can predict toxic responses in humans only if the response is not unique to humans [98]. Those compounds that are toxic to humans but relatively nontoxic to animals (i.e., thalidomide) are of greatest concern. The extrapolation of animal data to humans is likely to become even more complicated as molecular biology techniques continue to allow the more sophisticated characterization of specific human therapeutic targets (human enzymes, receptors) and the ultimate development of drugs specific for these targets.

Table 7.3 lists common undesirable drug effects seen in human studies; 76% of the findings are predictable from animal studies. Predictability is enhanced for those adverse effects that can be directly related to the pharmacologic mechanism of action of the compound. Adverse responses commonly referred to as dose- and time-related are relatively well predicted from animal studies. It is more difficult to extrapolate effects that are not dose or time related [239].

A small element of toxicity cannot be predicted until large-scale clinical studies are conducted [119]. This may be the result of a very low incidence of occurrence or idiosyncratic responses in a small subset of the patient population; however, considering the increased use of pharmaceutical agents and the relative infrequency of major incidence of human toxicity, the initial laboratory studies are clearly serving a valuable function [9]. A large majority of human drug exposures are free of toxicity and in good accordance with the results of animal toxicity studies [240]. The use of adequate test systems is critical to the predictive ability of animal toxicity evaluations. Cahn [21] reported that the cardiac effects of calcium antagonists (i.e., ectopic beats, ventricular tachycardia, and ventricular fibrillation) were seen in humans but were not described in long-term animal studies. These effects, however, were demonstrable in animals using appropriate functional evaluations not always included in routine toxicologic testing, and it is these kinds of clinical findings that contribute to the evolution of the science of regulatory toxicology. Oftentimes, toxicologically important endpoints, such as cardiac, pulmonary, or renal function, are not taken into consideration in the design of "routine" toxicology studies, which are frequently focused on changes in clinical signs and pathology. The toxicologist is challenged to consider potential adverse effects related to the pharmacodynamics of the test compound in the design of appropriate safety studies [174]. Furthermore, many new drug candidates are being targeted toward the aging population, where compromised hepatic or renal function may significantly modify the pharmacokinetics and thus toxicity of the compound. These examples emphasize the importance of using an adequate and appropriate test system to evaluate the toxicity of new drug candidates.

As with standard pharmaceuticals, no animal model is fully appropriate to evaluate the toxicity of highly specific human proteins [110]. Animal testing for biotechnology products is limited to the species showing the same pharmacologic response as humans, without showing signs of immunity [46,103]. This is only feasible when proteins are highly conserved across species. The production of neutralizing antibodies will limit the study duration and thus support for clinical trials. Administration of a highly specific human protein to laboratory animals for a sufficient duration to produce immune-complex disease will do nothing to reveal effects anticipated in clinical trials. Antigenicity of the test material can be a major complicating factor, in that the potential allergic etiology of all lesions developing in animals treated with human proteins must be considered [245]. Alternatively, nonclinical toxicology studies of biotechnology products may be less predictive of allergic responses that may occur in humans following chronic therapy [232]. The appropriate laboratory species for biotechnology product testing should demonstrate similar pharmacodynamics and adverse responses relative to humans. If an animal model demonstrating similar pharmacologic response to humans cannot be selected, species selection based on toxicity likely to be representative of that expected in humans may be acceptable [225]. The FDA does not currently require the study of recombinant proteins in primate models, but these animals demonstrate many similarities to humans at the molecular level and often turn out to be the most appropriate species for toxicity testing of NBEs [110].

## TOXICOKINETICS

Toxicokinetic analysis has become a standard component of the design, and a valuable tool in the interpretation, of the nonclinical safety profile of NCEs [54,69]. Various guidances that have been issued in recent years describe objectives and specific recommendations for toxicokinetic evaluation [133,134]. Well-designed toxicokinetic studies provide insight on exposure (typically expressed as drug concentration per unit time or AUC [area under the plasma concentration/time curve]) as well as other pharmacokinetic parameters such as clearance ($Cl_T$), volume of distribution ($Vd$), half-life ($T_{1/2}$), etc. These pharmacokinetic parameters are often further refined in specific studies to assess absorption, distribution, metabolism, and elimination (ADME).

Toxicokinetic parameters guide the assessment of dose proportionality, accumulation potential upon multiple doses, sex differences, and species differences. Figure 7.5 illustrates the relationship of dose vs. exposure after single and multiple doses. This example shows a linear, predictable increase in exposure across the range of doses examined. It is also noteworthy to point out that the saturation of systemic exposure may occur after oral dosing as a result of absorption-related processes. When saturation is reached, increasing the dose does not result in increases of exposure. Toxicokinetic analysis is useful in determining the dose at which saturation occurs. This may then be used to justify selection of a high dose in a toxicology study.

The majority of toxicokinetic assessments are designed to evaluate parent drug or metabolite concentrations in plasma [13]. These studies are usually an integrated part of the design of various toxicology studies, including acute, subchronic, chronic, reproductive/developmental, and carcinogenicity studies, regardless of route of administration or test species. An understanding of the internal exposure associated with dose-limiting toxicities or premonitory signs and symptoms that precede dose-limiting toxicities or adverse events provides the foundation for the establishment of a toxicokinetic/toxicodynamic (TK/TD) relationship of dose, exposure, and adverse events [175,246]. The role of $Cl_T$, absorption, or distribution changes in the relationship aid in the prediction of accumulation, saturation, or decreases in exposure observed after multiple doses.

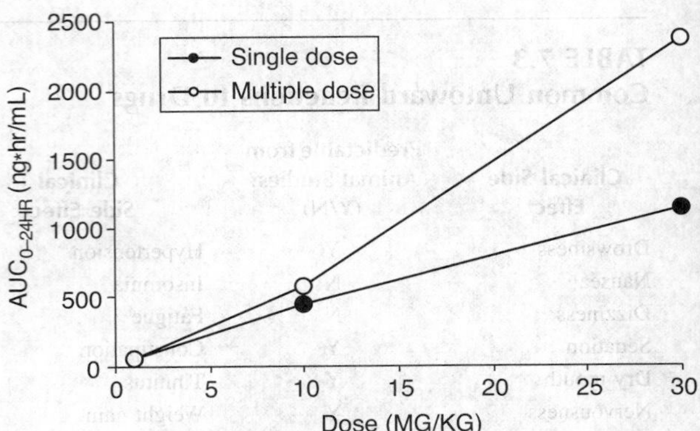


**FIGURE 7.5** Relationship of dose vs. exposure after single and multiple doses.

Similarities in the qualitative profiles of metabolism of xenobiotics across rat, mouse, dog, and nonhuman primates have been reported for decades [28,29]. Quantitative profiles, however, are often quite dissimilar and have been studied extensively; consequently, recent guidance [13] has been published to offer additional insight into the appropriate instances in which to quantify metabolites. Combined, these qualitative and quantitative assessments of an NCE early in the development cycle greatly aid in the selection of the appropriate preclinical species for safety assessment. This early assessment is often coupled with the previously mentioned ADME studies. Increasingly, various *in vitro* test systems are utilized for the initial assessment of some ADME properties. Specifically, the utilization of cellular, subcellular fractions, and tissue slices or segments to assess absorption and metabolism are the most widely employed test systems; for example, permeability across Caco-2 cell monolayers has been well established as a model for gut absorption [5].

Human and animal liver samples are utilized to create various *in vitro* test systems for metabolism. These systems offer the capability to study the rate, extent, and profile of metabolism in a comparative fashion across multiple species in a relatively short time frame. Subcellular fractions of hepatocytes are most commonly used and offer the advantages of speed, robustness, and capacity. In addition to containing the major drug metabolizing enzymes

(CYP450s and flavin-containing monooxygenases) and phase I (oxidative) capability present in subcellular fractions, liver slices and hepatocytes also possess greater capability to complete phase II (conjugative) metabolic transformations [205]. The intact cells are also thought to more closely mimic the intact functioning organ.

Both the activities and the abundance of individual isoforms of cytochrome P450s (CYP450s, the primary drug-metabolizing enzymes) have been extensively characterized in experimental animal species and humans. This level of characterization enables these systems to offer the potential for improvement of the selection of preclinical test species as well as decreased utilization of *in vivo* experiments. Predictions can also be made regarding the correlation of *in vitro* and *in vivo* estimates and the likelihood of clinical drug–drug interactions [27,143, 165,174,185].

Advances in technology, most notably liquid chromatography–tandem mass spectrometry (LC–MS/MS), have greatly enhanced the ability to quantify both parent drug and metabolite concentrations routinely in the sub-microgram-per-milliliter range [1]. This advance is often challenged by the trend to develop NCEs with increasingly higher potencies. The intended goal of toxicology studies is to dose animals in a fashion to obtain higher exposures than those predicted or observed in humans. Plasma concentration exposure assessments are now more widely used and accepted as a better estimate of safety multiples than comparisons of administered dose (e.g., mg/kg or $mg/m^2$). The assessment of systemic exposure in toxicity studies has been addressed by the ICH, and the objectives of toxicokinetic evaluation and specific recommendations are contained in ICH Topic S3A [133].

## TOXICOLOGY GUIDELINES

### Drug Development Timelines

Development time has become an important focus for the pharmaceutical industry. The available data indicate a fourfold increase in drug development time between the 1960s and the 1980s. From the early 1980s to 1996, however, mean drug development times have been relatively constant at about 10 to 12 years, with very wide variability [213]. The pharmaceutical industry has committed to increasing drug discovery and development efficiency, leading to a decrease in the mean drug development time to deliver innovative pharmaceuticals to patients more quickly and realize increased profitability to maintain aggressive research and development efforts. The FDA has also committed to increasing the development speed of safe and effective drugs through a program call the Critical Path Initiative, described in *Innovation/Stagnation: Challenge and Opportunity on the Critical Path to New Medical Products* [142].

New biological entities generally have had a shorter development time (approximately 6 to 9 years) than NCEs. The NBEs registered to date have generally been well-characterized natural molecules, and the shorter development time is probably related to a better understanding of their actions in humans. The introduction of analogs of natural proteins, some designed to be used at supraphysiologic levels, has led to an increase in development time for NBEs. From 1985 to 2004, the development times for NCEs has decreased approximately 22%, while the development time for NBEs has increased approximately 50% [178].

### Regulatory Guidelines for Toxicity Testing

In this section, the toxicology support packages for the registration of NCEs and NBEs are reviewed from slightly different perspectives. More detailed information on the specific studies conducted, their results, and their interpretation are included for the classical agents (omeprazole and zidovudine), because the majority of compounds currently in development would fall into this category and thus require similar testing strategies. Omeprazole was selected for discussion because it had a comprehensive toxicology package at the time of regulatory submission and represents an example of where additional mechanistic studies were critical in the approval process. Zidovudine (AZT) is discussed due to its proposed use in life-threatening disease where no adequate therapy was available. The rapid approval of the drug, in spite of significant toxicology findings and an abbreviated toxicology support package, demonstrates the inherent flexibility in the approval system even with regard to NCEs.

The discussion of specific human NBEs (gonadotropin-releasing hormone analogs, interferon, human insulin) is presented from a more philosophical perspective. Because these agents are naturally occurring and because the major limiting toxicity in animal studies (immunogenicity) is not applicable to clinical trials, the design of the toxicology package posed special issues that were considered on a case-by-case basis. Furthermore, the toxicologic profile was anticipated based on extensive clinical experience with less specific agents (animal-derived insulins) or a broad understanding of the physiological functions of the hormones. The following discussion of the NBEs poses questions, concerns, and general guidelines to be considered in the development of these agents.

### Acute, Subchronic, and Chronic Testing

Toxicity testing can be considered to be composed of several major types [45]. Acute (single-dose), subchronic (multiple dose; less than 6 months' duration), and chronic (multiple dose; greater than or equal to 6 months' duration) studies are intended to elucidate the target organs for toxicity and demonstrate dose–response relationships. They are useful for determining the mechanisms of toxic action and often provide important information for dose selection in other study types. A variety of endpoints are routinely evaluated in subchronic and chronic studies, including body weight, feed consumption, hematology, clinical chemistry, urinalysis, and gross and histologic pathology of numerous tissues. A list of common parameters assessed in subchronic and chronic studies is presented in Table 7.4; however, the toxicologist is continually challenged to modify study design to address the anticipated actions of the compound under investigation. This may result in the addition of certain parameters or tissues to be evaluated or a more comprehensive analysis of tissues (i.e., electron microscopic evaluation, immunohistochemistry). Furthermore, previous studies, or knowledge of the toxicity of other agents in the therapeutic class or those that have a similar structure may suggest alternative assessments, such as the determination of the propensity of the agent to induce hepatic microsomal enzymes, cause phospholipid accumulation, or result in peroxisome proliferation.

An assessment of bioavailability and pharmacokinetics is often an important endpoint of subchronic and chronic studies. As discussed previously, these data are critical to extrapolate toxicity findings to humans. Often, the toxicokinetic profile of the compound is determined early and late in the study so the potential for drug accumulation can be revealed. Alternatively, drug levels may be lower toward study termination or the metabolite profile may differ due to the induction of drug metabolizing enzymes. Tissues may also be collected for drug analysis, so levels in affected tissues can be related to the extent of the histopathologic findings. Finally, important dose–response relationships can be established, relative to both parent compound and metabolites, that may be critical in the interpretation of toxicity data.

Acute (single dose followed by a 2-week observation period) and subchronic (usually 2-week or 1-month studies) testing is required prior to the FHD. One-month studies in one rodent species (usually rat or mouse) and one nonrodent species (usually dog or primate) generally will support 1 to 2 weeks of dosing in humans. Where possible, the animal studies should be carried out using the same route of administration anticipated for use in patients. As an aside, it should be acknowledged that clinical studies may initially be conducted by the intravenous route, regardless of the desired ultimate route of administration, particularly for those molecules that are anticipated to show efficacy rapidly, to demonstrate the proof of concept of a new pharmacologic mechanism. Drug developers might thus avoid the time and expense associated with formulation development and maximization of the desired properties of the chemical if it has been demonstrated that the molecule or mechanism is ineffective. These clinical trials require the support of intravenous toxicology assessments. In these circumstances, it is extremely important to evaluate the risk associated with the potential for demonstrating toxicity in the intravenous study that may be irrelevant to the ultimate route of administration.

Phase II and III efficacy testing in patients is supported by longer term studies. Depending on the proposed duration of human exposure, toxicity studies to support phases II and III may be of 3-, 6-, or 9/12-months' duration. Two or more subchronic or chronic studies may be conducted simultaneously (e.g., 3-month and 6-month studies may be initiated at the same time) so patients can be placed on the trial earlier (upon completion of the 3-month study) and maintained on the trial longer (supported by the 6-month study) if the human efficacy and safety data support continued therapy. A potential problem with this approach is that dose selection for the more extended study may be found to be inadequate (doses either too low or too high) based on the findings of the shorter test.

Much discussion has surrounded the utility of 1-year studies. The FDA has been a strong proponent of the 1-year study approach [43], but Japan and the European Union (EU) have suggested that 1-year studies reveal little new information beyond that gained from 6-month studies. These data have been reviewed by Lumley et al. [162], who suggest that, for 154 compounds for which short-term (≤6 months) and long-term (>6 months) animal data are available, tests lasting longer than 6 months (excluding carcinogenicity studies) have not provided new substantive safety information. They also point out that, although new findings became evident after 6 months of treatment in 9 out of 75 cases, the data did not influence the decision of whether to continue the development of the compound. Parkinson et al. [180] have reported that long-term toxicity studies in dogs provide little new data when compared to 3-month dog studies in conjunction with short- and long-term rodent studies.

Based on the current ICH position, it is generally accepted that 6-month rodent and 9-month nonrodent multiple-dose studies are acceptable for a tripartite development plan [135]. Even so, it is strongly recommended that the sponsor have a written commitment from the appropriate reviewing division of the FDA on the acceptability of the 9-month vs. the 1-year nonrodent chronic toxicity study.

### Additional Toxicology Studies To Support Clinical Trials

Other tests conducted prior to initial clinical trials include mutagenicity studies and pharmacologic assessments. A variety of mutagenicity studies are currently employed that assess various types of DNA damage *in vitro* and *in vivo* in an attempt to predict the oncogenic potential of the compound under investigation. In pharmacologic

**TABLE 7.4**
**Parameters That Might Typically Be Assessed in a Subchronic/Chronic Toxicology Study**

| | | |
|---|---|---|
| **Live Phase** | | |
| Body weight | Clinical observations | Electrocardiogram (large animal) |
| Feed consumption | Ophthalmology | Physical examination |
| Efficiency of food utilization (g body weight gained per 100 g feed consumed) | | |
| **Hematology** | | |
| Erythrocyte count | Mean corpuscular hemoglobin | Thrombocyte count |
| Hemoglobin | Mean corpuscular hemoglobin concentration | Activated partial thromboplastin time |
| Packed cell volume | Total leukocyte count | Prothrombin time |
| Mean corpuscular volume | Leukocyte differential | M/E ratio (bone marrow smears) |
| **Clinical Chemistry** | | |
| Glucose | Gamma glutamyltransferase | Cholesterol |
| Blood urea nitrogen | Creatinine phosphokinase | Triglycerides |
| Creatinine | Calcium | Total protein |
| Total bilirubin | Inorganic phosphorus | Albumin |
| Alkaline phosphatase | Sodium | Globulin |
| Alanine transaminase | Potassium | Albumin/globulin ratio |
| Aspartate transaminase | Chloride | |
| **Urinalysis** | | |
| Color | Protein | Ketones |
| Clarity | Glucose | Bilirubin |
| Specific gravity | Occult blood | Urobilinogen |
| pH | | |
| **Organ Weights** | | |
| Kidneys | Ovaries | Adrenals |
| Liver | Testes | Thyroids (with parathyroids) |
| Heart | Prostate | Brain |
| **Histopathology** | | |
| Kidney | Stomach | Skin |
| Urinary bladder | Duodenum | Skeletal muscle |
| Liver | Jejunum | Bone |
| Gallbladder | Ileum | Bone marrow |
| Heart | Cecum | Adrenal |
| Aorta | Colon | Thyroid |
| Trachea | Rectum | Parathyroid |
| Lung | Ovary | Pituitary |
| Spleen | Uterus | Cerebrum |
| Lymph node | Cervix | Cerebellum |
| Thymus | Vagina | Brain stem |
| Salivary gland | Testis | Spinal cord |
| Pancreas | Epididymis | Sciatic nerve |
| Tongue | Prostate | Eye |
| Esophagus | Mammary gland | Harderian gland |
| **Other** | | |
| Blood levels of parent compound/metabolites P450 content | Hepatic microsomal enzyme activity/ cytochromes P450 content | Peroxisome proliferation Tissue phospholipid phosphorus concentration |

screening, the ability of the compound to produce toxicities or "side effects" based on its pharmacologic mechanism of action is assessed; for example, an agent that is shown to bind to β-receptors *in vitro* might be anticipated to influence cardiac function in subsequent toxicity and clinical testing. Mutagenicity and pharmacology safety studies are often employed as very early screens in the evaluation of potential drug candidates to select one of a group of structurally related compounds that would be least likely to result in carcinogenicity and most likely to demonstrate the specific desired pharmacologic activity. The types and utility of these studies are further described below. Finally, special studies might be conducted prior to initial clinical testing to address specific issues, such as irritation testing of an agent proposed for topical use in the patient population.

## Reproductive and Developmental Toxicity Studies

The thalidomide incident raised a great deal of concern relative to predictive testing for developmental toxicity, as well as fertility effects in both males and females. Although regulations have differed substantially among countries, worldwide harmonized guidelines for reproductive toxicity testing have been established [136,137]. The ultimate goal of these studies is to assess reproductive risk to adults, as well as to the developing individual, at all stages from conception to sexual maturity. Traditionally, animal studies have been conducted in three segments: in adults, treatment pre-mating through mating in the male and pre-mating through either implantation or lactation in the female (segment I); in pregnant animals, treatment during organogenesis (segment II or teratology studies); and in pregnant/lactating animals, treatment from the completion of organogenesis through lactation (segment III, peri- and postnatal study). Although guidelines addressing treatment regimens have been rather similar throughout the world, the required endpoints measured in both the adult and developing organism have varied widely, and this is an area where much duplicative testing has occurred to support worldwide registration.

The harmonized ICH guidelines [136,137] stress the need for flexibility in testing for reproductive and developmental toxicity and challenge the toxicologist to custom design a combination of studies that will reveal potential effects on all of the parameters considered in the classical segment I, II, and III studies. For treated adults, these include development and maturation of gametes, mating behavior, fertilization, implantation, parturition, and lactation. In the developing organism, where the maternal animal may be exposed to the drug candidate from prior to mating through lactation, assessments of early embryonic development, major organ formation, fetal development and growth, postnatal development and growth (including behavioral assessments), and attainment of full reproductive function are required. These evaluations might be carried out as one comprehensive study with interim assessments, or they might be segmented into several treatment components. Thus, the new guidelines have not diminished the extent of evaluation but allow

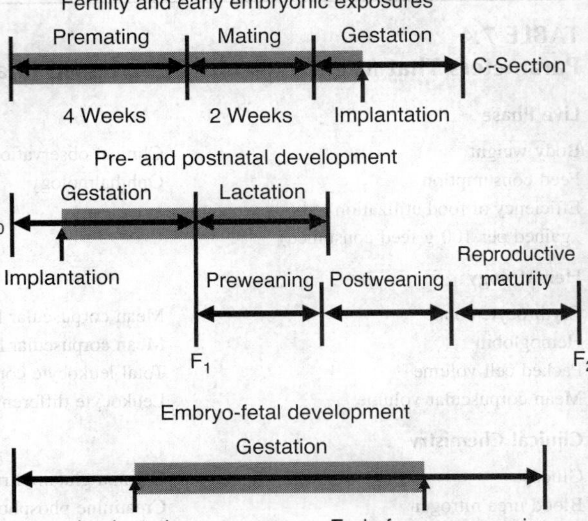


**FIGURE 7.6** The three-study design proposed for the assessment of reproductive and developmental toxicity for a standard pharmaceutical. (From ICH, *Reproductive Toxicity: Male Fertility Studies*, Topic S5B, Step 5, ICH Harmonized Tripartite Guideline. International Conference on Harmonisation of Technical Requirements for Registration of Pharmaceuticals for Human Use, Geneva, Switzerland 1996.)

flexibility in study design based on what is already known of the compound under investigation. The harmonized guideline suggests a three-study design that is likely to provide all of the developmental toxicity data necessary to support product registration, assuming no untoward toxicity (Figure 7.6). Should toxicity be demonstrated, further mechanistic studies would be conducted to clarify effects and determine whether the responsible mechanisms would be applicable to humans. The results from previous subchronic and chronic studies (e.g., evidence suggesting an effect on spermatogenesis upon histopathologic examination of the testes) are critical in the design of an appropriate reproduction package.

Women of child-bearing potential are generally first recruited into phase II trials, and several countries require that efficacy be demonstrated in male patients prior to recruitment of women of child-bearing potential into trials, regardless of the outcomes of the nonclinical reproductive studies. However, it should be noted that the FDA has encouraged the inclusion of women of child-bearing potential in early clinical trials, especially in the case of drugs intended to treat life-threatening conditions or in the study of disease states that more commonly afflict women. Typically, prior to the inclusion of women of child-bearing potential into clinical trials, studies are conducted to evaluate effects on organogenesis (segment II) in two species (usually the rabbit and the rodent species selected for the subchronic and chronic studies). Female fertility assessments are also usually undertaken prior to longer term treatment or addition of significant numbers of patients. Despite the availability of these data, clinical protocols frequently require that women of child-bearing potential use reliable contraceptive intervention. Early studies in men are supported by histological evaluation of the testes at the conclusion of the subchronic and chronic nonclinical studies, and specific animal studies to examine drug effects on male fertility are not required until phase III. Unless there are concerns regarding a specific chemical class or mechanism of action, the more sophisticated analyses of peri- and postnatal development and behavior (segment III studies) are often conducted in conjunction with the phase III clinical trials. Frequently, as is the case with other toxicity evaluations, special studies may be conducted to determine the mechanisms of observed reproductive effects in an effort to assess whether these findings are meaningful to humans and whether use of the compound should be restricted depending on the reproductive status of the patient (i.e., the drug should not be administered to pregnant women). Finally, the inclusion of female patients in clinical trials may be allowed following a more limited assessment of reproductive or developmental parameters if the compound under development is shown to be efficacious or is thought to provide distinct advantages over available therapies in the treatment of life-threatening disease.

## Carcinogenicity Studies

Among the final toxicity studies to be conducted to support the registration of chronic-use therapeutics are the carcinogenicity bioassays. These lifetime (2-year) studies are usually conducted in two rodent species (normally, rat and mouse). Selection of the top dose for the carcinogenicity bioassays has been an intensely debated topic of discussion in the international harmonization process. The international consensus has based the selection of the top dose for carcinogenicity bioassays on any one of the following [130]:

- Maximum tolerated dose (MTD; see discussion below)
- Area under the plasma concentration–time curve (AUC) rodent-to-human ratio of 25-fold, which applies when no genotoxicity is identified
- Saturation of absorption
- Dose-limiting pharmacodynamics (i.e., hypotension, hypoglycemia, decreased blood clotting time)
- Use of a limit dose (1500 mg/kg), which applies when the maximum recommended human dose is ≤500 mg/day and the AUC (rodent:human) is ≥10

The middle and low doses should also be selected to provide additional information to use in the risk assessment. The consideration of dose linearity, saturation of metabolic pathways, margin of safety, pharmacodynamics, specific animal physiology, threshold effects, and progression of toxic effects should be included in the selection of the middle and low doses for carcinogenicity evaluations.

The use of a MTD as the top dose in a carcinogenicity study is a subject of some controversy. The MTD has been classically defined as the dose that causes no more than a 10% decrease in body weight and does not produce mortality, clinical signs of toxicity, or pathologic lesions that would be predicted to shorten the natural life span of an experimental animal for any reason other than the induction of neoplasms [201]. Because the MTD is often based on a body weight effect, Hoffman et al. [126] have proposed an improved evaluation of growth data in rodent toxicology studies using a powerful statistical analysis method that reduces the number of false positives and provides a comprehensive summary of compound-related effects. The MTD is suggested to produce a level of toxicity indicative of sufficient chemical challenge to define chronic toxic manifestations [122]. Many regulatory bodies default to the use of the MTD as the maximum dose in the rodent bioassays. A major concern with the MTD approach is that metabolic saturation, as discussed above, may occur at high doses, leading to abnormal metabolism [172] or, in the case of inhaled therapeutics, abnormal clearance [170]. Chemicals administered at the MTD in animal bioassays tend to induce mitogenesis as a result of cell death due to frank intoxication, with the target tissues differing among species and sexes [108]. This stimulation of cell proliferation, a natural recovery process in response to severe toxicologic insult that does not normally occur at reasonable multiples of human exposure levels, can account for the carcinogenic response of nongenotoxic compounds [40]. Thus, the fact that a chemical is a carcinogen at MTD levels in rodents may provide little meaningful information relative to low-dose risk assessment in humans [2]. MTDs for chronic toxicity studies are usually estimated based on the results of subchronic toxicity studies; however, because compound distribution and disposition may be affected by the dose or duration of treatment, this may be a very crude estimate [172,176].

The choice of a MTD is a critical aspect of chronic toxicity evaluations [207]. Cell proliferation indices in subchronic toxicology studies may provide a useful estimation of an appropriate MTD by determining the highest dose that does not result in the phenomenon [20]. The use of kinetic parameters, ($C_{max}$, AUC related to dose) would better predict the dose at which saturation (nonlinearity) might occur and therefore provide a better estimate of the MTD in a particular species. Also, changes in the urinary metabolite profile, in relation to dose, may be a good way of indicating metabolic overload and could aid in more accurately selecting upper dose levels in toxicology studies [233]. A systemic exposure-based alternative to the MTD for carcinogenicity studies has been presented by Contrera et al. [44]. This alternative was discussed as being suitable for nongenotoxic compounds with low rodent toxicity and similar metabolism in rodents and humans. The relevance to human risk assessment of a carcinogenicity finding for nongenotoxic compounds was addressed by Silva Lima and van der Lann [200]. They discussed the main causes of carcinogenicity from nongenotoxic compounds: chronic cell injury, immunosuppression, increased secretion of trophic hormones, receptor activation, and other mechanisms such as CYP450 induction. An overall approach to addressing the human relevance of the mode of animal carcinogenicity (i.e., genotoxic or nongenotoxic) has been addressed by Cohen et al. [42].

The utility of using two rodent species has also been an active area for discussion. The ICH has indicated that the rat would be preferable to the mouse for conducting carcinogenicity studies [129]. The rat seems to have been given a preference because background mechanistic data are usually available for rats (not mice), studies of metabolic disposition are more often carried out in rats than in mice, and mouse carcinogenicity studies are dominated by liver tumors of questionable relevance to humans. A review of the European Regulatory Database has concluded that studies in the mouse add little to the ability to detect carcinogenic risk from pharmaceuticals [45,220]; however, the review found that carcinogenicity studies in two rodent species are necessary to identify trans-species tumorigens. NCEs active across species are considered to pose a relatively greater risk to humans than NCEs positive in only one specie. The conduct of a study using an alternative *in vivo* carcinogenicity model along with a standard bioassay in one species was considered to be an acceptable alternative for assessing carcinogenic potential [35].

The use of alternative models for carcinogenicity assessment (see below) meets the desire of the FDA to have an assessment in two species and offers the advantages of using fewer animals, being of shorter duration, and being capable of enhancing the predictability of the rodent bioassay [46]. A number of transgenic animal models are currently being evaluated as alternatives to the 2-year bioassay [6,34,106]; however, it is early yet to evaluate the validity of alternative models of carcinogenic potential to the risk assessment process, and it would be advisable to discuss the selection of alternative models with the appropriate regulatory agency prior to study initiation, as the use of alternative models is not routinely accepted by the FDA.

Within the FDA, it is highly recommended that protocols, dose justification documents, and supporting data be submitted to the Carcinogenicity Assessment Committee (CAC) for evaluation prior to initiation of carcinogenicity studies. The CAC provides consultation on study designs, ensures consistency and quality in the analysis and interpretation of animal carcinogenicity studies across industry and within the Agency, and monitors scientific developments to ensure that scientific standards of design and interpretation are upheld.

Since 1982, the number of NBEs presented for registration has risen; even so, these NBEs represent a relatively small number of molecules. Due, in part, to this lack of practical experience, the safety programs for the NBEs have been designed on a case-by-case basis. The ICH has provided two guidelines that address carcinogenicity studies with products of biotechnology [129,138]. It is generally acknowledged that carcinogenicity studies are not appropriate for biotechnology products given essentially as replacement therapy at physiological levels, especially when clinical experience exists (e.g., insulins, calcitonin, pituitary-derived growth hormone) [116,129,138]. Product-specific assessment of carcinogenic potential may be necessary depending on clinical dosing regimen, patient population, and biological activity of the product. For products that have the potential to induce cell proliferation (e.g., growth factors), an *in vitro* evaluation of receptor expression in cells relevant to the patient population may be conducted. If these data indicate a need for further evaluation of carcinogenic potential, 2-year studies in a single rodent species should be considered.

Long-term carcinogenicity evaluations with endogenous peptides and proteins or their analogs are generally indicated when [129]:

- Significant differences in biological effects to the natural substances are found.
- Modifications lead to significant changes in structure compared to the natural substance.
- Therapeutic exposure levels exceed those that normally occur in the systemic circulation or in tissues.

A specific example of studies recommended for the analog of a naturally occurring decapeptide, gonadotropin-releasing hormone (GnRH), is presented below [188].

Regulatory guidelines as to when and why carcinogenicity studies should be conducted with naturally occurring substances and their analogs exist, as do opinions from industry, academia, and regulatory agencies regarding the propriety of conducting these studies [116]. For each NBE under development, the existing opinions and guidance must be considered, a reasonable plan to establish safety must be developed, and a discussion with the appropriate regulatory agency should be held to test the plan. Safety evaluation of NBEs is still very much a case-by-case consideration. The pharmaceutical industry must be careful not to overinterpret the position that carcinogenicity studies are not usually appropriate for biotechnology products.

| | Early Discovery Nonclinical Testing | Phase I | Phase II | Phase III | FDA Review | Phase IV |
|---|---|---|---|---|---|---|
| Years | 3.5–6.5 | 1–1.5 | 2 | 3–3.5 | 2.5–1.5 | |
| Test population | Nonclinical laboratory studies | <100 healthy volunteers | <300 patients | <3000 patients | | |
| Purpose | Efficacy and safety | Kinetics dose range safety | Efficacy safety | Confirm efficacy long-term safety | Data review | Post-marketing testing required by FDA (could be related to line extension strategy) |
| Success rate | 100,000 NCEs | 100 NCEs enter trials (0.1%) | | | 10–20 approved (0.01–0.02%) | |

Short term studies ≤2 weeks, 1 or 2 species, alternatives detect serious adverse effects

File IND

Genotox., safety pharmacology, definitive studies ≤1 month, 2 species, safety evaluation in support of FHD

Definitive studies ≤ 6 months (9 months to 1 year nonrodent), support extended clinical trials, carcinogenicity studies or alternatives

File NDA

Carcinogenicty studies or other studies dictated by clinical experience

Fertility and teratology studies conducted before inclusion of women of child bearing potential (WCBP) in clinical trials

**FIGURE 7.7** Schematic of the drug development and approval process in the U.S. Similar processes are employed for worldwide pharmaceutical testing. (From Gordon, C. V. and Wierenga, D. E., *The Drug Development and Approval Process*, Pharmaceutical Manufacturers Association, Washington, D.C., 1991.)

## NEW CHEMICAL ENTITIES

The extent and types of safety testing of synthetic organic pharmaceutical agents in animal models depends on a variety of factors, including the potential duration of treatment of patients (e.g., short-term, antibiotics; chronic, antihypertensives), route of administration, pharmacologic mechanism of action, proposed patient population, and clinical experience with other agents in that therapeutic class. Furthermore, the design of animal toxicity studies that occur later in development must carefully consider the results of previous tests in animals and humans relative to bioavailability, unanticipated toxic responses, and relevance of species selected.

Generally, toxicity testing in animals can be considered in three phases: testing to support FHD (single- and multiple-dose; phase I), testing to support longer term and broader efficacy studies (weeks to months; phase II), and testing to support final registration and, if appropriate, chronic treatment (phase III) (Figure 7.7). Although the great majority of testing is performed prior to registration, special studies may be requested by regulatory agencies during the review and approval processes. Following widespread clinical use of a new agent, further testing may be appropriate to examine potential mechanisms of action for unanticipated side effects that become evident in the increasing patient population or subpopulations. These may occur due to genetic differences, environmental factors, age, patient history, existence of other diseases or pathologies, and drug interactions. Other tests may be considered if new formulations of the drug are developed, if the drug is suggested for new indications, or if it will be used in patient populations that were not anticipated during initial development (e.g., pediatrics).

The primary purpose of initial clinical testing (support for FHD) is to determine the toxicity and pharmacokinetics (and oral bioavailability, if appropriate) of the drug candidate in humans following one or several doses. Usually, the drug is administered to humans at doses below the anticipated efficacious dose, and doses are escalated until a satisfactory multiple over the anticipated efficacious dose is achieved or toxicity becomes evident. Unless the drug candidate has known, serious toxicity, as is the case with many oncolytics, it is usually first tested in a limited male, nonpatient population that is under constant observation. As indicated above, the FDA supports the early inclusion of women in clinical trials for new therapies, especially those to be used in the treatment of life-

**TABLE 7.5**
**International Guidelines for the Duration of Animal Toxicology Studies Necessary To Support Clinical Trials of Various Duration**

| Clinical Trial Duration | Toxicology Duration To Support Phase I and II (EU) and Phase I, II, and III (Japan) | | Toxicology Duration To Support Phase III (EU) and Marketing (All Regions) | |
|---|---|---|---|---|
| | Rodents | Non-Rodents | Rodents | Non-Rodents |
| Single dose | 2 weeks | 2 weeks | — | — |
| ≤2 weeks | 2 weeks | 2 weeks | 1 month | 1 month |
| 1 month | 1 month | 1 month | 3 months | 3 months |
| ≤3 months | 3 months | 3 months | 6 months | 3 months |
| >3 months | — | — | 6 months | Chronic |
| ≤6 months | 6 months | 6 months | — | — |
| >6 months | 6 months | Chronic | — | — |

*Source:* Data from ICH Topic M3(M) (see ICH website, http://www.ich.org).

threatening diseases [84,86]. Because of this interest, studies of developmental toxicity, which usually occur later, may be moved to a much earlier point in the drug development process [179].

When designing animal studies to support FHD, a major consideration in dose selection must be the anticipated margin of safety between animals and humans. Ideally, doses in animal studies should provide exposure to the compound well in excess of what is anticipated at the highest doses to be tested in humans. As discussed previously, a comparison of these doses on a milligram-per-kilogram basis is no longer considered to provide adequate information in this respect due to potential species differences in absorption and rates and routes of metabolism; thus, a good estimate of the pharmacokinetic behavior of the agent in animals is an important goal of nonclinical testing. No firm guideline exists regarding what should be considered an adequate margin of safety; however, a smaller margin between the potentially efficacious dose and a toxic dose is tolerated for those compounds under development for life-threatening diseases, particularly if they are expected to offer a distinct therapeutic advantage over other agents currently marketed in the class.

Based on the activities of the ICH, the designs and goals of clinical trials are similar throughout the world. Table 7.5 shows the recommendations for the duration of animal tests relative to proposed human exposure to NCEs. It is possible to discuss the nonclinical and clinical programs with regulatory bodies and, depending on the characteristics and proposed use of the chemical, modify these recommendations.

The duration of animal tests necessary to support a specific duration of clinical trials for NBEs is much more flexible [138]. Short-term clinical trials for life-threatening conditions may be supported by 2-week nonclinical toxicology studies. Subchronic clinical trials can be supported with toxicology studies of 1 to 3 months' duration. Clinical trials to support chronic therapy can be supported with 6-month toxicology studies.

Regulators throughout the world have recognized that resources could be used more efficiently, and efficacious and safe drugs could be made available more rapidly, if guidelines for nonclinical testing and registration were comparable across countries. The ICH has developed a comprehensive set of safety guidelines to harmonize the regulatory requirements of the EU, United States, and Japan. The ICH Expert Working Groups (EWG) has considered appropriate guidelines for all of the various types of toxicity tests, including acute and subchronic testing, chronic and carcinogenicity testing, reproduction and developmental toxicity studies, and mutagenicity testing. The required duration of animal studies to support human exposure has also been addressed. A good deal of collaboration has occurred between regulatory agencies and pharmaceutical companies in the development of the safety guidelines.

Dorato and Buckley [70] have provided a more complete discussion of the role of the ICH process in drug development. Table 7.6 shows the ICH guidelines that address the various nonclinical studies required to support clinical trials and registrations in the three major regions (EU, United States, and Japan) and to qualify impurities. A proposal on nonclinical studies to support pharmaceutical excipient development has been made by Baldrick [12]. Selected Internet websites providing information on the design of toxicology studies to support clinical trials are shown in Table 7.7

As mentioned previously, the speed with which safe and effective new drugs enter clinical trials and eventually the market is a focal topic for regulatory agencies and the pharmaceutical industry. In the past, the FDA [85] and the European Committee for Human Medicinal Products

**TABLE 7.6**
**ICH Guidelines for the Conduct of Nonclinical Studies**

| Topic | Topic Number[a] | Title and Contents |
|---|---|---|
| Toxicity testing | S4 | *Single Dose and Repeat Dose Toxicity Tests (Step 5):* Recommendation to abandon $LD_{50}$ determination; reduction in duration of longest-term dose toxicity study in rodents from 12 to 6 months |
| | S4A | *Repeat-Dose Toxicity Tests in Non-Rodents (Step 5):* Reduction of duration of repeat dose toxicity studies in nonrodents from 12 to 9 months |
| Carcinogenicity studies | S1A | *Need for Carcinogenicity Studies of Pharmaceuticals (Step 5):* Definition of circumstances requiring carcinogenicity studies, taking into account known risks, indications, and duration of exposure |
| | S1B | *Testing for Carcinogenicity in Pharmaceuticals (Step 5):* Need for studies in two species; Alternatives to 2-year rodent bioassay |
| | S1C | *Dose Selection for Carcinogenicity Studies in Pharmaceuticals (Step 5):* Criteria for selection of high dose |
| | S1C(R) | *Addendum to S1C: Addition of a Limit Dose and Related Notes (Step 5)* |
| Genotoxicity studies | S2A | *Genotoxicity: Specific Aspects of Regulatory Tests (Step 5):* Specific guidance for *in vitro* and *in vivo* tests plus glossary of terms |
| | S2B | *Genotoxicity: Standard Battery Tests (Step 5):* Identification of a standard set of assays; Extent of confirmatory experimentation |
| Reproductive toxicology | S5A | *Detection of Toxicity to Reproduction for Medicinal Products (Step 5):* Specific guidance for testing reproductive toxicity |
| | S5B(M) | *Maintenance of the ICH Guideline on Toxicity to Male Fertility: An Addendum to the Guideline on Detection of Toxicity to Reproduction for Medicinal Products* |
| Toxicokinetics and pharmacokinetics | S3A | *Toxicokinetics: Guidance on the Assessment of Systemic Exposure in Toxicity Studies (Step 5):* Integration of kinetic information into toxicity testing |
| | S3B | *Pharmacokinetics: Guidance for Repeat Dose Tissue Distribution Studies (Step 5):* Need for tissue distribution studies when appropriate data cannot be derived from other sources |
| Biotechnology products | S6 | *Preclinical Safety Evaluation of Biotechnology-Derived Pharmaceuticals (Step 5):* Nonclinical safety studies, use of animal models of disease and other alternative methods, need for genotoxicity and carcinogenicity studies, impact of antibody formation |
| Joint safety/efficacy (multidisciplinary) | M3(M) | *Maintenance of the ICH Guideline on Nonclinical Safety Studies for the Conduct of Human Clinical Trials for Pharmaceuticals (Step 5):* Principles for development of nonclinical testing strategies (addresses full range of studies to support clinical trials for NCEs) |
| Pharmacology studies | S7A | *Safety Pharmacology Studies for Human Pharmaceuticals (Step 5)* |
| | S7B | *The Nonclinical Evaluation of the Potential for Delayed Ventricular Repolarization (QT Interval Prolongation) by Human Pharmaceuticals (Step 3)* |
| Immunotoxicology studies | S8 | *Immunotoxicology Studies for Human Pharmaceuticals (Step 3)* |
| Impurities in new drug substances | Q3A(R) | *Impurities in New Drug Substances (Step 5)* |
| Impurities in new drug products | Q3B(R) | *Impurities in New Drug Products (Step 5)* |

[a] The most recent information on ICH guidelines can be found on the ICH website (http://www.ich.org).

CHMP, formerly the Committee for Proprietary Medicinal Products [CPMP]) [36] have published guidelines that provide an exception to the minimum 2-week rodent and nonrodent study requirements of the ICH and allow for single-dose toxicity studies in support of targeted clinical trials. The FDA guidance [85] was intended to facilitate choosing compounds to enter phase I human studies. The CHMP [36] guidance was intended to facilitate early characterization of pharmacokinetic properties or receptor selectivity profiles using, for example, positron emission tomography (PET). The clinical trials could test several closely related pharmaceuticals with the intent of choosing the preferred candidate or formulation for further development. The CHMP guidance recommended a maximum dose of 100 μg. Following these attempts, the FDA, in keeping with their Critical Path Initiative [142], published a draft guidance on exploratory investigational new drug (IND) studies [90]. The FDA draft guidance clarifies the

**TABLE 7.7**
**Internet Websites Providing Information on the Design and Expectations of Nonclinical Toxicology Studies**

| | |
|---|---|
| http://www.fda.gov/cder/guidance | Access to guidance documents representing the Agency's current thinking on a particular subject relating to the drug development process; includes access to adopted and draft ICH guidelines |
| http://www.eudra.org/humandocs/humans/swp.htm | Safety Working Party (SWP) documents covering aspects of safety evaluation in Europe; includes access to adopted and draft ICH guidelines |
| http://www.ich.org | The process and the adopted and draft safety guidelines of the International Conference on Harmonization of Technical Requirements for Registration of Pharmaceuticals for Human Use |
| http://www.eudra.org | The European Agency for the Evaluation of Medicinal Products (EMEA) |
| http://www.cmr.org | The Centre for Medicines Research International (CMR), a not-for-profit organization funded by the worldwide research-based pharmaceutical industry to provide unique data and expert analysis to address technical, medical, economic, regulatory, and policy issues in the discovery, development, and safe use of medicines |

nonclinical and clinical approaches to the conduct of exploratory studies that occur very early in phase I, involve very limited human exposure, and have no therapeutic intent (e.g., screening). The dosing duration in the clinical study is expected to be ≤7 days. The flexibility available for drug development and examples of the use of the exploratory IND approach are presented. This is a proposal of an approach that could facilitate the development of safe and effective drugs. At this time, it is in the comment stage. The content of the final proposal is yet to be established.

## Specific Agents

As mentioned previously, two NCEs (omeprazole and zidovudine) were selected for discussion due to their different target patient populations, which drove the design of customized toxicology testing strategies. This discussion is limited primarily to the data that supported their initial approvals (and, in the case of zidovudine, includes post-approval commitments) to demonstrate how disease state and early toxicology findings can influence subsequent development strategy.

### Omeprazole (Prilosec®)

Omeprazole (Figure 7.8) is a substituted benzimidazole that is a potent inhibitor of $H^+/K^+$ ATPase (proton pump) at the secretory surface of the gastric parietal cell, thereby inhibiting gastric acid secretion [192,224]. Omeprazole is indicated for the short-term (4- to 12-week) treatment of active duodenal and gastric ulcer, gastroesophageal reflux disease (GERD), severe erosive esophagitis, and maintenance of healing of erosive esophagitis, as well as the long-term treatment of pathological hypersecretory conditions such as the Zollinger–Ellison syndrome. It is also approved for use with clarithromycin and amoxicillin for the treatment of patients with *Helicobacter pylori* infection and duodenal ulcer disease. The recommended dosage for short-term indications is 20 to 40 mg daily (approximately 0.4 to 0.8 mg/kg in a 50-kg individual). For long-term indications, the recommended initial dose is 60 mg daily; however, doses up to 120 mg three times daily have been administered [183]. Table 7.8 lists the toxicology studies that were submitted to the FDA [82] to support the U.S. registration of omeprazole. The content of the toxicology package suggests that the intravenous route may have also been a considered route for therapy.

The results of acute, subchronic, and chronic studies suggested that the toxicology profile of omeprazole was generally unremarkable [30,73,82,120]. The acute toxicity of the compound in rats and mice was low, as demonstrated by the oral $LD_{50}$ values (the dose that kills 50% of the animals tested), generally in excess of 4 g/kg. Multiple dose studies in rats were conducted at doses up to 414 mg/kg/day for 3 months and up to 138 mg/kg/day for 6 months. No consistent effects on body weight or feed consumption were reported in those studies. Treatment-related findings that occurred at high doses in these studies included decreases in several erythrocytic parameters and decreases in plasma glucose and triiodothyronine. The latter finding was ascribed to a reduction in the peripheral conversion of thyroxine to triiodothyronine. Increased liver and kidney weights were observed in both studies as well as in the 24-month rat oncogenicity study. Elevated

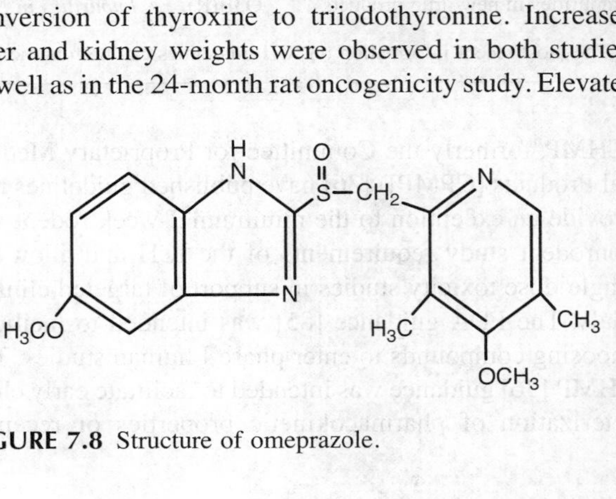

FIGURE 7.8 Structure of omeprazole.

**TABLE 7.8**
**Summary of Toxicology Studies Conducted To Support the Registration of Omeprazole in the United States**

| |
|---|
| *Acute toxicology* |
| Oral study in mice |
| Intravenous study in mice |
| Oral study in rats |
| Intravenous study in rats |
| Oral study in dogs |
| *Subchronic toxicology* |
| 2-Week intravenous study in rats |
| 1-Month intravenous study in rats |
| 1-Month intravenous study in dogs |
| 3-Month oral study in rats |
| 3-Month oral study in mice |
| 3-Month oral study in dogs |
| 3-Month oral study in dogs with 3-month recovery |
| *Chronic toxicology* |
| 6-Month oral study in rats |
| 3-Month and 6-month oral studies in rats with 2-week to 6-month recovery |
| 2-Year study in female rats to examine gastrin-dependent variables |
| 1-Year oral study in dogs with 4-month recovery |
| 5-Year oral study in dogs (ongoing at time of application) |
| *Genetic toxicology* |
| Ames *Salmonella* test with/without metabolic activation |
| Mouse lymphoma forward mutation assay |
| Mouse micronucleus test |
| Mouse chromosome aberrations |
| Rat liver DNA damage assay |
| *Reproductive and developmental toxicology* |
| Segment I oral fertility in rats |
| Segment II oral teratology in rats |
| Segment II oral teratology in rabbits |
| Segment III oral perinatal and postnatal in rats |
| Segment III extended oral perinatal and postnatal in rats |
| *Carcinogenicity studies* |
| 78-Week oral study in mice |
| 104-Week oral study in rats |
| 104-Week oral study in female rats |

kidney weights were correlated with an apparent exacerbation of the progress of chronic nephropathy that normally occurs in aging Sprague–Dawley rats.

Dogs were treated with omeprazole for 3 months at doses up to 138 mg/kg/day or 12 months at doses up to 28 mg/kg/day. Clinical chemistry findings were generally unremarkable, although, as observed in the rat, some decreases in hematology parameters and plasma triiodothyronine were noted. The most significant nonclinical finding in both rats and dogs was a reversible gastric mucosal cell hyperplasia with increases in mucosal thickness and folding. In the 6-month study in rats, omeprazole induced a dose-related eosinophilia of the zymogen granules of the pepsinogen-secreting chief cells, with slight atrophy of these cells occurring at the high dose. Slight chief cell atrophy was also observed in dogs given the high dose of omeprazole for 3 or 12 months. To characterize these gastric changes more rigorously, a rather extensive reversibility study was conducted in rats in which animals were treated with either 0 or 138 mg/kg/day omeprazole for 14 days or 1, 3, or 6 months. Other groups of animals were treated with that dose of omeprazole for 3 or 6 months, followed by recovery periods of 14 days or 1, 3, or 6 months. This study demonstrated the time dependency and complete reversibility of the gastric lesions in rats. A 3-month recovery period following 3 months of treatment in dogs showed that the slight chief cell atrophy observed at 3 months was reversible, and a 4-month recovery period following 12 months of treatment in dogs demonstrated the reversibility of mucosal hyperplasia and chief cell atrophy, although increased mucosal folding was still evident.

No teratologic findings were observed in the rat at omeprazole doses up to 138 mg/kg/day administered on days 6 to 15 of pregnancy. The two highest doses tested in rabbits (approximately 70 and 140 mg/kg), administered during days 6 to 18 of pregnancy, resulted in maternal toxicity as evidenced by anorexia and reduced water intake. Signs were sufficiently severe that treatment of animals at the high dose was discontinued on day 14. Fetal mortality was increased in conjunction with maternal toxicity, but fetal development was unaffected by maternal omeprazole treatment. The major finding of the fertility and peri- and postnatal studies was a decrease in the weight gain of pups of maternal animals given the high dose of 138 mg/kg/day during late pregnancy and lactation. This correlated with a decrease in maternal body weight and feed consumption during late lactation. Whether the decrease in pup weight gain was the result of the decrease in maternal feed consumption or whether it may have been a direct effect on offspring via the breast milk transfer of the compound is not known.

In the mouse oncogenicity study, animals were treated with up to 138 mg/kg/day omeprazole for 18 months. A decrease in survival was noted at the high dose, but no neoplasia was observed in any organ. Different results, however, were obtained in the rat oncogenicity study, in which animals were treated with 13.8, 43, or 138 mg/kg/day omeprazole for 24 months. Enterochromaffin-like (ECL) cell hyperplasia, progressing to ECL cell carcinoids, occurred in dose-related fashion in these animals, with males being affected at doses of 43 and 138 mg/kg/day and females being affected at all dose levels. These positive findings resulted in the temporary suspension of the clinical trial program. The carcinoids were characterized as "end-of-life" tumors, as the first was discovered at 82 weeks of treatment in an animal that had died prematurely. Carcinoid tumors were not identified as the cause of death in any animals, and no metastases were found. A 2-year study was repeated in females in an attempt to define a dose at which ECL cell carcinoids did not occur; however, carcinoid formation again occurred in a dose-related fashion, including at the lowest dose tested (1.7 mg/kg/day).

A major question that must be addressed following positive results in a carcinogenicity bioassay is whether tumorigenesis was the direct result of chemical insult or could be related to the pharmacologic mechanism of action of the compound. Furthermore, whether the model is appropriate for extrapolation of these findings to humans requires evaluation. For example, at this late stage in the development of a compound, sufficient pharmacokinetic data should be available in both the animal species tested and humans to determine whether a finding might be restricted to a species that metabolizes the compound quite differently from humans. If this is the case, further mechanistic studies can be designed to support or refute the applicability of the findings.

A number of *in vivo* and *in vitro* mutagenicity studies were conducted with omeprazole (Table 7.8). An initial mouse micronucleus test with omeprazole administered to animals at a high dose of 5000 mg/kg produced equivocal results, with slight increases in the mean numbers of micronucleated cells compared to controls (approximately 2-fold, compared to 30-fold following a 0.4-mg/kg dose of the positive control, mitomycin C). It was noted that the dose of 5000 mg/kg was not well tolerated. A second mouse micronucleus test conducted with a maximum dose of approximately 800 mg/kg did not show evidence of mutagenic potential. All other mutagenicity tests conducted produced negative results, suggesting that the tumorigenesis observed in the 2-year rat studies was not the result of a genotoxic action of omeprazole or its metabolites.

Mechanistic studies in dogs and rats, combined with correlative data from clinical trials, ultimately provided the information to support the safe use and registration of omeprazole. At the doses selected for the toxicity studies, the sustained decrease in luminal pH of the stomach resulting from the inhibition of gastric acid secretion by omeprazole caused a substantial increase in the release of gastrin into the blood. In fed rats, gastrin levels in plasma normally range from 150 to 200 pg/mL. Omeprazole administered at doses of 13.8 to 138 mg/kg/day increased plasma gastrin concentrations to 1000 to 3000 pg/mL. Gastrin has a trophic effect on the gastric mucosa and results in the hyperplasia of several cell types, including ECL cells, and consequent mucosal thickening. These data suggest that omeprazole does not inherently cause ECL cell hyperplasia and resulting carcinoid formation. Indeed, in antrectomized dogs (the major source of gastrin is surgically removed), high doses of omeprazole for 1 year resulted in neither hypergastrinemia nor mucosal hyperplasia [73]. Similarly, antrectomy in rats prevents the hypergastrinemia and ECL cell hyperplasia associated with omeprazole treatment [152].

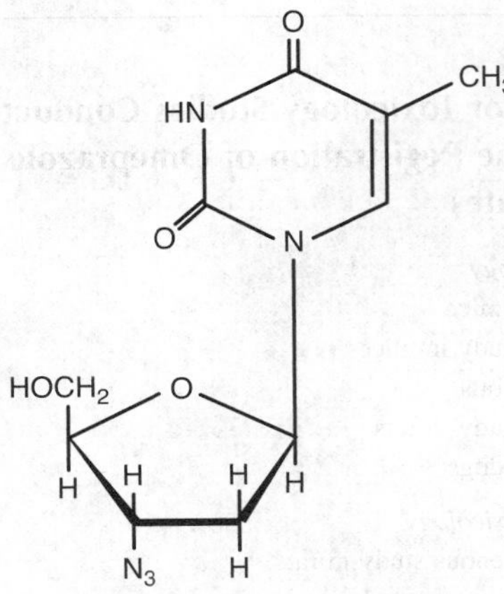

**FIGURE 7.9** Structure of zidovudine.

The course of development of omeprazole demonstrates the importance of conducting mechanistic studies to elucidate the significance of findings in animal safety studies and whether the effects can meaningfully be extrapolated to humans. Clearly, a close collaboration between the toxicologist and clinician during advancing human trials is critical to resolve questions related to human safety. The extensive clinical experience with omeprazole has confirmed its safe and effective use for treatment of the described indications [15,183]. In a group of patients who required continuous treatment with 40 mg of omeprazole for up to 4 years, no evidence was found for dysplastic or neoplastic changes [51]. Over 12,000 endoscopic biopsies further supported the clinical safety of omeprazole relative to its potential for causing hyperplastic changes.

It should be noted that development of omeprazole probably took years longer than anticipated due to the need to conduct the supplementary mechanistic and carcinogenicity nonclinical studies, as well as additional clinical investigations. Clearly, in these situations, the sponsor needs to evaluate whether the additional time and expense required for approval are supported by the market need. In this case, in 2001, 12 years after approval and prior to being approved for over-the-counter distribution, Prilosec was the world's second biggest selling drug.

## Zidovudine (Retrovir®/AZT)

Zidovudine (azidothymidine [AZT]; Figure 7.9) inhibits viral RNA-dependent DNA polymerase (reverse transcriptase) and, thus, viral replication. Furthermore, as a thymidine analog, zidovudine becomes incorporated into growing strands of DNA by viral reverse transcriptase and inhibits the further addition of nucleotides. It is intended

**TABLE 7.9**
**Summary of Toxicology Studies Submitted for Initial FDA Review To Support the Registration of Zidovudine in the United States**

| |
|---|
| *Acute toxicology* |
| Intravenous study in mice |
| Intravenous study in rats |
| *Subchronic toxicology* |
| 2-Week intravenous study in dogs with 2-week recovery |
| 2-Week oral study in rats |
| 2-Week oral study in dogs |
| 2-Week oral study in monkeys |
| 1-Month intravenous study in rats with 2-week recovery |
| 3-Month oral study in rats with 2-week recovery |
| 3-Month oral study in monkeys with 6-week recovery |
| 6-Month oral study in rats with 2-month recovery |
| *Reproductive and developmental toxicology* |
| Segment II oral teratology in rats |
| Segment II oral teratology in rabbits |
| *Genetic toxicology* |
| Mouse lymphoma assay |
| Ames *Salmonella* test with/without metabolic activation |
| Cell transformation assay |
| *In vivo* cytogenetic study in rats |
| *In vitro* cytogenetic study in human lymphocytes |
| *Summary of toxicology studies planned or in progress at the time of initial FDA review of zidovudine* |
| 6-Month oral study in monkeys |
| Segment I oral reproduction/fertility study in rats |
| Segment III oral peri- and postnatal study in rats |
| Segment II oral teratology study in rabbits |
| 1-Year oral study in rats |
| 1-Year oral study in monkeys |
| Oral carcinogenicity study in rats |
| Oral carcinogenicity study in mice |

for use in the management of adult patients with human immunodeficiency virus (HIV) infection when antiretroviral therapy is warranted [184]. It is also indicated for the prevention of maternal–fetal HIV transmission during gestation and labor and in the neonate after birth. The recommended dose for adults is 600 mg/day in divided doses in combination with other antiretroviral agents. Zidovudine is also available for intravenous infusion in patients with advanced disease and for use in women during labor and delivery. In spite of the intended long-term use of the compound, it was approved with a minimal toxicology package due to the serious nature of the disease and lack of alternative efficacious therapies.

The studies listed in the upper portion of Table 7.9 were either submitted as part of the original NDA in December of 1986 or as amendments to the application shortly thereafter [81]. At the time of the initial pharmacology/toxicology review, a variety of chronic toxicity studies were still underway or planned. The FDA commentary indicated that: "Nonclinical toxicity data submitted in support of the application include results of studies in rats, dogs, and cynomolgus monkeys. FDA guidelines would have prescribed more extensive nonclinical testing than that reported thus far; however, the urgency for developing an anti-AIDS drug has been so great that clinical testing has preceded the usual/customary nonclinical testing. For example, while data from a 6-month *clinical* study are available, results for the supporting 6-month *nonclinical* toxicity studies have not yet been submitted" [emphasis added]. An approvable letter issued by the FDA in March of 1987, less than 4 months following submission of the NDA, stipulated the timing for the conduct of these outstanding studies. Comprehensive reviews of the acute, subchronic, chronic, genetic, carcinogenic, reproductive, and developmental toxicity studies have been published [9,10,113].

Zidovudine demonstrated relatively low acute toxicity with intravenous median lethal doses (MLDs) of greater than 750 mg/kg in rats and mice. The most consistent findings in the subchronic and chronic studies with rats, dogs, and monkeys were effects on hematologic parameters. In rats given two divided doses of zidovudine at approximately 50, 150, or 500 mg/kg/day orally for 3 or 6 months, reversible decreases were observed in red blood cell counts and hemoglobin concentration, primarily in the mid- and high-dose groups. The severity of these effects appeared to progress slightly between 3 and 6 months of treatment. No remarkable histopathology was noted in these studies. A subsequent 1-year study, submitted well following the initial approval of zidovudine, revealed a toxicity profile similar to that observed in the 3- and 6-month studies. The severity of anemia did not progress between 6 and 12 months of exposure, and effects were again reversible following discontinuation of treatment.

Dogs were more sensitive to zidovudine treatment. In a 2-week study in which animals were administered 125 to 500 mg/kg/day orally in divided doses, leukopenia, thrombocytopenia, and decreases in erythroid values were observed at all dose levels. Cytostatic effects were observed in the small intestine at the high dose and were also evidenced by slight to moderate non-dose-related lymphoid depletion and mild to marked dose-related bone marrow hypocellularity at all dose levels. No cytostatic effects were observed at similar or higher doses in either rats or monkeys. Studies revealed that zidovudine was metabolized almost identically in monkeys and humans, and as a result the continued nonclinical development of the drug was conducted in the monkey rather than the dog. The species differences in metabolism were not of sufficient magnitude, however, to account for the much greater sensitivity of the dog to zidovudine, and the design of subsequent nonclinical and clinical studies continued to respect the significant findings in this species.

Monkeys responded to a 2-week treatment at divided doses of 125 to 500 mg/kg/day with a slight reduction in hemoglobin concentrations in one animal at the low dose and in both monkeys given the high dose. In a 3-month monkey study, at divided doses of 35 to 300 mg/kg/day, dose-related decreases in erythron parameters were noted as early as day 15 of treatment and progressed to live-phase termination. Platelet counts were also increased. Values returned to normal during the 6-week recovery phase of the study. Subsequent (post-approval) 6- and 12-month studies were conducted in the monkey. In addition to the findings in the 3-month study, bone marrow cytology revealed changes consistent with the hematology findings, and marginal decreases in white blood cell counts were observed at the 300-mg/kg/day dose. All findings were again reversible.

Teratology studies were also carried out in rats and rabbits prior to approval of zidovudine. Divided doses up to 500 mg/kg/day resulted in no evidence for teratogenicity in either species, but non-treatment-related low fertility rates and mortalities in the rabbit study prompted the FDA to request that a second study be initiated prior to drug approval. Effects were limited to an increase in fetal resorptions and an associated decrease in fetal body weights at the maternally toxic high dose. The potential use of zidovudine in pregnant women to inhibit transplacental HIV transmission prompted additional *in vitro* and *in vivo* reproductive and developmental toxicity studies following the initial approval of the compound. These subsequent studies demonstrated that zidovudine is embryotoxic in rats at doses that are not overtly maternally toxic. The postnatal survival, growth, and development of offspring from zidovudine-treated rats were unaffected following several treatment regimens. In general, exposure levels associated with the effects observed in the reproductive and developmental toxicity studies were significantly higher than those observed clinically.

No evidence for mutagenicity by zidovudine was observed in the Ames *Salmonella* study either with or without mammalian metabolic activation. The compound was weakly mutagenic in the mouse lymphoma assay without metabolic transformation at concentrations of 4000 and 5000 μg/mL; it was also weakly mutagenic with metabolic activation at concentrations greater than or equal to 1000 μg/mL. A positive response was obtained in the mammalian cell transformation assay at concentrations of 0.5 μg/mL or greater. In an *in vitro* cytogenetic assay in human lymphocytes, zidovudine caused structural chromosomal abnormalities at concentrations equal to or greater than 3 μg/mL; however, in an *in vivo* rat assay, no chromosomal abnormalities were noted following the intravenous administration of doses up to 300 mg/kg (plasma levels over 400 μg/mL). Subsequent *in vivo* micronucleus studies in mice and rats revealed dose-related increases in micronucleated erythrocytes, reflecting chromosome breakage or mitotic spindle damage.

Carcinogenicity studies in mice were initiated using single daily doses of 30, 60, or 120 mg/kg. These doses were reduced to 20, 30, or 40 mg/kg at 3 months of treatment due to treatment-related anemia. Rats were dosed with 80, 220, or 600 mg/kg/day, with the high dose being reduced to 450 then 300 mg/kg. As expected, hematologic changes were observed, but no deaths or morbidities occurred that were considered treatment related in either study. In the mouse study, one benign vaginal neoplasm occurred at 30 mg/kg, and 5 malignant and 2 benign neoplasms occurred at 40 mg/kg. Two vaginal neoplasms occurred at the high dose of the rat study. In both cases, the tumors were late occurring and nonmetastasizing. An eloquent argument was put forth suggesting that these vaginal tumors resulted from high local exposure to zidovudine due to the retrograde flow of urine containing high levels of the excreted compound into the vagina. An additional lifetime mouse study to support this hypothesis was conducted in which animals were administered zidovudine intravaginally. Thirteen vaginal squamous cell carcinomas occurred in animals receiving the highest concentration in that study, supporting the contention that systemic exposure to the drug was unlikely to be responsible for the neoplasia observed in the oral studies.

A variety of adverse reactions have been documented in patients receiving zidovudine. Due to the wide range of symptoms associated with the opportunistic infections seen in patients with autoimmune deficiency syndrome (AIDS), it is difficult to assess which adverse reactions are clearly the result of zidovudine therapy. The animal safety studies, however, were highly predictive of the major hematologic toxicities of zidovudine described in humans: granulocytopenia and severe anemia. Similar to the earlier discussion regarding the development of omeprazole, additional mechanistic studies were conducted with zidovudine to explain toxicity findings, even though the drug was intended for the treatment of a potentially fatal disease.

Although the toxicology support package for zidovudine ultimately responded to existing guidelines for registration of a chronic-use pharmaceutical in the United States, its development history demonstrates that the approval system allows considerable flexibility in cases where the market for a life-threatening disease is clearly not satisfied (i.e., antivirals, antifungals, oncolytics). This type of development strategy can only occur with close collaboration between the submitter and regulatory agency and after careful consideration of the risk/benefit assessments.

**TABLE 7.10**
**Safety Testing of Biotechnology Products**

| Category | Requirements |
|---|---|
| 1 | Identity, purity, pharmacology, safety pharmacology |
| 2 | Category 1, plus detailed pharmacologic activity (human, animal), relationship of plasma concentration and antibody titer (human, animal, *in vitro*), tolerance, selected toxicological testing |
| 3 | Categories 1 and 2, plus studies guided by indication, studies guided by duration of treatment |

*Source:* Bass, R. and Scheibner, E., *Arch. Toxicol.*, 11(Suppl.):182–190, 1987. With permission.

## NEW BIOLOGICAL ENTITIES

### SPECIAL AGENTS

The development of highly purified species-specific protein pharmaceutical agents, made possible through advances in rDNA technology, presents a significant challenge to toxicologists. The major question raised is "What nonclinical toxicology evaluations should be conducted to ensure safety in human clinical trials?" The major issue is the testing of these specific proteins in nonhomologous animal species, where the possibility of immunogenicity, not applicable to the clinical trial, exists [209]. As no universally accepted methods or procedures for the nonclinical evaluation of NBEs exist, decisions of appropriate nonclinical study designs are made on a case-by-case basis. The general consensus is that nonclinical toxicity evaluations with species-specific proteins reveal little more than enhanced pharmacodynamic activity rather than predicting the potential for adverse effects. Furthermore, the toxicity observed in animal studies may be the result of an immunologic response to the foreign protein. Toxicology studies with NBEs should demonstrate that the product has no adverse effects other than those specifically related to pharmacodynamics and that safety for the expected clinical dose range, rather than exaggerated toxicity (MTD), should be demonstrated [14,182,209].

Regulatory agencies have placed great emphasis on chemical characterization of the NBE as a means of establishing that it is identical with the naturally occurring protein (manufacturing contaminant issues aside). Establishing this identity has allowed for appropriate modification of toxicology requirements and abbreviation of the toxicology support package; however, NBEs are being developed that contain amino acid sequences that have been purposefully manipulated to differ from the naturally occurring protein to result, for example, in a prolonged duration of action over the naturally occurring agent. These molecules may require a more comprehensive toxicology package, such as those established for NCEs (see above).

Safety testing of NBEs can be presented in three categories (Table 7.10). The reasonably clear-cut time sequence of nonclinical and clinical studies established with NCEs is not often feasible with NBEs. The interactions between toxicologists and clinicians are important in addressing suspected adverse reactions in the clinical trials [14]. Nonclinical toxicology evaluations of NBEs should be designed according to the risks anticipated from the type of product, the contaminant profile, and the intended clinical use [14]. Major questions and differences of opinion will continue to exist relative to the evolution of nonclinical toxicology testing strategies of NBEs. The major questions will arise concerning appropriate species [67,209], the need to conduct genetic toxicology studies [14,209], the conduct of reproductive toxicology assessments [209], and the need for classical carcinogenicity studies. As examples of NBEs, we have chosen to discuss toxicology support for the registration of the gonadotropin-releasing hormone analogs, interferon, and human insulin. The development of interferon has provided a great deal of guidance for nonclinical toxicity testing of NBEs. The pharmacologic effects of insulin are well known from extensive clinical experience. This experience has aided the relatively rapid approval of rDNA insulin products and has allowed the chemical characterization of test material to play a major role in supporting a more limited toxicology profile. Human insulin, therefore, provides an example of a NBE that was approved rapidly, in approximately 5 months [153].

#### Gonadotropin-Releasing Hormone Analogs

Gonadotropin-releasing hormone analogs are either agonists or antagonists of the receptor for the naturally occurring hypothalamic decapeptide, GnRH. The chemical modifications either increase the biological activity and duration of action or affect the solubility, potency, and kinetics of the molecule. GnRH analogs were first introduced for the treatment of cancer (e.g., prostatic carcinoma), and their toxicologic assessment was less complete than usually recommended for new drugs. Since their introduction, the use of GnRH analogs has expanded into treatment of non-life-threatening conditions, and they now are expected to have to undergo the same rigorous toxicology evaluation as other new drugs [188]. In the case of GnRH analogs, the FDA has allowed the multiple-dose

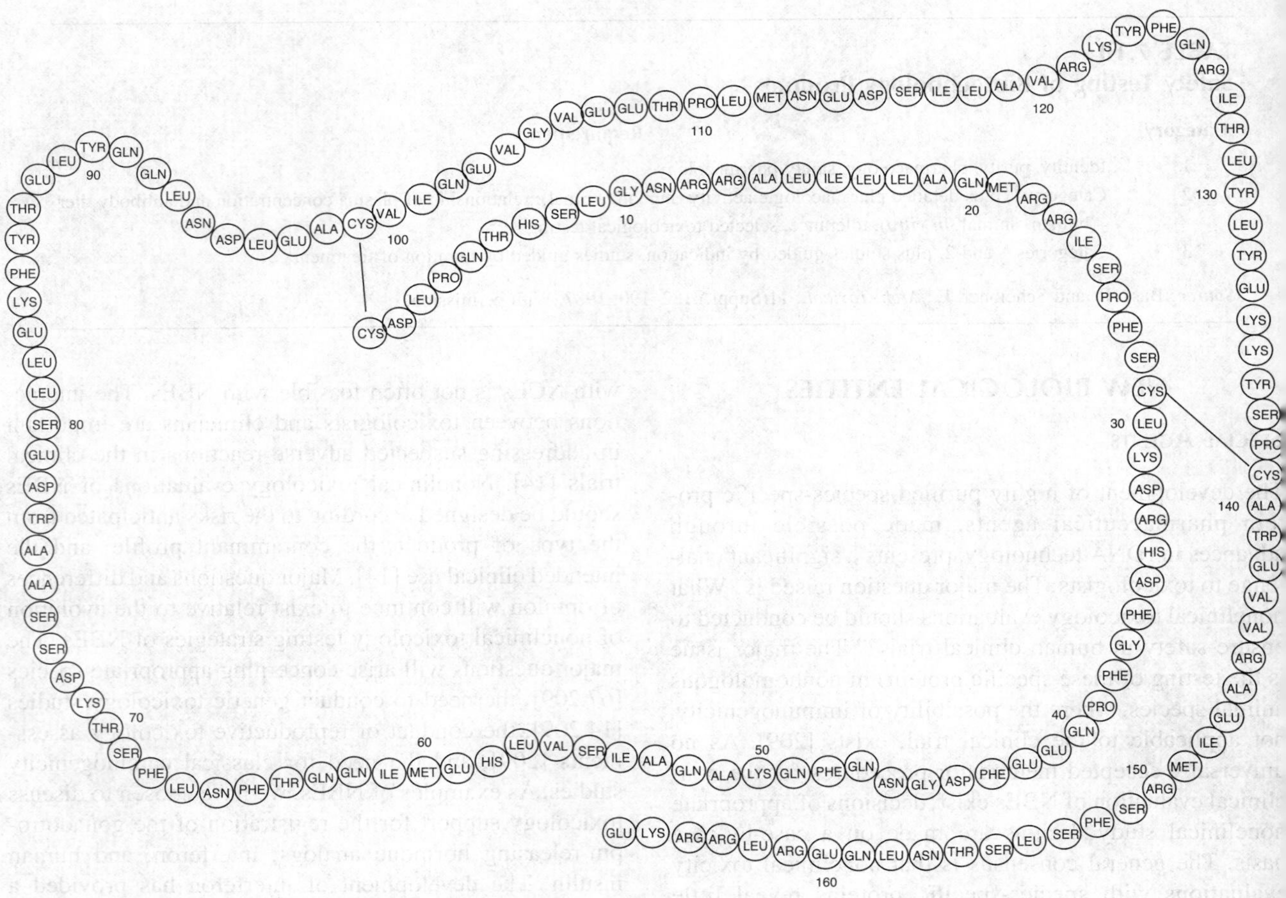


**FIGURE 7.10** Consensus sequence of human leukocyte interferons. (Adapted from Stebbing, N. and Weck, P.K., in *Recombinant DNA Products: Insulin, Interferon and Growth Hormone*, Bollon, A.P., Ed., CRC Press, Boca Raton, FL, 1984, pp. 75–114.)

toxicity studies to be conducted at a multiple of human exposure (30- to 50-fold) rather than at doses that define the toxic limits of the compound. Due to the chronic nature of therapy and the chemical dissimilarity with native GnRH, the FDA has recommended that both rat and mouse carcinogenicity studies be conducted. As is the case with multiple-dose toxicity studies, the FDA has allowed that the MTD need not be used but some multiple of the human clinical exposure must be used to set the top dose in the carcinogenicity studies (e.g., 15- to 50-fold). The full toxicity profile recommended for GnRH analogs includes single-dose acute toxicity (rodent and non-rodent), repeat dose toxicity studies through 6 months in rodents and 9 months in nonrodents, genetic toxicology, developmental toxicology, and carcinogenicity [188].

## Interferon

Interferons (IFNs) are classified as IFN-α (leukocyte), -β (fibroblast), or γ (immune) (Figure 7.10). IFN-α consists of a family of at least 14 highly homologous species. The amino acid sequence homology of the IFN-α subtypes has been reported to be 52 to 75% [202, 203,216]. The biologi activities of IFNs include antiviral, anticellular, and immu nomodulatory activities [216]. The properties of interferon and their potential uses have been reviewed by Bocci [16]

The adverse clinical experiences reported with the us of IFNs include fever, chill/rigor, headache, tremor, nausea vomiting, myalgia, anxiety, fatigue, malaise, anorexia, con fusion, local inflammation, cardiovascular toxicity, hepa totoxicity, and abnormal electroencephalograms (EEGs [92,197,198,203,222]. The most commonly reporte adverse effects are fever, fatigue, and leukopenia [203] The effects that cause the most distress in clinical subject are those related to central nervous system (CNS) depres sion [222]. The toxicity seen with very pure and single clone IFN preparations is almost identical to that reporte for the less pure, more heterogeneous preparations of IFNs The responses reported, particularly the influenza-like syn drome, therefore, are likely intrinsically related to IFN and not to a contaminant or impurity [203,222].

The species specificity of highly purified human IFN implies that classical animal (nonhomologous) efficac and toxicity models are not applicable in evaluation c

**TABLE 7.11**
**Recommendations for Interferon Testing by the French Ministry of Social Affairs**

| Toxicologic Test | Recommendation |
|---|---|
| Acute | Two species, both sexes in one species; two routes; 2-week observation |
| Subchronic | Two species, rodent and primate; 3 months of daily injections |
| Reproduction | Segments I, II, and III |
| Mutagenicity | *In vivo* and *in vitro* clastogenesis |
| Carcinogenicity | Not required |
| Pyrogenicity | Rabbit |
| Safety pharmacology | *In vivo* cardiopulmonary, isolated organs; neurobehavioral studies |
| Cell culture | Cytostatic and cytotoxic effects |

*Source:* Zbinden, G., in *Preclinical Safety of Biotechnology Products Intended for Human Use*, Graham, C. E., Ed., Alan R. Liss, New York, 1987. pp. 143–159. With permission.

these materials. Nonclinical safety testing of IFNs has not identified an appropriate animal model [96,121,189,194, 204,245], supporting the recommendation that the routine safety tests applied to NCEs should not be applied haphazardly [209] to NBEs. Yet, given the traditional significance and predictive nature of nonclinical toxicology evaluations and acknowledging the lack of generally accepted and validated nonclinical animal models for the testing of these entities, drug regulatory agencies have published safety testing guidelines that place NBEs on a level with conventional drugs relative to the comprehensive requirements for animal safety studies [99,144]. A representative example of these guidelines and requirements is given in Table 7.11.

Interferon-α2a (Roferon®-A) is a commercially available NBE identical to one of the 15 subtypes of human leukocyte IFN [217]. At the time that clinical trials were initiated with this drug, considerable clinical data were available from studies with other leukocyte IFNs to indicate the types of adverse reactions, described above, that might be expected [203,222]. The species specificity of recombinant IFN-α2a has led to production of neutralizing antibodies in rodent and nonrodent species [217]. This has impaired the ability of toxicology studies to detect the expected adverse clinical signs in common toxicology species.

Acute, single-dose toxicology studies (Table 7.12) were conducted in a variety of species in an attempt to disclose any unexpected acute toxicity related to the clinical dosage form (excipients, active ingredients). No mortalities were noted in the species tested. The $LD_{50}$ of IFN-α2a was determined to be >22.8 × 106 units/kg i.v. These studies were conducted at multiples of a single clinical dose (3 × $10^6$ units/kg), ranging from 10- to 167-fold. Multiple-dose toxicology studies were conducted over a range of 5 to 26 weeks at 3- to 78-fold the weekly clinical dose (9 × $10^6$ units/kg) (Table 7.13). A low frequency of treatment-

**TABLE 7.12**
**Acute Toxicology Studies Conducted with Interferon-2α**

| Species | Route[a] | Dose (units × $10^6$/kg) | Clinical Multiple[b] |
|---|---|---|---|
| Mouse | iv | 30,250 | 83× |
| | im | 30,500 | 167× |
| | sc | 30 | 10× |
| Rat | iv and im | 30,100 | ≤33× |
| | sc | 30 | 10× |
| Rabbit | iv and im | 100 | 33× |
| | im and sc | 30 | 10× |

[a] iv, intravenous; im, intramuscular; sc, subcutaneous.

[b] Recommended clinical dose = 3 × $10^6$ units/kg im or sc, three times weekly.

**TABLE 7.13**
**Multiple-Dose Toxicology Studies Conducted with Interferon-2α**

| Species | Route[a] | Duration (weeks) | Dose (units × $10^6$/kg) | Clinical Multiple[b] |
|---|---|---|---|---|
| Mouse | im | 5 | 0, 1.4, 2.8, 5.7 | ≤4× |
| Rat | im and iv | 5 | 0, 1, 10, 100 | ≤78× |
| Rat | im | 26 | 0, 7.5, 15, 30 | ≤23× |
| *Saimiri sciureus* | im | 2 | 0, 2.5 | 2× |
| *Macaca mulatta* | im | 4 | 0, 2.5, 10, 25 | ≤19× |
| *Macaca fasicularis* | im | 13 | 0, 2, 10 | ≤3× |

[a] iv, intravenous; im, intramuscular.

[b] Recommended clinical dose = 3 × $10^6$ units/kg three times weekly (9 × $10^6$ units/kg/week).

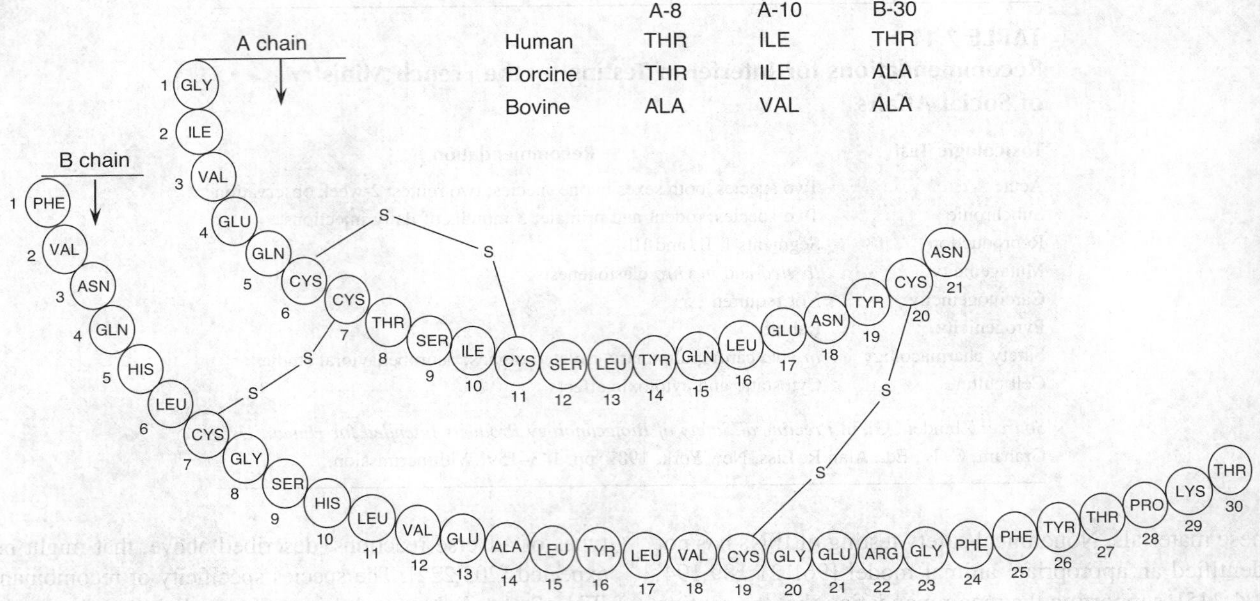


**FIGURE 7.11** Sequence of biosynthetic human insulin (BHI).

related adverse findings was reported: slight weight loss in rats; a slight, reversible increase in platelets and total leukocytes in mice; a slight decrease in hemoglobin and hematocrit in squirrel monkeys; dose-dependent anorexia and weight loss in *Macaca mulatta*; and transient anorexia in *M. fasicularis*. In studies longer than 2 weeks, neutralizing antibodies developed in rabbits, guinea pigs, and *M. fasicularis* [217]. These results were expected and may have affected the signs of toxicity. Reproductive studies carried out in *M. mulatta* indicated that a dose-dependent increase in abortion was related to the administration of IFN-α2a.

## Insulin

The nonclinical toxicity of biosynthetic human insulin (BHI) (Figure 7.11) was evaluated in an unconventional way. The use of graded increments of dose representing multiples of the projected clinical exposure was not feasible due to the pharmacologic effect of hypoglycemia caused by the insulin molecule. Because the pharmacologic effects of insulin were well known, a primary goal of the toxicology evaluations was to determine whether BHI contained potentially toxic contaminants or impurities (e.g., *Escherichia coli* proteins, endotoxins) that are introduced as a result of the synthetic process. Toxicology studies on BHI were conducted simultaneously with purified porcine pancreatic insulin (PPI) as a positive control, at doses previously established to produce hypoglycemia but not mortality. The doses selected for toxicology studies were varied according to species sensitivity, route of administration, and duration of treatment. In acute, single-dose toxicity studies (Table 7.14), the minimal lethal dose of BHI to rats and mice was >10 units/kg s.c. Dogs given single doses of 2 units/kg BHGI s.c. showed the expected hypoglycemia, but no toxicity. No toxic effects were seen in either rats or dogs given BHI s.c. or i.v. for 1 month (Table 7.15). Chronic toxicity, reproductive toxicity, and carcinogenicity studies were not conducted due to the extensive clinical experience with animal-derived insulin and the extensive chemical analysis of the NBE, establishing its identical nature with natural human insulin. BHI was negative in a genetic toxicology screen composed of bacterial mutation, DNA repair, and sister chromatic exchange evaluations. BHI was also not pyrogenic. Overall, BHI did not induce any effects different from those induced by PPI, and all effects seen were extensions of known insulin pharmacology. Investigations demonstrating the virtual absence of endogenous *E. coli* proteins and

**TABLE 7.14**
**Acute Toxicology Studies Conducted with Biosynthetic Human Insulin**

| Species | Route[a] | Dose (units/kg) | Clinical Multiple[b] |
|---|---|---|---|
| Mouse | sc | 10 | 40× |
| Rat | sc | 10 | 40× |
| Dog | sc | 2 | 8× |
| Monkey | iv | 0.1 | — |

[a] sc, subcutaneous; iv, intravenous.
[b] Anticipated clinical dose = 0.24 units/kg/day, sc.

**TABLE 7.15**
**Multiple Dose Toxicology Studies Conducted with Biosynthetic Human Insulin**

| Species | Route[a] | Duration (weeks) | Dose (units/kg) | Clinical Multiple[b] |
|---|---|---|---|---|
| Rat | sc | 4 | 2.4 | 10× |
| Dog | sc | 4 | 2.0 | 8× |
| Dog | iv | 4 | 0.1 | — |

[a] sc, subcutaneous; iv, intravenous.
[b] Anticipated clinical dose = 0.24 units/kg/day, sc.

the absence of antigenic response in rats and guinea pigs sensitized with *E. coli* polypeptides further addressed the safety of the rDNA-derived human insulin product.

## Special Issues

### No-Observed-Adverse-Effect Level

The no-observed-adverse-effect level (NOAEL) is an important concept in the evaluation of potentially toxic agents [22,156]. The use of the NOAEL in the development of pharmaceuticals has been reviewed by Dorato and Engelhardt [71]. The NOAEL for pharmaceuticals may be defined as the "highest dose/exposure that does not cause important increases in the frequency or severity of adverse effects between exposed and control groups based on careful biological and statistical analysis. While minimum toxic effects or pharmacodynamic responses may be observed at this dose, they are not considered to be adverse to human health or as precursors to serious adverse events with continued duration of exposure" [71]. This approach fits very well with the ICH position that the effect to be determined is the toxicologically relevant effect; that is, the effect that may endanger human health [125].

In addition to the more traditional approach to NOAEL determination, alternatives have been discussed. The benchmark dose approach claims to be a more powerful statistical tool than traditional NOAEL approaches [94]. The term *hormesis* may be considered to describe a response that is stimulatory at low doses and inhibitory at high doses; however, no universally accepted definition of hormesis relative to its use in safety assessment exists. Hormesis theory has been proposed as a method to improve toxicology risk assessment [23–26]. As a perspective, Axelrod et al. [8] have presented an argument that the existing toxicology data do not support a universal extension of the hormesis concept to regulatory policy. The last 40 years of drug safety evaluations support the more traditional NOAEL methodology as an appropriate approach to risk/benefit for pharmaceuticals. Toxicologists should be open, however, to alternative approaches and challenges to established practices.

## Immunotoxicology

Immunotoxicology can be defined as the discipline concerned with the study of adverse effects on the immune system as a result of exposure to xenobiotics [62]. The development of immunotoxicology, since the 1970s, has been reviewed by Koller [150]. It is not the purpose of this section to review in detail the specific evaluations conducted to define immunotoxicity [60,62,214,234], but to discuss the use of these evaluations in a hazard assessment tier approach. Adverse responses of the immune system are known to occur secondary to malnutrition, radiation exposure, neonatal thymectomy, and exposure to certain drugs and chemicals [64,221]. Historically, few chemicals have been shown to be immunosuppressive in toxicology evaluations, probably because the lymphoid organs and the immune system, in general, have been poorly examined. It would be desirable, therefore, to establish an effective tier approach to detect immunotoxicity in standard subchronic and chronic toxicology studies and also to evaluate the functional nature of the changes observed as a result of drug exposure. It is presumed that a functional change detected in the immune system is predictive of adverse health effects [234]. It must be remembered that a critical function must be depressed beyond a defined, minimal point (reserve capacity) to indicate a health risk [163]. The tier approach is encouraged because it more carefully directs the use of resources, and a single immune function assay may not comprehensively characterize the myriad of potential toxic effects on the immune system [62]. Specific immune function tests for increasing the capability of toxicology studies to reveal effects on lymphoid tissue and to evaluate more fully the risk of chemical exposure by determining the functional significance of the responses observed have been reviewed [60]. It is known that acute and chronic effects of drug exposure on the immune system can result in three principal undesirable effects: immunosuppression or enhancement, autoimmunity, and allergic reactions [163]. Immunosuppression has also been related to an increased incidence of neoplasia, although the relationship between immunosuppression and carcinogenesis is not a direct one [64]. An early consensus meeting held by the National Institute for Environmental Health Sciences (1979) resulted in the development of a list of relevant immunologic parameters for evaluating chemically induced immunotoxicity. This immunology screening panel has been reviewed [60] and includes pathotoxicology, hematology, host resistance, radiometric delayed hypersensitivity, lymphoproliferation, humoral immunity, and evaluation of bone marrow progenitor cells.

One of the first guidelines for immunotoxicology testing was that developed by the EU in the late 1970s. The focus of these guidelines was to evaluate the potential risk of chemical exposure by evaluating the functional signif-

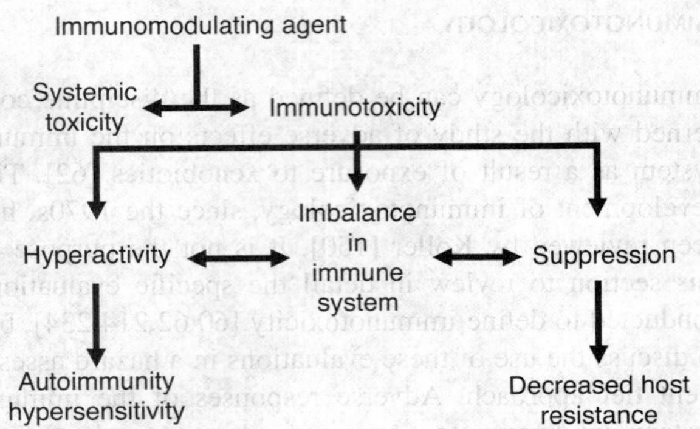


**FIGURE 7.12** Potential toxic responses of immunomodulating agents. (Adapted from Falchetti, R. et al., in *Current Problems in Drug Toxicology*, Zbinden, G. et al., Eds., John Libbey Eurotext, London, 1983, pp. 248–263.)

icance of any histopathological or hematological effects seen on lymphoid organs in routine toxicity studies [176]. The intention was to pursue the significance of these effects with specific function tests as necessary. It is known that immunotoxicity following drug exposure may take the form of changes in lymphoid tissue organ weights or histology or changes in bone marrow or peripheral leukocytes [62]. Norbury [177], however, pointed out that the evaluation of drug effects on the immune system is related to immune responsiveness and is not simply a single-point examination of lymphoid tissue using histopathology and hematology. Histopathologic changes are generally not believed to be sensitive indicators of drug-induced immunotoxicity, are seen only at fairly high dose levels, and do not necessarily equate with functional immune alterations [62,177,234]. The route and time of exposure relative to the maturational development of the immune system are important considerations in designing an immunotoxicity protocol [32,61].

The application of nonspecific immunotherapy for bacterial and viral diseases has led to an increased level of importance in the determination of immunotoxic effects. The standard acute, subchronic, and chronic/oncogenic studies that form the basis of toxicologic evaluations should be complemented with specific evaluations useful in determining functional effects on the immune system, especially if the agent in question is a known immune modulator. Immune system function results from a balance of the activities of various cellular components and their soluble factors [191], and an alteration in any factor could result in an imbalance of the entire system [78]. The effects of immunomodulating agents, therefore, could result in either enhancement (e.g., hypersensitivity, autoimmunity) or suppression (e.g., decreased host resistance) (Figure 7.12).

Several tier approaches to immunotoxicity testing have been proposed [59,163]; for example, the National Toxicology Program (NTP) has proposed an immunotoxicology testing strategy that includes a limited number of functional and host resistance assays [59]. The two-tier approach consists of a screen (tier 1), which represents a limited effort that includes the assessment of cell-mediated immunity, humoral immunity, and immunopathology. Tier 1 provides little information on the specificity of an observed immune defect or its relevance to the host; however, it can detect an immune alteration resulting from drug exposure (163). Tier 2 represents an in-depth evaluation, initiated only if functional changes are seen in tier 1 at otherwise nontoxic doses [163]. The in-depth immune function and host resistance evaluations provide information on the mechanisms of the immunotoxicity and aid risk assessment. Luster et al. [163] have reported that no compound evaluated to date has been found to produce an effect in tier 2 without demonstrating some effect in tier 1. The NTP procedure for detection of immune alterations following chemical or drug exposure in rodents is shown in Table 7.16. The concept of performing functional tests is critical to defining potential mechanisms of the toxic response and their applicability to humans. An inter-

## TABLE 7.16
## NTP Immunotoxicology Procedure

| | Tier 1 |
|---|---|
| Immunopathology | Hematology (complete and differential blood count) |
| | Organ weights (spleen, thymus, kidney, liver) |
| | Body weight |
| | Cellularity (spleen) |
| | Histology (spleen, thymus, lymph node) |
| Humoral-mediated immunity | Plaque-forming cells |
| Cell-mediated immunity | Lymphocyte response to mitogens |
| Nonspecific immunity | Natural killer cell activity |

| | Tier 2 |
|---|---|
| Immunopathology | Quantitation of splenic B and T cells |
| Humoral-mediated immunity | IgG response to sheep RBCs |
| Cell-mediated immunity | Delayed hypersensitivity |
| Nonspecific immunity | Macrophage function |
| Host resistance | Syngenic tumor cells (tumor incidence) |
| | *Listeria monocytogenes* (mortality) |
| | Influenza (mortality) |
| | *Plasmodium yoelii* (parasitemia) |

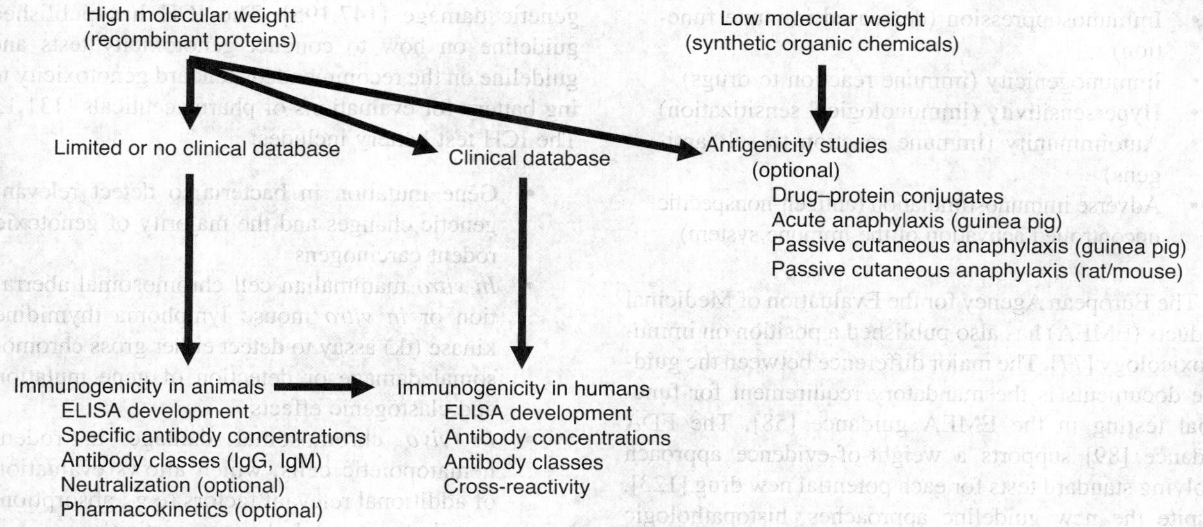


FIGURE 7.13 Proposed approach for addressing immunogenicity and antigenicity issues with established and novel biotechnology products and NCEs (pers. comm., D. Wierda, Eli Lilly and Company, Greenfield, IN, 1992).

national collaborative effort has focused on the evaluation of limited pathology or enhanced pathology evaluations to better understand potential immunotoxicity [212]. The enhanced pathology approach (e.g., weight determination, examination of additional lymphoid organs, grading of changes in the principal compartments of lymphoid tissue) was determined to provide an advantage in revealing effects on the immune system.

The direction for pharmaceutical development is to include tests of potential immune system involvement in the traditional toxicologic evaluations for subchronic and chronic toxicity. Due to the sensitivity of the immune system to toxicants that could adversely affect the critical balance of the various immune factors and the adverse health effects that could ensue, it is extremely important to define any potential interaction of a new drug and immune system function [234].

As can be anticipated from the previous discussion, immunogenicity is a major scientific issue relative to the development of biotechnology products. Concern has been raised over comparison of the recombinant protein and the naturally occurring protein, as animal models are thought to be inadequate to assess the chemically subtle, but potentially immunologically significant, differences in the human response to these molecules. It has been assumed that a recombinant protein, designed for human use, would produce a number of adverse effects (e.g., the production of neutralizing antibodies) in experimental animals. It has now become clear, through chronic exposures in nonclinical studies, that some low-molecular-weight human proteins are not immunogenic in animals or are only weakly so. They have also been observed not to produce neutralizing antibodies. In the case where antibodies to human proteins have been detected in nonclinical studies, they do not necessarily cause expected immunopathology or neutralization activity. The rhesus monkey has been shown to predict the relative immunogenicity of several recombinant proteins in humans [67] and may serve as a good model.

A further question is "Should all recombinant DNA products be routinely screened in animals prior to their introduction into humans?" The major reason for conducting immunotoxicity evaluations in experimental animals is to detect those compounds that could induce anaphylaxis or anaphylactoid reactions in humans [227]. New molecules, previously minimally tested in animals, such as enkephalins, would have a greater potential risk than well-known molecules, such as insulin. An approach to testing recombinant proteins as well as NCEs for immunogenicity or antigenicity has been suggested based on the extent of the clinical database and the existing regulatory requirements (Figure 7.13). Although studies in animals seem well justified for poorly characterized chemicals, it remains an open question whether or not regulatory agencies will accept an existing extensive clinical database as justification for not performing immunogenicity evaluations. Again, the chemical characterization of the recombinant product relative to the natural material will have some bearing on this debate.

The importance of defining the potential interaction between a new drug and the immune system is emphasized in the new regulatory approaches addressing immunotoxicology. The FDA has issued a guidance for industry [89] that identifies five major areas of immunotoxicology:

- Immunosuppression (decreased immune function)
- Immunogenicity (immune reaction to drugs)
- Hypersensitivity (immunological sensitization)
- Autoimmunity (immune reaction to self antigens)
- Adverse immunostimulation (antigen-nonspecific uncontrolled activation of the immune system)

The European Agency for the Evaluation of Medicinal Products (EMEA) has also published a position on immunotoxicology [77]. The major difference between the guidance documents is the mandatory requirement for functional testing in the EMEA guidance [58]. The FDA guidance [89] supports a weight-of-evidence approach involving standard tests for each potential new drug [123]. Despite the new guideline approaches, histopathologic evaluation of lymphoid organs and tissues has been important in identifying potential immunotoxicology [151].

The ICH has agreed to harmonize the immunotoxicology guidance among the United States, EU, and Japan (Table 7.6). They have published a guidance to address nonclinical testing for immunosuppression induced by low-molecular-weight drugs [141]. Although the guidance will not specify how each immunotoxicity study is performed, it will address the general approach to immunotoxicity testing for new and marketed drugs.

## Genetic Toxicology

Genotoxicity has been defined as the ability of either a chemical or physical agent to damage DNA, resulting in a mutation [37]. An important element of toxicology is the early identification of potentially hazardous substances. Because the actions of toxicants are ultimately exerted at the cellular level, isolated cell systems represent an important model for identifying toxic effects. *In vitro* assays allow a greater control over xenobiotic metabolism (e.g., addition of enzymes or inhibitors) and facilitate mechanistic studies that could not be performed *in vivo* [145]. *In vitro* tests generally provide a reasonable approximation of the potential for an agent to have an effect on genetic material; *in vivo* procedures provide a better test for the potential for genetic alterations to occur in the intact organism [149]. The short-term *in vitro* tests for genotoxicity, potentially predictive of *in vivo* carcinogenicity, are among the most important techniques available for the rapid determination of potential severe undesirable effects of compounds selected for development. They are also useful in the prioritization of compounds to be studied in the more extended and expensive *in vivo* toxicology studies. A number of assays are available to evaluate the potential for genotoxicity. The majority opinion is in favor of a battery approach to identify potential genotoxic activity because different assays assess different types of genetic damage [147,195]. The ICH has published a guideline on how to conduct genotoxicity tests and a guideline on the recommended standard genotoxicity testing battery for evaluations of pharmaceuticals [131,132]. The ICH test battery includes:

- Gene mutation in bacteria to detect relevant genetic changes and the majority of genotoxic rodent carcinogens
- *In vitro* mammalian cell chromosomal aberration or *in vitro* mouse lymphoma thymidine kinase (tk) assay to detect either gross chromosomal damage or detection of gene mutation and clastogenic effects
- *In vivo* chromosomal damage in rodent hematopoietic cells, which allows evaluation of additional relevant factors (e.g., absorption, distribution, metabolism, excretion)

This battery may be expanded when appropriate, such as when compounds with structural alerts are negative in the three standard tests [132].

Excellent reviews of methods to study the genotoxic potential and issues concerning nongenotoxic but carcinogenic chemicals are available [18,19,228,229]. *In viv* exposure assays have the advantage of an intact metaboli system to effectively assay those compounds that must be activated (metabolized to a reactive entity) to achieve a effect. The *in vitro* assays may be conducted with or without the addition of a postmitochondrial supernatant from livers of rats treated with polychlorinated biphenyls t maximally induce drug-metabolizing enzyme activitie and thus enhance the detection of indirect-acting agents

*In vitro* genotoxicity assays are often used in the early drug discovery process aimed at selecting drug candidate for further development. The definitive *in vitro* genotox icity evaluations of mutation and chromosomal damag are generally submitted prior to the FHD. The complet battery of recommended tests should be submitted prio to phase II clinical development.

Conducting genotoxicity screens on NBEs has bee an area of much discussion. The ICH has recognized tha the standard genotoxicity testing battery may not be appli cable to NBEs [138]. Despite this, many pharmaceutica companies conduct genotoxicity evaluations on NBEs pri marily to evaluate process impurities, to meet perceive regulatory expectations, or to meet specific regulator agency requests [116].

## Safety Pharmacology

*Safety pharmacology* involves establishing the pharmaco logic profile of new drug candidates by evaluating th pharmacodynamics related to the therapeutic indicatio and by evaluating the pharmacodynamics on other orga systems not related to the therapeutic indication [49

These studies are usually conducted at doses well below those used to establish the toxicology profile. It is not necessary or even desirable to produce frank toxicity to establish a valid pharmacologic profile.

Pharmacologic profiling was initially focused on guiding the synthetic chemist in the discovery of new pharmacologically active chemicals, rather than on the detection of adverse drug effects in humans [238]. In reviewing the common adverse drug findings in humans, Zbinden [238] pointed out that some responses are easily detected in both humans and experimental animals (e.g., sedation, anorexia, body weight changes, tremor, tachycardia), and some responses are only detectable in humans (e.g., tinnitus, vertigo, nausea, headache). In any event, when one considers the nature of the functional disturbances encountered in both nonclinical and clinical testing, it becomes evident that pharmacologic profiling is critical to the safety evaluation of potential therapeutic agents [230,231]. Every chemical that enters the body has the potential for creating effects that may or may not be related to its pharmacologic activity. Antihistamines, for example, produce sedation, which is related to their pharmacologic effects, and anticholinergic responses, which are not [230]. Pharmacologic profiling can help identify the potential incidence of effects unrelated to the known pharmacologic activity. Pharmacologic profiling also provides crucial information for the selection of NCEs during the early discovery process, for the design of toxicology studies, and for the approach to safety monitoring appropriate for clinical trials [159,186].

The Japanese Ministry of Health and Welfare (MHW) has published guidelines for general pharmacology studies [72]. These studies are designed to characterize effects and potency and to determine mechanism. The guidelines include studies to determine effects on general activity and behavior, the CNS, the autonomic nervous system and smooth muscle, the respiratory and cardiovascular systems, the gastrointestinal tract, and renal excretion.

The ICH guidelines now include select pharmacology guidance (Table 7.6). ICH Topic S7A [139] addresses the definition, objectives, and scope of safety pharmacology studies, as well as studies required prior to phase I clinical trials and prior to marketing approval. The safety pharmacology core battery has been built based on previously existing regional draft guidances [37] and now includes the following vital organ systems:

- Central nervous system
- Cardiovascular system
- Respiratory system

Follow-up studies are recommended to provide a greater depth of understanding of the effects observed in the core battery or from clinical trials on pharmacovigilance: renal/urinary system, autonomic nervous system, gastrointestinal system, dependency potential, skeletal muscle, and immune and endocrine functions.

The EU Committee for Propriety Medicine Products (CPMP) had published a *Points To Consider* document on QT interval prolongation [48]. The impact of QT prolongation and related occurrence of Torsades de Pointes with the use of quinolones has been reviewed by Frothingham [101]. The ICH Topic S7B(R) now provides recommendations for nonclinical studies to address the potential for QT interval prolongation [140]. It also provides guidance on integrated risk assessments. The prolongation of the QT interval is one of the few single physiological responses that could end the development of a potential new pharmaceutical [112].

## Measure of Exposure

The relationship of administered dose to toxicologic response is not always a simple correlation. Traditionally, the administered dose (mg/kg) has been the most commonly used expression to compare toxicologic responses between species. The value of the administered dose term as the most appropriate comparator with toxicologic response, however, has been questioned in scientific and regulatory circles. It is becoming increasingly well recognized that both beneficial and toxic effects of therapeutic agents are dependent on the quantity of material reaching the target site [169,214,243].

Knowledge of the level of exposure is critical for understanding not only safety but also efficacy. Fully elucidating this knowledge base provides the most robust estimates of safety multiples. In its simplest representation, the safety multiple can be calculated by dividing the highest exposure observed without the presence of an adverse effect by the exposure needed to elicit efficacy. In practice, much more complexity is encountered in making this estimate. Variables such as exposure comparisons across species, the absence of exposure determinations at the target site of activity or toxicity, and nonlinearity of the dose/exposure–response complicate the assessment.

Measures of exposure are also useful in establishing nonlinearity in kinetics, which is important in explaining toxic responses seen in particular species [169]. It seems more rational to establish an upper dose in toxicology studies based on linearity of kinetics rather than at the MTD, as it is often the case that the MTD falls in the range of nonlinear kinetics, saturating normal metabolic processes. Thus, an animal treated at the MTD may be exposed to much higher levels of parent drug or toxic metabolites than would be observed at meaningful multiples of the clinical dose.

The exposure differences often observed across species can sometimes be explained by known differences in rates of absorption and/or metabolic clearance across species. Figure 7.14 illustrates the general relationship of clearance

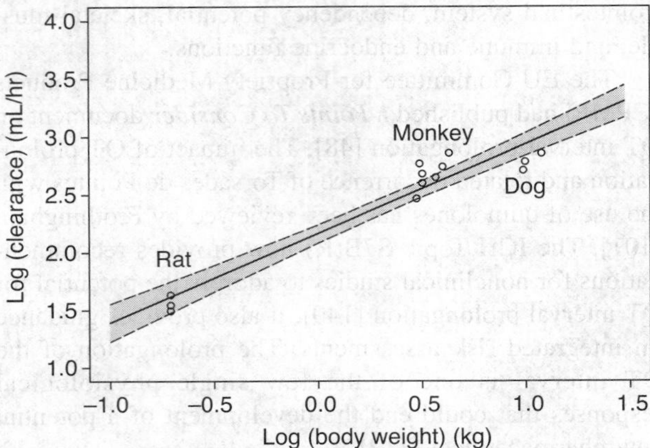


**FIGURE 7.14** The general relationship of clearance to body weight for dog, rat, and monkey.

to body weight for dog, rat, and monkey. From this generic depiction, it is apparent how a lower administered dose on a milligram-per-kilogram basis produces a higher exposure in dog compared to rat. This relationship can be extrapolated further to humans, who, in general, would have even slower rates of clearance. The relationship of administered dose to delivered dose remains a central issue in the interpretation of toxicology data. The measurement of plasma concentrations of parent compound and metabolites represents a partial resolution to this problem. There are limitations, however, in using plasma concentration as a relevant measure of exposure for those compounds that are tissue sequestered [168]. It has been noted that, although many therapeutic agents achieve tissue levels proportional to plasma concentration, some continue to accumulate in tissue with continued dosing.

Levels of exposure of test compounds or their metabolites at sites of action can be assessed by measuring tissue kinetics. Lovastatin is an example of a compound that exerts its primary pharmacology and toxicity at the same site. In this case, the pharmacokinetics of both lovastatin and its active metabolite have been well characterized [118]. An increasingly more common method is the utilization of quantitative whole body autoradiography. This technique uses radioactive drug ($^{14}C$ or $^{3}H$) to assess concentrations in tissues across the entire body (Figure 7.15). Concentration time profiles can be generated that subsequently can be used to derive various pharmacokinetic parameters. A shortcoming of this technique is its inability to distinguish parent compound from metabolite [34]. Perhaps the most critical aspect of the role of the toxicologist is to complete the multifaceted exposure comparisons, couple this knowledge with that of the observed effects of the test compound, and render an informed opinion on the overall safety profile of the test compound under study.

## Clinical Trials in Pediatric Populations

The FDA has found that most products indicated for treatment of diseases that occur in both adults and children have little clinical trial support for pediatric use. As a result, a new regulation requiring pediatric studies for certain NCEs and NBEs has been proposed [87]. The CPMP has also concluded that specific age-dependent differences in pharmacokinetics, pharmacodynamics, growth process and development, and specific pathology

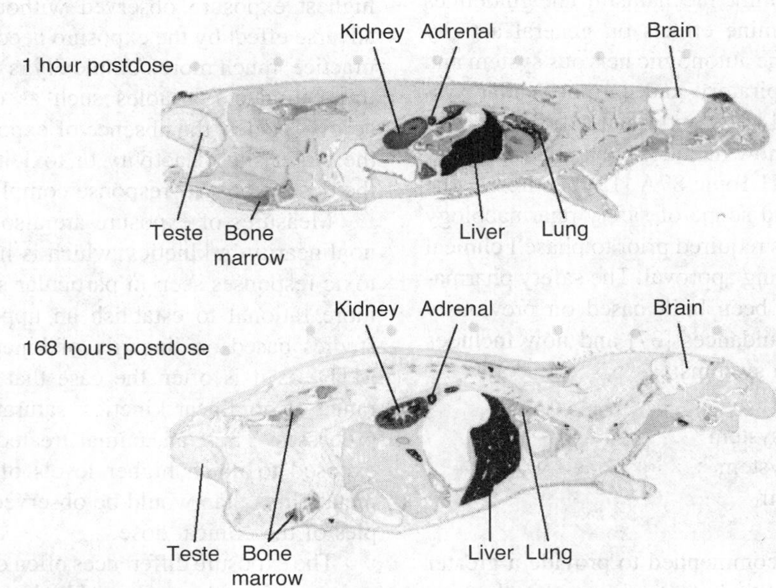


**FIGURE 7.15** Phosphor images of whole-body sections from male Fischer 344 rats after receiving a single dose of [$^{14}$C]-test compound.

require that therapeutic agents be tested in the target age group [47]. The ICH has recommended that pediatric clinical trials be supported by repeated-dose toxicity studies of an appropriate duration, all developmental toxicity studies, and the full battery of genotoxicity tests. These studies should be concluded before the pediatric clinical trials begin [128]. Due to the potentially long duration of treatment, carcinogenicity studies must be considered prior to the initiation of long-term pediatric clinical trials. The performance of nonclinical studies in juvenile animals may also be necessary if previous toxicology evaluations and human safety data are not sufficient or raise a cause for concern.

## Nonclinical Evaluation of Anticancer Drugs

As discussed earlier, the development of drugs for life-threatening diseases such as cancer and AIDS requires a modification of the approach established for the standard safety assessment of NCEs or NBEs. The treatment of cancer usually includes the use of potent agents designed to halt cell replication. The therapeutic index for these agents is often small. Due to the life-threatening nature of the disease, a greater tolerance of drug toxicity and a shorter nonclinical testing strategy are generally accepted for therapeutic agents in this class, as serious drug effects are often less threatening than the targeted disease; however, due to the greater intrinsic toxicity of the agents in this class, the early clinical trials are often conducted in patients rather than normal volunteers, as with other therapeutic agents.

The history of anticancer drug development has been extensively reviewed [63,114,215]. The basic approach for development of anticancer drugs includes:

- Establishment of a safe clinical trial entry dose
- Determination of potential dose-limiting target-organ toxicity
- Evaluation of reversibility of effects
- Determination of MTD in animals
- Determination of dose schedule toxicity

The use of nonclinical studies has been useful in accurately predicting MTDs in humans and safe starting doses for clinical trials [114]. The CPMP has provided a specific note for guidance on the nonclinical evaluation of anticancer agents [50]. Safety pharmacology is generally required prior to phase I studies, as are the determination of $C_{max}$ (maximum plasma concentration of the drug) and AUC at the animal MTD. Other kinetic parameters are expected to be determined prior to phase II/III testing. Determination of the single-dose MTD in rodents and the approximate MTD in nonrodents, using a relevant route of exposure, is also expected prior to phase I. Repeated-dose toxicity studies in two rodent species are expected prior to phase I clinical studies. Longer term repeat dose studies, in a rodent and nonrodent species, that are equal in duration to the clinical trial but less than 6 months, are expected prior to phase II/III. Genotoxicity testing is not necessary prior to phase I/II, but the genotoxicity battery is expected to be conducted prior to phase II/III. Because cytotoxic anticancer agents are known to have an adverse effect on reproduction, developmental toxicity studies are not required but are encouraged. The development of anticancer drugs and other therapies for life-threatening diseases has unique characteristics based on the nature of the disease and the inherent toxicity of the therapeutic agents.

## Alternative Methods for Carcinogenicity Determination

The testing for carcinogenic potential has relied primarily on the rodent bioassay. Recently, through the ICH process, the rat has been identified as either the most acceptable or most relevant model for the 2-year bioassay. In addition to the rat bioassay, an alternative short-term method of carcinogenicity evaluation is recommended (ICH S1B). These approaches may include studies in transgenic mice (e.g., $p53^{+/-}$ heterozygous gene-deficient mouse or Tg.AC mouse) or the use of a neonatal rodent tumorigenicity model.

To evaluate and verify the available alternative models, the International Life Sciences Institute (ILSI) has initiated a collaborative effort among industry, academic, and government laboratories to study chosen chemicals in the alternative models and evaluate the results in light of the known bioassay data. Current information is insufficient to guide us in the choice of suitable alternative models for carcinogenicity evaluation. It is highly recommended that any deviation from the standard two species (mouse and rat) carcinogenicity bioassay be discussed with the FDA prior to implementation.

When fully validated, the proposed transgenic animal model may be used as follows [34]:

- To confirm results in equivocal 2-year rodent bioassays
- To set priorities for 2-year carcinogenicity bioassays
- As an alternative to the mouse 2-year bioassay, in conjunction with the rat 2-year bioassay
- To assess carcinogenic potential of new genotoxic contaminants or degradants in a drug product after 2-year bioassays are completed

In addition, the use of transgenic animals may support weight-of-evidence decisions; they are relatively short term and are generally less expensive than the 2-year rodent bioassay.

Several transgenic mouse models are available to complement the rat 2-year bioassay. The Tg.AC transgenic mouse, the heterozygous $p53^{+/-}$ gene-deficient mouse, and the Tg/ras$H_2$ mouse are discussed as examples. It must be remembered that these models are not fully validated.

The Tg.AC mouse model [211] examines chemically initiated skin as a target for tumorigenesis. The Tg.AC line may be able to differentiate carcinogens from noncarcinogens but may not be able to differentiate genotoxic carcinogens from those that cause only tumor promotion activity. This model may only be useful in combination with other transgenic animal models [196]. The heterozygous $p53^{+/-}$-deficient mouse model is based on rendering mice heterozygous for the p53 tumor suppressor gene [106]. These animals are at elevated risk for tumor development. The model has been proposed to best approximate humans at risk for heritable forms of cancer. This model may be able to detect genotoxic carcinogens in a 6-month period. The Tg/ras$H_2$ transgenic mouse contains multiple copies of the human c-Ha-*ras* gene. It has been found to respond well to both genotoxic and nongenotoxic carcinogens [164].

The neonatal mouse assay has been available longer than the transgenic animal models. The detailed protocol for this 1-year study has been previously reviewed [97]. Neonates are treated with the test compound on days 8 and 15 of age and then observed to 1 year of age. At that time, the animals are evaluated for tumor production. This assay is sensitive to direct-acting carcinogens, primarily those that work through formation of covalently bound DNA adducts.

The assessment of carcinogenicity in the drug development process is at a crossroads. The rodent bioassay has been used for over 25 years and has provided useful data, although it is not a perfect system and has received much criticism. The investigation of alternatives to the standard bioassay in two rodent species has been encouraged by the ICH. Conducting a 2-year study in the rat and an alternative study in the mouse may provide an acceptable transition. The available alternative models generally accepted as useful in carcinogenicity testing include the Tg/ras$H_2$ transgenic mouse and the $p53^{+/-}$ knockout mouse. The Tg/ras$H_2$ model is preferred for testing nongenotoxic compounds, and the $p53^{+/-}$ knockout model is preferred for testing genotoxic compounds. The general utility of available alternative models of carcinogenicity testing has been reviewed by MacDonald et al. [164]. Alternative assay results should not be considered on their own but should be included in a weight-of-evidence approach for risk assessment [164]. A historical perspective on the industry's experience with alternative carcinogenicity models has been provided by Ashton et al. [7]. Much work remains to be done before the alternative models are validated and fully useful as potential replacements in carcinogenicity risk assessments. The alternative assays are relative newcomers full of promise but short on experience. Undue enthusiasm about their ability to dramatically improve our carcinogenicity risk assessment process is not warranted. An alternative approach to carcinogenic risk evaluation, focusing on chemical exposure rather than the 2-year rodent bioassay, has been proposed by Cohen [41].

## Development of Stereoisomeric Drugs

The development of individual enantiomers is largely guided by regulatory expectations established by the FDA in 1992 [83]. Further guidance has also been published regarding the development of stereoisomers intended for animal use [88]. Differences in both the primary pharmacology and the toxicology of stereoisomers are well known. Differences in the pharmacokinetic profiles of enantiomers are also well known. These differences are usually manifested in rates of clearance or metabolism; for example, the enantiomers of the hypnotic zopiclone exhibit an approximate threefold difference in human clearance after oral administration [93]. Some nonsteroidal anti-inflammatory drugs also exhibit stereoselective clearance. S-Ibuprofen is cleared more rapidly than its R+ enantiomer. Furthermore, in this example, interconversion of the R-isomer to the S-isomer is a major route of clearance of the R-isomer [208].

Specific guidance is offered regarding the evaluation of the pharmacokinetic and toxicologic profiles of enantoimers vs. racemic mixtures. An assessment of the disposition of the isomers must include study of the potential of isomers to interconvert in both preclinical and clinical circumstances. Previously discussed *in vitro* metabolism test systems can be quite useful for this evaluation as well. Development of single enantiomers after the racemate has been evaluated is specifically addressed in the cited guidance. A 3-month study of the enantiomer in both rodent and nonrodent species and an assessment of the developmental toxicology profile have been suggested. In addition to the safety profile, the potential for interconversion of enantiomers should also be monitored.

## Peroxisome Proliferator-Activated Receptor Agonists

Peroxisome proliferator-activated receptors (PPARs) are members of the nuclear hormone receptor superfamily of ligand-actuated transcription factors. PPARs are thought to play a role in cellular functions, including lipid metabolism, cell proliferation, differentiation, adipogenesis, and inflammatory signaling. PPAR agonists are being developed for several indications, such as dyslipidemia, type 2 diabetes, and obesity. An analyst's view of recent development issues with PPAR agonists indicates a series of delays and setbacks [167]. The FDA has provided guidance for the nonclinical safety evaluation of PPAR agonists, in part related to the PPAR subtype.

These recommendations have been presented at scientific meetings held in Washington, D.C. [74] and in Monaco [75,76]. This is a developing area where changes in the regulatory guidance have occurred, and we should expect additional changes as more information on specific agents becomes available. As of this writing, PPAR agonists are thought of as multispecies, multistrain, multisex, and multisite carcinogens in rodents. Common tumor types observed include hemangiosarcomas (mice), transitional cell carcinomas of the urinary tract (rats), lipoma/liposarcoma, and sarcomatous tumors at multiple sites (e.g., kidney, stomach, uterus, skin). PPAR agonists are not currently considered genotoxic; however, the high incidence of positive carcinogenicity findings with this class raises safety concerns. The FDA has provided guidance that clinical studies of 6 months' duration or longer cannot be conducted until the 2-year rodent carcinogenicity studies are completed and an adequate margin of safety has been demonstrated. Two-year carcinogenicity studies in mice, rats, and, when appropriate, hamsters are recommended. Due to the prevalence of positive tumor findings with PPAR agonists, alternative assays of carcinogenicity, in lieu of the 2-year mouse study, are not recommended by the FDA. For PPAR agonists of the alpha subtype, 2-year carcinogenicity studies in hamsters rather than mice have been recommended. The appropriate safety assessment of PPAR agonists is a developing area, requiring close attention by toxicologists. The potentially large therapeutic benefits of the agents continue to drive intense interest in the PPAR platform.

## CONCLUSION

The next decade promises to be another exciting one for the toxicologist. Physicians and patients continue to expect more efficacious and safer medications, more quickly and less expensively. Basic research efforts in biochemistry, physiology, and pharmacology have allowed the more precise characterization of receptors and the normal and perturbed sequelae of receptor binding, thus continuing to stimulate the development of more specific, potent modulators of cellular functions. The etiologies of human diseases are becoming better understood, thanks to the technologic ability to elucidate their characteristics at the molecular level. This has resulted in the potential to therapeutically modify the disease process at its origin: the human genome. These molecular approaches have already resulted in the development of agents that are highly species specific, and the use of these techniques to elucidate normal and pathological cellular function will only continue to escalate. It is unlikely that the classical tools of toxicology will be sufficient to ensure the human safety of the highly specific potential therapeutics forthcoming from these sophisticated technologies. The development of NBEs has already challenged the established norms of safety assessment.

Consider the example of an agent that has shown selectivity for modifying the activity of a human-specific enzyme critical to a pathological process. Although traditional animal studies are likely to be predictive of toxicity that is unrelated to the pharmacology of the compound, they will not be useful for the prediction of adverse findings relative to the action of the drug at the enzyme, which will occur only in humans. Thus, not only may data generated from classical animal studies be inadequate to predict toxic responses in humans, but also the information may be irrelevant or misleading. A major concern is whether modification of this enzyme activity in the only responsive species (humans) might result in unanticipated, severe toxicity. How can this best be predicted prior to the initiation of clinical trials?

The future direction of discovery research suggests that industrial and regulatory toxicologists will need to collaborate more closely in the design of the toxicology studies to support registration, and ultimately these may have to be considered on a case-by-case basis. Indeed, the current guidelines resulting from global harmonization efforts repeatedly emphasize the need for defending the scientific rationale supporting the design of proposed toxicologic assessments. Although these guidelines are viewed as much more flexible than the country-specific regulations of the past, they also place a greater burden on toxicologists relative to defending the relevance of their studies. The era of "checking the tox box" has, thankfully, come to a close.

Another challenge on the horizon concerns the need to improve the efficiency of the drug development process without compromising its quality. Currently, the drug development and approval processes are taking longer and costing more than ever before. More stringent regulatory requirements have resulted in the conduct of more studies (in both animals and humans) that take more time and cost more money. These costs are passed on to the patient, who ultimately must also compensate for the resultant decreased market life, due to patent length restrictions, of the registered product. Furthermore, the use of large numbers of experimental animals remains a concern from both the ethical and financial points of view. One approach to solving these dilemmas is to ensure that the toxicology studies conducted meet the needs of the regulatory agency, the physician, and ultimately the patient. As suggested earlier, this can be most efficiently accomplished by early and routine interactions between the industry and these customers, especially in cases where the agent under development represents a unique therapeutic approach.

The application of omics technologies, such as genomics (toxicopharmaconomics), proteonomics, and metabonomics, are expected to enhance the toxicology toolbox for evaluation of drug safety. Proteonomics, the analysis of protein expression patterns, and metabonomics, the evaluation of metabolite profiles, are relative newcomers to drug safety evaluation. Toxicogenomics (or pharmacogenomics) has, perhaps, the greatest potential to affect nonclinical safety assessment. Toxicogenomics focuses on the study of differential gene expression (DGE) as an adaptation to chemical or environmental stress. A basic assumption is that DGE underlies all drug induced toxic events, with the possible exception of very rapid cell death [123,151]. A goal of toxicogenomics is to identify gene expression patterns that predict potential toxicity [79,190]. The identification of a causal relationship between DGE and delayed manifestations of frank toxicity can facilitate early drug development. It is anticipated that toxicogenomics will be increasingly integrated into all phases of drug development, particularly in mechanistic and predictive toxicology and biomarker identification [117]. The FDA has published a guidance document on submission of pharmacogenetic data [91]. Within the guidance, the FDA acknowledges that the toxicology database required to support clinical trials and marketing of drugs is well established. Any proposal for substitution of new animal genomics safety testing will involve the international scientific and drug development committees [91].

Finally, the major role of the toxicologist as a mechanistic scientist will continue to be enhanced. For the reasons discussed above, the design of toxicology studies and the interpretation of toxicology data will become increasingly more sophisticated, requiring a broad knowledge base in a variety of other scientific disciplines. Elucidation of the mechanisms responsible for observed toxicities would improve the ability to achieve the traditional, ultimate purpose of the discipline of toxicology: appropriate extrapolation of these data to humans. Achievement of this goal will surely become more challenging but also more exciting in the future.

## ACKNOWLEDGMENT

The authors gratefully acknowledge and appreciate the efforts of Mrs. Arlene Adkins in the preparation of this document.

## REFERENCES


1. Ackermann, B. L., Berna, M. J., and Murphy, A. T. (2002): Recent advances in the use of LC/MS/MS for quantitative high throughput bioanalytical support of drug discovery. *Curr. Top Med. Chem.*, 2:53–66.
2. Ames, B. N., and Gold, L. S. (1990): Too many rodent carcinogens: mitogenesis increases mutagenesis. *Science*, 249:970–971.
3. Anacher, D. E. (2002): A toxicologist's guide to biomarkers of hepatic response. *Human Exp. Toxicol.*, 21:253–262.
4. Ankier, S. I., and Warrington, S. J. (1989): Research and development of new medicines. *J. Int. Med. Res.*, 17:407–416.
5. Artursson, P., Paam, K., and Luthman, K. (1996): Caco-2 monolayers in experimental and theoretical predictions of drug transport. *Adv. Drug Delivery*, 22:67–84.
6. Ashby, J. (1996): Alternatives to the 2-species bioassay for the identification of potential human carcinogens. *Human Exp. Toxicol.*, 15(3):183–202.
7. Ashton, G. A., Griffiths, S. A., and McAuslane, J. A. N (1999): *Industry Experience with Alternative Models for the Carcinogenicity Testing of Pharmaceuticals*, Centre for Medicines Research International, Surrey.
8. Axelrod, D. et al. (2004): Hormesis: an inappropriate extrapolation from the specific to the universal. *Int. J Occup. Environ. Health*, 10:335–339.
9. Ayers, K. M. et al. (1996): Nonclinical toxicology studies with zidovudine: genetic toxicity tests and carcinogenicity bioassays in mice and rats. *Fundam. Appl. Toxicol.* 32:148–158.
10. Ayers, K. M. et al. (1996): Nonclinical toxicology studies with zidovudine: acute, subacute, and chronic toxicity in rodents, dogs, and monkeys. *Fundam. Appl. Toxicol.* 32:129–139.
11. Baker, S. B. deC., and Davey, D. G. (1970): The predictive value for man of toxicological tests of drugs in laboratory animals. *Br. Med. Bull.*, 26(3):208–211.
12. Baldrick, P. (2000): Pharmaceutical excipient development: the need for preclinical guidance. *Regul. Toxicol Pharmacol.*, 32:210–218.
13. Ballie, T. A. et al. (2002): Drug metabolites in safety testing. *Toxicol. Appl. Pharmacol.*, 182:188–196.
14. Bass, R., and Scheibner, E. (1987): Toxicological evaluation of biotechnology products: a regulatory viewpoint *Arch. Toxicol.*, 11(Suppl.):182–190.
15. Berlin, R. G. (1991): Omeprazole: gastrin and gastric data *Digest. Dis. Sci.*, 36:1501–1502.
16. Bocci, V. (1992): Physicochemical and biological properties of interferons and their potential uses in drug delivery systems. *Crit. Rev. Ther. Drug Carrier Syst.*, 9(2) 91–133.
17. Brent, R. L. (1980): The prediction of human diseases from laboratory and animal tests for teratogenicity, carcinogenicity, and mutagenicity. In: *Controversies in Therapeutics* edited by L. Lasagna. W.B. Saunders, Philadelphia, PA pp. 131–150.
18. Brusick, D. (1989): Genetic toxicology. In: *Principles and Methods of Toxicology*, 2nd ed., edited by A. W. Hayes Raven Press, New York, pp. 407–434.
19. Butterworth, B. (1989): Nongenotoxic carcinogens in the regulatory environment. *Reg. Toxicol. Pharmacol* 9:244–256.
20. Butterworth, B. E. et al. (1991): The rodent cancer test: an assay under siege. *CIIT Activities*, 11(9):1–6.
21. Cahn, J. (1983): Forecasting of cardiac side effects: vincamine and calcium antagonists, a comparative study in animals and man. In: *Current Problems in Drug Toxicology*, edited by G. Zbinden, J. Y. Detaille, and G. Mazue John Libbey Eurotext, London, pp. 90–94.
22. Calabrese, E. J., and Baldwin, L. A. (1994): Improved method for selection of the NOAEL. *Regul. Toxicol. Pharmacol.*, 19:48–50.
23. Calabrese, E. J. (2004): Hormesis: basic generalizable central to toxicology and a method to improve risk assessment process. *Int. J. Occup. Environ. Health*, 10:466–46

24. Calabrese, E. J. (2005): Paradigm lost, paradigm found: the re-emergence of hormesis as a fundamental dose response model in the toxicological sciences. *Environ. Pollut.*, 138:379–412.
25. Calabrese, E. J., and Baldwin, L. A. (1998): Hormesis as a biological hypothesis. *Environ. Health Perspec.*, 106(1):357–362.
26. Calabrese, E. J., and Baldwin, L. A. (2003): Toxicology rethinks its central belief. *Nature*, 42:691–692.
27. Caldwell, G. W. et al. (2004): Allometric scaling of pharmacokinetic parameters in drug discovery: can human CL, Vss, and $t_{1/2}$ be predicted from *in vivo* rat data? *Eur. J. Drug Metab. Pharmacokinet.*, 2:133–143.
28. Caldwell, J. (1981): The current status of attempts to predict species differences in drug metabolism. *Drug Metab. Rev.*, 12(2):221–237.
29. Campbell, D. B., and Ings, R. M. (1988): New drug approaches to the use of pharmacokinetics in toxicology and drug development. *Hum. Toxicol.*, 7:469–479.
30. Carlsson, E. et al. (1986): Pharmacology and toxicology of omeprazole, with special reference to the effects on the gastric mucosa. *Scand. J. Gastroent.*, 118(Suppl.):31–38.
31. Cavagnaro, J. A. (2002): Preclinical safety evaluation of biotechnology-derived pharmaceuticals. *Nature Rev.*, 1:469–475.
32. Chan, P. K., O'Hara, G. P., and Hayes, A. W. (1981): Principles and methods for acute and subchronic toxicity. In: *Principles and Methods of Toxicology*, edited by A. W. Hayes, Raven Press, New York, pp. 1–51.
33. Chance, R. E., Kroeff, E. P., and Hoffman, J. A. (1981): Chemical, physical and biological properties of recombinant human insulin. In: *Insulins, Growth Hormone, and Recombinant DNA Technology*, edited by J. L. Gueriguian. Raven Press, New York, pp. 71–84.
34. Chay S. H., and Pohland R. C. (1994): Comparison of quantitative whole-body autoradiographic and tissue dissection techniques in the evaluation of the tissue distribution of [$^{14}$C]daptomycin in rats. *J. Pharm. Sci.*, 83:1294–1299.
35. Chien, R. E., Ed. (1979): *Issues in Pharmaceutical Economics*, Lexington Books, Lanham, MD.
36. CHMP (2004): *Position Paper on Nonclinical Safety Studies To Support Clinical Trials with a Single Microdose*, CPMP/SWP/02 Rev. 1, Committee for Human Medicinal Products, Clermont-Ferrand France.
37. Choy, W. N. (1996): Principles of genetic toxicology. *Drug Chem. Toxicol.*, 19(3):149–160.
38. Clemento, A. (1999): New and integrated approaches to successful accelerated drug development. *Drug Inf. J.*, 33:699–710.
39. Cluff, L. E. (1980): Is drug toxicity a problem of great magnitude? Yes! In: *Controversies in Therapeutics*, edited by L. Lasagna. W.B. Saunders, Philadelphia, PA, pp. 44–50.
40. Cohen, S., and Ellwein, L. B. (1990): Cell proliferation in carcinogenesis. *Science*, 249:1007–1011.
41. Cohen, S. M. (2004): Human carcinogenic risk evaluation: an alternative approach to the two-year rodent bioassay. *Toxicol. Sci.*, 80:225–229.
42. Cohen, S. M. et al. (2004): Evaluating the human relevance of chemically induced animal tumors. *Toxicol. Sci.*, 78:181–186.
43. Contrera, J. F. et al. (1993): Adverse drug reactions. *Toxicol. Rev.*, 12(1):63–76.
44. Contrera, J. F. et al. (1995): A systemic exposure-based alternative to the maximum tolerated dose for carcinogenicity studies of human therapeutics. *J. Am. Coll. Toxicol.*, 14(1):1–10.
45. Contrera, J. F., Jacobs, A. C., and DeGeorge, J. J. (1997): Carcinogenicity testing and the evaluation of regulatory requirements for pharmaceuticals. *Reg. Toxicol. Pharmacol.*, 25:130–145.
46. Contrera, J. F. (1998): Transgenic animals: refining the two-year rodent carcinogenicity study. *Lab. Animal*, 27(2):30–33.
47. CPMP (1997): *Note for Guidance on Clinical Investigation of Medicinal Products in Children*, CPMP/EWP/462/95. Committee for Proprietary Medicinal Products, European Medicines Agency, London.
48. CPMP (1997): *Points To Consider: The Assessment of the Potential for QT Interval Prolongation by Non-Cardiovascular Medicinal Products*, CPMP/986/96. Committee for Proprietary Medicinal Products, European Medicines Agency, London.
49. CPMP (1998): *Note for Guidance on Safety Pharmacology Studies in Medicinal Product Development (Draft of Preliminary Consultation)*, CPMP/SWP/872/98. Committee for Proprietary Medicinal Products, European Medicines Agency, London.
50. CPMP (1998): *Note for Guidance on the Preclinical Evaluation of Anticancer Medicinal Products*, CPMP/SWP/997/96. Committee for Proprietary Medicinal Products, European Medicines Agency, London.
51. Creutzfeldt, W., and Lamberts, R. (1991): Is hypergastrinaemia dangerous to man? *Scand. J. Gastroenterol.*, 180(Suppl.):179–191.
52. D'Agnolo, G. (1983): The control of drugs obtained by recombinant DNA and other biotechnologies. In: *Current Problems in Drug Toxicology*, edited by G. Zbinden, J. Y. Detaille, and G. Mazué. John Libbey Eurotext, London, pp. 241–247.
53. D'Aguanno, W. (1973): *Drug Toxicity Evaluation: Pre-Clinical Aspects, FDA Introduction to Total Drug Quality.* DHEW Publ. No. (FDA)74-3006, U.S. Food and Drug Administration, Washington, D.C., pp. 35–40.
54. Dahlem A., Allerheiligen, S. A., and Vodicnik, M. J. (1995): Concomitant toxicokinetics: techniques for and interpretation of exposure data obtained during the conduct of toxicology studies. *Toxicol. Pathol.*, 23(2):170–178.
55. Dayan, A. D. (1981): The troubled toxicologist. *TIPS*, 2(11):1–4.
56. Dayan, A. D. (1986): Preclinical safety studies on genetically engineered medicine for man. *BIBRA J.*, 5(3):12–15.
57. Dayan, A. D. (1988): Risk assessment of biotechnology products. *Hum. Toxicol.*, 7(1):50–52.
58. Dean, J. H. (2004): A brief history of immunotoxicology and a review of the pharmaceutical guidance for drugs. *Int. J. Toxicol.*, 23:83–90.

59. Dean, J. H. et al. (1982): Procedures available to examine the immunotoxicity of chemicals and drugs. *Pharmacol. Rev.*, 34:137–148.
60. Dean, J. H. et al. (1989): Immune system: evaluation of injury. In: *Principles and Methods of Toxicology*, 2nd ed., edited by A. W. Hayes. Raven Press, New York, pp. 741–760.
61. Dean, J. H., Luster, M. I., and Boorman, G. A. (1982): Methods and approaches for assessing immunotoxicity: an overview. *Environ. Health Perspect.*, 43:27–29.
62. Dean, J. H., and Vos, J. G. (1986): An introduction to immunotoxicology assessment. In: *Immunotoxicology of Drugs and Chemicals*, edited by J. Descotes. Elsevier, New York, pp. 3–31.
63. DeGeorge, J. J. et al. (1998): Regulatory considerations for preclinical development of anticancer drugs. *Cancer Chemother. Pharmacol.*, 41:173–185.
64. Descotes, G., Mazue, G., and Richey, P. (1982): Drug immunotoxicological approaches with some selected medical products: cyclophosphamide, methylprednisolone, betamethasone, cefoxitine, minor tranquillizers. *Toxicol. Lett.*, 13:129–138.
65. Diener, R. M. (1997): Safety assessment of pharmaceuticals. In: *Comprehensive Toxicology*, Vol. 2, *Toxicology Testing and Evaluation*, edited by I. G. Sipes, C. A. McQueen, and J. Gandolfi. Elsevier, New York, pp. 269–290.
66. DiMasi, J. A. (1994): Risks, regulation, rewards in new drug development in the United States. *Reg. Toxicol. Pharmacol.*, 19:228–235.
67. DiMasi, J. A. et al. (1991): Cost of innovation in the pharmaceutical industry. *J. Health Econ.*, 10:107–142.
68. DiMasi, J. A., Hansen, R. W., and Grabowski, H. G. (2003): The price of innovation: new estimates of drug development costs. *J. Health Econ.*, 22:151–185.
69. Dixit, R. et al. (2003): Toxicokinetics and physiologically based toxicokinetics in toxicology and risk assessment. *J. Toxicol. Environ. Health B Crit. Rev.*, 6(1):1–40.
70. Dorato, M. A., and Buckley, L. A. (2005): Toxicology in the drug development process. In *Current Protocols in Pharmacology*, edited by S. J. Enna, M. Williams, J. W. Ferkany, T. Kenakin, R. D. Porsolt, and J. P. J. Sullivan. John Wiley & Sons, New York.
71. Dorato, M. A., and Engelhardt, J. A. (2005): The no-observed-adverse-effect-level (NOAEL) in drug safety evaluations: use, issues, definitions. *Regul. Toxicol. Pharmacol.*, 42(3):265–274.
72. Anon. (1991): *Drug Registration Requirements in Japan*, 4th ed., Yakuji Nippo, Ltd., Tokyo, pp. 69–73.
73. Ekman, L. et al. (1985): Toxicological studies on omeprazole. *Scand. J. Gastroent.*, 108(Suppl.):53–69.
74. El-Hage, J. (2004): Preclinical and clinical safety assessments for PPAR agonists. In: *Proc. DIA 40th Annual Meeting*, Washington, D.C.
75. El-Hage, J. (2005): Peroxisome proliferation activated receptor (PPAR) agonists: nonclinical/clinical toxicity and safety assessments. In: *Proc. Third Int. Symp. on PPARS: Efficacy and Safety*, Monte Carlo, Monaco.
76. El-Hage, J. (2005): Peroxisome proliferation activated receptor (PPAR) agonists: Carcinogenicity findings and regulatory recommendations. In: *Proc. Third Int. Symp. on PPARs: Efficacy and Safety.* Monte Carlo, Monaco.
77. EMEA (2000): *Note for Guidance on Repeated Dose Toxicity*, CPMP/SWP/1042/99, corr. Appendix B. European Medicines Agency, London.
78. Falchetti, R. et al. (1983): Toxicological evaluation of immunomodulating drugs. In: *Current Problems in Drug Toxicology*, edited by G. Zbinden, J. Detaille, and G. Mazue. John Libbey Eurotext, London, pp. 248–263.
79. Farr, S., and Dunn, R. T. (1999): Concise review: gene expression applied to toxicology. *Toxicol. Sci.*, 50:1–9.
80. FDA (1987): *Good Laboratory Practice for Nonclinical Laboratory Studies*, Final Rule, 21CFR58. U.S. Food and Drug Administration, Washington, D.C.
81. FDA (1989): *Summary Basis of Approval for Zidovudine*. U.S. Food and Drug Administration, Washington, D.C.
82. FDA (1990): *Summary Basis of Approval for Omeprazole*. U.S. Food and Drug Administration, Washington, D.C.
83. FDA (1992): *FDA's Policy Statement for the Development of New Stereoisomeric Drugs* (revised 2001). U.S. Food and Drug Administration, Washington, D.C. (www.fda.gov/cder/guidance/stereo).
84. FDA (1993): Guideline for the study and evaluation of gender differences in the clinical evaluation of drugs. U.S. DHHS *Federal Register* Notice, July 22, 58FR39406.
85. FDA (1996): *Guidance for Industry: Single Dose Toxicity Testing for Pharmaceuticals*. U.S. Food and Drug Administration, Washington, D.C.
86. FDA (1997): Investigational new drug applications; proposed amendment to clinical hold regulations for products intended for life-threatening diseases. U.S. DHHS *Federal Register* Notice, Sept. 24, 62FR49946.
87. FDA (1997): Regulations requiring manufacturers to assess the safety and effectiveness of new drugs and biological products in pediatric patients. U.S. DHHS *Federal Register* Notice, July 24, 21CFR201,312,314,601.
88. FDA (2002): *Guidance for Industry: Development of Supplemental Applications for Approved New Animal Drugs*. Center for Veterinary Medicine, U.S. Food and Drug Administration, Washington, D.C. (www.fda.gov/cvm/Guidance/guide82).
89. FDA (2002): *Guidance for Industry: Immunotoxicology Evaluation of Investigational New Drugs*. U.S. Food and Drug Administration, Washington, D.C. (http://www.fda.gov/cder/guidance/index.htm).
90. FDA (2005): *Draft Guidance, Guidance for Industry, Investigators, and Reviewers: Exploratory IND Studies*, 70FR19764. U.S. Food and Drug Administration, Washington, D.C.
91. FDA (2005): *Guidance for Industry: Pharmacogenomic Data Submission*. U.S. Food and Drug Administration, Washington, D.C.
92. Fent, K., and Zbinden, G. (1987): Toxicity of interferon and interleukin. *TIPS*, 8:100–105.
93. Fernandez, C. et al. (1993): Pharmacokinetics of zopiclone and its enantiomers in caucasian young healthy volunteers. *Drug Metab. Disp.*, 21(6):1125–1128.
94. Filipsson, A. F. et al. (2003): The benchmark dose method: review of available models and recommendations for application in health risk assessment. *Crit. Rev. Toxicol.*, 33(5):505–542.

95. Findlay, G., and Kermani, F., Eds. (2000): *The Pharmaceutical R&D Compendium: CMR International/SCRIP's Complete Guide to Trends in R&D*. CMR International and PJB Publishers, Ltd., Surrey, U.K.
96. Finter, N. B., Woodrouffe, J., and Priestman, T. J. (1982): Monkeys are insensitive to pyrogenic effects of human alpha-interferons. *Nature (Lond.)*, 298:301.
97. Flammang, J. J. et al. (1997): Neonatal mouse assay for tumorigenicity: alternative to the chronic rodent bioassay. *Reg. Toxicol. Pharmacol.*, 26:230–240.
98. Fletcher, A. P.(1978): Drug safety tests and subsequent clinical experience. *J. Roy. Soc. Med.*, 71:693–696.
99. French Ministry of Social Affaires (1984): *Recommendation concernant le protocole toxicologigue des interferons pour l'obtention d'une autorisation de mise sur le marché*. Direction de la Pharmacie et du Medicament, Sous-Direction des Affaires Techniques et Scientifigues.
100. Friedmann, N. (1985): Thymopentin: safety overview. *Surv. Immunol. Res.*, 4(Suppl. 1):139–148.
101. Frothingham, R. (2001): Rates of Torsades de Pointes associated with ciprofloxacin, ofloxacin, levofloxacin, gatifloxacin and moxifloxacin. *Pharmacotheraphy*, 21(12):1468–1472.
102. Gad, S. C., and Chengelis, C. P. (1995): Human health products: drugs and medicinal devices. In: *Regulatory Toxicology*, edited by C. P. Chengelis, J. F. Holson, and S. C. Gad, Raven Press, New York, pp. 9–49.
103. Galbraith, W. M. (1987): Safety evaluation of biotechnology-derived products. In: *Preclinical Safety of Biotechnology Products Intended for Human Use*, edited by C. E. Graham. Alan R. Liss, New York, pp. 3–14.
104. Galloway, J. A., and Chance, R. E. (1984): Human insulin rDNA: from rDNA through the FDA, In: *Proceedings of the Second World Conference on Clinical Pharmacology and Therapeutics*, edited by L. Lemberger and M. M. Reidenberg. ASPET, MD, pp. 503–520.
105. Goeddel, D. V. et al. (1979): Expression in *Escherichia coli* of chemically synthesized genes for human insulin. *Proc. Natl. Acad. Sci. USA*, 76:106–110.
106. Goldsworthy, T. L. et al. (1994): Transgenic animals in toxicology. *Fund. App. Toxicol*., 22:8–19.
107. Gordon, C. V. and Wierenga, D. E. (1991): *The Drug Development and Approval Process*, Orphan Drugs in Development, Pharmaceutical Manufacturers Association, Washington, D.C.
108. Gori, G. B. (1991): Are animal tests relevant in cancer risk assessment? A persistent issue becomes uncomfortable. *Reg. Toxicol. Pharmacol.*, 13:225–227.
109. Goyan, J. (1981): Introduction. In: *Insulins, Growth Hormone and Recombinant DNA Technology*, edited by J. L. Gueriguian. Raven Press, New York, p. xviii.
110. Graham, C. E. (1987): Overview: the industry position. In: *Preclinical Safety of Biotechnology Products Intended for Human Use*, edited by C. E. Graham. Alan R. Liss, New York, pp. 183–187.
111. Grahame-Smith, D. G. (1982): Preclinical toxicological testing and safeguards in clinical trials. *Eur. J. Clin. Pharmacol.*, 22:1–6.
112. Gralinski, M. R. (2000): The assessment of potential for QT interval prolongation with new pharmaceuticals. Impact on Drug Development. *J. Pharmacol. Toxicol. Methods*, 43:91–99.
113. Greene, J. A. et al. (1996): Nonclinical toxicology studies with zidovudine: reproductive toxicity studies in rats and rabbits. *Fundam. Appl. Toxicol.*, 32:140–147.
114. Grieshaber, C. K. (1991): Prediction of human toxicity of new antineoplastic drugs from studies in animals. In: *The Toxicity of Anticancer Drugs*, edited by G. Powis and M. P. Hacker. Pergamon Press, New York, pp. 10–26.
115. Griffin, P. J. (1986): Predictive value of animal toxicity studies. In: *Long-Term Animal Studies: Their Predictive Value for Man*, edited by S. R. Walker and A. D. Dayan. MTP Press, Lancaster, PA, pp. 107–116.
116. Griffiths, S. A. et al. (1998): Non-clinical safety evaluation of products of biotechnology: industrial strategies. *CMR Int. Rep.*, 5–6.
117. Guerreiro, N. et al. (2003): Toxicogenomics in drug development. *Toxicol. Pathol.*, 31:471–479.
118. Halpin, R. A. et al. Biotransformation of Lovastatin V. (1993): Species differences in *in vivo* metabolite profiles of mouse, rat, dog, and human. *Drug Metab. Disp.*, 21(6): 1003–1006.
119. Hanley, T., Udall, V., and Weatherall, M. (1970): An industrial view of current practice in predicting drug toxicity. *Br. Med. Bull.*, 26(3):203–207.
120. Hansson, E., Havu, N., and Carlsson, E. (1986): Toxicology studies with omeprazole. *Scand. J. Gastroenterol.*, 118(Suppl.):89–91.
121. Harada, Y. (1987): Problems presented by animal toxicity studies. In: *Preclinical Safety of Biotechnology Products Intended for Human Use*, edited by C. E. Graham. Alan R. Liss, New York, pp. 127–142.
122. Haseman, J. K. (1985): Issues in carcinogenicity testing: dose selection. *Fundam. Appl. Toxicol.*, 5:66–78.
123. Hastings, K. L. (2002): Implications of the new FDA/CDER immunotoxicology guidance for drugs. *Int. Immunopharmacol.*, 2:1613–1618.
124. Hayes, A. H., Jr. (1990): Safety considerations in product development. *Drug Safety*, 5(Suppl. 1):24–26.
125. Hess, R. (1991): Repeated dose toxicity: industry perspective. In: *First International Conference on Harmonization*, edited by P. F. D'Arcy and D. W. G. Harron. Brussels, Belgium, p. 197.
126. Hoffman, W. P., Ness, D. K., and van Lier, R. B. L. (2002): Analysis of rodent growth data in toxicology studies. *Toxicol. Sci.*, 66:313–319.
127. Homburger, F. (1987): The necessity of animal studies in routine toxicology [comments]. *Toxicology*, 1(5): 245–255.
128. ICH (1997): *Non-Clinical Safety Studies for the Conduct of Human Clinical Trials for Pharmaceuticals*, ICH Topic M3, Step 5, ICH Harmonized Tripartite Guideline. International Conference on Harmonisation of Technical Requirements for Registration of Pharmaceuticals for Human Use, Geneva, Switzerland.
129. ICH (1995): *Need for Carcinogenicity Studies of Pharmaceuticals*, Topic S1A, Step 5. ICH Harmonized Tripartite Guideline. International Conference on Harmonisation of Technical Requirements for Registration of Pharmaceuticals for Human Use, Geneva, Switzerland.

130. ICH (1995): *Dose Selection for Carcinogenicity Studies in Pharmaceuticals*, Topic S1C(R), Step 5, ICH Harmonized Tripartite Guideline. International Conference on Harmonisation of Technical Requirements for Registration of Pharmaceuticals for Human Use, Geneva, Switzerland.
131. ICH (1996): *Genotoxicity: Specific Aspects of Regulatory Tests*, Topic S2A, Step 5, ICH Harmonized Tripartite Guideline. International Conference on Harmonisation of Technical Requirements for Registration of Pharmaceuticals for Human Use, Geneva, Switzerland.
132. ICH (1997): *Genotoxicity: Standard Battery of Tests*. Topic S2B, Step 5, ICH Harmonized Tripartite Guideline. International Conference on Harmonisation of Technical Requirements for Registration of Pharmaceuticals for Human Use, Geneva, Switzerland.
133. ICH (1995): *Toxicokinetics: Guidance on the Assessment of Systemic Exposure in Toxicity Studies*, Topic S3A, Step 5, ICH Harmonized Tripartite Guideline. International Conference on Harmonisation of Technical Requirements for Registration of Pharmaceuticals for Human Use, Geneva, Switzerland.
134. ICH (1994): *Pharmacokinetics: Guidance for Repeat Dose Tissue Distribution Studies. Need for Tissue Distribution Studies When Appropriate Data Cannot Be Derived from Other Sources*, Topic S3B, Step 5, ICH Harmonized Tripartite Guideline. International Conference on Harmonisation of Technical Requirements for Registration of Pharmaceuticals for Human Use, Geneva, Switzerland.
135. ICH (1997): *Draft Guideline for Duration of Chronic Toxicity Testing*, Topic S4A, Step 3, ICH Harmonized Tripartite Guideline. International Conference on Harmonisation of Technical Requirements for Registration of Pharmaceuticals for Human Use, Geneva, Switzerland.
136. ICH (1996): *Detection of Toxicity to Reproduction from Medicinal Products*, Topic S5A, Step 5, ICH Harmonized Tripartite Guideline. International Conference on Harmonisation of Technical Requirements for Registration of Pharmaceuticals for Human Use, Geneva, Switzerland.
137. ICH (1996): *Reproductive Toxicity: Male Fertility Studies*, Topic S5B, Step 5, ICH Harmonized Tripartite Guideline. International Conference on Harmonisation of Technical Requirements for Registration of Pharmaceuticals for Human Use, Geneva, Switzerland.
138. ICH (1997): *Preclinical Safety Evaluation of Biotechnology-Derived Pharmaceuticals*, Topic S6, Step 5, ICH Harmonized Tripartite Guideline. International Conference on Harmonisation of Technical Requirements for Registration of Pharmaceuticals for Human Use, Geneva, Switzerland.
139. ICH (2000): *Safety Pharmacology Studies for Human Pharmaceuticals*, Topic S7A, Step 5, ICH Harmonized Tripartite Guideline. International Conference on Harmonisation of Technical Requirements for Registration of Pharmaceuticals for Human Use, Geneva, Switzerland.
140. ICH (2004): *The Nonclinical Evaluation of the Potential for Delayed Ventricular Repolarization (QT Interval Prolongation) by Human Pharmaceuticals*, Topic S7B(R), Step 3, ICH Harmonized Tripartite Guideline. International Conference on Harmonisation of Technical Requirements for Registration of Pharmaceuticals for Human Use, Geneva, Switzerland.
141. ICH (2004): *Immunotoxicology Studies for Human Pharmaceuticals*, Topic S8, Step 3, ICH Harmonized Tripartite Guideline. International Conference on Harmonisation of Technical Requirements for Registration of Pharmaceuticals for Human Use, Geneva, Switzerland.
142. FDA (2004): *Innovation/Stagnation: Challenge and Opportunity on the Critical Path to New Medicinal Products*. U.S. Food and Drug Administration, Washington, D.C.
143. Ito, K., and Houston, J. B. (2005): Prediction of human drug clearance from *in vitro* and preclinical data using physiologically based and empirical approaches. *Pharm. Res.*, 22(1):103–112.
144. JMHW (1984): *Notification on Application Data for rDNA Drugs*, Notification No. 243, Pharmaceutical Affairs Bureau, Japanese Ministry of Health and Welfare, Tokyo.
145. Jenssen, D., and Romet, L. (1990): Studies of metabolism mediated mutagenicity *in vitro*. *Altern. Lab. Animals*, 18:243–250.
146. Karch, F. E. (1980): Is drug toxicity a problem of great magnitude? Probably not. In: *Controversy in Therapeutics*, edited by L. Lasagna. W.B. Saunders, Philadelphia, PA, pp. 51–57.
147. Kier, L D. (1985): Use of the Ames test in toxicology. *Reg. Toxicol. Pharmacol.*, 5:59–64.
148. Klaassen, C. D., and Doull, J. (1980): Evaluation of safety: toxicologic evaluation. In: *Toxicology: The Basic Science of Poisons*, 2nd ed., edited by J. Doull, C. D. Klaassen, and M. O. Amdur. Macmillan, New York, pp. 11–27.
149. Kluwe, W. M. (1995): The complementary roles of *in vitro* and *in vivo* tests in genetic toxicology assessment. *Reg. Toxicol. Pharmacol.*, 22:268–272.
150. Koller, L. D. (2001): A perspective on the progression of immunotoxicology. *Toxicology*, 160:105–110.
151. Kuper, C. F. et al. (2000): Histopathologic approaches to detect changes indicative of immunotoxicity. *Toxicol. Pathol.*, 28(3):454–466.
152. Larsson, H. et al. (1986): Plasma gastrin and gastric enterochromaffin-like cell activation and proliferation: studies with omeprazole and ranitidine in intact and antrectomized rats. *Gastroenterology*, 90:391–399.
153. Lasagna, L. (1986): Clinical testing of products prepared by biotechnology. *Reg. Toxicol. Pharmacol.*, 6:385–390.
154. Lasagna, L. (1987): Predicting human drug safety from animal studies: current issues. *J. Toxicol. Sci.*, 12:439–450
155. Lemberger, L. (1987): Early clinical evaluation in man: the buck stops here. *Xenobiotica*, 17(3):267–273.
156. Lewis, R. W. et al. (2002): Recognition of adverse and nonadverse effects in toxicity studies. *Toxicol. Path.* 30(1):66–74.
157. Litchfield, J. T. (1961): Forecasting drug effects in man from studies in laboratory animals. *JAMA*, 177:104–108.
158. Lumley, C. E. (1994): General pharmacology, the international regulatory environment, and harmonization guidelines. *Drug Dev. Res.*, 32:223–232.
159. Lumley, C. E., and Walker, S. R. (1988): Investigation of the relationship between animal and clinical data. *Abstr 29th Congress Eur. Soc. Toxicol.*, 188.
160. Lumley, C. E., and Walker, S. R. (1985): A toxicology databank based on animal safety evaluation studies of pharmaceutical compounds. *Hum. Toxicol.*, 4:447–460.

161. Lumley, C. E., and Walker, S. R. (1985): The value of chronic animal toxicology studies of pharmaceutical compounds: a retrospective analysis. *Fundam. Appl. Toxicol.*, 5:1007–1024.
162. Lumley, C. E., Parkinson, C., and Walker, S. R. (1992): An international appraisal of the minimum duration of chronic animal toxicity studies. *Hum. Exp. Toxicol.*, 11:155–162.
163. Luster, M. I. et al. (1988): Development of a testing battery to assess chemical-induced immunotoxicity: National Toxicology Program's Guidelines for immunotoxicity evaluation in mice. *Fundam. Appl. Toxicol.*, 10:2–19.
164. MacDonald, J. et al. (2004): The utility of genetically modified mouse assays for identifying human carcinogens: a basic understanding and path forward. *Toxicol. Sci.*, 77:188–194.
165. Mahmood, I. (2005): Interspecies scaling of biliary excreted drugs: a comparison of several methods. *J. Pharm. Sci.*, 94(4):883–892.
166. Martial, J. A. et al. (1979): Human growth hormone: complementary DNA cloning and expression in bacteria. *Science*, 205:602–607.
167. Melinkova, I. (2005): From the analyst's couch: raising HDL cholesterol. *Nat. Rev.*, 4:185.
168. Monro, A. M. (1990): Interspecies comparisons in toxicology: the utility and futility of plasma concentrations of the test substance. *Regul. Toxicol. Pharmacol.*, 12:137–160.
169. Monro, A. (1992): What is an appropriate measure of exposure when testing drugs for carcinogenicity in rodents? *Toxicol. Appl. Pharmacol.*, 112:171–181.
170. Morrow, P. E. (1992): Dust overloading of the lungs: update and appraisal. *Toxicol. Appl. Pharmacol.*, 113:1–12.
171. Mullin, R. (2003): Drug developments cost about $1.7 billion. *C&E News*, 81(50):8.
172. Munro, I. C. (1977): Considerations in chronic toxicity testing: the chemical, the dose, the design. *J. Environ. Pathol. Toxicol.*, 1:183–197.
173. Nagata, S. et al. (1980): Synthesis in *E. coli* of a polypeptide with human leukocyte interferon activity. *Nature (Lond.)*, 284:316–320.
174. Nagilla, R., and Ward, K. W. (2004): A comprehensive analysis of the role of correction factors in the allometric predictivity of clearance from rat, dog, and monkey to humans. *J. Pharm. Sci.*, 93(10):2522–2534.
175. Nakayama, Y. et al. (2005): Simulation of the toxicokinetics of trichloroethylene, methylene chloride, styrene and *n*-hexane by a toxicokinetics/toxicodynamics model using experimental data. *Environ. Sci.*, 12(1):21–32.
176. National Toxicology Program (1984): Report of the NTP *Ad Hoc* Panel on Chemical Carcinogenesis Testing and Evaluation. U.S. Department of Health and Human Services, Washington, D.C.
177. Norbury, K. C. (1982): Immunotoxicology in the pharmaceutical industry. *Environ. Health Perspect.*, 43:53–59.
178. TCSDD (2005): *Outlook*. Tufts Center for the Study of Drug Development, Boston, MA.
179. Parkinson, C. Thomas, K. E., and Lumley, C. E. (1997): Reproductive toxicity testing of pharmaceutical compounds to support the inclusion of women in clinical trials. *Hum. Exp. Toxicol.*, 16:239–246.
180. Parkinson, C., Lumley, C. E., and Walker, S. R. (1995): The value of information generated by long-term toxicity studies in the dog for the nonclinical safety assessment of pharmaceutical compounds. *Fund. Appl. Toxicol.*, 25:115–123.
181. Peck, H. M. (1968): An appraisal of drug safety evaluation in animals and the extrapolation of results to man. In: *Importance of Fundamental Principles in Drug Evaluation*, edited by D. E. Tedeschi, and R. E. Tedeschi. Raven Press, New York, pp. 450–471.
182. Petricciani, J. C. (1983): An overview of safety and regulatory aspects on new biotechnology. *Reg. Toxicol. Pharmacol.*, 3:428–433.
183. *Physicians' Desk Reference* (1999): Medical Economics Data, Montvale, NJ, pp. 584–587.
184. *Physicians' Desk Reference* (1999): Medical Economics Data, Montvale, NJ, pp. 1202–1210.
185. Poggesi, I. (2004): Predicting human pharmacokinetics from preclinical data. *Curr. Opin. Drug Discov. Dev.*, 7(1):100–111.
186. Proakis, A. G. (1994): Regulatory consideration on the role of general pharmacology studies in the development of therapeutic agents. *Drug Dev. Res.*, 32: 233–236.
187. Purchase, I. F. H. et al. (1998): Workshop overview: scientific and regulatory challenges to the reduction, refinement, and replacement of animals in toxicity testing. *Toxicol. Sci.*, 43:86–101.
188. Raheja, K. L., and Jordan, A. (1994): FDA recommendations for preclinical testing of gonadotropin-releasing hormone (GnRH) analogues. *Reg. Toxicol. Pharmacol.*, 19:168–175.
189. Ronneberger, H., and Hilfenhaus, J. (1983): Toxicity studies with human fibroblast interferon. *Arch. Toxicol. Suppl.*, 6:391–394.
190. Rosenblum, I. Y. (2003): Toxicogenomic applications to drug risk assessment. *Environ. Health Perspec.*, 111(15):A804–A805.
191. Rumjanek, V. M., Hanson, J. M., and Morley, J. (1982): Lymphokines and monokines. In: *Immunopharmacology*, edited by P. Sirois and M. Pleszczymski. Elsevier Press, Amsterdam, pp. 267–285.
192. Sachs, G. et al. (1988): Gastric H,K-ATPase as therapeutic target. *Annu. Rev. Pharmacol. Toxicol.*, 28: 269–284.
193. Schein, P. S. et al. (1970): The evaluation of anti-cancer drugs in dogs and monkeys for the prediction of qualitative toxicities in man. *Clin. Pharmacol. Ther.*, 11: 3–40
194. Schellebens, H., de Reus, A., and von den Meide, P. H. (1984): The chimpanzee as a model to test side effects of human interferons. *J. Med. Primatol.*, 13: 235–245.
195. Schreiner, C. A. (1983): Application of short-term tests to safety testing of industrial chemicals. *Ann. N.Y. Acad. Sci.*, 407:367–373.
196. Schwetz, B., and Gaylor, D. (1997): New directions for predicting carcinogenesis. *Mol. Carc.*, 20:275–279.
197. Scott, G. M. (1982): Interferon: pharmacokinetics and toxicity. *Phil. Trans. R. Soc. Lond.*, B299:91–107.

198. Scott, G. M. (1983): The toxic effects of interferon in man. *J. Interferon Res.*, 5:85–114.
199. Segre, G. (1983): New toxicological problems and proposed solutions: an introduction. In: *Current Problems in Drug Toxicology*, edited by G. Zbinden, J. Y., Detaille, and G. Mazué. John Libbey Eurotext, London, pp. 239–240.
200. Silva Lima, B., and van der Lann, J. W. (2000): Mechanisms of nongenotoxic carcinogenesis and assessment of human hazard. *Reg. Toxicol. Pharmacol.*, 32:135–143.
201. Sontag, J. M., Page, N. P., and Safiotti, V. (1976): *Guidelines for Carcinogen Bioassays in Small Rodents*, DHHS Publ. (NIH) 76–801. National Cancer Institute, Bethesda, MD.
202. Stebbing, N. (1984): Pharmacological assessment of interferons for clinical use. In: *Proceedings of the Second World Conference on Clinical Pharmacology and Therapeutics*, edited by L. Lemberger and M. M. Reidenberg. American Society for Pharmacology and Experimental Therapy, Bethesda, MD, pp. 521–534.
203. Stebbing, N., and Weck, P. K. (1984): Preclinical assessment of biological properties of recombinant DNA derived human interferons. In: *Recombinant DNA Products: Insulin, Interferon and Growth Hormone*, edited by A. P. Bollon. CRC Press, Boca Raton, FL, pp. 75–114.
204. Stebbing, N. et al. (1983): Antiviral effects of bacteria derived human leukocyte interferons against encephalomyocarditis virus infection of squirrel monkeys. *Arch. Viral.*, 76:365–372.
205. Stevens, J. C., Fayer, J. L., and Cassidy, K. C. (2001): Characterization of 2-[[4-[[2-(1H-tetrazol-5-ylmethyl)phenyl]methoxy]phenoxy]methyl] quinoline *n*-glucuronidation by *in vitro* and *in vivo* approaches. *Drug Metab. Disp.*, 29(3):289–295.
206. Stoll, R. E. (1987): The preclinical development of biotechnology-derived pharmaceuticals: the PMA perspective. In: *Preclinical Safety of Biotechnology Products Intended for Human Use*, edited by C. E. Graham. Alan R. Liss, New York, pp. 169–171.
207. Swenberg, J. A. (1995): Bioassay design and MTD setting: old methods and new approaches. *Reg. Toxicol. Pharmacol.*, 21:44–51.
208. Tan, S. C. et al. (2002): Stereoselectivity of ibuprofen metabolism and pharmacokinetics following the administration of the racemate to healthy volunteers. *Xenobiotica*, 32(8):683–697.
209. Teelmann, K., Hohbach, C., and Lehmann, H., The International Working Group (1986): Preclinical safety testing of species-specific proteins produced with recombinant DNA-techniques. *Arch. Toxicol.*, 59:195–200.
210. Temkin, C. L. et al. (1941): *Four treatments of Theophrastus von Hohemheim called Paraceleus*. The Johns Hopkins Press, Baltimore, MD (translated from the original German, with introductory essays).
211. Tennant, R. W., Spalding, J., and French, J. F. (1996): Evaluation of transgenic mouse bioassays for identifying carcinogens and noncarcinogens. *Mutation Res.*, 365:119–127.
212. The ICICIS Group Investigators (1998): Report of validation study of assessment of direct immunotoxicity in the rat. *Toxicol.*, 125:183–201.
213. Spence, C., Ed. (1997): *The Pharmaceutical R&D Compendium: CMR International/SCRIP's Complete Guide to Trends in R&D*. CMR International and PJB Publishers, Ltd., Surrey, U.K.
214. Thomas, P. T. (1990): Approaches used to assess chemically induced impairment of host resistance and immune function. *Toxic Subst. J.*, 10:241–278.
215. Tomaszewski, J. E., and Smith, A. C. (1997): Safety testing of antitumor agents, in: *Comprehensive Toxicology*, Vol. 2, *Toxicity Testing and Evaluation*, edited by P. D. Williams and G. H. Hottendorf. Elsevier, New York, pp. 299–309.
216. Trotta, P. P. (1986): Preclinical biology of alpha interferons. *Semin. Oncol.*, 13(3):3–12.
217. Trown, P. W., Wills, R. J., and Kamm, J. J. (1986): The preclinical development of Roferon®-A. *Cancer*, 57(8):1648–1656.
218. Tsang, L., and Beers, D. (2003): Legal and scientific considerations in nonclinical assessment of biotechnology products. *Drug Inf. J.*, 37:397–406.
219. Anon. (1992): U.S. biotechnology policy [editorial]. *Nature*, 356:1–2.
220. Van Oosterhoot, J. P. J. et al. (1997): The utility of two rodent species in carcinogenic risk assessment of pharmaceuticals in Europe. *Reg. Toxicol. Pharmacol.*, 25:6–17.
221. Vos, J. G. (1977): Immune suppression as related to toxicology. *CRC Crit. Rev. Toxicol.*, 5:67–101.
222. Wagstaff, J. et al. (1984): A phase I toxicity study of human rDNA interferon in patients with solid tumors. *Cancer Chemother. Pharmacol.*, 13:100–105.
223. Waife, S. O., and Lasagna, L. (1985): From DNA to NDA: the impact of recombinant DNA technology on new drug development. *Reg. Toxicol. Pharmacol.*, 5:212–224.
224. Wallmark, B. (1986): Mechanism of action of omeprazole. *Scand. J. Gastroent.*, Suppl. 118:11–16.
225. Weissinger, J. (1989): Nonclinical pharmacologic and toxicologic considerations for evaluating biologic products. *Reg. Toxicol. Pharmacol.*, 10:255–263.
226. Weissinger, J. (1990): Pharmacology and toxicology of novel drug delivery systems: regulatory issues. *Drug Safety*, 5(Suppl. 1):107–113.
227. Wierda, D. (2006): Personal communication. Eli Lilly and Company, Greenfield, IN.
228. Williams, G. M., and Weisburger, J. H. (1991): Chemical carcinogenesis. In: *Toxicology: The Basic Science of Poisons*, 4th ed., edited by M. O. Amdur, J. Doull, J., and C. D. Klaassen. Pergamon Press, New York, pp. 127–200.
229. Williams, G. M., Dunkel, V. C., and Ray, V. A., Eds. (1983): Cellular systems for toxicity testing. *Ann. N.Y. Acad. Sci.*, 407:1–482.
230. Williams, P. (1990): The role of pharmacological profiling in safety assessment. *Reg. Toxicol. Pharmacol.*, 12(3):238–252.
231. Williams, P. D. et al. (1991): General pharmacology of a new potent 5-hydroxytryptamine antagonist. *Arzneim. Forsch.*, 41(1):189–195.
232. Wilson, A. B. (1987): The toxicology of the end products from biotechnology processes. *Arch. Toxicol.*, 11:194–199.
233. Wolf, F. J. (1980): Effect of overloading pathways on toxicity. *J. Environ. Pathol. Toxicol.*, 3:113–134.

234. Yoshida, S., Golub, M. S., and Gershwin, M. E. (1989): Immunological aspects of toxicology: premises not promises. *Reg. Toxicol. Pharmacol.*, 9:56–80.
235. Zapp, J. A., Jr. (1977): Extrapolation of animal studies to the human situation. *J. Toxicol. Environ. Health*, 2:1425–1433.
236. Zbinden, G. (1964): The problem of the toxicologic examination of drugs in animals and their safety in man. *Clin. Pharmacol. Ther.*, 5:537–545.
237. Zbinden, G. (1966): The significance of pharmacologic screening tests in the preclinical safety evaluation of new drugs. *J. New Drugs*, 6:1–7.
238. Zbinden, G. (1976): A look at the world from inside the toxicologist's cage. *Eur. J. Clin. Pharmacol.*, 9:333–338.
239. Zbinden, G. (1978): Application of basic concepts to research in toxicology. *Pharmacol. Rev.*, 30(4):605–616.
240. Zbinden, G. (1982): Current trends in safety testing and toxicological research. *Naturwissenschaften*, 69:255–259.
241. Zbinden, G. (1986): Acute toxicity testing, public responsibility and scientific challenges. *Cell Biol. Toxicol.*, 2(3):325–335.
242. Zbinden, G. (1987): Biotechnology products intended for human use, toxicological targets and research strategies. In: *Preclinical Safety of Biotechnology Products Intended for Human Use*, edited by C. E. Graham. Alan R. Liss, New York, pp. 143–159.
243. Zbinden, G. (1989): Improvement of predictability of subchronic and chronic toxicity studies. *J. Toxicol. Sci.*, 14(Suppl. 3):3–21.
244. Zbinden, G. (1990): Effects of recombinant human alpha-interferon in a rodent cardiotoxicity model. *Toxicol. Lett.*, 50:25–35.
245. Zbinden, G. (1990): Safety evaluation of biotechnology products. *Drug Safety*, 5(Suppl. 1):58–64.
246. Zhong, W. Z, Williams, M. G., and Branstetter, D. G. (2000): Toxicokinetics in drug development: an overview of toxicokinetic application in the development of PNU-101017: an anxiolytic drug candidate. *Curr. Drug Metab.*, 1(3):243–254.
247. Zwickl, C. M. et al. (1991):Comparison of the immunogenicity of recombinant and pituitary human growth hormone in rhesus monkeys. *Fundam. Appl. Toxicol.*, 16:275–287.

# Notes

# 8 Statistics for Toxicologists

*Shayne C. Gad*

## CONTENTS

## INTRODUCTION

This chapter has been written for both practicing and student toxicologists and pathologists as a practical guide to the common statistical problems encountered in toxicologic pathology and the methodologies that are available to solve them. The chapter has been enriched by the inclusion of discussions of why a particular procedure or interpretation is recommended, by the clear enumeration of the assumptions that are necessary for a procedure to be valid, and by discussion of problems drawn from the actual practice of toxicology and toxicologic pathology.

Since 1960, the field of toxicology has become increasingly complex and controversial in both its theory and practice. Much of this change is due to the evolution of the field. As in all other sciences, toxicology started as a descriptive science. Living organisms, be they human or otherwise, were dosed with or exposed to chemicals or physical agents and the adverse effects that followed were observed. But, as a sufficient body of descriptive data was accumulated, it became possible to infer and study underlying mechanisms of action—to determine in a broader sense why adverse effects occurred. Toxicology has thus entered a later state of development, the mechanistic stage, where active contributions to the field encompass both descriptive and mechanistic studies.

Studies continue to be designed and executed to generate increased amounts of data. Genomics and proteomics have even accentuated this process. The resulting problems of data analysis have then become more complex, and toxicology has drawn more deeply from the well of available statistical techniques. Statistics has also been very active and growing during the last 35 years, to some extent, at least, because of the parallel growth of toxicology. These simultaneous changes have led to an increasing complexity of data and, unfortunately, to the introduction of numerous confounding factors that severely limit the utility of the resulting data in all too many cases.

A major difficulty is the very real necessity to understand the biological realities and implications of a problem as well as to understand the peculiarities of toxicological data before procedures are selected and employed for analysis. These characteristics include the following.

1. It is necessary to work with a relatively small sample set of data collected from the members of a population (laboratory animals, cultured cells, and bacterial cultures) that is not actually our population of interest (that is, humans or a target animal population).
2. Frequently, data are obtained from a sample that was censored on a basis other than by the investigator's as design. By censoring, of course, we mean that not all data points were collected as might be desired. This censoring could be the result of either a biological factor (the test animal being dead or too debilitated to manipulate) or a logistic factor (equipment being inoperative or a tissue being missed in necropsy).
3. The conditions under which our experiments are conducted are extremely varied. In pharmacology (the closest cousin to at least classical toxicology), the possible conditions of interaction of a chemical or physical agent with a person are limited to a small range of doses via a single route over a short course of treatment to a defined patient population. In toxicology, however, all these variables (dose, route, time span, and subject population) are determined by the investigator.
4. The time frames available to solve our problems are limited by practical and economic factors. This frequently means that there is not time to repeat a critical study if the first attempt fails, so a true iterative approach is not possible.

The training of most toxicologists in statistics remains limited to a single introductory course that concentrates on some theoretical basics. As a result, the armentarium of statistical techniques of most toxicologists is limited, and the tools that are usually present (*t*-tests, chi-square, analysis of variance, and linear regression) are neither fully developed nor well understood. It is hoped that this chapter will help change this situation.

As a point of departure toward this objective, it is essential that any analysis of study results be interpreted by a professional who firmly understands three concepts: (1) the difference between biological significance and statistical significance, (2) the nature and value of different types of data, and (3) causality. For the first concept, we should consider the four possible combinations of these two different types of significance, for which we find the relationship shown below:

| | | Statistical Significance | |
|---|---|---|---|
| | | **No** | **Yes** |
| **Biological Significance** | **No** | Case I | Case II |
| | **Yes** | Case III | Case IV |

Cases I and IV give us no problems, for the answers are the same statistically and biologically, but cases II and III present problems. In case II (the "false positive"), we have a circumstance where there is a statistical significance in the measured difference between treated and control groups, but there is no true biological significance to the finding. This is not an uncommon happening, for example, in the case of clinical chemistry parameters. This is called a *type I* error by statisticians, and the probability of this happening is called the $\alpha$ (alpha) level. In case III (the "false negative"), we have no statistical significance, but the differences between groups are biologically or

**TABLE 8.1**
**Approximate Total Sample Sizes for Comparisons Using the *t*-Test and Equal Group Sizes**

| | $\beta = 0.1$ | | $\beta = 0.2$ | |
|---|---|---|---|---|
| $\Delta/\sigma$ | $\alpha = 0.05$ | $\alpha = 0.10$ | $\alpha = 0.05$ | $\alpha = 0.10$ |
| 0.25 | 672 | 548 | 502 | 396 |
| 0.50 | 168 | 138 | 126 | 98 |
| 0.75 | 75 | 62 | 56 | 44 |
| 1.00 | 42 | 34 | 32 | 24 |
| 1.25 | 28 | 22 | 20 | 16 |
| 1.50 | 18 | 16 | 14 | 12 |

*Note:* $\Delta$ is the difference in the treatment group means, and $\sigma$ is the standard deviation.

toxicologically significant. This is called a *type II* error by statisticians, and the probability of such an error happening by random chance is called the $\beta$ (beta) level. An example of this second situation is when we see a few of a very rare tumor type in treated animals. In both of these latter cases, numerical analysis, no matter how well done, is no substitute for professional judgment. Along with this, however, one must have a feeling for the different types of data and for the value or relative merit of each. Note that the two error types interact, and in determining sample size we need to specify both $\alpha$ and $\beta$ levels. Table 8.1 demonstrates this interaction in the case of the *t*-test.

The reasons why biological and statistical significance are not identical are multiple, but a central one is certainly causality. Through our consideration of statistics, we should keep in mind that just because a treatment and a change in an observed organism are seemingly or actually associated with each other this does not "prove" that the former caused the latter. Although this fact is now widely appreciated for correlation (for example, the fact that the number of storks' nests found each year in England is correlated with the number of human births that year does not mean that storks bring babies), it is just as true in the general case of significance. Timely establishment and proof that treatment causes an effect require an understanding of the underlying mechanism and proof of its validity. At the same time, it is important that we realize that not finding a good correlation or suitable significance associated with a treatment and an effect likewise does not prove that the two are not associated—that a treatment does not cause an effect. At best, it gives us a certain level of confidence that, under the conditions of the current test, these items are not associated.

These points will be discussed in greater detail in the "Assumptions" sections for each method, along with other common pitfalls and shortcomings associated with the method. To help in better understanding the discussion to come, terms frequently used throughout this chapter should first be considered. These are presented in Table 8.2.

Each measurement we make—each individual piece of experimental information we gather—is called a *datum*; however, we gather and analyze multiple pieces at one time, the resulting collection being called *data*. Data are collected on the basis of their association with a treatment

**TABLE 8.2**
**Some Frequently Used Terms and Their General Meanings**

| Term | Meaning |
|---|---|
| 95% confidence interval | A range of values (above, below, or above and below) the sample (mean, median, mode, etc.) that has a 95% chance of containing the true value of the population (mean, median, mode); also called the fiducial limit equivalent to $p < 0.05$ |
| Bias | Systemic error as opposed to a sampling error; for example, selection bias may occur when each member of the population does not have an equal chance of being selected for the sample |
| Degrees of freedom | The number of independent deviations; usually abbreviated df |
| Independent variables | Also known as predictors or explanatory variables |
| *p*-value | Another name for significance level; usually 0.05 |
| Power | The effect of the experimental conditions on the dependent variable relative to sampling fluctuation. When the effect is maximized, the experiment is more powerful. Power can also be defined as the probability that there will not be a type II error (1-beta); conventionally, power should be at least .07 |
| Random | Each individual member of the population having the same chance of being selected for the sample |
| Robust | Having inferences or conclusions little affected by departure from assumptions |
| Sensitivity | The number of subjects experiencing each experimental condition divided by the variance of scores in the sample |
| Significance level | The probability that a difference has been erroneously declared to be significant, typically 0.05 and 0.01, corresponding to 5% and 1% chance of error |
| Type I error (false positives) | Concluding that there is an effect when there really is not an effect; its probability is the alpha level |
| Type II error (false negatives) | Concluding that there is no effect when there really is an effect; its probability is the beta level |

(intended or otherwise) as an effect (a property) that is measured in the experimental subjects of a study, such as body weights. These identifiers (that is, treatment and effect) are termed *variables*. Our treatment variables (those that the researcher or nature control and which can be directly controlled) are termed *independent*, and our effect variables (such as weight, life span, and number of neoplasms) are termed *dependent* variables; their outcome is believed to dependent on the treatment being studied.

All the possible measures of a given set of variables in all the possible subjects that exist is termed the *population* for those variables. Such a population of variables cannot be truly measured; for example, one would have to obtain, treat, and measure the weights of all the Fischer-344 rats that were, are, or ever will be. Instead, we deal with a representative group: a *sample*. If our sample of data is appropriately collected and of sufficient size, it serves to provide good estimates of the characteristics of the parent population from which it was drawn.

## Bias and Chance

Any toxicological study aims to determine whether a treatment elicits a response. An observed difference in response between a treated and control group need not necessarily be a result of treatment. There are, in principle, two other possible explanations: *bias*, or systematic differences other than treatment between the groups, and *chance*, or random differences. A major objective of both experimental design and analysis is to try to avoid bias. Wherever possible, treated and control groups to be compared should be alike with respect to all other factors. Where differences remain, these should be corrected for in the statistical analysis. Chance cannot be wholly excluded, as identically treated animals will not respond identically. Although even the most extreme difference might in theory be due to chance, a proper statistical analysis will allow the experimenter to assess this possibility. The smaller the probability of a false positive, the more confident the experimenter can be that the effect is real. Good experimental design improves the chance of picking up a true effect with confidence by maximizing the ratio between *signal* and *noise*.

## Hypothesis Testing and Probability (*p*) Values

A relationship of treatment to some toxicological endpoint is often stated to be "statistically significant ($p < 0.05$)." What does this really mean? A number of points have to be made. First, statistical significance need not necessarily imply biological importance, if the endpoint under study is not relevant to the animal's wellbeing. Second, the statement will usually be based only on the data from the study in question and will not take into account prior knowledge. In some situations, such as when one or two of a very rare tumor type are seen in treated animals, statistical significance may not be achieved but the finding may be biologically extremely important, especially if a similar treatment was previously found to elicit a similar response. Third, the $p$ value does not describe the probability that a true effect of treatment exists; rather, it describes the probability of the observed response, or one more extreme, occurring on the assumption that treatment actually had no effect whatsoever. A $p$ value that is not significant is consistent with a treatment having a small effect not detected with sufficient certainty in this study. Fourth, there are two types of $p$ values. A one-tailed (or one-sided) $p$ value is the probability of getting by chance a treatment effect in a specified direction as great as or greater than that observed. A two-tailed $p$ value is the probability of getting, by chance alone, a treatment difference in either direction that is as great as or greater than that observed. By convention, $p$ values are assumed to be two-tailed unless the contrary is stated. Where one can rule out in advance the possibility of a treatment effect except in one direction (which is unusual), a one-tailed $p$ value should be used. Often, however, two-tailed tests are to be preferred, and it is certainly not recommended to use one-tailed tests and *not* report large differences in the other direction. In any event, it is important to make it absolutely clear whether one- or two-tailed tests have been used.

It is a great mistake, when presenting results of statistical analyses, to mark, as do some laboratories, results simply as significant or not significant at one defined probability level (usually $p < 0.05$). This poor practice does not allow the reader any real chance to judge whether or not the effect is a true one. Some statisticians present the actual $p$ value for every comparison made. Although this gives precise information, it can make it difficult to assimilate results from many variables. One practice we recommend is to mark $p$ values routinely using plus signs to indicate positive differences (and minus signs to indicate negative differences) as follows: +++ $p$, 0.001; ++ $0.001 \leq p < 0.01$; + $0.01\ p < 0.05$ (+ $0.05 \leq p < 0.1$). This highlights significant results more clearly and also allows the reader to judge the whole range from "virtually certain treatment effect" to "some suspicion." Note that, when using two-tailed tests, bracketed plus signs indicate findings that would be significant at the conventional $p < 0.05$ level using one-tailed tests but are not significant at this level using two-tailed tests. In interpreting $p$ values it is important to realize they are only an aid to judgment to be used in conjunction with other available information. One might validly consider a $p < 0.01$ increase as chance when it was unexpected, occurred only at a low dose level with no such effect seen at higher doses, and was evident in only one subset of the data. In contrast, a $p < 0.05$ increase might be convincing if it occurred in the top dose and was for an endpoint one might have expected to be increased from known properties of the chemical or closely related chemicals.

## Multiple Comparisons

When a $p$ value is stated to be <0.05, this implies that, for that particular test, the difference could have occurred by chance less than 1 time in 20. Toxicological studies frequently involve making treatment–control comparisons for large numbers of variables and, in some situations, also for various subsets of animals. Some statisticians worry that the larger the number of tests, the greater the chance of picking up statistically significant findings that do not represent true treatment effects. For this reason, an alternative "multiple comparisons" procedure has been proposed in which, if the treatment was totally without effect, then 19 times out of 20 *all* the tests should show nonsignificance when testing at the 95% confidence level. Automatic use of this approach cannot be recommended. Not only does it make it much more difficult to pick up any real effects, but also there is something inherently unsatisfactory about a situation where the relationship between a treatment and a particular response depends arbitrarily on which other responses happened to be investigated at the same time. It is accepted that in any study involving multiple endpoints that inevitably a gray area will exist between those showing highly significant effects and those showing no significant effects, where there is a problem distinguishing chance and true effects. However, changing the methodology so the gray areas all come up as nonsignificant can hardly be the answer.

## Estimating the Size of the Effect

It should be clearly understood that a $p$ value does not give direct information about the size of any effect that has occurred. A compound may elicit an increase in response by a given amount, but whether a study finds this increase to be statistically significant will depend on the size of the study and the variability of the data. In a small study, a large and important effect may be missed, especially if the endpoint is imprecisely measured. In a large study, on the other hand, a small and unimportant effect may emerge as statistically significant.

Hypothesis testing tells us whether an observed increase can or cannot be reasonably attributed to chance but not how large it is. Although much statistical theory relates to hypothesis testing, current trends in medical statistics are toward confidence interval estimation, with differences between test and control groups expressed in the form of a best estimate, coupled with the 95% confidence interval (CI). Thus, if one states that treatment increases response by an estimated 10 units (95% CI, 3–17 units), this would imply a 95% chance that the indicated interval includes the true difference. If the lower 95% confidence limit exceeds zero, this implies that the increase is statistically significant at $p < 0.05$ using a two-tailed test. One can also calculate, for example, 99% or 99.9% confidence limits, corresponding to testing for significance at $p < 0.01$ or $p < 0.001$. In screening studies of standard design, the tendency has been to concentrate mainly on hypothesis testing; however, presentation of the results in the form of estimates with confidence intervals can be a useful adjunct for some analyses and is very important in studies aimed specifically at quantifying the size of an effect.

Two terms refer to the quality and reproducibility of our measurements of variables. The first, *accuracy*, is an expression of the closeness of a measured or computed value to its actual or true value in nature. The second, *precision*, reflects the closeness or reproducibility of a series of repeated measurements of the same quantity. If we arrange all of our measurements of a particular variable in order as points on an axis marked as to the values of that variable and if our sample were large enough, the pattern of distribution of the data in the sample would begin to become apparent. This pattern is a representation of the frequency distribution of a given population of data—that is, of the incidence of different measurements, their central tendency, and dispersion.

The most common frequency distribution—and one we will talk about throughout this chapter—is the normal (or Gaussian) distribution. The normal distribution is such that two thirds of all values are within one standard deviation of the mean (or average value for the entire population) and 95% are within 1.96 standard deviations of the mean. Symbols used are $\mu$ for the mean and $\sigma$ for the standard deviation.

In all areas of biological research, optimal design and appropriate interpretation of experiments require that the researcher understand both the biological and technological underpinnings of the system being studied and of the data being generated. From the point of view of the statistician, it is vitally important that the experimenter both know and be able to communicate the nature of the data, and understand its limitations. One classification of data types is presented in Table 8.3.

The nature of the data collected is determined by three considerations: (1) the biological source of the data (the system being studied), (2) the instrumentation and techniques being used to make measurements, and (3) the design of the experiment. The researcher has some degree of control over each of these, least over the biological system (the researcher normally has a choice of only one of several models to study) and most over the design of the experiment or study. Such choices, in fact, dictate the type of data generated by a study.

Statistical methods are based on specific assumptions. Parametric statistics (those that are most familiar to the majority of scientists) have more stringent underlying assumptions than do nonparametric statistics. Among the underlying assumptions for many parametric statistical methods (such as the analysis of variance) is that the data are continuous. The nature of the data associated with a

**TABLE 8.3**
**Types of Variables (Data) and Examples of Each Type**

| Classified By | Type | Example |
|---|---|---|
| Scale | Scalar | Body weight |
| Continuous | Ranked | Severity of a lesion |
| Discontinuous | Scalar | Weeks until the first observation of a tumor in a carcinogenicity study |
| | Ranked | Clinical observations in animals |
| | Attribute | Eye colors in fruit flies |
| | Quantal | Dead/alive or present/absent |
| Frequency distribution | Normal | Body weights |
| | Bimodal | Some clinical chemistry parameters |
| | Others | Measures of time to incapacitation |

variable (as described above) imparts a value to that data, the value being the power of the statistical tests that can be employed.

Continuous variables are those that can at least theoretically assume any of an infinite number of values between any two fixed points (such as measurements of body weight between 2.0 and 3.0 kg). Discontinuous variables, meanwhile, are those that can have only certain fixed values, with no possible intermediate values (such as counts of 5 and 6 dead animals, respectively).

Limitations on our ability to measure constrain the extent to which the real-world situation approaches the theoretical, but many of the variables studied in toxicology are in fact continuous. Examples of these are lengths, weights, concentrations, temperatures, periods of time, and percentages. For these continuous variables, we may describe the character of a sample with measures of central tendency and dispersion that we are most familiar with: the mean, denoted by the symbol $\bar{x}$ and also called the arithmetic average, and the standard deviation (SD), which is denoted by the symbol $\sigma$ and is calculated as being equal to:

$$\sqrt{\frac{\sum x^2 - \frac{\left(\sum x\right)^2}{N}}{N-1}}$$

where $x$ is the individual datum and $N$ is the total number of data in the group. Contrasted with these continuous data, however, we have discontinuous (or discrete) data, which can only assume certain fixed numerical values. In these cases, our choice of statistical tools or tests is, as we will find later, more limited.

## PROBABILITY

Probability is simply the frequency with which, in a sufficiently large sample, a particular event will occur or a particular value be found. Hypothesis testing, for example, is generally structured so the likelihood of a treatment group being the same as a control group (the so-called *null hypothesis*) can be assessed as being less than a selected low level (very frequently 5%), which implies that we are 1.0 – $\alpha$ (that is 1.0 – 0.05, or 95%) sure that the groups are *not* equivalent.

# FUNCTIONS OF STATISTICS

Statistical methods may serve to perform any combination of three possible tasks. The one we are most familiar with is hypothesis testing—that is, determining if two (or more) groups of data differ from each other at a predetermined level of confidence. A second function is the construction and use of models that may be used to predict future outcomes of chemical–biological interactions. This is most commonly seen in linear regression or in the derivation of some form of correlation coefficient. Model fitting allows us to relate one variable (typically, a treatment or independent variable) to another. The third function, reduction of dimensionality, continues to be less commonly utilized than the first two. This final category includes methods for reducing the number of variables in a system while only minimally reducing the amount of information, thus making a problem easier to visualize and understand. Examples of such techniques are factor analysis and cluster analysis. A subset of this last function is the reduction of raw data to single expressions of central tendency and variability (such as the mean and standard deviation). There is also a special subset of statistical techniques that is part of both the second and third functions of statistics. This is data transformation, which includes such things as the conversion of numbers to log or probit values.

## DESCRIPTIVE STATISTICS

Descriptive statistics are used to summarize the general nature of a dataset. As such, the parameters describing any single group of data have two components. One of these describes the location of the data, and the other gives a measure of the dispersion of the data in and about this location. Often overlooked is the fact that the choice of which parameters are used to give these pieces of information implies a particular type of distribution for the data.

Most commonly, location is described by giving the (arithmetic) mean and dispersion by giving the *standard deviation* (SD) or the *standard error of the mean* (SEM). The calculation of the first two of these has already been described. If we again denote the total number of data in a group as *N*, then the SEM would be calculated as:

$$\text{SEM} = \frac{\text{SD}}{\sqrt{N}}$$

The use of the mean with either the SD or SEM implies, however, that we have reason to believe that the sample of data being summarized is from a population that is at least approximately normally distributed. If this is not the case, then we should instead use a set of statistical descriptions that do not require a normal distribution. These are the *median* (for location) and the *semiquartile distance* (for a measure of dispersion). These somewhat less familiar parameters are characterized as follows.

When all the numbers in a group are arranged in a ranked order (that is, from smallest to largest), the *median* is the middle value. If the group has an odd number of values, then the middle value is obvious; for example, in the case of 13 values, the seventh largest is the median. When the number of values in the sample is even, the median is calculated as the midpoint between the ($N$/2)th and the ([$N$/2] + 1)th number; for example, in the series of numbers 7, 12, 13, 19, the median value would be the midpoint between 12 and 13, which is 12.5.

The standard deviation and the standard error of the mean are related to each other but yet are quite different. The SEM is quite a bit smaller than the SD, making it very attractive to use in reporting data. This size difference is because the SEM actually is an estimate of the error (or variability) involved in measuring the means of samples and not an estimate of the error (or variability) involved in measuring the data from which means are calculated. This is implied by the *central limit theorem*, which tells us three major things:

- The distribution of sample means, which will be approximately normal regardless of the distribution of values in the original population from which the samples were drawn
- The mean value of the collection
- The standard deviation of the collection of all possible means of samples of a given size, called the standard error of the mean, which depends on both the standard deviation of the original population and the size of the sample

The SEM should be used only when the uncertainty of the estimate of the mean is of concern, which is almost never the case in toxicology; rather, we are concerned with an estimate of the variability of the population, for which the standard deviation is appropriate.

When all the data in a group are ranked, a quartile of the data contains one ordered quarter of the values. Typically, we are most interested in the borders of the middle two quartiles, $Q_1$ and $Q_3$, which together represent the semiquartile distance and which contain the median as their center. Given that there are *N* values in an ordered group of data, the upper limit of the fourth quartile ($Q$) may be computed as being equal to the [($jN \div 1$)/4th] value. Once we have used this formula to calculate the upper limits of $Q_1$ and $Q_3$, we can then compute the semiquartile distance (which is also called the *quartile deviation* and as such is abbreviated as QD) with the formula QD = $(Q_3 - Q_1)/2$. For example, for the 15-value dataset 1, 2, 3, 4, 4, 5, 5, 5, 6, 6, 6, 7, 7, 8, 9, we can calculate the upper limits of $Q_1$ and $Q_3$ as:

$$Q_1 = \frac{1(15+1)}{4} = \frac{16}{4} = 4$$

$$Q_2 = \frac{3(15+1)}{4} = \frac{48}{4} = 12$$

The 4th and 12th values in this dataset are 4 and 7, respectively. The semiquartile distance can then be calculated as:

$$QD = \frac{7-4}{2} = 1.5$$

There are times when it is desired to describe the relative variability of one or more sets of data. The most common way of doing this is to compute the *coefficient of variation* (CV), which is calculated simply as the ratio of the standard deviation to the mean, or:

$$CV = \frac{SD}{\overline{X}}$$

A CV of 0.2 or 20% thus means that the standard deviation is 20% of the mean. In toxicology, the CV is frequently between 20 and 50% and may at times exceed 100%.

## Sampling

Sampling—the selection of which individual data point will be collected, whether in the form of selecting which animals to collect blood from or to remove a portion of a diet mix from for analysis—is an essential step upon which all other efforts toward a good experiment or study are based. Three assumptions about sampling are common to most of the statistical analysis techniques that are used in toxicology: The sample is collected without bias, each member of a sample is collected independently of the others, and members of a sample are collected with replacements. Precluding bias, both intentional and unintentional, means that at the time of selection of a sample to measure, each portion of the population from which that selection is to be made has an equal chance of being selected. Independence means that the selection of any portion of the sample is not affected by and does not affect the selection or measurement of any other portion. Finally, sampling with replacement means that, in theory, after each portion is selected and measured, it is returned to the total sample pool and thus has the opportunity to be selected again. This is a corollary of the assumption of independence. Violation of this assumption (which is almost always the case in toxicology and all the life sciences) does not have serious consequences if the total pool from which samples are drawn is sufficiently large (say, 20 or greater) that the chance of reselecting that portion is small anyway.

The four major types of sampling methods are *random*, *stratified*, *systematic*, and *cluster*. Random is by far the most commonly employed method in toxicology. It stresses the fulfillment of the assumption of avoiding bias. When the entire pool of possibilities is mixed or randomized, then the members of the group are selected in the order in which they are drawn from the pool.

Stratified sampling is performed by first dividing the entire pool into subsets or strata, then doing randomized sampling from each strata. This method is employed when the total pool contains subsets that are distinctly different but in which each subset contains similar members. An example is a large batch of a powdered pesticide in which it is desired to determine the nature of the particle size distribution. Larger pieces or particles are on the top, progressively smaller particles have settled lower in the container, and at the very bottom the material has been packed and compressed into aggregates. To determine a timely representative answer, proportionally sized subsets from each layer or strata should be selected, mixed, and randomly sampled. This method is used more commonly in diet studies.

In systematic sampling, a sample is taken at set intervals (such as every fifth container of reagent or taking a sample of water from a fixed sample point in a flowing stream every hour). This is most commonly employed in quality assurance or (in the clinical chemistry lab) in quality control.

In cluster sampling, the pool is already divided into numerous separate groups (such as bottles of tablets), and we select small sets of groups (such as several bottles of tablets) then select a few members from each set. What one gets then is a cluster of measures. Again, this is a method most commonly used in quality control or in environmental studies when the effort and expense of physically collecting a small group of units is significant.

In classical toxicology studies sampling arises in a practical sense in a limited number of situations. The most common of these are as follows:

- Selecting a subset of animals or test systems from a study to make some measurement (which either destroys or stresses the measured system, or is expensive) at an interval during a study; this may include such cases as doing interim necropsies in a chronic study or collecting and analyzing blood samples from some animals during a subchronic study
- Analyzing inhalation chamber atmospheres to characterize aerosol distributions with a new generation system
- Analyzing diet in which test material has been incorporated
- Performing quality control on an analytical chemistry operation by having duplicate analyses performed on some materials
- Selecting data to audit for quality assurance purposes

We have now become accustomed to developing exhaustively detailed protocols for an experiment or study prior to its conduct. *A priori* selection of statistical methodology (as opposed to the *post hoc* approach) is as significant a portion of the process of protocol development and experimental design as any other and can measurably enhance the value of the experiment or study (see Table 8.4). Prior selection of statistical methodologies is essential for proper design of other portions of a protocol such as the number of animals per group or the sampling intervals for body weight. Implied in such a selection is the notion that the toxicologist has both an in-depth knowledge of the area of investigation and an understanding of the general principles of experimental design, for the analysis of any set of data is dictated to a large extent by the manner in which the data are obtained.

The four basic statistical principles of experimental design are *replication*, *randomization*, *concurrent (local) control*, and *balance*. In abbreviated form, these may be summarized as follows:

- *Replication*—Any treatment must be applied to more than one experimental unit (animal, plate of cells, litter of offspring, etc.). This provides more accuracy in the measurement of a response than can be obtained from a single observation, because underlying experimental errors tend to cancel each other out. It also supplies an estimate of the experimental error derived from the variability among each of the measurements taken (or

**TABLE 8.4**
**Rules for Form Design and Preparation**

1. Forms should be used when some form of repetitive data must be collected. They may be either paper or electronic.
2. If only a few (two or three) pieces of data are to be collected, they should be entered into a notebook and not onto a form. This assumes that the few pieces are not a daily event, with the aggregate total of weeks/months/years ending up as lots of data to be pooled for analysis.
3. Forms should be self-contained but should not try to repeat the content of the standard operating procedures or method descriptions.
4. Column headings on forms should always specify the units of measurement and other details of entries to be made. The form should be arranged so sequential entries proceed down a page, not across. Each column should be clearly labeled with a heading that identifies what is to be entered in the column. Any fixed part of entries (such as °C) should be in the column header.
5. Columns should be arranged from left to right so there is a logical sequential order to the contents of an entry as it is made. An example would be date/time/animal number/body weight/name of the recorder. The last item for each entry should be the name or unique initials of the individual who made the data entry.
6. Standard conditions that apply to all the data elements to be recorded on a form or the columns of the form should be listed as footnotes at the bottom of the form.
7. Entries of data on the form should not use more digits than are appropriate for the precision of the data being recorded.
8. Each form should be clearly titled to indicate its purpose and use. If multiple types of forms are being used, each should have a unique title or number.
9. Before designing the form, carefully consider the purpose for which it is intended. What data will be collected, how often, with what instrument, and by whom? Each of these considerations should be reflected in some manner on the form.
10. Those things that are common or standard for all entries on the form should be stated as such once. These could include such things as instrument used, scale of measurement (°C, °F, or K), or the location where the recording is made.

replicates). In practice, this means that an experiment should have enough experimental units in each treatment group (that is, a large enough $N$) so that reasonably sensitive statistical analysis of data can be performed. The estimation of sample size is addressed in detail later in this chapter.

- *Randomization*—This is practiced to ensure that every treatment has its fair share of extreme high and extreme low values. It also serves to allow the toxicologist to proceed as if the assumption of independence is valid; that is, there is no avoidable (known) systematic bias in how one obtains data.
- *Concurrent control*—Comparisons between treatments should be made to the maximum extent possible between experimental units from the same closely defined population; therefore, animals used as a control group should come from the same source, lot, age, etc. as test group animals. Except for the treatment being evaluated, test and control animals should be maintained and handled in exactly the same manner.
- *Balance*—If the effect of several different factors is being evaluated simultaneously, the experiment should be laid out in such a way that the contributions of the different factors can be separately distinguished and estimated. There are several ways of accomplishing this using one of several different forms of design, as will be discussed below.

The four basic experimental design types used in toxicology are the *randomized block*, *Latin square*, *factorial design*, and *nested design*. Other designs that are used are really combinations of these and are rarely employed in toxicology. Before examining these four basic types, however, we must first examine the basic concept of blocking.

Blocking is, simply put, the arrangement or sorting of the members of a population (such as all of an available group of test animals) into groups based on certain characteristics that may (but are not sure to) alter an experimental outcome. Such characteristics that may cause a treatment to give a differential effect include genetic background, age, sex, and overall activity levels, among others. The process of blocking then acts (or attempts to act) so each experimental group (or block) is assigned its fair share of the members of each of these subgroups.

We should now recall that randomization is aimed at spreading out the effect of undetectable or unsuspected characteristics in a population of animals or some portion of this population. The merging of the two concepts of randomization and blocking leads to the first basic experimental design the randomized block. This type of design requires that each treatment group have at least one member of each recognized group (such as age), the exact members of each block being assigned in an unbiased (or random) fashion.

The second type of experimental design assumes that we can characterize treatments (whether intended or otherwise) as belonging clearly to separate sets. In the simplest case, these categories are arranged into two sets that may be thought of as rows (for, say, source litter of test animal

**TABLE 8.5**
**Sample Table**

| | Age (weeks) | | | |
|---|---|---|---|---|
| **Source Litter** | **6–8** | **8–10** | **10–12** | **12–14** |
| 1 | *A* | *B* | *C* | *D* |
| 2 | *B* | *C* | *D* | *A* |
| 3 | *C* | *D* | *A* | *B* |
| 4 | *D* | *A* | *B* | *C* |

with the first litter as row 1, the next as row 2, etc.) and the secondary set of categories as columns (for, say, our ages of test animals, with 6 to 8 weeks as column 1, 8 to 10 weeks as column 2, and so on). Experimental units are then assigned so each major treatment (control, low dose, intermediate dose, etc.) appears once and only once in each row and each column. If we denote our test groups as A (control), B (low), C (intermediate), and D (high), such an assignment would appear as shown in Table 8.5.

The third type of experimental design is the factorial design, which has two or more clearly understood treatments, such as exposure level to test chemical, animal age, or temperature. The classical approach to this situation (and to that described under the Latin square) is to hold all but one of the treatments constant and at any one time to vary just that one factor. In the factorial design, however, all levels of a given factor are combined with all levels of every other factor in the experiment. When a change in one factor produces a different change in the response variable at one level of a factor than at other levels of this factor, there is an interaction between these two factors that can then be analyzed as an interaction effect.

The last of the major varieties of experimental design is the nested design, where the levels of one factor are nested within (or are subsamples of) another factor; that is, each subfactor is evaluated only within the limits of its single larger factor.

Another concept that is essential to the design of experiments in toxicology is *censoring*. Censoring is the exclusion of measurements from certain experimental units, or indeed of the experimental units themselves, from consideration in data analysis or inclusion in the experiment at all. Censoring may occur either prior to initiation of an experiment (where, in modern toxicology, this is almost always a planned procedure), during the course of an experiment (when they are almost universally unplanned, resulting from such as the as the death of animals on test), or after the conclusion of an experiment (when usually data are excluded because of being identified as some form of outlier).

In practice, *a priori* censoring in toxicology studies occurs in the assignment of experimental units (such as animals) to test groups. The most familiar example is in the common practice of assignment of test animals to acute, subacute, subchronic, and chronic studies, where the results of otherwise random assignments are evaluated for body weights of the assigned members. If the mean weights are found not to be comparable by some preestablished criterion (such as a 90% probability of difference by analysis of variance), then members are reassigned (censored) to achieve comparability in terms of starting body weights. Such a procedure of animal assignment to groups is known as *censored randomization*.

The first precise or calculable aspect of experimental design encountered is determining sufficient test and control group sizes to allow one to have an adequate level of confidence in the results of a study (that is, in the ability of the study design with the statistical tests used to detect a true difference—or effect—when it is present). The statistical test contributes a level of power to such a detection. Remember that the power of a statistical test is the probability that a test results in rejection of a hypothesis (say, $H_0$) when some other hypothesis (say, $H$) is valid. This is considered the power of the test with respect to the (alternative) hypothesis $H$.

If there is a set of possible alternative hypotheses, the power, regarded as a function of $H$, is termed the *power function* of the test. When the alternatives are indexed by a single parameter $\theta$, simple graphical presentation is possible. If the parameter is a vector $\theta$, then one can visualize a *power surface*.

If the power function is denoted by $\beta(\theta)$ and $H_0$ specifies $\theta = \theta_0$, then the value of $\beta(P)$—the probability of rejecting $H_0$ when it is in fact valid—is the significance level. The power of a test is greatest when the probability of a type II error is the least. Specified powers can be calculated for tests in any specific or general situation.

Some general rules to keep in mind are:

- The more stringent the significance level, the greater the necessary sample size. More subjects are needed for a 1%-level test than for a 5%-level test.
- Two-tailed tests require larger sample sizes than one-tailed tests. Assessing two directions at the same time requires a greater investment.
- The smaller the critical effect size, the larger the necessary sample size. Subtle effects require greater efforts.
- Any difference can be significant if the sample size is large enough.
- The larger the power required, the larger the necessary sample seize. Greater protection from failure requires greater effort. The smaller the sample size, the smaller the power (i.e., the greater the chance of failure).
- The requirements and means of calculating the necessary sample size depend on the desired (or practical) comparative sizes of test and control groups.

This number ($N$) can be calculated, for example, for equal-sized test and control groups using the formula:

$$N = \frac{(t_1 + t_2)^2}{d^2} S$$

where $t_1$ is the one-tailed $t$ value with $(N - 1)$ degrees of freedom corresponding to the desired level of confidence, $t_2$ is the one-tailed $t$ value with $(N - 1)$ degrees of freedom corresponding to the probability that the sample size will be adequate to achieve the desired precision, and $S$ is the sample standard deviation, derived typically from historical data and calculated as:

$$S = \sqrt{\frac{1}{N-1} \sum (V_1 - V_2)^2}$$

A number of aspects of experimental design are specific to the practice of toxicology. Before we look at a suggestion for step-by-step development of experimental designs, these aspects should first be considered as follows:

1. Frequently, the data gathered from specific measurements of animal characteristics are such that there is wide variability in the data. Often, such wide variability is not present in a control or low-dose group, but in an intermediate dosage group variance inflation may occur; that is, a large standard deviation may be associated with the measurements from this intermediate group. In the face of such a set of data, the conclusion that there is no biological effect based on a finding of no statistically significance effect might well be erroneous.
2. In designing experiments, a toxicologist should keep in mind the potential effect of involuntary censoring on sample size. In other words, although a study might start with five dogs per group, this provides no margin should any die before the study is ended and blood samples are collected and analyzed. Just enough experimental units per group frequently leaves too few at the end to allow meaningful statistical analysis, and allowances should be made accordingly in establishing group sizes.
3. It is certainly possible to pool the data from several identical toxicological studies. For example, if we performed an acute inhalation study where only three treatment group animals survived to the point at which a critical measure (such as analysis of blood samples) was performed, we would not have enough data to perform a meaningful statistical analysis. We could then repeat the protocol with new control and treatment group animals from the same source. At the end, after assuring ourselves that the two sets of data are comparable, we could combine (or pool) the data from survivors of the second study with those from the first. The costs of this approach, however, would then be both a greater degree of effort expended (than if we had performed a single study with larger groups) and increased variability in the pooled samples (decreasing the power of our statistical methods).
4. Another frequently overlooked design option in toxicology is the use of an unbalanced design—that is, of different group sizes for different levels of treatment. There is no requirement that each group in a study (control, low-dose, intermediate-dose, and high-dose) have an equal number of experimental units assigned to it. Indeed, there are frequently good reasons to assign more experimental units to one group than to others, and all the major statistical methodologies have provisions to adjust for such inequalities, within certain limits. The two most common uses of the unbalanced design have larger groups assigned to either the highest dose, to compensate for losses due to possible deaths during the study, or to the lowest dose, to give more sensitivity in detecting effects at levels close to an effect threshold or more confidence to the assertion that no effect exists.
5. We are frequently confronted with the situation where an undesired variable is influencing our experimental results in a nonrandom fashion. Such a variable is called a *confounding variable*; its presence makes the clear attribution and analysis of effects at best difficult and at worst impossible. Sometimes such confounding variables are the result of conscious design or management decisions, such as the use of different instruments, personnel, facilities, or procedures for different test groups within the same study. Occasionally, however, such confounding variables are the result of unintentional factors or actions, in which case the variable is referred to as a *lurking variable*. Examples of such variables are almost always the result of standard operating procedures being violated (e.g., water not being connected to a rack of animals over a weekend, a set of racks not being cleaned as frequently as others, or a contaminated batch of feed being used).
6. Finally, some thought must be given to the clear definition of what is meant by experimental unit and concurrent control. The experimental unit in toxicology encompasses a wide variety of

possibilities. It may be cells, plates of microorganisms, individual animals, litters of animals, etc. The importance of clearly defining the experimental unit is that the number of such units per group is the $N$ that is used in statistical calculations or analyses and critically affects such calculations. The experimental unit is the unit that receives treatments and yields a response that is measured and becomes a datum. What this means in practice is that, for example, in reproduction or teratology studies where we treat the parental generation females and then determine results by counting or evaluating offspring, the experimental unit is still the parent; therefore, the number of litters, not the number of offspring, is the $N$ [8]. A true concurrent control is one that is identical in every manner with the treatment groups except for the treatment being evaluated. This means that all manipulations, including gavaging with equivalent volumes of vehicle or exposing to equivalent rates of air exchanges in an inhalation chamber, should be duplicated in control groups just as they occur in treatment groups.

The goal of the four principles of experimental design (*replication*, *randomization*, *concurrent control*, and *balance*) is statistical efficiency and the economizing of resources. The single most important initial step in achieving such an outcome is to clearly define the objective of the study and have a clear statement of what questions are being asked.

## EXPERIMENTAL DESIGN

Toxicological experiments generally have a twofold purpose. The first question is whether or not an agent results in an effect on a biological system. The second question, never far behind, is how much of an effect is present. It has become increasingly desirable that the results and conclusions of studies aimed at assessing the effects of environmental agents be as clear and unequivocal as possible. It is essential that every experiment and study yield as much information as possible and that the results of each study have the greatest possible chance of answering the questions it was conducted to address. The statistical aspects of such efforts, so far as they are aimed at structuring experiments to maximize the possibilities of success, are called *experimental design*.

Ten facets of any study may affect its ability to detect an effect of a treatment. The first six concern minimizing the role of chance; the last four relate to avoidance of bias:

- *Choice of species and strain*—Ideally, the responses of interest should be rare in untreated control animals but should be reasonably readily evoked by appropriate treatments. Some species or specific strains, perhaps because of inappropriate diets [1], have high background tumor incidences that make increases both difficult to detect and difficult to interpret when detected.
- *Dose levels*—This is a very important and controversial area. In screening studies aimed at hazard identification, it is normal, to avoid requiring huge numbers of animals, to test at dose levels higher than those to which humans will be exposed but not so high that marked toxicity occurs. A range of doses is usually tested to guard against the possibility of misjudgment of an appropriate high dose and that the metabolic pathways at the high doses differ markedly from those at lower doses and, perhaps, to ensure that no large effects occur at dose levels in the range to be used by humans. In studies aimed more at risk estimation, more and lower doses may be tested to obtain fuller information on the shape of the dose–response curve.
- *Number of animals*—This is obviously an important determinant of the precision of the findings. The calculation of the appropriate number depends on: (1) the critical difference (i.e., the size of the effect it is desire to detect); (2) the false-positive rate (i.e., the probability of an effect being detected when none exists; equivalent to the $\alpha$ level or type I error); (3) the false-negative rate, (i.e., the probability of no effect being detected when one of exactly the critical size exists; equivalent to the $\beta$ level or type II error); and (4) some measure of the variability in the material. Tables relating numbers of animals required to obtain values of critical size $\alpha$ and $\beta$ are given in Lee [2], and software (e.g., nQUERY ADVISOR) is also available for this purpose. As a rule of thumb, to reduce the critical difference by a factor $n$ for a given $\alpha$ and $\beta$, the number of animals required will have to increased by a factor of $n^2$.
- *Duration of the experiment*—It is obviously important not to terminate the study too early for fatal conditions, which are normally strongly age related. Less obviously, going on for too long in a study can be a mistake, partly because the last few weeks or months may produce relatively few extra data at a disproportionate cost and partly because diseases of extreme old age may obscure the detection of tumors and other conditions of more interest. For nonfatal conditions, the ideal is to kill the animals when the average prevalence is around 50%.

- *Accuracy of determinations*—This is of obvious importance. Although good laboratory practices (GLPs) and advances in technology have improved the situation here, it is necessary for those taking part in the study to be diligent.
- *Sampling*—Sampling is an essential step upon which any meaningful experimental result depends. Sampling may involve selection of the individual data points that will be collected, determining which animals tissue samples will be collected from, or taking a sample of a diet mix for chemical analysis. Three assumptions about sampling are common to most of the statistical analysis techniques used in toxicology. The assumptions are that the sample is collected without bias, each member of a sample is collected independently of the others, and members of a sample are collected with replacements.
- *Stratification*—To detect a treatment difference with accuracy, it is important that the groups being compared are as homogeneous as possible with respect to other known causes of the response. In particular, suppose that there is another known important cause of the response for which the animals vary, so the animals are a mixture of hyper- and hypo-responders from this cause. If the treated group has a higher proportion of hyper-responders, it will tend to have a higher response even if treatment has no effect. Even if the proportion of hyper-responders is the same as in the controls, it will be more difficult to detect an effect of treatment because of the increased between-animal variability. Given that this other factor is known, it will be sensible to take it into account in both the design and analysis of the study. In the design, it can be used as a blocking factor so animals at each level are allocated equally (or in the correct proportion) to control and treated groups. In the analysis, the factor should be treated as a stratifying variable, with separate treatment–control comparisons made at each level and the comparisons combined for an overall test of difference. This is discussed later, where we refer to the factorial design as one example of the more complex designs that can be used to investigate the separate effect of multiple treatments.
- *Balance*—If the effect of several different factors is being evaluated simultaneously, then the experiment should be laid out in such a way that the contributions of the different factors can be separately distinguished and estimated. There are several ways to accomplish this using different forms of design, as will be discussed later. It is important to recognize that mathematical comparisons are best when group sizes are similar. It may be tempting to place more animals in the treated group to "see" the effect, but such an action weakens statistical analysis of the experiment.
- *Randomization*—Random allocation of animals to treatment groups is a prerequisite of good experimental design. If not carried out, one can never be sure whether treatment–control differences are due to treatment or to confounding by other relevant factors. The ability to randomize easily is a major advantage animals experiments have other over epidemiology. Although randomization eliminates bias (as least in expectation), simple randomization of all animals may not be the optimal technique for producing a sensitive test. If there is another major source of variation (e.g., sex or batch of animals), it will be better to carry out stratified randomization (i.e., separate randomizations within each level of the stratifying variable). The need for randomization applies not only to the allocation of animals to the treatment but also to anything that can materially affect the recorded response. The same random number that is used to apply animals to treatment groups can be used to determine cage position, order of weighting, order of bleeding for clinical chemistry, order of sacrifice at terminations, and so on.
- *Adequacy of control group*—Although, on occasion, historical control data can be useful, a properly designed study demands that a relevant concurrent control group be included with which results for the test group can be compared. The principle that like should be compared with like, apart from treatment, demands that control animals should be randomized from the same source as the treatment animals. Careful consideration should also be given to the appropriateness of the control group. Thus, in an experiment involving treatment of a compound in a solvent, it would often be inappropriate to include only an untreated control group, as any differences observed could only be attributed to the treatment–solvent combination. To determine the specific effects of the compound, a comparison group

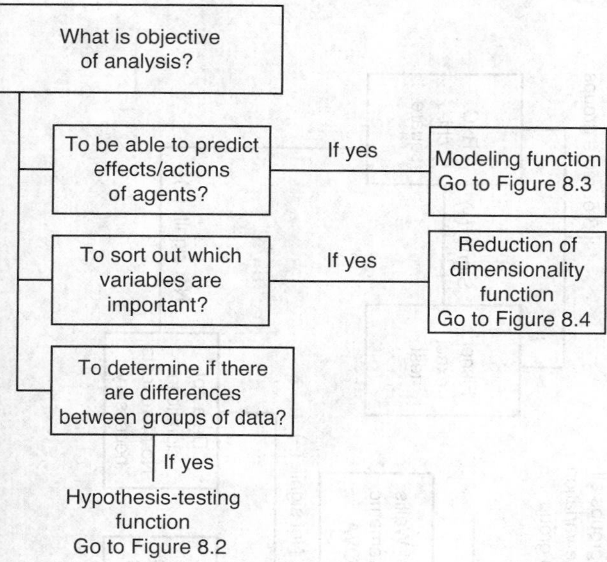


**FIGURE 8.1** Overall decision tree for selecting statistical procedures.

given the solvent only, by the same route of administration, would be required. It is not always generally realized that the position of the animal in the room in which it is kept may affect the animal's response. An example is the strong relationship between the incidence of retinal atrophy in albino rats and closeness to the lighting source. Systematic differences in cage position should be avoided, preferably via randomization.

For the reader who would like to further explore experimental design, a number of more detailed texts are available that include more extensive treatments of the statistical aspects of experimental design [3–8].

## GENERALIZED METHODOLOGY SELECTION

One approach for the selection of appropriate techniques to employ in a particular situation is to use a decision-tree method. Figure 8.1 is a decision tree that leads to the choice of one of three other trees to assist in technique selection, with each of the subsequent trees addressing one of the three functions of statistics that was defined earlier in this chapter. Figure 8.2 illustrates the selection of hypothesis-testing procedures, Figure 8.3 modeling procedures, and Figure 8.4 the reduction of dimensionality procedures. For the vast majority of situations, these trees will guide the user to the choice of the proper technique. The tests and terms used in these trees will be explained subsequently.

## GENERAL CONSIDERATIONS AND DATA CHARACTERIZATION FOR STATISTICAL ANALYSIS

### VARIABLES TO BE ANALYZED

Although some pathologists still regard their discipline as providing qualitative rather than quantitative data, it is abundantly clear that pathology, when applied to routine screening of animal toxicity and carcinogenicity studies, has to be quantitative to at least some degree so statistical statements can be made about possible treatment effects. Inevitably, there will be some descriptive text that will not be appropriate for statistical analysis. However, the main objective of the pathologist should be to provide information on the presence or absence (with severity grade or size where appropriate) of a list of conditions, consistently recorded from animal to animal by well-defined criteria, that can be validly used in a statistical assessment.

Given that statistical analysis is worth doing and data are available that would be analyzed, should one then analyze all the endpoints recorded? Some arguments have been put forward against analyzing all the endpoints studies, none of which really holds water. One argument is that some endpoints are not of interest. Perhaps the study is essentially a carcinogenicity study, so non-neoplastic endpoints are not considered to be background pathology and almost *per se* unrelated to treatment. In our view, this is illogical. If the pathologist has gone to the trouble of recording the data, then surely, in general, they ought to be analyzed; otherwise, why record them in the first place? After all, the costs of the statistical analyses are much less than those of doing the study and the pathology. While one might justify failure to analyze non-neoplastic data where tumor analysis has already shown that the compound is clearly carcinogenic and no longer of market potential, the general rule ought to be to analyze everything that has been specifically investigated.

Another argument put forward against doing multiple analyses is that it may yield many chance significant *p* values that have to be explained away. This seems to us a poor reason for not exploring the data fully. A detailed look at the data can only aid interpretation, provided that one is not hide-bound by the false argument that statistical significance necessarily equates with biological importance and definitely indicates a true effect of treatment.

Another reason not to analyze might be that visual inspection of summary tables reveals no suspicion of an effect for some endpoint. This seems to be, in this age of rapid and efficient computer programs, totally the wrong way to organize things. If the data are held on a computer, it is much better and quicker to do the actual analysis than to do the inevitably subjective, unreliable, and slow prescreening process. In any case, where substantial differences in survival between groups exist, it

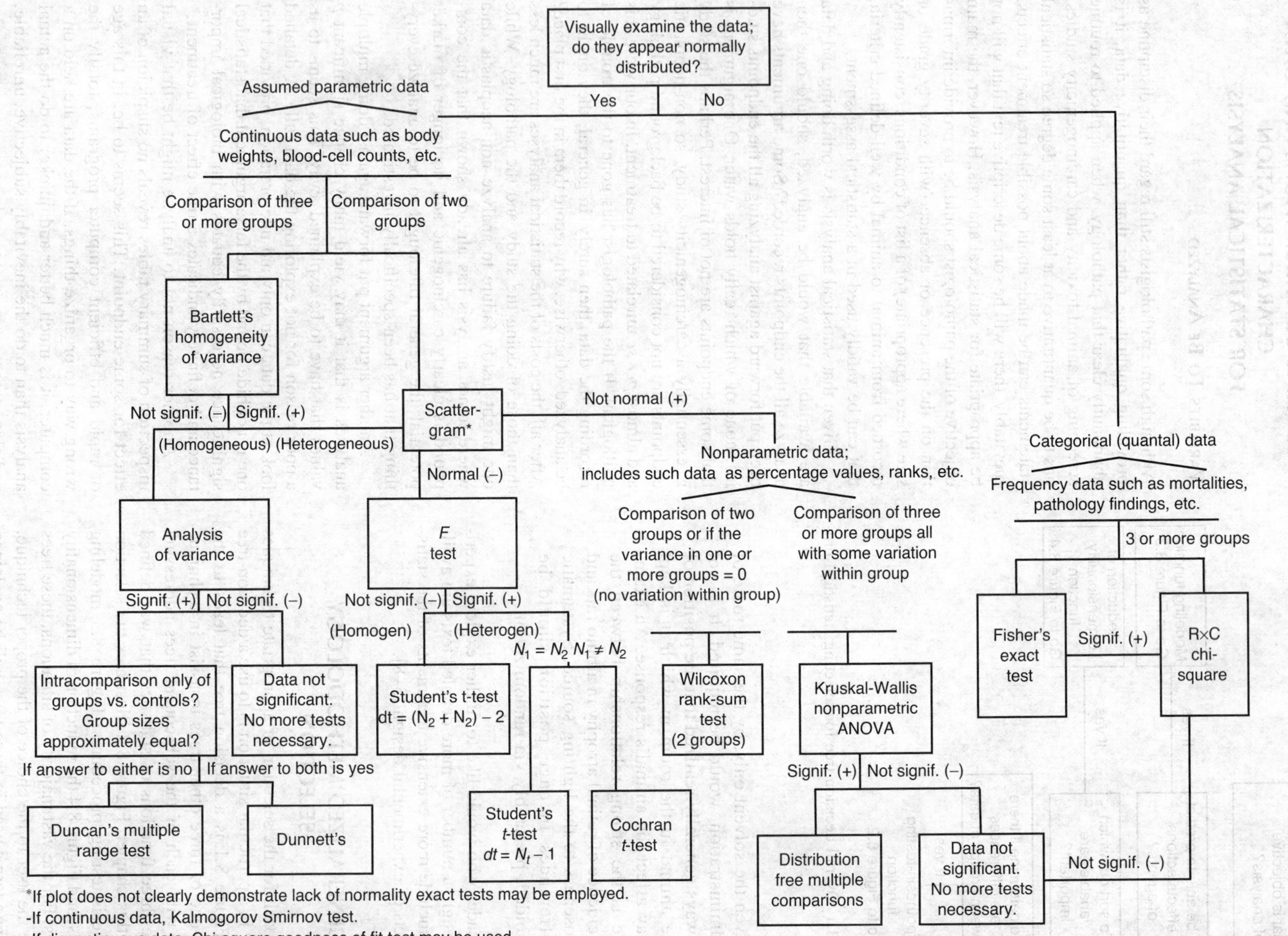


*If plot does not clearly demonstrate lack of normality exact tests may be employed.
-If continuous data, Kalmogorov Smirnov test.
-If discontinuous data, Chi-square goodness of fit test may be used.

**FIGURE 8.2** Decision tree for selecting hypothesis-testing procedures.

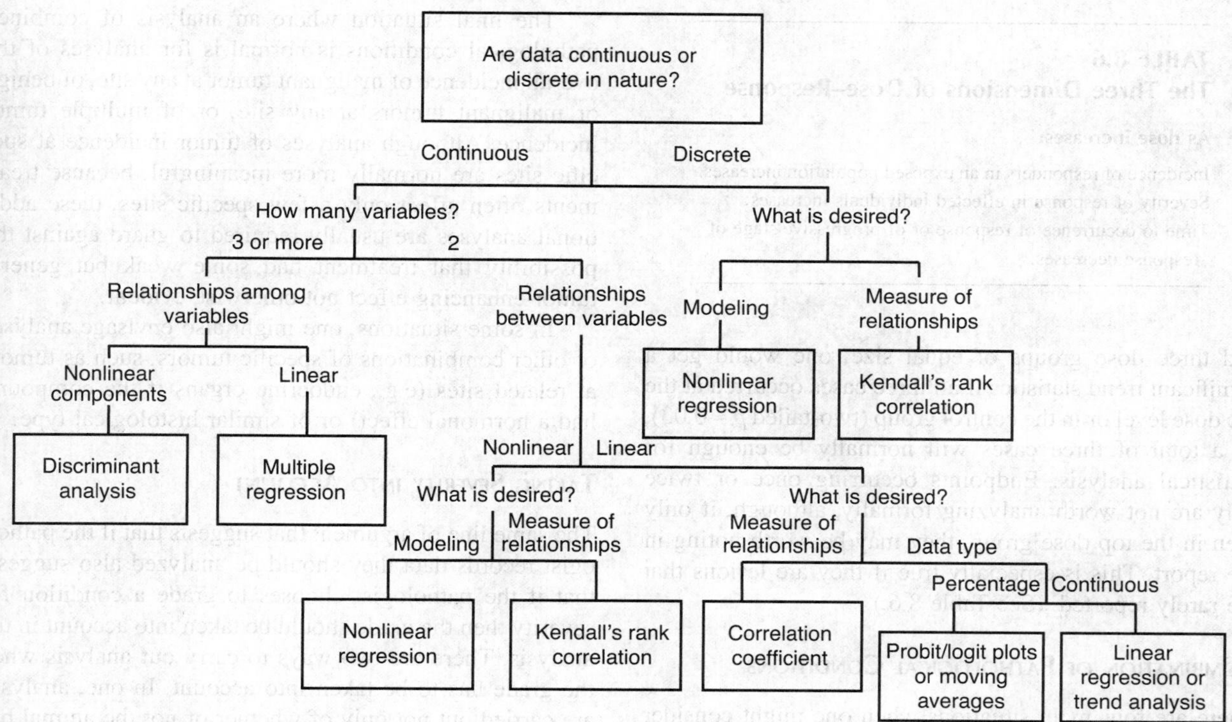


**FIGURE 8.3** Decision tree for selecting modeling procedures.

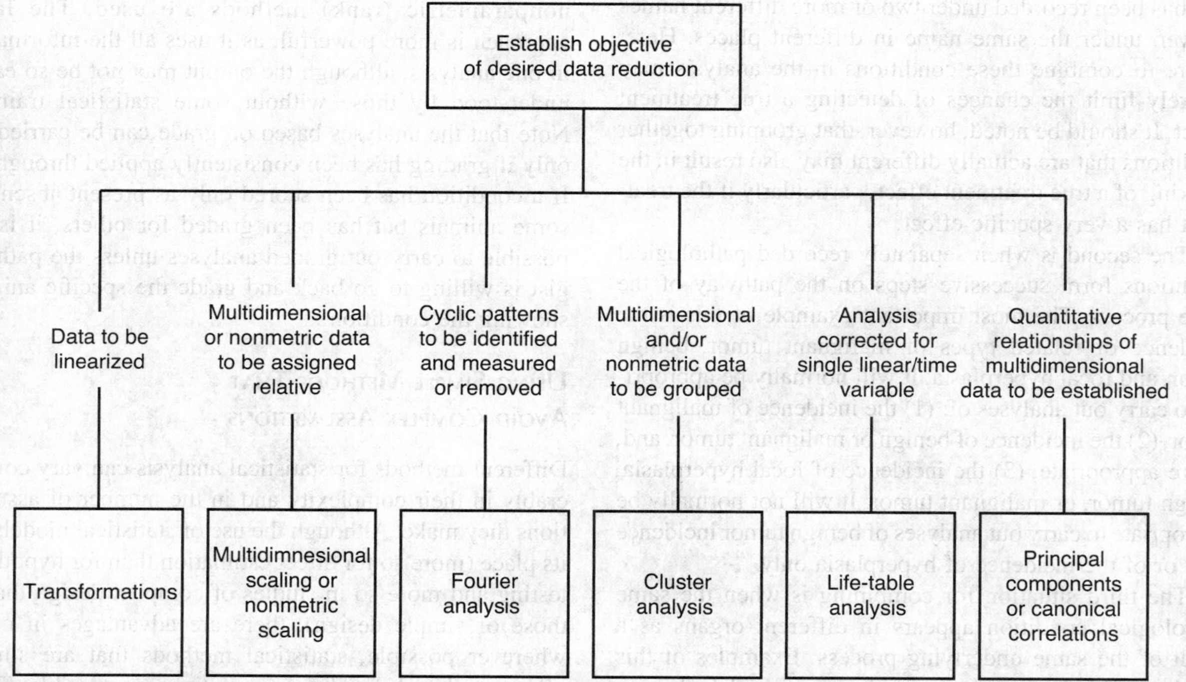


**FIGURE 8.4** Decision tree for selection of reduction of dimensionality procedures.

s very difficult to form a reliable view by inspection of non-age-adjusted frequencies on whether an effect might or might not have occurred.

A final, more valid, reason is that some endpoints occur only very rarely. One should, however, be clear what "very rarely" means. For a typical study with a control

**TABLE 8.6**
**The Three Dimensions of Dose–Response**

| **As dose increases:** |
|---|
| Incidence of responders in an exposed population increases. |
| Severity of response in effected individuals increases. |
| Time to occurrence of response or of progressive stage of response decreases. |

and three dose groups of equal size, one would get a significant trend statistics if all three cases occurred at the top dose level or in the control group (two-tailed $p \approx 0.03$), so a total of three cases will normally be enough for statistical analysis. Endpoints occurring once or twice only are not worth analyzing formally, although, if only seen in the top dose group, they may be worth noting in the report. This is especially true if they are lesions that are rarely reported. (See Table 8.6.)

## Combination of Pathological Conditions

There are four main situations when one might consider combining pathological conditions in a statistical analysis. The first is when essentially the same pathological condition has been recorded under two or more different names or even under the same name in different places. Here, failure to combine these conditions in the analysis may severely limit the chances of detecting a true treatment effect. It should be noted, however, that grouping together conditions that are actually different may also result in the masking of a true treatment effect, particularly if the treatment has a very specific effect.

The second is when separately recorded pathological conditions form successive steps on the pathway of the same process. The most important example of this is the incidence of related types of malignant tumor, benign tumor, and focal hyperplasia. It will normally be appropriate to carry out analyses of: (1) the incidence of malignant tumor, (2) the incidence of benign or malignant tumor, and, where appropriate, (3) the incidence of focal hyperplasia, benign tumor, or malignant tumor. It will not normally be appropriate to carry out analyses of benign tumor incidence only or of the incidence of hyperplasia only.

The third situation for combining is when the same pathological condition appears in different organs as a result of the same underlying process. Examples of this are the multicentric tumors (such as myeloid leukemia, reticulum cell sarcoma, and lymphosarcoma) or certain non-neoplastic conditions (such as arteritis/periarteritis and amyloid degeneration). Here, analysis will normally be carried out of only the incidence at any site, although in some situations site-specific analyses might be worth carrying out.

The final situation where an analysis of combined pathological conditions is normal is for analyses of the overall incidence of malignant tumor at any site, of benign or malignant tumors at any site, or of multiple tumor incidence. Although analyses of tumor incidence at specific sites are normally more meaningful, because treatments often affect only a few specific sites, these additional analyses are usually required to guard against the possibility that treatment had some weak but general tumor-enhancing effect not otherwise evident.

In some situations, one might also envisage analyses of other combinations of specific tumors, such as tumors at related sites (e.g., endocrine organs if the compound had a hormonal effect) or of similar histological type.

## Taking Severity into Account

The same line of argument that suggests that if the pathologist records data they should be analyzed also suggests that if the pathologist chooses to grade a condition for severity then the grade should be taken into account in the analysis. There are two ways to carry out analysis when the grade has to be taken into account. In one, analyses are carried out not only of whether or not the animal has a condition but also of whether or not the condition is at least grade 2, at least grade 3, etc. In the other approach nonparametric (rank) methods are used. The latter approach is more powerful, as it uses all the information in one analysis, although the output may not be so easily understood by those without some statistical training. Note that the analyses based on grade can be carried out only if grading has been consistently applied throughout. If a condition has been scored only as present/absent for some animals but has been graded for others, it is not possible to carry out graded analyses unless the pathologist is willing to go back and grade the specific animals showing the condition.

## Using Simple Methods That Avoid Complex Assumptions

Different methods for statistical analysis can vary considerably in their complexity and in the number of assumptions they make. Although the use of statistical models has its place (more so for effect estimation than for hypothesis testing and more so in studies of complex design than in those of simple design), there are advantages in using, wherever possible, statistical methods that are simple, robust, and make as few assumptions as possible. There are three reasons for this. First, such methods are more generally understandable to the toxicologist. Second, hardly ever enough data exist in practice to validate any given formal model fully. Third, even if a particular model is known to be appropriate, the loss of efficiency in using appropriate simpler methods is often only very small.

**TABLE 8.7**
**Dose–Response Effect on Time to Death**

| | Control | Exposed | Combined |
|---|---|---|---|
| Early deaths | 1/20 (5%) | 18/90 (20%) | 19/110 (17%) |
| Late deaths | 24/80 (30%) | 7/10 (70%) | 31/90 (34% |
| Total | 25/100 (25%) | 25/100 (25%) | 50/200 (25%) |

The methods we advocate for routine use for the analysis of tumor incidence tend, therefore, not to be based on the use of formal parametric statistical models. For example, when studying the relationship of treatment to incidence of a pathological condition and wishing to adjust for other factors (in particular, age at death) that might otherwise bias the comparison, methods involving stratification are recommended, rather than a multiple regression approach or time-to-tumor models. Analysis of variance (ANOVA) methods can be useful in the case of continuously distributed data for estimating treatment effects; however, they involve underlying assumptions (normally distributed variables, variability equal in each group). If these assumptions are violated, nonparametric methods based on the rank of observations, rather than their actual value, may be preferable for hypothesis testing.

### Using All of the Data

Often information is available about the relationship between treatment and a condition of interest for groups of animals differing systematically in respect of some other factor. Obvious examples are males and females, differing times of sacrifice, and differing secondary treatments. Although it will be necessary, in general, to look at the relationship within levels of this other factor, it will also be advisable to try to come to some assessment of the relationship over all levels of the other factor and where a combined inference is not sensible, but in far more situations this is not the case, and using all the data in one analysis allows a more powerful test of the relationship under study. Some scientists consider that conclusions for males and females should always be drawn separately, but there are strong statistical arguments for a joint analysis.

### Combining, Pooling, and Stratification

Suppose, in a hypothetical study of a toxic agent that induces tumors that do not shorten the lives of tumor-bearing animals, the data regarding the number of animals with tumor out of the number examined are as shown in Table 8.7. It can be seen that if the time of death is ignored and the *pooled* data are studied, the incidence of tumors is the same in each group, resulting in the false conclusion that treatment had no effect. Looking within each time of death, however, an increased incidence in the exposed group can be seen. An appropriate statistical method would *combine* a measure of difference between the groups based on the early deaths and a measure of difference based on the late deaths, and conclude correctly that incidence, after adjustment for time of death, is greater in the exposed groups.

In this example, time of death is the stratifying variable, with two strata—early deaths and late deaths. The essence of the methodology is to make comparisons only within strata (so one is always comparing like with like, except with respect to treatment) and then to combine the differences over strata. Stratification can be used to adjust for any variable or, indeed, combinations of variables.

Some studies are of factorial design, in which combinations of treatments are tested. The simplest such design is one in which four equal-sized groups of animals receive: (1) no treatment, (2) treatment A only, (3) treatment B only, and (4) treatments A and B. If one is prepared to assume that any effects of the two treatments are independent, one can use stratification to enable more powerful tests to be conducted of the possible individual treatment effects. Thus, to test for effects of treatment A, for example, one conducts comparisons in two strata, the first consisting of groups 1 and 2 not given treatment B and the second consisting of groups 3 and 4 given treatment B. Results combined from the two strata are based on twice as many animals and are therefore markedly more likely to detect possible effects of treatment A than is a simple comparison of groups 1 and 2. There is also the possibility of identifying interactions, such as synergism and antagonism, between the two treatments.

### Multiple Control Groups

In some routine long-term screening studies, the study design involves five groups of (usually) 50 animals of each sex, three of which are treated with successive doses of a compound and two of which are untreated controls. Assuming that there is no systematic difference between the control groups (e.g., the second control group is in a different room or from a different batch of animals), it will be normal to carry out the main analyses with the control groups treated as a single group of 100 animals. It will usually be a sensible preliminary precaution to carry out additional analyses comparing incidences in the two control groups.

### Trend Analysis, Low-Dose Extrapolation, and NOEL Estimation

Although comparisons of individual treated groups with the control group are important, a more powerful test of a possible effect of treatment will be to carry out a test for a dose-related trend. This is because most true effects of treatment tend to result in a response that increases (or

**TABLE 8.8**
**Lethality Incidence**

| | Control | Exposed |
|---|---|---|
| Early deaths | 0/20 | 0/20 |
| Middle deaths | 1/10 | 9/10 |
| Late deaths | 20/20 | 20/20 |
| Total | 21/50 | 29/50 |

decreases) with increasing dose and because trend tests take into account all the data in a single analysis. In interpreting the results of trend tests, it should be noted that a significant trend does not necessarily imply an increased risk at lower doses, nor, conversely, does a lack of increase at lower doses necessarily indicate evidence of a threshold (i.e., a dose below which no increase occurs).

Note that the testing for trend is seen as a more sensitive way of picking up a possible treatment effect than simple pairwise comparisons of treated and control groups. Attempting to estimate the magnitude of effects at low doses, typically below the lowest positive dose tested in the study, is a much more complex procedure and is heavily dependent on the assumed functional form of the dose–response relationship.

Such low-dose extrapolation is typically only conducted for tumors believed to be cause by a genotoxic effect that some, but by no means all, scientists believe has no threshold. For other types of tumors and for many non-neoplastic endpoints, a threshold cannot be estimated directly from data at a limited number of dose levels, a no-observed-effect level (NOEL) can be estimated by finding the highest dose level at which there is no significant increase in effects.

## Need for Age Adjustment

When marked differences in survival occur between treated groups, it is widely recognized that there is a need for an age adjustment (i.e., an adjustment for age at death or onset). This is illustrated in the example above, where, because of the greater number of deaths occurring early in the treated group, the true effect of treatment disappears if no adjustment is made. Thus, a major purpose of age adjustment is to avoid bias. It is not so generally recognized, however, that, even where there are no survival differences, age adjustment can increase the power to detect between-group differences. This is illustrated in Table 8.8. Here, treatment results in a somewhat earlier onset of a condition that occurs eventually in all animals. Failure to age adjust will result in a comparison of 29/50 with 21/50, which is not statistically significant. Age adjustment will essentially ignore the early and late deaths, which contribute no comparative statistical information, and will be based on the comparison of 9/10 with 1/10, which is statistically significant. Here, age adjustment sharpens the contrast, rather than avoiding bias, by avoiding the dilution of data capable of detecting treatment effects with data that are of little or no value for this purpose.

## Need To Take Context of Observation into Account

It is now widely recognized that age adjustment cannot properly be carried out unless the context of observation is taken into account. Three contexts are relevant, the first two relating to the situation where the condition is only observed at death (e.g., an internal tumor) and the third where it can be observed in life (e.g., a skin tumor). In the first context the condition is assumed to have caused the death of the animal (i.e., to be *fatal*). In this case, the incidence rate for a time interval and a group is calculated by:

$$\frac{\text{Number of animals dying in interval because of lesion}}{\text{Number of animals alive at start of interval}}$$

In the second context, the animal is assumed to have died of another cause (i.e., the condition is *incidental*). In this case, the rate is calculated by:

$$\frac{\text{Number of animals dying in interval with lesion}}{\text{Total number of animals dying in interval}}$$

In the third context, where the condition is *visible*, the rate is calculated by:

$$\frac{\text{Number of animals getting condition in interval}}{\text{Number of animals without condition at start of interval}}$$

A problem with the method of Peto et al. [9], which takes context of observation into account, is that some pathologists are unwilling or feel unable to decide whether, in any given case, a condition is fatal or incidental. A number of points should be made here. First, where there are marked survival differences, it may not be possible to conclude reliably whether a treatment is beneficial or harmful unless such a decision is made. This is well illustrated by the example in Peto et al. [9], where assuming that all pituitary tumors were fatal resulted in the (false) conclusion that *N*-nitrosodimethylamine (NDMA) was carcinogenic, and assuming that they were all incidental resulted in the (false) conclusion that NDMA was protective. Using, correctly, the pathologist's best opinion as to which were and which were not likely to be fatal resulted in an analysis that (correctly) concluded that NDMA had no effect. If the pathologist in this case had been unwilling to make a judgment as to fatality, believing it to be unreliable, no conclusion could have been reached. This state of affairs would, however, be a fact of life and *not* a position reached because an inappropriate statistical method was being used.

Although it will normally be a good routine for the pathologist to ascribe "factors contributory to death" for each animal that was not part of a scheduled sacrifice, it is in fact not strictly necessary to determine the context of observation for all conditions at the outset. An alternative strategy is to analyze under differing assumptions: (1) no cases fatal, (2) all cases occurring in descendents fatal, and (3) all cases of same defined severity occurring in decedents fatal, with, under each assumption, other cases incidental.

If the conclusion turns out the same under each assumption or if the pathologist can say, on general grounds, that one assumption is likely to be a close approximation to the truth, it may not be necessary to know the context of observation for the condition in question for each individual animal. Using the alternative strategy might result in a saving of the pathologist's time by only having to make a judgment for a limited number of conditions where the conclusion seems to hang on correct knowledge of the context of observation.

Finally, it should be noted that, although many non-neoplastic conditions observed at death are never causes of death, it is, in principle, as necessary to know the context of observation for non-neoplastic conditions as it is for tumors.

## Experimental and Observational Units

In many situations, the animal is both the experimental unit and the observational unit, but this is not always so. To determine treatment effects by the methods of the next section, it is important that each experimental unit provides only one item of data for analysis, as the methods all assume that individual data items are statistically independent. In many feeding studies, where the cage is assigned to a treatment, it is the cage, rather than the animal, that is the experimental unit. In histopathology, observations for a tissue are often based on multiple sections per animal, so the section is the observational unit. Multiple observations per experimental unit should be combined in some suitable way into an overall average for that unit before analysis.

## Missing Data

In many types of analysis, animals with missing data are simply removed from the analysis; however, in some situations this can be an inappropriate thing to do. One situation is when carrying out an analysis of a condition that is assumed to have caused the death of the animal. Although an animal dying at week 83 for which the section was unavailable for microscopic examination cannot contribute to the group comparison at week 83, one knows that it did not die because of any condition in previous weeks, so it should contribute to the denominator of the calculations in all previous weeks.

Another situation is when histopathological examination of a tissue is not carried out unless an abnormality is seen post mortem. In such an experiment, one might have the following data for that tissue:

- *Control group*—50 animals, 2 abnormal post mortem, 2 examined microscopically, 2 with tumor of specific type
- *Treated group*—50 animals, 15 abnormal post mortem, 15 examined microscopically, 14 with tumor of specific type

Ignoring animals with no microscopic sections, one would compare 2/2 = 100% with 14/15 = 93% and conclude that treatment nonsignificantly decreased incidence. This is likely to be a false conclusion, and it would be better here to compare the percentages of animals that had a post mortem abnormality that turned out to be a tumor—that is, 2/50 = 4% with 14/50 = 28%. Unless some aspect of treatment made tumors much easier to detect at post mortem, one could then conclude that treatment did have an effect on tumor incidence.

Particular care has to be taken in studies where the procedures for histopathological examination vary by group. In a number of studies conducted in recent years, the protocol has demanded full microscopic examination of a given tissue list in decedents in all groups and in terminally killed controls in high-dose animals. In other animals, terminally killed low- and mid-dose animals, microscopic examination of a tissue is only conducted if the tissue is found to be abnormal at post mortem. Such a protocol is designed to save money but leads to difficulty in comparing the treatment groups validly. Suppose, for example, that responses in terminally killed animals are 8/20 in the controls, 3/3 (with 17 unexamined) in the low-dose animals, and 5/6 (with 14 unexamined) in the mid-dose animals. Is one supposed to conclude that treatment at the low and mid-doses increased response, based on a comparison of the proportions examined microscopically (40%, 100%, and 83%), or that it decreased response, based on the proportion of animals in the group (40%, 15%, and 25%)? It could well be that treatment had no effect but some small tumors were missed at post mortem. In this situation, a valid comparison can only be achieved by ignoring the low- and mid-dose groups when carrying out the comparison for the age stratum "terminal kill." This, of course, seems wasteful of data, but these are data that cannot be usefully used due to the inappropriate protocol.

## USE OF HISTORICAL CONTROL DATA

In some situations, particularly where incidences are low, the results from a single study may suggest an effect of treatment on tumor incidence but be unable to demonstrate it conclusively. The possibility of comparing results in the treated groups with those of control groups from other studies is then often raised; thus, a nonsignificant incidence of 2 cases out of 50 in a treated group may seem much more significant if no cases have been seen in, say, 1000 animals representing controls from 20 similar studies. Conversely, a significant incidence of 5 cases out of 50 in a treated group as compared with 0 out of 50 in the study controls may seem far less convincing if many other control groups had incidences around 5 out of 50. While not understating the importance of looking at historical control data, it must be emphasized that there are a number of reasons why variation between studies may be greater than variation within a study. Differences in diet, in duration of the study, in intercurrent mortality, and in who the study pathologist is may all contribute. Statistical techniques that ignore this and carry out simple statistical tests of treatment incidence against a pooled control incidence may well give results that are seriously in error and are likely to overstate statistical significance considerably.

### METHODS FOR DATA EXAMINATION AND PREPARATION

The data from toxicology studies should always be examined before any formal analysis. Such examinations should be directed to determining if the data are suitable for analysis and, if so, what form the analysis should take (see Figure 8.2). If the data as collected are not suitable for analysis or if they are only suitable for low-powered analytical techniques, one may wish to use one of the many forms of data transformation to change the data characteristics so they are more amenable to analysis. For data examination, two major techniques are presented here: the *scattergram* and *Bartlett's test*. Likewise, for data preparation, two techniques are presented: *randomization* (including a test for randomness in a sample of data) and *transformation*. Exploratory data analysis (EDA) is presented and briefly reviewed later. This is a broad collection of techniques and approaches to probe data—that is, to both examine and perform some initial, flexible analysis of the data.

### SCATTERGRAM AND BARTLETT'S TEST

Two of the major points to be made throughout this chapter are (1) the use of the appropriate statistical tests, and (2) the effects of small sample sizes (as is often the case in toxicology) on our selection of statistical techniques. Frequently, simple examination of the nature and distribution of data collected from a study can also suggest patterns and results that were unanticipated and for which the use of additional or alternative statistical methodology is warranted. It was these points that caused the author to consider a section on scattergrams and their use essential for toxicologists.

Bartlett's test may be used to determine if the values in groups of data are homogeneous. If they are, this (along with the knowledge that they are from a continuous distribution) demonstrates that parametric methods are applicable. But, if the values in the (continuous data) groups fail Bartlett's test (i.e., are heterogeneous), we cannot be secure in our belief that parametric methods are appropriate until we gain some confidence that the values are normally distributed. With large groups of data, we can compute parameters of the population (kurtosis and skewness, in particular) and from these parameters determine if the population is normal (with a certain level of confidence). If our concern is especially marked, we can use a chi-square goodness-of-fit test for normality. But, when each group of data consists of 25 or fewer values, these measures or tests (kurtosis, skewness, and chi-square goodness-of-fit) are not accurate indicators of normality. Instead, in these cases, we should prepare a scattergram of the data and then evaluate the scattergram to estimate if the data are normally distributed. This procedure consists of developing a histogram of the data and examining the histogram to gain a visual appreciation of the location and distribution of the data.

The abscissa (or horizontal scale) should be in the same scale as the values and should be divided so the entire range of observed values is covered by the scale of the abscissa. Across such a scale we then simply enter symbols for each of our values. Example 1 shows such a plot. Example 1 is a traditional and rather limited form of scatterplot, but such plots can reveal significant information about the amount and types of association between the two variables, the existence and nature of outliers, the clustering of data, and a number of other two-dimensional factors [10,11].

Current technology allows us to add significantly more graphical information to scatterplots by means of graphic symbols (letters, faces, or different shapes, such as squares, colors, etc.) for the plotted data points. One relatively simple example of this approach is shown in Figure 8.5, where the simple case of dose (in a dermal study), dermal irritation, and white blood cell count are presented. This graph quite clearly suggests that as dose (variable $x$) is increased, dermal irritation (variable $y$) also increases; as irritation becomes more severe, white blood cell count (variable $z$), an indicator of immune system involvement suggesting infection or persistent inflammation, also increases. There is no direct association of variables $x$ and $z$, however. Cleveland and McGill [12] presented an excellent, detailed overview of the expanded capabilities of the scatterplot, and the interested reader should refer to that article. Cleveland later expanded this to a book [13]. Tufte [14] has also expanded on this.

Suppose we have the two data sets below:
Group 1 — 4.5, 5.4, 5.9, 6.0, 6.4, 6.5, 6.9, 7.0, 7.1, 7.0, 7.4, 7.5, 7.5, 7.5, 7.6, 8.0, 8.1, 8.4, 8.5, 8.6, 9.0, 9.4, 9.5, and 10.4
Group 2 — 4.0, 4.5, 5.0, 5.1, 5.4, 5.5, 5.6, 6.5, 6.5, 7.0, 7.4, 7.5, 7.5, 8.0, 8.1, 8.5, 8.5, 9.0, 9.1, 9.5, 9.5, 10.1, 10.0, and 10.4

Both of these groups contain 24 values and cover the same range. From them, we can prepare the following scattergrams:

**Group 1**

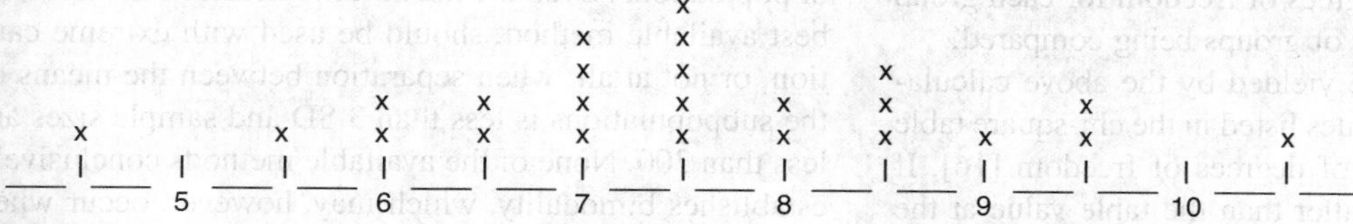


**Group 2**

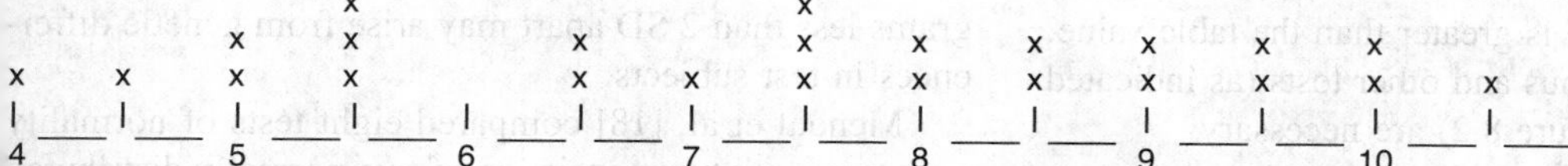


Group 1 can be seen to approximate a normal distribution (bell-shaped curve); we can proceed to perform the appropriate parametric tests with such data. But, group 2 clearly does not appear to be normally distributed; in this case, the appropriate nonparametric technique must be used.

**EXAMPLE 1**

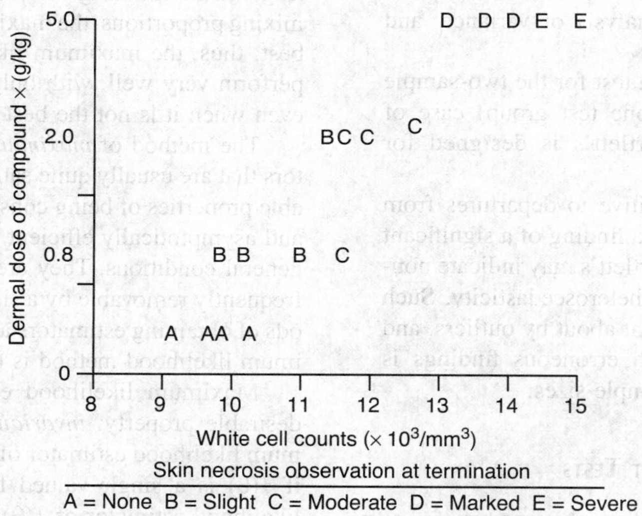


**FIGURE 8.5** Exploratory data analysis.

## Bartlett's Test for Homogeneity of Variance

Bartlett's test [15] is used to compare the variances (values reflecting the degree of variability in datasets) among three or more groups of data, where the data in the groups are continuous sets (such as body weights, organ weights, red blood cell counts, or diet consumption measurements). It is expected that such data will be suitable for parametric methods (normality of data is assumed), and Bartlett's is frequently used as a test for the assumption of equivalent variances.

Bartlett's is based on the calculation of the corrected $\chi^2$ (chi-square) value by the formula:

$$\chi^2_{corr} = 2.3026 \frac{\sum \mathrm{df}\left(\log_{10}\left[\frac{\sum \mathrm{df}(S^2)}{\sum \mathrm{df}}\right]\right) - \sum\left[\mathrm{df}(\log_{10} S^2)\right]}{1 + \frac{1}{3(K-1)}\left[\sum \frac{1}{\mathrm{df}} - \frac{1}{\sum \mathrm{df}}\right]}$$

where $S^2$ = variance, equal to:

$$\frac{\dfrac{n\sum X^2 - \left(\sum X\right)^2}{n}}{n-1}$$

$X$ = individual datum within each group; $n$ = number of data within each group; df = degrees of freedom for each group = $(N - 1)$; and $K$ = number of groups being compared.

The corrected $\chi^2$ value yielded by the above calculations is compared to the values listed in the chi-square table according to the numbers of degrees of freedom [16]. If the calculated value is smaller than the table value at the selected $p$ level (traditionally 0.05), the groups are accepted to be homogeneous and the use of ANOVA is assumed proper. If the calculated $\chi^2$ is greater than the table value, the groups are heterogeneous and other tests (as indicated by the decision tree in Figure 8.2) are necessary.

### Assumptions and Limitations

- Bartlett's test does not test for normality but rather homogeneity of variance (also called equality of variances or homoscedasticity).
- Homoscedasticity is an important assumption for Student's $t$-test, analysis of variance, and analysis of covariance.
- The $F$-test is actually a test for the two-sample (that is, control and one test group) case of homoscedasticity. Bartlett's is designed for three or more samples.
- Bartlett's is very sensitive to departures from normality. As a result, a finding of a significant chi-square value in Bartlett's may indicate non-normality rather than heteroscedasticity. Such a finding can be brought about by outliers, and the sensitivity to such erroneous findings is extreme with small sample sizes.

## Statistical Goodness-of-Fit Tests

A goodness-of-fit test is a statistical procedure for comparing individual measurements to a specified type of statistical distribution; for example, a normal distribution is completely specified by its arithmetic mean and variance (the square of the standard deviation). The null hypothesis, that the data represent a sample from a single normal distribution, can be tested by a statistical goodness-of-fit test. Various goodness-of-fit tests have been devised to determine if the data deviate significantly from a specified distribution. If a significant departure occurs, it indicates only that the specified distribution can be rejected with some assurance. This does not necessarily mean that the true distribution contains two or more subpopulations. The true distribution may be a single distribution, based on a different mathematical relationship (e.g., log-normal). In the latter case, logarithms of the measurement would not be expected to exhibit by a goodness-of-fit test a statistically significant departure from a log-normal distribution.

Everitt and Hand [17] recommended the use of a sample of 200 or more to conduct a valid analysis of mixtures of populations. Even the maximum likelihood method, the best available method, should be used with extreme caution, or not at all, when separation between the means of the subpopulations is less than 3 SD and sample sizes are less than 300. None of the available methods conclusively establishes bimodality, which may, however, occur when separation between the two means (modes) exceeds 2 SD. Conversely, inflections in probits or separations in histograms less than 2 SD apart may arise from genetic differences in test subjects.

Mendal et al. [18] compared eight tests of normality to detect a mixture consisting of two normally distributed components with different means but equal variances. Fisher's skewness statistic was preferable when one component comprised less than 15% of the total distribution. When the two components comprised more nearly equal proportions (35 to 65%) of the total distribution, the Engelman and Hartigan test [19] was preferable. For other mixing proportions, the maximum likelihood ratio test was best; thus, the maximum likelihood ratio test appears to perform very well, with only small loss from optimality, even when it is not the best procedure.

The method of *maximum likelihood* provides estimators that are usually quite satisfactory. They have the desirable properties of being consistent, asymptotically normal and asymptotically efficient for large samples under quite general conditions. They are often biased, but the bias is frequently removable by a simple adjustment. Other methods of obtaining estimators are also available, but the maximum likelihood method is the most frequently used.

Maximum likelihood estimators also have another desirable property: *invariance*. Let us denote the maximum likelihood estimator of the parameter $\theta$ by $\hat{\sigma}$. Then if $f(\theta)$ is a single-valued function of $\theta$, the maximum likelihood estimator of $f(\theta)$ is $f(\hat{\sigma})$. Thus, for example,

$$\hat{\sigma} = \left(\hat{\sigma}^2\right)^{1/2}$$

The principle of maximum likelihood tells us that we should use as our estimate that value which maximizes the likelihood of the observed event.

These maximum likelihood methods can be used to obtain *point estimates* of a parameter, but we must remember that a point estimator is a random variable distributed in some way around the true value of the parameter. The true parameter value may be higher or lower than our estimate. It is often useful therefore to obtain an interval within which we are reasonably confident the true value will lie, and the generally accepted method is to construct what are known as *confidence limits*.

The following procedure will yield upper and lower 95% confidence limits with the property that, when we say that these limits include the true value of the parameter, 95% of all such statements will be true and 5% will be incorrect:

1. Choose a (test) statistic involving the unknown parameter and no other unknown parameter.
2. Place the appropriate sample values in the statistic.
3. Obtain an equation for the unknown parameter by equating the test statistic to the upper 2.5% point of the relevant distribution.
4. The solution of the equation gives one limit.
5. Repeat the process with the lower 2.5% point to obtain the other limit.

One can also construct 95% confidence intervals using unequal tails (for example, using the upper 2% point and the lower 3% point). We usually want our confidence interval to be as short as possible, however, and with a symmetric distribution such as the normal or $t$, this is achieved using equal tails. The same procedure very nearly minimizes the confidence interval with other nonsymmetric distributions (for example, chi-square) and has the advantage of avoiding rather tedious computation.

When the appropriate statistic involves the square of the unknown parameter, both limits are obtained by equating the statistic to the upper 5% point of the relevant distribution. The use of two tails in this situation would result in a pair of nonintersecting intervals. When two or more parameters are involved, it is possible to construct a region within which we are reasonably confident the true parameter values will lie. Such regions are referred to as *confidence regions*. The implied interval for $p_1$ does not form a 95% confidence interval, however, nor is it true that an 85.7375% confidence region for $p_1$, $p_2$, and $p_3$ can be obtained by considering the intersection of the three separate 95% confidence intervals, because the statistics used to obtain the individual confidence intervals are not independent. This problem is obvious with a multiparameter distribution such as the multinomial, but it even occurs with the normal distribution because the statistic that we use to obtain a confidence interval for the mean and the statistic that we use to obtain a confidence interval for the variance are not independent. The problem is not likely to be of great concern unless a large number of parameters is involved.

## Randomization

Randomization is the act of assigning a number of items (plates of bacteria or test animals, for example) to groups in such a manner that there is an equal chance for any one item to end up in any one group. This is a control against any possible bias in assignment of subjects to test groups. A variation on this is censored randomization, which ensures that the groups are equivalent in some aspect after the assignment process is complete. The most common example of a censored randomization is one in which it is ensured that the body weights of test animals in each group are not significantly different from those in the other groups. This is done by analyzing group weights both for homogeneity of variance and by analysis of variance after animal assignment, then rerandomizing if there is a significant difference at some nominal level, such as $p \leq 0.10$. The process is repeated until there is no significant difference.

There are several methods for actually performing the randomization process. The three most commonly used are card assignment, use of a random number table, and use of a computerized algorithm. For the card-based method, individual identification numbers for items (plates or animals, for example) are placed on separate index cards. These cards are then shuffled and placed one at a time in succession into piles corresponding to the required test groups. The results are a random group assignment.

The random number table method requires only that one have unique numbers assigned to test subjects and access to a random number table. One simply sets up a table with a column for each group to which subjects are to be assigned. We start from the head of any one column of numbers in the random table (each time the table is used, a new starting point should be utilized). If our test subjects number less than 100, we utilize only the last two digits in each random number in the table. If they number more than 99 but less than 1000, we use only the last three digits. To generate group assignments, we read down a column, one number at a time. As we come across digits that correspond to a subject number, we assign that subject to a group (enter its identifying number in a column), proceeding to assign subjects to groups from left to right and filling one row at a time. After a number is assigned to an animal, any duplication of its unique number is ignored. We use as many successive columns of random numbers as we may need to complete the process.

The third (and now most common) method is to use a random number generator that is built into a calculator or computer program. Procedures for generating these are generally documented in user manuals.

One is also occasionally required to evaluate whether a series of numbers (such as an assignment of animals to test groups) is random. This requires the use of a randomization test, of which there are a large variety. The chi-square test can be used to evaluate the goodness-of-fit to a random assignment. If the result is not critical, a simple sign test will work. For the sign test, we first determine the middle value in the numbers being checked for randomness. We then go through a list of the numbers assigned to each group, scoring each as a "+" (greater than our middle number) or a "–" (less than our middle number). The number of pluses and minuses in each group should be approximately equal.

## Transformations

If our initial inspection of a dataset reveals it to have an unusual or undesired set of characteristics (or to lack a desired set of characteristics), we have a choice of three courses of action. We may proceed to select a method or test appropriate to this new set of conditions or abandon the entire exercise or transform the variables under consideration in such a manner that the resulting transformed variates ($X'$ and $Y'$, for example, as opposed to the original variates $X$ and $Y$) meet the assumptions or have the characteristics that are desired.

The key to all this is that the scale of measurement of most (if not all) variables is arbitrary. Although we are most familiar with a linear scale of measurement, there is nothing that makes this the correct scale on its own, as opposed to a logarithmic scale (familiar logarithmic measurements are pH values or earthquake intensity [Richter scale]). Transforming a set of data (converting $X$ to $X'$) is really as simple as changing a scale of measurement.

There are at least four good reasons to transform data:

1. Normalize the data, making them suitable for analysis by our most common parametric techniques such as analysis of variance (ANOVA). A simple test of whether a selected transformation will yield a distribution of data that satisfies the underlying assumptions for ANOVA is to plot the cumulative distribution of samples on probability paper (commercially available paper that has the probability function scale as one axis). One can then alter the scale of the second axis (that is, the axis other than the one that is on a probability scale) from linear to any other (logarithmic, reciprocal, square root, etc.) and see if a previously curved line indicating a skewed distribution becomes linear to indicate normality. The slope of the transformed line gives us an estimate of the standard deviation. And, if the slopes of the lines of several samples or groups of data are similar, we accordingly know that the variance of the different groups are homogenous.
2. Linearize the relationship between a paired set of data, such as dose and response. This is the most common use in toxicology for transformations and is demonstrated in the "Probit/Logit Transforms and Regression" section.
3. Adjust data for the influence of another variable. This is an alternative in some situations to the more complicated process of analysis of covariance. A ready example of this usage is the calculation of organ weight to body weight ratios in *in vivo* toxicity studies, with the resulting ratios serving as the raw data for an analysis of variance performed to identify possible target organs.
4. Make the relationships between variables clearer by removing or adjusting for interactions with third, fourth, etc. uncontrolled variables that influence the pair of variables of interest.

Common transformations are presented in Table 8.9.

## Exploratory Data Analysis

Over the past 20 years, an entirely new approach has been developed to get the most information out of the increasingly larger and more complex datasets that scientists are faced with. This approach involves the use of a very diverse set of fairly simple techniques that comprise exploratory data analysis (EDA). As expounded by Tukey [20], there are four major ingredients to EDA:

- *Displays*—These visually reveal the behavior of the data and suggest a framework for analysis. The scatterplot (presented earlier) is an example of this approach.
- *Residuals*—These are what remain of a set of data after a fitted model (such as a linear regression) or some similar level of analysis has been removed.
- *Reexpressions*—These involve questions of what scale would serve to best simplify and improve the analysis of the data. Simple transformations, such as those presented earlier in this chapter, are used to simplify data behavior (for example, linearizing or normalizing) and clarify analysis.
- *Resistance*—This is a matter of decreasing the sensitivity of analysis and summary of data to misbehavior so the occurrence of a few outliers, for example, will not complicate or invalidate the methods used to analyze the data. For example, in summarizing the location of a set of data, the median (but not the arithmetic mean) is highly resistant.

These four ingredients are utilized in a process falling into two broad phases: an exploratory phase and a confirmatory phase. The exploratory phase isolates patterns in and features of the data and reveals them, allowing an inspection of the data before there is any firm choice of actual hypothesis testing or modeling methods has been made.

**TABLE 8.9**
**Common Data Transformations**

| Transformation | How Calculated[a] | Example of Use |
|---|---|---|
| Arithmetic | $x' = x/y$ or $x' = x + c$ | Organ weight/body weight |
| Reciprocals | $x' = 1/x$ | Linearizing data, particularly rate phenomena |
| Arcsine (also called angular) | $x' = \text{arcsine}\sqrt{x}$ | Normalizing dominant lethal and mutation rate data |
| Logarithmic | $x' = \log x$ | pH values |
| Probability (probit) | $x' = \text{probability } x$ | Percentage responding[b] |
| Square roots | $x' = \sqrt{x}$ | Surface area of animal from body weights |
| Box Cox | $x' = (x^v - 1)v$ for $v \neq 0$<br>$x' = \ln x$ for $v = 0$ | A family of transforms for use when one has no prior knowledge of the appropriate transformation to use |
| Rank transformations | Depends on nature of samples | As a bridge between parametric and nonparametric statistics |

[a] $x$ and $y$ are original variables; $x'$ and $y'$ transformed values; $C$ stands for a constant.
[b] Plotting a double reciprocal (i.e., $1/x$ vs. $1/y$) will linearize almost any dataset, so will plotting the log transforms of a set of variables.

Confirmatory analysis allows evaluation of the repro-lucibility of the patterns or effects. Its role is close to that of classical hypothesis testing but also often includes steps such as: (1) incorporating information from an analysis of another, closely related set of data, and (2) validating a result by assembling and analyzing additional data. These techniques are in general beyond the scope of this text; however, Velleman and Hoaglin [21] and Hoaglin et al. [22] present a clear overview of the more important methods, along with codes for their execution on a microcomputer (they have also now been incorporated into Minitab). A short examination of a single case of the use of these methods, however, is in order.

Toxicology has long recognized that no population—animal or human—is completely uniform in its response to any particular toxicant. Rather, a population is composed of a (presumably normal) distribution of individuals, some resistant to intoxication (hypo-responders), the bulk responding close to a central value (such as an $LD_{50}$), and some being very sensitive to intoxication (hyper-responders). This population distribution can, in fact, result in additional statistical techniques. The sensitivity of techniques such as ANOVA is reduced markedly by the occurrence of outliers (extreme high or low values, including hyper-and hypo-responders), which, in fact, serve to markedly inflate the variance (standard deviation) associated with a sample. Such variance inflation is particularly common in small groups that are exposed or dosed at just over or under a threshold level, causing a small number of individuals in the sample (who are more sensitive than the other members) to respond markedly. Such a situation is displayed in Figure 8.6, which plots the mean and standard deviations of methemoglobin levels in a series of groups of animals exposed to successively higher levels of a hemolytic agent.

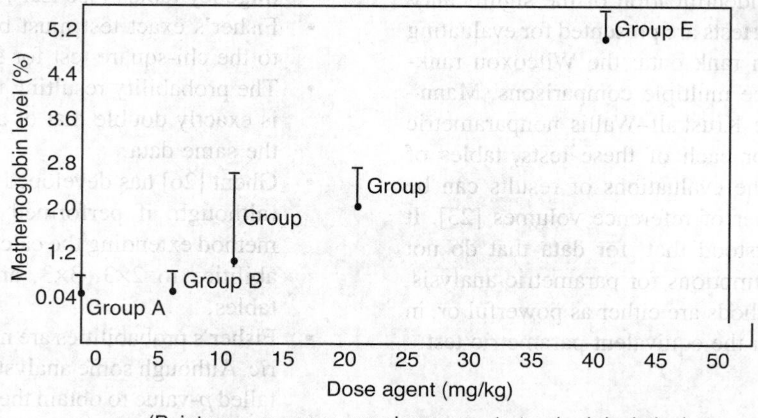


**FIGURE 8.6** Variance inflation.

Though the mean level of methemoglobin in group C is more than double that of the control group (A), no hypothesis test will show this difference to be significant because it has such a large standard deviation associated with it. Yet, this inflated variance exists because a single individual has such a marked response. The occurrence of the inflation is certainly an indicator that the data need to be examined closely. Indeed, all tabular data in toxicology should be visually inspected for both trend and variance inflation.

A concept related (but not identical) to resistance and exploratory data analysis is that of *robustness*. Robustness generally implies insensitivity to departures from assumptions surrounding an underlying model, such as normality. When summarizing the location of data, the median, though highly resistant, is not extremely robust, but the mean is both nonresistant and nonrobust.

## HYPOTHESIS TESTING OF CATEGORICAL AND RANKED DATA

Categorical (or contingency table) presentations of data can contain any single type of data, but generally the contents are collected and arranged so they can be classified as belonging to treatment and control groups, with the members of each of these groups then classified as belonging to one of two or more response categories (such as tumor/no tumor or normal/hyperplastic/neoplastic). For these cases, two forms of analysis are presented: Fisher's exact test (for the 2×2 contingency table) and the R×C (R, row; C, column) chi-square test (for large tables). It should be noted, however, that versions of both of these tests permit the analysis of any size of contingency table.

The analysis of rank data—what is generally called *nonparametric statistical analysis*—is an exact parallel of the more traditional (and familiar) parametric methods. There are methods for the single-comparison case (just as Student's *t*-test is used) and for the multiple-comparison case (just as analysis of variance is used) with appropriate *post hoc* tests for exact identification of the significance with a set of groups. Four tests are presented for evaluating statistical significance in rank data: the Wilcoxon rank-sum test, distribution-free multiple comparisons, Mann–Whitney U test, and the Kruskall–Wallis nonparametric analysis of variance. For each of these tests, tables of distribution values for the evaluations of results can be found in any of a number of reference volumes [23]. It should be clearly understood that, for data that do not fulfill the necessary assumptions for parametric analysis, these nonparametric methods are either as powerful or, in fact, more powerful than the equivalent parametric test.

### FISHER'S EXACT TEST

Fisher's exact test should be used to compare two sets of discontinuous, quantal (all or none) data. Small sets of such data can be checked by contingency data tables, such as those of Finney et al. [24]. Larger sets, however, require computation. These include frequency data such as incidences of mortality or certain histopathological findings. Thus, the data can be expressed as ratios. These data do not fit on a continuous scale of measurement but usually involve numbers of responses classified as either negative or positive—that is, a contingency table situation [15].

The analysis is started by setting up a 2×2 contingency table to summarize the numbers of positive and negative responses, as well as the totals of these, as follows:

| | Positive | Negative | Total |
|---|---|---|---|
| **Group I** | $A$ | $B$ | $A + B$ |
| **Group II** | $C$ | $D$ | $C + D$ |
| **Totals** | $A + C$ | $B + D$ | $A + B + C + D = N_{\text{total}}$ |

Using the above set of symbols, the formula for *P* appears as follows:*

$$P = \frac{(A+B)!(C+D)!(A+C)!(B+D)!}{N!A!B!C!D!}$$

The exact test produces a probability (*P*) that is the sum of the above calculation repeated for each possible arrangement of the numbers in the above cells (that is, *A*, *B*, *C*, and *D*) showing an association equal to or stronger than that between the two variables. The *P* resulting from these computations will be the exact one- or two-tailed probability depending on which of these two approaches is being employed. This value tells us if the groups differ significantly (with a probability less than 0.05, say) and the degree of significance.

#### Assumptions and Limitations

- Tables are available that provide individual exact probabilities for small sample size contingency tables (see Zar [25]).
- Fisher's exact test must be used in preference to the chi-square test for small cell sizes.
- The probability resulting from a two-tailed test is exactly double that of a one-tailed test from the same data.
- Ghent [26] has developed and proposed a good (although, if performed by hand, laborious) method extending the calculation of exact probabilities to 2×3, 3×3, and R×C contingency tables.
- Fisher's probabilities are not necessary symmetric. Although some analysts will double the one-tailed *p*-value to obtain the two-tailed result, this method is usually overly conservative.

* *A*! is *A* factorial; for example, for 4!, this would be (4)(3)(2)(1) = 24

## 2×2 Chi-Square Test

Although Fisher's exact test is preferable for analysis of most 2×2 contingency tables in toxicology, the chi-square test is still widely used and is preferable in a few unusual situations (particularly if cell sizes are large yet only limited computational support is available). The formula is simply:

$$\chi^2 = \frac{(O_1 - E_1)^2}{E_1} = \frac{(O_2 - E_2)^2}{E_2}$$

$$= \sum \frac{(O_i - E_i)^2}{E_i}$$

where the $O$'s are observed numbers (or counts) and $E$'s are expected numbers. The common practice in toxicology is for the observed figures to be test or treatment group counts. The expected figure is calculated as:

$$E = \frac{\text{(column total)(row total)}}{\text{grand total}}$$

for each box or cell in a contingency table. Our degrees of freedom are $(R-1)(C-1) = (2-1)(2-1) = 1$. Looking at a chi-square table for 1 degree of freedom, we see that this is greater than the test statistic at 0.05 (3.84) but less than that at 0.01 (6.64) so $0.05 > p > 0.01$.

### Assumptions and Limitations

*Assumptions*

1. Data are univariate and categorical.
2. Data are from a multinomial population.
3. Data are collected by random, independent sampling.
4. Groups being compared are of approximately same size, particularly for small group sizes.

*When to use*

1. The data are of a categorical (or frequency) nature.
2. The data fit the assumptions above.
3. Goodness-of-fit to a known form of distribution is being tested.
4. Cell sizes are large.

*When not to use*

1. The data are continuous rather than categorical.
2. Sample sizes are small and very unequal.
3. Sample sizes are too small (for example, when total $N$ is less than 50 or if any expected value is less than 5).
4. Any 2×2 comparison is being performed (use Fisher's exact test instead).

## R×C Chi-Square Test

The R×C chi-square test can be used to analyze discontinuous (frequency) data as in the Fisher's exact test or the 2×2 chi-square test; however, in the R×C test we wish to compare three or more sets of data. An example would be comparison of the incidence of tumors among mice on three or more oral dosage levels. We can consider the data as positive (tumors) or negative (no tumors). The expected frequency for any box is equal to (row total)(column total)/($N_{\text{total}}$). As in the Fisher's exact test, the initial step is to set up a table (this time a R×C contingency table):

| | Positive | Negative | Total |
|---|---|---|---|
| Group I | $A_1$ | $B_1$ | $A_1 + B_1 = N_1$ |
| Group II | $A_2$ | $B_2$ | $A_2 + B_2 = N_2$ |
| | ↓ | ↓ | ↓ |
| Group R | $A_R$ | $B_R$ | $A_R + B_R = N_R$ |
| Total | $N_A$ | $N_B$ | $N_{\text{total}}$ |

Using these symbols, the formula for chi-square ($\chi^2$) is:

$$\chi^2 = \frac{N_{tot}^2}{N_A N_B N_K}\left(\frac{A_1^2}{N_1} + \frac{A_2^2}{N_2} + \ldots + \frac{A_K^2}{N_K} - \frac{N_A^2}{N_{tot}}\right)$$

The resulting $\chi^2$ value is compared to table values [16] according to the number of degrees of freedom, which is equal to $(R-1)(C-1)$. If $\chi^2$ is smaller than the table value at the 0.05 probability level, the groups are not significantly different. If the calculated $\chi^2$ is larger, there is some difference among the groups, and 2×R chi-square or Fisher's exact tests will have to be compared to determine which groups differ from which other groups.

### Assumptions and Limitations

1. The test is based on data being organized in a table with *cells* (in the table below, *A*, *B*, *C*, and *D* are cells):

| | | Columns (C) | | |
|---|---|---|---|---|
| | | Control | Treated | Total |
| Rows (R) | No Effect | $A$ | $B$ | $A + B$ |
| | Effect | $C$ | $D$ | $C + D$ |
| Total | | $A + C$ | $B + D$ | $A + B + C + D$ |

2. None of the expected frequency values is less than 5.0.
3. The chi-square test is always one tailed.
4. Without the use of some form of correction, the test becomes less accurate as the differences between group sizes increases.
5. The results from each additional column (group) are approximately additive. Due to this characteristic, the chi-square test can be readily used for evaluating any R×C combination.

6. The results of the chi-square calculation must be a positive number.
7. The test is weak with either small sample sizes or when the expected frequency in any cell is less than 5 (this latter limitation can be overcome by pooling, or combining, cells).
8. Test results are independent of order of cells, unlike Kolmogorov–Smirnov.
9. The test can be used to test the probability of validity of any distribution.

### Wilcoxon Rank-Sum Test

The Wilcoxon rank-sum test is commonly used for the comparison of two groups of nonparametric (interval or not normally distributed) data, such as those which are not measured exactly but rather as falling within certain limits (for example, how many animals died during each hour of an acute study). The test is also used when there is no variability (variance = 0) within one or more of the groups we wish to compare [15].

The data in both groups being compared are initially arranged and listed in order of increasing value, then each number in the two groups must receive a rank value. Beginning with the smallest number in either group (which is given a rank of 1.0), each number is assigned a rank. If there are duplicate numbers (called *ties*), then each value of equal size will receive the median rank for the entire identically sized group; thus, if the lowest number appears twice, both figures receive a rank of 1.5. This, in turn, means that the ranks of 1.0 and 2.0 have been used and that the next highest number has a rank of 3.0. If the lowest number appears three times, then each is ranked as 2.0 and the next number has a rank of 4.0; thus, each tied number gets a median rank. This process continues until all of the numbers are ranked. Each of the two columns of ranks (one for each group) is totaled, giving the sum of ranks for each group being compared. As a check, we can calculate the value:

$$\frac{(N)(N+1)}{2}$$

where $N$ is the total number of data in both groups. The result should be equal to the sum of the sum of ranks for both groups.

The sums of rank values are compared to table values [27] to determine the degree of significant differences, if any. These tables include two limits (an upper and a lower) that are dependent on the probability level. If the number of data is the same in both groups ($N_1 \neq N_2$), then the lesser sum of ranks (smaller $N$) is compared to the table limits to find the degree of significance. Normally, the comparison of the two groups ends here and the degree of significant difference can be reported.

### Distribution-Free Multiple Comparison

The distribution-free multiple comparison test should be used to compare three or more groups of nonparametric data. These groups are then analyzed two at a time for any significant differences [28]. The test can be used for data similar to those compared by the rank-sum test. We often employ this test for reproduction and mutagenicity studies (such as comparing survival rates of offspring of rats fed various amounts of test materials in the diet).

As shown in Example 2, two values must be calculated for each pair of groups: the difference in mean ranks and the probability level value against which the difference will be compared. To determine the difference in mean ranks, we must first arrange the data within each of the groups in order of increasing values, then we must assign rank values, beginning with the smallest overall figure. Note that this ranking is similar to that in the Wilcoxon test except that it applies to more than two groups. The ranks are then added for each of the groups. As a check, the sum of these values should equal:

$$\frac{N_{tot}(N_{tot}+1)}{2}$$

where $N_{tot}$ is the total number of figures from all groups. Next, we can find the mean rank ($R$) for each group by dividing the sum of ranks by the numbers in the data ($N$) in the group. These mean ranks are then taken in those pairs that we want to compare (usually each test group vs the control), and the differences are found ($|R_1 - R_2|$). This value is expressed as an absolute figure; that is, it is always a positive number.

The second value for each pair of groups (the probability value) is calculated from the expression:

$$z\left[\frac{a}{K}(K-1)\right]\sqrt{\frac{N_{tot}(N_{tot}+1)}{12}}\sqrt{\frac{1}{N_1}\frac{1}{N_2}}$$

where $a$ is the level of significance for the comparison (usually 0.05, 0.01, 0.001, etc.), $K$ is the total number of groups, and $z$ is a figure obtained from a normal probability table and determining the corresponding $z$-score.

The result of the probability value calculation for each pair of groups is compared to the corresponding mean difference $|R_1 - R_2|$. If $|R_1 - R_2|$ is smaller, then there is no significant difference between the groups. If it is larger, then the groups are different, and $|R_1 - R_2|$ must be compared to the calculated probability values for $a = 0.01$ and $a = 0.001$ to find the degree of significance.

## Example 2

Consider the set of data provided below (ranked in increasing order), which could represent the proportion of rats surviving given periods of time during diet inclusion of a test chemical at four dosage levels (survival index).

| I 5.0 mg/kg | | II 2.5 mg/kg | | III 1.25 mg/kg | | IV 0.0 mg/kg | |
|---|---|---|---|---|---|---|---|
| % Value | Rank | % Value | Rank | % Value | Rank | % Value | Rank |
| 40 | 2.0 | 40 | 2.0 | 50 | 5.5 | 60 | 9.0 |
| 40 | 2.0 | 50 | 5.5 | 50 | 5.5 | 60 | 9.0 |
| 50 | 5.5 | 80 | 12.0 | 60 | 9.0 | 80 | 12.0 |
| 100 | 17.5 | 80 | 12.0 | 100 | 17.5 | 90 | 14.0 |
| | | 100 | 17.5 | 100 | 17.5 | 100 | 17.5 |
| | | | | | | 100 | 17.5 |
| Sum of ranks | 27.0 | | 49.0 | | 55.0 | | 79.0 |

$$N_{\text{I}} = 4,\ N_{\text{II}} = 5,\ N_{\text{III}} = 5,\ N_{\text{IV}} = 6 \quad N_{tot} = 20$$

Check sums of ranks = 210, (20 × 21)/2 = 210

Mean ranks ($R$):

$$R_1 = \frac{27.0}{4} = 6.75 \quad R_2 = \frac{49.0}{5} = 9.80$$

$$R_3 = \frac{55.0}{5} = 11.00 \quad R_4 = \frac{79.0}{6} = 13.17$$

| Comparison Groups | $R_1 - R_2$ | Probability Test Values |
|---|---|---|
| 5.0 vs. 0.0 | 6.42 | $(0.05/4(3)) = Z_{0.00417} = 2.637$ $\sqrt{\frac{(20)(21)}{12}}\sqrt{\frac{1}{4}+\frac{1}{6}} = 10.07$ |
| 2.5 vs. 0.0 | 3.37 | $(0.05/4(3)) = Z_{0.00417} = 2.637$ $\sqrt{\frac{(20)(21)}{12}}\sqrt{\frac{1}{5}+\frac{1}{6}} = 9.45$ |
| 1.25 vs. 0.0 | 2.17 | $(0.05/4(3)) = Z_{0.00417} = 2.637$ $\sqrt{\frac{(20)(21)}{12}}\sqrt{\frac{1}{5}+\frac{1}{6}} = 9.45$ |

Because each of the $|R_1 - R_2|$ values is smaller than the corresponding probability calculation, the pairs of groups compared are not different at the 0.05 level of significance.

### Assumptions and Limitations

1. As with the Wilcoxon rank-sum test, too many tied ranks inflate the false positive.
2. Generally, this test should be used as a *post hoc* comparison after Kruskall–Wallis.

## Mann–Whitney U Test

This is a nonparametric test in which the data in each group are first ordered from lowest to highest values, then the entire set (both control and treated values) is ranked, with the average rank being assigned to tied values. The ranks are then summed for each group and U is determined according to

$$U_t = n_c n_t + \frac{n_t(n_t+1)}{2} - R_t$$

$$U_c = n_c n_t + \frac{n_c(n_c+1)}{2} R_c$$

where $n_c$ and $n_t$ are the sample sizes for the control and treated groups, respectively, and $R_c$ and $R_t$ are the sums of ranks for the control and treated groups, respectively. For the level of significance for a comparison of the two groups, the larger value of $U_c$ or $U_t$ is used. This is compared to critical values as found in tables [29].

With the above discussion and methods in mind, we can now examine the actual variables that we encounter in teratology studies. These variables can be readily divided into two groups: measures of lethality and measures of teratogenic effect [30]. Measures of lethality include: (1) corpora lutea per pregnant female, (2) implants per pregnant female, (3) live fetuses per pregnant female, (4) percentage of preimplantation loss per pregnant female, (5) percentage of resorptions per pregnant female, and (6) percentage of dead fetuses per pregnant female. Measures of teratogenic effect include: (1) percentage of abnormal fetuses per litter, (2) percentage of litters with abnormal fetuses, and (3) fetal weight gain. As demonstrated in Example 3, the Mann–Whitney U test is employed for the count data, but which test should be employed for the percentage variables should be decided on the same grounds as described later under reproduction studies.

## Example 3

In a 2-week study, the levels of serum cholesterol in treatment and control animals are successfully measured and assigned ranks as below:

| Treatment | | Control | |
|---|---|---|---|
| Value | Rank | Value | Rank |
| 10 | 1 | 19 | 4 |
| 18 | 3 | 28 | 13 |
| 26 | 10.5 | 29 | 14.5 |
| 31 | 16 | 26 | 10.5 |
| 15 | 2 | 35 | 19 |
| 24 | 8 | 23 | 7 |
| 22 | 6 | 29 | 14.5 |
| 33 | 17 | 34 | 18 |
| 21 | 5 | 38 | 20 |
| 25 | 9 | 27 | 12 |
| Sum of ranks | 77.5 | | 132.5 |

The critical value for one-tailed $p \leq 0.05$ is $U \geq 73$. We then calculate:

$$U_t = (10)(10) + \frac{10(10+1)}{2} - 77.5$$
$$= 100 + \frac{110}{2} - 77.5 = 77.5$$
$$U_c = (10)(10) + \frac{10(10+1)}{2} - 132.5 = 2.5$$

Because 77.5 is greater than 73, these groups are significantly different at the 0.05 level.

### Assumptions and Limitations

1. It does not matter whether the observations are ranked from smallest to largest or *vice versa*.
2. This test should not be used for paired observations.
3. The test statistics from a Mann–Whitney test are linearly related to those of Wilcoxon. The two tests will always yield the same result. The Mann–Whitney is presented here for historical completeness, as it has been much favored in reproductive and developmental toxicology studies; however, it should be noted that the authors do not include it in the decision tree for method selection (Figure 8.2).

## Kruskal–Wallis Nonparametric ANOVA

The Kruskal-Wallis nonparametric one-way analysis of variance should be the initial analysis performed when we have three or more groups of data which are by nature nonparametric (not a normally distributed population, data of a discontinuous nature, or all the groups being analyzed not being from the same population) but not a categorical (or quantal) nature. Commonly, these will be either rank-type evaluation data (such as behavioral toxicity observation scores) or reproduction study data. The analysis is initiated [31] by ranking all the observations from the combined groups to be analyzed. Ties are given the average rank of the tied values (that is, if two values which would tie for 12th rank and therefore are ranked 12th and 13th then both would be assigned the average rank of 12.5).

The sum of ranks of each group ($r_1, r_2, \ldots, r_k$) is computed by adding all the rank values for each group. The test value $H$ is then compute as:

$$H = \frac{12}{n(n+1)} \sum r_1^2/n_1 + r_2^2/n_2 + \cdots + r_k^2/n_k - 3(n+1)$$

where $n_1, n_2, \ldots, n_k$ are the number of observations in each group. The test statistic is then compared with a table of $H$ values. If the calculated value of $H$ is greater than the table value for the appropriate number of observations in each group, there is significant difference between the groups, but further testing (using the distribution-free multiple comparisons method) is necessary to determine where the difference lies (as demonstrated in Example 4).

### Example 4

As part of a neurobehavioral toxicology study, righting reflex values (whole numbers ranging from 0 to 10) were determined for each of five rats in each of three groups. The values observed and their ranks are as follows:

| Control Group | | 5 mg/kg Group | | 10 mg/kg Group | |
|---|---|---|---|---|---|
| Reflex Score | Rank | Reflex Score | Rank | Reflex Score | Rank |
| 0 | 2 | 1 | 5 | 4 | 11 |
| 0 | 2 | 2 | 7.5 | 4 | 11 |
| 0 | 2 | 2 | 7.5 | 5 | 13 |
| 1 | 5 | 3 | 9 | 8 | 14.5 |
| 1 | 5 | 4 | 11 | 8 | 14.5 |
| Sum of ranks | 16 | | 40 | | 64 |

From these, the $H$ value is calculated as:

$$H = \frac{12}{15(15+1)}\left[\frac{16^2}{5} + \frac{40^2}{5} + \frac{64^2}{5}\right] - 3(15+1)$$
$$= \frac{12}{240}\left[\frac{(256+1600+4096)}{5}\right] - 48$$
$$= \frac{1}{20}(1190.4) - 48$$
$$= 59.52 - 48$$
$$= 11.52$$

Consulting a table of values for $H$, we find that for the case where we have three groups of five observations each, the test values are 4.56 (for $p = 0.10$), 5.78 ($p = 0.05$), and 7.98 (for $p = 0.01$). As our calculated $H$ is greater than the $p = 0.01$ test value, we have determined that there is a significant difference between the groups at the level of $p < 0.01$ and would now have to continue to a multiple comparisons test to determine where the difference is.

### Assumptions and Limitations

1. The test statistic $H$ is used for both small and large samples.
2. When we find a significant difference, we do not know which groups are different. It is not correct to then perform a Mann–Whitney U test on all possible combinations; rather, a multiple comparison method must be used, such as the distribution-free multiple comparisons.
3. Data must be independent for the test to be valid.
4. Too many tied ranks will decrease the power of this test and lead to increased false-positive levels.

5. When $k = 2$, the Kruskal–Wallis chi-square value has 1 degree of freedom. This test is identical to the normal approximation used for the Wilcoxon rank-sum test. As noted in previous sections, a chi-square with 1 degree of freedom can be represented by the square of a standardized normal random variable. In the case of $k = 2$, the $H$ statistic is the square of the Wilcoxon rank-sum $z$-test (without the continuity correction).
6. The effect of adjusting for tied ranks is to slightly increase the value of the test statistic, $H$; therefore, omission of this adjustment results in a more conservative test.

## LOG-RANK TEST

The log-rank test is a statistical methodology for comparing the distribution of time until the occurrence of the event in independent groups. In toxicology, the most common event of interest is death or occurrence of a tumor, but it could just as well be liver failure, neurotoxicity, or any other event that occurs only once in an individual. The elapsed time from initial treatment or observation until the *event* is the *event time*, often referred to as *survival time*, even when the event is not death.

The log-rank test provides a method for comparing risk-adjusted event rates, useful when test subjects in a study are subject to varying degrees of opportunity to experience the event. Such situations arise frequently in toxicology studies due to the finite duration of the study, early termination of the animal, or interruption of treatment before the event occurs. Examples where use of the log-rank test might be appropriate include comparing survival times in carcinogenity bioassay animals that are given a new treatment with those in the control group or comparing times to liver failure for several dose levels of a new NSAID where the animals are treated for 10 weeks or until cured, whichever comes first.

If every animal were followed until the event occurrence, the event times could be compared between two groups using the Wilcoxon rank-sum test; however, some animals may die or complete the study before the event occurs. In such cases, the actual time of the event is unknown because the event does not occur while under study observation. The event times for these animals are based on the last known time of study observation and are referred to as *censored* observations because they represent the lower bound of the true, unknown event times. The Wilcoxon rank-sum test can be highly biased in the presence of the censored data.

The null hypothesis tested by the log-rank test is that of equal event time distributions among groups. Equality of the distributions of event times implies similar event rates among groups, not only for the clinical trial as a whole but also for any arbitrary time point during the trial. Rejection of the null hypothesis indicates that the event rates differ among groups at one or more time points during the study.

The idea behind the log-rank test for the comparison of two life tables is simple: If there were no difference between the groups, the total deaths occurring at any time should split between the two groups at that time. So, if the numbers at risk in the first and second groups in, say, the sixth month were 70 and 30, respectively, and 10 deaths occurred in that month, then we would expect:

$$10 \times \frac{70}{70+30} = 7$$

of these deaths to have occurred in the first group, and

$$10 \times \frac{30}{70+30} = 3$$

of the deaths to have occurred in the second group.

A similar calculation can be made at each time of death (in either group). By adding together for the first group the results of all such calculations, we obtain a single number, the extent of exposure ($E_1$), which represents the expected number of deaths in that group if the two groups had the same distribution of survival times. An extent of exposure ($E_2$) can be obtained for the second group in the same way. Let $O_1$ and $O_2$ denote the actual total numbers of deaths in the two groups. A useful arithmetic check is that the total number of deaths ($O_1 + O_2$) must equal the sum of the extents of exposure ($E_1 + E_2$). The discrepancy between the $O$'s and $E$'s can be measured by the quantity:

$$x^2 = \frac{\left(|O_1 - E_1| - \frac{1}{2}\right)^2}{E_1} + \frac{\left(|O_2 - E_2| - \frac{1}{2}\right)^2}{E_2}$$

For rather obscure reasons, $x^2$ is known as the log-rank statistic. An approximate significance test of the null hypothesis of identical distributions of survival time in the two groups is obtained by referring $x^2$ to a chi-square distribution on 1 degree of freedom. This is demonstrated in Example 5.

### Example 5

In a study of the effectiveness of a new monoclonal antibody to treat specific cancer, the times to reoccurrence of the cancer in treated animals in weeks were as follows:

| **Control Group** | | | **Treatment Group** | | |
|---|---|---|---|---|---|
| 1 | 5 | 11 | 6 | 10 | 22 |
| 1 | 5 | 12 | 6 | 11 | 23 |
| 2 | 8 | 12 | 6 | 13 | 25 |
| 2 | 8 | 15 | 6 | 16 | 32 |
| 3 | 8 | 17 | 7 | 17 | 32 |
| 4 | 8 | 22 | 9 | 19 | 34 |
| 4 | 11 | 23 | 10 | 20 | 35 |

The table provided below presents the calculations for the log-rank test applied to these times. A chi-square value of 13.6 is significant at the $p < 0.001$ level.

*Illustration:*

$$t = 23,\quad 2 \times \frac{6}{7} = 1.7143,\quad 2 \times \frac{1}{7} = 0.2857$$

*Test of significance:*

$$x^2 = \frac{\left(|O_1 - E_1| - \tfrac{1}{2}\right)^2}{E_1} + \frac{\left(|O_2 - E_2| - \tfrac{1}{2}\right)^2}{E_2}$$

$$= \frac{\left(|9 - 19.2| - \tfrac{1}{2}\right)^2}{19.2} + \frac{\left(|21 - 10.8| - \tfrac{1}{2}\right)^2}{10.8} = 13.6$$

*Estimate of relative risk:*

$$\theta = \frac{9/19.2}{21/10.8} = 0.24$$

The log-rank test as presented by Peto et al. [32] uses the product-limit life-table calculations rather than the actuarial estimators shown above. The distinction is unlikely to be of practical importance unless the grouping intervals are very coarse.

Peto and Pike [33] suggest that the approximation in treating the null distribution of $\chi^2$ as a chi-square is conservative, so it will tend to understate the degree of statistical significance. In the formula for $\chi^2$ we have used the continuity correction of subtracting 1/2 from $|O_1 - E_1|$ and $|O_2 - E_2|$ before squaring. This is recommended by Peto et al. [32] when, as in nonrandomized studies, the permutational argument does not apply. Peto et al. [32] give further details of the log-rank test and its extension to comparisons of more than two treatment groups and to tests that control for categorical confounding factors.

### Assumptions and Limitations

1. The endpoint of concern is, or is defined so it is, right censored; that is, once it happens, it does not recur. Examples are death or a minimum or maximum value of an enzyme or physiologic function (such as respiration rate).
2. The method makes no assumptions on distribution.
3. Many variations of the log-rank test for comparing survival distributions exist. The most common variant has the form:

$$\chi^2 = \frac{(O_1 \quad E_1)^2}{E_1} + \frac{(O_2 \quad E_2)^2}{E_2}$$

where $O_i$ and $E_i$ are computed for each group, as in the formulas given previously. This statistic also has an approximate chi-square distribution with 1 degree of freedom under $H_0$. A continuity correction can also be used to reducing the numerators by 1/2 before squaring. Use of such a correction leads to even

**Log-Rank Calculation for Tumor Data**

| | At Risk | | | Relapses | | | Extent of Exposure | | |
|---|---|---|---|---|---|---|---|---|---|
| **Time ($t$)** | **T** | **C** | **Total** | **T** | **C** | **Total** | **T** | **C** | **Total** |
| 1 | 21 | 21 | 42 | 0 | 2 | 2 | 1.0000 | 1.0000 | 2 |
| 2 | 21 | 19 | 40 | 0 | 2 | 2 | 1.0500 | 0.9500 | 2 |
| 3 | 21 | 17 | 38 | 0 | 1 | 1 | 0.5526 | 0.4474 | 1 |
| 4 | 21 | 16 | 37 | 0 | 2 | 2 | 1.1351 | 0.8649 | 2 |
| 5 | 21 | 14 | 35 | 0 | 2 | 2 | 1.2000 | 0.8000 | 2 |
| 6 | 20.5 | 12 | 32.5 | 3 | 0 | 3 | 1.8923 | 1.1077 | 3 |
| 7 | 17 | 12 | 29 | 1 | 0 | 1 | 0.5862 | 0.4138 | 1 |
| 8 | 16 | 12 | 28 | 0 | 4 | 4 | 2.2857 | 1.7143 | 4 |
| 10 | 14.5 | 8 | 22.5 | 1 | 0 | 1 | 0.6444 | 0.3556 | 1 |
| 11 | 12.5 | 8 | 20.5 | 0 | 2 | 2 | 1.2295 | 0.7705 | 2 |
| 12 | 12 | 6 | 18 | 0 | 2 | 2 | 1.3333 | 0.6667 | 2 |
| 13 | 12 | 4 | 16 | 1 | 0 | 1 | 0.7500 | 0.2500 | 1 |
| 15 | 11 | 4 | 15 | 0 | 1 | 1 | 0.7333 | 0.2667 | 1 |
| 16 | 11 | 3 | 14 | 1 | 0 | 1 | 0.7857 | 0.2143 | 1 |
| 17 | 9.5 | 3 | 12.5 | 0 | 1 | 1 | 0.7600 | 0.2400 | 1 |
| 22 | 7 | 2 | 9 | 1 | 1 | 2 | 1.5556 | 0.4444 | 2 |
| 23 | 6 | 1 | 7 | 1 | 1 | 2 | 1.7143 | 0.2857 | 2 |
| Total | | | | 9 | 21 | 30 | 19.2080 | 10.7920 | 30 |
| | | | | | ($O_1$) | ($O_2$) | ($E_1$) | ($E_2$) | |

further conservatism and may be omitted when sample sizes are moderate or large.

4. The Wilcoxon rank-sum test could be used to analyze the event times in the absence of censoring. A generalized Wilcoxon test, sometimes called the *Gehan test*, based on an approximate chi-square distribution, has been developed for use in the presence of censored observations. Both the log-rank and the generalized Wilcoxon tests are nonparametric tests, and they require no assumptions regarding the distribution of event times. When the event rate is greater early in the trial than toward the end, the generalized Wilcoxon test is the more appropriate test because it gives greater weight to the earlier differences.
5. Survival and failure times often follow the exponential distribution. If such a model can be assumed, a more powerful alternative to the log-rank test is the likelihood ratio test. This parametric test assumes that event probabilities are constant over time; that is, the chance that a patient becomes event positive at time $t$ given that he is event negative up to time $t$ does not depend on $t$. A plot of the negative log of the event times distribution showing a linear trend through the origin is consistent with exponential event times.
6. Life tables can be constructed to provide estimates of the event time distributions. Estimates commonly used are known as the Kaplan–Meier estimates.

## HYPOTHESIS TESTING: UNIVARIATE PARAMETRIC TESTS

Jnivariate case data (where each datum is defined by one reatment and one effect variable) from normally distributed populations generally have a higher information value associated with them, but the traditional hypothesis esting techniques (which include all the methods lescribed in this chapter) are generally neither resistant or robust. All the data analyzed by these methods are lso, effectively, continuous; that is, at least for practical urposes, the data may be represented by any number and each such data number has a measurable relationship to other data numbers.

### TUDENT'S *t*-TEST (UNPAIRED *t*-TEST)

airs of groups of continuous, randomly distributed data re compared via this test. We can use this test to compare hree or more groups of data, but they must be intercompared by examination of two groups taken at time and re preferentially compared by analysis of variance (ANOVA). Usually, this means comparison of a test group vs. a control group, although two test groups may be compared as well. To determine which of the three types of $t$-tests described in this chapter should be employed, the $F$-test is usually performed first. This will tell us if the variances of the data are approximately equal, which is a requirement for the use of the parametric methods. If the $F$-test indicates homogeneous variances and the numbers of data within the groups ($N$) are equal, then the Student's $t$-test is the appropriate procedure [15]. If the $F$ is significant (the data are heterogeneous) and the two groups have equal numbers of data, the modified Student's $t$-test is applicable [34].

The value of $t$ for Student's $t$-test is calculated using the formula:

$$t = \frac{\bar{X}_1 - \bar{X}_2}{\sum D_1^2 + \sum D_2^2}\sqrt{\frac{N_1 N_2}{N_1 + N_2}(N_1 + N_2 - 2)}$$

where

$$\sum D^2 = \frac{N\sum X^2 - \left(\sum X\right)^2}{N}$$

The value of $t$ obtained from the above calculations is compared to the values in a $t$-distribution table according to the appropriate number of degrees of freedom (df). If the $F$ value is not significant (i.e., variances are homogeneous), the df = $N_1 + N_2 - 2$. If the $F$ is significant and $N_1 = N_2$, then the df = $N - 1$. Although this case indicates a nonrandom distribution, the modified $t$-test is still valid. If the calculated value is larger than the table value at $p = 0.05$, it may then be compared to the appropriate other table values in order of decreasing probability to determine the degree of significance between the two groups. Example 6 demonstrates this methodology.

#### Example 6

Suppose we wish to compare two groups (at test and control group) of dog weights following inhalation of a vapor. First, we would test for homogeneity of variance using the $F$-test. Assuming that this test gave negative (homogeneous) results, we would perform the $t$-test as follows:

| | Test Weight | | Control Weight | |
|---|---|---|---|---|
| **Dog** | $X_1$ **(kg)** | $X_1^2$ | $X_2$ **(kg)** | $X_2^2$ |
| 1 | 8.3 | 68.89 | 8.4 | 70.56 |
| 2 | 8.8 | 77.44 | 10.2 | 104.04 |
| 3 | 9.3 | 86.49 | 9.6 | 92.16 |
| 4 | 9.3 | 86.49 | 9.4 | 88.36 |
| Sums | $\Sigma X_1 = 35.7$ | $\Sigma X_1^2 = 319.31$ | $\Sigma X_2 = 37.6$ | $\Sigma X_2^2 = 355.12$ |
| Means | 8.92 | | 9.40 | |

The difference in means = 9/40 = 8.92 = 0.48.

$$\sum D_1^2 = \frac{4(319.31)-(35.7)^2}{4} = \frac{2.75}{4} = 0.6875$$

$$\sum D_2^2 = \frac{4(355.12)-(37.6)^2}{4} = \frac{6.72}{4} = 1.6800$$

$$t = \frac{0.48}{\sqrt{0.6875+1.6800}}\sqrt{\frac{4(4)}{4+4}}(4+4) = 1.08$$

The table value for $t$ at the 0.05 probability level for (4 + 4 – 2), or 6 degrees of freedom, is 2.447; therefore, the dog weights are not significantly different at $p = 0.05$.

### Assumptions and Limitations

1. The test assumes that the data are univariate, continuous, and normally distributed.
2. Data are collected by randomly sampling.
3. The test should be used when the assumptions in 1 and 2 are met and there are only two groups to be compared.
4. Do not use when the data are ranked, when the data are not approximately normally distributed, or when more than two groups are to be compared. Do not use for paired observations.
5. This is the most commonly misused test method, except in those few cases where one is truly only comparing two groups of data and the group sizes are roughly equivalent. It is not valid for multiple comparisons (because of resulting additive errors) or where group sizes are very unequal.
6. The test is robust for moderate departures from normality, and, when $N_1$ and $N_2$ are approximately equal, it is robust for moderate departures from homogeneity of variances.
7. The main difference between the $z$-test and the $t$-test is that the $z$-statistic is based on a known standard deviation ($\sigma$) while the $t$-statistic uses the sample standard deviation ($s$) as an estimate of $\sigma$. With the assumption of normally distributed data, the variance $\sigma^2$ is more closely estimated by the sample variance $s^2$ as $n$ gets large. It can be shown that the $t$-test is equivalent to the $z$-test for infinite degrees of freedom. In practice, a large sample is usually considered to be $n \geq 30$.

## Cochran *t*-Test

The Cochran test should be used to compare two groups of continuous data when the variances (as indicated by the $F$-test) are heterogeneous and the numbers of data within the groups are not equal ($N_1 \neq N_2$). This is the situation, for example, when the data, although expected to be randomly distributed, are found not to be [3]. Two $t$ values are calculated for this test: the observed $t$ ($t_{obs}$) and the expected $t$ ($t'$). The observed $t$ is obtained by:

$$t_{obs} = \frac{\bar{X}_1 - \bar{X}_2}{\sqrt{W_1 + W_2}}$$

where $W$ = SEM² (standard error of the mean squared) = $S^2/N$, where variance $S$ can be calculated from:

$$S = \frac{N\sum X^2 - \left(\sum X\right)^2}{\frac{N}{N-1}}$$

The value for $t'$ is obtained from:

$$t' = \frac{t_1'W_1 + t_2'W_2}{W_1 + W_2}$$

where $t_1'$ and $t_2'$ are values for the two groups taken from the $t$-distribution table corresponding to ($N$ – 1) degrees of freedom (for each group) at the 0.05 probability level (or such level as one may select).

The calculated $t_{obs}$ is compared to the calculated $t'$ value (or values, if $t'$ values were prepared for more than one probability level). If $t_{obs}$ is smaller than a $t'$, the groups are not considered to be significantly different at that probability level. This procedure is shown in Example 7

### Example 7

Using the red blood cell count comparison from the discussion of the $F$-test (with $N_1 = 5$, $N_2 = 4$), the following results were determined:

$$\bar{X}_1 = \frac{37.60}{5} = 7.52 \quad W_1 = \frac{0.804}{5} = 0.1608$$

$$\bar{X}_2 = \frac{29.62}{4} = 7.52 \quad W_2 = \frac{0.025}{4} = 0.1608$$

(Note that $S^2$ values of 0.804 and 0.025 are calculated using the formula set forth in the section on Bartlett's test.)

$$t_{obs} = \frac{7.52 - 7.40}{\sqrt{0.1608 + 0.0062}} = 0.29$$

From the $t$-distribution table we use $t_1$ = 2.776 (df = 4 and $t_2$ = 3.182 (df = 3) for the 0.05 level of significance there is no statistical difference at $p$ = 0.05 between the two groups.

### Assumptions and Limitations

1. The test assumes that the data are univariate, continuous, and normally distributed and that group sizes are unequal.
2. The test is robust for moderate departures from normality and very robust for departures from equality of variances.

### *F*-Test

This is a test of the homogeneity of variances between two groups of data [15]. It is used in two separate cases. The first is when Bartlett's indicates heterogeneity of variances among three or more groups (i.e., it is used to determine which pairs of groups are heterogeneous). Second, the *F*-test is the initial step in comparing two groups of continuous data that we would expect to be parametric (two groups not usually being compared using ANOVA), the results indicating whether the data are from the same population and whether subsequent parametric comparisons would be valid. The *F* is calculated by dividing the larger variance ($S_1^2$) by the smaller one ($S_2^2$). $S^2$ is calculated as:

$$S = \frac{\frac{N\sum X^2 - \left(\sum X\right)^2}{N}}{N-1}$$

where *N* is the number of data in the group and *X* represents the individual values within the group. Frequently, $S^2$ values may be obtained from ANOVA calculations. Use of this is demonstrated in Example 8. The calculated *F* value is compared to the appropriate number in an *F*-value table for the appropriate degrees of freedom ($N - 1$) in the numerator (along the top of the table) and in the denominator (along the side of the table). If the calculated value is smaller, it is not significant, and the variances are considered homogeneous (and the Student's *t*-test would be appropriate for further comparison). If the calculated *F* value is greater, *F* is significant and the variances are heterogeneous (and the next test would be a modified Student's *t*-test if $N_1 = N_2$ or the Cochran *t*-test if $N_1 \neq N_2$; see Figure 8.2 to review the decision tree).

#### Example 8

If we wished to compare the red blood cell counts (RBC) of rats receiving a test material in their diet with the RBCs of control rats we might obtain the following results:

| Test Weight | | Control Weight | |
|---|---|---|---|
| $X_1$ (kg) | $X_1^2$ | $X_2$ (kg) | $X_2^2$ |
| 8.3 | 68.89 | 8.4 | 70.56 |
| 8.8 | 77.44 | 10.2 | 104.04 |
| 9.3 | 86.49 | 9.6 | 92.16 |
| 9.3 | 86.49 | 9.4 | 88.36 |
| $\Sigma X_1 = 35.7$ | $\Sigma X_1^2 = 319.31$ | $\Sigma X_2 = 37.6$ | $\Sigma X_2^2 = 355.12$ |
| 8.92 | | 9.40 | |

| Test RBC | | Control RBC | |
|---|---|---|---|
| $X_1$ | $X_1^2$ | $X_2$ | $X_2^2$ |
| 8.23 | 67.73 | 7.22 | 52.13 |
| 8.59 | 73.79 | 7.55 | 57.00 |
| 7.51 | 56.40 | 7.53 | 56.70 |
| 6.60 | 46.56 | 7.32 | 53.58 |
| 6.67 | 44.49 | | |
| $ÂX_1 = 37.60$ | $ÂX_1^2 = 285.97$ | $ÂX_2 = 29.62$ | $ÂX_2^2 = 219.41$ |

Variance for $X_1$:

$$S_1^2 = \frac{\frac{5(285.97) - (37.60)^2}{5}}{5-1} = 0.804$$

Variance for $X_2$:

$$S_3^2 = \frac{\frac{4(219.41) - (29.62)^2}{4}}{4-1} = 0.025$$

$$F = \frac{0.804}{0.025} = 32.16$$

From a table for *F* values, for 4 (numerator) vs. 3 (denominator) degrees of freedom, we read the limit of 9.12 at the 0.05 level. As our calculated value is larger (and, therefore, significant), the variances are heterogeneous and the Cochran *t*-test would be appropriate for comparison of the two groups of data.

#### Assumptions and Limitations

1. This test could be considered as a two group equivalent of the Bartlett's test.
2. If the test statistic is close to 1.0, the results are (of course) not significant.
3. The test assumes normality and independence of data.

### Analysis of Variance

Analysis of variance (ANOVA) is used for the comparison of three or more groups of continuous data when the variances are homogeneous and the data are independent and normally distributed. A series of calculations are required for ANOVA, starting with the values within each group being added ($\Sigma X$) and then these sums being added ($\Sigma\Sigma X$). Each figure within the groups is squared, and these squares are then summed ($\Sigma X^2$) and these sums added ($\Sigma\Sigma X^2$). Next the correction factor (CF) can be calculated from the following formula.

$$CF = \frac{\left(\sum_1^K \sum_1^N X\right)^2}{N_1 + N_2 + \cdots + N_k}$$

where *N* is the number of values in each group, and *K* is the number of groups. The total sum of squares ($SS_{total}$) is then determined as follows:

$$SS_{total} = \sum_1^K \sum_1^N X^2 - CF$$

In turn, the sum of squares between groups ($SS_{bg}$) is found from:

$$SS_{bg} = \frac{\left(\sum X_1\right)^2}{N_1} + \frac{\left(\sum X_2\right)^2}{N_2} + \cdots + \frac{\left(\sum X_k\right)^2}{N_k} - CF$$

The sum of squares within group ($SS_{wg}$) is then the difference between the last two figures, or:

$$SS_{wg} = SS_{total} - SS_{bg}$$

Now, there are three types of degrees of freedom to determine. The first, total degrees of freedom, is the total number of data within all groups under analysis minus one ($N_1 + N_2 + \ldots + N_k - 1$). The second figure (the degrees of freedom between groups) is the number of groups minus one ($K - 1$). The last figure (the degrees of freedom within groups, or error df) is the difference between the first two figures ($\text{df}_{total} - \text{df}_{bg}$).

The next set of calculations requires determination of the two mean squares ($MS_{bg}$ and $MS_{wg}$). These are the respective sum of square values divided by the corresponding df figures ($MS = SS/\text{df}$). The final calculation is that of the $F$ ratio. For this, the $MS$ between groups is divided by the $MS$ within groups ($F = MS_{bg}/MS_{wg}$). A table of the results of these calculations (using data from Example 9 at the end of this section) would appear as follows:

| | df | SS | MS | F |
|---|---|---|---|---|
| Between groups | 3 | 0.04075 | 0.01358 | 4.94 |
| Within groups | 12 | 0.03305 | 0.00275 | |
| Total | 15 | 0.07380 | | |

For interpretation, the $F$ ratio value obtained in the ANOVA is compared to a table of $F$ values. If $F \leq 1.0$, the results are not significant and comparison with the table values is not necessary. The degrees of freedom for the greater mean square ($MS_{bg}$) are indicated along the top of the table. Then read down the side of the table to the line corresponding to the degrees of freedom for the lesser mean square ($MS_{wg}$). The figure shown at the desired significance level (traditionally 0.05) is compared to the calculated $F$ value. If the calculated number is smaller, there is no significant differences among the groups being compared. If the calculated value is larger, there is some difference but further (*post hoc*) testing will be required before we know which groups differ significantly.

### Example 9

Suppose we want to compare four groups of dog kidney weights, expressed as percentage of body weights, following an inhalation study. Assuming homogeneity of variance (from Bartlett's test), we could complete the following calculations:

| | 400 ppm | 200 ppm | 100 ppm | 0 ppm |
|---|---|---|---|---|
| | 0.43 | 0.49 | 0.34 | 0.34 |
| | 0.52 | 0.48 | 0.40 | 0.32 |
| | 0.43 | 0.40 | 0.42 | 0.33 |
| | 0.55 | 0.34 | 0.40 | 0.39 |
| $\Sigma X$ | 1.93 | 1.71 | 1.56 | 1.38 |

$\Sigma\Sigma X = 1.93 + 1.71 + 1.56 + 1.38 = 6.58$

Next, these figures are squared:

| | 400 ppm | 200 ppm | 100 ppm | 0 ppm |
|---|---|---|---|---|
| | 0.1849 | 0.2401 | 0.1156 | 0.1156 |
| | 0.2704 | 0.2304 | 0.1600 | 0.1024 |
| | 0.1849 | 0.1600 | 0.1764 | 0.1089 |
| | 0.3025 | 0.1156 | 0.1600 | 0.1521 |
| $\Sigma X^2$ | 0.9427 | 0.7461 | 0.6120 | 0.4790 |

$\Sigma\Sigma X^2 = 0.9427 + 0.7461 + 0.6120 + 0.4790 = 2.7798$

$$CF = \frac{(6.58)^2}{4+4+4+4} = 2.7060$$

$$SS_{total} = 2.7798 - 2.7060 = 0.0738$$

$$SS_{bg} = \frac{(1.93)^2}{4} + \frac{(1.71)^2}{4} + \frac{(1.56)^2}{4} + \frac{(1.38)^2}{4} - 2.7060 = 0.04075$$

$$SS_{wg} = 0.07380 - 0.04075 = 0.03305$$

The total degrees of freedom (df) = 4 + 4 + 4 + 4 − 1 = 15.

$$\text{df}_{bg} = 4 - 1 = 3$$

$$\text{df}_{wg} = 15 - 3 = 12$$

$$MS_{bg} = \frac{0.4075}{3} = 0.01358$$

$$MS_{wg} = \frac{0.03305}{12} = 0.00275$$

$$F = \frac{0.01358}{0.00275} = 4.94$$

Going to a table of $F$ values, we find the 3 $\text{df}_{bg}$ (greater mean square) and 12 $\text{df}_{wg}$ (lesser mean square), and the 0.05 value of $F$ is 3.49. Because our calculated value is greater, there is a difference among groups at the 0.05 probability level. To determine where the difference is further comparisons by a *post hoc* test will be necessary

### Assumptions and Limitations

1. What is presented here is the workhorse of toxicology—the one-way analysis of variance. Many other forms exist for more complicated experimental designs.
2. The test is robust for moderate departures from normality if the sample sizes are large enough; unfortunately, this is rarely the case in toxicology.

3. ANOVA is robust for moderate departures from equality of variances (as determined by Bartlett's test) if the sample sizes are approximately equal.
4. It is not appropriate to use a *t*-test (or a two groups at a time version of ANOVA) to identify where significant differences are within the design group. A multiple-comparison *post hoc* method must be used.

## Post Hoc Tests

There is a wide variety of *post hoc* tests available to analyze data after finding significant results in an ANOVA. Each of these tests has advantages and disadvantages, proponents and critics. Four of the tests are commonly used in toxicology and will be presented or previewed here. These are Dunnett's *t*-test and Williams' *t*-test. Two other tests that are available in many statistical packages are Turkey's method and the Student–Newman–Keuls method [25]. If ANOVA reveals no significance it is not appropriate to proceed to perform a *post hoc* test in hopes of finding differences. To do so would only be another form of multiple comparisons, increasing the type I error rate beyond the desired level.

### Duncan's Multiple Range Test

Duncan's test [35] is used to compare groups of continuous and randomly distributed data (such as body weights or organ weights). The test normally involves three or more groups taken one pair at a time. It should only follow observation of a significant $F$ value in the ANOVA and can serve to determine which group (or groups) differs significantly from which other group (or groups). There are two alternative methods of calculation. The selection of the proper one is based on whether the number of data ($N$) are equal or unequal in the groups.

### Groups with Equal Numbers of Data ($N_1 = N_2$)

Two sets of calculations must be carried out: (1) the determination of the difference between the means of pairs of groups, and (2) the preparation of a probability rate against which each difference in means is compared (as shown in the first of the two examples in this section). The means (averages) are determined (or taken from the ANOVA calculation) and ranked in either decreasing or increasing order. If two means are the same, they take up two equal positions (thus, for four means we could have ranks of 1, 2, 2, and 4 rather than 1, 2, 3, and 4). The groups are then taken in pairs, and the differences between the means ($\bar{X}_1 - \bar{X}_2$), expressed as positive numbers, are calculated. Usually, each pair consists of a test group and the control group, although multiple tests groups may be compared if so desired. The relative rank of the two groups being compared must be considered. If a test group is ranked 2 and the control group is ranked 1, then we say that there are two places between them; if the test group is ranked 3, then there would be three places between it and the control.

To establish the probability table, the standard error of the mean (SEM) must be calculated as presented earlier or as:

$$\sqrt{\frac{\text{error mean square}}{N}} = \sqrt{\frac{\text{mean square within group}}{N}}$$

where $N$ is the number of animals or replications per dose level. The mean square within groups ($MS_{wg}$) can be calculated from the information given in the ANOVA procedure (refer to the earlier section on ANOVA). The SEM is then multiplied by a series of table values [36,37] to set up a probability table. The table values used for the calculations are chosen according to the probability levels (note that the tables have sections for 0.05, 0.01, and 0.001 levels) and the number of means apart for the groups being compared and the number of error degrees of freedom. The error df is the number of degrees of freedom within the groups. This last figure is determined from the ANOVA calculation and can be taken from ANOVA output. For some values of df, the table values are not given and should thus be interpolated. Example 10 demonstrates this case.

#### Example 10

Using the data given in Example 9 (4 groups of dogs, with 4 dogs in each group), we can make the following calculations:

| | Rank 1 | 2 | 3 | 4 |
|---|---|---|---|---|
| Concentration (ppm) | 0 | 100 | 200 | 400 |
| Mean kidney weight ($\bar{X}$) | 0.345 | 0.390 | 0.428 | 0.482 |

| Groups Compared | $(\bar{X}_1 - \bar{X}_2)$ | No. of Means Apart | Probability |
|---|---|---|---|
| 2 vs. 1 (100 vs. 0 ppm) | 0.045 | 2 | $p > 0.05$ |
| 3 vs. 1 (200 vs. 0 ppm) | 0.083 | 3 | $p > 0.05$ |
| 4 vs. 1 (400 vs. 0 ppm) | 0.137 | 4 | $0.01 > p > 0.001$ |
| 4 vs. 2 (400 vs. 100 ppm) | 0.092 | 3 | $0.05 > p > 0.01$ |

The mean square within groups from the ANOVA example was 0.00275; therefore, the SEM = $\sqrt{0.00275} = 0.02622$. The error df ($df_{wg}$) was 12, so the following table values are used:

| No. of Means Apart | Probability Levels 0.05 | 0.01 | 0.001 |
|---|---|---|---|
| 2 | 3.082 | 4.320 | 6.106 |
| 3 | 3.225 | 4.504 | 6.340 |
| 4 | 3.313 | 4.662 | 6.494 |

When these are multiplied by the SEM we get the following probability table:

| No. of Means Apart | Probability Levels 0.05 | 0.01 | 0.001 |
|---|---|---|---|
| 2 | 0.0808 | 0.1133 | 0.1601 |
| 3 | 0.0846 | 0.1161 | 0.1662 |
| 4 | 0.0869 | 0.1212 | 0.1703 |

## Groups with Unequal Numbers of Data ($N_1 \neq N_2$)

This procedure is very similar to that discussed above. As before, the means are ranked and the differences between the means are determined ($\bar{X}_1 - \bar{X}_2$). Next, weighting values ($a_{ij}$ values) are calculated for the pairs of groups being compared in accordance with:

$$a_u = \sqrt{\frac{2N_iN_j}{(N_i+N_j)}} = \sqrt{\frac{2N_1N_2}{(N_1+N_2)}}$$

This weighting value for each pair of groups is multiplied by ($\bar{X}_1 - \bar{X}_2$) for each value to arrive at a $t$ value. This is the $t$ that will later be compared to a probability table.

The probability table is set up as earlier except that instead of multiplying the appropriate table values by SEM, SEM² is used. This is equal to $\sqrt{MS_{wg}}$.

For the desired comparison of two groups at a time, either ($\bar{X}_1 - \bar{X}_2$) value (if $N_1 = N_2$) is compared to the appropriate probability table. Each comparison must be made according to the number of places between the means. If the table value is larger at the 0.05 level, the two groups are not considered to statistically different. If the table value is smaller, the groups are different and the comparison is repeated at lower levels of significance; thus, the degree of significance may be determined. We might have significant differences at 0.05 but not at 0.01, in which case the probability would be represented at $0.05 > p > 0.01$. Example 11 demonstrates this case.

### Example 11

Suppose that the 400 ppm level from the above example had only 3 dogs, but that the mean for the group and the mean square within groups were the same. To continue Duncan's we would calculate the weighting factors as follows:

100 ppm vs. 0 ppm,

200 ppm vs. 0 ppm $N_1 = 4$; $N_2 = 4a_r = \sqrt{\frac{2(4)(4)}{4+4}} = 2.00$,

400 ppm vs. 0 ppm $N_2 = 3$; $N_4 = 4a_r = \sqrt{\frac{2(3)(4)}{3+4}} = 1.852$,

400 ppm vs. 100 ppm

Using the ($\bar{X}_1 - \bar{X}_2$) from the above example we can set up the following tables:

| Concentration (ppm) | No. of Means Apart | ($\bar{X}_1 - \bar{X}_2$) | $a_u$ | ($\bar{X}_1 - \bar{X}_2)a_u$ |
|---|---|---|---|---|
| 100 vs. 0 | 2 | 0.045 | 2.000 | 2.000(.045) = .090 |
| 200 vs. 0 | 3 | 0.083 | 2.000 | 2.000(.083) = .166 |
| 400 vs. 0 | 4 | 0.137 | 1.852 | 1.852(.137) = .254 |
| 400 vs. 100 | 3 | 0.092 | 1.852 | 1.852(.092) = .170 |

Next we calculate SEM² as being $0.00275 = 0.05244$. This is multiplied by the appropriate table values chosen for 11 degrees of freedom (df$_{wg}$ for this example). This gives the following probability table:

| No. of Means Apart | Probability Levels 0.05 | 0.01 | 0.001 |
|---|---|---|---|
| 2 | 0.1632 | 0.2303 | 0.3291 |
| 3 | 0.1707 | 0.2401 | 0.3417 |
| 4 | 0.1753 | 0.2463 | 0.3501 |

Comparing the $t$ values with the probability table values we get the following:

| Comparison | Probability |
|---|---|
| 100 ppm vs. 0 ppm | $p > 0.05$ |
| 200 ppm vs. 0 ppm | $p > 0.05$ |
| 400 ppm vs. 0 ppm | $0.01 > p > 0.001$ |
| 400 ppm vs. 100m | $0.05 > p > 0.01$ |

## Assumptions and Limitations

1. Duncan's assures a set alpha level or type I error rate for all tests when means are separated by no more than ordered step increases. Preserving this alpha level means that the test is less sensitive than some others, such as the Student–Newman–Keuls. The test is inherently conservative and not resistant or robust.

## Scheffe's Multiple Comparisons

Scheffe's is another *post hoc* comparison method fo groups of continuous and randomly distributed data. I also normally involved three or more groups [38,39]. It i widely considered a more powerful significance test tha Duncan's. Each *post hoc* comparison is tested by compar ing an obtained test value ($F_{contr}$) with the appropriat critical $F$ value at the selected level of significance—th table $F$ value multiplied by ($K$ – 1) for an $F$ with ($K$ – 1 and ($N$ – $K$) degrees of freedom, where $K$ is the numbe of groups being compared. $F_{contr}$ is computed as follows

a. Compute the mean for each sample (group).
b. Denote the residual mean square by $MS_{wg}$.
c. Compute the test statistic as:

$$F_{contr} = \frac{(C_1\bar{X}_1 + C_2\bar{X}_2 + \cdots + C_k\bar{X}_k^{12})}{(K-1)MS_{wg}(C_1^2/n_1 + \cdots + C_k^2/n_k)}$$

where $C_k$ is the comparison number such that the sum of $C_1, C_2, \ldots, C_k = 0$. This is demonstrated in Example 12.

### Example 12

At the end of a short-term feeding study, the following body weight changes were recorded:

| | Group 1 | Group 2 | Group 3 |
|---|---|---|---|
| | 10.2 | 12.2 | 9.2 |
| | 8.2 | 10.6 | 10.5 |
| | 8.9 | 9.9 | 9.2 |
| | 8.0 | 13.0 | 8.7 |
| | 8.3 | 8.1 | 9.0 |
| | 8.0 | 10.8 | |
| | | 11.5 | |
| Totals | 51.6 | 76.1 | 46.6 |
| Means | 8.60 | 10.87 | 9.32 |

$MS_{wg} = 1.395$

To avoid logical inconsistencies with pairwise comparisons, we compare the group having the largest sample mean (group 2) with that having the smallest sample mean (group 1), then with the group having the next smallest sample mean, and so on. As soon as we find a nonsignificant comparison in this process (or no group with a smaller sample mean remains), we replace the group having the largest sample mean with that having the second largest sample mean and repeat the comparison process.

Accordingly, our first comparison is between groups 2 and 1. We set $C_1 = -1$, $C_2 = 1$, and $C_3 = 0$ and calculate our test statistic:

$$F_{contr} = \frac{(10.87 - 8.60)^2}{(3-1)(1.395)(1/6 + 1/7)} = 5.97$$

The critical region for $F$ at $p \leq 0.05$ for 2 and 11 degrees of freedom is 3.98; therefore, these groups are significantly different at this level. We next compare groups 2 and 3 using $C_1 = 0$, $C_2 = 1$, and $C_3 = -1$:

$$F_{contr} = \frac{(10.87 - 9.32)^2}{(3-1)(1.395)(1/7 + 1/5)} = 2.51$$

This is less than the critical region value, so these groups are not significantly different.

### Assumptions and Limitations

1. The Scheffe procedure is robust to moderate violations of the normality and homogeneity of variance assumptions.
2. It is not formulated on the basis of groups with equal numbers (as one of Duncan's procedures is), and if $N_1 \neq N_2$ there is no separate weighting procedure.
3. It tests all linear contrasts among the population means (the other three methods confine themselves to pairwise comparison, except they use a Bonferroni-type correlation procedure).
4. The Scheffe procedure is powerful because of it robustness, yet it is very conservative. The type I error (the false-positive rate) is held constant at the selected test level for each comparison.

## DUNNETT'S *t*-TEST

Dunnett's *t*-test [40,41] has as its starting point the assumption that what is desired is a comparison of each of several means with one other mean and only one other mean; in other words, that one wishes to compare each and every treatment group with the control group but not compare treatment groups with each other. The problem here is that, in toxicology, one is frequently interested in comparing treatment groups with other treatment groups. However, if one does want only to compare treatment groups vs. a control group, then Dunnett's is a useful approach. In a study with $K$ groups (one of them being the control), we will wish to make $(K - 1)$ comparisons. In such a situation, we want to have a $P$ level for the entire set of $(K - 1)$ decisions (not for each individual decision). The Dunnett's distribution is predicated on this assumption. The parameters for utilizing a Dunnett's table, such as found in his original article, are $K$ (as above) and the number of degrees of freedom for the mean square within groups ($MS_{wg}$). The test value is calculated as:

$$t = \frac{|T_j - T_i|}{\sqrt{2MS_{wg/n}}}$$

where $n$ is the number of observations in each of the groups; the mean square within group ($MS_{wg}$) value is as we have defined it previously; $T_j$ is the control group mean; and $T_i$ is the mean of, in order, each successive test group observation. Note that one uses the absolute value of the positive number resulting from subtracting $T_i$ from $T_j$. This is to ensure a positive number for our final $t$. Example 13 demonstrates this test, again with the data from Example 9.

### Example 13

Assume that the means, $N$ values, and sums for the groups previously presented in Example 3 are:

| | Control | 100 ppm | 200 ppm | 400 ppm |
|---|---|---|---|---|
| Sum ($\Sigma X$) | 1.38 | 1.56 | 1.71 | 1.93 |
| $N$ | 4 | 4 | 4 | 4 |
| Mean | 0.345 | 0.39 | 0.4275 | 0.4825 |

The $MS_{wg}$ was 0.00275, and our test $t$ for 4 groups and 12 degrees of freedom is 2.41. Substituting in the equation, we calculate our $t$ for the control vs. the 400 ppm to be:

$$= \frac{|0.345 - 0.4825|}{\sqrt{2(0.00275)/4}} = \frac{0.1375}{\sqrt{0.001375}} = \frac{0.1375}{0.037081} = 3.708$$

which exceeds our test value of 2.41, showing that these two groups are significantly different at $p \leq 0.05$. The values for the comparisons of the control vs. the 200 and 100 ppm groups are then found to be 2.225 and 1.214, respectively. Both of these are less than our test value; therefore, the groups are not significantly different.

### Assumptions and Limitations

1. Dunnett's seeks to ensure that the type 1 error rate will be fixed at the desired level by incorporating correction factors into the design of the test value table.
2. Treated group sizes must be approximately equal.

## Williams' t-Test

Williams' $t$-test [42,43] is popular, although its use is quite limited in toxicology. It is designed to detect the highest level (in a set of dose/exposure levels) at which there is no significant effect. It assumes that the response of interest (such as change in body weights) occurs at higher levels but not at lower levels and that the responses are monotonically ordered so $X_0 \leq X_1 \ldots \leq X_k$. This is, however, frequently not the case. The Williams technique handles the occurrence of such discontinuities in a response series by replacing the offending value and the value immediately preceding it with weighted average values. The test also is adversely affected by any mortality at high dose levels. Such moralities "impose a severe penalty, reducing the power of detecting an effect not only at level $K$ but also at all lower doses" [43, p. 529]. Accordingly, it is not generally applicable in toxicology studies.

## Analysis of Covariance

Analysis of covariance (ANCOVA) is a method for comparing sets of data that consist of two variables (treatment and effect, with our effect variable being the *variate*), when a third variable (the *covariate*) exists that can be measured but not controlled and which has a definite effect on the variable of interest. In other words, it provides an indirect type of statistical control, allowing us to increase the precision of a study and to remove a potential source of bias. One common example of this is in the analysis of organ weights in toxicity studies. Our true interest here is the effect of our dose or exposure level on the specific organ weights, but most organ weights also increase (in the young, growing animals most commonly used in such studies) in proportion to increases in animal body weight. As we are not here interested in the effect of this covariate (body weight), we measure it to allow for adjustment. We must be careful before using ANCOVA, however, to ensure that the underlying nature of the correspondence between the variate and covariate is such that we can rely on it as a tool for adjustments [44,45].

Calculation is performed in two steps. The first is a type of linear regression between the variate $Y$ and the covariate $X$. This regression, performed as described under the linear regression section, gives us the model:

$$Y = a_1 + BX + e$$

which in turn allows us to define adjusted means ($\bar{Y}$ and $\bar{X}$) such that $\bar{Y}_{1a} = \bar{Y}_1 - (\bar{X}_1 - X^*)$.

If we consider the case where $K$ treatments are being compared such that $K = 1, 2, \ldots, K$ and we let $X_{ik}$ and $Y_{ik}$ represent the predictor and predicted values for each individual $i$ in group $k$, we can let $X_k$ and $Y_k$ be the means. Then, we define the between-group (for treatment) sum of squares and cross products as:

$$T_{xx} = \sum_{k-1}^{K} n_k \left(\bar{X}_K - \bar{X}\right)^2$$

$$T_{yy} = \sum_{k-1}^{K} n_k \left(\bar{Y}_K - \bar{Y}\right)^2$$

$$T_{xy} = \sum_{k-1}^{K} n_k \left(\bar{X}_K - \bar{X}\right)\left(\bar{Y}_K - \bar{Y}\right)$$

In a like manner, within-group sums of squares and cross products are calculated as:

$$\sum xx = \sum_{k=1}^{K} \sum_{i} \left(X_{ik} - X_k\right)^2$$

$$\sum yy = \sum_{k=1}^{K} \sum_{i} \left(Y_{ik} - Y_k\right)^2$$

$$\sum xy = \sum_{k=1}^{K} \sum_{i} \left(X_{ik} - X_k\right)\left(Y_{ik} - Y_k\right)$$

where $i$ indicates the sum from all the individuals within each group, and $f$ = total number of subjects minus number of groups. Also,

$$S_{xx} = T_{xx} + \sum xx$$

$$S_{yy} = T_{yy} + \sum yy$$

$$S_{xy} = T_{xy} + \sum xy$$

With these in hand, we can then calculate the residual mean squares of treatments ($St^2$) and error ($Se^2$):

$$St^2 = \frac{Tyy - \frac{S^2xy}{Sxx} + \frac{\sum^2 xy}{\sum xx}}{lc - 1}$$

$$Se^2 = \frac{\left(\sum yy - \frac{\sum^2 y}{\sum xx}\right)}{f - 1}$$

These can be used to calculate an $F$ statistic to test the null hypothesis that all treatment effects are equal:

$$F = \frac{St^2}{Se^2}$$

The estimated regression coefficient of Y or X is:

$$B = \frac{\sum xy}{\sum xx}$$

The estimated standard error for the adjusted difference between two groups is given by:

$$Sd = Se \frac{1}{n_j} + \frac{1}{n_i} + \frac{\left(X_i - X_j\right)}{\sum xx}$$

where $n_i$ and $n_j$ are the sample sizes of the two groups. A test of the null hypothesis that the adjusted differences between the groups is zero is provided by:

$$t = \frac{Y_1 - Y_0 - B\left(X_1 - X_0\right)}{Sd}$$

The test value for $t$ is then obtained from the $t$-table with $f - 1$ degrees of freedom. Computation is markedly simplified if all the groups are of equal size, as demonstrated in Example 14.

### Example 14

An ionophere was evaluated as a potential blood-pressure-reducing agent. Early studies indicated that there was an adverse effect on blood cholesterol and hemoglobin levels, so a special study was performed to evaluate this specific effect. The hemoglobin (Hgb) level covariate was measured at the start of the study along with the percent changes in serum triglycerides between the start of the study and at the end of the 13-week study. Was there a difference in effects of the two ionopheres?

| Ionophere A | | Ionophere B | |
|---|---|---|---|
| Hgb | Serum Triglyceride (% Change) | Hgb | Serum Triglyceride (% Change) |
| $x$ | $y$ | $x$ | $y$ |
| 7.0 | 5 | 5.1 | 10 |
| 6.0 | 10 | 6.0 | 15 |
| 7.1 | −5 | 7.2 | −15 |
| 8.6 | −20 | 6.4 | 5 |
| 6.3 | 0 | 5.5 | 10 |
| 7.5 | −15 | 6.0 | −15 |
| 6.6 | 10 | 5.6 | −5 |
| 7.4 | −10 | 5.5 | −10 |
| 5.3 | 20 | 6.7 | −20 |
| 6.5 | −15 | 8.6 | −40 |
| 6.2 | 5 | 6.4 | −5 |
| 7.8 | 0 | 6.0 | −10 |
| 8.5 | −40 | 9.3 | −40 |
| 9.2 | −25 | 8.5 | −20 |
| 5.0 | 25 | 7.9 | −35 |
| | | 5.0 | 0 |
| | | 6.5 | −10 |

To apply ANCOVA using Hgb as a covariate, we first obtain some summary results from the data as follows:

| | Ionophere A (Group 1) | Ionophere B (Group 2) | Combined |
|---|---|---|---|
| $\Sigma x$ | 112.00 | 119.60 | 231.60 |
| $\Sigma x^2$ | 804.14 | 821.64 | 1625.78 |
| $\Sigma y$ | −65.00 | −185.00 | −250.00 |
| $\Sigma y^2$ | 4575.00 | 6475.00 | 11050.00 |
| $\Sigma xy$ | −708.50 | −1506.50 | −2215.00 |
| $\bar{x}$ | 7.000 | 6.6444 | 6.8118 |
| $\bar{y}$ | −4.625 | −10.2778 | −7.3529 |
| $n$ | 16 | 18 | 34 |

We compute for the ionophere A group ($i = 1$):

$$S_{xx(1)} = 804.14 - (112)^2/16 = 20.140$$
$$S_{yy(1)} = 4575.00 - (-65)^2/16 = 4310.938$$
$$S_{xy(1)} = -708.50 - (112)(-65)/16 = -253.500$$

Similarly for the ionophere B group ($i = 2$), we obtain:

$$S_{xx(2)} = 26.964$$
$$S_{yy(2)} = 4573.611$$
$$S_{xy(2)} = -277.278$$

Finally, for the combined data (ignoring groups), we compute:

$$S_{xx} = 48.175$$
$$S_{yy} = 9211.765$$
$$S_{xy} = -512.059$$

The sums of squares can now be obtained as:

$$TOT(SS) = 9211.8$$

$$SSE = \frac{(20.140 + 26.964)(4310.938 + 4573.611) - [(-253.500) - 277.28]^2}{(20.140 + 26.964)} = 2903.6$$

$$SSG = \frac{(48.175)(9211.765) - (-512.059)^2}{48.175} - 2903.6 = 865.4$$

$$SSC = (4310.938 + 473.611) - 2903.6 = 5980.9$$

and the ANCOVA summary table can be completed as follows:

| Source | df | SS | MS | F |
|---|---|---|---|---|
| Treatment | 1 | 865.4 | 865.4 | 9.2[a] |
| X (Hgb) | 1 | 5980.9 | 5980.9 | 63.8 |
| Error | 31 | 2903.7 | 93.7 | |
| Total | 33 | 9211.8 | | |

[a] Significant ($p < 0.05$); critical $F$ value = 4.16.

The $F$ statistics are formed as the ratios of effect mean squares (MS) to the MSE (93.7). Each $F$ statistic is compared with the critical $F$ value with 1 upper and 31 lower degrees of freedom. The critical $F$ value for $\alpha = 0.05$ is 4.16. The significant covariate effect ($F = 63.8$) indicates that the triglyceride response has a significant linear relationship with Hgb. The significant $F$ value for treatment indicates that the mean triglyceride response adjusted for hemoglobin effect differs between treatment groups.

### Assumptions and Limitations

1. The underlying assumptions for ANCOVA are fairly rigid and restrictive. The assumptions include the following:
   a. The slopes of the regression lines of a $Y$ and $X$ are equal from group to group. This can be examined visually or formally (i.e., by a test). If this condition is not met, ANCOVA cannot be used.
   b. The relationship between $X$ and $y$ is linear.
   c. The covariate $X$ is measured without error. Power of the test declines as error increases.
   d. There are no unmeasured confounding variables.
   e. The errors inherent in each variable are independent of each other. Lack of independence effectively (but to an immeasurable degree) reduces sample size.
   f. The variances of the errors in groups are equivalent between groups.
   g. The measured data that form the groups are normally distributed. ANCOVA is generally robust to departures from normality.
2. Of the seven assumptions above, the most serious are the first four.

## MODELING

The mathematical modeling of biological systems, restricted even to the field of toxicology, is an extremely large and vigorously growing area. Broadly speaking, modeling is the principal conceptual tool by which toxicology seeks to develop as a mechanistic science. In an iterative process, models are developed or proposed, tested by experiment, reformulated, and so on in a continuous cycle. Such a cycle could also be described as two related types of modeling: *explanatory* (where the concept is formed) and *correlative* (where data are organized and relationships derived). An excellent introduction to the broader field of modeling of biological systems can be found in Gold [46].

In toxicology, modeling is of prime interest in seeking to relate a treatment variable with an effect variable and, from the resulting model, predict effects at exact points where no experiment has been done (but in the range where we have performed experiments, such as determining $LD_{50}$ values), to estimate how good our prediction is, and, occasionally, simply to determine if a pattern of effects is related to a pattern of treatment.

For use in prediction, the techniques of linear regression, probit/logit analysis (a special case of linear regression), moving averages (an efficient approximation method), and nonlinear regression (for doses where data cannot be made to fit a linear pattern) are presented. For evaluating the predictive value of these models, both the correlation coefficient (for parametric data) and Kendall's rank correlation (for nonparametric data) are given. And, finally, the concept of trend analysis is introduced and a method presented.

When we are trying to establish a pattern between several data points (whether this pattern is in the form of a line or a curve), what we are doing is *interpolating*. It is possible for any given set of points to produce an infinite set of lines or curves that pass near (for lines) or through (for curves) the data points. In most cases, we cannot actually know the real pattern, so we apply a basic principle of science: Occam's razor. We use the simplest explanation (or, in this case, model) that fits the facts (or data). A line is, of course, the simplest pattern to deal with and describe, so fitting the best line (linear regression) is the most common form of model in toxicology.

### Linear Regression

Foremost among the methods for interpolating within a known data relationship is regression, which involves the fitting of a line or curve to a set of known data points on a graph and interpolating (estimating) this line or curve in areas where we have no data points. The simplest of these regression models is that of linear regression (which is valid when increasing the value of one variable changes the value of the related variable in a linear fashion, either positively or negatively). This is the case we will explore here, using the method of least squares.

Given that we have two sets of variables, $x$ (say, mg/kg of test material administered) and $y$ (say, percentage of animals so dosed that die), it is necessary to solve for $a$ and $b$ in the equation $Y_i = a + bx_i$, where the uppercase $Y_i$ is the fitted value of $y_i$ at $x_i$ and we wish to minimize $(y_i - Y_i)^2$. So, we solve the equations:

$$a = \bar{y} - b\bar{x}$$

$$b = \frac{\sum x_1 y_1 \quad n\bar{x}\bar{y}}{\sum x_1^2 - n\bar{x}^2}$$

where $a$ is the $y$ intercept, $b$ is the slope of the time, and $n$ is the number of data points. Use of this is demonstrated in Example 13.

Note that, in actuality, dose–response relationships are often not linear and instead we must use either a transform (to linearize the data) or a nonlinear regression method [47]. Note also that we can use the correlation test statistic (to be described in the correlation coefficient section) to determine if the regression is significant (and, therefore, valid at a defined level of certainty. A more specific test for significance would be the linear regression analysis of variance [31]. To do so we start by developing the appropriate ANOVA table and then proceed to perform the linear regression portion of the ANOVA as shown in Example 15.

### Example 15

From a short-term toxicity study we have the following results:

| Dose Administered (mg/kg) $x_1$ | $x_1^2$ | Percent Animals Dead (%) $y_1$ | $x_1y_1$ |
|---|---|---|---|
| 1 | 1 | 10 | 10 |
| 3 | 9 | 20 | 60 |
| 4 | 16 | 18 | 72 |
| 5 | 25 | 20 | 100 |
| Sums $x_1$ = 13 | $x_1^2$ = 51 | $y_1$ = 68 | $x_1y_1$ – 242 |

$$\bar{x} = 3.25 \quad \text{and} \quad \bar{y} = 17$$

$$a = 17 - (2.4)(3.25) = 9.20$$

$$b = \frac{242-(4)(3.25)(17)}{51-(4)(10.5625)} = \frac{21}{8.75} = 2.40$$

We therefore see that our fitted line is $Y = 9.2 + 2.4X$. These ANOVA table data are then used as shown in Example 12.

**Linear Regression Analysis of Variance**

| Source of Variation (1) | Sum of Squares (2) | Degrees of Freedom (3) | Mean Square (=2/3) (4) |
|---|---|---|---|
| Regression | $b_1^2\left(\sum x_1^2 - n\bar{x}^2\right)$ | 1 | By division |
| Residual | By difference | $n-2$ | By division |
| Total | $\sum y_1^2 - n\bar{y}^2$ | $n-1$ | |

We then calculate $F_{1,n-2}$ = (regression mean square)/(residual mean square). This is demonstrated in Example 16.

### Example 16

We desire to test the significance of the regression line in Example 11:

$$\sum y_1^2 = 10^2 + 20^2 + 18^2 + 20^2$$

$$\text{Regression } SS = (2.4)^2\left[51 - 4(3.25)^2\right] = 50.4$$

$$\text{Total } SS = 1224 - 4(17^2) = 68.0$$

$$\text{Residual } SS = 68.0 - 50.4 = 17.6$$

$$F_{1,2} = 50.4/8.8 = 5.73$$

This value is not significant at the 0.05 level; therefore, the regression is not significant. A significant $F$ value (as found in an $F$ distribution table for the appropriate degrees of freedom) indicates that the regression line is an accurate prediction of observed values at that confidence level. Note that the portion of the total sum of squares explained by the regression is called the *coefficient of correlation*, which in the above example is equal to $0.86^2$ (or 0.74). Calculation or the correlation coefficient is described later in this chapter.

Finally, we might wish to determine the confidence intervals for our regression line; that is, given a regression line with calculated values for $Y_i$ given $x_i$, within what limits may we be certain (with, say, a 95% probability) what the real value of $Y_i$ is? If we denote the residual mean square in the ANOVA by $s^2$, then the 95% confidence limits for $a$ (denoted by $A$, the notation for the true, as opposed to estimated, value for this parameter) are calculated as:

$$t_{n-2} = \frac{a-A}{\sqrt{\dfrac{s^2\left(\sum x^2\right)}{n\sum x_1^2 - n^2\bar{x}^2}}}$$

$$= \frac{9.2-A}{\sqrt{\dfrac{8.8(51)}{4(51)-(16)(10.562)}}} = \frac{9.2-A}{\sqrt{\dfrac{448}{35.008}}}$$

$$9.2 - A = -15.405$$

$$A = 9.2 - 15.405$$

### Assumptions and Limitations

1. All the regression methods are for interpolation, not extrapolation; that is, they are valid only in the range for which we have data—the experimental region—not beyond.
2. The method assumes that the data are independent and normally distributed, and it is sensitive to outliers. The *x*-axis (or horizontal) component plays an extremely important part in developing the least-squares fit. All points have equal weight in determining the height of a regression line, but extreme *x*-axis values unduly influence the slope of the line.
3. A good fit between a line and a set of data (that is, a strong correlation between treatment and response variables) does not imply any causal relationship.
4. It is assumed that the treatment variable can be measured without error, that each data point is independent, that variances are equivalent, and that a linear relationship does not exist between the variables.
5. The many excellent texts on regression, which is a powerful technique, include Draper and Smith [48] and Montgomery et al. [49], which are not overly rigorous mathematically.

## Probit/Log Transforms and Regression

As we noted in the preceding section, dose–response problems (among the most common interpolation problems encountered in toxicology) rarely are straightforward enough to make a valid linear regression directly from the raw data. The most common valid interpolation methods are based on probability (*probit*) and logarithmic (*log*) value scales, with percentage responses (death, tumor incidence, etc.) being expressed on the probit scale and doses ($Y_i$) expressed on the log scale. There are two strategies for such an approach. The first is based on transforming the data to these scales, then doing a weighted linear regression on the transformed data. (If one does not have access to a computer or a high-powered programmable calculator, the only practical strategy is not to assign weights.) The second requires the use of algorithms (approximate calculation techniques) for the probit value and regression process and is extremely burdensome to perform manually.

Our approach to the first strategy requires that we construct a table with the pairs of values of $x_i$ and $y_i$ listed in order of increasing values of $Y_i$ (percentage response). Beside each of these columns a set of blank columns should be left so the transformed values may be listed. We then simply add the columns described in the linear regression procedure. Log and probit values may be taken from any of a number of sets of tables and the rest of the table is then developed from these transformed $x_i$ and $y_i$ values (denoted as $x_i'$ and $y_i'$). A standard linear regression is then performed (see Example 17). The second strategy we discussed has been broached by a number of authors [50–53]. All of these methods, however, are computationally cumbersome. It is possible to approximate the necessary iterative process using the algorithms developed by Abramowitz and Stegun [54], but even this merely reduces the complexity to a point where the procedure may be readily programmed on a small computer or programmable calculator.

### Example 17

(See table below.) Our interpolated log of the $LD_{50}$ is 1.000539, calculated by using $Y = -0.200591 - 0.240226x$, where $x$ equals 5.000 (the probit of 50%) in the regression equation. When we convert this log value to its linear equivalent, we get an $LD_{50}$ of 10.0 mg/kg. Finally, our calculated correlation coefficient is $r = 0.997$. A goodness-of-fit of the data using chi-square may also be calculated.

### Assumptions and Limitations

1. The probit distribution is derived from a common error function, with the mid-point (50% point) moved to a score of 5.00.
2. The underlying frequency distribution becomes asymptotic as it approaches the extremes of the range; that is, in the range of 16 to 84% the corresponding probit values change gradually. The curve is relatively linear, but beyond this range the values change ever more rapidly as they approach either 0% or 100%. In fact, there are no values for either of these numbers.
3. A normally distributed population is assumed, and the results are sensitive to outliers.

| Percentage of Animals Killed ($x_1$) | Probit of $x_1 = x_1'$ | Dose of Chemical (mg/kg) ($y_1$) | Log of $y_1 = y_1'$ | $(x_1')^2$ | $x'_1y'_1$ |
|---|---|---|---|---|---|
| 2 | 2.9463 | 3 | 0.4771 | 8.6806 | 1.40568 |
| 10 | 3.7184 | 5 | 0.6990 | 13.8264 | 2.59916 |
| 42 | 4.7981 | 10 | 1.0000 | 23.0217 | 4.79810 |
| 90 | 6.2816 | 20 | 1.3010 | 39.4585 | 8.17223 |
| 98 | 7.2537 | 30 | 1.4771 | 52.6162 | 10.4190 |
| | $\Sigma x_1' = 24.9981$ | | $\Sigma y_1' = 4.9542$ | $\Sigma(x_1')^2 = 137.6034$ | $\Sigma x_1'y_1' = 27.68974$ |

## Moving Averages

An obvious drawback to the interpolation procedures we have examined to date is that they do take a significant amount of time (although they are simple enough to be done manually), especially if the only result we desire is an $LD_{50}$, $LC_{50}$, or $LT_{50}$. The method of moving averages [55,56] gives a rapid and reasonable accurate estimate of this median effective dose ($m$) and the estimated standard deviation of its logarithm. Such methodology requires that the same number of animals be used per dosage level and that the spacing between successive dosage exposure levels be geometrically constant (e.g., levels of 1, 2, 4, and 8 mg/kg or 1, 3, 9, and 27 ppm). Given this and access to a table for the computation of moving averages one can readily calculate the median effective dose with the formula (illustrated for dose):

$$\log m = \log D + d(K - 1)/2 + df$$

where $m$ is the median effective dose or exposure; $D$ is the lowest dose tested; $d$ is the log of the ratio of successive doses/exposures; and $f$ is a table value taken from Gad [23] for the proper $K$ (the total number of levels tested minus 1). Example 18 demonstrates the use of the this method and the new tables.

### Example 18

As part of an inhalation study we exposed 4 groups of 5 rats each to levels of 20, 40, 80, and 160 ppm of a chemical vapor. These exposures killed 0, 1, 3, and 5 animals, respectively. From the $N = 5$, $K = 3$ tables on the $r$ value 0, 1, 3, 5 line, we get an $f$ of 0.7 and an $\alpha_f^4$ of 0.31623. We can then calculate the $LC_{50}$ to be:

$$\begin{aligned} \text{Log}LC_{50} &= 1.30130 + 0.30103(2)/2 + 0.30103(0.7) \\ &= 1.30103 + 0.51175 \\ &= 1.81278 \\ \therefore LC_{50} &= 65.0 \text{ ppm with 95\% confidence intervals of } \pm 2.179\, d\sigma_f \text{ or } \pm 2.179(0.30103) \times (0.31623) \\ &= \pm 0.20743 \end{aligned}$$

Therefore, the log confidence limits are 1.81278 ± 0.20743 = 1.60535 to 2.02021; on the linear scale, 40.3 to 104.8 ppm.

### Assumptions and Limitations

1. A common misconception is that the moving-average method cannot be used to determine the slope of the response curve. This is not true. Weil has published a straightforward method for determining slope in conjunction with a moving-average determination of the $LD_{50}$ [57].
2. The method also provides confidence intervals.

## Nonlinear Regression

More often than not in toxicology we find that our data demonstrate a relationship between two variables (such as age and body weight) that is not linear; that is, a change in one variable (e.g., age) does not produce a directly proportional change in the other (e.g., body weight), but some form of relationship between the variables is apparent. If understanding such a relationship and being able to predict unknown points is of value, we have a pair of options available to us. The first, which was discussed and reviewed earlier, is to use one or more transformations to linearize our data and then to make use of linear regression. This approach, although most commonly used, has a number of drawbacks. Not all data can be suitably transformed. Sometimes the transformations necessary to linearize the data require a cumbersome series of calculations, and the resulting linear regression is not always sufficient to account for the differences among sample values; there might be significant deviations around the linear regression line (that is, a line may still not give us a good fit to the data or do an adequate job of representing the relationship between the data). In such cases, we have available a second option—fitting the data to some nonlinear function such as some form of the curve. This is, in general form, nonlinear regression and may involve fitting data to an infinite number of possible functions, but most often we are interested in fitting curves to a polynomial function of the general form:

$$Y = a + bx + cx^2 + dx^2 + \ldots$$

where $x$ is the independent variable. As the number of powers of $x$ increases, the curve becomes increasingly complex and will be able to fit a given set of data increasingly well. Generally in toxicology, however, if we plot the log of a response (such as body weight) vs. a linear scale of our dose or stimulus, we get one of four types of nonlinear curves [16]:

- *Exponential growth*, where $\log Y = A(B^x)$, such as the growth curve for the log phase of a bacterial culture
- *Exponential decay*, where $\log Y = A(B^{-x})$, such as a radioactive decay curve
- *Asymptotic regression*, where $\log Y = A - B(p^x)$, such as a first-order reaction curve
- *Logistic growth curve*, where $\log Y = A/(1 + Bp^x)$, such as a population growth curve.

In all of these cases, $A$ and $B$ are constant and $p$ is a log transform. These curves are illustrated in Figure 8.7.

All four types of curves are fit by iterative processes; that is, best-guess numbers are initially chosen for each of the constants and, after a fit is attempted, the constants are modified to improve the fit. This process is repeated until an acceptable fit has been generated. Analysis of

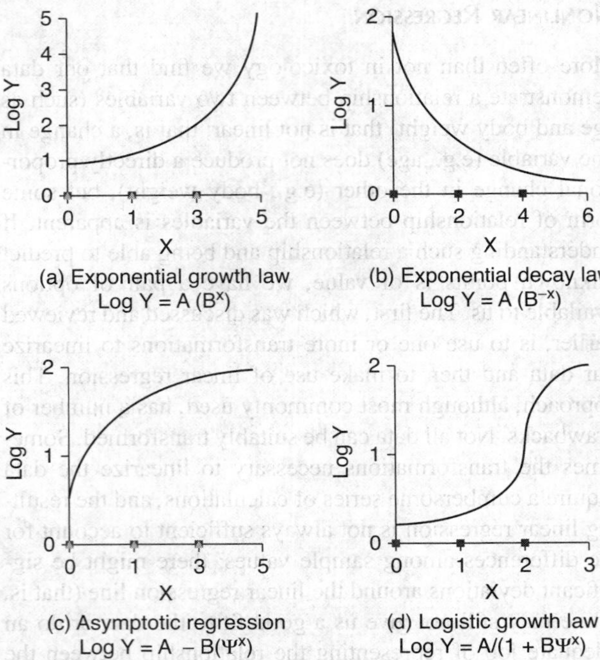


**FIGURE 8.7** Common curvilinear curves.

variance or covariance can be used to objectively evaluate the acceptability of it. Needless to say, the use of a computer generally accelerates such a curve-fitting process.

### Assumptions and Limitations

1. The principle of using least squares may still be applicable in fitting the best curve, if the assumptions of normality, independence, and reasonably error-free measurement of response are valid.
2. Growth curves are best modeled using a nonlinear method.

## Correlation Coefficient

The correlation procedure is used to determine the degree of linear correlation (direct relationship) between two groups of continuous (and normally distributed) variables; it will indicate whether there is any statistical relationship between the variables in the two groups. For example, we may wish to determine if the liver weights of dogs on a feeding study are correlated with their body weights. Thus, we will record the body and liver weights at the time of sacrifice and then calculate the correlation coefficient between these pairs of values to determine if there is some relationship. A formula for calculating the linear correlation coefficient ($r_{xy}$) is as follows:

$$r_{xy} = \frac{N\sum XY - \left(\sum X\right)\left(\sum Y\right)}{\sqrt{N\sum X^2 - \left(\sum X\right)^2}\sqrt{N\sum Y^2 - \left(\sum Y\right)^2}}$$

where $X$ is each value for one variable (such as the dog body weights in the above example), $Y$ is the matching value for the second variable (the liver weights), and $N$ is the number of pairs of $X$ and $Y$. Once we have obtained $r_{xy}$ it is possible to calculate $t_r$, which can be used for more precise examination of the degree of significant linear relationship between the two groups. This value is calculated as follows:

$$t_r = \frac{r_{zy}\sqrt{N-2}}{\sqrt{1-r_{zy}^2}}$$

This calculation is also equivalent to $r$ = sample covariance/$(S_xS_y)$, as was seen earlier under ANCOVA.

The value obtained for $r_{xy}$ can be compared to table values [16] for the number of pairs of data involved minus two. If the $r_{xy}$ is smaller (at the selected test probability level, such as 0.05), the correlation is not significantly different from zero (no correlation). If $r_{xy}$ is larger than the table value, there is a positive statistical relationship between the groups. Comparisons are then made at lower levels of probability to determine the degree of relationship (note that if $r_{xy}$ equals either 1.0 or −1.0 there is complete correlation between the groups). If $r_{xy}$ is a negative number and the absolute is greater than the table value, there is an inverse relationship between the groups; that is, a change in one group is associated with a change in the opposite direction in the second group of variables. Both computations are demonstrated in Example 19.

Because the comparison of $r_{xy}$ with the table values may be considered a somewhat weak test, it is perhaps more meaningful to compare the $t_r$ value with values in a $t$-distribution table for $(N-2)$ degrees of freedom (df), as is done for the Student's $t$-test. This will give a more exact determination of the degree of statistical correlation between the two groups. Note that this method examines only possible linear relationships between sets of continuous, normally distributed data.

### Example 19

If we computed the dog body weight vs. dog liver weight for a study we could obtain the following results:

| Dog | Body Weight (kgj) $X$ | $X^2$ | Liver Weight (g) $Y$ | $Y^2$ | $XY$ |
|---|---|---|---|---|---|
| 1 | 8.4 | 70.56 | 243 | 59049 | 2041.2 |
| 2 | 8.5 | 72.25 | 225 | 50625 | 1912.5 |
| 3 | 9.3 | 86.49 | 241 | 58081 | 2241.3 |
| 4 | 9.5 | 90.25 | 263 | 69169 | 2498.5 |
| 5 | 10.5 | 110.25 | 256 | 65536 | 2688.0 |
| 6 | 8.6 | 73.96 | 266 | 70756 | 2287.6 |
| Sum | $\Sigma X$ = 54.8 | $\Sigma X^2$ = 503.76 | $\Sigma Y$ = 1494 | $\Sigma Y^2$ = 373216 | $\Sigma XY$ = 13669.1 |

$$r_{xy} = \frac{6(13669.1)-(54.8)(1494)}{\left(\sqrt{6(503.76)-(54.8)^2}\right)\left(\sqrt{6(373216)-(1494)^2}\right)}$$
$$= 0.381$$

The table value for six pairs of data (read beside the ($N$ – 2) value, or 6 – 2 = 4) is 0.811 at a 0.05 probability level. Thus, there is a lack of statistical correlation (at $p = 0.05$) between the body weights and liver weights for this group of dogs. The $t_r$ value for these data would be calculated as follows:

$$t_r = \frac{0.381\sqrt{6-2}}{\sqrt{1-(0.381)^2}}$$

The value for the $t$-distribution table for 4 degrees of freedom at the 0.05 level is 2.776; therefore, this again suggests a lack of significant correlation at $p = 0.05$.

## Assumptions and Limitations

1. A strong correlation does not imply that a treatment causes an effect.
2. The distances of data points from the regression line are the portions of the data not explained by the model. These are called *residuals*. Poor correlation coefficients imply high residuals, which may be due to many small contributions (variations of data from the regression line) or a few large ones. Extreme values (outliers) greatly reduce correlation.
3. $X$ and $Y$ are assumed to be independent.
4. Feinstein [58] has provided a fine discussion of the difference between correlation (or association of variables) and causation.

## Kendall's Coefficient of Rank Correlation

Kendall's rank correlation, represented by τ (tau), should be used to evaluate the degree of association between two sets of data when the nature of the data is such that the relationship may not be linear. Most commonly, this is when the data are not continuous or normally distributed. An example of such a case is when we are trying to determine if there is a relationship between the length of hydra and their survival time (in hours) in a test medium, as is presented in Example 18. Both of our variables here are discontinuous, yet we suspect a relationship exists. Another common use is in comparing the subjective scoring done by two different observers.

Tau is calculated at $\tau = N/n(n-1)$ where $n$ is the sample size and $N$ is the count of ranks, calculated as $N = 4(^nC_i) - n(n-1)$, with the computing of $^nC_i$ being demonstrated in the example. If a second variable $Y_2$ is exactly correlated with the first variable $Y_1$, then the variates $Y_2$ should be in the same order as the $Y_1$ variates; however, if the correlation is less than exact, the order of the variates $Y_2$ will not correspond entirely to that of $Y$. The quantity $N$ measures how well the second variable corresponds to the order of the first. It has maximum value of $n(n-1)$ and a minimum value of $-n(n-1)$.

A table of data is set up with each of the two variables being ranked separately. Tied ranks are assigned as demonstrated earlier under the Kruskall–Wallis test. From this point, disregard the original variates and deal only with the ranks. Place the ranks of one of the two variables in rank order (from lowest to highest), paired with the rank values assigned for the other variable. If one (but not the other) variable has tied ranks, order the pairs by the variables without ties [15]. The most common way to compute a sum of the counts is also demonstrated in Example 20. The resulting value of tau will range from –1 to +1, as does the familiar parametric correlation coefficient, $r$.

### Example 20

During the validation of an *in vitro* method, it was noticed that larger hydra seem to survive longer in test media than do small individuals. To evaluate this, 15 hydra of random size were measured (mm), then placed in test media. How many hours each individual survives was recorded over a 24-hour period. These data are presented below, along with ranks for each variable.

| Length | Rank ($R_1$) | Survival | Rank ($R_2$) |
|---|---|---|---|
| 3 | 6.5 | 19 | 9 |
| 4 | 10 | 17 | 7 |
| 6 | 15 | 11 | 1 |
| 1 | 1.5 | 25 | 15 |
| 3 | 6.5 | 18 | 8 |
| 3 | 6.5 | 22 | 12 |
| 1 | 1.5 | 24 | 14 |
| 4 | 10 | 16 | 6 |
| 4 | 10 | 15 | 5 |
| 2 | 3.5 | 21 | 11 |
| 5 | 13 | 13 | 3 |
| 5 | 13 | 14 | 4 |
| 3 | 6.5 | 20 | 10 |
| 2 | 3.5 | 23 | 13 |
| 5 | 13 | 12 | 2 |

We then arrange this based on the order of the rank of survival time (there are no ties here). We then calculate our counts of ranks. The conventional method is to obtain a sum of counts $C_i$, as follows: Examine the first value in the column of ranks paired with the ordered column. In the case below, this is rank 15. Count all ranks subsequent to it that rank greater than 15. There are 14 ranks following the 2 and all of them are less than 15; therefore, we count a score of $C_1 = 0$. We repeat this process for each subsequent rank of $R_1$, giving us a final score of 1. By this point, it is obvious that our original hypothesis—that larger hydrae live longer in test media than do small individuals—was in error.

| $R_2$ | $R_1$ | Following $R_2$ Ranks Greater Than $R_1$ | Counts ($C_i$) |
|---|---|---|---|
| 1 | 15 | — | $C_1 = 0$ |
| 2 | 13 | — | $C_2 = 0$ |
| 3 | 13 | — | $C_3 = 0$ |
| 4 | 13 | — | $C_4 = 0$ |
| 5 | 10 | — | $C_5 = 0$ |
| 6 | 6.5 | 10 | $C_6 = 0$ |
| 7 | 10 | — | $C_7 = 0$ |
| 8 | 6.5 | — | $C_8 = 0$ |
| 9 | 6.5 | — | $C_9 = 0$ |
| 10 | 6.5 | — | $C_{10} = 0$ |
| 11 | 3.5 | 6.5 | $C_{11} = 0$ |
| 12 | 6.5 | — | $C_{12} = 0$ |
| 13 | 3.5 | — | $C_{13} = 0$ |
| 14 | 1.5 | — | $C_{14} = 0$ |
| 15 | 1.5 | — | $C_{15} = 0$ |
| | | | $C_i = 1$ |

Our count of ranks ($N$) is then calculated as:

$$\begin{aligned} N &= 4(1) - 15(15-1) \\ &= 4 - 15(14) \\ &= -206 \end{aligned}$$

We can then calculate tau as:

$$= \frac{-206}{15(15-1)} = \frac{-206}{210} = -0.9810$$

In other words, there is a strong negative correlation between our variables.

## Assumption and Limitation

1. This is a very robust estimator that does not assume normality, linearity, or minimal error of measurement.

## Trend Analysis

Trend analysis is a collection of techniques utilized by toxicology since the mid-1970s [59]. The actual methodology dates back to the mid-1950s [60]. Trend analysis methods are a variation on the theme of regression testing. In the broadest sense, the methods are used to determine whether a sequence of observations taken over an ordered range of a variable (most commonly time) exhibits some form of pattern of change (an increase or upward trend) associated with another variable of interest (in toxicology, some form or measure of dosage an exposure). Trend corresponds to sustained and systematic variations over a long period of time. It is associated with the structural causes of the phenomenon in question—for example, population growth, technological progress, new ways of organization, or capital accumulation.

The identification of trend has always posed a serious statistical problem. The problem is not one of mathematical or analytical complexity but of conceptual complexity. This problem exists because the trend as well as the remaining components of a time series are latent (nonobservables) variables so, therefore, assumptions must be made on their behavioral pattern. The trend is generally thought of as a smooth and slow movement over the long term. The concept of "long" in this connection is relative, and what is identified as a trend for a given series span might well be part of a long cycle once the series is considerably augmented. Often, a long cycle is treated as a trend because the length of the observed time series is shorter than one complete face of this type of cycle. The ways in which data are collected in toxicology studies frequently serve to complicate trend analysis, as the length of time for the phenomena underlying a trend to express themselves is frequently artificially censored.

To avoid the complexity of the problem posed by a statistically vague definition, statisticians have resorted to two simple solutions: One consists of estimating trend and cyclical fluctuations together (the *trend cycle*); the other consists of defining the trend in terms of the series length (the *longest nonperiodic movement*).

## Trend Models

Within the large class of models identified for trend, we can distinguish two main categories: deterministic trends and stochastic trends. Deterministic trend models are based on the assumption that the trend of a time series can be approximated closely by simple mathematical functions of time over the entire span of the series. The most common representation of a deterministic trend is by means of polynomials or of transcendental functions. The time series from which the trend is to be identified is assumed to be generated by a nonstationary process where the nonstationarity results from a deterministic trend. A classical model is the regression or error model [67], where the observed series is treated as the sum of a systematic part or trend and a random part or irregular. This model can be written as:

$$Z_t = Y_t + U_t'$$

where $U_t$ is a purely random process; that is, $U_t \sim$ i.i.d. (0 $2/u$) (independent and identically distributed with expected value 0 and variance $\sigma(2/u)$).

Trend tests are generally described as $k$-sample tests of the null hypothesis of identical distribution against an alternative of linear order; in other words, if sample $I$ has distribution function $F_i$, $i = 1$, then the null hypothesis $H_0$: $F_1 = F_2 - \ldots = F_k$ is tested against the alternative: $H_1$: $F_1 \geq F_2 \geq \ldots = F_k$ (or its reverse), where at least one of the inequalities is strict. These tests can be thought of as

| Month of Study | Control: Total X Animals with Tumors | Control: Change ($X_{A-B}$) | Low Doses: Total Y Animals with Tumors | Low Doses: Change ($Y_{A-B}$) | Low Doses: Compared to Control ($Y - X$) | High Doses: Total Z Animals with Tumors | High Doses: Change ($Z_{A-B}$) | High Doses: Compared to Control ($Z - X$) |
|---|---|---|---|---|---|---|---|---|
| 12 (A) | 1 | NA | 0 | NA | NA | 5 | NA | NA |
| 13 (B) | 1 | 0 | 0 | 0 | 0 | 7 | 2 | (+)2 |
| 14 (C) | 3 | 2 | 1 | 1 | (−)1 | 11 | 4 | (+)2 |
| 15 (D) | 3 | 0 | 1 | 0 | 0 | 11 | 0 | 0 |
| 16 (E) | 4 | 1 | 1 | 0 | (−)1 | 13 | 2 | (+)1 |
| 17 (F) | 5 | 1 | 3 | 2 | (+)1 | 14 | 1 | 0 |
| 18 (G) | 5 | 0 | 3 | 0 | 0 | 15 | 1 | (+)1 |
| 19 (H) | 5 | 0 | 5 | 2 | (+)2 | 18 | 3 | (+)3 |
| 20 (I) | 6 | 1 | 6 | 1 | 0 | 19 | 1 | 0 |
| 21 (J) | 8 | 2 | 7 | 1 | (−)1 | 22 | 3 | (+)1 |
| 22 (K) | 12 | 4 | 9 | 2 | (−)2 | 26 | 4 | 0 |
| 23 (L) | 14 | 2 | 12 | 3 | (+)1 | 28 | 2 | 0 |
| 24 (M) | 18 | 4 | 17 | 5 | (+)1 | 31 | 3 | (−)1 |
| | | | | Sum of signs | 4 + | | Sum of signs | 6 + |
| | | | | $Y - X$ | 4 − | | $Z - X$ | 1 − |
| | | | | | = 0 (no trend) | | | = 5 |

special cases of tests of regression or correlation in which association is sought between the observations and its ordered sample index. They are also related to analysis of variance except that the tests are tailored to be powerful against the subset of alternatives $H_1$ instead of the more general set $\{F_1 \neq F_j, \text{ some } i \neq j\}$.

Different tests arise from requiring power against specific elements or subsets of this rather extensive set of alternatives. The most popular trend test in toxicology is currently that presented by Tarone in 1975 [59] because it is the one used by the National Cancer Institute (NCI) in the analysis of carcinogenicity data. A simple, but efficient alternative is the Cox and Stuart test [66] which is a modification of the sign test. For each point at which we have a measure (such as the incidence of animals observed with tumors) we form a pair of observations—one from each of the groups we wish to compare. In a traditional NCI bioassay this would mean pairing control with low dose and low dose with high dose (to explore a dose-related trend) or each time period observation in a dose group (except the first) with its predecessor (to evaluate a time-related trend). When the second observation in a pair exceeds the earlier observation, we record a plus sign for that pair. When the first observation is greater than the second, we record a minus sign for that pair. A preponderance of plus signs suggests a downward trend, while an excess of minus signs suggests an upward trend. A formal test at a preselected confidence level can then be performed.

More formally put, after having defined what trend we want to test for, we first match pairs as $(X_1 - X_{1+c})$, $(X_2, X_{2+c})$, …, $(X_{n'-c}, X_{n'})$ where $c = n'/2$ when $n'$ is even and $c = (n' + 1)/2$ when $n'$ is odd (where $n'$ is the number of observations in a set). The hypothesis is then tested by comparing the resulting number of excess positive or negative signs against a sign test table such as are found in Beyer. We can, of course, combine a number of observations to allow ourselves to actively test for a set of trends, such as the existence of a trend of increasing difference between two groups of animals over a period of time. This is demonstrated in Example 21.

## Example 21

In a chronic feeding study in rats, we tested the hypothesis that, in the second year of the study, there was a dose-responsive increase in tumor incidence associated with the test compound. We utilize a Cox–Stuart test for trend to address this question (see table this page). All groups start the second year with an equal number of animals. Reference to a sign table is not necessary for the low-dose comparison (where there is no trend) but clearly shows the high dose to be significant at the $p \leq 0.5$ level.

## Assumptions and Limitations

1. Trend tests seek to evaluate whether there is a monotonic tendency in response to a change in treatment; that is, the dose–response direction is absolute. As the dose goes up, the incidence of tumors increases. Thus, the test loses power rapidly in response to the occurrences of reversals—for example, a low-dose group with a decreased tumor incidence. Methods are available [62] that smooth the bumps of reversals in long data series. In toxicology, however, most data series are short (that is, there are only a few dose levels).

Tarone's trend test is most powerful at detecting dose-related trends when tumor onset hazard functions are proportional to each other. For more power against other dose-related group differences, weighted versions of the statistic are also available [63,64]. In 1985, the U.S. *Federal Register* recommended that the analysis of tumor incidence data be carried out with a Cochran–Armitage trend test [65,66]. The test statistic of the Cochran–Armitage test is defined as the term:

$$T_{CA} = \sqrt{\frac{N}{(N-r)r}}, \frac{\sum_{i=0}^{k}\left(R_1 - \frac{n_1}{N}r\right)d_1}{\sqrt{\sum_{i=0}^{k}\frac{n_i}{N}d_i^2 - \left(\sum_{i=0}^{k}\frac{n_i}{N}d_1\right)^2}}$$

with dose scores $d_i$. Armitage's test statistic is the square of this term ($T_{CA}^2$). As one-sided tests are carried out for an increase of tumor rates, the square is not considered. Instead, the above-mentioned test statistic presented by Portier and Hoel [67] is used. This test statistic is asymptotically standard normal distributed. The Cochran–Armitage test is asymptotically efficient for all monotone alternatives [59], but this result only holds asymptotically. Tumors are rare events, so the binominal proportions are small. In this situation, approximations may become unreliable. As a result, exact tests that can be performed using two different approaches—conditional and unconditional—are considered. In the first case, the total number of tumors $r$ is regarded as fixed; thus, the null distribution of the test statistic is independent of the common probability $p$. The exact conditional null distribution is a multivariate hypergeometric distribution. The unconditional model treats the sum of all tumors as a random variable, and the exact unconditional null distribution is a multivariate binomial distribution. The distribution depends on the unknown probability.

## METHODS FOR THE REDUCTION OF DIMENSIONALITY

Techniques for the reduction of dimensionality are those that simplify the understanding of data, either visually or numerically, while causing only minimal reductions in the amount of information present. These techniques operate primarily by pooling or combining groups of variables into single variables but may also entail the identification and elimination of low-information-content (or irrelevant) variables. Descriptive statistics (calculations of means, standard deviations, etc.) are the simplest and most familiar form of reduction of dimensionality. Here, we first need to address classification, which provides the general conceptual tools for identifying and quantifying similarities and differences between groups of things that have more than a single linear scale of measurement in common (for example, which have both been determined to have or lack a number of enzyme activities). We will then consider two collections of methodologies that combine graphic and computational methods: multidimensional/nonmetric scaling and cluster analysis. Multidimensional scaling (MDS) is a set of techniques for quantitatively analyzing similarities, dissimilarities, and distances between data in a display-like manner. Nonmetric scaling is an analogous set of methods for displaying and relating data when measurements are nonquantitative (the data are described by attributes or ranks). Cluster analysis is a collection of graphic and numerical methodologies for classifying things based on the relationships between the values of the variables that they share. The final pair of methods for reduction of dimensionality that will be tackled in this chapter are Fourier analysis and life-table analysis. Fourier analysis seeks to identify cyclic patterns in data and then either analyze the patterns or the residuals after the patterns are taken out. Life-table analysis techniques are directed toward identifying and quantitating the time course of risks (such as death or the occurrence of tumors).

### CLASSIFICATION

Classification is both a basic concept and a collection of techniques that are necessary prerequisites for further analysis of data when the members of a set of data are (or can be) each described by several variables. At least some degree of classification (which is broadly defined as the dividing of the members of a group into smaller groups in accordance with a set of decision rules) is necessary prior to any data collection. Whether formally or informally, an investigator has to decide which things are similar enough to be counted as the same and develop rules for governing collection procedures. Such rules can be as simple as "measure and record body weights only of live animals on study" or as complex as that demonstrated by the expanded weighting classification presented in Example 22. Such a classification also demonstrates that the selection of which variables to measure will determine the final classification of data.

**Example 22**

| | |
|---|---|
| Is animal of desired species? | Yes/No |
| Is animal member of study group? | Yes/No |
| Is animal alive? | Yes/No |
| Which group does animal belong to? | |
| Control | |
| Low dose | |
| Intermediate dose | |
| High dose | |
| What sex is the animal? | Male/Female |
| Is the measured weight within an acceptable range? | Yes/No |

Classifications of data have two purposes [68,69]: *data simplification* (also known as *descriptive function*) and *prediction*. Simplification is necessary because there is a limit to both the volume and complexity of data that the human mind can comprehend and deal with conceptually. Classification allows us to attach a label (or name) to each group of data, to summarize the data (that is, assign individual elements of data to groups and to characterize the population of the group), and to define the relationships between groups (that is, develop a taxonomy).

Prediction, meanwhile, is the use of summaries of data and knowledge of the relationships between groups to develop hypotheses as to what will happen when further data are collected (as when more animals or people are exposed to an agent under defined conditions) and as to the mechanisms that cause such relationships to develop. Indeed, classification is the prime device for the discovery of mechanisms in all of science. A classic example of this was Darwin's realization that there were reasons (the mechanisms of evolution) behind the differences and similarities in species that had caused Linaeus to earlier develop his initial modern classification scheme (or taxonomy) for animals.

To develop a classification, one first sets bounds wide enough to encompass the entire range of data to be considered but not unnecessarily wide. This is typically done by selecting some global variables (variables every piece of datum has in common) and limiting the range of each so it just encompasses all the cases on hand. Then one selects a set of local variables (characteristics that only some of the cases have, such as the occurrence of certain tumor types, enzyme activity levels, or dietary preferences) that serves to differentiate between groups. Data are then collected, and a system for measuring differences and similarities is developed. Such measurements are based on some form of measurement of distance between two cases (*x* and *y*) in terms of each single variable scale. If the variable is a continuous one, then the simplest measure of distance between two pieces of data is the Euclidean distance, $d(x,y)$, defined as:

$$d(x,y) = \sqrt{(x_i - y_i)^2}$$

For categorical or discontinuous data, the simplest distance measure is the matching distance, defined as:

$$d(x,y) = \text{Number of times } x_i \neq y_i$$

After we have developed a table of such distance measurements for each of the local variables, some weighting factor is assigned to each variable. A weighting factor seeks to give greater importance to those variables that are believed to have more relevance or predictive value. The weighted variables are then used to assign each piece of data to a group. The actual act of developing numerically based classifications and assigning data members to them is the realm of cluster analysis and will be discussed later in this chapter. Classification of biological data based on qualitative factors has been well discussed; Gordon [69] and Glass [70] do an excellent job of introducing the entire field and mathematical concepts.

Relevant examples of the use of classification techniques range from the simple to the complex. Schaper et al. [71] developed and used a very simple classification of response methodology to identify those airborne chemicals that alter the normal respiratory response induced by $CO_2$. At the other end of the spectrum, Kowalski and Bender [72] developed a more mathematically based system to classify chemical data (a methodology they termed *pattern recognition*).

## Statistical Graphics

The use of graphics in one form or another in statistics is the single most effective and robust statistical tool and, at the same time, one of the most poorly understood and improperly used. Graphs are used in statistics (and in toxicology) for one of four major purposes. Each of the four is a variation on the central theme of making complex data easier to understand and use. These four major functions are exploration, analysis, communication and display of data, and graphical aids. Exploration (which may be simply summarizing data or trying to expose relationships between variables) is determining the characteristics of datasets and deciding on one or more appropriate forms of further analysis, such as the scatterplot. Analysis is the use of graphs to formally evaluate some aspect of the data, such as whether there are outliers present or if an underlying assumption of a population distribution is fulfilled. As long ago as 1960 [73], some 18 graphical methods for analyzing multivariate data relationships had been developed and proposed.

Communication and display of data are the most commonly used functions of statistical graphics in toxicology, whether for internal reports, presentations at meetings, or formal publications in the literature. When communicating data, graphs should not be used to duplicate data that are presented in tables but rather to show important trends or relationships in the data. Although such communication is most commonly of a quantitative compilation of actual data, it can be also be used to summarize and present the results of statistical analysis. The fourth and final function of graphics is one that is largely becoming outdated as microcomputers become more widely available. Graphical aids to calculation include nomograms (the classic example in toxicology of a nomogram is that presented by Litchfield and Wilcoxon for determining median effective doses) and extrapolating and interpolating data graphically based on plotted data.

There are many forms of statistical graphics (a partial list, classified by function, is presented in Table 8.10), and a number of these (such as scatterplots and histograms)

**TABLE 8.10**
**Forms of Statistical Graphics (by Function)**

| Exploration | | |
|---|---|---|
| **Data Summary** | **Two Variables** | **Three or More Variables** |
| Box and whisker plot | Autocorrelation plot | Biplot |
| Histogram | Cross-correlation plot | Cluster trees |
| Dot-array diagram | Scatterplot | Labeled scatterplot |
| Frequency polygon | Sequence plot | Glyphs and metroglyphs |
| Ogive | | Face plots |
| Stem and leaf diagram | | Fourier plots |
| | | Similarity and preference maps |
| | | Multidimensional scaling displays |
| | | Weathervane plot |
| **Analysis** | | |
| **Distribution Assessment** | **Model Evaluation and Assumption Verification** | **Decision Making** |
| Probability plot | Average vs. standard deviation | Control chart |
| *Q–Q* plot | Component-plus-residual plot | Cusum chart |
| *P–P* plot | Partial residual plot | Half-normal plot |
| Hanging histogram | Residual plots | Ridge trace |
| Rootagram | | Youden plot |
| Poissonness plot | | |
| **Communication and Display of Data** | | |
| **Quantitative Graphics** | **Summary of Statistical Analyses** | **Graphical Aids** |
| Line chart | Means plot | Confidence limits |
| Pictogram | Sliding reference distribution | Graph paper |
| Pie chart | Notched box plot | Power curves |
| Contour plot | Factor space/response | Nomographs |
| Stereogram | Interaction plot | Sample-size curves |
| Color map | Contour plot | Trilinear coordinates |
| Histogram | Predicted response plot | |
| | Confidence region plot | |

can be used for each of a number of possible functions. Most of these plots are based on a Cartesian system (that is, they use a set of rectangular coordinates), and our review of construction and use will focus on these forms of graphs.

Construction of a rectangular graph of any form starts with the selection of the appropriate form of graph followed by the laying out of the coordinates (or axes). Even graphs that are going to encompass multivariate data (that is, more than two variables) generally have as their starting point two major coordinates. The vertical axis or ordinate (also called the *y*-axis) is used to present an independent variable. Each of these axes is scaled in the units of measure that will most clearly present the trends of interest in the data. The range covered by the scale of each axis is selected to cover the entire region for which data are presented. The actual demarking of the measurement scale along an axis should allow for easy and accurate assessment of the coordinates of any data point, yet should not be cluttered.

Actual data points should be presented by symbols that present the appropriate indicators of location, and if they represent a summaries of data from a normal data population it would be appropriate to present a symbol for the mean and some indication of the variability (or error) associated with that population, commonly by using error bars, which present the standard deviation (or standard error) from the mean. If, however, the data are not normal or continuous it would be more appropriate to indicate location by the median and present the range or semiquartile distance for variability estimates. The symbols that are used to present data points can also be used to present a significant amount of additional information. At the simplest level, clearly distinct symbols (circles, triangles, squares, etc.) are very commonly used to provide a third dimension of data (most commonly the treatment group). But, by clever use of symbols, all sorts of additional information can be presented. Using a method such as Chernoff's faces [74], in which faces are used as symbols of the data points (and various aspects of the faces

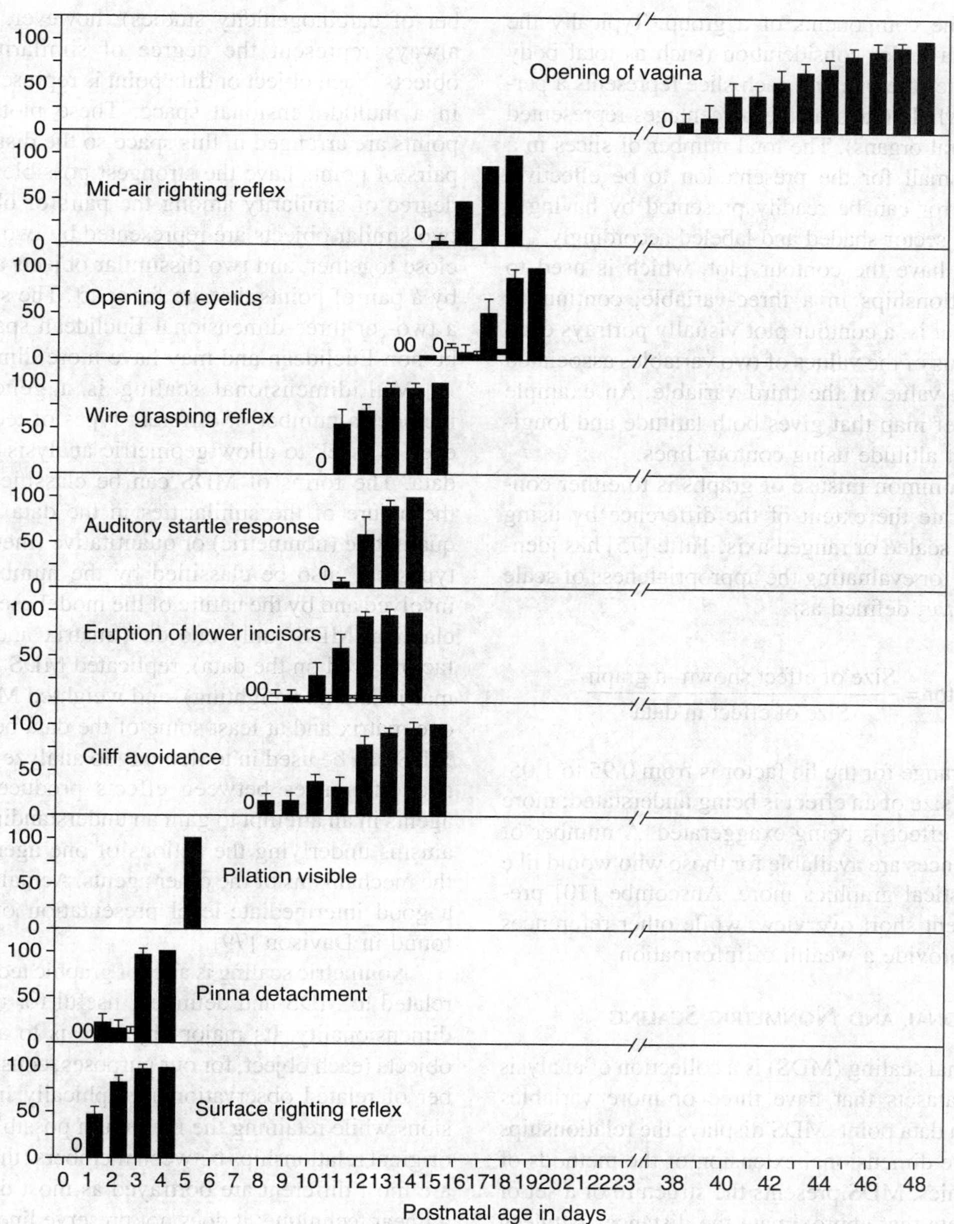


**FIGURE 8.8** Acquisitions of postnatal development landmarks in rats.

present additional data, such as the presence or absence of eyes denoting presence or absence of a secondary pathological condition), it is possible to present a large number of different variables on a single graph.

The three other forms of graphs that are commonly used in toxicology are histograms, pie charts, and contour plots. Histograms are graphs of simple frequency distribution. Commonly, the abscissa is the variable of interest (such as lifespan or litter size) and is generally shown as classes or intervals or measurements (such as age ranges of 0 to 10, 10 to 20, etc. weeks). The ordinate, meanwhile, is the incidence or frequency of observations. The result is a set of vertical bars, each of which represents the incidence of a particular set of observations. Measures of error or variability about each incidence are reflected by some form of error bar on top of or in the frequency bars, as shown in Figure 8.8. The size of class intervals may be unequal (in effect, one can combine or pool several small class intervals), but it is proper in such cases to vary the width of the bars to indicate differences in interval size.

Pie charts are the only common form of quantitative graphic technique that is not rectangular; rather, the figure is presented as a circle out of which several slices are delimited. The only major use of the pie chart is in presenting a breakdown of the components of a group. Typically the entire set of data under consideration (such as total body weight) constitutes the pie, and each slice represents a percentage of the whole (such as the percentages represented by each of several organs). The total number of slices in a pie should be small for the presentation to be effective. Variability or error can be readily presented by having a subslice of each sector shaded and labeled accordingly.

Finally, we have the contour plot, which is used to depict the relationships in a three-variable, continuous data system. That is, a contour plot visually portrays each contour as a locus of the values of two variables associated with a constant value of the third variable. An example would be a relief map that gives both latitude and longitude of constant altitude using contour lines.

The most common misuse of graphs is to either conceal or exaggerate the extent of the difference by using inappropriately scaled or ranged axis. Tufte [75] has identified a statistic for evaluating the appropriateness of scale size, the *lie factor*, defined as:

$$\text{Lie factor} = \frac{\text{Size of effect shown in graph}}{\text{Size of effect in data}}$$

An acceptable range for the lie factor is from 0.95 to 1.05. Less means the size of an effect is being understated; more means that the effect is being exaggerated. A number of excellent references are available for those who would like to pursue statistical graphics more. Anscombe [10] presents an excellent short overview, while other references [14,75,76,77] provide a wealth of information.

## Multidimensional and Nonmetric Scaling

Multidimensional scaling (MDS) is a collection of analysis methods for datasets that have three or more variables making up each data point. MDS displays the relationships of three or more dimensional extension of the methods of statistical graphics. MDS presents the structure of a set of objects from data that approximate the distances between pairs of the objects. The data, called *similarities*, *dissimilarities*, *distances*, or *proximities*, must be in such a form that the degree of similarities and differences between the pairs of the objects (each of which represents a real-life data point) can be measured and handled as a distance (remember the discussion of measures of distances under classifications). Similarity is a matter of degree; small differences between objects cause them to be similar (a high degree of similarity), while large differences cause them to be considered dissimilar (a small degree of similarity).

In addition to the traditional human conceptual or subjective judgments or similarity, data can be an objective similarity measure (the difference in weight between a pair of animals) or an index calculated from multivariate data (the proportion of agreement in the results of a number of carcinogenicity studies); however, the data must always represent the degree of similarity of pairs of objects. Each object or data point is represented by a point in a multidimensional space. These plots or projected points are arranged in this space so the distances between pairs of points have the strongest possible relation to the degree of similarity among the pairs of objects. That is, two similar objects are represented by two points that are close together, and two dissimilar objects are represented by a pair of points that are far apart. The space is usually a two- or three-dimensional Euclidean space, but it may be non-Euclidean and may have more dimensions.

Multidimensional scaling is a general term that includes a number of different types of techniques; however, all seek to allow geometric analysis of multivariate data. The forms of MDS can be classified according to the nature of the similarities in the data [78]. It can be qualitative (nonmetric) or quantitative (metric MDS). The types can also be classified by the number of variables involved and by the nature of the model used; for example, classical MDS (only one data matrix and no weighting factors used on the data), replicated MDS (more than one matrix and no weighting), and weighted MDS (more than one matrix and at least some of the data being weighted). MDS can be used in toxicology to analyze the similarities and differences between effects produced by different agents in an attempt to gain an understanding of the mechanisms underlying the actions of one agent to determine the mechanisms of the other agents. Actual algorithms and a good intermediate-level presentation of MDS can be found in Davison [79].

Nonmetric scaling is a set of graphic techniques closely related to MDS and definitely useful for the reduction of dimensionality. Its major objective is to arrange a set of objects (each object, for our purposes, consisting of a number of related observations) graphically in a few dimensions while retaining the maximum possible fidelity to the original relationships between members (that is, values that are most different are portrayed as most distant). It is not a linear technique; it does not preserve linear relationships (i.e., $A$ is not shown as twice as far from $C$ as $B$, even though its value difference may be twice as much). The spacings (interpoint distances) are kept such that, if the distance of the original scale between members $A$ and $B$ is greater than that between $C$ and $D$, then the distances on the model scale will likewise be greater between $A$ and $B$ than between $C$ and $D$. Figure 8.5, presented earlier, uses a form of this technique to add a third dimension by using letters to present degrees of effect on the skin. This technique functions by taking observed measures of similarity or dissimilarity between every pair of $M$ objects, then finding a representation of the objects as points in Euclidean space such that the interpoint distances in some sense match the observed similarities or dissimilarities by means of weighting constants.

## Cluster Analysis

Cluster analysis is a quantitative form of classification. It serves to help develop decision rules and then use these rules to assign a heterogeneous collection of objects to a series of sets. This is almost entirely an applied methodology (as opposed to theoretical). The final result of cluster analysis is one of several forms of graphic displays and a methodology (set of decision classifying rules) for the assignment of new members into the classifications. The classification procedures used are based on either density of population or distance between members. These methods can serve to generate a basis for the classification of large numbers of dissimilar variables, such as behavioral observations and compounds with distinct but related structures and mechanisms [80,81], or to separate tumor patterns caused by treatment from those caused by old age [27].

The five types of clustering techniques are [82]:

- Hierarchical techniques—Classes are subclassified into groups, with the process being repeated at several levels to produce a tree that gives sufficient definition to groups.
- Optimizing techniques—Clusters are formed by optimization of a clustering criterion. The resulting classes are mutually exclusive; the objects are clearly partitioned into sets.
- Density or mode-seeking techniques—Clusters are identified and formed by locating regions in a graphic representation that contains concentrations of data points.
- Clumping techniques—These are variations of density-seeking techniques in which assignment to a cluster is weighted on some variables so clusters may overlap in graphic projections.
- Others—These methods do not clearly fall into the other classes.

Romesburg [83] provides an excellent step-by-step guide to cluster analysis.

## Fourier or Time Analysis

Fourier analysis [84] is most frequently a univariate method used for either simplifying data (which is the basis for its inclusion in this chapter) or for modeling. It can, however, also be a multivariate technique for data analysis. In a sense, it is like trend analysis; it looks at the relationship of sets of data from a different perspective. In the case of Fourier analysis, the approach is to resolve the time dimension variable in the dataset. At the simplest level, it assumes that many events are periodic in nature and if we can remove the variation in other variables because of this periodicity (by using Fourier transforms) then we can better analyze the remaining variation from other variables. The complications to this are that (1) there may be several overlying cyclic time-based periodicities, and (2) we may be interested in the time-cycle events for their own sake.

Fourier analysis allows one to identify, quantitate, and (if we wish) remove the time-based cycles in data (with their amplitudes, phases, and frequencies) by use of the Fourier transform:

$$nJ_i = x_i \exp(-iw_i t)$$

where $n$ is length; $J$ is the discrete Fourier transform for that case; $x$ is actual data; $i$ is the increment in the series; $w$ is the frequency; and $t$ is time. A graphic example of the use of Fourier analysis in toxicology is provided in Figure 8.9.

## Life Tables

Chronic *in vivo* toxicity studies are generally the most complex and expensive studies conducted by a toxicologist. Answers to a number of questions are sought in such a study—notably, if a material results in a significant increase in mortality or in the incidence of tumors in those animals exposed to it. But, we are also interested in the time course of these adverse effects (or risks). The classic approach to assessing these age-specific hazard rates is the use of life tables (also called *survivorship tables*).

It may readily be seen that during any selected period of time ($t_i$) we have a number of risks competing to affect an animal. There are risks of natural death, death induced by a direct or indirect action of the test compound, and death due to such occurrences of interest of tumors [85]. Also, we are indeed interested in determining if (and when) the last two of these risks become significantly different than the natural risks (defined as what is seen to happen in the control group). Life-table methods enable us to make such determinations as the duration of survival (or time until tumors develop) and the probability of survival (or of developing a tumor) during any period of time.

We start by deciding the interval length ($t_i$) we wish to examine within the study. The information we gain becomes more exact as the interval is shortened, but as interval length is decreased, the number of intervals increases and calculations become more cumbersome and less indicative of time-related trends because random fluctuations become more apparent. For a 2-year or lifetime rodent study, an interval length of a month is commonly employed. Some life-table methods, such as the Kaplan–Meyer, have each new event (such as a death) define the start of a new interval.

Having established the interval length, we can tabulate our data [86]. We begin by establishing the following columns in each table, with a separate table being established for each group of animals (i.e., by sex and dose level):

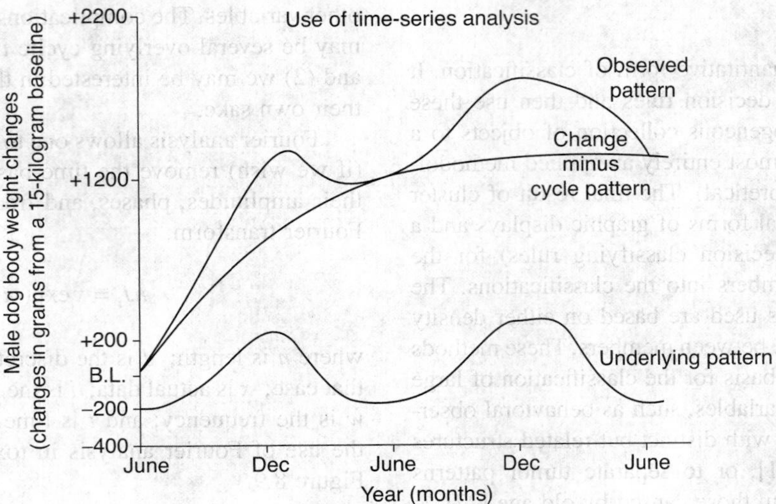


**FIGURE 8.9** Use of time-series analysis.

- The interval of time selected ($t_i$)
- The number of animals in the group that entered that interval of the study alive ($l_i$)
- The number of animals withdrawn from the study during the interval (such as those taken for an interim sacrifice or that may have been killed by a technician error) ($\omega_i$)
- The number of animals that died during the interval ($d_i$)
- The number of animals at risk during the interval $l_i = l_i - 1/2\omega_i$, or the number at the start of the interval minus one half the number withdrawn during the interval
- The proportion of animals that died ($D_i = d_i/l_i$)
- The cumulative probability of an animal surviving until the end of that interval of study: $P_i = 1 - D_i$, or 1 minus the number of animals that died during that interval divided by the number of animals at risk
- The number of animals dying until that interval ($M_i$)
- The number of animals found to have died during the interval ($m_i$)
- The probability of dying during the interval of the study: $c_i = 1 - (M_i + m_i/l_i)$, or the total number of animals dead until that interval plus the animals discovered to have died during that interval divided by the number of animals at risk through the end of that interval;
- The cumulative proportion surviving ($p_i$) is equivalent to the cumulative product of the interval probabilities of survival (i.e., $p_i = p_1 \cdot p_2 \cdot p_3 \cdot \ldots \cdot p_x$)
- The cumulative probability of dying ($C_i$), equal to the cumulative product of the interval probabilities to that point (i.e., $C_i = c_1 \cdot c_2 \cdot c_3 \cdot \ldots \cdot c_x$)

With such tables established for each group in a study (as shown in Example 23), we may now proceed to test the hypothesis that each of the treated groups has a significantly shorter duration of survival or that the treated groups died more quickly (note that plots of total animals dead and total animals surviving will give one an appreciation of the data but can lead to no statistical conclusions).

Now, for these two groups, we wish to determine effective sample size and to compare survival probabilities in the interval months 14–15. For the exposure group we compute sample size as:

$$l_{g14-15} = \frac{0.8400(1-0.8400)}{(0.0367)^2} = 99.7854$$

Likewise, we get a sample size of 98.1720 for the control group. The standard error of difference for the two groups here is:

$$\mathrm{SD} = \sqrt{0.0367^2 + 0.0173^2} = 0.040573$$

The probability of survival differences is $P_D = 0.9697 - 0.8400 = 0.1297$. Our test statistic is then 0.1297/0.040573 = 3.196. From our $z$ value table we see that the critical values are:

$$p \leq 0.05 = 1.960$$
$$p \leq 0.01 = 2.575$$
$$p \leq 0.001 = 3.270$$

As our calculated value is larger than all but the last of these, we find our groups to be significantly different at the 0.01 level ($0.01 > p > 0.001$). A multiplicity of methods is available for testing significance in life tables.

**Example 23**

| Interval (months) $l_i$ | Alive at Beginning of Interval $l_i$ | Animals Withdrawn $w_i$ | Died During Interval $d_i$ | Animals at Risk $l_i$ | Proportion of Animals Dead $D_i$ | Probability of Survival $P_i$ | Cumulative Proportion Surviving $P_i$ | Standard Error of Survival $S_i$ |
|---|---|---|---|---|---|---|---|---|
| **Test Level 1** | | | | | | | | |
| 8–9 | 109 | 0 | 0 | 109 | 0 | 1.0000 | 1.0000 | 0.0000 |
| 9–10 | 109 | 0 | 2 | 109 | 0.0184 | 0.9816 | 0.9816 | 0.0129 |
| 10–11 | 107 | 0 | 0 | 107 | 0 | 1.0000 | 0.9816 | 0.0128 |
| 11–12 | 107 | 10 | 0 | 102 | 0 | 1.0000 | 0.9816 | 0.0128 |
| 12–13 | 97 | 0 | 1 | 97 | 0.0103 | 0.9897 | 0.9713 | 0.0162 |
| 13–14 | 96 | 0 | 1 | 96 | 0.0104 | 0.9896 | 0.9614 | 0.0190 |
| 14–15 | 95 | 0 | 12 | 95 | 0.1263 | 0.8737 | 0.8400 | 0.0367 |
| 15–16 | 83 | 0 | 2 | 83 | 0.0241 | 0.9759 | 0.8198 | 0.0385 |
| 16–17 | 81 | 0 | 3 | 81 | 0.0370 | 0.9630 | 0.7894 | 0.0409 |
| 17–18 | 78 | 20 | 1 | 68 | 0.0147 | 0.9853 | 0.7778 | 0.0419 |
| 18–19 | 57 | 0 | 2 | 57 | 0.0351 | 0.6949 | 0.7505 | 0.0446 |
| **Control Level** | | | | | | | | |
| 11–12 | 99 | 0 | 1 | 99 | 0.0101 | 0.9899 | 0.9899 | 0.0100 |
| 12–13 | 98 | 0 | 0 | 98 | 0 | 1.0000 | 0.9899 | 0.0100 |
| 13–14 | 98 | 0 | 0 | 98 | 0 | 1.0000 | 0.9899 | 0.0100 |
| 14–15 | 98 | 0 | 2 | 98 | 0.0204 | 0.9796 | 0.9697 | 0.0172 |
| 15–16 | 96 | 0 | 1 | 96 | 0.0104 | 0.9896 | 0.9596 | 0.0198 |
| 16–17 | 95 | 0 | 0 | 95 | 0 | 1.0000 | 0.9596 | 0.0198 |
| 17–18 | 95 | 20 | 2 | 85 | 0.0235 | 0.8765 | 0.9370 | 0.0249 |
| 18–19 | 73 | 0 | 2 | 73 | 0.0274 | 0.9726 | 0.9113 | 0.0302 |

with (as is often the case) the power of the tests increasing as does the difficulty of computation [59,87–89].

We begin our method of statistical comparison of survival at any point in the study by determining the standard error of the K interval survival rate as [90]:

$$S_K = P_k \sqrt{\sum_{1}^{k} \left( \frac{D_i}{1'_x - d_x} \right)}$$

We can also determine the effective sample size ($l_1$) in accordance with:

$$l_1 = \frac{P(1-P)}{S^2}$$

We can now compute the standard error of difference for any two groups (1 and 2) as:

$$S_D = \sqrt{S_1^2 + S_2^2}$$

The difference in survival probabilities for the two groups is then calculated as:

$$P_D = P_1 - P_2$$

We can then calculate a test statistic as:

$$t' = \frac{P_D}{S_D}$$

This is then compared to the $z$ distribution table. If $t' > z$ at the desired probability level, it is significant at that level. Example 23 illustrates the life-table technique for mortality data. With increasing recognition of the effects of time (both as age and length of exposure to unmeasured background risks), life-table analysis has become a mainstay in chronic toxicology. An example is the reassessment of the $ED_{01}$ study [91] that radically changed interpretation of the results and understanding of underlying methods when adjustment for time on study was made. The increased importance of and interest in the analysis of survival data have not been restricted to toxicology, but rather have encompassed all the life sciences. Those with further interest should consult Lee [92] or Elandt-Johnson and Johnson [93], both general in their approach to the subject.

## MULTIVARIATE METHODS

In a chapter of this kind, an in-depth explanation of the available multivariate statistical techniques is an impossibility; however, as the complexity of problems in toxicology increases, we can expect to confront more frequently data that are not univariate but rather multivariate (or multidimensional). For example, a multidimensional study might be one in which the animals are being dosed with two materials that interact. Suppose we measure body weight, tumor incidence, and two clinical chemistry values for test material effects and interaction. Our dimensions—or variables—are now $A$ = dose $x$; $B$ = dose $y$; $W$ = body weight; $C$ = tumor incidence; $D$ and $E$, which are levels of clinical chemistry parameters; and possibly also $t$ (length of dosing).These situations are particularly common in chronic studies [94]. Although we can continue to use multiple sets of univariate techniques as we have in the past, we risk significant losses of power, efficiency, and information when doing so, as well as an increased possibility of error [95].

Here we will also look briefly at the workings and uses of each of the most commonly employed multivariate techniques, together with several examples from the literature of their employment in toxicology and the other biological sciences. We will group the methods according to their primary function: hypothesis testing (are these significant or not?), model fitting (what is the relationship between these variables or what would happen if a population would be exposed to $x$?), and reduction of dimensionality (which variables are most meaningful?). It should be noted (and will soon be obvious), however, that most multivariate techniques actually combine several of these functions.

The most fundamental concept in multivariate analysis is that of a multivariate population distribution. At this point, it is assumed that the reader is familiar with the univariate random variable and with such standard distributions as the normal distribution. Here, we extend these to the multivariate normal distribution.

Multivariate data are virtually never processed and analyzed other than by computer. One must first set up an appropriate database file and then enter the data, coding some of them to meet the requirements of the software being utilized; for example, if only numerical data are analyzed, sex may have to be coded as 1 for male and 2 for females. Having recorded the data, it is then essential to review for suspect values and errors of various kinds. There are many different types of suspect values, and it is helpful to distinguish among them:

- *Outliers*—These are defined to be observations that appear to be inconsistent with the rest of the data. They may be caused by gross recording or entering errors, but it is important to realize that an apparent outlier may occasionally be genuine and indicate a non-normal distribution or valuable data point.
- *Inversions*—A common type of error occurs when two consecutive digits are interchanged at the recording, coding, or entering stage. The error may be trivial if, for example, 56.74 appears as 56.47, but it may generate an outlier if 56.74 appears as 65.74.
- *Repetitions*—At the coding or entering stage, it is quite easy to repeat an entire number in two successive rows or columns of a table, thereby omitting one number completely.
- *Values in the wrong column*—It is also easy to get numbers into the wrong columns.
- *Other errors and suspect values*—Many other types of errors are possible, including misrecording of data of a minor nature.

The general term used to denote procedures for detecting and correcting errors is *data editing*. This includes checks for completeness, consistency, and credibility. Some editing can be done at the end of the data entry stage. In addition, many routine checks can be made by the computer itself, particularly those for gross outliers. An important class of such checks are range tests. For each variable an allowable range of possible values is specified, and the computer checks that all observed values lie within the given range. Bivariate and multivariate checks are also possible; for example, one may specify an allowable range for some functions of two of more variables. Checks called "if–then" checks are also possible; for example, if both age and date of birth are recorded for each animal, then one can check that the answers are consistent. If the date of birth is given, then one can deduce the corresponding age. In fact, in this example, the age observation is redundant. It is sometimes a good idea to include one or two redundant variables as a check on accuracy. Various other general procedures for detecting outliers are described by Barnett and Lewis [96].

When a questionable value or error is detected, the toxicologist must decide what to do about it. One may be able to go back to the original data source and check the observation. Inversions, repetitions, and values in the wrong column can often be corrected in this way. Outliers are more difficult to handle, particularly when they are impossible to check or have been misrecorded in the first place. It may be sensible to treat them as missing values and try to insert a value guessed in an appropriate way (e.g., by interpolation or by prediction from other variables). Alternatively, the value may have to be left as unrecorded and then either all observations for the given individual will have to be discarded or one will have to accept unequal numbers of observations for the different variables. With a univariate set of observations, the analysis usually begins with the calculation of two summary statistics—namely, the mean and standard deviation. In the multivariate case, the analysis usually begins with the calculation of the mean and standard deviation for each variable and, in addition, the correlation coefficient for each pair of variables is usually calculated. These summary statistics are vital in providing a preliminary look at the data.

The sample mean of the $j$th variable is given by:

$$\bar{x}_j = \sum_{r=1}^{n} x_{rj}/n$$

and the sample mean vector, **x**, is given by $x^T = [x_1, x_2, \ldots, x_n]$. If the observations are a random sample from a population with mean $\bar{x}$, then the sample mean vector **x** is usually the point estimate of $x$, and this estimate can easily be shown to be unbiased. The standard deviation of the $j$th variable is given by:

$$S_j = \sqrt{\left[\sum_{r=1}^{n}\left(x_{rj} - \bar{x}_j^2\right)/(n-1\right]}$$

The correlation coefficient of variables $i$ and $j$ is given by:

$$r_{ij} = \sum_{r=1}^{n}\left(x_{ri} - \bar{x}_j\right)\left(x_{rj} - \bar{x}_j\right)/(n-1)s_i s_j$$

These coefficients can be conveniently assembled in the sample correlation matrix ($R$), which is given by:

$$R = \begin{pmatrix} 1 & r_{12} & \cdots & r_{1n} \\ r_{21} & 1 & \cdots & r_{2n} \\ \vdots & & & \\ r_n^1 & r_n^2 & \cdots & 1 \end{pmatrix}$$

Note that the diagonal terms are all unity.

The interpretation of means and standard deviations is straightforward. It is worth looking to determine if, for example, some variables have much higher scatter than others. It is also worth looking at the form of the distribution of each variable and considering whether any of the variables must be transformed; for example, the logarithmic transformation is often used to reduce positive skewness and produce a distribution that is closer to normal. One may also consider the removal of outliers at this stage.

Three significant multivariate techniques have hypothesis testing as their primary function: *MANOVA*, *MANCOVA*, and *factor analysis*. Multivariate analysis of variance (MANOVA) is the multidimensional extension of the ANOVA process we explored before. It can be shown to have grown out of Hotelling's $T^2$ [97], which provides a means of testing the overall null hypothesis that two groups do not differ in their means on any of $p$ measures. MANOVA accomplishes its comparison of two (or more) groups by reducing the set of $p$ measures on each group to a simple number applying the linear combining rule $W_i = w_j X_{ij}$ (where $w_j$ is a weighting factor) and then computing a univariate $F$-ratio on the combined variables. New sets of weights ($w_j$) are selected in turn until the set that maximizes the $F$-ratio is found. The final resulting maximum $F$-ratio (based on the multiple discriminant functions) is then the basis of the significance test. As with ANOVA, MANOVA can be one-way or higher order, and MANOVA has as a basic assumption a multivariate normal distribution. When Gray and Laskey [98] used MANOVA to analyze the reproductive effects of manganese in the mouse, it allowed the identification of significant effects at multiple sites. Witten et al. [99] utilized MANOVA to determine the significance of the effects of dose, time, and cell division in the action of abrin on the lymphocytes.

Multivariate analysis of covariance (MANCOVA) is the multivariate analog of analysis of covariance. As with MANOVA, it is based on the assumption that the data being analyzed are from a multivariate normal population. The MANCOVA test utilizes the two residual matrices using the statistic and is an extension of ANCOVA with two or more uncontrolled variables (or covariables). A detailed discussion can be found in Tatsuoka [100].

Factor analysis is not just a technique for hypothesis testing; it can also serve a reduction of dimensionality function. It seeks to separate the variance unique to particular sets of values from that common to all members in that variable system and is based on the assumption that the intercorrelations among the $n$ original variables are the result of there being some smaller number of variables (factors) that explain the bulk of variation seen in the variables. All of the several approaches to achieving these end results seek a determination of what percentage of the variance of each variable is explained by each factor (a factor being one variable or a combination of variables). The model in factor analysis is $y = Af + xz$, where $y$ is an $n$-dimensional vector of observable responses; $A$ is a factor loading an $n \times q$ matrix of unknown parameters; $f$ is a $q$-dimensional vector of common factor; and $z$ is an $n$-dimensional vector of unique factor.

Used for the reduction of dimensionality, factor analysis is said to be a linear technique because it does not change the linear relationships between the variables being examined. Joung et al. [101] used factor analysis to develop a generalized water-quality index that promises suitability across the United States and has appropriate weightings for ten parameters. Factor analysis promises great utility as a tool for developing models in risk analysis where a number of parameters act and interact.

Now, we move on to two multivariate modeling techniques: *multiple regression* and *discriminant analysis*. Multiple regression and correlation seeks to predict one (or a few) variable from several others. It assumes that the available variables can be logically divided into two (or more) sets and serves to establish maximal linear (or some other scale) relationships among the sets. The linear model for the regression is simply:

$$Y = b_0 + b_1X_1 + b_2X_2 + \ldots + b_pX_p$$

where $Y$ is the predicted value, and the $b$ values are set to maximize correlations between $X$ and $Y$ and $Y$ and $Y$ (the actual observations). The $X$'s are independent of predictor variables, and the $Y$'s are dependent variables or outcome measures. One of the outputs from the process will be the coefficient of multiple correlation, which is simply the multivariate equivalent of the correlation coefficient ($r$).

Schaeffer et al. [102] have neatly demonstrated the utilization of multiple regression in studying the contribution of two components of a mixture to its toxicologic action, using quantitative results from an Ames test as an end point. Paintz et al. [103] similarly used multiple regression to model the quantitative structure–activity relationships of a series of 14 1-benzoyl-3-methyl-pyrazole derivatives.

Discriminant analysis has for its main purpose finding linear combinations of variables that maximize the differences between the populations being studied, with the objective of establishing a model to sort objects into their appropriate populations with minimal error. At least four major questions are, in a sense, being asked of the data:

- Are there significant differences among the $K$ groups?
- If the groups do exhibit statistical differences, how do the central masses (or centroids, the multivariate equivalent of means) of the populations differ?
- What are the relative distances among the $K$ groups?
- How are new (or at this point unknown) members allocated to establish groups? How do you predict the set of responses of characteristics of an as yet untried exposure case?

The discriminant functions used to produce the linear combinations are of the form:

$$D_I = d_{i1}X_i + d_{i2}Z_2 + \ldots + d_{ip}Z_p$$

where $D_I$ is the score on the discriminant function $I$; the $d$ values are weighting coefficients; and the $Z$ values are standardized values of the discriminating variables used in the analysis.

It should be noted that discriminant analysis can also be used for the hypothesis testing function by the expedient of evaluating how well it correctly classifies members into proper groups (say, control, treatment 1, treatment 2, etc.). Taketomo et al. [104] used discriminant analysis in a retrospective study of gentamycin nephrotoxicity to identify patient risk factors (that is, variables that contributed to a prediction of a patient being at risk).

Finally, we introduce four techniques whose primary function is the reduction of dimensionality: *canonical correlation analysis*, *principal components analysis*, *biplot analysis*, and *correspondence analysis*. Canonical correlation analysis provides the canonical $R$, an overall measure of the relationship between two sets of variables (one set consisting of several outcome measures, the other of several predictor variables). The canonical $R$ is calculated on two numbers for each subject:

$$W_i = \Sigma w_j X_{ij} \text{ and } V_i = \Sigma v_i Y_{ij}$$

where the X values are predictor variables; the Y values are outcome measures; and $W_j$ and $V_j$ are canonical coefficients.

MANOVA can be considered a special case of canonical correlation analysis. Canonical correlation can also be used in hypothesis testing for testing the association of pairs of sets of weights, each with a corresponding coefficient of canonical correlation, each uncorrelated with any of the preceding sets of weights, and each accounting for successively less of the variation shared by the two sets of variables. For example, Young and Matthews [105] used canonical correlation analysis to evaluate the relationship between plant growth and environmental factors at 12 different sites.

The main purpose of principal components analysis is to describe, as economically as possible, the total variance in a sample in a few dimensions; that is, one wishes to reduce the dimensionality of the original data while minimizing the loss of information. It seeks to resolve the total variation of a set of variables into linearly independent composite variables that successively account for the maximum possible variability in the data. The fundamental equation is $Y = AZ$, where $A$ is a matrix of scaled eigenvectors, $Z$ is the original data matrice, and $Y$ represents the principal components. The concentration here, as in factor analysis, is on relationships within a single set of variables. Note that the results of principal components analysis are affected by linear transformations. Cremer and Seville [106] used principal components analysis to compare the difference in blood parameters resulting from each of two separate pyrethroids. Henry and Hidy [107], meanwhile, used principal components analysis to identify the most significant contributors to air quality problems.

The biplot display [108] of multivariate data is a relatively new technique that promises wide applicability to problems in toxicology. It is, in a sense, a form of exploratory data analysis, used for data summarization and description. The biplot is a graphical display of a matrix $Y_{nmx}$ of $N$ rows and $M$ columns by means of row and column markers. The display carries one marker for each row and each column. The "bi" in biplot refers to the joint display of rows and columns. Such plots are used primarily for inspection of data and for data diagnostics when such data are in the form of matrices. Shy-Modjeska et al. [109] illustrated this usage in the analysis of aminoglycoside renal data from beagle dogs, allowing the simultaneous display of relationships among different observed variables and presentation of the relationship of both individuals and treatment groups to these variables.

Correspondence analysis is a technique for displaying the rows and columns of a two-way contingency table as points in a corresponding low-dimensional vector space. As such, it is equivalent to simultaneous linear regression (for contingency table data, such as tumor incidences, which is a very common data form in toxicology), and it can be considered a special case of canonical correlation analysis. The data are defined, described, and analyzed in a geometric framework. This is particularly attractive to such sets of observations in toxicology as multiple endpoint behavioral scores and scored multiple tissue lesions.

A number of good surveys of multivariate techniques are available that are not excessively mathematical [110–112]. More rigorous mathematical treatments on an introductory level are also available [113]. Most of the techniques we have described are available in the better computer statistical packages.

## META-ANALYSIS

Meta-analysis, meaning "analysis among" (and actually entailing analysis of multiple existing analyses), is being used increasingly in biomedical research to try to obtain a qualitative or quantitative synthesis of the research literature on a particular issue. The technique is usually applied to the synthesis of several separate but comparable studies to yield a single answer. Though dating back to the 1930s [114], only recently has it become popular. The process of systematic reviews and meta-analysis has three main components: (1) systematic review and selection of studies plus (2) quantitative and (3) qualitative analyses [115–117].

### Selection of Studies for Analysis: Systematic Reviews

The issue of study selection is perhaps the most problematic for those investigators doing meta-analysis. The criteria for selection may vary from project to project; however, several factors concerning selection must be addressed before analyses commence. Each choice made by the investigator must be weighed carefully as to the likely effect of selection bias vs. the perceived bias that the selection was designed to remove:

- *Use of gray literature*. The current dogma among many scientists is that only peer-reviewed literature is valuable for inclusion in reviews and, therefore, by inference systematic reviews. Should studies be limited to those that are peer reviewed or published? It is well known that negative studies, or those that report little or no benefit from following a particular course of action, are less likely to be published than positive studies; therefore, the published literature may be biased toward studies with positive results, and a synthesis of these studies would give a biased estimate of the impact of pursuing some courses of action. When a systematic review is planned, a plethora of industrial, academic, and government research papers have often been prepared that deal with the issue under consideration. Unfortunately, access to this gray literature is limited, although search engines are now available that can be used to attempt to discover this unpublished information. These studies may give a less biased report on the topic in question; however, some of these unpublished studies may be of lower quality than peer-reviewed materials. Sometimes poor research methods can produce reported results that underestimate the impact, hence providing an opposite bias to that described earlier.
- *Peer review*. As mentioned above, peer review is considered the primary method for quality control in scientific publishing. Should publications in a systematic review and meta-analysis be limited to peer-reviewed articles and, if so, what journals should be included or excluded? The choice of journal may be used as another filter based on the rigor of review and editor latitude given to fill the journal. Some investigators recommend that only those studies that are published in peer-reviewed publications be considered in meta-analysis. Although this may seem an attractive option, it might produce an even more highly biased selection of studies for systematic review.
- *Quality control*. Peer review is not the only method of providing quality control and quality assured data for meta-analysis. Additional quality control and assurance criteria may be used to select the best and most reliable data during systematic review. A rhetorical question we could ask is "Should studies be limited to those that meet additional quality control criteria?" If investigators, undertaking a systematic review, impose an additional set of criteria before including a study in the meta-analysis, the average quality of the studies used should be improved. Contrary to the quality issue is the concern about selection bias. In fact, by placing specific quality filters on data, the investigators may introduce more bias than created by the poor quality data. Moreover, different investigators may use different criteria for a valid study and therefore select a different group of studies for meta-analysis. The result is a possible conflicting output of the meta-analysis.

- *Study design*. Some investigators insist that systematic reviews be limited to randomized controlled studies. Such a limitation produces a variant of the potential bias described earlier. At one time, rigid quality standards were more likely to be met by randomized controlled studies than by observational studies, but this is no longer necessarily the case. Observational methods are currently used to evaluate naturally occurring effects, particularly those that are uncommon. It is quite possible that more important issues, such as combining data from studies performed in different laboratories and using different strains of a single animal species, may result in more systematic error than the study design.
- *Methodology*. Different methodologies can cause differing degrees of systematic bias on output data. This begs the question "Should studies selected for use in meta-analysis be limited to those using identical methods?" This limitation would mean using only separately published studies from the same laboratory in a limited time frame for which the methods were comprehensively monitored and determined to be identical. In practice, application of this filter would massively reduce the number of studies that could be used in the meta-analysis whose power would therefore be decreased greatly. Accordingly, the user must understand the inherent differences between studies and exercise caution and judgment in selecting and rejecting them for use.

### Pooled (Quantitative) Analysis

The main purpose of meta-analysis is to provide a quantitative assessment of the similarity of responses in a number of studies. The goal is to develop better overall estimates of the degree of benefit achieved by specific exposure and dosing techniques based on the combining, or pooling, of estimates found in the existing studies of the interventions. This type of meta-analysis is sometimes called a *pooled analysis* because the analyst pools the observations of many studies and then calculates parameters such as risk or odds ratios from the pooled data. Because of the many decisions regarding inclusion or exclusion of studies, different meta-analyses might reach very different conclusions on the same topic. Even after the studies are chosen, many other methodological issues are involved in choosing how to combine means and variances (e.g., what weighting methods should be used). Pooled analysis should report relative risks and risk reductions as well as absolute risks and risk reductions.

### Methodological (Qualitative) Analysis

Sometimes the question to be answered is not how much toxicity is induced by the use of a particular exposure but whether there is any biologically significant toxicity. In this case, a qualitative meta-analysis may be done, in which the quality of the research is scored according to a list of objective criteria. The analyst then examines the methodologically superior studies to determine whether the question of toxicity is answered consistently by them. This qualitative approach has been referred to as *methodological analysis* or *quality scores analysis*. In some cases, the methodologically strongest studies agree with one another and disagree with the weaker studies. These weaker studies may be consistent with one another.

## BAYESIAN INFERENCE

Sensitivity and specificity of a test are important to characterize to understand the accuracy and precision of data generated. Once a researcher decides to use a certain test to diagnose an illness, two important questions require answers. First, if the test results are positive, what is the probability that the researcher has uncovered the condition of interest? Second, if the test results are negative, what is the probability that the patient does not have the disease? Bayes' theorem provides a method to answer these two questions. The English clergyman after whom it is named first described the theorem centuries ago [118]. It is one of the most imposing statistical formulas in the biomedical sciences. Put in symbols more meaningful for researchers such as toxicologists and pathologists [119], the formula is as follows:

$$P(D \mid T+) = \frac{p(T+ \mid D+)p(D+)}{[p(T+ \mid D+)p(D+)] + [p(T+ \mid D-)p(D-)]}$$

where $p$ denotes probability, $D+$ means that the animal has the effect in question, $D-$ means that the animal does not have the effect, $T+$ means that a certain diagnostic test for the effect is positive, $T-$ means that the test is negative, and the vertical line (|) means "conditional upon" what immediately follows.

Most researchers, who have to address sensitivity, specificity, and predictive values, often do not wish to use Bayes' theorem; however, this is a useful formula. Closer examination of the equation reveals that Bayes' theorem is merely the formula for the positive predictive value.

The numerator of Bayes' theorem describes cell *a*, the true positive results, in a 2×2 table. The probability of being in cell *a* is equal to the prevalence multiplied by the sensitivity, where $p(D+)$ is the prevalence (i.e., the probability of being in the affected column) and $p(T+|D+)$ is the sensitivity (i.e., the probability of being in the top row, given the fact of being in the affected column). The denominator of Bayes' theorem consists of two terms, the first of which again describes cell $a$, the true positive results, and the second of which describes cell $b$, the false-positive error rate. This rate can be represented by $p(T+|D-)$, which is multiplied by the prevalence of unaffected animals, or $p(D-)$. True positive results ($a$) divided by true positive plus false-positive results ($a + b$) gives $a/(a + b)$, the positive predictive value.

In genetics, a simpler formula for Bayes' theorem is sometimes used. The numerator is the same, but the denominator is $p(T+)$. This makes sense because the denominator in $a/(a + b)$ is equal to all of those who have positive test results, whether they are true positive or false-positive results.

## BAYES' THEOREM IN THE EVALUATION OF SAFETY ASSESSMENT STUDIES

In a population with a low prevalence of a particular toxicity, most of the positive results in a screening program for that lesion or effect would be falsely positive. Although this does not automatically invalidate a study or assessment program, it does raise some concerns about cost effectiveness, which can be explored using Bayes' theorem.

An example to illustrate Bayes' theorem is a study employing an immunochemical stain-based test to screen tissues for a specific effect. This test uses small amounts of antibody, and the presence of an immunologically bound stain is considered a positive result. If the sensitivity and specificity of the test and the prevalence of biochemical effect are known, Bayes' theorem can be used to predict what proportion of the tissues with positive test results will have true positive results (i.e., truly showing the effect).

Figure 8.10 shows how the calculations are made. If the test has a sensitivity of 96% and if the true prevalence is 1%, then only 13.9% of tissues predicted showing a positive test result actually will be true positives. Pathologists and toxicologists can quickly develop a table that lists different levels of test sensitivity, test specificity, and effect prevalence and shows how these levels affect the proportion of positive results that are likely to be true positive results. Although this calculation is fairly straightforward and is extremely useful, it seldom has been used in the early stages of planning for large studies or safety assessment programs.

**PART 1 Initial data**

Sensitivity of immunological stain = 96% = 0.96
False-negative error rate of the test = 04% = 0.04
Specificity of the test = 94% = 0.94
False-positive error rate of the test = 06% = 0.06
Prevalence of effect in the tissues = 01% = 0.01

**PART 2 Use Bayes' theorem**

$$p(D+|T+) = \frac{p(T+|D+)p(D+)}{[p(T+|D+)p(D+)]+[p(T+|D-)p(D-)]}$$

$$= \frac{\text{(Sensitivity)(Prevalence)}}{[\text{(Sensitivity)(Prevalence)}] + \text{(False-Positive Error Rate)}(1 - \text{Prevalence})}$$

$$= \frac{(0.96)(0.01)}{[(0.96)(0.01)]+[(0.06)(0.99)]}$$

$$= \frac{0.0096}{0.0096 + 0.0594} = \frac{0.0096}{0.0690} = 0.139 = 13.9\%$$

**PART 3 Use of a 2×2 table, with numbers based on the assumption that 10,000 tissues are in the study**

**True Disease Status**

| **Test Result** | **Number Affected** | **Not Affected** | **Total** |
|---|---|---|---|
| Positive | 96 (96%) | 594 (6%) | 690 (7%) |
| Negative | 4 (4%) | 9306 (94%) | 9310 (93%) |
| Total | 100 (100%) | 9900 (100%) | 10,000 (100%) |

Positive predictive value = 96/690 = 0.139 = 13.9%.

**FIGURE 8.10** Use of Bayes' theorem or a 2×2 table to determine the positive predictive value of a hypothetical tuberculin-screening program.

## BAYES' THEOREM AND INDIVIDUAL ANIMAL EVALUATION

Uncertainty concerning the exact cause of death of an animal is a problem that faces most toxicologic pathologists. Suppose a toxicologic pathologist is uncertain about an animal's cause of death and has a positive test result, such as in the example given earlier. Even if the toxicologic pathologist knows the sensitivity and specificity of the test in question, interpretation is still problematic. In order to calculate the positive predictive value, it is necessary to know the prevalence of the particular true tissue/effect that the test is designed to detect. The prevalence is thought of as the expected prevalence in the population from which the animal comes. The actual prevalence is usually not known, but usually an estimate is attempted.

An example of such a situation is when a pathologist evaluates a male primate observed to have fatigue and signs of kidney stones. No other clinical signs of parathyroid disease are detected on physical examination. The toxicologic pathologist considers the possibility of hyperparathyroidism and arbitrarily decides that its prevalence is perhaps 2%, reflecting that in 100 such primates probably only 2 of them would have the disease. This probable disease prevalence is referred to as the *prior probability*,

**PART 1 Initial data**

Sensitivity of the first test = 90%= 0.90
Specificity of the first test = 95%= 0.95

**PART 2 Use Bayes' theorem**

$$p(D+|T+) = \frac{p(T+|D+)p(D+)}{[p(T+|D+)p(D+)]+[p(T+|D-)p(D-)]}$$

$$= \frac{(0.90)(0.02)}{[(0.90)(0.02)]+[(0.05)(0.98)]}$$

$$= \frac{0.018}{0.018+0.049} = \frac{0.018}{0.067} = 0.269 = 27\%$$

**PART 3 Use of a 2×2 table**

**True Disease Status**

| **Test Result** | **Number Affected** | **Not Affected** | **Total** |
|---|---|---|---|
| Positive | 18 (90%) | 49 (5%) | 67 (6.7%) |
| Negative | 2 (10%) | 931 (95%) | 933 (93.3%) |
| Total | 20 (100%) | 980 (100%) | 1000 (100.0%) |

Positive predictive value = 18/67 = 0.269 = 27%.

**FIGURE 8.11** Use of Bayes' theorem or a 2×2 table to determine posterior probability and positive predictive values.

reflecting the fact that it is estimated prior to the performance of laboratory tests. This probability is based on the estimated prevalence of a particular pathology among primates with similar signs and symptoms. Although the toxicologic pathologist believes that the probability of hyperparathyroidism is low, he considers the serum calcium concentrations to rule out the diagnosis. Somewhat to his surprise, the results of the test are positive, with an elevated level of calcium of 12.2 mg/dL. The pathologist could order other tests for parathyroid disease, but some test results may be positive and some negative for a number of reasons.

Under these circumstances, Bayes' theorem could be used to make a second estimate of probability, referred to as the *posterior probability*, reflecting the fact that this determination is made after the test results are known. Calculation of the posterior probability is based on the sensitivity and specificity of the test that was performed, which in this case was elevated serum calcium, and on the prior probability, which in this case was set at 2%. If the serum calcium test had a 90% sensitivity and a 95% specificity, a false-positive error rate of 5% would be expected. Note that specificity plus the false-positive error rate always equals 100%.

When this information is used in the Bayes' equation, as shown in Figure 8.11, the result is a posterior probability of 27%. This means that the animal in question is now within a group of primates with a significant possibility of parathyroid disease. In Figure 8.11, note

**PART 1 Initial data**

Sensitivity of the first test = 95%= 0.95
Specificity of the first test = 98%= 0.94
Prior probability of disease = 27% = 0.27

**PART 2 Use Bayes' theorem**

$$p(D+|T+) = \frac{p(T+|D+)p(D+)}{[p(T+|D+)p(D+)]+[p(T+|D-)p(D-)]}$$

$$= \frac{(0.95)(0.27)}{[(0.95)(0.27)]+[(0.02)(0.73)]}$$

$$= \frac{0.257}{0.257+0.049} = \frac{0.257}{0.272} = 0.9449^{a} = 94\%$$

**PART 3 Use of a 2×2 table**

**True Disease Status**

| **Test Result** | **Number Affected** | **Not Affected** | **Total** |
|---|---|---|---|
| Positive | 256 (95%) | 15 (2%) | 271 (27.1%) |
| Negative | 13 (5%) | 716 (98%) | 729 (72.9%) |
| Total | 269 (100%) | 731 (100%) | 1000 (100.0%) |

Positive predictive value = 256/271 = 0.9446[a] = 94%.

[a]The slight difference in the results for the two approaches is due to rounding errors. It is not important biologically.

**FIGURE 8.12** Use of Bayes' theorem or a 2×2 table to determine second posterior probability and second positive predictive values.

that the result is the same when a 2×2 table is used (i.e. 27%). This is true because the probability based on the Bayes' theorem is identical to the positive predictive value.

In light of the 27% posterior probability,the pathologist decides to order a parathyroid hormone radioimmunoassay, even though this test is expensive. If the radioimmunoassay had a sensitivity of 95% and a specificity of 98% and the results turned out to be positive, the Bayes theorem could again be used to calculate the probability of parathyroid disease. This time, however, the posterior probability for the first test (27%) would be used as the prior probability for the second test. The result of the calculation, as shown in Figure 8.12, gives a new probability of 94%. Thus, the primate in all probability did have hyperparathyroidism.

The reader may be wondering why the posterior probability increased so much the second time. One reason was that the prior probability was considerably higher in the second calculation compared to the first (27% vs. 2%) based on the fact that the first test yielded positive results. Another reason was that the specificity of the second test was high (98%), which markedly reduced the false-positive error rate and therefore increased the positive predictive value.

**TABLE 8.11**
**Classification of Data Commonly Encountered in Toxicology**

| Type of Data | Examples |
|---|---|
| Continuous normal | Body weights |
| | Food consumption |
| | Organ weights: absolute and relative |
| | Mouse ear swelling test (MEST) measurements |
| | Pregnancy rates |
| | Survival rates |
| | Crown–rump lengths |
| | Hematology (some) |
| | Clinical chemistry (some) |
| Continuous but not normal | Hematology (some; WBC) |
| | Clinical chemistry (some) |
| | Urinalysis |
| Scalar data | Neurobehavioral signs (some) |
| | PDI scores |
| | Histopathology (some) |
| Count data | Resorption sites |
| | Implantation sites |
| | Stillborns |
| | Hematology (some; reticulocyte counts, Howel–Jolly, WBC differentials) |
| Categorical data | Clinical signs |
| | Neurobehavioral signs (some) |
| | Ocular scores |
| | GP sensitization scores |
| | Mouse ear swelling test (MEST) sensitization |
| | Counts |
| | Fetal abnormalities |
| | Dose/mortality data |
| | Sex ratios |
| | Histopathology data (most) |

## Assumptions and Limitations

1. Test results must be independent of each other. This also means that the population remaining after one test must have the same proportional response to the following tests as the original population did.
2. If the calculations are done on an iterative basis, care must be taken to correct for cumulative round-off errors.

## DATA ANALYSIS APPLICATIONS IN TOXICOLOGY

Having reviewed basic principles and provided a set of methods for statistical handling of data, the remainder of this chapter addresses the practical aspects and difficulties encountered in day-to-day toxicological work. As a starting point, we present in Table 8.11 an overview of data types actually encountered in toxicology, classified by type (as presented earlier). It should be stressed, however, that this classification is of the most frequent measure of each sort of observation (such as body weight) and will not always be an accurate classification. There are now common practices in the analysis of toxicology data, although they are not necessarily the best. They are discussed in the remainder of this chapter, which seeks to review statistical methods on a use-by-use basis and to provide a foundation for the selection of alternatives in specific situations.

### MEDIAN LETHAL AND EFFECTIVE DOSES

For many years, the starting point for evaluating the toxicity of an agent was to determine its $LD_{50}$ or $LC_{50}$, which are the dose or concentration, respectively, of a material at which half of a population of animals would be expected to die. These figures are analogous to the $ED_{50}$ (effective dose for half a population) used in pharmacologic activities and are derived by the same means. To calculate either of these figures we need, at each of several dosage (or exposure) levels, the number of animals dosed and the number that died. If we seek only to establish the median effective dose in a range-finding test, then four or five animals per dose level, using Thompson's method of moving averages, is the most efficient methodology and will give a sufficiently accurate solution. With two dose levels, if the ratio between the high and low dose is two or less, even total or no mortality at these two dose levels will yield an acceptably accurate medial lethal dose, although a partial mortality is desirable. If, however, we wish to estimate a number of toxicity levels ($LD_{10}$, $LD_{90}$) and are interested in more precisely establishing the slope of the dose/lethality curve, the use of at least ten animals per dosage level with the probit/log regression technique is the most common approach. Note that in the equation $Y_i = a + bx_1$, $b$ is the slope of the regression line, and that our method already allows us to calculate 95% confidence intervals about any point on this line. The confidence interval at any one point will be different from the interval at other points and must be calculated separately. Additionally, the nature of the probit transform is such that toward the extremes—$LD_{10}$ and $LD_{90}$, for example—the confidence intervals will balloon. That is, they become very wide. Because the slope of the fitted line in these assays has a very large uncertainty, in relation to the uncertainty of the $LD_{50}$ itself (the midpoint of the distribution), much caution must be used with calculated $LD_x$ values other than $LD_{50}$. The imprecision of the $LD_{35}$, a value close to the $LD_{50}$, is discussed by Weil [120], as is that of the slope of the log dose–probit line [121]. Debanne and Haller [122] recently reviewed the statistical aspects of different methodologies for estimating a median effective dose.

There have been questions for years as to the value of $LD_{50}$ and the efficiency of the current study design (which uses large numbers of animals) in determining it. As long ago as 1953, Weil et al. [123] presented forceful arguments that an estimate having only minimally reduced precision could be made using significantly fewer animals. More recently, the last few years have seen an increased level of concern over the numbers and uses of animals in research and testing and have produced additional arguments against existing methodologies for determining the $LD_{50}$ or even the need to make the determination at all [124]. In response, several suggestions for alternative methodologies have been advanced [125–127].

## Body and Organ Weights

Among the sets of data commonly collected in studies where animals are dosed with (or exposed to) a chemical are body weight and the weights of selected organs; in fact, body weight is frequently the most sensitive indication of an adverse effect. How to best analyze this and in what form to analyze the organ weight data (as absolute weights, weight changes, or percentages of body weight) have been the subject of a number of articles [128–131].

Both absolute body weights and rates of body weight change (calculated as changes from a baseline measurement value that is traditionally the animal's weight immediately prior to the first dosing with or exposure to the test material) are almost universally best analyzed by ANOVA followed, if called for, by a *post hoc* test. Even if the groups were randomized properly at the beginning of a study (no group being significantly different in mean body weight from any other group and all animals in all groups within two standard deviations of the overall mean body weight), there is an advantage to performing the computationally slightly more cumbersome (compared to absolute body weights) changes in body weight analysis. The advantage is an increase in sensitivity, because the adjustment of starting points (the setting of initial weights as a zero value) acts to reduce the amount of initial variability. In this case, Bartlett's test is performed first to ensure homogeneity of variance and the appropriate sequence of analysis follows. With smaller sample sizes, the normality of the data becomes increasingly uncertain and nonparametric methods such as Kruskal–Wallis may be more appropriate [25].

The analysis of relative (to body weight) organ weights is a valuable tool for identifying possible target organs [126]. How to perform this analysis is still a matter of some disagreement, however. Weil [129] presented evidence that organ weight data expressed as percentages of body weight should be analyzed separately for each sex. Furthermore, because the conclusions from organ weight data of males differed so often from those of females, data from animals of each sex should be used in this measurement. Others [130,132–134] have discussed in detail other factors that influence organ weights and must be taken into account.

The two competing approaches to analyzing relative organ weights call for either of the following [131]:

- Calculate organ weights as a percentage of total body weight (at the time of necropsy) and analyzing the results by ANOVA.
- Analyze results by ANCOVA with body weights as the covariates, as discussed previously.

A number of considerations should be kept in mind when these questions are addressed. First, one must keep a firm grasp on the difference between biological significance and statistical significance. In this particular case we are especially interested in examining organ weights when an organ weight change is not proportional to changes in whole body weights. Second, we are now required to detect smaller and smaller changes while still retaining a similar sensitivity (i.e., the $p < 0.05$ level). Several devices are available to attain the desired increase in power. One is to use larger and larger sample sizes (number of animals) and the other is to utilize the most powerful test we can; however, the use of even currently employed numbers of animals is being vigorously ques tioned, and the power of statistical tests must, therefore, now assume an increased importance in our considerations.

The biological rationale behind analyzing both absolute body weight and the ratio of organ weight to body weight (this latter as opposed to a covariance analysis of organ weights) is that, in the majority of cases, except for the brain, the organs of interest in the body change weight (except in extreme cases of obesity or starvation) in proportion to total body weight. We are particularly interested in detecting cases where this is not so. Analysis of actual data from several hundred studies (unpublished data) has shown no significant difference in rates of weight change of target organs (other than the brain) compared to total body weight for healthy animals in those species commonly used for repeated dose studies (rats, mice, rabbits, and dogs). Furthermore, it should be noted that analysis of covariance is of questionable validity in analyzing body weight and related organ weight changes, because a primary assumption is the independence of treatment—that the relationship of the two variables is the same for all treatments [135]. Plainly, in toxicology this is not true.

In cases where the differences between the error mean squares are much greater, the ratio of $F$ ratios will diverge in precision from the result of the efficiency of covariance adjustment. These cases are where either sample sizes are much larger or where the differences between means themselves are much larger. This latter case is one that does not occur in the designs under discussion in any manner that would leave analysis of covariance as a valid approach, because group means start out being very similar and cannot diverge markedly unless there is a treatment effect. As we have discussed earlier, a treatment effect invalidates a prime underpinning assumption of analysis of covariance. Shirley and Newnham [136] have argued the case for ANCOVA, but without providing answers to arguments presented above.

## Clinical Chemistry

Several clinical chemistry parameters are commonly determined from the blood and urine collected from animals in chronic, subchronic, and occasionally acute toxicity studies. In the past (and still, in some places), the accepted practice has been to evaluate these data using univariate–parametric methods (primarily $t$-tests and/or ANOVA); however, this can be shown not to be the best approach on a number of grounds.

First, such biochemical parameters are rarely independent of each other, and our interest often is not focused on just one of the parameters; rather, there are batteries of the parameters associated with toxic actions at particular target organs. For example, increases in creatinine phosphokinase (CPK), γ-hydroxybutyrate dehydrogenase (γ-HBDH), and lactate dehydrogenase (LDH), occurring together, are strongly indicative of myocardial damage. In such cases, we are not just interested in a significant increase in one of these, but in all three. Table 8.12 gives a brief overview of the association of various parameters with actions at particular target organs. A more detailed coverage of the interpretation of such clinical laboratory tests can be found in other references [137–140].

Similarly, the serum electrolytes (sodium, potassium, and calcium) interact with each other; a decrease in one is frequently tied, for instance, to an increase in one of the others. Furthermore, the nature of the data (in the case of some parameters), either because of the biological nature of the parameter or the way in which it is measured, is frequently either not normally distributed (particularly because of being markedly skewed) or not continuous in nature. This can be seen in some of the reference data for experimental animals in Mitruka and Rawnsley [141] or Weil [142] in, for example, creatinine, sodium, potassium, chloride, calcium, and blood.

## Hematology

Much of what we said about clinical chemistry parameters is also true for the hematologic measurements made in toxicology studies. Which test to perform should be evaluated by use of a decision tree until one becomes confident as to the most appropriate methods. Keep in mind that sets of values and (in some cases) population distribution vary not only between species but also between the commonly used strains of species and that control or standard values will drift over the course of only a few years.

Again, the majority of these parameters are interrelated and highly dependent on the method used to determine them. Red blood cell (RBC) count, platelet counts, and mean corpuscular volume (MCV) may be determined using a device such as a Coulter counter to take direct measurements, and the resulting data are usually stable for parametric methods. The hematocrit, however, may actually be a value calculated from the RBC and MCV values and, if so, is dependent on them. If the hematocrit is measured directly, instead of being calculated from the RBC and MCV, it may be compared by parametric methods. (See Table 8.13.)

Hemoglobin is directly measured and is an independent and continuous variable However, and probably because at any one time a number of forms and conformations (oxyhemoglobin, deoxyhemoglobin, methemoglobin, etc.) of hemoglobin are actually present, the distribution seen is not typically a normal one but rather may be a multimodal one. Here, a nonparametric technique such as the Wilcoxon or multiple rank-sum analysis is called for.

Consideration of the white blood cell (WBC) and differential counts leads to another problem. The total WBC is, typically, a normal population amenable to parametric analysis, but differential counts are normally determined by counting, manually, one or more sets of 100 cells each.

**TABLE 8.12**
**Association of Changes in Biochemical Parameters with Actions at Particular Target Organs**

| Parameter | Blood | Heart | Lung | Kidney | Liver | Bone | Intestine | Pancreas | Notes |
|---|---|---|---|---|---|---|---|---|---|
| Albumin | | | | ↓ | ↓ | | | | Produced by the liver; very significant reductions indicate extensive liver damage |
| ALP | | | | | ↑ | ↑ | ↑ | | Elevations usually associated with cholestasis; bone alkaline phosphatase tends to be higher in young animals |
| Bilirubin (total) | ↑ | | | | ↑ | | | | Usually elevated due to cholestasis, due to either obstruction or hepatopathy |
| BUN | | | | ↑ | ↓ | | | | Estimates blood filtering capacity of the kidneys; does not become significantly elevated until the kidney function is reduced 60–75% |
| Calcium | | | | ↑ | | | | | Can be life threatening and result in acute death |
| Cholinesterase | | | | ↑ | ↓ | | | | Found in plasma, brain, and RBC |
| CPK | | ↑ | | | | | | | Most often elevated due to skeletal muscle damage but can also be produced by cardiac muscle damage; can be more sensitive than histopathology |
| Creatinine | | | | ↑ | | | | | Also estimates blood filtering capacity of kidney as BUN does |
| Glucose | | | | | | | | ↑ | Alterations other than those associated with stress uncommon and reflect an effect on the pancreatic islets or anorexia |
| GGT | | | | | ↑ | | | | Elevated in cholestasis; this is a microsomal enzyme, and levels often increase in response to microsomal enzyme induction |
| HBDH | | ↑ | | | ↑ | | | | — |
| LDH | | ↑ | ↑ | ↑ | ↑ | | | | Increase usually due to skeletal muscle, cardiac muscle, or liver damage; not very specific |
| Protein (total) | | | | ↓ | ↓ | | | | Absolute alterations usually associated with decreased production (liver) or increased loss (kidney); can see increase in case of muscle wasting (catabolism) |
| SGOT | | ↑ | | ↑ | ↑ | | | ↑ | Present in skeletal muscle and heart and most commonly associated with damage to these |
| SGPT | | | | | ↑ | | | | Elevations usually associated with hepatic damage or disease |
| SDH | | | | | ↑↓ | | | | Liver enzyme that can be quite sensitive but is fairly unstable; samples should be processed as soon as possible |

*Note:* ↑, increase in chemistry values; ↓, decrease in chemistry values; ALP, alkaline phosphatase; BUN, blood urea nitrogen; CPK, creatinine phosphokinase; GGT, gamma glutamyl transferase; HBDH, hydroxybutyric dehydrogenase; LDH, lactic dehydrogenase; RBC, red blood cells; SDH, sorbitol dehydrogenase; SGOT, serum glutamic oxaloacetic transaminase (also called AST [aspertate amino transferase]); SGPT, serum glutamic pyruvic transaminase (also called ALT [atanine amino transferase]).

The resulting relative percentages of neutrophils are then reported as either percentages or are multiplied by the total WBC count with the resulting count being reported as the absolute differential WBC. Such data, particularly in the case of eosinophils (where the distribution does not approach normality), should usually be analyzed by nonparametric methods. It is widely believed that relative (%) differential data should not be reported because they are likely to be misleading.

Finally, it should always be kept in mind that it is rare for a change in any single hematologic parameter to be meaningful. Rather, because these parameters are so interrelated, patterns of changes in parameters should be expected if a real effect is present, and analysis and interpretation of results should focus on such patterns of changes. Classification analysis techniques often provide the basis for a useful approach to such problems.

## Histopathologic Lesion Incidence

The last 20 years have seen increasing emphasis placec on histopathological examination of tissues collected fron animals in subchronic and chronic toxicity studies. It i not true that only those lesions that occur at a statisticall significantly increased rate in treated or exposed animal are of concern, for in some cases a lesion may be of sucl a rare type that the occurrence of only one or a few sucl in treated animals raises a red flag. It is true, however, tha in most cases a statistical evaluation is the only way t determine if what we see in treated animals is significantl worse than what has been seen in control animals. And although cancer is not our only concern, this category o lesions is that of greatest interest.

Typically, comparison of incidences of any one typ of lesion between controls and treated animals are mad

**TABLE 8.13**
**Some Probable Conditions Behind Hematological Changes**

| Parameter | Elevation | Depression |
|---|---|---|
| Red blood cells | Vascular shock<br>Excessive diuresis<br>Chronic hyposia<br>Hyperadrenocorticism | Anemias: blood loss, hemolysis, low RBC production |
| Hematocrit | Increased RBC<br>Stress<br>Shock: trauma, surgery<br>Polycythemia | Anemias<br>Pregnancy<br>Excessive hydration |
| Hemoglobin | Polycythemia (increase in production of RBC) | Anemias<br>Lead poisonings |
| Mean cell volume | Anemias<br>B12 deficiency | Iron deficiency |
| Mean corpuscular hemoglobin | Reticulocytosis | Iron deficiency |
| White blood cells | Bacterial infections<br>Bone marrow stimulation | Bone marrow depression<br>Cancer chemotherapy<br>Chemical intoxication<br>Splenic disorders |
| Platelets | — | Bone marrow depression<br>Immune disorder |
| Neutrophils | Acute bacterial infections<br>Tissue necrosis<br>Strenuous exercise<br>Convulsions<br>Tachycardia<br>Acute hemorrhage | — |
| Lymphocytes | Leukemia<br>Malnutrition<br>Viral infections | — |
| Monocytes | Protozoal infections | — |
| Eosinophils | Allergy<br>Irradiation<br>Pernicious anemia<br>Parasitism | — |
| Basophils | Lead poisoning | — |

*Sources:* Hayes, A.W., Ed., *Principles and Method of Toxicology*, 4th ed., Taylor & Francis, Philadelphia, PA, 2001; Minckler, J. et al., *Pathology: An Introduction*, Mosby, St. Louis, MO, 1971; Thomas, H.C., *Handbook of Automated Electronic Clinical Analysis*, Reston Publishing, Reston, VA, 1979.

ısing the multiple 2×2 chi-square test or Fisher's exact est with a modification of the numbers of animals as the lenominators. Too often, experimenters exclude from conideration all those animals (in both groups) that died prior o the first animals being found with a lesion at that site. 'he special case of carcinogenicity bioassays will be disussed in detail in the next chapter.

An option that should be kept in mind is that, frequently, pathologist can not only identify a lesion as present but lso grade those present as to severity. This represents a ignificant increase in the information content of the data hat should not be given up by performing an analysis based nly on the perceived quantal nature (present/absent) of the lata. Quantal data, analyzed by chi-square or Fisher's exact ests, are a subset (the 2×2 case) of categorical or continency table data. In this case, it also becomes ranked (or rdinal) data; the categories are naturally ordered (for example, no effect < mild lesion < moderate lesion < severe esion). This gives a 2×R table if there are only one treatnent and one control group, or an N×R (multiway) table if here are three or more groups of animals.

The traditional method of analyzing multiple, crosslassified data has been to collapse the N×R contingency table over all but two of the variables, following this with the computation of some measure of association between these variables. For an $N$-dimensional table this results in $N(N-1)/2$ separate analyses. The result is crude, giving away information and even (by inappropriate pooling of data) yielding a faulty understanding of the meaning of data. Though computationally more laborious, a multiway (N×R table) analysis should be utilized.

## Reproduction

The reproductive implications of the toxic effects of chemicals are being increasingly important. Because of this, reproduction studies, together with other closely related types of studies (such as teratogenesis, dominant lethal, and mutagenesis studies), are now commonly companion to chronic toxicity studies. One point that must be kept in mind with all reproduction-related studies is the nature of the appropriate sampling unit. What is the appropriate $N$ in such a study: the number of individual pups, the number of litters, the number of pregnant females? Fortunately, it is now fairly well accepted that the first case (using the number of offspring as the $N$) is inappropriate [130]. The real effects in such studies actually occur in the female that was exposed to the chemical or is mated to a male that was exposed. What happens to her and to the development of the litter she is carrying is biologically independent of what happens to every other female or litter in the stud. This cannot be said for each offspring in each litter; for example, the death of one member of a litter can and will be related to what happens to every other member. Also, the effect on all of the offspring might be similar for all of those from one female and different or lacking for those from another.

As defined by Oser and Oser [143], four primary variables are of interest in a reproduction study. First is the fertility index (FI), which may be defined as the percentage of attempted matings (i.e., each female housed with a male) that resulted in pregnancy, pregnancy being determined by a method such as the presence of implantation sites in the female. Second is the gestation index (GI), which is defined as the percentage of mated females, as evidenced by a vaginal plug being dropped or a positive vaginal smear, that deliver viable litters (i.e., litters with at least one live pup). Two related variables that may also be studied are the mean number of pups born per litter and the percentage of total pups per litter that are stillborn. Third is the viability index (VI), which is defined as the percentage of offspring born that survive at least 4 days after birth. The last in this four-variable system is the lactation index (LI), which is the percentage of animals per litter that survive 4 days and also survive to weaning. In rats and mice, this is classically taken to be 21 days after birth. An additional variable that may reasonably be included in such a study is the mean weight gain per pup per litter.

Given that our $N$ is at least 10, we may test each of these variables for significance using a method such as the Wilcoxon–Mann–Whitney U test or the Kruskal–Wallis nonparametric ANOVA. If $N$ is less than 10, we cannot expect the central limit theorem to be operative and should use the Wilcoxon sum of ranks (for two groups) or the Kruskal–Wallis nonparametric ANOVA (for three or more groups) to compare groups.

## Developmental Toxicology

When the primary concern of a reproductive/developmental study is the occurrence of birth defects or deformations (terata, either structural or functional) in the offspring of exposed animals, the study is one of developmental toxicology (teratology). In the analysis of the data from such a study, we must consider several points. First is sample size. Earlier, a method to estimate sufficient sample size was presented. The difficulties with applying these methods here revolve around two points: (1) selecting a sufficient level of sensitivity for detecting an effect, and (2) factoring in how many animals will be removed from study (without contributing a datum) by either not becoming pregnant or not surviving to a sufficiently late stage of pregnancy. Experience generally dictates that one should attempt to have 20 pregnant animals per study group if a pilot study has provided some confidence that the pregnant test animals will survive the dose levels selected. Again, it is essential to recognize that the litter, not the fetus, is the basic independent unit for each variable.

A more fundamental consideration, alluded to in the section on reproduction, is that as we use more animals, the mean of means (each variable will be such in a mathematical sense) will approach normality in its distribution. This is one of the implications of the central limit theorem; even when the individual data are not normally distributed, their means will approach normality in their distribution. At a sample size of 10 or greater, the approximation of normality is such that we may use a parametric test (such as a $t$-test or ANOVA) to evaluate results. At sample sizes less than 10, a nonparametric test (Wilcoxon rank-sum or Kruskal–Wallis nonparametric ANOVA) is more appropriate. Other methodologies have been suggested [144,145] but do not offer any prospect of widespread usage. One nonparametric method that is widely used is the Mann–Whitney U test, which was described earlier. Williams and Buschbom [146] further discuss some of the available statistical options and their consequences, and Rai and Ryzin [147] have recommended a dose responsive model.

## Dominant Lethal Assay

The dominant lethal study is essentially a reproduction study that seeks to study the endpoint of lethality to the fetuses after implantation and before delivery. The proper identification of the sampling unit (the pregnant female) and the design of an experiment so a sufficiently large sample is available for analysis are the primary statistical considerations. The question of sampling unit has been adequately addressed in earlier sections. Sample size is of concern here because the hypothesis-testing techniques that are appropriate with small samples are of relatively low power, as the variability about the mean in such cases is relatively large. With sufficient sample size (e.g., from 30 to 50 pregnant females per dose level per week [148]) variability about the mean and the nature of the distribution allow sensitive statistical techniques to be employed.

The variables that are typically recorded and included in analysis are (for each level/week): (1) the number of pregnant females, (2) live fetuses per pregnancy, (3) total implants per pregnancy, (4) early fetal deaths (early resorptions)per pregnancy, and (5) late fetal deaths per pregnancy.

A wide variety of techniques for analysis of these data have been used. Most common is the use of ANOVA after the data have been transformed by the arc sine transform [149]. Beta binomial [150,151] and Poisson distributions [152] have also been attributed to these data, and trans forms and appropriate tests have been proposed for use in each of these cases (in each case with the note that the transforms serve to stabilize the variance of the data). With sufficient sample size, as defined earlier in this section, the Mann–Whitney U test is recommended for use here. Smaller sample sizes necessitate the use of the Wilcoxon rank-sum test.

### Diet and Chamber Analysis

Earlier we presented the basic principles and methods for sampling. Sampling is important in many aspects of toxicology, and here we address its application to diet preparation and the analysis of atmospheres from inhalation chambers. In feeding studies, we seek to deliver doses of a material to animals by mixing the material with their diet. Similarly, in an inhalation study we mix a material with the air the test animals breathe. In both cases, we must then sample the medium (food or atmosphere) and analyze these samples to determine what levels or concentrations of material were actually present and to assure ourselves that the test material is homogeneously distributed. Having an accurate picture of these delivered concentrations, and how they varied over the course of time, is essential on a number of grounds:

1. The regulatory agencies and sound scientific practice require that analyzed diet and mean daily inhalation atmosphere levels be ±10% of the target level.
2. Excessive peak concentrations, because of the overloading of metabolic repair systems, could result in extreme acute effects that would lead to results in a chronic study that are not truly indicative of the chronic low-level effects of the compound but rather of periods of metabolic and physiologic overload. Such results could be misinterpreted if true exposure or diet levels were not maintained at a relatively constant level.

Sampling strategies are not just a matter of numbers (for statistical aspects) but also of geometry, so the contents of a container or the entire atmosphere in a chamber can be truly sampled; it is also a matter of time, with regard to the stability of the test compound. The samples must be both randomly collected and representative of the entire mass of what one is trying to characterize. In the special case of sampling and characterizing the physical properties of aerosols in an inhalation study, some special considerations and terminology apply. Because of the physiologic characteristics of the respiration of humans and of test animals, our concern is very largely limited to those particles or droplets that are of a respirable size. Unfortunately, respirable size is a complex characteristic based on aerodynamic diameter, density, and physiological characteristics. Although particles with an aerodynamic diameter of less than 10 microns are generally agreed to be respirable in humans (that is, they can be drawn down to the deep portions of the lungs), 3 microns in aerodynamic diameter is a more realistically value. Typically, it then becomes a matter of calculating measures of central tendency and dispersion statistics, with the identification of those values that are beyond acceptable limits [50].

### Genotoxicity

In the last 25 years, a wide variety of tests [153] for genotoxicity have been developed and brought into use. These tests give us a quicker and less expensive (although not as conclusive) way of predicting whether a material of interest is a mutagen, and possibly a carcinogen, than do longer term, whole-animal studies. How to analyze the results of the multitude of tests (Ames, DNA repair, micronucleus, chromosome aberration, cell transformation, and sister chromatid exchange, to name a few) is an extremely important question. Some workers in the field hold that it is not possible (or necessary) to perform statistical analysis, that the tests can simply be judged to be positive or not positive on the basis of whether or not they achieve a particular increase in the incidence of mutations in the test organism. Quantitations of potency are complicated by the fact that we are dealing with a nonlinear phenomenon; although low doses of most genotoxicants produce a linear response curve with increasing dose, the curve will flatten out (and even turn into a declining curve) as the higher doses provoke an acute response.

Several concepts different from those we have previously discussed need to be examined, for our concern has now shifted from how a multicellular organism acts in response to one of a number of complex actions to how a mutational event is expressed, most frequently by a single cell. Given that we can handle much larger numbers of experimental units in systems that use smaller test organisms, we can seek to detect both weak and strong mutagens.

Conducting the appropriate statistical analysis and utilizing the results of such an analysis properly must begin with an understanding of the biological system involved and, from this understanding, the correct model and hypothesis must be developed. We begin such a process by considering each of five interacting factors [154,155]:

1. $\alpha$, which is the probability of our committing a type I error (saying an agent is mutagenic when it is not, equivalent to our $p$ in such earlier considered designs as the Fisher's exact test)—false positive
2. $\beta$, which is the probability of our committing a type II error (saying an agent is not mutagenic when it is)—false negative

3. Δ, our desired sensitivity in an assay system (such as being able to detect an increase of 10% in mutations in a population)
4. σ, the variability of the biological system and the effects of chance errors
5. $n$, the single necessary sample size to achieve each of these (we can, by our actions, change only this portion of the equation) as $n$ is proportional to σ/α, β, and Δ

The implications of this are, therefore, that: (1) the greater σ is, the larger $n$ must be to achieve the desired levels of α, β, and Δ; (2) the smaller the desired levels of α, β, and/or Δ (if $n$ is constant), the larger our σ is.

What are the background mutation level and the variability in our technique? As any good genetic or general toxicologist will acknowledge, matched concurrent control groups are essential. Fortunately, with these test systems, large $n$ values are readily attainable, although there are other complications to this problem, which we will consider later. An example of the confusion that would otherwise result is illustrated in the intralaboratory comparisons on some of these methods done to date, such as that reviewed by Weil [156].

New statistical tests based on these assumptions and upon the underlying population distributions have been proposed, along with the necessary computational background to allow one to alter one of the input variables [α, β, or Δ]. A set that shows particular promise is that proposed by Katz [157,158] in his two articles. He described two separate test statistics: Φ, for when we can accurately estimate the number of individuals in both the experimental and control groups, and θ, for when we do not actually estimate the number of surviving individuals in each group and we can assume that the test material is only mildly toxic in terms of killing the test organisms. Each of these two test statistics is also formulated on the basis of only a single exposure of the organisms to the test chemicals. Given this, then we may compute:

$$\Phi = \frac{a(M_E - 0.5) - Kb(M_C + 0.5)}{\sqrt{Kab(M_E + M_C)}}$$

where $a$ and $b$ are the number of groups of control ($C$) and experimental ($E$) organisms, respectively; $K = N_E/N_C$, where $N_C$ and $N_E$ are the numbers of surviving microorganisms; $M_E$ and $M_C$ are the numbers of mutations in the experimental and control groups; and $\mu_e$ and $\mu_c$ are the true (but unknown) mutation rates (as $\mu_c$ gets smaller, $N$ must increase). We may compute the second case as:

$$\theta = \frac{a(M_E - 0.5) + (M_C + 0.5)}{ab(M_E - M_C)}$$

with the same constituents.

In both cases, at a confidence level for $I$ of 0.05, we accept that $\mu_c = \mu_e$ if the test statistic (either Φ or θ) is less than 1.64. If it is equal to or greater than 1.64, we may conclude that we have a mutagenic effect (at $\alpha = 0.05$).

In the second case (θ, where we do not have separate estimates of population sizes for the control and experimental groups), if $K$ deviates widely from 1.0 (if the material is markedly toxic), we should use more containers of control organisms (tables for the proportions of each to use given different survival frequencies may be found in Katz [158]). If different levels are desired, tables for θ and Φ may be found in Kastenbaum and Bowman [159].

An outgrowth of this is that the mutation rate per surviving cells ($\mu_c$ and $\mu_e$) can be determined. It must be remembered that, if the control mutation rate is so high that a reduction in mutation rates can be achieved by the test compound, these test statistics must be adjusted to allow for a two-sided hypothesis [160]. The α levels may likewise be adjusted in each case or tested for, if we want to assure ourselves that a mutagenic effect exists at a certain level of confidence (note that this is different from disproving the null hypothesis). It should be noted that numerous specific recommendations have been made for statistical methods designed for individual mutagenicity techniques, such as that of Bernstein et al. [161] for the Ames test.

## Behavioral Toxicity

A brief review of the types of studies or experiments conducted in the area of behavioral toxicology and the classification of these into groups is in order. Although a small number of studies do not fit into the following classification, the great majority may be fitted into one of the following four groups. Many of these points have been covered in earlier articles [80,162].

Observational score-type studies are based on observing and grading the response of an animal to its normal environment or to a stimulus that is imprecisely controlled. This type of result is generated by one of two major sorts of studies. Open-field studies involve placing an animal in the center of a flat, open area and counting each occurrence of several types of activities (grooming, moving outside a designated central area, rearing, etc.) or timing until the first occurrence of each type of activity. The data generated are scalar of either a continuous or discontinuous nature but frequently are not of a normal distribution. Tilson et al. [163] presented some examples of this sort.

Observational screen studies involve a combination of observing behavior and evoking a response to a simple stimulus, the resulting observation being graded as normal or as deviating from normal on a graded scale. Most of the data so generated are rank in nature, with some portions being quantal or interval. Irwin [164] and Gad [162] have presented schemes for the conduct of such studies which became the basis of the commonly used functional

**TABLE 8.14**
**Functional Observational Battery Parameters Showing Significant Differences Between Treated and Control Groups**

| | Rats (18-Crown-6 Animals Given 40 mg/kg i.p.) | | | | |
|---|---|---|---|---|---|
| **Parameter** | **Control Sum of Ranks** | $N_C$ | **18-Crown-6 Treated Sum of Ranks** | $N_T$ | **Observed Difference in Treated Animals (Compared to Controls)** |
| Twitches | 55.0 | 10 | 270.0 | 15 | Involuntary muscle twitches |
| Visual placing | 55.0 | 10 | 270.0 | 15 | Less aware of visual stimuli |
| Grip strength | 120.0 | 10 | 205.0 | 15 | Considerable loss of strength, especially in hind limbs |
| Respiration | 55.0 | 10 | 270.0 | 15 | Increased rate of respiration |
| Tremors | 55.0 | 10 | 270.0 | 15 | Marked tremors |

*Note:* All parameters above are significant at $p < 0.05$

observational battery. Table 8.14 gives an example of the nature (and of one form of statistical analysis) of such data generated after exposure to one material.

The second type of study is one that generates rates of response as data. The studies are based on the number of responses to a discrete controlled stimulus or are free of direct connection to a stimulus. The three most frequently measured parameters are licking of a liquid (milk, sugar water, ethanol, or a psychoactive agent in water), gross locomotor activity (measured by a photocell or electromagnetic device), or level pulling. Examples of such studies has been published by Annau [165] and Norton [166]. The data generated are most often of a discontinuous or continuous scalar nature and are often complicated by underlying patterns of biological rhythm.

The third type of study generates a variety of data classified as error rate. These are studies based on animals learning a response to a stimulus or memorizing a simple task (such as running a maze or a Skinner-box type of shock-avoidance system). These tests or trials are structured so animals can pass or fail on each of a number of successive trials. The resulting data are quantal, although frequently expressed as a percentage.

The final major type of study is one that results in data that are measures of the time to an endpoint. They are based on animals being exposed to or dosed with a toxicant and the time taken for an effect to be observed is measured. The endpoint is usually failure to continue to be able to perform a task and can, therefore, be death, incapacitation, or the learning of a response to a discrete stimulus. Burt [167] and Johnson et al. [168] present data of this form. The data are always of a censored nature—that is, the period of observation is always artificially limited as in measuring time-to-incapacitation in combustion toxicology data, where animals are exposed to the thermal decomposition gases to test materials for a period of 30 minutes. If incapacitation is not observed during these 30 minutes. it is judged not to occur. The data generated by these studies are continuous, discontinuous, or rank in nature. They are discontinuous because the researcher may check or may be restricted to checking for the occurrence of the endpoint only at certain discrete points in time. On the other hand, they are rank if the periods to check for occurrence of the endpoint are far enough apart, in which case one may actually only know that the endpoint occurred during a broad period to time—but not where in that period.

There is a special class of test which should also be considered at this point—the behavioral teratology or reproduction study. These studies are based on dosing or exposing either parental animals during selected periods in the mating and gestation process or pregnant females at selected periods during gestation. The resulting offspring are then tested for developmental defects of a neurological and behavioral nature. Analysis is complicated by a number of facts:

1. The parental animals are the actual targets for toxic effects, but observations are made on offspring.
2. The toxic effects in the parental generation may alter the performance of the mother in rearing its offspring, which in turn can lead to a confusion of prenatal and postnatal effects.
3. Different capabilities and behaviors develop at different times.

A researcher can, by varying the selection of the animal model (species, strain, sex), modify the nature of the data generated and the degree of dispersion of these data. In behavioral studies particularly, limiting the within-group variability of data is a significant problem and generally should be a highly desirable goal.

Most, if not all, behavioral toxicology studies depend on as least some instrumentation. Very frequently overlooked here (and, indeed, in most research) is that instru-

**TABLE 8.15**
**Overview of Statistical Testing in Behavioral Toxicology (Tests Commonly Used vs. Tests Most Frequently Appropriate)**

| Type of Observation | Most Commonly Used Procedures | Suggested Procedures |
|---|---|---|
| Observational scores | Student's *t*-test or one-way ANOVA | Kruskal–Wallis nonparametric ANOVA or Wilcoxon rank-sum |
| Response rates | Student's *t*-test or one-way ANOVA | Kruskal–Wallis ANOVA or one-way ANOVA |
| Error rates | ANOVA followed by a *post hoc* test | Fisher's exact, R×C chi-square, or Mann–Whitney U test |
| Times to endpoint | Student's *t*-test or one-way ANOVA | ANOVA then a *post hoc* test or Kruskal–Wallis ANOVA |
| Teratology and reproduction | ANOVA followed by a *post hoc* test | Fisher's exact test, Kruskal–Wallis ANOVA, or Mann–Whitney U test |

*Note:* That these are the most commonly used procedures was established by an extensive literature review that is beyond the scope of this chapter. The reader need only, however, look at the example articles cited in the text of this chapter to verify this fact.

mentation, by its operating characteristics and limitations, goes a long way toward determining the nature of the data generated by it. An activity monitor measures motor activity in discrete segments. If it is a "jiggle cage" type of monitor, these segments are restricted so only a distinctly limited number of counts can be achieved in a given period of time and then only if they are of the appropriate magnitude. Likewise, the technique can also readily determine the nature of the data. In measuring response to pain, for example, one could record it as a quantal measure (present or absent), a rank score (on a scale of 1 to 5 for decreased to increased responsiveness, with 3 being normal), or as scalar data (by using an analgesia meter that determines either how much pressure or heat is required to evoke a response).

Study design factors are probably the most widely recognized of the factors that influence the type of data resulting from a study. Number of animals used, frequency of measures, and length of period of observation are three obvious design factors that are readily under the control of the researcher and directly help to determine the nature of the data.

Finally, it is appropriate to review each of the types of studies currently utilized in behavioral toxicology according to the classification presented at the beginning of this section, in terms of which statistical methods are used now and what procedures should be recommended for use. The recommendations, of course, should be viewed with a critical eye. They are intended with current experimental design and technique in mind and can only claim to be the best when one is limited to addressing the most common problems from a library of readily and commonly available and understood tests. Table 8.15 summarizes this review and recommendation process.

## Carcinogenesis

In the experimental evaluation of substances for carcinogenesis based on experimental results in a nonhuman species at some relatively high dose or exposure level, an attempt is made to predict the occurrence and level of tumorogenesis in humans at much lower levels. An entire chapter could be devoted to examining the assumptions involved in this undertaking and review of the aspects of design and interpretation of animal carcinogenicity studies. Such is beyond the scope of this effort. The reader is referred to Gad [23] for such an examination. The single most important statistical consideration in the design of carcinogenicity bioassays in the past was based on the point of view that what was being observed and evaluated was a simple quantal response (cancer occurred or it did not) and that a sufficient number of animals needed to be used to have reasonable expectations of detecting such an effect. Though the single fact of whether or not the simple incidence of neoplastic tumors is increased due to an agent of concern is of interest, a much more complex model must now be considered. The time-to-tumor, patterns of tumor incidence, effects on survival rate, and age at first tumor all must now be included in a model.

### Bioassay Design

As presented earlier in the section on experimental design the first step that must be taken is to clearly state the objective of the study to be undertaken. Carcinogenicity bioassays have two possible objectives. The first objective is to detect possible carcinogens. Compounds are evaluated to determine if they can or cannot induce a statistically detectable increase in tumor rates over background levels, and only by happenstance is information generated that is useful in risk assessment. Most older studies have such detection as their objective. Current thought is that at least two species must be used for detection, although the necessity of a second species (the mouse) is increasingly questioned. The second objective for a bioassay is to provide a range of dose–response information (with tumor incidence being the response) so a risk assessment may be performed. Unlike detection, which requires only one treatment group with adequate survival times (to allow expression of tumors), dose–response requires at least three treatment groups with adequate survival. We will

**TABLE 8.16**
**Sample Size Required To Obtain a Specified Sensitivity at $p < 0.05$ Treatment Group Incidence**

| Background Tumor Incidence | $P^a$ | 0.95 | 0.90 | 0.80 | 0.70 | 0.60 | 0.50 | 0.40 | 0.30 | 0.20 | 0.10 |
|---|---|---|---|---|---|---|---|---|---|---|---|
| 0.30 | 0.90 | 10 | 12 | 18 | 31 | 46 | 102 | 389 | — | — | — |
| | 0.50 | 6 | 6 | 9 | 12 | 22 | 32 | 123 | — | — | — |
| 0.20 | 0.90 | 8 | 10 | 12 | 18 | 30 | 42 | 88 | 320 | — | — |
| | 0.50 | 5 | 5 | 6 | 9 | 12 | 19 | 28 | 101 | — | — |
| 0.10 | 0.90 | 6 | 8 | 10 | 12 | 17 | 25 | 33 | 65 | 214 | — |
| | 0.50 | 3 | 3 | 5 | 6 | 9 | 11 | 17 | 31 | 68 | — |
| 0.05 | 0.90 | 5 | 6 | 8 | 10 | 13 | 18 | 25 | 35 | 76 | 464 |
| | 0.50 | 3 | 3 | 5 | 6 | 7 | 9 | 12 | 19 | 24 | 147 |
| 0.01 | 0.90 | 5 | 5 | 7 | 8 | 10 | 13 | 19 | 27 | 46 | 114 |
| | 0.50 | 3 | 3 | 5 | 5 | 6 | 8 | 10 | 13 | 25 | 56 |

[a] $P$ is the power for each comparison of treatment group with background tumor incidence.

shortly look at the selection of dose levels for this case; however, given that the species is known to be responsive, only one species of animal needs to be used for this objective.

To address either or both of these objectives, three major types of study designs have evolved. First is the classical skin-painting study, usually performed in mice. A single, easily detected endpoint (the formation of skin tumors) is evaluated during the course of the study. Although dose–response can be evaluated in such a study (dose usually being varied by using different concentrations of test material in volatile solvent), most often detection is the objective of such a study. Although others have used different frequencies of application of test material to vary dose, there are data to suggest that this only serves to introduce an additional variable [169]. Traditionally, both test and control groups in such a test consist of 50 to 100 mice of one sex (males being preferred because of their very low spontaneous tumor rate). This design is also used in tumor initiation/promotion studies.

The second common type of design is the original National Cancer Institute (NCI) bioassay. The announced objective of these studies was detection of moderate to strong carcinogens, although the results have also been used in attempts at risk assessment. Both mice and rats were used in parallel studies. Each study used 50 males and 50 females at each of two dose levels (high and low) plus an equal-sized control group. The National Toxicology Program (NTP) has subsequently moved away from this design because of a recognition of its inherent limitations. More animals per group and more dose groups are now used.

Finally, the standard industrial toxicology design uses at least two species (usually rats and mice) in groups of no fewer that 100 males and females each. Each study has three dose groups and at least one control. Frequently, additional numbers of animals are included to allow for interim terminations and histopathological evaluations. In both this and the NCI design, a long list of organs and tissues are collected, processed, and examined microscopically. This design seeks to address both the detection and dose–response objectives with a moderate degree of success.

Selecting the number of animals to use for dose groups in a study requires consideration of both biological (expected survival rates, background tumor rates, etc.) and statistical factors. The prime statistical consideration is reflected in Table 8.16. It can be seen in this table that, if, for example, we were studying a compound that caused liver tumors and were using mice (with a background or control incidence of 30%), we would have to use 389 animals per sex per group to be able to demonstrate that an incidence rate of 40% in treatment animals was significant compared to the controls at the $p = 0.05$ level.

Perhaps the most difficult aspect of designing a good carcinogenicity study is the selection of the dose levels to be used. At the start, it is necessary to consider the first underlying assumption in the design and use of animal cancer bioassays—the need to test at the highest possible dose for the longest practical period. The rationale behind this assumption is that, although humans may be exposed at very low levels, detecting the resulting small increase (over background) in the incidence of tumors would require the use of an impractically large number of test animals per group.

This point is illustrated by Table 8.16, where, for example, only 46 animals (per group) are needed to show a 10% increase over a zero background (that is, a rarely occurring tumor type), but 770,000 animals (per group) would be needed to detect a .1% increase above a 5% background. As we increase dose, however, the incidence of tumors (the response) will also increase until it reaches the point where a modest increase (say 10% over a reasonably small background level (say 1%) could be detected using an accept-

**TABLE 8.17**
**Average Number of Animals Needed To Detect a Significant Increase in the Incidence of an Event (e.g., Tumors, Anomalies) Over Background Incidence (Control) at Expected Incidence Levels Using the Fisher Exact Probability Test ($p = 0.05$)**

| Background Incidence (%) | Expected Increase in Incidence (%) 0.01 | 0.1 | 1 | 3 | 5 | 10 |
|---|---|---|---|---|---|---|
| 0 | 46,000,000[a] | 460,000 | 4600 | 511 | 164 | 46 |
| 0.01 | 46,000,000 | 460,000 | 4600 | 511 | 164 | 46 |
| 0.1 | 47,000,000 | 470,000 | 4700 | 520 | 168 | 47 |
| 1 | 51,000,000 | 510,000 | 5100 | 570 | 204 | 51 |
| 5 | 77,000,000 | 770,000 | 7700 | 856 | 304 | 77 |
| 10 | 100,000,000 | 1,000,000 | 10,000 | 1100 | 400 | 100 |
| 20 | 148,000,000 | 1,480,000 | 14,800 | 1644 | 592 | 148 |
| 25 | 160,000,000 | 1,600,000 | 16,000 | 1840 | 664 | 166 |

[a] Number of animals needed in each group, controls as well as treated.

ably small-sized group of test animals (in Table 8.17 we see that 51 animals would be needed for this example case). There are, however, at least two real limitations to the highest dose level. First, the test rodent population must have a sufficient survival rate after receiving a lifetime (or 2 years) of regular doses to allow for meaningful statistical analysis. Second, we really want the metabolism and mechanism of action of the chemical at the highest level tested to be the same as at the low levels where human exposure would occur. Unfortunately, toxicologists usually must select the high dose level based only on the information provided by a subchronic or range finding study (usually 90 days in length), but selection of either too low or too high a dose will make the study invalid for detection of carcinogenicity and may seriously impair the use of the results for risk assessment.

There are several solutions to this problem. One of these has been the rather simplistic approach of the NTP Bioassay Program, which is to conduct a 3-month range-finding study with sufficient dose levels to establish a level that significantly (10%) decreases the rate of body weight gain. This dose is defined as the maximum tolerated dose (MTD) and is selected as the highest dose. Two other levels, generally one half MTD and one quarter MTD, are selected for testing as the intermediate and low dose levels. In many earlier NCI studies, only one other level was used.

The dose range-finding study is necessary in most cases, but the suppression of body weight gain is a scientifically questionable benchmark when dealing with establishment of safety factors. Physiologic, pharmacologic, or metabolic markers generally serve as better indicators of systemic response than body weight. A series of well-defined acute and subchronic studies designed to determine the chronicity factor and to study the onset of pathology can be more predictive for dose setting than body weight suppression. Also, the NTP's MTD may well be at a level where the metabolic mechanisms for handling a compound at real-life exposure levels have been saturated or overwhelmed, bringing into play entirely artifactual metabolic and physiologic mechanisms [170]. The regulatory response to questioning the appropriateness of the MTD as a high dose level [171] has been to acknowledge that occasionally an excessively high dose is selected but to counter by saying that using lower doses would seriously decrease the sensitivity of detection.

## DATA ANALYSIS APPLICATIONS IN TOXICOLOGIC PATHOLOGY

Having reviewed basic principles and provided a set of methods for the statistical handling of data, the remainder of this chapter addresses the practical aspects and difficulties encountered in preclinical safety assessment in the field of toxicologic pathology. Analyses of pathology data are well defined, although they may not necessarily use the best methods available. The use of statistical methodology is discussed in the remainder of this chapter. The aim of this section is to review statistical methods on a use-by-use basis and to provide a foundation for the selection of alternatives in specific situations. Meta-analyses and Bayesian approaches are not addressed in detail but should be kept in mind.

### Body and Organ Weights

Body weight and the weights of selected organs are usually collected in studies where animals are dosed with, or exposed to, a chemical. In fact, body weight is frequently the most sensitive indication of an adverse treatment effect. How to analyze these data best and in what form to analyze the organ weight data, such as absolute weights, weight changes, or percentages of body weight, have been the subject of great discussion.

Both absolute body weights and rates of body weight change are best analyzed by ANOVA followed, if called for, by a *post hoc* test. Body weight change is usually calculated as changes from a baseline measurement value, which is traditionally the animal's weight immediately prior to the first dosing with or exposure to the test material. To standardize body weight, no group should be significantly different in mean body weight from any other group, and all animals in all groups should lie within two standard deviations of the overall mean body weight. Even if the groups were randomized properly at the beginning of a study, there is an advantage to performing the computationally slightly more cumbersome changes in body weight analysis. The advantage of this calculation is an increase in sensitivity because the adjustment of starting points (i.e., the setting of initial weights as a zero value) reduces the amount of initial variability. In this case, Bartlett's test is performed first to ensure the homogeneity of variance and the appropriate sequence of analysis follows.

If sample sizes are small or normality of data is uncertain, nonparametric methods, such as Kruskal-Wallis, may be more appropriate. The analysis of relative organ weights is a valuable tool for identifying possible target organs. How to perform this analysis is still a matter of some disagreement. Organ weight data, expressed as percentages of body weight, should be analyzed separately for each sex. Often, conclusions from organ weight data of males differ from those of females; hence, separating these data by gender should always be done. Other factors, such as the effect of stage of the reproductive cycle on uterine weight, may also influence organ weights. These factors must be taken into account both in the stratification of animals and in the interpretation of results.

The two alternative approaches to analyzing relative organ weights call for either calculating organ weights as a percentage of total body weight at the time of necropsy and analyzing the results by ANOVA or analyzing results by ANCOVA, with body weights as the covariates as discussed previously. A number of considerations should be kept in mind when this choice is made. First, one must recognize the difference between biological significance and statistical significance. By evaluating relative body weight, the significance of a weight change that is not proportional to changes in whole body weights must be determined. Second, the toxicologic pathologist now must interpret small changes while still retaining a similar sensitivity (i.e., the $p < 0.05$ level).

Several tools can be used to increase the power of the analysis. One is to increase the sample size by increasing the number of animals, and the other is to utilize the most powerful test available that is appropriate to the data. The number of animals used in the groups is currently under debate with respect to power of detecting a significant change. The power of statistical tests is important in the consideration of animal numbers.

In the majority of cases, except for the brain, the organs of interest change weight in proportion to total body weight, except in extreme cases of obesity or starvation. This change is the biological rationale behind analyzing both absolute body weight and the ratio of organ weight to body weight. Analyses are designed to detect cases where this relative change does not occur. Analysis of data from several hundred studies has shown no significant difference in rates of weight change of target organs, other than the brain, compared to total body weight for healthy animals in rats, mice, rabbits, and dogs used for repeated dose studies. The analysis of covariance is of questionable validity in analyzing body weight and related organ weight changes, as a primary assumption is the independence of treatment. In toxicologic pathology, the assumption that the relationship of the two variables is the same for all treatments is not true.

In cases where the differences between the error mean squares are much greater during the analysis, the ratio of $F$ ratios will diverge in precision from the result of the efficiency of covariance adjustment. These cases occur where either sample sizes are large or where the differences between means themselves are great. This latter case is one that does not occur in the designs under discussion in any manner that would leave analysis of covariance as a valid approach because group means are very similar at the beginning of the experiment and cannot diverge markedly unless there is a treatment effect. As discussed earlier, a treatment effect invalidates a prime underpinning assumption of analysis of covariance.

## Clinical Chemistry

A number of clinical chemistry parameters are commonly determined on the blood and urine collected from animals in chronic, subchronic, and, occasionally, acute toxicity studies. In the past, and currently in some places, the accepted practice has been to evaluate these data using univariate–parametric methods, primarily $t$-tests and ANOVA; however, this is not the best approach. First, biochemical parameters are rarely independent of each other, and the focus of inquiry is rarely limited to only one of the parameters. Instead, several parameters can change when toxicity is seen in specific organs. For example, simultaneous elevations of creatinine phosphokinase, γ-hydroxybutyrate dehydrogenase, and lactate dehydrogenase are strongly indicative of myocardial damage. In such a case, the clinical importance of these findings is not limited to a significant elevation in one of these enzymes; all three must be considered together. Detailed coverage of the interpretation of such clinical laboratory tests can be found elsewhere.

Second, interaction occurs among parameters; therefore, each parameter is not independent. For example, serum electrolytes (sodium, potassium, and calcium) interact such that a decrease in one is frequently tied to an increase in one of the others. Finally, either because of the biological nature of the parameter or the way in which it is measured, data are frequently skewed or not continuous. This skewness and discontinuous nature of data can be seen in some of the reference data for experimental animals (e.g., creatinine, sodium, potassium, chloride, calcium, and blood).

## CARCINOGENESIS

Inferences about the potential human carcinogenicity of substances are based on experimental results obtained from a nonhuman species given the substance at a high dose or exposure level. The aim of this procedure is to predict the possibility and probability of occurrence of tumorogenesis in humans at much lower levels. An entire textbook could be devoted to examining the assumptions involved in this undertaking and review of the aspects of design and interpretation of animal carcinogenicity studies. Such detail is beyond the scope of this chapter. The reader is referred to Gad [23] for more detail.

In the past, the single most important statistical consideration in the design of carcinogenicity bioassays was based on a simple quantal response: cancer did or did not occur. Experiments were designed so a sufficient number of animals were used so as to have a reasonable expectation of detecting an effect if one occurred. Although the primary objective was to determine whether the incidence of tumors was increased following exposure to the test article of interest, a much more complex model should now be considered to answer other questions pertinent to the extrapolation of experimental results in animals to make inferences about risks to human health.

The time to tumor, patterns of tumor incidence, effects on survival rate, and age at first tumor can now be evaluated. The rationale for including these factors lies in concerns associated with likely planned or unplanned exposure of humans to xenobiotic and naturally occurring substances, and relatively small increases in the incidence of tumors over background would require the use of an impracticably large number of test animals per group.

To illustrate this point, examine the data provided in Table 8.17. Here, only 46 animals per group are required to show a 10% increase over a zero background, where the background included a rarely occurring tumor type. To detect a .1% increase above a 5% background, 770,000 animals per group would be needed! As dose increases the incidence of tumors, the response will also increase. This increase occurs until it reaches the point where a modest increase (e.g., 10%) over a reasonably small background level (e.g., 1%) could be detected using an acceptably small-sized group of test animals. Table 8.17 shows that 51 animals would be needed for such a situation. It can be seen that the number of animals required to demonstrate a 1/100,000 increase above a 25% background incidence would be very large.

At least two potential difficulties often occur in the group given the highest dose. First, mortality can be higher than other groups; a sufficient number of rodents must survive to the end of the study to allow for meaningful statistical analysis. Second, toxicologic pathologists must select the high dose level based only on the information provided by a subchronic or range-finding study, usually 90 days in length. To predict carcinogenic effects across species, it is necessary that the metabolism and mechanism of action of the chemical at the highest level tested are the same as at the low levels where human exposure would occur. Unfortunately, selection of a dose that is too low may make the study invalid for the detection of carcinogenicity, and selection of a dose that is too high, where toxicokinetics result in different metabolism, may seriously impair the use of the results for risk assessment.

## REFERENCES


1. Roe, D. (1989): *Handbook on Drug and Nutrient Interactions: A Problem Oriented Reference Guide*, 4th ed. American Dietetic Association, Chicago.
2. Lee, P. (1993): *Bayesian Statistics*. Oxford University Press, London.
3. Cochran, W. G. and Cox, G. M. (1975): *Experimental Designs*. John Wiley & Sons, New York, pp. 100–102.
4. Diamond, W. J. (1981): *Practical Experimental Designs*. Lifetime Learning Publications, Belmont, CA.
5. Federer, W. T. (1955): *Experimental Design*. Macmillan, New York.
6. Hicks, C. R. (1982): *Fundamental Concepts in the Design of Experiments*. Holt, Rinehart and Winston, New York.
7. Kraemer, H. C. and Thiemann, G. (1987): *How Many Subjects? Statistical Power Analysis in Research*. Sage Publications, Newbury Park, CA, p. 27.
8. Myers, J. L. (1972): *Fundamentals of Experimental Designs*. Allyn & Bacon, Boston.
9. Peto, R. et al. (1980): Guidelines for simple, sensitive significance tests for carcinogenic effects in long-term animal experiments, in *Long-Term and Short-Term Screening Assays for Carcinogens: A Critical Appraisal*, World Health Organization, Geneva.
10. Anscombe, F. J. (1973): Graphics in statistical analysis. *Am. Statist*., 27:17–21.
11. Chambers, J. M., Cleveland, W. S., Kleiner, B., and Tukey, P. A. (1983): *Graphical Methods for Data Analysis*. Wadsworth, Belmont, CA.
12. Cleveland, W. S. and McGill, R. (1984): Graphical perception: theory, experimentation, and application to the development of graphical methods. *J. Am. Stat. Assoc*. 79:531–554.
13. Cleveland, W. S. (1985): *The Elements of Graphing Data*. Wadsworth, Monterey, CA.

14. Tufte, E. R. (1990): *Envisioning Information*. Graphics Press, Cheshire, CT.
15. Sokal, R. R. and Rohlf, F. J. (1994): *Biometry*, 3rd ed. W.H. Freeman, San Francisco, CA.
16. Snedecor, G. W. and Cochran, W. G. (1980): *Statistical Methods*, 7th ed. Iowa State University Press, Ames, pp. 470–471.
17. Everitt, B. S. and Hand, D. J. (1981): *Finite Mixture Distributions*. Chapman & Hall, New York.
18. Mendell, N. R., Finch, S. J., and Thode, H. C. Jr. (1993): Where is the likelihood ratio test powerful for detecting two component normal mixtures? *Biometrics*, 49:907–915.
19. Engelman, L. and Hartigan, J. A. (1969): Percentage points of a test for clusters. *J. Am. Stat. Assoc.*, 64:1647–1648.
20. Tukey, J. W., (1977): *Exploratory Data Analysis*. Addison-Wesley, Reading, PA.
21. Velleman, P. F. and Hoaglin, D. C. (1981): *Applications, Basics and Computing of Exploratory Data Analysis*, Duxbury Press, Boston.
22. Hoaglin, D. C., Mosteller, F., and Tukey, J. W. (1983): *Understanding Robust and Explanatory Data Analysis*, John Wiley & Sons, New York.
23. Gad, S. C. (2005): *Statistics and Experimental Design for Toxicologists*, 4th ed., Taylor & Francis, Boca Raton, FL.
24. Finney, D. J., Latscha, R., Bennet, B. M., and Hsu, P. (1963): *Tables for Testing Significance in a 2×2 Contingency Table*. Cambridge University Press, Cambridge, U.K.
25. Zar, J. H. (1974): *Biostatistical Analysis*. Prentice-Hall, Englewood Cliffs, NJ, pp. 50, 151–161, 518–542.
26. Ghent, A. W. (1972): A method for exact testing of 2×2, 2×3, 3×3 and other contingency tables employing binomiate coefficients. *Am. Midland Nat.*, 88:15–27.
27. Beyer, W. H. (1976): *Handbook of Tables for Probability and Statistics*, CRC Press, Boca Raton, FL, 409–413.
28. Hollander, M. and Wolfe, D. A. (1973): *Nonparametric Statistical Methods*. John Wiley & Sons, New York, pp. 124–129.
29. Siegel, S. (1956): *Nonparametric Statistics for the Behavioral Sciences*. McGraw-Hill, New York.
30. Gaylor, D. W. (1978): Methods and concepts of biometrics applied to teratology. In: *Handbook of Teratology*, Vol. 4, edited by J. G. Wilson and F. C. Fraser. Plenum Press, New York, pp. 429–444.
31. Pollard, J. H. (1977): *Numerical and Statistical Techniques*. Cambridge University Press, Cambridge, U.K., pp. 170–173.
32. Peto, R., Pike, M. C., Armitage, P., Breslow, N. E., Cox. D. R., Howard, S. V., Kantel, N., McPherson, K., Peto, J., and Smith, P. G. (1977): Design and analysis of randomized clinical trials requiring prolonged observations of each patient. II. Analyses and examples. *Br. J. Cancer*, 35:1–39.
33. Peto, R. and Pike, M. C. (1973): Conservatism of approximation $\sigma(0 - e)^2/e$ in the log rank test for survival data on tumour incidence data. *Biometrics*, 29:579–584.
34. Dixon, W. J. (1994): *BMD–Biomedical Computer Programs*. University of California Press, Berkeley.
35. Duncan, D. B. (1955): Multiple range and multiple *F* tests. *Biometrics*, 11:1–42.
36. Harter, A. L. (1960): Critical values for Duncan's new multiple range test. *Biometrics*, 16:671–685.
37. Beyer, W. H. (1976): *Handbook of Tables for Probability and Statistics*, CRC Press, Boca Raton, FL.
38. Scheffe, H. (1959): *The Analysis of Variance*. Wiley, New York.
39. Harris, R. J. (1975): *A Primer of Multivariate Statistics*. Academic Press, New York, pp. 96–101.
40. Dunnett, C. W. (1955): A multiple comparison procedure for comparing several treatments with a control. *J. Am. Stat. Assoc.*, 50:1096–1121.
41. Dunnett, C. W. (1964): New tables for multiple comparison with a control. *Biometrics*, 16:671–685.
42. Williams, D. A. (1971): A test for differences between treatment means when several dose levels are compared with a zero dose control. *Biometrics*, 27:103:117.
43. Williams, D. A. (1972): The comparison of several dose levels with a zero dose control. *Biometrics*, 28:519–531.
44. Anderson, S., Auquier, A., Hauck, W. W., Oakes, D., Vandaele, W., and Weisburg, H. I. (1980): *Statistical Methods for Comparative Studies*. John Wiley & Sons, New York.
45. Kotz, S. and Johnson, N. L. (1982): *Encyclopedia of Statistical Sciences*, Vol. 1. John Wiley & Sons, New York, pp. 61–69.
46. Gold, H. J. (1977): *Mathematical Modeling of Biological System: An Introductory Guidebook*, John Wiley & Sons, New York.
47. Gallant, A. R. (1975): Nonlinear regression. *Am. Statist.*, 29:73–81.
48. Draper, N. R. and Smith, H. (1998): *Applied Regression Analysis*, 3rd ed. John Wiley & Sons, New York.
49. Montgomery, D. C., Peck, E. A., and Vining, G. (2001): *Introduction to Linear Regression Analysis*, 3rd ed. John Wiley & Sons, New York.
50. Bliss, C. I. (1935): The calculation of the dosage-mortality curve. *Ann. Appl. Biol.*, 22:134–167.
51. Finney, D. K. (1977): *Probit Analysis*, 3rd ed. Cambridge University Press, Cambridge, U.K.
52. Litchfield, J. T. and Wilcoxon, F. (1949): A simplified method of evaluating dose effect experiments. *J. Pharmacol. Exp. Ther.*, 96:99–113.
53. Prentice, R. L. (1976): A generalization of the probit and logit methods for dose response curves. *Biometrics*, 32:761–768.
54. Abramowitz, M. and Stegun, I. A. (1964): *Handbook of Mathematical Functions*. National Bureau of Standards, Washington, D.C., pp. 925–964.
55. Thompson, W. R. and Weil, C. S. (1952): On the construction of tables for moving average interpolation. *Biometrics*, 8:51–54.
56. Weil, C. S. (1952): Tables for convenient calculation of median-effective dose ($LD_{50}$ or $ED_{50}$) and instructions in their use. *Biometrics*, 8:249–263.
57. Weil, C. S. (1983): Economical $LD_{50}$ and slope determinations. *Drug Chem. Toxicol.*, 6:595–603.
58. Feinstein, A. R. (1979): Scientific standards vs. statistical associations and biological logic in the analysis of causation. *Clin. Pharmacol. Therap.*, 25:481–492.
59. Tarone, R. E. (1975): Tests for trend in life table analysis. *Biometrika*, 62:679–682.

60. Cox, D. R. and Stuart, A. (1955): Some quick tests for trend in location and dispersion. *Biometrics*, 42:80–95.
61. Anderson, T. W. (1971) *The Statistical Analysis of Time Series*. John Wiley & Sons, New York.
62. Dykstra, R. L. and Robertson, T. (1983): On testing monotone tendencies, *J. Am. Stat. Assoc.*, 78:342–350.
63. Breslow, N. (1988): Comparison of survival curves. In: *Cancer Clinical Trials: Methods and Practice*, edited by M. F. Buse, M. J. Staguet, and R. F. Sylvester. Oxford University Press, London, pp. 381–406.
64. Crowley, J. and Breslow, N. (1984): Statistical analysis of survival data. *Annu. Rev. Public Health*, 5:385–411.
65. Armitage, P. (1955) Tests for linear trends in proportions and frequencies. *Biometrics*, 11:375–386.
66. Cochran, W. F. (1954) Some models for strengthening the common $\chi^2$ tests. *Biometrics*, 10:417–451.
67. Portier, C. and Hoel, D. (1984) Type I error of trend tests in proportions and the design of cancer screens. *Comm. Stat. Theory Meth.*, A13:1–14.
68. Hartigan, J. A. (1983): Classification. In: *Encyclopedia of Statistical Sciences*, Vol. 2, edited by S. Katz and N. L. Johnson. John Wiley & Sons, New York.
69. Gordon, A. D. (1981) *Classification*. Chapman & Hall, New York.
70. Glass, L. (1975): Classification of biological networks by their qualitative dynamics. *J. Theor. Biol.*, 54:85–107.
71. Schaper, M., Thompson, R. D., and Alarie, Y. (1985): A method to classify airborne chemicals which alter the normal ventilatory response induced by $CO_2$. *Toxicol. Appl. Pharmacol.*, 79:332–341.
72. Kowalski, B. R. and Bender, C. F. (1972): Pattern recognition: a powerful approach to interpreting chemical data. *J. Am. Chem. Soc.*, 94:5632–5639.
73. Anderson E. (1960): A semigraphical method for the analysis of complex problems. *Technometrics*, 2:387–391.
74. Chernoff, H. (1973): The use of faces to represent points in *K*-dimensional space graphically. *J. Am. Stat. Assoc.*, 68:361–368.
75. Tufte, E. R. (1983): *The Visual Display of Quantitative Information*. Graphics Press, Cheshire, CT.
76. Schmid, C. F. (1983): *Statistical Graphics*. John Wiley & Sons, New York.
77. Tufte, E. R. (1997): *Visual Explanations*. Graphics Press, Cheshire, CT.
78. Young, F. W. (1985): Multidimensional scaling. In: *Encyclopedia of Statistical Sciences*, Vol. 5, edited by S. Katz and N. L. Johnson. John Wiley & Sons, New York, pp. 649–659.
79. Davison, M. L. (1983): *Multidimensional Scaling*. John Wiley & Sons, New York.
80. Gad, S. C. (1984): Statistical analysis of behavioral toxicology data and studies. *Arch. Toxicol. Suppl.*, 5:256–266.
81. Gad, S. C., Reilly, C., Siino, K. M., and Gavigan, F. A. (1985): Thirteen cationic ionophores: neurobehavioral and membrane effects. *Drug Chem. Toxicol.*, 8(6):451–468.
82. Everitt, B. (2001): *Cluster Analysis*, 4th ed. Oxford University Press, London.
83. Romesburg, H. C. (1984): *Cluster Analysis for Researchers*. Lifetime Learning Publications, Belmont, CA, pp. 45–58.
84. Bloomfield, P. (1976): *Fourier Analysis of Time Series: An Introduction*. John Wiley & Sons, New York.
85. Hammond, E. C., Garfinkel, L., and Lew, E. A. (1978): Longevity, selective mortality, and competitive risks in relation to chemical carcinogenesis. *Environ. Res.*, 16:153–173.
86. Cutler, S. J. and Ederer, F. (1958): Maximum utilization of the life table method in analyzing survival. *J. Chron. Dis.*, 8:699–712.
87. Salsburg, D. (1980): The effects of life-time feeding studies on patterns of senile lesions in mice and rats. *Drug Chem. Toxicol.*, 3:1–33.
88. Cox, D. R. (1972): Regression models and life-tables. *J. Roy. Stat. Soc.*, 34B:187–220.
89. Haseman, J. K. (1977): Response to use of statistics when examining life time studies in rodents to detect carcinogenicity. *J. Toxicol. Environ. Health*, 3:633–636.
90. Garrett, H. E. (1947): *Statistics in Psychology and Education*. Longmans, Green and Co., New York, pp. 215–218
91. SOT $ED_{01}$ Task Force (1981): Reexamination of the $ED_{01}$ study: adjusting for time on study. *Fundam. Appl. Toxicol.*, 1:8–123.
92. Lee, E. T. (1980): *Statistical Methods for Survival Data Analysis*. Lifetime Learning Publications, Belmont, CA.
93. Elandt-Johnson, R. C. and Johnson, N. L. (1980): *Survival Models and Data Analysis*. John Wiley & Sons, New York
94. Schaffer, J. W., Forbes, J. A., and Defelice, E. A. (1967) Some suggested approaches to the analysis of chronic toxicity and chronic drug administration data. *Toxicol. Appl Pharmacol.*, 10:514–522.
95. Davidson, M. L. (1972): Univariate versus multivariate tests in repeated-measures experiments. *Psych. Bull.* 77:446–452.
96. Barnett, V. and Lewis, T. (1994): *Outliers in Statistical Data*, 3rd ed. John Wiley & Sons, New York.
97. Hotelling, H. (1931): The generalization of Student's ratio *Ann. Math. Stat.*, 2: 360–378.
98. Gray, L. E. and Laskey, J. W. (1980): Multivariate analysis of the effects of manganese on the reproductive physiology and behavior of the male house mouse. *J. Toxicol. Environ Health*, 6:861–868.
99. Witten, M., Bennet, C. E., and Glassman, A. (1981): Studies on the toxicity and binding kinetics of abrin in normal and Epstein Barr virus-transformed lymphocyte culture. I Experimental results. *Exp. Cell. Biol.*, 49:306–318.
100. Tatsuoka, M. M. (1971): *Multivariate Analysis*. John Wiley & Sons, New York.
101. Joung, H. M., Miller, W. M., Mahannah, C. N., and Guitjens, J. C. (1979): A generalized water quality index based on multivariate factor analysis. *J. Environ. Qual.* 8:95–100.
102. Schaeffer, D. J., Glave, W. R., and Janardan, K. G. (1982) Multivariate statistical methods in toxicology. III. Specifying joint toxic interaction using multiple regression analysis. *J. Toxicol. Environ. Health*, 9:705–718.
103. Paintz, M., Bekemeier, H., Metzner, J., and Wenzel, U (1982): Pharmacological activities of a homologous series of pyrazole derivatives including quantitative structure–activity relationships (QSAR). *Agents Actions* 10(Suppl.):47–58.

104. Taketomo, R. T., McGhan, W. F., Fushiki, M. R., Shimada, A. and Gumpert, N. F. (1982): Gentamycin nephrotoxicity application of multivariate analysis. *Clin. Pharm.*, 1:554–549.
105. Young, J. E. and Matthews, P. (1981): Pollution injury in southeast Northumberland, England, U.K.—the analysis of field data using economical correlation analysis. *Environ. Pollut. B. Chem. Phys.*, 2:353–366.
106. Cremer, J. E. and Seville, M. P. (1982): Comparative effects of two pyrethroids dietamethrin and cismethrin, on plasma catecholamines and on blood glucose and lactate. *Toxicol. Appl. Pharmacol.*, 66:124–133.
107. Henry, R. D. and Hidy, G. M. (1979): Multivariate analysis of particulate sulfate and other air quality variables by principal components. *Atmos. Environ.*, 13:1581–1596.
108. Gabriel, K. R. (1981): Biplot display of multivariate matrices for inspection of data and diagnosis. In: *Interpreting Multivariate Data*, edited by V. Barnett. John Wiley & Sons, New York, pp. 147–173.
109. Shy-Modjeska, J. S., Riviere, J. E., and Rawldings, J. O. (1984): Application of biplot methods to the multivariate analysis of toxicological and pharmacokinetic data. *Toxicol. Appl. Pharmacol.*, 72:91–101.
110. Atchely, W. R. and Bryant, E. H. (1975): *Multivariate Statistical Methods: Among Groups Covariation*. Dowden, Hutchinson and Ross, Stroudsburg, PA.
111. Bryant, E. H. and Atchely, W. R. (1975): *Multivariate Statistical Methods: Within-Groups Covariation*. Dowden, Hutchinson and Ross, Stroudsburg, PA.
112. Seal, H. L. (1964): *Multivariate Statistical Analysis for Biologists*. Methuen, London.
113. Gnanadesikan, R. (1977): *Methods for Statistical Data Analysis of Multivariate Observations*. John Wiley & Sons, New York.
114. Tippett, L. C. (1931): *The Methods of Statistics*. Williams & Norgate, London.
115. Sacks, H. S., Berrier, J., Reitman, D., Ancona-Berk, V. A., and Chalmers, T. C. (1987): Meta-analyses of randomized controlled trials. *N. Engl. J. Med.*, 316(8):450–455.
116. Thacker, S. B. (1988): Meta-analysis: a quantitative approach to research integration. *JAMA*, 259:1685–1689.
117. Sutton, A. J., Abrams, K. R., Jones, D. R., Sheldon, T. A., and Song, F. (2000): *Methods for Meta-Analysis in Medical Research*. John Wiley & Sons, New York.
118. Bayes, T. (1763). An essay towards solving a problem in the doctrine of chances, *Phil. Trans. Roy. Soc.*, 53:370–418.
119. Goodman, S. (2001). *What Can Bayesian Analysis Do for Us?* Presented to USFDA Oncologic Drugs Advisory Committee, Pediatric Subcommittee, on November 28, 2001 (http://www.fda.gov/ohrms/dockets/ac/01/slides/3803s1_05A_Goodman/index.htm.
120. Weil, C. S. (1972): Statistics vs. safety factors and scientific judgment in the evaluation of safety for man. *Toxicol. Appl. Pharmacol.*, 21:459–472.
121. Weil, C. S. (1975): Toxicology experimental design and conduct as measured by inter-laboratory collaboration studies. *J. Assoc. Off. Anal. Chem.*, 58:687–688.
122. Debanne, S. M. and Haller, H. S. (1985): Evaluation of statistical methodologies for estimation of median effective dose. *Toxicol. Appl. Pharmacol.*, 79:274–282.
123. Weil, C. S., Carpenter, C. P., and Smith, H. I. (1953): Specifications for calculating the median effective dose. *Am. Indust. Hyg. Assoc. J.*, 14:200–206.
124. Zbinden, G. and Flury-Roversi, M. (1981): Significance of the $LD_{50}$ test for the toxicological evaluation of chemical substances. *Arch. Toxicol.*, 47:77–99.
125. DePass, L. R., Myers, R. C., Weaver, E. V., and Weil, C. S. (1984): An assessment of the importance of number of dosage levels, number of animals per dosage level, sex and method of $LD_{50}$ and slope calculations in acute toxicity studies. In: *Alternate Methods in Toxicology*. Vol. 2. *Acute Toxicity Testing: Alternate Approaches*, edited by A. M. Goldberg. Mary Ann Liebert, New York.
126. Gad, S. C., Smith, A. C., Cramp, A. L., Gavigan, F. A., and Derelanko, M. J. (1984): Innovative designs and practices for acute systemic toxicity studies, *Drug Chem. Toxicol*, 7:423–434.
127. Bruce, R. D. (1985): An up-and-down procedure for acute toxicity testing. *Fund. Appl. Toxicol.*, 5:151–157.
128. Jackson, B. (1962): Statistical analysis of body weight data. *Toxicol. Appl. Pharmacol.*, 4:432–443.
129. Weil, C. S. (1962): Applications of methods of statistical analysis to efficient repeated-dose toxicological tests. I. General considerations and problems involved—sex differences in rat liver and kidney weights. *Toxicol. Appl. Pharmacol.*, 4:561–571.
130. Weil, C. S. (1970): Selection of the valid number of sampling units and a consideration of their combination in toxicological studies involving reproduction, teratogenesis or carcinogenesis. *Food Cosmet. Toxicol.*, 8: 177–182.
131. Weil, C. S. and Gad, S. C. (1980): Applications of methods of statistical analysis to efficient repeated-dose toxicologic tests. 2. Methods for analysis of body, liver and kidney weight data. *Toxicol. Appl. Pharmacol.*, 52:214–226.
132. Weil, C. S. (1973): Experimental design and interpretation of data from prolonged toxicity studies. In: *Proc. 5th Int. Congress on Pharmacology*, Vol. 2. Beacon Press, San Francisco, CA, pp. 4–12.
133. Boyd, E. M. and Knight, L. M. (1963): Postmortem shifts in the weight and water levels of body organs, *Toxicol. Appl. Pharmacol.*, 5:119–128.
134. Boyd, E. M. (1972): *Predictive Toxicometrics*. Williams & Wilkins, Baltimore, MD.
135. Ridgemen, W. J. (1975): *Experimentation in Biology*. John Wiley & Sons, New York, pp. 214–215.
136. Shirley, E. A. C. and Newman, J. F. (1954): A distribution-free method for analysis of covariance, *Appl. Statist.*, 3:158–162.
137. Gad, S. C. and Chengelis, C. P. (1992): *Animal Models in Toxicology*. Marcel Dekker, New York.
138. Loeb, W. F. and Quimby, F. W. (1999): *The Clinical Chemistry of Laboratory Animals*, 2nd ed. Taylor & Francis, Philadelphia, PA.
139. Harris, E. K. (1978): Review of statistical methods of analysis of series of biochemical test results. *Ann. Biol. Clin.*, 36:194–197.
140. Martin, H. F., Gudzinowicz, B. J., and Fanger, H. (1975): *Normal Values in Clinical Chemistry*. Marcel Dekker, New York.

141. Mitruka, B. M. and Rawnsley, H. M. (1977): *Clinical Biochemical and Hematological Reference Values in Normal Animals*. Masson, New York.
142. Weil, C. S. (1982): Statistical analysis and normality of selected hematologic and clinical chemistry measurements used in toxicologic studies. *Arch. Toxicol., Suppl.*, 5:237–253.
143. Oser, B. L. and Oser, M. (1956): Nutritional studies in rats on diets containing high levels of partial ester emulsifiers. II. Reproduction and lactation. *J. Nutr.*, 60:429.
144. Kupper, L. L. and Haseman, J. K. (1978): The use of a correlated binomial model for the analysis of certain toxicological experiments. *Biometrics*, 34:69–76.
145. Nelson, C. J. and Holson, J. F. (1978): Statistical analysis of teratologic data: problems and advancements. *J. Environ. Pathol. Toxicol.*, 2:187–199.
146. Williams, R. and Buschbom, R. L. (1982): *Statistical Analysis of Litter Experiments in Teratology*. Battelle, Columbus, OH.
147. Rai, K. and Ryzin, J. V. (1985): A dose–response model for teratological experiments involving quantal responses. *Biometrics*, 41:1–9.
148. Bateman, A. T. (1977): The dominant lethal assay in the male mouse. In: *Handbook of Mutagenicity Test Procedures*, edited by B. J. Kilbey, M. Legator, W. Nichols, and C. Ramel. Elsevier, New York, pp. 325–334.
149. Mosteller, F. and Youtz, C. (1961): Tables of the Freeman–Tukey transformations for the binomial and Poisson distributions. *Biometrika*, 48:433–440.
150. Aeschbacher, H. U., Vautaz, L., Sotek, J., and Stalder, R. (1977): Use of the beta binomial distribution in dominant-lethal testing for "weak mutagenic activity," Part 1. *Mutat. Res.*, 44:369–390.
151. Vuataz, L. and Sotek, J. (1978): Use of the beta-binomial distribution in dominant-lethal testing for "weak mutagenic activity," Part 2. *Mutat. Res.*, 52:211–230.
152. Dean, B. J. and Johnston, A. (1977): Dominant lethal assays in the male mice: evaluation of experimental design, statistical methods and the sensitivity of Charles River (CD1) mice. *Mutat. Res.*, 42:269–278.
153. Kilbey, B. J., Legator, M., Nicholas, W., and Ramel, C. (1977): *Handbook of Mutagenicity Test Procedures*. Elsevier, New York, pp. 425–433.
154. Grafe, A. and Vollmar, J. (1977): Small numbers in mutagenicity tests. *Arch. Toxicol.*, 38:27–34.
155. Vollmar, J. (1977): Statistical problems in mutagenicity tests. *Arch. Toxicol.*, 38:13–25.
156. Weil, C. S. (1978): A critique of the collaborative cytogenetics study to measure and minimize interlaboratory variation. *Mutat. Res.*, 50:285–291.
157. Katz, A. J. (1978): Design and analysis of experiments on mutagenicity. I. Minimal sample sizes. *Mutat. Res.*, 50:301–307.
158. Katz, A. J. (1979): Design and analysis of experiments on mutagenicity. II. Assays involving micro-organisms. *Mutat. Res.*, 64:61–77.
159. Kastenbaum, M. A. and Bowman, K. O. (1970): Tables for determining the statistical significance of mutation frequencies. *Mutat. Res.*, 9:527–549.
160. Ehrenberg, L. (1977): Aspects of statistical inference in testing genetic toxicity. In: *Handbook of Mutagenicity Test Procedures*, edited by B. J. Kilbey, M. Legator, W. Nichols, and C. Ramel. Elsevier, New York, pp. 419–459.
161. Bernstein, L., Kaldor, J., McCann, J., and Pike, M. C. (1982): An empirical approach to the statistical analysis of mutagenesis data from the *Salmonella* test. *Mutat. Res.*, 97:267–281.
162. Gad, S. C. (1982): A neuromuscular screen for use in industrial toxicology. *J. Toxicol. Environ. Health*, 9:691–704.
163. Tilson, H. A., Cabe, P. A., and Burne, T. A. (1980): Behavioral procedures for the assessment of neurotoxicity. In: *Experimental and Clinical Neurotoxicology*, edited by P. S. Spencer and N. H. Schaumburg. Williams & Wilkins, Baltimore, MD, pp. 758–766.
164. Irwin, S. (1968): Comprehensive observational assessment. In: Systematic, quantitative procedure for assessing the behavioral and physiologic state of the mouse. *Psychopharmacologia*, 13:222–257.
165. Annau, Z. (1972): The comparative effects of hypoxia and carbon monoxide hypoxia on behavior. In: *Behavioral Toxicology*, edited by B. Weiss and V. G. Laties. Plenum Press New York, pp. 105–127.
166. Norton, S. (1973): Amphetamine as a model for hyperactivity in the rat. *Physiol. Behav.*, 11:181–186.
167. Burt, G. S. (1972): Use of behavioral techniques in the assessment of environmental contaminants. In: *Behavioral Toxicology*, edited by B. Weiss and V. G. Laties. Plenum Press, New York, pp. 241–263.
168. Johnson, B. L., Anger, W. K., Setzer, J. V., and Xinytaras C. (1972): The application of a computer controlled time discrimination performance to problems. In: *Behavioral Toxicology*, edited by B. Weiss and V. G. Laties. Plenum Press, New York, pp. 129–153.
169. Wilson, J. S. and Holland, L. M. (1982): The effect of application frequency on epidermal carcinogenesis assays. *Toxicology*, 24:45–53.
170. Gehring, P. J. and Blau, G. E. (1977): Mechanisms of carcinogenicity: dose response. *J. Environ. Pathol. Toxicol.*, 1:163–179.
171. Haseman, J. K. (1985): Issues in carcinogenicity testing dose selection. *Fund. App. Toxicol.*, 5:66–78.

# 9 Quantitative Extrapolations in Toxicology

*Joseph V. Rodricks, David W. Gaylor, and Duncan Turnbull*

## CONTENTS



## INTRODUCTION

The purpose of this chapter is to describe the scientific basis for extrapolation of toxicity findings in laboratory animals to predict outcomes in human populations. Such extrapolation is often said to comprise two components, one qualitative in nature, the second quantitative. Qualitative extrapolations generally concern the nature of the toxic response (are the specific toxicity endpoints observed in test animals also expected in similarly exposed human beings?), while quantitative extrapolations concern issues such as the magnitude and duration of the dose at which human beings and test animals are expected to be at equal risk of toxicity. Although both types of extrapolation are important, scientific understanding of the basis for quantitative extrapolations across species is more limited than is the basis for qualitative extrapolations. At the same time, because results from animal toxicity studies have become such important determinants of regulatory and other public health protection activities and because the latter typically require the establishment of quantitative limits on human exposure, questions regarding quantitative extrapolations have come to be seen as ultimately of greater significance than those that are purely qualitative in character. Thus, as in many other areas in which science plays a significant role in the development of social policies, the questions that are of most importance tend to be those about which science has the least clear answers. It is therefore no surprise that there is so much public skepticism about the predictions of toxicologists and risk assessors.

The scientific basis for quantitative extrapolations across species is not, however, as feeble as it is sometimes made out to be by those who would minimize the importance of toxicology, and science in general, in regulatory and public health decisions. The principal purpose of this chapter is to describe that basis as it is understood today and to point to the many active areas of research that are devoted to furthering that understanding. As has been the case in the evolution of other areas of science, the greatest understanding comes with the ability to provide quantitatively accurate, empirical descriptions of physical and biological phenomena and to build from these generally applicable, predictive models of those phenomena. It is true, of course, that quantitative understanding never provides a complete description of these phenomena—the qualitative aspects will always be necessary to complete the picture.

Following brief discussions of the historical context of and need for interspecies extrapolation and a description of the problem, the chapter continues with five sections on different aspects of the problem of extrapolation and concludes with a discussion of the overall strategy for scientifically based, interspecies extrapolation.

## HISTORICAL CONTEXT

The use of experimental animals to study biological phenomena arose in the mid-nineteenth century, but their modern use in toxicology had its origin during the third decade of the twentieth century, when several investigators began to study the effects of vitamins, minerals, and other food constituents. At about the same time efforts were initiated to identify and breed species and strains of laboratory animals whose genetic and physiological characteristics and nutritional requirements could be sufficiently well defined so they could be reliably used in controlled experiments. By the mid-1930s, a number of government and industrial laboratories in the United States, Europe, and Japan had begun the fairly routine use of laboratory animals to study occupational chemicals and, soon thereafter, reports of studies of food additives, pesticides, and pharmaceutical agents began to appear in the literature. These early efforts to use laboratory animals to investigate chemical toxicity were no doubt motivated by the belief that responses in animals were useful indicators of potential human responses, but there was little explicit discussion of how data from these studies should be used for that purpose.

Two FDA scientists, Arnold Lehman and O. Garth Fitzhugh, were perhaps the first toxicologists to deal explicitly with the issue. In a short but famous paper published in 1954, the two scientists described the basis for the belief that results from animal studies could be used qualitatively to predict responses in humans, but that quantitative predictions were more problematic. To deal with this problem, Lehmann and Fitzhugh postulated that "average" humans would likely respond to a chemical exposure at a lower dose than would a group of experimental animals, and that within the human population some individuals would respond at lower doses than would the "average" person. In modern parlance, these authors recognized that the *variability* in response at a given dose was likely to be much greater in a highly diverse human population than it is in a group of inbred and otherwise homogenous and healthy experimental animals. They further recognized that, if toxicology data from experimental animals were to be used to establish protective limits* (what Lehman and Fitzhugh called "safe levels") for humans, it would be necessary to "adjust" the experimental results. From their discussion and review arose the concept of "safety factors": A factor of 10 was proposed to adjust (downward) the animal dose (specifically, the no-observed-adverse-effect level [NOAEL]) to estimate the NOAEL for the "average" human and another factor of 10 to estimate the NOAEL for the "most sensitive" members of the human population. They offered the term *acceptable daily intake* (ADI) as their notion of a "safe level" of chronic chemical exposure for the general population, and the ADI was to be obtained by dividing experimental NOAELs from chronic animal toxicology studies by a factor of 100 ($10 \times 10$). This system, though modified

* "Protective limits" is our term, used through the chapter as a description of any quantitative measure that is derived from toxicology or epidemiology data and that is intended to establish the upper limit on exposure that is thought to be without significant risk of toxicity to humans. It is recognized that it is not possible to provide absolute assurance that such limits will protect every person in a population; moreover, exposure greater than these limits may pose no risk to many persons. The term is used here as a simple, one-phrase description of a concept that has in practice unfortunately attracted many different names (see below).

in several significant ways, remains in place today. It is used to establish various protective limits for chemical exposures in the general population. It is interesting that in the 1954 Lehman–Fitzhugh paper, an attempt is made to find an empirical basis for the two factors of 10, but the authors recognized that the database available for such an analysis was extremely limited [48].

Although the Lehman–Fitzhugh approach to quantitative extrapolation recognized the phenomena of inter- and intraspecies variability, it assumed implicitly that no toxic response was likely to occur in any individual unless exposures exceeded some threshold dose. The problem was to define that threshold for a large and diverse human population when the only significant data available arose from experimental studies. During the 1940s and 1950s, an influential body of scientists working in the area of experimental carcinogenesis espoused the view that this particular class of toxic agents behaved biologically in ways that called into question the viability of the threshold concept [54]. Government policies incorporated this view, which was until the 1970s captured by the phrase "no safe level." By that time, regulators saw that such a policy provided little useful guidance for decision-making and turned to the scientific literature to identify specific methodologies that could be applied to animal carcinogenicity data to estimate low-dose risks to humans for substances that might act by "no-threshold" mechanisms [54]. The concept of safety also took a turn at this time, as scientists and decision-makers recognized that safety could be defined only in relative terms. In an influential 1983 study issued by the National Research Council, the notion arose that decisions regarding levels of risk that were sufficiently small to ensure protection of the public health properly fell within the domain of *risk management*: Scientists have the task of assessing toxic risks and describing how their magnitudes change with exposure, but policymakers have the task of deciding how much risk reduction is needed to protect public health and how any needed risk reduction is to be achieved [54,57].

As these developments regarding the uses of toxicology data in decision-making evolved, so did the work of experimental toxicologists. Since the 1950s, animal studies have provided increasingly complete data on the effects of chemical exposures. Thus, increasing amounts of information have been developed on the influences of dose, duration, and routes of exposure; on the roles of chemical kinetics and metabolism; and on the biological and molecular mechanisms underlying the production of toxicity. Experimentalists continue to find useful ways to examine the influences of exogenous compounds on a greater variety of targets, including complex systems of the body. Alongside the work of toxicologists must now be placed developments in the field of epidemiology, because these offer increased possibilities for testing hypotheses generated in the toxicology laboratory.

These various developments, which will emerge more completely in the discussions to follow, place greater demands on the risk managers who must use the results of risk assessments in the formulation of health protection policies. At the same time, they provide better tools for risk assessment—of which interspecies extrapolation is a highly important component—and are proving to be of value in improving the scientific basis for risk-based decision-making. This is, after all, what toxicology is about.

## THE NEED FOR EXTRAPOLATION

### Experimental and Epidemiological Methods and Their Limits

In the best of all possible worlds there would be no need for any form of extrapolation, or, second-best, well-founded predictive models would be available to extrapolate from one set of conditions to another. In the area of predictive toxicology, we now live in a third-best world, and perhaps our goal is to achieve the second-best. The best world is probably not within our reach.

Why is extrapolation necessary? The answer to this question may seem so obvious that it is unnecessary to offer an explicit discussion. But, because so many observers see science as a purely empirical subject in which all forms of extrapolation are merely speculative (a highly naive view of science) or as a subject in which extrapolation is justified only when well-supported predictive models have come to be available (a more credible and, indeed, a proper view of science), the question of "Why engage in extrapolation?" is not as easy to answer as it might appear to be.

Given that there is a social need, expressed in many federal and state laws, to protect people from the toxic properties of chemicals in the environment and in foods, consumer products, medicines, and so on, it is necessary to rely on one or both of the two basic methods available for acquiring toxicological data—the epidemiological and the experimental. Both have strengths and limits (Table 9.1), but neither method is capable of providing direct measurements of toxic risks that are applicable in all situations of potential interest. Animal studies allow us to understand toxicity characteristics of a chemical before human exposure is allowed to occur, whereas the epidemiological method generally does not. They also allow much more thorough examination of the effects of chemical exposure, under a much wider variety of conditions, than do the methods of epidemiology. They usually provide better information on dose–response characteristics and also allow causal relations to be more readily established. They suffer, of course, one large disadvantage, in that they do not reveal responses directly in the species of interest; thus, extrapolation from animal data is necessary

**TABLE 9.1**
**Comparison of Epidemiology and Animal Studies for Identifying Toxic Properties**

| | Epidemiology Studies | Animal Studies |
|---|---|---|
| Opportunity to conduct study | Often not possible | Generally possible |
| Opportunity to obtain information prior to human exposure | No | Yes |
| Time requirements | Years to decades | Weeks to years |
| Species of interest | Yes[a] | No |
| Cause-effect determination | Difficult | Not as difficult |
| Opportunity to obtain quantitative dose–response data | Not frequently | Always |

[a] Note that epidemiology studies may not provide data on both sexes or on all relevant subgroups of the human population.

in many if not most cases if anything at all is to be said about potential human risk and protection of human populations.

Both methods, experimental and epidemiological, suffer from additional limitations. Most importantly, the information they yield, even under the best conditions, is restricted to that portion of the dose–response curve that is within the detection power of the method used; in both cases, the size of the population that can be studied, along with some other aspects of that population, is the principal determinant of detection power. Thus, empirical dose–response data will be limited to the relatively "high-dose, high-risk" portion of the dose–response curve. Most conditions about which toxicity dose–response data are sought concern exposures and doses that fall outside of (well below) the observable range, and therefore outside of the area of direct measurement.

It is also important to emphasize that, although epidemiological data derive from studies in humans, there always remains the question of the representativeness of the studied population for the population whose risk is being assessed. The typical problem concerns the use of information obtained in occupational cohorts for assessing toxic risk in the general population.

Other extrapolation issues arise and cannot be avoided, because it is often necessary to reach decisions in the absence of complete data. The most common problems arise when it is necessary to assess risk for a given route of exposure when data are available only for another route and when the assessment concerns risks associated with chronic exposures when only relatively short-term data on toxicity are available. Thus, if toxicity and epidemiology data are to be used at all to assess human risk and to establish protective limits, extrapolations will be necessary, sometimes several types.

## Using Epidemiological and Experimental Data: Distinguishing "Prudence" from "Science"

When engaging in any form of extrapolation for which a reliable, well-established model or empirical basis is not available, judgments must be made that, strictly speaking go beyond the realm of pure science [54]. If such judgments are not introduced, then, as described in the foregoing, no useful conclusions can be reached. Because of the social need to provide conclusions, however tentative and uncertain, judgments must be introduced, whether they concern qualitative or quantitative issues. The National Research Council, in its 1983 report on risk assessment in the federal government and also in its 1994 report on the same subject, recommended that regulatory agencies use the best available science but that the agencies should also adopt guidelines that would specify what judgments they would make (these have come to be loosely called "defaults") to fill knowledge and data gaps. The NRC's notion was that these "defaults" would be specified in advance, in the form of guidelines, and that they would be applied consistently, in specific risk assessments, to avoid case-by-case manipulations and to ensure explicitness in the assessment. The NRC also recommended flexibility in the use of "defaults" so chemical specific data, if reliable, could be used in specific cases to replace one or more generic "defaults."

The choice of "defaults" is a difficult topic, having both scientific and policy components. The EPA guidelines for carcinogen risk assessment can be consulted for information on the choice of defaults [90]. The emphasis in this chapter is on the scientific basis for extrapolation and its limits, and the regulatory defaults will be discussed only in passing. It is important, however, to attempt to distinguish what is in the area of well-established science from what is a "science-policy" choice. The truth is that all science is accompanied by uncertainty and that there is no sharp distinction between what is known with such high certainty that little or no judgment is called for and that which is insufficiently certain to stand on its own.

## DESCRIPTION OF THE PROBLEM

### Defining Extrapolation

The several types of extrapolations necessary to assess risks of toxicity in human populations from data obtained in experimental animals are described in the next five sections of this chapter, along with what are called the "empirical" and "biological" bases for each. Both qualitative and quantitative aspects are described, with an emphasis on the latter.

- *Cross-species* extrapolations are those pertaining to the attempt to describe expected toxic responses and their relationship to dose in human populations based on responses and their relationship to doses observed experimentally.
- *Within-species* extrapolations are those pertaining to the attempt to describe the expected variability in response within human populations, based either on observations in limited segments of that population or on responses predicted for limited segments of that population from observations in experimental animals.
- *Cross-dose* extrapolations are those pertaining to the attempt to describe toxic responses and their relationship to dose for the range of the dose–response relationship that falls below the range that is subject to direct measurement and within the range expected to be experienced by the population that is the subject of the risk assessment.
- *Cross-route* extrapolations are those pertaining to the attempt to describe expected toxic responses and their relationships to dose in populations exposed by one route, based on responses and their relations to dose observed when exposure occurs by another route.
- *Cross-time* extrapolations are those pertaining to the attempt to describe toxic responses and their relationships to dose for various exposure durations, based on responses and their relationships to dose observed over different exposure durations.

In each case, the "empirical" and "biological" bases for such extrapolations will be summarized. By "empirical" is meant an analysis based on comparisons made in more limited circumstances of actual observations relevant to the extrapolation being considered; for example, compilations and comparisons of human and animal dose–response data for several carcinogens are available, and these comparisons allow at least limited generalizations to be made regarding cross-species extrapolations. By "biological" bases is meant the use of basic knowledge in biology to provide support for particular forms of extrapolation. Both the empirical and biological bases are limited, but often inferences drawn from them converge and thus may provide support for some aspects of extrapolation.

The emphasis in this chapter is on quantitative extrapolations and on the types of scientific data and theories that are regarded as the basis for such extrapolations. As will be noted in the concluding sections, practical applications of many of the issues discussed in the following are not yet generally available; the thrust of the discussion thus concerns the directions necessary to improve the scientific basis for human risk assessment.

## CROSS-SPECIES EXTRAPOLATION

### Defining the Problem

When extrapolating toxicological data from one species to another, it is necessary to consider the various factors that may differentially affect the response of different species to the exposure of interest. As a first approximation, of course, for both cross-species and within-species extrapolation, it is common to normalize the toxicant dose to the body weight, and express it as mg/kg body weight per day. This approach assumes that it is the concentration of the toxicant in the body or at the target site (or some measure directly proportional to it) that determines the magnitude of the toxic effect. Although this appears to address at least one of the differences between species (body size), there are many examples in the literature showing that direct extrapolation simply on the basis of body weight is not accurate. Other aspects of differences among species need to be considered.

Two broad, overlapping classes of factors may affect differential responses: those related to the "effective dose" of the toxicant (determined in part by its pharmacokinetic and metabolic behavior) and those related to the inherent sensitivity of different species. In the absence of differences in inherent sensitivity (often called *pharmacodynamic* differences*), it is generally assumed that different species will show similar responses when exposed to the same effective dose of the toxicant. The effective dose

* The terms "pharmacokinetic" and "pharmacodynamic" derive from the pharmacological sciences. They have been retained by many toxicologists to describe the behavior of substances having toxic effects. Some have advocated the terms "toxicokinetics" and "toxicodynamics" but, although those terms seem more descriptive, they have not achieved widespread use.

may in turn be influenced by species differences in body size, lifespan, anatomy and physiology, and pharmacokinetics and metabolism. For a direct-acting toxicant that follows first-order kinetics, the effective dose may be simply proportional to the administered dose. For a chemical with a more complex pattern of kinetics and metabolism to generate a reactive metabolite, pharmacokinetic modeling may be needed to accurately predict the effective dose.

Two general approaches have been used to evaluate methods for interspecies extrapolation: empirical and biologically based. Empirical approaches rely on collection of data from multiple species exposed to the same substances under comparable conditions and evaluation of the data to identify consistent relationships among species that permit the prediction of the magnitude of the response in one species (typically humans) based on data from another species. This approach has the advantage of simplicity but does not readily allow consideration of chemical-specific factors that may affect *interspecies scaling*. Biologically based approaches make use of scientific knowledge of factors that may influence interspecies scaling, such as anatomical and physiological parameters and especially pharmacokinetics, metabolism, and mechanism of action.

## Empirical Approaches

Empirical approaches to cross-species scaling have been studied since the end of the nineteenth century, when Rubner [72] noted that oxygen utilization and caloric expenditure scaled among dogs of different sizes approximately on the basis of body surface area more closely than simply on body mass. Subsequently, this work was extended to other species and other parameters, including the development of criteria for selecting doses for cancer chemotherapeutic drugs [33,63]. These authors noted that, for a number of chemotherapeutic drugs, the $LD_{10}$ in mice, rats, and hamsters (treated daily for 5 days); the maximum tolerated dose in dogs and monkeys (the highest daily dose that killed no animals); and the maximum tolerated dose used clinically in humans were more closely comparable when the doses were expressed on a mg/m$^2$ body surface area basis than when expressed on the basis of body weight. Because of the difficulty in accurately measuring body surface area, as an approximation based on the relationship between the mass and surface area of a sphere, this surface area scaling is generally approximated as (body weight)$^{2/3}$.

A later reevaluation of these and other toxicity data by Travis and White [83], however, suggested that (body weight)$^{3/4}$ scaling gave a better correlation. Travis and White [83] reanalyzed the toxicity data in mouse, rat, dog, Rhesus monkey, and human on 14 chemotherapeutic drugs described by Freireich et al. [33] and similar data in mouse, hamster, rat, dog, monkey, and human on an additional 13 chemotherapeutic drugs. Based on all of these datasets and the use of a multiple linear regression model, the exponent of body weight giving the best correlation ($r^2 = 0.96$) among species was 0.73, with 95% confidence bounds of 0.69 to 0.77.

More recently, Rhomberg and Wolff [69] reported an analysis of cross-species scaling for acute oral $LD_{50}$ values covering ten mammalian species and a wide range of chemical types. They used data from the NIOSH Registry of Toxic Effects of Chemical Substances (RTECS) database, which contains information on over 135,000 substances. Several thousand of these had oral $LD_{50}$ data from at least two species to permit comparisons. Overall, based on the log–log scatter plots of $LD_{50}$ values in pairs of species (e.g., rat vs. mouse; mouse vs. rabbit) and log–log plots of mean ratios of $LD_{50}$ values vs. body weight for all species-pair comparisons, body weight scaling appeared to give a better correlation among species than did body surface area (body weight)$^{2/3}$ or (body weight)$^{3/4}$. The authors suggested that the difference between the apparent optimal scaling procedure for acute lethality (mg/kg) and that for repeat-dose lethality (mg/kg$^{2/3}$ or mg/kg$^{3/4}$) may be related to differences in the mechanism of lethality. With single acute doses, the lethal dose may depend on the level of defense capacities or reserves that are proportional to body mass; with repeated exposure, survival may, to a larger extent, be a function of repair or replacement rate, which may show scaling patterns more like those for basal metabolic rate or other rates. As discussed below, these rates tend to scale across species as a function of body surface area or (body weight)$^{3/4}$.

Empirical approaches have also been used extensively in attempts to identify the most appropriate approach for scaling of cancer data between species [2,20,21,35]. In one of the most extensive of these studies, Allen et al. [2] evaluated 23 chemicals for which data permitted quantitative estimation of cancer potency in humans and animals. As their measure of "potency," Allen et al. [2] calculated *risk-related doses* (RRDs), defined as the average daily dose per kg body weight that would be expected to result in an extra cancer risk of 25% over a lifetime. Allen et al. examined several different ways of expressing the dose, different scaling procedures, different subsets of animal data (restricted by experimental design, route of exposure, species, sex, tumor types), different statistical measures (maximum likelihood estimates and lower confidence limits on RRD), and different ways of considering results from multiple studies of the same chemicals (median response, most sensitive species/sex combination). When the logarithms of the RRDs derived from epidemiological studies were plotted against the logarithms of the predicted RRDs from animal studies, the scaling procedure giving the strongest correlation (by a slight margin) between humans and animals was scaling on the basis of body weight. However, there were wide variations in the apparent relative potency for individual chemicals; given the uncertainties in the experimental and epidemiological data, it was not possible to rule out surface area scaling—(body weight)$^{2/3}$—or some intermediate procedure, such as (body weight)$^{3/4}$, as being appropriate. Wide variations in apparent relative potency were also reported in the other comparisons [20,21,35]. These studies considered mostly potency comparisons between rats and mice. On average in these studies, rats appeared to be slightly more sensitive than mice when compared on a body-weight basis, as would be expected if the correct scaling was on the basis of (body weight)$^{2/3}$ or (body weight)$^{3/4}$. Because of the small sample sizes and wide variation for individual chemicals, however, these comparisons are of limited predictive value.

The U.S. Environmental Protection Agency (USEPA) has published an extensive discussion on the selection of a default scaling procedure for cancer risk assessment based on these and other considerations [82]. Its conclusion was that, although there was considerable uncertainty in the selection of any generic scaling procedure, the balance of evidence supported the use of (body weight)$^{3/4}$ as a general procedure when no chemical-specific data provide support for an alternative procedure.

## Biologically Based Approaches

### *Allometry*

The work of Rubner [72] and others [1,43,44] on the relationship between body size and metabolism, mentioned previously, led to the study of allometry, the relationships between body weight and various biological and physiological parameters. A large number of biological and physiological parameters have been found to show a relation to body weight of the form:

$$Y = aW^b$$

where $Y$ is the biological parameter of interest (e.g., organ size, blood flow rate, basal metabolic rate), and $a$ and $b$ are constants relating $Y$ to body weight ($W$).

When the exponent $b$ in the allometric equation is 1.0, cross-species scaling is based on relative body weights. When it is 0.67, the biological or physiological measure is said to scale on the basis of relative surface areas.

In examining a wide number of measurements, measures of size (e.g., organ weights, blood volumes) tended to show an exponent ($b$) of approximately 1.0, while measures of rates (e.g., ventilation rate, basal metabolic rate, drug clearance rate) typically showed values closer to 0.67 [1]. There was considerable variation in the best estimate of the exponent ($b$) for different parameters, and other authors have suggested that the value of $b$ for metabolic and other rates falls closer to 0.75 [11,64,75]. Figure 9.1, for example shows a plot of basal metabolic rate vs. body weight for warm-blooded animals ranging from mouse to elephant [7]. The slope of this line corresponds to a $b$ value of 0.76.

### *Pharmacokinetics*

Parameters affecting pharmacokinetic differences across species (organ sizes, physiological and metabolic rates) are also subject to allometric scaling. As a result, provided a chemical follows the same pharmacokinetic pathways in different species, pharmacokinetic scaling will also tend to follow allometric scaling. Travis et al. [84] have suggested that for toxicants that are metabolically inactivated scaling on the basis of (body weight)$^{3/4}$ provides the most accurate scaling across species. For reactive chemicals that are spontaneously inactivated, body weight scaling may provide a more accurate scaling metric.

All of the above discussion has centered on determining a generic procedure for extrapolating among species. As noted above, one of the most critical factors in cross-species extrapolation is the effective dose of the toxicant at the site of action. This is the purview of pharmacokinetic modeling. The use of pharmacokinetic modeling, particularly physiologically based pharmacokinetic (PBPK) modeling in exposure estimation for risk assessment has been extensively discussed [55,83,84] and is continually being updated and improved. For example, a series of increasingly comprehensive PBPK models have been applied to trichloroethylene over the past 10 to 15 years [39,78].

Physiologically based pharmacokinetic modeling uses physiological parameters, such as breathing rates, blood flow rates, and tissue volumes, to describe metabolic processes in physiologically realistic tissue groups (compartments), such as liver, lung, fat muscle, and other organs that are connected by venous and arterial blood flow. Each compartment is described mathematically by a set of differential equations. Such models can be used to relate exposure concentrations (e.g., in air, water, or food) to concentrations of the parent compound or its metabolites in different tissues, including the target tissue. A major strength of PBPK models is that one can use the same model developed for a chemical in one species to predict chemical transport and metabolism in other species (including humans) by substituting species-specific physiological, biochemical, and metabolic parameters in the model.

## Pharmacodynamics

Pharmacokinetics deals with the movement of chemicals in the body. Pharmacodynamics examines the interactions of chemicals and their metabolites with tissue constituents, their biological and physiological effects, and mechanisms of action. Even if one uses allometric scaling or pharmacokinetic modeling to identify an equivalent dose of a chemical at a target site, there may be differences in

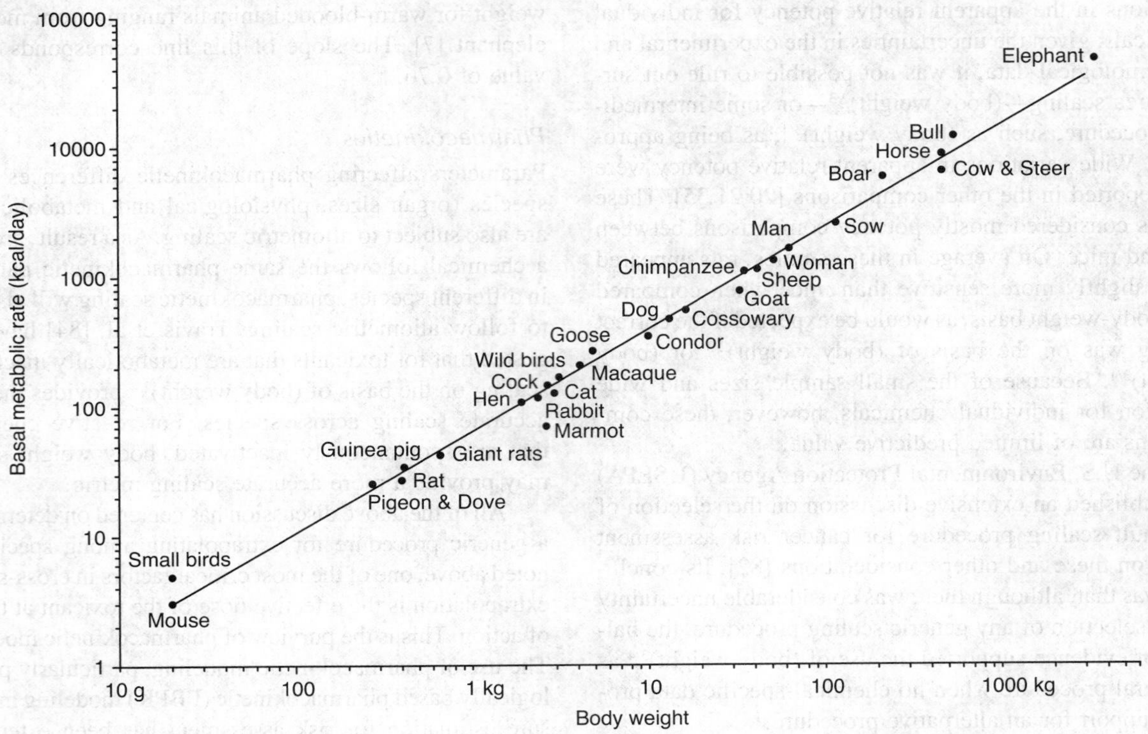


**FIGURE 9.1** Benedict's 1938 mouse-to-elephant graph showing the relationship between species body weight and basal metabolic rate. (From Benedict, F. G., *Vital Energetics: A Study in Comparative Basal Metabolism*. Carnegie Institute of Washington, Washington, D.C., 1938.)

response in different species because of differences in pharmacodynamics or susceptibility. Thus, for example, a diverse group of chemicals, including unleaded gasoline, decalin, and *d*-limonene, cause nephrotoxicity and renal tumors in male rats by a species- and sex-specific mechanism involving binding to $\alpha_{2\mu}$-globulin [10,80]. This phenomenon does not occur in female rats or in other species (including humans) that do not produce $\alpha_{2\mu}$-globulin, nor does it occur in male Black Reiter rats that also lack $\alpha_{2\mu}$-globulin [26].

A number of chemicals interact with glutathione and cause toxic effects when tissue glutathione levels are depleted. Knowledge of differences among species in the constitutive levels of glutathione in different tissue and its rate of synthesis can assist in predicting differences among species in sensitivity to such chemicals.

In some cases, data from *in vitro* studies can be used to assist in interspecies extrapolation; for example, chemicals that cause peroxisome proliferation in rodent liver have been studied in isolated hepatocytes to investigate differences among species in susceptibility [29,60]. Use of data from studies in isolated cells to predict effects *in vivo*, however, adds its own set of uncertainties, including those related to the representativeness of the cell or tissue samples used for the studies [59].

## Application of Allometric Scaling

To apply allometric scaling in practice for converting a specific animal dose (in mg/kg body weight) to an equivalent human dose (also in mg/kg) the animal dose is multiplied by the ratio of animal to human body weight to the $(1 - b)$ power. Thus, for scaling on the basis of (body weight)$^{3/4}$, the equivalent human dose is calculated as:

$$\text{Human dose (mg/kg)} = \text{animal dose (mg/kg)} \times \left(\frac{\text{animal body weight}}{\text{human body weight}}\right)^{1/4}$$

This scaling factor is derived as follows. For (body weight)$^b$ scaling:

$$\frac{D_h}{(W_h)^b} = \frac{D_a}{(W_a)^b}$$

where $D_h$ is the human dose (mg), $D_a$ is the animal dose (mg), $W_h$ is the human body weight (kg), and $W_a$ is the animal body weight. Rearranging this equation and dividing both sides by the human body weight gives:

$$\frac{D_h}{W_h} = \frac{D_a}{W_h} \times \left(\frac{W_h}{W_a}\right)^b$$

Multiplying the top and bottom of the right side by the animal body weight ($W_a$) gives:

$$\frac{D_h}{W_h} = \frac{D_a}{W_a} \times \frac{(W_h)^b \times W_a}{(W_a)^b \times W_h}$$

$$= \frac{D_a}{W_a} \times \frac{(W_a)^{1-b}}{(W_h)^{1-b}}$$

$$= \frac{D_a}{W_a} \times \left(\frac{W_a}{W_h}\right)^{1-b}$$

Examples of the application of these relationships are illustrated in Table 9.2.

**TABLE 9.2**
**Cross-Species Scaling Factors Based on Allometric Scaling**[a]

| Allometric Scaling | Mouse–Human Scaling Factor | Rat–Human Scaling Factor |
|---|---|---|
| $W^{1.0}$ | $(0.035/70)^0 = 1.0$ | $(0.3/70)^0 = 1.0$ |
| $W^{0.75}$ | $(0.035/70)^{0.25} = 0.150$ | $(0.3/70)^{0.25} = 0.256$ |
| $W^{0.67}$ | $(0.035/70)^{0.33} = 0.0814$ | $(0.3/70)^{0.33} = 0.165$ |

[a] Assumes human body weight of 70 kg, mouse body weight of 35 g, and rat body weight of 300 g. The scaling factor indicates the dose for a human (mg/kg/day) that is equivalent to 1 mg/kg/day for a mouse or rat; for example, with $W^{0.75}$ scaling, a dose of 1 mg/kg/day in a mouse is equivalent to 0.150 mg/kg/day for a 70-kg human.

### Current Trends

Currently, extensive work is underway to improve our ability to perform cross-species extrapolation, particularly in the areas if pharmacokinetic and pharmacodynamic modeling. As with other types of extrapolation, cross-species extrapolation is aided by an understanding of the mechanism of action of the toxicant under consideration. Because of the limitations of studying toxicants in humans, efforts are also underway to use isolated human cells to address some pharmacodynamic questions and, hence, improve the accuracy of extrapolation.

The increasing availability of biomonitoring data on a variety of chemicals in the general population [16] has encouraged the use of blood or tissue levels of chemicals in humans for interspecies extrapolation. In a recent risk assessment of perfluorooctanoic acid (PFOA), for example, Butenhoff et al. [12] used serum PFOA level as a basis for cross-species extrapolation from rats and monkeys to humans, assuming equal risk at equal steady-state serum PFOA level. Butenhoff et al. [12] reported that the use of serum levels of this nonmetabolized chemical was useful for minimizing uncertainty in cross-species extrapolation. Support for this view comes from the fact that rats, dogs, and monkeys all showed similar toxic effects at similar serum PFOA levels [12].

## WITHIN-SPECIES EXTRAPOLATION: VARIABILITY IN HUMAN RESPONSE

### Defining the Problem

The term "extrapolation" is, in this context, used somewhat differently than it is in the four other areas discussed in this chapter. The human population is without question highly *heterogenous* with respect to all of the many differences that affect response to a given dose of a given chemical. If the distribution of toxic responses at a specified dose were known with accuracy, it would be possible to specify a dose at which a specified fraction of the population would be at a given risk of toxicity. Moreover, if the distribution of threshold doses for the population were known, it would be possible to specify doses at which, for example, the vast majority of individuals would not be at risk; the specific fraction at the tail of the distribution that might remain at risk would be selected as a matter of policy (such a selection would be necessary on the assumption that any plausible distribution of population thresholds would not include an identifiable dose at which there were absolutely zero responders).

In practice, such population distributions are not available, and the traditional default approach has been, first, to estimate a threshold dose for some hypothetical person described as "average" with respect to responsiveness, and, second, to divide this threshold estimate by a factor, usually 10, that is thought to lead to an estimate of the threshold dose somewhere near the tail of the underlying, but unknown, distribution at which the "most sensitive" members of population are expected to be found. This estimate is usually considered protective for the "most sensitive" individuals and, necessarily, for all less sensitive individuals. Factors other than 10 have been used if there is some reason to believe that the data supporting a threshold represent individuals more or less sensitive than the hypothetical "average." This relatively crude procedure thus involves extrapolation from so-called "average" to "sensitive" individuals. Interestingly, no methodological tradition within the realm of carcinogen risk assessment is specifically designed to account for population variability [57].

**TABLE 9.3**
**Major Sources of Interindividual Variability in Responses to Chemical Exposures**

| Source | Cause of Variability | Some Major Influences |
|---|---|---|
| Uptake | Differences in contact with and absorption of chemical from its environmental sources | Age<br>Diet<br>Smoking<br>Health status of skin, respiratory tract, and gastrointestinal tract |
| Pharmacokinetics and metabolism | Differences in distribution, metabolism, and elimination, leading to different target site concentrations | Age<br>Gender<br>Health status<br>Other exposure (dietary, drug, chemicals)<br>Genetic polymorphisms<br>Pregnancy |
| Response at target sites (pharmacodynamics) | Differences in biological response at a given target site concentration | Age<br>Gender<br>Health status<br>Nutritional status<br>Hormonal status<br>Pregnancy<br>Immune status<br>Genetic polymorphisms |

The traditional approach to deriving protective doses offers no insight into the degree of population protection provided. Whether it is underprotective or extraordinarily overprotective depends on where on the true (but unknowable) distribution of threshold doses the protective dose happens to lie.

Movement away from the "default" approach toward a more scientifically rigorous one that provides some insight into the degree of protection from toxic risk provided at various doses depends on the development and incorporation of scientific information relevant to the question of population variability. The factors known to influence variability are described in the next subsection, and this is followed by a review of some empirical data on this question.

## Biological Bases for Interindividual Variability

Three major sources of interindividual variability in response to a chemical exposure can be described. First, individuals vary in intake of a chemical from the environment; second, pharmacokinetic and metabolic behaviors of chemicals may vary among individuals; and, third, interindividual variability may exist with respect to the response at the target site to a given dose (concentration × time) of a toxicologically active compound (pharmacodynamic differences). These three influences lead to variability in the size of the dose (concentration × time) of active compound (administered compound or, more often, a metabolite thereof) that reaches the target site and the magnitude of the response to that dose, even when all individuals are exposed to identical concentrations of a chemical in their environment. In addition, variability exists with respect to the intake of a chemical because of differences in the nature and extent of human contacts with the environmental media in which the chemical is present. In Table 9.3 are listed some of the major contributors to interindividual variability in toxic response.

Despite the substantial empirical support for the fact of human variability, large uncertainty exists regarding its magnitude. In any given exposure situation, some factors may serve to increase the relative responsiveness of some individuals, but in other exposure situations these same individuals may be at less relative risk. Thus, for example, infants lacking metabolic capacity, which does not fully develop until about one year of age, may be less susceptible to substances requiring metabolic activation yet be more sensitive to other substances because of their less than fully functional immune systems. The number of factors influencing responsiveness is so large and variable within the human population and the cumulative direction of their effects (to increase or decrease sensitivity) so unpredictable in any given exposure situation that no attempt to derive a generally applicable model of variability based on biological understanding of each of the factors known to influence it has proved successful. Instead, empirical evidence that captures the cumulative effects of all important influences has been generally regarded as of more value [57].

## Empirical Approaches to Understanding Interindividual Variability

Little empirical data supported Lehmann and Fitzhugh's original 1954 proposal to extrapolate from "average" to "sensitive" individuals by the incorporation of the assumption of a 10-fold difference in sensitivity, but this assumption has become the standard "default" uncertainty factor typically used by regulators when there is no known basis for another factor. Although it has never been explicitly described, the adoption of the 10-fold factor suggests that the total variability in response in the human population ranges over about 100-fold, assuming a symmetrical distribution of responses about the average. Of course, the location of the "least" and "most" sensitive individuals on the actual distribution is unknown, so it is not possible to describe the actual range.

By the early 1980s sufficient empirical information had accumulated to allow limited analysis of variability. Dourson and Stara's 1983 review [27] of a number of datasets concluded that a 10-fold factor was likely to reflect a wider range of interindividual variability than could be documented for the vast majority of chemicals. This analysis suggested that the 10-fold factor was adequately protective. Review of differences in human metabolism of chemicals [13] has found that a 10-fold factor covered the total range of variability for 80 to 95% of the population; this finding suggests that the range from "average" to "sensitive" is significantly less than 10-fold. $LD_{50}$ ratios of adults to newborns for 238 chemicals have been evaluated [76] as a measure of intraspecies variability. Although it was found that the median ratio reflected only a 2.6-fold variability and that 86% of the ratios were less than 10, the fact that most of the data derived from experimental animal studies casts some doubt on its applicability to humans. In a review published in 1996, the authors concluded, based on evaluations of the type just described, that [28]:

> In general, the default value of 10 for interhuman variability appears to be protective when starting from a median response, or by inference, from a NOAEL assumed to be from an average group of humans. However, when NOAELs are available in a known sensitive human subpopulation, or if human toxicokinetics or toxicodynamics are known with some certainty, this default value should be adjusted or replaced accordingly.

Some authors have proposed examining variability separately for factors influencing delivery of target site dose (uptake, pharmacokinetics, metabolism) and those affecting response. One review suggested that variability in the former factor was generally larger than it was for the latter and proposed that the 10-fold factor be subdivided in factors of 4 (pharmacokinetics) and 2.5 (pharmacodynamics) [67].

The limited empirical analysis supporting the factor of 10, or its subfactors, is perhaps reassuring. Little effort has been devoted to developing more complete descriptions of variability distributions. Heterogeneity in response might be derived by treating human data, where available, as animal data are often treated [68]. Probit plots have, for example, been developed using data derived from studies in Iraq of neurobehavioral outcomes in humans exposed *in utero* to methylmercury. A probit plot is one useful way of describing the variability in thresholds among individuals in a population. If it is assumed that the distribution is lognormal in character, then a plot of probit against log dose yields a straight line, the slope of which reflects variability. Steep slopes reveal less variability than do shallower ones. (Lognormal distributions arise when the factors contributing to variability act multiplicatively—addition of the logarithms of variables is identical to multiplying the variables themselves). It has been suggested that some responses to methylmercury show a probit slope as low as 1, corresponding to a geometric standard deviation of 10 [68]. It is inferred from this that 95% of the population would have thresholds spread over a range of 10,000-fold in dosage—from 100-fold lower to 100-fold higher than the threshold dose for individuals at the median. Such estimates provide the type of description of variability that could increase the level of risk information provided to decision-makers, because they allow more explicit analysis of the areas of the distribution that might be selected as the focus of regulation. At the same time, numerous difficulties attend the use of such statistical methods, not least of which is their failure to incorporate data on biological mechanisms. It nevertheless suggests the possibility of a more quantitative direction for this aspect of risk assessment.

# QUANTITATIVE STRATEGIES FOR HIGH- TO LOW-DOSE EXTRAPOLATION

## Introduction

In the production and use of chemicals, it is necessary to consider the health risks and benefits. Both the degree of risks and benefits depend on the amount of chemical present. For some chemicals (e.g., genotoxic carcinogens), even trace amounts may pose some risk. For other relatively nontoxic chemicals, large doses may be required before adverse health effects result. In either case, the goal is to eliminate or at least minimize the occurrence of adverse health effects. This generally entails establishing a dose–response curve that relates the incidence of disease (i.e., the proportion of individuals that develop a disease to the dose of a chemical). This provides a method for estimating doses associated with low probability of disease.

In the case of cancer, regulatory decisions regarding exposures to carcinogens are generally made to limit the estimated lifetime probability of cancer (risk) to less than one in 10,000 and often to less than one in 1,000,000 [70]. This creates a difficult problem. For studies conducted in laboratory animals, it would require tens of thousands of animals to estimate the incidence of cancer with precision at doses producing cancer risks below 1 in 10,000. Resources simply are not available to conduct such studies. Instead, experiments are conducted in animals at doses high enough to elicit potential toxic effects that can be observed in a moderate number, generally 50 or less, of animals per dose group. Such studies generally must produce incidence rates in excess of 10% to achieve statistical significance; this limit is 1000 to 100,000 times greater than the risks that are the subject of regulation. Often, these studies require doses that are tens, hundreds, or even thousands of times higher than human doses. In cancer studies, the highest dose generally is selected that is anticipated not to cause death, other than due to tumors, and does not cause average body weight losses greater than 10% compared to control animals. The use of high doses necessitates extrapolation to estimate the incidence of adverse health effects at human dose levels.

Occasionally, human data are available to access risk. Often these studies are based on occupational exposures that are higher than those experienced by the general public. Hence, extrapolation to lower doses is required, albeit much less than from animal studies. Toxicity studies for drugs often are conducted from near-human dose levels up to perhaps 100 times higher. In such studies, some extrapolation to lower dose levels may be required.

A question almost always arises from toxicological studies about extrapolating the results from high doses to lower doses experienced by humans. In the following sections, the biological and empirical basis for low-dose extrapolation is summarized. Finally, the uncertainties and future directions for low-dose extrapolation are discussed.

In the 1970s and early 1980s, many mathematical models were developed and proposed for use in low-dose extrapolation, some with a theoretical, biological basis, others more empirically based (providing a good mathematical fit to experimental data, with useful statistical properties). These include one-hit, multihit, multistage, Weibull, probit, and log-logistic models [54,58]. With some datasets, different models gave very different estimates of low-dose risk. In the face of uncertainty, regulatory agencies, in consideration of their role of ensuring public health, adopted models for low-dose extrapolation that were conservative; that is, they erred on the side of tending to overestimate risk [32,85]. More recently, there has been less emphasis on the specific mathematical model used for low-dose extrapolation and more emphasis on understanding the mode of action of the carcinogen and the implications of that mode of action for low-dose human risk [90]. In this section, we discuss current approaches used by regulatory agencies, particularly the USEPA, for low-dose extrapolation.

## Noncancer Endpoints

### Threshold Doses

For noncancer endpoints, it is generally assumed that small doses of chemicals can be tolerated without any adverse health effects. Experimental data often are compatible with the existence of a threshold dose; however, experimental data can only demonstrate that an effect is likely to be within certain limits. For example, when no adverse effect is seen among a group of 100 animals, it can be stated with 95% confidence that the true incidence is likely to be less than 3%. Threshold doses generally cannot be estimated with precision even in animal studies with homogeneous animals. Estimation of threshold doses for a heterogeneous human population is even more problematic.

The general approach to safety assessment for noncancer effects is to establish an acceptable daily intake (reference dose) based on dividing an experimental no-observed-adverse-effect level (NOAEL) by a series of safety (uncertainty) factors to allow for extrapolation from animals to humans, when necessary, and to allow for protection of sensitive individuals in a population [6,89]. Ideally, this results in a reference dose (RfD) that is below the threshold dose for most individuals, resulting in negligible risks. Because the RfD based on animal data is generally a factor of 100 or more below doses that produce adverse effects in bioassays, it is presumed that risks are negligible at these lower doses.

The safety assessment process described above makes no explicit use of a dose–response curve. No estimate of the risk at the NOAEL is made nor extrapolation to lower risks at lower doses. Gaylor [34] demonstrated that estimated risks of embryo/fetal deaths and malformations at the NOAEL varied from 0 to 4.5% for typical bioassays. Leisenring and Ryan [49] showed that risks at the NOAEL could be as high as 20% for quantal (incidence) data from typical bioassays. Recognizing the wide variation in risks at the NOAEL and the fact that smaller sample sizes result in larger NOAELs, Crump [22] suggested that, rather than the NOAEL, the point of departure for RfDs should be a benchmark dose (BD) corresponding to a specified low level (1 to 10%) of risk. A lower confidence limit for the BD is used to allow for experimental variation, and an additional uncertainty factor is introduced to account for the point of departure being associated with a low level of risk. The USEPA has endorsed the use of the BD methodology [89] and has employed it in several recent evaluations (e.g., antimony trioxide, benzene, beryllium, chloroform, 1,1-dichloroethylene, 1,3-dichloropropene, ethylene glycol monobutyl ether, hexachlorocyclopentadiene, methyl ethyl ketone, methylmercury, methylnaphthalene, methylene diphenyl diisocyanate, phosphoric acid, tributyl tin oxide, 1,1,1,2, tetrafluoroethane—all available at http://www.epa.gov/iris/index.html).

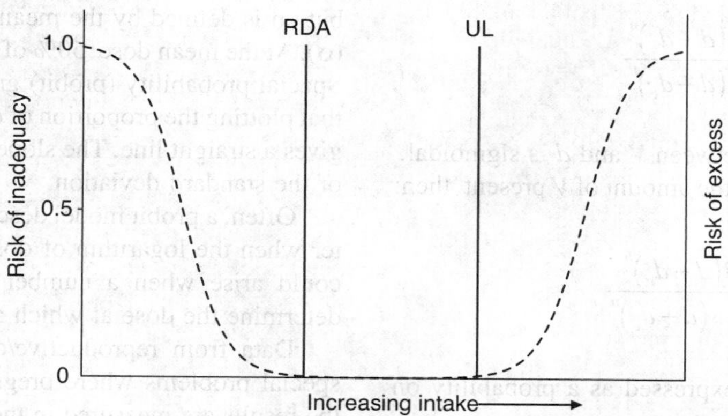


**FIGURE 9.2** U-shaped dose–response curve seen with essential nutrients.

The whole concept of threshold doses is challenged by the additivity-to-background argument [25]. For adverse health effects that occur at some background frequency in the human population that may be the result of chemical exposure, either endogenous or exogenous, threshold doses have apparently been surpassed for some individuals. Other individuals are, presumably, just below their threshold; hence, the addition of any chemical dose, no matter how small, will have an additional effect if it augments an existing toxic chemical pathway [25]. At low doses, the relationship between added health risk and dose is approximately linear [25]. For this to be true, however, the added chemical dose must not be a toxic effect that has a mechanism of action independent of existing mechanisms.

## U-Shaped Dose–Response

With most toxicants, the incidence of adverse effects increases with increasing dose once the threshold (if it exists) is exceeded. For some essential vitamins and minerals, however, a U-shaped dose–response curve may occur (Figure 9.2). At deficient levels, health risks increase, but health risks may also occur at excessive levels (e.g., vitamin A and iron) [41,42]; hence, an optimum dose or range exists that minimizes risk, with risk increasing at doses both below and above the optimum level.

Some evidence suggests a phenomenon (or group of phenomena) that can also display a U-shaped response—so-called *hormetic* [14,15]. These are dose–response relationships in which a small initial benefit (e.g., improvement of growth) is seen at low dose levels, followed at higher doses by impairment. This has been attributed to “a modest overcompensation to a disruption in homeostasis” [15]. Such phenomena have been seen in a variety of organisms, but their applicability to humans is uncertain.

Clearly, U-shaped dose–response curves require special attention. Risk assessment low-dose extrapolation procedures that only consider high dose risk may result in recommended doses that produce risks at low doses, particularly for essential or beneficial nutrients.

## Quantitative Extrapolation of Noncancer Effects

Although most low-dose extrapolation of noncancer effects uses the NOAEL/safety factor approach described above, some attempts have been made to extrapolate quantitatively to lower doses. The Michaelis–Menten equation is used widely in enzyme kinetics to relate the velocity ($V$) of an enzyme-mediated reaction to the substrate dose ($d$), where $V$ increases rapidly for small doses and then levels off, approaching a maximum rate ($V_{max}$) at high doses:

$$V = \frac{V_{max} d}{K + d}$$

where $K$ is the dose at which $V$ equals one half of the maximum value ($V_{max}$). $V_{max}$ and $K$ are generally estimated by a double reciprocal plot of $1/V$ vs. $1/d$ that gives a linear relationship with an intercept of $1/V_{max}$ and slope of $K$ times the intercept:

$$\frac{1}{V} = \frac{1}{V_{max}} + \frac{k}{V_{max}}\left(\frac{1}{d}\right)$$

In the event that an endogenous dose is present that is equivalent to $d_e$, then:

$$\frac{1}{V} = \frac{V_{max}\left(d + d_e\right)}{K + \left(d + d_e\right)}$$

Now, nonlinear regression procedures are required to estimate the parameters. The Hill equation is a generalization where:

$$V = \frac{V_{max}\left(d+d_e\right)^n}{K+\left(d+d_e\right)^n}$$

If $n > 1$, the relationship between $V$ and $d$ is sigmoidal.

If risk is proportional to the amount of $V$ present, then:

$$\text{Risk} = \frac{B\left(d+d_e\right)^n}{K+\left(d+d_e\right)^n}$$

where $B \le 1$ when risk is expressed as a probability on the scale of 0 to 1.

Even though there may be a biological basis for this dose–response curve, estimates of $B$, $K$, $d_e$, and $n$ may be quite imprecise unless the numbers of animals are adequate over a wide range of doses. Also, the relationship between risk and $V$ is crucial.

For developmental effects, Gaylor and Chen [36] have proposed that birth defects may be related to decreased fetal weight. It was assumed that fetal growth was exponential and that the growth rate constant was affected by chemical exposure during gestation. They suggested two models that fit a number of dose–response data about equally well for predicting the incidence ($P$) of a variety of structural malformations:

$$P = 1-\exp\left[-\left(b_0+b_1d^k\right)\right]$$

and

$$P = 1-\exp\left[-\left(b_0+b_1d+b_2d^2+\ldots+b_kd^k\right)\right]$$

where $d$ is the daily dose, and $b_0$, $b_1$, ..., $b_k$ and $k$ are estimated from the data.

Leroux et al. [50] developed a mathematical model to describe aspects of the dynamic process of organogenesis based on branching models of cell kinetics. The biological information incorporated in the model includes timing and rates of dynamic cell processes such as differentiation, migration, growth, and replication. The dose–response models produced can explain patterns of malformation rates as a function of both dose and time of exposure.

The probit model has a long history in biology [30]. It belongs to a class of tolerance distributions. The probit model is based on the Gaussian (normal) bell-shaped distribution. The tolerance distribution describes the relative probability that an individual will respond (suffer an adverse health effect) at a dose $d$. If a large number of factors act in an additive manner to determine the dose at which an individual responds, then by the central limit theorem a normal distribution is obtained. Integrating (summing the relative probabilities) up to a dose $d$ gives the proportion (probability) of individuals that develop an adverse effect at or below the dose $d$. The normal distribution is defined by the mean ($\mu$) and standard deviation ($\sigma$). At the mean dose, 50% of the individuals are affected. Special probability (probit) graph paper is available such that plotting the proportion of effected individuals vs. dose gives a straight line. The slope of the line is the reciprocal of the standard deviation.

Often, a probit model describes experimental data better when the logarithm of dose is used. This distribution could arise when a number of factors multiplicatively determine the dose at which an individual responses.

Data from reproductive/developmental studies pose special problems where pregnant animals are dosed and the results are measured in the offspring/fetuses. Correlation of results among offspring/fetuses within a litter must be considered. Kodell et al. [46] assume that the incidence of adverse effects for the offspring/fetuses of a litter behave according to a binomial distribution and the probability of adverse effects varies among litters according to a beta distribution. Further, it is assumed that the probability of an adverse effect may be a function of the size $s$ (number of offspring/fetuses in a litter). The expected probability of an adverse effect for an offspring/fetus in a litter of size $s$ at dose $d$ is:

$$P(d,s) = 1-\exp\left\{-\left[b_0+b_1\left(s-\bar{s}\right)\right]\right\}$$

for $d$ less than or equal to a threshold dose of $d_0$ and $\bar{s}$ is the average litter size across all dose groups in a bioassay. For doses above the threshold dose of $d_0$:

$$P(d,s) = 1-\exp\left(-\left\{\begin{matrix}\left[b_0+b_1\left(s-\bar{s}\right)\right]+\left[b_3+b_4\left(s-\bar{s}\right)\right] \\ \times\left(d-d_0\right)^k\end{matrix}\right\}\right)$$

where the $b \ge 0$ and $k \ge 1$ are estimated from the data, and

$$\left[b_0+b_1\left(s-\bar{s}\right)\right] \ge 0 \text{ and } \left[b_3+b_4\left(s-\bar{s}\right)\right] \ge 0 \text{ for all } s$$

Kupper et al. [47] proposed a model for reproductive/developmental data of the form:

$$P(d,s) = b_0+b_1s+\left[1-b_0-b_1s\right] / \left\{1+\exp\left[b_3+b_4s-b_5\log\left(d-d_0\right)\right]\right\}$$

Ryan [73] discussed multivariate models that simultaneously consider two or more biological effects, such as the proportion of malformed fetuses in a litter and the proportion of dead/resorbed fetuses in a litter.

Microbial risk assessment is an area currently receiving more attention than in the past. Haas [40] proposed that the probability of infection could be described as a function of the number $N$ of colony-forming units by:

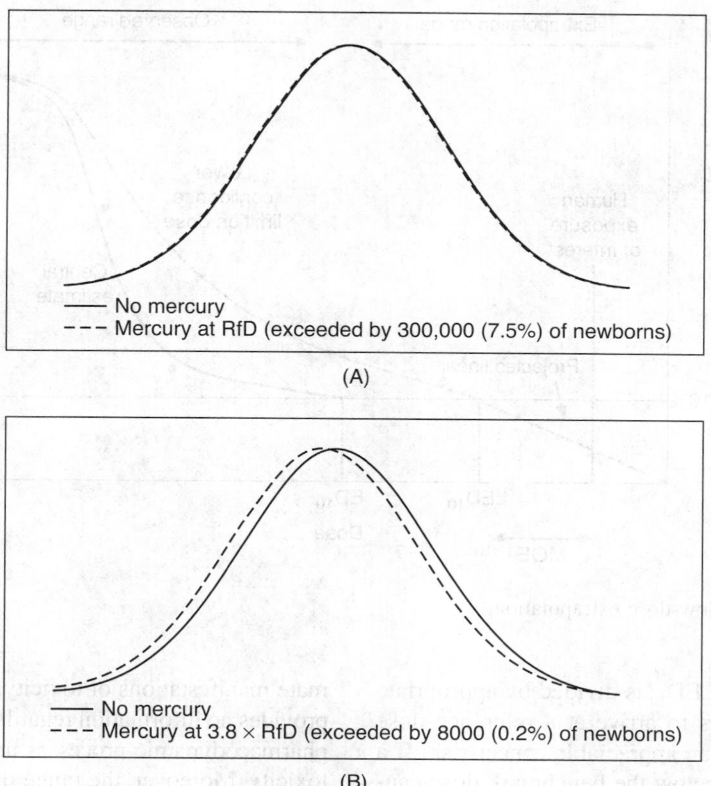


**FIGURE 9.3** Predicted distribution of performance in children not exposed to and in children whose mothers were exposed to methylmercury at the RfD (A) or at 3.8 times the RfD (B). (From Clewell, H.J. and Crump, K.S., *Risk Analysis*, 25, 285–289, 2005. With permission.)

$$P = 1 - [1 + (N/\beta)]^{-\alpha}$$

where $\alpha$ and $\beta$ are estimated from dose–response data.

Low-dose extrapolation using empirical models is suspect. The model may fit adequately in the experimental dose range, but there is little assurance of accuracy below the experimental dose range. However, linear extrapolation can serve as an upper-bound estimate of risk for dose–responses that are convex (curve upward).

In addition to examining changes in the proportions of a quantal (or quantized) effect at low doses, it is also possible to estimate the effect of a particular (low) dose on the distribution of a continuous variable (such as body weight). An interesting recent example of this appears in a paper by Clewell and Crump [18]. These authors used dose–response information used by the USEPA to derive the reference dose (RfD) for methylmercury to illustrate the change in the distribution of the predicted performance in a neuropsychological function test (Boston naming test) of children of mothers unexposed to methylmercury compared to those whose mothers were exposed at the RfD, or a multiple of the RfD. Figure 9.3 shows that, at the RfD, the distribution of performance is almost indistinguishable from the of the unexposed population, and even at 3.8 times the RfD, the difference is small (a change in the mean score of 1.6%).

## Cancer

More attention has been devoted to dose–response modeling for cancer than noncancer effects for risk assessment. As noted earlier, considerable effort was expended in the 1970s trying to establish the best mathematical model for low-dose extrapolation of cancer data. More recently, greater emphasis has been placed on understanding the mode of action of the carcinogen under study and how that mode of action is likely to influence the shape of the dose–response curve at low doses. Recognizing that current methodology generally does not permit precise estimates of cancer risks below 10%, the latest USEPA [90] carcinogen cancer risk assessment guidelines suggest estimating the dose associated with a 10% tumor risk ($ED_{10}$) as a point of departure for low-dose extrapolation. To account for experimental variation, a lower confidence limit ($LED_{10}$) is calculated for this benchmark dose. If evaluation of the likely mode of action reveals that a nonlinear (threshold-like) dose–response is likely below

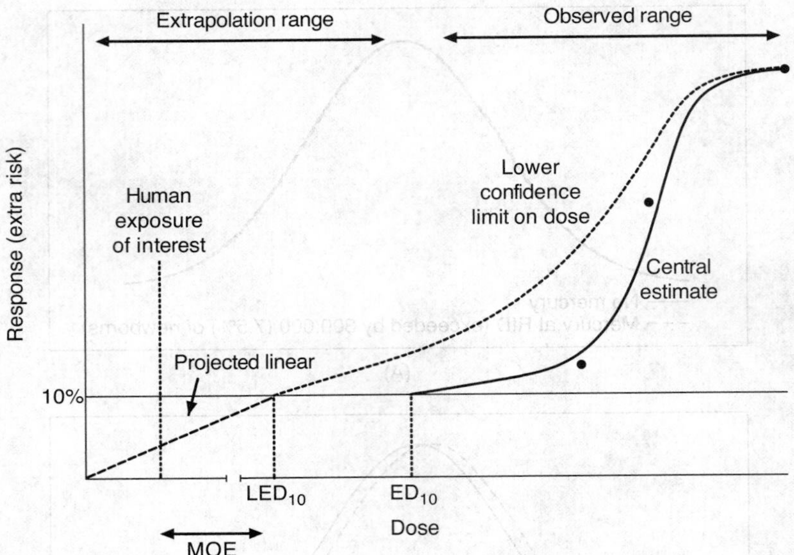


**FIGURE 9.4** Illustration of low-dose extrapolation.

this benchmark dose, the $LED_{10}$ is divided by appropriate uncertainty (safety) factors to arrive at a reference dose that is not likely to cause an appreciable cancer risk. If a nonlinear dose–response below the benchmark dose cannot be justified, the regulatory default is linear extrapolation from the $LED_{10}$ to 0 (Figure 9.4). This gives a low-dose cancer potency slope factor of $0.10/LED_{10}$. This often is similar to the value of $q_1{}^*$ that is estimated from the former default, the linearized multistage model of carcinogenesis [85]. In the rare cases where the mode of action is well understood and sufficient adequate data exist, a chemical-specific model may be developed to predict cancer risk at doses below the $LED_{10}$ [90].

## Mechanistic Toxicology and Cross-Dose Extrapolation

Several lines of experimental evidence seem to support the view that both pharmacokinetic and pharmacodynamic processes leading to toxicity may change as the magnitude of the dose changes. Gathering the evidence to support such changes in specific cases often presents significant experimental challenges, but much current work in toxicology is devoted to this topic. Several authors have referred to "dose-dependent transitions in toxicity mechanisms" when describing such phenomena [77]. Uncovering the existence of such transitions can, in principle, lead to more complete and accurate descriptions of dose–response relationships than can be obtained using standard animal toxicology experiments.

The dose–response relationship typically observed through experimental animal studies provides information on the relationship between administered doses and ultimate manifestations of toxicity. That observed relationship provides no information regarding the pharmacokinetic and pharmacodynamic processes underlying the production of toxicity. Moreover, the range of observable toxicity is typically well above the range of human exposure. Mechanistic studies (including both pharmacokinetic and pharmacodynamic processes) may explain at the most basic level the shape of the observed relationship and also provide details regarding the relationship close to or within the range of human exposures. This type of understanding would, in principle, eliminate many of the uncertainties in extrapolation from high (experimental) to low (human) doses.

Mechanistic transitions can be explained by the fact that many pharmacokinetic and pharmacodynamic processes may become saturable as dose increases, and some may be induced. Pharmacokinetic processes that are saturable or inducible include absorption, distribution, protein binding, metabolism, and elimination. Interactions with receptors and many other pharmacodynamic processes are similarly saturable or inducible. The cumulative effects of such processes at a given dose presumably account for the nature and extent of the toxic response at that dose. It is clear that in most cases the processes affected and the direction of the effect will change as doses change, so it is plain that the concept of dose-dependent transitions in toxicity mechanisms is a concept that fits well with current understanding of pharmacokinetic and pharmacodynamic processes [19].

Recent publications that arose from a series of workshops held under the auspices of the International Life Sciences Institute contain case studies that illustrate dose-dependent transitions in toxicity mechanisms and also provide comment on the use of the knowledge in risk assessment. There is also available in these publications useful commentary on the types of studies that are required to document the existence of such transitions [77].

## CROSS-ROUTE EXTRAPOLATION

### Defining the Problem

It is often necessary, particularly in the regulatory setting, to make some estimate of the likely toxic effect associated with exposure to a chemical by one route of exposure (e.g., inhalation) when toxicity data only from studies of exposure by a different route (e.g., oral) are available. Such extrapolation is subject to considerable uncertainty and often is not attempted. Currently, for example, the USEPA rarely uses oral data to derive inhalation RfCs or inhalation data to derive oral RfDs. In particular, chemicals quite frequently show route-specific toxic effects, especially at or near the site of administration (in the respiratory tract by inhalation, in the upper gastrointestinal tract by oral administration, or on the skin by dermal exposure). Such local effects may be indicative of a potential effect at a local site by another route of administration; for example, a chemical that causes local dermal irritation when applied to the skin will very likely also cause local irritation when inhaled or ingested. Quantitative extrapolation of such effects, however, is generally not possible because of difficulties in identifying the effective dose of the toxicant at the target site and particularly because of differences in the sensitivity of various tissues.

For systemic effects at distant sites, a crude cross-route adjustment can be performed (if absorption data are available) by simply normalizing the dose to a body-weight basis and adjusting for relative absorption by the different routes of exposure. This procedure can be quite misleading, however. As an example, if the chemical is subject to a significant first-pass effect in the liver (either activating a protoxicant or inactivating a direct-acting toxicant), then the ingestion and inhalation routes may show substantially different responses even with the same total absorbed dose. Clearly, if relevant metabolic and pharmacokinetic data are available, a preferred approach would be the development of a pharmacokinetic model that would take these factors into consideration.

## CROSS-TIME EXTRAPOLATION: QUANTITATIVE STRATEGIES FOR ADJUSTING FOR DIFFERENCES IN EXPOSURE DURATIONS

### Introduction

Dosing animals with chemicals in toxicological studies may vary from a single administration to continuous exposure over the lifetime of the animal. Likewise, human exposure may vary from a single episode to continuous lifetime exposure. Often, a 2-year rodent lifetime is considered equivalent to a 70-year human lifetime; hence, an exposure of an animal for one year is assumed equivalent to a 35-year exposure in humans. Generally, the durations of animal studies are chosen to mimic likely human exposure conditions; however, resources are not available to test chemicals at all of the possible human exposure conditions. As a result, statistical techniques have been devised to estimate the effects of short-term exposures from studies conducted with long-term exposures and *vice versa*.

### Biological Models

Generally, estimates of cancer risk are based on the average daily lifetime exposure; that is, the total dose is divided by the number of days in a lifetime. This is a plausible approach for a genotoxic carcinogen where it is assumed that carcinogenesis is a stochastic process. In such a case, the probability of a biological event is proportional to the number of molecules of the chemical available to interact with biological matter. To predict biological effects for different durations of exposures, it is necessary for dose–response models to contain a time or age element.

The multistage model of carcinogenesis does contain an element of time. Crump and Howe [24] provided estimates of risk as a function of age and duration of exposure. Kodell et al. [45] showed that the use of the average daily lifetime dose may overestimate risk but never underestimates the cancer risk by more than a factor of $k$ (generally 3 to 6) for a $k$-stage model. For example, based on the average daily lifetime dose, the estimated risk for exposure for one tenth of a lifetime would be one tenth of the risk for continuous lifetime exposure at that daily dose. According to Kodell et al. [45], this estimated risk should be multiplied by a factor of 3 to 6 to allow for exposure during a sensitive age.

For the Moolgavkar–Venson–Knudson two-stage clonal expansion model of carcinogenesis, Chen et al. [17] showed that the use of the average daily lifetime dose generally does not underestimate risk by more than a factor of ten. For exposures over one tenth of a lifetime, the estimated risk would be the same as for a continuous lifetime exposure. For exposures for a fraction ($f$) less than one tenth of a lifetime, the estimated risk is taken to be less than 10 times the upper limit on the estimated lifetime risk with continuous exposure. For the extreme of only a single exposure to $N$ milligrams of a carcinogen, the average daily lifetime (75-year) dose for a 70-kg person is $N/70 \times 75 \times 365 = 5 \times 10^{-7} \times N$ mg/kg. If $q_1{}^*$ is the estimated upper bound of the cancer risk per mg/kg body weight per lifetime daily exposure, then the estimated risk from a single exposure is $5 \times 10^{-6} \times q_1{}^*$, where a factor of 10 is included to allow for exposure at a sensitive age.

### Empirical Models

For noncancer effects, extrapolation of subchronic to chronic exposures is generally accompanied by an uncertainty factor of 10 [6]. That is, it is assumed that an effect observed with a subchronic exposure is not likely to occur at less than one tenth that dose for a chronic exposure. Swartout [79] compared NOAELs and LOAELs for subchronic and chronic exposures for about 100 cases. The median ratio of subchronic to chronic doses producing equivalent effects was 2 with a 95th percentile of 17. On the average, with a chronic exposure one half of the dose for a subchronic exposure produced the same biological effect. For 5% of the cases, the chronic dose was less than 1/17th of the subchronic dose for the same biological effect. The conventional default factor of 10 for subchronic to chronic extrapolation covered about 89% of the cases. Pieters et al. [62] conducted a similar study for 149 cases and obtained median ratio of subchronic doses for similar effects of 1.7 with a 95th percentile of 29.

Haber's rule has been used extensively to make small extrapolations between durations of exposure. Haber's rule states that equal biological effects are expected for equal exposures of concentration ($c$) times duration ($t$); that is, equal values of $c \times t$ are expected to produce equal biological effects. For example, if the exposure duration is doubled, the concentration would have to be halved to obtain the same biological effect. A generalization of Haber's rule was given by ten Berge et al. [82], where values of $c^n \times t$ are expected to produce equal biological effects. Estimation of the exponent $n$ requires dose–response data collected for different durations of exposure. For several data sets, ten Berge et al. [82] observed that $n$ varied from about 1 to 3 and tends to center around $n = 2$. In the absences of duration–dose–response data, the recommended extrapolation to different durations of exposure is calculated on the basis of $c^2 \times t$. For example, if the exposure time is increased by a factor of 4, the concentration needs to be halved to obtain an equivalent biological effect. To be conservative, it is recommended that $c^3 \times t$ be used when extrapolating from long to shorter exposure times and $c \times t$ be used when extrapolating from short to longer exposure durations.

## AN OVERALL STRATEGY FOR SCIENTIFICALLY BASED INTERSPECIES EXTRAPOLATION

To assess risks to human populations from exposures to potentially toxic substances, based on data from experimental studies, always requires cross-species and within-species extrapolations, almost always requires cross-dose extrapolations, and often requires cross-route and cross-time extrapolations. The need for specific types of extrapolation depends on (1) the specific risk situation under assessment, and (2) the nature of the data available for that assessment. Embarking upon a risk assessment thus requires, at the outset, a careful delineation of the problem to be evaluated. Once this is accomplished, efforts are made to collect all data that might be relevant to the risk question at hand. A matching of the data available with the problem to be assessed allows identification of the types of extrapolation that will have to be undertaken.

At the current stage of development of the toxicological sciences, most extrapolations are undertaken using the so-called "default" approaches discussed earlier. Increasingly, however, attempts are being made to search for the types of information needed to avoid resort to such "defaults" and to use approaches with more fully developed scientific bases. Table 9.4 describes the types of inquiries that might be made to move toward a more purely science-based approach. It is assumed that the risk assessment problem to be addressed requires the use of animal toxicology data (no significant epidemiology data are available) and that all five forms of extrapolation will be required. Thus, for example, the assessment might involve chronic, general population exposure to a drinking-water or food contaminant for which the only available toxicity data involves gavage or even inhalation exposures over 90 days in one or more species of experimental animals. Before resorting to the usual "defaults" for each of these types of required extrapolation and assuming that they are simply scientific uncertainties that cannot be overcome, it is now expected that toxicologists will inquire more fully, along the lines outlined in Table 9.4, into the possibility that alternative, data-based approaches can be developed. At the same time, it must be recognized that development of the data necessary for science-based extrapolations will necessarily introduce new uncertainties that have to be accommodated. Thus, although most would agree that reliable pharmacokinetic data can provide useful information on interspecies differences, it is likely that scientists will disagree on just how complete such data need to be before they can be used in risk assessment. Regulators typically display a high degree of skepticism about the incorporation of such data and tend to remain close to the usual "defaults," not because they dispute their relevance but because they question their completeness. Thus, risk assessors must work to reach consensus not only on the types of data needed to improve risk assessments but also on the difficult question of how complete they must be before they can be used for important public health decisions.

## QUESTIONS

1. Why is it necessary to extrapolate?
2. How is a dose–response curve selected?
3. Does a threshold dose always exist? Ever exist?
4. Is the risk zero at the no-observed-adverse-effect level? Explain.
5. When does linear extrapolation to lower doses overestimate risk?

**TABLE 9.4**
**Overall Strategy for Science-Based Interspecies Extrapolation: The Search for Information Necessary To Improve the Scientific Basis for Human Risk Assessment Based on Experimental Data**

| Type of Extrapolation | Type of Inquiry |
|---|---|
| Cross-species | Can data be found for quantitative cross-species comparisons of target site doses and their relationships to administered doses? |
| | Can PBPK models be developed to accomplish above? |
| | Can mechanistic data be developed to estimate differences in target site responsiveness across species? |
| Within species | Can quantitative estimates be developed, on a chemical-specific basis, of the ranges of human variability in uptake, pharmacokinetic and metabolic handling, and target-site responsiveness? |
| | Can methods be developed to integrate the above sources of variability? |
| Cross-dose | Can the data necessary to apply biologically based models for low-dose extrapolation be identified or developed? |
| Cross-route | Can pharmacokinetic data be developed to permit accurate assessments of inter-route differences? |
| Cross-time | Can empirical data be found to support extrapolations from one exposure duration to another? Are there biologically based mechanistic considerations to guide such extrapolations? |

6. Describe the basis for a Probit model.
7. Construct a numerical example that illustrates Haber's rule.
8. When is a safety factor not an uncertainty factor?
9. What factors affect interspecies differences in response?
10. What factors account for variability in response within species?
11. What is the difference between variability and uncertainty?

## REFERENCES


1. Adolph, E. F. (1949): Quantitative relations in the physiological constitutions of mammals. *Science*, 109:579.
2. Allen, B. C., Shipp, A. M., Crump, K. S., Kilian, B., Hogg, M. L., Tudor, J., and Keller, B. (1987): *Investigation of Cancer Risk Assessment Methods: Summary*, EPA/600/6-87/007a, NTIS PB88-127105, U.S. Environmental Protection Agency, Washington, D.C.
3. Armitage, P. and Doll, R. (1961): Stochastic models for carcinogenesis. In: *Proc. of the 4th Berkeley Symp. on Mathematical Statistics and Probability*, Vol. 4. University of California Press, Berkeley, pp. 19–38.
4. Bailer, A. J. and Portier, C. J. (1988): Effects of treatment-induced mortality and tumor-induced mortality on tests for carcinogenicity in small samples. *Biometrics* 44:417–431.
5. Baird, S. J. S., Cohen, J. T., Graham, J. D., Shlyakhter, A. I., and Evans, J. S. (1996): Noncancer risk assessment: a probabilistic alternative to current practice. *Human Ecol. Risk Assess.*, 2:79–102.
6. Barnes, D. G. and Dourson, M. (1988): Reference dose (RfD): description and use in health risk assessments. *Reg. Toxicol. Pharmacol.*, 8:471—486.
7. Benedict, F. G. (1938): *Vital Energetics: A Study in Comparative Basal Metabolism*. Carnegie Institute of Washington, Washington, D.C.
8. Berkson, J. (1944): Application of the logistic function to bio-assay. *J. Am. Statist. Assoc.*, 39:357–365.
9. Bieler, G. S. and Williams, R. L. (1993): Ratio estimates, the delta method, and quantal response tests for increased carcinogenicity. *Biometrics*, 49:793–801.
10. Borghoff, S. J., Short, B. G., and Swenberg, J. A. (1990): Biochemical mechanisms and pathobiology of alpha 2μ-globulin nephropathy. *Ann. Rev. Pharmacol. Toxicol.*, 30:349–367.
11. Boxenbaum, H. and Ronfeld, R. (1980): Interspecies pharmacokinetic scaling and the Dedrick plots. *Am. J. Physiol.*, R768–R774.
12. Butenhoff, J. L., Gaylor, D. W., Moore, J. A., Olsen, G. W., Rodricks, J., Mandel, J. H., and Zobel, L. R. (2004). Characterization of risk for general population exposure to perfluorooctanoate. *Regul. Toxicol. Pharmacol.*, 39: 363–380.
13. Calabrese, E. J. (1985): Uncertainty factors and interindividual variation. *Regul. Toxicol. Pharmacol.*, 5:190–196.
14. Calabrese, E. J. (2005): Toxicological awakenings: the rebirth of hormesis as a central pillar of toxicology. *Toxicol. Appl. Pharmacol.*, 204:1–8.
15. Calabrese, E. J. and Baldwin, L. A. (2003): Hormesis: the dose–response revolution. *Annu. Rev. Pharmacol. Toxicol.*, 43:175–197.
16. CDC. (2003): *Second National Report on Human Exposure to Environmental Chemicals*, NCEH Pub. No. 02-0716. Centers for Disease Control and Prevention, Atlanta, GA (http://www.cdc.gov/exposurereport/2nd/).
17. Chen, J. J., Kodell, R. L., and Gaylor, D. W. (1988): Using the biological two-stage model to assess risk from short-term exposures. *Risk Anal.*, 8:223–230.

18. Clewell, H. J. and Crump, K. S. (2005): Quantitative estimates of risk for noncancer endpoints. *Risk Anal.*, 25:285–289.
19. Clewell, H. J., Andersen, M. E., and Barton, H. A. (2002): A consistent approach for the application of pharmacokinetic modeling in cancer and noncancer risk assessment. *Environ. Health Perspect.*, 110:85–93.
20. Crouch, E. (1983): Uncertainties in interspecies extrapolations of carcinogenicity. *Environ. Health Perspect.*, 50:321–327.
21. Crouch, E. and Wilson, R. (1979): Interspecies comparison of carcinogenic potency. *J. Toxicol. Environ. Health*, 5:1095–1118.
22. Crump, K. S. (1984): A new method for determining allowable daily intakes. *Fund. Appl. Toxicol.*, 4: 854-871.
23. Crump, K. S. (1996): The linearized multistage model and the future of quantitative risk assessment. *Human Exp. Toxicol.*, 15:787–798.
24. Crump, K. S. and Howe, R. B. (1984): The multistage model with a time-dependent dose pattern: applications to carcinogen risk assessment. *Risk Anal.*, 4:163–179.
25. Crump, K. S., Hoel, D. G., Langley, C. H., and Peto, R. (1976): Fundamental carcinogenic processes and their implications for low dose risk assessment. *Cancer Res.*, 36:2973–2979.
26. Dietrich, D. R. and Swenberg, J. A. (1991): NCI–Black–Reiter (NBR) male rats fail to develop renal disease following exposure to agents that induce $\alpha$-2$\mu$-globulin ($\alpha_{2\mu}$) nephropathy. *Fundam. Appl. Toxicol.*, 16:749–762.
27. Dourson, M. L. and Stara, J. F. (1983): Regulatory history and experimental support of uncertainty (safety) factors. *Regul. Toxicol. Pharmacol.*, 3:224–238.
28. Dourson, M. L., Felter, S. P., and Robinson, D. (1996): Evolution of science-based uncertainty factors. *Regul. Toxicol. Pharmacol.*, 24:108–120.
29. Elcombe, C. R., Bell, D. R., Elias, E., Hasmall, S. C., and Plant, N. J. (1996): Peroxisome proliferators: species differences in response of primary hepatocyte cultures. *Ann. N.Y. Acad. Sci.*, 804:628–635.
30. Finney, D. J. (1964): *Probit Analysis*, 2nd ed. Cambridge University Press, Cambridge, U.K.
31. Food and Drug Administration (FDA). (1971): Advisory Committee on Protocols for Safety Evaluation. Panel on Carcinogenesis Report on Cancer Testing in the Safety Evaluation of Food Additives and Pesticides. *Toxicol. Appl. Pharmacol.* 20:419–438.
32. Food and Drug Administration (FDA). (1987): Sponsored compounds in food-producing animals; criteria and procedures for evaluating the safety of carcinogenic residues; animal drug safety policy. *Fed. Reg.*, 52:49572–49590.
33. Freireich, E. J., Gehan, E. A., Rall, D. P., Schmidt, L. H., and Skipper, H. E. (1966): Quantitative comparison of toxicity of anticancer agents in mouse, rat, hamster, dog, monkey, and man. *Cancer Chemother. Rep.*, 50:219–244.
34. Gaylor, D. W. (1992): Incidence of developmental defects at the no observed adverse effect level (NOAEL). *Reg. Toxicol. Pharmacol.*, 15:151–160.
35. Gaylor, D. W. and Chen, J. J. (1986): Relative potency of chemical carcinogens in rodents. *Risk Anal.*, 6:283–290.
36. Gaylor, D. W. and Chen, J. J. (1993): Dose–response models for developmental malformations. *Teratology*, 47:291–297.
37. Gaylor, D. W. and Zheng, Q. (1996): Risk assessment of nongenotoxic carcinogens based upon cell proliferation/death rates in rodents. *Risk Anal.*, 16:221–225.
38. Gaylor, D. W., Chen, J. J., and Sheehan, D. M. (1993): Uncertainty in cancer risk estimates. *Risk Anal.*, 13:149–154.
39. Greenberg, M. S., Burton, G. A., Jr., and Fisher, J. W. (1999). Physiologically based pharmacokinetic modeling of inhaled trichloroethylene and its oxidative metabolites in B6C3F$_1$ mice. *Toxicol. Appl. Pharmacol.*, 154:264–278.
40. Haas, C. N. (1983): Estimation of risk due to low doses of microorganisms: a comparison of alternative methodologies. *Am. J. Epidemiol.*, 118:573–582.
41. Institute of Medicine (IOM). (1998): *Dietary Reference Intakes: A Risk Assessment Model for Establishing Upper Intake Levels for Nutrients*. Food and Nutrition Board, Institute of Medicine. National Academy Press, Washington, D.C. (http://books.nap.edu/catalog/6432.html).
42. Institute of Medicine (IOM). (2000): *Dietary Reference Intakes for Vitamin A, Vitamin K, Arsenic, Boron, Chromium, Copper, Iodine, Iron, Manganese, Molybdenum, Nickel, Silicon, Vanadium, and Zinc*. Food and Nutrition Board, Institute of Medicine. National Academy Press, Washington, D.C. (http://books.nap.edu/catalog/10026.html).
43. Kleiber, M. (1932): Body size and metabolism. *Hilgardia*, 6:315–353.
44. Kleiber, M. (1947): Body size and metabolic rate. *Physiol. Rev.*, 27:511–541.
45. Kodell, R. L., Gaylor, D. W., and Chen, J. J. (1987): Using average lifetime dose rate for intermittent exposures to carcinogens. *Risk Anal.*, 7:339–345.
46. Kodell, R. L., Howe, R. B., Chen, J. J., and Gaylor, D. W. (1991): Mathematical modeling of reproductive and developmental toxic effects for quantitative risk assessment. *Risk Anal.*, 11:583–590.
47. Kupper, L., Portier, C., Hogan, M., and Yamamoto, E. (1986): The impact of litter effects on dose–response modeling in teratology. *Biometrics*, 42:85–98.
48. Lehman, A. J. and Fitzhugh, O. G. (1954): 100-fold margin of safety. *Assoc. Food Drug Off. U.S. Q. Bull.*, 18:33–35.
49. Leisenring, W. and Ryan, L. (1992): Statistical properties of the NOAEL. *Reg. Toxicol. Pharmacol.*, 15:161–171.
50. Leroux, B. G., Leisenring, W., Moolgavkar, S. H., and Faustman, E. M. (1996): A biologically based dose–response model for developmental toxicology. *Risk Anal.*, 16:449–458.
51. Lewis, S. C. (1993): Reducing uncertainty with adjustment factors: improvements in non-cancer risk assessment. *Fundam. Appl. Toxicol.*, 20:2–4.
52. Moolgavkar, S. H. and Knudson, A. G. (1981): Mutation and cancer: a model for human carcinogenesis. *J. Natl. Cancer Inst.*, 66:1037–1052.
53. Moolgavkar, S. H. and Venzon, D. J. (1979): Two-event models for carcinogenesis: Incidence curves for childhood and adult tumors. *Math. Biosci.*, 47:55–77.
54. National Research Council (NRC). (1983): *Risk Assessment in the Federal Government: Managing the Process* National Academy Press, Washington, D.C.

55. National Research Council (NRC). (1987): *Drinking Water and Health Volume 8: Pharmacokinetics in Risk Assessment*. National Academy Press, Washington, D.C.
56. National Research Council (NRC). (1993): *Issues in Risk Assessment*. National Academy Press, Washington, D.C.
57. National Research Council (NRC). (1994): *Science and Judgment in Risk Assessment*. National Academy Press, Washington, D.C.
58. Occupational Safety and Health Administration (OSHA). (1980): Identification, classification and regulation of potential occupational carcinogens. *Fed. Reg.*, 45:5001–5296.
59. Paine, A. J. (1996): Validity and reliability of *in vitro* systems in safety evaluation. *Environ. Toxicol. Pharmacol.* 2:207–212.
60. Perrone, C. E., Shao, L., and Williams, G. M. (1998): Effect of rodent hepatocarcinogenic peroxisome proliferators on fatty acyl-CoA oxidase, DNA synthesis, and apoptosis in cultured human and rat hepatocytes. *Toxicol. Appl. Pharmacol.*, 150:277–286.
61. Peto, R., Pike, M. C., Day, N. E., Gray, R. G., Lee, P. N., Parish, S., Peto, J., Richards, S., and Wahrendorf, J. (1980): Guidelines for simple, sensitive significance tests for carcinogenic effects in long-term animal experiments. *IARC Monographs*, Suppl. 2. International Agency for Research on Cancer, Lyon, France.
62. Pieters, M. N., Kramer, H. J., and Slob, W. (1998): Evaluation of the uncertainty factor for subchronic-to-chronic extrapolation: statistical analysis of toxicity data. *Reg. Toxicol. Pharmacol.*, 27:108–111.
63. Pinkel, D. (1958): The use of body surface area as a criterion of drug dosage in cancer chemotherapy. *Cancer Res.*, 18:853–856.
64. Prothero, J. W. (1980): Scaling of blood parameters in mammals. *Comp. Biochem. Physiol. A*, 67:649–657.
65. Purchase, I. F. H. and Auton, R. T. (1995): Thresholds in chemical carcinogenesis. *Reg. Toxicol. Pharmacol.*, 22:199–205.
66. Renwick, A. G. (1991): Safety factors and establishment of acceptable daily intake. *Food Add. Contam.* 8(2): 135–150.
67. Renwick, A. G. (1993): Data-derived safety factors for the evaluation of food additives and environmental contaminants. *Food Add. Contam.*, 10(3):275–305.
68. Rees, D. C. and Hattis, D. (1994): Developing quantitative strategies for animal to human extrapolation. In: *Principles and Methods of Toxicology*, 3rd ed., edited by A. W. Hayes. Raven Press, New York, pp. 275–315.
69. Rhomberg, L. R. and Wolff, S. K. (1998): Empirical scaling of single oral lethal doses across mammalian species based on a large database. *Risk Anal.*, 18:741–753.
70. Rodricks, J. V., Brett, S., and Wrenn, G. (1987): Significant risk decisions in federal regulatory agencies. *Reg. Toxicol. Pharmacol.*, 7:307–320.
71. Rodricks, J. V., Rudenko, L., Starr, T. B., and Turnbull, D. (1997): Risk assessment. In: *Comprehensive Toxicology*, Vol. I, *General Principles*, edited by J. Bond. Pergamon Press, New York.
72. Rubner, M. (1883): Ueber den einfluss der körpergrösse auf stoff und kraft wechsel. *Z. Biol.*, 1919:535–562.
73. Ryan, L. (1992): Quantitative risk assessment for developmental toxicity. *Biometrics*, 48:163–174.
74. Schlosser, P. M., Lilly, P. D., Conolly, R. B., Janszen, D. B., Kimbell, J. S. (2003): Benchmark dose risk assessment for formaldehyde using airflow modeling and a single-compartment, DNA-protein cross-link dosimetry model to estimate human equivalent doses. *Risk Anal.*, 23(3): 473–487.
75. Schmidt-Nielson, K. (1984): *Scaling: Why Is Animal Size So Important?* Cambridge University Press, Cambridge, U.K.
76. Sheehan, D. and Gaylor, D. W. (1990): Analyses of the adequacy of safety factors. *Teratology*, 41:590–591.
77. Slikker, Jr., W., Andersen, M. E., Bogdanffy, M. S., Bus, J. S., Cohen S. D. et al. (2004): Dose-dependent transitions in mechanisms of toxicity. *Toxicol. Appl. Pharmacol.*, 201: 203–225
78. Stenner, R. D., Merdink, J. L., Fisher, J. W., and Bull, R. J. (1998): Physiologically-based pharmacokinetic model for trichloroethylene considering enterohepatic recirculation of major metabolites. *Risk Anal.*, 18:261–269.
79. Swartout, J. (1996): *Subchronic-to-Chronic Uncertainty Factor for the Reference Dose*, Abstract F2.03. Society for Risk Analysis Annual Meeting, New Orleans, LA.
80. Swenberg, J. A., Short, B., Borghoff, S. Strasser, J., and Charbonneau, M. (1989): The comparative pathobiology of alpha 2μ-globulin nephropathy. *Toxicol. Appl. Pharmacol.* 97:35–46.
81. Tan, Y.M., Butterworth, B.E., Gargas, M. L., and Conolly, R. B. (2003): Biologically motivated computational modeling of chloroform cytolethality and regenerative cellular proliferation. *Toxicol Sci.*, 75(1):192–200.
82. ten Berge, W. F., Zwart, A., and Appelman, L. M. (1986): Concentration–time mortality response relationship of irritant and systemically acting vapours and gases. *J. Hazard. Mater.*, 3:301–309.
83. Travis, C. C. and White, R. K. (1988): Interspecific scaling of toxicity data. *Risk Anal.*, 8:119–125.
84. Travis, C. C., White, R. K., and Ward, R. C. (1990): Interspecies extrapolation of pharmaco-kinetics. *J. Theor. Biol.*, 142:285–304.
85. U.S. Environmental Protection Agency (USEPA). (1986): Guidelines for carcinogen risk assessment. *Fed. Reg.*, 51:33992–34003.
86. U.S. Environmental Protection Agency (USEPA). (1992): Draft report: a cross-species scaling factor for carcinogen risk assessment based on equivalence of mg/kg$^{3/4}$/day. *Fed. Reg.* 57:24152–24173.
87. U.S. Environmental Protection Agency (USEPA). (2000): *Benchmark Dose Technical Guidance Document*, External Review Draft. U.S. Environmental Protection Agency, Washington, D.C. (http://www.epa.gov/ncea/bnchmrk/bmds_peer.htm).
88. U.S. Environmental Protection Agency (USEPA). (2001): *Toxicological Review of Chloroform (CAS No. 67-66-3) In Support of Summary Information on the Integrated Risk Information System (IRIS)*, EPA/635/R-01/001. U.S. Environmental Protection Agency, Washington, D.C. (http://www.epa.gov/iris/toxreviews/0025-tr.pdf).

89. U.S. Environmental Protection Agency (USEPA). (2002): *A Review of the Reference Dose and Reference Concentration Processes*, EPA/630/P-02/002F, prepared for the Risk Assessment Forum, U.S. Environmental Protection Agency, Washington, D.C. (http://cfpub.epa.gov/ncea/raf/recordisplay.cfm?deid=55365).
90. U.S. Environmental Protection Agency (USEPA). (2005): *Guidelines for Carcinogen Risk Assessment*. *Risk Assessment Forum*, EPA/630/P-03/001B. U.S. Environmental Protection Agency, Washington, D.C. (http://cfpub.epa.gov/ncea/raf/recordisplay.cfm?deid=116283).
91. Zheng, Q., Lutz, W. K., and Gaylor, D. W. (1997): A carcinogenesis model describing mutational events at the DNA adduct level. *Math. Biosci.*, 144:23–44.

# 10 The Practice of Exposure Assessment

*Dennis J. Paustenbach and Amy K. Madl*

## CONTENTS

## INTRODUCTION

Health risk assessment is the process wherein toxicology data from animal and human epidemiologic studies are evaluated, a mathematical formula is applied to predict the response at low doses, and information about the degree of exposure is used to predict quantitatively the likelihood that a particular adverse response will be seen in a specific human population [1–3]. More simply, risk assessment is a process by which scientists evaluate the potential for adverse health effects from exposure to naturally occurring or synthetic agents [4]. Regulatory agencies have used the risk assessment process for nearly 50 years, most notably the U.S. Food and Drug Administration (USFDA) [5]; however, the difference between assessments performed in the 1950s and 1960s and those performed in the 1980s and 1990s is that dose-extrapolation models, quantitative exposure assessments, and quantitative descriptions of uncertainty have been added to the process [6]. Because of increased ability to measure and predict exposures, the availability of quantitative computer software programs, and better quantitative methods for estimating the low-dose response (such as physiologically based pharmacokinetic [PBPK] models), risk assessments conducted today provide more accurate risk estimates than in the past [3,7,8].

Since 1980, most environmental regulations and some occupational health standards have, at least in part, been based on health risk assessments [3,9,10]. They include standards for pesticide residues in crops, drinking water, ambient air, and food additives, as well as exposure limits for contaminants found in indoor air, consumer products, and other media. Risk managers increasingly rely on risk assessment to decide whether a broad array of risks are significant or trivial—an important task because more than 400 of the about 2000 chemicals routinely used in industry have been identified as carcinogens in various animal studies [11,12]. In theory, the results of risk assessments in the United States should influence virtually all regulatory decisions involving so-called "toxic agents" [13–15].

The risk assessment process has four parts: hazard identification, dose–response assessment, exposure assessment, and risk characterization [11]. Although progress has been made over the past 20 years with regard to how to conduct and interpret toxicology and epidemiology studies (e.g., hazard identification) and scientists believe that they are doing a better job of dose–response extrapolation than in the past, most significant advances in the risk assessment process have occurred in the field of exposure assessment [16–18].

Since about 1995, an increasing number of environmental scientists have embraced the view that "toxicology data are important, but they do not mean much without quantitative information about human exposure" [19]. For this reason, each year since the toxicology community has shown increasing interest in understanding the exposure assessment field [20,21]. Fortunately, a significant amount of research has been conducted to identify better values for many exposure parameters, and major improvements have been made in applying these exposure factors to various scenarios. This chapter is intended to familiarize toxicologists, risk assessors, and others with this evolving field.

## BASIC CONCEPTS

### Description of Exposure Assessment

Exposure assessment is the step that quantifies the intake of an agent resulting from contact with various environmental media (e.g., air, water, soil, food) [3,8,22,23]. Exposure assessments can address past, current, or future exposures, although uncertainties can become significant when attempting to anticipate what might have happened or what will happen [8,24–30]. Researchers have used a variety of methods to approximate historical and future exposures, including using geographic location, job history, historical records, biomonitoring, and estimates from mathematical models, as a proxy for exposure [31–36].

Exposure assessment in various forms dates back at least to the early twentieth century, and perhaps earlier, particularly in the fields of epidemiology [37,38], industrial hygiene [39,40], and health physics [41]. Exposure assessment combines elements of all three disciplines and relies upon aspects of biochemical toxicology (to estimate delivered dose), atmospheric sciences, anthropometry, analytical chemistry, food sciences, physiology, environmental modeling, and others [42].

Fundamentally, an exposure assessment describes the nature and size of the various populations exposed to a chemical agent and the magnitude and duration of their exposures [43,44]. It determines the degree of contact a person has with a chemical and estimates the magnitude of the absorbed dose [45]. Several factors need to be considered when estimating that dose, including characteristics of the contaminated media, exposure duration, route of exposure, chemical bioavailability from the contaminated media (e.g., soil), and, sometimes, the unique characteristics of the tested population (e.g., hairless mice absorb a greater percent of chemical than other mice). By definition, *duration* is the period of time over which the person is exposed. An *acute* exposure generally involves one contact with the chemical, usually for less than a day. An exposure is considered *chronic* when it takes place over a substantial portion of the person's lifetime. Exposures of intermediate duration are usually referred to as *subchronic* [42].

Knowledge of the chemical concentration in an environmental medium is essential to determine the magnitude of the absorbed dose. This information is usually obtained by analytical measurements of samples of the contaminated medium (air, water, soil, sediment, food, or house dust). Estimates can also be made using mathematical models, such as models relating air concentrations at various distances from a point of release (e.g., a smoke stack) to such factors as release rate, weather conditions, distance, and stability of the agent [46,47]. Needless to say, a significant number of factors must be considered to quantitatively evaluate a typical or complex contaminated site (Figure 10.1).

In general, since about 1995, our ability to perform exposure assessments has matured to a degree that they will usually possess less uncertainty than other steps in the risk assessment. Admittedly, many factors should be considered when estimating exposure; for example, it is a complicated procedure to understand the transport and distribution of a chemical that has been released into the environment [48]. Nonetheless, a number of studies have shown that scientists can now do an adequate job of quantifying chemical concentrations in various media and the resulting uptake by exposed persons [8,49,50]. No doubt, in the coming years, these estimates will be confirmed or rejected as larger amounts of biomonitoring data become available. For those who wish to question the claim that exposures can generally be estimated should consider the confidence of toxicologists in the results of animal bioassays that label a chemical a possible human carcinogen even though tumors were observed only at the maximum tolerated dose (MTD). Similarly, consider the confidence in dose extrapolation when three equally valid statistical models can yield risk estimates that are 1000-fold different at a typical environmental dose.

The primary routes of human exposure to chemicals in the ambient environment are dust and vapor inhalation, dermal contact with contaminated soils or dusts, and ingestion of contaminated food, water, house dust, or soil. In the workplace, the predominant exposure route usually is inhalation, followed by dermal uptake and, to a lesser extent, dust ingestion due to hand-to-mouth contact [7]. Uncertainty in environmental exposure assessment can be greater than in an occupational exposure assessment; however, many workplaces can experience large fluctuations in airborne concentrations, and the work practices of various employees may differ significantly, so it is difficult to measure dermal uptake and incidental ingestion [40,51–54].

Scientists in the field of radiological health were the first to quantitatively estimate human uptake of environmental contaminants [55,56]; thus, the published literature in health physics can be a source of valuable information when conducting assessments of chemical contaminants [57]. Work conducted after World War II has provided

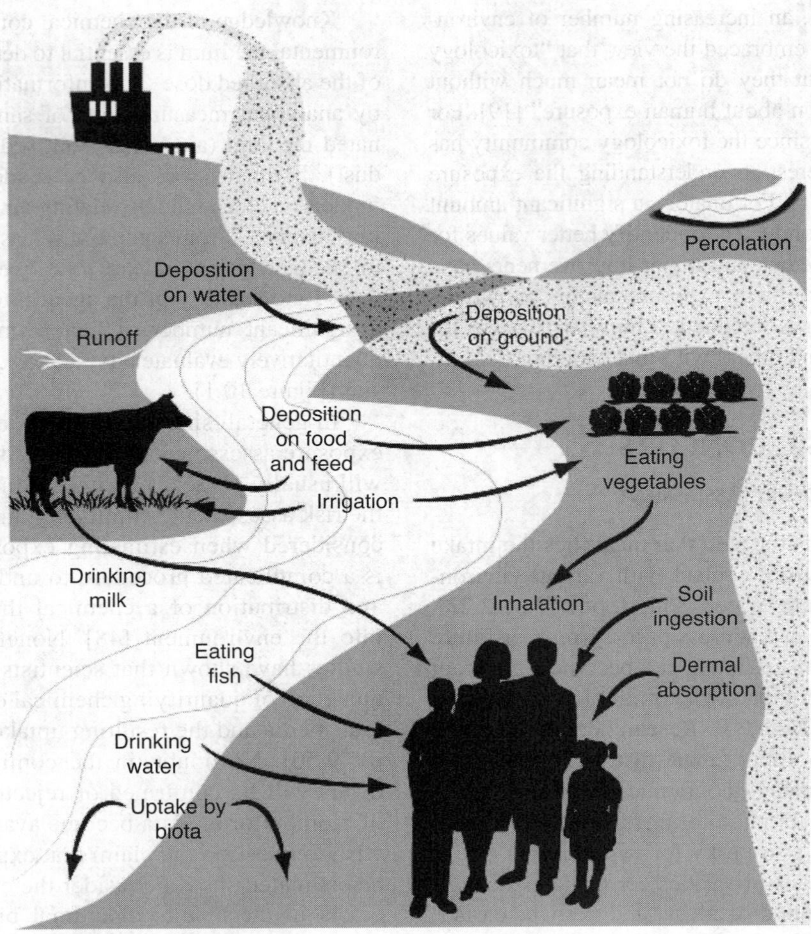


**FIGURE 10.1** Exposure pathways.

numerous methodologies for estimating the human uptake of environmental contaminants [57]. These have been refined over the past decade [26,58–60]. The availability of information on the degree of exposure associated with various scenarios has increased dramatically in recent years, as evidenced by the recent U.S. Environmental Protection Agency (USEPA) *Exposure Factors Handbook* and *Child-Specific Exposure Factors Handbook*, documents containing nearly 2000 pages of combined information on exposure assessment [44,61–64].

The practice of exposure assessment, at least for regulatory purposes, has changed over time; for example, beginning in the late 1970s, the U.S. regulatory policy encouraged or mandated the use of conservative approaches when conducting exposure assessments. This was codified in the USEPA's original document entitled *Risk Assessment Guidance for Superfund* (so-called RAGS) [65]. At that time, standardization of exposure assessments used to satisfy regulatory agencies was considered prudent because it guaranteed that risks would not be underestimated in order to ensure protection of public health. Beginning about 1985, concern was expressed that repeated use of conservative exposure factor assumptions was producing unrealistically high estimates of exposure and that the cost of achieving the recommended cleanup levels was becoming enormous [52,66–73]. Thus, to evaluate the accuracy of many of the factors used in these assessments, changes in the process began to occur. Even in 2006, government agencies such as the Office of Management and Budget (OMB) developed new guidelines and urged risk assessors to do an even better job at eliminating compounding conservatism in their assessments.

Around 1990, risk assessors began to apply Monte Carlo techniques to eliminate the possibility of compounding conservatism and to increase the transparency of the analyses. Application of Monte Carlo techniques to exposure assessment has dramatically improved our understanding of the certainty of exposure estimates thereby altering the field permanently [36,71,74–85]. The USEPA and other agencies have now embraced this approach, which is well described in several guidelines, as well as the document known as RAGS3A (a process for conducting probabilistic risk assessment) [82,86,87].

## What Is Exposure?

Over the years, the terminology used in published exposure assessment literature has been inconsistent. Although there is reasonable agreement that human exposure means contact with the chemical or the agent [23,65,88], no widespread agreement has yet been reached as to whether this means contact with (1) the visible exterior of the person (skin and openings into the body, such as mouth and nostrils), or (2) the so-called *exchange boundaries* where absorption takes place (skin, lungs, gastrointestinal tract) [42]. The differing definitions have led to some ambiguity in the use of terms and units for quantifying exposure; for example, the terms *dose*, *uptake*, and *intake* have often been used loosely.

Some scientists find it helpful to think of the human body as having a hypothetical outer boundary that separates the inside of the body from the outside [42]. The outer boundary of the body consists of the skin and openings into the body, such as the mouth, nostrils, or punctures and lesions in the skin. In most exposure assessments, chemical exposure is defined as contact of the chemical with some part of this boundary. An exposure assessment is the quantitative or qualitative evaluation of that contact. It describes the intensity, frequency, and duration of contact and often quantifies the rate at which the chemical crosses the boundary (chemical intake or uptake rates), the route of the chemical across the boundary (exposure route, such as dermal, oral, or respiratory), the resulting amount of chemical actually crossing the boundary (dose), and the amount of chemical absorbed (internal dose) [23,89]. A very workable quantitative definition of exposure is to think of it as "the product of (concentration), (time), and (duration), or rate of transport of toxicant (mg/min)" [51].

Depending on the purpose of the exposure assessment, the numerical output of these analyses may be an estimate of either exposure or dose. If an exposure assessment is being done as part of a risk assessment in support of an epidemiologic study, for example, sometimes only qualitative exposure levels are all that can be provided. In these situations, categories such as low-, medium-, and high-level exposure may be used (although strongly discouraged by the authors). In contrast, a greater portion of the assessments of environmental or occupational exposure conducted in recent years have attempted to quantitatively predict the absorbed dose (mg/kg/day) and, occasionally, the circulating blood level or the concentration of the toxicant in the target organ [90–96].

## Concepts of Exposure, Intake, Uptake, and Dose

The process of a chemical entering the body can be described in two steps—contact (exposure) followed by actual entry (crossing the boundary). Absorption, by crossing the boundary, leads to the availability of an amount of chemical to biologically significant sites within the body (target tissue dose). Although the description of contact with the outer boundary is simple conceptually (e.g., mg benzene per $cm^2$ skin), estimating the degree to which a chemical crosses this boundary is somewhat more complex [97].

In the early 1990s, some scientists described the transport of chemicals into the body as involving two separate steps: intake and uptake. Intake involved physically moving the chemical in question through an opening in the outer boundary (usually the mouth or nose), typically via inhalation, eating, or drinking. Normally, the chemical was contained in a medium such as air, food, water, or dust/soil. Here, the key question was the mass inhaled or ingested. Uptake, in contrast to intake, involved absorption of the chemical through the skin or across other barriers.

Today, most scientists tend to lump intake and uptake together, referring to the amount of chemical entering the body as *intake* or the *absorbed dose*. Some chemicals are absorbed completely, so systemic absorption is the same as that when the chemical is eaten or comes in contact with the skin. In other cases, the chemical is often contained in a carrier medium, and the medium itself typically is not absorbed at the same rate as the contaminant of interest, so estimates of the amount of chemical crossing the boundary cannot be made directly. For example, benzene on the surface of a contaminated soil particle will move quickly through the skin, but benzene in the center of the soil particle may never completely reach the surface and, therefore, is not bioavailable and may never enter the bloodstream. Of course, for many inorganic chemicals such as arsenic or lead in soil, bioavailability can be very low as the chemical is bound to the interstices of the soil particle. In this case, absorption can be very low. In short, if a chemical cannot be released, it has no bioavailability; consequently, because there is no absorbed dose, the chemical does not pose a risk.

Dermal absorption is an example of direct uptake across the outer boundary of the body. A chemical uptake rate is the amount of chemical absorbed per unit of time. In this process, mass transfer occurs by diffusion, so uptake will depend on the concentration gradient across the boundary, permeability of the barrier, and other factors [89,98,99]. Chemical uptake rates can be expressed as a function of the exposure concentration, permeability coefficient, surface area exposed, or as flux [7].

## Bioavailability

The study of the bioavailability of chemicals in various media began around 1980 and continues to be an important area of research [50,100–118]. Most studies are of oral bioavailability, although the dermal and inhalation bioavailability of chemicals on various media have also been studied. Bioavailability has been a bit confusing due to a lack of standard terminology [119]. The review paper by Ruby et al. [100] is probably the most authoritative one on this topic, although the text by Chen et al. [119] is also a valuable resource. We suggest that the following definitions be used in future assessments:

- *Bioavailability*—Oral bioavailability is defined as the fraction of an administered dose that reaches the central (blood) compartment from the gastrointestinal tract. Bioavailability defined in this manner is commonly referred to as *absolute bioavailability* and is equal to the oral absorption fraction.
- *Relative bioavailability*—Relative bioavailability refers to comparative bioavailabilities of different forms of a substance or for different exposure media containing the substance (e.g., bioavailability of a metal from soil relative to its bioavailability from water), expressed in this document as a *relative absorption factor* (RAF).
- *Relative absorption factor*—The RAF also describes the ratio of the absorbed fraction of a substance from a particular exposure medium relative to the fraction absorbed from the dosing vehicle used in the toxicity study for that substance; the term *relative bioavailability adjustment* (RBA) is also used to describe this factor.
- *Bioaccessibility*—The oral bioaccessibility of a substance is the fraction that is soluble in the gastrointestinal environment and is available for absorption. The bioaccessible fraction is not necessarily equal to the RAF (or RBA) but depends on the relation between results from a particular *in vitro* test system and an appropriate *in vivo* model.

Both *in vitro* and *in vivo* tests are available for evaluating bioavailability, and many different approaches have been suggested over the past 25 years [50,100–106,108–123]

As noted by Ruby et al. [100], a number of *in vitro* tests have been used to characterize the oral bioavailability of various chemicals in various media. Simple extraction tests have been used for several years to assess the degree of metals dissolution in a simulated gastrointestinal tract environment. The predecessor of these systems was developed originally to assess the bioavailability of iron from food in nutrition studies. In these systems, various metal salts or soils containing metals are incubated in low-pH solution for a period intended to mimic residence time in the stomach. The pH is then increased to near neutral, and incubation continues for a period intended to mimic residence time in the small intestine. Enzymes and organic acids are added to simulate gastric and small-intestinal fluids. The fraction of lead, arsenic, or other metals that dissolve during the stomach and small-intestinal incubations represents the fraction that is bioaccessible (i.e., is soluble and available for absorption).

A number of *in vivo* tests have also been used with varying success; for example, gastrointestinal absorption of lead in humans varies with the age, diet, and nutritional status of the subject, as well as with the chemical species and the particle size of lead that is administered. Age is a well-established determinant of lead absorption; adults typically absorb 7 to 15% of lead ingested from dietary sources, whereas estimates of lead absorption from dietary sources in infants and children range from 40 to 53%. For the purpose of modeling exposure to lead in soil, the USEPA currently assumes that the absolute bioavailability of lead in diet and water is 50% and that the absolute bioavailability of lead in soil is 30% for children. This corresponds to a soil RAF of 0.60 (60%) for the bioavailability of soil lead relative to lead in water (i.e., RAF = 0.3/0.5) [100].

The results of bioavailability studies must be considered in virtually all assessments involving human exposure [50,100,105,112,114,115,118,119]. Often, the effects in uptake will be minor, but in other cases one may find that insignificant quantities of a chemical are absorbed even though the applied dose or exposure is quite high [120,124].

## Applied Dose or Potential Dose

Applied dose has been defined as the amount of chemical available at the absorption barrier (skin, lung, gastrointestinal tract) [42]. It is useful to know the applied dose if a relationship can be established between it and the internal dose, a relationship that can sometimes be established experimentally. This can be estimated either through modeling or by direct measurement. For example, years ago, some researchers analyzed phenol concentrations in the blood of volunteers over time after the volunteers placed their hands in containers of nitrobenzene or benzene in an attempt to quantify the flux rate [125,126]. Usually, it is difficult to measure the applied dose directly, as many of the absorption barriers are internal to the human and not localized in such a way as to make measurement easy. An approximation of applied dose can be made, however using the concept of potential dose [42]. Potential dose is simply the amount of chemical that is ingested or inhaled or the amount of chemical contained in material applied to the skin. It is a useful term or concept in those instances when there is a measurable amount of chemical in a particular medium. The potential dose for ingestion and inhalation is analogous to the administered dose in a dose-response experiment.

For the dermal route, potential dose is the amount of chemical applied or the amount of chemical in the medium applied (e.g., soil deposited on the skin). Note that because all of the chemical in the soil particulate is no contacting the skin, this differs from exposure (the concentration in the particulate times the duration of contact) and applied dose (the amount in the layer actually touching the skin) [42].

As previously noted, the amount of chemical that reaches the exchange boundaries of the skin, lungs, or gastrointestinal tract may often be less than the potential dose if the material is only partly bioavailable and therefore only partially absorbed. For example, only about 0.001 to 1.0% of dioxins or polycyclic aromatic hydrocarbons (PAHs) on fly ash in contact with the skin are likely to penetrate [123]. When bioavailability data are available, adjustments to the potential dose should be made to convert it to the absorbed or internal dose [119,123].

## Internal Dose

The amount of chemical that has been absorbed and is available for interaction with biologically significant receptors (e.g., target organs) is called the *internal dose*. Estimating internal dose can be difficult but it is one of the primary objectives of a good exposure assessment [127,128]. Transport models are available to assist in this process [129]. Once absorbed, the chemical can be metabolized, stored, excreted, or transported within the body. The amount transported to an individual organ, tissue, or fluid of interest is termed the *delivered dose* [96,130,131]. The dose delivered to the target organ may be only a small part of the total internal dose but, by definition, it is the most relevant. For example, although 1 mg of PCB may be absorbed into the body, at any given time the amount in the liver (the target organ) may only be 0.001 mg. The time course over which that 0.001 mg is delivered is often equally important to understand. Work to refine the techniques used to estimate delivered dose has been among the most exciting areas of exposure assessment research in recent years. Currently, the best approach to estimate delivered dose is to measure blood or to use physiologically based pharmacokinetic (PB-PK) models [96,132–136]. Recent research efforts have involved the use of PBPK models and data on polymorphisms in metabolic enzymes to understand the disposition of environmental toxicants in potentially susceptible human populations [131].

The *biologically effective dose* (BED), or the amount that actually reaches cells, sites, or membranes where adverse effects occur [137], may represent only a fraction of the delivered dose, but it is obviously the best one for predicting adverse effects. Understanding BED is the ultimate goal of exposure assessment. Regrettably, thus far, toxicologists have rarely been able to estimate BED or measure it for most chemicals [42].

Currently, most risk assessments dealing with environmental chemicals (as opposed to pharmaceutical assessments) rely on dose–response relationships based on the potential (administered) dose or the internal dose,

**TABLE 10.1**
**Examples of PBPK Models for Toxic Materials**

| | |
|---|---|
| Benzene | Lead |
| Benzo(*a*)pyrene | Methanol |
| Butoxyethanol | Methoxyethanol (2-ME) |
| Butoxyethanol | Methyl ethyl ketone (MEK) |
| Carbon tetrachloride | Nickel |
| Chlorfenvinphos | Nicotine |
| Chloroform | Parathion |
| Chlorpentafluorobenzene | Physostigmine |
| *cis*-Dichlorodiammine platinum | PBB |
| Dichloroethane | PCBs |
| Dichloromethane | Styrene |
| Dieldrin | Toluene |
| Diisopropylfluorophosphate | TCDF |
| Dimethyloxazolidine dione | TCDD (dioxin) |
| Dioxane | Tetrachloroethylene |
| Ethylene oxide | Trichloroethane |
| Ethyoxy ethanol (2-EE) | Trichloroethylene |
| Formaldehyde | Trichlorotribluoroethane |
| Hexane | Vinyl chloride |
| Hexavalent chromium | Vinylidene fluoride |
| Kepone | Xylene |

*Note:* This table is an expansion of one presented in a paper by Leung and Paustenbach [138].

because our understanding of how to estimate the delivered dose or the biologically effective dose is insufficient for most chemicals. In general, the best method currently available for estimating the dose to the target organ is to use PB-PK models. These have been developed for nearly 100 high-volume industrial chemicals (Table 10.1) [138].

Often, it is more convenient in risk assessment to refer to dose rates, or the amount of a chemical dose (applied or internal) per unit time (e.g., mg/day), or as dose rates on a per-unit-body-weight basis (e.g., mg/kg/day). Most exposure data found in the various editions of the USEPA's *Exposure Factors Handbook* and *Child-Specific Exposure Factors Handbook* and other guidance documents are presented as dose rates (e.g., grams of fish consumed each day) rather than as absorbed dose [44,61,62,64,139,140].

## Exposure and Dose Relationships

Depending on the purpose of the exposure assessment and the mechanism of action of the chemical, different estimates of exposure and dose may have to be calculated. Often, estimates of uptake will be presented in units so the dose metric will be the same as that used in the toxicology study, which is not always informative.

When risk is a function of time of exposure, exposure or dose profiles can be very useful. In these profiles, the exposure concentration or dose is plotted as a function of time [141]. Concentration and time are used to depict

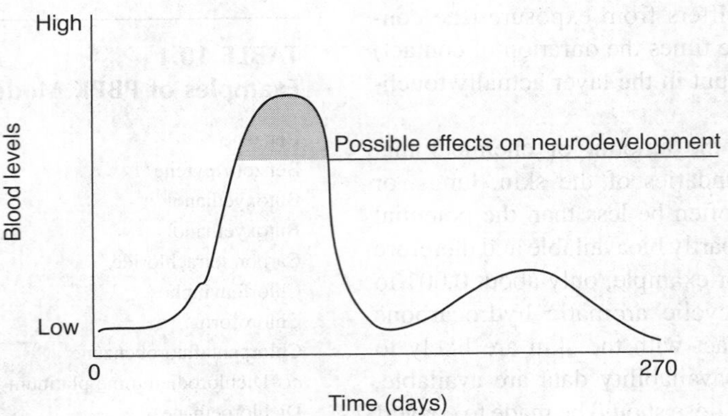


**FIGURE 10.2** Time course of exposure to a developmental toxicant. Note that the shaded portion represents the blood concentration of toxicant that is necessary to offer some probability that an inverse effect might occur.

exposure, and amount and time characterize dose. Such profiles are important for use in risk assessment where the severity of the effect depends on the pattern by which the exposure occurs, rather than on the total (integrated) exposure. For example, a developmental toxicant may only produce effects if exposure occurs during a particular stage of development [97]. As shown in Figure 10.2, the time above a certain dose rate (the shaded portion) presents an increased risk to the fetus of certain birth defects. Similarly, a single acute exposure to very high contaminant levels may induce adverse effects, even if the average exposure is much lower than apparent no-effect levels [97]. To understand hazards posed by most chemicals, it is important to consider their pharmacokinetics; for example, for a chemical that has a long biologic half-life, internal exposure continues long after the chemical is ingested because blood levels remain high until the substance is metabolized or eliminated. Conversely, for others, the chemical is inhaled, absorbed, metabolized, and excreted in less than an hour after exposure.

In general, it is necessary to consider the time elements of exposure assessment relative to the risk posed by the exposure. Standard approaches to time averaging to estimate long-term daily exposure concentration, in some cases, result in substantial underprediction of short-term variations in exposure. Similarly, the use of short-term measurements may overpredict long-term exposures [142,143]. It is useful to understand the relationship between the biological half-lives of toxicants and the subsequent critical time element of the exposure. Indeed, the appropriate consideration of these elements should dictate the averaging times for both the exposure limits and exposure assessment [53,54,144–146].

If a material causes its biological damage quickly and is gone from the body in a short time, then how we test its toxicity is critical; for example, consider a material with a half-life of a few minutes in the body. If we were to test it by spreading or apportioning the daily dose of this material over 24 hours using inhalation, we would achieve very different results than if the animal absorbed the same quantity in a couple of one-hour inhalation exposures. The same dose of this quick-acting material would do much more damage amassed in a bolus dose administered over a few minutes or even an hour or two than if absorbed over 24 hours.

Dosing times in animal studies, then, should be commensurate with the biological half-life. We often need to measure the exposure over an appropriately short period of time where the worst-case exposure may occur. The same logic also holds for the dermal (topically applied) and oral (normally ingested) exposures in that they appear to occur in time frames that are comparable to what we would expect in the environment. Bolus dosing by gavage or injection would, of course, be worst case. Thus, if a material causes its biological damage and is gone from the body in a relatively short period of time, then a long-term measure of exposure will generally be unnecessary.

On the other hand, if the biological half-life of the compound is longer than a few days, then relatively high spikes of exposure over a day or two are not particularly significant from a health impact perspective. What is important from a chronic toxicity perspective for these types of compounds is, of course, the weighted average over a significantly long time period. As such, it would only seem appropriate to use an annual average exposure when dealing with a compound with a very long (greater than 90 days) half-life in the body and no evidence of acute toxicity at high short-term dose rate [145,147,148].

Integrated or aggregate exposure is the sum total of exposure to a chemical via all routes of exposure (and all media). It is now commonplace to add as many as 6 to 10 different exposure sources per route (e.g., DDT in different fruits and vegetables) and 3 exposure routes (e.g., DDT via food, air, and dermal contact) [149,150]. Modeling software is available that characterizes doses from exposure to multiple chemicals, by multiple routes, and from multiple sources [149]. The units of aggregate exposure are concentration times duration. Aggregate exposure has been considered in complex assessments over the past 10 years (e.g., incinerators), but it came to the fore with passage of the Food Quality Protection Act (FQPA) in 1996. An increasing number of guidance documents from regulatory agencies and examples from researchers of how to perform these assessments have been published over the last several years [59,149,151]. Some research has been devoted to understanding aggregate environmental exposures, and many attempts have been made to determine the source contribution for different chemicals. These efforts have been the basis for studies such as the National Human Exposure Assessment Survey (NHEXAS), Children's Total Exposure to Persistent Pesticides and Other Persistent Organic Pollutants (CTEPP), Center for Health Assessment of Mothers and Children of Salinas (CHAMACOS), and Minnesota Children's Pesticides Exposure Study (MNCPES) [152–160].

Integrated exposure is the total area under the curve (AUC) of the exposure profile. An exposure profile (a picture of the exposure concentration over time) is particularly useful when trying to understand occupational exposure because it contains more information than a numerical estimate of the integrated exposure, including the duration and periodicity of exposure, the peak exposure, and the shape of the area under the time–concentration curve. The risk posed by most systemic toxicants with chronic effects is best understood by evaluating the blood concentration vs. time relationship.

A common way to characterize exposure is the time-weighted average (TWA). This is particularly relevant when conducting an assessment of a carcinogenic chemical in the workplace. In cancer risk assessments, the time over which exposure is integrated is usually 70 years [42]. A TWA dose rate is the total dose divided by the time period of dosing, usually expressed in units of mass per unit time, or mass/time normalized to body weight (e.g., mg/kg/day). TWA dose rates such as the lifetime average daily dose (LADD) are used in dose–response equations to estimate lifetime risk.

## Measures of Dose

For risk assessment purposes, dose estimates should be expressed in a manner that can be compared with available dose–response data from animal or human studies. For example, if data on human exposure is in mg of lead per deciliter of blood (mg/dL), it would be best to use the blood concentrations in an animal study to predict the risk to humans. Frequently, dose–response relationships are based on potential dose (called *administered dose* in animal studies), although dose–response relationships are sometimes based on internal dose. These differences must be accounted for. The measure of dose selected should be based on the mode of action of the adverse effect [23,128, 141,161–163]. For example, to assess a nasal irritant, the airborne concentration of the chemical is a relevant dose and an even better dose metric would be the milligrams of chemical contacting a square centimeter of nasal mucosa.

Doses may be expressed in several different ways. Solving Equation 10.1, for example, gives the dose rate over the time period of interest. The dose-per-unit time is the dose rate, which has units of mass/time. The most common dose measure is average daily dose (ADD), which is used to predict or assess the noncarcinogenic effects of a chemical:

$$ADD = [C \cdot IR \cdot B]/[BW \cdot AT] \quad (10.1)$$

where $ADD$ is the potential average daily dose, $C$ is the mean exposure concentration, $IR$ is the ingestion rate, $B$ is the bioavailability, $BW$ is the body weight, and $AT$ is the time period over which the dose is averaged (days). A typical calculation follows.

### Example Calculation 1: Determining the Average Daily Dose

A typical American eats a certain amount of lettuce over a lifetime (about 2000 kg). Assume that on any given week the maximum quantity ingested is 0.5 kg, and the maximum on any one day is 0.04 kg/day. Assume that the typical aldrin residue is 4 mg/kg on all lettuce ingested over the person's lifetime. What is the ADD of aldrin for the maximum week? Assume the oral bioavailability of aldrin in lettuce is 90%. We are given:

$C$ = 4 mg/kg (aldrin)
$IR$ = 0.5 kg
$B$ = 0.9
$BW$ = 70 kg
$AT$ = 7 days

Therefore:

$ADD$ = [C · IR · B]/[BW · AT]
$ADD$ = [4 mg/kg · 0.5 kg · 0.9]/[70 kg · 7 days]
$ADD$ = 0.004 mg/kg/day

When the primary health risk posed by a chemical is cancer or another chronic effect, then the biological response is usually described in terms of lifetime probabilities (e.g., the increased risk of developing cancer during a 70-year lifetime is 2 in 100,000). In these circumstances, even though exposure does not occur over the entire lifetime, doses are usually presented as LADDs [42]. The LADD takes the form of Equation 10.2, with lifetime ($LT$) replacing the averaging time ($AT$):

$$LADD_{pot} = [C \cdot IR \cdot B]/[BW \cdot LT] \quad (10.2)$$

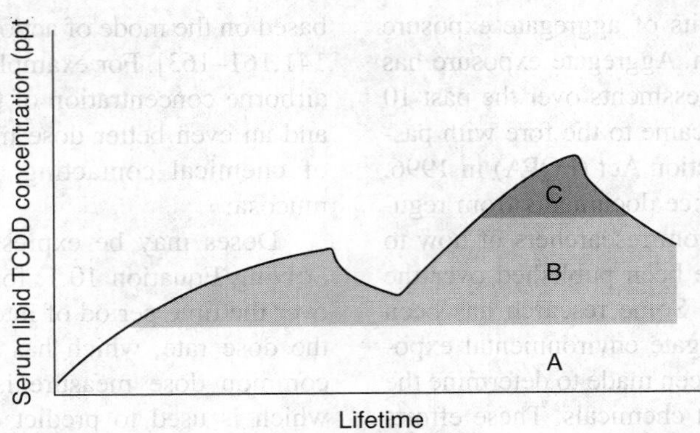


**FIGURE 10.3** Theoretical concentration vs. time curve for TCDD illustrating one possible relationship between AUC and response. This figure illustrates the possible combination of AUC and thresholds for production of various responses: area A, no effect; area B, enzyme induction occurs; area C, significant increased cell proliferations. (From Aylward, L.L. et al., *Environ. Sci. Technol.*, 30(12):3534–3543, 1996. With permission.)

### Example Calculation 2: Determining the Lifetime Average Daily Dose

What is the LADD in Example 1 involving aldrin in lettuce? Assume that the maximum reasonable lifetime uptake of lettuce (99% person) is 14,000 kg. We are given:

$C$ = 4 mg/kg (aldrin in lettuce)
$IR$ = 14,000 kg
$B$ = 0.9
$BW$ = 70 kg
$LT$ = 70 yr = 25,550 days

where:

$$LADD = \left[\overline{C} \cdot \overline{IR} \cdot \overline{B}\right] / \left[BW \cdot LT\right]$$

Then:

$$LADD = [4 \cdot 14{,}000 \cdot 0.9]/[70 \cdot 25{,}550]$$
$$LADD = 0.028 \text{ mg/kg/day}$$

Although other measures of chronic dose may be more appropriate for predicting the hazard posed by specific chronic toxicants, such as an area under the blood-concentration curve or the peak target tissue concentration, the LADD is the most common dose metric used in carcinogen risk assessment (Figure 10.3).

## CONCEPTUAL APPROACHES TO EXPOSURE ASSESSMENT

### Quantifying Exposure

Certain methods, such as environmental sensors and geographic information systems (GISs), can be used to derive information about external environmental exposures and the personal activity patterns that influence the magnitude, frequency, duration, and pathways of exposure. Other methods, such as biologic sensors, toxicogenomics, and body burden assays, are more frequently being used to derive measurements of internal biologic exposure [164]. Although exposure assessments are conducted for a variety of reasons, the process of estimating exposure can be generally approached using one of the following three methods [42]:

- *Direct measurement*—The exposure can be measured at the point of contact (the outer boundary of the body) while it is taking place, measuring the exposure concentration and time of contact and integrating them (point-of-contact measurement). An example is the measurement of the amount of contaminated soil on an exposed hand of someone digging a hole to plant a tree. The relevant exposure information would be contaminant concentration in soil (µg/g), surface area of the hand in contact with the soil (100 $m^2$), and time of exposure (2 hours).
- *Exposure scenario*—Sometimes one is concerned about an exposure that may or may not occur so a hypothetical exposure scenario is developed. In these assessments, specific data cannot actually be collected, but relevant information can be found. For example, if an incinerator were built, it would not be known today how much of each chemical in the airborne emissions would reach the various compartments in the environment (food, soil, sediment, surface water), but one could describe what would be likely to occur (a scenario).
- *Biomonitoring*—Sometimes historical exposure can be estimated based on the amount of chemical in the body or being eliminated in breath, urine, or feces. In recent years, doses have been reconstructed through internal indicators (biomarkers, body burden, excretion

levels) for persistent organics and several metals. Among the most common historical examples are lead in blood, urinary mercury, volatile hydrocarbons in the breath, and dioxins in blood fat.

These three approaches to exposure quantification (or dose) are independent because each is based on different assumptions or data. The fact that they are independent measures is useful in verifying or validating the results of the various approaches. Each of the three has strengths and weaknesses; using them in combination can considerably strengthen the credibility of an exposure assessment [12,23,165]. For example, results of the exposure assessment would be validated if one could mathematically predict the absorbed dose per day of a chemical, estimate the resulting blood concentrations, and confirm these estimates by sampling the blood of the exposed population [50].

### Estimates Based on Direct Measurement

Point-of-contact or direct exposure assessment evaluates the exposure as it occurs. Measuring chemical concentrations at the interface between the person and the environment as a function of time yields an exposure profile. The best known example of point-of-contact measurement is the radiation dosimeter. This small badge-like device measures radiation exposure as it occurs and provides an integrated exposure estimate for the period of time over which the measurement has been taken [42]. The Total Exposure Assessment Methodology (TEAM) studies conducted by the USEPA also makes use of direct measurements [166]. In the TEAM studies, a small pump with a collector and an absorbent was attached to a person's clothing to measure his or her exposure to airborne solvents or other pollutants as it occurred, just as has been done in industrial hygiene studies of the past 60 years [167]. In both of these examples, the measurements were collected at the interface between the person and the environment while exposure was occurring.

The area of exposure assessment known as *agricultural hygiene* has developed very sophisticated techniques for estimating the uptake (absorption) of chemicals during the mixing and application of pesticides [168–170]. Macroscale technologies such as laser-based and infrared-radiation-based sensors are currently being used for assessing population exposures to sulfur and nitric oxides in industrial-stack effluents. Other microscale sensors, including personal dosimeters, are being used to monitor levels of carbon dioxide, carbon monoxide, volatile organic compounds, pesticides, and PAHs in the workplace, households, and personal environments [164]. Recent efforts have focused on automated "lab-on-a-chip" sensing devices for detection environmental agents [164,171].

A common limitation in exposure assessment is the lack of information about patterns of physical activity and behavior that affect the likelihood of exposure, the frequency and duration of exposure, and the uptake and distribution of environmental agents in the body [164]. GIS approaches have been used for developing individual metrics for exposure to pesticides, drinking water contaminants, and air pollutants, such as nitric oxide, sulfur dioxide, and particulates [164,172–174]. Only recently, researchers have used GIS to derive personal exposure estimates by linking information about personal activity and behavioral patterns with environmental data [164,174–176]. Interestingly, personal dosimetry devices are being developed to measure individual variables related to activity such as motion, temperature, pressure, energy use, respiratory function, and heart rate [164,177–184].

Providing that the measurement devices are accurate, the direct measurement method likely gives the most accurate exposure value for the period of time over which the measurement was taken. It is often expensive, however, to use these techniques to evaluate persons in the community, and measurement devices and techniques do not currently exist for all chemicals (at least at ambient concentrations).

### Estimates Based on Exposure Scenarios

Using the exposure scenario approach, the assessor attempts to estimate or predict chemical concentrations in a medium or location and link this information with the time that individuals or populations are in contact with the chemical. An exposure scenario is the set of assumptions describing how this contact takes place. This is, by far, the most common approach to exposure assessment, and such an approach is necessary when trying to predict the impact of events that may occur in the future, such as building a new manufacturing facility or introducing a new pesticide or herbicide [27,185–187].

The first step to building a scenario is to determine the concentration of the contaminated media. This is typically accomplished indirectly by measuring, modeling, or using existing data on concentrations in the media of concern, rather than at the point of contact (e.g., pesticide residues on food or metal emissions on residential soils). Often, we assume that the concentration in the bulk medium is the same as the concentration at the point of exposure. This can be a source of potential error and should be discussed in the uncertainty analysis. For example, over the past 20 years, most assessments of the hazard posed by contaminated soil were based on soil samples collected in the top 6 inches of soil, even though most persons were exposed routinely to the surface soil (usually the top 2.5 inches). Arguments can be made in either direction about the appropriateness of this assumption.

The next step in conducting an exposure scenario is to estimate the contact time, identify who is likely to be exposed, and then develop estimates of the exposure frequency and duration. Like chemical concentration characterization, this is usually done indirectly using demographic data, survey statistics, behavior observation, activity diaries, activity models or, in the absence of more substantive information, assumptions about behavior [63,82].

Chemical concentration and population characterizations are ultimately combined in an exposure scenario. One of the major problems in evaluating dose equations is that the limiting assumptions used to derive them (e.g., steady-state assumptions) do not always hold true. Two approaches to this problem are available: (1) to evaluate the exposure or dose equation under conditions when the limiting assumptions do hold true, or (2) to build a dynamic model that accounts for both accumulation and degradation. The microenvironment method, which is typically used to evaluate air exposures, is an example of the first approach. This method evaluates segments of time and location when the assumption of constant concentration is approximately true and then sums the time segments to determine the total exposure for the respiratory route, effectively removing some of the uncertainty [188]. In occupational hygiene, this is done by combining time–motion data with short-term air concentration data. While estimates of exposure concentration and time of contact may be estimated in some situations, the concentration and time-of-contact estimates can be measured for each microenvironment. This avoids much of the error due to summing average values in cases where concentration and time of contact vary widely.

In the second approach, a computer model can efficiently predict dose if enough data are available [38,47,189]. When conducting modeling, various tools are used to describe uncertainty caused by parameter variation (i.e., Monte Carlo analysis), which may be necessary in some assessments. Monte Carlo techniques are rarely helpful when assessing individuals or small populations because actual "by person" data will often be available.

## Estimating Exposure Using Biological Monitoring

Exposure can often be estimated after it has taken place. One important factor is whether the biological half-life of the chemical is sufficiently long to allow for accurate measurement. If a total dose is known or can be reconstructed, and information about intake and uptake rates is available, an average past exposure rate can be estimated [141,190–194]. Dose reconstruction relies on measuring biological fluids (blood, urine), hair, nails, or feces after exposure, intake and uptake have already occurred and using these measurements to back-calculate dose [141]; however, data on body burden levels or biomarkers cannot be used directly unless a relationship can be established between these levels (or biomarker indications) and internal dose.

Biological monitoring can be used to evaluate the amount of a chemical in the body by measuring one or more parameters (Table 10.2). In general, if these measurements can be made and the biologic half-life is acceptable then past exposure estimates can be reasonably accurate. Not all of these can be measured for every chemical [42]:

- The concentration of the chemical itself in biological tissues or sera (e.g., blood, urine, breath, hair, adipose tissue)
- The concentration of the chemical's metabolites
- The biological effect that occurs as a result of human exposure to the chemical (e.g., alkylated hemoglobin or changes in enzyme induction)
- The amount of a chemical or its metabolites bound to target molecules

The results of biomonitoring can be used to estimate chemical uptake during a specific interval, if background levels do not mask the marker and the relationship between uptake and the selected marker is known [195]. The sampling time for biomarkers is often critical. Establishing a correlation between exposure and measurement of the marker, including pharmacokinetics, is necessary to properly back-calculate historical exposure [42].

The strengths of this method are that it demonstrates that exposure and absorption of the chemical have actually taken place; also, theoretically, it can give a good indication of past exposure. The drawbacks are that it will not work for every chemical because of interferences or the reactive nature of the chemical or because the biological half-life of the agent is too short; also, data relating internal dose to exposure are necessary. Additionally, it may be expensive.

For those chemicals where biological monitoring can be used to estimate past exposure, the information obtained can be invaluable for conducting retrospective exposure assessments that can be used in epidemiology studies. Some examples of chemicals for which past exposure can reliably be estimated include several metals, as well as numerous large organic chemicals (e.g., DDT, chlordane, dioxin, PBB, PCB) [141].

# INFORMATION UPON WHICH EXPOSURE ASSESSMENTS ARE BASED

Comprehensive exposure assessment of a complex scenario may require several hundred exposure factors to estimate the various chemical concentrations in one of several dozen different media. Among the most complex exposure assessments are those that address the risks posed by airborne

**TABLE 10.2**
**Examples of Types of Measurements To Characterize Exposure-Related Media and Parameters**

| Type of Measurement (Sample) | Usually Attempts to Characterize | Examples | Typical Information Needed To Characterize Exposure |
|---|---|---|---|
| Breath | Total internal dose for individuals or population (usually indicative of relatively recent exposures) | Measurement of volatile organic compounds (VOCs), alcohol (usually limited to volatile compounds) | 1. Relationship between individuals and population; exposure history (i.e., steady-state or not); pharmacokinetics (chemical half-life); possible storage reservoirs within the body<br>2. Relationship between breath content and body burden |
| Blood | Total internal dose for individuals or population (may be indicative of either relatively recent exposures to fat-soluble organics or long-term body burden for metals) | Lead studies, pesticides, heavy metals (usually best for soluble compounds, although blood lipid analysis may reveal lipophilic compounds) | 1. Same as above<br>2. Relationship between blood content and body burden |
| Adipose | Total internal dose for individuals or population (usually indicative of long-term averages for fat-soluble organics) | National Human Adipose Tissue Survey (NHATS) substances, dioxin, PCBs (usually limited to lipophilic compounds) | 1. Same as above<br>2. Relationship between adipose content and body burden |
| Nails, hair | Total internal dose for individuals or population (usually indicative of past exposure in weeks to months range; can sometimes be used to evaluate exposure patterns) | Heavy metal studies (usually limited to metals) | 1. Same as above<br>2. Relationship between nails, hair content, and body burden |
| Urine | Total internal dose for individuals or population (usually indicative of elimination rates); time from exposure to appearance in urine may vary, depending on chemical | Studies of tetrachloroethylene and trichloroethylene | 1. Same as above<br>2. Relationship between urine content and body burden |

*Source:* USEPA, *Lognormal Distribution in Environmental Applications*, EPA/600/R-97/006, U.S. Environmental Protection Agency, Washington, D.C., 1997.

emissions from combustors (Figure 10.4) [58,188,196]. To estimate the concentration, numerous dispersion models, as well as fate and transport models, may be required. In addition, the assessor may need to search the literature to identify relevant studies from as many as ten related fields of research. Sometimes, hundreds of published papers and government guidance documents must be evaluated, used, and cited. In short, the exercise can be formidable, especially for assessments involving food chain contamination. Equally difficult and highly complex exposure assessments are those that attempt to estimate the uptake of fish by various members of the angling public [78].

## Obtaining Data on Intake and Uptake

The numerous editions of the *Exposure Factors Handbook* and *Child-Specific Exposure Factors Handbook* present statistical data on many of the factors used to assess exposure, including intake rates and provide citations for primary references [44,61–64]. Today, this series of publications represents the most comprehensive single source of exposure assessment information. Some of the many intake factors in the various volumes include:

- Drinking-water consumption rates
- Breast-milk ingestion rates for infants
- Consumption rates for homegrown fruits, vegetables, beef, and dairy products
- Consumption rates for recreationally caught fish and shellfish
- Rates of hand-to-mouth and object-to-mouth activity for children
- Incidental soil ingestion rates
- Pulmonary ventilation rates
- Surface area of various parts of the human body
- Body weight for various age groups

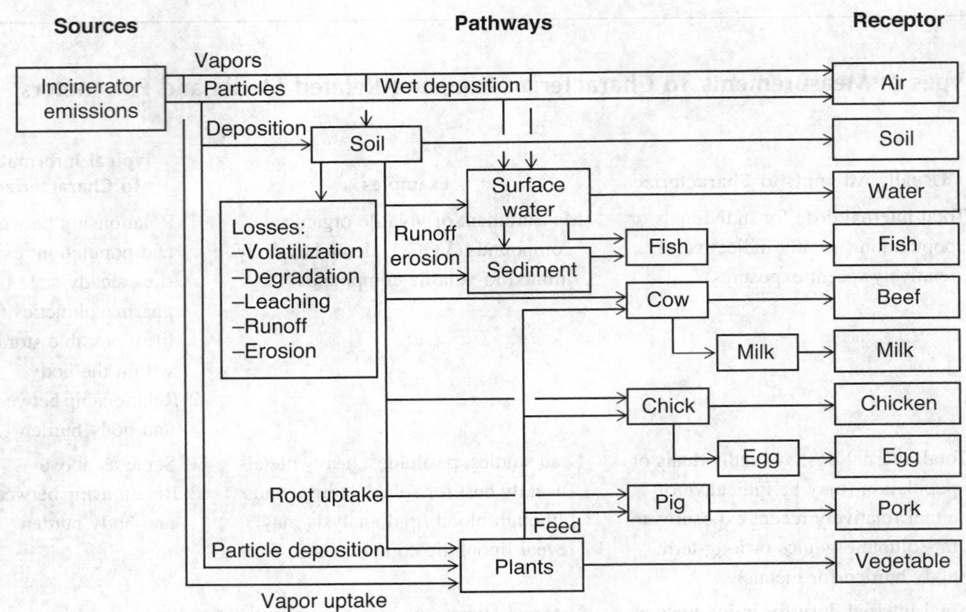


**FIGURE 10.4** The USEPA's conceptual approach to dealing with direct and indirect exposure pathways as illustrated by assessments of incinerator emissions.

- Duration and frequency in different locations and microenvironments
- Duration and use of consumer product use
- Duration of lifetime

Table 10.3 presents examples of some of the standard or default exposure factors used in risk assessment.

The *Exposure Factors Handbook* is updated routinely to include new research data on previously discussed factors. It also provides default parameter values that can be used when site-specific data are not available. Obviously, general default values should not be used in place of known, valid data that are more relevant to the assessment being conducted. The *Exposure Factors Handbook*, although substantial, may not contain all available information on exposure factors or relevant studies, so a supplemental literature search should be conducted to ensure that pertinent literature has been identified. As will be discussed later, if a probabilistic or Monte Carlo assessment is to be conducted, the documents titled *Risk Assessment Guidance (RAGS3A) for Conducting Probabilistic Risk Assessment* [86] and *Options for Development of Probability Distributions for Exposure Factors* [87] should be consulted.

## Concentration Measurements in Environmental Media

Sometimes, the published data are inadequate to conduct a proper site-specific assessment. In these cases, concentration data can be gathered by conducting a new field study or by evaluating data from past field studies to estimate concentrations. Media measurements taken close to the point of contact are preferable to measurements far removed geographically or temporally. As the distance from the point of contact increases, the certainty of the data at the point of contact usually decreases, and the obligation for the assessor to show relevance of the data to the assessment at hand becomes greater. For example an outdoor air measurement, no matter how close it is taken to the point of contact, cannot by itself adequately characterize indoor exposure [197].

Concentrations often vary considerably from place to place, seasonally, and over time due to changing emission and use patterns [46,52,198]. This needs to be considered not only when designing studies to collect new data, but also when evaluating the applicability of existing measurements as estimates of exposure concentrations in a new assessment. It is of particular concern when the measurement data will be used to extrapolate to long time periods such as lifetimes. Transport and dispersion models are frequently used to help answer these questions [189].

The exposure assessor is likely to encounter several different types of measurements. One type often used to understand concentration trends is outdoor fixed-location monitoring. This measurement is used by the USEPA and other groups to provide a record of pollutant concentration at one place over time. Nationwide air and water monitoring programs have been established so baseline values in these environmental media can be documented. Although it is not practical to set up a national monitoring network to gather data for a particular exposure

**TABLE 10.3**
**Some Standard Default Assumptions Used in Exposure Assessment**

| Variable | Assumption |
|---|---|
| Drinking water | 2 L/day (adult, RME) |
| | 1.4 L/day (adult, average) |
| | 1.0 L/day (child) |
| | 0.1 L/day (incidental ingestion during swimming) |
| Soil (Ingestion) | 200 mg/day (child, average) |
| | 800 mg/day (child, 90th percentile) |
| | 100 mg/day (adult) |
| Food | 2000 gm/day (adult, total) |
| Beef (home-grown) | 44 g/day (average) |
| | 75 g/day (RME) |
| | 100 g/day (all sources) |
| Dairy (home-grown) | 160 g/day (average) |
| | 300 g/day (RME) |
| | 400 gm/day (all sources) |
| Fruit (home-grown) | 28 g/day (average) |
| | 42 g/day (RME) |
| | 140 g/day (all sources) |
| Vegetables (home-grown) | 50 g/day (average) |
| | 80 g/day (RME) |
| | 200 gm/day (all sources) |
| Sport fish | 30 g/day (average) |
| | 140 g/day (RME) |
| Inhalation | 10 $m^3$/day (average 8-hr shift.) |
| | 20 $m^3$/day (adult, average) |
| | 30 $m^3$/day (RME) |
| Body weight | 13.2 kg (2–5 yr) |
| | 20.8 kg (6 yr) |
| | 70 kg (adult, average) |
| Lifespan | 70 yr |
| Exposed skin area | 0.2 $m^2$ (adult, average) |
| | 0.53 $m^2$ (adult, RME) |
| | 1.94 $m^2$ (male bathing) |
| | 1.69 $m^2$ (female bathing) |
| Showering | 7 min (average; 5-min. shower uses 40 gallons) |
| | 12 min (90th percentile) |
| Residence time | 9 yr (average) |
| | 30 yr (RME) |

assessment, data from existing networks can be evaluated for relevance to an exposure assessment. These data are often far removed from the point of contact. Adapting data from previous studies usually presents specific challenges.

Indoor air contaminant concentrations can vary as much or more than those in outdoor air [16,199–203]; consequently, indoor exposure is best represented by measurements taken at the point of contact. However, because pollutants such as carbon monoxide can exhibit substantial penetration to the indoor environment, both indoor and outdoor sources of the contaminant should be considered [21,204–206].

Contaminant concentrations in food and drinking-water measurements can also be measured. General characterization of these media, such as market-basket studies (where representative diets are characterized), shelf studies (where foodstuffs are taken from store shelves and analyzed), or drinking-water-quality surveys, are usually far removed from the point of contact for an individual but may be useful in evaluating exposure concentrations for a large population. Measurements of tap water or foodstuffs in a home and how they are used are closer to the point of contact. In evaluating the relevance of data from previous studies, variation in the distribution systems must be considered, as well as the space–time proximity [42].

Consumer or industrial product analysis is sometimes done to characterize the chemical concentrations in products. Product formulations can change substantially over time, similar products do not necessarily have similar formulations, and regional differences in product formulation can also occur. These factors should be considered when determining the relevance of extant data and when setting up sampling plans to gather new data [42].

Another type of concentration measurement is the microenvironmental measurement. Rather than using measurements to characterize the entire medium, this approach defines specific zones in which the concentration in the medium of interest is thought to be relatively homogenous and then characterizes the concentration in that zone [23,207]. Typical microenvironments include the home or parts of the home, office, automobile, or other indoor settings. Microenvironments can also be divided into time segments (e.g., kitchen during the day, kitchen during the night). This approach can produce measurements that are closely linked with the point of contact, both in location and time, especially when new data are generated for a particular exposure assessment. The more specific the microenvironment, however, the greater the burden on the exposure assessor to establish that the measurements are representative of the population of interest.

The concentration measurement that provides the closest link to the actual point of contact is personal monitoring. In virtually all cases, if available, this information should be the basis of exposure assessments of individuals. An obvious exception is the work environment where lapel sampling is conducted while the person is wearing a respirator; thus, in this case, personal sampling would not reflect genuine exposure.

## Models and Their Role

Often, the most critical element of an exposure assessment is estimating pollutant concentrations at exposure points. This is usually carried out by combining field data and modeling results. In the absence of field data, this process often relies on the results of mathematical models of aerial dispersion (i.e., ISCLT) or of water movement (i.e., MODFLOW) [208–212]. The USEPA's Science Advisory Board and others have recommended that modeling be ideally linked with monitoring data in regulatory assessments, although this is not always possible.

A modeling strategy has several aspects, including setting objectives, selecting the appropriate model, obtaining and installing the code, calibrating and running the computer model, and validating results. Many of these aspects are analogous to the quality assurance and quality control measures applied to measurements. Regardless of whether models are extensively used in an assessment or whether a formal modeling strategy is documented in the exposure assessment plan, when computer simulation models, such as fate and transport models, or exposure models are used in exposure assessments, the assessor must be aware of the performance characteristics of the model and state how the exposure assessment requirements are satisfied by the model [189].

The site must be characterized if models are to be used to simulate pollutant behavior at a specific site. Site characterization for any modeling study includes examining source characterization, dimensions and topography of the site, location of receptor populations, meteorology, soils, geohydrology, and ranges and distributions of chemical concentrations. For exposure models that simulate both chemical concentration and time of exposure (through behavior patterns), data on these two parameters must be evaluated [23,27,213]. Criteria are provided by the USEPA for selecting surface water models and groundwater models, respectively; the reader is referred to these documents for further details [212]. Similar selection criteria exist for air dispersion models [211].

A primary consideration in selecting a model is whether to perform a screening study or a detailed evaluation. A screening study makes a preliminary evaluation of a site or a general comparison between several sites. It may be generic to a type of site (e.g., an industrial segment or a climatic region) or may pertain to a specific site for which sufficient data are not available to properly characterize the site. Screening studies can help direct data collection at the site by providing an indication of the level of detection and quantification that would be required and the distances and directions from a point of release where chemical concentrations might be expected to be highest.

An example of a screening-level modeling effort would be to estimate the amount of lead deposited by an incinerator onto local crops using a basic air dispersion model, without considering local geographical or weather conditions. The next level of complexity would consider the presence of mountains, their proximity to the stack, the local weather patterns, and the number of atmospheric inversions per year. A higher level of analysis could incorporate yet other, more subtle factors.

The value of the screening-level analysis is that it is simple to perform and may indicate that no significant contamination exists. Screening-level models are frequently used to get a first approximation of the concentrations that may be present. Often, these models use very conservative assumptions; that is, they tend to overpredict concentrations or exposures. If the results of the conservative screening procedure predict concentrations or exposures at less than a predetermined no-concern level, then more detailed analysis is probably not necessary. If the screening estimates are above that level, refinement of the assumptions or a more sophisticated model is necessary to generate a more realistic estimate [42].

Screening-level models also help the user conceptualize the physical system, identify important processes, and locate available data. The assumptions used in the preliminary analysis should represent conservative conditions, such that the predicted results overestimate potential conditions, limiting false negatives. If the limited field measurements or screening analyses indicate that a contamination problem may exist, then a detailed modeling study may be useful. In contrast, the purpose of the detailed evaluation is to use the best data available to make the best estimate of spatial and temporal chemical distributions of a specific site. Detailed studies typically require higher quality data and more sophisticated models.

## Accounting for Background Concentrations

Background exposure to xenobiotics, especially environmentally persistent ones, can occur due to natural or anthropogenic sources [215]. In most exposure assessments, background soil concentrations are the focus of attention, but the same issue can be relevant when evaluating sediments, ambient air, groundwater, and vegetation (food stuffs). At some sites, it is important that these so-called "background" concentrations be accounted for because removing the quantity of toxicant due to humans may, in fact, not appreciably change the concentrations or be sufficient to reduce the risk to acceptable levels. For example, naturally occurring concentrations of lead, arsenic, and cadmium in some locations may be higher than cleanup levels established by various regulatory agencies [214,216]. The exposure assessor should try to determine local background concentrations by gathering data from nearby locations clearly unaffected by the site under investigation or by referring to published works that have assessed this issue [217].

## Description of Background Levels

When assessing soils, background levels can be viewed in at least four different ways [208]:

- *Pristine levels*—Some would like to equate background levels with those associated with the "pristine" state—that is, soils or landscapes unaffected by human activity. This rather idealistic situation probably no longer exists; even in Antarctica, mercury and dioxin concentrations can be detected in some media. Toxic elements mainly associated with the solid phase of some natural material (such as soil dust, plant or volcanic ash, vegetable matter) are relatively mobile in a global sense. For example, Nriagu [218] has suggested that about 40 million tons of heavy metals have been dispersed atmospherically over the many centuries of human activity. Increases in pollutant metal concentrations have been measured up to 60 km from smelters, and automotive lead (fine particles) has been measured in soils and rainfall up to about 50 km downwind from major cities. Soil contamination up to 50 to 100 m from highways by automotive lead (coarser particles) is an example of short-range transport and contrasts with transport of toxic metals on a continental or global scale (e.g., contamination of the Greenland ice sheets from the northern United States, mercury in the Florida Everglades due to aerial releases in South America, the snows of the New Zealand Alps contaminated by soil dust from inland Australia) [208].
- *Normal levels*—One might ask: "Are soils contaminated at farms in the higher rainfall areas of the Appalachians, which used to receive automotive exhaust particulates from the New York metropolitan area 200 km away?" These soils are not pristine, but the chemical concentrations are perfectly safe for growing food, raising farm animals, and residential living. Soils from such areas would have a range of what is often considered normal background values. To most exposure assessors, this mosaic of normal soils, affected only by the minor pollution of everyday activities associated with modern rural and urban life, should be the basis for defining background values. Statistically, this range of normal background values would constitute a single log-normally distributed population. Obviously, one needs to exclude the outliers or hot spots due to a geochemical anomaly or localized pollution arising from industrial emissions, disposal of waste products, or intensive (excessive) use of farm chemicals [208].
- *Historically polluted regions*—Local community and regulatory policies often affect what are defined as background levels. A community with highly developed environmental consciousness may insist on very low, possibly unreasonable, reference values. Some densely populated areas with historically derived pollution, perhaps from former mining activities, may sustain apparently healthy populations who pragmatically must accept higher background level values. The cities of Philadelphia, Baltimore, and New York and parts of Japan, for example, may fit in this category.
- *Geochemical variation*—Background levels of some potentially toxic elements may vary among geographical regions because of differences in soil type. The resulting concentrations are often referred to as *naturally occurring levels*.

An important factor is the composition of rocks and sediments that weather from soils. Some extreme examples are high concentrations of nickel, cobalt, and chromium in igneous rocks such as basalts that cover extensive areas in western Victoria and Tasmania, Australia, as well as high concentrations of boron in soils on marine sediments in the Riverina, Mallee, and Wimmera districts of South Australia and Victoria, Eyre Peninsula, and parts of western Australia.

Some regulatory agencies have provided written guidance describing how to select soil or sediment cleanup values that account for background chemical concentrations. These have varied significantly, but within the past 5 years there appears to be some convergence regarding the definition of background, how to measure it, and how it should affect exposure assessment calculations.

## ESTIMATING UPTAKE VIA THE SKIN

When attempting to predict chemical risks in the environmental or occupational setting, the dermal exposure route should nearly always be assessed. In most evaluations of hazardous waste sites and ambient air or water contaminants, this is not a major route of exposure. Although the uptake of chemicals via the skin has generally been overlooked in most workplace exposure assessments, it probably represents a substantial portion of the exposure for many occupations. Even though gloves are more frequently used than in years past and training has increased on the possible hazards of dermal exposure, ample evidence still indicates that, in order to conduct a complete exposure assessment, this route deserves attention [89,90,219,220].

In addition to the risks associated with systemic toxicity due to uptake via the skin, it is sometimes necessary to evaluate the allergic contact dermatitis (ACD) hazard. In recent years, techniques have been developed to quantitatively predict the likelihood of illication and induction of ACD [221,222,551]. Some regulatory agencies are concerned with ACD and have developed cleanup standards based on this health endpoint.

In the workplace, a worker's skin frequently comes into contact with solvents or chemicals mixed in water (aqueous materials). In most environmental settings where persons can be exposed to contaminated soil, dust, or water, dermal update must be assessed. Fortunately, a good deal of research has been conducted to understand the rate at which chemicals pass through the skin. Percutaneous absorption of neat chemicals (e.g., the pure liquid) was often studied in humans until the late 1970s [126, 223–229]. Because of the potential toxicity of many chemicals and improved laboratory techniques, *in vivo* human studies have been largely supplanted by experiments with laboratory animals, *in vitro* studies, or by using athymic rodents grafted with human skin [230]. Historical research has shown that, in general, chemical penetration of the human skin is similar to that of a pig or monkey and much slower than that of the rat and rabbit [231]; thus, for many chemicals, there is some level of confidence that the rate of dermal uptake of a chemical by humans can be inferred from animal data.

Beginning in the 1980s, *in vitro* studies using human skin began to be conducted on a more routine basis. In these studies, a piece of excised skin is attached to a diffusion apparatus with a top chamber to hold the applied chemical and a temperature-controlled bottom chamber containing saline or other fluids (plus a sampling port to withdraw fractions for analysis) [232]. Although human forearm skin is optimal, it is difficult to obtain, so abdominal or breast skin is commonly used. Generally, a properly conducted *in vitro* test can be a reasonably good predictor of the absorption rate *in vivo* [233,234]; however, due to the fragile nature of the technique, these studies must be carefully interpreted [235]. Often, depending on the conditions of the test, the results are not applicable to humans.

Aside from neat liquids and exposure to contaminated water, dermal exposures can also occur through contact with dust or dirt on surfaces and by way of contact with soil or dust-bound contaminants [197]. Surface-to-skin transfer of contaminants is a complex process; for example, a pesticide can be transferred to skin during contact with any contaminated exposure medium. Once on the skin, pesticide residues and contaminated particles can be transferred back to the contaminated surface during subsequent contact or due to loss by dislodgement or washing, or they can be transferred into the body by percutaneous absorption or hand-to-mouth activity [236]. Few studies have directly estimated soil loading on human skin, and only one of them attempted to measure dermal contact of contaminated equipment by workers [237–246]; however, recent efforts using a fluorescent tracer as a surrogate for pesticide residues and house dust particles have proved informative in understanding the parameters that affect residue transfer from surface to skin, skin to other objects, and skin to mouth [236,247]. In these studies, controlled transfer experiments were conducted by varying contact parameters with each trial. The mass of a tracer transferred was measured and the contact surface area estimated using video imaging techniques. Parameters evaluated included surface type, surface loading, contact motion, pressure, duration, and skin condition. Results have shown that surface loading and skin condition are among the important parameters for characterizing residue transfers [236].

The available studies probably provide sufficient data to generate point estimates of soil adherence and, perhaps, can provide a reasonable probability density function (PDF) for most persons exposed to contaminated soils. The degree of representativeness of the data to the general population is difficult to assess [248]. Recently, a couple of studies measured the adherence of soil to multiple skin surfaces (hands, forearms, lower legs, faces, and feet under ambient and recreational conditions [239,241]. Dermal loading on the hands was found to vary over five orders of magnitude and to be dependent on the type of activity. Differences between pre- and post-activity adherence demonstrated the episodic nature of dermal contact with soil; however, due to the activity-dependent nature of soil exposure, data from these studies must be interpreted for their relevance to the type of activity, frequency, duration, and otherwise site-specific nature of exposure. The various studies involving contaminated soil are informative for providing an estimate of exposure; however, they are probably a couple of orders of magnitude greater than what might be expected in a chemical plant. Nonetheless, this work serves as a starting point for bracketing potential exposure to dusts in the workplace.

Recently, a reasonable level of research has investigated exposure to house dust. The basis for this concern has been increasing evidence that controlling exposure to house dust, especially in homes located near sites with considerable surface soil contamination, is more important for reducing the health hazard than remediating the soil [197]. Numerous papers in recent years have shown that in-house exposure to toxics is much greater than that encountered due to ambient (so-called environmental) contamination [21,201,249,250]. In household dust studies, the source of the dust sample can provide a historical, as well as recent, collection of potential exposures from dust inside the home. Undisturbed surfaces (e.g., top of a refrigerator, attic dust) can be indicative of materials deposited over a long period of time, whereas frequently cleaned surfaces (e.g., kitchen countertops) will be indicative of the most recent deposits [251].

Along these lines, and of particular interest to those who study indoor exposure, is the recent work to develop standardized approaches for collecting wipe samples and estimating the amount of dust loading on the palm of the hand [252,253]. Using fluoroscein-tagged Arizona test dust as a possible surrogate for house dust, particle transfers to both wet and dry skin were quantified for contact events with stainless steel, vinyl, and carpeted surfaces that had been preloaded with the tagged test dust [247]. With these tests, researchers found that: (1) only about one third of the projected hand surface typically came into contact with the smooth test surfaces; (2) the fraction of particles transferred to the skin decreased as the surface roughness increased, with carpeting transfer coefficients averaging only one tenth those of stainless steel; (3) hand dampness significantly increased the particle mass transfer; and (4) consecutive presses decreased the particle transfer by a factor of three as the skin surface became loaded, requiring about 100 presses to reach an equilibrium transfer rate [247]. Although dermal absorption of toxicants in house dust will almost always pose a relatively low dermal uptake hazard, the uptake of toxicants due to hand-to-mouth contact can be substantial [197].

## Quantitative Description of Dermal Absorption

For the purposes of risk assessment, percutaneous absorption is defined as the transport of externally applied chemicals through cutaneous structures and the extracellular medium to the bloodstream [23,254]. In many settings, such as agricultural workers, platers, mechanics, and others, dermal uptake is the primary route of exposure. The simplest way to describe the rate of skin absorption is to apply Fick's first law of diffusion at steady state [255,256]:

$$J = dQ/dt = D \cdot k \nabla C / e \approx K_p \cdot C \qquad (10.3)$$

where $J = dQ/dt$ is the chemical flux or rate of chemical absorbed (mg/cm$^2$/hr), $D$ is the diffusivity in the stratum corneum (cm$^2$/hr), $k$ is the stratum corneum/vehicle partition coefficient of the chemical (unitless), $\nabla C$ is the concentration gradient (mg/cm$^3$), $e$ is the thickness of the stratum corneum (cm), $K_p$ is the permeability coefficient (cm/hr), and $C$ is the applied chemical concentration (mg/cm$^3$). The concentration gradient is equal to the difference between the concentration above and below the stratum corneum. Because the concentration below is small compared to the concentration above, $\nabla C$ can be approximated as equal to the applied chemical concentration. From the above equation, it can be seen that the rate of absorption is directly proportional to the applied concentration. The diffusivity represents the rate of migration of the chemical through the stratum corneum. Because the stratum corneum has a non-negligible thickness, there is a period of transient diffusion (lag time), during which the transfer rate rises to reach a steady state. In these studies, the steady state is maintained indefinitely, provided the system remains constant. Depending on the type of chemical, the lag time can range from minutes to days [89]. From an exposure assessment standpoint, if the exposure duration is shorter than the lag time, it is unlikely that there will be any significant systemic absorption [255,257].

The partition coefficient ($K_p$) is one of the key parameters that influences the degree to which a chemical penetrates the skin [255,257–260]. Fatty chemicals tend to accumulate in the stratum corneum. Conversely, the stratum corneum is an effective barrier for hydrophilic substances, which tend to have low skin absorption rates. Because stratum corneum/vehicle partition coefficients are difficult to measure, the three parameters ($D$, $k$, and $e$) are combined to give an overall permeability coefficient ($K_p$). It is noteworthy that Equation 10.3 only approximates most *in vivo* exposure situations because true steady-state conditions are rarely attained. In spite of its limitations, this equation has yielded satisfactory estimations of the actual absorption rates of chemicals for many situations (Table 10.4).

**TABLE 10.4**
**Human Cutaneous Permeability Coefficient Values for Some Industrial Chemicals in Aqueous Medium**

| | MW | $K_{ow}$ | Observed | Calculated[a] |
|---|---|---|---|---|
| Organic chemicals | | | | |
| 2-Amino-4-nitrophenol | 154.13 | 21.38 | 0.00066 | 0.019 |
| 4-Amino-2-nitrophenol | 154.13 | 9.12 | 0.0028 | 0.0081 |
| Aniline | 93.12 | 7.94 | 0.041[b] | 0.091 |
| Benzene | 78.11 | 134.90 | 0.11 | 0.39 |
| *p*-Bromophenol | 173.02 | 389.05 | 0.036 | 0.25 |
| Butane-2,3-diol | 90.12 | 0.12 | <0.00005 | 0.0009 |
| *n*-Butanol | 74.12 | 7.59 | 0.0025 | 0.024 |
| 2-Butanone | 72.10 | 1.94 | 0.0045 | 0.007 |
| Carbon disulfide | 76.14 | 100.00 | 0.54[b] | 0.3 |
| Chlorocresol | 142.58 | 1258.93 | 0.055 | 1.31 |
| *S*-Chlorophenal | 128.56 | 147.91 | 0.033 | 0.19 |
| *p*-Chlorophenal | 128.56 | 257.04 | 0.036 | 0.34 |
| Chloroxylenal | 156.61 | 1621.81 | 0.059 | 1.35 |
| *m*-Cresol | 108.13 | 100.00 | 0.015 | 0.18 |
| *o*-Cresol | 108.13 | 100.00 | 0.016 | 0.18 |
| *p*-Cresol | 108.13 | 85.11 | 0.018 | 0.15 |
| Decanol | 158.28 | 37153.52 | 0.08 | 30.11 |
| 2,4-Dichlorophenol | 163.01 | 1995.26 | 0.06 | 1.5 |
| 1,4-Dioxane | 88.10 | 0.38 | 0.00043 | 0.0016 |
| Ethanol | 46.07 | 0.49 | 0.0008 | 0.0036 |
| 2-Ethoxyethanol | 90.12 | 0.29 | 0.0003 | 0.0013 |
| Ethylbenzene | 106.16 | 1412.54 | 1.215[b] | 2.65 |
| Ethylether | 74.12 | 6.76 | 0.016 | 0.022 |
| *p*-Ethylphenol | 122.17 | 549.54 | 0.035 | 0.79 |
| Heptanol | 116.20 | 257.04 | 0.038 | 0.41 |
| Hexanol | 102.17 | 107.15 | 0.028 | 0.21 |
| Methanol | 32.04 | 0.17 | 0.0016 | 0.0026 |
| Methyl hydroxybenzoate | 152.15 | 91.20 | 0.0091 | 0.082 |
| β-Naphthol | 144.16 | 691.83 | 0.028 | 0.7 |
| 3-Nitrophenol | 139.11 | 100.00 | 0.0056 | 0.11 |
| 4-Nitrophenol | 139.11 | 81.28 | 0.0056 | 0.09 |
| Nitrosodiethanolamine | 134.13 | 0.13 | 0.0000055 | 0.0005 |
| Nonanol | 144.26 | 2951.21 | 0.06 | 2.99 |
| Octanol | 130.22 | 933.25 | 0.061 | 1.19 |
| Pentanol | 88.15 | 36.31 | 0.006 | 0.091 |
| Phenol | 94.11 | 32.36 | 0.0082 | 0.074 |
| Propanol | 60.09 | 2.00 | 0.0017 | 0.0088 |
| Resorcinol | 110.11 | 6.03 | 0.00024 | 0.011 |
| Styrene | 104.14 | 891.25 | 0.635[b] | 1.72 |
| Thymol | 150.21 | 1995.26 | 0.053 | 1.84 |
| Toluene | 92.13 | 489.78 | 1.01 | 1.15 |
| 2,4,6-Trichlorophenol | 197.46 | 2344.23 | 0.059 | 1.02 |
| 3,4-Xylenol | 122.16 | 169.82 | 0.036 | 0.25 |
| Inorganic chemicals | | | | |
| Cobalt chloride | 129.84 | — | 0.0004 | — |
| Lead acetate | 325.29 | — | 0.0000042[b] | — |
| Mercuric chloride | 271.50 | — | 0.00093 | — |
| Nickel chloride | 129.60 | — | 0.001 | — |
| Nickel sulfate | 154.75 | — | <0.000009 | — |
| Silver nitrate | 169.87 | — | <0.00035[b] | — |
| Sodium chromate | 161.97 | — | 0.0021[b] | — |

[a] Permeability coefficients were calculated using equations presented in Leung and Paustenbach [192].

[b] All of the observed permeability coefficients were obtained by using *in vitro* techniques except those denoted with the superscript "b," which were determined *in vivo*.

## Pharmacokinetic Models for Estimating the Uptake of Chemicals in Aqueous Solution

Pharmacokinetic models predict the uptake of a chemical through the skin based on fundamental thermodynamics. Several different models have been proposed. For example, a four-compartment pharmacokinetic model was developed in 1989 [260]. This model, which uses first-order rate constants, describes chemical movement through the compartments representing the various skin structures. It has been used successfully to predict the chemical disposition in the skin and plasma as a function of their physicochemical properties; when an input rate constant to the skin surface is added to the model, it can be used to assess vehicle effects. A similar model that treats the barrier membrane as a series of spaces filled with immiscible liquids has also been developed [258]; its advantage is that it allows examination of non-steady-state conditions where Fick's law does not apply.

Under an infinite-dose situation where the amount of a chemical lost by penetration is too small to alter the applied concentration (e.g., where one is swimming), the rate of absorption is essentially linear once steady state has been reached. In the finite-dose system, however, the chemical solution is applied as a thin film and the concentration decreases as penetration proceeds (e.g., a splash). All other model parameters being the same, penetration is reduced under finite-dose conditions. This is because the chemical concentration is continuously reduced over time, resulting in a decrease in the gradient across the stratum corneum. These modeling results indicate that the mechanism by which fluxes are affected must be considered when extrapolating to non-steady-state conditions.

Although classic pharmacokinetic modeling like that described by Guy et al. [257] can provide a good mathematical description of the disposition of chemicals, it does not depict exactly the biological processes in the intact animal. Fortunately, due to recent improvements in available computer hardware and software, pharmacokinetic methods based on physiological principles are now feasible alternatives for analysis of *in vivo* skin penetration studies. These so-called PB-PK models realistically describe the disposition of the chemical in the intact animal in terms of rates of blood flow, permeability of membranes, and partitioning of chemicals into tissues [258,261]. Characterizing dermal absorption in terms of actual anatomical, physiological, and biochemical parameters facilitates extrapolations to the real species of interest, humans.

In 1991, a PB-PK model was developed to describe percutaneous absorption of volatile organic contaminants in dilute aqueous solutions [262]. The exposure scenario modeled was either hand or full-body immersion into a vessel of solute-contaminated water. Modeling results suggested that chemical uptake in aqueous solutions is most markedly influenced by epidermal blood flow rates, followed by epidermal thickness and lipid content of the stratum corneum. In general, thicker and fattier skin provides a better barrier to dermal penetration of chemicals. These are precisely the principles under which barrier creams offer their protection for increasing the effective thickness and lipophilicity of the skin. This model also predicted that the dose of some volatile organic chemicals in water absorbed through the skin during a 20-minute bath may be equivalent to the amount inhaled [262].

Among the most complex and best validated of the various models for dermal uptake of liquids is that developed by McDougal et al. [98,263–276]. These authors have successfully predicted dermal uptake rates for humans for more than a dozen chemicals based on animal data. One advantage of dermal PB-PK models over traditional *in vivo* methods is their ability to accurately describe nonlinear biochemical and physical processes. For example, describing skin penetration based on blood concentrations or excretion rates as percent absorbed assumes that all processes have a simple linear relationship with the exposure concentration. This is often not the case. The kinetics become nonlinear when the absorption, distribution, metabolism, and excretion (ADME) of a chemical is saturated at high exposure concentrations. This model and other models developed since then are generally reliable methods for estimating dermal uptake for certain classes of chemicals.

## Factors Used To Estimate Dermal Uptake

Many factors need to be quantitatively accounted for to estimate the likely systemic uptake of a chemical that comes into contact with the skin, either as a liquid or when present in soil or dust [7,277].

### Dermal Bioavailability

The typical media of concern for assessing cutaneous contact to environmental chemicals, in contrast with occupational exposure, are house dust, soil, fly ash, and sediment. In the workplace, dermal uptake is due to direct contact with liquids and contact with surfaces contaminated with dirt or liquids. A number of parameters can influence the degree of cutaneous bioavailability of chemicals in complex matrices. These may include aging (time following contamination), soil type (e.g., silt, clay, sand), type and concentration of co-contaminants (e.g., oil and other organics), and the concentration of the chemical contaminant in the media [123]. The bioavailability of a chemical in soil will usually be affected by its physicochemical properties. Large molecular weight chemicals tend to bind to soil or dust and be less water soluble, while smaller molecules will frequently be water soluble, less tightly bound, and relatively bioavailable [99,119]. The cutaneous bioavailability of perhaps 20 to 30 chemicals in soils has been determined in animals [50,104,113,116, 118–120,123,278–281]. These studies show that different media and different chemicals can yield dramatically different cutaneous bioavailabilities. The results of these studies, for example, produce values of bioavailability for different chemicals that range from 0.001 to 3% for chemicals in soil. Dioxin and lead are considered classic case studies because many human populations have been exposed to these chemicals in dust or soil yet the blood levels can be very high or very low depending on the bioavailability.

### Skin Surface Area

An abundance of information is available regarding the surface area of different portions of the body. One simple approach is to use the "rule of nines" for estimating the surface area of the human body [282]: The head and neck are 9%, upper limbs are each 9%, lower limbs are each 18%, and the front or back of the trunk is 18% [69]. The USEPA has estimated an exposed surface area (arms, hands, legs, and feet) of 2900 $cm^2$ for children 0 to 2 years old; 3400 $cm^2$ for children 2 to 6 years old; and 2940 $cm^2$ for adults (an adult is assumed to wear pants, an open-neck short-sleeved shirt, shoes, and no hat or gloves) [42]. When assessing chemical exposure in the ambient environment, most of the necessary surface area information can be found in the USEPA's *Exposure Factors Handbook* and *Child-Specific Exposure Factors Handbook* [63,64]. Table 10.5 presents the skin surface areas commonly used when conducting exposure assessments [282]. A distribution plot of skin area vs. body weight has been developed [283].

### Soil Loading on the Skin

A key factor to consider when estimating dermal uptake is the soil-to-skin adherence rate. Values of 0.5 to 0.6 mg/$cm^2$ and 0.2 to 2.8 mg/$cm^2$ have been reported for adults and children, respectively, although it is important to carefully consider site-specific information when conducting these assessments [52,237,243,284]. Work by Finley et al. [238], Kissel et al. [241], and Holmes et al. [239] has built on prior studies to show that dermal loading can vary significantly among different activities and different people. Based on data collected in past studies, in 1992 the USEPA suggested a default soil-to-skin adherence rate of 0.2 mg/$cm^2$ (median) and 1.0 mg/$cm^2$ (95th percentile) for an adult. The *Exposure Factors Handbook* and *Child-Specific Exposure Factors Handbook* give considerable attention to this topic [63,64]. One approach to improving dermal uptake calculations is to use area-weighted adherence factors as recently suggested by the USEPA. When assessing the risk to large populations, Monte Carlo techniques are often useful in characterizing distributions of exposure.

**TABLE 10.5**
**Representative Surface Areas of the Human Body (Adult Male)**

| Body Portion | Area ($cm^2$) |
|---|---|
| Whole body | 18,000 |
| Head and neck | 1620 |
| Head | 1260 |
| Back of head | 320 |
| Neck | 360 |
| Back of neck | 90 |
| Torso | 6480 |
| Back | 2520 |
| Chest | 2520 |
| Sides | 1440 |
| Upper limbs | 3240 |
| Upper arms (elbow–shoulder) | 1440 |
| Lower arms (elbow–wrist) | 1080 |
| Hands | 720 |
| Palms | 360 |
| Upper arms (back of) | 360 |
| Lower arms (back of) | 270 |
| Lower limbs | 6480 |
| Thighs | 3240 |
| Lower legs (knee–ankle) | 2160 |
| Feet | 1080 |
| Soles of feet | 540 |
| Thighs (back of) | 810 |
| Lower legs (back of) | 540 |
| Perineum | 180 |

### Interpreting Wipe Samples

In some workplaces, wipe sampling has been conducted historically to assess the degree of surface contamination. Hospitals were among the first occupational settings, as long ago as 1940, to rely on this method to determine microbial levels in operating rooms. In pharmaceutical manufacturing, wipe sampling has been used as an indicator of hygienic conditions since the 1960s. The health physics profession has utilized wipe samples extensively as an indicator of the need for better housekeeping and decontamination; this group performed most of the early work in quantifying the relationship of wipe sample concentrations to dermal and oral uptake.

Over the years, few papers have discussed how to collect and interpret wipe samples [241,285–289]. Beginning in 1995, greater attention has been paid to the risk posed by contaminated house dust, and more study of this topic has been conducted. When the primary effect of a chemical is skin discoloration, allergic contact dermatitis, or chloracne, wipe sampling is nearly always the preferred approach for assessing the acceptability of the workplace (rather than relying on air samples). Beginning in the 1980s, a substantial number of wipe samples were collected in office buildings contaminated with dioxins and furans after electrical transformer fires to estimate the potential human exposure [290]. This approach was recently used to assess exposures to dust generated from the September 11, 2001, terrorist attacks [251,291–293]. No doubt, the two studies that will be considered landmark in their importance will be the ones conducted of dioxin contaminated sites: one in Midland, Michigan, and one in New Zealand. The Midland assessment is being conducted by the University of Michigan and, due to the magnitude of the study and the rigor of the design, it should provide a wealth of information regarding the relationships among outdoor soil, indoor dust, and blood levels (by dioxin congener).

Although wipe sampling data have generally been used as an indicator of cleanliness [285], these data can also be used to estimate systemic uptake of a contaminant if the degree of skin contact with the contaminated surfaces and the bioavailability of the chemical (in that media) are known. Although historical wipe sampling methods were rather imprecise, they were useful for obtaining a rough estimate of the possible exposure, which could be refined later by other means, such as biological monitoring.

If one knows that wipe sampling results are representative of what comes into contact with the hands (i.e., actually able to be absorbed), then the procedures for converting wipe sample data to estimates of systemic uptake are straightforward [294]. For example, if one knows the number of times a surface (e.g., valve handle, instrument controller, or drum) is touched, the surface area of the hand touching these items (usually the palm), and the percutaneous chemical absorption rate, then the uptake can be estimated using wipe sample information. The best wipe sampling data are those collected in a reliable and consistent manner, with a focus on the mass per unit area. Hand wash or hand wipe sampling is often more representative of the true hazard than surface wipe samples [295]. The bulk of what is known about how to best evaluate this matter comes from the relatively robust studies of agricultural workers that has been conducted over the past 20 years by hygienists who have specialized in this area (e.g., Kissel, Poppendorf, Spears, Knarr, Knack, and Fenske).

Until recently, no standardized approaches existed for conducting wipe sampling. Differences in the use of wetting agent (acetone, methylene chloride, water, saline, isopropanol, and ethanol) and sampling media (paper, cotton, and synthetic fibers) produced drastically different results. In some procedures, especially those that used methylene chloride (in which the paint was concurrently stripped by the solvent), the chemical in the paint matrix was assumed to be bioavailable (a completely unreasonable assumption). Clearly, much of the previous work, which measured the amount of chemical released following aggressive scrubbing of the contaminated surfaces with detergent or solvent, did not reflect a realistic exposure scenario. Thus, there has been a need for standard techniques that attempt to mimic the conditions in which a hand comes into contact with a contaminated surface [289]. Some of the techniques have been developed by hygienists involved in agricultural exposure assessment [169,296–298].

In an attempt to fill this need, fairly sophisticated work to standardize these procedures has been conducted by researchers at Rutgers University. A couple of their wipe sampling procedures and devices have been patented [252]. They have also developed a dry contact sampling device that offers promise for determining the hazard from surface dusts [253]. The implications from recent wipe sampling research are that: (1) a minimum number of samples is needed to have statistical confidence; (2) the pressure applied to the cloth during sample collection should be standardized; (3) neat solvent should not be used as a collection media; (4) the size of the sample area must be sufficient to collect enough contaminant for quantification; and (5) the technique should be validated by using glove analyses. More research will be devoted to this topic in the coming years as it has become quite clear that in-home vs. outdoor exposures are generally much better predictors of risk because such a large fraction of the day is spent indoors (and because rugs, upholstery, drapes, toys, and other objects tend to accumulate toxicants over time) [299–301].

## Estimating the Dermal Uptake of Chemicals in Soil

One of the most frequently occurring exposure scenarios involving environmental exposures is that of contaminated soil [52]. Unfortunately, dermal uptake of chemicals found on soil has infrequently been evaluated experimentally [89]. A model to estimate the amount of a chemical in soil that crosses the stratum corneum into the underlying tissue layer has been developed [99]. To differentiate this absorptive process from bioavailability, which also includes transport into blood, McKone [99] refers to the percentage of available chemical as an *uptake fraction*. The approach is based on the fugacity concept, which measures the tendency of a chemical to move from one phase to another. Because the skin has a fat content of about 10% and soil has an organic carbon content of 1 to 4%, a chemical in soil placed on the skin will move from the soil to the underlying adipose layers of the skin. However, this transfer depends on the period of time between deposition on the skin and removal by evaporative processes. The mass-transfer coefficients of the soil-to-skin layer and the soil-to-air layer define the rate at which these competing processes occur.

Results of this model suggest that the chemical uptake fraction in soil varies with the exposure duration, soil deposition rate, and physical properties of the chemical and is particularly sensitive to the values of the $K_{ow}$, as well as the mass or depth of soil deposited on the skin. When the amount of soil on the skin is low (<1 mg/cm$^2$), a high uptake fraction, approaching unity in some cases, is predicted. With higher soil loading (20 mg/cm$^2$), an uptake of only 0.5% is predicted. Because of the diverse variations of the uptake fraction with soil loading, results obtained from experiments with a single soil loading should be applied with caution to human soil-exposure scenarios.

The dermal uptake of chemicals in soil is a complex process, but its behavior is predictable if the controlling factors are accounted for and quantified [89,99]. In situations involving a relatively thin layer of a chemical on the skin, a few generalizations can be made. First, for chemicals with a high $K_{ow}$ and a low air–water partition coefficient, it is reasonable to assume 100% uptake in 12 hours. Second, for chemicals with an air–water partition coefficient greater than 0.01, the uptake fraction is unlikely to exceed 40% in 12 hours. Third, for chemicals with an air–water partition coefficient greater than 0.1, one can expect less than 3% uptake in 12 hours. In most occupational settings, contaminated soil will rarely be in contact with the skin for greater than 4 hours before it is washed off; consequently, this should be accounted for when attempting to predict systemic uptake.

## Dermal Uptake of Contaminants in Soil

To estimate chemical uptake, one needs to know the percutaneous absorption rate, the exposed skin area, the chemical concentration, and the exposure duration [302]. One scenario would be a thin film of chemical on the skin. For this finite-dose scenario, Equation 10.4 is useful:

$$\text{Uptake (mg)} = (C)(A)(x)(f)(t) \qquad (10.4)$$

where $C$ is the concentration of the chemical (mg/cm$^2$), $A$ is the skin surface area (cm$^2$), $x$ is the thickness of the film layer (cm), $f$ is the absorption rate (percent per hour), and $t$ is the duration of exposure (hour).

Another scenario would be an excess amount of a chemical on the skin (i.e., infinite dose). In this case, the thickness of the chemical layer is not calculated and steady-state kinetics are assumed. For a chemical in an aqueous or gaseous media:

$$\text{Uptake (mg)} = (C)(A)(K_p)(t)(d) \qquad (10.5)$$

where $d$ is the distribution factor, and $K_p$ is the permeability coefficient (cm/hr).

For a neat liquid chemical:

$$\text{Uptake (mg)} = (A)(J)(t) \qquad (10.6)$$

where $J$ is the flux of chemical (mg/cm$^2$/hr). The USEPA has suggested using the following equation for estimating percutaneous absorption of chemicals in soil [277]:

$$\text{Uptake (mg)} = (C)(A)(r)(B) \qquad (10.7)$$

where $C$ is the concentration of the chemical in soil (mg/g), $A$ is the skin surface area (cm$^2$), $r$ is the soil-to-skin adherence rate (g/cm$^2$), and $B$ is the cutaneous bioavailability (unitless).

### Example Calculation 3: Skin Uptake of a Chemical in Soil

A person gardens with soil contaminated on average with 250 ng dioxin per gram of soil (250 ppb). Assuming that the person's hands and lower arms are in contact with the soil, the soil loading is equal to 0.2 mg/cm$^2$, and the cutaneous bioavailability of dioxin in soil is 1% [123], what is the plausible uptake of dioxin by this person (using Equation 10.7)? Assume that the person washes his or her hands every 4 hours and the exposed area of skin is 1800 cm$^2$.

$$\text{Uptake (ng)} = (C)(A)(r)(B)$$

where:

$C$ = 250 ng/g
$A$ = 1,800 cm$^2$
$r$ = 0.2 mg/cm$^2$
$B$ = 0.01

By substitution:

$$\text{Uptake} = \left(\frac{250\ \text{ng TCDD}}{\text{g soil}}\right)\left(\frac{0.2\ \text{mg soil}}{\text{cm}^2\ \text{skin}}\right)\left(\frac{1\ \text{g}}{10^3\ \text{mg}}\right) \times (1800\ \text{cm}^2\ \text{skin})(0.01) = 0.9\ \text{ng TCDD}$$

*Note:* A preferred method for performing this calculation, if data are available, is to use a flux rate (ng/cm$^2$/hr) for the chemical. Assume that the rate is 500 ng/cm$^2$/hr:

$$\text{Uptake (ng)} = (C)(J)(A)(t)$$

where:

$J$ = 500 ng/cm$^2$/hr
$t$ = 4 hr

By substitution:

$$\text{Uptake} = \left(\frac{250\ \text{ng TCDD}}{1\ \text{g soil}}\right)\left(\frac{1\ \text{g}}{10^9\ \text{mg}}\right)(1800\ \text{cm}^2\ \text{skin}) \times (4\ \text{hr})\left(\frac{500\ \text{ng soil}}{\text{cm}^2-\text{hr}}\right) = 0.9\ \text{ng TCDD}$$

### Uptake of Chemicals in an Aqueous Matrix

Published estimates of dermal uptake of chemicals in water have generally focused on evaluating workplace or environmental exposure. A number of different scenarios has been evaluated [75,303–307]; for example, with regard to the possible uptake of a chemical present in water, the amount of chloroform absorbed through the skin by a man showering for 4 hours in water containing 1 ppb chloroform has been estimated [304,305]. About 10 years ago, it was recognized that, in the indoor environment, dermal exposure to volatile chemicals present in drinking water will rarely represent the vast majority of the hazard. Specifically, it was found that inhalation exposure due to the release of vapors from liquids to which people were in close contact was a greater contributor to the dose [308]; for example, comparisons have been made of the chloroform concentration in exhaled breath after a shower to that after an inhalation-only exposure [304,305].

### Example Calculation 4: Skin Uptake of a Chemical From Water

A person has filled his swimming pool with shallow well water contaminated with 0.002 mg/mL (2 ppb) toluene. What is the plausible dermal uptake of toluene while swimming in the contaminated water for half an hour? Assume that 18,000 cm² of skin are exposed and the $K_p$ is 1.01 cm/hr. From Equation 10.5:

$$\text{Uptake} = (C)(A)(K_p)(t)(d)$$

where:

$C$ = 0.002 mg/mL
$A$ = 18,000 cm²
$K_p$ = 1.01 cm/hr
$t$ = 0.5 hr
$d$ = distribution factor (1 mL of water covers 1 cm³)

By substitution:

$$\text{Uptake} = (0.002 \text{ mg/mL})(18{,}000 \text{ cm}^2)(1.01 \text{ cm/hr}) \times (0.5 \text{ hr})(1 \text{ mL water per } 1 \text{ cm}^3) = 18 \text{ mg}$$

### Percutaneous Absorption of Liquid Solvents

Whereas the percutaneous absorption of chemical solutes generally proceeds by simple diffusion, the skin uptake of neat chemical liquids is not necessarily exclusively governed by Fick's law. Consequently, the uptake of neat liquid through the skin needs to be estimated using direct *in vivo* skin contact techniques. Table 10.6 presents the percutaneous absorption rates of some neat industrial liquid solvents that have been determined in human volunteer studies.

**TABLE 10.6**
**Absorption Rates of Some Neat Industrial Liquid Chemicals in Human Skin *In Vivo***

| Chemical (mg/cm²/hr) | Absorption Rate |
|---|---|
| Aniline | 0.2–0.7 |
| Benzene | 0.24–0.4 |
| 2-Butoxyethanol | 0.05–0.68 |
| 2-(2-Butoxyethoxy)ethanol | 0.035 |
| Carbon disulfide | 9.7 |
| Dimethylformamide | 9.4 |
| Ethylbenzene | 22–23 |
| 2-Ethoxyethanol | 0.796 |
| 2-(2-Ethyoxyethoxy)ethanol | 0.125 |
| Methanol | 11.5 |
| 2-Methoxyethanol | 2.82 |
| 2-(2-Methoxyethoxy)ethanol | 0.206 |
| Methyl butyl ketone | 0.25–0.48 |
| Nitrobenzene | 2 |
| Styrene | 9–15 |
| Toluene | 14–23 |
| Xylene (mixed) | 4.5–9.6 |
| *m*-Xylene | 0.12–0.15 |

### Example Calculation 5: Skin Uptake of a Neat Liquid Chemical

Due to carelessness or a leak, the inside of a glove becomes contaminated with 2-methoxyethanol. How much can be absorbed if a worker wears the contaminated glove on one hand for half an hour? Assume the surface area of exposed skin is 360 cm² and the flux rate is 2.82 mg/cm²/hr. From Equation 10.6:

$$\text{Uptake} = (A)(J)(t)$$

where:

$A$ = 360 cm²
$J$ = 2.82 mg/cm²/hr
$t$ = 0.5 hr

By substitution:

$$\text{Uptake} = (360 \text{ cm}^2)(2.82 \text{ mg/cm}^2\text{/hr})(0.5 \text{ hr}) = 508 \text{ mg}$$

To understand the relative hazard from skin exposure vs. inhalation exposure, the quantity of 2-methoxyethanol absorbed by the same worker via inhalation for 8 hours

**TABLE 10.7**
**Percutaneous Absorption Rates for Chemical Vapors *In Vivo***

| Chemical | Skin Update in Combined Exposure (%)[a] | Permeability Coefficient $K_p$ (cm/hr) | |
|---|---|---|---|
| | | Rat | Human |
| Styrene | 9.4 | 1.75 | 0.35–1.42 |
| *m*-Xylene | 3.9 | 0.72 | 0.24–0.26 |
| Toluene | 3.7 | 0.72 | 0.18 |
| Perchloroethylene | 3.5 | 0.67 | 0.17 |
| Benzene | 0.8 | 0.15 | 0.08 |
| Halothane | 0.2 | 0.05 | — |
| Hexane | 0.1 | 0.03 | — |
| Isoflurane | 0.1 | 0.03 | — |
| Methylene chloride | — | 0.28 | — |
| Dibromomethane | — | 1.32 | — |
| Bromochloromethane | — | 0.79 | — |
| Phenol | — | — | 15.74–17.59 |
| Nitrobenzene | — | — | 11.1 |
| 1,1,1-Trichloroethane | — | — | 0.01 |

[a] In combined exposure, rats are simultaneously absorbing chemical vapors by inhalation and by whole-body absorption through the skin.

(10 m³ of air inhaled), assuming a Threshold Limit Value (TLV®) of 16 mg/cm³, can be estimated and compared to the dose due to inhalation exposure. Assume an 80% inhalation uptake efficiency:

$$\text{Inhalation uptake} = (16\ \text{mg/m}^3)(10\ \text{m}^3)(0.8) = 128\ \text{mg}$$

Thus, the uptake of 2-methoxyethanol following 30 minutes of skin exposure of a single hand can be as much as 4 times that from inhalation for 8 hours at the TLV concentration, a presumably safe level of exposure. From this example, it is clear that the cutaneous route of entry can, in some situations, significantly contribute to the total absorbed dose, especially in the occupational setting.

## Percutaneous Absorption of Chemicals in the Vapor Phase

Until the 1990s, it was generally assumed that the plausible dose resulting from vapors absorbed through the skin was too low to pose a hazard. Only a few studies have examined this issue [226,254,306]. A few clinical reports have encouraged some limited *in vivo* research to evaluate the absorption of several chemicals in the gaseous phase through the human skin (Table 10.7). A chamber system to measure the whole-body percutaneous absorption of chemical vapors in rats has been described by McDougal et al. [98], which has produced some interesting results [268]. In this system, chemical flux across the skin is determined from the chemical concentration in blood during exposure by using a PB-PK model. In most cases, vapor absorption through the skin amounts to less than 10% of the total dose received from a combined skin and inhalation exposure. Despite the good agreement between the rat and human in the relative ranking of the permeability coefficients among the chemicals studied, the rat skin appears to be two to four times more permeable than the human skin. These observations are consistent with previously reported data [123,126,228,233,235].

It is generally not necessary to account for the contribution from percutaneous uptake of vapors when the observed effect level is used as a guideline for acceptable exposure, because uptake via this route is usually inherent in the data; that is, the studies of animals or humans from which data were collected were usually exposed via inhalation (whole body), so dermal uptake of the vapor occurred. Although good work practices and the law prohibit situations where persons are placed in atmospheres that are life-threatening were it not for a supplied air respirator, sometimes in emergency situations airline (supplied air) respirators or self-contained breathing apparatus (SCBA) are worn in environments containing chemical concentrations 10-fold to 1000-fold greater than the TLV. In these cases, it is can be useful to account for vapor uptake through either exposed or covered skin.

Although nearly all data on vapor absorption involve bare skin, the role of clothing in preventing skin uptake has occasionally been evaluated. For example, a study of workers wearing denim clothing indicated no decreased uptake of phenol vapors [228] but found a 20% and 40% reduction in uptake of nitrobenzene and aniline vapor respectively [125,223]. Although standard clothing may slightly decrease the amount of a chemical transferred from air through the skin, it can be a significant source of continuous exposure if the clothing has been contaminated.

### Example Calculation 6: Skin Uptake of a Chemical Vapor

Assume that a person needs to repair a leaking pump, so he enters a room wearing an airline respirator. Assume the room contains 500 mg/m$^3$ nitrobenzene (100 times the current TLV) and it takes 30 minutes to repair the pump. How much nitrobenzene might be absorbed through the skin? The head, neck, and upper limbs are assumed to be exposed (surface area = 4,860 cm$^2$), and the rest of the body (surface area = 13,140 cm$^2$) is covered with clothing. Assume the percutaneous $K_p$ of nitrobenzene is 11.1 cm/hr, and that the clothing has reduced the skin uptake rate of vapors by about 20% [125]:

$$\text{Uptake} = (C)(A)\left(K_p\right)(t)$$

$$\begin{aligned}\text{Uptake through exposed skin} &= \left(500\ \text{mg/m}^3\right)\left(4860\ \text{cm}^2\right)\\ &\times (11.1\ \text{cm/hr})(0.5\ \text{hr})\\ &\times \left(1\ \text{m}^3/106\ \text{cm}^3\right)\\ &= 13.5\ \text{mg}\end{aligned}$$

$$\begin{aligned}\text{Uptake through clothing} &= \left(500\ \text{mg/m}^3\right)\left(13{,}140\ \text{cm}^2\right)\\ &\times (11.1\ \text{cm/hr})(0.8)\\ &\times (0.5\ \text{hr})\left(1\ \text{m}^3/106\ \text{cm}^3\right)\\ &= 29\ \text{mg}\end{aligned}$$

$$\text{Total uptake} = 13.5 + 29 = 42.5\ \text{mg}$$

From this example, it is clear that if one enters an environment containing a high concentration of an airborne contaminant, even if a supplied-air respirator is worn, the degree of skin uptake of the vapor may be worthy of evaluation to ensure that the worker is protected. In this example, uptake following one day of inhalation exposure at the TLV (5 mg/m$^3$) results in a 50-mg uptake (10 m$^3$ × 5 mg/m$^3$). These kinds of calculations sometimes have to be conducted in difficult work environments that are in a state of alert (e.g., submarines, chemical plants during emergency situations).

## ESTIMATING INTAKE VIA INGESTION

If the appropriate information is available, estimating the intake of various chemicals due to ingestion is a relatively straightforward exercise. In general, one is concerned with the ingestion of the following media: drinking water, other liquids, food, soil, and house dust. Drinking water contamination may occur because of soil contamination from leaking underground storage tanks, landfills, or hazardous waste sites, as well as discharges from contaminated streams or water transport systems. Nearly all foods in Western society contain a number of intentional and unintentional chemicals, including pesticide residues, naturally occurring chemicals, and food additives that serve as preservatives or enhancers of taste or visual appeal. Soils are ingested as a result of eating incompletely washed vegetables, hand-to-mouth contact, and through direct ingestion by children. Soils are also ingested when particles too large to reach the lower respiratory tract are inhaled (and then are swallowed). House dust contaminated with a number of chemicals can be ingested due to contact with foods, toys, upholstery, carpet, and hand-to-mouth activities [197,316].

### Estimating Intake of Chemicals in Drinking Water

Estimating the magnitude of the potential dose of toxics from drinking water requires knowledge of the amount of water ingested, the chemical concentrations in the water, and the chemical bioavailability in the gastrointestinal tract. The amount of water ingested per day varies with each person and is usually related to the amount of physical activity. A good deal of literature has addressed the amount of water ingested by persons engaged in different kinds of activities [63,209,210].

Currently, the USEPA suggests that, when little is known about the specifics of exposure, a value of 2 L/day for adults and 1 L/day for infants (body weight of less than 10 kg) should be used as the default value. These rates include drinking water consumed in the form of juices and other beverages. Numerous studies cited in the USEPA's *Exposure Factors Handbook* and *Child-Specific Exposure Factors Handbook* have generated data on drinking water intake rates [61,62,64]. In general, these sources support the USEPA's use of 2 L/day for adults and 1 L/day for children as upper-percentile tap-water intake rates. Many of the studies have reported fluid intake rates for both total fluids and tap water. Total fluid intake is defined as consumption of all types of fluids including tap water, milk, soft drinks, alcoholic beverages, and water intrinsic to purchased foods. Total tap water is defined as water consumed directly from the tap as a beverage or used to prepare foods and beverages (e.g., coffee, tea, frozen juices, soups). Data for both consumption categories are presented in numerous publications. Table 10.8 presents typical information reported from these studies [42,64].

All currently available studies on drinking water intake are based on short-term survey data. Although short-term data may be suitable for obtaining mean intake values that are representative of both short- and long-term consumption patterns, upper-percentile values may be different for short-term and long-term data because there is

**TABLE 10.8**
**Summary of Tap Water Intake by Age**

| | Intake (mL/d) | | Intake (mL/kg/d) | |
|---|---|---|---|---|
| **Age Group** | **Mean** | **10th–90th Percentiles** | **Mean** | **10th–90th Percentiles** |
| Infants (<1 yr) | 302 | 0–649 | 43.5 | 0–100 |
| Children (1–10 yr) | 736 | 286–1,294 | 35.5 | 12.5–64.4 |
| Teens (11–19 yr) | 965 | 353–1,701 | 18.2 | 6.5–32.3 |
| Adults (20–64 yr) | 1366 | 559–2,268 | 19.9 | 8.0–33.7 |
| Adults (65+ yr) | 1459 | 751–2,287 | 21.8 | 10.9–34.7 |
| All ages | 1193 | 423–2,092 | 22.6 | 8.2–39.8 |

generally more variability in short-term surveys. Most of the currently available drinking water surveys are based on recall, which may be a source of uncertainty in the estimated intake rates because of the subjective nature of this type of survey technique [42,63,64]. However, recently researchers have looked for ways to better characterize exposures that persons might have experienced as a result of ingestion of contaminated drinking water. These efforts have included the use of geographic information system (GIS) software to integrate fate and transport modeling of chemicals in groundwater with the geocoded study population, develop input data for the simulation model, and assign individual exposures to chemicals of interest by linking results of the model to the census block group of residence [174,317].

To estimate the intake of toxics via direct ingestion of drinking water, the calculation is straightforward:

$$\text{Intake} = (V)(C)(B)$$

where $V$ is the volume of water (L/day), $C$ is the concentration of chemical in water (µg/L), and $B$ is the bioavailability (unitless).

One of the more interesting observations of the past 20 years is that ingestion of contaminated drinking water is sometimes not the primary route of exposure to the toxicant in drinking water. Uptake of volatile chemicals via inhalation can be nearly as great in some homes as ingestion, which is the result of the presence of these chemicals in air due to showering, off-gases from the dishwasher, and other opportunities for volatilization of the chemical [254,268,304–306,308].

## Importance of Soil Ingestion When Estimating Human Exposure

Between 1980 and 1995, predicted risks associated with the ingestion of contaminated soil were the primary drivers for remediating many (if not most) hazardous waste sites. As discussed by Paustenbach et al. [52], there was no better example than the site in Times Beach, Missouri. Billions of dollars can be required to clean up these kinds of sites to levels that would not pose a significant risk if children actually ate significant quantities of contaminated soil. Because of the expense of remediation, a good deal of research has been conducted over the past 20 years to attempt to quantitatively understand this route of exposure.

Clearly, the ingestion of soil and house dust is a potential source of human exposure to toxicants [16,318,319]. The potential for contaminant exposure via this source is greater for children because they are more likely to ingest greater quantities of soil than adults. Inadvertent soil ingestion among children may occur through the mouthing of objects or hands [64]. Mouthing behavior is considered to be a normal phase of childhood development. Adults may also ingest soil or dust particles that adhere to food, cigarettes, or their hands. Deliberate soil ingestion is defined as pica and is considered to be relatively uncommon [64]. Because normal, inadvertent soil ingestion is more prevalent and data for individuals with pica behavior are limited, the focus of most exposure assessments is on normal levels of soil ingestion that occur as a result of mouthing or unintentional hand-to-mouth activity [42,52,71,320].

Mouthing activities by children are generally accepted as normal and commonplace; for example, Barltrop [321] estimated that almost 80% of all children at age 1 year exhibited mouthing tendencies. Such activities are potential exposure routes to trace amounts of soil and/or dust adhering to fingers, hands, and objects placed in the mouth. The available data indicate that soil exposure occurs through several indirect routes:

- Soil contributes to house dust (e.g., by local dust deposition, by mud and dirt carried in by shoes and pets).
- House dust (fine particles) adheres to objects and to children's hands.
- Children ingest dust particles when sucking and mouthing objects and fingers.

Obviously in some situations, exposure may be direct (a :hild playing outdoors may eat dirt directly). In other sit-uations, oral exposure may occur via contamination of domestic water supplies or contamination of vegetable pro-duce grown onsite. However, the content and concentration of dusts in the indoor environment, which may represent he most important source of indirect exposure to soil, must e better understood [197,208]. Considerable efforts have een made recently through large-scale studies, such as the National Children's Study, NHEXAS, and MNCPES, to etter characterize indoor exposures to chemicals adhered o household dust, particularly pesticide, lead, and allergen exposures to children [152,322–324]. Several researchers ave concluded that the hazards posed by the majority of ousehold pesticides are better detected by dust sampling han by air sampling [316,319,325–327]. Studies designed o characterize children's exposure to pesticides indicate hat the largest number of pesticides and the highest con-centrations are found in household dust compared with air, oil, and food [328,329]. Other recent research efforts have nvolved better characterization of personal activities using uestionnaires, videotaping, wireless-coupled infrared echnologies, and personal digital assistants (PDAs) to uantify dermal and ingestion exposures of microactivities, uch as hand-to-mouth and object-to-mouth activities 155,157,160,330,331,365].

Many studies have been conducted to estimate the mount of soil ingested by children. Most of the early tudies attempted to quantify the amount of soil ingested y measuring the amount of dirt present on children's ands and making generalizations based on behavior. More recently, soil intake studies have been conducted using a methodology that measures trace elements in eces. These measurements are used to estimate the mount of soil ingested over a specified period of time.

## Studies of Soil Ingestion

In light of the importance of soil ingestion for estimating uman exposure to contaminated soil, several literature urveys have been undertaken to identify the typical mount of soil consumed by children and adults 42,44,52,208,320]. Research evaluating lead uptake by hildren from ingestion of contaminated soil, paint chips, ust, and plaster provides the best source of information. Walter et al. [332] estimated that a normal child typically ngests very small quantities of dust or dirt between the ges of 0 to 2 years, the largest quantities between 2 and years, and nearly insignificant amounts thereafter. In the lassic text by Cooper [333], it was noted that the desire f children to eat dirt or place inedible objects in their mouths becomes established in the second year of life and isappears more or less spontaneously by the age of 4 to years. A study by Charney et al. [334] also indicated hat mouthing tends to begin at about 18 months and continues through 72 months, depending on several fac-tors such as nutritional and economic status, as well as race. Work by Sayre et al. [335] indicated that ages 2 to 6 years are the important years, but that "intensive mouth-ing diminishes after 2 to 3 years of age."

An important distinction that is often blurred is the difference between the ingestion of very small quantities of dirt due to mouthing tendencies and the disease known as pica. Children who intentionally eat large quantities of dirt, plaster, or paint chips (1 to 10 g/day) and conse-quently are at greater risk of developing health problems can be said to suffer from pica. This disease is known as *geophasia* if the craving is for dirt alone. Geophasia, rather than pica, is generally of greatest concern in areas with contaminated soil.

Duggan and Williams [336] have summarized the lit-erature on the amount of lead ingested through dust and dirt. In their opinion, a quantity of 50 μg of lead was the best estimate for daily ingestion of dust by children, assuming, on the high side, that an average lead concen-tration of 1000 ppm would indicate a soil and dust inges-tion rate of 50 mg/day. Lepow et al. [337] estimated an ingestion rate equal to 100 to 250 mg/day (specifically, 10 mg ingested 10 to 25 times a day). Barltrop and col-leagues [321,338] also estimated that the potential uptake of soils and dusts by a toddler is about 100 mg/day. In a Dutch study, the amount of lead on hands ranged from 4 to 12 ng. By assuming maximum lead concentrations of 500 ng/g (concentrations were typically lower) and com-plete ingestion of the contents adsorbed to a child's hand on 10 separate occasions, the amount of ingested dirt would equal 240 mg. Thus, in order to eat 10,000 mg of soil per day, the rate suggested by the Centers for Disease Control and Prevention (CDC), children would have to place their hands into their mouths 410 times a day, a rate that seems improbable [339,340].

A report by the National Research Council [339] that addressed the hazards of lead suggested a soil/dust inges-tion rate of 40 mg/day. Day et al. [341] measured the amount of dirt transferred from children's hands (age range from 1 to 3 years) to a sticky sweet and estimated that a daily intake of 2 to 20 sweets would lead to a dirt intake of 10 to 1000 mg/day. Bryce-Smith [342] estimated 33 mg/day. In its document addressing lead in air, the USEPA assumed that children ate 50 mg/day of household dust, 40 mg/day of street dust, and 10 mg/day of dust derived from their parents' clothing (i.e., a total of 100 mg/day).

Kimbrough et al. [320] estimated the ingestions of soil at Times Beach, Missouri, based on unpublished observa-tions about children's behavior and hand–mouth activity. A few years later, Kimbrough noted that their estimate of up to 10,000 mg/day was clearly not close to reality, and her personal estimate would be closer to 50 mg/day [208]. LaGoy [343] based his soil ingestion estimates on a review of the literature, in particular using empirical data derived by Binder et al. [344] and Van Wijnen et al. [345]. Similarly, Paustenbach et al. [213] based his estimates on a review of the literature, including the mass-balance quantitative study conducted by Calabrese et al. in 1989 [346].

De Silva [347,348] adopted a different approach that may overcome some of the uncertainties inherent in the assumptions of the above indirect studies by applying a slope factor increase of 0.6 μg/dL in children's blood lead levels for each 1000-ppm increase in soil lead (this factor was developed by Barltrop et al. [338] following his work on blood lead levels in children from villages on old mining sites). De Silva then deduced that an increase of 0.6 μg/dL in blood indicates an extra oral intake of 3.75 μg lead per day, based on a USEPA calculation that an increase of 1.0 μg lead per day in children's diets produces an increase of 0.16 in the blood lead level [211]. With a soil lead value of 1000 ppm, 3.75 mg of soil would contain 3.75 μg of lead, suggesting that 3.75 mg/day (say, 4 mg) of soil was ingested by the children. However, the slope factor used here may not be the most appropriate, as mining soil wastes typically have larger sized particles which tends to decrease lead bioavailability compared with soil contaminated by lead smelter activity and therefore reduces the slope factor.

A major step forward beyond estimating soil ingestion using indirect measurements was the attempt to study tracer elements found in soil with elements measured in the urine and feces of children. Several studies have been conducted thus far that have used this approach [309,310, 344,346–352]. One early tracer study evaluated the amount of soil eaten by 24 hospitalized and nursery school children by analyzing the amount of aluminum, titanium, and acid soluble residue in the feces of children ages 2 to 4 years [345]. They found an average of 105 mg/day of soil in the feces of nursery children and 49 mg/day in hospitalized children. Even with the limited number of children in the study, the difference between the two groups was significant ($p < 0.01$). If the value for the hospitalized children is assumed to be the background level because these substances are taken in from non-soil sources (e.g., diet and toothpastes), the estimated average amount of soil ingested by the nursery school children would be 56 mg/day. This value is in the lower range of estimates in the literature and supports the use of 100 mg/day as a reasonable daily average uptake of soil by toddlers (ages 2 to 4 years or 1.5 to 3.5 years).

Calabrese et al. have completed two major studies [309,310,346,349,350,353]. In the first, they quantitatively evaluated 6 different tracer elements in the stools of 65 school children ages 2 to 4 years. They attempted to evaluate children from diverse socioeconomic backgrounds. This study was more definitive than prior investigations because they analyzed the children's diets, assayed for the presence of tracers in the diapers, assayed house dust and surrounding soil, and corrected for the pharmacokinetics of the tracer materials.

In the second study, soil ingestion estimates were obtained from a stratified, simple random sample of 64 children ages 1 to 4 years residing on a superfund site in Montana [354]. The study was conducted during the month of September for 7 consecutive days. Soil ingestion was estimated by each soil tracer via traditional methods as well as by an improved approach using five trace elements (Al, Si, Ti, Y, and Zr), called the *best tracer method* (BTM), which corrects for error due to misalignment of trace input and output, as well as error occurring from ingestion of tracers from nonfood, nonsoil sources, while being insensitive to the particle size of the soil/dust ingested. According to the BTM, the median soil ingestion was less than 1 mg/day, and the upper 95% was 160 mg/day. No significant age- (1 year vs. 2 vs. 3) or sex-related differences in soil ingestion were observed. These estimates are lower than estimates observed in the first study, which was conducted in New England during September and October.

Based on the series of early papers by Calabrese and others [309,310,346–350], a few generalizations can be made. The first two studies were difficult to conduct and interpret. Second, only children from a single climate were studied, and it can be expected that rates vary with the amount of time spent indoors and outdoors. Third, only a handful of children have been studied (less than 500), so it is not possible to characterize the percentage of children who might tend to ingest large quantities of soil or house dust. Fourth, the relevant amount of soil or house dust ingested indoors vs. outdoors is not known yet. In most cases, the contaminant concentrations in dust can be quite different when found in a carpet vs. the yard [197]. This was demonstrated in a recent study that involved evaluation of aggregate daily exposures, contributions of specific pathways of exposure, and temporal variation in exposure to chlorpyrifos from a collection of indoor air, carpet dust, exterior soil, and duplicate diet samples [160]. Chlorpyrifos concentrations in each medium and self-reported rates of time spent inside at home, time and frequency of contact with carpet, frequency of contact with soil, and weights of the duplicate diet samples were used to derive exposure to chlorpyrifos from each medium, as well as average daily aggregate exposure. Although it was found that inhalation of indoor air and ingestion of solid food accounted for almost all (97.9%) exposure to chlorpyrifos on average, the authors did report significant differences in average chlorpyrifos concentrations in exterior soil and carpet dust [160]. Specifically, a chlorpyrifos concentration of 204 ng/kg was reported for exterior soil, whereas carpet dust showed a pesticide concentration of 238 ng/kg [160]. Fifth, despite some degree of uncertainty in the results of the various studies, it appears that a best estimate of soil intake for most children resides in the are

**TABLE 10.9**
**Values for Childhood and Adult Soil Ingestion Rates Used in Health Risk Assessments Conducted Between 1984 and 2005**

| Author | Age | Solid and Dust (mg/d) |
|---|---|---|
| Barltrop [338] | 2–6 | 100 |
| Lepow et al. [337] | 2–6 | 100–250 |
| Day et al. [341] | 2–6 | 10–1000 |
| Kimbrough et al. [320] | 0–9 months | 0 |
| | 9–18 months | 1000 |
| | 1.5–3.5 | 10,000 |
| | 3.5–5 | 1000 |
| | 5+ | 100 |
| Hawley [593] | 0–2 | Negligible |
| | 2–6 | 90 |
| | 6–18 | 21 |
| | 18–70 | 57 |
| La Goy [594] | 1–6 | 500 (maximum) |
| | 1–6 | 100 (average) |
| Calabrese et al. [346] | 1–4 | 27–85 (mean) |
| | | 9–16 (median) |
| Paustenbach [69] | 2–4 | 25–50 |
| | Adults | 2–5 |
| De Silva [348] | Children | ~4 |
| USEPA [63] | Children | 200 |
| Calabrese and Stanek [357] | Children | 30-60 (best estimate) |

of 10 to 25 mg/day. It appears that perhaps 1 to 5% of the children may ingest much larger amounts during certain days or weeks (e.g., 2000 mg/day), but these tendencies do not occur on a chronic basis.

The issue of how much soil and house dust children eat, as well as the percent of children who are engaged in these activities, remains an active area of research [345–347,351–353,355,356]. Work by Calabrese and Stanek [357] suggests that prior work yielded reasonable results for purposes of risk assessment. Most of the values discussed here are presented in Table 10.9. As discussed previously, another area of research impacting exposure assessments of contaminated soil which has been and continues to be actively pursued is the bioavailability of the contaminant in the soil matrix [50,101,102,104,108, 109,113,116–119,281,358].

## What Is the Significance of Pica?

There appears to be some confusion in the literature over what constitutes pica. Pica can be defined as the "habitual ingestion of substances not normally regarded as edible," but some authors have included mouthing and sucking activities in their definitions [359]. Others appear to assume that all children with pica necessarily must be habitual soil eaters. In fact, pica behavior may be generalized to the ingestion of many different (nonfood) substances or may be specific to one substance such as paper, soap, or earth. It is likely that repetitive pica behavior specifically for dirt (habitual geophagia) rarely occurs in the general population in most industrialized countries [360,361].

Pica should, therefore, be considered a normal temporary phenomenon in some children. In the general population, the prevalence of both mouthing and pica and the range of articles ingested have been shown to decrease with age [321]. In the 1-year-old age group, 78% of the children mouthed objects and 35% ingested them; this behavior decreased at the age of 4 years, when 33% were mouthing and only 6% had pica.

It is also relevant to note that, in certain circumstances, pica for soil may be culturally determined, such as eating clay, which is high in silicon and aluminum, for its medical properties in the relief of stomach discomfort and diarrhea by some Aborigines, or the custom of eating earth during pregnancy in certain cultures [208]. Some women, for example, in the southern portions of the United States have a craving for and eat certain clays during pregnancy.

Pica may be associated with physical disorders, including iron deficiency; however, it has been debated whether pica represents a cause or an effect of these deficiencies. Pica can also be associated with mental illness. It has also been reported that 25% of institutionalized mentally handicapped adults indulged in pica of one kind or another (including bizarre objects ranging from rags and string to rocks, insects, and feces) [360].

Calabrese and Stanek [357] have indicated that, in their studies, they have observed great variability in soil ingestion by children. They have noted, for example, that some children are highly variable in their soil ingestion activities, displaying little propensity for soil ingestion on one day but ingesting copious amounts the next day. Although there has not been any concerted focus on the soil pica child, the available data indicate that some children ingest over 50 g of soil on particular days. They note that, although it is true that some children will ingest large amounts of soil, it is far from certain whether soil pica is behavior that only a small subgroup displays over limited number of years (e.g., 1 to 6) or whether most children on occasion display this behavior or some combination of both behavioral patterns. Clearly, additional work is needed to understand this topic.

### Soil Ingestion by Adults

For most persons beyond the ages of 5 to 6 years, the daily uptake of dirt due to intentional ingestion is generally thought to be quite low. With the exception of some lower income persons who eat clays due to tradition or mineral deficiency, adults will not usually intentionally ingest dirt or soil; however, there are two other important ways in which adults eat dirt: incidental hand-to-mouth contact and through dust on vegetables. It has been shown that most soil ingested from crops comes from leafy vegetables. Interestingly, investigations at nuclear weapons trials have shown that particles exceeding 45 μm are seldom retained on leaves. Further, superficial contamination by smaller particles is readily lost from leaves, usually by mechanical processes or rain and certainly by washing [362]. As a result, unless the soil contaminant is absorbed into the plant, superficial contamination of plants by dirt will rarely present a health hazard [52,363].

The estimated deposition rate of dust from ambient air in rural environments is about 0.012 μg/cm$^2$-day, assuming that rural dust contains about 300 μg/g of lead (the substance for which these data were obtained). The USEPA has estimated that even at relatively high air concentrations (0.45 mg/m$^3$ total dust), it is unlikely that surface deposition alone can account for more than 0.6 to 1.5 μg lettuce per g dust (2 to 5 μg/g lead) on the surface of lettuce during a 21-day growing period [52]. These data suggest that daily ingestion of dirt and dust by adults due to eating vegetables is unlikely to exceed about 0 to 5 mg/day even if all of the 137 g of leafy and root vegetables, sweet corn, and potatoes consumed by adult males each day were replaced by family garden products.

In its document on lead, the USEPA uses worst-case assumptions to estimate that persons could take up to 100 μg of lead each day due to unwashed vegetables. The actual uptake by adults from vegetables should actually be much less and is probably negligible, because the USEPA's estimate assumes that all of the suspended dust is contaminated; persons do not wash the vegetables; garden vegetables are eaten throughout the year, rather than only during the growing season; and persons actually replace most vegetables with their own garden products.

With respect to the second route—unwashed vegetables—only a very limited amount of work has been conducted. It has been suggested that the primary route of uptake will be through accidental ingestion of dirt on the hands, which may be of special concern to smokers who tend to have more frequent hand-to-mouth contact. It is true that, before the importance of this route of entry was recognized, persons who worked in lead factories between 1890 and 1920 probably received a large portion of their body burden of lead due to poor hygiene; however, such conditions are now rare in the United States and most developed countries.

Some persons have evaluated the exposure experience of agricultural workers who apply or work with pesticide dusts. Due to the frequency and degree of pesticide exposure during its manufacture or application, these data do not appear to be appropriate surrogates for estimating soil uptake from the hands of persons who live on or near sites having contaminated soil. In addition, most of the published studies on pesticides involve liquids such as the organophosphates, rather than "soil-like" particles. Exposure studies of persons who apply granular pesticides might be more useful for defining upper-bound estimates of dermal exposure than estimates based on dusty workplaces [364].

At least one study has been conducted to specifically address soil uptake by adults involved in remediating waste sites [187,355,357]. The results suggest that the amount of soil eaten by these workers is much less than the default value of 100 mg/day suggested by the USEPA in a number of guidance documents or risk assessments.

### Estimating the Intake of Chemicals Via Food

Without question, the information necessary to accurately estimate the ingestion of xenobiotics via foods is one of the most complex of all exposure calculations. The hundreds of different possible foods and dozens of different chemicals that can be present as a pesticide residue and background concentrations of various chemicals in soil make this a formidable task.

The methodology for estimating uptake via ingestion must account for the quantity of food ingested each day the concentration of contaminant in the ingested material and the bioavailability of the contaminant in the media

Over the past 20 years, a significant amount of work has been directed at understanding these exposure factors. Specifically, an entire volume of the USEPA's *Exposure Factors Handbook* (Vol. II) is devoted to this topic [63]. In addition, it is also addressed in the USEPA's *Child-Specific Exposure Factors Handbook* [64].

The approach to estimating uptake via foods was first applied in the late 1940s by the Food and Drug Administration and did not change appreciably through 2000 [366]; however, because of the passage of the Food Quality Protection Act of 1996, which significantly amended the U.S. laws that regulate pesticides (e.g., Federal Insecticide, Fungicide, and Rodenticide Act [FIFRA] and the Federal Food, Drug, and Cosmetic Act), the methodology for estimating uptake of chemicals from foods has changed dramatically [367]. Specifically, the FQPA established a stringent health-based standard ("a reasonable certainty of no harm") for pesticide residues in foods to assure protection from unacceptable pesticide exposure and to strengthen health protections from pesticide risks for sensitive populations. In addition, the FQPA required the USEPA to consider the cumulative effects on human health that may result from exposure to mixtures of pesticides [367]. In response, the USEPA Office of Pesticide Programs, in consultation with the FIFRA scientific advisory panel, has developed guidelines for the cumulative risk assessment of pesticides that share a common mechanism of toxicity [87]. The approach is conceptually similar to methods developed by the USEPA for estimating exposure to mixtures of dioxins and dibenzofurans using toxicity equivalence factors to normalize the toxicity (e.g., binding to the aryl hydrocarbon receptor) of each member of the group with respect to that of a single chemical [368]. Specifically, the FQPA requires that all pesticide residues from foods be added together in a prescribed manner based on target organ, with the goal of understanding the total daily dose of all residual pesticides in the diet [369–371]. Then, if necessary, the pesticide manufacturers are expected to calculate the necessary residue level that their chemical may have in a particular food so the total dose does not exceed a fraction of the acceptable daily intake (ADI). Because of the hundreds of foods and dozens of residues, this presents a formidable challenge.

Ingestion of contaminated fruits and vegetables is a potential pathway of human exposure to toxic chemicals. Fruits and vegetables may become contaminated with toxic chemicals by several different pathways. Ambient air pollutants may be deposited on or absorbed by plants or dissolved in rainfall or irrigation waters that contact the plants. Plant roots may also absorb pollutants from contaminated soil and groundwater. The addition of pesticides, soil additives, and fertilizers may also result in food contamination [63]. Formulas are available to predict the concentration of chemicals from the soil that have deposited from the air and remain after treatment with a pesticide.

The primary information source on consumption rates of fruits and vegetables among the U.S. population is the U.S. Department of Agriculture's (USDA) Nationwide Food Consumption Survey (NFCS) and the USDA Continuing Survey of Food Intakes by Individuals (CSFII) [366,372–374]. Data from the NFCS have been used in various studies to generate consumer-only and per-capita intake rates for individual fruits and vegetables, as well as total fruits and total vegetables. CSFII data from the 1989–1991 survey have been analyzed by the USEPA to generate per-capita intake rates for various food items and food groups [63,372–375].

Consumer-only intake is defined as the quantity of fruits and vegetables consumed by individuals who ate these food items during the survey period. Per-capita intake rates are generated by averaging consumer-only intakes over the entire population of users and nonusers. In general, per-capita intake rates are appropriate for use in exposure assessment for which average dose estimates for the general population are of interest, because they represent both individuals who ate the foods during the survey period and individuals who may eat the food items at some time but did not consume them during the survey period. Total fruit intake refers to the sum of all fruits consumed in a day, including canned, dried, frozen, and fresh fruits. Likewise, total vegetable intake refers to the sum of all vegetables consumed in a day, including canned, dried, frozen, and fresh vegetables.

Intake rates may be presented on either an as-consumed or dry-weight basis. As-consumed intake rates (g/day) are based on the weight of food in the form in which it is consumed. In contrast, dry-weight intake rates are based on the weight of food consumed after the moisture content has been removed; therefore, when calculating exposures based on ingestion, the unit of weight used to measure the contaminant concentration in the produce must be understood. Intake data from the individual NFCS and CSFII components are based on "as eaten" (i.e., cooked or otherwise prepared) forms of the food items or groups; thus, no corrections are required to account for changes in portion sizes from cooking losses [373,374,376].

Estimating source-specific exposures to toxic chemicals in fruits and vegetables may also require information on the amount of fruits and vegetables exposed to or protected from contamination as a result of cultivation practices, the physical nature of the food product itself (e.g., those having protective coverings that are removed before eating would be considered protected), or the amount grown beneath the soil (e.g., most root crops such as potatoes). The percentages of foods grown above and below ground will be useful when the contaminant concentrations in foods are estimated from concentrations in soil, water, and air; for example, vegetables grown below ground would more likely be contaminated by soil pollut-

**TABLE 10.10**
**Summary of Default Exposure Factor Recommendations and Confidence Ratings for U.S. Citizens**

| Exposure Factor | Recommendation | Confidence Rating |
|---|---|---|
| Drinking-water intake rate | 21 mL/kg/d or 1.4 L/d (average) | Medium |
| | 34 mL/kg/d or 2.3 L/d (90th percentile) | Medium |
| | (Percentiles and distribution also included; means and percentiles also included for pregnant and lactating women) | |
| Total fruit intake rate | 3.4 g/kg/d (per capita, average) | Medium |
| | 12.4 g/kg/d (per capita, 95th percentile) | Low |
| | (Percentiles also included; means presented for individual fruits) | |
| Total vegetable intake rate | 4.3 g/kg/d (per capita, average) | Medium |
| | 10 g/kg/d (per capita, 95th percentile) | Low |
| | (Percentiles also included; means presented for individual vegetables) | |
| Total meat intake rate | 2.1 g/kg/d (per capita, average) | Medium |
| | 5.1 g/kg/d (per capita, 95th percentile) | Low |
| | (Percentiles also included; percentiles also presented for individual meats) | |
| Total dairy intake rate | 8.0 kg/d (per capita, average) | Medium |
| | 29.7 g/kg/d (per capita, 95th percentile) | Low |
| | (Percentiles also included; means presented for individual dairy products) | |
| Grain intake | 4.1 g/kg/d (per capita average) | High |
| | 10.8 g/kg/d (per capita 95th percentile) | Low in long-term |
| | (Percentiles also included) | Upper percentiles |
| Breast-milk intake rate | 742 mL/d (average) | Medium |
| | 1033 mL/d (upper percentile) | Medium |
| Fish intake rate | | |
| General population | 20.1 g/d (total fish) average | High |
| | 14.1 g/d (marine) average | High |
| | 6.0 g/d (freshwater/estuarine) average | High |
| | 63 g/d (total fish) 95th percentile long-term | Medium |
| | (Percentiles also included) | |
| Serving size | 129 g (average) | High |
| | 326 g (95th percentile) | High |
| Recreational marine anglers | 27 g/d (finfish only) | Medium |
| Recreational freshwater | 8 g/d (average) | Medium |
| | 25 g/d (95th percentile) | Medium |
| Native American subsistence population | 70 g/d (average) | Medium |
| | 170 g/d (95th percentile) | Low |

ants, but leafy above-ground vegetables would more likely be contaminated by the deposition of air pollutants on plant surfaces. Some examples of various exposure factors and confidence ratings for liquids and food are presented in Table 10.10 [63].

Individual average daily intake rates calculated from NFCS and CSFII data are based on averages of reported individual intakes over 1 day or 3 consecutive days. Such short-term data are suitable for estimating mean average daily intake rates representative of both short-term and long-term consumption; however, the *distribution* of average daily intake rates generated using short-term data (e.g., 3 day) do not necessarily reflect the long-term distribution of average daily intake rates. The distributions generated from short-term and long-term data will differ to the extent that each individual's intake varies from day to day; the distributions will be similar to the extent that individuals' intakes are constant from day to day [63].

The variation in day-to-day intake rates among individuals will be greatest for food items or groups that are highly seasonal and for items or groups that are eaten year-around but are not typically eaten every day. For these foods, the intake distribution generated from short-term data will not reflect long-term distribution. On the other hand, for broad categories of foods (e.g., vegetables) that are eaten on a daily basis throughout the year with minimal seasonality, the short-term distribution may be a reasonable approximation of the true long-term distribution although it will show somewhat more variability.

Other relevant fruit and vegetable intake studies include the USEPA's Dietary Risk Evaluation System (DRES), Office of Pesticide Programs (OPP). The OPP uses the DRES (formerly the Tolerance Assessment System) to assess the dietary risk of pesticide use as part of the pesticide registration process [373,374,376]. It sets tolerances for specific pesticides on raw agricultural commodities based on estimates of dietary risk. These estimates are calculated using pesticide residue data for the food item of concern and relevant consumption data. Intake rates are based primarily on the USDA 1977–1978 NFCS, although intake rates for some food items are based on estimations from production volumes or other data (e.g., some items were assigned an arbitrary value of 0.000001 g/kg/day) [63]. The OPP has calculated per-capita intake rates of individual fruits and vegetables for 22 subgroups of the population (age, regional, and seasonal) by determining the composition of NFCS food items and disaggregating complex food dishes into their component raw agricultural commodities (RACs) [63,377].

The advantage of using these data is that complex food dishes have been disaggregated to provide intake rates for a very large number of fruits and vegetables. These data are also based on the individual body weights of the respondents; therefore, using these data to calculate toxic chemical exposure may provide more representative estimates of potential dose per unit body weight. Because the data are based on the NFCS short-term dietary recall, however, the same limitations discussed previously for other NFCS data sets also apply here. In addition, consumption patterns may have changed since the data were collected in 1977–1978. The OPP is in the process of translating consumption information from the USDA CSFII 1989–1991 survey to be used in DRES [63,377].

The USDA has also conducted a study entitled *Food and Nutrient Intakes of Individuals in One Day in the U.S.* [372,377]. The USDA calculated mean intake rates for total fruits and total vegetables using NFCS data from 1977/1978 and 1987/1988 and CSFII data from 1994/1995 [63,372,377]. Mean per-capita total intake rates are based on intake data for 1 day from the 1977/1978 and 1987/1988 USDA and NFCS, respectively. Data from both surveys are presented in the *Exposure Factors Handbook* to demonstrate that, although the 1987–1988 survey had fewer respondents, the mean per-capita intake rates for all individuals agree with the earlier survey. Also, slightly different age classifications were used in the two surveys, providing a wider range of age categories from which exposure assessors may select appropriate intake rates. The age groups used in this dataset are the same as those used in the 1987/1988 NFCS. Information for per-capita intake rates and consumer-only intake rates for various ages of individuals is also available. Intake rates for consumers only were calculated by dividing the per-capita consumption rate by the fraction of the population using vegetables or fruits in a day [63]. The advantages of using these data are that they provide intake estimates for all fruits, all vegetables, or all fats combined. Again, these estimates are based on 1-day dietary data that may not reflect usual consumption patterns [63].

Children's exposure from food ingestion may differ from that of adults because of differences in the type and amounts of food eaten and intake per unit body weight [64]. Recent information on consumption rates of foods among children is available from the USDA's NFCS and the USDA'S CSFIIs. Data from the 1989/1991 and 1994/1996 CSFIIs have been used in various studies to generate children's per-capita intake rates for both individual foods and the major food groups. The Supplemental Children's Survey to the 1994/1996 CSFII was conducted in response to the Food Quality Protection Act (FQPA) [378,379], which required the USDA to provide data from a larger sample of children for use by the USEPA in estimating exposure to pesticide residues in the diets of children. The 1998 survey adds intake data from 5559 children from birth through 9 years of age to the intake data collected from 4253 children of the same age who participated in the CSFII 1994/1996 [378,379].

### Intake of Fish and Shellfish

Contaminated finfish and shellfish are potential sources of human exposure to toxic chemicals. Pollutants are carried in surface waters but also may be stored and accumulated in sediments as a result of complex physical and chemical processes. Consequently, various aquatic species can be exposed to pollutants and may become sources of contaminated food [63]. Accurately estimating exposure to various chemicals in a population that consumes fish from a polluted water body requires an estimation of caught-fish intake rates by fishermen and their families. Commercially caught fish are marketed widely, making the prediction of an individual's consumption from a particular commercial source difficult. Because the catch of recreational and subsistence fishermen is generally not diluted in this way, these individuals and their families represent the population that is most vulnerable to exposure by intake of contaminated fish from a specific location [63].

Over the years, fish consumption survey data have been collected using a number of different approaches that need to be considered when interpreting the survey results [380]. Generally, surveys are either "creel" studies in which fishermen are interviewed while fishing or broader population surveys using mailed questionnaires or phone interviews. Both data types can be useful for exposure assessment purposes, but somewhat different applications and interpretations are needed. In fact, creel study results have often been misinterpreted because of inadequate knowledge of survey principles [63,381,382].

The typical survey seeks to draw inferences about a larger population from a smaller sample of that population. The larger population from which the survey sample is taken and to which the survey results are generalized denotes the target population of the survey. To generalize from the sample to the target population, the probability of being sampled must be known for each member of the target population. This probability is reflected in weights assigned to each survey respondent, with weights being inversely proportional to sampling probability. When all members of the target population have the same probability of being sampled, all weights can be set to 1 and essentially ignored [383,384].

In a mail or phone study of licensed anglers, the target population generally involves all licensed anglers in a particular area, and in these studies the sampling probability is essentially equal for all target population members. In a creel study, the target population is anyone who fishes at the locations being studied; generally in a creel study, the probability of being sampled is not the same for all members of the target population. For example, if the survey is conducted for one day at a site, then it will include all persons who fish there daily, but only about 1/7th of the people who fish there weekly, 1/30th of the people who fish there monthly, etc. In this example, the probability of being sampled (or inverse weight) is seen to be proportional to the frequency of fishing. However, if the survey involves interviewers who revisit the same site on multiple days and persons who are only interviewed once for the survey, then the probability of being in the survey is not proportional to frequency; in fact, it increases less proportionally with greater frequency of fishing. If the same site is surveyed every day of the survey period with no re-interviewing, all members of the target population would have the same probability of being sampled, regardless of fishing frequency, implying that the survey weights should all equal 1 [383,385].

On the other hand, if the survey protocol calls for individuals to be interviewed each time an interviewer encounters them (i.e., without regard to whether they were previously interviewed), then the inverse weights will again be proportional to fishing frequency, no matter how many times interviewers revisit the same site. Note that, when individuals can be interviewed multiple times, the results of each interview are included as separate records in the database, and the survey weights should be inversely proportional to the expected number of times that an individual's interviews are included in the database [63,383,385].

Fish and shellfish exposure assessments are among the most complicated of all assessments [386]. A significant portion of the *Exposure Factors Handbook* addresses this topic [63]. Recently, fairly complex methods, including Monte Carlo modeling, have been applied to resolve many of the difficulties estimating exposure of anglers and their families [78,380].

## Aggregate Exposure and the FQPA

Pesticides are regulated under the Federal Insecticide, Fungicide, and Rodenticide Act (FIFRA) and the Federal Food, Drug, and Cosmetic Act (FFDCA). In 1996, Congress passed the Food Quality Protection Act (FQPA), which amended both FIFRA and FFDCA [367]. These laws mandated the USEPA to register pesticides and set tolerances based on a safety determination, a reasonable certainty that use of a given pesticide or consumption of raw agricultural commodity of processed foods that contain the pesticide and its residues will cause no harm to human health or the environment. The USEPA evaluates risks posed by the use and usage of each pesticide to make a determination of safety. Based on this determination, the USEPA regulates pesticides to ensure that use of the chemical is not unsafe.

In the past, the USEPA evaluated the safety of pesticides based on a single-chemical, single-exposure pathway scenario; however, FQPA requires that the Agency consider aggregate exposure in its decision-making process. Section 408(a)(4)(b)(2)(ii) of FFDCA specifies that, with respect to a tolerance, there must be a determination "that there is a reasonable certainty that no harm will result from aggregate exposure to the pesticide chemical residue, including all anticipated dietary exposures and all other exposures for which there is reliable information." Section (b)(2)(C)(ii)(I) states that "there is a reasonable certainty that no harm will result to infants and children from aggregate exposure to the pesticide chemical residues." *Aggregate dose* is defined as the amount of a single substance available for interaction with metabolic processes or biologically significant receptors from multiple routes of exposure. *Aggregate risk* is defined as the likelihood of the occurrence of an adverse health effect resulting from all routes of exposure to a single substance. Conversely, *cumulative risk* is defined as the likelihood of the occurrence of an adverse health effect resulting from all routes of exposure to a group of substances sharing a common mechanism of toxicity.

As shown in Figure 10.5, the most basic concept underlying all aggregate exposure assessments is that exposure occurs to an individual. The integrity of the data concerning this exposed individual must be maintained throughout the aggregate exposure assessment. In other words, each of the individual subassessments must be linked back to the same person [387]. Because exposures are based on that received by a single individual, aggregate exposure assessments must agree in time, place, and demographic characteristics. The individual's temporal (i.e., exposures via all pathways agree in time), spatial (i.e., exposures via all pathways agree in place/location) and demographic (i.e., exposures via all pathways agree in age, gender, and ethnicity and other demographic characteristics) characteristics are then used to develop a dis

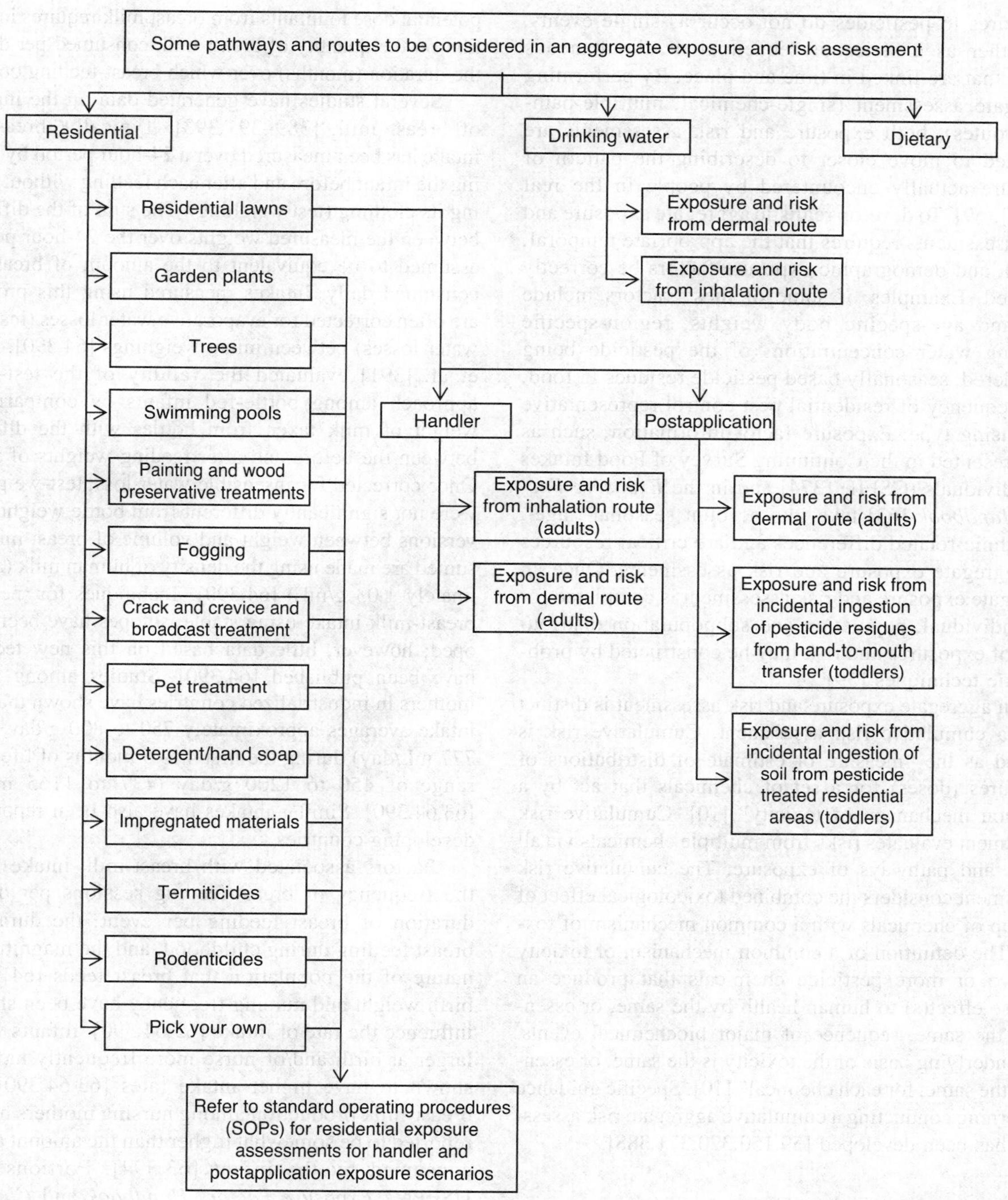


FIGURE 10.5 Factors to consider in an aggregate exposure assessment of a pesticide. (From USEPA, *Guidance for Performing Aggregate Exposure and Risk Assessments*. Office of Pesticide Programs, U.S. Environmental Protection Agency, Washington, D.C., 1999.)

ribution of total exposure to (many) individuals in a population of interest [369].

Each of these parameters has imbedded attributes that must be matched to create a reasonable assessment. Some of these imbedded attributes include:

- Time (duration, daily, seasonally)
- Place (location and type of home, urbanization, watersheds, region)
- Demographics (age, gender, reproductive status, ethnicity, personal preference)

Exposures to pesticides do not occur as single events, but rather as a series of sequential or simultaneous events that are linked in time and place. By performing aggregate assessment (single-chemical, multiple-pathways/routes), both exposure and risk assessments are expected to move closer to describing the pattern of exposure actually encountered by people in the real world [369]. To develop realistic aggregate exposure and risk assessments requires that the appropriate temporal, spatial, and demographic exposure factors be correctly assigned. Examples of some of these factors include sex- and age-specific body weights, region-specific drinking water concentrations of the pesticide being considered, seasonally based pesticide residues in food, and frequency of residential pest control representative of housing type. Exposure factor information, such as that presented in the Continuing Survey of Food Intakes by Individuals (CSFII) [374] and in the *Exposure Factors Handbook* [63] take into account seasonal-, age-, and ethnic-related differences and are critical resources for aggregate exposure and risk assessments. Once an aggregate exposure and risk assessment is completed for one individual, population and subpopulation distributions of exposures and risk may be constructed by probabilistic techniques [59].

An aggregate exposure and risk assessment is distinct from a cumulative risk assessment. Cumulative risk is defined as the "measure or estimate of distributions of exposures (doses) for a set of chemicals that act by a common mechanism of toxicity" [10]. Cumulative risk assessment evaluates risks from multiple chemicals via all routes and pathways of exposure. The cumulative risk assessment considers the combined toxicological effect of a group of chemicals with a common mechanism of toxicity. The definition of a common mechanism of toxicity is "two or more pesticide chemicals that produce an adverse effect(s) to human health by the same, or essentially the same, sequence of major biochemical events. The underlying basis of the toxicity is the same, or essentially the same, for each chemical" [10]. Specific guidance concerning conducting a cumulative aggregate risk assessment has been developed [59,150,370,371,388].

## Breast Milk

Breast milk is a potential source of exposure to toxic substances for nursing infants. Lipid-soluble chemical compounds accumulate in body fat and may be transferred to breast-fed infants in the lipid portion of breast milk. Because nursing infants obtain most (if not all) of their dietary intake from breast milk, they are especially vulnerable to exposures to these compounds. In fact, the peak body burdens of certain chemicals (such as dioxin) can reach their lifetime peak (μg/kg) on the last day of nursing at age 12 to 24 months. Estimating the magnitude of the potential dose to infants from breast milk requires information on the quantity of breast milk consumed per day and the duration (months) over which breast-feeding occurs.

Several studies have generated data on the ingestion of breast milk [389–391,393]. Typically, breast-milk intake has been measured over a 24-hour period by weighing the infant before and after each feeding without changing its clothing (test weighing). The sum of the difference between the measured weights over the 24-hour period is assumed to be equivalent to the amount of breast milk consumed daily. Intakes measured using this procedure are often corrected for evaporative water losses (insensible water losses) between infant weighings [64,390]. Neville et al. [391] evaluated the validity of the test-weight approach among bottle-fed infants by comparing the weight of milk taken from bottles with the difference between the before and after feeding weights of infants. Once corrected for insensible water loss, test-weight data were not significantly different from bottle weights. Conversions between weight and volume of breast milk consumed are made using the density of human milk (approximately 1.03 g/mL) [64,390]. Techniques for measuring breast-milk intake using stable isotopes have been developed; however, little data based on this new technique have been published [64,390]. Studies among nursing mothers in industrialized countries have shown that infant intake averages approximately 750 to 800 g/day (728 to 777 mL/day) during the first 4 to 5 months of life, with a range of 450 to 1200 g/day (437 to 1165 mL/day) [63,64,390]. Similar intakes have also been reported for developing countries.

Factors associated with breast-milk intake include the frequency of breast-feeding sessions per day, the duration of breast feeding per event, the duration of breast feeding during childhood, and the magnitude and nature of the population that breast feeds [64]. Infant birth weight and nursing frequency have been shown to influence the rate of intake [63,64,390]. Infants who are larger at birth and/or nurse more frequently have been shown to have higher intake rates [63,64,390]. Also breast-milk production among nursing mothers has been reported to be somewhat higher than the amount actually consumed by the infant [63,394]. Portions of the USEPA's *Exposure Factors Handbook* and *Child-Specific Exposure Factors Handbook* address this topic, and a few published papers have offered some novel approaches [63,64,82]. Some examples of breast-milk intake rates are presented in Table 10.11.

Information on the fat content of breast milk is also needed for estimating dose from breast-milk residue concentrations that have been indexed to lipid content [63,64,393,395]. Environmental chemicals with high lipid solubility are likely to be found in breast milk. These chemicals include polyhalogenated compounds, organochlorine insecticides, and polybrominated diphe

**TABLE 10.11**
**Values for Daily Intakes of Breast Milk**

| Age (months) | Number of Infants Surveyed | Intake (mL/day) Mean | Range of Means | Upper Percentile[a] | Refs. |
|---|---|---|---|---|---|
| 1 | 77 | 702 | 600–747 | 1007[b] | Pao et al [597]; Butte et al. [392]; Neville et al. [391]; Dewey and Lönnerdal [595] |
| 3 | 140 | 759 | 702–833 | 1025[b] | Pao et al. [597]; Butte et al. [392]; Neville et al. [391]; Dewey and Lönnerdal [595]; Dewey et al. [596] |
| 6 | 85 | 765 | 682–896 | 1059[b] | Paoet al. [597]; Neville et al. [391]; Dewey and Lönnerdal [595]; Dewey et al. [596] |
| 9 | 62 | 622 | 600–627 | 1038 | Neville et al. [391]; Dewey et al. [596] |
| 12 | 51 | 427 | 391–435 | 900 | Neville et al. [391]; Dewey et al. [596] |
| 12-month TWA | | 688 | | Range 900–1059 (middle of the range, 980) | |
| 1- to 6-month TWA | | 742 | | 1033 | |

[a] Upper percentile is reported (mean + 2 standard deviations) except as noted.
[b] Middle of the range; TWA, time-weighted average.

nylethers. For many long-lived environmental chemicals in lactating women, breast milk may be a major route of elimination [396]. These fat-soluble chemicals are incorporated into the milk as it is synthesized, and they must be measured in accordance with the fat content of the milk to allow for meaningful comparisons within an individual and among populations [397].

The lipid content of breast milk varies according to the length of time that an infant nurses, and it increases from the beginning to the end of a single nursing session. The lipid portion accounts for approximately 4% of human breast milk (39 ± 4.0 g/L) [63,64,395]. Most studies evaluating chemical exposures through breast milk have not accounted for the fact that women's stores of lipophilic chemicals in adipose tissue and breast milk are depleted over the duration of lactation [398]. A number of factors can affect contaminant levels in breast milk, including the health of the mother during pregnancy and during the lactation period, the presence and levels of other xenobiotics that may alter metabolism, changes in body mass index during pregnancy and lactation, diet, other factors that may mobilize fat, parity and length of previous lactation, number of children being breast-fed at one time, maternal age, and maternal body mass index [397]. In addition, variation in the time of breast-milk sampling (including time postpartum and time of day), the age of the mother, and the number of previously breast-fed children can also influence measured levels [398].

Whereas the study of prescription drugs has provided a basis for understanding the governing principles behind transfer of chemicals through breast milk [399], the research of environmental chemical exposure through breast milk has increased dramatically over the last few decades. One of the earliest reports of the measurement of an environmental chemical in breast milk was by Laug and colleagues in 1951 [400], who reported that the breast milk from 32 women from the general population of Washington, D.C., contained DDT (1,1,1-trichloro-1,2-*bis*(4-chlorophenyl)ethane) at an average concentration of 0.13 ppm. Since then, a number of studies have evaluated trends of exposure to environmental chemicals through breast milk. Studies have measured chordane, dieldrin/aldrin, heptachlor, hexachlorobenzene, hexachlorocyclohexane, dioxins and furans, polychlorinated biphenyls, polybrominated diphenylethers, and metals in breast milk. The data that exist suggest that bans and restrictions in recent decades on the use of many of the persistent organic pollutants have led to a dramatic decline in the levels of many of these chemicals in breast milk [401].

## ESTIMATING UPTAKE VIA INHALATION

Estimating intake via inhalation depends on only a few exposure factors, such as inhalation rate, airborne chemical concentration, bioavailability, and, if it is a particle, particle size. In general, uncertainty in estimates of intake via inhalation is among the smallest of all exposure calculations. Inhalation rates are known to vary directly with the amount of physical activity of the persons being evaluated. The default value used by the USEPA and others is 20 $m^3$/day. When conducting occupational exposure assessments, it is common to assume that workers inhale about 10 $m^3$ in an 8-hour workday [42].

Airborne chemical concentrations are obtained through either direct measurement or modeling. The form of the chemicals in the air will be a gas (includes vapors) or particles (dusts or fumes). Generally, it is assumed that virtually all of the vapors or gases will be absorbed if inhaled [42,226,402]. This may not be the case for volatile chemicals, if the concentration in the blood is approaching steady state. In those cases, a significant fraction of the inhaled vapors will be present in the exhaled breath and, therefore, not absorbed [126].

It is usually assumed that, if particles enter the lower respiratory tract, they will eventually be absorbed unless the chemical is highly insoluble. Generally, it is assumed that particles less than 150 μm are inhalable, but virtually all particles greater than 10 μm (by weight) will be captured in the upper respiratory tract (nose and throat) and then ingested. It has often been assumed that particles less than 10 μm will be captured in the lower respiratory tract and nearly 100% of these will eventually be absorbed. Notably, for some insoluble chemicals, such as silica, beryllium, and ambient particulate matter, the adverse effect is related to the particle size, so this must be taken into account.

In recent years, the focus on respirable particles has moved toward those in the 0.1- to 2.5-μm range. Many epidemiology studies have indicated that, on days when the airborne concentration of these particles is very high, a significant increase can occur in the number of persons admitted to the hospital. It is not yet known whether certain very small particles are primarily responsible for producing this response or whether other contaminants (e.g., sulfur dioxide or ozone) must be present to exacerbate the response. Nonetheless, the study of these very small particles, as well as those much smaller (called *ultrafine particles*), has probably been among the most exciting areas of research in the environmental sciences over the past several years [403–406].

### Various Inhalation Rates

A significant amount of research has been done to correlate various inhalation rates with different tasks and body weights. Most studies on this subject have been summarized in the most recent USEPA *Exposure Factors Handbook* [44]. Data are available for dozens of different levels of physical activity and the distributions for several populations are presented. A number of equations have been proposed for predicting the inhalation rate based on body weight [63]. The *Exposure Factors Handbook* and other sources provide a number of tables that relate physical activity with inhalation rate (see Table 10.12).

### Bioavailability of Airborne Chemicals

Because the mass of chemicals inhaled is usually quite small and because most particles less than 10 μm in diameter are thought to be fairly easily absorbed, it is generally assumed that particles are 100% bioavailable after they are trapped in the lower lung. Likewise, it is generally assumed that most vapors and gases are completely absorbed (100% bioavailable) if they reach the lower respiratory tract. Both are conservative assumptions that should be reassessed on a case-by-case basis.

**TABLE 10.12**
**Daily Inhalation Rates Estimated from Daily Activities**

| Subject | Inhalation Rate (IR) Resting ($m^3$/hr) | Light Activity ($m^3$/hr) | Daily Inhalation Rate (DIR) ($m^3$/d) |
|---|---|---|---|
| Adult man | 0.45 | 1.2 | 22.8 |
| Adult woman | 0.36 | 1.14 | 21.1 |
| Child (10 yr) | 0.29 | 0.78 | 14.8 |
| Infant (1 yr) | 0.09 | 0.25 | 3.76 |
| Newborn | 0.03 | 0.09 | 0.78 |

## ROLE OF UNCERTAINTY ANALYSIS

Exposure assessment uses a wide array of information sources and techniques. Even when actual exposure-related measurements exist, assumptions or inferences will still be required. Most likely, data will not be available for all aspects of the exposure assessment, and these data may be of questionable or unknown quality. In these situations, the exposure assessor will have to rely on a combination of professional judgment, inferences based on analogy with similar chemicals and conditions, estimation techniques, and the like. The net result is that the exposure assessment will be based on a number of assumptions with varying degrees of uncertainty [42].

The decision analysis literature has focused on the importance of explicitly incorporating and quantifying scientific uncertainty in risk assessments [385,386]. Reasons for addressing uncertainties in exposure assessments include [42]:

- Uncertainty information from different sources and of different quality must be combined.
- A decision must be made about whether and how to expend resources to acquire additional information (e.g., production, use, and emissions data; environmental fate information; monitoring data; population data) to reduce the uncertainty.
- So much empirical evidence exists that biases may occur, resulting in so-called best estimates that are not very accurate. Even when all that is needed is a best-estimate answer, the quality of the answer may be improved by incorporating a frank discussion of uncertainty into the analysis.

- Exposure assessment is an iterative process. The search for an adequate and robust methodology to handle the problem at hand may proceed more effectively and to a more certain conclusion if the associated uncertainty is explicitly included and it can be used as a guide in the process of refinement.
- A decision is rarely made on the basis of a single piece of analysis. Further, it is rare for there to be one discrete decision; a process of multiple decisions spread over time is the more common occurrence. Chemicals of concern may go through several levels of risk assessment before a final decision is made. During this process, decisions may be made based on exposure considerations. An exposure analysis that attempts to characterize the associated uncertainty allows the user or decision-maker to do a better evaluation in the context of the other factors being considered.
- Exposure assessors have a responsibility to present not just numbers but also a clear and explicit explanation of the implications and limitations of their analyses. Uncertainty characterization helps to achieve this.

Essentially, constructing scientifically sound exposure assessments and analyzing uncertainty go hand-in-hand. The reward for analyzing uncertainties is knowing that the results have integrity or that significant gaps exist in available information that can make decision-making a tenuous process.

## Variability vs. Uncertainty

Although some authors treat variability as a specific component of uncertainty, the USEPA and others advise risk assessors (and, by analogy, the exposure assessor) to distinguish between variability and uncertainty [12,407]. Specifically, uncertainty represents a lack of knowledge about factors affecting exposure or risk, whereas variability arises from true heterogeneity across people, places, or time. In other words, uncertainty can lead to inaccurate or biased estimates, whereas variability can affect the precision of the estimates and the degree to which they can be generalized.

Variability and uncertainty can complement or confound one another. The National Research Council [12] has drawn an instructive analogy based on estimating the distance between the Earth and the moon. Prior to fairly recent technological developments, it was difficult to accurately measure this distance, resulting in measurement uncertainty. Because the moon's orbit is elliptical, the distance is a variable quantity. If only a few measurements were taken without knowledge of the elliptical pattern, then either of the following incorrect conclusions might be reached:

- The measurements were faulty, thereby ascribing to uncertainty what was actually caused by variability.
- The moon's orbit was random, thereby not allowing uncertainty to shed light on seemingly unexplainable differences that are in fact variable and predictable.

A more fundamental error in the above situation might be to incorrectly estimate the true distance and assume that a few observations were sufficient. This latter pitfall—treating a highly variable quantity as if it were invariant or only uncertain—is most relevant to the exposure or risk assessor [42].

Now consider a situation that relates to exposure, such as estimating the average daily dose by one exposure route—ingestion of contaminated drinking water. Suppose that it is possible to measure an individual's daily water consumption (and concentration of the contaminant) exactly, thereby eliminating uncertainty in the measured daily dose. The daily dose still has an inherent day-to-day variability because of changes in the individual's daily water intake or concentration of the contaminant in the water [42].

Clearly, it is impractical to measure the individual's dose every day. For this reason, the exposure assessor may estimate the average daily dose (ADD) based on a finite number of measurements, in an attempt to average out the day-to-day variability. The individual has a true (but unknown) ADD, which has now been estimated based on a sample of measurements. Because the individual's true average is unknown, it is uncertain how close the estimate is to the true value; thus, the variability across daily doses has been translated into uncertainty in the ADD. Although the individual's true ADD has no variability, the estimate of the ADD has some uncertainty [42].

The preceding discussion pertains to the ADD for one person. Now consider a distribution of ADDs across individuals in a defined population (e.g., the general U.S. population). In this case, variability refers to the range and distribution of ADDs across individuals in the population. By comparison, uncertainty refers to the exposure assessor's state of knowledge about that distribution or about parameters describing the distribution (e.g., mean, standard deviation, general shape, various percentiles) [42].

As noted by the National Research Council [42], the realms of variability and uncertainty have fundamentally different ramifications for science and judgment. For example, uncertainty may force decision-makers to judge how probable it is that exposures have been overestimated or underestimated for every member of the exposed population, whereas variability forces them to cope with the certainty that different individuals are subject to exposures both above and below any of the exposure levels chosen as a reference point.

## Types of Variability

Variability in exposure is related to an individual's location, activity, and behavior or preferences at a particular point in time, as well as pollutant emission rates and physicochemical processes that affect concentrations in various media (e.g., air, soil, food, and water). The variations in pollutant-specific emissions or processes and in individual locations, activities, or behaviors are not necessarily independent of one another. For example, both personal activities and pollutant concentrations at a specific location might vary in response to weather conditions or between weekdays and weekends [42].

At a more fundamental level, three types of variability can be distinguished:

- Variability across locations (spatial variability)
- Variability over time (temporal variability)
- Variability among individuals (interindividual variability)

Spatial variability can occur both at regional (macroscale) and local (microscale) levels; for example, fish intake rates can vary depending on the region of the country. Higher consumption may occur among populations located near large bodies of water such as the Great Lakes or coastal areas. As another example, outdoor pollutant levels can be affected at the regional level by industrial activities and at the local level by activities of individuals. In general, higher exposures tend to be associated with closer proximity to the pollutant source, whether it is an industrial plant or related to a personal activity such as showering or gardening. In the context of exposure to airborne pollutants, the concept of a *microenvironment* has been introduced to denote a specific locality (e.g., a residential lot or a room in a specific building) where the airborne concentration can be treated as homogeneous (i.e., invariant) at a particular point in time.

Temporal variability refers to variations over time, whether long or short term. Seasonal fluctuations in weather, pesticide applications, use of woodburning appliances, and fraction of time spent outdoors are examples of longer term variability. Examples of shorter term variability are differences in industrial or personal activities on weekdays vs. weekends or at different times of the day.

Interindividual variability can be either of two types: (1) human characteristics such as age or body weight, and (2) human behaviors such as location and activity patterns. Each of these variabilities, in turn, may be related to several underlying phenomena that vary; for example, the natural variability in human weight is due to a combination of genetic, nutritional, and other lifestyle or environmental factors. Variability arising from independent factors that combine multiplicatively generally will lead to an approximately log-normal distribution across the population, or across spatial/temporal dimensions [30,42,408].

## Monte Carlo Analysis

Among the most significant advances in exposure assessment of the past 10 to 15 years is the application of Monte Carlo or other probabilistic analyses to environmental health issues [24,82,409–411]. Monte Carlo analysis has existed as an engineering analytical tool for many years, but the development of specialized computer software, such as Crystal Ball (Decisioneering; Boulder, CO) and @RISK (Palisades Corp., Newfield, NY), has allowed its application to new areas. As discussed previously, one criticism of many exposure assessments has been a reliance on overly conservative assumptions about exposure, as well as the problem of how to properly account for the highly exposed (but usually small) populations [74,412,413]. The Monte Carlo technique offers an approach to addressing this issue. Since about 1990, at least 30 risk assessments have been published in the peer-reviewed literature that rely on this technique.

The probabilistic or Monte Carlo model accounts for the uncertainty in select parameters evaluating the range and probability of plausible exposure levels. Instead of specifying input parameters as single values, this model allows for consideration of the probability distributions. The Monte Carlo statistical simulation is a statistical model in which the input parameters to an equation are varied simultaneously. The values are chosen from the parameter distributions, with the frequency of a particular value being equal to the relative frequency of the parameter in the distribution. The simulation involves the following three steps:

- The probability distribution of each equation parameter (input parameter) is characterized, and the distribution is specified for the Monte Carlo simulation. If the data cannot be fit to a distribution, the data are bootstrapped into the simulation, meaning that the input values are randomly selected from the actual data set without a specified distribution.
- For each iteration of the simulation, one value is randomly selected from each parameter distribution, and the equation is run. Many iterations are performed, such that the random selections for each parameter approximate the distribution of the parameter. For each dose equation, 5000 iterations are typically performed.
- Each iteration of the equation is evaluated and saved; hence, a probability distribution of the equation output (possible doses) is generated.

This technique generates distributions that describe the uncertainty associated with the risk estimate (resultant doses). The predicted dose for every 50th percentile to 95th percentile of the exposed population and the true

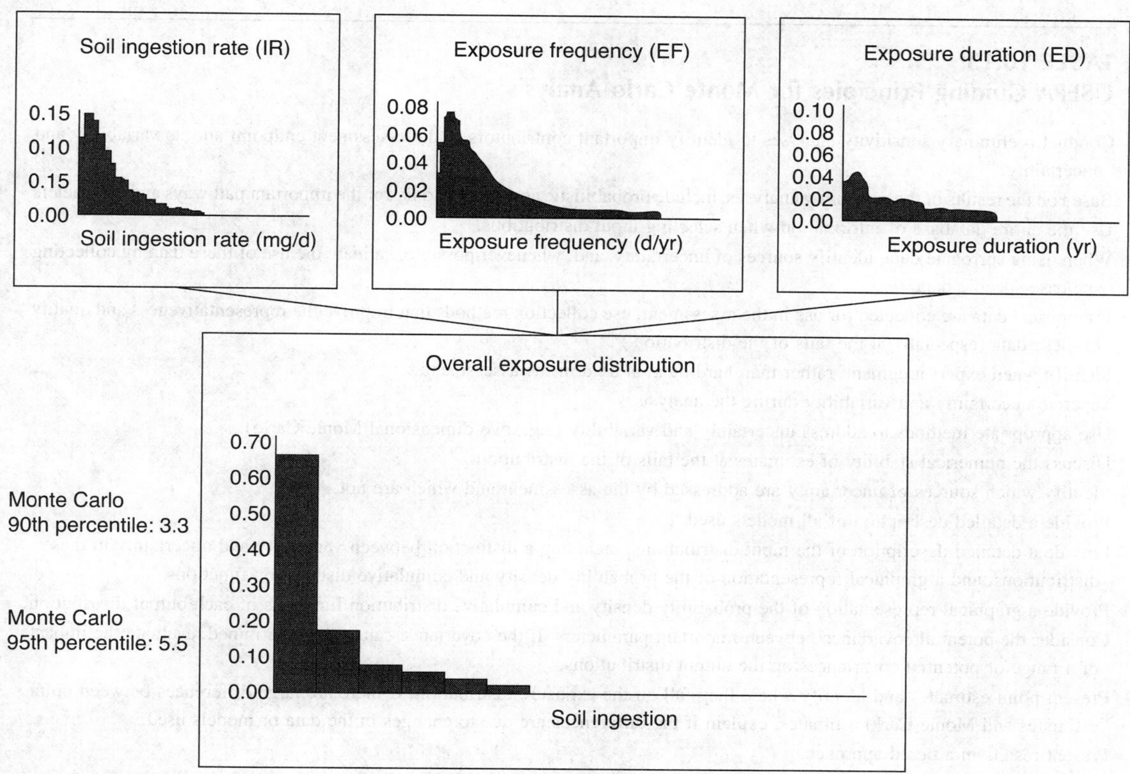


**FIGURE 10.6** Example of how probability density functions (distributions) for three different related exposure factors are combined to form a distribution for the amount of soil ingested by a population of children. The Monte Carlo technique allows the risk assessor to account for the variability in many exposure parameters within a population and then produce a distribution that characterizes the entire population.

mean are calculated. Using these models, the assessor is not forced to rely solely on a single exposure parameter or the repeated use of conservative assumptions to identify the plausible dose and risk estimates. Instead, the full range of possible values and their likelihood of occurrence are incorporated into the analysis to produce the range and probability of expected exposure levels [36,79–81,140, 411,412].

The methodology is illustrated in the following examples. The first example is developing an understanding of the time required to go shopping. Time spent shopping each month (minutes) is estimated by the product of two parameters: the number of trips per month and the total time spent in the store (minutes). Total time spent in the store is the sum of time spent shopping and time spent waiting in line. Using Monte Carlo techniques, a distribution of likely values is associated with each of these parameters. These distributions depend on the detail of information available to characterize each parameter; for example, the distribution compares all of the information, such as those days when the line at the check-out counter is short, as well as those when the line is long. It is noteworthy that each parameter has a different distribution: log-normal, Gaussian, and square. Total time spent shopping is then calculated repeatedly by combining parameter values that are randomly selected from these distributions. The result is a distribution of likely time spent shopping each month. Using this technique, information concerning each parameter is carried along to the final estimate.

The second example, which directly applies to toxicologists, is to build a distribution that describes the various soil ingestion rates for children. As shown in Figure 10.5, the three pertinent distributions are the basis for constructing the overall exposure distribution. Most of the variables used in an exposure assessment actually exist as ranges, rather than single point values. For example, the common assumption that adult body weight is 70 kg will be replaced in a Monte Carlo analysis by the appropriate distribution (i.e., normal) of body weights (including maximum, minimum, mean, and standard deviation). Using this approach, virtually every exposure variable, whether physiological, behavioral, environmental, or chronological, can be replaced with a probability distribution [36,76, 77,79–81,140,238,384,412–421]. Because no population (or individual) is exposed to a single concentration; breathes, eats, or drinks at a single rate; or is exposed for the same length of time, it is not appropriate to assess

**TABLE 10.13**
**USEPA Guiding Principles for Monte Carlo Analysis**

Conduct preliminary sensitivity analyses to identify important contributors to the assessment endpoint and its variability and uncertainty.

Based on the results of the sensitivity analyses, include probabilistic assessments only for the important pathways and parameters.

Use the entire database of information when selecting input distributions.

When using surrogate data, identify sources of uncertainty, and, whenever possible, validate the use of these data by collecting site/case specific data.

If empirical data are collected for use in the assessment, use collection methods that improve the representativeness and quality of these data (especially at the tails of the distribution.)

Identify when expert judgment, rather than hard data, is used in the assessment.

Separate uncertainty and variability during the analysis.

Use appropriate methods to address uncertainly and variability (e.g., two-dimensional Monte Carlo).

Discuss the numerical stability of estimates at the tails of the distribution.

Identify which sources of uncertainty are addressed by the assessment and which are not.

Provide a detailed description of all models used.

Provide a detailed description of the input distributions, including a distinction between variability and uncertainty in these distributions, and a graphical representation of the probability density and cumulative distribution functions.

Provide a graphical representation of the probability density and cumulative distribution functions of each output distribution.

Consider the potential covariance between important parameters. If the covariance cannot be determined, evaluate the impact of a range of potential covariances on the output distributions.

Present point estimates and identify where they fall on the exposure distribution. If there are large differences between point estimates and Monte Carlo estimates, explain if the differences are due to changes in the data or models used.

Present results in a tiered approach.

them as such. To be protective, high values are employed, resulting in the problems of compounding conservatisms mentioned previously [66,67,74].

The probabilistic analysis addresses the main deficiencies of the point estimate approach because it imparts more information to risk managers and the public and uses all of the available data [423,425–427]. The range of values (i.e., the distribution) for all the variables used in an exposure assessment is determined (e.g., normal, log-normal, uniform, triangular) and combined into a distribution of distributions. Because of the extrapolations involved and the assumptions made, the area of single greatest uncertainty in risk assessment is associated with the dose–response evaluations.

It should also be clear that, in addition to exposure variables, data forming the basis of the toxicological criteria—carcinogenic potency factors (CPFs) and reference doses (RfDs)—are also amenable to Monte Carlo-style analysis where a robust database exists [422,424,426–438]. As with exposure variables, the advantage to this approach is that it allows all data to be used (and weighted appropriately, where necessary), thus avoiding reliance on a single experiment or endpoint.

Probabilistic analyses have in recent years been recognized in regulatory guidance [82,86,87], and the USEPA's Risk Assessment Forum has published a document of principles for conducting Monte Carlo analyses (Table 10.13) [82].

Like traditional exposure analysis, one challenge to performing a Monte Carlo analysis properly is having appropriate distributions for use in the analysis. Numerous studies on individual variables have been published in the risk assessment literature [71,75,283 439–440,442,443], and the impact on the distributions employed on the outcome has also been discussed [30,444–448]. It should be pointed out that these techniques can be combined with other advanced risk assessment methods (e.g., PB-PK modeling) to further reduce uncertainty in risk estimates [83,449]. Recently two-dimensional Monte Carlo analyses and probabilistic approaches, in general, have been developed that take into account both variability and uncertainty and have been utilized as a method to quantify uncertainty in sensitivity analyses [85,450–454]. Information appropriate to probabilistic analyses can often be found in published papers in fields quite distant from the environmental sciences.

## Case Study Using Monte Carlo Technique

An example might be useful [441]. Assume that person are likely to be exposed to contaminated drinking wate at the maximum contaminant limit (MCL). Concern ha been raised that these regulatory limits are not sufficientl protective, and that certain federal and state regulator programs (e.g., Resource Conservation and Recovery Act

**TABLE 10.14**
**Risks Calculated for Exposure to Four Halogenated Solvents in Water Using Probabilistic Analysis at the MCL Level and for the 50th and 95th Percentile Exposure**

| Chemical | 50th Percentile Risk | 95th Percentile Risk | MCL Risk |
|---|---|---|---|
| Tetrachloroethylene | 0 | 0.000005 | 0.000007 |
| Chloroform | 0.000009 | 0.00014 | 0.000017 |
| Bromoform | 0.000002 | 0.000016 | 0.000023 |
| Vinyl chloride | 0.000005 | 0.000029 | 0.000054 |

are justified in requiring groundwater remediation to levels below that of drinking water standards. To test this supposition, it is necessary to evaluate the possible incremental cancer risk of exposure via tap-water ingestion, dermal contact with water while showering, inhalation of indoor vapors, and ingestion of produce irrigated with groundwater using a probabilistic approach. Probability density functions for each exposure variable (e.g., water ingestion, skin surface area, fraction of exposed skin, showering time, inhalation rate, air exchange and water use rates, exposure time) are then identified and used in the appropriate exposure equation to calculate dose and risk.

Some have suggested that the Latin hypercube (LHC) approach offers some advantages to traditional approaches for identifying the correct number of iterations. Often, one can reach convergence sooner with LHC than the Monte Carlo option in @RISK or Crystal Ball. In addition, LHC is more reproducible (to the hundredth decimal place). The Monte Carlo option requires more iterations to reach convergence.

The results of such an analysis are presented in Table 10.14 [441]. The risk associated with exposure to water at the current maximum contaminant level for four different contaminants, as well as the 50th and 95th percentiles of exposure as determined by the probabilistic analysis, are shown. At the 50th percentile level (the best estimate), the risk ranges from $6\times10^{-7}$ (tetrachloroethylene) to $9\times10^{-6}$ (chloroform); at the 95th percentile (the upper-bound risk), these risks range from $4\times10^{-6}$ (tetrachloroethylene) to $1.5\times10^{-4}$ (chloroform). These values can be compared to the point estimate risks calculated for the MCLs, which range from $7\times10^{-6}$ (tetrachloroethylene) to $5.4\times10^{-5}$ (vinyl chloride). For the 50th percentile (average) person, all calculated risks are within the range of acceptable risks adopted by regulatory authorities for Superfund sites ($1\times10^{-4}$ to $1\times10^{-7}$). For the 95th percentile person (upper bound), the risks are still mostly below the $1\times10^{-4}$ benchmark risk level generally used to separate acceptable from unacceptable risks. For tetrachloroethylene, these results are 30 (50th percentile) to 3 (risk at the MCL) times below the reasonable maximum exposure (RME) risk of $2\times10^{-5}$ developed by combining the 95th percentile values for each exposure variable using standard USEPA risk assessment methodologies. This point estimate is greater than the 99th percentile of risk and is consistent with statements regarding the conservatism of the RME approach. These results suggest that chemical residues in drinking water at the MCL levels will be health protective and that remedial goals based on *de minimis* requirements ($1\times10^{-6}$) might be unnecessarily low [441].

In terms of estimates for the RME individual which often serve as the basis for regulatory decisions, several observations on the utility of probabilistic assessment can be made. First, exposure assessments that incorporate two to three direct exposure pathways usually show that the 95th percentile probabilistic estimates are three to five times below the traditional RME estimates. Second, for multiple-pathway assessments that contain several indirect exposure pathways, the 95th percentile probabilistic estimates can be as much as an order of magnitude below the RME estimates. Third, when the number of distributions used in the exposure assessment is 10 or more, the difference between the 50th and 95th percentile estimates may be between 5 and 10. Finally, in such assessments, the difference between the RME estimates and the 95th percentile probabilistic estimates can be as high as 100. In the probabilistic approach to estimating exposure and risk, the complete range of potential risks can be illustrated along with the likelihood estimates and estimates of uncertainty associated with such risks. Although the availability and confidence of distributions for exposure variables differ, risk assessors ought to take advantage of this and similar approaches in their risk assessments to advance and improve the process. Additionally, because the highest degree of uncertainty in risk assessment tends to be the CPFs, attention ought to be directed to applying probabilistic analysis to the development of toxicity criteria in a similar manner [431,438].

## Sensitivity Analysis

In addition to establishing exposure and risk distributions, probabilistic analysis can also identify variables with the greatest impact on the estimates and illuminate uncertainties associated with exposure variables through sensitivity analysis [455–459]. Sensitivity analysis is the study of how the uncertainty in the output of a model (numerical or otherwise) can be apportioned to different sources of uncertainty in the model input. The sensitivity analysis is hence considered by some as a prerequisite for model building in any setting and in any field where models are used [454]. Sensitivity analysis can help in identifying critical control points, prioritizing additional data collection, and verifying and validating a model [451]. This provides some insight into the confidence that resides in exposure and risk estimates and has two important results. First, it identifies the inputs that would benefit most from additional research to reduce uncertainty and improve risk estimates. Second, assuming that a thorough assessment has been conducted, it is possible to phrase the results in more accessible terms, such as: "The risk assessment of PCBs in smallmouth bass is based on a large amount of high-quality reliable data, and we have high confidence in the risk estimates derived. The analysis has determined that 90% of the increased cancer risk could be eliminated through a ban on carp and catfish, but there is no appreciable reduction in risk from extending such a ban to bass and trout" [437]. Such a description provides all stakeholders with considerably more information than a simple point estimate of risk based on a traditional exposure and risk assessment.

If the most sensitive exposure variables are based on limited or uncertain data, confidence in these estimates will be poor. Robust datasets, on the other hand, lead to increased confidence in the resulting estimates. In the above example involving smallmouth bass, sensitivity is defined as the ratio of the relative change in risk produced by a unit relative to change in the exposure variables used. A Gaussian approximation (the product of the normalized sensitivity and the standard deviation of the distribution) of intake was used to allow both sensitivity and uncertainty to be gauged. In this case, the true mean of each distribution was chosen as the baseline point value, and the differential value for each variable was calculated by increasing this value by 10%. For each variable, the differential value was substituted, the risks recalculated, and the baseline value replaced [441]. Sensitivity was calculated using the following formula:

$$\text{Sensitivity} = \frac{\left|Risk_{baseline} - Risk_{10\%}\right|}{\left|X_{baseline} - X_{10\%}\right|} \times [\sigma]$$

where $X_{baseline}$ and $X_{10\%}$ are the baseline and differential values, respectively, for variable $X$, and $\sigma$ is the standard deviation for the distribution of variable $X$. The sensitivity of each variable relative to one another is assessed by summing the unitless sensitivity values and determining the relative percent that each variable contributes to the total.

**TABLE 10.15**
**Results of Sensitivity Analysis for Tetrachloroethylene Exposure in Household Water**

| Exposure Variable | Sensitivity (unitless) | Percentage Rank (%) |
|---|---|---|
| Shower exposure time | 0.000004 | 55.0 |
| Exposure duration | 0.000001 | 20.0 |
| Plant–soil partition factor | 0 | 8.4 |
| Water ingestion rate | 0 | 4.6 |
| Surface area of exposed skin | 0 | 4.4 |
| Body weight | 0 | 3.8 |
| Dermal permeability constant | 0 | 1.8 |
| Skin fraction contacting water | 0 | 1.5 |

Table 10.15 identifies the most important variables in the probabilistic analysis for tetrachloroethylene. In this case, the most sensitive exposure variables in household exposure to tap water are exposure time in the shower and exposure duration. Relatively small changes in these variables will result in relatively large changes in the risk estimates. Because these estimates are based on actual time-use studies and census information, this suggests that a high level of confidence can be placed on this estimate, particularly if site-specific data are being used. If the critical variables (in terms of sensitivity) were not based on robust data, this would suggest that the risk assessment could be improved by additional research on these exposure variables. It is interesting that the form of the distribution chosen for the variables is less important than the validity of the data [72]. When the empirical distribution of the tap-water ingestion rate was substituted with a lognormal distribution [460,461], the resultant change in risk estimates was less than 1%.

In this case, the value of the sensitivity analysis is that it allows input variables to be ranked in order of importance and confidence in the output to be established to a higher degree than previously possible. As pointed out by the USEPA, "Where possible, exposure assessors should report variability in exposures as numerical distributions and should characterize uncertainty as probability distributions. They need to identify clearly where they are using point estimates for 'bounding' potential exposure variables or estimates; these point estimates should not be misconstrued to represent, for example, the upper 95th percentile when information on the actual distribution is lacking" [437]. As noted by the USEPA, such explicit presentation of the data reduces the temptation to use the exposure assessment process for veiling policy judgments [462].

**TABLE 10.16**
**Effect of Matrix and Aging on the Bioavailability of Lead from Soil**

| Treatment | | Tibial Lead | | Relative Lead Absorption (%) |
|---|---|---|---|---|
| Lead Acetate (ppm diet) | Soil Lead (ppm) | ppm | Standard Deviation | |
| — | — | 0.3 | 0.3 | — |
| — | 11.3 | 0 | — | — |
| 50 | — | 247 | 10 | 100 |
| 50 | 11.3 | 130 | 30 | 53 |
| — | 706 | 40 | 6 | 16 |
| — | 995 | 108 | 26 | 44 |
| — | 1080 | 37 | 7.3 | 15 |
| — | 1260 | 53.6 | 7 | 22 |
| — | 10,420 | 173 | 22 | 70 |

## EVOLVING RESEARCH IN EXPOSURE ASSESSMENT

The field of exposure assessment will continue to benefit from ongoing research efforts. The following are some fruitful areas of ongoing research.

### Bioavailability

Areas of applied research that will improve the practice of exposure assessment include bioavailability, speciation, chemical fate, and the role of biological monitoring. For nearly 20 years, consideration of the bioavailability of a chemical in a various media has become an increasingly important aspect of the exposure assessment process [50,100,116,118,119,463]. Alexander [464] has shown that a variety of organic chemicals in soil lose the ability to interact with biological receptors over time, despite the fact that the chemical concentration in soil remains largely the same. The alteration in bioavailability extends across the various routes of exposure, as well [104–106,109–111,113, 117,120,123,278–280,463,465,466]. Inorganic compounds, even those posing a potentially significant degree of hazard (i.e., cyanide), can react similarly [467–470]. These losses in hazard potential are presumably due to irreversible chemical interactions with soil constituents. Table 10.16 indicates that the bioavailability of lead added to soil is immediately halved and that it is further reduced over time [471]. This would suggest that an assumption of 100% bioavailability of this compound (and many others) from soil is erroneous. It is also clear that the environmental media in which the compound occurs will influence its uptake into the body [465]. The USEPA recognized this fact when it developed two RfDs for manganese depending on whether it occurred in solid matrices (e.g., food or soil) or water [10]. One simple method to improve bioavailability estimates is to conduct extractions under more biologically relevant conditions.

Bench-scale extraction experiments in simulated gastric fluids or sweat can be used to inexpensively and accurately measure how readily environmental residues can be released from the media in which they occur [100, 122,472]. As with inhalation or ingestion of vapors or solutions, both the release and absorption rates of agents from an environmental matrix (e.g., soil) across biological membranes must be incorporated into the risk assessment when such data are available and generated when absent. This requirement is particularly an issue when assessing dermal exposure. The problem for materials in aqueous solutions is less problematic than from solid matrices [89]. For liquids, permeability constants expressed in terms of agent weight per unit area per time (mg/cm$^2$/min) have been developed for a number of agents, and *in vivo* and *in vitro* techniques or mathematical models exist to develop similar flux rates if needed [231,232,235,255, 257,262,280,466,474–476]. From soil, however, the typical approach in many risk assessments has been to assume a constant percent absorbed from soil adhered to skin as a default. For volatiles, an absorption rate of 25% has been used. For semivolatiles and inorganics, absorption rates of 10% and 1% have been used, respectively.

Some experimental data for absorption are available for a few agents (e.g., PCBs, DDT, dioxin, benzo(*a*)-pyrene), suggesting that the simple assumption of a constant percentage absorbed may overestimate or underestimate the dose depending on the agent, co-contaminants, soil type, exposure duration, and similar considerations [123,278,280,466]. The impact of this default approach results in an instantaneous dermal dose being assumed, regardless of whether the soil remains in contact with the skin for one minute or one day. This assumption, together with the questionable route-to-route adjustment of toxicity criteria from oral to dermal previously discussed, results in the dermal absorption of agents from soil, which arguably should present a minor exposure and risk in most cases, being a major driver in the risk assessment of soil-bound contaminants.

## Chemical Fate

Risk assessors ought to incorporate information on the fate of chemicals in the environment in their exposure estimates, whenever possible [468]. Many organic compounds tend to degrade over time and may disappear from exposed surfaces relatively quickly or otherwise change [69,477]. As suggested above, inorganic compounds may also undergo changes in the environment over time that affect their fates [470,471]. Influencing factors include degradation by sunlight, soil and water microbes, evaporation, and chemical interactions. The resultant changes can dramatically alter the outcome of exposure assessments [50,467]. For example, most criticism of incinerators has focused on the inhalation risk of dioxin emitted from the stacks. As it turns out, the environmental half-life of dioxin (as a vapor) is only 90 minutes because of photolytic degradation. In contrast, the half-life for dioxin in soil or fly ash is 50 to 500 years. The focus of concern is often not the main risk issue when environmental fate is considered because levels and availability change over time [7]. Incorporation of half-life data into risk assessments can have substantial benefits for improving the understanding of the potential exposures and risks associated with a specific situation [2,478]. In a similar manner, the risk from persistent contaminants (i.e., DDT) in fish has usually been assessed using results from the analysis of raw fish fillets in combination with assumptions about the size and number of fish meals. The effects of cleaning and cooking on these residues are not typically considered but they have been shown to be reduced substantially in many cases (i.e., 50% or greater) [479,480]. Because many of these risk assessments form the basis of fish advisories or bans with potentially significant economic repercussions, it is obviously important to make these exposure estimates as accurate as possible. Additionally, because of the known health benefits of fish consumption, making recommendations against eating fish based on theoretical risk needs to be rigorously defended.

## Biomarkers and Molecular Epidemiology

There is general agreement among the scientific community that diseases that contribute the greatest public health burden to society result from complex interactions between genetic and environmental factors, such as chemical pollutants, nutrition, lifestyle, and infectious agents [164,482–484]. The field of epidemiology is a critical field for understanding these interactions. The cornerstone of exposure assessment in epidemiologic studies is the development of the exposure metric, the estimate of exposure for each individual of the study population [164].

The past decade has witnessed a dramatic increase in the level of research activity, derivation of theoretical constructs, and development of practical applications for the direct measurement of biological events or responses that result from human exposure to xenobiotics [195,485]. These measurements, conveniently grouped under the descriptor "biological markers" or "biomarkers," reflect molecular and cellular alterations that occur along the temporal and mechanistic pathways connecting ambient exposure to a toxicant and eventual disease. As such, an almost limitless array of biomarkers is theoretically available for assessment, and only a minute fraction of these has been recognized and investigated to date [485–487].

The term *biomarker* is a general term for specific measurements of an interaction between a biological system and an environmental agent [156,488]. Biomarkers can be broadly grouped into several categories: biomarkers of internal exposure, biomarkers of early biological effects, susceptibility biomarkers, genomic biomarkers, and biomarkers of health risk [489]. The International Program of Chemical Safety defines three classes of biomarkers [488]:

- The *biomarker of exposure* is defined as an exogenous substance or its metabolite or the product of an interaction between a xenobiotic agent and some target molecule or cell that is measured in a compartment within an organism.
- The *biomarker of effect* is a measurable biochemical, physiological, behavioral, or other alteration within an organism that, depending on the magnitude, can be recognized as being associated with an established or possible health impairment or disease.
- The *biomarker of susceptibility* is an indicator of an inherent or acquired ability of an organism to respond to the challenge of exposure to a specific xenobiotic substance.

Some events that can technically be classified as biomarkers of chemical exposure (e.g., hematological changes following high levels of exposure to lead or benzene, acetylcholinesterase inhibition by organophosphates) have been measured for decades. However, the recent surge of interest in this field has been driven by technical advances in analytical chemistry and molecular genetic techniques and by the recognition that classical toxicology and epidemiology may not be able to alone resolve critical questions regarding causation of environmentally induced disease [195].

Epidemiology relies on the inference of associations between exposure and response variables. Typically, the measurements of response in epidemiologic studies reflect late-stage endpoints of morbidity, mortality, body-weight decrease, tumor development, and tissue pathology [164,490,491]. Defining risk at a late stage in the disease process provides little opportunity to intervene and redirect the outcome. It is clearly more desirable to identify early changes in biologic processes that can serve as predictive markers of exposure, of early effect or of susceptibility [164,492]. Biomarkers are an important component of the emerging discipline of molecular epidemiology, which seeks to expand the capabilities and overcome the limitations of classical epidemiology by incorporating biological measurements collected in exposed humans [195,493].

Early efforts at utilizing biomarkers to make quantitative estimates of exposure and to predict human cancer risk were made by Ehrenberg and Osterman-Golkar [494]. Using ethylene oxide as a model xenobiotic, these investigators explored the use of macromolecular reaction products (e.g., hemoglobin adducts) as internal dosimeters. By employing hemoglobin adduction data, they predicted the level of ambient ethylene oxide that would correspond to a tumorigenic dose of $\gamma$-radiation, which they termed the "rad-equivalent dose." Seminal work in the area of biomarkers as applied to the molecular epidemiology of cancer was performed by Perera and Weinstein [495], who proposed the use of such techniques to identify environmental contributors to human cancer incidence. Important early applications of biomarkers to characterize environmental and occupational exposure have also been explored by several other groups in the United States and abroad.

As presented in the original NRC report, biomarkers of internal dose reflect the absorbed fraction of a xenobiotic (i.e., the amount of material that has successfully crossed physiological barriers to enter the organism) [492]. Consequently, the magnitude of the biomarker accounts for bioavailability and is influenced by numerous parameters such as route of exposure, physiological characteristics of the receptor, and chemical characteristics of the xenobiotic. Generally, simple measurement of xenobiotic levels in biological media (blood, tissue, urine) can provide data on internal dose, and this is referred to as *biomonitoring* [497–499]. As employed in most studies, biomonitoring indicates the presence of the substance or marker in the body at a single point in time, corresponding to when the specimen was taken, but such data alone provide no information on the source, pathway, the magnitude, the frequency, or the duration of exposure [500,501]. Biomonitoring of exposure involves measurement of the concentration of a chemical in a given biologic matrix during or after absorption, distribution, metabolism, and excretion (ADME), and its concentration level depends on the amount of the chemical that has been absorbed into the body, the pharmacokinetics (ADME) of the chemical, and the exposure scenario, including the time sequence of exposure and time since last exposure [502,503]. Ideally, to link the dose with adverse health outcomes, measurements of the biologically effective dose (the dose at the target site that causes an adverse health effect) are preferred [501,502].

Biomarkers reflect internal dose (in terms of proximity to downstream events in the sequence) and could include the measurement of a metabolite in selected biological media, particularly if such metabolite is active or critical to the toxic effects seen [195]. A biomarker should be biologically relevant, sensitive, and specific (i.e., valid) and should be readily accessible, inexpensive, and technically feasible. Analytical, metabolic, and source specificity are important aspects to consider when identifying an appropriate biomarker, where the analytical specificity refers to the capability of the analytical method to exclusively measure the chemical (parent or metabolite) of interest, metabolic specificity means that the chemical measured is derived exclusively from the parent chemical of interest, and source specificity indicates the source of the chemical in the body [504]. This combination of requirements is rarely achieved, and some tradeoff is inevitable in order to obtain useful biomarker data in a timely manner. A few promising examples are presented in Table 10.17. The validation process for a biomarker involves determining the relationship between the biological parameter measured and both upstream and downstream events in the continuum; that is, the dose–response curve must be characterized [195].

For the majority of chemicals of interest, occupational or environmental sources may not be the only source of exposure. The identification of the exposure depends on the concentration difference in the exposed individual or group, in comparison to the general population [504]. Over the last decade, the CDC has expanded its biomonitoring efforts to better characterize potential trends in chemical exposures to the general population [156]. In March 2001, the CDC released the *National Report on Human Exposure to Environmental Chemicals*, which provided summary analyses for blood and urine samples obtained in 1999 from the National Health and Nutrition Examination Survey (NHANES 99+) and enhanced information from previous NHANES. In January 2003, the CDC released the *Second National Report on Human Exposure to Environmental Chemicals*, which presented exposure information and separate analyses by age, sex, and race/ethnicity on 116 environmental chemicals in people who had blood and urine samples taken during 1999 and 2000 [107,505]. The national exposure information identified which chemicals get into Americans in measurable quantities, determines whether exposure levels are higher among population subgroups, determines how many Americans have levels of chemicals above recognized health threshold levels, establishes reference ranges that define general population expo-

**TABLE 10.17**
**Biomarkers Examined for Selected Occupational and Environmental Chemicals**

| Chemical | Biomarker: Exposure | Biomarker: Effect | Biomarker: Susceptibility |
|---|---|---|---|
| PAH | DNA adducts[a] | *hprt* mutation | GST-M1 |
| | Hb adducts | *gpa* mutation | NAT-2 |
| | SA adducts | *fes* oncogene activation | CYP1A1 |
| | Urinary 1-HP[a] | *ras* p21 level | CYP2A2 |
| | Sister chromatid exchange (SCE) | DNA single-strand breaks | |
| | SCE (high-frequency cells) | Chromosomal aberrations | |
| | | Micronuclei | |
| 1,3-Butadiene | Hb adducts[a] | *hprt* mutation | |
| | Sister chromatid exchange (SCE) | Chromosomal aberrations | |
| | Urinary metabolites | Micronuclei | |
| | | *ras* oncogene activation | |
| Acrylamide | Hb adducts[a] | | |
| | Urinary metabolites | | |

[a] Biomarkers for which cumulative data indicate best correlation with ambient exposure.

sure so unusual exposures can be recognized, assesses the effectiveness of public health efforts to reduce population exposure to selected chemicals, and tracks over time trends in U.S. population exposure [506].

These data have been proven to be very useful in understanding national trends of chemical exposure, as well as the effectiveness of laws intended to restrict environmental emissions. The biomonitoring of lead and persistent organic pollutants is probably the best example of how biomonitoring can confirm the reduction of human exposures as a result of restrictions of environmental emissions (Figure 10.7). Blood lead measurements in the population were important in identifying lead in gasoline as a significant source of human lead exposure and documenting the reduction in blood lead levels in the population as a result of removing lead from gasoline and other products in the

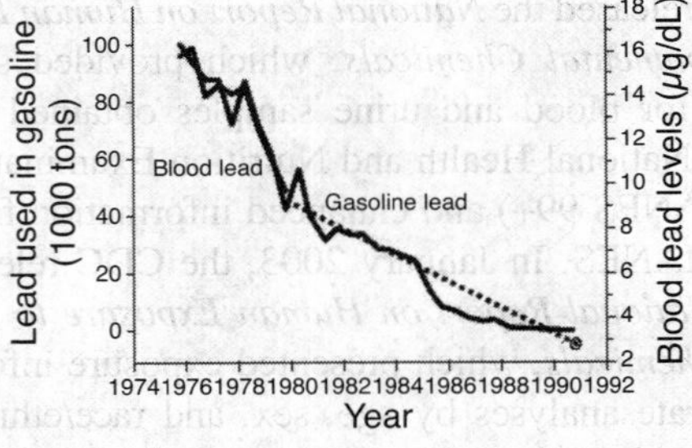


**FIGURE 10.7** Blood lead levels (μg/dL) in relation to the use of lead in gasoline. The NHANES III dataset illustrates how human lead levels in the United States continued a steep decline over the course of two decades following the removal of lead from gasolines used for passenger vehicles. (From CDC, *Third National Report on Human Exposure to Environmental Chemicals*, Centers for Disease Control and Prevention, 2005, http://www.cdc.gov/exposurereport/3rd/.)

United States [506]. In the United States, the mean lead blood concentration in children during the consecutive phases of NHANES II, III, and IV in 1976–1980, 1988–1991, and 1991–1994 were 150, 36, and 27 μg/L, respectively [156,507]. Serum cotinine levels in the early 1990s indicated more widespread exposure to environmental tobacco smoke (ETS) in the United States than previously thought; additional measurements in 1999 and 2000 documented major declines in exposure to ETS as a result of public health actions in the 1990s [506]. The results of biological monitoring for evaluating the background contamination or the trends regarding contamination have also been used to compare internal exposure to organochlorine compounds or the trends in the concentration of dioxins in breast milk [156,508–510].

Persistent and nonpersistent chemicals can react with biomolecules such as DNA, hemoglobin, or fatty acids to form biomolecular adducts. By using these adducts as a surrogate for exposure, a greater length of time (dependent on the life of the adducts) can pass after exposure before measurements are collected [502]. Biomarkers with half-lives of 7 days or longer exhibit physiologic dampening of fluctuations in external contaminant levels and can offer advantages when compared to short-lived biomarkers of exposures assessed by air monitoring [511]. Ehrenberg and co-workers first proposed using hemoglobin (Hb) adducts to monitor the internal dose of alkenes and epoxides such as ethylene oxide over two decades ago [494]. This methodology has since evolved into a widely used and highly sensitive technique for quantitating N-terminal Hb adducts of a variety of xenobiotic metabolites in human blood. Hb adducts have been employed as internal exposure biomarkers for aromatic amines, nitrosamines, polycyclic aromatic hydrocarbons, and other compounds [195].

Toxicogenomics-based methods have recently been used in laboratory settings to develop biomarkers of exposure, early biologic response, and susceptibility [164]. Toxicogenomics is a broad field that seeks to define, on a global basis, the levels, activities, regulation, and interaction of genes, mRNA transcripts (transcriptomics), proteins (proteomics), and metabolites (metabolomics) in a biologic sample or system [164]. The approaches have been used for classifying exposures to a variety of chemicals and drugs (e.g., hydrazine, 2-bromoethanamine, lead, acetate, cadmium, and acetaminophen) based on mechanism of action and dose and for classifying health outcomes for cardiovascular disease and cancer based on disease status and severity [164,512–525]. The primary basis of classification and discovery in these studies is the molecular signature. Once the discriminating elements of the molecular signature are identified, biologic function can be inferred by mapping components to known biologic pathways and verifying functionality in follow-up studies [164]. One of the greatest challenges with this technology is that background levels of expression and variability for mRNA transcripts, proteins, and metabolites in human tissues are currently not known but must be defined if toxicogenomic methods are to be used to assess personal exposures in epidemiologic studies. Expression levels are expected to vary widely because of differences in diet, lifestyle, health status, and genetic predisposition [164]. Despite the enormous promise of toxicogenomics for advancing our understanding of the relationship between environmental exposure and disease, the challenge has been, and will continue to be, the development of genetic and biologic markers that are predictive of adverse health outcomes in both experimental and human studies [164].

Research over the last decade has expanded the use of biomarkers in applying human biological monitoring data of exposure to individual disease and susceptibility information [195,526,527]. The advancements in analytical methods have allowed researchers to measure markers in a variety of biological specimens, including serum, cord blood, urine, feces, hair, bone, teeth, breast milk, saliva, and exhaled breath [502,508,528–531]. In addition, recent technical advances have also allowed researchers to use biomarkers to address questions about the effectiveness of personal protection equipment and engineering controls in preventing exposures in the workplace [529], the contribution of different exposure pathways on biomarker patterns [532], the effect of genetic polymorphisms on biomarkers of mutagenicity [527,533], the impact of personal activities on individual exposure [154,508], and the correlation of biomarkers of exposure and disease risk [531,534].

Recent research efforts, such as the National Human Exposure Assessment Survey (NHEXAS) [154,535], the Children's Total Exposure to Persistent Pesticides and Other Persistent Organic Pollutants (CTEPP) [152], and other USEPA programs [153], have focused on developing databases of exposures of human populations to a wide range of pollutants in air, water, food, soil, and indoor/residential environments over a wide range of space and time scales. For example, the University of California, Berkeley Center for Health Assessment of Mothers and Children of Salinas (CHAMACOS), is collecting biomarkers in farming communities for pesticides and other important pollutants from mothers and their newborn children from conception through early childhood [153]. Data from NHEXAS has been used to understand lead, phenanthrene, naphthol, and chlorpyrifos environmental concentrations in outdoor air, soil, indoor air, dust, dermal, water, beverages, and diet solids in relation to urine and blood concentrations [154]. In particular, these data were used to evaluate how personal activities and lifestyle factors—for example, seasonal differences, ventilation (window, central air/heat), home environment (paint, cement, carpet, fireplace), pesticide use, garden care, smoking, gas grill use, vacuuming method, schedule (work at home), and personal hygiene—may contribute to exposure to the specific chemicals within the home environment [154].

Another important research question to be addressed with biomarker research is whether PBPK models are broadly applicable as tools for relating dose biomarkers to measures of population exposure and health risk [158]. This approach was recently evaluated in the USEPA dioxin reassessment, in which PBPK models were used to evaluate the reasonablenesss of their earlier estimated cumulative dietary intake of dioxin compounds [536,537]. Wallace et al. [538,539] assessed the utility of using exhaled breath for estimating exposure and body burden for volatile organic compounds (VOCs) based on PBPK models. Chinnery and Gleason [540] and McKone [475] used PBPK models of chloroform applied to breath samples reported by Jo et al. [304,305] to determine the relative contribution of inhalation and dermal exposure routes for adults showering with water containing residual chloroform from disinfection. More recently, Aylward et al. [541] used serial measurements of serum lipid 2,3,7,8-tetrachlorodibenzo-*p*-dioxin (TCDD) concentrations in 36 adults from Seveso, Italy, and 3 patients from Vienna, Austria, to model the distribution and elimination of dioxins. The measurement of chemicals and biomarkers has revolutionalized the field of exposure assessment. New challenges will involve the interpretation of these data for minimizing exposure and health risks, as well as effectively communicating the risk trade-offs to the general public.

## Statistical and Analytical Issues

Despite the use of precise and reproducible analytical methods, we often do not have enough data regarding chemical concentrations to estimate exposure with great certainty. Due to resource availability, over the past 15 to 20 years it has often been the case that a single round of analytical results or samples collected for other purposes serves as input and the surrogate for long-term or lifetime exposure [542]. As noted previously, chemical concentrations vary over both time and space which makes the task of dose estimation all the more difficult. For example, to use the (estimated) average dose to predict the typical lifetime dose may seriously overestimate or underestimate the actual dose. Additionally, the average dose may be less important in the biological scheme of things than peak exposures or exposures at specific times (i.e., developmental effects) and ought to be considered as such in the evaluation of exposure [163]. Techniques do exist for estimating long-term exposure from short-term data [543–545], but the reliability of these estimates is uncertain. Similarly, a variety of mathematical or bench-scale models exist that have been used to estimate exposure in the absence of measurements or long-term monitoring data [546]. As has been noted on several occasions, "all models are wrong, but some are useful," and risk assessors should carefully evaluate mesoscale and microscale models, as well as model outputs, for relevance and accuracy. Often, field measurements can serve as useful and relatively inexpensive reality checks for model results.

Equally important in exposure assessment are the statistics used to analyze field data. Environmental data are most often log-normally distributed. Under such conditions, a geometric average is generally assumed to be a better measure of the central tendency of data than the arithmetic mean [547]. Despite this, the arithmetic mean (and the 95% upper confidence limit of the arithmetic mean) is typically used to identify environmental concentrations for use in exposure assessment. Because the advances in analytical chemistry have improved our ability to measure trace amounts of chemicals in different media and identify potential sources in some situations, less reliance should be placed on the use of mathematical models to predict the distribution of chemical and physical agents in the environment, and actual field data should be collected.

Another important issue in exposure assessment is how the analytical limit of detection (LOD) is handled in calculations. An agent reported as a nondetect may be treated as a numerical zero or occurring at the LOD or some fraction of the LOD, typically one half of the LOD or the LOD divided by the square root of 2, for purposes of calculating statistics. The manner in which censured data is assessed may affect the outcome of the risk assessment process [548–550,552–555]. As an example, analysis of highly contaminated samples or samples containing interfering substances may result in high LODs. Under such conditions and in the absence of additional analysis, assuming that nondetects are present at one half the LOD could result in the exposure assessment and subsequent risk assessment being driven by compounds that are not truly present in the environmental media. When such an approach is used on a site that may be only 2 to 10% contaminated (based on surface area), the predicted severity of the level of contamination will be much higher than what actually exists [547]. In these cases, it is often appropriate to insert a value lower than one half the LOD when conducting exposure assessment calculations.

The practical result of these decisions can be illustrated by considering the following 11 data points resulting from analysis of field samples: Nondetect (ND), ND, ND, ND, ND, 5, 6, 6, 8, 55, and 500 ppm. The results are log-normally distributed, as expected. The detection limit is 0.05 ppm, and nondetects are assumed to be present at one half the detection limit (0.025 ppm). Using these assumptions, the arithmetic mean of the data set is 52.7 ppm, and the geometric mean is 1.3 ppm. The practical consequence of choosing one descriptor over the other may be to misidentify or mischaracterize the dose and ultimately the risk, which will influence regulatory decisions involving remediation and regulation.

## CLOSING THOUGHTS

The field of exposure assessment has evolved significantly over the past 20 years. We have learned a great deal about where people are exposed to xenobiotics and the relative degree of exposure. Not that long ago, most of our concerns were about industrial chemicals in our water, ambient air, and the soil. Today, we know that indoor exposure to particles, vapors, and gases in the home (influenced by smoking) often represents the predominant source of exposure for most persons. Aggregate exposure assessment and biomonitoring has changed the field of exposure assessment tremendously and is moving research to look at more complex and less obvious sources and pathways of exposure. A greater portion of our work in the future will undoubtedly focus on better understanding the individual contribution that environmental and indoor sources have to personal exposures [164,204,206], differential exposures among susceptible populations (e.g., children, elderly, disease-compromised) [556], and biological markers that identify biologic events early in the exposure–disease continuum.

It is the authors' personal view that, of the four portions of a risk assessment, exposure assessment has made the biggest improvement in quality over the 25-year history of health risk assessment. Often, exposure assessments will contain less uncertainty than other steps in a risk assessment, especially the dose–response portion. Admittedly, the number of factors to consider is large when estimating exposure, and it is a complicated procedure to understand the transport and distribution of a chemical that has been released into the environment. Nonetheless, the available data indicate that scientists can do an adequate job of quantifying the concentration of the chemicals in the various media and the resulting uptake by exposed persons if they account for all the factors that should be considered.

We have learned at least 11 significant lessons about conducting exposure assessments in recent years; had we not had to learn through experience, avoiding these lessons could potentially have saved the United States hundreds of millions of dollars and thousands of person-years of work. The first lesson is that experience has shown that, in our attempts to be prudent, we placed too much emphasis on the so-called maximally exposed individual (MEI) [68,70,71,75]. Often, the results of those analyses were misinterpreted by the public or misrepresented by some scientists or lawyers; as a result, poor decisions were made by risk decision-makers.

Second, as we have learned how to accurately characterize the risks of exposure for about 95% of the population, more emphasis has been placed on evaluating the various special groups (e.g., Eskimos, subsistence fishermen, dairy farmers) [78,313,394,508,557–560] and potentially susceptible populations (e.g., children, fetuses, elderly) [152,155,159,322,326,328–331,502,556,561–572]. Although the risk for these populations must be understood, the typical levels of exposure for the majority of the population should be the initial focus of the assessment. Risk managers need to understand the size of the exposed populations and the risks.

The third lesson is a variation of the second: Do not allow the repeated use of conservative assumptions to dictate the results of the assessment [473]. In recent years, many investigators have addressed this issue and have demonstrated its importance. Monte Carlo techniques can generally be successful in addressing this problem.

Fourth, we have learned that risk managers and the public want to understand the statistical confidence in our estimates of risk. Sensitivity analyses can yield important information about the critical exposure variables [451–454, 548,549,554,573]. Further, most risk assessments can benefit from analyses of both variability and uncertainty. Without these, risk managers are not fully informed.

Fifth, we have improved our techniques for statistically handling samples that have no detectable amount of a contaminant. Frequently, regulatory agencies have used the limit of detection (LOD) of the analysis or one half the LOD in the exposure calculations, relying on the premise that the contaminant might be present at that level. We learned that when such an approach is used (without reflection) on a site that may be only 2 to 10% contaminated (based on surface area), the impact of a few samples on the results could lead us to improper conclusions about the level of risk to persons who live there or nearby.

Sixth, we have gained a significant degree of confidence in our ability to estimate historical exposures: so-called dose-reconstruction or restrospective risk assessments. Over the past 20 years, these assessments have been used in epidemiology studies to understand the likely exposure to workers and those in the community nearly 40 to 50 years ago based on estimated chemical usage and emission data, measured data, and models [28,29,36,193, 194,219,574–579].

Seventh, we now understand the need to quantitatively account for indirect pathways of exposure. For example, the uptake of a contaminant in water by humans due to ingestion is obvious (and direct), but the uptake of the same contaminant by garden vegetables due to watering or uptake via the inhalation of volatile contaminants from the water while showering is an example of indirect pathways that had not always been evaluated in assessments. Perhaps the most important indirect route of exposure, which had not been considered before 1986 when regulating airborne nonvolatile chemicals, is the ingestion of particulate emissions that have deposited onto soil and plants and are subsequently eaten by grazing animals [58]. Much additional research in this area will be conducted and it will probably change our views on many chemicals.

Eighth, we have learned that the exposure patterns of children are unlike those of adults [152,155,159,322,326, 328–331,502,556,561–572]. As some have said in more ways than one, children are not miniature adults! Their intake of certain foods, percentage of time outdoors, proximity to carpets, and inhalation rates per body weight are all different.

The ninth lesson learned is to use biological monitoring to validate or confirm the predicted degree of human exposure. Over the past 15 years, analytical chemists have increased their ability to detect very small quantities of dozens of chemicals in blood, urine, hair, feces, breath, and fat [33,154,156,158,397,398,489,500,502,504,506,508, 511, 526–534,541,560,580–583]. For many chemicals, these data represent a direct indicator of recent exposure and, in some cases (such as PCBs and dioxins), chronic exposure. Validation of our exposure assessments should be one of the major areas of study during the next few decades (both through biomonitoring and molecular epidemiology).

Tenth, it has become clear that, in most cases, the most significant risks due to exposure to chemicals occurs in the workplace. Even though great strides have been made in industrial hygiene over the past 50 years, the doses to which persons can legally be exposed are much greater (often by a factor of 100) than those to which most persons not in those occupations will ever be exposed.

Eleventh, and perhaps most important, we have learned that (for most persons) exposures to chemicals and bacteria in the home pose a greater risk than to those in the ambient air or through the ingestion of water. Many fine studies conducted in the 1970s through the current day continue to show that in-home exposures to most chemicals are 2 to 20 times greater than exposures in the ambient environment [21,201,203,206,249,311,316,319, 402,421,584–592].

We have come a long way in a short time. Several professional societies, including the International Society of Exposure Analysis (ISEA), Society for Risk Analysis (SRA), American Industrial Hygiene Association (AIHA),

Air and Waste Management Association (AWMA), American Chemical Society (ACS), Society of Toxicology (SOT), and International Society for Regulatory Toxicology and Pharmacology (ISRTP), among others, have all placed an emphasis on improving the practice of exposure assessment. All indications are that the information we have gained has significantly improved the quality of recent risk assessments, and it can be expected that due to better exposure assessments, future decisions by risk managers will be much better informed.

## QUESTIONS

1. What are the routes of exposure normally considered in an exposure assessment?
   *Answer:* Dermal, inhalation, and ingestion
2. What is the definition of exposure assessment?
   *Answer:* Exposure assessment is the step in the risk assessment that quantifies the uptake of an agent resulting from contact with various media (e.g., air, water, soil, and food). These assessments can address past, current, or future anticipated exposures.
3. When estimating uptake through the skin, what are the factors to be considered?
   *Answer:* Percutaneous absorption rate, surface area of exposed skin, the chemical concentration, exposure duration, and interspecies scaling factor (if data were not collected using human skin).
4. What is a PBPK model, and why are PBPK models considered an important improvement over traditional toxicological methods?
   *Answer:* A physiologically based pharmacokinetic (PBPK) model is a quantitative description of the absorption, distribution, metabolism, and excretion (ADME) of a chemical in living organisms (fish, laboratory animal, or human). These models are usually capable of scaling-up animal data to predict the behavior of the toxicologically important substance (parent or metabolite) in humans, thus representing a major improvement over traditional qualitative or semiquantitative approaches.
5. Uncertainty analyses are an important component of exposure assessments. In these analyses, uncertainty is contrasted with variability. What is the difference between these two terms? Give an example.
   *Answer:* Uncertainty represents a lack of knowledge about factors affecting exposure. For example, if one can precisely measure a particular value, such as the amount of chicken eaten by a specific person on a particular day, then there would be no uncertainty in the measurement. On the other hand, if one wanted to understand the ingestion of chicken during adulthood, this would vary from day to day and week to week; thus, this would represent variability. To understand the degree of variability, measurements would be necessary. The three most common forms of variability are the variability across locations, variability over time, and variability among individuals.
6. Over the past 10 years, Monte Carlo or probabilistic techniques have become an important and useful component of exposure assessment. Describe the technique and discuss what is learned from their application.
   *Answer:* Monte Carlo techniques attempt to describe the uncertainty in select exposure parameters without having to make a particular measurement during every event over a lifetime. For example, these techniques allow one to estimate with confidence the daily ingestion of water by a typical adult male without having to collect every glass of water drunk by a person (or group of persons) over that person's lifetime. The technique generates distributions that describe the uncertainty associated with the risk estimate. By using this approach, the assessor is not forced to rely solely on a single exposure parameter or the repeated use of conservative assumptions to identify the possible dose and risk estimates for a population of persons.

## REFERENCES


1. Commission on Life Sciences (CLS). (1983): *Risk Assessment in the Federal Government: Managing the Process*. The National Academies Press, Washington, D.C.
2. Paustenbach, D.J., Ed. (1989): *The Risk Assessment of Environmental Hazards: A Textbook of Case Studies*. John Wiley & Sons, New York.
3. Paustenbach, D. J. (1995): The practice of health risk assessment in the United States (1975–1995): how the U.S. and other countries can benefit from that experience. *Hum. Ecol. Risk Assess.*, 1(1):29–79.
4. Society of Toxicology. (2000): *Risk Assessment: What's It All About?* Society of Toxicology, Reston, VA.
5. Lehmann, A. J and Fitzhugh, O. G. (1954): 100-fold margin of safety. *Q. Bull. Assoc. U.S. FDA*, 18:33.
6. Center for Risk Analysis (1994): *Historical Roots of Health Risk Assessment*. Harvard University, School of Public Health, Cambridge, MA.
7. Paustenbach, D. J., Leung, H. W., and Rothrock, J. (1999) Health risk assessment. In: *Occupational Skin Disease* edited by R. Adams. W.B. Saunders, Philadelphia, PA.
8. Paustenbach, D. J., Ed. (2002): *Human and Ecological Risk Assessment: Theory and Practice*. John Wiley & Sons New York.
9. Carnegie Commission on Science. (1993): *Risk and the Environment: Improving Regulatory Decision Making*. The Carnegie Corporation, New York.

10. USEPA. (1998): *Integrated Risk Information System (IRIS)*. U.S. Environmental Protection Agency, Washington, D.C.
11. Committee on Risk Perception and Communication, National Research Council. (1989): *Improving Risk Communication*. The National Academies Press, Washington, D.C.
12. Committee on Risk Assessment of Hazardous Air Pollutants, Commission on Life Sciences, National Research Council. (1994): *Science and Judgment in Risk Assessment*. The National Academies Press, Washington, D.C.
13. Stern, P. C. and Fineberg, H. V., Eds. (1996): *Understanding Risk: Informing Decisions in a Democratic Society*. The National Academies Press, Washington, D.C.
14. Presidential/Congressional Commission on Risk Assessment and Risk Management (CRAM). (1997): *Framework for Environmental Health Risk Management*, Final Report, Vol. I. U.S. Government Printing Office, Washington, D.C.
15. Presidential/Congressional Commission on Risk Assessment and Risk Management (CRAM). (1997): *Risk Assessment and Risk Management in Regulatory Decision-Making*, Final Report, Vol. 1. U.S. Government Printing Office, Washington, D.C.
16. Roberts, J. W., Budd, W. T., Chuang, J., and Lewis, R. G. (1993): *Chemical Contaminants in House Dust*, EPA/600/A-93/215. U.S. Environmental Protection Agency, Washington, D.C.
17. Wallace, L. A. (1998): The Weslowski Lecture, personal correspondence.
18. Belzer, R. B. (2002): Exposure assessment at a crossroads: the risk of success. *J. Expo. Anal. Environ. Epidemiol.*, 12(2):96–103.
19. Rhomberg, L. R. (1997): A survey of methods for chemical risk assessment among federal regulatory agencies. *Hum. Ecol. Risk Assess.*, 3:1029–1196.
20. Ott, W. R. (1995): Human exposure assessment: the birth of a new science. *J. Expo. Anal. Environ. Epidemiol.*, 5(4):449–472.
21. Ott, W. R. and Roberts, J. W. (1998): Everyday exposure to toxic pollutants. *Sci. Am.*, 278(2):86–91.
22. ATSDR. (1995): *Public Health Assessment Guidance Manual*. Agency for Toxic Substances and Disease Registry, U.S. Department of Health and Human Services, Washington, D.C.
23. USEPA. (1992): *Supplemental Guidance to RAGS: Calculating the Concentration Term*, OSWER Directive 9285.7-081. U.S. Environmental Protection Agency, Office of Solid Waste and Emergency Response, Washington, D.C.
24. Duan, N. and Mage, D. T. (1997): Combination of direct and indirect approaches for exposure assessment. *J. Expo. Anal. Environ. Epidemiol.*, 7(4):439–470.
25. Georgopoulos, P. G. and Lioy, P. J. (1994): Conceptual and theoretical aspects of human exposure and dose assessment. *J. Exp. Anal. Environ. Epidemiol.*, 4:253–285.
26. Paustenbach, D. J., Jernigan, J. D., Bass, R., Kalmes, R., and Scott, P. (1992): A proposed approach to regulating contaminated soil: identify safe concentrations for seven of the most frequently encountered exposure scenarios. *Regul. Toxicol. Pharmacol.*, 16(1):21–56.
27. Paustenbach, D. J., Meyer, D. M., Sheehan, P. J., and Lau, V. (1991): An assessment and quantitative uncertainty analysis of the health risks to workers exposed to chromium contaminated soils. *Toxicol. Indust. Health*, 7(3):159–196.
28. Ripple, S. R. (1992): Looking back: the use of retrospective health risk assessment. *Environ. Sci. Technol.*, 26:1270–1277.
29. Stewart, P. and Herrick: R. F., (1991): Issues in performing retrospective exposure assessment. *Appl. Occup. Environ. Hyg.*, 6:421–427.
30. Hoffman, F. O. and Hammonds, J. S. (1994): Propagation of uncertainty in risk assessments: the need to distinguish between uncertainty due to lack of knowledge and uncertainty due to variability. *Risk Anal.*, 14(5):707–712.
31. Brody, J. G., Vorhees, D. J., Melly, S. J., Swedis, S. R., Drivas, P. J., and Rudel, R. A. (2002): Using GIS and historical records to reconstruct residential exposure to large-scale pesticide application. *J. Expo. Anal. Environ. Epidemiol.*, 12(1):64–80.
32. Johansen, K., Tinnerberg, H., and Lynge, E. (2005): Use of history science methods in exposure assessment for occupational health studies. *Occup. Environ. Med.*, 62(7):434–441.
33. Mage, D. T., Allen, R. H., Gondy, G., Smith, W., Barr, D. B., and Needham: L. L. (2004): Estimating pesticide dose from urinary pesticide concentration data by creatinine correction in the Third National Health and Nutrition Examination Survey (NHANES-III). *J. Expo. Anal. Environ. Epidemiol.*, 14(6):457–465.
34. Rull, R. P. and Ritz, B. (2003): Historical pesticide exposure in California using pesticide use reports and land-use surveys: an assessment of misclassification error and bias. *Environ. Health Perspect.*, 111(13):1582–1589.
35. Ramachandran, G. (2001): Retrospective exposure assessment using Bayesian methods. *Ann. Occup. Hyg.*, 45(8):651–667.
36. Williams, P. R. and Paustenbach, D. J. (2003): Reconstruction of benzene exposure for the Pliofilm cohort (1936–1976) using Monte Carlo techniques. *J. Toxicol Environ. Health A*, 66(8):677–781.
37. Eisenbud, M. (1978): *Environment, Technology and Health: Human Ecology in Historical Perspective*. New York University Press, New York.
38. Lynch, J. R. (1985). Measurement of worker exposure. In: *Patty's Industrial Hygiene and Toxicology*, edited by L. J. Cralley and L. V. Cralley. Wiley Interscience, New York.
39. McCord, C.P. (1943): *Industrial Hygiene for Engineers*. Martin Press, Chicago, IL.
40. Paustenbach, D. J. (1990): Health risk assessment and the practice of industrial hygiene. *Am. Indust. Hyg. Assoc. J.*, 51(7):339–351.
41. Upton, A. C. (1988): Evolving perspectives on the concept of dose in radiobiology and radiation protection. *Health Phys.*, 55(4):605–614.
42. USEPA. (1992): Guidelines for exposure assessment: notice. *Fed. Reg.*, 57(104):22888–22938.
43. USEPA. (1988): Proposed guidelines for exposure-related measurements. *Fed. Reg.*, 53(232):48830–48853.
44. USEPA. (1996): *Exposure Factors Handbook*. Vol. I. *General Factors: Review Draft*, EPA 600/P-95/002A. Office of Health and Environmental Assessment, U.S. Environmental Protection Agency, Washington, D.C.
45. Committee on Advances in Assessing Human Exposure to Airborne Pollutants, National Research Council. (1991): *Human Exposure Assessment for Airborne Pollutants: Advances and Opportunities*. The National Academies Press, Washington, D.C.

46. Scott, P. K., Finley, B. L., Sung, H. M., Schulze, R. H., and Turner: D. B. (1997): Identification of an accurate soil suspension/dispersion modeling method for use in estimating health-based soil cleanup levels of hexavalent chromium in chromite ore processing residues. *J. Air Waste Manag. Assoc.*, 47(7):753–765.
47. Zannetti, P. (1992): Particle modeling and its application for simulating air pollution phenomena. In: *Environmental Modeling*, edited by P. Melli. Computational Mechanics Publications, Southampton, U.K.
48. Lorber, M. (2001): Indirect exposure assessment at the United States Environmental Protection Agency. *Toxicol. Indust. Health*, 17(5–10):145–156.
49. Finley, B. L. and Paustenbach, D. J. (1997): Using applied research to reduce uncertainty in health risk assessment: five case studies involving human exposure to chromium in soil and groundwater. *J. Soil Contam.*, 6:649–705.
50. Paustenbach, D. J., Hays, S. M., Sururi, S., and Underwood, P. (1997): Comparing the predicted uptake of TCDD using exposures calculations with the actual uptake: a case study of resident of Time Beach, Missouri. Paper presented at the 17th Int. Symp. on Chlorinated Dioxins and Related Compounds, Indianapolis, IN.
51. Jayjock, M., Lynch, J., and Nelson, D., Eds. (2000): *Risk Assessment Principles for the Industrial Hygienist*. ACGIH Press, Cincinnati, OH.
52. Paustenbach, D. J., Shu, H. P., and Murray, F. J. (1986): A critical examination of assumptions used in risk assessments of dioxin contaminated soil. *Regul. Toxicol. Pharmacol.*, 6(3):284–307.
53. Lapare, S., Brodeur, J., and Tardif, R. (2003): Contribution of toxicokinetic modeling to the adjustment of exposure limits to unusual work schedules. *AIHA J.*, 64(1):17–23.
54. Verma, D. K. (2000): Adjustment of occupational exposure limits for unusual work schedules. *AIHA J.*, 61(3):367–374.
55. Romney, E. M., Lindberg, N. G., Hawthorne, H. A., Brystrom, B. B., and Larson, K. H. (1963): Contamination of plant foliage with radioactive nuclides. *Annu. Rev. Plant Physiol.*, 14:271–279.
56. ICRP. (1994): *Human Respiratory Tract Model for Radiological Protection*, Publ. No. 66. International Commission on Radiological Protection, Tarrytown, NY.
57. Baes, C. F. I., Sharp, R. D., Sjoreen, A., and Shor, W. R. (1984): *A Review and Analysis of Parameters for Assessing Transport of Environmental Released Radionuclides Through Agriculture*, ORNL-5786. U.S. Department of Energy, Oak Ridge National Laboratory, Oak Ridge, TN.
58. Fries, G. F. and Paustenbach, D. J. (1990): Evaluation of potential transmission of 2,3,7,8-tetrachlorodibenzo-*p*-dioxin-contaminated incinerator emissions to humans via foods. *J. Toxicol. Environ. Health*, 29(1):1–43.
59. ILSI. (1998): *Aggregate Exposure Assessment*. International Live Science Institute, Washington, D.C.
60. McKone, T. E. and Bogen, K. T. (1991): Predicting the uncertainties in risk assessment. *Environ. Sci Technol.*, 25:16–74.
61. USEPA. (1996): *Exposure Factors Handbook*. Vol. III. *Activity Factors: SAB Review Draft*, EPA/600/P-95/002P. Office of Health and Environmental Assessment, Office of Research and Development, U.S. Environmental Protection Agency, Washington, D.C.
62. USEPA. (1996): *Exposure Factors Handbook*. Vol. II. *Food Ingestion Factors: SAB Review Draft*, EPA/600/P-95/002Bb. Office of Health and Environmental Assessment, Office of Research and Development, U.S. Environmental Protection Agency, Washington, D.C.
63. USEPA. (1997): *Exposure Factors Handbook* (update to the May 1989 edition), EPA/600/P-95/0002Fa. U.S. Environmental Protection Agency, Washington, D.C.
64. USEPA. (2002): *Child-Specific Exposure Factors Handbook*, EPA/600/P-00/002B. Office of Research and Development, National Center for Environmental Assessment, U.S. Environmental Protection Agency, Washington, D.C.
65. USEPA. (1989): *Risk Assessment Guidance for Superfund*, EPA/540/1-89/002. Office of Emergency and Remedial Response, Washington, D.C.
66. Cullen, A. C. (1994): Measures of compounding conservatism in probabilistic risk assessment. *Risk Anal.*, 14(4):389–393.
67. Maxim, D. (1989): Problems associated with the use of conservative assumptions in exposure and risk analysis. In: *The Risk Assessment of Environmental and Human Health Hazards: A Textbook of Case Studies*, edited by D. J. Paustenbach. John Wiley & Sons, New York, pp. 526–560.
68. Nichols, A. L. and Zeckhauser, R. J. (1988): The perils of prudence: how conservative risk assessments distort regulation. *Regul. Toxicol. Pharmacol.*, 8(1):61–75.
69. Paustenbach, D. J. (1989): A survey of environmental risk assessment. In: *The Risk Assessment of Environmental and Human Health Hazards: A Textbook of Case Studies*, edited by D. J. Paustenbach. John Wiley & Sons, New York, p. 139.
70. Wilson, M. D., McCormick, W. P., and Hinton, T. G. (2004): The maximally exposed individual: comparison of maximum likelihood estimation of high quantiles to an extreme value estimate. *Risk Anal.*, 24(5):1143–1151.
71. Copeland, T. L., Paustenbach, D. J., Harris, M. A., and Otani, J. (1993): Comparing the results of a Monte Carlo analysis with EPA's reasonable maximum exposed individual (RMEI): a case study of a former wood treatment site. *Regul. Toxicol. Pharmacol.*, 18(2):275–312.
72. Finley, B. L., Scott, P. K., and Paustenbach: D. J. (1993): Evaluating the adequacy of maximum contaminant levels as health-protective cleanup goals: an analysis based on Monte Carlo techniques. *Regul. Toxicol. Pharmacol.* 18(3):438–455.
73. Goldstein, B. D. (1989): The maximally exposed individual. *Environ. Forum*, Nov.–Dec.:13–16.
74. Burmaster, D. E. and Harris, R. H. (1993): The magnitude of compounding conservatisms in Superfund risk assessments. *Risk Anal.*, 13:131–134.
75. Finley, B. L. and Paustenbach, D. J. (1994): The benefits of probabilistic exposure assessment: three case studies involving contaminated air, water, and soil. *Risk Anal.* 14(1):53–73.
76. Thompson, K. M. and Burmaster, D. E. (1991): Parametric distributions for soil ingestion by children. *Risk Anal.* 11:339–342.
77. Thompson, K. M., Burmaster, D. E., and Crouch, E. A. (1992): Monte Carlo techniques for quantitative uncertainty analysis in public health risk assessments. *Risk Anal.*, 12(1):53–63.

78. Wilson, A. L., Price, P., and Paustenbach, D. (2000): An assessment of the risk of DDT and PCB in fish from Palos Verdes shelf. In: *Human and Ecological Risk Assessment: Theory and Practice*, edited by D. Paustenbach. John Wiley & Sons, New York.
79. Burmaster, D. E. and von Stackelberg, K. (1991): Using Monte Carlo simulations in public health risk assessments: estimating and presenting full distributions of risk. *J. Expo. Anal. Environ. Epidemiol.*, 1(4):491–512.
80. Burmaster, D. E. and Anderson, P. D. (1994): Principles of good practice for the use of Monte Carlo techniques in human health and ecological risk assessments. *Risk Anal.*, 14(4):477–481.
81. Smith, R. L. (1994): Use of Monte Carlo simulation for human exposure assessment at a superfund site. *Risk Anal.*, 14(4):433–439.
82. USEPA. (1997): *Guiding Principles for Monte Carlo Analysis*, EPA/630/R-97/001. Risk Assessment Forum, Office of Research and Development, U.S. Environmental Protection Agency, Washington, D.C.
83. Simon, T. (1997): Combining physiologically based pharmacokinetics modeling with Monte Carlo simulation to derive an acute inhalation guidance value for trichloroethylene. *Regul. Toxicol. Pharmacol.*, 26:257–270.
84. USEPA. (1996): *Summary Report for the Workshop on Monte Carlo Analysis*, EPA/630/R-96/010. Office of Research and Development, U.S. Environmental Protection Agency, Washington, D.C.
85. Vose, D. (1996): *Quantitative Risk Analysis: A Guide to Monte Carlo Simulation Modeling*. John Wiley & Sons, New York.
86. USEPA. (1999): *Risk Assessment Guidance (RAGS3A) for Conducting Probabilistic Risk Assessment*. U.S. Environmental Protection Agency, Washington, D.C.
87. USEPA. (2000): *Options for Development of Probability Distributions for Exposure Factors*, EPA/600/R-00/058. Office of Research and Development, National Center for Environmental Assessment, U.S. Environmental Protection Agency, Washington, D.C.
88. Allaby, M. (1989): *A Dictionary of the Environment*, 3rd ed. New York University Press, New York.
89. Leung, H. W. and Paustenbach, D. J. (1994): Techniques for estimating the percutaneous absorption of chemicals due to environmental and occupational exposure. *Appl. Occup. Environ. Hyg.*, 9(3):187–197.
90. Paustenbach, D. J. (1988): Assessment of the developmental risks resulting from occupational exposure to select glycol ethers within the semiconductor industry. *J. Toxicol. Environ. Health*, 23(1):29–75.
91. Reitz, R. H., Gargas, M. L., Andersen, M. E., Provan, W. M., and Green, T. L. (1996): Predicting cancer risk from vinyl chloride exposure with a physiologically based pharmacokinetic model. *Toxicol. Appl. Pharmacol.*, 137(2):253–267.
92. Clewell, H. J. et al. (2002): Review and evaluation of the potential impact of age- and gender-specific pharmacokinetic differences on tissue dosimetry. *Crit. Rev. Toxicol.*, 32(5):329–389.
93. Clewell, R. A. and Gearhart, J. M. (2002): Pharmacokinetics of toxic chemicals in breast milk: use of PBPK models to predict infant exposure. *Environ. Health Perspect.*, 110(6):A333–A337.
94. Cox, Jr., L. A. and Ricci, P. F. (1992): Reassessing benzene cancer risks using internal doses. *Risk Anal.*, 12(3):401–410.
95. Dennison, J. E., Bigelow, P. L., and Andersen, M. E. (2004): Occupational exposure limits in the context of solvent mixtures, consumption of ethanol, and target tissue dose. *Toxicol. Indust. Health*, 20(6–10):165–175.
96. Simmons, J. E., Evans, M. V., and Boyes, W. K. (2005): Moving from external exposure concentration to internal dose: duration extrapolation based on physiologically based pharmacokinetic derived estimates of internal dose. *J. Toxicol. Environ. Health A*, 68(11–12):927–950.
97. USEPA. (1992): *Exposure Assessment Guidelines*. U.S. Environmental Protection Agency, Washington, D.C.
98. McDougal, J. N., Jepson, G. W., Clewell, 3rd, H. J., Gargas, M. L., and Andersen, M. E. (1990): Dermal absorption of organic chemical vapors in rats and humans. *Fundam. Appl. Toxicol.*, 14(2):299–308.
99. McKone, T. E. (1990): Dermal uptake of organic chemicals from a soil matrix. *Risk Anal.*, 10(3):407–419.
100. Ruby, M. V. et al. (1999): Advances in evaluating the oral bioavailability of inorganics in soil for use in human health risk assessment. *Environ. Sci. Technol.*, 33(21):3697–3705.
101. Barriuso, E., Koskinen, W. C., and Sadowsky, M. J. (2004): Solvent extraction characterization of bioavailability of atrazine residues in soils. *J. Agric. Food Chem.*, 52(21): 6552–6556.
102. Braida, W. J., White, J. C., and Pignatello, J. J. (2004): Indices for bioavailability and biotransformation potential of contaminants in soils. *Environ. Toxicol. Chem.*, 23(7):1585–1591.
103. Burger, J., Diaz-Barriga, F., Marafante, E., Pounds, J., and Robson, M. (2003): Methodologies to examine the importance of host factors in bioavailability of metals. *Ecotoxicol. Environ. Safety*, 56(1):20–31.
104. Casteel, S. W. et al. (1997): Bioavailability of lead to juvenile swine dosed with soil from the Smuggler Mountain NPL site of Aspen, Colorado. *Fundam. Appl. Toxicol.*, 36(2):177–187.
105. Caussy, D. (2003): Case studies of the impact of understanding bioavailability: arsenic. *Ecotoxicol. Environ. Safety*, 56(1):164–173.
106. Caussy, D., Gochfeld, M., Gurzau, E., Neagu, C., and Ruedel, H. (2003): Lessons from case studies of metals: investigating exposure, bioavailability, and risk. *Ecotoxicol. Environ. Safety*, 56(1):45–51.
107. CDC. (2003): *Second National Report on Human Exposure to Environmental Chemicals*, 03–0572. National Center for Environmental Health, U.S. Centers for Disease Control and Prevention, Atlanta, GA.
108. Echevarria, G., Massoura, S. T., Sterckeman, T., Becquer, T., Schwartz, C., and Morel, J. L. (2006): Assessment and control of the bioavailability of nickel in soils. *Environ. Toxicol. Chem.*, 25(3):643–651.
109. Ehlers, L. J. and Luthy, R. G. (2003): Contaminant bioavailability in soil and sediment. *Environ. Sci. Technol.*, 37(15):295A–302A.
110. Grabowski, L. A., Houpis, J. L., Woods, W. I., and Johnson, K. A. (2001): Seasonal bioavailability of sediment-associated heavy metals along the Mississippi river floodplain. *Chemosphere*, 45(4–5):643–651.

111. Hunt, J. R. (2003): Bioavailability of iron, zinc, and other trace minerals from vegetarian diets. *Am. J. Clin. Nutr.*, 78(3, Suppl.):633S–639S.
112. Janssen, C. R., Heijerick, D. G., De Schamphelaere, K. A., and Allen, H. E. (2003): Environmental risk assessment of metals: tools for incorporating bioavailability. *Environ. Int.*, 28(8):793–800.
113. Ng, J. C., Kratzmann, S. M., Qi, L., Crawley, H., Chiswell, B., and Moore, M. R. (1998): Speciation and absolute bioavailability: risk assessment of arsenic-contaminated sites in a residential suburb in Canberra. *Analyst*, 123(5):889–892.
114. Peijnenburg, W., Sneller, E., Sijm, D., Lijzen, J., Traas, T., and Verbruggen, E. (2004): Implementation of bioavailability in standard setting and risk assessment. *Environ. Sci.*, 11(3):141–149.
115. Peijnenburg, W., Sneller, E. Sijm, D. Lijzen, J. Traas, T. and Verbruggen, E. (2004): Implementation of bioavailability in standard setting and risk assessment: suggestions based on a workshop with emphasis on metals. *Arh. Hig. Rada Toksikol.*, 55(4):273–278.
116. Schoof, R. A. and Nielsen, J. B. (1997): Evaluation of methods for assessing the oral bioavailability of inorganic mercury in soil. *Risk Anal.*, 17(5):545–555.
117. Young, A. L., Giesy, J. P., Jones, P. D., and Newton, M. (2004): Environmental fate and bioavailability of Agent Orange and its associated dioxin during the Vietnam War. *Environ. Sci. Pollut. Res. Int.*, 11(6):359–370.
118. Paustenbach, D. J., Bruce, G. M., and Chrostowski, P. (1997): Current views on the oral bioavailability of inorganic mercury in soil: implications for health risk assessments. *Risk Anal.*, 17(5):533–544.
119. Chen, W., Hrudey, S. E., and Rousseaux, C. (1996): *Bioavailability in Environmental Risk Assessment*, CRC Press, Boca Raton, FL.
120. Umbreit, T. H., Hesse, E. J., and Gallo, M. A. (1986): Acute toxicity of TCDD contaminated soil from an industrial site. *Science*, 232:497–499.
121. Jayjock, M. A., Hazelton, G. A., Lewis, P. G., and Wooder, M. F. (1996): Formulation effect on the dermal bioavailability of isothiazolone biocide. *Food Chem. Toxicol.*, 34(3):277–282.
122. Ruby, M. V., Davis, A., Kempton, J. H., Drexter, J. W., and Bergstrom, P. D. (1992): Lead bioavailability under simulated gastric conditions. *Environ. Sci. Technol.*, 26:1242–1248.
123. Shu, H. et al. (1988): Bioavailability of soil-bound TCDD: dermal bioavailability in the rat. *Fundam. Appl. Toxicol.*, 10(2):335–343.
124. van den Berg, M., Olie, K., and Hutzinger, O. (1984): Uptake and selective retention in rats of orally administered chlorinated dioxins and PCDF from fly ash. *Chemosphere*, 13:531–544.
125. Piotrowski, J. (1967): Further investigations on the evaluation of exposure to nitrobenzene. *Br. J. Indust. Med.*, 24(1):60–65.
126. NIOSH. (1977): *Exposure Tests for Organic Compounds in Industrial Toxicology.* National Institute for Occupational Safety and Health, Cincinnati, OH.
127. Ramsey, J. D. and Andersen, M. E. (1984): A physiologically based description of the inhalation pharmacokinetics of styrene is rats and humans. *Toxicol. Appl. Pharmacol.*, 73:159–175.
128. USEPA. (1996): Draft guidelines for carcinogenic risk assessment. *Fed. Reg.*, 61(79):17960–18011.
129. McKone, T. E. and Bogen, K. T. (1992): Uncertainties in health-risk assessment: an integrated case study based on tetrachloroethylene in California groundwater. *Regul. Toxicol. Pharmacol.*, 15(1):86–103.
130. Clewell, 3rd, H. J. (1995): The application of physiologically based pharmacokinetic modeling in human health risk assessment of hazardous substances. *Toxicol. Lett.*, 79(1–3):207–217.
131. Haber, L. T., Maier, A., Gentry, P. R., Clewell, H. J., and Dourson, M. L. (2002): Genetic polymorphisms in assessing interindividual variability in delivered dose. *Regul. Toxicol. Pharmacol.*, 35(2, Pt. 1):177–197.
132. Andersen, M. E. et al. (1991): Physiologically based pharmacokinetic modeling with dichloromethane, its metabolite, carbon monoxide, and blood carboxyhemoglobin in rats and humans. *Toxicol. Appl. Pharmacol.*, 108(1):14–27.
133. Lilly, P. D., Andersen, M. E., Ross, T. M., and Pegram: R. A. (1998): A physiologically based pharmacokinetic description of the oral uptake, tissue dosimetry, and rate of metabolism of bromodichloromethane in the male rat. *Toxicol. Appl. Pharmacol.*, 150(2):205–217.
134. Marino, D. J. et al. (2006): Revised assessment of cancer risk to dichloromethane. Part I. Bayesian PBPK and dose–response modeling in mice. *Regul. Toxicol. Pharmacol.*, 45(1):44–54.
135. Merrill, E. A. et al. (2003): PBPK predictions of perchlorate distribution and its effect on thyroid uptake of radioiodide in the male rat. *Toxicol. Sci.*, 73(2):256–269.
136. Merrill, E. A. et al. (2005): PBPK model for radioactive iodide and perchlorate kinetics and perchlorate-induced inhibition of iodide uptake in humans. *Toxicol. Sci.*, 83(1):25–43.
137. Committee on Geosciences, Environment and Resources, National Research Council. (1990): *Human Exposure Assessment for Airborne Pollutants: Advances and Applications*. National Academy Press, Washington, D.C.
138. Leung, H. W. and Paustenbach, D. J. (1995): Physiologically based pharmacokinetic and pharmacodynamic modeling in health risk assessment and characterization of hazardous substances. *Toxicol. Lett.*, 79(1–3):55–65.
139. AIHC. (1994): *Exposure Factors Sourcebook*. American Industrial Health Council, Washington, D.C.
140. Finley, B. L., Proctor, D. M., Scott, P. K., Harrington, N. Paustenbach, D. J., and Price, P. (1994): Recommended distributions for exposure factors frequently used in health risk assessment. *Risk Anal.*, 14(4):533–553.
141. Aylward, L. L., Hays, S. M., Karch, N. J., and Paustenbach D. J. (1996): Relative susceptibility of animals and humans to the cancer hazard posed by exposure to 2,3,7,8-tetrachlorodibenzo-*p*-dioxin using internal measures of dose *Environ. Sci. Technol.*, 30(12):3534–3543.
142. Lorenzana, R. M., Troast, R., Klotzbach, J. M., Follansbee M. H., and Diamond, G. L. (2005): Issues related to time averaging of exposure in modeling risks associated with intermittent exposures to lead. *Risk Anal.*, 25(1):169–178
143. Wallace, L. and Williams, R. (2005): Validation of a method for estimating long-term exposures based on short term measurements. *Risk Anal.*, 25(3):687–694.

144. Paustenbach, D. J. (2000). Pharmacokinetics and unusual work schedules. In: *Patty's Industrial Hygiene and Toxicology*, 5th ed., edited by R. L. Harris. OEM Press, Beverly Farms, MA.
145. Roach, S. A. (1966): A more rational basis for air sampling programs. *Am. Indust. Hyg. Assoc. J.*, 27(1):1–12.
146. Pastino, G. M., Kousba, A. A., Sultatos, L. G., and Flynn, E. J. (2003): Derivation of occupational exposure limits based on target blood concentrations in humans. *Regul. Toxicol. Pharmacol.*, 37(1):66–72.
147. Rappaport, S. M. (1985): Smoothing of exposure variability at the receptor: implications for health standards. *Ann. Occup. Hyg.*, 29(2):201–214.
148. Rappaport, S. M. and Spear, R. C. (1988): Physiological damping of exposure variability during brief periods. *Ann. Occup. Hyg.*, 32(1):21–33.
149. Price, P. S. and Chaisson, C. F. (2005): A conceptual framework for modeling aggregate and cumulative exposures to chemicals. *J. Expo. Anal. Environ. Epidemiol.*, 15(6):473–481.
150. USEPA. (2001): *General Principles for Performing Aggregate Exposure and Risk Assessment*. Office of Pesticide Programs, U.S. Environmental Protection Agency, Washington, D.C.
151. Petersen, B. J. (2003): Methodological aspects related to aggregate and cumulative exposures to contaminants with common mechanisms of toxicity. *Toxicol. Lett.*, 140–141:427–435.
152. Bradman, A. and Whyatt, R. M. (2005): Characterizing exposures to nonpersistent pesticides during pregnancy and early childhood in the National Children's Study: a review of monitoring and measurement methodologies. *Environ. Health Perspect.*, 113(8):1092–1099.
153. Castorina, R., Bradman, A., McKone, T. E., Barr, D. B., Harnly, M. E., and Eskenazi, B. (2003): Cumulative organophosphate pesticide exposure and risk assessment among pregnant women living in an agricultural community: a case study from the CHAMACOS cohort. *Environ. Health Perspect.*, 111(13):1640–1648.
154. Egeghy, P. P., Quackenboss, J. J., Catlin, S., and Ryan, P. B. (2005): Determinants of temporal variability in NHEXAS: Maryland environmental concentrations, exposures, and biomarkers. *J. Expo. Anal. Environ. Epidemiol.*, 15(5):388–397.
155. Freeman, N. C. et al. (2001): Quantitative analysis of children's microactivity patterns: the Minnesota Children's Pesticide Exposure Study. *J. Expo. Anal. Environ. Epidemiol.*, 11(6):501–509.
156. Jakubowski, M. and Trzcinka-Ochocka, M. (2005): Biological monitoring of exposure: trends and key developments. *J. Occup. Health*, 47(1):22–48.
157. Moschandreas, D. J. et al. (2001): On predicting multi-route and multimedia residential exposure to chlorpyrifos and diazinon. *J. Expo. Anal. Environ. Epidemiol.* 11(1):56–65.
158. Sohn, M. D., McKone, T. E., and Blancato, J. N. (2004): Reconstructing population exposures from dose biomarkers: inhalation of trichloroethylene (TCE) as a case study. *J. Expo. Anal. Environ. Epidemiol.*, 14(3):204–213.
159. Clayton, C. A., Pellizzari, E. D., Whitmore, R. W., Quackenboss, J. J., Adgate, J., and Sefton, K. (2003): Distributions, associations, and partial aggregate exposure of pesticides and polynuclear aromatic hydrocarbons in the Minnesota Children's Pesticide Exposure Study (MNCPES). *J. Expo. Anal. Environ. Epidemiol.*, 13:100–111.
160. Pang, Y., D. L. MacIntosh, D. E. Camann, and P. B. Ryan: Analysis of aggregate exposure to chlorpyrifos in the NHEXAS-Maryland investigation. *Environ. Health Perspect* 110(3):235–240. (2002):
161. Andersen, M. E., Clewell, 3rd, H., and Krishnan, K. (1995): Tissue dosimetry, pharmacokinetic modeling, and interspecies scaling factors. *Risk Anal.*, 15(4):533–537.
162. Andersen, M. E. and Conolly, R. B. (1998): Mechanistic modeling of rodent liver tumor promotion at low levels of exposure: an example related to dose–response relationships for 2,3,7,8-tetrachlorodibenzo-*p*-dioxin. *Hum. Exp. Toxicol.*, 17(12):683–690.
163. Andersen, M. E., MacNaughton, M. G., Clewell, H. J., and Paustenbach, D. J. (1987): Adjusting exposure limits for long and short exposure periods using a physiological pharmacokinetic model. *Am. Indust. Hyg. Assoc. J.*, 48(4):335–343.
164. Weis, B. K. et al. (2005): Personalized exposure assessment: promising approaches for human environmental health research. *Environ. Health Perspect.*, 113(7): 840–848.
165. U.S. Office of Science and Technology Policy. (1993): *Researching Health Risks*. Office of Technology Assessment, U.S. Department of Health and Human Services, Washington, D.C.
166. USEPA. (1987): *The Total Exposure Assessment Methodology (TEAM) Study: Summary and Analysis*, EPA/600/6–87/002a. Office of Research and Development, U.S. Environmental Protection Agency, Washington, D.C.
167. ACGIH. (1998): *Industrial Hygiene Instruments Handbook*. ACGIH Press, Cincinnati, OH.
168. Knaak, J. B., Dary, C. C., Patterson, G., and Blancato J. N. (2001): The worker hazard posed by reentry into pesticide-treated foliage: reassessment of reentry intervals using foliar residue transfer-percutaneous absorption PB-PK/PD models, with emphasis on isofenphos and parathion. In: *Human and Ecological Risk Assessment: Theory and Practice*, edited by D. J. Paustenbach. John Wiley & Sons, New York.
169. Knaak, J. B., Iwata, Y., and Maddy K. T. (1989): The worker hazard posed by reentry into pesticide-treated foliage: development of safe reentry times, with emphasis on chlorthiophos and carbosulfan. In: *The Risk Assessment of Environmental Hazards: A Textbook of Case Studies*, edited by D. J. Paustenbach. John Wiley & Sons, New York, pp. 797–842.
170. Kissel, J. and Fenske, R. (2000): Improved estimation of dermal pesticide dose to agricultural workers upon reentry. *Appl. Occup. Environ. Hyg.*, 15(3):284–290.
171. Hood, L., Heath, J. R., Phelps, M. E., and Lin, B. (2004): Systems biology technologies enable predictive and preventive medicine. *Science*, 306(5696):640–643.
172. Bellander, T. et al. (2001): Using geographic information systems to assess individual historical exposure to air pollution from traffic and house heating in Stockholm. *Environ. Health Perspect.*, 109(6):633–639.
173. Kunzli, N. (2005): Unifying susceptibility, exposure, and time: discussion of unifying analytic approaches and future directions. *J. Toxicol. Environ. Health A*, 68(13–14): 1263–1271.

174. Nuckols, J. R., Ward, M. H., and Jarup, L. (2004): Using geographic information systems for exposure assessment in environmental epidemiology studies. *Environ. Health Perspect.*, 112(9):1007–1015.
175. Hellstrom, L., Jarup, L., Persson, B., and Axelson, O. (2004): Using environmental concentrations of cadmium and lead to assess human exposure and dose. *J. Expo. Anal. Environ. Epidemiol.*, 14(5):416–423.
176. Jarup, L. (2004): Health and environment information systems for exposure and disease mapping, and risk assessment. *Environ. Health Perspect.*, 112(9):995–997.
177. Balbatun, A., Louka, F. R., and Malinski, T. (2003): Dynamics of nitric oxide release in the cardiovascular system. *Acta Biochim. Pol.*, 50(1):61–68.
178. Cao, Y., Lee Koo, Y. E., and Kopelman, R. (2004): Poly(decyl methacrylate)-based fluorescent PEBBLE swarm nanosensors for measuring dissolved oxygen in biosamples. *Analyst*, 129(8):745–750.
179. Jianrong, C., Yuqing, M., Nongyue, H., Xiaohua, W., and Sijiao, L. (2004): Nanotechnology and biosensors. *Biotechnol. Adv.*, 22(7):505–518.
180. Kalinowski, L., Dobrucki, I. T., and Malinski, T. (2004): Race-specific differences in endothelial function: predisposition of African Americans to vascular diseases. *Circulation*, 109(21):2511–2517.
181. Koo, Y. E., Cao, Y., Kopelman, R., Koo, S. M., Brasuel, M., and Philbert, M. A. (2004): Real-time measurements of dissolved oxygen inside live cells by organically modified silicate fluorescent nanosensors. *Anal. Chem.*, 76(9):2498–2505.
182. Miljanic, S., Knezevic, Z., Stuhec, M., Ranogajec-Komor, M., Krpan, K., and Vekic, B. (2003): Energy dependence of new thermoluminescent detectors in terms of HP(10) values. *Radiat. Prot. Dosimetry*, 106(3):253–256.
183. Mo, J. W. and Smart, W. (2004): Lactate biosensors for continuous monitoring. *Front. Biosci.*, 9:3384–3391.
184. Salimi, A., Compton, R. G., and Hallaj, R. (2004): Glucose biosensor prepared by glucose oxidase encapsulated sol–gel and carbon-nanotube-modified basal plane pyrolytic graphite electrode. *Anal. Biochem.*, 333(1):49–56.
185. Nessel, C. S., Butler, J. P., Post, G. B., Held, J. L., Gochfeld, M., and Gallo, M. A. (1991): Evaluation of the relative contribution of exposure routes in a health risk assessment of dioxin emissions from a municipal waste incinerator. *J. Expo. Anal. Environ. Epidemiol.*, 1(3):283–307.
186. Paustenbach, D. J., Rinehart, W. E., and Sheehan, P. J. (1991): The health hazards posed by chromium-contaminated soils in residential and industrial areas: conclusions of an expert panel. *Regul. Toxicol. Pharmacol.*, 13(2):195–222.
187. Proctor, D., Zak, M. A., and Finley, B. (1997): Resolving uncertainties associated with the construction worker soil ingestion rate: a proposal for risk-based remediation goals. *Hum. Ecol. Risk Assess.*, 3:299–304.
188. Price, P. S., Su, S. H., Harrington, J. R., and Keenan, R. E. (1996): Uncertainty and variation in indirect exposure assessments: an analysis of exposure to tetrachlorodibenzo-*p*-dioxin from a beef consumption pathway. *Risk Anal.*, 16(2):263–277.
189. Calabrese, E. J. and Kostecki, P. T. (1992): *Risk Assessment and Environmental Fate Methodologies*. Lewis Press, Ann Arbor, MI.
190. Goodman, M. et al. (2000): Epidemiologic study of pulmonary obstruction in workers occupationally exposed to ethyl and methyl cyanoacrylate. *J. Toxicol. Environ. Health A*, 59(3):135–163.
191. Plato, N., Krantz, S., Gustavsson, P., Smith, T. J., and Westerholm, P. (1995): A cohort study of Swedish man-made mineral fiber (MMMF) production workers. Part I. Fiber exposure assessment in the rock/slag wool production industry 1938–1990. *Scand. J. Work Environ. Health*, 21:345–352.
192. Sathiakumar, N. et al. (1998): Mortality from cancer and other causes of death among synthetic rubber workers. *Occup. Environ. Med.*, 55(4):230–235.
193. Smith, T. J., Hammond, S. K., and Wong, O. (1993): Health effects of gasoline exposure: I. Exposure assessment for U.S. distribution workers. *Environ. Health Perspect.*, 101(6):13–21.
194. Stewart, P., Lees, P. S. J., and Francis, M. (1996): Quantification of historical exposures in occupational cohort studies. *Scand. J. Work Environ. Health*, 22:405–414.
195. DeCaprio, A. P. (1997): Biomarkers: coming of age for environmental health and risk assessment. *Environ. Sci. Technol.*, 31(7):1837–1848.
196. USEPA. (1997): *Methodology for Assessing Health Risks Associated with Multiple Exposure Pathways of Combustor Emissions*, NCEA-C-0238. National Center for Environmental Assessment, U.S. Environmental Protection Agency, Washington, D.C.
197. Paustenbach, D. J., Finley, B. L., and Long, T. F. (1997): The critical role of house dust in understanding the hazards posed by contaminated soils. *Int. J. Toxicol.*, 16:339–362
198. Scott, P. K., Finley, B. L., Harris, M. A., and Rabbe, D. E (1997): Background air concentrations of Cr(VI) in Hudson County, New Jersey: implications for setting health-based standards for Cr(VI) in soil. *J. Air Waste Manag Assoc.*, 47(5):592–600.
199. Wallace, L. et al. (1987): The "TEAM" study: personal exposures to toxic substances in air, drinking water, and breath of 400 residents of New Jersey, North Carolina, and North Dakota. *Environ. Res.*, 43:290–307.
200. Wallace, L. et al. (1987): Emissions of volatile organic compounds from building materials and consumer products. *Atmos. Environ.*, 21:385–393.
201. Wallace, L., Pellizzari, E., and Wendel, C. (1991): Total volatile organic concentrations in 2700 personal, indoor and outdoor air samples collected in the U.S. EPA TEAM studies. *Indoor Air*, 4:465–477.
202. Wallace, L. A. (1986): The Total Exposure Assessment Methodology (TEAM) study: direct measurement of personal exposures through air and water for 600 residents of several U.S. cities. In: *Pollutants in a Multimedia Environment* edited by Y. Cohen. Plenum Press, New York, pp. 289–315
203. Wallace, L. A., Pellizzari, E., Hartwell, T., Sparacino, C. Sheldon, L. S., and Zelon, H. (1985): Personal exposures indoor–outdoor relationships and breath levels of toxic air pollutants measure for 355 persons in New Jersey. *Atmos Environ.*, 19:1651–1661.

204. Payne-Sturges, D. C., Burke, T. A., Breysse, P., Diener-West, M., and Buckley, T. J. (2004): Personal exposure meets risk assessment: a comparison of measured and modeled exposures and risks in an urban community. *Environ. Health Perspect.*, 112(5):589–598.
205. Sarnat, J. A., Schwartz, J., Catalano, P. J., and Suh, H. H. (2001): Gaseous pollutants in particulate matter epidemiology: confounders or surrogates? *Environ. Health Perspect.*, 109(10):1053–1061.
206. Weisel, C. P. et al. (2005): Relationship of Indoor, Outdoor and Personal Air (RIOPA) study: study design, methods and quality assurance/control results. *J. Expo. Anal. Environ. Epidemiol.*, 15(2):123–137.
207. Price, P., Scott, P., Wilson, N. D., and Paustenbach, D. J. (1998): An empirical approach for deriving information on total duration of exposure from information on historical exposure. *Risk Anal.*, 18:611–619.
208. El Saadi, O. and Langley, A. (1994): *The Health Risk Assessment and Management of Contaminated Sites*, Proceedings of a National Workshop on the Health Risk Assessment and Management of Contaminated Sites, South Australian Health Commission, Adelaide.
209. USEPA. (1983–1989): *Methods for Assessing Exposure to Chemical Substances*, EPA/560/5–85/002, NTIS PB86-107067. Exposure Evaluation Division, Office of Toxic Substances, U.S. Environmental Protection Agency, Washington, D.C.
210. USEPA. (1985): *Development of Statistical Distributions or Ranges of Standard Factors Used in Exposure Assessments*, EPA/600/8-85/010. Office of Health and Environmental Assessment, U.S. Environmental Protection Agency, Washington, D.C.
211. USEPA. (1986): *Guidance on Air Quality Models (Rev.)*, EPA/450/2-78/027R. Office of Air Quality Planning and Standards, U.S. Environmental Protection Agency, Research Triangle Park, NC.
212. USEPA. (1987): *Selection Criteria for Mathematical Models Used in Exposure Assessments: Surface Water Models*, EPA/600/8–87/042, NTIS PB88-139928/AS. Office of Research and Development, Office of Health and Environmental Assessment, U.S. Environmental Protection Agency, Washington, D.C.
213. Paustenbach, D. J., Wenning, R. J., Lau, V., Harrington, N. W., Rennix, D. K., and Parsons, A. H. (1992): Recent developments on the hazards posed by 2,3,7,8-tetrachlorodibenzo-*p*-dioxin in soil: implications for setting risk-based cleanup levels at residential and industrial sites. *J. Toxicol. Environ. Health*, 36(2):103–149.
214. Dragun, J. (1998): *The Soil Chemistry of Hazardous Materials*. Amherst Scientific, Amherst, MA.
215. Travis, C. C. and Hester, S. T. (1990): Background exposures to chemicals: what is the risk? *Risk Anal.*, 10:463–466.
216. Dragun, J. and Chiasson, A. (1991): *Elements in North American Soil*. Hazardous Materials Control Resources Institute, Greenbelt, MD.
217. Ott, W. R. (1994): *Environmental Statistics and Data Analysis*. CRC Press, Boca Raton, FL.
218. Nriagu, J. (1979): *Heavy Metals in the Environment*. John Wiley & Sons, New York.
219. Paustenbach, D. J. et al. (1992): Reevaluation of benzene exposure for the Pliofilm (rubberworker) cohort (1936–1976). *J. Toxicol Environ. Health*, 36(3):177–231.
220. Tinkle, S. S. et al. (2003): Skin as a route of exposure and sensitization in chronic beryllium disease. *Environ. Health Perspect.*, 111(9):1202–1208.
221. Jayjock, M. A. (1998): Risk assessment of contact allergens. *Am. J. Contact Dermat.*, 9(3):155–161.
222. Nethercott, J. R. et al. (1994): A study of chromium induced allergic contact dermatitis with 54 volunteers: implications for environmental risk assessment. *Occup. Environ. Med.*, 51(6):371–380.
223. Dutkiewicz, T. and Piotrowski, J. (1961): Experimental investigations on the quantitative estimation of aniline absorption in man. *Pure Appl. Chem.*, 3:319–323.
224. Dutkiewicz, T. and Tyras, H. (1967): A study of the skin absorption of ethylbenzene in man. *Br. J. Indust. Med.*, 24(4):330–332.
225. Feldmann, R. J. and Maibach, H. I. (1974): Percutaneous penetration of some pesticides and herbicides in man. *Toxicol. Appl. Pharmacol.*, 28(1):126–132.
226. Krivanek, N. D., McLaughlin, M., and Fayweatherm W. E. (1978): Monomethylformamide levels in human urine after repetitive exposure to dimethylformamide vapor. *J. Occup. Med.*, 20(3):179–182.
227. Mraz, J. and Nohova, H. (1992): Percutaneous absorption of *N,N*-dimethylformamide in humans. *Int. Arch. Occup. Environ. Health*, 64(2):79–83.
228. Piotrowski, J. K. (1971): Evaluation of exposure to phenol: absorption of phenol vapour in the lungs and through the skin and excretion of phenol in urine. *Br. J. Indust. Med.*, 28(2):172–178.
229. Stewart, R. D. and Dodd, H. C. (1964): Absorption of carbon tetrachloride, trichloroethylene, tetrachloroethylene, methylene chloride, and 1,1,1-trichloroethane through the human skin. *Am. Indust. Hyg. Assoc. J.*, 25:439–446.
230. Klain, G. J. and Black, K. E. (1990): Specialized techniques: congenitally athymic (nude) animal models. In: *Methods for Skin Absorption*, edited by B. W. Kemppainen and W. G. Reifenrath. CRC Press, Boca Raton, FL, pp. 165–174.
231. Bartek, M. J., LaBudde, J. A., and Maibach: H. I. (1972): Skin permeability *in vivo*: comparison in rat, rabbit, pig and man. *J. Invest. Dermatol.*, 58(3):114–123.
232. Frantz, S. W. (1990): Instrumentation and methodology for *in vitro* skin diffusion cells. In: *Methods for Skin Absorption*, edited by B. W. Kemppainen and W. G. Reifenrath. CRC Press, Boca Raton, FL, pp. 35–59.
233. Bronaugh, R. L., Stewart, R. F. Congdon, E. R. and Giles, Jr., A. L. (1982): Methods for *in vitro* percutaneous absorption studies. I. Comparison with in vivo results. *Toxicol. Appl. Pharmacol.*, 62(3):474–480.
234. Scott, R. C., Batten, P. L., Clowes, H. M., Jones, B. K., and Ramsey, J. D. (1992): Further validation of an *in vitro* method to reduce the need for *in vivo* studies for measuring the absorption of chemicals through rat skin. *Fundam. Appl. Toxicol.*, 19(4):484–492.
235. Barber, E. D., Teetsel, N. M., Kolberg, K. F., and Guest, D. (1992): A comparative study of the rates of *in vitro* percutaneous absorption of eight chemicals using rat and human skin. *Fundam. Appl. Toxicol.*, 19(4):493–497.

236. Cohen Hubal, E. A., Suggs, J. C., Nishioka, M. G., and Ivancic, W. A. (2005): Characterizing residue transfer efficiencies using a fluorescent imaging technique. *J. Expo. Anal. Environ. Epidemiol.*, 15(3):261–270.
237. Driver, J. H., Konz, J. J., and Whitmyre, G. K. (1989): Soil adherence to human skin. *Bull. Environ. Contam. Toxicol.*, 43(6):814–820.
238. Finley, B. L., Scott, P. K., and Mayhall, D. A. (1994): Development of a standard soil-to-skin adherence probability density function for use in Monte Carlo analyses of dermal exposure. *Risk Anal.*, 14(4):555–569.
239. Holmes, K. K., Kissel, J. C., and Richter, K. Y. (1996): Investigation of the influence of oil on soil adherence to skin. *J. Soil Contam.*, 5(4):301–308.
240. Johnson, J. E. and Kissel, J. C. (1996): Prevalence of dermal pathway dominance in risk assessment of contaminated soils: a survey of Superfund risk assessment, 1989–1992. *Hum. Ecol. Risk Assess.*, 2:356–365.
241. Kissel, J. C., Richter, K. Y., and Fenske R. A. (1996): Field measurement of dermal soil loading attributable to various activities: implications for exposure assessment. *Risk Anal.*, 16(1):115–125.
242. Lepow, M. L., Bruckman, L., Gillette, M., Markowitz, S., Robino, R., and Kapish, J. (1975): Investigations into sources of lead in the environment of urban children. *Environ. Res.*, 10(3):415–426.
243. Que Hee, S. S., Peace, B., Clark, C. S., Boyle, J. R., Bornschein, R. L., and Hammond, P. B. (1985): Evolution of efficient methods to sample lead sources, such as house dust and hand dust, in the homes of children. *Environ. Res.*, 38(1):77–95.
244. Roels, H. A., Buchet, J. P., Lauwenys, R. R., Claeys-Thoreau, F., Lafontaine, A., and Verduyn, G. (1980): Exposure to lead by oral and pulmonary routes of children living in the vicinity of a primary lead smelter. *Environ. Res.*, 22:81–94.
245. Sheppard, S. C. and Evenden, W. G. (1994): Contaminant enrichment and properties of soil adhering to skin, *J. Environ. Qual.*, 23:604–613.
246. Marlow, D., Sweeney, M. H., and Fingerhut, M. (1990): Estimating the amount of TCDD absorbed by workers who manufactured 2,4,5-T. In: *Proc. Tenth Int. Conf. on Chlorinated Dioxins and Related Compounds*, Bayreuth, Germany.
247. Rodes, C. E., Newsome, J. R., Vanderpool, R. W., Antley, J. T., and Lewis R. G. (2001): Experimental methodologies and preliminary transfer factor data for estimation of dermal exposures to particles. *J. Expo. Anal. Environ. Epidemiol.*, 11(2):123–139.
248. Burmaster, D. E. and Thompson, K. M. (1997): Estimating exposure point concentrations for surface soils for use in deterministic and probabilistic risk assessments. *Hum. Ecol. Risk Assess.*, 3:363–384.
249. Wallace, L. A., Pellizzari, E., Hartwell, T., Whitmore, R. W., Sparacino, C., and Zelon, H. (1986): Total Exposure Assessment Methodology (TEAM) study: personal exposure, indoor–outdoor relationships, and breath levels of volatile organic compounds in New Jersey. *Environ. Int.*, 12:369–387.
250. Wallace, L. A. (1989): The Total Exposure Assessment Methodology (TEAM) study: an analysis of exposures, sources, and risk associated with four chemicals. *J. Am. Coll. Toxicol.*, 8:883–895.
251. Lioy, P. J., Freeman, N. C. and Millette, J. R. (2002): Dust: a metric for use in residential and building exposure assessment and source characterization. *Environ. Health Perspect.*, 110(10):969–983.
252. Lioy, P. J., Wainman, T., and Weisel, C. (1993): A wipe sampler for the quantitative measurement of dust on smooth surfaces: laboratory performance studies. *J. Expo. Anal. Environ. Epidemiol.*, 3(3):315–330.
253. Lioy, P. J., Yiin, L. M., Adgate, J., Weisel, C., and Rhoads, G. G. (1998): The effectiveness of a home cleaning intervention strategy in reducing potential dust and lead exposures. *J. Expo. Anal. Environ. Epidemiol.*, 8(1):17–35.
254. USEPA. (1999): *Risk Assessment Guidelines for Dermal Assessment*. U.S. Environmental Protection Agency, Washington, D.C.
255. Surber, C., Wilhelm, K. P., Maibach, H. I., Hall, L. L., and Guy, R. H. (1990): Partitioning of chemicals into human stratum corneum: implications for risk assessment following dermal exposure. *Fundam. Appl. Toxicol.*, 15(1): 99–107.
256. Wepierre, J. and Marty, J. P. (1979): Percutaneous absorption of drugs. *Trends Pharmacol. Sci.*, 1:23–26.
257. Guy, R. H., Hadgraft, J., and Maibach, H. I. (1982): A pharmacokinetic model for percutaneous absorption. *Int J. Pharmacol.*, 11:119–129.
258. Anderson, B. D., Higuchi, W. I., and Raykar, P. V. (1988) Heterogeneity effects on permeability-partition coefficient relationships in human stratum corneum. *Pharm. Res.* 5(9):566–573.
259. Flynn, G. L. (1990): Physicochemical determinants of skin absorption. In: *Principles of Route-to-Route Extrapolation for Risk Assessment*, edited by T. R. Gerrity and C. J Henry. Elsevier, New York, pp. 93–127.
260. Gargas, M. L., Burgess, R. J., Voisard, D. E., Cason, G H., and Andersen, M. E. (1989): Partition coefficients of low-molecular-weight volatile chemicals in various liquids and tissues. *Toxicol. Appl. Pharmacol.*, 98(1):87–99.
261. Andersen, M. E., Clewell, H. J. I., Gargas, M. L., Smith F. A., and Reitz, R. H. (1987): Physiologically-based pharmacokinetics and the risk assessment for methylene chloride. *Toxicol. Appl. Pharmacol.*, 87:185–205.
262. Shatkin, J. A. and Brown, H. S. (1991): Pharmacokinetics of the dermal route of exposure to volatile organic chemicals in water: a computer simulation model. *Environ. Res.* 56(1):90–108.
263. McDougal, J. N. (1996): Physiologically-based pharmacokinetic modeling. In: *Dermatology*, edited by F. N. Marzulli and H. I. Maibach. Taylor & Francis, Washington D.C.
264. Islam, M. S., Zhao, L., Zhou, J., Dong, L., McDougal, J. N., and Flynn, G. L. (1996): Systemic uptake and clearance of chloroform by hairless rats following dermal exposure I. Brief exposure to aqueous solutions. *Risk Anal* 16(3):349–357.

265. Islam, M. S., Zhao, L., Zhou, J., Dong, L., McDougal, J. N., and Flynn, G. L. (1999): Systemic uptake and clearance of chloroform by hairless rats following dermal exposure: II. Absorption of the neat solvent. *Am. Ind. Hyg. Assoc. J.*, 60(4):438–443.
266. Jepson, G. W. and McDougal, J. N. (1997): Physiologically based modeling of nonsteady state dermal absorption of halogenated methanes from an aqueous solution. *Toxicol. Appl. Pharmacol.*, 144(2):315–324.
267. Jepson, G. W. and McDougal, J. N. (1999): Predicting vehicle effects on the dermal absorption of halogenated methanes using physiologically based modeling. *Toxicol. Sci.*, 48(2):180–188.
268. Mattie, D. R., Bates, Jr., G. D., Jepson, G. W., Fisher, J. W., and McDougal, J. N. (1994): Determination of skin:air partition coefficients for volatile chemicals: experimental method and applications. *Fundam. Appl. Toxicol.*, 22(1):51–57.
269. Mattie, D. R., Grabau, J. H., and McDougal, J. N. (1994): Significance of the dermal route of exposure to risk assessment. *Risk Anal.*, 14(3):277–284.
270. McDougal, J. N. and Boeniger, M. F. (2002): Methods for assessing risks of dermal exposures in the workplace. *Crit. Rev. Toxicol.*, 32(4):291–327.
271. McDougal, J. N. and Robinson, P. J. (2002): Assessment of dermal absorption and penetration of components of a fuel mixture (JP-8). *Sci. Total Environ.*, 288(1–2):23–30.
272. McDougal, J. N., Jepson, G. W., Clewell, 3rd, H. J., and Andersen, M. E. (1985): Dermal absorption of dihalomethane vapors. *Toxicol. Appl. Pharmacol.*, 79(1):150–158.
273. McDougal, J. N., Jepson, G. W., Clewell, 3rd, H. J., MacNaughton, M. G., and Andersen, M. E. (1986): A physiological pharmacokinetic model for dermal absorption of vapors in the rat. *Toxicol. Appl. Pharmacol.*, 85(2):286–294.
274. McDougal, J. N. and Jurgens-Whitehead, J. L. (2001): Short-term dermal absorption and penetration of chemicals from aqueous solutions: theory and experiment. *Risk Anal.*, 21(4):719–726.
275. Morgan, D. L. et al. (1991): Dermal absorption of neat and aqueous volatile organic chemicals in the Fischer 344 rat. *Environ. Res.*, 55(1):51–63.
276. Rogers, J. V. and McDougal, J. N. (2002): Improved method for *in vitro* assessment of dermal toxicity for volatile organic chemicals. *Toxicol. Lett.*, 135(1–2):125–135.
277. USEPA. (1992): *Dermal Exposure Assessments: Principles and Applications*, EPA/600/8–91/011B. Exposure Assessment Group, Office of Health and Environmental Assessment, U.S. Environmental Protection Agency, Washington, D.C.
278. Skrowronski, G. A., Turkall, R. M., and Abdel-Rahman, M. S. (1988): Soil absorption alters bioavailability of benzene in dermally exposed male rats. *Am. Indust. Hyg. Assoc. J.*, 49:506–511.
279. Umbreit, T. H., Hesse, E. J., and Gallo, M. A. (1986): Comparative toxicity of TCDD contaminated soil from Times Beach, Missouri, and Newark, New Jersey. *Chemosphere*, 15:121–2124.
280. Wester, R. C., Maibach, H. I., Sedik, L., Melendres, J., Wade, M., and DiZio, S. (1993): Percutaneous absorption of pentachlorophenol from soil. *Fundam. Appl. Toxicol.*, 20(1):68–71.
281. Fouchecourt, M. O., Arnold, M., Berny, P., Videmann, B., Rether, B., and Riviere, J. L. (1999): Assessment of the bioavailability of PAHs in rats exposed to a polluted soil by natural routes: induction of EROD activity and DNA adducts and PAH burden in both liver and lung. *Environ. Res.*, 80(4):330–339.
282. Snyder, W. S. et al. (1975): *Report of the Task Group on Reference Man*. Pergamon Press, Oxford, U.K., pp. 134–144
283. Burmaster, D. E. (1998): Lognormal distributions for skin area as a function of body weight. *Risk Anal.*, 18(1):27–32.
284. CDHS. (1986): *Development of Applied Action Levels for Soil Contact: A Scenario for the Exposure of Humans to Soil in a Residential Setting*. California Department of Health Services, Sacramento, CA.
285. Caplan, K. (1993): The significance of wipe samples. *Am. Indust. Hyg. Assoc. J.*, 53(2):70–75.
286. ECETOC. (1993): *Strategy for Assigning a "Skin Notation,"* revised ECETOC Document No. 31. European Center for Ecotoxicology and Toxicology of Chemicals, Brussels, Belgium.
287. ECETOC. (1993): *Percutaneous Absorption*, Monograph No. 20. European Center for Ecotoxicology and Toxicology of Chemicals, Brussels, Belgium.
288. Fenske, R. A. (1993): Dermal exposure assessment techniques. *Ann. Occup. Hyg.*, 37(6):687–706.
289. McArthur, B. (1992): Dermal measurement and wipe sampling methods: a review. *Appl. Occup. Environ. Hyg.*, 7:599–606.
290. Michaud, J. M., Huntley, S. L., Sherer, R. A., Gray, M. N., and Paustenbach, D. J. (1994): PCB and dioxin re-entry criteria for building surfaces and air. *J. Expo. Anal. Environ. Epidemiol.*, 4(2):197–227.
291. Lioy, P. J. et al. (2002): Characterization of the dust/smoke aerosol that settled east of the World Trade Center (WTC) in lower Manhattan after the collapse of the WTC 11 September, 2001. *Environ. Health Perspect.*, 110(7): 703–714.
292. Tang, K. M., Nace, Jr., C. G., Lynes, C. L., Maddaloni, M. A., LaPosta, D., and Callahan, K. C. (2004): Characterization of background concentrations in upper Manhattan, New York, apartments for select contaminants identified in World Trade Center dust. *Environ. Sci. Technol.*, 38(24):6482–6490.
293. Yiin, L. M. et al. (2004): Comparisons of the dust/smoke particulate that settled inside the surrounding buildings and outside on the streets of southern New York City after the collapse of the World Trade Center, September 11, 2001, *J. Air Waste Manag. Assoc.*, 54(5):515–528.
294. Brouwer, D. H. and van Hemmen, J. J. (1992): Elements of a sampling strategy for dermal exposure assessment [abstract]. In: *Workshop on Occupational Skin Exposure to Chemical Substances*, International Occupational Hygiene Association, Brussels, Belgium, December 7.

295. Fenske, R. A. and Lu, C. (1994): Determination of handwash removal efficiency: incomplete removal of the pesticide chlorpyrifos from skin by standard handwash techniques. *Am. Indust. Hyg. Assoc. J.*, 55(5):425–432.
296. Lavy, T. L., Shepard, J. S., and Bouchard, D. C. (1980): Field worker exposure and helicopter spray pattern of 2,4,5-T. *Bull. Environ. Contam. Toxicol.*, 24(1):90–96.
297. Lavy, T. L., Walstad, J., Flynn, R., and Mattice, J. (1982): (2,4-Dichlorophenoxy)acetic acid exposure received by aerial application crews during forest spray operations. *J. Agric. Food Chem.*, 30(2):375–381.
298. Popendorf, W. J. and Leffingwell, J. T. (1982): Regulating OP pesticide residues for farmworker protection. *Residue Rev.*, 82:125–201.
299. Wickens, K., Lane, J., Siebers, R., Ingham, T., and Crane, J. (2004): Comparison of two dust collection methods for reservoir indoor allergens and endotoxin on carpets and mattresses. *Indoor Air*, 14(3):217–222.
300. Foarde, K. and Berry, M. (2004): Comparison of biocontaminant levels associated with hard vs. carpet floors in nonproblem schools: results of a year long study. *J. Expo. Anal. Environ. Epidemiol.*, 14(Suppl. 1):S41–S48.
301. Yiin, L. M. et al. (2002): Comparison of techniques to reduce residential lead dust on carpet and upholstery: the New Jersey assessment of cleaning techniques trial. *Environ. Health Perspect.*, 110(12):1233–1237.
302. Costa, M., Zhitkovich, A., Harris, M., Paustenbach, D. J., and Gargas, M. (1997): DNA–protein cross-links produced by various chemicals in cultured human lymphoma cells. *J. Toxicol. Environ. Health*, 50(5):433–449.
303. Byard, J. (1989): Hazard assessment of 1,1,1-trichloroethane in groundwater. In: *The Risk Assessment of Environmental and Human Health Hazards: A Textbook of Case Studies*, edited by D. J. Paustenbach. John Wiley & Sons, New York, pp. 331–334.
304. Jo, W. K., Weisel, C. P., and Lioy, P. J. (1990): Routes of chloroform exposure and body burden from showering with chlorinated tap water. *Risk Anal.*, 10(4):575–580.
305. Jo, W. K., Weisel, C. P., and Lioy, P. J. (1990): Chloroform exposure and body burden from showering with chlorinated tap water. *Risk Anal.*, 10:575–580.
306. Kezic, S., Mahieu, K., Monster, A. C., and de Wolff, F. A. (1997): Dermal absorption of vaporous and liquid 2-methoxyethanol and 2-ethoxyethanol in volunteers. *Occup. Environ. Med.*, 54(1):38–43.
307. Scow, K., Wechsler, A. E., Stevens, J., Wood, M., and Callahan, M. A. (1979): *Identification and Evaluation of Waterborne Routes of Exposure from Other Than Food and Drinking Water*, EPA/440/4-79/016. U.S. Environmental Protection Agency, Washington, D.C.
308. Kerger, B. D., Schmidt, C. E., and Paustenbach, D. J. (2000): Assessment of airborne exposure to trihalomethanes from tap water in residential showers and baths. *Risk Anal.*, 20(5):637–651.
309. Calabrese, E. J. and Stanek, 3rd, E. J. (1991): A guide to interpreting soil ingestion studies. II. Qualitative and quantitative evidence of soil ingestion. *Regul. Toxicol. Pharmacol.*, 13(3):278–292.
310. Calabrese, E. J., Stanek, E. J., Gilbert, C. E., and Barnes, R. M. (1990): Preliminary adult soil ingestion estimates: results of a pilot study. *Regul. Toxicol. Pharmacol.*, 12(1):88–95.
311. Clayton, C. A. et al. (1993): Particle Total Exposure Assessment Methodology (PTEAM) study: distributions of aerosol and elemental concentrations in personal, indoor, and outdoor air samples in a southern California community. *J. Expo. Anal. Environ. Epidemiol.*, 3(2):227–250.
312. Fiserova-Bergerova, V., Pierce, J. T., and Droz, P. O. (1990): Dermal absorption potential of industrial chemicals: criteria for skin notation. *Am. J. Indust. Med.*, 17(5):617–635.
313. Fitzgerald, E. F., Hwang, S. A., Brix, K. A., Bush, B., Cook, K., and Worswick, P. (1995): Fish PCB concentrations and consumption patterns among Mohawk women at Akwesasne. *J. Expo. Anal. Environ. Epidemiol.*, 5(1):1–19.
314. Gomez, M. R. (2000): Exposure assessment must stop being local. *Appl. Occup. Environ. Hyg.*, 15(1):15–20.
315. Jarabek, A. M., Menache, M. G., Overton, Jr., J. H., Dourson, M. L., and Miller, F. J. (1990): The U.S. Environmental Protection Agency's inhalation RfD methodology: risk assessment for air toxics. *Toxicol. Indust. Health*, 6(5):279–301.
316. Butte, W. and Heinzow, B. (2002): Pollutants in house dust as indicators of indoor contamination. *Rev. Environ. Contam. Toxicol.*, 175:1–46.
317. Reif, J. S., Burch, J. B., Nuckols, J. R., Metzger, L., and Anger, W. K. (2003): Neurobehavioral effects of exposure to trichloroethylene through a municipal water supply. *Environ. Res.*, 9:248–258.
318. Roberts, J. W. et al. (1992): Human exposure to pollutants in the floor dust of homes and office. *J. Exp. Anal. Environ. Epidemiol. Suppl.*, 1:127–146.
319. Roberts, J. W. and Dickey, P. (1995): Exposure of children to pollutants in house dust and indoor air. *Rev Environ. Contam. Toxicol.*, 143:59–78.
320. Kimbrough, R. D., Falk, H., Stehr, P., and Fries, G. (1984): Health implications of 2,3,7,8-tetrachlorodibenzodioxin (TCDD) contamination of residential soil. *J. Toxicol. Environ. Health*, 14(1):47–93.
321. Barltrop, D. (1966): The prevalence of pica. *Am. J. Dis Child.*, 112(2):116–123.
322. Fenske, R. A., Bradman, A., Whyatt, R. M., Wolff, M. S. and Barr, D. B. (2005): Lessons learned for the assessment of children's pesticide exposure: critical sampling and analytical issues for future studies. *Environ. Health Perspect.* 113(10):1455–1462.
323. Vojta, P. J. et al. (2002): First National Survey of Lead and Allergens in Housing: survey design and methods for the allergen and endotoxin components. *Environ. Health Perspect.*, 110(5):527–532.
324. McCauley, L. A. et al. (2001): Work characteristics and pesticide exposures among migrant agricultural families: a community-based research approach. *Environ. Health Perspect.*, 109(5):533–538.
325. Whitmore, R. W., Immerman, F. W., Camann, D. E., Bond, A. E., Lewis, R. G., and Schaum, J. L. (1994): Non-occupational exposures to pesticides for residents of two U.S. cities. *Arch. Environ. Contam. Toxicol.*, 26:47–59.
326. Fenske, R. A., Lu, C., Barr, D., and Needham, L. (2002): Children's exposure to chlorpyrifos and parathion in an agricultural community in central Washington State. *Environ. Health Perspect.*, 110(5):549–553.

327. Roberts, J. W. et al. (1991): Development and field testing of a high volume sampler for pesticides and toxics in dust. *J. Expo. Anal. Environ. Epidemiol.*, 1(2):143–155.
328. Lewis, R. G., Fortmann, R. C., and Camann, D. E. (1994): Evaluation of methods for monitoring the potential exposure of small children to pesticides in the residential environment. *Arch. Environ. Contam. Toxicol.*, 26(1):37–46.
329. Simcox, N. J., Fenske, R. A., Wolz, S. A., Lee, I. C., and Kalman, D. A. (1995): Pesticides in household dust and soil: exposure pathways for children of agricultural families. *Environ. Health Perspect.*, 103(12):1126–1134.
330. Gilliland, F. et al. (2005): Air pollution exposure assessment for epidemiologic studies of pregnant women and children: lessons learned from the Centers for Children's Environmental Health and Disease Prevention Research. *Environ. Health Perspect.*, 113(10):1447–1454.
331. Ozkaynak, H., Whyatt, R. M., Needham, L. L., Akland, G., and Quackenboss, J. (2005): Exposure assessment implications for the design and implementation of the National Children's Study. *Environ. Health Perspect.*, 113(8):1108–1115.
332. Walter, S. D., Yankel, A. J., and Lindern, I. H. (1980): Age-specific risk factors for lead absorption in children. *Arch. Environ. Health*, 35:53–58.
333. Cooper, M. (1957). *Pica*. Charles C Thomas, Springfield, IL, pp. 60–74.
334. Charney, E., Sayre, J., and Coulter, M. (1980): Increased lead absorption in inner city children: where does the lead come from? *Pediatrics*, 65(2):226–231.
335. Sayre, J., Charney, E., Vostal, J., and Pless, B. (1974): House and hand dust as a potential source of childhood lead exposure. *Am. J. Dis. Child.*, 127:167–170.
336. Duggan, M. J. and Williams, S. (1977): Lead in dust in city streets. *Sci. Total Environ.*, 7(1):91–97.
337. Lepow, M. L., Bruckman, L., Rubino, R. A., Markowtiz, S., Gillette, M., and Kapish, J. (1974): Role of airborne lead in increased body burden of lead in Hartford children. *Environ. Health Perspect.*, 7:99–102.
338. Barltrop, D., Strehlow, C. D., Thornton, I., and Webb, J. S. (1975): Absorption of lead from dust and soil. *Postgrad. Med. J.*, 51(601):801–804.
339. NRC. (1974): *Lead in the Environment*. National Academy Press, Washington, D.C.
340. Paustenbach, D. J. (1987): Assessing the potential environmental and human health risks of contaminated soil. *Comments Toxicol.*, 1:185–220.
341. Day, J. P., Hart, M., and Robinson, M. S. (1975): Leas in urban street dust. *Nature (Lond.)*, 253:343–345.
342. Bryce-Smith, D. (1974): Lead absorption in children. *Phys. Bull.*, 25:178–187.
343. LaGoy, P. K. (1987): Estimated soil ingestion rates for use in risk assessment. *Risk Anal.*, 7(3):355–359.
344. Binder, S., Sokal, D., and Maughan, D. (1986): Estimating soil ingestion: the use of tracer elements in estimating the amount of soil ingested by young children. *Arch. Environ. Health*, 41(6):341–345.
345. Van Wijnen, J. H. et al. (1990): Estimated soil ingestion by children. *Environ. Res.*, 51:147–162.
346. Calabrese, E. J. et al. (1989): How much soil do young children ingest: an epidemiologic study. *Regul. Toxicol. Pharmacol.*, 10(2):123–137.
347. de Silva, P. E. (1991): *Assessment of Heath Risk to Residents of Contaminated Sites*. Occupational Health Services Report to Gas and Fuel Corporations, AMCOSH, Werribbee, Victoria, Canada.
348. de Silva, P. E. (1994): How much soil do children ingest: a new approach. *Appl. Occup. Environ. Hyg.*, 9:40–43.
349. Calabrese, E. J. et al. (1996): Methodology to estimate the amount and particle size of soil ingested by children: implications for exposure assessment at waste sites. *Regul. Toxicol. Pharmacol.*, 24(3):264–268.
350. Stanek, E. J. and Calabrese, E. J. (1991): A guide to interpreting soil ingestion studies. I. Development of a model to estimate the soil ingestion detection level of soil ingestion studies. *Regul. Toxicol. Pharmacol.*, 13:263–277.
351. Stanek, E. J. and Calabrese, E. J. (1995): Daily estimates of soil ingestion in children. *Environ. Health Perspect.*, 103:276–285.
352. Stanek, E. J. and Calabrese, E. J. (1995): Improved soil ingestion estimates for use in site evaluations using the best tracer method. *Hum. Ecol. Risk Assess.*, 1:133–157.
353. Stanek, E. J., Calabrese, E. J., and Xu, L. (1997): Soil ingestion in adults: results of a second pilot study. *Ecotoxicol. Environ. Safety*, 36:249–257.
354. Calabrese, E. J., Stanek, 3rd, E. J., Pekow, P., and Barnes, R. M. (1997): Soil ingestion estimates for children residing on a superfund site. *Ecotoxicol. Environ. Safety*, 36(3):258–268.
355. Stanek, E. J. and Calabrese, E. J. (1995): Soil ingestion estimates for use in site evaluation based on the best tracer method. *Hum. Ecol. Risk Assess.*, 1:133–156.
356. Stanek, E. J. and Calabrese, E. J. (1998): Prevalence of soil mouthing/ingestion among healthy children aged 1 to 6. *Soil Contam.*, 2:27–42.
357. Calabrese, E. J. and Stanek, E. J. (1998): Soil ingestion in children and adults: a dominant influence in site-specific risk assessment. *Environ. Law Reporter*, 28(10):660–710.
358. Schoof, R. A. (2004): Bioavailability of soil-borne chemicals: method development and validation. *Hum. Ecol. Risk Assess.*, 10(4):637.
359. Lourie, R. S. and Cayman, E. M. (1963): Why children eat things that are not food. *Children*, 10:143–146.
360. Danford, D. E. (1982): Pica and nutrition. *Annu. Rev. Nutr.*, 2:303–322.
361. Taylor, E. R. (1983): How much soil do children eat? In: *The Health Risk Assessment and Management of Contaminated Sites*, edited by O. El Saadi and A. Langley. South Australian Health Commission, Adelaide, pp. 72–77.
362. Russell, R. S. (1966): Entry of radioactive materials into plants. In: *Radioactivity and Human Diet*, edited by R. S. Russell. Pergamon Press, New York.
363. Martin, W. E. (1964): Loss of Sr-90, Sr-89, and I-131 from fallout of contaminated plants. *Radiat. Bot.*, 4:174–183.
364. Knarr, R. D., Cooper, G. L., Brian, E. A., Kleinschmidt, M. G., and Graham, D. G. (1985): Worker exposure during aerial application of a liquid and a granular formulation of Ordram Selective Herbicide to rice. *Arch. Environ. Contam. Toxicol.*, 14(5):523–527.
365. Zartarian, V. G., Ferguson, A. C., and Leckie, J. O. (1998): Quantified mouthing activity data from a four-child pilot field study. *J. Expo. Anal. Environ. Epidemiol.*, 8(4):543–553.

366. USDA. (1972): *Food Consumption: Households in the United States, Seasons, and Year 1965–1966*. U.S. Department of Agriculture, Washington, D.C.
367. Food Quality Protection Act of 1996, Public Law 104-170, 1996.
368. USEPA. (1989): *Interim Procedures for Estimating Risks Associated with Exposures to Mixtures of Chlorinated Dibenzo-p-Dioxins and -Dibenzofurans (CDD and CDFs) and 1989 Update*, EPA/625/3-89/016. Risk Assessment Forum, U.S. Environmental Protection Agency, Washington D.C.
369. USEPA. (1999): *Guidance for Performing Aggregate Exposure and Risk Assessments*. Office of Pesticide Programs, U.S. Environmental Protection Agency, Washington, D.C.
370. USEPA. (2002): *Guidance on Cumulative Risk Assessment of Pesticide Chemicals That Have a Common Mechanism of Toxicity*. Office of Pesticide Programs, Office of Prevention, Pesticides, and Toxic Substances, U.S. Environmental Protection Agency, Washington, D.C.
371. USEPA (2002): *Revised Cumulative Risk Assessment of Organophosphate Pesticides*. Office of Pesticide Programs, Office of Prevention, Pesticides, and Toxic Substances, U.S. Environmental Protection Agency, Washington, D.C.
372. USDA. (1980): *Food and Nutrient Intakes of Individuals in One Day in the United States, Spring 1977: Nationwide Food Consumption Survey 1977–1978*, Preliminary Report 2. U.S. Department of Agriculture, Washington, D.C.
373. USDA (1992): *Food and Nutrient Intakes of Individuals in One Day in the United States, 1987–1988: Nationwide Food Consumption Survey 1987–1988*, Report 87-1-1. U.S. Department of Agriculture, Washington, D.C.
374. USDA. (1992): *Continuing Survey of Food Intakes by Individuals (CSFII) 1988–1991*. U.S. Department of Agriculture, Washington, D.C.
375. USEPA. (1984): *An Estimation of the Daily Food Intake Based on Data From the 1977–1978. USDA Nationwide Food Consumption Survey*, EPA/520/1–84/015. Office of Pesticide Programs, Office of Prevention, Pesticides, and Toxic Substances, U.S. Environmental Protection Agency, Washington, D.C.
376. Pao, E. M., Fleming, K. H., Guenther, P. M., and Mickle, S. J. (1982): *Food Commonly Eaten by Individuals: Amount Per Day and Per Eating Occasion*, Home Economics Report No. 44. U.S. Department of Agriculture, Beltsville, MD.
377. White, S. B., Peterson, C. A., Clayton, C. A., and Duncan, D. P. (1983): *The Construction of a Raw Agricultural Commodity Consumption Database*, Interim Report No. 1. Office of Pesticide Programs, Office of Prevention, Pesticides, and Toxic Substances, U.S. Environmental Protection Agency, Washington, D.C.
378. USDA. (1998): *Continuing Survey of Food Intakes by Individuals (CSFII): 1994–96*. U.S. Department of Agriculture, Washington, D.C.
379. USDA. (1999): *Food and Nutrient Intakes by Children 1994–96*. Food Surveys Research Group, Beltsville Human Nutrition Research Center, U.S. Department of Agriculture, Beltsville, MD.
380. Finley, B. L. et al. (2003): The Passaic creel/angler survey: expert panel review, findings, and recommendations. *Hum. Ecol. Risk Assess.*, 9:829–855.
381. Price, P., Su, S. H., and Gray, M. N. (1994): The effect of sampling bias on estimates of angler consumption rates in creel surveys. *J. Exp. Anal. Environ. Epidemiol.*, 4:355–372.
382. Puffer, H. W., Azen, S. P., Duda, M. J., and Young, D. R. (1981): *Consumption Rates of Potentially Hazardous Marine Fish Caught in the Metropolitan Los Angeles Area*, EPA Grant R807.120010. U.S. Environmental Protection Agency, Washington, D.C.
383. Ebert, E. S., Price, P. S., and Keenan, R. E. (1994): Selection of fish consumption estimates for use in the regulatory process. *J. Exp. Anal. Environ. Epidemiol.*, 4:373–394.
384. Ruffle, B., Burmaster, D. E., Anderson, P. D., and Gordon, H. D. (1994): Lognormal distributions for fish consumption by the general U.S. populations. *Risk Anal.*, 14(4): 395–404.
385. Roseberry, A. M. and Burmaster: D. E. (1991): A note: estimating exposure concentration of lipophilic organics chemicals to humans via finfish. *J. Exp. Anal. Environ. Epidemiol.*, 1:513–521.
386. Murray, D. M. and Burmaster, D. E. (1994): Estimated distribution for average daily consumption of total and self-caught fish for adults in Michigan angler households. *Risk Anal.*, 14:513–520.
387. USEPA. (1999): *Guidance for Performing Aggregate Exposure and Risk Assessments Under the Food Quality Protection Act. (Draft)*. Office of Pesticide Programs, Office of Prevention, Pesticides, and Toxic Substances, U.S. Environmental Protection Agency, Washington, D.C.
388. USEPA. (1999): *Guidance for Identifying Pesticide Chemicals and Other Substances That Have a Common Mechanism of Toxicity*. Office of Pesticide Programs, Office of Prevention, Pesticides, and Toxic Substances, U.S. Environmental Protection Agency, Washington, D.C.
389. Kohler, L., Meeuwisse, G., and Mortensson, W. (1984): Food intake and growth of infants between six and twenty-six weeks of age on breast milk, cow's milk formula, or soy formula. *Acta Paediatr. Scand.*, 73(1):40–48.
390. Institute of Medicine. (1991): *Nutrition During Lactation* The National Academies Press, Washington, D.C.
391. Neville, M. C. et al. (1988): Studies in human lactation: milk volumes in lactating women during the onset of lactation and full lactation. *Am. J. Clin. Nutr.*, 48(6): 1375–1386.
392. Butte, N. F., Garza, C., Smith, E. O. et al. (1984): Human milk intake and growth in exclusively breast-fed infants J. Pediatr., 104:187–195.
393. Arcus-Arth, A., Krowech, G., and Zeise, L. (2005): Breast milk and lipid intake distributions for assessing cumulative exposure and risk. *J. Expo. Anal. Environ. Epidemiol.* 15(4):357–365.
394. Ebert, E. S., Harrington, J. R., Boyle, J. R., Knight, J., and Keenan, R. E. (1993): Estimating consumption of fresh water fish among Maine anglers. *North Am. J. Fisheries Manag.*, 13:737–745.
395. Smith, A. E. (1987): Infant exposure assessment for breast milk dioxins and furans derived from waste incineration emissions. *Risk Anal.*, 7(3):347–353.

396. Gallenberg, L. A. and Vodicnik, M. J. (1989): Transfer of persistent chemicals in milk. *Drug Metab. Rev.*, 21:277–317.
397. Needham, L. L. and Wang, R. Y. (2002): Analytic considerations for measuring environmental chemicals in breast milk. *Environ. Health Perspect.*, 110(6):A317–A324.
398. LaKind, J. S., Berlin, C. M., and Naiman, D. Q. (2001): Infant exposure to chemicals in breast milk in the United States: what we need to learn from a breast milk monitoring program. *Environ. Health Perspect.*, 109(1):75–88.
399. Knowles, J. A. (1965): Excretion of drugs in milk: a review. *J. Pediatr.*, 66(6):1068–1082.
400. Laug, E. P., Kunze, F. M., and Prickett, C. S. (1951): Occurrence of DDT in human fat and milk. *Arch. Indust. Hyg.*, 3:245–246.
401. Solomon, G. M. and Weiss, P. M. (2002): Chemical contaminants in breast milk: time trends and regional variability. *Environ. Health Perspect.*, 110(6):A339–A347.
402. Andelman, J. B. (1985): Human exposures to volatile halogenated organic chemicals in indoor and outdoor air. *Environ. Health Perspect.*, 62:313–318.
403. Krewski, D. et al. (2003): Overview of the reanalysis of the Harvard Six Cities Study and American Cancer Society Study of Particulate Air Pollution and Mortality. *J. Toxicol. Environ. Health A*, 66(16–19):1507–1551.
404. Pope, C. A. et al. (2002): Lung cancer, cardiopulmonary mortality, and long-term exposure to fine particulate air pollution. *JAMA*, 287(9):1132–1141.
405. Sarnat, J. A., Schwartz, J., and Suh, H. H. (2001): Fine particulate air pollution and mortality in 20 U.S. cities. *N. Engl. J. Med.*, 344(16):1253–1254.
406. Villeneuve, P. J., Goldberg, M. S., Krewski, D., Burnett, R. T., and Chen, Y. (2002): Fine particulate air pollution and all-cause mortality within the Harvard Six-Cities Study: variations in risk by period of exposure. *Ann. Epidemiol.*, 12(8):568–576.
407. Carrington, C. D. and Bolger, P. M. (1998): Uncertainty and risk assessment. *Hum. Ecol. Risk Assess.*, 4:253–258.
408. Hoffman, F. O. and Hammonds, J. S. (1992): *An Introductory Guide to Uncertainty Analysis in Environmental and Health Risk Assessment*, ES/ER/TM-35. Martin Marietta Energy Systems, Oak Ridge, TN.
409. Morgan, M. G. and Henrion, M. (1990): *Uncertainty: A Guide to Dealing with Uncertainty in Quantitative Risk and Policy Analysis*. Cambridge: Cambridge University Press, Cambridge, U.K.
410. USEPA. (1995): *Guidance for Risk Characterization*. Science Policy Council, U.S. Environmental Protection Agency, Washington, D.C.
411. USEPA. (2000): *Human Health Risk Assessment for Diazinon*. Health Effects Division, Office of Pesticide Programs, Office of Prevention, Pesticides, and Toxic Substances, U.S. Environmental Protection Agency, Washington, D.C.
412. Anderson, P. D. and Yuhas, A. L. (1996): Improving risk management by characterizing reality: a benefit of probabilistic risk assessment. *Hum. Ecol. Risk Assess.*, 2:55–58.
413. Burmaster, D. E. and Huff, D. A. (1997): Using lognormal distributions and lognormal probability plots in probabilistic risk assessments. *Hum. Ecol. Risk Assess.*, 3:223–234.
414. Burmaster, D. E. and Maxwell, N. I. (1991): Time- and loading-dependence in the McKone model for dermal uptake of organic chemicals from a soil matrix. *Risk Anal.*, 11(3):491–497.
415. Burmaster, D. E. and Thompson, K. M. (1995): Backcalculating cleanup targets in probabilistic risk assessments when the acceptability of cancer risk is defined under different risk management policies. *Hum. Ecol. Risk Assess.*, 1(1):101–120.
416. Glickman, T. S. (1986): A methodology for estimating time-of-day variations in the size of a population exposed to risk. *Risk Anal.*, 6(3):317–324.
417. Israeli, M. and Nelson, C. B. (1992): Distribution and expected time of residence for U.S. households. *Risk Anal.*, 12:65–72.
418. Murray, D. M. and Burmaster, D. E. (1992): Estimated distributions for total body surface area of men and women in the United States. *J. Expo. Anal. Environ. Epidemiol.*, 2(4):451–461.
419. Price, P. et al. (1996): Monte Carlo modeling of time-dependent exposures using a microexposure event approach. *Risk Anal.*, 16:339–348.
420. Taylor, A. C., Evans, J. S., and McKone, T. E. (1993): The value of animal test information in environmental control decisions. *Risk Anal.*, 12:403–412.
421. Trowbridge, P. R. and Burmaster, D. E. (1997): A parametric distribution for the fraction of outdoor soil in indoor dust. *J. Soil Contam.*, 6:161–168.
422. Allen, B., Gentry, R., Shipp, A., and Van Landingham, C. (1998): Calculation of benchmark doses for reproductive and developmental toxicity observed after exposure to isopropanol. *Regul. Toxicol. Pharmacol.*, 28(1):38–44.
423. Frey, H. C. and Rhodes, D. S. (1998): Characterization and simulation of uncertainty frequency distributions: effects of distribution choice, variability, uncertainty, and parameter dependence. *Hum. Ecol. Risk Assess.*, 4:423–469.
424. Beck, B. D. and Cohen, J. T. (1997): Risk assessment for criteria pollutants versus other noncarcinogens: the difference between implicit and explicit conservatism. *Hum. Ecol. Risk Assess.*, 3:671–626.
425. Mertz, C. K., Slovic, P., and Purchase, I. F. (1998): Judgments of chemical risks: comparisons among senior managers, toxicologists, and the public. *Risk Anal.*, 18(4):391–404.
426. Baird, S. J. S., Cohen, J. T., Graham, J. D., Shlyakhter, A. I., and Evans, J. S. (1996): Noncancer risk assessment: a probabilistic alternative to current practice. *Hum. Ecol. Risk Assess.*, 2:79–102.
427. Boyce, C. P. (1998): Comparison of approaches for developing distributions for carcinogenic potency factors. *Hum. Ecol. Risk Assess.*, 4:527–589.
428. Cox, L. A. J. (1996): More accurate dose–response estimation using Monte Carlo uncertainty analysis: the data cube approach. *Hum. Ecol. Risk Assess.*, 2:150–175.
429. Crouch, E. A. (1996): Uncertainty distributions for cancer potency factors: combining epidemiological studies with laboratory bioassays—the example of acrylonitrile. *Hum. Ecol. Risk Assess.*, 2:130–149.
430. Crouch, E. A. (1996): Uncertainty distributions for cancer potency factors: laboratory animal carcinogenicity and interspecies extrapolation. *Hum. Ecol. Risk Assess.*, 2:103–129.

431. Evans, J. S., Graham, J. D., Gray, G. M., and Sielken, Jr., R. L. (1994): A distributional approach to characterizing low-dose cancer risk. *Risk Anal.*, 14(1):25–34.
432. Evans, J. S. et al. (1994): Use of probabilistic expert judgment in uncertainty analysis of carcinogenic potency. *Regul. Toxicol. Pharmacol.*, 20(1, Pt. 1):15–36.
433. Hill, R. A. and Hoover, S. M. (1997): Importance of the dose–response model form in probabilistic risk assessment: a case study of health effects from methylmercury in fish. *Hum. Ecol. Risk Assess.*, 3:465–481.
434. Shlyakhter, A. I., Goodman, G., and Wilson, R. (1992): Monte Carlo simulation of rodent carcinogenicity bioassays. *Risk Anal.*, 12:73–82.
435. Sielken, Jr., R. L. (1989): Useful tools for evaluating and presenting more science in quantitative cancer risk assessment. *Tox. Subst. J.*, 9:353–404.
436. Sielken, Jr., R. L. and Stevenson, D. E. (1997): Opportunities to improve quantitative risk assessment. *Hum. Ecol. Risk Assess.*, 3:479–490.
437. Sielken, Jr., R. L. and Valdez-Flores, C. (1996): Comprehensive realism's weight-of-evidence based distributional dose–response characterization. *Hum. Ecol. Risk Assess.*, 2:175–193.
438. Velazquez, S. F., McGinnis, P. M., Vater, S. T., Stiteler, W. S., Knauf, L. A., and Schoeny, R. S. (1994): Combination of cancer data in quantitative risk assessments: case study using bromodichloromethane. *Risk Anal.*, 14:285–292.
439. Burmaster, D. E. (1998): A lognormal distribution for time spent showering. *Risk Anal.*, 18:33–36.
440. Copeland, T. L., Holbrow, A. M., Otani, J. M., Connor, K. T., and Paustenbach, D. J. (1994): Use of probabilistic methods to understand the conservatism in California's approach to assessing health risks posed by air contaminants. *Air Waste*, 44(12):1399–1413.
441. Gargas, M. L., Finley, B. L., Paustenbach, D. J., and Long, T. F. (1999): Environmental risk assessment: theory and practice. In: *General and Applied Toxicology*, edited by B. Ballantyne, T. Marrs, and T. Syverson. Macmillan, London, pp. 1749–1809.
442. Sedman, R., Funk, L. M., and Fountain, R. (1998): Distribution of residence duration in owner occupied housing. *J. Expo. Anal. Environ. Epidemiol.*, 8(1):51–58.
443. Smith, A. E., Ryan, P. B., and Evans, J. S. (1992): The effects of neglecting correlations when propagating uncertainty and estimating population distribution of risk. *Risk Anal.*, 12:467–474.
444. Bukowski, J., Korn, L. R., and Wartenberg, D. (1995): Correlated inputs in quantitative risk assessment: the effects of distributional shape. *Risk Anal.*, 15:215–219.
445. Cooper, J. A., Ferson, S., and Ginzberg, L. (1996): Hybrid processing of stochastic and subjective uncertainty data. *Risk Anal.*, 16:785–792.
446. Haas, C. N. (1997): Importance of the distributional form in characterizing inputs to Monte Carlo risk assessments. *Risk Anal.*, 17:107–113.
447. Hamed, M. M. and Bedient, P. B. (1997): On the effect of probability distributions of input variables in public health risk assessment. *Risk Anal.*, 17(1):97–105.
448. Hattis, D. B. and Burmaster, D. E. (1994): Assessment of variability and uncertainty distributions for practical risk analyses. *Risk Anal.*, 17:97–105.
449. Cronin, W. J. T., Oswald, E. J., Shelley, M. L., Fisher, J. W., and Flemming, C. D. (1995): A trichloroethylene risk assessment using a Monte Carlo analysis of parameter uncertainty in conjunction with physiologically-based pharmacokinetic modeling. *Risk Anal.*, 15(5): 555–565.
450. Cullen, A. C. and Frey, C. (1998): *Probabilistic Techniques in Exposure Assessment*. Plenum Press, New York.
451. Frey, H. C. and Patil, S. R. (2002): Identification and review of sensitivity analysis methods. *Risk Anal.*, 22(3):553–578.
452. Greenland, S. (2001): Sensitivity analysis, Monte Carlo risk analysis, and Bayesian uncertainty assessment. *Risk Anal.*, 21(4):579–583.
453. Helton, J. C. and Davis, F. J. (2002): Illustration of sampling-based methods for uncertainty and sensitivity analysis. *Risk Anal.*, 22(3):591–622.
454. Saltelli, A. (2002): Sensitivity analysis for importance assessment. *Risk Anal.*, 22(3):579–590.
455. Bogen, K. T. and Spear, R. C. (1987): Integrating uncertainty and interindividual variability in environmental risk assessment. *Risk Anal.*, 7(4):427–436.
456. Iman, R. L. and Helton, J. C. (1991): The repeatability of uncertainty and sensitivity analyses for complex probabilistic risk assessments. *Risk Anal.*, 11:591–606.
457. Rai, S. N. and Krewski, D. (1998): Uncertainty and variability analysis in multiplicative risk models. *Risk Anal.*, 18(1):37–45.
458. Robinson, R. B. and Hurst, B. T. (1997): Statistical quantification of the sources of variance in uncertainty analysis. *Risk Anal.*, 17:447–454.
459. Shlyakhter, A. I. (1994): An improved framework for uncertainty analysis: accounting for unsuspected errors. *Risk Anal.*, 14:441–447.
460. Roseberry, A. M. and Burmaster, D. E. (1992): Lognormal distributions for water intake by children and adults. *Risk Anal.*, 12:99–104.
461. Ershow, A. G. and Cantor, K. P. (1989): *Total Tapwater Intake in the United States: Population-Based Estimates of Quantities and Sources*. Life Sciences Research Office Federated American Societies for Experimental Biology Bethesda, MD.
462. Graham, J. D. et al. (1992): The role of exposure databases in risk assessment. *Arch. Environ. Health*, 47:408–420.
463. Lucier, G. W. and Schecter, A. (1998): Human exposure assessment and the National Toxicology Program. *Environ Health Perspect.*, 106(10):623–627.
464. Alexander, M. How toxic are chemicals in soil? *Environ Sci Technol* 29:2713–2717. (1995):
465. Ruoff, W. L., Diamond, G. L., Velazquez, S. F., Stiteler W. M., and Gefell, D. J. (1994): Bioavailability of cadmium in food and water: a case study on the derivation of relative bioavailability factors for inorganics and their relevance to the reference dose. *Regul. Toxicol. Pharmacol.*, 20(2):139–160.

466. Wester, R. C., Bucks, D. A. W., and Maibach, H. (1993): Percutaneous absorption of contaminants from soil. In: *Health Risk Assessment: Dermal and Inhalation Exposure and Absorption of Toxicants*, edited by G. M. Wang, J. B. Knaak, and H. Maibach. CRC Press, Boca Raton, FL.
467. Davis, A., Bloom, N. S., and Que Hee, S. S. (1997): The environmental geochemistry and bioaccessibility of mercury in soils and sediments: a review. *Risk Anal.*, 17(5):557–569.
468. Davis, A., Drexter, J. W., Ruby, M. V., and Nicholson, A. (1993): Micromineralogy of mine waste in relation to lead bioavailability, Butte, Montana. *Environ. Sci Technol.*, 27:1415–1425.
469. Davis, A., Ruby, M. V., and Bergstrom, P. D. (1992): Bioavailability of arsenic and lead from the Butte Montana, mining district. *Environ. Sci. Technol.*, 26:461–468.
470. Shifrin, N. S., Beck, B. D., Gauthier, T. D., Chapnick, S. D., and Goodman, G. (1996): Chemistry, toxicology and human health risks of cyanide compounds in soils at former manufactured gas plant sites. *Regul. Toxicol. Pharmacol.*, 23:106–116.
471. Chaney, R. L., Sterrett, S. B., and Mielke, H. W. (1984): The potential for heavy metal exposure from urban gardens and soils. In: *Proc. Symp. Heavy Metals in Urban Gardens*, edited by J. R. Preer, Agriculture Experiment Station, University of the District of Columbia, Washington, D.C.
472. Horowitz, S. B. and Finley, B. L. (1993): Using human sweat to extract chromium from chromite ore processing residue: applications to setting health-based cleanup levels. *J. Toxicol. Environ. Health*, 40(4):585–599.
473. Bogen, K. T. (1994): A note on compounded conservatisms. *Risk Anal.*, 14:379–382.
474. Bogen, K. T., Keating, G. A. Meissner, S., and Vogel, J. S. (1998): Initial uptake kinetics in human skin exposed to dilute aqueous trichloroethylene *in vitro*. *J. Expo. Anal. Environ. Epidemiol.*, 8(2):253–271.
475. McKone, T. E. (1993): Linking a PBPK model for chloroform with measured breath concentrations in showers: implications for dermal exposure models. *J. Expo. Anal. Environ. Epidemiol.*, 3(3):339–365.
476. Wester, R. C. and Noonan, P. K. (1980): Relevance of animal models for percutaneous absorption. *Int. J. Pharmacol.*, 7:99–110.
477. ACS. (1983): *Fate of Chemicals in the Environment*, ACS Symp. Ser. 225. American Chemical Society, Washington, D.C.
478. Borgert, C. J., Roberts, S. M., Harbison, R. D., and James, R. C. (1995): Influence of soil half-life on risk assessment of carcinogens. *Regul. Toxicol. Pharmacol.*, 22(2): 143–151.
479. Morgan, J. N., Berry, M., and Graves, R. L. (1997): Effects of commonly used cooking practices on total mercury concentration in fish and their impact on exposure assessments. *J. Exp. Anal. Environ. Epidemiol.*, 7:119–133.
480. Wilson, N. D., Shear, N. M., Paustenbach, D. J., and Price, P. S. (1998): The effect of cooking practices on the concentration of DDT and PCB compounds in the edible tissue of fish. *J. Expo. Anal. Environ. Epidemiol.*, 8(3):423–440.
481. Thomas, K. W., Sheldon, L. S., Pellizzari, E., Handy, R. W., Roberds, J. M., and Berry, M. (1997): Testing duplicate diet sample collection methods for measuring personal dietary exposures to chemical contamination. *J. Exp. Anal. Environ. Epidemiol.*, 7:17–36.
482. Doll, R. and Peto, R. (1981): The causes of cancer. *J. Natl. Cancer Inst.*, 66:1191–1308.
483. Hemminki, K., Lonnstedt, I., Vaittinen, P., and Lichtenstein, P. (2001): Estimation of genetic and environmental components in colorectal and lung cancer and melanoma. *Genet. Epidemiol.*, 20(1):107–116.
484. World Cancer Research Fund Panel. (1997): *Food Nutrition and the Prevention of Cancer: A Global Perspective*. American Institute for Cancer Research, Washington D.C.
485. Hattis, D. B. (1986): The promise of molecular epidemiology for quantitative risk assessment. *Risk Anal.*, 6(2):181–193.
486. Holdway, D. A. (1996): The role of biomarkers in risk assessment. *Hum. Ecol. Risk Assess.*, 2:263–267.
487. McMillan, A., Whittemore, A. S., Silvers, A., and DiCiccio, Y. (1994): Use of biological markers in risk assessment. *Risk Anal.*, 14(5):807–813.
488. IPCS. (1993): *Biomarkers and Risk Assessment: Concepts and Principles*, Vol. 155. International Programme of Chemical Safety, World Health Organization, Geneva.
489. Au, W. W., Lee, E., and Christiani, D. C. (2005): Biomarker research in occupational health. *J. Occup. Environ. Med.*, 47(2):145–153.
490. Bocchetta, M. and Carbone, M. (2004): Epidemiology and molecular pathology at crossroads to establish causation: molecular mechanisms of malignant transformation. *Oncogene*, 23(38):6484–6491.
491. Maier, A., Savage, Jr., R. E., and Haber, L. T. (2004): Assessing biomarker use in risk assessment: a survey of practitioners. *J. Toxicol. Environ. Health A*, 67(8–10):687–695.
492. Committee on Biologic Markers of the National Research Council. (1987): Biological markers in environmental health research. *Environ. Health Perspect.*, 74:3–9.
493. Wolfe, D. A. (1996): Insights on the utility of biomarkers for environmental impact assessment and monitoring. *Hum. Ecol. Risk Assess.*, 2:245–250.
494. Ehrenberg, L. and Osterman-Golkar, S. (1980): Alkylation of macromolecules for detecting mutagenic agents. *Teratog. Carcinog. Mutagen*, 1(1):105–127.
495. Perera, F. P. and Weinstein, I. B. (2000): Molecular epidemiology: recent advances and future directions. *Carcinogenesis*, 21(3):517–524.
496. CDC. (2005): *Third National Report on Human Exposure to Environmental Chemicals*, Centers for Disease Control and Prevention (http://www.cdc.gov/exposurereport/3rd/).
497. Ashley, D. L., Bonin, M. A., Cardinali, F. L., McCraw, J. M., and Wooten, J. V. (1994): Blood concentrations of volatile organic compounds in a nonoccupationally exposed U.S. population and in groups with suspected exposure. *Clin. Chem.*, 40(7, Pt. 2):1401–1404.
498. Ashley, D. L., Bonin, M. A., Cardinali, F. L., McCraw, J. M., and Wooten, J. V. (1996): Measurement of volatile organic compounds in human blood. *Environ. Health Perspect.*, 104(Suppl. 5):871–877.

499. Que Hee, S. (1993): *Biological Monitoring*. Wiley, New York.
500. Borgert, C. J. (2005): Understanding human biomonitoring. *Regul. Toxicol. Pharmacol.*, 43(2):215–218.
501. Pirkle, J. L., Needham, L. L., and Sexton, K. (1995): Improving exposure assessment by monitoring human tissues for toxic chemicals. *J. Expo. Anal. Environ. Epidemiol.*, 5(3):405–424.
502. Barr, D. B., Wang, R. Y., and Needham, L. L. (2005): Biologic monitoring of exposure to environmental chemicals throughout the life stages: requirements and issues for consideration for the National Children's Study. *Environ. Health Perspect* 113(8):1083–1091.
503. Sexton, K., Callahan, M. A., and Bryan, E. F. (1995): Estimating exposure and dose to characterize health risks: the role of human tissue monitoring in exposure assessment. *Environ. Health Perspect.*, 103(Suppl. 3):13–29.
504. Aitio, A. and Kallio, A. (1999): Exposure and effect monitoring: a critical appraisal of their practical application. *Toxicol. Lett.*, 108(2–3):137–147.
505. Duggan, A. et al. (2003): Di-alkyl phosphate biomonitoring data: assessing cumulative exposure to organophosphate pesticides. *Regul. Toxicol. Pharmacol.*, 37(3):382–395.
506. Pirkle, J. L., Osterloh, J., Needham, L. L., and Sampson, E. J. (2005): National exposure measurements for decisions to protect public health from environmental exposures. *Int. J. Hyg. Environ. Health*, 208(1–2):1–5.
507. ATSDR. (1999): *Toxicological Profile for Lead*. Agency for Toxic Substances and Disease Registry, U.S. Department of Health and Human Services, Atlanta, GA.
508. Hwang, S. A., Yang, B. Z., Fitzgerald, E. F., Bush, B., and Cook, K. (2001): Fingerprinting PCB patterns among Mohawk women. *J. Expo. Anal. Environ. Epidemiol.*, 11(3):184–192.
509. Schmidt, K. et al. (1997): Internal exposure to hazardous substances of persons from various continents: investigations on exposure to different organochlorine compounds. *Int. Arch. Occup. Environ. Health*, 69:399–406.
510. Van Leeuwen, F. R. X. and Malish, R. (2002): Results of the third round of the WHO-coordinated study on the levels of PCBs, PCDDs, and PCFFs in human milk. *Organohalogen Compounds*, 56:311–316.
511. Symanski, E. and Greeson, N. M. (2002): Assessment of variability in biomonitoring data using a large database of biological measures of exposure. *AIHA J.*, 63(4):390–401.
512. Brindle, J. T. et al. (2002): Rapid and noninvasive diagnosis of the presence and severity of coronary heart disease using 1H-NMR-based metabonomics. *Nat. Med.*, 8(12):1439–1444.
513. Chevalier, R. L. (2004): Biomarkers of congenital obstructive nephropathy: past, present and future. *J. Urol.*, 172(3):852–857.
514. Choi, W. W. L., Lewis, M. M., Lawson, D., Yin-Geon, Q., Birdsong, G. G., and Cotsonis, G. A. (2004): Angiogenic and lymphangiogenic microvessel density in breast carcinomas: correlation with clinicopathologic parameters and the VEGF-family gene expression. *Mod. Pathol.*, 18(1):143–152.
515. Chung, C. H., Bernard, P. S., and Perou, C. M. (2002): Molecular portraits and the family tree of cancer. *Nat. Genet.*, 32(Suppl.):533–540.
516. Coen, M., Lenz, E. M., Nicholson, J. K., Wilson, I. D., Pognan, F., and Lindon, J. C. (2003): An integrated metabonomic investigation of acetaminophen toxicity in the mouse using NMR spectroscopy. *Chem. Res. Toxicol.*, 16(3):295–303.
517. Griffin, J. L., Walker, L. A., Shore, R. F., and Nicholson, J. K. (2001): Metabolic profiling of chronic cadmium exposure in the rat. *Chem. Res. Toxicol.*, 14(10):1428–1434.
518. Hamadeh, H. K., Bushel, P. R., Jayadev, S., Martin, K., DiSorbo, O., and Sieber, S. (2002): Gene expression analysis reveals chemical-specific proteins. *Toxicol. Sci.*, 67:219–231.
519. Holleman, A. et al. (2004): Gene-expression patterns in drug-resistant acute lymphoblastic leukemia cells and response to treatment. *N. Engl. J. Med.*, 351(6):533–542.
520. Holmes, E., Nicholson, J. K., and Tranter, G. (2001): Metabonomic characterization of genetic variations in toxicological and metabolic responses using probabilistic neural networks. *Chem. Res. Toxicol.*, 14(2):182–191.
521. Kimura, J. et al. (2004): Th1 and Th2 cytokine production is suppressed at the level of transcriptional regulation in Kawasaki disease. *Clin. Exp. Immunol.*, 137(2):444–449.
522. Petricoin, E. F., Ornstein, D. K., and Liotta, L. A. (2004): Clinical proteomics: applications for prostate cancer biomarker discovery and detection. *Urol. Oncol.*, 22(4):322–328.
523. Robertson, D. G., Reily, M. D., Sigler, R. E., Wells, D. F., Paterson, D. A., and Braden, T. K. (2000): Metabonomics: evaluation of nuclear magnetic resonance (NMR) and pattern recognition technology for rapid *in vivo* screening of liver and kidney toxicants. *Toxicol. Sci.*, 57(2):326–337.
524. Tallman, M. S. et al. (2004): Effects of all-*trans* retinoic acid or chemotherapy on the molecular regulation of systemic blood coagulation and fibrinolysis in patients with acute promyelocytic leukemia. *J. Thromb. Haemost.*, 2(8):1341–1350.
525. Troyer, D. A., Mubiru, J., Leach, R. J., and Naylor, S. L (2004): Promise and challenge: markers of prostate cancer detection, diagnosis and prognosis. *Dis. Markers* 20(2):117–128.
526. Kawasaki, T., Kono, K., Dote, T., Usuda, K., Shimizu, H. and Dote, E. (2004): Markers of cadmium exposure in workers in a cadmium pigment factory after changes in the exposure conditions. *Toxicol. Indust. Health* 20(1–5):51–56.
527. Kuljukka-Rabb, T. et al. (2002): The effect of relevant genotypes on PAH exposure-related biomarkers. *J. Expo Anal. Environ. Epidemiol.*, 12(1):81–91.
528. Barbosa, Jr., F., Tanus-Santos, J. E., Gerlach, R. F., and Parsons, P. J. (2005): A critical review of biomarkers used for monitoring human exposure to lead: advantages, limitations, and future needs. *Environ. Health Perspect.* 113(12):1669–1674.

529. Coble, J., Arbuckle, T., Lee, W., Alavanja, M., and Dosemeci, M. (2005): The validation of a pesticide exposure algorithm using biological monitoring results. *J. Occup. Environ. Hyg.*, 2(3):194–201.
530. Harkins, D. K. and Susten, A. S. (2003): Hair analysis: exploring the state of the science. *Environ. Health Perspect.*, 111(4):576–578.
531. Swenberg, J., Gorgeiva, N., and Ham, A. (2002): Linking pharmacokinetics and biomarker data to mechanism of action in risk assessment. *Hum. Ecol. Risk Assess.*, 8(6):1315.
532. Chang, H. Y., Tsai, C. Y., Lin, Y. Q., Shih, T. S., and Lin, Y. C. (2004): Urinary biomarkers of occupational *N,N*-dimethylformamide (DMF) exposure attributed to the dermal exposure. *J. Expo. Anal. Environ. Epidemiol.*, 14(3):214–221.
533. Godderis, L. et al. (2004): Influence of genetic polymorphisms on biomarkers of exposure and genotoxic effects in styrene-exposed workers. *Environ. Mol. Mutagen*, 44(4):293–303.
534. Brandt-Rauf, P. W., Luo, J., and Cheng, T. (2002): Molecular biomarkers and epidemiologic risk assessment. *Hum. Ecol. Risk Assess.*, 8(6):1295.
535. Sexton, K., Kleffman, D. E., and Callahan, M. A. (1995): An introduction to the National Human Exposure Assessment Survey (NHEXAS) and related phase I field studies. *J. Expo. Anal. Environ. Epidemiol.*, 5(3):229–232.
536. Pinsky, P. F. and Lorber, M. N. (1998): A model to evaluate past exposure to 2,3,7,8-TCDD. *J. Expo. Anal. Environ. Epidemiol.*, 8(2):187–206.
537. USEPA. (2001): *Exposure and Human Health Reassessment of 2,3,7,8-Tetrachlorodibenzo-p-Dioxin (TCDD) and Related Compounds (Draft)*, EPA/600/P-001. Office of Research and Development, National Center for Environmental Assessments, U.S. Environmental Protection Agency, Washington, D.C.
538. Buckley, T. J., Prah, J. D., Ashley, D., Zweidinger, R. A., and Wallace, L. A. (1997): Body burden measurements and models to assess inhalation exposure to methyl tertiary butyl ether (MTBE). *J. Air Waste Manag. Assoc.*, 47(7):739–752.
539. Wallace, L. A. and Pellizzari, E. D. (1995): Recent advances in measuring exhaled breath and estimating exposure and body burden for volatile organic compounds (VOCs). *Environ. Health Perspect.*, 103(Suppl. 3):95–98.
540. Chinnery, R. and Gleason, K. A. (1993): A compartment model for the prediction of breath concentration and absorbed dose of chloroform after exposure while showering. *Risk Anal.*, 13:51–62.
541. Aylward, L. L. et al. (2005): Concentration-dependent TCDD elimination kinetics in humans: toxicokinetic modeling for moderately to highly exposed adults from Seveso, Italy, and Vienna, Austria, and impact on dose estimates for the NIOSH cohort. *J. Expo. Anal. Environ. Epidemiol.*, 15(1):51–65.
542. Graham, J. D., Green, L., and Roberts, M. (1988): *In Search of Safety: Chemicals and Cancer Risks.* Harvard University Press, Cambridge, MA.
543. Buck, R. J., Hammerstrom, K. A., and Ryan, P. B. (1995): Estimating long-term exposures from short-term measurements. *J. Expo. Anal. Environ. Epidemiol.*, 5(3):359–373.
544. Buck, R. J., Hammerstrom, K. A., and Ryan, P. B. (1997): Bias in population estimates of long-term exposure from short-term measurements if individual exposure. *Risk Anal.*, 17:455–465.
545. Slob, W. (1996): A comparison of two statistical approaches to estimate long-term exposure distributions from short-term measurements. *Risk Anal.*, 16:195–200.
546. Stanek, E. J., Calabrese, E. J., and Xu, L. (1998): A caution for Monte Carlo risk assessment of long term exposures based on short term exposure data. *Hum. Ecol. Risk Assess.*, 4:409–422.
547. Crump, K. S. (1998): On summarizing group exposures in risk assessment: is an arithmetic mean or a geometric mean more appropriate? *Risk Anal.*, 18(3):293–297.
548. Gilbert, R. O. (1987): *Statistical Methods for Environmental Pollution Monitoring*. Wiley, New York.
549. Haas, C. N. and Scheff, P. A. (1990): Estimation of averages in truncated samples. *Environ. Sci. Technol.*, 24:912–919.
550. Helsel, D. R. (1990): Less than obvious: statistical treatment of data below the detection limit. *Environ. Sci. Technol.*, 24:1766–1774.
551. Horowitz, S. B. and Finley, B. L. (1994): Setting health-protective soil concentrations for dermal contact allergens: a proposed methodology. *Regul. Toxicol. Pharmacol.*, 19(1):31–47.
552. Parkin, T. B., Melsinger, J. J., Chester, S. T., Starr, J. L., and Robinson, J. A. (1988): Evaluation of statistical estimation methods for lognormally distribution variables. *Soil Sci. J.*, 52:323.
553. Perkins, J. L., Cutter, G. N., and Cleveland, M. S. (1990): Estimating the mean, variance, and confidence limits from censored (limit of detection), lognormally-distributed exposure data. *Am. Indust. Hyg. Assoc. J.*, 51:416–419.
554. Rappaport, S. M. and Selvin, S. (1987): A method for evaluating the mean exposure from a lognormal distribution. *Am. Indust. Hyg. Assoc. J.*, 48(4):374–379.
555. Travis, C. C. and Land, M. L. (1990): Estimating the mean of data sets with nondetectable values. *Environ. Sci. Technol.*, 24:961–962.
556. Inoue, K. et al. (2004): Perfluorooctane sulfonate (PFOS) and related perfluorinated compounds in human maternal and cord blood samples: assessment of PFOS exposure in a susceptible population during pregnancy. *Environ. Health Perspect.*, 112(11):1204–1207.
557. Richardson, G. M. and Currie, D. J. (1993): Estimating fish consumption rates for Ontario Amerindians. *J. Expo. Anal. Environ. Epidemiol.*, 3(1):23–38.
558. Stern, A. H., Korn, L. R., and Ruppel, B. E. (1996): Estimation of fish consumption and methylmercury intake in the New Jersey population. *J. Expo. Anal. Environ. Epidemiol.*, 6(4):503–525.
559. Coad, S. and Newhook, R. C. (1992): PCP exposure for the Canadian general population: a multimedia analysis. *J. Expo. Anal. Environ. Epidemiol.*, 2(4):391–413.

560. Ayotte, P., Muckle, G., Jacobson, J. L., Jacobson, S. W., and Dewailly, E. (2003): Assessment of pre- and postnatal exposure to polychlorinated biphenyls: lessons from the Inuit Cohort Study. *Environ. Health Perspect.*, 111(9):1253–1258.
561. Kimmel, C. A., Collman, G. W., Fields, N., and Eskenazi, B. (2005): Lessons learned for the National Children's Study from the National Institute of Environmental Health Sciences/U.S. Environmental Protection Agency Centers for Children's Environmental Health and Disease Prevention Research. *Environ. Health Perspect.*, 113(10):1414–1418.
562. Kimmel, G. L. (2005): An overview of children as a special population: relevance to predictive biomarkers. *Toxicol. Appl. Pharmacol.*, 206(2):215–218.
563. Needham, L. L., Barr, D. B., and Calafat, A. M. (2005): Characterizing children's exposures: beyond NHANES. *Neurotoxicology*, 26(4):547–553.
564. Ginsberg, G., Hattis, D., Russ, A., and Sonawane, B. (2004): Physiologically based pharmacokinetic (PBPK) modeling of caffeine and theophylline in neonates and adults: implications for assessing children's risks from environmental agents. *J. Toxicol, Environ. Health A*, 67(4):297–329.
565. Charnley, G. and Putzrath, R. M. (2001): Children's health, susceptibility, and regulatory approaches to reducing risks from chemical carcinogens. *Environ. Health Perspect* 109(2):187–192.
566. Adgate, J. L. et al. (2001): Measurement of children's exposure to pesticides: analysis of urinary metabolite levels in a probability-based sample. *Environ. Health Perspect.*, 109(6):583–590.
567. Koch, D., Lu, C., Fisker-Andersen, J., Jolley, L., and Fenske, R. A. (2002): Temporal association of children's pesticide exposure and agricultural spraying: report of a longitudinal biological monitoring study. *Environ. Health Perspect.*, 110(8):829–833.
568. Campbell, J. R., Rosier, R. N., Novotny, L., and Puzas, J. E. (2004): The association between environmental lead exposure and bone density in children. *Environ. Health Perspect.*, 112(11):1200–1203.
569. Haley, V. B. and Talbot, T. O. (2004): Geographic analysis of blood lead levels in New York State children born 1994–1997. *Environ. Health Perspect.*, 112(15): 1577–1582.
570. Wang, R. Y., Needham, L. L., and Barr, D. B. (2005): Effects of environmental agents on the attainment of puberty: considerations when assessing exposure to environmental chemicals in the National Children's Study. *Environ. Health Perspect.*, 113(8):1100–1107.
571. Eskenazi, B. et al. (2005): Methodologic and logistic issues in conducting longitudinal birth cohort studies: lessons learned from the Centers for Children's Environmental Health and Disease Prevention Research. *Environ. Health Perspect.*, 113(10):1419–1429.
572. Clayton, C. A., Pellizzari, E. D., Whitmore, R. W., Quackenboss, J. J., Adgate, J., and Sefton, K. (2003): Distributions, associations, and partial aggregate exposure of pesticides and polynuclear aromatic hydrocarbons in the Minnesota Children's Pesticide Exposure Study (MNCPES). *J. Expo. Anal. Environ. Epidemiol.*, 13(2):100–111.
573. Travis, C. C., White, R. K., and Ward, R. C. (1990): Interspecies extrapolation of pharmacokinetics. *J. Theor. Biol.*, 142(3):285–304.
574. Widner, T. (2000): Dose reconstruction for radionuclides and chemicals released from the federal nuclear facility in Oak Ridge, Tennessee. In: *Human and Ecological Risk Assessment: Theory and Practice*, edited by D. J. Paustenbach. John Wiley & Sons, New York.
575. Hallock, M. F., Smith, T. J., Woskie, S. R., and Hammond, S. K. (1994): Estimation of historical exposures to machining fluids in the automotive industry. *Am. J. Indust. Med.*, 26(5):621–634.
576. Madl, A. K. and Paustenbach, D. J. (2002): Airborne concentrations of benzene and mineral spirits (Stoddard solvent) during cleaning of a locomotive generator and traction motor. *J. Toxicol. Environ. Health A*, 65(23): 1965–1979.
577. Madl, A. K. and Paustenbach, D. J. (2002): Airborne concentrations of benzene due to diesel locomotive exhaust in a roundhouse. *J. Toxicol. Environ. Health A*, 65(23): 1945–1964.
578. Mangold, C., Clark, K., Madl, A., and Paustenbach, D. (2006): An exposure study of bystanders and workers during the installation and removal of asbestos gaskets and packing. *J. Occup. Environ. Hyg.*, 3(2):87–98.
579. Paustenbach, D. J., Madl, A. K., Donovan, E., Clark, K., Fehling, K., and Lee, T. C. (2006): Chrysotile asbestos exposure associated with removal of automobile exhaust systems (ca. 1945–1975) by mechanics: results of a simulation study. *J. Expo. Sci. Environ. Epidemiol.*, 16(2): 156–171.
580. Hanninen, O. O. et al. (2004): The EXPOLIS study: implications for exposure research and environmental policy in Europe. *J. Expo. Anal. Environ. Epidemiol.*, 14(6): 440–456.
581. Koo, H. J. and Lee, B. M. (2005): Human monitoring of phthalates and risk assessment. *J. Toxicol. Environ. Health A*, 68(16):1379–1392.
582. Olsen, G. W. et al. (2003): An occupational exposure assessment of a perfluorooctanesulfonyl fluoride production site: biomonitoring. *AIHA J.*, 64(5):651–659.
583. Swan, S. H. et al. (2003): Semen quality in relation to biomarkers of pesticide exposure. *Environ. Health Perspect.*, 111(12):1478–1484.
584. Daisey, J. M., Hodgson, A. T., Fish, W. J., Mendell, M. J. and Ten Brinke, J. (1994): Volatile organic compounds in 12 California office buildings: classes, concentrations, and sources. *Atmos. Environ.*, 28(22):3557–3562.
585. Gesell, T. F. and Prichard, H. M. (1980): The contribution of radon in tap water to indoor radon concentrations. In *Natural Radiation Environment III*, edited by T. F. Gesell and W. M. Lowder. U.S. Department of Energy, Washington, D.C., pp. 1347–1363.
586. Jenkins, P. L., Phillips, T. J., Mulberg, E. J., and Hui, S P. (1992): Activity patterns of Californians: use of and proximity to indoor pollutant sources. *Atmos. Environ.* 26A(12):2141–2148.
587. Krieger, R. I., Ross, J. H., and Thongsinthusak, T. (1992) Assessing human exposures to pesticides. *Rev. Environ Contam. Toxicol.*, 128:1–15.

588. Lioy, P. J., Waldman, J. M., Buckley, T. J., Butler, J. P., and Pietarinen, C. (1990): The personal, indoor, and outdoor concentrations of PM-10 measured in an industrial community during the winter. *Atmos. Environ.*, 24B(1): 57–60.
589. Lioy, P. L., Waldman, J. M., Greenberg, A., Harkov, R., and Pietarinen, C. (1988): The Total Human Environmental Exposure Study (THEES) to benzo(*a*)pyrene: comparison of the inhalation and food pathways. *Arch. Environ. Health*, 43(4):304–312.
590. Wallace, L. A. (2000): Real-time monitoring of particles, PAH, and CO in an occupied townhouse. *Appl. Occup. Environ. Hyg.*, 15:19.
591. McBride, S. J. et al. (1999): Investigations of the proximity effect for pollutants in the indoor environment. *J. Expo. Anal. Environ. Epidemiol.*, 9(6):602–621.
592. Conner, J. M., Oldaker, G. B. I., and Murphy, J. J. (1990): Method for assessing the contribution of environmental tobacco smoke to respirable particles in indoor microenvironments. *Environ. Technol.*, 11:189–196.
593. Hawley, J. K. (1985): Assessment of health risk from exposure to contaminated soil. *Risk Anal.*, 5:289–302.
594. LaGoy, P. K. (1987): Estimated soil ingestion rates for use in risk assessment. *Risk Anal.*, 7:355–359.
595. Dewey, K. G. and Lönnerdal, B. (1983): Milk and nutrient intake of breast-fed infants from 1 to 6 months: relation to growth and fatness. *J. Pediatr. Gastroenterol. Nutr.*, 2:497–506.
596. Dewey, K. G. et al. (1991): Adequacy of energy intake among breast-fed infants in the DARLING study: relationships to growth, velocity, morbidity, and activity levels. *J. Pediatr.*, 119:538–547.
597. Pao, E. M., Hines, J. M., and Roche, A. F. (1980): Milk intakes and feeding patterns of breast-fed infants. *J. Am. Diet. Assoc.*, 77:540–545.

# Notes

# 11 Epidemiology for Toxicologists

*Ralph R. Cook*

## CONTENTS



## INTRODUCTION

The search for scientific "truth" regarding the causes of human disease is a laborious multistep process, a winnowing of a large number of postulated hypotheses down to the few that can be supported with data derived from testing and observation. Success depends on the replication of results, coherence of evidence from many different fields, and, ultimately, an understanding of the underlying biological mechanisms of action. In evaluating the potential human health effects of chemical exposures, three major sources of scientific information are used by the courts, various government agencies, and the larger scientific community: experimental laboratory research, controlled clinical investigations, and observational epidemiology studies. These three are not mutually exclusive in method or thought; nonetheless, each makes a unique contribution toward understanding the etiologies of human disease and each has certain inherent limitations. Ultimately, the determination of causation depends on the demonstration of a meaningful elevated risk for the disease among those with the "exposure" and a biological explanation for the excess. The former can only be obtained via epidemiology studies; the latter usually comes from an interplay of information derived from experimental laboratory research and controlled clinical investigations.

Toxicology is one of the key experimental disciplines. In toxicology, investigators are able to carefully control the exposures of genetically homogeneous groups of animals, some purposefully bred so they exhibit a marked predilection for specific diseases. Toxicologists also have the opportunity of evaluating each and every subject to the same exquisite detail; therefore, at least in theory, the experimental method can provide comprehensive results unperturbed by extraneous variables. It is increasingly being recognized, however, that this approach does not automatically eliminate, or reduce to a level of insignificance, all forms of technical bias [66,104], nor does it protect against personal bias, a problem that can plague all types of research irrespective of the affiliations of the investigators [76,80,124]. Toxicology also has become stylized to the extent that the format of research results is both predictable and quantitative. This has made it very convenient for information derived from toxicology experiments to be used in quantitative risk assessments and related regulations. Paradoxically, this convenience facilitates a major violation of scientific principles: an extrapolation beyond the data to make inferences about health effects not only for levels of exposure that were not administered but also for species that were not studied.

Clinical investigators also administer measured doses according to a predetermined schedule, but do so to humans, thereby eliminating the need to extrapolate between species. Other technical biases, at least in theory, are minimized by randomly assigning potential study participants to either the exposed group or the unexposed controls; however, humans are not passive participants in health research. At the very least, they must consent to be studied. Some do not, and key characteristics of those who do and those who do not may be markedly different, quite possibly to the extent of compromising the utility of the initial randomization. Furthermore, specific subgroups (e.g., only men or only those who are patients of a single clinician) may be eligible for inclusion in the research. For all of these reasons, "controlled" clinical trials are, at best, quasi-experimental research. This means that care must be exercised in extrapolating their findings too broadly.

On the plus side, clinical research can collect data both on objective signs of pathology and on more subjective symptoms, data that animal research cannot provide. On the negative side, it may not be feasible to obtain an equivalent and comprehensive dataset for each study subject. For example, histopathology of all the organs is only available on those few of the deceased whose next of kin allow an autopsy to be performed.

The two major strengths of observational epidemiology research are that, one, it studies humans and, two, it deals with the effects of real exposures—actual levels, durations, and patterns of exposure to individual agents and to mixtures. If epidemiology studies are well done, they furnish results that reasonably can be extended to larger populations. Unfortunately, epidemiologists are often forced to handle exposure as a qualitative variable (either as a yes/no or some variation of high, medium, and low). This can limit the utility of the research results for those who require quantitative information. In addition, because the research is observational in the sense that the investigators simply observe natural experiments and do not exercise control over the key variables, epidemiologists routinely must grapple with a number of technical biases that are largely transparent to those in the other two fields. If these biases (in particular, selection, misclassification, and confounding) are not adequately addressed during the study design or data analysis, the study results may be unduly imprecise and important associations missed. Alternatively, the results may be relatively precise but precisely inaccurate, thereby leading to interpretations that are incorrect. Some of these problems can be exacerbated if the epidemiologist utilizes secondary sources of data (data originally gathered for purposes other than the specific research project, possibly even for reasons unrelated to research), especially if the methods for the original data collection process were poorly documented.

One other point differentiates epidemiology from the other two fields. Although acceptance into the American College of Epidemiology provides an imprimatur of professional competence, in practice epidemiology requires no graduate degree, certification, or licensure. Some enter the field with strong statistical skills; others have extensive training in human biology. The formal graduate degree programs require a proficiency in both, but anyone with any level of education can gather a dataset, analyze it, report the results, and call their efforts epidemiology research. There is no law against it, and no professional body condemns the practice. Epidemiologists can be very egalitarian. The rule of thumb for the consumer of epidemiology results therefore, especially if a report appears in other than the peer-reviewed literature, is *caveat emptor*.

Epidemiology has been defined as the study of the distribution and determinants of disease in humans [81]. Although commonly used, this definition is incomplete. Although epidemiologists certainly search for the factors associated with human disease, they also attempt to identify both interventions that likely will benefit those who are at risk for getting the condition (perhaps because of unique patterns of exposure to combinations of putative agents or a genetic predisposition for reacting adversely to such exposures) and treatments that will help control or cure any significant pathology once it occurs. They also, implicitly or explicitly, try to determine which agents do not cause a specific disease, which interventions will not be successful, and which treatments are not effective.

As with toxicology and clinical research, good epidemiology is an amalgam of subject specific knowledge and methods. And, just as clinical specialties and areas of expertise in toxicology have evolved with the growing complexities of each of those two fields, epidemiology is divided into a number of overlapping subgroups: occupational, environmental, reproductive, cardiovascular, cancer, infectious disease, molecular, genetic, nutritional, medical device, clinical, etc. Some of these are subdivided still further; for example, AIDS is a subcategory of viral which, in turn, is a subset of infectious disease. Although certain knowledge and techniques may be unique to a subgroup, many concepts are common across the discipline.

The primary objective of this chapter is to introduce those concepts that span the field so toxicologists might become better consumers of the epidemiology literature. Table 11.1 provides an outline of the major topics: data, measures of disease frequency, measures of risk and association, methods, and issues. All of these are interrelated so the order of presentation is somewhat arbitrary, but the first three set the stage for the last one, in particular the key issues impacting validity—selection, misclassification, and confounding. Each of these in turn is presented to show how it can bias the measures of association. The intent is to make the readers more sensitive to possible flaws so they might determine for themselves how well the investigators recognized a potential problem and addressed it during study design, data collection, or data analysis—and during data interpretation.

Most of the major points are illustrated with recent examples from the epidemiology literature or, to a lesser extent, with toxicology or clinical references. In part, this was done to emphasize the point that all science—experimental, quasi-experimental, and observational—is based on assumptions that may not be correct. Some of these assumptions are relatively innocuous. Conversely, some are so important that, if violated, the science is severely flawed and any policies based on that science also are likely flawed. Those are the two poles of a continuum, but validity is not a dichotomous variable like pregnancy. The challenge for the consumer of scientific reports is to determine which provide useful information, which do not, and which lie somewhere in between—intriguing enough to warrant additional research but not strong enough to merit intervention. This winnowing equates to first identifying potential problems that may have compromised the validity of the research; second, determining the probability to which the problems occurred; and, third, estimating the impact, if any, they had upon the results.

**TABLE 11.1**
**Major Topics**

- Data
  - Prevalence
  - Incidence
- Measures of disease frequency
  - Prevalence rate
  - Incidence rate
- Measures of risk and association
  - Absolute risk
  - Relative risk
  - Standardized mortality ratio (SMR)
  - Proportional mortality ratio (PMR)
  - Rate difference
  - Attributable risk
- Methods
  - Cohort
  - Case control
  - Cross-sectional
  - Case studies or case series
- Issues
  - Selection
  - Misclassification
  - Confounding
  - Chance
  - Causation

This chapter is by design a limited overview. For those who wish to have a more detailed presentation of the field, the list of references includes a number of recently published textbooks [22,52,68,104,107]. Note that, throughout this chapter, key "terms of art" are highlighted. Most can be found in Last's *Dictionary of Epidemiology*, an invaluable resource to any technical library [78]. Additionally, at the end of the chapter, Appendix A, Guidelines for Good Epidemiology Practices for Occupational and Environmental Epidemiologic Research (GEPs), provides an analog of the toxicology good laboratory practices (GLPs). This is followed by Appendix B, which provides an outline that can be used for critiquing epidemiology reports.

## DATA AND MEASURES OF DISEASE FREQUENCY

Data, information, and knowledge are related but not equivalent terms. Data are the raw materials gathered by the investigator during the course of an investigation. Information is analyzed data. Knowledge is meaningful information that can be used to predict or solve problems.

By way of analogy, data are bricks, information the wall, and knowledge the building. Just as good bricks can be put together poorly to build an unstable wall, so can data be valid but aggregated in a fashion that produces useless information. Pseudo-knowledge may be a function of either bad data or useless information, just as a structurally unsound building may result from poor quality bricks or an unstable wall.

## Prevalence and Incidence

For the epidemiologist, the two general types of data are *prevalence* and *incidence*. Prevalence is what is observed at a single point, a snapshot of what is prevalent, what exists, at a specific point in chronological or biological time. For example, the number of toxicologists currently employed by the federal regulatory agencies is prevalence data. Incidence is the number of incidents, of new events, that develop over time. Because it represents a delta (a change), incidence data have to be gathered at two or more points in time. In a sense, if prevalence is a snapshot, incidence is a movie. The number of toxicologists hired by the federal agencies in any given year is an example of incidence data. They did not work for the agencies at the start of the year (observation point one), but they did work for the government later in the year (observation points two or more).

Parenthetically, when epidemiologists speak of prevalence data they are usually referring to *point prevalence*, but they may mean *period prevalence*. Period prevalence is a combination of what exists at the beginning as well as what occurs during a specified period. The number of toxicologists who were employed by the federal agencies at any time during a given year is period prevalence data. It includes those who were working at the beginning of the year (point prevalence data) and those who were hired during the year (incidence data). Period prevalence may or may not be the same as the number of those who were employed at the end of the year (more point prevalence data) because some toxicologists may have left government employment during the period of observation.

Whether data are period prevalence or incidence can sometimes be difficult to discern because both refer to events occurring during a span of time. The key is whether the data represent a combination of existing and new events (period prevalence) or just new events (incidence). Unless otherwise noted, the term "prevalence" is used in this chapter as a synonym for "point prevalence."

The difference between prevalence and incidence data is important for at least three reasons. One, incidence data can be used to evaluate cause and effect; prevalence data usually cannot, at least not without additional assumptions. Two, because prevalence data can be gathered at a single point in time, it is much easier to obtain; therefore, many reports in the medical literature are based on prevalence data. Three, the medical literature often incorrectly uses the two terms interchangeably; as a consequence, reports that use valid prevalence data to develop nonsense information about cause and effect appear in even the most prestigious journals.

Although the two are different, they are related. Prevalence is a function of both the incidence ($I$) and the duration ($D$) of the disease ($P = I \times D$). What this means is that a chemical may not cause a disease, may not increase the incidence of the disease, but it may still be associated with a higher prevalence of the condition. Whether that is good news or bad depends on the circumstances; for example, the incidence of diabetes may be quite stable in a population, but, if that population is given access to a chemical called insulin, then the prevalence of the condition likely will increase dramatically. It will increase because the insulin extends the duration of the disease by allowing more of the afflicted to live longer.

Conversely, the prevalence of minor birth defects (prevalence because the events are measured at a single point in biological time: birth) could be lower among live children born to women exposed to some agent, not because the agent prevents the development of minor defects *in utero* but because the agent causes major malformations, including some among those fetuses who happen to have minor defects. If the major malformations lead to early spontaneous abortions, the *incidence* of minor defects might be quite stable, but the duration *in utero* of those with both types of congenital defects would be shortened and fewer newborns with minor problems would be observed at birth. The *prevalence* of minor defects among live births would be lower.

Figure 11.1 and Figure 11.2 illustrate these points. In both, a group of six patients (A through F) is observed for two years. In Figure 11.1, the condition is time limited; it spontaneously resolves, it is cured through some treatment, or the patient dies. The point prevalence at the initial baseline observation (year 0) is one. Two additional cases subsequently occur and all three resolve before year's end. At the end of year 1, the point prevalence is zero, the incidence is two, and the period prevalence is three. During the following year, two more cases develop and one patient dies (patient D) of an unrelated cause before the end of the year. At year's end, therefore, the point prevalence is zero, the incidence is two, and the period prevalence is also two.

In Figure 11.2, the condition is chronic, perhaps because, like diabetes, it has been extended through treatment. Note that the incidence is exactly the same as in the previous example. It is two in each year; however, the extended duration has impacted both measures of prevalence. The point prevalence at the time of the three observations is respectively one, three, and four. The period prevalence for the first year is three. For the second, it is five even though patient D died before year's end. Each of these

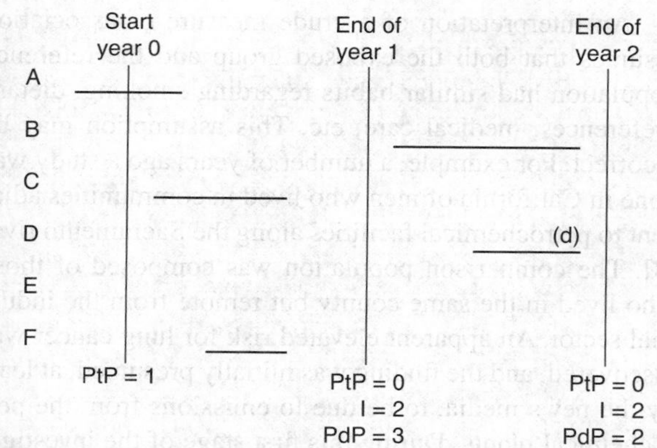


**FIGURE 11.1** Prevalence vs. incidence (time-limited condition).

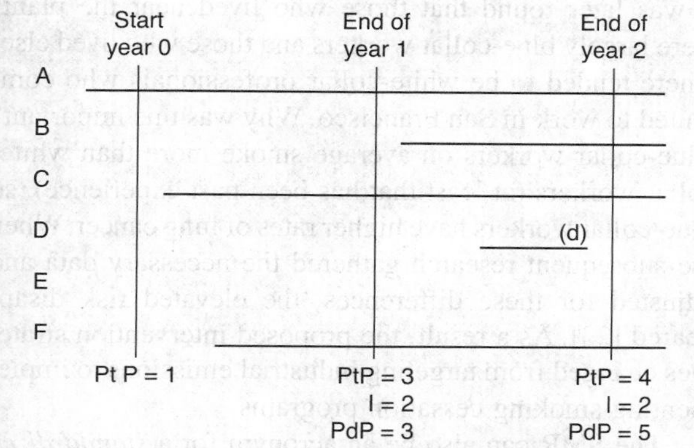


**FIGURE 11.2** Prevalence vs. incidence (condition chronic).

three measures of disease provides valuable information, but using either type of prevalence data for interpretations about cause and effect depends on assumptions about *incidence time* (i.e., when the health event actually occurred) and disease duration that are often untestable or incorrect.

## Rates

Technically, incidence and prevalence refer to numerator data; however, in both the epidemiology and medical literature, these terms often are used interchangeably with, respectively, *incidence rate* and *prevalence rate*. An incidence rate is the number of new events of a disease in a defined population that occur during some specified period of time. A prevalence rate is the number of cases of disease observed in a defined population at a point in time. In both, the numerator is a subset of the denominator. Obviously, errors of count in either the numerator or the denominator can impact the accuracy of a rate; nonetheless, in some technical reports, the former may not be a subset of the latter, and the description as to how either was compiled may be less than clear.

Rather than presenting a rate as the actual numerator (the exact number of new events observed) in comparison to the actual denominator (the precise count of the group under study) at or during the period of observation, for convenience a rate is usually given as the number of cases per 100 or per 1000 or per 10,000. For example, if the study group had 486 persons and 5 new events occurred during a 12-month period of observation, then the incidence rate might be presented 1.0 per 100 per year (5 divided by 486 times 100) or, alternatively, as 10.3 per 1000 per year.

As opposed to toxicology, in epidemiology the study groups can be either *fixed* or *open* (sometimes referred to as *dynamic*). In a fixed study group, those included are defined at the start and followed over time. If no losses occur during the period of study, the group may be called a *closed population*. In an open study group, individuals may be added or lost during the time of study. Just the events that occur and just the time that passes during the period each individual was under observation are counted. This so-called *person–time experience* assumes that observing 10 people for 1 year is the same as observing 1 person for 10 years. In some situations, the assumption is appropriate; in others, it may not be. The determination of which is which depends on the underlying biological model.

# MEASURES OF RISK AND ASSOCIATION

## Absolute Risk

It is an immutable fact of life that we are all going to get ill at some time and ultimately we are all going to die. On a personal level, the questions for each of us are by what disease and when? Epidemiologists are also interested in those questions, but they are particularly interested in whether the disease occurs more frequently or more severely in association with some type of exposure. In other words, when it comes to identifying the causes of disease, what is at issue is whether the *absolute risk* for a specific disease among the exposed is greater than the absolute risk of that same disease in the unexposed.

## Relative Risk

An incidence rate provides a measure of absolute risk. The ratio of the incidence rates in two different groups is a *rate ratio*, *risk ratio*, or *relative risk*, a key measure of association between exposure and disease. If the relative risk (*RR*) is appreciably greater than 1 among those with a particular exposure, it is evidence that the agent may be causing the disease. May be. Conversely, if the relative risk is below 1, the agent may be protecting against the disease. May be. And, if the relative risk approximates 1, there may be no meaningful association between the two variables. Once again, may be. "May be" is an important caveat in all three situations because how well the *apparent relative risk* (the number derived as the result of a particular investigation) corresponds to the *true relative risk* (the actual underlying biological truth) depends not only on the statistical stability of the estimate of relative risk but also on how well the potential technical biases of selection, misclassification, and confounding were controlled in the study design, during data collection, and by data analyses.

## Standardized Mortality Ratio

In cohort mortality studies, the measure of association may be provided as a *standardized mortality ratio* (SMR). Because this is simply the ratio of the number of deaths observed in the study group to the number that would have been expected if the study group had the same death rate as a reference (i.e., standard) population, it is sometimes presented as a ratio of *observed to expected* deaths (O/E). By convention, this measure of association is given as a percentage, but the interpretations parallel those of the relative risk. A SMR of 150 is analogous to a RR of 1.5, a SMR of 75 to a RR of 0.75, and a SMR of 100 to a RR of 1.0. Because the observed number of deaths occurs in discrete increments and the number of expected deaths is for all intents and purposes a continuous variable (i.e., the expected deaths might be a biologically impossible number such as 1.27365…), by convention many epidemiologists will not calculate a SMR if the number of observed deaths is less than 2 [22]. They may simply provide the two numbers (the observed and the expected) or just give a confidence interval. Sometimes, they will do neither and merely indicate that the numbers were too small to be meaningful

Note that the "controls" (the comparison group) in a SMR analysis are essentially those in a hypothetical group statistically constructed from the reference population to have approximately the same age, race, and gender characteristics as the exposed group. For many occupational studies, the mortality experience of U.S. white males is used as a reference, even if a small number of those in the occupational cohort are of a different race or ethnic group, the assumption being that the calculations of expected deaths will be adequate. Using a SMR approach also means that the investigator is assuming that no one in the reference population was exposed to the agent of interest. If the exposure is relatively rare among those in the reference population, the assumption is probably reasonable because the mortality experience of those few who were exposed would have had very little impact on the population statistics. On the other hand, if the exposure is relatively common (e.g., something like chlorinated drinking water), then the assumption may be unreasonable and another type of study would have to be done to obtain valid information.

An interpretation of a crude measure of association assumes that both the exposed group and the reference population had similar habits regarding smoking, dietary preferences, medical care, etc. This assumption may be incorrect. For example, a number of years ago a study was done in California of men who lived in communities adjacent to petrochemical facilities along the Sacramento river [8]. The comparison population was composed of those who lived in the same county but remote from the industrial sector. An apparent elevated risk for lung cancer was discovered, and the finding was initially presumed, at least by the news media, to be due to emissions from the petrochemical plants. During this first stage of the investigation, however, no attempt had been made to control for the effects of cigarette smoking. For efficiency, that activity had been deferred to subsequent stages of the research. It was later found that those who lived near the plants were largely blue-collar workers and those who lived elsewhere tended to be white-collar professionals who commuted to work in San Francisco. Why was this important? Blue-collar workers on average smoke more than white-collar workers (at least that has been past experience), so blue-collar workers have higher rates of lung cancer. When the subsequent research gathered the necessary data and adjusted for these differences, the elevated risk disappeared [23]. As a result, the proposed intervention strategies changed from targeting industrial emissions to implementing smoking cessation programs.

The SMR can also be an acronym for a *standardized morbidity ratio*. Instead of death being the outcome of interest, it is illness, but the calculations and the resultant interpretations are basically the same. So, too, are the underlying assumptions. If the assumptions were violated to the degree that the study results were affected, then the reader should look for confirmation elsewhere.

## Proportional Mortality Ratio

Neither the RR or the SMR should be confused with the *proportional mortality* (or *morbidity*) *ratio* (PMR). The PMR is a measure of the relative importance of an individual category of disease *among those with any disease*. As such, both numbers in the ratio are numerator data. Although it is a convenient measure to obtain, it must be used with caution in etiologic research because it compares proportions and not rates. It makes the assumption that a higher proportion of a particular disease is the same as an increased frequency of that disease. Because a PMR calculation works like a teeter-totter, that assumption may be invalid. Although a higher proportion of disease A may be due to an increased incidence of disease A, it also may simply be a function of a lower frequency (and therefore a lower proportion) of some other condition, disease B. For example, a higher PMR for cancer among an occupational group with a certain exposure may mean that more of those with the exposure were developing (and dying) from cancer than those in the standard population, but it is also consistent with the interpretation that those with the exposure were *not* dying more often from cancer; they were just dying less often from noncancer events. In other words, in a PMR analysis an apparently "adverse" finding may be spurious (i.e., solely a function of the *healthy worker effect*) [85].

### Rate Difference and Attributable Risk

With two incidence rates, it is possible to calculate not only a rate ratio but also a *rate difference*. If the association between the exposure and the disease is truly causal, the rate difference provides a measure of the excess burden of disease that an exposed population might expect to experience as a result of the exposure. Stated another way, it represents the amount of the disease that would have never occurred if the exposure had been prevented. In such situations, it is may be called an *attributable risk*, *attributable risk percent*, *attributable fraction* or a number of other related terms as derived for just the exposed group or for the general population as a whole. Unfortunately, some will calculate a risk difference and use the term "attributable" even when causation has not been established.

Note that the two measures—the rate ratio and the rate difference—provide very different information. The higher a RR is above one, the greater the likelihood that a true cause-and-effect relationship exists, but a high RR for a very rare disease among a few individuals with a unique exposure may be of *de minimis* concern from a public health perspective whereas a lower RR for a relatively common condition might equate to an enormous number of cases. By way of example, it is generally accepted that excess exposures to vinyl chloride monomer cause angiosarcoma of the liver. The RR for this association is quite high, but the total number of excess cases, worldwide, approximates 100. By way of contrast, the RR for heart disease among cigarette smokers is only about 1.5, but the rate difference equates to a large number of cases—many, many orders of magnitude more than 100. This is because both the disease and the exposure are relatively common. From a public health perspective, it is much more important to control the excess risk of disease related to smoking than it is the risk associated with vinyl chloride monomer. Yet, for the purpose of establishing a cause for the disease, it took many fewer epidemiology studies to establish an etiologic association between vinyl chloride monomer and liver angiosarcomas than it did for cigarettes and cardiovascular problems.

Rate ratios and rate differences are derived from research in which two groups are defined based on exposure status, and the disease patterns of each are followed forward in time. On occasion, it is easier to get groups based on whether they do or do not have a specific disease and then collect data on previous exposures. For example, it may be more convenient to identify all those who developed lung cancer during some period, possibly via the use of data from a tumor registry, and identify a comparable group of healthy individuals from the general population, perhaps by means of random digit dialing within the same area codes as the cases. Gathering data on previous exposures from those in each group (or from their next of kin) would allow the calculation of an *odds ratio* (OR), the odds of having been exposed to a particular agent given one had the disease vs. the odds of having been exposed to that same agent among the healthy controls. If a study is done properly, the OR will approximate the RR; for example, if the RR for getting lung cancer among cigarette smokers is 10, the OR of having been a cigarette smoker among those with lung cancer also will be about 10. For simplicity, the rest of the text will focus predominately on two measures of association: the relative risk and the odds ratio.

## METHODS

### Cohort

Over the years, epidemiologists have developed a variety of methods to evaluate cause and effect. The most intuitively obvious, and the most analogous to the approach used in toxicology, is the *cohort* study. A cohort is simply a group with some common characteristic (e.g., gender, ethnic background, health behavior, or exposure to a particular chemical or medicine). In a cohort study, the health experiences of at least two cohorts are compared, one with an exposure to the agent of interest and one without. Ideally, multiple cohorts, each with a different level of exposure, are identified so the effects of low, medium, and higher levels of exposure can be assessed. Irrespective of the number of groups, conceptually exposure status is determined first, and health data—on subsequent mortality, morbidity, blood cholesterol levels, whatever—are then gathered forward in time. If the exposure status is determined in the present and the health data are then gathered into the future, the term *prospective cohort study* is used.

Prospective cohort studies, for all of their advantages, may not be the method of choice in preliminary investigations of the causes of disease, especially if the disease has a long latency. As opposed to toxicologists who dose animals of species with relatively short lifespans (a standard chronic feeding study of mice takes 2 years), epidemiologists examine a long-lived species, humans; therefore, if they only did prospective cohort studies of chronic disease, then they likely would complete very few projects during their professional careers. To overcome this problem, epidemiologists will often use historical records—personnel files, medical archives, industrial

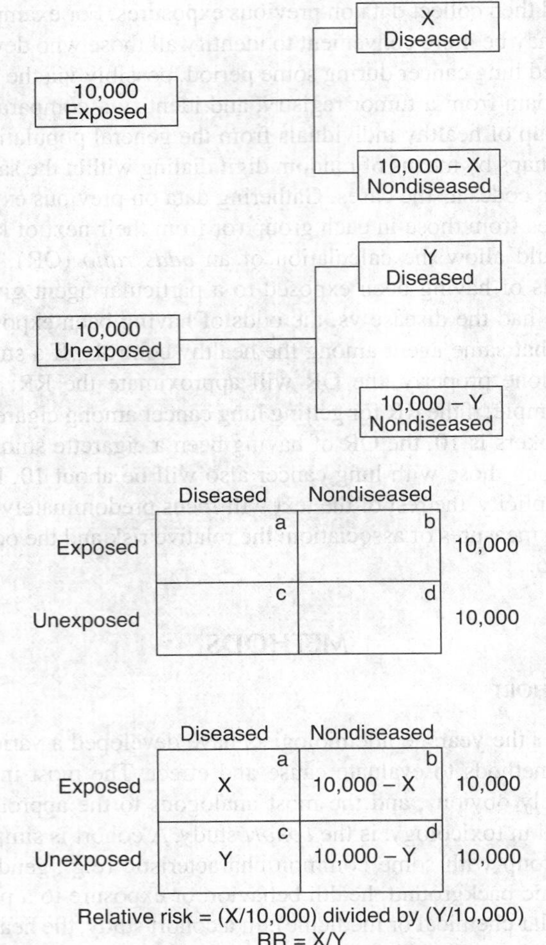


**FIGURE 11.3** Illustration of the cohort method.

hygiene reports, etc.—to define their exposed and unexposed study groups at some arbitrary date in the past. They will then gather health data on each individual in the study groups from that point up to the present. These are sometimes called *retrospective cohort studies*, or, to differentiate them from the case control method, which also gathers data on former events, they may be labeled *nonconcurrent prospective studies* or *historical prospective studies* or even *retrospective prospective studies*. Irrespective of whether the starting point for a cohort study is at the present or in the past, the results are based on incidence data presented as relative risks (and, if appropriate, risk differences).

Figure 11.3 illustrates how this is done. Two groups of healthy individuals are identified at a point in time. One group is selected because they have (or had) a known or presumptive exposure to a specific agent; the second because they don't have (and ideally never had) the exposure. The health experience of those in each group is then compiled in an equivalent fashion over some defined period of time. This health experience is converted into incidence rates and the rates compared by means of a relative risk. It is possible to calculate relative risks for all health events combined (e.g., total causes of mortality) or for any number of distinct outcomes (e.g., just deaths due to angiosarcoma of the liver). When the cohort approach is used in exploratory data analysis, it can be considered an exposure in search of a disease, a hypothesis-generating exercise. If it targets just one or a limited number of specific associations of *a priori* concern, it is akin to hypothesis testing. Many epidemiology studies are a combination of both and it may be difficult for the reader to discern which associations were of concern at the beginning of the research and which were simply serendipitous findings [3]. At times, it is possible to make this determination only by reviewing the original study protocol, if there was one.

In the example, each group at the start of the study had 10,000 individuals; therefore, the marginals for the 2×2 table are both 10,000. During the period of study, $X$ individuals in the exposed group were observed to have developed the disease (cell *a*) while the remainder (10,000 – $X$) did not (cell *b*). The incidence rate for the exposed is $X$ divided by 10,000 (10,000 being the totals of those in cells *a* and *b*). Among the unexposed, $Y$ developed the same disease (cell *c*) and 10,000 – $Y$ did not (cell *d*). The incidence rate among the unexposed is therefore $Y$ divided by 10,000. Dividing $X$ over 10,000 by $Y$ over 10,000 gives the relative risk. Because both groups had the same denominator, this particular RR simplifies to $X/Y$. In real life, that seldom happens.

Hypothetically, the investigators might have found that 50 individuals among the exposed developed the disease and only 10 among the unexposed (Figure 11.4). After plugging these numbers into the table, the resultant calculations would produce a RR of 5. The exposed had 5 times the risk of developing the disease as did the unexposed, assuming there was no selection, misclassification, or confounding bias and the finding was not a chance occurrence.

In a cohort study, those in both groups must be free of the condition at the start of the investigation. This implies that no one in either the exposed or the unexposed group is eligible until they are first examined and determined to be disease free. In other words, the first step of any prospective incidence study is, conceptually, a *cross-sectional* or *prevalence study*. The data from this cross-sectional study, even though they are collected on two or more cohorts, cannot be used to make interpretations concerning etiology. They are prevalence data.

In actual practice, it may be impossible to determine baseline health status; for example, in a nonconcurrent cohort morbidity study, an investigator cannot go back in time to examine the study participants in any of the groups. Even when the study has a prospective orientation, it may

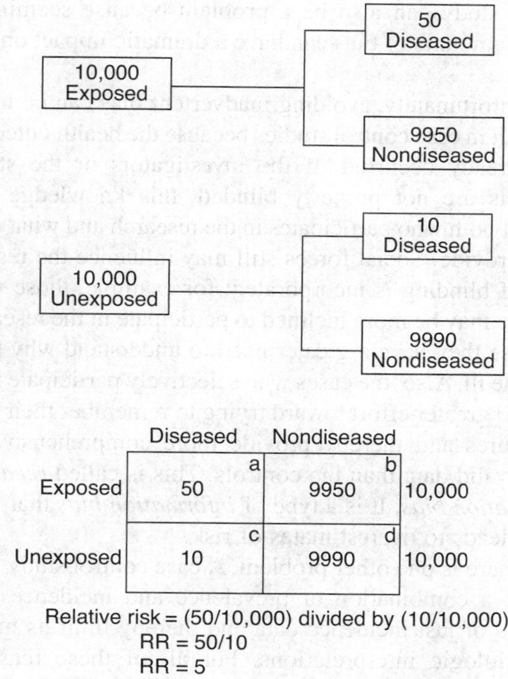


**FIGURE 11.4** Illustration of the cohort method; RR = 5.

not be feasible to examine those in the control group if a SMR-type approach is used because that would mean everyone in the standard group (e.g., the U.S. white male population) would have to be examined, a logistical impossibility. Nevertheless, if the natural history of the health condition is well understood, adjustments can be made to overcome this problem. With diseases of long latency, the investigators might simply ignore the health data from the first couple of years. For mortality research, a person might be presumed living at the start of the study if he or she was then employed, paying taxes, or receiving retirement benefits.

A cohort study can be a very labor-intensive process. Exposure histories have to be compiled and validated. Study subjects (or their next of kin) may have to be traced and contacted and data obtained on personal habits, hobbies, and a host of other variables. Medical records must then be collected and coded. Many things can complicate the process.

The first major obstacle is simply finding the study subjects. In our society, it is not unusual for someone to change their residence multiple times during his or her lifetime. Women may leave the workforce, get married, and, in the process, assume a new last name. Conversely, someone may have a name so common that it is very difficult to determine which "John Miller" is the correct study subject and which is not.

A second major obstacle is finding comparable health data on each individual. The amount of medical information can vary from person to person simply because of differences in healthcare-seeking behavior. The study subjects may have many different physicians, each providing a different level of care, possessing diverse diagnostic skills, and having office records with unique formats. Many states and municipalities have disparate rules governing access to government records such as death certificates. In addition, litigation and regulations may obstruct the process of data collection [5,7,32].

## Case Control

Even if all of these obstacles can be satisfactorily addressed, it means that a great deal of effort may be required to gather a lot of data that produces relatively little useful information. In the hypothetical example, 20,000 individuals were tracked to identify the 60 who actually got the disease. To overcome the inefficiencies of cohort studies, epidemiologists developed the *case control* method. With case control studies, the past exposures of those with some disease are compared to the past exposures of those who do not have the disease; for example, smoking histories might be compared between men who do and do not have lung cancer. Due to the fact that data are gathered from the past, a case control study may be referred to as a *retrospective study* or a *trohoc* ("cohort" spelled backwards) to differentiate it from a retrospective cohort study.

Because the study participants for a case control investigation are first determined in the disease axis of the 2×2 table and data are then gathered on exposure status to fill in each of the four cells, it makes no sense to calculate incidence rates or relative risks. Instead, a different measure of association is used based on the odds of past exposure. The odds of past exposure are calculated respectively for the disease group and for the nondiseased control group. These odds are then compared to develop an *odds ratio* (OR). Because the magnitude of the OR in well-done case control research closely approximates that of the underlying RR, it allows the interpretations of a case control study to parallel those of the cohort method: An OR appreciably above 1 suggests a causal association between the disease and the exposure, an OR appreciably below 1 suggests protection, and a ratio near 1, plus or minus, suggests no association between the exposure and the disease.

If we go back to the hypothetical example, there were 60 new cases of disease in this closed population (Figure 11.5). Knowing the age, gender, race, and perhaps other key characteristics of the diseased, 60 "matched" nondiseased individuals could be randomly selected from the remainder of the 20,000. These 120 would constitute the bottom marginals of the 2×2 table, 60 in each column. Data could then be collected on past exposure. In this particular example, 50 of the diseased group would end

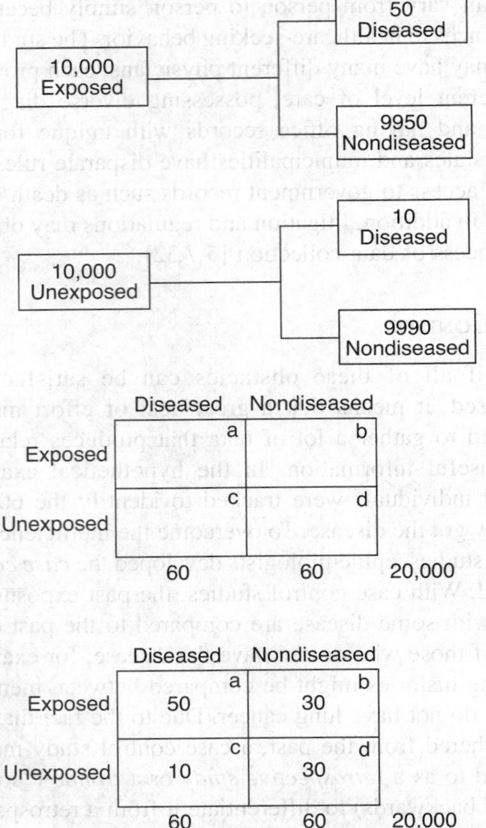


**FIGURE 11.5** Illustration of the case control method; OR = 5.

up in cell *a* and 10 in cell *c*. Among the nondiseased, approximately 30 would end up in cell *b* and another 30 in cell *d*. Parenthetically, the nondiseased in each exposure category were approximately the same, 9950 and 9990; therefore, random sampling of 60 from the aggregate 19,940, should select equal numbers from each group: 30 and 30. With data in all four cells, the odds of exposure among the patients would be 50 to 10 (5 to 1) and the odds of exposure among the controls would be 30 to 30 (1 to 1), giving an odds ratio or OR of 5. Conveniently, the calculations for case control studies often simplify to a *cross-product ratio* (*ad* divided by *bc*). In the example, 50 times 30 divided by 30 times 10 simplifies to 5.

For this case control study, an evaluation of just 120 individuals provided the same information as a study of 20,000. In the context of the time, effort, and cost, gathering data on such a limited number of study subjects can be a tremendous advantage. Paradoxically, the small size of the study can also be a problem because seemingly minor amounts of bias can have a dramatic impact on the OR.

Unfortunately, avoiding inadvertent bias can be more difficult in case control studies because the health outcome has already occurred. If the investigators or the study subjects are not properly blinded, this knowledge can impact both who participates in the research and what data they provide. Social forces still may influence the results even if blinding is incorporated; for example, those with disease may be more inclined to participate in the research because they have a greater need to understand why they became ill. Also, the cases who selectively participate may expend greater effort toward trying to remember their past exposures and thereby provide more comprehensive or more valid data than the controls. This is called *recall* or *rumination bias*. It is a type of *information bias* that very often leads to overestimates of risk.

There is one other problem. A case control study may utilize a combination of prevalence and incidence data instead of just incidence data and thereby limit its utility for etiologic interpretations. For all of these reasons, results from case control studies are considered lesser evidence than those derived from cohort research.

Even with its limitations, a case control approach can be very attractive. Because the two groups are initially defined on disease status, data on any number of exposures can be collected. As a consequence, a large number of different associations can be evaluated simultaneously and rapidly reduced to just a few that deserve further study. For that reason, an exploratory case control study can be considered a disease in search of an exposure, the mirror image of the hypothesis-generating exercise done in a cohort mode.

A case control study also can focus in depth on just one disease exposure association, testing a hypothesis derived from case reports or other types of research with much greater sophistication than might be feasible in a cohort study. In certain situations, it can be advantageous to use the cohort and the case control approach in series to generate a relatively small and well-defined number of hypotheses. Such a *nested case control study* can combine the strengths of both methods; for example, the cohort approach could be used to identify a cluster of disease within a broadly defined group, perhaps all those ever employed at a multiple-chemical-manufacturing facility, and a case control study could then be implemented within the larger cohort not only to narrow the focus to those few agents that appear to be most important for that particular disease but also to do so with proper adjustments for confounding. This integrated approach, therefore, can achieve both efficiency and rigor.

Cohort and case control studies are sometimes referred to as *analytic* research, in contrast to other types of epidemiology investigations that are simply *descriptive* of time, place, and person. In theory, the term "analytic" should be restricted to those studies designed to test *a priori* hypotheses, but in practice it is often used more broadly to refer to any cohort or case control research, regardless of whether it generates or tests hypotheses. That is unfortunate, because it blurs the distinction between these two important stages of discovery and the role each plays in the search for the causes of human disease.

Understanding disease etiology depends on a complex, iterative course of inquiry called the *scientific method* [130]. To quote Hazen [67], "The scientific method is an elegant process for learning about the natural world, but it is neither intuitive nor obvious." This method can be idealized as a cycle of observation (data collection), synthesis (data analyses), hypothesis (reasoned conjecture based on the interpretations of patterns derived during the data analyses, often as interpreted in the context of other information), and prediction. The prediction then has to be tested (sometimes referred to as the hypothesis testing stage of scientific inquiry) with a new round of observation and synthesis, providing results that, given they replicate the original findings, reinforce the initial hypothesis. Alternatively, the results might not support the hypothesis, in whole or in part; therefore, the original hypothesis might be dismissed outright or modified and retested by means of a new round of prediction, observation, and synthesis. This goes on until there is some level of consensus that a provisional truth has been identified. As a rule of thumb, the more provocative the association, the more imperative the need to replicate the findings.

It is important to recognize that the prediction must be constructed in a form that is both unambiguous and refutable; therefore, although the hypothesis may be stated as "exposure to agent *X* is associated with an increased risk to disease *Y*," the prediction has to be stated in the null (i.e., exposure to agent *X* will not be followed by an increased risk of disease *Y*) and the null refuted. Refuting the null lends support for the hypothesis, but in science the default is always the null; consequently, a theory of cause and effect may be disproved by an unfulfilled prediction, but it can never be completely proved—thus the caveat about "provisional truth." Alternatively, an anomaly (an exception to a prediction) may lead to new insights; for example, it may suggest a prediction that restricts causal actions of agent *X* to higher levels of exposure.

It is also important to recognize that at the core of any scientific inquiry there is always a paradigm, a prevailing expectation about the workings of the natural world; for example, the current paradigm regarding dose–response for carcinogens is linear, nonthreshold. That paradigm is being challenged. Calabrese [21] has shown that for a large number of agents both attributes of the current paradigm are wrong, particularly at the lower ranges of exposure. That knowledge suggests that our current approach to risk assessment and risk management have to be reevaluated.

## Case Studies or Case Series

Hypotheses for analytic epidemiology may originate from toxicology studies or from epidemiology investigations, but many evolve from clinical observations and are published in the form of *case studies* or *case series*. Although based a great deal on intuition, a case study is a time-honored way for a physician to develop new theories about the causes of human disease. It has been said with some justification that every human carcinogen was first identified by an astute clinician who published his findings in the form of a case study or case series. Nonetheless, that does not mean case studies can be used to unerringly identify new etiologic associations. Although the theories derived from case studies are not always wrong, history teaches that they are seldom right [9,34,101]. Determining which is which depends on data developed by others using experimental, quasi-experimental, and observational research. If we go back to the 2×2 table, we can see why.

To test a hypothesis about a new cause for human disease (to identify an elevated risk in analytic epidemiology research), data are needed in all four cells of the 2×2 table—data that are properly defined on both variables. Case studies tend to focus just on those in one of the four cells: cell *a*, the exposed with disease. Very little if any data are gathered by the clinician on those in each of the other three cells. Furthermore, those from whom data are gathered are a *convenience sample*. They probably are not a representative sample of any well-defined group, especially not a representative sample of the healthy—irrespective of their exposure history. They are not because physicians tend to direct their efforts toward diagnosing and treating those with medical problems.

In Figure 11.6, examples a through c, the three 2×2 tables represent the three possible types of association. In the first, the 50 in cell *a* translates to a RR of 5; in the second, the 30 to a RR of 1; and in the third, the 20 to a RR of 0.5. The three relative risks have very different meanings. Although it is conceivable that any clinician practicing in a community might become suspicious if a cluster of 3 or so patients came to him with the same rare disease and all had a similar exposure history, based on the information available he would not be able to determine whether the cluster was a subset of those in cell *a* from Figure 11.6a, Figure 11.6b, or Figure 11.6c. Most clusters, however provocative, are meaningless [102]. Furthermore, additional case reports do not satisfy the need for replication and confirmation. Once a testable hypothesis has been formulated, additional case reports proposing the same hypothesis contribute nothing.

By way of example, in the silicone breast implant controversy, it was originally hypothesized that women who received this medical device were at increased risk

Example a. Relative risk is 5

| | Diseased | Nondiseased | |
|---|---|---|---|
| Exposed | 50 | 9950 | 10,000 |
| Unexposed | 10 | 9990 | 10,000 |
| | 60 | 19,940 | 20,000 |

Example b. Relative risk is 1

| | Diseased | Nondiseased | |
|---|---|---|---|
| Exposed | 30 | 9950 | 10,000 |
| Unexposed | 30 | 9990 | 10,000 |
| | 60 | 19,940 | 20,000 |

Example c. Relative risk is 0.5

| | Diseased | Nondiseased | |
|---|---|---|---|
| Exposed | 20 | 9950 | 10,000 |
| Unexposed | 40 | 9990 | 10,000 |
| | 60 | 19,940 | 20,000 |

A physician sees three patients with the condition and all three were exposed to the same chemical, leading him to conclude that the disease in all three was caused by the chemical exposure. Is he correct?

**FIGURE 11.6** Case studies and case series.

for breast cancer. The theory was based on clinical observations, and concern was increased because of an animal toxicology study that demonstrated an Oppenheimer effect, the tumorigenic properties of foreign bodies as observed in rodents [13,49,89]. As a result of subsequent research, both experimental and observational, something between Figure 11.6b and Figure 11.6c is now thought to most closely approximate the association between silicone breast implants and human breast cancer. It is being theorized that these medical devices or the materials from which they were constructed offer some type of protective effect against breast cancer [16,126]. The current data-based theory is in exact opposition to the hypothesis originally derived from the case reports. Interestingly, no action has been taken on this information. Why? Probably because, even though the epidemiology study results have been reasonably consistent and demonstrate coherence with the findings of the experimental animal research and the public health implications of such an association could be profound considering both the frequency and the life-threatening characteristics of the cancer, the underlying biological mechanisms of protection have not been identified.

## ISSUES

Peer review is an imperfect process. Even the most prestigious journals publish findings that are wrong. As a consequence, everything must be read with a degree of healthy skepticism [96]. This can be difficult enough within a single field, but it is truly a daunting task when a scientist tries to evaluate the merit of work from a different discipline. If a toxicologist understands the basics of data, measures of disease frequency, measures of association, and methods, the epidemiology literature can be screened fairly rapidly using the mantra of *selection*, *misclassification*, *confounding*, *chance*, and *causation*. Consultation with an epidemiologist or biostatistician might still prove necessary, but only for the smaller number of studies.

The order of this mantra is important. If obvious technical biases are related to selection, misclassification, or confounding, it may make very little sense to spend time trying to evaluate the merit of the investigators' statistical machinations, much less to assume the findings of statistical significance have biological meaning. It is no accident that the scientific literature has a highly stylized format: some variation of abstract, introduction, methods and materials, results, discussion, and conclusion. This format allows the reader to rapidly focus on the key components of the work. If the authors provide a one-sided presentation of the topic in the introduction, supply insufficient detail regarding their methods and materials, or do not critique their own work in the discussion—pointing out the potential biases of selection, misclassification, and confounding and how they were addressed—the reader should exercise extreme caution before accepting either the results or the conclusions, even as provisional truth.

## Selection

In epidemiology, *bias* is used to denote a deviation from the truth but not necessarily to imply that the deviation was intentional [3,132]. *Selection bias* refers to errors that are related to systematic differences between those who are and are not selected in a study. Even if the data gathered are valid for the examined, it may be inappropriate to use any information derived from the data for purposes of extrapolation to a larger population; for example, the results of a study of hormone replacement therapy among women cannot logically be extended to men. In epidemiology research, various types of selection bias can be introduced by the study subjects, the investigators, or even traditional medical practice and other social forces.

*Self-selection* occurs in both clinical research and some epidemiology studies. It is well recognized that those who participate in controlled clinical investigations, those who actually sign informed consents, may not be representative of the general population; therefore, even with randomization of treatment, care must be taken before extending the study results too broadly. A similar problem occurs in observational studies in which some type of active participation, some type of action on the part of the study subjects, is required. For example, informed consent is required for any epidemiology study in which biological samples are collected. Usually, the more invasive the procedure, the more disinclined are the potential subjects to participate and the greater the potential for bias; however, in other situations, this bias may be less obvious or, paradoxically, so obvious that it is largely overlooked. As an example, how many times have you received a questionnaire in the mail and, rather than filling it out, tossed it away? By doing so, you introduced a potential *participation bias* into that investigator's work, the potential for which may not be acknowledged in the final report.

In certain types of observational research, self-selection is not a problem. Projects that can be conducted without the active cooperation of the subjects often are able to achieve close to 100% follow-up; for example, occupational cohort mortality studies that utilize personnel records and industrial hygiene reports to identify the exposed and death certificates to document the cause of death can be conducted with little or no self-selection [90].

WANTED

Kansas farmers suffering from non-Hodgkins lymphoma especially those exposed to herbicides such as 2,4-D. Needed to interview for research article. Please send name, address, and phone number to Box P7629 Classified Dept., The Star, 1729 Grand, K.C., MO 64108.

**FIGURE 11.7** Recruitment of study subjects.

The same arguably holds for some studies that utilize medical records, but only if the medical records relate to the total health experience of a well-defined population. Such is the case in certain countries with socialized medicine in which all the hospital and clinic records are available for the entire citizenry. In the United States, such opportunities are rare and even those few are disappearing rapidly. The Mayo Clinic is a world renowned referral center, providing both state-of-the-art medical treatment and highly sophisticated research on the underlying mechanisms of disease. In addition, it serves most of the primary medical needs for those who live in the relatively isolated community of Rochester, MN, and shares medical records by agreement with the few other primary care facilities that operate in that area [53,54]. Having access to the total health experience of those in the community has allowed the Mayo epidemiologists to focus some of their research just on the residents and thereby to conduct high-quality, population-based epidemiology research that has minimal self-selection or referral bias. Recently, ostensibly for reasons of privacy and confidentiality, the state legislature passed a law requiring study-specific informed consent from all study subjects before any of their data may be utilized for research purposes, even if the patients had previously expressed a willingness to have their medical records used for any such activities [75]. This action by the Minnesota legislature, although undoubtedly politically expedient, will not only complicate the logistics of future research at Mayo Clinic but may also unfortunately compromise the validity of the work.

Either intentionally or not, investigators can introduce selection bias when they decide who to study, especially if they make a greater effort to get participation among the exposed than the unexposed, or the diseased than the healthy. Figure 11.7 is an advertisement that appeared in a Kansas paper in the late 1980s. It apparently was placed by investigators who wished to identify more subjects for a research project and thereby improve its statistical power. What they presumptively did not recognize was that by recruiting simultaneously on both health outcome

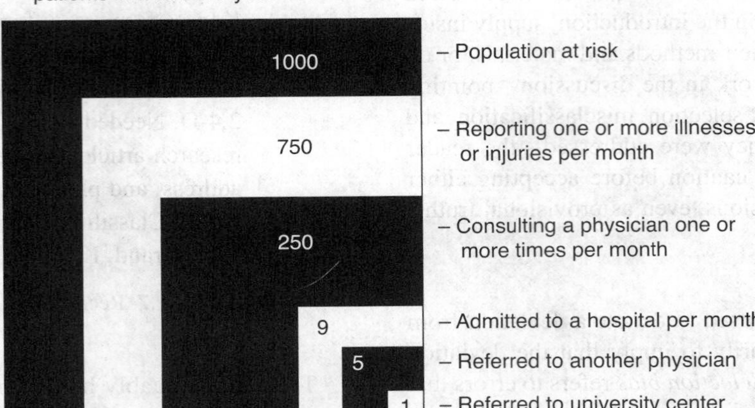


**FIGURE 11.8** Berkson's bias: potential selection bias by referral. (From Gehlbach, S.H., *Interpreting the Medical Literature*, McGraw-Hill, New York, 1993. With permission.)

(non-Hodgkin's lymphoma) and exposure (2,4-D) they would introduce a significant selection bias into their work, one potentially so severe as to invalidate any of their findings.

More recently, a study was published in the *Journal of the American Medical Association* of children with esophageal dysfunction who had been born to mothers with silicone breast implants [79]. Once again, the key study subjects had been selected on the dual characteristics of health outcome and exposure. The investigators characterized their work as a case control study and indicated they had findings that were supportive of a cause-and-effect association. The fact that their report was little more than a case series was missed during the peer-review process and corrected later in the form of an obscure errata, and apparently only then because of *ad hoc* peer review (i.e., a series of highly critical letters to the editor) [28,39,41,43,93].

Figure 11.8 illustrates the dynamic that leads to *Berkson's bias*, a particular type of selection bias that occurs as a result of the patterns of referral, either self-referral or physician referral [55]. Although there is some merit in asserting that the 250 individuals who initially consulted a physician represent those with the more definitive illness among the 1000 in the population at risk and thus are legitimate subjects for etiologic research, it is less likely that the same thing can be said about the 5 referred to a specialist or the 1 who finally ended up at a university center. Patients seen by specialists or at tertiary referral centers include a disproportionate number whose disease is complicated, obscure, or atypical. In our chemophobic society, these patients also may be referred because of a suspicion that the condition is related to what Peter Huber has called the latest *terror du jour* [70]. A spuriously elevated relative risk will predictably be found in any research in which the study subjects are selected on the joint characteristics of the condition of interest and the putative agent of concern.

Even if no formal study is conducted, the specialist may develop a marked suspicion concerning the presumptive cause for the condition and then act on that presumption. Once it becomes known in the community that a physician or a referral center is interested in patients with a particular condition, especially when it occurs in conjunction with exposure to a specific agent, additional referrals or self-referrals further compromise the value of the sample for etiologic research [121]. Ironically, the more caring the physician in the sense of being more willing to provide therapy to those who have been unsuccessfully treated or refused treatment by others, the more that physician becomes a magnet for these patients.

The 1995 publication by Robinson and colleagues entitled "Analysis of Explanted Silicone Implants: A Report of 300 Patients" illustrates a number of potential selection biases [99]. Among the 300 women who Dr. Robinson explanted over the course of 3 years, 214 (71.3%) reportedly had "disruption" (defined as frank rupture of an implant or severe silicone bleed). Interestingly, these authors noted that there was "virtually no difference in the disruption rates between those patients relating symptoms to their implants and those who did not (71.8% vs. 70.9%)", suggesting that health complaints were not a consequence of implant status. Nonetheless, they extrapolated from this sample to predict that most implants will lose their integrity somewhere between 8 and 14 years and recommended that all gel-filled implants be removed "preferably before 8 years from implantation."

Robinson et al. based their rates and their interpretations and formulated a policy of explantation on data from a denominator of 300, but that was not the group they

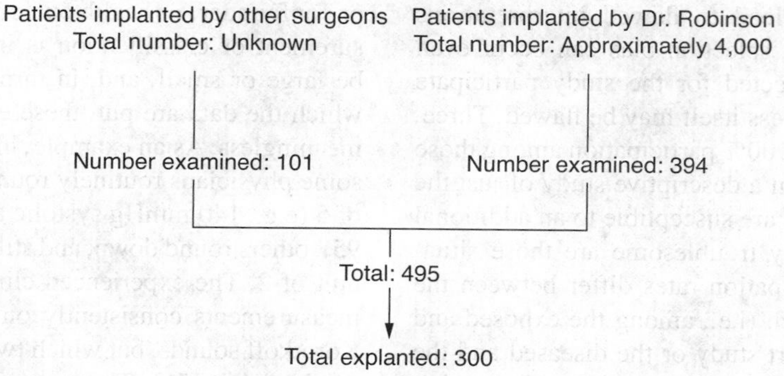


**FIGURE 11.9** Convenience sample. (From Robinson, O.G. et al., *Ann. Plastic Surg.*, 34, 1–7, 1995. With permission.)

actually studied (Figure 11.9). According to the paper, Dr. Robinson saw 495 women who would have been eligible for this investigation, 101 who had been implanted by other surgeons and 394 of his own patients. The 300 were drawn from the 495, but note that, even if he had studied all 495, he still would not have been able to develop rates that were free of potential selection bias. Even with 100% participation of his sample, he would not have been able to develop rates that meaningfully could be extrapolated back to a larger group. That's because the 495 were a convenience sample, an ill-defined and likely highly biased sample of the larger population from whence they came. The larger population included all of Dr. Robinson's implant patients and, by implication, all the breast implant patients of the other 15 to 20 plastic surgeons who practiced concurrently in the same community [1]. Court records indicate that Dr. Robinson implanted approximately 4000 women. and it is quite possible that at least some of the surgeons in his community implanted comparable numbers [98].

So what can we make of the Robinson information? The data collected for this report were prevalence data. Although gathered over a 3-year period, for the individual study subjects they were obtained at a single point in biological time—the time of surgical explantation. Because they had prevalence data, the researchers could not differentiate between events that occurred at the time of surgical implantation, during the period the implant was within the body, or at explantation. Their interpretation, therefore, that implant failure is a function of the aging of the device, presumptively related to biological degradation of the silicone elastomer shell, required assumptions (e.g., the incidence time of rupture was just before explantation) that were not adequately addressed in this research.

Based on the work of others, at least some of those assumptions appear to be incorrect. Rapaport et al. [94] found that an appreciable number of implant ruptures occur secondary to micropunctures caused by needles or other medical devices used during the implant procedure. Others have done work that expands on this observation [14]. Brandon and colleagues [15], using lot-matched controls, reported that the material properties of the silicone shell are not affected by implantation for time periods up to 21 years and concluded "that the silicone elastomer undergoes little or no change during implantation." Slavin and Goldwyn [116] noted that approximately 25% of the implant ruptures they observed occurred during the explant procedure. At least two other mechanisms contribute to implant ruptures *in vivo*: closed capsulotomies (manual compression of the breast to rupture the tissue capsule surrounding the medical device) and so-called "fold flaws" (disruption of the elastomer by excessive flexing at the site of folds in the shell). Both involve mechanical trauma. Obviously, different approaches might better be used to prevent, control, or otherwise address implant ruptures caused by different mechanisms.

Setting aside the questions of the validity of the data and the causes of implant rupture, if the 300 who were explanted are a representative sample of the ever-implanted, then it is quite possible that a high proportion of implanted women have "disrupted" implants. Further, if "disruption" equates to implant rupture, either overt or occult, it suggests that there may be a high rupture rate for these medical devices, at least for those brands and models favored by Dr. Robinson and his colleagues [25]. On the other hand, if the 214 with disrupted implants are the majority of those in the numerator of a true rate, especially if disruption does not equate to rupture, then it is likely that the actual rupture rate is quite low, quite possibly a single-digit phenomenon. Of course, if neither scenario is correct, then the information is invalid and has no utility at all. Furthermore—and in spite of the question about rupture rates—if these authors are correct in their observation that there is a lack of association between implant integrity and health outcome, a conclusion reached independently by others, then is it good public health policy to expose all implanted women to the predictable risks of explant surgery [19,137]? Probably not.

A number of lessons can be learned from this report: One, not understanding the difference between prevalence and incidence data can lead to flawed interpretations [29,30,31,57,58,59]. Two, selection bias can occur even when 100% of those selected for the study participate because the selection process itself may be flawed. Three, anytime there is less than 100% participation among those originally selected, even in a descriptive study of just the exposed group, the results are susceptible to an additional selection bias. Particularly troublesome are those situations in which the participation rates differ between the groups in analytic research (i.e., among the exposed and the unexposed for a cohort study or the diseased and the healthy in a case control study) because this suggests that the reasons for participation may not have been equivalent and therefore there may have been a spurious correlation between health outcome and exposure among one group or the other. The consequence of selection bias is an incorrect measure of association, possibly an underestimate of risk but often an overestimate. Complicating the situation still further, the dynamics of selection bias can change over time as a result of a well-publicized environmental controversy, a lawsuit, a provocative news program, or any number of other things; thus, different types of selection bias can wax and wane. Four, flawed studies can lead to flawed policies, policies that ironically may put those whom they are designed to protect at greater risk.

In evaluating the literature, the reader needs to ask two questions related to selection bias: Was the sample that the investigators were attempting to study truly representative of some larger group? Were the researchers successful in getting participation from all or a large majority of those they sought to study? An individual epidemiology report probably will have little or no value if the answer to either question is "no." The operative term in the previous sentence is "probably." It is important to note that not every potential selection bias is real; therefore, not every study with less than 100% participation should be dismissed as meaningless. The question is how does one determine whether or not a study with less than optimal participation provides relatively unbiased results. Usually, one cannot make that determination from the single study. The question can only be addressed in the context of the larger body of literature. If the results of the potentially flawed study are comparable with those of other work in which selection bias is a lesser concern, the consistency suggests a cross-validation of findings. On the other hand, if the results of multiple studies are markedly different, it raises concern that the findings of one or more of the reports are biased.

## MISCLASSIFICATION

*Measurement* or *misclassification bias*, also called *information bias*, is systematic error arising from the inaccurate measurement or inappropriate classification of subjects on the study variables—either exposure (to the putative agent or confounder) or health outcome. At some level, all measurement or classification is inaccurate. The errors may be large or small, and, in turn, depending on the use to which the data are put, these errors may be important or meaningless. As an example, in measuring blood pressure some physicians routinely round up to the next increment of 5 (e.g., 140 mmHg systolic and 90 diastolic or 145 and 95), others round down, and still others record to the closet unit of 2. The experienced clinician tends to make these measurements consistently on the same two of the five Korotkoff sounds, but which two may vary from physician to physician [51]. These variations from the true blood pressure probably have very little importance in the clinical setting if the patient is consistently measured and treated by the same physician, but they could be very important if treatment is provided by multiple physicians. They also could be important if the clinical data were used to judge the relative efficacy of a variety of treatments as administered by different physicians.

Misclassification can be introduced into an epidemiology study by the study subjects, the measurement tool, the observer, or even, after the fact, by the consumer of the research findings; for example, Edwards and associates [40] conducted interviews to gather data on alcohol consumption. They observed that men reported significantly lower age-adjusted mean levels of alcohol use when a third party was present during the interview (probably the spouse, in most instances). Conversely, study subjects may over-report specific conditions. Cautioning against placing too much reliance on self-reported data, Star et al. [122] noted that the self-reported diagnosis of rheumatoid arthritis could be confirmed in only about 20% of elderly women, a finding replicated more broadly across age strata by Sanchez-Guerrero and colleagues [108] in a larger study of nurses.

Over-reporting or under-reporting by study subjects may be a function of a number of factors unrelated to the biology of the disease [63]. For example, medical students tend to develop the symptoms of the latest disease they are studying even to the extent that some male students reportedly have complained of sympathetic labor pains during their obstetrics training! To address such *reporting bias*, Turner and associates [131] emphasized the importance of double-blinding in clinical trials of pain medications, positing that even inadvertent clues of voice inflection or facial expression by an unblinded investigator could influence how a patient might report his or her symptoms.

An epidemiologist might use various techniques to avoid or reduce the potential for either purposeful or unintentional misreporting. Concealing the intent of the research from the study subjects is one, but such blinding of subjects is increasingly difficult to use in a climate of mandated informed consent and almost instantaneous dissemination of news about the latest health controversy. Another approach is to add a dummy health variable whose association with the exposure is biologically implausible; for example, a query about dental caries could be incorporated into a study evaluating the effects of exercise on angina. If a strong correlation exists between the frequency with which the study subjects reported the dummy variable and the health outcome of concern, one should suspect a misreporting problem. In such a situation, it may be necessary to validate the reports—perhaps, if feasible, by examining a subset of the respondents or via review of medical records that predate the controversy or by use of a biological marker such as saliva cotinine for cigarette smoking [77,100,135].

At a minimum, the processes by which the data were collected should be well defined. Even then, there could be problems. It is well recognized by the seasoned researcher that mechanical or electronic instruments of assessment periodically must be calibrated to ensure a consistency of measurement over time. To achieve validity, they must be calibrated to an external standard. The application of other data collection tools such as questionnaires may be less than rigorous. With survey instruments, the order in which the questions are posed can be important. Even if the questionnaire is not open ended, the words themselves may have alternative connotations for different ethnic or racial groups. To the extent possible, epidemiology research should use tools whose strengths and limitations are well recognized or should incorporate a validation pilot into the research project.

*Diagnostic bias*, a type of observer bias, occurs when a physician's diagnosis is influenced by his or her knowledge of certain exposures or surrogates of such exposures. In a study of eosinophilia–myalgia syndrome, Wagner et al. [136] found up to a sixfold increase in diagnosing the condition when physicians were told the patients had ingested L-tryptophan even though use of this dietary supplement was not part of the definition for the condition. In unpublished work, Cook submitted a series of chest x-rays to a board-certified radiologist and resubmitted the same x-rays about a month later. As is customary, a short description of each patient accompanied his respective radiograph. During one submission, the patients were identified as office workers, during the other as pipefitters. When presented as pipefitters, they were more frequently diagnosed as having asbestosis.

Even laboratory tests that logically should be free of this bias may not be. In attempting to satisfy himself of the utility of a new laboratory test that purported to distinguish between those who did and did not have a particular environmental exposure, Young [138] found a stronger correlation between test positivity and a history of exposure than the correlation with actual exposure, even when some of the histories had been fabricated. He concluded that: "Since the test costs at least $350, it is probably wise for surgeons to advise their patients to find a better way to spend their money."

*Conflict of interest* is another form of observer bias. It occurs when special interests of the investigators unintentionally or intentionally compromise observer objectivity. Journal requirements for disclosure of the authors affiliations and financial interests are an attempt to control this bias, or at least make the readers aware of its potential, but this implies that affiliation and money are the only threats to objectivity [45,60,83,103]. Others are much more insidious: power, prestige, position, promotion, social philosophy, and a need to publish being just a few. The best protection against observer bias, whatever the cause, is for the investigators to be aware that it might exist and design it out at the project's inception. This equates to a well thought out protocol, review and approval of the protocol by an appropriate third party, slavish adherence to the approved procedures, and possibly even independent oversight of the conduct of the research. In other words, it requires good epidemiology practices.

Misclassification can even occur after a study has been published! In the Robinson et al. [99] article, silicone breast implants were classified as "disrupted" if, at time of explantation, the device shell was broken or it was simply judged subjectively that an excess of what appeared to be silicone gel surrounded the intact implant. Others, referencing the published paper, erroneously have implied that all the "disrupted" implants were ruptured [19,57]. Familiarity with the literature is the best protection against this form of misclassification bias. For those new to a field or an issue, it may be necessary to go back to the original sources before accepting the conclusions of a literature review or a meta-analysis.

## Sensitivity and Specificity

Consistency (i.e., precision) of measurement, although important, does not ensure the absence of measurement bias, whatever its underlying cause. As illustrated in Figure 11.10, it is possible to be precise and precisely incorrect. What is more important is accuracy (i.e., the validity of the data). The key measures of validity are *sensitivity* and *specificity*. Sensitivity is a measure of how well the test identifies a true condition (disease or exposure). Specificity is a measure of how well it documents a true noncondition (the absence of disease or exposure).

It is actually more complicated than that. A number of tests used in medicine, such as blood cholesterol or antinuclear antibody status, do not clearly separate the normal individuals from those who are abnormal [68,127]. The distribution of values in each group overlap (Figure 11.11). In such situations, the operational diagnostic breakpoint between the two can be somewhat arbitrary. It can be set to identify all the true abnormals (point a; all the true positives) but only by accepting a certain number of false positives—incorrectly labeling some normal indi-

**FIGURE 11.10** Precision and accuracy.

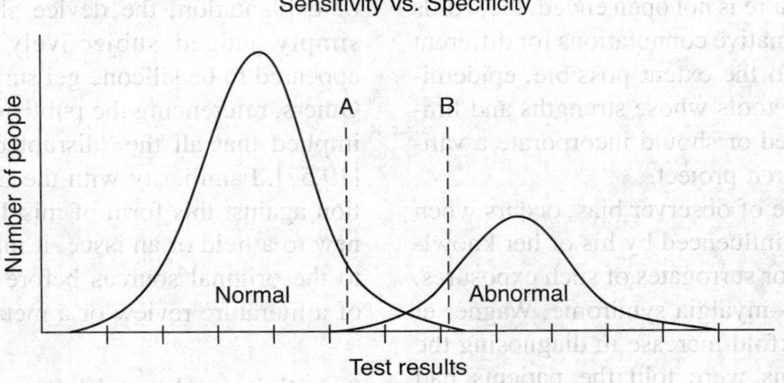


**FIGURE 11.11** Sensitivity vs. specificity.

viduals as abnormal. Or, the breakpoint could be moved to correctly identify all the true normals (point b), but only at the expense of getting more false negatives—misclassifying some of the abnormal patients. In other words, where the breakpoint is set can impact the sensitivity and specificity of a test, and if sensitivity is improved it may mean the specificity has been compromised and *visa versa*. For example, one could arbitrarily declare all chemicals human carcinogens. Such an error on the side of caution would certainly correctly label all the true carcinogens and would guarantee a sensitivity of 100%, but the specificity of such a strategy would be abysmal because the number of false positives would be huge.

Note that, to determine the sensitivity and specificity of a test, its results must be compared to those of a *gold standard*, an accepted test or procedure that reliably determines the presence or absence of the condition. Why then is it necessary to have the new test? Why not just use the gold standard? The new test may be needed because the gold standard is not as useful a tool in the clinical setting. It might be more expensive, inconvenient, invasive, or dangerous.

Paradoxically, data gathered by means of a gold standard actually might have little utility for etiologic research in and by itself but still may be important for the development of tools that can be used in such investigations. For example, explantation (the surgical removal of a medical device such as a breast implant) is the gold standard for implant rupture [19]; however, for both technical and ethical reasons, it can only be used to collect prevalence data. Other noninvasive techniques such as mammography, ultrasound, and magnetic resonance imaging (MRI) can collect incidence data and at lesser risk to the patient, but the relative validity of each can only be established via the gold standard of explantation.

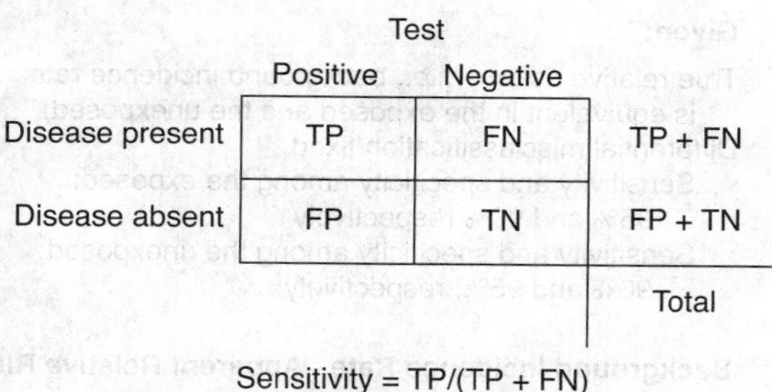


Sensitivity = TP/(TP + FN)
Specificity = TN/(FP + TN)

| | Test Positive | Test Negative | |
|---|---|---|---|
| Disease present | 90 | 10 | 100 |
| Disease absent | 90 | 810 | 900 |
| | | | 1000 |

Sensitivity = 90/100 or 90%
Specificity = 810/900 or 90%

**FIGURE 11.12** Sensitivity and specificity.

Given that there is a suitable gold standard, Figure 11.12 shows how these measures are calculated. In the figure, TP is true positive, FN is false negative, FP is false positive, and TN is true negative. Sensitivity is obtained by dividing TP by (TP + FN) and specificity by dividing TN by (FP + TN). For this particular example, the sensitivity and specificity are both 90%, quite good for most clinical tests [101]. Sensitivity reflects how well, given that the condition is actually present, the test detects the condition. Conversely, specificity is a measure of how well, given that the condition is really absent, the test does not erroneously document its presence.

## Predictive Value Positive and Predictive Value Negative

In real life, whether or not the condition is actually present is unknown before the test is performed. That is the reason for doing the test! For most investigations, what is of greatest interest is the predictive capabilities of a test: how well, given the test result is positive or negative, it respectively predicts the presence or absence of the condition. These measures, *predictive value positive* (PV+) and *predictive value negative* (PV–), can be obtained by making calculations in the vertical axis of the 2×2 table (Figure 11.13). In this example, the PV+ is 50% and the PV– is 98.8%. What this indicates is that, among those who are diagnosed as having a illness on the basis of an abnormal test result, only 50% of them are truly diseased, but among those whose test result was in the normal range 98.8% are actually healthy.

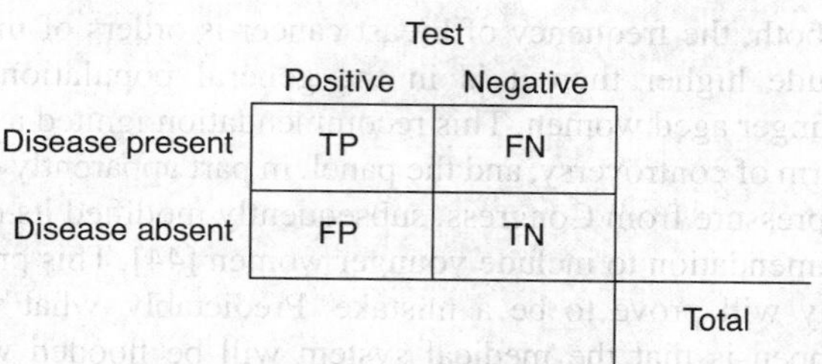


Predictive value positive = TP/(TP + FP)
Predictive value negative = TN/(FN + TN)

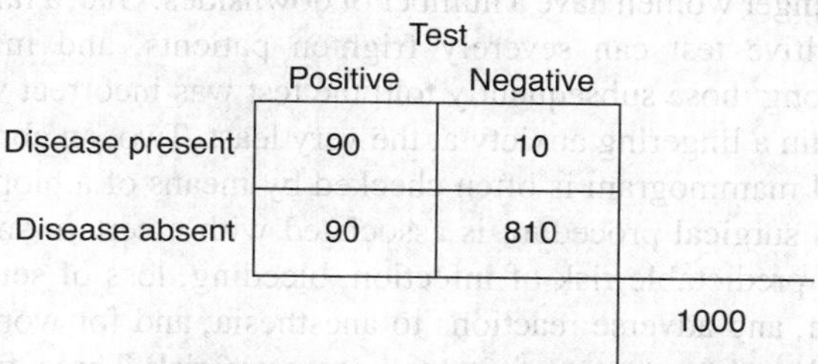


Predictive value positive = 90/180 or 50%
Predictive value negative = 810/820 or 98.8%

**FIGURE 11.13** Predictive value positive and predictive value negative

**Given:**
Sensitivity = 90%
Specificity = 90%

| Disease Frequency | PV+ | PV– |
|---|---|---|
| 1 in 10 | 50% | 98.8% |
| 1 in 100 | 8.3% | 99.9% |
| 1 in 1000 | 0.9% | 99.99% |

**FIGURE 11.14** Predictive values as a function of disease frequency.

The two sets of measures are related but not equivalent. Although the sensitivity and specificity are relatively stable attributes of a test, the predictive values vary widely as a function of the background frequency of the condition being studied. Figure 11.14 illustrates this point. For a given test, the less frequent the condition, the lower the PV+ and the higher the PV–. When the disease frequency drops to 1 in a 1000, the PV+ is less than 1%. In other words, if used as a screening tool the test would label patients as abnormal incorrectly more than 99 times out of 100. This interplay between the underlying validity of a test and the relative frequency of the condition being studied not only impacts epidemiology, but it also has important implications for medical surveillance (and the government regulations that mandate, fund, or otherwise support such procedures) as well as toxicology, in particular as it impacts risk assessment.

A blue-ribbon panel of experts, for example, recently recommended that routine mammography screening be restricted to women over 50 or those in high-risk groups.

In both, the frequency of breast cancer is orders of magnitude higher than it is in the general population of younger aged women. This recommendation ignited a fire storm of controversy, and the panel, in part apparently due to pressure from Congress, subsequently modified its recommendation to include younger women [44]. This probably will prove to be a mistake. Predictably, what will happen is that the medical system will be flooded with false positives [42].

False-positive breast cancer screening tests among younger women have a number of downsides. One, a false-positive test can severely frighten patients, and many among those subsequently told the test was incorrect will retain a lingering anxiety at the very least. Two, an abnormal mammogram is often checked by means of a biopsy; this surgical procedure is associated with a certain small but predictable risk of infection, bleeding, loss of sensation, and adverse reactions to anesthesia, and for women without breast cancer it is an unnecessary risk. Three, false positives put a strain on our medical care delivery system and misdirect limited resources. Four, procedures that produce false results cost money, a financial burden that must be born by the individual patient in the form of direct payments and by society at large in the form of increased insurance premiums and higher taxes. Five, and arguably of greatest importance, some young women after one, two, or more false-positive reports may lose confidence in the procedure. A certain number of these will drop out of the mammography program and never re-enroll. This means that they will not get the screening test later when they would benefit from it most.

The results of toxicology studies are not immune to this problem, in part because their results are routinely extrapolated to humans. If a high dose of an agent is found to cause tumors among rodents, it is current policy to assume that it will cause some form of cancer in humans at lower levels of exposure. Unfortunately, although the sensitivity of toxicology research is quite high (but not perfect), its predictive value positive for extrapolations between different species of rodents is low, on the order of 50% in a study of various chemicals purposefully selected because of their presumed carcinogenicity [73]. Arguably, its predictive value positive is even lower for humans, especially for chemicals being tested simply to satisfy a mandated protocol.

## Nondifferential and Differential Misclassification

Descriptive epidemiology research, such as that done by Robinson and colleagues [99], focuses on a single group. Nonetheless, misclassification obviously can produce erroneous information. In analytic epidemiology, the problem is compounded because data are gathered on and compared between two or more groups. This can lead to errors that are either *nondifferential* or *differential*.

**Given:**

True relative risk = 1 (i.e., background incidence rate is equivalent in the exposed and the unexposed)
Differential misclassification fixed
Sensitivity and specificity among the exposed: 95% and 90% respectively
Sensitivity and specificity among the unexposed: 90% and 95%, respectively

| Background Incidence Rate | Apparent Relative Risk |
|---|---|
| 10 per 100 per year | 1.9 |
| 1 per 100 per year | 5.5 |
| 1 per 1000 per year | 9.2 |

**FIGURE 11.15** Risk estimates in the presence of differential misclassification.

With *nondifferential mismeasurement or misclassification*, error is equivalent across all study groups; for example, an investigator may wish to compare the effects of growth between two groups with different dietary habits. As the health outcome, height might be assessed and recorded to the nearest inch. Even if the measurements were made carefully, two individuals with identical recorded values could easily vary in height by a half inch or more. In spite of the mislabeling, if the measurements were conducted consistently, the rank order of the various study subjects by height, short to tall, would be reasonably accurate. Furthermore, this would be the situation irrespective of the exposure group to which a particular individual might belong. As a consequence, meaningful comparisons could be made between the two groups. What the nondifferential misclassification might do is add a degree of statistical variability to the data and thereby bias the measure of association (the RR or OR) toward the null, to a greater or lesser extent depending on the magnitude of the bias.

With *differential misclassification bias*, the relative invalidity of the data varies by study group. This may give rise to unpredictable shifts of the RR or OR either *toward* or *away from* the null; that is, it may generate marked under- or over-estimates of relative risk depending on the size and direction of the differential error. It can even generate large measures of apparent excess risk where none really exists. In addition, this error is magnified as a function of the underlying frequency of the condition. Figure 11.15 illustrates these points. In this example, the respective sensitivities and specificities were quite high, but a slightly better job of identifying problems was done among the employees in plant A, where there was a better sensitivity but also a concomitant decrement in the measure of specificity. Although the *true relative risk* was 1 (i.e., there was no excess risk in either group), the *apparent relative risks* among the employees in plant A were quite high.

Does this happen in real life? Yes. Dr. Irving Selikoff justifiably has been recognized as one of the icons of modern occupational medicine. In 1968, he and his associates published a paper in the *Journal of the American Medical Association* entitled "Asbestos Exposure, Smoking and Neoplasia," in which they reported an excess risk of lung cancer among insulation workers [110]. Quoting from the article, "A copy of the death certificate was obtained for each of the 94 deaths. In addition, we examined hospital records, postmortem findings (41 cases), as well as the surgical and pathological reports when surgery was performed (39 cases). We also re-examined histologic specimens. It was found that the death certificate was inaccurate in 14 instances." For comparison numbers (i.e., expected deaths by cause), they used "United States 1964 life tables for white males." In their well-meaning attempts to be thorough, they introduced a significant bias into their study because, by using multiple sources to diagnose the exposed and only one source (life tables based on death certificates) to determine cause of death among the unexposed, they compromised their study results with differential misclassification. As a consequence, they overestimated the true relative risk.

Differential misclassification continues to contaminate research and clinical practice. Propelled by a series of case reports in the late 1980s and early 1990s, a cascade of systemic diseases was alleged to have been caused by silicone breast implants [111,114,120]. Many of these theories have been evaluated in case control and cohort studies and found wanting [36,125]. In spite of this, some still argue that women with implants are at higher risk to something called atypical connective tissue disease (ACTD), siliconosis, or systemic silicone-related disease (SSRD) [113,117]. One of the problems with this alleged condition is that no one can define it well enough so it can be rigorously studied to determine whether it occurs uniquely or more frequently among women with breast implants. Investigators who have tried have reported that the condition is essentially the same as fibromyalgia or chronic fatigue syndrome, and the risks appear to be equivalent in women who both do and do not have breast implants [24]. However, those who allege a unique disease does exist and waits to be discovered criticize this work as "studying the wrong disease" [118,139].

To address this presumptive shortcoming, one physician proposed that epidemiologists utilize a set of criteria that he developed to diagnose SSRD in his medical practice [119]. It is based on a series of inclusionary, exclusionary, and relatively nonspecific clinical criteria. By definition, one of the two inclusionary criteria is required to make a diagnosis: either "current or past silicone gel-filled breast implants" or a "local disease" such as capsular contracture or implant rupture. Obviously, neither of the latter could occur unless a woman had an implant. They are therefore surrogates for the exposure of interest, silicone breast implants.

Restricting a diagnosis of SSRD to those with the joint characteristics of exposure and health outcome means that two women with exactly the same signs and symptoms, one with silicone breast implants and the other without, could never be classified as having the same condition. By means of the inclusionary criteria, Dr. Solomon ensured, however inadvertently, that any epidemiology study that used his definition would be biased by essentially a 100% differential misclassification. In theory, any etiologic research based on his definition would produce a spurious elevated relative risk that approached infinity.

Similar biases occur in toxicology research, both nondifferential and differential. The traditional acceptance of tumors, benign and malignant, as a surrogate for cancer is one form of misclassification. In well-conducted studies, it probably is nondifferential, but any time the methods for disease determination differ between the exposed animals and the controls, it could be differential. For example, if more histopathological slides are made or read for the exposed animals than the controls, it is more likely that small occult tumors will be found among the exposed. This is differential misclassification, one that would introduce an over-estimate of risk.

The reader of scientific reports can garner clues regarding the potential for misclassification bias, both nondifferential and differential, from the "Methods and Materials" sections of these articles. The variables, both exposure and disease, should be well defined and equivalent throughout. If they are not, the reasons for the differences should be discussed. The techniques for data collection should be reasonable and applied consistently across all groups. It is important to be particularly vigilant if a study report does not discuss the potential for misclassification bias and how it was addressed. This suggests that the authors were either naïve regarding the problem or chose to ignore it. In either situation, the potential for bias could be high.

## Confounding

A potential *confounder* is a determinant for the disease in question, an alternative "cause" whose influence may confound or confuse the results of an epidemiology study. It can either be an agent itself or a surrogate for that agent. Age, for example, is a surrogate for a constellation of biological, environmental, and social factors that individually and in aggregate are associated with increased risks to certain diseases. The same can be said for race and ethnic background. One of the two major characteristics of a confounder is that it is itself a "cause" (i.e., a predictor) for the disease under study. Different agents have different effects. None is a universal confounder.

A confounder is not the same as an *effect modifier*, although an agent, depending on the study, can be one, both, or neither. Effect modification produces a nonuniformity of effect across various levels of the effect modifier [104]; for example, the consequences of exposure to pathogenic organisms varies by immunization status.

In addition to being an alternative cause, the other major attribute of a confounder is that it must be unequally distributed across study groups; that is, *confounding* occurs only when a determinant of the outcome of interest (a confounder) is unequally distributed among the exposed and the unexposed in a cohort study or among the diseased and the nondiseased in a case control study. As with the biases related to selection and misclassification, the degree of differential distribution determines the direction and magnitude of the error. In addition, the relative potency of the confounder can also, to a greater or lesser extent, influence the apparent relative risk or odds ratio.

Cigarette smoking, for example, is one of the major determinants of lung cancer. Any epidemiology study investigating the carcinogenic potential of a particular agent *vis-à-vis* lung cancer has to take this into consideration and has to control for smoking. If not, the greater the proportion of the exposed who are or were smokers, the greater will be the overestimate of actual risk. On the other hand, if more controls smoked, the true risk to the putative agent will be underestimated.

Smoking is also associated with mortality due to cardiovascular disease, but not to the degree to which it causes lung cancer. In other words, equivalent amounts of unequal distribution between the two study groups may not have the same impact on the measures of risk for different conditions because the potency, the biological activity of a confounder, varies from disease to disease. With lung cancer, smoking equates to a relative risk of perhaps 10, whereas for cardiovascular mortality the relative risk lies closer to 1.5; for still other diseases, it has a RR that approximates 1 (no effect).

Very few diseases have only one etiology. Even a rare malignancy such as angiosarcoma of the liver has a number of alternative causes aside from vinyl chloride monomer [46]. Agents with high potency are relative easy to discern. It is those with lesser biological activity that are more difficult to identify. An indeterminate number of the latter undoubtedly have not yet been discovered. Theoretically, because all of the causes for the various diseases are unknown, some level of confounding may occur in any epidemiology study (and any toxicology study for that matter). In addition, it is highly likely that there are synergistic and antagonistic actions between various agents, both exogenous and endogenous, further complicating the picture.

Evidence is also growing that indicates that effects seen at high dose may be reversed at low dose, at least for some agents, a phenomenon that makes interpretation of the dose–response curve more challenging [20,21]. It also means that some of the basic assumptions (defaults, if you will) inherent to the current approaches to risk assessment are wrong. As a consequence, policies based on many of the risk assessments that incorporate these defaults may actually negatively impact public health.

In experimental studies, the number of variables is purposefully kept to a minimum and ostensibly all of them are under the control of the investigators. Those who conduct observational studies of humans do not have the same advantages. The number of variables is limited only by life itself. Each participant in an epidemiology study has his or her own unique genetic makeup and own unique pattern of extraneous exposures (e.g., diet, medications, personal habits). Although either or both may be only weak confounders for a particular health outcome under investigation, they may be one reason why epidemiology research, especially any single study, has difficulty in reliably identifying putative agents with lesser biological potency, with true relative risks less than 3 or so [129]. This is because even in the absence of selection and misclassification biases the signal may be swamped by the noise of uncontrolled confounding. In epidemiology, the signal-to-noise ratio is improved via more research, especially more targeted research. As the exposure–disease associations become more focused, the relative risks should increase in size. If they do not, be suspicious of claims of causation. Also be suspicious of etiologic interpretations based on one study unless supporting evidence is available.

## Control of Confounding

Confounding can be addressed through study design or data analysis, or a combination of both. As an example, if smoking is a confounder for a particular disease (i.e., those who smoke get the disease more frequently than those who do not smoke, but those who do not smoke still get the disease), confounding by smoking can be dealt with via a technique called *subject category restriction*—that is, by restricting the study subjects (both those exposed to the putative agent and the controls) to just those who never smoked. This design strategy simplifies the analysis and interpretation of the data, but it also restricts how broadly the results can be extrapolated. If only nonsmokers are studied, the results derived from the sample usually only apply to the larger population of nonsmokers. Comparable information about smokers must come from another study restricted to exposed and unexposed individuals, all of whom smoked.

Alternatively, if a certain number of subjects is being evaluated and it is known that a proportion of those in the exposed and unexposed groups were smokers, controlling for confounding could be attempted at the analysis stage of the research, possibly by means of a *stratified data analysis* whereby different strata of smokers are analyzed and the results combined across strata. With the advent of high-speed computers, ever more sophisticated statistical techniques have been developed to control confounding but most of these incorporate assumptions that may or may not be valid depending upon the circumstances—and because they involve complex calculations within a "black box" it is often impossible for the reader (and perhaps even the investigators) to assess the relative impact of the various assumptions on the results.

*Matching* of potential confounders (e.g., age, race, gender, smoking) is an intuitively attractive way of addressing confounding that combines elements of both study design and data analysis; however, it is not a panacea [104]. Not only may it be difficult to do properly, but it also places certain constraints on the types of information that can be developed. Also, it may lead to overmatching (i.e., to matching on surrogates of exposure or health outcome) [55].

No matter what method is used to prevent or control confounding, decisions about which specific potential confounders might be important must be made at the stage of protocol development, if for no other reason than to ensure that adequate data are collected. Obviously, it would be impossible to control for smoking during the analysis stage of the research if no data concerning cigarette smoking had been collected.

Confounding is not restricted to epidemiology research. It also occurs in toxicology; for example, Hart and associates [65] have explored the impact of food intake in laboratory animals. They noted that animals fed *ad libitum* have poorer health and longevity than those whose diet has been restricted. The total caloric load appears to play a role, but trace contaminants may also be important. As reported recently by Paolini and colleagues [91], most standardized diet formulations used by cancer research laboratories worldwide "contain the well-known mutagenic carcinogenic element manganese at the same level and, in some cases, at an even higher level (up to ninefold) compared to that used to study the carcinogenicity of manganese itself." Obviously, the more animals eat, the higher their caloric load and the higher their dose of this carcinogen; however, the amount ingested could be an unintended consequence of the experiment (e.g., ever larger amounts of the test chemical mixed with the food may make the food less and less palatable). For those experiments in which ingestion varied by dose level of the experimental agent, it is quite possible that the results reflect a measure of confounding and perhaps effect modification. Paolini et al. [91] also summarized a number of problems with using historical controls; for example, "B6C3F$_1$ mice have a higher natural incidence of tumors than humans, and this incidence has also changed over time, increasing in excess of 50% over a period of just 10 years."

Although it is impossible to control for all possible confounders in any single study, the reader of epidemiology reports should determine whether attempts were made to control those factors that likely would have had the greatest impact on the results. As with other types of potential bias, a paper can offer a number of clues as to how well this issue was or was not properly managed. If confounding was ignored or obviously inadequately addressed, be skeptical of the information. Look for confirmation in other work that did try to minimize confounding.

## CHANCE

Within the mantra of selection, misclassification, confounding, chance, and causation, the rubric *chance* covers all things mathematical and statistical and some that are methodological. For example, did the investigators add, subtract, multiply, and divide properly? Were the numbers of subjects consistently the same in the abstract, results, discussion, and tables? With more complex statistical procedures, especially those conducted in the mode of exploratory data analysis, it is possible for even the most seasoned epidemiologist to inadvertently lose part of a dataset or to ignore a key assumption and thereby produce erroneous results. If numbers are inconsistent within a report, do the authors explain why? And, if they do, does the explanation seem appropriate or does it smack of gerrymandering or numerology? If either of the latter, look for confirmation of the results elsewhere. Or look for a correction published as an errata in a subsequent issue of the journal.

Also, determine how the data were aggregated for analysis. Does it make sense, in particular, biological sense? By way of example, in 2001 investigators affiliated with the Food and Drug Administration published a study entitled "Silicone Gel Breast Implant Rupture, Extracapsular Silicone, and Health Status in a Population of Women" [18]. This paper has been represented as demonstrating a causal association between "leaking silicone gel implants" and fibromyalgia [140]. The original article, its conclusions, and subsequent interpretations have been severely criticized for a variety of reasons [12,41].

For the purposes of this discussion, the point of interest is how the investigators chose to make their key comparison. Through the use of explant surgery, they determined the prevalence of implant status among women with breast implants (no women without implants were included in the research). Three categories of exposure were defined: extracapsular rupture (obvious silicone adjacent to or remote from the outside of the tissue capsule that surrounds every implant), intracapsular rupture (silicone outside the medical device but apparently confined inside the tissue capsule), and intact implants. The authors reported a statistically significant excess (OR, 2.8; 95% CI, 1.2–6.3) of self-reported symptoms consistent with fibromyalgia (FM), but only when they compared the complaints of women who had extracapsular rupture with those in the aggregate group who either had intracapsular rupture *or* intact implants.

It was pointed out in subsequent letters to the editor that the strategy made no biological sense. As one critic noted, "…if an association exists between implant status and

FM, one would hypothesize … the true gradient would be: intact < intracapsular rupture < extracapsular rupture." Reformatting the data presented in the paper so ORs not reported in the published paper could be calculated, he noted that the odds ratio between fibromyalgia and ruptured status reached statistical significance largely as a consequence of the way the comparison group was structured [12]:

> The OR between FM and extracapsular rupture (compared to intact devices) is 1.88 and not statistically significant. The OR between FM and intracapsular rupture (compared to intact devices) is 0.50 and also not statistically significant. The OR for any rupture versus intact devices is 0.87 … the largest difference in FM risk is between extracapsular and intracapsular rupture. If a gradient in risk exists, these data seem to suggest a gradient for FM that is: intracapsular rupture < intact < extracapsular rupture … suggesting that intracapsular rupture may protect women against FM!

The term "statistical significance" is used by both epidemiologists and toxicologists. It means that, within some acceptable measure of statistical "wobble," two findings were not equivalent, a measure of association such as the relative risk was different than 1, a trend was found, two variables were highly correlated, etc. It is not the same as biological significance, because it does not speak to the underlying validity of the data. As a consequence, if it is apparent that the dataset in a research study is likely biased by selection, misclassification, or confounding, it may make very little sense to analyze the data or to accept any information resulting from a data analysis.

Statistical significance does not equate to cause and effect, even in situations where the underlying data may be valid. For example, Vojandi et al. [134] published a study in 1992 that compared the results of a large number of tests of immune function among women who had breast implants for more then 10 years with those of a sex- and age-matched control group composed of women who did not have these medical devices. In this study, they identified a number of differences ($p < 0.001$) between the exposed and the unexposed and concluded that "these immunological abnormalities in individuals who underwent silicone breast augmentation indicate a mechanism of tissue injury to these patients causing autoimmune diseases or syndromes."

Their data may have been valid, but their inference regarding breast implants was not. In the study, all of the implanted women had "symptomatology in relation to the musculoskeletal and nervous system" and all the unimplanted women did not. In the context of the 2×2 table, they only collected data for two of the four cells: *a* and *d*, the exposed/diseased and the unexposed/healthy, respectively. As a consequence, they could not disentangle the two variables and determine whether the implants "caused" the disease. They could not calculate a relative risk or odds ratio and therefore could not determine whether women with breast implants were more likely to get disease. At best, what their study could do was to appraise the efficacy of their test battery for differentiating between those who did and did not have disease, irrespective of exposure [27]. But, even for that, the exercise was of little utility because the "disease" was so poorly defined and the battery of tests so broad.

## *p* Values and Confidence Intervals

Increasingly, epidemiologists are moving away from the use of *p values* and toward *confidence intervals* (CI) [47]. Although useful, *p* values can obscure important characteristics of the underlying dataset. By itself, a *p* value less than 0.05 or 0.01 suggests that a finding deviates from the null (e.g., a RR that differs from 1), but not whether the result is higher or lower, nor does it necessarily provide insight regarding statistical power. With confidence intervals, one set of numbers representing the range of values that are consistent with the data observed (e.g., the 95% confidence interval) not only provides an indication of where the point estimate of risk lies relative to the null but also gives the reader a sense of the underlying variability of the data and, therefore, the *statistical power* of the study to detect a problem given one exists. If 1 lies within a 95% CI, it indicates that the finding is not statistically significantly different from 1. If the lower value of a 95% CI is greater than 1, the estimate of risk is statistically significantly elevated. If the upper value is less than 1, it is statistically significantly decreased. Furthermore, the width of a confidence interval is an indication of the power of that study, at least for that particular outcome. If narrow, the power of the study is large; conversely, if wide, the power is low.

It is important to note that a study result may have a wide confidence interval and still be valid. Statistical power and study validity are not equivalent concepts. One addresses precision, the other accuracy. In fact, a result from a small study relatively unbiased by selection, misclassification, and confounding may be more valid than the result from a larger study that has a narrower confidence interval. Although the former may have limited utility in and by itself to support or refute causation as a consequence of its low power, when combined with the results of other studies of comparable quality it may prove to be very valuable. This is the rationale underlying *meta-analysis*.

### Meta-Analysis

Meta-analysis refers to the use of statistical tools to combine the results of different studies. Originally, it was confined to randomized controlled clinical trials, to combining results of multiple small studies of the equivalent design (i.e., those with identical dosing regimens and comparable, well-defined outcomes). It is increasingly being used to aggregate the findings of multiple epidemiology studies, even when their results were derived by means of disparate methods (e.g., cohort and case control studies), the sample sizes varied by orders of magnitude, the categories of exposure differed, and the disease outcomes were similar but not equivalent [11]. Although some decry the use of meta-analysis for this purpose, others view it as an important adjunct to the traditional, more subjective literature review. Done properly, meta-analysis not only promises an aggregate quantitative measure of risk that has a narrower confidence interval than each individual study, but it also facilitates the identification of any studies that may be outliers, perhaps because of various types of technical bias or differences in study design.

Meta-analysis is not the same as *data pooling*. Whereas meta-analysis depends on the research results as obtained from epidemiology reports, pooling refers to the aggregation of the actual raw data from many different studies and the subsequent analysis of this larger, single dataset. Conceptually, pooling has some advantages over meta-analysis, but in practice it also has a number of disadvantages, a major one being access to the data. Unlike meta-analysis, where the results have been distributed publicly via the scientific journals, data are not as readily available. In part, this is because of concerns related to protecting the privacy of individual study subjects and the confidentiality of their data [6].

The validity of a meta-analysis is dependent on the validity of the studies included in the exercise. To address this problem, some have suggested that *a priori* rules must be established with respect to which studies to include or exclude. Unfortunately, these rules may reflect the personal biases of the one doing the meta-analysis. For that reason, a type of sensitivity analysis is arguably a better approach [92]. In this type of analysis, the results of all available studies are first evaluated together and then various combinations are used to better understand how the different methods, number of study subjects, classifications of exposure, or definitions of health outcome may have influenced the calculations. It can even be used to compare and contrast the results of different studies that may have different types of bias and to explore whether potential bias is a likely explanation for why one or just a few of the studies seem to be outliers. If a comprehensive sensitivity analysis is conducted and the results published, readers also have the opportunity to make their own interpretations, something that can be difficult to do with the traditional literature review or even with pooling.

One particular type of bias to which both literature reviews and meta-analyses are particularly susceptible is *publication bias*. Publication bias is a type of selection bias. It refers to the tendency of authors to submit and editors preferentially to accept studies with provocative findings [3,37,74]. This has also been referred to as *positive results bias* and can be exacerbated by a *hot stuff bias* [106]. The publication of "I had a patient like that, too" case reports is an example of the latter. Such a flurry of case reports following the initial announcement of an interesting finding in either a medical journal or the popular press can give undue credibility to hypothesized associations, even if they are not real. A number of different approaches can be used to assess the possibility of publication bias, but the best way to avoid it is to aggressively search for pertinent research reports, including those in the form of dissertations, abstracts, and publications in obscure journals [96].

## Exploratory Data Analysis and Multiple Comparisons Bias

To the general public, all findings of statistical significance have basically the same merit. They either accept them as exact and correct or, when faced with apparent contradictions, become frustrated with science. The late Senator Muskie, following an exhaustive series of federal hearings in which various experts testified about a complex environmental issue, epitomized that frustration when he reportedly said that he sorely wished to meet a one-armed scientist, someone who did not always say, "On the one hand this, but on the other hand that."

Scientific discovery is not a destination. It is a journey with many side trips along the way. It starts with a hypothesis, a theory whose genesis may be any number of things ranging from the subjective (clinical observations which seem unusual for intuitive reasons) to the super quantitative (statistically significant findings derived during *exploratory data analysis* of a large medical dataset, such as the health claim files of a private insurance company or of Medicare/Medicaid). Before these findings can be accepted as even provisional truth, they have to be confirmed by additional research, preferably well-focused hypothesis testing research.

In both hypothesis generating and hypothesis testing exercises, the same statistical tools and the same levels of statistical significance may be used; yet, the findings of the former do not carry the same interpretive weight as those from the latter [115]. That's because the former, in addition to uncontrolled confounding, are subject to a *multiple comparisons bias* [128].

The statistical tests used in health research factor in both a type I and a type II error. A type I is the error of rejecting a null hypothesis, of concluding that a difference exists when, in truth, it does not. By convention, the alpha level (the probability of a type I error) is usually set at 0.05 (which equates to a 95% CI). This means that a certain predictable number of statistically significant findings are incorrect, about 1 in 20. The greater the number of comparisons, the greater is the number of spurious associations that may be found (i.e., the larger is the multiple comparisons bias). Various techniques have been developed to address this bias, the simplest perhaps being the *Bonferroni correction* in which the putative alpha is divided by the total number of comparisons, and the "corrected alpha" is used to determine the presence or absence of statistical significance [84]. For example, if the study alpha level was preset at 0.05 and ten comparisons were made, a Bonferroni-corrected 95% CI would, in essence, be a 99.5% CI.

In many studies in which a large number of comparisons are made, the authors will do a Bonferroni correction or some analogous procedure and report the confidence intervals with and without the adjustment.In others, they they will not, but they will indicate the total number of comparisons and thus allow the reader to develop his or her own opinions about the merit of the findings. In still others, it may be difficult for the reader to recognize the potential for a multiple comparisons bias, especially if investigators practice surreptitious data dredging—engage in exploratory statistical analyses of large and diverse datasets but selectively report only those results which support their own pet theories [88,109,128]. Because few comparisons are presented, the reader is given the erroneous impression that only those few were considered and therefore they must have been of some *a priori* concern. This approach can be particular attractive to quasi-scientific advocacy groups who recognize the publicity value of a statistically significant cluster.

## *Post Hoc* Reasoning

The latter is but one of a number of variations on the theme of purposefully biased science [72,87]. In another, the investigators simply scan a dataset and determine which hypotheses they wish to test. Or, they may gerrymander the dataset and thereby construct an artificial cluster. In either case, by having foreknowledge of what the cluster is and where it is located in the dataset, the investigators can reduce the total number of actual statistical procedures and, therefore, even with "overly conservative" corrections for multiple comparisons, claim to have refuted the null hypothesis. The nefarious may even point to a hypothesis in a protocol that predated the formal statistical analysis. Although the work seems to fit the scientific method, giving the results an aura of biological credibility, the findings are a product of *post hoc reasoning*. They are worthless. Using this approach, statistically significant clusters can even be generated from a table of random numbers.

Investigators who are guilty of *post hoc* reasoning are sometimes derisively called *Texas sharpshooters* [62]. In most target shooting, one shoots at a bull's eye. The Texas sharpshooter first shoots at the side of the barn (perhaps from very close to the building) and then draws the bull's eye around the holes. By doing so, he claims his marksmanship is both precise and accurate.

You seem to be in fine health
but let's run a few tests.
I'm sure we can find something wrong with you.

**FIGURE 11.16** Multiple comparisons bias in clinical medicine.

If clusters of disease are the catalyst for an epidemiology study, they can introduce another form of self-fulfilling reasoning into the research. It occurs when an investigator stumbles upon a cluster of disease, perhaps in an occupational group, and then uses the cluster to develop a hypothesis about one or more of the chemicals to which the group was exposed and also to test this hypothesis; that is, the cluster is incorporated into any subsequent analytic research. If the disease is rare, it is quite possible that an elevated relative risk will be found in the formal epidemiology study even if no new cases are discovered in the expanded cohort. Although the additional research in this situation may be designed, initiated, and conducted after the theory was developed, it will not be an independent test of the hypothesis [3,26,48].

In summary, even the most precise results may be wrong, a consequence of simple mathematical errors, technical bias, or less innocent intent. Although exploratory data analysis is a valuable tool, more is not always better. This maxim applies equally well to epidemiology, toxicology and clinical medicine (Figure 11.16). To be interpreted properly, the results of tests must be put in the context of the size of the dataset, the number of tests that were performed, the body of information that is already available, and, if possible, the mind-set of the investigators at the inception of the research. The latter may be obvious from the introduction of the paper or from the protocol, but sometimes it can only be surmised.

## Causation

Even when selection, misclassification, and confounding are minimal, identification of the causes of human disease is not simply an exercise of calculating which exposure–disease associations are statistically significant. It is a thoughtful process based on the preponderance of evidence and a logical ordering of that information. Sir Bradford Hill, a British statistician/epidemiologist, presented his criteria for determining causation in the mid-1960s and subsequently refined them for his textbook [69]. They are still in wide use. In interpreting data, he noted that an

**TABLE 11.2**
**Hill Criteria for Causation**

| |
|---|
| Strength of the association |
| Consistency |
| Specificity |
| Temporal relationship |
| Biological gradient |
| Biological plausibility |
| Coherence of the evidence |
| Experiment |
| Reasoning by analogy |

investigator must deal with two basic problems: *significance* (the statistical reliability of a finding) and *inference* (the interpretations one might make from such a finding). With the former, he cautioned against either over- or under-interpreting the importance of statistical significance—noting that, if absent, "chance is a not unlikely reason" for an apparent difference, for an apparent association, or for an apparent elevated relative risk, but, if present, "chance is still a possible, though unlikely, explanation." He also advised that conclusions related to a new finding have to be "more guarded" and stress the "limitations" of the data (size of the sample, potential for bias, etc.). As for inference, he offered nine criteria for differentiating between "causation or merely association" when faced "with a clear and significant association between some form of sickness and some feature of the environment" (Table 11.2).

His first criterion was *strength of the association*—in other words, the size of the relative risk or odds ratio. Obviously, not every statistically significant relative risk is meaningful, but the larger the number, the less likely any observed association is simply the result of random error or the consequence of selection, misclassification, and confounding. The question is, "How large is large enough?" For isolated findings, seasoned epidemiologists are reluctant to accept relative risks of less than 3 or 4 [129].

Sir Bradford's second criterion was *consistency*, the finding of similar relative risks for the same condition and exposure in different epidemiology studies conducted by different investigators on different groups of participants. In part, this is important because it is unlikely that the equivalent errors would be replicated in all the studies; therefore, a finding that is consistent across many studies is more likely true. It logically follows that a summary measure of risk as derived from consistent findings will more likely reflect the underlying biological truth than the results of any single study. As mentioned earlier, meta-analysis provides such a summary measure. It is a way of teasing out a signal from the cacophony of noise that is inherent to epidemiology. If statistically significant, the findings of multiple small studies may be biologically important, but also meaningful can be the absence of elevated risks in study after study after study, or as de Grasse Tyson has emphasized, "Null results matter, too" [34]. Although it is theoretically impossible to prove the negative, when multiple studies fail to identify an association between disease and a particular exposure pragmatic scientists conclude proof of causation is lacking and move on.

As his third criterion, he offered *specificity*, elevated risks to a single or small number of well-defined health problems. When many disparate conditions are attributed to an agent, at some point it becomes questionable whether any of them are a likely consequence of exposure. The need for specificity also applies to the disease itself. No meaningful body of etiologic research can be conducted to determine if a condition occurs more frequently among the exposed if the "disease" cannot be defined because, perforce, each individual study would be evaluating a different outcome. The same holds for exposure. Although the initial stages of investigation may incorporate broader categories of disease such as "pulmonary disease" and mixtures of chemicals, knowledge comes with focus.

Sir Bradford's fourth criterion dealt with the *temporal relationship* of the exposure and the disease or, as he put it, "Which is the cart and which is the horse?" In cross-sectional or prevalence research, it is often impossible to make this determination. Conditions with long latency or those whose signs and symptoms wax and wane over time can further complicate the picture [131]. Nonetheless, if the condition occurs before the exposure, it cannot have been caused by the exposure.

His fifth criterion was *biological gradient*; that is, if small doses cause harm, do larger doses cause greater harm? Parenthetically, this is not a variation of the assumption inherent to quantitative risk assessment (i.e., if large doses are associated with health problems, lesser doses cause lesser problems) [20,21]. Something akin to linear extrapolation back through zero exposure must be assumed for the latter. Such an assumption is not required for the former.

*Biological plausibility* was presented as a sixth criterion. This he implicitly categorized as one of the lesser tier of criteria because "what is biologically plausible depends on the biological knowledge of the day." Some consider this necessary to prove causation; that is, the underlying mechanisms of action must be understood before cause and effect can be accepted. For many, it is too stringent a requirement. They are satisfied if a meaningful association is found for a risk factor even if the exact causal agent and the process by which it works is unknown. In a sense, biological plausibility also is a lesser criterion because it is subordinate to other criteria. For example, a biologically plausible explanation for a disease excess is meaningless if there is no disease excess.

The seventh criteria addressed the *coherence of the evidence*, the amalgamation of what is known concerning the natural history and biology of the disease, the presumptive actions of the etiologic agent, the results of experimental research on animals, and the contributions of other types of information. The evidence can come from within a single study or across studies from many different disciplines. Cigarette smoking, for example, is associated with both an increase in lung cancer and an excess risk for a constellation of other diseases, in part because smoke is a mixture of noxious agents. Although lung cancer may be the outcome of interest in a particular study (say, one evaluating the impact of low levels of smoking), an increase in both lung cancer and the other pertinent diseases would add coherence to any evidence of harm. As for multidisciplinary evidence, the decrease of mammary tumors among methyl-nitroso-urea-exposed animals implanted with silicone-gel-filled devices adds credibility to the epidemiology findings of lower breast cancer risk among women with silicone breast implants [126].

The next attribute was *experiment*, but not necessarily in the context of a laboratory experiment. He also considered the removal of the presumptive etiologic agent a type of experiment. If a problem resolves following such removal, it may provide support for cause and effect, but even this is not absolute proof. Diseases wax and wane. If the putative exposure is removed at the apex of disease severity, resolution may take place coincidentally, and the condition, in the absence of exposure, may return at a later date. If that were the case, it would suggest that the original "experiment" was incomplete and therefore lent fallacious support to conclusions about cause and effect. To eliminate this possibility, there must be adequate follow-up of the patients following removal of the putative agent. Even then, a number of other things can confound such experiments. Humans react to subliminal clues, exhibiting both placebo and nocebo effects, and these can present as either subjective symptoms or more objective signs of disease [123]. Resolution of a condition may be related to concomitant treatment. Alternatively, its original presentation and subsequent resolution can be due to malingering [112].

Sir Bradford's ninth and final criterion was *reasoning by analogy*; that is, if agent *X* can cause disease *Y* perhaps a material similar to *X* can cause a disease comparable to *Y*. Some have argued that because new environmental immunologically mediated diseases such as eosinophilia myalgia secondary to L-tryptophan exposure are still being identified, it is possible (they imply probable) that silicone is also associated with a new disease. Eosinophilia myalgia has a relatively short latency, however, and it has a characteristic clinical presentation. Both of these attributes are missing with silicone. If there is an epidemic of a unique autoimmune disease caused by silicone, it has not yet been discovered. If this is because it is a disease of long latency and therefore the epidemic has not yet occurred, one has to ask the question, "What then is the basis for the legal controversy?"

## Legal Causation

At one time, courts tended to disregard epidemiology as simply a statistical exercise that provided information of little probative value; however, within the last 10 to 15 years, it has become key to the legal theory of causation as used in the particular type of litigation that deals with tort or product liability [10]. Epidemiology research helps establish not only whether an agent is causally associated with a particular disease but also whether the association supports a finding of "*more likely than not*." This equates to an attributable risk percent (AR%) of greater than 50% and, with knowledge of the relative risk (RR), can be calculated with the following formula: AR% = (RR – 1)/RR. For example, a relative risk of 3 would equate to an attributable risk percent of 67%.

As mentioned earlier, the various calculations regarding attributable risk have no meaning until causation for human disease is established, until an acceptable number of the Hill criteria have been satisfied. In theory, therefore, an exposure–disease association has four characteristics that must be demonstrated before a claim of causation logically can be accepted in legal deliberations: (1) The putative agent must be a known cause of the disease; (2) the causal relationship must be more likely than not; (3) the plaintiff must have been exposed to the agent in adequate quantity and for sufficient duration; and (4) the plaintiff must have developed the appropriate disease after the exposure. The first two deal with *general causation*. The last two pertain to *specific causation*. In tort liability cases, the plaintiff has the burden to prove all four, at least in theory. Trials are emotional events and jury deliberations can sometimes be more influenced by the subjective than the objective.

Prior to the 1993 Daubert decision, juries were the triers of fact and judges basically functioned as the umpires of the proceedings [33,50,95]. They made rulings regarding process but few about content. The Daubert case changed that [33]. After a series of appeals that went all the way to the Supreme Court, judges were given the additional responsibility of serving as "gatekeepers." Juries retained the role of triers of fact, but judges were charged with determining which body of facts were relevant and reliable vs. which were simply junk science—which testimony would assist the jury in their deliberations, and "whether the 'probative value' of the testimony substantially outweighed the risks of prejudice, confusion or wasted time" [71]. In practice, this means federal judges now must decide which expert witnesses can and cannot testify and what opinions they will be permitted to convey to the jury. Many state courts are also moving toward a process based on the Daubert principles.

Some courts have done an impressive job in rendering judgments that have included sophisticated legal arguments well infused with scientific principles [86]. Others have accomplished the same result with the help of outside experts employed directly by the court, an option acknowledged in the Daubert decision [64]. Still others at the state level have yet to apply the Daubert principles, in part because some judges feel uncomfortable with their new role and in part because the new rules technically pertain to just the federal judiciary [38].

Lawyers and judges, even at the federal level, are still exploring the limits of the gatekeeper function and how certain statistical and epidemiologic thought might be translated into legal concepts. As an example, statistical significance means that a finding has a lower confidence limit above 1; that is, there is some assurance that the estimate of risk is different than 1. The legal notion of "more likely than not" requires a relative risk above 2, but it is unclear whether the key finding, to be admissible, has to be statistically significantly different than 1 or statistically significantly different than 2. When there is just one or a limited number of epidemiology studies, the latter makes more sense; however, the former is not inconsistent with epidemiology opinion when a large number of reasonably valid studies have similar results.

### Clinical Causation

Neither epidemiology causation (what Sir Bradford Hill called "medical causation") nor legal causation should be confused with *clinical causation*. The primary goal of clinical medicine is diagnosis and treatment. In a sense, the major reason for a diagnosis is to predict which treatment will most successfully reverse, eliminate, or control a patient's troublesome symptoms or signs of pathology. If the diagnosis is correct, the resulting treatment works and the patient is well served. If not, the patient likely gets no better, possible may get worse, or even may develop additional adverse outcomes as a result of the inappropriate therapy.

Experienced clinicians are adept at the technique of *differential diagnosis*. Through the use of various signs, symptoms, and test results, and factoring in the risks inherent to alternative treatments, they identify the most probable diagnoses, weigh the merits of each, and use the resultant information to help select a treatment that likely will be most successful. If that particular treatment does not work, they move on to the next most likely diagnosis and a different treatment and, if that does not work, to still another, continually balancing benefit and risk.

When clinicians speak of searching for the "cause" of a patient's problems, they usually are referring to identifying the most likely diagnosis, quite possibly one whose underlying mechanisms of action are unknown. Arguably, knowledge regarding the underlying cause of a particular disease is only important in the clinical setting if it materially impacts treatment decisions—for example, if a specific type of bacterial pneumonia is more efficaciously treated by a particular antibiotic—and the underlying causes are not initially discovered by the process of differential diagnosis. Such knowledge is derived from experimental animal research, controlled clinical investigations, and observational epidemiology studies. Contrary to what some physicians have asserted, differential diagnosis, no matter how sophisticated, does not obviate the need for etiologic research [2,56,61,97]. As the many programs of the National Institutes of Health demonstrate, research regarding cause and effect and that related to diagnosis and treatment are complementary but not equivalent.

Parenthetically, proper diagnoses are made by means of pattern recognition, by what Margolis has called "habits of the mind" [82]. Within the context of clinical causation, this has a number of implications. One, the more extensive a physician's training and experience, the larger the number of mental templates he acquires against which he can compare the next patient's combination of signs, symptoms, and test results; thus, even if the underlying etiology for a condition is unknown (i.e., the condition is idiopathic), a physician may develop successful strategies for treating the syndrome. Two, this knowledge, no matter how prodigious, is always finite. Physicians recognize this. They specialize so they might concentrate their energy on developing in-depth knowledge within one sector of medical practice, and even within that specialty they refer patients to their peers, a tacit acknowledgment that another physician may be better suited to diagnose and treat a particular individual. Three, because the number of templates increases as a direct result of experience, the more seasoned the clinician, the greater the clinician's ability to diagnose and, paradoxically, the greater the potential for a multiple comparisons bias. The latter is reflected in case reports.

## CONCLUSION

As de Grasse Tyson noted in his recent essay, *Certain Uncertainties*, "The frontier of science is a messy place" [35]. As a consequence, to the uninitiated science appears to provide contradictory and therefore unreliable findings, regardless of whether the research is experimental, quasi-experimental, or observational, but perhaps more so for the latter (Figure 11.17). Part of the reason for the apparent inconsistencies is related to technical bias (selection, misclassification, and confounding), but part is due to over-interpretation of the findings of any single study, either by the study investigators or by the consumers of research reports.

One of the primary goals of any scientist should be the elimination of bias from his or her research. The first step is to acknowledge that various types of bias exist, the second is to understand how they occur, and the third is to develop methods and procedures to avoid, minimize,

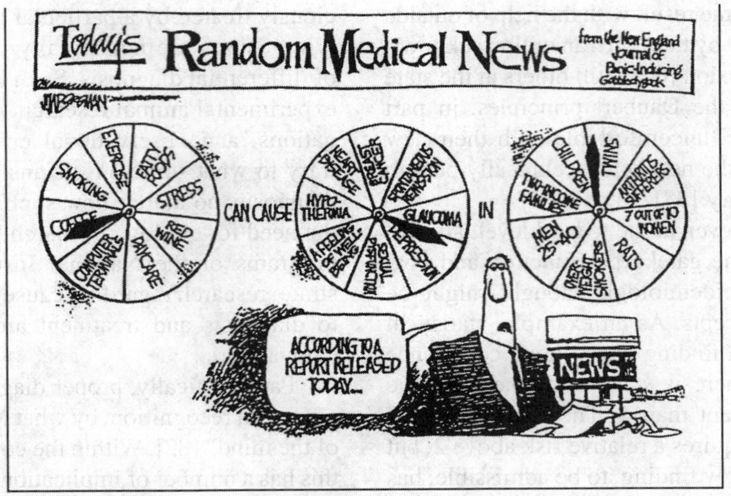


**BY JIM BORGMAN**

**FIGURE 11.17** Today's random medical news. (Copyright 1997, *The Cincinnati Enquirer*. Reprinted with permission of King Features Syndicate.)

or control bias. Over the years, well-trained epidemiologists have found ways to address potential error and improve the validity of their research. The same can be said for toxicologists and clinical investigators. The scientific method has been core to all of these endeavors.

The scientific method is one of the major discoveries in human history [17,133]. It has allowed mankind to gain a more objective view of the universe, to better understand the workings of the atom, and to successfully identify the causes of disease and with that understanding to treat, control, and even eliminate some of the major scourges that once were endemic worldwide. Unfortunately, the scientific method can be laborious, inconvenient, time consuming, and expensive. The temptation to take shortcuts can be great, but history teaches that shortcuts often just lead to further confusion [105]. Identifying the truth can be difficult enough even in the best of circumstances, but it is impossible with biased data, inappropriate methods, or muddled logic.

One of the responsibilities of the technical journals is to screen research papers and determine which have sufficient rigor in data, methods, and interpretation to warrant publication; however, scientific investigation is a human endeavor and peer review an imperfect process, so flawed studies still get published, even in the best of journals. For that reason, the ultimate judgment regarding the value of any single report or group of reports may have to be made by the consumer: the epidemiologist, toxicologist, physician, lawyer, judge, newspaper reporter, or other member of the general public less well versed in the scientific method. This chapter provides a conceptual framework whereby such winnowing of fact from fancy might be accomplished. Within a basic understanding of epidemiology data, measures of association and methods, it is based on the mantra of selection, misclassification, confounding, chance, and causation.

## QUESTIONS

### Exercise One

Among 50 employees of a chemical company producing ethylmethyl chicken wire, the company physician identified 2 last year with lung cancer. In the general population the incidence rate of lung cancer is 5 per 1000 per year What is the relative risk for lung cancer among the chemical company employees? Does this finding support the legal concept of "general causation"? What about "medical causation" as defined by Sir Bradford Hill?

*Answer*

Based on the information provided, the presumptive incidence rate for lung cancer among the employees was 2 per 50 per year, which is the same as 4 per 100 or 40 pe 1000 per year. Because the incidence rate in the genera population is 5 per 1000 per year, the apparent relative risk is 40 divided by 5, or 8.0.

To support the legal concept of "general causation," two conditions must be satisfied. One, the putative agent must be a known cause of the condition. Two, the association must be "more likely than not." In other words, the relative risk must be above 2 and statistically significant In this situation, we do not know whether ethylmethy chicken wire is generally accepted as a known cause fo lung cancer, in part because we have no knowledge of the results of any other experimental or observational researc on this chemical. Furthermore, we do not know whethe

the finding in this particular study is statistically significantly greater than 2 or even whether it is significantly above 1. Calculating the 95% confidence interval could be done, but, with due consideration for the small number of cases that were observed, it is quite possible that this was a chance occurrence.

General causation aside, even if this had been the first report of lung cancer associated with the chemical, the data provided are insufficient to conclude that ethylmethyl chicken wire should be considered a potential risk factor for this type of malignancy. For example, we do not know how the two diagnoses of lung cancer were made. Were they based on x-ray or confirmed by biopsy? The former is much less likely to be correct than the latter. We also do not know whether the two lung cancers developed before or after first exposure. Although determining the exact incident time of a malignancy is usually impossible, if the company had been producing the chemical for just a short time or the two employees had just recently been hired, the known latency of lung cancer would suggest that the condition predated any possible putative exposure. Moreover, who made the original diagnosis and when? If the diagnoses had been made earlier by the employees' personal health providers and only "identified" later by the company physician during the course of a routine clinic visit, then the cancers could have been *prevalence* and not *incidence* cases.

In addition to more information about those in the numerator, what about the denominator? Did this company have only 50 employees? If there were considerably more and only 50 had been seen at the company's medical clinic, it is quite possible that a selection bias could have artificially inflated the incidence rate for lung cancer. If such were the case, the relative risk was seriously elevated.

Furthermore, to calculate the incidence rate for the plant population, we had to assume that the two who developed the condition, given they were diagnosed accurately, actually were exposed. If they worked in a part of the company remote from the production facilities (e.g., in accounting or sales), then an incorrect assumption that they were exposed could have introduced a *misclassification bias* into the calculations, one that resulted in another spurious elevation in relative risk.

Finally, the comparison incidence rate, the absolute risk among the unexposed (5 per 1000 per year) was that of the general population. That means that there were other causes for lung cancer, such as smoking. Furthermore, the term "general population" suggests that the rate was based on the experience of both men and women. The two genders have distinctly different incidence rates of lung cancer. If the distribution of either the attribute of gender or smoking was not similar in both the exposed and controls, then *confounding* could have biased the results.

### Exercise Two

There are two hospitals in the same city: a smaller one with 30 births a month and a larger one with 300 births a month. At the end of the year, which hospital is likely to have experienced more months with more than 60% male births?

*Answer*

The sample size is smaller for the smaller hospital. As a consequence, the variability of the data is greater; thus, there is a greater chance that the smaller hospital will have more months with more than 60% male births.

### Exercise Three

In a study of chemicals associated with site-specific neoplasia in rodents, Huff and colleagues [73] reported that "25 chemicals were carcinogenic to the liver in both rats and mice, 9 chemicals caused liver cancer only in rats, 53 caused liver cancer only in mice, and in 226/313 studies no chemically related liver tumors were observed in either rats or mice." They also stated that "the overall interspecies concordance in liver carcinogenicity is 80% (251/313)." That means that, if the mouse bioassay were used as a screening test of the carcinogenic potential of chemicals for rats, then it would have a sensitivity of 73.5% and a specificity of 81.0%. In this example, what were the predictive value positive and predictive value negative of the mouse bioassay?

*Answer*

The easiest way to solve this problem is to set up a 2 × 2 table similar to that found in Figure 11.13. If the outcome in rats is the "truth" that we wish to predict, then there were 25 true positives (TP), 9 false negatives (FN), 53 false positives (FP), and 226 true negatives (TN). The *predictive value positive* was TP/(TP + FP), or 25/78; in other words, less than a third (32.1%) of the chemicals that were positive in the mouse bioassay were actually carcinogenic to rats. The *predictive value negative* was TN/(TN + FN), or 226/235; the mouse bioassay correctly predicted which chemicals would not cause liver tumors in rats 96.2% of the time.

### Exercise Four

In Exercise Three, 34 of the 313 chemicals tested caused liver tumors in rats (about 1 in 10). If, among additional chemicals to be tested randomly, only 1 in 100 would actually cause liver tumors in humans, how predictive would be the results of the mouse bioassay? Calculate the predictive value positive, the predictive value negative, and the concordance. Assume the same level of sensitivity and specificity for the mouse bioassay as found with rats. For convenience, also assume that 3400 chemicals were tested.

*Answer*

Among the 3400 chemicals tested, 34 are actually human carcinogens. With a sensitivity of 73.5%, that would mean that 25 would be true positives (TP) and 9 would be false negatives (FN). With a specificity of 81% among the remaining 3366 chemicals, approximately 2726 would be true negatives (TN) and 640 would be false positives (FP). Using these numbers, a 2 × 2 table can be set up and the predictive values and concordance calculated. The *predictive value positive* would be TP/(TP + FP), or 25/665 (3.8%); of the 665 positive tests in mice, less than 4% of them would correctly predict what would happen in humans. The *predictive value negative* would be TN/(TN + FN), or 2726/2735 (99.7%); in other words, of the 2735 negative mouse bioassay studies, 99.7% would correctly predict that the chemicals would not produce cancer in humans. Concordance would be (TP + TN)/(TP + FN + TN + FP), or 2751/3400 (80.9%), basically the same as that found by Huff for his interspecies study.

## REFERENCES


1. *Allison v. McGhan Medical Corp*. (No. 1:93-CV-2051-RLV) (N.D. Ga. November 3, 1998).
2. ASPRS/PSEF. (1996): *American Society of Plastic and Reconstructive Surgery and the Plastic Surgery Education 1996 Combined Roster*. ASPRS/PSEF, Arlington Heights, IL.
3. Anderson, B. (1990): *Methodological Errors in Medical Research: An Incomplete Catalog*. Blackwell Scientific, Oxford.
4. Angell, M. (1997): *Science on Trial: The Clash of Medical Evidence and the Law in the Breast Implant Case*. W.W. Norton, New York.
5. Anon. (1997): Informed consent litigation could severely hamper epidemiologic research. *Epidemiol. Monitor*, 18(8):1–3.
6. Anon. (1999): OMB explains how it intends to implement new requirements for release of research data collected under federal grant dollars. *Epidemiol. Monitor*, 20(3):7–10.
7. Anon. (1997): Pharmacoepidemiologists moving to protect access to medical record information. *Epidemiology Monitor*, 18(6):1–3.
8. Austin, D.F. (1979): Preliminary Report, Cancer Incidence Rates, Industrial and Non-Industrial Areas of Contra Costa County (unpublished). California Department of Health Services, Emeryville, CA.
9. Bender, A. P., Williams, A. N., Johnson, R. A., and Jagger, H. G. (1990): Appropriate publish health responses to clusters: the art of being responsibly responsive. *Am. J. Epidemiol*., 132:S48-52.
10. Black, B. (1990): Matching evidence about clustered health events with tort law requirements. *Am. J. Epidemiol*., 132:S79–S86.
11. Blair, A., Burg, J., Foran, J., Gibb, H., Greenland, S., Morris, Raabe, G., Savitz, D., Teta, J., Wartenberg, D., Wong, O., and Zimmerman, R. (1995): Guidelines for application of meta-analysis in environmental epidemiology. *Regul. Toxicol. Pharmacol*., 22:189–197.
12. Bowlin, S. J. (2001): Silicone gel breast implants (letter to the editor). *J. Rheumatol*., 28:2760–2761.
13. Brand, K. G., Johnson, K. H., and Buoen, L. C. (1976): Foreign body tumorigenesis. *CRC Crit. Rev. Toxicol*., 4:353–394.
14. Brandon, H. J., Young, V. L., Jerina, K. L., Wolf, C., and Schorr, M. W. (1997): Diagnosis of breast implant failure mechanisms. In: *Proc. of the 13th European Conf. on Biomaterials*, Goteborg, Sweden, Sept. 4–7, 1997.
15. Brandon, H. J., Young, V. L., Wolf, C., and Jerina, K. L. (1997): Long-term material stability of explanted breast implants. In: *Proc. of the 66th Annual Scientific Meeting of the ASPRS Plastic Surgery Forum*, Vol. XX, pp. 215–216.
16. Brinton, L. A., Malone, K. E., Coates, R. J., Schoenberg, J. B., Swanson, C. A., Daling, J. R., and Stanford, J. L. (1996): Breast enlargement and reduction: results from a breast cancer case-control study. *Plast. Reconstr. Surg*., 97:269–275.
17. Bronowski, J. (1956): *Science and Human Values*. Harper & Row, New York, pp. 33–35.
18. Brown, S. L. Pennello, G., Berg, W. A., Soo, M. S., and Middleton, M. S. (2001): Silicone gel breast implant rupture, extracapsular silicone, and health status in a population of women. *J. Rheumatol*., 28:996–1003.
19. Brown, S. L., Silverman, B. G., and Berg, W. A. (1997): Rupture of silicone-gel breast implants: causes, sequelae and diagnosis. *Lancet*, 350:1531–1537.
20. Calabrese, E. J. and Baldwin, L. A. (1997): A quantitatively based methodology for the evaluation of chemical hormesis. *Hum. Ecol. Risk Asses*., 3:545–554.
21. Calabrese, E. J. and Blain R. (2005): The occurrence of hormetic dose responses in the toxicological literature, the hormesis database: an overview. *Toxicol. Appl. Pharmacol*., 202:289–301.
22. Checkoway, H., Pearce, N., and Crawford-Brown, D (1989): *Research Methods in Occupational Epidemiology* Oxford University Press, New York.
23. Cheevers, J. (1981): CC industry may not be cancer culprit *Contra Costa Times*, October 20, 1981.
24. Chow, H. Y., Cash, J. M., Calabrese, H. H., and Wilke, W S. (1996): Patients with chronic fatigue syndrome (CFS and silicone-associated disease (SAI) are similarly disabled. *Arthritis Rheum*., 38(Suppl. 9):S52.
25. Collis, N. and Sharpe, D. T. (1998): Rupture of silicone gel breast implants. *Lancet*, 351:520.
26. Cook, R. R. (1981): Dioxin, chloracne and soft tissue sarcoma. *Lancet*, 1:618–619.
27. Cook, R. R. (1993): But is it significant? *Ann. Plast Surg*. 31:94–95.
28. Cook, R. R. (1994): Sclerodermalike esophageal disease in children breast-fed by mothers with silicone breast implants. *JAMA*, 272:767–768.
29. Cook, R. R., Curtis, J. M., Perkins, L. L., and Hoshaw, S J. (1998): Rupture of silicone-gel breast implants. *Lancet* 351:520–521.
30. Cook, R. R., Hoshaw, S. J., and Perkins, L. L. (1998): Failure of silicone gel breast implants: analysis of literature data fo 1652 explanted prostheses. *Plast. Reconstr. Surg*., 101:1162
31. Cook, R. R., Hoshaw, S. J., and Perkins, L. L. (1999) Failure of silicone gel breast implants. *Plast. Reconstr* 103:1091–1092.

32. Cook, R. R., Tirey, S. L., Spadacene, N. W., and Woodbury, M. (1994): Access to data for epidemiological studies. In: *Environmental Epidemiology: Effects of Environmental Chemicals on Human Health*, edited by W. M. Draper. American Chemical Society, Washington, D.C., pp. 231–244.
33. *Daubert v. Merrell Dow Pharmaceuticals, Inc.*, 509 U.S. 579, 1993.
34. de Grasse Tyson, N. (1998): Belly up to the error bar. *Natural History*, 11:70–74.
35. de Grasse Tyson, N. (1998): Certain uncertainties. *Natural History*, 10:86–88.
36. Diamond, B. A., Hulka, B. S., Kerkvliet, N. I., and Tugwell, P. (1998): *Silicone Breast Implants in Relation to Connective Tissue Diseases and Immunologic Dysfunction: A Report by a National Science Panel to the Honorable Sam C. Pointer, Jr., Coordinating Judge for the Federal Breast Implant Multi-District Litigation. In re: Silicone Breast Implants Products Liability Litigation* (MDL 926) (No. CV 92-10000-S) (N.D. Ala. November 17, 1998).
37. Dickersin, K. (1990): The existence of publication bias and risk factors for its occurrence. *JAMA*, 263:1385–1389.
38. *Dow Chemical Company v. Mahlum*, 970 P.2d 98 (Nev. Supreme Court, 1998).
39. Editors. (1994): Correction: incorrect study design in abstract. *JAMA*, 272:770.
40. Edwards, S. L., Slattery, M. L., and Ma, K. (1998): Measurement errors stemming from non-respondents present at in-person interviews. *Ann. Epidemiol.*, 8:272–277.
41. Ehrlich, G. E. (2001): Silicone gel breast implants (letter to the editor). *J. Rheumatol.*, 28:2760.
42. Elmore, J. G., Barton, M. B., Moceri, V. M., Polk, S., Arena, P. J., and Fletcher, S. W. (1998): Ten-year risk of false positive screening mammograms and clinical breast examinations. *N. Engl. J. Med.*, 338:1089–1096.
43. Epstein, W. A. (1994): Sclerodermalike esophageal disease in children breast-fed by mothers with silicone breast implants. *JAMA*, 272:768.
44. Ernster, V. L. (1997): Mammography screening for women 40 through 49: a guidelines saga and a clarion call for informed decision making. *Am. J. Public Health*, 87: 1103–1106.
45. Fairweather, W. E., Higginson, J., and Beauchamp, T. L., Eds. (1991): *Ethics in Epidemiology*. Pergamon Press, New York.
46. Falk, H., Herbert, J., Crowley, S., Ishak, K. G., Thomas, L. B., Popper, H., and Caldwell, G. (1981): Epidemiology of hepatic angiosarcoma in the United States: 1964–1974. *Environ. Health Perspect.*, 40:107–113.
47. Feinstein, A. R. (1998): *P*-values and confidence intervals: two sides of the same unsatisfactory coin. *J. Clin. Epidemiol.*, 51:355–360.
48. Fingerhut, M. A., Halperin, W. E., Marlow, D. A., Piactelli, D. A., Honchar, P. A., Sweeney, M. A., Griefe, A. L., Dill, P. A., Steenland, K., and Suruda, A. J. (1991): Cancer mortality in workers exposed to 2,3,7,8-tetrachlorodibenzo-*p*-dioxin. *N. Engl. J. Med.*, 324:212–218.
49. FDA. (1991): *Background Information on the Possible Health Risks of Silicone Breast Implants*. U.S. Food and Drug Administration, Rockville, MD (rev. Feb. 8, 1991).
50. Foster, K. R. and Huber, P. W. (1997): *Judging Science: Scientific Knowledge and the Federal Courts*. MIT Press, Cambridge, MA.
51. Fraser, G. E. (1986): *Preventive Cardiology*. Oxford University Press, New York.
52. Friedman, G. D. (1994): *Primer in Epidemiology*, 4th ed. McGraw-Hill, New York.
53. Gabriel, S. E., Woods, J. E., O'Fallon, W. M., Beard, C. M., Kurland, L. T., and Melton, L. J., III. (1997): Complications leading to surgery after breast implantation. *N. Engl. J. Med.*, 336:677–682.
54. Gabriel, S. E., O'Fallon, W. M., Kurland, L. T., Beard, C. M., Woods, J. E., and Melton, L. J., III. (1994): Risk of connective-tissue diseases and other disorders after breast implantation. *N. Engl. J. Med.*, 330:1697–702.
55. Gehlbach, S. H. (1993): *Interpreting the Medical Literature*. McGraw-Hill, New York.
56. Gershwin, E. (1997): Testimony in *Spitzfaden v. Dow Corning Corporation* (No. CV 92-2589) (LA. Civ. Dist. Ct., April 22, 1997).
57. Goldberg, E. P., Widenhouse, C., Marotta, J., and Martin, P. (1997): Failure of silicone gel breast implants: analysis of literature data for 1652 explanted prostheses. *Plast. Reconstr. Surg.*, 100:281–284.
58. Goldberg, E. P., Widenhouse, C., Marotta, J., and Martin, P. (1998): Failure of silicone gel breast implants: analysis of literature data for 1652 explanted prostheses. *Plast. Reconstr. Surg.*, 101:1163–1164.
59. Goldberg, E. P., Widenhouse, C., Marotta, J., and Martin, P. (1999): Failure of silicone gel breast implants. *Plast. Reconstr. Surg.*, 103:1092.
60. Goldwyn, R. M. (1997): Financial disclosure is not full disclosure. *Plast. Reconstr. Surg.*, 99:2034–2035.
61. Gorman C. (1999): The web of deceit. *Time*, 153(5):76.
62. Grufferman, S. (1982): Hodgkin's disease. In: *Cancer Epidemiology and Prevention*, edited by D. Schottenfeld and J. F. Fraumeni. W.R. Saunders, Philadelphia, PA, p. 734.
63. Hahn, R. A. (1997): The nocebo phenomenon: concept, evidence and implications for public health. *Prevent. Med.*, 26:607–611.
64. *Hall v. Baxter Healthcare Corporation*, 947 F. Supp. 1387 (D. Or., 1996).
65. Hart, R. W., Neumann, D. A., and Robertson, M. (1995): *Dietary Restriction: Implications for the Design and Interpretation of Toxicity and Carcinogenicity Studies*. ILSI Press, Washington, D.C.
66. Haseman, J. K., Huff, J. E., Rao, G. N., and Eustis, S. L. (1989): Sources of variability in rodent carcinogenicity studies. *Fundam. Appl. Toxicol.*, 12:793–804.
67. Hazen, R. M. (2001): *The Joy of Science*. The Teaching Company, Chantilly, VA.
68. Hennekens, C. H., and Buring, J. E. (1987): *Epidemiology in Medicine*, edited by S. L. Mayrent. Little, Brown and Co., Boston.
69. Hill, A. B. (1971): *Principles of Medical Statistics*, 9th ed. Oxford University Press, New York.
70. Huber, P. (1997): The health scare industry. *Forbes*, October 6, 1997:15.
71. Huber, P. (1998): Joiner, Scheffer and Kumbo: refining the standards of evidence. *Civil Justice Memo*, 35:1–5.

72. Huff, D. (1954): *How To Lie with Statistics*. W.W. Norton, New York.
73. Huff, J., Cirvello, J., Haseman, J., and Bucher, J. (1991): Chemicals associated with site-specific neoplasia in 1394 long-term carcinogenesis experiments in laboratory rodents. *Environ. Health Perspect.*, 93:247–270.
74. Ioannidis, J. P. A. (1998): Effect of the statistical significance of results on the time to completion and publication of randomized efficacy trials. *JAMA*, 279:281–286.
75. Jacobsen, S. J., Xia, Z., Campion, M. E., Darby, C. H., Plevak, M. F., and Melton, L. J. (1997): Authorization for research use of medical records: who declines. In: *Proc. of The American College of Epidemiology Annual Scientific Sessions* (Abstract A-3), Cambridge, MA, Sept. 21–23, 1997.
76. Kohn, A. (1986): *False Prophets: Fraud and Error in Science and Medicine*. Basil Blackwell, Oxford.
77. Kvien, T. K., Glennas, A., Knudsrod, O. G., and Smedstad, L. M. (1996): The validity of self-reported diagnosis of rheumatoid arthritis: results from a population survey followed by clinical examinations. *J. Rheumatol.*, 23:1866–1871.
78. Last, J. M. (1995): *A Dictionary of Epidemiology*, 3rd ed. Oxford University Press, New York.
79. Levine, J. J. and Ilowite, N. T. (1994): Sclerodermalike esophageal disease in children breast-fed by women with silicone breast implants. *JAMA*, 271:213–216.
80. Lock, S. (1988): Misconduct in medical research: does it exist in Britain? *Br. Med. J.*, 297:1531–1535.
81. MacMahon, B. and Pugh, T. F. (1970): *Epidemiology: Principles and Methods*. Little, Brown and Co., Boston.
82. Margolis, H. (1993): *Paradigms and Barriers: How Habits of the Mind Govern Scientific Beliefs*. University of Chicago Press, Chicago.
83. Marshall, E. (1997): Journals joust over conflict-of-interest. *Science*, 276:524.
84. Matthews, D. E. and Farewell, V. (1985): *Using and Understanding Medical Statistics*. Karger, Basel.
85. McMichael, A. J. (1976): Standardized mortality ratios and the 'healthy worker effect': scratching beneath the surface. *J. Occup. Med.*, 18:165–168.
86. *Merrell Dow Pharmaceuticals, Inc. v. Havner*. 953 S.W. 2d 706, (Tex., 1997).
87. Michael III, M., Boyce, W. T., and Wilcox, A. J. (1984): *Biomedical Bestiary: An Epidemiologic Guide to Flaws and Fallacies in the Medical Literature*. Little, Brown and Co., Boston.
88. Mills, J. L. (1993): Data torturing. *N. Engl. J. Med.*, 329:1196–1199.
89. Moore, G. E. and Palmer, W. N. (1977): Money causes cancer: Ban it! *JAMA*, 238:397.
90. Olsen, G. W., Lacy, S. E., Bodner, K. M., Chau, M., Arceneaux, T. G., Cartmill, J. B., Ramlow, J. M., and Boswell, J. M. (1997): Mortality from pancreatic and lymphopoietic cancer among workers in ethylene and propylene chlorohydrin production. *Occup. Environ. Med.*, 54:592–598.
91. Paolini, M., Biagi, G. L., and Cantelli-Forti, G. (1997): A hidden paradox in carcinogenesis bioassays. *J. Natl. Cancer Inst.*, 89:736.
92. Perkins, L. L., Clark, B. D., Klein, P. J., and Cook, R. R. (1995): A meta-analysis of breast implants and connective tissue diseases. *Ann. Plast. Surg.*, 35:561–570.
93. Placik, O. J. (1994): Sclerodermalike esophageal disease in children breast- fed by mothers with silicone breast implants. *JAMA*, 272:768–769.
94. Rapaport, D. P., Stadelmann, W. K., and Greenwald, D. P. (1997): Incidence and natural history of saline-filled implant deflations: comparison of blunt-tipped versus cutting and tapered needles. *Plast. Reconstr. Surg.* 100:1028–1032.
95. Reed, M. E. (1997): *Daubert* and the breast implant litigation: how is the judiciary addressing the science? *Plast Reconstr. Surg.*, 100:1322–1326.
96. Riegleman, R. K. and Hirsch, R. P. (1996): *Studying a Study and Testing a Test: How To Read the Health Science Literature*, 3rd ed. Little, Brown and Co., Boston.
97. Roberts, H. J. (1988): Reactions attributed to aspartame containing products: 551 cases. *J. Appl. Nutr.*, 40:85–94.
98. Robinson, O. G. (1994): Deposition testimony, *In re: Silicone Breast Implants Product Liability Litigation*, (MDL 926) (No. CV 92-P-10000-S) (N.D. Ala, March 12, 1994).
99. Robinson, O. G., Bradley, E. L., and Wilson, D. S. (1995): Analysis of explanted silicone implants: a report of 300 patients. *Ann. Plast. Surg.*, 34:1–7.
100. Roht, L. H., Vernon, S. W., Weir, F. W., Pier, S. M., Sullivan, P., and Reed, L. J. (1985): Community exposure to hazardous waste disposal sites: assessing reporting bias. *Am. J. Epidemiol.*, 122:418–433.
101. Rothman, K. J. (1987): Clustering of disease. *Am. J. Public Health*, 77:13–15.
102. Rothman, K. J. (1990): A sobering start for the cluster busters' conference. *Am. J. Epidemiol.*, 132:S6–S13.
103. Rothman, K. J. (1993): Conflict of interest: the new McCarthyism in science. *JAMA*, 269:2782–2784.
104. Rothman, K. J. and Greenland, S. (1998): *Modern Epidemiology*, 2nd ed. Lippincott–Raven, Philadelphia, PA.
105. Rousseau, D. L. (1992): Case studies in pathological science. *Am. Sci.*, 80:54–62.
106. Sackett, D. L. (1979): Bias in analytic research. *J. Chron Dis.*, 32:51–63.
107. Sackett, D. L., Haynes, R. B., Guyatt, G. H., and Tugwell P. (1991): *Clinical Epidemiology: A Basic Science for Clinical Medicine*, 2nd ed. Little, Brown and Co., Boston.
108. Sanchez-Guerrero J., Colditz, G. A., Karlson, E. W Hunter, D. J., Speizer, F. E., and Liang, M. H. (1995): Silicone breast implants and the risk of connective tissue diseases and symptoms. *N. Engl. J. Med.*, 332:1666–1670
109. Schneiderman, M. A. (1994): More on torturing data. *N Engl. J. Med.*, 330:861–862.
110. Selikoff, I. J., Hammond, E. C., and Churg, J. (1968): Asbestos exposure, smoking and neoplasia. *JAMA*, 204:106–112
111. Shoaib, B. O., Patten, B. M., and Calkins, D. S. (1994): Adjuvant breast disease: an evaluation of 100 symptomatic women with breast implants for silicone fluid injection. *Keio J. Med.*, 43:79–87.
112. Shorter, E. (1992): *From Paralysis to Fatigue: A History of Psychosomatic Illness in the Modern Era*. The Free Press, New York.
113. Silverman, S., Borenstein, D., Solomon, G., Espinoza, L and Colin, M. (1996): Preliminary operational criteria for systemic silicone related disease (SSRD). *Arthritis Rheum.*, 39(Suppl. 9):S51.

114. Silverstein, M. J., Handel, N., Gamagami, P., Waisman, J. R., Gierson, E. D., Rosser, R. J., Steyskal, R., and Colburn, W. (1988): Breast cancer in women after augmentation mammoplasty. *Arch. Surg.*, 123:681–685.
115. Skrabanek, P. (1994): The emptiness of the black box. *Epidemiology*, 5:553–555.
116. Slavin, S. A. and Goldwyn, R. M. (1995): Silicone gel implant explantation: reasons, results and admonitions. *Plast. Reconstr. Surg.*, 95:63–9.
117. Solomon, G. (1993): Clinical and serologic features of 176 women with silicone implants: evidence for a novel disease, siliconosis. *Arthritis Rheum.*, 36(Suppl. 9):S117.
118. Solomon, G., Espinoza, L., and Silverman, S. (1994): Breast implants and connective-tissue disease. *N. Engl. J. Med.*, 331:1231.
119. Solomon, G. E. (1996): Operational criteria for systemic silicone related disease (SSRD), declaration submitted *In re: Breast Implant Litigation*, (No. 92-182-JO-LEAD (E.D. N.Y., August 2, 1996).
120. Spiera, H. (1988): Scleroderma after silicone augmentation mammoplasty. *JAMA*, 260:236–238.
121. Spiera, H. and Kerr, L. D. (1993): Scleroderma following silicone implantation: a cumulative experience of 11 cases. *J. Rheumatol.*, 20:958–961.
122. Star, V. L., Scott, J. C., Sherwin, R., Lane, N., Nevitt, M. C., and Hochberg, M. C. (1996): Validity of self-reported rheumatoid arthritis in elderly women. *J. Rheumatol.*, 23:1862–1865.
123. Staudenmayer, H. (1999): *Environmental Illness: Myth and Reality*. Lewis, London.
124. Steimle, S. (1998): Will Germany's Good Scientific Practice Guidelines prevent fraud? *J. Natl. Cancer Inst.*, 90:1694–1695.
125. Sturrock, R. D., Batchelor, J. R., Harpwood, V., Long, D. R., Milward, T. M., Silman, A. J., and Sloane, J. P. (1998): *Silicone Gel Breast Implants: The Report of the Independent Review Group*. Medical Device Agency of the British Department of Health, London.
126. Su, C. W., Dreyfuss, D. A., Krizek, T. J., and Leoni, K. J. (1995): Silicone breast implants and the inhibition of cancer. *Plast. Reconstr. Surg.*, 96:513–520.
127. Tan, E. M., Feltkamp, T. E. W., Smolen, J. S., Butcher, B., Dawkins, R. et al. (1997): Range of antinuclear antibodies in 'healthy' individuals. *Arthr. Rheum.*, 40:1601–11.
128. Tannock, I. F. (1996): False-positive results in clinical trials: multiple significance tests and the problem of unreported comparisons. *J. Natl. Cancer Inst.*, 88:206–207.
129. Taubes, G. (1995): Epidemiology faces its limits. *Science*, 269:164–169.
130. Trefil, J. S. and Hazen, R. M. (1997): *The Sciences: An Integrated Approach*. John Wiley & Sons, New York.
131. Turner, J. A., Deyo, R. A., Loeser, J. D., Von Korff, M., and Fordyce, W. E. (1994): The importance of placebo effects in pain treatment and research. *JAMA*, 271:1609–1614.
132. Ungar, W. (1998): Bias: it's everywhere! *Pharmacoepidemiol. Drug Safety*, 7:425–427.
133. Van Doren, C. (1991): The invention of the scientific method. In: *A History of Knowledge*. Ballantine Books, New York, pp. 184–212.
134. Vojandi, A., Campbell, A., and Brautbar, N. (1992): Immune functional impairment in patients with clinical abnormalities and silicone breast implants. *Toxicol. Indust. Health*, 8:415–429.
135. Wagenknecht, L. E., Burke, G. L., Perkins, L. L., Haley, N. J., and Friedman, G. D. (1982): Misclassification of smoking status in the CARDIA study: a comparison of self-report with serum cotinine levels. *Am. J. Public Health*, 82:33–36.
136. Wagner, K. R., Elmore, J. G., and Horwitz, R. I. (1996): Diagnostic bias in clinical decision making: An example of L-tryptophan and the diagnosis of eosinophilia–myalgia syndrome. *J. Rheumatol.*, 23:2079–2085.
137. Young, V. L., Elliott, L. F., Peters, W. J., and Lassus, C. (1997): Panel discussion: management of displaced breast implants. *Aesth. Surg. J.*, 17:247–253.
138. Young, V. L. (1996): Testing the test: an analysis of the reliability of the silicone sensitivity test (SILS) in detecting immune-mediated responses to silicone breast implants. *Plast. Reconstr. Surg.*, 97:681–683.
139. Zuckerman, D. (1999): Uncertainty about breast implants' safety won't stop thousands from trying them. *San Jose Mercury News*, February 1, 1999.
140. Zuckerman, D. (2003): A split decision on breast implants. *The Washington Post*, October 26, 2003.


## APPENDIX A. GUIDELINES FOR GOOD EPIDEMIOLOGY PRACTICES FOR OCCUPATIONAL AND ENVIRONMENTAL EPIDEMIOLOGIC RESEARCH

The Guidelines for Good Epidemiology Practices for Occupational and Environmental Epidemiologic Research are included in this text by courtesy of the Chemical Manufacturers Association (CMA). They were developed by the CMA Epidemiology Task Group as part of the Epidemiology Resource and Information Center (ERIC) Pilot Project and, prior to publication in 1991, were modified following review and comment by an *ad hoc* panel of epidemiologists from academia, various government agencies, and the private sector. Although they do not have the force of law, they have been recognized by a number of groups and are analogous to the toxicology good laboratory practices (GLPs). The Guidelines for Good Epidemiology Practices (GEPs) were developed in part to provide an alternative to the GLPs, one that would appropriately address the issues confronted by epidemiologists conducting non-experimental research.

The Guidelines for Good Epidemiology Practices address the conduct of studies generally undertaken to answer questions about human health in relation to the workplace or the environment. The GEPs propose minimum practices and procedures that should be considered to help ensure the quality and integrity of data used in epidemiologic research and to provide adequate documentation of the research methods. Epidemiologic studies often evolve through a number of stages that precede the development of a protocol (e.g., proposals, feasibility studies, and measurement instrument validation). Although the GEPs are intended to address all activities that begin with protocol development, it was the opinion of the Task Group that adherence to the spirit of the guidelines would prove beneficial for those activities preceding protocol development as well as more informal investigations such as health hazard assessments/evaluations or small cluster investigations.

A copy of the original guidelines as published by ERIC can be obtained from the CMA. The complete document includes an introduction, eight sections, and three appendices. Only the eight sections are presented here, and they are in a slightly abridged form. A more detailed discussion of these GEPs can be found in the proceedings of a conference published in December 1991 issue of the *Journal of Occupational Medicine*.

## I. Organization and Personnel

### A. Organizational Structure

The organization or individual conducting the research shall be fully responsible for the operation and performance of the research. The organization shall be a legal entity with a governing body that sets policy and that is fully responsible for the administrative aspects of the organization and its related research activities. The relationship, roles, and responsibilities of the organizations and/or individuals sponsoring or conducting the study should be carefully defined in writing.

### B. Personnel

Personnel engaged in epidemiologic research and related activities shall have the education, training, and/or experience necessary to competently perform the assigned functions. The organization shall maintain a current summary of training and experience of these personnel. A job description for each individual engaged in or supervising activities shall be maintained and updated periodically.

## II. Facilities, Resource Commitment, and Contractors

### A. Facilities

Adequate physical facilities shall be provided to all those engaged in epidemiologic research and related activities. Sufficient resources (e.g., office space, relevant equipment, and office/professional supplies) shall be available to ensure timely completion of all studies. Suitable storage facilities shall be available to maintain research materials in a safe and secure environment.

### B. Resource Commitment

Sufficient commitment shall be made at the beginning of each study to ensure its timely and proper completion

### C. Contractors

For the purposes of ensuring and documenting the contractor's conformance with the Guidelines for Good Epidemiology Practices, it is recommended that the study sponsor have the right during the course of the study, and for a reasonable period following completion of the study, to inspect the contractor's facilities, including equipment, technical records, and records relating to the work conducted under the sponsor's contract.

## III. Protocol

Each study shall have a written protocol. This protocol must be approved before the study begins. The protocol should include the following:

A. A descriptive title
B. The names, titles, degrees, addresses, and affiliations of the study director, principal investigator, and all co-investigators
C. The name and address of the sponsor
D. An abstract of the protocol
E. The proposed study tasks and milestones, including study approval date (date protocol signed by all signatories), study start date (first date that the protocol is implemented), periodic progress review dates, and estimated completion date
F. A statement of the research objectives, specific aims, and rationale. The statement should identify the immediate purpose of the investigation; for example, it might indicate whether the study will be exploratory data analysis, hypothesis testing, or a combination of both.
G. A critical review of the relevant literature to evaluate applicable findings. This should include pertinent animal and human experiments, clinical studies, vital statistics, and previous epidemiologic studies. The literature review should be in sufficient depth to identify potential confounders and effect modifiers and to determine areas where new knowledge is needed
H. A description of the research methods, including:
   1. The overall research design and the reasons for choosing the proposed study design
   2. The data sources for exposure, health status, and risk factors
   3. Clear definitions of health outcomes, exposure and other measured risk factors as well as selection criteria, as appropriate, for exposed and nonexposed persons, morbidity and mortality cases, and referent groups
   4. The project's study size and, if appropriate, statistical power

5. The methods to be used in assembling the study data, including a description of, or reference to, methods used to control, measure, or reduce various forms of error (e.g., bias due to selection, misclassification, interviewer, or confounding) and its impact on the study. Pretesting procedures for research instruments and any manuals and formal training to be provided to interviewers, abstractors, coders, or data entry personnel should also be described or referenced.
6. The procedures for handling data in the analysis
7. The methods for data analysis
8. The major limitations of the study design, data sources, and analytic methods
9. The criteria for interpreting the results

I. A description of plans for protecting human subjects
J. The quality assurance and quality control procedures for all phases of the study; as appropriate, a certification and/or qualifications of any supporting laboratory or research groups
K. A description of plans for disseminating and communicating study results
L. The resources required to conduct the study
M. The bibliographic references
N. Addenda, as appropriate (e.g., informed consent forms, questionnaires, and representative samples of other documents to be used in the study)
O. A data protocol review and approval sign-off sheet for the study director, principal investigator, co-investigators, and all reviewers
P. The dated amendments to the protocol

## V. Review and Approval

### A. Scientific Review

The study protocol shall receive appropriate scientific review by qualified persons who are not part of the investigative team to ensure that the study is designed to address the objectives of the research and that the protocol is written according to the Guidelines for Good Epidemiology Practices. The nature and the circumstances of this review shall be documented.

### B. Ethical Review

The ethical aspects of each study protocol shall be reviewed by an institutional review board or other comparable review procedure. This review should consider:

1. Obligations to research subjects
2. Obligations to society
3. Obligations to funders and employers
4. Obligations to colleagues

### C. Administrative Review

The administrative aspects of the study protocol shall receive appropriate review and written approval by sponsors, contractors, and associated third parties to ensure that sufficient resources are available to complete the study in a timely and proper fashion.

## V. Study Conduct

While the study director shall be responsible for the overall research program, the principal investigator shall be responsible for the individual research project, including the day-to-day conduct of the study, interpretation of the study data, and preparation of a final report. These responsibilities extend to all aspects of the study including periodic reporting of study progress as well as quality assurance. In some situations, the study director and the principal investigator may be the same person. To ensure the proper conduct of the study, personnel shall adhere to sound research principles and practices established according to the protocol. A protocol must be approved before the study begins. The study shall be conducted in accordance with the protocol; all deviations from the protocol shall be properly documented and authorized by the principal investigator. If a decision is made not to complete a research project, the reasons for that decision shall be put in writing, dated, and signed by the responsible party (i.e., the individual who makes the decision to terminate the study).

### A. Protection of Human Subjects

Procedures for protecting human subjects shall be followed. Confidential information about study subjects shall be protected using established procedures. If stipulated by the study protocol and/or required by an institutional review board, each study subject shall be informed about the purpose of the study and any risks associated with participating in the study. Written consent, if required, shall be obtained from each study subject before he/she participates in the study. Written consent shall include at a minimum:

1. The purpose of the research or study
2. The names, addresses, and phone numbers of personnel available to answer questions about the research and the rights of study subjects
3. The expected duration of a subject's participation
4. The eligibility requirements for study participation
5. The possible benefits of the study results to the study subject or others
6. A statement on the voluntary nature of participation in the study and the right of the study subject to discontinue participation at any time
7. A statement of confidentiality of records identifying the study subject, including reasonable exceptions

to absolute confidentiality (e.g., sharing of information with the study subject's personal physician or as required by court order)
8. A description of any foreseeable risks or discomforts to the study subject
9. A statement of the availability of the results

## B. Data Collection and Verification

All data collected for the study should be recorded directly, accurately, promptly, and legibly. The individuals responsible for the integrity of the data, computerized and hard copy, shall be identified. All procedures used to verify and promote the quality and integrity of the data shall be outlined in writing. A historical file of these procedures shall be maintained, including all revisions and the dates of such revisions. Any changes in data entries shall be documented.

## C. Analysis

All data management and statistical analysis programs and packages used in the analyses should be documented. All dated versions used in research shall be kept with accompanying documentation.

## D. Study Report

Completed studies shall be summarized in a final report that accurately and completely presents the study objectives, methods, results, and the principal investigator's interpretation of the findings. Although the content and length of any technical publication based on the research may be subject to requirements of the particular journal, if a more comprehensive report is written it should include:

1. A descriptive title
2. An abstract
3. The purpose (objectives) of the research as stated in the protocol
4. The names, titles, degrees, addresses and affiliations of the study director, principal investigator, and all the co-investigators
5. The name and address of the sponsor
6. The dates on which the study was initiated and completed
7. An introduction with background, purpose, and specific aims of the study
8. A description of the research methods, including:
   a. The selection of study subjects and controls
   b. The data collection methods
   c. The transformations, calculations or operations on the data
   d. The statistical methods used in data analyses
9. A description of circumstances that may have affected the quality or integrity of the data
10. A summary of the data analyses, including sufficient tables, graphs, and illustrations to present the pertinent data and to reflect the analyses performed
11. A statement of the conclusions drawn from the analyses of the data
12. A discussion of the implications of the study results
13. A list of references
14. A statement describing the location where all source data and the final report are stored
15. A dated study report review sign-off sheet for the study director, principal investigator, co-investigators, and reviewers and/or auditors

# VI. Communication

Each organization shall predetermine procedures under which communications of the intent, conduct, results, and interpretations of an epidemiologic study will occur, including what function individuals associated with the research will fulfill. These individuals should include the principal investigator, study director, and/or the sponsor. This procedure may be documented in the form of a standard operating procedure, in the study protocol, or through contractual agreement. Government agencies shall be informed of study results in a manner that complies with applicable regulatory requirements. To the extent possible, scientific peers shall be informed of study results by publication in the scientific literature or via presentations at scientific conferences, workshops or symposia. As feasible, all study subjects shall be informed of the study results and any interpretations of the study findings and conclusions. Information about the study results should be presented in language appropriate to the audience.

# VII. Archiving

Physically secure archives must be designated for the orderly storage and expedient retrieval of all study related material. An index shall be prepared to identify the archived contents and their location, and to identify by name and location any material that by their general nature are not retained in a specific study archive. Access to the archives shall be controlled and limited to authorized personnel only. Special procedures may be necessary to ensure that confidential information about study subject is protected. Individual study archives should contain, or refer to, the following:

A. The original signed and dated study protocol and all approved modifications
B. The original signed and dated final report of the study

C. All source data and, where feasible, biological specimens. A printed sample of the master computer data files with reference to the location of the machine readable master.

D. Documentation adequate to identify and locate all computer programs and statistical procedures used, including version numbers where appropriate

E. Copies of computer printouts, including relevant execution code, that form the basis of any tables, graphs, discussions, or interpretations in the final report. Any manually developed calculations shall be documented on a work sheet and similarly retained.

F. Correspondence pertaining to the study, standard operating procedures, informed consent releases, copies of all relevant representative material, copies of signed institutional review board and other external reviewer reports, and copies of all quality assurance reports and audits. As appropriate, this would include questionnaires, the name, make and model numbers of relevant measurement instruments, calibration information and procedures.

G. Original documents for the certain research materials that may be unique to the study such as laboratory notebooks and coder modification records

## VIII. Quality Assurance

Written procedures shall be established to ensure the quality of the data used in a study. These procedures shall address data collection and completeness, coding and computer input, storage and retrieval, and data validation and analysis. Any deviations from the GEPs shall be explained and documented in the final report. An individual who is not part of the investigative team should be assigned as a study quality assurance auditor. This individual shall, no less than annually, review study compliance with the written quality assurance procedures. The study quality assurance auditor shall prepare a written summary of the audit. The principal investigator should respond in writing to the audit report, including any remedial actions taken. Quality assurance activities shall address the preceding sections of these guidelines as well as monitor conformance with established standard operating procedures.

# APPENDIX B. OUTLINE FOR CRITIQUING EPIDEMIOLOGY REPORTS

We live in the Information Age, in a time when we all are inundated with the latest news from places near, far, and unknown: tips from financial experts about opportunities that seem to good to be true, conflicting health reports from scientists of varied repute, and "facts" from advocates for this cause or the other. The Internet can now deliver the compiled knowledge of humanity, uncensored and almost instantaneously. A major problem for most of us is not accessing information; it is determining what to accept and what to ignore.

For the scientist, the problem is particularly acute. Technical journals are proliferating at an astonishing rate. This phenomenon has occurred in part because of the explosion of discovery based on new technologies and the related desire of investigators to share their finding with their colleagues, both those within a particular specialty and those in allied fields—and to share their opinions with the courts, the regulatory agencies, funding organizations, and the general public. It also has occurred simply as a function of the increase in the absolute number of scientists and the needs of each to be published.

Unfortunately, peer review is an imperfect process and that means a certain percentage of published articles are fatally flawed, even those that appear in the most respected journals. In addition, scientific discourse increasingly is taking place on the Internet, a process of information exchange largely devoid of any semblance of peer review. Therefore, scientific reports, irrespective of source, must be read with a degree of healthy skepticism.

Critiquing research reports requires a combination of subject specific knowledge and an understanding of the strengths and shortcomings of the available biological tools and statistical techniques. For anyone trying to evaluate information from the field of epidemiology, it also requires an appreciation of the methods used in observational research, some of which may seem counterintuitive (e.g., case control studies) and others more rigorous than they actually are (e.g., case reports, case series, and ecologic studies). For those outside the field, that makes critically assessing the worth of epidemiology research much more difficult; nonetheless, it is not impossible, at least to the degree of winnowing the dissonance of claims and counterclaims down to the few that seem to have a legitimate message.

Fortunately, all scientific articles conceptually share a common format, a format that places a burden upon the authors to not only share the results of their research but also to explain the process by which they were obtained. The authors are also expected to discuss the potential limitations of their work. Some are remiss in this endeavor, in whole or in part, and the burden falls upon the reader.

Over the years, I have found that the quickest way to review an epidemiology report is through the use of the mantra of selection, misclassification, confounding, chance, and causation. For those who routinely must evaluate medical or public health reports in greater depth, I recommend the book by Riegelman and Hirsch, *Studying a Study and Testing a Test: How To Read the Health Science Literature* [96]. They use a slightly different format based on assignment, assessment, analysis, interpretation, and extrapolation. They also provide a number of excellent examples to illustrate key points. For those with an intermediate need, consider the following 11-step outline. No study is perfect, but the outline will assist you in determining which ones may have merit.

- *Background information*—Does the introduction provide a balanced presentation of the biological issues? Is there an adequate differentiation between fact and conjecture? Is the reader directed to other references which, given the reader is so inclined, will provide a more in-depth presentation of the issue?
- *Basic design of the research*—What study design was used: case report, case series, cross-sectional, case control, cohort, clinical trial, ecologic, or hybrid? Does the paper present the study design coherently and discuss its strengths and weaknesses? Was there another study method that could have been used more effectively? If so, what was it and why? What type of data were used: incidence, prevalence or a combination of both? Was the study conducted in an ethical fashion?
- *Objective of the study*—Was the objective of the study clearly stated? Did the study attempt to generate or test hypotheses, or was it a combination of both? Is there a sense that the authors clearly separated one logic from the other? Was the problem important enough to have been studied?
- *Sampling strategy*—What sampling strategy was used: random, judgmental, convenience, cluster, or other? Will the sampling strategy allow the results to be extrapolated back to the target population from whom the sample was drawn? If a cluster was used to initiate the study, were those in the cluster included from this particular research? How was the comparison group identified? Was the sample size large enough to provide meaningful information? Did the author discuss the power of the study? Did they need to? Is the group being evaluated independent of other groups that have been studied?
- *Risk of bias*—What was the potential for selection bias. What was the potential for misclassification bias? How did the authors address these technical problems? Were any potential biases likely equivalent or different among the exposed and unexposed or the diseased and non-diseased? What were the likely impacts of any of these technical biases on the point estimates of risk and the variability of the data?
- *Estimate of exposure/independent variable(s)*—Was each independent variable well defined? What methods were used to estimate exposure? Did they provide direct or surrogate measures of exposure? Were the variables adequate; that is, were they reliable and valid? Were they precise? Were they meaningful? If an association was found, were the exposures well enough defined so that control strategies could be developed? Was there a possibility for differential misclassification? How probable was it? If probable, how might it have influenced the risk estimates or the confidence intervals? And how might they, in turn, have influenced the authors' interpretations?
- *Health indicators of response/dependent variable(s)*—Were the health outcomes well defined or nebulous? Were they defined with acceptable criteria? Was there a possibility for differential misclassification? If so, how probable was it? If probable, how might it have influenced the risk estimates or the confidence intervals? And how might they, in turn, have influenced the authors' interpretations?
- *Identification and control of confounders/covariables*—What were the confounders that were considered? Were they the important ones? Were they controlled and, if so, how? During study design, data collection, data analysis, or interpretation? Were they controlled for all the results, or at least all the results upon which the major interpretations were based? Were there other obvious potential confounders that were ignored (e.g., age, race, gender, socioeconomic class, or alternative exposures of merit)? What were the likely impacts of these other confounders on the risk estimates or confidence intervals or interpretations?
- *Analysis of data*—How were the data analyzed? Was there any evidence of gerrymandering of data or groups or selective use of results? What statistical tests were used? Were they simple to understand? Were they unnecessarily complex? Were they appropriate? What assumptions were violated? If important assumptions were potentially violated, was the impact of violating them discussed in the paper? How were the results presented: point estimates and $p$ values, "statistically significant," "significant," point estimates and confidence intervals, correlation coefficients, etc.? Were the calculations correct? Did the numbers in the tables add up? Were the findings presented in sufficient detail to allow the reader to come to his own conclusions? Were the findings internally consistent? How did they handle the issue of multiple comparisons: make adjustments, ignore the problem, or rationalize it away?
- *Interpretation of results*—Were the objectives of the study met? Were the interpretations made within the constraints of the original study objectives, or did they ask one question and answe

another? Do the study design, data, and results support the interpretations and conclusions? Do they make sense? How strong are they? Do they support cause and effect in terms of biological plausibility, temporal relationship, consistency with other studies by other investigators (if there were any), coherence of the total body of evidence, specificity, strength of the association, and dose–response—or are they simply not inconsistent with cause and effect? Are there alternative explanations for the results because of a high potential for one or more technical biases or with other biological paradigms? Were these alternatives also discussed, even briefly? What is the public health significance of the findings? Do the summary and the abstract fairly represent the study results, or do they only provide the most provocative findings?

- *Suggested further research*—What new information, pro or con, did the study produce? What new hypotheses were developed or existing hypotheses refined by this research? What useful information was generated? What additional research is needed? If multiple additional research projects were suggested, are they feasible? In what order should they be done?

# Notes

# 12 Principles of Pathology for Toxicology Studies

*Steven R. Frame and Peter C. Mann*

## CONTENTS



## INTRODUCTION

Anatomical pathology findings often define the critical outcomes of the hazard identification process, including primary target organ effects, no-observed-effect levels, adversity of exposure, and interspecies relevance of exposure. As such, the anatomical pathology evaluation is essential to the identification and characterization of target organ toxicity.

Pathology is defined as "the medical science, and specialty practice, concerned with all aspects of disease, but with special reference to the essential nature, causes, and development of abnormal conditions, as well as the structural and functional changes that result from the disease processes" [1]. This definition encompasses the traditional role of the pathologist in identifying morphological changes in tissues at the gross and microscopic levels. It also describes the more comprehensive and often more complex aspects of pathology. These include identifying and characterizing the nature (e.g., inflammatory, degenerative, or disturbances of growth) of abnormal findings, as well as their cause and development (pathogenesis).

Toxicologic pathology is the science that integrates the disciplines of pathology and toxicology and is concerned with the effects of potentially noxious substances [2,3]. The Society of Toxicologic Pathologists defines a toxicologic pathologist as follows: "Any person who is a toxicologic pathologist by virtue of training, experience, and/or scientific contributions to the field, and is actively involved in safety assessment, teaching, or research in toxicologic pathology or the administration of these activities [4]." The role of toxicologic pathologists is to identify pathological changes, to determine the etiology and significance of those changes, and to clearly and accurately report their conclusions to other scientists. This role requires not only an understanding of normal and abnormal tissue morphology but also a basic understanding of general ("whole animal") physiology and clinical medicine. This understanding necessarily contains a comparative component, as the pathology evaluation usually encompasses multiple laboratory animal species and is typically used as one component of human risk or safety assessment. In addition, the pathology assessment must integrate other relevant data such as in-life study parameters (e.g., clinical signs, body weight changes), clinical pathology findings, and metabolism and pharmacokinetic data; therefore, the pathologist must draw from a broad spectrum of disciplines in the biological and medical sciences.

The gold standard of the pathology evaluation in toxicity studies has been the examination of paraffin-embedded, hematoxylin- and eosin-stained tissue sections. These standardized methods are time tested, and accurate diagnosis and reporting of findings derived from such routinely prepared specimens will continue to play a critical role in the hazard identification process. A wide array of new technologies is continually arising from advances in computer technology and molecular biology. Examples include the various "-omics" (genomics, proteinomics, metabinomics), genetically modified animal models, and special microscopy (e.g., digital, laser capture, confocal). Many of these technologies are, or will become, critical tools in drug discovery and development, as well as in the hazard identification and risk assessment of xenobiotics; however, accurate morphological diagnoses using standard pathology methods will likely remain important guideposts directing the practical application of these newer technologies.

This chapter provides: (1) an overview of standard pathology procedures commonly used in toxicity studies; (2) a discussion of important considerations in the evaluation, interpretation, and reporting of pathology findings; and (3) a discussion of quality assurance practices in pathology, including the pathology peer review and pathology working groups. Comprehensive discussions of general pathology and organ-specific toxicologic pathology are provided in a number of standard texts devoted to these subjects [4–9].

## PATHOLOGY PROCEDURES

The primary goals of the pathology examination are to identify and collect all gross lesions, to collect all tissues as listed in the study protocol, to trim and process all required tissues for microscopic evaluation, and to diagnose and report all lesions accurately. Perhaps most importantly, all of these procedures must be done in a consistent manner and in accordance with standardized procedures.

### The Necropsy

Necropsy refers to the examination of a body after death [10]. Although the necropsy is often one of the shortest phases in a toxicology study, it is one of the most critical. Procedures typically performed during the necropsy for toxicology bioassays include the gross examination, determination of organ weights, and collection of tissue for microscopic examination. Other procedures that commonly occur at necropsy include terminal blood and urine collection, preparation of bone marrow smears, and collection of samples for biochemical or molecular biology procedures.

The necropsy represents the beginning of data generation in the post mortem phase of the study and is the link between in-life findings and histopathological findings [11]. Because improper or incomplete necropsy examination can negatively impact an entire study, it is essential that the necropsy be performed by highly trained technicians operating under the supervision of a qualified pathologist [12]. Errors introduced during necropsy may produce tissue artifacts that confound the subsequent histopathological evaluation. Sources of artifact include freezing rather than refrigerating animals that die on study, autolysis and other tissue artifacts resulting from prolonged intervals between death and tissue fixation, and inadequate tissue fixation, such as occurs with immersion fixation of tissues that are too large or from the use of improper or compromised fixative.

Necropsy observations may provide the first evidence of target organ effects, including carcinogenicity and cause of death. In addition, intercurrent disease or procedural factors that could complicate study interpretation may also be discovered at necropsy. Because most aspects of the necropsy are not reproducible events within a study, errors occurring during this phase of the study generally cannot be corrected retrospectively. This underscores the need for careful planning and conduct of the necropsy; therefore prior to and during the necropsy, the necropsy team must have the following documents available for review [11,12]

- Study protocol, including amendments
- Clinical records
- Standard operating procedures (SOPs)

The study protocol sets forth the study objectives and study-specific procedures and takes precedence over other documents. The clinical findings for each animal should also be available to the necropsy technician to ensure that any unusual findings observed in life are identified and collected for microscopic evaluation. SOPs provide study personnel with information on the conduct of specific procedures that may be performed during the necropsy. Technicians should be knowledgeable about the procedures documented in the SOPs and consult them if questions arise. Inconsistencies in euthanasia procedures, tissue dissection and retrieval, descriptive terminology, or tissue fixation at necropsy can introduce variables that may significantly complicate subsequent evaluation of organ weight data and microscopic changes. For this reason, institutional procedures, as documented in SOPs, must be in place to standardize tissue collection and weighing, description of gross findings, and tissue trimming. These SOPs are required by regulatory agencies and contribute to an efficient and comprehensive necropsy [3]. Every animal presented for necropsy should have an individual necropsy record for recording body and organ weights, gross findings, and tissues collected. Euthanasia date, time, and method and signature lines for all personnel involved in the necropsy of the animal should also be included [13].

## Euthanasia Procedures

A number of pathology parameters may be influenced by the method of euthanasia and the choice of euthanasia agent; for example, barbiturate euthanasia agents are known to cause pooling of blood in the spleen, resulting in gross enlargement (splenomegaly) and congestion of the spleen in dogs. If the spleen were anticipated to be a primary or secondary target of a test compound, the use of an alternative method of euthanasia might be considered, or at least the known effects of the barbiturate on the spleen would have to be considered in interpreting the gross and organ weight data. The choice of euthanasia procedure may also produce specific histopathological changes that should not be confused with treatment-related changes; for example, carbon dioxide asphyxiation may produce focal, acute alveolar hemorrhage in the lung (Figure 12.1).

Methods of euthanasia should adhere to the recommendations of the American Veterinary Medical Association Panel on Euthanasia [14] and the *Guide for Care and Use of Laboratory Animals* [15]. The euthanasia procedure should seek to minimize pain and distress, should be easy to perform consistently, and should minimize tissue artifacts. The selection of specific agents and methods for euthanasia will depend on the species involved and the objectives of the study. Some common methods of euthanasia are given in Table 12.1. Generally, inhalant or noninhalant chemical agents are preferable to physical methods such as decapitation [15]. Although the selection of specific agents and methods may be species and protocol dependent, all methods of euthanasia should be reviewed and approved by the institutional animal care and use committee (IACUC). To limit post mortem autolysis and the potential introduction of confounding artifacts, the interval between death and necropsy should be minimized. Prolonged intervals between euthanasia and necropsy may produce significant alterations in organ weights and histology; for example, increased liver weights (both absolute and relative to body weight) and microscopic vacuolation of the liver may occur with a delay of only 25 minutes between euthanasia and necropsy [16]. Dissection should begin no longer than 5 minutes after euthanasia, and dissection time should not exceed about 20 minutes.

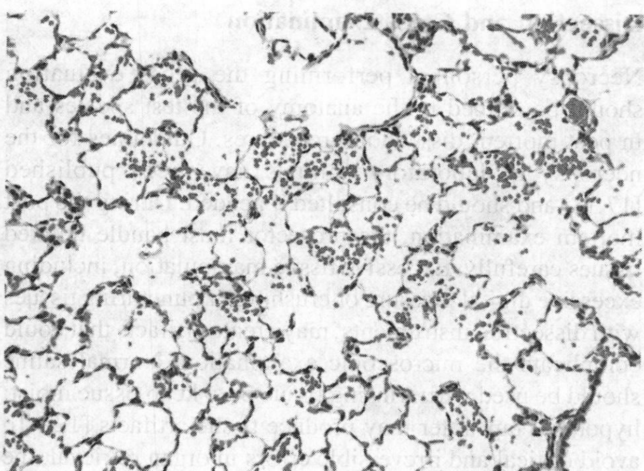

**FIGURE 12.1** Focal acute alveolar hemorrhage in the lung of a rat euthanized by $CO_2$ anesthesia and exsanguination.

**TABLE 12.1**
**Methods of Euthanasia**

| |
|---|
| *Asphyxiation* |
| Carbon dioxide |
| *Anesthesia* |
| Isoflurane |
| Sodium pentobarbital |
| Methoxyflurane |
| Halothane |
| Chloroform |
| *Cervical dislocation* |
| *Decapitation* (guillotine) |

## Dissection and Gross Examination

Necropsy personnel performing the gross evaluation should be trained in the anatomy of the test species and in post mortem dissection procedures. Guidelines for the necropsy of laboratory animals have been published [17,18] and should be consulted if needed. During the post mortem examination, the prosector must handle unfixed tissues carefully. Excessive tissue manipulation, including excessive digital pressure or crushing or puncturing tissues with dissection instruments, may create artifacts that could complicate the microscopic examination. Normal saline should be used when rinsing tissue or to keep tissue moist; hypotonic tap water may produce tissue artifacts [12]. To avoid critical and irreversible errors in organ retrieval, the protocol should be strictly followed and consulted as necessary with regard to the tissues and organs that are to be collected. Table 12.2 provides a list of tissues recommended by the Society of Toxicologic Pathology (STP) for subchronic and chronic toxicity studies [19]. This list is a minimum core list for all types of repeat-dose studies. Additional tissue may be added based on exposure route, species or strain of test animal, or known targets.

Detection and an accurate description of gross lesions are essential aspects of the necropsy examination, as the gross examination guides subsequent tissue trimming and histopathology. Gross identification and retrieval of lesions is particularly important for nonprotocol tissue, which would otherwise not be examined microscopically. To ensure consistency between prosectors, both within and across studies, gross observations should be identified according to standard descriptive terms. Abnormal gross findings should generally be described using some or all of the following criteria: location, number, size, color, consistency, distribution, and any special features that characterize the lesion. A list of selected terms that can be used in the gross description is given in Table 12.3. The gross descriptions of tissue changes should be concise and descriptive (rather than diagnostic), with special attention given to consistency throughout the necropsy. An example of a gross description is given in Figure 12.2.

## Tissue Fixation

Tissues may be preserved by immersion, inflation, or perfusion [12]. A wide array of fixatives and fixation procedures is available. The most common fixation method is immersion fixation in neutral buffered formalin; however, no universal fixative exists because no one fixative is perfect for all applications. The specific fixation procedure should be determined by the study objectives and study protocol. Prior to immersion fixation, tissues should be trimmed to approximately 0.5-cm thickness. Optimally, tissues are placed in formalin for at least 24 to 48 hours at a 10:1 volume ratio of fixative to tissue; however, ratios as low as 3:1 are adequate if tissues are properly prepared for fixation. Formalin provides relatively rapid fixation, is easy to use, and is inexpensive, but formalin is potentially toxic and thus requires proper ventilation and disposal. Also, tissue artifacts, such as retinal detachment, may occur secondary to tissue shrinkage. Nevertheless, for most routine studies and for most tissues, 10% neutral buffered formalin is typically the fixative of choice. Other fixatives may be used for specific tissue or procedures. A common example is the use of modified Davidson's fixative for fixation of the eye and testes [20]. Glutaraldehyde-based fixatives are often employed for ultrastructural studies. Inflation is the preferred method for fixation of the lung and may also be used for hollow organs, including the urinary bladder, stomach, and intestines [12]. Perfusion may be used in special target organ toxicity studies; for example, whole body intravascular perfusion with Karnovsky fixative is commonly employed in neurotoxicity studies [21].

**TABLE 12.2**
**STP-Recommended Core List of Tissues To Be Examined Histopathologically in Repeat-Dose Toxicity Studies (For All Species Where Applicable)**

| | |
|---|---|
| Adrenal gland | Peripheral nerve |
| Aorta | Pituitary |
| Bone with bone marrow[a] | Prostate |
| Brain | Salivary gland |
| Cecum | Seminal vesicle |
| Colon | Skeletal muscle |
| Duodenum | Skin |
| Epididymis | Spinal cord |
| Esophagus | Spleen |
| Eye | Stomach |
| Gallbladder | Testis |
| Harderian gland | Thymus |
| Heart | Thyroid gland |
| Ileum | Trachea |
| Jejunum | Urinary bladder |
| Kidney | Uterus |
| Liver | Vagina |
| Lung | Gross lesions |
| Lymph node(s) | Tissue masses |
| Mammary gland[b] | Tissues relevant to route of exposure[c] |
| Ovary | Tissues unique to the species or strain |
| Pancreas | Known target tissues |
| Parathyroid gland | |

[a] For nonrodents, either rib or sternum; for rodents, femur including articular cartilage.

[b] Females only.

[c] Such as nose and larynx for inhalation studies.

*Source:* Adapted from Bregman, C.L. et al., *Toxicol. Pathol.*, 31, 252, 2003. With permission.

**TABLE 12.3**
**Gross Lesion Description**

| **General Location** | **Number** | **Consistency** | **Special Characteristics** |
|---|---|---|---|
| Cutaneous | Single | Brittle | Area |
| Subcutaneous | Two | Caseous | Adhesion |
| Peritoneal | Four | Fibrinous | Circumscribed |
| Abdominal | Greater than *x* | Firm | Depressed |
| Thoracic | Multiple | Friable | Distended |
| Cranial | Size | Fluctuant | Flat |
| Sacral | Small | Gelatinous | Irregular |
| Lumbar | Enlarged | Granular | Layered |
| Cervical | Increased in size | Greasy | Linear |
| Axillary | Decreased in size | Gritty | Lobulated |
| Inguinal | Exact measurement | Hard | Macule |
| | | Mucoid | Mass |
| **Specific Location** | **Color** | Oily | Nodule |
| Ventral | Black | Rough | Oval |
| Dorsal | Blue | Rubbery | Papillary |
| Lateral | Brown | Scaly | Papule |
| Medial | Clear | Soft | Pedunculated |
| Distal | Cloudy | Thin | Perforated |
| Proximal | Dark | Viscous | Pitted |
| Deep | Green | Watery | Plaque |
| Hilus | Grey | | Polypoid |
| Wall | Mottled | **Distribution** | Prominent |
| Lumen | Opaque | Focal | Umbilicated |
| Mucosa | Pale | Multifocal | Raised |
| Superficial | Pink | Diffuse | Round |
| Serosa | Purple | Patchy | Spherical |
| Cortex | Red | Bilateral | |
| Medulla | Tan | Symmetrical | |
| Parenchyma | Transparent | Confluent | |
| Peripheral | Translucent | Unilateral | |
| Margin/Edge | White | All Lobes | |
| Anterior | Yellow | Random | |
| Posterior | | | |
| Right | | | |
| Left | | | |
| Cranial | | | |
| Caudal | | | |

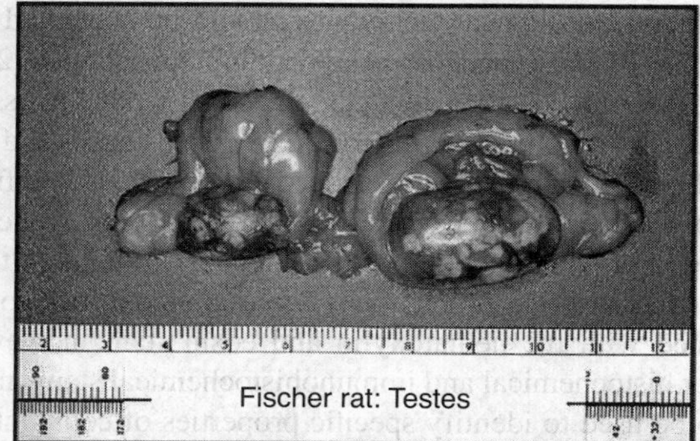


**FIGURE 12.2** Gross description: Testes. Masses, bilateral, multiple, firm, tan, 0.2 to 0.5 cm in diameter.

## Organ Weight Determinations

Organ weight changes can be sensitive indicators of target organ toxicity, and significant changes in organ weights may occur in the absence of changes in other pathology parameters [22]; for example, increased liver weight associated with hepatic cytochrome P450 induction is a common finding in toxicology studies. Liver weight increases of up to 20% relative to controls may occur without microscopic evidence of hepatocellular hypertrophy or changes in serum chemistries [23]. Similarly, modest dose-related changes in kidney weight commonly occur in toxicology studies without histopathological evidence of cellular alteration [24].

As with other necropsy procedures, the collection of organ weights must be done using standardized methods to ensure consistency and avoid artifactual weight changes. More consistent results for organ weight determinations can be attained if the animal is bled out prior to weighing organs [12]. Care should also be taken to remove extraneous tissue and blood clots and to prevent tissue dehydration. For some small tissues, such as rodent adrenal or thyroid glands, weighing tissue after fixation may help minimize artifacts associated with the handling of fresh tissue [3]. To minimize weighing errors during necropsy, a range of expected normal weights for each organ (matched to the species, strain, sex, and age of the test animal) should be available. Any weights that are outside this range should be verified before recording. As with many other endpoints collected during the necropsy, organ weights cannot be reproduced at a later date, so consistency and accuracy are critical to generating meaningful organ weight data.

## Histology Procedures

All changes to tissue that occur after fixation are, in effect, artifact. The goal in processing slides for microscopic (or ultrastructural) examination is to control the artifactual change so it is consistent across organs and across animals [13]. As noted for necropsy procedures, clearly written SOPs for histology procedures and strict adherence to these SOPs are essential for consistency in histological slide presentation. It is also important that each batch of tissues processed includes tissues from animals in all study groups to avoid apparent compound effects that are actually the result of variation in processing, embedding, or staining.

### Tissue Trimming

The first step in processing tissues for microscopic evaluation is tissue trimming. Trimming should be performed by trained technicians with knowledge of gross anatomy and medical terminology and an understanding of the meaning of gross observations made during necropsy. The technician should have the study protocol, as well as the gross findings, available prior to trimming. Tissue trimming is yet another procedure where inconsistencies may result in biased sampling and poor comparability across different groups within a study or across many different studies. Thus, for each organ, SOPs should be available that outline standard trimming procedures to be followed, including the plane of trim (e.g., transverse or longitudinal) and, for larger organs such as the liver or lung, the number of sections and the specific areas to be trimmed. Guides for tissue trimming in rodents have been published in an attempt to standardize tissue trimming across laboratories [25–27]. Tissues should be trimmed to a maximum of 0.3 cm for processing. Smaller tissues can be embedded intact. When trimming masses, adjacent normal tissue should be included where possible, and multiple sections should be taken of masses that are large or variable in appearance.

**Dehydration**
- Necessary because paraffin is not miscible with water
- Graded alcohols most commonly used method

↓

**Clearing**
- Clearing reagents must be miscible with both the dehydrant (alcohol) and paraffin
- Commonly used reagents: xylene, toluene, d-Limonene
- As the alcohol is removed, the tissue clears
- Must be carefully regulated to avoid excess hardening of the tissue

↓

**Impregnation**
- Complete removal of clearing reagents by substitution with paraffin
- Temperature of paraffin baths is critical; temperatures > 5°C above the melting point of 56–58°C will cause excessive hardening of the tissue
- Vacuum removes air, gases, and remaining clearing reagent and draws paraffin into all areas of the tissue

↓

**Embedding**
- Orientation of tissue in melted paraffin
- Provides a firm medium for tissue sectioning
- Orientation and location of tissue in paraffin block must be consistent across animals

**FIGURE 12.3** Procedures for tissue processing and embedding in paraffin.

### Tissue Processing, Staining, and Embedding

Following fixation and trimming, tissue is most commonly processed and embedded in paraffin for microscopic evaluation. The steps involved in processing to paraffin are given in Figure 12.3. Embedding media composed of acrylic or epoxy resins are commonly used for high-resolution light microscopy and electron microscopy. Textbooks on histological techniques should be consulted on the use of these and other embedding procedures [28]. Care should be taken to maintain consistency in tissue processing, as artifacts may be introduced that could be confused with compound-induced lesions [29]. Paraffin-embedded tissue blocks are routinely cut with a microtome to produce sections with a thickness of about 4 to 6 μm (microns). Paraffin tissue sections are most commonly stained with the hematoxylin and eosin stain; however, other histochemical and immunohistochemical stains may also be used to identify specific properties of cells, intracellular structures, or microorganisms [28,30]. Examples of some histochemical special stains used to identify specific structures or materials are given in Table 12.4.

**TABLE 12.4**
**Selected Histochemical Stains Used in the Histopathologicial Evaluation**

| Stain | Uses |
|---|---|
| Hematoxylin and eosin (HE) | Most commonly used stain for routine histopathology |
| | Stains nucleus basophilic (blue) and cytoplasm eosinophilic (pink to red) |
| Periodic acid-Schiff (PAS) | Stains PAS positive structures magenta; examples include: |
| | Glycogen and mucin |
| | Basement membranes |
| | Microorganisms, including some fungi and protozoa |
| Masson's trichrome | Stains collagen blue and muscle red |
| Oil red O | Stains lipids red; requires fresh smears or cryostat sections |
| Perl's iron stain | Stains iron (hemosiderin) blue |
| Luxol fast blue | Stains myelinated nerve fibers blue |
| Von Kossa's | Stains calcium deep purple |
| Toluidene Blue | Stains mast cell granules violet |

## THE HISTOPATHOLOGY EXAMINATION

### Overview

Histopathology is the study of morphological changes in tissues at the light-microscopic level [31]. Histopathological findings frequently form the basis of the no-observed-adverse-effect level (NOAEL) in toxicology studies and are a critical part of hazard identification and risk assessment of pharmaceuticals, chemicals, biologics, and medical devices. The Society of Toxicologic Pathology has published a best practices guideline for toxicologic histopathology [31], which is aimed at identifying and defining the fundamental elements of the histopathological examination and appropriate techniques to minimize observer bias. Some of the recommendations given in this guideline are summarized in Table 12.5, and selected topics are discussed in more detail below.

**TABLE 12.5**
**Some Fundamental Elements of the Histopathological Evaluation**

**Information that should be available to the pathologist prior to the microscopic evaluation**

The nature of the test substance and the class of compounds to which it belongs
Results of previous studies with the test compound in the same or different species, including target organ effects
All details of the experimental design (study protocol)
In-life data, including clinical signs, body weight, and nutritional data
Clinical pathology data (hematology, clinical chemistries, urinalysis), as well as hormone and enzyme induction data
Gross findings and organ weight data for individual animals

**The process of histopathological evaluation**

Assess specimen quality (reflects adequacy of tissue collection, fixation, trimming, processing, and staining)
Ensure that appropriate sections of tissues or organs are present on the slide and request recuts if necessary
Use concise, standardized diagnostic nomenclature and diagnostic criteria for tabular summaries
Use detailed free text as needed to better define complex lesions
Evaluate tissues either animal by animal (allows a more comprehensive assessment of animal's health) or organ by organ (provides a more focused examination of changes and greater consistency in severity grading)
Use a severity grading system that is definable, reproducible and meaningful
In carcinogenicity studies:
- Distinguish hyperplasia, dysplasia, and neoplasia
- Classify neoplasms as benign or malignant and as primary or metastatic
- Provide evaluation of cause of death

**Procedures that may be used to enhance accuracy and consistency of the histopathology evaluation**

Informal reevaluation of specific changes in specific tissues; may be conducted using masking techniques
Peer review by a second pathologist of defined subsets of animals and tissues, as well as study conclusions
Review by a pathology working group (PWG) consisting of experts for the target tissues of interest

*Source:* Adapted from Crissman, J.W. et al., *Toxicol. Pathol.*, 32, 126, 2004.

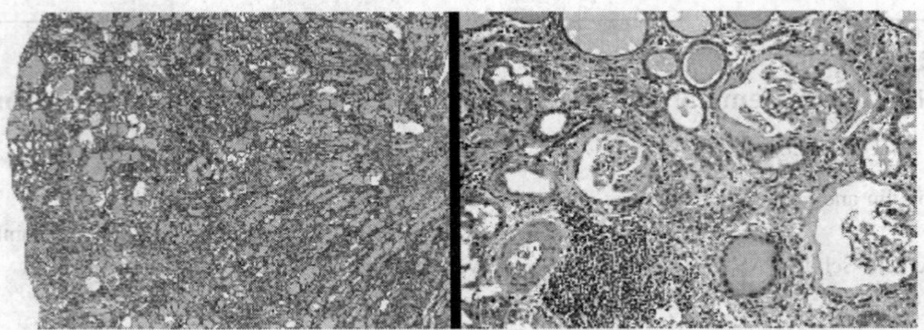

**Morphologic diagnosis based on classic descriptive pathology:**
Kidney, nephropathy, progressive, chronic, characterized by basement membrane thickening, tubular degeneration and regeneration, tubular dilatation, proteinaceous casts, mixed interstitial inflammatory infiltrate, interstitial fibrosis, mesangial proliferation, dilatation of Bowman's space, proliferation of parietal cells, glomerular adhesions, and sclerotic glomeruli

**Morphological diagnosis for tabular summaries:**
Nephropathy, chronic, progressive, severe

**FIGURE 12.4** Morphological diagnoses in classic and toxicologic pathology.

### Qualifications and Responsibilities of the Pathologist

The histopathological evaluation is the responsibility of the pathologist who must clearly communicate the results of that evaluation not only to other pathologists but also to toxicologists and other scientists. Unlike the approach used in classic diagnostic pathology, which is characterized by lengthy detailed descriptions of each morphological abnormality, toxicologic pathologists must record their findings in a manner that allows for meaningful tabular summaries of the data. A comparison of the diagnostic and descriptive approaches used in classic and toxicologic pathology is given in Figure 12.4. Using the pathology narrative, the pathologist must also ensure proper interpretation of the summarized microscopic findings; thus, histopathology is interpretive as well as descriptive [31,32].

A uniform accreditation standard for toxicologic pathologists does not exist, and the basic requirements to qualify as a toxicologic pathologist vary by country; however, a general consensus is that practicing toxicologic pathology requires formal training in the biomedical field and postgraduate training in toxicologic pathology. A review of the regional standards for the training and accreditation of toxicologic pathologists has been conducted by the International Federation of Societies of Toxicologic Pathologists [4].

One of the most challenging aspects of the histopathological evaluation in toxicologic pathology is achieving consistency among pathologists both within and across studies, as well as over time. Bucci [3] noted that, other than inaccuracy, inconsistency was the most undesirable characteristic of toxicology data. Histopathological diagnoses typically include some degree of subjectivity; therefore, although the diagnostic terms used by different pathologists for the same lesion should be comparable, they are not expected to always be identical [31,33]. Even though no two pathologists can be expected to produce identical findings (diagnoses, lesion grades) across all tissues examined in a study, qualified pathologists should identify the same treatment-related lesions and the no-observed-adverse effect level for those lesions.

### Procedures for the Histopathological Examination

In most circumstances, best results are achieved when all tissues from a study are evaluated by one pathologist [31]; however, in some circumstances, such as large studies or critical studies with short time lines, it may be necessary for two or more pathologists to read different subsets of tissues. Most commonly this involves different pathologists evaluating tissue from males and females. Peer review by a single pathologist is important in achieving consistency in studies where more than one pathologist has evaluated a study. The order of slide review among study groups may vary based on study type and personal preference of the pathologist. For some studies, especially those with small group sizes, all controls may be evaluated first to establish the spectrum of "normal." This is followed by the evaluation of high and intermediate dose groups as required by the protocol. Alternatively, the histopathological evaluation may proceed through alternating subsets of animals from each group.

### Coded (Blinded, Masked) Histopathology Evaluations

The practice of conducting the initial histopathological evaluation with the pathologist having no knowledge of treatment status of individual animals is a controversial and much debated issue [34,35]. The controversy exist primarily between pathologists and nonpathologists, as most practicing toxicologic pathologists do not endorse coded evaluations for the *initial* slide review [34,36]. The position of the Society of Toxicologic Pathology on coded evaluations is as follows [37]:

> The Society of Toxicologic Pathologists unequivocally supports *open* or *non-blinded* microscopic evaluation of tissues from experimental animals obtained from toxicology studies. The Society supports the long standing diagnostic pathology practice that a pathologist, when making a microscopic evaluation of tissues, must have access to all available information about the animals from which the tissues were derived. Over the years, this method has been proven as an efficient and effective way of generating accurate data in a setting where time, costs, and productivity must all be considered.

Similarly, the American College of Veterinary Pathologists (ACVP) stated the following position [38]:

> In the opinion of the ACVP, such procedure [coded initial evaluation] is not appropriate for the routine evaluation of slides from toxicology studies.

The foremost objection to coded slide evaluation in the initial histopathological examination is loss of knowledge of the range of normal that exists in known controls. The pathologist uses the concurrent control to establish a baseline for what is expected for a given species, strain, sex, and age of animal. Without this baseline, subtle differences between treated and control groups may be difficult to detect. Knowledge of the treatment group also allows the pathologist to assess the spectrum of related morphological changes and determine the most appropriate diagnostic terminology, including combining related diagnoses when indicated. The use of multiple diagnoses for a single disease process may obscure treatment-related effects. In addition to these considerations, the additional procedures required to code and decode data and the additional effort necessary for the pathologist to record essentially all observations (including those well within "normal") would increase study costs and timing; thus, the disadvantages of coded evaluations for the initial evaluation are both scientific and economic [36,38].

Although coded evaluations are not recommended for the initial slide review, a coded reexamination of selected target organs is commonly undertaken by toxicologic pathologists to confirm subtle changes or to clarify slight treatment-related effects on the incidence or severity of common background lesions.

## Diagnostic Nomenclature

The histopathological diagnosis is the primary means of communicating the results of the microscopic evaluation. Standardized terminology should be chosen and should clearly communicate the important aspects of tissue changes. Internationally recognized standards for diagnostic nomenclature, such as the *Standardized System of Nomenclature and Diagnostic Criteria (SSNDC): Guides for Toxicologic Pathology* [39] have been developed. A number of toxicologic pathology reference texts also provide guidance on diagnoses and diagnostic criteria [7–9,40].

The construction of a morphological diagnosis is typically hierarchical and includes the topography or site (organ or tissue), the major pathological process, and qualifiers. Qualifiers may specify subsites within the organ, distribution, duration, character, and severity [12,33,41]. An example is given in Figure 12.5. Qualifiers are used as needed to make distinctions that are toxicologically relevant. Except for the severity grade, different terminology for topography, process, or qualifiers defines a separate and distinct diagnosis; thus, the hierarchical approach is highly flexible and allows for an almost unlimited number of diagnoses. Overuse of this flexibility, however, by selecting different combinations of diagnostic terms to describe similar lesions can potentially obscure a treatment-related effect or create the appearances of an effect where none is present. In Figure 12.5, for example, some possible diagnoses might include the following:

- Liver. Necrosis, acute, centrilobular, moderate
- Liver. Necrosis, moderate (no duration or distribution qualifier)
- Liver. Necrosis, acute, coagulative, centrilobular, moderate (character qualifier added)

Although describing the same primary process, each of these diagnoses as constructed would be summarized and tabulated separately when, instead, a single diagnosis would be more appropriate; however, liver necrosis with a distinctly different distribution may be indicative of a different pathogenesis for the necrosis and thus should be diagnosed separately from the centrilobular lesion presented in Figure 12.5. As an example, subcapsular necrosis of the liver has been reported to occur following drug-induced liver microsomal enzyme induction with associated hepatomegaly and compression of hepatocytes adjacent to the liver capsule [23]. It is the responsibility of the pathologist to appropriately group lesions of similar morphology, location, and pathogenesis under a single diagnosis that best allows for detection of treatment-related changes.

Neoplasia has been defined as an abnormal mass of tissue, the growth of which exceeds and is uncoordinated with that of normal tissue and persists in the same excessive manner after cessation of the stimulus that evoked the change [43]. For neoplastic lesions, the diagnosis should indicate whether the neoplasm is benign or malignant and whether it is primary to the tissue being examined or is metastatic. Tumor multiplicity and their bilateral or

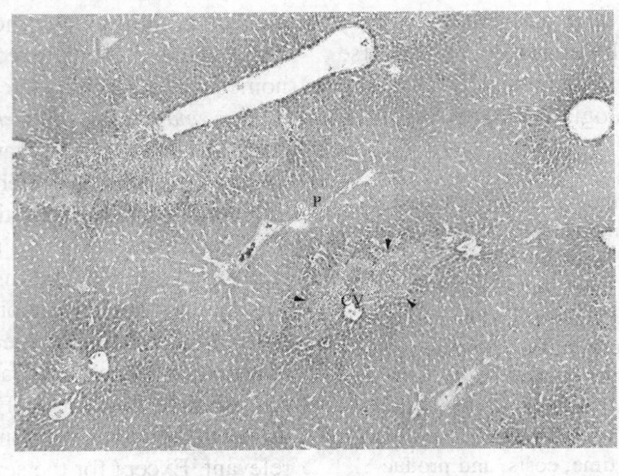


| | | Qualifiers | | |
|---|---|---|---|---|
| **Organ/Tissue** | **Primary Pathological Process** | **Duration** | **Distribution** | **Severity** |
| Liver | Necrosis | Acute | Centrilobular | Moderate |

**FIGURE 12.5** Liver from a Sprague–Dawley rat. Acute coagulative necrosis of hepatocytes (arrowheads) is centered about centra veins (CV); hepatocytes in portal areas (P) are unaffected.

unilateral presence in paired organs should also be noted. Lesions occurring secondary to the neoplasm, such as inflammation and necrosis, are generally not recorded, as these secondary changes seldom provide useful information.

Although there are exceptions, most neoplasms are classified as to their predicted biological behavior and their cell of origin. To designate biological behavior, the suffix "-oma" typically indicates that the lesion is benign, while "carcinoma" and "sarcoma" indicate malignant neoplasms of epithelial or mesenchymal origin, respectively. For histiogenesis, prefixes such as "adeno-" (glandular tissue) and "fibro-" (fibrous or connective tissue) are used to designate the cell or tissue of origin. Thus, mammary gland adenocarcinoma refers to a malignant neoplasm of glandular tissue, specifically of the mammary gland. Selected examples of taxonomy of neoplasms are given in Table 12.6 [42].

As noted earlier, the classification of proliferative lesions as hyperplasia, benign neoplasia, or malignant neoplasia is based on *predictive* biological behavior. This prediction is based on historical correlation of certain microscopic features of neoplasms to clinical behavior; for example, neoplasms are diagnosed as malignant if there is evidence of invasion or metastasis or if histological features of the neoplasm have been shown historically to correlate with invasion or metastasis. It is important to remember, however, that these are operational terms based on evaluation at one point in time, and the future progression of the lesion cannot be definitively predicted unless metastasis has already occurred. Furthermore, the morphological distinction between hyperplasia and neoplasia or benign and malignant neoplasia is not always clear; thus, although the designations hyperplasia, benign neoplasia or malignant neoplasia have practical utility, they do represent simplifications of a complex process [43].

In toxicologic pathology, diagnostic criteria for proliferative lesions that are relatively objective and reasonably predictive of biological behavior are ideal. At some points along the morphological continuum for some proliferative lesions, however, clear features that differentiate, for example, hyperplasia from adenoma are difficult to identify. In these cases, the size of the lesion may be the central (albeit not the only) criterion used to differentiate hyperplasia from adenoma. Although size should not be the only diagnostic criterion, the use of size as a central feature for some lesions adds consistency to the application of diagnostic terms and thus facilitates comparison across studies and with historical incidence data. It is important to recognize that in these cases, size may not be a reliable predictor of biological behavior; for example, for proliferative lesions of pancreatic acinar cells, the two-dimensional size of the lesion has been proposed as an important criterion, with lesions less than 5 mm in diameter being considered hyperplasia and those greater than 5 mm being considered as benign neoplasms [44]. The use of size as one of the main criteria in this instance facilitates consistency in diagnosis across studies and laboratories, and some relationship between size and biological behavior for proliferative pancreatic acinar cell lesions has been suggested [44]. However, the differential diagnosis between hyperplastic acinar cell lesions approaching 5 mm in diameter and adenomas slightly greater than 5 mm in diameter is somewhat arbitrary and not necessarily reflective of meaningful differences between the respective lesions.

**TABLE 12.6**
**Selected Taxonomy of Neoplasia**

| Tissue | Benign Neoplasia[a] | Malignant Neoplasia[b] |
|---|---|---|
| *Epithelium* | | |
| Squamous | Squamous cell papilloma | Squamous cell carcinoma |
| Transitional | Transitional cell papilloma | Transitional cell carcinoma |
| Glandular | — | — |
| Liver cell | Hepatocellular adenoma | Hepatocellular carcinoma |
| Islet cell | Islet cell adenoma | Islet cell adenocarcinoma |
| *Connective tissue* | | |
| Adult fibrous | Fibroma | Fibrosarcoma |
| Embryonic fibrous | Myxoma | Myxosarcoma |
| Cartilage | Chondroma | Chondrosarcoma |
| Bone | Osteoma | Osteosarcoma |
| Fat | Lipoma | Liposarcoma |
| *Muscle* | | |
| Smooth muscle | Leiomyoma | Leiomyosarcoma |
| Skeletal muscle | Rhabdomyoma | Rhabdomyosarcoma |
| Cardiac muscle | Rhabdomyoma | Rhabdomyosarcoma |
| *Endothelium* | | |
| Lymph vessels | Lymphangioma | Lymphangiosarcoma |
| Blood vessels | Hemangioma | Hemangiosarcoma |
| *Lymphoreticular* | | |
| Thymus | (Not recognized) | Thymoma |
| Lymph nodes | (Not recognized) | Lymphosarcoma (malignant lymphoma) |
| *Hematopoietic* | | |
| Bone marrow | (Not recognized) | Leukemia |
| | | Granulocytic |
| | | Monocytic |
| | | Erythroleukemia |
| *Neural tissue* | | |
| Nerve sheath | Neurilemmoma | Neurogenic sarcoma |
| Glioma | Glioma | Malignant glioma |
| Astrocytes | Astrocytoma | Malignant astrocytoma |
| Embryonic cells | (Not recognized) | Neuroblastoma |

[a] "-oma," benign neoplasm.
[b] "Sarcoma," malignant neoplasm of mesenchymal origin; "carcinoma," malignant neoplasm of epithelial origin.

*Source:* Haschek, W.M. and Rousseaux, C.G., Eds., *Fundamentals of Toxicologic Pathology*, Elsevier, San Diego, CA, 1998, p. 16. With permission.

## Severity Grading of Lesions

Severity grading is the semiquantitative application of a defined severity score to specific lesions [33,45]. Although a wide array of morphometic methods is available to quantitate changes in tissue, semiquantitative rating systems, if adequately defined and consistently applied, are fully capable of detecting treatment-related changes. Comparisons of results from quantitative and semiquantitative methods often show no relevant difference between the two methods in identifying a treatment-related effect or determining the no-observed-effect level for that effect; thus, the routine histopathological evaluation is typically conducted using semiquantitative methods [12,45].

Severity grading is used as a diagnosis qualifier, primarily for nonneoplastic lesions, and is especially useful in identifying treatment-related effects that are not clearly incidence based; for example, a treatment-related effect may manifest only as increased severity of a common spontaneous lesion, such as chronic nephropathy. Lesion grading, however, may not be applicable to some nonneoplastic lesions, such as cysts. Severity grading is subjective, and systems for grading may vary among pathologists; therefore, reproducibility of results from severity

**TABLE 12.7**
**Some Commonly Used Severity Grading Schemes**

**Grading Scheme I**

0 = Not present
1 = Minimal (<1%)
2 = Slight (1–25%)
3 = Moderate (26–50%)
4 = Moderately severe/high (51–75%)
5 = Severe/high (76–100%)

| **Grading Scheme II** | **A** | **B** |
|---|---|---|
| Grade 1 = Minimal | (<10%) | (0–25%) |
| Grade 2 = Mild | (10–39%) | (26–50%) |
| Grade 3 = Moderate | (40–79%) | (51–75%) |

**Grading Scheme III**

Grade 1 = Minimal
Grade 2 = Slight (same as mild)
Grade 3 = Moderate
Grade 4 = Marked (same as severe)
Grade 5 = Massive (same as very severe)

*Source*: Shackleford, C. et al, *Toxicol. Pathol.*, 30, 93, 2002. With permission.

grading requires that the grading scheme be clearly defined. Clear definitions not only provide the reviewer with an image of the spectrum of changes observed for a particular lesion but also aid in the peer review process. There are no standardized guidelines for lesion grading; however, grading schemes most commonly use four or five severity grades, designated by descriptive terms or numerical grade, to denote the extent of tissue involvement and/or the degree of tissue damage. Some commonly used grading schemes are given in Table 12.7.

### Diagnostic Drift

Diagnostic drift refers to gradual changes in nomenclature or application of severity grading scales that may occur in a single study group or across several groups in a single study or when several studies are compared. The use of multiple terms or many different qualifiers to diagnose different morphological changes that are essentially the same is one source of diagnostic drift. Terminology and severity grading may also change over the course of evaluating the study as the pathologist becomes better aware of the full spectrum of treatment-related effects. Diagnostic drift cannot be appreciated by observing a single event but rather requires numerous data points separated by time. It is more commonly a problem in large studies containing large numbers of animals and tissues that must be evaluated over a relatively long period of time. Diagnostic drift is a source of variation that, if severe enough, may falsely create or mask treatment-related changes or may complicate determination of the no-observed-effect level. A slide evaluation method used to minimize diagnostic drift is to evaluate replicates of animals across all groups—for example, five controls, five high-dose, five low-dose, five mid-dose, and so on. If diagnostic drift is clearly identified in a study, the tissue affected should be reevaluated. In these circumstances, a coded evaluation of the specific tissue and lesion in question may be beneficial [12,31,33].

## EVALUATION OF PATHOLOGY DATA

Interpretation of pathology findings requires a comprehensive assessment of gross, organ weight, and histopathological data, as well as in-life and clinical pathology findings. As noted previously, careful attention to the quality and consistency of the processes used to generate pathology data will minimize confounding factors, such as tissue sampling bias or tissue artifacts, that may mimic histopathological lesions. The pathology findings must, in turn, be considered in the context of other study factors such as study design (dose, duration, route of exposure, post-exposure recovery), test animal (species, strain, age, sex, mode of death), and animal husbandry (group vs. individual housing, *ad libitum* vs. diet optimization, caging, bedding) [46].

As with other data generated in toxicology studies, the evaluation of pathology data is primarily concerned with identifying changes that are due to treatment and determining if those changes are adverse. Although the focus of routine bioassays typically involves comparison of endpoints across distinct treatment groups, the evaluation of factors such as statistically significant differences between group means or percent change of a treated group mean from control should not occur at the expense of a careful examination of individual animal data. This is especially important in shorter term studies, which typically have group sizes of ten or fewer animals, and for endpoints whose measurements are inherently imprecise. Outlined below are some of the issues that frequently must be addressed in the evaluation of pathology findings in routine toxicology studies. Some additional issues that are most commonly encountered in carcinogenicity studies are discussed in several reviews on the subject. These include general texts on design and interpretation of long-term studies [47], application of statistics [48–50], guidelines for combining neoplasms [51], assessment of hyperplastic lesions in carcinogenicity studies [52], and the use of historical control data in carcinogenicity studies [53].

### Assessing Cause–Effect and Adversity of Pathology Findings

A number of factors should be considered when assessing whether differences between treated and control groups are due to chance or represent an effect of the test article

**TABLE 12.8**
**Discriminating Factors for Assessing Cause–Effect Relationships and Adversity of Pathology Findings**

**Discriminating factors for assessing cause–effect relationship**
There is no obvious dose response.
The group change is due to an outlier in one or more animals.
The measurement of the endpoint is inherently imprecise.
The change is within normal biological variation (historical control or reference values).
There is a lack of biological plausibility (e.g., the difference is inconsistent with class effects, mode of action, or what is known or expected of the test material).

**Discriminating factors for assessing adversity**
The effect causes no alteration in the general function of the test organism or of the organs/tissues affected.
The effect is adaptive.
The effect is transient (i.e., resolves in the course of treatment vs. reversibility, which refers to resolution with cessation of treatment).
The severity of the effect is limited (below thresholds of concern).
The effect is isolated and independent. Changes in other parameters usually associated with the effect of concern are not observed.
The effect is not a precursor (i.e., not part of a continuum of changes known to progress with time to an established adverse effect).
The effect is secondary to other adverse effects.
The effect is a consequence of the experimental model (e.g., stress associated with restraint or reactions to physical properties of the test substance, such as taste or odor).

*Source:* Adapted from Lewis, R.W. et al., *Toxicol. Pathol.*, 30, 66, 2002.

nd if treatment-related effects are adverse. Lewis et al. 54] have proposed a list of factors that can be considered n determining if changes between treated and control roups are due to the test material and if the changes are dverse. These are summarized in Table 12.8, but note hat none of the factors listed in Table 12.8 should be onsidered in isolation; rather, in assessing cause–effect r adversity, these general factors should be considered in ombination with specific information for a given study, uch as study design and known effects of the test article rom previous studies (weight of evidence approach). For xample, as for other endpoints in a study, dose–response s an important factor in determining if differences bserved in pathology endpoints between controls and reated groups are likely to represent true effects or are ue to chance. An otherwise expected dose–response may ot occur in some circumstances even when the response bserved is treatment related; for example, test article ffects at lower doses may be obscured or absent at higher oses due to overt toxicity, including lethality. Or, dose-ependent mechanisms of action, wherein different (even pposite) mechanisms of action may exist for a compound epending on dose, may also alter an expected dose esponse [54].

Outliers are extreme deviations in an individual find-ng from the group norm, as well as from historical values 54]. In the context of assessing test article-related vs. hance effects, the term *outlier* also assumes the deviation om norm is not due to the test article. Outliers that are etermined to be due to technical errors or that occur econdary to disease states unrelated to test article admin-stration do not reflect a group effect of the test article and should be removed from the analysis. Statistical outliers, however, may represent low-incidence occurrences of compound-related effects; therefore, as previously noted, a weight-of-evidence approach must be taken when determining whether to include or exclude putative outliers.

Although concurrent study controls are the first and best reference for comparison to treated groups, historical data can be a valuable tool in assessing causality (or adversity) of apparent treatment-related effects in a study. A robust historical control dataset for a given parameter may provide a better insight into the true incidence and variability of a lesion within an untreated population. As noted by Lewis et al. [54], however, the use of historical data "should not be seen as a convenient device for discounting unwanted or difficult findings." The fact that an altered value for a given parameter falls within historical values would not, in isolation, indicate a chance finding but instead may be more indicative of nonadversity of the effect. For assessing causality, historical data may be helpful in identifying aberrant values within concurrent controls or in assessing potential compound-related effects whose incidence within the control population is very low or very high and variable [54].

Several definitions of adversity have been proposed [54–56]. Although these definitions vary, common elements in defining an adverse effect include some treatment-related change (e.g., morphological, biochemical, developmental) that alters the function of an organ or system and/or alters the ability to respond to additional environmental challenges. Evaluating the adversity of an effect that has been determined to be due to exposure to a test article is complicated by a number of factors. First, the terminology associated with adverse effects is varied, with words and terms such as *adverse*, *toxicologically significant* (or *relevant*), and *biologically significant* (or *relevant*) being used interchangeably by some or having distinct meanings by others. In addition, most definitions of adversity are not accompanied by criteria for determining adversity. Finally, determining the adversity of effects usually requires case-by-case expert judgment that often precludes rote approaches, but some of the factors given in Table 12.8 may be considered in a weight-of-evidence approach to determining adversity.

### Primary vs. Secondary Effects

Some changes in organs or tissues that are observed following exposure to a test compound may be the result of primary target effects on some other organ or tissue or due to general systemic toxicity. These secondary effects are generally not considered adverse (Table 12.8); however, in some cases, such as massive iron accumulation in the spleen secondary to hemolysis, the secondary response may produce adverse effects in the affected organ.

An example of secondary effects commonly encountered in toxicology studies is changes that may occur in some organ weight parameters in association with a decrease in body weight. Failure to consider the relationship between body weight and the associated weight change that may occur in some organs may lead to misinterpretation of organ weight findings. A common example is failure to consider the effects of body weight decrements on the organ weight/body weight ratio. The absolute organ weight of some organs, such as brain and testes, is relatively unaffected by modest decrements in body weight. In these cases, because the numerator (organ weight) is constant but the denominator (body weight) is decreased, the organ weight relative to body weight for these organs is increased in association with treatment-related decreases in body weight. Absent other evidence of primary effects, this increase in organ weight relative to body weight should not be interpreted as a primary pathological change in the affected organ. In contrast, many organs, most notably the liver, decrease in weight with decreased body weight, so the ratio of liver weight to body weight may remain normal relative to controls under conditions of decreased body weight. An increase in the liver weight relative to body weight, even in association with decrements in body weight, may indicate a primary weight increase in the liver [57,58].

Bailey et al. [22] investigated the effects of body weight changes on organ weights and ratios of organ weight to body or brain weight. The organ weight relative to body weight was the most appropriate parameter to evaluate organ weight effects in liver and thyroid gland, and the organ/brain weight ratio was most appropriate for adrenal gland and ovary. For other organs, alternative methods, such as analysis of covariance, were recommended. Whereas these results focused on the rat, the body and organ weight correlations observed were considered to be generally applicable to other species [22].

Many histopathological changes may also be observed secondary to severe systemic toxicity, and these may complicate the interpretation of compound-related effects; for example, atrophy and weight loss in lymphoid organs, especially the thymus, may occur in response to general stress, which may be produced at high doses in routine bioassays. Differentiating primary immunomodulating effects from a high-dose generalized response to stress may be problematic; however, careful examination of the dose response for lymphoid organ changes may help in the evaluation. Immunosuppressive drugs often produce dose-related effects on lymphoid organs at doses not associated with other significant effects. In contrast, lymphoid changes occurring secondary to stress are expected to be limited to high doses and to occur in association with other signs of toxicity such as weight loss or general clinical suppression [59].

Another common secondary microscopic observation is atrophy of female reproductive organs due to nonspecific general toxicity associated with stress or reduced food intake, which in turn result in reduced gonadotropin secretion [60]; thus, it may be difficult to distinguish some primary effects in female reproductive organs from nonspecific secondary effects at doses of a compound that produce severe stress or body weight effects. Nevertheless, making such distinction between primary and secondary effects on reproductive organs can have important implications for reproductive hazard classification. As discussed previously, the significance of organ weight or histopathological changes must be considered in the context of other study findings, as well as any other information known about the test compound or its class.

### Effects Associated with Extreme Doses or Severe Cytotoxicity

Some histopathological changes, including neoplasia, may occur only at high doses of a test compound that overwhelm normal physiological defense mechanisms or that produce severe cytotoxicity with resultant regenerative hyperplasia. Although these effects certainly represent adverse findings to the test species, their relevance to the hazard identification process is questionable, and it is important for the pathologist to fully characterize and contrast these findings with those observed at lower, more relevant doses.

One example of histopathological changes due to excessive doses is the constellation of lung lesions that have been reported to occur in rats following chronic exposure to very high concentrations of a number of different particulates that are poorly soluble and of low inherent

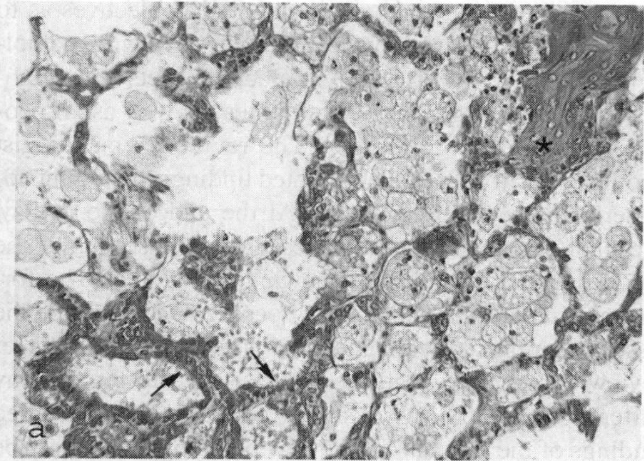


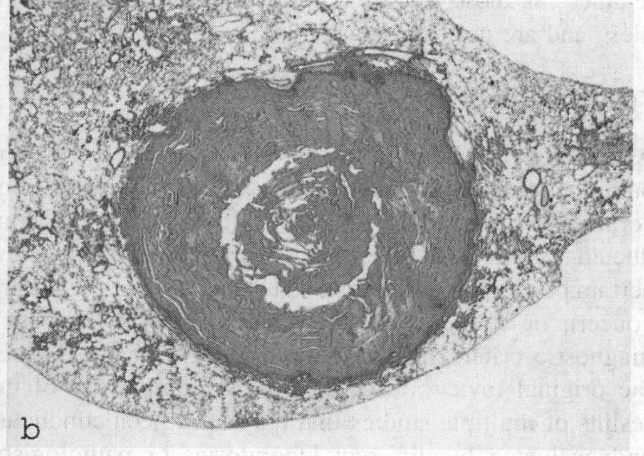


IGURE 12.6 Lung from a Sprague–Dawley rat exposed by nhalation to high concentrations of a dust. (a) Lesions are charcterized by marked accumulation of dust-laden macrophages n alveolar ducts and alveoli, with interstitial inflammation, yperplasia of type II pneumocytes, and bronchiolarization arrows) and focal squamous metaplasia (*) of alveolar ducts. b) With chronic exposure, proliferative keratinizing squamous esions, including proliferative keratin cysts, may develop.

oxicity ("lung overload") [61,62]. Microscopically, lung hanges are characterized by marked accumulation of paricle-laden macrophages in alveolar spaces, interstitial nflammation, hyperplasia of type II pneumocytes, and ronchiolarization and squamous metaplasia of alveolar ucts. With increased duration of exposure, squamous netaplasia and the formation of large cystic, keratinizing quamous lesions or squamous neoplasms may occur in ne lung (Figure 12.6). The pathogenesis of these lesions thought to involve alteration in macrophage clearance nd persistent inflammation due to the large surface dose f dust in the lungs. At lower dust concentrations that are ot associated with the marked inflammatory and adaptive esponses, proliferative squamous lesions are not observed. This finding underscores the role of high dust concentraons or surface dose and the associated inflammatory and roliferative changes in the pathogenesis of the lesions.

It is important to determine if proliferative lesions such as hyperplasia or neoplasia are due to a direct effect of the compound or are a response to a primary degenerative or necrotic event leading to regenerative hyperplasia. If hyperplasia can be clearly associated with tissue toxicity, then exposures that do not produce the primary toxic event are unlikely to produce cancer in the affected tissue [52].

### Assessing Cause of Death

Determining the cause of death in individual animals may be an important consideration in some toxicity studies. The pathologist is responsible for determining the cause of death or morbidity in animals that die prior to the scheduled necropsy and should attempt to identify a cause of death whenever possible. The pathologist should also determine if overall mortality and differences in mortality among groups is the result of the test compound. The assignment of cause of death should be included as part of the pathology peer review [63].

Determining cause of death requires a thorough knowledge of the systemic pathology of the test species and professional judgment. The pathologist should have all study data for the individual animal available, including clinical observations and clinical pathology, as well as pathology data, to make this judgment [63,64]. The cause of death for each animal should be based on the primary disease process judged to have lead to morbidity or mortality. The World Health Organization defines cause of death as "the disease or injury which initiated the train of events leading directly to death, or the circumstances of accident or violence which produced fatal injury [65]." If there are several potential causes of death, the one judged to have most likely lead to mortality should be chosen; however, more than one cause of death may be assigned if it is determined that multiple disease processes contributed significantly to morbidity or death. If the cause of death cannot be determined from the information available, then this should also be recorded [63].

## QUALITY ASSURANCE AND THE PATHOLOGY PEER REVIEW

Because pathology data often provides important information used in making regulatory decisions on the health hazard and risk to humans of many drugs, industrial chemicals, and pesticides, clearly defined procedures must be in place to ensure the accuracy of pathology data, including the generation, interpretation, and reporting of the data. These quality review procedures usually include a quantitative pathology data review (data audit or pathology materials review) and a pathology peer review. In addition, review of pathology findings for some studies may include a pathology working group review [12,66,67]. These aspects of the pathology quality assurance are discussed in more detail below.

## Pathology Data Review

The purpose of the pathology data review is to ensure the quality of the materials and procedures used to generate histopathological data, as well as other pathology data. The data review is most commonly conducted by a quality assurance unit. The pathology materials reviewed include individual animal necropsy records, histology laboratory worksheets, fixed tissues, blocks, and microscopic slides. A complete (100%) inventory of slides, blocks, and bags of remaining wet tissue should be conducted to ensure proper identification and labeling, and all slides and blocks should be matched. A random subset of slides should be examined macroscopically and microscopically to evaluate slide quality, including cover slipping, tissue placement, staining, and presence or absence of artifacts. A random subset of residual wet tissues should also be examined to verify the animal identification number. In addition, residual wet tissue may also be examined by a pathologist to ensure that no additional gross lesions are present that were either not identified at necropsy or were not trimmed. Items that may be discovered in the data review include untrimmed lesions, incorrectly identified animals, and slides that do not correspond to their respective blocks [12,66].

## The Pathology Peer Review

The pathology peer review is a procedure whereby a second pathologist reviews a subset of tissues and other data from the initial pathology evaluation. The primary purpose of the peer review is to verify the accuracy of toxicologically significant microscopic findings; however, it is not intended to corroborate every detail of every microscopic finding in a study [68]. A pathology peer review serves to ensure the integrity of the pathology evaluation, encourages consistency in the application of diagnostic criteria and terminology, and provides a method of continuing education for participants [66,69]. Documented histopathology peer reviews also increase the confidence of regulatory agencies in the pathology portion of the report [70].

The histopathology peer review may be informal (undocumented consultations) or formal, and the formal peer review may be prospective or retrospective. The prospective peer review is conducted prior to finalization of the study and is conducted by an informed reviewer, its procedures are included in the protocol, and the results of the peer review are documented in the final report. A retrospective peer review, such as that conducted by a pathology working group, generally occurs after the data are finalized. The results of a retrospective peer review should be documented in a separate report [66].

A formal peer review is conducted by an independent pathologist whose objectives are to ensure that diagnoses are accurate, that lesions are diagnosed consistently across animals and groups, and that generally accepted diagnostic criteria and nomenclature are followed [66]. Approaches may vary between study types and study objectives as to the sampling size to be evaluated by the reviewing pathologist, but the review typically includes subsets from high-dose and control groups of both sexes, as well as all neoplasms and all target organs. The reviewing pathologist should confirm all treatment-related findings, including no-observed-adverse-effect levels. At the end of the review the final diagnosis should represent the consensus of the study pathologist and the reviewing pathologist. The formal peer review should be fully documented to include the tissues examined, the diagnosis of both the study and peer review pathologists, and the actions taken to resolve any differences [66,68]. Worksheets containing the detailed findings of the primary and review pathologist need not be retained, as these are the equivalent of "pathology work files" and are not raw data [68].

## The Pathology Working Group

A pathology working group (PWG) is a panel of expert pathologists assembled to review a specific question concerning study results. A PWG review is typically conducted for finalized studies and may be convened to review certain pivotal or controversial studies that raise regulatory concern or studies that have critical effects for which diagnostic criteria and terminology have changed since the original review; also, they allow comparison of the results of multiple studies that may have been conducted and evaluated by different laboratories or pathologists. The PWG may also be used to address differences between a study and reviewing pathologist that could not be resolved during the peer review process. Sources of disagreement may include unfamiliarity with a lesion, use of different criteria for tumor classification, use of different thresholds for diagnosis of lesions (especially non-neoplastic aging lesions), use of different terminology for the same lesion, diagnostic drift, and varying pathology reporting system data input and reporting requirements. The technical aspects of the PWG have been reviewed by Mann [66] and are summarized in Table 12.9. The U.S. Environmental Protection Agency has also outlined the procedures and documentation necessary for the results of a PWG to be considered in place of the original reading (Table 12.10) [71].

## QUESTIONS

1. What documents should be available to the necropsy team prior to and during the necropsy, and what information is provided in each of these documents?
2. What are the disadvantages of coded or blinded histopathological evaluations, and what are some circumstances where a blinded evaluation would be indicated?

**TABLE 12.9**
**Pathology Working Group: Technical Approach**

**Composition**

Three to five experienced toxicologic pathologists
A chairperson
- Usually nonvoting
- Organizes and presents material so as to resolve issues in an unbiased and scientifically sound manner
- Anticipates and seeks resolution of potential problems with the data that might affect interpretation of the study

The study pathologist and reviewing pathologist
- May attend as panel members

**Procedures**

The PWG slide review is coded; members have no knowledge of treatment group or previous diagnosis.
PWG reviews representative slides of the target tissue containing the potential treatment-related changes in question.
PWG reviews all slides for which different diagnoses were recorded between the peer review and study pathologists.
PWG provides a consensus diagnosis for each slide examined:
- Consensus diagnosis is based on majority vote.
- Discussion and reexamination may be required.
- The final consensus diagnosis is recorded by the chairperson.
- No changes are made to the consensus diagnosis once the slides are uncoded.

After examination and uncoding of slides, the PWG members may be asked to utilize their expertise to discuss the biological significance of their findings

**The PWG Report**

The report is assembled by the PWG chairperson.
The PWG narrative summary includes the following:
- Incidence tables
- Comments on the diagnostic terminology used
- Morphological descriptions of the lesions examined
- An evaluation of the study pathologist's report
- Comments on the quality of the histological preparations and tissue availability
- The opinions of the PWG

*Source:* Adapted from Mann, P.C., *Toxicol. Pathol.*, 24, 650, 1996.

**TABLE 12.10**
**Pathology Working Group Procedures Required by the U.S. Environmental Protection Agency**

For any target tissue reevaluated, all slides containing that tissue in all dose groups, as well as the controls, must be reread by the peer review pathologist. This is to include the following:
- Slides previously classified by the study pathologist as within normal limits
- Slides having tumors, hyperplasia, hypertrophy, foci of cellular alteration, or other non-neoplastic lesions

The pathology reports from both the study and peer review pathologists and the original slides are to be submitted to a PWG.

The PWG will review, as a minimum, all slides about which there were significantly differing diagnoses between the study and peer review pathologists.

A detailed pathology report should be provided that presents the following:
- The PWG findings, including the original diagnosis and the new diagnosis for each slide read
- A comment column to note any discrepancies, missing slides, etc.

*Source:* Adapted from USEPA, *Requests for Reconsiderations of Carcinogenicity Peer Review Decisions Based on Changes in Pathology Diagnoses*, PR Notice 94-5, U.S. Environmental Protection Agency, Washington, D.C., 1994.

3. What is diagnostic drift, what are some factors that may cause diagnostic drift, and how can it be minimized?
4. What are some discriminating factors to consider in a weight-of-evidence approach to determine if effects are treatment related and if they are adverse?
5. What are the primary purpose and key elements of the pathology peer review?

## REFERENCES


1. *Steadman's Medical Dictionary*, 27th ed., Lippincott Williams & Wilkins, Hagerstown, MD, 2000.
2. Rousseaux, C. G., Haschek, W. M., and Wallig, M. A., Toxicologic pathology: an introduction, in *Handbook of Toxicologic Pathology*, 2nd ed., Haschek, W. M., Rousseaux, C. G., and Wallig, M. A., Eds., Academic Press, San Diego, CA, 2002, chap. 1.
3. Bucci, T. J., Basic techniques, in *Handbook of Toxicologic Pathology*, 2nd ed., Haschek, W. M., Rousseaux, C. G., and Wallig, M. A., Eds., Academic Press, San Diego, CA, 2002, chap. 8.
4. International Federations of Societies of Toxicologic Pathologists, Report of the IFSTP Professional Standards Subcommittee (PSSC), *Toxicol. Pathol.*, 31, 562, 2003.
5. Slauson, D. O. and Cooper, B. J., *Mechanisms of Disease: A Textbook of Comparative General Pathology*, 3rd ed., Mosby, St. Louis, MO, 2002.
6. Greaves, P., *Histopathology of Preclinical Toxicity Studies*, 2nd ed., Elsevier, Amsterdam, 2000.
7. Boorman, G. A., Eustis, S. L., Elwell, M. R., Montgomery, C. A., and MacKenzie, W. F., Eds., *Pathology of the Fischer Rat: Reference and Atlas*. Academic Press, San Diego, CA, 1990.
8. Haschek, W. M., Rousseaux, C. G., and Wallig, M. A., Eds, *Handbook of Toxicologic Pathology*, 2nd ed., Academic Press, San Diego, CA, 2002.
9. Maronpot, R. R., Boorman, G. A., and Gaul, B. W., Eds, *Pathology of the Mouse: Reference and Atlas*, Cache River Press, Vienna, IL, 1999.
10. Anderson, D. M., *Dorland's Illustrated Medical Dictionary*, 29th ed., W.B. Saunders, Philadelphia, PA, 2000.
11. Black, H. E., A manager's view of the 'musts' in a quality necropsy, in *Managing Conduct and Data Quality of Toxicology Studies: Sharing Perspectives and Horizons*, Hoover, B. K. et al., Eds., Princeton Scientific, Princeton, NJ, 1986, p. 249.
12. Hardisty, J. F. and Eustis, S. L., Toxicologic pathology: a critical stage in study interpretation, in *Progress in Predictive Toxicology*, Clayton, D. B. et al., Elsevier, Amsterdam, 1990, chap. 3.
13. Mann, P. C., Hardisty, J. F., and Parker, M. D., Managing pitfalls in toxicologic pathology, in *Handbook of Toxicologic Pathology*, 2nd ed., Haschek, W. M., Rousseaux, C. G., and Wallig, M. A., Eds., Academic Press, San Diego, CA, 2002, chap. 9.
14. American Veterinary Medical Association, 2000 Report of the AVMA Panel on Euthanasia, *J. Am. Vet. Med. Assoc.*, 218, 669, 2000.
15. National Research Council, *Guide for Care and Use of Laboratory Animals*, National Academy Press, Washington, D.C., 1996.
16. Li, X. et al., Morphogenesis of postmortem hepatocyte vacuolation and liver weight increases in Sprague–Dawley rats, *Toxicol. Pathol.*, 31, 682, 2003.
17. Feldman, D. B. and Seely, J. C., Eds., *Necropsy Guide. Rodents and Rabbits*, CRC Press, Boca Raton, FL, 1988.
18. Olds, R. J. and Olds, J. R., *A Colour Atlas of the Rat-Dissection Guide*, Wolfe Medical, London, 1979.
19. Bregman, C. L. et al., Recommended tissue list for histopathologic examination in repeat-dose toxicity and carcinogenicity studies: a proposal of the Society of Toxicologic Pathology (STP), *Toxicol. Pathol.*, 31, 252, 2003.
20. Latendresse, J. R. et al., Fixation of testes and eyes using a modified Davidson's fluid: comparison with Bouin's fluid and conventional Davidson's fluid, *Toxicol. Pathol.*, 30 524, 2002.
21. Fix, A. S. and Garman, R. H., Practical aspects of neuropathology: a technical guide for working with the nervous system, *Toxicol. Pathol.*, 28, 122, 2000.
22. Bailey, S. A., Zidell, R. H., and Perry R. W., Relationships between organ weight and body/brain weight in the rat what is the best analytical endpoint? *Toxicol. Pathol.*, 32 448, 2004.
23. Amacher, D. E. et al., The relationship among microsomal enzyme induction, liver weight and histological change in rat toxicology studies, *Food Chem. Toxicol.*, 36, 831, 1998
24. Greaves, P., *Histopathology of Preclinical Toxicity Studies* 2nd ed., Elsevier, Amsterdam, 2000, chap. 9.
25. Ruehl-Fehlert, C. et al., Revised guides for organ sampling and trimming in rats and mice, Part 1, *Exp. Toxicol. Pathol.* 55, 91, 2003.
26. Kittel, B. et al., Revised guides for organ sampling and trimming in rats and mice, Part 2, *Exp. Toxicol. Pathol* 55, 413, 2004.
27. Morawietz, G. et al., Revised guides for organ sampling and trimming in rats and mice, Part 3, *Exp. Toxicol. Pathol.* 55, 433, 2004.
28. Bancroft, J. D. and Gamble, M., *Theory and Practice of Histological Techniques*, 5th ed., Churchill Livingstone New York, 2001.
29. Thompson, S. W. and Luna, L. G., *An Atlas of Artifact Encountered in the Preparation of Microscopic Tissue Sections*, Charles C Thomas, Springfield, IL, 1978.
30. Prophet, E. B. et al., Eds., *Armed Forces Institute of Pathology: Laboratory Methods in Histotechnology*, America Registry of Pathology, Washington, D.C., 1992.
31. Crissman, J. W. et al., Best practices guideline: toxicologi histopathology, *Toxicol. Pathol.*, 32, 126, 2004.
32. Morgan, K. T. and Eustis, S. L., Criteria for classificatio of neoplasms for pathologists and statisticians, in *Carcinogenicity: The Design, Analysis, and Interpretation of Long-Term Animal Studies*, Grice H. C. and Ciminera, J. L., Eds., ILSI Monographs, Springer-Verlag, New York 1988, chap. 10.
33. Herbert, R. A. et al., Nomenclature, in *Handbook of Toxicologic Pathology*, 2nd ed., Haschek, W. M., Rousseaux C. G., and Wallig, M. A., Eds., Academic Press, San Diego CA, 2002, chap. 7.

34. Dodd, D. C., Blind slide reading or the uninformed versus the informed pathologist, *Comm. Toxicol.*, 2, 81, 1988
35. Temple, R. et al., The case for blind side reading, *Comm. Toxicol.*, 2, 99, 1988.
36. Goodman, D. G., Factors affecting histopathologic interpretation of toxicity-carcinogenicity studies, in *Carcinogenicity: The Design, Analysis, and Interpretation of Long-Term Animal Studies*, Grice H. C. and Ciminera, J. L., Eds., ILSI Monographs, Springer-Verlag, New York, 1988, chap. 14.
37. Society of Toxicologic Pathologists, Society of Toxicologic Pathologists' position paper on blind slide reading [editorial], *Toxicol. Pathol.*, 14, 493, 1986.
38. Prasse, K. et al., Microscopic evaluation of veterinary pathology slides, *Toxicol. Appl. Pharmacol.*, 83, 184, 1986.
39. *Standardized System of Nomenclature and Diagnostic Criteria (SSNDC): Guides for Toxicologic Pathology*, STP/ARP/AFIP, Washington, D.C.
40. Mohr, U., Ed., *International Classification of Rodent Tumors* [series], Scientific Publ. No. 122, International Agency for Research on Cancer, Oxford University Press, Oxford.
41. Glaister, J. R., General pathology, in *Principals of Toxicologic Pathology*. Taylor & Francis, London, 1986, chap. 2.
42. Haschek, W. M. and Rousseaux, C. G., Chemical carcinogenesis, in *Fundamentals of Toxicologic Pathology*, Haschek, W. M. and Rousseaux, C. G., Eds., Academic Press, San Diego, CA, 1998, chap. 2.
43. Eustis, S. L., The sequential development of cancer: a morphological perspective, *Toxicol. Lett.*, 49, 267, 1989.
44. Hansen, J. F. et al., Proliferative and other selected lesions of the exocrine pancreas in rats. GI-6, in *Guides for Toxicologic Pathology*, STP/ARP/AFIP, Washington, D.C., 1995.
45. Shackleford, C. et al., Qualitative and quantitative analysis of nonneoplastic lesions in toxicology studies, *Toxicol. Pathol.*, 30:93-96, 2002.
46. Wolf, D. C. and Mann, P. C., Confounders in interpreting pathology for safety and risk assessment., *Toxicol. Appl. Pharmacol.*, 202, 302, 2005.
47. Grice H. C. and Ciminera, J. L., Eds., *Carcinogenicity: The Design, Analysis, and Interpretation of Long-Term Animal Studies*, ILSI Monographs, Springer-Verlag, New York, 1988.
48. Gad, S. C. and Rousseaux, C. G., Use and misuse of statistics in the design and interpretation of toxicity studies, in *Handbook of Toxicologic Pathology*, 2nd ed., Haschek, W. M., Rousseaux, C. G., and Wallig, M. A., Eds, Academic Press, San Diego, CA, 2002, chap. 15.
49. Peto Analysis Working Group of the STP, Draft recommendations on classification of rodent neoplasms for Peto analysis, *Toxicol Pathol.*, 29, 265, 2001.
50. Elwell, M. et al., The Society of Toxicologic Pathology's recommendations on statistical analysis of rodent carcinogenicity studies, *Toxicol. Pathol.*, 30, 415, 2002.
51. McConnell E. E. et al., Guidelines for combining neoplasms for evaluation of rodent carcinogenesis studies, *J. Natl. Cancer Inst.*, 76, 283, 1986.
52. Boorman, G. et al., Society of Toxicologic Pathology position on assessment of hyperplastic lesions in rodent carcinogenicity studies, *Toxicol. Pathol.*, 32, 124, 2004.
53. Haseman, J. K., Huff, J., and Boorman, G. A., Use of historical control data in carcinogenicity studies in rodents, *Toxicol. Pathol.*, 12, 126, 1984.
54. Lewis, R. W. et al., Recognition of adverse and nonadverse effects in toxicity studies, *Toxicol Pathol*, 30, 66, 2002.
55. Higgins, I. T. T., What is an adverse health effect? *J. Air Pollut. Control Assoc.*, 33, 661, 1983.
56. Organisation for Economic Co-operation and Development (OECD), *Guidance Notes for Analysis and Evaluation of Repeat-Dose Toxicity Studies*, OECD Series on Testing and Assessment No. 32 and OECD Series on Pesticides No. 10, 2001, chap. 1.
57. Feron, V. J. et al., An evaluation of the criterion 'organ weight' under conditions of growth retardation, *Food Cosmet. Toxicol.*, 11, 85, 1973.
58. Oishi, S., Oishi, H., and Hiraga, K., The effect of food restriction for 4 weeks on common toxicity parameters in male rats, *Toxicol. Appl. Pharmacol.*, 47, 15, 1979.
59. Greaves, P., *Histopathology of Preclinical Toxicity Studies*, 2nd ed., Elsevier, Amsterdam, 2000, chap. 3.
60. Yuan, Y.-D. and Foley, G. I., Female reproduction system, in *Handbook of Toxicologic Pathology*, 2nd ed., Haschek, W. M., Rousseaux, C. G., and Wallig, M. A., Eds., Academic Press, San Diego, CA, 2002, chap. 43.
61. Carlton, W. W., 'Proliferative keratin cyst,' a lesion in the lungs of rats following chronic exposure to *para*-aramid fibrils, *Fundam. Appl. Toxicol.*, 23, 304, 1994.
62. Borm, P. J., Schins, R. P., and Albrecht, C., Inhaled particles and lung cancer. Part B. Paradigms and risk assessment, *Int. J. Cancer*, 110, 3, 2004.
63. Long, G., Recommendations to guide determining cause of death in toxicity studies, *Toxicol. Pathol.*, 32, 269, 2004.
64. Ettlin, R. A., Stirnimann, P., and Prentice, D. E., Causes of death in rodent toxicity and carcinogenicity studies, *Toxicol. Pathol.*, 22, 165, 1994.
65. WHO, Rules and guidelines for mortality and morbidity coding, in *International Statistical Classification of Diseases and Related Health Problems*, ICD-10, World Health Organization, Geneva, 1993.
66. Mann, P. C., Pathology peer review from the perspective of an external peer review pathologist, *Toxicol. Pathol.*, 24, 650, 1996.
67. Boorman, G. A. et al., Quality review procedures necessary for rodent pathology databases and toxicogenomic studies: the National Toxicology Program experience, *Toxicol. Pathol.*, 30, 88, 2002.
68. Society of Toxicologic Pathology, Commentary: documentation of pathology peer review; position of the Society of Toxicologic Pathologists, *Toxicol. Pathol.*, 25, 655, 1997.
69. Ward, J. M. et al., Peer review in toxicologic pathology, *Toxicol Pathol.*, 23, 226, 1995.
70. USEPA, *Pathology Raw Data Definition As It Relates to Pathology Data Trails and Independent Pathology Peer Review System*, PR Notice 87-10, U.S. Environmental Protection Agency, Washington, D.C., 1987.
71. USEPA, *Requests for Reconsiderations of Carcinogenicity Peer Review Decisions Based on Changes in Pathology Diagnoses*, PR Notice 94-5, U.S. Environmental Protection Agency, Washington, D.C., 1994.

# Notes

# 13 The Information Infrastructure of Toxicology

*Philip Wexler, Pertti (Bert) Hakkinen, Patricia Nance, Ann Parker, and Jacqueline Patterson*

## CONTENTS



## SCOPE OF TOXICOLOGY

The U.S. Society of Toxicology (SOT) developed the following definition of toxicology in 2005:

> Toxicology is the study of the adverse effects of chemical, physical or biological agents on living organisms and the ecosystem, including the prevention and amelioration of such adverse effects. Discussion: Toxicity is the adverse end product of a series of events that is initiated by exposure to chemical, physical or biological agents. Toxicity can manifest itself in a wide array of forms, from mild biochemical malfunctions to serious organ damage and death. These events, any of which may be reversible or irreversible, include absorption, transport, metabolism to more or less toxic metabolites, excretion, interaction with cellular macromolecules and other modes of toxic action. Toxicology integrates the study of all of these events, at all levels of biological organization, from molecules to complex ecosystems. The broad scope of toxicology, from the study of fundamental mechanisms to the measurement of exposure, including toxicity testing and risk analysis, requires an

> extensively interdisciplinary approach. This approach utilizes the principles and methods of other disciplines, including molecular biology, chemistry (analytical, organic, inorganic and biochemistry), physiology, medicine (veterinary and human), computer science and informatics.

This definition reflects the multidisciplinary nature of toxicology. Indeed, toxicology data and information can be distributed via a number of resources such as paper and electronic journals and books, databases, websites, professional societies, trade associations, and government organizations. Data are generated from laboratory animal or *in vitro* studies (e.g., cell culture), or via *in silico* methods (i.e., using computer programs to estimate the toxicity of chemicals). In addition, limited ethically responsible human studies can provide a degree of toxicology and safety information.

Finding the best ways to keep current with the vast amounts of literature and other information associated with the many aspects of toxicology is both a challenge and an opportunity. An earlier review article examined the web's expanding information role in the intersecting disciplines of toxicology and risk analysis [2]. Significant advances have been made over the years in ease of access to toxicological information, and 24-hour-a-day access is now expected and, in the vast majority of cases, readily available. The vast array of websites, in addition to Internet mailing lists, provides numerous opportunities for training and information sharing.

## SCOPE OF INFORMATION RESOURCES AND THE ROLE OF THE WEB

Information in most established scientific disciplines is widespread and, typically, diffuse, despite continuing efforts to integrate, coordinate, and consolidate it. In the not so distant past, things used to be simpler. In the predigital era, information was either oral or written and on a physical support such as paper, clay tablets, parchment, or vellum. Oral information referred largely to the collegial network–colleagues communicating with each other in face-to-face discussions or via telephone calls as a means of diffusing knowledge. Written information took on a discrete number of forms–textbooks and other monographs, technical journals, newsletters, dissertations, abstracting and indexing services, etc. With the advent of computers and, particularly, computer databases, the nature of information began to change, and with the onset and flourishing of the web, information and its infrastructure have taken on a level of complexity hardly dreamt of even by those already conversant with an earlier era of database searching.

Today, the lines demarcating one kind of information from another are increasingly blurring, as more and more packaged and synthesized information, not to mention raw data, and the invariable unsubstantiated musings are all democratically finding their ways onto the web. Although bibliographic information and summaries of standard technical documentation as reflected in books and journals already have a robust history of online access, the goal of *full* text digital availability has lagged behind. It will likely just be a matter of time before such documentation becomes routinely digitized and more readily accessible. Elsevier's Science Direct (http://www.sciencedirect.com) is one such collection of full-text literature. Authors seem to be increasingly accepting of the concept of online publication [10].

Online libraries are in our future. Project Gutenberg (http://www.gutenberg.org/), begun in 1971, is a producer of over 13,000 free electronic books on the Internet. It consists largely of older literary works in the public domain in the United States, but a scientific counterpart is inevitable. The issue of who will ultimately pay for this, and how remains to be resolved and will be explored a bit more at the end of this chapter. More recently, in 2004, the company that created the web search engine Google™ announced that it is working with libraries at the University of Michigan, Harvard University, Stanford University, Oxford University, and the New York Public Library to digitize books in their collections and make them accessible online. The list of libraries has since expanded. This massive scanning project will bring millions of volumes of printed books into the Google Print database (http://print.google.com/) for around-the-clock and around-the-world free access. Interestingly, a group of European libraries has made plans to move forward with a similar complementary effort to put European literature online (http://europa.eu.int/information_society/activities/digital_libraries).

Virtually every organization of relevance to toxicology, as reflected here, has a web presence. Activities that would once have required reams of paper to describe and would have resulted in a distribution nightmare are usually consolidated on websites where the user comes to the information instead of *vice versa*. Consider government agencies such as the U.S. Environmental Protection Agency (EPA) and the U.S. Food and Drug Administration (FDA), whose extensive websites are rife with information. The challenge here, if anything, is how to manage and efficiently navigate (and then digest) the large amount of information and data. Utilization of new *push technologies*, which will be discussed later, offers an alternative approach to delivering information.

The formerly strict lines between multimedia and computerized representation have been long erased. Digital audio or video transmissions via the Internet (i.e., streaming media) are commonplace, and as bandwidth increases quality will improve. Computer technologies used for entertainment hold great potential for educational applications. Indeed, many websites provide audio and video access to live meetings, and instructional videos are increasingly incorporating tools such as Flash™ and ActionScript™ to visually enhance learning.

## MONOGRAPHS

This section provides a selective list of books in five areas dealing with toxicology, as well as an "other" category. The five main areas considered are general toxicology, industrial/occupational toxicology, environmental toxicology, clinical toxicology, and risk assessment. The "other" category is comprised of books on a variety of topics, such as risk communication, forensic toxicology, and more. This list offers mostly recent books published since 2000 but also includes older books that are considered classics in the field.

### General Toxicology

Ballantyne, B., Marrs, T., and Syversen, T. (1999): *General and Applied Toxicology*, 2nd ed., Macmillan, London.

Derelanko, M. J. and Hollinger, M. A. (2001): *Handbook of Toxicology*, 2nd ed., CRC Press, Boca Raton, FL.

Gilbert, S. G. (2004): *A Small Dose of Toxicology: The Health Effects of Common Chemicals*, CRC Press, Boca Raton, FL.

Hardman, J. G., Limbird, L. E., and Gilman, A. G. (2001): *Goodman & Gilman's The Pharmacological Basis of Therapeutics*, 10th ed., McGraw-Hill, New York.

Harris, R., Bingham, E., Cohrssen, B., and Powell, C. (2000): *Patty's Industrial Hygiene and Toxicology*, 5th ed., Wiley, New York.

Hayes, A. W. (2007): *Principles and Methods of Toxicology*, 5th ed., Taylor & Francis, Boca Raton, FL.

Illing, P. (2001): *Toxicity and Risk*, CRC Press, Boca Raton, FL.

Kent, C. (1998): *Basics of Toxicology*, Wiley, New York.

Klaassen, C. D. (2003): *Casarett & Doull's Essentials of Toxicology*, McGraw-Hill, New York.

Klaassen, C. D. (2001): *Casarett & Doull's Toxicology: The Basic Science of Poisons*, 6th ed., McGraw-Hill, New York.

Lewis, R. A. (1998): *Lewis' Dictionary of Toxicology*, CRC Press, Boca Raton, FL.

Lu, F. C. and Kacew, S. (2002): *Lu's Basic Toxicology: Fundamentals, Target Organs and Risk Assessment*, 4th ed., Taylor & Francis, London.

Timbrell, J. A. (2001): *Introduction to Toxicology*, 3rd ed., CRC Press, Boca Raton, FL.

Vettorazzi, G. (2001): *The ITIC International Dictionary of Toxicology*, ITIC Press, San Sebastián, Spain.

Wexler, P. (2005): *Encyclopedia of Toxicology*, Elsevier, Oxford.

Wexler, P., Hakkinen, P. J., Kennedy, G., and Stoss, F. (2000): *Information Resources in Toxicology*, 3rd ed., Academic Press, San Diego, CA (4th edition to be published by Elsevier in 2008).

Williams, P. L. and Burson, J. L. (2000): *Principles of Toxicology: Environmental and Industrial Applications*, 2nd ed., Wiley, New York.

Woolley, A. (2003): *A Guide to Practical Toxicology: Evaluation, Prediction, and Risk*, Taylor & Francis, London.

### Clinical Toxicology

Barile, F. A. (2003): *Clinical Toxicology: Principles and Mechanisms*, American Association for Clinical Chemistry (AACC) Press, Washington, D.C.

Dart, R. C. (2003): *Medical Toxicology*, 3rd ed., Lippincott Williams & Wilkins, Philadelphia, PA.

Delaney, K. A., Ling, L. J., Erickson, T., and Ford, M. D. (2000): *Clinical Toxicology*, W.B. Saunders, Philadelphia, PA.

Goldfrank., L., Flomenbaum, N., Lewin, N., Howland, M. A., Hoffman, R., and Nelson, L. (2002): *Goldfrank's Toxicologic Emergencies*, 7th ed., McGraw-Hill Professional, New York.

Gupta, S. K., Kaleekal, T., and Peshin, S. S. (2003): *Emergency Toxicology: Management of Common Poisons*, Narosa Publishing House, New Delhi.

Haddad, L. M., Shannon, M. W., and Winchester, J. F. (1998): *Clinical Management of Poisoning and Drug Overdose*, 3rd ed., W.B. Saunders, Philadelphia, PA.

Rossoff, I. S. (2001): *Encyclopedia of Clinical Toxicology: A Comprehensive Guide and Reference*, Parthenon, New York.

Sullivan, J. B. and Krieger, G. R. (2001): *Clinical Environmental Health and Toxic Exposures*, 2nd ed., Lippincott Williams & Wilkins, Philadelphia, PA.

### Environmental Toxicology

Baird, C. (1998): *Environmental Chemistry*, 2nd ed., W.H. Freeman, New York.

Crosby, D. G. (1998): *Environmental Toxicology and Chemistry*, Oxford University Press, Oxford.

Hoffman, D. J., Rattner, B. A., Burton, G. A., and Cairns, J. (2002): *Handbook of Ecotoxicology*, 2nd ed., CRC Press, Boca Raton, FL.

Landis, W. G. and Yu, M. (2004): *Introduction to Environmental Toxicology: Impacts of Chemicals Upon Ecological Systems*, 3rd ed., CRC Press, Boca Raton, FL.

Manahan, S. E. (2004): *Environmental Chemistry*, 8th ed., CRC Press, Boca Raton, FL.

Moriarty, F. (1999): *Ecotoxicology: The Study of Pollutants in Ecosystems*, 3rd ed., Academic Press, London.

Newman, M. C., and Unger, M. A. (2002): *Fundamentals of Ecotoxicology*, 2nd ed., CRC Press, Boca Raton, FL.

Rand, G. M. and Petrocelli, S. R. (1985): *Fundamentals of Aquatic Toxicology*, Hemisphere Publishing, Washington, D.C.

Schüürmann, G. and Markert, B. (1997): *Ecotoxicology: Ecological Fundamentals, Chemical Exposure, and Biological Effects*, Environmental Science and Technology: A Wiley–Interscience Series of Texts and Monographs, Wiley, New York.

Streit, B. and Braunbeck, T. (1997): *Encyclopedic Dictionary of Ecotoxicology and Environmental Chemistry*, Taylor & Francis, New York.

Suter, G., Efroymson, R., Sample, B., and Jones, D. (2000): *Ecological Risk Assessment for Contaminated Sites*. CRC Press, Boca Raton, FL.

U.S. Environmental Protection Agency. (2000): *Ecological Risk Assessment: Federal Guidelines*, ABS Consulting, Houston, TX.

Walker, C. H., Hopkin, S. P., Sibley, R. M., and Peakall, D. B. (2001): *Principles of Ecotoxicology*, 2nd ed., CRC Press, Boca Raton, FL.

Wright, D. A., Welbourn, P., Campbell, G. C., Harrison, R. M., and deMora, S. J. (2001): *Environmental Toxicology*, Cambridge University Press, Cambridge, U.K.

Yen, T. F. (2005): *Environmental Chemistry: Chemistry of Major Environmental Cycles*, World Scientific, Hackensack, NJ.

Zakrzewski, S. F. (2002): *Environmental Toxicology*, 3rd ed., Oxford University Press, Oxford.

## Industrial Hygiene/Occupational Health


American Conference of Governmental Industrial Hygienists (ACGIH), Cincinnati, OH, http://www.acgih.org; numerous relevant publications, including their TLVs and BEIs and documentation for deriving them.

American Industrial Hygiene Association (AIHA), Reston, VA, http://www.aiha. org; numerous relevant publications, including their *Emergency Response Planning Guidelines and Workplace Environmental Exposure Level Handbook*.

Berger, E. H., Ward, W. D., Royster, J. C., and Morrill, L. H. (1986): *Noise and Hearing Conservation Manual*, American Industrial Hygiene Association, Reston, VA.

Boleij, J. S., Buringh, E., Heederik, D., and Kromhout, H. (1995): *Occupational Hygiene of Chemical and Biological Agents*, Elsevier, New York.

Burke, R. (2002): *Hazardous Materials Chemistry for Emergency Responders*, 2nd ed., CRC Press, Boca Raton, FL.

DiBerardinis, L. J. (1998): *Handbook of Occupational Safety and Health*, 2nd ed., Wiley, New York.

Greenberg, M., Hamilton, R., Philips, S., and McCluskey, G. J. (2003): *Occupational, Industrial, and Environmental Toxicology*, 2nd ed., Mosby, St. Louis, MO.

Franklin, C. and Worgan, J. (2005): *Occupational and Residential Exposure Assessment for Pesticides*, Wiley, Hoboken, NJ.

Friend, M. A. (2003): *Fundamentals of Occupational Safety and Health*, 3rd ed., Government Institutes, Rockville, MD.

Hathaway, G. and Proctor, N. H. (2004): *Proctor & Hughes' Chemical Hazards of the Workplace*, 5th ed., Wiley, New York.

Lewis, R. J. (2004): *Sax's Dangerous Properties of Industrial Materials*, 11th ed., Wiley, New York.

McCunney, R. J. (2003): *A Practical Approach to Occupational and Environmental Medicine*, 3rd ed., Lippincott Williams & Wilkins, Philadelphia, PA.

Plog, B. A. and Quinlan, P. J. (2001): *Fundamentals of Industrial Hygiene*, 5th ed., National Safety Council, Itasca, IL.

Que Hee, S. (1993): *Biological Monitoring: An Introduction*, Wiley, New York.

Stellman, J. M. (1998): *Encyclopedia of Occupational Health and Safety*, 4th ed., International Labor Office, Geneva.

Williams, P. L. and Burson, J. L. (2000): *Principles of Toxicology: Environmental and Industrial Applications*, 2nd ed., Wiley, New York.


## Risk and Risk Assessment


Asante-Duah, K. (2002): *Public Health Risk Assessment for Human Exposure to Chemicals*, Springer, New York.

Benjamin, S. L. and Belluck, D. A. (2001): *A Practical Guide to Understanding, Managing, and Reviewing Environmental Risk Assessment Reports*, CRC Press, Boca Raton, FL.

Bradley, J. (2002): *Elimination of Risk in Systems: Practical Principles for Eliminating and Reducing Risk in Complex Systems*, Tharsis, Canada.

Byrd, D. M. and Cothern, C. R. (2000): *Introduction to Risk Analysis: A Systematic Approach to Science-Based Decision Making*, Government Institutes, Rockville, MD.

Chavas, J. P. (2004): *Risk Analysis in Theory and Practice*, Academic Press, London.

Covello, V. T. and Merkhoher, M. W. (1993): *Risk Assessment Methods: Approaches for Assessing Health and Environmental Risks*, Plenum Press, New York.

Cox, L. A. (2001): *Risk Analysis: Foundations, Models, and Methods*, Kluwer Academic, London.

Haimes, Y. Y. (2004): *Risk Modeling, Assessment, and Management*, 2nd ed., Wiley–Interscience, New York.

Hyatt, N. (2003): *Guidelines for Process Hazards Analysis (PHA, HAZOP), Hazards Identification, and Risk Analysis*, CRC Press, Boca Raton, FL.

Knopman, D., Lockwood, J. R., Cecchine, G., Willis, H., and Macdonald, J. (2004): *Unexploded Ordnance: A Critical Review of Risk Assessment Methods*, RAND Corp., Santa Monica, CA.

Lachin, J. M. (2000): *Biostatistical Methods: The Assessment of Relative Risks*, Wiley–Interscience, New York.

Lundgren, R. E. and McMakin, A. H. (2004): *Risk Communication: A Handbook for Communicating Environmental, Safety, and Health Risks*, 3rd ed., Battelle Press, Columbus, OH.

McDaniels, T. and Small, M. (2004): *Risk Analysis and Society: Interdisciplinary Perspectives*, Cambridge University Press, Cambridge, U.K.

Morgan, M. G., Fischhoff, B., Bostrom, A., and Atman, C. J. (2001): *Risk Communication: A Mental Models Approach*, Cambridge University Press, Cambridge, U.K.

Paustenbach, D. (2002): *Human and Ecological Risk Assessment: Theory and Practice*, Wiley, New York.

Ropeik, D. and George, G. (2002): *Risk: A Practical Guide for Deciding What's Really Safe and What's Really Dangerous in the World Around You*, Houghton Mifflin, New York.

Slovic, P. (2000): *The Perception of Risk*, Earthscan Publications, London.

Wilson, R., Edmund, A., and Crouch, C. (2001): *Risk–Benefit Analysis*, 2nd ed., Harvard University Press, Boston, MA.


## Other


Calabrese, E. (1994): *Biological Effects of Low-Level Exposures: Dose–Response Relationships*, CRC Press, Boca Raton, FL.

Fenton, J. (2001): *Toxicology: A Case-Oriented Approach*, CRC Press, Boca Raton, FL.

Harris, J. (2000): *Criminal Poisoning: Investigational Guide for Law Enforcement, Toxicologists, Forensic Scientists, and Attorneys*, Trestrail Humana Press, Totowa, NJ.

Haschek, W. M. and Rousseaux, C. G. (1998): *Fundamentals of Toxicologic Pathology*, Academic Press, San Diego, CA.

Holladay, S. D. (2004): *Developmental Immunotoxicology*, CRC Press, Boca Raton, FL.

Plumlee, K. H. (2003): *Clinical Veterinary Toxicology*, Mosby, St. Louis, MO.

Levine, B. (2003): *Principles of Forensic Toxicology*, 2nd ed., American Association for Clinical Chemistry (AACC) Press, Washington, D.C.

Lide, D. R. (2003): *Handbook of Chemistry and Physics*, 84th ed., CRC Press, Boca Raton, FL.

Reddy, M., Yang, R. S., Andersen, M. E., and Clewell, H. J. (2005): *Physiologically Based Pharmacokinetic Modeling: Science and Applications*, Wiley–Interscience, Hoboken, NJ.

Roder, J. D. (2001): *Veterinary Toxicology: The Practical Veterinarian*, Butterworth–Heinemann, Philadelphia, PA.

Walker, J. D. (2003): *QSARs for Pollution Prevention, Toxicity Screening, Risk Assessment, and Web Applications*, SETAC Press, Pensacola, FL.

## JOURNALS AND NEWSLETTERS

Great strides have been made in the early twenty-first century to provide more immediate and easy access to scientific publications. Journals covering many areas of toxicology and related disciplines that affect toxicological research are published by various groups, including commercial publishers, societies, and government agencies. Traditional print journals are giving way to electronic format, which is speeding the transfer of new research and findings to scientists around the world. Free online journals and publishing services go even further to share results with an even wider audience.

Scientific journals covering research and other topics in toxicology have been with us since at least the mid-nineteenth century. Among the successors to key early journals begun in the 1930s and 1940s are *Archives of Toxicology*, *Pharmacology and Toxicology*, *Toxicology and Applied Pharmacology*, and *Eksperimentalnaia i Klinicheskaia Farmakologiia*.

Today, dozens of journals focus on toxicology, either broadly or on special subareas, such as biomarkers, carcinogenesis, and metabolism. In addition, many others focus on applications of toxicology in drug development or in settings such as the ambient environment or the workplace. Other journals cover research and analysis that crosses multiple disciplines and address issues—for example, in environmental or public health.

The most significant advancement in the last decade has been the proliferation of access via the Internet and other electronic media. Many journals now provide subscribers and others with early notification of publications (tables of contents). These services alert readers to the latest research quickly and can often be tailored to the user's particular interests. Some journals provide direct access to abstracts online for paid subscribers and the general public, and many now provide full-text online for subscribers or for single article purchase. Further innovations in access to journal literature are described later in this chapter.

Many journals now encourage authors to submit their articles electronically to facilitate faster peer review and acceptance. Journal peer reviews are often handled through e-mail or the Internet, and the use of these technologies has shortened review time considerably. Some journals release accepted papers in manuscript form almost immediately upon acceptance. These papers have not yet been copyedited or formatted for publication and are eventually replaced with the official publication version; however, they are considered published and are citable.

To keep current with available journals is difficult. Names change, as do publishers. New journals are born and others cease publication. The Internet is the easiest way to find a journal of interest, as most publishers have websites for their journals, where one can find information on contents, subscriptions, submissions, and sometimes access to abstracts and full text of articles. *NewJour* is a comprehensive and free listing of electronic journals and newsletters on the Internet and is updated frequently (http://gort.ucsd.edu/newjour/NewJourWel.html). In addition, one can access *PubList* (http://www.publist.com/), a free service that provides searches of a database of over 150,000 magazines, journals, newsletters, and other periodicals from around the world; *Ulrich's Periodicals Directory*, a bibliographic database with information on more than 260,000 print and electronic periodicals (http://www.bowker.com/brands/ulrichs.htm); and the U.S. Library of Congress, which has online catalogs of its extensive holdings, including journals (http://www.loc.gov/).

Table 13.1 provides some key toxicology journals, largely in English, and their publishers. These journals cover toxicology and the related areas of occupational and environmental health, ecotoxicology, risk analysis, and medicine. Readers can easily locate current websites for these journals or their publishers via a standard Web search engine, such as Google™. It should be emphasized that this is a selective list of journals. Compared to journals, newsletters offer a generally less formal but sometimes more compelling means of keeping up to date in a subject. A few of the more prominent online toxicology newsletters are listed in Table 13.2.

## ORGANIZATIONS

An extensive overview of toxicology-related organizations was published by Hakkinen, Stoss, Behrendt, and Wexler (2000). Many different types of organizations are associated with toxicology. These include professional societies, trade associations, government organizations, nongovernmental organizations (NGOs), centers or departments of universities, and companies. Key examples of these types of organizations are described below, together with key publications and websites providing compilations of this type of information. The reader is urged to exercise caution when considering data and advice from the Web, and should carefully consider the source. The authors of this chapter are not responsible for the information provided by the organizations noted, including the content of their sites. Further, the authors do not necessarily endorse any products or services mentioned by the organizations.

**TABLE 13.1**
**Selected Toxicology Journals and Publishers**

| Title | Publisher |
|---|---|
| *Adverse Drug Reaction Bulletin* | Lippincott Williams & Wilkins |
| *Adverse Drug Reactions and Toxicological Reviews* | Adis International |
| *American Journal of Industrial Medicine* | Wiley |
| *Annals of Occupational Hygiene* | Oxford University Press |
| *Annals of the ICRP* | Elsevier |
| *Annual Review of Pharmacology and Toxicology* | Annual Reviews |
| *Aquatic Toxicology* | Elsevier |
| *Archives of Environmental Contamination and Toxicology* | Springer |
| *Archives of Environmental Health* | Heldref Publications |
| *Archives of Toxicology* | Springer-Verlag |
| *Bulletin of Environmental Contamination and Toxicology* | Springer-Verlag |
| *Carcinogenesis* | Oxford University Press |
| *Cell Biology and Toxicology* | Springer |
| *Chemical Research in Toxicology* | ACS Publications |
| *Chemico Biological Interactions* | Elsevier |
| *Chemosphere* | Elsevier |
| *Comparative Biochemistry and Physiology. Part C: Pharmacology, Toxicology, and Endocrinology* | Elsevier |
| *Contact Dermatitis* | Blackwell |
| *Critical Reviews in Toxicology* | Taylor & Francis |
| *Cutaneous and Ocular Toxicology* | Taylor & Francis |
| *Drug and Chemical Toxicology* | Taylor & Francis |
| *Ecotoxicology* | Springer |
| *Environmental Health Perspectives* | U.S. National Institute of Environmental Health Sciences |
| *Environmental Research* | Elsevier |
| *Environmental Toxicology* | Wiley |
| *Environmental Toxicology and Chemistry* | SETAC Press (Society for Toxicology and Chemistry) |
| *Environmental Toxicology and Pharmacology* | Elsevier |
| *Experimental and Toxicologic Pathology* | Elsevier |
| *Food and Chemical Toxicology* | Elsevier |
| *Free Radical Biology and Medicine* | Elsevier |
| *Human and Experimental Toxicology* | Hodder Arnold |
| *Inhalation Toxicology* | Taylor & Francis |
| *International Journal of Toxicology* | Taylor & Francis |
| *Journal of Analytical Toxicology* | Preston Publications |
| *Journal of Applied Toxicology* | Wiley |
| *Journal of Exposure Analysis and Environmental Epidemiology* | Nature Publishing Group |
| *The Journal of Occupational and Environmental Hygiene* | Taylor & Francis |
| *Journal of Occupational and Environmental Medicine* | Lippincott Williams& Wilkins |
| *Journal of Pharmacological and Toxicological Methods* | Elsevier |
| *Journal of Toxicology. Clinical Toxicology* | Taylor & Francis |
| *Journal of Toxicology and Environmental Health* | Taylor & Francis |
| *Molecular Carcinogenesis* | Wiley |
| *Mutagenesis* | Oxford University Press |
| *NeuroToxicology* | Elsevier |
| *Neurotoxicology and Teratology* | Elsevier |
| *Regulatory Toxicology and Pharmacology* | Elsevier |
| *Reproductive Toxicology* | Elsevier |
| *Risk Analysis* | Blackwell Publishing |
| *Science of the Total Environment* | Elsevier |
| *Teratogenesis, Carcinogenesis, Mutagenesis* | Wiley |
| *Teratology* | Wiley |
| *Toxicologic Pathology* | Taylor & Francis |
| *Toxicological Reviews* | Adis International |
| *Toxicological Sciences* | Oxford University Press |
| *Toxicology* | Elsevier |
| *Toxicology and Applied Pharmacology* | Elsevier |
| *Toxicology and Industrial Health* | Hodder Arnold |
| *Toxicology in Vitro* | Elsevier |
| *Toxicology Letters* | Elsevier |
| *Toxicology Mechanisms and Methods* | Taylor & Francis |
| *Toxicon* | Elsevier |
| *Xenobiotica* | Taylor & Francis |

**TABLE 13.2**
**Online Toxicology Newsletters**

| Newsletter | URL |
|---|---|
| *Biological Effects of Low Level Exposures (BELLE)* | http://www.belleonline.com/toc.html |
| *CIIT Centers for Health Research Newsletter* | http://www.ciit.org/news_events/activities.asp |
| *Environmental Toxicology Newsletter (UC Davis)* | http://extoxnet.orst.edu/newsletters/ucdnl.htm |
| *Greenfacts.org* | http://greenfacts.org/ |
| *Harvard Center for Risk Analysis* | http://www.hcra.harvard.edu/risk.html |
| *National Institute of Occupational Safety and Health (NIOSH) E-News* | http://www.cdc.gov/niosh/enews/ |
| *Society for Risk Analysis (SRA) Newsletter* | http://www.sra.org/newsletter.php |
| *Society of Toxicology (SOT) Communiqué Newsletter* | http://www.toxicology.org/Information/publications/communique.html |
| *Trends in Risk & Remediation (Gradient)* | http://www.gradientcorp.com/coinfo/trends.html |

## PROFESSIONAL SOCIETIES

Professional societies associated with toxicology have been reviewed by Kehrer and Mirsalis [5] and Patterson et al. [7].

### American College of Toxicology (ACT)

The mission of the American College of Toxicology (http://www.actox.org) is to educate and lead professionals in industry, government, and related areas of toxicology by actively promoting the exchange of information and perspectives on the current status of safety assessment and the applications of new developments in toxicology. ACT maintains an outstanding collection of toxicological-related links to its website. The ACT newsletter is available on the website, as are announcements of upcoming meetings.

### International Association of Forensic Toxicologists (TIAFTnet)

The aims of TIAFT (http://www.cbft.unipd.it/tiaft/) are to promote cooperation and coordination of efforts among members and to encourage research in forensic toxicology. The website contains meeting information, grants, and new developments, including the newest compilation of blood or plasma levels of more than 800 substances and active metabolites.

### International Society of Exposure Analysis (ISEA)

The International Society of Exposure Analysis (http://www.iseaweb.org/) was established in 1989 to foster and advance the science of exposure analysis related to environmental contaminants, both for human populations and ecosystems. The membership promotes communication among all disciplines involved in exposure analysis, recommends exposure analysis approaches to address substantive or methodological concerns, and works to strengthen the impact of exposure assessment on environmental policy. ISEA's newsletters are available from this website, as are announcements for upcoming ISEA meetings.

### International Union of Toxicology (IUTOX)

The International Union of Toxicology (http://www.iutox.org) members include SOT and over other societies around the world. IUTOX now has 47 national/ regional Society members representing over 20,000 toxicologists from industry, academia and government as members. Its purpose is to foster international scientific cooperation among national and other groups of toxicologists and promote worldwide acquisition, dissemination, and utilization of toxicology knowledge. The IUTOX website provides links to the sites of all available member societies (http://www.iutox.org/toxlinks.asp).

### Society for Risk Analysis (SRA)

The Society for Risk Analysis (http://www.sra.org) is a multidisciplinary group designed to provide opportunities to exchange information, ideas, and methodologies for risk analysis and risk problem solving. The Society publishes a peer-reviewed journal, *Risk Analysis*, which provides a focal point for new developments in risk analysis for scientists from a wide range of disciplines. The journal covers health risks; engineering, mathematical, and theoretical aspects of risks; and social and psychological aspects of risk such as risk perception, acceptability, economics, and ethics. The Society's newsletter and annual meeting abstracts are available from the website.

### Society of Forensic Toxicologists (SOFT)

The Society of Forensic Toxicologists (http://www.soft-tox.org/) is an organization composed of practicing forensic toxicologists and those interested in the discipline for the purpose of promoting and developing forensic toxicology. The website contains meeting information, education, research, employment opportunities, publications, and related links.

### Society of Toxicology (SOT)

The Society of Toxicology (http://www.toxicology.org/) is perhaps the best-known professional society associated with toxicology and has a global, albeit mostly U.S., membership. SOT is an organization of scientists who practice toxicology in many areas. The Society holds annual meetings, publishes a journal and a member newsletter, and sponsors continuing education courses. The Society addresses toxicological issues through several of its specialty sections. It maintains an excellent collection of toxicology-related websites of various organizations, along with other sites of interest for toxicologists and for people considering a career in, or just seeking information about, toxicology (http://www.toxicology.org/AI/CRAD/careerguide.asp).

## Trade Associations

Many trade associations are active in toxicology efforts and evaluations. Listed below are a number of broad chemical trade associations. Many other more specialized groups represent more specific areas.

### American Chemistry Council (ACC)

The American Chemistry Council (http://www.americanchemistry.com/), formerly the Chemical Manufacturers Association, is the voice of the U.S. chemical industry. The ACC represents the chemical industry on public policy issues, coordinates the industry's research and testing programs (e.g., the Long-Range Research Initiative [LRI]), and administers the industry's environmental, health, and safety performance improvement initiative, known as Responsible Care®. Members include corporations in the chemical and chemical-using (e.g., consumer product) industries.

### British Industrial Biological Research Association (BIBRA)

The British Industrial Biological Research Association (http://www.bibra-information.co.uk) members include British chemical manufacturers. Its activities include the development of summaries (toxicology profiles) of the data for numerous chemicals. Further, BIBRA can keep members of its Information Advisory Service up to date with new developments or provide tailored service to a company.

### European Chemical Industry Council (Cefic)

The European Chemical Industry Council (http://www.cefic.be) is both the forum and the voice of the European chemical industry. It represents, directly or indirectly, chemical companies that account for nearly a third of world chemical production. It has numerous working groups related to chemical safety and the regulation of chemicals, and sponsors research related to toxicology and other risk assessment.

### European Centre for Ecotoxicology and Toxicology of Chemicals (ECETOC)

The focus of ECETOC (http://www.ecetoc.org/) includes the manufacture, processing, handling, and use of chemicals. ECETOC also cooperates with governmental agencies and other organizations concerned with the effects of chemicals on health and the environment.

### International Council of Chemical Associations (ICCA)

The International Council of Chemical Associations (http://www.icca-chem.org/) is the global voice of the chemical industry, representing chemical manufacturers and producers. It accounts for more than 75% of chemical manufacturing operations. ICCA promotes and coordinates Responsible Care® and other voluntary chemical industry initiatives and has a central role in the exchange of information within the international industry and in the development of position statements on matters of policy. It is also the main channel of communication between the industry and various international organizations that are concerned with health, environment, and trade-related issues.

## Government Organizations

Government organizations serving as toxicology information resources are numerous and include not only the U.S. Environmental Protection Agency [8] and other federal agencies [1] but also state and local governments.

### Environment Canada and Health Canada

The Canadian Environmental Protection Act (CEPA) requires the establishment of a Priority Substances List (PSL) to identify substances that are of priority for assessing whether environmental exposure to them poses a risk to the health of Canadians or to the environment. Over 60 Priority Substances have been evaluated, and some of the assessment documents containing tolerable intakes, tolerable concentrations, and tumorigenic doses and concentrations for these substances are available through their website (http://www.hc-sc.gc.ca/ewh-semt/contaminants/existsub/index_e.html).

### European Union, European Commission, Joint Research Centre, Institute for Health and Consumer Protection, European Chemicals Bureau (ECB)

Through its Existing Chemicals work area, the European Bureau (http://ecb.jrc.it/) conducts data collection, priority setting, and risk assessment of existing chemicals. Risk assessment documents of Existing Chemicals are available in the Existing Chemicals section of the ECB website, along with information about the International Uniform Chemical

Information Database (IUCLID), the European Union System for the Evaluation of Substances (EUSES), and the Harmonized Electronic Data Set (HEDSET). Through its New Chemicals work area, the ECB is responsible for notification of new chemical substances to be placed on the European Market. EUSES is an integrated modeling system that uses a single framework for comparing the potential risks of different substances released to multiple environmental media (water, soil, and air) and multiple human exposure pathways (inhalation, ingestion, and dermal). EUSES can be used for indirect human exposures and for consumer product and worker exposures. This website also provides access to the forthcoming REACH (Registration, Authorisation, and Regulation of Chemicals) regulatory framework. The main access to REACH information will eventually be through the website of the European Chemicals Agency (http://ec.europa.edu/echa/).

The European Commission's Joint Research Centre, Institute for Health and Consumer Protection, also includes the European Centre for the Validation of Alternative Methods (ECVAM). ECVAM is "the reference centre, at an international level, for the development, scientific and regulatory acceptance of alternative testing methods aimed at replacing, reducing or refining the use of laboratory animals and to be applied in different fields in the biomedical sciences." The Scientific Information Service (SIS; http://ecvam-sis.jrc.it/index.html) is a database that provides factual and evaluated information on advanced non-animal test development and validation for toxicology assessments coming from a wide range of international information sources.

## National Institute of Public Health and the Environment (RIVM), The Netherlands

The knowledge and expertise of RIVM (http://www.rivm.nl/en/) in the fields of health, nutrition, and environmental protection (including the conduct of research, monitoring, modeling, and risk assessment) are used primarily for advising the Dutch government. As an example of its efforts, RIVM develops human toxicological risk limits (maximum permissible risks, or MPRs). Reports for a variety of chemicals based on chemical assessments are compiled in the framework of the Dutch governmental program on risks in relation to soil quality. These MPR values are published in RIVM reports, many of which can be downloaded from the Publications section of this site.

## U.S. Centers for Disease Control and Prevention (CDC), Department of Health and Human Services

The Agency for Toxic Substances and Disease Registry (ATSDR; http://www.atsdr.cdc.gov) is an agency within the CDC. Its functions include public health assessments of waste sites, health consultations concerning specific hazardous substances, health surveillance and registries, response to emergency releases of hazardous substances, applied research in support of public health assessments, information development and dissemination, and education and training concerning hazardous substances. ATSDR develops toxicological profiles (http://www.atsdr.cdc.gov/toxpro2.html) for hazardous substances found at National Priority List sites and for the Department of Defense and the Department of Energy for substances related to federal sites. Within these documents, ATSDR develops minimal risk values (MRLs; see http://www.atsdr.cdc.gov/mrls.html). Also available is ATSDR's ToxFAQs™ (http://www.atsdr.cdc.gov/toxfaq.html), a series of summaries about hazardous substances and their health effects. Information for this series is excerpted from the ATSDR Toxicological Profiles and Public Health Statements. The ATSDR Science Corner (http://www.atsdr.cdc.gov/cx.html) is a user-friendly gateway to environmental health information and resources.

The National Institute for Occupational Safety and Health (NIOSH; http://www.cdc.gov/niosh/homepage.html) is another part of the CDC and is the federal agency responsible for conducting research and making recommendations for the prevention of work-related injury and illness. Its objectives include the conduct of research to reduce work-related illnesses and injuries, the promotion of safe and healthy workplaces through interventions, recommendations and capacity building, and the enhancement of global workplace safety and health through international collaborations. The NIOSH website provides access to information on numerous health and safety topics, including emerging issues such as nanotechnology.

## U.S. Consumer Product Safety Commission (CPSC)

The Consumer Product Safety Commission (http://www.cpsc.gov/) protects the U.S. public against risk of injury or harm from consumer products. Among its efforts, the CPSC evaluates and develops standards and guidelines for safety issues and regulations for labeling and packaging. It has also performed exposure-related research as part of the development of risk assessments for exposures to various chemicals in consumer products.

## U.S. Department of Energy (DOE)

The Department of Energy's Center for Risk Excellence (http://riskcenter.doe.gov) develops and implements policy practices, guidance, tools, support, and training that result in risk-based decisions that protect both human health and the environment. Its website offers information and links to sources of information on toxicology and risk assessment tutorials, fact sheets for chemicals, and risk tools. Particularly useful are the Risk Assessment Tools and Models (http://

rais.ornl.gov/cre_tools.shtml). The DOE's Risk Assessment Information System (RAIS; http:/rais.ornl.gov/) is a website developed to provide a service-oriented environmental risk assessment expert system. RAIS provides tools for performing basic risk assessment activities, such as preliminary remediation goals, toxicity values and profiles (including the EPA's IRIS and HEAST), federal and state guidelines, human health risk models, and ecological benchmarks.

### U.S. Environmental Protection Agency (EPA)

The mission of the EPA, founded in 1970, is to protect human health and the environment. Several of its components are of particular relevance to toxicology. The National Center for Environmental Assessment (NCEA; http://www.epa.gov/ncea/), for example, serves as the EPA national resource center for the overall process of human health and ecological risk assessments, including the integration of hazard, dose–response, and exposure data and models to produce risk characterizations. The Office of Pollution Prevention and Toxics (OPPT; http://www.epa.gov/opptintr/) promotes pollution prevention, safer chemicals, risk reduction, and public understanding of risks. Its Office of Research and Development (ORD; http://www.epa.gov/ord/) is the scientific research arm of EPA. ORD's research helps provide the solid underpinning of science and technology for EPA. It has developed the Environmental Information Management System (EIMS; http://www.epa.gov/eims/eims.html) to organize and provide access to EPA resources through a searchable database.

Numerous databases accessible through the EPA's website present a wealth of relevant and reliable data. The Integrated Risk Information System (IRIS; http://www.epa.gov/iris) is the EPA's consensus database of information on human health effects that may result from exposure to various chemicals found in the environment, including values such as oral reference doses (RfDs) and inhalation reference concentrations (RfCs) for noncarcinogenic health effects, as well as oral slope factors and oral and inhalation unit risks for carcinogenic effects. The Toxics Release Inventory (TRI; http://www.epa.gov/tri) is a database containing information focused on the estimated numbers of pounds of certain toxic chemicals released to the environment, augmented by source reduction and recycling data. Both IRIS and TRI are accessible via the EPA's website and the National Library of Medicine's TOXNET system (http://toxnet.nlm.nih.gov). Recent informatics efforts in genomics at EPA and NIEHS are discussed further later in this chapter.

### U.S. Food and Drug Administration (FDA)

The Food and Drug Administration (http://www.fda.gov/) is an agency consisting of a number of centers within the Department of Health and Human Services with the mission of protecting the public health of Americans by helping safe and effective products reach the market in a timely way, monitoring products for continued safety after they are in use, and helping the public get accurate, science-based information. The Center for Biologics Evaluation and Research (CBER; http://www.fda.gov/opacom/factsheets/justthefacts/) regulates biological products for disease prevention and treatment. The Center for Devices and Radiological Health (CDRH; http://www. fda.gov/opacom/factsheets/justthefacts/5cdrh.html) ensures that new medical devices are safe and effective before they are marketed; monitors devices throughout the product life cycle, including a nationwide postmarket surveillance system; and ensures that radiation-emitting products (e.g. microwave ovens, television sets, cell phones, and laser products) meet radiation safety standards. The Center for Drug Evaluation and Research (CDER; http://www.fda.gov/opacom/factsheets/justthefacts/) evaluates all new prescription and over-the-counter drugs before they are sold and serves as a consumer watchdog for marketed drugs to be sure they continue to meet the highest standards. The Center for Food Safety and Applied Nutrition (CFSAN; http://www.fda.gov/opacom/factsheets/justthefacts/ 2cfsan.html) is responsible for the safety of the entire food supply, except for meat, poultry, and some egg products that are regulated by the U.S. Department of Agriculture. In addition, CFSAN has developed rapid methods for the detection of microbial and viral food contaminants and works closely with public and private sector partners to operate systems for rapid identification and control of outbreaks of foodborne diseases. The Center for Veterinary Medicine (CVM; http://www.fda.gov/opacom/factsheets/justthefacts/) helps ensure that animal food products are safe and evaluates the safety and effectiveness of drugs used to treat companion animals (e.g. dogs, cats, and horses). The mission of the National Center for Toxicological Research (NCTR; http://www.fda.gov/nctr/index.html) is to conduct peer-reviewed scientific research that supports and anticipates the FDA's current and future regulatory needs. This involves fundamental and applied research specifically designed to define biological mechanisms of action underlying the toxicity of products regulated by the FDA. This research is aimed at understanding critical biological events in the expression of toxicity and at developing methods to improve the assessment of human exposure, susceptibility, and risk.

### U.S. National Institutes of Health (NIH)

The major information component of the NIH is the National Library of Medicine (NLM). The TOXNET system (http://toxnet.nlm.nih.gov) is a group of databases managed by the NLM's Toxicology and Environmental Health Information Program (TEHIP), situated within its Specialized Information Services (SIS) Division. The data banks and biblio graphic files are built, maintained, and funded by several federal agencies. Its databases are described later in this chapter. Further, NLM's Specialized Information Services (SIS) provides online toxicology training via its Toxicology Tutor I, II, and III (http://sis.nlm.nih.gov/enviro/toxtutor.html) and provides extensive information about resources related to toxicology and environmental health education (http://sis.nlm.nih.gov/enviro/envirohealthlinks.html), including academic program directories, continuing education and tutorials, distance learning, etc.

Another NIH institute particularly relevant to toxicology is the National Institute of Environmental Health Sciences (NIEHS; http://www.niehs.nih.gov). Its website links to resources from the National Toxicology Program (NTP). The NTP consists of the relevant toxicology activities of the NIEHS, NIOSH, and NCTR. NTP's goals are to provide toxicological evaluations on substances of public health concern, develop and validate test methods, develop approaches and generate data that strengthen the scientific basis for risk assessments, and communicate program plans and results to all stakeholders including governmental agencies, the medical and scientific communities, and the public. The NTP website also provides access to NTP testing information and study results, the NTP Report on Carcinogens (RoC), chemical health and safety information, special reports, and announcements, as well as links to the websites for the NTP Center for the Evaluation of Alternative Toxicological Methods (NICEATM) and the Center for the Evaluation of Risks to Human Reproduction (CERHR).

### Other Organizations

Readers should take note that, as with all the lists in this chapter, the compilation of organizations is highly selective with regard to a number of variables, including geography. There are important toxicology related groups and activities (for example, in Asia, Africa, and Latin America) which, because of space limitations, could not be included here.

## International Organizations

### Intergovernmental Forum on Chemical Safety (IFCS)

The Intergovernmental Forum on Chemical Safety (http://www.who.int/ifcs/) was established to coordinate and monitor international and regional action related to sound chemicals management and to identify and build consensus on chemicals assessment and management priorities. The IFCS is an overarching, participatory forum where governments meet with intergovernmental and nongovernmental organizations to discuss chemical safety issues and provide policy guidance for the sound management of chemicals, to be implemented by national governments and organizations. The World Health Organization is the administering agency for the IFCS and its Secretariat.

### International Agency for Research on Cancer (IARC)

The International Agency for Research on Cancer (http://www.iarc.fr/) is part of the World Health Organization, and its mission is to coordinate and conduct research on the causes of human cancer and the mechanisms of carcinogenesis, as well as to develop scientific strategies for cancer control. IARC is involved in both epidemiological and laboratory research and disseminates information through publications, meetings, courses, and fellowships. Cancer databases and other resources at IARC include the Monographs Database (a complete list of agents, mixtures, and exposures, all evaluated with their classifications) and the Cancer Epidemiology Database among others.

### International Consumer Products Health and Safety Organization (ICPHSO)

The International Consumer Products Health and Safety Organization (http://www.icphso.org/) is dedicated to health and safety issues related to consumer products manufactured and marketed in the global marketplace. ICPHSO includes a broad range of health and safety professionals and interested consumers, and sponsors workshops to inform and educate manufacturers, importers, distributors, retailers, and others of their product safety responsibilities.

### International Labour Organization (ILO)

The International Labour Organization (http://www.ilo.org/) is the United Nations (UN) specialized agency seeking the promotion of social justice and internationally recognized human and labor rights. Of particular interest to toxicologists is its International Occupational Safety and Health Information Centre (CIS), the knowledge management arm of the InFocus Programme on Safety and Health at Work and the Environment (SafeWork) (http://www.ilo.org/public/english/protection/safework/index.htm). The goal of the CIS is to ensure that workers and everyone concerned with their protection have access to the facts they need to prevent occupational injuries and diseases.

### International Programme on Chemical Safety (IPCS)

The IPCS (http://www.who.int/pcs/index.htm) is a joint program of the World Health Organization, the International Labour Organization, and the United Nations Environmental Programme. The purpose of IPCS is to establish the scientific basis for safe use of chemicals and to strengthen national capabilities and capacities for chemical safety. The organization evaluates chemical risks to human health and the environment and develops methodologies for these evaluations. It publishes a variety of high-quality, peer-reviewed monographs, such as the Environmental Health Criteria documents. The OECD/IPCS Database on Hazard/Risk Assessment Methodologies (http://appli1.oecd.org/ehs/ipcs.nsf, a searchable database) is the outcome of a joint project of the OECD and the IPCS and includes extensive listings and links to hazard/risk assessment methodologies for industrial chemicals and pesticides.

The IPCS Harmonization Project (http://www.who.int/ipcs/methods/harmonization/en/) is the result of the IPCS having taken the lead to globally harmonize approaches to the assessment of risk from exposure to chemicals. The issue areas include cancer, mutagenicity, reproductive and developmental toxicity, non-neoplastic effect, neurotoxicity, immunotoxicity, and terminology. IPCS INCHEM (http://www.inchem.org) provides free access to hundreds of publications about chemicals from the international organizations that cooperate with the IPCS. These publications are produced and peer reviewed by teams of experts from around the world and are intended for a wide range of professionals concerned about the safe use of chemicals. IPCS INTOX (http://www.intox.org) is a computerized poisons information package designed to assist poison centers, health ministries, and other related institutions to efficiently manage information related to poisoning, national product registration, and chemical incidents. The INTOX package consists of a database and collection of documents on a variety of potentially poisonous substances.

### Strategic Approach to International Chemicals Management (SAIC)

The SAIC (http://www.chem.unep.ch/saicm), adopted by the International Conference on Chemicals Management (ICCM) in 2006, is a policy framework for international action on chemical hazards. It supports the goal of ensuring that, by the year 2020, chemicals are produced and used in ways that minimize significant adverse impacts on the environment and human health.

## Nongovernmental Organizations (NGOs)

### CIIT Centers for Health Research

CIIT (http://www.ciit.org/) is a private, not-for-profit research organization created by the U.S. chemical industry to conduct independent research on the effects of chemicals. CIIT research is concentrated in four major areas: (1) toxicological and physiological studies to assess responses of the organism and cell to chemical exposures; (2) high-throughput genomics with *in vivo*, *in vitro*, and *ex vivo* preparations to catalog and evaluate tissue and cell responses; (3) computational analysis of biological data using simulation and bioinformatics to interpret these responses; and (4) quantitative modeling of dynamic systems that predict dose–response behavior of these biological responses under realistic exposures. The major focus of research currently underway at CIIT is on reproductive and developmental biology and respiratory biology. In implementing a systems biology approach to health effects research, CIIT is expanding its research capabilities in genomics technology and developing a functional genomics program.

### Health Effects Institute (HEI)

The Health Effects Institute (http://www.healtheffects.org/) is an independent, nonprofit corporation chartered in 1980 to provide high-quality, impartial, and relevant science on the health effects of pollutants from motor vehicles and from other sources in the environment. Supported jointly by the U.S. Environmental Protection Agency and industry, HEI has funded hundreds of studies and published over 100 research reports and several special reports that provide important research findings on the health effects of a variety of pollutants, including carbon monoxide, methanol and aldehydes, nitrogen oxides, diesel exhaust, ozone, and, most recently, particulate air pollution.

### International Life Sciences Institute (ILSI)

The International Life Sciences Institute (http://rsi.ilsi.org/) is a nonprofit, worldwide scientific research foundation seeking to improve the well being of the general public through the pursuit of sound and balanced science. Its goal is to further the understanding of scientific issues relating to nutrition, food safety, toxicology, risk assessment, and the environment.

### Toxicology Excellence for Risk Assessment (TERA)

Toxicology Excellence for Risk Assessment (http://www.tera.org) is an independent nonprofit research and education organization dedicated to the best use of toxicity data in risk assessment. TERA develops risk values; improves methods for human health risk assessment through their research program; sponsors expert review of risk assessments, risk values, methods, and research through their independent peer review and peer consultation programs; provides technical support, training courses, and risk communication to diverse groups through their education program; and compiles and distributes peer-reviewed risk values to the international user community through the International Toxicity Estimates for Risk (ITER) database described later in this chapter.

## University-Affiliated Organizations

### Extension Toxicology Network (EXTOXNET)

EXTOXNET (http://ace.orst.edu/info/extoxnet) is a effort of the University of California at Davis, Oregon State University, Michigan State University, Cornell Un versity, and the University of Idaho. EXTOXNET provides information on pesticides, discussion of concepts in toxicology and environmental chemistry, and fact sheets. This information has been developed by toxicologists and chemists within the Extension Service of these universities with the goal of developing unbiased information in a format understandable by the nonexpert.

### Harvard Center for Risk Analysis (HCRA)

The Harvard Center for Risk Analysis (http://www.hcra.harvard.edu) attempts to provide a big-picture overview of public health by comparing and ranking a wide range of hazards, analyzing the results of dealing with those hazards in various ways, and developing sound scientific data identifying policy choices that are most likely to achieve the greatest health, safety, and environmental benefits with the most efficient use of finite resources. The Center's current research programs focus on the areas of motor vehicle safety, medical technology, environmental health, food, agriculture, and children's health. The Center has a newsletter entitled *Risk in Perspective*, which discusses results of research projects.

### The Johns Hopkins University

Under the management of The Johns Hopkins Center for Alternatives to Animal Testing (CAAT), a diverse group of organizations serve on the Altweb Project Team, many of which maintain their own websites that provide key links from and to Altweb (http://altweb.jhsph.edu/). The intent of Altweb is to be the online clearinghouse for resources, information, and news about alternatives to animal testing and to serve as the most comprehensive resource on animal alternatives for scientists, educators, veterinarians, and individuals throughout the world. This effort is designed to bring together government agencies, the academic community, animal protection groups, and private industry to encourage the use of alternative methods.

## DATABASES AND WEBSITES

The Internet plays a vital role in the provision of information resources. We have access to sources, including full-text journal articles, online that in the past were only available by physically visiting a library. The following websites and databases are intended to serve as a selective list and do not approach the scope of a comprehensive compilation.

### Cal/EPA–OEHHA Toxicity Criteria Database

(http://www.oehha.ca.gov/risk/chemicaldb)—This toxicity criteria database is maintained by the California Office of Environmental Health Hazard Assessment (OEHHA). The database provides California Public Health goals, chronic reference exposure levels, cancer classification, cancer potency values, and Proposition 65 No Significant Risk Levels.

### Canadian Centre for Occupational Health and Safety (CCOHS)

(http://www.ccohs.ca/)—The Canadian Centre for Occupational Health and Safety (CCOHS) promotes a safe and healthy working environment by providing publications, fact sheets, pocket guides, Workplace Hazardous Materials Information System (WHMIS) criteria, and Web information service databases. The Web databases include RTECS, MSDS plus CHEMINFO, CHEMpendium, and Occupational Safety and Health (OSH) references and legislation regulations. Some of the databases are restricted to subscribers.

### Dialog

(http://www.dialog.com)—Dialog is a vast database collection with full-text access to most of the search results. The database offers comprehensive, global coverage of biomedical research, chemicals, computer science, energy and environment, health industry, mechanical and civil engineering, medical practice, medical devices, pharmaceuticals, software, therapy and treatment breakthroughs, and drug interactions. Dialog database content is available through Telnet and desktop (Windows®) software or through a Web browser format with DialogWeb. Dialog is available by subscription or on a pay-per-use basis with a credit card through the DialogWeb browser. Below are examples of a few of the databases that can be searched within Dialog:

- *BIOSIS Previews*® contains citations from *Biological Abstracts*® (BA) and *Biological Abstracts/Reports, Reviews, and Meetings*® (BA/RRM; formerly *BioResearch Index*®), the major publications of BIOSIS®. Together, these publications constitute the major English- language service providing comprehensive worldwide coverage of research in the biological and biomedical sciences.
- *CANCERLIT*® is produced by the International Cancer Research DataBank Branch (ICRDB) of the U.S. National Cancer Institute. The database consists of bibliographic records referencing cancer research publications dating from 1963 to the present.
- *EMBASE* has long been recognized as an important, comprehensive index of the world's literature on human medicine and related disciplines.
- *SciSearch*®, a cited reference science database, is an international, multidisciplinary index to the literature of science, technology, biomedicine, and related disciplines produced by

Thomson. It contains all of the records published in the *Science Citation Index*® (SCI®), plus additional records in engineering technology, physical sciences, agriculture, biology, environmental sciences, clinical medicine, and the life sciences.

- *ToxFile* covers the toxicological, pharmacological, biochemical, and physiological effects of drugs and other chemicals: adverse drug reactions, chemically induced diseases, carcinogenesis, mutagenesis, teratogenesis, environmental pollution, waste disposal, radiation, and food contamination are typical areas of coverage.

### ESIS (European Chemical Substances Information System)

(http://ecb.jrc.it/esis)—This database is a wealth of information for European chemical information. The following areas can be searched: European Inventory of Existing Commercial Substances (EINECS), which contains general chemical information such as CAS Registry Numbers, EINECS numbers, substance names, chemical formulas, structure, classification, and labeling (risk and safety phrases, danger, etc.); ELINCS European List of Notified Chemical Substances (ELINCS); high production volume chemicals (HPVCs) and low production volume chemicals (LPVCs), including EU producer/importer lists; IUCLID Chemical Data Sheets (in Adobe and OECD format); priority lists; and risk assessment processes and tracking systems in relation to Council Regulation (EEC) 793/93, also known as Existing Substances Regulation (ESR).

### EXICHEM

(http://webdomino1.oecd.org/ehs/exichem.nsf)—The EXICHEM database is maintained and updated by the Organization for Economic Cooperation and Development (OECD). It is a pointer system on current, planned, and completed activities on existing chemicals in OECD member countries and other relevant bodies. It was created to provide information for the OECD member countries on who is doing what with which chemicals (e.g., information gathering, testing, and evaluation).

### Haz-Map®

(http://hazmap.nlm.nih.gov/)—This occupational toxicology database is designed primarily for health and safety professionals but also for consumers seeking information about the health effects of exposure to chemicals at work. It contains approximately 1000 chemicals or biological agents and links jobs and hazardous tasks with occupational diseases and their symptoms. This association indicates an increased risk for significant exposure and subsequent disease.

### ILPI–MSDS (Finding Material Safety Data Sheets on the Internet)

(http://www.ilpi.com/msds/index.html)—This website is a comprehensive MSDS resource maintained by Interactive Learning Paradigms, Inc. (ILPI). The website includes everything from links to material safety data sheets to OSHA regulations and interpretations.

### MedWeb

(http://www.medweb.emory.edu/MedWeb/)—MedWeb is a catalog of biomedical and health-related websites maintained by the staff of the Robert W. Woodruff Health Sciences Center Library at Emory University. The website includes resources in toxicology, biological and physical sciences, pharmacology, clinical practice, and publications

### National Pesticide Information Center (NPIC)

(http://npic.orst.edu)—The National Pesticide Information Center is a cooperative agreement between Oregon State University and the U.S. Environmental Protection Agency. The website contains information about pesticide safety and use, publications, fact sheets, health databases, regulations, and many pesticide links.

### National Service Center for Environmental Publications (NSCEP)

(http://www.epa.gov/ncepihom/index.htm)—The National Service Center for Environmental Publications maintains and distributes EPA publications in hardcopy, CD-ROM, and other multimedia formats. The current publication inventory includes over 7000 titles, which are free to all requestors.

### NIOSH Databases and Information Resources

(http://www.cdc.gov/niosh/database.html)—This National Institute for Occupational Safety and Health website contains a vast array of occupational health and safety resources. The most popular databases include the International Chemical Safety Cards (ICSC), NIOSH Pocket Guide to Chemical Hazards, Immediately Dangerous to Life and Health (IDLH) values, and NIOSHTIC-(a bibliographic database of occupational safety and health publications).

### Pesticide Action Network (PAN) Pesticide Database

(http://www.pesticideinfo.org)—The PAN Pesticide Database is a project of Pesticide Action Network North America (PANNA) and has been supported by grants from the EPA and a number of foundations. The database is a diverse array of information on pesticides from many di ferent sources that provides human toxicity (chronic and acute), ecotoxicity, and regulatory information for about 5400 pesticide active ingredients and their transformation products, as well as adjuvants and solvents used in pesticide products. References to data sources are provided.

### Preliminary Remediation Goals (PRGs)

(http://www.epa.gov/region09/waste/sfund/prg/index.html) —Preliminary Remediation Goals are tools for evaluating and cleaning up contaminated sites in the U.S. Superfund and the Resource Conservation and Recovery Act programs. They are risk-based concentrations that are intended to assist risk assessors and others in initial screening-level evaluations of environmental measurements. The PRGs contained in the Region 9 PRG table are generic. They are calculated without site-specific information; however, they may be re-calculated using site specific data.

### PubMed®

(http://pubmed.gov)—PubMed, a service of the National Library of Medicine, includes over 15 million citations for biomedical articles back to the 1950s. These citations are from MEDLINE and additional life science journals. PubMed includes links to many sites providing full-text articles and other related resources.

### Registry of Toxic Effects of Chemical Substances (RTECS®)

(http://www.mdli.com/; http://ccinfoweb.ccohs.ca/rtecs/search.html)—The Registry of Toxic Effects of Chemical Substances helps users find critical toxicological information by providing citations on over 150,000 chemical substances from more than 2500 sources. The RTECS® database was previously maintained by the National Institute for Occupational Safety and Health and is now licensed through MDL Information Services, Inc. It is available by subscription via the web or by CD-ROM from MDL or licensed dealers such as CCOHS.

### RxList: The Internet Drug Index

(http://www.rxlist.com/)—The Internet Drug Index provides general descriptions and information on clinical pharmacology, indications and doses, precautions, side effects, interactions, overdosage, and contradictions.

### ScienceDirect

(http://www.sciencedirect.com/)—Since its launch in 1997, ScienceDirect has evolved from a web database of Elsevier journals to one of the world's largest providers of scientific, technical, and medical literature. Access to full text articles is available to subscribers or on a pay-per-view basis.

### Scirus

(http://www.scirus.com/srsapp/)—Scirus is a comprehensive science-specific internet search engine. Driven by the latest search engine technology, Scirus searches over 167 million science-specific web pages.

### Scopus

(http://www.scopus.com/)—Scopus is a navigation tool covering the world's largest collection of abstracts, references, and indexes of scientific, technical, and medical literature. Updated daily, it includes the abstracts and cited references of over 14,000 titles from more than 4000 international publishers. Seamless links to full-text articles and other library resources make Scopus quick, easy, and comprehensive. Access is only available through subscribing libraries.

### Scorecard

(http://www.scorecard.org/)—Environmental Defense's Scorecard is an Internet service that provides information about chemical releases in the United States. It provides information about local air pollution, including interactive maps that can be accessed by Zip Code; information on toxic chemicals released by manufacturing facilities; and information about the health risks of air pollution. Scorecard ranks and compares pollution in areas across the United States and profiles 6800 chemicals and shows locations in which they are used, as well as their hazards. Scorecard integrates over 400 scientific and governmental databases to generate its profiles.

### TOXNET® (Toxicology Data Network)

(http://toxnet.nlm.nih.gov; as a link on http://sis.nlm.nih.gov)—TOXNET® is a group of databases covering toxicology, hazardous chemicals, environmental health, and related areas. It is managed by the Toxicology and Environmental Health Information Program (TEHIP) in the Division of Specialized Information Services of the National Library of Medicine. TOXNET provides free access and easy searching of the following databases:

- Chemical Carcinogenesis Research Information System (CCRIS)—A scientifically evaluated and fully referenced data bank developed and maintained by the National Cancer Institute (NCI), CCRIS contains over 9200 chemical records with carcinogenicity, mutagenicity, tumor promotion, and tumor inhibition test results. Data are derived from studies cited in primary journals, current awareness tools, NCI reports, and other special sources. Test results have been reviewed by experts in carcinogenesis and mutagenesis.

- ChemIDplus—A database providing access to structure and nomenclature authority databases used for the identification of chemical substances cited in National Library of Medicine databases, ChemIDplus contains over 368,000 chemical records, of which over 235,000 include chemical structures. ChemIDplus is searchable by name, synonym, CAS Registry Number, molecular formula, classification code, locator code, and structure.
- Developmental and Reproductive Toxicology/Environmental Teratology Information Center (DART®/ETIC)—A bibliographic database covering literature on reproductive and developmental toxicology, DART®/ETIC is managed by the National Library of Medicine and funded by the EPA, the National Institute of Environmental Health Sciences, and the NLM. DART®/ETIC contains references to reproductive and developmental toxicology literature published since 1965.
- GENE-TOX (Genetic Toxicology)—This toxicology database was created by the EPA and contains genetic toxicology test results on over 3000 chemicals. Selected literature was reviewed by scientific experts for each of the test systems under evaluation; the results are represented in GENE-TOX.
- Hazardous Substances Data Bank (HSDB®)—This factual database focuses on the toxicology of over 5000 potentially hazardous chemicals, but provides information in the areas of emergency handling procedures, industrial hygiene, environmental fate, human exposure, detection methods, and regulatory requirements. The data are fully referenced and peer reviewed by a scientific review panel composed of expert scientists.
- Integrated Risk Information System (IRIS; http://www.epa.gov/iris)—This database is managed by the National Center for Environmental Assessment of the EPA and contains carcinogenic and noncarcinogenic health-risk information on over 540 chemicals. These chemical files contain descriptive and quantitative information about oral reference doses (RfDs) and inhalation reference concentrations (RfCs) for chronic noncarcinogenic health effects and hazard identification, as well as oral slope factors and oral and inhalation unit risks for carcinogenic effects. IRIS risk assessment data has been scientifically reviewed by EPA scientists and represents EPA consensus. IRIS is widely used in the EPA for risk-based decision-making.
- International Toxicity Estimates for Risk (ITER; http://www.tera.org/iter)—This database contains human health risk values and cancer classifications for over 640 chemicals of environmental concern from multiple organizations worldwide in support of human health risk assessments. It is compiled by Toxicology Excellence for Risk Assessment (TERA) and provides a comparison of international risk assessment information in a side-by-side tabular format. ITER explains differences in risk values derived by different organizations and contains links to the source documentation. It is the only database that includes risk information from independent parties whose risk values have undergone peer review through TERA's ITER Peer Review Program.Program. A risk notification system is being designed to complement ITER with information related to risk values under development.
- LactMed (Drugs and Lactation Database)—Containing over 500 records, LactMed is a peer-reviewed and fully referenced database of drugs to which breastfeeding mothers may be exposed. Among the data included are maternal and infant levels of drugs, possible effects on breastfed infants and on lactation, and alternative drugs to consider.
- Toxics Release Inventory (TRI)—This series of databases describes the releases of toxic chemicals into the environment annually for the 1987 to 2002 reporting years. TRI is mandated by the Emergency Planning and Community Right-to-Know Act and is based on data submitted to the EPA from industrial facilities throughout the United States. Information is included on over 650 chemicals and chemical categories. Pollution prevention data are also reported by each facility for each chemical.
- TOXLINE®—A bibliographic database providing comprehensive coverage of the biochemical, pharmacological, physiological, and toxicological effects of drugs and other chemicals from 1965 to the present. TOXLINE contains over 3.5 million citations, almost all with abstracts and/or index terms and CAS Registry Numbers. TOXLINE references are drawn from sources grouped into two major parts, TOXLINE Core and TOXLINE Special, both of which offer a variety of search and display capabilities.

The TOXNET® "Multi-Database" option allows for simultaneous searching of HSDB, IRIS, ITER, CCRIS, GENE-TOX, and LactMed.

### Other Resources

The NLM's Environmental Health and Toxicology pages (http://sis.nlm.nih.gov/enviro.html) contain links to additional relevant databases in areas such as occupational safety and health, household products, drugs and lactation, and dietary supplements. A new search engine, ToxSeek®, offers powerful tools for accessing toxicological information on the Web. ToxLearn is the successor to ToxTutor, an online tutorial that trains users in the basic principles of toxicology. The International Union of Pure and Applied Chemistry's (IUPAC) toxicology glossary is also made available by NLM, as are programs geared towardthe public and schoolchildren (e.g., Tox Town®, ToxMystery).

### TSCATS (Toxic Substance Control Act Test Submission Database)

(http://www.syrres.com/esc/tscats.htm; also available as a subfile in NLM's TOXLINE database)—TSCATS was developed by Syracuse Research Corporation (SRC) for the EPA in 1985. It is a central system for the collection, maintenance, and dissemination of information on unpublished technical reports submitted by industry to the EPA under the Toxic Substances Control Act (TSCA). Studies on over 8000 chemicals are categorized into three broad subject areas (health effects, environmental effects, and environmental fate).

### Other Tools

- *Chemfinder* (http://chemfinder.cambridgesoft.com/) is a compilation of free and subscription databases for chemical information including chemical structures, physical properties, and hyperlinks.
- *General search engines* include http://www.altavista.com; http://www.excite.com; http://www.google.com; http://www.infoseek.com; http://www.lycos.com; http:// www.msn.com; http://www.webcrawler.com; http://www.yahoo.com.
- *Meta-search engines* include http://www.dogpile.com; http://www.info.com; http://www.monstercrawler.com.

## NEW DEVELOPMENTS, RECENT ISSUES

One might think there is no more room for new technologies to expand the scope and power of information organization and retrieval, but it would be a flawed assumption. Periodically, new ways of managing information overload are developed. Some, inevitably, fall by the wayside, but others offer us opportunities to reconsider the way we have been processing and displaying information all along.

### Open Access Publishing

It used to be understood, and was rarely questioned, that the fruits of scientific research, in the form of original data and information, were for sale; however, as a crisis evolved in the pricing of serials, globally networked computers became commonplace, and users began particularly to question the ethics of requiring paid subscriptions to access to data generated by taxpayer dollars, a movement to open access to literature began to take hold, despite continued resistance by some publishers [6]. The Bethesda Statement on Open Access Publishing (http://www.earlham.edu/~peters/fos/bethesda.htm) has recently defined open access publishing and formulated the following two conditions that must be met for a publication to be considered open access:

1. The author(s) and copyright holder(s) grant(s) to all users a free, irrevocable, worldwide, perpetual right of access to, and a license to copy, use, distribute, transmit and display the work publicly and to make and distribute derivative works, in any digital medium for any responsible purpose, subject to proper attribution of authorship, as well as the right to make small numbers of printed copies for their personal use.
2. A complete version of the work and all supplemental materials, including a copy of the permission as stated above, in a suitable standard electronic format is deposited immediately upon initial publication in at least one online repository that is supported by an academic institution, scholarly society, government agency, or other well-established organization that seeks to enable open access, unrestricted distribution, interoperability, and long-term archiving (for the biomedical sciences, PubMed Central is such a repository).

Note that: (1) open access is a property of individual works, not necessarily journals or publishers, and (2) community standards, rather than copyright law, will continue to provide the mechanism for enforcement of proper attribution and responsible use of the published work, as they do now.

#### BioMed Central

Covering over 100 journals, BioMed Central (http://www.biomedcentral.com/) serves as an independent publishing house supporting free, immediate, and permanent open access to peer-reviewed biomedical research. Some of the journals require a subscription to view additional content, such as reviews. Among the journals with a focus on toxicology-related disciplines are Environmental Health-A Global Access Science Source (http://www.ehjournal.net/), Particle and Fibre Toxicology (http://www.particleandfibretoxicology.com/), and Journal of Carcinogenesis (http://www.carcinogenesis.com/).

### Public Library of Science

The Public Library of Science (http://www.plos.org) is a nonprofit organization of scientists and physicians and is another publication outlet for peer-reviewed open access technical literature. As of Spring 2005, they have published online and in print, PLoS series in biology, medicine, genetics, computational biology, and pathogens. A unique feature of these journals is the inclusion of a synopsis article written for the general public with each research article.

### Environmental Health Perspectives

*Environmental Health Perspectives* (http://ehp.niehs.nih.gov/) is a well-regarded monthly journal of peer-reviewed research, published by the National Institute of Environmental Health Sciences (NIEHS) and available free online.

### National Institutes of Health Policy

Certainly consequential for toxicology, is a recently announced U.S. National Institutes of Health policy designed to accelerate the public's access to published articles resulting from NIH-funded research. It calls on scientists to release to the public, manuscripts from research supported by NIH as soon as possible and within 12 months of final publication. These peer-reviewed, NIH-funded research publications will be available in a web-based archive to be managed by the National Library of Medicine. The online archive will increase the public's access to health-related publications at a time when demand for such information is increasing. PubMed Central (http://www.pubmedcentral.nih.gov), a part of the NIH's National Library of Medicine, serves as the NIH's digital repository of full-text, peer-reviewed biomedical, behavioral, and clinical research journals. It is a publicly accessible, stable, permanent, and searchable electronic archive.

Open access publishing is still a work in progress and modifications will clearly be made as the scientific community, publishers, legal experts, and the public continue discussions on the best means (and there will likely be several) to implement it. At least one website, FreeMedicalJournals.com (http://freemedicaljournals.com/), is tracking freely available, online, and full-text medical journals.

## Advance Access

Advance access (unrelated to open access *per se*) is a means of publishing papers online in manuscript form soon after they have been accepted for publication and considerably sooner than they appear as print publications. *Toxicological Sciences*, the official journal of the U.S. Society of Toxicology, published by Oxford University Press adheres to advance access publication. Unique codes known as Digital Object Identifiers are generated for individual articles and serve as a means of identifying different versions (including the final version) of the same article.

## Push Technology

Repeatedly searching the web for information on a particular topic takes time; however, a set of technologies collectively referred to as *push technology* can deliver information directly to one's computer on a daily basis or even as soon as the information is identified. Some websites push user-specified information that a user has specified they want, such as newspaper stories about a chemical or government organization (e.g., see Google's Alerts; http://www.google.com/alerts), or even job openings in a particular field of science and even in a particular country or region of the world (e.g., see the *New Scientist* job alerts; http://www.newscientistjobs.com/). The profile acts as a filter and is stored either on the client's machine (client-based filter) or on the push vendor's server (server-based filter). Traditional push technologies are evolving into forms such as Really Simple Syndication (RSS) that allow for more precise and efficient distribution channels.

## Mapping and Other Visualizations

The old saw, "A picture is worth a thousand words," just will not go away and, often, for good reason. Although certain pictures cannot capture the complexity that requires a verbal explanation and need for an additional "thousand words" to be fully understood, pictures, in addition to presenting a visual image of the thing itself, are useful in enhancing, crystallizing, or enhancing concepts that words and numbers alone cannot. TOXMAP, for example, is a tool developed by the National Library of Medicine to go hand-in-hand with the EPA's Toxic Release Inventory (TRI), a database accessible via the National Library of Medicine's TOXNET system (http://toxnet.nlm.nih.gov). TOXMAP helps users create maps showing where chemicals are released into the air, water, and ground. It identifies the releasing facilities, color-codes release amounts for a single year, and provides multiple-year chemical release trends.

Although there are scattered image files of toxicological microphotographs, poisonous plants and animals, intact organisms with toxic damage, etc., no consolidated online library of toxicological images exists yet, although this would be a very useful project and is being pursued by at least one U.S. government agency. One already available website of images is Clinical Toxinology Resources (http://www.toxinology.com), which is a searchable database of over 6000 images designed to meet the needs of anyone seeking information on venomous and poisonous organisms and poisonous plants throughout the world. This website was developed by the Toxinology Department of the Women's and Children's Hospital, Adelaide, and the Department of Paediatrics at the University of Adelaide.

### Computational Toxicology and Toxicoinformatics

A branch of toxicological information apart from the presentation of standard data and scientific literature is gradually developing. The advent of the "-omics" disciplines (genomics, proteomics, metabonomics) has given a boost to toxicology, and they offer new and still evolving approaches to understand the risks posed by many chemicals to human health and the environment. These new biological disciplines, when in the service of toxicology, have been collectively referred to as *toxicogenomics*. Developing mathematical and computer models to predict toxic outcomes and better understand mechanisms of action is known as *computational toxicology* or, sometimes, *toxicoinformatics* [4].

Several major efforts within U.S. government agencies are underway to utilize these technologies. In addition to the National Institute of Environmental Health Sciences' National Center for Toxicogenomics (http://www.niehs.nih.gov/nct/), the EPA has recently established a National Center for Computational Toxicology (http://www.epa.gov/comptox/), and the FDA's National Center for Toxicological Research now includes a Center for Toxicoinformatics (http://www.fda.gov/nctr/science/centers/toxicoinformatics/).

New databases and other tools are being developed to make sense of the enormous amounts of data being generated. The EPA's Distributed Structure-Searchable Toxicity (DSSTox) Database Network (http://www.epa.gov/NHEERL/dsstox/), for example, provides a community forum for publishing standard format, structure-annotated chemical toxicity data files for open public access. One of its goals is to facilitate development of improved models for predicting toxicity based on chemical structure.

Another relevant and evolving database is the Chemical Effects in Biological Systems (CEBS) Knowledge Base (http://www.niehs.nih.gov/cebs-df/) from the National Institute of Environmental Health Sciences. The stated goals of the CEBS are to:

- Create a reference toxicogenomic information system of studies on environmental chemicals/stressors and their effects.
- Develop relational and descriptive compendia on toxicologically important genes, groups of genes, SNPs, mutants, and their functional phenotypes that are relevant to human health and environmental disease.
- Create a toxicogenomics knowledge base to support hypothesis-driven research.

### International Information Coordination

The inevitable forward march of globalization requires interaction between toxicologists at an international level. The International Union of Toxicology (IUTOX) has been aware of this for some time [9]. The U.S. National Library of Medicine is in the process of building an online World Library of Toxicology, Chemical Safety, and Environmental Health. This is being designed as a portal to sources of information from specific countries and international groups and is intended to foster cooperation and collaboration in research and other activities and to minimize duplication of effort.

## CONCLUSION

Information and data generated by advances in the toxicological sciences, in line with that of other scientific disciplines, continues to grow at a phenomenal rate. The increasing complexity derives from the interdisciplinary nature of toxicology and the societal manifestations of the science.

## QUESTIONS

1. Identify some of the key web-based databases in toxicology, environmental health, and risk assessment.
2. What organizations consider toxicology from a global perspective?
3. Discuss such issues as mapping and computational toxicology with regard to the future direction of toxicological information; entertain pluses and minuses and possible outcomes for information controversies, such as open access publishing.

## REFERENCES


1. Brinkhuis, R. P. (2001): Toxicology information from U.S. government agencies. *Toxicology*, 157:25–49.
2. Hakkinen, P. J. (2001): Global toxicology and risk analysis: roles of the Internet and World Wide Web. *Toxicology*, 160:59–63.
3. Hakkinen, P. J., Stoss, F. W., Behrendt, B., and Wexler, P. (2000): Organizations. In: *Information Resources in Toxicology*, 3rd ed., edited by P. J. Hakkinen, G. L. Kennedy, Jr., and F. W. Stoss. Academic Press, San Diego, CA, pp. 439–489.
4. Kavlock, R. et al. (2005): Computational toxicology: framework, partnerships, and program development; September 29–30, 2003; Research Triangle Park, North Carolina. *Reprod. Toxicol.*, 19(3):265–280.
5. Kehrer, J. P. and Mirsalis, J. (2001): Professional toxicology societies: web based resources. *Toxicology*, 157:67–76.
6. Liesegang, T. J., Schachat, A. P., and Albert, D. M. (2005): The open access initiative in scientific and biomedical publishing: fourth in the series on editorship. *Am. J. Ophthalmol.*, 139(1):156–167.

7. Patterson, J., Hakkinen, P. J., and Wullenweber, A. E. (2002): Human health risk assessment: selected Internet and World Wide Web resources. *Toxicology*, 173:123–143.
8. Poore, L. M., King, G., and Stefanik, K. (2001): Toxicology information resources at the Environmental Protection Agency. *Toxicology*, 157:11–23.
9. Schou, J. S. and Hodel, C. M. (2003): The International Union of Toxicology (IUTOX): history and its role in information on toxicology. *Toxicology*, 190(1–2):117–124.
10. Schroter, S., Barratt, H., and Smith, J. (2004): Author's perceptions of electronic publishing: two cross sectional surveys. *Br. Med. J.*, 328(7452):1350–1353.

# Part 2

## Agents

# 14 Foodborne Toxicants

*Chada S. Reddy and A. Wallace Hayes*

## CONTENTS

The complex chemical milieu of unspoiled foods in their pure and conventional form is safe in normal human populations. Major components of such foods are macro- and micronutrients essential for human and animal survival; however, adverse effects can occur in individuals as a result of chronic overindulgence in certain foods, greater sensitivity, the presence of larger amounts of chemical defense agents/contaminants in foods subjected to drought and pest attack, intentional introduction of additives, increased consumption of herbal and nonherbal dietary supplements, or chemicals introduced during certain food production, processing, cooking, storage, and serving conditions. An excellent summary of these classes of agents has been presented by Deshpande [68]. The simultaneous presence of dietary components capable of enhancing or protecting against the adverse effects of other food toxicants has, without a doubt, contributed to the lack of correlations between experimental animal data and epidemiological data from humans. This review attempts to summarize our current understanding of the multitude of such agents and their toxic or beneficial effects on human and animal health.

## NATURAL TOXICANTS

Acute intoxications from chemicals naturally occurring in human diet derived from plants appear limited to individuals and selected classes of compounds such as protein allergens and fava bean glycosides [170]. Food derived from animal sources, however, can contain toxic components or contaminants that have caused acute intoxications in healthy human populations. Milk sickness and quail poisoning (coturnism) are examples of acute intoxication resulting from the consumption of meat containing plant toxicants and of the regional nature of certain intoxications. Just in the United States, more than 75 million acute illnesses, 600,000 hospitalizations, and up to 5000 deaths are caused annually by naturally occurring foodborne infectious agents alone. The true magnitude of problems pertinent to food safety in the modern world begins to appear in focus only when one considers the fact that many more acute infections as well as the long-term or delayed human effects resulting from a wide array of other plant and animal toxins are either unrecognized or unreported. This review attempts to summarize current knowledge pertinent to this bewildering array of food toxicants and the conditions they cause in human beings and animals.

### TOXICANTS IN FOODS OF PLANT ORIGIN

Foods of plant origin account for most (>70%) of the world's supply of protein. Although plants with obvious toxic effects have been excluded from human diet by trial and error, deleterious (toxic as well as antinutritive) effects from the following groups of compounds are deemed significant for human health.

#### Alkaloids

Alkaloids are nitrogenous heterocyclic organic compounds that protect plants against herbivorous consumption and attack by insects, parasites, and competitors. Major alkaloi groups of concern from the standpoint of human consump-:ion include pyrrolizidines, xanthines, and solanines. Oth-ers—such as piperidines from *Conium* and tobacco, quin-olizidines from *Lupinus*, and indolizidines from *Astragalus*, *Swainsona*, and red clover—are mainly consumed by graz-ng animals and can be potentially transferred to humans hrough milk. For a review of the toxic effects of alkaloids n humans and animals, see Cheeke [45].

*Pyrrolizidine alkaloids* (PAs) are a group of more than 250 geographically ubiquitous plant metabolites posing a najor threat to animal health by their presence in plants ncluding *Senecio*, *Crotalaria*, and *Heliotropium*, among others. Human exposure and possible health effects result rom the wide use of coltsfoot (*Tussilago*), comfrey (*Sym-phytum*), and petasites (*Petasites*) as herbal remedies, foods e.g., salads), and tea; from the contamination of food grains with seeds from PA-containing plants; from the consumption of honey derived from pansy ragwort (*Sene-cio* sp.) and Patterson's curse (*Echium* sp.); and by the consumption of milk from animals grazing the alkaloid-containing plants mentioned above [58].

Huxtable [115] reviewed human intoxications with PAs among which the acyclic diesters and macrocyclic diesters such as retronecine, senecionine, and petasitenine re more toxic [107]. Highly reactive pyrrole derivatives of PA and their hydrolysis products formed by the action of mixed-function oxidases are considered to be respon-ible for the toxic effects of PA [180]. Typically, high mortality with hepatic venoocclusive disease character-zed by occlusion of small branches of hepatic vein lead-ng to ascites/edema and renal disease with reduced uri-ary output occurs mostly in children. Survivors often manifest cirrhosis. Histologically, endothelial prolifera-on and medial hypertrophy lead to occlusion of small epatic veins which then leads to centrilobular congestion esulting in sinusoidal widening and blood pooling. Necrosis and fibrosis ultimately results. Certain dehydro A's (monocrotaline, in particular) are known also to nduce similar occlusive changes in pulmonary arterioles eading to pulmonary hypertension and right ventricular ypertrophy and ultimately to right heart congestive fail-re. Impairment of serotonin and norepinephrine clear-nce by endothelial cells appears to contribute to pulmo-ary hypertension [115].

Many PAs and their pyrrole metabolites are bifunc-onal alkylating agents cross-linking to macromolecules, ncluding DNA [107,115], thus accounting for their mutagenicity and carcinogenicity in experimental animals 107]. At least six species of plants (*Senecio longilobus*, *etasites japonicus*, *Tussilago farfara*, *Symphytum offici-ale*, *Farfugium japonicum*, and *Senecio cannabifolis*) and ght PAs have been shown to induce one or more of the ollowing types of cancer: hepatic carcinoma, heman-ioendothelial sarcoma in the liver, liver cell adenoma, holangiosarcoma, astrocytoma, squamous cell carcinoma of the skin, pulmonary adenoma, adenocarcinoma of the small intestines, adenomyoma of the ileum, and rhab-domyosarcoma [107].

Recently, the National Toxicology Program (NTP) showed that the oral exposure to the pyrrolizidine alkaloid riddelliine, in addition to producing non-neoplastic lesions in the liver and kidney of male and female rats and mice and in the lung and arteries (multiple tissues) of female mice, increased the incidences of hemangiosar-coma in the liver, hepatocellular adenoma, and mononu-clear cell leukemia in male and female rats; of heman-giosarcoma in the liver in male mice; and of alveolar/bronchiolar neoplasms in female mice [198]. Alterations in hepatocyte/endothelial vascular endothelial growth factor (VEGF) synthesis, in KDR/flk-1 activation, and in cell-cycle modulators such as K-ras, beta-catenin, and p53 have been proposed as mechanisms leading to sus-tained endothelial cell proliferation and thus to heman-giosarcoma [109,196].

Pyrrolizidine alkaloids also seem to have protective effects against hepatocellular neoplasms in male and female mice by an unknown mechanism and against gastric/duode-nal ulcers by decreasing acid and increasing gastrin, epi-dermal growth factor, and mucus production [193,287]. PAs such as heliotrine are developmentally toxic, inducing lower jaw hypoplasia, musculoskeletal defects involving ribs, and general growth retardation [107]. Clarifying the causal rela-tionship of foodborne PAs to aberrant human development was a recent report [225] of hepatic venoocclusive disease in a preterm neonate (who died shortly after a caesarean delivery) exhibiting hepatomegaly and ascites in association with the presence of high amounts of pyrrolizidine alkaloids in the herbal mixture used for cooking the family meal and in neonatal liver.

*Solanum alkaloids*, including predominantly solanine, chaconine, and tomatine, are found in potato, eggplant, and tomato (species of *Solanum* genus), among others. Reviews on the biosynthesis, occurrence, and toxicology of solanum alkaloids include those by Sharma and Salunkhe [262] and Keeler et al. [130]. Potatoes, espe-cially sprouted, greened, blighted, injured, or spoiled, appear to have raised the most concern relating to possible alkaloid intoxication, with over 200 cases of human poi-sonings from potato ingestion documented [195]. Signs of intoxication in humans, some of which may be related to the irritant, estrogenic, and cholinesterase-inhibiting activity of the alkaloids, appear at >20 mg alkaloid per 100 g of tuber and include headache, vomiting, diarrhea, neurological signs, debilitation, and death. Prolonged exposure at lower doses of these alkaloids causes increased liver-to-body weight ratios [85] and antiandro-genic effects [98] in animals. Certain of these alkaloids have apoptotic/anticarcinogenic [164], antithrombotic, cardioprotective, and antibacteremic shock effects [239,308]. A combination of anticholinergic and antagonistic actions against tumor necrosis factor alpha (TNF$\alpha$)-induced elevation of [$Ca^{2+}$]i and plasminogen activator inhibitor likely account for these actions.

An earlier connection between the consumption of blighted potatoes and the incidence of anencephaly–spina bifida (ASB) in humans appears to be a false alarm [252]. Exposure of rats to the certain alkaloids, including solanine, solasodine, choconine, and cytochalasins B, D, and E, resulted in minor skeletal to major facial and central nervous system abnormalities in animals, whereas the exposure to others such as tomatidine was without effect [252]. Aglycones of many commonly occurring alkaloids also show adverse effects on gestational weight gain and fetal body weights [85]. Vitamin D glycoside contained in *Solanum malacoxylon* is excreted in milk and crosses the placenta leading to skeletal alterations in the offspring despite increased Ca and P levels [97]. Although normal levels of glycoalkaloids in domestic potatoes conform to the U.S. Department of Agriculture (USDA) guideline of 20 mg/100 g of tuber, glycoalkaloid content can be increased several fold due to exposure to light, immature tubers, wounding of potatoes, and stresses such as fungal attack [29]. Baking, boiling, or microwaving does not destroy solanine or chaconine in potatoes. Protection of tubers from sunlight, $\gamma$-irradiation, soaking in water under controlled conditions, dipping damaged potatoes in emulsified water, treating potatoes with sprout inhibitors during storage, waxing and heating, dipping in oils (corn, olive, or mineral), spraying tubers with lecithin (such as PAM®) or simply spray-rinsing tubers with an aqueous solution of an edible surfactant (Tween® 85) appear to be some simple methods to prevent glycoalkaloid formation during storage [262].

Three major *xanthine alkaloids*—caffeine, theobromine, and theophylline—are found as major components of coffee (*Coffee arabica*), cocoa (*Theobroma cocas*), and tea (*Thea sinensis*), respectively. Caffeine, in addition, is added to many beverages, foods, and medications [76]. Caffeine-related adverse effects begin when 0.5 to 1.0 g of caffeine (10 cups of coffee) is ingested by an adult; fatalities occur in children at 5 g and in adults at 5 to 10 g [61,76]. Caffeine and other methylxanthines inhibit phosphodiesterase, leading to intracellular accumulation of cyclic AMP; block adenosine receptors; and cause increased release of $Ca^{2+}$ from the terminal cisternae of sarcoplasmic reticulum [61]. Major effects of xanthines involve CNS stimulation (hyperesthesia to convulsions), emesis, cardiovascular effects (cardiac stimulation to arrhythmias), diuresis, and smooth muscle effects leading to decreased vascular resistance and bronchodilation [61]. In addition, caffeine enhances gastric secretion of acid and pepsin.

In habitual coffee drinkers, cessation of caffeine consumption results in a withdrawal syndrome characterized by headache, fatigue, drowsiness, depression, difficulty concentrating, irritability, and lack of clarity in thinking [121]. Caffeine increases serum homocysteine, a risk factor for cardiovascular disease [301], and induces bone loss in postmenopausal women [224]. Caffeine and theobromine are mutagenic in bacterial systems and can potentiate DNA damage caused by other genotoxins but are neither directly carcinogenic in animals nor associated with human cancer [10,61]. Caffeine actually appears to protect against certain cancers, type 2 diabetes, preeclampsia of pregnancy, and development of Parkinsonism [17,132,243].

Caffeine is teratogenic in experimental animals, causing mostly limb and facial defects [252]. Although high caffeine consumption during pregnancy may increase the risk of spontaneous abortion and low-birth-weight babies, no correlation exists between caffeine consumption and birth defects in humans [132,273]. Greater danger, however, appears to be associated with combined consumption of caffeine with other vasoactive herbal ingredients and medications. Oral exposure to caffeine (30 mg/kg) along with ephedrine (25 mg/kg) in rats (1.4- and 12-fold, respectively, above average human exposure), resulted in the death of rats within 4 to 5 hours, accompanied by massive interstitial hemorrhage and degeneration and necrosis of myofibers in myocardium of the left ventricle and interventricular septum [206]. Caffeine, in addition, appears to enhance alcohol-mediated acetaminophen hepatotoxicity [69]. The Canadian government [202] issued guidelines to limit daily caffeine consumption to 400 mg (~8 cups of coffee) in healthy adults, 300 mg (~ cups of coffee) in women of reproductive age, and 2.. mg/kg in children. The U.S. Food and Drug Administration (FDA) issued a warning to pregnant women to limit coffee consumption [61].

## Allergens

Food allergies are a group of disorders characterized by an exaggerated immunologic response to a component of food. Although all foods are capable of eliciting an allergic reaction, most food allergies are associated with only eight foods or food groups, many of which show cross-reactivity with latex allergens. Allergens in foods are generally heat- and acid-stable proteins, glycoproteins, or peptides with a molecular weight often between 5000 and 70,000. The prevalence of food allergies is about 4%, is age dependent (up to 8% of children under 3 years and 1.5% of adults are affected), and is on the rise [215,244,245]. Milk (casein, $\beta$-lactoglobulin) and eggs (ovomucoid, ovalbumin) are the most commonly incriminated agents [13 Allergens belonging to the cupin and prolamin superfamily of proteins and the proteins of the defense system are the most widespread groups of plant allergens and are present in peanuts (Aha I and II, aglutinin) and treenut wheat (globulins, glutenine), soy proteins (in formulae fish, shrimp, crustaceans, tomatoes, fruits, chocolate, and certain beverages [15,34,36,311].

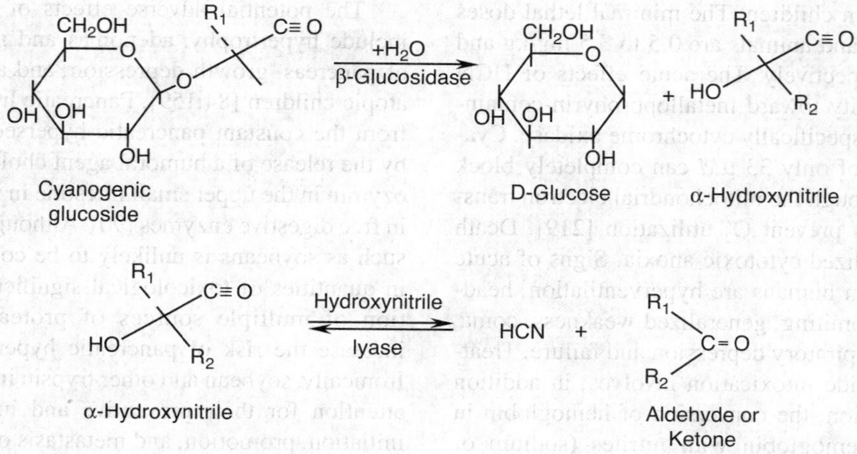

FIGURE 14.1 Enzymatic hydrolysis of cyanogenic glycosides. Initially, the glycoside is hydrolyzed by a β-glucosidase releasing glucose and α-hydroxynitrile. The hydroxynitrile dissociates either enzymatically or nonenzymatically to yield HCN and the corresponding aldehyde or ketone.

Normally, the gut-associated lymphoid tissue renders systemic antibodies nonresponsive to food allergens via signaling involving various cytokines of the immature dendritic cells, T-cells, and Treg cells [268]. A breakdown in this signaling is thought to predispose individuals to food allergies. Although multiple tissues of the body can be affected by a single allergen, the skin (eczema and urticaria) and the respiratory tract (e.g., rhinitis, pneumonitis, asthma) account for 90% of food allergies [216]. Also suggestive of allergic reaction to food are abdominal distress, vomiting, diarrhea, hypotension and shock secondary to hypovolemia, and nervous system involvement as indicated by headaches, convulsions, and behavioral problems [216]. Rapid-onset allergies are dependent on the reaction of the antigens with circulating antibodies of the IgE class (reagin) and the eventual release of vasoactive substances such as serotonin and histamine, leading to life-threatening anaphylaxis. Non-IgE type food allergies are dependent on specifically sensitized lymphocytes that are attracted to the site of antigen exposure by lymphokines released by already existing T-lymphocytes and require several hours or days to fully manifest.

Recent evidence suggests a role for food allergies in autoimmune disorders (celiac disease), juvenile or insulin-dependent diabetes mellitus, migraine, and arthritis in children [36,134]. Idiosyncratic reactions such as lactose intolerance, a result of genetic deficiency of lactase leading to luminal accumulation of lactic acid and osmotic diarrhea, must be differentiated from immunologically mediated allergic reactions. Although recent characterization of food allergens and an understanding of the immunopathogenesis of the associated disorders are likely to lead to novel diagnostic and immunotherapeutic approaches (such as the successful anti-IgE therapy for peanut allergy) [245], avoidance of known sources of allergens still remains the mainstay of current therapy for food allergies.

## Cyanogens

Cyanogenic glycosides, which release highly toxic hydrocyanic acid upon hydrolysis, are derived not only from plants (more than 2000) but also from fungi, bacteria, and even members of the animal kingdom [193]. Although cassava, sweet potatoes, yam, maize, millets, bamboo, sugarcane, peas, beans, almond kernel, lemon, lime, apple, pear, cherry, apricot, prune, and plum constitute sources for humans, poisonings are mainly associated with the consumption of improperly processed cassava in Africa, Asia, and Latin America [193,219]. Among more than 20 glycosides identified, only four (i.e., amygdalin, dhurrin, linamarin, and lotaustralin) appear to be of toxicologic importance. Cyanogenic lipids, although of unknown toxicological significance, are also present in plants and yield carbonyl compound and hydrogen cyanide (HCN) upon hydrolysis [67].

Hydrolysis of the glycoside is triggered by physical disruption (e.g., mastication, trampling) or stress (e.g., drought, cooking, frost) and is catalyzed by β-glucosidase and hydroxynitrile lyase, which are present within the plant or in bacteria in the gastrointestinal tract of humans and animals [219]. Figure 14.1 presents the scheme of breakdown leading to the formation of glucose and hydroxynitrile from the glycoside followed by breakdown of hydroxynitrile into carbonyl compounds and HCN. Rhodanese catalyzes the conversion of HCN to thiocyanate in the presence of thiosulfate [219].

Animals have often been acutely poisoned by young sorghum and arrow grass. Young bamboo shoots and tea made from peach leaves are examples of dietary sources of HCN poisoning in children. The minimal lethal doses of HCN in humans and animals are 0.5 to 3.5 mg/kg and 2 to 10 mg/kg, respectively. The acute effects of HCN result from its affinity toward metalloporphyrin-containing enzymes, more specifically cytochrome oxidase. Cyanide concentration of only 33 µ*M* can completely block electron transfer through the mitochondrial electron transport chain and thus prevent $O_2$ utilization [219]. Death results from generalized cytotoxic anoxia. Signs of acute cyanide poisoning in humans are hyperventilation, headache, nausea and vomiting, generalized weakness, coma, and death due to respiratory depression and failure. Treatment of acute cyanide intoxication involves, in addition to artificial respiration, the conversion of hemoglobin in the blood to methemoglobin with nitrites (sodium or amyl). Methemoglobin competes with cytochrome oxidase for HCN and forms cyanmethemoglobin. Coadministration of sodium thiosulfate will convert free cyanide present in the blood to thiocyanate, which is eliminated. As free cyanide in the blood decreases, additional cyanide dissociates from the cyanmethemoglobin and is subsequently eliminated [46].

Tropical ataxic neuropathy (TAN), characterized by myelopathy, bilateral optical atrophy, deafness, and polyneuropathy (konzo, an irreversible upper motoneuron paralytic disease of women and children) [23]; goiter; epigastric burning pain; dizziness; and abdominal distension/vomiting [2] have been linked to longer term consumption of cassava diets in Africa and other tropical countries [210]. These diets were also poor in protein and sulfur-containing amino acids which can detoxify HCN to thiocyanoalanine and subsequently to inert 2-amino-4-thiazolidine carboxylic acid [219]. Although the chronic effects of cyanogen exposure were earlier thought to be due to thiocyanates, recent evidence suggests that both linamarin and cyanide may be responsible for konzo in humans and animals [23,272]. In addition, cyanogenic glycosides exert antitumor promoting activity [89].

## Enzyme Inhibitors

Although plant and animal foods contain inhibitors of proteases, amylases, and lipases, only the inhibitors of proteases may pose some hazard to human health, if any. Kunitz inhibitor, the major protease inhibitor of soybeans, is a heat-labile protease capable of inhibiting trypsin and to a lesser degree chymotrypsin and other proteases [129]. The cationic form, the major active human trypsin, is only weakly inhibited, whereas the anionic form is fully inhibited [158]. In addition, a heat-stable, lower molecular weight Bowman–Birk inhibitor and others are also present in raw soybeans. Egg white, milk, beans, peas, cereal grains, alfalfa, sunflower, and potatoes also have been shown to contain one or more protease inhibitors [118].

The potential adverse effects of protease inhibitors include hypertrophy, adenomas and nodular hyperplasia of pancreas, growth depression, and allergic reactions in atopic children [84,159]. Pancreatic hypertrophy is likely from the constant pancreatic hypersecretion necessitated by the release of a humoral agent cholecystokinin pancreozymin in the upper small intestine in response to a deficit in free digestive enzymes [90]. Although any single source such as soybeans is unlikely to be consumed by humans in quantities of toxicological significance, the consumption of multiple sources of protease inhibitors may increase the risk of pancreatic hypertrophy and cancer. Ironically, soybean and other trypsin inhibitors are gaining attention for their preventive and inhibitory effects on initiation, promotion, and metastasis of cancer induced by many agents in multiple tissues [62,84]. Recent studies suggest that Bowman–Burk inhibitor-induced specific proteasomal inhibition leads to alterations in cell-cycle proteins and to cell-cycle arrest, thus accounting for the anticancer effects of these agents [47]. The contribution of other mechanisms such as inhibition of enzymes involved in oxygen free-radical formation and the induction of amino acid deprivation in cancer cells by protease inhibition is yet to be investigated.

## Estrogens and Plant Sterols

Hundreds of species of plants contain plant sterols and estrogenic isoflavonoids (e.g., genistein, glycetein, daidzein) or their glycosides (genistin, glyectin, daidzin) coumestans (e.g., coumestrol, 4-*O*-methylcoumestrol) and lignans [3,277]. Phytoestrogens, although capable of causing infertility in animals grazing heavily on estrogen (coumestan)-containing forages (subterranean clover alfalfa), have not been proven to cause human problems. Genistein from soybeans appears to be the plant estrogen of most (if any) significance in human health [6]. Zearalenone and zearalenol, two major resorcylic acid lactone estrogens, are produced in corn in response to infection by toxigenic strains of the fungus *Fusarium roseum* and are discussed along with other mycotoxins. Human infants can be exposed to 4 mg/kg body weight or more of isoflavones from soy-based formula [6]. Although phytoestrogens bind to multiple steroid (estrogen, progesterone, and androgen) receptors, they are considered to be selective estrogen receptor modulators (SERMs). Due to their lower potency (500 to 10,000 times) compared to estradiol, phytoestrogens can actually impede the action of endogenous estrogen and at higher doses also induce antigonadotropic effects at hypothalamic, pituitary, and gonadal levels in both sexes [6]. In addition to the above effects, genistein inhibits protein tyrosine kinases associated with a number of growth factors and other enzymes with roles in cell proliferation and differentiation [6].

Reported effects of phytoestrogens in animals include infertility in sheep fed subterranean clover and cattle consuming alfalfa [6], as well as feminization of males following developmental exposure [52]. Recent reports of detrimental effects of high levels of phytoestrogens in soy-based products on learning, memory, and anxiety behaviors in male but not female, rats [154] suggest the need for caution in feeding male infants soy formula. Although phytoestrogens appear to be noncarcinogenic when given orally [299], some (genistein, coumestrol, quercetin, zearalenone, resveratrol, and some metabolites of daidzein) are genotoxic, and all exhibit pro-apoptotic effects *in vitro* [25,279]. In humans, reversible changes in menstrual cycle and follicle-stimulating hormone (FSH) and luteinizing hormone (LH) surges in premenopausal women appear to result from soy consumption, but no developmental or infertility problems were noted in populations consuming large quantities of phytoestrogens [6]. Phytoestrogens may actually exert antioxidant activity and may protect humans against coronary heart disease; cancer of breast, prostate, and colon; obesity; and postmenopausal osteoporosis [4,28,154]. Average human adults are only exposed to 102 μg dietary estrogenic equivalents (reflecting both potency and exposure) daily compared to 3.35 mg/day from estrogen replacement therapy and 16.7 mg/day from oral contraceptives [242]. Their exposure in high soy consumers, especially infants, and thus their longer term effects are likely to be significant.

Phytosterols (campesterol and sitosterol, and their 5α-saturated stanols) are normal dietary components (200 to 300 mg/day) that chemically resemble and thus interfere with absorption of dietary cholesterol and together reportedly decrease the incidence of coronary heart disease by 20 to 25% [99]. Paradoxically, high levels of plasma phytosterols alone may be associated with increased coincidence of coronary heart disease. Because plant stanol esters reduce absorption and serum concentrations of both cholesterol and plant sterols, beneficial supplementation of human diets over the long term may be better accomplished with the use of plant stanols alone.

## Glucosinolates

Glucosinolates are a group of more than 100 flavor-imparting thioglucoside compounds found at up to 60 mg/g in all crucifers such as broccoli, cabbage, Brussels sprouts, cauliflower, calabrese, turnip, radish, horseradish, mustard, and rapeseed and related plants. Common names of some important glucosinolates include sinigrin, progoitrin, epiprogoitrin, glucobrassicin, and neoglucobrassicin. Not only the parent glucosinolates but also their products of plant and human digestive tract bacterial myrosinase (thioglucosidase) hydrolysis—isothiocyanates, nitriles, oxazolidinethione (OZT), and thiocyanate ions—contribute to their biological effects [302].

Although evidence is lacking in humans, thiocyanate ion inhibits the uptake of iodine by the thyroid, leading to iodine-reversible hyperplasia and hypertrophy of the thyroid (cabbage and legume goiter) and growth suppression in animals. OZT also inhibits thyroxine synthesis and induces goiter (brassica seed goiter) in rats by inhibiting the incorporation of iodine into precursors of thyroxine [179]. This condition is not reversible by iodine supplementation. In addition to goiter, epiprogoitrin and progoitrin also induce liver and kidney enlargement and death at 2.6% in the diet via their nitrile metabolites [295]. Bile duct hyperplasia, hepatocyte necrosis, and megalocytosis of renal tubular epithelium were also seen in these animals [288].

Isothiocyanates are embryocidal and cause fetal weight reduction [29]. Isothiocyanates and certain glucosinolates (e.g., sinigrin) are mutagenic in the Ames assay whereas thiocyanates are not [29]. Recent evidence indicates that the desulfo precursors of glucosinolates may also be carcinogenic [309]. Higher intake of cruciferous vegetables in humans and animals, however, may exert an anticarcinogenic effect attributable to the formation of isothiocyanates (at least seven), indoles, indole-3-carbinol, 3-indoleacetonitrile, and 3,3′-diindolylmethane [302]. Anticarcinogenic effects of glucosinolates may result from their induction of phase II enzymes (quinone reductase) in the gastrointestinal tract and liver [284] and stimulation of apoptosis [270].

## Lectins (Phytohemagglutinins)

Lectins are high-molecular-weight (100,000 to 150,000), heat-labile proteins, lipoproteins, or glycoproteins (up to >10% of total seed protein) detected in over 800 edible plant species, of which 600 belong to the leguminosae (e.g., beans, peas). In addition, lectins are also present in animals such as sponges, crustaceans, mollusks, fish blood, amphibian eggs, and even mammalian tissue [67]. Interactions of animal lectins such as annexin and galectin with animal cell proteins such as Bcl-2 and synexin and subsequent triggering of signaling cascades give them an ability to regulate various endogenous functions involving glycoprotein and cell (normal and tumor) recognition, adhesion, and clearance; signal transduction; extracellular glycoprotein trafficking; mitogenesis; apoptosis; and immune functions [192,294].

Upon ingestion, plant lectins survive digestion by the gastrointestinal enzymes and bind to membrane glycosyl groups of the cells lining the digestive tract. This leads to nonspecific inhibition of digestion and active and passive absorption of many nutrients (e.g., amino acids, fats, vitamins, minerals, thyroxine) across the intestinal mucosa, alters the bacterial flora, modulates the immune status of the digestive tract, damages the luminal membranes of the epithelium, and induces necrosis of intestinal epithelial cells [133,298]. These effects account for growth suppression and possibly goiter after long-term oral exposure to high levels [118]. Mortality following acute systemic lectin exposure is associated with damage to the liver [116] and other organs. The most toxic lectin—ricin from castor bean (lethal dose in humans of 1 to 10 μg/kg body weight)—can cause severe intestinal epithelial cell necrosis and death from multiple-organ damage [296]. Recent evidence suggests that lipid peroxidation mediated by reactive oxygen species may be involved in ricin-induced thyroid damage [247]. The lectin portion (B chain) of the ricin dipeptide binds the galactosyl residues on the surface of intestinal epithelial cells and facilitates the intracellular uptake of the enzymatic (RNA-specific *N*-glycosidase) A chain via clathrin-dependent as well as clathrin-independent endocytosis. The A chain then enters golgi and the endoplasmic reticulum (ER), inhibits protein synthesis, and causes cell death [249,296]. Less toxic lectins may act by the same mechanism to stimulate protein synthesis, mitogen activation, and immune stimulation.

## Lipids

Although lipids are essential for normal development, growth, and cellular function, their overconsumption has been associated with weight gain, obesity, cardiovascular disease, and a condition known as metabolic syndrome which, in addition to above, is characterized by increased propensity for type 2 (insulin-resistant) diabetes. Factors such as the departure from established food-use patterns, the use of new lipids in human diets, or inborn errors of metabolism (due to gene polymorphism) act in concert with lipid overconsumption to induce hyperlipidemia, cardiovascular disease, hypertension, and obesity. Recent evidence implicates a positive feedback circuit between high dietary fat and increase in brain galanin (GAL), a feeding stimulant peptide, in the onset of human overeating syndrome and obesity [153]. Mechanistically, circulating lipids interact directly (as fatty acids) or indirectly (via biosynthetic intermediates such as prostaglandins, leukotrienes, etc., or via the interaction of lipids and their derivatives such as diacylglycerol) with a variety of signal-transduction pathways and transcription factors such as peroxisome proliferator-activated receptor (PPAR), liver X receptor, hepatocyte nuclear factor 4, carbohydrate-response-element-binding protein, farnesoid X receptor, and sterol-regulatory-element-binding protein (SREBP) [233]. These interactions lead to alterations in the expression of genes (such as adipocytokines) that mediate cellular responses involved in inflammation, a prerequisite to insulin resistance and diabetes. Paradoxically, dietary supplementation with monounsaturated fatty acids (oleic and omega- or alpha-3 fatty acids) and conjugated linoleic acids can prevent or reverse the onset of insulin resistance/metabolic syndrome by altering membrane fluidity and signaling and reducing adipose tissue TNFα and subsequent alterations in SREBP [230,233].

Erucic acid (*cis*-13-docosanoic acid) is predominantly a component of rape (*Brassica napus* and *B. campestris*) and mustard (*B. hirta* and *B. juncea*) seeds. Canada, Argentina, Mexico, China, India, Pakistan, Japan, and several European countries are the major producers and users of these oils. Growth suppression, myocardial fatty infiltration, mononuclear cell infiltration, and fibrosis were observed in weanling rats fed erucic acid at levels supplying greater than 20% of the dietary calories. In addition, ducklings showed hydropericardium and cirrhosis, and guinea pigs developed splenomegaly and hemolytic anemia [181]. Organ-specific inhibition of glutamate oxidation and adenosine triphosphate (ATP) synthesis in cardiac mitochondria [110] could be mechanistically involved in the pathogenesis of these lesions. In humans, however, although the long-term use of Lorenzo's oil (oleic acid and erucic acid) in the treatment of adrenoleukodystrophy or adrenomyeloneuropathy leads to thrombocytopenia and lymphopenia [291], adverse effects from the dietary consumption of erucic acid have not been reported.

Refsum disease is a genetic peroxisomal fatty acid oxidase and catalase deficiency resulting in an inability of the affected individuals to convert phytanic acid (3,7,11,15-tetramethylhexadecanoic acid, a product of chlorophyll metabolism in the rumen) from dairy products and ruminant fats to α-hydroxyphytanic acid in preparation for further oxidation. This results in accumulation of lipids containing phytanic acid in many tissues and a disorder characterized by poor physical and mental growth, blindness, deafness, and other neurologic signs [49]. Elimination of dairy and ruminant fats from the diet of these individuals results in partial remission.

Cyclopropene fatty acids such as sterculic acid (C19) and malvalic acid (C18) are natural components of oils from plants of the order Malvales, most important of which are cotton and kapok seeds. Cyclopropene fatty acids have been incriminated in the pink discoloration of egg white and reduced egg production in cottonseed-fed laying hens, growth suppression and impaired female reproduction in rats, and increased saturated fatty acids (possibly causing atherosclerosis) in the tissues of pigs and other animals [181]. They are themselves carcinogens and markedly increase the carcinogenicity of aflatoxin in trout [10].

Increasing the consumption of polyunsaturated fatty acids in the diet to lower blood cholesterol, although beneficial in decreasing the incidence of coronary disease, has raised concerns for adverse effects such as increasing the total triglyceride levels and induction of vitamin E deficiency [181,230]. Carrol [42] demonstrated a strong correlation between dietary fat and age-adjusted mortality rates from breast and intestinal cancer. Pancreatic cancer was found to be enhanced by a diet containing 20% corn oil but not by one containing 18% hydrogenated coconut oil and 2% corn oil [234]. Unsaturated fatty acids are easily oxidized during cooking to a variety of mutagens, enals and other aldehydes, and alkoxy and hydroperoxy radicals [10]. Lipid oxidation products alter signal-transduction pathways [283] and thus enhance cell proliferation and promote carcinogenesis. Lipid-induced inhibition of immune responses and enhanced formation of some of the known tumor promoters such as prostaglandins and bile acids also have been reported [42]. Interestingly, Hayasu et al. [103] showed that oleic and linoleic acids may, in fact, be antimutagenic. The overall effect of dietary fats may depend on the ratio of beneficial fatty acids to those of the causative fatty acids for each effect. Until a clear understanding of the role of dietary fat in human disease is obtained, prevention of weight gain, obesity, type 2 diabetes, and cardiovascular disease appears to be best achieved by a diet low in fat and sugars and high in fiber and protein.

## Oxalates and Phytates

Certain plants, including spinach, rhubarb, beet leaves, tea, and cocoa, contain high levels (0.2 to 2.0% on a fresh weight basis) of oxalic acid. Cattle and sheep have been poisoned following ingestion of the toxic plants *Halogeton* and *Sarcobatus* (grease wood). Toxic signs result from binding of the oxalic acid to serum calcium, leading to hypocalcemia, coagulation defects, and tetany. Degeneration and necrosis of kidneys and vasculature from $Ca^{2+}$ oxalate deposition may result in severe cases. A subpopulation of urinary-stone-forming patients who are "hyperabsorbers" of oxalates, absorbing more than the normal to 8% of the 150- to 250-mg/day of dietary oxalate intake, may benefit from a reduction in dietary oxalate intake [177]. It is possible that antibiotic-therapy-induced loss of the intestinal oxalate metabolizer *Oxalobacter formigenes* may contribute to elevated body and urinary oxalate, thus increasing the risk of recurrent calcium oxalate kidney stone formation [276]. Either directly or via their increased production of reactive oxygen species, oxalates alter lipid signaling pathways, leading to changes in renal membrane characteristics and damage that promotes oxalate crystal enucleation and growth into stones [253]. Chronic oxalate consumption interferes with absorption of calcium, iron, magnesium, and copper and inhibits succinate dehydrogenase and carbohydrate metabolism [211]. Approximately 2.5 kg of tomato or 0.5 kg of spinach leaves must be consumed to approach a lethal dose (5 g or more) of oxalates.

Phytic acid, the hexaphosphoric ester of myo-inositol (IP6), is present at high levels (up to 1.5 g%) in the bran and germ of wheat, followed by other cereals, nuts, seeds, spices, and legumes [119]. Phytates bind di- and trivalent metals in the order of $Cu^{2+} > Zn^{2+} > Co^{2+} > Mn^{2+} > Fe^{3+}$ $Ca^{2+}$, causing mineral deficiencies (especially of $Ca^{2+}$ and $Fe^{3+}$) in developing countries that are heavily reliant on cereals as the exclusive source of protein. Inclusion of phytase, an enzyme that releases phosphate from plant phytic acid, in animal feeds ensures phosphate utilization and reduces environmental phosphate pollution from animal production. Supplementation with minerals and vitamin D can antagonize most effects of oxalates and phytates [119]. By altering inositol second-messenger pathways, phytates (especially in combination with inositol) inhibit the cell cycle, increase malignant cell differentiation and reversion to normal cells, and are anticarcinogenic in animals and humans for prostate and possibly breast, colon, liver, prostate, and skin cancer; leukemia; and sarcomas [265].

## Plant Phenolics

Plant phenolics comprise a group of several thousand substituted phenolic compounds occurring in trace amounts as esters or glycoconjugates and play essential roles in plant growth and development, defense, symbiosis, pollen development and male fertility, polar auxin transport, protection against ultraviolet radiation, and cell-cycle regulation. They are widespread constituents of fruits, vegetables, cereals, dry legumes, chocolate, and beverages such as tea, coffee, or wine. Acute human and animal poisonings are mostly caused by phenolics that are either uncommon or are present as contaminants in human food and which include coumarins, aflatoxins, and gossypol. Phenolics common in human foods belong to three general classes: nonflavonoids (gallic, syringic, caffeic, and other acids), flavonoids (flavones such as tangeritin, flavonols such as kaempferol and quercetin, isoflavones such as coumestrol, aurones, chalcones, and anthocyanin pigments), and polyphenols (tannins and lignin). Common human foods rich in flavonoids include coffee, chocolate, tea extract, green and black tea, pomegranate juice, grape juice and grape extracts, virgin olive oil, red wine, and soy proteins.

Polyphenols are widely distributed and present in relatively larger amounts in cereals, millets, legumes, and fruits. Deleterious effects of long-term exposure to both hydrolyzable (polyphenolic acid) and condensed (polyflavonoid) tannins include reductions in the digestibility of foods and feeds, protein utilization, and body weight gain; damage to and sloughing of the mucosal lining of the gastrointestinal tract; and cancer of the mouth and esophagus [107,229]. In contrast to mild acute effects in humans, livestock losses can exceed $10 million annually, attributable to the toxic effects of hydrolyzable oak tannins consumed when other forages are unavailable [267]. An epidemiological correlation exists between the high consumption of condensed tannins (sorghums and dark beer prepared from them, tea, red wines, and areca nuts) and high rates of human oral and esophageal cancer [67].

Parenteral exposure to tannins reportedly has led to a high incidence of liver and other tumors in rodents [107]. On the other hand, negative associations between tea drinking and stomach cancer [278] and coffee consumption and kidney cancer [117] also exist. Polyphenols, however, are not directly damaging to the DNA [45], and experimental evidence suggests an anticarcinogenic effect of penta-*O*-gallyl-beta-D-glucose and epigallocatechin gallate, two green tea tannins [87].

Flavonoids and nonflavonoids exert no less than 40 different physiologic and pharmacologic actions, thus accounting for their therapeutic and extensive health food use. These actions include antiandrogenic, anticoagulant, antihistaminic, antihypercholesterolemic, antiinflammatory, antinutritional (inhibit protein digestibility and nonheme iron absorption leading to iron deficiency), antioxidant, antiproliferative, antipruritic, antipyretic, antirheumatic, antiseptic, antithrombogenic, antithyroid, antitumor, apoptotic, estrogenic, and vasoactive effects. In addition, polyphenols alter the bioavailability and thus the biologic effects of certain drugs, including benzodiazepines, terfenidine, and cyclosporine. Many if not all of these actions are based on their ultraviolet-absorbing, chelating, oxidative phosphorylation uncoupling, and oxidant/antioxidant properties. In addition, their induction of P450-mediated enzymes and alteration of enzymes (phospholipases, ATPases, cyclooxygenases, lipoxygenases, protein kinases), oncogenes, and other signaling components critical for cell survival and proliferation [45,82] contribute to an array of opposing effects in many systems.

Although human consumption of flavonoids alone can be greater than 1 g/day [119], the toxicological implications of exposure to flavonoids and other simple phenolics arise from their life-time exposure. High intake of flavonoid supplements in humans and the experimental feeding of high levels in the diet have caused acute renal failure possibly due to hemolysis, liver failure, contact dermatitis, and anemia [160,186]. Endocrine disruption involving inhibition of thyroid peroxidase and thyroid hormone biosynthesis by flavonoids leading to increased thyroid weight (goiter) and decreased plasma levels of thyroid hormones is of particular concern for babies exposed to high doses of isoflavones through soy feeding. Their estrogenic (at moderate doses) and antiandrogenic (at higher levels) activities are of little concern in adults at normal intake (0.2 to 5 mg/day from a Western diet and 20 to 120 mg/day from an Asian diet), but their antiluteinizing hormone effect at levels present in soy-based infant formula can have adverse effects on the sexual maturation of male infants, who normally exhibit luteinizing hormone secretion between birth and 6 months of age.

Many flavonoids, including the most abundant quercetin, kaempferol, myricetin, hesperetin, naringenin, wogonin, and norwogonin, as well as their glycosides, are mutagenic in bacterial or mammalian systems [107,168]. Although polyphenols such as caffeic acid, quercetin, and green-tea catechins are known to induce tumors in the forestomach, colon, kidney, and highly oxidative tissues in rats and mice, reports of their human carcinogenic effects are scanty, likely due to the efficient repair of quercetin quinone methide–DNA adducts [24,74,186]. A preponderance of evidence from animal and *in vitro* systems, however, points to the preventive effects of flavonoids and polyphenols against cardiovascular diseases, cancers, neurodegenerative diseases, diabetes, and osteoporosis [250]. Clinical studies on biomarkers of oxidative stress, cardiovascular disease risk factors, and tumor or bone resorption biomarkers, however, have often led to contradictory results, likely due to differences in type and levels of phenolics consumed. Phenolics, for example, are known to exert antioxidant effects at low doses and prooxidant effect at higher doses [317]. A recent study associated an increased intake of flavones and flavonols, but not other flavonoids, with decreased incidences of breast cancer [33].

The anticarcinogenic effects of phenolics in animals appear to involve both initiation and progression phases of cancer and a combination of mechanisms, including inhibition of metabolic enzymes leading to reduced levels of reactive intermediates, induction of detoxifying enzymes such as glutathione *S*-transferase, reduced formation of oxidation products, alteration of the activity of protein kinases and oncogenes that stimulate cell proliferation, increased apoptosis, reduced expression of matrix metalloproteinases involved in metastasis, and inhibition of angiogenesis [21,45,107,123,161,313]. Protective effects of flavonoids on the cardiovascular system have been shown more consistently in both animals and human beings. The mechanisms for such effects appear to involve reduction of low-density lipoprotein (LDL) oxidation (antiatherogenic effect) and platelet aggregation (antithrombotic), vasodilation, relaxation of cardiovascular smooth muscle, and their antiinflammatory and antihypercholesterolemic (decrease in low-density lipoprotein, increase in high-density lipoprotein [HDL]) effects among others [82,111,173].

Gossypol (1,1,6,6,7,7-hexahydroxy-5,5-diisopropyl-3,3-dimethyl [2,2-binaphthalene]-8,8-dicarboxaldehyde), a yellow phenolic pigment in cottonseed, can bind to proteins and minerals and reduce the biological availability of iron and lysine [118]. Similar to other phenolics, free gossypol (>60 ppm) inhibits oxidative phosphorylation and causes a myriad of other effects leading to acute toxicity in animals on a high cottonseed diet. In general, higher doses cause cardiac failure associated with liver and lung (pulmonary edema) damage, whereas chronic exposure leads to general malnutrition and reproductive effects [45]. Signs of gossypol toxicity include loss of appetite and body weight; rough hair coat; edematous flui in body cavities, lungs, and pericardium giving rise to gasping; hemorrhagic degenerative changes in liver; and necrosis of cardiac myocytes [321]. Changes in plasma $K^{2+}$ (increase in calves and decrease in humans) may be responsible for gossypol toxicity. Olive discoloration of egg yolk and decreased egg hatchability occur in poultry [45]. Male antifertility effects of gossypol in mammals are only partially reversible and include reduced sperm production as well as motility during the late stages of spermatogenesis likely caused by mitochondrial damage [222] or inhibition of protein kinases [286]. Gossypol is not mutagenic in the Ames test [45] but appears to induce genetic damage (dominant lethal mutations) in rats and may be both an initiator and a promoter of carcinogenesis [10]. In rat lymphocytes, gossypol induces DNA breaks secondary to cytotoxicity [221].

Gossypol and polyphenol (tannin) toxicity can be prevented by the addition of iron, supplemental protein, vitamins E and K, and alkalinizing agents such as sodium hydroxide. In addition, non-ionic detergents such as Tween® 80, methyl donors such as choline and methionine, and dehulling and peeling of grains and fruits have been shown to counteract the toxic effects of tannins [66,267]. A glandless (gossypol-free) variety of cottonseed is expected to eliminate gossypol toxicity in animals but appears to be more susceptible to insect attack and has yet to gain popularity.

## Proteins, Peptides, and Amino Acids

The average American dietary protein supplies 15% of total calories. In general, it appears that long-term consumption of high levels of protein, especially animal-derived and in amounts that supply >45% of daily caloric needs, has been associated with weakness, nausea, diarrhea, diabetes, renal glomerular sclerosis, Crohn's disease, and osteoporosis due to increased $Ca^{2+}$ loss from bones in response to acidosis [26,54,266,315]. Huang et al. [114] showed that soybean proteins extracted with 20% ethanol can inhibit thyroid hormone (TH) receptor (TR) binding to the TR element in TH-regulated genes and hypothesized that soy protein rather than other components may account for the hypocholesterolemic and hypolipidemic and thus cardioprotective effects of soybeans. Feeding high levels of soy protein, however, markedly increased pancreatic weights and reduced spleen weights in both male and female rats, possibly due to the presence of active residual trypsin inhibitors (known to induce hypertrophy and hyperplasia of the acinar cells) and soy fiber, respectively. Products such as D-amino acids and lysinoalanine formed during alkaline/heat treatment of proteins such as casein, lactalbumin, soy protein isolate, or wheat proteins can reduce the digestibility of other dietary proteins [95].

Protein toxicants such as allergens, hemagglutinins (lectins), and enzyme inhibitors have already been discussed. Certain microbial protein toxins are discussed in subsequent sections. Toxic peptides from mushrooms are discussed below.

### *Mushroom Peptides*

Among the approximately 5000 species of mushrooms that exist in nature, up to 300 have been shown to be safely edible; however, the ingestion of 50 to 100 types of mushrooms (generally collected from the wild) can lead to toxic and occasionally lethal (12 known) consequences [31,77, 146,312]. Ninety percent of these poisonings occur in individuals under 19 years of age. One or more of the following are often involved: cyclopeptide, orellanine, monomethylhydrazine, disulfiram-like hallucinogenic indoles, muscarinic, isoxazole, and GI-specific irritant toxins (Table 14.1). Cultivated mushrooms, for the most part, are safe.

Cyclopeptide toxicants are thermostable and are comprised of amatoxin (Figure 14.2), phallotoxin, and verotoxin groups, the latter two producing effects only at high doses. Approximately half of a mature cap of *Amanita verna* ("destroying angel," common in the United States) or *A. phalloides* ("green death cap," in Europe), which contain amanitin, can be lethal in an adult [182]. Clinical effects that begin to appear after a 12-hour latency period include epigastric tenderness, intense and cramping abdominal pain, nausea, vomiting, severe secretory diarrhea (possibly bloody), and hepatomegaly, as well as secondary acid–base disturbances, electrolyte abnormalities, hypoglycemia, dehydration, and hypotension. This is followed by elevation of liver enzymes (AST and ALT) and bilirubin, coagulopathy, hypoglycemia, acidosis, hepatic encephalopathy, hepatorenal syndrome, multiple organ failure (including pancreas, adrenal, and testes), disseminated intravascular coagulation, mesenteric thrombosis, convulsions, and death 6 to 16 days after ingestion [31]. Fatalities are common (10 to 30%) even following intensive symptomatic care, which includes fluid replacement, activated charcoal hemoperfusion, forced diuresis, etc. Penicillin therapy (by an unknown mechanism) and, in countries other than France and the United States, the use of silibinin (from the milk thistle plant, *Silybum marianum*), which prevents hepatocyte uptake of amatoxins, have produced beneficial effects in direct relationship with the speed of onset of therapy [182,303]. A return toward normal glucose, factor V, and fibrinogen is prognostic of recovery [182] and may take several weeks to months. Amatoxins act by binding to and inhibiting RNA polymerase II and thus mRNA and protein synthesis leading to cell necrosis [31,303].

The effects of phallotoxins include swelling of the liver due to engorgement of hepatic sinusoids with blood and depletion of blood in the peripheral circulation leading to shock. Hepatocyte damage occurs due to a reduction

**TABLE 14.1**
**Mushroom-Induced Syndromes**

| Syndrome | Mushroom Species | Toxic Compound(s) | Effects | Mechanism | Prevention/ Treatment |
|---|---|---|---|---|---|
| | | | **Rapid Onset** | | |
| Gastrointestinal | *Chlorophyllum molybditis* *Entoloma lividum* *Omphalotus olearius* *Paxillus involutus* *Trichodoma pardinum* | Many unknown | Emesis, diarrhea | Unknown | Cooking/fluid replacement |
| Parasympathetic | *Inocyte* sp. *Clitocybe* sp. *Omphalotus illudens* *Amanita* sp. | Muscarine and related | Increased salivation, lacrimation, and urination; diarrhea; dyspnea; sweating; bradycardia; tremors, etc. | Parasympathetic stimulation | Avoid/atropine |
| CNS syndrome | *Psilocybe* sp. *Panaeolus* sp. *Copelandia* sp. *Gymnopilus* sp. | Psilocybin, psilocin | Hallucinations involving all sensations, hyperthermia, convulsions, coma, death | Serotonin agonist | Avoid/diazepam and cooling |
| | *Amanita pantheria* *A. muscaria* | Ibotenic acid, muscinol, stizolobic and stizolobinic acid | Alternating depression and neuromuscular stimulation | Stimulation of bicuculim-reactive post-synaptic receptors | Avoid/diazepam and respiration |
| Alcohol sensitization | *Coprinus* sp. *Clitocybe claviceps* *Boletus luridus* *Verpa bohemica* | Coprine and others | Nausea, vomiting, headache, hypotension, tingling, palpitations, tachycardia, testicular damage, etc. | Inhibit acetaldehyde dehydrogenase | Avoid mushroom and alcohol; supportive |
| | | | **Delayed Onset** | | |
| Headache | *Gyromitra esculenta* (false morel) *Gyromitra* sp. | Gyromitrin, monomethylhydrazine, etc. | Fatigue, head and body ache, vomiting, liver damage, death, carcinogenic | Interfere with pyridoxine? | Cook or dry; do not inhale vapors |
| Nephropathy | *Cortinarius* sp. | Orellanine, cortinarin | Polydypsia, oliguria, nausea, head and body aches, chills, etc.; renal tubular and liver necrosis; death | Membrane damage from oxygen-derived free radicals (similar to Paraquat) | Hemodialysis |
| Carcinogenic | *Agaricus bisporus* (edible) | Agaritine, hydrazines | Lung tumors | Genotoxic | Cooking |
| Hepatotoxic | *Amanita phalloides* (Europe) *A. virosa* (U.S.) *Galerina* sp. *Lepiota* sp. | Amatoxins, phallotoxins, virotoxins | Emesis and diarrhea, increase in serum enzymes, decrease in glucose and clotting factors, hepatic and renal damage, jaundice, coma, death | Inhibit RNA polymerase; enhance G-actin polymerization into F-actin; inhibit F-actin depolymerization | Correct glucose and clotting effects; decontaminate; penicillin and silibinin; supportive; liver transplant |

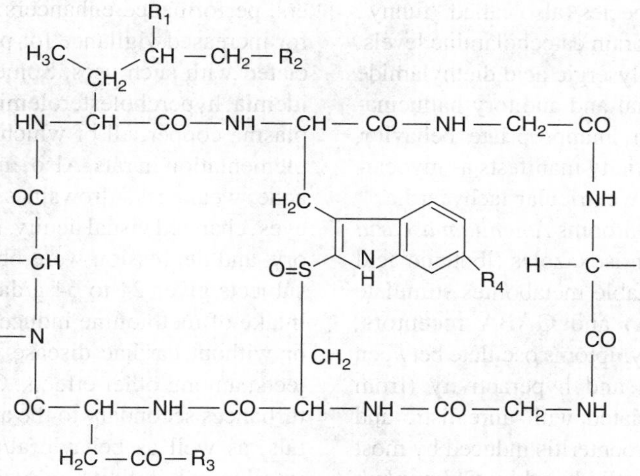

| | $R_1$ | $R_2$ | $R_3$ | $R_4$ |
|---|---|---|---|---|
| α-Amanitin | •OH | •OH | $•NH_2$ | •OH |
| β-Amanitin | •OH | •OH | •OH | •OH |
| γ-Amanitin | •CH | •H | $•NH_2$ | •OH |
| ε-Amanitin | •OH | •H | •OH | •OH |
| Amanin | •OH | •OH | •OH | •H |
| Amanullin | •H | •H | $•NH_2$ | •OH |
| Amaninamide | •OH | •OH | $•NH_2$ | •H |

**FIGURE 14.2** The structures of amatoxins.

in cellular G-actin concentration resulting from the combined effect of stimulated G-actin polymerization into F-actin and inhibition of F-actin depolymerization leading to a loss of membrane elasticity and thus to cell surface vesiculation [303].

The toxic syndromes produced by these and other (some nonpeptide) classes of mushroom toxicants are summarized in Table 14.1 and discussed in detail by Berger and Guss [32].

Some degree of initial gastritis (vomiting) and enteritis (diarrhea) is a common feature of most mushroom poisonings. Delayed (for up to 3 weeks) acute renal failure (interstitial fibrosis and acute tubular necrosis manifesting as polyuria followed by oliguria) is caused by the heat-stable bipyridyl orellanine (2,2′-bipyridine-3,3′,4,4′-tetrol-1,1-dioxide or 3,3′,4,4′-tetrahydroxy-2,2′-bipyridine-*N*,*N*′-dioxide) present in *Cortinarius orellanus*, *C. speciosissimus*, *C. splendens*, and *C. gentiles* (common in Scandinavia and Europe) or *Amanita smithiana* (common in the U.S. pacific northwest). Toxicosis manifests as weakness, lassitude, and headache and, in severe cases, hepatic failure developing over several days, resulting in hypoglycemia, delirium, and seizures progressing to coma and death. It is caused, especially in isoniazid-sensitive individuals, by a hydrolysis product (*N*-monomethylhydrazine) of the volatile nonpeptide toxin gyromitrin, which is present in false morels (represented by *Gyromitra* species, particularly *Gyromitra esculenta* in Europe). Inhibition of GABA synthesis and induction of pyridoxine deficiency appear to be the contributory mechanisms.

A disulfiram (antabuse)-like syndrome characterized by headache, paresthesias of the hands and feet, metallic taste, facial flushing, palpitations, tachycardia, orthostatic hypotension, chest pain, nausea, vomiting, and diaphoresis occurs if alcohol is consumed within 72 hours after consumption of the otherwise safe mushrooms *Coprinus atramentarius* (common lawn mushroom) and *Clitocybe clavipes*. The syndrome lasts for up to 2 days and is induced by the amino acid coprine ($N^5$-[1-hydroxycyclopropyl]–L-glutamine). 1-Aminocyclopropanol hydrochloride, a hydrolytic product of coprine, inhibits aldehyde dehydrogenase, leading to accumulation of acetaldehyde from alcohol metabolism.

A cholinergic syndrome characterized by salivation, lacrimation, abdominal pain, diarrhea, emesis, perspiration, and occasionally miosis, rhinorrhea, flushed skin, bradycardia, and hypotension is induced by a heat-stable parasympathomimetic compound, muscarine, present in several lawn and park mushrooms of the genera *Inocybe* or *Clitocybe*. Muscarine activates acetylcholine receptors on the heart, apocrine glands, and smooth muscle.

Psilocybin (*O*-phosphoryl-4-hydroxy-*N*,*N*-dimethyltryptamine), present in at least 75 mushrooms belonging to the genera *Psilocybe* and *Paneolus*, as well as *Conocybe*,

*Gymnophilus*, and *Stropharia* species (also called "funny" or "magic" mushrooms), alters brain catecholamine levels, especially serotonin, leading to lysergic acid diethylamide (LSD)-like signs including visual and auditory hallucinations, confusion, disorientation, inappropriate behavior, and mydriasis. Rarely, cardiotoxicity manifests as myocardial infarction and serious supraventricular tachycardia.

The beautifully colored mushrooms *Amanita muscaria* and *Amanita pantherina* contain isoxazoles (ibotenic acid and muscimol) whose thermostable metabolites stimulate *N*-methyl-D-aspartate (NMDA) and GABA receptors, inducing a syndrome in which symptoms oscillate between various degrees of depression and hyperactivity (from obtundation to delirium) associated with unrealistic and bizarre behavior. Although gastroenteritis induced by most common mushrooms is mild, that induced by *Chlorophyllum molybditis* (found in lawns, fields, and open woods in southern and midwestern parts of the United States) can be severe, requiring immediate medical attention.

Other identified human conditions associated with mushroom production, commerce, and consumption include hypersensitivity to edible mushrooms in certain populations; hypersensitive allergic alveolitis and other pulmonary allergic changes in mushroom workers from spores of certain edible mushrooms (mushroom worker's lung); hemolytic reactions following consumption of mushrooms belonging to the genera *Gyromitra*, *Boletus*, and *Paxillus*; and dermatitis (allergic) from contact with one or more species of the genera *Boletus*, *Lactarius*, *Calvaria*, and *Agaricus*. Treatment of mushroom poisoning is mostly supportive.

Extracts and isolated metabolites from mushrooms can either enhance or suppress innate and acquired immunity, leading to beneficial effects such as increased disease resistance, anticancer activity, suppression of autoimmune (T[H]1 type T-cell-mediated) and allergic (T[H]2 type T-cell-mediated) diseases. Mechanistically, low-molecular-weight metabolites affect apoptosis-, angiogenesis-, metastasis-, cell-cycle-related signaling, and high-molecular-weight components (polysaccharides or polysaccharide–protein complexes) enhance innate and cell-mediated immune responses, leading to altered mitogenic response, T-cell differentiation, and activation of immune effector cells such as lymphocytes, macrophages, and natural killer cells [167]. Additional beneficial effects of mushrooms include inhibition of clotting, and reductions in blood cholesterol and pressure [32].

### Amino Acids

Adverse effects from amino acids, reviewed by Garlick [92], appear to be restricted to very high parenteral doses or diets with low protein levels. Recent increases in the consumption of amino acid dietary supplements, flavorings (glutamate as monosodium glutamate [MSG] and aspartate and phenylalanine in aspartame), health promoters, performance enhancers, and behavior modifiers call for increased vigilance for potential adverse effects associated with such uses. Some examples include hyperlipidemia, hypercholesterolemia, enlarged liver, and reduced plasma copper, all of which are reversed by copper supplementation in rats. Also, increases in urinary zinc, headache, weakness, drowsiness, nausea, anorexia, painful eyes, changed visual acuity, mental confusion, poor memory, and depression were observed in overweight human subjects given 24 to 64 g/day of histidine. The excessive intake of methionine induces hyperhomocystinemia with or without cardiac disease in both rats and human subjects, among other effects. Ocular lesions and visual disturbances secondary to the accumulation of tyrosine crystals, as well as behavioral/perceptual and performance, intellectual deficits in neonatal babies and animals appear following *in utero* and/or neonatal exposure to high-tyrosine diets/formulas. The therapeutic use of greater than ten times the required dose of amino acids when given on empty stomach, can lead to adverse effects such as gastric distress (essential amino acids); nausea febrile reaction or headache (methionine, isoleucine, and threonine); and disorientation (methionine and tryptophan) in mental patients treated with monoamine oxidase inhibitors [102].

Monosodium glutamate has long been used as a flavor enhancer in commercially processed foods. MSG as well as other acidic amino acids, but not basic or neutral amino acids, produced lesions in rats and mice in the arcuate nucleus of the hypothalamus, retina, lateral geniculate nucleus, and in other brain areas devoid of the blood–brain barrier [208]. Numbness of the neck and back, weakness and palpitations, the typical signs of so-called Chinese restaurant syndrome, were later found not to be associated with dietary MSG [285].

Hypoglycin A (β-methylene cyclopropyl alanine) and its γ-glutamyl conjugate, hypoglycin B, are components of the fruit of the plant *Blighia sapida* ("ackee" in Jamaica and "isin" in Nigeria). Consumption of this fruit in the unripened stage has been associated with hypoglycemia resulting from inhibition of gluconeogenesis involving inhibition of fatty acyl-CoA dehydrogenases and thus β-oxidation of fatty acids by cyclopropylacetyl CoA (a metabolite of hypoglycin A). Signs of intoxication include vomiting, convulsions, hypothermia, coma, and even death. Pretreatment with clofibrate (stimulator of peroxisomal fatty acid oxidases) has been shown to prevent many but not all signs, lesions, and biochemical effects [297].

Koa haoli (*Leucaena leucocephala*), a legume found in Hawaii, and other legume species belonging to the Mimosidae family have potentially high nutritive value for animals and humans [201]; however, use of these legumes is precluded in ruminants by the goitrogenic effect of the metabolite (3,4-dihydroxypyridine) of an unusual amino acid, mimosine (3-*N*-(3-hydroxypyridone-4)-2-amino propionic acid), present in this plant. Mimosine also causes reversible destruction of the hair follicle matrix (loss of hair), reduced bone strength and mineral composition in poultry, and growth depression in both ruminants and nonruminants. The ability of mimosine to chelate Zn and Mg, reduce plasma thyroid and other hormone levels [220], and inhibit a large number of enzymes leading to DNA synthesis inhibition and cell-cycle arrest [122,157] explains many of the effects.

Djenkolic acid, an amino acid that is structurally similar to cystine, is present in the djenkol bean (*Pithocolobium lobatum*) found in Sumatra and Java. It cannot substitute for cystine nor can it be totally metabolized but it can crystallize in the kidney, causing hematuria and crystalluria [157].

Favism is a hemolytic disease (accompanied by jaundice and hemoglobinuria) in persons genetically deficient in glucose-6-phosphate dehydrogenase (G6PD) and thus in NADPH and glutathione content. It results from the consumption of the amino acid 3,4-dihydroxyphenylalanine and the pyrimidine aglycones (divicine and isouramil) of the glycosides vicine and convicine in broad beans (*Vicia faba*) found mainly in the Mediterranean region and in the Middle East [48]. Ohga et al. [207] observed the beneficial effects of human heptaglobin administration in managing this crisis.

The etiology of the neurologic disease characterized by posterior sensory ataxia in cattle consuming cycads may be an amino acid, β-*N*-methylamino-L-alanine. When incorporated into structural animal proteins, certain seleno-amino acids (e.g., methylselenocysteine, selenocystathionine, selenocysteine, and selenomethionine) that are found in plants that grow on high-selenium soils [157] may produce defective hair and hooves, which are eventually lost during longer term exposure in livestock. In human beings, a syndrome characterized by abdominal distress, nausea, vomiting, diarrhea, and loss of scalp and body hair had been reported following the consumption of coco de mono (*Lecythis ollaria*) nuts containing high levels of selenocystathionine [16].

The amino acids L-2,4-diaminobutyric acid (DABA), -*N*-oxalyl-L-2,3-diaminopropionic acid (ODAP), -cyanoalanine, and 4-glutamylcyanoalanine, and related homologs, are present in the seeds of several species of *Lathyrus* and *Vicia sativa* in the Indian subcontinent; they have been implicated in the pathogenesis of neurolathyrism, a syndrome characterized by muscular rigidity, weakness, paralysis of leg muscles, and death following long-term, high-level consumption of *L. sativus* seeds [296]. The mechanism of action appears to involve irreversible binding of ODAP to the glutamate receptor and enhanced release or reduced reuptake of glutamine at relevant nerve terminals leading to vascular degeneration and necrosis of neurons [212]. In certain individuals, amino acids such as β-aminopropionitrile and the dipeptide (*N*-glutamyl) aminopropionitrile, as well as certain urides, hydrazides, and hydrazines, from the green parts of *Lathyrus* and other plants lead to osteolathyrism characterized by bone deformities and reductions in the tensile strength of aorta [101] resulting from the irreversible inhibition of lysyl oxidase and interference with cross-linking of collagen [314].

Creeping indigo (*Indigofera endecapylla*), a tropical forage, contains a nitric oxide synthase inhibitor, indospicine, that causes liver damage in sheep, rats, and mice by inhibiting the incorporation of arginine, the amino acid it resembles, into protein [157,214]. 3-Nitropropionic acid, a neurotoxin capable of inhibiting mitochondrial succinate dehydrogenase and thus cellular respiration, is also present [7].

## Saponins

The saponins are bitter-tasting steroidal (C27) or mono-, di-, tri-, and sesquiterpenoid (C30) glycosides from plants, fish, and sponges capable of reducing surface tension, hemolyzing red blood cells, and causing toxic effects in cold-blooded animals. Their occurrence, biological effects, and relevance to food, agriculture, and medicine are reviewed by Walker and Yamazaki [306,307].

So far, d-limonene and other saponins from citrus oils; ginseng saponins; medicagenic acid and hederosides in alfalfa and *Hedera helix*, respectively; and oleanolic and ursolic acid in a variety of food, medicinal, and other plants, as well as their aglycones (sapogenins), have been studied to some extent. Their analgesic, antiatherosclerotic, anticarcinogenic, anticholinergic, antihypercholesterolemic, antihyperglycemic, antiinflammatory, antitubercular, cardioprotective, diuretic, and hepatoprotective effects are likely to encourage increased dietary, supplemental, and medicinal utilization of saponin-containing plants such as ginseng [163,171,223,316]. The mechanisms of protection involve $Ca^{2+}$-antagonistic and vasodilatory/venoconstrictive, immunomodulatory, bile-acid-binding, antiproliferative, membrane-permeabilizing, antioxidant, and anti-cytochrome P450 effects [163,223]. Feeding high levels of saponin from a variety of sources, however, results in a lower growth rate, increased serum LDH and GOT associated with hepatocellular necrosis, and increased BUN, hematuria, and proteinuria associated with renal tubular necrosis in animals [136,200]. Several steroidal and nonsteroidal saponins from pasture weeds such as *Hypericum perforatum* and *Narthecium ossifragum*, vines such as *Tribulus terrestris*, and tropical grasses such as *Brachiaria* and *Panicum* sp. cause primary or hepatogenic photosensitization [44]. Alpha-hederin, a saponin that induces metallothionein in maternal tissues, appears to induce visceral and skeletal defects in offspring born to exposed mothers by possibly reducing zinc availability to the fetus [71]. Similar to the effects of phenolics, the beneficial effects of saponins can be derived from daily doses present in a balanced diet.

## Vaso- and Psychoactive Substances

High levels of amines such as tyramine and its methyl derivatives octopamine, dopamine, epinephrine, norepinephrine, histamine, serotonin, and others are present in cheese, yeast products, fermented foods, beer, wine, pickled herring, snails, chicken liver, coffee, broad beans, chocolate, cream products, and plants such as pineapple, banana, plantain, and avocado [166]. Moderate amounts of cheese and yeast products commonly contain the dose of 10 mg of tyramine required to cause severe hypertensive crises in individuals treated with nonselective monoamine oxidase (MAO) inhibitors for disorders of mood [22]. Inhibition of MAO leads to a combined vasopressor effect of unmetabolized biogenic as well as dietary amines. In addition, tyramine enhances release of catecholamines that are present in supranormal amounts in the adrenal medulla [22]. Palpitations, migraine headaches, and in some instances intracranial bleeding and death may ensue. Use of selective (MAO-A or -B) inhibitors for therapy appears not to sensitize individuals to dietary tyramine [138].

Psychoactive substances include central nervous system stimulants such as xanthines (caffeine, theophylline, and theobromine present in coffee, tea, and cocoa), depressants such as alcohol and high doses of atropine (from jimsonweed and henbane), and hallucinogens such as myristicin from nutmeg, psilocybin and psilocin from mushrooms, and nondietary consumption of cocaine and lysergic acid derivatives by drug cults [180]. Chronic overindulgence in xanthine-containing beverages may lead to restlessness, disturbed sleep, myocardial stimulation reflected as premature systoles and tachycardia (palpitations), and tremors. Herbs containing toxic psychoactive agents include California poppy, catnip, cinnamon, hops, hydrangea, juniper, kola nut, nutmeg, periwinkle, thornapple, and wild lettuce [29]. The essential oils of coffee and the tannins in tea may cause diarrhea and constipation, respectively [61]. Caffeine is not mutagenic by itself nor does it enhance the mutagenic effects of other compounds in mammalian cells; it may actually be anticarcinogenic [50].

## Vitamins and Antivitamins

Vitamin A (retinol), vitamin D, and pyridoxine have low safety margins (ten times the recommended daily allowance [RDA]) and should be used on a longer term basis only under medical supervision [175]. Others with safety ratios of 50 to 100 relative to the RDA are generally safe. Therapeutic uses of vitamin A for night blindness, steatorrhea, hyperkeratosis, acne vulgaris, certain immune disorders, and cancer, along with daily consumption of carotenoids and vitamin A in plant and animal tissues (especially the liver), account for the total vitamin A exposure in humans [209]. Oral doses of 18,000 to 60,000 IU/day and 100,000 IU/day can cause hypervitaminosis A in infants and adults, respectively, with premature epiphyseal closure and retardation of long bone growth in children and headaches, blurred vision, fatigue, hair loss, drying and flaking skin, pruritis, nose bleeds, anemia, and liver and spleen enlargement in adults [209]. Therapy of acute promyelocytic leukemia with tretinoin (all-*trans*-retinoic acid) as an adjunct to antineoplastic therapy induces retinoic acid syndrome (fatal leukocytosis, body weight gain, respiratory distress, cardiac and renal failure) in 25% of cases [81]. Excess vitamin A (retinoids) is teratogenic and can cause craniofacial, thymic, heart, and CNS malformations subsequent to interaction with cellular retinoic acid or retinol binding proteins [64]. The Teratology Society recommends that the oral doses of vitamin A not exceed 6000 IU/day during pregnancy.

Vitamin D, following hydroxylation at $C_{25}$ and $C_1$ in the liver and kidney, respectively, functions to facilitate the action of parathyroid hormone to release $Ca^{2+}$ from the bone and to promote its intestinal absorption and inhibit its renal loss. Ergosterol (a plant steroid) converted to ergocalciferol (vitamin $D_2$) by ultraviolet light endogenous dehydrocholesterol (in skin) converted to cholecalciferol (vitamin $D_3$) by sunlight, and vitamin-D-fortified milk are predominant sources of vitamin D for humans [209]. The recommended dose is 400 IU/day excessive exposure to vitamin D from 1000 to 3000 IU/day in infants and 10,000 to 500,000 IU/day in adults has resulted in toxicosis. The recent popularity of vitamin D and calcium supplements such as shark cartilage especially in cancer patients already having a certain degree of hypercalcemia, has resulted in many cases of symptomatic hypercalcemia [143]. Poisonings in dogs and cats have resulted from accidental consumption of insecticidal vitamin D packages. Toxic signs of vitamin D in humans, which are similar to those seen in laboratory animals, are a result of hypercalcemia leading to extraskeletal calcifications, especially blood vessel walls and kidneys, which in turn lead to hypertension, renal failure, and cardiac insufficiency [209]. Recent evidence suggests that supplementation with vitamin $D_3$ may lead to an increase in LDL, may negate putative cardioprotection by hormone replacement therapy in postmenopausal women [105], and may exacerbate intimal hyperplasia in balloon-damaged rat arteries [144].

Long-term supplemental use of vitamin E can result in coagulopathy (by reduced iron utilization and increased vitamin K requirement) as well as tumor promotion (both stage I and II); the long-term use of pyridoxine can result in photosensitivity reaction and sensory neuropathy [175,190,209].

Table 14.2 lists examples of antivitamin factors in foods, the effects of which, in general, only manifest in individuals with already low levels of the vitamin in question.

**TABLE 14.2**
**Examples of Anti-Vitamin Factors in Natural Foods**

| Vitamin | | Antagonist(s) | Mechanism | Effect(s) | Source(s) |
|---|---|---|---|---|---|
| A | | Lipoxidase | Oxidizes β-carotene; | Lower blood Vit A level | Soybeans |
| | | Citral | inhibits retinoic acid (P450) dehydrogenase | Endothelial damage cardiovascular disease? | Oranges |
| B | Thiamine | Thiaminase, tannins, and *ortho*-catechols | Inactivates thiamin | Neurologic syndrome | Bracken fern, other plants, and seafood |
| | Riboflavin | Hypoglycin A | — | Vomiting, sickness | Ackee plum |
| | Niacin | Leucine? | Irreversible binding | Pallagra | Cereal crops |
| | Biotin | Avidin | — | — | Egg white |
| | Pyridoxine | Linatine, agaritine | Hydrazine metabolites condense with the vitamin | | Flaxseed, Shiitake mushroom |
| | Pantothenic acid | Unknown | Unknown | — | Pea seedlings |
| | B12 | Unknown | Unknown | — | Soybeans |
| C | | Ascorbic acid oxidase | Oxidizes vitamin C | Normally none | Fruits and vegetables |
| D | | β-Carotene, plant steroids (some) | Reduce absorption | Rickets and osteomalacia | Green leafy vegetables |
| | | Unknown | Unknown | — | Soybeans |
| E | | β-Tocopherol oxidase | Oxidizes vitamin E | Muscular dystrophy | Kidney beans and others |
| | | Polyunsaturated fatty acids | Increases viatmin E demand | Liver necrosis | Vegetable oils, beans |
| K | | Diconmarol | Inhibits epoxide reductase | Hemorrhages | Sweet clover |

## Miscellaneous Plant Toxicants

A common human intoxication known as milk sickness was one of the most dreaded diseases from the Colonial times to the early nineteenth century in an area extending from North Carolina and Virginia to the midwestern United States [156]. The disease manifested as weakness, nausea and vomiting, constipation, tremors, prostration, delirium, and even death. It resulted from the consumption of dairy products made from milk derived from cows (even healthy ones) grazing on white snakeroot (*Eupatorium rugosum*) or rayless goldenrod (*Haplopappus heterophillus*). The causative agent appears to be tremetol, an unsaturated alcohol, in combination with a resin acid [156]. Other plant toxins excreted through milk that pose toxic hazards for children and nursing animals include pyrrolizidine, piperidine, and quinolizidine alkaloids; sesquiterpene lactones of bitterweed and rubberweed; and glucosinolates [213]. Animals grazing on high-selenium forage may excrete high levels of selenium in milk and contribute to chronic selenium toxicosis in the offspring [213]. Current processing methods have kept these conditions in check for the most part. In cattle, consumption of 5 to 10 of snakeroot causes weakness and trembling of various groups of muscles, labored respiration, and death.

Fool's parsley (*Aethusa cynapium*) and other members of the Umbelliferae family contain highly toxic acetylene derivatives (e.g., aethusin) that are responsible for many human poisonings [158]. Carotatoxin and other acetylene compounds in carrots are neurotoxic [29] but are not likely to cause problems in humans.

Purple mint (*Perilla frustescens*), widely distributed in the United States and Japan, is used as a flavoring agent, for medicinal purposes, and as animal feed. The presence, in mint, of a ketone-substituted furan capable of causing acute pulmonary emphysema and other lung lesions in cattle and other animals [260] raises questions about the safety of these practices in humans and animals.

Cycads, the palm-like plants adapted for adverse climatic conditions of the tropical and subtropical areas of the world, are still used (seeds and stem) as a source of starch in Guam, Kenya, Amami Oshima, Miyako Island, and southern Japan by small groups of people [178]. Adverse effects result from incomplete extraction of toxicants, including cycasin and β-*N*-methylamino-L-alanine (BMAA), during preparation of the flour. Neurologic conditions include a paralytic disease (amyotrophic lateral sclerosis [ALS]) and Parkinsonism–dementia (PD) among the native Chamorro in Guam consuming cycad. Gait disturbances, motor weakness, and paralysis in cattle grazing on cycads; Parkinsonian features and degenerative changes in CNS motor neurons in monkeys; and, very importantly, Alzheimer's dementia (AD) in human beings appear to be related to cycad toxicants, especially BMAA [178,197, 273]. In this regard, it is important to note that recently BMAA has been shown to be produced by potentially all Cyanobacteria [56] in many parts of the ecosystem, including the root tissues of cycad trees. BMAA also accumulates in cycad seeds and is biomagnified by seed-eating flying foxes which, in turn, are consumed by the Chamorros. Because cyanobacteria function as primary producers in many food chains, the likelihood that this environmental neurotoxin could be a major etiological factor for Alzheimer's, worldwide would be a significant breakthrough. Although the presence of BMAA in the brains of Chamorros who have died of neurodegenerative ALS/PD syndrome, as well as Canadians with Alzheimer's dementia [197] is consistent with this hypothesis, the recent inability to reproduce these results by Montine et al. [194] and to produce lesions of AD in experimental animals fed BMAA adds some uncertainty to this theory. Mechanistically, attenuation of the cycad-induced neurotoxic syndrome by AP7 and MK801, two selective antagonists of *N*-methyl-D-aspartate receptor and its associated ion channel, suggests a role for the excitatory neurotransmitters in the causation of ALS/PD, other motor-system diseases (e.g., Huntington's chorea, Parkinson's disease, and olivopontocerebellar atrophy), and Alzheimer's disease [273]. Other effects of cycasin, its aglycone, or cycad flour include hepatic necrosis; subserosal hemorrhages; accumulation of yellow fluid in serosal cavities; benign and malignant tumors in the liver, kidney, lungs, and gastrointestinal tract (mainly colon); neuroteratologic effects in offspring; death in experimental animals; and mutagenic effects in a variety of *in vitro* and *in vivo* systems. Interestingly, cycasin in neither toxic nor carcinogenic when given parenterally to conventional rats or when given either orally or parenterally to germ-free rats, suggesting that the intestinal flora mediate cycasin toxicity. Bacterial β-glucosidase hydrolyzes cycasin to the active carcinogen methylazoxymethanol (MAM), which produces hepatomas in rats. MAM spontaneously breaks down to methyldiazonium hydroxide, which methylates hepatic DNA, RNA, and enzymes [178]. Certain cycad glycosides inhibit aromatase and may be useful in the treatment of estrogen-dependent cancer [139].

## Herbal Supplements

Herbal supplements are becoming increasingly popular both as alternatives and as aids to conventional agents in the therapy of ailments, as well as in health promotion and the prevention of disease. Regardless of their effectiveness and although most such preparations appear safe at suggested doses, excessive use of certain preparations can have adverse effects. In general, these effects represent a combination of effects of all the ingredients (many discussed above) in the crude herbal preparation. The effects of some commonly used herbal supplements are summarized by Cupp [59] and are presented in Table 14.3.

## Toxicants in Foods of Animal Origin

The discussion on plant toxicants has included certain agents from this category (e.g., lipids, vasoactive agents in cheese, plant-derived toxicants in milk). Others such as bacterial toxins in meats are discussed in a subsequent section. Many region-specific diseases such as coturnism (quail poisoning seen as rhabdomyolysis without neurological signs) are associated with geographical and cultural eating patterns and are beyond the scope of this review. Residues in meats and milk, from pesticides, antibiotics, and growth promotants including hormones such as estrogenic substances are addressed briefly in the section on food additives and in other chapters. Natural toxicant hazards in foods of animal origin are mostly limited to those derived from marine sources.

### Marine Toxins in Food

Of the many marine organisms capable of containing toxins (>1200 species), only a few are involved in food poisoning. Modern transportation and recent increases in the frequency and intensity of toxic algal blooms have led to an increase in the incidence as well as the spread of seafood poisoning inland. Toxicants may be produced by the fish itself, by the marine plankton or algae consumed by the fish with or without the aid of certain marine bacteria. Brett [35], Leftley and Hannah [151], and Russel and Dart [241] presented a detailed discussion of the toxicology of fishborne toxins.

*Shellfish poisoning* is one of several (amnesic, digestive, neurotoxic, paralytic) disease entities resulting from the consumption of shellfish (e.g., clams, crustaceans, lobsters, mussels, oysters, scallops) that have ingested toxic marine algae, especially certain dinoflagellates. The shellfish are toxic during seasons of heavy algal bloom (such as red tide) and may contain 200 organisms per milliliter or more. Toxicity increases in proportion to the concentration of algae and disappears within 2 weeks after the toxic plankton has disappeared from the waters [240].

Saxitoxin, neosaxitoxin, and gonyautoxins are the most potent of the more than 20 toxins present in paralytic shellfish poison, produced by the dinoflagollates belonging to *Alexandrium*, *Gymnodinium*, *Guanyalax* and *Pyrodinium* species. Saxitoxin blocks the action potential in nerves and muscles by preferential blockade of inward flow of sodium ions with no effect on the flow of potassium or chloride ions [126]. Consumption of 1 mg of the toxin (in 1 to 5 mussels or clams weighing 150 g each) can be mildly toxic, whereas 4 mg can be fatal if not treated vigorously. Toxic symptoms begin as numbness of the lips, tongue, and fingertips within minutes after eating. Numbness then extends to the legs, arms, and neck and is followed by general muscular incoordination, which progresses to respiratory paralysis and death. Decreased heart rate and contractile force, headache, dizziness, increased sweating, and thirst may also be noted. Boiling in bicarbonate-treated water and discarding the broth is suggested as a means of preventing shellfish poisoning [100].

Diarrheic shellfish poisoning occurs globally from the consumption of shellfish (mussels, cockles, scallops, oysters, cockles, whelks, and green crabs) contaminated by one of several species of *Dinophysis* and containing a combination of okadaic acid (OA), dinophysis toxins (DPTs), pectenotoxins, and yessotoxins [35]. Both OA and DPTs are powerful inhibitors of protein phosphotases and potent tumor promoters [138]. Whether protein phosphotase inhibition leads to the observed increase in the permeability of intestinal epithelial cells exposed to OA and thus its diarrheic effect is unknown.

Neurotoxic shellfish poisoning is characterized by nausea, vomiting, diarrhea, chills, headache, muscle weakness and pain, and eye and nasal irritation, as well as, in severe cases, paresthesia, difficulty in breathing, double vision, dysphonea, dysphagia, tachycardia, and convulsions. It has been reported along the Gulf of Mexico, the eastern coast of Florida, and in New Zealand following the consumption of shellfish (mussels, oysters, and whelks) or inhalation of airborne blooms containing a heavy load of *Gymnodinium breve* or similar organisms. The lipophilic polyether toxin brevitoxin promotes $Na^+$ influx and thus depolarization by its action on site 5 of the voltage-dependent $Na^+$ channels [35,151].

Amnetic shellfish poisoning, characterized by short-term and sometimes permanent memory loss associated with gastrointestinal signs and hallucinatory state, has been reported mostly from the coastal areas in North America, Canada, France, Portugal, and the United Kingdom. Neuronal degeneration and necrosis in hippocampus, coma, and death result in severe cases. A water-soluble, acidic, non-protein amino acid, domoic acid (and its isomers), is produced by the diatom *Pseudonitzschia* species in king scallops (at lower levels in blue mussels, queen scallops, crab, razor fish, anchovies, sardines, mackerel, jack smelt, albacore, sand dabs, krill, and humpback whales). Domoic acid acts as a competitive glutamate antagonist at various sites and has been ascribed the etiological role [35,151].

Azaspiracids, a new class of algal toxins potentially produced in mussels and other shellfish throughout northern Europe by *Protoperidinium* species, act by unknown mechanisms to induce gastrointestinal symptoms that include nausea, vomiting, severe diarrhea, and stomach cramps lasting for up to 5 days. In mice, necroses in the lamina propria of the small intestine, thymus, and spleen; fatty changes in the liver; chronic interstitial pneumonia; and lung tumors are observed [35].

Between 300 and 400 tropical reef and semipelagic species of edible marine animals, including barracudas, groupers, sea basses, snappers, surgeon fishes, parrotfishes, jacks, wrasses, eels, and certain gastropods, accumulate toxins in their liver and other viscera that are capable of causing ciguatera poisoning, at an estimated rate of 20,000 to 50,000 cases per year worldwide [35,162]. The intoxication, common in the South Pacific and the Caribbean, appears to follow the spatial and temporal pattern of the distribution of a photosynthetic dinoflagellate *Gamblerdiscus toxicus*, which is consumed by the smaller herbivorous fish and in turn by the ciguatoxic fish [240]. Ciguatoxins, a group of 23 colorless and heat-stable lipophilic polyethers (molecular weight of 1100), appear to play a major role in intoxication with some contribution from the water soluble maitotoxin [151]. Ciguatoxins increase membrane permeability to sodium ions, causing depolarization of nerves. In addition, ciguatoxin inhibits subsequent inactivation of open $Na^+$ channels and possesses anticholinesterase activity in experimental animals [151,240]. Maitotoxin, on the other hand, inactivates voltage-dependent and receptor-mediated $Ca^{2+}$ channels, leading to high intracellular $Ca^{2+}$ and cell death [151]. Ciguatoxicosis is the most common marine toxicosis n humans manifesting as tingling of the lips, tongue, and throat followed by numbness, nausea, vomiting, abdominal pain, diarrhea, pruritis, bradycardia, dizziness, muscle and joint pain, and ataxia. Severe cases exhibit paresis of the legs and infrequently death due to cardiovascular or respiratory failure [151,240]. Prevention of ciguatera poisoning is difficult, although extensive evisceration of fish may help.

*Pufferfish* (fugu fish) *poisoning*, known to occur as far back as 2000 to 3000 B.C. in China and Japan, results from the consumption of tetrodotoxin present in the liver and ovaries of pufferfish, ocean sunfishes, porcupine fishes, blue-ringed octopus, and certain amphibians of the family *Salamandridae* [127,240]. Toxin accumulation is greatest just prior to spawning in the spring. Tetrodotoxin (TTx), with a cyclic hemilactal structure, is highly lethal ($LD_{50}$, 10 mg/kg) to all vertebrates and is active after boiling for 1 hour but is inactivated under alkaline conditions [86]. Tetrodotoxin prevents the increase in the early $Na^+$ permeability in both motor and sensory neuronal membranes similar to that of saxitoxin [240]. In humans, numbness of the lips, tongue, fingers, and arms; muscular paralysis and ataxia; hypotension; and respiratory paralysis leading to death progress rapidly beginning 30 to 60 minutes after consumption of 1 to 2 mg of tetrodotoxin (1 to 10 g of roe or liver). Although current treatment is only symptomatic, experimental evidence [43,232] indicates that anti-TTx antibodies or 4-aminopyridine may be effective in antagonizing the cardiorespiratory toxic effects of TTx. Training of personnel in proper evisceration techniques and licensing of fugu restaurants is of the essence.

*Scombroid poisoning* is the most widespread fish-borne intoxication resulting from the consumption of inadequately preserved abalone, amberjack, bluefish, tuna, mackerel, mahi mahi, and sardines in which histamine and saurine are produced as a result of bacterial scombrotoxic action [162]. Scombroid fish apparently has a sharp or peppery taste. Signs of intoxication include

**TABLE 14.3**
**Toxicology of Commonly Used Herbal Supplements**

| Name | Uses | Pharmacology | Adverse Effects |
|---|---|---|---|
| Aloe | Heal wounds, burns, ulcers, frostbite, and dry skin; laxative | Inhibits gastric acid secretion, antiinflammatory and antiviral activity; increases cytotoxic T-cell number and function; reduces production of inflammatory prostaglandins and leukocyte migration to wound site | Bloat, cramps, melanosis coli, acute renal failure, dermatologic lesions, and death from consumption of multiple herbal preparations including aloe |
| Borage | Source of ω-6- fatty acids; aids in therapy of diabetic neuropathy, arthritis, and hypertension | Promotes formation of antiinflammatory and immunoregulatory prostaglandin $E_1$; inhibits T-lymphocyte proliferation; augments baroreceptor-mediated vascular resistance | Constipation |
| Calamus | Fever and gastrointestinal (GI) distress in Asia | Acts as a sedative, muscle relaxant, negative ino- and chronotropic | Contact dermatitis; mutagenic and carcinogenic |
| Cascara Sagrada® | Laxative | Promotes active water secretion into the GI lumen and inhibits water reabsorption | Respiratory allergy; mutagenic |
| Cat's claw | Antiviral, anticancer aid, GI diseases such as leaky bowel | Antimutagenic (mechanism unknown) | Acute allergic interstitial nephritis |
| Chamomile | Antiinflammatory | Benzodiazepine-like | Anaphylaxis |
| Chaparral | Antioxidant, anticancer | Inhibits cellular respiration | Contact dermatitis, hepatopathy, cystic nephropathy, renal cell carcinoma |
| Coltsfoot | Antihistamines, decongestants, expectorants; endocrine rejuvenator | Demulcent, respiratory stimulant; increases ciliary motion; cardiac stimulant | Pyrrolizidine alkaloids (PAs) induce hepatic venoocclusive disease, liver damage, fibrosis, and cirrhosis, several types of liver cancer |
| Comfrey | Cosmetic use to remove skin oils; used to prevent kidney stones; used to treat injuries (e.g., burns) and peptic ulcer | Increases prostaglandin synthesis in stomach | Effects similar to coltsfoot (from PAs) |
| Cranberry | Prevention and relief of urinary infection and stones | Reduces urinary pH; reduces bacterial adherence to urinary mucosa; antiviral | Diarrhea and weight gain |
| Dong Quai | Menstruation and menopause | Calcium channel blockade, estrogenic, vasodilating, antispasmodic, sedative | Photosensitization, carcinogenic |
| Echinacea | Immunostimulant | Macrophage activation | GI signs; allergic cross-sensitivity to chamomile, daisies, ragweed, etc. |
| Ephedra (Ma Huang) | Asthma, cough, hypotension, pain, venereal disease, weight loss | α- and β- adrenergic and CNS stimulant | Stroke, psychosis, renal calculi, cardiomyopathy, amphetamine production |

| | | | |
|---|---|---|---|
| Feverfew | Migraine, arthritis | Inhibition of phospholipase $A_2$ and prostaglandins | Contact dermatitis, allergies, GI signs |
| Garlic | Antiatherosclerotic; prevents heart attack, stroke, and cancer | Reduces cholesterol level and platelet aggregation; $G_1$-phase arrest of tumor cells; macrophage stimulation | Allergy (asthma, dermatitis, diarrhea, rhinitis), contact dermatitis (burn) |
| Ginger | Prevent/relieve nausea (motion and morning sicknesses), anti-inflammatory | Inhibition of thromboxanes/prostaglandins and platelet aggregation, positive inotrope on heart | Contains mutagenic and antimutagenic compounds. |
| *Ginkgo biloba* | Improve circulation and mental acuity, impotence, vertigo | Antioxidant; inhibits platelet aggregation; blocks serotonin; lowers corticosteroid synthesis | Neurotoxic food poisoning, cerebral and eye hemorrhages |
| Ginseng | Physical, mental and sexual invigoration | Stimulates hypothalamo–pituitary–adrenal axis; CNS stimulant; vasodilatory; antiplatelet aggregatory; immunostimulant; antineoplastic effect | Ginseng abuse syndrome (GAS): hypertension, nervousness, sleeplessness, skin lesions, and diarrhea; CNS excitation and arousal |
| Hawthorn | Hypertension, arteriosclerosis, heart failure, angina, vascular and other diseases | Caffeine-like increase in heart contractility and coronary blood flow secondary to phosphodiesterase inhibition; antiarrhythmic; antioxidant; hypocholesterolemic | Mutagenic |
| Kava | Relief of anxiety and stress | GABA-ergic and antiadrenergic | Photosensitization, liver damage, EKG changes, sudden death |
| Licorice | Addison's disease and a variety of digestive and respiratory diseases in China; flavors tobacco products in the United States | Antiinflammatory; mineralocorticotropic (inhibits 11β-hydroxy steroid dehydrogenase); antiviral | Hypokalemia, hypertension, rhabdomyolysis, renal failure, pseudoaldosteronism, pulmonary edema |
| Pokeweed | Cathartic, emetic, narcotic, gargle, and other products | Irritant; antimicrobial, mitogenic and immunostimulant; antiinflammatory; hemagglutination, vagal/sympathetic, and neuro stimulation | Gastroenteritis, hemolysis, anemia, muscle weakness and spasms, tremors, respiratory and cardiac arrest |
| Sassafras | Tonic, stimulant, blood purifier; used to treat many conditions | Antiinflammatory and antimicrobial activities | Vomiting, flushing, hypertension, hallucinations, paralysis; hepatocarcinogen |
| Saw palmetto | Aids in treatment of benign prostate hypertrophy; improves sexual function | Immunostimulant, antiandrogenic, antiestrogenic, inflammatory (inhibits prostaglandin synthesis) | Hypertension, rare liver damage, GI signs, reduced libido, urine retention |
| Scullcap | Sedative, anxiolytic, spasmolytic | Inhibits muscle contractility, decreases tumor cell viability, ameliorates myelosuppression from antineoplastic chemotherapy | Acute hepatitis with or without fibrosis |
| Senna | Laxative | Promotes active water secretion into the GI lumen and inhibits water reabsorption | Mild morphologic changes; mutagenic |
| St John's wort | Antidepressant | Cholinergic, serotonergic, melatonergic | Phototoxicity of skin and cutaneous axons; hypomania |
| Valerian | Headache, fatigue, insomnia | GABA-mediated sedative/hypnotic; muscle relaxation; positive inotrophy and negative chronotropy in heart | CNS depression, liver damage |

nausea, vomiting, diarrhea, epigastric distress, flushing of the face, throbbing headache, and burning of the throat followed by numbness and urticaria. Severe cases may lead to cyanosis and respiratory distress and, rarely, to death. These signs appear within 2 hours of the meal and disappear in 16 hours [240]. The disease readily responds to antihistamine treatment.

Prorocentrolides, pinnatoxins, and spirolides are thought to activate calcium channels, and some were implicated in over 2500 cases of illness in Japan following consumption of the bivalve *Pinna pectinata*. Certain other compounds from algae may be of therapeutic potential, including amphidinolides and carbenolides with cytotoxic activity against tumor cells, zooxanthellotoxins with vasoconstrictor activity, and gambierdic acids and gonodiomin with antifungal properties (35).

## FOOD CONTAMINANTS

Whereas some naturally occurring toxicants and food additives either impart resistance to plants against pests or help preserve or enhance the nutritional quality of the diet, respectively, the biological and manmade industrial chemical contaminants only increase the risk of foodborne illness and deserve a much broader margin of safety in their control than food additives. The FDA, in consultation with other federal agencies, establishes legal action levels (i.e., the maximal level allowed in foods and feeds) of a contaminant based on economic considerations and technological feasibility [104].

### Bacterial Infections and Intoxications

Foods contaminated with microbial agents are a major known source of human disease estimated to afflict tens of millions of people and cost $22 billion annually in the United States alone [8]. With a few exceptions, these can be prevented by adequate cooking and proper cooling, storage, and reheating of cooked foods in clean containers [176]. Bacterial foodborne disease may result from the consumption, in food, of either bacteria (e.g., *Salmonella* sp. and *Clostridium perfringens*) that can cause disease by multiplying in the intestinal mucosa, where they may elaborate toxins (enterotoxins), or a sufficient amount of preformed microbial toxins (staphylococcal enterotoxins and botulinum toxin). In addition to the above well-known etiologies, genetic changes in bacteria that increase virulence, changes in eating habits, altered food production and distribution systems, increased numbers of immunocompromised food consumers, and improved detection systems have led to the identification of other pathogens such as *Escherichia coli*, *Listeria*, and *Yersinia* as causing foodborne illness.

*Clostridium perfringens* frequently causes foodborne infections that subsequently lead to sporulation of the organism in the large intestine. The enterotoxin, released during sporulation of the bacteria, is capable of causing fluid accumulation in the intestines. The α-toxin, possessing lethal, necrotizing, and hemolytic activities, is also produced by certain types. Among the five distinct types of *C. perfringens* (type A through E), type A is almost always involved in foodborne gastroenteritis and associated signs in humans in the United States. Only meat and fish products are capable of providing all the amino acids and growth factors required for the growth of *C. perfringens*. Roast beef, beef stew, gravy, and meat pies for type A and pork, other meats, and fish for type C are frequently involved [37]. Typically, foods involved are cooked at 100°C for less than an hour and are subsequently kept warm or slowly cooled. Spores that survive the heat shock multiply faster in the food than those not subjected to heat treatment, and they elaborate the enterotoxin in the gut. The enterotoxin appears to form ion-permeable channels in the cell membrane leading to movement of extracellular calcium and water into the cells, resulting in cell death [281]. Entry of the toxin into the bloodstream leads to the release of potassium from hepatocytes, hyperkalemic cardiac failure, and death [281]. Due to the ubiquitous distribution of the organism in soil and in the gastrointestinal tract of humans and animals, preventing contamination is difficult. Multiplication and toxin production can be inhibited by heating food to proper temperature (165 to 212°F), prompt and effective cooling, and avoiding prolonged reheating before consumption.

*Staphylococcus aureus* is probably the leading cause of foodborne disease worldwide. The organisms are Gram-positive, nonmotile, non-spore-forming cocci that occur ubiquitously in the environment. Although humans are the leading source of food contamination by way of nasal discharge and infected cuts and wounds, the organism can be present in milk derived from mastitic cows and meat derived from arthritic poultry [189]. Baked ham, poultry, fish and shellfish, meat and potato salads, cream-filled bakery goods, and high-protein leftover foods are frequently involved in such intoxication [37]. The multiplication of *S. aureus* in raw food products is inhibited by the presence of other spoilage organisms. As a result, most cooked products subsequently contaminated by infected handlers and stored at a warm temperature for several hours before consumption are capable of causing intoxication. The causative agent is one of more than six immunologically distinct heat-stable enterotoxic protein (molecular weight, 26,000 to 34,000) whose secretion is regulated by chromosomes during growth (A, D, and E) or by plasmids (B and C). In addition, *S. aureus* also produces many other substances such as coagulase, DNase, hemolysins, lipases, fibrinolysin, and hyaluronidase that are toxic to one or more animal species. Although all strains of *S. aureus* are potentially pathogenic, the enterotoxin production is closely related to the presence of coagulase and DNase. Signs and symptoms begin 1 to 6 hours after consumption of contaminated food and include nausea, salivation, vomiting, retching, occasional diarrhea, abdominal cramps, sweating, dehydration, and weakness followed by recovery in 1 to 3 days. Severe cases may show fever, chills, drop in blood pressure, and prostration [189]. Preventive measures effective against *S. aureus* food intoxication include educating food handlers regarding hygienic practices to reduce post-cooking contamination of high-protein foods and eliminating prolonged storage of cooked foods at room temperature before consumption.

*Botulism* is a neurotoxic syndrome caused by the consumption of improperly cooked and stored foods containing one of seven (A through G) heat-labile neurotoxins produced by *Clostridium botulinum*. It is a ubiquitous, anaerobic, Gram-positive, motile rod capable of forming heat-resistant spores. High moisture, a pH above 4.6, and prolonged anaerobic storage are required for sufficient toxin production [189]. Common foods involved are home-canned fruits and vegetables such as beans, corn, leafy vegetables, and especially peppers, all of which contain toxins A and B. Nonpoultry meats contain toxin B, whereas cheese and other dairy products contain toxin A. Type E is isolated mostly from fish products [189]. Types C and D, causing botulism in animals and birds, do not affect humans. Outbreaks of the infection, however, are often from more unusual sources such as chili peppers, tomatoes, and improperly handled baked potatoes wrapped in aluminum foil. Although the FDA approved the use of botulinum toxin type A for the treatment of eye muscle disorders, cervical dystonia (a neurological movement disorder causing severe neck and shoulder contractions), and frown lines between the eyebrows, it is also considered an agent of bioterrorism because of its fatal effects in aerosolized form and its use in weapons by rogue states.

Botulinum toxins are stable in the acid pH of the stomach, where they are protected from gastric juice and pepsin by a nontoxic component of the toxin molecule. Once in the duodenum, the toxin is activated by trypsin with no change in molecular size and is subsequently absorbed into lymphatics. The toxin irreversibly binds to the myoneural junction and, acting as a zinc endopeptidase, degrades peptides involved in the release of acetylcholine (ACh), thus inhibiting its release at the peripheral cholinergic nerve endings [189]. Signs and symptoms of botulism usually appear within 12 to 24 hours (range, 2 hours to 6 days) following consumption of the toxin-containing food. Initial signs of nausea, vomiting, and diarrhea are followed later by predominantly neurologic signs including headache, dizziness, blurred or double vision, loss of light reflex, weakness of facial muscles, and pharyngeal paralysis (difficulty in speech and swallowing). Fever is absent. Sensory reflexes and mental alertness are intact. Paralysis of the respiratory muscles leads to failure of respiration and death, usually in 3 to 10 days [269]. Foodborne botulism can be prevented by proper canning techniques, boiling vegetables for at least 3 minutes before serving, and discarding all swollen and damaged canned products after boiling. The control of botulism cases involves the use of monovalent (E), bivalent (A and B), or polyvalent (A, B, and E) antitoxin; the recall of all involved commercial products; proper reporting; and epidemiologic investigation. Boiling for 3 minutes or heating at 80°C for 30 minutes destroys the preformed toxin, whereas the use of salt, the antimicrobial compound nisin, polyphosphates, smoke, spices, lactic acid, and nitrite can inhibit the growth of *C. botulinum* and thus toxin formation [189]. If the nitrite content of cured meats and fish as well as fermented sausages is reduced from current levels as a means of decreasing the levels of carcinogenic dietary nitrosamines, it is conceivable that the incidence of botulism from the consumption of such foods will increase unless suitable replacements are found.

Foodborne disease outbreaks involving *Bacillus cereus* have occurred in Northern and Eastern Europe. A diarrheal illness involving a wide variety of meats and vegetables, various desserts, fish, pasta, milk, and ice cream (similar to that of *Clostridium perfringens*), and a vomiting illness involving flour-based foods such as cereals and fried rice served in Chinese restaurants (similar to that of *Staphylococcus aureus*) are both apparently caused by this organism [189]. At least seven toxins, including the heat-stable (121°C for 90 minutes) emetic toxin cereulide and the enterotoxins hemolysin BL and its nonhemolytic homologue, contribute to the syndrome [189,255]. Enterotoxin appears to disrupt cell membranes, leading to increased permeability whereas the mechanism of emetic toxin is unknown.

*Salmonella* sp. consists of over 2200 serotypes possessing somatic O, flagellar H, and capsular Vi antigens, of which 50 serotypes commonly occur. *S. typhi*, *S. paratyphi*, and *S. sendai* are adapted to human hosts who serve as sole carriers for those organisms. *S. typhimurium* and *S. enteritidis* are the two most common disease-causing agents in the United States. Feces of infected humans, domestic and wild animals, and birds serve as sources of contamination in a variety of meat and milk products and more recently raw fruits and vegetables, causing severe gastrointestinal signs along with fever, septicemia, shock, and sequelae of embolism including pneumonia, meningitis, and abortion. Mortality is rare and occurs in very young, very old, and immunocompromised patients. Some individuals develop a chronic condition called Reiters syndrome, which manifests as painful joints, irritated eyes, and painful urination. Enteritis can result from bacterial multiplication within the mucosa as well as from enterotoxins secreted by some serotypes. *Salmonella*-free birds can be raised by vaccination and by raising the birds in *Salmonella*-free environments using *Salmonella*-free pelleted feed. Thorough cooking of meats, pasteurization of milk and dairy and egg products, prevention of cross-contamination between cooked and raw products, and testing, isolation, and treatment of carrier animals and food-handling personnel are all extremely important in controlling the incidence of this most common foodborne disease [78]. In this regard, evidence of the emergence of antibiotic-resistant strains such as *S. typhimurium* DT104 in the United Kingdom and United States suggests that future research must be directed at understanding the mechanisms of microbial adaptation to stresses if we are to better control such infections.

Four species of *Shigella* (i.e., *S. dysenteriae*, *S. flexneri*, *S. boydii*, and *S. sonnei*) cause an estimated 165 million cases of *Shigella* diarrhea (acute bacillary dysentery) annually, 99% of which occur in developing countries and 69% in children under 5 years of age, resulting in 1.1 million deaths [203]. Young children in daycare and custodial institutions are most susceptible. *Shigella* enteritis is characterized by fever, mucohemorrhagic diarrhea, abdominal cramps, and tenesmus. *S. flexneri* can lead to Reiters syndrome (see *Salmonella* discussion) and eventually to chronic arthritis. *S. dysenteriae* type 1 produces Shiga toxin and can lead to life-threatening hemolytic uremic syndrome (HUS), the same complication that develops in some cases of infection with enterohemorrhagic *Escherichia coli*. Shigellosis is highly contagious and produces a large number of secondary cases in each outbreak that involve persons in contact with infected patients. Contamination of food by unhygienic food handlers and the consumption of raw vegetables raised in contaminated soils are two main contributors to the incidence. Therapy with antibiotics and fluids and prevention through thorough handwashing are known effective strategies against the spread of infections.

*Campylobacter jejuni* and others in this genus (e.g., *C. sputorum*) are the leading cause of bacterial diarrhea, accounting for up to 2.4 million cases in the United States and up to 14% of diarrheal illness worldwide. Many aspects of the illness, including the symptoms that rarely require treatment, are similar to those of salmonellosis [9]. Reactive arthritis, inflammation of urethra and conjunctiva, and Guillain–Barré syndrome (paralysis of limbs and weakness of respiratory muscles) have been described as sequelae in occasional cases [9].

*Escherichia coli*, a close relative of the genus *Shigella*, has recently raised concern as a fatal foodborne disease agent. Of the more than 160 serotypes (based on O, H, or capsular K antigen), 43 can induce gastroenteritis sometimes associated with life-threatening HUS in humans [231]. In North America, HUS is the most common cause of acute kidney failure in children, who are particularly prone to this complication. Pneumonia, meningitis, thrombotic and thrombocytopenic purpurea, bladder and kidney infections, and septicemia may also result from *E. coli* infections. Based on virulence factors (which bestow the organism with an ability to attack, invade, and produce toxin in the host cells) located in the plasmids, five virotypes have been identified as pathogenic: enterotoxigenic *E. coli* (ETEC), enteroaggregative *E. coli* (Eagg EC), enteropathogenic *E. coli* (EPEC), enterohemorrhagic *E. coli* (EHEC), and enteroinvasive *E. coli* (EIEC). Serogroup O 157:H7, which belongs to the EHEC group and produces a Shiga-like toxin, may be the most common serotype causing nausea, vomiting, watery (bloody?) diarrhea, and HUS mainly traceable to the consumption of contaminated beef products [231]. Less frequently, unpasteurized milk and juices; ham, turkey, salami, and cheese sandwiches; dry fermented sausage; salad; and nonchlorinated water have been involved. The genome of *E. coli* O157:H7 is about 70% homologous with and 30% larger than that of the harmless serotype K12, suggesting that further study of differences could lead to a better understanding of the virulence and pathogenicity and tissue predilection of *E. coli* O157:H7. Once in the intestines *E. coli* produces Shiga-like toxins SL1 and SL2 which act similar to cholera toxin and ricin, by receptor binding and entry into a vesicular pathway, followed by release and translocation of the enzymatic A1 domain of the A subuni into the target cell cytosol. Covalent modification of intra cellular targets then leads to the activation of adenylate cyclase and a sequence of events culminating in ion fluxes the secretion of serotonin and prostaglandins, and alter ation in the expression of genes, leading to intestinal cel death and life-threatening diarrheal disease [63,231].

Several species and serotypes of *Vibrio*, especially th serogroups O1 and O139, have been responsible for larg epidemics of cholera worldwide. Cholera toxin is struc turally and mechanistically similar to the heat-labil *Escherichia coli* enterotoxin [63]. Although consumptio of sewage-contaminated drinking water is the predomi nant source of major epidemics, foodborne vibriosis ca result from the consumption of fecal-contaminated food such as vegetables, fish, and pork products. Consumptio of raw vegetables fertilized with untreated sewage and o shellfish harvested from sewage-contaminated estuarie are also common sources. Symptoms of intoxication ar severe diarrhea characterized by watery stool (ofte referred to as rice-water stool), associated with muscl cramps, hypovolemia, hypotension, shock, and metaboli acidosis due to the loss of bicarbonate and poor tissu perfusion. Therapy mainly involves oral or intravenous (i extremely severe cases) rehydration therapy.

*Listeriosis*, in addition to being transmitted by othe routes, is an emerging foodborne disease resulting fro the consumption of *Listeria monocytogenes*-contaminate soft cheeses, milk and other milk products, poultry, me (especially deli meats and frankfurters), coleslaw, an other products (e.g., salads) derived from contaminated vegetables. Food products are contaminated by contact with soil, feces, discharges, and urine from infected animals and humans. The clinical foodborne disease occurs primarily in pregnant women, neonates, and older and immunocompromised populations. It is characterized by gastrointestinal or flu-like symptoms within 12 hours of exposure followed by bacteremia leading to abortions, stillbirths, or premature births in pregnant women; meningitis, respiratory distress, and skin nodules in neonates; and meningitis-related signs in adults [53]. The disease can be treated with antibiotics and other supportive measures. Prevention involves improvement of sanitation of the environment and equipment and education to identify and avoid contaminated food products.

## Viral Foodborne Illnesses

In the United States, as in all industrialized countries, nearly every person will have viral gastroenteritis at least once; 610,000 hospitalizations and more than 4000 deaths occur annually. The advent of polymerase chain reaction (PCR), microarray, and proteomic virus detection techniques is beginning to allow the realization that most of the foodborne diarrheal illnesses that failed detection in the past are likely of viral origin. Although viruses can neither grow nor produce toxins in foods, they can induce foodborne illness by their mere presence either in fresh produce or in processed food contaminated by fecal material. As reviewed by Clark and McKendrick [51] and Leach [150] and described below, noroviruses, hepatitis virus A and E, rotaviruses, and astroviruses are the major culprits contributing to up to two thirds of all foodborne microbial illnesses. Their highly infectious nature and survival in pH and temperature extremes renders prevention by education, hygiene, and immunization (such as for hepatitis A) especially important with regard to protecting against these illnesses.

### Norovirus

The ability of asymptomatic individuals to shed viral particles and the stability of the virus render noroviruses the etiological agent responsible for most cases of foodborne gastroenteritis in the United States and worldwide, accounting for from 60 to 93% of all viral gastroenteritis cases. Infections occur following the ingestion of airborne or foodborne (shellfish, water) viral particles. Outbreaks occur mostly in hospitals, in nursing homes, and on cruise ships. The expression of carbohydrates belonging to the ABH histo-blood group antigens that allow intestinal cell attachment by noroviruses renders these individuals more susceptible to virus infection. Generally mild clinical features of acute infection include fever, nausea, vomiting, diarrhea, abdominal cramps, headache, and myalgia with a more severe and sometimes fatal course in patients with immunosuppression (cancer or post-transplantation chemotherapy) and a chronic course in normal but stressed individuals. Secondary attacks occur at a high rate, resulting in high rates of transmission and large outbreaks. The high level of norovirus genetic variability in response to changing environment will likely pose considerable challenges to disease control by vaccination akin to the situation with influenza virus.

### Rotavirus

Infection with this virus is the foremost cause of severe gastroenteritis of young children under 5, resulting in over 2 million hospitalizations and up to 600 000 deaths per year worldwide. Fever, nausea, vomiting, diarrhea, abdominal cramps, headache, and myalgia are also common features, with the severity increasing in immunocompromized individuals (e.g., HIV, solid organ transplantation, bone marrow transplantation). Rotavirus diarrhea results from a combination of cell-damage-induced, malabsorptive, viral, enterotoxic-peptide-mediated secretory and enteric-nervous-system-mediated hypermotility components. Rotavirus also enters extraintestinal sites, including the blood, central nervous system, liver, spleen, and kidney. Whether this explains the occasionally reported sudden death, convulsions, and biliary atresia in children is under investigation.

### Hepatitis A and E

In 2001, 10,609 cases (approximately 45,000 clinical cases and 93,000 new infections when accounting for underreporting) of hepatitis A virus infection were reported in the United States (3.77 cases per 100,000), with young adult men (ages 25 to 39 years) at higher risk. Hepatitis A is transmitted primarily by the fecal–oral route and through contaminated foods or drinks, especially uncooked fruits and vegetables and shellfish collected from contaminated habitat. After ingestion and absorption, the virus replicates in the liver and is excreted in bile, reaching the highest concentrations in the stool within 2 weeks, at which time the risk of transmission is highest. Twenty percent of children less than 3 years old and 75% of adolescents and adults show symptoms of fever, malaise, abdominal pain, and jaundice lasting for 2 months or longer, culminating in fulminant liver failure with the highest case-fatality rates occurring in adults older than 50 years. No specific therapy is available. Good hand hygiene, effective public water sanitation, and food hygiene are also important. Immunoglobulin (from pooled plasma) and two inactivated hepatitis A vaccines offer a high degree of short-term and long-term protection, respectively [150]. These measures appear to be paying off as evidenced by steadily declining cases each year.

Hepatitis E outbreaks are rare in the United States and are related to consumption of contaminated drinking water during travel to endemic areas such as South Asia and North Africa, where the mortality rate, especially in pregnant patients, can be high (15 to 25%).

### Spongiform Encephalopathies (Prion Diseases)

Spongiform encephalopathies (TSEs) include Creutzfeldt–Jakob disease (CJD), Gerstmann–Straussler–Scheinker syndrome, fatal familial insomnia, and Kuru in humans; scrapie in sheep; transmissible mink encephalopathy; chronic wasting disease in deer; and bovine spongiform encephalopathy (BSE) in animals. These prion diseases are showing increasing evidence of transmissibility across species. They are degenerative disorders of the nervous system characterized pathologically by spongiform degeneration of the grey matter, neuronal death, astrocytosis, and the accumulation of protease-resistant prion proteins (PrPSc), conformational variants of the normal nerve cell membrane prion protein (PrPc). Most cases of CJD are sporadic with unknown mode of transmission, 10 to 15% of cases are inherited, and a small number have been transmitted by medical procedures. The BSE epidemic in the United Kingdom; its presence in Japan, North America, and most member states of the European Union; evidence that BSE prions have infected humans in the United Kingdom, causing variant Creutzfeldt–Jakob disease (vCJD); and the historic implication that consumption of infected human tissue is involved in the spread of Kuru together strongly suggest a risk of transmission of BSE to humans via animal foods/products on a global basis [172]. Further, the potential BSE prion infection in sheep flocks and high prevalence of chronic wasting disease in wild cervids in the United States suggest a human risk from dietary exposure to prions from sheep, deer, and elk. Surprisingly, cytosolic accumulation of PrPc itself, but neither overexpression of PrPSc nor loss of PrPc function, appears to mediate neurotoxic mechanisms. The fact that targeted deletion of PrPc prevented the onset of neurodegeneration and reversed early pathology in mice infected with prions [172] confirms this mechanism and suggests that either PrPc may be nonfunctional or its loss of function can be compensated and that prevention or treatment of such diseases is feasible.

Other foodborne microbial agents that contribute significantly to the gastroenteritis toll around the world with increasing severity and duration in children, elderly, and immunocompromized individuals include *Yersinia* sp. due to the consumption of improperly cooked chitterlings (porcine large intestines) by people of African origin during major holidays; *Cryptosporidium* sp. due to contaminated water and unpasteurized apple cider; *Cyclospora* infection from contaminated water and fresh berries; *Brucella* sp. arising from unpasteurized milk and meats from infected cattle, sheep, goats, and their products; and viruses such as astroviruses, enteric adenovirus, severe-acute-respiratory-syndrome-inducing coronavirus (SARS-CoV), toroviruses, human parechovirus, picobirnaviruses, cytomegalovirus, and herpes simplex virus.

### Prevention and Control of Microbial Food Hazards

The National Animal Health Monitoring System has stepped up efforts to monitor food animal and poultry health on the farm and thus develop strategies to deal with potential increases in existing as well as emerging foodborne disease threats. The Food Safety Inspection Service (FSIS) of the USDA began implementing a Hazard Analysis and Critical Control Point (HACCP) system for pathogen reduction in 1996 for all slaughter and processing operations. The HACCP directs each processing unit to conduct a hazard analysis, identify critical control points at which a safety hazard can be prevented, establish limits at each point, develop monitoring procedures and corrective action when limits are exceeded, and implement recordkeeping that will allow subsequent verification by the FSIS for compliance [108]. Data for the *Escherichia coli* and *Salmonella* burdens of carcasses are used as evidence of fecal and enteric pathogen reduction. These two programs together with recent advances in the establishment of microbial genomic sequences and the development of PCR, DNA microarray, and proteomic methods will provide means for the rapid detection and identification of contaminating organisms and thus for minimizing the incidence of foodborne disease from animal foods in human populations. The FDA's recent approval of low-dose irradiation of red meats to control pathogens coupled with the previously approved irradiation of poultry for pathogen reduction; pork for the control of trichinae; fruits, vegetables, and grains for insect control; and spices, seasonings, and dry enzymes used in food processing for microbial reduction [14] should not only contribute further to the prevention of foodborne disease caused by microbial pathogens but also help in increasing the shelf life of such products without undesirable organoleptic, toxicological, or nutritional changes. Irradiation has yet to be approved for pathogen control of seafood products and is unsuitable for dairy products because of the development of off-flavors and discoloration. In the final analysis, however, the keys to minimizing microbial foodborne illness are at the food preparer or consumer level in the form of hygienic processing, canning and packaging; choosing reliable and clean food and water sources and processing aids; cooking at the right temperature; avoiding cross contamination (of cooked foods with raw); hygienic service (e.g., exclusion of infected food handlers from work) and prompt and appropriate storage and reheating.

The potential person-to-person spread of microbial diseases via the medium of food and the modern-day global nature of human travel and movement of food require global harmonization of efforts in the prevention of the spread of foodborne disease agents. A recent collaborative effort among the World Health Organization (WHO), the Food and Agriculture Organization (FAO), and the World Organization for Animal Health (WOAH) resulted in an agreement to develop reporting and surveillance methods for the incidence of the diseases at national and international levels; develop international animal health standards for foodborne disease agents that do not cause clinical disease in animals; to study farm ecology (environmental survival, multiplication, and spread and colonization in the animal) of foodborne pathogens; harmonization of foodborne disease investigation, diagnostic methodology, quality control; and uniform application of a risk-based farm-to-table approach when developing food safety standards [169].

## MYCOTOXINS

From the standpoint of human and animal health, toxigenic molds belonging to the genera *Aspergillus*, *Fusarium*, and *Penicillium* have received the most attention due to their frequent occurrence in food and feed commodities. Unfavorable conditions such as drought and damage to seeds by insects or mechanical harvesting can enhance fungal toxin (mycotoxin) production during both growth and storage, thus making mycotoxicoses a problem of both developing as well as developed countries. Although more than 100 mycotoxins have been identified throughout the world, the following discussion is limited to those with known public health significance. This subject has been reviewed by Bennet and Klich [30].

### Aflatoxins

The aflatoxins are a group of highly substituted coumarins containing a fused dihydrofuran moiety (Figure 14.3). They are produced predominantly by the molds *Aspergillus flavus* and *A. parasiticus*. Four major aflatoxins, designated $B_1$, $B_2$, $G_1$, and $G_2$ (based on blue or green fluorescence under ultraviolet light), are produced in varying quantities in a variety of produce (e.g., peanuts, various other nuts, cottonseed, corn, cereal grains, figs) that have not been adequately dried at harvest and are stored at relatively high temperatures [40]. Human exposure can occur from the consumption of these products as well as from the tissues and milk ($AFM_1$, a hydroxylated metabolite) of food animals consuming them.

Aflatoxin $B_1$ ($AFB_1$), the most potent and most commonly occurring aflatoxin, is acutely toxic ($LD_{50}$, 0.3 to 0.9 mg/kg) to all species of animals, birds, and fishes [55]. The sensitivity of the animals varies depending on the balance between the metabolic activation (cytochrome P450) and protection (glutathione synthesis) mechanisms. Effects of $AFB_1$ in animals are predominantly in liver and include death without signs or with signs of anorexia, depression, ataxia, dyspnea, anemia, and acute hemorrhages from body orifices. In subchronic cases, icterus, hypoprothrombinemia, hematomas, and gastroenteritis are common. Chronic aflatoxicosis is characterized by bile-duct proliferation, periportal fibrosis, icterus, and cirrhosis of the liver and is associated with loss of weight, as well as reduced resistance to disease (immune suppression); it is more prevalent in domestic animals and is also likely to occur in humans [211]. Prolonged exposure to low levels of $AFB_1$ in animals also leads to hepatoma, cholangiocarcinoma, or hepatocellular carcinoma and other tumors [40].

The National Research Council [205] reviewed epidemiological studies and concluded that the risk of primary hepatocellular carcinoma from $AFB_1$ exposure may be one in 10,000 in the United States; however, in populations infected with hepatitis B, the risk may be 10 to 100 times higher. $AFB_1$ is mutagenic following metabolic activation in many systems including HeLa cells, *Bacillus subtilis*, *Neurospora crossa*, and *Salmonella typhimurium* [40]. $AFB_1$ is partly metabolized by the cytochrome P450 system in the liver into a variety of reactive products (e.g., $AFB_1$ 8,9-epoxide) to form adducts with proteins and DNA [113]. The DNA lesions lead to inactivation of the tumor suppressor gene p53 due to G→T transversions of codon 249 and can explain a high proportion of liver cancer in areas with high aflatoxin exposures [148]. Such effect biomarkers as well as exposure biomarkers (e.g., $AFB_1$–DNA or $AFB_1$–albumin adducts) can be used to assess the effectiveness of preventive strategies [131]. By way of protein adducts, $AFB_1$ inhibits many enzymes involved in DNA synthesis, DNA-dependent RNA polymerase activity, messenger RNA synthesis, and protein synthesis [112], which may be related to several lesions and signs of aflatoxicosis, including fatty liver (failure to mobilize fats from the liver), coagulopathy (inhibition of prothrombin synthesis), and reduced immune function.

Other less widespread human clinical syndromes in which aflatoxins have been implicated include childhood cirrhosis in India, possibly Reye's syndrome in many parts of the world, and rarely acute hepatitis (aflatoxicosis) in India, Taiwan, and certain countries in Africa [261]. Widespread concern regarding the toxic effects of aflatoxins in humans and animals and the possible transfer of residues from animal tissues and milk to humans has led to regulatory actions governing the interstate as well as global transport and consumption of aflatoxin-contaminated food and feed commodities. Action levels for total aflatoxins in corn and other feed commodities used to feed mature, nonlactating animals range from 100 to 300 ppb. For milk, the action level is 0.5 ppb. For other commodities destined

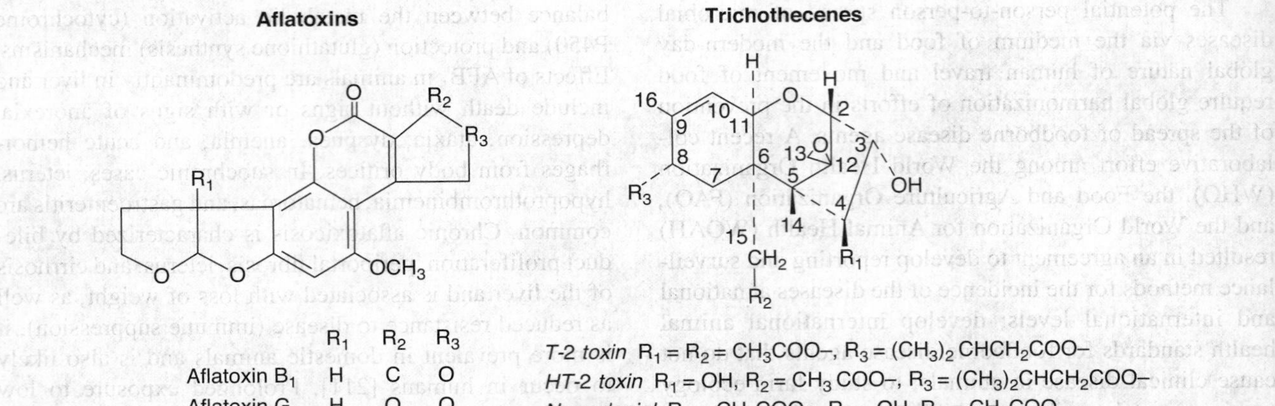

**Aflatoxins**

| | $R_1$ | $R_2$ | $R_3$ |
|---|---|---|---|
| Aflatoxin $B_1$ | H | C | O |
| Aflatoxin $G_1$ | H | O | O |
| Aflatoxin $M_1$ | OH | C | O |
| Aflatoxicol | H | C | OH |

**Trichothecenes**

*T-2 toxin* $R_1 = R_2 = CH_3COO–$, $R_3 = (CH_3)_2CHCH_2COO–$
*HT-2 toxin* $R_1 = OH$, $R_2 = CH_3COO–$, $R_3 = (CH_3)_2CHCH_2COO–$
*Neosolaniol* $R_1 = CH_3COO–$, $R_3 = OH$, $R_2 = CH_3COO–$
*Diacetoxyscirpenol* $R_1 = R_2 = CH_3COO–$, $R_3 = H$
*Monoacetoxyscirpenol* $R_1 = OH$, $R_2 = CH_3COO–$, $R_3 = H$

Zearalenone

Ochratoxin A

**FIGURE 14.3** The structure of the mycotoxins, aflatoxins, trichothecenes, ochratoxin A, and zearalenone.

for human consumption and interstate and potential global commerce, the action limit is 20 ppb (per FDA Compliance Policy Guides 7106.10, 7120.26, and 7126.23).

## Epipolythiodioxopiperazines

Epipolythiodioxopiperazines (ETPs) are toxic secondary metabolites made only by fungi [91]. The best-known ETP is gliotoxin, which appears to be a virulence factor associated with invasive aspergillosis of immunocompromised (e.g., organ transplant) patients, most likely due to its own immunosuppressive effects. In addition, it is anticarcinogenic, due probably to its proapoptotic effect. The toxicity of ETPs is due to the presence of a disulfide bridge that can inactivate proteins via reaction with thiol groups and to the generation of reactive oxygen species by redox cycling. Inhibition of NFκB, mitochondrial ATP synthase, adenine nucleotide transporter, and farnesyltransferase may have relevance to its immunosuppressive and apoptotic effects. Gliotoxin was detected in the sera of mice as well as human patients suffering from *Aspergillus fumigatus* infection, a rare demonstration that a mycotoxin (gliotoxin) is produced in the infected organs of human patients of aspergillosis at a significant level. Another member of this group, sporidesmin, induces facial eczema in cattle, sheep, and other animals secondary to free-radical-mediated liver damage. Chaetomin is produced by *Chaetomium globosum*; some chaetomin isolates infect skin and nails and cause deadly systemic infection in immunocompromised humans.

## Ergot Alkaloids

*Ergotism*, which is now rare, was first associated with the consumption of scabrous (ergotized) grain in the mid-sixteenth century. Subsequent studies led to the identification of *Claviceps purpurea* as the fungal agent invading rye, oats, wheat, and Kentucky bluegrass and *C. paspa* as the agent invading Dallis grass. Lysergic acid derivatives (the peptides and the amine alkaloids of ergot) were identified as the causative agents of the gangrenous and nervous forms of the disease. The gangrenous form, resulting from a predominance of alkaloids with α-adrenergic (ergotoxine) and vasopressor (ergotamine) action [141], typically manifests as prickly and intense heat and cold sensations in the limbs and swollen, inflamed, necrotic, and gangrenous extremities that eventually slough off. Convulsive ergotism, characterized by CNS signs, numbness, cramps, severe convulsions, and death, as well as abortions in animals, results from antiserotonin or adrenergic effects in the central nervous system and the uterotonic effects of many of these alkaloids combined [141].

### Fumonisins

*Fusarium moniliforme* Sheldon is a common fungal contaminant of cereals, especially corn, around the world. Contamination of corn by *F. moniliforme* as well as its major metabolites, fumonisins $B_1$ and $B_2$, can induce one of several human and animal diseases, such as leukoencephalomalacia (LEM) in horses; pulmonary edema in swine; renal and hepatotoxicosis in horses, swine, and rats; and hepatocarcinogenic effects in rats [75]. Recent evidence suggests that $FB_1$ increases chromosomal aberrations in primary rat hepatocytes [135] and developmental effects in the offspring secondary to hepatotoxicity in pregnant mice [228]. The consumption of high levels of fumonisins in home-grown corn has been associated with higher incidences of human esophageal cancer in certain regions of South Africa, China, northern Italy, and the United States (e.g., Charleston, SC). Fumonisins induce neural tube defects in animals, and their presence in corn products is potentially linked to a cluster of anencephaly and spina bifida cases in Texas [30,174]. Although the mechanisms of toxic and carcinogenic effects are not clearly understood, inhibition of spingolipid biosynthesis [304], enhancement of lipid peroxidation [1], elevated secretion of tumor necrosis factor-alpha [73], depletion of glutathione levels [125], elevated nitric oxide synthesis [236], induction of protein kinase C translocation via its action on phorbol ester binding site [320], and inhibition of protein serine/threonine phosphotases [88] are among the effects invoked to explain some or all of the effects of $FB_1$.

### Ochratoxins

The ochratoxins, a group of seven isocoumarin derivatives linked with phenylalanine by an amide bond, are produced by *Aspergillus ochraceus* and *Penicillium verrucosum* (among others) in barley, corn, wheat, oats, rye, green coffee beans, peanuts, wine, coca, dried fruits, certain grape wines, and tissues (e.g., pork) and blood from contaminated animals [30,258]. In experimental animals, ochratoxin A (OTA) produces predominantly renal proximal tubular lesions and liver degeneration. The acute oral $LD_{50}$ of OTA ranges from between 0.2 mg/kg for the dog and 59 mg/kg in mice. An association has been clearly established between a diet with high levels of OTA and nephropathy in humans and swine in the Balkan countries and swine in Denmark and the United States [140,165]. Signs include lassitude, fatigue, anorexia, abdominal (epigastric or diffuse) pain, and severe anemia followed by signs of renal damage. Reduced concentrating ability, reduced renal plasma flow, and decreased glomerular filtration occur sequentially, accompanied by gross and microscopic renal changes such as necrosis, fibrosis with some tubular regeneration, glomerular hyalinization, and interstitial sclerosis. Death results from uremia. Ochratoxins are teratogens and probable carcinogens (IARC class 2B) [258] that induce hepatomas and renal adenomas secondary to genotoxic effects in mice [128]. Relevant cellular effects that mediate the effects of OTA include alterations in enzymes involved in glucose metabolism, ATP synthesis, anion transport, lipid peroxidation, prostaglandins, and extracellular signal-regulated kinases [57,142,185,256].

### Psoralens

Psoralens are furocoumarin compounds that have been used in repigmenting achromatic skin lesions in an acquired disease called vitiligo, in some suntan lotions, and in drugs used to treat psoriasis [41]. Abuse of such compounds can result in dermatitis following exposure to the sun along with nausea, vomiting, vertigo, and mental excitation. A phototoxic dermatitis in celery pickers has also been linked to the presence of psoralens (8-methoxypsoralen, 5-methoxypsoralen, and trimethylpsoralen) in stalks infected with *Sclerotinia sclerotiorum* (pink rot), *S. rolfsii*, *Rhizoctonia solani*, or *Erwinia aroideae* or in celery stalks soaked in 5% NaCl [261]. Fig, parsley, parsnip, lime, and clove also contain psoralens. 8-Methoxypsoralen appears to undergo epoxidation of the furan ring similar to aflatoxins and may thus react with DNA in a similar fashion. Treatment with 8-methoxypsoralen and ultraviolet light led to squamous cell carcinomas of the ear in mice [41].

Unlike other photosensitizing agents, psoralens seem to act by photoreacting with DNA and to a lesser extent with RNA. The mechanism of psoralen photosensitivity appears to involve intercalation and cross-linking of psoralen in the DNA which occurs in three steps: (1) reversible intercalation of psoralen between two pyrimidines on opposing sides of the helix; (2) formation of a monoadduct, with the 5,6 double bond of the pyrimidine following absorption of 1 quantum of ultraviolet light; and (3) cross-link formation by absorption of a second quantum of ultraviolet light and linking of the monoadduct to the 5,6 double bond of thymidine [257]. In general, an excellent correlation exists between photoadduct formation and photosensitization of psoralens.

## Trichothecenes

Trichothecenes are a group of 12,13-epoxy trichothecenes produced by *Fusarium poae*, *F. tricinctum*, *F. graminearum*, *F. nivale*, *F. solani*, *Myrothecium roridum*, and *Stachybatrys atra*, among others, in cereal grains, including wheat. The group of macrocyclic trichothecenes includes satratoxins, verrucarins, and roridins and is produced mainly by *Stachybotrys* sp. in hay. Although more toxic, this group does not pose a significant human health threat due to its lack of prevalence. Group A trichothecenes (T-2 toxin, diacetoxyscirpenol) contain a side chain and are relatively less polar compared to group B (nivalenol, deoxynivalenol [DON], fusarenon). A two-volume treatise on trichothecene toxins and their role in human and animal health is available [27].

Most trichothecenes of health significance are produced by *Fusarium* sp. Common to all toxic syndromes are the characteristic signs of alimentary toxic aleukia (ATA) caused by T-2 toxin and related trichothecenes, including radiometric damage such as irritation and necrosis of skin and mucous membranes, hemorrhage, destruction of thymus and bone marrow, hematologic changes, nervous disturbances, necrotic angina, and shock [27]. Feed refusal, vomiting, and immune suppression are common problems caused by DON-contaminated wheat and corn in farm animals, especially swine and dogs and possibly humans [211,217]. Paradoxically, prolonged nivalenol and deoxynivalenol exposure induces autoimmune-like effects similar to human IgA nephropathy [237]. Trichothecenes (T-2 toxin) can cause fetal death, abortions, and teratogenic effects [27]. Although several trichothecenes are genotoxic in bacterial, yeast, and cell culture systems [135,289], they exhibit no initiator or promoter effect in whole animal systems [145].

The metabolism of trichothecenes occurs rapidly through deacetylation and hydroxylation and subsequent glucuronidation in the liver and kidneys [27,237], thus posing little problem of residues in meats from contaminated animals. At the molecular level, DON and other trichothecenes disrupt normal cell function by inhibiting protein synthesis via binding to the ribosome and by activating critical cellular kinases involved in signal transduction related to proliferation, differentiation, and apoptosis [217]. In addition, they also affect serotonergic pathways in the brain and induce expression of a number of cytokines [237]. Recently, the European Commission Scientific Committee on Food (SCF) and the Joint FAO/WHO Expert Committee on Food Additives (JECFA) established tolerable daily intakes (TDIs) of 1, 0.7 and 0.06 μg/kg body weight for DON, nivalenol, and the sum of T-2 and HT-2, respectively [254].

## Zearalenone

Zearalenone and zearalenol are nonsteroidal estrogenic contaminants (produced by *Fusarium roseum*) in grains such as corn, wheat, sorghum, barley, and oats. Zearalenone induces effects consistent with those produced by excessive steroidal estrogens (i.e., anabolic and uterotropic activities and regulation of serum gonadotropins). Although swine appear to be the most sensitive and exhibit signs of hyperestrogneic syndrome (i.e., changes in serum luteinizing hormone, swollen and edematous vulva, hypertrophic myometrium, vaginal cornification and prolapse, and infertility) [211], human exposure to zearalenone and its metabolites by way of cereal products can also be significant. The high frequency of premature menarche in Puerto Rico is suspected to be a result of high levels of zearalenone and similar compounds in the diet. Recommended safe daily human consumption of zearalenone appears to be 0.05 μg/kg body weight [30].

The mode of action of zearalenone involves interaction with estrogen receptors, translocation of the receptor-zearalenone complex to the nucleus, combination with chromatin receptors, selective RNA transcription leading to biochemical effects including increased water and lowered lipid content in muscle, and increased permeability of the uterus to glucose, RNA, and protein precursors [93]. Available evidence indicates that rapid conversion of zearalenone and zearalanol to conjugated metabolites that are excreted in urine and feces makes consumption of meat and milk from animals receiving Ralgro® an insignificant risk to humans.

Zearalenone is genotoxic in bacterial systems [94], forms DNA adducts in female mouse tissues, and induces hepatocellular adenomas in female mice [218]. The carcinogenic risk to humans and whether or not potentiative interaction exists between the adverse effects of zearalenone and those of dietary or endogenous estrogens as well as the xenoestrogens in the environment are unknown at the present time.

## Other Mycotoxins

A number of other mycotoxins (Table 14.4) have been identified either as contaminants in foods destined for human consumption or as metabolites of fungi isolated from human foods [39]. Although some of these have been associated with outbreaks of domestic animal diseases, the link between human consumption and disease is either emerging or nonexistent. Other mycotoxins have been shown to induce toxic and lethal effects in laboratory animals with no association between consumption of these toxins by animals or humans and a disease syndrome. Several of these (e.g., cytochalasins and secalonic acid D) have been used as research tools to expand our understanding of normal as well as abnormal cellular responses to xenobiotics [227]. Although it is difficult to assess the total significance of consumption of mycotoxins in human foods, it is easy to conceive that such a task requires extensive research into hundreds of known and potentially large number of as yet unknown mycotoxins. In spite of the vast number of toxic metabolites, a reduction in mycotoxin levels in foods and feeds and the prevention of mycotoxicoses in humans and animals can be achieved for the most part by avoiding stress in crops and damage to seeds by pests and by mechanical harvesting. Rapid postharvest drying and avoiding conditions that promote mold growth during storage are equally important.

## Pesticides

Pesticides are essential in agriculture and their use in the United States is regulated by the U.S. Environmental Protection Agency (EPA). Although acute intoxication with a pesticide usually results from accidental or suicidal ingestion, careless storage, or improper use, consumption of the residues of dietary pesticides over a lifetime can have deleterious effects, including cancer, endocrine disruption, and reproductive and immune system effects. The toxic effects of individual pesticides are presented elsewhere in this book.

The National Monitoring Program for Food and Feed, comprised of three federal surveillance programs (i.e., the Total Diet Study of market foods by the FDA, nationwide monitoring of unprocessed food and feed by the FDA, and analysis of meat and poultry by the USDA) monitors residues of organochlorines, organiphosphates, carbamates, and very infrequently herbicides and inorganic pesticides such as arsenic and bromide in various agricultural products. Despite the ban on the use of DDT and other persistent pesticides beginning in 1972, based on data on food items in the FDA's Total Diet Study and on fruits and vegetables from the USDA's Pesticide Data Program, persistent organic pollutant (POP) residues continue to be detected in virtually all categories of foods. Up to five have been found to be present in a single food, the most common POPs being dieldrin and DDT metabolites [251]. A recent study estimated that, in many plant materials that contain pesticide levels below those currently recommended as minimal risk levels (MRLs) of free pesticides typically in the range of 0.05 to 1 mg/kg), the actual total exposure far exceeds these values when one considers the bound fraction, basic chemical characterization, digestibility, and bioavailability in the target animal rather than the rat, a commonly used experimental animal [248].

The regulation of these residues by various U.S. regulatory agencies shows a lack of harmonization and thus consistency of risk estimation. Consumption of a single serving of fish, a glass of milk, or all items in a full day's diet (worst-case scenario) contaminated, for example, with DDT residues equal to the amount permitted by the FDA's action level would expose an adult to up to 50, 10, or 90 times and a child to up to 300 times the daily exposure considered "safe" by the health-based standards (RfDs and MRLs) set by the EPA and the Agency for Toxic Substances and Disease Registry (ATSDR) [251]. Although this is likely contributed to by the historically higher estimates of exposure and thus risk by the EPA [72] and the earlier "zero risk" policy mandated by the Delaney Clause, new authority derived from the Food Quality Protection Act of 1996 (FQPA; FFDCA 408(b)(2)(D)(v) and (vi)) to apply a science-based "reasonable certainty of no harm" principle and more reasonable estimation of exposures would bring these standards closer together. Additional major provisions of the FQPA include consideration of aggregate (drinking water, residential, and dietary) and cumulative exposure to a pesticide and other substances with common mechanisms of toxicity; consideration of children's special sensitivity and exposure to pesticides; use of an extra 10-fold safety factor in addition to the traditional 100-fold safety factor, unless, on the basis of reliable data, a different level is determined to be safe for children; an explicit determination that a tolerance (legal residue limit) is safe for children; and the development and application of a screening and testing program for chemicals with the potential to disrupt the endocrine process. The benefits from these changes to human, especially child, health have yet to be determined.

## Toxic Metals

A high proportion of the total daily exposure to metals by the general population occurs from their natural presence in foods. Beverages, water, air, and contact with metal-containing consumer products contribute to the rest. Because children consume more calories per unit body weight and have a higher absorption rate than adults, they are at a higher risk than adults. Industrial and agricultural uses of metal products pose a hazard of food-contamination associated with their use, storage, accidental spillage, and improper disposal. The recent decline in the use of heavy-metal-based pesticides, including herbicides, makes acute poisonings from dietary toxic metals less likely. A decline in the use of containers with metal coatings that dissolve during food manufacture, cooking, and storage has also contributed to the decline in acute toxicities associated with metals. Foodborne intoxications from metals are mostly limited to long-term consumption of water and food products from environments that contain naturally high levels of metals (e.g., arsenic, selenium, and fluoride) or that are contaminated by mining, smelting, and industrial discharge (e.g., methylmercury and Minamata disease). Most problems associated with water- and foodborne metal intoxications are preventable by adequate testing of drinking water and of soil prior to growing foods that accumulate metals from the soil. The reader is referred elsewhere in this edition (Chapter 17) for information on the toxic effects of metals.

**TABLE 14.4**
**Miscellaneous Mycotoxins**

| Mycotoxin | Major Producing Organisms | Source of Fungi | Principal Toxic Effects |
|---|---|---|---|
| Alternariol and alternariol methyl ether | *Alternaria* sp. | Sorghum, peanuts, wheat | Highly teratogenic to mice; cytotoxic to HeLa cells; lethal to mice |
| Altenuene, altenuisol | *Alternaria* sp. | Peanuts | Cytotoxic to HeLa cells |
| Altertoxin I | *Alternaria* sp. | Sorghum, peanuts, wheat | Cytotoxic to HeLa cells; lethal to mice |
| Ascladiol | *Aspergillus clavatus* | Wheat flour | Lethal to mice |
| Austamide and congeners | *Aspergillus ustus* | Stored foodstuffs | Toxic to ducklings |
| Austadiol | *Aspergillus ustus* | Stored foodstuffs | Toxic to ducklings |
| Austin | *Aspergillus ustus* | Peas | Lethal to chicks |
| Austocystins | *Aspergillus ustus* | Stored foodstuffs | Toxic to ducklings; cytotoxic to monkey kidney epithelial cells |
| Chaetoglobosins | *Penicillium aurantiovirens*<br>*Chaetomium globosum* | Pecans | Toxic to chicks; cytotoxic to HeLa cells |
| Citreoviridin | *Penicillium citreoviride* | Rice | Neurotoxic, producing convulsions in mice |
| Citrinin | *Penicillium viridicatum*<br>*Penicillium citrinum* | Corn, barley | Nephrotoxic in swine |
| Cyclopiazonic acid | *Penicillum cyclopium* | Ground nuts, meat products | Nephrotoxic, enterotoxic |
| Cytochalasins | *Aspergillus clavatus*<br>*Phoma* sp.<br>*Phomopsis* sp.<br>*Hormiscium* sp.<br>*Helminthosporium dematioideum*<br>*Metarrhizium anisopliae* | Rice, potatoes, kodo millet, pecans, tomatoes | Cytotoxic to HeLa cells; teratogenic to mice and chickens |
| Diplodiatoxin | *Diplodia maydis* | Corn | Nephrotoxic and enterotoxic to cattle and sheep |
| Emodin | *Aspergillus wentii* | Chestnuts | Lethal to chicks |
| Fumigaclavines | *Aspergillus fumigatus* | Silage | Enterotoxic to chicks |
| Kojic acid | *Aspergillus flavus* | Squash, spices | Lethal to mice |
| Malformins | *Aspergillus niger* | Onions, rice | Lethal to rats |
| Maltoryzine | *Aspergillus oryzae* | Malted barley | Hepatotoxic; causes paralysis |
| Moniliformin | *Fusarium moniliforme* | Corn | Cardiotoxic in rodents |
| Oosporein (chaetomidin) | *Chaetomium trilaterale* | Peanuts | Lethal to chicks |

## FOOD ADDITIVES

The increasing demand for food by an ever-increasing world population and for ready-to-eat foods resulting from changes in lifestyles in developed societies has necessitated the use of chemical additives to help preserve, nutritionally fortify, and process the foods. Concerns of adulteration (masking of low-quality food by chemical additives) and toxic effects from chronic dietary chemical exposure have led to the passage of the Food and Drug Act of 1906; the Food, Drug, and Cosmetic Act of 1938; the Miller Pesticide Amendment of 1954; the Food Additive Amendment of 1958; the color additive amendment of 1960; the animal drug amendment of 1968; the Food, Drug, and Cosmetic Act (FDCA) of 1976; and the Food Quality Protection Act (FQPA) of 1996. The term *food additive* is defined in these Acts as "any substance the intended use of which results or may reasonably be expected to result, directly or indirectly, in its becoming a component or otherwise affecting the characteristic of any food (including any substance intended for use in producing, manufacturing, packing, processing, preparing, treating, packaging, transporting or holding food and including any source of radiation intended for any such use), if such substance is not generally recognized among experts qualified by scientific training and experience to evaluate its safety as having been adequately shown through scientific procedures (or, in the case of substances used in food prior to January 1, 1958, through either scientific procedures or experience based on common use in food) to be safe under the conditions of its intended use."

Food additives fall into two broad categories: direct and indirect. Direct additives are intentionally added to food and are justified by their ability to preserve, improve the nutritional/organoleptic quality, or to aid in the production and processing of foods. Some of the more than 30 functional classes of direct additives are antioxidants, inhibitors of bacterial and mold growth, vitamins and minerals, color and antifoaming agents. Among the approximately 3000

**TABLE 14.4 (cont.)**
**Miscellaneous Mycotoxins**

| Mycotoxin | Major Producing Organisms | Source of Fungi | Principal Toxic Effects |
|---|---|---|---|
| Paspalamines | *Claviceps paspali* | Dallis grass | Neurotoxic to cattle and horses; causes paspalum staggers |
| Patulin | *Penicillium urticae* | Apple juice | Lethal to mice; mutagenic; teratogenic to chicks; pulmonary effects in dogs; carcinogenic to rats |
| Penicillic acid | *Penicillium* sp. | Corn, dried beans | Lethal to mice; mutagenic; carcinogenic to rats |
| PR toxin | *Penicillium roqueforti* | Mixed grains | Hepatotoxic and nephrotoxic to rats; abortion in cattle |
| Roseotoxin B | *Trichothecium roseum* | Corn | Toxic to mice and ducklings |
| Rubratoxins | *Penicillium rubrum* | Corn | Causes hemorrhage in animals; hepatotoxic to cattle |
| Secalonic acids | *Aspergillus aculeatus* *Penicillium oxalicum* | Rice, corn | Lethal, cardiotoxic, lung irritant, and teratogenic to mice |
| Slaframine | *Rhizoctonia leguminicola* | Red clover | Salivation and lacrimation in horses and cattle |
| Sporidesmins | *Pithomyces chartarum* | Pasture grasses | Hepatotoxic; causes photosensitization in ruminants |
| Sterigmatocystin | *Aspergillus flavus* | Mammals | Mutagen, carcinogen, and hepatotoxic to mammals |
| Tenuazonic acid | *Alternaria* sp. | Grains, nuts | Lethal to mice |
| Terphenyllins | *Aspergillus candidus* | Wheat flour | Hepatoxic to mice; cytotoxic to HeLa cells |
| Tremorgenic mycotoxins | | | |
| Fumitremorgens A, B | *Aspergillus fumigatus* | Rice | Neurotoxic (prolonged tremors and convulsions) |
| Paxilline | *Penicillium paxilli* | Pecans | Neurotoxic (prolonged tremors and convulsions) |
| Penitrems A, B, and C | *Penicillium cyclopium* | Peanuts, meat products, cheese | Neurotoxic (prolonged tremors and convulsions) to cattle, sheep, dogs, and horses |
| Tryptoquivalines | *Aspergillus clavatus* | Rice | Neurotoxic (prolonged tremors and convulsions) |
| Verruculogen (TR-1) | *Penicillium verruculosum* | Peanuts | Neurotoxic (prolonged tremors and convulsions) |
| Unidentified toxin(s) | *Aspergillus terrus* *Balansia epichloe* *Epichloe typhina* *Fusarium tricinctum* *Others* | Fescue grass | Gangrene (Fescue foot); summer slump syndrome; fat necrosis and agalactia in cattle |
| Xanthoascin | *Aspergillus candidus* | Wheat flour | Hepatotoxic and cardiotoxic to mice |

*Source:* Condensed and modified from Busby, Jr., W.F. and Wogan, G.N., Psoralens, in *Mycotoxins and Nitroso Compounds: Environmental Risks*, Vol. 2, Shank, R.C., Ed., CRC Press, Boca Raton, 1981.

lirect additives, more than 600 are generally recognized as safe (GRAS) and about 150 are sanctioned as safe prior to September 6, 1958. Their number is likely to increase. They are exempt from regulation as food additives unless the scientific review of these substances warrants reclassification in the future or are used at levels higher than accepted to achieve the intended purpose. Sucrose, corn syrup, dextrose, and salt (all on the GRAS list) account for ~93%, by weight, of all the food additives used. Indirect additives are chemicals that gain their way into foods unintentionally or unavoidably during some phase of production, processing, storage, or packaging. Components of packaging containers and materials that migrate into foods fall under this category. In addition, the presence of pesticide residues in crops and processed foods and residues from animal drugs in milk, meat, and eggs is allowed based on their potential health risk as balanced against the benefits of their use. In Europe and the rest of the world, food additive use is governed by similar standards set by the European Union and the World Health Organization.

## Safety Assessment of Food Additives

Foods in their natural forms are assumed safe unless they are "ordinarily injurious." The presence of avoidable contaminants posing a risk of injury to health renders the food unsafe and subject to recall, whereas the presence of unavoidable contaminants posing such risk is considered unsafe if it exceeds tolerance levels set by the FDA (or EPA) or action levels (informal and not subject to law) set by the FDA. Intentional additives other than those GRAS or those that have been sanctioned prior (e.g., colors) are subject to regulations described above and can only be used at levels posing a risk at or above the level considered acceptable.

The safety evaluation of direct food additives involves establishment of a no-observed-adverse-effect level (NOAEL) in experimental animals followed by, for non-carcinogenic effects, establishment of acceptable daily intake (ADI) in the total diet for human beings using a suitable safety factor (usually 100). For carcinogens, an

**TABLE 14.5**
**Test Schedule Recommended by the FDA for Additives at Various Concern Levels**

| Concern Level I | Concern Level II | Concern Level III |
|---|---|---|
| Short-term (at least 28 days) feeding study in a rodent species | Subchronic feeding study in a rodent species | Chronic (at least 1 yr) feeding study in a rodent species |
| Short-term test for carcinogenic potential | Subchronic feeding study in a non-rodent species | Chronic (at least 1 yr) feeding study in a non-rodent species |
| | Multiple (at least two) generation reproduction study with teratology in a rodent species | Multiple (at least two) generation reproduction study with teratology in a rodent species |
| | Short-term test for carcinogenic potential | Short-term test for carcinogenic potential |
| | | Carcinogenicity studies in two rodent species (test 1 can be a part of this) |

additional factor of 10 is used for unusual susceptibilities such as in children. For unavoidable contaminants, the levels of tolerance or tolerable daily intakes and, for cumulative chemicals, provisionally tolerable weekly intake (PTWI) levels are set to limit the quantity of the agent in each commodity based on a risk–benefit cost analysis.

The FDA-recommended safety testing approach is based on the concept of *level of concern* [38], which depends on the level of exposure, structural correlation with known toxic compounds (if no toxicity data are available), and existing toxicologic data. Subjective categorization of additives into concern levels I, II, or III (level III being of highest concern) are made for compounds contributing <0.05 ppm, 0.05 to 1.0 ppm, and >1.0 ppm, respectively, to the total diet. Also, low, medium, or high toxicity or a structural similarity to compounds with low, medium, or high toxicity places a compound in concern levels I, II, or III, respectively. Furthermore, formation of active metabolites would place the compound in level III. Table 14.5 lists recommended toxicologic tests for each level of concern. This testing scheme allows compounds producing effects only at high levels and those with lower levels of human exposure to be tested less extensively. Protocols for testing color additives and indirect food additives are similar to direct food additive testing. Safety testing of animal drugs and feed additives, however, is more complex because an additional animal species (target animal) is involved. Also, it is necessary to understand the impact of the target animal metabolism on the diversity and toxic potential of the metabolites to humans, to develop residue detection methods and elimination strategies in the target animal species, and to set maximum allowable residues (tolerance) of the parent compound and metabolites in tissues of the target animal.

Until recently, the use of carcinogenic food additives has been strictly regulated by a special anticancer clause, the Delaney Clause, in the Food Additive Amendment which prohibited the use of these additives "if they are found to induce cancer when ingested by man or animal, or if found, after tests which are appropriate to the evaluation of safety of such substances, to induce cancer in man or animal" [80]. Recent developments in the understanding of the mechanisms of carcinogenesis required changes as reflected in the FQPA of 1996. Currently, the regulation of pesticide residues has become the responsibility of the EPA and thus is not a food additive issue for the FDA. The EPA can approve pesticide applications if it concludes that there is "a reasonable certainty that no harm will result from its aggregate exposure." This would translate to no greater than a risk of 1 in 1 million lifetime risk for carcinogens and levels with a 100-fold safety margin to the NOAEL for threshold-limited effects. For already approved carcinogenic pesticides, the EPA is allowed to retain tolerance posing greater than negligible risk (1 in 1 million) if the pesticide either protects consumers from a greater health risk or is necessary to avoid disruption of adequate, wholesome and economical food supply [187]. In addition, the FQPA directs the EPA to take higher intakes for certain pesticides in children into consideration in setting tolerances and to apply an additional safety factor of 10 for threshold effects in calculating the ADI.

All new animal drugs and feeds containing them must receive premarket clearance from the Bureau of Veterinary medicine of the FDA. Antibiotics and steroidal as well as nonsteroidal growth promotants are used in feeds for the prevention of disease or for growth promotion in the raising of 60 to 100% of food-producing animals. Two major concerns with the use of antibiotics in animal feeds are (1) the development of resistance in enteric bacteria in animals that could be transferred via plasmids to pathogenic bacteria in the gut of animals and thus to humans through meat or milk making antibiotics currently used in human medicine ineffective and (2) the presence of carcinogenic residues in meat or residues that form carcinogenic nitrosamines following reaction with nitrite in the meat. Increasing pressure on the FDA is expected to result in the potential exclusion of all antibiotics as feed additives.

Similarly, concerns about the use of hormonal and other growth promotants in feeds are related to possible chronic toxic effects, mainly carcinogenicity. The synthetic estrogen diethylstilbestrol (DES) was banned as a feed additive due to its carcinogenic effects and the lack of a method sensitive enough to detect residues of health significance (causing >1 lifetime cancer in 1,000,000). Steroids approved for feed additive use in one or more animal species include estradiol, progesterone, testosterone, melengestrol acetate, and zearalanol. Other additives commonly used in animal production include monensin, iodides (e.g., EDDI), phenothiazine, and thiabendazole. Carcinogenic animal drugs can be approved for use by the FDA if an adequate withdrawal period is recommended between the last dose of the additive and slaughter to allow residues to fall below those judged to be capable of inducing greater than negligible (1 lifetime cancer in 1,000,000) incidence of cancer.

A brief discussion on major classes of additives is presented below. Further details on various classes and examples of food additives, their intended function, and their toxic effects can be obtained from Deshpande [68]. Toxic reactions associated with the use of selected additives that have been restricted, banned, or are currently being critically reviewed are also presented below.

## Direct Food Additives

### Flavoring Agents

Flavoring agents are by far the largest (>2000) and most chemically diverse group of small-molecular-weight (<300) food additives used in small quantities to specifically impart new flavor to or modify or mask the existing flavor of food. Because only small amounts (<2%) of the total dietary flavoring agents are added (98% of the total quantity of flavoring agents consumed are already present in foods), the flavoring agents in normal daily use appear very safe [79,305]. To reduce the time and cost burden involved in a systematic assessment of the safety of the large number of agents involved, a joint FAO/WHO Expert Committee on Food Additives established a new approach [79]. The estimated intake of individual agent or a combination of agents in a natural mixture is compared with the appropriate intake threshold of toxicological concern (ITTC) for the entire structurally similar class of agents to determine whether or not the intake represents a safety concern. The human ITTCs have been determined to be 1800, 540, and 90 μg/day for three structural classes of agents based on the fifth-percentile no-observed-effect level (NOEL) from animal experiments, adjusted by the safety factor of 100. The second threshold of 1.5 μg/day is a pragmatic and conservative value for individual agents derived from the linear extrapolation from animal experiments and represents an intake posing lifetime cancer risk of <1 in 1 million and is conservative enough to cover all toxicological responses. If an estimated intake of an agent is <1.5 μg/day, it would be considered to be of no safety concern. All agents with intakes of >1.5 μg/day are subject to safety evaluation by standard methods. For a mixture of compounds as encountered in natural flavor complexes, if the exposures for all compounds in each structural class add up to less than the appropriate threshold for that class, the mixture is considered safe. Using this approach, the committee failed to identify any among a large number of the flavoring agents evaluated as being of safety concern.

### Food Colors

Color is a quality of foods that makes them visually acceptable and aids in their recognition. Foods containing added colors include candy and confections; bakery goods; soft drinks; cereals; dairy products such as butter, ice cream, and sherbet; margarine; snack foods; jams and jellies; and dessert powders. Following the passage of the Color Additive Amendment of 1960, 20 natural colors (including preparations such as dried algae meal, annatto extract, beet powder, grape-skin extract, fruit juice, paprika, caramel, carrot oil, cochineal extract, ferrous gluconate, iron oxide, and turmeric) were exempted from certification, whereas all the synthetic colors including the ones approved prior to the Amendment were required to be retested if questions regarding their safety arose. A provisional certification was given to those in use that required further testing. Currently, seven certified synthetic colors have unlimited uses (FD&C colors Blue No. 1, Red No. 3, Red No. 40, and Yellow No. 5 are permanently listed whereas FDB Blue No. 2, Green No. 3, and Yellow No. 6 are provisionally listed; according to good manufacturing practices); one permanently listed color (Citrus Red No.2) is used only for coloring the skins of oranges at 2 ppm. Several colors, including Green 1, Green 2, Orange B, Red 2, Red 4, and Violet 1, were delisted due to concerns of their carcinogenicity and other chronic toxic effects. A controversy linking food colors to allergies and hyperkinesis in children remains unresolved.

### Selected Food Additives

*Aspartame (Nutrasweet®)*

The use of aspartame (*L*-aspartyl-L-phenylalanine methyl ester) as a sweetener in a large variety of food products increases the likelihood that acceptable daily intake is exceeded. A previously suspected association between aspartame consumption and anecdotal reports of hypertension, headache, dizziness, and seizures in adults and hyperkinesia in children appears to be unsupported. Although nitrosation products of aspartame are moderately mutagenic [263], the question of the role of aspartame in the recent increase of lymphoma of the brain is still unresolved. The metabolism of aspartame to aspartic acid, phenylalanine, methanol, and formaldehyde led to the suggestion that patients with phenylketonuria should avoid aspartame-containing products. For the rest of the population, the ADI of 40 mg/kg/day appears safe.

### *Butylated Hydroxyanisole (BHA) and Toluene (BHT)*

The synthetic phenolic antioxidant BHA has been GRAS and used for decades as an antioxidant to retard the autoxidation of lipids and prevent rancidity in foods. Together with ascorbic acid, α-tocopherol, gallate esters, and BHT, it fulfills almost 100% of the antioxidant requirements of foods. BHA may be a rodent carcinogen involving squamous cells in the forestomach but not the human equivalent glandular or other cells. The carcinogenicity of BHA appears to involve *O*-demethylation of BHA to tertiary butyl hydroquinone (TBHQ), oxidation of TBHQ to tertiary butyl semiquinone and tertiary butylquinone (TBQ), conjugation of TBQ with glutathione (GSH), and formation of DNA-reactive oxygen species including the hydroxyl radical [300]. However, the genotoxicity of BHA has not been demonstrated. Because humans lack forestomach, evidence for direct genotoxic effect is lacking, and because human doses are well below doses inducing nongenotoxic effects, it is highly unlikely that BHA carcinogenicity in rodents is relevant to the safety of BHA in human foods. High doses (0.5 to 1.0 g/kg) of BHT, on the other hand, can form an electrophilic metabolite and cause renal and hepatic damage in male and female rats, whereas lower doses increase liver weight and inhibit hepatic enzymes. BHT is neither genotoxic nor carcinogenic and, in fact, may be anticarcinogenic [147]. Acceptable daily intakes of BHA and BHT are, respectively, 0.5 and 0.05 mg/kg/day [68].

### *Cyclamates*

Sodium and calcium cyclamate were introduced as nonnutritive sweeteners in 1950 and were included in the 1959 GRAS list. Significant consumption of cyclamates in low-calorie foods and drinks followed. The subsequent demonstration of a link between enhanced bladder carcinogenicity or cocarcinogenicity and the consumption of cyclamates led to the removal of cyclamates from the GRAS list and thus from food additive use in the United States, despite a lack of evidence in human beings [5] and many other animal studies that failed to show carcinogenic effects. Cyclamates, however, are still used in many other countries.

### *Monosodium Glutamate (MSG)*

Monosodium glutamate first appeared on the GRAS list in 1958 and has been used both as a seasoning agent and a flavor enhancer. The demonstration of lesions in the retina and the lateral arcuate and geniculate nucleus in MSG-exposed neonatal rats and mice led to a voluntary discontinuation of its use in infant foods in the United States. This appears justified, as the neonatal effects of MSG may last through adulthood [280]. A link has been suggested between fetal and newborn glutamate-exposure-induced destruction of arcuate nucleus neurons in mice resulting in permanently elevated plasma leptin levels that fail to adequately counterregulate food intake, leading to obesity and metabolic syndrome [106]. In adults, the MSG symptom complex (headache, muscle tightness, numbness/tingling, general weakness, and flushing, among others) occurs in a third of the population at a threshold dose of 2.5 g of MSG [318].

### *Nitrates, Nitrites, and Nitrosamines*

Leafy vegetables contribute 99% of the total daily dietary intake of nitrates (100 to 150 mg/day) [65]. Their use to cure meats (to give characteristic flavor and pink color, to prevent rancidity, and to prevent growth of the spores of *Clostridium botulinum*) contributes <0.1 mg/day [65]. Nitrates can be reduced endogenously by microbial systems to nitrites, which then oxidize the hemoglobin to methemoglobin (heme iron from ferrous to ferric state). Methemoglobin, being unable to combine with oxygen, can accumulate in sufficient quantities to lead to anoxia. The use of water with high (>30 mg/L) nitrate (from soils, fertilizers, etc.) in making baby formula and foods, spinach with high nitrate content, and occasionally meats with high levels of added nitrates and nitrites have resulted in life-threatening methemoglobinemia in humans, especially children. The consumption by animals of plants high in nitrates has caused significant economic loss for owners. In adult humans, however, the daily intake of nitrate and nitrite amount to <69% and 0.7% of the ADI of 3.64 and 0.135 mg/kg/day [65].

Nitrite reacts with secondary amines to form a variety of *N*-nitrosamines that is present in foods, pharmaceuticals, cosmetics, agricultural chemicals, tobacco, and tea. *In vivo*, nitrosamines are converted to unstable hydroxy alkyl compounds which subsequently form reactive alkyl carbonium ions capable of alkylating DNA [300]. Nitrosamines are mutagens and rodent carcinogens that produce cancer in a variety of organs, including the liver, respiratory tract, kidney, urinary bladder, esophagus, stomach, lower gastrointestinal tract, and pancreas [300]. Nitrite itself may promote carcinogenesis. Because they are not added to foods, however, nitrosamines are not subject to the restrictions of the Delaney Clause. Inhibition of nitrosamine formation in foods by ascorbate, cysteine, gallic acid, tannins, sodium sulfite, and sodium erythorbate prompted the FDA to suggest that one of these compounds be concurrently added to meats during curing to reduce nitrite added from 200 to 120 ppm. Such a practice, however, is ill advised until a suitable additive is found to deal with the threat of *Clostridium botulinum* growth effectively at reduced nitrate levels and consumers accept the ensuing changes in the appearance and organoleptics of meat cured with this combination. Perhaps a change to vegetables containing lower levels of nitrates to reduce nitrate intake is a less dangerous option.

### *Olestra*

Olestra is a sucrose polyester of eight long-chain fatty acids; it is a nonabsorbable (thus non-calorie-contributing) substitute for conventional fats in use since 1996 in certain snacks (e.g., potato chips, crackers, tortilla chips) as a means to reduce fat intake and thus control obesity. Although it also reduces cholesterol, the FDA considered it insignificant. Animal and human experiments from Proctor & Gamble as well as independent studies revealed major problems with its use, such as enteritis characterized by intestinal cramps and loose stools, as well as inhibition of the absorption of dietary lipid-soluble vitamins (A, D, E, and K) and of carotenoids such as alpha-carotene, beta-carotene, lutein, zeaxanthin, and lycopene [292]. Due to the purported roles of carotenoids (lutein and zeaxanthin) in the prevention of macular degeneration and cancer, whether or not olestra consumption will increase the incidence of these conditions is an open question. Although the FDA-required addition of vitamins A, D, E, and K may address some issues, recent evidence indicates that a carotenoid deficit cannot be reversed by such dietary supplementation [290]. In light of the more than 20,000 public complaints (more than for all other food additives combined) and that no NOEL or ADI have been set since the lowest dose tested in humans caused adverse effects, relaxation of the earlier required consumer warning label relating to the gastrointestinal effects angered consumer advocacy groups such as the Center for Science in the Public Interest. This decision, along with the facts that olestra failed to gain approval in Canada, the United Kingdom, and other countries and that olestra-containing products have declined in popularity in the United States, appears to reflect the FDA's gullibility in accepting industry-funded postmarket research findings strongly associated with the researchers' financial relationships with the product manufacturer [155].

### *Saccharin*

Saccharin is used as a sweetener in soft drinks and for table-top uses. It has been in use since the beginning of the twentieth century and survived an FDA ban via a Congressional moratorium and an attempted ban of its use due to suspected weak bladder-cancer-promoting activity [199] at relatively high doses. A dose of 300 mg/day in adult humans does not pose a health threat. Its average intake is 7.1 mg/day in the United States and 15.0 mg/day in Europe, with per capita intake reaching as high as 25 mg/day in certain subpopulations [118].

### *Safrole*

Safrole and other alkenylbenzene compounds (β-asarone, methyleugenol, estragole, and isosafrole) are active components of many spice flavors. Sassafras, which contains high levels of safrole, has been used as a flavoring agent in sarsaparilla root beer. Safrole consumption has also occurred in the form of sassafras oil and sassafras tea, the latter still occurring to a limited extent in the United States. A total dose of only 0.5 to 1.5 mg of safrole orally or intraperitoneally to infant male mice caused high liver tumor incidence. Dihydrosafrole, a synthetic safrole, caused esophageal tumors in rats, and some of the other natural alkenylbenzenes are also carcinogenic [188]. These findings resulted in the FDA ban on the use of safrole, sassafras, and sassafras oil from commercial use in foods, including root beer, in 1960. A metabolite, 1-hydroxy sulfate ester, is apparently the ultimate carcinogen, forming adducts with guanine and adenine [188].

## Other Chemical Additives

Other chemical additives prohibited from use due to a potential risk or to lack of demonstration of safety include calamus and its derivatives in 1968 (containing aklenylbenzene flavoring agents); coumarin flavoring compounds in 1953; chlorofluorocarbon propellants in self-pressurized containers in 1978 due to their role in the dissolution of the Earth's ozone layer, which results in increased skin cancer risk from ultraviolet radiation; diethyl pyrocarbonate (DEPC), an antimicrobial agent in beers and juices (cold pasteurization) and a ferment inhibitor in 1972 due to the presence of the carcinogen urethane in DEPC-treated products; and dulcin, a sweetener, in 1950 due to liver and bladder cancer in rats. On the other hand, the safety of some recently approved additives such as the fat substitute olestra (long-chain fatty acid esters with sugar) continues to be debated. Table 14.6 lists intentional food additives based on their current safety concerns as assessed by the Center for Science in the Public Interest.

## Indirect Food Additives

## Packaging Materials

Packaging is an essential part of food processing that aids in the preservation of the wholesomeness of foods by preventing: (1) contamination or destruction by dirt, microorganisms, insects, and rodents; (2) loss or gain of moisture, odors, flavors, or aroma; and (3) deterioration from air, light, heat, and contaminating gases. Other functions served by packaging include assembling a variety of items, convenient handling, labeling, and finally sales promotion. A variety of materials ranging from metal foils to complex plastic substances are in use. Examples of package modifications employing chemical additives are oleoresinous coating with or without suspended ZnO, which is used in the preservation of acid foods that do (e.g., seafood) or do not (e.g., cherries) produce sulfides; stabilizers to prevent degradation of plastic when exposed to heat and light; and hot-melt adhesives used to glue multilayered packages (e.g., tea, hydrated soups, potato chips). A complete list of additives approved for use in packaging is included in Deshpande [68].

**TABLE 14.6**
**Classification of Common Intentional Food Additives Based on Safety Concerns**

| Class | Reason | Examples |
|---|---|---|
| Reduce intake | Nontoxic as recommended but may be toxic or nutritionally undesirable in large amounts | Caffeine, corn syrup, dextrose (corn sugar, glucose), high-fructose corn syrup, hydrogenatated starch hydrolysate, hydrogenated vegetable oil, invert sugar, lactitol, maltitol, mannitol, polydextrose, salatrim, salt, sorbitol, sugar, tagatose |
| Try to avoid | May pose risk; requires more testing | Citrus red 2, red 40, brominated vegetable oil (BVO), butylated hydroxyanisole (BHA), butylated hydroxytoluene (BHT), heptyl paraben, quinine |
| Sensitive people must avoid | Allergic or other reactions; unsafe in sensitive people | Yellow 5, artificial and natural flavoring, aspartame (Nutrasweet), beta-carotene, caffeine, carmine, cochineal, casein, gum tragacanth, hydrolyzed vegetable protein (HVP), lactose, monosodium glutamate (MSG), mycoprotein, quinine, sodium bisulfite, sulfites, sulfur dioxide |
| Generally unsafe | Unsafe as recommended or poorly tested | Acesulfame potassium, blue 1, blue 2, green 3, red 3, yellow 6, aspartame (Nutrasweet), cyclamate, olestra (Olean™), potassium bromate, propyl gallate, saccharin, sodium nitrite, sodium nitrate, stevia |

*Source:* Compiled from information provided by the Center for Science in the Public Interest; http://www.cspinet.org/reports/chemcuisine.htm.

To approve new packaging material, the FDA requires extraction studies involving one or more of aqueous (8% alcohol), alcoholic (50% alcohol), or lipid solvents (corn oil or triglycerides) followed by toxicity testing, depending on the extent of extraction (>1 ppm requiring extensive testing including chronic toxicity). The National Science Foundation (NSF) estimated that as many as 3000 chemicals may enter foods indirectly from the process of packaging itself [104]. A safety review by the FDA has resulted in banning the adhesive Flectol H, polyurethane resins, curing agents, food packaging adhesives containing 4,4-methylenebis (2-chloroanaline), and the synthetic chemicals mercaptoimidazoline and 2-mercaptoimidazoline used in the production of rubber articles [104]. The use of polyvinylchloride for the packaging of liquors has been banned in the United States. Among the packaging-derived contaminants likely to be encountered in U.S. diets, benzene and vinyl chloride are known carcinogens; acrylonitrile, 1,3-butadiene, epichlorohydrin, formaldehyde, propylene oxide, and styrene oxide are probable carcinogens; and 2,4-diaminololuene, dibutyl- and diethylhexylphthalates, dimethylformamide, 1,4-dioxane, ethylacrylate, phenyl glycidil ether, styrene, and toluene diisocyanate are possible carcinogens [118,205].

## Toxic Factors Produced During Processing

Food processing is aimed at improving the quality of foodstuffs, ensuring safety, and enhancing the ease of preparation. This requires various chemical and physical treatments of food that may result in (1) partial or complete destruction or removal of nutrients, (2) inferior digestibility or utilization of nutrients, and (3) the generation of new and potentially harmful chemicals. The first two effects can be overcome by nutritional supplementation. The third effect represents a need for appropriate toxicological investigation. In addition, similar products can be formed during storage due to continuous effects of heat, humidity, light, oxygen, and catalysts present in foods.

The formation of cross-linked amino acid side chains such as lysinoalanine, ornithinoalanine, and lanthionine as well as racemization of amino acids to D-analogs appears to take place during alkali treatment, for example, of soybean protein for preparing imitation meat [83]. These products, especially lysinoalanine, have been shown to cause nephrocytomegaly (enlarged nuclei and cytoplasm) of the pars recta cells. Nonenzymatic browning reactions (Maillard reactions) occurring during the heating of foods (drying, frying, roasting, baking, and broiling) involve chemical interactions between amino acids and reducing sugars (aldoses and ketoses) forming mutagenic reductones, furans, amino-carbonyls, pyrazines, and other premelanoid secondary amine derivatives (Amadori and Heyns' products) that have been proposed to inhibit growth, impair reproduction, damage liver, cause allergies, play a role in aging, and induce lens lesions [118]. In general, high-protein foods appear to possess more mutagenic activity compared with foods rich in carbohydrates or fats. Pyrrolysis of proteins and amino acids at high temperatures (300°C or more) is known to yield a series of heterocyclic compounds that can be metabolized to mutagenic products and were positive in one or more rodent species for carcinogenicity [118,282]. Using estimates of various heterocyclic amines ingested and cancer potencies in animal studies, Layton et al. [149] estimated that only 0.25% of human colorectal cancer ma be due to these compounds. Certain cooking practices such as frying high-nitrite foods (e.g., cured bacon) results in the formation of nitrosamines, the carcinogenic effects of which have already been discussed.

A variety of polycyclic aromatic hydrocarbons (PAHs) are formed in foods by pyrrolysis during cooking or by their prior contact with petroleum or coal-tar products. Although the carcinogenic effects of PAHs are known, the contribution of dietary PAHs to cancer in humans is likely to be insignificant.

Fats (polyunsaturated) undergo three basic changes during storage and heat treatments: autoxidation, thermal oxidation, and thermal polymerization. Autoxidation occurs at below 100°C in the presence of enzymes (lipoxygenases) or upon exposure to light and results in the generation of hydroperoxides via a free-radical or singlet oxygen mechanism, leading to rancidity [118]. Hydroperoxides can be degraded into alkanes, aldehydes, and ketones, among others. Termination of peroxidative reactions generally involves scavenging of the radicals or their polymerization into nonreactive products. Lipid hydroperoxides, at subtoxic levels, can stimulate signal transduction mediated by $Ca^{2+}$ and protein phosphorylation by acting as second messengers in pathways involved in cell proliferation, chemotaxis, apoptosis, and other cellular mechanisms [283]. High levels of rancid fats (5% or more of the diet) can cause decreased food consumption, diarrhea, weight loss, leukopenia, and hair loss. Hydroperoxides and their products (e.g., hydroxynonenal, melonyldialdehyde) can disrupt gap-junctional communication, can form DNA adducts, and are mutagenic and carcinogenic, increasing the incidence of tumors and atherosclerosis [118,319]. Components of heated oils that fail to form adducts with urea, especially the cyclic monomeric fatty acids followed by polymers of fatty acids, appear to be toxic. Toxic effects include, in addition to those described above, hepatomegaly and carcinogenicity.

Yeast-fermented foods and beverages such as yogurt, cider, malt beverages, bread, soy sauce, wine, and sake, in addition to the psychoactive and vasoactive amines discussed earlier, contain mutagenic and carcinogenic ethyl carbamate (urethane) derived in the presence of heat and light from arginine, asparagine, cyanogenic glycosides, or ethanol in the fermented commodity. Levels of <10 ppb or less for soft drinks and <30 to 400 ppb for various alcoholic beverages have been recommended as acceptable by the FAO/WHO and the Canadian government, respectively [119]. The major fermentation product consumed by humans is ethanol, which, in addition to death from toxic effects, contributes to human deaths from occupational as well as automobile accidents. The toxic effects of ethanol can manifest in many organ systems but display major involvement of the central nervous system (dependence and depression), the developing fetus (fetal alcohol syndrome of mental deficiency and microcephaly), and the liver (hepatomegaly followed by cirrhosis). Mechanisms of ethanol toxicosis may involve direct effects of alcohol, effects of its metabolite acetaldehyde, ethanol-induced malnutrition, ethanol-induced endotoxin release by intestinal bacteria that stimulates the release of reactive chemicals by Kupfer's cells, ethanol-induced potentiation of other hepatotoxic agents, or a combination of these. An International Agency for Research on Cancer (IARC) expert panel considered ethanol to be a human carcinogen that causes tumors of the oral cavity, pharynx, esophagus, and liver [271].

The toxic effects of processed food as a whole, however, cannot be estimated by simply adding up the toxic, mutagenic, and carcinogenic potentials of the products present in it. This is due to the fact that chemical derivatives that both enhance as well as antagonize the myriad of toxic effects of other dietary components are formed during processing [205]. At present, these chemicals and their interactions with each other, for the most part, are unknown. As a result, the overall adverse effects of cooked foods can only be determined reliably based on the assessment of risk from the complex milieu of the product in question.

## Irradiation Products

Irradiation as a means to kill foodborne bacteria has been studied in the United States and Europe since the late nineteenth century, although major efforts only materialized in the1950s [274]. After many false starts—due to the conflicting results of adverse effects from feeding irradiated foods to animals (enlarged atria, premature death, and smaller litter size, since shown to be due to vitamin deficiency) and changing regulatory demands of both the U.S. Congress (that classified irradiation as a food additive) and the FDA (required irradiated products to be shown noncarcinogenic in long-term studies)—approval of low-dose gamma radiation for disinfecting wheat and inhibiting sprouting in potatoes was granted by the FDA in 1963 and 1964. The Joint Expert Committee on the Wholesomeness of Irradiated Foods of the International Atomic Energy Agency determined and the Codex Alimentarius Commission (the United Nations body that establishes international food standards) accepted that radiolytic products produced in any food exposed to low and medium doses of radiation were safe for human consumption and endorsed as wholesome all foods irradiated up to the medium dose of 1 million rads. The Bureau of Foods Irradiated Foods Committee (BFIFC) of the FDA subsequently announced that "all foods irradiated at doses not to exceed 100 krads and spices and vegetable flavorings irradiated at slightly higher doses are wholesome and safe for human consumption." The FDA subsequently approved low-dose irradiation of pork in 1985 to control the *Trichinella spiralis* parasite; to disinfect dry herbs,

**TABLE 14.7**
**Examples of Biotechnology-Derived Crops Approved Globally and Their Desired Traits**

| Phenotypic Trait | Example Crops |
|---|---|
| Fatty acid composition modified | Canola, soybean, |
| Fertility restoration | Canola, chicory |
| Herbicide tolerance | Alfalfa, bent grass, canola, carnation, chicory, cotton, linseed, lentil, maize, rice, soybean, sugar beet, sunflower, tobacco, wheat |
| Insect resistance | Cotton, maize, potato |
| Lepidopteran resistance | Cotton, maize, tomato |
| Modified color | Carnation |
| Nicotine reduction | Tobacco |
| Increased shelf life | Carnation, melon, tomato |
| Virus resistance | Papaya, potato, squash |

*Source:* Compiled from information provided by AGBIOS, Merrickville, Ontario, Canada (www.agbios.com).

spices, seeds, teas, and vegetable seasonings; to inhibit ripening and sprouting of fruits and vegetables; to control *Salmonella* and other foodborne bacteria in poultry in 1990; and to control foodborne pathogens in beef in 1997.

Although the FDA appears convinced of the safety of irradiated foods and the ability of irradiation to reduce the growing threat of deadly foodborne illnesses, many negatives still linger. These include unusual colors, textures, and flavors of irradiated foods; the presence of radiolytic products such as 2-alkylcyclobutanones (2-ACBs), radiolytic products of triglycerides, and their potential for tumor promotion [226]; requirement for the addition of new and costly nuclear processing facilities containing highly radioactive sources; and issues of worker and transportation safety, the disposal of radioactive waste, and its image as a target for terrorism. The net result of efforts over the last century amounted to the irradiation of only about 0.002% of the fruits, vegetables, and poultry consumed annually in the United States by the end of the 1990s. Public acceptance of irradiated foods will only occur when the public is convinced of its cost effectiveness and its greater effectiveness and safety compared to existing preservation methods.

## BIOTECHNOLOGY-DERIVED FOODS

Modern biotechnology, unlike classical breeding techniques that involve the transfer of large portions of genomic DNA between organisms (plants, animals, and microorganisms), is more selective, as it involves the transfer of a specific, known gene from one organism to another. Foods derived from biotechnology (BT foods) offer such benefits as more economical and healthier foods that are more nutritive and hypoallergenic, a reduction in pesticide use, and a reduction in the loss of habitat [152]. These goals are accomplished by the insertion of genes that lead, for example, to the synthesis and purification of essential nutrients (such as riboflavin) and processing aids (such as calf chymosin used in cheese manufacture) on a large scale, to the expression of biological insecticidal proteins (*Bacillus thuringiensis* endotoxin), and to altered plant characteristics (drought and herbicide resistance) and nutrient composition (e.g., starch in potato). Unintended effects can result from altered levels of nutrients or antinutrients in the food; increased or silenced expression of genes for preexisting normal, toxic, or pharmacologically active substances; and new gene-product-induced alterations in host metabolic pathways in the consumer as well as effects on other plant and animal life sharing the environment with the altered plant. The Animal and Plant Health Inspection Service (APHIS) of the U.S. Department of Agriculture (USDA) is responsible for ensuring that the growth of genetically engineered plants does not harm the agricultural environment. The EPA is responsible for ensuring the human and environmental safety of pesticidal substances engineered into plants, and the FDA is responsible for ensuring that foods derived through genetic engineering are as safe as their original counterpart (i.e., "substantially equivalent"). In 2003, the Codex Alimentarius Commission (a worldwide regulatory body) adopted international guidelines for biotech food safety that are consistent with the FDA's approach. Table 14.7 presents a summary of biotechnology-derived crops approved globally with the list expanding rapidly.

For the most part, foods undergoing single genetic alteration are likely to be substantially equivalent to the host species in their nutritional value (with respect to micro-, macro-, and antinutrients) and safety as long as the selected gene is not a known toxin or allergen and does not induce wide-ranging chemical changes. An attempt to nutritionally enhance soybean using the expression of a Brazil nut protein was curtailed because of the allergenicity of the Brazil nut protein. The introduction of the gene for *Bacillus thuringiensis* (BT) delta-endotoxin

**TABLE 14.8**
**Stability in Gastric Fluid and No-Observed-Effect Level for Novel Biotechnology-Derived Food Proteins in Animals**

| Protein | Crop | Abundance (% Total Protein) | Stability in Gastric Juice (sec) | NOEL (mg/kg Body Weight) |
|---|---|---|---|---|
| Cry1Ac | Cotton, tomato | <0.01 | 30 | 4200 |
| Cry1Ab | Maize | <0.01 | 30 | 4000 |
| Cry2Aa | Cotton | <0.01 | <15 | 3000 |
| Cry2Ab | Maize, cotton | <0.01 | <15 | 3700 |
| Cry3A | Potato | <0.01 | <15 | 5200 |
| CP4 EPSPS | Soybean, maize, cotton, canola, sugarbeet | <0.1 | <15 | 572 |
| mzEPSPS | Maize | <0.05 | <15 | 350 |
| NPTII | Cotton, potato, tomato | <0.01 | 10 | 5000 |
| GUS | Sugarbeet | 0.01 | <15 | 100 |
| GOX | Canola | <0.01 | <15 | 100 |
| ACCdeaminase | Tomato | NA | NA | 602 |

*Note*: NA, not available.

*Source:* Compiled from information provided by AGBIOS, Merrickville, Ontario, Canada (www.agbios.com).

nto corn raised concerns regarding the toxic and allergenic effects of BT-toxin on humans consuming such corn, on the migratory Monarch butterflies, and on the nontarget insects consuming the pollen from such corn. BT-toxins (e.g., Cry1Ab protein) disrupt intestinal ion flow by opening cellular pores following selective binding to sites localized on the brush-border intestinal epithelium of only the Lepidopteran insect but not nontarget species. Although recent studies have discounted the effects of BT-corn pollen on the Monarch butterfly [259] and highlighted the unlikelihood of toxic effects in mammals due to the easy digestibility of BT-toxins and the specificity of their adverse effects towards Lepidopterans [264], BT-corn (StarLink™) failed to gain U.S. regulatory approval (although it is approved in Europe, Asia, and Central America) for human consumption due to the uncertainty involved in current methodology for the assessment of the allergenicity of the protein. Data presented in Table 14.8 on the NOELs and gastric stability of certain novel proteins in biotechnology-derived foods illustrate their safety, with safety factors ranging from >1000 for CP4 to 2.6 million for Cry3A. Pleiotropic changes are likely to be detected because of their detrimental effects on the survival of the altered plant or the resulting phenotypic or chemical analytical alterations. The careful selection of transgenes and the ability of conventional safety testing protocols to detect adverse effects of certain biotechnology-derived foods, leading to discontinuation of their production, illustrate the effectiveness of the current biotechnology-derived food production/testing strategy, but public apprehension against their use in the United States has yet to be overcome. The acceptance of many therapeutic recombinant peptide hormones such as insulin and of milk produced by cows receiving recombinant bovine somatotropin (rbST) in the United States and 30 other countries suggests, however, that consumer education can overcome such fears.

In addition to the toxic effects, the protein nature of the gene products introduced into biotechnology-derived foods raises concerns about the allergenicity of such proteins (known allergens, cross-reactive allergens, or new proteins), a justified major concern for risk assessment. Currently, all genes introduced into food crops undergo a series of tests designed to assess their allergenicity by their source (allergenic or nonallergenic), the similarity of their amino acid sequences to known allergens, and their stability with regard to digestion by proteases from the stomach [18], and efforts are underway to develop an appropriate animal model for optimal biotechnology-derived allergen testing. These and breakthroughs in biotechnology, such as posttranscriptional gene silencing to prevent protein accumulation (e.g., Gly m Bd 30 K protein, a major soy allergen), the alteration of an allergen's secondary or tertiary structure (e.g., reduction of disulfide bonds of wheat and milk allergens by *Escherichia coli* thioredoxin), and modification of the primary amino acid sequence of genes encoding allergens (e.g., shrimp tropomyosin), are expected not only to reduce allergenicity of foods but also to yield modified proteins useful in desensitization of allergic patients in the near future [152].

## NONHERBAL DIETARY SUPPLEMENTS

Although herbal products and supplements have been used by herbalists and laypersons for centuries, the use of nonherbal dietary supplements has gained popularity only relatively recently. Because they are neither foods nor drugs used for therapeutic purposes, FDA regulation and information regarding their safety and toxic effects in animals and users at levels recommended for use and in actual use by consumers are sparse. This information for some commonly encountered supplements, as summarized by Cupp [60], is presented in Table 14.9.

## CARCINOGENS AND MUTAGENS IN FOODS

Cancer, a disease of most public concern for the past half a century, is a multistage process involving initiation (induction of DNA damage thus resulting in a transformed cell), promotion (a nongenotoxic effect leading to rapid multiplication of the transformed cell and thus establishment of a cancerous lesion), and progression. Multistage models involving a sequence of multiple genetic events with the incidence increasing in proportion to the exponent of time seem to fit most human cancers [205]. Naturally occurring food toxicants provide examples of both initiators and promoters, as up to 70% of cancer deaths can be attributed to dietary factors [70]. Examples of likely dietary carcinogens as reviewed by the National Research Council [205] and others are provided in Table 14.10. This list includes carcinogens derived from natural products both by commercial processing (alcohol) and by biotransformation in the body (allylisothiocyanate and nitrosamines), as well as initiators (e.g., aflatoxins, furocoumarins, pyrrolizidine alkaloids) and promoters (e.g., phorbol esters, fat, caffeine). In addition, residues of synthetic chemicals can be present in foods subsequent to accidental contact or intentional use to increase production. An added dimension to diet is the formation of animal carcinogens during cooking, such as nitrosamines, aromatic hydrocarbons, heterocyclic amines (aminocarbolines, imidazoquinolines, imidazoquinoxalines, and imidazopyridines); fat oxidation products; and acrylamides inducing cancer of the liver, stomach, intestines, zymbal and clitoral glands, skin, and oral cavity, among others [120,205]. The IARC classifies this group of carcinogens as possibly (2A) or probably (2B) carcinogenic to humans. Coffee, in addition to caffeine, is known to yield several carcinogens, including caffeic acid, catechol, furfural, hydrogen peroxide, and hydroquinone, during roasting and brewing [11]. Carcinogenic natural pesticides are present in all classes of plant foods, including fruits, vegetables, and spices [11].

Balancing this bewildering array of toxins and carcinogens, in almost every food item is another group of chemicals capable of antagonizing these effects. Dietary antimutagens and anticarcinogens, whose mechanisms of action are not always understood, feature a wide variety of chemical structures (Table 14.11). As with mutagens, multiple species of antimutagens and anticarcinogens appear to be present in each dietary component (at least five are known in soybeans and three or more in broccoli). Interactions between carcinogens and anticarcinogens are complicated, as indicated by the study of indole-3-carbinol, a component of cruciferous vegetables known to inhibit mammary and forestomach neoplasia in rodents. When given as a pretreatment, indole carbinol reduced the carcinogenicity of aflatoxin $B_1$, whereas exposure to indole carbinol after the carcinogen exposure resulted in an increase in aflatoxin carcinogenicity [20].

## NATURAL VS. SYNTHETIC CHEMICALS

The widely held belief that naturally (free of synthetic chemicals) grown foods are inherently safer than those grown with the aid of synthetic chemicals is flawed. Certain natural chemicals in the human diet (such as indole carbinol in cruciferous vegetables) interact with the same receptor (Ah) with which dioxin (TCDD), one of the most feared synthetic toxicants, interacts [12]. An EPA reference dose—a dose estimated to produce 1 cancer in 1 million individuals (6 fg/kg/day)—of TCDD is comparable to 5 mg of indole carbinol per 100 g of broccoli or cabbage, a level of exposure not unrealistic. The EPA banned Alar™ based on a worst-case scenario of risk estimation in response to public outcry resulting from less-than-objective reporting by the media and a passive attitude by knowledgeable academicians and researchers [235]. This ban suggests that regulatory agencies may also subscribe to this misconception.

The National Research Council [205] considers natural carcinogens to be at least as potent and, considering the extent of exposure, more potent compared to synthetic carcinogens. The fact that disproportionately fewer natural chemicals have been tested so far suggests a greater need to test natural chemicals, a daunting task considering the fact that toxicants in all classes of foods are not known and that exposures to those that are known are wide ranging or unknown. Adding to this complexity is the fact that simultaneous exposure to two initiators, an initiator and a promoter, or two promoters can lead to additive, multiplicative and supramultiplicative carcinogenic responses in experimental settings [137]. At low exposure levels, such as those occurring in natural foods, the differences are lost such that the overall risk of a mixture becomes only additive [204]. Furthermore, the safety assessment of human dietary ingredients is an almost impossible task due to the presence of protective (e.g., anticarcinogens) agents in the same mixture, the interactive effects among themselves, and their combined antagonistic effects against the effects of the mixture of carcinogens and other toxicants also present. Testing various crude solvent extracts of each dietary ingredient or even selected total diets (composed of average daily per capita amounts of each of the common dietary ingredients in the U.S., for example) may reduce the amount of testing. Interactions between components of various extracts and between extractable and nonextractable components will still have to be estimated or further tested.

## FOODBORNE BIOTERRORISM

Attacks with biological agents are appealing to organizations with limited resources such as terrorist and radical groups that intend to scare the masses rather than inflict mass casualties. Biological and toxic agents can be grown inexpensively but are difficult to weaponize for aerosol dispersal, making large-scale bioterrorist attacks unlikely. Bioterrorism involving local or regional food and water supplies is a more practical alternative. The Centers for Disease Control and Prevention [238] groups biological warfare agents into:

- Category A—Easily disseminated or transmitted from person to person and capable of high mortality rates; has the greatest impact on public health systems and the civilian psyche. Examples include variola (smallpox), *Bacillus anthracis* (anthrax), *Yersinia pestis* (plague), *Clostridium botulinum* (botulism), *Francisella tularensis* (tularemia), and filo- and arenaviruses (viral hemorrhagic fever).
- Category B—Moderate dissemination and morbidity and lower mortality rates. Examples include *Coxiella burnetii* (Q fever), *Brucella* (brucellosis), *Burkholderia mallei* (glanders), *B. pseudomallei* (melioidosis), alphaviruses (encephalitis), *Rickettsia prowazekii* (typhus), toxins (toxicoses), *Chlamidia psittaci* (psittacosis), food safety threats (e.g., *Salmonella*, *Escherichia coli*), and water safety threats (e.g., *Vibrio*, *Cryptosporidium*).
- Category C—Emerging pathogens currently limited by availability and difficulty in production. Examples include Nipah virus (encephalitis), huntavirus (pulmonary syndrome), tickborne hemorrhagic fever virus, yellow fever, and multidrug-resistant tuberculosis.

Most biological agents are unstable in the environment. They are destroyed by public water treatment methods, boiling water, and cooking food, and they cause only short-term vomiting and diarrhea. Also, they would require large amounts to overcome dilution. Nevertheless, the many steps involved in centralized food processing and the rapid and wide distribution of foods still present a window of vulnerability for the intentional introduction of biological agents (organisms and toxins, most likely ones including botulinum toxin, *Salmonella*, *Shigella*, *Escherichia coli*, and *Vibrio cholerae*) into food products. Recent reviews of bioterrorism agents include those of Karwa et al. [124] and Meinhardt [184].

Compared to conventional chemical weapons, toxins are generally difficult to produce in large quantities. They are nonvolatile, more toxic by weight, dermally inactive, odorless, tasteless, immunogenic, and slow acting [124]. Among the toxins used as aerosols, agents that are highly toxic but difficult to produce may be more of a threat in a closed-space delivery system, while those that are stable and easily produced and delivered are likely to be used as open-air weapons. Some toxins are also effective when ingested, and others are dermally active. Food bioterrorism, in a manner similar to that of a nonterroristic foodborne disease, involves large numbers of people within a geographical area consuming the same contaminated food products and exhibiting signs characteristic of the agent involved within a short time frame (hours for toxins and up to 72 hours for microorganisms) after consumption. Following the September 11, 2001, terrorist attack on the World Trade Center in New York, under the authority of the Public Health Security and Bioterrorism Preparedness and Response Act passed in June 2002, the FDA developed four new regulations that address the registration of all (domestic and foreign) food facilities, prior notification of importation of food shipments, establishment and maintenance of records of receipts and shipments by all processors, and administrative detention of suspect food, in preparation for dealing with foodborne terrorism events. More recently, the FDA added Food Security Preventive Measures Guidance [293] by listing security and testing measures that ensure the physical and chemical safety of milk and food products.

Table 14.12 lists toxicants that could potentially be employed as foodborne or waterborne terrorism agents. Of these, the marine toxins (saxitoxin and tetrodotoxin) are difficult to produce and are considered only remote threats in bioterrorism. As discussed below, the botulinum toxin, staphylococcal enterotoxin B, ricin, and trichothecene mycotoxins have been stockpiled or allegedly used in warfare and terrorism in the past and are most likely to be used in future bioterrorism.

### BOTULINUM TOXIN

In addition to its well-known involvement as a foodborne toxic agent and its approved therapeutic and cosmetic uses, the botulinum toxin, produced by the bacterium *Clostridium botulinum*, was once stockpiled, experimented with, or used by the United States, Russia, Iraq, and the Aum Shinrikyo sect in Japan [124]. Its absorption via the lung also renders it a potential threat by aerosol dispersion (particle size 0.1 to 0.3 $\mu g$), the most efficient means of attack; 1 kg of toxin is capable of causing the death of 1.5 million people. Wein and Liu [310] recently

**TABLE 14.9**
**Adverse Effects of Common Dietary Supplements**

| Supplement | Uses | Mechanisms | Adverse Effects |
|---|---|---|---|
| Androstenedione, etc. | Increases muscle mass and strength; alternative to Viagra | Significant conversion to testosterone and estrogen by 17-β-hydroxysteroid dehydrogenase and aromatase only at >300 mg/day | Increased libido in females, decreased libido/feminization in males; occasional acute priapism; pancreatic and/or prostate cancer (?); reduced HDL |
| Chitosan | Controls obesity and hyperlipidemia | Binds to and prevents lipid absorption | Block absorption of lipid-soluble vitamins and micronutrients (?) |
| Chromium picolinate | Promotes insulin sensitivity and maintains glucose levels; burns fat; increases muscle mass | Cofactor for insulin; influences glucose, fat, and protein metabolism | Rhabdomyolysis, hepato- and nephrotoxicity, learning deficits in infants; carcinogen |
| Coenzyme Q10 (ubiquinone) | Improves heart, immune, and many cellular functions and lipid and energy status | Mitochondrial electron transport, antioxidant, free-radical scavenger | Abnormal renal and liver function |
| Colloidal silver | Immunostimulant and antiinflammatory used in infectious disease, cancer, diabetes, allergy, etc. | Unknown | Permanent skin discoloration (argyria); teratogen (ear, face, and neck anomalies) |
| Creatine monohydrate | Enhances strength and sprint activities; used in neuromuscular diseases and treatment of inborn errors of metabolism | Creatine gets phosphorylated and, in turn, donates it to ADP to aid ATP synthesis; improves energy efficiency | Isolated reports of glomerulosclerosis and interstitial nephritis |
| Dehydroepiandrosterone (DHEA) | Anabolic; hormone replacement; anti-aging; antineurodegenerative | Converted to many steroids in the body; has estrogenic and androgenic actions | Increased libido in females, decreased libido/feminization in males; hepatitis; reduced HDL |
| Dimethylglycine | Used as a performance enhancer and in autism; anticonvulsant; vaccine adjuvant | Unknown | Mutagenic (requires activation) (?) |
| Fish oil | Promotes cardiovascular, psychiatric, and joint health | Vasodilation by NO release; increases membrane fluidity; antiarrhythmic, antithrombotic, and antiinflammatory | Fish taste, diarrhea, hypervitaminosis A (myalgia, nausea, vomiting, pruritis, angular chelitis), bleeding tendency, high LDL, cancer |
| γ-OH butyric acid, γ-butyrolactone, and 1,4-butanediol | Date-rape drug (BD), euphoriant, sexual enhancer, sleep aid; anabolic | Neurotransmitter, dopaminergic, depressant; inhibits energy metabolism | Vomiting, hypothermia, hypernatremia; cardiac, respiratory, and CNS depressant |
| Germanium | Used to treat arthritis, cardiovascular diseases, cancer, depression, infections, cirrhosis, etc. | Unknown | Nausea, vomiting, nephro- and neurotoxicity, myelosuppression; teratogenic in animals (skeletal and eye defects) |

| | | | |
|---|---|---|---|
| Glucosamine and chondroitin | Improves cartilage and joint strength | Increases cartilage proteoglycan synthesis | Gastroenteritis, angioedema, insulin resistance, skin lesions |
| Huperzine | Prevents the onset and improves signs of dementias | Inhibits cerebral acetylcholinesterase, inhibits NMDA receptors and thus glutamate-mediated neurodegeneration | Nausea, vomiting, diarrhea, hyperactivity, dizziness |
| Hydrazine sulfate | Used to treat anorexia of cancer and cachexia | Inhibits phosphoenolpyruvate carboxykinase; involved in gluconeogenesis by cancer cells that leads to cachexia in cancer patients | Nausea, vomiting, hepatorenal damage, paresthesias; alters sperm morphology; teratogenic (bone and soft tissue defects; carcinogenic |
| 5-Hydroxytryptophan | Used to treat serotonin deficits; to promote weight loss; as a sleep aid; for pain relief; to balance mood; to treat depression and aggressiveness | Serotonergic action | Hypersalivation, nausea, vomiting, diarrhea, hypotension, seizures with hyperthermia |
| Melatonin | Sleep aid, jet-lag aid; aphrodisiac | Controls sleep and circadian rhythm; immunostimulant; antioxidant; free-radical scavenger | GI signs; cardiac and CNS stimulant; sleep disturbances; allergic reaction; headache |
| Methylsulfonymethane | Antiinflammatory and other uses | Donates sulfur for cysteine and methionine synthesis | Nausea, diarrhea, headache |
| Pyruvate | Inhibits fat storage; promotes fat loss and endurance | Increases energy expenditure, reduces lipid synthesis | Diarrhea, gastric and intestinal gas; possible sodium overload |
| Red yeast rice | Reduce total and LDL cholesterol | Inhibits HMG-CoA (similar to statin drugs) | Anaphylactoid reaction (respiratory); rhabdomyolysis, hepatotoxicity |
| *S*-Adenosylmethionine (SAM) | Psychiatric, liver, and joint health | Methyl donor for neurotransmitter, nucleic acid, phospholipids, and protein synthesis | GI complaints, mood elevation and mania, insomnia, hyperhomocystinemia |
| Shark cartilage | Cancer therapy aid | Inhibits angiogenesis (endothelial proliferation) | Nausea, vomiting, diarrhea, hyperglycemia, weakness, hepatotoxicity. |
| L-Tryptophan | Used to treat insomnia, seasonal depression, premenstrual mood swings, stress | Serotonin precursor | Dizziness, GI signs, insomnia palpitations, confusion |
| Vanadyl sulfate | Used to treat insulin-independent diabetes and hyperlipidemia; chemoprevention | Decreases plasma glucose levels by possibly increasing insulin sensitivity | Diarrhea, irregular respiration, ataxia, hind-limb paralysis |

**TABLE 14.10**
**Carcinogens and Potential Carcinogens in the Diet**

| Carcinogen/Mutagen | Major Foods Containing the Chemical |
|---|---|
| Alcohol | Grains and fruits |
| Allylisothiocyanate | Cabbage, collard greens, Brussels sprouts, mustard |
| Caffeic acid, caffeine, and theobromine | Coffee, cocoa, fruits, vegetables |
| Cyclopropene fatty acids | Cottonseed oil, kapok, okra |
| Fat (unsaturated and cholesterol-containing) | Vegetable and animal fats |
| Flavonoids (e.g., quercetin) | Vegetables, tea, coffee |
| Furocoumarins (psoralen) | Celery, figs, parsley, parsnips |
| Gossypol | Cottonseed oil |
| Hormones (e.g., estrogen, testosterone, progestins) | Meats as residues, supplements |
| Hydrazines (agaritine, gyromitrin) | Mushrooms |
| D-Limonene | Citrus juices |
| Methylazoxymethanol, cycasin | Cycads |
| Mycotoxins (aflatoxins, fumonisins, ochratoxin A, sterigmatocystin) | Corn, cottonseed, peanuts, wheat and other grains |
| Nitrosamines | Beets, celery, spinach, meat preserved in nitrite |
| Phorbol esters | Croton oil, other Euphorbaceae (herbal teas) |
| Polyphenols (tannic acid) | Beverages (tea, cider, cocoa, red wine), fruits |
| Ptaquiliside | Bracken fern |
| Pyrrolizidine alkaloids | Herbs, herbal teas, honey |
| Safrole, estragole, methyleugenol, piperine, etc. | Nutmeg, other spices, black pepper |
| Processed food carcinogens/mutagens | Processed/cooked (e.g., irradiated, fried, overheated) foods |
| Acrylamide | |
| Amino acid pyrrolysates (Trp-P-1, Trp-P-2, Glu-P-1, Glu-P-2) | |
| Carbolines | |
| Coffee-derived mutagens/carcinogens | |
| Fat oxidation products | |
| Imidazoquinolines and quinoxalines | |
| Maillard reaction products | |
| Polyaromatic hydrocarbons | |
| Radiolytic:2-alkylcyclobutanones (2-ACBs) | |

estimated, however, that less than 1 g of botulinum toxin introduced into the milk supply at some point during processing (between milking on the farm and bottling in the processing plant) can result in 100,000 casualties in the absence of testing or detection. Their findings suggest that direct contamination of the food supply may be even more dangerous, although more difficult. Other products such as juices and other beverages that are subject only to pasteurization temperatures before consumption are also candidates for such attack. Signs mainly include vision disturbances, dysphagia, and dysphonia early, followed by descending paralysis, hypotension, and respiratory failure as early as 24 hours after exposure. Treatment involves activated charcoal, respiratory support, and administration of the antitoxin. Early symptomatic detection can prevent up to two thirds of the casualties, whereas rapid ELISA testing to detect the toxin at each point in the sequence of events (e.g., between milking and bottling) prevents nearly all cases. Thus, employment of security measures and testing at each point of production, collection, processing, and transport of foods such as those proposed in the Food Security Preventive Measures Guidance for milk and other food products by the FDA [293] are the ultimate safeguards against terrorism involving food products.

## Ricin

Audi et al. [19] reviewed the bioterror potential of ricin, a lectin from castor bean (*Ricinus communis*). Its recent discovery at a South Carolina post office, a White House mail center, and a U.S. senator's office and its still unknown origin are of concern. Ricin is one of the most easily produced and potent toxins introduced through oral, inhalation, or parenteral exposure. It is highly lethal, especially when inhaled, which makes it highly attractive for bioterrorists. The most likely scenarios of ricin use include aerosol release into the environment or adulteration of food and beverages. Ingestion of ricin leads to nausea, vomiting, diarrhea, and abdominal pain beginning within 12 hours and progressing to hypotension, liver failure, renal dysfunction, and death due to multiple organ failur

**TABLE 14.11**
**Important Antimutagens and Anticarcinogens Naturally Occurring in Foods**

| Class/Subclass | Examples | Foods Containing Them |
|---|---|---|
| Alkaloids | Indole-3-carbinol, caffeine | Broccoli, cabbage, cauliflower, coffee |
| Amino acids | Cysteine and tryptophan, curcumin | Many plants and animals |
| Arylheptanoids | — | Turmeric |
| Benzenoids | Gingerol, paradol | Ginger root and related plants |
| Cyclitols | Myoinostol, phytic acid | Wheat, other cereals, nuts, and meats (?) |
| Estrogens | Sitosterol | Soybeans, alfalfa, etc. |
| Fatty acid derivatives | Conjugated linoleic and arachidonic acid | Vegetable oils |
| Fiber | Acid-soluble, neutral, etc. | Fruits and vegetables, cereal bran |
| Minerals | Se, $Ca^{2+}$ | Crops grown on Se-containing soils, milk, meat |
| Phenolics | | |
| Phenolic acids | Gallic and protocatechuic acids | Many fruits and vegetables |
| Phenyl propanoids | Caffeic, cinnamic, chlorogenic and ferrulic acids, enginol, myristicin | Broccoli, other vegetables |
| Flavones | Apigenin, myricetin, quercetin, robinetin, rutin | Fruits, herbs, and vegetables |
| Isoflavones | Biochanin A, genistein, daidzein, etc. | Soybeans and others |
| Polyphenols | | |
| Lignins | Sesamin | Sesame seed |
| Tannins | Ellagic and tannic acids, epigallo-catechin-gallate | Chinese green tea, other teas, cereals, legumes, and fruits |
| Protease inhibitors | Antipain, elastatinal | — |
| Porphyrins | Chlorophyll, chlorophyllin, cytochrome C, hemin, hemoglobin, myoglobin | Green leafy vegetables, meats |
| Sulfur-containing compounds | Benzyl isothiocyanate, cysteanine, diallyl sulfide and disulfide, glutathione, isothiocyanate, phenethyl, sinigrin, sulforaphane | Broccoli, cabbage, cauliflower, and others |
| Terpenoids | | |
| Monoterpenes | Carveol, limonene, menthols | Citrus fruits, grapes, mint, other plants, wine |
| Diterpenes | Cafestol, kahweol | Coffee, variety of plants, sponges, corals, etc. |
| Triterpenes | Glycerrhetinic acid, its glycoside, limonin, oleanolic acid, and ursolic acid | Citrus fruits, and a variety of medicinal plants |
| Sesquiterpenes | Nerolidol | Medicinal plants and herbs |
| Unidentified | Unknown | Beef, cabbage, germinating wheat, mushrooms, etc. |
| Vitamins | | |
| Carotenoids | Canthaxanthin, β-carotene, fucoxanthin | Fresh green leafy vegetables |
| Others | Vitamins A, C, E, and riboflavin | Fruits and vegetables (fresh), meats, fish |

r cardiovascular collapse. Inhalational exposure produces ough, dyspnea, arthralgias, and fever and may progress o respiratory distress and death, with few other organ ystem manifestations. Ricin analysis at federal laborato- ies and supportive measures are the only aids to diagnosis nd treatment.

## TAPHYLOCOCCAL ENTEROTOXIN B

n addition to the earlier discussion of enterotoxin as a oodborne toxicant, staphylococcal enterotoxin can be nass produced easily from cultures of *Staphylococcus ureus* and is stable as an aerosol. Inhalational exposure esults in its binding to the major histocompatibility com- lex that stimulates T-cells, leading to the massive release f cytokines. This produces signs associated with intersti- al pulmonary edema, including fever, myalgia, cough, chest tightness, dyspnea, headache, and vomiting [124], as well as signs of toxic shock syndrome (hypotension, shock). The low mortality and large amounts of toxin required to produce effects make this toxin less desirable compared to others.

## TRICHOTHECENE TOXINS

The fungal toxins (e.g., aflatoxin $B_1$, fumonisin $B_1$, ochratoxin A, and trichothecenes) have the ability to induce effects immediately upon contact and to produce lethality at levels of only a few milligrams; therefore, T-2 toxin and other trichothecenes have been scrutinized more closely for use as the ideal biologic warfare agent (i.e., one that is lethal, easy and inexpensive to produce, and stable for aerosol dispersal over wide areas; has no antidote or vaccine; and can spread from person to person) [124,275].

**TABLE 14.12**
**Toxins with Potential for Use in Foodborne Terrorism**

| Toxin Type | Examples | Source | Syndrome |
|---|---|---|---|
| Bacterial | Botulinum toxins | *C. botulinum* | *Inhalation:* neurologic (descending paralysis) |
| | *Clostridium perfringens* toxins | *C. perfringens* | *Ingestion:* gastroenteritis |
| | Staphylococcal enterotoxin B (SEB) | *S. aureus* | *Ingestion:* gastroenteritis<br>*Inhalation:* toxic shock and pulmonary edema |
| Fungal | Aflatoxin | *Aspergillus flavus* | *Inhalation:* pulmonary edema and hemorrhage<br>*Ingestion:* hemorrhagic gastroenteritis |
| | T-2 toxin | *Fusarium* spp. | *Dermal:* blistering<br>*Inhalation:* tracheobronchitis and hemoptysis |
| Algal | Anatoxin A | Blue-green algae | *Ingestion:* paralysis |
| | Microcyctin | Blue-green algae | *Ingestion:* paralysis and liver damage |
| Marine | Saxitoxin | Dinoflagellate | *Ingestion:* neurotoxin |
| | Tetrodotoxin | Pufferfish | *Ingestion:* neurotoxin |
| Plant | Ricin | Castor bean | *Inhalation:* respiratory distress |
| | Abrin | Precatory bean | *Ingestion:* gastroenteritis and shock, similar to ricin |

*Source:* Data from Meinhardt [184] and Karwa et al. [124].

All allegations of their use—by the United States against North Korea and China in 1952 and by the Soviet Union to attack Hmong tribesmen in Laos and Kampuchia (as "yellow rain") in 1981 and later in Afghanistan—have either remained unsubstantiated or have been disproved. Victims in the "yellow rain" incident appear to have exhibited signs similar to those expected from trichothecene intoxication (blistering of the skin, corneal injury, wheezing, cough, tracheobronchitis, and hemoptysis), and leaf samples from the area contained traces of trichothecenes. Subsequently, these allegations were negated by reports that the material in the so-called "yellow rain" is likely a mass defecation by swarms of Asian honeybees and the trace levels of trichothecenes likely reflected natural production in this area. The U.S. military still considers these agents as serious bioweapons, as evidenced by the clearance of a reactive skin decontamination lotion by the FDA in 2003 for use by the military to remove or neutralize chemical warfare agents and T-2 fungal toxin from the skin. Additional information on the history of bioweapon use, the mechanisms of action, and signs of intoxication of trichothecenes can be found in Karwa et al. [124], Stark [275], and earlier discussion in this chapter.

### *Vibrio cholerae*

Among the agents causing diarrhea by secreting enterotoxins, *Vibrio cholerae* causes the most severe seasonal disease epidemics, mostly in Asia, Africa, and Latin America. Its main mode of transmission is via contaminated water and food—the same media that could be used by terrorists [184] and has apparently been used by the Japanese in World War II in China [124]. Serogroup O1 (which has two main serotypes, Inaba and Ogawa, and two biotypes, classica and El Tor) and the most recently identified serogroup, 0139 Bengal, cause the most severe disease. The cholera toxin (CT) is similar to the heat-labile enterotoxin secreted by *Escherichia coli* and causes diarrhea by the same mechanism: excessive net secretion of electrolytes and water from the upper fifth of the small intestine. The review by Sanchez and Holmgren [246] provides details on the factors that trigger and mechanisms involved in the secretion of CT, the mode of action of CT, and the immunology of the disease. Interestingly, the presence of lytic cholera phages in environmental waters appears to reduce the presence of *V. cholerae*, thus providing a mechanism whereby the emergence and duration of cholera epidemics can be naturally controlled. Recently developed mixed serotype and biotype inactivated *V. cholerae* O1 and attenuated classical *V. cholerae* O1 Inaba vaccines appear to be superior to the earlier vaccines due to improved local (gut) immunity; however the immunity lasts only up to 6 months.

For details on disease caused by enterotoxigenic *Escherichia coli* and *Shigella* species, please refer to the discussion on foodborne bacterial diseases earlier in this chapter.

### Conventional Chemical Weapons

Although the contamination of foods with conventional chemical weapons—including nerve agents (e.g., tabun, sarin, soman, and especially VX), cyanide, incapacitating agents (BZ and Agent 15), vesicants (e.g., mustards, phosgene, lewisite), and choking agents (e.g., phosgene, chlorine, bromine)—and with radionuclides is possible, the volatile nature of the former (with the exception of VX

**TABLE 14.13**
**Measures To Minimize Consumption of Foodborne Toxicants/Carcinogens**

| |
|---|
| Purchase foods from reputable retailers/producers. |
| Carefully check and avoid obviously spoiled foods. |
| Consume foods raw, fresh, or boiled; minimize cooking over open flame or frying. |
| Be sure food handlers and consumers are aware of proper hygiene during the handling of foods. |
| Properly process foods before consumption: evisceration of fugu fish, proper extraction of cassava, washing to eliminate external contamination. |
| Avoid direct contact of food with wood- or fossil-fuel-derived smoke. |
| Avoid direct contact of food with the flame during cooking. |
| Cook bacon, etc., in the microwave instead of griddle or frying pan; add antioxidants to frying oil and other foods to minimize formation of oxidation products. |
| Reduce nitrate and nitrite levels and add ascorbic acid before curing meat products. |
| Avoid overcooking, cooking at a high temperature, or a large weight loss of cooked product. |
| Avoid cross-contamination of cooked foods with raw foods. |
| Avoid leaving cooked food at room temperature for prolonged periods; refrigerate leftovers promptly after each meal. |
| Include in the daily diet food items containing agents that minimize carcinogen activation (e.g., isoflavones in soy, berries, grapes). |
| Avoid chronic overindulgence on any single food type at the expense of others in a balanced meal. |
| Avoid consuming excessively hot food or drink. |
| Promote research on ways to minimize levels of harmful chemicals in foods (e.g., use of biotechnology). |

nd the restricted availability of the latter makes threats with such agents involving foods less likely. Examples of agents and other sign-specific treatments available to handle such emergencies include the FDA-approved ATNAA (atropine/pralidoxime) autoinjector to treat nerve gas intoxication; new dosage forms of AtroPen (atropine) autoinjectors for use in children and adolescents to deal with nerve agents; ThyroSafe (potassium iodide) tablets to protect the thyroid from general radiation exposures; Prussian blue to inhibit the absorption of radioactive cesium and thallium; and pentetate calcium trisodium (Ca-DTPA) and pentetate zinc sodium (Zn-DTPA) to increase the elimination of internal contamination with plutonium, americium, or curium, which are found in the fallout from nuclear detonation and waste from nuclear power plants [183].

## STRATEGIES TO DEAL WITH DIETARY HAZARDS

The best but not a perfect approach to a more realistic risk appraisal may involve testing pelleted, dehydrated, edible, whole products, such as meat, fruits, or vegetables, individually or in combination in animals at doses reflecting human intake. Such an approach not only reduces the cost of testing compared to individual chemical testing strategy but also provides data more relevant to natural exposure to a complex diet. At the same time, transgenic plants with improved nutritional quality, the capability to withstand processing, resistance to spoilage, lower levels of toxic compounds using antisense or other recombinant DNA technology to inactivate genes regulating biosynthesis and metabolism), and reduced susceptibility for fungal infestation should be developed and lower risk associated with consumption of such plants confirmed prior to extensive consumer use. A reduction in fungal susceptibility can be achieved by lowering the levels of micro- or macronutrients in the plant necessary for fungal growth or other appropriate techniques [205]. Success in risk reduction is also likely if preservation, processing, and storage methods to reduce the levels or the effects of natural toxicants in foods are developed and put in practice. Examples of such processes include methods to process cassava root to extract and neutralize cyanide; evisceration of fugu fish to eliminate tetrodotoxicosis; waxing, heating, dipping in corn oil, spraying with lecithin, and immersing potatoes in dilute detergents to reduce their glycoalkaloid content; and the prudent addition of antioxidants to reduce the formation of harmful agents such as lipid oxidation products and nitrosamines. Last, but not least, of the strategies in successfully dealing with natural dietary chemicals involves public education in avoiding and reducing exposure to natural dietary hazards. Table 14.13 lists recommendations to reduce the consumption of carcinogens and mutagens in the diet.

## CONCLUSION

Although much progress has occurred in our understanding of the identification and management of foodborne hazards, large gaps exist in our knowledge in the areas of mechanisms of pathogenesis of known human intoxications associated with foods; interactions between multiple toxicants present simultaneously, between toxicants and nutritional components, and between toxicants and antitoxicants (including antimutagens and anticarcinogens) in foods; methods of extrapolating realistic human health risks from animal data; and the development of safer plant varieties and processing and cooking methodologies that minimize toxic hazards to consumers. Because natural dietary toxicants are at least as toxic as the synthetic ones and their exposure is greater in quantity and consistency than the synthetic toxicants, U.S. and worldwide research resources should be shifted toward achieving a realistic balance in the study of health hazards, more toward natural dietary components. Current testing of purified individual food toxicants in animals is inadequate and must be replaced by feeding realistic levels of such compounds in the complex milieu of the product in which the toxicant is naturally present (smoked meats, for example) or even perhaps in the total human diet. Realistically speaking, although this task is impossible because of the vastly variable composition of individual food ingredients as well as that of total human diet, we can edge closer to this goal by designing a diet containing various dietary ingredients (vegetables, fruits, grain, dairy products, and meats) at a level equal to percentages of their average per capita human consumption. Education of the consumers to minimize dietary risks using practicable methods similar to those cited above and to shatter the myths that "natural is healthy" and "manmade or synthetic is toxic" needs to be vigorously pursued. An educated populace is less likely to be unduly alarmed and is more likely to accept prudent regulatory actions resulting from realistic scenarios of risk estimation. Finally, the application of newer molecular methodologies (such as PCR) to confirm intoxication from bacterial and other biotoxins and intensified activities of national animal health monitoring system combined with more rigorous application of HACCP will lead to significant reduction in currently widespread incidence of microbial diseases from food sources.

## QUESTIONS

1. Design a Hazard Analysis Critical Control Point system for a foodborne microbial agent of your choice. Can this system, in principle, be used to control other toxic (both natural and synthetic) hazards? If not, is it possible to modify the HACCP system or to develop a similar system to suit such needs?
2. Using the information in this chapter and other available resources (including your imagination), propose a strategy to test natural dietary toxicants in a way that allows more realistic extrapolation to humans than is currently used.
3. Do synthetic chemicals or biotechnology products pose a greater hazard than natural foodborne chemicals in your informed opinion? If so, are we doing all we can to keep them out of our food supply? Is there more that can be done to achieve this goal? If not, how can we use our resources in the right context of food safety?
4. Most foodborne toxicants have both beneficial and adverse effects. Using all available information, calculate the average daily consumption for a toxicant or group that is safe and yet allows most beneficial effects.
5. Study the methodology required to weaponize a bioterrorism agent, and determine which of the currently known toxic agents is ideal for this purpose. Describe the effects of this toxic agent on a regional scale and ways to control and prevent the effects of such attacks.

## REFERENCES


1. Abado-Becognee, K. et al. (1998): Cytotoxicity of fumonisin B1: implication of lipid peroxidation and inhibition on protein and DNA syntheses. *Arch. Toxicol.* 72(4):233–236.
2. Abuye, C., Kelbassa, U., and Wolde-Gebriel, S. (1998): Health effects of cassava consumption in South Ethiopia. *East Afr. Med. J.*, 75:166–170.
3. Adams, H. R. (1989): Phytoestrogens. In: *Toxicants of Plant Origin*. Vol. IV. Phenolics, edited by P. R. Cheeke. CRC Press, Boca Raton, FL, pp. 23–51.
4. Adlercreutz, H. (1995): Phytoestrogens: epidemiology and a possible role in cancer protection. *Environ. Health Perspect.*, 103(Suppl. 7):103–112.
5. Ahmed, F. E. and Thomas, D. B. (1992): Assessment of the carcinogenicity of the non-nutritive sweetener cyclamate. *Crit. Rev. Toxicol*, 22:81–118.
6. Aldridge, D. and Tahourdin, C. (1998): Natural oestrogenic compounds. In: *Natural Toxicants in Foods*, edited by D. H. Watson. CRC Press, Boca Raton, FL, pp. 54–83.
7. Alston, T. A., Mela, L., and Bright, H. G. (1977): 3-Nitropropionate, the toxic substance of Indigofera, is a suicide activator of succinate dehydrogenase. *Proc. Natl. Acad. Sci. (USA)*, 74:3767–3771.
8. Altekruse, S. F., Swerdlow, D. L., and Wells, S. J. (1998): Factors in the emergence of foodborne diseases. *Vet. Clin. N. Am. Food Anim. Pract.*, 14:1–15.
9. Altekruse, S. F., Swerdlow, D. L., and Stern, N. J. (1998): *Campylobacter jejuni*. *Vet. Clin. N. Am. Food Anim. Pract.* 14:31–40.
10. Ames, B. N. (1983): Dietary carcinogens and anticarcinogens: oxygen radicals and degenerative diseases. *Science* 221:1256–1264.
11. Ames, B. N. and Gold, L. S. (1990): Too many rodent carcinogens: mitogenesis increases mutagenesis. *Perspec. Sci.*, 249:970–971.
12. Ames, B. N., Profet, M., and Gold, L.S. (1990): Nature's chemicals and synthetic chemicals: comparative toxicology. *Proc. Natl. Acad. Sci. (USA)*, 87:7782–7786.
13. Anderson, J. A. (1997): Milk, eggs, and peanuts: food allergies in children. *Am. Fam. Phys.*, 56:1365–1374.
14. Andrews, L. S., Ahmedna, M., Grodner, R. M., Liuzzo, J. A., Murano, P. S., Murano, E. A., Rao, R. M., Shane, S. and Wilson, P. M. (1998): Food preservation using ionizing radiation. *Rev. Environ. Contam. Toxicol.*, 154:1–53.

15. Angus, F. (1998): Nut allergens. In: *Natural Toxicants in Food*, edited by D. H. Watson. CRC Press, Boca Raton, FL, pp. 84–104.
16. Aronow, L. and Kerdel-Vegas, F. (1965): Selino-cystathionine, a pharmacologically active factor in the seeds of *Lecythis ollaria*: cytotoxic and depilatory effects. *Nature (Lond.)*, 205:1185–1186.
17. Ascherio, A. et al. (2001): Prospective study of caffeine consumption and risk of Parkinson's disease in men and women, *Ann. Neurol.*, 50:56–63.
18. Astwood, J. D. et al. (2003): Food biotechnology and genetic engineering. In: *Food Allergy*, 3rd ed., edited by A. Metcalfe, A. Sampson, and A. Simon. Blackwell Scientific, Oxford, pp. 51–70.
19. Audi, J., Belson, M., Patel, M., Schier, J., and Osterloh, J. (2005): Ricin poisoning: a comprehensive review. *JAMA*, 294:2342–2351.
20. Bailey, G., Goeger, D., Hendricks, J., Nixon, J., and Pawlowski, N. (1985): Indole-3-carbinol promotion and inhibition of aflatoxin $B_1$ carcinogenesis in rainbow trout [abstract]. *Proc. Am. Assoc. Cancer Res.*, 26:115.
21. Balasubramanian, S. and Govindaswamy, S. (1996): Inhibitory effect of dietary flavonol, quercetin, on 7,12-cimethylbenzanthracine-induced hamster baccal pouch carcinogenesis. *Carcinogenesis*, 17:877–879.
22. Baldessarini, R. J. (1985): Drugs and the treatment of psychiatric disorders. In: *Goodman and Gilman's The Pharmacological Basis of Therapeutics*, 7th ed., edited by A. G. Gilman, L. Goodman, T. W. Rall, and F. Murad. Macmillan, New York, pp. 387–445.
23. Banea-Muyambu, J. P., Tylleskar, T., Gitebo, N., Mtadi, N., Gebre-Medhim, M., and Rosling, A. (1997): Geographical and seasonal association between linamarin and cyanide exposure from cassava and the upper motor neuron disease Konzo in former Zaire. *Trop. Med. Int. Health*, 2:1143–1151.
24. Barotto, N. N., Lopez, C. B., Eynard, A. R., Fernandez-Zapico, M. D., and Valentich, M. A. (1998): Quercetin enhances pre-tumorous lesions in the NMU model of rat pancreatic carcinogenesis. *Cancer Lett.*, 129:1–6.
25. Bartholomew, R. M. and Ryan, D. A. (1980): Lack of mutagenicity of some phytoestrogens in the *Salmonella*/mammalian microsome assay. *Mutat. Res.*, 78: 317–321.
26. Barzel, U. S. and Massey, L. K. (1998): Excess dietary protein can adversely affect bones. *J. Nutr.*, 128:1051–1053.
27. Beasley, V. R. (1989): *Trichothecene Mycotoxicosis: Pathophysiologic Effects*, Vols. 1 and 2. CRC Press, Boca Raton, FL.
28. Beck, V., Rohr, U., and Jungbauer, A. (2005): Phytoestrogens derived from red clover: an alternative to estrogen replacement therapy? *J. Steroid Biochem. Mol. Biol.*, 94:499–518.
29. Beier, R. C. (1990): Natural pesticides and bioactive components in foods. *Rev. Environ. Contam. Toxicol.*, 113:47–137.
30. Bennett, J. W. and Klich, M.( 2003): Mycotoxins. *Clin. Microbiol. Rev.*, 16:497–516.
31. Berger, K. J. and Guss, D. A. (2005): Mycotoxins revisited, Part I. *J. Emerg. Med.*, 28:53–62.
32. Berger, K. J. and Guss, D. A. (2005): Mycotoxins revisited, Part II. *J. Emerg. Med.*, 28:175–183.
33. Bosetti, C. et al. (2005): Flavonoids and breast cancer risk in Italy. *Cancer Epidemiol. Biomark. Prev.*, 14:805–808.
34. Breiteneder, H. and Radaur, C. (2004): A classification of plant food allergens. *J. Allergy Clin. Immunol.*, 113:821–830.
35. Brett, M. M. (2003): Food poisoning associated with biotoxins in fish and shellfish. *Curr. Opin. Infect. Dis.*, 16:461–465.
36. Bruggink, T. (1997): Food allergy and food intolerance. In: *Food Safety and Toxicity*, edited by J. DeVries. CRC Press, Boca Raton, FL, pp. 183–194.
37. Bryan, F. L. (1979): Infections and intoxications caused by other bacteria. In: *Foodborne Infections and Intoxications*, 2nd ed., edited by H. Riemann and F. L. Bryan. Academic Press, New York, pp. 212–298.
38. Bureau of Foods (1982): *Toxicological Principles for the Safety Assessment of Direct Food Additives and Color Additives Used in Food*. U.S. Food and Drug Administration, Washington, D.C.
39. Busby, Jr., W. F. and Wogan, G. N. (1979): Foodborne mycotoxins and alimentary mycotoxicoses. In: *Foodborne Infections and Intoxications*, 2nd ed., edited by H. Riemann and F. L. Bryan. Academic Press, New York, pp. 519–610.
40. Busby, Jr., W. F. and Wogan, G. N. (1981): Aflatoxins. In: *Mycotoxins and Nitroso Compounds: Environmental Risks*, Vol. 2, edited by R. C. Shank. CRC Press, Boca Raton, FL, pp. 3–28.
41. Busby, Jr., W. F. and Wogan, G. N. (1981): Psoralens. *Mycotoxins and Nitroso Compounds: Environmental Risks*, Vol. 2, edited by R. C. Shank. CRC Press, Boca Raton, FL, pp. 105–119.
42. Carrol, K. K. (1982): Dietary fat and its relationship to human cancer. In: *Carcinogens and Mutagens in the Environment*, Vol. 1, edited by H. F. Stich. CRC Press, Boca Raton, FL, pp. 31–38.
43. Chang, F. C. T., Spriggs, D. L., Benton, B. J., Ketter, S. J., and Capucio, B. R. (1997): 4-Aminopyridine reverses saxitoxin and tetrodotoxin-induced cardiorespiratory depression in chronically instrumented guinea pigs. *Fundam. Appl. Toxicol.*, 38:75–88.
44. Cheeke, P. R. (1996): Biological effects of feed and forage saponins and their impacts on animal production. *Adv. Exp. Med. Biol.*, 405:377–385.
45. Cheeke, P. R. (1989): Toxicants of plant origin. In: *Phenolics*, Vols. 1–4. CRC Press, Boca Raton, FL.
46. Chen, K. K. and Rose, C. L. (1952): Nitrite and thiosulfate therapy in cyanide poisoning. *JAMA*, 149:113–119.
47. Chen, Y. W., Huang, S. C., Lin-Shiau, S. Y., and Lin, J. K. (2005): Bowman–Birk inhibitor abates proteasome function and suppresses the proliferation of MCF7 breast cancer cells through accumulation of MAP kinase phosphatase-1. *Carcinogen*, 26:1296–1306.
48. Chevion, M., Mager, J., and Claser, G. (1983): Favism producing agents. In: *Handbook of Naturally Occurring Food Toxicants*, edited by M. Rechcigl, Jr. CRC Press, Boca Raton, FL, pp. 63–79.
49. Chow, C. W., Poulos, A., Fellenberg, A. J., Christodoulon, J., and Danks, D. M. (1992): Autopsy findings in two siblings with infantile Refsum's disease. *Acta Neuropathol.*, 83:190–195.

50. Chung, F. L., Wang, M., Rivenson, A., Iatropoulous, M. J., Reinhardt, J. C., Pittman, B., Ho, C. T., and Amin, S. G. (1998): Inhibition of lung carcinogenesis by black tea in Fischer rats treated with a tobacco-specific carcinogen: caffeine as an important constituent. *Cancer Res.*, 58(18):4096–4101.
51. Clark, B. and McKendrick, M. (2004): A review of viral gastroenteritis. *Curr. Opin. Infect. Dis.*, 17:461–469.
52. Clarkson, T. B. (1995): Estrogenic soybean isoflavones and chronic disease. *Trends Endocr. Metab.*, 6:11–16.
53. Cooper, J. and Walker, R. D. (1998): Listeriosis. *Vet. Clin. N. Am. Food Anim. Pract.*, 14:113–125.
54. Cordain, L., Miller, J. B., Eaton, S. B., Mann, N., Holz, H., and Speth, J. D. (2000): Plant–animal substance ratios and macronutrient energy estimations in world wide hunter–gatherer diets. *Am. J. Clin. Nutr.*, 32:741–749.
55. Coulombe, Jr., R. A. (1991): Aflatoxins. In: *Mycotoxins and Phytoalexins*, edited by R. P. Sharma and D. K. Salunkhe. CRC Press, Boca Raton, FL, pp. 103–143.
56. Cox, P. A. et al. (2005): Diverse taxa of cyanobacteria produce beta-*N*-methylamino-L-alanine, a neurotoxic amino acid. *Proc. Natl. Acad. Sci. USA*, 10:5074–5078.
57. Creppy, E. E., Baudrimont, I., and Betbeder, A. M. (1995): Prevention of nephrotoxicity of ochratoxin A, a food contaminant. *Toxicol. Lett.*, 82–83:869–877.
58. Crews, C. (1998): Pyrrolizidine alkaloids. In: *Natural Toxicants in Food*, edited by D. H. Watson. CRC Press, Boca Raton, FL, pp. 11–28.
59. Cupp M.J. (2000): *Toxicology and Clinical Pharmacology of Herbal Products*. Humana Press, Totowa, NJ, pp. 1–302.
60. Cupp, M. J. and Tracy, T. S. (2003): *Dietary Supplements: Toxicology and Clinical Pharmacology*, Humana Press, Totowa, NJ, pp. 1–399.
61. Daly, J. W. (1993): Mechanism of action of caffeine. In: *Coffee, Caffeine, and Health*, edited by S. Garattini. Raven Press, New York, pp. 97–150.
62. DeClerk, Y. A. and Inven, S. (1994): Protease inhibitors: role and potential therapeutic use in human cancer. *Eur. J. Cancer*, 30A:2170–2180.
63. De Haan, L. and Hirst, T. R. (2004): Cholera toxin: a paradigm for multi-functional engagement of cellular mechanisms. *Mol. Memb. Biol.*, 21:77–92.
64. Dencker, L., Gustafson, A. L., Annerwall, E., Busch, C., and Eriksson, U. (1991): Retinoid-binding proteins in craniofacial development. *J. Craniofac. Genet. Dev. Biol.*, 11(4):303–314.
65. Derks, H. J. G. M., Groen, C., Olling, M., and Zeilmaker, M. J. (1997): Extrapolation of toxicity data in risk assessment. In: *Food Safety and Toxicity*, edited by J. DeVries. CRC Press, Boca Raton, FL, pp. 241–254.
66. Deshpande, S. S., Sathe, S. K., and Salunkhe, D. K. (1984): Chemistry and safety of plant polyphenols. In: *Nutritional and Toxicological Aspects of Food Safety*, edited by M. Friedman. Plenum Press, New York, pp. 457–495.
67. Deshpande, S. S. and Sathe, S. K. (1991): Toxicants in plants. In: *Mycotoxins and Phytoalexins*, edited by R. P. Sharma and D. K. Salunkhe. CRC Press, Boca Raton, FL, pp. 671–730.
68. Deshpande, S. S. (2002): *Handbook of Food Toxicology*. Marcel Dekker, New York, pp. 1–880.
69. DiPetrillo, K., Wood, S., Kostrubsky, V., Chatfield, K., Bement, J., Wrighton, S., Jeffery, E., Sinclair, P., and Sinclair, J. (2002): Effect of caffeine on acetaminophen hepatotoxicity in cultured hepatocytes treated with ethanol and isopentanol. *Toxicol. Appl. Pharmacol.*, 185:91–97.
70. Doll, R. and Peto, R. (1981): The causes of cancer: quantitative estimates of avoidable risks of cancer in the United States today. *J. Natl. Canc. Inst.*, 66:1193–1308.
71. Duffy, J. Y., Baines, D., Overmann, G. J., Keen, C. L., and Daston, G. P. (1997): Repeated administration of alpha-hederin results in alterations in maternal zinc status and adverse developmental outcome in the rat. *Teratology* 56(5):327–334.
72. Duggan, A., Charnley, G., Chen, W., Chukwudebe, A. Hawk, R., Krieger, R. I., Ross, J., and Yarborough, C (2003): Di-alkyl phosphate biomonitoring data: assessing cumulative exposure to organophosphate pesticides. *Reg Toxicol. Pharmacol.*, 37:382–395.
73. Dugyala, R. R., Sharma, R. P., Tsunoda, M., and Riley, R. T (1998): Tumor necrosis factor-alpha as a contributor in fumonisin $B_1$ toxicity. *J. Pharmacol. Exp. Ther.*, 285: 317–324.
74. Dunnick, J. K. and Hailey, J. R. (1992): Toxicity and carcinogeniality studies of quercetin, a natural component of foods. *Fundam. Appl. Toxicol.*, 19:423–431.
75. Dutton, M. F. (1996): Fumonisins, mycotoxins of increasing importance: their nature and their effects. *Pharmacol Ther.*, 70:137–161.
76. Ellenhorn, M. J. and Barceloux, D. G. (1988): *Medical Toxicology: Diagnosis and Treatment of Human Poisoning* pp. 508–514, 606–613, Elsevier, New York.
77. Ellenhorn, M. J., Schonwald, S., Ordog, G., and Wasserberger, J. (1997): Natural toxins: plants, mycotoxins mushrooms. In: *Ellenhorn's Medical Toxicology*, edited by M. J. Ellenhorn, S. Schonwald, G. Ordog, and J. Wasserberger. Williams & Wilkins, Philadelphia, PA, pp 1880–1896.
78. Ekperigen, H. E. and Nagaraja, K. V. (1998): *Salmonella Vet. Clin. N. Am. Food Anim. Pract.*, 14:17–29.
79. FAO/WHO. (2005): Evaluation of certain food additives *World Health Org. Tech. Rep. Ser.*, 928:1–156.
80. FDA. (1976): Federal Food, Drug, and Cosmetic Act, as Amended. U.S. Food and Drug Administration, Washington, D.C.
81. Fenaux, P. and DeBotton, S. (1998): Retinoic acid syndrome: recognition, prevention, and management. *Drug Safety*, 18(4):273–279.
82. Formica, J. V. and Regelson, W. (1995): Review of the biology of quercetin and related bioflavonoids. *Food Chem. Toxicol.*, 33:1061–1080.
83. Friedman, M., Gumbmann, M. R., and Masters, P. M (1984): Protein-alkali reactions: chemistry, toxicology, and nutritional consequences. In: *Nutritional and Toxicological Aspects of Food Safety*, edited by M. Friedman. Plenum Press, New York, pp. 367–412.
84. Friedman, M. and Brandon, D. L. (2001): Nutritional and health benefits of soy proteins. *J. Agric. Food Chem* 49:1069–1086.
85. Friedman, M., Henika, P. R., Mackey, B. E. (2003): Effect of feeding solanidine, solasodine and tomatidine to non-pregnant and pregnant mice. *Food Chem. Toxicol.*, 41:61–71.

86. Fuhrman, F. A. (1983): Toxic constituents of animal foodstuffs: eggs of fishes and amphibians. In: *Handbook of Naturally Occurring Food Toxicants*, edited by M. Rechcigl, Jr. CRC Press, Boca Raton, FL, pp. 301–311.
87. Fujiki, H. et al. (1992): Anticarcinogenic effects of penta-*O*-galloyl-beta-D-glucose and epigallocatechin gallate. *Prev. Med.*, 21:503–509.
88. Fukuda, H., Shima, H., Vesonder, R. F., Tokuda, H., Nishino, H., Katoh, S., Tamura, S., Sugimura, T., and Nagao, M. (1996): Inhibition of protein serine/threonine phosphatases by fumonisin $B_1$, a mycotoxin. *Biochem. Biophys. Res. Commun.*, 220(1):160–165.
89. Fukuda, T., Ito, H., Mukainaka, T., Tokuda, H., Nishino, H., and Yoshida, T. (2003): Anti-tumor promoting effect of glycosides from *Prunus persica* seeds. *Biol. Pharmaceut. Bull.*, 26:271–273
90. Gallaher, D. and Schneeman, B. O. (1984): Nutritional and metabolic response to plant inhibitors of digestive enzymes. *Adv. Exp. Med. Biol.*, 177:299–320.
91. Gardiner, D. M., Waring, P., Howlett, B. J. (2005): The epipolythiodioxopiperazine (ETP) class of fungal toxins: distribution, mode of action, functions and biosynthesis. *Microbiology*, 151:1021–1032.
92. Garlick, P. J. (2004): The nature of human hazards associated with excessive intake of amino acids. *J. Nutr.*, 134(6 Suppl.):1633S–1639S; discussion, 1664S–1672S.
93. Gentry, P. A. (1986): Comparative biochemical changes associated with mycotoxicosis other than aflatoxicosis and trichothecene toxicosis. In: *Diagnosis of Mycotoxicoses*, edited by J. R. Richard and J. R. Thurston. Martinus Nijhoff, Dordrecht, Netherlands, pp. 125–139.
94. Ghedira-Chekir, L. et al. (1998): Induction of a SOS repair system in lysogenic bacteria by zearalenone and its prevention by vitamin E. *Chemico-Biol. Interact.*, 113:15–25.
95. Gilani, G. S., Cockell, K. A., and Sepehr, E. (2005): Effects of antinutritional factors on protein digestibility and amino acid availability in foods. *J. AOAC Int.*, 88:967–987.
96. Godon, K. A. H. and Houstead, J. (1998): Transmissible spongiform encephalopathies in food animals, *Vet. Clin. N. Am. Food Anim. Pract.*, 14(1):49–70.
97. Gorniak, S. L. et al. (2003) Assessment of the perinatal effects of maternal ingestion of *Solanum malacoxylon* in rats. *Reprod. Toxicol.*, 17:67–72.
98. Gupta, R. S. and Dixit, V. P. (2002): Effects of short-term treatment of solasodine on cauda epididymis in dogs. *Indian J. Exp. Biol.*, 40:169–73.
99. Gylling, H. and Miettinen, T.A. (2005): The effect of plant stanol- and sterol-enriched foods on lipid metabolism, serum lipids and coronary heart disease. *Ann. Clin. Biochem.*, 42:254–263.
00. Halstead, B. W. (1978): *Poisonous and Venomous Marine Animals of the World*. Darwin Press, Princeton, NJ.
01. Haque, A. et al. (1997): Evidence of osteolathyrism among patients suffering from neurolathyrism in Bangladesh. *Natural Toxins*, 5(1):43–46.
02. Harper, A. E. (1973): Amino acids of nutritional importance. In: *Toxicants Occurring Naturally in Foods*, 2nd ed., edited by Committee on Food Protection, National Research Council. National Academy of Sciences Press, Washington, D.C., pp. 130–152.
103. Hayatsu, S., Arimoto, K., Togawa, K., and Mokita, M. (1981): Inhibitory effects of the ether extract of human feces on activities of mutagens: Inhibition of oleic and linoleic acids. *Mutat. Res.*, 81:287–293.
104. Hayes, J. R. and Campbell, T. C. (1986): Food additives and contaminants. In: *Casarett and Doull's Toxicology: The Basic Science of Poisons*, 3rd ed. edited by C. D. Klaassen, M. O. Amdur, and J. Doull. Macmillan, New York, pp. 771–800.
105. Heikkinen, A. M., Tuppurainen, M. T., Niskanen, L., Komulainen, M., Penttila, I., and Saarikoski, S. (1997): Long-term vitamin $D_3$ supplementation may have adverse effects on serum lipids during postmenopausal hormone replacement therapy. *Eur. J. Endocrinol.*, 137:495–502.
106. Hermanussen, M. and Tresguerres, J. A. (2003): Does the thrifty phenotype result from chronic glutamate intoxication? A hypothesis. *J. Perinat. Med.*, 31:489–495.
107. Hirono, I. (1987): *Naturally Occurring Carcinogens of Plant Origin: Toxicology, Pathology, and Biochemistry*, Elsevier, New York, pp. 1–227.
108. Hogue, A. T., White, P. L., and Heninover, J. A. (1998): Pathogen reduction and hazard analysis and critical control point (NACCP) systems for meat and poultry. *Vet. Clin. N. Am. Food Anim. Pract.*, 14(1):151–164.
109. Hong, H. L. et al. (2003): Chemical-specific alterations in ras, p53, and beta-catenin genes in hemangiosarcomas from B6C3F$_1$ mice exposed to *o*-nitrotoluene or riddelliine for 2 years, *Toxicol. Appl. Pharmacol.*, 191:227–34.
110. Houtsmuller, U. M. T., Struijk, C. B., and Van Der Beek, A. (1970): Decrease in rate of ATP synthesis of isolated rat heart mitochondria induced by dietary erucic acid. *Biochim. Biophys. Acta*, 218:564–566.
111. Howard, B. V. and Kritchevsky, D. (1996): Phytochemicals and cardiovascular disease: a statement for health care professionals from the American Heart Association. *Circulation*, 95:2591–1593.
112. Hsieh, D. P. H. (1979): Basic metabolic effects of mycotoxins. In: *Interactions of Mycotoxins in Animal Production*, National Academy of Sciences, Washington, D.C., pp. 43–55.
113. Hsieh, D. P. H. (1986): Genotoxicity of mycotoxins. In: *New Concepts and Developments in Toxicology*, edited by P. L. Chambers, P. Gebring, and F. Sakai. Elsevier, New York, pp. 251–259.
114. Huang, W. et al. (2005): Soy protein isolate increases hepatic thyroid hormone receptor content and inhibits its binding to target genes in rats. *J. Nutr.*, 135:1631–1635.
115. Huxtable, R. J. (1989): Human health implications of pyrrolizidine alkaloids and herbs containing them. In: *Toxicants of Plant Origin*. Vol. I. *Alkaloids*, edited by P. R. Cheeke. CRC Press, Boca Raton, FL, pp. 41–86.
116. Ikeguonu, F. I. and Bassir, O. (1977): Effects of phytohemagglutinins from immature legume seeds on the function and enzyme activities of the liver and on the organs of the rat. *Toxicol. Appl. Pharmacol.*, 40:217–226.
117. Jacobsen, B. K. and Bjelke, E. (1982): Coffee consumption and cancer: a prospective study [abstract]. In: *Proc. of the 13th Int. Cancer Congress*, Seattle, WA.
118. Janssen, M. M. T. (1997): Antinutritives; Food contaminants; Food additives; Nutrients. In: *Food Safety and Toxicity*, edited by J. De Vries. CRC Press, Boca Raton, FL, pp. 39–98.

119. Janssen, M. M. T., Put, H. M. T., and Nout, M. J. R. (1997): Natural toxins. In: *Food Safety and Toxicity*, edited by J. De Vries. CRC Press, Boca Raton, FL, pp. 7–38.
120. Jagerstad, M. and Skog, K. (2005): Genotoxicity of heat-processed foods. *Mutat. Res.*, 574:156–172.
121. Juliano, L. M. and Griffiths, R. R. (2004): A critical review of caffeine withdrawal: empirical validation of symptoms and signs, incidence, severity, and associated features. *Psychopharmacology*, 176:1–29.
122. Kalejta, R. F. and Hamlin, J. L. (1997): The dual effect of mimosine on DNA replication. *Exp. Cell. Res.*, 231:173–183.
123. Kanadaswami, C. et al. (2005): The antitumor activities of flavonoids. *In Vivo*, 19:895–909.
124. Karwa, M., Currie, B., and Kvetan, V. (2005): Bioterrorism: preparing for the impossible or the improbable. *Crit. Care Med.*, 33(Suppl.):S75–S95.
125. Kang, Y. J. and Alexander, J. M. (1996): Alterations of the glutathione redox cycle status in fumonisin $B_1$-treated pig kidney cells. *J. Biochem. Toxicol.*, 11:121–126.
126. Kao, C. Y. (1967): Comparison of the biological actions of tetrodotoxin and saxitoxin. In: *Animal Toxins*, edited by F. E. Russell and P. R. Saunders. Pergamon Press, Oxford, pp. 109–114.
127. Kao, C. Y. (1966): Tetrodotoxin, saxitoxin, and their significance in the study of excitation phenomena. *Pharmacol. Rev.*, 18:997–1049.
128. Kanisawa, M. and Suzuki, S. (1978): Induction of renal and hepatic tumors in mice by ochratoxin A, a mycotoxin. *Gann Jpn. J. Cancer Res.*, 69:599–600.
129. Kassell, B. (1970): Inhibitors of proteolytic enzymes. *Methods Enzymol.*, 19:839–906.
130. Keeler, R. F., Baker, D. C., and Gaffield, W. (1991): Solanum alkaloids. In: *Mycotoxins and Phytoalexins*, edited by R. P. Sharma and D. K. Salunkhe. CRC Press, Boca Raton, FL, pp. 607–636.
131. Kensler, T. W., Groopman, J. D., and Roebuck, B. D. (1998): Use of aflatoxin adducts as intermediate endpoints to assess the efficacy of chemopreventive interventions in animals and man. *Mutat. Res.*, 402:165–172.
132. Khoury, J. C. et al. (2004): Consequences of smoking and caffeine consumption during pregnancy in women with type 1 diabetes. *J. Maternal-Fetal Neonat. Med.*, 15:44–50.
133. King, T. P., Pusztai, A., and Clarke, E. M. W. (1980): Kidney bean lectin-induced lesions in rat small intestine. I. Light microscopic studies. *J. Comp. Pathol.*, 90:585–593.
134. Kitts, D., Yuan, Y., Joneja, J., Scott, F., Szilagyi, A., Amiot, J., and Zarkadas, M. (1997): Adverse reactions to food constituents: allergy, intolerance, and autoimmunity. *Can. J. Physiol. Pharmacol.*, 75:241–254.
135. Knasmuller, S., Bresgen, N., Kassie, F., Mersch-Sundermann, V., Gelderblom, W., Zohrer, E., and Eckl, P. M. (1997): Genotoxic effects of three fusarium mycotoxins, fumonisin $B_1$, moniliformin, and vomitoxin, in bacteria and in primary cultures of rat hepatocytes. *Mutat. Res.*, 391:39–48.
136. Kobayashi, M., Suzuki, K., Nagasawa, S., and Mimaki, Y. (1993): Purification of toxic saponins from narthecium asiaticum maxim. *J. Vet. Med. Sci.*, 55:401–407.
137. Kodell, R. L., Krewski, D., and Zielinski, J. M. (1991): Additive and multiplicative risks in the two-stage clonal expansion model of carcinogenesis. *Risk Anal.*, 11:483–490.
138. Korn, A., Wagner, B., Moritz E., and Dingemanse, J. (1996): Tyramine pressor sensitivity in healthy subjects during combined treatment with moclobemide and selegiline. *Eur. J. Clin. Pharmacol.*, 49:273–278.
139. Kowalska, M. T., Itzhak, Y., and Puett, D. (1995): Presence of aromatase inhibitors in cycads. *J. Ethnopharmacol.* 47:113–116.
140. Krogh, P., Hald, B., Plestina, R., and Ceovic, S. (1977): Balkan nephropathy and food-borne ochratoxin A: preliminary results of a survey of foodstuffs. *Acta Pathol. Microbiol. Scand. Sect. B* 85:238–240.
141. Kunkel, D. B. and Jallo, D. S. (1990): Ergot. In: *Clinical Management of Poisoning and Drug Overdose*, 2nd ed. edited by L. M. Haddad and J. F. Winchester. W.B. Saunders, Philadelphia, PA, pp. 1401–1406.
142. Kuramochi, G., Gekle, M., and Silbernagle, S. (1997): Derangement of pH homeostasis in the renal papilla: ochratoxin A increases pH in vasa recta blood. *Nephron* 76(4):472–476.
143. Lagman, R. and Walsh, D. ( 2003): Dangerous nutrition? Calcium, vitamin D, and shark cartilage nutritional supplements and cancer-related hypercalcemia. *Support Care Cancer*, 11:232–235.
144. Lamawansa, M. D., Wysocki, S. J., House, A. K., and Norman, P. E. (1996): Vitamin D3 exacerbates intimal hyperplasia in balloon-injured arteries. *Br. J. Surg.* 83:1101–1103.
145. Lambert, L. A., Hines, F. A., and Eppleyl, R. M. (1995): Lack of initiation and promotion potential of deoxynivalenol for skin. *Food Chem. Toxicol.*, 33:217–222.
146. Lampe, K. F. (1983): Mushroom poisoning. In: *Handbook of Naturally Occurring Food Toxicants*, edited by M. Rechcigl, Jr. CRC Press, Boca Raton, FL, pp. 193–212.
147. Lanigan, R. S. and Yamarik, T. A. (2002): Final report on the safety assessment of BHT. *Int. J. Toxicol.*, 21:19–94.
148. Lasky, T. and Magder, L. (1997): Hepatocellular carcinoma p53 G→T transversions at codon 249: the fingerprint of aflatoxin exposure? *Environ. Health Perspect.*, 105(4): 392–397.
149. Layton, D. W., Bogen, K. T., Knize, M. G., Hatch, F. T., Johnson, V. M., and Felton, J. S. (1995): Cancer risk of heterocyclic amines in cooked foods: an analysis and implications for research. *Carcinogenesis*, 16:39–52.
150. Leach CT. (2004): Hepatitis A in the United States. *Pediatr. Infect. Dis. J.*, 23(6):551–552.
151. Leftley, J. W. and Hannah, F. (1998): Phycotoxins in seafood. In: *Natural Toxicants in Food*, edited by D. H. Watson. CRC Press, Boca Raton, FL, pp. 182–224.
152. Lehrer, S. B. and Bannon, G. A. (2005): Risks of allergic reactions to biotech proteins in foods: perception and reality. *Allergy*, 60:559–564.
153. Leibowitz, S. F. (2005): Regulation and effects of hypothalamic galanin: relation to dietary fat, alcohol ingestion, circulating lipids and energy homeostasis. *Neuropeptides* 39:327–332.

154. Lephart, E. D., Setchell, K. D., Handa, R. J., and Lund, T. D. (2004): Behavioral effects of endocrine-disrupting substances: phytoestrogens. *ILAR J.*, 45:443–454.
155. Levine, J., Gussow, J. D., Hastings, D., and Eccher, A. (2003): Authors' financial relationships with the food and beverage industry and their published positions on the fat substitute olestra. *Am. J. Public Health*, 93:664–669.
156. Lewis, W. H. and Elvin-Lewis, M. P. F. (1977): *Medical Botany: Plants Affecting Human Health*. John Wiley & Sons, New York, p. 57.
157. Liener, I. E. (1980): Miscellaneous toxic factors. In: *Toxic Constituents of Plant Foodstuffs*, 2nd ed., edited by I. E. Liener. Academic Press, New York, pp. 429–467.
158. Liener, I. E. and Kakade, M. L. (1980): Protease inhibitors. In: *Toxic Constituents of Plant Foodstuffs*, 2nd edition, edited by I. E. Liener. Academic Press, New York, pp. 7–71.
159. Liener, I. E. (1995): Possible adverse effects of soybean anti-carcinogens. *J. Nutr.*, 125(3, Suppl.):744–750.
160. Lin, J. K., Chen, Y. C., Huang, Y. T., and Lin-Shiau, S. Y. (1997): Suppression of protein kinase C and nuclear oncogene expression as possible molecular mechanisms of cancer chemoprevention by apigenin and curcumin. *J. Cell. Biochem.*, 28–29(Suppl.):39–48.
161. Lin, J. L. and Ho, Y. S. (1994): Flavonoid-induced acute nephropathy. *Am. J. Kidney Dis.*, 23(3):433–440.
162. Lipp, E. K. and Rose, J. B. (1997): The role of seafood in food borne diseases in the United States of America. *Rev. Sci. Tech.*, 16:620–640.
163. Liu, J., Liu, Y., Bullock, P., and Klaassen, C. D. (1995): Suppression of liver cytochrome P450 by alpha-hederin: relevance to hepatoprotection. *Toxicol. Appl. Pharmacol.*, 134(1):124–131.
164. Liu, L. F. et al. (2004): Action of solamargine on human lung cancer cells: enhancement of the susceptibility of cancer cells to TNFs, *FEBS Lett.*, 577:67–74.
165. Lloyd, W. E., Daniels, G. N., and Stahr, H. M. (1985): Cases of nephrotoxic mycotoxicoses in cattle and swine in the United States. In: *Trichothecenes and Other Mycotoxins*, edited by J. Lacey. John Wiley & Sons, New York, pp. 545–548.
166. Lovenberg, W. (1973): Some vaso- and psychoactive substances in food. In: *Toxicants Occurring Naturally in Foods*, 2nd ed., edited by Committee on Food Protection, National Research Council. National Academy of Sciences Press, Washington, D.C., pp. 170–188.
167. Lull, C., Wichers, H. J., and Savelkoul, H. F. (2005): Anti-inflammatory and immuno-modulating properties of fungal metabolites. *Mediat. Inflamm.*, 2005:63–80.
168. MacGregor, J. T. (1984): Genetic and carcinogenic effects of plant flavonoids: an overview. In: *Nutritional and Toxicological Aspects of Food Safety*, edited by M. Friedman. Plenum Press, New York, pp. 497–526.
169. MacKenzie, A. A., Allard, D. G., Perez, E., and Hathaway, S. (2004): Food systems and the changing patterns of food-borne zoonoses. *Rev. Sci. Tech. Off. Int. Epiz.*, 23:677–684
170. Mager, J., Chevion, M., and Claser, G. (1980): Favism. In: *Toxic Constituents of Plant Foodstuffs*, 2nd ed., edited by I. E. Liener. Academic Press, New York, pp. 266–294.
171. Malinow, M. R., Bardana, Jr., E. J., Pirofsky, B., Craig, S., and McCluagblin, P. (1982): Systemic lupus erythematosus-like syndrome in monkeys fed alfalfa sprouts: role of a non-protein amino acid. *Science*, 216:415–417.
172. Mallucci, G. and Collinge, J. (2004): Update on Creutzfeldt–Jakob disease. *Curr. Opin. Neurol.*, 17:641–647.
173. Manach, C., Mazur, A., and Scalbert, A. (2005): Polyphenols and prevention of cardiovascular diseases. *Curr. Opin. Lipidol.*, 16:77–84.
174. Marasas, W. F. (1995): Fumonisins and their implications for human and animal health. *Nat. Toxins*, 3:193–198.
175. Marks, J. (1989): The safety of the vitamins: an overview. *Int. J. Vit. Nutr. Res.*, 30(Suppl.):12–20.
176. Marth, E. H. (1981): Food-borne hazards of microbial origin. In: *Food Safety*, edited by H. R. Roberts. John Wiley & Sons, New York, pp. 15–65.
177. Massey, L. K. (2003): Dietary influences on urinary oxalate and risk of kidney stones. *Front. Biosci.*, 8:S584–594.
178. Matsumoto, H. (1983): Cycasin. In: *Handbook of Naturally Occurring Food Toxicants*, edited by M. Rechcigl, Jr. CRC Press, Boca Raton, FL, pp. 43–61.
179. Matsumoto, T., Itoh, H., and Akiba, Y. (1968): Goitrogenic effects of 5-vinyl-2-oxazolidinethione, a goitrogen in rapeseed, in growing chicks. *Poultry Sci.*, 47:1323–1330.
180. Mattocks, A. R. (1986): *Chemistry and Toxicology of Pyrrolizidine Alkaloids*. Academic Press, New York.
181. Mattson, F. H. (1973): Potential toxicity of food lipids. In: *Toxicants Occurring Naturally in Foods*, 2nd ed., edited by Committee on Food Protection, National Research Council. National Academy of Sciences Press, Washington, D.C., pp. 189–209.
182. McPartland, J. M., Vigaly, R. J., and Cubeta, M. A. (1997): Mushroom poisoning. *Am. Fam. Phys.*, 55:1797–1811.
183. Meadows, M. (2004): The FDA and the fight against terrorism. *FDA Consumer*, 38:20–27.
184. Meinhardt, PL. (2005): Water and bioterrorism: preparing for the potential threat to U.S. water supplies and public health, *Ann Rev. Public Health*, 26: 213–237.
185. Meisner, H. and Cimbala, M. (1985): Effect of ochratoxin A on gene expression in rat kidneys. In: *New Concepts and Developments in Toxicology*, edited by P. L. Chambers, P. Gehring, and F. Sakai. Elsevier, New York, pp. 261–271.
186. Mennen, L. I., Walker, R., Bennetau-Pelissero, C., and Scalbert, A. (2005): Risks and safety of polyphenol consumption. *Am. J. Clin. Nutr.*, 81(1, Suppl.):326S–329S.
187. Merril, R. A. (1997): Food safety regulation: reforming the Delaney Clause. *Ann. Rev. Public Health*, 18:313–340.
188. Miller, J. A., Miller, E. C., and Phillips, D. H. (1982): The metabolic activation and carcinogenicity of alkenylbenzenes that occur naturally in many spices. In: *Carcinogens and Mutagens in the Environment*, Vol. 1, edited by H. F. Stich. CRC Press, Boca Raton, FL, pp. 93–96.
189. Miller, I., Gray, D., and Kay, H. (1998): Bacterial toxins found in foods. In: *Natural Toxicants in Food*, edited by D. H. Watson. CRC Press, Boca Raton, FL, pp. 105–146.
190. Mitchel, R. E. and McCann, R. (1993): Vitamin E is a complete tumor promoter in mouse skin. *Carcinogen*, 14(4):659–662.

191. Miura, K., Nakajima, Y., Yamanaka, N., Terao, K., Shibato, T., and Ishino, S. (1998): Induction of apoptosis with fusarenon-X in mouse thymocytes. *Toxicology*, 127(1–3):195–206.
192. Mody, R., Joshi, S., and Chaney, W. (1995): Use of lectins as diagnostic and therapeutic tools for cancer. *J. Pharmacol. Toxicol. Methods*, 33:1–10.
193. Montgomery, R. D. (1980): Cyanogens. In: *Toxic Constituents of Plant Foodstuffs*, 2nd ed., edited by I. E. Liener. Academic Press, New York, pp. 143–160.
194. Montine, T. J., Li, K., Perl, D. P., and Galasko, D. (2005): Lack of beta-methylamino-L-alanine in brain from controls, AD, or Chamorros with PDC. *Neurology*, 65: 768–769.
195. Morris, S. C. and Lee, T. H. (1984): The toxicity and teratogenicity of Solanaceae glycoalkaloids, particularly those of the potato (*Solanum tuberosum*): a review. *Food Technol. Aust.*, 36:118–124.
196. Moyer, C., Allen, D., Basabe, A., Maronpot, R. R., and Nyska, A. (2004): Analysis of vascular endothelial growth factor (VEGF) and a receptor subtype (KDR/flk-1) in the liver of rats exposed to riddelliine: a potential role in the development of hemangiosarcoma, *Exp. Toxicol. Pathol.*, 55:455–465.
197. Murch, S. J., Cox, P. A., Banack, S. A., Steele, J. C., and Sacks, O. W. (2004): Occurrence of beta-methylamino-L-alanine (BMAA) in ALS/PDC patients from Guam. *Acta Neurol. Scand.*, 110:267–269.
198. National Toxicology Program. (2003): Toxicology and carcinogenesis studies of riddelliine (CAS No. 23246-96-0) in F344/N rats and B6C3F$_1$ mice (gavage studies), *NTP Tech. Rep. Ser.*, (508):1–280.
199. Nakanishi, K., Hagiwara, A., Shibata, M., Imaida, K., Tetematsu, M., and Ito, N. (1980): Dose response of saccharin induction of urinary bladder hyperplasia in Fisher 344 rats pretreated with *N*-butyl *N*- (4-hydroxybutyl) nitrosamine. *J. Natl. Cancer Inst.*, 65: 1005–1010.
200. Nakhla, H. B., Mohamed O. S., Abu, I. M., Fatuh, A. L., and Adam, S. E. (1991): The effect of *Trigonella foenum graecum* (fenugreek) crude saponins on Hisex-type chicks. *Vet. Hum. Toxicol.*, 33(6):561–564.
201. NAS (1977): *Leucaena, Promising Forage, and Tree Crop for the Tropics*. National Academy of Sciences Press, Washington, D.C.
202. Nawrot, P., Jordan, S., Eastwood, J., Rotstein, J., Hugenholtz, A., and Feeley, M. (2003): Effects of caffeine on human health, *Food Addit. Contam.*, 20:1–30.
203. Niyogi, S. K. (2005): Shigellosis. *J. Microbiol.*, 43: 133–143.
204. National Research Council. (1988): Complex mixtures: methods for *in vivo* toxicity testing. National Academy Press, Washington, D.C.
205. National Research Council. (1996): *Carcinogens and Anticarcinogens in the Human Diet*. National Academy Press, Washington, D.C., pp. 1–417.
206. Nyska, A. et al. (2005): Acute hemorrhagic myocardial necrosis and sudden death of rats exposed to a combination of ephedrine and caffeine. *Toxicol. Sci.*, 83: 388–396.
207. Ohga, S., Higashi, E., Nomura, A., Matsuzaki, A., Hirono, A., Miwa, S., Fujii, H., and Ueda, K. (1995): Haptoglobin therapy for acute favism: a Japanese boy with glucose-6-phosphate dehydrogenase Guadalajara. *Br. J. Haematol.*, 89:421–423.
208. Olny, J. W. (1982): The toxic effects of glutamate and related compounds in the retina and the brain. *Retina* 2:341–359.
209. Omaye, S. T. (1984): Safety of megavitamin therapy. In: *Nutritional and Toxicological Aspects of Food Safety*, edited by M. Friedman. Plenum Press, New York, pp. 169–203.
210. Osuntokun, B. O. (1973): Ataxic neuropathy associated with high cassava diets in West Africa. In: *Chronic Cassava Toxicity*, edited by B. Nestel and R. MacIntyre. International Development Research Center, Ottawa, Canada, pp 127–138.
211. Osweiler, G. D., Carson, T. L., Buck, W. B., and Van Gelder G. A. (1985): *Clinical and Diagnostic Veterinary Toxicology*, Kendall-Hunt Publishing, Dubuque, IA.
212. Padmanaban, G. (1980): Lathyrogens. In: *Toxic Constituents of Plant Foodstuffs*, 2nd ed., edited by I. E. Liener Academic Press, New York, pp. 239–263.
213. Panter, K. E. and James, L. F. (1990): Natural plant toxicants in milk: a review. *J. Anim. Sci.*, 68:892–904.
214. Pass, M. A., Arab, H., Pollitt, S., and Hegarty, M. P. (1996) Effects of the naturally occurring arginine analogues indospicine and canavanine on nitric oxide mediated functions in aortic endothelium and peritoneal macrophages *Nat. Toxins*, 4(3):135–140.
215. Pearl, E. R. (1997): Food allergy. *Lippincott's Primary Care Pract.*, 1:154–167.
216. Perlman, F. (1980): Allergens. In: *Toxic Constituents of Plant Foodstuffs*, 2nd ed., edited by I. E. Liener. Academic Press, New York, pp. 295–327.
217. Pestka, J. J. and Smolinski, A. T. (2005): Deoxynivalenol toxicology and potential effects on humans. *J. Toxicol Environ. Health Part B: Crit. Rev.*, 8:39–69.
218. Pfohl-Leszkowicz, A., Chekir-Ghedira, L., and Bacha H (1995): Genotoxicity of zearalenone, and estrogenic mycotoxin: DNA adduct formation in female mouse tissues *Carcinogen*, 16(10):2315–2320.
219. Poulton, J. E. (1983): Cyanogenic compounds in higher plants and their toxic effects. In: *Handbook of Natural Toxins*. Vol. 1. *Plant and Fungal Toxins*, edited by R. F. Keeler and A. T. Tu. Marcel Dekker, New York, pp. 117–160.
220. Puchala, R., Pierzynowski, S. G., Sahlu, T., and Hart, S. P (1996): Effects of mimosine administered to a perfused area of skin in Angora goats. *Br. J. Nutr.*, 75(1):69–79.
221. Quintana, P.J., de Peyster, A., Klatzke, S., and Park, H. J (2000): Gossypol-induced DNA breaks in rat lymphocytes are secondary to cytotoxicity. *Toxicol Lett.*, 117:85–94.
222. Randel, R. D., Chase, C. C., and Wyse, S. J., (1992): Effect of gossypol and cottonseed products on reproduction in mammals. *J. Anim. Sci.*, 70:1628–1638.
223. Rao, A. V. and Sung, M. K. (1995): Saponins as anticarcinogens. *J. Nutr.*, 125(Suppl. 1):717S–724S.
224. Rapuri, P. B. et al. (2001): Caffeine intake increases the rate of bone loss in elderly women and interacts with vitamin D receptor genotypes [see comment]. *Am. J. Clin Nutr.*, 74:694–700.

225. Rasenack, R., Muller, C., Kleinschmidt, M., Rasenack, J., and Wiedenfeld, H. (2003): Veno-occlusive disease in a fetus caused by pyrrolizidine alkaloids of food origin. *Fetal Diag. Ther.*, 18:223–225.
226. Raul, F., Gosse, F., Delincee, H., Hartwig, A., Marchioni, E., Miesch, M., Werner, D., and Burnouf, D. (2002): Food-borne radiolytic compounds (2-alkylcyclobutanones) may promote experimental colon carcinogenesis. *Nutr. Cancer*, 44:189–191.
227. Reddy, C.S. (2005): Alterations in protein kinase A signaling and cleft palate: a review, *Hum. Exp. Toxicol.*, 24:235–242.
228. Reddy, R. V., Johnson, G., Rottinghaus, G. E., Casteel, S. W., and Reddy, C. S. (1996): Developmental effects of fumonisin B in mice. *Mycopathology*, 134:161–166.
229. Reed, J. D. (1995): Nutritional toxicology of tannins and related polyphenols in forage legumes. *J. Anim. Sci.*, 43:1516–1528.
230. Riccardi, G., Giacco, R., and Rivellese, A. A. (2004): Dietary fat, insulin sensitivity and the metabolic syndrome. *Clin. Nutr.*, 23:447–456.
231. Riemann, H. P. and Oliver, D. O. (1998): *Escherichia coli* O157:H7, *Vet. Clin. N. Am. Food Anim. Pract.*, 14(1): 41–48.
232. Rivera, V. R., Pol, M. A., and Bignami, G. S. (1995): Prophylaxis and treatment with a monoclonal antibody of tetrodotoxin poisoning in mice. *Toxicon*, 33:1231–1237.
233. Roche, H. M. (2004): Dietary lipids and gene expression. *Biochem. Soc. Trans.*, 32:999–1002.
234. Roebuck, B. D., Yeager, Jr., J. D., Longnecker, D. S., and Wilpone, S. A. (1981): Promotion by unsaturated fat of azaserine-induced pancreatic carcinogenesis in the rat. *Cancer Res.*, 41:3961–3966.
235. Rosen, J. D. (1990): Much ado about Alar. *Issues Sci. Technol.*, VIII:85–90.
236. Rotter, B. A. and Oh, Y. N. (1996): Mycotoxin fumonisin $B_1$ stimulates nitric oxide production in a murine macrophage cell line. *Nat. Toxins*, 4(6):291–294.
237. Rotter, B. A., Prelusky, D. B., and Pestka, J. J. (1996): Toxicology of deoxynivalenol (vomitoxin). *J. Toxicol. Environ. Health*, 48(1):1–34.
238. Rotz, L. D. et al. (2002): Public health assessment of potential biological terrorism agents. *Emerg. Infect. Dis.*, 8:225–230.
239. Ruan, Q. R. et al. (2001): Anisodamine counteracts lipopolysaccharide-induced tissue factor and plasminogen activator inhibitor-1 expression in human endothelial cells: contribution of the NF-kappa B pathway, *J. Vasc. Res.*, 38:13–19.
240. Russell, F. E. (1986): Toxic effects of animal toxins. In: *Casarett and Doull's Toxicology: The Basic Science of Poisons*, 3rd ed., edited by C. D. Klaassen, M. O. Amdur, and J. Doull. Macmillan, New York, pp. 706–756.
241. Russell, F. E. and Dart, R. C. (1991): Toxic effects of animal toxins. In: *Casarett and Doull's Toxicology: The Basic Science of Poisons*, 4th ed., edited by M. O. Amdur, J. Doull, and C. D. Klaassen. Pergamon Press, New York, pp. 753–803.
242. Safe, S. H. (1995): Environmental and dietary estrogens and human health: is there a problem? *Environ. Health Perspect.*, 103:346–351.
243. Salazar-Martinez, E., Willett, W. C., Ascherio, A., Manson, J. E., Leitzmann, M. F., Stampfer, M. J., Hu, F. B. (2004): Coffee consumption and risk for type 2 diabetes mellitus, *Ann. Intern. Med.*, 140:1–8.
244. Sampson, H. A. (1997): Food allergy. *JAMA*, 278:1888–1894.
245. Sampson, H. A. (2004): Update on food allergy. *J. Allergy Clin. Immunol.*, 113:805–819.
246. Sanchez, J. and Holmgren, J. (2005): Virulence factors, pathogenesis and vaccine protection in cholera and ETEC diarrhea. *Curr. Opin. Immunol.*, 17:388–398.
247. Sandani, G. R., Soman, C. S., Deodhar, K. K., and Nadharni, G. D. (1997): Reactive oxygen species involvement in ricin-induced thyroid toxicity in the rat. *Hum. Exp. Toxicol.*, 16:254–256.
248. Sandermann, Jr., H. (2004): Bound and unextractable pesticidal plant residues: chemical characterization and consumer exposure. *Pest. Manag. Sci.*, 60:613–623.
249. Sandvig, K. and Van Deurs, B. (1997): Endocytosis, intracellular transport and cytotoxic action of shiga toxin and ricin. *Physiol. Rev.*, 76:949–966.
250. Scalbert, A., Manach, C., Morand, C., Remesy, C., and Jimenez, L. (2005): Dietary polyphenols and the prevention of diseases. *Crit. Rev. Food Sci Nutr.*, 45:287–306.
251. Schafer, K. S. and Kegley, S. E. (2002): Persistent toxic chemicals in the U.S. food supply. *J. Epidemiol. Commun Health*, 56:813–817.
252. Schardein, J. L. (1985): *Chemically Induced Birth Defects*. Marcel Dekker, New York, pp. 709–716.
253. Scheid, C. R., Cao, L. C., Honeyman, T., and Jonassen, J. A. (2004): How elevated oxalate can promote kidney stone disease: changes at the surface and in the cytosol of renal cells that promote crystal adherence and growth. *Front. Biosci.*, 9:797–808.
254. Schlatter, J. (2004): Toxicity data relevant for hazard characterization. *Toxicol. Lett.*, 153:8389.
255. Schoeni, J. L. and Wong, A. C. (2005): Bacillus cereus food poisoning and its toxins. *J. Food Prot.*, 68:636–648.
256. Schramek, H., Wilflingseder D., Pollack, V., Freudinger, R., Mildenberger, S., and Gekle, M. (1997): Ochratoxin A-induced stimulation of extracellular signal-regulated kinases 1/2 is associated with Madin–Darby canine kidney-C7 cell dedifferentiation. *J. Pharmacol. Exp. Ther.*, 283(3):1460–1468.
257. Scott, B. R., Pathak, M. A., and Mohn, G. R. (1976): Molecular and genetic basis of furocoumarin reactions. *Mutat. Res.*, 39:29–74.
258. Scudamore, K. A. (1998): Mycotoxins. In: *Natural Toxins in Foods*, edited by D. H. Watson. CRC Press, Boca Raton, FL, pp. 147–181.
259. Sears, M. K. et al. (2001): Impact of Bt corn pollen on monarch butterfly populations: a risk assessment. *PNAS*, 98:11937–11942.
260. Selman, I. E., Wiseman, A., Breeze, R. G., and Pirie, H. M. (1976): Fog fever in cattle: various theories on its etiology. *Vet. Rec.*, 99:181–184.
261. Shank, R. C. (1981): Environmental toxicoses in humans. In: *Mycotoxins and Nitroso Compounds: Environmental Risks*, Vol. 1, edited by R. C. Shank. CRC Press, Boca Raton, FL, pp. 107–140.

262. Sharma, R. P. and Salunkhe, D. K. (1989): Solanum glycoalkaloids. In: *Toxicants of Plant Origin*. Vol. I. *Alkaloids*, edited by P. R. Cheeke. CRC Press, Boca Raton, FL, pp. 179–236.
263. Shephard, S. E. (1993): Mutagenic activity of peptides and artificial sweetener aspartame after nitrosation. *Food Chem. Toxicol*., 31:323–329.
264. Siegel, J. P. (2001): The mammalian safety of *Bacillus thuringiensis*-based insecticides. *J. Invert. Pathol*., 77, 13–21.
265. Singh, R. P. and Agarwal, R. (2005): Prostate cancer and inositol hexaphosphate: efficacy and mechanisms. *Anticancer Res*., 25:2891–2903.
266. Shoda, R., Matsueda, K., Yamamoto, S., and Umeda, N. (1996): Epidemiologic analysis of Crohn's disease in Japan: increased dietary intake of n-6 polyunsaturated fatty acids and animal protein relates to increased incidence of Crohn's disease. *Am. J. Clin. Nutr*., 63:741–745.
267. Singleton, V. L. and Kratzer, F. H. (1973): Plant phenolics. In: *Toxicants Occurring Naturally in Foods*, 2nd ed., edited by Committee on Food Protection, National Research Council. National Academy of Sciences Press, Washington, D.C., pp. 309–345.
268. Smith, D. W. and Nagler-Anderson, C. (2005): Preventing intolerance: the induction of non-responsiveness to dietary and microbial antigens in the intestinal mucosa, *J. Immunol*., 174:3851–3857.
269. Smith, L. D. (1977): *Botulism: The Organism, Its Toxins, the Disease*. Charles C Thomas, Springfield, IL.
270. Smith, T. K., Lund, E. K., and Johnson, I. K. (1998): Inhibition of dimethyl-hydrazine induced aberrant crypt foci and induction of apoptosis in rat colon following oral administration of the glucosinolate, sinigrain. *Carcinogens*, 19:267–273.
271. Snyder, R. and Andrews, L. S. (1996): The effects of solvents and vapors. In: *Casarett and Doull's Toxicology: The Basic Science of Poisons*, 5th ed., edited by C. D. Klaassen. McGraw-Hill, New York, pp. 737–771.
272. Soto-Blanco, B., Maiorka, P. C., and Gorniak, S. L. (2002): Neuropathologic study of long term cyanide administration to goats. *Food Chem. Toxicol*., 40:1693–1698.
273. Spencer, P. D., Nunn, P. B., Hugon, J., Ludolph, A. C., Ross, S. M., Roy, D. N., and Robertson, R. C. (1987): Guam amyotrophic lateral sclerosis–Parkinsonism–dementia linked to a plant-excitant neurotoxiri. *Science*, 237:517–522.
274. Spiller J. (2004): Radiant cuisine: the commercial fate of food irradiation in the United States. *Technol. Culture*, 45(4):740–763.
275. Stark, A. A. (2005): Threat assessment of mycotoxins as weapons: molecular mechanisms of acute toxicity. *J. Food Prot*., 68:1285–1293.
276. Stewart, C. S., Duncan, S. H., Cave, D. R. (2004): Oxalobacter formigenes and its role in oxalate metabolism in the human gut. *FEMS Microbiol. Lett*., 230:1–7.
277. Stob, M. (1983): Estrogens. In: *Handbook of Naturally Occurring Food Toxicants*, edited by M. Rechcigl, Jr. CRC Press, Boca Raton, FL, pp. 81–100.
278. Stocks, P. (1970): Cancer mortality in relation to national consumption of cigarettes, solid fuel, tea, and coffee. *Br J. Cancer*, 24:215–225.
279. Stopper, H., Schmitt, E., and Kobras, K. (2005): Genotoxicity of phytoestrogens. *Mutat. Res*., 574:139–155.
280. Stricker-Krongrad, A., Burlet, C., and Beck, B. (1998): Behavioral deficits in monosodium glutamate rats: specific changes in structure and behavior. *Life Sci*., 62: 2127–2132.
281. Sugimoto, N., Horiguchi, Y., and Matsuda, M. (1996): Mechanism of action of *Clostridium perfringens* enterotoxin. In: *Natural Toxins II*, edited by B. R. Singh and A Tu. Plenum Press, New York, pp. 257–269.
282. Sugimura, T. (1986): Past, present, and future of mutagen in cooked foods. *Environ. Health Perspect*., 67:5–10.
283. Suzuki, Y. J., Forman, H. J., and Sevanian, A. (1997) Oxidants as stimulators of signal transduction. *Free Radical Biol. Med*., 22:269–285.
284. Talaley, P. and Zhang, Y. (1996): Chemoprotection agains cancer by isothiocyanates and glucosinolates. *Biochem Soc. Transact*., 24:806–810.
285. Tarasoff, L. and Kelly, M. F. (1993): Monosodium L glutamate: a double-blind study and review. *Food Chem Toxicol*., 31(2):1019–1035.
286. Teng, C. S. (1995): Gossypol-induced apoptotic DNA fragmentation correlates with inhibited protein kinase C activity in speratocytes. *Contraception*, 52:389–395.
287. Toma, W., Trigo, J. R., de Paula, A. C., and Brito, A. R (2004): Modulation of gastrin and epidermal growth facto by pyrrolizidine alkaloids obtained from *Senecio brasiliensis* in acute and chronic induced gastric ulcers, *Can. J Physiol. Pharmacol*., 82:319–325.
288. Tookey, H. L., Van Etten, C. H., and Daxenbichler, M. E (1980): Glucosinolates. In: *Toxic Constituents of Plan Foodstuffs*, 2nd ed., edited by I. E. Liener. Academic Press New York, pp. 103–142.
289. Tsuda, S., Kosaka, Y., Murakami, M., Matsuo, H., Matsusaka, N., Taniguchi, K., and Sasaki, Y. F. (1998): Detectio of nivalenol genotoxicity in cultured cells and multipl mouse organs by the alkaline single-cell gel electrophoresis assay. *Mutat. Res*., 415(3):191–200.
290. Tulley, R. T. et al. (2005): Daily intake of multivitamin during long-term intake of olestra in men prevents decline in serum vitamins A and E but not carotenoids. *J. Nutr* 135:1456–1461.
291. Unkrig, C. J., Schroeder, R., Scharf, R. E., and Aubourg P. (1994): Lorenzo's oil and lymphocytopemia (letter *New Engl. J. Med*., 330:577.
292. U.S. FDA. (1996): Healthful snacks for the chip-and-di crowd: olestra approved with special labeling. *FDA Consumer*, Vol. 30.
293. U.S. FDA. (2003): *Dairy Farms, Bulk Milk Transporter Bulk Milk Transfer Stations and Fluid Milk Processors Food Security Preventive Measures Guidance*, U.S. Foo and Drug Administration, Washington, D.C.
294. Van Damme, E. J., Barre, A., Rouge, P., and Peumans, W J. (2004): Cytoplasmic/nuclear plant lectins: a new stor *Trends Plant Sci*., 9:484–489.

295. Van Etten, C. H. and Tookey, H. L. (1983): Glucosinolates. In: *Handbook of Naturally Occurring Food Toxicants*, edited by M. Rechcigl, Jr. CRC Press, Boca Raton, FL, pp. 15–30.
296. Van Genderen, H. (1997): Adverse effects of naturally occurring non-nutritive substances. In: *Food Safety and Toxicity*, edited by J. DeVries. CRC Press, Boca Raton, FL, pp. 147–162.
297. Van Hoff, F., Hue, L., Vamecq, J., and Sherratt, H. S. (1985): Protection of rats by clofibrate against the hypoglycaemic and toxic effects of hypoglycin and pent-4-enoate: an ultrastructural and biochemical study. *Biochem. J.*, 229:387–297.
298. Vasconcelos, I. M. and Oliveira, J. T. (2004): Antinutritional properties of plant lectins. *Toxicon*, 44:385–403.
299. Verdeal, K., Brown, R. R., Richardson, T., and Ryan, D. S. (1980): Affinity of phytoestrogens for the estradiol-binding proteins and effect of coumestrol on growth of 7,12–dimethylbenz (A) anthracene-induced rat mammary tumors. *J. Natl. Cancer Inst.*, 64:285–290.
300. Verhagen, H. (1997): Adverse effects of food additives. In: *Food Safety and Toxicity*, edited by J. DeVries. CRC Press, Boca Raton, FL, pp. 121–132.
301. Verhoef, P., Pasman, W. J., Van Vliet, T., Urgert, R., and Katan, M. B. (2002): Contribution of caffeine to the homocysteine-raising effect of coffee: a randomized controlled trial in humans. *Am. J. Clin. Nutr.*, 76:1244–1248.
302. Verkerk, R., Dekker, M., and Jongen W. M. F. (1998): Glucosinolates. In: *Natural Toxicants in Foods*, edited by D. H. Watson. CRC Press, Boca Raton, FL, pp. 29–53.
303. Vetter, J. (1998): Toxins of *Amanita phalloides*. *Toxicon*, 36:13–24.
304. Voss, K. A. et al. (1995): Subchronic toxicity of fumonisin $B_1$ to male and female rats. *Food Addit. Contam.*, 12(3):473–478.
305. Waddell, W. J. (2002): Thresholds of carcinogenicity of flavors. *Toxicol. Sci.*, 68:275–279.
306. Walker, G. R. and Yamazaki, K. (1996): Saponins in food and agriculture. *Adv. Exp. Med. Biol.*, 404:1–422.
307. Walker, G. R. and Yamazaki, K. (1996): Saponins in traditional and modern medicine. *Adv. Exp. Med. Biol.*, 405:1–576.
308. Wang, L. Z. et al. (2001): Verapamil, cyproheptadine, and anisodamine antagonized $[Ca^{2+}]i$ elevation induced by TNFalpha in a single endothelial cell, *Acta Pharmacol. Sinica*, 22:918–922.
309. Weil, M. J., Zhang, Y., and Nair, M. G. (2004): Colon cancer proliferating desulfosinigrin in wasabi (*Wasabia japonica*). *Nutr. Cancer*, 48:207–213.
310. Wein, L. M. and Liu, Y. (2005): Analyzing a bioterror attack on the food supply: the case of botulinum toxin in milk, PNAS, 102:9984–9989.
311. Whitley, B. D., Holmes, A. R., Shepherd, M. G., and Ferguson, M. M. (1991): Peanut sensitivity as a cause of burning mouth. *Oral Surg Oral Med. Oral Pathol* 72:671–674.
312. Wieland, T. and Faulstich, H. (1983): Peptide toxins from Amanita. In: *Handbook of Natural Toxins, Vol 1, Plant and Fungal Toxins*, edited by R. F. Keeler and A. T. Tu. Marcel Dekker, New York, pp. 117–160.
313. Williamson, G., Faulkner, K., and Plumb, G. W. (1998): Glucosinolates and phenolics and antioxidants from plant foods. *Eur. J. Cancer Prev.*, 7:17–21.
314. Wilmarth, K. R. and Froines, J. R. (1992): *In vitro* and *in vivo* inhibition of lysyl oxidase by aminopropionitriles. *J. Toxicol. Environ. Health*, 37:411–423.
315. Wolever, T. M., Hamad, S., Gittelsohn, J., Gao, J., Hanley, A. J., Harris, S. B., and Zinman, B. (1997): Low dietary fiber and high protein intake associated with newly diagnosed diabetes in a remote aboriginal community. *Am. J. Clin. Nutr.*, 66:1470–1474.
316. Xu, R., Zhao, W., Xu, J., Shao, B., and Qin, G. (1996): Studies on bioactive saponins from Chinese medicinal plants. *Adv. Exper. Med. Biol.*, 404:371–382.
317. Yamanaka, N., Oda, O., and Nagao, S. (1997): Prooxidant activity of caffeic acid, dietary non-flavonoid phenolic acid, on $Cu^{2+}$ induced low density lipoprotein oxidation, *FEBS Lett.*, 405:186–190.
318. Yang, W. H., Drouin, M. A., Herbert, M., Mao, Y., and Karsh, J. (1997): The monosodium glutamate symptom complex: assessment in double-blind placebo-controlled, randomized study. *J. Allergy Clin. Immunol.*, 99:757–762.
319. Esterbauer, H. (1993): Cytotoxicity and genotoxicity of lipid peroxidation products. *Am. J. Clin. Nutr.*, 57:779–786.
320. Yeung, J. M., Wang, H. Y., and Prelusky, D. B. (1996): Fumonisin $B_1$ induces protein kinase C translocation via direct interaction with diacylglycerol binding site. *Toxicol. Appl. Pharmacol.*, 141(1):178–184.
321. Zelski, R. Z., Rothwell, J. T., Moore, R. E., and Kennedy, D. J. (1995): Gossypol toxicity in preruminant calves. *Aust. Vet. J.*, 72:394–398.

# Notes

# 15 Solvents and Industrial Hygiene

*David L. Dahlstrom and Margaret Buckalew*

## CONTENTS

As used in this chapter, the term *solvent* generally refers to organic liquids whose common use is to dissolve or mix with other organic compounds. Common organic solvents in use by industry include various:

- Aliphatic hydrocarbons (e.g., *n*-hexane, gasoline, petroleum naphtha, mineral spirits)
- Aromatic hydrocarbons (e.g., benzene, toluene, xylene, phenols, naphthalene)
- Alcohols and glycols (e.g., methanol, ethanol, allyl alcohol, ethylene chlorohydrin, ethylene glycol, cellosolve)
- Carbon disulfide
- Chlorinated hydrocarbons (e.g., methylene chloride, trichloroethylene, methyl chloroform, perchloroethylene, ethylene dichloride)
- Cyclic hydrocarbons (e.g., cyclohexane)
- Ethers and epoxides (e.g., diethyl ether dioxane, isopropyl ether, dimethyl ether)
- Esters (ethyl formate, vinyl acetate, 2-nitropropane)
- Ketones and aldehydes (e.g., formaldehyde, acetaldehyde, acetone, methyl ethyl ketone)
- Aromatic nitro- and amino-containing solvents (e.g., dinitrobenzene, toluidines, diphenylamines, 2-naphthylamine, aniline)

Today, in both industry and the home, the use of solvents is widespread (see Figure 15.1). Billions of pounds are produced and used annually. Solvents are in use as common degreasing agents, paint thinners and removers, dry-cleaning agents, chemical intermediates, extractants, and carrier vehicles for paints, varnishes, and industrial coatings. They are used in industry to make waxes, paints, varnishes, lacquers, pharmaceuticals, plastics, pesticides, rubber goods, synthetic textiles, adhesives, shoe polish, floor cleaners, and many other major products in everyday use.

Industrial hygienists and other health and safety (H&S) professionals need to know about the basic properties and behavior of organic solvents in common use because most are environmental and occupational toxicants that are injurious to varying degrees when used without proper and adequate administrative and engineering controls. The common industrial hygiene principles of anticipation, recognition, evaluation, and control of hazards represented by the use of toxic materials within the workplace have done much to better characterize and minimize the potential adverse effects that may relate to elevated concentrations of solvents and their vapors. General statements of toxic effects upon human health due to solvent exposure are difficult to make, especially when one considers the numerous ways in which exposure to organic solvents occurs in industry, the variation of solvent mixtures and vapors produced, the variability of exposures, and the range in relative health and age of exposed individuals. The overriding problem for the industrial hygienist and occupational health specialist is not so much identifying an exposure to an individual solvent but determining at what concentration, frequency, duration, and routes of entry (oral, inhalation, skin and eye contact) a dose of the solvent/mixture will result in a harmful effec (e.g., dermatitis, systemic injury, irritation, narcosis, neuropathy, carcinogenicity).

## PROPERTIES OF SOLVENTS

Key to the recognition and evaluation of solvents and the potential hazards they represent is understanding their respective chemical, physical, and toxicological properties. The following description of key properties should be considered when attempting to evaluate and control the risks (hazard × toxicity) associated with work involving organic solvents. It should also be recognized that most solvents used in the workplace often contain small amounts of impurities and that these impurities can affec the properties inherent in the particular solvent.

### Boiling Point

The temperature at which the vapor pressure of a liqui becomes equal to the pressure of the surrounding atmosphere is the liquid's *boiling point*. In typical use, a pressure of 1 atmosphere (760 mmHg or 14.7 psi) is considered standard atmospheric pressure at sea level. This atmospheric pressure will vary, however, depending o elevation above or below sea level and should be take into account when estimating boiling temperatures. A point for recognition is that, like many liquids, solvent begin to evaporate at temperatures below their boilin point. As the temperature of the liquid is increased, the volume of vapor evolving will increase. Also, the lower the boiling point of a solvent, the higher the relative vapor pressure will be and the more readily the solvent will volatilize into the surrounding air [11,12].

## VAPOR PRESSURE

The *vapor pressure* of a liquid is a measure of the pressure exerted (usually in millimeters of mercury, or mmHg) by the vapor of the liquid at a given temperature above its liquid form. As noted in our discussion of boiling point, the vapor pressure of a liquid will increase as the temperature of the liquid is increased; therefore, the amount of vapor above the surface of a liquid entering an open environment is dependant on the surface area of the liquid, the temperature, and the atmospheric pressure. Another factor to note is that the higher the vapor pressure of the liquid, the greater is its propensity to evaporate [11,12]. An industrial hygienist or other H&S professional is able to use the vapor pressure of a liquid to calculate what the parts per million (ppm) concentration of the liquid will be at its point of saturation or equilibrium within a given environment. This is performed by dividing the vapor pressure of the liquid by the total atmospheric barometric pressure and multiplying by 1,000,000:

$$\frac{\text{Vapor pressure of the liquid}}{\text{Total atmospheric barometric pressure}} \times 1{,}000{,}000$$
$$= \text{PPM of liquid}$$

## EVAPORATION RATE

The rate by which a liquid evaporates is dependent on a number of intrinsic properties of the liquid and various external factors; therefore, the *evaporation rate* of a liquid or solvent is relative to the rate at which it evaporates in comparison with a reference liquid or solvent under identical conditions. Most frequently, the rate at which a specific volume of a solvent evaporates is compared to the rate at which a specific volume ethyl ether evaporates at the same temperature and atmospheric pressure. Because ethyl ether as a reference is given an evaporation rate of 1.0, those solvents with an evaporation rate of less than 1.0 evaporate more quickly than ethyl ether and those whose evaporation rate is greater than 1.0 evaporate more slowly [11,12].

## FLASH POINT

The lowest temperature of a liquid at which a vapor is given off in sufficient concentration that the vapor/air mixture above the surface of the liquid will spread a flame away from the flame source through the vapor/air mixture is identified as the *flash point* of the liquid. The point to recognize here is that a solvent vapor/air mixture that may be below the lower flammability limit of the solvent may burn in the immediate area surrounding the ignition source without spreading; however, if the solvent vapor/air mixture is within the flammable range, then the flame will spread through it when an ignition source is provided [11,12]. The flash point of a solvent is the temperature below which a solvent will not give off vapors sufficient to be ignited. It can be used, in part, as a criterion to define under what temperature a solvent may be safely stored in open containers and used. Do note, however, that solvents introduced into the atmosphere as sprays or mists can be ignited below the flash point of the solvent. The flash point can be defined based on the temperature at which a solvent within a "closed-cup" creates enough vapor sufficient to flash when an ignition source is introduced. Flash point values are most often reported based upon the results of this "closed-cup" test.

## FLAMMABLE (EXPLOSIVE) RANGE

The range of concentrations of a gas or vapor that, when mixed with air and exposed to an ignition source, will ignite or rapidly combust (explode) is recognized as the flammability (explosivity) range for the particular gas or vapor. The flammability (explosivity) range is inclusive of all concentration levels of a gas or vapor from the lowest concentration (lower flammability limit [LFL] or lower explosive limit [LEL]) to the highest concentration (upper flammability limit [UFL] or upper explosive limit [UEL]) a vapor or gas can be made to combust. The range of concentrations between the LFL/LEL and the UFL/UEL is commonly reported in percent concentration (by volume) of the gas or vapor in air, and they are typically based on an ambient temperature of 68 degrees Fahrenheit (°F). When considering the LFL/LEL and UFL/UEL of a gas or vapor in conditions other than that of an ambient temperature of 68°F and 1 standard atmosphere, one should recognize that the general effect of increasing the temperature or pressure is to lower the lower limits and raise the upper limits. Conversely, a lowering of the temperature or pressure will serve to raise the lower limits and lower the upper limit [11,12]. Figure 15.2 provides a graphical depiction of the relationships among flammable/explosive limits, vapor pressure, and flash points [11,12].

## SPECIFIC GRAVITY

The ratio of the weight of a given volume of a substance (e.g., solvent), at a given temperature (usually given as 75°F) and atmospheric pressure, to the weight of an equal volume of water at the same temperature and atmospheric pressure is known as the *specific gravity* of the substance.

**Aliphatic hydrocarbons**

Straight or branched chains of carbon and hydrogen.

Hexane* — 50 ppm
Heptane — 400 ppm
VM&P Naphtha — 300 ppm

**Cyclic hydrocarbons**

Ring structure saturated and unsaturated with hydrogen.

Cyclohexane — 300 ppm
Turpentine — 100 ppm

**Esters**

Formed by interaction of an organic acid with an alcohol.

Ethyl acetate — 400 ppm
Isopropyl acetate — 250 ppm

**Aromatic hydrocarbons**

Contain a 6-carbon ring structure with one hydrogen per carbon bound by energy from several resonant forms.

Benzene — 0.5 ppm
Toluene — 50 ppm
Xylene — 100 ppm

**Alcohols**

Contain a single hydroxyl group.

Methanol — 200 ppm
Ethanol — 1000 ppm
Isopropanol — 400 ppm

**Ketones**

Contain a double bonded carbonyl group, C=O, with two hydrocarbon groups on the carbon.

Methyl ethyl ketone — 200 ppm
Acetone — 500 ppm
Methyl isobutyl ketone — 50 ppm

*TLV®, American Conference of Governmental Industrial Hygienists (ACGIH) Threshold Limit Value (TLV-TWA), 1998.

**FIGURE 15.1** Classes of organic solvents. (From AAI, *Handbook of Organic Industrial Solvents*. Alliance of American Insurers Schaumburg, IL, 1997. With permission.)

If the specific gravity of a solvent is less than 1, it will float on top of the water; if it is greater than 1, it will tend to sink, depending on its solubility. In conditions where a solvent is spilled into a water body, the additional concern of its ability to catch fire should it reach an ignition source should be considered for a solvent with a specific gravity less than 1; however, skimming the surface of the water body above a certain depth would be a feasible remediation measure. Similar considerations for remediation methods would come into play should the specific gravit

**Halogenated hydrocarbon**

A halogen atom has replaced one or more hydrogen atoms on the hydrocarbon.

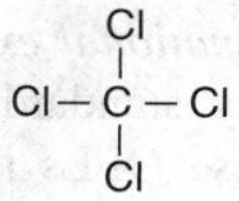

Carbon tetrachloride — 5 ppm
Methyl chloroform — 350 ppm
Chloroform — 10 ppm

**Glycols**

Contain double hydroxyl groups.

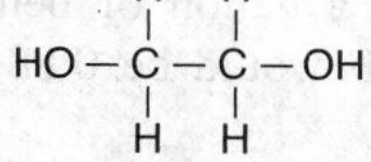

Ethylene glycol — 100 mg/m$^3$ (ceiling)
Hexylene glycol — 25 ppm (ceiling)

**Aldehydes**

Contain the double-bonded carbonyl group, C=O, with only one hydrocarbon group on the carbon.

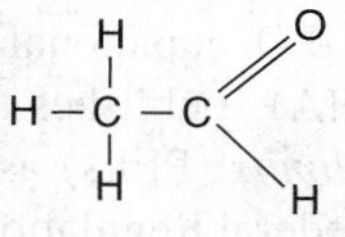

Acetaldehyde — 25 ppm (Ceiling)
Formaldehyde — 0.3 ppm (Ceiling)

**Nitro-hydrocarbons**

Contain an $NO_2$ group.

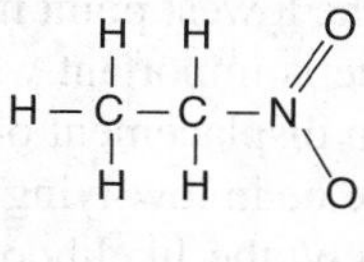

Nitroethane — 100 ppm
Nitromethane — 20 ppm

**Ethers**

Contain the C–O–C linkage.

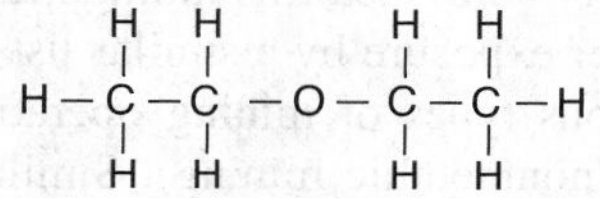

Ethyl ether — 400 ppm
Isopropyl ether — 250 ppm

**FIGURE 15.1 (cont.)**

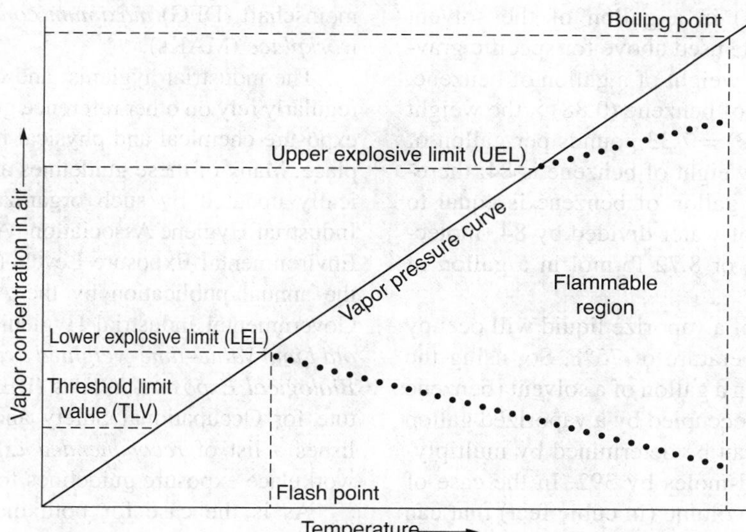


**FIGURE 15.2** Diagram of vapor pressure vs. temperature showing relationships among upper and lower flammable (explosive) limits, flammable and nonflammable regions, Threshold Limit Value, boiling point, flash point, and vapor pressure curve. This diagram shows what happens to a vapor/air mixture as concentrations and temperature vary.

of the solvent (depending on its solubility) be greater than 1 [11,12]. Another way to consider this value is by the following example. If the weight of a gallon of water at 75°F is 8.33 pounds and the specific gravity of benzene is 0.88, then the weight of a gallon of benzene at an ambient temperature of 75°F would be 0.88 × 8.33, or 7.33 pounds.

## Vapor Density

The ratio of weight of a given volume of vapor or gas to the weight of an equal volume of air at the same temperature and atmospheric pressure is known as the *vapor density* of the vapor or gas. Vapors or gases with a vapor density less than 1 (lighter than air) will rise into the air, whereas those with a vapor density greater than 1 (heavier than air) will tend to sink to the lowest point in the immediate surroundings. This value is important when considering the potential for oxygen displacement or fire/explosion should an ignition source be in low-lying areas when a solvent is spilled. Conversely, the likelihood of vapors rising and representing an acute respiratory hazard when the vapor density is less than 1 is equally worthy of consideration [11,12].

## Vapor Volume

The ability to calculate the number of cubic feet of a vapor resulting from the evaporation of a known volume (in gallons) of a solvent can be quite useful when considering the potential for various adverse effects to occur. The process by which this is performed involves the use of the specific gravity of the solvent, its molecular weight, and the weight of a gallon of water (8.33 pounds), as well as calculating the weight of the number of pound-moles (lb-mol) in a gallon of the solvent [11,12]. As in the example used above for specific gravity and calculation of the weight of a gallon of benzene, the specific gravity (SG) of benzene (0.88) × the weight of a gallon of water (8.33) = 7.33 pounds per gallon of benzene. The molecular weight of benzene is 84; therefore, a pound-mole of a gallon of benzene is equal to 0.88 (SG) × 8.33 lb/gal of water divided by 84 (molecular weight of benzene), or 8.72 lb-mol in a gallon of benzene.

Now, a pound-mole of a vaporize liquid will occupy 392 cubic feet at a temperature of 75°F. So, using the number of pound-moles in a gallon of a solvent (benzene in this case), the volume occupied by a vaporized gallon of the liquid (benzene) can be determined by multiplying the number of pound-moles by 392. In the case of a gallon of benzene, the volume (in cubic feet) that can be occupied by the vaporization of a gallon of benzene is calculated by following the process above: (8.72 × 392) or 3417.24 cubic feet, a pretty good size area to be potentially affected.

## Workplace Exposure Limits

In the case of evaluating workplace exposure doses, the industrial hygienist and other H&S professionals rely on various *occupational exposure limits* (OELs) established by regulatory standards and reference guidelines. The intent of these OELs is to establish levels of exposure deemed to be protective of worker health and wellbeing. Around the world, many government agencies and scientific organizations have issued OELs as government-regulated standards or guidelines. OELs are intended for use solely for the protection of worker health and wellbeing within the workplace and should not be applied to the assessment of exposures occurring outside the occupational environment.

In the United States, the U.S. Department of Labor, through the Occupational Safety and Health Administration (OSHA), publishes a list of regulated *permissible exposure limits* (PELs) as threshold limits as part of the Code of Federal Regulations (CFR), Titles 1910 and 1926. These PELs establish by regulation the averaged airborne concentrations of over 400 chemicals to which the "normal" worker may be exposed over an 8-hour day, 40 hours a week, over a working lifetime (40 to 45 years) without suffering adverse impact to health. Likewise, the Mine Safety and Health Administration (MSHA) regulates miner exposure by a similar list of PELs as they apply to various types of mining operations (e.g., coal, metallic and nonmetallic mining). Similarly, the Control of Substances Hazardous to Health (COSHH) regulations within the United Kingdom regulate workplace exposure levels by specifying *maximum exposure limits* (MELs) and *occupational exposure standards* (OESs). German workforce exposures are regulated by the Deutsche Forschungsgemeinschaft (DFG) *maximum concentration values in the workplace* (MAKs).

The industrial hygienist and other H&S professionals regularly rely on other reference guidelines when assessing exposure chemical and physical hazards within the workplace. Many of these guidelines are published and periodically updated by such organizations as the American Industrial Hygiene Association (AIHA) and its Workplace Environmental Exposure Levels (WEEL) Committee and the annual publication by the American Conference of Governmental Industrial Hygienists (ACGIH) of *Threshold Limit Value–time-weighted averages* (TLV–TWAs) and *Biological Exposure Indices* (BEIs®). The National Institute for Occupational Safety and Health (NIOSH) publishes a list of *recommended exposure limits* (RELs) as workplace exposure guidelines for many chemicals.

As is the case for both the OSHA PELs and the ACGIH TLVs, these exposure limits values represent a *time-weighted average* (TWA) of the airborne concentration of specific chemicals within a workplace. The TWA allows for periodic excursions above the PEL or TLV for short periods of time, provided there are compensating exposures below the PEL or TLV during the workday. Although the PEL or TLV are considered to represent the airborne concentration of a workplace chemical to which most workers can be exposed 8 hours per day and 40 hours per week throughout their working lifetime without suffering adverse effects to health and wellbeing, they should not be looked upon as being absolute values between hazard and nonhazard or an index of relative toxicity.

To illustrate the TWA concept, consider a worker who is degreasing metal parts at two different workstations using trichloroethylene (50 ppm TLV). The employee spends 240 minutes at Station 1 with an average exposure of 30 ppm, followed by 45 minutes with no exposure (lunch), then 195 minutes at Station 2 with an exposure of 60 ppm. Using the formula:

$$\frac{(C_1T_1)+(C_2T_2)+(C_3T_3)+\ldots+(C_nT_n)}{480\text{ minutes}} = X\text{ ppm}$$

where $C$ denotes the concentration of the chemical, and $T$ refers to the respective time period of use of that chemical, the 8-hour TWA for the scenario above can be calculated as follows [6]:

$$\frac{\left[\begin{array}{l}(30\text{ ppm}\times 240\text{ min})+(0\text{ ppm}\times 45\text{ min})\\ +(60\text{ ppm}\times 195\text{ min})\end{array}\right]}{480\text{ minutes}} = 39\text{ ppm}$$

The TLV–TWA of 50 ppm for trichloroethylene exposure has not been exceeded.

One may want to evaluate the employee's work practices to consider whether the chemical in use can also be absorbed through the skin or eyes, as well as other aspects of the operation (e.g., provision of local exhaust ventilation) to reduce the exposure as much as possible (refer to the sections on exposure controls and personal protective equipment). The TLV for a mixture of airborne chemicals can be calculated by using the formula [2]:

$$\frac{C_1}{T_1}+\frac{C_2}{T_2}+\cdots+\frac{C_n}{T_n}=1$$

provided the components of the mixture have similar toxicological effects and the workplace air is analyzed for each component. If the calculated results indicate a value greater than 1, then the TLV of the mixture measured has been exceeded. As an example, suppose a worker was exposed to a mixture of 25 ppm *n*-hexane (TLV = 50 ppm) and 200 ppm VM&P naphtha (TLV = 300 ppm ) during the shift. The calculated TLV is 25 ppm/50 ppm + 200 ppm/300 ppm = 0.5 + 0.67 = 1.17. The threshold limit has been exceeded, and action should be taken to reduce this exposure. For other examples of TLVs for mixtures, refer to the most current annual version of the ACGIH TLV handbook [4].

Two other categories related to evaluation of the airborne chemical exposure assessment of OELs (e.g., PELs or TLVs) described in the literature are the *short-term exposure limit* (STEL) and *ceiling limit* (C). A STEL is a 15-minute exposure limit (above the specified 8-hour TWA) to which a worker can be continuously exposed without suffering debilitating effects (e.g., irritation, chronic or irreversible tissue damage, narcosis). The STEL should not be exceeded at any time during the 15-minute period of exposure. Allowances are made for up to four STEL excursions per day, as long as there are at least 60 minutes between exposure periods and the 8-hour TLV–TWA is not exceeded. The STEL is supplementary to the TWA limit and should not be used for any other purposes, such as workplace engineering design. Methanol is an example of a solvent having both an 8-hour TLV–TWA (200 ppm) and a STEL (250 ppm). A ceiling limit is the airborne concentration of an applicable chemical in air that should not be exceeded at any time, even instantaneously. Isophorone, a solvent used in some printing inks, has a ceiling limit of 5 ppm [4]. For certain compounds, a *skin notation* (S) has been added to indicate a possible significant contribution to overall exposure from absorption through the skin, mucous membranes, or the eyes. Both methanol and propanol have skin notations.

When considering useful methods of worker medical surveillance and exposure monitoring, biological exposure indices of specific chemicals are useful. BEIs, as provided in the ACGIH's annual publication of Threshold Limit Values and in other sources, provide levels of the specific chemical or its metabolites, as measured primarily from the exhaled air, blood, and urine of workers exposed to the airborne contaminant, for a worker with inhalation exposure to the same chemical. BEIs are a measure of the amount of chemical or its metabolites in the body and may be useful when evaluating the possibility of skin absorption, effectiveness of personal protective equipment, or nonoccupational exposure. BEIs are strictly related to 8-hour exposures, 40 hours a week, and to the specified timing for the collection of the sample. For example, the BEI determinant for *n*-hexane exposure is 2,5-hexanedione measured in urine (5 mg/g creatinine) collected at the end of the shift. In the case of altered work schedules, BEIs may be extrapolated based on pharmacokinetic and pharmacodynamic considerations [4].

The AIHA WEEL Committee develops *workplace environment exposure limits* (WEELs) for chemical agents in common use in the workplace that have no other current OEL guidelines established by other organizations. WEELs are expressed as an average concentration measured over a time period, as different time periods of exposure measurement are specified depending on the properties of the particular chemical. The skin notation for a WEEL is used in the same manner as the ACGIH TLV.

The NIOSH recommended exposure limits are expressed as TWAs or as ceiling limits, or both. These recommended limits are published as criteria documents and are revised periodically as new research information regarding the particular chemical becomes available. The RELs are applicable to worker exposure assessments for up to a 10-hour day and are intended to provide the maximum possible health protection for all workers against acute and chronic effects of exposure. Likewise, the RELs provide skin notations for specific chemicals, where applicable.

The OSHA permissible exposure limits are national consensus standards incorporated by law as federal OSHA standards in Section 1910.1000 of Title 29 of the Code of Federal Regulations, Table Z-2. These PELs are derived from the 1968 ACGIH TLVs and certain air-quality standards recognized as *maximum allowable concentrations* (MACs) of the American National Standards Institute. The PELs represent the legal allowable concentrations of airborne contaminants within workplaces regulated by OSHA. Although the ACGIH revises some TLVs each year, the PELs remain as created unless changes are made in the law. A number of revisions and additions have been made to the PEL list since 1970. The PELs identify chemical TWAs, ceiling values, and skin notations.

The National Institute for Occupational Safety and Health has also identified those airborne concentrations of specific chemicals it considers to be *immediately dangerous to life or health* (IDLH). The IDLH nomenclature addresses extremely hazardous conditions. It is the estimated maximum concentration of an airborne contaminant from which a worker can escape (e.g., after failure of respiratory protection) without losing his or her life or suffering permanent health impairment.

An area of focus within the occupational health and safety community that is becoming more and more common in today's workplace involves the evaluation of potential adverse effects upon workers resulting from chemical exposures during extended work shifts (e.g., 10-hour or 12-hour days). To compensate for the higher accumulated doses and reduced recovery times caused by the longer work periods, adjustments to the exposure limits must be made. For a discussion of this subject, refer to the article by Paustenbach [104].

## INDUSTRIAL HYGIENE SAMPLING METHODOLOGY

In the workplace, the industrial hygienist utilizes many different sampling techniques and methodologies to identify and measure the various chemical, biological, radiological, and physical hazards to worker safety and health. In the case of airborne chemical, biological, or radiological hazards, the concentration of the particular workplace hazard is measured either in the general area where the work is being performed or in the immediate area of personal dose intake. The most accurate method of evaluating the amount of a particular airborne contaminant a worker may inhale is to collect measured samples over a defined time period from an area approximately one foot in radius from the worker's nose called the *breathing zone*. This type of sample is recognized as a *personal sample*. If, on the other hand, the industrial hygienist's objective is to identify and evaluate the types and concentrations of chemicals being put into the general workplace air from various points of generation (e.g., degreasing stations, paint spray booths, mixing kettles, metal forming and cutting lines), the technique used is the collection of measured samples over time within the general work environment close to the operation in question. This type of sample is generally termed an *area sample*.

A number of methods are available to collect and evaluate solvent (airborne contaminant) concentrations, depending on the nature of the operation, the solvents of interest, and the objectives of the measurement process. The primary categories of industrial hygiene sampling include *active sampling*, *direct measurement*, and *passive dosimetry*. The qualitative and quantitative value of the measurements obtained from each of these sampling methods is dependant on the level of accuracy required, the urgency in obtaining results, the cost of sample collection and analysis, and the level of difficulty for collecting the samples. As chemical detection capabilities and instrument sensitivities have improved, the trend in evaluating the workplace environment in recent years has moved toward the use of direct-reading instruments (e.g., colorimetric detector tubes and handheld instruments) and passive dosimetry (e.g., organic vapor badges). This is due, in large part, to the immediate feedback and ease of use associated with these methods. For additional information on industrial hygiene sampling, refer to the ACGIH text, *Air Sampling Instruments* [1].

### ACTIVE SAMPLING

Active sampling involves the use of a battery-powered sampling pump to draw a measured volume of contaminated air onto a suitable collection medium over a defined period of time which is then analyzed in the laboratory to accurately quality or quantify the amount of material collected. The sampling pump requires field or laboratory calibration to ensure that it is drawing air across and through the collection medium at the desired flow rate. The collection medium is an integral part of the sampling train, being connected to the sampling pump via tubing. The complete sampling train is then positioned in the immediate area of the process (point of generation) or is attached to the worker, with the sampling pump on the hip and collection medium located in the worker's breathing zone (typically the shirt collar). When completed, the collection medium is provided to an accredited industrial hygiene analytical laboratory where the chemicals of interest are extracted and analyzed. Various types of collection media are available for solvents, depending on such factors as the polarity and complexity of the chemical being evaluated. Examples of the variety of sorbant media utilized include:

- Activated charcoal for sampling solvents such as chlorinated hydrocarbons, gasoline, many alcohols, and ketones
- Silica gel for amines, methanol, phenols, and aldehydes
- Chemically treated media, including filters for toluene diisocyanates, naphthylamines, and toluidines

### Direct Measurement

Direct measurement devices allow solvent (airborne contaminant) concentrations to be measured within the ambient workplace with nearly instantaneous results. This approach is of great assistance when an area or operation requires immediate or high-frequency assessment. For these purposes, direct-reading instrumentation is useful in a number of applications, including identifying potential process leak locations, determining high-exposure areas and occurrences, evaluating the effectiveness of engineering controls, or for continuous monitoring applications. This type of sampling method is referred to as a *grab sample* and represents the presence and general concentration of the air contaminant only at that location and point in time. The results obtained should be considered semiquantitative or semiqualitative, due to larger acceptable standard error and method limitations.

A wide variety of direct-reading devices and instruments is available for measuring solvent concentrations. The measurement methods generally incorporate the use of colorimetric detector (or indicator) tubes and badges or direct-reading instruments. Colorimetric detector tubes and badges contain reagent-containing media that react with airborne solvent vapor (categorically or specifically) to produce a color change. The intensity of color change or the length of stain is compared with a calibration scale to determine the concentration (generally within ±25%) of the solvent vapor. The use of these tubes generally incorporate the use of a calibrated bellows-type pump or battery-powered pump; they are easy to use and relatively inexpensive and may be used for short (several minutes) or longer (hours) sampling intervals. Disadvantages include possible interfering compounds, lower accuracy, and some subjectivity in the readings. Direct-reading instruments are preferred over detector tubes when multiple readings are desired. Types of direct-reading instruments suitable for measuring solvents include analyzers with flame ionization and infrared detectors, combustible gas/vapor meters, photoionization detectors, and portable gas chromatographs. Direct-reading instruments can be either handheld (for portability) or fixed (for continuous area monitoring).

### Passive Dosimetry

Area or personal sampling that utilizes passive dosimeters (e.g., organic vapor monitors) incorporates the principles of molecular diffusion into the collection media rather than a sampling pump. This detection and measurement technology is particularly well suited for personal sampling because these devices [19]:

- Are lightweight
- Are unobtrusive
- Require no external power source
- Require no calibration
- Can be used to obtain short-term or full-shift exposures

Organic vapor monitors are accurate (within an accepted standard error and limitations) and can be used to sample for many industrial solvents. Analysis generally involves chemical desorption and gas chromatographic methods and is similar to that for the charcoal tubes mentioned above.

## EXPOSURE CONTROLS

Worker overexposure to solvents may be avoided through the selection and implementation of proper administrative policies and practices, management programs (e.g., employee training, education, housekeeping, waste disposal, job safety analysis, medical surveillance), workplace planning (e.g., process design and location, equipment design, materials storage, chemicals used), and the use of engineering controls (e.g., supply and exhaust ventilation), where necessary. The selection of the appropriate worker exposure control methods will depend on the nature of the hazard, how the potential hazard enters the work space, and the routes of anticipated exposure. Specific controls may be mandated by federal health and safety regulations (as in the case of benzene and vinyl chloride) or when exposure levels exceed established occupational exposure limits (e.g., PELs and TLVs). The workplace controls of first choice should always focus on engineering and management/administrative practices. The incorporation and use of appropriate personal protective equipment within the workplace should be viewed as a control method of last resort when no other practical means are available to control worker exposure. Control of solvent exposure is generally not addressed by a single control measure and is most often achieved by the application of a combination of these methods. The preferred approach in controlling solvent exposure is through the implementation of effective administrative/management programs and engineering controls.

### Administrative and Management Control

The first line of defense in the control of employee health and safety results from the implementation of effective workplace operations policies, and procedures. These policies and procedures set the expectations for personnel practices, workplace behavior, product quality, and management commitment. It is the measure of management commitment, and adherence to the policies and procedures specified defines the level of success in workplace safety, hazards management, employee health, staff and line integration, and community wellbeing. Typical management programs that serve to enhance workplace safety and employee wellbeing include:

- Employee education (hazard communications, hazard awareness, emergency procedures, frequent safety meetings, proper use of personal protective equipment)
- Employee training (e.g., process procedures, materials use, equipment operation and maintenance, standard operating procedures, safe work practices)
- Community outreach programs (e.g., community right-to-know practices)

### Engineering Controls

Engineering controls play a key role in the safe use of production equipment, the workplace environment, and employee wellbeing. To accomplish these objectives, significant consideration must be given to the design and integration of applicable engineering controls in the workplace layout, selection and use of production equipment, chemical and materials usage, process design and operating procedures, ventilation system configuration, fire suppression, utilities, etc. Engineering controls should figure prominently in the control of process operations, including the use of solvents and other chemical hazards. In this regard, useful engineering controls would include practices to:

- Eliminate use of the solvent in the process.
- Substitute a less toxic solvent in the process.
- Isolate the process from surrounding operations.
- Enclose the process to minimize worker exposure.
- Use effective local exhaust or dilution ventilation to eliminate or minimize hazardous levels of solvent vapors from the worker's breathing zone.
- Revise processes to minimize or eliminate process hazards by changing from manual to mechanical systems, wet methods from dry methods, water-based cleaners from organic solvent-based ones, etc.

### Elimination

Organic solvent-based processes can often be redesigned to eliminate the use of a particular solvent in the operation. Eliminating the solvent is considered the best approach to controlling worker exposure. Such options should be evaluated during the initial design of the process and can be instituted whenever safer process methods become available. Over the years, the practice of solvent elimination has been utilized in various organic solvent-based processes such as metals degreasing, cleaning, printing, painting, and treatment. These initiatives are the result of a combination of health and safety awareness, materials cost control, good business practice, and government regulation. Industry has recognized that the elimination of organic solvents may have additional benefits as well, including the reduction of hazardous air pollutant emissions into the ambient environment, the cost savings associated with decreases in waste treatment and disposal costs, and significant reductions in personal protective equipment purchases. Examples of organic solvent elimination include replacing chlorinated solvent degreasers with water-based detergent or subcritical carbon dioxide systems (discussed later); replacing solvent-based paints with water-based paints; improving flux application systems in circuit-board manufacturing to eliminate the need for cleaning with chlorinated compounds; and using water-based or vegetable oil-based inks to eliminate solvent-based inks.

### Substitution

When the elimination of solvents from a process is not practical, it is often possible to replace the more toxic solvent with another of lower toxicity or less hazardous physical properties (e.g., higher flash point, lower vapor pressure). Often, substitutions may be made within a chemical series by retaining the active group; for example, substitution of butyl cellosolve for methyl cellosolve may be advantageous. The general group also can be retained such as in the substitution of aromatic naphtha for toluene or toluene for benzene. Substituting a solvent with similar polar characteristics but different toxicity, such as ethanol for methanol, may also be possible. Other examples include replacing perchloroethylene with citrus-based products in metal degreasing, substituting isocyanate-containing coatings with toluene-based materials, and replacing formaldehyde used in preserving laboratory specimens with glycol-based compounds.

## Isolation and Enclosure

A process can sometimes be enclosed or automated to isolate the worker from the hazards of operation. When total enclosure of a solvent-based process is not possible, the operation can be separated from adjacent areas to minimize the number of workers potentially exposed to the vapor. The isolation by enclosure of a solvent-based process usually requires the introduction of local exhaust or dilution ventilation (see below) to prevent or minimize the accumulation of toxic concentrations of vapors within the workspace or process enclosure (fire/explosion hazard). Examples of isolation are found in most manufacturing environments; for example, manual painting in automotive assembly plants and manual metal-plating operations have been replaced with robotic systems. These automated processes often can be operated and monitored from remote locations.

## Process Revision

The revision of production processes remains a viable engineering control. Often, an engineering cost–benefit analysis that includes consideration of potential decreases in employee health costs, medical surveillance testing, the use of personal protective equipment, and the return on investment for new equipment will determine the practicality of the option whenever newer techniques and production equipment become available. Two examples of process revision related to the potential for exposure to solvents in industrial practice are seen in the spray-painting practice. Such revisions in process include replacing spray-painting with paint dipping and replacing compressed-air spray-painting with electrostatic methods, resulting in a decrease in the volume of paint overspray into the workplace.

# WORKPLACE VENTILATION

When the engineering control methods discussed above are not feasible or available, the incorporation of mechanical ventilation methods for the control of worker exposure to airborne contaminants such as solvent vapors is appropriate. Within the occupational workplace setting, this involves the balanced delivery of a sufficient supply of uncontaminated air into the work area (dilution ventilation) or direct removal of contaminated air by both general and local exhaust ventilation methods.

## Dilution Ventilation

The incorporation of mechanical ventilation in the workplace to manage worker exposures to airborne contaminants can, in some situations, be accomplished by the introduction of sufficient fresh air in specific work locations to dilute vapors to acceptable levels. Typical applications for dilution ventilation are controlling heat exposures (as in foundries) or regulating humidity and odor. If dilution ventilation is to be used to reduce the concentration of solvent vapors in the ambient air, at least four conditions must be met [3]:

1. The concentrations of solvent vapor generated must be relatively low or the air volume necessary for dilution will become so costly and inefficient as to be impractical.
2. The worker must be positioned a sufficient distance from the immediate source of solvent vapor generation to ensure that the established PEL, TLV–TWA, STEL, and ceiling limit are not exceeded.
3. The solvent must have a relatively low toxicity rating.
4. The solvent vapor must be released into the work environment at a uniform rate.

## Local Exhaust Ventilation

In contrast to dilution ventilation, *local exhaust ventilation* (LEV) functions to remove the solvent vapors (air contaminant) at their point of generation. In most instances, the use of local exhaust ventilation proves to be more effective in protecting the worker from exposure and less expensive to operate because lower air volumes and smaller fans are required. LEV systems can consist of a canopy hood or slotted capture collection system (depending on the vapor density of the solvent), duct work, a suction fan, and an optional filtration system for contaminant removal prior to discharge to the outside environment. Figure 15.3 shows a typical local exhaust ventilation system [6]. The decision of whether or not to install LEV for solvent exposure control is based on a number of factors, including:

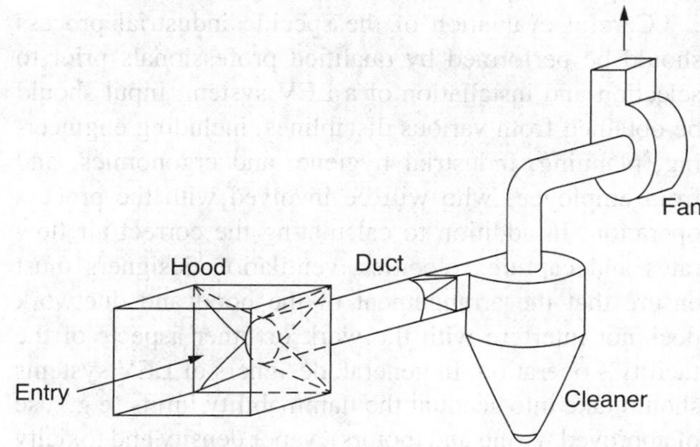


**FIGURE 15.3** Typical local exhaust system components. (From DiNardi, S. R., Ed., *The Occupational Environment: Its Evaluation and Control*, AIHA Press, Fairfax, VA, 1997. With permission of the American Industrial Hygiene Association.)

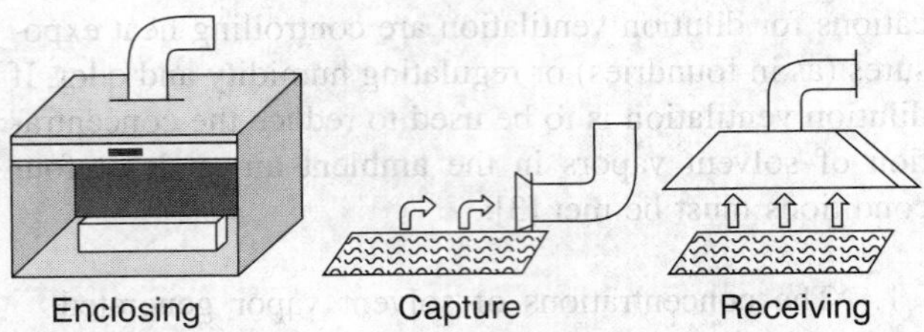


**FIGURE 15.4** Three types of local exhaust ventilation hoods.

- Lack of more cost-effective controls
- Volume and toxicity rating of the solvent vapor generated
- Regulatory requirements
- Good management practice

In LEV systems, the contaminated air is exhausted to the outside ambient environment either directly or by passing the air stream through some variety of contaminant collection or filtration system. The three types of LEV hood designs for solvent vapor control are *enclosing*, *exterior* (or capture), and *receiving*. Figure 15.4 provides an illustration of each [6].

Enclosing hoods partially or completely enclose the process so the point of contaminant generation is located inside the hood. Enclosing the process as much as possible increases the effectiveness and efficiency of LEV systems. Examples of enclosing systems include laboratory chemical fume hoods and spray-paint hoods. Exterior hoods, also called *capture* or *slotted hoods*, are located near the point of contaminant generation but do not enclose it. Examples of exterior hoods are slot-type hoods used on vapor degreasing processes and flexible hoods used to exhaust solvent-based mixing processes. Receiving hoods are typically canopy-type hoods used for exhausting hot processes (e.g., ovens and detergent baths). They are generally less suitable for solvent operations such as metal cleaning and degreasing.

Careful evaluation of the specific industrial process should be performed by qualified professionals prior to selection and installation of a LEV system. Input should be obtained from various disciplines, including engineering, planning, industrial hygiene, and ergonomics, and from employees who will be involved with the process operation. In addition to calculating the correct air flow rates and capture velocities, ventilation designers must ensure that the arrangement of the hood and ductwork does not interfere with the work or other aspects of the facility's operation. In general, designers of LEV systems should take into account the flammability limits (e.g., use of approved wiring and motors), vapor density and toxicity of the solvent, the anticipated concentration and volume of vapor generated, possible interfering air currents in the room, whether access to the work area is needed, and, the amount of airflow or capture velocity required to adequately exhaust the contaminant [6]. *Capture velocity* is the air velocity at any point in front of the hood or at the hood opening that is necessary to overcome surrounding air currents to capture the contaminated air at the point of generation by causing it to flow into the hood. Recommended capture velocities for solvents vary between 50 and 500 feet per minute depending on the nature of the operation, toxicity of the vapor, and conditions of solvent dispersion into the air [3].

Figure 15.5, Figure 15.6, and Figure 15.7 provide design detail principles of local exhaust ventilation, including hood nomenclature and design considerations. For further reading on local exhaust ventilation systems, refer to the fundamental text *Industrial Ventilation: A Manual of Recommended Practice*, published by the ACGIH [3].

## Personal Protective Equipment

If engineering or administrative controls discussed above are not feasible or do not provide adequate protection, *personal protective equipment* (PPE) must be used to minimize exposures. PPE should always be considered a last resort and managed carefully by qualified individuals. This is due to a number of limiting factors associated with PPE, which include the following:

- PPE does not eliminate the hazard. When PPE (e.g., air purifying respirator and elastomeric gloves) is used to control solvent (workplace) exposures, the PPE functions as a barrier that separates the worker from the airborne and liquid hazards present. Should this PPE be compromised, either by tearing or degradation, the worker will be directly exposed to the hazard.
- The selection and use of the correct PPE are relative to its effectiveness. Because no single respirator type, filtration cartridge or canister, or glove/barrier clothing elastomer is effective in all conditions of solvent use, it is mandatory that the worker and the industrial hygienist/OHS specialist work together to ensure that the PPE selected remains effective against the hazards it is intended to protect against. It is therefore recommended that the employer select only that PPE that is approved by NIOSH whenever possible.
- The safe use and care of the PPE selected for protection of the worker require proper worker training and are mandated by federal OSHA regulations, as specified in Title 29 CFR 1910 (General Industry), 1915 (Shipyards), 1917 (Marine Terminals), 1918 (Longshoring), and 1926 (Construction). It is therefore incumbent upon management to ensure that proper PPE-

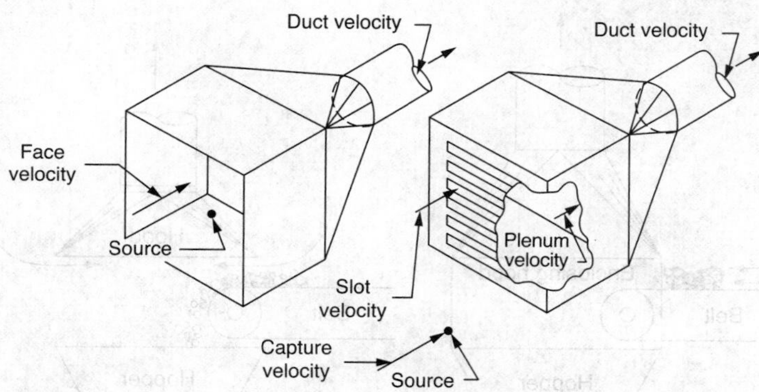


**FIGURE 15.5** Principles of exhaust hoods. (From ACGIH, *Industrial Ventilation: A Manual of Recommended Practice*, 23rd ed., American Conference of Governmental Industrial Hygienists, Cincinnati, OH, 1998. With permission.)

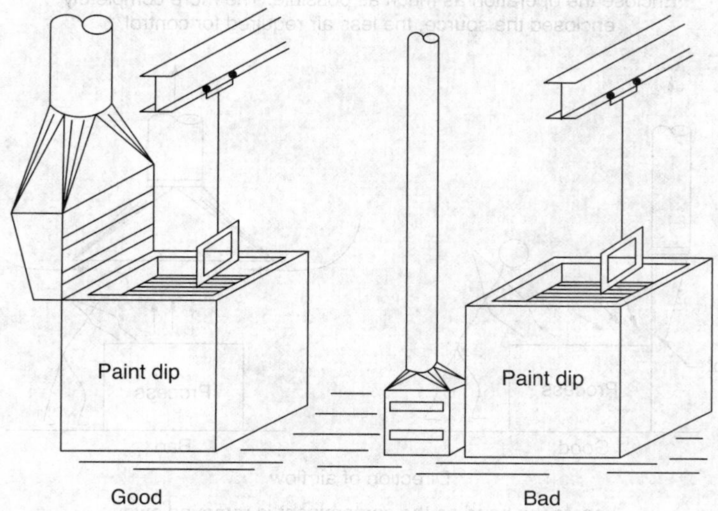


**FIGURE 15.6** Principles of exhaust hoods. (From ACGIH, *Industrial Ventilation: A Manual of Recommended Practice*, 23rd ed., American Conference of Governmental Industrial Hygienists, Cincinnati, OH, 1998. With permission.)

use programs and procedures are written and available to the respective employees and that the employees are sufficiently trained to know when and how to properly use the PPE available to them. Likewise, it is incumbent upon each employee to participate in such training programs and to use the PPE made available to them in the proper and intended way.

- The use of PPE may provide a false sense of security to the user. Depending on the frequency of use and the level of training the worker receives, the PPE user may believe that the PPE selected will provide complete protection under all circumstances, leading them to enter into circumstances for which the PPE is not intended or effective. This type of mistake is most commonly evident in chemical spill or emergency incident situations when the volume and concentration of solvent (chemical) vapors and liquid are sufficient to quickly compromise the effectiveness of the standard workplace PPE.

The types and variety of PPE used within the workplace are numerous, making the proper selection and use of this equipment a necessary, sometime difficult task for management and the worker alike. The PPE selected for use in a particular job setting should match not only the level and type of hazard to be confronted (based on the chemical and physical properties of the chemicals in use, the workers level of direct contact with the hazard [solvent], and the nature of the work to be performed) but also the worker's activity, vision, and dexterity needs while wearing the PPE. Often, the worker must confront a combination of hazards and chemical mixtures during job performance, making PPE selection and use decisions complex; consequently, it is not only advisable for management to consult with respective PPE manufacturers

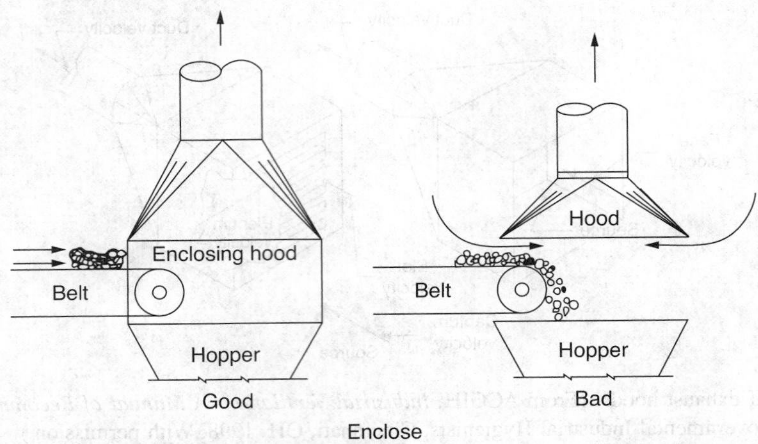
Enclosing hood
Belt
Hopper
Good
Hood
Belt
Hopper
Bad
Enclose
Enclose the operation as much as possible. The more completely enclosed the source, the less air required for control.

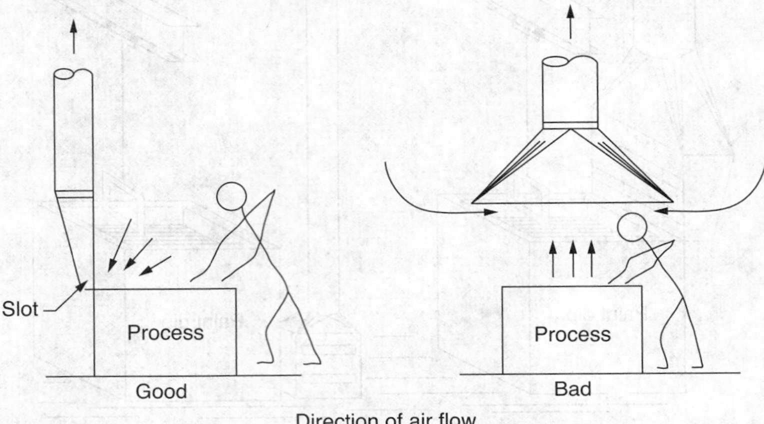
Slot
Process
Good
Process
Bad
Direction of air flow
Locate the hood so the contaminant is removed away from the breathing zone of the operator.

**FIGURE 15.7** Principles of exhaust hoods. (From ACGIH, *Industrial Ventilation: A Manual of Recommended Practice*, 23rd ed. American Conference of Governmental Industrial Hygienists, Cincinnati, OH, 1998. With permission.)

regarding the intended use of PPE within the workplace but also for the worker to be ever vigilant while using PPE to ensure that the integrity of the PPE remains protective against the hazards confronted.

The primary categories of PPE in common use within the workplace include respirators, clothing (suits, gloves, foot coverings, sleeves and aprons), eye, and face protection. The combined use of PPE representing these categories is dependent on the nature of the hazard, the frequency and duration of exposure, and the nature of the work to be performed.

## Respirators

Respiratory protection is used to provide the wearer with clean breathing air while working in the presence of airborne contaminants at levels that may prove to be injurious to health or wellbeing. Often, the use of respiratory equipment in the workplace is employed to protect the worker from intermittent exposures that can occur during process operations or during emergency repair and maintenance; however, respirators may also be the only feasible method of protection for exposures that may occur during normal work operations.

When workplace engineering (e.g., local exhaust or dilution ventilation) or other control methods are not feasible, the employer should implement an effective respiratory protection program, including worker training and medical surveillance (as applicable), and provide workers with respiratory protection appropriate to the hazard. The goal is to ensure that the appropriate type of respiratory protection is selected and used correctly. OSHA's Respiratory Protection Standard for General Industry, as established in 29 CFR 1910.134, specifies the employer and employee responsibilities when respiratory equipment is to be used within the workplace. This standard mandates that the employer develop and implement a written respiratory protection program that describes when respiratory equipment is to be used, what respiratory equipment is to be used, which employees are qualified to use the respiratory protection specified, and how the respiratory equipment is to be used and maintained. Elements of the program include respirator selection, user training and fit testing, medical approval, and specific instructions for cleaning and maintenance (refer to the 2005 updated standard). Only approved respirators (e.g., by the National Institute for Occupational Safety and Health) should selected for use. NIOSH approval program requirements are specified in 42 CFR Part 84.

The two major categories of respirators are air-purifying and atmosphere-supplying. Air-purifying respirators for chemical and particulate (e.g., solvents) exposure provide protection to the user by removing the hazardous contaminant from the stream of the air prior to its inhalation by the user. This air-cleaning process is accomplished by drawing the contaminated air through specially prepared cartridges or canisters containing various filtering or sorbent materials (e.g., activated charcoal). Air-purifying respirators are available in various facepiece configurations, including quarter-, half-, and full-face mounts, with or without eye protection. They are designed to be disposable or for reuse and can be powered or standard air-purifying devices. It is important to recognize that air-purifying respirators *do not* provide an independent source of breathing air and should not be used in environments that are oxygen deficient or when the airborne contaminants present do not have good warning properties (e.g., noticeable taste, odor, minor irritating effects below the established occupational exposure limit) to indicate cartridge or canister overloading, improper facepiece fit, or damage to the respirator. Air-purifying respirators should never be worn for protection against airborne chemicals with poor warning properties [6]. Because people vary greatly in their ability to detect odors, other methods such as cartridge replacement schedules or visible end-of-service-life indicators are being developed by various groups to ensure greater safety when air-purifying equipment is used. It is also important to recognize that, because people have different facial configurations, no one make or model of respirator will provide a sufficient face-to-facepiece seal for all users.

Atmosphere-supplying respirators provide the user with an independent source of clean breathing air separate from the local environment. Examples of atmosphere-supplying respirators include air-line devices and self-contained breathing apparatus (SCBA). Whereas air-line devices are designed to provide the user with an independent source of air for extend periods of time, the SCBA is designed to provide the user with an independent source of breathing air ranging from 5 minutes (emergency escape only) to 60 minutes, depending on user activity and equipment configuration.

Other factors that must be considered when selecting respiratory protection are the nature of the hazard, including oxygen deficiency, concentration of the airborne contaminant, and adequacy of warning properties. The expertise of the industrial hygienist is necessary for proper identification and assessment of the airborne contaminant levels, flammable limit status, and ambient oxygen concentration; evaluating the configuration and location of the work area in relation to an available area of clean air; implementing appropriate and periodic workplace and user monitoring procedures during respirator use; and determining the proper respiratory equipment for use.

### Protective Clothing

Protective clothing is used to protect the user from dermal contact and exposure to chemicals by forming a barrier between the skin and the hazard (e.g., solvent). The proper use of protective equipment, including clothing, eye, face, hand, and foot protection, is mandated by OSHA standards found in 29 CFR 1910 Subpart I and should be referred to by employer and employee alike.

Protective clothing includes gloves, laboratory coats, rubber aprons, chemical resistant suits, and boots. Various configurations of chemical-resistant clothing can be selected, depending on the nature and concentration of the solvent (hazard) of concern, the kind of work activity and time period to be performed, and the level of protection required. Chemical-resistant elastomers used as barrier coatings for protective clothing include neoprene, nitrile, natural, or butyl rubber; polyvinyl chloride or polyvinyl alcohol; and Viton™. Some operations may require only partial protection (such as a protective apron, sleeves, or leggings), while others may require the use of full-body enclosures (such as those used by emergency response workers). It must be recognized that the type of protective clothing configuration, barrier material, and level of protection chosen can significantly affect the mobility, vision, and manual dexterity of the worker. Depending on the characteristics of the work environment, the use of protective clothing, especially encapsulating suits, can present potential heat stress hazards and may therefore require the close monitoring of workers and workplace conditions.

The degree of protection afforded by a given type of chemical-protective clothing is related to three primary performance factors: *permeation*, *degradation*, and *penetration*. Permeation is the ability of a chemical to pass through the molecular configuration of a protective barrier (e.g., the clothing or glove) and is defined on the basis of the permeation rate for the particular material being challenged. Degradation of a material results from a reduction in one or more of the physical properties of protective clothing or gloves due to direct contact with a chemical and is defined by the degradation rate for the material with respect to the specific chemical challenge. Penetration is measured as the rate of flow of a chemical through physical aspects of the clothing or glove, such as zippers, seams, pores, or imperfections in the material. Manufacturers of protective clothing determine product-specific performance data via laboratory tests conducted in accordance with methods established by the American Society for Testing and Materials (ASTM). All three factors should be considered when choosing protective clothing, because data for some but not all of these factors may not correlate with a given type of clothing and target chemical; for example, a glove may have acceptable degradation ratings for use with a specific chemical, but the chemical may readily permeate the material. No single glove or type of protective clothing provides adequate protection against every hazard. Furthermore, a glove type from one manufacturer often has different performance data from the same glove type produced by another manufacturer. Another general source of information is *Chemical Protective Clothing*, published by the American Industrial Hygiene Association [5]. This two-volume set provides the data required to select and use chemical protective clothing. Included in the document is a discussion of permeation theory, testing methods, and available vendors.

### Eye and Face Protection

Eye and face protection is used to prevent injuries that may occur while handling or transporting solvents and other chemical liquids, vapors, fumes, or particulates. Two types of protective eyewear in common use to prevent exposure are chemical splash goggles and face shields. Chemical splash goggles are designed to completely enclose the eyes (as opposed to safety glasses, which are designed to prevent physical injuries that may result from an object striking the eye). Some goggles may also prevent vapor exposure to the eye in addition to contact with the liquid. Face shields are often worn in conjunction with goggles to protect the face and neck. Face shields and goggles that meet recognized safety standards bear the engraving of ANSI 2-87, which indicates that the device has passed safety performance tests conducted by the American National Standards Institute. OSHA regulates the safe use of eye and face protection within the workplace as describe in 29 CFR Part 133.

## ABSORPTION OF SOLVENTS AND INHALATION EXPOSURE

The chemical property of solubility (lipid or water) is a key factor in determining how a solvent, whether by ingestion, dermal exposure, or inhalation, partitions itself into blood and tissues. When considering solvent exposure via the inhalation route, the rate of solvent uptake and the subsequent equilibrium concentration in tissues are also dependent on individual activity, breathing rate, and the minute volume of blood flow through the lung and other organs.

Solvents that are highly soluble in water-based systems, such as blood and tissues, are absorbed very readily into the system by the inhalation route, causing blood concentrations to rise rapidly. The driving force is the difference in concentration of the solvent between inspired air and blood. The amount of solvent diffusing through the alveolar capillary membrane is dependent on the air–blood partition coefficient. Tissue equilibrium concentrations with solvents such as xylene, styrene, and acetone, which are highly soluble in blood and tissues, are not limited by pulmonary ventilation because the tissues act as a sink for the inhaled solvent. As pulmonary ventilation is increased, the blood and tissue concentrations continue to rise. The limiting factor in attaining the tissue equilibrium concentration is the blood flow through the tissues and the blood–tissue partition coefficient.

Solvents such as methyl chloroform, methylene chloride, trichloroethylene, and toluene, which have lower solubilities in blood and tissues, reach equilibrium rapidly because of low solubility or low blood–air partition coefficients. Tissue concentrations also will reach equilibrium rapidly because of low tissue–blood partition coefficients. In this case, the limiting factor in tissue concentration is the solubility of the solvent in the tissue and the individual's pulmonary ventilation rate. To achieve a higher concentration in tissues and blood, pulmonary ventilation must increase, allowing more solvent to enter the blood and a new blood–tissue equilibrium to be reached [13,14].

## DERMAL UPTAKE OF SOLVENTS

The opportunity for solvents to enter the body via contact with the skin is enhanced, in part, due to the large surface area of the skin (18 $ft^2$). Fortunately, the barrier properties of the skin associated with filamentous proteins and lipids of the stratum corneum naturally inhibit penetration by harmful non-lipid-soluble substances. Disruption of this barrier, however, by injury, illness, or removal of lipids can facilitate passage of these materials; for example treatment of the skin with polar organic solvents, detergents, and some surfactants can remove the lipids, thereby increasing the skin's permeability.

Penetration of the skin by a solvent depends on a number of other factors such as the thickness of the skin layers, the integrity of the skin, the concentration gradient of solvent on either side of the epithelium, and a number of physical constants. In addition, the degree to which the skin hydrates can increase absorption by affecting its permeability. Movement of water-soluble compounds may be impeded, however, when the stratum corneum is highly hydrated. Although hair follicles and sweat glands comprise only a small proportion of the skin's surface area they, too, provide pathways for solvent penetration.

Solvents can denature the lipids in the skin, resulting in drying and irritating effects, cellular hyperplasia, and swelling; for example, the careless use of solvents without the use of barrier creams or proper hand and arm protection frequently leads to cases of dermatitis in the workplace. In studies on the effect of solvents on the lipid barrier in the skin, the ability of a solvent to penetrate the skin is dependent on the polarity of the solvent and the surface charge of the skin. Results comparing penetration or removal of skin lipids by several solvents indicate that ethanol, the solvent with the greatest polarity, extracts the most lipids, followed by acetone and ether [17]. Treatment of the skin with solvents can also enhance the penetration rate of other compounds. In a study using excised human skin, the effect of several solvents, including dimethyl sulfoxide (DMSO), dimethylacetamide, formamide, and diethylformamide, on the penetration rate of sarin was examined [86]. The result of this study demonstrate that solvent pretreatment of the skin increases the rate of sarin transport across the skin barrier by a factor of 10 to 100 over that of sarin alone on control skin.

In studies using toluene, xylene, and styrene vapors to assess the rate of skin penetration of these aromatic solvents in human volunteers, human volunteers were exposed to 300 ppm or 600 ppm concentrations of the solvent vapors for 3.5 hours in a dynamic exposure situation. The subjects wore full-face respirators to prevent pulmonary absorption of the solvents via the inhalation route. Each subject exercised for a 10-minute period, sufficient to make the subjects perspire and to raise the skin temperature about 0.5°C. Perspiration and warm skin temperature enhance the hydration of the skin and subsequent percutaneous absorption. After termination of exposure, these solvents displayed biphasic elimination from the blood into exhaled air with a short half-life of about 1 hour and a much longer half-life of approximately 10 hours. Xylene and styrene showed a slight delay in excretion in exhaled air after percutaneous exposure when compared with similar exposure via the inhalation route. Delayed excretion after dermal exposure may be accounted for by a slow release from the skin after termination of exposure.

Overall percutaneous absorption of the xylene, toluene, and styrene concentrations corresponded to only about 0.1% of the amount estimated to be absorbed by the pulmonary route. This observation indicates a very small absorption potential for these solvents by the percutaneous route. When the percutaneous absorption of xylene vapor is compared to earlier work with xylene liquid, the vapor displays an approximately tenfold greater efficiency in penetrating the skin than does the liquid. According to Riihimäki and Pfäffli [116], it is not uncommon to observe greater penetration with vapor exposure because liquid solvents remove the lipids from the stratum corneum and thus interfere with absorption. Additionally, exercise promoted the absorption of solvents because of the warm hydrated skin. In general, percutaneous absorption of solvent vapors would not contribute significantly to the total blood concentrations of these solvents.

## TOXICOLOGY OF SELECTED SOLVENTS

This section provides insight into various solvents of occupational concern due to their propensity to produce neurotoxic, reproductive, or carcinogenic effects in humans. Examples are provided for these as well as less toxic, alternative solvents. Generally speaking, acute exposure to high levels of solvents can result in temporary or long-term alterations of central nervous system (CNS) function.

### Effects of Acute Solvent Exposure on the Central Nervous System

Although varying widely in chemical structure and physical properties, solvents produce a rather stereotypical set of toxicological manifestations upon acute exposure [10], the significance of which is dependent on dose concentration, duration, and frequency. Most commonly, acute exposure is evidenced by a varied level of CNS dysfunction and, if exposure is sufficiently severe, narcosis. The systemic toxicity of solvents is observable either throughout the body or in an organ with selective vulnerability distant from the point of entry of the chemical, as with solvents and peripheral neuropathies. Exposure to certain solvents can be associated with some temporary alteration of cognitive and psychomotor function following short-term exposures at or near the TLV. Exposure to greater concentrations may provoke such symptoms as headache, dizziness, ataxia, euphoria, drowsiness, lightheadedness, disorientation, confusion, tremors, and nausea. Exposure to potentially lethal (IDLH) levels of solvents can result in stupor, loss of consciousness, coma, respiratory depression, and abnormal cardiac function.

### Other Toxic Effects of Solvent Exposure

Exposure to solvents at concentrations too low to induce many of the acute symptoms cited above is of special concern with regard to neurotoxicity, because the capacity of nervous tissues for post-toxicity regeneration is limited and repeated insults may lead to cumulative damage. The subtlest symptoms of chronic solvent exposure include relatively mild alterations of mood and behavior not accompanied by quantifiable evidence of dysfunction on neurobehavioral tests [18,57]. Although dose–response and causal relationships have been difficult to study in the absence of animal models, symptoms of chronic solvent exposure may include increased irritability, decreased span of attention, and loss of interest in daily activities. More severe damage to the nervous system, both central and peripheral, occurs upon repeated exposure to certain solvents such as carbon disulfide and *n*-hexane, as discussed later in this chapter.

Numerous neurobehavioral and functional tests have been used to detect such changes in both clinical and experimental settings [69,112]. Whether the acute effects of solvents play a role in determining the pathogenesis of toxic lesions observed after chronic exposure to the same solvents is uncertain; however, current thought is that the acute effects on the nervous system are mediated through nonspecific interactions of solvents with the cell membrane—that is, increases in membrane fluidity or functional alteration of cell surface receptors—while the effects of chronic exposure are mediated by specific biochemical actions of solvents.

It is well known that neurotoxic chemicals can have a negative impact on sensory function. Often, the symptoms reported following chemical exposure to such chemicals are related to the senses [48]. Toluene, xylene, styrene, trichloroethylene, and carbon disulfide are examples of solvents associated with adverse affects on the auditory system [98]. In the industrial environment, workers are often exposed to solvents as well as high levels of noise, which is known to damage the inner ear and result in hearing loss. In recent years, evidence has emerged from workplace studies and animal experimentation that the combined effects of noise and ototoxic solvents may increase the susceptibility to hearing loss [68,98]. In one animal study, rats were exposed to toluene, to noise, or to toluene followed by noise and then their auditory functions were tested. Results showed that rats exposed to toluene followed by noise exhibited a decrease in auditory sensitivity greater than the sum of the effects of toluene and noise alone [98]. The risk for hearing loss may be increased by factors other than noise, such as drugs or other chemicals, and can also be influenced by heredity and aging [98]. It is important to take all of these factors into account when evaluating hearing loss in the workplace.

## CHEMICAL INTERACTIONS AFFECTING TOXICITY

It is not unusual for a toxic chemical to enter the body and interact with another toxic substance or with a medical drug. The result of this interaction of chemical agents, drugs, etc., can be mediated in several ways through effects on absorption, protein binding, biotransformation, or excretion of the drugs or chemicals. The possible results of chemical interactions within the body are listed below:

1. *Additive effect* (e.g., 2 + 2 = 4)—The effect observed when the combination of two chemicals having independent toxicities results in a combined toxic effect equal to the addition of the two. *Example:* Two organophosphates simultaneously used as pesticides will depress cholinesterase levels equal to the additive concentrations of each agent.
2. *Synergistic effect* (e.g., 2 + 1 = 20)—The toxicological effect that occurs when chemicals having independent toxicity produce a toxic response significantly greater in effect than the additive sum of the two substances observed individually. *Example:* The hepatotoxicity of carbon tetrachloride is significantly increased in the presence of ethyl alcohol.
3. *Potentiation* (e.g., 2 + 0 = 10)—The effect that is observed when one substance acts as a catalyst in the presence of another substance to enhance the toxicity of the second substance. *Example:* A steady diet of corn oil potentiates the effect of incomplete carcinogens.
4. *Antagonism* (e.g., 2 + 2 = 1)—A substance recognized as an antagonist is one whose effect tends to decrease the adverse effect of a second substance. There are various types of antagonist actions. Functional antagonism occurs when two chemicals produce opposing physiological effects and result in an overall no-net-effect exposure. Chemical antagonism occurs when the interaction of the two chemicals results in an interference with the normal chemical transformation of the chemicals so less toxic agent is available. Dispositional antagonism occurs when absorption, distribution, or excretion of the chemical is altered; therefore, less of the chemical mixture reaches the target tissue. Receptor antagonism occurs when competition for the same receptor results in less of either chemical reaching the receptor. *Example:* Many of the principles of antagonism are used in the design of antidotes in clinical toxicology or for the poisoning of humans.

The body eliminates foreign chemicals or chemical in higher concentrations than normal by xenobiotic metab olism. A xenobiotic is any chemical or foreign substanc that is found in the body of an organism that is not normall produced by or expected to be present in the organisn [149]. This consists of the deactivation and the secretio of xenobiotics and happens mostly at the liver. Secretio routes are urine, feces, breath, and sweat. Hepatic enzyme are responsible for the metabolism of xenobiotics, by firs activating them (oxidation, reduction, hydrolysis, an hydration of the xenobiotic) and then conjugating the activ secondary metabolite with glucuronic or sulfuric acid, o glutathione, followed by excretion in bile or urine. A example of a group of enzymes involved in xenobioti metabolism is the hepatic microsomal cytochrome P450s

Organs that catalyze relatively few types of chemica biotransformation reactions or have low rates of xenobioti metabolism, such as lung, nasal mucosa, and testes, may also be target organs for the toxicity of some solvents; for example, ethylene glycol monomethyl ether and related glycol ethers are recognized testicular toxicants [58,95], as are hexane and the hexane metabolite 2,5-hexanedione [22]. In recent years, special attention has been paid to the potential susceptibility of the tissues lining the upper airways (e.g., the nasal mucus) to solvent-induced toxicity. These tissues are known to have high levels of certain xenobiotic-metabolizing enzymes and, in addition, are generally exposed to high solvent concentrations relative to the lung and other organs. In particular, solvent esters such as propylene glycol monomethyl ether acetate, dimethylphthalate, and diethylsuccinate are enzymatically transformed by nasal carboxylesterase to yield acidic products that may accumulate to toxic levels in the nasal mucosa [95,126,133]. Certain solvents require no metabolism to adversely affect the tissues of the upper respiratory tract; vapors or aerosols of aldehydes cause local tissue damage to the nasal epithelium [97], presumably due to the activity of these solvents in forming protein–protein and protein–DNA cross-links [61].

Organs that receive a high percentage of the cardiac output are exposed to greater dose concentrations of absorbed toxicants than poorly perfused tissues. A major determinant of target-organ selectivity for the toxicity of solvents is how the solvent is metabolized, often referred to as *xenobiotic metabolism*. Whereas pharmacokinetics defines the quantity of solvent reaching a particular organ or tissue after absorption, metabolism may yield products with increased toxic potential relative to the parent chemical. Thus, well-perfused organs with high capacities for specific types of biotransformation reactions, mainly those catalyzed by cytochrome P450, are common targets for solvent-induced toxicity. In particular, the liver is vulnerable to the toxicity of many solvents, due to its high capacity for xenobiotic metabolism. Many common hepatotoxic solvents yield toxic intermediates or end products upon biotransformation (e.g., carbon tetrachloride [113], chloroform [109], and trichloroethylene [16]); however, some solvents, such as ethanol, may exert their hepatotoxic effects indirectly by altering the cellular reduction–oxidation balance during metabolism, thereby disrupting normal liver function and structure [117].

The kidney, as a filtering and concentrating organ of excretion, receives not only untransformed solvents but also the products of hepatic metabolism of solvents. These biotransformation products (e.g., conjugates of trichloroethylene) may be more toxic than the parent chemical and produce renal-specific toxicity [81]. The ion-transport and solute-concentrating functions of renal tubules also contribute to the vulnerability of the kidney to certain chemical toxicants [82]. In addition, the biochemical peculiarities of certain species and genders may play a major role in bringing about solvent-induced renal toxicity. A notable example is the susceptibility of the male rat to renal toxicity caused by exposure to 1,4-dichlorobenzene, Stoddard solvent, VM&P naphtha, and other hydrocarbon solvents. This has been attributed to the male-rat-specific abundance of the low-molecular-weight protein $\alpha_{2\mu}$-globulin, which acts as a carrier for lipophilic molecules [128]. $\alpha_{2\mu}$-Globulin is normally degraded in renal tubule lysosomes, and binding to a solvent ligand slows degradation of the protein so the $\alpha_{2\mu}$-globulin–hydrocarbon complex is sequestered by lysosomes [77]. The sequestered protein apparently disrupts lysosomal function and cytotoxicity results when large amounts of $\alpha_{2\mu}$-globulin accumulate [121]. No apparent counterpart to this type of nephrotoxicity exists in species other than the rat [7].

The human body has evolved a rapid and effective biotransformation system to eliminate nonpolar xenobiotics (e.g., benzene, benzo(*a*)pyrene, carbon tetrachloride) to protect it from the buildup of the xenobiotic agent within the body and from the partitioning of the xenobiotic agent into the lipid membrane layer. This partitioning effect often results in a disruption of fluidity of the membrane and results in toxicity. Nonpolar xenobiotics are transformed in the body into polar substances through the process of bioactivation to permit and enhance direct elimination or by further polar modification. The bioactivation process occurs through oxidation, reduction, or hydrolysis. Bioactivation of a xenobiotic agent becomes a danger to the cell when it overwhelms or creates an imbalance between bioactivation and detoxification processes, causing toxicity. Bioactivation often involves insertion or addition of oxygen molecules, or the transfer of an electron to create a reactive or unstable compound that subsequently rearranges to become a reactive compound.

## Solvent Mixtures

Humans are often exposed to multiple chemicals at work or in the home. An example, as reported by Worksafe Australia (the Australian National Occupational Health and Safety Commission), involved solvent exposure and health effects observed in spray-paint apprentices [143]. This study identified 32 different solvents contained in 20 thinner products used by the painters. Within this group, six different categories of solvents were represented: alcohols, aromatic hydrocarbons, esters, glycol ethers, ketones, and mixtures. Of significance was the fact that the workers commonly perceived the thinners to be equivalent and safe to use. This underscores the need for chemical communication programs to inform workers about the potential hazards of working with mixtures of chemicals.

As previously discussed, exposure to multiple chemicals, either simultaneously or sequentially, may alter the toxicological impact of the individual chemicals upon the body of the receptor, leading to a modification in toxic effects as might be predicted by simply summing their individual toxic behaviors. Thus, a combination of certain chemicals may affect (positively, negatively, or not at all) the absorption, distribution, metabolism, and excretion of the chemical mixture within the body of the receptor [74]. The study of chemical interactions has evolved most extensively in the area of therapeutic drugs. Although some information exists on interactions of industrial chemicals, most research on chemical toxicity to date has dealt with single, pure chemicals. These single chemical studies are important because they allow researchers to gather fundamental knowledge about the mechanisms of toxicity under conditions that are well controlled; however, much research remains in the evaluation of potential health effects associated with exposures to multiple chemical compounds [148].

## Glycol Ethers

Glycol ethers represent an important category of solvents that are widely used in mixtures for industrial and consumer applications. They are grouped as ethylene glycol, propylene glycol, or butylene glycol with the ether component of the molecule containing methyl, ethyl, propyl, butyl, or higher molecular weight moieties [52]. Additional members of this class of compounds are the corresponding acetate esters. The miscibility of glycol ethers with water and many organic compounds make them ideally suited as solvents in oil–water compositions. Production capacity of the ethylene-based ethers in 1992 exceeded 1 billion pounds, with the coatings (paint) industry being the major consumer [52]. In addition to coatings, glycol ethers are found in many household goods such as brake fluids, waxes, cleaners, dyes, detergents, degreasers, and inks. In particular, 2-butoxyethanol has been formulated into hundreds of consumer products [27].

The current ACGIH TLV–TWAs and German MAKs for three widely used glycol ethers—2-methoxyethanol (ME), 2-ethoxyethanol (EE), and 2-butoxyethanol (BE)—are 5 ppm, 5 ppm, and 20 ppm, respectively. The NIOSH RELs for ME and EE are significantly lower at 0.1 ppm and 0.5 ppm, respectively. All have skin notations. ACGIH bases their limit for ME on possible blood, reproductive, and CNS effects. For EE and BE, effects on reproduction and the blood are considered, respectively. The TLV and REL for propylene glycol monomethyl ether (PGME) are both 100 ppm and are based on potential irritation and CNS effects [2,4,101].

The commonly encountered glycol ethers are colorless liquids with mild odors. The primary routes of exposure in the industrial environment are inhalation and skin absorption [52]. Outside of the workplace, some cases of accidental or intentional ingestion of products containing glycol ethers by children and adults are reported [27]. In general, the ethylene glycol ethers exhibit low acute oral toxicity [52]. Experiments in rats have shown that the methyl, ethyl, and butyl ethers are readily absorbed through the skin [118]. As the molecular weights of the glycol ethers increase, the potential for inhalation exposure and skin absorption decreases. Because the methyl and ethyl ethers of ethylene glycol and their acetates have demonstrated adverse reproductive, embryotoxic, teratogenic, and developmental effects in animal studies [58–60,99,100], their use in consumer products has declined [52].

Metabolically, the monoalkyl ethers of ethylene glycol are converted to their respective alkoxyacetic acids via the actions of alcohol dehydrogenase [27]. Many of the observed adverse effects caused by ethylene glycol ethers in animals, such as hemotoxicity (e.g., 2-butoxyethanol) and testicular toxicity, are attributed to these toxic metabolites. Whereas rat erythrocytes have demonstrated vulnerability to the hemolytic effects of 2-butoxyacetic acid (from BE), human erythrocytes have been shown to be much less susceptible to these effects [27]. PGME and its acetate (PGMEA) are relatively innocuous compounds when compared to the ethylene glycol ethers discussed earlier. Overexposure to PGME has been associated only with increased liver weight and CNS depression. Studies have shown that EE and PGME are metabolized by different routes and the types of metabolites produced are responsible for the marked differences in toxicity; for example, methoxyacetic acid is the primary metabolite of EE, and propylene glycol is the main biotransformation product of PGME and PGMEA [94,96].

Investigators have studied the potential interaction of ethanol and EE due to the recognition of similar metabolic pathways and the likelihood of concomitant exposure to ethanol due to personal behaviors in some individuals [100]. When dose levels of EE are presented to rats alone or in combination with ethanol, researchers have noted an apparent increase the duration of pregnancy. Exposure to dose levels of EE during gestational days 7 to 13 resulted in observation of a decrease in certain behavioral tests such as rotorod performance; however, when test animals were exposed to dose levels of EE and also consumed ethanol, the behavioral deficits observed were diminished. When dose levels of EE were administered alone during late gestation, the motor activity levels of pups were depressed and performance during avoidance conditioning trials was retarded. Observers note that the combined administration of EE and ethanol appears to generate a synergistic effect on the behavioral deficits induced by EE and to depress both activity and learning. Also, it has been observed that ethanol during late gestation altered the neurochemical effects of EE.

In summary, concomitant exposure to ethanol and EE can have differential effects depending on the stage of gestation. Ethanol administration during the early period of gestation tended to improve both the behavioral and neurochemical effects of EE to approximately 50% of the response produced by EE alone. In the late stage of ges tation, however, the combination of ethanol with EE exaggerated the effects of EE alone. These scientific observations indicate that the possibility exists for ethanol-induced exaggeration of the potential toxic effects of EE exposure in pregnant workers.

Retrospective epidemiological studies of workers exposed to ME and EE report evidence of adverse effects on the male reproductive system, with increased frequency of reduced sperm counts [145]. Evaluation of sperm production in humans and several other animal species indicate that the output of human sperm is about one fourth that of other mammals when compared on a per-gram tissue basis. This finding suggests that humans may be more susceptible to occupational toxicants than predicted by laboratory animals [130]. As is the case with many widely used chemicals with potentially harmful effects, substitutes are being considered. PGMEA and ethyl-3-propionate have been identified as useful and less-toxic alternatives to ethylene glycol ether solvents [23].

## Benzene

Benzene has been used extensively over the years as a raw material in the manufacturing of polymers, detergents, pesticides, dyes, plastics, and resins and as a solvent for waxes, oils, natural rubber, and other compounds [93,123]. In addition, benzene is a component of gasoline and is generally present at low levels throughout the ambient environment [64]. Exposure to benzene in the workplace is primarily through inhalation, although skin absorption may also contribute to the overall body burden. OSHA regulates benzene specifically by standards established in 29 CFR 1910.1028 and recognizes benzene as an occupational carcinogen [102]. The OSHA PEL for benzene is 1 ppm and the STEL is 5 ppm [102]. The 2005 ACGIH TLV–TWA and STEL for benzene are 0.5 ppm and 2.5 ppm (skin notation), respectively. The ACGIH designates benzene as a confirmed human carcinogen [2,4]. The NIOSH REL and STEL for benzene are 0.1 and 1 ppm, respectively. NIOSH identifies benzene as an occupational carcinogen [2,4]. The ACGIH BEI for benzene is 25 pg of the metabolite *S*-phenylmercapturic acid per gram of creatinine in urine, as measured at the end of the work shift [4].

Due to its high lipid solubility, acute exposure to benzene can depress the CNS to the point of narcosis. Headache, dizziness, nausea, and vomiting are all features of benzene overexposure. Exposure to benzene at high concentrations can lead to blurring of vision, unconsciousness, convulsions, ventricular irregularities, and respiratory failure. Death as a result of exposure to extremely high concentrations of benzene may occur because of respiratory failure or cardiac arrhythmias [123,144]. Concomitant exposure to benzene and high concentrations of catecholamines can sensitize the heart and lead to ventricular fibrillation.

Benzene is hematotoxic and carcinogenic following repeated exposure to high concentrations [90]. Numerous rodent studies have shown that benzene can also cause cytogenetic damage *in vivo* [64]. In addition, examination of the chromosomes of humans exposed to high levels of benzene reveal an elevated rate of chromosomal aberrations that persist after cessation of exposure [47]. Chronic exposure to benzene leads to a progressive depression of bone marrow function [83]. Epidemiological studies demonstrate that blood dyscrasias such as pancytopenia, aplastic anemia, and acute myelogenous leukemia can develop in humans as a result of this exposure [90,123]. Furthermore, some clinical investigations indicate that it may take several years after the termination of exposure for benzene-induced leukemia to appear [138].

Enzymes linked to the metabolic activation of benzene and its metabolites are the cytochrome P450 monooxygenases and myeloperoxidase [90]. The major metabolic pathway for benzene appears to be oxidation to a phenol, which is then converted to a sulfate conjugate and excreted in urine. Other hydroxylated metabolites include hydroquinone and catechol. Benzene metabolism can be affected by interactions of benzene with its metabolites or other compounds; for example, experiments in mice suggest that benzene can inhibit the oxidation of phenol. Furthermore, animal and human studies have demonstrated that coexposure to toluene may significantly alter the formation of benzene metabolites. Finally, treatment with ethanol induces benzene and phenol metabolism in the liver, resulting in higher levels of active metabolites [90].

The actual mechanism of benzene-induced leukemia is not known. Potential mechanisms for benzene-induced bone marrow disease include metabolism of the parent compound to phenols and other metabolites, in particular, quinone-type metabolites such as catechol, quinol, and pyrogallol, which could react with chromosomes and interfere with mitosis. Another possibility could be the depletion of sulfur available for glutathione detoxification, thereby leading to interaction of toxic intermediates with critical elements of the bone marrow. Another suggested mechanism involves transfer of benzene metabolites from the liver to the bone marrow [123]. Researchers have investigated the metabolism and binding of radioisotope-labeled benzene in the isolated hind limb of rats in which benzene was administered directly into the bone marrow space. Metabolites of benzene were found covalently bound to macromolecules in the bone marrow, indicating that the bone marrow has the potential of metabolizing benzene to reactive intermediates [67]. The fact that benzene or benzene metabolites have been shown to inhibit the multiplication of erythrocyte precursor cells in the bone marrow may imply an additional mode of action [76]. The potential for benzene to induce leukemia in experimental animals has been difficult to demonstrate. In a 2-year carcinogenicity study, rats and mice fed benzene in corn oil developed dose-related leukopenia and tumors in multiple organs, but the study failed to show benzene-associated leukemia [64].

## Toluene

Toluene is a flammable solvent that has been used extensively in the chemical, rubber, paint, and drug industries. It is also useful as a solvent for paints, inks, lacquers, dyes, and other compounds and as an additive for gasoline. Sources of toluene in the ambient environment include manufacturing plants, automobile emissions, gasoline evaporation, and cigarette smoke [31,93]. Various exposure limits and biological indicators of exposure apply to toluene. The ACGIH TLV–TWA and German MAK for toluene are 50 ppm (skin notation) and 50 ppm, respectively. The NIOSH REL is 100 ppm as a time-weighted average and the STEL is 150 ppm. The established IDLH for toluene is 500 ppm. The current OSHA PEL is 200 ppm, with a 300-ppm ceiling limit as a 10-minute peak per 8-hour work shift [2,4]. The ACGIH BEIs are 0.05 mg of toluene per liter of venous blood, collected before the last shift of the work week; 1.6 g of hippuric acid per gram of creatinine in the urine, collected at the end of the shift; and, 0.5 mg of *o*-cresol per liter of urine, collected at the end of the shift [2,4,101]. Toluene in expired air has also been evaluated to determine its usefulness as an indicator of exposure. Analysis of expired air in toluene-exposed workers revealed that the toluene concentration was correlated to the exposure environment, representing approximately 15 to 20% of the environmental concentration [28].

The principal toxic effect of toluene is injury to the nervous system. Toluene is most rapidly absorbed by inhalation, followed by ingestion and skin contact. A substantial amount of inhaled toluene is retained in the body. The toxicity of toluene is similar to that of benzene except that it does not exhibit the hematopoietic effects characteristic of benzene. Toluene is an eye and skin irritant, and animal studies indicate that its acute oral toxicity is less than that of other alkylbenzenes [31]. In humans, acute effects of toluene exposure can resemble alcoholic intoxication by first stimulating and later depressing the central nervous system.

Exposure to high concentrations of toluene, as seen in cases of solvent abuse (e.g., glue sniffing), may cause death by sensitizing the myocardium [115,144]. In chronic abusers of toluene, irreversible neurological toxicity and reversible renal damage have also been reported [129,142]. Symptoms associated with the intentional inhalation of high concentrations of toluene include euphoria, mild tremors, unsteady gait, and changes in behavior. Encephalographic examination of these individuals has shown abnormalities indicative of cerebellar atrophy [71]. Toluene is a lipid-soluble compound that readily crosses the placenta and, as such, may pose a teratogenic risk in cases of high exposure, as with intentional abuse. A pattern of teratogenicity, like that of fetal alcohol syndrome (described in the section on ethanol), is prevalent in human studies relating to excessive *in utero* exposure to toluene. Simultaneous abuse of alcohol and toluene may heighten the risks [142].

Toluene is metabolized to benzoic acid, which is subsequently conjugated with glycine or glucuronic acid to form hippuric acid or benzoylglucuronates, respectively. These conjugates, as well as another metabolite, *o*-cresol, are excreted in the urine [75]. In human studies, ethanol has been shown to inhibit the metabolism of toluene at blood ethanol concentrations of 21 mmol/L [42]. Test results indicate that the concentration of toluene in the alveolar air of the toluene/ethanol-exposed group can be significantly higher than that of the toluene control group. In these studies, hippuric acid and *o*-cresol excretion is significantly reduced as compared to controls. Additionally, during the 24 hours following the last exposure, excretion of both hippuric acid and *o*-cresol was about 40 to 50% of that excreted by subjects who received only toluene. These results suggest that ethanol may alter the metabolism of inhaled toluene and prolong its elimination from the body; therefore, the possibility of ethanol consumption should be considered during biological monitoring, as ethanol intake could lead to an underestimation of the actual toluene exposure [42].

In contrast to the above observations, pretreatment of rats with phenobarbital (PB) indicates that the metabolism of toluene can be enhanced to form benzoic acid. The pretreatment did not, however, appear to effect the rate of conjugation of benzoic acid with glycine to form hippuric acid. The hippuric acid concentration in the urine of PB pretreated rats was about three times that of rats receiving toluene only. In addition, the toluene concentration in the blood of the PB pretreated group was only about half that in the toluene-exposed rats. Not only did the phenobarbital pretreatment enhance metabolism of toluene to benzoic acid (with subsequent conversion to hippuric acid), but it also reduced the blood concentration of toluene, thus shortening the sleeping time induced by the narcotic effect of toluene [66].

At present, the mechanisms of the neurotoxic effect of toluene are not well understood. Some experimental work with rats indicate that exposure to 30,000 ppm of toluene for a few minutes reduced the concentration of tryptophan and tyrosine in plasma by about 50% and 20%, respectively, compared to controls. Tryptophan and tyrosine are known to be precursors of the neurotransmitters noradrenaline, dopamine, and 5-hydroxytryptamine. The reason for the decrease in the precursors is unknown but it is speculated to be an alteration in the hepatic uptake or utilization of these amino acids [141]. A potential factor in toluene-induced neurotoxicity is the production of reac tive oxygen species that can result in cell damage. Experiments using rats suggest that benzaldehyde, a metabolite of toluene, accelerates the production of these reactive oxygen species within the nervous system and may also contribute to the overall neurotoxicity [87].

## *n*-Hexane

*n*-Hexane is a flammable liquid and one of the most toxic of the alkanes. It is an excellent organic solvent that has been used in industrial applications such as printing, low-temperature thermometers, adhesives, extractions, and cleaning processes [30,62]. The primary routes of exposure in the industrial setting are by inhalation and skin contact. The ACGIH TLV–TWA, NIOSH REL, and German MAK are all 50 ppm. ACGIH and the German MAK recognize *n*-hexane with a skin notation. NIOSH recognizes the IDLH concentration for *n*-hexane as1100 ppm (10% of the lower explosive limit). ACGIH set the TLV based on possible neuropathy, CNS effects, and irritation [2,4,101]. Acute toxic responses after accidental ingestion include nausea, gastrointestinal irritation, and CNS effects. Inhalation overexposure leads to dizziness, a sense of euphoria, and numbness of the extremities. Exposure to high concentrations causes vertigo and a marked anesthetic effect. Hexane is also an irritant to the skin upon dermal exposure [30].

Many cases of polyneuropathy in workers exposed to *n*-hexane have been noted, with the earliest occurring in Japan [147]. The severity of symptoms in the Japanese workers varied directly with degree and duration of exposure, and in some cases there was incomplete recovery [62]. Polyneuropathy has also been reported in cases of solvent abuse [30]. The neurotoxic effect of *n*-hexane has characteristically been a progressive motor or sensorimotor neuropathy with symptoms usually reported after several months of exposure [62]. In cases from occupational exposure, symptoms have often been sensory, with numbness and paresthesia in the distal extremities, most notably the feet or hands. Improvement of symptoms is noted after cessation of exposure, and mild cases can recover completely.

Hexane is readily absorbed in laboratory animals and has an affinity for tissues high in lipid content [24]. It is rapidly metabolized to hydroxylated compounds prior to being converted to a keto- form [72,84]. 2,5-Hexanedione and methyl *n*-butyl ketone are the metabolites suspected of being responsible for the production of neurotoxicity.

The mechanism of 2,5-hexanedione-induced neuropathy is not known but several hypotheses have been presented [39]. These include a reduction in energy production in the axon resulting in disruption of axonal transport, alteration of protein structure, and inadequate proteolysis of neurofilaments in the nerve terminal. 2,5-Hexanedione has been shown to interact with glyceraldehyde-3,5-dehydrogenase and phosphofructokinase, inhibiting their glycolytic properties and resulting in decreased energy production and possible disruption of axonal flow. Reaction of 2,5-hexanedione with lysine amine moieties to form pyrrole adducts and modification of neurofilament or axonal skeletal proteins is also an attractive hypothesis [40]. Modification of the proteins may lead to cross-linking of the neurofilaments, which could cause difficulty in neurofilament passage through narrow regions of the axon, such as the node of Ranvier, and therefore an accumulation of proteins at the site of constriction. Possible biophysical membrane changes as a result of 2,5-hexanedione may influence the degeneration of the axon. 2,5-Hexanedione binding and inactivation of calcium-dependent proteases that are important for degradation of neurofilament proteins are the last mechanisms mentioned that might lead to accumulation of neurofilaments. Although none of the mechanisms mentioned fully answers all of the questions concerning *n*-hexane-induced neurotoxicity, these hypotheses offer some contributions to the understanding of the toxic response. It may be that several mechanisms act in parallel to produce the neurotoxic effects.

Repeated exposure of rats to *n*-hexane not only produces the characteristic pattern of neurotoxicity but also results in testicular lesions [146]. The testicular effects are linked to disruption of the cytoskeleton of Sertoli cells. Secondary effects, caused by a loss in functional spermatogonial cells, are seen in affected tubules. Acute exposure led to reversible effects but inhalation or oral exposures of 2 to 5 weeks led to irreversible effects. Although the neurotoxic effect of *n*-hexane is observed in humans, the testicular effect seen in rats has not been well documented in humans.

## Methyl *n*-Butyl Ketone

Industrial uses of methyl *n*-butyl ketone (2-hexanone, MBK) as a solvent or cosolvent (e.g., with methyl ethyl ketone) include the manufacture of adhesives, lacquers, vinyl coatings, printing inks, oils, varnish removers, and other materials [25,73]. Occupationally, the principal routes of exposure to MBK are via inhalation and skin contact with the liquid. The OSHA PEL for MBK is 100 ppm. Since 1998, the ACGIH has identified the TLV–TWA for MBK at 5 ppm (skin notation) to protect against possible neuropathy. The German MAK is also 5 ppm. The NIOSH REL and IDLH for MBK are 1 ppm and 1600 ppm, respectively [2,4,101].

Methyl *n*-butyl ketone demonstrates a low acute oral toxicity. The inhalation of high vapor concentrations of MBK can result in eye and respiratory tract irritation followed by CNS depression and narcosis [131]. MBK easily penetrates the skin, and inhalation exposure yields approximately 80 to 85% pulmonary retention. In addition, MBK is widely distributed in the tissues, the highest concentrations being found in the blood and the liver [25]. Chronic exposure to low doses may produce degenerative axonal changes, primarily in the peripheral nerves and long spinal cord tracts [124,125,131]. Depending on the route of administration, a number of metabolites in varying amounts can be detected in the blood. The primary neurotoxic metabolite, as with *n*-hexane, is 2,5-hexanedione. Other metabolites identified following oral, intraperitoneal, or respiratory exposures include 2-hexanol and 5-hydroxy-2-hexanediol [25].

Since the 1970s, MBK has been considered a neurotoxic agent after instances of neurotoxicity were reported in the printing and painting industries [9,91]. Inhalation appears to be the primary route of exposure, with the severity of the toxicity being proportional to the extent of exposure. The characteristic disorder associated with methyl *n*-butyl ketone exposure begins several months after chronic exposure commences. Symptoms include weight loss and distal sensory neuropathy marked by a tingling sensation in the hands or feet. The muscular weakness that develops usually involves the hands and feet, but in severe cases may extend to the legs and thighs. The sensory loss is symmetrical and a moderate reduction of nerve conduction velocity is found in peripheral nerves.

## Carbon Disulfide

Carbon disulfide ($CS_2$) is a toxic and highly flammable solvent in extensive use in the manufacture of rayon, soil disinfectants, carbon tetrachloride, and electronic vacuum tubes. It is commonly used as a solvent in industrial hygiene analytical procedures. Other applications include its use as a fumigant for grain and a corrosion inhibitor [21,93].

Inhalation and skin contact are the main routes of occupational exposure. Because the sense of smell is quickly fatigued and sensitized to carbon disulfide's characteristic rotten-egg odor, this warning property is not useful in judging exposure. The ACGIH TLV–TWA exposure limit is 10 ppm with skin notation. The NIOSH REL is 1 ppm with a STEL/ceiling limit of 10 ppm and skin notation. The OSHA PEL–TWA is 20 ppm with a ceiling limit of 30 ppm as a 30-minute peak over an 8-hour work shift [2,4,101]. The ACGIH TLV–TWA was set to protect against cardiovascular, central nervous system, and neuropathic effects. NIOSH has established an IDLH value of 500 ppm [101]. In addition to these levels, proposals in the literature have suggested lowering the occupational exposure limit to 4 ppm to prevent neurological sequelae [63]. The BEI recommended by the ACGIH is 5 mg of the metabolite 2-thiothiazolidine-4-carboxylic acid (TEA) per gram of creatinine in urine, measured at the end of the work shift [4].

Acute exposure to high concentrations of carbon disulfide can result in restlessness, euphoria, nausea, vomiting, headache, mucous membrane irritation, unconsciousness, and fatal convulsions. Chronic exposure can lead to abnormalities such as irritability, hallucinations, auditory and visual disturbances, and weight loss [21,55,78, 93,137]. Distal sensorimotor neuropathy is the most common chronic effect associated with $CS_2$ exposure. This has been confirmed in experimental animals as a neurofilamentous axonopathy that affects long axons in the CNS and peripheral nervous system [36,54]. Peripheral neuropathy takes place only after frequent and prolonged exposures to $CS_2$ and is characterized by a loss of distal sensory and motor function. The condition can progress more proximally with continued exposure. Chronic exposure to $CS_2$, as well as hexane, 2-hexanone, and their metabolite 2,5-hexanedione, results in large swellings of the distal axons, which are filled with neuron filaments. Continued exposure causes axonal degeneration distal to the axonal swellings [34,54]. In addition to these effects, encephalopathy, detected by neurological examination and neuropsychological testing, has been reported. Evidence suggests that exposure to $CS_2$ accelerates the rate of atherosclerosis [54]. In addition, an investigation to determine a possible association between $CS_2$ exposure and ischemic heart disease mortality found that the relationship is meaningful only for workers exposed to high levels for many years. Price has suggested a safe level of between 15 and 20 ppm [111]. Approximately 70 to 90% of absorbed $CS_2$ is metabolized and excreted in the urine. The remaining 10 to 30% is exhaled in the breath unchanged. In addition to TTCA, mentioned above, other metabolites found in workers' urine include 2-mercapto-2-thiazolin-5-one and thiocarbamide [63,107,108,135,136].

In a study of rayon production workers with long-term exposure to $CS_2$ at concentrations well above the TLV, evidence of neuropathy was observed in a significant number of workers and consisted of distal sensory loss, altered tendon reflexes, reduced muscle power, and reduction in nerve conduction velocity. These abnormalities persisted for up to 10 years after removal from exposure and were considered to be permanent impairments in nervous system physiology [35].

## Methanol

Synthetic methanol (or methyl alcohol, wood alcohol) production exceeded 1 billion pounds in 2005. The largest use of methanol is in the production of methyl *t*-butyl ether (MTBE), an additive in gasoline. It is also utilized as a denaturant for ethanol, a raw material in the production of numerous other chemicals such as formaldehyde and acetic acid, and as a solvent or antifreeze in paints and strippers, cleaners, and windshield washer compounds [43].

The major routes of exposure to methanol in the industrial environment are through inhalation and dermal contact. The ACGIH TLV–TWA of 200 ppm (250 ppm STEL) is based on potential ocular toxicity and CNS effects. The OSHA PEL, German MAK, and NIOSH REL are all set at 200 ppm. NIOSH has further established an IDLH value of 6000 ppm for methanol, and the ACGIH and NIOSH have added skin notations as indications that skin absorption can be a contributor to the overall body burden. The ACGIH BEI is 15 mg methanol per liter of urine, collected at the end of the work shift [2,4,101].

Most information regarding methanol toxicity in humans is gathered from acute exposures, primarily from ingestion, but adverse health effects from inhalation and dermal exposures have been reported [80]. In one NIOSH study, teachers' aides reported headaches, blurred vision, and other symptoms following inhalation exposure to methanol used in duplicating machines. Concentrations at the site were about 2 to 15 times the current REL. Adverse effects have also been reported following skin applications of methanol for various purposes, although inhalation may have also contributed to these exposures [80].

Methanol is readily absorbed following oral, inhalation, or dermal exposure and is distributed throughout the body according to the water content of the tissues [80]. Ingestion of as little as 2 teaspoonfuls may cause toxicity, whereas the fatal dose in humans is between 2 and 8 oz. [53]. In the absence of medical treatment, a dose of between 4 and 10 mL of methanol taken internally can lead to blindness [114], and, depending on the amount of methanol ingested, mild to severe CNS depression can occur. A latent period, commonly 12 to 24 hours, usually ensues followed by severe abdominal pain, difficult breathing, blurred vision, and pain in the eyes, among other symptoms. Visual impairment or total blindness can occur within days depending on individual susceptibility and the time when treatment began [80]. Metabolic acidosis due to formic acid production is thought to be the cause of the delayed symptoms and ocular toxicity [114].

Metabolism of methanol in the liver accounts for a high percentage of absorbed methanol in both nonhuman primates and rats. Lesser amounts are excreted unchanged in the urine and breath. Metabolism is important not because of its primary role in clearance but because of the connection between its metabolites and the acute toxic effects mentioned above. Methanol is oxidized by the catalase–peroxidative system in rats, rabbits, and guinea pigs and an alcohol dehydrogenase system in humans and primates. The metabolic sequence proceeds from methanol to formaldehyde to formic acid (formate) and finally to carbon dioxide and water. Formic acid is metabolized in both rats and primates via a folate-dependent pathway. Rats are able to utilize this pathway more efficiently than primates, allowing for a more rapid conversion to carbon dioxide. Because the process is slower in humans and primates, high doses of methanol cause a buildup of formate in tissues, including the eye, resulting in the observed toxicity [114]. Administration of ethanol has been used in treating methanol poisoning because ethanol inhibits the oxidation of methanol by competing for the same metabolic pathway. Prompt hemodialysis (able to remove both methanol and formate), coupled with concurrent administration of ethanol and bicarbonate, has been successful in many poisoning cases [53].

## Ethanol

Ethanol (ethyl alcohol, grain alcohol) is produced in large quantities and is utilized extensively as a solvent in industry, in numerous consumer preparations, and as an additive to gasoline (gasohol). It is used industrially as a raw material in the production of pharmaceuticals, plastics, perfumes, cosmetics, and other compounds. Other applications include products such as hairsprays, mouthwashes, cleaning products, and drug formulations [53,80]. Denaturants (e.g., methanol) are added to the alcohol in a number of these products to discourage ingestion. Synthesis from ethylene represents the largest source of ethanol; smaller amounts are made from the fermentation of natural materials [80].

Human exposure to ethanol is primarily through ingestion of alcoholic beverages and inhalation of ethanol vapors from industrial processes and consumer products. Percutaneous absorption appears to be much less important [80]. OSHA, ACGIH, and NIOSH have established the exposure limit of 1000 ppm for ethanol [2,4]. The German MAK is 500 ppm. The NIOSH IDLH of 3300 ppm is set because of safety concerns (10% of the lower explosive limit) rather than toxicological considerations [101].

Although there is no clear evidence that ethanol is carcinogenic in animals, it has been shown to be a tumor promoter. Additionally, the International Agency for Research on Cancer (IARC) has classified alcoholic beverages as a Group 1 carcinogen based on the occurrence of a variety of tumors in humans that have been causally related to ingestion of these beverages [80]. An unfortunate occurrence associated with chronic maternal consumption of large amounts of alcohol is a pattern of congenital abnormalities commonly called *fetal alcohol syndrome*. Effects may include growth retardation, microcephaly, mental deficiency, facial abnormalities, and poor coordination. Children who have been affected may display a few or many of the features characteristic of the syndrome [33,80,110].

Ethanol is a CNS depressant that is capable of inducing all stages of anesthesia. It is readily absorbed by the gastrointestinal tract and the lungs and is distributed throughout the body water [53]. Absorption can be delayed, however, by food in the stomach. Subjects exposed to 5000 to 10,000 ppm of ethanol vapor experienced eye irritation and coughing [114]. Individuals with tolerance to alcohol experienced headache, drowsiness, and stupor when exposed to concentrations of 9400 to 13,200 mg/m$^3$ (5000 to 7000 ppm) for a period of 110 minutes [114]. Ingestion of approximately 1 liter of an alcoholic beverage (655% ethanol) within several minutes can result in death [53]. Individuals with blood alcohol levels of approximately 0.05 to

0.15% (50 to 150 mg/dL) may exhibit decreased inhibitions, poor coordination, blurred vision, and slowed reaction time. Increasing blood levels to 0.15 to 0.30% can result in slurred speech, visual impairment, hypoglycemia, and staggering. At 0.3 to 0.5% blood alcohol content (severe intoxication), symptoms can include poor muscular coordination, hypothermia, vomiting and nausea, and convulsions. In adults, coma and death are typically associated with levels exceeding 0.5% [80,114]. The wide ranges reported above reflect the differences in tolerance and susceptibility of individuals to the effects of alcohol.

Like methanol, ethanol is metabolized primarily (about 90%) by the liver. Elimination from the body by urinary excretion and pulmonary exhalation is minimal [80]. Oxidation of ethanol to acetaldehyde occurs via alcohol dehydrogenase within the cytosol. Acetaldehyde is then converted to acetic acid by the action of aldehyde dehydrogenase. Both enzymes utilize oxidized nicotinamide adenine dinucleotide (NAD) as a cofactor [53]. Following release to the blood, acetic acid is metabolized to carbon dioxide and water in the peripheral tissues [80]. Alternative, but less active, metabolic pathways have been demonstrated in humans and other species. These include catalase and microsomal ethanol-oxidizing systems [26,79]. Adults metabolize ethanol at a rate of about 7 to 10 g/hr. This rate remains essentially constant for each individual within a wide range of exposure. Metabolic rates are higher for chronic alcoholics and children [80,114].

The interaction of ethanol with other hepatotoxins is well known. Ethanol pretreatment has been shown to increase the toxicity of carbon tetrachloride, chloroform, trichloroethylene, dimethylnitrosamine, chlorpromazine, and other compounds [127]. The induction of cytochrome P450 isozymes may be responsible for their metabolic effects [80].

## Methylene Chloride

Methylene chloride (dichloromethane) is widely used in a number of diverse applications, including the manufacture of polyurethane foams, pharmaceuticals production, boat building, paint stripping, vapor degreasing, extraction of caffeine from coffee and tea, and in various consumer products. Its high volatility; good solvent properties for fats, oils, and other compounds; and relatively good water solubility compared to other chlorinated compounds have made it quite valuable [103,132].

Due to the high vapor pressure of methylene chloride, the primary route of human exposure is through inhalation; however, dermal contact can be significant, depending on the application. The ACGIH TLV–TWA of 50 ppm was set to protect against CNS effects and anoxia. In addition, the ACGIH has designated methylene chloride as a confirmed animal carcinogen but also states that available epidemiological studies do not confirm an increased risk of cancer in exposed humans [4]. NIOSH recommends that methylene chloride be regarded as a potential occupational carcinogen [101]. OSHA regulates methylene chloride in the workplace under 29 CFR 1910.1052 [103]. OSHA considers methylene chloride a potential human carcinogen and has reduced the PEL for methylene chloride from 500 ppm to 25 ppm, with a STEL of 125 ppm (15 min) and an action level of 12.5 ppm that triggers certain requirements [103]. The current German MAK is 100 ppm [2].

The primary acute hazards associated with exposure to methylene chloride are due to its narcotic effect and can result in CNS depression and eye, skin, and respiratory tract irritation. In addition, one of the products of methylene chloride metabolism is carbon monoxide, which can impair health in a manner similar to direct exposure to carbon monoxide. The resulting carboxyhemoglobin levels reduce the supply of oxygen to the heart and may aggravate preexisting heart disease [103].

Metabolism of methylene chloride can proceed via two pathways, one by a route involving cytochrome P450 mixed-function oxidase (MFO) and the other by a route utilizing glutathione *S*-transferase (GST). Carbon dioxide is an end-product in both systems, but carbon monoxide is only produced via the MFO route. At low concentrations, the MFO system appears to dominate, but at higher concentrations (above 300 to 500 ppm) the glutathione pathway increases in a disproportionate manner [132].

Methylene chloride was shown in a 1986 National Toxicology Program inhalation study to produce lung and liver tumors in male and female mice and benign mammary tumors in male and female rats [56]. Recent research has suggested that mice may be uniquely sensitive at high exposures to methylene-chloride-induced lung and liver cancer [56]. The tumors appear to be caused by a genotoxic mechanism involving metabolites of the GST pathway. The particular metabolites responsible are not found in high concentrations in lung or liver tissue in humans or rats.

In a study to determine the effects of alcohols and toluene upon methylene chloride-induced carboxyhemoglobin in the rat and monkey, it was shown that ethanol, methanol isopropanol, and toluene inhibited the formation of carboxyhemoglobin. In addition, neither the rat nor the monkey demonstrated the methanol potentiation of carboxyhemoglobin that has been reported to occur in humans [32].

A study of the pharmacokinetics of [$^{14}$C]-methylene chloride in rats at 50, 500, and 1500 ppm for 6 hours showed that metabolic processes were saturated above the 50-ppm exposure concentration. At 48 hours after exposure, approximately 95% of the body burden attributable to the 50-ppm exposure was metabolized, in contrast to 69% and 45% at 500 and 1500 ppm, respectively [89]. In addition, the production of carboxyhemoglobin reached a steady-state range of 10 to 13% regardless of the exposure concentration suggesting that the CO metabolic pathway was saturated.

Tetrachloroethylene (perchloroethylene) is another solvent in which patterns of elimination are altered when metabolic pathways become saturated [105]. In a study comparing oral and inhalation exposure of rats to [$^{14}$C]-tetrachloroethylene, it was found that, with increasing dose, metabolism was saturated, resulting in more of the parent compound being eliminated unchanged at 72 hours after exposure [105]. These results with methylene chloride and tetrachloroethylene indicate that just increasing the exposure concentration does not always increase the body burden in a linear manner. Such information may be useful for safety evaluations to avoid the overestimation of body burden.

## NONTRADITIONAL SOLVENTS

Given the negative health and environmental impacts created by some of the more widely used solvents, a great deal of effort has gone into finding suitable replacements. The following compounds are examples of nontraditional materials that show promise as replacement solvents.

### D-LIMONENE

d-Limonene is a naturally occurring monocyclic terpene found in citrus peel oils, spices, evergreens, and human milk [140]. It is considered to have low acute toxicity and is listed as generally recognized as safe (GRAS) as a food additive by the U.S. Food and Drug Administration (21 CFR 182.60). It has found wide application as a solvent in numerous cleaning and degreasing applications, replacing more toxic and environmentally undesirable chlorinated solvents, glycol ethers, xylene, and chlorofluorocarbons (CFCs) [46]. Skin contact with d-limonene may cause irritation and sensitization (attributed to the oxidation product d-limonene oxide) [140]. d-Limonene has been shown to produce hyaline droplet nephropathy and renal tubular tumors in male rats; however, these effects are attributed to the unique presence of $\alpha_{2\mu}$-globulin in the male rat and are not deemed relevant to other species, including humans [45]. Among the attributes of d-limonene are its antimicrobial, antiviral, antifungal, and antilarval properties [29]. d-Limonene and related monoterpenes have also demonstrated chemopreventive and chemotherapeutic efficacy in experimental cancer-therapy models [37]. Based on similar metabolic pathways in rats and humans and the therapeutic successes in rodents, it has been suggested that d-limonene may be an efficacious chemotherapeutic agent for human malignancies [37].

### CARBON DIOXIDE

Carbon dioxide ($CO_2$) is a gas under standard temperature and pressure conditions. It can be converted, however, to the liquid and supercritical phases by increasing pressure and temperature. The critical point of carbon dioxide is 31°C and 73 atm. Below this point, $CO_2$ can be maintained in a liquid state (e.g., 65 atm and 25°C), whereas above 31°C no amount of pressure can be applied to liquefy it (supercritical phase) [65]. In either of these dense phases, $CO_2$ exhibits good solvent properties. Beneficial characteristics include liquid-like density, gas-like diffusivity, and low surface tension. In particular, liquid $CO_2$ acts like a hydrocarbon solvent, it has good homogenizing properties (immiscible liquids form a single phase when mixed with $CO_2$), and it is a good solvent for many aliphatic hydrocarbons and most small aromatic hydrocarbons. Other chemical groups such as halocarbons, esters, ketones, and low-molecular-weight alcohols also exhibit good solvency in $CO_2$ [65]. Since the mid-1970s, supercritical $CO_2$ technology has been employed in the food, beverage, pharmaceutical, and perfume industries. Applications include the production of spice extracts, natural dyes, decaffeinated coffee and tea, plant extracts, active substances from drugs, and volatile oils [20,38,88]. It has also been used in wastewater treatment, chemical analysis, and at times as an aerosol propellant. More recently, liquid $CO_2$ has found favor as an alternative for metal parts degreasing and as a solvent for dry-cleaning clothes [38,70].

One such $CO_2$ degreasing system is being used in a pen manufacturing operation to replace perchloroethylene. It consists primarily of two separate systems: a hot oil pretreatment process and an automated system that employs liquid carbon dioxide in a pressure vessel. The application is to degrease and remove chips from ball points after machining. The hot oil unit is used to displace fatty esters contained in machining oil and to remove chips in the point cavity. The automated unit then removes oil from the points using liquid carbon dioxide. The carbon dioxide and oil are separated in a recycling system, and the carbon dioxide is used again during the next cleaning cycle.

Advantages of $CO_2$ usage over conventional solvents are numerous. Carbon dioxide is nonflammable, noncorrosive, nonreactive, nontoxic, inexpensive, and plentiful. Products obtained are solvent free. Selective separations are possible. Finally, environmental problems are eliminated, because the gas is recovered for future use. One of the disadvantages of $CO_2$ systems involves the relatively high start-up costs for equipment; however, these may be recouped through improved productivity and reduced costs for waste disposal, for example.

### IONIC LIQUIDS

Ionic systems, which are made up of salts that are liquid at room temperature, are finding applications in a number of chemical processes. Ionic liquids have good solvent properties for many inorganic, organic, and, polymeric materials and, in some cases, these compounds can serve as both catalyst and solvent [49]. Research has indicated that partitioning of organic solvents between an ionic liquid and water corresponds closely with that found for molecular organic solvents and water; thus, ionic liquids have the potential to replace the toxic, flammable, and volatile organic compounds currently used in liquid–liquid separations [50]. The room-temperature ionic compounds, such as 1-butyl-3-methylimidazolium hexafluorophosphate and 1-butylpyridinium nitrate, consist of nitrogen-containing organic cations and inorganic anions. Their physical and chemical properties can be altered according to the choice of ions. Advantages compared to conventional organic solvents include low volatility and relative ease of recycling [50]. Other potential uses include removal of organic contaminants from wastewater, soil cleanup, replacement of corrosive mineral acids in refinery processes, and spent nuclear fuel treatment [49]. The safety and toxicological profiles of these compounds have yet to be thoroughly developed; therefore, caution must be exercised before they are put into general use.

## OPPORTUNITIES IN THE TOXICOLOGICAL EVALUATION OF SOLVENTS

Human exposure to solvents is quite common in today's society. These exposures frequently involve multiple chemicals that are found in numerous products such as cleaning agents, paint thinners, and fuels. Although most toxicological research to date has dealt with single chemicals, questions remain about the long-term health effects associated with low-level exposures to multiple chemicals and the sensitivity of the toxicological endpoints that are currently being relied upon. Development of innovative experimental protocols and new quantitative mechanistic approaches to the study of chemical interactions may be beneficial in this regard [74,148]. Economic concerns and the desire for less toxic and more environmentally friendly chemicals have resulted in the introduction of numerous alternative compounds into the marketplace. In some cases, little may be known about the health and environmental impacts of these materials; examples include the ionic liquids discussed above. It is therefore essential that sufficient toxicological and environmental data be gathered before replacements are introduced on a wide scale. Research has shown that many neurotoxic chemicals are capable of adversely affecting the sensory function. Minor changes in vision or hearing, for example, can dramatically alter job performance and the overall quality of life. While most reports to date have dealt with changes in the visual system, additional investigations into the effects of solvents on hearing, taste, and smell would provide important new information on this subject [48].

## ACKNOWLEDGMENTS

The authors wish to thank Leslie Bienenfeld for her assistance in editing the chapter. The authors also gratefully acknowledge the significant contributions of the previous writers, Paul H. Ayres, W. David Taylor, Michael J. Olson, Robert C. Spiker, Jr., and Gary B. Morris.

## QUESTIONS

1. You are a toxicologist with industrial hygiene responsibilities in a large manufacturing company. Your boss has just told you that the solvent the factory is using to degrease metal parts will be banned by the EPA within the next 6 months. Your job is to lead a team of employees, who have a vested interest in the current solvent, in coming up with a suitable alternative material. What are your considerations in recommending a replacement? Explain.
2. Assume that the solvent chosen above will be used in six locations in the factory. You surmise that some sort of ventilation will be required to protect the employees. What factors must you take into account in recommending the proper system?
3. One of your employees has begun using a solvent mixture containing xylene and toluene. To ensure the safety of the worker, you have conducted personal air monitoring throughout the day and have come up with the following sampling times and monitoring results: 0800–1000, 60 ppm xylene and 25 ppm toluene; 1000–1200, 92 ppm xylene and 45 ppm toluene; 1200–1300, no exposure because employee left for lunch; 1300–1600, 110 ppm xylene and 47 ppm toluene. Calculate the TWA exposure for each chemical. Assume that there is no dermal exposure and that the toxic effects contributed by each solvent are additive. Has the TLV–TWA been exceeded?
4. Match each solvent or metabolite with the appropriate fact listed below:

| Solvent/Metabolite | Fact |
|---|---|
| 1. d-Limonene | a. Antidotal in methanol poisonings |
| 2. Carbon disulfide | b. Associated with bone marrow disease in humans |
| 3. 2,5-Hexanedione | c. Potentially useful in cancer therapy |
| 4. Toluene | d. Teratogen and embryotoxin |
| 5. Methanol | e. Metabolism produces carboxyhemoglobin |
| 6. 2-Butoxyacetic acid | f. Frequently "sniffed" to obtain euphoric effect |
| 7. Ethylene glycol monomethyl ether | g. Used in rayon production |
| 8. Benzene | h. A few milliliters can lead to blindness |
| 9. Ethanol | i. Primary causative agent in polyneuropathy |
| 10. Methylene chloride | j. Produces hemolytic effects in rats |

## REFERENCES


1. ACGIH. (1995): *Air Sampling Instruments for Evaluation of Atmospheric Contaminants*, 8th ed. American Conference of Governmental Industrial Hygienists, Inc., Cincinnati, OH.
2. ACGIH. (2005): *Guide to Occupational Exposure Values: 2005*. American Conference of Governmental Industrial Hygienists, Inc., Cincinnati, OH.
3. ACGIH. (1998): *Industrial Ventilation: A Manual of Recommended Practice*, 23rd ed. American Conference of Governmental Industrial Hygienists, Inc, Cincinnati, OH.
4. ACGIH. (2005) *TLVs® and BEIs®: Threshold Limit Values for Chemical Substances and Physical Agents, and Biological Exposure Indices*. American Conference of Governmental Industrial Hygienists, Inc., Cincinnati, OH.
5. AIHA. (1990): *Chemical Protective Clothing*, edited by I. S. Johnson and K. J. Anderson. American Industrial Hygiene Association, Akron, OH.
6. AIHA. (1997): *The Occupational Environment: Its Evaluation and Control*, edited by S. R. DiNardi. AIHA Press, Fairfax, VA.
7. Alden, C. L. (1986): A review of unique male rat hydrocarbon nephropathy. *Toxicol. Pathol*., 14:109–111.
8. Allen, N. (1979): Solvents and other industrial organic compounds. In: *Handbook* of *Clinical Neurology Intoxications of the Nervous System*, Part 1(36), edited by P. J. Vinken and G. W. Bruyn. Elsevier/North-Holland, New York, pp. 361–389.
9. Allen, N., Mendell, J. R., Billmaier, D. J., Fontaine, R. E., and O'Neill, J. (1975): Toxic polyneuropathy due to methyl *n*-butyl ketone. *Arch. Neurol*., 32:209–218.
10. Anger, W. K. (1986): Workplace exposures. In: *Neurobehavioral Toxicology*, edited by Z. Annau. The Johns Hopkins University Press, Baltimore, MD, pp. 331–347.
11. AAI. (1987): *Handbook of Organic Industrial Solvents*. Alliance of American Insurers, Schaumburg, IL.
12. AAI. (1988): *Handbook of Hazardous Materials*, Alliance of American Insurers Schaumburg, Illinois.
13. Astrand, I. (1975): Uptake of solvents in the blood and tissues of man. *Scand. J. Work Environ. Health*, 1:199–218.
14. Astrand, I. (1985): Uptake of solvents from the lungs. *Br. J. Indust. Med*., 42:217–218.
15. Baelum, J., Anderson, I., Lundqvist, G. R., Molhave, L., Pedersen, O. F., Vaeth, M., and Wyon, D. P. (1985): Response of solvent-exposed printers and unexposed controls to six-hour toluene exposure. *Scand. J. Work Environ. Health*, 11:271–280.
16. Baerg, R. D. and Kimberg, D. V. (1970). Centrilobular hepatic necrosis and acute renal failure in 'solvent sniffers.' *Ann. Intern. Med*., 73:713–720.
17. Bahl, M. K. (1985): ESCA studies on skin lipid removal by solvents and surfactants. *J. Soc. Cosmet. Chem*., 36:287–296.
18. Baker, B. L. (1988): Organic solvent neurotoxicity. *Annu. Rev. Public Health*, 9:223–232.
19. Bamberger, R. L., Esposito, G. G., Jacobs, B. W., Podolak, G. E., and Mazur, J. F. (1978): A new personal sampler for organic vapors. *Am. Indust. Hyg. Assoc. J*., 39:701–798.
20. Basta, N. and McQueen, S. (1985): Supercritical fluids: still seeking acceptance. *Chem. Eng*., 92:14–17.
21. Beliles, R. P. and Beliles, E. M. (1993): Phosphorus, selenium, telfurium, and sulfur. In: *Patty's Industrial Hygiene and Toxicology*, Vol. 11A, 4th ed., edited by G. D. Clayton and F. E. Clayton. John Wiley & Sons, New York, pp. 818–822.
22. Boekelheide, K. (1987): 2,5-Hexanedione alters microtubule assembly. 1. Testicular atrophy, not nervous system toxicity, correlates, with enhanced tubulin polymerization. *Toxicol. Appl. Pharmacol*., 88:370–382.
23. Boggs, A. (1989): Comparative risk assessment of casting solvents for positive photo resist. *Appl. Indust. Hyg*., 4:81–87.
24. Bohlen, P., Schlunegger, U. P., and Lauppi, E. (1973): Uptake and distribution of hexane in rat tissues. *Toxicol. Appl. Pharmacol*., 2S:242–249.
25. Bos, P. M., deMik, G., and Bragt, P. C. (1991): Critical review of the toxicity of methyl *n*-butyl ketone: risk from occupational exposure. *Am. J. Indust. Med*., 20:115–194.
26. Bradford. B. U., Seed, C. B., Handler, J. A., Forman, D. T., and Thurman, R. G. (1993): Evidence that catalase is a major pathway of ethanol oxidation *in vivo*: dose–response studies in deer mice using methanol as a selective substrate. *Arch. Biochem. Biophys*., 303:172–176.
27. Browning, R. G. and Curry, S. C. (1994): Clinical toxicology of ethylene glycol monoalkyl ethers. *Human Exp. Toxicol*., 13:325–335.
28. Brugnone, F., Perbellini, L., Gaffuri, E., and Apostoli, P. (1980): Biomonitoring of industrial solvent exposures in workers' alveolar air. *Int. Arch. Occup. Environ. Health*., 47:245–261.
29. Cavender, F. (1994): Alicyclic hydrocarbons: limonene. In: *Patty's Industrial Hygiene and Toxicology*, Vol. IIB, 4th ed., edited by G. D. Clayton and F. E. Clayton. John Wiley & Sons, New York, pp. 1282–1283.
30. Cavender, F. (1994): Aliphatic hydrocarbons: hexanes. In: *Patty's Industrial Hygiene and Toxicology*, Vol. IIB, 4th ed., edited by G. D. Clayton and F. E. Clayton. John Wiley & Sons, New York, pp. 1233–1234.
31. Cavender, F. (1994): Aromatic hydrocarbons: toluene. In: *Patty's Industrial Hygiene and Toxicology*, Vol. IIB, 4th ed., edited by G. D. Clayton and F. E. Clayton. John Wiley & Sons, New York, pp. 1326–1332.
32. Ciuchta, H. P., Savell, G. M., and Spiker, R. C. (1979): The effects of alcohols and toluene upon methylene-chloride-induced carboxyhemoglobin in the rat and monkey. *Toxicol. Appl. Pharmacol*., 49:347–354.
33. Clarren, S. L. and Smith, D. W. (1978): The fetal alcohol syndrome. *N. Engl. J. Med*., 198:1063–1067.
34. Colombi, A., Maroni, M., Picchi, O., Rota, E., Castano, P., and Foa, V. (1981): Carbon disulfide neuropathy in rats: a morphological and ultrastructural study of degeneration and regeneration. *Clin. Toxicol*., 18:1463–1474.
35. Corsi, G., Maestrelli, P., Picotti, G., Manzoni, S., and Negrin, P. (1983): Chronic peripheral neuropathy in workers with previous exposure to carbon disulphide. *Br. J. Indust. Med*., 40:209–211.
36. Costa, L. G. and Manzo, L. (1998): Biological monitoring of occupational neurotoxicants. In: *Occupational Neurotoxicology*, edited by L. G. Costa and L. Manzo. CRC Press, Boca. Raton, FL, p. 90.

37. Crowell, P. L., Elson, C. E, Bailey, H. H., Elegbode, A., Haag, J. D., and Gould, M. N. (1994): Human metabolism of the experimental cancer therapeutic agent d-limonene. *Cancer Chemother. Pharmacol.*, 35:31–37.

38. Darvin, C. H. and Hill, E. A. (1996): Demonstration of liquid $CO_2$ as an alternative for metal parts cleaning. *Precision Cleaning*, 4(9):25–32.

39. DeCaprio, A. P. (1985): Molecular mechanisms of diketone neurotoxicity. *Chem. Biol. Interact.*, 54:257–270.

40. DeCaprio, A. P. and O'Neill, E. A. (1985): Alterations in rat axonal cytoskeletal proteins induced by *in vitro* and *in vivo* 2,5-hexanedione exposure. *Toxicol. Appl. Pharmacol.*, 78:235–247.

41. DiVincenzo, G. D., Hamilton, M. L., Kaplan; C. J., Krasavage, W. J., and O'Donoghue, J. L. (1978): Studies on the respiratory uptake and excretion and the skin absorption of methyl *n*-butyl ketone in humans and dogs. *Toxicol. Appl. Pharmacol.*, 44:593–604.

42. Dossing, M., Baelum, J., Hansen, S. H., and Lundqvist, G. R. (1984): Effect of ethanol, cimetidine, and propranolol on toluene metabolism in man. *Int. Arch. Occup. Environ. Health*, 54:309–315.

43. EPA. (1994): *Chemical Summary for Methanol*, EPA 749-F-94-013a. Office of Pollution Prevention and Toxics, U.S. Environmental Protection Agency, Washington, D.C., pp. 1–9.

44. Fiserova-Bergcrova, V. and Diaz, M. L. (1986): Determination and prediction of tissue–gas partition coefficients. *Int. Arch. Occup. Environ. Health*, 58:75–87.

45. Flamm, W. G. and Lehman-McKeeman, L. D. (1991): The human relevance of the renal tumor-inducing potential of d-limonene in male rats: implications for risk assessment. *Reg. Toxicol. Pharmacol.*, 13:70–86.

46. Florida Chemical Co., Inc. (1997): *d-Limonene Product Data Sheet*. Winter Haven, FL.

47. Forni, A. M., Cappcllini, A., Pacifico, E., and Vigliani, E. C. (1971): Chromosome changes and their evolution in subjects with past exposure to benzene. *Arch. Environ. Health*, 23:385–391.

48. Fox, D. S. (1998): Sensory system alterations following occupational exposure to chemicals. In: *Occupational Neurotoxicology*, edited by L. G. Costa and L. Manzo, CRC Press, Boca Raton, FL, pp. 169–184.

49. Freemantle, M. (1998): Designer solvents. *Chem. Eng. News*, 13:32–37.

50. Freemantle, M. (1998): Ionic liquids show promise for clean separation technology. *Chem. Eng. News*, 34:12.

51. Gargas. M. L., Burgess, R. J., Voisard, D. J., Cason, G. H., and Andersen, M. E. (1989): Partition coefficients of low molecular weight volatile chemicals in various liquids and tissues. *Toxicol. Appl. Pharmacol.*, 98:87–99.

52. Gingell, R. et al. (1994): Glycol ethers and other selected glycol derivatives. In: *Patty's Industrial Hygiene and Toxicology*, Vol. IID, 4th ed., edited by G. D. Clayton and F. E. Clayton. John Wiley & Sons, New York, pp. 2761–2966.

53. Gosselin. R. E., Smith, R. P., and Hodge, H. C. (1984): Ethyl alcohol and methyl alcohol In: *Clinical Toxicology of Commercial Products*, Section III, 5th ed., Williams & Wilkins, Baltimore, MD, pp. 166–171, 275–279.

54. Graham. D. G., Amarnath, V., Valentine, W. M., Pyle, S. I., and Anthony, D. C. (1995): Pathogenic studies of hexane and carbon disulfide neurotoxicity. *Crit. Rev. Toxicol.*, 25(2):91–112

55. Grasso, P., Sharratt, M., Davies, D. M., and Irvine, D. (1984): Neurophysiological and psychological disorders and occupational exposure to organic solvents. *Food Chem. Toxicol.*, 22:819–852.

56. HSIA. (1998): *Methylene Chloride White Paper*. Halogenated Solvents Industry Alliance, Washington, D.C., pp. 1–6.

57. Hanninen, H. (1985): Twenty-five years of behavioral toxicology within occupational medicine: a personal account *Am. Indust. Med.*, 7:19–30.

58. Hardin, B. D. (1983): Reproductive toxicity of the glycol ethers. *Toxicology*, 27:91–102.

59. Hardin, B. D., Bond, G. P., Sikov, M. R., Andrew, F. D. Beliles, R. P., and Niemeier, R. W. (1981): Testing of selected workplace chemicals for teratogenic potential *Scand. J. Work Environ. Health*, 7:66–75.

60. Hardin, B. D., Niemeier, R. W., Smith, R. J., Kuczuk, M H., Mathinos, P. R., and Weaver, T. F. (1982): Teratogenicity of 2-ethoxyethanol by dermal application. *Drug Chem. Toxicol.*, 5:277–294.

61. Heck, H. d'A., Casanova, M., and Starr, T. B. (1990): Formaldehyde toxicity: new understanding. *CRC Crit. Rev Toxicol.*, 20:397–426.

62. Herskowitz, A., Ishii, N., and Schaumburg, H. (1971): *n* Hexane neuropathy. *N. Engl. J. Med.*, 285:82–85.

63. Hoet, P. and Lauwerys, R. (1998): Biological monitoring of occupational neurotoxicants. In: *Occupational Neurotoxicology*, edited by L. G. Costa and L. Manzo. CRC Press, Boca Raton, FL, pp. 57–58.

64. Huff, J. E., Haseman, J. K., DeMarini, D. M., Eustis, S Maronpot, R. R., Peters, A. C., Persing, R. L., Chrisp, C E., and Jacobs, A. C. (1989): Multiple-site carcinogenicit of benzene in Fischer 344 rats and B6C3F mice. *Environ Health Perspect.*, 82:125–163.

65. Hyatt, J. A. (1984): Liquid and supercritical carbon dioxid as organic solvents. *J. Org. Chem.*, 49:5097–5101.

66. Ikeda, M. and Ohtsuji, H. (1971): Phenobarbital-induce protection against toxicity of toluene and benzene in th rat. *Toxicol. Appl. Pharmacol.*, 20:30–43.

67. Irons, R. D., Dent, J. G., Baker, T. S., and Rickert, D. E (1980): Benzene is metabolized and covalently bound i bone marrow *in situ*. *Chem. Biol. Interact.*, 30:241–245.

68. Johnson, A. and Nylen, P. (1995): Effects of industria solvents on hearing. *Occup. Med.*, 10(3):623–640.

69. Johnson, B. L., Ed. (1990): *Advances* in *Neurobehaviora Toxicology: Applications in Environmental and Occupational Health*. Lewis, Chelsea, MI.

70. Kaplan, K. (1597): A new spin on dry cleaning. *Los Angeles Times*, September 8, 1997.

71. Knox, J. W. and Nelson, J. R. (1966): Permanent encephalopathy from toluene inhalation. *N. Engl. J. Med* 273:1494–1496.

72. Kramer, A., Standinger, H., and Ullrich, V. (1974): Effec of *n*-hexane inhalation on the monooxygenase system i mice liver microsomes. *Chem. Biol. Interact.*, 8:11–18.

73. Krasavage, W. J., O'Donoghue, J. L., DiVincenzo, G. D., and Terhaar, C. J. (1980): The relative neurotoxicity of methyl *n*-butyl ketone, *n*-hexane, and their metabolites. *Toxicol. Appl. Pharmacol.*, 52:433–441.
74. Krishnan, K., Andersen, M. E., Clewell III, H. I., and Yang, R. S. H. (1994): Physiologically based pharmacokinetic modeling of chemical mixtures. In: *Toxicology of Chemical Mixtures*, edited by R. S. H. Yang. Academic Press, San Diego, CA, pp. 399–433.
75. Laham, S. (1970): Metabolism of industrial solvents. *Indust. Med.*, 39:61–64.
76. Lee, E. W., Kocsis, J. J., and Snyder, R. (1974): Acute effects of benzene on S9Fe incorporation into circulating erythrocytes. *Toxicol. Appl. Pharmacol.*, 22:431–436.
77. Lehman-McKeeman, L. D., Rivera-Torres, M. I., and Caudill, D. (1990): Lysosomal degradation of $\alpha_{2\mu}$-globulin and $\alpha_{2\mu}$-globulin–xenobiotic conjugates. *Toxicol. Appl. Pharmacol.*, 103:539–548.
78. Lewey, F. H. (1941): Neurological, medical, and biochemical signs and symptoms indicating chronic industrial carbon disulphide absorption. *Ann. Intern. Med.*, 15:869–883.
79. Lieber, C. S. and DeCarli, L. M. (1970): Hepatic microsomal ethanol-oxidizing system. *J. Biol. Chem.*, 245: 2505–2512.
80. Lington, A. W. and Bevan, C. (1994): Alcohols. In: *Patty's Industrial Hygiene and Toxicology*, Vol. IID, 4th ed., edited by G. D. Clayton and F. E. Clayton. John Wiley, New York, pp. 2585–2622.
81. Lock, E. A. (1988): Studies on the mechanism of nephrotoxicity and nephrocarcinogenicity of halogenated alkenes. *CRC Crit. Rev. Toxicol.*, 19:23–42.
82. Lock, E. A. and Ishmael, J. (1985): Effect of the organic acid transport inhibitor probenicid on renal cortical uptake and proximal tubular toxicity of hexachloro-1,3-butadiene and its conjugates. *Toxicol. Appl. Pharmacol.*, 81:3242.
83. Longacre, S. L., Kocsis, J. J., and Snyder, R. (1981): Influence of strain differences in mice on the metabolism and toxicity of benzene. *Toxicol. Appl. Pharmacol.*, 60:398–409.
84. Lu, A. Y. H., Strobel, H. W., and Coon, M. J. (1970): Properties of a solubilized form of the cytochrome P450-containing mixed-function oxidase of liver microsomes. *Mol. Pharmacol.*, 6:213–220.
85. Maron, S. H. and Prutton, C. F. (1965): *Principles of Physical Chemistry*, 4th ed. Macmillan, New York, pp. 215–216, 285.
86. Matheson, Jr., L. E., Wurster, D. E., and Ostrenga, J. A. (1979): Sarin transport across excised human skin. II. Effect of solvent pretreatment on permeability. *J. Pharm. Sci.*, 11:1410–1413.
87. Mattia, C. J., LeBel, C. P., and Bandy, S. C. (1991): Effects of toluene and its metabolite on cerebral reactive oxygen species generation. *Biochem. Pharmacol.*, 42:879–882.
88. McHugh, M. A. (1986): Extraction with supercritical liquids. In: *Recent Developments in Separation Science*, Vol. 9, edited by N. Li and J. Cala. CRC Press, Boca Raton, FL, pp. 75–105.
89. McKenna, M. J., Zempel, J. A., and Braun, W. H. (1982): The pharmacokinetics of inhaled methylene chloride in rats. *Toxicol. Appl. Pharmacol.*, 65:1–10.
90. Medinsky, M. A., Schlosser, P. M., and Bond, J. A. (1994): Critical issues in benzene toxicity and metabolism: the effect of interactions with other organic chemicals an risk assessment. *Environ. Health Perspect.*, 102(9):119–124.
91. Mendell, J. R., Saida, K., Ganansia, M. F., Jackson, D. B., Weiss, H., Gardier, R. W., Chrisman, C., Allen, N., Couri, D., O'Neill, J. J., Marks, B. H., and Hetland, L. B. (1974): Toxic polyneuropathy produced by methyl *n*-butyl ketone. *Science*, 185:787–789.
92. Menger, F. M., Goldsmith, D. I., and Mandell, L. (1972): *Organic Chemistry: A Concise Approach*. W.A. Benjamin, Menlo Park, CA, p. 450.
93. Budavari, S., Ed. (1996): *Merck Index*, 12th ed. Merck, Whitehouse Station, NJ.
94. Miller, R. R., Hermann, E. A., Langvardt, P. W., McKenna, M. J., and Schwetz, B. A. (1983): Comparative metabolism and disposition of ethylene glycol monomethyl ether and propylene glycol monomethyl ether in male rats. *Toxicol. Appl. Pharmacol.*, 67:229–237.
95. Miller, R. R., Hermann, E. A., Young, J. T., Calhoun, L. L., and Kastl, P. E. (1984): Propylene glycol monomethyl ether acetate (PGMEA) metabolism, disposition, and short-term vapor inhalation toxicity studies. *Toxicol. Appl. Pharmacol.*, 75:521–530.
96. Miller, R. R., Hermann, E. A., Young, J. T., Landry, T. D., and Calhoun, L. L. (1984): Ethylene glycol monomethyl ether and propylene glycol monomethyl ether: metabolism, disposition, and subchronic inhalation toxicity studies. *Environ. Health Persp.*, 52:233–239.
97. Monteir-Riviere, N. A. and Popp, I. A. (1986): Ultrastructural evaluation of acute nasal toxicity in the rat respiratory epithelium in response to formaldehyde gas. *Fund. Appl. Toxicol.*, 6:251–262.
98. Morata, T. C. and Dunn, D. E. (1994): Occupational exposure to noise and ototoxic organic solvents. *Arch. Environ. Health*, 49:359–365.
99. Nelson, B. K., Setzer, J. V., Brightwell, W. S., Mathinos, P. R., Kuauk, M. H., Weaver, T. E., and Goad, P. T. (1984): Comparative inhalation teratogenicity of four glycol ether solvents and an amino derivative in rats. *Environ. Health Persp.*, 57:261–271.
100. Nelson, B. K., Brightwell, W. S., Setzer, I. V., and O'Donohue, T.L. (1984): Reproductive toxicity of the industrial solvent zethoxvethanol in rats and interactive effects of ethanol. *Environ. Health Persp.*, 57:255–259.
101. NIOSH. (2005): *Pocket Guide to Chemical Hazards*. National Institute for Occupational Safety and Health, Cincinnati, OH, pp. 1–454.
102. OSHA. (2005) *Benzene*, 29 CFR 1910.1028. Occupational Safety and Health Administration, Washington, D.C.
103. OSHA. (1997): *Methylene Chloride*, 29 CFR 1910.1052. Occupational Safety and Health Administration, Washington, D.C.
104. Paustenbach, D. J. (1994): Occupational exposure limits, pharmacokinetics, and unusual work schedules. In: *Patty's Industrial Hygiene and Toxicology*, Vol. IIIA, 4th ed., edited by G. D. Clayton and F. E. Clayton. John Wiley & Sons, New York, pp. 191–348.

105. Pegg, D. G., Zempel, J. A., Braun, W. H., and Watanabe, P. G. (1979): Disposition of tetrachloro($^{14}$C)ethylene following oral and inhalation exposure in rats. *Toxicol. Appl. Pharmacol.*, 51:465–474.
106. Perbellini, L., Bruguone, F., Caretta, D., and Maranelli, G. (1985): Partition coefficients of some industrial aliphatic hydrocarbons (C5–C7) in blood and human tissues. *Br. J. Indust. Med.*, 42:162–167.
107. Pergal, M., Vukojevic, N., and Djuric, D. (1972): Isolation and identification of thiocarbamide. *Arch. Environ. Health*, 25:42–44.
108. Pergal, M., Vukojevic, N., Cirin-Popov, N., Djuric, D., and Bojovic, T. (1972): Carbon disulfide metabolites excreted in the urine of exposed workers. *Arch. Environ. Health*, 25:38–41.
109. Pohl, L. R. (1979): Biochemical toxicology of chloroform. In: *Reviews in Biochemical Toxicology*, Vol. 1, edited by E. Hodgson, J. R. Bend, and R. M. Philpot. Elsevier/North Holland, New York, pp. 79–107.
110. Ratt, G. E. (1982): Alcohol and the developing fetus. *Br. Med. Bull.*, 38:48–53.
111. Price, B., Bergman, T. S., Rodriquez, M., Henrich, R. T., and Moran. E. J. (1997): A review of carbon disulfide exposure data and the association between carbon disulfide exposure and ischemic heart disease mortality. *Reg. Toxicol. Pharmacol.*, pp. 119–128.
112. Rafales. L. S. (1986): Assessment of locomotor activity. In: *Neurobehavioral Toxicology*, edited by Z. Annau, pp. 54–68. The Johns Hopkins University Press, Baltimore, MD.
113. Recknagel, R. O. (1967): Carbon tetrachloride hepatotoxicity. *Pharmacol. Rev.*, 19:145–208.
114. Reese, E. and Kimbrough, R. D. (1993): Acute toxicity of gasoline and some additives. *Environ. Health Prospect.*, 101(Suppl. 6):115–131.
115. Reinhardt, C. F., Mullin, L. S., and Maxfield, M. E. (1973): Epinephrine-induced cardiac arrhythmia potential of some common industrial solvents. *J. Occup. Med.*, 15:953–955.
116. Riihimäki, V. and Pfäffli, P. (1978): Percutaneous absorption of solvent vapors in man. *Scand. J. Work Environ. Health*, 4:73–85.
117. Rubin, E. and Lieber, C. S. (1972): The effects of ethanol on the liver. In: *International Review of Experimental Pathology*, edited by G. W. Richter and M. A. Epstein. Academic Press, San Diego, CA, pp. 177–232.
118. Sabourin, P. I., Medinsky, M. A., Thurmond, F., Birnbaum, L. S., and Henderson, R. F. (1992): Effect of dose on the disposition of methoxyethanol, ethoxyethanol, and butoxyethanol administered dermally to male F344/N rats. *Fund. Appl. Toxicol.*, 19:124–132.
119. Sato, A. and Nakajima, T. (1979): Partition coefficients *of* some aromatic hydrocarbons and ketones in water, blood, and oil. *Br. J. Indust. Med.*, 36:231–234.
120. Scheflan, L. and Jacobs, M. B. (1953): *The Handbook of Solvents*. Van Nostrand Reinhold, New York, p. 728.
121. Short, B. G., Burnett, V. L., Cox, M. G., Bus, J. S., and Swenberg, J. A. (1987): Site-specific renal cytotoxicity and cell proliferation in male rats exposed to petroleum hydrocarbons. *Lab. Invest.*, 57:564–577.
122. Snyder, C. A., Goldstein, B. D., Sellakumar, A., Wohan, S. R., Bromberg, L., Erlichman, M. N., and Laskin, S (1978): Hematotoxicity of inhaled benzene to Sprague–Dawley rats and AKR mice at 300 ppm. *J. Toxicol Environ. Health*, 4:605–618.
123. Snyder, R. and Kocsis, J. J. (197.5): Current concepts of chronic benzene toxicity. *CRC Crit. Rev. Toxicol.* 3:265–288.
124. Spencer, P. S. and Schaurnburg, H. H. (1977): Ultrastructural studies of the dying-back process. IV. Differential vulnerability of PNS and CNS fibers in experimental central–peripheral distal axonopathies. *J. Neuropathol. Exp Neurol.*, 36:300–320.
125. Spencer, P. S., Schaumburg, H. H., Raleigh, R. L., and Terhaar, C. J. (1975): Nervous system degeneration produced by the industrial solvent methyl *n*-butyl ketone *Arch. Neurol.*, 32:219–222.
126. Stott, W. T. and McKenna, M. J. (1985): Hydrolysis of several glycol ether acetates and acrylate esters by nasal mucosal carboxylesterase *in vitro*. *Fund. Appl. Toxicol.* 5:399–404.
127. Strubelt, O. (1980): Interaction between ethanol and othe hepatotoxic agents. *Biochem. Pharmacol.*, 29:1445–1449
128. Swenberg, J. A., Short, B., Borghoff, S., Strasser, J., and Charbormeau, M. (1989): The comparative pathobiology of $\alpha_{2\mu}$-globulin nephropathy. *Toxicol. Appl. Pharmacol.* 97:35–46.
129. Taher, S. M., Anderson, R. J., McCartney, R., Popovtzer M. M., and Schrier, R. W. (1974): Renal tubular acidosi associated with toluene 'sniffing.' *N. Engl. J. Med* 290:765–768.
130. Thomas, J. A. and Ballantyne, B. (1990): Occupationa reproductive risk: sources, surveillance, and testing. *J Occup. Med.*, 32:547–554.
131. Topping. D. C., Moreott, D. A., David. R. M., and O'Donohue. J. L. (1994): Ketones. In: *Putty's Industrial Hygien and Toxicology*, Vol. IIC, 4th ed., edited by G. D. Clayto and F. E. Clayton. John Wiley & Sons, New York, pp 1739–1787.
132. Torkelson, T. R. (1994): Halogenated aliphatic hydrocarbons containing chlorine, bromine, and iodine. In: *Party Industrial Hygiene and Toxicology*, Vol. IIE, 4th ed., edite by G. D. Clayton and F. E. Clayton. John Wiley & Son New York, pp. 4034–4045.
133. Trela, B. A. and Bogdanffy, M. S. (1991): Carboxylesteras dependent cytotoxicity of dibasic esters (DBE) in rat nasa explants. *Toxicol. Appl. Pharmacol.*, 107:285–301.
134. Van Dolah, R. W. (1965): Flame propagation, extinguishment, and environmental effects on combustion. *Fire Technol.*, 2:138–145.
135. van Doom, R., Delbressine, L. P. C., Leijdekkers, C. M Vertin, P. G., and Hendenon, P. H. (1981): Identificatio and determination of 2-thiothiazolidine-4-carboxylic aci in urine of workers exposed to carbon disulfide. *Arch Toxicol.*, 475–458.
136. van Doorn, R., Leijdekkers, C. P. M. J. M., Henderson, T., Vanhoome, M., and Vertin, P. G. (1981): Determinatic of thio compounds in urine of workers exposed to carbo disulfide. *Arch. Environ. Health*, 36:289–297.

137. Vigliani, E. C. (1950): Clinical observations an carbon disulfide intoxication in Italy. *Indust. Med. Surg.*, 19:240–242.
138. Vigliani, E. C. and Fomi, A. (1976): Benzene and leukemia. *Environ. Res.*, 11:122–127.
139. Vincent, I. H. (1998): International occupational exposure standards: a review and commentary. *Am. Indust. Hyg. Assoc. J.*, 59:729–742.
140. Von Burg, R. (1995): Toxicology update: limonene. *J. Appl. Toxicol.*, 15(6):495–499.
141. Voog, L. and Eriksson, T. (1984): Toluene-induced decrease in rat plasma concentrations of tyrosine and tryptophan. *Acta Pharmacol. Toxicol.*, 54151–54153.
142. Wilkins-Haug, L. (1997): Teratogen update: toluene. *Teratology*, 55:145–151.
143. Winder, C. and Ng, S. K. (1995): The problem of variable ingredients and concentration in solvent thinners. *Am. Indust. Hyg. Assoc. J.*, 56:1225–1228.
144. Winek, C. L. and Collom, W. D. (1971): Benzene and toluene fatalities. *J. Occup. Med.*, 13:259–261.
145. WHO. (1990): *2-Methoxyethanol, 2-Ethoxyethanol, and Their Acetates*, Environmental Health Criteria 115. International Program on Chemical Safety, World Health Organization, Geneva.
146. WHO. (1991): *n-Hexane*, Environmental Health Criteria 122. International Program on Chemical Safety, World Health Organization, Geneva.
147. Yamada, S. (1964): An occurrence of polyneuritis by *n*-hexane in the polyethylene laminating plants. *Jpn. J. Indust. Health*, 6:192–194.
148. Yang, R. S. H. (1994): Introduction to the toxicology of chemical mixtures. In: *Toxicology* of *Chemical Mixtures*, edited by R. S. H. Yang. Academic Press, San Diego, CA, pp. 1–10.
149. Wikipedia. (2005). http://en.wikipedia.org/wiki/Xenobiotic.

# Notes

# 16 Crop Protection Chemicals: Mechanisms of Action and Hazard Profiles

*Charles B. Breckenridge and James T. Stevens*

## CONTENTS

## INTRODUCTION

The use of chemicals to control pests dates back more than 1000 years B.C. to the Chinese, who discovered that sulfur was effective as a fumigant; in the sixteenth century, they discovered that arsenic could be used as an insecticide [220]. Tobacco leaf (nicotine) and the seed of Strychnos nux vomica (strychnine) were used as rodenticides in the eighteenth century [252], and the insecticidal active botanicals rotenone, derived from the root of Derris eliptica, and pyrethrum, from the flowers of chrysanthemums, were used as insecticides in the mid-1800s. The Bordeaux mixture (copper sulfate, lime, calcium hydroxide, and water) was introduced in France for the control of mildew in grapes in 1880 [220]. Paris Green (copper arsenite) and later calcium arsenite were used extensively at the turn of the century to control the Colorado potato beetle [252].

The era of modern agriculture, which began after World War II, depended on (1) the introduction of highly mechanized farming practices; (2) the use of fertilizers, whose production was diverted from a large munitions manufacturing capacity that had developed during the war (3) the use of pesticides aimed at controlling pests; an (4) optimizing yield, especially in monoculture stapl crops such as corn, soybeans, rice, and wheat. The dis covery of more efficacious (e.g., low use rate) and selec tive (e.g., tolerance to beneficial plants, insects, and ani mals) pesticides has largely been based on the use o screening methods and, more recently, combinatoria chemistry coupled with high-throughput screening tech niques to discover new classes of biologically active ingre dients. The drive to discover new pesticides comes fron the following business imperatives: (1) cost effectiveness which confers competitive advantage; (2) societal pressur for improved safety; and (3) the development of pest resis tance (pests evolve over generations by the selection c polymorphic forms that have developed a tolerance to th pesticide). The incorporation of genes that confer pesticid tolerance to relatively inexpensive, and comparativel safe, nonselective herbicides in key crops is a recent devel opment. More recent still is the insertion of genes int plants to produce the insecticidal protein delta endotoxi (derived from *Bacillus thuringiensis*), which is thought to create a lytic pore 1 to 2 nm in diameter in the mid-gut of the insect [310]. On the immediate horizon is the use of genetic engineering, perhaps combined with chemically induced changes in plant metabolism to create functional food and fiber exhibiting traits that have enhanced nutritional value, facilitate processing, or have other desirable attributes.

In this chapter, we have characterized the hazard of pesticides that either have significant economic value or are representative of a group of active ingredients with a specific biochemical mode of action in targeted species. The modes of action, where known, and the hazard profiles of organic pesticides, such as those described in the first paragraph or others that have subsequently been discovered, have been included for the sake of comparison with those of synthetic pesticides because of the recent increase in popularity of organically grown food and the use of natural pesticides [313].

Pesticides are grouped according to their modes of action in targeted species. This method of classifying fungicides [233], herbicides [253], and insecticides [260] has been developed by agronomists to assist growers in preventing the development of resistance in targeted species (see Table 16.70). Grouping pesticides according to their mode of action is better informed of potential biological outcomes and, in some instances, provides data directly relevant to the toxicologist because the mode of action in the pest species may be more or less conserved in mammals. This is also relevant to pesticide registration because in 1996 the U.S. Environmental Protection Agency (EPA) was directed by Congress to conduct cumulative risk assessments on chemicals that share a common mechanism of action in mammals. In the guidance developed by the EPA [489], they considered the initial grouping of chemicals based on one of the following four criteria: structural similarity, the mechanism underlying the effect of the pesticide on target species, the general mechanism of mammalian toxicity, or a specific toxic effect in mammals.

In this chapter, we have elected to organize chemicals by their mechanism of action in the targeted species of fungi, insects, or plants. Chemical structures are provided but structure–activity relationships are not discussed extensively. Hazard profiles for members of a pest-based "common mechanism class" have the potential to reveal whether there should be an animal-based common mechanism group, although such judgments are usually based on mode-of-action studies in animals. Even if it is established that a group of chemicals belong to a common mechanism class, this still leaves unaddressed the important question as to whether the modes of action elaborated in the target species or in animals models are relevant to humans, and, if so, how doses should be scaled between species.

## HAZARD CHARACTERIZATION OF PESTICIDES

### Federal Insecticide, Fungicide and Rodenticide Act

The Federal Insecticide, Fungicide, and Rodenticide Act (FIFRA) was passed by the U.S. Congress in 1947 [370]. The legislation was administered by the U.S. Department of Agriculture (USDA) and remained primarily a labeling requirement. FIFRA has been amended several times and its registration provisions strengthened [202]. Pesticide use in the United States is also regulated under the Federal Food, Drug, and Cosmetic Act (FFDCA), which was amended in 1954 (Section 408, Miller Amendment) to require the establishment of pesticide tolerances on food [371]. An additional amendment in 1958 (Section 409) created the requirement to establish tolerances for food additives present in processed foods [371]. Section 409 of the FFDCA contains the Delaney Clause, which forbad the use of carcinogens as food additives. The EPA applied Section 409 of the FFDCA to those circumstances when pesticide residues found in processed food were greater than those found in the raw agricultural commodity. Under such circumstances, food additive tolerances are also required, and such additives must not be carcinogenic.

### Food Quality Protection Act of 1996

In 1996, the Food Quality Protection Act (FQPA) reauthorized the FFIFRA provisions and required tolerances to be reassessed as part of reregistrations [7]. FQPA amendments to the FFDCA [371] and FIFRA [370] directed the EPA to consider a number of factors when assessing risk as part of the tolerance setting procedure [7]. FQPA provides for a single, health-based standard and eliminates the problem posed by having different standards for pesticide concentration in raw and processed foods. FQPA required that, in the process of setting tolerances for pesticide residues in food, the EPA must evaluate the aggregate risk arising from exposure to pesticides from all routes of exposure, including oral, dermal, and inhalation exposure. Occupational exposure assessment, however, remained outside the jurisdiction of FQPA. FQPA also directed the EPA to consider pesticides having a common mechanism of toxicity and to evaluate the cumulative effect of exposure to pesticides sharing a common mechanism. Finally, the Agency was charged with developing techniques for evaluating the potential for pesticides to affect the endocrine system. Most of these provisions reflect concerns that children may be more susceptible to chemicals than adults, thereby taking into account key recommendations of a National Academy of Sciences report, *Pesticides in the Diets of Infants and Children* [330]. Under FQPA, the EPA assumes an extra tenfold uncertainty factor to account for increased susceptibility of children including effects of *in utero* exposure, unless data suggest otherwise.

## STUDY REQUIREMENTS

Toxicology testing guidelines have been promulgated by the EPA [372]; the Japanese Ministry of Agriculture, Forestry, and Fisheries (MAFF) [320]; the European Community (EC) [140,155]; and the Organization for Economic Cooperation and Development (OECD) [335]. These guidelines have been harmonized among the various regulatory authorities [224,225,320,335,473,544] and are revised or enhanced [473] as new testing procedures are developed (Table 16.1).

Acute toxicity studies are conducted by administering the chemical by the oral, dermal, or inhalation route to estimate the dose that is expected to causes mortality in 50% ($LD_{50}$) of the animals. Studies are also conducted to evaluate the irritation potential of chemicals when applied to skin and eyes. The potential of chemicals to cause allergic reactions when applied to skin (i.e., skin sensitization) is also determined. Acute oral and inhalation studies are usually conducted in rats, dermal and eye irritation studies are typically conducted in rabbits, and the sensitization study is carried out in guinea pigs. The results from these studies are used to establish the precautionary language (Table 16.2) used on product labels for crop protection chemicals [361].

Oral toxicity studies are conducted in rats, mice, or dogs fed diets containing the pesticide for various durations of time (28 days, 90 days, 1 year) or for the lifetime of the animal (24 months for rats, 18 months for mice). Animals are randomly assigned to either a control group or one of several treatment groups, each comprised of 10 to 50 rats or mice and 4 to 6 dogs. Typically, each study has at least four groups: a control group and three groups of animals that receive low, medium, and high concentrations of the pesticide in their respective diets. The high-dose group is typically administered a maximally tolerated dose [229] or, if the pesticide is nontoxic, a maximum limited dose of 1000 mg/kg/day. Lower doses are established to have minimal to moderate effects, and one dose ideally should have no effect. Typically, clinical symptoms and effects on survival, body weight, food consumption, blood chemistry, hematological, and urinary parameters are evaluated on multiple occasions during the in-life phase of the study. At study termination, individual organs are weighed, and gross and microscopic examinations are conducted on approximately 50 tissues per animal. The effects of the pesticide are described (hazard identification) and the lowest observed-effect level (LOEL), no-observable-effect level (NOEL), and no-observed-adverse-effect level (NOAEL) are determined.

Dermal toxicity is evaluated by applying the chemical to the skin for 6 hours a day for 21 days in rats or 28 days in rabbits. Developmental toxicity studies are conducted to evaluate the potential of the pesticide to affect the development of offspring, and birth defects are evaluated in both the rat and rabbit. In addition to developmental toxicity studies, a reproduction study is conducted in rats. This study involves feeding diets containing the chemical to young adult male and female rats for approximately 3 months prior to mating. The females are allowed to produce a litter of offspring that are then reared to adulthood. The animals are fed diets containing the chemical during this entire period of time. After reaching sexual maturity second-generation animals are allowed to mate and produce a second litter of offspring ($F_2$ generation) that are administered the pesticide until they reach adulthood at approximately 90 days of age. A variety of toxicological and reproductive parameters are assessed to determine the effect of the pesticide on neonatal development and reproductive function in young males and females animals.

The mutagenic potential of pesticides typically is assessed by evaluating its possible interaction with (1) genes (gene mutation tests), (2) the chromosome (clastogenic tests), and (3) directly with DNA (classified as other tests). The carcinogenic potential of pesticides is evaluated in mice and rats typically fed the chemical in their diet for 18 months and 24 months, respectively. The concentration of the chemical administered in the diet in chronic studies is generally selected based on the results from a 90-day feeding study [229]. Approximately 50 tissues from each animal are examined for the presence of tumors or other evidence of tissue damage. Pesticides are considered to be potential human carcinogen if (1) they significantly increase the incidence of any tumor above the incidence observed in concurrent or historical control animals, (2) they increases the incidence of rare or malignant tumors, and (3) they shorten the latency to tumor development period.

To determine whether a chemical is likely to be carcinogenic in humans, regulatory agencies around the world [259,410,544] conduct weight-of-evidence assessments using methods similar to those described in the EPA cancer classification scheme [410,544]. In a weight-of-evidence assessment, the carcinogenic potential of the chemical is evaluated by considering the results from animal studies, including but not limited to the details of the tumor response seen in animal bioassays (dose–response, including an evaluation of the evidence for nonlinearity, structure–activity considerations, details on mode of action and a description of key events, temporal and dose congruity, biological plausibility, and alternative modes of action including but not limited to an assessment of the genotoxic potential. Results from epidemiological investigations are weighted heavily, if such data are available. The outcome of this assessment classifies the pesticide a

**TABLE 16.1**
**Series 870 Health Effects Test Guidelines**

| OPPT No. | Name | Existing Numbers OPPT | OPP | OECD |
|---|---|---|---|---|
| **Group A—Acute Toxicity Test Guidelines** | | | | |
| 870.1000 | Acute toxicity testing-background | None | None | None |
| 870.1100 | Acute oral toxicity | 798.1175 | 81-1 | 401 |
| 870.1200 | Acute dermal toxicity | 798.1100 | 81-2 | 402 |
| 870.1300 | Acute inhalation toxicity | 798.1150 | 81-3 | 403 |
| 870.2400 | Acute eye irritation | 798.4500 | 81-4 | 405 |
| 870.2500 | Acute dermal irritation | 798.4470 | 81-5 | 404 |
| 870.2600 | Skin sensitization | 798.4100 | 81-6 | 406 |
| **Group B—Subchronic Toxicity Test Guidelines** | | | | |
| 870.3100 | 90-Day oral toxicity in rodents | 798.2650 | 82-1 | 408 |
| 870.3150 | 90-Day oral toxicity in nonrodents | None | 82-1 | 409 |
| 870.3200 | 21/28-Day dermal toxicity | None | 82-2 | 410 |
| 870.3250 | 90-Day dermal toxicity | 798.2250 | 82-3 | 411 |
| 870.3465 | 90-Day inhalation toxicity | 798.2450 | 82-4 | 413 |
| 870.3700 | Prenatal developmental toxicity study | 798.4900 | 83-3 | 414 |
| 870.3800 | Reproduction and fertility effects | 798.4700 | 83-4 | 416 |
| **Group C—Chronic Toxicity Test Guidelines** | | | | |
| 870.4100 | Chronic toxicity | 798.3260 | 83-1 | 452 |
| 870.4200 | Carcinogenicity | 798.3300 | 83-2 | 451 |
| 870.4300 | Combined chronic toxicity/carcinogenicity | 798.3320 | 83-5 | 453 |
| **Group D—Genetic Toxicity Test Guidelines** | | | | |
| 870.5100 | Bacterial reverse mutation test | 798.5100, 798.5265 | 84-2 | 471, 472 |
| 870.5140 | Gene mutation in *Aspergillus nidulans* | 798.5140 | 84-2 | None |
| 870.5195 | Mouse biochemical specific locus test | 798.5195 | 84-2 | None |
| 870.5200 | Mouse visible specific locus test | 798.5200 | 84-2 | None |
| 870.5250 | Gene mutation in *Neurospora crassa* | 798.5250 | 84-2 | None |
| 870.5275 | Sex-linked recessive lethal test in *Drosophila* | 798.5275 | 84-2 | 477 |
| 870.5300 | *In vitro* mammalian cell gene mutation test | 798.5300 | 84-2 | 476 |
| 870.5375 | *In vitro* mammalian chromosome aberration test | 798.5375 | 84-2 | 473 |
| 870.5380 | Mammalian spermatogonial chromosomal aberration | 798.5380 | 84-2 | 483 |
| 870.5385 | Mammalian bone marrow chromosomal aberration test | 798.5385 | 84-2 | 475 |
| 870.5395 | Mammalian erythrocyte micronucleus test | 798.5395 | 84-2 | 474 |
| 870.5450 | Rodent dominant lethal assay | 798.5450 | 84-2 | 478 |
| 870.5460 | Rodent heritable translocation assays | 798.5460 | 84-2 | None |
| 870.5500 | Bacterial DNA damage or repair tests | 798.5500 | 84-2 | None |
| 870.5550 | Unscheduled DNA synthesis in mammalian cells | 798.5550 | 84-2 | 482 |
| 870.5575 | Mitotic gene conversion in *Saccharomyces cerevisiae* | 798.5575 | 84-2 | 481 |
| 870.5900 | *In vitro* sister chromatid exchange assay | 798.5900 | 84-2 | 479 |
| 870.5915 | *In vivo* sister chromatid exchange assay | 798.5915 | 84-2 | None |
| **Group E—Neurotoxicity Test Guidelines** | | | | |
| 870.6100 | Acute and 28-day delayed neurotoxicity of organophosphorus substances | 798.6450, 798.6540, 798.6560 | 81-7, 82-5, 82-6 | 418, 419 |
| 870.6200 | Neurotoxicity screening battery | 798.6050, 798.6200, 798.6400 | 81-8, 82-7, 83-1 | 424 |
| 870.6300 | Developmental neurotoxicity study | None | 83-6 | None |
| 870.6500 | Schedule-controlled operant behavior | 798.6500 | 85-5 | None |
| 870.6850 | Peripheral nerve function | 798.6850 | 85-6 | None |
| 870.6855 | Neurophysiology: sensory evoked potentials | 798.6855 | None | None |
| **Group F—Special Studies Test Guidelines** | | | | |
| 870.7200 | Companion animal safety | None | None | None |
| 870.7485 | Metabolism and pharmacokinetics | 798.7485 | 85-1 | 417 |
| 870.7600 | Dermal penetration | None | 85-3 | None |
| 870.7800 | Immunotoxicity | None | 85-7 | None |

*Source:* U.S. EPA, *Pesticide Assessment Guidelines*, OPPTS Harmonized 870 Health Effects Test Guidelines Series, U.S. Environmental Protection Agency, Washington, D.C., 1998.

## TABLE 16.2
## USEPA Acute Toxicology Classification Scheme

| Toxicology Category | Signal Word | Oral $LD_{50}$ (mg/kg) | Dermal $LD_{50}$ (mg/kg) | Inhalation $LC_{50}$ (mg/L) | Eye Irritation | Skin Irritation |
|---|---|---|---|---|---|---|
| I | Danger[a] | Up to 50 | Up to 200 | Up to 0.2 | Corrosive; corneal opacity not reversed in 7 days | Corrosive |
| II | Warning | From 50 through 500 | From 200 through 2000 | From 0.2 through 2.0 | Corneal opacity reversed in 7 days; irritation persisting 7 days | Severe irritation at 72 hours |
| III | Caution | From 500 through 5000 | From 2000 through 5000 | From 2.0 through 20 | No corneal opacity; irritation reversed within 7 days | Moderate irritation at 72 hours |
| IV | Caution | Greater than 5000 | Greater than 5000 | Greater than 20 | No irritation | Mild or slight irritation at 72 hours |

[a] The word "Poison" is used on the label if the "Danger" category is based on oral, dermal, or inhalation toxicity.

*Source:* Stevens, J. T. et al., in *Primer on Regulatory Toxicology*, Chenzelis, C. et al., Eds., RavenPress, New York. With permission.

## TABLE 16.3
## Schemes for the Classification of Carcinogens

| U.S. Environmental Protection Agency Original Classification Scheme [147] | | | | International Agency for Research on Cancer 2005 Guideline [266] | |
|---|---|---|---|---|---|
| Carcinogen Category | Criteria for Classification | Risk Character RfD | Risk Character $Q_1$* | Descriptive Characterization | |
| A–Human | Sufficient evidence in humans | | X | Carcinogenic to humans | Group 1: The agent is carcinogenic to humans |
| B–Probable Human | | | | | |
| B1 | Limited evidence in humans<br>Sufficient evidence in animals (two species with tumors) | | X | Likely to be carcinogenic to humans | Group 2A: The agent is probably carcinogenic to humans |
| B2 | Inadequate human evidence<br>Sufficient evidence in animals | | X | | |
| C–Possible Human | No evidence in humans<br>Limited evidence in animals | X | X | Suggestive evidence of carcinogenic potential | Group 2B: The agent is possibly carcinogenic to humans |
| D–Not Classifiable | Inadequate animal or human evidence | X | | Inadequate information to assess carcinogenic potential | Group 3: The agent is not classifiable as to its carcinogenicity to humans |
| E–Not a Human Carcinogen | Sufficient animal testing with no evidence of carcinogenicity and human experience | X | | Not likely to be carcinogenic to humans | Group 4: The agent is probably not carcinogenic to humans |

one of several categories shown in Table 16.3 for the EPA and International Agency for Research on Cancer (IARC).

## FUNGICIDES

The Fungicide Resistance Action Committee (FRAC) has classified fungicides according to their mechanism of action in fungi [233]. In Table 16.4, pesticides are grouped according to their assigned FRAC code. Not all FRAC codes have been included in this chapter, and some commercially important members of the selected FRAC code may have been omitted. Table 16.4 provides a summary of the mechanism of action of pesticides in fungi for the selected FRAC code. It also provides conclusions as to whether or not the identified mechanisms of action in fungi are likely conserved in humans and whether any other basis exists for establishing a common mechanism of toxicity group, as set forth by FQPA and elaborated by EPA [489

## ACYLALANINES AND OXAZOLIDINONES (FRAC CODE A1)

The acylanine fungicides are represented here by metalaxyl and by mefenoxam, which is the active R enantiomer of metalaxyl. The oxazolidinones are represented by oxadixyl. These chemicals inhibit RNA synthesis in fungi by interfering with RNA polymerase I [233]. The activity of mefenoxam and oxadixyl is restricted to the control of downy mildew and late blight in a broad spectrum of crops. Data suggest that structural features of ribosome gene promoters are conserved from plants to humans, but specific base sequences are not [663]; therefore, it is unlikely that a pesticide selected to interfere with RNA polymerase I in fungi would have the same effect in mammalian cells. Furthermore, because RNA polymerase I is central to cell replication and growth [685], it is unlikely that an effect on RNA polymerase I in animal bioassays would go unnoticed unless the chemicals were poorly absorbed or rapidly metabolized and eliminated. The hazard profiles for metalaxyl and oxadixyl are unremarkable, and they provide no indication of any effect on the cell cycle, although specific studies have not been conducted. (See Table 16.5.)

## BENZIMIDAZOLES AND THIOPHANATES (FRAC CODE B1)

The benzimidazole (benomyl and thiabendazole) and thiophanate (thiophanate-methyl) fungicides have broad-spectrum activity, as indicated by a mode of action involving the inhibition of mitosis by preventing polymerization of β-tubulin, one of the constituent building blocks of microtubules [233]. Microtubules play a critical role in both the plant and animal kingdoms during mitosis [679,681] and provide the basis for a cytoarchitecture that permits intracellular transport of molecules using the motor proteins dynein and kinesin [646,679,681]. Recently, it has been suggested that microtubules may provide the network for laying down cellulose structural support in plants [676]. Whereas the α- and β-tubulin subunits that make up microtubules are conserved between kingdoms, microtubules in plants are fundamentally different from those found in fungi and animals because plant microtubules do not arise from discrete organizing centers such as the centrosome or the spindle pole body, as they do in fungi and animals [700]. During mitosis in fungi and animals cells, γ-tubulin serves as a microtubule nucleating factor, playing a role in assembling the α- and β-tubulin dimmers that are embedded in and grow out of a pair of centrioles and associated proteins. Recent data suggest that γ-tubulin may also play a role in microtubule nucleation in plant cells [657]. Based on new and developing information and considering that a number of isoforms of β-tubulin have been described in mammalian cells [698], it is not known, but is certainly possible, that the benzimidazole and thiophanate fungicides could interfere with microtubule formation in animal cells. The structure, use, and hazard profiles of benzimidazole and thiophanate fungicides are presented in Table 16.6. The hazard characteristics of benomyl, thiabendazole, and thiophanate-methyl have been reviewed by the EPA and by a Joint Meeting of the FAO/WHO Panel of Experts on Pesticide Residues in Food (JMPR) [262,289,534]. Benomyl has been classified by the EPA as a Category C carcinogen (possible human carcinogen) based on the occurrence of treatment-related mouse liver tumors in a mouse carcinogenicity study [508]. The EPA revoked all tolerances for benomyl on July 17, 2002, and sales of the product in the channels of trade were discontinued after December, 31, 2002 [508].

## CARBOXAMIDES (FRAC CODE C2)

Carboxamides are represented here by flutolanil, which is a systemic fungicide with protective and curative action. It is used to control sheath blight, white mold, and snow blight in rice, cereals, sugar beet, and other crops. The structure, use, and hazard profile of flutolanil are given in Table 16.7. Carboxamides inhibit mitochondrial respiration by blocking electron transport at the succinate dehydrogenase stage (complex II) in the Krebs cycle (see Figure 16.1 and Figure 16.2). A recent characterization of the crystalline structure of succinate dehydrogenase in *Escherichia coli* [665,703] has permitted structural modeling of the active sites of the enzyme [652,667]. Future work may permit a determination of where carboxamides bind to succinate dehydrogenase of fungi and address the question of whether homologous binding sites are present in human mitochondria. If it is assumed that carboxamides bind to human mitochondrial targets, one would predict that they would be highly toxic. Flutolanil is remarkably nontoxic, which belies any effect as an inhibitor of mitochondrial respiration in mammals.

## METHOXYACRYLATE AND OXIMINOACETATE (STROBILURINS; FRAC CODE C3)

The first member of the strobilurin family (strobilurin A) was isolated from *Oudemansiella mucidu* found growing on beech trees [3,327] and from the pine cone fungus *Strobilurus tenacellus* [654]. The strobilurins are selective, yet they have broad fungicidal activity that includes protective, eradicant, and antisporant effects. The strobilurins inhibit mitochondrial respiration by binding to the $Q_0$ site on cytochrome *b* (Figure 16.2) thereby blocking electron transfer between cytochrome *b* and cytochrome $c_1$ in the Krebs cycle (Figure 16.1), resulting in a disruption of adenosine triphosphate (ATP) synthesis [687]. The crystalline structure of the cytochrome $bc_1$ complex has been elucidated alone [644] and bound to its substrate cyto-

**TABLE 16.4**
**Fungicides Listed According To FRAC Classification System**

| Chemical Group | Common Name | FRAC MOA (Code) | Effect at Target (FRAC Code) | MOA Conserved in Mammals? | Other Common MOA in Mammals |
|---|---|---|---|---|---|
| Acylalanines<br>Oxazolidinones | Mefenoxam<br>Oxadixyl<br>(Table 16.5) | Nucleic acid synthesis inhibition (A) | Complex II: succinate dehydrogenase (SDG) (C2) | Unlikely | None known |
| Benzimidazoles<br>Thiophanates | Benomyl<br>Thiabendazole<br>Thiophanate-methyl<br>(Table 16.6) | Mitosis and cell division (B) | β tubulin assembly during mitosis | Possible | None known |
| Carboxamides | Flutolanil<br>(Table 16.7) | Respiration (C) | Complex II: succinate dehydrogenase (C2) | Likely but the active ingredient does not appear to reach target in mammals | Glutathione activity in liver mitochondria |
| Methoxyacrylates<br>Oximinoacetate | Azoxystrobin<br>Picoxystrobin<br>Trifloxystobin<br>(Table 16.8) | Respiration (C) | Inhibits cytochrome $bc_1$ (ubiquinol oxidase) at the outside quinine site (C3) | Likely but the active ingredient does not appear to reach target in mammals | None known |
| Organotin chemicals | Triphenyl tin acetate/hydroxide<br>(Table 16.9) | Respiration (C) | Inhibitors of oxidative phosphorylation (C6) | Likely but the active ingredient does not appear to reach target in mammals | None known |
| Anilinopyrimidines | Cyprodinil<br>Pyrimethanil<br>(Table 16.10) | Inhibit amino acid and protein synthesis (D) | Methionine biosynthesis and protease secretion inhibition (D1) | Unlikely; monogastric mammals do not synthesize methionine | None known |
| Phenylpyrroles | Fenpiclonil<br>Fludioxonil<br>(Table 16.11) | Signal transduction (E) | MAP kinases involved in osmotic signal transduction (E2) | Possible, as mammalian analogs of HOG1 exist | None known |
| Dicarboxyimides | Iprodione<br>Vinclozolin<br>(Table 16.12) | Lipids and membrane synthesis (F) | NADH cytochrome *c* reductase in lipid peroxidation (F1) | Likely | Share a common metabolite 3,5-dichloro-aniline [501]; potential antiandrogenic |

| | | | | | |
|---|---|---|---|---|---|
| Triazole<br>Imidazoles | Cyproconazole<br>Difenconazole<br>Fenbuconazole<br>Hexaconazole<br>Myclobutanil<br>Propiconazole<br>Tebuconazole<br>Triadimefon<br>Triadimenol<br>Imazalil<br>Prochloraz<br>(Tables 13 to 15) | Sterol biosynthesis in membranes (G) | C14 demethylase in sterol biosynthesis (G1) | Yes | Cyp 19 (aromatase) inhibition [664] |
| Dithiocarbamates<br>Ethylenebisdithio-carbamates | Ferbam<br>Thiram<br>Ziram<br>(Table 16.16)<br>Mancozeb<br>Maneb<br>Zineb<br>(Table 16.17) | Multi-site contact activity (M3) | Interferes with oxygen uptake and inhibits sulfur containing enzymes<br>Breaks down cyanide which then reacts with sulfhydryl groups in cells | Possible | Cholinesterase inhibition (ziram and metam sodium); neurotoxic (distal peripheral neuropathy); potential to chelate cations; mancozeb, maneb, and metiram share a common metabolite, ethylenethio-urea [505] |
| Phthalimides | Captan<br>(Table 16.18) | Multi-site contact activity (M4) | Preferential react with protein sulfhydryl groups; enzyme inhibitor | Likely | None known |
| Chloronitriles | Chlorothalonil<br>(Table 16.19) | Multi-site contact activity (M5) | Conjugates with cellular thiols | Likely | None known |

*Source:* FRAC, *FRAC Code List 2: Fungicides Sorted by Mode of Action*, Fungicide Resistance Action Committee, 2005 (http://www.frac.info/frac/index.htm).

**TABLE 16.5**
**Structure, Use, and Hazard Profiles for Acylalanine (Metalaxyl, Mefenoxam) and Oxazolidinone (Oxadixyl) Fungicides (FRAC Code A1)**

| Fungicide | Structure | Principal Use/Crop | Application Rate |
|---|---|---|---|
| Mefenoxam (metalaxyl-M, Ridomil Gold®) [123] | 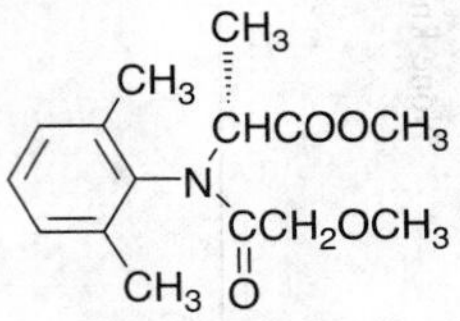 | Used on alfalfa, apples, asparagus, avocadoes, berries, citrus, cole crops, cotton, cucurbits, hops, peanuts, stone fruit, soybeans, sugar beets, tobacco and vegetables. | 100–1000 g active ingredient per hectare |
| Oxadixyl (Anchor®, Sandofan®) [139] | 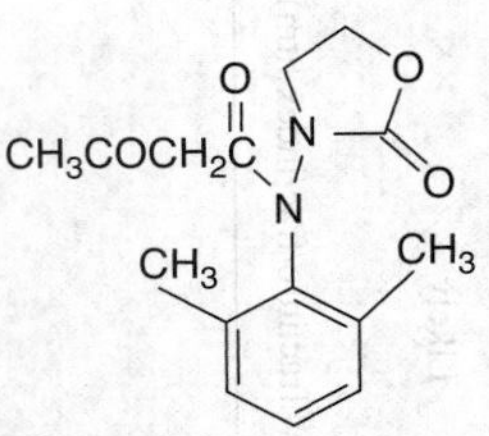 | Used on vines, maize, potatoes, tobacco, hops, sunflowers, citrus, fruit, vegetables, and as seed dressing for cotton, peas, and sunflowers | 200–300 g active ingredient per hectare |

| Fungicide | Irritation | | $LD_{50}$ (mg/kg) | | $LC_{50}$ (mg/L) | Sensitization Potential | Signal Word |
|---|---|---|---|---|---|---|---|
| | Eye | Skin | Oral | Dermal | Inhalation | | |
| Mefenoxam | Severe irritant | Slight irritant | 490 | >2000 | >2.3 | Negative | Warning |
| Oxadixyl | Non-irritant | Non-irritant | 1860 | >2000 | >5.6 | Negative | Caution |

| Fungicide | Species/Study | NOEL (mg/kg/day) | Toxicity Study | Hazard Indicator |
|---|---|---|---|---|
| Mefenoxam [384,390,439] | Rat/2-year | 13 | Mutagenicity | Not mutagenic |
| | Dog/52-week | 8 | Developmental | Not teratogenic |
| | Mouse/18-month | 38 | Reproductive | No evidence |
| | RfD (Based on the 6-month dog study) | 0.08 | Oncogenicity | No evidence |
| Oxadixyl [139,512] | Rat/2-year | 250 | Mutagenicity | Not mutagenic |
| | Dog/52-week | 500 | Developmental | Not teratoric |
| | Mouse/18-month | — | Reproductive | Not reproductive toxin |
| | RfD | — | Oncogenicity | — |

chrome *c* [672], and details regarding binding to the dimeric structure have been proposed [692]. The $Q_0$ binding site for stigmatellin, an analog of the strobilurins, has been identified by crystallization of cytochrome $bc_1$ in the presence of an excess of the inhibitor [644]. The amino acid sequence of the $Q_0$ binding site appears to be highly conserved across species, including mammals. Tests carried out on 14 strobilurins on mitochondrial enzyme preparations from fungi, housefly, rat, and maize showed no selectivity [687]. The structure, use, and hazard profiles for three economically important members of this class are presented in Table 16.8. The fact that the strobilurins are relatively nontoxic to animals suggests that they do not reach these mitochondrial targets.

## Organotin Compounds (FRAC Code C6)

The organometallic fungicides, which are represented here by triphenyltin, are limited in their spectrum for disease control but are effective as protective, curative, and anti sporulants in the treatment of early and late blight, scab leaf blotch, and powdery mildew [81]. Trialkyltins affec mitochondrial respiration by (1) disrupting the membran potential by exchanging halide ions for hydroxyl ion across the membrane; (2) binding to ATP synthase thereby reducing ATP production; and (3) causing mito chondrial swelling, especially for the more lipophilic com pounds. They are also capable of causing cellular lysis *i vitro* [683] and in aquatic organisms [702]. Of thes effects, the effect on ATP synthase is the most specifi (see Figure 16.2).

The ATP synthases are enzymes that make up tw rotary motors. The membrane-embedded $F_0$ motor convert energy from a transmembrane electrochemical ($Na^+$) grad ent into torque, which is transmitted through a commo shaft to the water-exposed $F_1$ motor, where it drives syn thesis of ATP from adenosine diphosphate (ADP) and pho

**TABLE 16.6**
**Structure, Use, and Hazard Profiles for Benzimidazole (Benomyl, Thiabenazole) and Thiophanate Fungicides (FRAC Code B1)**

| Fungicide | Structure | Principal Use/Crop | Application Rate (g active ingredient per hectare) |
|---|---|---|---|
| Benomyl (Benlate®) [28] | $CONH(CH_2)_3CH_3$; $NHCO_2CH_3$ | Used against Ascomycetes and Basidomycetes in cereals, grapes, pome fruit, stone fruit, rice, and vegetables | 140–550<br>Tree: 550–1100<br>Storage: 25–200 g/hL |
| Thiabenazole (Mertech®) [181] | H, N, S | Used for the control of *Aspergillus*, *Botrytis*, and others in vegetables, bananas, cereals, cabbage, stone fruit, citrus fruit, and hops | 0.2–2.2 g/L |
| Thiophanate-methyl (Topsin-M®) [184] | $NHCSNHCO_2CH_3$; $NHCSNHCO_2CH_3$ | Used for eyespot on cereals, scab and rot on apples and pears, and powdery mildew on pome fruit, stone fruit, vegetables, strawberries, and vines | 30–50 |

| Fungicide | Irritation: Eye | Irritation: Skin | $LD_{50}$ (mg/kg): Oral | $LD_{50}$ (mg/kg): Dermal | $LC_{50}$ (mg/L): Inhalation | Sensitization Potential | Signal Word |
|---|---|---|---|---|---|---|---|
| Benomyl | Moderately irritating | Slight irritant | >10,000 | >5000 | >2.0 | Not sensitizer | Caution |
| Thiabendazole | Non-irritating | Non-irritant | 3100 | — | 0.07 | Not sensitizer | Caution |
| Thiophanate-methyl | Mildly irritating | Mildly irritating | >5000 | >10000 | 1.7 | Not sensitizer | Caution |

| Fungicide | Species/Study | NOEL (mg/kg/day) | Toxicity Study | Hazard Indicator |
|---|---|---|---|---|
| Benomyl [289,374,508] | Rat/2-year oral | >125 | Mutagenicity | Not mutagenic |
| | Dog/52-week oral | 12.5 | Developmental | Teratogenic in mice |
| | Mouse/18-month oral | 40 | Reproductive | Not reproductive toxin |
| | RfD (Based on the rat reproduction study | 0.05 | Oncogenicity | C with RfD (liver tumors in mice) |
| Thiabendazole [262,520] | Rat/2-year oral | 40 | Mutagenicity | Not mutagenic |
| | Dog/52-week oral | — | Developmental | Not teratogenic |
| | Mouse/18-month oral | — | Reproductive | Not reproductive toxin |
| | ADI (human study) | 0.035 | Oncogenicity | Not oncogenic |
| Thiophanate-methyl [292,534] | Rat/2-year oral | 8.0 | Mutagenicity | Not mutagenic |
| | Dog/52-week oral | 50 | Developmental | Not teratogenic |
| | Mouse/18-month oral | 23 | Reproductive | Not reproductive toxin |
| | RfD (based on 2-year rat study) | 0.08 | Oncogenicity | Not oncogenic |

hate [656,693]. More is known about the operation of the $F_1$ motor than about the membrane-bound $F_0$ motor, the presumptive site of action of the organotin compounds [645,678,697]; however, interference with the production of ATP by organotin compounds would have significant toxicological consequences if they reach and bind to ATP synthase.

The structure and mammalian toxicity profile of triphenyltin (fentin) are presented in Table 16.9. Triphenyltin hydroxide is moderately acutely toxic (oral $LD_{50}$ = 140 mg/kg/day) but is highly toxic following longer term administration (acceptable daily intake [ADI] = 0.0005 mg/kg/day). It has been classified by the EPA as a Category B2 (probable human) carcinogen, based on mouse liver and rat pituitary and testicular tumors. It has not been possible to predict toxicity across species based on structure alone because some substituted organotin compounds are more toxic to some species than to others [683,702].

## Anilinopyrimidines (FRAC Code D1)

The spectrum of activity for anilinopyrimidines, which are represented here by cyprodinil and pyrimethanil, is limited

**TABLE 16.7**
**Structure, Use, and Hazard Profile for Carboxamide (Flutolanil) Fungicides (FRAC Code C1)**

| Fungicide | Structure | Principal Use/Crop | Application Rate (g active ingredient per hectare) |
|---|---|---|---|
| Flutolanil (Folistar®, Moncut®) [91,400] | (structure: CONH-linked benzene rings bearing $CF_3$ and $OCH(CH_3)_2$) | Used on peanuts rice, cereals, sugar beets, fruits, and vegetables | 300–1000 (peanuts) |

| Irritation | | $LD_{50}$ (mg/kg) | | $LC_{50}$ (mg/L) | Sensitization | |
|---|---|---|---|---|---|---|
| Eye | Skin | Oral | Dermal | Inhalation | Potential | Signal Word |
| Slightly irritating | Non-irritating | >10,000 | >5000 | >6.0 | Negative | Caution |

| Species/Study | NOEL (mg/kg/day) | Toxicity Study | Hazard Indicator |
|---|---|---|---|
| Rat/2-year oral | 87 | Mutagenicity | Not mutagenic |
| Dog/2-year oral | 50 | Developmental | Not teratogenic |
| Mouse/18-month oral | 735 | Reproductive | Not reproductive toxin |
| RfD (NOEL = 64 in rat reproduction study; UF = 300) | 0.2 | Oncogenicity | E (not oncogenic) |

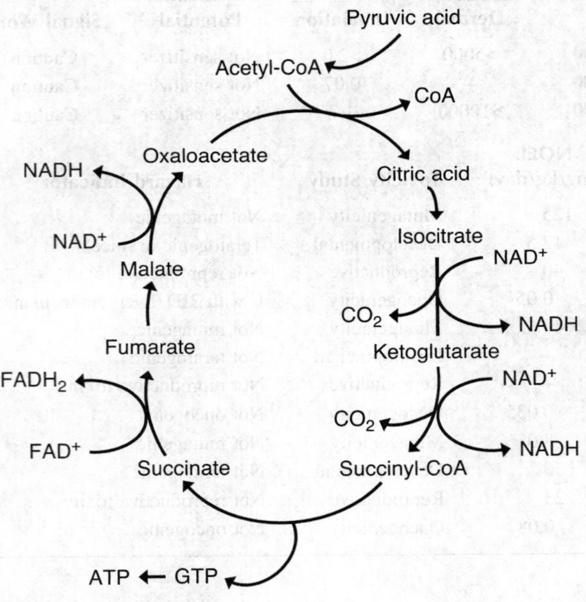


**FIGURE 16.1** Kreb cycle.

to ascomycetes and deuteromycetes [59,162]. The anilinopyrimidines are used to control gray mold on vines, fruit, vegetables, and ornamentals and leaf scab on pome fruit. The anilinopyrimidine fungicides inhibit methionine biosynthesis [659] but not likely through an action of cystathionine β-lyase as originally proposed [658, 660, 675]. Cystanthionine γ-synthase and methionine synthase are other possible targets of the anilinopyrimidine fungicides, but the actual target in plants is not likely relevant to humans. Methionine is one of four essential sulfur-containing amino acids in the aspartate pathway (the others being lysine, threonine, and isoleucine) that humans and other monogastric animals cannot synthesize *de novo* [666,684]. The structure, use, and hazard profiles of the anilopyrimidine fungicides cyprodinil and pyrimethanil are presented in Table 16.10. The hazard profiles for these chemicals suggest minimal risk to humans.

## Phenylpyrroles (FRAC Code E2)

It has been proposed that the phenylpyrrole fungicides represented here by fenpiclonil and fludioxonil, interfere with the two-component, MAP-kinase-mediated osmotic signal transduction pathway shown in Figure 16.3 [669]. *Neurospora crassa*, incubated with fenpicolnil or fludioxonil, respond by accumulating glycerol, which causes the fungi to rupture [339,340]. Os-2 mutants do not accumulate glycerol when treated with phenylpyrroles [669,706] indicating that if the activation of the osmotic signal transduction pathway is blocked then the fungicidal action of the phenylpyrroles is prevented. Although the exact molecular target of the phenylpyrrole fungicides within the osmotic signal transduction pathway has not been convincingly demonstrated [704,706], evidence suggests that at least one component of the two-component pathway (Figure 16.3) is conserved in vertebrate cells [670,671,699]; therefore, it could be conservatively assumed that the phenylpyrroles may affect this signaling pathway in mammals [232,249]. The structure, use, and hazard profiles of fenpiclonil and fludioxonil are presented in Table 16.11. Fen

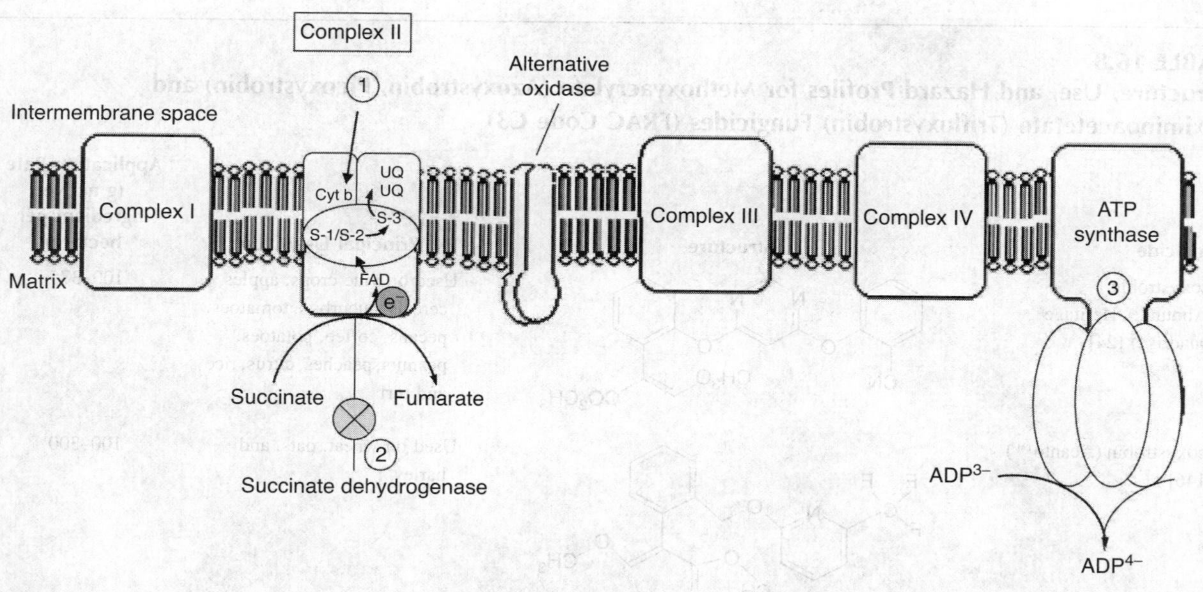


FIGURE 16.2 Mitocondrial respiration.

piclonil and fludioxonil are not acutely toxic, and they do not exhibit a remarkable repeat-dose toxicity profile. Fludioxonil has been classified as a Category D carcinogen (nonclassifiable in regard to carcinogenicity).

## DICARBOXIMIDES (FRAC CODE F1)

It has been proposed that the dicarboximides cause lipid peroxidation in membranes by inhibiting NADH cytochrome *c* reductase, which is part of the mitochondrial respiratory chain (Figure 16.1 and Figure 16.2) present in all eukaryotic cells [312]. The dicarboximides may also inhibit spore germination, and because of reports of resistance development for the phenylpyrroles and the dicarboximides, it has been suggested that the dicarboximides may also have effects on the osmotic signal transduction pathway discussed for the phenylpyrroles [4,206,341,704]. The dicarboximides have a narrow spectrum of activity limited to *Botrytis*, *Sclerotinia*, *Monilinia*, and *Alternaria*. The dicarboximides are used to treat diseases in turf, strawberries, stone fruit, peanuts, and vines. The structure, use, and hazard profiles of the dicarboximide fungicides iprodione and vinclozolin are presented in Table 16.12.

Iprodione interferes with androgen synthesis [291] and caused an elevated incidence of interstitial cell tumors in male rats at a concentration of 1600 ppm in the diet. Vinclozolin is metabolized in animals to the antiandrogenic metabolites, 2-1[(3,5-dichlorophenyl) carbamoyl]oxyl-2-methyl-3-butenoic acid and 3,5′-dichloro-2-hydroxy-2methylbutyl-3-enanilide. These metabolites are presumed to cause infertility in male rats [298]. It has been proposed that this effect is due to a feminization of the outer genital organs of males exposed to the metabolites during development [453].

The EPA has not established a common mechanism of toxicity group for the dicarboximide fungicides based on their antiandrogenic potential. For cancer risk assessment, however, the EPA has applied a linear low-dose method for evaluating the cumulative risk resulting from exposure to the metabolite 3,5-dichloroaniline (3,5-DCA), which is common to the dicarboximide fungicides iprodione, procymidone, and vinclozolin. The EPA has assumed that this terminal, plant, animal, and environmental metabolite is mutagenic and carcinogenic because of a predicted structure–activity relationship with *p*-chloroaniline [501], which caused an increased incidence of sarcoma in the spleen of male Fischer 344 rats [331].

## DEMETHYLASE INHIBITORS (FRAC CODE G1)

The sterol biosynthesis inhibitors or, more precisely, the sterol demethylase inhibitor (DMI) group is comprised of imidazole, piperazine, pyridine, pyrimidine, and triazole fungicides. DMI inhibitors affect fungi by inhibiting the synthesis of ergosterol [233]. Ergosterol synthesis inhibition results in an accumulation of methylated ergosterol derivatives, which due to their bulkier structure cannot be packed correctly into the lipid bilayer of the fungal

**TABLE 16.8**
**Structure, Use, and Hazard Profiles for Methoxyacrylate (Azoxystrobin, Picoxystrobin) and Oximinoacetetate (Trifloxystrobin) Fungicides (FRAC Code C3)**

| Fungicide | Structure | Principal Use/Crop | Application Rate (g active ingredient per hectare) |
|---|---|---|---|
| Azoxystrobin (Abound®, Heritage®, Quadris®) [24] | 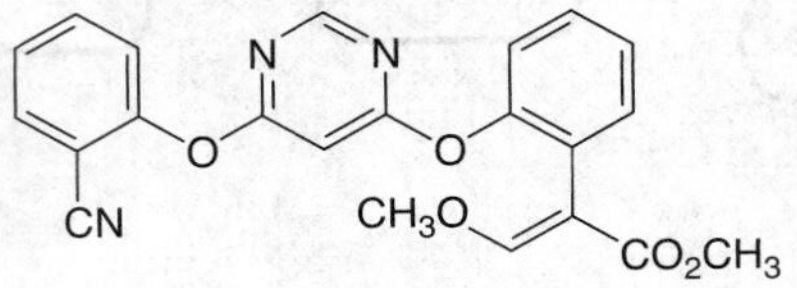 | Used on vine crops, apples, cereals, cucurbits, tomatoes, pecans, coffee, potatoes, peanuts, peaches, citrus, rice and turf | 100–375 |
| Picoxystrobin (Acanto™) [146] | 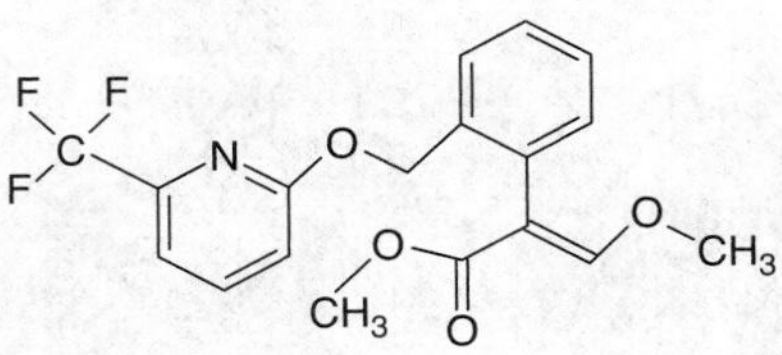 | Used on wheat, oats, and barley | 100–300 |
| Trifloxystrobin (Flint®, Stratego™) [190] | 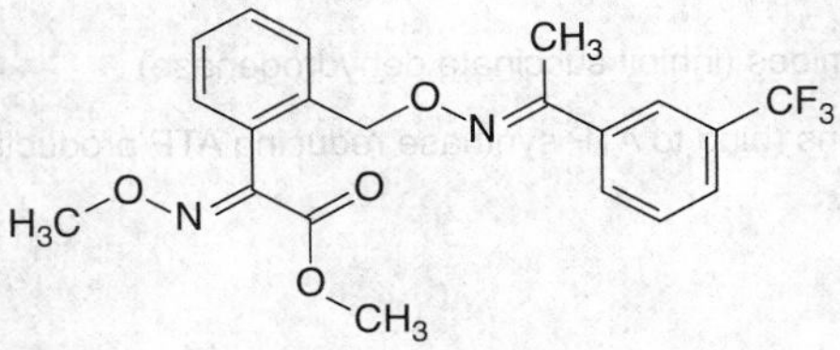 | Used on cucurbits and fruiting vegetables, pome fruit, stone fruit, grapes, hops, and pistachio | 50–187.5 |

| Fungicide | Irritation: Eye | Irritation: Skin | $LD_{50}$ (mg/kg): Oral | $LD_{50}$ (mg/kg): Dermal | $LC_{50}$ (mg/L): Inhalation | Sensitization Potential | Signal Word |
|---|---|---|---|---|---|---|---|
| Azoxystrobin | Slight irritant | Slight irritant | >5000 | >2000 | >0.7 | No positive | Caution |
| Picoxystrobin | Non-irritant | Non-irritation | >5000 | >2000 | 4.59 | Not sensitizer | Caution |
| Trifloxystrobin | Mild irritant | Mild irritant | >5000 | >2000 | 4.65 | Strong | Caution |

| Fungicide | Species/Study | NOEL (mg/kg/day) | Toxicity Study | Hazard Indicator |
|---|---|---|---|---|
| Azoxystrobin [413] | Rat/2-year | 18 | Mutagenicity | No evidence |
| | Dog/52-week | 25 | Developmental | Not teratogenic |
| | Mouse/18-month | 381 | Reproductive | No evidence |
| | RfD | 0.18 | Oncogenicity | E (no evidence) |
| Picoxystrobin [223] | Rat/2-year | 12.2 | Mutagenicity | No evidence |
| | Dog/52-week | — | Developmental | Not teratogenic |
| | Mouse/18-month | — | Reproductive | No evidence |
| | ADI | 0.043 | Oncogenicity | Not likely human carcinogen |
| Trifloxystrobin [498] | Rat/2-year | 9.8 | Mutagenicity | No evidence |
| | Dog/52-week | 5 | Developmental | Not teratogenic |
| | Mouse/18-month | 39.4 | Reproductive | No evidence |
| | RfD | 0.038 | Oncogenicity | Not likely human carcinogen |

membrane. These membrane alterations hinder the uptake and storage of nutrients, resulting in cell death. Effects of DMIs on fungal lipid, nucleic acid, and protein synthesis are likely secondary to their effect on cell membranes. DMI fungicides also inhibit 14-α-demethylase (P450 CYP51) in mammalian cells, where 14-α-demethylase catalyzes the conversion of lanosterol to zymosterol, a precursor to cholesterol and all the mammalia sex steroids. In addition to inhibiting 14-α-demethylase the DMI fungicides have the potential to inhibit the P45 CYP19 enzyme aromatase, which catalyzes the conver sion of testosterone to 17β-estradiol and the conversio of androstenedione to estrone (Figure 16.4). The EPA ha not determined if a common mechanism grouping of th

**TABLE 16.9**
**Hazard Profile for Triphenyltin (FRAC Code C6)**

| Structure | Principal Use/Crop | Application Rate (g active ingredient per hectare) |
|---|---|---|
| 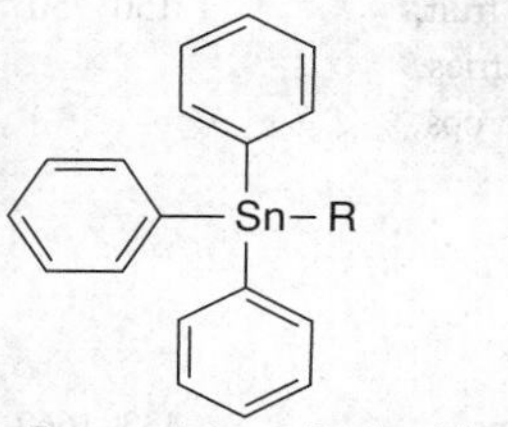 Triphenyl–Sn–R; R = acetate or hydroxide | Used on potatoes, celery, onions, sugar beets, peanuts, beans, wheat, coffee, and pecans | 160–240 |

| Fungicide | Irritation: Eye | Irritation: Skin | $LD_{50}$ (mg/kg): Oral | $LD_{50}$ (mg/kg): Dermal | $LC_{50}$ (mg/L): Inhalation | Sensitization Potential | Signal Word |
|---|---|---|---|---|---|---|---|
| Triphenyltin acetate | Severe irritant | Non-irritant | 140 | 450 | 0.044 | Positive | Danger |
| Triphenyltin hydroxide | Severe irritant | Slight irritant | 110 | 1600 | 0.060 | Negative | Danger |

| Fungicide | Species/Study | NOEL (mg/kg/day) | Toxicity Study | Hazard Indicator |
|---|---|---|---|---|
| Triphenyltin hydroxide [81,638] | Rat/2-year | <0.3 | Mutagenicity | No evidence |
| | Dog/52-week | 0.2 | Developmental | Not teratogenic |
| | Mouse/18-month | 1.4E | Reproductive | No evidence |
| | ADI | 0.0005 | Oncogenicity | B2 (liver tumors in mice; pituitary and testicular tumors in rats) |

DMI fungicides can be made based on these potential common molecular targets, although it has conducted an aggregate dietary risk assessment for three DMI fungicides based on the toxicity of the common metabolite 1,2,4-triazole [569].

The use and structures of the most prominent DMI fungicides are provided in Table 16.13, and their hazard profiles appear in Table 16.14 and Table 16.15. High doses of the DMI fungicides cause a treatment-related increased incidence of mouse liver tumors for the majority of these chemicals (hexaconazole and myclobutanol are the exceptions). These effects are apparently not mediated through a direct genotoxicity mechanism. It has been suggested that a phenobarbital-like induction of P450 enzymes is responsible for the liver tumor response seen in mice and that a similar response is unlikely to occur in humans [530]. In addition to the carcinogenic effect, evidence of developmental and reproductive system effects have been noted for several of the DMI fungicides, including cyproconazole [386] and triadimenol [269]. The mechanism of action has not been established but it has been postulated that some of these effects may be secondary to effects of the azole fungicides on steroid biosynthesis [643].

## Inorganic Fungicides (FRAC Code M1 and M2)

Inorganic chemicals such as sulfur were used before 1000 B.C. [220], and elemental sulfur and copper (hydroxide, oxychloride, and sulfate) are still used as fungicides today. The mode of action of the inorganic fungicides is protective or preventative; they exert their effects by the inhibiting of mitochondrial respiration [175]. The inorganic fungicides are relatively ineffective as they must be applied at high use rates ranging from 1000 to 10,000 g/hectare. Severe eye irritation is seen with copper hydroxide [51], whereas copper oxychloride and copper sulfate are not eye irritants [52,53]. Elemental sulfur is considered practically nontoxic to humans and animals [175].

## Dithiocarbamates and Ethylenebisdithiocarbamates (FRAC Code M3)

The dithiocarbamates are broad-spectrum protective fungicides having multiple sites of action [82,185,194]. They are used to control scab on pome fruit, blue mold on tobacco, rust on ornamentals, and numerous diseases on

**TABLE 16.10**
**Structure, Use and Hazard Profiles for Anilinopyrimidine Fungicides (FRAC Code D1)**

| Fungicide | Structure | Principal Use/Crop | Application Rate (g active ingredient per hectare) |
|---|---|---|---|
| Cyprodinil (Vangard®) [59] | | Used on cereals, grapes, pome fruit, stone fruit, almonds, strawberries, vegetables, vegetables, field crops, and as a seed dressing | 150–750 |
| Pyrimethanil (Mythos®, Scala®) [162] | | Used on pome fruit, vine crops, vegetables and ornamentals | 333–1667 |

| Fungicide | Irritation: Eye | Irritation: Skin | $LD_{50}$ (mg/kg): Oral | $LD_{50}$ (mg/kg): Dermal | $LC_{50}$ (mg/L): Inhalation | Sensitization Potential | Signal Word |
|---|---|---|---|---|---|---|---|
| Cyprodinil | Minimal irritant | Slight irritant | 2796 | >2000 | >1.2 | Positive | Caution |
| Pyrimethanil | Slight irritant | Non-irritant | >4149 | >5000 | >1.98 | Not positive | Caution |

| Fungicide | Species/Study | NOEL (mg/kg/day) | Toxicity Study | Hazard Indicator |
|---|---|---|---|---|
| Cyprodinil [459] | Rat/2-year | 3.75 | Mutagenicity | No evidence |
| | Dog/52-week | 65.6 | Developmental | Not teratogenic |
| | Mouse/18-month | 16.1 | Reproductive | No evidence |
| | RfD | 0.038 | Oncogenicity | E (No evidence) |
| Pyrimethanil [448] | Rat/2-year | 20 | Mutagenicity | No evidence |
| | Dog/52-week | 30 | Developmental | Not teratogenic |
| | Mouse/18-month | 211 | Reproductive | No evidence |
| | RfD | 0.2 | Oncogenicity | C with RfD (thyroid tumors in rats |

vegetables. These agents interfere with oxygen uptake and may bind to sulfur-containing enzymes. The dithiocarbamates are applied at rates of 500 to over 10,000 g/hectare. Ferbam, thiram, and ziram are the commercially important chemicals in this group. The ethylenebisdithiocarbamate (EBDC) fungicides (mancozeb, maneb, and zineb) have a broad spectrum of activity, although their fungicidal mode of action is primarily protective. Their mechanism of action is to form cyanide, which reacts with thiol compounds within cells [120,121,193].

The EPA has considered including the dithiocarbamates in a common mechanism group [325,505] based on their potential to (1) generate carbon disulfide, which is a potential cause of distal peripheral neuropathy [505]; (2) form the common metabolite ethylenethiourea (ETU), which has potential carcinogenic effects; (3) chelate physiologically important polyvalent cation such as copper, zinc, lead, or cadmium, with potentia neurotoxicity resulting from nervous system sequestration of heavy metals; and (4) inhibit acetyl cholinest erase [325]. In their final decision, the EPA conclude that the available evidence suggests that neuropathology induced by the treatment of rats with dithiocarbamates could not be linked to the formation of carbo disulfide. A common mechanism grouping for mancozeb, maneb, and metiram was supported based on thei ability to form the common metabolite ETU. It was als concluded that two dithiocarbamate pesticides, ziran and metam sodium, share a common mechanism fo acetylcholinesterase inhibition [505].

The structure, use, and hazard profiles for dithiocar bamates are given in Table 16.16. Ferbam, thiram, an

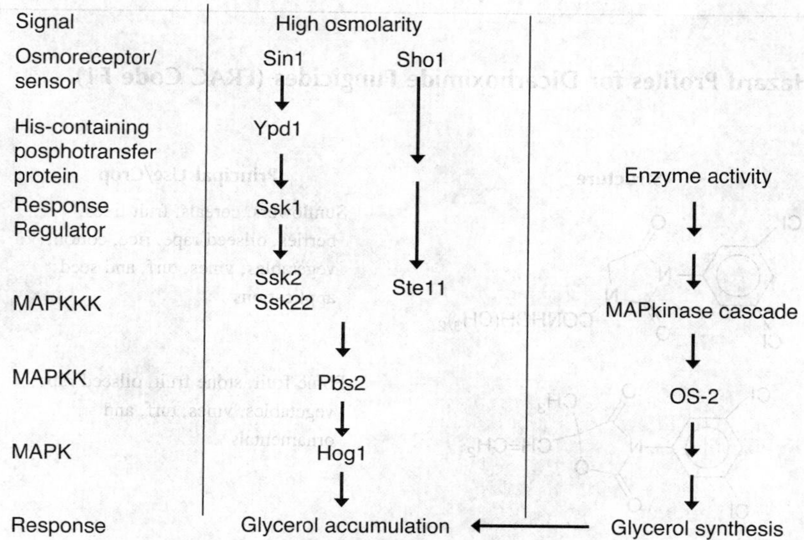


FIGURE 16.3 Osmotic signal transduction.

**TABLE 16.11**
**Structure, Use, and Hazard Profiles for Phenylpyrrole Fungicides (FRAC Code E2)**

| Fungicide | Structure | Principal Use/Crop | Application Rate |
|---|---|---|---|
| Fenpiclonil [80] | | Used for seed application on cereals and peas; potato seed dressing | 20 g active ingredient per 100 kg<br>20–50 g active ingredient per ton |
| Fludioxonil (Maxim®, Scholar®) [85] | | Used for seed application on rice and on grapes, stone fruit, vegetables, field crops, turf, and ornamentals | 2.5–10 g active ingredient per 100 kg |

| Fungicide | Irritation: Eye | Irritation: Skin | $LD_{50}$ (mg/kg): Oral | $LD_{50}$ (mg/kg): Dermal | $LC_{50}$ (mg/L): Inhalation | Sensitization Potential | Signal Word |
|---|---|---|---|---|---|---|---|
| Fenpiclonil | Non-irritant | Non-irritant | >5000 | >2000 | >1.5 | Negative | Caution |
| Fludioxonil | Slight irritant | Non-irritant | >5000 | >2000 | >2.6 | Negative | Caution |

| Fungicide | Species/Study | NOEL (mg/kg/day) | Toxicity Study | Hazard Indicator |
|---|---|---|---|---|
| Fenpiclonil [80] | Rat/2-year | 1.25 | Mutagenicity | No evidence |
| | Dog/52-week | 100 | Developmental | Not teratogenic |
| | Mouse/18-month | 20 | Reproductive | No evidence |
| | ADI | 0.013 | Oncogenicity | No evidence |
| Fludioxonil [530] | Rat/2-year | 50 | Mutagenicity | Clastogenic (*in vitro*) |
| | Dog/52-week | 3.3 | Developmental | Not teratogenic |
| | Mouse/18-month | 143 | Reproductive | No evidence |
| | RfD | 0.03 | Oncogenicity | D with RfD |

**TABLE 16.12**
**Structure, Use, and Hazard Profiles for Dicarboximide Fungicides (FRAC Code F1)**

| Fungicide | Structure | Principal Use/Crop | Application Rate (g active ingredient per hectare) |
|---|---|---|---|
| Iprodione (Rovral®) [113] | Cl, O, N, N, $CONHCH(CH_3)_2$, Cl, O | Sunflowers, cereals, fruit trees, berries, oilseed rape, rice, cotton, vegetables, vines, turf, and seed applications | 500–12,000 |
| Vinclozolin (Roilan®, Flotilla®) [192] | Cl, O, $CH_3$, N, $CH{=}CH_2$, O, Cl, O | Pome fruit, stone fruit, oilseed rape, vegetables, vines, turf, and ornamentals | 300–430 |

| Fungicide | Irritation | | $LD_{50}$ (mg/kg) | | $LC_{50}$ (mg/L) | Sensitization | Signal |
|---|---|---|---|---|---|---|---|
| | Eye | Skin | Oral | Dermal | Inhalation | Potential | Word |
| Iprodione | Mild irritant | Non-irritant | 4468 | >2000 | >5.2 | Negative | Cautior |
| Vinclozolin | Minimal irritant | Minimal irritant | >15,000 | >5000 | 29.1 | Positive | Caution |

| Fungicide | Species/Study | NOEL (mg/kg/day) | Toxicity Study | Hazard Indicator |
|---|---|---|---|---|
| Iprodione | Rat/2-year | 6.0 | Mutagenicity | No evidence |
| [291,375,467] | Dog/52-week | 4.2 | Developmental | Not teratogenic |
| | Mouse/18-month | 1870 | Reproductive | No evidence |
| | RfD (UF = 300) | 0.04 | Oncogenicity | B2 (liver, testes) |
| Vinclozolin | Rat/2-year | 1.2 | Mutagenicity | No evidence |
| [293,298,453, | Dog/52-week | 2.4 | Developmental | Not teratogenic |
| 501,502] | Mouse/18-month | 21 | Reproductive | Antiandrogenic metabolite |
| | RfD | 0.012 | Oncogenicity | B2 with RfD (multiple benign tumors in rats) |

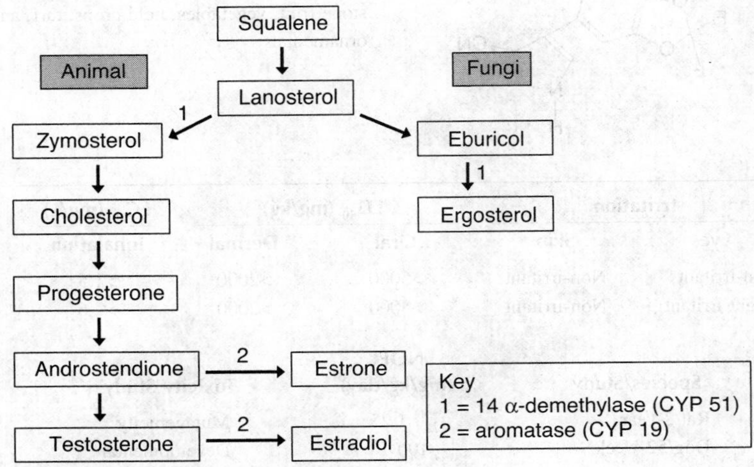


**FIGURE 16.4** Steroid biosynthesis in fungi, and animals.

ziram have significant acute toxicity, especially by inhalation. Both ferbam and ziram affect spermatazoa in mice, and thiram is teratogenic at high doses. No evidence suggests that any of these fungicides are carcino genic, although positive mutagenic studies wer reported for thiram. The structure, use, and hazard pro files for mancozeb, maneb, and zineb and their common metabolite ethylenethiourea are provided in Table 16.17. Both mancozeb and maneb are classified as Category B2 (probable human) carcinogens [241], based on the formation of mouse liver tumors and thyroid follicular cell tumors in rats. Although zineb was not found to be oncogenic in the rat or mouse, it was observed to produce non-neoplastic hyperplasia of the follicular cells of the thyroid in rats [437]. All three fungicides are transformed in animals to the metabolite ethylenethiourea, which inhibits thyroid peroxidase and causes progressive lesions in the thyroid follicular cells, often leading to tumor formation [283,284,286]. The EPA has regulated the risk associated with exposure to ETU using a cancer slope factor ($Q_1^*$) of 0.06 $(mg/kg/day)^{-1}$ [437].

## Phthalimides (FRAC Code M4)

The phthalimide fungicides represent a relatively old group of synthetic chemicals of which only captan remains in use. Captan has a broad spectrum of activity that is attributed to the formation of thiophosgene [5]. Thiophosgene has the potential to react with thiol groups, most likely at point of contact with the mucosa of the stomach, as shown in Figure 16.5. Folpet, another member of this class of fungicides, is also capable of producing thiophosgene. The structure, use, and hazard profiles for captan are given in Table 16.18. Captan has been shown to bind to DNA *in vitro* but not *in vivo*. Captan has been classified by the EPA as a Category B2 (probable human) carcinogen based on gastrointestinal-tract tumors in the mouse [241].

## Chloronitriles (FRAC Code M5)

The chloronitrile fungicides, represented here by chlorothalonil, have a broad spectrum of fungicidal activity and are considered protective. Chlorothalonil controls fungal infection by binding to sulfur-containing enzymes [44]. The structure, use, and hazard profiles of chlorothalonil are given in Table 16.19. Chlorothalonil is a severe eye irritant, a moderate skin irritant, and a potential skin sensitizer. The EPA has classified chlorothalonil as a likely human carcinogen based on kidney and forestomach tumors in both rats and mice [493]. The occurrence of kidney tumors was preceded by a pronounced hyperplasia in the proximal tubules. It has been proposed that these preneoplastic changes are due to the formation of nephrotoxic thiol metabolites of chlorothalonil [632]. It has also been suggested that β-lyase catalyzes the conversion of the cysteine conjugate to the ultimate toxiphore (see Figure 16.6). Because the activity of GST (glutathione *S*-transferase) and β-lyase in the human kidney is about 10% that of the rat kidney, it is likely that humans are less susceptible to chlorthalonil than are rats.

# INSECTICIDES

Table 16.20 provides the list of insecticides reviewed in this chapter, grouped into categories according to mode of action [260] as defined by the Insecticide Resistance Action Committee (IRAC). The majority of the insecticidal modes of action described below are relevant to humans, although a few exceptions are noted in Table 16.20 and are discussed in the appropriate section below.

## Carbamates: AChE Inhibitors (IRAC Code 1A)

Acetylcholine is an excitatory neurotransmitter substance that is released from synaptic vesicles that are found within boutons of the presynaptic neuron (Figure 16.7). Neurotransmitter release is triggered by voltage-dependent calcium ion influx across the presynaptic membrane in response to an action-potential-driven depolarization (sodium current) of the membrane. Once the neurotransmitter is released, it diffuses into the synaptic cleft where it binds to a postsynaptic receptor, triggering an excitatory postsynaptic potential (EPSP).

The carbamate insecticides inhibit the enzyme acetylcholinesterase (AChE), which is found in the postsynaptic cleft and catalyzes the cleavage of acetylcholine into its constituent components acetyl and choline, thereby inactivating the neurotransmitter. Cholinesterase inhibition, which is defined as the percent reduction in AChE activity, results in a prolonged stimulation of the cholinergic receptors leading to a hyperexcitatory state downstream from the site of action. This translates into an intense activation of the autonomic nervous system, which, depending on the severity of AChE inhibition, results in piloerection, salivation, tremor, convulsion, respiratory arrest and death. Carbamates affect neurotransmission in an almost identical fashion to the organophosphorus insecticides except that carbamate insecticides carbamylate AChE whereas the organophophorus insecticides phosphorylate AChE, as shown in Figure 16.8.

Carbamate insecticides exist as esters of carbamic acid, typically having an aryl (ring) substituent as the leaving group. Carbamates react with the serine group on acetylcholinesterase to yield a carbamylation of the serine hydroxyl group. A hydroxylated leaving group is generated. The carbamylation of AChE is reversible, unlike the phosphorylation of the AChE by organophosphate insecticides. The carbamylated complex will typically hydrolyze in minutes. The organophosphorus insecticide (OP) reacts with acetylcholinesterase at a serine hydroxyl group within the active site of the enzyme. In this reaction, the hydroxyl group is phosphorylated, yielding a leaving group. Reactivation of the enzyme can take many hours or even days.

**TABLE 16.13**
**Structure and Use Profiles for Demethylase Inhibitors (FRAC Code G1)**

| Fungicide | Structure | Principal Use/Crop |
|---|---|---|
| Cyproconazole (Alto®) [58] | | Cereal, sugar beets, fruit trees, vines, coffee, turf, bananas, and vegetables for application of rust, powdery mildew, *Septoria*, *Venturia*, and others; application rates are 60–100 g/ha |
| Difenoconazole (Dividend®) [69] | | Seed application, grapes, fruit trees, potatoes, sugar beets, oilseed rape, banana, ornamentals, and vegetables for treating a variety of fungal diseases; application rates are 30–125 g/ha |
| Fenbuconazole (Indar®) [77] | | Cereals, fruit trees, vines, beans, sugar beets, rice, bananas, ornamentals, tree nuts, and vegetables; application rates are 30–75 g/ha |
| Hexaconazole (Amizol®) [98] | | Vine, coffee, bananas, peanuts, and vegetables for treating a variety of fungal diseases; application rates are 15–250 g/ha |
| Myclobutanil (Rally®, Nova®) [134] | | Seed application, grapes, fruit trees, rice, cotton, barley, wheat, maize, grass seed, ornamentals, and vegetables for treating a variety of fungal diseases; application rates are typically 30–60 g/ha |

The pesticidal modes of action of the carbamate and organophosphorus insecticides are preserved in mammals. The EPA has determined that a separate common mechanism grouping exists for the *N*-methyl carbamate and the organophosphorus insecticides [474] based on the differences shown in Figure 16.8. The structure and use of some representative carbamate insecticides are given in Table 16.21, and their hazard profiles appear in Table 16.22. Aldicarb is the most acutely toxic of the carbamates selected for inclusion in this chapter, with an oral $LD_{50}$ below 1 mg/kg and a dermal $LD_{50}$ of 20 mg/kg. Carbofuran, methomyl, and propoxur have been classified by the EPA as Category C (possible human carcinogens; aldicarb is classified as Category D (nonclassifiable in regard to carcinogenicity) [241].

## Organophosphorus Insecticides: AChE Inhibitors (IRAC Code 1B)

Organophosphorus insecticides (OPs) vary tremendously in chemical structure and properties. The OPs are classified into the following groups depending on the position-

**TABLE 16.13 (cont.)**
**Structure and Use Profiles for Demethylase Inhibitors (FRAC Code G1)**

| Fungicide | Structure | Principal Use/Crop |
|---|---|---|
| Propiconazole (Tilt®) [155] | | Wheat, rice, coffee, bananas, peanuts, stone fruit, maize, and turf for treating a variety of fungal diseases (rate: 24 to 110 g active ingredient per ha) |
| Tebuconazole Folicur® [176] | | Seed application, cereals, coffee, fruit trees, grapes, grass seed, oilseed rape, soybeans, sugar beets, bananas, ornamentals, turf, and vegetables for treating a variety of fungal diseases; application rates are 200–375 g/ha |
| Triadimefon (Bayleton®) [186] | | Cereals, corn, fruit trees, vines, berries, sugar cane, tobacco and vegetables for treating a variety of fungal diseases; application rates are 125–500 g/ha |
| Triadimenol (Baytan®) [187] | | Seed application, cereals, fruit trees, hops, vines, and vegetables for treating a variety of fungal diseases; application rates are 100–500 g/ha |
| Imazalil (Fungaflor®) [102] | | Seed, fruit trees, potatoes, bananas, vegetables, ornamentals, and cereals for treating a variety of fungal diseases (rate: 4–5 g active ingredient per 100 kg seed) |
| Prochloraz (Sportak®) [149] | | Citrus, tropical fruit (dip), beets, oilseed rape, mushrooms, ornamentals, and cereals (seed application) (rate: 400–600 g active ingredient per ha) |

ing of the central phosphorus atom: phosphates, phosphonates, phosphorothionates, phosphorodithioates, and phosphoroamidothioates. Examples of OPs within these different subgroups are presented in Table 16.23. The toxicological profiles for the OPs are presented in Table 16.24. Of the OPs reviewed here, azinphos-methyl is the most acutely toxic, and malathion is the least toxic. Acephate causes liver tumors in mice, and delayed neurotoxicity was seen is studies on dichlrovos. Monocrotophos has the lowest chronic reference dose.

## Cyclodiene Organochlorines: GABA Antagonists (IRAC Code 2A and 2B)

In both insects and mammals, chloride channel-blocking insecticides cause hyperexcitability, convulsions, and death [9]. Overstimulation of neuronal pathways in the central nervous system (CNS) results from blocking the action of the inhibitory neurotransmitter γ-aminobutyric acid (GABA). Normally, when GABA is released from the presynaptic nerve terminal, it binds to a

**TABLE 16.14**
**Hazard Profiles for Demethylase Inhibitors (FRAC Code G1)**

| | Irritation | | $LD_{50}$ (mg/kg) | | $LC_{50}$ (mg/L) | | |
|---|---|---|---|---|---|---|---|
| **Fungicide** | **Eye** | **Skin** | **Oral** | **Dermal** | **Inhalation** | **Sensitization Potential** | **Signal Word** |
| Cyproconazole | Non-irritant | Non-irritant | >1020 | >2000 | 5.7 | Negative | Caution |
| Difenoconazole | Moderate irritant | Slight irritant | 1453 | >2000 | 3.3 | Negative | Caution |
| Fenbuconazole | Non-irritant | Non-irritant | >2000 | >5000 | >2.1 | Negative | Caution |
| Hexaconazole | Mild irritant | Non-irritant | 2189 | >2000 | >5.9 | Positive | Caution |
| Myclobutanil | Irritant | Non-irritant | >1600 | >5000 | >5.0 | Positive | Danger |

| Fungicide | Species/Study | NOEL (mg/kg/day) | Toxicity Study | Hazard Indicator |
|---|---|---|---|---|
| Cyproconazole [386] | Rat/2-year | 2.2 | Mutagenicity | Clastogenic (CHO) |
| | Dog/52-week | 1.0 | Developmental | Teratogenic in rabbit |
| | Mouse/18-month | 1.8 | Reproductive | No evidence |
| | RfD | 0.01 | Oncogenicity | B2 (Mouse liver tumors in both sexes) |
| Difenoconazole [425] | Rat/2-year | 1.0 | Mutagenicity | No evidence |
| | Dog/52-week | 3.4 | Developmental | Not teratogenic |
| | Mouse/18-month | 4.7 | Reproductive | No evidence |
| | RfD | 0.01 | Oncogenicity | C with RfD ( Mouse liver tumors in both sexes) |
| Fenbuconazole [388] | Rat/2-year | 3.0 | Mutagenicity | No evidence |
| | Dog/52-week | 3.8 | Developmental | Not teratogenic |
| | Mouse/18-month | 1.4 | Reproductive | E (No evidence) |
| | RfD | 0.03 | Oncogenicity | C with RfD (Mouse liver tumors -both sexes/ thyroid tumors–male rats) |
| Hexaconazole [402] | Rat/2-year | 0.5 | Mutagenicity | No evidence |
| | Dog/52-week | 2.0 | Developmental | Not teratogenic |
| | Mouse/18-month | 4.7 | Reproductive | No evidence |
| | RfD | 0.005 | Oncogenicity | C with $Q_1$* (Male rat Leydig cell tumor) |
| Myclobutanil [440] | Rat/2-year | 2.5 | Mutagenicity | No evidence |
| | Dog/52-week | 3.1 | Developmental | Not teratogenic |
| | Mouse/18-month | 13.7 | Reproductive | Testicular atrophy |
| | RfD | 0.025 | Oncogenicity | E (No evidence) |

postsynaptic receptor protein containing an intrinsic chloride ion channel. When GABA binds to its receptor, the chloride channel is opened, and chloride ions flow across the postsynaptic membrane. This increase in chloride permeability hyperpolarizes (makes more negative) the membrane, resulting in an inhibitory postsynaptic potential (IPSP), which has a dampening effect on neuronal excitation, making it less likely that a postsynaptic action potential will occur. Attenuation of GABA-mediated neuronal inhibition leads to hyperexcitation of downstream neuronal pathway because GABA neuronal pathways are inhibitory [9]. The structure, use, and hazard profiles for the cyclodiene organochlorines, represented here by endosulfan, lindane (γ-HCH), and fipronil, are given in Table 16.25. They are moderately acutely toxic (lindane is an eye irritant) and have relatively low chronic references doses. They are not mutagenic, developmental, or reproductive toxins and are not carcinogenic.

## Organochlorines: Sodium Channel Modulators (IRAC Code 3)

The organochlorines are one of the oldest groups of synthetic insecticides, dating back to the early 1940s [318]. These lipophilic compounds are environmentally stable and persistent, and many (e.g., dieldrin, endrin, and DDT) have been banned in the United States; however, more biodegradable materials such as lindane and endosulfan still have limited use today in some countries. Fipronil, which is an arylheterocycle with a similar mode of action has improved selective toxicity toward insects. The organochlorine insecticides induce repetitive action potentials by slowing the kinetics of sodium channel activation and inactivation (closing), resulting in prolonged tail currents that cause a state of hyperstimulation of the CNS [329] Because DDT is highly lipophilic and because of the relatively small size of the insect and its lower body temperature, DDT more readily reaches its target (the sodium

**TABLE 16.15**
**Hazard Profiles for Demethylase Inhibitors (FRAC Code G1) (Continued)**

| Fungicide | Irritation | | $LD_{50}$ (mg/kg) | | $LC_{50}$ (mg/L) | Sensitization Potential | Signal Word |
|---|---|---|---|---|---|---|---|
| | Eye | Skin | Oral | Dermal | Inhalation | | |
| Propiconazole | Mild irritant | Slight irritant | 1517 | >6000 | >5.8 | Negative | Caution |
| Tebuconazole | Mild irritant | Non-irritant | >3933 | >5000 | >0.37 | Negative | Caution |
| Triadimefon | Non-irritant | Non-irritant | >363 | >2000 | >3.6 | Positive | Warning |
| Triadimenol | Non-irritant | Non-irritant | >1100 | >5000 | >0.9 | NA | Caution |
| Imazalil | Non-irritant | Mild irritant | >227 | 4200 | 16 | Negative | Warning |
| Prochloraz | Irritant | Mild irritant | 1600 | 3000 | 0.42 | Negative | Caution |

| Fungicide | Species/Study | NOEL (mg/kg/day) | Toxicity Study | Hazard Indicator |
|---|---|---|---|---|
| Propiconazole [155,406,569] | Rat/2-year | 3.6 | Mutagenicity | No evidence |
| | Dog/26-week oral | 1.3 | Developmental | Not teratogenic |
| | Mouse/18-month | 15 | Reproductive | No evidence |
| | RfD | 0.013 | Oncogenicity | C with RfD (mouse liver tumors in males) |
| Tebuconazole [450] | Rat/2-year | 7.4 | Mutagenicity | No evidence |
| | Dog/52-week | 3.0 | Developmental | Teratogenic in rat |
| | Mouse/18-month | 2.9 | Reproductive | No evidence |
| | RfD | 0.03 | Oncogenicity | C with RfD (mouse liver tumors in both sexes) |
| Triadimefon [268,407,569] | Rat/2-year | 16.4 | Mutagenicity | No evidence |
| | Dog/2-year oral | 11.4 | Developmental | Not teratogenic |
| | Mouse/18-month | 40 | Reproductive | No evidence |
| | RfD (52-wk dog study; UF = 300) | 0.04 | Oncogenicity | C with RfD (mouse liver tumors in both sexes) |
| Triadimenol [269,569] | Rat/2-year | 7.0 | Mutagenicity | No evidence |
| | Dog/52-week | 3.75 | Developmental | Teratogenic in rat |
| | Mouse/18-month | 30 | Reproductive | No evidence |
| | ADI | 0.038 | Oncogenicity | C with RfD (liver tumors in female mice) |
| Imazalil [276] | Rat/2-year | 5.0 | Mutagenicity | No evidence |
| | Dog/52-week | 2.5 | Developmental | Not teratogenic |
| | Mouse/18-month | 40 | Reproductive | No evidence |
| | ADI | 0.025 | Oncogenicity | C-Q*(Mouse liver) |
| Prochloraz [267] | Rat/2-year | 1.9 | Mutagenicity | No evidence |
| | Dog/52-week | 0.9 | Developmental | Not teratogenic |
| | Mouse/18-month | 11.7 | Reproductive | Decreased litter size |
| | ADI | 0.009 | Oncogenicity | C-Q* (mouse liver tumors in both sexes) |

channel in the nervous system), so it has a greater effect (approximately 500- to 4500-fold) in insects than in mammals [329]. The structure, use, and hazard profiles for DDT and methoxychlor are provided in Table 16.26. DDT is moderately acutely toxic, and DDT and methoxychlor both have weak estrogenic activity [205].

## Pyrethroids: Sodium Channel Modulators (IRAC Code 3)

The pyrethroid insecticides, typically esters of chrysanthemic acid, were isolated from the flowers of chrysanthemum. Synthetic pyrethroid chemistry and insecticidal effects for type I pyrethroids are rather broadly defined, and they include pyrethroids containing des-cyano-3-phenoxybenzyl or other alcohols [674]. Many of the older nonphenoxybenzyl, type I compounds (e.g., pyrethrins, allethrin, and tetramethrin) are unstable in the environment, and this characteristic prevented their use in row crops. Introduction of the phenoxybenzyl (e.g., permethrin) and halogenated (e.g., tefluthrin) alcohols improved chemical stability and allowed the use of pyrethroids on row crops. The characteristic clinical signs seen in mammals following exposure to type I pyrethroids [696] include the occurrence of fine tremors, hyperexcitability, and myoclonus (T-syndrome). The type II pyrethroids are more narrowly defined in terms of their chemical structure. They contain an α-cyano-3-phenoxybenzyl alcohol, which increases insecticidal activity about tenfold [674]. Clinical signs seen in mammals following exposure to type II pyrethroids [696] include sinuous writhing (choreoathetosis), salivation, hyperactivity, and clonic/tonic convulsions (CS syndrome).

**TABLE 16.16**
**Structure, Use, and Hazard Profiles for the Dithiocarbamates Fungicides (FRAC Code M3)**

| Fungicide | Structure | Principal Use/Crop | Application Rate (g active ingredient per hectare) |
|---|---|---|---|
| Ferbam (Metam®) [82] | $[(CH_3)_2-N-C(=S)-S^-]_3\,Fe^{3+}$ | Pome fruit, peaches, and tobacco | 300–500 |
| Thiram (Vitavax®) [185] | $(CH_3)_2N-C(=S)-S-S-C(=S)-N(CH_3)_2$ | Seed dressing | 13–18 g/100 lb seed |
| Ziram (Attivar®) [194] | $CH_3-N(CH_3)-C(=S)-S-Zn-S-C(=S)-N(CH_3)-CH_3$ | Pome fruit, stone fruit, nuts, vines, vegetables, and ornamentals | 1550–2760 |

| Fungicide | Irritation | | $LD_{50}$ (mg/kg) | | $LC_{50}$ (mg/L) | Sensitization Potential | Signal Word |
|---|---|---|---|---|---|---|---|
| | Eye | Skin | Oral | Dermal | Inhalation | | |
| Ferbam | Mild irritant | Slight irritant | >4000 | >4000 | 0.4 | Weak positive | Warning |
| Thiram | Slight irritant | Irritant | >1800 | >2000 | >0.1 | Positive | Warning |
| Ziram | Severe irritant | Non-irritant | 270 | >2000 | 0.06 | Positive | Danger |

| Fungicide | Species/Study | NOEL (mg/kg/day) | Toxicity Study | Hazard Indicator |
|---|---|---|---|---|
| Ferbam [294,541] | Rat/2-year | 12.0 | Mutagenicity | No evidence |
| | Dog/52-week | 5.0 | Developmental | Not teratogenic |
| | Mouse/18-month | — | Reproductive | Effects on sperm in mice |
| | ADI | 0.003 (interim) | Oncogenicity | No evidence |
| Thiram [278,535] | Rat/2-year | 1.2 | Mutagenicity | Positive Ames, and SCE |
| | Dog/ 2-year oral | 0.84 | Developmental | Teratogenic in mice, and hamster at high doses |
| | Mouse/18-month | 3.0 | Reproductive | No evidence |
| | ADI | 0.008 | Oncogenicity | No evidence |
| Ziram [295,524] | Rat/2-year | <2.5 | Mutagenicity | Clastogenic |
| | Dog/52-week | 1.6 | Developmental | Not teratogenic |
| | Mouse/18-month | 3.0 | Reproductive | Effects on sperm in mice |
| | ADI (UF = 1000] | 0.003 | Oncogenicity | No evidence |

The distinction between type I and type II pyrethroids was confirmed in a recent comparative acute neurotoxicity study aimed at providing a detailed description of clinical signs at the time of peak effect using a modern functional observational battery (FOB) of tests in rats treated with minimally effective or maximally tolerated doses of type I (bifenthrin, *S*-bioallethrin, permethrin, pyrethrin, resmethrin, tefluthrin) and type II (β-cyfluthrin, cypermethrin, deltamethrin, esfenvalerate, fenpropathrin, λ-cyhalothrin) pyrethroids. The FOB data, which were subjected to a principal components/factor analysis, confirmed that two major factors (T and CS) accounted for more that 90% of the variability in the group means. Dose-responsiveness was observed with marginally effective doses clustering with the control groups near the origin, whereas groups receiving larger doses tended to be deployed along the T or CS axis (Figure 16.9).

Despite clear differences in the profiles of mammalian clinical signs between the type I and type II pyrethroids, it is generally believed that the pyrethroids exert their effects by modifying the kinetic characteristics of the sodium channel function, largely based on results from studies conducted *ex vivo* [650] and by the observation that the development of knockdown resistance (kdr-associated gene mutations) confer decreased sensitivity of insects to both DDT and the pyrethroid insecticides [217,689].

In a comparative study where the $NA_V1.8$ mammalian sodium channel was expressed in *Xenopus* oocytes [653], the kinetics of the sodium channel response to electrical

**TABLE 16.17**
**Hazard Profiles for Dithiocarbamates Fungicides Mancozeb, Maneb, and Zineb (FRAC Code M3)**

| Fungicide | Structure | Principal Use/Crop | Application Rate (g active ingredient per hectare) |
|---|---|---|---|
| Mancozeb (Dithane®, Manzate®) [120] | $\left[ S-\overset{S}{\overset{\|}{C}}-\underset{H}{\underset{\|}{N}}-CH_2-CH_2-\underset{H}{\underset{\|}{N}}-\overset{S}{\overset{\|}{C}}-S-Mn \right]_x (Zn)_y$ | Potatoes, tomatoes, fruits, vegetables, cereals, vines, ornamental, and tobacco | 1500–2000 |
| Maneb (Kypman®) [121] | $\left[ S-\overset{S}{\overset{\|}{C}}-\underset{H}{\underset{\|}{N}}-CH_2-CH_2-\underset{H}{\underset{\|}{N}}-\overset{S}{\overset{\|}{C}}-S-Mn \right]_x$ | Potatoes, tomatoes, vegetables, apples, pears, cereals, ornamentals, vines, and tobacco | 450–3600 |
| Zineb (Kypzin®) [193] | $\left[ S-\overset{S}{\overset{\|}{C}}-\underset{H}{\underset{\|}{N}}-CH_2-CH_2-\underset{H}{\underset{\|}{N}}-\overset{S}{\overset{\|}{C}}-S-Zn \right]_x$ | Oilseed rape, berries, apples, pears, stone fruit, citrus fruit, bananas, currants, olives, celery, vegetables, and vines | 2250 |
| Ethylenethiourea (common metabolite) [325,437,505] | Cyclic structure: NH–C(=S)–$S^-$ / NH–C(=S)–$S^-$ with ethylene bridge, $Mn^{2+}$ | — | — |

| Fungicide | Irritation: Eye | Irritation: Skin | $LD_{50}$ (mg/kg): Oral | $LD_{50}$ (mg/kg): Dermal | $LC_{50}$ (mg/L): Inhalation | Sensitization Potential | Signal Word |
|---|---|---|---|---|---|---|---|
| Mancozeb | Severe irritant | Slight irritant | >5000 | >5000 | 5.14 | Positive | Danger |
| Maneb | Moderate irritant | Slight irritant | 6750 | >5000 | 7.38 | Positive | Warning |
| Zineb | Mild irritant | Slight irritant | >5200 | >6000 | NA | Negative | Caution |

| Fungicide | Species/Study[a] | NOEL (mg/kg/day) | Toxicity Study | Hazard Indicator |
|---|---|---|---|---|
| Mancozeb [283,437,546] | Rat/2-year | 4.8 | Mutagenicity | Equivocal evidence |
| | Dog/52-week | 7.0 | Developmental | Teratogenic at high doses |
| | Mouse/18-month | 17 | Reproductive | No evidence |
| | ADI | 0.031 | Oncogenicity | B2 (thyroid tumors in rats of both sexes) |
| Maneb [284,437,438,547] | Rat/2-year | 5.0 | Mutagenicity | No evidence |
| | Dog/52-week | 6.4 | Developmental | Not teratogenic |
| | Mouse/18-month | 11 | Reproductive | No evidence |
| | ADI | 0.031 | Oncogenicity | B2 (liver tumors in mice of both sexes; thyroid tumors in rats) |
| Zineb [286] | Rat/2-year | <25 | Mutagenicity | No evidence |
| | Dog/52-week | 50 | Developmental | Not teratogenic |
| | Mouse/18-month | No adequate study | Reproductive | No evidence |
| | ADI | 0.031 | Oncogenicity | No evidence |

[a] ADI based on ethylenethiourea as a metabolite common to mancozeb, maneb, metiram, and zineb; see also Mulkey, M. E., *Determination of Whether Dithiocarbamate Pesticides Share a Common Mechanism of Toxicity*, U.S. Environmental Protection Agency, Washington, D.C., 2001 (http://www.epa.gov/pesticides/cumulative/dithiocarb.pdf).

stimulation differed among 11 pyrethroids. Multidimensional scaling was used to quantify and display the magnitude of the difference between chemicals with respect to the kinetics of activation, fast inactivation, and tail current decay parameters (see Figure 16.10). The results show that, in general, the pyrethroids cluster (i.e., are

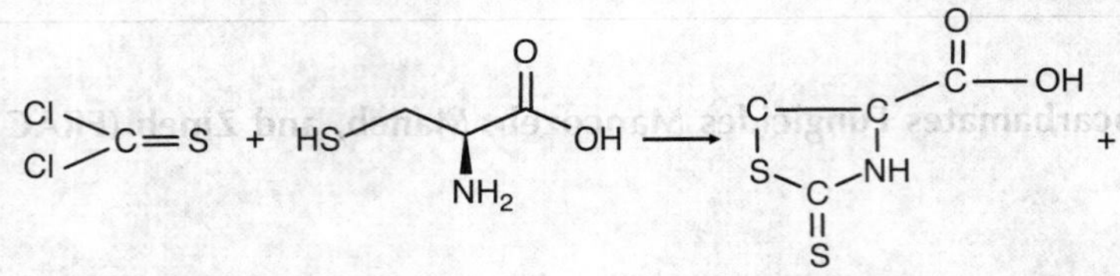

Thiophosgene + Cysteine ⟶ 2-Thioxothiazolidine 4-carboxylase (TCC) + 2HCl

**FIGURE 16.5** Proposed toxiphore for captan.

**TABLE 16.18**
**Structure, Use, and Hazard Profiles for the Multisite Contact Phthalimide Fungicide Captan (FRAC Code M4)[a]**

| Structure | Principal Use/Crop | Application Rate (g active ingredient per hectare) |
|---|---|---|
| 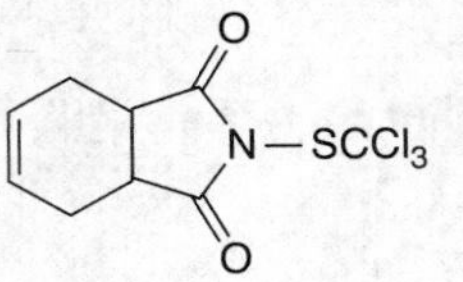 N—$SCCl_3$ | Used on stone fruit, citrus, almonds, vegetables, potatoes, tomatoes, oilseed rape, berries, and ornamentals | 2000–4800 |

| Irritation | | $LD_{50}$ (mg/kg) | | $LC_{50}$ (mg/L) | Sensitization | Signal |
|---|---|---|---|---|---|---|
| Eye | Skin | Oral | Dermal | Inhalation | Potential | Word |
| Corrosive | Mild irritant | 9000 | >4500 | 5.8 | Positive | Danger |

| Species/Study | NOEL (mg/kg/day) | Toxicity Study | Hazard Indicator |
|---|---|---|---|
| Rat/2-year | 25 | Mutagenicity | Positive *in vitro* |
| Dog/66-week oral | 60 | Developmental | Positive in monkey and hamster |
| Mouse/18-month | — | Reproductive | No evidence |
| RfD (based on rat reproduction study) | 0.13 | Oncogenicity | B2 (G.I. tract tumors, mouse; kidney tumors, rat) |

[a] See references 5, 38, 290, and 526.

similar) with respect to sodium channel kinetic parameters along the type I/type II classification. Exceptions were noted for bifenthrin (type I) and fenvalerate and fenpropathrin (type II), which appear intermediated between the predominant type I/type II grouping. Similar results have been seen in comparative studies of the effect of pyrethroid insecticides on functional characteristics of the calcium [691] and chloride [649] channels (data not shown).

The voltage-gated sodium channel is formed by transmembrane proteins [217], which in insects is comprised of an α-subunit that has four repeated homologous domains (I to IV), each having six members (S1 to S6) spanning the membrane and connected to each other by intracellular and extracellular loops of amino acids (Figure 16.11). The S5 and S6 sections of each domain line the ion channel pore, with S4 being the voltage sensor element [651]. Nine sodium channel isoforms have been identified in mammals [661], compared to only one in insects [217]. There is evidence of structural and functional homology between mammalian sodium channel isoforms and the insect sodium channel [677], although there are exceptions. For example, Usherwood et al. [695] showed that specific mutations that result in resistance to the pyrethroids do not confer resistance to DDT, contrary to expectation. Gilles et al. [237] reported that α-like toxin from scorpion venom, which inhibits sodium current inactivation in insects, bound to receptor site three of the insect sodium channel but did not bind to rat brain synaptosomes. Gordon et al. [242] reported that scorpion toxin binds to homologous but not identical receptor sites in rat brain and insect sodium channels.

The biological basis for a common mechanism of toxicity for type I and type II pyrethroids, including a discussion of the sodium, calcium, and GABA-gated

**TABLE 16.19**
**Structure, Use, and Hazard Profile for the Contact Chloronitrile Fungicide Chlorothalonil (FRAC Code M5)[a]**

| Structure | Principal Use/Crop | Application Rate (g active ingredient per hectare) |
|---|---|---|
| CN, Cl, Cl, Cl, Cl, Cl (chlorothalonil ring structure) | Used on pome fruit, stone fruit, citrus, cane fruit, vegetables, corn, ornamentals, mushrooms, tobacco, soya, and turf | 1000–2500 |

| Irritation | | $LD_{50}$ (mg/kg) | | $LC_{50}$ (mg/L) | Sensitization Potential | Signal Word |
|---|---|---|---|---|---|---|
| Eye | Skin | Oral | Dermal | Inhalation | | |
| Severe irritant | Mild irritant | >10000 | >10000 | 0.093 | Negative | Danger |

| Species/Study | NOEL (mg/kg/day) | Toxicity Study | Hazard Indicator |
|---|---|---|---|
| Rat/2-year | 2.0 | Mutagenicity | No evidence |
| Dog/52-week | 150 | Developmental | Not teratogenic |
| Mouse/18-month | 5.35 | Reproductive | No evidence |
| ADI2 | 0.03 (JMPR) | Oncogenicity | Likely: forestomach tumors, mice; kidney tumors, rats $Q_1^* = 7.6 \times 10^{-2}$ (mg/kg/day)$^{-1}$ |
| RfD (noncancer) | 0.02 | | |
| RfD (cancer) | 0.015 | | |

[a] See references 44, 277, 493, and 632.

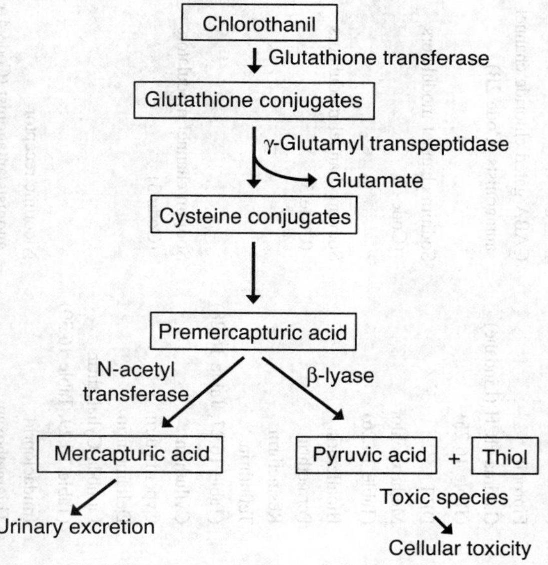


**FIGURE 16.6** Proposed mechanism of action for chlorothalonil.

chloride channels, has yet to be considered by regulatory authorities [342,351,357]. The structures and use of some of economically important non-cyano (type I) pyrethroids are given in Table 16.27, and their hazard profiles appear in Table 16.28. Comparable information for type II (cyano-substituted) pyrethroids is given in Table 16.29 and Table 16.30.

The synthetic pyrethroids are generally effective as insecticides in the low grams per hectare range. In general, the mammalian acute toxicity of the type II pyrethroids is greater than type I pyrethroids. Because the pyrethroids are highly lipophilic, the acute oral $LD_{50}$ can differ by up to 40-fold depending on the lipophilic characteristics and the volume of the vehicle used to administer the chemical. For example, deltamethrin's $LD_{50}$ is reported to range from 128 mg/kg to greater than 5000 mg/kg [226]. Because pyrethroids are 500 to 4500 times less toxic to mammals than to insects, they have been safely used with relatively few reports of human poisoning [11]. Aside from acute neurotoxic, the hazard profile for the pyrethroids is unremarkable, although a treatment-related increase in the incidence of lung and liver tumors has been reported for permethrin [562].

## NICOTINE AND NEONICOTINOIDS: ACETYLCHOLINE RECEPTOR AGONISTS (IRAC CODE 4A)

Nicotine has been used as a contact insecticide since the middle of the eighteenth century [136]. Nicotine mimics the

**TABLE 16.20**
**Insecticides Listed According to IRAC Classification System**

| Chemical Group | Common Name | IRAC MOA (Code) | Effect in Target Species | Biochemical Target Conserved in Mammals? | Other Biochemical Target(s) in Mammals? |
|---|---|---|---|---|---|
| Carbamates | Aldicarb<br>Carbaryl<br>Carbofuran<br>Methomyl<br>Propoxur<br>(Table 16.21, Table 16.22) | Inhibits acetylcholine esterase (AChE) (Code 1A) | Depression of AChE activity in the peripheral and central nervous systems | Yes | Common mode of action grouping based on AChE inhibition |
| Organophosphates | Acephate<br>Azinphos-methyl<br>Chlorpyrifos<br>Diazinon<br>Dichlorvos<br>Malathion<br>Monocrotophos<br>(Table 16.23, Table 16.24) | Inhibits acetylcholine esterase (AChE) (Code 1B) | See Code A above | Yes | Common mode of action grouping based on AChE inhibition |
| Cyclodiene | Endosulfan<br>(Table 16.25) | GABA gated chloride channel antagonists (Code 2A) | Blocks inhibitory circuits leading to hyperexcitation | Likely | Possible |
| Phenylpyrazoles<br>Organochlorine | Fipronil<br>Gamma-HCH (Lindane)<br>(Table 16.25) | GABA gated chloride channel antagonists (Code 2B) | Blocks inhibitory circuits leading to hyperexcitation | Likely | Possible |
| Organochlorines | DDT<br>Methoxychlor<br>(Table 16.26) | Sodium channel modulators (Code 3) | Delays sodium ion channel inactivation leading to hyperexcitation | Likely | Possible |
| Non-cyano pyrethroids | Bioallethrin<br>Permethrin<br>Resmethrin<br>Tefluthrin<br>(Table 16.27, Table 16.28) | Sodium channel modulators (Code 3) | Delays sodium ion channel inactivation leading to hyperexcitation | Likely | Common mode of action grouping yet to be determined |
| Cyano-pyrethroids | Cyfluthrin<br>Cypermethrin<br>Deltamethrin<br>Lambda-Cyhalothrin<br>(Table 16.29, Table 16.30) | Sodium channel modulators (Code 3) | Delays sodium ion channel inactivation leading to hyperexcitation | Likely | Common mode of action grouping yet to be determined |
| Neonicotinoids | Imidacloprid<br>Thiamethoxam<br>(Table 16.31) | Nicotinic receptor agonists/antagonist (Code 4A) | Activates nicotinic receptors leading to hyperexcitation | Likely | Common mode of action grouping to be determined |

| | | | | | |
|---|---|---|---|---|---|
| Nicotine | Nicotine (Table 16.31) | Nicotinic receptor agonists/antagonist (Code 4A) | Activates nicotinic receptors leading to hyperexcitation | Yes | Common mode of action grouping yet to be determined |
| Spinosyns | Spinosad (Table 16.32) | Allosteric nicotinic receptor agonists (Code 5) | Likely activates nicotinic receptors leading to hyperexcitation | Likely | Common mode of action grouping yet to be determined |
| Avermectins<br>Milbemycins | Abamectin<br>Emamectin benzoate<br>Milbemycin<br>(Table 16.32) | Chloride channel activators (Code 6) | Simulates a GABA like activation of the chloride channel leading to hyperexcitation | Likely | Common mode of action grouping yet to be determined |
| Sesquiterpenoids<br>Carbamate<br>Alkoxypyrimidine | Methoprene (Code 7A)<br>Fenoxycarb (Code 7B)<br>Pyriproxfen (Code 7C)<br>(Table 16.33) | Juvenile hormone mimics (Code 7) | Mimics JH3; interferes with molting | Possible | None known |
| Azomethine pyridines | Pymetrozine (Table 16.33) | Selective feeding blocker (Code 9B) | Affects feeding behavior | Unlikely | None known |
| Phenyltetrazine<br>Carboxamide | Clofentezine<br>Hexythiazox<br>(Table 16.34) | Unknown or non-specific mode of action (Code 10A) | Inhibits mite growth | Unknown | None known |
| *Bacillus thuringiensis* toxin | *Bacillus thuringiensis* sp. *aizawai*<br>*B. thuringiensis* sp. *kurstaki*<br>(Table 16.35) | Microbial disruption of the insect midgut membrane (Code 11) | Produces toxins that bind to protein receptors in the midgut of susceptible insects and subsequently forms pores in the insect midgut epithelium [236] | Unlikely | None known |
| Benzylureas | Diflubenzuron<br>Teflubenzuron<br>(Table 16.36) | Type O inhibitors of chitin biosynthesis (Code 15) | Blocks chitin biosynthesis; interferes with molt | Unlikely | None known |
| Aminotriazine | Cyromazine (Table 16.34) | Molting disruptor; also has larvacide activity (Code 17) | Is larvacidal in feed through systems and disrupts moulting | Unlikely | None known |
| Diacylhydrazine | Tebufenozide (18A) (Table 16.37) | Ecdysone agonist/ molting disruptor (Code 18) | Interferes with molt | Possible | None known |
| Formamidine | Amitraz (Table 16.38) | Octopaminergic agonist (Code 19) | Affects the CNS; yohimbine is an antidote in mammals | Yes; homologous receptor is the $\alpha_2$-adrenergic receptor | Inhibits synthesis of monoamine oxidase and prostaglandin $E_2$ [264] |
| Pyrrole | Chlorfenapyr (pro-pesticide requiring oxidative *N*-dealkylation to the NH derivative) | Uncouples oxidative phosphorylation from electron transport (Code 13) | Disrupts proton gradient by transporting protons across the mitochondrial membrane | Yes | None known |
| Amidinohydrazone | Hydramethylnon | blocks electron trans-port in mitrocondrial complex III (Code 20) | Coupling site II (cytochrome $bc_1$) | Yes | |

**TABLE 16.20 (cont.)**
**Insecticides Listed According to IRAC Classification System**

| Chemical Group | Common Name | IRAC MOA (Code) | Effect in Target Species | Biochemical Target Conserved in Mammals? | Other Biochemical Target(s) in Mammals? |
|---|---|---|---|---|---|
| Pyridazinone | Pyridaben | Blocks electron transport in mitrocondrial complex I (Code 21) | — | Yes | |
| — | Rotenone (Table 16.39, Table 16.40) | blocks electron transport in mitrocondrial complex I (Code 21) | — | Yes | |
| Benzodioxole | Piperonyl butoxide (Table 16.38) | Inhibits P450 dependent monoxygenases (Code 27A) | Inhibits methyl farnesoate expoxidase, which catalyzes synthesis of juvenile hormone III<br>Inhibits mixed-function oxidases, thus reducing the biodegradation of permethrin (synergism) | Yes | None known |

*Source:* IRAC, *Mode of Action Classification, Version 5*. Insecticide Resistance Action Committee, 2005 (http://www.irac-online.org/).

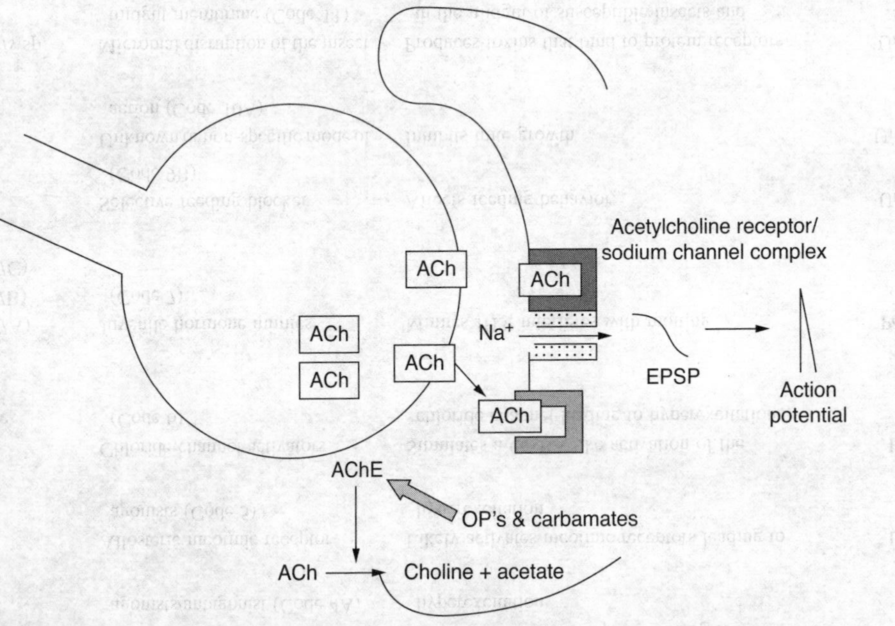


**FIGURE 16.7** Pre/postsynaptic events associated with acetylcholine-mediated neurotransmission.

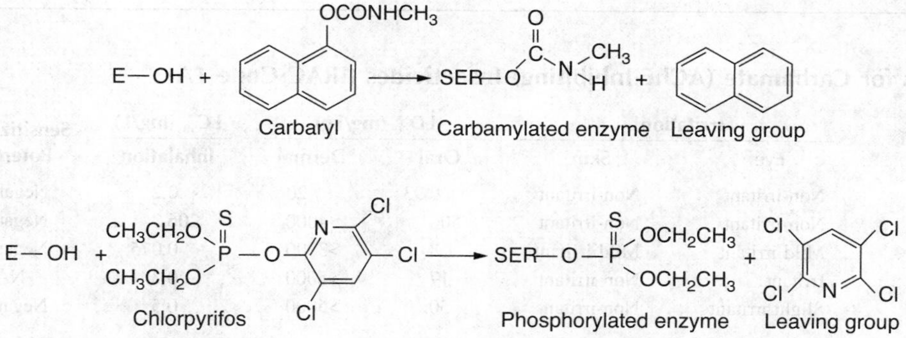

$$E{-}OH + \text{Carbaryl} \longrightarrow SER{-}O{-}C(=O){-}NH{-}CH_3 + \text{naphthalene}$$

Carbaryl (OCONHCH$_3$) — Carbamylated enzyme — Leaving group

$$E{-}OH + (CH_3CH_2O)_2P(=S){-}O{-}\text{(trichloropyridinyl)} \longrightarrow SER{-}O{-}P(=S)(OCH_2CH_3)_2 + \text{trichloropyridinol}$$

Chlorpyrifos — Phosphorylated enzyme — Leaving group

**FIGURE 16.8** Interaction of organophosphate, and carbamate insecticides with AChE.

## TABLE 16.21
## (Structure, and Use Profiles for Carbamate, AChE Inhibiting Insecticides (IRAC Code 1A)

| Insecticide | Structure | Principal Use/Crop | Application Rate (g active ingredient per hectare) |
|---|---|---|---|
| Aldicarb (Temik®) [17] | 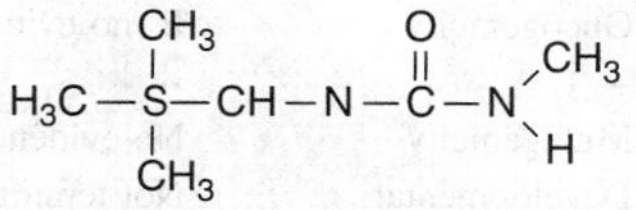 $H_3C{-}S(CH_3)_2{-}CH{-}N{-}C(=O){-}N(CH_3)H$ | Controls chewing, and sucking insects in vegetables and various crops | 350–5600 |
| Carbaryl (Sevin®) [39] | 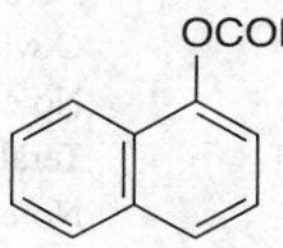 1-naphthyl OCONHCH$_3$ | Controls chewing, and sucking insects in vegetables and various crops | 250–2000 |
| Carbofuran (Furadan®) [40] | 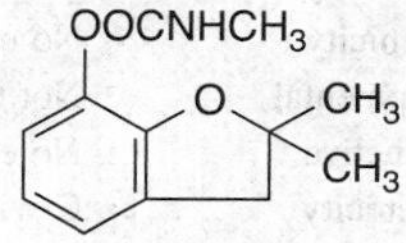 OOCNHCH$_3$ (2,2-dimethyl-dihydrobenzofuranyl, CH$_3$, CH$_3$) | Controls soil dwelling and foliar feeding insects in food crops | 260–2050 |
| Methomyl (LanoxC®) [124] | 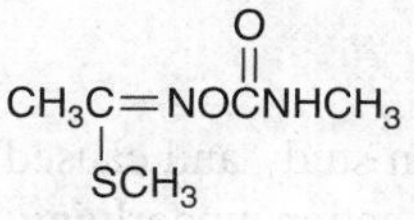 $CH_3C(SCH_3){=}NOC(=O)NHCH_3$ | Controls chewing, and sucking insects in vegetables, food crops, and turf | 120–2000 |
| Propoxur (Aprocarb®) [156] | 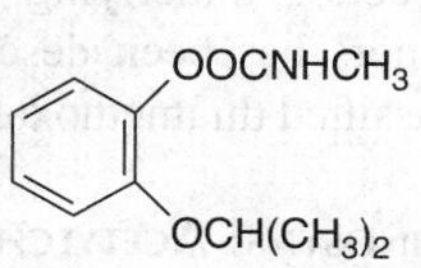 OOCNHCH$_3$, OCH(CH$_3$)$_2$ (phenyl) | Controls cockroaches, flies, fleas, ants, and mosquitoes | 1200 |

action of acetylcholine, which is a major excitatory neurotransmitter in the central nervous system. Nicotine, which acts as a ligand to the postsynaptic nicotinic acetylcholine receptor (nAchR), activates an intrinsic cation channel resulting in the depolarization of the postsynaptic cell due to an influx of sodium. The resulting excitatory postsynaptic potential triggers an action potential if the degree of membrane depolarization is sufficient. Persistent activation of the nicotinic acetylcholine receptors results in an overstimulation of the cholinergic neurotransmission system, resulting in hyperexcitation, convulsions, paralysis, and death.

The neonicotinoids, represented here by imidacloprid and thiamethoxam, are nicotine-like agonists that are used as insecticides. These chemicals, which are absorbed by plants either following foliar application or when applied as seed treatment, are effective in controlling piercing and sucking insects such as aphids, leafhoppers, and whiteflies [366]. Whereas the neonicotinoids act as ligands on

**TABLE 16.22**
**Hazard Profiles for Carbamate (AChE-Inhibiting) Insecticides (IRAC Code 1A)**

| Insecticide | Irritation | | $LD_{50}$ (mg/kg) | | $LC_{50}$ (mg/L) | Sensitization Potential | Signal Word |
|---|---|---|---|---|---|---|---|
| | Eye | Skin | Oral | Dermal | Inhalation | | |
| Aldicarb | Non-irritant | Non-irritant | 0.93 | 20 | 0.2 | Negative | Danger |
| Carbaryl | Non-irritant | Non-irritant | 500 | >4000 | 206 | Negative | Caution |
| Carbofuran | Mild irritant | Mild irritant | 8 | >3000 | 0.075 | Negative | Danger |
| Methomyl | Irritant | Non-irritant | 17 | >5000 | 0.3 | NA | Danger |
| Propoxur | Slight irritant | Non-irritant | 50 | >5000 | 0.5 | Negative | Warning |

| Insecticide | Species/Study | NOEL (mg/kg/day) | Toxicity Study | Hazard Indicator |
|---|---|---|---|---|
| Aldicarb [288,373] | Rat/2-year | 0.3 | Mutagenicity | No evidence |
| | Dog/104-week | 0.1 | Developmental | Not teratogenic |
| | Mouse/18-month | 0.3 | Reproductive | No evidence |
| | ADI | 0.003 | Oncogenicity | D |
| | RfD | 0.001 | | |
| Carbaryl [228,263,527] | Rat/2-year | 200 | Mutagenicity | No evidence |
| | Dog/52-week | 1.43 | Developmental | Not teratogenic |
| | Mouse/18-month | — | Reproductive | No evidence |
| | Human | 0.01 | Oncogenicity | E (no evidence) |
| | RfD | 0.01 | — | |
| Carbofuran [265,556] | Rat/2-year | 20 | Mutagenicity | No evidence |
| | Dog/2-year oral | 10 | Developmental | Not teratogenic |
| | Mouse/18-month | 20 | Reproductive | No evidence |
| | RfD | 0.002 | Oncogenicity | C with RfD (mouse liver tumors in both sexes) |
| Methomyl [273,470] | Rat/2-year | 200 | Mutagenicity | No evidence |
| | Dog/52-week | 200 | Developmental | Teratogenic in mice |
| | Mouse/18-month | 500 | Reproductive | No evidence |
| | ADI | 0.02 | Oncogenicity | C with RfD (liver tumors, female mice) |
| Propoxur [274,446] | Rat/2-year | 5.0 | Mutagenicity | No evidence |
| | Dog/52-week | 1.25 | Developmental | Not teratogenic |
| | Mouse/18-month | 40 | Reproductive | No evidence |
| | ADI | 0.01 | Oncogenicity | C with RfD |

homologous receptors in insects and vertebrates, the affinity of the neonicotinoid for the insect nicotinic acetylcholine receptor is reported to be 5- to 3500-fold greater than that observed in vertebrates [365,366]. The molecular basis for differences in affinity has been proposed [315,366], and molecular design aimed at achieving greater selectivity for insects has been discussed [297]. Differences in absorption, distribution, metabolism, and elimination are also expected to play a role in insect selectivity. A common mechanism grouping for neonicotinoid insecticides has not yet been proposed but appears to be supported by what is known about their mode of action.

The structure, use, and hazard profiles for nicotine, imidacloprid, and thiamethoxam are presented in Table 16.31. Imidacloprid and thiamethoxam are moderately acutely toxic by the oral route and are much less toxic to mammals than is nicotine. Imidacloprid is not a developmental toxin or a carcinogen. High doses of thiamethoxam caused testicular effects in the multigeneration reproduction study and caused liver tumors in the mouse. A mode of action underlying the occurrence of the mouse liver tumors has been described [246,338], and the EPA has classified thiamethoxam as not likely a human carcinogen.

## Spinosyns: Acetylcholine Receptor Agonists (IRAC Code 5)

Spinosad, which is a fermentation-produced macrolid, was initially derived from the soil actinomycete *Saccharopolyspora spinasa* and is comprised of syinosyn A and D (Figure 16.12). It is highly toxic to lepidopteran, dipteran, and some coleopteran insects [347]. Symptoms seen in insects include CNS hyperexcitation, involuntary muscle contraction, and tremor which ultimately result in neuromuscular fatigue and paralysis [347,348]. Spinosyns are believed to be nAchR agonists although the experimental evidence supporting this proposed mode of action is inconclusive [343]. In addition to the proposed effects mediated by binding to nAchR, the

**TABLE 16.23**
**Structures, and Use Profiles for the Organophosphate (AChE-Inhibiting) Insecticides (IRAC Code 1B)**

| Insecticide | Structure | Principal Use/Crop | Application Rate (g active ingredient per hectare) |
|---|---|---|---|
| Acephate (Amithene®) [14] | Phosphoramidothiate | Control sucking, and chewing insects | 500–1000 |
| Azinphos-methyl (Guthion®) [23] | Phosphorodithionate | Control of sucking, and chewing insects | — |
| Chlorpyrifos (Lorsban®) [45] | Phosphorothionate | Control of sucking, chewing, and boring insects | 300–600 |
| Diazinon (Spectracide®) [64] | Phosphorothionate | Control of sucking and chewing insects, and mites | 400–800 |
| Dichlorvos (Vapona®) [67] | Phosphonate | Control of sucking and chewing insects, and spider mites in household sprays, etc. | 100 |
| Malathion (Acimal®) [119] | Phosphorodithionate | Control of sucking and chewing insects | 500–1250 |
| Monocrotophos (Monocron®) [133] | Phosphate | Control of sucking, chewing, and boring insects and spider mites | — |

spinosyns caused a dose-responsive reduction in the response to γ-aminobutyric acid (GABA) in isolated small-diameter cockroach neurons [576], an effect that is probably mediated through the chloride channel/chloride current. The hazard profile for spinosad (shown in Table 16.32) suggests that the mode of action is highly selective for insects. Spinosad is not very acutely toxic to mammals (oral $LD_{50}$ = 3738), and it is not neurotoxic [449].

## Avermectins and Milbemycin: Chloride Channel Activators (IRAC Code 6)

The avermectins and milbemycin are a group of closely related 16-membered macrocyclic lactones isolated from *Streptomyces avermitilis* and *Streptomyces hygroscopicus*, respectively [364]. The chemical structures for abamectin, emamectin-benzoate, and milbemycin are

**TABLE 16.24**
**Hazard Profiles for the Organophosphate, AChE Inhibiting, Insecticides (IRAC Code 1B)**

| Insecticide | Irritation | | $LD_{50}$ (mg/kg) | | $LC_{50}$ (mg/L) | Sensitization | Signa |
|---|---|---|---|---|---|---|---|
| | Eye | Skin | Oral | Dermal | Inhalation | Potential | Wor |
| Acephate | — | Non-irritant | 866 | >2000 | >15 | Negative | Cauti |
| Azinphos-methyl | Mild irritant | Non-irritant | 6–19 | 150 | 0.15 | Positive | Dange |
| Chlorpyrifos | Non-irritant | Non-irritant | 2680 | >2000 | >0.67 | Negative | Cauti |
| Diazinon | Non-irritant | Non-irritant | 1250 | >2150 | 2.33 | Negative | Cauti |
| Dichlorvos | Irritant | Irritant | 50 | 90 | 0.34 | Negative | Dange |
| Malathion | NA | NA | 1000E | 4100 | >5.2 | NA | Cauti |
| Monocrotophos | Non-irritant | Non-irritant | 18 | 130 | 0.08 | NA | Dang |

| Insecticide | Species/Study | NOEL (mg/kg/day) | Toxicity Study | Hazard Indicator |
|---|---|---|---|---|
| Acephate [270,553] | Rat/2-year | 0.5 | Mutagenicity | No evidence |
| | Dog/26-week oral | 0.75 | Developmental | Not teratogenic |
| | Mouse/18-month | — | Reproductive | No evidence |
| | Human | 0.3 | Oncogenicity | C (mouse liver tumor) |
| | ADI (human; UF = 10) | 0.03 | Neurotoxicity | Not neurotoxic |
| Azinphos-methyl [275,555] | Rat/2-year | 0.86 | Mutagenicity | Effects *in vitro;* not *in vivo* |
| | Dog/52-week oral | 0.74 | Developmental | No evidence |
| | Mouse/18-month | 0.88 | Reproductive | Effects on fertility |
| | Human | 0.005 | Oncogenicity | E (no evidence) |
| | ADI | 0.005 | Neurotoxicity | Not neurotoxic |
| Chlorpyrifos [266,509] | Rat/2-year | 0.1 | Mutagenicity | No evidence |
| | Dog/13-week oral | 10 | Developmental | Not teratogenic |
| | Mouse/18-month | 3.9 | Reproductive | No evidence |
| | Human | 0.1 | Oncogenicity | No evidence |
| | ADI (UF = 10)[a] | 0.01 | Neurotoxicity | Not delayed neurotoxin |
| Diazinon [280,529] | Rat/2-year | 0.07 | Mutagenicity | No evidence |
| | Dog/2-year oral | 0.02 | Developmental | Not teratogenic |
| | Mouse/18-month | — | Reproductive | No evidence |
| | Human | 0.025 | Oncogenicity | No evidence |
| | ADI (UF = 10)[a] | 0.002 | Neurotoxicity | Not delayed neurotoxin |
| Dichlorvos [281,558] | Rat/2-year | 2.4 | Mutagenicity | May be mutagenic |
| | Dog/52-week | — | Developmental | Not teratogenic |
| | Mouse/18-month | 10 | Reproductive | No evidence |
| | Human/21-day | 0.04 | Oncogenicity | No evidence |
| | ADI (UF = 10)[a] | 0.004 | Neurotoxicity | Delayed neuropathy |
| Malathion [561,637] | Rat/2-year | <1.2 | Mutagenicity | No evidence |
| | Dog/52-week | — | Developmental | Not teratogenic |
| | Mouse/18-month | | Reproductive | Effects on litter size |
| | Human/56-day | 0.34 | Neurotoxicity | Not delayed neurotoxin |
| | ADI | 0.02 | Oncogenicity | No evidence |
| Monocrotophos [287] | Rat/2-year | 0.025 | Mutagenicity | No evidence |
| | Dog/52-week | 0.0125 | Developmental | Not teratogenic |
| | Mouse/18-month | — | Reproductive | No evidence |
| | ADI | 0.0006 | Oncogenicity | No evidence |

[a] An uncertainty factor of 10 was applied as human data were available.

shown in Figure 16.13. These potent acaracides cause signs of ataxia, paralysis, and death, but the hyperexcitation typically found with most other insecticides is absent [352]. Although a number of pharmacologic effects of ivermectin have been described [369], it is generally agreed that the principal mode of action of this class is an activation of chloride ion current by a GABA-like opening of the chloride channel [10,200]. In addition, opening of a glutamate-gated chloride channel has also been implicated [231,311]. The toxicity profiles fo

**TABLE 16.25**
**Structure, Use, and Hazard Profiles for Organochlorine, GABA-Gated, Chloride-Channel Antagonists (IRAC Code 2A, and 2B)**

| Insecticide | Structure | Principal Use | Application Rate (g active ingredient per hectare) |
|---|---|---|---|
| Endosulfan [76] | 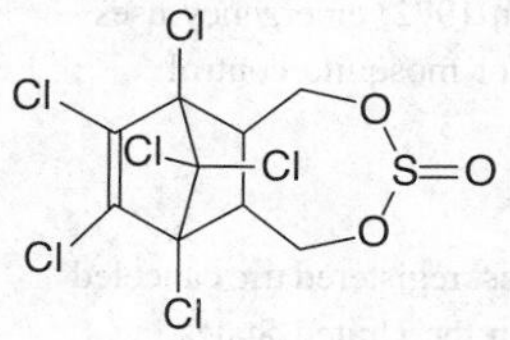 | Used to control sucking, chewing, and boring insects in a variety of crops, including fruit, vines, vegetables, cotton, and cereal | 1000–2500 |
| Fipronil [83] | 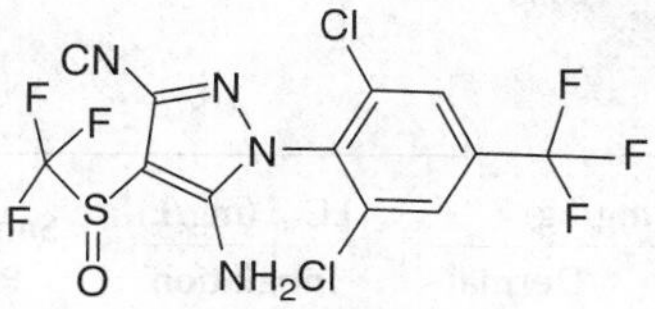 | Used to control thrips, corn root worms, and termites | 100–200 |
| Gamma-HCH (Lindane) [97] | 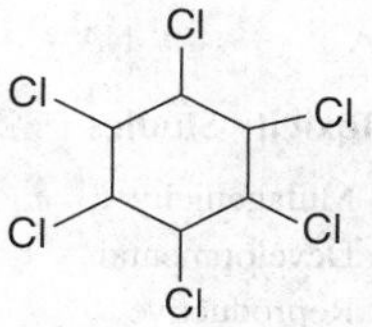 | Used to control soil-inhabiting insects, public-health pests, and animal ectoparasites | 250–750 |

| Insecticide | Irritation: Eye | Irritation: Skin | $LD_{50}$ (mg/kg): Oral | $LD_{50}$ (mg/kg): Potential | $LC_{50}$ (mg/L): Inhalation | Sensitization Potential | Signal Word |
|---|---|---|---|---|---|---|---|
| Endosulfan | Non-irritant | Non-irritant | 70 | 359 | >0.034 | Negative | Danger |
| Fipronil | Non-irritant | Non-irritant | 97 | >2000 | 0.68 | Negative | Warning |
| Gamma-HCH | Irritant | Irritant | >88 | >900 | 1.6 | Negative | Warning |

| Insecticide | Species/Study | NOEL (mg/kg/day) | Toxicity Study | Hazard Indicator |
|---|---|---|---|---|
| Endosulfan [271,519] | Rat/2-year | 0.60 | Mutagenicity | No evidence |
| | Dog/52-week | 0.57 | Developmental | Not teratogenic |
| | Mouse/18-month | 0.84 | Reproductive | No evidence |
| | ADI | 0.006 | Oncogenicity | No evidence |
| Fipronil [399,431] | Rat/2-year | 0.20 | Mutagenicity | No evidence |
| | Dog/52-week | 0.30 | Developmental | Not teratogenic |
| | Mouse/18-month | 0.50 | Reproductive | No evidence |
| | RfD | 0.0002 | Oncogenicity | No evidence |
| | | | Neurotoxicity | Not neurotoxic |
| Gamma-HCH (lindane) [272,560] | Rat/2-year | 0.75 | Mutagenicity | No evidence |
| | Dog/52-week | 1.6 | Developmental | Not teratogenic |
| | Mouse/18-month | — | Reproductive | No evidence |
| | ADI | 0.008 | Neurotoxicity | Not neurotoxic |
| | | | Oncogenicity | No evidence |

abamectin, emamectin benzoate, and milbemycin are given in Table 16.32. Abamectin and emamectin are neurotoxic in mammals, which exhibit hyperexcitability, tremors, incoordination, ataxia, and coma-like sedation [673]. Much of the early hazard evaluation for abamectin and emamectin was conducted in the wild-type CF-1 mouse [359,673], which has been found to be heterozygous for P-glycoprotein [279,304,305,482]. The toxic

**TABLE 16.26**
**Structure, Use, and Hazard Profiles for Organochlorine Sodium-Channel Modulators (IRAC Code 3)**

| Insecticide | Structure | Principal Use | Application Rate (g active ingredient per hectare) |
|---|---|---|---|
| Dichlorodiphenyl-trichloroethane (DDT) [62] | Cl–C₆H₄–CH(CCl₃)–C₆H₄–Cl | Banned in the United States in 1972; emergency uses for mosquito control | — |
| Methoxychlor [126] | CH₃O–C₆H₄–CH(CCl₃)–C₆H₄–OCH₃ | Last registered use canceled in the United States in 2003 | — |

| Insecticide | Irritation: Eye | Irritation: Skin | $LD_{50}$ (mg/kg): Oral | $LD_{50}$ (mg/kg): Dermal | $LC_{50}$ (mg/L): Inhalation | Sensitization Potential | Signal Word |
|---|---|---|---|---|---|---|---|
| DDT | NA | Non-irritant | 113 | >2000 | NA | NA | NA |
| Methoxychlor | NA | NA | 3460 | NA | NA | NA | NA |

| Insecticide | Species/Study | NOEL (mg/kg/day) | Toxicity Studies | Hazard Indicator |
|---|---|---|---|---|
| DDT [228,318,353,358] | Rat/2-year | 6.25 | Mutagenicity | Possibly mutagenic |
| | Dog/2-year oral | <5 | Developmental | NA |
| | Mouse/18-month | <8.33 | Reproductive | Weak estrogenic properties |
| | RfD 2 | 0.01 | Oncogenicity | Equivocal |
| Methoxychlor [126,264,635] | Rat/2-year | 20 | Mutagenicity | Not mutagenic |
| | Dog/52-week | 21 | Developmental | Not teratogenic |
| | Mouse/18-month | 28 | Reproductive | Weak estrogenic properties |
| | RfD2 | NA | Oncogenicity | Probably not carcinogenic |

effects of abamectin were reduced in studies using animals having a fully intact P-glycoprotein blood–brain barrier, supporting the idea that differential expression of P-glycoprotein, which is a substrate for the avermectins, might account for differences in selectivity among species [633]. The importance of an intact blood–brain barrier in humans has been considered [221]. Milbemycin is less acutely toxic than the avermectins and it is not a developmental toxin in animal bioassays.

## Juvenile Hormone Mimics and Selective Feeding Blocker (IRAC Codes 7 and 9)

Juvenile hormones modulate an extraordinarily broad range of morphological and physiological processes during larval development and metamorphosis [214], in addition to having effects on various aspects of adult reproduction and behavior [639]. Juvenile hormones are terpenoid-based compounds (Figure 16.14) that from an evolutionary point of view could be precursors to steroids and retinoids, which are also terpenoid derivatives [302]. The failure to identify classical nuclear receptors for juvenile hormones in spite of decades of effort [296] may be attributed to the possibility that juvenile hormones signal through membrane receptors [701]. Because juvenile hormones are capable of binding to a large number of proteins, they may exert their effects by binding to G-coupled membrane receptors which could trigger a cascade of intracellular MAP-kinase signaling molecules such as have been described for gonadotropin-releasing hormone in mammals [328].

The juvenile hormone mimics (JHMs) are compounds that bear a structural resemblance to the juvenile hormones in insects, which are lipophilic sesquiterpenoids containing an epoxide and methyl ester groups (see Figure 16.14). These chemicals mimic the action of juvenile hormones and affect a number of physiological processes, such as molting and reproduction. Exposure to JHMs at molting causes death by producing mixed larval/pupal or larval/adult morphologies. The efficacy of these compounds is greatest when normal juvenile hormone titers are low—namely, in the last larval or early pupal stages [214,296,570].

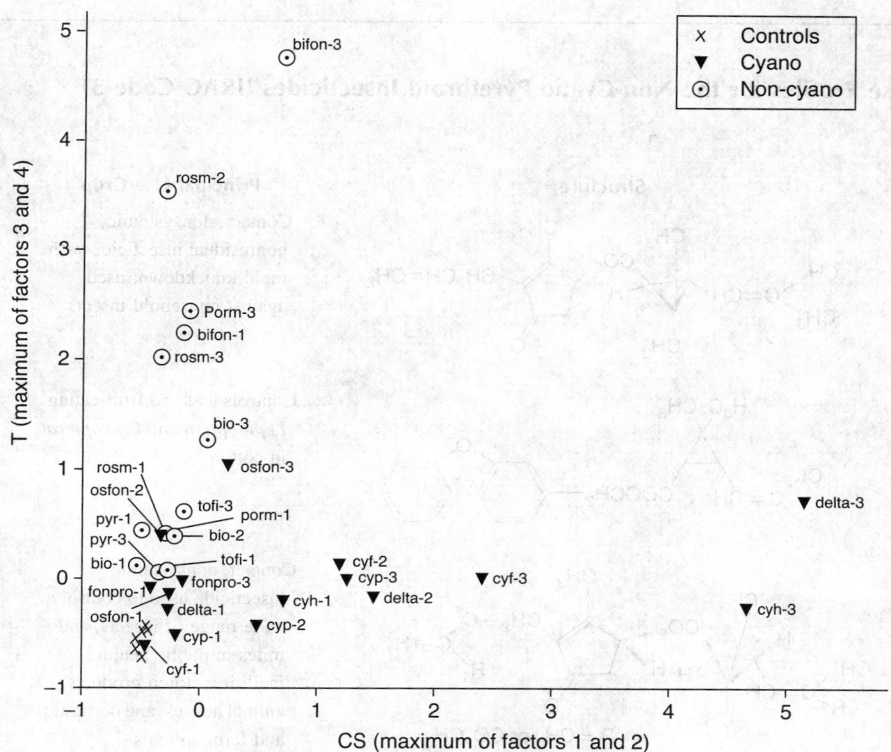


**FIGURE 16.9** Results from a principal components/factor analysis of FOB data.

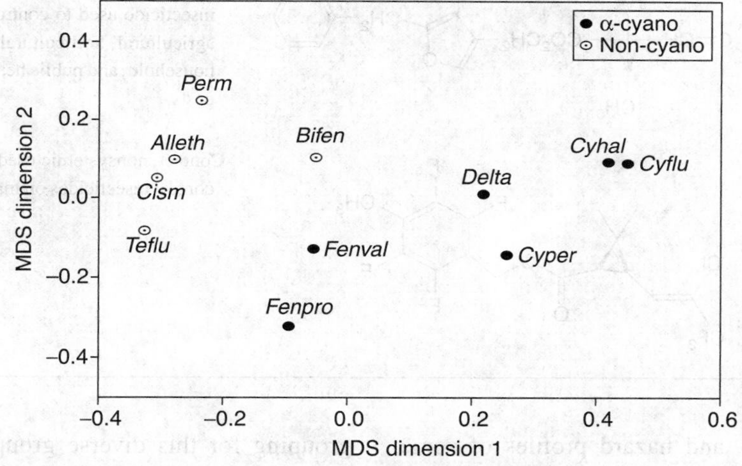


**FIGURE 16.10** Multidimensional scaling of sodium channel kinetic parameters.

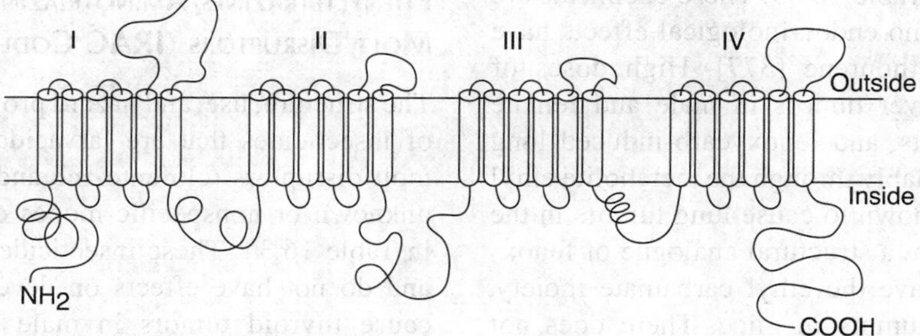


**FIGURE 16.11** Schematic of the transmembrane voltage-gated sodium channel.

**TABLE 16.27**
**Structure and Use Profiles for the Non-Cyano Pyrethroid Insecticides (IRAC Code 3)**

| Insecticide | Structure | Principal Use/Crop | Application Rate (g active ingredient per hectare) |
|---|---|---|---|
| Bioallethrin [31] | 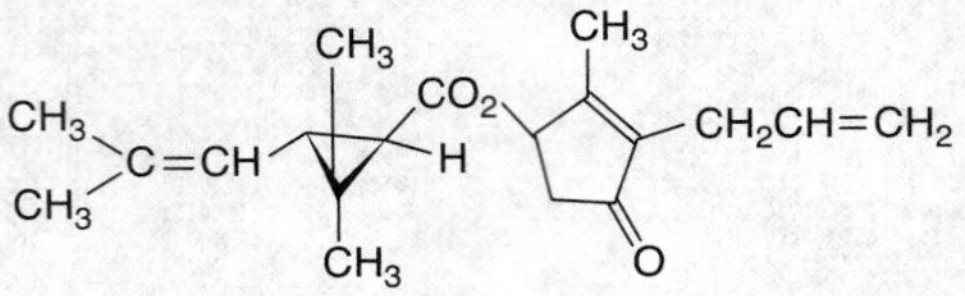 | Contact, nonsystemic, nonresidual insecticide with rapid knockdown; used against household insects | — |
| Permethrin (Ambush®) [144] | 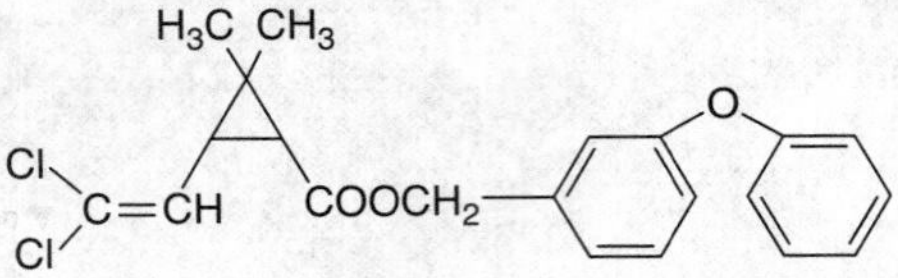 | Controls leaf and fruit-eating *Lepidoptera* and *Coleoptera* in cotton | 25–200 |
| Pyrethrins (pyrethrum) [159] | 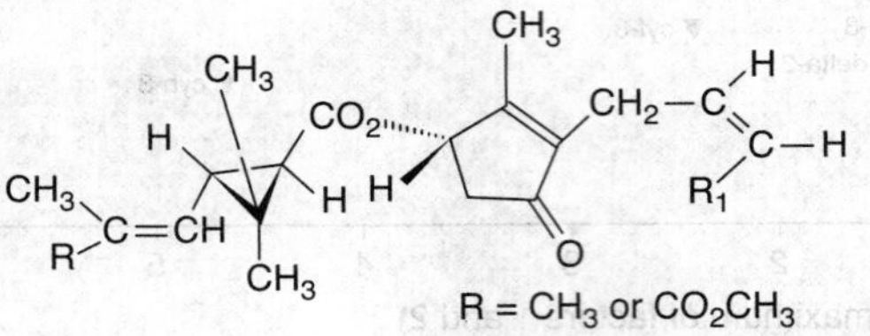 R = $CH_3$ or $CO_2CH_3$<br>$R_1$= CH=$CH_2$ or $CH_3$ or $CH_2CH_3$ | Contact, nonsystemic insecticide used to control a wide range of insects, and mites in public health facilities, stored products, animal houses, and domestic and farm animals | 1400–2800 |
| Resmethrin (Crossfire®) [166] | 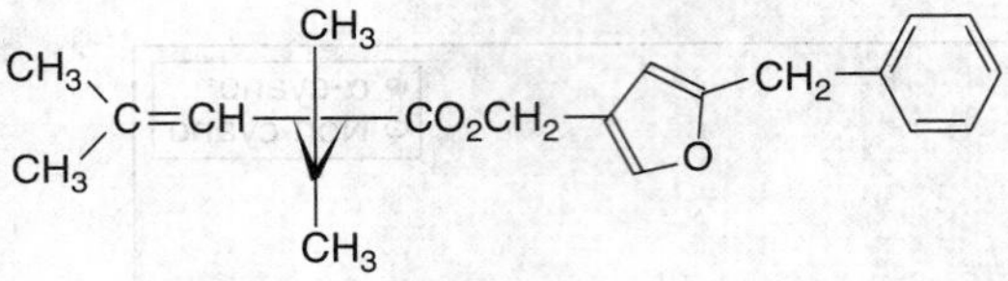 | Contact, nonsystemic insecticide used to control agricultural, horticultural, household, and public health pests | 4–10 |
| Tefluthrin (Force™) [179] | 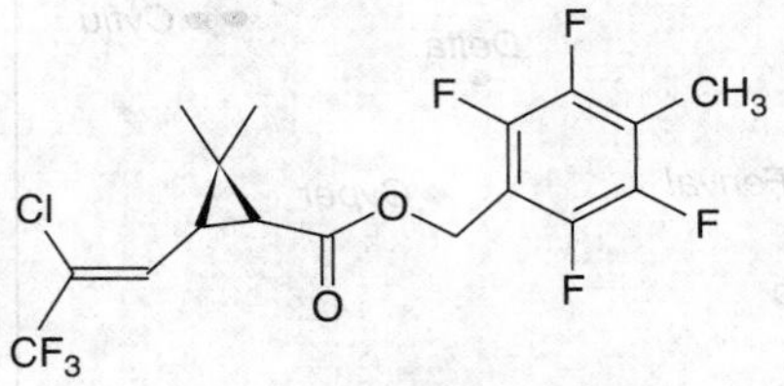 | Contact, nonsystemic used to control insecticides on maize | 35–60 |

The structures, uses, and hazard profiles of some juvenile hormone mimics (methoprene, fenoxycarb, and pyriproxifen) and the feeding inhibitor pymetrazine [251] are presented in Table 16.33. These chemicals are not acutely toxic, and no endocrinological effects have been reported for methoprene [377]. High doses of pymetrozine caused liver tumors in male and female mice and in female rats, and fenoxycarb-induced lung tumors in mice, presumably through the metabolite ethyl carbamate, which is known to cause lung tumors in the mouse [694]. Difolatan, a structural analogue of fenoxycarb that does not have the ethyl carbamate moiety, does not cause lung tumors in mice. There does not appear to be a basis for creating a common mechanism grouping for this diverse group of chemicals because juvenile hormone is unique to insects.

## Phenyltetrazines/Aminotriazines: Larvacides/Molt Disruptors (IRAC Codes 10A and 17)

The structure, use, and hazard profiles for a diverse group of insecticides that are larvacidal (cyromazine) and/or molt disruptors (clofentezine and hexythiazox) through unknown or nonspecific modes of action are presented in Table 16.34. These insecticides are not acutely toxic and do not have effects on development. Clofentezine cause thyroid tumors in male rats, and hexythiazox causes liver tumors.

**TABLE 16.28**
**Hazard Profiles for Non-Cyano Pyrethroids (IRAC Code 3)**

| Insecticide | Irritation: Eye | Irritation: Skin | $LD_{50}$ (mg/kg): Oral | $LD_{50}$ (mg/kg): Dermal | $LC_{50}$ (mg/L): Inhalation | Sensitization Potential | Signal Word |
|---|---|---|---|---|---|---|---|
| Bioallethrin | Non-irritant | Non-irritant | 53.8 | >2000 | 2.5 | Negative | Warning |
| Permethrin | Non-irritant | Non-irritant | 430 | >2000 | >0.68 | Moderate | Warning |
| Pyrethrin | — | — | 1030 | >1500 | 3.4 | Sensitizer | Warning |
| Resmethrin | Non-irritant | Non-irritant | >2500 | >3000 | 9.49 | Negative | Caution |
| Tefluthrin | — | Slight irritant | 21.8 | 316 | 0.037 | Sensitizer | Danger |

| Insecticide | Species/Study | NOEL (mg/kg/day) | Toxicity Study | Hazard Indicator |
|---|---|---|---|---|
| Bioallethrin [342,416] | Rat/2-year | 5 | Mutagenicity | Not mutagenic |
| | Dog/52-week | 1.5 | Developmental | Not teratogenic |
| | Mouse/18-month | 2.5 | Reproductive | Not reproductive toxin |
| | ADI | 0.04 | Oncogenicity | Not carcinogenic |
| | Acute NOEL | 1 | Neurotoxicity | Not studies available |
| Permethrin [285,562] | Rat/2-year | 5.0 | Mutagenicity | No evidence |
| | Dog/52-week | 5.0 | Developmental | Not teratogenic |
| | Mouse/18-month | 7.1 | Reproductive | No evidence |
| | RfD | 0.05 | Oncogenicity | C with RfD (lung, and liver tumors in female mice) |
| | Acute NOEL | — | Neurotoxicity | Neurotoxic |
| Pyrethrins [565] | Rat/2-year | 25 | Mutagenicity | Not mutagenic |
| | Dog/52-week | — | Developmental | Not teratogenic |
| | Mouse/18-month | 143 | Reproductive | Not a reproductive toxin |
| | ADI | 0.125 | Oncogenicity | Not carcinogenic |
| | Acute NOEL | — | Neurotoxicity | — |
| Resmethrin [227,566] | Rat/2-year | <25 | Mutagenicity | Not mutagnic |
| | Dog/6-month oral | 10 | Developmental | Not teratogenic |
| | Mouse/18-month | 50 | Reproductive | Not a developmental toxin |
| | ADI | 0.1 | Oncogenicity | Not carcinogenic |
| | Acute NOEL | — | Neurotoxicity | Not neurotoxic |
| Tefluthrin [451] | Rat/2-year | 4.6 | Mutagenicity | Not mutagenic |
| | Dog/52-week | 0.5 | Developmental | Delayed development |
| | Mouse/18-month | 3.4 | Reproductive | Not a reproductive toxin |
| | ADI | 0.005 | Oncogenicity | Not carcinogenic |
| | Acute NOEL | 0.5 | Neurotoxicity | Neurotoxic |

## DELTA-ENDOTOXINS DERIVED FROM *BACILLUS THURINGIENSIS* (IRAC CODE 11)

*Bacillus thuringiensis* (Bt) forms a crystalline inclusion body during sporulation that contains a number of insecticidal proteins [310]. When consumed by the insect, the inclusion body is dissolved in the midgut and releases delta-endotoxins. Mixtures of different delta-endotoxins are usually present in the inclusion body, and individual toxin proteins are designated with the prefix "cry-." These proteins contain a few hundred to over 1000 amino acids. After they are ingested, delta-endotoxins are cleaved to an active form by proteases within the midgut. The active toxins bind specifically to the membranes of the midgut epithelia and alter their ion permeability by forming a cation channel or pore. Ion movements through this pore disrupt potassium and the pH gradients and lead to lysis of the epithelium, gut paralysis, and death [236,310,629]. Delta-endotoxins derived from Bt have been used as insecticides over a 30-year period, and more recently plants have been genetically engineering to express delta-endotoxins. Considering that these bacterial strains of Bt are found in nature and that there is no evidence of adverse effects on human health resulting from the use of delta-endotoxin as an insecticide [414,455–457], the EPA requires only a limited toxicological assessment of new Bt products. Acute oral toxicity, *in vitro* digestibility, and infectivity/pathogenicity studies are required along with an evaluation of amino acid homology [457]. (See Table 16.35.)

**TABLE 16.29**
**Structure and Use Profiles for the α-Cyano Pyrethroid Insecticides (IRAC Code 3)**

| Insecticide | Structure | Principal Use/Crop | Application Rate (g active ingredient per hectare) |
|---|---|---|---|
| Cyfluthrin (Baythroid®) [55] | | Contact, nonsystemic insecticide used on cereals, cotton, fruit, and vegetables | 14–56 |
| Cypermethrin (Ammo™) [57] | | Contact, nonsystemic insecticide used on fruits, vines, coffee, cerials, ornamentals | 110–670 |
| Deltamethrin (Crackdown™) [63] | | Contact, nonsystemic insecticide used on cereals, citrus, cotton, grapes, cotton, maize, oilseed rape, and soya beans | 2.5–12.5 |
| Lambda-Cyhalothrin (Karate®) [56] | | Controls broad spectrum of chewing and piercing insects on cereals, hops, ornamentals, vegetables, and cotton | 2–5 |

## Benzoylureas: Chitin Synthesis Inhibitors (IRAC Code 15)

The benzoylureas, represented here by diflubenzuron and teflubenzuron, block molting in insects by preventing the formation of a new cuticle exoskeleton, which is comprised of about 50% chitin. Chitin is a polysaccharide comprised of *N*-acetylglucosamine. It has been proposed that polymerization is blocked by the benzoylureas by (1) the inhibition of chitin synthetase or its biosynthesis, (2) the inhibition of proteases or their biosynthesis, or (3) the inhibition of a membrane transport step involving UDP-*N*-acetylglucosamine [368]. These biochemical pathways are not present in mammals [319]. Diflubenzuron and teflubenzuron are not toxic to mammals. They are highly lipophilic and therefore tend to bioconcentrate in fat, which for some members of this class (lufenuron) has led to the expression of delayed toxicity in animal studies, apparently due to tissue bioaccumulation. The structure, use, and hazard profiles of diflubenzuron and teflubenzuron are presented in Table 16.36.

## Diacyhydrazine: Ecdysone Agonists (IRAC Code 18A)

The hormone 20-hydroxyecdysone (20E) is synthesized by the insect from cholesterol or phytosteroids, which it obtains from its diet; insects do not possess the biochemical pathways required to synthesize the steroid nucleus [254]. The ecdysone titer in hemolymph rises progressively up until the time of molt and then falls precipitously. Ablation of the eyestalk, the source of this hormone in crustaceans, results in a failure of molting behavior and no change in ecdysteroid concentration in hemolymph [198]. The ecdysteroid hormone 20E apparently mediates this effect on molting and reproductive function by binding to the nuclear ecdysteroid receptor (EcR) [345]. The ecdysone agonists, represented here by tebufenozide (see Figure 16.15), are lethal to lepidopteran pests by inducing premature molting [214]. The toxicity of tebufnozide (RH-5992) to insect larvae is proportional to its binding affinity to EcR proteins [354]. EcR is comprised of at least two proteins, which are gene products of the EcR and USP (ultraspiracle) genes. These genes are members of the steroid hormone receptor superfamily and

**TABLE 16.30**
**Hazard Profiles for α-Cyano Pyrethroids (IRAC Code 3)**

| Insecticide | Irritation: Eye | Irritation: Skin | $LD_{50}$ (mg/kg): Oral | $LD_{50}$ (mg/kg): Dermal | $LC_{50}$ (mg/L): Inhalation | Sensitization Potential | Signal Word |
|---|---|---|---|---|---|---|---|
| Cyfluthrin | Mild irritant | Non-irritant | 500 | >5000 | 0.5 | Negative | Warning |
| Cypermethrin | Irritant | Irritant | 247 | >4920 | 2.5 | Weak sensitizer | Warning |
| Deltamethrin | Mild irritant | Non-irritant | >5000 | >2000 | 2.2 | Negative | Caution |
| Lambda cyhalothrin | Non-irritant | Mild irritant | 56 | 632 | 0.60 | Negative | Danger |

| Insecticide | Species/Study | NOEL (mg/kg/day) | Toxicity Study | Hazard Indicator |
|---|---|---|---|---|
| Cyfluthrin [423] | Rat/2-year | 50 | Mutagenicity | Not mutagenic |
| | Dog/52-week | 5 | Developmental | Not teratogenic |
| | Mouse/18-month | 200 | Reproductive | Not a reproductive toxin |
| | ADI | 0.02 | Oncogenicity | Not carconogenic |
| | Acute NOEL | 20 | Neurotoxicity | Not neurotoxic |
| Cypermethrin [517,539] | Rat/2-year | 7.5 | Mutagenicity | Not mutagenic |
| | Dog/52-week | 5 | Developmental | Not teratogenic |
| | Mouse/18-month | 14 | Reproductive | Not a reproductive toxin |
| | ADI | 0.05 | Oncogenicity | Not carcinogenic |
| | Acute NOEL | 5 | Neurotoxicity | Neurotoxic |
| Deltamethrin [226,342,424] | Rat/2-year | 1 | Mutagenicity | Not mutagenic |
| | Dog/2-year oral | 1 | Developmental | Not teratogenic |
| | Mouse/18-month | 12 | Reproductive | Not a reproductive toxin |
| | ADI | 0.01 | Oncogenicity | Not carcinogenic |
| | Acute NOEL | 1 | Neurotoxicity | No studies available |
| Lambda-cyhalothrin [469] | Rat/2-year | 2.5 | Mutagenicity | No evidence |
| | Dog/52-week | 0.1 | Developmental | Not teratogenic |
| | Mouse/18-month | 14.2 | Reproductive | No evidence |
| | ADI | 0.001 | Oncogenicity | D (not classifiable) |
| | Acute NOEL | 0.5 | Neurotoxicity | Neurotoxic |

are the insect homolog of the vertebrate retinoid X receptor gene [300]. As such, it is possible that ecdysone agonists could have affinity for vertebrate receptor proteins. The structure, use, and hazard profiles of tebufenozide are presented in Table 16.37. It is not acutely toxic nor is it a developmental or reproductive toxin.

## Octopaminergic Agonists and Monoamine Oxidase Inhibitors (IRAC Codes 19 and 27)

Octopamine is an excitatory neurotransmitter in insects. The octopaminergic agonists amitraz and chlordimeform, both members of the formamidine class of insecticides, are selective for parasitic mites and ticks and some *Lepidoptera* and *Homoptera* species [256]. Recent studies in vertebrates have shown that amitraz and chlordimeform cause sympathomimetic effects [324], apparently by binding to α-2-adrenergic receptors [2,204]. It has been suggested that amines such as the octopamies could signal through G-protein-coupled receptors [308]. Chlordimeform has been shown to modify function in the hypothalamic–pituitary–gonadal axis in male [238] and female rats [239,240]. Human poisoning associated with amitraz exposure has been reported [641]. Piperonyl butoxide inhibits methyl farnesoate expoxidase, which catalyzes the synthesis of juvenile hormone III, thereby affecting molting behavior and reproductive function in insects [690]. Piperonyl butoxide also inhibits P450 monooxygenases, thereby retarding the metabolism of the pyrethroid insecticides and serving to prolong their period of effective action. The structure, use, and hazard profiles for amitraz and piperonyl butoxide are provided in Table 16.38. Amitraz is moderately acutely neurotoxic and has effects on development and reproductive function. It is also an animal carcinogen. Piperonyl butoxide is not acutely toxic nor does it have reproductive or developmental effects [563].

## Respiratory Inhibitors and Uncouplers (IRAC Codes 13, 20, and 21)

Compounds that disrupt energy metabolism have been identified from both natural and synthetic sources. An important natural product is rotenone, which is derived from the roots of *Derris* and *Lonchocatus* [306] and the

## TABLE 16.31
## Hazard Profiles for Nicotine and the Neonicotinoid Insecticides (IRAC Code 4)

| Insecticide | Structure | Principal Use | Application Rate (g active ingredient per hectare) |
|---|---|---|---|
| Imidacloprid (Admire®, Provado®) [110] | 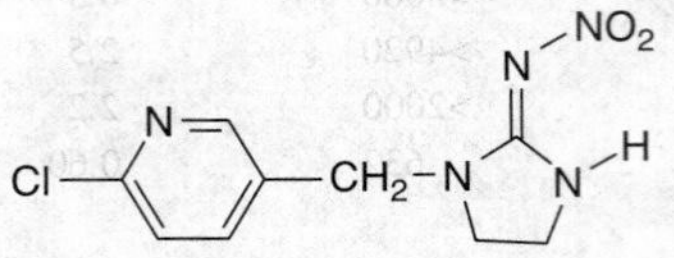 | Used to control sucking insects including aphids, thrips, and whiteflies | 25–100 |
| Thiamethoxam (Actara®, Cruiser®, Platinum®) [182] | 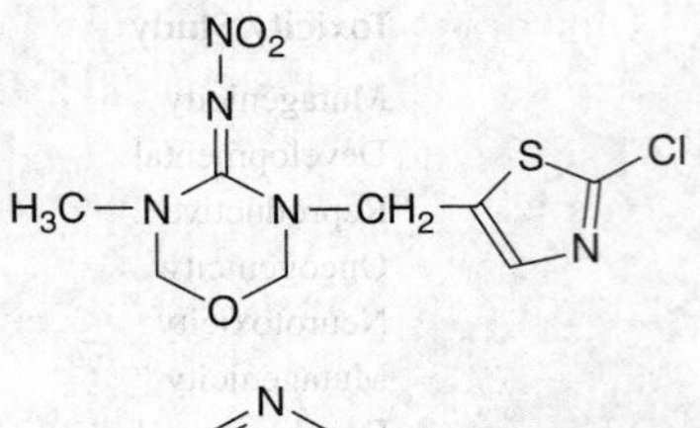 | Used to control sucking insects including ricehoppers, asphids, thrips, and whiteflies | 10–200 |
| Nicotine (Nico® Soap) [136] | 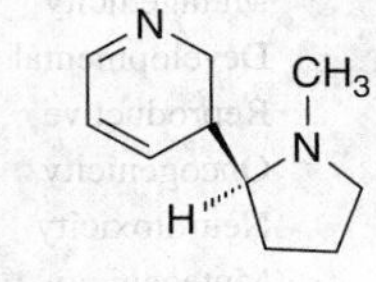 | Used to control sucking insects including ricehoppers, asphids, thrips, and whiteflies | Limited use |

| Insecticide | Irritation: Eye | Irritation: Skin | $LD_{50}$ (mg/kg): Oral | $LD_{50}$ (mg/kg): Dermal | $LC_{50}$ (mg/L): Inhalation | Sensitization Potential | Signal Word |
|---|---|---|---|---|---|---|---|
| Imidacloprid | Non-irritant | Non-irritant | 424 | >5000 | 0.07 | Negative | Warning |
| Thiamethoxam | Non-irritant | Non-irritant | 1563 | >2000 | >3.72 | Negative | Caution |
| Nicotine | Irritant | Mild irritant | 50 | 50 | NA | Negative | Danger |

| Insecticide | Species/Study | NOEL (mg/kg/day) | Toxicity Study | Hazard Indicator |
|---|---|---|---|---|
| Imidacloprid [326,435] | Rat/2-year | 5.7 | Mutagenicity | No evidence |
| | Dog/2-year oral | 41 | Developmental | Not teratogenic |
| | Mouse/18-month | 208 | Reproductive | No evidence |
| | RfD | 0.057 | Oncogenicity | E (no evidence) |
| Thiamethoxam [246,338,497,551] | Rat/2-year | 21.0 | Mutagenicity | No evidence |
| | Dog/52-week | 4.05 | Developmental | Not teratogenic |
| | Mouse/18-month | 2.63 | Reproductive | Testicular effects |
| | RfD2 | 0.0006 | Oncogenicity | Unlikely (mouse liver tumors) |
| Nicotine [136] | Rat/2-year | — | Mutagenicity | No evidence |
| | Dog/52-week | | Developmental | Not teratogenic |
| | Mouse/18-month | | Reproductive | — |
| | RfD | | Oncogenicity | — |

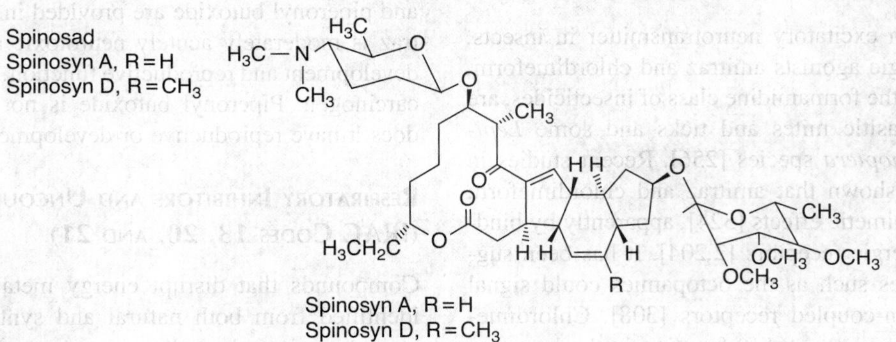

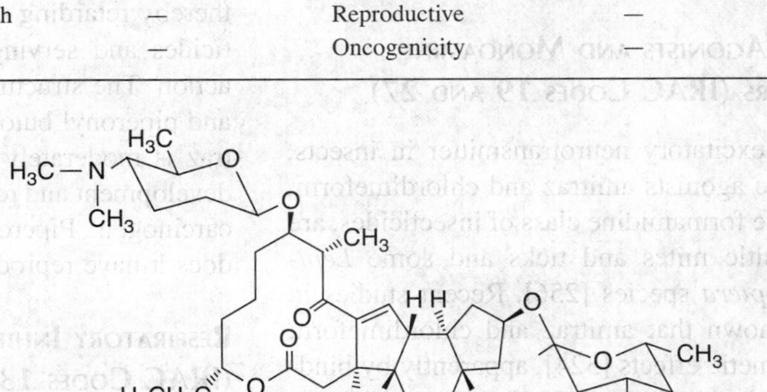

**FIGURE 16.12** Chemical structure of spinosad.

**TABLE 16.32**
**Hazard Profiles for Spinosad, the Avermectins, and Milbemycin (IRAC Codes 5 and 6)**

| Insecticide | Irritation: Eye | Irritation: Skin | $LD_{50}$ (mg/kg): Oral | $LD_{50}$ (mg/kg): Dermal | $LC_{50}$ (mg/L): Inhalation | Sensitization Potential | Signal Word |
|---|---|---|---|---|---|---|---|
| Spinosad [171] | Non-irritant | Non-irritant | 3738 | >5000 | >5.18 | Negative | Caution |
| Abamectin [12] | Mild irritant | Non-irritant | 13.6 | >2000 | 5.73 | Negative | Danger |
| Emamectin benzoate [75] | Severe irritant | Non-irritant | 76 | >2000 | 2.12 | Negative | Danger |
| Milbemycin [131] | Mild irritant | Non-irritant | 456 | >5000 | 1.9 | Negative | Caution |

| Insecticide | Species/Study | NOEL (mg/kg/day) | Toxicity Study | Hazard Indicator |
|---|---|---|---|---|
| Spinosad [347,348,449] | Rat/2-year | 5.0 | Mutagenicity | No evidence |
| | Dog/26-week oral | 2.7 | Developmental | Not teratogenic |
| | Mouse/18-month | 7.5 | Reproductive | No evidence |
| | RfD | 0.027 | Neurotoxicity | Not neurotoxic |
| | | | Oncogenicity | E (no evidence) |
| Abamectin [279,303] | Rat/2-year | 1.5 | Mutagenicity | No evidence |
| | Dog/26-week oral | 0.25 | Developmental | Teratogenic (rabbit, mouse) |
| | Mouse/18-month | 4.0 | Reproductive | No evidence |
| | RfD (Based on rat reproduction study; UF = 1000) | 0.00012 | Neurotoxicity | Neurotoxic in rat and dog |
| | | | Oncogenicity | E (no evidence) |
| Emamectin Benzoate [487] | Rat/2-year | 0.25 | Mutagenicity | No evidence |
| | Dog/26-week oral | 0.25 | Developmental | Not teratogenic |
| | Mouse/18-month | 2.5 | Reproductive | No evidence |
| | RfD (based on 15-day neurotoxicity in CF-1 rats) | 0.00083 | Neurotoxicity | Neurotoxicity exhibited in rodents and dogs |
| | Mouse (UF = 900) | | Oncogenicity | E (no evidence) |
| Milbemycin [352,633] | Rat/2-year | 6.81 | Mutagenicity | Not mutagenic |
| | Dog/26-week oral | 3 | Developmental | Not teratogenic |
| | Mouse/18-month | 18.9 | Reproductive | Not a reproductive toxin |
| | RfD | 0.03 | Neurotoxicity | Not neurotoxic |
| | | | Oncogenicity | Not carcinogenic |

leaves of some species of *Tephrosia* [261]. The synthetic compounds in this structurally diverse group include the pyrrole chlorfenapyr and the amidinohydrazones hydramethylnon, pyridaben, and rotenone. Disruption of energy metabolism occurs in the mitochondria and usually takes the form of either an inhibition of the electron transport system or an uncoupling of the transport system from ATP production. Inhibition of the electron transport system blocks the production of ATP and causes a decrease in oxygen consumption by the mitochondria. These uncouplers act on coenzyme Q oxidoreductase in the electron transport chain or the cytochrome $bc_1$ complex [255]. The electron transport system functions normally, but the production of ATP is uncoupled from the electron transport process due to a dissipation of the proton gradient across the inner mitochondrial membrane (see Figure 16.2). In the presence of uncouplers, oxygen consumption increases, but no ATP is produced [256]. The disruption of energy metabolism and the subsequent loss of ATP results in a slowly developing toxicity, and the effects of all of these insecticides include inactivity, paralysis, and death. It is expected that this mode of action would be preserved in mammalian systems if the chemical reaches its enzyme target; thus, chemicals that interfere with mitochondrial respiration are expected to have similar effects in mammalian species. The structure, use, and hazard profiles for these insecticides that inhibit mitochondrial respiration (chorfenapyr, hydramethylnon, pyridaben, and rotenone) are given in Table 16.39 and Table 16.40.

## Pheromones

Pheromones are chemical attractants secreted by special glands of insects to assist them in identifying or locating members of the opposite gender [317]. The EPA has defined pheromones as chemicals produced by arthropods (insects, arachnids, or crustaceans) that modify the behavior of other individuals of the same species [389]. The EPA has registered 17 arthropod pheromone active ingredients, 11 of which are lepidopteran pheromones [389]. The information submitted covered compounds that were from 6- to 16-carbon unbranched alcohols, acetates, and aldehydes that are volatile. Fewer data are required for the registration of pheromones. The available data on lepidopteran and

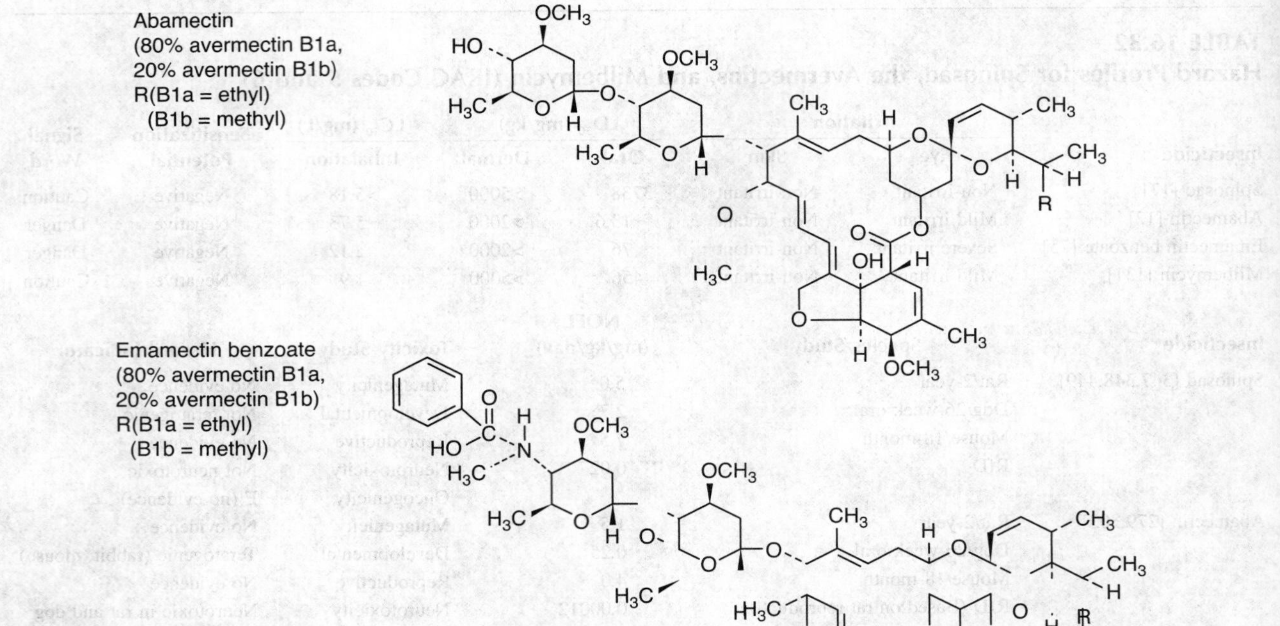

Abamectin
(80% avermectin B1a,
20% avermectin B1b)
R(B1a = ethyl)
(B1b = methyl)

Emamectin benzoate
(80% avermectin B1a,
20% avermectin B1b)
R(B1a = ethyl)
(B1b = methyl)

Milbemycin D : $R_5$ = —OH $R_{25} = CH(CH_3)_2$
Milbemycin $A_3/A_4$ : $R_5$ = —OH $R_{25} = CH_3$ and $C_2H_5$
Milbemycin A3/A4 5-Oxime : $R_5$ = NOH $R_{25} = CH_3$ and $C_2H_5$
Milbemycin : $R_5$ = —OH $R_{25}$ = $R_{23}$ = $NOCH_3$

**FIGURE 16.13** Chemical structure for the avermectins, and mibemycin.

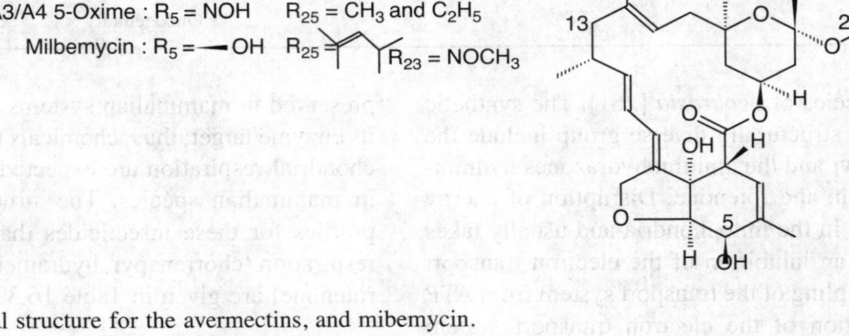

Methoprene $(CH_3)_2C(OCH_3)CH_2CH_2CH_2CH(CH_3)CH_2-CH=C(CH_3)-CH=CH-CO_2CH(CH_3)_2$

Juvenile hormone 3

**FIGURE 16.14** Comparison of the structure of methoprene to juvenile hormone 3.

**TABLE 16.33**
**Structure, Use, and Hazard Profiles for Juvenile Hormone Mimics Methoprene, Fenoxycarb, and Pyriproxifen (IRAC Code 7) and the Selective Feeding Blocker Pymetrozine (IRAC Code 9)**

| Insecticide | Structure | Principal Use | Application Rate (g active ingredient per hectare) |
|---|---|---|---|
| Methoprene (Apex®) [125] | 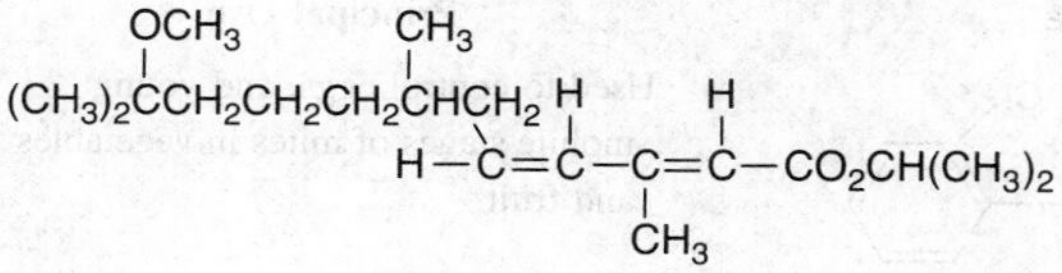 $(CH_3)_2C(OCH_3)CH_2CH_2CH_2CH(CH_3)CH_2-CH=C(CH_3)-CH=CH-CO_2CH(CH_3)_2$ | Prevents metamorphosis to viable adults; used in public health, food-handling facilities, mushroom houses | 11,300 |
| Fenoxycarb [79] | 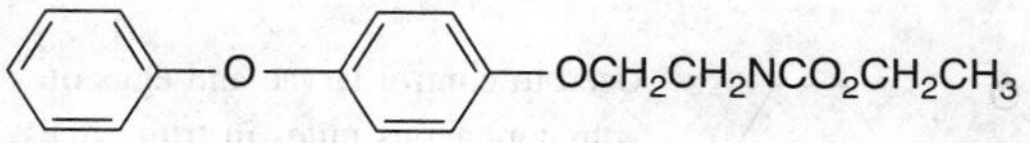 $C_6H_5-O-C_6H_4-OCH_2CH_2NCO_2CH_2CH_3$ | Used for control of fire ants, other ants, and other public health insect pests | 25–50 |
| Pymetrozine (Sterling®) [158] | 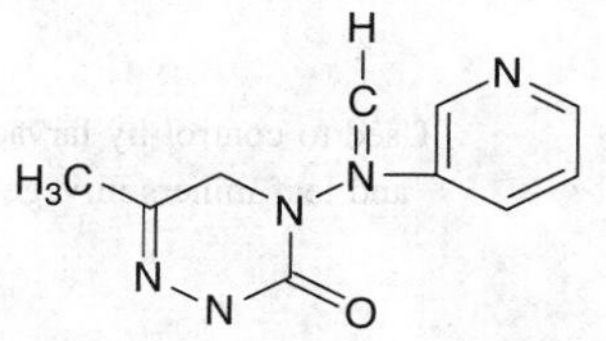 | Used to control aphids and whiteflies in vegetables, ornamentals, cotton, and field crops | 150–300 |
| Pyriproxyfen (Knack®) [164] | 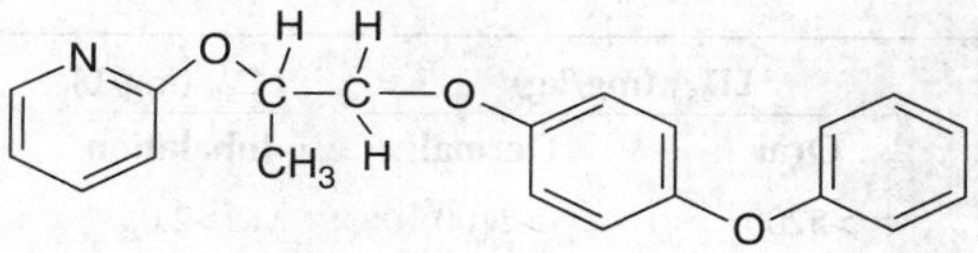 | Use to control public health insect pests | 25–50 |

| Insecticide | Irritation: Eye | Irritation: Skin | $LD_{50}$ (mg/kg): Oral | $LD_{50}$ (mg/kg): Dermal | $LC_{50}$ (mg/L): Inhalation | Sensitization Potential | Signal Word |
|---|---|---|---|---|---|---|---|
| Methoprene | Non-irritant | Non-irritant | 34,600 | 3500 | 210 | Negative | Caution |
| Fenoxacarb | Slight irritant | Non-irritant | >10,000 | >2000 | 4.4 | Negative | Caution |
| Pymetrozine | Non-irritant | Non-irritant | >5820 | >2000 | >1.8 | Negative | Caution |
| Pyriproxyfen | Non-irritant | Non-irritant | >5000 | >2000 | >3.1 | Negative | Caution |

| Insecticide | Species/Study | NOEL (mg/kg/day) | Toxicity Study | Hazard Indicator |
|---|---|---|---|---|
| Methoprene [377] | Rat/2-year | 5000 | Mutagenicity | Not mutagenic |
| | Dog/52-week | — | Developmental | Not treratogenic |
| | Mouse/18-month | 2500 | Reproductive | Not reproductive toxin |
| | ADI | 0.1 | Oncogenicity | Not carcinogenic |
| Fenoxycarb [429,570] | Rat/2-year | 10 | Mutagenicity | Not mutagenic |
| | Dog/52-week | 25 | Developmental | Not teratogenic |
| | Mouse/18-month | 5 | Reproductive | Not reproductive toxin |
| | RfD (based on $Q_1$*) | 0.0000007 | Oncogenicity | Lung/liver in mice, C with $Q_1$* of $5.6 \times 10^{-2}$ $(mg/kg/day)^{-1}$ |
| Pymetrozine [251,476,477] | Rat/2-year | 3.7 | Mutagenicity | No evidence |
| | Dog/52-week | 0.57 | Developmental | Not teratogenic |
| | Mouse/18-month | — | Reproductive | No evidence |
| | RfD | 0.0057 | Oncogenicity | No evidence |
| Pyriproxyfen [479,491] | Rat/2-year | 35 | Mutagenicity | No evidence |
| | Dog/52-week | 100 | Developmental | Not teratogenic |
| | Mouse/18-month | 85 | Reproductive | No evidence |
| | RfD | 0.35 | Oncogenicity | E (no evidence) |

**TABLE 16.34**
**Structure, Use, and Hazard Profiles for Phenyltetrazine (Clofentezine, Hexythiazox; IRAC Code 10A) and Aminotriazine(Cyromazine; IRAC Code 17) Larvicides/Growth and Moulting Disruptors with an Unknown or Nonspecific Mode of Action**

| Insecticide | Structure | Principal Use | Application Rate (g active ingredient per hectare) |
|---|---|---|---|
| Clofentezine (Apollo®) [49] | 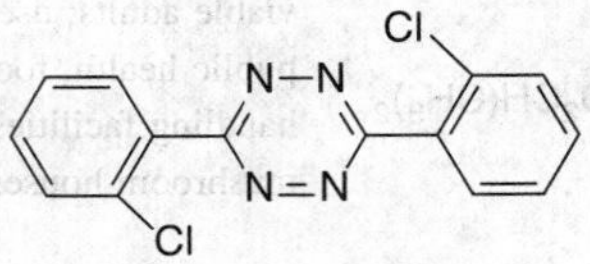 | Used to control eggs, and young mobile stages of mites in vegetables and fruit | 100–400 |
| Hexythiazox (Nissorun®) [99] | 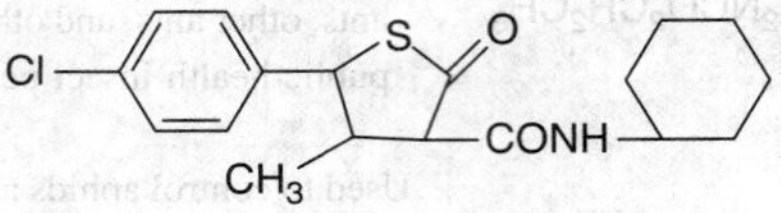 | Used to control larvae and eggs of phygophagous mites in fruit, vines, cotton, and vegetables | 150–300 |
| Cyromazine (Trigard®) [60] | 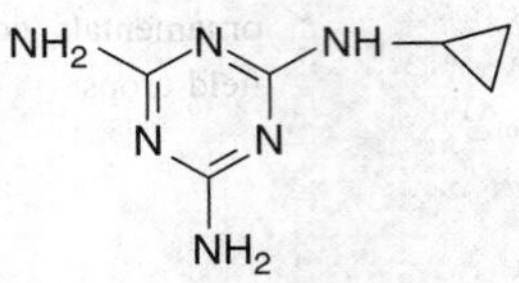 | Used to control fly larvae in manure and leaf miners in vegetables | 75–450 |

| Insecticide | Irritation: Eye | Irritation: Skin | $LD_{50}$ (mg/kg): Oral | $LD_{50}$ (mg/kg): Dermal | $LC_{50}$ (mg/L): Inhalation | Sensitization Potential | Signal Word |
|---|---|---|---|---|---|---|---|
| Clofentezine | Non-irritant | Non-irritant | >5200 | >2100 | >2.0 | Weak positive | Caution |
| Hexythiazox | Mild irritant | Non-irritant | >5000 | >5000 | >2.0 | Negative | Caution |
| Cyromazine | Non-irritant | Mild irritant | 2029 | >1370 | >2.7 | Negative | Caution |

| Insecticide | Species/Study | NOEL (mg/kg/day) | Toxicity Study | Hazard Indicator |
|---|---|---|---|---|
| Clofentezine [483,484] | Rat/2-year | 2.0 | Mutagenicity | No evidence |
| | Dog/26-week oral | 1.25 | Developmental | Not teratogenic |
| | Mouse/18-month | 7.1 | Reproductive | No evidence |
| | RfD | 0.012 | Oncogenicity | C with Q* (thyroid tumors in male rats) |
| Hexythiazox [465] | Rat/2-year | 21.5 | Mutagenicity | No evidence |
| | Dog/26-week oral | 2.5 | Developmental | Not teratogenic |
| | Mouse/18-month | 37.5 | Reproductive | No evidence |
| | RfD | Q* = 0.039 (mg/kg/day)$^{-1}$ | Oncogenicity | C with Q* (based on liver tumors ) |
| Cyromazine [486,494] | Rat/2-year | 1.8 | Mutagenicity | No evidence |
| | Dog/26-week oral | 0.75 | Developmental | Not teratogenic |
| | Mouse/18-month | 6.5 | Reproductive | No evidence |
| | RfD | 0.008 | Oncogenicity | E (no evidence) |

other arthropod pheromones, including several aromatic pheromones, have shown no acute mammalian toxicity at the limit dose levels tested. The acute toxicity profile generally reveals oral and dermal $LD_{50}$ values of greater than 5000 mg/kg and 2000 mg/kg, respectively [452]. Acute inhalation $LC_{50}$ values generally are greater than 5 mg/L. Eye and skin irritation potentials fall in the mild or not irritating range, and there is no evidence of skin sensitization potential. Because small amounts of the pheromone are present inside bait stations, human contact is minimal therefore, the full data package required for conventiona pesticides is waived by the EPA [452].

## HERBICIDES

The modes of action (MOAs) of herbicides used in crop protection have been classified by the Herbicide Resistance Action Committee (HRAC) into groups [253] as

**TABLE 16.35**
**Structure, Use, and Hazard Profiles for Delta Endotoxin Derived from *Bacillus thuringiensis* subspp. *aizawai* and *kurstaki* (IRAC Code 11)**

| Insecticide | Structure of Delta–Endotoxin Protein | Principal Use | Application Rate (g active ingredient per hectare) |
|---|---|---|---|
| *Bacillus thuringiensis* subspp. *aizawai* and *kurstaki* [25,26,236,310,414,455–457] | 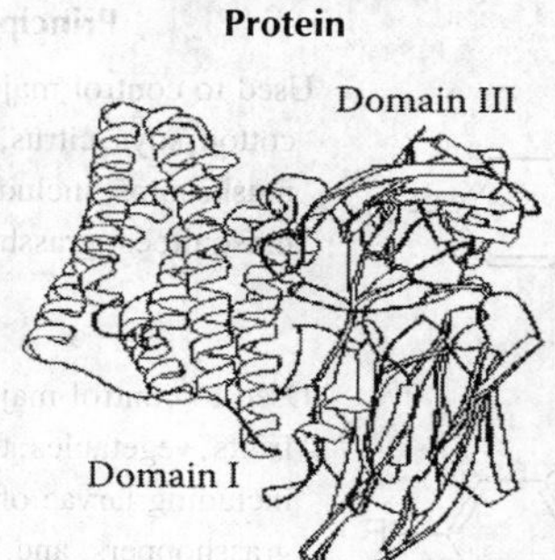 | Control of caterpillars of the *Lepidoptera* (butterflies, moths, and corn root worm) but also mosquito larvae and blackflies that vector river blindness in Africa | 1121 |

| Insecticide | Irritation | | LD$_{50}$ (mg/kg) | | LC$_{50}$ (mg/L) | Sensitization Potential | Signal Word |
|---|---|---|---|---|---|---|---|
| | Eye | Skin | Oral | Dermal | Inhalation | | |
| *aizawai* subsp. | — | — | N.i. | — | — | — | Caution |
| *kurstaki* subsp. | No infectivity | No infectivity | N.i. | N.i. | 5.4 | — | Caution |

| Insecticide | Species/Study | NOEL (mg/kg/day) | Toxicity Study | Hazard Indicator |
|---|---|---|---|---|
| *aizawai* subsp | Rat/2-year | 8.4 | Mutagenicity | Waived |
| | Dog/52-week | — | Developmental | Waived |
| | Mouse/18-month | — | Reproductive | Waived |
| | RfD | — | Oncogenicity | Waived |
| *kurstaki* subsp. | Rat/2-year | — | Mutagenicity | Waived |
| | Dog/52-week | — | Developmental | Waived |
| | Mouse/18-month | — | Reproductive | Waived |
| | ADI | — | Oncogenicity | Waived |

presented in Table 16.41. It has been suggested that the MOA for 60% of herbicides introduced during the period from 1960 to 2000 involve biochemical pathways specific to chloroplasts or plant signaling hormones [572]. Whether any of the molecular targets of herbicides in plants have homologous targets in animals will be considered in the following section.

## Acetyl–CoA Carboxylase Inhibitors (HRAC Code A)

The aryloxyphenoxypropionate and cyclohexanedione herbicides inhibit acetyl coenzyme A (acetyl–CoA) carboxylase (ACCase), although the exact binding site of these herbicides on this enzyme has not yet been determined [212]. They block the synthesis of fatty acids essential for the production of plant lipids, which are vital to the integrity of cell membranes and the formation of cuticle waxes during new plant growth (see Figure 16.16). Injury is slow to develop (7 to 10 days) and appears first on new leaves emerging from the whorl of the grass plant. The herbicide is taken up by the foliage and moves in the phloem to areas of new growth [219]. ACCase exists in two forms in plants: (1) the prokaryote form, found in broadleaf dicotyledonous plants, which are 400 to 6000 times more tolerant to these herbicides; and (2) the eukaryote form, found in perennial and annual grasses, which are more susceptible. The prokaryotic form of ACCase is heterodimeric, comprised of four gene products: the biotin carboxyl carrier (BCC), biotin carboxylase (BCase), and α- and β-subunits of carboxyltransferase (CTase). The eukaryotic, homodimeric form is a single 220- to 230-dDa polypeptide comprised of linked BCC, BCase, and CTase domains [212,574]. Differences in tolerance among the moncotyledonous plants are attributed to differences in the rate of detoxification among the subspecies [195–197]. Resistance development is attributed the acquisition of mutations that (1) increase the expression of ACCase, (2) alter binding of the herbicide to ACCase, or (3) increase expression of enzymes involved in herbicide metabolism [212]. Bio-

**TABLE 16.36**
**Structures, Use, and Hazard Profiles for the Benzoylurea Chitin Synthesis Inhibitors Diflubenzuron and Teflubenzuron (IRAC Code 15)**

| Insecticide | Structure | Principal Use | Application Rate (g active ingredient per hectare) |
|---|---|---|---|
| Diflubenzuron (Amilin®) [71] | 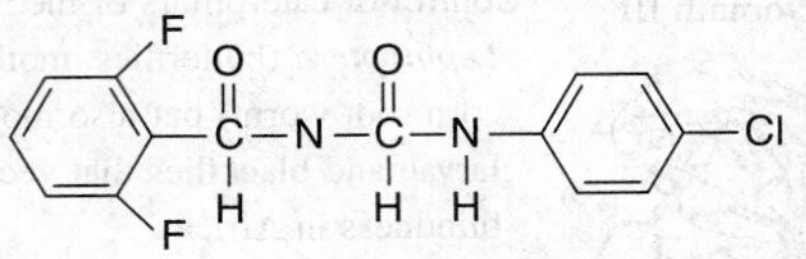 | Used to control major insect pests in cotton, soya, citrus, tea, vegetables, and mushrooms, including larvae of flies, mosquitoes, grasshoppers, and locust | 25–75 |
| Teflubenzuron (Nomolt®) [178] | 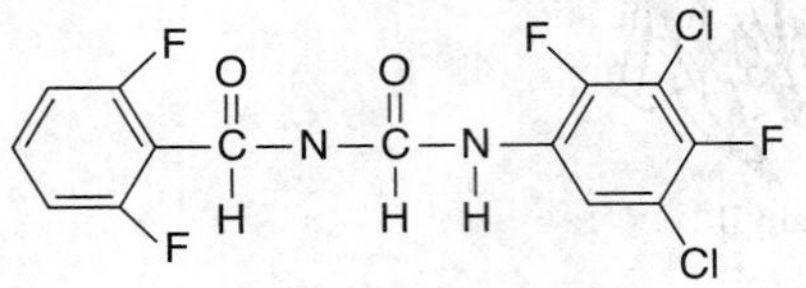 | Used to control major insect pests in fruits, vegetables, tobacco, and cotton, including larvae of flies, mosquitoes, grasshoppers, and locust | 30–60 |

| Insecticide | Irritation: Eye | Irritation: Skin | $LD_{50}$ (mg/kg): Oral | $LD_{50}$ (mg/kg): Dermal | $LC_{50}$ (mg/L): Inhalation | Sensitization Potential | Signal Word |
|---|---|---|---|---|---|---|---|
| Diflubenzuron | Non-irritant | Non-irritant | >4640 | >10,000 | >35 | Negative | Caution |
| Teflubenzuron | Non-irritant | Non-irritant | >5000 | >2000 | >3.1 | Negative | Caution |

| Insecticide | Species/Study | NOEL (mg/kg/day) | Toxicity Study | Hazard Indicator |
|---|---|---|---|---|
| Diflubenzuron [426,461,462] | Rat/2-year | 2.0 | Mutagenicity | No evidence |
| | Dog/52-week | 2.0 | Developmental | Not teratogenic |
| | Mouse/18-month | 2.0 | Reproductive | No evidence |
| | RfD | 0.02 | Oncogenicity | E (no evidence) |
| Teflubenzuron [178] | Rat/2-year | 4.8 | Mutagenicity | No evidence |
| | Dog/52-week | 3.2 | Developmental | Not teratogenic |
| | Mouse/18-month | 2.1 | Reproductive | No evidence |
| | ADI (based on 18-month mouse study (UF = 200) | 0.01 | Oncogenicity | No evidence |

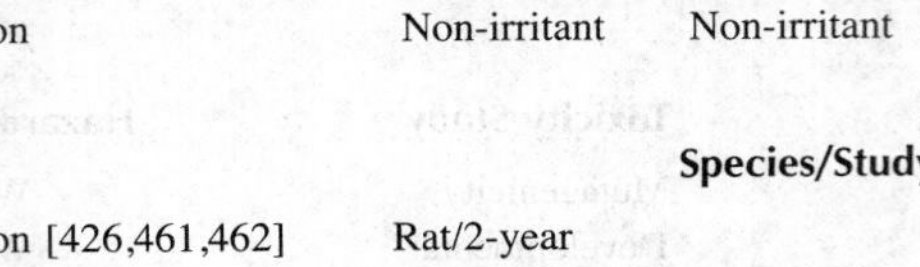

**FIGURE 16.15** Comparison of the structure of ecdysone 20 to tebufenozide.

chemical pathways for fatty acid synthesis are conserved in mammalian species (Figure 16.16). ACCase in mammals is like the eukaryotic form. No specific data indicate that the aryloxyphenoxypropionate (AOPP) or cyclohex anedione (CHD) herbicides alter fatty-acid synthesis in animal studies, although drugs have been developed to block this pathway; for example, the rate-limiting enzyme for sterol synthesis (HMG–CoA reductase) is inhibited by simvastatin, a drug designed to reduce cholesterol biosyn thesis.

## Aryloxyphenoxypropionates (HRAC Code A)

The structure, use, and hazard profiles for six aryloxyphe noxypropionate herbicides are presented in Table 16.42 and Table 16.43. These herbicides are generally no acutely toxic. Clodinafop-propargyl and haloxyfop have been identified as peroxisomal proliferators in the rodent The relevance of peroxisomal proliferation to humans i determined on a case-by case basis using a framework assessment of the weight of evidence [299].

**TABLE 16.37**
**Structure, Use, and Hazard Profile for the Diacylhydrazine Ecdysone Agonist Tebufenozide (IRAC Code 18A)**

| Structure | Principal Use | Application Rate (g active ingredient per hectare) |
|---|---|---|
| | Used for control lepidopteran larvae on rice, fruit, row crop, nut crops, vegetables, and vines. | 56 |

| Insecticide | Irritation: Eye | Irritation: Skin | $LD_{50}$ (mg/kg): Oral | $LD_{50}$ (mg/kg): Dermal | $LC_{50}$ (mg/L): Inhalation | Sensitization Potential | Signal Word |
|---|---|---|---|---|---|---|---|
| Tebufenozide | Non-irritant | Non-irritant | >5000 | >5000 | 4.5 | Negative | Caution |

| Insecticide | Species/Study | NOEL (mg/kg/day) | Toxicity Study | Hazard Indicator |
|---|---|---|---|---|
| Tebufenozide [177,480,570] | Rat/2-year | 4.8 | Mutagenicity | No evidence |
| | Dog/52-week | 1.8 | Developmental | Not teratogenic |
| | Mouse/18-month | 143 | Reproductive | No evidence |
| | RfD | 0.018 | Oncogenicity | E (no evidence) |

**TABLE 16.38**
**Structure, Use, and Hazard Profiles for the Octopaminergic Agonist Amitraz (IRAC Code 19) and the P450 Monooxygenase Inhibitor Piperonyl Butoxide (IRAC Code 27)**

| Insecticide | Structure | Principal Use | Application Rate (g active ingredient per hectare) |
|---|---|---|---|
| Amitraz [19] | | Nonsystemic, with contact and respiratory action; expels tick, mites, scale insects, whitefly, asphids, and others | 1120–3360 |
| Piperonyl butoxide [147] | | Inhibits insects' MFO, increasing the efficacy of the applied insecticide | 5382 |

| Insecticide | Irritation: Eye | Irritation: Skin | $LD_{50}$ (mg/kg): Oral | $LD_{50}$ (mg/kg): Dermal | $LC_{50}$ (mg/L): Inhalation | Sensitization Potential | Signal Word |
|---|---|---|---|---|---|---|---|
| Amitraz | Non-irritant | Non-irritant | 531 | >200 | 2.4 | Negative | Danger |
| Piperonyl butoxide | Non-irritant | Non-irritant | 4570 | >2000 | >5.9 | Positive | Caution |

| Insecticide | Species/Study | NOEL (mg/kg/day) | Toxicity Study | Hazard Indicator |
|---|---|---|---|---|
| Amitraz [204,324,395,641] | Rat/2-year | 2.5 | Mutagenicity | Not mutagenic |
| | Dog/52-week | 0.25 | Developmental | Delayed development in rabbits |
| | Mouse/18-month | 3.75 | Reproductive | Effects on fecunity |
| | ADI | 0.003 | Oncogenicity | C with $Q_1^* = 5 \times 10^{-2}$ $(mg/kg/day)^{-1}$ |
| Piperonyl butoxide [563] | Rat/2-year | 30 | Mutagenicity | Not mutagenic |
| | Dog/52-week | 16 | Developmental | Not teratogenic |
| | Mouse/18-month | 30 | Reproductive | Not a reproductive toxin |
| | ADI | 0.2 | Oncogenicity | C (ileocecal, liver tumors at high doses) |

**TABLE 16.39**
**Structure and Use Profiles for the Mitochondrial Respiration Inhibiting Insecticides Chlorfenapyr (IRAC Code 13), Hydramethynon (IRAC Code 20), and Pyridaben and Rotenone (IRAC Code 21)**

| Insecticide | Structure | Principal Use | Application Rate (g active ingredient per hectare) |
|---|---|---|---|
| Chlorfenapyr (Pirate®) [42] | 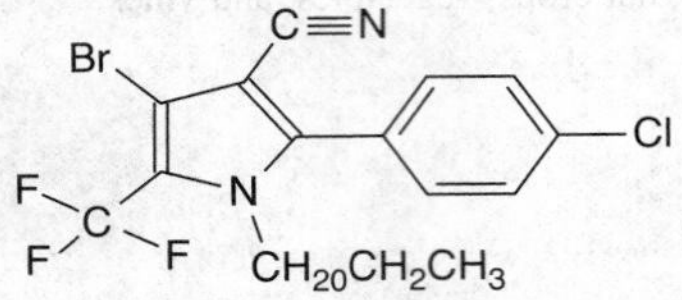 | Used to control many insects and mites in cotton, vegetables, citrus, vines, and soya beans | 1177 |
| Hydramethylnon (Amdro®) [100] | 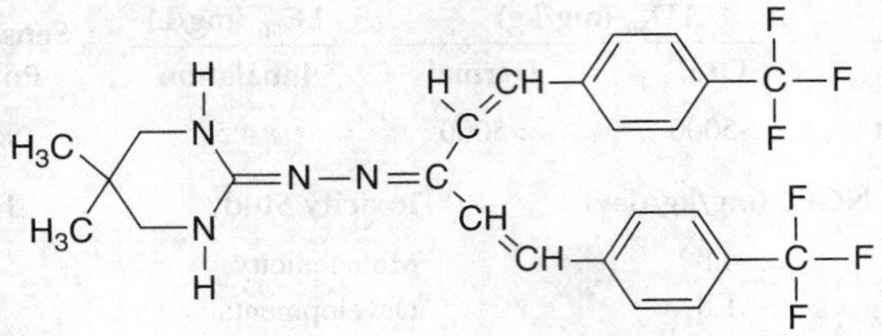 | Used to control agricultural and household Formicidae | 16 |
| Pyridaben (Poseidon®) [160] | 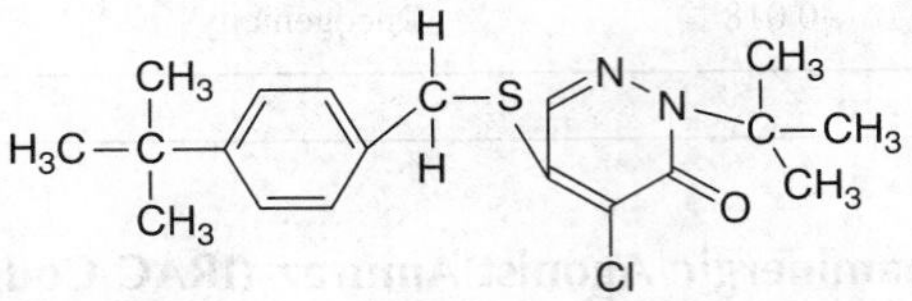 | Used to control acarids on field crops, fruits, vegetables, and ornamentals | 100–300 |
| Rotenone [168] | 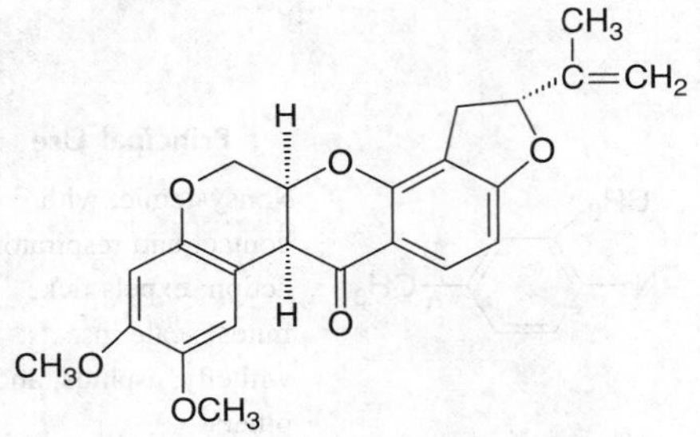 | Used to control aphids, thrips, suckers, moths, beetles, and spider mites in fruits and vegetables | 280–420 |

## Cyclohexanediones (HRAC Code A)

The structure, use, and hazard profiles for the cyclohexanedione herbicides clethoxydim and sethoxydim are presented in Table 16.44. Both clethodim and sethoxydim have limited toxicity in acute and repeat dose toxicity studies. They are not mutagenic, carcinogenic, or developmental or reproductive toxins.

## Acetolactate Synthase Inhibitors (HRAC Code B)

The acetolactate synthase (ALS) inhibitors, which are comprised of the sulfonyurea, imidazolinone, triazolopyridimidine, and pyriidinyl thiobenzoate classes of herbicides interact with the acetolactate synthase enzyme, thereby blocking the biosynthesis of branched-chain amino acids, valine, leucine, and isoleucine, as illustrated in Figure 16.17 [688]. The binding site is considered to be a vestigial quinine binding site on the enzyme [350]. Because the biochemical pathway for the synthesis of branched-chain amino acids does not exist in monogastric animals, this herbicidal mode of action is not relevant to humans. In fact, some researchers have taken advantage of this selectivity to design antituberculosis drugs [243].

## Sulfonylureas (HRAC Code B)

Sulfonylurea (SU) herbicides belong to a class of compounds comprised of three distinct components: an aryl group linked to a nitrogen-containing hetrocycle via a sulfonylurea bridge. Sulfonylurea herbicides inhibit root and shoot growth in rapidly growing plants by suppressing

**TABLE 16.40**
**Hazard Profiles for Mitochondrial Respiration Inhibiting Insecticides Chlorfenapyr (IRAC Code 13), Hydramethynon (IRAC Code 20), and Pyridaben and Rotenone (IRAC Code 21)**

| | Irritation | | $LD_{50}$ (mg/kg) | | $LC_{50}$ (mg/L) | | |
|---|---|---|---|---|---|---|---|
| **Insecticide** | **Eye** | **Skin** | **Oral** | **Dermal** | **Inhalation** | **Sensitization Potential** | **Signal Word** |
| Chlorfenapyr | Moderate irritant | Non-irritant | 441 | >2000 | 1.9 | Negative | Warning |
| Hydramethylnon | Mild irritant | Non-irritant | 817 | >2000 | 2.9 | Negative | Caution |
| Pyridaben | Slight irritant | Non-irritant | 820 | >2000 | 0.66 | Negative | Caution |
| Rotenone | — | — | 39.5 | — | — | — | — |

| Insecticide | Species/Study | NOEL (mg/kg/day) | Toxicity Study | Hazard Indicator |
|---|---|---|---|---|
| Chlorfenapyr [418] | Rat/2-year | 2.9 | Mutagenicity | No evidence |
| | Dog/52-week | 4.0 | Developmental | Not teratogenic |
| | Mouse/18-month | 2.8 | Reproductive | No evidence |
| | RfD | 0.03 | Oncogenicity | E (no evidence) |
| | | | Neurotoxicity | Not neurotoxic |
| Hydramethylnon [255,466] | Rat/2-year | 50 | Mutagenicity | No evidence |
| | Dog/52-week | 1.0 | Developmental | Not teratogenic |
| | Mouse/18-month | 25 | Reproductive | No evidence |
| | ADI | 0.01 | Oncogenicity | C with RfD (lung, and liver tumors in mice) |
| | | | Neurotoxicity | Not neurotoxic |
| Pyridaben [447] | Rat/2-year | 1.13 | Mutagenicity | No evidence |
| | Dog/52-week | <0.5 | Developmental | Not teratogenic |
| | Mouse/18-month | 2.78 | Reproductive | No evidence |
| | ADI | 0.005 | Oncogenicity | E (no evidence) |
| | | | Neurotoxicity | Not neurotoxic |
| Rotenone [261,459,567] | Rat/2-year | 7.5 | Mutagenicity | Not mutagenic |
| | Dog/52-week | 0.4 | Developmental | Not teratogenic |
| | Mouse/18-month | 10.7 | Reproductive | Not a reproductive oxin |
| | ADI | — | Oncogenicity | Not carcinogenic |

cell division [4]. Initial research conducted on *Escherichia coli* and *Salmonella typhimurium* and later confirmed in plants and yeasts indicate that the herbicidal activity is due to the inhibition of acetolactate synthase, an enzyme necessary for the biosynthesis of branched-chain amino acids in bacteria, fungi, and higher plants. A large number of sulfonylurea herbicides have been developed for commercial use in North America and Europe. The structure, use, and applications rates are provided in Table 16.45 and Table 16.46 for the most commonly used SUs.

Sulfonylurea herbicides generally are not acutely toxic or irritating to the skin and eye nor are they mutagenic, developmentally toxic, or oncogenic. Their hazard profiles are given in Table 16.47. Various target organs have been identified at high doses in chronic studies in rodents and dogs, including bone marrow, liver, kidney, testes, and the peripheral and central nervous systems. Tumor incidence was elevated above control levels in the liver (primisulfuron) at doses that exceeded the maximum tolerated dose. An earlier appearance of mammary tumors has also been observed in female Sprague–Dawley rats (prosulfuron, tribenuron). A unitary mode of action underlying the effects of this class of chemical on mammalian systems is not discernable. The diversity of the effects observed in various target organs is attributed to specific functional groups and not to the defining characteristic of the class, the sulfonylurea bridge.

An alternate mode of action for the SUs is derived from the fact that a sulfonylurea receptor protein in pancreatic β-cell plays an important role in glucose regulation. An ATP-sensitive potassium ion channel ($K_{ATP}$) has been identified in B-cells of the pancreas. The ultrastructure of the $K_{ATP}$ channel is unique among $K^+$ ion channels in that it is comprised of two proteins: a sulfonylurea receptor protein (SUR), which belongs to the family of ABC (ATP cassette) transporter proteins, and a smaller protein (Kir6.2), which belongs to a family of inward-rectifying potassium current proteins. Four Kir6.2 subunits are constitutively expressed with four SUR subunits to make up the selective $K^+$ pore [322]. $K_{ATP}$ channels containing the SUR1 isoform can be blocked by sulfonylureas [647] and opened with diazoxide. SUR1 is thus critically involved in regulation of $K_{ATP}$ channel activity. It is proposed that an elevation in blood glucose concentration leads to an

**TABLE 16.41**
**Hebicides Listed According to HRAC Classification System**

| Chemical Group | Common Name | HRAC MOA (Code) | Effect in Target Species | MOA Conserved in Mammals? | Other Common MOA in Mammals |
|---|---|---|---|---|---|
| Aryloxyphenoxy-propionates (AOPP) | Clodinafop propargyl<br>Diclofop-methyl<br>Fenoxaprop-p-ethyl<br>Fluazifop-p-butyl<br>Haloxyfop<br>Propaquizafop<br>(Table 16.42, Table 16.43) | Inhibits acetyl–CoA carboxylase (ACCase) (Code A) | Blocks membrane lipid formation by inhibiting the synthesis of fatty acids | Yes | Rodent peroxsomal proliferation; not likely relevant to humans |
| Cyclohexanediones (CHD) | Clethodim<br>sethoxydim<br>(Table 16.44) | Inhibits acetyl–CoA carboxylase (ACCase) (Code A) | Blocks membrane lipid formation by inhibiting the synthesis of fatty acids | Yes | None identified |
| Sulfonylureas (SU) | Bensulfuron-methyl<br>Chlorimuron-ethyl<br>Chlorsulfuron<br>Halosulfuron<br>Imazosulfuron<br>Metsulfuron-methyl<br>Nicosulfuron<br>Oxasulfuron<br>Primisulfuron-methyl<br>Prosulfuron<br>Rimsulfuron<br>Sulfometuron-methyl<br>Sulfosulfuron<br>Thifensulfuron-methyl<br>Triasulfuron<br>Tribenuron-methyl<br>(Table 16.45, Table 16.46, Table 16.47) | Acetolactate synthetase (ALS) or AHAS (acetohydroxyacid synthase) inhibition (Code B) | Blocks synthesis of branched-chain amino acids: leucine and valine (ALS inhibitors) and isoleucine (AHAS inhibitors) | No | Binding to sulfonylurea receptors in pancreatic cells results in release of insulin and decrease in glucose levels; SU herbicides do not appear to be effective (i.e., potent) SU receptor binders, as no interference with glucose regulation has been reported for this class of herbicides |
| Imidazolinones | Imazameth (Imazapic)<br>Imazamethabenz-methyl<br>Imazamox<br>Imazapyr<br>Imazaquin<br>Imazethapyr<br>(Table 16.48, Table 16.49) | ALS and AHAS inhibition (Code B) | Blocks branched-chain amino acid synthesis | No | None identified |

| Triazolopyrimidines | Flumetsulam (Table 16.50) | ALS and AHAS inhibition (Code B) | Blocks branched-chain amino acid synthesis | No | None identified |
|---|---|---|---|---|---|
| Pyrimidinylthio-benzoates | Pyriminobac-methyl (Table 16.50) | ALS and AHAS inhibition (Code B) | Blocks branched-chain amino acid synthesis | No | None identified |
| Triazines<br>Triazinones | Atrazine<br>Cyanazine<br>Propazine<br>Simazine<br>Ametryn<br>Prometryn<br>Prometon<br>Metribuzin<br>(Table 16.51, Table 16.52) | Inhibits photosynthesis at photosystem II (Code C1) | Binds to the D1 trans-membrane protein that binds plastoquinone; resistance develops in plants that acquire single-point mutations in the stomal D1 loop, usually a Ser264 to Gly mutation [213] | No | Chloro-s-triazines belong to a common mechanism class based on effects on the hypothalamic-pituitary gonadal axis [513] |
| Uracils<br>Pyridazinones | Bromacil<br>Terbacil<br>Norflurazon<br>(Table 16.53) | Inhibits photosynthesis at photosystem II (Code C1) | See Code C1 above | No | None known |
| Ureas | Diuron<br>Fluometuron<br>Linuron<br>(Table 16.54) | Inhibits photosynthesis at photosystem II (Code C2) | Target similar to that described for Code C1 | No | None known |
| Amide | Propanil | Inhibits photosynthesis at photosystem II (Code C2) | Target similar to that described for Code C1 | No | None known |
| Nitriles | Dichlobenil<br>Ioxynil<br>(Table 16.55) | Inhibits photosynthesis at photosystem II (Code C3) | Target similar to that described for Code C1 | | |
| Benzothiadiazinone<br>Phenylpyridazine | Bentazone<br>Pyridate<br>(Table 16.56) | Inhibits photosynthesis at photosystem II (Code C3) | Target similar to that described for Code C1 | No | None known |
| Bipyridylium | Diquat<br>Paraquat<br>(Table 16.57) | Inhibits photosynthesis at photosystem I (Code D) | Accepts electrons from one of the iron–sulfur carriers in photosystem I to form reactive supraoxide, peroxide and hydroxyl radials | No | Capacity to undergo redox recycling in specific tissues |
| Diphenylethers | Acifluorfen<br>Fomesafen<br>Lactofen<br>Oxyfluorfen<br>(Table 16.58) | Protoporphyrinogen oxidase (PPO) inhibitors (Code E) | Inhibits PPO, the final enzyme in the tetra-pyrrole biosynthetic pathway before it branches to chlorophyll or heme; protoporphyrin IX accumulates | Yes; effects on the heme biosynthetic pathway in mammals | Peroxisome proliferation and potential binding to PPARγ receptors [203,299] |

**TABLE 16.41 (cont.)**
**Hebicides Listed According to HRAC Classification System**

| Chemical Group | Common Name | HRAC MOA (Code) | Effect in Target Species | MOA Conserved in Mammals? | Other Common MOA in Mammals |
|---|---|---|---|---|---|
| *N*-Pyenylphlhalimide<br>Thiadiazole<br>Triazolinone | Flumiclorac-pentyl<br>Fluthiacet-methyl<br>Carfentrazone-ethyl<br>(Table 16.59) | Protoporphyrinogen oxidase (PPO) inhibitors (Code E) | Inhibits PPO, the final enzyme in the tetra-pyrrole biosynthetic pathway before it branches to chlorophyll or heme | Yes; effects on the heme biosynthetic pathway in mammals | None known |
| Oxadiazole<br>Pyrimidindione | Oxadiazon<br>Butafenacil<br>(Table 16.60) | Protoporphyrinogen oxidase (PPO) inhibitors (Code E) | Inhibits PPO, the final enzyme in the tetra-pyrrole biosynthetic pathway before it branches to chlorophyll or heme | Yes; effects on the heme biosynthetic pathway in mammals | None known |
| Pyridazinones | Norflurazon<br>Fluridone (Table 16.61) | Phytoene desaturase (PDS) inhibitors (Code F1) | Bleaching inhibition of carotenoid biosynthesis at the PDS step. | No | None known |
| Isoxazole<br>Triketone | Isoxaflutole<br>Sulcotrione<br>Mesotrione<br>(Table 16.62) | Inhibits 4-hydroxyl-phenypyruvate-dioxygenase (4-HPPD) (Code F2) | Bleaching inhibition of carotenoid biosynthesis at the HPPD step | Yes | HPDD inhibition leads to an increase in tyrosine in animal models for some members of this class |
| Triazole<br>Isoxazolidinone | Amitrole<br>Clomazone<br>(Table 16.63) | Target unknown (Code F3) | Bleaching inhibition of carotenoid biosynthesis; site of binding unknown | Unknown | The triazole, amitrole may have an effect on cholesterol synthesis |
| Glycine | Glyphosate<br>(Table 16.64) | Inhibits 5-enol-pyruvylshikimate-3-phosphate snythase (EPSP) (Code G) | Blocks synthesis of aromatic amino acids, phenylalanine, and tyrosine | No | None known |
| Phosphinic acid | Glufosinate ammonium<br>(Table 16.64] | Inhibits glutamine synthetase (Code H) | Blocks synthesis of glutamine | Yes | None known |
| Carbamate | Asulam (Table 16.64] | Inhibits dihydro-pteroate synthase (DHP) (Code I) | Blocks synsthesis of the vitamin folate | Yes | None known |
| Dinitroaniline | Benfluralin<br>Pendimethalin<br>Trifluralin<br>(Table 16.65) | Inhibits microtubule assembly (Code K1) | Binds to α- or β-tubulin monomers, thereby blocking tubulin polymerization | Yes | None known, but this class has a common structural homology to aniline |
| Chloroacetamide | Alachlor<br>Acetochlor<br>Metolachlor<br>Dimethenamid<br>(Table 16.66) | Inhibits very-long-chain fatty acid (VLCFA) synthesis in cell walls (Code K3) | Inhibits cell division by blocking the formation of VLCFAs | Unlikely | Alachlor and acetochlor share a common mechanism based on their ability to form a reactive quininone imine [230,245] |

| | | | | | |
|---|---|---|---|---|---|
| Benzamide | Isoxaben (Table 16.67) | Inhibits cellulose synthesis (Code L) | Blocks the creation of new plant cell walls | No | None known |
| Thiocarbamate | Butylate<br>Molinate<br>(Table 16.67) | Inhibits lipid synthesis (Code N) | May block the synthesis of long-chain fatty acids by inhibiting fatty acid synthetase complex and specific elongases | — | — |
| Phenoxycarboxylic acid<br>Benzoic acid<br>Pyridine carboxylic acid<br>Quinoline carboxylic acid | 2,4-Dichlorophenoxy acetic acid (2,4-D)<br>Dicamba<br>Picloram<br>Quinclorac<br>(Table 16.68) | Auxin (indole acetic acid) hormone mimic (Code O) | May bind to the auxin receptor, inducing ACC synthase activity, leading to cyanide accumulation and tissue necrosis [247] | Unlikely | None known |
| Pyrazolium<br>Organoarsenical | Difenzoquat<br>Monosodium methanearsonate (MSMA)<br>(Table 16.69) | Herbicides with an unknown mode of action (Code Z) | — | Unknown | None known |

*Source:* HRAC, *Classification of Herbicides According to Mode of Action*, Herbicide Resistance Action Committee, 2005 (http://www.plantprotection.org/hrac/Bindex.cfm?doc=moa2002.htm).

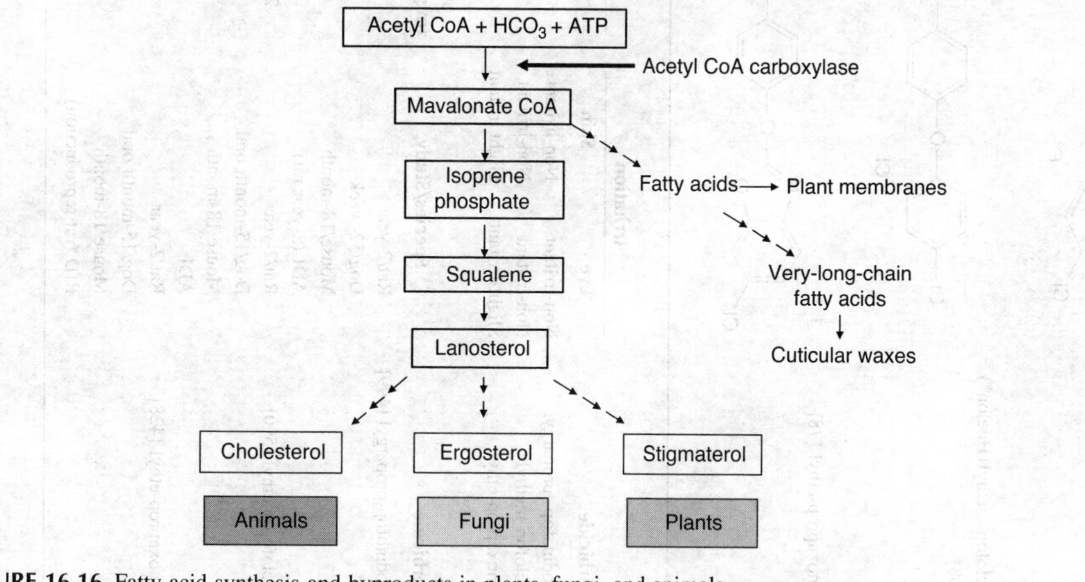


**FIGURE 16.16** Fatty acid synthesis and byproducts in plants, fungi, and animals.

**TABLE 16.42**
**Structure, Use, and Hazard Profiles of Acetyl–CoA Carboxylase (ACCase)-Inhibiting Aryloxyphenoxypropionate (AOPP) Herbicides (HRAC Code A)**

| Herbicide | Structure | Principal Use/Crop | Application Rate (g active ingredient per hectare) |
|---|---|---|---|
| Clodinafop-propargyl (Discover®) [48] | 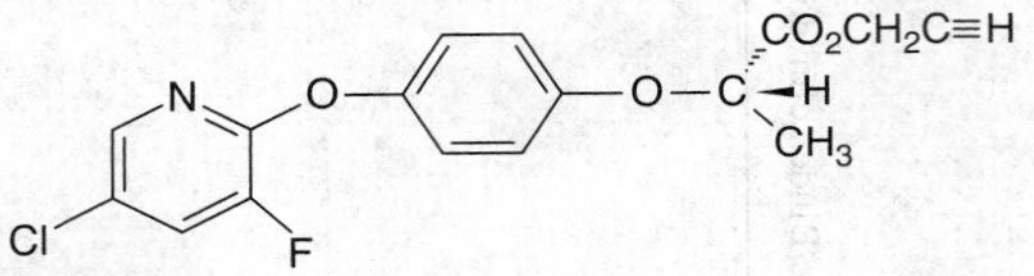 | Cereals | 30–60 |
| Diclofop-methyl (Hoelon®) [68] | 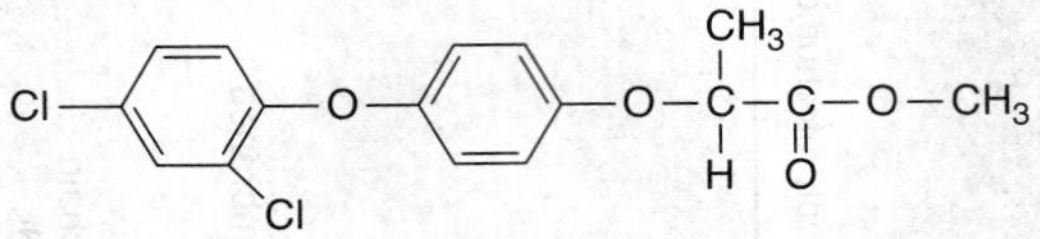 | Cereals | 840–1680 |
| Fenoxaprop-ethyl [78] | 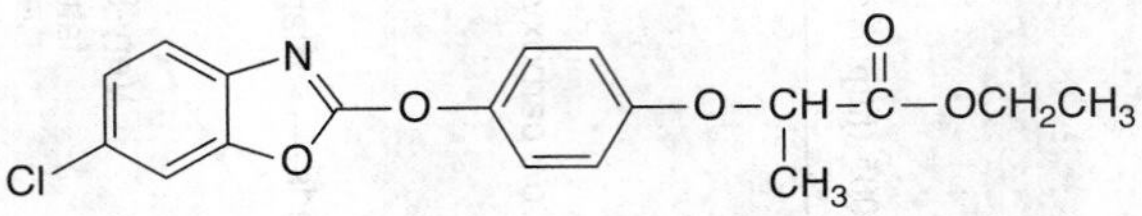 | Cereals, soybeans, and turf | 37.5–111 |

| Herbicide | Irritation | | $LD_{50}$ (mg/kg) | | $LC_{50}$ (mg/L) | Sensitization Potential | Signal Word |
|---|---|---|---|---|---|---|---|
| | Eye | Skin | Oral | Dermal | Inhalation | | |
| Clodinafop- propargyl | Non-irritant | Non-irritant | 1829 | >2000 | 2.325 | Positive | Caution |
| Diclofop-methyl | Non-irritant | Non-irritant | 2020 | >5000 | >3.83 | NA | Caution |
| Fenoxaprop-ethyl | Slight irritant | Slight irritant | 2565 | >2000 | >0.511 | Negative | Caution |

| Herbicide | Species/Study | NOEL (mg/kg/day) | Toxicity Study | Hazard Indicator |
|---|---|---|---|---|
| Clodinofop-propargyl [499] | Rat/2-year | 0.35 | Mutagenicity | Not a mutagen |
| | Dog/52-week | 3.3 | Developmental | Not a developmental toxin |
| | Mouse/18-month | 1.2 | Reproductive | Not a reproductive toxin |
| | ADI (2-year rat) | 0.004 | Oncogenicity | Peroxisomal proliferator (mouse liver tumors |
| Diclofop-methyl [590] | Rat/2-year | 20 | Mutagenicity | Not a mutagen |
| | Dog/15-month oral | 8.0 | Developmental | Not a developmental toxin |
| | Mouse/18-month | NA | Reproductive | Not a reproductive toxin |
| | ADI | 0.001 | Oncogenicity | NA |
| Fenoxaprop-ethyl [428] | Rat/2-year | 1.5 | Mutagenicity | Not a mutagen |
| | Dog/15-month oral | 0.375 | Developmental | Not a developmental toxin |
| | Mouse/18-month | 5.7 | Reproductive | Not a reproductive toxin |
| | RfD (rat reproduction) | 0.0025 | Oncogenicity | C (pending; adrenal tumors) |

increased rate of glucose metabolism in pancreatic β-cells and the consequent alteration in the intracellular ratio of ATP/ADP, resulting in inhibition of $K_{ATP}$ channels. The subsequent depolarization of the β-cell plasma membrane activates voltage-sensitive $Ca^{2+}$ channels and the ensuing influx of $Ca^{2+}$ initiates insulin secretion [333]. It appears that the sulfonylurea herbicides have a rather low affinity to SUR because there is no evidence of an effect on glucose regulation in the animal studies.

## Imidazolinones (HRAC Code B)

The structure, use, and hazard profiles for the imidazolinones imazameth and imazamethabenzmethyl are pro

**TABLE 16.43**
**Structures, Use, and Hazard Profiles of Acetyl–CoA Carboxylase Inhibiting Aryloxyphenoxypropionate (AOPP) Herbicides (HRAC Code A)**

| Herbicide | Structure | Principal Use/Crop | Application Rate (g active ingredient per hectare) |
|---|---|---|---|
| Fluazifop-p-butyl (Fusilade®) [84] | 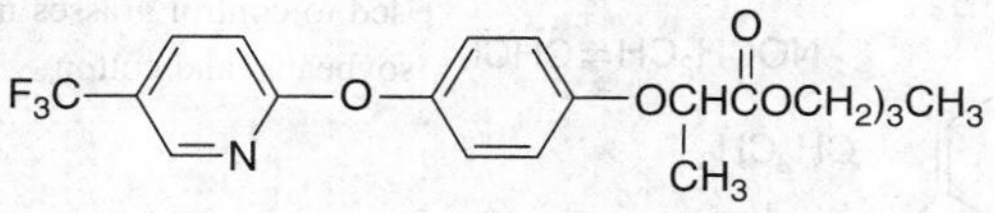 $F_3C$–(pyridyl, N)–O–(phenyl)–$OCH(CH_3)C(=O)O(CH_2)_3CH_3$ | Cotton, fruit, and soybeans | 125–375 |
| Haloxyfop (Galant™) [96] | 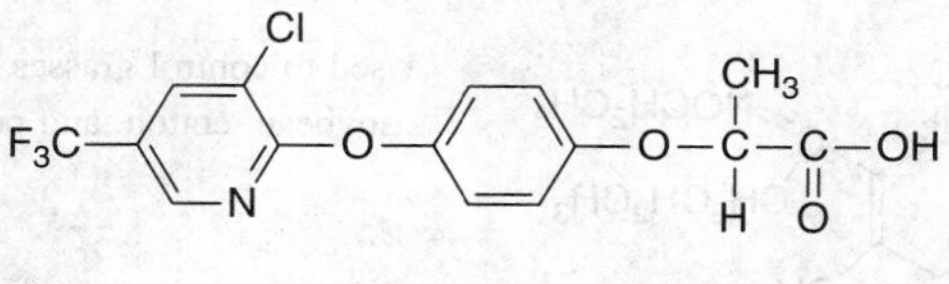 $F_3C$–(3-Cl-pyridyl, N)–O–(phenyl)–O–C(H)($CH_3$)–C(=O)–OH | Cotton, soybeans, sunflowers, oilseed rape | 104–208 |
| Propaquizafop (AGIL®) [153] | 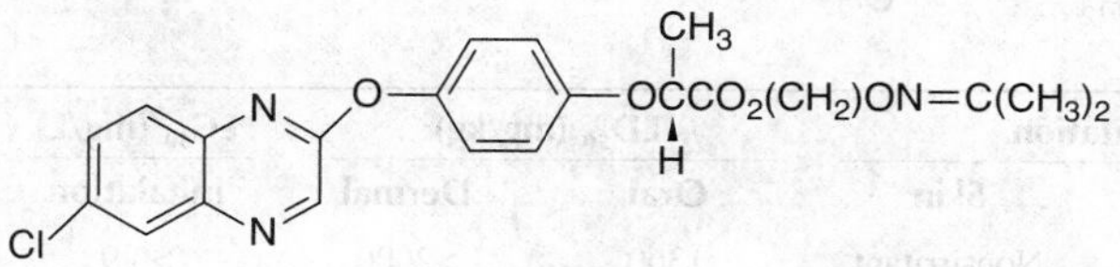 Cl–(quinoxalinyl, N, N)–O–(phenyl)–$OC(H)(CH_3)CO_2(CH_2)ON{=}C(CH_3)_2$ | Soybeans, cotton, sunflower, sugar beets, potatoes, oilseed rape, vegetables, peanuts, tobacco | 60–200 |

| Herbicide | Irritation: Eye | Irritation: Skin | $LD_{50}$ (mg/kg): Oral | $LD_{50}$ (mg/kg): Dermal | $LC_{50}$ (mg/L): Inhalation | Sensitization Potential | Signal Word |
|---|---|---|---|---|---|---|---|
| Fluazifop-p-butyl | Mild irritant | Slight irritant | 4096 | >2420 | >5.24 | Negative | Caution |
| Haloxyfop | Moderate irritant | Non-irritant | 518 | >5000 | NA | Negative | Caution |
| Propaquizafop | Moderate irritant | Non-irritant | >5000 | >2000 | 2.5 | Possibly positive | Caution |

| Herbicide | Species/Study | NOEL (mg/kg/day) | Toxicity Study | Hazard Indicator |
|---|---|---|---|---|
| Fluazifop-p-butyl [542] | Rat/2-year | 0.5 | Mutagenicity | Negative |
| | Dog/52-week | 5.0 | Developmental | Delayed skeletal ossification |
| | Hamster/18-month | 12.1 | Reproductive | Not a reproductive toxin |
| | RfD | 0.0074 | Oncogenicity | Not likely |
| Haloxyfop [195,600] | Rat/2-year | 0.065 | Mutagenicity | No evidence |
| | Dog/52-week | 0.5 | Developmental | Not a developmental toxin |
| | Mouse/18-month | 0.6 | Reproductive | No evidence |
| | ADI | 0.0003 | Oncogenicity | Peroxisomal proliferator |
| Propaquizafop [153] | Rat/2-year | 1.5 | Mutagenicity | — |
| | Dog/52-week | 20 | Developmental | — |
| | Mouse/18-month | 1.5 | Reproductive | — |
| | ADI | 0.015 | Oncogenicity | — |

vided in Table 16.48. These ALS inhibitors are relatively nontoxic, even at high doses, with no evidence of mutagenic, developmental, or oncogenic effects.

## TRIAZOLOPYRIMIDINES (HRAC CODE B)

The triazolopyrimidine ALS inhibitors include imazamox, mazapyr, imazaquin, and imazethapyr. As with the imidazolinones, these chemicals also have excellent hazard profiles (Table 16.49). No evidence of significant target organ toxicity, mutagenic, developmental, or oncogenic potential has been realized even at doses that approximate the limit dose of 1000 mg/kg/day.

## PYRIMIDINYLTHIOBENZOATES (HRAC CODE B)

Members of the pyrimidinylthiobenzoate class of ALS-inhibiting herbicides, flumetsulam and pyriminobac-methyl (Table 16.50) are slightly less well tolerated in mammalian systems than other ALS inhibitors as evidence

**TABLE 16.44**
**Structure, Use, and Hazard Profiles of Acetyl–CoA Carboxylase (ACCase)-Inhibiting Cyclohexanedione (CHD) Herbicides (HRAC Code A)**

| Herbicide | Structure | Principal Use/Crop | Application Rate (g active ingredient per hectare) |
|---|---|---|---|
| Clethodim (Select®) [47] | 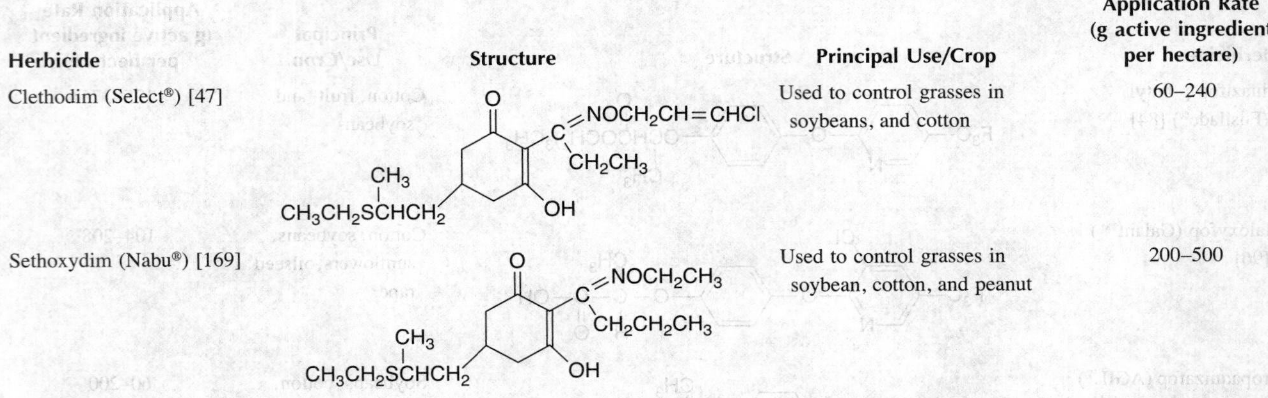 | Used to control grasses in soybeans, and cotton | 60–240 |
| Sethoxydim (Nabu®) [169] | | Used to control grasses in soybean, cotton, and peanut | 200–500 |

| Herbicide | Irritation: Eye | Irritation: Skin | $LD_{50}$ (mg/kg): Oral | $LD_{50}$ (mg/kg): Dermal | $LC_{50}$ (mg/L): Inhalation | Sensitization Potential | Signal Word |
|---|---|---|---|---|---|---|---|
| Clethodim | NA | Non-irritant | 1360 | >2000 | >3.9 | Negative | Caution |
| Sethoxydim | Non-irritant | Non-irritant | 2676 | >5000 | 6.1 | Negative | Caution |

| Herbicide | Species/Study | NOEL (mg/kg/day) | Toxicity Study | Hazard Indicator |
|---|---|---|---|---|
| Clethodim [420,587] | Rat/2-year | 19 | Mutagenicity | Not a mutagen |
| | Dog/52-week | 1 | Developmental | Not a developmental toxin |
| | Mouse/18-month | 28 | Reproductive | Not a reproductive toxin |
| | ADI | 0.01 | Oncogenicity | No evidence |
| Sethoxydim [195–197,550,620] | Rat/2-year | 17.2 | Mutagenicity | Not a mutagen |
| | Dog/52-week | 8.9 | Developmental | Not a developmental toxin |
| | Mouse/18-month | 14 | Reproductive | Not a reproductive toxin |
| | ADI | 0.14 | Oncogenicity | No evidence |

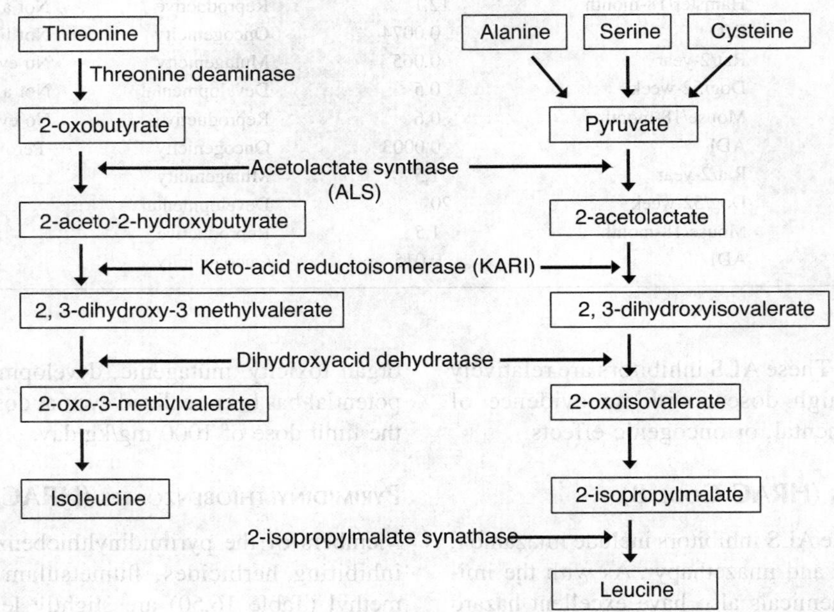


**FIGURE 16.17** Branched chain amino acid synthesis acetolactate synthase (ALS).

**TABLE 16.45**
**Structure and Use Profiles of the Acetolactate Synthase (ALS)- and Acetohyroxyacid Synthase (AHAS)-Inhibiting Sulfonylurea Herbicides (HRAC Code B)**

| Chemical/Common Name | Structure | Principal Use/Crop | Application Rate (g active ingredient per hectare) |
|---|---|---|---|
| Bensulfuron-methyl (Londax®) [29] | $CO_2CH_3$; $CH_2SO_2NHCONH$; $OCH_3$; N; N; $OCH_3$ | Rice | 46–60 |
| Chlorimuron-ethyl (Classic®) [43] | $CO_2CH_2CH_3$; $SO_2NHCONH$; Cl; N; N; $OCH_3$ | Soybeans, peanuts | 9–13 |
| Chlorsulfuron (Glean®) [46] | $SO_2NHCONH$; Cl; $OCH_3$; N; N; N; $CH_3$ | Cereals, IWC | 9–140 |
| Halosulfuron-methyl (Permit®) [95] | $COOCH_3$; Cl; N; N; $CH_3$; $SO_2NHCONH$; $OCH_3$; N; N; $OCH_3$ | Cereals, corn sorghum, turf | 18–35 |
| Imazosulfuron (Sibatito®, Takeoff®) [109] | N; N; Cl; $SO_2NHCONH$; $CH_3$; N; N; $CH_3$ | Cereals, rice, turf | 75–1000 |
| Metsulfuron-methyl (Ally®, Escort®) [130] | $CO_2CH_3$; $SO_2NHCONH$; Cl; $OCH_3$; N; N; N; $CH_3$ | Cereals | 4–7.5 |
| Nicosulfuron (Accent®) [135] | N; $SO_2NHCONH$; $CON(CH_3)_2$; $OCH_3$; N; N; $OCH_3$ | Corn | 35–70 |

**TABLE 16.46**
**Structure and Use Profiles of the Acetolactate Synthase (ALS)- and Acetohyroxyacid Synthase (AHAS)-Inhibiting Sulfonylurea Herbicides (HRAC Code B)**

| Chemical/Common Name | Structure | Principal Use/Crop | Application Rate (g active ingredient per hectare) |
|---|---|---|---|
| Oxasulfuron (Expert®) [140] | 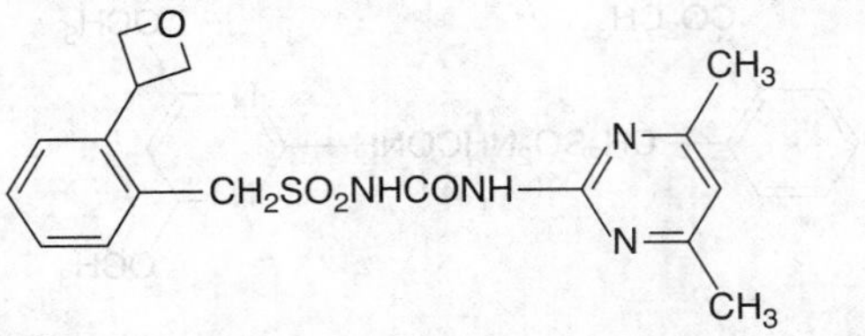 O; $CH_3$; N; $CH_2SO_2NHCONH$; N; $CH_3$ | Soybeans | 60–90 |
| Primisulfuron-methyl (Beacon®) [148] | 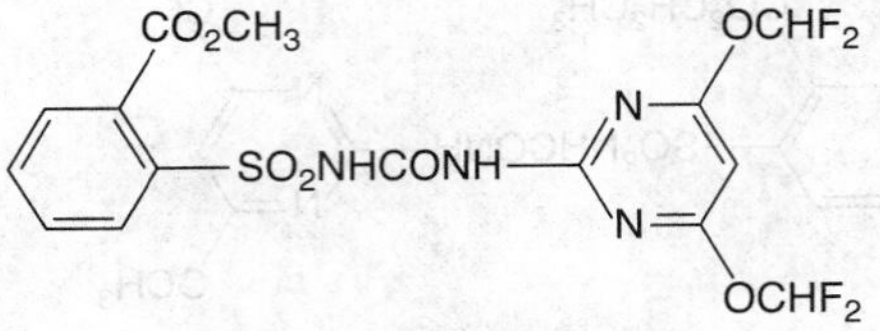 $CO_2CH_3$; $OCHF_2$; N; $SO_2NHCONH$; N; $OCHF_2$ | Corn | 20–40 |
| Prosulfuron (Peak®) [157] | 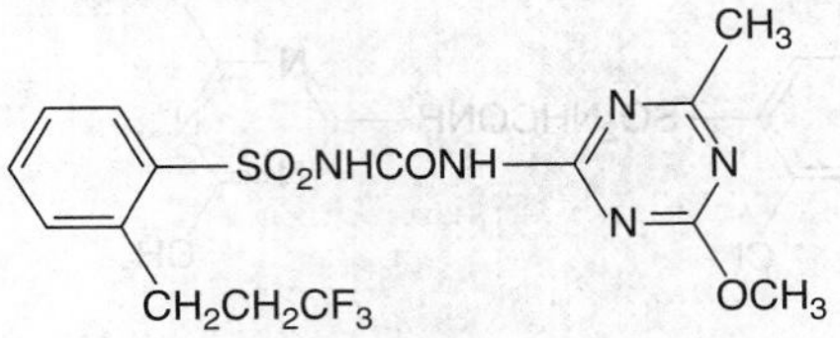 $CH_3$; N; $SO_2NHCONH$; N; N; $CH_2CH_2CF_3$; $OCH_3$ | Cereals, corn, sorghum, pasture | 10–40 |
| Rimsulfuron (Matrix®) [167] | 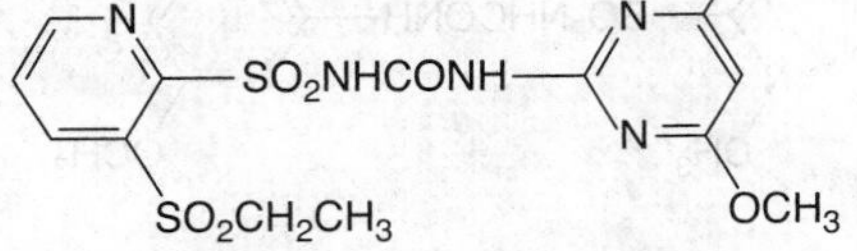 $OCH_3$; N; N; $SO_2NHCONH$; N; $SO_2CH_2CH_3$; $OCH_3$ | Corn, tomatoes, and potatoes | 15 |
| Sulfometuron-methyl (Oust®) [173] | 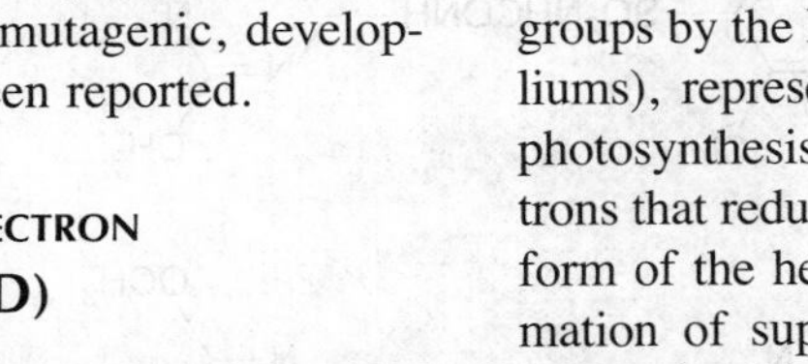 $CO_2CH_3$; $CH_3$; N; $SO_2NHCONH$; N; $CH_3$ | IWC | 26–420 |

by lower NOELs; however, the hazard profiles for these chemicals are still favorable, as no mutagenic, developmental, or oncogenic effects have been reported.

## Inhibition of Photosynthetic Electron Transport (HRAC Code C and D)

Photosynthesis is a process unique to plants whereby light energy captured by chlorophyll is converted to electrochemical energy through an electron transport chain to produce NADPH (photosystem I) or ATP (photosystem II). Herbicides that interfere with electron transport in the photosynthetic pathways have been grouped into tw groups by the HRAC [253]. Group D herbicides (bipyridy liums), represented here by paraquat and diaquat, bloc photosynthesis at the photosystem I stage by capturing elec trons that reduce the herbicide (Figure 16.18). The reduce form of the herbicide is then oxidized, leading to the for mation of supraoxides and hydrogen peroxide and ulti mately hydroxyl radicals, which damage cellular compo nents, thus affecting unsaturated membrane lipids an resulting in fatty acid peroxidation, loss of membrane semi permeability, desiccation, and cell death [355]. Group C (triazines, triazolinones, uracils, pyridazones, and phenyl

**TABLE 16.46 (cont.)**
**Structure and Use Profiles of the Acetolactate Synthase (ALS)- and Acetohyroxyacid Synthase (AHAS)-Inhibiting Sulfonylurea Herbicides (HRAC Code B)**

| Chemical/Common Name | Structure | Principal Use/Crop | Application Rate (g active ingredient per hectare) |
|---|---|---|---|
| Sulfosulfuron) [174] | $SO_2CH_2CH_3$; $SO_2NHCONH$; $OCH_3$; N; $OCH_3$ | Cereal (wheat), IWC | 10–35 |
| Thifensulfuron-methyl (Pinnacle®, Harmony®) [183] | $CO_2CH_3$; S; $SO_2NHCONH$; $CH_3$; $OCH_3$; N; $CH_3$ | Cereals, corn, soybean, pastures | 9–60 |
| Triasulfuron (Amber®, Logran®) [188] | $SO_2NHCONH$; $OCH_2CH_2Cl$; $OCH_3$; N; $CH_3$ | Cereal (wheat), IWC | 5–10 |
| Tribenuron-methyl (Express®) [189] | $CO_2CH_3$; $SO_2NHCONH$; $CH_3$; $OCH_3$; N; $CH_3$ | Cereal (wheat) | 7.5–30 |

carbamates), C2 (ureas and amides), and C3 (nitriles, benzolthiadiazinone, and phenylpyridazines) herbicides all affect photosystem II (Figure 16.18). When electron transport is interrupted by group C herbicides, and light continues to fall on the chloroplast, the energy level of chlorophyll is raised from a singlet to a triplet state, which itself damages cell membrane lipids or creates reactive oxygen species that interact with cellular lipids, proteins, and nucleic acids [208,573].

Aside from creating reactive oxygen species, photosynthesis inhibitors also block food-producing processes in susceptible plants by limiting the availability of NADPH and ATP to enter into the so-called dark reaction (Calvin cycle) where $CO_2$ is fixed and carbohydrates are produced. The reduction in carbohydrate synthesis may result in a slow starvation of the plant [655]. Signs of injury include yellowing (chlorosis) of leaf tissue followed by death (necrosis) of the tissue. Pre-emergent- or early post-emergent-applied herbicides such as the triazines are taken up into the plant via the roots or foliage and move in the xylem to the plant leaves. As a result, signs of injury first appear on older leaves or along leaf margins. Foliar-applied photosynthetic inhibitors generally remain in the foliar portion of the treated plant, with movement from foliage to roots being negligible [248].

### Triazines and Triazinone (HRAC Code C1)

The s-triazines, phenylureas, and uracil herbicides all inhibit photosynthetic electron transport (Hill reaction) in photosystem II [662] by binding to the D1 protein [213] and blocking the mobile electron carrier plastoquinone [208]. The most common mechanism of resistance to s-triazines is a mutation of the psbA gene, which encodes the D1 protein, whereby glycine is substituted

**TABLE 16.47A**
**Hazard Profiles of the Acetolactate Synthase (ALS)- and Acetohyroxyacid Synthase (AHAS)-Inhibiting Sulfonylurea Herbicides (HRAC Code B)**

| Herbicide | Irritation | | $LD_{50}$ (mg/kg) | | $LC_{50}$ (mg/L) | Sensitization | Signal |
|---|---|---|---|---|---|---|---|
| | Eye | Skin | Oral | Dermal | Inhalation | Potential | Word |
| Bensulfuron | Non-irritant | Non-irritant | >5000 | >2000 | >7.5 | NA | Caution |
| Chlorimuron | Non-irritant | Non-irritant | 4102 | >2000 | >5.0 | Negative | Caution |
| Chlorsulfuron | Slight irritant | Non-irritant | 5545 (male) | 2500 | >5.9 | Negative | Caution |
| Halosulfuron | NA | NA | 8866 | >2000 | NA | NA | Caution |
| Imazosulfuron | Non-irritant | Non-irritant | >5000 | >2000 | >2.4 | Negative | Caution |
| Metsulfuron | Mod. irritant | Mild irritant | >5000 | >2000 | >5.0 | Negative | Caution |

| Herbicide | Species/Study | NOEL (mg/kg/day) | Toxicity Study | Hazard Indicator |
|---|---|---|---|---|
| Bensulfuron [581] | Rat/2-year | 37.5 | Mutagenicity | No evidence |
| | Dog/52-week | 227 | Developmental | Not teratogenic |
| | Mouse/18-month | 455 | Reproductive | Not reproductive toxin |
| | ADI2 | 0.2 | Oncogenicity | No evidence |
| Chlorimuron [528,585] | Rat/2-year | 12.5 | Mutagenicity | No evidence |
| | Dog/52-week | 6.25 | Developmental | Not teratogenic |
| | Mouse/18-month | 180 | Reproductive | No evidence |
| | ADI2 | 0.02 | Oncogenicity | No evidence |
| Chlorsulfuron [510,538,586] | Rat/2-year | 5 | Mutagenicity | No evidence |
| | Dog/52-week | 50 | Developmental | Not teratogenic |
| | Mouse/18-month | 71 | Reproductive | No evidence |
| | RfD 2 | 0.05 | Oncogenicity | No evidence |
| Halosulfuron [464] | Rat/2-year | 50 | Mutagenicity | No evidence |
| | Dog/52-week | 10 | Developmental | Not teratogenic |
| | Mouse/18-month | 430 | Reproductive | No evidence |
| | ADI | 0.1 | Oncogenicity | No evidence |
| Imazosulfuron [109] | Rat/2-year | 106 | Mutagenicity | No evidence |
| | Dog/52-week | 75 | Developmental | Teratogenic in mice |
| | Mouse/18-month | NA | Reproductive | No evidence |
| | RfD or ADI | NA | Oncogenicity | No evidence |
| Metsulfuron-methyl [609] | Rat/2-year | 25 | Mutagenicity | No evidence |
| | Dog/52-week | 12.5 | Developmental | Not teratogenic |
| | Mouse/18-month | 710 | Reproductive | No evidence |
| | ADI (Germany) | 0.0125 | Oncogenicity | No evidence |

for serine at amino acid 264 in the stromal loop of the D1 protein [213]. While the molecular targets relating to inhibition of the Hill reaction do not exist in mammalian systems, a common mode of action of toxicity for the chloro-*s*-triazines has been defined based on effects of the hypothalamic–pituitary–gonadal axis [513]. The specific molecular target underlying this mode of action has not been identified; however, it is likely to relate to the formation of a good leaving group by the chlorine atom, as indicated by reactivity with glutathione to form glutathione conjugates as part of the detoxification pathway or reactions with sulfhydral groups to form adducts to proteins [250,316]. The structure and use of the symmetrical triazines and the asymmetrical triazine or triazinone metribuzin are presented in Table 16.51, and the hazard profiles are giving in Table 16.52.

The triazines are generally not acutely toxic. The symmetrical chloro-*s*-triazines—atrazine, cyanazine, propazine, and simazine—induce an earlier onset or an increase in the incidence of mammary tumors in lifetime feeding studies in Sprague–Dawley female rats [363,381]. The mode of action underlying the occurrence of these tumors in female Sprague–Dawley rats has been described and is not considered relevant to humans [554]. The IARC has classified atrazine and simazine as not classifiable as to carcinogenicity to humans [259]. The EPA has also classified atrazine and simazine as not likely to be carcinogenic to humans [554,568].

**TABLE 16.47B**
**Hazard Profiles of the Acetolactate Synthase (ALS)- and Acetohyroxyacid Synthase (AHAS)-Inhibiting Sulfonylurea Herbicides (HRAC Code B)**

| Herbicide | Irritation: Eye | Irritation: Skin | $LD_{50}$ (mg/kg): Oral | $LD_{50}$ (mg/kg): Dermal | $LC_{50}$ (mg/L): Inhalation | Sensitization Potential | Signal Word |
|---|---|---|---|---|---|---|---|
| Nicosulfuron | Mod. irritant | NA | >5000 | >2000 | 5.47 | Negative | Caution |
| Oxasulfuron | Non-irritant | Non-irritant | >5000 | >2000 | 5.08 | Negative | Caution |
| Primisulfuron | Slight irritant | Non-irritant | >5050 | >2010 | >4.8 | Negative | Caution |
| Prosulfuron | Non-irritant | Non-irritant | 986 | >2000 | >5.0 | Negative | Caution |
| Rimsulfuron | Mod. irritant | Non-irritant | >5000 | >2000 | >5.4 | Negative | Caution |
| Sulfometuron | Slight irritant | Slight irritant | >5000 | >2000 | >11 | Negative | Caution |

| Herbicide | Species/Study | NOEL (mg/kg/day) | Toxicity Study | Hazard Indicator |
|---|---|---|---|---|
| Nicosulfuron [532,612] | Rat/2-year | 1000 | Mutagenicity | No evidence |
| | Dog/52-week | 125 | Developmental | Not teratogenic |
| | Mouse/18-month | 1070 | Reproductive | No evidence |
| | ADI | 1.25 | Oncogenicity | No evidence |
| Oxasulfuron [140] | Rat/2-year | 8.3 | Mutagenicity | No evidence |
| | Dog/52-week | 1.3 | Developmental | Not teratogenic |
| | Mouse/18-month | 1.5 | Reproductive | No evidence |
| | ADI | 0.0026 | Oncogenicity | No evidence |
| Primisulfuron [616] | Rat/2-year | 13 | Mutagenicity | No evidence |
| | Dog/52-week | 25 | Developmental | Not teratogenic |
| | Mouse/18-month | 45 | Reproductive | Testicular degeneration |
| | ADI | 0.13 | Oncogenicity | D (liver tumor in male mice at doses > MTD) |
| Prosulfuron [157] | Rat/2-year | 8.6 | Mutagenicity | No evidence |
| | Dog/52-week | 1.9 | Developmental | Not teratogenic |
| | Mouse/18-month | 80 | Reproductive | No evidence |
| | ADI | 0.019 | Oncogenicity | D (mammary tumors in male rats; early onset) |
| Rimsulfuron [495] | Rat/2-year | 11.8 | Mutagenicity | No evidence |
| | Dog/52-week | 1.6 | Developmental | Not teratogenic |
| | Mouse/18-month | 351 | Reproductive | No evidence |
| | RfD | 0.016 | Oncogenicity | No evidence |
| Sulfometuron [622] | Rat/2-year | 2.5 | Mutagenicity | No evidence |
| | Dog/52-week | 5.0 | Developmental | Teratogenic in two species |
| | Mouse/18-month | 140 | Reproductive | No evidence |
| | ADI | 0.025 | Oncogenicity | No evidence |

## Uracils and Pyridazinones (HRAC Code C1)

The structure, use, and hazard profiles for two uracils (bromacil and terbacil) and the pyridazinone herbicide norflurazon are given in Table 16.53. The acute toxicity of bromacil, terbacil, and norflurazon is unremarkable. These herbicides are not mutagenic, teratogenic, or reproductive toxins; however, bromacil and norflurazon have been classified as Category C (possible human) carcinogens based on mouse liver tumors.

## Ureas (HRAC Code C2)

The structure, use, and toxicity for the urea class of photosynthesis-inhibiting herbicides diuron, fluometuron, and linuron are provided in Table 16.54. Diuron, linuron, and fluometuron have limited acute toxicity. Fluometuron caused hemosiderosis in the spleen in repeat-dose studies. Diuron and linuron have been classified by the EPA as either known or likely human carcinogens as defined in the EPA 1996 classification scheme [544] or as Category C (possible human) carcinogens based on an earlier scheme [410]. The cancer classification of fluometuron is pending; a slight elevation in the incidence of lymphoma was noted in the rat study on fluometuron.

## Nitriles (HRAC Code C3)

The structure and use of the nitriles dichlobenil, and ioxynil are presented along with their hazard profiles in Table 16.55. Ioxynil is minimally toxic. Dichlobenil was classified as a Category C carcinogen based on mouse liver tumors.

**TABLE 16.47C**
**Hazard Profiles of the Acetolactate Synthase (ALS)- and Acetohyroxyacid Synthase (AHAS)-Inhibiting Sulfonylurea Herbicides (HRAC Code B)**

| Herbicide | Irritation: Eye | Irritation: Skin | $LD_{50}$ (mg/kg): Oral | $LD_{50}$ (mg/kg): Dermal | $LC_{50}$ (mg/L): Inhalation | Sensitization Potential | Signal Word |
|---|---|---|---|---|---|---|---|
| Sulfosulfuron | Non-irritant | Slight irritant | >5000 | >5000 | NA | Negative | Caution |
| Thifensulfuron | Slight irritant | Non-irritant | >5000 | >2000 | >7.9 | Negative | Caution |
| Triasulfuron | Slight irritant | Non-irritant | >5000 | >2000 | >5.1 | Negative | Caution |
| Tribenuron | Slight irritant | Non-irritant | >5000 | >2000 | >5.0 | Positive | Caution |

| Herbicide | Species/Study | NOEL (mg/kg/day) | Toxicity Study | Hazard Indicator |
|---|---|---|---|---|
| Sulfosulfuron [496] | Rat/2-year | 24.4 | Mutagenicity | Not mutagenic |
| | Dog/52-week | 100 | Developmental | Not teratogenic |
| | Mouse/18-month | 93.4 | Reproductive | Not a developmental toxin |
| | ADI | 0.24 | Oncogenicity | Likely human carcinogen |
| | | | Neurotoxicity | Not neurotoxic |
| Thifensulfuron [624] | Rat/2-year | 2.6 | Mutagenicity | No evidence |
| | Dog/52-week | 19 | Developmental | Not teratogenic |
| | Mouse/18-month | 1070 | Reproductive | No evidence |
| | ADI | 0.026 | Oncogenicity | No evidence |
| Triasulfuron [625] | Rat/2-year | 32.1 | Mutagenicity | No evidence |
| | Dog/52-week | 33 | Developmental | Not teratogenic |
| | Mouse/18-month | 1.2 | Reproductive | No evidence |
| | ADI | 0.012 | Oncogenicity | No evidence |
| Tribenuron [626] | Rat/2-year | 1.25 | Mutagenicity | No evidence |
| | Dog/52-week | 8.2 | Developmental | Not teratogenic |
| | Mouse/18-month | 30 | Reproductive | No evidence |
| | ADI | 0.011 | Oncogenicity | C (Mammary tumors in female rats; early onset) |

## Benzothiadiazoles and Phenylpyridazine (HRAC Code C3)

These two classes of herbicides are represented by bentazone and pyridate, respectively. Bentazone and pyridate are not toxic in either acute- or repeat-dose studies. (See Table 16.56.)

## Bipyridyliums (HRAC Code D)

Diquat and paraquat, which are bipyridylium photosynthesis inhibitors, are unlike the HRAC Code C1–C3 herbicides in that the bipyridyliums inhibit electron flow in photosystem I. The structure, use, and hazard profiles for diquat and paraquat are provided in Table 16.57. Diquat is less acutely toxic than paraquat, but they both are moderately toxic in long-term studies. These bipyridyliums are not mutagenic, teratogenic, or carcinogenic, nor are they reproductive toxicants; however, both diquat and paraquat are capable of undergoing redox recyling as they are reduced by electron donors, and they then undergo oxidization as they react with oxygen to form reactive oxygen species [355,356]. This redox potential is believed to account for tissue damage seen in the lung of animals treated with paraquat [442] and cataractogenic effects in diquat-treated animals at low to moderate doses [387]. The EPA concluded that, although both diquat and paraquat are capable of generating oxygen free radicals, their effects are unlikely to be additive because of differences in tissue distribution and hence target organ selectivity [442].

## Protoporphyrinogen Oxidase Inhibitors (HRAC Code E)

Protoporphyrinogen-oxidase inhibitors block the biosynthesis of chlorophyll by inhibiting protoporphyrinogen oxidase (PPO) found in chloroplasts and mitochondria in plants. A similar action in animals interferes with the biosynthesis of heme and cytochrome P450 enzymes (Figure 16.19). It is not uncommon to find evidence of anemia in rodents, especially rats, exposed to PPO-inhibiting herbicides. In addition to effects on heme synthesis, it is theorized that light- and oxygen-dependent peroxidation of cell membrane lipids may lead to cell lysis and death, particularly in organs where protoporphyrin IX forms or bioconcentrates as a result of PPO inhibition [207]. Such a

**TABLE 16.48**
**Structure, Use, and Hazard Profiles of the Acetolactate Synthase (ALS)-Inhibiting Imidazolinone Herbicides (HRAC Code B)**

| Chemical/Common Name | Structure | Principal Use/Crop | Application Rate (g active ingredient per hectare) |
|---|---|---|---|
| Imazameth (Cadre®) [105] | (structure: $CH_3$, COOH, CH($CH_3$)$_2$, $CH_3$, N, O) | Soybeans, peanuts, sugarcane | 70–110 |
| Imazamethabenz-methyl (Assert®) [103] | (structure: $CH_3$, $COOCH_3$, CH($CH_3$)$_2$, $CH_3$, N, O) | Wheat, barley, sunflower | 250–700 post |

| Herbicide | Irritation: Eye | Irritation: Skin | $LD_{50}$ (mg/kg): Oral | $LD_{50}$ (mg/kg): Dermal | $LC_{50}$ (mg/L): Inhalation | Sensitization Potential | Signal Word |
|---|---|---|---|---|---|---|---|
| Imazameth | NA | Non-irritant | >5000 | >5000 | 2.38 | NA | Caution |
| Imazamethabenz-methyl | Slight irritant | Non-irritant | >5000 | >2000 | >5.8 | Negative | Caution |

| Herbicide | Species/Study | NOEL (mg/kg/day) | Toxicity Study | Hazard Indicator |
|---|---|---|---|---|
| Imazameth (imazapic) [403] | Rat/2-year | 1029 | Mutagenicity | No evidence |
| | Dog/52-week | <137 | Developmental | Not teratogenic |
| | Mouse/18-month | 1134 | Reproductive | No evidence |
| | RfD (UF = 300) | 0.5 | Oncogenicity | No evidence |
| Imazamethabenz-methyl [531,601] | Rat/2-year | 12.5 | Mutagenicity | No evidence |
| | Dog/52-week | 6.25 | Developmental | Not teratogenic |
| | Mouse/18-month | 19.5 | Reproductive | No evidence |
| | ADI | 0.06 | Oncogenicity | No evidence |

hypothesis is consistent with experimental observations that liver damage and liver tumor formation, particularly in mice, often result from high-dose exposure to PPO inhibitors. It has also been postulated that liver damage and the subsequent tumor response seen in animals following high-dose exposure to PPO inhibitors may result from peroxisome proliferative effects of these herbicides. Furthermore, it has been postulated that there may a linkage between peroxisome proliferation and binding to PPARγ receptors, discussed in more detail below [203,299].

## Diphenyl Ethers (HRAC Code E)

The diphenyl ether PPO inhibitors are represented here by acifluorfen, formesafen, lactofen, and oxyfluorfen. Their structure, use, and hazard profiles are presented in Table 16.58. Lactofen is a severe eye irritant; otherwise, the acute toxicity of the PPO inhibitors is not remarkable. Acifluorfen and lactofen are classified as Category B2 (probable human) carcinogens based on an increased incidence of liver and stomach tumors. Oxyfluorfen and formesafen are classified as Category C (possible human) carcinogens based on an increased incidence of liver tumors. A mode of action underlying liver tumors commonly seen in this class of herbicide has been proposed [203,299].The key events leading to tumor expression are redrawn in Figure 16.20 from the paper by Klaunig et al. [299]. It has been proposed that diphenyl ethers serve as ligands to the PPARγ receptor, thereby activating genes involved in peroxisome proliferation (key event 2a), regulation of the cell cycle, suppression of apoptosis (key event 2b), and lipid metabolism (key event 2c). Suppres-

**TABLE 16.49**
**Structures, Use, and Hazard Profiles of the Acetolactate Synthase (ALS) Inhibiting Imidazolinone Herbicides (HRAC Code B) (Continued)**

| Chemical/Common Name | Structure | Principal Use/Crop | Application Rate (g active ingredient per hectare) |
|---|---|---|---|
| Imazamox (Raptor®) [104] | 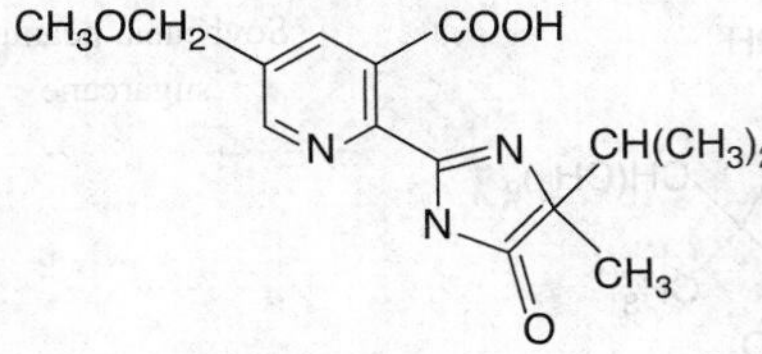 $CH_3OCH_2$, COOH, N, N, N, $CH(CH_3)_2$, $CH_3$, O | Soybeans, legumes | 34–43 |
| Imazapyr (Arsenal®) [106] | 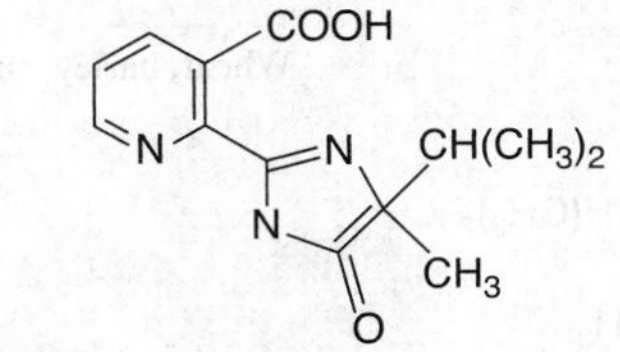 COOH, N, N, N, $CH(CH_3)_2$, $CH_3$, O | IWC | 250–1700 |
| Imazaquin (Scepter®) [107] | 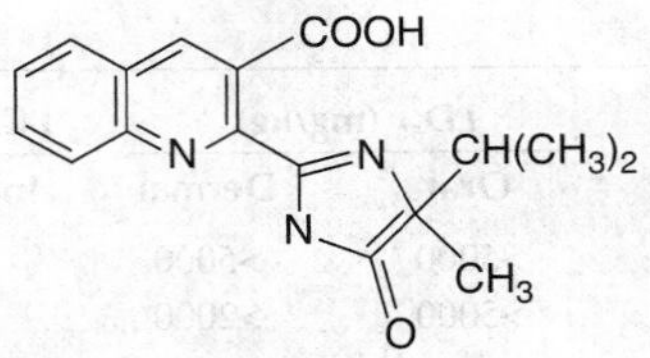 COOH, N, N, N, $CH(CH_3)_2$, $CH_3$, O | Soybeans | 70–140 pre-plant, PPI, pre, post |
| Imazethapyr (Pursuit®) [108] | 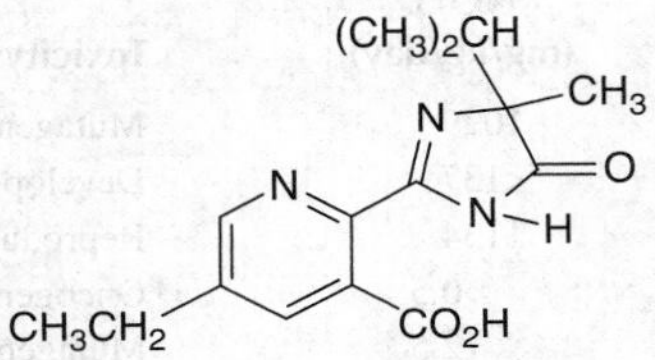 $(CH_3)_2CH$, $CH_3$, N, O, N, N–H, $CH_3CH_2$, $CO_2H$ | Soybeans, corn, legumes, peanuts | 130–260 early pre-plant, PPI, pre, post |

| Herbicide | Irritation: Eye | Irritation: Skin | $LD_{50}$ (mg/kg): Oral | $LD_{50}$ (mg/kg): Dermal | $LC_{50}$ (mg/L): Inhalation | Sensitization Potential | Signal Word |
|---|---|---|---|---|---|---|---|
| Imazamox | Mild irritant | Non-irritant | >5000 | >4000 | >6.3 | Negative | Caution |
| Imazapyr | Irreversible | Non-irritant | >5000 | >2000 | >1.3 | Negative | Danger |
| Imazaquin | Non-irritant | Slight irritant | >5000 | >2000 | >5.7 | Negative | Caution |
| Imazethapyr | Slight irritant | Slight irritant | >5000 | >2000 | >2.6 | Negative | Caution |

| Herbicide | Species/Study | NOEL (mg/kg/day) | Toxicity Study | Hazard Indicator |
|---|---|---|---|---|
| Imazamox [434] | Rat/2-year | 1068 | Mutagenicity | No evidence |
| | Dog/52-week | 1165 | Developmental | Not teratogenic |
| | Mouse/18-month | — | Reproductive | No evidence |
| | RfD | 3.0 | Oncogenicity | E (no evidence) |
| Imazapyr [559,602] | Rat/2-year | 500 | Mutagenicity | No evidence |
| | Dog/52-week | 250 | Developmental | Not teratogenic |
| | Mouse/18-month | 1500 | Reproductive | No evidence |
| | ADI | 2.5 | Oncogenicity | No evidence |
| Imazaquin [545,603] | Rat/2-year | 500 | Mutagenicity | No evidence |
| | Dog/52-week | 25 | Developmental | Not teratogenic |
| | Mouse/18-month | 150 | Reproductive | No evidence |
| | ADI | 0.25 | Oncogenicity | No evidence |
| Imazethapyr [604] | Rat/2-year | 500 | Mutagenicity | No evidence |
| | Dog/52-week | 25 | Developmental | Not teratogenic |
| | Mouse/18-month | 750 | Reproductive | No evidence |
| | ADI | 0.25 | Oncogenicity | No evidence |

**TABLE 16.50**
**Structures, Use, and Hazard Profiles of Acetolactate Synthase (ALS)-Inhibiting Triazolopyrimidine and Pyrimidinylthiobenzoate Herbicides (HRAC Code B)**

| Chemical/ Common Name | Structure | Principal Use/Crop | Application Rate (g active ingredient per hectare) |
|---|---|---|---|
| Flumetsulam (Broadstrike®) [86] | [structure] | Corn, soybeans | 25–78 |
| Pyriminobac-methyl (Prosper®) [163] | [structure] | Cotton | 30–60 |

| Herbicide | Irritation | | $LD_{50}$ (mg/kg) | | $LC_{50}$ (mg/L) | Sensitization Potential | Signal Word |
|---|---|---|---|---|---|---|---|
| | Eye | Skin | Oral | Dermal | Inhalation | | |
| Flumetsulam | Slight irritant | Non-irritant | >5000 | >2000 | >5.9 | Negative | Caution |
| Pyriminobac-methyl | Slight irritant | Slight irritant | >5000 | >2000 | >5.5 | NA | Caution |

| Herbicide | Species/Study | NOEL (mg/kg/day) | Toxicity Study | Hazard Indicator |
|---|---|---|---|---|
| Flumetsulum [594] | Rat/2-year | 35 | Mutagenicity | No evidence |
| | Dog/52-week | 100 | Developmental | Not teratogenic |
| | Mouse/18-month | 32 | Reproductive | No evidence |
| | ADI | 0.32 | Oncogenicity | No evidence |
| Pyriminobac-methyl [163] | Rat/2-year | 0.9 | Mutagenicity | No evidence |
| | Dog/52-week | — | Developmental | Not teratogenic |
| | Mouse/18-month | 8.1 | Reproductive | No evidence |
| | ADI | 0.009 | Oncogenicity | No evidence |

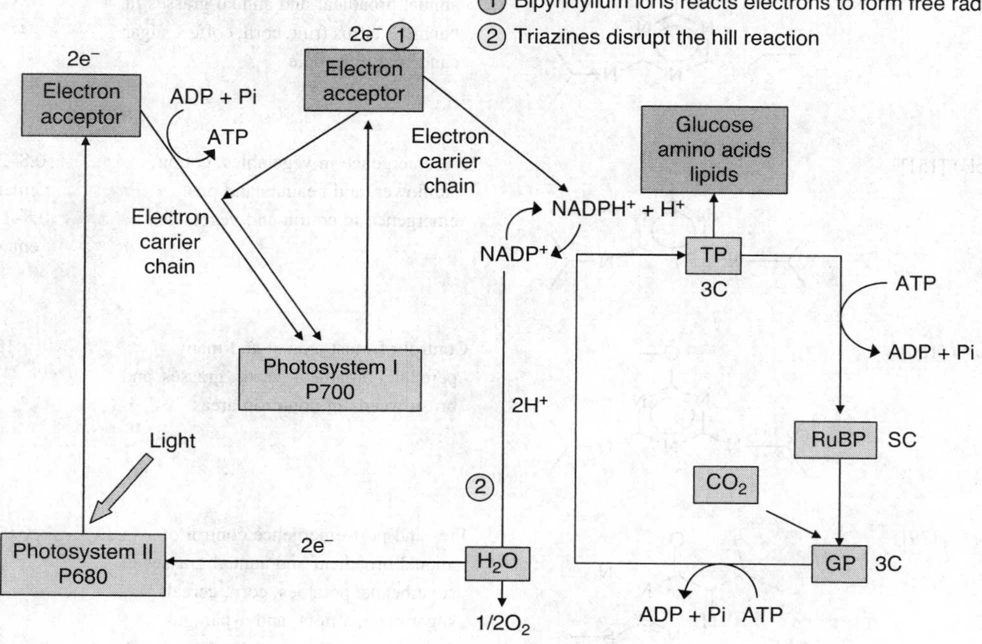


FIGURE 16.18 Photosynthesis molecular targets.

**TABLE 16.51**
**Structures, and Use Profiles of the Photosynthesis Inhibiting Triazines, and Triazinones Herbicides (HRAC Code C1)**

| Herbicide | Structure | Principal Crops /Use | Application Rate (kg active ingredient per hectare) |
|---|---|---|---|
| Atrazine (Aatrex®) [22] | 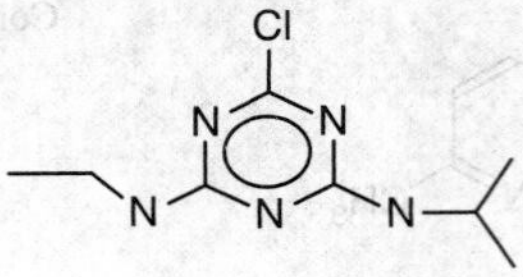 | Pre- and post-emergence control of annual broadleaf and annual grasses in corn, sorghum, sugar cane, and pineapple | 1.5–2.5 |
| Cyanazine (Bladex®) [54] | 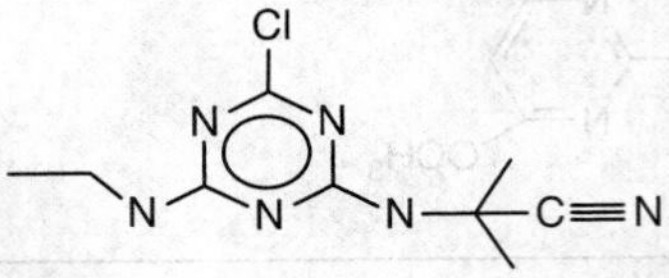 | Pre-emergence in broad beans, corn, and peas and post-emergence in barley and wheat | 1–3 (pre-emergence) 0.26–0.33 (post-emergence) |
| Propazine (Milo-Pro®) [154] | 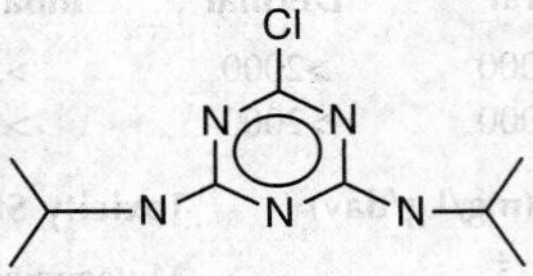 | Pre- and post-emergence control of annual broadleaf and annual grasses in sorghum, carrots, chervil, and parley | 0.5–3 |
| Simazine (Princep®) [170] | 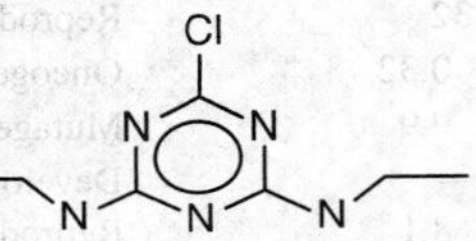 | Pre- and post-emergence control of annual broadleaf and annual grasses in pome fruit, stone fruit, citrus, vines, corn, sorghum, sugar cane, and pineapple | 1.5–3 |
| Ametryn (Evik®) [18] | 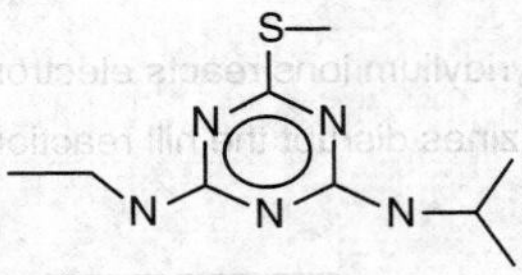 | Pre- and post-emergence control of annual broadleaf and annual grasses in bananas, citrus fruit, corn, coffee, sugar cane, and pineapple | 2–4 |
| Prometryn (Caparol®) [151] | 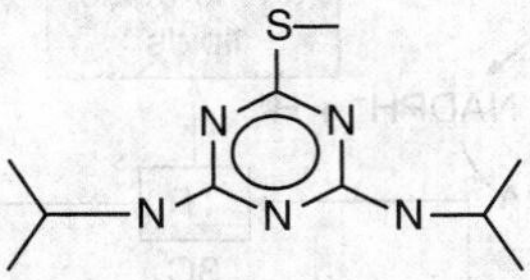 | Pre-emergence in vegetables, cotton, sunflower, and peanuts and post-emergence in cotton and vegetables | 0.8–2.5 (pre-emergence) 0.8–1.5 (post-emergence) |
| Prometon (Pramitol®) [150] | 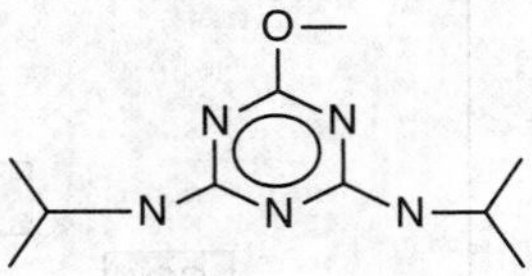 | Control of most annual and many perennial broadleaf weeds, grasses, and brush weeds in non-crop areas | 10–20 |
| Metribuzin (Sencor®) [129] | 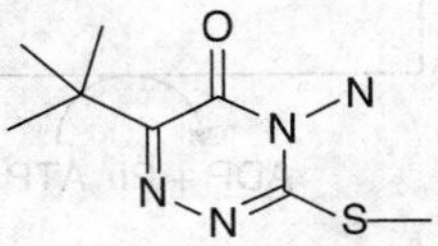 | Pre- and post-emergence control of annual broadleaf and annual grasses in soya beans, potatoes, corn, cereals, sugar cane, alfalfa, and asparagus | 0.07–1.45 |

**TABLE 16.52A**
**Hazard Profiles of the Photosynthesis Inhibiting Triazines, and Triazinones Herbicides (HRAC Code C1)**

| Herbicide | Irritation | | $LD_{50}$ (mg/kg) | | $LC_{50}$ (mg/L) | Sensitization | Signal |
|---|---|---|---|---|---|---|---|
| | Eye | Skin | Oral | Dermal | Inhalation | Potential | Word |
| Atrazine | Non-irritant | Non-irritant | 3090 | >3100 | >5.0 | Positive | Caution |
| Cyanazine | Non-irritant | Non-irritant | 182 | >2000 | >5.3 | Negative | Warning |
| Propazine | Mild irritant | Non-irritant | >7000 | >3100 | >2.0 | Negative | Caution |
| Simazine | Non-irritant | Mild irritant | >5000 | >3100 | >5.5 | Negative | Caution |

| Herbicide | Species/Study | NOEL (mg/kg/day) | Toxicity Study | Hazard Indicator |
|---|---|---|---|---|
| Atrazine | Rat/2-year | 0.5 | Mutagenicity | No evidence |
| [1,222,316,360,381, | Dog/52-week | 3.75 | Developmental | Not teratogenic |
| 554,580,628] | Mouse/18-month | 1.2 | Reproductive | No evidence |
| | ADI | 0.005 | Oncogenicity | Not likely carcinogenic in humans |
| Cyanazine [381,588] | Rat/2-year | 12 | Mutagenicity | No evidence |
| | Dog/52-week | 25 | Developmental | Teratogenic in rat, and rabbit |
| | Mouse/18-month | 1.4 | Reproductive | No evidence |
| | ADI | NA | Oncogenicity | Category C with Q* (based on mammary tumors in female Sprague–Dawley rats) |
| Propazine [445] | Rat/2-year | 5.8 | Mutagenicity | No evidence |
| | Dog/52-week | 1.3 | Developmental | Not teratogenic |
| | Mouse/18-month | 15 | Reproductive | No evidence |
| | RfD | 0.02 | Oncogenicity | Category C with Q* (based on mammary tumors in female Sprague–Dawley rats) |
| Simazine [381,568,621] | Rat/2-year | 0.5 | Mutagenicity | No evidence |
| | Dog/52-week | 7.5 | Developmental | Not teratogenic |
| | Mouse/18-month | 5.7 | Reproductive | No evidence |
| | ADI | 0.005 | Oncogenicity | Not likely carcinogenic in humans |

**TABLE 16.52B**
**Hazard Profiles of the Photosynthesis-Inhibiting Triazine and Triazinone Herbicides (HRAC Code C1)**

| Herbicide | Irritation | | $LD_{50}$ (mg/kg) | | $LC_{50}$ (mg/L) | Sensitization | Signal |
|---|---|---|---|---|---|---|---|
| | Eye | Skin | Oral | Dermal | Inhalation | Potential | Word |
| Ametryn | Non-irritant | Non-irritant | 1160 | >2020 | >5.1 | Positive | Caution |
| Prometryn | Slight irritant | Non-irritant | 4550 | >2020 | >5.1 | Negative | Caution |
| Prometon | Irritant | Mild irritant | 1518 | >2020 | >3.2 | Negative | Warning |
| Metribuzin | Non-irritant | Non-irritant | 1090 | >20000 | >0.65 | Negative | Caution |

| Herbicide | Species/Study | NOEL (mg/kg/day) | Toxicity Study | Hazard Indicator |
|---|---|---|---|---|
| Ametryn [537,578] | Rat/2-year | 2.5 | Mutagenicity | No evidence |
| | Dog/52-week | 10 | Developmental | Not teratogenic |
| | Mouse/18-month | 1.5 | Reproductive | No evidence |
| | RfD | 0.025 | Oncogenicity | E (no evidence) |
| Prometryn [405,475,618] | Rat/2-year | 37 | Mutagenicity | No evidence |
| | Dog/106-week oral | 3.7 | Developmental | Not teratogenic |
| | Mouse/102-week oral | 1.0 | Reproductive | No evidence |
| | RfD (based on 2-year dog study; UF = 100) | 0.037 | Oncogenicity | E (no evidence) |
| Prometon [617] | Rat/2-year | 1.0 | Mutagenicity | No evidence |
| | Dog/52-week | 5.0 | Developmental | Not teratogenic |
| | Mouse/18-month | 70 | Reproductive | No evidence |
| | RfD | — | Oncogenicity | No evidence |
| Metribuzin [472, 608] | Rat/2-year | 5.0 | Mutagenicity | No evidence |
| | Dog/104-week oral | 2.5 | Developmental | Not teratogenic |
| | Mouse/18-month | 120 | Reproductive | No evidence |
| | RfD (based on 2-year dog study; UF = 100) | 0.025 | Oncogenicity | No evidence |

**TABLE 16.53**
**Structures, Use, and Hazard Profiles for the Photosynthesis-Inhibiting Uracil and Pyridazinone Herbicides (HRAC Code C1)**

| Herbicide | Structure | Principal Use/Crop | Application Rate (g active ingredient per hectare) |
|---|---|---|---|
| Bromacil (Hyvar®) [34] | 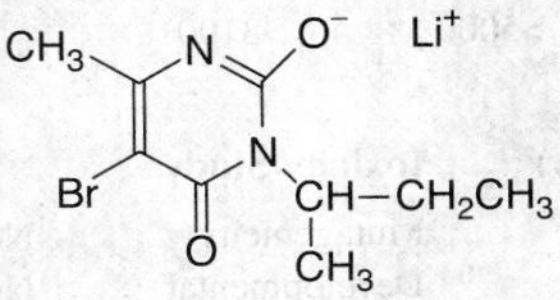 | Used to control grasses, broadleaf weeds, and brush in non-cropland areas | 1500–15,000 |
| Terbacil (Sinbar®) [180] | 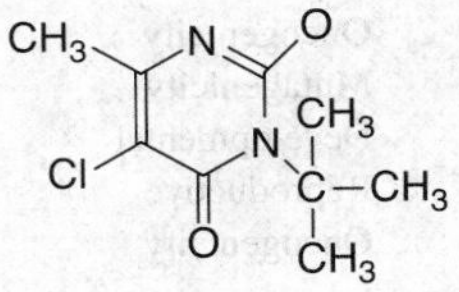 | Used to control grasses and broadleaf weeds in nut trees, mint, alfalfa, and fruits | 500–8000 |
| Norflurazon (Predict®) [137] | 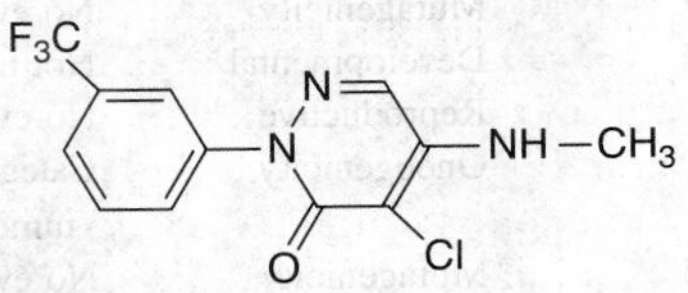 | Used to control broadleaf weeds and sedges in fruits, nuts, and berries; also used on right of ways | 500–4000<br>9000 (right of ways) |

| Herbicide | Irritation | | $LD_{50}$ (mg/kg) | | $LC_{50}$ (mg/L) | Sensitization Potential | Signal Word |
|---|---|---|---|---|---|---|---|
| | Eye | Skin | Oral | Dermal | Inhalation | | |
| Bromacil | Mild irritant | Mild irritant | 5175 | >5000 | >4.8 | Positive | Caution |
| Terbacil | Mild irritant | Non-irritant | 1255 | >5000 | >4.4 | Negative | Caution |
| Norflurazon | Non-irritant | Non-irritant | 9000 | >20,000 | NA | Negative | Caution |

| Herbicide | Species/Study | NOEL (mg/kg/day) | Toxicity Study | Hazard Indicator |
|---|---|---|---|---|
| Bromacil [396,583] | Rat/2-year | 2.5 | Mutagenicity | No evidence |
| | Dog/52-week | 15.6 | Developmental | Not teratogenic |
| | Mouse/18-month | — | Reproductive | No evidence |
| | RfD | 0.1 | Oncogenicity | C (liver tumors; male mice) |
| Terbacil [623] | Rat/2-year | 2.5 | Mutagenicity | No evidence |
| | Dog/104-week | 1.25 | Developmental | Not teratogenic |
| | Mouse/18-month | 7.1 | Reproductive | No evidence |
| | ADI | 0.013 | Oncogenicity | E (no evidence) |
| Norflurazon [441,533,613] | Rat/2-year | 19 | Mutagenicity | No evidence |
| | Dog/26-week | 1.6 | Developmental | Not teratogenic |
| | Mouse/18-month | 41 | Reproductive | No evidence |
| | RfD | 0.02 | Oncogenicity | C (liver tumors in mice) |

sion of apoptosis coupled with a stimulation of cell proliferation allows cells with mutation to be selected for clonal expansion (key event 7), leading to preneoplastic foci and tumors. Peroxisome proliferation may cause oxidative stress (key event 5) and cell death, thereby further stimulating cell turnover [299].

## *N*-PHENYLPHTHALIMIDES, THIADIAZOLES, AND TRIAZOLINONES (HRAC CODE E)

The structure, use, and hazard profiles for the *N*-phenylphthalimide (flumiclorac-pentyl), thiadiazole (fluthiacet-methyl), and triazolinone (carfentrazone-ethyl) PPC

**TABLE 16.54**
**Structures, Use, and Hazard Profiles for the Photosynthesis-Inhibiting Urea Herbicides (HRAC Code C2)**

| Herbicide | Structure | Principal Use/Crop | Application Rate (g active ingredient per hectare) |
|---|---|---|---|
| Diuron (Diumate®) [74] | Cl; Cl–; O; NH–C–N; $CH_3$; $CH_3$ | Used to control many annual weeds at a lower rate and perennials at a higher rate in nuts, berries, spices, and cereals | 10,000–30,000 |
| Fluometuron (Cotoran®) [88] | $F_3C$; O; NH–C–N; $CH_3$; $CH_3$ | Used to control broadleaf weeds and grasses | 1000–1500 |
| Linuron (Lorox®) [117] | Cl; Cl–; O; NH–C–N; $CH_3$; $OCH_3$ | Used to control broadleaf weeds in vegetables and cereals | 250–2240 |

| Herbicide | Irritation: Eye | Irritation: Skin | $LD_{50}$ (mg/kg): Oral | $LD_{50}$ (mg/kg): Dermal | $LC_{50}$ (mg/L): Inhalation | Sensitization Potential | Signal Word |
|---|---|---|---|---|---|---|---|
| Diuron | Mild irritant | Non-irritant | 3400 | 2000 | >2.5 | Negative | Caution |
| Fluometuron | Slight irritant | Non-irritant | 6416 | >10,000 | >2.0 | Negative | Caution |
| Linuron | Non-irritant | Non-irritant | 1090 | >20,000 | >0.65 | Negative | Caution |

| Herbicide | Species/Study | NOEL (mg/kg/day) | Toxicity Study | Hazard Indicator |
|---|---|---|---|---|
| Diuron [427,522,593] | Rat/2-year | <1.02 | Mutagenicity | No evidence |
| | Dog /104-week | 0.625 | Developmental | Not teratogenic |
| | Mouse/18-month | >50 (LDT) | Reproductive | No evidence |
| | RfD (UF = 300] | 0.002 | Oncogenicity | Known/likely (liver, mice; bladder, rats) |
| Fluometuron [596] | Rat/2-year | 0.55 | Mutagenicity | No evidence |
| | Dog/52-week | 10 | Developmental | Not teratogenic |
| | Mouse/18-month | 1.3 | Reproductive | No evidence |
| | ADI2 | 0.0055 | Oncogenicity | Classification pending |
| Linuron [607] | Rat/2-year | 2.5 | Mutagenicity | No evidence |
| | Dog/104-week oral | 0.77 | Developmental | Not teratogenic |
| | Mouse/18-month | 21 | Reproductive | No evidence |
| | RfD | 0.008 | Oncogenicity | C (interstitial cell tumors in male rat) |

nhibitors are presented in Table 16.59. Flumiclorac-pen-yl, which is applied at a rate of 30 to 60 g/hectare, is not oxic in acute or repeat-dose studies. The thiadiazole luthiacet-methyl, which is also applied at a low rate of 4 o 15 g/hectare, has a low reference dose based on results rom the chronic mouse study. Fluthiacet-methyl caused iver tumors in mice and pancreatic tumors in rats. The eference dose (RfD) for carfentrazone-ethyl (0.03 ng/kg/day) is based on the chronic rat study where evi-lence of porphyrin deposits were seen in the liver at high loses [458].

## Oxadiazole and Pyrimidindione Herbicides (HRAC Code E)

The structure, use, and hazard profiles for the oxadiazole (oxadiazon) and pyrimidindione (butafenacil) PPO inhibiting herbicides are presented in Table 16.60. Butafenacil has an unremarkable hazard profile, whereas oxadiazon has an extremely low reference dose based on hepatotoxicity and hemolytic anemia seen at high doses. Oxadiazon has been classified by the EPA as a likely human carcinogen.

**TABLE 16.55**
**Structure, Use, and Hazard Profile for the Photosynthesis-Inhibiting Nitrile Herbicides (HRAC Code C3)**

| Herbicide | Structure | Principal Use/Crop | Application Rate (g active ingredient per hectare) |
|---|---|---|---|
| Dichlobenil (Acme®) [66] | 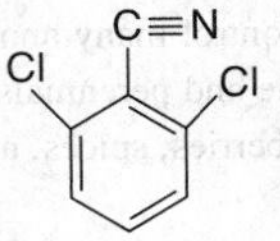 2,6-dichlorobenzonitrile (C≡N; Cl; Cl) | Used to control annual, biennial broadleaf weeds and grasses in orchards, at industrial sites, under asphalt, and in non-crop areas | 2700–8100 |
| Ioxynil (Totril®) [112] | 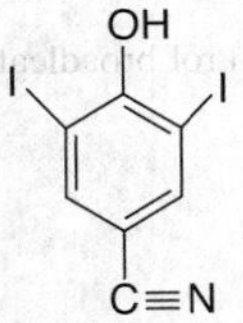 4-hydroxy-3,5-diiodobenzonitrile (OH; I; I; C≡N) | Used for control of selected weeds in fall-planted small grains | 350–490 |

| Herbicide | Irritation: Eye | Irritation: Skin | $LD_{50}$ (mg/kg): Oral | $LD_{50}$ (mg/kg): Dermal | $LC_{50}$ (mg/L): Inhalation | Sensitization Potential | Signal Word |
|---|---|---|---|---|---|---|---|
| Dichlobenil | Non-irritant | Non-irritant | >1000 | >2000 | >0.25 | Negative | Warning |
| Ioxynil | Non-irritant | Mild irritant | 110 | 1050 | >0.40 | Negative | Warning |

| Herbicide | Species/Study | NOEL (mg/kg/day) | Toxicity Study | Hazard Indicator |
|---|---|---|---|---|
| Dichlobenil [589] | Rat/2-year | 2.5 | Mutagenicity | No evidence |
| | Dog/52-week | 1.25 | Developmental | Not teratogenic |
| | Hamster/18-month oral | 10 | Reproductive | No evidence |
| | RfD | 0.013 | Oncogenicity | C (liver tumors in female rat) |
| Ioxynil [605] | Rat/2-year | 0.5 | Mutagenicity | No evidence |
| | Dog/30-week oral | 1.0 | Developmental | Not teratogenic |
| | Mouse/18-month | <1.5 | Reproductive | No evidence |
| | ADI | 0.005 | Oncogenicity | No evidence |

## Bleaching Herbicides (HRAC Code F1)

The bleaching herbicides disrupt the synthesis of carotenoid pigments, which protect chlorophyll pigments from photo-oxidation in strong light (see Figure 16.21). In the absence of carotenoids, chlorophyll is destroyed and turns white; thus, the leaves of the plant have a bleached appearance. The pyridazinone triketone and isoxazole bleaching herbicides are considered here. The pyridazinones inhibit carotenoid biosynthesis at the phytoene desaturase step [686], whereas the triketones, which were initially identified in the bottle-brush plant [307,321], and isoxazoles inhibit 4-hydroxyphenylpyruvate dioxygenase (HPDD) [337]. The site of action for the triazoles and isoxazolidinones is unknown [253]. Of these modes of action, only the HPDD inhibition is relevant to humans because blocking HPDD results in an increased incidence of tyrosinemia in animals [507].

## Pyridazinones (HRAC Code F1)

The structure, use, and hazard profiles for norflurazon and fluridone are given in Table 16.61. Neither norflurazon nor fluridone is acutely toxic; the repeat-dose profile for fluridone is unremarkable. Norflurazon is classified as a Category C (possible human) carcinogen based on mouse liver tumors.

## Triketones and Isoxazoles (HRAC Code F2)

The structure, use, and hazard profiles for sulcotrione and mesotrione (triketone) and isoxaflutole (isoxazole) are presented in Table 16.62. Sulcotrione is not acutely toxic, but no publicly available data are available to assess developmental, reproductive, or chronic toxicity or carcinogenic potential. Isoxaflutole is not acutely toxic, but it causes developmental and neurotoxic effects and liver tumors in mice and rats. Mesotrione is not mutagenic, carcinogenic, or neurotoxic. It has a low reference dose based on effects on tyrosine and their sequella in the rat. Mice and dogs are less sensitive and more similar to humans than are rats with respect to inhibition of HPDD *in vitro*.

**TABLE 16.56**
**Structures, Use, and Hazard Profile for the Photosynthesis-Inhibiting Benzothiadiazole and Phenylpyridazine Herbicides (HRAC Code C3)**

| Herbicide | Structure | Principal Use/Crop | Application Rate (g active ingredient per hectare) |
|---|---|---|---|
| Bentazone (Basagran®) [30] | 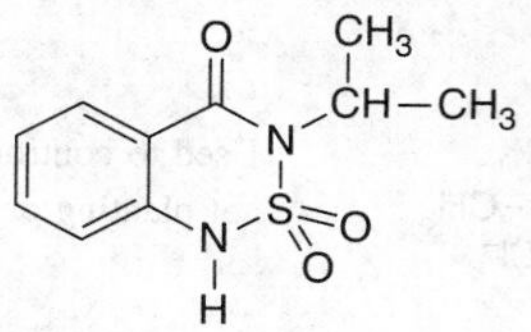 | Used to control annual broadleaf weeds in soybeans, peas, peanuts, and cereals | 1000–2240 |
| Pyridate [161] | 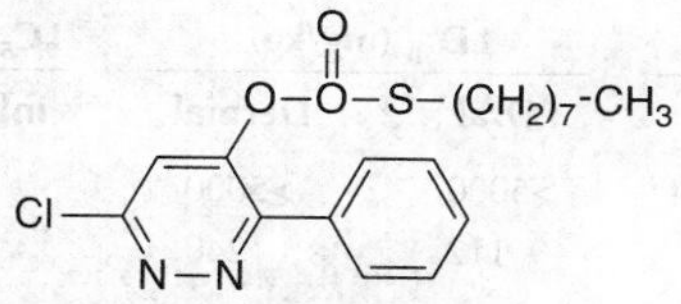 | Used to control annual broadleaf weeds in cereals, turf, and vegetables | 900 |

| Herbicide | Irritation | | $LD_{50}$ (mg/kg) | | $LC_{50}$ (mg/L) | Sensitization Potential | Signal Word |
|---|---|---|---|---|---|---|---|
| | Eye | Skin | Oral | Dermal | Inhalation | | |
| Bentazon | Moderate irritant | Moderate irritant | 1100 | >2500 | 5.1 | — | Caution |
| Pyridate | Non-irritant | Moderate irritant | 4690 | >2000 | >4.7 | Positive | Caution |

| Herbicide | Species/Study | NOEL (mg/kg/day) | Toxicity Study | Hazard Indicator |
|---|---|---|---|---|
| Bentazone [382,582] | Rat/2-year | 17.5 | Mutagenicity | No evidence |
| | Dog/52-week | 3.2 | Developmental | Not teratogenic |
| | Mouse/18-month | 50 | Reproductive | No evidence |
| | RfD | 0.03 | Oncogenicity | E (no evidence) |
| Pyridate [478,619] | Rat/2-year | 10.8 | Mutagenicity | No evidence |
| | Dog/104-week oral | 20 | Developmental | Not teratogenic |
| | Mouse/18-month | <48 | Reproductive | No evidence |
| | RfD | 0.11 | Oncogenicity | E (no evidence) |

## Triazoles and Isoxazolidinones (HRAC Code F3)

The chemical structure, use, and hazard profiles for amitrole (triazole) and clomazone (isoxazolidinone) are presented in Table 16.63. These compounds are not acutely toxic. Clomazone has an unremarkable repeated-dose toxicity profile. Amitrole has been classified as a Category B2 (probable human) carcinogen based on thyroid tumors in rats and liver tumors in mice [241]. These tumor responses could be related to a triazole-related action on the liver (see triazole fungicides for a detailed discussion of this mode of action and its relevance to humans).

## EPSP Synthase, Glutamine Synthase, and Dihydropteroate (DHP) Synthase Inhibitors (HRAC Codes G, H, and I, Respectively)

Glyphosate is a nonselective herbicide that inhibits 5-enolpyruvlshikimate-3-phosphate synthase (EPSPS) in the shikimate pathway (Figure 16.22) of plant plastids [234]. Inhibition of EPSPS blocks the biosynthesis of the aromatic amino acids phenylalanine (Phe), tyrosine (Tyr), and tryptophan (Trp) and secondary products important for plant growth and development, including lignans, alkaloids, flavonoids, and benzoic acids [209]. This pathway does not exist in mammals; therefore, this mode of action is not relevant to humans.

Glutamine synthetase is the initial enzyme in the pathway that assimilates inorganic nitrogen into organic compounds in plants. It is the pivotal enzyme in nitrogen metabolism in plants that, in addition to assimilating ammonia, recycles ammonia produced by other processes including photorespiration and deamination reactions (Figure 16.23). Glutamine synthetase is found in analogous pathways in mammals and plays a similar role in recycling nitrogen. Glufosinate (DL-phosphinothricin) is a close structural analog of glutamic acid and is considered to be the active pesticidal component in the natural tripeptide phosphinothricin-alanyl-alanine (bialaphos), first discovered in *Streptomcyces viridochronogenes* [218].

**TABLE 16.57**
**Structure, Use, and Hazard Profiles for the Photosynthesis-Inhibiting Bipyridylum Herbicides (HRAC Code D)**

| Herbicide | Structure | Principal Use/Crop | Application Rate |
|---|---|---|---|
| Diquat (Weedtrine®) [73] | 2Br⁻ (bipyridylium dication) | Used to control algae in ponds, lakes, and drainage ditches | 400–1000 g active ingredient per hectare |
| Paraquat (Cyclone®) [142] | $H_3C-^{+}N$ … $N^{+}-CH_3$ 2Cl⁻ | Used to control existing vegetation at planting or no till | 400–1000 g active ingredient per hectare |

| Herbicide | Irritation | | $LD_{50}$ (mg/kg) | | $LC_{50}$ (mg/L) | Sensitization Potential | Signal Word |
|---|---|---|---|---|---|---|---|
| | Eye | Skin | Oral | Dermal | Inhalation | | |
| Diquat | Non-irritant | Slight irritant | >5000 | >5000 | >6 | Negative | Caution |
| Paraquat | NA | Irritant | 112 | 240 | — | Negative | Warning |

| Herbicide | Species/Study | NOEL (mg/kg/day) | Toxicity Study | Hazard Indicator |
|---|---|---|---|---|
| Diquat [387,398,592] | Rat/2-year | 0.6 | Mutagenicity | No evidence |
| | Dog/52-week | 0.5 | Developmental | Not teratogenic |
| | Mouse/18-month | 3.5 | Reproductive | No evidence |
| | RfD | 0.005 | Oncogenicity | E (no evidence) |
| Paraquat [394,442,614] | Rat/2-year | 1.25 | Mutagenicity | No evidence |
| | Dog/52-week | 0.45 | Developmental | Not teratogenic |
| | Mouse/18-month | 1.87 | Reproductive | No evidence |
| | RfD | 0.0045 | Oncogenicity | E (no evidence) |

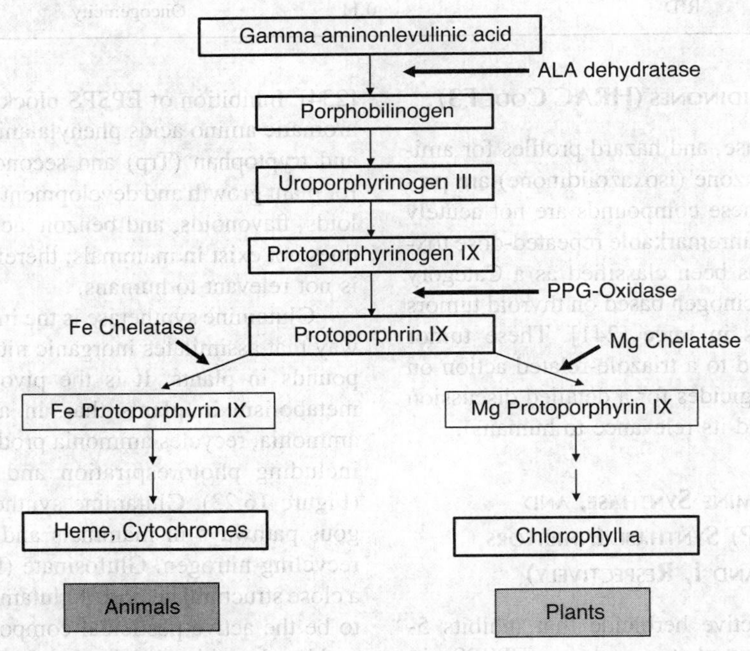


**FIGURE 16.19** Protoporphrin biosynthetic pathway.

**TABLE 16.58**
**Structures, Use, and Hazard Profile for the Protoporphyrin-Inhibiting Diphenylether Herbicides (HRAC Code E)**

| Herbicide | Structure | Principal Use/Crop | Application Rate (g active ingredient per hectare) |
|---|---|---|---|
| Acifluorfen (Scepter®) [15] | $F_3C$, Cl, O, COOH, $NO_2$ | Used to control annual broadleaf weeds in peanuts, beans, and rice | 200–600 |
| Fomesafen (Flosil®) [92] | $F_3C$, O, Cl, $CONHSO_2CH_3$, $NO_2$ | Used to control annual broadleaf weeds in soybeans | 200–400 |
| Lactofen (Cobra®) [116] | $F_3C$, O, Cl, $CO_2CHCO_2CH_2CH_3$, $CH_3$, $NO_2$ | Used to control annual broadleaf weeds in cereals, potatoes, soya, and rice | 70–220 |
| Oxyfluorfen (Goal®) [141] | $F_3C$, O, Cl, $OCH_2CH_3$, $NO_2$ | Used to control annual broadleaf weeds in conifers, vegetables, nuts, and vine crops | 250–2240 |

| Herbicide | Irritation: Eye | Irritation: Skin | $LD_{50}$ (mg/kg): Oral | $LD_{50}$ (mg/kg): Dermal | $LC_{50}$ (mg/L): Inhalation | Sensitization Potential | Signal Word |
|---|---|---|---|---|---|---|---|
| Acifluorfen | Non-irritant | Moderate irritant | 1450 | >2000 | >6.9 | Negative | Caution |
| Fomesafen | Moderate irritant | Mild irritant | 1250 | >1000 | 4.97 | Negative | Caution |
| Lactofen | Severe irritant | Non-irritant | >5000 | 2000 | — | — | Danger |
| Oxyfluorfen | Moderate irritant | Non-irritant | >5000 | >5000 | — | Negative | Caution |

| Herbicide | Species/Study | NOEL (mg/kg/day) | Toxicity Study | Hazard Indicator |
|---|---|---|---|---|
| Acifluorfen [411,412,577] | Rat/2-year | 25 | Mutagenicity | No evidence |
| | Dog/52-week | NA | Developmental | Not teratogenic |
| | Mouse/18-month | 38 | Reproductive | No evidence |
| | RfD (rat reproduction) | 0.013 | Oncogenicity | B2 (liver/stomach tumors) |
| Fomesafen [433] | Rat/2-year | 0.25 | Mutagenicity | No evidence |
| | Dog/52-week | 1.0 | Developmental | Not teratogenic |
| | Mouse/18-month | 1.0 | Reproductive | No evidence |
| | RfD | 0.0025 | Oncogenicity | CQ* (liver tumors in mice) |
| Lactofen [404,500] | Rat/2-year | 25 | Mutagenicity | No evidence |
| | Dog/52-week | 5.0 | Developmental | Not teratogenic |
| | Mouse/18-month | 1.5 | Reproductive | No evidence |
| | RfD (mouse; UF = 1000) | 0.002 | Oncogenicity | B2 (liver/stomach tumors) |
| Oxyfluorfen [393,516] | Rat/2-year | 2.0 | Mutagenicity | No evidence |
| | Dog/2-year oral | 2.5 | Developmental | Not teratogenic |
| | Mouse/18-month | 0.3 | Reproductive | No evidence |
| | RfD | 0.003 | Oncogenicity | C (liver tumors in mice) |

Glufosinate inhibits glutamine synthase, resulting in an accumulation of ammonium and the inhibition of photosynthesis [209].

Folic acid, or its coenzyme form, serves as an intermediate carrier of hydroxymethyl, formyl, or methyl groups in enzyme-mediated reactions leading to the synthesis of amino acids, purines, and pyrimidines (Figure 16.24). Dihydropteroate synthase (DHP) catalyzes the first step of the folic acid biosynthetic pathway [210]. Asulam inhibits DHP synthase, thereby blocking folic acid synthesis, which is needed for the formation of purine nucleotides required for cell division [210]. Asulam and other members of this class are structural analogs of 4-aminobenzoic acid and likely serve as substrates for DDP synthase because the administration of 4-aminobenzoic acid reverses the phytotoxicity seen in plants and microbes

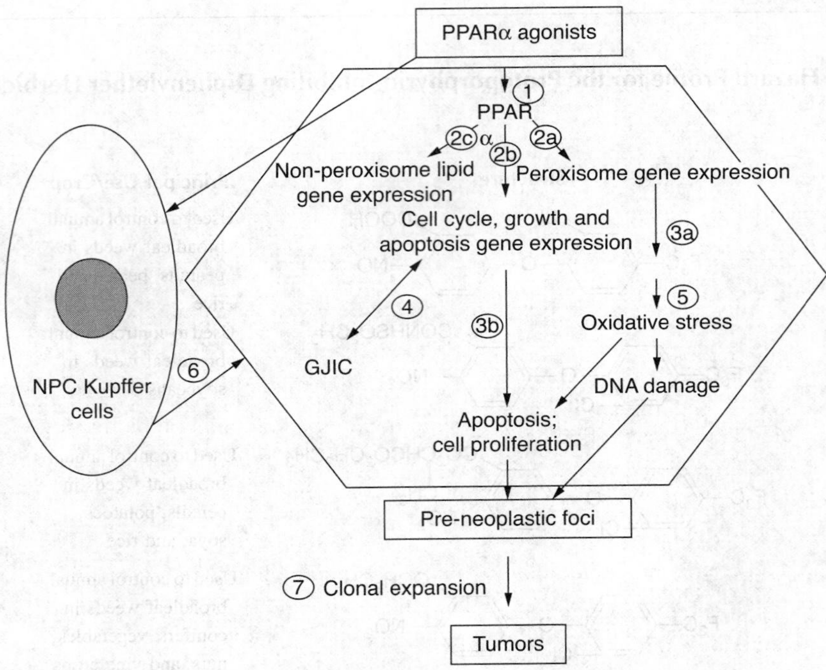


**FIGURE 16.20** Proposed mode of action for PPARα agonists.

treated with DHP synthase inhibitors [210]. The structures, uses, and hazard profiles for representative EPSP synthase, glutamine synthetase, and dihydropteroate synthase inhibitors are given in Table 16.64.

Glyphosate, glufosinate-ammonium, and asulam are not acutely toxic. Glyphosate and glufosinate-ammonium are not toxic in repeat-dose studies, and neither chemical is a developmental toxin, mutagen, or carcinogen. Asulum is not mutagenic, teratogenic, or a reproductive toxin, but a statistically significant increase in thyroid and adrenal gland tumors was observed in the male rat. A margin-of-exposure approach was used to assess carcinogenic risk, likely because of the structural similarity of asulam to products occurring naturally in plants [385].

## Dinitroaniline Microtubule Assembly Inhibitors (HRAC Code K1)

Several groups of herbicides including the dinitroanilines (e.g., benfluralin, pendimethalin, and trifluralin) phosphoroamidates (amirprophos-methyl, butamiphos), pyridines (dithiopyr, thiazopyr), benzamides (propyzamide, tebutam), and benzoic acid (DCPA) bind to tubulin monomers and prevent microtubule polymerization. This mode of action is considered relevant to humans (see the discussion on microtubule formation in the section on benzimidazole and thiophanate fungicides). The structure, use, and hazard profiles of the three commercially important dinitroaniline microtubule assembly inhibitors are given in Table 16.65. Benfluralin, pendimethalin, and trifluralin are not acutely toxic. Benfluralin caused liver and thyroid tumors in rats at doses that exceeded the maximum tolerated dose and therefore was considered a suggestive human carcinogen; a margin-of-exposure approach was used for cancer risk assessment [525]. Pendimethalin, which caused an increased incidence of thyroid tumors was classified as a Category C (possible human) carcinogen using a margin-of-exposure approach for cancer risk assessment [444]. Trifluralin caused thyroid, bladder, and kidney tumors and its carcinogenic risk was regulated using $Q_1$* [408,536]. These tumorigenic effects are unlikely related to the herbicidal mode of action of this class.

## Chloroacetamide Inhibitors of Very-Long-Chain Fatty Acid Synthesis (HRAC Code K3)

Boger and Matthes [2002] reviewed the evidence suggesting that chloroacetamides block the formation of very-long-chain saturated fatty acids (VLCFAs) by inhibiting the fatty acid elongase [648], shown in Figure 16.25 [648]. Vertebrates, including humans, have the biochemical mechanisms necessary to synthesize long-chain fatty acids including the enzyme long-chain fatty acid acyl elongase [258]. Humans, however, lack the desaturase enzymes (not shown in Figure 16.25) that produce the health-beneficial very-long-chain polyunsaturated fatty acids synthesized by plants and fish [367]. It is plausible that the chloroacetamides could perturb fatty acid synthesis in mammals, but there is direct no evidence of this in animal studies.

## TABLE 16.59
**Structures, Use, and Hazard Profiles for the Protoporphyrin-Inhibiting *N*-Phenylphthalimide, Thiadiazole, and Triazolinone Herbicides (HRAC Code E)**

| Chemical/Common Name | Structure | Principal Use/Crop | Application Rate (g active ingredient per hectare) |
|---|---|---|---|
| Flumiclorac-pentyl (Resource®) [87] | $OCH_2COO(CH_2)_4CH_3$; Cl | Used to control broadleaf weeds in soybeans and corn | 30–60 |
| Fluthiacet-methyl (Action®) [90] | F; Cl; $SCH_2COOCH_3$ | Used to control annual broadleaf weeds in corn, soybeans, and cereals | 4–15 |
| Carfentrazone-ethyl (Affinity®, Aurora®) [41] | $F_2HC$; $H_3C$; Cl; $CH_2CHCO_2H$; F; Cl | Used to control annual broadleaf weeds in cereals | 9–35 |

| Herbicide | Irritation: Eye | Irritation: Skin | $LD_{50}$ (mg/kg): Oral | $LD_{50}$ (mg/kg): Dermal | $LC_{50}$ (mg/L): Inhalation | Sensitization Potential | Signal Word |
|---|---|---|---|---|---|---|---|
| Flumiclorac–pentyl | Slight irritant | Non-irritant | >5000 | >2000 | >5.9 | Negative | Caution |
| Fluthiacet-methyl | Non-irritant | Non-irritant | >5000 | >2000 | >5.0 | NA | Caution |
| Carfentrazone-ethyl | Minimal irritant | Non-irritant | 5143 | >4000 | >5.0 | Negative | Caution |

| Herbicide | Species/Study | NOEL (mg/kg/day) | Toxicity Study | Hazard Indicator |
|---|---|---|---|---|
| Flumiclorac-pentyl [543,595] | Rat/2-year | 35 | Mutagenicity | No evidence |
| | Dog/52-week | 100 | Developmental | Not teratogenic |
| | Mouse/18-month | 32 | Reproductive | No evidence |
| | RfD | 0.32 | Oncogenicity | E (no evidence) |
| Fluthiacet-methyl [432,488] | Rat/2-year | 2.1 | Mutagenicity | No evidence |
| | Dog/52-week | 30 | Developmental | Not teratogenic |
| | Mouse/18-month | 0.1 | Reproductive | No evidence |
| | RfD | 0.001 | Oncogenicity | Likely carcinogen (mouse liver tumors; rat pancreatic tumors) |
| Carfentrazone-ethyl [458] | Rat/2-year | 3.0 | Mutagenicity | No evidence |
| | Dog/52-week | 50 | Developmental | Not teratogenic |
| | Mouse/18-month | 10 | Reproductive | No evidence |
| | RfD | 0.03 | Oncogenicity | E (no evidence) |

Aside from the herbicidal mode of action of the chloroacetamides, alachlor, acetochlor, and butachlor have been identified as sharing a common mechanism of toxicity [504]. Alachlor, acetochlor, and butachlor all undergo dealkylation to form aniline and a reactive quinine imine (see Figure 16.26) which is thought to be the ultimate carcinogenic moiety responsible for nasal epithelial adenomas and carcinomas found in rats [230]. A sulfoxide metabolite of acetochlor was found in the plasma of rats treated with acetochlor and bioconcentrated in the nasal epithelial tissue. Nasal epithelial tissue from humans apparently does not support the metabolic conversion of the sulfoxide to the sulfoxide of the quinine imine, whereas the rat made this conversion [245].

**TABLE 16.60**
**Structure, Use, and Hazard Profiles for the Protoporphyrinogen Oxidase (PPO)-Inhibiting Oxadiazole and Pyrimidindione Herbicides (HRAC Code E)**

| Chemical/Common Name | Structure | Principal Use/Crop | Application Rate (g active ingredient per hectare) |
|---|---|---|---|
| Oxadiazon (Ronstar®) [138] | | Used for control of bindweed and annual broadleaf weeds in flowers, fruit trees, bushes, sunflowers, and onions | 1000–4000 |
| Butafenacil (Inspire™) [36] | | Used as a low-rate cotton defoliant | 50–100 |

| Herbicide | Irritation: Eye | Irritation: Skin | $LD_{50}$ (mg/kg): Oral | $LD_{50}$ (mg/kg): Dermal | $LC_{50}$ (mg/L): Inhalation | Sensitization Potential | Signal Word |
|---|---|---|---|---|---|---|---|
| Oxadiazon | Slight irritant | Negligible irritant | >5000 | >2000 | >2.77 | Negative | Caution |
| Butafenacil | Mildly iritating | Non-irritant | >5000 | >4000 | >5.1 | Negative | Caution |

| Herbicide | Species/Study | NOEL (mg/kg/day) | Toxicity Study | Hazard Indicator |
|---|---|---|---|---|
| Oxadiazon [523] | Rat/2-year | 0.36 | Mutagenicity | Not mutagenic |
| | Dog/52-week | — | Developmental | Not teratogenic |
| | Mouse/18-month | 10 | Reproductive | — |
| | RfD | 0.0036 | Oncogenicity | Likely to be carcinogenic $Q_1^* = 7.11 \times 10^{-2}$ $(mg/kg/day)^{-1}$ |
| Butafenacil [36] | Rat/2-year | 3.76 | Mutagenicity | Not mutagenic |
| | Dog/52-week | 100 | Developmental | Not teratogenic |
| | Mouse/18-month | 1.71 | Reproductive | Not a reproductive toxin |
| | RfD | 0.012 | Oncogenicity | Not carcinogenic |

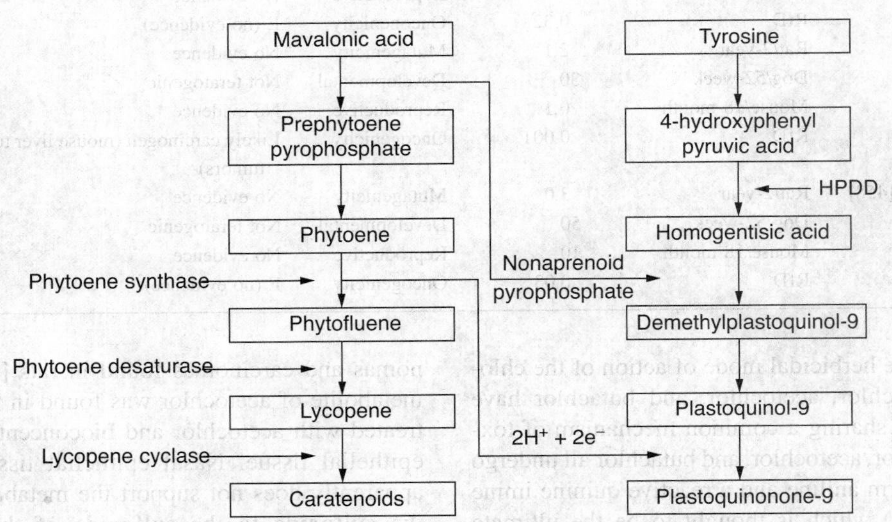


**FIGURE 16.21** Carotenoid pigment, and chlorophyll biosynthetic pathways.

**TABLE 16.61**
**Structure, Use, and Hazard Profiles for the Photobleaching Pyridazinones Herbicides (HRAC Code F1)**

| Chemical/ Common Name | Structure | Principal Use/Crop | Application Rate (g active ingredient per hectare) |
|---|---|---|---|
| Norflurazon (Evital®) [137] | $F_3C$-phenyl, N, N, $NHCH_3$, O, Cl | Fruit tree, nut, and vine crops; soybean; peanut; cotton; ornamentals; IWC | 500–3360 |
| Fluridone (Sonar®) [89] | $CH_3$, N, $F_3C$-phenyl, O, phenyl | Aquatic herbicide | 2240 (0.075–0.15 mg/L) |

| Herbicide | Irritation | | $LD_{50}$ (mg/kg) | | $LC_{50}$ (mg/L) | Sensitization Potential | Signal Word |
|---|---|---|---|---|---|---|---|
| | Eye | Skin | Oral | Dermal | Inhalation | | |
| Norflurazon | Non-irritant | Non-irritant | >9000 | >20,000 | >0.2 | Negative | Caution |
| Fluridone | Slight | Non-irritant | >10,000 | >5000 | >4.12 | Negative | Caution |

| Herbicide | Species/Study | NOEL (mg/kg/day) | Toxicity Study | Hazard Indicator |
|---|---|---|---|---|
| Norflurazon [533,613] | Rat/2-year | 19 | Mutagenicity | No evidence |
| | Dog/52-week | 1.5 | Developmental | Not teratogenic |
| | Mouse/18-month | 41 | Reproductive | No evidence |
| | RfD | 0.02 | Oncogenicity | C (mouse liver tumors) |
| Fluridone [597] | Rat/2-year | 8.0 | Mutagenicity | No evidence |
| | Dog/52-week | 11.4 | Developmental | Not teratogenic |
| | Mouse/18-month | 11.6 | Reproductive | No evidence |
| | RfD | 0.08 | Oncogenicity | E (no evidence) |

The structure, use, and hazard profiles for alachlor, acetochlor, metolachlor, and dimethenamid are provided in Table 16.66. All four chloracetamide herbicides are potential skin sensitizers. Alachlor and acetochlor both exhibit mutagenic potential and significant oncogenic potential in both rats and mice. Dimethenamid also has exhibited weak genotoxicity and a tumor response in the female rat. Metolachlor showed a weak oncogenic response in the liver of the female rat. A common mechanism of action has been proposed by the EPA for alachlor and acetochlor based on the carcinogenic potential of the quinine imine reactive intermediate [504].

## Cellulose and Lipid Synthesis Inhibitors (HRAC Code L and N)

Isoxaben is a member of the benzamide class of herbicides that block the incorporation of glucose into cellulose [571]. The mechanism by which isoxaben has this effect is not well understood, but it is unlikely relevant to humans. Quinclorac also blocks the radiolabeled uptake of glucose into cellulose in certain monocot species [301]; however, it has also been suggested that quinclorac activates the auxin receptor, which has been tentatively identified [668], leading to increased ACC (1-aminocyclopropane-1-carboxylic acid) synthase activity that results in the production of ethylene but also increased amounts of cyanide (Figure 16.27). Ethylene triggers biochemical changes leading to senescence in plants but cyanide, which is formed in grasses treated with quinclorac, is directly phytotoxic [247]. This pathway is not present in mammals, although TIR1 has homology with the human SKP2 (S-phase kinase-associated protein) and its corresponding gene [346].

The herbicidal mode of action for the thiocarbamate herbicides butylate and molinate is through the inhibition of fatty acid synthesis. Inhibition is achieved not by blocking ACCase (Figure 16.16) but rather by inhibiting long-chain fatty acid synthesis (Figure 16.25). Some of the thiocarbamate herbicides (EPTC, molinate,

**TABLE 16.62**
**Structures, Use, and Hazard Profiles for the Photobleaching Triketone and Isoxazole Herbicides (HRAC Code F2)**

| Chemical/Common Name | Structure | Principal Use/Crop | Application Rate (g active ingredient per hectare) |
|---|---|---|---|
| Sulcotrione (Mikito®) [172] | | Corn, sugar cane | 200–300 |
| Mesotrione (Callisto™) [122] | | Controls broadleaf and some grass weeds in maize | 70–225 |
| Isoxaflutole [115] | | Corn | 75–140 |

| Herbicide | Irritation: Eye | Irritation: Skin | $LD_{50}$ (mg/kg): Oral | $LD_{50}$ (mg/kg): Dermal | $LC_{50}$ (mg/L): Inhalation | Sensitization Potential | Signal Word |
|---|---|---|---|---|---|---|---|
| Sulcotrione | Mild irritant | Non-irritant | >5000 | >4000 | >1.6 | Positive | Caution |
| Mesotrione | Mild irritant | Slight irritant | >5000 | >5000 | >5.19 | Negative | Caution |
| Isoxaflutole | Mild irritant | Minimal | >5000 | >2000 | >5.3 | Negative | Caution |

| Herbicide | Species/Study | NOEL (mg/kg/day) | Toxicity Study | Hazard Indicator |
|---|---|---|---|---|
| Sulcotrione [172] | Rat/2-year | — | Mutagenicity | No evidence |
| | Dog/52-week | — | Developmental | Not teratogenic |
| | Mouse/18-month | — | Reproductive | NA |
| | RfD orADI | — | Oncogenicity | NA |
| Mesotrione [321,507] | Rat/2-year | 0.16[a] | Mutagenicity | Not mutagenic |
| | Dog/52-week | 10 | Developmental | Not teratogenic |
| | Mouse/18-month | 56.2 | Reproductive | Not a reproductive toxin |
| | RfD | 0.007 | Oncogenicity | Not carcinogenic |
| | | | Neurotoxicity | Not a neurotoxin |
| Isoxaflutole [337,436,468] | Rat/2-year | 2.0 | Mutagenicity | No evidence |
| | Dog/52-week | 45 | Developmental | Developmental toxicity |
| | Mouse/18-month | 3.2 | Reproductive | No evidence |
| | RfD | 0.002 | Neurotoxicity | Neurotoxicity |
| | | | Oncogenicity | Likely to be a carcinogen (liver tumors in both sexes of rats and mice) |

[a] NOAEL.

pebulate, and cycloate) have been determined to belong to a common mechanism grouping with the carbamate insecticides, based on their ability to inhibit acetyl cholinesterase [505]. However, the most sensitive toxicological endpoint of this class of herbicide is sciatic nerve degeneration, and it is the NOELs for these endpoints that are used to assess acute and chronic risk [515]. The EPA did not establish a common mechanism grouping for the thiocarbamate herbicides based on neuropathology because they concluded that the proposed mechanism underlying this response [631] was not adequately understood [514].

**TABLE 16.63**
**Structure, Use, and Hazard Profiles for the Photobleaching Triazole and Isoxazolidinone Herbicides (HRAC Code F3)**

| Herbicide | Structure | Principal Use/Crop | Application Rate (g active ingredient per hectare) |
|---|---|---|---|
| Amitrole (Amizol®) [20] | | Fruit trees, grapes, olives, ornamentals, cereal, IWC, aquatic plants | 1000–5000 |
| Clomazone (Command®) [50] | | Soybeans, peas peppers | 560–1700 |

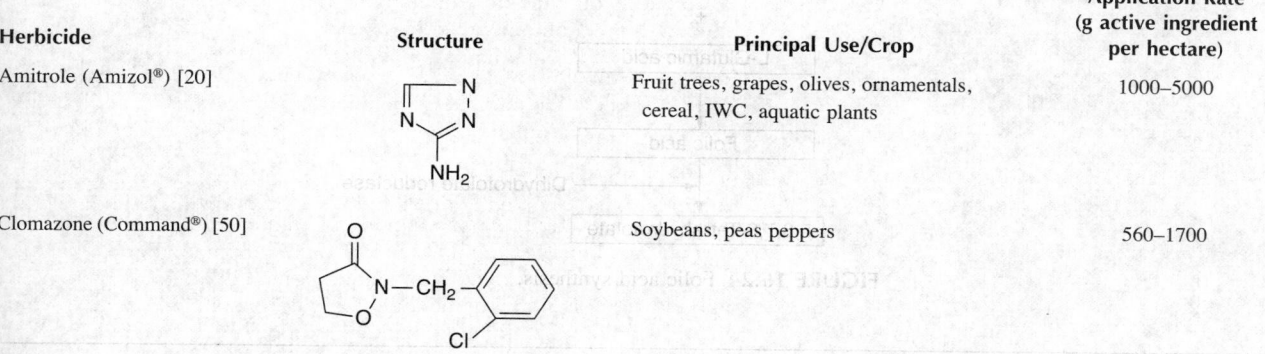

| Herbicide | Irritation: Eye | Irritation: Skin | $LD_{50}$ (mg/kg): Oral | $LD_{50}$ (mg/kg): Dermal | $LC_{50}$ (mg/L): Inhalation | Sensitization Potential | Signal Word |
|---|---|---|---|---|---|---|---|
| Amitrole | Slight irritant | Slight irritant | >5000 | >2000 | — | — | Caution |
| Clomazone | Non-irritant | Minimal | 2077 | >2000 | 4.23 (female) | Negative | Caution |

| Herbicide | Species/Study | NOEL (mg/kg/day) | Toxicity Study | Hazard Indicator |
|---|---|---|---|---|
| Amitrole [492] | Rat/2-year | 0.5 | Mutagenicity | No evidence |
| | Dog/52-week | NA | Developmental | Not teratogenic |
| | Mouse/18-month | 1.4 | Reproductive | No evidence |
| | RfD | Q* = 1.13 $(mg/kg/day)^{-1}$ | Oncogenicity | B2 (thyroid tumor in both sexes of rats and mice; liver tumor in mice) |
| Clomazone [485] | Rat/2-year | 4.3 | Mutagenicity | No evidence |
| | Dog/52-week | 12.5 | Developmental | Not teratogenic |
| | Mouse/18-month | 143 | Reproductive | No evidence |
| | Rfd | 0.043 | Oncogenicity | E (no evidence) |

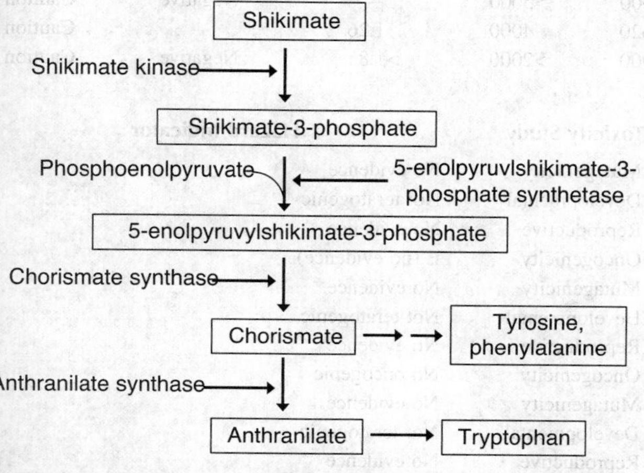


FIGURE 16.22 Aromatic amino acid biosynthetic pathway.

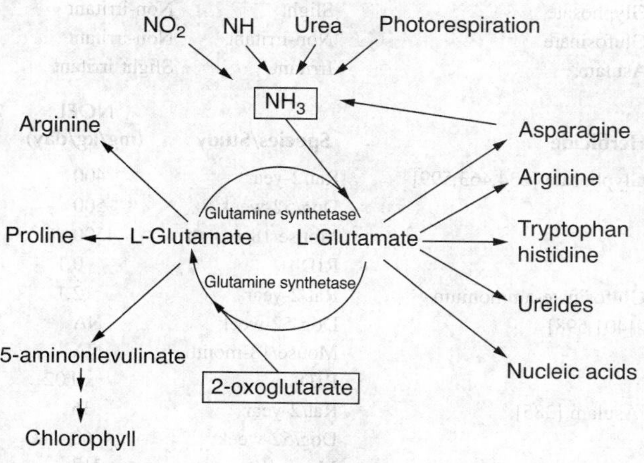


FIGURE 16.23 Nitrogen fixation and recycling in plants.

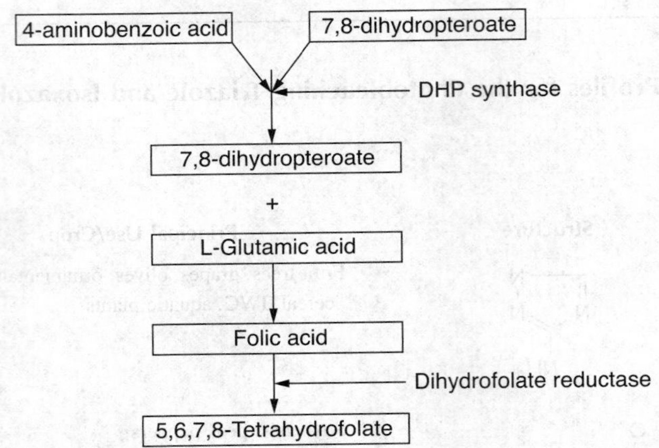


**FIGURE 16.24** Folic acid synthesis.

**TABLE 16.64**
**Structure, Use, and Hazard Profiles for EPSP Glutamine- and DHP Synthase-Inhibiting Glycine and Phosphinic Acid Herbicides (HRAC Codes G, H & I)**

| Herbicide | Structure | Principal Use/Crop | Application Rate (g active ingredient per hectare) |
|---|---|---|---|
| Glyphosate (Roundup®) EPSP synthase inhibitor [94] | $HO-C(=O)-CH_2NHCH_2P(=O)(OH)-OH$ | Corn, soybeans, IWC | 1500–4300 |
| Glufosinate-ammonium (Finale®) glutamine synthase inhibitor [93] | $CH_3P(=O)(O)-CH_2-CH_2-CH(NH_2)CO_2H\ \ NH_4^+$ | Fruit trees, grapes, rubber, palm ornamentals, vegetables, IWC | 400–1500 |
| Asulam (Asulux®) [21] | $H_2N-C_6H_4-SO_2NHCO_2CH_3$ | Sugar cane, alfalfa, banana, coffee, tea, cocoa, pasture, forestry | 1000–10,000 |

| Herbicide | Irritation: Eye | Irritation: Skin | $LD_{50}$ (mg/kg): Oral | $LD_{50}$ (mg/kg): Dermal | $LC_{50}$ (mg/L): Inhalation | Sensitization Potential | Signal Word |
|---|---|---|---|---|---|---|---|
| Glyphosate | Slight | Non-irritant | 5600 | >5000 | — | Negative | Caution |
| Glufosinate | Non-irritant | Non-irritant | 1620 | 4000 | 1.26 | — | Caution |
| Asulam | Irritant | Slight irritant | >5000 | >2000 | >1.8 | Negative | Caution |

| Herbicide | Species/Study | NOEL (mg/kg/day) | Toxicity Study | Hazard Indicator |
|---|---|---|---|---|
| Glyphosate [234,463,599] | Rat/2-year | 400 | Mutagenicity | No evidence |
| | Dog/52-week | 500 | Developmental | Not teratogenic |
| | Mouse/18-month | 4500 | Reproductive | No evidence |
| | RfD | 0.1 | Oncogenicity | E (no evidence) |
| Glufosinate-ammonium [401,598] | Rat/2-year | 2.1 | Mutagenicity | No evidence |
| | Dog/52-week | NA | Developmental | Not teratogenic |
| | Mouse/18-month | NA | Reproductive | No evidence |
| | RfD | 0.02 | Oncogenicity | No oncogenic |
| Asulam [385] | Rat/2-year | 36 | Mutagenicity | No evidence |
| | Dog/52-week | 60 | Developmental | Not teratogenic |
| | Mouse/18-month | 713 | Reproductive | No evidence |
| | RfD orADI | 0.36 | Oncogenicity | C (thyroid and adrenal gland tumors in male rats |

**TABLE 16.65**
**Structures, Use, and Hazard Profiles for the Dinitroaniline Microtuble Assembly Inhibiting Herbicides (HRAC Code K1)**

| Herbicide | Structure | Principal Use/Crop | Application Rate (g active ingredient per hectare) |
|---|---|---|---|
| Benfluralin (Balan®, Benefin®) [27] | 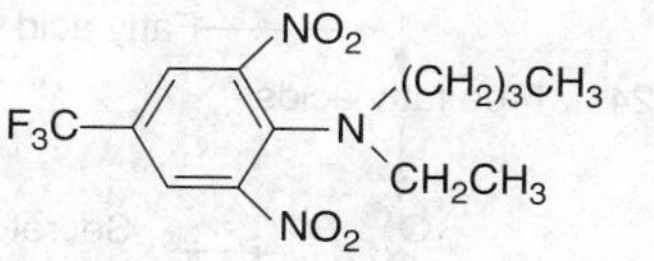 $F_3C$–(ring with two $NO_2$)–$N(CH_2CH_3)((CH_2)_3CH_3)$ | Alfalfa, clover, lettuce, and tobacco | 1260–1680 |
| Pendimethalin (Prowl®) [143] | 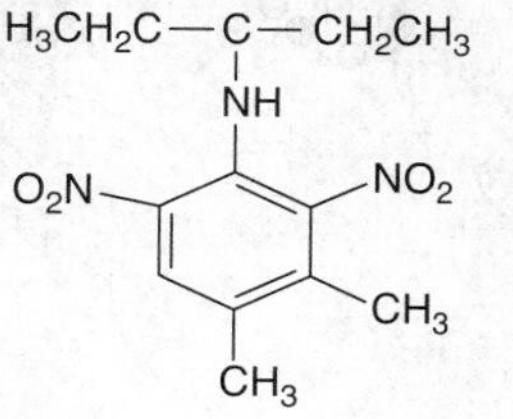 $H_3CH_2C$–C–$CH_2CH_3$ / NH / ring with $O_2N$, $NO_2$, $CH_3$, $CH_3$ | Corn, sorghum, rice, soybeans, cotton, potatoes, tobacco, sugarcane, beans, onions, and sunflower | 560–3360 |
| Trifluralin (Treflan®) [191] | 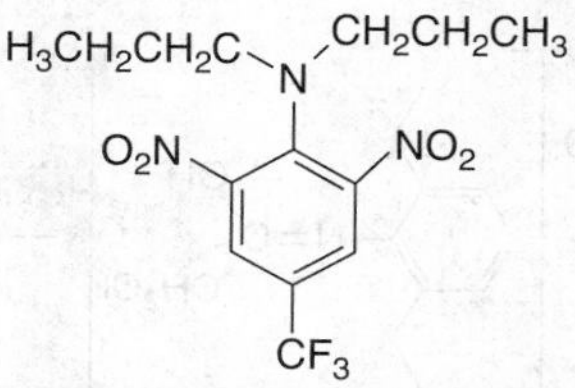 $H_3CH_2CH_2C$–N–$CH_2CH_2CH_3$ / ring with $O_2N$, $NO_2$, $CF_3$ | Alfalfa, asparagus, beans, carrots, celery, cole crops, cucurbits, onions, okra, peas, peppers, potatoes, sunflower, tomatoes, wheat, barley, flax soybeans corn, sorghum, and ornamentals | 500–1000 |

| Herbicide | Irritation | | $LD_{50}$ (mg/kg) | | $LC_{50}$ (mg/L) | Sensitization Potential | Signal Word |
|---|---|---|---|---|---|---|---|
| | Eye | Skin | Oral | Dermal | Inhalation | | |
| Benfluralin | Slight irritant | Slight irritant | >10,000 | >5000 | >2.3 | Positive | Caution |
| Pendimethalin | Slight irritant | Non-irritant | 1050 | >5000 | 320 (nominal) | Negative | Caution |
| Trifluralin | Slight irritant | Non-irritant | >5000 | >5000 | >4.8 | Positive | Caution |

| Herbicide | Species/Study | NOEL (mg/kg/day) | Toxicity Study | Hazard Indicator |
|---|---|---|---|---|
| Benfluralin [525] | Rat/2-year | 0.5 | Mutagenicity | No evidence |
| | Dog/52-week | 25 | Developmental | Not teratogenic |
| | Mouse/18-month | 6.5 | Reproductive | No evidence |
| | RfD | 0.005 | Oncogenicity | Suggestive (liver and throid tumors at doses > MTD) |
| Pendimethalin [443,444,615] | Rat/2-year | 10.0 | Mutagenicity | No evidence |
| | Dog/104-week | 12.5 | Developmental | Not teratogenic |
| | Mouse/18-month | 75 | Reproductive | No evidence |
| | RfD | 0.13 | Oncogenicity | C with RfD (thyroid follicular cell adenomas) |
| Trifluralin [408,536,627] | Rat/2-year | 2.5 | Mutagenicity | No evidence |
| | Dog/52-week | 2.4 | Developmental | Not teratogenic |
| | Mouse/18-month | 7.5 | Reproductive | No evidence |
| | RfD | 0.024 | Oncogenicity | C with Q* (bladder, kidney, thyroid tumors) |

The structure, use, and hazard profiles of the cellulose synthesis inhibitor isoxaben and quinclorac and the lipid synthesis inhibitors represented by butylate and molinate are given in Table 16.67. Isoxaben exhibits low acute toxicity and is only moderately toxic in repeat-dose studies (chronic RfD = 0.05 mg/kg/day), but it is a developmental toxin at maternally toxic doses. There was a positive micronucleus test, and adrenal gland and liver tumors occurred. The toxicity profile for butylate is unremarkable; it was not neurotoxic, nor did it have the effects on gonadal parameters that were described for molinate [631], although it was positive in the skin

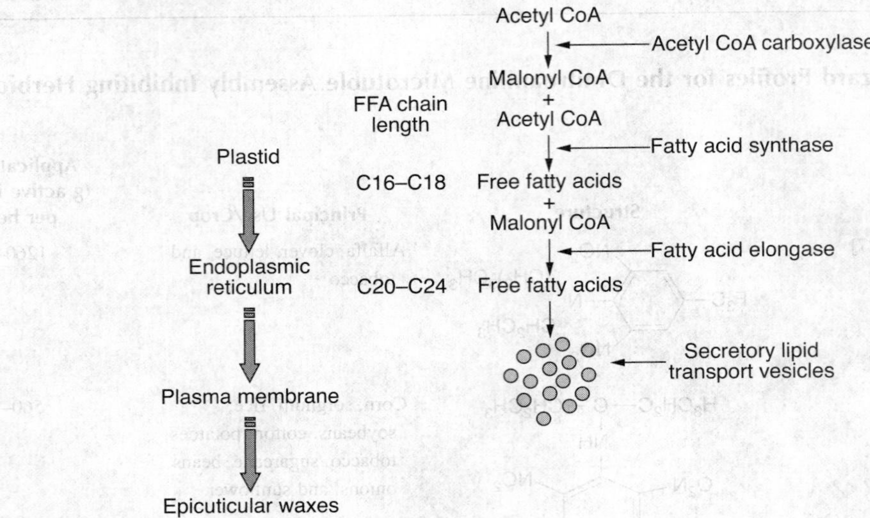


**FIGURE 16.25** Long-chain fatty acid synthesis in plants.

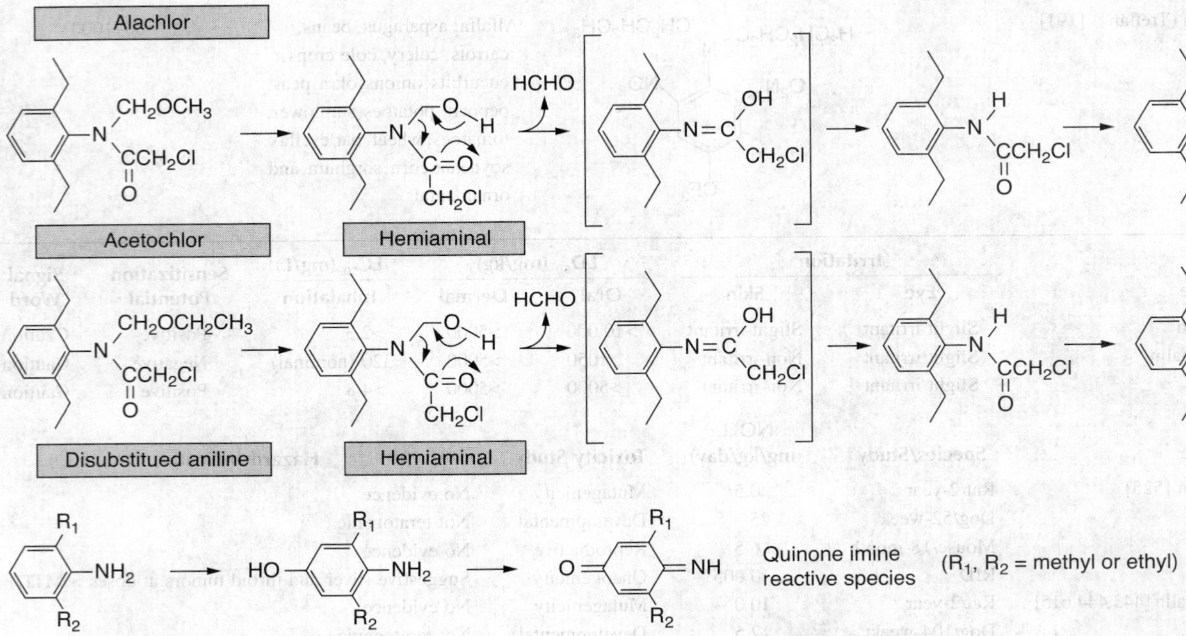

Alachlor

Acetochlor

Hemiaminal

Disubstitued aniline

Quinone imine reactive species

($R_1$, $R_2$ = methyl or ethyl)

**FIGURE 16.26** Chloroacetamide common mechanism grouping.

sensitization study. Molinate is acutely toxicity by inhalation ($LC_{50}$ = 0.003 mg/L); it has effects on sperm morphology in male rats and delays reproductive development in female rats. It was weakly positive in a mouse lymphoma study, but all other mutagenicity studies were negative [515]. An increased incidence of kidney tumors was found in rats at high doses. The EPA classified the oncogenic potential of molinate as "suggestive evidence for carcinogenicity but not sufficient to assess human carcinogenic potential," and a margin-of-exposure approach was recommended for risk assessment [511].

## Synthetic Auxin Mimics (Phenoxy, Benzoic, and Pyridine Acids) (HRAC Code O)

Indole 3-acetic acid, which is the plant hormone auxin plays a critical role in regulating plant cell growth and differentiation by binding to its receptors [199,215,216]. The mechanism of auxin signaling has been the subject of intense research efforts [309,336], and progress has been made in identifying the critical receptor proteins, the arabidopsis TIR1-related F-box proteins AFB1, AFB2, and AFB3 [215,216,668], which are involved in the acti

## TABLE 16.66
## Structure, Use, and Hazard Profiles for the Chloroacetamide Inhibitors of Very-Long-Chain Fatty Acid Synthesis (HRAC Code K3)

| Herbicide | Structure | Principal Use/Crop | Application Rate (g active ingredient per hectare) |
|---|---|---|---|
| Alachlor (Lasso®) [16] | 2,6-diethylphenyl–N($CH_2OCH_3$)($COCH_2Cl$) | Corn, beans, peanuts, sorghum, soybeans, sunflowers, and ornamentals | 1500–4500 |
| Acetochlor (Surpass®) [13] | 2-methyl-6-ethylphenyl–N($CH_2OCH_2CH_3$)($COCH_2Cl$) | Corn, soybeans, sorghum, and wheat | 900–3360 |
| Metolachlor (Dual®) [128] | 2-ethyl-6-methylphenyl–N($CH(CH_3)CH_2OCH_3$)($COCH_2Cl$) | Corn, soybeans, and sorghum, cucurbits, onions, peas, pecans, peppers, potatoes, and sugar beets | 1250–6200 |
| Dimethenamid (Frontier®) [72] | 2,4-dimethylthien-3-yl–N($CH(CH_3)CH_2OCH_3$)($COCH_2Cl$) | Corn and soybeans | 850–1440 |

| Herbicide | Irritation: Eye | Irritation: Skin | $LD_{50}$ (mg/kg) Oral | $LD_{50}$ (mg/kg) Dermal | $LC_{50}$ (mg/L) Inhalation | Sensitization Potential | Signal Word |
|---|---|---|---|---|---|---|---|
| Alachlor | Non-irritant | Non-irritant | 930 | 13,300 | >1.04 | Positive | Caution |
| Acetochlor | Slight irritant | Non-irritant | 2148 | 4166 | >3.0 | Positive | Caution |
| Metolachlor | Non-irritant | Minimal irritant | >2780 | >10000 | >1.75 | Positive | Caution |
| Dimethenamid | Slight irritant | Non-irritant | 1570 | >2000 | >5.0 | Positive | Caution |

| Herbicide | Species/Study | NOEL (mg/kg/day) | Toxicity Study | Hazard Indicator |
|---|---|---|---|---|
| Alachlor [230,454,504] | Rat/2-year | 2.5 | Mutagenicity | Positive (UDS) |
| | Dog/52-week | 1.0 | Developmental | Not teratogenic |
| | Mouse/18-month | 16.6 | Reproductive | No evidence |
| | RfD | 0.01 | Oncogenicity | C/RfD (nasal, rats; lung, mice) |
| Acetochlor [245,380,504,552] | Rat/2-year | 8.0 | Mutagenicity | Positive (CHO, UDS, mouse lymphoma) |
| | Dog/52-week | 2.0 | Developmental | Not teratogenic |
| | Mouse/18-month | 13 | Reproductive | No evidence |
| | RfD | 0.02 | Oncogenicity | B2 (liver, thyroid, nasal tumors in rats; lung tumors in mice) |
| Metolachlor [391,504] | Rat/2-year | 15 | Mutagenicity | No evidence |
| | Dog/52-week | 10 | Developmental | Not teratogenic |
| | Mouse/18-month | 120 | Reproductive | No evidence |
| | RfD | 0.1 | Oncogenicity | C/RfD ( liver tumors in female rat) |
| Dimethenamid [397] | Rat/2-year | 5.0 | Mutagenicity | Weak positive (CHO, UDS) |
| | Dog/52-week | 9.6 | Developmental | Not teratogenic |
| | Mouse/18-month | 40 | Reproductive | No evidence |
| | RfD | 0.05 | Oncogenicity | C/RfD (liver, ovary; female rat) |

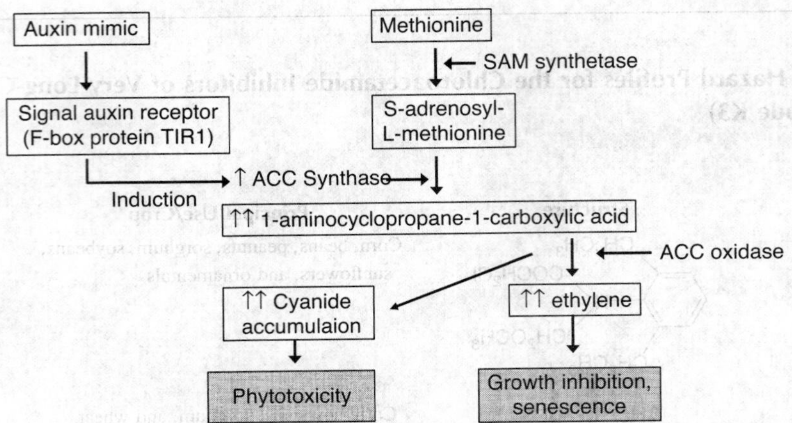


**FIGURE 16.27** Auxin signal transduction.

vation of these signaling pathways. Auxin binding mediates the association of a set of transcriptional repressor proteins (AUX/IAA proteins; 29 members in arabidopsis) with another protein (SCF) complex. The SCF complex mediates polyubiquitination of the repressor proteins, which are then targeted for degradation by proteasome. With the repressor proteins removed, various sets of genes are induced, including those involved with cell elongation and cell division [199]. Grossman has suggested that the herbicidal action of the auxin mimics is through auxin-induced ACC synthase upregulation which results in ethylene-mediated senescence as well as cyanide-mediated phytotoxicity [247,664]. These pathways are not present in mammalian systems although homologs of the F-box proteins have been identified in mammals [346]. Table 16.68 provides the structure, use, and hazard profiles for the auxin mimics 2,4-D (phenoxy acid), dicamba (benzoic acid), and clopyralid and picloram (pyridine acids). The hazard profile for 2,4-D is unremarkable. Dicamba, clopyralid, and picloram are all eye irritants, and dicamba and picloram are potential skin sensitizers; otherwise, the hazard profiles for these chemicals are unremarkable.

### Herbicides with Unknown Mechanisms of Action (HRAC Code Z)

The structures, uses, and hazard profiles of difenzoquat and monosodium methanearsonic acid (MSMA) are presented in Table 16.69. Difenzoquat is labeled with the signal word "Danger" because it is corrosive to the eyes but otherwise is toxicologically unremarkable. MSMA, an organic arsenical, is a mild skin and eye irritant. MSMA caused decrease fertility in the rat reproduction study and is classified as a Category B2 (probable human) carcinogen based on bladder tumors in the rat.

## CONCLUSIONS AND QUESTIONS

According to a Byzantine proverb, "He who has bread may have many problems, but he who lacks it has only one." In a world where the current population is burgeoning, one might say that collectively we face more than 6.6 billion individual problems, ultimately leading to only one question: How will we sustain and possibly increase food production in the face of declining arable land and the future demand to convert a part of biomass production into energy production? A corollary question is: How will we produce food locally where it is needed or redistribute it from regions of high productivity to other needy parts of the world? All this must be done in the face of strident attacks, at least in the developed countries, against the basic sciences and the economy of food production that up to now have permitted the diversion of human resources away from growing food to other economic, political, and cultural activities.

So, how will we overcome these apparently insurmountable challenges? Perhaps we should look to the past 60 years, the period of time that might be considered the period of modern agriculture, for clues as to what has worked and what has failed. The most noticeable advances have included:

- The development of strains of plants that are resistant to disease or result in a greater yield using conventional breeding techniques or, more recently, molecular marker-assisted breeding methods
- The introduction of mechanized farming methods that resulted in a reduction in the reliance on human or animal labor
- The utilization of cropping practices and fertilizers to enhance productivity

## TABLE 16.67
## Structure, Use, and Hazard Profile for Benzamide, Quinoline Carboxylic Acid, and Thiocarbamate Inhibitors of Cellulose or Lipid Synthesis (HRAC Codes L and N)

| Herbicide | Structure | Principal Use/Crop | Application Rate (g active ingredient per hectare) |
|---|---|---|---|
| Isoxaben (Gallery®) [114] | $OCH_3$, $OCH_3$; —CO—NH— isoxazole (O—N) —$C(CH_2CH_3)_2CH_3$ | Turf, ornamentals, non-bearing fruit, and nut trees, and conifers | 50–1000 |
| Quinclorac (Facet®) [165] | $CO_2H$, Cl, N, Cl (quinoline) | Soyabean, rice | 250–750 |
| Butylate (Sutan®) [37] | $H_3C-H_2C-S-C(=O)-N(CH_2-CH(CH_3)CH_3)_2$ | Corn | 3000–4000 |
| Molinate (Ordram®) [132] | $H_3C-H_2C-S-C(=O)-N$ (azepane ring) | Rice | 2000–4000 |

| Herbicide | Irritation: Eye | Irritation: Skin | $LD_{50}$ (mg/kg): Oral | $LD_{50}$ (mg/kg): Dermal | $LC_{50}$ (mg/L): Inhalation | Sensitization Potential | Signal Word |
|---|---|---|---|---|---|---|---|
| Isoxaben | Moderate irritant | Slight irritant | >10,000 | >2000 | >2.68 | Negative | Caution |
| Quinclorac | Non-irritant | Non-irritant | 2680 | >2000 | >5.2 | — | Caution |
| Butylate | Non-irritant | Mild irritant | 4659 | 1659 | 4.64 | Positive | Caution |
| Molinate | Moderate irritant | Mild irritant | 720 | ~4000 | 0.003 | Negative | Danger |

| Herbicide | Species/Study | NOEL (mg/kg/day) | Toxicity Study | Hazard Indicator |
|---|---|---|---|---|
| Isoxaben [606] | Rat/2-year | 5.0 | Mutagenicity | Positive micronucleus test |
| | Dog/52-week | 10.0 | Developmental | Positive at maternal toxic doses |
| | Mouse/18-month | 14.3 | Reproductive | Not a reproductive toxin |
| | RfD orADI | 0.05 | Oncogenicity | C (adrenal, and liver tumors) |
| Quinclorac [247] | Rat/2-year | 675 | Mutagenicity | Not mutagenic |
| | Dog/52-week | 33 | Developmental | Not teratogenic |
| | Mouse/18-month | 42 | Reproductive | Not a reproductive toxin |
| | RfD orADI | — | Oncogenicity | Not carcinogenic |
| Butylate [379,584] | Rat/2-year | 50 | Mutagenicity | No evidence |
| | Dog/52-week | 5.0 | Developmental | Not teratogenic |
| | Mouse/18-month | 20 | Reproductive | No evidence |
| | RfD orADI | 0.05 | Oncogenicity | E (no evidence) |
| Molinate [511,514,515,610,631] | Rat/2-year | 0.3 | Mutagenicity | Weak positive, mouse lymphoma |
| | Dog/52-week | 1.0[a] | Developmental | Not teratogenic |
| | Mouse/18-month | 1.0 | Reproductive | Abnormal sperm morphology; delayed vaginal opening |
| | RfD orADI | 0.001 | Oncogenicity | Suggestive (kidney tumors in rats) MOE approach |

[a] LOAEL.

**TABLE 16.68**
**Structure, Use, and Hazard Profiles for the Phenoxy, Benzoic, and Pyridine Acids Herbicides That Mimic Indole Acetic Acid (Auxin) (HRAC Code O)**

| Herbicide | Structure | Principal Use/Crop | Application Rate (g active ingredient per hectare) |
|---|---|---|---|
| 2,4-D Wedare® [61] | Cl–C6H3(Cl)–OCH2COOH | Turf, cereals, sorghum, corn, soybeans, asparagus, and fruit trees | 280–2240 |
| Dicamba (Banvel®) [65] | C6H2(Cl)2(OCH3)CO2H | Corn, turf, sorghum, cereals, pastures, and asparagus | 100–400 |
| Clopyralid (Reclam®) [50a] | Cl2-pyridine–COH2 | Sugar beets, corn, grass seed, conifers, and pasture | 70–560 |
| Picloram (Tordon®) [145] | Cl3(NH2)-pyridine–COH2 | Industrial weed control, forestry, pasture, and range land | 35–1120 |

| Herbicide | Irritation: Eye | Irritation: Skin | $LD_{50}$ (mg/kg): Oral | $LD_{50}$ (mg/kg): Dermal | $LC_{50}$ (mg/L): Inhalation | Sensitization Potential | Signal Word |
|---|---|---|---|---|---|---|---|
| 2,4-D | Severe irritant | Moderate irritant | 639 | >2000 | 1.8 | Negative | Warning |
| Dicamba | Corrosive | Non-irritant | 1851 | >2000 | >9.6 | Positive | Danger |
| Clopyralid | Severe irritant | Slight irritant | 4300 | >2000 | 1.3 | Negative | Warning |
| Picloram | Moderate irritant | Non-irritant | 4012 | >2000 | >0.035 | Positive | Danger |

| Herbicide | Species/Study | NOEL (mg/kg/day) | Toxicity Study | Hazard Indicator |
|---|---|---|---|---|
| 2,4-D [481,540] | Rat/2-year | 5.0 | Mutagenicity | No evidence |
| | Dog/52-week | 1.0 | Developmental | Not teratogenic |
| | Mouse/18-month | 1.0 | Reproductive | No evidence |
| | RfD orADI | 0.01 | Oncogenicity | D (Not classifiable) |
| Dicamba [460,557] | Rat/2-year | 125 | Mutagenicity | Positive (*B. subtilis*; UDS) |
| | Dog/52-week | 60 | Developmental | Not teratogenic |
| | Mouse/18-month | 108 | Reproductive | No evidence |
| | RfD orADI | 0.6 | Oncogenicity | D (not classifiable) |
| Clopyralid [421] | Rat/2-year | 50 | Mutagenicity | No evidence |
| | Dog/52-week | 100 | Developmental | Not teratogenic |
| | Mouse/18-month | 500 | Reproductive | No evidence |
| | RfD orADI | 0.5 | Oncogenicity | E (no evidence) |
| Picloram [490] | Rat/2-year | 20 | Mutagenicity | No evidence |
| | Dog/52-week | 35 | Developmental | Not teratogenic |
| | Mouse/18-month | 500 | Reproductive | No evidence |
| | RfD orADI | 0.2 | Oncogenicity | E (no evidence) |

**TABLE 16.69**
**Structure, Use, and Hazard Profiles for Organoarsenical and Pyrazolium Herbicides with an Unknown Mode of Action (HRAC Code Z)**

| Herbicide | Structure | Principal Use/Crop | Application Rate (g active ingredient per hectare) |
|---|---|---|---|
| Difenzoquat (Avenge®) [70] | [structure: 1,2-dimethyl-3,5-diphenylpyrazolium; N–N⁺, $H_3C$, $CH_3$] | Barley and wheat | 700–1120 |
| Monosodium methanearsonate (MSMA) (Drexar®) [127] | $H_3C-As(=O)(OH)-O^-$ $Na^+$ | Controls broadleaf weeds in noncrop areas, cotton, and turf | 2220–2770 |

| Herbicide | Irritation: Eye | Irritation: Skin | $LD_{50}$ (mg/kg): Oral | $LD_{50}$ (mg/kg): Dermal | $LC_{50}$ (mg/L): Inhalation | Sensitization Potential | Signal Word |
|---|---|---|---|---|---|---|---|
| Difenzoquat | Corrosive | Moderate irritant | 373E | >2000 | 0.5 | Negative | Danger |
| MSMA | Mild irritant | Mild irritant | 1059E | >2000 | >6.0 | — | Caution |

| Herbicide | Species/Study | NOEL (mg/kg/day) | Toxicity Study | Hazard Indicator |
|---|---|---|---|---|
| Difenzoquat [383,591] | Rat/2-year | 25 | Mutagenicity | — |
| | Dog/52-week | 20 | Developmental | Not teratogenic |
| | Mouse/18-month | 75 | Reproductive | No evidence |
| | RfD | 0.20 | Oncogenicity | E (no evidence) |
| MSMA [392,611] | Rat/2-year | 3.2 | Mutagenicity | NA |
| | Dog/52-week | — | Developmental | Not teratogenic |
| | Mouse/18-month | — | Reproductive | Decreased fertility |
| | RfD | 0.01 | Oncogenicity | B2 (bladder fibrosarcomas) |

- The use of agricultural chemicals to enhance yield by reducing loss to disease and the destruction by pests
- The development of highly effective pesticides that control pests with a near-pharmacologic level of potency (i.e., grams active ingredient per hectare)

One must also acknowledge, however, that there have been some failures to achieve the main goal of pesticide science—namely, to achieve the selectivity of effect and thereby to prevent unwanted effects on animals or the environment. Thus, pesticide development is about one thing and one thing only: selectivity. The mission of pesticide scientists is to find chemicals that control pests at some reasonable and economically achievable dose, ultimately by interfering with a fundamental life process of the pest, without having any effect on beneficial plants, insects, or mammals, including humans.

In reviewing the more that 50 different modes of actions discussed in this chapter, one cannot help but be amazed at the ingenuity of the researchers who have spent their lives in this hunt for better and more effective pesticides. These men and women have achieved success mainly using relatively crude biological screening tools combined with ingenious chemical synthesis strategies, which, with luck and a lot of persistence, have led to the discovery of pesticide candidates that have then undergone extensive optimization to finally become useful agricultural tools. In hindsight, once the molecular mechanisms have been described, it is incredible that the specificity of effect was achieved using what could be best described as an intelligent random search process. Even today, with high-throughput screening methods and the use of combinatorial chemistry, the crop protection industry has not turned to rational design, largely because the science still is not there to build a molecule from the ground up based on what is currently known about molecular targets.

Using traditional methods, then, it may still be possible to increase selectivity, lower use rates, and identify chemicals with new modes of actions. The incorporation of genes that create tolerance in beneficial plants and the expression of pesticides by the species that are to be protected are examples from the recent past. In the future, it is likely that plants and animals will continue to be genetically modified or bred using more sophisticated selection

**TABLE 16.70**
**Index of Fungicides, Insecticides, and Herbicides**

| Fungicides | | | |
|---|---|---|---|
| **Common Name** | **Chemical Class** | **MOA** | **Table No.** |
| Azoxystrobin | Methoxyacrylate | Inhibit mitochondrial respiration | 16.8 |
| Benomyl | Benzimidazole | Inhibit mitosis and cell division | 16.6 |
| Captan | Phthalimide | Reacts with sulfhydryl groups | 16.18 |
| Chlorothalonil | Chloronitriles | Conjugates with cellular thiols | 16.19 |
| Cyproconazole | Triazole | Inhibit sterol biosynthesis | 16.13, 16.14, 16.15 |
| Cyprodinil | Anilinopyrimidine | Inhibit methionine biosynthesis | 16.10 |
| Difenconazole | Triazole | Inhibit sterol biosynthesis | 16.13, 16.14, 16.15 |
| Fenbuconazole | Triazole | Inhibit sterol biosynthesis | 16.13, 16.14, 16.15 |
| Fenpiclonil | Phenylpyrrole | Inhibit osmotic signal transduction | 16.11 |
| Ferbam | Dithiocarbamate | Multi-site contact activity | 16.16 |
| Fludioxonil | Phenylpyrrole | Inhibit osmotic signal transduction | 16.11 |
| Flutolanil | Carboxamide | Inhibit mitochondrial respiration | 16.7 |
| Hexaconazole | Triazole | Inhibit sterol biosynthesis | 16.13, 16.14, 16.15 |
| Imazalil | Imidazole | Inhibit sterol biosynthesis | 16.13, 16.14, 16.15 |
| Iprodione | Dicarboxyimide | Inhibit lipid synthesis | 16.12 |
| Mancozeb | Ethylenebisdithio-carbamates | Reacts with sulfhydryl groups | 16.17 |
| Maneb | Ethylenebisdithio-carbamates | Reacts with sulfhydryl groups | 16.17 |
| Mefenoxam | Acylalanine | Inhibit nucleic acid synthesis | 16.5 |
| Myclobutanil | Triazole | Inhibit sterol biosynthesis | 16.13, 16.14, 16.15 |

methods to achieve improved traits that create greater health benefits or facilitate the preprocessing of food and fiber. Targeted delivery of pesticides, whether derived from natural sources or that are the result of chemical optimization of what has been designed by nature, may be another way to move toward solutions that have lower environmental impact. Whatever the outcome, the next 50 years should prove to be critical, given current population predictions and the emerging environmental and energy crises.

## QUESTIONS

1. Farmers must contend with some 80,000 plant diseases, 30,000 species of weeds, 1000 species of nematodes, and more than 10,000 species of insects. Today, national and international agricultural organizations estimate that as much as 45% of the world's crops continues to be lost to these types of hazards. In the United States alone, about $20 billion worth of crops (one tenth of production) is lost each year. What do you think would be the status of our national food production capacity without the use of pesticides?
2. Who ensures that pesticides can be used without unacceptable hazard to the consumer to protect food crops and maximize yields?
3. How stringent are the testing requirements for the registration of a pesticide when compared to those for products used in the household and yard, industrial chemicals, or even pharmaceuticals?
4. Has the introduction of pesticides into your food supply had a positive or negative impact on your life?

## REFERENCES


1. Abel, E.L., Opp, S.M., Verlinde, C.L.M.J., Bammler, T.K., and Eaton, D.L. (2004): Characterization of atrazine biotransformation by human and murine glutathione S-transferase. *Toxicol. Sci.*, 80:230–238.
2. Altobelli, D., Martire, M. Maurizi, S., and Preziosi, P. (2001): Interaction of formamidine pesticides with the presynaptic $\alpha_2$-adrenoceptor regulating [$^3$H]noradrenaline release from rat hypothalamic synaptosomes. *Toxicol Pharmacol.*, 172:179–185.
3. Anke, T., Hecht, H.J., Schramm, G., and Steglich, W. (1979): Antibiotics from Basidiomycetes. IX. Oudemansin, an antifungal antibiotic from *Oudemansiella mucida* (Schrader ex Fr.) Hoehnel (Agaricales), *J. Antibiot.*, 32:1112–1117.
4. Avenot, H., Simoneau, P., Iacomi-Vasilescu, B., and Bataille-Simoneau, N. (2005): Characterization of mutations in the two-component histidine kinase gene AbNIK1 from *Alternaria brassicicola* that confer high dicarboximide and phenylpyrrole resistance. *Curr. Genet.*, 47:234–243.
5. Bernard, B.K. and Gordon, E.B. (2000): An evaluation of the common mechanism approach to the Food Quality Protection Action: captan and four related fungicides, a practical example. *Int. J. Toxicol.*, 19:43–61.

**TABLE 16.70 (cont.)**
**Index of Fungicides, Insecticides, and Herbicides**

**Fungicides (cont.)**

| Common Name | Chemical Class | MOA | Table No. |
|---|---|---|---|
| Oxadixyl | Oxazolidinone | Inhibit nucleic acid synthesis | 16.5 |
| Picoxystrobin | Methoxyacrylate | Inhibit mitochondrial respiration | 16.8 |
| Prochloraz | Imidazole | Inhibit sterol biosynthesis | 16.13, 16.14, 16.15 |
| Propiconazole | Triazole | Inhibit sterol biosynthesis | 16.13, 16.14, 16.15 |
| Pyrimethanil | Anilinopyrimidine | Inhibit methionine biosynthesis | 16.10 |
| Tebuconazole | Triazole | Inhibit sterol biosynthesis | 16.13, 16.14, 16.15 |
| Thiabendazole | Benzimidazole | Inhibit mitosis and cell division | 16.6 |
| Thiophanate-methyl | Thiophanate | Inhibit mitosis and cell division | 16.6 |
| Thiram | Dithiocarbamate | Multi-site contact activity | 16.16 |
| Triadimefon | Triazole | Inhibit sterol biosynthesis | 16.13, 16.14, 16.15 |
| Triadimenol | Triazole | Inhibit sterol biosynthesis | 16.13, 16.14, 16.15 |
| Trifloxystobin | Oximinoacetate | Inhibit mitochondrial respiration | 16.8 |
| Triphenyltin (fentin) | Organotin | Inhibits mitochondrial respiration | 16.9 |
| Vinclozolin | Dicarboxyimide | Inhibit lipid synthesis | 16.12 |
| Zineb | Ethylenebisdithio-carbamates | Reacts with sulfhydryl groups | 16.17 |
| Ziram | Dithiocarbamate | Multi-site contact activity | 16.16 |

**Insecticides**

| Common Name | Chemical Class | MOA | Table No. |
|---|---|---|---|
| Abamectin | Avermectin | Chloride channel activator | 16.32 |
| Acephate | Organophosphate | Inhibit acetylcholinesterase | 16.23, 16.24 |
| Aldicarb | Carbamate | Inhibit acetylcholinesterase | 16.21, 16.22 |
| Amitraz | Formamidine | Octopaminergic agonist | 16.38 |
| Azinphos-methyl | Organophosphate | Inhibit acetylcholinesterase | 16.23, 16.24 |
| Bioallethrin | Non-cyano pyrethroid | Sodium channel modulator | 16.27, 16.28 |
| Carbaryl | Carbamate | Inhibit acetylcholinesterase | 16.21, 16.22 |
| Carbofuran | Carbamate | Inhibit acetylcholinesterase | 16.21, 16.22 |
| Chlorfenapyr | Pyrrole | Uncouple oxidative phosphorylation | 16.39, 16.40 |
| Chlorpyrifos | Organophosphate | Inhibit acetylcholinesterase | 16.23, 16.24 |
| Clofentezine | Phenyltetrazine | Mite growth inhibitor | 16.34 |
| Cyfluthrin | Cyano pyrethroid | Sodium channel modulator | 16.29, 16.30 |
| Cypermethrin | Cyano Pyrethroid | Sodium channel modulator | 16.29, 16.30 |
| Cyromazine | Aminotriazine | Molt disrupter, larvacide | 16.34 |
| DDT | Organochlorine | Sodium channel modulator | 16.26 |
| Delta-endotoxin | Source: Bacillus thuringiensis | Create ion pores in insect mid-gut | 16.35 |
| Deltamethrin | Cyano Pyrethroid | Sodium channel modulator | 16.29, 16.30 |
| Diazinon | Organophosphate | Inhibit acetylcholinesterase | 16.23, 16.24 |
| Dichlorvos | Organophosphate | Inhibit acetylcholinesterase | 16.23, 16.24 |
| Diflubenzuron | Benzylurea | Inhibit chitin biosynthesis | 16.36 |
| Emamectin benzoate | Avermectin | Chloride channel activator | 16.32 |
| Endosulfan | Cyclodiene | GABA antagonist | 16.25 |
| Fenoxycarb | Carbamate | Juvenile hormone mimic | 16.33 |
| Fipronil | Phenylpyrazole | GABA antagonist | 16.25 |
| Hexythiazox | Carboxamide | Mite growth inhibitor | 16.34 |
| Hydramethylnon | Amidinohydrazone | Inhibit mitochondrial respiration | 16.39, 16.40 |
| Imidacloprid | Neonicotinoid | Nicotinic receptor agonist/antagonist | 16.31 |
| Lambda Cyhalothrin | Cyano Pyrethroid | Sodium channel modulator | 16.29, 16.30 |
| Lindane (Gamma-HCH) | Organochlorine | GABA antagonist | 16.25 |
| Malathion | Organophosphate | Inhibit acetylcholinesterase | 16.23, 16.24 |
| Methomyl | Carbamate | Inhibit acetylcholinesterase | 16.21, 16.22 |
| Methoprene | Sesquiterpenoid | Juvenile hormone mimic | 16.33 |
| Methoxychlor | Organochlorine | Sodium channel modulator | 16.26 |
| Milbemycin | Milbemycin | Chloride channel activator | 16.32 |

**TABLE 16.70 (cont.)**
**Index of Fungicides, Insecticides, and Herbicides**

**Insecticides (cont.)**

| Common Name | Chemical Class | MOA | Table No. |
|---|---|---|---|
| Monocrotophos | Organophosphate | Inhibit acetylcholinesterase | 16.23, 16.24 |
| Nicotine | Nicotine | Nicotinic receptor agonist/antagonist | 16.31 |
| Permethrin | Non-Cyano Pyrethroid | Sodium channel modulator | 16.27, 16.28 |
| Piperonyl butoxide | Benzodioxole | Inhibit P450 monooxygenases | 16.38 |
| Propoxur | Carbamate | Inhibit acetylcholinesterase | 16.21, 16.22 |
| Pymetrozine | Azomethine Pyridine | Selective feeding blocker | 16.33 |
| Pyridaben | Pyridazinone | Inhibit mitochondrial respiration | 16.39, 16.40 |
| Pyriproxfen | Alkoxypyrimidine | Juvenile hormone mimic | 16.33 |
| Resmethrin | Non-Cyano Pyrethroid | Sodium channel modulator | 16.27, 16.28 |
| Rotenone | Source: Derris root | Inhibit mitochondrial respiration | 16.39, 16.40 |
| Spinosad | Spinosyn | Nicotinic receptor agonist/antagonist | 16.32 |
| Tebufenozide | Diacylhydrazine | Ecdysone agonist | 16.37 |
| Teflubenzuron | Benzylurea | Inhibit chitin biosynthesis | 16.36 |
| Tefluthrin | Non-Cyano Pyrethroid | Sodium channel modulator | 16.27, 16.28 |
| Thiamethoxam | Neonicotinoid | Nicotinic receptor agonist/antagonist | 16.31 |

**Herbicides**

| Common Name | Chemical Class | MOA | Table No. |
|---|---|---|---|
| Acetochlor | Chloroacetamide | Inhibit VLCFA | 16.66 |
| Acifluorfen | Diphenylethers | Inhibit protoporphyrinogen oxidase | 16.58 |
| Alachlor | Chloroacetamide | Inhibit VLCFA | 16.66 |
| Ametryn | Triazine | Inhibit Photosystem II | 16.51, 16.52 |
| Amitrole | Triazole | Inhibit carotenoid biosynthesis | 16.63 |
| Asulam | Carbamate | Inhibit dihydro-pteroate synthase | 16.64 |
| Atrazine | Triazine | Inhibit photosystem II | 16.51, 16.52 |
| Benfluralin | Dinitroaniline | Inhibit microtubule assembly | 16.65 |
| Bentazone | Benzothiadiazinone | Inhibit photosystem II | 16.56 |
| Bensulfuron-methyl | Sulfonylurea | Inhibit acetolactate synthetase | 16.45, 1646, 16.47 |
| Bromacil | Uracil | Inhibit photosystem II | 16.53 |
| Butafenacil | Pyrimidindione | Inhibit protoporphyrinogen oxidase | 16.60 |
| Butylate | Thiocarbamate | Inhibit lipid synthesis | 16.67 |
| Carfentrazone-ethyl | Triazolinone | Inhibit protoporphyrinogen oxidase | 16.59 |
| Chlorimuron-ethyl | Sulfonylurea | Inhibit acetolactate synthetase | 16.45, 16.46, 16.47 |
| Chlorsulfuron | Sulfonylurea | Inhibit acetolactate synthetase | 16.45, 16.46, 16.47 |
| Clethodim | Cyclohexanedione | Inhibit acetyl–CoA carboxylase | 16.44 |
| Clodinafop propargyl | Aryloxyphenoxy-propionate | Inhibit acetyl–CoA carboxylase | 16.42, 16.43 |
| Clomazone | Isoxazolidinone | Inhibit carotenoid biosynthesis | 16.63 |
| Cyanazine | Triazine | Inhibit photosystem II | 16.51, 16.52 |
| 2,4-D | Phenoxycarboxylic acid | Auxin hormone mimic | 16.68 |
| Dicamba | Benzoic acid | Auxin hormone mimic | 16.68 |
| Dichlobenil | Nitrile | Inhibit photosystem II | 16.55 |
| Diclofop-methyl | Aryloxyphenoxy-propionate | Inhibit acetyl–CoA carboxylase | 16.42, 16.43 |
| Difenzoquat | Pyrazolium | Unknown mode of action | 16.69 |
| Dimethenamid | Chloroacetamide | Inhibit VLCFA | 16.66 |
| Diuron | Urea | Inhibit photosystem II | 16.54 |
| Diquat | Bipyridylium | Inhibit photosystem I | 16.57 |
| Fenoxaprop-p-ethyl | Aryloxyphenoxy-propionate | Inhibit acetyl–CoA carboxylase | 16.42, 16.43 |
| Fluazifop-p-butyl | Aryloxyphenoxy-propionate | Inhibit acetyl–CoA carboxylase | 16.42, 16.43 |
| Fluthiacet-methyl | Thiadiazole | Inhibit protoporphyrinogen oxidase | 16.59 |
| Flumetsulam | Triazolopyrimidine | Inhibit acetolactate synthetase | 16.50 |
| Flumiclorac-pentyl | *N*-Pyenylphlhalimide | Inhibit protoporphyrinogen oxidase | 16.59 |
| Fluometuron | Urea | Inhibit photosystem II | 16.54 |
| Fluridone | Pyridazinone | Inhibit phytoene desaturase | 16.61 |

**TABLE 16.70 (cont.)**
**Index of Fungicides, Insecticides, and Herbicides**

| Herbicides (cont.) | | | |
|---|---|---|---|
| **Common Name** | **Chemical Class** | **MOA** | **Table No.** |
| Fomesafen | Diphenylethers | Inhibit protoporphyrinogen oxidase | 16.58 |
| Glufosinate | Phosphinic acid | Inhibit glutamine synthetase | 16.64 |
| Glyphosate | Glycine | Inhibit EPSPS | 16.64 |
| Halosulfuron | Sulfonylurea | Inhibit acetolactate synthetase | 16.45, 16.46, 16.47 |
| Haloxyfop | Aryloxyphenoxy-propionate | Inhibit acetyl–CoA carboxylase | 16.42, 16.43 |
| Imazameth (imazapic) | Imidazolinone | Inhibit acetolactate synthetase | 16.48, 16.49 |
| Imazamethabenz-methyl | Imidazolinone | Inhibit acetolactate synthetase | 16.48, 16.49 |
| Imazamox | Imidazolinone | Inhibit acetolactate synthetase | 16.48, 16.49 |
| Imazapyr | Imidazolinone | Inhibit acetolactate synthetase | 16.48, 16.49 |
| Imazaquin | Imidazolinone | Inhibit acetolactate synthetase | 16.48, 16.49 |
| Imazethapyr | Imidazolinone | Inhibit acetolactate synthetase | 16.48, 16.49 |
| Imazosulfuron | Sulfonylurea | Inhibit acetolactate synthetase | 16.45, 16.46, 16.47 |
| Ioxynil | Nitrile | Inhibit photosystem II | 16.55 |
| Isoxaben | Benzamide | Inhibit cellulose synthesis | 16.67 |
| Isoxaflutole | Isoxazole | Inhibit HPPD | 16.62 |
| Lactofen | Diphenylethers | Inhibit protoporphyrinogen oxidase | 16.58 |
| Linuron | Urea | Inhibit photosystem II | 16.54 |
| Mesotrione | Triketone | Inhibit HPPD | 16.62 |
| Metolachlor | Chloroacetamide | Inhibit VLCFA | 16.66 |
| Metribuzin | Triazinone | Inhibit photosystem II | 16.51, 16.52 |
| Metsulfuron-methyl | Sulfonylurea | Inhibit acetolactate synthetase | 16.45, 16.46, 16.47 |
| Molinate | Thiocarbamate | Inhibit lipid synthesis | 16.67 |
| MSMA | Organoarsenical | Unknown mode of action | 16.69 |
| Nicosulfuron | Sulfonylurea | Inhibit acetolactate synthetase | 16.45, 16.46, 16.47 |
| Norflurazon | Pyridazinone | Inhibit phytoene desaturase | 16.61 |
| Oxadiazon | Oxadiazole | Inhibit protoporphyrinogen oxidase | 16.60 |
| Oxasulfuron | Sulfonylurea | Inhibit acetolactate synthetase | 16.45, 16.46, 16.47 |
| Oxyfluorfen | Diphenylethers | Inhibit protoporphyrinogen oxidase | 16.58 |
| Paraquat | Bipyridylium | Inhibit photosystem I | 16.57 |
| Pendimethalin | Dinitroaniline | Inhibit microtubule assembly | 16.65 |
| Picloram | Pyridine carboxylic acid | Auxin hormone mimic | 16.68 |
| Primisulfuron-methyl | Sulfonylurea | Inhibit acetolactate synthetase | 16.45, 16.46, 16.47 |
| Prometon | Triazine | Inhibit photosystem II | 16.51, 16.52 |
| Prometryn | Triazine | Inhibit photosystem II | 16.51, 16.52 |
| Propanil | Amide | Inhibit photosystem II | 16.55 |
| Propaquizafop | Aryloxyphenoxy-propionate | Inhibit acetyl–CoA carboxylase | 16.42, 16.43 |
| Propazine | Triazine | Inhibit photosystem II | 16.51, 16.52 |
| Prosulfuron | Sulfonylurea | Inhibit acetolactate synthetase | 16.45, 16.46, 16.47 |
| Pyridate | Phenylpyridazine | Inhibit photosystem II | 16.56 |
| Pyriminobac-methyl | Pyrimidinylthio-benzoate | Inhibit acetolactate synthetase | 16.50 |
| Quinclorac | Quinoline carboxylic acid | Auxin hormone mimic | 16.68 |
| Rimsulfuron | Sulfonylurea | Inhibit acetolactate synthetase | 16.45, 16.46, 16.47 |
| Sethoxydim | Cyclohexanedione | Inhibit acetyl–CoA carboxylase | 16.44 |
| Simazine | Triazine | Inhibit photosystem II | 16.51, 16.52 |
| Sulcotrione | Triketone | Inhibit HPPD | 16.62 |
| Sulfometuron-methyl | Sulfonylurea | Inhibit acetolactate synthetase | 16.45, 16.46, 16.47 |
| Sulfosulfuron | Sulfonylurea | Inhibit acetolactate synthetase | 16.45, 16.46, 16.47 |
| Terbacil | Uracil | Inhibit photosystem II | 16.53 |
| Thifensulfuron-methyl | Sulfonylurea | Inhibit acetolactate synthetase | 16.45, 16.46, 16.47 |
| Triasulfuron | Sulfonylurea | Inhibit acetolactate synthetase | 16.45, 16.46, 16.47 |
| Tribenuron-methyl | Sulfonylurea | Inhibit acetolactate synthetase | 16.45, 16.46, 16.47 |
| Trifluralin | Dinitroaniline | Inhibit microtubule assembly | 16.65 |

6. Beyer, E. M., Jr., Duffy, M. J., Hay, J. V., and Schlueter, D. D. (1988): Sulfonylurea. In: *Herbicides: Chemistry, Degradation, and Mode of Action*, edited by P. C. Kearney and D. D. Kaufman, pp. 117–189. Marcel Dekker, New York.
7. Bliley, R. (1996): *Food Quality Protection Act of, 1996*. 104 Congress, 2nd Session, Report 104-669, Part 2, pp. 1–89. U.S. Government Printing Office, Washington, D.C.
8. Bloomquist, J. R. (1993): Neuroreceptor mechanisms in pyrethroid mode of action and resistance. *Rev. Pestic. Toxicol.*, 2:185–226.
9. Bloomquist, J. R. (1993): Toxicology, mode of action, and target site-mediated resistance to insecticides acting on chloride channels. *Mini Rev. Comp. Biochem. Physiol.*, 106C:301–314.
10. Bloomquist, J. (1996) Ion channels as targets for insecticides. *Ann. Rev. Entomol.*, 41:163–190.
11. Bradberry, S.M., Cage, S.A., Proudfoot, A.T., and Vale, A. (2005): Poisoning due to pyrethroids. *Toxicol. Rev.*, 24 (2):93–106.
12. BCPC. (2003): Abamectin. In: *A World Compendium: The Pesticide Manual*, 13th ed., edited by C. D. S. Tomlin, pp. 3–4. British Crop Protection Council, Alton Hampshire, U.K.
13. BCPC. (2003): Acetochlor. In: *A World Compendium: The Pesticide Manual*, 13th ed., edited by C. D. S. Tomlin, pp. 8–9. British Crop Protection Council, Alton Hampshire, U.K.
14. BCPC. (2003): Acephate. In: *A World Compendium: The Pesticide Manual*, 13th ed., edited by C. D. S. Tomlin, pp. 5–6. British Crop Protection Council, Alton Hampshire, U.K.
15. BCPC. (2003): Acifluorfen-sodium. In: *A World Compendium: The Pesticide Manual*, 13th ed., edited by C. D. S. Tomlin, pp. 11–12. British Crop Protection Council, Alton Hampshire, U.K.
16. BCPC. (2003): Alachlor. In: *A World Compendium: The Pesticide Manual*, 13th ed., edited by C. D. S. Tomlin, pp. 17–19. British Crop Protection Council, Alton Hampshire, U.K.
17. BCPC. (2003): Aldicarb. In: *A World Compendium: The Pesticide Manual*, 13th ed., edited by C. D. S. Tomlin, pp. 20–21. British Crop Protection Council, Alton Hampshire, UK.
18. BCPC. (2003): Ametryn. In: *A World Compendium: The Pesticide Manual*, 13th ed., edited by C. D. S. Tomlin, pp. 25–26. British Crop Protection Council, Alton Hampshire, U.K.
19. BCPC. (2003): Amitraz. In: *A World Compendium: The Pesticide Manual*, 13th ed., edited by C. D. S. Tomlin, pp. 29–30. British Crop Protection Council, Alton Hampshire, U.K.
20. BCPC. (2003): Amitrole. In: *A World Compendium: The Pesticide Manual*, 13th ed., edited by C. D. S. Tomlin, pp. 30–31. British Crop Protection Council, Alton Hampshire, U.K.
21. BCPC. (2003): Asulam. In: *A World Compendium: The Pesticide Manual*, 13th ed., edited by C. D. S. Tomlin, pp. 35–36. British Crop Protection Council, Alton Hampshire, U.K.
22. BCPC. (2003): Atrazine. In: *A World Compendium: The Pesticide Manual*, 13th ed., edited by C. D. S. Tomlin, pp. 39–41. British Crop Protection Council, Alton Hampshire, U.K.
23. BCPC. (2003): Azinphos-methyl. In: *A World Compendium: The Pesticide Manual*, 13th ed., edited by C. D. S. Tomlin, pp. 49–50. British Crop Protection Council, Alton Hampshire, U.K.
24. BCPC. (2003): Azoxystrobin. In: *A World Compendium: The Pesticide Manual*, 13th ed., edited by C. D. S. Tomlin, pp. 52–53. British Crop Protection Council, Alton Hampshire, U.K.
25. BCPC. (2003): *Bacillus thuringiensis. In: A World Compendium: The Pesticide Manual*, 13th ed., edited by C. D. S. Tomlin, pp. 55–58. British Crop Protection Council, Alton Hampshire, U.K.
26. BCPC. (2003): *Bacillus thuringiensis delta endotoxins. In: A World Compendium: The Pesticide Manual*, 13th ed., edited by C. D. S. Tomlin, pp. 59. British Crop Protection Council, Alton Hampshire, U.K.
27. BCPC. (2003): Benfluralin. In: *A World Compendium: The Pesticide Manual*, 13th ed., edited by C.D.S. Tomlin, pp. 66–67. British Crop Protection Council, Alton Hampshire, U.K.
28. BCPC. (2003): Benomyl. In: *A World Compendium: The Pesticide Manual*, 13th ed., edited by C. D. S. Tomlin, pp. 70–72. British Crop Protection Council, Alton Hampshire, U.K.
29. BCPC. (2003): Bensulfuron-methyl. In: *A World Compendium: The Pesticide Manual*, 13th ed., edited by C.D.S. Tomlin, pp. 73–74. British Crop Protection Council, Alton Hampshire, U.K.
30. BCPC. (2003): Bentazone. In: *A World Compendium: The Pesticide Manual*, 13th ed., edited by C.D.S. Tomlin, pp 77–78. British Crop Protection Council, Alton Hampshire U.K.
31. BCPC. (2003): Bioallethrin. In: *A World Compendium: The Pesticide Manual*, 13th ed., edited by C. D. S. Tomlin, pp. 91–92. British Crop Protection Council, Alton Hampshire U.K.
32. BCPC. (2003): Borax. In: *A World Compendium: The Pesticide Manual*, 13th ed., edited by C. D. S. Tomlin, pp 102–103. British Crop Protection Council, Alton Hampshire, U.K.
33. BCPC. (2003): Bordeaux mixture. In: *A World Compendium: The Pesticide Manual*, 13th ed., edited by C. D. S Tomlin, pp. 102–103. British Crop Protection Council Alton Hampshire, U.K.
34. BCPC. (2003): Bromacil. In: *A World Compendium: The Pesticide Manual*, 13th ed., edited by C. D. S. Tomlin, pp 106–107. British Crop Protection Council, Alton Hampshire, U.K.
35. BCPC. (2003): Bromoxynil. In: *A World Compendium The Pesticide Manual*, 13th ed., edited by C. D. S. Tomlin pp. 111–113. British Crop Protection Council, Alton Hampshire, U.K.
36. BCPC. (2003): Butafenacil. In: *A World Compendium: The Pesticide Manual*, 13th ed., edited by C. D. S. Tomlin, pp 120. British Crop Protection Council, Alton Hampshire U.K.

37. BCPC. (2003): Butylate. In: *A World Compendium: The Pesticide Manual*, 13th ed., edited by C. D. S. Tomlin, pp. 128–129. British Crop Protection Council, Alton Hampshire, U.K.
38. BCPC. (2003): Captan. In: *A World Compendium: The Pesticide Manual*, 13th ed., edited by C. D. S. Tomlin, pp. 133–134. British Crop Protection Council, Alton Hampshire, U.K.
39. BCPC. (2003): Carbaryl. In: *A World Compendium: The Pesticide Manual*, 13th ed., edited by C. D. S. Tomlin, pp. 135–136. British Crop Protection Council, Alton Hampshire, U.K.
40. BCPC. (2003): Carbofuran. In: *A World Compendium: The Pesticide Manual*, 13th ed., edited by C. D. S. Tomlin, pp. 139–140. British Crop Protection Council, Alton Hampshire, U.K.
41. BCPC. (2003): Carfentrazone-ethyl. In: *A World Compendium: The Pesticide Manual*, 13th ed., edited by C. D. S. Tomlin, pp. 143–144. British Crop Protection Council, Alton Hampshire, U.K.
42. BCPC. (2003): Chlorfenapyr. In: *A World Compendium: The Pesticide Manual*, 13th ed., edited by C. D. S. Tomlin, pp. 154–155. British Crop Protection Council, Alton Hampshire, U.K.
43. BCPC. (2003): Chlorimuron-ethyl. In: *A World Compendium: The Pesticide Manual*, 13th ed., edited by C. D. S. Tomlin, pp. 161–162. British Crop Protection Council, Alton Hampshire, U.K.
44. BCPC. (2003): Chlorothalonil. In: *A World Compendium: The Pesticide Manual*, 13th ed., edited by C. D. S. Tomlin, pp. 169–170. British Crop Protection Council, Alton Hampshire, U.K.
45. BCPC. (2003): Chlorpyrifos. In: *A World Compendium: The Pesticide Manual*, 13th ed., edited by C. D. S. Tomlin, pp. 173–174. British Crop Protection Council, Alton Hampshire, U.K.
46. BCPC. (2003): Chlorsulfuron. In: *A World Compendium: The Pesticide Manual*, 13th ed., edited by C. D. S. Tomlin, pp. 176–177. British Crop Protection Council, Alton Hampshire, U.K.
47. BCPC. (2003): Clethodim. In: *A World Compendium: The Pesticide Manual*, 13th ed., edited by C. D. S. Tomlin, pp. 185–186. British Crop Protection Council, Alton Hampshire, U.K.
48. BCPC. (2003): Clodinafop-propargyl. In: *A World Compendium: The Pesticide Manual*, 13th ed., edited by C. D. S. Tomlin, pp. 186–187. British Crop Protection Council, Alton Hampshire, U.K.
49. BCPC. (2003): Clofentezine. In: *A World Compendium: The Pesticide Manual*, 13th ed., edited by C. D. S. Tomlin, pp. 189–190. British Crop Protection Council, Alton Hampshire, U.K.
50. BCPC. (2003): Clomazone. In: *A World Compendium: The Pesticide Manual*, 13th ed., edited by C. D. S. Tomlin, pp. 191. British Crop Protection Council, Alton Hampshire, U.K.
50a. BCPC. (2003): Clopyralid. In: *A World Compendium: The Pesticide Manual*, 13th ed., edited by C. D. S. Tomlin, pp. 194–195. British Crop Protection Council, Alton Hampshire, U.K.
51. BCPC. (2003): Copper hydroxide. In: *A World Compendium: The Pesticide Manual*, 13th ed., edited by C. D. S. Tomlin, p., 202–203. British Crop Protection Council, Alton Hampshire, U.K.
52. BCPC. (2003): Copper oxychloride. In: *A World Compendium: The Pesticide Manual.*, 13th ed., edited by C. D. S. Tomlin, pp. 203–204. British Crop Protection Council, Alton Hampshire, U.K.
53. BCPC. (2003): Copper sulfate. In: *A World Compendium: The Pesticide Manual.*, 13th ed., edited by C. D. S. Tomlin, pp. 205–206. British Crop Protection Council, Alton Hampshire, U.K.
54. BCPC. (2003): Cyanazine. In: *A World Compendium: The Pesticide Manual*, 13th ed., edited by C. D. S. Tomlin, pp. 215–216. British Crop Protection Council, Alton Hampshire, U.K.
55. BCPC. (2003): Cyfluthrin. In: *A World Compendium: The Pesticide Manual*, 13th ed., edited by C. D. S. Tomlin, pp. 226–227. British Crop Protection Council, Alton Hampshire, U.K.
56. BCPC. (2003): Lambda-Cyhalothrin. In: *A World Compendium: The Pesticide Manual*, 13th ed., edited by C. D. S. Tomlin, pp. 233–235. British Crop Protection Council, Alton Hampshire, U.K.
57. BCPC. (2003): Cypermethrin. In: *A World Compendium: The Pesticide Manual*, 13th ed., edited by C. D. S. Tomlin, pp. 237–239. British Crop Protection Council, Alton Hampshire, U.K.
58. BCPC. (2003): Cyproconazole. In: *A World Compendium: The Pesticide Manual*, 13th ed., edited by C.D.S. Tomlin, pp. 248–249. British Crop Protection Council, Alton Hampshire, U.K.
59. BCPC. (2003): Cyprodinil. In: *A World Compendium: The Pesticide Manual.*, 13th ed., edited by C. D. S. Tomlin, pp. 249–251. British Crop Protection Council, Alton Hampshire, U.K.
60. BCPC. (2003): Cyromazine. In: *A World Compendium: The Pesticide Manual.*, 13th ed., edited by C. D. S. Tomlin, pp.251–252. British Crop Protection Council, Alton Hampshire, U.K.
61. BCPC. (2003):2,4–D. In: *A World Compendium: The Pesticide Manual.*, 13th ed., edited by C. D. S. Tomlin, pp. 254–258. British Crop Protection Council, Alton Hampshire, U.K.
62. BCPC. (2003): DDT. In: *A World Compendium: The Pesticide Manual.*, 13th ed., edited by C. D. S. Tomlin, pp. 267–268. British Crop Protection Council, Alton Hampshire, U.K.
63. BCPC. (2003): Deltamethrin. In: *A World Compendium: The Pesticide Manual.*, 13th ed., edited by C. D. S. Tomlin, pp. 271–272. British Crop Protection Council, Alton Hampshire, U.K.
64. BCPC. (2003): Diazinon. In: *A World Compendium: The Pesticide Manual.*, 13th ed., edited by C. D. S. Tomlin, pp. 277–278. British Crop Protection Council, Alton Hampshire, U.K.
65. BCPC. (2003): Dicamba. In: *A World Compendium: The Pesticide Manual.*, 13th ed., edited by C. D. S. Tomlin, pp. 278–280. British Crop Protection Council, Alton Hampshire, U.K.

66. BCPC. (2003): Dichlobenil. In: *A World Compendium: The Pesticide Manual*., 13th ed., edited by C. D. S. Tomlin, pp. 281–282. British Crop Protection Council, Alton Hampshire, U.K.
67. BCPC. (2003): Dichlorvos. In: *A World Compendium: The Pesticide Manual*., 13th ed., edited by C. D. S. Tomlin, pp. 291–292. British Crop Protection Council, Alton Hampshire, U.K.
68. BCPC. (2003): Diclofop-methyl. In: *A World Compendium: The Pesticide Manual*., 13th ed., edited by C. D. S. Tomlin, pp. 293–295. British Crop Protection Council, Alton Hampshire, U.K.
69. BCPC. (2003): Difenoconazole. In: *A World Compendium: The Pesticide Manual*., 13th ed., edited by C. D. S. Tomlin, pp. 304–305. British Crop Protection Council, Alton Hampshire, U.K.
70. BCPC. (2003): Difenzoquat metilsulfate. In: *A World Compendium: The Pesticide Manual*., 13th ed., edited by C. D. S. Tomlin, pp. 306–307. British Crop Protection Council, Alton Hampshire, U.K.
71. BCPC. (2003): Diflubenzuron. In: *A World Compendium: The Pesticide Manual*., 13th ed., edited by C. D. S. Tomlin, pp. 308–310. British Crop Protection Council, Alton Hampshire, U.K.
72. BCPC. (2003): Dimethenamid. In: *A World Compendium: The Pesticide Manual*., 13th ed., edited by C. D. S. Tomlin, pp. 320–322. British Crop Protection Council, Alton Hampshire, U.K.
73. BCPC. (2003): Diquat dibromide. In: *A World Compendium: The Pesticide Manual*., 13th ed., edited by C. D. S. Tomlin, pp. 341–342. British Crop Protection Council, Alton Hampshire, U.K.
74. BCPC. (2003): Diuron. In: *A World Compendium: The Pesticide Manual*., 13th ed., edited by C. D. S. Tomlin, pp. 347–348. British Crop Protection Council, Alton Hampshire, U.K.
75. BCPC. (2003): Emamectin benzoate. In: *A World Compendium: The Pesticide Manual*., 13th ed., edited by C. D. S. Tomlin, pp. 359–360. British Crop Protection Council, Alton Hampshire, U.K.
76. BCPC. (2003): Endosulfan. In: *A World Compendium: The Pesticide Manual*.13th ed., edited by C. D. S. Tomlin, pp. 363–364. British Crop Protection Council, Alton Hampshire, U.K.
77. BCPC. (2003): Fenbuconazole. In: *A World Compendium: The Pesticide Manual*.13th ed., edited by C. D. S. Tomlin, pp. 402–403. British Crop Protection Council, Alton Hampshire, U.K.
78. BCPC. (2003): Fenoxaprop-p-ethyl. In: *A World Compendium: The Pesticide Manual*.13th ed., edited by C. D. S. Tomlin, pp. 414–415. British Crop Protection Council, Alton Hampshire, U.K.
79. BCPC. (2003): Fenoxycarb. In: *A World Compendium: The Pesticide Manual*.13th ed., edited by C. D. S. Tomlin, pp. 416–417. British Crop Protection Council, Alton Hampshire, U.K.
80. BCPC. (2003): Fenpiclonil. In: *A World Compendium: The Pesticide Manual*.13th ed., edited by C. D. S. Tomlin, pp. 417–418. British Crop Protection Council, Alton Hampshire, U.K.
81. BCPC. (2003): Fentin (triphenyl tin): In: *A World Compendium: The Pesticide Manual*, 13th ed., edited by C. D. S. Tomlin, pp. 425–427. British Crop Protection Council, Alton Hampshire, U.K.
82. BCPC. (2003): Ferbam. In: *A World Compendium: The Pesticide Manual*, 13th ed., edited by C. D. S. Tomlin, pp. 430–431. British Crop Protection Council, Alton Hampshire, U.K.
83. BCPC. (2003): Fipronil. In: *A World Compendium: The Pesticide Manual*, 13th ed., edited by C. D. S. Tomlin, pp. 433–435. British Crop Protection Council, Alton Hampshire, U.K.
84. BCPC.(, 2003): Fluazifop-p-butyl. In: *A World Compendium: The Pesticide Manual*, 13th ed., edited by C. D. S. Tomlin, pp. 444–446. British Crop Protection Council, Alton Hampshire, U.K.
85. BCPC. (2003): Fludioxonil. In: *A World Compendium: The Pesticide Manual*, 13th ed., edited by C. D. S. Tomlin, pp. 452–453. British Crop Protection Council, Alton Hampshire, U.K.
86. BCPC. (2003): Flumetsulam. In: *A World Compendium: The Pesticide Manual*, 13th ed., edited by C. D. S. Tomlin, pp. 459–460. British Crop Protection Council, Alton Hampshire, U.K.
87. BCPC.(, 2003): Flumiclorac-pentyl. In: *A World Compendium: The Pesticide Manual*, 13th ed., edited by C. D. S Tomlin, pp. 460–461. British Crop Protection Council Alton Hampshire, U.K.
88. BCPC. (2003): Fluometuron. In: *A World Compendium. The Pesticide Manual*, 13th ed., edited by C. D. S. Tomlin pp. 463–464. British Crop Protection Council, Alton Hampshire, U.K.
89. BCPC. (2003): Fluridone. In: *A World Compendium: The Pesticide Manual*, 13th ed., edited by C. D. S. Tomlin, pp 476–477. British Crop Protection Council, Alton Hampshire, U.K.
90. BCPC. (2003): Fluthiacet-methyl. In: *A World Compendium: The Pesticide Manual*, 13th ed., edited by C. D. S Tomlin, pp. 485–486. British Crop Protection Council Alton Hampshire, U.K.
91. BCPC. (2003): Flutolanil. In: *A World Compendium: The Pesticide Manual*, 13th ed., edited by C. D. S. Tomlin, pp 486–487. British Crop Protection Council, Alton Hampshire, U.K.
92. BCPC. (2003): Fomesafen. In: *A World Compendium: The Pesticide Manual*, 13th ed., edited by C. D. S. Tomlin, pp 492–493. British Crop Protection Council, Alton Hampshire, U.K.
93. BCPC. (2003): Glufosinate-ammonium. In: *A World Compendium: The Pesticide Manual*, 13th ed., edited by C. D. S. Tomlin, pp. 511–512. British Crop Protection Council Alton Hampshire, U.K.
94. BCPC. (2003): Glyphosate. In: *A World Compendium: The Pesticide Manual*, 13th ed., edited by C. D. S. Tomlin, pp 513–516. British Crop Protection Council, Alton Hampshire, U.K.
95. BCPC. (2003): Halosulfuron-methyl. In: *A World Compendium: The Pesticide Manual*, 13th ed., edited by C. D. S Tomlin, pp. 523–524. British Crop Protection Council Alton Hampshire, U.K.

96. BCPC. (2003): Haloxyfop. In: *A World Compendium: The Pesticide Manual*, 13th ed., edited by C. D. S. Tomlin, pp. 524–526. British Crop Protection Council, Alton Hampshire, U.K.
97. BCPC. (2003): Gamma-HCH (Lindane): In: *A World Compendium: The Pesticide Manual*, 13th ed., edited by C. D. S. Tomlin, pp. 528–530. British Crop Protection Council, Alton Hampshire, U.K.
98. BCPC. (2003): Hexaconazole. In: *A World Compendium: The Pesticide Manual*, 13th ed., edited by C. D. S. Tomlin, pp. 533–534. British Crop Protection Council, Alton Hampshire, U.K.
99. BCPC. (2003): Hexythiazox. In: *A World Compendium: The Pesticide Manual*, 13th ed., edited by C. D. S. Tomlin, pp. 540–541. British Crop Protection Council, Alton Hampshire, U.K.
100. BCPC. (, 2003): Hydramethylnon. In: *A World Compendium: The Pesticide Manual*, 13th ed., edited by C. D. S. Tomlin, pp. 541–542. British Crop Protection Council, Alton Hampshire, U.K.
101. BCPC. (, 2003): Hydrogen Cyanide. In: *A World Compendium: The Pesticide Manual*, 13th ed., edited by C. D. S. Tomlin, pp. 541–542. British Crop Protection Council, Alton Hampshire, U.K.
102. BCPC. (2003): Imazalil. In: *A World Compendium: The Pesticide Manual*, 13th ed., edited by C. D. S. Tomlin, pp. 549–550. British Crop Protection Council, Alton Hampshire, U.K.
103. BCPC. (2003): Imazamethabenz-methyl. In: *A World Compendium: The Pesticide Manual*, 13th ed., edited by C. D. S. Tomlin, pp. 551–552. British Crop Protection Council, Alton Hampshire, GU34, 2QD UK
104. BCPC. (2003): Imazamox. In: *A World Compendium: The Pesticide Manual*, 13th ed., edited by C. D. S. Tomlin, pp. 552–553. British Crop Protection Council, Alton Hampshire, U.K.
105. BCPC. (2003): Imazapic (imazameth): In: *A World Compendium: The Pesticide Manual*, 13th ed., edited by C. D. S. Tomlin, pp. 554–555. British Crop Protection Council, Alton Hampshire, U.K.
106. BCPC. (2003): Imazapyr. In: *A World Compendium: The Pesticide Manual*, 13th ed., edited by C. D. S. Tomlin, pp. 555–556. British Crop Protection Council, Alton Hampshire, U.K.
107. BCPC. (2003): Imazaquin. In: *A World Compendium: The Pesticide Manual*, 13th ed., edited by C. D. S. Tomlin, pp. 557–558. British Crop Protection Council, Alton Hampshire, U.K.
108. BCPC. (2003): Imazethapyr. In: *A World Compendium: The Pesticide Manual*, 13th ed., edited by C. D. S. Tomlin, pp. 558–560. British Crop Protection Council, Alton Hampshire, U.K.
109. BCPC. (2003): Imazosulfuron. In: *A World Compendium: The Pesticide Manual*, 13th ed., edited by C. D. S. Tomlin, pp. 560–561. British Crop Protection Council, Alton Hampshire, U.K.
110. BCPC. (2003): Imidacloprid. In: *A World Compendium: The Pesticide Manual*, 13th ed., edited by C. D. S. Tomlin, pp. 562–564. British Crop Protection Council, Alton Hampshire, U.K.
111. BCPC. (2003): Indol-3ylacetic acid (auxin): In: *A World Compendium: The Pesticide Manual*, 13th ed., edited by C. D. S. Tomlin, pp. 570–571. British Crop Protection Council, Alton Hampshire, U.K.
112. BCPC. (2003): Ioxynil. In: *A World Compendium: The Pesticide Manual*, 13th ed., edited by C. D. S. Tomlin, pp. 574–576. British Crop Protection Council, Alton Hampshire, U.K.
113. BCPC. (2003): Iprodione. In: *A World Compendium: The Pesticide Manual*, 13th ed., edited by C. D. S. Tomlin, pp. 579–580. British Crop Protection Council, Alton Hampshire, U.K.
114. BCPC. (2003): Isoxaben. In: *A World Compendium: The Pesticide Manual*, 13th ed., edited by C. D. S. Tomlin, pp. 587–588. British Crop Protection Council, Alton Hampshire, U.K.
115. BCPC. (2003): Isoxaflutole. In: *A World Compendium: The Pesticide Manual*, 13th ed., edited by C. D. S. Tomlin, pp. 589–590. British Crop Protection Council, Alton Hampshire, U.K.
116. BCPC. (2003): Lactofen. In: *A World Compendium: The Pesticide Manual*, 13th ed., edited by C. D. S. Tomlin, pp. 596–597. British Crop Protection Council, Alton Hampshire, U.K.
117. BCPC. (2003): Linuron. In: *A World Compendium: The Pesticide Manual*, 13th ed., edited by C. D. S. Tomlin, pp. 599–600. British Crop Protection Council, Alton Hampshire, U.K.
118. BCPC. (2003): Lufenuron. In: *A World Compendium: The Pesticide Manual*, 13th ed., edited by C. D. S. Tomlin, pp. 600–601. British Crop Protection Council, Alton Hampshire, U.K.
119. BCPC. (2003): Malathion. In: *A World Compendium: The Pesticide Manual*, 13th ed., edited by C. D. S. Tomlin, pp. 600–601. British Crop Protection Council, Alton Hampshire, U.K.
120. BCPC. (2003): Mancozeb. In: *A World Compendium: The Pesticide Manual*, 13th ed., edited by C. D. S. Tomlin, pp. 606–608. British Crop Protection Council, Alton Hampshire, U.K.
121. BCPC. (2003): Maneb. In: *A World Compendium: The Pesticide Manual*, 13th ed., edited by C. D. S. Tomlin, pp. 608–609. British Crop Protection Council, Alton Hampshire, U.K.
122. BCPC. (2003): Mesotrione. In: *A World Compendium: The Pesticide Manual*, 13th ed., edited by C. D. S. Tomlin, pp. 631–632. British Crop Protection Council, Alton Hampshire, U.K.
123. BCPC. (2003): Metalaxyl-M. (Mefenoxam) In: *A World Compendium: The Pesticide Manual*, 13th ed., edited by C. D. S. Tomlin, pp. 633–635. British Crop Protection Council, Alton Hampshire, U.K.
124. BCPC. (2003): Methomyl. In: *A World Compendium: The Pesticide Manual*, 13th ed., edited by C. D. S. Tomlin, pp. 650 - 651. British Crop Protection Council, Alton Hampshire, U.K.
125. BCPC. (2003): Methoprene. In: *A World Compendium: The Pesticide Manual*, 13th ed., edited by C. D. S. Tomlin, pp. 652 - 653. British Crop Protection Council, Alton Hampshire, U.K.

126. BCPC. (2003): Methoxychlor. In: *A World Compendium: The Pesticide Manual*, 13th ed., edited by C. D. S. Tomlin, pp. 654 - 655. British Crop Protection Council, Alton Hampshire, U.K.
127. BCPC. (2003): Methylarsonic acid (MSMA): In: *A World Compendium: The Pesticide Manual*, 13th ed., edited by C. D. S. Tomlin, pp. 656 - 658. British Crop Protection Council, Alton Hampshire, U.K.
128. BCPC. (2003): *S*-Metolachlor. In: *A World Compendium: The Pesticide Manual*, 13th ed., edited by C. D. S. Tomlin, pp. 669 - 670. British Crop Protection Council, Alton Hampshire, U.K.
129. BCPC. (2003): Metribuzin. In: *A World Compendium: The Pesticide Manual*, 13th ed., edited by C. D. S. Tomlin, pp. 675–676. British Crop Protection Council, Alton Hampshire, U.K.
130. BCPC. (2003): Metsulfuron-methyl. In: *A World Compendium: The Pesticide Manual*, 13th ed., edited by C. D. S. Tomlin, pp. 677–678. British Crop Protection Council, Alton Hampshire, U.K.
131. BCPC. (2003): Milbemectin (milbemycin): In: *A World Compendium: The Pesticide Manual*, 13th ed., edited by C. D. S. Tomlin, pp. 680–681. British Crop Protection Council, Alton Hampshire, U.K.
132. BCPC. (2003): Molinate. In: *A World Compendium: The Pesticide Manual*, 13th ed., edited by C. D. S. Tomlin, pp. 683 - 684. British Crop Protection Council, Alton Hampshire, U.K.
133. BCPC. (2003): Monocrotophos. In: *A World Compendium: The Pesticide Manual*, 13th ed., edited by C. D. S. Tomlin, pp. 684–685. British Crop Protection Council, Alton Hampshire, U.K.
134. BCPC. (2003): Myclobutanil. In: *A World Compendium: The Pesticide Manual*, 13th Ed., edited by C. D. S. Tomlin, pp. 688–689. British Crop Protection Council, Alton Hampshire, U.K.
135. BCPC. (2003): Nicosulfuron. In: *A World Compendium: The Pesticide Manual*, 13th Ed., edited by C. D. S. Tomlin, pp. 702–703. British Crop Protection Council, Alton Hampshire, U.K.
136. BCPC. (2003): Nicotine. *A World Compendium: The Pesticide Manual*., 13th ed., edited by C. D. S. Tomlin, pp. 703–704. British Crop Protection Council, Alton Hampshire, U.K.
137. BCPC. (2003): Norflurazon. *A World Compendium: The Pesticide Manual*., 13th ed., edited by C. D. S. Tomlin, pp. 711–712. British Crop Protection Council, Alton Hampshire, U.K.
138. BCPC. (2003): Oxadiazon. In: *A World Compendium: The Pesticide Manual*, 13th ed., edited by C. D. S. Tomlin, pp. 727–728. British Crop Protection Council, Alton Hampshire, U.K.
139. BCPC. (2003): Oxadixyl. In: *A World Compendium: The Pesticide Manual*, 13th ed., edited by C. D. S. Tomlin, pp. 728–729. British Crop Protection Council, Alton Hampshire, U.K.
140. BCPC. (2003): Oxasulfuron. In: *A World Compendium: The Pesticide Manual*, 13th ed., edited by C.D.S. Tomlin, pp. 731–732. British Crop Protection Council, Alton Hampshire, U.K.
141. BCPC. (2003): Oxyfluorfen. In: *A World Compendium: The Pesticide Manual*, 13th ed., edited by C.D.S. Tomlin, pp. 738–739. British Crop Protection Council, Alton Hampshire, U.K.
142. BCPC. (2003): Paraquat dichloride. In: *A World Compendium: The Pesticide Manual*, 13th ed., edited by C.D.S. Tomlin, pp. 742–743. British Crop Protection Council, Alton Hampshire, U.K.
143. BCPC. (2003): Pendimethalin. In: *A World Compendium: The Pesticide Manual*, 13th ed., edited by C.D.S. Tomlin, pp. 752–753. British Crop Protection Council, Alton Hampshire, U.K.
144. BCPC. (2003): Permethrin. In: *A World Compendium: The Pesticide Manual*, 13th ed., edited by C.D. S. Tomlin, pp. 758–760. British Crop Protection Council, Alton Hampshire, U.K.
145. BCPC. (2003): Picloram. In: *A World Compendium: The Pesticide Manual*, 13th ed., edited by C.D. S. Tomlin, pp. 782–785. British Crop Protection Council, Alton Hampshire, U.K.
146. BCPC. (2003): Picoxystrobin. In: *A World Compendium: The Pesticide Manual*, 13th ed., edited by C.D. S. Tomlin, pp. 786–787. British Crop Protection Council, Alton Hampshire, U.K.
147. BCPC. (2003): Piperonyl butoxide. In: *A World Compendium: The Pesticide Manual*, 13th ed., edited by C. D. S Tomlin, pp. 788–789. British Crop Protection Council, Alton Hampshire, U.K.
148. BCPC. (2003): Primisulfuron-methyl. In: *A World Compendium: The Pesticide Manual*, 13th ed., edited by C. D. S. Tomlin, pp. 800–801. British Crop Protection Council, Alton Hampshire, U.K.
149. BCPC. (2003): Prochloraz. In: *A World Compendium: The Pesticide Manual*., 13th ed., edited by C. D. S. Tomlin, pp. 802–804. British Crop Protection Council, Alton Hampshire, U.K.
150. BCPC. (2003): Prometon. In: *A World Compendium: The Pesticide Manual*., 13th ed., edited by C. D. S. Tomlin, pp. 810–811. British Crop Protection Council, Alton Hampshire, U.K.
151. BCPC. (2003): Prometryn. In: *A World Compendium: The Pesticide Manual*., 13th ed., edited by C. D. S. Tomlin, pp. 812–813. British Crop Protection Council, Alton Hampshire, U.K.
152. BCPC. (2003): Propanil. In: *A World Compendium: The Pesticide Manual*., 13th ed., edited by C. D. S. Tomlin, pp. 816–817. British Crop Protection Council, Alton Hampshire, U.K.
153. BCPC. (2003): Propaquizafop. In: *A World Compendium: The Pesticide Manual*, 13th ed., edited by C. D. S. Tomlin, pp. 818–819. British Crop Protection Council, Alton Hampshire, U.K.
154. BCPC. (2003): Propazine. In: *A World Compendium: The Pesticide Manual*, 13th ed., edited by C. D. S. Tomlin, pp. 821–822. British Crop Protection Council, Alton Hampshire, U.K.
155. BCPC. (2003): Propiconazole. In: *A World Compendium. The Pesticide Manual*, 13th ed., edited by C. D.S. Tomlin pp. 825–826. British Crop Protection Council, Alton Hampshire, U.K.

156. BCPC. (2003): Propoxur. In: *A World Compendium: The Pesticide Manual*, *1*t3h ed., edited by C. D. S. Tomlin, pp. 829–830. British Crop Protection Council, Alton Hampshire, U.K.
157. BCPC. (2003): Prosulfuron. In: *A World Compendium: The Pesticide Manual*, 13th ed., edited by C. D. S. Tomlin, pp. 836–837. British Crop Protection Council, Alton Hampshire, U.K.
158. BCPC. (2003): Pymetrozine. In: *A World Compendium: The Pesticide Manual*, 13th ed., edited by C. D. S. Tomlin, pp. 840–841. British Crop Protection Council, Alton Hampshire, U.K.
159. BCPC. (2003): Pyrethrins. In: *A World Compendium: The Pesticide Manual*, 13th ed., edited by C. D. S. Tomlin, pp. 849–852. British Crop Protection Council, Alton Hampshire, U.K.
160. BCPC. (2003): Pyridaben. In: *A World Compendium: The Pesticide Manual*, 13th ed., edited by C. D. S. Tomlin, pp. 854–855. British Crop Protection Council, Alton Hampshire, U.K.
161. BCPC. (2003): Pyridate. In: *A World Compendium: The Pesticide Manual*, 13th ed., edited by C. D. S. Tomlin, pp. 857–858. British Crop Protection Council, Alton Hampshire, U.K.
162. BCPC. (2003): Pyrimethanil. In: *A World Compendium: The Pesticide Manual*, 13th ed., edited by C. D. S. Tomlin, pp. 861–862. British Crop Protection Council, Alton Hampshire, U.K.
163. BCPC. (2003): Pyriminobac-methyl. In: *A World Compendium: The Pesticide Manual*, 13th ed., edited by C. D. S. Tomlin, pp. 863–864. British Crop Protection Council, Alton Hampshire, U.K.
164. BCPC. (2003): Pyriproxyfen. In: *A World Compendium: The Pesticide Manual*, 13th ed., C. D. S. Tomlin, pp. 864–865. British Crop Protection Council, Alton Hampshire, U.K.
165. BCPC. (2003): Quinclorac. In: *A World Compendium: The Pesticide Manual*, 13th ed., C. D. S. Tomlin, pp. 869–870. British Crop Protection Council, Alton Hampshire, U.K.
166. BCPC. (2003): Resmethrin. In: *A World Compendium: The Pesticide Manual*, 13th ed., edited by C. D. S. Tomlin, pp. 878–880. British Crop Protection Council, Alton Hampshire, U.K.
167. BCPC. (2003): Rimsulfuron. In: *A World Compendium: The Pesticide Manual*, 13th ed., edited by C. D. S. Tomlin, pp. 881–882. British Crop Protection Council, Alton Hampshire, U.K.
168. BCPC. (2003): Rotenone. In: *A World Compendium: The Pesticide Manual*, 13th ed., edited by C. D. S. Tomlin, pp. 882–883. British Crop Protection Council, Alton Hampshire, U.K.
169. BCPC. (2003): Sethoxydim. In: *A World Compendium: The Pesticide Manual*, 13th ed., edited by C. D. S. Tomlin, pp. 887–888. British Crop Protection Council, Alton Hampshire, U.K.
170. BCPC. (2003): Simazine. In: *A World Compendium: The Pesticide Manual*, 13th ed., edited by C. D. S. Tomlin, pp. 891–892. British Crop Protection Council, Alton Hampshire, U.K.
171. BCPC. (2003): Spinosad. In: *A World Compendium: The Pesticide Manual*, 13th ed., edited by C. D. S. Tomlin, pp. 898–900. British Crop Protection Council, Alton Hampshire, U.K.
172. BCPC. (2003): Sulcotrione. In: *A World Compendium: The Pesticide Manual*, 13th ed., edited by C. D. S. Tomlin, pp. 908–909. British Crop Protection Council, Alton Hampshire, U.K.
173. BCPC. (2003): Sulfometuron-methyl. In: *A World Compendium: The Pesticide Manual*, 13th ed., edited by C. D. S. Tomlin, pp. 912–913. British Crop Protection Council, Alton Hampshire, U.K.
174. BCPC. (2003): Sulfosulfuron. In: *A World Compendium: The Pesticide Manual*, 13th ed., edited by C. D. S. Tomlin, pp. 913–915. British Crop Protection Council, Alton Hampshire, U.K.
175. BCPC. (2003): Sulfur. In: *A World Compendium: The Pesticide Manual*, 13th ed., edited by C. D. S. Tomlin, pp. 916–917. British Crop Protection Council, Alton Hampshire, U.K.
176. BCPC. (2003): Tebuconazole. In: *A World Compendium: The Pesticide Manual*, 13th ed., edited by C. D. S. Tomlin, pp. 923–925. British Crop Protection Council, Alton Hampshire, U.K.
177. BCPC. (2003): Tebufenozide. In: *A World Compendium: The Pesticide Manual*, 13th ed., edited by C. D. S. Tomlin, pp. 926–927. British Crop Protection Council, Alton Hampshire, U.K.
178. BCPC. (2003): Teflubenzuron. In: *A World Compendium: The Pesticide Manual*, 13th ed., edited by C. D. S. Tomlin, pp. 932–933. British Crop Protection Council, Alton Hampshire, U.K.
179. BCPC. (2003): Tefluthrin. In: *A World Compendium: The Pesticide Manual*, 13th ed., edited by C. D. S. Tomlin, pp. 934–935. British Crop Protection Council, Alton Hampshire, U.K.
180. BCPC. (2003): Terbacil. In: *A World Compendium: The Pesticide Manual*, 13th ed., edited by C. D. S. Tomlin, pp. 937–938. British Crop Protection Council, Alton Hampshire, U.K.
181. BCPC. (2003): Thiabendazole. In: *A World Compendium: The Pesticide Manual*, 13th ed., edited by C. D. S. Tomlin, pp. 957–958. British Crop Protection Council, Alton Hampshire, U.K.
182. BCPC. (2003): Thiamethoxam. In: *A World Compendium: The Pesticide Manual*, 13th ed., edited by C. D. S. Tomlin, pp. 960–961. British Crop Protection Council, Alton Hampshire, U.K.
183. BCPC. (2003): Thifensulfuron-methyl. In: *A World Compendium: The Pesticide Manual*, 13th ed., edited by C. D. S. Tomlin, pp. 963–965. British Crop Protection Council, Alton Hampshire, U.K.
184. BCPC. (2003): Thiophanate-methyl. In: *A World Compendium: The Pesticide Manual*, 13th ed., edited by C. D. S. Tomlin, pp. 973–974. British Crop Protection Council, Alton Hampshire, U.K.
185. BCPC. (2003): Thiram. In: *A World Compendium: The Pesticide Manual*, 13th ed., edited by C. D. S. Tomlin, pp. 975–976. British Crop Protection Council, Alton Hampshire, U.K.

186. BCPC. (2003): Triadimefon. In: *A World Compendium: The Pesticide Manual*, 13th ed., edited by C. D. S. Tomlin, pp. 986–987. British Crop Protection Council, Alton Hampshire, U.K.
187. BCPC. (2003): Triadimenol. In: *A World Compendium: The Pesticide Manual*, 13th ed., edited by C. D. S. Tomlin, pp. 987–989. British Crop Protection Council, Alton Hampshire, U.K.
188. BCPC. (2003): Triasulfuron. In: *A World Compendium: The Pesticide Manual*, 13th ed., edited by C. D. S. Tomlin, pp. 990–991. British Crop Protection Council, Alton Hampshire, U.K.
189. BCPC. (2003): Tribenuron-methyl. In: *A World Compendium: The Pesticide Manual*, 13th ed., edited by C. D. S. Tomlin, pp. 996–998. British Crop Protection Council, Alton Hampshire, U.K.
190. BCPC. (2003): Trifloxystrobin. In: *A World Compendium: The Pesticide Manual*, 13th ed., edited by C. D. S. Tomlin, pp. 1007–1008. British Crop Protection Council, Alton Hampshire, U.K.
191. BCPC. (2003): Trifluralin. In: *A World Compendium: The Pesticide Manual*, 13th ed., edited by C. D. S. Tomlin, pp. 1012–1014. British Crop Protection Council, Alton Hampshire, U.K.
192. BCPC. (2003): Vinclozolin. In: *A World Compendium: The Pesticide Manual*, 13th ed., edited by C. D. S. Tomlin, pp. 1027–1028. British Crop Protection Council, Alton Hampshire, U.K.
193. BCPC. (2003): Zineb. In: *A World Compendium: The Pesticide Manual*, 13th ed., edited by C. D. S. Tomlin, pp. 1032–1033. British Crop Protection Council, Alton Hampshire, U.K.
194. BCPC. (2003): Ziram. In: *A World Compendium: The Pesticide Manual*, 13th ed., edited by C. D. S. Tomlin, pp. 1034–1035. British Crop Protection Council, Alton Hampshire, U.K.
195. Burton, J.D., Gronwald, J.W, Somners, D.A. Connelly, J.A., Gengenbach, B.G, and Wise, D.L. (1987): Inhibition of acetyl-CoA carboxylase by the herbicides sethoxydim and haloxyfop. *Biochem. Biophys. Res. Commun.*, 148:1039–1044.
196. Campbell, J.R. and Penner, D. (1985a) Sethoxydim metabolism in monocotyledonous and dicotyledonous plants. *Weed Sci.*, 33:771–773.
197. Campbell, J.R. and Penner, D. (1985b) Retention, absorption, translocation and distribution of sethoxydim in monocotyledonous and dicotyledonous plants. *Weed Res.*, 27:179–186.
198. Chang, E.S. (1993): Comparative endocrinology of molting and reproduction: insects and crustaceans. *Annu. Rev. Entomol.*, 38:161–180.
199. Chen, J.-G. (2001): Dual auxin signaling pathways control cell elongation and division. *J. Plant Growth Reg.*, 20:255–264.
200. Clark, J. M., Scott, J. G., Campos, F., and Bloomquist, J. R. (1995): Resistance to avermectins: extent, mechanisms, and management implications. *Annu. Rev. Entomol.*, 40:1–30.
201. Cornes, D. (2005): Callisto: a very successful maize herbicide inspired by allelochemistry. In: *Proc. of the 13th Australian Agronomy Conf.*, September 10–15, 2006, Perth, WA, Australia (http://www.regional.org.au/ay/allelopathy/2005/2/7/2636_cornesd.htm).
202. Conner, J. D., Jr., Ebner, L. S., Landfair, S. W., O'Connor, C. III, Weinstein, K. W., and Jovanovich, A. P. (1991): *Pesticide Regulations Handbook*, 3rd ed. Executive Enterprises, New York, p. 1.
203. Corton, J. C., Lapinskas, P. J., and Gonzalez, F. J. (2000): Central role of PPARγ in the mechanism of action of hepatocarcinogenic peroxisome proliferators. *Mutat. Res.*, 448:139–151.
204. Costa, L. G., Wu, D. S., Olibet, G., and Murphy, S. D. (1989): Formamidine pesticides and alpha 2-adrenoceptors: studies with amitraz and chlordimeform in rats and development of a radioreceptor binding assay. *Neurotoxicol. Teratol.*, 11:405–411.
205. Crisp, T. M., Clegg, E. D., Cooper, R. L., Wood, W. P., Anderson, D. G., Baetcke, K. P., Hoffmann, J. L., Morrow, M. S., Rodier, D. J., Schaeffer, J. E., Touart, L. W., Zeeman, M. G., and Patel, Y. M. (1998): Environmental endocrine disruption: an effects assessment and analysis. *Rev. Environ. Health*, 106(1):11–56.
206. Cui, W. Beever, R. E., Parkes, S. L., and Templeton, M. D. (2004): Evolution of an osmosensing histidine kinase in field strains of *Botryotinia fuckeliana* (*Botrytis cinerea*) in response to dicarboximide fungicide usage. *Popul. Biol.*, 94 (10):1129–1135.
207. De Matteis, F. and Marks, G. S. (1996): Cytochrome P450 and its interaction with the heme biosynthesis pathway. *Can. J. Physiol. Pharmacol.*, 74:1–8.
208. Devine, M., Duke, S. O., and Fedtke, C., Eds. (1993): Herbicide inhibition of photosynthetic electron transport. In: *Physiology of Herbicide Action*, Prentice Hall, Englewood Cliffs, NJ, pp. 113–140.
209. Devine, M., Duke, S. O., and Fedtke, C., Eds. (1993): *Physiology of Herbicide Action*, Prentice Hall, Englewood Cliffs, NJ, pp. 251–294
210. Devine, M., Duke, S. O., and Fedtke, C. (1993): Other sites of herbicide action. In: *Physiology of Herbicide Action*, Prentice Hall, Englewood Cliffs, NJ, pp. 310–332.
211. Devine, M., Duke, S. O., and Fedtke, C., Eds. (1993): Herbicides with auxin activity. In: *Physiology of Herbicide Action*, Prentice Hall, Englewood Cliffs, NJ, pp 295–309.
212. Devine, M. D. (2002): Acetyl-CoA carboxylase inhibitors In: *Herbicide Classes in Development*, edited by P. Boger K. Wakabayashi, and K. Hirai. Springer-Verlag, Berlin, pp 103–113.
213. Devine, M. D. and Preston, C. (2000): The molecular basis of herbicide resistance. In: *Herbicides and Their Mechanisms of Action*, edited by A. H. Cobb and R. C. Kirkwood Sheffield Academic Press, Sheffield, U.K., pp. 72–104.
214. Dhadialla, T. S., Carlson, G. R., and Le, D. P. (1998): New insecticides with ecdysteroidal and juvenile hormone activity. *Annu. Rev. Entomol.*, 43:545–569.
215. Dharmasiri, N., Dharmasiri, S., and Estelle, M. (2005) The F-box protein TIR1 is an auxin receptor. *Nature*, 435 (26):441–445.
216. Dharmasiri, N., Dharmasiri, S., Weijers, D., Lechner, E. Yamada, M., Hobbie, L., Ehrismann, J. S., Jurgens, G. and Estelle, M. (2005): Plant development is regulated by a family of auxin receptor F box proteins. *Develop. Cell* 9:109–119.

217. Dong, K. (2003): Voltage-gated sodium channels as insecticide targets. In: *Chemistry of Crop Protection*, edited by G. Voss and G. Ramos. Wiley-VCH, Berlin, pp. 167–176.
218. Donn, G. and Kocher, H. (2002): Inhibitors of glutamine snythetase. In: *Herbicide Classes in Development*, edited by P. Boger, K. Wakabayashi, and K. Hirai. Springer-Verlag, Berlin, pp. 87–101.
219. Duke, S. O. and Kenyon, W. H. (1988): Polycyclic alkanoic acids. In: *Herbicides: Chemistry, Degradation, and Mode of Action*, edited by P. C. Kearney and D. D. Kaufman, pp. 71–116. Marcel Dekker, New York.
220. Ecobichon, D. J. (1993): Toxic effects of pesticides. In: *Casarett and Doull's Toxicology: The Basic Science of Poisons*, 4th ed., edited by J. Doull, C. D. Klaassen, and M. O. Amdur, pp. 565–621. Macmillan, New York.
221. Edwards, G. (2003): Ivermectin: does P-glycoprotein play a role in neurotoxicity? *Filaria J.*, 2(Suppl. 1):1–6.
222. Eldridge, J. C., Stevens, J. T., Wetzel, L. T., Tisdel, M. O., Breckenridge, C. B., McConnell, R. F., and Simpkins, J. W. (1996): Atrazine: mechanisms of hormonal imbalance in female SD rats. *Fundam. Appl. Toxicol.*, 24(12):2–5.
223. European Commission. (2003): *Picoxystrobin*, SANCO/10196/2003/Final, June 3, 2003. (http://ec.europa.eu/food/plant/protection/evaluation/newactive/list1_picoxystrobin.en.pdf).
224. European Economic Community. (1993): Commission Directive 93/67/EEC:1993. Laying Down the Principles for Assessment of Risks to Man and the Environment of Substances Notified in Accordance with Council Directive 67/548/EEC. July, 20, 1993.
225. European Economic Community. (1994): Commission Directive 94/79/EC of December, 1994 amending Council Directive 91/414/EEC concerning the placing of plant protection products on the market. *Off. J. Euro. Communities*, Dec. 31:L354/16.
226. Extension Toxicology Network. (1995): *Deltamethrin*. Pesticide Information Profiles, Oregon State University.
227. Extension Toxicology Network. (1996): *Resmethrin*. Pesticide Information Profiles, Extension Toxicology Network http://extoxnet.orst.edu/pips/resmethr.htm.
228. Extension Toxicology Network. (2006): *DDT (Dichlorodiphenyltrichloroethane)*. http://pmep.cce.cornell.edu/profiles/extoxnet/carbaryl-dicrotophos/ddt-ext.html.
229. Farber, T. M. (1987): *Pesticide Assessment Guidelines*. Subdivision F. *Position Document: Selection of a Maximum Tolerated Dose (MTD) in Oncogenicity Studies*. Toxicology Branch, Hazard Evaluation Division, Office of Pesticides Programs, U.S. Environmental Protection Agency, Washington, D.C.
230. Feng, P. C. et al. (1990): Metabolism of alachlor by rat and mouse liver and nasal turbinate tissues. *Drug Metab. Dispos.*, 18(3):373–377
231. Forrester, S. G., Beech, R. N., and Prichard, R. K. (2004): Agonist enhancement of macrocyclic lactone activity at a glutamate-gated chloride channel subunit from *Haemonchus contortus*. *Biochem. Pharmacol.*, 67:1019–1024.
232. Galcheva-Gargova, Z., Derijard, B., Wu, I.-H., and Davis, R. J. (1994): An osmosensing signal transduction pathway in mammalian cells. *Science*, 265:806–808.
233. FRAC. (2005): *FRAC Code List 2: Fungicides Sorted by Mode of Action*. Fungicide Resistance Action Committee (http://www.frac.info/frac/index.htm).
234. Geiger, D. R., and Fuchs, M. A. (2002): Inhibitors of aromatic amino acid biosynthesis (glyphosate): In: *Herbicide Classes in Development*, edited by P. Boger, K. Wakabayashi, and K. Hirai, pp. 59–85. Springer-Verlag, Berlin.
235. Gianessi, L. P. (1986): *A National Pesticide Usage Data Base*. Resources for the Future, Washington, D.C., pp. 1–14.
236. Gill, S. S., Cowles, E. A., and Pietrantonio, P. V. (1992): The mode of action of *Bacillus thuringiensis* endotoxins. *Ann. Rev. Entomol.*, 37:615–636.
237. Gilles, N., Krimm, I., Bouet, F., Froy, O., Gurevitz, M., Lancelin, J.-M., and Gordon, D. (2000): Structural implications on the interaction of scorpion α-like toxins with the sodium channel receptor site inferred from toxin iodination and pH-dependent binding. *J. Neurochem.*, 75(4), 1735–1745.
238. Goldman, J. M., Cooper, R. L., Laws, S. C., Rehnberg, G. L., Edwards, T. L., McElroy, W. K., and Hein, J. F. (1990): Chlordimeform-induced alterations in endocrine regulation within the male rat reproductive system. *Toxicol. Appl. Pharmacol.*, 104 (1):25–35.
239. Goldman, J. M., Cooper, R. L., Edwards, T. L., Rehnberg, G. L., McElroy, W. K., and Hein, J. F. (1991): Suppression of the luteinizing hormone surge by chlordimeform in ovariectomized, steroid-primed female rats. *Pharmacol. Toxicol.*, 68(2):131–136.
240. Goldman, J. M., Stoker, T. E., Perreault, S. D., Cooper, R. L., and Crider, M. A. (1993): Influence of the formamidine pesticide chlordimeform on ovulation in the female hamster: dissociable shifts in the leutinizing hormone surge and oocyte release. *Toxicol. Appl. Pharmacol.*, 121: 279–290.
241. Goldman, L. R. (1998): Chemicals and children's environment: what we don't know about risk. *Environ. Health Perspect.*, 106(Suppl. 3):875–880.
242. Gordon, D., Martin-Eauclaire, M.-F., Cestele, S., Kopeyan, C., Carlier, E., Khalifa, R. B., Pelhate, M., and Rochat, H. (1996): Scorpion toxins affecting sodium current inactivation bind to distinct homologous receptor sites on rat brain and insect sodium channels. *J. Biol. Chem.*, 271(14):8034–8045.
243. Grandoni, J. A., Marta, P. T., and Schloss, J. V. (1998): Inhibitors of branched-chain amino acid biosynthesis as potential anti-tuberculosis agents. *J. Antimicrob. Chemother.*, 42:475–482.
244. Graham, D. G., Amarnath, V., Valentine, W. M., Pyle, S. J., and Anthony, D. C. (1995): Pathogenic studies of hexane and carbon disulfide neurotoxicity. *Crit. Rev. Toxicol.*, 25(2):91–112.
245. Green, T., Lee, R., Moore, R. B., Ashby, Willis, G. A. Lund, V. J., and Clapp, M. J. L. (2000): Acetochlor-induced rat nasal tumors: further studies on the mode of action and relevance to humans. *Reg. Toxicol. Pharmacol.*, 32: 127–133.
246. Green, T., Toghill, A., Lee, R. Waechter, F., Weber, E., and Noakes, J. (2005): Thiamethoxam induced mouse liver tumors and their relevance to humans. Part 1. Mode of action studies in the mouse. *Toxicol. Sci.*, 86(1):36–47.

247. Grossman, K. (2000): The mode of action of quinclorac: a case study of a new auxin-type herbicide. In: *Herbicides and Their Mechanisms of Action*, edited by A. H. Cobb and R. C. Kirkwood. Sheffield Academic Press, Sheffield, U.K., pp. 181–214.

248. Gunsolus, G. L. and Curran, W. S. (1999): *Herbicide Mode of Action and Injury Symptoms*, http://www.mes.umn.edu/Documents/D/C/DC3832.htm.

249. Han, J., Lee, J.-D., Bibbs, L., and Ulevitch, R. J. (1994): A MAP kinase targeted by endotoxin and hyperosmolarity in mammalian cells. *Science*, 265:808–811.

250. Hamboeck, H. et al. (1981): The binding of *s*-triazine metabolites to rodent hemoglobins appear irrelevant to other species. *Molec. Pharmacol.*, 20:579–584.

251. Harrewijn, P. (1997): Pymetrozine, a fast-acting and selective inhibitor of aphid feeding: *in situ* studies with electronic monitoring of feeding behaviour. *Pesticides Sci.*, 49:130–140.

252. Hayes, W. J., Jr. (1991): Introduction. In *Handbook of Pesticide Toxicology*. Vol. I. *General Principles*, edited by W. J. Hayes, Jr., and E. R. Laws, Jr., pp. 1–37. Academic Press, San Diego, CA.

253. HRAC. (2005): *Classification of Herbicides According to Mode of Action*. Herbicide Resistance Action Committee, http://www.plantprotection.org/hrac/Bindex.cfm?doc=moa2002.htm.

254. Hoffman, K. H. and Lorenz, M. W. (1968): Recent advances in hormones in insect pest control. *Phytoparasitica*, 26(4):1–8.

255. Hollingshaus, J. (1987): Inhibition of mitochondrial electron transport by hydramethylnon: a new amidinohyrazone insecticide. *Pestic. Biochem. Physiol.*, 27:61–70.

256. Hollingworth, R. M. (1976): Chemistry, biological activity and uses of formamidine pesticides. *Environ. Health Perspect.*, 14:57–69.

257. Hollingworth, R. et al. (1994): New inhibitors of complex I of the mitochondrial electron transport chain with activity as pesticides. *Biochem. Soc. Trans. (Lond.)*, 22:230–233.

258. Horton, J. D., Goldstein, J. L., and Brown, M. S. (2002): SREBPs: activators of the complex program of cholesterol and fatty acid synthesis in the liver. *J. Clin. Invest.*, 109(9): 1125–1131.

259. IARC. (1998): *Evaluation or Re-evaluation of Some Agents Which Target Specific Organs in Rodent Bioassays*, Vol. 73, Working Group Monographs. International Agency for Research on Cancer, http://193.51.164.1l/pastandfuture/OCT98.html.

260. IRAC. (2005): *Mode of Action Classification, Version 5*. Insecticide Resistance Action Committee, http://www.irac-online.org/.

261. Irvine, J. E. and Freyre, R. H. (1959): Occurrence of rotenone in some species of the genus *Tephrosia*. *J. Agric. Food Chem.*, 7:106–107.

262. JECFA. (1997): *Thiabendazole (Thiabendazole)*, JECFA Monograph Series 31. Joint Expert Committee on Food Additives, World Health Organization, Geneva, pp. 1–23.

263. Joint Meeting of the FAO/WHO Panel of Experts on Pesticide Residues in Food (JMPR). (1974): *Carbaryl*, FAO Agricultural Studies No. 92, WHO Tech. Rep. Ser. No. 545, Evaluation of Some Pesticide Residues in Food, FAO/AGP/1974/M/ll.

264. Joint Meeting of the FAO/WHO Panel of Experts on Pesticide Residues in Food (JMPR). (1977): *Methoxychlor*. http://www.inchem.org/documents/jmpr/jmpmono/v077pr37.htm.

265. Joint Meeting of the FAO/WHO Panel of Experts on Pesticide Residues in Food (JMPR). (1983): *Carbofuran*. FAO Plant Production and Protection Paper 46. Food and Agriculture Organization of the United Nations and World Health Organization, Geneva.

266. Joint Meeting of the FAO/WHO Panel of Experts on Pesticide Residues in Food (JMPR). (1983): *Chlorpyrifos*. FAO Plant Production and Protection Paper 46. Food and Agriculture Organization of the United Nations and World Health Organization, Geneva.

267. Joint Meeting of the FAO/WHO Panel of Experts on Pesticide Residues in Food (JMPR). (1985): *Prochloraz*. FAO Plant Production and Protection Paper 56. Food and Agriculture Organization of the United Nations and World Health Organization, Geneva.

268. Joint Meeting of the FAO/WHO Panel of Experts on Pesticide Residues in Food (JMPR). (1985): *Triadimefon*. FAO Plant Production and Protection Paper 68. Food and Agriculture Organization of the United Nations and World Health Organization, Geneva.

269. Joint Meeting of the FAO/WHO Panel of Experts on Pesticide Residues in Food (JMPR). (1989): *Triadimenol*. FAO Plant Production and Protection Paper 99. Food and Agriculture Organization of the United Nations and World Health Organization, Geneva.

270. Joint Meeting of the FAO/WHO Panel of Experts on Pesticide Residues in Food (JMPR). (1990): *Acephate*. FAO Plant Production and Protection Paper 103. Food and Agriculture Organization of the United Nations and World Health Organization, Geneva.

271. Joint Meeting of the FAO/WHO Panel of Experts on Pesticide Residues in Food (JMPR). (1990) *Endosulfan*. FAO Plant Production and Protection Paper 99. Food and Agriculture Organization of the United Nations and World Health Organization Geneva.

272. Joint Meeting of the FAO/WHO Panel of Experts on Pesticide Residues in Food (JMPR). (1990): *Lindane*. FAO Plant Production and Protection Paper 99. Food and Agriculture Organization of the United Nations and World Health Organization, Geneva.

273. Joint Meeting of the FAO/WHO Panel of Experts on Pesticide Residues in Food (JMPR). (1990): *Methomyl*. FAO Plant Production and Protection Paper 99. Food and Agriculture Organization of the United Nations and World Health Organization, Geneva.

274. Joint Meeting of the FAO/WHO Panel of Experts on Pesticide Residues in Food (JMPR). (1990): *Propoxur*. FAO Plant Production and Protection Paper 99. Food and Agriculture Organization of the United Nations and World Health Organization, Geneva.

275. Joint Meeting of the FAO/WHO Panel of Experts on Pesticide Residues in Food (JMPR). (1991): *Azinphos-methyl*. FAO Plant Production and Protection Paper 111. Food and Agriculture Organization of the United Nations and World Health Organization, Geneva.

276. Joint Meeting of the FAO/WHO Panel of Experts on Pesticide Residues in Food (JMPR). (1992): *Imazalil*. FAO Plant Production and Protection Paper 111. Food and Agriculture Organization of the United Nations and World Health Organization, Geneva.
277. Joint Meeting of the FAO/WHO Panel of Experts on Pesticide Residues in Food (JMPR). (1993): *Chlorothalonil*. FAO Plant Production and Protection Paper 116. Food and Agriculture Organization of the United Nations and World Health Organization, Geneva.
278. Joint Meeting of the FAO/WHO Panel of Experts on Pesticide Residues in Food (JMPR). (1993): *Thiram*. FAO Plant Production and Protection Paper 116. Food and Agriculture Organization of the United Nations and World Health Organization, Geneva.
279. Joint Meeting of the FAO/WHO Panel of Experts on Pesticide Residues in Food (JMPR). (1994): *Abamectin*. FAO Plant Production and Protection Paper 127. Food and Agriculture Organization of the United Nations and World Health Organization, Geneva.
280. Joint Meeting of the FAO/WHO Panel of Experts on Pesticide Residues in Food (JMPR). (1994): *Diazinon*. FAO Plant Production and Protection Paper 122. Food and Agriculture Organization of the United Nations and World Health Organization, Geneva.
281. Joint Meeting of the FAO/WHO Panel of Experts on Pesticide Residues in Food (JMPR). (1994): *Dichlorvos*. FAO Plant Production and Protection Paper 122. Food and Agriculture Organization of the United Nations and World Health Organization, Geneva.
282. Joint Meeting of the FAO/WHO Panel of Experts on Pesticide Residues in Food (JMPR). (1994): *Fenpropathrin*. FAO Plant Production and Protection Paper 122. Food and Agriculture Organization of the United Nations and World Health Organization, Geneva.
283. Joint Meeting of the FAO/WHO Panel of Experts on Pesticide Residues in Food (JMPR). (1994): *Mancozeb*. FAO Plant Production and Protection Paper 122. Food and Agriculture Organization of the United Nations and World Health Organization, Geneva.
284. Joint Meeting of the FAO/WHO Panel of Experts on Pesticide Residues in Food (JMPR). (1994): *Maneb*. FAO Plant Production and Protection Paper 122. Food and Agriculture Organization of the United Nations and World Health Organization, Geneva.
285. Joint Meeting of the FAO/WHO Panel of Experts on Pesticide Residues in Food (JMPR). (1994): Permethrin. FAO Plant Production and Protection Paper 122. Food and Agriculture Organization of the United Nations and World Health Organization, Geneva.
286. Joint Meeting of the FAO/WHO Panel of Experts on Pesticide Residues in Food (JMPR). (1994): *Zineb*. FAO Plant Production and Protection Paper 122. Food and Agriculture Organization of the United Nations and World Health Organization, Geneva.
287. Joint Meeting of the FAO/WHO Panel of Experts on Pesticide Residues in Food (JMPR). (1995): *Monocrotophos*. FAO Plant Production and Protection Paper 133. Food and Agriculture Organization of the United Nations and World Health Organization, Geneva.
288. Joint Meeting of the FAO/WHO Panel of Experts on Pesticide Residues in Food (JMPR). (1996): *Aldicarb*. FAO Plant Production and Protection Paper 133. Food and Agriculture Organization of the United Nations and World Health Organization, Geneva.
289. Joint Meeting of the FAO/WHO Panel of Experts on Pesticide Residues in Food (JMPR). (1996): *Benomyl*. FAO Plant Production and Protection Paper 133. Food and Agriculture Organization of the United Nations and World Health Organization, Geneva.
290. Joint Meeting of the FAO/WHO Panel of Experts on Pesticide Residues in Food (JMPR). (1996): *Captan*. FAO Plant Production and Protection Paper 133. Food and Agriculture Organization of the United Nations and World Health Organization, Geneva.
291. Joint Meeting of the FAO/WHO Panel of Experts on Pesticide Residues in Food (JMPR). (1996): *Iprodione (addendum)*. FAO Plant Production and Protection Paper 133. Food and Agriculture Organization of the United Nations and World Health Organization, Geneva.
292. Joint Meeting of the FAO/WHO Panel of Experts on Pesticide Residues in Food (JMPR). (1996): *Thiophanate-methyl*. FAO Plant Production and Protection Paper 133. Food and Agriculture Organization of the United Nations and World Health Organization, Geneva.
293. Joint Meeting of the FAO/WHO Panel of Experts on Pesticide Residues in Food (JMPR). (1996): *Vinclozolin*. FAO Plant Production and Protection Paper 133. Food and Agriculture Organization of the United Nations and World Health Organization, Geneva.
294. Joint Meeting of the FAO/WHO Panel of Experts on Pesticide Residues in Food (JMPR). (1997): *Ferbam*. FAO Plant Production and Protection Paper 140. Food and Agriculture Organization of the United Nations and World Health Organization, Geneva.
295. Joint Meeting of the FAO/WHO Panel of Experts on Pesticide Residues in Food (JMPR). (1997): *Ziram*. FAO Plant Production and Protection Paper 140. Food and Agriculture Organization of the United Nations and World Health Organization, Geneva.
296. Jones, G. (1995): Molecular mechanisms of action of juvenile hormones. *Annu. Rev. Entymol.*, 40:147–169.
297. Kagabu, S. (2003): Molecular design of neonicotinoids: past, present and future. In: *Chemistry of Crop Protection*, edited by G. Vos and G. Ramos. Wiley-VCH, Berlin, pp. 191–212.
298. Kelce, W. R., Monosson, E., Gamcsik, M. P., Laws, S. C, and Gray, L. E., Jr. (1994): Environmental hormone disruptors: evidence that vinclozolin developmental toxicity is mediated by antiandrogenic metabolites. *Toxicol. Appl. Pharmacol.*, 126:276–285.
299. Klaunig, J. E., Babich, M. A., Baetcke, K. P., Cook, J. C., Corton, J. C., David, R. M., DeLuca, J. G., Lai, D. Y., McKee, R. H., Peters, J. M., Roberts, R. A., and Fenner-Crisp, P. A. (2003): PPARγ agonist-induced rodent tumors: mode of action and human relevance. *Crit. Rev. Toxicol.*, 33(6):655–780.
300. Koelle, M. R. Talbot, W. S., Segraves, W. A., Bender, M. T., Cherbas, P., and Hogness, D. S. (1991): The Drosophila EcR gene encodes an ecdysone receptor, a new member of the steroid receptor super-family. *Cell*, 67: 59–77.

301. Koo, S. J., Neal, J. C., and DiTomaso, J. M. (1996): 3,7-Dichloroquinolinecarboxylic acid inhibits cell-wall biosynthesis in maize roots. *Plant Physiol.*, 112: 1383–1389.
302. Kushiro, T., Nambara, E., and McCourt, P. (2003): The key to signaling. *Nature*, 422:122.
303. Lances, G. R. and Gordon, L. R. (1989): Toxicology. In: *Ivermectin and Abamectin*, edited by W. R. Campbell, pp. 89–112. Springer-Verlag, New York.
304. Lankas, G. R., Minsker, D. H., and Robertson, R. T. (1989): Effects of ivermectin on reproduction and neonatal toxicity in rats. *Food Chem. Toxicol.*, 27:523–529.
305. Lankas, G. R., Cartwright, M. E., and Umbenhauer, D. (1997): *p*-Glycoprotein deficiency in a subpopulation of CF-1 mice enhances avermectin-induced neurotoxicity. *Toxicol. Appl. Pharmacol.*, 143:357–365.
306. Larson, L. L. (1999): *Novel Organic and Natural Product Insect Management Tools*. National IPM Network, hppt://ipmworld.umn.edu/chapters/larson.htm.
307. Lee, D. L., Prisbylla, M. P. Cromartie, T. H. et al. (1997): The discovery and structural requirements of inhibitors of *p*-hydroxyphenylpyruvate dioxygenase. *Weed Sci.*, 45:601–609.
308. Lewin, A. H. (2006): Receptors of mammalian trace amines. *Am. Assoc. Adv. Sci.*, 8(1):E138–E145.
309. Leyser, O. (2006): Dynamic integration of auxin transport and signaling. *Curr. Biol.*, 16:R424–R433.
310. Li, J., Carroll, J., and Ellar, D. J. (1991): Crystal structure of insecticidal delta-endotoxin from *Bacillus thuringiensis* at, 2.5 Å resolution. *Nature,* 353:815–821.
311. Liu, J., Dent, J. A., Beech, R. N., and Prichard, R. K. (2004): Genomic organization of an avermectin receptor subunit from *Haemonchus contortus* and expression of its putative promoter region in *Caenorhabditis elegans*. *Molec. Biochem. Parasitol.*, 134:267–274.
312. Ludwig, B., Bender, E., Arnold, S., Huttemann, M., Lee, I., and Kadenbach, B. (2001): Cytochrome *c* oxidase and the regulation of oxidative phosphorylation. *Chembiochem.*, 2:392–403.
313. Magkos, F., Arvaniti, F., and Zampelas, A. (2006): Food: buying more safety or just peace of mind? A critical review of the literature. *Crit. Rev. Food Sci. Nutr.*, 46:23–56.
314. Matringe, M., Clair, D., and Scala, R. (1990): Effects of peroxidizing herbicides on protoporphyrin IX levels in non-chlorophyllous soybean cell culture. *Pestic. Biochem. Physiol.*, 36:300–307.
315. Matsuda, K., Buckingham, S. D., Kleier, D., Rauh, J. J., Grauso, M., and Sattelle, D. B. (2001): Neonicotinoids: insecticides acting on insect nicotinic acetylcholine receptors. *Trends Pharmacol. Sci.*, 22 (11):573–580.
316. McMullin, T. S., Brzezicki, J. M., Cranmer, B. K., Tessari, J. D., and Andersen, M. E. (2003): Pharmacokinetic modeling of disposition and time-course studies with [$^{14}C$] atrazine. *J. Toxicol. Environ. Health A*, 66:941–964.
317. Meister, R. T. (1997): *Farm Chemicals Handbook*. Meister, Willoughby, OH, p. C286.
318. Mellanby, K. (1992): *The DDT Story*. British Crop Protection Council, Farnham, Surrey, pp. 6–7.
319. Merzendorfer, H. (2006): Insect chitin synthases: a review. *J. Comp. Physiol. B*, 176:1–15.
320. MAFF. (1985): *Notification of the Director-General. Requirements for Safety Evaluation of Agricultural Chemicals*. Agricultural Production Bureau, Ministry of Agriculture, Forestry, and Fisheries, Japan.
321. Mitchell, G., Bartlett, D. W., Fraser, T. E. M., Hawkes, T. R., Holt, D. C., Townson, J. K. and Wichert, R. A. (2001): Mesotrione: a new selective herbicide for use in maize. *Pestic. Manage. Sci.*, 57:120–128.
322. Moreau, C., Jacquet, H., Prost, A-L., D'halan, N., and Vivaudou, M. (2000): The molecular basis of the specificity of action of the $K_{ATP}$ channel openers. *Eur. Molec. Biol. Org. J.*, 19(24):6644–6651.
323. Morishima, Y., Osabe, H., and Goto, Y. (1990): Action mechanism of DLH-1777, a novel, 4-pyridone-3-carboxamide herbicide: peroxidizing activity and accumulation of porphyrins. *J. Pestic. Sci.*, 15:553–559.
324. Moser, V. C., McDaniel, K. L., and Phillips, P. M. (1991): Rat strain and stock comparisons using a functional observational battery: baseline values and the effects of amitraz. *Toxicol. Appl. Pharmacol.*, 108:267–283.
325. Mulkey, M. E. (2001): *Determination of Whether Dithiocarbamate Pesticides Share a Common Mechanism of Toxicity*. U.S. Environmental Protection Agency, Washington, D.C. (http://www.epa.gov/pesticides/cumulative/dithiocarb.pdf).
326. Mullins, J. W. (1993): *Imidacloprid: A New Nitroguanidine Insecticide*. ACS Symp. Ser. 524: Newer Pest Control Agents and Technology with Reduced Environmental Impact. ACS, Washington, D.C.
327. Musilek, V., Cerna, J., Sasek, V., Semerdzieva, M., and Vondracek, M. (1969): Antifungal antibiotic of the Basidomycete *Oudemansiella mucida*. I. Isolation and cultivation of a producing strain, *Folia Microbiol. (Prague)*, 14:377–387
328. Naor, Z., Harris, D., and Shacham, S. (1998): Mechanism of GnRH signaling: combinatorial cross-talk of $Ca^{2+}$ and protein kinase C. *Front. Neuroendocrinol.*, 19:1–19.
329. Narahashi, T. (2000): Neuroreceptors, and ion channels as the basis for drug action: past, present and future. *J. Pharmacol. Therap.*, 294 (1):1–26.
330. National Research Council. (1993): *Pesticides in the Diets of Infants and Children*. National Academy Press, Washington, D.C.
331. NTP. (1979): *Bioassay of p-Chloroaniline for Possible Carcinogenicity*, CAS No. 106-47-8, Tech. Rep. No. 189. National Toxicology Program, Washington, D.C.
332. NTP. (2002): *Studies of Urethane, Ethanol and Urethane/Ethanol*, CAS No. 51-79-6, CAS No. 64-17-5, Tech. Rep. No. 510. National Toxicology Program, Washington, D.C.
333. Nestorowicz, A., Wilson, B. A., Schoor, K. P., Inoue, H., Glaser, B., Landau, H., Stanley, C. A., and Thornton., P. S. (1996): Mutations in the sulfonylurea receptor gene are associated with familial hyperinsulinism in Ashkenazi Jews. *Human Molec. Genet.*, 5 (11):1813–1822.
334. Nikolau, B. J., Wurtele, E. S., Caffrey, J., Chen, Y., Crane, V., Diez, T., Huang, J. -Y., Me Dowell, M. T, Shang, X. M., Song, J., Wang, X., and Weaver, L. M (1993): The biochemistry and molecular biology of acetyl-CoA carboxylase and other biotin enzymes. In: *Biochemistry and Molecular Biology of Membrane and Storage Lipids of Plants*, edited by N. Murata and C. Somerville, pp. 138–149. American Society of Plant Physiologists Press. Brentwood, TN.

335. OECD. (1981): *OECD Guideline for Testing of Chemicals*. Section 4. *Health Effects*. Organization for Economic Cooperation and Development, Paris, France.
336. Paciorek, T. and Frimi, J. (2006): Auxin signaling. *J. Cell Sci*., 119(7):1199–1202.
337. Pallett, K. E. (2000): The mode of action of isoxaflutole. In: *Herbicides and Their Mechanisms of Action*, edited by A. H. Cobb and R. C. Kirkwood, pp. 215–238. Sheffield Academic Press, Sheffield, U.K.
338. Pastoor, T., Rose, P., Lloyd, S., Peffer, R., and Green, T. (2005): Case study: weight of evidence evaluation of the human health relevance of thiamethoxam-related mouse liver tumors. *Toxicol. Sci*., 86 (1):56–60.
339. Pillonel, C. (2005):Evaluation of phenylaminopyrimidines as antifugal protein kinase inhibitors. *Pest Manage. Sci*. 61:1069–1076.
340. Pillonel, C. and Meyer, T. (1997): Effect of phenylpyrroles on glycerol accumulation and protein kinase activity of *Neurospora crassa*. *Pesticide Sci*., 49:229–236.
341. Ramesh, M. A., Laidlaw, R. D., Durrenberger, F., Orth, A. B., and Kronstad, J. W. (2001): The cAMP signal transduction pathway mediates resistance to dicarboximide and aromatic hydrocarbon fungicides in *Ustilago maydis*. *Fungal Genet. Biol*., 32:183–193.
342. Ray, D. E., Burr, S. A., and Lister, T. (2006): The effect of exposure to the pyrethroid, deltamethrin and *S*-bioallethrin on hippocampal inhibition and skeletal muscle hyperexcitability in rats. *Toxicol. Appl. Pharmacol*., 216: 354–362.
343. Raymond-Delpech, V., Matsuda, K., Sattelle, B. M., Rauh, J. J., and Sattelle, D. B. (2005): Ion channels: molecular targets of neuroactive insecticides. *Invertebrate Neurosci*., 5:119–133.
344. Retzinger, Jr., J. E. and Mallory-Smith, C. (1999): *Classification of Herbicides by Site of Action for Weed Resistance Management Strategies*. Oregon State University, Corvallis (http://www.css.orst.edu/weeds/Publications/siteaction.htm).
345. Riddiford, L. M., Cherbas, P., and Truman, J. W. (2001): Ecdysone receptors and their biological actions. *Vitam. Horm*., 60:1–17.
346. Ruegger, M., Dewey, E., Gray, W. M., Hobbie, L., and Turner, J. (1998): The TIR1 protein of Arabidopsis functions in auxin response and is related to human SKP2 and yeast Grr1p. *Genes Dev*., 12:198–207.
347. Salgado, V. (1998): Studies on the mode of action of spinosad: insect symptoms and physiological correlates. *Pestic. Biochem. Physiol*., 60:91–102.
348. Salgado, V. L., Sheets, J. J., Watson, G. B., and Schmidt, A. L. (1998): Studies on the mode of action of spinosad: the internal effective concentration and the concentration dependence of neural excitation. *Pestic. Biochem. Physiol*., 60:103–110.
349. Sandmann, G. (2002): Bleaching herbicides: action mechanism in carotenoid biosynthesis, structural requirements and engineering of resistance. In: *Herbicide Classes in Development*, edited by P. Boger, K. Wakabayashi, and K. Hirai, pp. 43–57. Springer-Verlag, Berlin.
350. Schloss, J. V., Ciskanik, L. M., and Van Dyk, V. E. (1988): Origin of the herbicide binding site of acetolactate synthase. *Nature*, 331:360–362.
351. Shafer, T. J. and Meyer, D. A. (2004): Effects of pyrethroids on voltage-sensitive calcium channels: a critical evaluation of strengths, weaknesses, data needs and relationship to assessment of cumulative neurotoxicity. *Toxicol. Appl. Pharmacol*., 196:303–318.
352. Shoop, W. L., Mrozik, H., and Fisher, M. H. (1995): Structure and activity of avermectins and milbemycins in animal health. *Vet. Parasitol*., 59:139–156.
353. Simmons, S. W. (1959): The use of DDT insecticides in human medicine. In: *DDT: The Insecticide Dichlorodiphenyl-Trichloroethane and Its Significance*, Vol. 2, edited by P. Miller, pp. 251–502. Birkhaeriger, Basel.
354. Smagghe, G., Dhadialla, T. S., and Lezzi, M. (2002): Comparative toxicity and ecdysone receptor affinity on non-steroidal ecdysone agonists and 20-hydroxyecdysone in *Chironomus tentans*. *Insect Biochem. Molec. Biol*., 32, 187–192.
355. Smith, L. L. (1986): The response of the lung to foreign compounds that produce free radicals. *Ann. Rev. Physiol*., 48:681–692.
356. Smith, L. L. (1987): Mechanism of paraquat toxicity in lung and its relevance to treatment. *Hum. Toxicol*., 6(1):31–36.
357. Soderlund, D. M. Clark, J. M., Sheets, L. P., Mullin, L. S., Piccirillo, V. J., Sargent, D., Stevens, J. T., and Weiner, M. L. (2002): Mechanism of pyrethroid neurotoxicity: implications for cumulative risk assessment. *Toxicology*, 171:3–59.
358. Solecki, R. (2000): *Pesticide Residues in Food: DDT (para,para′-Dichlorodiphenyltrichloroethane)* [addendum]. INCHEM, International Programme on Chemical Safety http://www.inchem.org/documents/jmpr/jmpmono/v00pr03.htm#_00032230.
359. Stevens, J. T. and Breckenridge, C. B. (2001): The avermectins: insecticidal and antiparasitic agents. In: *Handbook of Pesticide Toxicology*. Vol. 2. *Agents*, 2nd ed., edited by R. Krieger, pp. 1157–1167. Academic Press, New York.
360. Stevens, J. T., Breckenridge, C. B., Wetzel, L. T., Thakur, A. K., Liu, C, Werner, C, Luempert III, L. C., and Eldridge, J. C. (1999): A risk characterization for atrazine: oncogenicity profile. *J. Toxicol. Environ. Health A*, 56:69–109.
361. Stevens, J. T., Sumner, D. D., and Luempert, L. (1995): Agricultural chemicals: the impact of regulations under FIFRA on science and economics. In: *Primer on Regulatory Toxicology*, edited by C. Chenzelis, J. Holson, and S. Gad, pp. 133–163. Raven Press, New York.
362. Stevens, J. T., Werner, C, Breckenridge, C. B., and Sumner, D. D. (2007): Hazard assessment for selected symmetrical and asymmetrical triazine herbicide. In: *The Triazine Herbicides*, edited by H. M. LeBaron, J. McFarland, O. Burnside, and R. Clark, in press.
363. Stevens, J. T., Wetzel, L. T., Breckenridge, C. B., Gillis, J. H., Luempert III, L. G., and Eldridge, J. C. (1994): Hypothesis for mammary tumorigenesis in female Sprague–Dawley rats exposed to chloro-i-triazine herbicides. *J. Toxicol. Environ. Health*, 43(2):139–154.
364. Tanaka, Y. and Omura, S. (1993): Agroactive compounds of microbial origin. *Annu. Rev. Microbiol*., 47:57–87.
365. Tomizawa, M. and Casida, J. E. (2003): Selective toxicity of neonicotinoids attributed to specificity of insect and mammalian nicotinic receptors. *Annu. Rev. Entomol*., 48:339–364.

366. Tomizawa, M. and Casida, J. E. (2005): Neonicotinoid insecticide toxicology: mechanism of selective action. *Annu. Rev. Pharmacol. Toxicol.*, 45:247–268.
367. Truska, M., Wu, G., Vrinten, P., and Qiu, X. (2006): Metabolic engineering of plants to produce very long chain polyunsaturated fatty acids. *Transgenic Res.*, 15:131–137.
368. Tunaz, H. and Uygun, N. (2004): Insect growth regulators for insect pest control. *Turkish J. Agric. For.*, 28:377–387.
369. Turner, M. J. and Schaeffer, J. M. (1989): Mode of action of ivermectin. In: *Ivermectin and Abameclin*, edited by W. R. Campbell, pp. 73–88. Springer-Verlag, New York.
370. U. S. Congress. (1947): *Federal Insecticide, Fungicide and Rodenticide Act (FIFRA)*, Publ. No. 80-104, 61 Stat., 163. U.S. Congress, Washington, D.C., p. 1.
371. U.S. Congress. (1958): *Food Additive Amendments to the Federal Food, Drug, and Cosmetic Act (FFDCA)*, 409. Publ. No. 85-929, 72 Stat., 1785. U.S. Congress, Washington, D.C., p. 1.
372. U.S. EPA. (1982): *Pesticide Assessment Guidelines*. Subdivision F. *Hazard Evaluation: Human and Domestic Animals*, 540/9-82-025. U.S. Environmental Protection Agency, Washington, D.C. (available from NTIS, Springfield, VA).
373. U.S. EPA. (1983): *Aldicarb*, CASRN, 116-06-3. Integrated Risk Information System, U.S. Environmental Protection Agency, Washington, D.C.
374. U.S. EPA. (1987): *Benomyl*, CASRN, 17804-35-2. Integrated Risk Information System, U.S. Environmental Protection Agency, Washington, D.C.
375. U.S. EPA. (1988): *Iprodione*, CASRN 36734-19-7. Integrated Risk Information System, U.S. Environmental Protection Agency, Washington, D.C.
376. U.S. EPA. (1988): *Rotneone*, EPA Pesticide Fact Sheet. U.S. Environmental Protection Agency, Washington, D.C.
377. U.S. EPA. (1991): *Methoprene RED Fact Sheet*, updated June, 2001. U.S. Environmental Protection Agency, Washington, D.C.
378. U.S. EPA. (1992): *Indole-3-Butyric Acid: Reregistration Eligibility Decision (RED)*. U.S. Environmental Protection Agency, Washington, D.C.
379. U.S. EPA. (1993): *Butylate: Reregistration Eligibility Decision (RED)*, EPA-738-F-93-014. U.S. Environmental Protection Agency, Washington, D.C.
380. U.S. EPA. (1994): Acetochlor: pesticide tolerance. *Fed. Reg.*, 59(56):13654–13558.
381. U.S. EPA. (1994): *Atrazine, Simazine and Cyanazine: Notice of Initiation of Special Review*. U.S. Environmental Protection Agency, Washington, D.C.
382. U.S. EPA. (1996): *Bentazon: Reregistration Eligibility Decision (RED)*, EPA-738-R-94-029. U.S. Environmental Protection Agency, Washington, D.C.
383. U.S. EPA. (1994): *Difenzoquat: Reregistration Eligibility Decision (RED)*, EPA-738-R-94-014. U.S. Environmental Protection Agency, Washington, D.C.
384. U.S. EPA. (1994): *Metalaxyl: Reregistration Eligibility Decision (RED)*, EPA-738-R-94-017. U.S. Environmental Protection Agency, Washington, D.C.
385. U.S. EPA. (1995): *Asulam: Reregistration Eligibility Decision (RED)*, EPA-738-R-95-024. U.S. Environmental Protection Agency, Washington, D.C.
386. U.S. EPA. (1995): Cyproconazole: pesticide tolerance. *Fed. Reg.*, 60(153):40545–40548.
387. U.S. EPA. (1995): *Diquat Dibromide: Reregistration Eligibility Decision (RED)*, EPA-738-R-95-016. U.S. Environmental Protection Agency, Washington, D.C.
388. U.S. EPA. (1995): Fenbuconazole: pesticide tolerances. *Fed. Reg.*, 60(100):27419–27421.
389. U.S. EPA. (1995): Lepidopteran pheromones: tolerance exemption: final rule. *Fed. Reg.*, 60(168):45060–45062.
390. U.S. EPA. (1995): Metalaxyl: pesticide tolerance. *Fed. Reg.*, 60(244):65579–5581.
391. U.S. EPA. (1995): *Metolachlor: Reregistration Eligibility Decision (RED)*, EPA-738-R-95-006. U.S. Environmental Protection Agency, Washington, D.C.
392. U.S. EPA. (1995): Monosodium methanearsonate and disodium methanearsonate; toxic chemical release reporting; community right to know. *Fed. Reg.*, (http://www.epa.gov?fedrgstr/EPA-TRI/April/Day-20/pr-I3.html).
393. U.S. EPA. (1995): Oxyfluorfen: pesticide tolerance. *Fed. Reg.*, 60(187):49816–49818.
394. U.S. EPA. (1995): Paraquat: pesticide tolerance. *Fed. Reg.*, March 15.
395. U.S. EPA. (1996): *Amitraz: Reregistration Eligibility Decision (RED)*, List A, Case 0234. U.S. Environmental Protection Agency, Washington, D.C.
396. U.S. EPA. (1996): *Bromacil: Reregistration Eligibility Decision (RED)*, EPA-738-R-96-013. U.S. Environmental Protection Agency, Washington, D.C.
397. U.S. EPA. (1996): Dimethenamid: pesticide tolerance petition, notice of filing. *Fed. Reg.*, 61(62):10681–10684.
398. U.S. EPA. (1996): Diquat: pesticide tolerance. *Fed. Reg.*, 61(60):13474–13476.
399. U.S. EPA. (1996): *Fipronil: Pesticide Fact Sheet*, EPA 737-F-96-005. U.S. Environmental Protection Agency, Washington, D.C.
400. U.S. EPA. (1996): Flutolanil: pesticide tolerance. *Fed. Reg.*, 61(124):33041–3304.
401. U.S. EPA. (1996): Glufosinate-ammonium: pesticide tolerance petition, notice of filing. *Fed Reg.*, 61(223): 58684–58688.
402. U.S. EPA. (1996): Hexaconazole: pesticide tolerance. *Fed. Reg.*, 61 (70):15895–15896.
403. U.S. EPA. (1996): Imazameth: pesticide tolerance. *Fed Reg.*, 61(55):11311–11313.
404. U.S. EPA. (1996): Lactofen: pesticide tolerance. *Fed. Reg.*, 61(47):9399–9401.
405. U.S. EPA. (1996): *Prometryn: Reregistration Eligibility Decision (RED)*, EPA-738-R-95-033. U.S. Environmental Protection Agency, Washington, D.C.
406. U.S. EPA. (1996): Propiconazole: pesticide tolerances for emergency exemptions. *Fed. Reg.*, 61(220):58135-58140.
407. U.S. EPA. (1996): Triadimefon: pesticide tolerances for emergency exemptions. *Fed. Reg.*, 61(232): 63726–63726.
408. U.S. EPA. (1996): *Trifluralin: Reregistration Eligibility Decision (RED)*, EPA-738-R-95-040. U.S. Environmental Protection Agency, Washington, D.C.
409. U.S. EPA. (1996): *Office of Pesticide Programs Reference Dose Tracking Report*. Office of Prevention, Pesticides and Toxic Substances. U.S. Environmental Protection Agency, Washington, D.C., pp. 1–77

410. U.S. EPA. (1996): *Proposed Guidelines for Carcinogen Risk Assessment*, EPA/600/p-92/003c. Office of Research and Development, U.S. Environmental Protection Agency, Washington, D.C.
411. U.S. EPA. (1997): Acifluorfen: notice of filing of pesticide petitions. *Fed. Reg.*, 62(143):39967–39974.
412. U.S. EPA. (1997): *Sodium Acifluorfen: Reregistration Eligibility Decision (RED)*, Case 2605. U.S. Environmental Protection Agency, Washington, D.C.
413. U.S. EPA. (1997): Azoxystrobin: notice of filing of pesticide petitions. *Fed. Reg.*, 62(48):11441–11447. U.S. Environmental Protection Agency, Washington, D.C.
414. U.S. EPA. (1997): *Bacillus thuringiensis* subspecies tolworthi Cry9C: notice of filing of pesticide petitions. *Fed. Reg.*, 62(182):49224–49226.
415. U.S. EPA. (1997): Receipt of petition for determination of nonregulated status for genetically engineered corn. *Fed. Reg.*, 62(I56):43311–43312.
416. U.S. EPA. (1997): Bioallethrin: pesticide tolerances, final rule. *Fed. Reg.*, 62(128):62961–62970.
417. U.S. EPA. (1997) CGA329351: notice of filing of pesticide petitions. *Fed. Reg.*, 62(143):40080–40086.
418. U.S. EPA. (1997): Chlorfenapyr: pesticide tolerance, petition filing. *Fed. Reg.*, 62(24):5399–5540.
419. U.S. EPA. (1997): Chorothalonil: pesticide tolerance, petition filing. *Fed. Reg.*, 62(63):15700–15704.
420. U.S. EPA. (1997): Clethodim: pesticide petition, notice of filing. *Fed. Reg.*, 62(232):63942–63946.
421. U.S. EPA. (1997): Clopyralid: pesticide tolerance, emergency exemption. *Fed. Reg.*, 62(48):11360–11364.
422. U.S. EPA. (1997): Cloransulam-methyl: pesticide tolerance, emergency exemption. *Fed. Reg.*, 62(48): 11360–11364.
423. U.S. EPA. (1997): Cyfluthrin: pesticide tolerance, final rule. *Fed. Reg.*, 62(128):63010–63019.
424. U.S. EPA. (1997): Deltamethrin and tralomethrin: pesticide tolerance, final rule. *Fed. Reg.*, 62(128): 62993–63002.
425. U.S. EPA. (1997):Difenoconazole: notice of filing of pesticide petitions. *Fed. Reg.*, 62(143):40075–40080.
426. U.S. EPA. (1997): *Diflubenzuron: Reregistration Eligibility Decision (RED)*, EPA-738-R-97-008. U.S. Environmental Protection Agency, Washington, D.C.
427. U.S. EPA. (1997): Diuron: pesticide tolerance-petition filing. *Fed. Reg.*, 62(16):3685–3688.
428. U.S. EPA. (1997): Fenoxaprop-ethyl: notice of filing of pesticide petitions. *Fed. Reg.*, 62(180):48837–48842.
429. U.S. EPA. (1997): Fenoxycarb: FQPA assessment. *Fed. Reg.*, 62(195):52552–52558.
430. U.S. EPA. (1997): Fenpropathrin: pesticide tolerances, emergency exemptions. *Fed. Reg.*, 62(134): 37516–37522;
430a. U.S. EPA. (1997): Fenvalerate: pesticide tolerances. *Fed. Reg.*, 62(228):63019–63037.
431. U.S. EPA. (1997): Fipronil: notice of filing of pesticide petitions. *Fed Reg.*, 62(119):33641–33647.
432. U.S. EPA. (1997): Fluthiacet-methyl: pesticide tolerance, petition. *Fed Reg.*, 63(193):53660-53662.
433. U.S. EPA. (1997): Fomesafen: pesticide tolerance, emergency exemption, final rule. *Fed. Reg.*, 62(223):61639-6I645.
434. U.S. EPA. (1997): Imazamox: pesticide tolerance, final rule. *Fed. Reg.*, 62(105):29669–29673.
435. U.S. EPA. (1997): Imidacloprid: pesticide tolerance, petition filing. *Fed Reg.*, 62(38):8734–8734.
436. U.S. EPA. (1997): Isoxaflutole: pesticide tolerance, petition filing. *Fed. Reg.*, 62(38):8737–8740.
437. U.S. EPA. (1997): Mancozeb, maneb, and ethylenethiourea tolerances: notice of filing of pesticide petitions. *Fed. Reg.*, 62(148):41383–4l386.
438. U.S. EPA. (1997): Maneb: pesticide tolerances for emergency exemptions. *Fed. Reg.*, 62(185):49918–49925.
439. U.S. EPA. (1997): Mefenoxam: pesticide tolerance for emergency exemptions. *Fed. Reg.*, 62(149): 42019–42030.
440. U.S. EPA. (1997): Myclobutanil: pesticide tolerance for emergency exemptions. *Fed. Reg.*, 62(6):1284–1288.
441. U.S. EPA. (1997): Norflurazon: pesticide tolerance, petition filing. *Fed. Reg.*, 62(58):14423–14426.
442. U.S. EPA. (1997): *Paraquat Dichloride: Reregistration Eligibility Decision (RED)*, EPA-738-F-96-018. U.S. Environmental Protection Agency, Washington, D.C.
443. U.S. EPA. (1997): Pendimethalin: pesticide tolerance for emergency exemptions. *Fed. Reg.*, 62(100):28355–28361.
444. U.S. EPA. (1997): *Pendimethalin: Reregistration Eligibility Decision (RED)*, EPA-738-R-97-007. U.S. Environmental Protection Agency, Washington, D.C.
445. U.S. EPA. (1997): Propazine: pesticide tolerance, petition filing. *Fed. Reg.*, 63(193):53657–53660.
446. U.S. EPA. (1997): *Propoxur: Reregistration Eligibility Decision (RED)*, EPA-738-R-97-009. U.S Environmental Protection Agency, Washington, D.C.
447. U.S. EPA. (1997): Pyridaben: pesticide tolerance, petition filing. *Fed. Reg.*, 62(48):11450–11453.
448. U.S. EPA. (1997): Pyrimethanil: pesticide tolerance. *Fed. Reg.*, 62(231):63662–63669.
449. U.S. EPA. (1997): Spinosad: pesticide tolerance, final rule. *Fed. Reg.*, 62(38):8626–8632.
450. U.S EPA. (1997): Tebuconazole: pesticide tolerance, petition filing. *Fed. Reg.*, 62(43):10047–10050.
451. U.S. EPA. (1997): Tefluthrin: pesticide tolerance, final rule. *Fed. Reg.*, 62(228):62954–62961
452. U.S. EPA. (1997): *Trans-11-Tetradecenyl Acetate Technical Pheromone Pesticide Fact Sheet: Unconditional Registration*. U.S. Environmental Protection Agency, Washington, D.C.
453. U.S. EPA.1997): Vinclozolin: pesticide tolerance petition. *Fed. Reg.*, 62(53):13000–13005.
454. U.S. EPA. (1998): *Alachlor: Registration Eligibility Decision*. U.S. Environmental Protection Agency, Washington, D.C.
455. U.S. EPA. (1998): *Bacillus thuringiensis* variety *kurstaki*: notice of filing of pesticide petitions. *Fed. Reg.*, 63(67): 17174–1717.
456. U.S. EPA. (1998): *Bacillus thuringiensis: Reregistration Eligibility Decision (RED)*, EPA-738-F-98-001. U.S. Environmental Protection Agency, Washington, D.C.
457. U.S. EPA. (1998): *Bacillus thuringiensis: Reregistration Eligibility Decision (RED): Microbial Pesticides: Bacillus thuringiensis*, EPA738-R-98-004. U.S. Environmental Protection Agency, Washington, D.C.

458. U.S. EPA. (1998): Carfentrazone-ethyl: pesticide tolerances. *Fed. Reg.*, 63(189):52174–52180.
459. U.S. EPA. (1998): Cyprodinil: Novartis Crop Protection, Inc., approval of a pesticide product. *Fed. Reg.*, 63(l08):30749–30770.
460. U.S. EPA. (1998): Dicamba: notice of filing of pesticide petitions. *Fed. Reg.*, 63(2240):64481–64484.
461. U.S. EPA. (1998): Diflubenzuron: notice of filing of pesticide petitions. *Fed. Reg.*, 63(37):9528–9532.
462. U.S. EPA. (1998): Diflubenzuron: temporary pesticide tolerance. *Fed. Reg.*, 63(92):26481–26488.
463. U.S. EPA. (1998): Glyphosate: pesticide tolerance, final rule. *Fed. Reg.*, 63(195):54058–54066.
464. U.S. EPA. (1998): Halosulfuron-methyl: pesticide tolerance petition filing. *Fed Reg.*, 63(103):29401–29409.
465. U.S. EPA. (1998): Hexythiazox: pesticide tolerance petition filing. *Fed. Reg.*, 63(137):38644–38646.
466. U.S. EPA. (1998): *Hydramethylnon: Reregistration Eligibility Decision (RED)*, EPA-738-R-98-023, December. U.S. Environmental Protection Agency, Washington, D.C.
467. U.S. EPA. (1998): *Iprodione: Reregistration Eligibility Decision (RED)*, EPA-738-R-98-019. U.S. Environmental Protection Agency, Washington, D.C.
468. U.S. EPA.1998): Isoxaflutole: pesticide tolerances. *Fed. Reg.*, 63(184):50773–50784.
469. U.S. EPA. (1998): Lambda-cyhalothrin: pesticide tolerances. *Fed. Reg.*, 63(30):7291–7299.
470. U.S. EPA. (1998): *Methomyl: Reregistration Eligibility Decision (RED)*, EPA-738-R-98-021. U.S. Environmental Protection Agency, Washington, D.C.
471. U.S. EPA. (1998): Metolachor: pesticide tolerances for emergency exemptions. *Fed. Reg.*, 63(176):48586–48594.
472. U.S. EPA. (1998): *Metribuzin: Reregistration Eligibility Decision (RED)*, EPA-738-R-97-006. U.S. Environmental Protection Agency, Washington, D.C.
473. U.S. EPA. (1998): *Pesticide Assessment Guidelines*, OPPTS Harmonized 870 Health Effects Test Guidelines Series. U.S. Environmental Protection Agency, Washington, D.C.
474. U.S. EPA. (1998): Science policy on a common mechanism of toxicity: the organophosphate pesticides, *Fed. Reg.*, 64(24):5795–5799.
475. U.S. EPA. (1998): Prometryn: pesticide tolerances. *Fed. Reg.*, 63(37):9494–9499.
476. U.S. EPA. (1998): Pymetrozine: notice of filing of pesticide petitions. *Fed. Reg.*, 63(97):27723–27727.
477. U.S. EPA. (1998) Pymetrozine: notice of filing of pesticide petitions. *Fed. Reg.*, 63(194):53906–53909.
478. U.S. EPA. (1998): Pyridate: pesticide tolerances, final rule. *Fed. Reg.*, 63(194):53837–53844.
479. U.S. EPA. (1998): Pyriproxyfen: pesticide tolerances, final rule. *Fed. Reg.*, 63(128):33366–36373.
480. U.S. EPA. (1998): Tebufenozide: Rohm and Haas Company, notice of filing of pesticide tolerance. *Fed. Reg.*, 63(160):44439–44456.
481. U.S. EPA. (1999): 2,4-D: time-limited pesticide tolerances, final rule. *Fed. Reg.*, 64(46):11792–11799.
482. U.S. EPA. (1999): Avermectin: pesticide tolerances for emergency exemptions, final rule. *Fed. Reg.*, 64(66): 16843–16850.
483. U.S. EPA. (1999): Clofentezine: pesticide tolerance petition filing. *Fed. Reg.*, 64(18):4414–4418.
484. U.S. EPA. (1999): Clofentezine: pesticide tolerance petition filing. *Fed Reg.*, 64(74):19042-19050.
485. U.S. EPA. (1999): Clomazone: pesticide tolerance petition filing. *Fed. Reg.*, 64(32):8087–8090.
486. U.S. EPA. (1999): Cyromazine: pesticide tolerances for emergency exemptions, final rule. *Fed. Reg.*, 62(168): 45735-45741.
487. U.S. EPA. (1999): Emamectin benzoate: pesticide tolerance, final rule. *Fed. Reg.*, 64(96):27192–27200.
488. U.S. EPA. (1999): Fluthiacet-methyl: pesticide tolerance, final rule. *Fed. Reg.*, 64(7):18351–18357.
489. U.S. EPA. (1999): *Guidance for Identifying Pesticide Chemicals and Other Substances That Have a Common Mechanism of Toxicity*. U.S. Environmental Protection Agency, Washington, D.C.
490. U.S. EPA. (1999): Picloram: time-limited pesticide tolerances, final rule. *Fed. Reg.*, 64(2):418–425.
491. U.S. EPA. (1999): Pyriproxyfen: notice of filing of pesticide petitions. *Fed. Reg.*, 64(34):8638–8641.
492. U.S. EPA. (1999): *Amitrole Reregistration Eligibility Decision (RED)*. U.S Environmental Protection Agency, Washington, D.C.
493. U.S. EPA. (1999): *Chlorothalonil Reregistration Eligibility Decision (RED)*, EPA-738-R-99-004. U.S. Environmental Protection Agency, Washington, D.C.
494. U.S. EPA. (1999): Cyromazine: pesticide tolerances for emergency exemptions, final rule. *Fed. Reg.*, 62(168): 45735–45741.
495. U.S. EPA. (1999): Rimsulfuron; Pesticide tolerances for emergency exemptions, final rule. *Fed. Reg.*, 64(41): 10227–10233.
496. U.S. EPA. (1999): Sulfosulfuron: pesticide tolerances, final rule. *Fed. Reg.*, 64(96):27186–72192.
497. U.S. EPA. (1999): Thiamethoxam: notice of filing of pesticide petitions. *Fed. Reg.*, 64(86):24153–24160.
498. U.S. EPA. (1999): *Trifloxystrobin: New Chemical Registration*, Pesticide Fact Sheet. U.S. Environmental Protection Agency, Washington, D.C.
499. U.S. EPA. (2000): *Clodinafop-Propargyl*, Pesticide Fact Sheet. U.S. Environmental Protection Agency, Washington, D.C.
500. U.S. EPA. (2000): *Lactofen: Toxicology Evaluation*. U.S Environmental Protection Agency, Washington, D.C.
501. U.S. EPA. (2000): *Vinclozolin: Common Mechanism of Toxicity of Dicarboximide Fungicides*. U.S. Environmental Protection Agency, Washington, D.C.
502. U.S. EPA. (2000): *Vinclozolin: Reregistration Eligibility Decision (RED)*, EPA-738R-00-023, U.S. Environmental Protection Agency, Washington, D.C.
503. U.S. EPA. (2001): *A Determination of the Existence of a Common Mechanism of Toxicity and a Screening Level Cumulative Food Risk Assessment*. U.S. Environmental Protection Agency, Washington, D.C.
504. U.S. EPA. (2001): *Chloroacetanilides: The Grouping of a Series of Chloroacetanilide Pesticides Based on a Common Mechanism of Toxicity*. U.S. Environmental Protection Agency, Washington, D.C.

505. U.S. EPA. (2001): *Dithiocarbamate Pesticides: The Grouping of a Series of Dithiocarbamate Pesticides Based on a Common Mechanism of Toxicity*. U.S. Environmental Protection Agency, Washington, D.C.
506. U.S. EPA. (2001): Memorandum from Paul Lewis to Marcia Mulkey, SAP Report, 2001-11, November, 1. U.S. Environmental Protection Agency, Washington, D.C.
507. U.S. EPA. (2001): Mesotrione: pesticide tolerance, final rule. *Fed. Reg.*, 66(120):33187–33194.
508. U.S. EPA. (2002): *Benomyl: Reregistration Eligibility Decision (RED)*, EPA-738-R-02-011. U.S. Environmental Protection Agency, Washington, D.C.
509. U.S. EPA. (2002): *Chlorpyrifos: Reregistration Eligibility Decision (RED)*, EPA-738-R-01-007. U.S. Environmental Protection Agency, Washington, D.C.
510. U.S. EPA. (2002): Chlorsulfuron: pesticide tolerance, final rule. *Fed. Reg.*, 67(157):52866–52873.
511. U.S. EPA. (2002): *Molinate: Evaluation of the Carcinogenic Potential of Molinate (Second Review)*, HED Document No. 014407. U.S. Environmental Protection Agency, Washington, D.C.
512. U.S. EPA. (2002): Oxadixyl: proposed revocation of tolerances. *Fed. Reg.*, 67(25):5548–5552.
513. U.S. EPA. (2002): *Triazines: The Grouping of a Series of Triazine Pesticides Based on a Common Mechanism of Toxicity*. U.S. Environmental Protection Agency, Washington, D.C.
514. U.S. EPA. (2002): *Molinate: Assessment of Molinate by the Mechanism of Toxicity SARC*. U.S. Environmental Protection Agency, Washington, D.C.
515. U.S. EPA. (2002): *Molinate: Reregistration Eligibility Decision (RED)*. U.S. Environmental Protection Agency, Washington, D.C.
516. U.S. EPA. (2002): *Oxyfluorfen: Registration Eligibility Decision (RED)*, EPA-738-R-02-014. U.S. Environmental Protection Agency, Washington, D.C.
517. U.S. EPA. (2002): Zeta-cypermethrin and its Inactive R isomers: pesticide tolerances, final rule. *Fed. Reg.*, 66 (180):47979-47994.
518. U.S. EPA. (2002): Notice of filing pesticide petitions to establish a tolerance for certain pesticide chemicals in or on food. *Fed. Reg.*, 67(100):36178–36184.
519. U.S. EPA. (2002): *Endosulfan: Registration Eligibility Decision (RED)*, EPA-738-R-02-013. U.S. Environmental Protection Agency, Washington, D.C.
520. U.S. EPA. (2002): *Thiabendazole: Registration Eligibility Decision (RED)*, EPA-738-R-02-xxx. U.S. Environmental Protection Agency, Washington, D.C.
521. U.S. EPA. (2003): Butafencil: pesticide tolerance, final rule. *Fed. Reg.*, 68(182):54818–54827.
522. U.S. EPA. (2003): *Diuron: Reregistration Eligibility Decision (RED)*. U.S. Environmental Protection Agency, Washington, D.C.
523. U.S. EPA. (2003): *Oxadiazon: Reregistration Eligibility Decision (RED)*, EPA-738-R-04-003, List B, Case 2485. U.S. Environmental Protection Agency, Washington, D.C.
524. U.S. EPA. (2003): *Ziram: Reregistration Eligibility Decision (RED)*, List B, Case 2180. U.S. Environmental Protection Agency, Washington, D.C.
525. U.S. EPA. (2004): *Benfluralin: Reregistration Eligibility Decision (RED)*, EPA-738-R-04-012. U.S. Environmental Protection Agency, Washington, D.C.
526. U.S. EPA. (2004): *Captan: Amendment to the, 1999 Captan RED*. U.S. Environmental Protection Agency, Washington, D.C.
527. U.S. EPA. (2004): *Carbaryl: Interim Reregistration Eligibility Decision (IRED)*. U.S. Environmental Protection Agency, Washington, D.C.
528. U.S. EPA. (2004): *Chlorimuron-Ethyl: Human Health Risk Assessment*. U.S. Environmental Protection Agency, Washington, D.C.
529. U.S. EPA. (2004): *Diazinon: Interim Reregistration Eligibility Decision (IRED)*, EPA-738-R-04-006. U.S. Environmental Protection Agency, Washington, D.C.
530. U.S. EPA. (2004): Fludioxonil: pesticide tolerances, final rule. *Fed. Reg.*, 69(188):58084–58091.
531. U.S. EPA. (2004): *Imazamethabenz-methyl: HED Chapter for the Tolerance Reassessment Eligibility Decision (TRED)*. U.S. Environmental Protection Agency, Washington, D.C.
532. U.S. EPA. (2004): *Nicosulfuron: Report of the Food Quality Protection Act (FQPA) Tolerance Reassessment Progress and Risk Management Decision (TRED)*. U.S. Environmental Protection Agency, Washington, D.C.
533. U.S. EPA. (2004): *Norflurazon: Reregistration Eligibility Decision (RED)*, List B, Case 0229. U.S. Environmental Protection Agency, Washington, D.C.
534. U.S. EPA. (2004): *Thiophanate-Methyl: Reregistration Eligibility Decision (RED)*, List B, Case 2680. U.S. Environmental Protection Agency, Washington, D.C.
535. U.S. EPA. (2004): *Thiram: Reregistration Eligibility Decision (RED)*, EPA-738-R-04-012. U.S. Environmental Protection Agency, Washington, D.C.
536. U.S. EPA. (2004): *Trifluralin: Tolerance Reassessment Progress and Risk Management Decision (TRED)*, EPA-738-R-95-040. U.S. Environmental Protection Agency, Washington, D.C.
537. U.S. EPA. (2005): *Ametryn: Reregistration Eligibility Decision (RED)*, EPA 738-R-05-006. U.S. Environmental Protection Agency, Washington, D.C.
538. U.S. EPA. (2005): *Chlorsulfuron: Reregistration Eligibility Decision (RED)*, EPA-378-F-05-002. U.S. Environmental Protection Agency, Washington, D.C.
539. U.S. EPA. (2005): *Cypermethrin. Phase 2 HED Risk Assessment for the Reregistration Eligibility Decision (RED)*. U.S. Environmental Protection Agency, Washington, D.C.
540. U.S. EPA. (2005): *2,4-D: Reregistration Eligibility Decision (RED)*, EPA-738-R-05-002, List A. U.S. Environmental Protection Agency, Washington, D.C.
541. U.S. EPA. (2005): *Ferbam: Reregistration Eligibility Decision (RED)*, EPA-378-R-05-009. U.S. Environmental Protection Agency, Washington, D.C.
542. U.S. EPA. (2005): *Fluazifop-P-Butyl: Report of the Food Quality Protection Act (FQPA) Tolerance Reassessment Progress and Risk Management Decision (TRED)*, EPA-738-R-05-005. U.S. Environmental Protection Agency, Washington, D.C.

543. U.S. EPA. (2005): *Flumiclorac Pentyl: Report of the Food Quality Protection Act (FQPA) Tolerance Reassessment Progress and Risk Management Decision (TRED)*. U.S. Environmental Protection Agency, Washington, D.C.
544. U.S. EPA. (2005): *Guidelines for Carcinogen Risk Assessment: Risk Assessment Forum*, EPA-630/P-03/001F. U.S. Environmental Protection Agency, Washington, D.C., pp. 2-49–2-58.
545. U.S. EPA. (2005): *Imazaquin and Its Salts: HED Chapter of the Tolerance Reassessment Eligibility Decision (TRED)*. U.S. Environmental Protection Agency, Washington, D.C.
546. U.S. EPA. (2005): *Mancozeb: Reregistration Eligibility Decision (RED)*, EPA-738-R-04-012. U.S. Environmental Protection Agency, Washington, D.C.
547. U.S. EPA. (2005): *Maneb: Reregistration Eligibility Decision (RED)*, EPA-738-R-05-xxx. U.S. Environmental Protection Agency, Washington, D.C.
548. U.S. EPA. (2005): *Procymidone: Report of the Food Quality Protection Act (FQPA) Tolerance Reassessment Progress and Risk Management Decision (TRED)*. U.S. Environmental Protection Agency, Washington, D.C.
549. U.S. EPA. (2005): *Rotenone: Decisions on Critical Effects and Endpoint Selection*, results of the DED Hazard Science Policy Council, June, 28. U.S. Environmental Protection Agency, Washington, D.C.
550. U.S. EPA. (2005): *Sethoxydim: Reregistration Eligibility Decision (RED)*, List B, Case 2600. U.S. Environmental Protection Agency, Washington, D.C.
551. U.S. EPA. (2005): Thiamethoxam: pesticide tolerances for emergency exemptions, final rule. *Fed. Reg.*, 70(28): 7177–7182.
552. U.S. EPA. (2006): *Acetochlor: Report of the Food Quality Protection Act (FQPA) Tolerance Reassessment Progress and Risk Management Decision (TRED)*. U.S. Environmental Protection Agency, Washington, D.C.
553. U.S. EPA. (2006): *Acephate: Finalization of the Interim Reregistration Eligibility Decision (IRED)*, EPA-738-R-01-013. U.S. Environmental Protection Agency, Washington, D.C.
554. U.S. EPA. (2006): *Atrazine: Finalization of the Interim Reregistration Eligibility Decision and Completion of the Tolerance Reassessment and Reregistration Eligibility Process*. U.S. Environmental Protection Agency, Washington, D.C.
555. U.S. EPA. (2006): *Azinphos-Methyl: Finalization of the Interim Reregistration Eligibility Decision (IRED)*, Case 0235. U.S. Environmental Protection Agency, Washington, D.C.
556. U.S. EPA. (2006): *Carbofuran: Reregistration Eligibility Decision (RED)*, EPA-738-R-06-031, List A, Case 0101. U.S. Environmental Protection Agency, Washington, D.C.
557. U.S. EPA. (2006): *Dicamba and Associated Salts: Reregistration Eligibility Decision (RED)*, List B, Case 0065. U.S. Environmental Protection Agency, Washington, D.C.
558. U.S. EPA. (2006): *Dichlorvos (DDVP): Interim Reregistration Eligibility Decision (IRED)*, EPA-738-R-06-013. U.S. Environmental Protection Agency, Washington, D.C.
559. U.S. EPA. (2006): *Imazapyr: Reregistration Eligibility Decision (RED)*, EPA-728-R-06-007. U.S. Environmental Protection Agency, Washington, D.C.
560. U.S. EPA. (2006): *Lindane: Addendum to the Lindane Reregistration Eligibility Decision (RED) (July, 2006) and the Lindane RED (2002)*, HQ-OPP-2002-0202-0074; EPA-HQ-OPP-2002-02020027. U.S. Environmental Protection Agency, Washington, D.C.
561. U.S. EPA. (2006): *Malathion: Reregistration Eligibility Decision (RED)*, EPA-738-R-06-030. U.S. Environmental Protection Agency, Washington, D.C.
562. U.S. EPA. (2006): *Permethrin: Reregistration Eligibility Decision (RED)*, EPA-738-R-06-017. U.S. Environmental Protection Agency, Washington, D.C.
563. U.S. EPA. (2006): *Piperonyl Butoxide: Reregistration Eligibility Decision (RED)*, EPA-738-R-06-005, List B, Case 2525. U.S. Environmental Protection Agency, Washington, D.C.
564. U.S. EPA. (2006): *Propanil: Amendment to the Reregistration Eligibility Decision (RED) of Propanil (March, 2006) and the Propanil RED (September, 2003)*, EPA-738-R-06-017. U.S. Environmental Protection Agency, Washington, D.C.
565. U.S. EPA. (2006): *Pyrethrins: Reregistration Eligibility Decision (RED)*, EPA-738-R-06-004, List B, Case 2580. U.S. Environmental Protection Agency, Washington, D.C.
566. U.S. EPA. (2006): *Resmethrin: Reregistration Eligibility Decision (RED)*, EPA-738-R-06-003. U.S. Environmental Protection Agency, Washington, D.C.
567. U.S. EPA. (2006): *Rotenone: Phase 3 HED Chapter of the Reregistration Eligibility Decision Document (RED)*. U.S. Environmental Protection Agency, Washington, D.C.
568. U.S. EPA. (2006): *Simazine: Reregistration Eligibility Decision (RED)*, EPA-738-R-06-008. U.S. Environmental Protection Agency, Washington, D.C.
569. U.S. EPA. (2006): 1,2,4-Triazole, Triazole Alanine, Triazole Acetic Acid: Human Health Aggregate Risk Assessment in Support of Reregistration and Registration Actions for Triazole-Derivative Fungicide Compounds: Triadimefon, EPA-HQ-OPP-2005-0258; Triadimenol, EPA-HQ-OPP-2006-0038; Propiconazole, EPA-HQ-OPP-2005-0497. U.S. Environmental Protection Agency, Washington, D.C.
570. Valentine, B. J., Gurr, G. M., and Thwaite, W. G. (1996): Efficacy of the insect growth regulators tebufenozide and fenoxycarb on lepidopteran pest control in apples, and their compatibility with biological control for integrated pest management. *Austr. J. Exp. Agric*, 36:501–506.
571. Vaughn, K. C. (2002): Cellulose biosynthesis inhibitor herbicides. In: *Herbicide Classes in Development*, edited by P. Boger, K. Wakabayashi, and K. Hirai, pp. 139–150. Springer-Verlag, Berlin.
572. Wakabayashi, K. and Boger, P. (2002): Target sites for herbicides: entering the 21st century. *Pestic. Manage. Sci.*, 58:1149–1154.
573. Wakabayashi, K. and Boger, P. (2004): Phytotoxic sites of action for molecular design of modern herbicides. Part 1. The photosynthetic electron transport system. *Weed Biol. Manage.*, 4:8–18.
574. Wakabayashi, K. and Boger, P. (2004): Phytotoxic sites of action for molecular design of modern herbicides. Part 2. Amino acid, lipid and cell wall biosynthesis, and other targets for future herbicides. *Weed Biol. Manage.*, 4: 59–70

575. Ware, G. W. (1999): *An Introduction to Insecticides*. Radcliffe's IPM World Textbook Home Page, http://ipmworld.umn.edu/chapters/bloomq.htm.
576. Watson, G. B. (2001): Action of insecticidal spinosyns on γ-aminobutyric acid responses from small-diameter cockroach neurons. *Pestic. Biochem. Physiol.*, 71: 20–28.
577. WSSA. (1994): Acifluorfen. In: *Herbicide Handbook*, 7th ed., edited by W. H. Ahrens, pp. 5–7. Weed Science Society of America, Champaign, IL.
578. WSSA. (1994): Ametryn. In: *Herbicide Handbook*, 7th ed., edited by W. H. Ahrens, pp. 12–14. Weed Science Society of America, Champaign, IL
579. WSSA): (1994): Asulum. In: *Herbicide Handbook*, 7th ed., edited by W. H. Ahrens, pp. 18–19. Weed Science Society of America, Champaign, IL.
580. WSSA): (1994): Atrazine. In: *Herbicide Handbook*, 7th ed., edited by W. H. Ahrens, pp. 20–23. Weed Science Society of America, Champaign, IL.
581. WSSA. (1994): Bensulfuron. In: *Herbicide Handbook*, 7th ed., edited by W. H. Ahrens, pp. 28–30. Weed Science Society of America, Champaign, IL.
582. WSSA. (1994): Bentazon. In: *Herbicide Handbook*, 7th ed., edited by W. H. Ahrens, pp. 32–34. Weed Science Society of America, Champaign, IL.
583. WSSA. (1994): Bromacil. In: *Herbicide Handbook*, 7th ed., edited by W. H. Ahrens, pp. 37–39. Weed Science Society of America, Champaign, IL.
584. WSSA. (1994): Butylate. In: *Herbicide Handbook*, 7th ed., edited by W. H. Ahrens, pp. 43–45. Weed Science Society of America. Champaign, IL
585. WSSA. (1994): Chlorimuron. In: *Herbicide Handbook*, 7th ed., edited by W. H. Ahrens, pp. 56–58. Weed Science Society of America, Champaign, IL.
586. WSSA. (1994): Chlorsulfuron. In: *Herbicide Handbook*, 7th ed., edited by W. H. Ahrens, pp. 58–60. Weed Science Society of America, Champaign, IL.
587. WSSA. (1994): Clethodim. In: *Herbicide Handbook*, 7th ed., edited by W. H. Ahrens, pp. 62–64. Weed Science Society of America, Champaign, IL.
588. WSSA. (1994): Cyanazine. In: *Herbicide Handbook*, 7th ed., edited by W. H. Ahrens, pp. 72–74. Weed Science Society of America, Champaign, IL.
589. WSSA. (1994): Dichlobenil. In: *Herbicide Handbook*, 7th ed., edited by W. H. Ahrens, pp. 94–96. Weed Science Society of America, Champaign, IL.
590. WSSA. (1994): Diclofop. In: *Herbicide Handbook*, 7th ed., edited by W. H. Ahrens, pp. 101–103. Weed Science Society of America, Champaign, IL.
591. WSSA. (1994): Difenzoquat: In: *Herbicide Handbook, 7th* ed., edited by W. H. Ahrens, pp. 106–108. Weed Science Society of America, Champaign, IL.
592. WSSA. (1994): Diquat. In: *Herbicide Handbook* 7th ed., edited by W. H. Ahrens, pp. 108–110. Weed Science Society of America, Champaign, IL.
593. WSSA. (1994): Diuron. In: *Herbicide Handbook*, 7th ed., edited by W. H. Ahrens, pp. 113–115. Weed Science Society of America, Champaign, IL.
594. WSSA. (1994): Flumetsulam. In: *Herbicide Handbook*, 7th ed., edited by W. H. Ahrens, pp. 131–133. Weed Science Society of America, Champaign, IL.
595. WSSA. (1994): Flumiclorac-pentyl. In: *Herbicide Handbook*, 7th ed., edited by W. H. Ahrens, pp. 133–135. Weed Science Society of America, Champaign, IL
596. WSSA. (1994): Fluometuron. In: *Herbicide Handbook, 7th* ed., edited by W. H. Ahrens, pp. 135–137. Weed Science Society of America, Champaign, IL.
597. WSSA. (1994): Fluridone. In: *Herbicide Handbook*, 7th ed., edited by W. H. Ahrens, pp. 141–143. Weed Science Society of America, Champaign, IL.
598. WSSA. (1994): Glufosinate-ammonium. In: *Herbicide Handbook*, 7th ed., edited by W. H. Ahrens, pp. 147–149. Weed Science Society of America, Champaign, IL.
599. WSSA. (1994): Glyphosate. In: *Herbicide Handbook*, 7th ed., edited by W. H. Ahrens, pp. 149–152. Weed Science Society of America, Champaign, IL
600. WSSA. (1994): Haloxyfop. In: *Herbicide Handbook*, 7th ed., edited by W. H. Ahrens, pp. 153–156. Weed Science Society of America, Champaign, IL.
601. WSSA. (1994): Imazamethabenz: In: *Herbicide Handbook*, 7th ed., edited by W. H. Ahrens, pp. 159–161. Weed Science Society of America, Champaign, IL.
602. WSSA. (1994): Imazapyr. In: *Herbicide Handbook*, 7th ed., edited by W. H. Ahrens, pp. 161–163. Weed Science Society of America, Champaign, IL.
603. WSSA. (1994): Imazaquin. In: *Herbicide Handbook*, 7th ed., edited by W. H. Ahrens, pp. 163–166. Weed Science Society of America, Champaign, IL
604. WSSA. (1994): Imazethapyr. In: *Herbicide Handbook*, 7th ed., edited by W. H. Ahrens, pp. 166–168. Weed Science Society of America, Champaign, IL.
605. WSSA. (1994): Ioxynil. In: *Herbicide Handbook*, 7th ed., edited by W. H. Ahrens, pp. 168–171. Weed Science Society of America, Champaign, IL.
606. WSSA. (1994): Isoxaben. In: *Herbicide Handbook*, 7th ed., edited by W. H. Ahrens, pp. 173–175. Weed Science Society of America, Champaign, IL.
607. WSSA. (1994): Linuron. In: *Herbicide Handbook*, 7th ed., edited by W. H. Ahrens, pp. 177–179. Weed Science Society of America, Champaign, IL.
608. WSSA. (1994): Metribuzin. In: *Herbicide Handbook*, 7th ed., edited by W. H. Ahrens, pp. 200–203. Weed Science Society of America, Champaign, IL
609. WSSA. (1994): Metsulfuron. In: *Herbicide Handbook, 7th* ed., edited by W. H. Ahrens, pp. 203–205. Weed Science Society of America, Champaign, IL.
610. WSSA. (1994): Molinate. In: *Herbicide Handbook* 7th ed., edited by W. H. Ahrens, pp. 205–206. Weed Science Society of America, Champaign, IL.
611. WSSA. (1994): MSMA. In: *Herbicide Handbook*, 7th ed., edited by W. H. Ahrens, pp. 209–211. Weed Science Society of America, Champaign, IL.
612. WSSA. (1994): Nicosulfuron. In: *Herbicide Handbook*, 7th ed., edited by W. H. Ahrens, pp. 216–217. Weed Science Society of America, Champaign, IL.
613. WSSA. (1994): Norflurazon. In: *Herbicide Handbook*, 7th ed., edited by W. H. Ahrens, pp. 218–220. Weed Science Society of America, Champaign, IL.
614. WSSA. (1994): Paraquat. In: *Herbicide Handbook*, 7th ed., edited by W. H. Ahrens, pp. 226–228. Weed Science Society of America, Champaign, IL.

615. WSSA. (1994): Pendimethalin. In: *Herbicide Handbook*, 7th ed., edited by W. H. Ahrens, pp. 230–233. Weed Science Society of America, Champaign, IL.
616. WSSA. (1994): Primisulfuron. In: *Herbicide Handbook*, 7th ed., edited by W. H. Ahrens, pp. 238–240. Weed Science Society of America, Champaign, IL.
617. WSSA. (1994): Prometon. In: *Herbicide Handbook*, 7th ed., edited by W. H. Ahrens, pp. 243–244. Weed Science Society of America, Champaign, IL.
618. WSSA. (1994): Prometryn. In: *Herbicide Handbook*, 7th ed., edited by W. H. Ahrens, pp. 245–247. Weed Science Society of America, Champaign, IL.
619. WSSA. (1994): Pyridate. In: *Herbicide Handbook*, 7th ed., edited by W. H. Ahrens, pp. 256–258. Weed Science Society of America, Champaign, IL.
620. WSSA. (1994): Sethoxydim. In: *Herbicide Handbook*, 7th ed., edited by W. H. Ahrens, pp. 266–267. Weed Science Society of America, Champaign, IL.
621. WSSA. (1994): Simazine. In: *Herbicide Handbook*, 7th ed., edited by W. H. Ahrens, pp. 270–272. Weed Science Society of America, Champaign, IL.
622. WSSA. (1994): Sulfometuron. In: *Herbicide Handbook*, 7th ed., edited by W. H. Ahrens, pp. 274–276. Weed Science Society of America, Champaign, IL.
623. WSSA. (1994): Terbacil. In: *Herbicide Handbook*, 7th ed., edited by W. H. Ahrens, pp. 278–280. Weed Science Society of America, Champaign, IL
624. WSSA. (1994): Thifensulfuron. In: *Herbicide Handbook*, 7th ed., edited by W. H. Ahrens, pp. 282–283. Weed Science Society of America, Champaign, IL.
625. WSSA. (1994): Triasulfuron. In: *Herbicide Handbook*, 7th ed., edited by W. H. Ahrens, pp. 287–289. Weed Science Society of America, Champaign, IL.
626. WSSA. (1994): Tribenuron. In: *Herbicide Handbook*, 7th ed., edited by W. H. Ahrens, pp. 290–291. Weed Science Society of America, Champaign, IL.
627. WSSA. (1994): Trifluralin. In: *Herbicide Handbook*, 7th ed., edited by W. H. Ahrens, pp. 296–299. Weed Science Society of America, Champaign, IL.
628. Wetzel, L. T., Luempert III, L. C., Breckenridge, C. B., Tisdel, M. O., Stevens, J. T., Thakur, A. K., Extrom, P. J., and Eldridge, J. C. (1994): Chronic effects of atrazine on estrus and mammary tumor formation in female Sprague–Dawley and Fischer 344 rats. *J. Toxicol. Environ. Health*, 43(2):182–196.
629. Whalon, M. E. and Wingerd, B. A. (2003): Bt: mode of action and use. *Arch. Insect Biochem. Physiol*., 54, 200–211.
630. Whysner, J., Ross, P. M., and Williams, G. M. (1996): Phenobarbital mechanistic data and risk assessment: enzyme induction, enhanced cell proliferation and tumor progression. *Pharmacol. Therap*., 71(1/2):153–191.
631. Wickramaratne, G. A., Foster, J. R., Ellis, M. K., and Tomenson, J. A. (1998): Molinate: rodent reproductive toxicity and its relevance to humans, a review. *Regul. Toxicol. Pharmacol*., 27:112–118.
632. Wilkinson, C. F. and Killeen, J. C. (1996): A mechanistic interpretation of the oncogenicity of chlorothalonil in rodents and an assessment of human relevance. *Reg. Toxicol. Pharmacol*., 24:69–84.
633. Wolstenhome A. J. and Rogers, A. T. (2005): Glutamate-gated chloride channels and the mode of action of the avermectin/milbemycin anthelmintics. *Parasitology*, 131(Suppl.): S85–S95.
634. Wood N. J., Lambe, K. G., Myers, K. A., Tugwood, J. D., and Roberts, R. A. (1999): The peroxisome proliferator (PP) response element upstream of the human acyl CoA oxidase gene is inactive among a sample human population: significance for species differences in response to PPs. *Carcinogenesis*, 20(3):369–372
635. WHO. (1965): *Methoxychlor: Evaluation of the Toxicity of Pesticide Residues in Food*, FAO Meeting Rep. No. PL/1965/10/1. World Health Organization, Geneva (http://www.inchem.org/documents/jmpr/jmpmono/v065pr31.htm).
636. WHO. (1967): *WHO Expert Committee on Malaria*, Thirteenth Report, WHO Tech. Rep. Serv. No. 357. World Health Organization, Geneva.
637. WHO. (1977): *Malathion*, Data Sheets on Pesticides No. 29. World Health Organization, Geneva (http://www.inchem.org/documents/jmpr/jmpmono/v91pr02.htm).
638. WHO. (1999): *Triphenyltin Compounds*, Concise International Assessment Document 13. World Health Organization, Geneva (http://www.mindfully.org/Plastic/Stabilizers/Triphenyltin-Compounds-WHO1999. htm).
639. Wyatt, G. H. and Davey, K. G. (1996): Cellular and molecular actions of juvenile hormones. 2. Roles of juvenile hormones in adult insects. *Adv. Insect Physiol*., 26:1–155.
640. Yamane, D. and Andoh, A. (1989): Porphyrin synthesis involvement in diphenyl ether-like mode of action of TNPP-ethyl, a novel phenylpyrazole herbicide. *Pestic. Biochem Physiol*., 35:70–80
641. Yilmaz, H. L. and Yildizdas, D. R. (2003): Amitraz poisoning an emerging problem: Epidemiology, clinical features, management, and preventive strategies. *Arch. Disease in Children*. 88:130–134.
642. Yoshimi, A., Tsuda, M., and Tanaka, C. (2004): Cloning and characterization of the histidine kinase gene Dic1 from *Cochliobolus heterostrophus* that confers dicarboximide resistance and osmotic adaptation. *Molec. Gen. Genom*. 271:228–236.
643. Zarn, J. A. Bruschweiler, B. J., and Schlatter, J. R. (2003): Azole fungicides affect mammalian steroidogenesis by inhibiting sterol, 14α-demethylase and aromatase. *Environ. Health Perspect*., 111(3):255–261.
644. Zhang, Z., Huang, L., Shulmeister, V. M., Chi, Y. -I., Kim K. K., Hung, L. -W., Crofts, A. R., Berry, E. A., and Kim, S -H. (1998): Electron transfer by domain movement in cytochrome $bc_1$. *Nature*, 392:677–684.
645. Aldridge, W. N. (1958): The biochemistry of organotin compounds. *Biochem J*., 69:367–376
646. Asbury, C. L. (2005): Kinesin: World's tiniest biped. *Curr Opin. Cell. Biol*., 17:89–97.
647. Ashcroft, F. M. and Gribble, F. M. (1999): Correlating structure and function in ATP-sensitive K+ channels. *Trends Neurosci*., 21 (7):288–293.
648. Boger, P. and Matthes, B. (2002): Inhibitors of biosynthesis of very long chain fatty acids. In: *Herbicide Classes in Development*, edited by P. Boger, K. Wakabayashi, and K Hirai, pp. 115–137. Springer-Verlag, Berlin.

649. Burr, S. A. and Ray, D. E. (2004): Structure–activity and interaction effects of 14 different pyrethroids on voltage-gated chloride ion channels. *Tox. Sci.*, 77:341–346.

650. Casida, J. E. et al. (1983): Mechanisms of selective action of pyrethroid insecticides. *Ann. Rev. Pharmacol. Toxicol.*, 23:413–438.

651. Cattrell, W. A. (1995): Structure and function of voltage-gated ion channels. *Ann. Rev. Biochem.*, 64:463–531.

652. Cecchini, G., Maklashina, E., Yankovskaya, V., Iverson, T. M., and Iwata, S. (2003): Variation in proton donor/acceptor pathways in succinate: quinine oxidoreductases. *FEBS Lett.*, 545:31–38.

653. Choi, J-S. and Soderlund, D. M. (2006): Structure–activity relationships for the action of 11 pyrethroid insecticides on rat $Na_v$ 1.8 sodium channels expressed in *Xenopus* oocytes. *Toxicol. Appl. Pharmacol.*, 211:233–244.

654. Clough, J. M (1993): The strobilurins, oudemansins, and myxothiazols, fungicidal derivatives of β-methoxyacrylic acid. *Nat. Prod. Rep.*, 10:565–574.

655. Copping, L. G. and Hewitt, H. G. (1998): Herbicides. In: *Chemistry and Mode of Action of Crop Protection Agents*, pp. 17–45. Royal Society of Chemistry, Thomas Graham House, Cambridge, U.K.

656. Dimroth, P., von Ballmoos, C., and Meier, T. (2006): Catalytic and mechanical cycles in F-ATP synthases. *EMBO Rep.*, 7(3):276–282.

657. Eckardt, N., A. (2006): Function of γ-tubulin in plants. *Plant Cell*, 18:1327–1329.

658. Ejim, L. J. et al. (2004): Cystathionine β-lyase is important for virulence of *Salmonella enterica* Serovar *Typhimurium*. *Infect. Immun.*, 72(6):3310–3314.

659. Fritz, R., Lanen, C., Colas, V., and Leroux, P. (1997): Inhibition of methionine biosynthesis in *Botrytis cinerea* by the anilinopyrimidine fungicide pyrimethanil. *Pestic. Sci.*, 49:40–46.

660. Fritz, R. et al. (2003): Effect of the anilinopyrimidine fungicide pyrimethanil on the cystanthionine β-lyase of *Botrytis cinerea*. *Pestic. Biochem. Physiol.*, 77:54–65.

661. Goldin, A., (1999): Diversity of mammalian voltage-gated sodium channels. *Ann. NY. Acad. Sci.*, 88:38–50.

662. Good, N. E. (1961): Inhibitors of the Hill reaction. *Plant Physiol.*, 36(2):788–803.

663. Grummt, I. (2003): Life on a planet of its own: regulation of RNA polymerase I transcription in the nucleuolus. *Genes Dev.*, 17:1691–1702.

664. Hansen H. and Grossman K., (2000): Auxin-induced ethylene triggers abscisic acid and growth inhibition. *Plant Physiol.*, 124:1437ñ1448

665. Hederstedt, L. (2003): Complex II is complex too. *Science*, 299:671–672.

666. Hesse, H. et al. (2004): Current understanding of the regulation of methionine biosynthesis in plants. *J. Exp. Bot.*, 55(404):1799–1808.

667. Horsefield, R., Yankovskaya, V., Sexton, G. Whittingham, W., Shiomi, K., Omura, S., Byrne, B., Cecchini, G., and Iwata, S. (2006): Structural and computational analysis of the quinine-binding site of complex II (succinate-ubiquinone oxidoreductase): a mechanism of electron transfer and proton conduction during ubiquinone reduction. *J. Biol. Chem.*, 281(11):7309–7316.

668. Kepinski, S. and Leyser O. (2005): The Arabidopsis F-box protein TIR1 is an auxin receptor. *Nature*, 435(26): 446–451.

669. Kojima, K., Takano, Y., Yoshimi, A, Tanaka, C., Kikuchi, T., and Okuno, T., (2004): Fungicide activity through activation of a fungal signaling pathway. *Mol. Microbiol.*, 53(6):1785–1796.

670. Kultz, D. (2001): Evolution of osmosensory MAP kinase signaling pathways. *Am. Zool.*, 41:743–757.

671. Kultz, D. and Burg, M. (1998): Evolution of osmotic stress signaling via MAP kinase cascades. *J. Exp. Biol.*, 201:3015–3021.

672. Lange, C. and Hunte, C. (2002): Crystal structure of the yeast cytochrome $bc_1$ complex with its bound substrate cytochrome *c*. *Proc. Nat. Acad. Sci.*, 99(5):2800–2805.

673. Lankas, G. R., and Gordon, L. R. (1989): Toxicology. In: *Ivermectin and Abameclin*, edited by W. R. Campbell, pp. 89–112. Springer-Verlag, New York.

674. Lawrence, L. J. and Casida, J. E. (1982): Pyrethroid toxicology: mouse intracerebral structure–activity relationships. *Pestic. Biochem. Physiol.*, 18(1):9–14.

675. Leroux, P., Fritz, R., Debieu, D., Albertini, C. Lanen, C. Bach, J., Gredt, M., and Chapeland, F., (2002): Mechanisms of resistance to fungicides in field strains of *Botytis cinerea*. *Pestic. Manage. Sci.*, 58:876–888.

676. Lloyd, C. (2006): Microtubules make tracks for cellulose. *Science*, 312:1482–1483.

677. Loughney, K., Kreber, R., and Ganetzky, B. (1989): Molecular analysis of the para locus, a sodium channel gene in *Drosophila*. *Cell*, 58:1143–1154.

678. Matsuno-Yagi, A. and Hatefi, Y. (1993): Studies on the mechanism of oxidative phosphorylation. *J. Biochem. Chem.*, 268(9):6168–6173.

679. Mazumdar, M. and Misteli, T. (2005): Chromokinesins: multitalented players in mitosis. *Trends Biochem. Sci.*, 15(7):349–355.

680. Miki, H., Okada, Y., and Hirokawa, N. (2005): Analysis of the kinesin superfamily: insight into structure and function. *Trends Cell Biol.*, 15(9):467–476.

681. Moore, A. and Wordeman, L. (2004): The mechanism, function and regulation of depolymerizing kinesins during mitosis. *Trends Biochem. Sci.*, 14(10):537–546.

682. Motoyama, T., Ohira, T., Kadkura, K., Ichiishi, A., Fujimura, M., Yamaguchi, I., and Kudo, T. (2005): An Os-1 family of histidine kinase from a filamentous fungus confers fungicide-sensitivity to yeast. *Curr. Genet.*, 47:298–306.

683. Nicklin, S. and Robson, M. W. (1988) Organotins: toxicology and biological effects. *Appl. Organometal. Chem.*, 2, 487–508.

684. Ravanel, S., Gakiere, B., Job, D., and Douce, R. (1998): The specific features of methionine biosynthesis and metabolism in plants. *Proc. Natl. Acad. Sci.*, 95:7805–7812.

685. Russell, J. and Zomerdijk, J. C. B. M. (2005): RNA-polymerase-I-directed rDNA transcription, life and works. *Trends Biochem. Sci.*, 30(2):87–96.

686. Sandmann, G. (2002): Bleaching herbicides: action mechanism in carotenoid biosynthesis, structural requirements and engineering resistance. In: *Herbicide Classes in Development*, edited by P. Boger, K. Wakabayashi, and K. Hirai, pp. 43–57. Springer-Verlag, Berlin.

687. Sauter, H., Steglich, W., and Anke, T. (1999): Strobilurins: evolution of a new class of active substances. *Angew. Chem. Int. Ed.*, 38:1328–1349.
688. Shimizu, T., Nakayama, I., and Nagayama, K. (2002): Acetolactate synthase inhibitors. In: *Herbicide Classes in Development*, edited by P. Boger, K. Wakabayashi, and K. Hirai, pp. 1–41. Springer-Verlag, Berlin.
689. Soderlund, D. M. and Knipple, D. C. (2003): The molecular biology of knockdown resistance to pyrethroid insecticides. *Insect. Biochem. Mol. Biol.*, 33:563–577.
690. Staal, G. B. (1986): Antijuvenile hormone agents. *Ann. Rev. Entomol.*, 39:391–429.
691. Symington, S. B. (2005): The Action of T- and CS-Syndrome Pryrethroids on Voltage-Sensitive Calcium Channels in Rat Brain, Ph.D. dissertation. University of Massachusetts, Amherst, (http://gradworks.umi.com/31/63/3163712.html).
692. Trumpower, B. L. (2002): A concerted, alternating sites mechanism of ubiquinol oxidation by the dimeric cytochrome $bc_1$ complex. *Biochim. Biophys. Acta*, 1555: 166–173.
693. Ueno, H., Suzuki, T., Kinosita, K., and Yoshida, M. (2005): ATP-driven stepwise rotation of $F_0F_1$-ATP synthase. *Proc. Natl. Acad. Sci. USA*, 102(5):1333–1338.
694. U. S. DHHS, NTP. (2002): *Toxicology and Carcinogenicity Studies of Urethane, Ethanol, and Urethane/Ethanol in B6C3F$_1$ Mice, Drinking Water Studies*, Tech. Rep. No. 510, Publ. No. 02–4444. National Toxicology Program, U.S. Department of Health and Human Services, Washington, D.C.
695. Usherwood, P. N. R., Vais, H., Khambay, B. P. S., Davies, T. G. E., and Williamson, M. S. (2005): Sensitivity of the *Drosophila* para sodium channel to DDT is not lowered by the super-kdr mutation M918T on the IIS4-S5 linker that profoundly reduces sensitivity to permethrin and deltamethrin. *FEBS Lett.*, 579:6317–6325.
696. Verschoyle, R. D. and Aldridge, W. N. (1980): Structure–activity relationships of some pyrethroids in rats. *Arch. Toxicol.*, 45:325–329.
697. von Ballmoos, C., Brunner, J., and Dimroth, P. (2004): The ion channel of F-ATP synthase is the target of organotin compounds. *Proc. Natl. Acad. Sci. USA*, 101(31): 11239–11244.
698. Wang, D., Villansante, A., Lewis, S. A., and Cowan, N. J. (1986): The mammalian β-tubulin repertoire: Hematopoietic expression of a novel heterologous β-tubulin isotype. *J. Cell Biol.*, 103:1903–1910.
699. Waskiewicz, A. J. and Cooper, J. A. (1995): Mitogen and stress response pathways: MAP kinase cascades and phosphatase regulation in mammals and yeasts. *Curr. Opin. Cell Biol.*, 7:798–805.
700. Wasteneys, G. O. (2002): Microtubule organization in the green kingdom: chaos or self-order? *J. Cell Sci.*, 115: 1345–1354.
701. Wheeler, D. and Nijhout, H. F. (2003): A perspective for understanding the modes of juvenile hormone action as a lipid signaling system. *BioEssays*, 25:994–1001.
702. White, J. S. Tobin, J. M., and Cooney, J. J. (1999), Organotin compounds and their interactions with microorganisms. *Can. J. Microbiol.*, 45:541–554.
703. Yankovskaya, V., Horsefield, R., Tornroth, S., Luna-Chavez, C., Miyoshi, H., Leger, C., Byrne, B., Cecchini, G., and Iwata, S. (2003): Architecture of succinate dehydrogenase and reactive oxygen species generation. *Science*, 299:700–704.
704. Yoshimi, A., Kojima, K., Takano, Y., and Tanaka, C. (2005): Group III histidine kinase is a positive regulator of Hog1-type mitogen-activated protein kinase in filamentous fungi. *Eukaryot. Cell*, 4(11):1820–1828.
705. Yuh-Ru, J. L. and Liu, B. (2004): Cytoskeletal motors in Arabidopsis: sixty-one kinesins and seventeen myosins *Plant Physiol.*, 136:3877–3883.
706. Zhang, Y., Lamm, R., Pillonel, C., Lam, S., and Xu, J.-R (2002): Osmoregulation and fungicide resistance: the *Neurospora crassa* os-2 gene encodes a HOG1 mitogen-activated protein kinase homologue. *Appl. Environ Microbiol.*, 68(2):532–538.

# 17 Metals

*Jill C. Merrill,* Joseph J.P. Morton, and Stephen D. Soileau*

**CONTENTS**



* The views expressed in this chapter do not necessarily represent the views of the U.S. Food and Drug Administration.

## INTRODUCTION

Metals are elements generally characterized by ductility, luster, being electropositive with a tendency to lose electrons, and having the property of conducting heat and electricity; however, a number of the elements individually discussed in the body of this chapter are not true metals (e.g., arsenic, fluorine). The attempt was made to include elements that have physiological actions (both beneficial and toxic) by virtue of their chemical ionic form. Elements such as oxygen and sulfur, which are essential to life in some forms (e.g., water, amino acids) but which also exist in forms that are chemically reactive and hence toxic (e.g., hydroperoxide, sulfuric acid), are not covered in this chapter.

Metals can have a variety of physiological effects, and it is often possible to demonstrate the toxicity of any given metal in any given organ, provided that the dose is both high and prolonged (but not so high and prolonged that the primary target organ receives a fatal dose). Essential elements may be toxic at a dose that overwhelms homeostatic controls on absorption and excretion, and the mechanism of toxicity is commonly related to an essential physiological role of the metal (e.g., control of osmolarity for sodium consumption in excess of water intake, neurotransmission for potassium consumption in excess of water intake, redox reactions for iron intake in excess of protein binding capacity). Physiological actions of nonessential elements include substituting for essential elements in enzymatic reactions, energy metabolism, neurotransmission, structural components (bone), reacting covalently or noncovalently with enzymes, membranes, DNA, and stimulating the production of active oxygen species [257]. The variety of physiological effects makes it difficult to determine which action is responsible for toxicity in the most sensitive target organ. In some cases organs are most sensitive for a biochemical reason (e.g., thallium interferes with energy metabolism, and target organs are those with the highest energy requirement); in other cases, the most sensitive organ is simply the organ in which the accumulation is greatest (e.g., cadmium and uranium accumulate in the kidneys, which are the target organs). Metals can interact with each other either to enhance toxicity (e.g., by affecting the same target organ) or to reduce toxicity (e.g., by stimulating defense mechanisms); this must be particularly kept in mind for the interpretation of animal experiments (e.g., levels of calcium, iron, and zinc should be controlled in investigations of cadmium toxicity) and epidemiological studies (e.g., fluoride reduces the incidence of dental caries; therefore a population with the lowest fluoride exposure is likely to have the highest exposure to mercury and other metals used in dental restorations). The number of combinations of metals that could potentially be investigated is huge, and such studies are most useful either when a sensitive subpopulation is identified (e.g., individuals with insufficient intake of specific nutrients) or when a specific mechanism is revealed. Few treatments for metal toxicity are based on interfering with the mechanism of action; rather, measures are designed to reduce gastrointestinal absorption (from acute poisoning) by removing or binding the metal or are designed to speed elimination from the body (e.g., chelation therapy) [203]. Prevention of excessive exposure is generally the best way to reduce the potential for metal toxicity.

The variety of physiological effects that metals can have is also the reason that adverse effects can often be demonstrated in most organ systems. Reproductive, developmental, immunological, and neurological toxicity, which are often not investigated in routine bioassays, are endpoints of increasing concern. For metals in particular, which on general principles would be expected to at least have the potential for these types of toxicity, toxicological understanding should not be considered complete without some information on whether these systems might be the most sensitive.

## QUANTIFICATION OF TOXICOLOGICAL EFFECTS OF METALS

Consideration of the toxicity of metals must be quantitative because of the need to identify the most sensitive organ among all the systems that can be affected by the metal and also because metals are naturally occurring and ubiquitous. Exposure to any metal cannot be banned the way exposure to, for example, an organic pesticide or food additive can be banned; some elements are essential to life, and even for those that are not, with sufficiently sensitive analytic techniques their presence can be demonstrated in any given sample of food, water, soil, or air. The quantification of the toxic effects of metals must attempt to precisely identify the highest level that is not expected to cause undue adverse effects because in many cases the traditional approach of using a safety/uncertainty factor of 10 would quickly lead to calculated levels that are below those essential for health (e.g., zinc, molybdenum) or levels that are below background exposures from food or water and hence extremely costly to achieve (e.g., cadmium, arsenic).

An example of the need to quantify toxicity is in the U.S. Environmental Protection Agency (USEPA) program to address abandoned hazardous-waste sites. For each site, a quantitative risk assessment is performed to determine the need for and extent of remediation [668]. Essentially, this risk assessment calculates doses of contaminants based on the concentration in a medium (air, soil, food, water) and the intake of that medium (e.g., adults are assumed to ingest 100 mg of soil per day). This dose, to the maximally exposed person, is then compared to two quantitative toxicological values. The first is the reference dose (RfD) or reference concentration (RfC), which is the highest dose or concentration not thought to be associated with adverse noncancer health effects (the "threshold"). The derivation of RfDs and RfCs is described in Chapter 2 in this volume. RfDs quantify oral (and potentially dermal) toxicity, and RfCs quantify inhalation toxicity; separate values may be derived for acute, intermediate, and chronic exposure duration. The second toxicological value is the slope factor, which quantifies the cancer risk corresponding to a given lifetime dose. There are commonly separate slope factors for the inhalation and oral routes; risks from less-than-lifetime exposure are evaluated by dividing the duration of exposure by an (assumed) 70-year lifetime. Cleanup standards for a site are commonly set as the concentrations that would deliver a dose to the most exposed individual that is less than the RfD for each chemical and that results in an "acceptable" cancer risk (e.g., $10^{-4}$ to $10^{-6}$ incremental lifetime cancer risk). For metals, one important issue is whether they are present at the hazardous-waste site at levels exceeding the natural background level (which are not necessarily below what would pose an unacceptable risk to the most exposed individual). Another issue that arises is incorporation of uncertainty into derivation of cleanup standards. One example is chromium, commonly measured as total chromium, which leads to uncertainty because only the rarer hexavalent form, not the more common trivalent form, is considered a carcinogen [674]; considering all chromium detected at a hazardous site to be hexavalent will lead to an overestimate of the risk and the need for cleanup by an unknown amount. Another example is antimony, which has an oral RfD derived with a safety factor of 1000 [674]; it is likely that this safety factor is too conservative and that cleanup standards will be more stringent and therefore more expensive than needed. For both chromium and antimony, additional information (speciation at the site for chromium; better toxicology data for antimony) would allow risk assessments to determine more precisely acceptable levels, which would prevent the setting of potentially unnecessarily strict standards.

On the other hand, there may be situations where current standards of exposure are not strict enough. Human activities such as mining and smelting, fossil fuel burning and incineration, fertilizer-intensive agriculture, and other industrial processes have increased human exposure to many elements to levels far above those of the preindustrial environment. Lead and cadmium are two examples of metals for which the level of exposure deemed acceptable has dropped many times over the years, as concern about frank toxicity among workers was replaced by concern about more subtle signs of toxicity in workers which was in turn replaced by concern about even more subtle adverse effects in the general population exposed through environmental (including dietary) routes [104, 119]. Numerous examples of metals exist for which our knowledge of their toxicology primarily consists of information on frank toxicity in exposed workers and a few animal studies, very similar to the extent of information that was used to derive standards for lead and cadmium that we now know could cause substantial toxicity in the general population. One of the major reasons for the advances in knowledge about lead, cadmium, and a few other metals was the development of biomarkers of exposure (blood lead levels and urinary cadmium levels) that provide a way to quantify environmental exposure and thus allow studies linking exposure to health effects in the general population. Biomarkers of exposure commonly provide much more precise quantification of exposure than is possible by traditional means, particularly for the general population that may be exposed by several routes (food, air, water), all of which are variable in time and location. One important future direction for the investigation of metals toxicology is developing and validating biomarkers of exposure and using these biomarkers to investigate potential adverse effects in the general population.

For most metals, quantification of toxicological effects has not been done using human studies with validated biomarkers. Instead, the traditional methods are used: assembling the entire dataset, surveying the data to determine the most sensitive target organ (the organ exhibiting an adverse effect at the lowest dose), identifying the no-observed-adverse-effect level (NOAEL) or the lowest observed-adverse-effect level (LOAEL), and applying safety or uncertainty factors to derive a threshold below which no noncancer effects are expected to occur. For cancer risk assessment, the process involves determining a weight-of-evidence judgment as to whether the element has the potential to cause cancers in humans; for example, the USEPA uses classifications of Group A, known human carcinogen; Group B, probable human carcinogen; Group C, possible human carcinogen; and Group D, not classifiable as to human carcinogenicity. A separate step is to quantify the cancer risk associated with a given dose on the no-threshold assumption that any exposure carries some cancer risk, with a safety margin built in by using the most sensitive sex/species/organ carcinogenic response and by using the upper 95th confidence limit of the slope [668]. Several issues for the qualitative and quantitative evaluation of toxicity pertain particularly to metals and are outlined below.

## Essentiality

Recommended dietary allowances (RDAs) are defined as "the levels of intake of essential nutrients that, on the basis of scientific knowledge, are judged by the National Research Council (NRC) Food and Nutrition Board to be adequate to meet the known nutrient needs of practically all healthy persons" [483]. They are revised and published periodically by the NRC, which convenes expert committees to estimate the mean dietary requirement for the population based on deficiency studies, balance studies, nutritional intakes, bioavailability, interactions, and homeostatic regulatory mechanisms [515]. A normal Gaussian distribution for the range of requirements within the population and a coefficient of variation of 15% are generally assumed. The RDA is then set at two standard deviations above the mean. Statistically, the RDA represents the 97.5th percentile of the nutrient requirement in the healthy population [92]. The first RDAs were set during World War II, when food was rationed, and it was important to set minimum requirements to prevent frank deficiency diseases. Today, although preventing nutrient-deficient diseases is still important, public health concerns are directed toward defining the amounts of nutrients needed to ensure optimum health, provide excellent physiological and mental function, and prevent degenerative diseases [343]. For example, the 1989 recommendation for selenium is based on the amount required to support maximal activity of the selenium-dependent enzyme glutathione peroxidase and prevent cardiomyopathy [483], but recent epidemiological studies suggest selenium has cancer-preventive activity at levels significantly higher than that needed to support maximal activity of glutathione peroxidase [148,149,714]. A higher recommendation for selenium might be set if a reduction in cancer risk was chosen instead of preventing the disease process associated with frank selenium deficiency, cardiomyopathy [158]. This requires the use of a reconstructed RDA, which is termed the dietary reference intake (DRI) [321,457]. The new DRIs have four components: (1) the recommended dietary allowances (RDAs); (2) the estimated average requirements (EARs); (3) adequate intakes (AIs); and (4) tolerable upper intake levels (ULs). These reference values are defined and depicted graphically in Figure 17.1 [324]. The Food and Nutrition Board of the Institute of Medicine, National Academy of Sciences, has formed the Committee on the Scientific Evaluation of Dietary Reference Intakes to address these issues for the various nutrients.

## Routes of Exposure

Two major routes of exposure to metals are by inhalation and oral exposure. Inhalation of metals, particularly as fumes or dusts, commonly causes systemic effects on the lung, ranging from mild, self-limiting metal fume fever from acute exposure and benign pneumoconiosis from chronic exposure for some metals to severe chronic obstructive lung disease for others [496]. Standards for inhalation exposure to metals are developed by the American Conference of Governmental Industrial Hygienists (ACGIH)

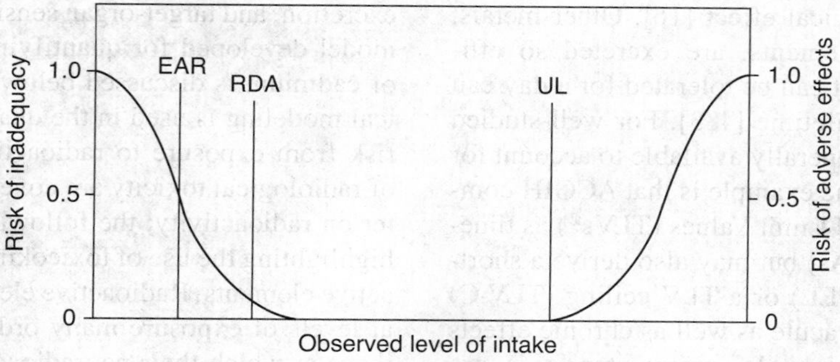


**FIGURE 17.1** Dietary reference intakes. (From Institute of Medicine, *Dietary Reference Intakes: Guiding Principles for Nutrition Labeling and Fortification*, National Academy Press, Washington, D.C., 2003, p. 3. With permission.)

based primarily on occupational data; the highest allowable standard (10 mg/m$^3$) pertains to dusts that are not chemically reactive but present a cumulative, physical burden on the lung that can be harmful from long-term exposure [34]. Oral exposure to many metals with known lung toxicity often has no adverse effects (although at high enough oral doses, most metals cause acute gastrointestinal irritation and distress [203]), possibly due in part to the faster turnover of gastrointestinal vs. lung cells. The carcinogenicity of metals can be dependent on the route of exposure. Several metals are considered to be carcinogenic by the inhalation route, but not by the oral route (e.g., cadmium, chromium, nickel). This classification is based on the observation of an increased rate of lung but not other forms of cancer among workers and experimental animals exposed by inhalation and on no observed increased rate among experimental animals exposed orally. These metals, however, could have weak rather than no oral carcinogenicity; on theoretical grounds, it could be argued that most mechanisms by which an element is carcinogenic to lung tissue could operate in other tissues as well, and certain metals such as arsenic are known to be both lung carcinogens and systemic carcinogens. The potential human oral carcinogenicity of metals, particularly those known to be inhalation carcinogens, is an area deserving further study, and, again, valid biomarkers of exposure would be very valuable for such studies.

Another point concerning route of exposure is that some metals, such as cadmium, are known to have very different toxicokinetic and toxicological properties by parenteral routes than by oral or inhalation routes. For cadmium at least, this is most likely due to the binding to metallothionein as a required step in oral or inhalation absorption, which is bypassed by parenteral exposure [16]. Cancers can be induced in experimental animals by implantation of solids (metals as well as other solids). This solid-state carcinogenicity is typically considered to be only marginally relevant to human exposures; however, solid-state carcinogenesis may be relevant in humans with implanted metal-containing prosthetic devices [693]. In general, studies using parenteral routes of exposure are often of limited use, unless the goal is to evaluate human parenteral exposure from medical procedures. A final point with respect to route of exposure is that most metals are considered not to be absorbed through the skin (with certain exceptions such as mercury and thallium); however, few data actually exist to substantiate this assumption, and further studies would be useful to quantify the dermal absorption of metals.

### Form

Some metals, such as mercury, exist in elemental, ionic, and organic forms, and each of these forms has a unique toxicity. Other metals, such as chromium, may exist in two or more valence states with different effects. Still others, such as nickel, may be primarily protein bound in food sources but primarily free ions in water, which may affect absorption. Finally, there are some metals, such as cadmium, that appear to have similar toxicological effects regardless of their form. Ideally, information would be available to do a separate quantification for each toxicologically distinct form; this would only be useful, of course, in situations where the form to which humans are to be exposed is actually known. Failing this ideal, the attempt is made to derive a standard for the most toxic form of a given metal; however, in many cases, this leads to overly stringent standards and in some cases lack of information on the most toxic form of a metal may lead to standards that are too lax.

### Duration of Exposure

The influence of duration of exposure on the quantitative, and even qualitative, toxicology of a metal depends on its toxicokinetics. Cadmium is an example of a cumulative toxin. To a relatively good approximation, the same total dose given over a week, a month, or a year will accumulate in the kidneys to the same extent and have the same physiological effect [16]. Other metals, particularly essential elements, are excreted so efficiently that any dose that can be tolerated for a day can also be tolerated for a lifetime [483]. For well-studied metals, information is generally available to account for duration of exposure. One example is that ACGIH commonly derives Threshold Limit Values (TLVs®) as time-weighted averages (TWAs) but may also derive a short-term exposure limit (STEL) or a TLV ceiling (TLV-C) for substances that have acute as well as chronic effects [34]. For less-studied metals, many standards are derived based on animal data with very little information on toxicokinetics, which means that extrapolation to durations of exposure other than those used in the study at hand are quite uncertain.

## Age at Exposure

Infants and young children may be particularly sensitive to toxic effects of metals both because they often absorb a greater fraction of ingested metals than older children or adults and because some developing systems (particularly the nervous system) are more sensitive to toxic effects than mature systems. Lead is an example of a metal that is known to be most deleterious to fetuses, infants, and toddlers [119,120]. The elderly are another group that may be more sensitive than healthy adults to the toxic effects of metals due to diminution of homeostatic and adaptive mechanisms. Definition of a safe level of exposure has an inherent uncertainty for metals lacking data on effects on infants and the elderly.

## Animal vs. Human Data

For well-studied metals, animal and human toxicity appear to be in general qualitative agreement, although there are some exceptions, such as the difficulty in demonstrating that arsenic is carcinogenic to experimental animals. Quantitative differences do occur, however; for example, gastrointestinal absorption of cadmium is about two to three times lower in experimental animals than in humans [16]. Use of a tenfold safety factor for animal-to-human extrapolation in this case would yield a toxicological value that is about three times more stringent than necessary. This emphasizes the importance of using human data whenever possible for the quantification of toxicity.

## Toxicokinetic Modeling

Toxicokinetic modeling is very useful for evaluating the toxicity of well-studied metals. Good models can integrate information on the effects of routes of exposure, chemical forms, age at exposure, duration of exposure, and interindividual variation on absorption, distribution, excretion, and target-organ sensitivity. The toxicokinetic model developed for quantifying the systemic toxicity of cadmium is discussed below. In addition, toxicological modeling is used in the quantification of the cancer risk from exposure to radioactive elements. Principles of radiological toxicity are covered in detail in the chapter on radioactivity; the following is a brief description highlighting the use of toxicokinetic modeling for radioactive elements. Radioactive elements can cause damage at levels of exposure many orders of magnitude below those at which their nonradioactive forms cause chemical damage because radioactive decay involves the release of a large amount of energy in the form of alpha particles, beta particles, and gamma rays. A single radioactive decay can initiate a cascade of events that creates a huge number of active oxygen species, which are thought to be the ultimate cause of radioactive damage. This damage can cause cell death at high levels of exposure; of greater concern is the possibility of mutations that can initiate or promote cancer or result in hereditary defects [667].

Quantification issues are addressed for individual metals in the remainder of this chapter. The reader should recognize that the USEPA continually reviews and revises toxicity values, and that the numbers presented here are simply the values that were specified in December 2004.

## Sources of Information

Numerous sources of information on the toxicity of metals are available, many of which are updated on a regular basis. The USEPA maintains the Integrated Risk Information System (IRIS), which has a summary of information (including RfD and RfC values) on numerous toxic chemicals, including a number of metals [674]. This database is regularly updated, and is available on the EPA's website (http://www.epa.gov). The ACGIH publishes annually a listing of all chemicals for which TLVs and biological exposure indices (BEIs) exist [35]. The documentation for the development of these values is also available, with the latest update of this publication occurring in 1996 [34]. The U.S. Agency for Toxic Substances and Disease Registry (ATSDR) has published many documents that summarize the toxicological effects of elements and chemicals; documents exist for many of the elements discussed in this chapter. Finally, the series *Patty's Industrial Hygiene and Toxicology*, contains a vast amount of information on the toxicity of metals (and other elements and compounds), with the major emphasis being on industrial exposures and effects [67,68,166,531]; other general texts on metals are useful sources, as well [445]. Treatments for exposures to metals and other compounds are described in detail in *Clinical Toxicology of Commercial Products* [256] and in *Medical Toxicology: Diagnosis and Treatment of Human Poisoning* [203].

## ESSENTIAL ELEMENTS

### CALCIUM

Calcium is essential both for the physical structure of bone and for normal physiological function (e.g., nerve conduction, muscle contraction, blood clotting, membrane permeability, enzyme activation, acetylcholine synthesis) [28]. The average healthy adult body contains about 1200 g of calcium, 99% of which is found in bone and teeth, with the remaining 1% in extracellular fluids, intracellular structures, and cell membranes. The average calcium content of the blood ranges from 9.0 to 10.5 mg/dL with tight physiological controls. Decreased body calcium leads to loss of bone mineralization, reduction of bone strength, and increased susceptibility to fractures [557] and may increase blood pressure [438], particularly among pregnant women [69]. Calcium deficiency is also associated with convulsions and tetany. The RDA for calcium was derived from the need to maintain skeletal calcium, using an estimated 200 to 250 mg/day obligatory loss and an oral absorption fraction of 30 to 40%, leading to a recommendation of 1200 mg/day for ages 11 to 24 and 800 mg/day for older age groups [483]. In 1997, the Food and Nutrition Board of the Institute of Medicine released their new DRI for calcium [321]. The AI of calcium was designed to maximize calcium retention to promote bone strength and prevent osteoporosis [101], and the recommended intakes are 1300 mg/day for ages 9 to 18, 1000 mg/day for ages 19 to 50, and 1200 mg/day for ages >51. The UL is 2.5 g/day.

Calcium is not a very toxic metal, but adverse effects may occur at intakes greater than 2000 mg/day [482]. Intestinal absorption of calcium decreases as intake increases; however, very large intakes of calcium can increase the calcium body burden [483] as well as interfere with the absorption of magnesium [595], zinc [226], and iron [277]. Very large chronic intakes are associated with hypercalcemia or hypercalciuria. Other symptoms of calcium excess include renal failure and soft tissue calcification. High-calcium diets could increase the risk of kidney stones in susceptible individuals and reduce the bioavailability of zinc and iron. Although excessive calcium intake from food and municipal water was previously seen mainly in individuals with conditions predisposing them to increased calcium absorption, such as parathyroidectomy [435], the consumption of calcium-fortified foodstuffs (e.g., sparkling water, breakfast cereal, orange juice) in addition to a diet containing generous amounts of dairy products could theoretically lead to levels of concern [703]. Education is needed to prevent both calcium excess in one population and calcium deficiency in another. The timing of calcium supplementation (with meals vs. at bedtime) may affect the risk of calcium oxalate stone formation, and a recent study suggests taking calcium supplements with meals to reduce the risk of stone formation [188]. A potential adverse effect associated with high levels of calcium supplementation is ingestion of heavy metals, such as arsenic, cadmium, and lead, which have been found to contaminate some calcium supplements [90,360,702]. With the increased interest in daily calcium supplementation as a preventive measure for colon cancer, osteoporosis, and hypertension, the possible contamination of these supplements is of concern. Manufacturers of calcium supplements are responsible for ensuring premarket safety evaluations of their products, and the heavy metal content varies accordingly; however, the USFDA monitors the postmarketing claims made for dietary supplements. In June 2003, they issued warning letters to 18 firms marketing coral calcium supplements with false and unsubstantiated claims of efficacy against diseases and conditions such as cancer, multiple sclerosis, lupus, and heart disease [93].

### CHLORIDE

Chloride is the principal extracellular inorganic ion. It is required for maintenance of fluid and electrolyte balance and for the production of gastric acid [483]. Dietary chloride deficiency is rare, but prolonged loss of electrolytes from vomiting, diarrhea, heavy sweating, and so forth can lead to hypochloremic metabolic alkalosis [483]. For both adult men and women, the AI of chloride is 2.3 g/day, and the UL is 3.6 g/day [325]. Reactive chlorine compounds (e.g., chlorine gas, hydrochloric acid, hypochlorite, chlorine dioxide) are irritating to the tissues they contact, but neutral chloride solutions are nontoxic [210]. Habitual excess intake of table salt may contribute to hypertension in susceptible individuals, and animal data suggest that the chloride ion may play a role as well as the sodium ion [472]. This question has more than theoretical implications because potassium chloride is widely used as a salt substitute by individuals seeking to restrict their sodium intake. The sparse human data on the association between chloride intake and blood pressure are generally negative, and more studies are needed before restriction of chloride intake, independent of sodium intake, could be suggested to have a beneficial effect in the general population. The USEPA has not derived any toxicity values for chloride [674].

### CHROMIUM

Chromium is a first-series transition metal; its name is derived from the Greek word for "color" because most chromium compounds are brightly colored. The only important chromium ore is chromite. Chromium is used as an alloy with other metals, and is also used for plating of metals [67]. Although chromium can have valences from –2 to +6, the most important valences are +3 and +6 [34]. Trivalent chromium is the most abundant form of chromium in the environment. Chromium (III) is an essential nutrient that plays a role in glucose metabolism [38,67]. Although $Cr^{3+}$ is poorly absorbed orally [189], absorption is greatly enhanced by the presence of the glucose tolerance factor, which forms a complex with $Cr^{3+}$ [593]. The AIs for men and women (ages 19 to 50 years) are 35 μg/day and 25 μg/day, respectively [323]. Chromium (III) is considered to be relatively nontoxic *in vivo* [67]; however, a case report indicated that chronic ingestion of high levels of chromium (III) picolinate induced reversible liver and kidney toxicity [135]. Mice exposed to chromium (III) acetate in drinking water for over 2 years did not show an increased incidence of tumors [583].

Hexavalent chromium is the most important valence from a toxicity standpoint. Unlike chromium (III), chromium (VI) is readily absorbed by all tissues. Because chromate ($CrO_4^{2-}$) is structurally similar to phosphate and sulfate [163], it readily enters all cells via the general anion channel protein. Chromium (VI) is acutely toxic, with most reports of human toxicity occurring as a result of accidental or intentional ingestion. The lethal oral dose of soluble chromates in humans is estimated to be in the range of 50 to 70 mg/kg. Symptoms of acute toxicity include vomiting and generalized gastrointestinal tract damage with gastrointestinal bleeding leading to cardiovascular shock. If the victim survives the initial toxic effects, liver necrosis, tubular necrosis of the kidney, and damage to the blood-forming tissues can occur [67]. In one case of acute poisoning, liver transplantation led to complete recovery of the patient [624]. Long-term occupational exposure to chromium has been associated with either low-molecular-weight proteinuria or elevated levels of proteins normally found in the urine [67,163]. Although animal studies have shown that parenteral administration of 15 mg/kg potassium chromate (6+) is nephrotoxic, chronic renal disease due to occupational or environmental exposure has not been reported [67].

Dermal exposure to potassium dichromate and other chromium compounds can lead to the development of a sensitization reaction. The resulting hypersensitivity results from chromium binding to proteins and becoming a hapten [179]. Prior to the implementation of appropriate industrial hygiene precautions, occupational inhalation exposure to $Cr^{6+}$ was associated with changes in the septal mucosa, ranging from irritation to septal perforation [67]; however, inhalation exposure rarely causes asthma [496].

The carcinogenicity of chromium in the respiratory tract has been well established, beginning when the first nasal tumors were described among Scottish chrome pigment workers in the late nineteenth century [498], and has been reviewed in the recent literature [67,154,163]. The mechanism of action is believed to be from a direct modification of DNA [513]. After hexavalent chromate enters a cell, it is rapidly reduced to $Cr^{3+}$. During the reduction process, unstable and reactive intermediates, including Cr(IV), Cr(V), hydroxide, thiyl, and organic (RS and R) radicals and active oxygen radicals are formed, and it is believed that these moieties are responsible for chromium carcinogenicity [154]. Because $Cr^{6+}$ is readily absorbed by all tissues, one could postulate that chromium-induced cancers should be noted in other organs. Although the evidence is not as strong, exposure to hexavalent chromium is associated with an increased incidence of many types of cancers [163].

Both trivalent and hexavalent chromium are fetotoxic and teratogenic when administered to rabbits. At 500 ppm in drinking water, both materials significantly reduced the number of implantation sites and viable fetuses. Malformations included dwarfism and kinky and short tail [201]. Oral administration of hexavalent chromium produced similar adverse effects in mice [650].

The USEPA has established an oral RfD for chromium (III) insoluble salts of 1.5 mg/kg/day, an oral RfD for chromium (VI) of 3 μg/kg/day, an inhalation RfC for chromic acid mists and dissolved Cr(VI) aerosols of $8 \times 10^{-3}$ μg/m$^3$, and an inhalation RfC for Cr(VI) particulates of 0.1 μg/m$^3$. The USEPA has classified chromium (III) as a Group D (not classifiable) carcinogen and chromium (VI) as a Group A (human) carcinogen (inhalation exposure) and Group D carcinogen (oral exposure).

## Cobalt

Cobalt is an essential component of vitamin $B_{12}$ (hydroxycobalamin) which is involved in intermediary metabolism, nucleic acid synthesis, and in single-carbon metabolism; it is required to prevent macrocytic megaloblastic anemia, atrophic gastritis, achlorhydria, neurologic degeneration and dementia [28]. Vitamin $B_{12}$ is synthesized by bacteria, fungi, and algae, but not by yeasts, plants, or animals [483]. Cobalt deficiency may develop in animals dependent on gut microflora for their vitamin $B_{12}$, such as ruminants, and in strict vegetarians consuming no animal products. The RDA for vitamin $B_{12}$ is 2.4 μg/day, and although cobalt is known to activate the enzyme arginase [639] the only recognized requirement for cobalt is as a component of vitamin $B_{12}$. Cobalt is a hard silvery metal widely distributed in rocks and soils and always occurs with nickel and usually with arsenic [13]. It is primarily used in the production of superalloys, as a drier in paints, in magnets, and in the production of prosthetic devices. Occupational exposure occurs in the hard metal industry, among cobalt blue dye plate painters and coal miners, and this exposure is reflected in elevated levels of cobalt in tissues and body fluids.

Cobalt can be toxic. For the general population, ingestion is the primary route of exposure [13]. Oral exposure to cobalt caused cardiomyopathy among individuals who drank excessive amounts of beer (8 to 25 pints/day) containing cobalt as a foam stabilizer [467]. This effect may have been potentiated by a combination of alcohol, preexisting heart damage, and poor diets associated with heavy alcohol consumption because anemic individuals have been exposed to higher levels of cobalt without a similar effect [467,596]. Cobalt can cause allergic dermatitis (eczema and urticaria, mainly of the hands) [29,684], and cross-reaction with nickel is frequent [572,681]. Inhalation exposure to cobalt alloyed to tungsten carbide (hard metal) is associated with hard metal disease, which is characterized by interstitial fibrosis and restrictive respiratory impairment [407]. The toxic mechanism of hard metal particles is thought to involve both cobalt sensitivity and the generation of oxygen radicals by the carbide particles [407,497]. Cobalt by itself has caused occupational asthma in diamond polishers, and the effect has been attributed to an immunologic mechanism with cobalt acting as a hapten [241]. The carcinogenicity of cobalt is uncertain. Animal studies are positive only for subcutaneous, intramuscular, or intratracheal administration, but not for inhalation, and the excess rates of lung cancer observed in men occupationally exposed to cobalt dust could be explained by simultaneous exposure to nickel, arsenic, and tobacco [391,508].

In a 1991 review of cobalt and its compounds, the International Agency for Research on Cancer (IARC) concluded evidence for carcinogenicity in humans was inadequate but sufficient evidence existed in experimental animals [329]. IARC concluded that cobalt and cobalt compounds are possibly carcinogenic to humans (Group 2B). Subsequent data indicate that hard metal dust is genotoxic in *in vitro* and *in vivo* systems [169]. Production of active oxygen species and inhibition of DNA repair are likely modes of action. Given the fact that epidemiological evidence links exposure to hard metal dust with an increased risk of lung cancer, De Boeck et al. [169] suggest that an IARC revision is advisable. In view of the fact that patients with cobalt-containing, metal-on-metal, total hip replacements were found to have blood cobalt concentrations up to 50 times higher than controls, it seems prudent to monitor these patients frequently to ensure that toxic effects of cobalt wear debris are detected early [401]. The USEPA has not derived toxicity values for cobalt. The ACGIH has adopted TLV–TWA values for cobalt carbonyl and cobalt hydrocarbonyl of 0.1 mg Co per m$^3$ [35].

## COPPER

Copper occurs naturally as the free metal and occurs in compounds in +1 or +2 valence states. Copper is incorporated into several enzymes involved in hemoglobin formation, carbohydrate metabolism, catecholamine biosynthesis, and cross-linking of collagen, elastin, and hair keratin [6]. These enzymes include cytochrome *c* oxidase, dopamine β-hydroxylase, ascorbic acid oxidase, and superoxide dismutase, as well as interaction with ceruloplasmin and metallothionein. Copper deficiency causes anemia, neutropenia, and impaired growth, particularly in children [483]. The ingestion of copper in foods is the primary source for copper intake. The intake from copper plumbing and unpolluted fresh water is not significant. The RDA for adult men and women is 900 μg/day, with an additional 100 μg/day during pregnancy and an additional 400 μg/day during lactation [323]. The USEPA action level for copper in tap water is 1.3 mg/L [153].

Copper is readily absorbed following oral ingestion, but homeostatic mechanisms limit further intake once requirements are met. Copper overload is normally further controlled by binding to metallothionein. Copper is either active or in transit, and little or no excess copper is normally stored [405]. Following absorption, copper is bound to albumin and transcuprein and is mainly deposited in liver hepatocytes with lesser amounts in the kidney. Biliary excretion is the major route, with small amounts secreted in the urine. Considering these homeostatic mechanisms following oral intake, absorption through the inhalation or dermal routes may allow toxic levels to pass unimpeded into the blood.

The consumption of water containing high levels of copper or suicide attempts with copper sulfate can result in vomiting, diarrhea, nausea, abdominal pain, hemolytic anemia, hepatic and renal neurosis, and death. Industrial exposure to copper fumes may occur, resulting in metal fume fever with dyspnea, chills, headache, and nausea [67]. The ACGIH has adopted TLV–TWA values for copper of 1 mg/m$^3$ for dusts and mists and 0.2 mg/m$^3$ for fumes [35]. The OSHA permissible exposure limit (PEL) differs, as it is 0.1 mg/m$^3$ for copper fume. Copper can be dermally absorbed from copper-containing topical products [388,537,538]. Dermal irritation and allergic contact dermatitis have been associated with copper jewelry, intrauterine contraceptive devices, and through occupational exposure to electroplating and copper-containing agricultural products [388].

Wilson's disease is one of several examples of toxicity involving copper in humans. This disease is due to an autosomal recessive disorder that affects normal copper homeostasis. The retention of hepatic copper is excessive and is accompanied by a decreased concentration of plasma ceruloplasmin, impaired biliary copper excretion, and hypercupremia resulting in hepatic and renal lesions and hemolytic anemia [6]. Tetrathiomolybdate, a new anticopper drug developed for Wilson's disease, is a promising antiangiogenic drug useful against cancer in animals and clinically. Speculation is that copper levels serve as a primitive angiogenesis and growth-signal regulator [96]. Menkes' disease is a multisystemic lethal disorder characterized by neurodegenerative symptoms and connective tissue manifestations. The disease is attributable to a deficiency of one or more copper-dependent enzymes [656].

## Fluorine

Fluorine, the most reactive of the elements, is a pale yellow gas with a pungent odor. The chief fluoride sources are fluorspar ($CaF_2$) and cryolite ($Na_3AlF_6$). Fluorine, hydrogen fluoride, and other fluorine compounds are used in a wide number of applications in the nuclear (in the synthesis of uranium hexafluoride), agrochemical (pesticides), drug (anticaries agents), and other industries [34,67]. Fluorine gas is a severe eye, mucosal, and skin irritant [34]. Hydrogen fluoride (HF) is a weak acid that causes severe burns on the skin and in the eye, either in aqueous solution or as the anhydrous acid [67]. In addition to causing dermal and ocular damage, hydrogen fluoride is readily absorbed through the skin. Once absorbed, fluoride complexes with calcium and causes hypocalcemia. If the hypocalcemia is severe, death can occur via cardiac arrhythmia. Hydrogen fluoride burns over as little as 2.5% of the body surface have caused fatalities, depending on the HF concentration [136,140,369].

Fluoride is incorporated into bones and tooth enamel, making teeth more resistant to caries. A fluoride deficiency has never been conclusively demonstrated in humans or animals, but goats fed <1 mg F per kg dry ration had reduced growth and survival [43]. The NRC classifies fluoride as a beneficial but not an essential element [483]. Fluoride replaces hydroxyl ions in enamel, yielding an apatite crystal that is more resistant to acid. Some studies suggest that fluoride supplements may also increase bone strength [522]. Fluoride in aqueous solutions is virtually 100% absorbed, while absorption of fluoride in bone meal may be as low as 40% [483]. The AIs for fluoride are set at 4 mg/day and 3 mg/day for adult men and women, respectively [321]. Although serious complications are rare (because of limitations set by the USFDA on the total amount of fluoride in an OTC anticaries drug product), acute fluoride toxicity can occur from accidental ingestion of fluoride-containing products. In his review of reported accidental fluoride poisoning cases, Whitford [700,701] proposed a "probably toxic dose" of 5 mg/kg, although toxicity has been reported at doses as low as 0.1 mg/kg [24].

Chronic ingestion of fluoride above 2 mg/day can cause mottled teeth in children, doses over 8 mg/day can cause osteosclerosis, and doses of 20 mg/day for 10 to 20 years can cause hypermineralization of bone, leading to crippling skeletal fluorosis and renal toxicity [434,483]. Fluoride increases bone mass but decreases its tensile strength and is apparently not a treatment for osteoporosis [58]. Case reports indicate that administration of sodium fluoride for treatment of osteoporosis can exacerbate rheumatoid arthritis, possibly by stimulating leukocytes and other mediators of the acute inflammatory response [194]. Human epidemiological studies have found no evidence that fluoride causes gastrointestinal, respiratory, reproductive, or developmental toxicity [102]. Skeletal and dental changes can be seen in rodents exposed to fluoride, as well as chronic stomach inflammation and ulcers [102,434]. The USEPA has derived an oral RfD for fluorine based on a no-observed-effect level (NOEL) for objectionable mottling of the teeth (dental fluorosis), which may occur in children drinking water with more than 1 ppm of fluoride, leading to a NOEL of 0.06 mg/kg/day in a 20-kg child drinking 1 L/day and ingesting 0.1 mg/kg/day of dietary fluoride [674]. The endpoint of dental fluorosis is not considered toxic or adverse. The ACGIH has adopted a TLV–TWA value for fluorides of 2.5 mg F/m$^3$, with a carcinogenicity classification of A4 (not classifiable as a human carcinogen) and a TLV–TWA value for fluorine of 1 ppm and a STEL value of 2 ppm [35].

The carcinogenicity of fluoride is debatable. No increase in tumors was found among mice exposed to 1.75 mg/kg/day of sodium fluoride in water for 30 months [349]. Sprague–Dawley rats had no statistically significant increase in tumors following 2 years of exposure to doses up to 25 mg/kg/day [434]. When the National Toxicology Program (NTP) conducted a 2-year drinking water study of sodium fluoride at doses up to 10 mg/kg/day, they found no evidence for carcinogenicity in female rats, male mice, or female mice and equivocal evidence of carcinogenicity in male rats [102]. The evidence in male rats consisted of an increase in bone osteosarcomas with a dose–response trend that was statistically significant but an incidence in the highest dose group that was not significantly elevated compared to controls [102]. Also, no osteosarcomas were found in female rats even though they accumulated fluoride in bones to the same extent as the male rats and they exhibited fluoride-induced osteosclerosis [102]. The USEPA has not yet evaluated fluoride for potential human carcinogenicity [674]. Whereas human epidemiology studies have generally been negative, the question of whether fluoride is a potential human carcinogen is still open, and more studies are needed to resolve this question of some public health importance [58,102,434].

## Iodine

Iodine is the heaviest of the halogens that are of industrial interest. In solid form, iodine takes the form of gray-black plates or granules. It volatilizes at room temperature, yielding a violet vapor [531]. The major sources of iodine are oil and natural gas brines; Japan's natural-gas-well brines are credited with as much as four fifths of the world's iodine reserve [531]. Topical iodine solutions (2% iodine and 2% sodium iodide in 50% alcohol, USP) have been used for decades as germicides and antiseptics [34]. When inhaled, iodine vapor can be intensely irritating to mucous membranes and affects the upper and lower portions of the pulmonary tract [32]. Flury and Zernik [225] reported that humans could work undisturbed at 0.1 ppm, could work with difficulty at 0.2 ppm, and could not work at 0.3 ppm. Topical application of iodine solutions can cause irritation, and strong solutions can cause burns [34]. Iodide is required for the synthesis of the thyroid hormones thyroxine and triiodothyronine. Iodide is efficiently absorbed, and excess iodide is excreted in the urine [483]. Deficiency of iodide causes hypothyroidism and goiter, and severe deficiency in the newborn may cause cretinism and mental retardation [215]. The RDA for iodine is 150 μg/day for adults, with an extra 70 μg/day during pregnancy and 140 μg/day during lactation [323]. Chronic absorption of high levels of iodide can lead to a condition known as iodism. This condition is characterized by sleeplessness, tremor, rapid heart rate, diarrhea, weight loss, conjunctivitis, rhinitis, and bronchitis. This syndrome is usually associated with long-term ingestion of iodide containing medications [34]. The USEPA and ACGIH have not derived toxicity values for iodide [35,674].

## IRON

Iron is a silver-white solid metal found mainly in combination with other elements as oxides, carbonates, sulfides, and silicates [67]. It exists in two stable oxidation states, oxidized ferric ($Fe^{3+}$) and reduced ferrous ($Fe^{2+}$), which accounts for its essentiality as a trace element and its crucial role in the oxygen and electron transport reactions of all living cells. Dietary iron is available as either heme or nonheme [28]. Heme iron is found in meats and is relatively well absorbed compared with nonheme iron, which is also found in meats, grains, and vegetables. Intestinal absorption of iron depends on iron status; 10% of the total iron (heme plus nonheme) is absorbed when iron status is normal, but up to 20% in deficiency states. Adequate intakes of vitamin C increase the intestinal absorption of nonheme iron by two- to fourfold [162], which may be of significance to the iron status of vegetarians. Iron is lost through the shedding of cells, sweat, nails, hair, blood loss, menstruation, and in the urine. Early symptoms of iron deficiency are nonspecific and include fatigue and weakness. This progresses to iron deficiency anemia, which is characterized by small red blood cells with low hemoglobin content (microcytic hypochromic anemia). These symptoms resolve after administration of iron. The RDA for women 19 to 50 years of age is 18 mg/day, but this increases to 27 mg/day during pregnancy [323]. The RDA for men of all ages and postmenopausal women is 8 mg/day. The UL for males and females 14 years and older is 45 mg/day.

Free iron is an oxygen-reactive substance that is highly toxic to cells and will enhance the formation of free radicals and peroxidation of membrane lipids [52,53,570]. Oxidative damage associated with elevated brain iron has been suggested as a risk factor for early age at onset for neurodegenerative diseases such as Alzheimer's disease and Parkinson's disease [62]. Humans are unable to eliminate excess iron and regulate body iron stores by limiting absorption [437]. Divalent iron is taken up by intestinal mucosa and converted to the trivalent form. The trivalent form is bound to transferrin [67,282], a glycoprotein with two-iron binding sites [419]. Iron is transported to the liver or spleen as transferrin where it is stored as ferritin, which has a large iron storage capacity and prevents iron from participating in the Fenton reaction [63]. Of the typical 4-g body iron stores found in adults, 66% is bound as hemoglobin, 10% as the protein myoglobin, a minute amount in iron-containing enzymes, and the rest as intracellular storage proteins. The physiological controls on this essential but potentially toxic metal can be overwhelmed, either by an acute large intake (e.g., accidental ingestion of dietary supplements by children [121]) or by chronic excessive intake (endogenous Sub-Saharan African populations with a probable genetic defect who consume beverages brewed in steel drums may develop pancreatic, hepatic, or renal toxicity from their accumulation of excessive iron) [255]. Inhalation exposure to iron dust or fumes has resulted in pulmonary siderosis; fibrosis does not develop, and the clinical course is benign [496]. Hepatotoxicity is typically seen in patients with iron overload and can progress from portal fibrosis to cirrhosis [621]. A gray-bronze hyperpigmentation of the skin caused by increased melanin and iron deposition usually resolves after iron removal. Free radical stress and lipid peroxidation have both been suggested as factors in the etiology of diabetes [512], and increased iron stores have been reported to contribute to the development of non-insulin-dependent diabetes [574]. An increased risk of infection by a number of microorganisms, including *Vibrio vulnificus*, *Listeria monocytogenes*, *Yersinia enterocoloitica*, *Escherichia coli*, and *Candida* species, may result from excessive iron intake, due to direct effects on the immune system or enhanced bacterial growth due to the increased availability of iron [254,284]. The possibility that excess body iron might play a role in atherosclerosis (the "iron hypothesis") was first proposed by J.L. Sullivan in 1981 [628], and epidemiologic evidence suggests that excess dietary iron is a coronary risk factor [573,658,659]. Although regular blood donation in middle-aged males is associated with a reduced risk of myocardial infarction [575], further clinical studies are required to test the iron hypothesis and determine potential treatments [600,715,716].

Iron poisoning is the most common fatal poisoning in children reported to poison control centers in the United States [409, 698]. Despite supplements being packaged in child-resistant packages and carrying warning labels, the public perception of their potential danger is low [315], and fatalities in children have recently increased [37,73,698]. Iron poisoning is characterized by four distinct clinical stages, but individual patients do not always experience each stage [203,262,441]: Stage I (initial period) occurs 0.5 to 2 hours after ingestion and is characterized by the onset of acute gastrointestinal symptoms (vomiting and diarrhea), but central nervous system (CNS) symptoms (lethargy and coma) may be present in severe cases. Stage II (apparent recovery) occurs 6 to 24 hours after ingestion, and the patient may appear to have improved. Stage III (recurrent period) occurs 12 to 48 hours after ingestion, when the patient may experience shock, gastrointestinal perforation, hepatic and renal failure, and metabolic acidosis. Stage IV (late period develops 2 to 6 weeks after ingestion, with symptoms of gastrointestinal scarring and small bowel obstruction. Mistaking stage II iron poisoning for complete recovery is a serious medical error [455] that has resulted in early discharge of iron-poisoned patients [203].

Gastrointestinal symptoms typically occur following the ingestion of 20 mg elemental iron per kg body weight, and doses greater than 60 mg/kg are often lethal. Treatment involves stabilizing vital functions, removing unabsorbed iron from the gastrointestinal tract, and intravenous administration of deferoxamine if symptoms are severe [37,441,455]. Deferoxamine is an iron chelator produced by *Streptomyces pilosus* that removes iron from transferrin and ferritin but not hemoglobin [413,640]. Current research is directed toward the design of an orally active, nontoxic, selective iron chelator [295,371]. The USFDA has issued regulations requiring all iron-containing products to carry a label stating the dangers of iron overdosage and unit dose packaging for products containing 30 mg or more per dosage unit [676]. The USEPA has not derived any toxicity values for iron [674]; the ACGIH has adopted TLV–TWA values for iron of 5 mg/m$^3$ for iron oxide dust and fume and 1 mg/m$^3$ for soluble iron salts [35]. Although the carcinogenicity of iron is still under debate [311], epidemiological reports suggest an elevated risk of colorectal cancer with increased iron exposure. Suggested mechanisms for a carcinogenic role of iron include induction of oxidative stress, enhanced tumor cell growth, and alteration of the immune system [180].

### Magnesium

Magnesium is essential to a large number of biochemical and physiological processes, including neuromuscular conduction in skeletal and cardiac muscle [67]. It is also an important structural component of bone [483]. Plasma concentrations of magnesium are regulated within a narrow range (0.65 to 1.0 m*M*), primarily by adjustments in the reabsorption of filterable magnesium in the loop of Henle and also by the passive buffering by bone magnesium [483]. Magnesium deficiency can occur secondarily to general malnutrition, alcoholism [558], or other disease states that affect gastrointestinal electrolyte absorption or excretion or renal cation reabsorption. Magnesium deficiency results in reduced levels of potassium and calcium, as well as symptoms of nausea, muscle weakness, irritability, and mental derangement [402,607]. The RDAs for magnesium are 420 mg/day for men and 320 mg/day for women, with an extra 30 to 80 mg/day (depending on maternal age) during pregnancy [321]. Average intakes of U.S. adults are not much above the RDAs; however, there is no definitive evidence of effects attributable to magnesium deficiency [483]. Vitamin D facilitates the absorption of magnesium [458]. Oral exposure to magnesium is not toxic, except in individuals with impaired renal function, who may experience nausea, vomiting, and hypotension, followed by CNS depression accompanied by a sharp drop in blood pressure and respiratory paralysis [67,483]. Magnesium salts are poorly absorbed orally and are commonly used as antacids or cathartics. Inhalation exposure to magnesium oxide can cause metal fume fever [67,496]. The USEPA has not derived toxicity values for magnesium; the ACGIH has adopted a TLV–TWA value for magnesium oxide fumes of 10 mg/m$^3$ (nuisance dust) [35].

### Manganese

Manganese is a silver-gray soft metal that occurs in ores mainly as oxides [67]. Manganese and its compounds are used in numerous products and applications including iron and steel alloys, dry-cell batteries, paints, inks, fertilizers, and fungicides [244]. Manganese is an essential trace metal that is a component of several mitochondrial enzymes, pyruvate carboxylase, and superoxide dismutase and activates a wide variety of enzymes (decarboxylases, transferases, hydrolases). It occurs in meats, poultry, nuts, grains, green leafy vegetables, and tea. Although outright manganese deficiency has not been observed in the human population, suboptimal manganese intake may be a concern [674]. In animals, manganese deficiency can cause impaired growth, skeletal abnormalities, and altered metabolism of carbohydrates and lipids [22]. The AI for manganese is set at 2.3 mg/day and 1.8 mg/day for adult men and women [323]. The USEPA has reviewed numerous human and animal studies and related information and concluded that an appropriate chronic oral reference dose for manganese is 10 mg/day (0.14 mg/kg/day) [674]. Only between 3 and 10% of dietary manganese is absorbed in normal adults, and total body stores are controlled by a complex homeostatic mechanism regulating absorption and excretion. Calcium, iron deficiencies, age, and other factors may increase manganese absorption [22,674].

Occupational inhalation exposure is the primary route for manganese toxicity. The primary toxic effect of occupational inhalation exposure is neurological damage [561]; however, inhalation exposure to manganese can also affect the lung directly, causing metal fume fever, pneumonitis, chronic obstructive lung disease, and pneumonia [22, 496]. Occupational exposure to manganese at levels of about 1 mg/m$^3$ may decrease male fertility [238,390]. The neurological effects of inhalation of manganese dusts, termed manganism, typically begin with weakness and lethargy and may progress to disturbances in speech and gait, a mask-like face, tremor, and possibly hallucinations and psychosis [561]. Symptoms may resemble Parkinson's disease, but there is only minimal response to L-dopa therapy. The pathobiochemical aspects of manganism involve the striatum and globus pallidus. Cell damage may be due to the autooxidation of dopamine with the formation of free radicals [682]. Manganese applied to the nasal cavity in rats is taken up in the olfactory receptor cells and transported along the primary neurons to the olfactory bulbs, with subsequent migration into most parts of the brain. This route circumvents the blood–brain barrier [190,291]. More subtle nonclinical neurological damage can be identified by neurobehavioral tests (e.g., reaction time, finger tapping, hand steadiness) in men chronically exposed to levels as low as 0.14 mg/m$^3$ for 1-35 years [335]. Impairment of speed and coordination of motor function are noted.

The USEPA has derived a recent inhalation RfC of 0.05 μg/m$^3$ based on studies of Roels et al. [561,674]. The previous RfC was 0.4 μg/m$^3$. An occupational exposure guideline for Mn using the benchmark method for subclinical effects was determined to range from 0.1 to 0.3 mg/m$^3$ (8-hr TWA) for the *respirable* particulate fraction only [151]. The ACGIH has also lowered the TLV–TWA value for Mn to 0.2 mg/m$^3$ for elemental and inorganic compounds [35]. Manganese cyclopentadienyl tricarbonyl (MMT) is a gasoline octane enhancer in use since 1970. The major combustion products of MMT are manganese particulates of manganese phosphate with some sulfates and a small amount of oxides. The TLV–TWA for MMT is 0.1 mg Mn/m$^3$, with a notation noting the potential for dermal absorption [35]. Little evidence exists to suggest that manganese has carcinogenic potential [240]. A 2-year bioassay of manganese sulfate monohydrate in the diet found no evidence of carcinogenicity to rats and equivocal evidence of carcinogenicity to mice [22]. The USEPA has classified manganese as a Group D carcinogen (not classifiable as to human carcinogenicity) based on inadequate evidence in humans and animals [674].

## Molybdenum

Molybdenum is a silver-white metal of the second transition series. The primary molybdenum-containing ore is molybdenite ($MoS_2$), with minor ores being powellite ($CaMoO_4$) and wulfenite ($PbMoO_4$). Metallic molybdenum is used in a number of important applications, such as in high-temperature and tool steel alloys and in missile and aircraft parts. Molybdenum disulfide is used as a dry lubricant or as a component in lubricants [34]. Molybdenum is a constituent of several enzymes, including aldehyde oxidase, xanthine oxidase, and sulfide oxidase [483]. Deficiency is extremely rare; one patient on total parenteral nutrition had disturbed sulfur and uric acid metabolism that resolved after molybdenum supplementation [483]. The RDA for molybdenum is set at 45 μg/day for both adult men and women [323]. High levels of molybdenum in herbage eaten by cattle caused diarrhea in cattle [217], which could be alleviated by the administration of copper salts [494]. Further study has shown an inverse relationship between molybdenum and copper. When molybdenum intake in cattle in increased, the concentration of utilizable copper in the liver decreases [67].

The acute oral toxicity of molybdenum compounds is related to their solubility. Molybdenum trioxide, calcium molybdate, and ammonium molybdate caused fatalities in rats when administered at doses from 1.2 to 6.0 g Mo per kg; conversely, administration of insoluble molybdenum disulfide to rats at concentrations as great as 6.0 g Mo per kg did not cause any fatalities [214]. The USEPA has derived an oral RfD for molybdenum of 5 μg/kg/day (350 μg/day for a 70-kg adult) based on an increase in urinary uric acid levels in humans exposed to 10 mg Mo per day in the diet with an uncertainty factor of 30 [674]. Rodent bioassays of molybdenum trioxide indicate that this compound is carcinogenic in rats and mice, causing an increased incidence of alveolar/bronchiolar adenoma or carcinoma (combined). Male rats and mice appear to be more sensitive to the carcinogenic effects of molybdenum trioxide [137].

Dental technicians exposed to the dust of vitallium alloy, which contains chromium, cobalt, and molybdenum, can develop pneumoconiosis that is clearly different from hard metal lung disease associated with cobalt exposure [496], and some data suggest that molybdenum inhalation can cause pneumoconoiosis [34]. The ACGIH has adopted TLV–TWA values for Mo of 10 mg/m$^3$ (inhalable) and 3 mg/m$^3$ (respirable) for insoluble compounds and 0.5 mg/m$^3$ (respirable) for soluble compounds [35].

## Phosphorus

Phosphorus is an essential component of bone and also participates in many important biochemical reactions [483]. Approximately 85% of the body store of phosphorus is in bone, with the rest as soluble phosphate ion and a component of a variety of biomolecules. Absorption of phosphate ranges from 50 to 70% when intake is adequate to 90% when intake is low [483]. Dietary phosphorus deficiency is rare but can occur following prolonged use of the antacid aluminum hydroxide, which binds phosphorus into an unavailable form. Symptoms of phosphorus deficiency include bone loss, weakness, anorexia, and pain. The RDA for phosphorus is 1250 mg/day for ages 9 to 18 and 700 mg/day for older age groups [321].

High-level phosphate intake in the forms of phosphate-fortified infant formulas, phosphoric acid in carbonated beverages, or purified amino acids may cause calcium loss, which can be adverse in situations of inadequate calcium intake; however, phosphorus in the form of complex proteins does not seem to have this effect [483,618]. Certain reactive forms of phosphate may be chemically irritating, but neutral phosphate solutions are essentially nontoxic [356]. Phosphorous as the free element does not occur in nature. It exists either as relatively nontoxic red phosphorous or toxic yellow (or white) phosphorous [178]. Toxic exposure to yellow phosphorous can occur through the oral, dermal, or respiratory routes. Rodenticides and insecticides containing yellow phosphorous have accounted for poisonings characterized initially by gastrointestinal burning and severe abdominal pain, vomiting, and diarrhea. Acute cardiovascular collapse may occur [317,535]. If the victim survives, a second stage of symptoms may occur up to several weeks later, resulting in systemic toxic effects on the liver, heart, kidneys, or central nervous system. Phosphorous can cause necrotic skin burns. The fumes are irritating to the respiratory tract, eyes, and skin. Phosphorous is converted to phosphates and excreted in the urine. The USEPA has derived a chronic oral RfD for elemental phosphorus based on studies of Condray [161], which found increased mortality in pregnant rats near the end of gestation at a dose of 0.075 mg/kg/day for 80 days prior to mating and during gestation. A NOAEL of 0.015 mg/kg/day was converted to an RfD of 0.02 μg/kg/day (1.4 μg/day for a 70-kg adult) with low confidence, using an uncertainty factor of 1000 (a factor of 10 for interspecies variation, 10 for intraspecies variation, and 10 for incomplete reproductive/developmental data and a less-than-adequate lifetime study), and a modifying factor of 1 [674]. The ACGIH has adopted TLV–TWA values for phosphorus of 0.2 ppm for phosphorus trichloride, 1 mg/m$^3$ for phosphorus pentasulfide, 0.1 ppm for phosphorus pentachloride, 0.1 ppm for phosphorus oxychloride, and 0.1 mg/m$^3$ for yellow phosphorus [35]. No data exist to suggest that phosphorus may have a carcinogenic potential.

## Potassium

Elemental potassium is a highly reactive soft metal with a silver-colored appearance and is not found in nature. Potassium compounds are common. Elemental potassium is even more reactive than sodium and must be stored under airtight anhydrous conditions, such as under xylene. Oxidation may form highly reactive superoxides on the surface of the metal which can detonate the bulk, causing spattering and skin and eye penetration [377]. Autoignition can occur at room temperature. Dermal and ocular thermal burns and liquefaction necrosis due to the formation of potassium hydroxide are the primary effects following exposure. Imbedded particles require surgical debridement. Water irrigation is contraindicated. Potassium is the principal cation of intracellular fluid, accumulating to a concentration about 30 times higher than in plasma. Potassium in plasma is involved in nerve transmission, muscle contraction, and blood pressure homeostasis. The gastrointestinal absorption of potassium is nearly complete; plasma concentrations are kept within a narrow range by regulation of urinary excretion and by depletion of body stores in cases of low potassium intake [483]. Dietary potassium deficiency is rare, but prolonged vomiting, diarrhea, or diuretic use may deplete potassium enough to cause weakness, anorexia, nausea, drowsiness, irrational behavior, and, in severe cases, potentially fatal cardiac arrhythmias [483]. Potassium appears to moderate the effect of increased sodium intake on elevating blood pressure, probably by affecting renal sodium excretion [472]. The AI for potassium is 4.7 g/day for both adult men and women [325]. Dietary potassium is not toxic if sufficient water is ingested and renal function is adequate to maintain homeostasis; symptoms of hyperkalemia from dehydration or acute renal failure are similar to those of hypokalemia, including muscle weakness, fatigue, and paralysis [80]. The USEPA has not derived any toxicity values for potassium.

## Selenium

Selenium is widely distributed in nature and found in combination with sulfides and other minerals [20,68]. It has semiconducting properties and is used in photocopying machines, light meters, and rectifiers. Selenium is used in agriculture as a component of fertilizers, pesticides, and animal feeds, and selenium sulfide is an active ingredient in antidandruff shampoo [20,34,227]. Although selenium has long been known to protect vitamin-E-deficient rats from liver necrosis [592], a specific biochemical role had not been elucidated until Rotruck et al. [566] demonstrated it to be an essential constituent of glutathione peroxidase. This enzyme protects polyunsaturated fatty acids in the cell membrane from oxidative damage caused by free radicals. Its identification in human erythrocytes established selenium as an essential trace element in human nutrition [49]. Selenium deficiency has been identified as the major causal factor in the potentially fatal cardiomyopathy (Keshan disease) affecting young children and women of child-bearing age in Keshan County of the People's Republic of China [355]. A diet based primarily on local produce grown in the selenium-poor soil resulted in a selenium deficiency, alleviated by supplementing the diet with selenium-fortified table salt [145]. Additional evidence for its essentiality in humans is provided by the observed cardiomyopathy seen in patients maintained on long-term total parenteral nutrition [344,410,680]. Selenium also plays an important role in the control of thyroid hormone [76], which is essential for normal growth, development, and metabolism. The selenoenzymes iodothyronine deiodinases are responsible for the activation of thyroxine (T4) to triiodothyronine (T3), and a selenium deficiency may cause reduced growth rates. The RDA for selenium was derived from the intake associated with a plateauing of plasma glutathione peroxidase activity in Chinese adult males (40 μg/day) [711], adjusted for differences in body weight between the reference Chinese and North American male, with an additional safety factor of 1.3 to account for individual variation [399,483]. The RDA for selenium is 55 μg/day for males and females, with additional recommendations of 5 μg/day for pregnancy and 15 μg/day for lactation [322]. The UL is 400 μg/day. Selenium is readily absorbed from the gastrointestinal tract, and the average U.S. diet typically provides 60 to 150 μg/day [620], which should be adequate to prevent cardiomyopathy in the general population.

Selenium toxicity has long been observed in cattle grazing on milk-vetch (legumes of *Astragalus* species) grown in the seleniferous soils of Wyoming and South Dakota [281,351,430,470]. Acute intoxication in livestock is known as "blind staggers" and is characterized by signs of CNS impairment (ataxia, impaired vision, disorientation) and respiratory distress. Chronic exposure to moderately toxic selenium levels is known as "alkali disease" and results in skin lesions with alopecia, hoof necrosis and loss, growth retardation, anemia, and cardiac atrophy. In humans, chronic sublethal selenium toxicity has been observed in individuals living in seleniferous areas and is characterized by hair or nail loss, thickened or brittle nails, garlicky breath, tooth decay [272,273], skin lesions, gastrointestinal disorders, and CNS abnormalities, including peripheral anesthesia, acroparesthesia, and pain in the extremities [584,617,711]. It has also been reported following the ingestion of superpotent selenium dietary supplements, and consumers need to be aware of its potential for toxicity [289,341]. The deterioration of keratinized tissue is thought to result from the replacement of sulfur with selenium in sulfur-containing amino acids. Acute selenium intoxication resulting from ingestion is rare in humans [236] but has been reported following suicidal, accidental, and homicidal exposure [112,374,420,569]. Symptoms include gastrointestinal disturbances due to the irritative properties of selenium, a characteristic garlicky breath from the exhalation of dimethyl selenide [440], formication of the nose, signs of rhinitis, neurological symptoms ranging from mild tremors to myoclonic jerks, and cardiovascular shock. Acute inhalation of hydrogen selenide has been reported to cause severe dyspnea with abnormal pulmonary function tests [581], and chronic inhalation of the gas leads to garlicky breath, gastrointestinal disturbances, dental caries, nail deformities, and conjunctivitis [25]. Chronic overexposure to selenium has been associated with the motor neuron disease amyotrophic lateral sclerosis [380,685]. Although selenium is known to be an avian teratogen [230,302,470], there is inconclusive evidence linking it to mammalian teratogenesis [339,602]; Yang et al. [710,711] did not observe teratogenesis in babies during epidemiological studies in seleniferous regions where malformed chicks hatched from local eggs.

The USEPA has established a chronic oral RfD for selenium using the study of Yang et al. [710] and corroborated by Longnecker et al. [412]. The NOAEL of 0.85 mg/day for selenium was converted to a dose of 0.015 mg/kg/day (based on an average adult body weight of 55 kg), and an RfD of 5 μg/kg/day was derived using an uncertainty factor of 3 (less than a full factor of 10 was used to account for sensitive individuals because of the availability of epidemiological data from two independent studies of moderate size) [674]. Confidence in this RfD is considered high [674]. The USEPA has not derived RfCs for selenium. The ACGIH has adopted a TLV–TWA value for selenium of 0.2 mg/m$^3$ [35].

Various animal models report a protective effect of pharmacologic levels of selenium against chemical carcinogenesis [150,334,549,642]. In 1969, Shamberger and Frost [601] reported an inverse relationship between cancer mortality rates in the U.S. and plant selenium levels as mapped by Kubota et al. [379]. Subsequent epidemiological studies have reported promising but inconclusive findings [160,314,370]. A nutritional prevention of cancer trial reported that 200 μg selenium daily did not protect against the development of recurrent nonmelanoma skin cancers but was inversely associated with mortality from total, prostate, and colorectal cancers [148,149]. Recent analyses of the complete results of the trial (13 years) suggest that the effect of selenium supplementation was strongest for an effect on prostate cancer [159]. Another study reports an inverse relationship between advanced prostate cancer and toenail selenium concentration [714], an indicator of past selenium intake. In the Western world, prostate cancer is the second leading cause of male cancer deaths [697], making it an important public health issue. The National Cancer Institute has sponsored a randomized, prospective, double-blind study designed to determine if selenium and/or vitamin E will decrease the risk of prostate cancer in healthy men [263,450,481]. This study, known as the Selenium and Vitamin E Cancer Prevention Trial (SELECT), is the largest-ever prostate cancer prevention trial (enrollment closed on June 24, 2004, with 35,534 participants) and is designed to confirm earlier findings suggesting that selenium and vitamin E may reduce the risk of developing prostate cancer. The study contains four arms: (1) placebo + placebo, (2) selenium + placebo, (3) vitamin E + placebo, and (4) selenium + vitamin E. This phase III trial will conclude in 2013.

Selenium sulfide has been shown to be a rodent carcinogen by the oral [479] but not dermal [478,480] route. A 2-year gavage bioassay of selenium sulfide by the National Toxicology Program produced evidence of carcinogenicity in male rats (liver), female rats (liver), and female mice (liver and lung), but not male mice [479].

The USEPA classifies selenium sulfide as a Group B2 carcinogen (probable human carcinogen), based on inadequate data from human studies and sufficient evidence from rodent studies; no quantitative risk assessment was performed [674]. Other selenium compounds are classified as Group D carcinogens (not classifiable as to carcinogenicity in humans) based on inadequate evidence in both humans and animals [674]. The suggested beneficial antioxidant effects of selenium and the potential widespread use of selenium supplements makes it important to gain a fuller understanding of selenium toxicology.

### Sodium

Sodium is a highly reactive soft metal with a silver appearance that is not found in the elemental form in nature [377]. Sodium compounds are ubiquitous in nature. Elemental sodium must be stored under airtight anhydrous conditions, such as under oil, to prevent oxidation, which can result in autoignition at room temperature. Superoxides may form, resulting in a violent explosion. Dermal and ocular thermal burns and liquefaction necrosis due to the formation of sodium hydroxide are the primary effects following sodium exposure. Explosion may cause particles to imbed in the skin and eye, requiring surgical debridement. Water irrigation is contraindicated. Sodium is the principal cation of extracellular fluid and the primary regulator of extracellular fluid volume. Sodium also regulates osmolarity, acid–base balance, and membrane potential and participates in active transport across cell membranes. Renal excretion of sodium maintains homeostasis over a wide range of intakes and losses, via aldosterone control of tubular excretion. Sodium deficiency is very uncommon but may occur after heavy and prolonged sweating, chronic diarrhea, or renal disease and constitutes a medical emergency. Dietary sodium is not toxic if sufficient water is ingested and renal function is adequate to maintain homeostasis [483]. Lifelong excess intake of sodium may predispose sensitive individuals to hypertension, and individuals diagnosed with high blood pressure are commonly advised to limit sodium intake to 1 to 2 g/day or less [472]. The AIs are 1.5 g/day for men and women ≤50 years of age, 1.3 g/day for ages 51 to 70, and 1.2 g/day for ages >70 [325]. At present, the public health benefit of restricting sodium intake in the general population is not firmly established [483]. The USEPA has not derived any toxicity values for sodium.

### Zinc

Zinc is a bluish-white, soft metal extracted from ore; it is used in alloys, for galvanizing iron to prevent corrosion and oxidation, and in numerous compounds, including those for use in cosmetics, pharmaceuticals, and dry-cell batteries [221]. At temperatures approaching its boiling point, zinc volatilizes and oxidizes to the white fume of zinc oxide [34]. Zinc is an essential trace element and is a required component of many enzymes [483]. Zinc is stored in bone and muscle but is not readily released from these stores during deficiency. Gastrointestinal absorption of zinc is higher when body stores are lower and is also higher from more refined diets. Zinc deficiency causes loss of appetite, growth retardation, and slow wound healing; no single enzyme function has been identified as being associated with these signs of zinc deficiency. Severe zinc deficiency causes hypogonadism and dwarfism, which are alleviated with zinc supplementation. The RDAs for zinc are 11 mg/day for adult men and 8 mg/day for adult women, with an additional 3 mg/day during pregnancy and 4 mg/day during lactation [323]. Inhalation exposure to zinc oxide fume can cause metal fume fever [34]. Zinc chloride fume is a corrosive material that has caused chemical pneumonitis, alveolar and bronchial obliteration, and death [21]. Zinc compounds are absorbed orally and excreted primarily in the feces. Zinc has low human toxicity by the oral route, but high levels can cause gastrointestinal distress [21]. Long-term oral intakes of zinc at levels of 18.5 to 25 mg/day can interfere with copper absorption, and intakes 10 to 30 times the RDA can impair immune responses and decrease serum high-density lipoprotein [483]. The USEPA has derived an oral RfD for zinc of 0.3 mg/kg/day [674]. No inhalation RfC has been derived for zinc. The USEPA has classified zinc as a Group D carcinogen (not classifiable as to human carcinogenicity) based on inadequate evidence in humans and animals. The ACGIH has adopted TLV–TWA values for zinc of 10 mg/m$^3$ for zinc oxide dust, 5 mg/m$^3$ for zinc oxide fume, and 1 mg/m$^3$ for zinc chloride fume [35].

A summary of quantitative values for essential elements is given in Table 17.1. Elements are listed in the order of highest to lowest dietary requirement.

## MAJOR TOXIC METALS

### Arsenic

Arsenic is a Group VA element of the periodic table, the 52nd most abundant element in the Earth's crust. Arsenic is refined from the minerals arsenopyrite and loellingite, or it can be prepared from the reduction of arsenic trioxide. The main use of arsenic in the United States is in the production of herbicides and other agricultural chemicals. Arsenic is also used in the semiconductor industry [35]. Although arsenic can exist in several valence states, the +3 and +5 states are the most prevalent, with arsenite (+3) being more toxic than arsenate (+5) [203]. Dietary consumption of arsenic is generally low. The typical daily American intake is 145 µg/day from both food and water [64]; however, consumption of seafood and grain can increase the amount of arsenic ingested [326,582,655].

**TABLE 17.1**
**Essential Elements**

| Element | Recommended Intake (mg/kg/day)[a] | Chronic Oral Toxicity RfD[b] (mg/kg/day) | Chronic Oral Toxicity Confidence | Chronic Inhalation Toxicity RfC[c] (μg/m³) | Chronic Inhalation Toxicity Confidence | Carcinogenicity Inhalation Slope Factor (risk (μg/m³)⁻¹) | Carcinogenicity Classification[d] |
|---|---|---|---|---|---|---|---|
| Potassium | 67 | — | — | — | — | — | |
| Chlorine | 33 | — | — | — | — | — | |
| Sodium | 21 | — | — | — | — | — | |
| Calcium | 14 | — | — | — | — | — | |
| Phosphorus | 10 | — | — | — | — | — | |
| Magnesium | 6 | — | — | — | — | — | |
| Iron | $1.1 \times 10^{-1}$ | — | — | — | — | — | |
| Copper | $1.3 \times 10^{-2}$ | — | — | — | — | — | D |
| Zinc | $1.6 \times 10^{-2}$ | $3 \times 10^{-1}$ | Medium | — | — | — | D |
| Manganese | $3.3 \times 10^{-2}$ | $1.4 \times 10^{-1}$ | Medium | $5 \times 10^{-2}$ | Medium | — | D |
| Fluorine | $5.7 \times 10^{-2}$ | $6 \times 10^{-2}$ (cosmetic) | High | — | — | — | — |
| | | $12 \times 10^{-2}$ (adverse) | High | — | — | — | — |
| Iodine | $2 \times 10^{-3}$ | — | — | — | — | — | — |
| Chromium | $5 \times 10^{-4}$ | 1.5 ($Cr^{3+}$) | Low | — | — | — | D ($Cr^{3+}$) |
| | — | $3 \times 10^{-3}$ ($Cr^{6+}$) | Low | $8 \times 10^{-3}$[e] | Low | $1.2 \times 10^{-2}$ ($Cr^{6+}$) | A ($Cr^{6+}$) |
| | | | | $1 \times 10^{-1}$[f] | Medium | — | — |
| Molybdenum | $6.4 \times 10^{-4}$ | $5 \times 10^{-3}$ | Medium | — | — | — | — |
| Selenium | $7.9 \times 10^{-4}$ | $5 \times 10^{-3}$ | High | — | — | — | B2 ($SeS_2$) D (all other) |

[a] For a 70-kg adult (30- to 50-year-old) male; see text for actual values.
[b] RfD, reference dose.
[c] RfC, reference concentration.
[d] Group A. known human carcinogen; Group B2, probable human carcinogen; Group D, not classifiable as to human carcinogenicity.
[e] Chromic acid mists and soluble Cr(VI) aerosols
[f] Chromium (VI) particulates

Arsenic is readily absorbed via the gut [67], and excretion occurs primarily in the urine [103,165,425,634]. Two processes are involved in the metabolism of arsenate and arsenite: (1) the interconversion of arsenate and arsenite, and (2) the conversion of these moieties to monomethyl arsonic acid and dimethyl arsinic acid. Because the methylated forms of arsenic are less toxic and because methylation results in lower tissue retention of inorganic arsenic, the methylation process is viewed as a detoxification mechanism [15].

Arsenic is believed to exert its toxic effects through at least two mechanisms, depending on its valence state. Arsenate inhibits ATP synthesis by uncoupling oxidative phosphorylation, whereas arsenite reacts with thiol groups on the active sites of many enzymes and tissue proteins, such as keratin (i.e., skin, nails and hair) [635]. Because of this reactivity with thiol groups, arsenic concentrates in the skin, hair, and nails. Mee's lines (horizontal white lines on the fingernails) appear in exposed individuals after the exposed nail bed grows to the exterior [446]. At one time, inorganic arsenic was widely used as a "criminal poison" because it was odorless and nearly tasteless. The lethal dose of arsenic trioxide can be as low as 0.2 g [256]. Acute toxicity is characterized by severe gastrointestinal symptoms, which occur from 30 minutes to several hours after ingestion. Eventually, severe gastrointestinal hemorrhaging occurs, leading to profound losses of fluid and electrolytes and resulting in collapse, shock, and death [256]. If the victim survives the initial toxic sequelae, jaundice, renal failure, and peripheral neuropathology can develop [203,256].

In cases of acute intoxication, chelation therapy can be very effective in reducing or preventing symptoms. The agent of choice is British anti-Lewisite (BAL), which is dimercaptopropanol [706]. D-Penicillamine is also effective as a chelating agent, but nephrotoxicity and optic neuritis can result from long-term use [203]; therefore, BAL remains the treatment of choice in arsenic poisoning

[256]. Arsenic is teratogenic when administered intraperitoneally in several species, but no reproductive or developmental adverse effects were noted when arsenic was administered orally in nonmaternally toxic doses [495]. Chronic ingestion of arsenic can be difficult to diagnose. Diarrhea and abdominal pain can occur, as well as hyperpigmentation, hyperketatosis, and numerous other skin- and hair-related disorders [51,111,203,306,638]. Peripheral vascular occlusive disease has also been linked to chronic exposure to high levels of arsenic in drinking water in Chile (Raynaud's phenomenon) [89] and in Taiwan (Blackfoot disease) [651]. Neurological changes have been associated with occupational inhalation exposure to inorganic arsenic by smelter workers [87,216,294]. Neurologic changes included peripheral neuropathy of sensory and motor neurons, as measured by motor and sensory deficits [468], and encephalopathy, as evidenced by hallucinations and other psychological disturbances [65].

Chronic exposure to arsenic in drinking water is associated with an increased incidence of cancer. Numerous studies have been conducted in Taiwan comparing residents in the Blackfoot disease endemic area with residents in areas with low levels of arsenic in drinking water. These studies have consistently shown an increase in the incidence of skin cancer and several internal cancers in areas with high arsenic consumption [141–144,651,652,707]. A similar study was conducted in Japan, which showed an association between high levels of ingested arsenic and lung and urinary tract cancer [654].

Chronic exposure to arsenic via the inhalation route is also associated with the development of tumors. Studies of smelter worker populations have shown strong associations between exposure and an increased incidence of lung cancer [50,206,394,553,644], as have studies of pesticide manufacturing workers [418,521), and case reports of lung cancer in arsenical pesticide applicators [565]. Whereas the carcinogenicity of arsenic is well established in humans, carcinogenicity in animal models has been more difficult to establish. Of the many animal carcinogenicity studies reviewed by the IARC in 1980 [655], only two gave positive results: one with subcutaneous/intravenous administration of sodium arsenite in mice in a multigenerational study [519] and one with intratracheal installation of copper and calcium arsenate in rats [337]. Later studies showed that both calcium arsenate and arsenic trioxide are carcinogenic when administered intratracheally to Syrian golden hamsters [336,532,533]. The mechanism of carcinogenicity is postulated to be related to the multi-step metabolism of pentavalent arsenic to dimethyl arsinic acid, during which free radicals are produced [67,635].

Arsine ($AsH_3$) is a gaseous form of arsenic that is formed whenever arsenic is in the presence of hydrogen [256] and as such can be generated in metal tanks storing acids that contain arsenic impurities [67]. The toxicity profile of arsine is different from all other arsenic compounds. The hallmark of arsine toxicity is hemolysis, sometimes followed by acute renal failure [256]. BAL and D-penicillamine are not effective treatments for arsine poisoning [256].

The USEPA has established an oral RfD of 0.3 μg/kg/day for inorganic arsenic compounds and considers it a Group A (human) carcinogen. The inhalation RfC for arsine is 0.05 μg/m$^3$ [674]. The ACGIH has adopted the following TLVs for arsenic compounds: arsenic, elemental arsenic, and inorganic compounds, as As, 0.01 mg/m$^3$ (A1 carcinogenicity notation; confirmed human carcinogen); arsine, 0.05 ppm [35].

## Cadmium

Cadmium is a soft, silver-white transition metal, often found in association with zinc ores and obtained primarily as a byproduct of zinc preparation [67]. It is used primarily in the production of nickel–cadmium batteries but also for pigments in plastics, ceramics, and glasses, as stabilizers for polyvinyl chloride, for coatings on steel and some nonferrous metals, and as a component of specialized alloys [16]. The toxicity of cadmium has been widely investigated, and cadmium has been shown to affect nearly every organ system if the dose is high enough [77]. Acute effects of cadmium depend on the route of exposure. Symptoms of acute inhalation exposure to cadmium develop 4 to 10 hours after exposure and initially simulate metal fume fever (fever, nausea, vomiting, headache, cough, dyspnea, nasopharyngeal irritation) but can progress to chemical pneumonitis and a potentially fatal pulmonary edema [77,82,85,195]. A fatal dose can be inhaled by exposed individuals who are unaware of either the presence of cadmium or its inhalation hazard [77,414]. Cadmium absorption following inhalation exposure is dependent on particle size and solubility [232] but has been reported to be as high as 90% [688]. Fatal doses have been estimated at 50 mg/m$^3$ for 1 hour [60,106] and 9 mg/m$^3$ for 5 hours [77]. Recovery following acute high-level exposure or chronic exposure at lower levels may be accompanied by pulmonary fibrosis [59,168,646]. Oral exposure to cadmium is rarely fatal because the associated gastrointestinal irritation leads to vomiting, eliminating most of the dose before absorption [55,105,415,509,608]. Gastrointestinal absorption is about 5% [333]; however, cadmium and iron are both absorbed through a common pathway involving the divalent metal transporter 1 (DMT1) [633]. Iron deficiency results in increased expression of intestinal DMT1 and is therefore associated with increased cadmium absorption [223,579].

Chronic inhalation or ingestion of cadmium results in kidney damage, characterized by tubular or glomerular dysfunction with proteinuria, low concentration capacity and decreased inulin clearance [232]. Increased urinary excretion of $\beta_2$-microglobulin, a low-molecular-weight protein normally reabsorbed in the proximal tubule, is an early indicator of renal dysfunction and should be regarded as an adverse effect because it is predictive of an increase in the age-related decline in the glomerular filtration rate [562]. Absorbed cadmium is first transported to the liver, where it stimulates the synthesis of metallothionein and is sequestered as cadmium-metallothionein. Small amounts of liver cadmium-metallothionein are released into the plasma following normal cell turnover, filtered with the primary urine, reabsorbed into the proximal tubular cells where lysosomes degrade the metallothionein portion, with the release of cadmium which then induces renal metallothionein synthesis. Renal damage results when the kidneys can no longer produce sufficient metallothionein to sequester the cadmium ion and prevent its interaction with critical macromolecules [258]. Free cadmium may inactivate metalloenzymes, activate calmodulin, and damage cell membranes through activation of oxygen [690]. Excess inhalation or ingestion exposure to cadmium leads to abnormalities of calcium metabolism, and susceptible individuals may develop a painful bone disease as first discovered in a cadmium-contaminated area in Japan (Toyama Perfecture) and termed *itai-itai* ("ouch-ouch") disease [474,653]. The disease is characterized by osteomalacia and osteoporosis with an increased tendency to spontaneous fracture and is associated with bone pain and renal tubular dysfunction. Cadmium has been shown to increase bone resorption and inhibit bone formation in both *in vivo* and *in vitro* systems [79]. It is hypothesized that cadmium causes prolonged urinary calcium loss leading to skeletal demineralization and an increased risk of fractures [579]. Although intravenous cadmium administration produces a hypertensive response in rats [546], no difference in blood pressure was found between high- and low-exposure workers after adjusting for age, weight, and cigarette smoking [16]. Cadmium-exposed populations are not reported to have elevated death rates associated with cardiovascular disease. Maternal and fetal toxicity of cadmium is well documented in rodents [16,523]. Elevated levels of cadmium in neonates is associated with a decreased birth weight [312], but further research is required to determine if developmental effects of cadmium are of concern at environmental levels. Tobacco plants (*Nicotiana* sp.) are known to concentrate cadmium independent of the soil content [579], and it is estimated one-pack/day smokers can absorb 1 to 3 g cadmium each day [400]. Cadmium oxide is reported to be highly bioavailable, and smokers are known to have higher levels of cadmium in their blood (4 to 5 times higher) and kidneys (2 to 3 times higher) than nonsmokers [579]. In 1919, Alsberg and Schwartze [30] reported testicular necrosis induced by cadmium. Cadmium-induced testicular damage leads to infertility in experimental animals and may damage the reproductive ability of cadmium-exposed workers [330]. Although the mechanism of cadmium-induced testicular toxicity is poorly understood, cadmium treatment is known to increase vascular permeability in rat testis [72], and one theory is that the damage is the result of testicular blood vessel toxicity. Other evidence suggests that cadmium-induced testicular toxicity is associated with oxidative damage through the production of reactive oxygen species [625]. In support of this theory, it has been reported that zinc-deficient rats are more susceptible to cadmium-induced testicular damage [520]; other researchers report that ascorbic acid or alpha-tocopherol supplementation protects rats from cadmium-induced testicular damage [270,271].

The USEPA has derived a chronic oral RfD for cadmium based on the highest level of human renal cadmium not associated with significant proteinuria [674]. Separate values were derived for food and water exposure, assuming 2.5% absorption of cadmium from food and 5% from water with a 0.01%-per-day excretion. A kidney concentration of 200 μg Cd per g wet human renal cortex is considered the NOAEL [674], and an uncertainty factor of 10 was used for intrahuman variability. The resulting RfD values for Cd are 0.001 mg/kg/day (food) and 0.0005 mg/kg/day (water); confidence in these values is considered high. No reference concentration values for chronic cadmium inhalation exposure were calculated. The Agency for Toxic Substances and Disease Registry has calculated chronic minimal risk levels (MRLs) for cadmium based on human studies with measured exposures [16]. The inhalation MRL was calculated from a NOAEL for renal effects in workers exposed to 0.0016 mg/m$^3$ [340]; adjusting for continuous lifetime exposure and using an uncertainty factor of 10 to account for sensitive members of the population, the chronic inhalation MRL is 0.0002 mg/m$^3$. A chronic oral MRL of 0.0007 mg/kg/day was calculated from a study in a Japanese population exposed to cadmium in rice [507]. The average nonsmoking American absorbs approximately 1 to 3 μg Cd per day from the diet [16], which is a level only two to four times lower than the oral MRL, indicating that there is not a large margin of safety with respect to cadmium toxicity, particularly given evidence that postmenopausal women and diabetics may be more sensitive to cadmium toxicity than members of the general population [104].

In its Ninth Report on Carcinogens, the National Toxicology Program upgraded cadmium and its compounds to "known to be human carcinogens," whereas it had previously reported them as "reasonably anticipated to be human carcinogens" [490]. The USEPA classifies cadmium as a Group B1 carcinogen (probable human carcinogen) [674], based on evidence in humans and animals [631,643]. An inhalation unit risk of $1.8 \times 10^{-3}$ (μg/m$^3$)$^{-1}$ was calculated from the study of Thun et al. [643]. Likewise, the World Health Organization's IARC has designated cadmium and cadmium compounds as carcinogenic to humans (Group 1) [330], based primarily on its role in lung cancer [622]. Cadmium's role in other human cancers is less clear. Prostate tumors have been reported in male rats after oral cadmium exposure [691], and several studies suggest a role for cadmium in human prostate cancer [2,56,692,699]. Occupational exposure to cadmium has been associated with an increased risk of renal cell carcinoma [310]. The mechanisms of cadmium carcinogenesis are not clearly understood [688,689]. Cadmium is a poor mutagen [688] but may act as an epigenetic or indirectly genotoxic carcinogen. Potential mechanisms include inhibition or faulty DNA repair, aberrant gene expression, and suppressed apoptosis [689,283], but further work is necessary to define cadmium's role as a carcinogen [689].

### Lead

Lead is a heavy, bluish-gray metal and, although it serves no biological purpose, is the most widely used nonferrous metal [17,218]. Lead and lead compounds have been used in many industrial applications, including batteries, ammunition, paints and varnishes, gasoline, pigments, radiation shields, medical equipment, solder, glass, and ceramic glazes [17]. Inhalation and ingestion are the main routes of exposure for inorganic lead [299]. Adults are primarily exposed occupationally [17] by inhalation, with 35 to 40% of inhaled lead dust or fumes being deposited in the lungs and extensive (95%) blood absorption [395]. Children are primarily exposed by ingestion and absorb 50% of an ingested dose through the gastrointestinal tract. In contrast, adults absorb 10% of an ingested dose, but gastrointestinal absorption will vary with particle size (inverse proportion), solubility, nutritional status, and fasting. Elimination is mainly in the urine, with lesser amounts in the feces, sweat, hair, and nails.

In adults, early symptoms are often nonspecific (fatigue, depression, sleep disturbance, anorexia, intermittent abdominal pain, nausea, constipation, diarrhea, and myalgia [299]. The blood lead level (PbB) is the single best diagnostic test for lead exposure [299]. Animal experimentation and many epidemiologic studies suggest that low increases in PbBs may elevate blood pressure [280,590,605], but the results are not definitive [293]. No consistent relationship between blood pressure and PbB was found after examination of the NHANES III dataset [176], and the subject remains controversial. Other researchers suggest that long-term lead accumulation, measured as bone lead, as opposed to PbB, which reflects recent exposure, may be associated with developing hypertension [147,308]. Reversible slowing of nerve conduction velocity has been observed at PbBs as low as 30 µg/dL [599], and adverse effects on reaction time, mood, and visual-motor coordination have been observed at 30 to 50 µg/dL [54,626]. Anemia is not seen until PbBs are in excess of 50 µg/dL [309]. Overt neurotoxicity (wrist drop) is reported at levels in excess of 80 µg/dL [253]. Chronic irreversible nephropathy requires high and sustained PbB exposure [259], but low-level lead exposure (PbB $<$ 10 µg/dL) is associated with renal impairment, as measured by an increase in serum creatinine [361]. Morphological alterations and decreases in sperm count, density, and motility have all been reported in heavily exposed males (PbB $>$ 40 µg/dL) [45,48,398,657]. Paternal occupational lead exposure has been reported to increase the risk of low birth weight and prematurity [404]. Children with prenatal exposure to lead are reported to have reduced academic performance [70]. Lead readily crosses the placenta to the fetus [357,560]; maternal PbBs in excess of 15 µg/dL are associated with low birth weights and preterm delivery, and PbBs in excess of 30 µg/dL are associated with spontaneous abortions [207,211,564].

Lead poisoning in children caused by the ingestion of lead paint was first noted in Australia and became recognized as a public health problem in the United States in the 1920s [406]. Early symptoms of chronic poisoning in children are often nonspecific, including headaches, anorexia, vomiting, and constipation, progressing to anemia with basophilic stippling of red cells, Burton's line, chronic nephritis, peripheral neuropathy (manifested as wrist or foot drop), and radiographs of long bones revealing lead deposits [511,529]. Frank encephalopathy (PbB $>$ 80 µg/dL) is characterized by ataxia, coma, convulsions, cerebral edema, and even death. The long-term neurologic consequences of childhood lead poisoning were recognized in 1943 when Byers and Lord [108] followed up 20 "cured" cases and found poor academic performance in all but one. In 1975, de la Burde and Choate [172] reported school failure due to learning and behavioral problems in asymptomatic lead-exposed children. Asymptomatic children in first and second grade who had elevated dentine lead levels scored lower on standardized tests, especially in areas measuring verbal performance and auditory processing, and were more likely to exhibit disruptive behavior relative to controls [492]. Reexamined 11 years later as adolescents, those with greater lead exposure were more at risk for dropping out of school, reading disability, absenteeism, poor hand–eye coordination, and low scholastic class standing [493,591,645].

Since 1970, the CDC has repeatedly reduced the action level for PbB from 70 to 10 µg/dL, and numerous federal laws have been enacted to reduce lead exposure [250]: the 1971 Lead-Based Paint Poisoning Prevention Act [660]; the USEPA's phase-out of lead in gasoline, starting in 1973 [666], with completion in 1995 [672]; the USEPA's ban on lead in plumbing, fixtures, fittings, and solder [661]; the U.S. Consumer Product Safety Commission's 1978 ban on the use of paint containing more than 0.06% lead by weight for interior/exterior residential surfaces, toys, and furniture [664]; the USFDA's ruling to eliminate lead-solder in food cans by December 1995 [675]; and the Residential Lead Based Paint Hazard Reduction Act of 1992 [662]. Results

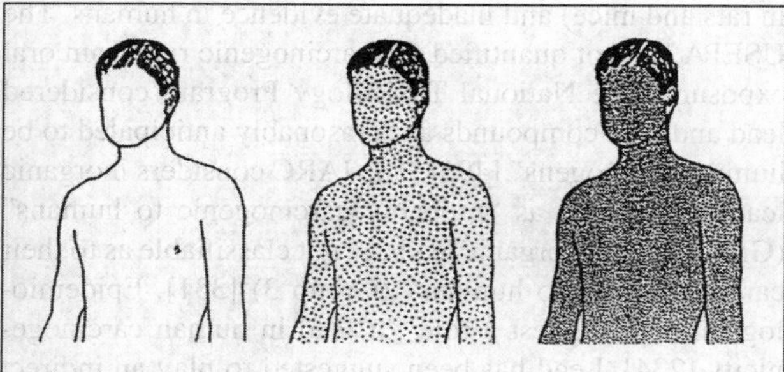

**FIGURE 17.2** Comparison of lead body burden (from left to right): Ancient people uncontaminated by industrial lead (1 dot); typical American (1000 dots); level associated with clinical lead poisoning (4000 dots). Each dot represents 40 μg Pb per 70-kg person. (From Patterson, C. et al., *Sci. Total Environ.*, 107, 205–236, 1991. With permission.

of two National Health and Nutrition Examination Surveys, NHANES II (1976 to 1980) and NHANES III (phase I, 1988 to 1991; phase II, 1991 to 1994), indicate a substantial decline in PbB [123,124,250,536]. Since the late 1970s, the average PbB in children 1 to 5 years of age has declined from 15 μg/dL to 2.7 μg/dL and the reduction of lead in gasoline and dietary sources (mainly by eliminating the use of lead-soldered cans for food and beverages) is believed to be responsible for this effect [97,536]. Although this decline is substantial, Patterson et al. [525] estimate that today's average American has a mean body burden of 40 mg industrial Pb per 70 kg, whereas analysis of pre-Colombian American Indian skeletons indicates that their mean body burden was 40 μg Pb per 70 kg (Figure 17.2). Based on bone/blood lead ratios, this approximates a PbB of 0.016 μg/dL which is 600-fold lower than the current level of concern (i.e., 10 μg/dL) [224] and places current guidelines much closer to lethal Pb levels than natural [610].

It is estimated that 890,000 (4.4%) U.S. preschool children have a PbB of 10 μg/dL or higher [123,124]. Further reductions in PbB will require primary prevention efforts to reduce exposure to lead remaining in housing and soil [385,536,611]. The CDC's 1997 lead poisoning prevention program recommends targeted screening and follow-up care for high-risk children (i.e., children who live in older homes, children from low-income families) [123,124]. Although controversial [423], the program reserves universal screening of young children to those meeting at least one of the following criteria: (1) child resides in a Zip Code where at least 27% of the housing predates 1950; (2) child receives public assistance for the poor; or (3) caretaker's response to a risk assessment questionnaire suggests that the child is at risk.

Although the CDC set 10 μg/dL as the blood lead level of concern in 1991 [119], this should not be interpreted to mean there are no adverse effects below this level [387]. Reduced academic performance is associated with blood lead levels below 5 μg/dL [386], and there appears to be no threshold for the toxic effects of lead on cognitive function [33,384, 387]. Some researchers have suggested further lowering the level of concern [74], and this issue was reviewed by the CDC's Advisory Committee on Childhood Lead Poisoning Prevention (ACCLPP) [132]. Based on ACCLPP recommendations, the CDC's blood lead level of concern remains at 10 μg/dL for the following reasons [134]:

- "No effective clinical interventions are known to lower the blood lead levels for children with levels less than 10 μg/dL or to reduce the risk for adverse developmental effects.
- Children cannot be accurately classified as having blood lead levels above or below a value less than 10 μg/dL because of the inaccuracy inherent in laboratory testing.
- Finally, there is no evidence of a threshold below which adverse effects are not experienced. Thus, any decision to establish a new level of concern would be arbitrary and provide uncertain benefits."

The elimination of elevated blood lead levels in children (defined as at or above 10 μg/dL) by 2010 is a national health objective of the Department of Health and Human Services [177].

The use of car radiators containing lead solder for the illegal distillation of alcohol ("moonshine") has long been associated with lead poisoning [209], and recent reports suggest that middle-aged men in rural settings continue to be at risk [204,526]. Excessive PbBs (>40 μg/dL) have been reported in automobile radiator repair mechanics, and "take-home" lead is a potential source of elevated PbB in their children [251], as has been reported in other lead-related industries, such as ceramics and furniture-stripping [129,352]. Apart from prenatal exposure, lead poisoning in infants has been reported from the use of traditional folk remedies, which are often known by their common names of azarcon, greta, and ghasard [114–117]; the use of lead-contaminated water to prepare formula [604]; the use of lead-soldered samovar (urn) for formula preparation [603]; and home renovation and repair [122,552]. Lead poisoning in children has been reported following the ingestion of foreign objects, including an imported clothing accessory [208], fishing sinkers [469], curtain weights [83]; and imported lead-contaminated food-related products (candy, food coloring) [125,417], as well as the use of imported ceramic dinnerware [131]. Child and adult lead poisoning has been reported following the use of imported traditional remedies [126,133,459], indicating that "culturally appropriate educational efforts are needed to inform persons of the potential health risks posed by these remedies" [133].

Elevated blood lead levels have been found in newly arrived refugee children, and refugee status is considered a risk factor for lead poisoning [237]. It is an important health issue for internationally adopted children [127], and the American Academy of Pediatrics recommends screening for elevated blood lead levels in children who have been adopted or emigrated from countries where lead poisoning is prevalent [33]. Enactment of the above-mentioned federal laws designed to reduce lead exposure has made fatal childhood lead poisoning infrequent in the United States; however, a 2-year-old child died of lead poisoning in 2000 [128], 10 years after the last reported childhood lead fatality [118]. Although the child and her family had recently arrived in the United States from Egypt, isotopic analysis indicated that the lead in her blood was more likely from the lead in the paint and dust of her older U.S. apartment than from soil and dust samples taken from her home in Egypt [128].

The dissolution of retained lead gunshot has resulted in lead poisoning [358,424], with rapid onset when the bullet lodges in contact with synovial fluid [448]. Lead in crystal leaches into alcoholic beverages, and lead contents as high as 21.5 mg/L have been reported in beverages stored in crystal decanters [261]. The lead content of various calcium supplements (bonemeal, dolomite, calcium carbonate) has been tested, and levels of supplementation providing 800 mg of calcium would also contain over 6 μg lead in over one quarter of the 70 different brands tested [91,702]. Although cases of lead intoxication by this route have not been reported, pregnant women and children are the populations most at risk from this source. Lead-contaminated heroin has been reported as a source of lead intoxication, and physicians need to be aware of this possibility [524].

The primary treatment for lead poisoning is prevention [387] and involves identifying sources and eliminating exposure; however, for children with venous PbBs > 45 μg/dL, the CDC [119] currently recommends chelation therapy. In 1991, the USFDA approved the use of *meso*-2,3-dimercaptosuccinic acid (DMSA; succimer) which is an effective oral chelating agent and is more specific for lead than $CaNa_2EDTA$ [505], the use of which is associated with the urinary loss of essential trace elements [530]. Results of a recent randomized controlled clinical trial of DMSA did not find a neurobehavioral benefit of chelation therapy for children with blood lead levels between 20 and 44 μg/dL [563]. Iron deficiency is associated with increased lead absorption [421] and should be treated in all cases [39].

The adverse health effects of lead exposure occur at PbBs so low there appears to be no threshold, and the USEPA considers it inappropriate to derive RfD values for lead [674]. The USEPA classifies lead as a Group B2 carcinogen (probable human carcinogen) based on sufficient evidence in animals (via injection or oral exposure in rats and mice) and inadequate evidence in humans. The USEPA has not quantified the carcinogenic risk from oral exposure. The National Toxicology Program considered lead and lead compounds as "reasonably anticipated to be human carcinogens" [491]. The IARC considers inorganic lead compounds as "probably carcinogenic to humans" (Group 2A) and organic lead as "not classifiable as to their carcinogenicity to humans" (Group 3) [331]. Epidemiological data suggest a role for lead in human carcinogenicity [234]. Lead has been suggested to play an indirect role in carcinogenicity by inhibiting DNA repair or otherwise enhancing the DNA damage of other genotoxic compounds [612].

## Mercury

Mercury is a silver-white fluid trace metal found in igneous and sedimentary rocks and in the form of the ore cinnabar (mercury sulfide) [67]. It is biologically nonessential and toxic to all organisms. Mercury may occur in the elemental form or as inorganic and organic compounds. Mercury and its compounds are used in electrical meters, in chloralkali production, in thermometers, and as antimicrobial preservatives in paints, cosmetics, and pharmaceuticals. Use in dry-cell batteries is now restricted due to environmental toxicity concerns following disposal [663,673]. Use in interior latex paints is prohibited [669]. Toxicity is related to the covalent binding of mercury to sulfhydryl groups, as well as to carboxyl, amide, amine, and phosphoryl groups, thereby inactivating cellular functions [110].

Acute elemental mercury ingestion is usually of no significance due to poor absorption from the gastrointestinal tract. Acute exposure to high concentrations of elemental mercury vapors are irritating to the respiratory tract. Chronic exposure to the vapors produces CNS toxicity, which includes muscle weakness and tremors, nervousness, memory loss, and anorexia. Inorganic mercury compounds generally demonstrate local irritant or corrosive activity. Acute ingestion may result in necrosis to the gastrointestinal tract and renal tubular necrosis. Chronic effects produce CNS toxicity similar to that noted for elemental mercury. Organic mercury compounds have been used for the treatment of syphilis and as diuretics but have been replaced by less toxic drugs. Organic mercury compounds, such as phenylmercuric acetate, thimerosal, and mercurochrome, are primarily used as antimicrobial preservatives in ophthalmic preparations, vaccines, and nasal sprays [110]. Contact dermatitis may occur to both inorganic and organic mercurials, with cross-sensitivity to each being reported [18,219]. Elemental mercury and its compounds are excreted in urine and feces and through respiration.

Dietary intake from agricultural products treated with mercurial fungicides and fish from mercury-pol-

**TABLE 17.2**
**Major Toxic Metals**

| | Chronic Oral Toxicity | | Chronic Inhalation Toxicity | | Carcinogenicity | | |
|---|---|---|---|---|---|---|---|
| Metal | RfD[a] (mg/kg/day) | Confidence | RfC[b] (μg/m$^3$) | Confidence | Oral Slope Factor (risk (μg/m$^3$)$^{-1}$) | Inhalation Slope Factor (risk (μg/m$^3$)$^{-1}$) | Classification[c] |
| Lead | —[d] | — | — | — | —[e] | — | B2 |
| Cadmium | $1 \times 10^{-3}$ (food) | High | — | — | — | $1.8 \times 10^{-3}$ | B1 |
| | $5 \times 10^{-4}$ (water) | — | — | — | — | — | — |
| Mercury | — | — | $3 \times 10^{-1}$ | Medium | — | — | D |
| Arsenic | $3 \times 10^{-4}$ | Medium | — | — | 1.5 | $4.3 \times 10^{-3}$ | A |

[a] RfD, reference dose.
[b] RfC, reference concentration.
[c] Group A, known human carcinogen; Group B1 and B2, probable human carcinogen; Group D, not classifiable as to human carcinogenicity.
[d] The USEPA considers it inappropriate to determine RfD values for lead.
[e] The USEPA does not quantify lead carcinogenicity.

luted water is the major route for toxicity of organic mercury compounds. Metallic and inorganic mercury can enter the air and water from rock and ore deposits, burning of fossil fuels, industrial and agricultural emissions, and trash disposal and incineration. Atmospheric fallout adds to water pollution. Inorganic mercury compounds may be methylated by the microflora of soil and water to form methylmercury. Through the food chain, edible fish can concentrate methylmercury to levels a thousand times greater than in the environment [110]. Methylmercury is neurotoxic, and the effects are both dose and time dependent [67]. Ataxia is an early symptom followed by slurred speech, weakness, vision and hearing loss, tremors, coma, and death. Well-documented poisonings from contaminated fish and grains occurred in Japan and Iraq [67,110]. Additionally, methylmercury is a well-known neuroteratogen [18].

The use of dental amalgam fillings has generated concern, both because exposure of dental workers may exceed occupational standards and because a variety of illnesses (e.g., multiple sclerosis, rheumatoid arthritis, leukemia) have been attributed to dental mercury exposure in the general population [44,235,381,687]. Contact dermatitis is experienced by some dental patients; for those in whom symptoms do not quickly subside with antihistamine treatment, replacement of fillings with non-mercury materials may alleviate immunological and dermatological symptoms [235,381]. Other diseases have not been firmly linked to mercury exposure, and at present replacement of mercury-containing fillings cannot be justified in nonallergic individuals, although the issue of whether or not mercury should continue to be used for new dental fillings is more controversial [41,235,381]. Additional research is needed before the potential for dental amalgams to cause harm, and the benefit of substituting more costly or less durable materials can be reasonably evaluated. Concern must include the potential risk of effects on the fetus [18]. Recent reviews support the safety of dental amalgam use [473,712].

The USEPA has derived a chronic inhalation exposure RfC for elemental mercury (vapor) of 0.3 μg/m$^3$ based on critical effects of hand tremor, memory disturbances and autonomic dysfunction (acrodynia) with a medium level of confidence [674]. There is no current chronic oral exposure RfD pending further review [674]. The ACGIH has adopted TLV–TWA values of 0.025 mg/m$^3$ for inorganic forms and metallic mercury, of 0.1 mg/m$^3$ for aryl mercury compounds, and 0.01 mg/m$^3$ for alkyl compounds [35]. These TLV–TWA values carry a skin notation which points out the potential for dermal absorption. The ACGIH and USEPA consider mercury as not classifiable as to human carcinogenicity based on no evidence of carcinogenicity in humans and inadequate evidence in animals [35,674].

A summary of some quantitative toxicity values for these four major toxic metals is given in Table 17.2. Metals are listed in approximate reverse order of toxicity (least toxic first).

## MINOR TOXIC METALS WITH RfDS

### Antimony

Antimony is a brittle silver-colored metal extracted from ores [67]. Compounds of antimony cover the full range of toxicity. Less toxic compounds have found use in cosmetic pigments (antimony sulfide) and medicinals (antimony potassium tartrate, tartar emetic). Stibine, the metal hydride of antimony, is a colorless, highly toxic gas used in the manufacture of semiconductors [67]. Ingestion of antimony compounds can cause gastrointestinal, cardiac, dermatological, hepatic, and neurological toxicity in humans and animals [9]. Mechanisms for these effects include binding to sulfhydryl groups and inhibiting protein and carbohydrate metabolism [436]. Acute inhalation exposure to antimony trichloride or antimony pentachloride may cause pneumonitis, but the injury may be caused by the chloride rather than by the antimony itself; acute exposure to antimony hydride can cause hemolysis but, again, the antimony itself may not be responsible [496]. Long-term inhalation exposure to antimony can cause benign pneumoconiosis [496] and may raise blood pressure [9]. Dermatological reactions to antimony (eczema, pustules) exhibit signs of an acute inflammatory response but do not appear to be an allergic reaction [436]. The USEPA has derived a chronic oral RfD for antimony based on the study of Schroeder et al. [589], which found changes in blood glucose and cholesterol levels in rats exposed to antimony in drinking water at a dose of 0.35 mg/kg/day. This LOAEL was converted to an RfD of 0.4 μg/kg/day using an uncertainty factor of 1000 (10 for interindividual variation, 10 for interspecies variation, and 10 for the use of a LOAEL rather than a NOAEL). Confidence in the RfD was considered low [674]; however, a recent study of potassium antimony tartrate on rats following 90-day exposure via drinking water gave a NOAEL level of 0.06 mg/kg/day [539]. Using a 100-fold safety factor, as above, 0.6 μg/kg/day approximates the RfD. The USEPA has not derived an inhalation RfC for antimony [674]. The ACGIH has adopted a TLV–TWA value for antimony of 0.5 mg/m$^3$ [35].

The carcinogenicity of antimony is uncertain. Mice given antimony potassium tartrate in drinking water at a dose of 0.88 mg/kg/day for 33 months had no increased incidence of lung, liver, or total tumors [349]. Workers exposed to antimony concentrations well over 5 mg/m$^3$ had an increased risk of lung cancer [186]; however, exposure to arsenic may have caused the excess [436]. Female rats exposed to >30 mg/m$^3$ had an increased incidence of lung tumors [269]. A chronic inhalation oncogenicity study in rats of antimony trioxide dust at doses less than 30 mg/m$^3$ did not show carcinogenicity [269]. Antimony has not been evaluated for human carcinogenic potential by the USEPA [674]. The ACGIH has classified antimony trioxide production as a suspect human carcinogen for which exposure levels should be as low as reasonably achievable [34,35].

### Barium

Barium is a silvery-white alkaline earth metal and is found in nature in combination with other elements [10]. The barium ion is highly reactive, and its toxicity is dependent on the solubility of the specific compound, with water-soluble forms (i.e., chloride, hydroxide, nitrate) being more toxic than insoluble forms (i.e., sulfate, carbonate). Barium compounds are used primarily as lubricating agents in drilling muds, but also in the manufacture of paints, bricks, tiles, glass, rubber, and pesticides. Barium sulfate is safely used medically as a contrast agent in x-ray diagnosis because it is not normally absorbed across the gastrointestinal lumen; however, in Brazil, the use of nonpharmaceutical-grade barium sulfate in the production of a single lot of Brand A contrast solution resulted in fatalities following radiologic examination [130]. Clinicians worldwide are advised to monitor patients for barium toxicity after the use of barium-containing contrast solution. Hospital staff familiar with the safe use of barium sulfate may fail to recognize other forms of barium as a potential toxic agent, and this has contributed to at least one fatality [191]. The general population is exposed by ingestion (i.e., food, drinking water) and inhalation. Some plants bioconcentrate barium from the soil, with Brazil nuts having very high concentrations (3000 to 4000 ppm) [66].

Occupational exposure to inhaled barium sulfate can cause a benign pneumoconiosis (baritosis) [185], which resolves with elimination of exposure. Acute ingestion of soluble $Ba^{2+}$ salts acts as a muscle poison, characterized by stimulation followed by paralysis [551]. Symptoms of poisoning start with the gastrointestinal muscles (nausea, vomiting, diarrhea) and progress to skeletal and cardiac muscle with ventricular fibrillation followed by death due to respiratory muscle paralysis [5,183,551,567]. The barium ion is thought to act as a potassium antagonist, producing an extracellular hypokalemia [567] relieved by intravenous infusion of potassium salts [5,183,551]; however, potassium infusion does not relieve the hypertension [183,567,674]. Prompt oral administration of 2 to 5% sodium sulfate to form the highly insoluble barium sulfate (1 g dissolved in 400,000 parts water) has been used to prevent absorption [5,551]. The USEPA has derived an oral RfD for barium based on two studies involving humans, one experimental [704] and one epidemiological [95], as well as on the subchronic and chronic rodent studies performed by the NTP [486]. Wones et al. [704] found a NOAEL for barium of 0.21 mg/kg/day (the highest dose tested) in healthy male volunteers exposed to barium in drinking water. Brenniman and Levy [95] found no convincing evidence of a difference in hypertension or other effects between two communities, one exposed to <0.2 mg Ba per L and the other to a mean of 7.3 mg Ba per L (0.20 mg Ba per kg per day). These very similar NOAELs were converted to an RfD of 0.07 mg/kg/day using an uncertainty factor of 3 (to account for database deficiencies and to protect sensitive individuals, a factor lower than 10 was considered appropriate because the supporting studies considered adult males, those likely to be most sensitive to barium's hypertensive effects) [674]. Confidence in the RfD was considered medium. The ACGIH has adopted

TLV–TWA values of 0.5 mg/m$^3$ for soluble barium compounds and 10 mg/m$^3$ for barium sulfate [35].

The NTP performed a 2-year rodent bioassay with barium chloride dihydrate in drinking water and found no carcinogenic effects in either rats or mice [486]. Under the USEPA's 1986 Guidelines for Carcinogen Risk Assessment, barium was classified as Group D (not classifiable as to human carcinogenicity), and under the Proposed Guidelines for Carcinogen Risk Assessment, barium is considered not likely to be carcinogenic to humans following oral exposure [674]. The carcinogenic potential of inhaled barium cannot be determined due to the lack of adequate animal inhalation studies.

## Beryllium

Beryllium is alkaline earth metal that is the lightest of the structural metals. Beryllium is a rare metal and is extracted primarily from bertrandite (beryllium–silicate ore) and beryl (beryllium–aluminum oxide–silicate ore). The primary uses of beryllium are as a structural metal in lightweight applications, in metal alloys, and in nuclear reactor technology, as beryllium is an excellent neutron reflector and moderator [34,67].

As beryllium is a rare metal, its toxic effects were not completely recognized until it became widely used in the 1940s. Although beryllium and its compounds can cause contact dermatitis, the primary target organ is the lung. Two types of beryllium-induced lung injury can occur: acute and chronic. The acute and frequently fatal syndrome resembles chemical pneumonitis and is associated with exposure to soluble forms of beryllium (e.g., beryllium sulfate and beryllium fluoride), where the concentration of airborne beryllium is greater than 0.1 mg/m$^3$ [199]. First reported in the early 1930s [67,679], this syndrome has been virtually eliminated in the workplace after 1950 because of controls limiting the concentration of these beryllium compounds in the air [11,200].

Chronic beryllium disease was first reported in 1946 as a delayed pneumonitis [279]. The disease (a.k.a. berylliosis) is characterized by granuloma formation, fibrosis, emphysema, and reductions in the vital capacity of the lung and total lung capacity [11]. The chronic disease has two forms, one that occurs during exposure and a second where the disease becomes evident 10 or more years after cessation of the exposure [67]. The mechanism of the delayed onset of the condition is not known. The disease has a strong immunological component; chelation treatment has little effect on the course of the disease, whereas corticosteroid treatment has been effective in disease suppression [67].

Beryllium dermatitis is a hypersensitivity reaction that is usually noted 1 to 2 weeks after exposure to soluble beryllium salts. Patch tests of individuals with soluble beryllium salts provoke a positive response. Beryllium can also induce dermal ulceration if particles of beryllium salts become imbedded in the skin [67]. The ulceration can be long lasting, and surgical intervention can be required to resolve the condition [641].

Beryllium compounds are also carcinogenic. The IARC reviewed the available literature and in 1993 published its findings that there was sufficient evidence in humans and animals for the carcinogenicity of beryllium [330]. In several retrospective epidemiologic studies conducted in the 1970s and 1980s, a consistent (but small) increase was noted in the incidence of lung cancer in workers exposed to beryllium. The decade of hire was one of the strongest correlates of lung cancer mortality.

The USEPA has established an oral RfD for beryllium of 2 μg/kg/day. The inhalation RfC is 0.02 μg/m$^3$. Beryllium is classified as a Group B1 (probable human) carcinogen [674]. The ACGIH has adopted TLV values for beryllium and compounds of 0.002 mg/m$^3$ as an 8-hour TWA and 0.01 mg/m$^3$ as a STEL/C. An A1 carcinogenicity notation (confirmed human carcinogen) is present [35].

## Boron

Boron is a metalloid. It is a solid element that, because of its high affinity for oxygen, always occurs in nature bound to oxygen in the form of inorganic borates [12,198,461]. Boron and associated compounds have many industrial applications, including the production of borosilicate glass, laundry bleaches (sodium perborate), wood preservatives, fire retardants, pesticides (cockroach control), fertilizers, cosmetics, and pharmaceuticals [12]. In 1875, Lister used boric acid as an antiseptic [408], but its effectiveness has since been discredited. The world's two largest borate deposits occur in the Mojave Desert (near Boron, California) and in Western Turkey [198]. Borates have long been known to be essential for plants, but a specific biochemical role remains to be determined [86]. Although boron deficiency has been reported in rats, chickens, and humans, as yet no requirement has been established in humans [181,483]. Nielsen [502,503] classifies it as an ultratrace element, and the World Health Organization's Expert Committee on Trace Elements in Human Nutrition concluded that it is "probably essential" in human nutrition [164]. As an essential plant nutrient, boron occurs naturally in fruits and vegetables, and in the adult American population the daily median boron intake is 1 mg/day [548]. Biochemical and physiological consequences of boron deprivation in humans suggest it affects calcium and magnesium metabolism [501]. Inadequate dietary boron (<0.2 mg/day) has been suggested as a factor contributing to osteoporotic bone loss [500].

Inorganic borates exhibit a low order of acute toxicity in mammals. The oral $LD_{50}$ for boric acid in male rats is 4.5 g/kg body weight [166]; however, inadvertent use of a 2.5% boric acid solution in preparation of infant formula has resulted in toxicity and death [705]. The oral lethal dose in adults has been reported as 15 to 20 g [609], but doses of 80 to 297 g have been tolerated in a single ingestion [686]. Nausea with vomiting and diarrhea (both a characteristic blue–green color) are common. Frequently the skin shows signs of erythema, desquamation (boiled lobster appearance), and exfoliation. Death generally occurs several days after ingestion and results from renal injury, circulatory collapse, and shock. Although boric acid solutions are no longer considered antiseptic, concentrated boric acid is still used as a household pesticide, and parents of young children need to recognize it as a potential poison [609,614]. Borates are not absorbed through intact skin [197], including that of newborns [233], but they are absorbed if the skin is damaged, abraded, or otherwise compromised [193]. The use of boric acid as a dusting powder during diapering has resulted in fatalities [249]. The Cosmetic Ingredient Review Expert Panel [78] reviewed the use of borates in cosmetics and concluded that "cosmetic formulations containing free sodium borate or boric acid at this concentration (5%) should not be used on infant or injured skin." Animal experiments indicate that chronic oral exposure to boric acid or borax is toxic to the male reproductive system, with testicular lesions being observed in rats, dogs, and mice [212,485]. Although the specific mechanism is unknown, the available data suggest a toxic effect on the Sertoli cell [213]. Boric acid is a developmental toxicant in all three mammalian species tested (rat, mouse, rabbit) [288,542,543], with the most sensitive endpoints being decreased fetal body weight and malformations/variations of the ribs [213]. The rat was the most sensitive species for developmental effects, with a NOAEL of 9.6 mg/kg/day [542]. At present, human data are insufficient to determine if boron causes male reproductive toxicity [320], but boric acid is considered a high-priority chemical for study with respect to human reproductive health [463].

The USEPA has recently revised their toxicity summary for boron and compounds [674]. The new chronic oral RfD for boron is 0.2 mg/kg-day with decreased fetal body weights being considered the critical effect [288,541, 542,544]. Confidence in this RfD is considered high. No inhalation RfC has been derived for boron due to an insufficient database. The ACGIH has adopted several TLV values for boron compounds: TWA values of 10 $mg/m^3$ for boron oxide, 5 $mg/m^3$ for sodium tetraborate decahydrate, and 1 $mg/m^3$ for sodium tetraborate pentahydrate and anhydrous, as well as ceiling values of 1 ppm for boron tribromide and boron trifluoride [35]. The NTP conducted a 2-year carcinogenesis bioassay in male and female $B6C3F_1$ mice and reported testicular atrophy and interstitial cell hyperplasia in males receiving 201 mg/kg/day but found no evidence of carcinogenicity [485]. Under the Draft Revised Guidelines for Carcinogen Risk Assessment, the data are considered inadequate for an assessment of the human carcinogenic potential of boron [674].

## Nickel

Nickel is member of the Group VIIIB series of transition metals. The three principal classes of nickel ores are sulfide, silicate, and arsenide. Nickel is used in a wide variety of applications, with 80% of the nickel in the United States being used in the production of nickel metal and alloys [67]. The essentiality of nickel in humans is debatable. Nickel deficiency can be experimentally induced in rats and larger mammals [41,483]. Some evidence suggests that nickel is essential for methyl metabolism and iron, calcium, and zinc absorption [41,527], but, if essential, the amount of nickel required would be amply met by the amount of nickel in a typical American diet [527].

Animal experiments have indicated that nickel compounds can be nephrotoxic, hepatotoxic, immunotoxic, and teratogenic [629]. In humans, nickel can cause allergic contact dermatitis, particularly in young women using nickel-containing earrings in pierced ears [107]; this is the most frequent disease among nickel workers. Statistical evaluations showed that up to 17% of all occupational allergies may be related to nickel occupational exposure [67]. Allergic asthma is rare [496], but case reports have been published [67]. Acute inhalation exposure to metallic nickel can cause metal fume fever [496]. Nickel carbonyl is a colorless, volatile liquid that is particularly hazardous. It has been estimated that exposure to 30 ppm nickel carbonyl for 30 minutes may be lethal in humans [36]. Acute inhalation exposure to this material can cause immediate and delayed toxic effects. Headache, dizziness and nausea are the immediate manifestations. Ten to 36 hours after exposure substernal pain, coughing, and dyspnea, consistent with chemical pneumonitis, are observed [34,67]. Sodium diethyldithiocarbamate (dithiocarb, a chelating agent) has been employed in the therapy of nickel-carbonyl-exposed workers [36]. Recovery is protracted, and is characterized by fatigue upon slight exertion [67]. Short-term exposure to 150 ppb $Ni(CO)_4$ can cause immediate, but not delayed, symptoms, whereas short-term exposure to concentrations on the order of a few parts per million can cause the more severe, delayed-type reactions [67]. The USEPA has derived a chronic oral RfD for soluble salts of nickel of 20 μg/kg/day. Neither oral nor inhalation RfDs have been established for nickel subsulfide, nickel refinery dust, or nickel carbonyl [674].

Inhalation of nickel compounds can cause lung cancer. The initial observation of excessive lung cancer and nasal tumors among nickel refinery workers was made as early as 1932 [67,466]. Since the initial observation, numerous epidemiologic studies have been conducted that show conclusively the association between occupational exposure to nickel refinery dust and nickel subsulfide and lung and nasal cancer [67]. The latency period for nickel-induced lung cancer was 13 to 14 years and that for nasal cancer was 15 to 24 years after first employment [34]. Nickel exposure has not been clearly associated with respiratory cancer in any of the industries using nickel [34]. The USEPA has classified nickel subsulfide and nickel refinery dust as Group A carcinogens (human carcinogen) on the basis of animal and epidemiologic carcinogenicity data; nickel carbonyl has been classified as a Group B2 carcinogen (probable human carcinogen) [674]. The IARC has classified nickel compounds as carcinogenic to humans [328]. The NTP published results on the carcinogenicity of inhaled nickel oxide [487], nickel subsulfide [488], and nickel sulfate hexahydrate [489] in rats and mice. There was no evidence of carcinogenicity of nickel sulfate hexahydrate in either species, clear evidence of carcinogenicity of nickel subsulfide in both species, and no, equivocal, or some evidence of carcinogenicity of nickel oxide, depending on species and sex. The ACGIH has adopted the following TLVs for nickel compounds: elemental/metal, 1.5 mg/m$^3$; soluble compounds, 0.1 mg/m$^3$; insoluble compounds, 0.2 mg/m$^3$; nickel carbonyl, 0.05 ppm; and nickel subsulfide, 0.1 mg/m$^3$ [35].

### Silver

Alloys of silver are used in jewelry, tableware, photographic materials, electronics, and dental products and as topical antibacterial agents for the treatment of burn wounds [67,305]. Silver is generally low in toxicity. It is absorbed following inhalation, ingestion, or topical application [34]. Accumulation of silver results in argyria, a blue-gray discoloration of the skin, mucous membranes, and eyes. Silver sulfadiazine used in the management of burn wound sepsis has resulted in argyria, ocular injury, leucopenia, and toxicity in kidney, liver, and neurological tissues. Silver may affect the immune system, and contact dermatitis has been observed following exposure to various silver compounds [7]. Toxicity has been attributed to the free silver ion released into solution and interaction with sulfhydryl, amino, carboxyl, and other groups on membrane or enzyme proteins [305]. Excretion from oral, respiratory, or topical exposure is primarily through the gastrointestinal tract [7]. Mucociliary escalator activity accounts for removal of silver following respiratory exposure. Silver is not considered to be a carcinogen or a reproductive or developmental toxicant [7,674]. The USEPA has derived a chronic oral RfD for silver of 5 µg/kg/day [674]. No inhalation RfC has been derived for silver. The ACGIH has adopted TLV–TWA values for silver of 0.1 mg/m$^3$ for the metal and 0.01 mg/m$^3$ for soluble compounds [35]. The ionic form is highly toxic to fish, but the ionic form is extremely low in the aquatic environment, and other more common forms of silver show only low to moderate toxicity [94,304].

### Strontium

Radiotoxicity is beyond the scope of this chapter, and the following discussion is limited to stable strontium. Strontium is a soft silvery alkaline earth metal which turns yellow upon formation of the oxide [67]. Its salts are used in the manufacture of color television screens, pyrotechnics (strontium and salts give a characteristic red color to flames), and electrical materials [671]. Over 99% of the typical body burden of 320 mg is found in bone, and there is no conclusive evidence it is an essential trace element in mammals [67,504]. Epidemiologic data suggest that drinking water containing strontium in the presence of fluoride (5 to 6 mg $Sr^{2+}$ and 1 mg $F^-$ per L) decreases the incidence of dental caries in children [167]. The gastrointestinal absorption of the strontium ion is poor [671]. Acute strontium toxicity is low; an oral $LD_{50}$ of 2250 mg/kg body weight has been reported for strontium chloride in rats [109]. Acute lethality is due to respiratory failure [156]. In contrast, bone is the target organ following chronic strontium toxicity. High doses inhibit calcification of the epiphyseal cartilage and cause deformities of long bones [615]. Strontium causes these effects by substituting for calcium in the hydroxyapatite crystal during calcification or displacing calcium from existing calcified bone. The metabolic basis of strontium's effect on calcium metabolism is thought to be inhibition of the renal synthesis of 1,25-dihydroxyvitamin $D_3$ [516]. Young animals (still growing) are more susceptible to the toxic effects of strontium than adults, with widening of the epiphyseal cartilage being observed at lower levels of dietary strontium [627]. Dietary calcium plays a protective role in strontium toxicity. Weanling rats maintained on diets containing 950 mg strontium per kg and 0.69% calcium for 4 weeks exhibited rachitic changes that were not seen in rats supplemented with 1.6% calcium [205]. In contrast to these toxic effects, pharmacologic treatment with low doses of strontium suppresses bone resorption [428], and strontium ranelate treatment has reduced the risk of vertebral fractures and increased bone mineral density in postmenopausal women with osteoporosis [439,451]. The USEPA has derived an oral RfD for strontium of 0.6 mg/kg/day, and confidence in this value is medium [674]. Stable strontium has not been adequately evaluated for carcinogenic potential. Strontium chromate is a human carcinogen (inhalation route), but this is due to the presence of hexavalent chromium, which is a genotoxic carcinogen [23].

### Thallium

Thallium is a soft, bluish-white metal widely but sparingly distributed in the Earth [67]. Historically it has been used to treat gout, venereal disease, dysentery, ringworm, and tuberculosis [138]. Thallium sulfate was widely used as a rodenticide, but its use was banned in the United States in 1972 [67]. It is still used as a rodenticide in other parts of the world, and rodent resistance to warfarin may cause this use to increase [353]. Thallium intoxication from contaminated heroin [547] and cocaine [319], presumably imported from areas where thallium is still used as a rodenticide, has been reported. It is currently used in the electronics industry and in the manufacture of prisms, costume jewelry, pigments, low-temperature thermometers, and infrared spectrometers.

Thallium is well absorbed following oral ingestion and causes severe gastrointestinal symptoms followed by painful paresthesia of the extremities (soles of the feet), motor paralysis, and death from respiratory failure [292,367,422,449,471]. Individuals surviving the acute phase develop scalp alopecia about 10 days after ingestion [292,471,678]. Mee's lines are often seen on fingernails 2 to 4 weeks after exposure [303]. Both alopecia and a painful peripheral neuropathy suggest thallium poisoning, but a definite diagnosis requires identifying elevated thallium levels in hair, nails, feces, saliva, blood or urine [303]. For an adult, the $LD_{50}$ has been calculated to be 8 to 12 mg/kg [471]. Treatment with oral Prussian blue (potassium ferric ferrocyanide) [303] interferes with the enterohepatic recirculation of thallium, trapping it in the intestines and promoting its elimination from the body.

In October 2003, the USFDA approved a Prussian blue formulation (Radiogardase®) for suspected internal contamination with thallium (www.fda.gov/cder/drug/infopage/prussian_blue). Prussian blue's crystal lattice makes it an effective ion exchanger for univalent cations in general [303], and its clinical use depends on the preferential binding of thallium over potassium [376]. Chelating agents (dithiocarb) have caused a redistribution of thallium to target organs, with an increase in toxicity [347]. The precise mechanism of thallium toxicity is unknown but may involve substitution of the thallous ion for potassium in the sodium/potassium ATPase pump or interference with sulfhydryl enzymes [471]. Presumably alopecia and Mee's lines result from thallium's interference with the formation of disulfide bonds [471]. Interference with tissue riboflavin with subsequent effects on metabolic pathways have also been suggested [113].

The USEPA has derived a chronic oral RfD for thallium based on its own 90-day study of rats exposed to aqueous thallium sulfate by gavage at doses up to 0.20 mg/kg/day [674]. Treatment-related effects on serum chemistry changes, alopecia, and lacrimation without histopathological changes were not considered adverse, and 0.2 mg/kg/day was considered the NOAEL [674]. This NOAEL was converted to an RfD of 0.09 μg/kg/day using an uncertainty factor of 3000 (10 for intraspecies extrapolation, 10 for interspecies variation, 10 for less than chronic exposure duration, and 3 for lack of reproductive and chronic toxicity data). Confidence in the RfD is considered low. No inhalation RfD has been derived for thallium. The ACGIH has adopted a TLV–TWA of 0.1 mg/m$^3$ for soluble thallium compounds [35]. Thallium is classified as Group D (not classifiable as to human carcinogenicity) based on two inadequate negative studies in humans and a lack of animal studies designed to examine carcinogenic endpoints [674]. Existing data do not indicate that thallium is mutagenic [396].

## Uranium

Uranium is a soft, malleable metal of the actinide series in the periodic table. The primary uranium ores are pitchblende (uranium oxide) and carnotite (uranium/vanadium containing mineral). Uranium is primarily used as nuclear fuel, but some minor applications include its use as a colorant in ceramics or glass or as depleted uranium (a by-product of the uranium enrichment process; >99% $^{238}U$) in armor-piercing projectiles [34]. Occupational exposure occurs in mining operations and in uranium enrichment (uranium hexafluoride), and the first significant military exposure occurred in "friendly fire" incidents in the 1991 Persian Gulf War [452].

Acute inhalation exposure to uranium hexafluoride can cause pneumonitis, but the injury may be caused by the fluoride rather than the uranium itself [496]. Chronic inhalation exposure of uranium dioxide dust at a concentration of 5 mg/m$^3$ produced no observable adverse effects in rats, dogs, or monkeys [392]. The kidney is the main target of uranium's chemical toxicity, with the target being the pars recta of the proximal tubules, the ascending limb of the loop of Henle, and collecting tubules [34]. Drinking water studies with uranyl nitrate hexahydrate indicated a LOAEL for renal damage of 0.96 mg uranyl nitrate per kg in the Sprague–Dawley rat and New Zealand white rabbit [242,243], with incomplete recovery 91 days after cessation of exposure in the rabbit [243]. An implantation study in the rat indicated that urinary mutagenicity might be used as a biomarker to detect exposure to internalized depleted uranium in potentially exposed soldiers [452]. The USEPA developed a chronic RfD of soluble salts of uranium of 3 μg/kg/day; no inhalation RfC has been derived for uranium [674].

Uranium isotopes are radioactive but only weakly due to their extremely long half-lives. Natural uranium has a radioactivity of about 0.7 pCi/μg [670]. No direct evidence exists that uranium is carcinogenic to humans or animals; however, based on the fact that uranium does emit ionizing radiation as it decays, the USEPA has classified uranium as a Group A carcinogen (known human carcinogen) and has proposed to quantify the cancer risk of uranium in drinking water using toxicokinetic modeling [670].

**TABLE 17.3**
**Minor Toxic Metals with RfDs**

| | Chronic Oral Toxicity | | Chronic Inhalation Toxicity | | Chronic Oral Toxicity | | |
|---|---|---|---|---|---|---|---|
| **Metal** | **RfD[a] (mg/kg/day)** | **Confidence** | **RfC[b] (μg/m³)** | **Confidence** | **Oral Slope Factor (risk (μg/m³)⁻¹)** | **Inhalation Slope Factor (risk (μg/m³)⁻¹)** | **Classification[c]** |
| Strontium | $6 \times 10^{-1}$ | Medium | — | — | — | — | — |
| Boron | $2 \times 10^{-1}$ | High | — | — | — | — | — |
| Barium | $7 \times 10^{-2}$ | Medium | — | — | — | — | D |
| Nickel | | | | | | | |
| Nickel refinery dust | — | — | | — | | $2.4 \times 10^{-4}$ | A |
| $Ni_3S_2$ | — | — | | — | | | A |
| $Ni(CO)_4$ | — | — | | | | — | B2 |
| Soluble nickel | $2 \times 10^{-2}$ | Medium | — | | | | — |
| Vanadium ($V_2O_5$) | $9 \times 10^{-3}$ | Low | — | — | | | — |
| Silver | $5 \times 10^{-3}$ | Low | | — | — | | D |
| Beryllium | $2 \times 10^{-3}$ | Low to medium | $2 \times 10^{-2}$ | Medium | — | $2.4 \times 10^{-3}$ | B1 |
| Uranium (soluble) | $3 \times 10^{-3}$ | Medium | — | — | — | — | — |
| Antimony | $4 \times 10^{-4}$ | Low | — | — | — | — | — |
| Thallium (salt) | $9 \times 10^{-5}$ | Low | — | — | — | — | D |

[a] RfD, reference dose.
[b] RfC, reference concentration.
[c] Group A, known human carcinogen; Group B1, probable human carcinogen; Group D, not classifiable as to human carcinogenicity.

## VANADIUM

Vanadium is a white to gray common trace metal that occurs in nature only in combination with oxygen, sodium, sulfur, and chloride. Vanadium deficiency can occur in laboratory animals on a very strict diet [42,483], and evidence suggests that vanadium helps regulate some phosphoryl transfer enzymes [527]. A requirement for vanadium extrapolated from animal experiments is 10 to 25 μg/day, whereas typical intake is 8 to 18 μg/day [527]; however, the NRC believes that there is only weak evidence that vanadium is essential and that any vanadium requirement would be met by naturally occurring levels [483].

Inhalation exposure to vanadium pentoxide can cause tracheobronchitis with persistent bronchial hyperreactivity and inflammation [14,34,496]. A greenish black discoloration of the tongue, gastrointestinal symptoms, neurotoxicity, and renal toxicity have also been reported in workers exposed to vanadium pentoxide. The short-term repeated inhalation in rats of vanadium metavanadate (8 hr/day for 4 days) at a concentration encountered by humans over a typical workweek altered pulmonary immune cell function and produced significant changes in the lungs themselves [155]. A 2-year inhalation study in rats and mice conducted by NTP found vanadium pentoxide to be a pulmonary carcinogen at and slightly above the OSHA permissible occupational exposure limit of 0.5 mg/m³ [554,588]. The progression of pathological lung changes with exposure and time is thought to affect the pattern and extent of vanadium lung deposition [184]. Oral exposure to vanadium can be toxic to the gastrointestinal, renal, and neurological systems. Vanadium is rapidly excreted in feces and urine following termination of exposure. The USEPA has derived a chronic oral RfD for vanadium pentoxide of 5 μg/kg/day (9 μg $V_2O_5$ per kg per day) which was derived using an uncertainty factor of 100 (10 for interindividual variation and 10 for interspecies variation); confidence in this RfD was considered low [674]. The USEPA has not derived an inhalation RfC for vanadium compounds. The ACGIH has adopted TLV–TWA values for vanadium of 1 mg/m³ for ferrovanadium dust and 0.05 mg/m³ for vanadium pentoxide [35].

The results of behavior testing show that oral sodium metavanadate in rats has resulted in significant reductions in both general activity and learning [577]. Mice given vanadyl sulfate in drinking water at doses up to 1 mg/kg/day for up to 33 months had no significant increase in tumor incidence [348,586]. The USEPA has not evaluated the potential human carcinogenicity of vanadium [674].

Table 17.3 presents the quantitative toxicity values derived by the USEPA for the metals listed in this section. The metals are listed in reverse order of toxicity (least to most toxic).

# MINOR TOXIC METALS WITHOUT RfDS

## ALUMINUM

Aluminum is the third most abundant element in the Earth's crust and is extracted from bauxite ore. Although aluminum is not an essential element, humans consume a substantial amount in the diet and are exposed to aluminum from a number of nondietary sources. On average, American adults consume 2 to 25 mg aluminum daily from food and beverages [265], with average amounts being 8.2 mg/day for males and 7.1 mg/day for females [528]. Aluminum is naturally present at low levels in most foods, but the primary source of dietary aluminum is from food additives. Over-the-counter antacids contain large amounts of aluminum hydroxide, and millions of consumers are dermally exposed to aluminum salts from the use of antiperspirants and deodorants [265].

Aluminum is generally considered to have a low order of toxicity, but it can cause reproductive toxicity when administered in high doses to experimental animals [252]. The potential role of aluminum in either causing Alzheimer's disease or in speeding its progression is highly controversial. Aluminum is certainly neurotoxic. In renal dialysis patients, excessive parenteral exposure to aluminum can cause a progressive, fatal neurological syndrome known as dialysis dementia [27]. In addition, injection of aluminum salts into the brain of rabbits leads to the development of neurofibrillary tangles, but not β-amyloid plaques, which are also indicators of Alzheimer's disease [368]. Some studies have found elevated levels of aluminum in some regions of the brain [378,648,694,708,713], whereas others have found no difference in aluminum levels between Alzheimer's and control brain tissue [338,429,649].

Epidemiologic studies do not show an association between aluminum exposure and the incidence of Alzheimer's disease. Canadian miners, between 1944 and 1979, were exposed to high concentrations of aluminum and aluminum oxide powder (McIntyre Powder) preceding each shift as a prophylactic treatment against silicotic lung disease. In an initial study of this population, there was no increased incidence of neurological disorders in exposed miners, but there was an increase in neurological impairment as measured by cognitive testing [556]. A follow-up study was conducted to address several methodological weaknesses in the initial study. No statistically significant differences were noted between exposed and non-exposed miners in either neurological disease or cognitive impairment incidence [555]. Similarly, an association was noted with aluminum in drinking water and the incidence of Alzheimer's disease [432]. A follow-up study (with methodological improvements) found no evidence of such an association [431]. Another recent study has shown no association between occupational exposure to aluminum and the incidence of Alzheimer's disease [260]. The ACGIH has adopted TLV–TWA values for aluminum of 10 mg/m$^3$ for metal dust and aluminum oxide, 5 mg/m$^3$ for pyro powders and welding fumes, and 2 mg/m$^3$ for soluble salts and alkyls [35].

## BISMUTH

Elemental bismuth is a soft lustrous metal that can occur naturally or in combined forms in ores. Bismuth is used in low-melting alloys. Insoluble bismuth salts are poorly absorbed orally or dermally. Excretion is primarily through the gastrointestinal tract. Bismuth compounds demonstrate a low order of toxicity [67]. They are used as cosmetics coloring agents and in pharmaceuticals, including those used for diarrhea, gastroesophageal reflux, and in ulcer therapy. Bismuth subsalicylate, used in ulcer therapy, has no substantial capacity to neutralize gastric acid but rather provides cytoprotection involving the enhanced secretion of mucus and $HCO^{3-}$, inhibition of pepsin activity, and the formation of bismuth–protein complexes that may afford a protective barrier against peptic digestion. Primary activity may be due to the antibacterial effect of bismuth compounds against the bacteria *Helicobacter pylori* in the gastrointestinal mucosa [99]. Toxic bismuth levels are generally not reached with normal use, although salicylism has been reported following the use of bismuth subsalicylate [683]; however, the potential for toxic levels of bismuth ion has been addressed during *H. pylori* therapy concurrent with agents causing hypochlorhydria [534]. Rats exposed to bismuth oxychloride in the diet for 2 years at doses up to 2.0 mg/kg/day were found to have no increased incidence of tumors [540]. The ACGIH has adopted a TLV–TWA value for bismuth telluride of 10 mg/m$^3$ [35].

## BROMINE

Bromine is a reddish-brown, noncombustible liquid. Although the Earth's crust contains a vast amount of bromine, the most readily recoverable sources of bromine are in salt lakes and brines. The largest use of bromine is in the production of fire retardants, as gasoline anti-knock agents, and in the agricultural chemical industry [531]. Bromine vapors are highly toxic; exposure to 1000 ppm bromine is rapidly fatal in humans, and brief exposures to 40 to 60 ppm are dangerous [531]. The symptoms of inhalation exposure include coughing, nosebleed, dizziness and headache, and abdominal pain and diarrhea; sometimes, a measles-like eruption on the trunk and extremities can occur [531]. Bromine vapor is extremely irritating to the eyes, skin, and mucous membranes and produces inflammatory lesions in the upper respiratory tract [26]. Prolonged contact with the skin causes ulceration [531].

Bromine may be an essential element. A bromine-deficient diet impaired the growth and reproductive success of goats; however, the evidence for the essentiality of bromine is weak, and no cases of human deficiency are likely to occur due to the widespread distribution of bromine in foods [40].

Excessive oral intake of bromide can cause neurological symptoms in humans (e.g., headache, lethargy, ataxia, disorientation), and high levels of bromine in the diet (400 to 1200 mg/kg) can cause CNS depression in mice [623]. Potassium bromate is a widely used form of bromine. Its primary use is as a conditioner for flour and dough [531]. Potassium bromate has been investigated for possible carcinogenic activity. Renal and other tumors were induced in male and female Fischer 344 rats and renal tumors in mice exposed to potassium bromate in drinking water in 2-year bioassays; tumor formation was dose dependent [173,327]. The IARC has classified potassium bromate as a Group 2B (possibly carcinogenic to humans) carcinogen. Conversely, mice and rats fed diets high in bread containing up to 75 mg/kg potassium bromate showed no increase in tumor incidence [222,246]. The lack of a carcinogenic effect could be attributable to the degradation of potassium bromate during the baking process [327].

The ACGIH has adopted TLV values for bromine gas of 0.1 ppm as an 8-hour TWA and 0.2 ppm as a STEL/C [35]. Although the USEPA has established a number of toxicity values for bromine-containing compounds, the USEPA has not derived any toxicity values for bromine [674].

## Cerium

Cerium is a lanthanous rare earth metal that is used in fireworks and cigarette lighter flints, in self-cleaning ovens, and as an abrasive for polishing glass [67]. Occupational inhalation exposure has been reported to cause a pneumoconiosis without pulmonary functional impairment [316]. Intravenous administration of cerium chloride produces severe hepatotoxicity in rats [476]. The USEPA has not derived toxicity values for cerium [674].

## Gallium

Gallium is a relatively rare metal that has found uses in diagnostic radiology [67], as an antineoplastic agent [88], and for the control of cancer-related hypercalcemia [146,187,695]. It is also used in the manufacture of alloys and semiconductor electronic devices. There is limited indications of occupationally related toxicity. Occupational exposure to $GaF_3$ fumes has resulted in a rash with subsequent reversible neurological effects consisting of muscular weakness [67]. Gallium is excreted in urine. Renal toxicity is noted in rats with the formation of precipitates of gallium complexed with calcium and phosphate [67]. No adverse effects were noted in a reproduction study conducted in male mice [157]. No occupational health standards have been established for gallium or its compounds [67]. The USEPA has not derived toxicity values for gallium.

## Germanium

Germanium is a Group IVA semiconducting metal. The pure metal has a metallic appearance but is very brittle, much like glass. Germanium is not found in the free state but always in combination with other elements, such as silver, copper, and arsenic. Germanium is used in the semiconductor industry (germanium was used in the first transistor), and it is often used in combination with other materials, such as arsenic and antimony, and alloyed with aluminum, gallium, and indium. It is also used in certain optic applications, as the pure metal is transparent to infrared radiation. Industrial exposures are to the dusts and fumes of germanium metal during extraction from ore and metal fumes from welding operations [67]. Germanium oxide and germanium sesquioxide have been used in elixirs for the treatment of cancer and AIDS [637].

In longer term oral animal studies, germanium and germanium oxide have been shown to be nephrotoxic [576], neurotoxic [362,433], and myotoxic [433]. The potential for germanium to induce lung injury is unclear; in one 4-week inhalation toxicity study of germanium powder in rats, histopathologic changes consistent with pulmonary toxicity were present [46], but a follow-up study using germanium dioxide showed no treatment-related histopathologic effects [47]. Germanium does not appear to be carcinogenic; in fact, certain germanium compounds appear to have antineoplastic activity [239]. In a lifetime feeding study in rats, animals receiving 5 ppm sodium germanate in water had a significantly lower incident of tumors than the control animals [349]. Because of this anticancer activity, germanium-containing elixirs have been sold, first in Japan and then in other countries, as a treatment for cancer and other diseases. To date, there have been at least 31 reported cases of toxicity associated with oral intake of germanium compounds, of which 9 were fatal [637]. Nephrotoxicity is the primary manifestation of germanium intoxication, although neurotoxicity and myotoxicity have been reported [632,637]. No TLV has been set for germanium or germanium oxide, but a TLV–TWA of 0.2 ppm has been set for germanium tetrahydride [35].

## Gold

Gold is a soft yellowish metal and belongs to Group IB of the periodic table. Its excellent heat and electrical conductivity and malleability have made it important in industrial applications [66]. Medically, it is used either orally or by intramuscular injection to slow the progression of rheumatoid arthritis, but treatment is associated with a high incidence of toxicity [318]. Adverse skin and mucous membrane effects (dermatitis, stomatitis, pruritus) are most frequent, with the incidence and severity being less for oral as opposed to parenteral treatment [647]. A mild proteinuria is the most common renal effect, but gold-induced nephrosis may occur. Aplastic anemia is relatively rare and has been associated with poor prognosis [709], which may improve with bone marrow transplantation [345]. Although traditionally regarded as inert, gold is being recognized as a common contact allergen [307]. In Sweden, it is second only to nickel [100], and results from the North American Contact Dermatitis Group rank it among the ten most common allergens in the United States [442]. Gold was named the Contact Allergen of the Year in 2001 by the *American Journal of Contact Dermatitis* [228]. Eyelid dermatitis was found in 7.5% of patients with a positive gold patch test reaction [229]. Gold allergy is more common in women than men and is linked to nickel and cobalt allergy. Gold hypersensitivity is characterized by late reactions, and failure to monitor the test site for a minimum of 3 weeks may result in false negatives. The USEPA has not derived toxicity values for gold [674].

### Hafnium

Hafnium is a gray metallic element with a silver-like luster; it is found in association with zirconium ores [34,202]. It has outstanding corrosion resistance and is used for this characteristic in atomic reactors, electronic components, and alloys. It has been evaluated for its antimicrobial properties as a coating on surgical implants [3]. Hafnium compounds show moderate toxicity in acute animal tests by several routes of administration [276]. Studies indicate concentration in the liver and skeleton. Hafnium is poorly absorbed orally [366], and the dust is considered to have relatively low toxicity. Workers exposed to 150 mg/m$^3$ of hafnium- and zirconium-containing dusts showed no adverse effects after 2 to 6 years [196]. The ACGIH has adopted a TLV–TWA value for hafnium of 0.5 mg/m$^3$ [35].

### Indium

Indium is a Group IIIA metal that is widely distributed in the Earth's crust. It is not found in the free state but most commonly in association with copper, zinc, and sulfur. Indium is used in the surface protection of metals and in many alloys because of its ability to increase hardness. Indium compounds are also used in the photovoltaic and semiconductor industry. Industrial exposures to indium occur during extraction and purification and in plating and the manufacture of certain electronic instruments [67]. Absorption of indium compounds is highly dependent on form; insoluble indium compounds are poorly absorbed and distributed, whereas soluble compounds, such as $InCl_3$ and $In_2(SO_4)_3$, are rapidly absorbed and distributed [67,717]. Consistent with these findings, soluble indium compounds are also more toxic than their insoluble counterparts. The acute lethal dose range for the soluble compound indium chloride in rabbits, rats, and dogs was 0.33 to 3.6 mg/kg [192], whereas the minimum lethal dose for insoluble indium oxide in rats was 955 mg/kg [4], and the oral and intraperitoneal $LD_{50}$ for insoluble indium phosphide was greater than 5 g/kg [346].

Indium compounds are toxic when inhaled. Copper indium diselenide, indium trichloride, and indium phosphide, when acutely administered intratracheally to rats at high doses (higher than would be expected in an industrial exposure), induced a persistent inflammatory response [84,464,630]. Copper indium diselenide was only slightly fibrogenic to the lung, and this corresponds with the limited solubility of this compound [465]. Subchronic inhalation of indium sesquioxide in rats induced a persistent inflammatory response; no fibrosis was noted [34]. Hamsters were treated once per week for 15 weeks with either indium arsenide or indium phosphide (dose = 7.5 mg arsenic or phosphorus) by intratracheal installation and were examined at the end of their lifespan. Adverse histopathologic findings were significantly higher in the treated groups [636]. Several studies have investigated the reproductive and developmental toxicity of indium compounds. Indium arsenide, administered intratracheally, reduced epididymal sperm counts in rats [518], but not hamsters [517]; intratracheal instillation of indium chloride in mice did not affect the reproductive performance of either males or females but it was fetotoxic [139]. Indium trichloride was a developmental toxin in mice and rats when administered intravenously [475]. The OSHA TLV–TWA for indium and its compounds is 0.1 mg/m$^3$ [35].

### Lithium

Lithium is a silvery white metal and the lightest solid element. Although it is used in batteries, for organic synthesis (Grignard reagent), in the space industry, as a swimming pool sanitizer, and in air conditioners, industrial intoxication has not been reported [397]. Lithium hydride in contact with water releases hydrogen gas (flammable), and it must be stored under airtight anhydrous conditions [67]. Inhalation exposure to lithium hydride can cause pulmonary edema, but the hydride and not the lithium is thought to be responsible [496]. The ACGIH has adopted a TLV–TWA value for lithium hydride of 0.025 mg/m$^3$ [35].

Oral lithium salts (lithium carbonate, lithium citrate) are widely used in the treatment of manic–depressive disorders, but routine serum monitoring is required because of the narrow therapeutic index [267,427]. The same lithium levels are almost without psychotropic effects in healthy individuals [427]. Effective treatment generally requires levels between 0.8 and 1.2 mEq/L, and toxic effects have been seen at serum levels above 1.5 mEq/L. Signs of lithium toxicity are primarily neurologic and range from fine tremors and muscle weakness in mild cases to dysarthria, hyperreflexia, coma, and collapse. Lithium therapy may produce lasting neurologic consequences [266]. Lithium has properties similar to sodium and substitution for body cations (sodium, potassium) may account for these effects [678]. There is no specific antidote, and treatment is based on limiting absorption and enhancing excretion [416]. Hemodialysis is used to enhance excretion [514]. Renal symptoms of intoxication include polyuria, polydipsia, and renal failure [514]. Lithium therapy during pregnancy has been associated with an increased risk of cardiac anomalies, and there is sufficient animal and human data to indicate lithium can cause developmental toxicity [460].

The Health Council of the Netherlands, acting on recommendations from the Committee for Compounds Toxic to Reproduction, has concluded that lithium carbonate and lithium chloride should be classified in Category 1 (substances known to cause developmental toxicity in humans) and be labeled with R61 (may cause harm to the unborn child) and R64 (may cause harm to breastfed babies) [286]. The USEPA has not derived toxicity values for lithium [674].

## Niobium

Niobium is a white-colored, soft metal found in ores in combination with tantalum and other elements [67]. Niobium is used in alloys and may find use in surgical implants and dental applications [67,75,506,598]. Organometallic niobium compounds have shown antitumor and anti-HIV activity *in vitro* and in mice [372,373,588]. Acute and chronic animal tests have been conducted on several niobium compounds [275,359,589]. Niobium is poorly absorbed from the gastrointestinal tract. Parenteral administration of niobium pentachloride results in decreased respiration, lethargy, and death. The compound is a moderate to severe skin irritant, with less irritation noted in rabbit eyes. Life-term studies of sodium niobate in mice and rats did not show carcinogenicity [349]. Occupational or general health standards have not been established for niobium in the United States. No reports of occupational health hazards from dust or fumes associated with forging or other fabrication techniques of niobium metal and alloys have been reported [67].

## Osmium

Osmium is a platinum group metal. Osmium tetroxide is a noncombustible, colorless to pale yellow solid, with a disagreeable chlorine-like odor. Osmium tetroxide is apparently formed quite readily from finely divided osmium metal by heating in air or even at room temperature [278]. Osmium is found in combination with platinum- and nickel-bearing ores. The major use of osmium is as osmium tetroxide, which is used as a biological stain for adipose tissues [34].

Metallic osmium and most of its other compounds are not considered highly toxic [34]; however, osmium tetroxide has been shown to be toxic in animals and in humans. The oral $LD_{50}$ for osmium tetroxide has been reported to be 14 mg/kg in the rat and 162 mg/kg in the mouse; the intraperitoneal $LD_{50}$ for the mouse was 14 mg/kg [67]. The reported $LC_{50}$ for the rat and mouse is 400 mg/m$^3$ [34]. Additionally, rabbits exposed for 30 minutes to osmium tetroxide at a concentration of 130 mg/m$^3$ died after 4 days from pulmonary edema [98]. Application of a drop of a 1% solution of osmium tetroxide to the rabbit eye caused severe corneal damage, permanent opacity, and superficial vascularization [98]. Toxic effects have also been reported on guinea pig bone marrow, although the route of administration, dose, and duration were not reported [278].

Osmium tetroxide-induced toxicity in humans has been reported in the early toxicology literature. Inhalation exposure to $OsO_4$ can cause irritation of the nose and throat which can persist for at least 12 hours [278]. Industrial exposure to osmium tetroxide concentrations ranging from 0.1 to 0.6 mg/m$^3$ induces lacrimation and disturbances in vision (i.e., the appearance of rings around lights). Other complaints included conjunctivitis, cough, and headache. Recovery usually occurred within a few days [443]. One human fatality resulting from inhalation of osmium tetroxide has been reported [443]. The exposure concentration was not reported; death was attributed to capillary bronchitis and pulmonary edema. The ACGIH has adopted a TLV–TWA of 0.0002 ppm and a STEL of 0.0006 ppm for osmium tetroxide, both measured as osmium [35].

## Platinum

Although platinum is relatively rare, it is found both as the pure metal and in combination with nickel, copper, and gold [67]. Platinum is used as a catalyst in the automotive, chemical, and pharmaceutical industries, and its nobility (resistance to oxidation) makes it important in the manufacture of laboratory equipment [559]. Metallic platinum is relatively inert, but the complex salts are frequent sensitizers, producing conjunctivitis, urticaria, dermatitis, and eczema following inhalation or dermal exposure [247]. A syndrome, previously known as "platinosis," is characterized by lacrimation, sneezing, rhinorrhea, cough, dyspnea, bronchial asthma (from chloroplatinates), and cyanosis [247]; however, this term suggests a pneumoconiosis and fibrosis which are not typical of platinum allergy syndrome, and the condition is better known as "allergy to platinum compounds containing reactive halogen ligands" [313]. The above-mentioned symptoms are elicited by either an immediate (type I) or delayed (type IV, within 24 hours) hypersensitivity reaction [247]. The platinum analog cisplatin has been used as a chemotherapeutic agent against various cancers, especially testicular and ovarian tumors, despite nephrotoxicity at therapeutic doses [247]. More recently, carboplatin has been used with comparable efficacy (for many types of cancer) and less toxicity, with thrombocytopenia being the major side effect [247,411]. The ACGIH has adopted TLV–TWA values of 1 mg/m$^3$ for platinum metal and 0.002 mg/m$^3$ for soluble salts [35]. The TLV for platinum salts protects against sensitization but does not offer protection to a previously sensitized individual. No increased risk of cancer has been reported from occupational exposure to platinum [247]. The USEPA has not evaluated the toxicity of platinum [674].

## Rhodium

Rhodium is a silver-white, hard metal that can form highly corrosive-resistant alloys and coatings used in electrical contacts, reflectors, and jewelry [34,247]. Unlike the related platinum compounds, rhodium compounds have not been found to be clinically active in cancer therapy. Antimalarial activity is reported for a rhodium–chloroquine complex [578]. Only a limited toxicity profile has been developed for rhodium and its compounds. Intravenously administered rhodium trichloride was moderately low in acute toxicity in rats and rabbits (approximately 200 mg/kg), with death possibly due to central nervous system depression [383]. Oral rhodium trichloride was low in toxicity ($LD_{50}$ > 500 mg/kg) [247]. A chronic feeding study showed slight carcinogenic activity in mice [585]. Rhodium has been reported to cause allergic contact urticaria [220]. The ACGIH has adopted a TLV–TWA for rhodium of 1 mg/m$^3$ for elemental and insoluble compounds, and 0.1 mg/m$^3$ for soluble rhodium compounds concurrent with the determination that elemental and rhodium compounds are not classifiable as human carcinogens [34,35]. The OSHA PELs for elemental, insoluble, and soluble rhodium compounds are 1/10 these levels [34].

## Tantalum

Tantalum is a gray, hard metal found in ores in combination with niobium and other metals. It is used in electric capacitors and as the carbide for tools, and it has found a wide range of uses in medical diagnostic and surgical implant applications [34,67,81]. Elemental tantalum and its principal oxide are essentially nontoxic *in vitro* and *in vivo* [506]. Occupational exposure to tantalum and its oxide has shown no overt adverse health effects [34]. Medical uses include tantalum gauze in the repair of hernias, implant plates and screws, and radiographic lung and bone markers [81]. The TLV–TWA for tantalum metal and oxide as dust is 5 mg/m$^3$ [35].

## Tellurium

Tellurium is placed in Group IVA of the periodic table. Tellurium has a number of industrial uses and is also found in a variety of food products (e.g., condiments, dairy products, nuts, fish) in high concentration. Pneumonitis and hemolytic anemia are prominent features of acute tellurium intoxication [31]. Tellurium hydride has been shown to be highly toxic, causing pulmonary irritation and intravascular hemolysis [696]. Acute oral or parenteral tellurium intoxication resulted in numerous symptoms, with hematuria noted in all animals treated [31,444]. Weanling rats fed 1% tellurium in the diet developed a peripheral neuropathy characterized by a transient demyelinating/remyelinating event [71,382]. There have been no reports of serious illness or death in workers exposed to tellurium and its compounds; however, absorbed tellurium is slowly metabolized to dimethyl telluride and is excreted in urine, sweat, and breath [175]. It is dimethyl telluride that is responsible for the "garlic breath" associated with tellurium exposure [389]. When an 18-month-old child ingested an unknown quantity of 1.7% tellurium dioxide in 60% HCl, there was no evidence of tellurium toxicity [297]. Two fatalities occurred after unintentional treatment with 2 g of sodium tellurite by ureteral catheter [354]. The autopsy revealed acute fatty degeneration and edema of the liver. The ACGIH has adopted a TLV–TWA value of 0.1 mg/m$^3$ for tellurium and compounds [35].

## Tin

Tin is a soft white metal that occurs in combination with other chemicals (e.g., chlorine, oxygen) [8]. It is alloyed with other metals to make pewter, solder, bronze, and a special cast bronze termed "bell metal" (up to 24% tin), which is noted for its tonal quality [67]. Most of the tin used in the United States is for plating steel cans. The fluoride is used in toothpaste, and the chloride is used to make frost-free windshields [580]. Organotins function as antimicrobials in agriculture and industry, as stabilizers in PVC plastics, and as marine antifouling agents [67]. Although Schwarz et al. [594], in 1970 reported a significant growth effect of dietary tin in weanling rats maintained on purified diets, this has not since been independently confirmed, and tin is not considered to be essential.

Inorganic tin compounds are poorly absorbed from the gastrointestinal tract (rats dosed orally absorbed 2.8% of Sn (II) and less than 1% of Sn (IV) [298]) and therefore are relatively nontoxic. The acute oral $LD_{50}$ for $SnCl_2$ is rats is 700 mg/kg [109], but after intravenous administration it is reported to be 100 mg Sn/kg [597]. Soluble salts of inorganic tin are gastric irritants producing nonspecific signs of nausea, vomiting, and diarrhea. Rats maintained on diets containing 0.3% or more as soluble inorganic tin salts (i.e., stannous chloride) experienced growth retardation and anemia [171]. Injected stannous chloride is a potent inducer of rat renal microsomal heme oxygenase, enhancing heme breakdown [350]. Diets supplemented with high levels of iron and copper protected rats from the anemia but did not alleviate growth depression [170]. Tin has adverse effects on the absorption and metabolism of the essential elements iron, copper, and zinc [170,171,264].

In contrast, organotins, especially the trialkyl derivatives, are highly toxic (rat acute oral $LD_{50}$ is 10 mg/kg) [363]. Triethyltin compounds are skin irritants and potent neurotoxins, producing a decrease in myelin content of the CNS and edema of the white matter [571]. Uncoupling of oxidative phosphorylation has been proposed as the mechanism of action [462]. Tributyltin (TBT) is known to masculinize the sex organs of female mollusks [332], resulting in a pseudohermaphroditism termed "imposex" [616]. Although the mechanisms by which TBT causes masculinization in mollusks is unknown, butyltins have been shown to inhibit human placental cytochrome P450 aromatase activity when measured *in vitro* [287] and to affect male sexual development in rats [268].

Acute inhalation exposure to tin can cause metal fume fever, and chronic exposure can cause a benign pneumoconiosis, stannosis [496]. The ACGIH has adopted TLV–TWA values of 2 mg/m$^3$ for the inorganic compounds (except tin hydride, $SnH_4$) and a TLV–TWA of 0.1 mg/m$^3$ for organic tin compounds [35]. Although the USEPA has not evaluated tin and tin-containing compounds for carcinogenicity, the NTP has performed a 2-year bioassay for stannous chloride in rats and mice with negative results in all but male rats, where the results were equivocal for thyroid C-cell tumors [484].

## Titanium

Titanium is a silver-gray colored metal that can occur naturally in several forms including titanium dioxide. Titanium is a component of several alloys and is used in surgical implants, where it is considered nontoxic [285]. Titanium dioxide, the most common oxide of titanium, is extensively used as a white pigment in paints, plastics, inks, and cosmetics [67,245]. Titanium dioxide is generally considered to be essentially nontoxic by the oral, dermal, and inhalation routes. A 2-year feeding study of titanium dioxide at maximum doses of 2.5 g/kg/day in rats and 6.4 g/kg/day in mice found no evidence of carcinogenicity [477], although the dose–response relationship was statistically significant for thyroid tumors in female rats and keratoacanthomas in male rats [248]. Toxicity from the more prevalent respiratory exposure to titanium dioxide has been investigated. A 2-year inhalation study was conducted in rats with acceptable results at a level of 10 mg/m$^3$; high levels produced squamous cell carcinomas, which are postulated to be the result of saturation of normal pulmonary clearance mechanisms [34,393]. Epidemiological findings and related information do not conclusively support a relationship between occupational exposure to titanium dioxide and pulmonary fibrosis, cancer, or other adverse health effects [34]. The ACGIH has adopted a TLV–TWA value of 10 mg/m$^3$ for titanium dioxide [35].

## Tungsten

Tungsten is a member of the third series of transition metals and occurs in nature in combination with iron, manganese, and calcium. The major use of tungsten is in cutting and wear-resistant materials [67]. Tungsten–iron shot is also in use as a less toxic replacement for lead shot [456]. In dogs, the bulk of inhaled tungsten oxide is rapidly excreted [1]. Although exposure to soluble tungsten compounds can be toxic in experimental animals [364,613], insoluble tungsten compounds have a low order of toxicity [174,231,365,454,456]. Male and female rats given sodium tungstenate in water for 2.5 years at doses of 0.25 and 0.29 mg/kg/day had no significant increase in tumor incidence [589]. Pulmonary fibrosis observed in men with inhalation exposure to cobalt-cemented tungsten carbide [453] has been attributed to cobalt [67]. Evaluation of workers with long-term exposure to tungsten or its insoluble compounds showed no development of pneumoconioses [34]. The tungsten ion antagonized the normal metabolic action of the molybdate ion, and therefore can inhibit molybdate-dependent enzymes [296,342,510]. The ACGIH has adopted TLV–TWA values of 5 mg tungsten per m$^3$ for insoluble compounds and 1 mg tungsten per m$^3$ for soluble compounds [35].

## Yttrium

Yttrium is a reactive rare earth lanthanide and is silvery white in color [67]. It is used as an alloying agent in stainless steels requiring high resistance to corrosion. When used in combination with zirconium, it improves the strength of magnesium castings. In the electronics industry, it is used as the matrix producing the red color in television tubes [67]. Yttrium chloride has been reported to cause granulomatous changes in the rat lung following intratracheal instillation [300], and the liver and spleen are reported to be the primary target organs following intravenous injection [301]. Despite a long history of industrial use, there are no definitive reports of adverse effects in workers [67]. The $LD_{50}$ for yttrium chloride following intraperitoneal injection in rats is 132 mg/kg body weight [152]. The USEPA has not derived toxicity values for yttrium. The ACGIH had adopted a TLV–TWA value of 5 mg/m$^3$ but reduced this to 1 mg/m$^3$ [35] based on a report that intratracheal administration of $Y_2O_3$ caused severe lung damage [67].

## Zirconium

Zirconium is a grayish white element of the second series of transition metals. The metal is produced from two main sources of ore, zircon ($ZrO \cdot SiO_2$), and baddeleyite ($ZrO_2$); hafnium is always associated with zirconium [35]. Zirconium is used for the cladding in nuclear fuel rods, and zirconium compounds are also used in foundry and sandblasting applications. Industrial exposure occurs during mining and purification operations, and in foundry and other industries. A significant percentage of the general population is exposed (dermally) to aluminum zirconium chlorohydrate complexes in commercially marketed antiperspirant products. Zirconium oxide has a low order of toxicity via the inhalation route in animals [619]; slight toxicity was noted in dogs when exposed to an airborne mist of zirconium chloride at 6 mg/m$^3$ for 2 months. Zirconium oxide and zirconium chloride exposure at 3.5 mg Zr per m$^3$ for 1 year had no measurable adverse effect on the animals exposed [67]. Similarly, in most studies of industrially exposed workers, no adverse effects have been associated with inhalation exposure to zirconium fumes or other zirconium compounds [274,426,550]; however, several cases of either fibrotic [61] or granulomatous [375,403] changes in the lung associated with inhalation exposure to zirconium compounds have been reported. Long-term exposure of mice to zirconium sulfate was not associated with increased tumor incidence [349].

Certain zirconium compounds, such as zirconium lactate, when applied to human skin [57,67,568,606] or the skin of experimental animals [545], can produce dermal granulomas of allergic origin. Aluminum zirconium chlorohydrate complexes, used as active ingredients in antiperspirants, do not appear to cause these granulomatous reactions, but, because of risk–benefit considerations, the USFDA and other global regulatory authorities banned the use of these materials in aerosolized drug and cosmetic products [665]. The ACGIH has adopted a TLV–TWA value of 5 mg/m$^3$ and a STEL value of 10 mg/m$^3$ for zirconium compounds [35].

# QUESTIONS

1. Margins of safety are normally calculated by determining a no-observed-adverse-effect level (NOAEL) in an animal species and then modifying the value by factors to account for inter- and intraspecies variability. USEPA reference doses (RfDs) are calculated using a similar approach. How does the oral RfD for the essential metal molybdenum show the limitations of this approach for calculating margins of safety?
   *Answer:* The estimated safe and adequate daily dietary intake for molybdenum is set at the current U.S. dietary intake of 75 to 250 μg/day, whereas the oral RfD for molybdenum is 350 μg/day, a value very close to current dietary intake. In fact, the USEPA had initially derived an RfD of 280 μg/day, essentially the same as current U.S. dietary intake levels. These calculations show that the indiscriminant use of safety factors (without consideration of other relevant information) can lead to overestimation of the toxicity of a substance.
2. Your 15-kg patient has ingested 10 tablets, each containing 324 mg ferrous sulfate. How much elemental iron did the child ingest? (Ferrous sulfate contains 20% elemental iron.)
   *Answer:* $10 \times 324 \times 0.20 = 648$ mg elemental iron; $648/15 = 43.2$ mg Fe per kg body weight. Inaccurate calculation of the dose of elemental iron ingested is a known problem in the management of iron poisoning.
3. Why is toenail selenium used instead of blood or hair selenium concentrations for estimating past selenium intake?
   *Answer:* Hair selenium content is difficult to accurately measure because of the common use of selenium-containing antidandruff shampoos. Blood selenium is not a good indicator of past selenium intake because the life of an erythrocyte is only 120 days.
4. Some metals elicit clinical neurotoxicity at certain workplace exposure levels; for example, manganism is often characterized by Parkinson-like symptoms. Describe tests useful in determining neurotoxicity prior to clinical symptoms so safe exposure levels may be established and corrective action taken.
   *Answer:* Nonclinical neurological injury can be evaluated by controlled neurobehavioral tests that evaluate reaction time, eye–hand coordination, and hand steadiness.

# ACKNOWLEDGMENT

The authors gratefully acknowledge the significant contributions of the previous author Fanny K. Ennever.

## REFERENCES


1. Aamodt, R. L. (1975): Inhalation of $^{181}W$ labeled tungstic oxide by six beagle dogs. *Health Phys.*, 28:733–742.
2. Abd Elghany, A., Schumacher, M. C., Slattery, M. L. et al. (1990): Occupation, cadmium exposure, and prostate cancer. *Epidemiology*, 1:107–115.
3. Abdullin, I. S., Mironov, N.M., and Garipova, G.I. (2004): Bactericidal and biologically stable coatings for medical implants and instruments. *Med. Tekh.*, 4:20–22.
4. Adamson, R. H., Canellos, G. P., and Sieber, S. M. (1975): Studies on the antitumor activity of gallium nitrate and other Group IIIa metal salts. *Cancer. Chemother. Rep.*, 59:599–610.
5. Agarwal, A. K., Ahlawat, S. K., Gupta, S. et al. (1995): Hypokalemic paralysis secondary to acute barium carbonate poisoning. *Trop. Doctor*, 25:101–103.
6. ATSDR. (1990): *Toxicological Profile for Copper*. Agency for Toxic Substances and Disease Registry, Atlanta, GA.
7. ATSDR. (1990): *Toxicological Profile for Silver*. Agency for Toxic Substances and Disease Registry, Atlanta, GA.
8. ATSDR. (1990): *Toxicological Profile for Tin*. Agency for Toxic Substances and Disease Registry, Atlanta, GA.
9. ATSDR. (1991): *Toxicological Profile for Antimony*. Agency for Toxic Substances and Disease Registry, Atlanta, GA.
10. ATSDR. (1992): *Toxicological Profile for Barium*. Agency for Toxic Substances and Disease Registry, Atlanta, GA.
11. ATSDR. (1992): *Toxicological Profile for Beryllium*. Agency for Toxic Substances and Disease Registry, Atlanta, GA.
12. ATSDR. (1992): *Toxicological Profile for Boron*. Agency for Toxic Substances and Disease Registry, Atlanta, GA.
13. ATSDR. (1992): *Toxicological Profile for Cobalt*. Agency for Toxic Substances and Disease Registry, Atlanta, GA.
14. ATSDR. (1992): *Toxicological Profile for Vanadium*. Agency for Toxic Substances and Disease Registry, Atlanta, GA.
15. ATSDR. (1993): *Toxicological Profile for Arsenic*. Agency for Toxic Substances and Disease Registry, Atlanta, GA.
16. ATSDR. (1999): *Toxicological Profile for Cadmium*. Agency for Toxic Substances and Disease Registry, Atlanta, GA (http://www.atsdr.cdc.gov/toxprofiles/tp5–c2.pdf).
17. ATSDR. (1993): *Toxicological Profile for Lead*. Agency for Toxic Substances and Disease Registry, Atlanta, GA.
18. ATSDR. (1993): *Toxicological Profile for Mercury*. Agency for Toxic Substances and Disease Registry, Atlanta, GA.
19. ATSDR. (1994): *Toxicological Profile for Magnesium*. Agency for Toxic Substances and Disease Registry, Atlanta, GA.
20. ATSDR. (1994): *Toxicological Profile for Selenium*. Agency for Toxic Substances and Disease Registry, Atlanta, GA.
21. ATSDR. (1994): *Toxicological Profile for Zinc*. Agency for Toxic Substances and Disease Registry, Atlanta, GA.
22. ATSDR. (1998): *Toxicological Profile for Manganese*, draft for public comment (update). Agency for Toxic Substances and Disease Registry, Atlanta, GA.
23. ATSDR. (2004): *Toxicological Profile for Strontium*. Agency for Toxic Substances and Disease Registry, Atlanta, GA.
24. Akiniwa, K. (1997): Re-examination of acute toxicity of fluoride. *Fluoride*, 30:89–104.
25. Alderman, L. C. and Bergin, J. J. (1986): Hydrogen selenide poisoning: an illustrative case with review of the literature. *Arch. Environ. Health*, 41:354–358.
26. Alexandrov, D. D. (1983): Bromine and Compounds. In: *Encyclopaedia of Occupational Health and Safety*, 3rd ed., Vol. 1, edited by L. Parmeggiani, pp. 326–329. International Labour Organization, Geneva.
27. Alfrey, A. C. (1993): Aluminum toxicity in patients with chronic renal disease. *Ther. Drug Metab.*, 15:593–597.
28. Allen, L. A. (1996): Nutritional products. In: *Handbook of Nonprescription Drugs*, pp. 361–392. American Pharmaceutical Association, Washington D.C.
29. Alomar, A., Conde-Salazar, L., and Romaguera, C. (1985): Occupational dermatoses from cutting oils. *Contact Derm.*, 12:129–138.
30. Alsberg, C.L. and Schwartze, E.W. (1919): Pharmacological action of Cd. *Pharmacology*, 13:504–510.
31. Amdur, M. L. (1958): Tellurium oxide, an animal study in acute toxicity. *A. M. A. Arch. Indust. Health*, 17:665–667.
32. Amdur, M. O. (1978): Respiratory response to iodine vapor alone and with sodium chloride aerosol. *J. Toxicol. Environ. Health*, 4:619–630.
33. American Academy of Pediatrics. (1998): Screening for elevated blood lead levels. *Pediatrics*, 101:1072–1078.
34. ACGIH. (1996): *Documentation of the Threshold Limit Values and Biological Exposure Indices*, 6th ed., American Conference of Governmental Industrial Hygienists, Cincinnati, OH.
35. ACGIH. (2004): *2004 TLVs and BEIs*, American Conference of Governmental Industrial Hygienists, Cincinnati, OH.
36. AIHA. (1968): *Hygienic Guide Series—Nickel Carbonyl*. American Industrial Hygiene Association, Fairfax, VA, pp. 304–307.
37. Anderson, A. C. (1994): Iron poisoning in children. *Curr. Opin. Pediatr.*, 6:289–294.
38. Anderson, R. A. (1997): Chromium as an essential nutrient for humans. *Reg. Toxicol. Pharmacol.*, 26:S35–S41.
39. Angle, C. R. (1993): Childhood lead poisoning and its treatment. *Annu. Rev. Pharmacol. Toxicol.*, 32:409–434.
40. Anke, M., Groppel, G., and Arnhold, W. (1990): Essentiality of the trace element bromine. *Acta Agronom. Hung.*, 39:297–303.
41. Anke, M., Groppel, G., and Krause, U. (1991): Essentiality of the toxic elements cadmium, arsenic, and nickel. In: *Trace Elements in Man and Animals*, Vol. 7, edited by B. Momcilovic, pp. 11-6–11-8. IMI, Zagreb, Croatia.
42. Anke, M., Groppel, G., and Krause, U. (1991): Essentiality of the toxic elements aluminum and vanadium. In: *Trace Elements in Man and Animals*, Vol. 7, edited by B. Momcilovic, pp. 11-9–11-11. IMI, Zagreb, Croatia.
43. Anke, M., Groppel, G., and Krause, U. (1991): Fluorine deficiency in goats. In: *Trace Elements in Man and Animals*, Vol. 7, edited by B. Momcilovic, pp. 26-28–26-27. IMI, Zagreb, Croatia.

44. Anneroth, G., Ericson, T., Johansson, I. et al. (1992): Comprehensive medical examination of a group of patients with alleged adverse effects from dental amalgams. *Acta Odontolog. Scand.*, 50:101–111.
45. Apostoli, P., Kiss, P., Porru, S., Bonde, J. P., Vanhoorne, M., and ASCLEPIOS (1998): Male reproductive toxicity of lead in animals and humans. *Occup. Environ. Med.*, 55:364–374.
46. Arts, J. H. E., Reuzel, P. G. J., Falke, H. E., and Beems, R. B. (1990): Acute and sub-acute inhalation toxicity of germanium metal powder in rats. *Food Chem. Toxicol.*, 28:571–579.
47. Arts, J. H. E., Til, H. P., Kuper, R., and Swennen, B. (1994): Acute and subacute inhalation toxicity of germanium dioxide in rats. *Food Chem. Toxicol.*, 32:1037–1046.
48. Assennato, G., Paci, C., Baser, M. E., Molinini, R., Candela, R. G., Altamura, B. M., and Giorgino, R. (1987): Sperm count suppression without endocrine dysfunction in lead-exposed men. *Arch. Environ. Health*, 42:123–127.
49. Awasthi, Y. C., Beutler, E., and Srivastava, S. K. (1975): Purification and properties of human erythrocyte glutathione peroxidase. *J. Biol. Chem.*, 250:5144–5149.
50. Axelson, O., Dahlgren, E., Jansson, C.-D., and Rehnlund, S. O. (1978): Arsenic exposure and mortality: a case referent study from a Swedish copper smelter. *Br. J. Ind. Med.*, 35:8–15.
51. Ayres, S., Jr., and Anderson, N. P. (1934): Cutaneous manifestations of arsenic poisoning. *Arch. Dermatol. Syphil.*, 30:33–43.
52. Bacon, B. R., and Britton, R. S. (1990): The pathology of hepatic iron overload: a free radical mediated process? *Hepatology*, 11:127–137.
53. Bacon, B. R., Tavill, A. S., Brittenham, G. M. et al. (1983): Hepatic lipid peroxidation *in vivo* in rats with chronic iron overload. *J. Clin. Invest.*, 71:429–439.
54. Baker, E. L., White, R. F., Pothier, L. J. et al. (1985): Occupational lead neurotoxicity: improvement in behavioral effects after reduction of exposure. *Br. J. Indust. Med.*, 42:507–516.
55. Baker, T. D. and Hafner, W. G. (1961): Cadmium poisoning from a refrigerator shelf used as an improvised barbecue grill. *Pub. Health Rep.*, 76:543–544.
56. Bako, G., Smith, E. S., Hanson, J. et al. (1982): The geographical distribution of high cadmium concentrations in the environment and prostate cancer in Alberta. *Can. J. Public Health*, 73:92–94.
57. Baler, G. R. (1965): Granulomas from topical zirconium in poison ivy dermatitis. *Arch. Dermatol.*, 91:145–148.
58. Banting, D. W. (1991): The future of fluoride: an update one year after the National Toxicology Program study. *J. Am. Dental Assoc.*, 123:86–91.
59. Barnhart, S. and Rosenstock, L. (1984): Cadmium chemical pneumonitis. *Chest*, 86:789–791.
60. Barret, H. M. and Card, B. Y. (1947): Studies on the toxicity of inhaled cadmium. II. The acute lethal dose cadmium oxide for man. *J. Indust. Hyg. Toxicol.*, 29:286–293.
61. Bartter, T., Irwin, R. S., Abraham, J. L., Dascal, A., Nash, G., Himmelstein, J. S., and Jederlinic, P. J. (1991): Zirconium compound-induced pulmonary fibrosis. *Arch. Intern. Med.*, 151:1197–1201.
62. Bartzokis, G., Tishler, T.A., Shin, I. S. et al. (2004): Brain ferritin iron as a risk factor for age at onset in neurodegenerative diseases. *Ann. N.Y. Acad. Sci.*, 1012:224–236.
63. Bast, A., Haenen, G. R., and Doelman, C. J. (1991): Oxidants and antioxidants: state of the art. *Am. J. Med.*, 91(Suppl. 3C):2–13.
64. Bates, M. N., Smith, A. H., and Hopenhayn-Rich, C. (1992): Arsenic ingestion and internal cancers: a review. *Am. J. Epidemiol.*, 135:462–476.
65. Beckett, W. S., Moore, J. L., Keogh, J. P., and Bleecker, M. L. (1986): Acute encephalopathy due to occupational exposure to arsenic. *Br. J. Indust. Med.*, 43:66–67.
66. Beliles, R. P. (1979): The lesser metals. In: *Toxicity of Heavy Metals in the Environment*, edited by F. W. Oehme, pp. 547–615. Marcel Dekker, New York.
67. Beliles, R. P. (1994): The metals. In: *Patty's Industrial Hygiene and Toxicology*, edited by G. D. Clayton and F. E. Clayton, pp. 1879–2352. John Wiley & Sons, New York.
68. Beliles, R. P. and Beliles, E. M. (1994): Phosphorus, selenium, tellurium, and sulfur. In: *Patty's Industrial Hygiene and Toxicology*, edited by G. D. Clayton and F. E. Clayton, pp. 783–829. John Wiley & Sons, New York.
69. Belizan, J. M., Villar, J., Zalazar, A. et al. (1983): Preliminary evidence of the effect of calcium supplementation on blood pressure in normal pregnant women. *Am. J. Obstet. Gynecol.*, 146:175–180.
70. Bellinger, D., Leviton, A., Waternaux, C., Needleman, H. and Rabinowitz, M. (1987): Longitudinal analyses of prenatal and postnatal lead exposure and early cognitive development. *N. Engl. J. Med.*, 316:1037–1043.
71. Berciano, M. T., Calle, E., Fernández, R., and Lafarga, M. (1998): Regulation of Schwann cell numbers in tellurium-induced neuropathy: apoptosis, supernumerary cells and internodal shortening. *Acta Neuropathol.*, 95:269–279.
72. Bergh, A.R. (1990): The acute vascular effects of cadmium in the testis do not require the presence of Leydig cells. *Toxicology*, 63:183–186.
73. Berkovitch, M., Matsui, D., Lamm, S. H. et al. (1994): Recent increases in numbers and risk of fatalities in young children ingesting iron preparations. *Vet. Hum. Toxicol.*, 36:53–55.
74. Bernard, S.M. (2003): Should the Centers for Disease Control and Prevention's childhood lead poisoning intervention level be lowered? *Am. J. Public Health*, 93:1253–1260.
75. Berry, J. P., Bertrand, F., and Galle, P. (1993): Selective intra-lysosomal concentration of niobium in kidney and bone marrow cells: a microanalytical study. *BioMetals*, 6:17–23.
76. Berry, M.J. and Larsen, P.R. (1992): The role of selenium in thyroid hormone action. *Endocr. Rev.*, 13:207–219.
77. Beton, D. C., Andrews, G. S., Davies, H. J. et al. (1966): Acute cadmium fume poisoning: five cases with one death from renal necrosis. *Br. J. Indust. Med.*, 23:292–301.
78. Beyer, K. H., Bergfeld, W. F., and Berndt, W. O. (1983): Final report on the safety assessment of sodium borate and boric acid. *J. Am. Coll. Toxicol.*, 2:87–125.
79. Bhattacharyya, M. H., Jeffery, E., and Silbergeld, E. K. (1996): Bone metabolism: effects of essential and toxic trace metals. In: *Toxicology of Metals*, edited by L. W. Chang, pp. 959–971. CRC Press, Boca Raton, FL.

80. Birch, N. J. and Karim, A. R., (1988): Potassium. In: *Handbook on Toxicology of Inorganic Compounds*, edited by N. G. Sieler and H. Sigel, pp. 543–553. Marcel Dekker, New York.
81. Black, J. (1994): Biological performance of tantalum. *Clin. Mater.*, 16:167–173.
82. Blanc, P. and Boushey, H. A. (1993): The lung in metal fume fever. *Seminars Resp. Med.*, 14:212–225.
83. Blank, E. and Howieson, J. (1983): Lead poisoning from a curtain weight. *J. Am. Med. Assoc.*, 249:2176–2177.
84. Blazka, M. E., Dixon, D. Haskins, E., and Rosenthal, G. J. (1994): Pulmonary toxicity to intratracheally administered indium trichloride in Fischer 344 rats. *Fundam. Appl. Toxicol.*, 22:231–239.
85. Blejer, H. P. (1966): Death due to cadmium oxide fumes. *Indust. Med. Surg.*, 35:363–364.
86. Blevins, D. G. and Lukaszewski, K. M. (1994): Proposed physiologic functions of boron in plants pertinent to animal and human metabolism. *Environ. Health Persp.*, 102(Suppl. 7):31–33.
87. Blom, S., Lagerkvist, B., and Linderholm, H. (1985): Arsenic exposure to smelter workers: clinical and neurophysiological studies. *Scand. J. Work Environ. Health*, 11:265–269.
88. Bockman, R. S., Wilhelm, F., Siris, E., Singer, F., Chausmer, A., Britton, R., Kotler, J., Bosco, J., Eyre, D. R., and Levenson, D. (1995): A multicenter trial of low dose gallium nitrate in patients with advanced Paget's disease of bone. *J. Clin. Endocrin. Metab.*, 80:595–602.
89. Borgoño, J. M., Vicent, P., Venturino, H., and Infante, A. (1977): Arsenic in the drinking water of the city of Antofagasta: epidemiological and clinical study before and after the installation of a treatment plant. *Environ. Health Perspec.*, 19:103–105.
90. Boulos, B. M. and von Smolinski, A. (1988): Alert to users of calcium supplements as antihypertensive agents due to trace metal contaminants. *Am. J. Hypertens.*, 1:137S–142S.
91. Bourgoin, B. P., Evans, D. R., Cornett, J. R. et al. (1993): Lead content in 70 brands of dietary calcium supplements. *Am. J. Public Health*, 83:1155–1160.
92. Bowman B. A. and Rishert, J. F. (1994): Comparison of the methodological approaches used in the derivation of recommended dietary allowances and oral reference doses for nutritionally essential elements. In: *Risk Assessment of Essential Elements*, edited by W. Mertz, C. O. Abernathy, and S. S. Olin, pp. 63–73. ILSI Press, Washington, D.C.
93. Brackett, R. E. (2004): Statement before Committee on Government Reform Subcommittee on human rights and wellness, U.S. House of Representatives, March 24 (http://www.fda.gov/ola/2004/dietarysupplements0324.html).
94. Brauner C. J. and Wood, C. M. (2002): Effect of long-term silver exposure on survival and ionoregulatory development in rainbow trout (*Oncorhynchus mykiss*) embryos and larvae, in the presence and absence of added dissolved organic matter. *Comp. Biochem. Physiol. C Toxicol. Pharmacol.*, 133(1–2):161–173.
95. Brenniman, G. R. and Levy, P. S. (1984): Epidemiological study of barium in Illinois drinking water supplies. In: *Advances in Modern Toxicology*, Vol. 9, edited by E. J. Calabrese, pp. 231–249. Princeton Scientific, Princeton, NJ.
96. Brewer, G.J. and Merajver, S.D. (2002): Cancer therapy with tetrathiomolybdate: antiangiogenesis by lowering body copper-a review. *Integr. Cancer Ther.*, 1(4):327–337.
97. Brody, D. J., Pirkle, J. L., Kramer, R. A. et al. (1994): Blood lead levels in the U.S. population. Phase 1 of the 3rd National health and Nutrition Examination Survey. *J. Am. Med. Assoc.*, 272:277–283.
98. Brunot, F. R. (1933): The toxicity of osmium tetroxide (osmic acid). *J. Indust. Hyg.*, 15:136–143.
99. Brunton, L. L. (1996): In: *Agents for Control of Gastric Acidity and Treatment of Peptic Ulcers*, edited by J. G. Hardman, L. E. Limberd, P. B. Molinoff, and R.W. Ruddon, pp. 901–915. McGraw-Hill, New York.
100. Bruze, M., Hedman, H., Bjorkner, B., and Moller, H. (1995): The development and course of test reactions to gold sodium thiosulfate. *Contact Derm.*, 33:386–391.
101. Bryant, R.J., Cadogan, J., and Weaver, C.M. (1999): The new dietary reference intakes for calcium: implications for osteoporosis. *J. Am. Coll. Nut.*, 18:406S–412S.
102. Bucher, J. R., Hejtmanick, M. R., Toft, J. D. et al. (1991): Results and conclusions of the National Toxicology Program's rodent carcinogenicity studies with sodium fluoride. *Int. J. Cancer*, 48:733–737.
103. Buchet, J. P., Lauwerys, R., and Roels, H. (1981): Urinary excretion of inorganic arsenic and its metabolites after repeated ingestion of sodium metaarsenite by volunteers. *Int. Arch. Occup. Environ. Health*, 48:111–118.
104. Buchet, J. P., Lauwerys, R., Roels, H. et al. (1990): Renal effects of cadmium body burden of the general population. *Lancet*, 336:699–702.
105. Buckler, H. M., Smith, W. D., and Rees, W. D. (1986): Self poisoning with oral cadmium chloride. *Br. Med. J.*, 292:1559–1560.
106. Bulmer, F. M. R., Rothwell, N. F., and Frankish, E. R. (1938): Industrial cadmium poisoning, a report of fifteen cases, including two deaths. *Can. Public Health J.*, 29:19–26.
107. Burrows, D. (1988): Mischievous metals—chromate, cobalt, nickel and mercury. *Clin. Exp. Dermatol.*, 14:266–272.
108. Byers, R. K. and Lord, E. E. (1943): Late effects of lead poisoning on mental development. *Am. J. Dis. Child.*, 66:471–494.
109. Calvery, H. O. (1942): Trace elements in food. *Food Res.*, 7:313–331.
110. Campbell, D., Gonzales, M., and Sullivan, Jr., J. B. (1992): Mercury. In: *Hazardous Materials Toxicology*, edited by J. B. Sullivan, Jr., and G. R. Krieger, pp. 824–833. Williams & Wilkins, Baltimore.
111. Carleton, A. B., Peters, R. A., and Thompson, R. H. S. (1948): The treatment of arsenical dermatitis with dimercaptopropanol (BAL). *Q. J. Med.*, 17:49–79.
112. Carter, R. F. (1966): Acute selenium poisoning. *Med. J. Aust.*, 1:525–528.
113. Cavanagh, J. B. (1991): What have we learnt from Graham Frederick Young? Reflections on the mechanism of thallium neurotoxicity. *Neuropath. Appl. Neurobiol.*, 17:3–9.
114. Centers for Disease Control and Prevention. (1981): Use of lead tetroxide as a folk remedy for gastrointestinal illness. *MMWR*, 30:546–547.

115. Centers for Disease Control and Prevention. (1983): Folk remedy-associated lead poisoning in Hmong children: Minnesota. *MMWR*, 32:555–556.
116. Centers for Disease Control and Prevention. (1983): Lead poisoning from Mexican folk remedies: California. *MMWR*, 32:554–555.
117. Centers for Disease Control and Prevention. (1984): Lead poisoning-associated death from Asian Indian folk remedies: Florida. *MMWR*, 33:638, 643–645.
118. Centers for Disease Control and Prevention. (1990): Fatal pediatric poisoning from leaded paint: Wisconsin, 1990. *MMWR*, 40:193–195.
119. Centers for Disease Control and Prevention. (1991): *Preventing Lead Poisoning in Young Children: A Statement by the Centers for Disease Control, October 1991*. Centers for Disease Control and Prevention, U.S. Department of Health and Human Services, Atlanta, GA.
120. Centers for Disease Control and Prevention. (1991): *Strategic Plan for the Elimination of Childhood Lead Poisoning*. Centers for Disease Control and Prevention, U.S. Department of Health and Human Services, Atlanta, GA.
121. Centers for Disease Control and Prevention. (1993): Toddler deaths resulting from ingestion of iron supplements: Los Angeles, 1992–1993. *MMWR*, 42:111–113.
122. Centers for Disease Control and Prevention. (1996): Children with elevated blood lead levels attributed to home renovation and remodeling activities: New York, 1993–1994. *MMWR*, 45:1120–1123.
123. Centers for Disease Control and Prevention. (1997): *Screening Young Children for Lead Poisoning: Guidance for State and Local Public Health Officials*. Centers for Disease Control and Prevention, U.S. Department of Health and Human Services, Atlanta, GA.
124. Centers for Disease Control and Prevention. (1997): Update: blood lead levels: United States, 1991–1994. *MMWR*, 46:141–146.
125. Centers for Disease Control and Prevention. (1998): Lead poisoning associated with imported candy and powdered food coloring: California and Michigan. *MMWR*, 47:1041–1043.
126. Centers for Disease Control and Prevention. (1999): Adult lead poisoning from an Asian remedy for menstrual cramps: Connecticut, 1997. *MMWR*, 48:27–29.
127. Centers for Disease Control and Prevention. (2000): Elevated blood lead levels among internationally adopted children: United States, 1998. *MMWR*, 49:97–100.
128. Centers for Disease Control and Prevention. (2001): Fatal pediatric lead poisoning: New Hampshire, 2000. *MMWR*, 50:457–459.
129. Centers for Disease Control and Prevention. (2001): Occupational and take-home lead poisoning associated with restoring chemically stripped furniture: California, 1998. *MMWR*, 50:246–248.
130. Centers for Disease Control and Prevention. (2003): Barium toxicity after exposure to contaminated contrast solution: Goias State, Brazil, 2003. *MMWR*, 52:1047–1048.
131. Centers for Disease Control and Prevention. (2004): Childhood lead poisoning from commercially manufactured French ceramic dinnerware: New York City, 2003. *MMWR*, 53:584–586.
132. Centers for Disease Control and Prevention. (2004): *Advisory Committee on Childhood Lead Poisoning Prevention (ACCLPP)*. Centers for Disease Control and Prevention, U.S. Department of Health and Human Services, Atlanta, GA (http://www.cdc.gov/nceh/lead/ACCLPP/acclpp_main.htm).
133. Centers for Disease Control and Prevention. (2004): Lead poisoning associated with ayurvedic medications. *MMWR* 53:582–584.
134. Centers for Disease Control and Prevention. (2004): *Why Not Change the Blood Lead Level of Concern At This Time?* Centers for Disease Control and Prevention, U.S Department of Health and Human Services, Atlanta, GA (http://www.cdc.gov/nceh/lead/spotLights/change-BLL.htm).
135. Cerulli J., Grabe D. W., Gauthier I., Malone M., and McGoldrick M. D. (1998): Chromium picolinate toxicity *Ann. Pharmacother*, 32:428–431.
136. Chan, K-M., Svancarek, W. P., and Creer, M. (1987): Fatality due to acute hydrofluoric acid exposure. *J. Toxicol. Clin Toxicol.*, 25:333–339.
137. Chan, P. C., Herbert, R. A., Roycroft, J. H., Haseman, J K., Grumbein, S. L., Miller, R. A., and Chou, B. J. (1998) Lung tumor induction by inhalation exposure to molybdenum trioxide in rats and mice. *Toxicol. Sci.*, 45:58–65.
138. Chandler, H. A. and Scott, M. (1986): A review of thallium toxicology. *J. Roy. Nav. Med. Serv.*, 72:75–79.
139. Chapin, R. E., Harris, M. W., Hunter, S., Davis, B. J. Collins, B. J., and Lockhart, A. C. (1995): The reproductive and developmental toxicity of indium in the Swiss mouse *Fundam. Appl. Toxicol.*, 27:140–148.
140. Chela, A., Reig, R., Sanz, P., Huguet, E., and Corbella, J (1989): Death due to hydrofluoric acid. *Am. J. Forensic Med. Pathol.*, 10:47–48.
141. Chen, C.-J., Chuang, Y.-C., Lin, T.-M., and Wu, H.-Y (1985): Malignant neoplasms among residents of a blackfoot disease-endemic area in Taiwan: high-arsenic artesian well water and cancers. *Cancer Res.*, 45:5895–5899.
142. Chen, C.-J., Chuang, Y.-C., You, S.-L., Lin, T.-M., and Wu H.-Y. (1986): A retrospective study on malignant neoplasms of bladder, lung and liver in blackfoot disease endemic area in Taiwan. *Br. J. Cancer*, 53:399–405.
143. Chen, C.-J. and Wang, C.-J. (1990): Ecological correlation between arsenic level in well water and age-adjusted mortality from malignant neoplasms. *Cancer Res* 50:5470–5474.
144. Chen, C.-J., Wu, M.-M., Lee, S.-S., Wang, J.-D., Cheng S.-H., and Wu, H.-Y. (1988): Atherogenicity and carcinogenicity of high-arsenic artesian well water. Multiple risk factors and related malignant neoplasms of Blackfoot disease. *Arteriosclerosis*, 8:452–460.
145. Cheng, Y.-Y. and Qian, P.-C., (1990): The effect of selenium-fortified table salt in the prevention of Keshan disease on a population of 1.05 million. *Biomed. Environ Sci.*, 3:422–428.
146. Chitambar, C.R. (2004): Gallium compounds as antineoplastic agents. *Curr. Opin. Oncol.*, 16(6):547–552.
147. Chu, N.F., Liou, S.H., Wu, T.N. et al. (1999): Reappraisal of the relation between blood lead concentration and blood pressure among the general population in Taiwan. *Occup Environ. Med.*, 56:30–33.

148. Clark, L. C., Combs, Jr., G. F., Turnbull, B. W. et al. (1996): Effects of selenium supplementation for cancer prevention in patients with carcinoma of the skin. *J. Am. Med. Assoc.*, 276:1957–1963 (published erratum appears in *J. Am. Med. Assoc.*, 277:1520, 1997).

149. Clark, L. C., Dalkin, B., Krongrad, A. et al. (1998): Decreased incidence of prostate cancer with selenium supplementation: results of a double-blind cancer prevention trial. *Br. J. Urol.*, 81:730–734.

150. Clayton, C. C. and Bauman, C. A. (1949): Diet and azo dye tumors: effect of diet during a period when the dye is not fed. *Cancer Res.*, 9:575–582.

151. Clewell, H. J., Lawrence, G. A., Calne, D. B. et al. (2003): Determination of an occupational exposure guideline for manganese using the benchmark method. *Risk Anal.*, 23(5):1031–46.

152. Cochran, K. W., Doull, J., Mazur, M., and DuBois, K. P. (1950): Acute toxicity of zirconium, columbium, strontium, lanthanum, cesium, tantalum and yttrium. *Arch. Indust. Health*, 1:637–650.

153. Code of Federal Regulations. Protection of the Environment (1997): National Primary Drinking Water Standards, Vol. 40, Part 141.80. U.S. Government Printing Office, Washington, D.C.

154. Cohen, M. D., Kargacin, B., Klein, C. B., and Costa M. (1993): Mechanisms of chromium carcinogenicity and toxicity. *CRC Crit. Rev. Toxicol.*, 23:255–281.

155. Cohen, M. D., Yang, Z., Zelikoff, J. T., and Schlesinger, R. B. (1996): Pulmonary immunotoxicity of inhaled ammonium metavanadate in Fischer 344 rats. *Fundam. Appl. Toxicol.*, 33:254–263.

156. Cole, V. V., Harned, B. K., and Hafkesbring, R. (1941): The toxicity of strontium and calcium. *J. Pharm. Exp. Ther.*, 71:1–5.

157. Colomina, J. M., Llobtet, J. M., Sirvent, J. J., Domingo, J. L., and Corbella, J. (1993): Evaluation of the reproductive toxicity of gallium nitrate in mice. *Food Chem. Toxicol.*, 31:847–851.

158. Combs, G. F. (1996): Should intakes with beneficial actions, often requiring supplementation, be considered for RDAs? *J. Nutr.*, 126:2373S–2376S.

159. Combs, G.F. (2004): Status of selenium in prostate cancer prevention. *Br. J. Cancer*, 91:195–199.

160. Comstock, G. W., Bush, T. L., and Helzlsouer, K. (1992): Serum retinol, beta-carotene, vitamin E, and selenium as related to subsequent cancer of specific sites. *Am. J. Epidemiol.*, 135:115–121.

161. Condray, J. R. (1985): *Elemental Yellow Phosphorus One-Generation Reproduction Study in Rats*. IR-82-215, IRD No. 401-189. Monsanto Company, St. Louis, MO.

162. Cook, J. D. and Monsen, E. R. (1977): Vitamin C, the common cold, and iron absorption. *Am. J. Clin. Nutr.*, 30:235–241.

163. Costa, M. (1997): Toxicity and carcinogenicity of Cr(VI) in animal models and humans. *Crit. Rev. Toxicol.*, 27:431–442.

164. Coughlin, J. R. (1996): Inorganic borates: chemistry, human exposure, and health and regulatory guidelines. *J. Trace Elem. Exp. Med.*, 9:137–151.

165. Crecelius, E. A. (1977): Changes in the chemical speciation of arsenic following ingestion by man. *Environ. Health Perspec.*, 19:147–150.

166. Culver, B. D., Smith, R. G., Brotherton, R. J. et al. (1994): Boron. In: *Patty's Industrial Hygiene and Toxicology*, edited by G.D. Clayton and F.E. Clayton, pp. 4411–4448. John Wiley & Sons, New York.

167. Curzon, M. E. J., Spector, P. C., and Iker, H. P. (1978): An association between strontium in drinking water supplies and low caries prevalence in man. *Arch. Oral Biol.* 23:317–321.

168. Davison, A. G., Fayers, P. M., Taylor, A. J. et al. (1988): Cadmium fume inhalation and emphysema. *Lancet*, 663–667.

169. De Boeck, M., Kirsch-Volders, M., and Lison, D. (2003): Cobalt and antimony: genotoxicity and carcinogenicity. *Mut. Res.*, 533:135–152.

170. De Groot, A. P. (1973): Subacute toxicity of inorganic tin as influenced by dietary levels of iron and copper. *Food Cosmetic. Toxicol.*, 11:955–962.

171. De Groot, A. P., Feron, V. J., and Til, H. P. (1973): Short-term toxicity studies on some salts and oxides of tin in rats. *Food Cosmetic. Toxicol.*, 11:19–30.

172. de la Burde, B. and Choate, M. S. (1975): Early asymptomatic lead exposure and development at school age. *J. Pediatr.*, 87:638–642.

173. DeAngelo A. B., George M. H., Kilburn S. R., Moore T. M., and Wolf D.C. (1998): Carcinogenicity of potassium bromate administered in the drinking water to male $B6C3F_1$ mice and F344/N rats. *Toxicol. Pathol.*, 26:587–594.

174. Delahant, A. B. (1955): An experimental study of the effects of rare metals on animal lungs. *A.M.A. Arch. Indust. Health*, 12:116–120.

175. DeMeio, R. H. (1947): Tellurium. II. Effect of ascorbic acid on the tellurium breath. *J. Indust. Hyg. Toxicol.*, 29:393–395.

176. Den Hond, E., Nawrot, T., and Staessen, J. A. (2002): The relationship between blood pressure and blood lead in NHANES III. *J. Hum. Hyperten.*, 16:563–568.

177. Department of Health and Human Services. (2004): *Healthy People 2010*. http://www.healthypeople.gov/Document/HTML/objectives/08–11.htm.

178. Desai, H. (1992): Phosphorus and phosphorus compounds. In: *Hazardous Materials Toxicology*, edited by J. B. Sullivan, Jr., and G. R. Krieger, pp. 937–939. Williams & Wilkins, Baltimore, MD.

179. Descotes, J. (1989): *Immunotoxicology of Drugs and Chemicals*, Elsevier, Amsterdam.

180. Deugnier, Y. (2003): Iron and liver cancer. *Alcohol*, 30:145–150.

181. Devirian, T. A. and Volpe, S. L. (2003): The physiological effects of dietary boron. *Crit. Rev. Food Sci. Nutr.*, 43(2):219–231.

182. Dietrich, K. N. (1991): Human fetal lead exposure: Intrauterine growth, maturation and postnatal neurobehavioral development. *Fundam. Appl. Toxicol.*, 16:17–19.152

183. Digenott, D., Rozsa, O., Levy, N., and Muammar, S. (1964): Hypokalemia in barium poisoning. *Lancet*, 2:343–344.

184. Dill, J. A., Lee, K. M., and Mellinger, K. H. (2004): Lung deposition and clearance of inhaled vanadium pentoxide in chronically exposed F344 rats and B6C3$F_1$ mice. *Toxicol. Sci.*, 77:6–18.
185. Doig, A. T. (1976): Baritosis: a benign pneumoconiosis. *Thorax*, 31:30–39.
186. Doll, R. (1985): Relevance of epidemiology to policies for the prevention of cancer. *Hum. Toxicol.*, 4:81–96.
187. Domingo, J. L., and Corbella, J. (1991): A review of the health hazards from gallium exposure. *Trace Elements in Med.*, 8:56–64.
188. Domrongkitchaiporn, S., Sopassathit, W., Stitchantrakul, W. et al. (2004): Schedule of taking calcium supplement and the risk of nephrolithiasis. *Kidney Int.*, 65: 1835–1841.
189. Donaldson, R. M. and Barreras, R. F. (1966): Intestinal absorption of trace quantities of chromium. *J. Lab. Clin. Med.*, 68:484–493.
190. Dorman, D.C., McManus B.E., Parkinson, C.U. et al. (2004): Nasal toxicity of manganese sulfate and manganese phosphate in young male rats following subchronic (13 week) inhalation exposure. *Inhal. Toxicol.*, 16(6–7): 481–488.
191. Downs, J. C., Milling, D., and Nichols, C. A. (1995): Suicidal ingestion of barium-sulfide-containing shaving powder. *Am. J. Forens. Med. Path.*, 16:56–61.
192. Downs, W. L., Scott, J. K., Steadman, L. T., and Maynard, E. A. (1959): *The Toxicity of Indium*. University of Rochester Atomic Energy Report UR-588, University of Rochester, Rochester, NY.
193. Draize, J. H. and Kelley, E. A. (1959): The urinary excretion of boric acid preparations following oral administration and topical applications to intact and damaged skin of rabbits. *Toxicol. Appl. Pharmacol.*, 1:267–276.
194. Duell, P. B. and Chesnut III, C. H. (1991): Exacerbation of rheumatoid arthritis by sodium fluoride treatment of osteoporosis. *Arch. Intern. Med.*, 151:783–784.
195. Dunphy, B. (1967): Acute occupational cadmium poisoning: a critical review of the literature. *J. Occup. Med.*, 9: 22–26.
196. Duverger-van Bogaert, M. and Lambotte-Vandepaer, M. (1988): Hafnium. In: *Handbook on Toxicity of Inorganic Compounds*, edited by H. G. Sieler and H. Sigel, pp. 313–318. Marcel Dekker, New York.
197. ECETOC. (1995): *Reproductive and General Toxicology of Some Inorganic Borates and Risks Assessment for Human Beings*, Tech. Rep. No. 63. European Center for Ecotoxicology and Toxicology of Chemicals, Brussels, Belgium.
198. ECETOC. (1997): *Ecotoxicology of Some Inorganic Borates*, Special Report No. 11. European Center for Ecotoxicology and Toxicology of Chemicals, Brussels, Belgium.
199. Eisenbud, M., Berghout, C. F., and Steadman, L. T. (1948): Environmental studies in plants and laboratories using beryllium: the acute disease. *J. Indust. Hyg. Toxicol.*, 30:282–285.
200. Eisenbud, M. and Lisson, J. (1983): Epidemiological aspects of beryllium-induced nonmalignant lung disease: a 30-year update. *J. Occup. Med.*, 25:196–202.
201. El-Tawil, O. S. Z. and Morgan, A. M. (2000): Assessment of the teratogenicity of trivalent and hexavalent chromium compounds in female rabbits. *Toxicologist*, 54:292.
202. Elinder, C. G. and Zenz, C. (1994): Other metals and their compounds: hafnium and its compounds. In: *Occupational Medicine*, edited by C. Zenz, O. B. Dickerson, and E. P. Horvath, Jr., pp. 595–616. Mosby, St. Louis, MO.
203. Ellenhorn, M. J. and Barceloux, D. G. (1988): *Medical Toxicology: Diagnosis and Treatment of Human Poisoning*. Elsevier, New York.
204. Ellis, T. and Lacy, R. (1998): Illicit alcohol (moonshine) consumption in West Alabama revisited. *S. Med. J.*, 91:858–860.
205. Engfeldt, B. and Hjertquist, S. O. (1969): Effect of strontium administration on bones and teeth of rats maintained on diets with different calcium contents. *Virchows Arch. Abt. A Path. Anat.*, 346:330–344.
206. Enterline, P. E. and Marsh, G. M. (1982): Cancer among workers exposed to arsenic and other substances in a copper smelter. *Am. J. Epidemiol.*, 116:895–911.
207. Ernhart, C. B. (1992): A critical review of low-level prenatal lead exposure in the human: effects on the fetus and newborn. *Reprod. Toxicol.*, 6:9–19.
208. Esernio-Jenssen, D., Donatelli-Guagenti, A., and Mofenson, H. C. (1996): Severe lead poisoning from an imported clothing accessory: "watch" out for lead. *Clin. Toxicol.*, 34:329–333.
209. Eskew, A. E., Crutcher, J. C., Zimmerman, S. L. et al (1961): Lead poisoning resulting from illicit alcohol consumption. *J. Forensic Sci.*, 6:337–350.
210. Ewers, U., Manojilovic, N., Hadnagy, W., and Grover, Y P. (1988): Chlorine. In: *Handbook on Toxicity of Inorganic Compounds*, edited by H. G. Sieler and H. Sigel, pp 223–237. Marcel Dekker, New York.
211. Fahim, M. S., Fahim, Z., and Hall, D. G. (1976): Effects of subtoxic lead levels on pregnant women in the State of Missouri. *Res. Commun. Chem. Pathol. Pharmacol.*, 13:309–331
212. Fail, P. A., George, J. D., Seely, J. C., Grizzle, T. B., and Heindel, J. J. (1991): Reproductive toxicity of boric acid in Swiss (CD-1) mice: assessment using the continuous breeding protocol. *Fundam. Appl. Toxicol.*, 17:225–239.
213. Fail, P. A., Chapin, R. E., Price, C. J. et al. (1998): General reproductive, developmental, and endocrine toxicity of boronated compounds. *Reprod. Toxicol.*, 12(1):1–18.
214. Fairhall, L. T., Dunn, R. C., Sharpless, N. E., and Pritchard E. A. (1945): *The Toxicity of Molybdenum*. Public Health Bull. No. 293, U.S. Government Printing Office, Washington, D.C.
215. Farwell, A. P. and Braverman, L. E. (1996): Thyroid and antithyroid drugs. In: *Goodman and Gilman's The Pharmacological Basis of Therapeutics*, edited by J. G. Hardman, L. E. Limbird, and A. G. Gilman, p 1392. McGraw Hill, New York.
216. Feldman, R. G., Niles, C. A., Kelly-Hayes, M., Sax, D. S., Dixon, W. J., Thompson, D. J., and Landau, E. (1979): Peripheral neuropathy in arsenic smelter workers. *Neurology*, 29:939–944.
217. Ferguson, W. S., Lewis, A. H., and Watson, S. J. (1938): Action of molybdenum in nutrition of milking cows. *Nature*, 141:553.

218. Fischbein, A. (1992): Occupational and environmental lead exposure. In: *Environmental and Occupational Medicine*, edited by W. N. Rom, pp. 735–758. Little, Brown & Co., Boston.
219. Fisher, A. A. (1986): Antiseptics and disinfectants. In: *Contact Dermatitis*, pp. 178–194. Lea & Febiger, Philadelphia.
220. Fisher, A. A. (1986): Contact urticaria. In: *Contact Dermatitis*, p. 698. Lea & Febiger, Philadelphia.
221. Fisher, D. (1992): Zinc. In: *Hazardous Materials Toxicology*, edited by J. B. Sullivan, Jr., and G. R. Krieger, pp. 865–868. Williams & Wilkins, Baltimore, MD.
222. Fisher, N., Hutchinson, J. B., Berry, R., Hardy, J., Ginocchio, A. V., and Waite, V. (1979): Long-term toxicity and carcinogenicity studies of the bread improver potassium bromate. I. Studies in rats. *Food Cosmet. Toxicol.*, 17:33–39.
223. Flanagan, P. R., McLellan, J. S., Haist, J. et al. (1978): Increased dietary cadmium absorption in mice and human subjects with iron deficiency. *Gastroenterology*, 74: 841–846.
224. Flegal, A. R. and Smith, D. R. (1992): Lead levels in preindustrial humans. *N. Engl. J. Med.*, 326:1293–1294.
225. Flury, F. and Zernik, F. (1931): *Schädliche gase dämpfe, nebel, rauch- und staubarten*, p. 309. Verlag von Julius Springer, Berlin, Germany.
226. Forbes, R. M. (1960): Nutritional interactions of zinc and calcium. *Fed. Proc.*, 19:643–647.
227. Foster, L. H. and Sumar, S. (1997): Selenium in health and disease: a review. *Crit. Rev. Food Sci. Nutr.*, 37:211–228.
228. Fowler, J.F. (2001): Gold. *Am. J. Cont. Dermat.*, 12(1):1–2.
229. Fowler, J.F., Taylor, J., Storrs, F. et al., (2001): Gold allergy in North America. *Am. J. Cont. Dermat.*, 12(1):3–5.
230. Franke, K. W., Moxon, A. L., Poley, W. E., and Tully, W. C. (1936): Monstrosities produced by the injection of selenium salts into hens' eggs. *Anat. Rec.*, 65:15–22.
231. Fredrick, W. G. and Bradley, W. R. (1946): Toxicity of some materials used in the manufacture of cemented tungsten carbide tools. *Indust. Med.*, 15:482–483.
232. Friberg, L. (1984): Cadmium and the kidney. *Environ. Health Perspect.*, 54:1–11.
233. Friis-Hansen, B., Aggerbeck, B., and Jansen, J. A. (1982): Unaffected blood boron levels in newborn infants treated with a boric acid ointment. *Food Chem. Toxicol.*, 20: 451–454.
234. Fu, H. and Boffetta, P. (1995): Cancer and occupational exposure to inorganic lead compounds: a meta-analysis of published data. *Occup. Environ. Med.*, 52:73–81.
235. Fung, Y. K. and Molvar, M. P. (1992): Toxicity of mercury from dental environment and from amalgam restorations. *Clin. Toxicol.*, 30:49–61.
236. Gasmi, A., Garnier, R., Galliot-Guilley, M. et al. (1997): Acute selenium poisoning. *Vet. Human Toxicol.*, 39: 304–308.
237. Geltman, P.L., Brown, M.J., and Cochran, J. (2001): Lead poisoning among refugee children resettled in Massachusetts, 1995 to 1999. *Pediatrics*, 108:158–162.
238. Gennart, J.-P., Buchet, J.-P., Roels, H., Ghyselen, P., Ceulemans, E., and Lauwerys, R. (1992): Fertility of male workers exposed to cadmium, lead, or manganese. *Am. J. Epidemiol.*, 135: 1208–1219.
239. Gerber, G. B. and Léonard, A. (1997): Mutagenicity, carcinogenicity and teratogenicity of germanium compounds. *Mutat. Res.*, 387:141–146.
240. Gerber, G.B., Leonard, A., and Hantson, P. (2002): Carcinogenicity, mutagenicity and teratogenicity of manganese compounds. *Crit. Rev. Oncol. Hematol.*, 42(1):25–34.
241. Gheysens, B., Auwerx, J., Van den Eeckhout, A., and Demedts, M. (1985): Cobalt-induced bronchial asthma in diamond polishers. *Chest*, 88:740–744.
242. Gilman, A. P., Villeneuve, D. C., Secours, V. E., Yagminas, A. P., Tracy, B. L., Quinn, J. M., Valli, V. E., Willes, R. J., and Moss, M. A. (1998): Uranyl nitrate: 28-day and 91-day toxicity studies in the Sprague–Dawley rat. *Toxicol. Sci.*, 41:117–128.
243. Gilman A. P., Moss M. A., Villeneuve D. C., Secours V. E., Yagminas A. P., Tracy B. L., Quinn J. M., Long G., and Valli V. E. (1998): Uranyl nitrate: 91-day exposure and recovery studies in the male New Zealand white rabbit. *Toxicol. Sci.*, 41:138–151.
244. Gilmore, Jr., D. A. and Bronstein, A. C. (1992): Manganese and magnesium. In: *Hazardous Materials Toxicology*, edited by J. B. Sullivan, Jr., and G. R. Krieger, pp. 896–901. Williams & Wilkins, Baltimore, MD.
245. Gilmore, Jr., D. A. and Bronstein, A. C. (1992): Titanium. In: *Hazardous Materials Toxicology*, edited by J. B. Sullivan, Jr., and G. R. Krieger, pp. 904–905. Williams & Wilkins, Baltimore, MD.
246. Ginocchio, A. V., Waite, V., Hardy, J., Fisher N., Hutchinson, J. B., and Berry, R. (1979): Long-term toxicity and carcinogenicity studies of the bread improver potassium bromate. 2. Studies in mice. *Food Cosmet. Toxicol.*, 17:41–47.
247. Goering, P. L. (1992): Platinum and related metals: palladium, indium, osmium, rhodium, and ruthenium. In: *Hazardous Materials Toxicology*, edited by J. B. Sullivan, Jr., and G. R. Krieger, pp. 874–881. Williams & Wilkins, Baltimore, MD.
248. Gold, L. S., Sawyer, C. B., Magaw, R. et al. (1984): A carcinogenic potency database of the standardized results of animal bioassays. *Environ. Health Persp.*, 58:9–319.
249. Goldbloom, R. B. and Goldbloom, A. (1953): Boric acid poisoning: report of four cases and a review of 109 cases from the world literature. *J. Pediatr.*, 43:631–643.
250. Goldman, L. R. (1998): Linking research and policy to ensure children's environmental health. *Environ. Health Perspect.*, 106:S857–S862.
251. Goldman, R. H., Baker, E. L., Hannan, M., and Kamerow, D. B. (1987): Lead poisoning in automobile radiator repair mechanics. *N. Engl. J. Med.*, 317:214–218.
252. Golub, M. S. and Domingo, J. L. (1996): What we know and what we need to know about developmental aluminum toxicity. *J. Toxicol. Environ. Health*, 48:585–597.
253. Gompertz, D. (1981): Assessment of risk by biological monitoring. *Br. J. Indust. Med.*, 38:198–201.
254. Gordeuk, V. R., McLaren, G. D., and Samowitz, W. (1994): Etiologies, consequences, and treatment of iron overload. *CRC Crit. Rev. Clin. Lab. Sci.*, 31:89–133.
255. Gordeuk, V. R., Mukiibi, J., Hasstedt, S. J. et al. (1992): Iron overload in Africa: interaction between a gene and dietary iron content. *N. Engl. J. Med.*, 326:95–100.

256. Gosselin, R. E., Smith, R. P., and Hodge, H. C. (1984): Arsenic. In: *Clinical Toxicology of Commercial Products*, edited by Gosselin, R. E., Smith, R. P., and Hodge, H. C., pp. III-42–III-47. Williams & Wilkins, Baltimore, MD.
257. Goyer, R. A. (1986): Toxic effects of metals. In: *Casarett and Doull's Toxicology: The Basic Science of Poisons*, 3rd ed., edition by C. D. Klaassen, M. O. Amdur, and J. Doull, pp. 582–635. Macmillan, New York.
258. Goyer, R. A., Miller, C. R., Zhu, S.-Y., and Victery, W. (1989): Non-metallothionein-bound cadmium in the pathogenesis of cadmium nephrotoxicity in the rat. *Toxicol. Appl. Pharmacol.*, 101:232–244.
259. Goyer, R. A. and Rhyne, B. C. (1973): Pathological effects of lead. *Int. Rev. Exp. Pathol.*, 12:1–77.
260. Graves, A. B., Rosner, D., Echeverria, D., Mortimer, J. A., and Larson, E. B. (1998): Occupational exposures to solvents and aluminum and estimated risk of Alzheimer's disease. *Occup. Environ. Med.*, 55:627–633.
261. Graziano, J. H. and Blum, C. (1991): Lead exposure from lead crystal. *Lancet*, 337:141–142.
262. Greengard, J. (1975): Iron poisoning in children. *Clin. Toxicol.*, 8:575–597.
263. Greenwald, P. (2004): Clinical trials in cancer prevention: current results and perspectives for the future. *J. Nutr.*, 134:3507S–3512S.
264. Greger, J. L. and Johnson, M. A. (1981): Effect of dietary tin on zinc, copper, and iron utilization by rats. *Food Cosmet. Toxicol.*, 19:163–166.
265. Greger, J. L. and Sutherland, J. E. (1997): Aluminum exposure and metabolism. *Crit. Rev. Clin. Lab. Sci.*, 34: 439–474.
266. Grignon, S. and Bruguerolle, B. (1996): Cerebellar lithium toxicity: a review of recent literature and tentative pathophysiology. *Therapeutics*, 51:101–106.
267. Groleau, G. (1994): Lithium toxicity. *Conc. Controv. Toxicol.*, 12:511–531.
268. Grote, K., Stahlschmidt, B., Talsness, C. E. et al. (2004): Effects of organotin compounds on pubertal male rats. *Toxicology*, 202:145–158.
269. Groth, D. H., Stettler, L. E., and Burg, J. R. (1986): Carcinogenic effects of antimony trioxide and antimony ore concentrate in rats. *J. Toxicol. Environ. Health*, 18: 607–626.
270. Gupta, R. S., Gupta, E. S., Dhakal, B. K. et al. (2004): Vitamin C and vitamin E protect the rat testes from cadmium-induced reactive oxygen species. *Mol. Cells*, 17:132–139.
271. Gupta, R. S., Kim, J., Gomes, C. et al. (2004): Effect of ascorbic acid supplementation on testicular steroidogenesis and germ cell death in cadmium-treated male rats. *Mol. Cell. Endocrinol.*, 221:57–66.
272. Hadjimarkos, D. M. (1965): Effect of selenium on dental caries. *Arch. Environ. Health*, 10:893–899.
273. Hadjimarkos, D. M., Storvick, C. A., and Remmert, L. F. (1952): Selenium and dental caries. *J. Pediatr.*, 40:451–455.
274. Hadjimichael, O. C. and Brubaker, R. E. (1981): Evaluation of an occupational respiratory exposure to a zirconium-containing dust. *J. Occup. Med.*, 23:543–547.
275. Haley, T. J., Komesu, N., and Raymond, K. (1962): Pharmacology and toxicology of niobium chloride. *Toxicol. Appl. Pharmacol.*, 4:385–392.
276. Haley, T. J., Raymond, K., Komesu, N., and Upham, H. C. (1962): The toxicologic and pharmacologic effects of hafnium salts. *Toxicol. Appl. Pharmacol.*, 4:238–246.
277. Hallberg, L., Brune, M., Erlandsson, M. et al. (1991): Calcium: effect of different amounts of nonheme- and heme-iron absorption in humans. *Am. J. Clin. Nutr.*, 53:112–119.
278. Hamilton, A. and Hardy, H. (1974): Osmium. In: *Industrial Toxicology*, pp. 155–156. Publishing Sciences Group, Acton, MA.
279. Hardy, H. L. and Tabershaw, I. R. (1946): Delayed chemical pneumonitis occurring in workers exposed to beryllium compounds. *J. Indust. Hyg. Toxicol.*, 28:197–211.
280. Harlan, W. R. (1988): The relationship of blood lead levels to blood pressure in the U.S. population. *Environ. Health Perspec.*, 78:9–13.
281. Harr, J. R. and Muth, O. H. (1972): Selenium poisoning in domestic animals and its relationship to man. *Clin. Toxicol.*, 5:175–186.
282. Hartman, R. S., Conrad, M. E., Hartman, R. E. et al. (1963): Ferritin-containing bodies in human small intestinal epithelium. *Blood*, 22:397–405.
283. Hartwig, A. (1998): Carcinogenicity of metal compounds: possible role of DNA repair inhibition. *Toxicol. Lett.* 102–103:235–239.
284. Hatchcock, J. N. and Rader, J. I. (1990): Macronutrient safety. *Ann. N.Y. Acad. Sci.*, 587:257–266.
285. Haug, R. H. (1996): Retention of asymptomatic bone plates used for orthognathic surgery and facial fractures *J. Oral Maxillofac. Surg.*, 54:611–617.
286. Health Council of the Netherlands, Committee for Compounds Toxic to Reproduction (2000): *Lithiumcarbonate and Lithiumchloride: Evaluations of the Effects on Reproduction, Recommendation for Classification*, Publ. No 2000/06OSH. Health Council of the Netherlands, The Hague (http://www.gr.nl/overig/pdf).
287. Heidrich, D. D., Steckelbroeck, S., and Klingmuller, D (2001): Inhibition of human cytochrome P450 aromatase activity by butyltins. *Steroids*, 66:763–769.
288. Heindel, J. J., Price, C. J., Field, E. A. et al. (1992): Developmental toxicity of boric acid in mice and rats. *Fundam Appl. Toxicol.*, 18:266–277.
289. Helzlsouer, K., Jacobs, R., and Morris, S. (1985): Acute selenium poisoning in the United States. *Fed. Proc.*, 44 1670.
290. Henderson, Y. and Haggard, H. W. (1943): Noxious gases p. 133. Reinhold Publishing, New York.
291. Henriksson, J. and Tjalve, H. (2000): Manganese uptake into the CNS via the olfactory pathway in rats affects astrocytes. *Toxicol. Sci.*, 55(2):392–398.
292. Herrero, F., Fernandez, E., Gomez, J. et al. (1995): Thallium poisoning presenting with abdominal colic, paresthesia, and irritability. *Clin. Toxicol.*, 33: 261–264.
293. Hertz-Picciotto, I. and Croft, J. (1993): Review of the relation between blood lead and blood pressure. *Epidemiol Rev.*, 15:352–373.

294. Heyman, A., Pfeifer, J. B., Willett, R. W., and Taylor, H. M. (1956): Peripheral neuropathy caused by arsenical intoxication. *N. Engl. J. Med.*, 254:401–408.
295. Hider, R. C., Choudhury, R., Rai, B. J. et al. (1996): Design of orally active iron chelators. *Acta Haematol.*, 95:6–12.
296. Higgins, E. S., Richert, D. A., and Westerfeld, W. W. (1956): Molybdenum deficiency and tungstate inhibition studies. *J. Nutr.*, 59:539–559.
297. Higgins T., Curry S., and Ruha A. (1999): Tellurium ingestion in an 18-month-old male. *Toxicol. Clin. Toxicol.*, 37:625.
298. Hiles, R. A. (1974): Absorption, distribution, and excretion of inorganic tin in rats. *Toxicol. Appl. Pharmacol.*, 27: 366–379.
299. Hipkins, K. L., Materna, B. L., Kosnett, M. J. et al. (1998): Medical surveillance of the lead exposed worker. *AAOHN J.*, 46:330–339.
300. Hirano, S., Kodama, N., Shibata, K., and Suzuki, K. T. (1990): Distribution, localization, and pulmonary effects of yttrium chloride following intratracheal instillation into the rat. *Toxicol. Appl. Pharmacol.*, 104:301–311.
301. Hirano, S., Kodama, N., Shibata, K, and Suzuki, K. T. (1993): Metabolism and toxicity of intravenously injected yttrium chloride in rats. *Toxicol. Appl. Pharmacol.*, 121: 224–232.
302. Hoffman, D. J., Ohlendorf, H. M., and Aldrich, T. W. (1988): Selenium teratogenesis in natural populations of aquatic birds in Central California. *Arch. Environ. Contam. Toxicol.*, 17:519–525.
303. Hoffman, R.S. (2003): Thallium toxicity and the role of Prussian blue in therapy. *Toxicol. Rev.*, 22(1):29–40.
304. Hogstrand, C. and Wood, C. M. (1998). Toward a better understanding of the bioavailability, physiology, and toxicity of silver in fish: implications for water quality criteria. *Environ. Toxicol. Chem.*, 17:547–561.
305. Hollinger, M. A. (1996): Toxicological aspects of topical silver pharmaceuticals. *CRC Crit. Rev. Toxicol.*, 26:255–260.
306. Holmquist, I. (1951): Occupational arsenical dermatitis: a study among employees at a copper ore smelting work including investigations of skin reactions to contact with arsenic compounds. *Acta Derm. Venereol.*, 31(Suppl. 26):1–214.
307. Hostynek, J. J. (1997): Gold: an allergen of growing significance. *Food Chem. Toxicol.*, 35:839–844.
308. Hu, H., Aro, A., Payton, M., Korrick, S., Sparrow, D., Weiss, S. T., and Rotnitzky, A. (1996): The relationship of bone and blood lead to hypertension. *JAMA*, 275: 1171–1176.
309. Hu, H., Watanabe, H., Payton, M., Korrick, S., and Rotnitzky, A. (1994): The relationship between bone lead and hemoglobin. *JAMA*, 272:1512–1517.
310. Hu, J., Mao, Y., and White, K. (2002): Renal cell carcinoma and occupational exposure to chemicals in Canada. *Occup. Med. (Lond.)*, 52:157–164.
311. Huang, X. (2003): Iron overload and its association with cancer risk in humans: evidence for iron as a carcinogenic metal. *Mutat. Res.*, 533:153–171.
312. Huel, G., Boudene, C., and Ibrahim, M. A. (1981): Cadmium and lead content of maternal and newborn hair: relationship to parity, birth weight and hypertension. *Arch. Environ. Health*, 36:221–227.
313. Hughes, E. G. (1980): Medical surveillance of platinum refinery workers. *J. Soc. Occup. Med.*, 30:27–30.
314. Hunter, D. J., Morris, J. S., Stampfer, M. J. et al. (1990): A prospective study of selenium status and breast cancer risk. *JAMA*, 264:1128–1131.
315. Huott, M. A. and Storrow, A. B. (1997): A survey of adolescents' knowledge regarding toxicity of over-the-counter medications. *Acad. Emerg. Med.*, 4:214–218.
316. Husain, M. H., Dick, J. A., and Kaplan, Y. S. (1980): Rare earth pneumoconiosis. *J. Soc. Occup. Med.*, 30:15–19.
317. Hussey, H. H. (1976): Phosphorus poisoning in children. *J. Am. Med. Assoc.*, 235:1366.
318. Insel, P. A., (1995): Analgesic-antipyretic and antiinflammatory agents and drugs employed in the treatment of gout. In: *Goodman & Gilman's The Pharmacological Basis of Therapeutics*, edited by J. G. Hardman, L. E. Limbird, and A. G. Gilman, pp. 617–657. McGraw-Hill, New York.
319. Insley, B.M., Grufferman, S., and Ayliffe, H.E. (1986): Thallium poisoning in cocaine abusers. *Am. J. Emerg Med.*, 4:545–548.
320. Institute for Evaluating Health Risks (1997): An assessment of boric acid and borax using the IEHR evaluative process for assessing human developmental and reproductive toxicity of agents. *Reprod. Toxicol.*, 11:123–160.
321. Institute of Medicine. (1997): *Dietary Reference Intakes for Calcium, Phosphorus, Magnesium, Vitamin D, and Fluoride*. National Academy Press, Washington, D.C.
322. Institute of Medicine. (2000): *Dietary Reference Intakes for Vitamin C, Vitamin E, Selenium, and Carotenoids*. National Academy Press, Washington, D.C.
323. Institute of Medicine. (2001): *Dietary Reference Intakes for Vitamin A, Vitamin K, Arsenic, Boron, Chromium, Copper, Iodine, Iron, Manganese, Molybdenum, Nickel, Silicon, Vanadium, and Zinc*. National Academy Press, Washington, D.C.
324. Institute of Medicine. (2003): *Dietary Reference Intakes: Guiding Principles for Nutrition Labeling and Fortification*. National Academy Press, Washington, D.C.
325. Institute of Medicine. (2004): *Dietary Reference Intakes for Water, Potassium, Sodium, Chloride, and Sulfate*. National Academy Press, Washington, D.C.
326. IARC. (1980): *IARC Monographs on the Evaluation of the Carcinogenic Risk of Chemicals to Humans: Some Metals and Metallic Compounds*, Vol. 23, pp. 39–141. International Agency for Research on Cancer, Lyon, France.
327. IARC. (1986): *IARC Monographs on the Evaluation of the Carcinogenic Risk of Chemicals to Humans: Potassium Bromate*, Vol. 40, International Agency for Research on Cancer, Lyon, France.
328. IARC. (1990): *IARC Monographs on the Evaluation of the Carcinogenic Risk of Chemicals to Humans: Chromium, Nickel and Welding*, Vol. 49, pp. 257–445. International Agency for Research on Cancer, Lyon, France.
329. IARC. (1991): *IARC Monographs on the Evaluation of the Carcinogenic Risk of Chemicals to Humans: Chlorinated Drinking-Water; Chlorination By-Products; Some Other Halogenated Compounds; Cobalt and Cobalt Compounds*, Vol. 52, pp. 1–544. International Agency for Research on Cancer, Lyon, France.

330. IARC. (1993): *IARC Monographs on the Evaluation of the Carcinogenic Risk of Chemicals to Humans: Beryllium, Cadmium, Mercury, and Exposures in the Glass Manufacturing Industry*, Vol. 58, pp. 41–237. International Agency for Research on Cancer, Lyon, France.
331. IARC. (2004): *IARC Monographs on the Evaluation of the Carcinogenic Risk of Chemicals to Humans: Inorganic and Organic Lead Compounds*, Vol. 87, pp. 10–17. International Agency for Research on Cancer, Lyon, France (http://monographs.iarc.fr/htdocs/announcements/vol87.htm).
332. International Programme on Chemical Safety. (1990): *Tributyltin Compounds*, Environmental Health Criteria 116. World Health Organization, Geneva (http://www.inchem.org/documents/ehc/ehc/ehc116.htm).
333. International Programme on Chemical Safety. (1992): *Cadmium*, Environmental Health Criteria 134. World Health Organization, Geneva (http:///www.inchem.org/documents/ehc/ehc/ehc134.htm).
334. Ip, C. (1985): Selenium inhibition of chemical carcinogenesis. *Fed. Proc.*, 44:2573–2578.
335. Iregren, A. (1990): Psychological test performance in foundry workers exposed to low levels of manganese. *Neurotoxicol. Teratol.*, 12:673–675.
336. Ishinishi, N., Yamamoto, A., Hisanaga, A., and Inamasu, T. (1983): Tumorigenicity of arsenic trioxide to the lung in Syrian golden hamsters by intermittent instillations. *Cancer Lett.*, 21:141–147.
337. Ivankovic, S., Eisenbrand, G., and Preussmann, R. (1979): Lung carcinoma induction in BD rats after a single intratracheal instillation of an arsenic-containing pesticide mixture formerly used in vineyards. *Int. J. Cancer*, 24: 786–788.
338. Jacobs, R. W., Duong, T., Jones, R. E., Trapp, G. A., and Scheibel, A. B. (1989): A reexamination of aluminum in Alzheimer's disease: analysis by energy dispersive x-ray microprobe and flameless atomic absorption spectrophotometry. *Can. J. Neurol. Sci.*, 16:498–503.
339. Jaffe, W. G. and Velez, F. B. (1973): Selenium intake and congenital malformations in humans. *Archiv. Latinoam. Nutr.*, 23:515–517.
340. Jarup, L., Elinder, C. G., and Spang, G. (1988): Cumulative blood-cadmium and tubular proteinuria: a dose–response relationship. *Int. Arch. Occup. Environ. Health*, 60: 223–229.
341. Jensen, R., Closson, W., and Rothenberg, R. (1984): Selenium intoxication: New York. *MMWR*, 33:157–158.
342. Johnson, J. L. and Rajagopalan, K. V. (1974): Molecular basis of the biological function of molybdenum. *J. Biol. Chem.*, 249:859–866.
343. Johnson, P. E. (1996): New approaches to establish mineral element requirements and recommendations: an introduction. *J. Nutr.*, 126:2309S–2311S.
344. Johnson, R. A., Baker, S. S., Fallon, J. T. et al. (1981): An occidental case of cardiomyopathy and selenium deficiency. *N. Engl. J. Med.*, 304:1210–1212.
345. Jones, G. and Brooks, P. M. (1996): Injectable gold compounds: an overview. *Br. J. Rheumatol.*, 35:1154–1158.
346. Kabe, I., Omae, K., Nakashima, H., Nomiyama, T., Uemura, T., Hosoda, K., Ishizuka, C., Yamazaki, K., and Sakurai, H. (1996): *In vitro* solubility and *in vivo* toxicity of indium phosphide. *J. Occup. Health*, 38:6–12.
347. Kamerbeek, H. H., Rauws, A. G., Ham, M. T. et al. (1971): Redistribution of thallium by treatment with sodium diethyldithiocarbamate. *Acta Med. Scand.*, 189:149–154.
348. Kanisawa, M. and Schroeder, H. A. (1967): Life term studies on the effect of arsenic, germanium, tin and vanadium on spontaneous tumors in mice. *Cancer Res.*, 27:1192–1195.
349. Kanisawa, M. and Schroeder, H. A. (1969): Life term studies on the effect of trace elements on spontaneous tumors in mice and rats. *Cancer Res.*, 29:892–895.
350. Kappas, A. and Maines, M. D. (1976): Tin: a potent inducer of heme oxygenase in kidney. *Science*, 192:60–62.
351. Katz, S. A. (1995): The toxicity/essentiality of dietary minerals: a review on some micronutrients prepared in honor of the award for life achievement to Doctor Krist Kostial. *Arh. Hig. Rada. Toksikol.*, 46:333–345.
352. Kaye, W. E., Novotny, T. E., and Tucker, M. (1987): New ceramics-related industry implicated in elevated blood lead levels in children. *Arch. Environ. Health*, 42:161–164.
353. Kazantzis, G. (1979): Thallium. In: *Handbook on the Toxicology of Metals*, edited by L. Friberg, G. F. Nordberg, V. B. Vouk, pp. 599–612. Elsevier/North Holland, New York.
354. Keall, J. H. H., Martin, N. H., and Tunbridge, R. E. (1946): A report of three cases of accidental poisoning by sodium tellurite: *Br. J. Indust. Med.*, 3:175–176.
355. Keshan Disease Research Group (1979): Epidemiologic studies on the etiologic relationship of selenium and Keshan disease. *Chin. Med. J.*, 92:477–482.
356. Kettrup, A. and Hüppe, U. (1988): Phosphorus. In: *Handbook on Toxicity of Inorganic Compounds*, edited by H. G. Sieler and H. Siegel, pp. 521–532. Marcel Dekker, New York.
357. Khera, A. K., Wibberley, D. G., and Dathan, J. G. (1980): Placental and stillbirth tissue lead concentrations in occupationally exposed women. *Br. J. Indust. Med.*, 37: 394–396.
358. Kikano, G. E. and Stange, K. C. (1992): Lead poisoning in a child after a gunshot injury. *J. Family Pract.*, 34: 498–504.
359. Kim, G.-S., Judd, D. A., Hill, C. L., and Schinazi, R. F. (1994). Synthesis, characterization, and biological activity of a new potent class of anti-HIV agents, the peroxoniobium-substituted heteropolytungstates. *J. Med. Chem.*, 37:816–820.
360. Kim, M., Kim, C., and Song, L. (2003): Analysis of lead in 55 brands of dietary calcium supplements by graphite furnace atomic absorption spectrometry after microwave digestion. *Food Addit. Contam.*, 20:149–153.
361. Kim, R., Rotnitzky, A., Sparrow, D., Weiss, S. T., Wager, C., and Hu, H. (1996): A longitudinal study of low-level lead exposure and impairment of renal function. *JAMA*, 275:1177–1181.
362. Kim, T. S. and Yim, S. Y. (1997): Peripheral nerve and muscle diseases, I. *Brain Pathol.*, 7:1117–1121.

363. Kimbrough, R. D. (1976): Toxicity and health effects of selected organotin compounds: a review. *Environ. Health Persp.*, 14:51–56.
364. Kinard, F. W. and van de Erve, J. (1940): Rat mortality following sodium tungstate injection. *Am. J. Med. Sci.*, 199:668–670.
365. Kinard, F. W. and van de Erve, J. (1943): Effect of tungsten metal diets in the rat. *J. Lab. Clin. Med.*, 28:1541–1543.
366. Kittle, C. F., King, E. R., and Brucer, M. (1951): The tissue distribution and excretion of radioactive hafnium mandelate in the rat. *J. Pharmacol. Exp. Therap.*, 101:21.
367. Klaassen, C. D. (1995): Nonmetallic environmental toxicants, In: *Goodman & Gilman's The Pharmacological Basis of Therapeutics*, 9th ed., edited by J. G. Hardman, L. E. Limbird, and A. G. Gilman, pp. 1673–1696. McGraw-Hill, New York.
368. Klatzo, I., Wesniewski, H., and Streicher, E. (1965): Experimental production of neurofibrillary degeneration. *J. Neuropath. Exp. Neurol.*, 24:187–199.
369. Kleinfeld, M. (1965): Acute pulmonary edema of chemical origin. *Arch. Environ. Health.*, 10:942–946.
370. Kok, F. J., de Bruijn, A. M., Hofman, A. et al. (1987): Is serum selenium a risk factor for cancer in men only? *Am. J. Epidemiol.*, 125:12–16.
371. Kontoghiorghes, G. J. (1995): Comparative efficacy and toxicity of desferrioxamine, deferiprone and other iron and aluminum chelating agents. *Toxicol. Lett.*, 80:1–18.
372. Köpf-Maier, P. and Klapötke, T. (1992): Antitumor activity of ionic mobecene and molybdenocene complexes in high oxidation states. *J. Cancer Res. Clin. Oncol.*, 118:216–221.
373. Köpf-Maier, P. and Köpf, H. (1994): Organometallic titanium, vandadium, niobium, molybdenum and rhenium complexes: early transition metal antitumour drugs. In: *Metal Compounds in Cancer Therapy*, edited by S. P. Fricker, pp. 109–146. Chapman & Hall, London.
374. Koppel, C., Baudisch, H., Beyer, K.-H. et al. (1986): Fatal poisoning with selenium dioxide. *Clin. Toxicol.*, 24:21–35.
375. Kotter, J. M. and Zieger, G. (1992): Sarkoidale Granulomatose nach mehrjähriger zircokoniumexposition, eine 'zirkoniumlunge.' *Pathologe*, 13:104–109.
376. Kravzov, J., Rios, C., Altagracia, M., Monroy-Noyola, A., and Lopez, F. (1993): Relationship between physiochemical properties of Prussian blue and its efficacy as antidote against thallium poisoning. *J. Appl. Toxicol.*, 13:213–216.
377. Krenzelok, E. P. (1992): Sodium and potassium. In: *Hazardous Materials Toxicology*, edited by J. B. Sullivan, Jr., and G. R. Krieger, pp. 797–799. Williams & Wilkins, Baltimore, MD.
378. Krishnan, S. S., Harrison, J. E., and Crapper McLachlan, D. R. (1987): Origin and resolution of the aluminum controversy concerning Alzheimer's neurofibrillary degeneration. *Biol. Trace Element Res.*, 13:35–42.
379. Kubota, J., Allaway, W. H., Carter, D. L. et al. (1967): Selenium in crops in the United States in relation to selenium-responsive diseases of animals. *J. Agric. Food Chem.*, 15:448–453.
380. Kurtzke, J. F. (1991): Risk factors in amyotrophic lateral sclerosis. *Adv. Neurol.*, 56:245–270.
381. Laine, J., Kalimo, K., Forssell, H., and Happonen, R. P. (1992): Resolution of oral lichenoid lesions after replacement of amalgam restorations in patients allergic to mercury compounds. *Br. J. Dermatol.*, 126: 10–15.
382. Lampert, P. W. and Garret, R. S. (1971): Mechanism of demyelination in tellurium neuropathy: electron microscope observations. *Lab. Invest.*, 25:380–388.
383. Landolt, R. R., Berk, H. W., and Russell, H. T. (1972): Studies on the toxicity of rhodium trichloride in rats and rabbits. *Tox. Appl. Pharmacol.*, 21:589–590.
384. Landrigan, P.J. (2000): Pediatric lead poisoning: is there a threshold? *Public Health Rep.*, 115:530–531.
385. Lanphear, B. P. (1998): The paradox of lead poisoning prevention. *Science*, 281:1617–1618.
386. Lanphear, B.P., Dietrich, K., Auinger, P. et al. (2000): Cognitive deficits associated with blood lead concentrations <10 microg/dL in U.S. children and adolescents. *Public Health Rep.*, 115:521–529.
387. Lanphear, B.P., Dietrich, K.N., and Berger, O. (2003): Prevention of lead poisoning in U.S. children. *Ambulat. Ped.*, 3:27–36.
388. Lansdown, A. B. G. (1995): Physiological and toxicological changes in the skin resulting from the action and interaction of metal ions. *CRC Crit. Rev. Toxicol.*, 25: 397–462.
389. Larner, A. J. (1995): Biological effects of tellurium: a review. *Trace Elem. Electrol.*, 12:26–31.
390. Lauwerys, R., Roels, H., Benet, P. et al. (1985): Fertility of male workers exposed to mercury vapor or to manganese dust: a questionnaire study. *Am. J. Indust. Med.*, 7: 171–176.
391. Lauwerys, R. R. (1989): Metals: epidemiological and experimental evidence for carcinogenicity. *Arch. Toxicol.*, 13(Suppl.):21–27.
392. Leach, L. J., Maynard, E. A., Hodge, H. C., Scott, J. K., Yuile, C. L., Sylvester, G. E., and Wilson, H. B. (1970): A five year inhalation study with uranium dioxide ($UO_2$) dust. I. Retention and biologic effect in the monkey, dog and rat. *Health Phys.*, 18:599–612.
393. Lee, K. P., Henry III, N. W., Trochimowicz, H. J., and Reinhardt, C. F. (1986): Pulmonary response to impaired lung clearance in rats following excessive $TiO_2$ dust deposition. *Environ. Res.*, 44:144–167.
394. Lee-Feldstein, A. (1983): Arsenic and respiratory cancer in man: follow-up of an occupational study. In: *Arsenic: Industrial, Biomedical, and Environmental Perspectives*, edited by W. H. Lederer and R. J. Fensterheim, pp. 245–265. Van Nostrand Reinhold, New York.
395. Leggett, R. W. (1993): An age-specific kinetic model of lead metabolism in humans. *Environ. Health Perspec.*, 101:598–616.
396. Leonard, A. and Gerber, G. B. (1997): Mutagenicity, carcinogenicity and teratogenicity of thallium compounds. *Mut. Res.*, 387:47–53.
397. Leonard, A., Hantson, Ph., and Gerber, G. B. (1995): Mutagenicity, carcinogenicity and teratogenicity of lithium compounds. *Mutat. Res.*, 339:131–137.

398. Lerda, D. (1992): Study of sperm characteristics in persons occupationally exposed to lead. *Am. J. Indust. Med.*, 22: 567–571.
399. Levander, O. A. (1991): Scientific rationale for the 1989 recommended dietary allowance for selenium. *Perspect. Pract.*, 91:1572–1576.
400. Lewis, G. P., Coughlin, L., Jusko, W. et al. (1972): Contribution of cigarette smoking to cadmium accumulation in man. *Lancet*, 1:291–292.
401. Lhotka, C., Szekeres, T., Steffan, I. et al. (2003): Four-year study of cobalt and chromium blood levels in patients managed with two different metal-on-metal total hip replacements. *J. Orthopaed. Res.*, 21:189–195.
402. Liebscher, D. H. and Liebscher, D. E. (2004): About the misdiagnosis of magnesium deficiency. *J. Am. Coll. Nutr.*, 23(6):730S–731S.
403. Liipo, K. K., Anttila, S. L., Taikina-aho, O., Ruodonen, E.-L., Toivonen, S. T., and Tuomi, T. (1993): Hypersensitivity pneumonitis and exposure to zirconium silicate in a young ceramic tile worker. *Am. Rev. Respir. Dis.*, 148:1089–1092.
404. Lin, S., Hwang, S. A., Marshall, E. G., and Marion, D. (1998): Does paternal occupational lead exposure increase the risks of low birth weight or prematurity? *Am. J. Epidemiol.*, 148:173–181.
405. Linder, M. C. and Hazegh, M. (1996): Copper biochemistry and molecular biology. *Am. J. Clin. Nutr.*, 63:797S–811S.
406. Lin-Fu, J. S. (1980): Lead poisoning and undue lead exposure in children: History and current status. In: *Low Level Lead Exposure: The Clinical Implications of Current Research*, edited by H. L. Needleman, pp. 5–16. Raven Press, New York.
407. Lison, D. (1996): Human toxicity of cobalt-containing dust and experimental studies on the mechanism of interstitial lung disease (hard metal disease). *CRC Crit. Rev. Toxicol.*, 26:585–616.
408. Lister, J. (1875): Recent improvements in the details of antiseptic surgery. *Lancet*, 603–605.
409. Litovitz, T. L., Holm, K. C., Bailey, K. M., and Schmitz, B. F. (1992): 1991 Annual report of the American Association of Poison Control Centers National Data Collection System. *Am. J. Emerg. Med.*, 10:452–505.
410. Lockitch, G., Taylor, G. P., Wong, L. T. K. et al. (1990): Cardiomyopathy associated with nonendemic selenium deficiency in a Caucasian adolescent. *Am. J. Clin. Nutr.*, 52:572–577.
411. Lokich, J. and Anderson, N. (1998): Carboplatin versus cisplatin in solid tumors: an analysis of the literature. *Ann. Oncol.*, 9:13–21.
412. Longnecker, M. P., Taylor, P. R., Levander, O. A. et al. (1991): Selenium in diet, blood, and toenails in relation to human health in a seleniferous area. *Am. J. Clin. Nutr.*, 53:1288–1294.
413. Lovejoy, F. H. (1982): Chelation therapy in iron poisoning. *J. Toxicol. Clin. Toxicol.*, 19:871–874.
414. Lucas, P. A., Jariwalla, A. G., Jones, J. H., Gough, J., and Vale, P. T. (1980): Fatal cadmium fume poisoning. *Lancet*, 205.
415. Lufkin, N. H. and Hodges, F. T. (1944): Cadmium poisoning, report of outbreak. *U.S. Nav. Med. Bull.*, 43:1273–1276.
416. Lydiard, R. B. and Gelenberg, A. J. (1982): Hazards and adverse effects of lithium. *Ann. Rev. Med.*, 333:327–344.
417. Lynch, R. A., Boatright, D. T., and Moss, S. K. (2000): Lead-contaminated imported tamarind candy and children's blood lead levels. *Public Health Rep.*, 115:537–543.
418. Mabuchi, K., Lilienfeld, A. M., and Snell, L. M. (1979): Lung cancer among pesticide workers exposed to inorganic arsenicals. *Arch. Environ. Health*, 34:312–320.
419. MacGillivray, R. T., Mendez, E., Sinha, S. K. et al. (1982): The complete amino acid sequence of human serum transferrin. *Proc. Natl. Acad. Sci. USA*, 79:2504–2508.
420. Mack, R. B. (1990): The fat lady enters stage left. *N.C. Med. J.*, 51:636–638.
421. Mahaffey-Six, K. and Goyer, R. A. (1972): The influence of iron deficiency on tissue content and toxicity of ingested lead in the rat. *J. Lab. Clin. Med.*, 79:128–136.
422. Malbrain, M. L., Lambrecht, G. L., Zandijk, E. et al. (1997): Treatment of severe thallium intoxication. *Clin. Toxicol.*, 35:97–100.
423. Manheimer, E. W. and Silbergeld, E. K. (1998): Critique of CDC's retreat from recommending universal lead screening for children. *Public Health Rep.*, 113:38–46.
424. Manton, W. I. (1994): Lead poisoning from gunshots: a five century heritage. *Clin. Toxicol.*, 32:387–389.
425. Mappes, R. (1977): Versuche zur ausscheidung von arsen im urin. *Int. Arch. Occup. Environ. Health*, 40:267–272.
426. Marcus, R. L., Turner, S., and Cherry, N. M. (1996): A study of lung function and chest radiograms in men exposed to zirconium compounds. *Occup. Med.*, 46: 109–113.
427. Marcus, W. L. (1994): Lithium: a review of its pharmacokinetics, health effects, and toxicology. *J. Environ. Pathol Toxicol. Oncol.*, 13:73–79.
428. Marie, P. J., Gabra, M.-T., Hott, M., and Miravet, L. (1985): Effect of low doses of stable strontium on bone metabolism in rats. *Miner. Electrolyte Metab.*, 11:5–13.
429. Markesbery, W. R., Ehmann, W. D., Hossain, T. I. M. Aluaddin, M., and Goodin, D. T. (1981): Instrumental neutron activation analysis of brain aluminum in Alzheimer disease and aging. *Ann. Neurol.*, 10:511–516.
430. Martin, J. L. and Gerlach, M. L. (1972): Selenium metabolism in animals. *Ann. N.Y. Acad. Sci.*, 192:193–199.
431. Martyn, C. N., Coggon, D. N., Inskip, H., Lacey, R. F. and Young, W. F. (1997): Aluminum concentrations in drinking water and risk of Alzheimer's disease. *Epidemiology*, 8:281–286.
432. Martyn, C. N., Osmond, C., Edwardson, J. A., Barker, D. J. P., Harris, E. C., and Lacey, R. F. (1989): Geographical relation between Alzheimer's disease and aluminum in drinking water. *Lancet*, 1:59–62.
433. Matsumuro, K., Izumo, S., Higuchi, I., Ronquillo, A. T. Takahashi, K., and Osame, M. (1993): Experimental germanium dioxide-induced neuropathy in rats. *Acta Neuropathol.*, 86:547–553.
434. Maurer, J. K., Cheng, M. C., Boysen, B. G., and Anderson R. L. (1990): Two-year carcinogenicity study of sodium fluoride in rats. *J. Natl. Cancer Inst.*, 82:1118–1126.
435. McAlister, N. H., Abrams, H. B., Schlosser, R., and Sturtridge, W. (1990): Unintentional self-intoxication with inorganic calcium. *J. Intern. Med.*, 228:193–195.

436. McCallum, R. 1. (1989): The industrial toxicology of antimony: the Ernestine Henry Lecture 1987. *J. R. Coll. Phys. Lond.*, 23:28–32.
437. McCance R. A. and Widdowson, E. M. (1937): Absorption and excretion of iron. *Lancet*, 2:680.
438. McCarron, D. A., Morris, C. D., and Cole, C. (1982): Dietary calcium in human hypertension. *Science*, 217:267–269.
439. McCaslin, F. E. and Janes, J. M. (1959): The effect of strontium lactate in the treatment of osteoporosis. *Mayo Clin. Proc.*, 34:329–334.
440. McConnell, K. P. and Portman, O. W. (1952): Excretion of dimethyl selenide by the rat. *J. Biol. Chem.*, 195: 277–282.
441. McGuigan, M. A. (1996): Acute iron poisoning. *Pediatr. Ann.*, 25:33–38.
442. McKenna, K. E., Dolan, O., Walsh, M. Y. et al. (1995): Contact allergy to gold sodium thiosulfate. *Contact Derm.*, 32:143–146.
443. McLaughlin, A. I. G., Milton, R., and Perry, K. M. A. (1946): Toxic manifestations of osmium tetroxide. *Br. J. Indust. Med.*, 3:183–186.
444. Mead, L. D. and Geis, W. J. (1901): Physiological and toxicological effects of tellurium compounds, with a special study of their influence on nutrition. *Am. J. Physiol.*, 5:104–149.
445. Medeiros, D. M., Wildman, R., and Liebes, R. (1997): Metal metabolism and toxicities. In: *Handbook of Human Toxicology*, edited by E. J. Massaro, pp. 149–188. CRC Press, New York.
446. Mees, R. A. (1919): The nails with arsenical polyneuritis. *JAMA*, 72:1337.
447. Meggs, W. J., Cahill-Morasco, R., Shih, R. D. et al. (1997): Effects of Prussian blue and *N*-acetylcysteine on thallium toxicity in mice. *Clin. Toxicol.*, 35:163–166.
448. Meggs, W. J., Gerr, F., Aly, M. H. et al. (1994): The treatment of lead poisoning from gunshot wounds with succimer (DMSA). *Clin. Toxicol.*, 32:377–385.
449. Meggs, W. J., Hoffman, R. S., Shih, R. D. et al. (1994): Thallium poisoning from maliciously contaminated food. *Clin. Toxicol.*, 32:723–730.
450. Meuillet, E., Stratton, S., and Cheruki, D. P. (2004): Chemoprevention of prostate cancer with selenium: an update on current clinical trials and preclinical findings. *J. Cell. Biochem.*, 91:443–458.
451. Meunier, P. J. et al. (2004): The effects of strontium ranelate on the risk of vertebral fracture in women with postmenopausal osteoporosis. *N. Engl. J. Med.*, 350:459–468.
452. Miller, A. C., Fuciarelli, A. F., Jackson, W. E., Ejnik, E. J., Emond, C., Strocko, S., Hogan, J., Page, N., and Pellmar, T. (1998): Urinary and serum mutagenicity studies with rats implanted with depleted uranium or tantalum pellets. *Mutagenesis*, 13:643–648.
453. Miller, C. W., Davis, M. W., Goldman, A., and Wyatt, J. P. (1953): Pneumoconiosis in the tungsten-carbide tool industry. *A.M.A. Arch. Indust. Hyg. Occup. Med.*, 8: 453–465.
454. Miller, J. W. and Sayers, R. R. (1941): The response of peritoneal tissue to industrial dusts. *U.S. Public Health Serv. Rep.*, 56(1):264–272.
455. Mills, K. C., and Curry, S. C. (1994): Acute iron poisoning. *Emerg. Med. Clin. North Am.*, 12:397–413.
456. Mitchell, R. R., Fitzgerald, S. D., Aulerich, R. J., Balander, R. J., Powell, D. C., Tempelman, R. J., Stickle, R. L., Stevens, W., and Bursian, S. J. (2001): Health effects following chronic dosing with tungsten–iron and tungsten–polymer shot in adult game-farm mallards. *J. Wildlife Dis.*, 37:451–458.
457. Monsen, E. R., (1996): New dietary reference intakes proposed to replace the Recommended Dietary Allowances. *J. Am. Diet. Assoc.*, 96:754–755.
458. Moon, J. (1994): The role of vitamin D in toxic metal absorption: a review. *J. Am. Coll. Nutr.*, 13:559–569.
459. Moore, C. and Adler, R. (2000): Herbal vitamins: lead toxicity and developmental delay. *Pediatrics*, 106: 600–602.
460. Moore, J. A. (1995): An assessment of lithium using the IEHR evaluative process for assessing human developmental and reproductive toxicity of agents. *Reprod. Toxicol.*, 9:175–210.
461. Moore, J. A. (1997): An assessment of boric acid and borax using the IEHR evaluative process for assessing human developmental and reproductive toxicity of agents. *Reprod. Toxicol.*, 11:123–160.
462. Moore, K. E. and Brody, T. M. (1961): Effect of triethyl tin on mitochondrial swelling. *Biochem. Pharmacol.*, 6:134–142.
463. Moorman, W. J., Ahlers, H. W., Chapin, R. E. et al. (2000): Prioritization of NTP reproductive toxicants for field studies. *Reprod. Toxicol.*, 14:293–301.
464. Morgan, D. L., Shines, C. J., Jeter, S. P. et al. (1997): Comparative pulmonary absorption, distribution, and toxicity of copper gallium diselenide, copper indium diselenide, and cadmium telluride in Sprague–Dawley rats. *Toxicol. Appl. Pharmacol.*, 147:399–410.
465. Morgan, D. L., Shines, C. J., Jeter, S. P., Wilson, R. E., Elwell, M. P., Price, H. C., and Moskowitz, P. D. (1995): Acute pulmonary toxicity of copper gallium diselenide, copper indium diselenide, and cadmium telluride intratracheally instilled into rats. *Environ. Res.*, 71: 16–24.
466. Morgan, J. G. (1958): Some observations on the incidence of respiratory cancer in nickel workers. *Br. J. Indust. Med.*, 15:224–234.
467. Morin, Y. and Daniel, P. (1967): Quebec beer-drinkers' cardiomyopathy: etiological considerations. *Can. Med. Assoc. J.*, 97:926–928.
468. Morton, W. E. and Caron, G. A. (1989): Encephalopathy: an uncommon manifestation of workplace arsenic poisoning? *Am. J. Indust. Med.*, 15:1–5.
469. Mowad, E., Haddad, I., and Gemmel, D. J. (1998): Management of lead poisoning from ingested fishing sinkers. *Arch. Pediatr. Adolesc. Med.*, 152:485–488.
470. Moxon, A. L. and Rhian, M. (1943): Selenium poisoning. *Physiol. Rev.*, 23:305–337.
471. Mulkey, J. P. and Oehme, F. W. (1993): A review of thallium toxicity. *Vet. Hum. Toxicol.*, 35:445–453.
472. Muntzel, M. and Drüeke, T. (1992): A comprehensive review of the salt and blood pressure relationship. *Am. J. Hypertens.*, 5:1S–42S.

473. Mutter, J., Naumann, J., Sadaghiani, C. et al. (2004): Amalgam studies: disregarding basic principles of mercury toxicity. *Int. J. Hyg. Environ. Health*, 207(4):391–397.
474. Nakada, T., Furuta, H., Koike, H. et al. (1989): Impaired urine concentrating ability in *itai-itai* (ouch-ouch) disease. *Int. J. Urol. Nephrol.*, 21:201–209.
475. Nakajima, M. et al. (2000): Comparative developmental toxicity study of indium in rats and mice. *Teratog. Carcinog. Mutagen.*, 20:219–227.
476. Nakamura, Y., Tsumura, Y., Tonogai, Y. et al. (1997): Differences in behavior among the chlorides of seven rare earth elements administered intravenously to rats. *Fundam. Appl. Toxicol.*, 37:106–116.
477. NCI. (1978): *Bioassay of Titanium Dioxide for Possible Carcinogenicity*, Tech. Rep. Ser. No. 97. National Cancer Institute, Bethesda, MD.
478. NCI. (1980): *Bioassay of Selenium Sulfide (Dermal Study) for Possible Carcinogenicity*, NCI Tech. Rep. Ser. No. 197, NTP No. 80-18. National Cancer Institute, Bethesda, MD.
479. NCI. (1980): *Bioassay of Selenium Sulfide (Gavage) for Possible Carcinogenicity*. NCI Tech. Report Ser. No. 194, NTP No. 80-17. National Cancer Institute, Bethesda, MD.
480. NCI. (1980): *Bioassay of Selsun (Trade Name) for Possible Carcinogenicity*. NCI Tech. Report Ser. No. 199, NTP No. 80-19. National Cancer Institute, Bethesda, MD.
481. NCI. (2005): *The SELECT Prostate Cancer Prevention Trial*, National Cancer Institute, Bethesda, MD (http://www.nci.nih.gov/select).
482. National Institutes of Health. (1994): Optimal calcium intake. *NIH Consens. Stat.*, 12:1–31.
483. National Research Council. (1989): *Recommended Dietary Allowances*, 10th ed. National Academy Press, Washington, D.C.
484. NTP. (1982): *Bioassay of Stannous Chloride for Possible Carcinogenicity*. Tech. Rep. No. 231. National Toxicology Program, Research Triangle Park, NC.
485. NTP. (1987): *Toxicology and Carcinogenesis Studies of Boric Acid (CAS No. 10043-35-3) in B6C3F$_1$ Mice*, Tech. Rep. Ser. 324. U.S. Department of Health and Human Services, National Institute of Health, National Toxicology Program, Research Triangle Park, NC.
486. NTP. (1994): *Toxicology and Carcinogenesis Studies of Barium Chloride Dihydrate (CAS No. 10326–27–9) in F344/N Rats and B6C3F$_1$ Mice (Drinking Water Studies)*, Tech. Rep. No. 432. U.S. Department of Health and Human Services, National Institute of Health, National Toxicology Program, Research Triangle Park, NC.
487. NTP. (1996): *Toxicology and Carcinogenesis Studies of Nickel Oxide in F344/N Rats and B6C3F$_1$ Mice*, Tech. Rep. Ser. No. 451, NIH Publ. No. 96–3367. U.S. Department of Health and Human Services, National Institute of Health, National Toxicology Program, Research Triangle Park, NC.
488. NTP. (1996): *Toxicology and Carcinogenesis Studies of Nickel Subsulfide in F344/N Rats and B6C3F$_1$ Mice*, Tech. Rep. Ser. No. 453, NIH Publ. No. 96-3369. U.S. Department of Health and Human Services, National Institute of Health, National Toxicology Program, Research Triangle Park, NC.
489. NTP. (1996): *Toxicology and Carcinogenesis Studies of Nickel Sulfate Hexahydrate in F344/N Rats and B6C3F$_1$ Mice*, Tech. Rep. Ser. No. 454, NIH Publ. No. 96-3370. U.S. Department of Health and Human Services, National Institute of Health, National Toxicology Program, Research Triangle Park, NC.
490. NTP. (2005): *Report on Carcinogens*, 11th ed. National Toxicology Program, Research Triangle Park, NC (http://ntp.niehs.nih.gov/ntp/roc/toc11.html).
491. National Toxicology Program Executive Committee Working Group for the Report on Carcinogens (RG2) (2005): *Review Summary*, http://ntp.niehs.nih.gov/ ntp/newhomeroc/roc11/Lead_RG2Summ.pdf.
492. Needleman, H. L., Gunnoe, C., Leviton, A. et al. (1979): Deficits of psychologic and classroom performance of children with elevated dentine lead levels. *N. Engl. J. Med.*, 300:689–695.
493. Needleman, H. L., Schell, A., Bellinger, D. et al. (1990): The long-term effects of exposure to low doses of lead in childhood. An 11-year follow-up report. *N. Engl. J. Med.*, 322:83–88.
494. Neilands, J. B., Strong, F. M., and Elvehjem, C. A. (1948): Molybdenum in the nutrition of the rat. *J. Biol. Chem.*, 172:431–439.
495. Nemec, M. D., Holson, J. F., Farr, C. H., and Hood, R. D (1998): Developmental toxicity assessment of arsenic acid in mice and rabbits. *Reprod. Toxicol.*, 12:647–658.
496. Nemery, B. (1990): Metal toxicity and the respiratory tract *Eur. Respir. J.*, 3:202–219.
497. Nemery, B., Lewis, C. P. L., and Demerts, M. (1994): Cobalt and possible oxidant-mediated toxicity. *Sci. Total Environ.*, 150:57–64.
498. Newman, D. (1890): A case of adeno-carcinoma of the left inferior turbinated body, and perforation of the nasal septum, in the person of a worker in chrome pigments. *Glasgow Med. J.*, 33:469.
499. Newton, P. E., Bolte, H. F., Daly, I. W., Pillsbury, B. D. Terrill, J. B., Drew, R. T., Ben-Dyke, R., Sheldon, A. W. and Rubin, L. F. (1994): Subchronic and chronic inhalation toxicity of antimony trioxide in the rat. *Fundam. Appl Toxicol.*, 22:561–576.
500. Nielsen, F. H. (1992): Facts and fallacies about boron. *Nutr Today*, 27:6–12.
501. Nielsen, F. H. (1994): Biochemical and physiologic consequences of boron deprivation in humans. *Environ. Health Persp.*, 102(Suppl. 7):59–63.
502. Nielsen, F. H. (1996): Evidence for the nutritional essentiality of boron. *Trace Elem. Exp. Med.*, 9 215–229.
503. Nielsen, F. H. (1996): How should dietary guidance be given for mineral elements with beneficial actions suspected of being essential? *J. Nutr.*, 126:2377S–2385S.
504. Nielsen, S.P. (2004): The biological role of strontium *Bone*, 35:583–588.
505. Nightingale, S.L. (1991): Succimer (DMSA) approved for severe lead poisoning. *JAMA*, 265:1802.
506. Niinomi, M. (2003): Fatigue performance and cyto-toxicity of low rigidity titanium alloy, Ti-29Nb-13Ta-4.6Zr. *Biomaterials*, 24(16):2673–2683.

507. Nogawa, K., Honda, R., Kido, T. et al. (1989): A dose–response analysis of cadmium in the general environment with special reference to total cadmium intake limit. *Environ. Res.*, 48:7–16.
508. Nordberg, G. (1994): Assessment of risks in occupational cobalt exposures. *Sci. Total Environ.*, 150:201–207.
509. Nordberg, G., Stenstrom, T., and Slorach, S. (1973): Cadmium poisoning caused by a cooled-soft-drink machine [in Swedish]. *Lakartidningen*, 70:601–604.
510. Notton, B. A. and Hewitt, E. J. (1971): The role of tungsten in the inhibition of nitrate reductase in spinach (*Spinacea oleracea* L.) leaves. *Biochem. Biophys. Res. Comm.*, 44:702–710.
511. Nye, L. J. J. (1929): An investigation of the extraordinary incidence of chronic nephritis in young people in Queensland. *Med. J. Austr.*, 2:145–169.
512. Oberley, L. (1988): Free radicals and diabetes. *Free Radic. Biol. Med.*, 5:113–124.
513. O'Brien, P. and Kortenkamp, A. (1995): The chemistry underlying chromate toxicity. *Transition Met. Chem.*, 20:636–642.
514. Okusa, M. D. and Crystal, L. J. T. (1994): Clinical manifestations and management of acute lithium intoxication. *Am. J. Med.*, 97:383–389.
515. Olin, S. S. (1998): Between a rock and a hard place: methods for setting dietary allowances and exposure limits for essential metals. *J. Nutr.*, 128:364S–367S.
516. Omdahl, J. L. and DeLuca, H. F. (1971): Strontium induced rickets: metabolic basis. *Science*, 174:949–951.
517. Omura, M., Hirata, M., Tanaka, A., Zhao, M., Makita, Y., Inoue, N., Gotoh, K., and Ishinishi, N. (1996): Testicular toxicity evaluation of arsenic-containing binary compound semi-conductors, gallium arsenide and indium arsenide, in hamsters. *Toxicol. Lett.*, 89:123–129.
518. Omura, M., Tanaka, A., Hirata, M., Zhao, M., Marita, Y., Gotoh, K., and Ishinishi, N. (1996): Testicular toxicity of gallium arsenide, indium arsenide, and arsenic oxide in rats by repetitive intratracheal instillation. *Fundam. Appl. Toxicol.*, 32:72–78.
519. Osswald, H. and Goerttler, K. (1971): Arsenic-induced leucoses in mice after diaplacental and postnatal application. *Verh. Dtsch. Gesellsch. Path.*, 55:289–293.
520. Oteiza, P. I., Adonaylo, V. N., and Keen, C. L. (1999): Cadmium-induced testes oxidative damage in rats can be influenced by dietary zinc intake. *Toxicology*, 137:13–22.
521. Ott, M. G., Holder, B. B., and Gordon, H. L. (1974): Respiratory cancer and occupational exposure to arsenicals. *Arch. Environ. Health*, 29:250–255.
522. Pak, C. Y. C., Sakhafe, K., Zerwekh, J. E., Parcel, C., Peterson, R., and Johnson, K. (1989): Safe and effective treatment of osteoporosis with intermittent slow release sodium fluoride: augmentation of vertebral bone. *J. Clin. Endocrinol. Metab.*, 68:150–159.
523. Parizek, J. (1965): The peculiar toxicity of cadmium during pregnancy. An experimental 'toxaemia of pregnancy' induced by cadmium salts. *J. Reprod. Fertil.*, 9:111–112.
524. Parras, F., Patier, J. L., and Ezpeleta, C. (1987): Lead-contaminated heroin as a source of inorganic-lead intoxication. *N. Engl. J. Med.*, 316:755.
525. Patterson, C., Ericson, J., Manea-Krichten, M., and Shirahata, H. (1991): Natural skeletal levels of lead in *Homo sapiens sapiens* uncontaminated by technological lead. *Sci. Total Environ.*, 107:205–236.
526. Pegues, D. A., Hughes, B. J., and Woernle, C. H. (1993): Elevated blood lead levels associated with illegally distilled alcohol. *Arch. Intern. Med.*, 153: 1501–1504.
527. Pennington, J. A. and Jones, J. W. (1987): Molybdenum, nickel, cobalt, vanadium, and strontium in total diets. *J. Am. Diet. Assoc.*, 87:1644–1650.
528. Pennington, J. A. and Shoen, S. A. (1995): Estimates of dietary exposure to aluminum. *Food Addit. Contam.*, 12:119–128.
529. Perlstein, M. A. and Attala, R. (1966): Neurologic sequelae of plumbism in children. *Clin. Pediatr.*, 5: 292–298.
530. Perry, Jr., H. M. and Perry, E. E. (1959): Normal concentrations of some trace metals in human urine: changes produced by ethylenediaminetetraacetate. *J. Clin. Invest.*, 38:1452–1463.
531. Perry, W. G., Smith, F. A., and Kent, M. B. (1994): The halogens. In: *Patty's Industrial Hygiene and Toxicology*, edited by G. D. Clayton and F. E. Clayton, pp. 4449–4521. John Wiley & Sons, New York.
532. Pershagen, G. and Bjorklund, N.-E. (1985): On the pulmonary tumorigenicity of arsenic trisulfide and calcium arsenate in hamsters. *Cancer Lett.*, 27:99–104.
533. Pershagen, G., Nordberg, G., and Björklund, N.-E. (1984): Carcinomas of the respiratory tract in hamsters given arsenic trioxide and/or benzo[*a*]pyrene by the pulmonary route. *Environ. Res.*, 34:227–241.
534. Phillips, R. H., Whitehead, M. W., Doig, L. A. et al. (2001): Is eradication of Helicobacter pylori with colloidal bismuth subcitrate quadruple therapy safe? *Helicobacter*, 6(2):151–156.
535. Pietras, R., Stavrakos, C., Gunnar, R. M., and Tobin, Jr., J. R. (1968): Phosphorus poisoning simulating acute myocardial infarction. *Arch. Intern. Med.*, 122:430–434.
536. Pirkle, J. L., Brody, D. J., Gunter, E. W. et al. (1994): The decline in blood lead levels in the United States. The National Health and Nutrition Examination Surveys (NHANES). *JAMA*, 272:284–291.
537. Pirot, F., Millet, J., Kalia, Y. N., and Humbert, Ph. (1996): *In vitro* study of percutaneous absorption, cutaneous bioavailability and bioequivalence of zinc and copper from five topical formulations. *Skin Pharmacol.*, 9:259–269.
538. Pirot, F., Panisset, F., Agache, P., and Humbert, P. (1996): Simultaneous absorption of copper and zinc through human skin *in vitro*. *Skin Pharmacol.*, 9: 43–52.
539. Poon, R., Chu, I., Lecavalier, P., Valli, V. E., Foster, W., Gupta, S., and Thomas, B. (1998). Effects of antimony on rats following 90-day exposure via drinking water. *Food Chem. Toxicol.*, 36:21–35.
540. Preussmann, R. and Ivankovic, S. (1975): Absence of carcinogenic activity in BD rats after oral administration of high doses of bismuth oxychloride. *Food Cosmet. Toxicol.*, 13:503–508.

541. Price, C.J., Field, E.A., Marr, M.C. et al. (1990): *Developmental Toxicity of Boric Acid (CAS No. 10043-35-3) in Sprague–Dawley Rats*, NTP Rep. No. 90-105 and Rep. Suppl. No. 90-105A). National Toxicology Program, Public Health Service, U.S. Department of Health and Human Services, Research Triangle Park, NC.

542. Price, C. J., Marr, M. C., and Myers, C. B. (1994): *Determination of the No-Observable-Adverse-Effect-Level (NOAEL) for Developmental Toxicity in Sprague–Dawley (CD) Rats Exposed to Boric Acid in Feed on Gestational Days 0 to 20, and Evaluation of Postnatal Recovery Through Postnatal Day 21*, Rep. No. 65C-5657-200. Research Triangle Institute, Research Triangle Park, NC.

543. Price, C. J., Marr, M. C., Myers, C. B. et al. (1991): *Final Report on the Developmental Toxicity of Boric Acid (CAS No. 10043-35-3) in New Zealand White Rabbits*, NIEHS/NTP Order PB92-129550. National Toxicology Program, Research Triangle Park, NC.

544. Price, C.J., Strong, P.L., Marr, M.C. et al (1996): Developmental toxicity NOAEL and postnatal recovery in rats fed boric acid during gestation. *Fund. Appl. Toxicol.*, 32:179–193.

545. Prior, J. T., Rustad, H., and Cronk, G. A. (1957): Pathological changes associated with deodorant preparations containing sodium zirconium lactate: an experimental study. *J. Invest. Dermatol.*, 29:449–463.

546. Puri, V. N. (1999): Cadmium induced hypertension. *Clin. Exp. Hypertens*. 21:79–84.

547. Questel, F., Dugarin, J., and Dally, S. (1996): Thallium-contaminated heroin. *Ann. Intern. Med.*, 124:616.

548. Rainey, C. J., Christensen, R. E., Nyquist, L. A. et al. (1999): Daily boron intake from the American diet. *J. Am. Diet. Assoc.*, 99(3):335–340.

549. Reddy, B. S., Rivenson, A., El-Bayoumy, K. et al. (1997): Chemoprevention of colon cancer by organoselenium compounds and impact of high- or low-fat diets. *J. Natl. Cancer Inst.*, 89:506–512.

550. Reed, C. E. (1956): A study of the effects on the lung of industrial exposure to zirconium dusts. *A.M.A. Arch. Indust. Health*, 13:578–580.

551. Reeves, A.L. (1979): Barium. In: *Handbook on the Toxicology of Metals*, edited by L. Friberg, G. F. Nordberg, V. B. Vouk, pp. 321–328. Elsevier, New York.

552. Reissman, D. B., Matte, D., Gurnitz, K. L. et al. (2002): Is home renovation or repair a risk factor for exposure to lead among children residing in New York City? *J. Urban Health: Bulletin of New York Acad. Med.*, 79:502–511.

553. Rencher, A. C., Carter, M. W., and McKee, D. W. (1977). A retrospective epidemiological study of mortality at a large western copper smelter. *J. Occup. Med.*, 19: 754–758.

554. Ress, N. B., Chou, B. J., Renne, N. B. et al. (2003): Carcinogenicity of inhaled vanadium pentoxide in F344/N rats and B6C3F$_1$ mice. *Toxicol. Sci.*, 74(2):287–296.

555. Rifat, S. L., Corey, P. N., and McLachlan, D. R. C. (1997): Neuropsychiatric disorders in a follow-up study of northern Ontario miners. *Am. J. Epidemiol.*, 145:S16.

556. Rifat, S. L., Eastwood, M. R., Crapper McLachlan, D. R., and Corey, P. N. (1990): Effect of exposure of miners to aluminum powder. *Lancet*, 336:1162–1165.

557. Riggs, B. L. and Melton, L. J. (1983): Evidence for two distinct syndromes of involutional osteoporosis. *Am. J. Med.*, 75:899–901.

558. Rivlin, R. S. (1994): Magnesium deficiency and alcohol intake: mechanisms, clinical significance and possible relation to cancer development (a review). *J. Am. Coll. Nutr.*, 13:416–423.

559. Rodgers, K. (1998): Platinum. In: *Immunotoxicology of Environmental and Occupational Metals*, edited by J. T. Zelikoff and P. T. Thomas, pp. 195–206. Taylor & Francis, Bristol, PA.

560. Roels, H. A., Hubermont, G., Buchet, J. P., and Lauwerys, R. (1978): Placental transfer of lead, mercury, cadmium, and carbon monoxide in women. *Environ. Res.*, 16:236–247.

561. Roels, H. A., Lauwerys, R., Buchet, J. P. et al. (1987): Epidemiological survey among workers exposed to manganese: effects on lung, central nervous system, and some biological indices. *Am. J. Indust. Med.*, 11:307–327.

562. Roels, H. A., Lauwerys, R. R., Buchet, J. P. et al. (1989): Health significance of cadmium induced renal dysfunction: a five year follow up. *Br. J. Indust. Med.*, 46:755–764.

563. Rogan, W. J., Dietrich, K. N., Ware, J. H. et al. (2001): The effect of chelation therapy with succimer on neurophysiological development in children exposed to lead. *N. Engl. J. Med.*, 344:1421–1426.

564. Rom, W. N. (1976): Effects of lead on female reproduction: a review. *Mt. Sinai J. Med.*, 43:542–552.

565. Roth, F. (1958): Über den Brochialkrebs Arsengeschädigter Winzer. *Virchows Arch.*, 331:119–137.

566. Rotruck, J. T., Pope, A. L., Ganther, H. E. et al. (1973): Selenium: biochemical role as a component of glutathione peroxidase. *Science*, 179:588–590.

567. Roza, O. and Berman, L. B. (1971): The pathophysiology of barium: hypokalemic and cardiovascular effects. *J. Pharmacol. Exp. Therap.*, 177: 433–439.

568. Rubin, L. Slepyan, A. H., Weber, L. F., and Neuhauser, I. (1956): Granulomas of the axillas caused by deodorants. *J. Am. Med. Assoc.*, 162:953–955.

569. Ruta, D. A. and Haider, S. (1989): Attempted murder by selenium poisoning. *Br. Med. J.*, 299:316–317.

570. Ryan, T. P. and Aust, S. D. (1992): The role of iron in oxygen-mediated toxicities. *CRC Crit. Rev. Toxicol.*, 22:119–141.

571. Rybak, L. P. (1992): Hearing: the effects of chemicals. *Otolaryngol. Head Neck Surg.*, 106:677–686.

572. Rystedt, I. and Fischer, T. (1983): Relationship between nickel and cobalt sensitization in hard metal workers. *Contact Derm.*, 9:195–200.

573. Salonen, J. T., Nyyssonen, K., Korpela, H. et al. (1992): High stored iron levels are associated with excess risk of myocardial infarction in eastern Finnish men. *Circulation* 86:803–811.

574. Salonen, J. T., Tuomainen, T. P., Nyyssonen, K. et al (1998): Relation between iron stores and non-insulin dependent diabetes in men: case-control study. *Br. Med J.*, 317:727.

575. Salonen, J. T., Tuomainen, T. P., Salonen, R. et al. (1998): Donation of blood is associated with reduced risk of myocardial infarction. The Kuopio Ischaemic Heart Disease Risk Factor Study. *Am. J. Epidemiol.*, 148:445–451.

576. Sanai, T., Okuda, S., Onoyama, K., Oochi, N., Takaichi, S., Mizuhira, V., and Fujishima, M. (1991): Chronic tubulointerstitial changes induced by germanium dioxide in comparison with carboxyethylgermanium sesquioxide. *Kidney Int.*, 40:882–890.
577. Sanchez, D. J., Colomina, M. T., and Domingo, J. L. (1998): Effects of vanadium on activity and learning in rats. *Phys. Behav.*, 63:345–350.
578. Sanchez-Delgado, R. A., Navarro, M., Perez, H., and Urbina, J. A. (1996): Toward a novel metal-based chemotherapy against tropical diseases. 2. Synthesis and antimalarial activity *in vitro* and *in vivo* of new ruthenium and rhodium-chloroquine complexes. *J. Med. Chem.*, 39: 1095–1099.
579. Satarug, S. and Moore, M. R. (2004): Adverse health effects of chronic exposure to low-level cadmium in foodstuffs and cigarette smoke. *Environ. Health Perspect.* 112:1099–1103.
580. Schafer, S. G. and Femfert, U. (1984): Tin: a toxic heavy metal? A review of the literature. *Reg. Tox. Pharmacol.*, 4:57–69.
581. Schecter, A., Shanske, W., Stenzler, A. et al. (1980): Acute hydrogen selenide inhalation. *Chest*, 77:554–555.
582. Schoof, R. A., Yost, L. J., Eickhoff, E. A., Crecelius, D. W., Meacher, D. M., and Menzel, D. B. (1999): A market basket survey of inorganic arsenic in food. *Food. Chem. Toxicol.*, 37:839–846.
583. Schroeder, H. A., Balassa, J. J., and Vinton, Jr., W. H. (1964): Chromium, lead, cadmium, nickel and titanium in mice: effect on mortality, tumors and tissue levels. *J. Nutr.*, 83:239–250.
584. Schroeder, H. A., Frost, D. V., and Balassa, J. J. (1970): Essential trace elements in man: selenium. *J. Chronic Dis.*, 23:227–243.
585. Schroeder, H. A. and Mitchener, M. (1971): Scandium, chromium (VI), gallium, yttrium, rhodium, palladium, indium in mice: effects on growth and life span. *J. Nutr.*, 101:1431–1437.
586. Schroeder, H. A. and Mitchener, M. (1975): Life-term effects of mercury, methyl mercury, and nine other trace metals on mice. *J. Nutr.*, 105:452–458.
587. Schroeder, H. A. and Mitchener, M. (1975): Life-term studies in rats: effects of aluminum, barium, beryllium, and tungsten. *J. Nutr.*, 105:421–427.
588. Schroeder, H. A., Mitchener, M., Balassa, J. J., Kanisawa, M., and Nason, A. P. (1968): Zirconium, niobium, antimony and fluorine in mice: effects on growth, survival and tissue levels. *J. Nutr.*, 95:95–101.
589. Schroeder, H. A., Mitchener, M., and Nason, A. P. (1970): Zirconium, niobium, antimony, vanadium and lead in rats: life term studies. *J. Nutr.*, 100:59–68.
590. Schwartz, J. (1988): The relationship between blood lead and blood pressure in NHANES II survey. *Environ. Health Perspect.*, 78:15–22.
591. Schwartz, J. (1994): Societal benefits of reducing lead exposure. *Environ. Res.*, 66:105–124.
592. Schwarz, K. and Foltz, C. M. (1957): Selenium as an integral part of factor 3 against dietary necrotic liver degeneration. *J. Am. Chem. Soc.*, 79:3292–3293.
593. Schwarz, K. and Mertz, W. (1959): Chromium (III) and the glucose tolerance factor. *Arch. Biochem. Biophys.*, 85:292–295.
594. Schwarz, K., Milne, D. B., and Vinyard, E. (1970): Growth effects of tin compounds in rats maintained in a trace element-controlled environment. *Biochem. Biophys. Res. Comm.*, 40:22–29.
595. Seelig, M. S. and Master, A. C. N. (1994): Consequences of magnesium deficiency on the enhancement of stress reactions: preventive and therapeutic implications (a review). *J. Am. Coll. Nutr.*, 13:429–446.
596. Seghizzi, P., D'Adda, F., Borleri, D., Barbic, F., and Mosconi, G. (1994): Cobalt cardiomyopathy. A critical review of literature. *Sci. Total Environ.*, 150:105–109.
597. Seifert, J. (1943): Intravenous injections of soluble tin compounds. *J. Lab. Clin. Med.*, 28:1344–1348.
598. Semlitsch, M., Staub, F., and Weber, H. (1985): Titanium–aluminum–niobium alloy, development for biocompatible, high strength surgical implants. *Biomed. Tech.*, 30:334–339.
599. Seppalainen, A. M., Hernberg, S., Vesanto, R., and Kock, B. (1983): Early neurotoxic effects of occupational lead exposure: a prospective study. *Neurotoxicology*, 4:181–192.
600. Shah, S. V. and Alam, M. G. (2003): Role of iron in atherosclerosis. *Kidney Dis.*, 41(S1):80–83.
601. Shamberger, R. J. and Frost, D. V. (1969): Possible protective effect of selenium against human cancer. *Can. Med. Assoc. J.*, 100:682.
602. Shamberger, R. J. (1971): Is selenium a teratogen? *Lancet*, 1316.
603. Shannon, M. (1998): Lead poisoning from an unexpected source in a 4-month old infant. *Environ. Health Perspec.*, 106:313–316.
604. Shannon, M. and Graef, J. W. (1989): Lead intoxication from lead-contaminated water used to reconstitute infant formula. *Clin. Ped.*, 28:380–382.
605. Shelkovnikov, S. A. and Gonick, H. C. (2001): Influence of lead on rat thoracic aorta contraction and relaxation. *Am. J. Hypertens.*, 14:873–878.
606. Shelley, W. B. and Hurley, H. (1958): The allergic origin of zirconium deodorant granulomas. *Br. J. Dermatol.*, 70:75–101.
607. Shils, M. E. (1988): Magnesium in health and disease. *Annu. Rev. Nutr.*, 8:429–460.
608. Shipman, D. L. (1986): Cadmium food poisoning in a Missouri school. *J. Environ. Health*, 49:89.
609. Siegel, E. and Wason, S. (1986): Boric acid toxicity. *Ped. Clin. N. Am.*, 33:363–367.
610. Silbergeld, E. K. (1996): Lead poisoning: the implications of current biomedical knowledge for public policy. *Maryland Med. J.*, 45:209–217.
611. Silbergeld, E. K. (1997): Preventing lead poisoning in children. *Ann. Rev. Public Health*, 18:187–210.
612. Silbergeld, E.K. (2003): Facilitative mechanisms of lead as a carcinogen. *Mutat. Res.*, 533:121–133.
613. Sivjakov, K. I. and Braun, H. A. (1959): The treatment of acute selenium, cadmium, and tungsten intoxication in rats with calcium disodium ethylenediaminetetraacetate. *Toxicol. Appl. Pharmacol.*, 1:602–608.
614. Skipworth, G. B., Goldstein, N., and McBride, W. P. (1967): Boric acid intoxication from 'medicated talcum powder.' *Arch. Dermatol.*, 95:83–86.

615. Skoryna, S. C. (1984): Metabolic aspects of the pharmacologic use of trace elements in human subjects with specific reference to stable strontium. In: *Trace Substances in Environmental Health*, edited by D. D. Hemphill, pp. 3–20. University of Missouri, St. Louis.
616. Smith, B.S. (1971): Sexuality in the American mud snail, *Nassarius obsoletus*. *Proc. Malacol. Soc. Lond.*, 39:377–378.
617. Smith, M. I., Franke, K. W., and Westfall, B. B. (1936): The selenium problem in relation to public health. *Public Health Rep.*, 51:1496–1505.
618. Spencer, H., Kramer, L., and Osis, D. (1988): Do protein and phosphorus cause calcium loss? *J. Nutr.*, 118–657–660.
619. Spiegl, C. J., Calkins, M. C., DeVoidre, J. J. et al. (1956): *Inhalation Toxicity of Zirconium Compounds*. I. *Short-Term Studies*. Atomic Energy Commission Project, Rep. No. UR-460. University of Rochester, Rochester, NY.
620. Stadtman, T. C. (1977): Biological function of selenium. *Nutr. Rev.*, 35:161–166.
621. Stal, P. (1995): Iron as a hepatotoxin. *Dig. Dis.*, 13: 205–222.
622. Stayner, L., Smith, R., Thun, M. et al. (1992): A dose–response analysis and quantitative assessment of lung cancer risk and occupational cadmium exposure. *Ann. Epidemiol.*, 2:177–194.
623. Sticht, G. and Käferstein, H. (1988): Bromine. In: *Handbook on Toxicity of Inorganic Compounds*, edited by H. G. Sieler and H. Siegel, pp. 143–154. Marcel Dekker, New York.
624. Stift, A., Friedl, J., and Laengle, F. (1998): Liver transplantation for potassium dichromate poisoning. *N. Engl. J. Med.*, 338:766–767.
625. Stohs, S. J., Bagchi, D., Hassoun, E. et al. (2001): Oxidative mechanism in toxicity of chromium and cadmium ions. *J. Environ. Pathol. Oncol.*, 20:77–88.
626. Stollery, B. T., (1996): Reaction time changes in workers exposed to lead. *Neurotoxicol. Teratol.*, 18:477–483.
627. Storey, E. (1961): Strontium "rickets": bone, calcium and strontium changes. *Austral. Ann. Med.*, 10:213–222.
628. Sullivan, J.L. (1981): Iron and the sex difference in heart disease risk. *Lancet*, 1:1293–1294.
629. Sunderman, F. W., Jr. (1988): Nickel. In: *Handbook on Toxicity of Inorganic Compounds*, edited by H. G. Sieler and H. Sigel, pp. 454–468. Marcel Dekker, New York.
630. Takebayashi, T., Omae, K., Oda, K., Uemura, T., Nomiyama, T., Ishizuka, C., and Sakurai, H. (1998): Toxicity of intratracheally-administered indium phosphide (InP) in rats. *J. Toxicol. Sci.*, 23(Suppl. II):403.
631. Takenaka, S., Oldiges, H., Konig, H. et al. (1983): Carcinogenicity of cadmium chloride aerosols in W rats. *J. Natl. Cancer Inst.*, 70:367–373.
632. Takeuchi, A., Yoshizawa, N., Oshima, S., Kubota, T., Oshikawa, Y., Akashi, Y., Oda., T., Niwa, H., Imazeki, N., Seno, A., and Fuse, Y. (1992): Nephrotoxicity of germanium compounds: report of a case and review of the literature. *Nephron*, 60:436–442.
633. Tallkvist, J., Bowlus, C. L., and Lonnerdal, B. (2001): DMT1 gene expression and cadmium absorption in human absorptive enterocytes. *Toxicol. Lett.*, 122: 171–177.
634. Tam, G. K., Charbonneau, S. M., Bryce, F., Pomroy, C., and Sandi, E. (1979): Metabolism of inorganic arsenic (74As) in humans following oral ingestion. *Toxicol. Appl. Pharmacol.*, 50:319–322.
635. Tamaki, S. and Frankenberger, Jr., W. T. (1992): Environmental biochemistry of arsenic. *Rev. Environ. Contam. Toxicol.*, 124:79–110.
636. Tanaka, A., Hisanaga, A., Hirata, M., Omura, M., Makita, Y., Inoue, N., and Ishinishi, N. (1996): Chronic toxicity of indium arsenide and indium phosphide to the lungs of hamsters. *Fukuoka Acta Med.*, 87:108–115.
637. Tao, S.-H. and Bolger, P. M. (1997): Hazard assessment of germanium supplements. *Reg. Tox. Pharmacol.*, 25:211–219.
638. Tay, C.-H. and Seah, C.-S. (1975): Arsenic poisoning from anti-asthmatic herbal preparations. *Med. J. Aust.*, 2:424–428.
639. Taylor A. and Marks, V. (1978): Cobalt: a review. *J. Hum. Nutr.*, 32:165–177.
640. Tenenbein, M. (1996): Benefits of parenteral deferoxamine for acute iron poisoning. *Clin. Toxicol.*, 34:485–489.
641. Tepper, L. B., Hardy, H. L., and Chamberlin, R. I. (1961): *Toxicity of Beryllium Compounds*, Elsevier, New York.
642. Thompson, H. J., and Becci, P. J. (1980): Selenium inhibition of *N*-methyl-*N*-nitrosourea-induced mammary carcinogenesis in the rat. *J. Natl. Cancer Inst.*, 1299–1301.
643. Thun, M. J., Schnorr, T. M., Smith., A. B. et al. (1985): Mortality among a cohort of U.S. cadmium production workers: an update. *J. Natl. Cancer Inst.*, 74:325–333.
644. Tokudome, S. and Kuratsune, M. (1976): A cohort study on mortality from cancer and other causes among workers at a metal refinery. *Int. J. Cancer*, 17:310–317.
645. Tong, S., Baghurst, P., McMichael, A. et al. (1996): Lifetime exposure to environmental lead and children's intelligence at 11–13 years: the Port Pirie cohort study. *Br. Med J.*, 312:1569–1575.
646. Townshend, R. H. (1982): Acute cadmium pneumonitis: a 17-year follow-up. *Br. J. Indust. Med.*, 39:411–412.
647. Tozman, E. C. S. and Gottlieb, N. L. (1987): Adverse reactions with oral and parenteral gold preparations. *Med Toxicol.*, 2:177–189.
648. Trapp, G. A., Miner, G. D., Zimmerman, R. L., Mastri, A R., and Heston, L. L. (1978): Aluminum levels in brain in Alzheimer's disease. *Biol. Psychiat.*, 13:709–718.
649. Traub, R. D., Rains, T. C., Garruto, R. M., Gadjusek, D C., and Gibbs, C. J. (1981): Brain destruction alone does not elevate brain aluminum. *Neurology*, 31:986–990.
650. Trivedi, B., Saxena, D. K., Murthy, R. C., and Chandra, S V. (1989): Embryotoxicity and fetotoxicity of orally administered hexavalent chromium in mice. *Reprod. Toxicol.*, 3:275–278.
651. Tseng, W. P. (1977): Effects and dose–response relationships of skin cancer and blackfoot disease with arsenic *Environ. Health Perspec.*, 19:109–119.
652. Tseng, W. P., Chu, H. M., How, S. W., Fong, J. M., Lin C. S., and Yeh, S. (1968): Prevalence of skin cancer in an endemic area of chronic arsenicism in Taiwan. *J. Natl Cancer Inst.*, 40:453–463.
653. Tsuchiya, K. (1969): Causation of ouch-ouch disease (*itai itai byo*): an introductory review. Part 1. Nature of the disease. *Keio J. Med.*, 18:181–194.

654. Tsuda, T., Babazono, A., Yamamoto, E. et al. (1995): Ingested arsenic and internal cancer: a historical cohort study followed for 33 years. *Am. J. Epidemiol.*, 141:198–209.
655. Tsuda, T., Inoue, I., Kojima, M., and Aoki, S. (1995): Market basket and duplicate portion estimation of dietary intakes of cadmium, mercury, arsenic, copper, manganese, and zinc by Japanese adults. *J. Assoc. Off. Anal. Chem. Int.*, 78:1363–1368.
656. Tümer, Z. and Horn, N. (1996): Menkes disease: recent advances and new insights into copper metabolism. *Ann. Med.*, 28:121–129.
657. Tuohimaa, P. and Wickmann, L. (1985): Sperm production of men working under heavy-metal or organic solvent exposure. In: *Occupational Hazards and Reproduction*, edited by K. Hemminki, M. Sorsa, and H. Vanio, pp. 73–80. Hemisphere, New York.
658. Tuomainen, T. P., Punnonen, K., Nyyssonen, K., and Salonen, J. T. (1998): Association between body iron stores and the risk of acute myocardial infarction in men. *Circulation*, 97:1461–1466.
659. Tzonou, A., Lagiou, P., Trichopoulou, A. et al. (1998): Dietary iron and coronary heart disease risk: a study from Greece. *Am. J. Epidemiol.*, 147:161–166.
660. U.S. Congress (1971): *Lead-Based Paint Poisoning Prevention Act*, Public Law 91-695. U.S. Government Printing Office, Washington, D.C.
661. U.S. Congress (1986): *Amendments to the Safe Drinking Water Act*, Public Law 99-339. U.S. Government Printing Office, Washington, D.C.
662. U.S. Congress (1992): *Residential Lead-Based Paint Hazard Reduction Act of 1992*, Public Law 102-550, Title X. U.S. Government Printing Office, Washington, D.C.
663. U.S. Congress (1996): *Mercury-Containing and Rechargeable Battery Management Act*, Public Law 104-142. U.S. Government Printing Office, Washington, D.C.
664. U.S. Consumer Product Safety Commission. (1977): Lead-containing paint and certain consumer products bearing lead-containing paint. *Fed. Reg.*, 42:44199–44201.
665. U.S. Department of Health, Education, and Welfare, Food and Drug Administration (1977): 21 CFR Parts 310.510 and 700.16, Final Rule. *Fed. Reg.*, 42:41374–41376.
666. USEPA. (1973): Control of lead additives in gasoline. *Fed. Reg.*, 38:33734–33741.
667. USEPA. (1989): *Office of Radiation Programs: Risk Assessment Methodology, Environmental Impact Statement for NESHAPS Radionuclides*. Vol. I. *Background Information Document*, EPA 520/1-89-005. U.S. Environmental Protection Agency, Washington, D.C.
668. USEPA. (1990): *Office of Emergency and Remedial Response: Risk Assessment Guidance for Superfund*. Vol. I. *Human Health Evaluation Manual (Part A)*, Interim final, EPA/540/1–89/002. U.S. Environmental Protection Agency, Washington, D.C.
669. USEPA. (1990). Pesticide products containing phenylmercury and other mercury compounds: receipt of requests for voluntary cancellation and amendments to delete uses. *Fed. Reg.*, 55:26754–26756.
670. USEPA. (1991): National primary drinking water regulations: radionuclides; proposed rule. *Fed. Reg.*, 56:33050–33127.
671. USEPA. (1992): *Health and Environmental Effects Document for Stable Strontium*, ECAO-CIN-G111. Office of Solid Waste and Emergency Response, Washington, D.C.
672. USEPA. (1996): Prohibition on gasoline containing lead or lead additives for highway use. *Fed. Reg.*, 61:3832–3838.
673. USEPA. (1997): *Implementation of the Mercury-Containing and Rechargeable Battery Management Act*. U.S. Environmental Protection Agency, Washington, D.C.
674. USEPA. (1998): *Integrated Risk Information System (IRIS)*. U.S. Environmental Protection Agency, Washington, D.C. (http://www.epa.gov/ngispgm3/iris/subst-fl.htm).
675. USFDA. (1995): Lead-soldered food cans. *Fed. Reg.*, 60:33106–33109.
676. USFDA. (1997): Iron-containing supplements and drugs: label warning statements and unit-dose packaging requirements; final rule. *Fed. Reg.*, 62:2217–2250.
677. USFDA. (2005): *Questions and Answers on Prussian Blue*. Center for Drug Evaluation and Research, U.S. Food and Drug Administration, Washington, D.C. (htpp://www.fda.gov/cder/drug/infopage/prussian_blue/Q&A.htm).
678. van der Voet, G. B. and de Wolff, F. A. (1996): Human exposure to lithium, thallium, antimony, gold, and platinum. In: *Toxicology of Metals*, edited by L. Magos and T. Suzuki, pp. 455–460. CRC Press, New York.
679. Van Ordstrand, H. S., Hughes, R., and Carmody, M. G. (1943): Chemical pneumonia in workers extracting beryllium oxide. *Cleve. Clin. Q.*, 10:10–18.
680. Van Rij, A. M., Thomson, C. D., McKenzie, J. M., and Robinson, M. F. (1979): Selenium deficiency in total parenteral nutrition. *Am. J. Clin. Nutr.*, 32:2076–2085.
681. Veien, N. K., Hattel, T., Justesen., O., and Norholm, A. (1987): Oral challenge with nickel and cobalt in patients with positive patch tests to nickel and/or cobalt. *Acta Derm. Venereol.*, 67:321–325.
682. Verity, M. A. (1997): Manganese neurotoxicity: pathobiochemical aspects. abstract. In: *Proc. of the 15th Int. Neurotoxicology Conf.*, Little Rock, AR.
683. Vernace, M. A., Bellucci, A. G., and Wilkes, B. M. (1994): Chronic salicylate toxicity due to consumption of over-the-counter bismuth subsalicylate. *Am. J. Med.*, 97: 308–309.
684. Vilaplana, J., Grimalt, F., Romaguera, C., and Mascaro, J. M. (1987): Cobalt content of household cleaning products. *Contact Derm.*, 16:139–141.
685. Vinceti, M., Guidetti, D., Pinotti, M. et al. (1996): Amyotrophic lateral sclerosis after long-term exposure to drinking water with high selenium content. *Epidemiology*, 7:529–532.
686. Von Burg, R. (1992): Boron, boric acid, borates and boron oxide. *J. Appl. Toxicol.*, 12:149–152.
687. Votaw, A. L. and Zey, J. (1991): Vacuuming a mercury-contaminated dental office may be hazardous to your health. *Dent. Assist.*, 60:27–29.
688. Waalkes, M.P. (2000): Cadmium carcinogenesis in review. *J. Inorg. Biochem.* 79:241–244.
689. Waalkes, M.P. (2003): Cadmium carcinogenesis. *Mutat. Res.*, 533:107–120.
690. Waalkes, M. P. and Goering, P. L. (1990): Metallothionein and other cadmium-binding proteins: recent developments. *Chem. Res. Toxicol.*, 3:281–288.

691. Waalkes, M. P. and Rehm, S. (1992): Carcinogenicity of oral cadmium in the male Wistar (WF/NCr) rat: Effect of chronic dietary zinc deficiency. *Fundam. Appl. Toxicol.*, 19:512–520.
692. Waalkes, M. P., and Rehm, S. (1994): Cadmium and prostate cancer. *J. Toxicol. Environ. Health*, 43:251–169.
693. Ward, J. J., Thronbury, D. D., Lemons, J. E., and Dunham, W. K. (1990): Metal-induced sarcoma: a case report and literature review. *Clin. Orthopaed. Relat. Res.*, 252:299–306.
694. Ward, N. I. and Mason, J. A. (1987): Neutron activation analysis techniques for identifying elemental status in Alzheimer's disease. *J. Radioanal. Nucl. Chem.*, 113:515–526.
695. Warrell, R. P. (1997): Gallium nitrate for the treatment of bone metastases. *Cancer*, 80(Suppl.):1680–1685.
696. Webster, S. H. (1946): Volatile hydrides of toxicological importance. *J. Indust. Hyg. Toxicol.*, 28:167–182.
697. Weir, H. K., Thun, M. J., Hankey, B. F. et al. (2003): Annual report to the nation on the status of cancer, 1975–2000, featuring the uses of surveillance data for cancer prevention and control. *J. Natl. Cancer Inst.*, 95:1276–1299.
698. Weiss, B., Alkon, E., Weindlar, F. et al. (1993): Toddler deaths resulting from ingestion of iron supplements: Los Angeles. *MMWR*, 42:111–113.
699. West, D. W., Slattery, M. L., Robison, L. M. et al. (1991): Adult dietary intake and prostate cancer risk in Utah: A case-control study with special emphasis on aggressive tumors. *Cancer Causes Control*, 2:85–94.
700. Whitford, G. M. (1987): Fluorides in dental products: safety considerations. *J. Dent. Res.*, 66:1056–1060.
701. Whitford, G. M. (1992): Acute and chronic fluoride toxicity. *J. Dent. Res.*, 71:1249–1254.
702. Whiting, S. J. (1994): Safety of some calcium supplements questioned. *Nutr. Rev.*, 52:95–97.
703. Whiting, S. J. and Wood, R. J. (1997): Adverse effects of high-calcium diets in humans. *Nutr. Rev.*, 55:1–9.
704. Wones, R. G., Stadler, B. L., and Frohman, L. A. (1990): Lack of effect of drinking water barium on cardiovascular risk factors. *Environ. Health Perspect.*, 85:355–359.
705. Wong, L. C., Heimbach, M. D., Truscott, D. R., and Duncan, B. D. (1964): Boric acid poisoning: Report of 11 cases. *Can. Med. Assoc. J.*, 90:1018–1023.
706. Woody, N. C. and Kometani, J. T. (1948): BAL in the treatment of arsenic ingestion of children. *Pediatrics*, 1:372–378.
707. Wu, M.-M., Kuo, T.-L., Hwang, Y.-H., and Chen, C.-J. (1989): Dose–response relation between arsenic concentration in well water and mortality from cancers and vascular diseases. *Am. J. Epidemiol.*, 130:1123–1132.
708. Xu, N., Majidi, V., Markesbery, W. R., and Ehmann, W. D. (1992): Brain aluminum in Alzheimer's disease using an improved GFAAS method. *Neurotoxicology*, 13:735–744.
709. Yan, A. and Davis, P. (1990): Gold induced marrow suppression: a review of 10 cases. *J. Rheumatol.*, 17:47–51.
710. Yang, G., Yin, S., Zhou, R. et al. (1989): Studies of safe maximal daily dietary Se-intake in a seleniferous area in China. II. Relation between Se-intake and the manifestation of clinical signs and certain biochemical alterations in blood and urine. *J. Trace Elem. Electrolytes Health Dis.*, 3:123–130.
711. Yang, G.-Q., Wang, S. Z., Zhou, R. H., and Sun, S. Z. (1983): Endemic selenium intoxication of humans in China. *Am. J. Clin. Nutr.*, 37:872–881.
712. Yip, H. K., Li, D. K., and Yau, D. C. (2003): Dental amalgam and human health. *Int. Dent. J.*, 53(6):464–8.
713. Yoshimasu, F., Yasui, M., Yoshiro, Y., Iwata, S., Gajdusek, C., Gibbs, C. J., and Chen, K-M. (1980): Studies on amyotrophic lateral sclerosis by neutron activation analysis. 2. Comparative study of analytical results on Guam PD, Japanese ALS and Alzheimer disease cases. *Folia Psychiat. Neurolog. Jpn.*, 34:75–82.
714. Yoshizawa, K., Willett, W. C., Morris, S. J. et al. (1998): Study of prediagnostic selenium level in toenails and the risk of advanced prostate cancer. *J. Natl. Cancer. Inst.*, 90:1219–1224.
715. Yuan, X.-M. and Wei, L. (2003): The iron hypothesis of atherosclerosis and its clinical impact. *Ann. Med.*, 35:578–591.
716. Zacharski, L. R. and Gerhard, G.S. (2003): Atherosclerosis: a manifestation of chronic iron toxicity? *Vascular Med.*, 8:153–155.
717. Zheng, W., Winter, S. M., Kattnig, M. J., Carter, D. E., and Sipes, I. G. (1994): Tissue distribution and elimination of indium in male Fischer 344 rats following oral and intratracheal administration of indium phosphide. *J. Toxicol. Environ. Health*, 43:483–494.

# 18 Ionizing Radiation

*Lorris G. Cockerham, Thomas L. Walden, Jr., Cham E. Dallas, G. Andrew Mickley, Jr., and Michael A. Landauer*

**CONTENTS**

## INTRODUCTION

The increasing use of radiation in the modern world and recent incidents of massive radiation exposure dictate that certain basic elements of radiation toxicity be addressed. Radiation toxicology is the study of the adverse effects of radiation on living organisms. It is a multidisciplinary science, borrowing freely from several of the basic sciences. The cytopathologic consequences of radiation exposure are similar to those induced in other types of cellular injury. Radiation-induced cell changes may result in death of the organism, death of the cells, modulation of physiological activity, or cancers that have no features distinguishing them from those induced by other types of cell injury. Electromagnetic radiation is divided into nonionizing and ionizing radiation according to the energy required to eject electrons from molecules [688]. Ionizing radiation, which may exhibit the properties of both waves and particles, has sufficient energy to produce ionization in matter. The ionizing radiations that exhibit corpuscular properties include alpha and beta particles, while those that behave more like waves of energy include x-rays and gamma rays. Radiation exposure comes from many sources and may be *directly ionizing* or *indirectly ionizing*. Directly ionizing radiation carries an electric charge that directly interacts, by electrostatic attraction or repulsion, with atoms in the tissue or medium exposed. Indirectly ionizing radiation is not electrically charged but results in production of charged particles by which its energy is absorbed. A characteristic of charged particles produced directly or indirectly is *linear energy transfer* (LET), the energy loss per unit of distance traveled, usually expressed in kiloelectron volts (keV) per micrometer (mm). The LET, depending on the velocity and charge of the particle, may vary from about 0.2 to more than 1000 keV/mm.

## DOSIMETRY AND EXPOSURE

### Systéme Internationale

The International Commission on Radiological Units and Measurement (ICRU) introduced the Systéme Internationale or SI units in 1980 to express radiation dose [310]. The gray (Gy), the SI unit for absorbed dose, corresponds to an energy absorption of 1 joule/kg or 100 rads. This concept of energy absorption is useful for determining absorbed doses of x-rays and gamma rays; however, determination of the absorbed dose in tissues exposed to fast neutron radiation involves more elaborate calculations. The absorbed dose of neutron radiation depends on the

**TABLE 18.1**
**Radiation Quantities and Units Used in Radiobiology**

| Unit or Quantity | Symbol | Application |
|---|---|---|
| Becquerel | Bq | SI quantity of radioactivity<br>Bq = 1 disintegration/s<br>Bq = $2.7 \times 10^{-11}$ Ci |
| Curie | Ci | Quantity of radioactivity<br>1 Ci = $3.7 \times 10^{10}$ dps<br>1 Ci = $3.7 \times 10^{10}$ Bq |
| Gray | Gy | SI unit of absorbed dose<br>1 Gy = 100 rad = 1 J/kg |
| Rad | rad | Unit of absorbed dose<br>1 rad = 0.01 Gy = 100 erg/g |
| Rem | rem | Unit of dose equivalent<br>rad × $Q$ × other modifying factor<br>1 rem = 0.01 Sv |
| Sievert | Sv | SI unit of dose equivalent<br>rad × $Q$ × other modifying factor<br>1 Sv = 100 rem |
| Linear energy transfer | LET | Energy deposition per unit of path length, usually in eV/micron |
| Relative biological effectiveness | RBE | Same effect from same dose of reference radiation used in radiobiology |
| Quality factor | Q | Biological effectiveness of radiations |
| Working level | WL | $1.3 \times 10^{-5}$ MeV α-energy/L air |
| Working level month | WLM | 1 WL × 170 hr |
| Electron volt | eV | Unit of energy<br>1 eV = $1.6 \times 10^{-12}$ ergs<br>1 eV = $1.6 \times 10^{-19}$ J |

*Source:* Data from Baisakhatov, R. and Khanson, K.P., *Radiobiologiya*, 11, 155–159, 1971; Landauer, M.R. et al., *J. Radiat. Res. (Tokyo)*, 38, 45–54, 1997.

transfer of energy from neutrons to directly ionizing particles in the tissue and is described by the kinetic energy released in the material. For general use, a quantity different from the rad or gray has been introduced: the dose equivalent. The dose equivalent allows for the relative effectiveness of a particular type of radiation. Gamma rays and x-rays are regarded as the standard, and a quality factor of 1 is multiplied times the dose to compute the dose equivalent; therefore, the dose equivalent (Seiverts) for x-rays and gamma rays is equal to the dose (grays). Neutrons, however, are thought to be roughly 10 times more effective in producing tissue damage than x-rays and therefore are assigned a quality factor of 10.

## Exposure Factors

Before discussing the effects of ionizing radiation, some of the factors that influence the toxicity of radiation should be reviewed. One of the major factors related to the exposure is the dose or total amount of radiation received (Table 18.1). The absorbed dose of radiation is the quotient dE/dm, where dE is the differential energy deposited into a differential mass (dm) [310]. The unit of absorbed dose, in the centimeter–gram–second (CGS) system, is the rad (radiation-absorbed dose), and 1 rad = 100 erg/g (i.e., a dose of 1 rad of ionizing radiation has been absorbed when 100 ergs of energy have been deposited in each gram of material) [664]. Another term commonly used, particularly in the field of radiation protection, is the rem (roentgen equivalent man). This unit was developed to enable radiation protection personnel to set standards of exposure (rem = rad × quality factor × distribution factor). The quality factor is a unit to equate the relative biological effectiveness (RBE) of one radiation to another, and the distribution factor attempts to compensate for the varying sensitivity of the different parts of the body. The roentgen, the amount of radiation required to produce one electrostatic unit of charge per cubic centimeter of air, is an older radiation exposure term that may still be seen in literature. This is a measure of only the actual ionizations produced by x-ray or gamma-ray irradiation in air.

A second factor influencing the toxicity of radiation is the dose rate: $D = dD/dt$, the differential dose with respect to time or, when there is no variability in dose $dE/dm$, then

**TABLE 18.2**
**Average Annual Effective Dose Equivalent of Ionizing Radiations**

| | Dose Equivalent | | Effective Dose Equivalent | |
|---|---|---|---|---|
| **Source** | **mSv** | **mrem** | **mSv** | **%** |
| *Natural* | | | | |
| Radon | 24.0 | 2400 | 2.0 | 55.0 |
| Cosmic | 0.27 | 27 | 0.27 | 8.0 |
| Terrestrial | 0.28 | 28 | 0.28 | 8.0 |
| Internal | 0.39 | 39 | 0.39 | 11.0 |
| Total natural | — | — | 3.0 | 82.0 |
| *Artificial* | | | | |
| Medical | | | | |
| X-ray diagnosis | 0.39 | 39 | 0.39 | 11 |
| Nuclear medicine | 0.14 | 14 | 0.14 | 4.0 |
| Consumer products | 0.10 | 10 | 0.10 | 3.0 |
| Occupational | 0.009 | 0.9 | <0.01 | <0.3 |
| Nuclear fuel cycle | <0.01 | <1.0 | <0.01 | <0.03 |
| Fallout | <0.01 | <1.0 | <0.01 | <0.03 |
| Miscellaneous | <0.01 | <1.0 | <0.01 | <0.03 |
| Total artificial | — | — | 0.63 | 18 |
| Total natural and artificial | — | — | 3.6 | 100 |

*Source:* BEIR V, *Health Effects of Exposure to Low Levels of Ionizing Radiation*. Committee on the Biological Effects of Ionizing Radiations, National Research Council, National Academy Press, Washington, D.C., 1990. With permission.

$D = D/t$. When reviewing experiments in radiation toxicology, the variables of total dose, dose rate, type of radiation, and variability of the model must be considered.

The exposure to radon ($^{222}Rn$) and radon daughters may be expressed, by convention, as the concentration of radon daughters measured in working levels (WLs), and cumulative exposures over time are measured in working-level months (WLMs) [68]. The WL is defined as any combination of radon daughters in 1 liter of air that results in the ultimate release of $1.3 \times 10^5$ MeV of potential alpha energy. This is approximately the alpha energy emitted by the radon daughters in equilibrium with 100 pCi of radon. The WLM is defined as exposure to this concentration for a working month of 170 hours.

## SOURCES

The quantity of radiation present could range from irreducible natural background levels to large-scale releases such as occurred at Chernobyl and Chelyabinsk in Eastern Europe. Biological damage may be detected at levels only slightly above the former, while the latter would result in extreme biological toxicity of unknown proportions. The sources of radiation may be broken into two major components: technologically induced or man-made radiation and natural radiation (Table 18.2).

### Technologic or Man-Made Radiation

#### Health Sciences

The use of man-made radiation in the health sciences is normally divided into three areas: (1) diagnostic x-ray examinations, (2) nuclear medicine, and (3) therapeutic radiation. In the more highly developed countries, exposure from medical sources may equal or exceed natural background radiation, but in undeveloped countries the relative contribution of medical irradiation may be only about 5% of the total exposure [510].

The use of x-rays in diagnostic examinations, including dental, represents the single largest man-made source of radiation exposure in the U.S. population. The Bureau of Radiological Health of the Food and Drug Administration estimated that approximately 65% of the people in the United States were exposed to x-rays for medical and dental diagnostic examinations in 1970. The mean active bone marrow dose to adults was 103 mrads, with the 65 and older age group receiving the highest per capita dose [67]. Dental x-ray examinations are the most common of the diagnostic examinations, and approximately 30% of the total diagnostic examinations are received on an outpatient basis [510].

The use of radiopharmaceuticals in nuclear medicine has almost doubled over a ten-year period. It is estimated that up to 12 million doses of radiopharmaceuticals are given each year in the United States for diagnostic purposes [67]. However, the per capita effective dose equivalent from these procedures in the United States is only about 140 FSv [510]. Radiation therapy has been used almost exclusively for the treatment of malignant neoplasms. The high absorbed dose (50 to 70 Gy) required in most malignant conditions leads to *nonstochastic* or direct effects such as cell death; therefore, some of the normal tissue surrounding the neoplasm may be exposed and incur some long-range risk. The risk, however, is usually eclipsed by the immediate benefits normally associated with increased life expectancy.

## Nuclear Weapons

The first atomic weapon was detonated in 1945 on a New Mexico desert north of Alamagordo. Since that day, hundreds of test explosions have been conducted by the United States, the Soviet Union, the United Kingdom, India, France, and the People's Republic of China. Between 1945 and 1984, the total estimated yield of all atmospheric nuclear explosions was approximately 546 megatons [510]. A 1-megaton (MT) explosion equals the explosive force of 1 million tons of TNT.

The radioactive fallout from nuclear explosions may be divided into three portions depending on the yield and height of the burst. The larger, intensively radioactive particles fall out close to the site within hours. Slightly smaller particles behave somewhat like aerosols and are dispersed into the troposphere, where they will stay for a matter of months. The fallout from this portion remains in bands around the Earth at the latitude of the detonation. The third portion penetrates the stratosphere and its particles are deposited worldwide over a period of months to years [231]. Most of the radioactive fallout is downwind from the explosion, and up to 70% is in the larger particle portion, returning to the Earth close to the detonation site within hours. The intensity of the radioactivity varies inversely with distance from the site of explosion. With a steady wind, the pattern of accumulated dose of radioactivity assumes the shape of nested cigar-shaped contours, each contour denoting a particular dose.

A 1-MT thermonuclear weapon detonating at ground level with a steady wind of approximately 15 miles per hour would produce a fallout radioactivity dose rate of 400 rem in 24 hours in an area of approximately 400 square miles. At a dose rate of 2 rem per year, more than 20 times the maximum recommended by the U.S. Environmental Protection Agency (EPA), an area of 1200 square miles would remain unfit for use for a year, and more than 20,000 square miles would be uninhabitable for a month [238].

In a nuclear explosion, over 400 radioactive isotopes are released into the biosphere. Among these, about 40 radionuclides are considered potentially hazardous. Of particular interest are those isotopes whose organ specificity and long half-lives present a danger of irreversible damage or induction of malignant alterations. Both early and delayed fallout result in the deposition of radioactive material in the environment [131]. The annual average whole-body fallout rate in the United States, approximately 45 $\mu$Sv (4.5 mrem), was projected to stay at this level through the year 2000 [67,510].

## Dirty Bombs

A radiological dispersal device (RDD) is designed to spread radioactive material passively as an aerosol or actively as the result of an explosion [11,172]. The RDD is a radiological device that causes the purposeful dissemination of radioactive material across an area without a nuclear detonation. It is usually a simple device, very inexpensive by most standards and easy to construct. Probably the most familiar of the RDDs is the dirty bomb, typically a conventional explosive such as dynamite or trinitrotoluene (TNT) or a fertilizer-based truck bomb that has been packed or covered with radioactive material such as cesium-137 ($^{137}$Ce) or cobalt-60 ($^{60}$Co) [118,317]. Such a dirty bomb may be the terrorist weapon of choice and can be produced by anyone with access to industrial or medical radioisotopes [751]. A possible dirty bomb scenario has been presented: "A stick of cobalt, an inch thick and a foot long, is taken from among hundreds of such sticks at a food irradiation plant. It is blown up with just 10 pounds of explosives in a 'dirty bomb' at the lower tip of Manhattan, with a one-mile-per-hour breeze blowing. Some 1000 square kilometers in three states is contaminated, and some areas of New York City become uninhabitable for decades" [420]. The dirty bomb has been called Osama bin Ladin's "formula for fear," because plans he left behind in Afghanistan have been found [752]. In March 2002, information came out of Pakistan showing that *Al Qaeda* had plans to build a dirty bomb and smuggle it into the United States [751].

Why would anyone want to use a dirty bomb, as the bomb maker faces death from handling the radioactive material, and fatalities from the blast would be nothing like those caused by a crude nuclear weapon [390,449]? Killing people is not the main goal of a dirty bomb; rather, it is the contamination of a large area with radioactivity in a nested cigar-shaped pattern resembling the fallout following the detonation of a nuclear weapon or a nuclear accident. The extent of the contaminated area depends on the amount of radioactive and explosive material used in the dirty bomb, the altitude of the detonation above ground level, and the winds at the time of detonation. Even so, only a few people would be killed in the actual blast, and the number affected by the radioisotope fallout would likely be limited, with long-term effects appearing as radiation sickness, high rates of cancer, increased infant mortality and sickness associated with a compromised immune system [118,402,752].

Actually, the greatest concern immediately following the detonation of a dirty bomb is that more people would probably be injured in the evacuation than in the actual event itself [752]. The terror induced in people not realizing the level of danger they are exposed to would lead to a general panic, disrupting communications and all forms of transportation. Highways, commuter trains, and air terminals would be jammed. Even hospitals in the area would be overflowing with people who may only think that they have been exposed to the radiation. Rather than considering a dirty bomb to be a weapon of mass destruction (WMD), it probably should be called a "weapon of mass disruption" [118,402].

## Nuclear Power Production

When radiation exposure from nuclear power production is mentioned, most persons immediately think of nuclear power reactors and the environmental dispersion of radionuclides, particularly krypton-85, tritium, carbon-14, and iodine-129; however, exposure from nuclear power production should also include mining, uranium fuel fabrication, and waste storage and disposal [510]. Although uranium mines increase the amount of uranium and its decay products, along with radon and its daughters, the environmental risks from the radioactive emissions from uranium mines is insignificant [231,336]. Mill tailings, however, may represent a significant source of environmental radiation due to the emanation of $^{222}$Rn, dispersion of the tailings by wind and water, and by the use of mill tailings in building construction. About 1000 land-based nuclear reactors have been constructed and operated at some time throughout the world. Some of the reactors were built for research or the production of radioisotopes and plutonium. Approximately 200 naval vessels throughout the world are powered by nuclear reactors; yet, the environmental release from nuclear operations in the United States results in a dose rate for the average person of less than 1 mrem/yr [67].

## Accidents

Although the environmental release of radionuclides from nuclear reactor operations is approximately 1 mrem/yr per person, malfunctions can develop and accidents can happen [67]. The contents of the reactor at the time of an accident and the amount of contaminant, including its physical and chemical properties, depend on the reactor type, its application, and the duration of operation [714]. Not all of the nearly 800 nuclides produced in reactors are radioactive, and of these only 54 are considered significant in risk assessment [231]. With core damage, the severity of the accident and therefore the risk depend on the radioactivity (mainly as $^{131}$I and $^{137}$Cs) being released to the environment.

Since 1952, 14 reactor accidents have involved core damage. One, the Windscale, U.K., Atomic Energy works accident, was the first time radioactive material was released from a reactor accident. In October 1957, a plutonium production reactor located on the coast of Cumbria in northwest England released approximately 740 TBq $^{131}$I, 22 TBq $^{137}$Cs, 8.8 TBq $^{210}$Po, and 3 TBq $^{89}$Sr [152,231]. The core in the no. 1 pile of the two air-cooled, graphite-moderated, natural uranium reactors was partially consumed by fire, releasing the fission products onto the seashore and foothills southwest of the Cumbrian Mountains, over much of England and parts of northern Europe. As an aftermath of this accident, the village of Seascale had four fatal leukemia cases in children that were under 20 years of age between 1950 and 1980. Based on statistics, only 0.5 cases would have been expected [743].

On March 28, 1979, the worst accident in the history of U.S. commercial nuclear power generation occurred on Three Mile Island (TMI) in Pennsylvania [336]. Even though the accident at TMI-2 released the radionuclides $^{131}$I, $^{133}$Xe, and $^{135}$Xe into the environment, the collective dose equivalent to the population from the release was less than 1% of the dose accrued from natural background radiation in a year [714]. As with the Windscale accident, the radionuclide identified as being of principal concern was $^{131}$I, but in this case only 1 TBq of $^{131}$I was released, and the fission products at TMI were retained within the vessel [151,152]. Although radiation exposure to the plant workers and the public was insignificant, the nuclear power industry was set back almost a decade. Even orders for the construction of new nuclear plants were canceled [231].

The largest airborne dispersion of radionuclides thus far occurred from the explosion and ensuing 10-day fire at the graphite-moderated reactor of unit no. 4 of the Chernobyl nuclear power station in the former Soviet Union (now Ukraine) on April 26, 1986. It has been estimated that this has been the single most costly industrial accident in history [152,231,360,714]. One revealing evaluation of the economic loss from this accident has been the estimate that from 8 to 10 annual budgets for the Republic of Belarus will be consumed to appropriately address the effects of this disaster, just for the needs generated in that country alone [117]. Another more conservative estimate puts the costs of modestly dealing with the Chernobyl nuclear accident at 45 billion dollars, not including, of course, the human and ecological toll [711]. The health care needs, for example, are only now becoming apparent [186].

Fallout from the Chernobyl nuclear disaster was very widespread, with a large amount of a variety of radionuclides distributed throughout the northern hemisphere [30,253,410,790]. In Eastern Europe and Scandinavia, it has been estimated that approximately $4 \times 10^{18}$ Becquerels (Bq) were dispersed by the accident [365,637,729]. The radioactive cloud moved outside of the area of the Soviet

Union in the first few days after the accident, and the event only became publicized when radioactive levels in Sweden became elevated, reaching 14 times background levels [151]. The importance of meteorological conditions at the time and place of nuclear accidents or thermonuclear detonations was clearly evidenced in the dispersion patterns of the fallout. The majority of the ensuing environmental radioactivity now resides in the newly created nation of Belarus (immediately north of the Chernobyl nuclear complex), which now has the distinction of approximately 30% of its territory incurring significant contamination from the accident, including 20% of the forests and 18% of the farmland [365]. Ukraine and Russia have most of the remainder of the inventory that was released. The most prominent radionuclides released, in terms of quantity and widespread geographic dispersion, were $^{137}$Cs, $^{90}$Sr, and $^{131}$I. Although less widespread, significant levels of $^{239}$Pu and various transuranics were also distributed in the areas around the reactor complex. In and around the 30-km Chernobyl exclusion zone, about 2000 km$^2$ were significantly contaminated with hot particles representing fuel particles of nonvolatile radionuclides such as strontium, plutonium, and americium [386].

The International Atomic Energy Agency (IAEA) report published by the Soviet Union in 1987 determined a radiation dose for the two most highly contaminated areas as: (1) a 106,340-man-Sv collective 50-year dose for a population of 10.1 million in Belarus, and (2) a 80,660-man-Sv collective 50-year dose for a population of 29.8 million in Ukraine [795]. Although many east European scientists in post-Soviet scientific circles challenge these estimates as being too low, they do suggest a significant radiation dose to a large number of people in the contaminated areas of these two countries. Indeed, newer estimates are that the 1 million people in the most contaminated areas of Belarus and Ukraine will accumulate between 150,000 and 200,000 Sv (as much as the total dose for both countries in the 1987 report), with the total population dose approaching 1,000,000 Sv [502]. Many villages surrounding the Chernobyl reactor area had to be abandoned, along with the entire city of Pripyat, which formerly housed over 50,000 people. This makes Pripyat the first sizable city in history to be abandoned solely on the basis of radioactive fallout. It has been estimated that some of the inhabitants of the abandoned villages have accumulated at least 35 rem per person [502]. In Belarus alone, approximately 2.2 million people live in the areas significantly contaminated by the accident, including 800,000 children [154]. At least 135,000 people were evacuated just from the Chernobyl exclusion zone established in the areas immediately around the reactor complex [738,837].

One result of the Chernobyl accident has been that the people and the ecosystem of the contaminated areas of Belarus, Ukraine, and Russia have become a living laboratory of the consequences of widespread radioactive contamination. In the ecosystem, deposition of the radionuclides from the reactor fire and dispersion tended to be very patchy, with variation of over 100% in soil and sediment samples taken only meters apart [368]. Uptake of radionuclides into wildlife around the reactor was very high, with radiocesium concentrations averaging 18,000 Bq/g in one rodent species [419] and up to 200 Bq/g in fish [368]. Food contamination from environmental radioactivity resulted in leafy vegetables reaching levels of 10 μCi/kg, and iodine levels in milk were commonly measured at 1 μCi/L [231]. As humans began moving back into the Chernobyl exclusion zone over time, evaluation of the irradiated dose of these illegal settlers has concluded that as of 2003 the radionuclide intake from potable water was insignificant, beta irradiation was insignificant, and the primary health risk was from external gamma irradiation and consumption of radioactive foods [55].

The ecotoxicological effects from the Chernobyl accident have also provided an unprecedented observation of the widespread dispersion of radionuclides in the environment. Evaluations of the blood cell DNA of fish from the radioactively contaminated aquatic habitats near the reactor revealed abnormal DNA distributions, hyperdiploidy, and cell-cycle perturbations, although there were no gross physical malformations [187,188,452,757]. An even more extensive evaluation of rodents in Chernobyl-contaminated areas has been published in the last decade. A reduction in fertility and various other physiological disorders were reported in some mammalian species in the years immediately following the accident [262,423]. Cytogenetic and other mutagenic effects were observed in rodents from sites ranging from the Chernobyl power plant vicinity to Sweden [174,175,290,722]. At least some of the variation in response observed can be attributed to a species difference in sensitivity, such as the high radioresistance reported for *Clethrionomys glareolus* from Chernobyl-contaminated areas [358,419]. Species differences in oxidative stress enzyme response were also found in rodents from these areas, with *C. glareolus* showing radioresistance relative to another species [344], despite a much higher deposition of radionuclides measured internally [143]. Changes in reproductive parameters such as litter size, pregnancy rate, and newborn mortality rate were reported for rodents kept over multiple generations in a Chernobyl vivarium [713]. Sufficient radioactivity was released in the accident to kill about 400 hectares of pine forest, and more than 1 million square meters of ground was bulldozed and buried [503].

The human health effects of the Chernobyl nuclear accident are only now becoming evident, with additional reports appearing each year. A relatively high incidence of thyroid cancer has been documented in Belarus, Ukraine, and Russia [86,272,615]. The incidence of leukemia is still under investigation, but a statistically relevant increase relative to the accident has not yet been substantiated [62,289,385,723,841,848]. An increased frequency of chromosomal aberrations has been detected in the blood cells of people living in the contaminated areas or those who worked as liquidators in the cleanup after the accident [696,745]. An increased mutation rate at human minisatellites was found in children born in a contaminated region of Belarus relative to a control population [222,223]. These minisatellites provide perhaps the only currently available system for the efficient monitoring of germline mutation in humans, and it was concluded that the damage was probably not due to DNA damage induced directly at the minisatellites but from radiation-induced damage at other sites in the genome [223].

Somatic minisatellite mutation events were present in a subset of radiation-induced but not sporadic thyroid cancers, suggesting that this type of genomic instability may play a role in radiation-induced tumorigenesis in the thyroid gland [593]. There was an increased frequency in both lymphocyte micronuclei (an assay of chromosomal integrity) and somatic mutations in erythrocytes at the glycophorin A locus of residents of Chernobyl-contaminated cities in Belarus, and these effects were significantly correlated with radiocesium content [457,642]. Somatic mutation responses were also found in the glycophorin A locus of erythrocytes from Chernobyl workers from the Baltic countries [80,81] and in immigrants to Israel [845]. There has been some indication that children exposed to low LET radiation due to Chernobyl may have an increase in cataract formation [199].

After long-term observation of 180,000 Chernobyl cleanup workers, a group of 259 people was identified as being at high risk for immunohematological effects [62]. After 13 years of follow-up with acute radiation sickness convalescents after Chernobyl, mature circumferent T-cell deficiency returned to normal levels over time, proving a complete restitution of T-cell compartment in the fixing of an immune-deficient state [149]. An investigation of four generations of mice permanently exposed to Chernobyl fallout radionuclides found a 30 to 70% change in immune function between matched control and exposed animals [687,695]. Among children in Belarus, increases in endocrine and dermatologic diseases, digestive organ diseases, autoimmune thyroiditis, and chronic tonsillitis and adenoiditis were reported after the accident [460].

Another serious radiation accident occurred in the central Brazilian plateau state of Goias but went virtually unnoticed by most of the world [176]. The Instituto Goiano de Radioterapia, a private radiotherapy clinic in Goiânia, Brazil, ceased operation in 1985, leaving a $^{137}$Cs radiotherapy unit in an insecure situation in an abandoned treatment room. In September 1987, the 50.9-TBq $^{137}$Cs source was removed from the protective housing of the therapy unit. With the later rupture of the container, the $^{137}$Cs became widely dispersed throughout the city's population of 1 million. Exposure to the cesium chloride resulted in 4 deaths, 28 other cases of acute radiation sickness, and 3500 m$^3$ of radioactive waste. The four persons who died received estimated doses ranging from 4.5 to 6.0 Gy. Other than massive accidents with widespread contamination such as occurred at Chernobyl, this is one of the most serious radiation accidents that has ever occurred.

## Nuclear Waste Management

The disposal of radioactive waste is a dilemma faced by a technologically advanced society. One of the basic demands of such a growing society is the availability of convenient and inexpensive sources of energy. As the demand for energy increases, the reliance on nuclear power will increase, as will the production of radioactive waste. Radioactive waste is classified by its physical and chemical properties as well as its source [231,276]. Three general categories of radioactive waste are (1) low level, (2) transuranic, and (3) high level. Low-level radioactive waste (LLRW) includes residues from laboratory research, medical institutions, uranium mill tailings, and waste generated in the cleanup of uranium, radium, and thorium processing plants. LLRW is further subdivided into classes A, B, and C, depending on the concentration, energy levels, half-lives, and sources of the radionuclides in the waste. Radionuclides found in LLRW include $^{241}$Am, $^{14}$C, $^{242}$Cm, $^{60}$Co, $^{137}$Cs, $^{129}$I, $^{241}$Pu, $^{226}$Ra $^{90}$Sr, $^{99}$Tc, $^{230}$Th, and $^{235}$U. Because the national inventory of LLRW waste is growing at a rate of $10^5$ m$^3$/yr (30% from medical institutions), management of the waste is facing a crisis in storage and disposal.

Transuranic (TRU) wastes are materials containing radionuclides with atomic numbers greater than uranium, such as americium, curium, and plutonium. These wastes originate mainly as by-products in the production and fabrication of plutonium for military purposes. Resulting from an industrial process involving transuranic materials, the TRU wastes are predominantly contaminated with $^{238}$Pu and $^{239}$Pu. These wastes tend to be water soluble and pose a distinct health hazard because they can contaminate a variety of physical forms, ranging from absorbent papers and rubber to discarded tools.

The most radioactivity and the highest concentration of radionuclides associated with nuclear wastes are found in spent fuel from civilian nuclear power reactors and in the reprocessing of civilian and military spent fuel. Typical radionuclides found in this high-level radioactive waste (HLW) are $^{60}$Co, $^{137}$Cs, $^{239-242}$Pu, $^{106}$Ru, and $^{90}$Sr. Because of the high hazard duration (>$10^5$ years) associated with these nuclides, large quantities (80 million gallons in 1982) of highly toxic liquid and solid HLW must be isolated from the environment for thousands of years.

Currently, much of the HLW is stored at temporary sites in concrete-encased steel tanks a few meters below the surface of the ground. Considering the finite lifetime of the steel tanks (15 to 40 years), it is clear that these wastes must be transferred to other containers or sites in the future. Several options for the permanent repository of radioactive waste have been considered [231,291], and the method currently in favor is in deep underground mined cavities.

The worst example of nuclear waste management to date is the experience of the Soviet nuclear weapons production complex, MAYAK, which started in 1948 on the Techa River in the Ural mountains of the Soviet Union (about 60 miles from the city of Chelyabinsk). The city constructed nearby to support the facility was originally named Chelyabinsk-65 (it is now called Ozersk), and this nuclear contamination area is usually now referred to simply as Chelyabinsk. Over $2 \times 10^7$ Ci were released over time into the surrounding area, making this the largest release of radionuclides at a single site in history [501,781]. Most of this inventory was released in 1957 in an explosion in a fuel reprocessing plant, referred to afterwards as the Kyshtym accident. The workers in the MAYAK facility were reported to have had high levels of radioactive exposure, exceeding 1 Gy annually for 25% of the radiochemical workers in the first 5 years of operation, with 11% of the workers overall receiving 6.3 Gy over the first decade [289]. This resulted in some deaths from chronic radiation lung injury [414] as well as elevated lung cancer deaths and leukemia in the workers [415]. In the people living in the villages along the Techa river and in the areas contaminated by the Kyshtym accident, there were also reports of high internal radionuclide doses [289] and an increased incidence of leukemia [416].

## Natural Environmental Radiation

Natural background radiation is the greatest contributor to radiation exposure in the world. In most countries, natural background radiation contributes slightly more than half of the absorbed radiation dose [510]. Relative contributions to the total absorbed dose may range from 42% in highly developed countries to 94% in most developing countries. Exposure to natural sources of irradiation is unavoidable, and life has evolved under a continuous exposure of ionizing radiation. This background radiation has three components: (1) cosmic radiation (external), (2) terrestrial radiation (external), and (3) naturally occurring radionuclides (internal).

### Cosmic and Solar Radiation

Cosmic and solar radiation originate predominately from galactic sources and consist mostly of high-energy protons and alpha particles [67,510,664]. Cosmic radiation at the Earth's surface varies with altitude, geomagnetic latitude, and solar modulation [253,336]; for example, in the United States, 48% of the population lives at sea level to 152.5 m and receives a dose rate of approximately 27 mrem/yr (0.27 μSv/yr), while in Leadville, Colorado (altitude 3200 m), the residents receive about 125 mrem/yr. This effect of altitude becomes increasingly important to passengers and crews of high-flying aircraft. It is estimated that cabin attendants and crew members receive approximately 160 mrem/yr above that received at sea level [231]. The cosmic rays are reduced by the Earth's atmosphere, resulting in a shielding effect. This shielding effect decreases with altitude, with cosmic ray exposure doubling every 1500 meters above the Earth's surface [510].

Above the Earth's atmosphere, the radiation consists of two main components. One is the dose from highly energetic cosmic radiation geomagnetically trapped in the Earth's magnetic field. The second component is received beyond the Earth's magnetic field and is due to background cosmic radiation of about 85% protons and 14% alpha particles. Astronauts traveling into outer space must traverse two belts of geomagnetically trapped radiation, the primary cosmic radiation, radiation from solar flares, and directed beams of gamma rays emitted by certain quasars and pointed directly at Earth [244]. Within the United States, the effect of latitude on cosmic radiation dose rate is less than 10%, with an average dose rate at sea level of about 270 μSv/yr [510]; however, in the United Kingdom the annual dose rate varies from about 280 μSv a year in the south of England to 310 μSv a year in the north of Scotland [253]. The dose rate variation with latitude depends primarily upon the variations in the earths magnetic field, with which cosmic radiation interacts [67].

### Terrestrial Radiation

Terrestrial radiation levels and rates from natural background sources are functions of geographic location and living habits. In most areas on Earth, the terrestrial radiation level varies within relatively narrow limits, but in certain regions of Brazil, China, France, Italy, Madagascar, and Nigeria the terrestrial radiation substantially exceeds the normal range [231,468,510]. For example, a person sunbathing on some beaches along the Atlantic coast of Brazil may receive as much as 17.5 cGy/yr from the sand alone [510]. Meanwhile, the exposure from the fine monazite particles of the soil in the Dong-anling and Tongyou regions of China would run between 18 cGy and 20 cGy per year [468].

The conterminous United States may be divided into three general radiation regions [67]. The Atlantic and Gulf coastal plains receive an average of 23 mrem/yr, while levels in the Colorado plateau area may be as high as 140 mrem/yr. The average terrestrial level for the remainder of the United States is only 46 mrem/yr, with an estimated national average of 40 mrem/yr.

The terrestrial radiation rate varies with the type of soil in the area and the naturally occurring radionuclide content of the soil. Approximately 70 of the 340 nuclides found in nature are radioactive [231]. These radionuclides have existed on the Earth's crust since its formation and are known as primordial radionuclides. These primordial radionuclides have half-lives comparable to the age of the universe and are the source of terrestrial radiation [510].

Three distinct chains of primordial radioactive elements are found in the Earth's crust and account for much of the terrestrial radiation exposure [510]: (1) the uranium series, (2) the thorium series, and (3) the actinium series. Uranium, the origin of the actinium series, is found in various quantities in rocks and soils. The uranium isotopes are alpha emitters and therefore do not contribute to the gamma background radiation. The presence of uranium in soils and in fertilizers leads to its presence, via the food chain, in plant and animal tissues. At equilibrium, an adult human male may be expected to have a uranium content of 100 to 125 μg. The thorium ($^{232}$Th) decay series may also move through the food chain, but, due to its relative insolubility and low specific gravity, it is present in biological materials only in insignificant amounts [231]. Thorium may be found in silty clay and peaty soils and in such vegetables as potatoes, corn, carrots, beans, and squash; however, the principal source of human exposure is inhalation of soil particles. Thorium is removed very slowly from bone, and its concentration increases with age.

Radium-226 ($^{226}$Ra), an alpha emitter originating in the uranium decay series, is present in varying amounts in all rocks, soils, and water and is of special importance, along with its daughter products [510]. $^{226}$Ra, with a half-life of 1622 years, decays to radon ($^{222}$Rn), a noble gas radionuclide with a half-life of 3.8 days. Radon, to be discussed later, also emits alpha particles but adds to the gamma radiation level of the environment through its gamma-emitting descendants.

Radium is very similar to calcium and is absorbed by plants from the soil like calcium. It then passes through the food chain to humans, where 70 to 90% is concentrated in bone. The amount of $^{226}$Ra moving through the food chain depends on its content in the soils and its rate of absorption by plants. This rate of absorption by plants is related to the amount of exchangeable calcium in the soil. Brazil nuts, because of their tendency to concentrate barium, another chemical very similar to radium, may have a $^{226}$Ra content approximately 1000 times greater than the average diet.

## Radionuclides

Internal radiation results from naturally occurring radionuclides contained within the body and contributes approximately 11 to 17% of the average radiation exposure of the population [69,739]. While some of the radioactive emitters may be freely dispersed throughout the body, others are concentrated in specific organs, and all of the emitted decay energy is absorbed locally [664,856]. The deposition of naturally occurring radionuclides such as bismuth, carbon, hydrogen, lead, polonium, potassium, radium, radon, thorium, and uranium results primarily from the inhalation and ingestion of these materials in air, food, and water [67].

In a terrestrial ecosystem, radionuclides such as $^{210}$Ra, $^{226}$Ra, and $^{222}$Rn that occur in the soil or are deposited in the soil are incorporated metabolically into plants [231]. In addition to root absorption, plants are contaminated by direct foliar deposition. Foliar deposition is potentially a major source of food chain radionuclide contamination, as the radionuclide may be absorbed metabolically by the plant or transferred directly to animals consuming or coming in direct contact with the foliage. Individual radionuclides pass from the roots or the leaves to the remainder of the plant. Mean $^{232}$Th concentrations of 0.018 ± 0.022 pCi/kg have been found in the edible portions of 25 vegetables, including beans, carrots, corn, potatoes, and squash. Flora near the summit of the Morro do Ferro, a hill in the state of Minas Gerais, Brazil, have absorbed so much $^{228}$Ra that they can easily be autoradiographed [231].

Atmospheric radionuclides are eventually deposited on surface waters as well as on the soil; therefore, the atmosphere is coupled to soils, surface waters, and subsurface aquifers. Radionuclides are eventually transported into streams or subsurface aquifers. Those that appear in deep underground aquifers may eventually reach surface waters and become incorporated into the biosphere.

Rivers, estuaries, and coastal waters are major receptors of effluent radionuclides from industrial plants and cities. These waters are of special importance because of their high biological activity and productivity. Phytoplankton in these relatively shallow waters convert mineral resources in the aquatic environment into food for higher organisms. Zooplankton, the basic food of several higher trophic levels, uses phytoplankton as its source of nourishment. Certain bottom-dwelling fish and animals also use phytoplankton as a source of nourishment.

Once the radionuclides have settled in an aquatic system over time, they tend to accumulate in the bottom sediments. In a collection pond built to contain a significant radioactive spill, 85% of the primary radionuclide present (radiocesium) was irreversibly bound up in the sediments, with some of the rest of the radionuclides available for remobilization from the sediment and subsequent exchange with the water column [838,839]. This accumulation in the sediments is an important factor in determining uptake and deposition in various aquatic species, with bottom-dwellers tending to accumulate more radionuclides than some other organisms in the water column.

The importance of radionuclides in marine and fresh water foods depends, in part, on where the radionuclide is located in the organism. A radionuclide is a higher risk if it concentrates in parts of an organism consumed by higher organisms, such as humans, than if it is deposited in a portion that is not eaten. The radionuclides of cobalt ($^{60}Co$) and zinc ($^{65}Zn$) concentrate in edible tissues, while those of radium ($^{226}Ra$) and strontium ($^{90}Sr$), although concentrated by clams, oysters, scallops, and certain crabs, are stored in the shell, which is not ordinarily consumed [231].

Uptake and retention of radionuclides are influenced by the portal of entry, chemistry and solubility, metabolism, and particle size. Internal contamination normally occurs via three principal routes of entry: inhalation, ingestion, and skin absorption. Of the three, inhalation is the biggest problem. Direct ingestion from contaminated food is also a problem. Gastrointestinal (GI) exposure depends on transit time through the gut, and absorption depends on the solubility of the radionuclide. Contamination of skin with radionuclides is of less consequence, as the skin forms a formidable barrier; however, contamination of an open wound may result not only in continuous radiation of the surrounding tissue but also in the introduction of the radionuclide into the rest of the body.

Regardless of the portal of entry, the radionuclide passes throughout the body of the animal and into the milk, flesh, internal organs, and eggs. When the radioactive material enters the body, it becomes an internal emitter. It will continue to radiate the body until it is excreted by some physiologic process, mainly through urine and feces, or until its radioactivity decays [131]. The time it takes an organism to eliminate half of the radionuclide is known as the biologic half-life, and the time necessary for a radionuclide to decay to half of its activity is the physical half-life (Table 18.3). If the biologic and physical half-lives are known for a particular radionuclide, the effective half-life may be calculated [510].

### Radon

Radon ($^{222}Rn$), the short-lived radionuclide decay product of $^{226}Ra$, accounts for approximately 60% of the effective dose equivalent from internal emitters [510]. As seen in Table 18.2, radon and its decay products or progeny ($^{222}Rn$, $^{214}Bi$, $^{214}Pb$, $^{214}Po$, $^{218}Po$) contribute 55% of the total average annual effective dose equivalent of 3.6 mSv. Since 1974, international concern has been centered on radon and radon progeny as indoor air pollutants that concentrate in nearly airtight homes and office buildings resulting from efforts directed toward energy conservation. These energy-efficient homes cause exposure to all segments of the population in which much lower air concentrations of radon progeny may be inhaled during life-span exposures (that is, at much lower dose rates than those that have occurred in uranium mining populations).

**TABLE 18.3**
**Half-Lives of Some Biologically Significant Radionuclides**

| | Half-Life | | |
|---|---|---|---|
| Nuclide | Physical | Biological (Whole Body) | Effective |
| $^{241}Am$ | 458 yr | 100 yr | 100 yr |
| $^{14}C$ | 5730 yr | 40 d | 40 d |
| $^{137}Cs$ | 30 yr | 70 d | 70 d |
| $^{131}I$ | 8 d | 138 d | 8 d |
| $^{55}Fe$ | 657 d | 2000 d | 494 d |
| $^{32}P$ | 14 d | 260 d | 14 d |
| $^{239}Pu$ | 24,000 yr | 180 yr | 180 yr |
| $^{24}Na$ | 15 hr | 11 d | 14 hr |
| $^{90}Sr$ | 28 yr | 36 yr | 16 yr |
| $^{3}H$ | 12 yr | 12 d | 12 d |
| $^{235}U$ | $7.1 \times 10^8$ yr | 20 d | 15 d |
| $^{65}Zn$ | 245 d | 400 d | 152 d |

*Source:* Data from Brizzee and Ordy [102], d'Avella, D. et al. [196], Landauer et al. [437], and Munro [575].

In addition, the low dust concentrations in such buildings with very low air changeover rates result in greatly increased fractions of radon progeny that are unattached to air carrier aerosols, causing proportionately higher radiological doses in the basal-cell epithelium of the conducting airways of the lungs. Measurements of air concentrations in Colorado Plateau uranium miners showed less than 2% unattached radon progeny, but, in regions of quiet air such as in private homes, levels became an order of magnitude higher; 81% of the attached RaA may become unattached upon decay [319]. During the years 1940 to 1988, many models were developed to calculate the radiation dose to the lungs as a whole or to selected regions of the respiratory tract. Dose conversion factors of rad/WLM show considerable increases above unity as the unattached fraction of radon progeny increases.

To define the role of attachment of radon progeny to aerosols and the incidence and site of radon progeny-induced respiratory carcinoma, a series of experimental studies used specific pathogen-free Wistar rats [756]. Groups of 32 rats received inhalation exposure 84 hours per week to 900-WL radon progeny attached to 15 mg/m$^3$ carnotite uranium ore dust. Exposures lasted for 150 days and the animals were then held for life-span carcinogenesis studies. These animals showed 60% incidence of squamous carcinoma or adenocarcinoma in the periphery of the respiratory tract following these prolonged inhalation exposures to radon progeny that were 98 to 99% attached to uranium ore dust aerosol, but the animals showed no tumors of the nasal pharynx. In marked contrast, matched groups of rodents that received exposure to radon progeny with only room air aerosol (from 10 to 25%

unattached, as is likely to occur in private homes) displayed less than 6% squamous carcinoma in the peripheral lung but 100% nasal squamous metaplasia and several cases of squamous carcinoma in the nasal pharynx, plus 22% squamous metaplasia in the major conducting airways such as the bronchi and first generations of secondary bronchi [756]. Additional studies in the same laboratory involved rats exposed to several concentrations of radon progeny attached to uranium ore dust and were designed to determine the effect of exposure rate and unattached fraction of radon progeny on the nature and incidence of pulmonary carcinoma [179]. Groups of 32 or 48 male, specific-pathogen-free Wistar rats received inhalation exposures to radon progeny in groups having unattached percentages of 1.6 and 10, the latter representing exposure conditions that might occur in minimally ventilated dwellings. Pulmonary neoplasms included epidermoid carcinoma, adenocarcinoma, adenosquamous carcinoma, mesothelioma, and adenoma. Exposure regimes included background level and 250, 500, and 1000 WL, with total exposures of 640 or 2560 WLM. Percentages of lung tumors resulting from these exposures ranged from none in the case of animals receiving background exposures to laboratory air to 47% in those animals at the lower dose rate—that is, 500 WL, with the total received exposure of 2560 WLM. Pathological evaluation resulting from the dose rate exposure study indicated an increase in the risk of pulmonary lung tumors as the exposure rate *decreased*.

Epidemiological studies of the incidence of bronchiogenic carcinoma among uranium miners in the Colorado Plateau, as well as in several European uranium mining studies, show considerable consistency in the relationship of the risk of bronchial carcinoma per WLM [319]. Studies using human volunteers who inhaled significant concentrations of radon progeny attached to uranium mine aerosols in the Colorado Plateau underground mines showed 90 to 100% attached radon progeny deposited in the regions of bronchi and subsegmental bronchi [178]. Because of the efficient deposition of unattached RaA in the brachia and major bronchi, the dose to bronchial epithelium from unattached RaA can be greater than three times that of unattached progeny per unit concentration in the atmosphere. Values for the dose conversion factor of rad/WLM in this region of the respiratory tract are 0.5 for underground miners. Analyses of exposures of individuals in the general population indicate values of 0.7 rad/WLM for men, 0.6 for women, 1.2 for children, and 0.6 for infants [319]. The unit WLM is defined only in terms of potential alpha-energy from radon progeny per liter of air. There can be very significant changes in the magnitude of this dose with factors of up to two-fold higher due to differing characteristics of inhaled atmospheres. These significant differences are not accounted for when using WLM alone as a unit or exposure.

In 1984, the focus of attention of carcinogenesis resulting from radon progeny inhalation by humans shifted from the miners of the Colorado Plateaus to the discovery of unusually high radon levels in a home built upon a geological formation called the Reading Prong in Pennsylvania. It soon became evident that each state had different problems associated with radon exposure to the general population. Although Pennsylvania had about 22,000 homes on its section of the Reading Prong, more than 250,000 homes were located on the New Jersey Reading Prong [591].

To explain the current concern regarding radon and radon progeny levels in the home, a further definition of the working levels is necessary. Because it is far simpler to measure radon, the parent, rather than the individual radon progeny, even though the latter contribute 95 to 98% of the dose to the respiratory epithelium, the working level can be defined according to its original estimate—that is, 100 pCi of radon per liter of air, which at 100% equilibrium with its progeny will give, by definition, 1.0 WL. In 1984, the National Council on Radiation Protection and Measurement Report No. 74 recommended an action level of 8 pCi per liter environmental exposures (i.e., exposures to the general population). Because only 50% of equilibrium is generally assumed for radon progeny $^{218}Po$, $^{214}Pb$, $^{214}Bi$, and $^{214}Po$ (i.e., one half the concentration of the parent $^{222}Rn$), this corresponds to an equivalent of 0.04 WL.

In 1986, to provide a more conservative position, the EPA issued a citizen's guide to radon recommending 4 pCi/L as an action level [794]. The recommended caution level by the EPA is based on linear extrapolations of high-dose exposures over 5 to 30 years of a small mining population on the Colorado Plateau [794]; the equating of lung cancer risk from a specified level of radon progeny to a given level of cigarette smoking is particularly difficult due to the close synergism between the incidence of bronchogenic carcinoma in the uranium miner population and high levels of cigarette smoking by these men. The wide limits of uncertainty found in the lung cancer risk estimate associated with a given total exposure level and the importance of dose rate factors demonstrated by recent animal and epidemiological studies have resulted in risk estimate calculations by agencies within the United States as well as in Canada and Europe that suggest possible lower risk per WLM or pCi/L exposure level. The action level for homes recommended by the EPA of 4 pCi/L is low when compared with that recommended by Canada and Finland [139]. It is likely that countries of the European Common Market will follow the example of these two countries with action levels at 8 or 20 pCi/L.

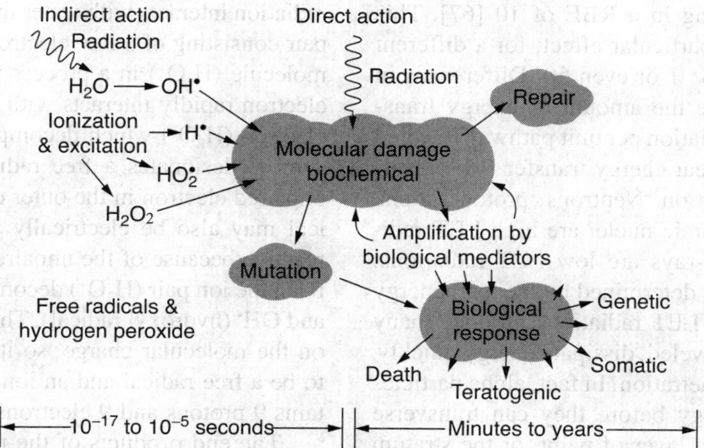


**FIGURE 18.1** Schematic for radiation injury to biological systems. The small quantity of radiation interacts with matter by ionization and excitation, producing both direct and indirect damage; the latter occurs primarily through free radical attack. Because free radicals induce a chain reaction of free radical injury, the initial event is magnified. The molecular and biochemical damage in turn stimulate further amplification through the release of biological mediators and hormones. At each of these points, actions can be taken (i.e., radioprotection, radiosensitization) that will enhance or suppress the biological expression of injury. (From Grosch, D.S. and Hopwood, L.E., *Biological Effects of Radiation*, Academic Press, New York, 1979; Salter, C.A., *Mil. Med.*, 166, 17–18, 2001. With permission.)

## RADIATION BIOCHEMISTRY

Radiation toxicity represents a dynamic interaction between radiation physical constraints and olecular damage, which are further amplified by biological processes resulting in injury (Figure 18.1). Biochemistry provides this important link, through which a quantity of energy sufficient to raise the temperature of a liter of water by 0.1°C, an amount less than the caloric energy of a candy bar, can elicit toxicity resulting in death or the beneficial therapy of a malignant tumor. Diverse extremes and yet mechanistically similar at the molecular level, the initial injurious events occur in $10^{-17}$ to $10^{-5}$ seconds, may be irreversible, and may require seconds to years for expression. The expression of injury for some cancers is so slow that it may not be observed over the individual's lifetime. The four processes mediating and amplifying radiation toxicity are ionization and excitation, molecular injury, biochemical damage, and amplification and expression of injury, as illustrated in Figure 18.1. The first three will be discussed in this section, followed by amplification and expression of biological injury in the remaining sections. The latter sections emphasize the important roles of biological mediators and hormones in this process, as well as delineating specific risk concerns such as cancer.

The four basic ways in which ionizing radiation may interact with an atom are direct and indirect ionization, excitation, and pair production. Ionization occurs through the impartation of sufficient energy to eject an electron out of its orbit, resulting in formation of an ion pair: the positively charged atom and the negatively charged electron. If the ejected electron is also imparted with additional energy, then it too may interact with other atoms to produce additional ion pairs, and this is an indirect ionization process. With excitation, the energy imparted to the electron is sufficient to raise it to a higher electron orbital but not to escape. Ionization does not occur, but the energy may be sufficient to break chemical bonds. Between 20 and 35 electron volts (eV) are required to produce an ion pair, but hydrogen bonds can be broken by 5 eV [820,826]. Finally, photon radiation with energies greater than 1.02 MeV is capable of interacting with the electromagnetic field of the nucleus to convert the energy into the formation of positron and an electron, or pair production. Neutrons, alpha particles, and other particulate radiations also produce ionization and excitation. Further, in the process of nuclear activation, the neutron is absorbed by the nucleus, forming a radioactive isotope.

Although differing types of radiation may be equal in their initial energy, they differ in the pattern in which the radiation is imparted to matter and as a result differ in their biological responsiveness. These differences are quantified for their relative biological effectiveness (RBE), determined as the dose of a given type of radiation (neutron, gamma, x-ray, etc.) to produce a reference biological effect divided by the dose of a standard radiation, usually 250-kilovolt potential (kVP) x-rays to produce the same effect. This is illustrated by the threshold for cataract formation in humans of 2 Gy for x-rays and 0.2 Gy for neutron exposures, resulting in a RBE of 10 [67]. This RBE applies only to this particular effect; for a different response the RBE might be 1 or even 50. Differences in effectiveness arise because the amount of energy transferred by a unit dose of radiation per unit pathway traveled through matter, termed linear energy transfer (LET), varies with the type of radiation. Neutrons, protons, alpha and beta particles, and atomic nuclei are high LET radiations, and gamma and x-rays are low LET radiations. The depth of penetration is determined by the total energy and the LET factor. High LET radiations produce many ionizations per distance traveled, dissipate energy quickly, and have low depths of penetration. In fact, alpha particles dissipate all of their energy before they can transverse through a sheet of paper, a layer of paint, or the stratum corneum of the skin. The consequence of such a concentrated delivery to a biological system is that more ionizations and therefore more damage are concentrated in a smaller area, making it more difficult to repair the damage [68,820,821,826]. Higher RBEs for neutrons (up to 46 for life shortening and 80 for cancer transformation) are elicited when lower doses and dose rates are used because of the better ability of cells to repair injury to low LET radiation [144,311,588]. Low-dose neutron exposure is more carcinogenic for mammary tissue than x-rays, and a low environmental or accidental exposure to neutron could therefore be more damaging than the total exposure dose alone might indicate [69]. In general, the LET increases from electrons to neutrons to alpha particles. The RBE increases with increasing LET up to 100 to 125 keV/μm [69,311]. Of particular concern, low-dose and dose-rate exposures to high LET neutron radiation may be more effective at causing cancers than similar exposures to low LET radiation.

Radiation interacts with matter by direct and indirect processes to form ion pairs, some of which may be free radicals. These ion pairs rapidly interact with themselves and other surrounding molecules to produce free radicals. Both the indirect and direct activities of ionizing radiation lead to molecular damage which is then translated into biochemical damage. Biochemical damage may then be amplified and expressed as biological injury in one of three basic processes, including damage to the DNA that may become expressed, stimulation and release of biological mediators, and alteration of nutritional vascular support.

## Radiolysis of Water

Radiation may impart energy to matter in one of two primary ways: either directly through ionizations or indirectly by transfer of energy and formation of free radicals. The most abundant molecule in living systems is water. It accounts for about 55% of the mass in humans. Ionizing radiation interacts with water molecules to form an ionized pair consisting of a free electron ($e^-$) and an ionized water molecule ($H_2O^+$) in a process termed radiolysis. The free electron rapidly interacts with water to form the hydrated electron ($H_2O^-$) which decomposes to $OH^-$ and $H^\bullet$. The • symbol designates a free radical, a molecule having an unpaired electron in the outer electron shell. The free radical may also be electrically neutral but remains highly reactive because of the unpaired electron. The second ion from the ion pair ($H_2O^+$) decomposes to $H^+$ (hydrogen ion) and $OH^\bullet$ (hydroxyl radical). The ionic designation depends on the molecular charge, so it is possible for a molecule to be a free radical and an ion. The hydroxyl radical contains 9 protons and 9 electrons and is electrically neutral.

The end products of the radiolysis of water without oxygen are $H^\bullet$, $OH^\bullet$, $H^+$, and $OH^-$. Of these, $H^\bullet$ and $OH^\bullet$ are the most important and comprise 55% of the initial relative yield [68]. Both are highly reactive and have half-lives of $10^{-11}$ seconds. This allows for the initial impact of the radiation ionization event, which may have missed the biological target, membrane, or DNA to diffuse away from the initial site and produce damage by free radical attack of a nearby molecule. The two primary factors influencing the formation of free radicals are the presence of oxygen and the LET. Radiolysis in the presence of oxygen produces the hydroperoxy radical ($HO_2^\bullet$), the hydroperoxy ion ($HO_2^-$), and hydrogen peroxide ($H_2O_2$). These chemical entities are powerful oxidizing agents with longer half-lives on the order of $10^{-10}$, and they may diffuse even farther from the initial site of ionization [68,342,826]. The type of radical species formed is also governed by how closely or how rapidly radicals are formed, as they may interact to form $H_2$, $H_2O$, and $H_2O_2$, neutralizing the radical attack before reaching the biological target; for example, $H^\bullet$ reacts with $OH^-$ to form $H_2O$. High LET radiations produce more ionizing events closer together, with a greater chance of free radical neutralization; hence, direct effects tend to predominate with high LET. The net effect of the above processes is that biological material irradiated in a dry state and in the presence of oxygen is more resistant to injury than when the reverse conditions are present. Interaction of a free radical with a biological molecule (RH) results in the formation of an organic free radical and stabilization of the initial radical:

$$OH^\bullet + RH \rightarrow H_2O + R^\bullet$$

Although yields vary depending on experimental and physiological parameters, the radical yield for gamma radiation on golden hamster embryo cells irradiated in a frozen living state determined by electron spin resonance was 12% $H^+$, 72% OH radicals, and 16% organic radicals [549].

**TABLE 18.4**
**Effects of Ionizing Radiation on Macromolecules**

| | |
|---|---|
| Amino acids | Liberation of ammonia, hydrogen sulfide, pyruvic acid, carbon dioxide, and hydrogen |
| Carbohydrates | Cleaveage of glycosidic bonds; depolymerization of individual monomers; oxidation of terminal alcohols to aldehydes |
| DNA | Degradation with base loss or modification; breakage of hydrogen bonds or sugar– phosphate bonds; cross-linking of DNA:DNA or DNA:protein; strand breakage; formation of guanyl, thymidyl, and sugar radicals |
| Lipids | Peroxidation; bond rearrangement; conjugated diene formation; aldehyde formation; β-scission; cross-linking; increased microviscosity |
| Proteins | Degradation and modification of amino acids; chain scission; cross-linkage; denaturation and changes in molecular weight and solubility |
| Thiols | Oxidation; reduction; radical formation; cross-linkage |

## The Oxygen Effect

The most effective sensitizer of biological tissues is oxygen [311]. Biological material and living systems irradiated in the presence of oxygen are more susceptible to injury than when irradiation occurs without oxygen. This response, known as the oxygen effect, was first observed in 1909 by Gottwald Schwartz, who noted that pressure applied to the forearm during irradiation reduced the resulting erythema [820]. He was not aware that lack of oxygen *per se* was responsible. The effectiveness of oxygen in modifying the response to radiation is expressed as the oxygen enhancement ratio (OER), the radiation dose in the absence of oxygen required to observe a given response divided by the radiation dose required to observe the response in the presence of oxygen. OERs for low LET radiations are generally 2.8 to 3 and tend to decrease as the RBE increases, up to a point between 100 and 200 keV/μm. The radiosensitizing effects of oxygen are particularly important in radiotherapy, as tumors may have hypoxic centers because of poor vascular supply, tumor compression from the outer cells, or altered metabolism. Therapeutic attempts to address this issue have tried to increase tumor oxygenation (through increased vascularization, vasodilation, hyperbaric oxygen, or increased oxygen delivery to tissues using intravascular perfluorocarbons) or to use electronegative compounds that mimic oxygen [311]. Hypoxia decreases the radiosensitivity of tissues but increases the sensitivity to hyperthermic treatments. The oxygen concentration must be reduced below 2.0% to see any appreciable protection, which increases rapidly with decreasing oxygen concentrations below 0.5% (3 mm partial pressure of oxygen, compared to 40 mm in venous blood and 60 to 80 mm in oxygenated arterial blood), at which point the OER is about 2 [311]. Oxygen results in the formation of hydroperoxy and hydrogen peroxide radicals, which are more damaging because of their longer half-lives. Free radical damage is fixed through the reaction of oxygen leading to the formation of peroxy and hydroperoxy organic products that are more resistant to biochemical repair processes [305,311]. The presence of oxygen can also enhance the therapeutic efficacy of some chemotherapeutic agents, while some agents including misonidazole, mitomycin C, doxorubicin, and tirapazamine are more effective in hypoxic environments [311,770,771]. Ethanol, narcotics, leukotrienes, and some thiols are radioprotective in part through their ability to induce hypoxia [278,820]. Ethanol and narcotics suppress the respiratory center in the central nervous system, resulting in hypoxia. Exaproxiral (RSR13) has been shown to radiosensitize tumors by increasing the whole blood P50, resulting in increased oxygen unloading by hemoglobin with increased tissue oxygenation [758]. Motexafin gadolinium enhances radiation sensitization by increasing reactive oxygen species and cell death [511]. When given with whole brain radiation in patients with brain metastases from lung cancer, neurological progression was delayed, but there was no difference in overall survival.

## Effects of Radiation on Macromolecules

Radiation-induced modification of biomolecules can be divided into structural degradation and decomposition, cross-linking of molecules, and breakage of chemical bonds [16,305,342,814,819,820,826]. The individual responses for the different classes of macromolecules are presented in Table 18.4. Bond breakage occurs through energy transfer, ionization transfer, or electron transfer. Cross-linking can occur between similar and also different classes of molecules: protein–protein, lipid–lipid, protein–DNA, etc. Structural and conformational changes may alter or eliminate biochemical activity directly and expose internal sites to radical attack [146,273,826]. The DNA bases

**TABLE 18.5**
**Steps in Lipid Peroxidation**

| | |
|---|---|
| Initiation | Enzymatic |
| (A) | LH + ionizing radiation → LH• + H• |
| (B) | LH + OH• → L• + $H_2O$ |
| (C) | LH + R• → L• + RH |
| Propagation | L• + $O_2$ → LOO• |
| | LOO• + LH (or RH) →→ L• (or R•) + $LO_2H$ |
| Termination | LOO• + LOO• (or ROO•) |
| | L• + LOO• |
| | LO• + LO• |
| | L• + L• |
| | L• + R• |
| | L• + H•L• + free radical scavenger |

*Note:* L, lipid; R, other organic molecules.

*Source:* Data from Ibuki, Y. and Goto, R., *Biol. Pharm. Bull.*, 23, 1094–1096, 2000; Miyachi, Y. et al., *Neurosci. Lett.*, 175, 92–94, 1994.

are protected from free radical attack because of their position in the center of the helix but are exposed by strand breaks, breakage of hydrogen bonds, and unwinding processes. Conformation is also critical for enzyme activity and structural proteins. The response of macromolecules *in vitro* may vary from the effects *in vivo* due to molecular interaction and biological repair processes.

## Lipid Peroxidation

Radiation injury of lipids *in vitro* and *in vivo* occurs primarily through peroxidation by free radical attack at the double bonds and carbonyls [650,819,820]. Lipid peroxidation, as illustrated in Table 18.5, consists of three phases: initiation, propagation, and termination. Lipid peroxidation is important to homeostasis and is associated with the formation of lipid mediators, including the prostaglandins and leukotrienes, as well as lipid degradation (Figure 18.2). Peroxidation may be mediated by three possible initial events [650,820] (Table 18.5). Initiation may occur enzymatically, as with cyclooxygenase and lipoxygenase enzymes, and requires molecular oxygen and ferric cofactors. Radiation-induced lipid peroxidation is initiated by direct or indirect ionization or by free radical attack. The primary radical species involved are the hydroxyl radical and superoxide [228,650].

Lipid peroxidation is affected by the lipid structure and composition, presence of oxygen and antioxidants, pH, temperature, and conditions of irradiation [228,650, 820]. Increasing the number of double bonds in the lipid carbon backbone enhances its susceptibility to free radical oxidation in solution; for example, arachidonic acid possesses four double bonds and is more sensitive than linoleic acid, which has three double bonds [560]. The greater the lipid concentration, the greater the likelihood that the free radicals propagate the chain reaction by attacking another lipid molecule. On the other hand, with increased rate of radical formation, the risk of radical interaction and neutralization increases; therefore, unlike other radiation-induced molecular injury processes, greater lipid peroxidation product yield is obtained with lower radiation doses and exposure rates.

Lipid peroxidation is a chain reaction in which interaction of the lipid radical with another organic molecule results in conversion of that organic molecule to the free radical state and propagation of damage. Alternatively, the lipid radical may terminate the reaction by one of several different processes as outlined in Table 18.5. It may react with another free radical or with a free radical scavenger. The primary free radical scavengers in biological systems are vitamins A and E and the thiols, and their membrane concentrations influence cellular radiosensitivity [412,820,821]. In addition, at least eight enzyme systems are associated with detoxification and repair of free radical injury, not including DNA-specific repair enzymes [427,820]. These include glutathione transferase, NADPH-dependent glutathione reductase, selenium-dependent glutathione peroxidase, selenium-independent glutathione peroxidase, ferric superoxide dismutase, manganese superoxide dismutase, and copper–zinc superoxide dismutase and catalase. Metallothionein is another protein with free radical scavenging ability, as one third of its amino acids are cysteine residues. Synthesis of this protein is enhanced under stress situations, including lipid peroxidation and radiation injury [820], and can be induced by zinc supplementation. Another thiol-containing protein, thioredoxin, participates in the reduction of a number of important DNA enzymes and transcription factors, including ribonucleotide reductase [264], and is a radioprotectant. It is itself reduced by NADPH-dependent thioredoxin reductase, a selenoprotein. Vitamin E is an important free radical scavenger that is located primarily in the cell membrane. Once modified or activated by free radical attack, it is detoxified and renewed by specific enzymatic pathways. Vitamin E interacts with organic radicals to form a stable organic alcohol [412]. In this process, vitamin E is converted to an excited state that is restored by interaction with vitamin C. Vitamin C, in turn, is renewed through the action of an NADH-dependent enzyme. Interestingly, melatonin, the pineal gland hormone, has been demonstrated to scavenge free radicals [654,812], and pretreatment of human lymphocytes irradiated *in vitro* reduced subsequent micronuclei damage formation [811]. Other studies have shown protection of whole animal survival [812].

Thiols are molecules containing free or potential sulfhydryl groups ($SH^-$) in their structure. Examples include the amino acids methionine and cysteine and the complex lipid thiol ether leukotriene $C_4$. The most abundant non

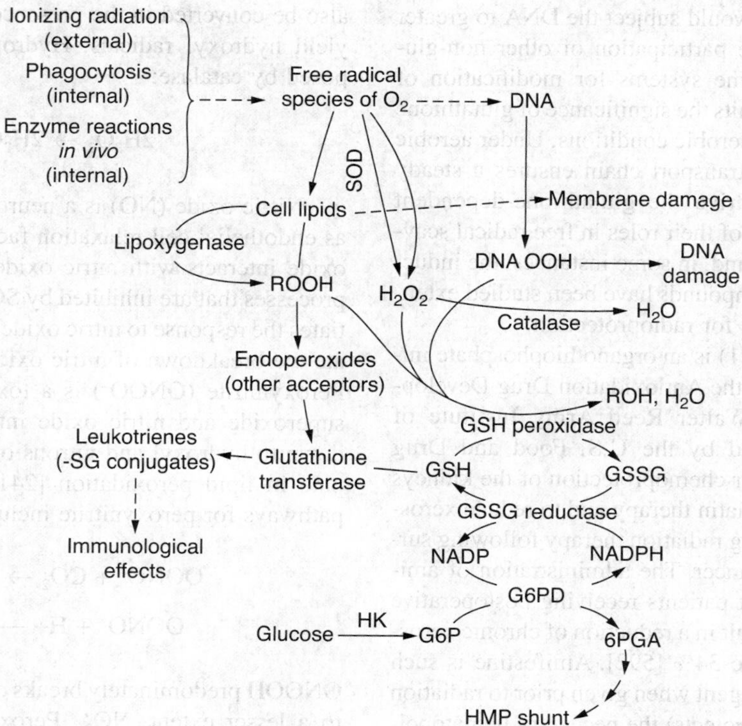


**FIGURE 18.2** Enzymatic repair of peroxide damage: Reactive oxygen species formed by enzymatic and nonenzymatic processes may induce oxidative attack of DNA, lipids, and other biomolecules. Repair of this damage can be mediated by glutathione (GSH, reduced glutathione) or glutathione-dependent peroxidases. As shown, oxidized glutathione (GSSH, glutathione disulfide) can be reduced or renewed, and most of the system can be maintained as long as a sufficient supply of reducing power (NADPH, nicotinamide adenosine diphosphate) generated by the metabolism of glucose is available. Abbreviations: DNA, deoxyribonucleic acid; G6PD, glucose-6-phosphate dehydrogenase; HK, hexokinase; HMP, hexose monophosphate shunt; ROOH, organic peroxide; SOD, superoxide dismutase. (From Jagoe, C.H. et al., *Ecotoxicology*, 7, 202–210, 1998. With permission.)

protein thiol, the tripeptide glutathione (GSH), is present intracellularly at 1 to 3 m*M*. Thiols act as cofactors for some enzymatic processes and participate in radical scavenging and detoxification processes. Their activities are dependent on cellular concentration, location, synthesis, and catabolism. During the reaction of a thiol with a free radical, donation of hydrogen by the reduced sulfhydryl of the thiol to the organic radical results in a form of chemical repair:

$$GSH + R^{\bullet} \rightarrow RH + GS^{\bullet}$$

Two glutathione radicals (GSC) may then react to form a disulfide (GSSG), terminating the radical chain reaction. This disulfide, or oxidized glutathione, is regenerated to yield two GSH molecules by the action of NAPDH-dependent glutathione reductase (Figure 18.2). Thiols may also affect responses to irradiation through formation and destruction of disulfide bridges. Glutathione may form a disulfide bridge with the free sulfhydryl group of another molecule, perhaps in an enzyme active site, masking and protecting the sulfhydryl from radical injury. Organic peroxides are repaired through the actions of selenium-dependent and independent glutathione peroxidases [427,820].

The hydrogen donation reducing power for these processes is provided by NADPH generated through the hexose monophosphate shunt; therefore, anything interfering with the hexose monophosphate shunt, glucose utilization, or the synthesis of $NAD^+$, $NADP^+$, or glutathione may ultimately influence free radical scavenging and tissue injury [78,427]. In fact, hypoxic cells depleted of glutathione become more sensitive to radiation [78,541,820]. This has implications for the radiotherapy of tumor cells, although sensitization has not been consistently observed for aerobic cells [820]. The difficulty arises in that glutathione concentrations must be reduced below 5% for sensitization to be observed [541,820]; yet, even at these low levels, glutathione-dependent enzymes usually maintain activity because they retain glutathione. The depletion of glutathione levels by use of specific glutathione synthetase inhibitors, such as buthionine sulfoximine, may have a greater effect on DNA damage as glutathione synthesis does not take place in the nucleus [40]. Glutathione must diffuse into the nucleus. As a result, glutathione depletion in the nucleus would subject the DNA to greater free radical damage. The participation of other non-glutathione-dependent enzyme systems for modification of oxidative damage also limits the significance of glutathione during irradiation under aerobic conditions. Under aerobic conditions, the electron transport chain ensures a steady supply of NADPH to drive the glutathione-dependent repair enzymes. Because of their roles in free radical scavenging, chemical repair, and, in some instances, the induction of hypoxia, thiol compounds have been studied extensively as potential agents for radioprotection.

Amifostine (WR-2721) is an organothiophosphate initially developed through the Antioxidation Drug Development Program of the Walter Reed Army Institute of Research; it is approved by the U.S. Food and Drug Administration (FDA) for chemoprotection of the kidneys in patients receiving cisplatin therapy and to reduce xerostomia in patients receiving radiation therapy following surgery for head or neck cancer. The administration of amifostine to head and neck patients receiving postoperative radiation therapy can result in a reduction of chronic xerostomia from 57% down to 34% [592]. Amifostine is such a potent radioprotective agent when given prior to radiation exposure that it raises (protects) the necessary hematopoietically lethal dose of radiation by 2.5 to 2.7 [867]. This compound has also been used in clinical trials to minimize radiation injury to normal tissues of patients receiving radiation therapy [101,144,213,311,408,592,730]. Amifostine is FDA approved for intravenous administration. It has been given in animals intravenously, subcutaneously, and intraperitoneally. The potential side effects include hypotension, nausea, vomiting, and hypocalcemia. Studies demonstrate that WR-2721 is preferentially taken up by and therefore protects normal tissues, as opposed to tumor tissues [101,592,868]. Central nervous system protection is minimal, as it does not cross the blood–brain barrier.

Superoxide is produced physiologically by several enzyme systems, is released by activated neutrophils, and is formed during the radiolysis of water in the presence of oxygen [820]. It is removed through conversion to hydrogen peroxide by the action of superoxide dismutase (SOD):

$$2H^+ + 2O_2 \rightarrow H_2O_2 + O_2$$

Treatment with SOD has been shown to protect macromolecules, cells *in vitro*, and animals. Superoxide is more likely to be transformed to hydrogen peroxide by superoxide dismutase rather than with nitric oxide (to form peroxynitrate, $ONOOO^-$) because of the higher cellular concentration of superoxide dismutase [532]. Hydrogen peroxide may be transformed via myleoperoxidase to yield hypochlorous acid ($HOCl^-$) which can then form hydroxyl radicals through superoxide interaction. Hydrogen peroxide is itself a potent oxidizing agent that can also be converted in the presence of ferric compounds to yield hydroxyl radicals. Hydrogen peroxide is decomposed by catalase:

$$2H_2O_2 \rightarrow 2H_2O + O_2$$

Nitric oxide (NO) is a neurotransmitter and also acts as endothelial cell relaxation factor, a vasodilator. Superoxide interacts with nitric oxide to form $NO_2$ and $NO_3$, processes that are inhibited by SOD. As such, SOD potentiates the response to nitric oxide. Cytochrome *c* can mediate the breakdown of nitric oxide in mitochondria [623]. Peroxynitrite ($ONOO^-$) is a toxic oxidant product from superoxide and nitric oxide interactions that can break down to hydroxyl and nitrous oxide radicals that participate in lipid peroxidation [241,532]. Potential reaction pathways for peroxynitrite include:

$$OONO^- + CO_2 \rightarrow ONOOCO_2^-$$

$$OONO^- + H+ \rightarrow ONOOH$$

ONOOH predominately breaks down to $NO_3^-$ (nitrate) and, to a lesser extent, $NO_2^{\bullet}$. Peroxynitrite can also interact with carbon dioxide to form $ONOOCO_2^-$, which can breakdown to form nitrate ($NO_3^-$) or nitrogen dioxide and carbonate radical ($CO_3^{\bullet}$). Nitrogen dioxide ($NO_2^{\bullet}$) can react with superoxide to form $ONOOO^-$, but it is more reactive toward thiols [245,532]. Nitrite dioxide ($NO_2^{\bullet}$) has a diffusion path length of 0.2 μm in cytosol and 0.8 μm in plasma [245]. In cytosol, the major reaction is formation of *S*-nitrosothiols, while in plasma urates also provide a source for nitration [245,532]. Nitric-oxide-catalyzed *S*-nitrosylation of metallothionein results in zinc cofactor release and inactivation [532] and may regulate zinc-dependent enzymes and signaling pathways. $NO_2^{\bullet}$ can also interact with the OH group on tyrosine in proteins (nitration) [245,532]. Tyrosine is oxidated to a tyrosine phenoxyl radical that can then interact with a similar radical to form a 3,3′-dityrosine cross-link or with nitrite oxide or nitrite dioxide to form 3-nitrotyrosine. Tyrosine phenoxyl radicals may also interact with superoxide, resulting in free tyrosine. Protein tyrosine nitrations have been observed in cells irradiated *in vivo* and in specific proteins including manganese-oxide-dependent superoxide dysmutase [445,532]. Tyrosine radicals have a long half-life. Long-lived radicals have been detected in proteins following irradiation of albumin in solution [862] and cells *in vivo* [426,445,862]. The half-life of long-lived protein radicals is 20 hours, and levels can be reduced by the addition of vitamin C to cell cultures following irradiation [862]. H-added phenylalanine has also been detected as a long-lived radical in cells irradiated *in vivo* [426]. Suppression of the long-lived protein radicals reduces mutagenicity but not cell lethality [426,862].

There is much interest in the biological roles of nitric oxide. Nitrite synthetase has three isozymes. Two are regulated by calcium levels; the third is induced usually by macrophages [791] and is also stimulated in the rat colon and ileum following radiation exposure [472]. The increase in nitric oxide is thought to be associated with radiation-induced ileal dysfunction [472]. Exposure of mice to 7 Gy of gamma irradiation has been reported to stimulate the L-arginine-dependent production of nitric oxide from the terminal guanidine group in mouse liver, lung, brain, and spleen [813]. Treatment of mice with a specific inhibitor of nitric oxide synthetase, or with DEA/NO, a nitric-oxide-releasing agent results in increased animal survival of the irradiated mice. Interestingly, both agents have been shown to increase the hypoxic fraction of the mouse bone marrow, and this may be related to alterations in regional blood flow (induction of acute hypoxia). Protection is most effective when these compounds are given close to and before irradiation.

Because of its similarities to oxygen, nitric oxide has also been evaluated as a radiosensitizing agent to hypoxic cells *in vitro*, with a sensitization enhancement ratio of 2.4 [542]. It is thought that, under hypoxic conditions, the free radical nitric oxide interacts with carbon-based free radicals to fix radiation-induced free radical injury in a manner similar to oxygen.

## Effects on Amino Acids, Peptides, and Proteins

The primary effects of radiation on amino acids and proteins in solution are provided in Table 18.4 [16,450,820]. Radiation-induced breakage of hydrogen bonds and disulfide bridges, or cross-linkage formation, can affect conformation and therefore activity and function. *In vivo*, nitration of the hydroxyl on tyrosine can be observed when nitrogen reactive species are present [791]. The radiation dose to inactivate proteins in the dried state is proportional to the molecular weight; that is, a larger target requires more radiation. This relationship has established radiation-inactivation of proteins as an accepted method of determining molecular weights. Further alterations may occur from the moderation of radiation synthesis, although, in general, protein synthesis is not affected by radiation doses within the lethal range for humans [16,604]. RNA synthesis usually decreases following irradiation of radiosensitive tissues [16,820]. Radiation effects on inducible enzyme systems depend on the particular system and vary between species and sexes. Some drug detoxification enzymes are reduced in males in association with decreased testosterone synthesis. Chronic radiation exposure has been shown to induce the Hsp70 heat shock protein in mouse lung [509], while an exposure of 0.25 Gy to cells *in vitro* induces PBP74/mortalin/Grp75, another member of the heat shock protein family [682]. Induction of heat shock proteins has been related to increased radioresistance. Irradiation of histones can result in protein-bound stable hydroperoxides, which can be released later, generating DNA and RNA base damage [467].

## Effects on Carbohydrates

The basic effects on carbohydrates in solution are shown in Table 18.4. Radiation may cause depolymerization of glycogen, and the α-glycosidic linkages found in glycogen and cellulose are more radiosensitive than β-D-glycosidic linkages that might be found in bacterial cell walls [820]. *In vivo*, the effects of radiation on carbohydrates are dominated by alterations in metabolic processes. Radiation promotes glycogenesis and gluconeogenesis during the first several days after irradiation primarily as a result of hormonal influences and alterations in metabolic enzymes [16,142,820]. Insulin and adrenocorticoid release stimulates an increase in blood glucose, thus providing a source for the glycogen synthesis, which is further supplemented by the shunting of amino acids released by tissue injury. Decreases in hexokinase, aldolase, and pyruvate kinase and an increase in transketolase are observed [142,820]. Additional NAPDH reducing power is obtained by the shunting of carbohydrates through the pentose phosphate pathway. Malondialdehyde, formaldehyde, and acetylaldehyde can be released following irradiation of fructose, sucrose, and glucose in solution [234]. Formation of these organic aldehydes from carbohydrates was found to be concentration and pH dependent.

## Effects on Nucleic Acids and DNA

Ionizing radiation exposure of DNA may cause degradation of bases and sugars, breakage of hydrogen or sugar-phosphate bonds, or cross-linking (see Table 18.4 and Figure 18.3) [146,273,305,342,814,826]. The incidences for the individual types of damage are base damage > single-strand breaks > DNA-protein cross-links > double-strand breaks [826]. In base damage, thymine sensitivity > cytosine > adenine > guanine [263], and there can be more than one type of injury [791]. At radiation doses of 200 kGy, it is proposed that the two primary radicals formed in DNA are the thymine anion and the guanine cation [341,814,815]. Radicals may be of a charged anion, deprotonated cation, H-addition radicals, H-abstraction radical, and opened sugar ring nature [341]. DNA damage to the carbohydrate produces strand breakage, glycosylic bond breakage, and formation of precursors to malonaldehyde [305,815,826]. DNA cross-linkages occur between DNA and DNA, DNA and proteins, or DNA and lipids. DNA to DNA cross-links may form between adjacent bases on the same strand or between different strands. The most common DNA protein adducts would be between DNA and the associated histones and

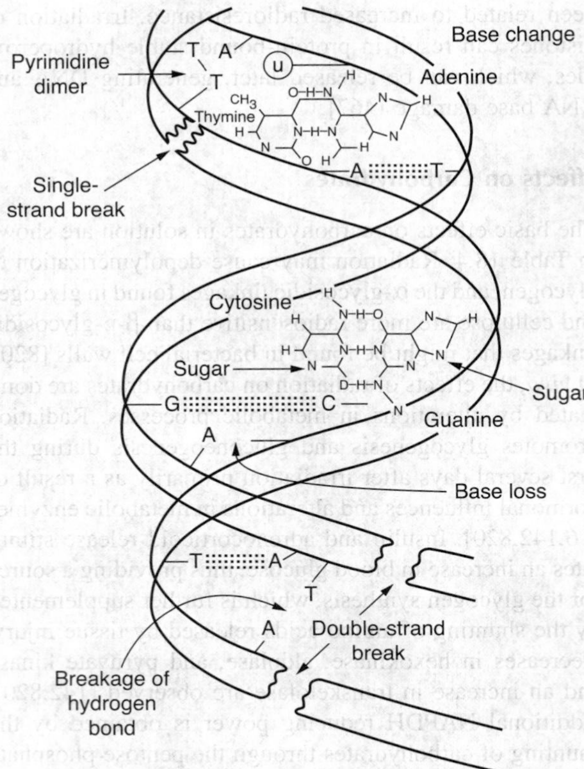


**FIGURE 18.3** Radiation damage to deoxyribonucleic acid. The basic categories of molecular damage—degradation, bond breakage, and cross-linking—that occur in biomolecules following radiation exposure are illustrated for this double-stranded segment of DNA. The dotted lines represent hydrogen bonding between the bases—three between cytosine (C) and guanine (G) and two between adenine (A) and thymine (T). (From Jagoe, C.H. et al., *Ecotoxicology*, 7, 202–210, 1998. With permission.)

are more likely to involve tyrosine or lysine residues [605]. DNA–protein cross-linkage formation is more common in expanded chromatin than in compressed chromatin [146]. Intracellularly, DNA–protein cross-links are enhanced by glutathione depletion but are reduced in the presence of oxygen. Important factors controlling injury are the medium composition, DNA conformation, and presence of repair enzymes [820,826]; for example, metal ions are thought to induce structural changes inhibiting free radical access to key sites and to stabilize the double-strand helix [741].

Thymine glycol and 8-hydroxy-guanine are oxidative DNA base damage products that can result from agents inducing oxidative damage including radiation exposure [444,791]. Using a sensitive assay with $10^{-15}$-micromole detection limits, thymine glycol has been identified in the DNA of A549 human lung carcinoma cells irradiated *in vitro* with doses less than 5 cGy. The sensitivity at this low radiation dose was sufficient to detect 4.3 molecules of thymine glycol per 1 billion DNA bases. The responses to this assay, while linear, are dependent on the source of DNA and on irradiation of the DNA inside the cell (lower response rate, protective of DNA), as opposed to naked DNA in solution. The A549 cells were able to remove 50% of the thymidine glycol from their DNA within 2 hours and 80% by 4 hours [444]. Both thymine glycol and 8-hydroxy-guanine can be detected in the urine but may also be elevated through diet and as by-products of enzymatic activity [791].

Breakage of sugar–phosphate bonds leads to single-strand breaks (SSBs) and double-strand breaks (DSBs). There are two hypotheses regarding DSB formation [826]. In the first, hydroxyl radical attack on one strand produces a strand break through radical formation which attacks the opposite strand, producing a DSB. According to the second hypothesis, DSBs are produced by local, multiply damaged sites to regional areas of both strands, resulting in a DSB. Cell lethality is directly related to DSBs. A 1-Gy dose of ionizing radiation results in 63 to 70 DSBs per cell [68]. The importance of DSBs over SSBs in relation to radiation-induced lethality is illustrated by the fact that a radiation dose that kills 63% of the exposed cells produces 1000 SSBs but only 40 DSBs [826]. There would also be an estimated 440 locally multiply damaged sites. Exposure to hydrogen peroxide by comparison would require production of over 2.5 million SSBs per cell to kill 63% of the cells. Single-strand breaks are chemical assay phenomena that do not lead to mutational events or expression of injury because the remaining strand is intact, holding the two ends together for repair by DNA ligase. With double-strand breaks, there is no stable endpoint for repair, and the chromosomal material may be lost during the subsequent cell division or rejoined to a different chromosome, forming a chromosomal aberration.

Several types of chromosomal aberrations are observed following irradiation. These include rejoining to the original chromosome—normal, inversions, and translocations; terminal or interstitial deletions; and ring and dicentric formation. Irradiation of the chromosomes in G results in chromosomal aberrations, while exposure after DNA synthesis in $G_2$ results in chromatid aberrations. Chromosomal exchanges occurring during $G_1$ irradiation may either be reciprocal (two chromosomes) or nonreciprocal (three chromosomes). In human fibroblasts irradiated *in vitro*, 50% of the chromosomal exchanges were nonreciprocal [106]. The rate of chromosomal aberration is directly related to the radiation dose and is higher for higher LET irradiation and higher dose rates [68,791, 819,820]. Chromosomal aberrations were still evident in the lymphocyte chromosomes of the Japanese atomic bomb survivors 35 years after the bombing [651], and elevations have been identified in radiotherapy patients and in persons receiving occupational radiation exposure such as uranium miners [68,69,819]. The incidence of chromosomal aberrations in exposed human populations decreases with time from radiation exposure and can be influenced by prior irradiation [791].

Chromosomal aberrations have also been demonstrated to be elevated in the lymphocytes of astronauts who completed long-term space flights on the Russian MIR space station [294,599], but this was not correlated to the duration of the flights [294]. Irradiation of blood from astronauts obtained before and after space flight show a slight increase in aberration induction of the post-flight samples (increased sensitivity) [294]. This appears to be related in part to the time when the sample was obtained following the flight and may be affected by other factors (microgravity exposure in flight, radiation exposure during flight, decreased activity/confinement, or stress.

The assessment of chromosomal aberrations represents a reasonable means of biological dosimetry and can be detected by several methods, including phytohemagglutinin-stimulated peripheral blood lymphocytes; premature chromosomal condensation; micronuclei techniques; fluorescent *in situ* hybridization (FISH), or chromosome painting; and the comet assay [84,227,392,463,791,819,820]. The results are influenced by the scoring criteria, type of irradiation, percent of body irradiated, degree of time following radiation exposure, culture time, and age of the individual, as well as exposure to other environmental toxins or agents that might also produce aberrations. Assays also exist for the detection of somatic cell mutations including the hypoxanthine phosphoribosyl transferase locus (HPRT) and the glycophorin A locus. The latter codes for a glycoprotein found on the surface of red blood cells (similar to and in conjunction with the ABO blood grouping) and expressed by alleles M and N (can also have null, O). Increases in glycophorin A variants are dose dependent and have been elevated in atomic bomb survivors and in patients from the radiation accidents in Goiana, Brazil, and Chernobyl [440]. The assay requires 0.1 mL of blood and can be completed in less than a day. An example of a variant of somatic mutation would be the detection of an MO allele expression in a person who was genetically MN or MM).

The final expression of DNA damage is modified by repair processes. Repair may occur at several levels: nonspecific suicide enzymatic repair, excision/ligation repair, SOS repair (mismatch repair), single base repair as with DNA glycosylases, and methylation. Depending on the process, more than one enzyme may be required. Excision/ligation repair involves several enzymes: one to make a nick on the DNA strand containing the damaged site (presuming that radiation has not already provided that nick); an excinuclease that excises damaged bases; synthesis of the new sequences based on the complimentary strand by the action of a DNA polymerase; and, finally, rejoining of the two loose ends by a DNA ligase [305,446, 691,819,820,826]. The main repair enzyme in bacteria is DNA polymerase I; in mammals, it is DNA polymerase II. In addition, bacteria have a second category of repair enzymes called SOS-repair or error-prone repair. As their name implies, they play a major role in the repair of double-stranded breaks and are associated with a high error rate [305]. Similar mismatch repair genes that have been identified in humans are designated hMSH1, hMSH2, hMLH1, and PMS2 [252]. Cells with functional mismatch repair genes are more sensitive to ionizing radiation (cell survival) than mutants lacking active forms of these repair genes [252]. The hardest break to repair without errors is the double-strand break, and the repair depends on whether the two breaks are coincident, with possible loss of intervening information, or the two breaks are separated, with overlapping sequences to ensure proper rejoining and repair [68,826]. The closer the approximation of the two breaks, the greater the chance of loss.

**TABLE 18.6**
**Eight R's Influencing Biochemical Reactions to Ionizing Radiation Exposure**

- Radiation factors
  - Dose
  - Dose rate
  - Dose fractionation
  - Radiation quality (LET)
  - Internal vs. external irradiation
- Radioprotectors/radiosensitizers
- Reoxygenation
- Resilience
  - Age/health/individual
  - Vascular integrity
- Response
  - Nutritional/homeostasis
  - Biological mediators
  - Mutational
  - Circadian
- Repopulation
- Recovery
- Repair

*Source:* Data from Alman et al. [12], Bachofer and Gautereaux [48], Baetcke et al. [49], Gasteiger and Campbell [269], Samaan [686], Sambur et al. [687], and Schultz-Hector et al. [708].

## Mechanisms

The biochemical response of a biological system to ionizing radiation is influenced by eight major groups of factors as outlined in Table 18.6. Repair, reoxygenation, redistribution, and regeneration have been called the "four R's of radiotherapy" [846]. It is appropriate to continue the "R" series. The first group of factors relates to the radiation itself [68,820]. Most biochemical responses are dose dependent or at least may have a damage threshold [16,820]. In addition, the rate at which the radiation dose is delivered and whether the dose was delivered in fractions or in a single exposure affect the biological response and depend on the process and cell type in question. In general, increasing the dose rate over the range of 1 to 100 cGy/min results in an enhancement of the effectiveness of the radiation in producing injury secondary to overcoming compensating mechanisms of repair of sublethal damage and cellular proliferation [65,311].

Cells have specific repair processes for DNA and the ability to repair small amounts of damage to cell membranes; therefore, damage that occurs at rates below the cellular capacity will not be as cumulatively injurious. LET is another factor influencing damage. High LET radiation produces more ionizations per unit of matter and more localized damage, making repair difficult. Larger radiation doses given in single treatments tend to cause more injury than the same doses given in fractions or over a more prolonged period. Injury to early responding tissues (hematopoietic stem cells, intestinal crypt stem cells, and many tumors) is reduced by prolonging the duration of the exposure (or, in the case of cancer therapy, by extending the treatment over days or weeks), while the total dose and fraction size are important to the degree of an effect on late responding tissues (muscle, connective tissue, and nerve tissue) [311]. Internal exposure is also a consequence for alpha and beta emitters; for example, $^{214}$Po, $^{218}$Po, $^{222}$Rn, and $^{236}$U cause alpha exposure to lung alveoli [68,69].

A number of external factors combined with radiation may influence radiation-induced responses when present at the time of radiation [312,791]. Effects on pulmonary function and on carcinogen activity of radionuclides can be affected by the presence of smoking, dust, asbestos, and potentially genetic subtypes [791]. There are experimental reports of magnetic fields resulting in synergistic growth reduction on cells exposed *in vitro* and radioprotection of mice to whole body radiation [791], but the data are insufficient to extrapolate a harmful effect under general occupational exposures. Transmutation produces both radiation exposure and chemical instability. The transmutation of carbon-14 in a key structural position to nitrogen causes mutations in *Drosophila* and in mice [67]. Transmutation of tritium also causes mutations; however, the major effect of a transmutation is the radiation exposure [67,69]. The ability of beta emitters to produce localized irradiation has been successfully applied to the use of $^{89}$Sr and samarium-153 to treat painful bone metastases in patients with breast cancer, prostate cancer, and multiple myeloma. These compounds are incorporated into bone in a manner similar to calcium and are concentrated at sites of increased bone remodeling/growth activity, such as a fracture or a cancerous lesion in bone. The localized radionuclide then releases radiation to the surrounding area and tumor as it radioactively decays. Similar therapies are under development to bring radioactive nuclides in contact with soft tissue tumors by attaching the radionuclide to a monoclonal antibody directed to the tumor. The FDA has approved Y-90 ibritumomab tiuxetan and I-131 tositumomab for the treatment of refractory or relapsed CD20-positive non-Hodgkin's lymphoma. Each of these compounds contain a monoclonal antibody to the CD20 antigen found on the non-Hodgkin's lymphoma cells (and also on normal B-cells) which binds to the tumor and brings the radioactive particle in close contact. Yttrium-90 and I-131 both undergo beta decay. I-131 is also used in the treatment of thyroid cancer. Population exposure to nonmedical radioactive particles through the food chain is also a potential environmental problem. The incorporation of $^{90}$Sr and $^{131}$I represents a concern for human exposure from nuclear weapon fallout and nuclear power plant accidents, such as Chernobyl.

Biochemical responses may be modified by the presence of either radioprotective or radiosensitizing agents [278,820,822]. Substances may induce radioprotection by several mechanisms, including hypoxia, free radical scavenging, immunomodulation, hematopoietic and intestinal stem cell recovery, and modulation of the cell cycle [279]. Reversal of these same processes is radiosensitizing. Cells exposed to ultraviolet radiation or hyperthermia are more sensitive to ionizing radiation, while animals exposed to hypothermia are more resistant to ionizing radiation. Radiation may affect core body temperature through prostaglandin synthesis and histamine release [383]. Numerous biological processes, enzymatic rates, and fidelity are temperature dependent. Even the normal spontaneous depurination of DNA can be accelerated by elevations in temperature, and this could affect mutations. Irradiated tissues remain sensitive to extreme temperature variations. Cold temperatures can accelerate the degeneration of irradiated skin [753]. Dietary modifications may influence radiation responses through mechanisms including reduction of growth stimuli, free radical scavenging, and immune deficiency, while the malabsorption seen in a number of populations, in chronic illnesses (including cancer), and in alcoholics would be an adverse modifier [791]. Important toxic metals with environmental influences on humans include arsenic (lung cancer), antimony, beryllium (lung disease), cadmium (renal cancer), chromium (lung and stomach cancer), lead, mercury, nickel, vanadium, and zinc [326,478,791]. These may modify biochemical processes including enzymatic activity, free radical formation and damage repair, and the physiological half-life clearance of other radioactive metals (potentially increasing radiation exposure) [791]. Experiments with cadmium, beryllium and lead have shown sub- to superadditive effects with radiation in cells in culture and in mice, and air arsenic concentrations are related to lung cancers in miners, but studies have not shown clear superadditive effects with radiation in human populations [791].

An interesting observation that may ultimately prove useful in the development of a radioprotective agent for the medical, civil defense, and space environments is the paradoxical nature of many of the biological mediators such as the cytokines, prostaglandins, leukotrienes, histamines, and serotonin [312,589,819,821,822]. Their release following radiation exposure plays a key role in the biological amplification of injury; yet, when administered before radiation, they are the most potent "protectors" of the naturally occurring biological substances. The degree of protection afforded by a radioprotective agent is described by its dose reduction factor (DRF) or dose modifying factor (DMF). Both values are ratios greater than 1 for protection and represent the radiation exposure in the presence of a radioprotective agent to produce a given biological response divided by the radiation dose required without the agent. Radiosensitizing substances are described by the sensitizer enhancement ratio (SER). It is a positive value representing the dose of radiation required to produce a given biological endpoint without a protective agent divided by the radiation dose in the presence of the radioprotective agent. Many of the chemotherapeutic drugs (e.g., adriamycin, *bis*-chloroethyl nitrosourea, bleomycin, camptothecan, chorambucil, *cis*-platin, cyclophosphamide, cytosine arabinoside, doxorubicin, epirubicin, fludarabine, 5-fluorouracil, gemcitabine, hydroxyurea, interferon, methotrexate, mitomycin C, vincristine, taxol, tirapazimine, and topotecan) have additive or synergistic responses when given with radiation exposures [141,309, 311,312,443,791]. These medications act through a number of different mechanisms, including alkylation of DNA resulting in cross-links, effects on glutathione levels and free radical formation, antimetabolites (such as folate metabolism), increased DNA breaks, inhibition of DNA synthesis and repair (such as topotecan and camptothecan blockage of DNA replication forks during the S phase), and modification of oxidative stress and signaling pathways [312,443,791]. Genotoxic chemicals have been divided into three classes: activation-independent alkylating agents, metabolism-dependent alkylating agents, and free radical generators [791]. These may be beneficial for tumor therapy but may also adversely affect normal tissues such as heart (adriamycin), lung (bleomycin), bone marrow, and mucosal tissue (e.g., inner lining of the mouth and rectum). Significant advances in oncology have been made using combined modality therapy (chemotherapy and radiotherapy) to improve local control, disease-free survival, and even overall survival. Some patients with laryngeal, bladder, or rectal cancer are able to receive organ preservation therapies involving chemo- and radiation therapy without compromising survival. Patients with nasopharyngeal carcinomas who are treated by concomitant cisplatin/5-fluorouracil chemotherapy with radiation therapy have a greater overall survival than patients who receive an identical course of radiotherapy alone: 78% vs. 47% 3-year overall survival, respectively [9]. The chemotherapy is thought to reduce the tumor volume, radiosensitize the tumor, and provide systemic treatment to prevent distant metastases. Similar increased benefit has been demonstrated in patients with head and neck cancers who receive concurrent postoperative chemotherapy and radiation therapy.

For some drugs, because the mode of injury is similar, there may be a memory response equivalent to or resembling a fractionated radiation exposure rather than drug injury at one time and radiation injury at another. Caffeine and the cardiac glycosides digoxin and ouabain, as well as metoclopramide (an antiemetic agent), are also radiosensitizers. Drugs that produce free radical damage or alkylation of DNA or produce biological responses similar to ionizing radiation are described as radiomimetic [41,311]. Examples include azaserine, benzene, bleomycin, BCNU, chloranbucil, cyclophosphamide, dacarbazine, diethylthiocarbamate, furazolidone, hydrogen peroxide, melphalan, mitomycin-C, neocarzinostatin, nitrofurantoin, nitrosoureas, nitrosoguanidine, ozone, superoxide, streptonigrin, sulfur mustard, tetranodecanoyl phorbor acetate, thiotepa, and the trichothecene mycotoxin T-2 [41,311,312,791].

Alterations in cell populations influence biochemical responses by several mechanisms. Radiation may alter the population of the cell type responsible for producing a particular product, as in the reduction of estrogen that may occur following ovarian irradiation, or it may alter the population of a controlling or modifying cell. These processes may have dire consequences such as prostaglandin production by the bone marrow stromal support cells responsible for maintaining a microenvironment suitable for hematopoietic stem cell development. Studies of bone marrow populations in culture demonstrate that radiation doses that do not kill the hematopoietic stem cells may still halt stem cell progression by injuring the stromal support cells. When stromal synthesis of prostaglandins was inhibited by ionizing radiation treatment, the hematopoietic stem cell development stopped [279]. It resumed when the marrow stromal cells were treated with a 100-Gy dose, killing the stromal cells but stimulating endogenous prostaglandin release. Repopulation kinetics and cell–cell influences play important roles in some forms of radiation-induced congenital anomalies where loss, reduction, or inhibition of one cell or tissue type influences the development of surrounding tissues and organs.

Hormonal control of homeostasis is affected by ionizing radiation and at the same time may affect biochemical responses and even survival to ionizing radiation. Physiological processes may be influenced by electrolyte imbalances, polydipsia, and polyuria following irradiation [820,864]. As mentioned later in the chapter, stress hormone levels of ACTH, glucagon, cortisol, insulin, and growth hormone are altered by ionizing radiation exposure. Radiation acts as a general stressing agent on the body [16,819,820]. Estrogen and testosterone both induce radioprotection when administered to animals, and radiosensitivity has been shown to vary throughout the estrous cycle [819,820]. Correspondingly, testosterone acts as a growth stimulus for prostate cancer cells. The use of luteinizing hormone-releasing hormone agonists that reduce testosterone levels in combination with radiation therapy can improve the success of therapy in some stages of prostate cancer [634,660]. Circadian rhythm and season variations also modify responses. The spontaneous induction rate for congenital anomalies and the ability of radiation to induce anomalies are both enhanced during the winter months [819]. In patients receiving radiation therapy, the major indirect effects of radiation observed are fatigue (with or without an associated anemia), reactions of skin just outside the radiation fields (absorptive reactions), and nausea. The sensitivity of the individual to ionizing radiation is modified by the state of health, individual genetic and physiological variations, and the presence of combined injuries. Patients with medical histories of scleroderma and xeroderma pigmentosum have worse skin reactions to ionizing radiation. Although the lethal radiation dose for humans is approximately 4.5 Gy, a 5% lethality is estimated for a 2-Gy dose [864]. In large radiation exposures, the presence of combined injuries such as infections or pathophysiological injuries, including burns, wounds, or other trauma, plays an important factor in survival [232,791]. Other factors are more difficult to describe. For example, a study on the ability of $^{210}$Po to induce lung cancers in hamsters found a 0% incidence of lung cancers when 40 nCi of $^{210}$Po was introduced by intratracheal administration, but a 5% incidence of lung tumors if the same administration was followed by an injection of saline [453]. Radiation-induced gastrointestinal metaplasia in rats can be increased by administration of 1% sodium chloride solution in the diet and lowered in mice receiving a 10% sodium chloride solution, but neither affected the carcinogenic incidence [828].

Radiation induces a block in the $G_2$ phase of the cell cycle and a prolongation of the S phase [16,311,312,313,676,791,820]. Cells are more sensitive when irradiated during the mitosis and $G_2$ phases of the cell cycle and progressively less sensitive during early and late S phases, where preparation for and synthesis of DNA occurs. The $G_2$ delay is probably related to repair processes. Studies with inhibitors of protein synthesis indicate that radiation inhibits the synthesis of a protein or key proteins necessary to proceed through the cell cycle [820]. Decreased synthesis and phosphorylation of histones is observed after irradiation [608]. When mitosis resumes, there is usually an increase in the percentage of cells undergoing mitosis compared to the unirradiated population due to a partial synchronization and abortive mitoses. There may also be an overcompensation from stimulation of cells normally in the $G_0$ stage back into active cycling. Increasing the radiation dose proportionally increases the mitotic delay time up to a point at which the possibility of radiation-induced cell death becomes a more likely outcome. This delay in resumption of mitosis has significant consequences for repopulation kinetics and potential lethality. There are four major acute radiation lethality syndromes: instantaneous, hematopoietic, intestinal, and central nervous system death [864]. Of these, the hematopoietic and intestinal cell deaths are governed by repopulation of the respective stem cell compartments. Hematopoietic death results in humans from doses between 2 and 7 Gy, with lethality occurring between 2 weeks and 2 months after irradiation. Death is directly attributable to the mitotic inhibition/death of two stem cell populations, one responsible for hematopoiesis and one for marrow support. The formed components of blood, red and white blood cells and platelets have finite lifetimes, 4.5 days for platelets to 120 days for red blood cells. They are continually replaced by the bone marrow but, if marrow production is halted, maturation depletion occurs with hematopoietic progenitor cells in various stages of development. An acute aplastic anemia develops without the continued formation of new progenitors. The duration of mitotic delay is crucial in the recovery process, because a point might be reached at which the pathological consequences of the loss of platelets and white blood cells cannot be reversed by the onset of new hematopoiesis in time to avert death. In theory, survival or transplantation of a single hematopoietic stem cell (with proper stromal support) can repopulate the entire hematopoietic system resulting in survival. Bone marrow transplantation has been used to treat victims of radiation accidents. It is not necessary for the entire marrow to be mitotically inhibited for the acute hematopoietic syndrome to be initiated. Production can be suppressed below a threshold. With a loss of platelets and white blood cells, hemorrhaging and infectious processes lead to death.

The Department of Defense, through the Armed Forces Radiobiology Research Institute (AFFRI), has developed the Biodosimetry Assessment Tool software which can utilize blood lymphocyte counts to estimate potential radiation exposure doses for triage and casualty management in the event of a radiation accident or disaster [733]. The program is available on line through the AFFRI website (http://www.afrri.usuhs.mil). Radiation-induced cell death, mitotic delay, and inhibition also occur in the intestinal crypts of Lieberkuhn after doses greater than 1 Gy and progresses to an acute intestinal syndrome death between 3 and 7 days after irradiation. Intestinal villi cells continue to be sloughed off or die without replacement. This produces a decrease in villi height, with an associated decrease in absorptive area, followed by eventual denuding of the villi surface, loss of electrolytes, and loss of the protective barrier permitting infection and hemorrhaging. Attempts to transplant intestinal crypt cells have not been successful. A person may survive a hematopoietically ethal dose of radiation only to die instead (and at an earlier period) from the acute intestinal syndrome. Agents that are radioprotective for the hematopoietic system may not necessarily affect the intestinal crypt cells in the same manner.

Radiation is unique among environmental mutagens in hat it may affect any of the three stages of tumor devel-opment: initiation, promotion, and latency [67,68,311,312, 791,819,820]. Several chromosomal breakage syndromes, such as xeroderma pigmentosum and ataxia telangiectasia, are associated with deficiencies in DNA repair and are, herefore, more susceptible to radiation-induced muta-ional injury. Cancer induction has been related to the acti-vation of protooncogenes to oncogenic status. Oncogenes may influence the development of cancer by direct stimu-atory effects through altered or amplified gene products or by suppressive activity. In the latter case, the normal unction of the gene acts to suppress cancer, and inactiva-ion or loss leads to expression of transformation of cells nd carcinogenesis. The most studied human suppressor oncogene, p53, is associated with the induction of retino-blastoma, colon cancer, and lung cancers in humans. Dele-ions and point mutations in the p53 gene have been observed in some radiation-induced cancers, including, in one study, 7 of 19 radon-associated lung cancers from uranium miners [797], and 16 of 52 in another study [768]. nterestingly, 31% of the miners with lung cancer in the econd study had a specific transversion of AGG to ATG arginine to methionine) at codon 249 [768]. This codon was not mutated in miners from the first study.

Radiation-induced genetic injury may arise from chro-nosomal loss of a gene or an entire chromosome, dele-ions resulting in codon loss or frame-shifts, and point nutations involving base substitutions that produce amino cid substitutions or stop codons. Radiation-induced nutations have been observed in the p53, c-*myc*, c-*abl*, M, c-*fms*, K-*ras* and H-*ras* oncogenes. Analyses of radi-ation-induced gene inactivation *in vitro* suggest that nduced mutations are more likely to be expressed through DNA deletions or rearrangements than as point mutations 773]. Interestingly, the types of activated K-*ras* point nutations observed in neutron-radiation-induced murine hymic lymphomas were lower in yield and differed from he spectrum of *ras* point mutations-induced by gamma radiation [736]. They also lacked a characteristic point nutation of adenine for guanine in codon 12 that was resent in 87% of the gamma-radiation-induced *ras* muta-ons. Characteristic deletions have been observed for ther radiation-induced cancers, such as the deletions on hromosome 2 in low LET radiation-induced murine mye-oid leukemias [782].

Radiation exposure may alter both enzyme levels and he availability of substrates and cofactors; for example, he increase of prostaglandins (PGs) in mouse spleens following 2-Gy exposures was shown to be related to a decrease in the activity of the enzyme associated with PG degradation, 15-hydroxy PG dehydrogenase [823]. This effect was not observed in irradiated human colon tissues [263]. Prostaglandin synthesis has also been studied in pig skin following x-irradiation [874]. An increase in $PGE_2$ levels occurs within 12 hours after 10-Gy irradiation followed by decreases 24 to 48 hours after irradiation [874]. Interestingly, during this same period $PGF_{2\alpha}$ increases. Both changes were related to an increase in NADPH-dependent $PGE_2$ 9-keto reductase, an enzyme that converts $PGE_2$ to $PGF_{2\alpha}$. In general, $PGE_2$ antagonizes the actions of $PGF_{2\alpha}$, much like prostacyclin ($PGI_2$) antagonizes the platelet-clotting activity of thromboxane. Prostaglandins are synthesized from arachidonic acid by cyclooxygenase-1 (COX1) or cyclooxygenase-2 (COX2), an inducible enzyme that is elevated in certain cancers including breast, colon, and lung [533]. Inducible COX2 expression in the brains of irradiated mice results in corresponding increases in prostaglandin E2 and thromboxane [422,561] and is associated with radiation-induced edema in the brain. COX2 inhibitor administration reduced both the edema and reduced the levels of tumor necrosis factors (TNFs) and several other cytokines [422]. Dexamethasone, an antiinflammatory steroid, is routinely administered to patients receiving whole brain therapy for brain metastases to reduce pretreatment edema associated with the brain tumor as well as potential radiation-induced side-effects. Selective COX2 inhibitors administered to nude mice with murine or human tumors implanted in one of their legs resulted in a 3.64 factor enhancement of radiation-induced tumor growth delay [533]. Similar responses have been identified in other tumor cells *in vitro* and *in vivo* [533,583,800], and clinical trials using COX2 inhibitors in combination with radiotherapy and chemotherapy are in progress.

Changes in the tissue levels of the lipoxygenase pathway are also observed following radiation, including a decrease in leukotriene $B_4$ in the colon of ferrets [250] but elevated synthesis of leukotriene $C_4$ and prostaglandin $E_2$ in murine peritoneal macrophages [744]. Murine peritoneal macrophages also exhibit postradiation elevation of acetyltransferase and acetylhydrolase, enzymes involved in platelet activating factor (PAF) metabolism [744].

Exposure to ionizing radiation results in the induction of a number of immediate early genes and early response genes [7,17,18,66,79,84,230,246,313,584,606,638,676, 682,684,791]. More than 100 stress-inducible genes, or immediate early genes, are induced by ionizing radiation [638]. Induction of gene expression has been observed to occur in both a time- and dose-dependent manner following radiation exposure to cells *in vivo* and to animals exposed to whole body radiation [17,18,84,791]. Immediately after radiation, the messenger RNA expression of several genes, including BAX, BCL-2, CDKN1A (CIP1/

WAF1), and GADD45A, occurs, and they can be quantitated by a multiplex real-time reverse transcriptase polymerase chain reaction assay [17,19,84]. The responses were quantifiable between 2 and 50 cGy for CDKN1A, between 20 and 200 cGy for DDB2 [18], and between 25 and 100 cGy for BAX, DDB2, and GADD45A in another study [84]. The responses for CDKN1A [18] and DDB2 [18,84] expression were linear with dose. The responses of some inducible genes vary with repeated exposure. Microarray hybridization to assess the expression of multiple genes may provide a rapid quantifiable system for estimating radiation exposure in the event of a radiation accident [17] and has been examined from the white blood cells of patients receiving whole body radiation therapy [19]. Many of the early response genes can also be induced by oxidative stress and free radical production [313,746] and are associated with transcription (including c-*fos*, c-*jun*, and NF-κB), translation, cell cycle regulation, apoptosis, carcinogenesis, and immunosuppression [7,17,18,313]. Several are protooncogenes, including c-*fos*, c-*Ha-ras*, c-*jun*, and c-*myc* [313,638]. In addition, radiation exposure produces biological mediators and second messengers, such as the cytokines α-interferon, interleukin 1 (IL-1), interleukin 6 (IL-6), tumor necrosis factor α (TNFα), and transforming growth factor β (TGFβ) [28,52,313,675], which in turn can induce specific gene transduction [313]. Some genes are downregulated, producing postirradiation consequences. The decreased production of cyclin B following radiation exposure is an underlying factor in radiation-induced mitotic delay [208,313,353]. Cyclin A and topoisomerase II-α also decrease following radiation exposure [208]. Heme oxygenase 1 is an oxidative-stress-related protein responsible for the breakdown of heme, in the process releasing iron and carbon dioxide [194]; increased expression follows irradiation *in vitro* [677] and *in vivo* [194], and its expression in the kidneys of irradiated rats is related to angiotensin II stimulation.

A number of important biological mediators amplify or elicit inflammation, fibrosis, or biological endpoints. Among these mediators are hormones, histamine, serotonin, leukotrienes, prostaglandins, phospholipids, cyclic nucleotides, and cytokines. Each of these may have unique or cooperative interactions that may result in injury or death or may even be necessary for recovery processes. Histamines and prostaglandins are mediators of radiation-induced skin erythema [111,820]. The cytokines are a diverse class of proteins produced by lymphoid cells and other tissues that mediate cell–cell interactions in a hormonal manner. These proteins include the IL-1 through IL-12, TNF, granulocyte/macrophage colony-stimulating factor (GM-CSF), stem cell factor, TGFβ, and the interferons, to name a few. These molecules act through specific receptors and have important physiological and pathological functions and responses. Lung pneumonitis and fibrosis after irradiation are in part related to elevation of TGFβ and IL-1α, both of which are capable of stimulating collagen gene expression and fibrosis. Prolonged plasma elevation of TGFβ in patients receiving radiotherapy has been associated with increased risk of developing pulmonary toxicity [28,114,589].

Depending on the situation and the cytokine interactions, many of these compounds are able to elicit radioprotective properties or radiosensitization [589]. TNF, stem cell factor (SCF), IL-1, and IL-12 are radioprotective in mice when administered individually 18 to 24 hours prior to irradiation, while IL-6, interferons α and β, and TGFβ act as radiosensitizers [589]. The endpoint of study is important, as administration of IL-12 to mice results in radioprotection of hematopoietic tissue but gastrointestinal radiosensitization. Other cytokines, including granulocyte colony-stimulating factor (G-CSF), keratinocyte growth factor, and TGFβ, are being examined in terms of their wound-healing effects and are undergoing clinical trial tests in patients receiving radiotherapy to minimize the injury to normal tissues while not affecting the tumor tissues [786]. Other agents that modify cytokine activity, such as lisofylline reduction of inflammatory cytokines, are also being evaluated to minimize radiation toxicity [786]. G-CSF has been used to increase white blood cell production in patients who have become neutropenic to combinations of ionizing radiotherapy and chemotherapy. Thrombopoietin has been isolated and is being investigated in clinical trials. It, too, should have advantages in those patients with radiation-induced bone marrow suppression (but not ablation), but its use will have to await appropriate clinical trials and FDA approval.

Biological processes may also be altered through changes in receptor expression [820]. Prostaglandin, leukotriene, β-adrenergic, histamine, serotonin, alpha IIb β3 integrin, and [$^3$H]corticosterone receptors have been studied in irradiated cells or animals. Radiation exposure has no effect on the specific binding of leukotriene $C_4$ at doses up to 20 Gy [822]. Adrenergic receptors play important roles in homeostasis through the autonomic nervous system. Radiation induces a decrease in adrenergic receptors in both the rat and rabbit during the first week after irradiation [441,777]. In rats, this is followed by an increase in β-adrenergic receptors apparently associated with later developing, radiation-induced congestive heart failure [441]. Changes in receptor expression can be related to increased production or to increased use and destruction. An exposure of 2 Gy (a common daily fraction size in clinical radiation oncology) has been shown to increase the mRNA synthesis of the epidermal growth factor receptor in cells irradiated *in vitro* within the first 24 hours [698]. The serotonin$_3$, or 5-HT$_3$, receptor is active in radiation-induced emesis, and specific receptor antagonists (dolasetron, granisetron, and ondansetron ) have been successful in reducing emesis in radiotherapy patients

Administration of 5-$HT_3$ antagonists to ferrets reportedly stopped radiation-induced emesis within 30 seconds of administration [77]. Antagonist to neurokinin 1 (substance P) have also been used to treat or prevent radiation- and chemotherapy-induced nausea in animals [95], and one, aprepitant, has been approved by the FDA for chemotherapy-induced emesis. Some radiation-induced receptor changes may have dire consequences; for example, the increased expression of integrin receptors by B16 murine melanoma cells following irradiation is associated with increased lung colony forming ability, or metastasis [607]. Expression of the integrin receptor was detected within 15 minutes after irradiation, with downregulation occurring by 4 hours after irradiation. Cellular adhesion is controlled by several protein systems, including the integrins, which are a family of cell adhesion receptors [171] involved in cell adhesion and in transmembrane signaling processes. The adhesion of cells results in radioresistance through intregin-mediated cell signaling, and radiation exposure results in increased integrin subunit synthesis [171]. Integrin signaling involves focal adhesion kinase and integrin-linked protein kinase activity. Integrin-linked kinase can inhibit GSK-3B-dependent cyclin $D_1$ proteolysis, resulting in upregulation of cyclin $D_1$ and cell cycle inhibition [171]. Cyclin $D_1$ can be stimulated through *ras* and MAPK pathways [171].

Several important signal transduction pathways and cascades are variously activated by ionizing radiation with important resultant modifications of cellular physiology [380,394,699,746]. Among these are the stress-activated protein kinase (SAPK) pathway, mitogen-activated protein kinase (MAPK) pathway, protein kinase C, modifications of intracellular calcium with corresponding activation of calcium dependent enzymes, bioactive lipids synthesized or released by phospholipases, and the nuclear transcription factor NF-κB [204,394,699,700]. Activation of the MAPK pathway with corresponding stimulation of mitogenic proliferation is thought to account for the accelerated repopulation of irradiated cells which affects the sensitivity, radioprotection, and response of cells (including tumor cells) to chronic low doses, as might occur environmentally or accidentally (e.g., Chernobyl), or fractionated doses of ionizing radiation that are routinely used clinically [699]. The steps involved in the activation of this pathway by ionizing radiation are illustrated in Figure 18.4. This pathway can be activated by binding of the receptors on the membrane cell surface to their ligands (including TNF and other cytokines released or elevated by ionizing radiation) [394] or intracellularly through reactive oxygen intermediates [394,746], changes in intracellular calcium, and phosphorylated activation of the epidermal growth factor receptor [699]. The epidermal growth factor receptor (EGFR) can be activated by ionizing radiation with doses as low as 0.5 Gy and indirectly following radiation by TNFα released from the cell membrane [204].

When MAPK kinase has been activated, during the following period of 8 or more hours the pathway may respond to receptor activation but does not respond to repeat ionizing radiation induction [204]. The MAPK kinase pathway tends to inhibit apoptosis, while TNF can influence apoptosis by the activation of procaspases; however, the response in a particular cell can be influenced (as might be expected) by cell type, radiation dose, and interactions with other signaling pathways, with the sum governing the response. Nucleoside excision repair enzyme and x-ray cross complementing group 1 protein participate in DNA repair following radiation exposure and can be elicited by radiation-induced activation of the MAPK kinase pathway and indirect activation of the pathway by TNFα release in DU145 human prostate cancer cells *in vitro* [204,853]. Inhibitors that block MAPK resulted in increased radiosensitivity and DNA damage [853]. There is considerable research interest in modifying or blocking EGFR activity through antibodies (such as cetuximab) or by tyrosine kinase inhibitors (gefitinib), with resultant increased tumor radiosensitivity. Inhibition of EGFR phosphorylation *in vitro* has been shown to block ionizing radiation-induced cellular proliferation [699]. In addition, use of specific inhibitors of MAPK *in vitro* resulted in an enhancement of double-stranded DNA breaks and reproductive cell death [124]. NF-κB, a nuclear transcription factor, is one of the early response genes induced by ionizing radiation exposure [313,394]. Its regulation is dysfunctional in people with the genetic disease ataxia telangiectasia, and it is thought to account for their increased sensitivity to ionizing radiation [394]. An ataxia-telangiectasia-mutated (ATM) gene participates in DNA repair and recombination, suppression of apoptosis, and induction of p53 [676]. The mutated gene is deficient in these functions.

## MOLECULAR AND CELLULAR EFFECTS

### EFFECTS ON ENERGY SYSTEMS

Radiation exposure affects biological energy processes by specific and nonspecific interactions. Nonspecifically, absorptive functions in the stomach and colon may be altered, in addition to prolonged gastric emptying times [220], induction of nausea and vomiting, increased stool frequency, loose stools or diarrhea, fatigue, or loss of appetite. Radiation affects carbohydrate metabolism, as mentioned earlier, by the release of stress-related hormones and by reducing the levels of the thyroid hormones triiodothyronine and thyroxine [820]. The reduced thyroid activity is also reflected by an increase in thyroid-stimulating hormone (TSH). In humans receiving radiotherapy, a single dose of 7.5 Gy results in a 35% incidence of elevated TSH [734]. Subclinical hypothyroidism is more common (normal thyroid hormone level, elevated TSH),

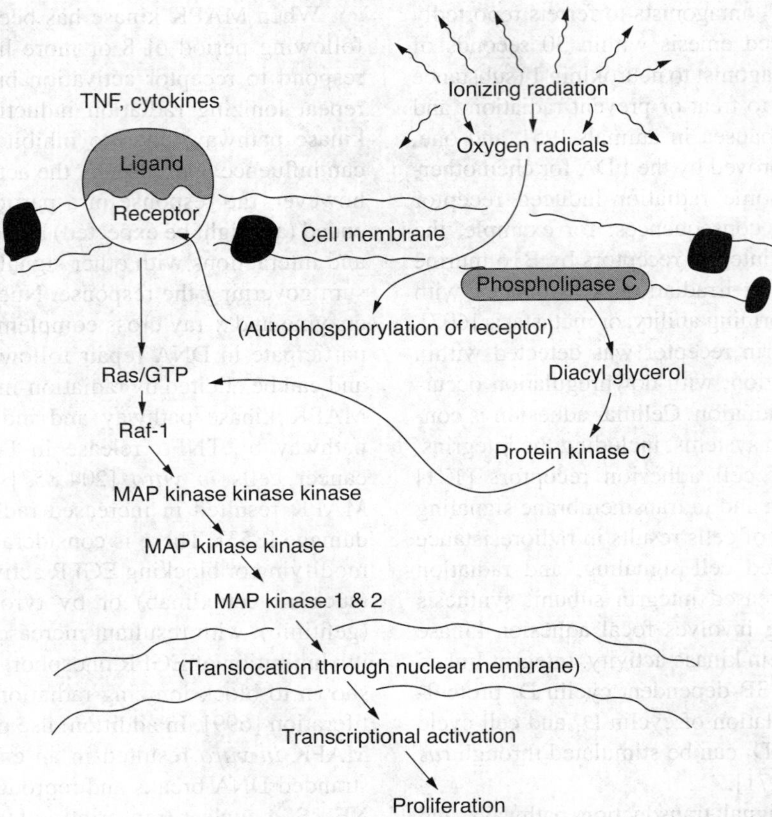


**FIGURE 18.4** Mitogen-activated protein (MAP) kinase signal transduction pathways can be stimulated in response to ionizing radiation. Radiation exposure results in signal transduction activation of the MAP kinase pathway. The MAP kinase pathway can normally be elicited by cytokines, including tumor necrosis factor (TNF). The best characterized receptor-mediated initiation has been the radiation-induced activation through the epidermal growth factor receptor (EGF) by autophosphorylation. The exact mechanism is under investigation and may involve reactive oxygen intermediates, intracellular calcium mobilization, or, indirectly other biological mediators (e.g., diacyl glycerol is released from the phospholipid membrane through the action of phospholipase C). The pathway involves activation of several kinases (phosphorylation activation) and can also be inhibited by phosphatase activities. The resultant pathway action is cell proliferation and differentiation. (Data from Herman and Panksepp [329], Nam et al. [584], and Patchen et al. [619].)

and, like clinical hypothyroidism, it responds to thyroid hormone (levothyroxine) supplementation.

Specifically, radiation exposure causes an uncoupling of nuclear and oxidative phosphorylation with corresponding decreases in oxygen-dependent adenosine triphosphate (ATP) synthesis [16,820]. It has been suggested as a mechanism for radiation-induced interphase death [409]. This effect has been observed in the rat spleen with doses as low as 1 Gy [799] and as early as 15 minutes after irradiation [50]. It is observed in the spleen, liver, and thymus but is not present in all tissues. The decrease in oxidative phosphorylation is preceded by a decrease in $NAD^+$ levels, and the decreases in ATP are preceded by reductions in adenosine monophosphate (AMP) and adenosine diphosphate (ADP). The losses are attributable to decreased synthesis, use of remaining stores, and increased leakage through the cell membrane. In nuclei, macrophages, and neutrophils, a specific membrane (NADPH oxidase) consumes molecular oxygen to produce superoxide but depletes NADPH in the process. The depletion of NADPH, coupled with radiation-induced inhibition of glucose-6-phosphate dehydrogenase, results in the uncoupling of oxidative phosphorylation.

Nicotinamide adenine dinucleotide (NAD) concentrations also decrease following irradiation. Decreases are observed within 15 minutes after irradiation and have been elicited with doses as low as 0.25 Gy although doses of 9 Gy or higher are usually required to observe changes [820]. This decrease is related to the activation of adenosine 5′-diphospho-5-β-D-ribosyl transferase (ADPRT) by double-stranded DNA breaks. ADPRT uses the ADP contained in NAD to form a poly(ADP-ribosyl) chain that attaches to histones and to several DNA repair enzymes, including DNA ligase II, $Ca^{2+}$- and $Mg^{2+}$-dependent endonuclease, and protein elongation factor 2 (PEF2). Although its specific function is not known, poly(ADP-ribosyl) therefore affects many biological processes including DNA repair and cell growth and differentiation. It is also important in some infectious processes since cholera and diphtheria toxins have ADPRT activity. The poly(ADP-ribosyl) binds to the histones, initiating a relaxation of chromatin and facilitating DNA repair. As long as double-stranded DNA damage is present, ADPRT will remain activated and the poly(ADP-ribosyl) chain will remain attached to the histones, preventing condensation of DNA necessary for metaphase. A suicide model of cell death has been proposed in which extensive DNA damage leads to activation of ADPRT with the subsequent use and depletion of NAD [75]. The decrease in NAD for electron transport results in a decrease in ATP synthesis, leading to a decrease in other biosynthetic and active transport processes with consequential interphase cell death. Most cells can maintain sufficient NAD concentrations to avoid this complication with radiation doses under 20 Gy. Because poly(ADP-ribosyl) participates in DNA repair, where inhibition would be radiosensitive, and in this cell death model, where inhibition would be protective, the use of ADPRT inhibitors such as caffeine, nicotinamide, and 3-aminobenzamine has produced mixed results [820].

Human immunodeficiency virus (HIV) activation is not a major response to ionizing radiation exposure; however, both ultraviolet radiation [875] and x-irradiation can stimulate *in vitro* replication of the HIV responsible for the acquired immunodeficiency syndrome (AIDS) in humans [594]. A 200% increase in viral replication was observed following a 1.5-Gy x-irradiation dose. The increase was thought to be mediated by a cAMP-dependent process. Much larger doses of either ultraviolet or x-irradiation are required for inactivation [328]. The $LD_{37}$ dose was 4.5 kGy. In another study, transcription of the HIV promoter could be activated in cells *in vivo* by doses as low as 25 cGy [235]. Irradiation of allograft tissues prior to transplant has been proposed as a method of inactivating potential latent HIV contamination; however, active virus could still be identified after irradiation of 50 kGy in one study [737], and the higher doses compromise tissue integrity for transplant. Radiotherapy remains an effective means of treating Kaposi sarcomas, lymphomas, and other radiosensitive tumors in AIDS patients [170]. Radiation can also cause activation of the herpes zoster virus responsible for chicken pox and shingles in humans. Reactivation of the virus (shingles) has been reported in 16% (up to 50%, in some studies) of patients receiving radiation therapy for Hodgkin's disease [299]. Most cases occurred within 8 months following completion of radiation therapy, and reactivation was most common in patients receiving chemotherapy and radiotherapy, and in children.

## The Target

The target for radiation-induced cellular injury depends on the endpoint examined. The membrane appears to be the target for some of the behavioral alterations following ionizing radiation, while the chromosome material is thought to be the target for cell lethality. This is supported by the results of microsurgical techniques on irradiated ameba to transfer the nucleus to unirradiated ameba and to transfer an unirradiated nucleus to the irradiated cytoplasm. Significantly greater lethality followed the irradiated nucleus. The unirradiated ameba that contained the irradiated nucleus behaved as if the entire nucleus had been irradiated, while the ameba with the irradiated cytoplasm containing the unirradiated nucleus required much higher doses to the cytoplasm for lethality [296].

In a classic set of experiments, Munro [575] irradiated either the cytoplasm or the nucleus of Chinese hamster fibroblast cells using a needle made of $^{210}Po$, an alpha emitter. When the needle remained in the cytoplasm, away from the nucleus, the alpha radiation doses of 250 Gy were dissipated in the cytoplasm with no effect; however, when the needle was placed close to the nucleus, cell lethality resulted from much smaller doses. Finally, cell lethality is directly proportional to the number of chromosomal aberrations observed after irradiation. They may be detected by several methods, including phytohemoagglutinin (PHA)-stimulated peripheral blood lymphocytes, premature chromosomal condensation, micronuclei techniques, and fluorescent *in situ* hybridization [820]. The result is influenced by radiation factors, degree of time following exposure to ionizing radiation, culture time, and age of the individual. The premature chromosomal condensation technique can be assayed in 2 hours, while the other assays require 48 to 52 hours to allow the stimulated lymphocytes to grow sufficiently for assay. The basic technique permits growth of the blood lymphocytes to provide a sufficient number of cells to conduct the assay. The complete assay time including culturing the blood lymphocytes takes 3 to 5 days.

Other experiments indicating that the DNA is the main target for radiation injury include those conducted with the halogenated pyrimidines 5-iododeoxyuridine and 5-bromodeoxyuridine [311,359]. These compounds are incorporated into the DNA and increase cellular radiosensitivity by increasing susceptibility to free radical attack (including the hydrated electron) and damage and by modification of DNA repair [311,359,375]. Cells in exponential growth are more sensitive than cells in plateau growth containing equivalent amounts of base substitution. Experiments incorporating $^{125}I$-iododeoxyuridine into cellular DNA or iodinated proteins into cell membranes further demonstrated that the nucleus, rather than the plasma membrane, is the target for the radiation-induced chromosomal instability that occurs several cell divisions to later after-radiation exposure [384].

## Bystander Effect

In cells *in vitro* and *in vivo*, bystander effects have been described in cells that were not directly irradiated, and these effects could be induced by low doses of high LET and low LET radiation [42,43,105,461,492,563,564,567, 568,716]. This has implications both for environmental radiation toxicology (low-dose alpha particle radiation effects are particularly important, given the role of environmental radon gas exposure as a etiological factor in lung cancer) and for therapeutic modification to enhance the radiation effects on tumor cells for cancer therapies. Bystander effects, which occur when a nonirradiated cell is influenced by an irradiated cell, can be induced or postulated by a number of mechanisms, including reactive-oxygen-induced radicals (also referred in the literature as reactive oxygen species) [531,852], reactive nitrogen species [531,716], carbon monoxide produced from radiation-induced heme oxygenase 1 [531,577], gap–junction interactions [43,531], and cellular mediators, including cytokines [469,531,586,610] and hormones. Irradiation of the cytoplasm of Chinese hamster ovary (CHO) cells *in vitro* had a minimal effect on cell lethality (90% survival) but increased the mutation rate at a specific CD59 test locus from 43 per 100,000 surviving cells to 125 in cells receiving 8 particle transversals [852]. The effects of cytoplasmic irradiation on the mutation rate could be eliminated by irradiation of cells in the presence of dimethylsulfoxide (DMSO), which is used as a free radical scavenger [852]. The mutant spectra were different when the nucleus was irradiated. The induction of sister chromatid exchanges in CHO cells in another experiment using plutoniun-238-induced alpha particles found an increased frequency with 31 millirads, although only 1% of cellular nuclei would have been exposed [580].

These observations have been further expanded to demonstrate a bystander response to nonirradiated cells. In an elegant experiment at the Gray Laboratory in England, charged helium ions were directed at specific glioma cells in a mixed culture also containing fibroblasts [716]. The cells were plated at a low density to minimized cell/cell contact (and hence gap junction interaction]. The radiation was sufficiently specific to target either the nucleus or the cytoplasm of single cells and resulted in DNA damage in bystanders cells assessed by micronuclei formation. A 36% increase in micronuclei detection was observed 50 hours later in surrounding nonirradiated glioma cells, and the proportion was the same whether the cytoplasms of one or ten cells were irradiated. The response was not cell specific, as a 78% increase in micronuclei detection in the surrounding unirradiated fibroblast cells occurred when the cytoplasm or nuclei of the glioma cells were targeted.

The bystander effect was prevented or eliminated by the presence of the nitric oxide scavenger cPTIO (2-(4-carboxyphenyl)-4,4,5,5-tetramethyl-imidazoline-1-oxy-3-oxide) during radiation or filipin, which modifies cytoplasmic membrane stability. Nitric oxide production doubled in the unirradiated bystander cells even when the cytoplasm of a single cell was irradiated [716]. Other studies have identified a nitric-oxide-related soluble factor released into the culture medium of irradiated cells. Conditioned culture medium from irradiated human glioblastoma cells *in vitro* resulted in p53 protein and nitric oxide synthetase expression in nonirradiated cells [492]. The effect was eliminated by treatment of the medium with nitric oxide scavengers and inhibitors to nitric oxide synthetase. Irradiation of the glioblastoma cells in medium from the previously irradiated cells as opposed to fresh medium resulted in a radioprotective effect—less cell killing thought mediated by the nitric-oxide-mediated induction.

Bystander effects are also mediated through direct cell-to-cell connections by gap junctions between the plasma membranes of adjacent cells [43,531]. Gap junctions are formed by connexin proteins (of which connexin 43 is the most common), and their organization, stability, and channeling functions are modified by phosphorylation [421]. Human diploid skin cells irradiated by alpha particles to such a low doses (0.16 cGy) that the nuclei of only 1% of cells would be transversed by the radiation (i.e., 99% not exposed) showed DNA micronuclei damage, and induction of p21waf expression in cells was increased 2.5-fold [43]. The expression was evident in more cells by staining than would be expected from an exposure of 1% of the cells to ionizing radiation, implicating a cellular communication process. The bystander effects could be reduced by lindane, which blocks gap junction interactions and was absent in cells lacking gap junctions.

## Cellular Responses

There are three basic categories of cellular damage and eight basic responses of cells to ionizing radiation. The basic categories of damage are sublethal damage, potentially lethal damage, and lethal damage. Potentially lethal damage is radiation-induced damage that would result in cell death if not repaired [82] and is affected by postradiation therapy or environment [311]. The basic responses of cells to ionizing radiation include no visible response mitotic delay, transformation, hyperplasia without accompanying cell division leading to giant cells, instant cell death, interphase cell death, reproductive cell death, and apoptosis.

The cell cycle describes the process of cells preparing for and undergoing cell division. Some cells are in a resting phase, termed $G_0$, and are not actively dividing, or preparing for division. Cells go through a $G_1$ (gap 1) growth phase prior to initiating the S phase (synthesis), during which the chromosomes are replicated. This is followed by a second gap ($G_2$), in between DNA synthesis and cell division (mitosis). Classical radiobiology studies of synchronized

**TABLE 18.7**
**Role of Cyclins in Radiation-Induced Cell-Cycle Response**

| Cyclin | Dependent Kinase | Cell-Cycle Phase | Radiation Effect |
|---|---|---|---|
| Cyclin A | cdk2 | S | $G_1$/S block |
| Cyclin B | cdk1 | $G_2$/M | $G_2$/M block |
| Cyclin C | cdk8, cdk9 | $G_1/G_2$, $G_2$/M | (?) |
| Cyclin D | cdk4, cdk6 | $G_0$, $G_1$ | $G_1$ block |
| Cyclin E | cdk2 | $G_1$/S | $G_1$ block |
| Cyclin G | cdk5 | $G_1$/S, $G_2$/M | $G_2$/M block |
| Cyclin H | cdk7 | $G_2$/M | (?) |

cells in culture have shown that cells tend to be most sensitive to ionizing radiation when irradiated during $G_2$ and mitosis and to be less sensitive when irradiated in the S phase. Because cells in mitosis and $G_2$ are more sensitive than the other stages of the cell cycle, cells that are rapidly cycling are more likely to be in a radiosensitive phase of the cell cycle than cells that are slowly dividing or cells that are removed from the cell cycle ($G_0$, resting phase). This can be particularly important in radiation therapy where differences in radiosensitivity are exploited to maximize tumor kill and patient cure. In fact, agents that affect cell cycle progression or synchronize cells (e.g., hydroxyurea) can be used for radiosensitization.

Molecular and cellular biology studies are providing insight into cell cycle regulation and how radiation is able to affect the cell cycle [313,578,676]. Two key checkpoints in the cell cycle can be inhibited by ionizing radiation: $G_1$/S and $G_2$/M. A third checkpoint also exists in the S phase [312,356,622]. Several sets of key proteins are involved in sensitizing, modifying, and eliciting the checkpoints [622]. The first set of proteins, called cyclins, are specific for particular phases of the cell cycle. They bind to and activate phosphorylating enzymes called cyclic-dependent kinases [622,640,676]. There may be several cyclins within a particular class, and there are specific inhibitors for the activated cyclin/cyclin-dependent kinase complexes. Seven different cyclin classes (A to H) have been identified (Table 18.7).

Radiation results directly and indirectly in the stimulation and accumulation of p53 (Figure 18.5), a tumor suppressor protein with DNA transcriptional and other regulatory roles. There appear to be several pathways, but evidence indicates that the two most important enzymes activated by DNA damage are ataxia-telangiectasia-mutated (ATM) and ataxia-telangiectasia and RAD3-related (ATR) proteins, which are phosphatidylinositol-3-phosphate-related serine and threonine phosphokinases [622,640]. These enzymes phosphorylate p53 (also seen

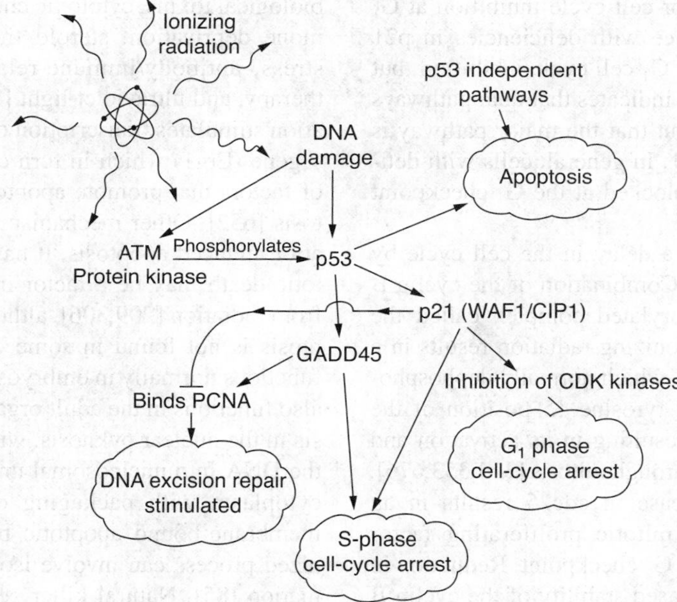


**FIGURE 18.5** Roles of p53 in response to ionizing radiation. Radiation or DNA damage results in activation of the Ataxia telangectasia gene (ATM), a protein kinase that phosphorylates and activates the tumor suppressor/transcriptional activator protein p53. The actions controlled by p53 are shown. p53 induces transcription of p21, a protein that subsequently inhibits kinases regulating cell cycle control. p53 increases levels of GADD45 protein, which can interact with PCNA (proliferating cell nuclear antigen) to stimulate DNA repair (as shown, p21 can inhibit PCNA binding and activity). p53 in turn, is important in mediating apoptosis, but some p53-independent pathways (ceramide signaling) also result in apoptosis.

in the literature as TP53) at serine 15, resulting in activation, and they also phosphorylate breast cancer-related gene 1 (BRCA1). Phosphorylation of BRCA1 at serine 1423 participates in $G_2$/M checkpoint control, phosphorylation at serine 1387 assists in the S phase checkpoint, and phosphorylation at serine 988 can also occur [356]. ATM can also regulate the S phase check point through phosphorylation of the Nijmegen breakage syndrome protein (NSB1) and the structure maintenance of chromosomes protein (SMC1) [640]. ATM phosphorylation of MDM2 at serine 395 inactivates MDM2-assisted degradation of p53, prolonging its biological half-life [640]. Activated p53 in turn induces transcription of the protein p21 (Cip1/waf1, also referred to as CDKN1A). Increased levels of p21 (CDKN1A) inhibit both cyclin E-cyclin-dependent kinase 2 (Cdk2) and cyclin A-cdk2 kinase with a resultant arrest in $G_1$ [224,313,676]. The cdk4/cyclin D kinase activity is also inhibited. In the presence of p21, the cyclin E–cdk2 complex remains in the unphosphorylated and inactivated form. Normally, cdk2 would phosphorylate and inactivate the retinoblastoma protein with resultant progression from $G_1$ phase to S phase. The retinoblastoma protein belongs to a family of three tumor suppressors that also exhibit histone modification enzyme activity and transcription regulation. Inactivation of the retinoblastoma (Rb) protein results in activation of E2F transcription factors controlling synthesis of proteins necessary for cell cycle progression [640]. Cells with p53 deficiencies or mutations are unable to elicit the inhibition of cyclin E/cdk2 activation or cell cycle inhibition at $G_1$ [311,313,676]. Irradiated mice with deficiencies in p21 ($p21^{-/-}$) are still able to elicit $G_1$ cell cycle inhibition, but to a lesser degree [202]. This indicates that dual pathways can result in $G_1$ inhibition, but that the major pathway is through p53 induction of p21. In general, cells with deficient p53 genes do not get blocked at the $G_1$ checkpoint after exposure to ionizing.

Radiation also produces a delay in the cell cycle by blocking the $G_2$ checkpoint. Combination of the cyclin B with cdk1 forms a phosphorylated complex called the mitotic proliferating factor. Ionizing radiation results in a decrease in the cdc25 protein, which normally dephosphorylates the threonine 14 and tyrosine 15 position of the mitotic proliferating factor, resulting in its activation and allowing the cell to proceed through mitosis [313,353,676]. The radiation-induced decrease in cdc25 results in an unactivated/phosphorylated mitotic proliferating factor and, in turn, inhibition at the $G_2$ checkpoint. Reduction in cyclin B transcription, decreased stability of the cyclin B mRNA, and increases in elongation factor 1-delta following irradiation have also been implicated in cell cycle blockage at $G_2$ [313,353,578,676]. In addition, transport of pathway factors out of the nucleus can also affect the cell cycle [622]. Two distinct $G_2$/M checkpoints have been identified: an ATM-dependent process and an ATM-independent process [622]. This is a simplified description of a more complicated interaction involving over- and under-expression of cyclin-dependent kinase inhibitors, upstream regulators, and other integrated pathways. Interestingly, overexpression of cyclins has been identified in several cancers [676].

If the reproductive potential of a cell has been eliminated but other functional components of a cell remain intact, the cell may continue to grow without dividing, forming a giant cell. In interphase, death occurs after an individual or organism receives a dose of radiation with so much resulting damage that the cell dies before it can undergo mitosis and potentially gets stuck in mitosis. The lymphocyte is considered a very radiosensitive cell in humans, among whom doses of 0.05 Gy have been shown to kill lymphocytes [864]. Reproductive death refers to the cellular processes by which a cell dies by inhibition in metaphase within several cell divisions after radiation exposure. Apoptosis, characterized by nuclear pyknosis, cytoplasmic condensation, and cellular phagocytosis with the appearance of apoptotic bodies, requires the synthesis of specific proteins and has been observed in rat thymocytes following exposure to ionizing radiation [855].

Following radiation exposure, some cells die by apoptotic death through activation of a genetically preprogrammed cascade. The apoptotic cascade is not specific to ionizing radiation but has been reported in cells exposed to other cellular stresses, including oxidative stress, increased cellular calcium, hyperthermia, viruses, microbiological toxins, cytotoxic chemotherapeutic agents, hormone deprivation, steroid therapy, cytokine-stimulated stress, antibody/immune-related stress, photodynamic therapy, and ultraviolet light [85,306,395,582,652]. Radiation stimulates transcription of the early growth response-1 gene (Erg1) which in turn can induce the transcription of factors that promote apoptosis and inhibitors of apoptosis [652]. Other mechanisms are also involved in radiation-induced apoptosis. It has been estimated that apoptotic death may be a factor in up to 25% of cells dying from radiation [209,306], although radiation-induced apoptosis is not found in some cell types [306]. Apoptosis functions normally in embryos to reduce cell numbers and also functions in the adult organism. It differs from necrosis in the nuclear pyknosis, with organized degradation of the DNA into nucleosomal units and condensation of the cytoplasm with packaging of apoptotic products into membrane-bound apoptotic bodies [85,393]. This organized process can involve isolated cells, without inflammation [85]. Natural killer cells utilize granzyme B and other apoptotic inducers to kill tumor cells [652].

Ionizing radiation initiates apoptosis by at least two mechanistic routes. The first is by initiation of DNA damage [85,306,395], and the second is mediated through the cell membrane action of sphingomyelinase releasing the second messenger ceramide [147,306]. The pathway mediated through DNA damage requires nonmutated/functional (wild-type) expression of the p53 protein, which controls production of Bax, an apoptotic inducing protein. Cells that have a mutated p53 protein (as many cancer cells have), do not undergo apoptotic death in response to radiation-induced DNA damage and p53 activation but may still be induced to undergo apoptosis through the membrane-induced (ceramide) pathway [147,306]. Bcl-2 and Bcl-$x_L$ are suppressor proteins of apoptosis that are thought to bind to each other and to Bax [85,306], forming either homodimers or heterodimers. The key events in apoptosis are speculated to occur either from homodimerization of Bax or from the decreased availability of Bcl-2 and Bcl-$x_L$ suppressors [85]. Membrane receptor-mediated events can inactivate Bad, another apoptotic promoter that would otherwise bind to and inactivate Bcl-$x_L$. Bcl-2 and Bcl-$x_L$ prevent the translocation of cytochrome *c* from the mitochondria [248].

In the cytoplasm, cytochrome *c* acts as an apoptotic protease activating factor, leading to the activation of cysteinyl-containing, aspartate-specific proteases (caspases, at least 11 types) that mediate the apoptotic cascade, including a caspase-activated DNase [248] and poly(ADP ribose) polymerase, as well as the release of endonuclease G [652]. Activation of caspases 3, 8, 9, and 10 (caspase 10 usually functions with caspase 8) is considered important in the apoptotic cascade, and further differences in activation pathways can be delineated based on whether the initiating processes are caspase 9 dependent, caspase 8 dependent, or caspase 3 and 9 independent (caspase 8 dependent) [307]. Among the ligands that can bind to receptors and activate the apoptosis pathway are the Fas ligand, tumor necrosis factor, and tumor necrosis factor-related apoptosis-inducing ligand (TRAIL) [652]. The membrane-mediated pathway for radiation-induced apoptosis is inhibited by protein kinase C activation [306]. There is interest in tumor therapy to enhance the radiosensitivity of tumor cells through stimulation of apoptosis, either by restoring a normal (wild-type) p53 protein or by inhibiting protein kinase C. Other pathways for apoptosis have been characterized that involve endoplasmic reticulum injury, Golgi apparatus damage, promyeloctic leukemia gene nuclear pathway, and nutritional factors [652]. Further, apoptosis is inhibited by a number of proteins, including the inhibitor of apoptosis proteins (IAPs) and Fas-associated, death-domain-like interleukin-1β converting enzyme inhibitory protein (FLIP), as well as Bcl-2 and Bcl-$x_L$, discussed above. Their levels, distribution, and the cell type involved all interact to influence or inhibit apoptosis [18,652].

The sensitivity of cells to ionizing radiation is influenced by several factors, but the general principles describing cellular radiosensitivity were described for cancer cells in 1906 [76] and are referred to as the Law of Bergonie and Triblondeau. The radiosensitivity of cells and tissues is related to the rate of cell division and the reproductive potential. Cells that are more rapidly dividing and have the potential for a number of successive generations will be more radiosensitive. There are some exceptions to these principles, such as the terminally differentiated lymphocytes, but it may be used to explain why rapidly dividing cells such as hematopoietic stem cells and intestinal crypt cells are more sensitive than nondividing tissues such as heart, skeletal muscle, and nerve cells. The law is usually presented as an explanation for the radiosensitivity of cancer tissues that are rapidly dividing and have unstable chromosome compositions. There may not be sufficient time to repair the chromosomal damage before the next cell division. The expression of injury in some biological systems may require long periods simply because the rate of cell division in a particular tissue is very slow. Even if radiation exposure produced so great a damage as to modify or prevent cell division, it would take years for the cell to reach the point where the division would occur and injury is expressed. The cell cycle time of some ocular cells is 3 years, explaining in part the long latency period for the expression of radiation-induced cataract formation [67].

Prokaryotes and insects also represent contradictions to the Law of Bergonie and Triblondeau. Mammals are much more sensitive; the $LD_{50}$ values for pigs, mice, and humans are 2.5 Gy, 6 to 8 Gy, and 4.5 Gy, respectively [33]. The $LD_{50}$ for *Escherichia coli* is between 20 and 50 Gy, compared to 100 to 1000 Gy for insects and 1000 Gy for ameba [33]. Plant sensitivities range from 4 to 400 Gy; for example, 75 to 90 Gy is lethal for the common chrysanthemum [49]. These differences are explained in part by application of the target theory, in that cells with large chromosome numbers and large nuclear to cytoplasmic ratios tend to be more radiosensitive, in part because they have larger critical target sizes [49]. The chrysanthemum has a polyploidy of 22. Some plant species such as pine trees have radiosensitivities similar to humans, making them ideal biological dosimeters surrounding nuclear facilities [296].

One last interesting aspect of radiobiology is that radiation exposure does not always result in harm [119,357,464,616,792]. Hormesis describes the beneficial responses to biologically harmful agents such as poisons, heavy metals, insecticides, and even chemotherapy agents, including adriamycin (a standard agent used in adjuvant combined chemotherapy regimens for breast cancer) [120]. Radiation has been used to improve the growth of seeds and yield, separately from the production of genetic mutations. It has even been reported to extend the life span of female mice, although it did so by sterilization, which eliminated the repeated stress of pregnancy. Increased life spans following low-dose radiation exposure [464,792] were also observed in crickets exposed to 0.5 to 2 Kr, house flies exposed to 100 to 150 Gy, and mice exposed

**TABLE 18.8**
**Radiation Dose Ranges and Associated Pathophysiological Events**

| | Pathophysiological Events | | |
|---|---|---|---|
| **Dose Range (cGy)** | **Prodromal Effects** | **Manifest Illness Effects** | **Survival** |
| 75–150 | Mild | Slight decrease in blood cell count | Virtually certain |
| 150–300 | Mild to moderate | Beginning symptoms of bone marrow damage | Probable (>90%) |
| 300–530 | Moderate | Moderate to severe bone marrow damage | Possible<br>Bottom third of range: $LD_{5/60}$<br>Middle third: $LD_{10/60}$<br>Top third: $LD_{50/60}$ |
| 530–830 | Severe | Severe bone marrow damage | Death within 3.5–6 weeks<br>Bottom half: $LD_{90/60}$<br>Top half: $LD_{99/60}$ |
| 830–1100 | Severe | Bone marrow pancytopenia and moderate intestinal damage | Death within 2–3 weeks |
| 1100–1500 | Severe | Combined gastrointestinal and bone marrow damage; hypotension | Death within 1–2.5 weeks |
| 1500–3000 | Severe gastrointestinal damage with upper half of range; early transient incapacitation (ETI); gastrointestinal death | | Death within 5–12 days |
| 3000–4500 | Gastrointestinal and cardiovascular damage | | Death within 2–5 days |

*Source:* Azzam, E.I. and Little, J.B., *Human Exp. Toxicol.*, 23, 61–65, 2004. With permission.

to 0.7 rad per day [792]. Increased embryo protection was observed when rainbow trout were exposed to 25 cGy [590]. Improvement or stimulation of the immune system has been observed, including mouse antibody response to tumor cells by 15 rad [20], enhancement of mouse lymphocyte antigen response by 2 cGy [356], and conconavalin-A-induced splenocyte proliferation [356] 4 hours after irradiation by an indirect effect [357]. Indirect effects and roles for reactive oxygen and nitrogen species [357] participate in the hormetic effects on macrophage activity. Studies have also shown a reduction in mutagenic and cancer induction [792]. Mechanisms for hormesis vary with the system under observation but include many of the mechanisms reviewed earlier in this chapter: inactivation of immediate response genes; cytokines, hormones, and other biological response-modifying agents; immune stimulation; and enhanced repair systems [464,792]. Hormetic responses including those to low-dose ionizing radiation exhibit biphasic patterns not consistent with the linear or linear-quadratic models with or without thresholds utilized for public protection recommendations [120,616]. As pointed out in Upton [792] and also reviewed in this chapter, epidemiological studies in human populations exposed to ionizing radiation (medical, occupational, exposure during war, and accidental environmental contamination such as the Chernobyl accident) suggest a risk for cancer development with exposure to low-dose ionizing radiation. Despite the potential beneficial stimulus of the hormetic effects, it is prudent to continue to apply as low as reasonably achievable (ALARA) standards to radiation exposure/protection.

## SOMATIC EFFECTS OF RADIATION

In higher animals, no simple, direct relationship exists between nuclear chromosome volume and sensitivity to ionizing radiation; rather, the effects of irradiation on specific organ systems are more critical [601].

### Acute Radiation Syndrome

Fewer than 25 documented fatalities worldwide, between 1946 and 1985, can be attributed to radiation accidents [510]. Although exposure of the whole body to lethal amounts of ionizing radiation is very rare, any discussion of the biological effects of ionizing radiation would be incomplete without mentioning acute radiation sickness and the acute radiation syndrome (ARS). Acute radiation sickness is manifest in characteristic clinical sequelae known as the ARS, a combination of syndromes determined primarily by the total radiation dose received, the rate the radiation is delivered, and how the radiation is distributed in the body [864]. Signs and symptoms of the ARS result from injury to bone marrow, gastrointestinal system, cardiovascular system, central nervous system, gonads, and skin. The variation in radiation sensitivity of these tissues causes the signs and symptoms of the ARS to occur in three successive phases: an initial prodromal phase, a later latent period, and the manifest illness phase. The length of each phase may vary directly with the radiation dose, and the time between each phase may vary indirectly with the dose, so at an extremely high dose of radiation the phases will blend, with the latent period disappearing completely (Table 18.8).

The prodromal phase may begin about 2 to 4 hours after doses of 3 to 5 Gy or within minutes after exposure to 45 Gy or higher. The initial prodromal phase is characterized by a combination of gastrointestinal and neuromuscular symptoms such as anorexia, nausea, vomiting, diarrhea, apathy, tachycardia, fever, headaches, insomnia, dizziness, and vertigo. The pathogenesis of the prodromal phase is not known, but several causal factors have been suggested, including direct radiation effects on the central and autonomic nervous systems, disturbance of the endocrine balance, and the production and release of various chemical mediators [211,821]. The latent period, which follows the prodromal phase, is relatively asymptomatic and is believed to be the time between initial cell damage and the interference of radiation with cell renewal in the affected organs [864].

The manifest illness phase of the ARS is classically divided into three major syndromes traditionally known as the hemopoietic, gastrointestinal, and central nervous system (CNS) syndromes [482]; however, the current view replaces the CNS syndrome with the neurovascular syndrome [864]. The hemopoietic syndrome may be encountered after exposure to 2 to 7 Gy, and 1 Gy or more can significantly damage the blood-forming capability of the body. Radiation kills the mitotically active hemopoietic precursors of red cells, white cells, and platelets. Pathophysiological consequences include increased susceptibility to infection, bleeding, anemia, and lowered immunity. Death usually results from hemorrhage and infection [482,510,864]. Radiation exposure above 7 Gy contributes to the gastrointestinal syndrome by inhibiting the renewal of the cells lining the digestive tract. At doses of 3 to 8 Gy, tight junctions between the epithelial cells are disrupted, allowing increased fluid and electrolyte loss and permitting the movement of bacterial endotoxins into the blood. At higher doses (10 to 15 Gy), denudation of the mucosa occurs and death results from dehydration, electrolyte imbalance, and septicemia [510,864]. The neurovascular syndrome is the least understood of the radiation-induced deaths. The syndrome is unique in that death occurs very quickly before damage to the gastrointestinal and hemopoietic systems becomes apparent. Readily obvious CNS signs and symptoms include disorientation, loss of muscular coordination, respiratory distress, apathy, prostration, convulsive seizures, and coma associated with death. Some researchers believe that 50 Gy is necessary for the neurovascular syndrome, and doses above 100 Gy are required for direct damage of the nervous system; however, ionizing radiation exposure modifies electroencephalographic activity [778], and *in vivo* exposure decreases hippocampal synaptic transmission and spike generation [343] at lower doses than anticipated. Details of nervous system effects are given later.

## Radiation-Induced Developmental Effects

Preimplantation *in utero* is never succeeded directly by postimplantation, the two being separated by the time it takes the placenta to develop into a functioning organ [553]. This interval of development takes 1 to 2 days in rodents and 10 to 14 days in humans; therefore, the traditional categories of prenatal development known as preimplantation, postimplantation, and fetal are replaced by conceptus, embryo, and fetus. The term *conceptus* is used for the stage of development lasting from ovulation until the placenta becomes a functioning organ. Even though there is some controversy about the maturation of an embryo to a fetus, most classifications of human developmental stages agree that this metamorphosis occurs in the last portion of the first trimester of gestation [335].

The developing conceptus and early embryo have a variety of rapidly dividing progenitor cells. These proliferating cells are much more sensitive to irradiation than differentiated, nondividing cells, making an organism more radiosensitive during its early stages of development than at any other stage of its life. The conceptus exhibits the lowest $LD_{50}$ of any stage of development and is easily killed by doses of ionizing radiation that would cause abnormalities at later stages of development [99,664]. Rather than a malformed organism developing to term, it dies in the conceptus stage due to irradiation-induced chromosomal damage [99]; thus, it is believed by some that the highest risk of irradiation during this stage is the death of the developing organism rather than teratogenesis [553].

In some animals, exposure to ionizing radiation *in utero* can result in anomalies in every organ system, and the concept has been formulated that irradiation, however small, can inflict damage to the embryo or fetus [664,668]. Irradiation-induced anomalies occurring during the middle stages of development may result in death of the organism or abnormal development of one or more organ systems; however, instead of lethality, morphological abnormalities are associated with irradiation during this time. Exposure during this period may result in gross malformations, growth retardation at term or as an adult, and structural pathology [510,869]. In the human, most major organogenesis occurs during the first trimester of pregnancy, with embryonic death and congenital abnormalities resulting from irradiation exposure during this period [664].

During late organogenesis and in the perinatal period, just before and just after birth, radiation damage tends to be functional rather than structural. Perinatal irradiation with x-rays and $\gamma$-rays (140 to 180 cGy) induces changes in tissue enzyme activity [23], hormone production [225,361], and hemopoiesis [287,552]. The major effects of perinatal *in utero* exposure in humans is seen in the developing central nervous system; neurological damage and behavioral changes are not always obvious in histological examination [99,155,534,551,597,664].

Susceptibility to radiation carcinogenesis is relatively high during prenatal development [514,665,818]. During the last four decades a major concern has been the risk of childhood leukemia and other neoplasms following irradiation *in utero* [69,74,99,287,510,630,819]. Studies of the effects of ionizing radiation on the fetus are extremely important, as there seems to be no biological reason to expect the fetus to be resistant [192]. The embryo may be 50 times more vulnerable than the adult to irradiation-induced leukemia [99], and the risk of a child dying of cancer before his or her 10th birthday may be increased 40 to 60% by *in utero* irradiation [559].

The developing conceptus, embryo, and fetus show high susceptibility to ionizing radiation, and the extent of injuries depends on the stage of development as well as the dose of radiation. Developmental anomalies are induced with doses much lower than previously used to demonstrate anomalies in adults [367,514,536,552]. Indeed, even antepartum dental radiography in pregnant women has been associated with an increased risk for infant low-term birth weight [350].

The risk of leukemia in children with Down's syndrome (trisomy 21) is estimated to be about 20 times higher than in the normal population, and there has been some evidence that preconceptional ionizing radiation is linked to the nondisjunction of chromosome 21 [769]. Clusters of cases of Down's syndrome have been alleged in Germany following the passage of the radioactive cloud from Chernobyl, and experimental results have demonstrated that ionizing radiation may induce nondisjunction in oogenesis and spermatogenesis [807]. Although effects on stillbirths have been inferred from spatial temporal analysis in Europe [697], it still cannot be concluded whether or not congenital malformations are associated with Chernobyl radiation anywhere.

Despite the long latency period of radiation-induced cancer, a dramatic increase of up to 100-fold in the number of childhood thyroid cancers has been observed in the heavily radiation-contaminated areas of Belarus and Ukraine and the Bryansk regions of Russia following the Chernobyl accident [39,680]. A strong relationship is indicated between the thyroid cancer and radiation from the Chernobyl cloud.

The *in utero* developing nervous system is particularly vulnerable to ionizing radiation, with defects of the eye and of spinal development being among some of the more common malformations encountered following early gestational exposure [109,110,283]. Later prenatal radiation exposure may result in dose-related abnormalities of the hippocampus with disorganized and loosely scattered neurons in the CA-1 and CA-3 regions and agenesis of the corpus callosum [529]. Perinatal irradiation, during neuronal migration and differentiation, resulted in delay of migration and severe reduction of neuronal tissue in the cerebral and cerebellar cortex and in the hippocampus [259,260,371]. Postnatal cephalic irradiation of newborn rat pups produced an increase (122%) in noradrenaline activity and a marked decrease [38%] in monoamine oxidase activity in the cerebellum [212]. The irradiation induced 60% reduction in cerebellar weight may account for some of the marked increase (223%) in noradrenaline concentration and increase (206%) in tyrosine hydroxylase activity.

## Direct Effects of Radiation on Reproductive Organs

During a period of approximately 20 years in the early part of the twentieth century, radiation was used in an attempt to increase fertility. Exposure normally was to 1.5 to 2.25 Gy over a period of 3 weeks. These levels apparently had little effect on fertility or on any later conceived children [510]; however, neonatal irradiation of rats [251] and hamsters [767] demonstrated impaired fertility in mature male and female animals. *In utero* irradiation of male rat pups resulted in atrophy of the testes, ventral prostates, and seminal vesicles with a complete disappearance of germinal cells from the testes [759]. Irradiation treatment of children 15 years of age and younger with Hodgkin's disease resulted in azoospermia in many of the mature males and ovarian injury in some of the mature females [609].

The safety of radiation levels is often questioned because the threshold for the effects of ionizing radiation on male reproduction are difficult to predict [704]. Although fully developed sperm cells and primary spermatocytes are relatively radioresistant, the quiescent and proliferating spermatogonial cells of the testis are highly sensitive to ionizing radiation [510,664,802,803]. Germ cell dysfunction is common following testicular irradiation, and a dose-dependent impairment of spermatogenesis with gradual recovery may be seen following doses of up to 6 Gy [318,405,735]. The effects of ionizing radiation on spermatogenesis are normally reversible, with the recovery of fertility predictable. Although an acute irradiation dose of 6 Gy to the testis is likely to produce permanent sterility, conception has occurred for males after years of either aspermic or hypospermic conditions following absorbed doses between 2.3 and 3.7 Gy [510,664]. Following the Chernobyl nuclear reactor disaster, however there has been a widespread fear of damage to the reproductive system, with implications for fertility problems and adverse effects on offspring. A pilot study of 18 salvage workers (liquidators) revealed incomplete genesis of sperm characterized by certain ultramorphological parameters of the sperm head [243]. The frequency of amorphous sperm head shape in the study group was significantly higher than in the local Ukrainian control group of 18 men.

Although the human testis is considered relatively resistant to the carcinogenic effects of radiation [50], occupational radiation exposure of the testis produces significant changes in serum gonadotropins and semen parameters. Testicular irradiation may be associated with elevated plasma levels of follicle-stimulating hormone (FSH) and luteinizing hormone (LH), reduced levels of androgen-binding protein (ABP) and testosterone, and reduced prostate and seminal vesicle weights [286,381, 405,505,632,820]. The decreased ABP levels and the increased FSH levels are associated with Sertoli cell dysfunction, although the FSH increase may be a secondary result of germ cell depletion rather than a direct effect of irradiation. The changes in LH and testosterone levels and the decrease in prostate and seminal vesicle weights are indicative of Leydig cell function impairment.

As a direct effect of ionizing radiation, the human prostate epithelial cells may also display malignant transformation after multiple exposure [424]. After a cumulative x-ray dose of 30 Gy, tumors characterized as poorly differentiated adenocarcinomas developed from prostate epithelial cells. This report may provide the first evidence of malignant transformation of human prostate epithelial cells resulting from direct exposure to ionizing radiation.

Although the mature female reproductive system has no proliferating stem cells, the oocytes are in follicles in various stages of development. Animal experiments indicate that radiosensitivity of the ova depends on the maturity of the follicle [664]. Irradiation depletion of the radiosensitive mature and intermediate follicles will result in periods of temporary sterility followed by fertility due to maturation of surviving immature follicles. Sensitivity varies between species, and temporary sterility can be produced in humans with doses as low as 1.5 Gy [820]. Cases have been reported of irradiated women receiving as much as 6.4 Gy becoming pregnant as long as 2 years later and delivering normal children [510]; however, the estimated dose required to produce permanent sterility in the female ranges between 6.25 and 30 Gy, depending on age of the subject [664], and sensitivity increases with the approach of menopause [820].

Ovarian failure is associated with whole body irradiation [181], abdominal irradiation [824,825], or radiation therapy for cervical carcinoma [498]. Pubertal failure or premature menopause was common in females irradiated in childhood [824,825]. Premenopausal women receiving radiation therapy may produce signs and symptoms associated with menopause, such as amenorrhea, dyspareunia, hot flashes, irritability, and loss of libido [236]. Breast cancer, the most frequent spontaneous malignancy diagnosed in women in the Western world, is increasing in incidence [679]. Exposure of the breast to ionizing radiation is now known to increase the risk of breast cancer, especially for younger women [493,494]. The risk of developing breast cancer is very high in women exposed to ionizing radiation before or during puberty, when the differentiation of terminal mammary end buds and alveolar structures is occurring [724]. In fact, the carcinogenic effects of ionizing radiation on the vestigial male breast may be quite similar to that seen in the prepubertal female breast [644]. The male breast displays an increase in risk for breast cancer with three or more radiographic examinations.

### The Lymphohematopoietic System and Immune Competency

One of the systems most sensitive to irradiation is the hematopoietic system, especially in the bone marrow system [566]. Because of the rapidly proliferating hematopoietic elements of the bone marrow, the hemopoietic syndrome may be encountered following irradiation of 100 cGy or more [558,664]. Approximately 50% of individuals exposed to 300 cGy ($LD_{50/60}$) will die within 2 months. The signs and symptoms result from radiation damage to the bone marrow, lymphatic organs, and immune system. The hematopoietic syndrome is characterized by a depression in the peripheral blood levels of mature erythrocytes, granulocytes, lymphocytes, monocytes, and platelets. Except for lymphocytes, the mature blood cells are relatively radioresistant and function normally in the peripheral blood after irradiation levels that will produce bone marrow damage.

Mature cells in peripheral blood have limited life spans, and the replacement of their functional cell types is dependent on the proliferation of the hematopoietic elements of the bone marrow. One type of stem cell, a pluripotent stem cell (PPSC), has the dual capability of self-perpetuation and differentiation and can meet the demands of the lymphohematopoietic and reticuloendothelial systems. The progeny of the PPSCs have specific functions in the body. Granulocytes are involved in activities against invasive bacteria and are related to the nonspecific immune response, while the cell-mediated and humoral responses of the lymphocytes are related to the specific immune response. The monocyte migrates into specific tissues and differentiates into a macrophage of the reticuloendothelial system. The platelet is a critical element of hemostasis, and thrombocytopenia may result in hemorrhage and purpura; thus, radiation damage to the PPSCs may seriously compromise all of these systems, resulting in hemorrhage, infection, and death.

Depression of the mature cells in circulating blood is dose dependent [510,558]. At radiation doses near the $LD_{50}$, changes in the small lymphocytes can be seen in one hour and lymphocytes may totally disappear from peripheral blood in 2 to 3 days [285,481,490,510,664]. Although lymphocyte depletion may be measured in hours, granulocytes and platelets are depleted over days, and depletion of erythrocytes may be measured in weeks.

The small lymphocytes in lymphoid tissue are some of the most radiosensitive cells in the body, and the small lymphocyte is the earliest to decrease in the peripheral blood following irradiation of humans or animals. The lymphocyte not only shows the most rapid reduction in number but is also the slowest to return to normal [490]. The reduction in lymphocytes leads to impaired specific immune responses and immunosuppression. Regeneration of both the T- and B-lymphocyte levels depends on the lymphoid stem cells, which in turn depend on the PPSCs of the bone marrow. Radiation-induced alterations in the nuclear material of lymphocytes might also have a significant impact on the function of this cell type. Chromosomal aberrations have been found in individuals exposed to radiation from both Chernobyl accident-related cleanup activities [745] and the subsequent fallout [696].

Radiation depresses nonspecific immune responses by reducing the levels of circulating monocytes and granulocytes [558]. Granulocytes serve as the first line of bactericidal defense at wounds, while the macrophage, a progeny of the monocyte, can phagocytize and catabolize foreign substances such as microorganisms and toxins. The macrophage is also involved in the humoral specific immune response by processing foreign substances and presenting them as antigens for recognition. Radiation-induced depression of both specific and nonspecific immune responses through the reduction of monocytes, granulocytes, and lymphocytes is potentially life threatening because of the enhanced susceptibility to opportunistic infections [232,621].

Ionizing radiation induces functional and quantitative abnormalities in the lymphoid cells of both humans and experimental animals [9,292]. Ionizing radiation was found to deplete both T and B murine cells in equal proportions in the spleen [292], but studies of the late effects of atomic-bomb radiation on the immune system showed alterations of the balance and interaction between T and B cells, with a decrease in the T-cell population and an increase in the B-cell population in the periphery [9]. Autoimmune deviations, both humoral and cellular, were observed in residents of areas contaminated from the Chernobyl accident and subjects participating in the clean-up of the accident [38,191,554,726]. Structural and functional changes seen in the kidney, thyroid, and crystalline lens of the eye well may be associated with significant changes in the humoral immunity systems related to ionizing radiation exposure.

Irradiation usually does not depress the platelet (thrombocyte) count significantly in a healthy individual [510,558]. The platelet's life span is 9 to 12 days, and depressions in circulating levels reach an initial nadir 10 days after irradiation. The individual will face problems similar to those of a patient with aplastic anemia: thrombocytopenia, capillary fragility, abnormal bleeding, and purpura. When the platelets reach a critical level, hemorrhage is likely to occur and may result in death of the individual; therefore, platelet transfusion becomes critical for the irradiated individual. Because of the relative radioresistance and long life span of erythrocytes (120 days), circulating blood levels fall slowly without complicating hemorrhage or infection [558]. In the presence of thrombocytopenia and hemorrhage, transfusions may be required. Erythrocyte recovery after irradiation normally follows granulocyte and platelet recovery.

Radiation at the $LD_{50/30}$ dose level (350 to 400 cGy in humans) may kill more than 99% of the critical cells in the hematopoietic tissue [510,558]. At this dose level cytological restoration of the bone marrow begins about 25 days after irradiation in humans. The PPSCs will self-replicate and differentiate to produce hematopoietic progenitor cells. The risk of death from hematopoietic radiation injury depends on the regeneration rate of bone marrow stem cells [354] and the level of medical treatment provided [710]. With bone marrow transplantation and heroic supportive therapy, survival is possible following whole body irradiation as high as 1200 cGy [134,339,510]. Although bone marrow transplantation certainly may be indicated, other clinical support regimens will stimulate hematopoietic regeneration, accelerate recovery, increase the $LD_{50/30}$, and enhance survival [476,477,618,619,620,709].

The most common neoplastic disorders of the hematopoietic system are the leukemias, and several different types of leukemia have been observed in a variety of experimental animals following exposure to ionizing radiation [658,851]. Leukemias account for about 32% of all cancers diagnosed in children, with about 85% of leukemias being classified as acute [871]. The acute leukemias in children may be related to parental occupational exposures to a number of carcinogenic substances or with prenatal and postnatal exposures to ionizing radiation.

Clinical evidence leads to the concept that ionizing radiation can cause leukemia by inducing DNA damage. Two hematopoietic cell lines were subjected to gamma irradiation to investigate the susceptibility of human cells to irradiation at the genetic recombination stage of leukemogenesis [201]. The irradiation induced the formation of fusion genes characteristic of leukemia in both cell lines. The cell lines studied showed differences in susceptibility and frequency at which the different fusion genes were formed. These differences in selectivity may help to explain the differences in risk development of some types of leukemia that have been observed following high doses of irradiation.

## Digestive Tract Dysfunction

The digestive tract includes the esophagus, small and large intestine, and rectum. The mucosa of the digestive tract undergoes continuous stress, and, for its functions t remain unimpaired, it must renew itself rapidly to replace lost cells. This fast turnover supported by a marked mitotic activity makes the digestive tract mucosa extremely radiosensitive. The duodenum is the most radiosensitive region of the digestive tract, followed by jejunum, ileum, esophagus, stomach, colon, and rectum, in order of decreasing radiosensitivity [63].

Surprisingly little has been written about radiation damage to the esophagus [796]. The mucosal cells are characterized by a rapid proliferation rate and a relatively high degree of radiosensitivity. Acute radiation injuries are quite symptomatic with submucosal congestion and leukocytic infiltration, followed by mucosal necrosis and sloughing. Healing occurs rapidly, and in animals with esophageal irradiation the mucosa appeared completely normal one year following 2.5-Gy exposure. Humans that received 7.3 to 7.6 Gy irradiation exhibited narrowing of the esophageal lumen, partial loss of the mucosa and muscularis, and widening of the submucosa for 2 to 8 months following exposure. In spite of the initial radiation injury, radiation therapy for esophageal carcinoma results in very low mortality.

At radiation doses of 7 to 50 Gy, injury to the gastrointestinal (GI) tract inhibits the renewal of the cell lining. The intestinal epithelial stem cell is the target of radiation damage, and the resulting decrease in mitotic activity leads to denudation of the intestinal mucosa, fluid and electrolyte imbalance, and bacteremia [300]. The symptoms of the GI syndrome include lethargy, emesis, diarrhea, dehydration, and sepsis. At doses of 3 to 8 Gy, temporary injury to the tight junctions between epithelial cells of the mucosal lining permits the escape of bacterial endotoxins into the bloodstream. As dose increases, the epithelial lining is more extensively depleted. With doses of 10 to 15 Gy, denudation of the mucosa exacerbates the loss of fluid and electrolytes. Beginning at about 12.5 Gy, early mortality occurs due to dehydration and electrolyte imbalance, with death occurring 4 to 5 days after exposure.

At doses of 1 Gy or more, irradiation of many mammals produces nausea and vomiting, signifying the prodronal phase of the acute radiation syndrome [63,221, 300,403]. Radiation-induced emesis, often accompanied by delayed gastric emptying, may be associated with areas of the brain known as the area postrema and the vomiting center, both located in the medulla [300]. Ablation of the area postrema has been observed to abolish radiation-induced emesis in some mammals [300], and zacopride, an antiemetic, inhibited radiation-induced emesis and suppression of gastric emptying in the monkey [221] and abolished radiation-induced emesis in ferrets [404]. Inhibition of radiation-induced emesis has been achieved also through the use of the selective 5-hydroxytryptamine ($5\text{-}HT_3$) receptor antagonists granisetron [77,353], ondansetron [330,639], and Y-25130 [255].

Functional alteration of the stomach by radiation includes a decrease in the production and secretion of HCl, pepsinogen, and mucus [63] and an increase in serum levels of pepsinogen and gastrin [834]. Histopathological alterations included vasodilation and edema indicative of increased microvascular permeability [112], and marked degenerative features, including atrophic mucosa and ulceration [63,97].

Intestinal mucosa cells originate from a single stem cell type located at the base of the intestinal crypts. As the cells proliferate and differentiate, they move to the tips of the villi, a journey of 3 to 5 days. The differentiated cells are shed continuously from the tips of the villi. Radiation damage to the cells of the crypt leads to the death of some cells and an arrest of mitotic activity of others [24,63,585]. The severely damaged crypt stem cells do not divide and replace the cells lost from the tips of the villi. This results in decreased absorption and allows a ready entry for intestinal flora into the systemic circulation [132,301,613]. Electrolyte transport is also altered in the jejunum and ilium following exposure to ionizing radiation, and this functional change may be related to the decreased mast cells and histamine [473,474].

Intestinal absorption may also be decreased due to vascular concrescence. In examinations of irradiated animals, the villous capillaries showed initial marked vasodilation followed by constriction, with many capillaries becoming totally nonpatent while the endothelial cells showed changes consistent with vascular damage [4,5,475]. These findings were consistent with functional changes seen in several studies reviewed by Cockerham and Hawkins [154].

The colon has a relatively high radiotolerance, possibly due to the long turnover of its cells, and the rectum may tolerate more than 50-Gy irradiation [63]. Post-irradiation pathological events are essentially the same as in the small intestine, with inhibition of mitosis, changes in cell morphology, and edema and vascular changes in the submucosal and serosal layers. An inflammatory response occurs within 24 hours, and a progressive degeneration leading to ulcerations occurs after high doses [112]. A late development of colorectal irradiation seen in murine studies was a dose-dependent decrease in compliance [488], possibly due to altered ratios of collagen isotypes, especially in the circular muscle layer and villi [489]. As with the other portions of the gastrointestinal tract, the loss of water and electrolyte imbalance due to diarrhea and the development of bacteremia are considered the most important factors in the gastrointestinal syndrome [300]. The rat colon also becomes unresponsive to neurally evoked electrolyte transport following exposure to ionizing radiation of 10 Gy [247]. This response also correlates with decreased mast cells and histamine.

## Cardiovascular Dysfunction

Cardiovascular dysfunction (CVD) has been defined as the inability of any element of the cardiovascular system to perform adequately upon demand. The maintenance of cardiovascular integrity is determined by changes in the (1) pumping action of the heart, (2) compliance of the vascular beds, (3) resistance of the peripheral circulation, (4) quantity of blood in the vascular system, and (5) viscosity of the blood. Failure of any of the mechanisms to respond properly may compromise the integrity of the entire cardiovascular system [154,323,324]. Exposure to supralethal doses of radiation has been shown to induce alterations in cardiovascular function in many species, including humans. The extent of the radiation-induced CVD and its etiology may vary with the species, level of exposure, and dose rate.

Radiation-induced CVD is surprisingly common [37] and may be manifest as circulatory shock. Although postirradiation hypotension does not occur with equal frequency in all species, it has been reported in rats, monkeys, and dogs [157,162,282,515]. However, evidence indicates that, even if sublethal doses of radiation induce a functional cardiovascular deficiency that manifests itself as early hypotension, the lesion may be masked during a period of circulatory deterioration because of cardiovascular reserve [271,579]. When the cardiovascular reserve is no longer capable of maintaining homeostasis the damage may then be recognized as radiation-induced shock [325].

Irradiated rats displayed compromised myocardial function with a decline in cardiac output and an increased left ventricular end-diastolic volume [708,859], which correlated with a decline in capillary density and focal degeneration of the myocardium. Cardiac performance after irradiation using an isolated working rat heart preparation showed a dose-dependent decrease in cardiac function and Frank–Starling curves suggesting a loss of contractile function of the myocardium [850]. Ultrastructural findings in the irradiated rat heart included intercalated disc damage and mitochondrial damage of the myocytes and swelling of the capillary endothelial cells and collapse of the capillaries [150]. Mediastinal irradiation of human patients damages endothelial cells with a loss of capillaries and ischemia leading to increases in collagen and fibrous tissue throughout the heart [37]. Long-term side effects of mediastinal irradiation include pericarditis, accelerated coronary artery disease, myocardial fibrosis, and valvular injury [37,122].

The response of the gastrointestinal microcirculation to radiation has received little attention, although this may be an important factor in the development of both the cardiovascular and gastrointestinal sub-syndromes. Here, as in other parts of the vascular system, the endothelial cell is one of the most radiosensitive cells [653]. The initial expression of radiation injury to the endothelial cells is an increased vascular permeability leading to changes in extracellular environment [35]. Following a single irradiation of 10 to 20 Gy, acute damage to endothelial cells may be detected in 1 to 5 days. Although not identical, radiation-induced endothelial damage in the lungs, kidney, myocardium, and intestine is similar and characterized by the plasma membrane becoming irregular with projections into the vascular lumen, followed by focal or generalized cytoplasmic swelling, which narrows the lumen and may obstruct it completely. Damaged endothelial cells may retract from the basement membrane, causing exposure of the membrane. This results in platelet adhesion followed by aggregation and the development of thrombosis and vascular occlusion. Damaged capillaries may be manifest by telangiectasia or may be replaced by a collagen scar formation [51,653]. When reviewing the response of any microcirculation to irradiation, the variables of total dose, dose rate, type of radiation, variability of the animal model, and postirradiation time of observation must be considered. Irradiation damage to the microcirculation may cause dilation or constriction and either an increase or decrease in blood flow, depending on the above factors [154].

A complication involving the endothelium of intracerebral vessels is the impaired integrity of the blood–brain barrier (BBB) following irradiation [323]. Functional alterations of the BBB are manifest in the endothelium by the activation of pinocytotic vesicular transport [195,784] and in astrocytes by glycogen deposition. The changes in BBB permeability seem to be the result of the intense vesicular response of the endothelium rather than opening of endothelial tight junctions or altered regional blood flow [195,785]. The opening of the BBB has been associated with cerebral vasogenic edema and ischemia [406,760]; however, Gobbel et al. [288] suggested that postirradiation, edema-induced vascular compression was not responsible for changes in regional cerebral blood flow observed in dogs.

A reduction in systemic blood pressure can reduce the driving force required to maintain cerebral blood flow and result in cerebral ischemia. The acute irradiation-induced hypotension in the monkey has a temporal, if not causal relationship with observed postirradiation reduction in regional cerebral blood flow [164]. However, the reduced cerebral blood flow seen in hippocampi and cortices of rat brains and the hypothalamus of humans 10 to 24 weeks postirradiation was probably due to the telangiectatic vessels, spreading edema, focal regions of necrosis, and hemorrhage observed in the brains [44,144,413,458].

Several biochemical mediators have been implicated in postirradiation CVD and reduced cerebral blood flow including histamine, serotonin, opiate peptides, platelet activating factor, eicosanoids, cyclic nucleotides, and catecholamines [211,323,820]. Evidence that further implicates histamine includes the finding that plasma histamine increases precipitously in dogs and monkeys after exposure to radiation [159,163]. Infusing histamine into humans resulted in decreased blood pressure and altered cerebral blood flow [12,721]. Pretreatment with antihistamines diminished the cardiovascular affects of irradiation in dogs and monkeys [158,163], further implicating histamine in postirradiation CVD. However, in another study infusing histamine into humans, Krabbe and Olesen [418] were unable to alter either blood pressure or cerebral blood flow. Likewise, the "histamine hypothesis" does not explain the postirradiation response of the rat [323]. Considering that the etiology of postirradiation CVD may vary with the species, perhaps the radiation-induced production or release of other intermediates such as serotonin [161] or free radicals [156] may account for the CVD.

## Radiation Effects on Bone, Cartilage, and Muscle

Mature bone is relatively radioresistant, and radionecrosis of bone is very rare [510,701]. Many of the effects from irradiation seen in bone may be attributed to a reduction in the number of blood vessels supplying the bone and a decreased blood flow. Radiation damage to bone may be apparent only after months or years following irradiation because of impaired progenitor cell proliferation [193,840]. If the vascular support of the bone can recover, mitotic activity may reappear within 2 weeks after a dose of less than 17.5 Gy; however, osteoradionecrosis, the characteristic late bone injury, often accompanied by osteomyelitis occurs at a minimum dose of 50 Gy [701].

Radiation-damaged adult bone characteristically displays a decreased ability to resist infection, increased susceptibility to fractures, and poor healing after damage [180,510]. The most common site for osteoradionecrosis and postirradiation complications is the mandible, probably due to its less than abundant blood supply. Clavicles and ribs have an increased incidence of fractures after radiation therapy for breast cancer, and tumor-induced fractures of long bones do not heal following radiotherapy of approximately 30 Gy [701]. Following surgical trauma of rat femur, a dose-dependent radiation-induced delay is seen in new bone formation [36].

Radiation effects on bone are age dependent. Developing bone is more sensitive to irradiation than adult bone, with the most pronounced effects being seen during organogenesis [510,701]; however, a reduction in number of blood vessels and decreased blood flow are not thought to be the primary causes of reduction in growth. Impaired progenitor cell proliferation may be the reason, as a single dose of 6 Gy has been demonstrated to produce a decrease in mitotic activity. Internal irradiation of bone by radionuclides may occur through occupational exposure or therapeutic administration [510]. A famous case of iatrogenic poisoning involved the use of Radithor, a patent medicine used as a metabolic stimulant and aphrodisiac [471]. Unfortunately, the victim died of radium poisoning after his skeleton accumulated a dose that may have been greater than 350 Sv.

Probably a more famous case of radioisotope poisoning involved the manufacture of watch dials painted with luminous compounds containing radium [510,674]. The radioisotopes $^{226}$Ra and $^{228}$Ra were ingested by the women workers when they tipped the brushes with their lips. Isotopes of radium, strontium, and calcium are considered volume seekers and ultimately are included in the matrix of bone, while plutonium and thorium isotopes are considered surface seekers and accumulate on the periosteum and endosteal surfaces of the bone. The accumulation of radium in the bones of the female dial painters correlated with fractures of long bones, coarsening of the trabecular pattern, bone infarcts, and aseptic osteonecrosis. Bone sarcomas and head carcinomas occurred among these women at a higher than normal rate. More than 3000 children of the 1495 women dial painters in the basic group were exposed continuously to an alpha- and gamma-enhanced radiation environment during their entire period of gestation [674], but no evidence exists to suggest that any effects of that exposure occurred.

Growing cartilage is more radiosensitive than growing bone [701]. Radiation doses exceeding 18 Gy cause permanent cessation of growth, but chondrocytes recover from irradiation less than 10 Gy. Children ages 6 years and under and during puberty are the most vulnerable to irradiation-induced growth depression. Uneven irradiation of the spine results in scoliosis, and doses up to 20 Gy result in major deformities. Pronounced growth retardation occurs above 35 Gy. Mature cartilage, like adult bone, is fairly radioresistant [510,701]. Doses of 60 to 70 Gy may be tolerated by mature cartilage if the irradiation is applied over 6 to 7 weeks; however, if the same dose is given in less than 6 weeks, radionecrosis of the cartilage is to be expected.

Although atrophy of muscle fibers may be seen following fractionated irradiation of 22 to 54 Gy, doses greater than 500 Gy are required to produce acute radionecrosis of skeletal muscle [510]. Recently, destructive alterations in muscle proteins were observed after $^{60}$Co gamma irradiation as low as 1.0 kGy, with a threefold decrease in elasticity occurring at doses of 15 kGy [411]. Ischemia from a radiation-damaged vascular supply may result in fibrosis, but the extent of muscle damage depends, in part, on whether the entire muscle was exposed or only a portion.

## Radiation Dermatosis

The effects of ionizing radiation on the skin range from erythema to necrosis. The regular sequence of change progresses, as the dose is increased, over two periods, one occurring within 70 to 120 days and the other from 4 months to years later [32]. The first period is characterized by erythema, pigmentation, epilation, dry desquamation, and moist desquamation, and the second period is characterized by atrophy, telangiectasia, fibrosis, and necrosis. Erythema, associated with an increased vascular permeability, appears after a single dose of 500 cGy or more and after multiple dose fractions when the total dose is 1200 cGy or more [32]. The first phase of erythema, presumably due to the release of vasoactive amines, usually occurs within the first 1 to 2 days and lasts for a week. The second phase begins at about 10 to 12 days, reaches a maximum at about 20 days, and lasts for 30 to 40 days. This second phase is due to vascular damage and increased blood flow [510]. The fading of erythema merges with increased pigmentation, which may be permanent or fade for days or weeks as dry desquamation proceeds [32,510]. This pigmentation is associated with an increased melanin content of the basal layer.

Some epilation may be noted at 10 days following irradiation with a single dose of 300 to 600 cGy [32,510]. The evolution and time course of epilation are not dose dependent, but epilation may be complete at 4 weeks with hair beginning to return in the second month and continuing for up to a year; however, a single dose of 700 cGy may cause permanent epilation [510]. Radiation epilation sensitivity varies with body area; the scalp and beard are the most sensitive, followed by chest, axillary, abdominal, eyebrow, eyelash, and pubic hair.

Dry desquamation, preceded by decreasing erythema and an increasing pigmentation, is characterized by a loss of epidermal cells accompanied by replacement. The cells may scale off or peel off in a sheet, leaving an intact, erythematous epidermal surface. The regenerative capacity usually exceeds the destructive capacity as long as the single dose does not exceed 2000 cGy or the multiple fraction total does not exceed 4500 cGy [32,510]. The reduced proliferative potential and regenerative capacity of irradiated skin cells may also manifest as an interference with wound healing at doses of 400 cGy or greater [173].

If the dose exceeds the levels allowing regeneration to occur (2400 cGy for single dose and 5000 cGy for multiple fraction total), then the epidermal cell population becomes depleted and the loss of epidermis allows serum leakage and a moist desquamation [32]. A bullous-type, moist desquamation may occur, with the small blisters tending to coalesce and rupture [510]. Blisters may even form beneath the basal layer, and the lesion may appear similar to a second- or third-degree thermal burn. The ruptured blisters may become infected and ulceration may occur. The ulceration is usually associated with a reduction in circulation due to obliterative arteriolar and small artery changes. Late skin damage may follow the early reactions by week to years or not at all. Alternatively, late skin damage may be manifest without an earlier reaction. The late reaction is, in part, dose dependent and may progress or remain static [32].

In the weeks or months following irradiation at a single dose level of 1700 to 2400 cGy or a multiple fraction total of 4500 to 5000 cGy, telangiectasia manifests as superficial, elongated, and dilated blood vessels [32]. Radiation-damaged endothelial cells are lost and the microvessels shorten, uncoil, and dilate. A loss of total microvasculature occurs as well as a decrease in functional vessels. The formation of telangiectatic vessels is dose dependent, and if the epidermal response is severe focal keratosis and dysplasia may be present.

Skin changes occurring months to years following irradiation may include increased induration, stiffening, and thickening of the dermis associated with increasing fibrosis [32,510]. Although the onset and formation of fibrosis are dose dependent, once it begins fibrosis is progressive, with a characteristic proliferation of the small arteries and arterioles. As the degree of fibrosis increases, so does the probability that necrosis will result.

Following a single dose of radiation that exceeds 2700 cGy or a multiple fraction total dose greater than 6000 cGy, the end stage of radiation dermatosis is a nonhealing necrosis [32]. Radiation-induced necrosis is associated with progressive loss of the dermal microvasculature and is the end stage to progressive fibrosis.

Chronic exposure to low-dose (0.015 cGy/sec) x-ray for 9 to 18 months (total doses equal 2.025 and 4.05 cGy) will produce a hyperkeratinization in the rat, along with a decrease in skin concentration of zinc and an increased concentration of iron [140]. The chronic sequelae following cutaneous radiation may include telangiectases, radiation keratoses, radiation ulcers, hemangiomas, splinter hemorrhages in the distal nail bed, lentiginous hyperpigmentation, and severe subcutaneous fibrosis [628]. This predominant involvement of the skin is sometimes described as the cutaneous radiation syndrome and can become the characteristic feature of chronic cutaneous irradiation.

## The Urinary System

The kidney is relatively radiosensitive compared to other abdominal organs and has a definite but low sensitivity to radiation carcinogenesis [67,69,510,854]. The time course for pathophysiological and histopathological changes is dose dependent; the fractionated tolerance dose (TD) for the human kidney is 20 to 23 Gy [510,854]. The kidneys are considered to be late-reacting organs, as the effects of radiation nephropathy appear months to years after exposure; however, pathologic changes in the endothelial cells of the renal microvasculature seen soon after exposure may have long-lasting effects, and later tubule and glomeruli degeneration may be secondary to renal ischemia

[854]. Functional changes seen in mice following fractionated irradiation with x-rays included decreased ethylenediaminetetraacetic acid (EDTA) clearance, increased urine output, and reduced hematocrit [750].

Reports of patients dying of renal failure and hypertension following therapeutic radiation include subacute changes of intimal necrosis, subendothelial thickening, fibrinoid thrombosis, atrophy of tubules, and replacement with collagen [510]. In some instances myointimal proliferation with foamy cells and sclerosis of the glomeruli may be seen as well.

No human data on acute radiation effects following exposure to single doses are available [510]; however, acute pathological changes seen in animals are hyperemia, increased capillary permeability, interstitial edema, and microvascular endothelial degeneration. These changes are usually followed by occlusive changes in the interlobular arteries and afferent arterioles, reducing blood flow to the nephron. Although in many cases the changes are transient, they may progress to severe diffuse endarteritis and necrotizing vasculitis, which may result in malignant hypertension [854].

Irradiation of pigs with a single $^{60}$Co gamma-ray dose of 7.8 Gy or higher resulted in a dose-dependent reduction in effective renal plasma flow (ERPF) and glomerular filtration rate (GFR) [661,662]. A normochromic normocytic anemia with a significant reduction in erythrocyte count and hematocrit and hemoglobin levels developed within 6 to 8 weeks following irradiation. Studies in pigs [663] and mice [747] indicate that the kidney fails to exhibit complete recovery in function following irradiation and that irradiation of a previously irradiated kidney is likely to lead to severe renal damage.

Chronic renal dysfunction develops 1 to 5 years after irradiation and involves a slow evolution of anemia, hypertension, and impairment of renal function [510]. The changes are progressive and irreversible, with the treatment usually being symptomatic. The pathology is an extension of that seen in subacute radiation-induced renal dysfunction. Chronic glomerulonephritis has been reported in subjects participating in the clean-up after the Chernobyl accident [191]. The renal pathology was associated with significant changes in the humoral immunity system manifested by increased serum levels of immunoglobulin M, immunoglobulin G, and circulating immune complexes; however, chronic ingestion of drinking water containing uranium (0.004 to 9 Fg/kg body weight) indicated that the proximal tubule, rather than the glomerulus, was the site of injury [870].

The urinary bladder is relatively more radioresistant than the kidney and can usually tolerate 55 to 60 Gy if the dose is fractionated [510]. Bladder complications are seen most often in humans following radiotherapy for cancer of the cervix, prostrate, or bladder [755]. The syndrome of acute radiation cystitis appears 4 to 6 weeks following treatment and includes dysuria, nocturia, and increased frequency. Edema, hyperemia, and partial desquamation of the mucosa may be seen. Acute pathological changes in humans are less well documented than those in animals.

Acute changes in the bladders of dogs following irradiation are more pronounced than those seen in rodents [755]. Acute to subacute pathophysiological changes in the rodent bladder include an increase in urination frequency, reduced bladder volume, decreased compliance of the bladder wall, and diminished pressure during micturition [214,465,513,748]. Re-irradiation tolerance for late bladder damage was inversely related to the first dose and independent of the interval between treatments [749].

Late-occurring or chronic pathologic alterations seen in the urinary bladders of dogs following irradiation are similar to those observed in humans. These include a small, shrunken, and contracted bladder with thick and fibrotic walls [510,755]. There may be multiple areas of edema and telangiectasia, and collagen may replace muscular tissue. Squamous metaplasia is common, and extensive mucosal ulceration may extend into and beyond the muscle layers. In women treated for cervical carcinoma, this ulceration can lead to vesicovaginal fistulas.

The most radioresistant portion of the urinary system is said to be the ureter [510]; however, others classify the ureter as radiosensitive, with ureteral fibrosis, stenosis, and obstruction following doses of 12.5 Gy [338,719]. Dogs have been shown to tolerate 17.5-Gy irradiation of the ureter, with early injury due to ulceration of the epithelium being seen at 25 Gy or above [281]; however, histologic evidence suggested that chronic injury of the canine ureter seen after 5 years was of vascular etiology.

## Radiation-Induced Hepatic Dysfunction

Early literature describing hepatic irradiation contained many contradictory reports concerning the radiosensitivity of the liver [372,510]. The hepatic cells are relatively radioresistant, and a marked capacity for regeneration is observed following destruction of a large portion of the liver. The liver is able to tolerate fractionated doses of 30 to 35 Gy over 3 to 4 weeks, but 35 Gy should not be exceeded [372]; however, following liver damage the regenerating portion is more radiosensitive [510]. Radiation-induced hepatic injury represents a continuum of clinical, pathological, and radiographic findings ranging from asymptomatic biochemical changes to fulminate, fatal hepatic failure [372]. Radiolesions in the liver are dose dependent in animals [94,302,303] and are primarily due to damage to the fine vasculature and connective tissue [274,275,372]. Characteristic changes following liver irradiation portray a nonspecific form of venous occlusive disease (VOD) resembling pathologically the Budd–Chiari syndrome [372,510].

Pathological changes that occur following irradiation of the liver to doses greater than 35 Gy may be divided into two stages. The acute phase may begin 2 to 6 weeks postirradiation and continue for 3 to 6 months. Clinical signs include hepatomegaly, ascites, jaundice, and elevated serum alkaline phosphatase and serum transaminases (SGOT and SGPT) [94,274,372,510]. A distinct decrease in hepatic biotransformation and tolerance to a wide variety of drugs may be seen within a few days after irradiation [569,849]. These clinical manifestations are usually associated with pathological changes that include sinusoidal congestion, occlusion of the central vein, disrupted intrahepatic blood flow, and parenchymal cell damage, atrophy and necrosis [275,372,510].

The late phase of pathological changes and hepatic dysfunction manifests itself more than 6 months after irradiation [372,510]. There is less hepatic congestion, but the signs and symptoms of portal hypertension and right-sided congestive heart failure are apparent. Pathological findings are those of a venoocclusive process leading to obstruction of hepatic outflow. The centrilobular veins may be obliterated with dense collagen, and periportal fibrosis is extensive. The liver appears shrunken and pale, and atrophy due to cell loss is evident. Serum phosphatase and transaminase levels may be slightly elevated or normal, but the serum albumin levels are decreased [372,569]. The postirradiation VOD seen in the liver is unique but is similar to conditions caused by various drugs [510].

## Radiation Pneumonitis and Pulmonary Fibrosis

The lung is relatively radiosensitive, and lesions in the lung are common after any irradiation. Involvement of even a small portion of the thorax results in some degree of pulmonary damage [510,783]. The lung's response to irradiation is biphasic [270]. The first phase, radiation pneumonitis, may vary in the postirradiation time of onset depending on the species, type of radiation, and dose [270,442,629,776]. A clinical threshold of 6 to 7 Gy and a maximum of 8 Gy in a single dose have been suggested for the development of radiation pneumonitis [510]. During the pneumonitis phase, functional changes are prominent, including respiratory distress, hypoxemia, increased bronchoalveolar lavage protein, impaired surfactant function, decrease in pulmonary blood flow, and pulmonary hypertension [270,629,776]. Alveolar epithelium and endothelial damage [442,629] may be associated with a radiation-induced production of free radicals; release of histamine, leukotriene, and prostaglandin; and the acceleration of lipid peroxidation [100,293,598]. Pulmonary blood flow problems were indicative of loss of fine vasculature, pulmonary hypertension, right ventricular hypertrophy, and radiation-induced heart failure [73,718]. The second phase of pulmonary injury, pulmonary fibrosis, is seen with or without the presence of pneumonitis [783]. Although many patients are asymptomatic during the fibrosis phase, functional changes include arterial hypoxia, a decreased lung volume, decreased compliance, and a reduced maximum breathing capacity. Endothelial and epithelial cell damage may contribute to the development of postirradiation pulmonary fibrosis [629], but the final picture includes replacement of septa by collagen, decreased total alveolar volume, reduced functional microvasculature, and atelectasis [510,783].

## Radiation Effects on Endocrine Function

Many discussions of radiation effects of the endocrine system are limited to the pituitary, thyroid, parathyroid, and adrenal glands [510]. Other endocrine glands less often considered are the pineal, pancreas, ovary, and testis. Although most endocrine glands are relatively radioresistant, with direct effects of radiation resulting from injury to the fine vasculature, endocrine abnormalities are relatively common following irradiation of the head and neck [510].

### Pineal Gland (Epiphysis)

Melatonin, a hormone that inhibits ovarian and testicular function, is synthesized and secreted by the pineal gland [180,820]. In several species, including humans, melatonin synthesis and secretion increase during the dark period of the day and are at lower levels during the daylight hours [268]. Pineal synthesis of melatonin in response to the light cycle is altered by radiation, with a decreased synthesis seen in rats following 3.5-Gy exposure [820]; however, melatonin has been shown to be radioprotective, with cellular destruction occurring postirradiation in other glands without melatonin [428,820]. Even so, the function of melatonin and the pineal in the radiation response of humans remains relatively obscure.

### Pituitary Gland (Hypophysis)

Physiologically, the pituitary gland is divisible into two distinct portions: (1) the anterior pituitary, or adenohypophysis, and (2) the posterior pituitary, or neurohypophysis. Most control of secretion by the pituitary comes from the hypothalamus by either nervous or hormonal signals. Secretion by the anterior pituitary is controlled by hormones secreted within the hypothalamus and transported to the anterior pituitary through the hypothalamic–hypophysial portal system of blood vessels. Secretion from the posterior pituitary is controlled by nerve fibers originating in the hypothalamus and terminating in the posterior pituitary. The hypothalamus receives signals from sources throughout the nervous system and uses this information to control secretion of the pituitary hormones [268,304].

Six important hormones are secreted by the anterior pituitary and play major roles in the control of metabolic functions throughout the body. These six hormones are: (1) growth hormone (GH), (2) adrenocorticotropic hormone (ACTH; corticotropin), (3) thyroid-stimulating hormone (TSH; thyrotropin), (4) prolactin (luteotropic hormone [LTH]), (5) follicle-stimulating hormone (FSH), and (6) luteinizing hormone (LH). The two hormones secreted in the posterior pituitary are antidiuretic hormone (ADH; vasopressin), which controls water excretion, and oxytocin, which helps deliver milk from the glands to the nipple and may help in delivery at the end of gestation. ADH and oxytocin are formed in the supraoptic and paraventricular nuclei of the hypothalamus and released from nerve endings in the posterior pituitary [268,304].

Hypothalamic–pituitary failure is a common complication of cranial and neck irradiation, and signs of endocrine deficiency may appear from 1 to 15 years after irradiation. Some researchers propose that radiation-induced alterations in pituitary function can be attributed to effects of radiation on the hypothalamus [686]. There is good evidence that the earliest irradiation damage to the hypothalamic–pituitary axis is at the level of the hypothalamus and that any subject receiving a total irradiation dose of 20 Gy or more to the axis is at risk of hypopituitarism [349,429,454]. In general, the direct effects of radiation on the hypothalamic–pituitary axis result in hypopituitarism manifested through alterations of the direct actions of the pituitary hormones or through their influence on other endocrine glands [455,510]. Radiation-induced pituitary dysfunction may result in loss of weight, loss of body hair, dry skin, slow pulse, low body temperature, dwarfism in children, genital atrophy, primary amenorrhea in females, failure of sexual development in males, and other dysfunctions associated with the endocrine system [510,686].

## Thyroid

Thyrotropin-releasing hormone (TRH), secreted by the hypothalamus, stimulates the release of thyroid-stimulating hormone (TSH) by pituitary thyrotrophs and the release of calcitonin from the C-cells of the thyroid. TSH stimulates all thyroidal functions associated with the production and release of triiodothyronine ($T_3$) and tetraiodothyronine ($T_4$; thyroxine). Through a classic feedback loop, excessive amounts of $T_3$ and $T_4$ suppress the release of TSH by the pituitary [396]. Calcitonin, from the C-cells or parafollicular cells in the thyroid gland, serves to lower the serum calcium and phosphate levels [268].

Radiation-induced injury of the hypothalamic–pituitary axis may be manifest in the development of hypothyroidism [429,454]; however, direct irradiation of the thyroid gland results in decreased production of $T_3$ and $T_4$ [820]. Estimates of irradiation doses required to produce hypothyroidism vary from 2 to 50 Gy, with thyroid ablation being a possible result of the larger doses. Data confirm the high incidence of thyroid dysfunction when the gland is included in the radiation field [29,240,316, 602,858]. Radiation hypothyroidism, through the feedback loop, will result in an increase in TSH secretion by the pituitary. This increase in TSH has been associated with increases in radiation-induced thyroid cancer [820]. Thyroid-stimulating hormone has also been studied in relation to radiation exposure in children, particularly after the Chernobyl nuclear disaster [289,641].

The latent period for radiation-induced thyroid cancer was in excess of 15 years following the exposure of the Japanese to the atomic bomb explosions at Hiroshima and Nagasaki. Increased numbers of thyroid cancer cases became evident within just 4 or 5 years following the Chernobyl nuclear accident, however, and continued to substantially increase over the next decade [86,272]. Another interesting aspect of the dynamics of thyroid cancer incidence in that accident was that geographic correlation of thyroid cancers was more closely related to the transportation corridors than to the isopleths of $^{131}$I distribution or to population density [615]. Recent analysis has validated that an increase in thyroid cancer can be related to Chernobyl radiation exposure even in nations that received relatively moderate levels of radioactivity from the accident [576].

Hypothyroidism following 10-Gy irradiation has been associated with a shift in the ratio of the alpha-myosin heavy chain to the beta-myosin heavy chain in the rat heart, a shift usually associated with overload of the heart or aging [820]. This change may correspond with a low resting ejection fraction and decreased response to exercise seen in cardiac scans following therapy for Hodgkin's disease [504]. Therapy for hyperthyroidism using $^{131}$I decreased basal calcitonin levels and may cause C-cell deficiency [60]; however, this may not be of extreme consequence as total thyroidectomy does not reduce the circulating level of the hormone to zero [268].

The thyroid gland of children is especially vulnerable to the carcinogenic action of ionizing radiation, as it has one of the highest risk coefficients of any organ and may be the only tissue with convincing evidence for risk at about 0.10 Gy [669]. This vulnerability is dramatically illustrated in the 100-fold increase in the number of childhood thyroid cancers observed in heavily contaminated areas of Belarus, Ukraine, and Bryansk regions of Russia following the accident at the Chernobyl nuclear power plant in 1986 [39,680].

## Parathyroid

In humans, usually four parathyroid glands are embedded in the poles of the thyroid gland. Chief cells in the parathyroid are sensitive to circulating levels of ionized calcium and act to secrete parathyroid hormone (PTH) in response to decreased $C^{++}$ levels. The PTH acts directly on bone to increase bone resorption and mobilize $Ca^{++}$. PTH also acts to depress plasma phosphate by increasing phosphate excretion [268]. Although parathyroid cells are relatively resistant to radiation, data confirm an association between hyperparathyroidism and radiation exposure [61,254,510]. The hyperparathyroidism may be secondary to a depletion of calcium that occurs in mammals following exposure to greater than 3-Gy radiation [820,821]. Other biological responses occurring as a result of irradiation-induced calcium loss include increased blood clotting, bone damage, and convulsions [820].

## Adrenal Glands

The endocrine functions of the adrenal glands are associated with the adrenal cortex, which produces over 30 different steroids. Only two of these corticosteroids are of exceptional importance to the endocrine function of the adrenal cortex: (1) aldosterone, the principal mineralocorticoid, and (2) cortisol, the principal glucocorticoid. The mineralocorticoids affect the levels of the electrolytes of the extracellular fluids, and the glucocorticoids serve to increase blood glucose concentration and affect protein and fat metabolism. Aldosterone secretion is regulated by: (1) the potassium ion concentration in extracellular fluid, (2) the renin–angiotensin system, (3) the quantity of sodium in the body, and (4) adrenocorticotropic hormone (ACTH). Secretion of cortisol is controlled almost entirely by ACTH [304].

Radiation-induced changes in the endocrine functions of the adrenal glands are difficult to document because stress factors, including radiation, result in increased release of ACTH from the pituitary [510,820]. ACTH acts on the adrenal cortex to stimulate the synthesis and release of aldosterone and cortisol. Even though irradiation of the hypothalamic–pituitary axis produces a defect in ACTH release, adrenal corticosterone levels may be normal or elevated following irradiation, indicating a possible hypersensitivity to the ACTH present or the presence of some other controlling factor [820].

The direct effects of radiation on the adrenal glands are manifest in three phases of activity, each associated with increases in plasma and adrenal corticosterone levels [820]. The first phase occurs early after irradiation, and the second peak of activity is associated with gastrointestinal damage. The third phase of activity is associated with hematopoietic injury. The increase in activity may be seen following absorbed doses of 15 to 35 Gy [510,820]; however, if the dose exceeds 35 Gy, normal steroidogenesis may occur under nonstress conditions, but the ability to respond to stress is impaired. Hypertrophy of the adrenal cortex has been demonstrated following irradiation with 15 Gy and higher [207,820].

## Pancreas

The pancreas has both endocrine and exocrine functions. The islets of Langerhans in the pancreas are associated with its endocrine function and act to secrete insulin and glucagon. Insulin is produced by the beta cells of the islets and glucagon by the alpha cells. The secretion of insulin and glucagon is controlled by the blood glucose concentration. An increase in blood glucose concentration stimulates insulin secretion and inhibits glucagon secretion. A decrease in blood glucose concentration has the opposite effect on both hormones. Insulin acts to increase glucose uptake by most tissues of the body and to stimulate glycogen synthesis. Glucagon stimulates the breakdown of hepatic glycogen and adipose tissue and also stimulates gluconeogenesis from amino acids. All of its actions increase blood glucose concentration [268]. The pancreas is relatively radioresistant compared to the surrounding structures such as the liver and small intestine [510]. The islet cells show more postirradiation changes than do the acinous cells, with the beta cells of the islets being more radioresistant than the alpha cells [207,560]. Decreases in insulin and glucagon have been observed following radiotherapy, and impaired insulin secretion and hypoglycemia were seen in rats 4 days after 10-Gy irradiation [689,820]. One month after irradiation, the insulin secretion impairment persisted and was accompanied by a reduced number of beta cells.

## Ovaries

The two types of ovarian hormones, estrogens and progestins, are secreted by the ovaries in response to follicle-stimulating hormone (FSH) and luteinizing hormone (LH) from the anterior pituitary. FSH and LH, in turn, are secreted by the anterior pituitary in response to luteinizing-hormone-releasing hormone (LHRH) from the hypothalamus. By far the most important of the estrogens is estradiol, secreted by the theca interna and granulosa cells of the ovarian follicles and by the corpus luteum. The most important progestin is progesterone, secreted by the corpus luteum [268,304]. Ovarian radiation severely reduces the formation of ovarian steroid hormones, even to the point of gonadal failure [455,825]. Estrogen decreases have been seen in humans following radiation doses of 6 to 100 Gy [455,820,825], and large doses produce premature menopause [820]. Persistently elevated gonadotrophin levels (FSH and LH) and amenorrhea are associated with reduced ovarian hormones [455,825]. Abdominal radiation of females in childhood has resulted in pubertal failure or premature menopause [824]. Neonatal irradiation of rats with 15 cGy produced a decrease in progesterone levels but not estradiol levels in adult animals.

## Testes

The testes secrete several hormones that are called androgens because of their masculinizing effects. Testosterone, considered to be the most significant testicular androgen, is formed by the interstitial cells of Leydig. As in the female, LHRH from the hypothalamus stimulates secretion of LH by the anterior pituitary. LH, in turn stimulates hyperplasia of the Leydig cells and the production of testosterone by these cells [268,304]. The most dramatic endocrine effect of irradiation of the testis is the increase in FSH and LH secretion from the anterior pituitary [505]. FSH levels have been used as an indication of damage to the germinal epithelium. Elevated LH levels associated with normal testosterone levels are indicative of Leydig cell damage [126,405,456,505,633]. This condition may occur at irradiation doses of 2 to 12 Gy because damage to the Leydig cells may be compensated by an increase in the number of cells (hyperplasia) in response to the elevated LH [820]. Higher doses of irradiation (20 to 30 Gy) have been shown to produce Leydig cell damage, decrease testosterone production, and increase secretion of LH [286,505,715]. Of course, sperm cells and the less differentiated cells giving rise to sperm cells are also known to be affected by radiation exposure; for example, ultramorphological sperm characteristics have been altered in workers cleaning up materials containing relatively high levels of radioactivity following the Chernobyl nuclear accident [243].

# Nervous System

## Radiogenic Effects on Sensory Functioning

Ionizing radiation can be sensed at extremely low levels [698]; for example, the olfactory response threshold to radiation is less than $1.0 \times 10^{-4}$ Gy. The visual system is sensitive to levels below $5.0 \times 10^{-6}$ Gy. Ionizing radiation has been shown to be as efficient as light in producing retinal activity (as assessed by the electroretinogram), and the visibility of ionizing radiation is now firmly established [397]. While visual system pathomorphology occurs only at high doses [258], this is not true of visual function disruption. Rats trained to a brightness discrimination task were unable to differentiate shades of gray after 3.6 Gy or to make sensitivity changes after 6 Gy of whole body x-rays [397]. Chimpanzees showed impaired accuracy and visual acuity on visual discrimination tests after about 4 Gy of gamma irradiation [659]. Kekcheyev [389] reported that 1 day after exposure to 0.3 to 1.0 Gy of x-rays, temporary decrements in scotopic visual sensitivity were observed in humans. Further, Lenoir [448] found long-term delays [20 to 36 days] in dark adaptation in patients exposed to 4 to 62 Gy of x-rays.

Most of the literature suggests that significant hearing changes, unlike vision changes, require massive doses (e.g., 10 to 70 Gy) of radiation [256]. Heinz [331], using fractionated head-only exposures of baboons to x-rays, found that 10, 12, and 15 Gy caused a long-term hearing deficit. The highest exposure produced a hearing loss of >90 dB that was not frequency specific. Lower doses of x-rays caused slowly developing, transient elevations in auditory reaction times. Vestibular function may be more radiosensitive than audition. Depression in vestibular function may exist at doses close to the $LD_{50}$, with higher doses producing longer lasting disruptions than low doses [31].

Although not much literature is available, several reports exist of olfactory, gustatory, and cutaneous sensory changes in patients exposed to therapeutic irradiation [256]. Altered taste perception was found in patients exposed to 36 Gy of x-rays, with a metallic taste being the most common report. Transient changes in taste and olfactory sensitivity were also reported in radiotherapy patients and rats [397]. Empirical evidence suggests radiogenic changes in pain perception. While gamma photons produce a dose-dependent analgesia in mice [772], data also suggest that x-rays or gamma rays do not alter the analgesic effects of morphine or the anesthetic effects of halothane in rats except under a narrow set of experimental conditions [115,216]. Miyachi and colleagues reported that the olfactory system of mice is important in detecting radiation [545,548] as well as in modulating radiation-induced analgesia [547].

## Radiogenic Pathology of the Adult Nervous System

A review of older radiobiology textbooks revealed the common belief that the adult CNS is relatively resistant to damage from ionizing radiation exposure [125]. This conclusion was derived, in part, from early clinical reports suggesting that radiation exposures, given to produce some degree of tumor control, produced no immediate morphological effects on the CNS [345]. This view was changed, however, when it was later shown that the latency period for the appearance of radiation damage in the CNS is simply longer than it is in other organ systems [470]. Later interest in the pathogenesis of delayed radiation necrosis in clinical medicine has produced a significant body of literature. Studies of radiation-induced brain damage in patients used computed axial tomography (CAT) technology to confirm CNS abnormalities that are not associated with tumor treatment but that occur because of the radiotherapy [340].

General, although not universal, agreement exists that there is a threshold dose below which no late radiation-induced morphological sequelae in the CNS occur. In laboratory animals, single doses of radiation up to 10 Gy produced no late morphological changes in the brain or spinal cord [327,451]. Necrotic lesions were observed in the forebrain white matter from doses of 15 Gy [121,129,391]. In humans, the "safe" dose has been a topic of considerable debate. Depending on the radiation field size, the threshold for CNS damage was estimated to be 30 to 40 Gy if the radiation was given in fractions [614]; spinal cord damage occurred with fractionated doses as low as 25 Gy [226]. The difference between a safe and a pathogenic radiation dose to the brain may be as small as 4.3 Gy [485].

Different topographical regions of the brain may vary in susceptibility to ionizing radiation [798]. The most sensitive area is the brain stem [35]. The cerebral cortex may be less sensitive than the subcortical structures [451] such as the hypothalamus [861], the optic chiasm, and the dorsal medulla [672]. Although radiation lesions occur more frequently in brain white matter [355,666,801], the radiosensitivity of white matter also appears to vary from region to region [451]. It may be that selective necrosis of white matter is due to the slow reproductive loss of glia or their precursors. The radiosensitivity of certain types of glial cells (beta astrocyte) is well recognized [655,656]. The earliest sign of their damage is widening of the nodes of Ranvier and segmental demyelination as early as 2 weeks after a dose of 5 to 60 Gy [491]. Cerebral RNA isolated from mice exposed to 0.1 or 2 Gy gamma radiation revealed modulation of the expression of 1574 genes, with 30% exhibiting dose-dependent variations [860].

The technique and endpoints selected to assess neuropathology can profoundly influence its detection. In proton-irradiated brain tissue stained with silver to detect degenerating neural elements, punctate brain lesions were found within 3 days after exposures as low as 2 Gy [761]. The lesions were not detectable with standard H and E stains. These effects are similar to a multi-infarction syndrome, in which the effects of small infarctions accumulate and may become symptomatic. Similarly, Philpott et al. [631] found that both the synaptic density and the spine length in area CA1 of the hippocampus were lower in mice irradiated with 0.005 or 0.5 Gy of $^{40}$Ar. Additionally, chronic or repetitive exposure (0.05 Gy of x-rays for 10 days) was found to be a stronger inducer of cellular and molecular changes in the hippocampus and frontal cortex than acute exposure (0.05 Gy for 1 day) in total-body irradiated mice [732].

The phenomenon of latent CNS radiation damage with doses above threshold has been well documented [74,125,678]. The long latent period has led to considerable speculation on the likely pathogenesis of late radiation lesions: (1) Radiation may act primarily on the vascular system, with necrosis secondary to edema and ischemia, and (2) radiation may have a primary effect on cells of the neural parenchyma, with vascular lesions exerting a minor influence [345].

The first evidence in support of a vascular hypothesis was obtained when canine brains that had been exposed to x-rays were examined [470]. It was suggested that delayed damage of capillary endothelial cells may occur, leading to a breakdown of the blood–brain barrier. This would result in vasogenic edema [196], the elevated pressure-impaired circulation of cerebral spinal fluid, and eventually neuronal and myelin degeneration [129,130]. The finding that hypertension accelerated the appearance of vascular lesions in the brain after irradiation with 10 to 30 Gy also supports a hypothesis of vascular pathogenesis [346]. The occlusive effects of radiation on arterial walls may cause a transient cerebral ischemia [334]. Sequential monkey-brain CAT scans revealed brain edema and hydrocephalus that accompanied hypoactivity and the animal's loss of alertness following 20 Gy of radiation [314]. Head-only exposure of rabbits to 4, 6, or 8 Gy of x-rays disturbed the blood–brain barrier permeability, which returned to normal after only 6 days [842]. The transient nature of the vascular phenomena may partially explain some of the behavioral deficits observed after exposure to intermediate or large doses of ionizing radiation [499,720].

Evidence of the direct action of radiation on the parenchymal cells of the nervous system, rather than the indirect effect through the vascular bed, was first provided when brain tissue in irradiated human patients was examined [600]. None of the brain lesions could be attributed to vascular damage because they were (1) predominantly in white matter and not codistributed with blood vessels, (2) not morphologically typical of ischemic necrosis, and (3) often found without any vascular effects [178,347,362, 625,872]. Thus, it appears that direct neuronal or glial mechanisms caused at least some of the observed radiogenic brain lesions.

### Alterations in Nervous System Physiology and Functioning

In addition to radiogenic changes in CNS morphology, a variety of changes in parameters of brain function were reported; for example, changes in brain metabolism were reported after very low (0.11 to 0.24 Gy) doses of ionizing radiation [229]. In a more detailed analysis with the $^{14}$C-2-deoxyglucose method of measuring local cerebral glucose utilization, a dose of 15 Gy of x-rays was administered to the rat brain [363]. Significantly lower rates of glucose use were found in 16 different rat brain structures at 4 days after irradiation and in 25 structures at 4 weeks. Although large radiogenic changes exist in the metabolism of particular brain nuclei, a weighted average rate for the irradiated brains was approximately 15% below that for the controls.

Researchers measured the functional sensitivity of some brain areas and the insensitivity of others [6,518]. The activation of behaviors through electrical stimulation of the lateral hypothalamus (but not of the sepal nucleus or substantia nigra) is still possible after 100 Gy [148,518]; however, years after clinical irradiations, dysfunctions of the hypothalamus are prominent even without evidence of hypothalamic necrosis [500]. Local subcortical changes may exist in the reticular formation and account for radiation-induced convulsability of the brain [670,671]. Similarly, postirradiation spike discharges are more likely to be observed in the hippocampal electroencephalograph (EEG) than in the cortical EEG [266]. This idea of selective neurosensitivity is further supported by experiments in which electrical recordings were made from individual nerve fibers after irradiation [277]. These data reveal a hierarchy of radiosensitivity in which gamma nerve fibers are more sensitive than beta fibers, and alpha nerve fibers are the least sensitive.

## Electrophysiology

Measures of electrophysiology illustrate changes in brain function after exposure to ionizing radiation. Several studies were reported in which cortical EEG changes were observed in humans and in animals following doses as low as 0.05 Gy [447]. Typically, an initial temporary increase in bioelectric amplitude was followed, within minutes, by a depression. Other investigations frequently required higher doses of radiation to observe changes in EEG; for example, changes were not seen in EEGs after 0.03- to 0.04-Gy x-rays, but significant alterations were observed after 2 Gy [308]. At a higher dose (15 Gy), monkey cortical EEG abnormalities consisted of the slowing of activity, with an increase in amplitude [672]. Spiking and patterns of grand mal seizure also occurred. A rapid onset of high-amplitude slow waves (delta waves) seemed to relate to periods of behavioral incapacitation [497]. Exposures to 4- to 6-Gy $^{60}$Co gamma radiation appeared to stimulate spontaneous activity in the neocortex, whereas exposures higher than 9 Gy inhibited all brain activities [557]. Many of the liquidators involved in the cleanup of the Chernobyl nuclear accident were reported to have abnormal EEGs [459,816,873].

The hippocampus shows significant changes in physiological activities after gamma irradiation, with even less than half of the 18-Gy threshold dose needed to produce changes in cortical activities [35,267]. Hippocampal spike discharges were first identified in cats [266] and later confirmed in rabbits [267]. This spiking developed soon after irradiation (2 to 4 Gy x-rays) when no other clinical signs of neurological damage or radiation sickness were present.

The apparent radiosensitivity of the hippocampus and its importance in critical functions, such as learning, memory, and motor performance [556], have led others to investigate the electrophysiology of this brain area. The firing of hippocampal neurons was found to be altered by exposure to 4 Gy of $^{60}$Co gamma radiation in rabbits [57]. In guinea pigs exposed to 5 or 10 Gy of x-rays, significant changes in hippocampal neuronal function were observed to be time, dose, and dose-rate dependent [625]. Higher doses (40- to 65-Gy x-rays) decreased the ability of hippocampal neurons to generate an action potential [626]. In addition, *in vitro* experiments suggested that spontaneous discharges of hippocampal pacemaker-like neurons were induced by x-rays and gamma rays at a dose of only 0.08 Gy [624]. These data suggest that hippocampal electrophysiology may be one of the most sensitive measures of functional brain changes after irradiation. Alterations in the thresholds and patterns for audiogenic and electroconvulsive seizures have been produced by exposing animals to ionizing radiations. Such effects are interpreted as reflecting gross changes in CNS reactivity. Early work with dogs showed that spontaneous seizures sometimes occurred following very large doses of radiation [470]. Later experiments confirmed that seizures can be induced by whole-body or head-only exposures to 30 to 250 Gy in a variety of species; for example, rats were exposed to 5 Gy of x-radiation, and the electroconvulsive shock (ECS) threshold was determined for 180 days after irradiation [670]. ECS thresholds were reduced in irradiated rats over the entire test period. Later studies [671] reported that considerably lower doses (perhaps as little as 0.01 Gy) also reduced the thresholds for ECS seizures and audiogenic seizures [535,635].

Unlike the CNS, peripheral nerves are quite resistant to the functional alterations produced by ionizing radiation. Most data indicate that peripheral nerves do not show any changes in electrophysiology with x-ray exposures below 100 Gy [690]. After higher doses, the action-potential amplitude and the conduction velocity temporarily increase but then gradually decrease [45–48]. Also, alpha and beta particles are more destructive to peripheral nerves than gamma rays or x-rays, and usually cause a monophasic depression of function without the initial enhancement of activity [261,269,857]. Perhaps the lowest dose of ionizing radiation ever found to produce an alteration in the function of peripheral nerves was reported in a study in which T-shaped preparations of isolated frog sciatic nerves were produced when the nerves were partially divided longitudinally [421]. Electrical stimulation was applied to the intact stem of the T, and electrical recordings were made from the ends of the two branches. A small segment of one of the branches was irradiated with 0.04 to 0.06 Gy of alpha particles, producing a definite decrease in action-potential amplitude and an increase in chronaxy. These results were remarkable because of the much higher doses required to affect these peripheral nerve functions in most other studies.

Paralysis of the hind limbs of animals can result from localized irradiation of the spinal cord. Rabbits developed this paralysis at 4 to 33 weeks after exposure of the upper thoracic region to 30 to 110 Gy of x-radiation at 2.5 Gy/day

[702]. The minimum single exposure found to produce paralysis at 5 months was 20 Gy [703]. As in other model systems, the interval between irradiation and the appearance of neurological symptoms decreased as dose increased; for example, 50 Gy of x-rays to the monkey midthoracic spinal cord produced immediate paraplegia, while 40 Gy was effective only after a latent period of about 5.5 months [164]. Some success in the amelioration of myelopathy of the cervical spinal cord in rodent models has been attributed to platelet-derived growth factor [22].

Radiation effects on the electrophysiology of the synapse were first studied using the cat spinal reflex [462,691–694]. These studies showed that excitatory synaptic transmission was significantly increased by x-ray exposures of 4 to 6 Gy. Synaptic transmission at the upper cervical ganglion of the cat was also facilitated 15 to 20 minutes after exposure to 8 Gy of x-rays [570]. Both monosynaptic and polysynaptic spinal reflexes were significantly augmented immediately after exposure to 5 Gy of x-radiation. Interestingly, significant augmentation of monosynaptic excitatory postsynaptic potentials (EPSPs) was found immediately after exposure to 6 to 12 Gy of x-rays, while inhibitory postsynaptic potentials (IPSPs) recorded from the same cell were not significantly affected by a 12-Gy exposure [691,694]. Similarly, polysynaptic EPSPs were significantly augmented as the dose increased, while the polysynaptic IPSPs were little influenced even by an exposure of 158 Gy. At higher doses (50 to 200 Gy), ionizing radiation may damage both synaptic and postsynaptic functioning, probably through different molecular mechanisms [778]. These radiogenic changes in synaptic transmission may be important factors underlying the complicated functional changes that occur in the CNS following radiation exposures.

## Neurochemistry

Ion flow across the neuronal semipermeable membrane is one of the most important mechanisms of postirradiation nervous transmission to be studied. In particular, the flow of sodium ions is believed to be involved in the control of neuronal excitability [128] and apparently can be disrupted after either a very high or very low dose of radiation. A study using the radioactive isotope $^{24}$Na compared the sodium intake across the membrane of the squid giant axon before and after exposure to x-rays [673]. A significant increase in sodium intake was found to occur during the initial hyperactive period induced by a dose of 500 Gy. These observations were confirmed, although a simultaneous decrease in the rate of sodium extrusion also occurred in a study of frog sciatic nerves that had been irradiated with 1500 to 2000 Gy of alpha particles [261]. As was described earlier, peripheral nerves may be less radiosensitive than CNS neurons and perhaps differ in their radiation response. In a study that used a different technique, the artificially stimulated uptake of sodium into brain synaptosomes was significantly reduced by an ionizing radiation exposure (high-energy electrons or gamma radiation) of 0.1 to 1000.0 Gy [574,847].

The brain has been described as a radiosensitive biochemical system [229], and many significant changes in brain neurochemistry have been observed after irradiation. An early study revealed that 1 to 2 days after an exposure to 3 Gy of x-radiation, neurosecretory granules in the hypophysial–hypothalamic system showed a transient increase in number over the controls [764]. A leaking of brain monoamines from the neuronal terminals of rats irradiated with 40 Gy of x-rays was also observed [185]. These changes in neuronal structure may correlate with radiogenic alterations of neurotransmitter systems.

Normal catecholamine functioning appears to be damaged following exposure to intermediate or high doses of ionizing radiation. After 100 Gy of $^{60}$Co gamma radiation, a transient disruption in dopamine functioning (similar in some ways to dopamine-receptor blockade) was demonstrated [351]. This radiogenic change in dopaminergic systems is further supported by the finding that a 30-Gy $^{60}$Co radiation exposure increased the ability of haloperidol (a dopamine-receptor-blocking drug) to produce cataleptic behavior [378]. Relatively low doses of $^{56}$Fe (0.1 to 1.0 Gy) also caused a profound reduction in $K^+$-stimulated dopamine release from perfused striatal slices of rat brain [376,377]. This decrement lasted as long as 180 days after exposure. Radiation-induced effects on dopamine have been correlated in time with behavioral deficits; however, other neuromodulators (such as prostaglandins) also seemed to influence dopaminergic systems to help produce some radiation-induced behavioral changes [378]. A transient reduction in the norepinephrine content of a monkey hypothalamus was observed on the day of exposure to 6.6 Gy of gamma radiation. Levels of this neurotransmitter returned to normal 3 days later [425]. Similar effects were reported elsewhere [806], but another study found no change in noradrenaline after 8.5 Gy of x-rays [373]. An increase in the catecholamine enzyme monoamine oxidase (MAO) was reported within 4 minutes of exposure and lasted for at least 3 hours [127].

A variety of functions involving the neurotransmitter acetylcholine (ACh) are significantly altered by exposure to ionizing radiation. ACh synthesis rapidly increased in the hypothalamus of the rat after as little as 0.02 Gy of beta radiation, but it was inhibited at only slightly higher radiation doses [229]. A dose of 4 Gy of $^{60}$Co gamma radiation produced a long-term increase in the rate of ACh synthesis in dogs [198]. Also, high-affinity choline uptake (a correlate of ACh turnover and release) slowly increased to 24% above control levels 15 minutes after irradiation with 100 Gy [351]. Choline uptake returned to normal by 30 minutes after exposure. Massive doses of gamma rays or x-rays (up to 600 Gy) were required to alter brain acetylcholinesterase activity [681], while much smaller doses depressed plasma acetylcholinesterase by 30% [466].

Exposure to large doses of ionizing radiation resulted in postirradiation hypotension in monkeys [108,163,324], with arterial blood pressure decreasing to less than 50% of normal [219]. Postirradiation hypotension also produced a decrease in cerebral blood flow immediately after a single dose of either 25 or 100 Gy of $^{60}Co$ gamma radiation [135,136,153,160]. This hypotension may be responsible for the early transient incapacitation (ETI; see later description) observed after a supralethal dose of ionizing radiation [108,137,788]. A study with untrained monkeys, whose postirradiation blood pressures were maintained by norepinephrine or other pressor drugs, showed that as long as arterial pressure was above a critical level the monkeys remained attentive and alert [534]; however, in a follow-up study on monkeys trained to perform a task, norepinephrine maintained blood pressure but did not consistently improve performance during the first 30 minutes after irradiation [789]. Other authors observed no close association between blood pressure and behavioral changes [497]. Further contrary evidence was obtained from experiments with the spontaneously hypertensive rat (SHR), in which exposure to ionizing radiation reduced the blood pressure of most rats to near-normal levels; however, the irradiated SHRs still showed a significant behavioral deficit after exposure to 100 Gy of high-energy electrons [519]. Finally, a significant association was found between the degree of hypotension and the frequency of early performance decrements (EPDs) [108]. Still, half the monkeys with a 50% drop in blood pressure did not show behavioral decrements; thus, even though the relationship between decreased blood pressure and impaired performance is intriguing, simple changes in blood pressure may not be sufficient to explain transient behavioral changes.

The massive release of histamine observed after exposure to a large dose of ionizing radiation was proposed as a mediator of radiogenic hypotension and EPDs [217]. Histamine was found to be a very active biogenic amine and putative neurotransmitter located in neurons and mast cells throughout the body, especially around blood vessels [215]. Attempts to alter the development of behavioral deficits by treating animals with antihistamines before exposure have been encouraging [93,218,219]. Monkeys pretreated with chlorpheniramine ($H_1$-receptor blocker) performed better and survived longer after irradiation than did controls [219]. Similar benefits were observed in irradiated rats [516]. Further, the use of diphenhydramine (a histamine $H_1$-receptor antagonist) inhibited radiation-induced cardiovascular dysfunction [15]. Because these antagonists produced only partial relief from radiation effects, it appears that the histamine hypothesis explained just a portion of the behavioral and physiological deficits observed after radiation exposure [123].

When most animal species were exposed to a sufficiently large dose of ionizing radiation, they exhibited lethargy, hypokinesia, and deficits in performance [138,397,518]. Because these behaviors seemed similar to those observed after a large dose of morphine, a role was proposed for endogenous opioids (endorphins) in the production of radiation-induced behavioral changes [206,364]. Endogenous morphine-like substances were thought to be released as a reaction to some [8,300,329] but not all [484] stressful situations. Like a sufficiently large injection of morphine, endogenous opioids produced lethargy, somnolence, and a reduction in behavioral responsiveness [387,484]. Cross-tolerance between endorphins and morphine was demonstrated for a variety of behavioral and physiological measures [107,762]. Because of the similarity of radiation- and opiate-induced symptoms, it is not surprising that endorphins are involved in some aspects of radiogenic behavioral change. Ionizing radiation produced dose-dependent analgesia in mice, and this radiogenic analgesia was reversed by the opiate antagonist naloxone [772]. In another experiment, morphine-induced analgesia of the rat was significantly enhanced 24 hours after neutron (but not gamma) irradiation, suggesting some combined delayed effects of endogenous and exogenous analgesics that may be radiation specific [115]. Ionizing radiation exposure also attenuated the naloxone-precipitated abstinence syndrome in morphine-dependent rats [184].

Further supporting the hypothesis that endorphins are involved in radiation-induced behavioral change, C57Bl/6J mice exhibited a stereotypic locomotor hyperactivity similar to that observed after morphine injection after receiving 10 to 15 Gy of $^{60}Co$ gamma radiation [520]. This radiogenic behavior was reversed by administering naloxone or by preexposing the mice to chronically stressful situations, a procedure that produces endorphin tolerance [521]. In addition, opiate-experienced mice reduced the self-administration of morphine after irradiation, suggesting that the internal production of an endorphin reduced the requirement for an exogenous opioid compound [522]. Biochemical assays also revealed changes in mouse brain beta-endorphin after exposure to ionizing radiation [523]. Rats and monkeys had enhanced blood levels of beta-endorphin after irradiation [14,190], and morphine-tolerant rats showed less performance decrement after irradiation than nontolerant subjects [524]. Further, naloxone given immediately before exposure to 100 Gy of high-energy electrons significantly attenuated the early behavioral deficits observed in rats [14]. Conversely, rats either underwent no change or were made more sensitive to radiation effects after chronic treatment with naloxone on a schedule that increased the number of endorphin receptors [565]; however, the manipulation of opioid systems did not produce total control over postirradiation performance deficits, thus these data do not suggest an exclusive role for endorphins in radiogenic behavioral change.

# BEHAVIORAL EFFECTS OF IONIZING RADIATION

## Behavioral and Neurophysiological Effects of Prenatal or Neonatal Radiation Exposure

The developing nervous system is significantly more radiosensitive than the adult nervous system because ionizing radiation, like other teratogenic agents, is apt to affect embryonic cells with high proliferative and metabolic activity [400,775]. Thus, prenatal exposure to ionizing radiation can cause either organogenic malformations (abnormal closure of the neural tube) or histogenic abnormalities (abnormal proliferation or migration of neurons) [382,649]. Because postnatal neurogenesis may be protracted [298], radiosensitivity of selected brain areas (e.g., the dentate gyrus) may well extend into the neonatal period and even young adulthood in some species [550]; for example, 4 Gy of $^{60}$Co gamma radiation caused a significant reduction in synaptic contacts made by hippocampal neurons from 7-day old rats [315]. Early postnatal exposure to $^{137}$Cs (5 to 25 Gy) produced a dose-dependent reduction in myelin synthesis [366]. Other cells important to the functioning of the CNS (e.g., glia) develop early in gestation and continue to divide in the mature organism. Glia guide neuronal migration in the fetal brain and subserve neuronal activity in the adult. The limit for the detection of morphological changes in glia was determined to be as low as 0.2 Gy x-rays [657].

Corresponding behavioral alteration may also be observed after relatively low doses of ionizing radiation if exposure occurs prenatally or soon after birth when many cells are actively dividing or migrating [133]. In fact, behavioral indicators were shown to be more sensitive indicators of radiogenic damage than were morphological assessments of brain development [382,611]. Schull and associates [706,707] estimated that survivors exposed *in utero* to the atomic bombings in Hiroshima and Nagasaki exhibited a diminution in intelligence score of 21 to 27 points per Gy. The highest risk of severe mental retardation occurred during the 8th to 15th week of gestation when radiation exposure coincided with the most rapid period of proliferation of neuronal elements and the migration of immature neurons to the cerebral cortex [612,705].

The behavioral results of prenatal irradiation are discussed in several in-depth reviews [98,397,400]. Most of the research focuses on motor performance. Prenatal irradiation can cause significant alterations in gait; for example, D'Amato and Hicks [189] observed a hopping locomotion in rats exposed to 1.5 Gy on gestation days 14 and 15. After some initial difficulty, these animals learned to traverse horizontal ladders by adapting their hopping gait to navigate the rungs. Norton and Kimler [595] also found that rats exposed to 1 Gy from a $^{137}$Cs source on gestation day 15 showed deficits in muscular endurance as measured by their ability to sustain their own weight by hanging from a rod. This behavioral deficit (and others) were correlated with reduced thickness of the cerebral cortex. Fractionation of the radiation dose resulted in less damage to the developing rat cerebral cortex, as measured by postnatal growth, behavioral tests, and morphological assessment [401,808]. Motor deficits were also observed in rats exposed to a low dose (0.6 Gy) on gestation day 16 [93] or in mice exposed to 0.35 Gy on days 11 or 12 after conception [56]. These animals had difficulty obtaining a reward if they were required to make motor responses in rapid succession (pressing a lever 4 times in 2 seconds). Administration of the radical reactive oxygen species (ROS) scavenger amifostine 30 minutes prior to exposure of rat pups to x-irradiation resulted in reducing cerebellar morphological damage and motor gait impairment [297]. Locomotor hyperactivity was reported after prenatal or perinatal radiation exposure [526]. In particular, locomotor activity was enhanced in mice irradiated with 1 Gy of $^{137}$Cs on gestation day 14 [539] and in rats exposed to 1.25 Gy of x-rays on gestation days 14 or 15 [596], as well as 2 Gy of x-rays on gestation day 17 [371]. Mice exposed to 1.0 Gy of $^{137}$Cs on day 14 of gestation exhibited higher levels of open-field activity at 19 to 20 months, but not at 6 to 7 or 12 to 13 months of age; thus, it was concluded that later behavioral changes of prenatally irradiated animals may depend on the age of testing [538].

Other behaviors were also affected by prenatal irradiation. Male mice irradiated with 2 Gy of $^{137}$Cs on gestation day 14 and tested at 100 to 135 days of age exhibited increased aggressiveness compared to controls [537]. When rats were exposed to 2 Gy x-irradiation on gestation day 17 and tested as adults, they showed enhanced performance on an active avoidance task (requiring movement in a shuttle box) [371,763] while passive avoidance (requiring a freezing response) was impaired [763]. Rats exposed to 1.5 Gy of $^{60}$Co gamma radiation on gestation day 15 exhibited a hyperresponse and delayed habituation on an acoustic startle test [540]. Mice receiving 0.1 to 0.5 Gy of x-rays [731] or 0.5 Gy of mixed neutron/gamma radiation [210] on gestation day 18 and tested in adulthood showed impairment on a spatial memory task [731]. Administration of D-amphetamine 10 minutes before testing as adults, however, alleviated neonatally induced x-irradiation-related deficits in short-term memory of rats [332]. Radiogenic deficits were also observed when animals (previously irradiated *in utero*) performed tasks with substantial cognitive components [102]. In 90-day-old squirrel monkeys exposed to gamma radiation (0.5 or 1 Gy) on gestation days 89 to 90, the correct responses in visual orientation, discrimination, and reversal learning tasks were significantly lower than those of controls. Decrements in reversal learning persisted undiminished in the irradiated subjects at 2 years of age.

Long-term neurocognitive deficits are common sequelae of cranial radiation therapy, particularly in children [2,21,25,34,333,740,805]. A dose of 24 Gy or more of radiation to the CNS of children under 5 years of age resulted in neurocognitive deficits that, it was reported, may not become apparent until 2 to 5 years after treatment [562]. Young children are most vulnerable due to toxicity to the developing brain. Mullenix and her colleagues used a neonatal rat model to investigate drugs to mitigate the effects of cranial radiation-induced behavioral deficits [572,573].

Significant changes in neurotransmitter levels can be measured in brains of rats exposed prenatally (e.g., 0.95 to 1.5 Gy on day 10, 12, or 15 of pregnancy). Marked changes in serotonin and serotonin receptors were found in several brain structures (e.g., hippocampus), and dopamine increased significantly in striatum. There were also significant increases in glutamate, glutamine, and gamma-aminobutyric acid (GABA) in cortex, hippocampus, striatum, and thalamus [205,530,555]. Administration of the NMDA receptor antagonist dizocilpine (MK-801), a glutamate blocker, before neonatal x-irradiation produced a dose-dependent behavioral protection in adult rats with radiation-induced hippocampal damage [527]. In a related study, dizocilpine administered 20 minutes after neonatal $^{60}$Co irradiation significantly reduced neuronal damage in rats 6 hours after exposure [10].

## Behavioral Effects of Adult Radiation Exposure

### Naturalistic Behaviors

Naturalistic behaviors (normal parts of an animal's response repertoire) may be altered by radiation exposure. In particular, spontaneous locomotor activity is of interest because this behavior is an important component of many other responses. Jones et al. [374] reported an immediate depression in rat volitional activity-wheel performance following an acute, whole-body dose of 2- to 7-Gy x-rays. Exposure of rats to 10-Gy high-energy electrons [237] or mice to 10-Gy $^{60}$Co gamma radiation [480] resulted in reductions in both horizontal and vertical activity within 1 hour of exposure. Radiation-induced hypoactivity has also been reported for a variety of species [525]. A biphasic locomotor response to radiation exposure (initial decrease, followed by partial recovery and then secondary hypoactivity) was reported for mice [432,433,804] and is consistent with the phasic postirradiation clinical symptomatology observed in humans [358].

Exposure to ionizing radiation also produces a dose-dependant reduction in food and water consumption, as well as nausea and vomiting (emesis) in a variety of species [256,257,397,404,495,779]. Taste-aversion learning (an association between a distinctive taste and radiation-induced malaise) appears to be an especially sensitive indicator of the effects of radiation on consummatory behaviors [644,645]. Taste aversions have been observed at very low doses following $^{56}$Fe exposure [646,647]. Diltiazem, a calcium channel blocker, prevented the onset of radiation-induced taste aversions in rats [571].

Another indicator of gastrointestinal malaise, vomiting, is more likely observed after irradiation with neutrons than gamma rays [220]. Increasing doses of radiation up to 10 Gy (neutron/gamma = 0.4) corresponded with the enhanced likelihood of vomiting in the monkey [528]. Above 10 Gy, however, the number of monkeys that vomited decreased with increasing dose. The $ED_{50}$ was approximately 4.5 Gy. Of the monkeys exposed to mixed neutron/gamma radiation (6 Gy) with a high neutron/gamma ratio of 0.85, 80% vomited in about 45 minutes [486].

The relationship between vomiting and performance decrement was shown to be complex [863], as irradiated animals rarely vomited during early behavioral incapacitations (see later discussions); for example, Franz [249] found no relationship between vomiting and early performance deficits in monkeys performing in a physical activity wheel and exposed to less than 50 Gy of mixed neutron/gamma radiation. Animals that were not incapacitated but received the same dose as the incapacitated animals vomited as expected [863,865]. Although these data are revealing, the relationship between radiation-induced vomiting and behavioral deficits remains to be fully elucidated. Serotonin receptor (5-$HT_3$) blocking agents were shown to be effective against radiation-induced vomiting [1,70,221,233] and to have relatively few behavioral side effects [72,73,91]. Treatment guidelines for radiation-induced nausea and vomiting have been described [239,348].

Social behaviors have not received much attention from radiobiologists. Miyachi and colleagues reported a paradoxical effect when the sexual [544] and aggressive [543,546] behaviors of mice were reduced following exposure to doses of 0.05 to 0.15 Gy of x-rays but not after higher doses of 0.25 to 0.35 Gy. Maier and Landauer [479,480] found that aggressive behaviors were surprisingly robust after 10 Gy (electron or $^{60}$Co gamma radiation) and persisted until radiogenic moribund behavior was evident.

### Motor Performance

Several studies revealed chronic deterioration on motivated motor performance tasks after doses of radiation at or below the $LD_{50}$; for example, Stapleton and Curtis [742] reported a long-term (42-week) progressive deterioration of forced wheel running behavior in mice exposed to a $LD_{50}$ dose of neutron radiation. Kimeldorf et al. [399] also found significant reductions in the motor capacity of rats swum daily to exhaustion before and after exposure to 3 to 10 Gy of x-rays. Rats performing a task where they had to press a bar 20 times rapidly to avoid footshock showed significant performance decrements following 7.5 Gy of $^{60}Co$ gamma radiation. This decrement persisted for the first 4 weeks after exposure [507]. Performance of a physically demanding motor task can alter survival after irradiation. Kimeldorf and Jones [398] reported that swimming to exhaustion before and after x-irradiation significantly reduced rat performance and lowered the $LD_{50}$ by about 2 Gy. Bogo [88,89] observed a similar phenomenon in rats performing a strenuous, shock-motivated motor task after irradiation.

## Learning, Memory, and Cognition

A number of studies suggest that learning can be altered by ionizing radiation exposure. Meyerson [512] conditioned rabbits to associate a light and tone stimulus with the respiratory reflex of apnea (cessation of breathing) produced by inhalation of ammonia vapor. A 15-Gy exposure to $^{60}Co$ gamma radiation produced an absent, or considerably reduced, conditioned apnea response to the light or tone. In contrast, the unconditioned apnea (normal response to ammonia inhalation) was enhanced after irradiation, suggesting that the performance capacity of the animal was intact. Others reported reduced maze-learning behavior after up to 10 Gy of x-rays [242]. Urmer and Brown [793] also found a temporary reduction in the ability of rats to reorganize previously learned material after 4-Gy $^{60}Co$ gamma radiation. In humans, adults receiving radiotherapy for gliomas, nasopharyngenal carcinoma, and head and neck tumors were reported to exhibit some degree of cognitive deficit following therapy [3,407,430]. An assessment of Chernobyl liquidators 9 to 12 years after the accident revealed significant neurocognitive deficits [265].

These data support the notion that radiation affects some components of learning, and they are consistent with other results [116] suggesting that radiogenic disruptions in behavior may not merely reflect defects in nonassociative factors. Although research in this field has found that there are postirradiation learning deficits, improved or unaltered learning capacity after irradiation was observed under some circumstances [525]. A 10-Gy dose of x-rays of the brain induced cognitive impairments that were associated with reduced cell proliferation in the dentate subgranular zone of the hippocampus [643,667].

Radiation exposure may also disrupt memory; for example, Wheeler and Hardy [835] found a significant retrograde amnesia in rats after 0.001 to 0.1 Gy of electron irradiation. Moderate doses (4.5 Gy) of acute gamma radiation to mice were reported to produce alterations in cognitive behavior (single-trial passive avoidance task), while low doses (1.5 Gy) caused no significant behavioral changes [487]. Exposure to $^{56}Fe$ particles produces cognitive deficits in rats that are similar to those observed in aged animals [203,379,727,728]. Human memory may be impaired by radiation exposure, as well. A few cases of acute retrograde amnesia were reported by people who survived the bombing of Hiroshima [370]. Five years after the attack, deficits in memory and intellectual capacity were noted in individuals experiencing radiation sickness [169]. Although there may be alternative explanations for these amnesias (e.g., psychological trauma), the data are consistent with Soviet literature, which reported memory deficits in patients undergoing therapeutic irradiations [369].

Exposure to ionizing radiation is known to alter performance on behavioral tasks requiring nondemanding physical movements and the involvement of functional cognitive processes, such as timing, decision making, or concept formation. For example, cynomolgus monkeys tested 2 to 3.5 months after a 20-Gy, head-only exposure to x-rays or gamma rays showed a deficit on a series of discrimination problems [659]. Cranial irradiation of rats with single doses of 20 or 25 Gy of x-rays produced delayed impairment of spatial learning and working memory [337]. Chimpanzees exhibited a chronic inability to perform an oddity discrimination task after whole body $^{60}Co$ gamma radiation of 4 Gy [659]. Highly trained complex behavior in laboratory rats can be disrupted with sublethal levels (4.5 to 6.75 Gy) of $^{60}Co$ gamma radiation [506–508,843,844]. Typically, radiation-induced disruptions in rates and patterns of responding become evident within 1 day of exposure, with a duration of 24 to 96 hours before returning to pre-exposure levels [508].

## Early, Transient Performance Deficits

Transient performance deficits were observed in animals and humans after a large, rapidly delivered dose of ionizing radiation. This response has been termed early transient incapacitation (ETI) [707]. An idealized, individual ETI profile is shown in the upper part of Figure 18.6. As shown, 5 to 10 minutes after radiation exposure, performance rapidly fell to near zero, followed by partial or total recovery 10 to 15 minutes later. Delayed ETIs also occurred about 45 minutes and 4 hours after irradiation. A less severe variant of ETI is early performance decrement (EPD), in which performance is significantly degraded rather than totally suppressed. ETI and EPD were presumed to occur only after supralethal radiation doses in which, following behavioral recovery, death occurred in hours or days; however, some data suggest that lower doses may also produce these effects [87].

Early transient performance decrements are modulated by several factors; for example, the radiation dose required to disrupt behavior is directly related to the requirements of the task being performed. Demanding/complex tasks and tasks that require rapid responding are most disrupted after radiation exposure [525]. Radiation dose and dose

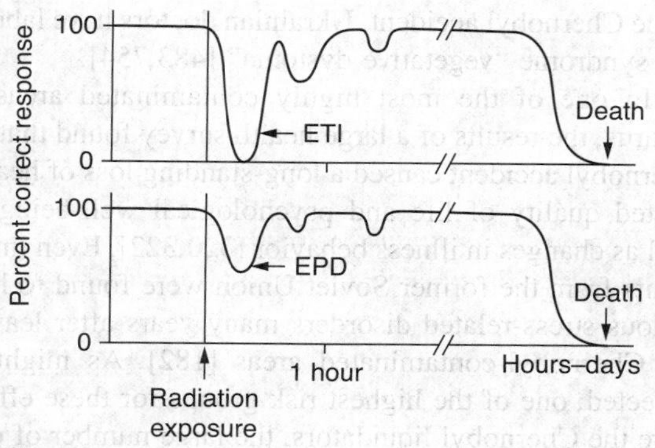


**FIGURE 18.6** Idealized performance time courses for acute radiation-induced behavioral decrement: As shown, soon after a sufficiently large dose of radiation, several animal species exhibit an early transient incapacitation (ETI, upper panel) or an early performance decrement (EPD, lower panel). Subsequent smaller transient deficits may occur approximately 45 minutes and 4 hours later. (Data from di Cicco et al. [210], Norton and Kimler [595], and Shigematzu [723].)

rate can also influence behavioral deficits. When 10 Gy of $^{60}$Co gamma radiation was given at dose rates from 0.3 to 1.8/minute, it produced 7 to 81% ETI [108].

The type of radiation can also differentially influence early transient behavioral deficits. The median effective doses required to disrupt rat performance on an accelerating rotating rod were 61 Gy for 18.6 MeV electrons, 81 Gy for 18.1 MVp Bremsstrahlung, 89 Gy for 1.25 MeV gamma photons, and 98 Gy for 1.67 MeV neutrons. Thus, in contrast to typical lethality data, electrons are significantly more effective than neutrons in producing motor deficits [90].

Several human accidents have involved very large doses of ionizing radiations sufficient to produce behavioral incapacitation. One of these exposures occurred in the early days of fissionable material production at the Los Alamos Scientific Laboratory and resulted in the fatal radiation injury of a worker known as Mr. K. The accident victim received a rapid total body dose of 45 Gy and an estimated upper abdominal dose of 120 Gy of mixed neutron and gamma radiations [725]. During the event, Mr. K either fell or was knocked to the floor. For a short period, he was apparently dazed as he turned his plutonium-mixing apparatus off and then back on again. He was able to run to another room but soon became ataxic and disoriented. He could not stand unaided, was incapacitated, and drifted in and out of consciousness for more than a half hour before he was rushed to a local hospital. Later, Mr. K regained consciousness and coherence. From 2 to 30 hours after the accident he showed significant behavioral recovery, at some points experiencing euphoria, although his clinical signs were grave. The few hours before his death (35 hours after the irradiation) were characterized by irritability, uncooperativeness, mania, and eventually coma. This case is consistent with the animal literature suggesting that a supralethal dose of radiation can produce early transient performance deficits. The physiological and behavioral symptoms as well as lethality associated with acute radiation effects in humans following exposure to radiation were summarized by Anno and his colleagues [26,27]. The medical management of acute radiation exposure is reviewed by Waselenko and colleagues [827].

The Chernobyl nuclear power plant accident in 1986 also produced behavioral deficits in personnel attempting to perform duties in high-radiation environments. A firefighter who fought the blaze of the burning reactor core suffered performance deficits and eventually had to withdraw because of radiation exposure; another individual exposed to an estimated 2.0 to 3.5 Gy during the accident was reported to have permanent headaches and vision impairment [83]. These human accidents add to the animal literature suggesting that sublethal doses of radiation can also induce performance decrements; however, children exposed to low doses of radiation from Chernobyl were reported to show no abnormal neurobehavioral or cognitive performance [53].

## Protection Against Radiogenic Behavioral Disruption

Few studies have attempted to normalize behavioral changes observed immediately (up to 24 hours) after irradiation; however, as described earlier, antihistamines (e.g., chlorpheniramine) [218,516] and opiate antagonists (e.g., naloxone) may offer behavioral radioprotection under certain circumstances [520]. Other data suggest that estrogens may reduce the intensity and duration of radiation-induced early transient behavioral deficits in castrated rats trained to perform an avoidance task [515]. Early studies administered *n*-decylaminoethanethiosulfuric acid to monkeys and reported some protection against ETI [717,787]. Radioprotectants that have been used to reduce the lethal effects of radiation have also been evaluated for behavioral toxicity [436,832]. A number of compounds are currently being examined for radioprotective efficacy [113,627,833]. To date, the only FDA-approved radioprotective drug is amifostine [417,765], formerly known as WR-2721. Although amifostine is available for clinical use in conjunction with radiotherapy [92,766,817], the side effects of the drug (emesis, hypotension) limit its use [694,831]. Unfortunately, WR-2721 and other phosphorothioates, at the best radioprotective doses, were behaviorally toxic alone (that is, they disrupted trained behavior or reduced locomotor activity) and potentiated rather than attenuated radiation-induced performance decrements [434,435,438,439,496,832]. In general, the behavioral toxicity of each of a variety of radioprotective compounds increases in parallel with the radioprotective properties [436,437].

## Psychological Factors of Radiation Exposure

Compared to what we know about the physiological changes brought about by ionizing radiation exposure, we know little about the psychological changes that may also be exhibited. The information we do have is derived from the Japanese atomic bomb experience, human radiation accidents, clinical radiation exposures, and selected animal studies. The data from animal studies reflect behavioral changes resulting from direct effects of radiation on nervous system functioning. The human data reflect additional social, cognitive, and cultural factors that modulate human emotional and psychological phenomena [64,517].

The animal data have shown, for example, that we can expect motivational changes after sufficiently large doses of ionizing radiation. Rats that initially exerted similar amounts of work to receive rewarding brain stimulation of several different brain nuclei, following exposure to 100 Gy of high-energy electrons, stopped working for stimulation of some nuclei (e.g., septum) but continued working for stimulation of the lateral hypothalamus [518]. Although these irradiated subjects maintained the capacity to perform bar pressing, selective modification of their motives occurred such that they chose to exert work to obtain a selected subset of the incentives that were previously all rewarding.

In addition to the physiologically mediated changes in psychological variables that might be revealed by animal studies, human perceptions, interactions, and expectations can combine to produce distinct changes in emotional responses after radiation exposure [166,685,866]. For example, the nuclear reactor accident in 1979 at Three Mile Island produced virtually no radiation exposure above background levels. Still, the perceived radiation hazard and the public's nuclear phobia evoked long-term emotional, behavioral, and physiological signs of stress [58,165,167,168]. The particular fear associated with potential radiation exposure from nuclear power plants seems to be heightened by the fact that ionizing radiation presents an invisible, unfamiliar, man-made (and therefore unnatural) hazard [617,683]. Interestingly, fear and dread of radiation are lessened when the radiation source is natural and the individual may encounter it in a familiar setting (e.g., his or her own basement), as is the case with radon [829].

Psychological symptoms that have been reported following radiation accidents have been quite dramatic. Fear, anxiety, stress, depression, neurasthenia, and hypochondria were reported as part of the clinical course of persons exposed to radiation during the Chernobyl nuclear power plant accident [104,183,284,388,780,809,830]. Deficits in memory, attention, and sensorimotor activities were also observed in these patients [145]. Many individuals and children, in particular, reported symptoms of fatigue, pallor, inattention, abdominal pain, and headache as a result of the Chernobyl accident. Ukrainian doctors have labeled this syndrome "vegetative dystonia" [483,754].

In one of the most highly contaminated areas of Belarus, the results of a large health survey found that the Chernobyl accident caused a long-standing loss of health-related quality of life and psychological well-being, as well as changes in illness behavior [320,322]. Even immigrants from the former Soviet Union were found to have various stress-related disorders many years after leaving the Chernobyl-contaminated areas [182]. As might be expected, one of the highest risk groups for these effects were the Chernobyl liquidators, the large number of people who were involved in cleaning up after the accident [809]. The clinical significance of this well-documented psychological stress due to the Chernobyl accident continues to be investigated [280,321,810].

In the 1987 accident involving a radiation therapy source in Goiania, Brazil, 79 people received measurable radiation contamination, of which 20 were hospitalized; however, 112,000 people were screened out of fear of exposure [96,200]. The survivors of these accidents, as well as those in Hiroshima and Nagasaki [54], all exhibit either a preoccupation with somatic symptoms, a fear about future health effects, neurasthenia (fatigue, weakness, dizziness, headaches), posttraumatic stress symptoms, or clinical depression. This observation led Bromet [103] to conclude that "radiation catastrophes share an important underlying dimension, namely, perception of risk to health, a perception that for many survivors can turn into an unresolvable fear."

## Summary

Radiation doses below the $LD_{50}$ (whole body) do not produce permanent sensory changes; however, transient alterations were reported in several modalities at doses from 1 to 5 Gy. High radiation doses can cause more permanent sensory and perceptual impairments. Radiogenic damage to mature brain morphology may occur after an exposure of less than 15 Gy and is an accepted finding at higher doses. The developing CNS is significantly more sensitive than the mature nervous system to radiation-induced changes in a variety of histological, morphological, and behavioral parameters; likewise, indicators of brain functioning (e.g., electrophysiology, neurochemistry) are more radiosensitive than are neuroanatomical endpoints.

Under many circumstances, exposure to ionizing radiations can significantly impede performance. This conclusion is supported by extensive research on experimental animals and limited human experiences. At low to intermediate doses of radiation (up to 10 Gy), performance deficits may be slow in developing and relatively long lasting. After high doses, the behavioral effects are often rapid (within minutes) and usually abate before the debilitation of chronic radiation sickness begins. These rapid effects can also occur at intermediate doses under certain circumstances.

Not all task performance is equally radiosensitive. Tasks with complex, demanding requirements are more easily disrupted than simple ones. The exception may be found in certain naturalistic behaviors (e.g., eating and drinking behavior) that are also quite radiosensitive. Postirradiation deficits in memory, cognitive ability, and motor performance have been reported. Radiation parameters such as dose, dose rate, and radiation quality can all influence the degree of performance decrement observed.

Many of the pharmacological compounds that protect animals from the lethal effects of ionizing radiation also have severe behavioral effects. In addition to the well-studied physiological and behavioral effects of ionizing radiations, psychological changes (e.g., motivational deficits, anxiety, depression) may also accompany radiation exposure or the threat of exposure.

## CONCLUSION

To assess the average exposure of residents of the United States to ionizing radiation, the National Council on Radiation Protection and Measurements obtained the collective effective dose equivalent from each of six main radiation source categories [587]. The collective effective dose equivalent is calculated by multiplying the average per capita effective dose equivalent by the estimated number of people exposed [47]. The average effective dose equivalent is then calculated by dividing the collective effective dose equivalent by the total U.S. population. The dose equivalent accounts for differences in relative biological effectiveness by multiplying the absorbed dose by the quality factor while the effective dose equivalent relates the dose equivalent to risk.

As seen in Table 18.2, natural radiation sources contribute 82% of the total average annual effective dose equivalent of 3.6 mSv. By far the largest contribution (55%) is made by radon and its decay products. Radon in domestic water supplies is also the chief contributor to radiation exposure from consumer products [34]. Although much is written about radiation exposure from nuclear power production and nuclear weapons testing fallout, their contributions are negligible compared to the importance of environmental radon, the largest source of human exposure to ionizing radiation.

Unfortunately, the exception to the rule occurred on April 26, 1986, when the graphite-moderated reactor of unit no. 4 of the Chernobyl nuclear power station of the former Soviet Union exploded, distributing a large amount of a variety of radionuclides throughout the northern hemisphere in what has been estimated to be the single most costly industrial accident in history [30,152,231,253,360, 410,714,790]. One result of the accident is that the contaminated areas of Belarus, Ukraine, and Russia have become a living laboratory of the consequences of radioactive contamination. The ecotoxicological effects of the Chernobyl accident have provided unprecedented observations of radionuclides in the environment, and the human health effects are only now becoming evident, with additional reports appearing each year. Many of these late reports have been included in this chapter because they are changing what is known about the toxicity of ionizing radiation.

## QUESTIONS

1. What processes are involved in the amplification of a single exposure of ionizing radiation to a final endpoint of biological expression (injury, cancer, or death)?
2. The most important sensitizer of biological tissues to ionizing radiation is ________.
3. Describe how radiation is thought to affect the cell cycle. In which phase of the cell cycle are cells more sensitive to ionizing radiation? In which phase are they more sensitive?
4. Why is a 20-cGy dose of neutron radiation able to result in cataract formation, but not a 20-cGy dose of x-radiation? What is the difference between a high LET radiation and a low LET radiation? How are RBE and LET related?
5. The Law of Bergonie and Tribondeau put forth in 1906 describes basically why some cells are more sensitive to ionizing radiation than other cells. Why are bone marrow cells and cancer cells sensitive to ionizing radiation?
6. What is the difference between *ionization* and *excitation*? What is a thiol, and how does it interact with a free radical?
7. What is the estimated lethal dose of ionizing radiation for humans? What are the lethality syndromes induced by ionizing radiation? Why can an organism tolerate a larger dose if it is protracted or fractionated, while a smaller, acute exposure dose can be lethal?
8. What is the role of apoptosis in response to radiation injury?
9. What have we learned about the effects of human populations living in the relatively highly contaminated areas around the Chernobyl reactor?
10. What is the reality of the actual incidence of birth defects from human exposure to environmental (not medical) radiation?
11. Discuss the behavioral alterations that can occur as a result of prenatal irradiation.
12. What is radiation-induced early transient incapacitation?
13. Explain the relatively high incidence of thyroid cancer in Belarus, Ukraine, and Russia in the 1990s.

14. How would a diet high in Brazil nuts contribute to a high body burden of radionuclides and where in the body would they be concentrated?
15. Explain the difference in the risk of consuming clams, oysters, or scallops from an area contaminated with $^{60}$Co and from one contaminated with $^{90}$Sr.
16. Explain why denudation of the gastrointestinal mucosa may not be seen with death occurring from the neurovascular syndrome associated with the acute radiation syndrome.
17. Why does the conceptus exhibit the lowest irradiation $LD_{50}$ of any stage of an organism's life?
18. How does irradiation depress both the specific and nonspecific immune responses?
19. Why is a dirty bomb not considered a weapon of mass destruction?

## REFERENCES


1. Aapro, M. (2004): Granisetron: an update on its clinical use in the management of nausea and vomiting. *Oncologist*, 9:673–686.
2. Abayomi, O. K. (1996): Pathogenesis of irradiation-induced cognitive dysfunction. *Acta Oncol.*, 35:659–663.
3. Abayomi, O. K. (2002): Pathogenesis of cognitive decline following therapeutic irradiation for head and neck tumors. *Acta Oncol.*, 41:346–351.
4. Abbas, B., Boyle, F. C., and Wilson, D. J. et al. (1990): Radiation induced changes in the blood capillaries of rat duodenal villi: a corrosion cast, light and transmission electron microscopical study. *J. Submicrosc. Cytol. Pathol. (Bologna)*, 22:63–70.
5. Abbas, B., Hume S. P., McCullough, J. S. et al. (1990): Early morphological changes in blood capillaries of mouse duodenal villi induced by x-irradiation. *J. Submicrosc. Cytol. Pathol. (Bologna)*, 22:609–614.
6. Abdullin, G. Z. (1962): *Study of Comparative Radiosensitivity of Different Parts of Brain in Terms of Altered Function*, Atomic Energy Commission TR-5141, OTS/Department of Commerce, Washington, D.C.
7. Ahmed, M. M. (2004): Regulation of radiation-induced apoptosis by early growth response-1 gene in solid tumors. *Curr. Cancer Drug Targets*, 4:43–52.
8. Akil, H., Madden, J., Patrick III, R. L., and Barchas, J. D. (1976): Stress-induced increase in endogenous opioid peptides: concurrent analgesia and its reversal by naloxone. In: *Opiates and Endogenous Opiate Peptides*, edited by H. W. Kosterlitz, p. 63. Elsevier, North-Holland, Amsterdam.
9. Akiyama, M. (1995): Late effects of radiation on the human immune system: an overview of immune response among the atomic-bomb survivors. *Int. J. Radiat. Biol.*, 68(5):497–508.
10. Alaoui, F., Pratt, J., Trocherie, S. et al. (1995): Acute effects of irradiation on the rat brain: protection by glutamate blockade. *Eur. J. Pharmacol.*, 276:55–60.
11. Allison, G. T. (2004): *Nuclear Terrorism: The Ultimate Preventable Catastrophe*. Henry Holt & Co., New York.
12. Alman, R. W., Rosenberg, M., and Fazekas, J. F. (1952): Effects of histamine on cerebral hemodynamics and metabolism. *AMA Arch. Neurol. Psychiat.*, 67:354–356.
13. Al-sarraf, M., LeBlanc, M., Giri, P. G. S. et al. (1998): Chemoradiation therapy versus radiation therapy in patients with advanced nasopharyngeal cancer: phase III randomized intergroup study 0099. *J. Clin. Oncol.*, 16:1310–1317.
14. Alter, W., Mickley, G. A., Catravas, G. et al. (1980): Role of histamine and beta-endorphin in radiation-induced hypotension and acute performance decrement in the rat. In: *Proc. of the 51st Annual Scientific Meeting of the Aerospace Medical Association*, pp. 225–226.
15. Alter, W. A., Catravas, G. N., Hawkins, R. N., and Lake, C. R. (1984): Effect of ionizing radiation of physiological function in the anesthetized rat. *Radiat. Res.*, 99:394–409.
16. Altman, K. I., Gerber, G. B., and Okada, S. (1970): *Radiation Biochemistry*. Academic Press, New York.
17. Amundson, S. A., Bittner, M., Meltzer, P. et al. (2001): Induction of gene expression as a monitor of exposure to ionizing radiation. *Radiat. Res.*, 156:657–661.
18. Amundson, S. A., Bittner, M., and Fornace, Jr., A. J. (2003): Functional genomics as a window on radiation stress signaling. *Oncogene.*, 22:5828–5833.
19. Amundson, S. A., Grace, M. B., McLeland, C. B. et al. (2004): Human *in vivo* radiation induced biomarkers: gene expression changes in radiotherapy patients. *Cancer Res.*, 64:6368–6371.
20. Anderson, R. E., Tokuda, S., Williams, W. L., and Spellman, C. W. (1986): Low dose irradiation permits immunization of A/J mice with subimmunogenic numbers of Sal cells. *Brit. J. Cancer*, 54:505–509.
21. Anderson, V., Godber, T., Smibert, E., and Ekert, H. (1997): Neurobehavioural sequelae following cranial irradiation and chemotherapy in children: an analysis of risk factors. *Pediatr. Rehabil.*, 1:63–76.
22. Andratschke, N. H., Nieder, C., Price, R. E. et al. (2004): Modulation of rodent spinal cord radiation tolerance by administration of platelet-derived growth factor. *Int. J. Radiat. Oncol. Biol. Phys.*, 60:1257–1263.
23. Andrew, F. D. and Lytz, P. S. (1981): Biochemical disturbances associated with developmental toxicity. In: *Developmental Toxicology*, edited by C. Kimmel and J. Buelke-Sam, pp. 145–165. Raven Press, New York.
24. Andrushchak, L. I., Gol'Dshmid, B. Y., Nikitchenko, V. V. et al. (1993): Morphological and ultrastructural changes in small intestine of rats under long-term constant action of low doses of ionizing radiation (Russia). *Tsitologiya I. Genetika.*, 27(6):13–19.
25. Andrykowski, M. A., Altmaier, E. M., Barnett, R. L. et al (1990): Cognitive dysfunction in adult survivors of allogeneic marrow transplantation: relationship to dose of total body irradiation. *Bone Marrow Transplant.*, 6:269–276.
26. Anno, G. H., Baum, S. J., Withers, H. R., and Young, R W. (1989): Symptomatology of acute radiation effects in humans after exposure to doses of 0.5–30 gy. *Health Phys.* 56:821–838.
27. Anno, G. H., Young, R. W., Bloom, R. M., and Mercier J. R. (2003): Dose response relationships for acute ionizing-radiation lethality. *Health Phys.*, 84:565–575.

28. Anscher, M. S., Kong, F.-M., and Jirtle, R. L. (1998): The relevance of transforming growth factor B1 in pulmonary injury after radiation therapy. *Lung Cancer*, 19:109–120.
29. Antonellia, A., Silvano, G., Bianchi, F. et al. (1995): Risk of thyroid nodules in subjects occupationally exposed to radiation: a cross-sectional study. *Occup. Environ. Med.*, 52(8):500–504.
30. Aoyama, M., Hirose, K., Inoue, H. et al. (1989): 30 years records of the radioactive fallout in Japan. *J. Radiat. Res. (Tokyo)*, 30:11.
31. Apanasenko, Z. I. (1967): *Combined Effect of Double Exposure to Vibration and Chronic Irradiation on the Functional State of Vestibular Apparatus*, NASA Technical Translation F-413, pp. 212–228. National Aeronautic and Space Administration, Washington, D.C.
32. Archambeau, J. O. (1987): Relative radiation sensitivity of the integumentary system: dose response of the epidermal, microvascular, and dermal populations. In: *Advances in Radiation Biology*. Vol. 12. *Relative Radiation Sensitivities of Human Organ Systems*, edited by J. T. Lett and K. I. Altman, pp. 147–203. Academic Press, San Diego, CA.
33. Arena, V. (1971): *Ionizing Radiation and Life*. Mosby, St. Louis, MO.
34. Armstrong, C. L., Gyato, K., Awadalla, A. W. et al. (2004): A critical review of the clinical effects of therapeutic irradiation damage to the brain: the roots of controversy. *Neuropsychol. Rev.*, 14:65–86.
35. Arnold, A., Bailey, P., and Harvey, R. A. (1954): Intolerance of primate brain stem and hypothalamus to conventional high energy radiations. *Neurology*, 4:575–585.
36. Arnold, M. and Kummermehr, J. (1988): Radiation induced damage to the regenerative capacity of surgically traumatized rat femur after single doses of x-rays. In: *Terrestrial Space Radiation and Its Biological Effects*, edited by P. D. McCormack, C. E. Swenberg, and H. B¸cker, pp. 475–486. Plenum Press, New York.
37. Arsenian, M. A. (1991): Cardiovascular sequelae of therapeutic thoracic radiation. *Prog. Cardiovasc. Dis.*, 33:299–311.
38. Asfandiyarova, N. S., Romadin, A. E., Kolcheva, N. G. et al. (1998): Immunity system in residents of territories contaminated with radionuclides after the Chernobyl accident (Russia). *Terapevticheskii Arkhiv.*, 70(1):55–59.
39. Astakhova, L. N., Anspaugh, L. R., Beebe, G. W. et al. (1998): Chernobyl-related thyroid cancer in children of Belarus: a case-control study. *Radiat. Res.*, 150:349–356.
40. Astor, M. B., Anderson, M. E., and Meister, A. (1988): Relationship between intracellular GSH levels and hypoxic cell radiosensitivity. *Pharmac. Ther.*, 39:115–121.
41. Auerbach, C. (1958): Radiomimetic substances. *Radiat. Res.*, 9:33–47.
42. Azzam, E. I. and Little, J. B. (2004): The radiation-induced bystander effect: evidence and significance. *Human Exp. Toxicol.*, 23:61–65.
43. Azzam, E. I., del Toledo, S. M., and Little, J. B. (2001): Direct evidence for the participation of gap junction-mediated intercellular communication in the transmission of damaged signals from α-particle irradiated to nonirradiated cells. *Proc. Natl. Acad. Sci. U.S.A.*, 98:473–478.
44. Babadzhanova, Sh.-A. and Busakov, B. S. (1997): Radiation as a risk factor for nerve system diseases. *Uzbekiston Tibbiet Zhurnali*, 0(5–7):41–43.
45. Bachofer, C. S. (1957): Enhancement of activity of nerves by x-rays. *Science*, 125:1140–1141.
46. Bachofer, C. S. and Gautereaux, M. E. (1959): X-ray effects on single nerve fibers. *J. Gen. Physiol.*, 42:723–735.
47. Bachofer, C. S. and Gautereaux, M. E. (1960): Bioelectric activity of mammalian nerves during x-irradiation. *Radiat. Res.*, 12:575–586.
48. Bachofer, C. S. and Gautereaux, M. E. (1960): Bioelectric response *in situ* of mammalian nerves exposed to x-rays. *Am. J. Physiol.*, 198:715–717.
49. Baetcke, K. P., Sparrow, A. H., Nauman, C. H., and Schwemmer, S. S. (1967): The relationship of DNA content to nuclear and chromosome volumes and to radiosensitivity ($LD_{50}$). *Proc. Natl. Acad. Sci. U.S.A.*, 58:533–540.
50. Baisakhatov, R. and Khanson, K. P. (1971): Comparison of the content of adenylic nucleotides and the activity of the process of oxidative phosphorylation in the rat thymus after total x-irradiation. *Radiobiologiya*, 11:155–159.
51. Baker, D. G. and Krochak, R. J. (1989): The response of the microvascular system to radiation: a review. *Cancer Invest.*, 7:287–294.
52. Barcellos-Hoff, M. H. (1993): Radiation-induced transforming growth factor β and subsequent extracellular matrix reorganization in the murine mammary gland. *Cancer Res.*, 53:3880–3886.
53. Bar Joseph, N., Reisfeld, D., Tirosh, E. et al. (2004): Neurobehavioral and cognitive performances of children exposed to low-dose radiation in the Chernobyl accident: the Israeli Chernobyl health effects study. *Am. J. Epidemiol.*, 160:453–459.
54. Barnaby, F. (1995): The effects of the atomic bombings of Hiroshima and Nagasaki. *Med. War.*, 11:1–9.
55. Baryakhtar, V. G., Sobotovich, E. V., Kulachinfskiy, A. et al. (2000): Evaluation of irradiation dose to illegal resettlers in the Chernobyl exclusion zone. In: *Acute and Remote Immunohematological Effects after the Chernobyl Accident in the Ukraine*, Special Issue, Vol. 1, pp. 77–84. Environmental Science and Pollution Research International.
56. Baskar, R. and Devi, P. U. (2000): Influence of gestational age to low-level gamma irradiation on postnatal behavior in mice. *Neurotoxicol. Teratol.*, 22:593–602.
57. Bassant, M. H. and Court, L. (1978): Effects of whole-body irradiation on the activity of rabbit hippocampal neurons. *Radiat. Res.*, 75:593–606.
58. Baum, A., Gatchel, R. J., and Schaeffer, M. A. (1983): Emotional, behavioral, and physiological effects of chronic stress at Three Mile Island. *J. Consult. Clin. Psychol.*, 51:565–572.
59. Baum, S. J., Anno, G. H., Young, R. W., and Withers, H. R. (1984): *Nuclear Weapon Effect Research at PSR-1983*. Vol. 10. *Symptomatology of Acute Radiation Effects in Humans after Exposure to Doses of 75 to 4500 rads (cGy) Free-in-Air*, DNA TR-85-50. Defense Nuclear Agency, Washington, D.C.
60. Bayraktar, M., Gedik, O., Akalin, S. et al. (1990): The effect of radioactive iodine treatment on thyroid C cells. *Clin. Endocrinol.*, 33:625–630.

61. Beard, C. M., Heath III, H., O'Fallon, W. M. et al. (1989): Therapeutic radiation and hyperparathyroidism: a case-control study in Rochester, Minn. *Arch. Intern. Med.*, 149:1887–1890.
62. Bebeshko, V. G., Bazyka, D. A., Chumak, A. A. et al. (2000): Short-term and long-term effects of radiation on laboratory animals and their progeny living in the Chernobyl nuclear power plant region. *Environ. Sci. Pollut. Res. Int.*, Special Issue (1):85–94.
63. Becciolini, A. (1987): Relative radiosensitivities of the small and large intestine. In: *Advances in Radiation Biology*, Vol. 12. *Relative Radiation Sensitivities of Human Organ Systems*, edited by J. T. Lett and K. I. Altman, pp. 83–128. Academic Press, San Diego, CA.
64. Becker, S. M. (2001): Psychosocial effects of radiation accidents. In: *Medical Management of Radiation Accidents*, edited by A. Igor, I. A. Gusev, A. K. Guskova, and F. A. Mettler, pp. 519–526. CRC Press, Boca Raton, FL.
65. Bedford, J. S. and Mitchell, J. B. (1973): Dose rate effects in synchronous mammalian cells in culture. *Radiat. Res.*, 54:316–327.
66. Beetz, A., Messer, G., Oppel, T. et al. (1997): Induction of interleukin 6 by ionizing radiation in a human epithelial cell line: control by corticosteroids. *Int. J. Radiat. Biol.*, 72:33–43.
67. BEIR III. (1980): *The Effects on Populations of Exposure to Low Levels of Ionizing Radiation*. Committee on the Biological Effects of Ionizing Radiations, National Research Council, National Academy Press, Washington, D.C.
68. BEIR IV. (1988): *Health Risks of Radon and Other Internally Deposited Alpha-Emitters*. Committee on the Biological Effects of Ionizing Radiations, National Research Council, National Academy Press, Washington, D.C.
69. BEIR V. (1990): *Health Effects of Exposure to Low Levels of Ionizing Radiation*. Committee on the Biological Effects of Ionizing Radiations, National Research Council, National Academy Press, Washington, D.C.
70. Belkacemi, Y., Ozsahin, M., Pene, F. et al. (1996): Total body irradiation prior to bone marrow transplantation: efficacy and safety of granisetron in the prophylaxis and control of radiation–induced emesis. *Int. J. Radiat. Oncol. Biol. Phys.*, 36:77–82.
71. Bengtsson, G. (1991): Introduction: present knowledge on the effects of radioactive contamination on pregnancy outcome. *Biomed. Pharmacother.*, 45:221–223.
72. Benline, T. A. and French, J. (1997): Anti-emetic drug effects on cognitive and psychomotor performance: granisetron vs. ondansetron. *Aviat. Space Environ. Med.*, 68:504–511.
73. Benline, T. A., French, J., and Poole, E. (1997): Anti-emetic drug effects on pilot performance: granisetron vs. ondansetron. *Aviat. Space Environ. Med.*, 68:998–1005.
74. Berg, N. O. and Lindgren, M. (1958): Time dose relationship and morphology of delayed radiation lesions of the brain of the rabbit. *Acta Radiol.*, 167:1–118.
75. Berger, N. A., Sims, J. L., Catino, D. M., and Berger, S. J. (1983): Poly(ADP-ribose)polymerase mediates the suicide response to massive DNA damage: studies in normal and DNA-repair defective cells. In: *ADP-Ribosylation, DNA Repair and Cancer*, edited by M. Miwa, O. Hayaisha, S. Shall, M. Smulson, and T. Sugimura, pp. 219–226. Science Society Press, Tokyo, Japan.
76. Bergonie, J. and Triblondeau, L. (1906): De quelques resultats de la radiotherapie et essai de fixation d'une technique rationannelle. *Comptes rendus des seances de l'acadenie des sciences*, 143:983–985; English translation by G. H. Fletcher (1959): Interpretation of some results of radiotherapy and an attempt at determining a logical technique of treatment. *Radiat. Res.*, 11:587–588.
77. Bermudez, J., Boyle, E. A., Miner, W. D., and Sanger, G. J. (1988): The anti-emetic potential of the 5-hydroxytryptamine$_3$ receptor antagonist BRL 43694. *Br. J. Cancer*, 58:644–650.
78. Biaglow, J. E., Varnes, M. E., Clark, E. P., and Epp, E. R. (1987): Role of glutathione and other thiols in cellular response to radiation and drugs. In: *Radiation Research: Proc. of the 8th Int. Congress of Radiation Research*, Vol. 2, edited by E. M. Fielden, J. F. Fowler, J. H. Hendry, and D. Scott, pp. 677–682. Taylor & Francis, New York.
79. Biard, D. S. F., Saintigny, Y., Maratrat, M. et al. (1997): Enhanced expression of the Kin17 protein immediately after low doses of ionizing radiation. *Radiat. Res.*, 147:442–450.
80. Bigbee, W. L., Jensen, R. H., Veideaum, T. et al. (1996): Glycophorin A biodosimetry in Chernobyl cleanup workers from the Baltic countries. *Br. Med. J.*, 312:1078–1079.
81. Bigbee, W. L., Jensen, R. H., Veidebaum, T. et al. (1997): Biodosimetry of Chernobyl cleanup workers from Estonia and Latvia using the glycophorin A *in vivo* somatic cell mutation assay. *Radiat. Res.*, 147:215–224.
82. Billen, D. (1987): Free radical scavenging and the expression of potentially lethal damage in x-irradiated repair deficient *Escherichia coli*. *Radiat. Res.*, 111:354–360.
83. Birioukov, A., Meurer, M., Peter, R. U. et al. (1993): Male reproductive system in patients exposed to ionizing irradiation in the Chernobyl accident. *Arch. Androl.*, 30:99–104.
84. Blakley, W. F., Miller, A. C., McLeland, C. B. et al. (2003): Radiation biodosimetry: applications for spaceflight. *Adv. Space Res.*, 31:1487–1493.
85. Blank, K. R., Rudolz, M. S., Kao, G. D. et al. (1997): The molecular regulation of apoptosis and implications for radiation oncology. *Int. J. Radiat. Biol.*, 71:455–466.
86. Bleuer, J. P., Averkin, Y. I., and Abelin, T. (1997): Chernobyl-related thyroid cancer: what evidence for role of short-lived iodines? *Environ. Health Perspect.*, 105(6):1483–1486.
87. Bogo, V., Franz, C. F., and Young, R. W. (1987): Effects of radiation on monkey visual discrimination performance. In: *Proc. of the 8th Int. Congress of Radiation Research*, edited by E. M. Fielden, J. F. Fowler, J. H. Hendry, and D. Scott, p. 259. International Association for Radiation Research, Edinburgh, Scotland.
88. Bogo, V. (1988): Radiation: behavioral implications in space. *Toxicology*, 49:299–307.
89. Bogo, V., Zeman, G. H., and Dooley, M. (1989): Radiation quality and rat motor performance. *Radiat. Res.*, 118:341–352.
90. Bogo, V., Dennison, B. A., and Mulvihill, M. (1989): Motor performance, radiation and mortality in rats. In: *Proc. of the 37th Annual Meeting of the Radiation Research Society*, Vol. 1, p. 139.
91. Bogo, V., Boward, C., and Fiala, N. et al. (1989): Zacopride: a nonbehaviorally toxic radiation antiemetic. In: *Proc. of the 60th Annual Scientific Meeting of the Aerospace Medical Association*.

92. Bohuslavizki, K. H., Brenner, W., Klutmann, S. et al. (1998): Radioprotection of salivary glands by amifostine in high-dose radioiodine therapy. *J. Nucl. Med.*, 39:1237–1242.
93. Bornhausen, M. (1986): Analysis of behavioral changes induced by prenatal irradiation. In: *Radiation Risks to the Developing Nervous System*, edited by H. Kriegel, W. Schmahl, G. B. Gerber, and F.-E. Stieve, pp. 283–293. Stuttgart, Gustav Fischer Verlag.
94. Bossola, M., Merrick, H. W., Eltaki, A. et al. (1990): Rat liver tolerance for partial resection and intraoperative radiation therapy: regeneration is radiation dose dependent. *J. Surg. Oncol.*, 45:196–200.
95. Bountra, C., Bunce, K., Dale, T. et al. (1993): Anti-emetic profile of a non-peptide neurokinin NK1 receptor antagonist, CP-99,994, in ferrets. *Eur. J. Pharmacol.*, 249:R3–R4.
96. Brandao-Mello, C. E., Oliveria, A. R., and Carvalho, A. B. (1991): The psychological effects of the Goiania radiation accident on the hospitalized victims. In: *The Medical Basis for Radiation-Accident Preparedness*. Vol. III. *The Psychological Perspective*, edited by R. C. Ricks, M. E. Berger, and F. M. O'Hara, Jr., pp. 121–129. Elsevier, New York.
97. Breiter, N., Trott, K. R., and Sassy, T. (1989): Effect of x-irradiation on the stomach of the rat. *Int. J. Radiat. Oncol. Biol. Phys.*, 17:779–784.
98. Brent, R. L. (1984): The effects of ionizing radiation, microwaves, and ultrasound on the developing embryo: clinical interpretations and applications of the data. *Curr. Prob. Pediatr.*, 14:1–87.
99. Brent, R. L., Beckman, D. A., and Jensh, R. P. (1987): Relative radiosensitivity of fetal tissues. In: *Advances in Radiation Biology*. Vol. 12. *Relative Radiation Sensitivities of Human Organ Systems*, edited by J. T. Lett and K. I. Altman, pp. 239–256. Academic Press, San Diego, CA.
100. Breuer, R., Tochner, Z., Conner, M. W. et al. (1992): Superoxide dismutase inhibits radiation-induced lung injury in hamsters. *Lung*, 170:19–29.
101. Brizel, D. M., Wasserman, T. H., Henke, M. et al. (2001): Phase III randomized trial of amifostine as a radioprotector in head and neck cancer. *J. Clin. Oncol.*, 19:1233–1244.
102. Brizzee, K. R. and Ordy, J. M. (1986): Effects of prenatal ionizing radiation on neural function and behavior. In: *Radiation Risks to the Developing Nervous System*, edited by H. Kriegel, W. Schmahl, G. B. Gerber, and F.-E. Stieve, pp. 255–282. Stuttgart, Gustav Fischer Verlag.
103. Bromet, E. J. (1998): Psychological effects of radiation catastrophes. In: *Effects of Ionizing Radiation: Atomic Bomb Survivors and Their Children* (1945–1995), edited by L. E. Peterson and S. Abrahamson. Joseph Henry Press, Washington, D.C.
104. Bromet, E. J., Gluzman, S., Schwartz, J. E., and Goldgaber D. (2002): Somatic symptoms in women 11 years after the Chernobyl accident: prevalence and risk factors. *Environ. Health Perspect.*, 110(Suppl. 4):625–629.
105. Brooks, A. (2004): Evidence for 'bystander' effects *in vivo*. *Human Exp. Toxicol.*, 23:67–70.
106. Brown, J. M. and Kovacs, M. S. (1993): Visualization of nonreciprocal chromosome exchanges in irradiated human fibroblasts by fluorescence *in situ* hybridization. *Radiat. Res.*, 136:71–96.
107. Brown, R. G. and Segal, D. S. (1980): Alterations in beta-endorphin-induced locomotor hyperactivity in morphine tolerant rats. *Neuropharmacology*, 19:619–621.
108. Bruner, A. (1977): Immediate dose-rate effects of $^{60}$Co on performance and blood pressure in monkeys. *Radiat. Res.*, 70:378–390.
109. Bruni, J. E., Persaud, T. V., Huang, W. et al. (1993): Postnatal development of the rat CNS following *in utero* exposure to a low dose of ionizing radiation. *Exp. Toxicol. Pathol.*, 45(4):223–231.
110. Bruni, J. E., Persaud, T. V., Froese, G. et al. (1994): Effects of *in utero* exposure to low dose ionizing radiation on development in the rat. *Histol. Histopathol.*, 9(1):27–33.
111. Bucky, G., Blank, F., and Distelheim, I. H. (1950): Influence of genz rays on histamine-induced manifestations. *Arch. Derm. Syph.*, 62:319–322.
112. Buell, M. G. and Harding, R. K. (1989): Proinflammatory effects of local abdominal irradiation on rat gastrointestinal tract. *Dig. Dis. Sci.*, 34:390–399.
113. Bump, E. and Malaker, K. (1998): *Radioprotectors: Chemical, Biological, and Clinical Perspectives*, CRC Press, Washington, D.C.
114. Burger, A., Loeffler, H., Bamburg, M., and Rodemann, H. P. (1998): Molecular and cellular basis of radiation fibrosis. *Int. J. Radiat. Biol.*, 73:401–408.
115. Burghardt, W. F. and Hunt, W. A. (1984): The interactive effects of morphine and ionizing radiation on the latency of tail withdrawal from warm water in the rat. In: *Proc. of the 9th Symp. on Psychology in the Department of Defense*, USAFA Tech. Rep. No. 84-2, edited by G. E. Lee and T. E. Ulrich, pp. 73–76. U.S. Air Force Academy, Colorado Springs, CO.
116. Burt, D. H. and Ingersoll, E. H. (1965): Behavioral and neuropathological changes in the rat following x-irradiation of the frontal brain. *J. Comp. Physiol. Psychol.*, 59:90–93.
117. Byelorussian SSR. (1991): *Byelorussia and Chernobyl*. The delegation of the Byelorussian SSR at the 45th Session of the United Nations General Assembly, a review, pp. 6–52. Minsk Belarus Publishers, Minsk, Belarus.
118. Byrnes, M. E., King, D. A., and Tierno, Jr., P. M. (2003): *Nuclear, Chemical, and Biological Terrorism. Emergency Response and Public Protection*. CRC Press, Boca Raton, FL.
119. Calabrese, E. J. and Baldwin, L. A. (2000): Radiation hormesis: the demise of a legitimate hypothesis. *Human Exp. Toxicol.*, 19:76–84.
120. Calabrese, E. J. and Baldwin, L. A. (2004): Hormesis: a generalizable and unifying hypothesis. *Crit. Rev. Toxicol.*, 31:353–424.
121. Calvo, W. (1993): Experimental radiation damage of the central nervous system. *Recent Results Cancer Res.*, 130:175–188.
122. Carlson, R. G., Mayfield, W. R., Normann, S., and Alexander, J. A. (1991): Radiation-associated valvular disease. *Chest*, 99:538–545.
123. Carpenter, D. O. (1979): *Early Transient Incapacitation: A Review with Considerations of Underlying Mechanisms*, AFRRI Scientific Report, SR 79-1, Armed Forces Radiobiology Research Institute, Bethesda, MD.

124. Carter, S., Auer. K. L., Reardon, D. B. et al. (1998): Inhibition of mitogen activated protein (MAP) kinase cascade protentiates cell killing by low dose ionizing radiation in A431 human squamous carcinoma cells. *Oncogene.*, 16:2787–2796.
125. Cassaret, G. W. (1980): *Radiation Histopathology.* CRC Press, Boca Raton, FL.
126. Castillo, L. A., Craft, A. W., Kernahan, J. et al. (1990): Gonadal function after 12-Gy testicular irradiation in childhood acute lymphoblastic leukaemia. *Med. Pediatr. Oncol.*, 18:185–189.
127. Catravas, G. N. and McHale, C. G. (1973): *Activity Changes of Brain Enzymes in Rats Exposed to Different Qualities of Ionizing Radiation*, AFRRI Scientific Report SR 73-19. Armed Forces Radiobiology Research Institute, Bethesda, MD.
128. Catterall, W. A. (1984): The molecular basis of neuronal excitability. *Science*, 223:653–661.
129. Caveness, W. F. (1977): Pathology of radiation damage to the normal brain of the monkey. *Natl. Cancer Inst. Monogr.*, 46:57–76.
130. Caveness, W. F. (1980): Experimental observations: delayed necrosis in normal monkey brain. In: *Radiation Damage to the Nervous System*, edited by H. A. Gilbert and A. R. Kagen, pp. 1–38. Raven Press, New York.
131. Cerveny, T. J. and Cockerham, L. G. (1986): Medical management of internal radionuclide contamination. *Med. Bull. U.S. Army, Europe*, 43:24–27.
132. Cerveny, T. J., MacVittie, T. J., and Young, R. W. (1989): Acute radiation syndrome in humans. In: *Medical Consequences of Nuclear Warfare*, edited by R. I. Walker and T. J. Cerveny, pp. 15–36. Office of the Surgeon General, TMM Publications, Falls Church, VA.
133. Chaillan, F. A., Devigne, C., Diabira, D. et al. (1997): Neonatal gamma-ray irradiation impairs learning and memory of an olfactory associative task in adult rats. *Eur. J. Neurosci.*, 9:884–894.
134. Champlin, R. (1988): Treatment for victims of nuclear accidents: the role of bone marrow transplantation. *Radiat. Res.*, 113:205–210.
135. Chapman, P. H. and Young, R. J. (1968): Effect of cobalt-60 gamma irradiation on blood pressure and cerebral blood flow in the *Macaca mulatta*. *Radiat. Res.*, 35:78–85.
136. Chapman, P. H. and Young, R. J. (1968): *Effect of Head Versus Trunk Fission-Spectrum Radiation on Learned Behavior in the Monkey*, U.S.A.F. SAM Tech. Rep. TR 68–80. School of Aerospace Medicine, Brooks Air Force Base, TX.
137. Chapman, P. H. and Young, R. J. (1968): Effect of high energy x-irradiation of the head on cerebral blood flow and blood pressure in the *Macaca mulatta*. *Aerosp. Med.*, 3:1316–1321.
138. Chaput, R. L. and Wise, D. (1970): Miniature pig incapacitation and performance decrement after mixed gamma-neutron irradiation. *Aerosp. Med.*, 41:290–293.
139. Charlton, D. E., Nikjoo, H., and Humm, J. L. (1989): Calculation of initial yields of single- and double-strand breaks in cell nuclei from electrons, protons and alpha particles. *Int. J. Radiat. Biol.*, 56:1–19.
140. Chaterjee, J., De, K., Basu, S. K., and Das, A. K. (1994): Low-level x-ray exposures on rat skin: hyperkeratinization and concomitant changes in biometal concentration. *Biol. Trace Element Res.*, 46(3):203–210.
141. Chen, A. Y., Okunieff, P., Pommier, Y., and Mitchell, J. B. (1997): Mammalian DNA topoisomerase I mediates the enhancement of radiation cytotoxicity by camptothecin derivatives. *Cancer Res.*, 57:1529–1536.
142. Cherkasova, L. S. and Mironova, T. M. (1976): Effects of ionizing radiation on enzymes of carbohydrate metabolism. *Radiobiologiya*, 16:657–664.
143. Chesser, R. K. et al. (2000): Concentrations and dose rate estimates of $^{134,137}$cesium and $^{90}$strontium in small mammals at Chernobyl, Ukraine. *J. Environ. Toxicol. Chem.*, 19:305–312.
144. Chieng, P. U., Huang, T. S., Chang, C. C. et al. (1991): Reduced hypothalamic blood flow after radiation treatment of nasopharyngeal cancer: SPECT studies in 34 patients. *Am. J. Neuroradiol.*, 12:661–665.
145. Chinkina, O. V. (1991): Psychological characteristics of patients exposed to accidental irradiation at the Chernobyl atomic-power station. In: *The Medical Basis for Radiation-Accident Preparedness.* Vol. III. *The Psychological Perspective*, edited by R. C. Ricks, M. E. Berger, and F. M. O'Hara, pp. 93–103. Elsevier, New York.
146. Chiu, S.-M., Xue, L.-Y., Friedman, L. R., and Olenik, N. L. (1992): Chromatin compaction and the efficiency of formation of DNA-protein cross-links in γ-irradiated mammalian cells. *Radiat. Res.*, 129:184–191.
147. Chmura, S. J., Nodzenski, E., Beckett, M. A. et al. (1997): Loss of ceramide production confers resistance to radiation-induced apoptosis. *Radiat. Res.*, 57:1270–1275.
148. Christensen, H. D., Flesher, A. M., and Haley, T. J. (1969): Changes in brain self-stimulation rates after exposure to x-irradiation. *J. Pharm. Sci.*, 58:128–129.
149. Chumak, A. A., Bazyka, D. A., Bieliaeva, N. V. et al. (2000): Immunological effects in acute radiation sickness convalescents: results of thirteen years of follow-up. *Int. J. Radiat. Med.*, 1(5):65–82.
150. Cilliers, G. D., Harper, I. S., and Lochner, A. (1989): Radiation-induced changes in the ultrastructure and mechanical function of the rat heart. *Radiother. Oncol.*, 16:311–326.
151. Clarke, R. H. (1987): Dose distributions in western Europe following Chernobyl. In: *Radiation and Health. The Biological Effects of Low-Level Exposure to Ionizing Radiation*, edited by R. Jones and R. Southwood, pp. 251–264. John Wiley & Sons, New York.
152. Clarke, R. H. (1989): Current radiation risk estimates and implications for the health consequences of Windscale, TMI and Chernobyl accidents. In: *Medical Response to Effects of Ionizing Radiation*, edited by W. A. Crosbie and J. H. Gittus, pp. 103–118. Elsevier, New York.
153. Cockerham, L. G. and Forcino, C. D. (1995): Effect of antihistamines, disodium cromoglycate (DSCG) or methysergide on post-irradiation cerebral blood flow and mean systemic arterial blood pressure in primates after 25 Gy whole-body, gamma irradiation. *J. Radiat. Res. (Tokyo)* 36:77–90.

154. Cockerham, L. G. and Hawkins, R. N. (1987): Radiation injury and the splanchnic circulation. In: *Pathophysiology of the Splanchnic Circulation*, Vol. II, edited by P. R. Kvietys, J. A. Barrowman, and D. N. Granger, pp. 55–66. CRC Press, Boca Raton, FL.
155. Cockerham, L. G. and Prell, G. D. (1989): Prenatal radiation risk to the brain. *Neurotoxicology*, 10:467–474.
156. Cockerham, L. G., Arroyo, C. M., and Hampton, J. D. (1988): Effects of 4-hydroxypyrazolo (3,4-d) pyrimidine (Allopurinol) on postradiation cerebral blood flow: implications of free radical involvement. *Free Radic. Biol. Med.*, 4:279–284.
157. Cockerham, L. G., Cerveny, T. J., and Hampton, J. D. (1986): Postradiation regional cerebral blood flow in primates. *Aviat. Space Environ. Med.*, 57:578–582.
158. Cockerham, L. G., Doyle, T. F., Donlon, M. A., and Gossett-Hagerman, C. J. (1985): Antihistamines block radiation-induced increased intestinal blood flow in canines. *Fundam. Appl. Toxicol.*, 5:597–604.
159. Cockerham, L. G., Doyle, T. F., Donlon, M. A., and Helgeson, E. A. (1984): Canine postradiation histamine levels and subsequent response to compound 48/80. *Aviat. Space Environ. Med.*, 55:1041–1045.
160. Cockerham, L. G., Doyle, T. F., Paulter, E. L., and Hampton, J. D. (1986): Disodium cromoglycate, a mast-cell stabilizer, alters postradiation regional cerebral blood flow in primates. *J. Toxicol. Environ. Health*, 18:91–101.
161. Cockerham, L. G., Forcino, T. C., Pellmar, T. C., and Smart, S. W. (1987): Effect of methysergide on postirradiation hypotension and cerebral ischemia. In: *Proc. of the Cerebral Hypoxia and Stroke Symposium*, August 22–24, Budapest, Hungary.
162. Cockerham, L. G., Hampton, J. D., and Doyle, T. F. (1986): Dose dependent radiation-induced hypotension in the canine. *Life Sci.*, 39:1543–1547.
163. Cockerham, L. G., Pautler, E. L., Carraway, R. E. et al. (1988): Effect of disodium cromoglycate (DSCG) and antihistamines on postirradiation cerebral blood flow and plasma levels of histamine and neurotensin. *Fundam. Appl. Toxicol.*, 10:233–242.
164. Cockerham, L. G., Prell, G. D., Cerveny, T. J. et al. (1991): Effects of aminoguanidine on pre- and postirradiation regional cerebral blood flow and systemic blood pressure in the primate. *Agents Actions*, 32:237–244.
165. Collins, D. L. (1991): Stress at Three Mile Island: altered perceptions, behaviors, and neuroendocrine measures. In: *The Medical Basis for Radiation-Accident Preparedness*. Vol. III. *The Psychological Perspective*, edited by R. C. Ricks, M. E. Berger, and F. M. O'Hara, pp. 71–79. Elsevier, New York.
166. Collins, D. L. (1992): Behavioral differences of irradiated persons associated with the Kyshtym, Chelyabinsk, and Chernobyl nuclear accidents. *Mil. Med.*, 157:548–552.
167. Collins, D. L. (2002): Human responses to the threat of or exposure to ionizing radiation at Three Mile Island, Pennsylvania, and Goiania, Brazil. *Mil. Med.*, 167:137–138.
168. Collins, D. L. and de Carvalho, A. B. (1993): Chronic stress from the Goiania $^{137}$Cs radiation accident. *Behav. Med.*, 18:149–157.
169. Committee for the Compilation of Materials on Damage Caused by the Atomic Bombs in Hiroshima and Nagasaki (1981): Psychological trends among A-bomb victims. In: *Hiroshima and Nagasaki: Physical, Mental and Social Effects of the Atomic Bombings*, translated by E. Ishikawa and D. Swain, pp. 485–500. Basic Books, New York.
170. Cooper, J. S. (1997): Classic and acquired immunodeficiency syndrome (AIDS)-related Kaposi's sarcoma. In: *Principles and Practice of Radiation Oncology*, 3rd ed., edited by C. A. Perez and L. W. Brady, pp. 745–762. Lippincott–Raven, Philadelphia, PA.
171. Cordes, N. and Meineke, V. (2004): Integrin signaling and the cellular response to ionizing radiation. *J. Molec. Histol.*, 35:327–337.
172. Couch, D. (2003): *The United States Armed Forces Nuclear, Biological, and Chemical Survival Manual*. Basic Books, Perseus Books Group, New York.
173. Cox, A. B., Lee, A. C., and Lett, J. T. (1988): Delayed effects of proton irradiation in the lens and integument: a primate model. In: *Terrestrial Space Radiation and Its Biological Effects*, edited by P. D. McCormack, C. E. Swenberg, and H. B¸cker, pp. 415–422. Plenum Press, New York.
174. Cristaldi, M., D'Arcangelo, E. D., Leradi, L. A. et al. (1990): $^{137}$Cs determination and mutagenicity tests in wild *Mus musculus domesticus* before and after the Chernobyl accident. *Environ. Pollut.*, 64:1–9.
175. Cristaldi, M., Leradi, L. A., Mascanzoni, D., and Mattei, T. (1991): Environmental impact of the Chernobyl accident: mutagenesis in bank voles from Sweden. *Int. J. Radiat. Biol.*, 59:31–40.
176. Croft, J. R. (1989): The Goi,nia accident. In: *Medical Response to Effects of Ionizing Radiation*, edited by W. A. Crosbie and J. H. Gittus, pp. 83–101. Elsevier, New York.
177. Crompton, M. R. and Layton, D. D. (1961): Delayed radionecrosis of the brain following therapeutic x-irradiation of the pituitary. *Brain*, 84:85–101.
178. Cross, F. T., Palmer, R. F., Busch, R. H. et al. (1981): Development of lesions in Syrian Golden hamsters following exposure to radon daughters and uranium ore dust. *Health Phys.*, 41:135–153.
179. Cross, F. T., Palmer, R. F., Filipy, R. E. et al. (1982): Carcinogenic effects of radon daughters, uranium ore dust and cigarette smoke in beagle dogs. *Health Phys.*, 42:33–52.
180. Currey, J. D., Foreman, J., Laketic, I. et al. (1997): Effects of ionizing radiation on the mechanical properties of human bone. *J. Orthop. Res.*, 15(1):111–117.
181. Cust, M. P., Whitehead, M. I., Powles, R., Hunter, M., and Milliken, S. (1989): Consequences and treatment of ovarian failure after total body irradiation for leukaemia. *Br. Med. J.*, 299:1494–1497.
182. Cwikel, J., Abdelgani, A., Goldsmith, J. R. et al. (1997): Two-year follow up study of stress-related disorders among immigrants to Israel from the Chernobyl area. *Environ. Health Perspect.*, 105(6):1545–1550.
183. Cwikel, J. (1997): Comments on the psychosocial aspects of the International Conference on Radiation and Health. *Environ. Health Perspect.*, 105(6):1607–1608.

184. Dafny, N. and Pellis, N. R. (1986): Evidence that opiate addiction is in part an immune response: immune system destruction by irradiation altered opiate withdrawal. *Neuropharmacology*, 25:815–818.

185. Dahlstrom, A., Haggendal, J., and Rosengren, B. (1973): The effect of Roentgen irradiation on monoamine containing neurons of the rat brain. *Acta Radiol. Ther. Phys. Biol.*, 12:191–200.

186. Dallas, C. E. (1993): Aftermath of the Chernobyl nuclear disaster: pharmaceutical needs in the Republic of Belarus. *Am. J. Pharmaceut. Educ.*, 57:182–185.

187. Dallas, C. E., Jagoe, C. H., Fisher, S. K. et al. (1995): Evaluation of genotoxicity in wild organisms due to the Chernobyl nuclear disaster. *Ecol. Indust. Regions*, 1:44–54.

188. Dallas, C. E., Lingenfelser, S. F., Lingenfelser, J. T. et al. (1998): Flow cytometric analysis of leukocyte and erthrocyte DNA in fish from Chernobyl-contaminated ponds in the Ukraine. *Ecotoxicology*, 7:211–219.

189. D'Amato C. J. and Hicks, S. P. (1980): Development of the motor system: effects of radiation on developing corticospinal neurons and locomotor function. *Exp. Neurol.*, 70:1–23.

190. Danquechin-Dorval, E., Mueller, G. P., Eng, R. R. et al. (1985): Effect of ionizing radiation on gastric secretion and gastric motility in monkeys. *Gastroenterology*, 89: 374–380.

191. Danylash, M. M., Voshchepynets, H. A., Urban, V. I., and Fekiishgazi, S. B. (1996): Immune status parameters and renal function in subjects with prior ionizing radiation exposure (Ukr). *Likars'Ka Sprava.*, 0(1–2):18–20.

192. Darby, S. C. and Weiss, H. A. (1995): Human studies in radiation leukaemogenesis. In: *Radiation Toxicology: Bone Marrow and Leukaemia*, edited by J. H. Hendry and B. I. Lord, pp. 335–353. Taylor & Francis, Bristol, PA.

193. Dare, A., Hachisu, R., Yamaguchi, A. et al. (1997): Effects of ionizing radiation on proliferation and differentiation of osteoblast-like cells. *J. Dental Res.*, 76(2):658–664.

194. Datta, P. K., Moulder, J. E., Fish, B. L. et al. (2001): Induction of heme oxygenase 1 in radiation neuropathy: role of angiotensin II. *Radiat. Res.*, 155:734–739.

195. d'Avella, D., Cicciarello, R., Albiero, F. et al. (1992): Quantitative study of blood–brain barrier permeability changes after experimental whole-brain radiation. *Neurosurgery*, 30:30–34.

196. d'Avella, D., Cicciarello, R., Angileri, F. F. et al. (1998): Radiation-induced blood-brain barrier changes: pathophysiological mechanisms and clinical implications. *Acta Neurochir. Suppl. (Wien)*, 71:282–284.

197. Davidoff, L. M., Dyke, C. G., Elsberg, C. A., and Tarlov, I. M. (1938): The effect of radiation applied directly to brain and spinal cord. I. Experimental investigations on Macaca rhesus monkeys. *Radiology*, 31:451–463.

198. Davydov, B. I. (1961): Acetylcholine metabolism on the thalamic region of the brain of dogs after acute radiation sickness. *Radiobiologiia*, 1:550–554.

199. Day, R., Gorin, M. B., and Eller, A. W. (1995): Prevalence of lens changes in Ukrainian children residing around Chernobyl. *Health Phys.*, 68:632–642.

200. de Carvalho, A. B. (1991): The psychological effects of the Goiania radiological accident on the emergency responders. In: *The Medical Basis for Radiation-Accident Preparedness*. Vol. III. *The Psychological Perspective*, edited by R. C. Ricks, M. E. Berger, and F. M. O'Hara, pp. 132–141. Elsevier, New York.

201. Deininger, M. W. N., Bose, S., Gora-Tybor, J. et al. (1998): Selective induction of leukemia-associated fusion genes by high-dose ionizing radiation. *Cancer Res.*, 58(3):421–425.

202. Deng, C., Zhang, P., Harper, J. W. et al. (1995): Mice lacking p21(CIP1/WAF1) undergo normal development but are defective in $G_1$ checkpoint control. *Cell*, 82:675–684.

203. Denisova, N. A., Shukitt-Hale, B., Rabin, B. M., and Joseph, J. A. (2002): Brain signaling and behavioral responses induced by exposure to (56)Fe-particle radiation. *Radiat. Res.*, 158:725–734.

204. Dent, P., Yacoub, A., Contessa, J. et al. (2003): Stress and radiation-induced activation of multiple intracellular signaling pathways. *Radiat. Res.*, 159:283–300.

205. Deroo, J., Gerber, G. B., and Maes, J. (1986): *Radiation Risks to the Developing Nervous System*, edited by H. Kriegel, W. Schmahl, G. B. Gerber, and F.-E. Stieve, pp. 211–219. Stuttgart, Gustav Fischer Verlag.

206. DeRyck, M., Schallert, T., and Teitelbaum, P. (1980): Morphine versus haloperidol catalepsy in the rat: a behavioral analysis of postural support mechanisms. *Brain Res.*, 201:143–172.

207. Deshmukh, B. D. and Suryawanshi, S. A. (1989): Effects of gamma irradiation on histomorphology of some endocrine glands of the rain quail, *Coturnix coromandelica* (Gmelin). *Indian J. Exp. Biol.*, 27:780–784.

208. de Toledo, S. M., Azzam, E. I., Gasmann, M. K., and Mitchell, R. E. (1995): Use of semiquantitative transcription polymerase chain reaction to study gene expression in normal human skin fibroblasts following low dose-rate irradiation. *Int. J. Radiat. Biol.*, 67: 135–143.

209. Dewey, W. C., Ling, C. C., and Meyn, R. E. (1995): Radiation-induced apoptosis: revelance to radiotherapy. *Int. J. Radiat. Oncol. Biol. Phys.*, 33:781–796.

210. di Cicco, D., Antal, S., and Ammassari-Teule, M. (1991): Prenatal exposure to gamma/neutron irradiation: sensorimotor alterations and paradoxical effects on learning. *Teratology*, 43:61–70.

211. Donlon, M. A. and Walden, Jr., T. L. (1988): The release of biologic mediators in response to acute radiation injury. *Comm. Toxicol.*, 2:205–216.

212. Dopico, A. M. and Zieher, L. M. (1993): Neurochemical characterization of the alterations in the noradrenertic afferents to the cerebellum of adult rats exposed to x-irradiation at birth. *J. Neurochem.*, 61(2):481–489.

213. Dorr, R. T. (1998): Radioprotectants: pharmacology and clinical applications of amifostine. *Semin. Radiat. Oncol.* 4(Suppl. 1):10–13.

214. D^rr, W. and Schultz-Hector, S. (1992): Early changes in mouse urinary bladder function following fractionated x irradiation. *Radiat. Res.*, 131:35–42.

215. Douglas, W. W. (1985): Histamine and 5-hydroxytryptamine (serotonin) and their antagonists. In: *Goodman and Gilman's The Pharmacological Basis of Therapeutics*, edited by A. G. Gilman, L. S. Goodman, T. W. Rall, and F. Murad, pp. 605–615. Macmillan, New York.
216. Doull, J. (1967): Pharmacological responses in irradiated animals. *Radiat. Res.*, 30:334–341.
217. Doyle, T. F. and Strike, T. A. (1977): Radiation-released histamine in the rhesus monkey as modified by mast cell depletion and antihistamine. *Experientia*, 33:1047–1049.
218. Doyle, T. F., Curran, C. R., and Turns, J. E. (1974): The prevention of radiation-induced early transient incapacitation of monkeys by an antihistamine. *Proc. Soc. Exp. Biol. Med.*, 145:1018–1024.
219. Doyle, T. F., Turns, J. E., and Strike, T. A. (1971): Effect of antihistamine on early transient incapacitation of monkeys subjected to 4000 rads of mixed gamma-neutron radiation. *Aerosp. Med.*, 42:400–403.
220. Dubois, A. (1988): Effect of ionizing radiation on the gastrointestinal tract. *Comm. Toxicol.*, 2:233–242.
221. Dubois, A., Fiala, N., Boward, C. A., and Bogo, V. (1988): Prevention and treatment of the gastric symptoms of radiation sickness. *Radiat. Res.*, 115:595–604.
222. Dubrova, Y. E., Nesterov, V. N., Krouchinsky, N. G. et al. (1996): Human minisatellite mutation rate after the Chernobyl accident. *Nature*, 380:683–686.
223. Dubrova, Y. E., Nesterov, V. N., Krouchinsky, N. G. et al. (1997): Further evidence for elevated human minisatelllite mutation rate in Belarus eight years after the Chernobyl accident. *Mutag. Mutat. Res. Fundam. Mol. Mech. Mutag.*, 38(2):267–278.
224. Dulic, V., Kaufmann, W. K., Wilson, S. J. et al. (1994): p53–dependent inhibtion of cyclin-dependent kinase activities in human fibroblast during radiation-induced $G_1$ arrest. *Cell*, 76:1013–1023.
225. Dygalo, N. N., Sakharov, D. G., and Shishkina, G. T. (1997): Corticosterone and testosterone in the blood of adult rats: the effects of low doses and the times of the action of ionizing radiation during intrauterine development (Russia). *Radiat. Biol. Radioecol.*, 37:377–381.
226. Dynes, J. B. and Smedal, M. J. (1960): Radiation myelitis. *Am. J. Roentgenol. Radium Ther. Nuc. Med.*, 83:78–87.
227. Edwards, A. A. (1997): The use of choromosomal aberrations in human lymphocytes for biological dosimetry. *Radiat. Res.*, 148:S39–S44.
228. Edwards, J. C., Cramp, W. C., Chapman, D., and Yatvin, M. B. (1984): The effects of ionizing radiation on biomembrane structure and function. *Prog. Biophys. Mol. Biol.*, 43:71–93.
229. Egana, E. (1962): Some effects of ionizing radiations on the metabolism of the central nervous system. *Int. J. Neurol.*, 3:631–647.
230. Ehrhart, E. J., Segarini, P., Tsang, M. L.-S. et al. (1997): Latent transforming growth factor B1 activation *in situ*: quantitative and functional evidence after low-dose gamma-irradiation. *FASEB J.*, 11:991–1002.
231. Eisenbud, M. (1987): *Environmental Radioactivity from Natural, Industrial, and Military Sources*, 3rd ed. Academic Press, New York.
232. Elliott, T. B., Brook, I., and Stiefel, S. M. (1990): Quantitative study of wound infection in irradiated mice. *Int. J. Radiat. Biol.*, 58:341–350.
233. Endo, T., Minami, M., Hirafuji, M. et al. (2000): Neurochemistry and neuropharmacology of emesis: the role of serotonin. *Toxicology*, 153:189–201.
234. Fan, X. (2003): Ionizing radiation induces formation of malondialdehyde, formaldehyde, and acetylaldehyde from carbohydrates and organic acid. *J. Agric. Food Chem.*, 24:5946–5949.
235. Faure, E., Cavard, C., Zider, A. et al. (1995): X-irradiation-induced transcription from HIV type 1 long term repeat. *AIDS Res. Human Retroviruses*, 11:41–43.
236. Feldman, J. E. (1989): Ovarian failure and cancer treatment: incidence and interventions for premenopausal women. *Oncol. Nurs. Forum*, 16:651–657.
237. Ferguson, J. L., Kandasamy, S. B., Harris, A. H. et al. (1996): Indomethacin attenuation of radiation-induced hyperthermia does not modify radiation-induced motor hypoactivity. *J. Radiat. Res. (Tokyo)*, 37:209–215.
238. Fetter, S. A. and Tsipis, K. (1981): Catastrophic releases of radioactivity. *Sci. Am.*, 244:41–47.
239. Feyer, P. C., Maranzano, E., Molassiotis, A. et al. (2004): Radiotherapy-induced nausea and vomiting (RINV): antiemetic guidelines. *Support Care Cancer*, 13(2):122–128.
240. Feyerabend, T., Kapp, B., Richter, E. et al. (1990): Incidence of hypothyroidism after irradiation of the neck with special reference to lymphoma patients: a retrospective and prospective analysis. *Acta Oncol.*, 29:597–602.
241. Fici, G. J., Althaus, J. S., and von Voigtlander, P. F. (1997): Effects of lazaroids and a peroxynitrite scavenger in a cell model of peroxynitrite toxicity. *Free Radic. Biol. Med.*, 22:223–228.
242. Fields, P. E. (1957): The effect of whole-body x-irradiation upon activity drum, straight runway, and maze performances of white rats. *J. Comp. Physiol. Psychol.*, 50:386–391.
243. Fischbein, A., Zabludovsky, N., Eltes, F. et al. (1997): Ultramorphological sperm characteristics in the risk assessment of health effects after radiation exposure among salvage workers in Chernobyl. *Environ. Health Perspect.*, 105(6):1445–1450.
244. Flam, F. (1992): Quasars: ablaze with gamma rays. *Science*, 256:311.
245. Ford, E., Hughes, M. N., and Wardman, P. (2002): Kinectics of the reactions of nitrogen dioxide with glutathione, cysteine, and uric acid at physiological pH. *Free Radical Biol. Med.*, 32:1314–1323.
246. Fornace, A. J. (1992): Mammalian genes induced by radiation: activation of genes associated with growth control. *Annu. Rev. Genet.*, 26:507–526.
247. Francois, A., Aigueperse, J., Gourmelon, P. et al. (1998): Exposure to ionizing radiation modifies neurally-evoked electrolyte transport and some inflammatory responses in rat colon *in vitro*. *Int. J. Radiat. Biol.*, 73(1):93–101.
248. Franke, T. F. and Lewis, C. C. (1997): A bad kinase makes good. *Nature*, 390:116–117.
249. Franz, C. G. (1985): Effects of mixed neutron-gamma total body irradiation on physical activity performance of rhesus monkeys. *Radiat. Res.*, 101:434–441.

250. Freeman, S. L., Hossain, M., and McNaughton, W. K. (2001): Radiation-induced acute intestinal inflammation differs following total-body versus abdominopelvic irradiation in the ferret. *Int. J. Radiat. Biol.*, 77:389–95.
251. Freud, A., Canfi, A., Sod-Moriah, U. A., and Chayoth, R. (1990): Neonatal low-dose gamma irradiation-induced impaired fertility in mature rats. *Isr. J. Med. Sci.*, 26: 611–615.
252. Fritzell, J. A., Narayanan, L., Baker, S. M. et al. (1997): Role of DNA mismatch repair in the cytotoxicity of ionizing radiation. *Cancer Res.*, 57:5143–5147.
253. Fry, F. A. (1987): Doses from environmental radioactivity. In: *Radiation and Health: The Biological Effects of Low-Level Exposure to Ionizing Radiation*, edited by R. Jones and R. Southwood, pp. 9–17. John Wiley & Sons, New York.
254. Fujiwara, S., Sposto, R., Ezaki, H. et al. (1992): Hyperparathyroidism among atomic bomb survivors in Hiroshima. *Radiat. Res.*, 130:372–378.
255. Fukuda, T., Setoguchi, M., Inaba, K. et al. (1991): The antiemetic profile of Y-25130, a new selective 5-$HT_3$ receptor antagonist. *Eur. J. Pharmacol.*, 196:299–305.
256. Furchtgott, E. (1963): Behavioral effects of ionizing radiations, 1955–61. *Psychol. Bull.*, 60:157–199.
257. Furchtgott, E. (1971): Behavioral effects of ionizing radiations. In: *Pharmacology and Biophysical Agents and Behavior*, edited by E. Furchgott, pp. 1–64. Academic Press, New York.
258. Furchtgott, E. (1975): Ionizing radiations and the nervous system. In: *Biology of Brain Dysfunction*, Vol. 3, edited by G. E. Galli, pp. 343–379. Plenum Press, New York.
259. Fushiki, S. (1997): Pathogenesis of the neuronal migration disorder, with special reference to the animal model of prenatal exposure to low-dose ionizing radiation (Japan). *No To Hattatsu*, 29(2):102–107.
260. Fushiki, S., Hyodo-Taguchi, Y., Kinoshita, C. et al. (1997): Short- and long-term effects of low-dose prenatal x-irradiation in mouse cerebral cortex, with special reference to neuronal migration. *Acta Neuropathol. (Berlin)*, 93(5): 443–449.
261. Gaffey, C. T. (1962): Bioelectric effects of high energy irradiation on nerve. In: *Response of the Nervous System to Ionizing Radiation*, edited by T. J. Haley and R. S. Snider, pp. 277–296. Academic Press, New York.
262. Gaichenko, V. A., Kryzhanovsky, V. I., and Stovbchaty, V. N. (1994): Post-accident state of the Chernobyl Nuclear Power Plant Alienated Zone faunal complexes. *Radiat. Biol. Ecol.*, Special Issue, 27–32.
263. Gal, D., Strickland, D. M., Lifshitz, S. et al. (1984): Effect of radiation on prostaglandin production by human bowel *in vitro*. *Int. J. Rad. Oncol. Biol. Phys.*, 10:653–657.
264. Gallegos, A., Berggren, M., Gasdaska, J. R., and Powis, G. (1997): Mechanisms of the regulation of thioredoxin reductase activity in cancer cells by the chemopreventive agent selenium. *Cancer Res.*, 57:4965–4970.
265. Gamache, G. L., Levinson, D. M., Reeves, D. L. et al. (2005): Longitudinal neurocognitive assessments of Ukrainians exposed to ionizing radiation after the Chernobyl nuclear accident. *Arch. Clin. Neuropsychol.*, 20:81–93.
266. Gangloff, H. (1962): Acute effects of x-irradiation on brain electrical activity in cats and rabbits. In: *Effects of Ionizing Radiation on the Nervous System: Proceedings*, pp. 123–138. International Atomic Energy Agency, Vienna.
267. Gangloff, H. and Haley, T. J. (1960): Effects of x-irradiation on spontaneous and evoked brain electrical activity in cats. *Radiat. Res.*, 12:694–704.
268. Ganong, W. F. (1989): *Review of Medical Physiology*, 14th ed. Appleton & Lange, San Mateo, CA.
269. Gasteiger, E. L. and Campbell, B. (1962): Alteration of mammalian nerve compound action potentials by beta irradiation. In: *Response of the Nervous System to Ionizing Radiation*, edited by T. J. Haley and R. S. Snider, pp. 597–605. Academic Press, New York.
270. Geist, B. J. and Trott, K. R. (1992): Radiographic and function changes after partial lung irradiation in the rat. *Strahlenther Onkol.*, 168:168–173.
271. Geist, B. J., Lauk, S., Bornhausen, M., and Trott, K. R. (1990): Physiologic consequences of local heart irradiation in rats. *Int. J. Radiat. Oncol. Biol. Phys.*, 18:1107–1113.
272. Gembicki, M., Stozharov, A. N., Arinchin, A. N. et al. (1997): Iodine deficiency in Belarusian children as a possible factor stimulating the irradiation of the thyroid gland during the Chernobyl catastrophe. *Environ. Health Perspect.*, 105(6):1487–1490.
273. George, A. M. and Cramp, W. A. (1987): The effects of ionizing radiation on structure and function of DNA. *Prog. Biophys. Molec. Biol.*, 50:121–169.
274. Geraci, J. P., Mariano, M. S., and Jackson, K. L. (1991): Hepatic radiation injury in the rat. *Radiat. Res.*, 125:65–72.
275. Geraci, J. P., Mariano, M. S., and Jackson, K. L. (1992): Radiation hepatology of the rat: microvascular fibrosis and enhancement of liver dysfunction by diet and drugs. *Radiat. Res.*, 129:322–332.
276. Gershey, E. L., Klein, R. C., Party, E., and Wilkerson, A. (1990): *Low-Level Radioactive Waste: From Cradle to Grave*. Van Nostrand Reinhold, New York.
277. Gersterner, H. B. (1956): Effect of high-intensity x-irradiation on the A group fibers of the frog sciatic nerve. *Am. J. Physiol.*, 184:333–337.
278. Giambarresi, L. and Jacobs, A. J. (1987): Radioprotectants. In: *Military Radiobiology*, edited by J. J. Conklin and R. I. Walker, pp. 265–301. Academic Press, San Diego, CA.
279. Gibson, D. P., DeGowin, R. L., and Knapp, S. A. (1982): Effect of x irradiation on release of prostaglandin E from marrow stromal cells in culture. *Radiat. Res.*, 89:537–545.
280. Giel, R. (1991): The psychosocial aftermath of two major disasters in the Soviet Union. *J. Traumatic Stress*, 4: 381–393.
281. Gillette, S. L., Gillette, E. L., Powers, B. E. et al. (1989): Ureteral injury following experimental intraoperative radiation. *Int. J. Radiat. Oncol. Biol. Phys.*, 17:791–798.
282. Gillette, S. M., Powers, B. E., Orton, E. C., and Gillette E. L. (1991): Early radiation response of the canine heart and lung. *Radiat. Res.*, 125:34–40.
283. Gilmore, S. A., Sims, T. J., Davies, D. L. et al. (1997): Microglial development is altered in immature spinal cord by exposure to radiation. *Int. J. Develop. Neurosci.* 15(1):1–14.

284. Ginzburg, H. M. (1993): The psychological consequences of the Chernobyl accident: findings from the International Atomic Energy Agency Study. *Public Health Rep.*, 108:184–192.
285. Girinsky, T., Baume, D., Socie, G. et al. (1991): Blood cell kinetics after a 385 cGy total body irradiation given to a CML patient for bone marrow transplantation. *Bone Marrow Transplant.*, 7:317–320.
286. Giwercman, A., von der Maase, H., Berthelsen, J. G. et al. (1991): Localized irradiation of testes with carcinoma *in situ*: effects on Leydig cell function and eradication of malignant germ cells in 20 patients. *J. Clin. Endocrinol. Metab.*, 73:596–603.
287. Gluzman, D. F., Moutet, A., Simmonet, M.-L. et al. (1994): Oncohematological aspects of ionizing radiation exposure on human embryo and fetus (Russia). *Eksperimental'Naya Onkologiya*, 16(4–6):279–287.
288. Gobbel, G. T., Seilhan, T. M., and Fike, J. R. (1992): Cerebrovascular response after interstitial irradiation. *Radiat. Res.*, 130:236–240.
289. Goldman, M. (1997): The Russian radiation legacy: its integrated impact and lessons. *Environ. Health Perspect.*, 105(6):1385–1392.
290. Goncharova, R. I. and Ryabokon, N. I. (1995): Dynamics of cytogenetic injuries in natural populations of bank vole in the Republic of Belarus. *Radiat. Protect. Dosim.*, 62:37–40.
291. Gonzales, S. (1982): Host rocks for radioactive-waste disposal. *Am. Sci.*, 70:191–200.
292. Goud, S. N. (1995): Effect of irradiation of lymphocyte proliferation and differentiation: potential of IL-6 in augmenting antibody responses in cultures of murine spleen cells. *Int. J. Radiat. Biol.*, 67(4): 461–468.
293. Graham, M. M., Evans, M. L., Dahlen, D. D. et al. (1990): Pharmacological alteration of the lung vascular response to radiation. *Int. J. Radiat. Oncol. Biol. Phys.*, 19:329–339.
294. Greco, O., Durante, M., Gialanella, G. et al. (2003): Biological dosimetry in Russian and Italian astronauts. *Adv. Space Research*, 31:1495–1503.
295. Grevert, D. and Goldstein, A. (1977): Some effects of naloxone on behavior in the mouse. *Psychopharmacology (Berlin)*, 53:111–113.
296. Grosch, D. S. and Hopwood, L. E. (1979): *Biological Effects of Radiation*. Academic Press, New York.
297. Guelman, L. R. et al. (2003): WR-2721 (amifostine, ethyol) prevents motor and morphological changes induced by neonatal x-irradiation. *Neurochem. Int.*, 42:385–391.
298. Gueneau, G., Baille, V., Dubos, M., and Court L. (1986): Protracted postnatal neurogenesis and radiosensitivity in the rabbit's dentate gyrus. In: *Radiation Risks to the Developing Nervous System*, edited by H. Kriegel, W. Schmahl, G. B. Gerber, and F.-E. Stieve, pp. 133–140. Stuttgart, Gustav Fischer Verlag.
299. Guinee, V. F., Guido, J. J., Pfalzgraf, K. A. et al. (1985): The incidence of herpes zoster in patients with Hodgkin's disease. *Cancer*, 56:642–648.
300. Gunter-Smith, P. J. (1987): Effect of ionizing radiation on gastrointestinal physiology. In: *Military Radiobiology*, edited by J. J. Conklin and R. Walker, pp. 135–151. Academic Press, San Diego, CA.
301. Gunter-Smith, P. J. (1989): Gamma radiation affects active electrolyte transport by rabbit ileum. II. Correlation of alanine and theophylline response with morphology. *Radiat. Res.*, 117:419–432.
302. Gupta, M. L. and Umadevi, P. (1990): Response of reptilian liver to external gamma irradiation. *Radiobiol. Radiother. (Berlin)*, 31:285–288.
303. Gupta, M. L. and Umadevi, P. (1990): Response of piscine liver to external gamma irradiation. *Radiobiol. Radiother. (Berlin)*, 31:289–292.
304. Guyton, A. C. (1986): *Textbook of Medical Physiology*, 7th ed. W.B. Saunders, Philadelphia, PA.
305. Hagen, U. (1989): Biochemical aspects of radiation biology. *Experientia*, 45:7–12.
306. Haimovitz-Friedmann, A., Kolesnick, R. N., and Fuks, Z. (1996): Modulation of the apoptotic response: potential for therapeutic applications in radiation oncology. *Semin. Radiat. Oncol.*, 6:273–283.
307. Hakem, R., Hakem, A., Duncan, G. S. et al. (1998): Differential requirement for caspase 9 in apoptotic pathway *in vivo*. *Cell*, 94:339–352.
308. Haley, T. J. (1962): Changes induced in brain activity by low doses of x-irradiation. In: *Effects of Ionizing Radiation on the Nervous System: Proceedings*, pp. 171–185. International Atomic Energy Agency, Vienna.
309. Hall, E. C. and Cox, J. D. (1989): Physical and biologic basis of radiation therapy. In: *Radiation Oncology: Rationale, Technique, Results*, 6th ed., edited by W. T. Moss and J. D. Cox, pp. 1–57. Mosby, St. Louis.
310. Hall, E. J. (1984): *Radiation and Life*, 2nd ed. Pergammon Press, New York.
311. Hall, E. J. (1994): *Radiobiology for the Radiologist*, 4th ed. Lippincott, New York.
312. Hall, E. J. (2000): *Radiobiology for the Radiologist*, 5th ed. Lippincott, Williams & Wilkins, Philadelphia, PA.
313. Hallahan, D. E. (1996): Radiation-medicated gene expression in the pathogenesis of the clinical radiation response. *Semin. Radiat. Oncol.*, 6:250–267.
314. Halpern, J., Kishel, S. P., Park, J. et al. (1984): Radiation-induced brain edema in primates, studies with sequential brain CAT scanning and histopathology. *Res. Commun. Chem. Pathol. Pharmacol.*, 45:463–470.
315. Hamdorf, G., Shahar, A., Cervos-Navarro, J. et al. (1992): Irradiation neurotoxicity assessed in organotypic cultures of rat hippocampus. *Neurotoxicology*, 13:165–170.
316. Hancock, S. L., Cox, R. S., and McDougall, I. R. (1991): Thyroid diseases after treatment of Hodgkin's disease. *N. Engl. J. Med.*, 325:599–605.
317. Hanley, C. J. (2004, June 19): Study: dirty bomb almost a sure thing. *Arkansas Democrat Gazette*, Little Rock, AR.
318. Hansen, P. V., Trykker, H., Svennekjaer, I. L., and Hvolby, J. (1990): Long-term recovery of spermatogenesis after radiotherapy in patients with testicular cancer. *Radiother. Oncol.*, 18:117–125.

319. Harley, N. H., Cross, F. T., and Stuart, B. O. (1984): *Evaluation of Occupational and Environmental Exposures to Radon and Radon Daughters in the United States*, NCRP Report No. 78. National Council on Radiation Protection and Measurement, Washington, D.C.
320. Havenaar, J., Rumyantzeva, G., Kasyanenko, A. et al. (1997): Health effects of the Chernobyl disaster: illness or illness behavior? A comparative general health survey in two former Soviet regions. *Environ. Health Perspect.*, 105(6):1533–1538.
321. Havenaar, J. M., van den Brink, W., Kasyanenko, A. P. et al. (1995): Mental health problems in the Gomel Region (Belarus): an analysis of risk factors in an area affected by the Chernobyl disaster. *Psychol. Med.*, 26:845–855.
322. Havenaar, J. M., de Wilde, E. J., van den Bout, J. et al. (2003): Perception of risk and subjective health among victims of the Chernobyl disaster. *Soc. Sci. Med.*, 56:569–572.
323. Hawkins, R. N. and Cockerham, L. G. (1987): Postirradiation cardiovascular dysfunction. In: *Military Radiobiology*, edited by J. J. Conklin and R. I. Walker, pp. 153–163. Academic Press, San Diego, CA
324. Hawkins, R. N. and Forcino, C. D. (1988): Effects of radiation on cardiovascular function. *Comments Toxicology*, 2:243–252.
325. Hawkins, R. N., Alter, W. A., Jr., Doyle, T. F., and Catravas, G. N. (1983): Radiation-induced cardiovascular dysfunction in the rhesus monkey. *Radiat. Res.*, 94:654.
326. Hayes, R. B. (1997): The carcinogenicity of metals in humans. *Cancer Causes Control*, 8:371–385.
327. Haymaker, W. (1962): Morphological changes in the nervous system following exposure to ionizing radiation. In: *Effects of Ionizing Radiation on the Nervous System*, pp. 309–358. International Atomic Energy Agency, Vienna.
328. Henderson, E. E., Tudor, G., and Yang, J. Y. (1992): Inactivation of the human immunodeficiency virus type 1 (HIV-1) by ultraviolet and x irradiation. *Radiat. Res.*, 131: 169–176.
329. Herman, B. H. and Panksepp, I. (1978): Effects of morphine and naloxone on separation distress and approach attachment: evidence for opiate mediation of social effect. *Pharmacol. Biochem. Behav.*, 9:213–220.
330. Hewitt, M., Cornish, J., Pamphilon, D., and Oakhill, A. (1991): Effective emetic control during conditioning of children for bone marrow transplantation using ondansetron, a 5-$HT_3$ antagonist. *Bone Marrow Transplant*, 7:431–433.
331. Hienz, R. D. (1992): *Effects of Ionizing Radiation on Auditory and Visual Thresholds*. DNA-TR-91-47. Defense Nuclear Agency, Alexandria, VA.
332. Highfield, D. A., Hu, D., and Amsel, A. (1998): Alleviation of x-irradiation-based deficit in memory-based learning by D-amphetamine: suggestions for attention deficit-hyperactivity disorder. *Proc. Natl. Acad. Sci. U.S.A.*, 95:5785–5788.
333. Hill, J. M., Kornblith, A. B., Jones, D. et al. (1998): A comparative study of the long term psychosocial functioning of childhood acute lymphoblastic leukemia survivors treated by intrathecal methotrexate with or without cranial radiation. *Cancer*, 82:208–218.
334. Hirata, Y., Matsukado, Y., Mihara, Y., and Kochi, M. (1985): Occlusion of the internal carotid artery after radiation therapy for the chiasmal lesion. *Acta Neurochir. (Wien.)*, 74:141–147.
335. Hoar, R. M. and Monie, I. W. (1981): Comparative development of specific organ systems. In: *Developmental Toxicology*, edited by C. A. Kimmel and J. Buelke-Sam, pp. 13–33. Raven Press, New York.
336. Hobbs, C. H. and McClellan, R. O. (1986): Toxic effects of radiation and radioactive materials. In: *Casarett and Doull's Toxicology*, 3rd ed., edited by C. D. Klaassen, M. O. Amdur, and J. Doull, pp. 669–705. Macmillan Publishing Company, New York.
337. Hodges, H., Katzung, N., Sowinski, P. et al. (1998): Late behavioural and neuropathological effects of local brain irradiation in the rat. *Behav. Brain Res.*, 91:99–114.
338. Hoekstra, H. J., Mehta, D. M., Oosterhuis, J. W. et al. (1990): The short- and long-term effect of single high-dose intra-operative electron beam irradiation of retroperitoneal structures: an experimental study in dogs. *Eur. J. Surg. Oncol.*, 16:240–247.
339. Hofer, M., Viklicka, S., Tkadlecek, L., and Karpfel, Z. (1989): Haemopoiesis in murine bone marrow and spleen after fractionated irradiation and repeated bone marrow transplantation, II. Granulopoiesis. *Folia Biol. (Praha)*, 35:418–428.
340. Hohwieler, M. L., Lo, T. C., Silverman, M. L., and Freiberg, S. R. (1986): Brain necrosis after radiotherapy for primary intracerebral tumor. *Neurosurgery*, 18:67–74.
341. Hole, E. D., Nelson, W. H., Sagstuen, E., and Close, D. M. (1992): Free radical formation in single crystals of 2′-deoxyguanosine 5′-monophosphate tetrahydrate disodium salts: an EPR/ENDOR study. *Radiat. Res.*, 129: 119–138.
342. Hollahan, Jr., E. V. (1987): Cellular radiation biology. In: *Military Radiobiology*, edited by J. J. Conklin and R. I Walker, pp. 87–110. Academic Press, Orlando, FL.
343. Hollinden, G. E. and Pellmar, T. C. (1989): Attenuation of synaptic transmission in hippocampal slices following whole animal exposure to ionizing radiation. *Soc. Neurosci. Abstr.*, 15:134.
344. Holloman, K., Dallas, C. E., Jagoe, C. H. et al. (1998): Superoxide dismutase and radiocesium activities in rodents from Chernobyl-contaminated areas in Ukraine. In: *Proc of SETAC 19th Annual Meeting: The Natural Connection. Environmental Integrity and Human Health*, November 15–19, Charlotte, NC, p. 161.
345. Hopewell, J. W. (1979): Late radiation damage to the central nervous system: a radiobiological interpretation. *Neuropathol. Appl. Neurobiol.*, 5:329–343.
346. Hopewell, J. W. and Wright, E. A. (1970): The nature of latent cerebral irradiation damage and its modification by hypertension. *Br. J. Radiol.*, 43:161–167.
347. Hopewell, J. W. and Wright, E. A. (1975): The effects of dose and field size on late radiation damage to the rat spinal cord. *Int. J. Radiat. Biol.*, 28:325–333.
348. Horiot, J. C. and Aapro, M. (2004): Treatment implications for radiation-induced nausea and vomiting in specific patient groups. *Eur. J. Cancer*, 40:979–987.

349. Huang, T. S., Chen, S. T., Lui, L. T. et al. (1990): Early effects of cranial irradiation on hypothalamic pituitary function. *Taiwan I Hsueh Hui Tsa Chih*, 89:541–547.

350. Hujoel, P. P., Bollen, A. M., Noonan, C. J., and del Aguila, M. A. (2004): Antepartum dental radiography and infant low birth weight. *JAMA*, 291(16):1987–1993.

351. Hunt, W. A., Dalton, T. K., and Darden, J. H. (1979): Transient alterations in neurotransmitter activity in the caudate nucleus of rat brain after a high dose of ionizing radiation. *Radiat. Res.*, 80:556–562.

352. Hunter, A. E., Prentice, H. G., Pothecary, K. et al. (1991): Granisetron, a selective 5-HT3 receptor antagonist, for the prevention of radiation induced emesis during total body irradiation. *Bone Marrow Transplant*, 7:439–441.

353. Hwang, A. and Muschel, R. J. (1998): Radiation and the $G_2$ phase of the cell cycle. *Radiat. Res.*, 150(Suppl.): S52–S59.

354. Hyer, M. and Nielsen, O. S. (1992): Influence of dose on regeneration of murine hematopoietic stem cells after total body irradiation and 5-fluorouracil. *Oncology*, 49:166–172.

355. Ibrahim, M. Z. M., Haymaker, W., Miquel, J., and Riopelle, A. J. (1967): Effects of radiation on the hypothalamus in monkeys. *Archiv fur Psychiatrie und Zeitschrift f. d. ges. Neurologie*, 210:1–15.

356. Ibuki, Y. and Goto, R. (1994): Enhancement of concanavalin A-induced proliferation of splenolymphocytes by low-dose-irradiated macrophages. *J. Radiat. Res.*, 35:83–91.

357. Ibuki, Y. and Goto, R. (2000): Enhancement of $O_2$ production from resident peritoneal macrophages by low dose *in vivo* gamma-irradiation. *Biol. Pharm. Bull.*, 23: 1094–1096.

358. Il'enko, A. I. and Krapivko, T. P. (1994): Radioresistance of populations of bank voles *Clethrionomys glareolus* in radionuclide-contaminated areas. *Doklady Biol. Sci.*, 336: 262–266.

359. Iliakis, G., Wang, Y., Pantelias, G. E., and Metzger, L. (1992): Mechanism of radiation sensitization by halogenated pyrimidines: effect of BrdU on repair of DNA breaks, interphase chromatin breaks, and potentially lethal damage in plateau-phase CHO cells. *Radiat. Res.*, 129: 202–211.

360. Imanaka, T., Seo, T., and Koide, H. (1988): Radioactivity release from the Chernobyl-4 accident and its cancer consequences. *J. Radiat. Res. (Tokyo)*, 29:80.

361. Inano, H., Suzuki, K., Ishii-Ohba, H. et al. (1989): Steroid hormone production in testis, ovary, and adrenal gland of immature rats irradiated *in utero* with $^{60}$Co. *Radiat. Res.*, 117:293–303.

362. Innes, J. R. and Carsten, A. (1961): Demyelination or malacic myelopathy. *Arch. Neurol.*, 4:190:199.

363. Ito, M., Patronas, N. J., Di Chiro, G. et al. (1986): Effect of moderate level x-radiation to brain on cerebral glucose utilization. *J. Comput. Assist. Tomogr.*, 10:584–588.

364. Iverson, S. D. and Iverson, L. L. (1981): *Behavioral Pharmacology*. Oxford University Press, New York.

365. Izrael, Yu. A., Petrov, V. A., Avdjushin, S. I. et al. (1987): Radioactive pollution of the natural environment in the zone of the accident of the Chernobyl Atomic Power Plant (Russia). *Meterologiya Hydrologiya*, 2:5–18.

366. Jacobs, A. J., Maniscalco, W. M., Parkhurst, A. B., and Finkelstein, J. N. (1986): *In vivo* and *in vitro* demonstration of reduced myelin synthesis following early postnatal exposure to ionizing radiation. *Radiat. Res.*, 105:97–104.

367. Jaenke, R. S. and Angleton, G. M. (1990): Perinatal radiation-induced renal damage in the beagle. *Radiat. Res.*, 122:58–65.

368. Jagoe, C. H., Chesser, R. K., Smith, M. H. et al. (1998): Radiocesium, mercury and lead in fish, and sediment radiocesium in waters near Chernobyl, Ukraine. *Ecotoxicology*, 7:202–210.

369. Jammet, H., Mathe, G., Pendic, B. et al. (1959): Study of six cases of accidental whole-body irradiation. *Rev. Fr. Etud. Clin. Biol.*, 4:210–225.

370. Janis, I. L. (1951): *Air War and Emotional Stress*. McGraw-Hill, New York.

371. Jensh, R. P., Eisenman, L. M., and Brent, R. L. (1995): Postnatal neurophysiologic effects of prenatal x-irradiation. *Int. J. Radiat. Biol.*, 67(2):217–227.

372. Jirtle, R. L., Anscher, M. S., and Alati, T. (1990): Radiation sensitivity of the liver. In: *Advances in Radiation Biology*. Vol. 14. *Relative Radiation Sensitivities of Human Organ Systems, Part II*, edited by K. I. Altman and J. T. Lett, pp. 269–311. Academic Press, San Diego, CA.

373. Johnsson, J. E., Owman, C. H., and Sjoberg, N. O. (1970): Tissue content of noradrenaline and 5-hydroxytryptamine in the rat after ionizing radiation. *Int. J. Radiat. Biol.*, 18:311–316.

374. Jones, D. C., Kimeldorf, D. J., Rubadeau, D. O. et al. (1954): Effects of x-irradiation on performance of volitional activity by the adult male rat. *Am. J. Physiol.*, 177: 243–250.

375. Jones, G. D. D., Ward, J. F., Limoli, C. L. et al. (1995): Mechanisms of radiosensitization in iododeoxyuridine-substituted cells. *Int. J. Radiat. Biol.*, 67:647–653.

376. Joseph, J. A., Erat, S., and Rabin, B. M. (1998): CNS effects of heavy particle irradiation in space: behavioral implications. *Adv. Space Res.*, 22:209–216.

377. Joseph, J. A., Hunt, W. A., Rabin, B. M., and Dalton, T. K. (1992): Possible 'accelerated striatal aging' induced by $^{56}$Fe heavy particle irradiation: implications for manned space flight. *Radiat. Res.*, 130:88–93.

378. Joseph, J. A., Kandasamy, S. B., Hunt, W. A. et al. (1988): Radiation-induced increases in sensitivity of cataleptic behavior to haloperidol: possible involvements of prostaglandins. *Pharmacol. Biochem. Behav.*, 29:335–341.

379. Joseph, J. A., Shukitt-Hale, B., McEwen, J., and Rabin, B. M. (2000): CNS-induced deficits of heavy particle irradiation in space: the aging connection. *Adv. Space Res.*, 25:2057–2064.

380. Jung, M. and Dritschilo, A. (1996): Signal transduction and cellular responses to ionizing radiation. *Semin. Radiat. Oncol.*, 6:268–272.

381. Kader, H. A. and Rostom, A. Y. (1991): Follicle stimulating hormone levels as a predictor of recovery of spermatogenesis following cancer therapy. *Clin. Oncol. (R. Coll. Radiol.)*, 3:37–40.

382. Kameyama, Y. and Hoshino, K. (1986): Sensitive phases of CNS development. In: *Radiation Risks to the Developing Nervous System*, edited by H. Kriegel, W. Schmahl, G. B. Gerber, and F.-E. Stieve, pp. 75–92. Stuttgart, Gustav Fischer Verlag.

383. Kandasamy, S. B., Hunt, W. A., and Mickley, A. G. (1988): Implications of prostaglandins and histamine $H_1$ and $H_2$ receptors in radiation-induced temperature responses of rats. *Radiat. Res.*, 114:42–53.
384. Kaplan, M. I. and Morgan, W. F. (1998): The nucleus is the target for radiation-induced chromosomal instability. *Radiat. Res.*, 150:382–390.
385. Karaoglou, A., Desmet, G., Kelly, G. N., and Menzel, H. G. (1995): The radiological consequences of the Chernobyl accident. In: *Proc. of the First Int. Conf.*, Minsk, Belarus. Office for Official Publications of the European Communities, Luxembourg.
386. Kashparov, V. A. (2003): Hot particles at Chernobyl. *Environ. Sci. Pollut. Res. Int.*, Special Issue, (1):21–30.
387. Katz, R. J., Carroll, B. J., and Baldright, G. (1978): Behavioral activation by enkephalins in mice. *Pharmacol. Biochem. Behav.*, 8:493–496.
388. Kaul, A., Landfermann, H., and Thieme, M. (1996): One decade after Chernobyl: summing up the consequences. *Health Phys.*, 71:634–640.
389. Kekcheyev, K. (1941): Changes in the threshold of achromatic vision of man by the action of ultrashort, ultraviolet and x-ray waves. *Probl. Fisiol. Optics*, 1:77–79.
390. Kelly, H. (2002): Testimony before the Senate Committee on Foreign Relations, Washington D.C.
391. Kemper, T. L., O'Neill, R., and Caveness, W. F. (1977): Effects of single dose supervoltage whole brain radiation in *Macaca mulatta*. *J. Neuropathol. Exp. Neurol.*, 36:916–940.
392. Kent, C. R. H., Eady, J. J., Ross, G. M., and Steel, G. G. (1995): The comet moment as a measure of DNA damage in the comet assay. *Int. J. Radiat. Biol.*, 67:655–660.
393. Kerr, J. F. R., Wyllie, A. H., and Currie, A. R. (1972): Apoptosis: a basic biological phenomenon with wide-ranging implications in tissue kinetics. *Br. J. Cancer*, 26:239–256.
394. Keyse, S. M. (1998): Protein phosphatases and the regulation of MAP kinase activity. *Cell Devel. Biol.*, 9:143–152.
395. Khodarev, N. N., Sokolova, I. A., and Vaughan, A. T. M. (1998): Mechanisms of induction of apoptotic DNA fragmentation. *Int. J. Radiat. Biol.*, 73:455–467.
396. Kim, J. H., Mandell, L. R., and Leeper, R. (1990): Radiation effects on the thyroid gland. In: *Advances in Radiation Biology.* Vol. 14. *Relative Radiation Sensitivities of Human Organ Systems, Part II*, edited by K. I. Altman and J. T. Lett, pp. 119–156. Academic Press, San Diego, CA.
397. Kimeldorf, D. J. and Hunt, E. L. (1965): *Ionizing Radiation: Neural Function and Behavior*. Academic Press, New York.
398. Kimeldorf, D. J. and Jones, D. C. (1951): The relationship of radiation dose to lethality among exercised animals exposed to Roentgen rays. *Am. J. Physiol.*, 167:626–632.
399. Kimeldorf, D. J., Jones, D. C., and Castanera, T. J. (1953): Effect of x-irradiation upon the performance of daily exhaustive exercise by the rat. *Am. J. Physiol.*, 174:331–335.
400. Kimler, B. F. (1998): Prenatal irradiation: a major concern for the developing brain. *Int. J. Radiat. Biol.*, 73:423–434.
401. Kimler, B. F., Vidal-Pergola, G. M., Peterson, S. L. et al. (1994): Effect of *in utero* radiation dose fractionation on rat postnatal development, behavior and brain structure: 3-hour interval. *Neurotoxicology*, 15:183–189.
402. King, G. (2004): *Dirty Bomb: Weapon of Mass Disruption*. Chamberlain Bros., Penguin Group, New York.
403. King, G. L. (1988): Characterization of radiation-induced emesis in the ferret. *Radiat. Res.*, 114:599–612.
404. King, G. L. and Landauer, M. R. (1990): Effects of Zacopride and BMY25801 (Batanopride) on radiation-induced emesis and locomotor behavior in the ferret. *J. Pharmacol. Exp. Ther.*, 253:1026–1033.
405. Kinsella, T. J. (1989): Effects of radiation therapy and chemotherapy on testicular function. *Prog. Clin. Biol. Res.*, 302:157–177.
406. Klatzo, I., Suzuki, R., Orzi, F. et al. (1984): Pathomechanisms of ischemic brain edema. In: *Recent Progress in the Study and Therapy of Brain Edema*, edited by K. G. Co and A. Raathmann, pp. 1–17. Plenum Press, New York.
407. Klein, M., Heimans, J. J., Aaronson, N. K. et al. (2002): Effect of radiotherapy and other treatment-related factors on mid-term to long-term cognitive sequelae in low-grade gliomas: a comparative study. *Lancet*, 360:1361–1368.
408. Kligerman, M. M., Liu, T., Scheffler, B. et al. (1992): Interim analysis of a randomized trial of radiotherapy of rectal cancer with/without WR-2721. *Int. J. Radiat. Oncol. Biol. Phys.*, 22:799–802.
409. Klouwen, H. M. and Betel, I. (1963): Radiosensitivity of nuclear ATP synthesis. *Int. J. Radiat. Biol.*, 6:441–461.
410. Koga, T., Morishima, H., Niwa, T., and Kawai, H. (1991): Tritium precipitation in European cities and in Osaka Japan, owing to the Chernobyl nuclear accident. *J. Radiat Res. (Tokyo)*, 32:267–276.
411. Kondakova, N. V., Lisakovskii, S. V., Sakharova, V. V. et al. (1994): Effect of ionizing radiation on human muscle tissue (Russia). *Voprosy Meditsinskoi Khimii*, 40(4):46–50
412. Konings, A. W. T. (1987): Role of membrance lipid composition in radiation-induced death of mammalian cells In: *Prostaglandin and Lipid Metabolism in Radiation Injury*, edited by T. L. Walden, Jr., and H. N. Hughes, pp 29–43. Plenum Press, New York.
413. Konoplyannikov, A. G. (1997): Molecular and cellular mechanisms of late radiation damages. *Radiatsionnaya Biologiya Radioekologiya*, 37(4):621–628.
414. Koshurnikova, N., Buldakov, L., Bysogolov, G. et al (1994): Mortality from malignancies of the hematopoietic and lympathic tissue among personnel of the first nuclear plant in the U.S.S.R. *Sci. Total Environ.*, 142:19–23.
415. Koshurnikova, N. A., Bysogolov, G. D., Bolotnikova, M G. et al. (1996): Mortality among personnel who worked at the MAYAK complex in the first years of its operation *Health Phys.*, 71:90–93.
416. Kossenko, M. M. (1996): Cancer mortality among Techa river residents and their offspring. *Health Phys.*, 71:77–82
417. Koukourakis, M. I. (2002): Amifostine in clinical oncology: current use and future applications. *Anticancer Drugs* 13:181–209.
418. Krabbe, A. A. and Olesen, J. (1982): Effect of histamine on regional cerebral blood flow in man. *Cephalalgia*, 2 15–18.
419. Krapivko, T. P. and Il'enko, A. I. (1988): First features of radioadaptation in a population of red-backed vole (*Clethrionomys glareolus*) in a radiation biogeocenosis *Doklady Akademii Nauk SSR*, 302(5):1272–1274.

420. Kristoff, J. D. (2004, March 10): A nuclear 9/11. *The New York Times*.
421. Krobel, W. and Kroem, G. (1959): Die wirkung geringer strahlungsdosen auf die signal-erzeugungs und fortleitungs-eigenshaftens-eigenschaften in froschnerven. *Atomkernergie*, 4:280–286.
422. Krykanides, S., Moore, A. H., Olschowka, J. A. et al. (2002): Cyclooxygenase-2 modulates brain inflammation-related gene expression in central vervous system radiation injury. *Brain Res. Mol. Brain. Res.*, 104:159–169.
423. Kryshev, I. I. (1992): *Radioecological Consequences of the Chernobyl Accident*. Nuclear Society International, Moscow.
424. Kuettle, M. R., Thraves, P. J., Jung, M. et al. (1996): Radiation-induced neoplastic transformation in human prostate epithelial cells. *Cancer Res.*, 56(1):5–10.
425. Kulinski, V. I. and Semenov, L. F. (1965): Content of catecholamines in the tissues of macaques during the early periods after total gamma irradiation. *Radiobiologiia*, 5:494–500.
426. Kumagai, J., Masui, K., Itagaki, Y. et al. (2003): Long-lived mutagenic radicals induced in mammalian cells by ionizing radiation are mainly localized to proteins. *Radiat. Res.*, 160:95–102.
427. Kumar, K. S., Vaishnav, Y. N., and Weiss, J. F. (1988): Radioprotection by antioxidant enzymes and enzyme mimetics. *Pharmac. Ther.*, 39:301–309.
428. Kundurovic, Z., Scepovic, M., Causevic, A., and Mornjakovic, Z. (1991): Histochemical aspects and fine structural characteristics of thyreocytes in pinealectomized and melatonin treated rats prior to irradiation. *Acta Med. Croatica*, 45:347–355.
429. Lam, K. S., Tse, V. K., Wang, C. et al. (1991): Effects of cranial irradiation on hypothalamic-pituitary function: a 5-year longitudinal study in patients with nasopharyngeal carcinoma. *Q. J. Med.*, 78:165–176.
430. Lam, L. C., Leung, S. F., and Chan, Y. L. (2003): Progress of memory function after radiation therapy in patients with nasopharyngeal carcinoma. *J. Neuropsychiatry Clin. Neurosci.*, 15:90–97.
431. Lampe, P. D. and Lau, A. F. (2004): The effects of connexin phosphorylation on gap junctional communication. *Int. J. Biochem. Cell Biology*, 36:1171–1186.
432. Landauer, M. R. (2002): Radiation-induced performance decrement. *Mil. Med.*, 167:128–130.
433. Landauer, M. R., Davis, H. D., Dominitz, J. A., and Weiss, J. F. (1987): Effects of acute gamma radiation exposure on locomotor activity of Swiss–Webster mice. *The Toxicologist*, 7:253.
434. Landauer, M. R., Davis, H. D., Dominitz, J. A., and Weiss, J. F. (1987): Dose and time relationships of the radioprotector WR-2721 on locomotor activity in mice. *Pharmacol. Biochem. Behav.*, 27:573–576.
435. Landauer, M. R., Davis, H. D., Dominitz, J. A., and Weiss, J. F. (1988): Long-term effects of radioprotector WR-2721 on locomotor activity and body weight of mice following exposure to ionizing radiation. *Toxicology*, 49:315–323.
436. Landauer, M. R., Davis, H. D., Kumar, K. S., and Weiss, J. F. (1992): Behavioral toxicity of selected radioprotectors. *Adv. Space Res.*, 12:273–283.
437. Landauer, M. R., McChesney, D. G., and Ledney, G. D. (1997): Synthetic trehalose dicorynomycolate (S-TDCM): Behavioral effects and radioprotection. *J. Radiat. Res. (Tokyo)*, 38:45–54.
438. Landauer, M. R., Walden, T. L., and Davis, H. D. (1990): Behavioral effects of radioprotective agents in mice: combination of WR-2721 and 16,16-dimethyl prostaglandin E2. In: *Frontiers in Radiation Biology*, edited by E. Riklis, pp. 199–207. VCH Publishers, New York.
439. Landauer, M. R., Castro, C. A., Benson, K. A. et al. (2001): Radioprotective and locomotor responses of mice treated with nimodipine alone and in combination with WR-151327. *J. Appl. Toxicol.*, 21:25–31.
440. Langlois, R. G., Ariyama, M., Kusunoki, Y. et al. (1993): Analysis of somatic cell mutations at the glycophorin A locus in atomic bomb survivors: a comparative study of assay methods. *Radiat. Res.*, 136:111–117.
441. Lauk, S., Bohm, M., Feiler, G. et al. (1989): Increased number of cardiac adrenergic receptors following local heart irradiation. *Radiat. Res.*, 119:157–165.
442. Law, M. P. and Ahier, R. G. (1989): Vascular and epithelial damage in the lung of the mouse after x rays or neutrons. *Radiat. Res.*, 117:128–144.
443. Lawrence, T. S., Blackstock, W. A., and McGinn, C. (2003): The mechansim of action of radiosensitization of conventional chemotherapeutic agents. *Semin. Radiat. Oncol.*, 13:13–21.
444. Le, X. C., Xing, J. Z., Lee, J. et al. (1998): Inducible repair of thymine glycol detected by an ultrasensitive assay for DNA damage. *Science*, 280:1066–1069.
445. Leach, J. K., Black, S. M., Schmidt-Ullrich, R. K., and Mikkelsen, R. B. (2002): Activation of constitutive nitric-oxide synthase activity is an early signaling event induced by ionizing radiation. *J. Biol. Chem.*, 277:15400–15406.
446. Leadon, S. A. (1996): Repair of DNA damage produced by ionizing radiation: a mini-review. *Semin. Radiat. Oncol.*, 6:295–305.
447. Lebedinsky, A. V., Grigoryev, U. G., and Demirchoglyan, G. G. (1958): On the biological effect of small doses of ionizing radiation. In: *Proc. of Second United Nations Int. Conf. on Peaceful Uses of Atomic Energy*, Vol. 22, pp. 17–28.
448. Lenoir, A. (1944): Adaptation and rontgenbestrahlung. *Radiol. Clin. (Basel)*, 13:264–276.
449. Levi, M., Nelson, R., and Yassif, J. (2002): Dirty bombs: response to a threat. *J. Fed. Am. Sci.*, 55:2
450. Liebster, J. and Kopoldova, J. K. (1964): The radiation chemistry of amino acids. *Adv. Radiat. Biol.*, 1:157–226.
451. Lindgren, M. (1958): On tolerance of brain tissue and sensitivity of brain tumors to irradiation. *Acta Radiol.*, 170: 5–75.
452. Lingenfelser, S. K., Dallas, C. E., Jagoe, C. H. et al. (1997): Variation in blood cell DNA content in *Carassius carrassius* from ponds near Chernobyl, Ukraine. *Ecotoxicology*, 6:187–203.
453. Little, J. B., McGrandy, R. B., and Kennedy, A. R. (1978): Interactions between polonium-210, alpha-radiation, benzo(*a*)pyrene, and 0.9% NaCl solution instillations in the induction of experimental lung cancer. *Cancer Res.*, 38:1929–1935.

454. Littley, M. D., Shalet, S. M., and Beardwell, C. G. (1990): Radiation and hypothalamic-pituitary function. *Baillieres Clin. Endocrinol. Metab.*, 4:147–175.
455. Littley, M. D., Shalet, S. M., and Beardwell, C. G. (1991): Radiation and the hypothalamic-pituitary axis. In: *Radiation Injury to the Nervous System*, edited by P. H. Gutin et al., pp. 303–324. Raven Press, New York.
456. Littley, M. D., Shalet, S. M., Morgenstern, G. R., Deakin, D. P. (1991): Endocrine and reproductive dysfunction following fractionated total body irradiation in adults. *Q. J. Med.*, 78:265–274.
457. Livingston, G. K., Jensen, R. H., Silberstein, E. B. et al. (1997): Radiobiological evaluation of immigrants from the vicinity of Chernobyl. *Int. J. Radiat. Biol.*, 72(6):703–713.
458. Lo, E. H., Frankel, K. A., Steinberg, G. K. et al. (1992): High-dose single-fraction brain irradiation: MRI, cerebral blood flow, electrophysiological, and histological studies. *Int. J. Radiat. Oncol. Biol. Phys.*, 22:47–55.
459. Loganovsky, K. N. and Yuryev, K. L. (2004): EEG patterns in persons exposed to ionizing radiation as a result of the chernobyl accident. Part 2. Quantitative EEG analysis in patients who had acute radiation sickness. *J. Neuropsychiatry Clin. Neurosci.*, 16:70–82.
460. Lomat, L., Galburt, G., Quastel, M. R. et al. (1997): Incidence of childhood disease in Belarus associated with the Chernobyl accident. *Environ. Health Perspect.*, 105:1529–1532.
461. Lorimore, S. A. and Wright, E. G. (2003): Radiation-induced genomic instability and bystander-effects: related inflammatory-type responses to radiation-induced stress and injury? A review. *Int. J. Radiat. Biol.*, 79:15–25.
462. Lott, J. R. (1962): Changes in ventral root potentials during x-irradiation of the spinal cord in the cat. In: *Effects of Ionizing Radiation on the Nervous System: Proceedings*, pp. 85–92. International Atomic Energy Agency, Vienna.
463. Lucas, J. N. (1997): Dose reconstruction for individuals exposed to ionizing radiation using chromosome painting. *Radiat. Res.*, 148:S33–S38.
464. Luckey, T. D. (1980): *Hormesis with Ionizing Radiation*. CRC Press, Boca Raton, FL.
465. Lundbeck, F., Uls, N., and Overgaard, J. (1989): Cystometric evaluation of early and late irradiation damage to the mouse urinary bladder. *Radiother. Oncol.*, 15:383–392.
466. Lundin, J., Clemedson, C. J., and Nelson, A. (1957): Early effects of whole-body irradiation on cholinesterase activity in guinea pig's blood with special regard to radiation sickness. *Acta Radiol.*, 48:52–64.
467. Luxford, C., Dean, R. T., and Davies, M. J. (2000): Radicals derived from histone hydroperoxides damage nucleobases in RNA and DNA. *Chem. Res. Toxicol.*, 13:665–672.
468. Luxin, W., Yongru, Z., Zufan, T. et al. (1990): Epidemiological investigation of radiological effects in high background radiation areas of Yangjiang, China. *J. Radiat. Res. (Tokyo)*, 31:119–136.
469. Lyer, R. and Lehnert, B. E. (2000): Factors underlying the cell growth-related bystander responses to $\alpha$-particles. *Cancer Res.*, 60:1290–1298.
470. Lyman, R. S., Kupalov, R. S., and Scholz, W. (1933): Effects of Roentgen rays on the central nervous system: results of large doses on the brains of adult dogs. *AMA Arch. Neurol. Psychiat.*, 29:56–87.
471. Macklis, R. M., Bellerive, M. R., and Humm, J. L. (1990): The radiotoxicology of Radithor: analysis of an early case of iatrogenic poisoning by a radioactive patent medicine. *JAMA*, 264:619–621.
472. MacNaughton, W. K., Aurora, A. R., Bhamra, J. et al. (1998): Expression, activity and cellular localization of inducible nitric oxide synthase in rat ileum and colon post-irradiation. *Int. J. Radiat. Biol.*, 74:255–264.
473. MacNaughton, W. K., Leach, K. E., Prud'Homme-LaLonde, L., and Harding, R. K. (1997): Exposure to ionizing radiation increases responsiveness to neural secretory stimuli in the ferret jejunum *in vitro*. *Int. J. Radiat. Biol.*, 72(2):219–226.
474. MacNaughton, W. K., Leach, K. E., Prud'Homme-LaLonde, L. et al. (1994): Ionizing radiation reduces neurally evoked electrolyte transport in ral ileum through a mast cell-dependent mechanism. *Gastroenterology*, 106(2): 324–335.
475. MacNaughton, W. K. and Prud'Homme-LaLonde, L. (1995): Exposure to ionizing radiation alters vasoreactivity in rat jejunum *ex vivo*. *Can. J. Physiol. Pharmacol.*, 73(6):699–705.
476. MacVittie, T. J., Monroy, R. L., Patchen, M. L., and Souza, L. M. (1990): Therapeutic use of recombinant human G-CSF (rhG-CSF) in a canine model of sublethal and lethal whole-body irradiation. *Int. J. Radiat. Biol.*, 57:723–736.
477. MacVittie, T. J., Monroy, R. L., Vigneulle, R. M. et al. (1991): The relative biological effectiveness of mixed fission-neutron-$\gamma$ radiation on the hematopoietic syndrome in the canine: effect of therapy on survival. *Radiat. Res.*, 128:S29–S36.
478. Magos, L. (1991): Epidemiological and experimental aspects of metal carcinogenesis: physiological properties, kinetics, and the active species. *Environ. Health Perspect.*, 95:157–89.
479. Maier, D. M. and Landauer, M. R. (1989): Effects of acute sublethal gamma radiation exposure on aggressive behavior in male mice: a dose–response study. *Aviat. Space Eviron. Med.*, 60:774–778.
480. Maier, D. M. and Landauer, M. R. (1990): Onset of behavioral effects in mice exposed to 10 Gy $^{60}$Co radiation. *Aviat. Space Environ. Med.*, 61:893–898.
481. Maier, D. M., Landauer, M. R., Davis, H. D., and Walden, T. L. (1989): Effect of electron radiation on aggressive behavior, activity, and hemopoiesis in mice. *J. Radiat. Res. (Tokyo)*, 30:255–265.
482. Maisin, J. R. (1988): Acute radiation syndromes in man. In: *Terrestrial Space Radiation and Its Biological Effects*, edited by P. D. McCormack, C. E. Swenberg, and H. Bücker, pp. 445–463. Plenum Press, New York.
483. Malysheva, O. A. and Shirinskii, V. S. (1998): Seasonal changes of secondary immunodeficiency in patients with vascular dystonia (Russia). *Klin. Med. (Mosk.)*, 76:34–36.
484. Margules, D. L. (1979): Beta-endorphin and endoloxone: hormones of the autonomic nervous system for the conservation of expenditure of bodily resources and energy in anticipation of famine or feast. *Neurosci. Biobehav. Rev.* 3:155–162.

485. Marks, J. E. and Wong, J. (1985): The risk of cerebral radionecrosis in relation to dose, time and fractionation. *Prog. Exp. Tumor Res.*, 29:210–218.
486. Martin, C., Roman, V., Agay, D., and Fatome, M. (1998): Anti-emetic effect of ondansetron and granisetron after exposure to mixed neutron and gamma irradiation. *Radiat. Res.*, 149:631–636.
487. Martin, C., Martin, S., Viret, R. et al. (2001): Low dose of the gamma acute radiation syndrome (1.5 Gy) does not significantly alter either cognitive behavior or dopaminergic and serotoninergic metabolism. *Cell Mol. Biol. (Noisy-le-grand)*, 47:459–465.
488. Martin, S., Vojnovic, B., and Murray, J. C. (1991): Determination of x-ray-induced damage to the murine colon using tissue compliance measurements. *Int. J. Radiat. Biol.*, 59:503–515.
489. Martin, S., Stratford, M. R. L., Watfa, R. R. et al. (1992): Collagen metabolism in the murine colon following x irradiation. *Radiat. Res.*, 130:38–47.
490. Maruyama, Y. and Feola, J. M. (1987): Relative radiosensitivities of the thymus, spleen, and lymphohemopoietic systems. In: *Advances in Radiation Biology*, Vol. 14. *Relative Radiation Sensitivities of Human Organ Systems, Part II*, edited by K. I. Altman and J. T. Lett, pp. 1–82. Academic Press, San Diego, CA.
491. Mastaglia, F. L., McDonald, W. I., Watson, J. V., and Yogendran, K. (1976): Effects of x-irradiation on the spinal cord: an experimental study of the morphological changes in central nerve fibers. *Brain*, 99:101–122.
492. Matsumoto, H., Hayashi, S., Hatashita, M. et al. (2001): Induction of radioresistance by a nitric oxide-mediated bystander effect. *Radiat. Res.*, 155:387–396.
493. Mattsson, A., Ruden, B.-I., Hall, P. et al. (1993): Radiation-induced breast cancer: long-term follow-up of radiation therapy for benign breast disease. *J. Nat. Cancer Inst. (Bethesda)*, 85(20):1679–1685.
494. Mattsson, A., Ruden, B.-I., Palmgren, J. et al. (1995): Dose- and time-response for breast cancer risk after radiation therapy for benign breast disease. *Br. J. Cancer*, 72(4):1054–1061.
495. Mattsson, J. L. and Yochmowitz, M. G. (1980): Radiation-induced emesis in monkeys. *Radiat. Res.*, 82:191–199.
496. McDonough, J. H., Mele, P. C., and Franz, C. G. (1992): Comparison of behavioral and radioprotective effects of WR-2721 and WR-36–89. *Pharmacol. Biochem. Behav.*, 42:233–243.
497. McFarland, W. L. and Levin, S. G. (1974): Electroencephalographic responses of 2500 rads of whole-body gamma-neutron radiation in the monkey *Macaca mulatta*. *Radiat. Res.*, 58:60–73.
498. McKay, M. J., Bull, C. A., Houghton, C. R., and Langlands, A. O. (1990): Persisting cyclical uterine bleeding in patients treated with radical radiation therapy and hormonal replacement for carcinoma of the cervix. *Int. J. Radiat. Oncol. Biol. Phys.*, 18:921–925.
499. McMahon, T. and Vahora, S. (1986): Radiation damage to the brain. *Neuropsychiatric Aspects*, 8:437–441.
500. Mechanick, J. I., Hochberg, F,H., and LaRocque, A. (1986): Hypothalamic dysfunction following whole brain irradiation. *J. Neurosurg.*, 65:490–494.
501. Medvedev, Z. A. (1979): *Nuclear Disaster in the Urals*. W.W. Norton, New York.
502. Medvedev, Z. A. (1990): *The Legacy of Chernobyl*, p. 187. W.W. Norton, New York.
503. Medvedev, Z. A. (1994): Chernobyl: eight years after. *TREE*, 9:369–371.
504. Mefferd, J. M., Donaldson, S. S., and Link, M. P. (1989): Pediatric Hodgkin's disease: pulmonary, cardiac, and thyroid function following combined modality therapy. *Int. J. Radiat. Oncol. Biol. Phys.*, 16:679–685.
505. Meistrich, M. L. and van Beek, M. E. A. B. (1990): Radiation sensitivity of the human testis. In: *Advances in Radiation Biology*, Vol. 14. *Relative Radiation Sensitivities of Human Organ Systems, Part II*, edited by K. I. Altman and J. T. Lett, pp. 227–268. Academic Press, San Diego, CA.
506. Mele, P. C., Franz, C. G., and Harrison, J. R. (1988): Effects of sublethal doses of ionizing radiation on schedule-controlled performance in rats. *Pharmacol. Biochem. Behav.*, 30:1007–1014.
507. Mele, P. C., Franz, C. G., and Harrison, J. R. (1990): Effects of ionizing radiation on fixed-ratio escape performance in rats. *Neurotoxicol. Teratol.*, 12:367–373.
508. Mele, P. C. and McDonough, J. H. (1995): Gamma radiation-induced disruption in schedule-controlled performance in rats. *Neurotoxicology*, 16:497–510.
509. Melkonyan, H. S., Ushakova, T. E., and Umansky, S. R. (1995): Hsp 70 gene expression in mouse lung cells upon chronic gamma-irradiation. *Int. J. Radiat. Biol.*, 68: 277–280.
510. Mettler, Jr., F. A. and Moseley, Jr., R. D. (1985): *Medical Effects of Ionizing Radiation*. Grune & Stratton, New York.
511. Meyers, C. A., Smith, J. A., Bezjak, A. et al. (2004): Neurocognitive function and progression in patients with brain metastases treated with whole brain radiation and motexafin gadolinium: results of a randomized phase III trial. *J. Clin. Oncol.*, 22:157–165.
512. Meyerson, F. G. (1958): Effect of damaging doses of gamma-radiation on unconditioned and conditioned respiratory reflexes. In: *Works of the Institute of Higher Nervous Activity, Pathophysiological Series*, Vol. 4. Izvestia Akademi, Moscow, pp. 25–41.
513. Michailov, M. C., Neu, E., Tempel, K. et al. (1991): Influence of x-irradiation on the motor activity of rat urinary bladder *in vitro* and *in vivo*. *Strahlenther Onkol.*, 167:311–318.
514. Michel, C. (1989): Radiation embryology. *Experientia*, 45:69–77.
515. Mickley, G. A. (1980): Behavioral and physiological changes produced by a supralethal dose of ionizing radiation: evidence for hormone-influenced sex differences in the rat. *Radiat. Res.*, 81:48–75.
516. Mickley, G. A. (1981): Antihistamine provides sex-specific radiation protection. *Aviat. Space Environ. Med.*, 52:247–250.
517. Mickley, G. A. (1991): Can animals serve as useful models for research on the psychological effects of radiation exposure? In: *The Medical Basis for Radiation-Accident Preparedness*. Vol. III. *The Psychological Perspective*, edited by R. C. Ricks, M. E. Berger, and F. M. O'Hara, pp. 25–38. Elsevier, New York.

518. Mickley, G. A. and Teitelbaum, H. (1978): Persistence of lateral hypothalamic-mediated behaviors after a supralethal dose of ionizing radiation. *Aviat. Space Environ. Med.*, 49:863–873.
519. Mickley, G. A., Teitelbaum, H., Parker, G. A. et al. (1982): Radiogenic changes in the behavior and physiology of the spontaneously hypertensive rat: evidence for a dissociation between acute hypotension and incapacitation. *Aviat. Space Environ. Med.*, 53:633–638.
520. Mickley, G. A., Stevens, K. E., White, G. A., and Gibbs, G. L. (1983): Endogenous opiates mediate radiogenic behavioral change. *Science*, 220:1185–1187.
521. Mickley, G. A., Sessions, G. R., Bogo, V., and Chantry, K. H. (1983): Evidence for endorphin-mediated cross-tolerance between chronic stress and the behavioral effects of ionizing radiation. *Life Sci.*, 33:749–754.
522. Mickley, G. A., Stevens, K. E., White, G. A., and Gibbs, G. L. (1983): Changes in morphine self-administration after exposure to ionizing radiation: evidence for the involvement of endorphins. *Life Sci.*, 33:711–718.
523. Mickley, G. A., Stevens, K. E., Moore, G. H. et al. (1983): Ionizing radiation alters beta-endorphin-like immunoreactivity in brain but not blood. *Pharmacol. Biochem. Behav.*, 19:979–983.
524. Mickley, G. A., Stevens, K. E., Burrows, J. M. et al. (1983): Morphine tolerance offers protection from radiogenic performance decrements. *Radiat. Res.*, 93:381–387.
525. Mickley, G. A., Bogo, V., and West, B. (1989): Behavioral and neurophysiological changes with exposure to ionizing radiation. In: *Textbook of Military Medicine*, edited by R. Zajtchuck, D. P. Jenkins, R. F. Bellamy, V. M. Ingram, R. I. Walker, and T. J. Cerveny, pp. 105–151. U.S. Army, Washington, D.C.
526. Mickley, GA., Ferguson, J. L., Mulvihill, M. A., and Nemeth, T. J. (1989): Progressive behavioral changes during the maturation of rats with early radiation-induced hypoplasia of *fascia dentata* granule cells. *Neurotoxicol. Teratol.*, 11:385–393.
527. Mickley, G. A., Ferguson, J. L., and Nemeth, T. J. (1992): Serial injections of MK 801 (Dizocilpine) in neonatal rats reduce behavioral deficits associated with x-ray-induced hippocampal granule cell hypoplasia. *Pharmacol. Biochem. Behav.*, 43:785–793.
528. Middleton, G. R. and Young, R. W. (1975): Emesis in monkeys following exposure to ionizing radiation. *Aviat. Space Environ. Med.*, 46:170–172.
529. Miki, T., Fukiu, Y., Hisano, S. et al. (1996): Histogenetic abnormalities of the hippocampus in prenatally gamma-irradiated rats. *Teratology*, 54(4):15A.
530. Miki, T., Sawada, K., Sun, X. Z. et al. (1999): Abnormal distribution of hippocampal mossy fibers in rats exposed to x-irradiation *in utero*. *Brain Res. Dev. Brain Res.*, 112:275–280.
531. Mikkelsen, R. (2004): Redox signaling mechanisms and radiation-induced bystanders effects. *Human Exp. Toxicol.*, 23:75–79.
532. Mikkelsen, R. B. and Wardmann, P. (2003): Biological chemistry of reactive oxygen and nitrogen and radiation-induced signal transduction mechanisms. *Oncogene.*, 22:5734–5754.
533. Milas, L. (2001): Cyclooxygenase-2 (COX2) enzyme inhibtors as potential enhancers of tumor radioresponse. *Semin. Radiat. Oncol.*, 11:290–299.
534. Miletich, D. J. and Strike, T. A. (1970): *Alteration of Postirradiation Hypotension and Incapacitation in the Monkey by Administration of Vasopressor Drugs*, AFRRI Scientific Report SR70-1. Armed Forces Radiobiology Research Institute, Bethesda, MD.
535. Miller, D. S. (1962): Effects of low level radiation on audiogenic convulsive seizures in mice. In: *Response of the Nervous System to Ionizing Radiation*, edited by T. J. Haley and R. S. Snider, pp. 513–531. Academic Press, New York.
536. Miller, R. W. (1990): Effects of prenatal exposure to ionizing radiation. *Health Phys.*, 59:57–61.
537. Minamisawa, T., Hirokaga, K., Sasaki, S., and Noda, Y. (1992): Effects of fetal exposure to gamma rays on aggressive behavior in adult male mice. *J. Radiat. Res. (Tokyo)*, 33:243–249.
538. Minamisawa, T. and Hirokaga, K. (1995): Long term effects of prenatal exposure to low level gamma rays on spontaneous circadian motor activity of male mice. *J. Radiat. Res. (Tokyo)*, 36:179–184.
539. Minamisawa, T. and Hirokaga, K. (1995): Long-term effects of prenatal exposure to low levels of gamma rays on open-field activity in male mice. *Radiat. Res.*, 144:237–240.
540. Mintz, M., Yovel, G., Gigi, A., and Myslobodsky, M. S. (1998): Dissociation between startle and prepulse inhibition in rats exposed to gamma radiation at day 15 of embryogeny. *Brain Res. Bull.*, 45:289–296.
541. Mitchell, J. B. and Russo, A. (1987): The role of glutathione in radiation and drug induced cytotoxicity. *Br. J. Cancer*, 55:96–104.
542. Mitchell, J. B., Wink, D. A., and DeGraff, W. et al. (1993): Hypoxic mammalian cell radiosensitization by nitric oxide. *Cancer Res.*, 53:5845–5848.
543. Miyachi, Y., Kasai, H., Ohyama, H., and Yamada, T. (1994): Changes of aggressive behavior and brain serotonin turnover after very low-dose x-irradiation of mice. *Neurosci. Lett.*, 175:92–94.
544. Miyachi, Y. and Yamada, T. (1994): Low-dose x-ray-induced depression of sexual behavior in mice. *Behav. Brain Res.*, 65:113–115.
545. Miyachi, Y., Koizumi, T., and Yamada, T. (1994): Immediate arousal response and adaptation to low-dose x-rays in mouse and its disappearance by olfactory bulbectomy and nitric oxide inhibitor. *Neurosci. Lett.*, 177:32–34.
546. Miyachi, Y. and Yamada, T. (1996): Head-portion exposure to low-level x-rays reduces isolation-induced aggression of mouse, and involvement of the olfactory carnosine in modulation of the radiation effects. *Behav. Brain Res.* 81:135–140.
547. Miyachi, Y. (1997): Analgesia induced by repeated exposure to low dose x-rays in mice, and involvement of the accessory olfactory system in modulation of the radiation effects. *Brain Res. Bull.*, 44:177–182.
548. Miyachi, Y. (2000): Disappearance of stress-induced hyperthermia following a low dose of x-irradiation involvement of the vomeronasal system in the modulation of the radiation-induced effects. *Br. J. Radiol.*, 73:51–57.

549. Miyazaki, T., Hayakawa, Y., Suzuki, K., and Watanabe, M. (1990): Radioprotective effects of dimethyl sulfoxide in golden hamster embryo cell exposed to gamma rays at 77 K. I. Radical formation as studied by electron spin resonance. *Radiat. Res.*, 124:66–72.
550. Mizumatsu, S., Monje, M. L., Morhardt, D. R. et al. (2003): Extreme sensitivity of adult neurogenesis to low doses of x-irradiation. *Cancer Res.*, 63:4021–4027.
551. Mole, R. H. (1986): Problems related to prenatal exposure of the nervous systems: history and perspective. In: *Radiation Risks to the Developing Nervous System*, edited by H. Kriegel, W. Schmahl, G. B. Gerber, and F.-E. Stieve, pp. 1–20. Gustav Fischer, Stuttgart, Germany.
552. Mole, R. H. (1990): Severe mental retardation after large prenatal exposures to bomb radiation: reduction in oxygen transport to fetal brain: a possible abscopal mechanism. *Int. J. Radiat. Biol.*, 58:705–711.
553. Mole, R. H. (1992): Expectation of malformations after irradiation of the developing human *in utero*: the experimental basis for predictions. In: *Advances in Radiation Biology*, Vol. 15. *Relative Radiation Sensitivities of Human Organ Systems, Part III*, edited by K. I. Altman and J. T. Lett, pp. 217–301. Academic Press, San Diego, CA.
554. Molostovov, G. S. and Shavrova, E. N. (1997): Immunophenotyping of peripheral blood lymphocytes in children and adolescents with Hashimoto's thyroiditis (Russia). *Vyestsi Akademii Navuk Byelarusi Syeryya Biyalahichnykh Navuk*, 0(1):93–100.
555. Momosaki, S., Sun, X. Z., Takai, N. et al. (2002): Changes in histological construction and decrease in $^3$H-QNB binding in the rat brain after prenatal x-irradiation. *J. Radiat. Res.* (Tokyo), 43:277–282.
556. Monje, M. L. and Palmer, T. (2003): Radiation injury and neurogenesis. *Curr. Opin. Neurol.*, 16:129–134.
557. Monnier, M. and Krupp, P. (1962): Action of gamma radiation on electrical brain activity. In: *Response of the Nervous System to Ionizing Radiation*, edited by T. J. Haley and R. S. Snider, pp. 607–617. Academic Press, New York.
558. Monroy, R. L. (1987): Radiation effects on the lymphohematopoietic system: a compromise in immune competency. In: *Military Radiobiology*, edited by J. J. Conklin and R. I. Walker, pp. 113–134. Academic Press, San Diego, CA.
559. Monson, R. R. and MacMahon, B. (1984): Prenatal x-ray exposure and cancer in children. In: *Radiation Carcinogenesis: Epidemiology and Biological Significance*, edited by J. D. Boice, Jr. and J. F. Faumeni, Jr., pp. 97–105. Raven Press, New York.
560. Mooibroek, J., Trieling, W. B., and Konings, W. T. (1982): Comparison of the radiosensitivity of unsaturated fatty acids, structured as micelles or liposomes, under different experimental conditions. *Int. J. Radiat. Biol.*, 42:601–609.
561. Moore, A. H., Olschowka, J. A., Williams, J. P. et al. (2004): Radiation-induced edema is dependent on cyclooxygenase 2 activity in mouse brain. *Radiat. Res.*, 161:153–160.
562. Moore, I. M., Kramer, J. H., Wara, W. et al. (1991): Cognitive function in children with leukemia: effect of radiation dose and time since irradiation. *Cancer*, 68: 1913–1917.
563. Morgan, W. F. (2003): Non-targeted and delayed effects of exposure to ionizing radiation. II. Radiation-induced genomic instability and bystander effects *in vivo*, clastogenic factors and transgenerational effects. *Radiat. Res.*, 159:581–596.
564. Morgan, W. F. (2003): Non-targeted and delayed effects of exposure to ionizing radiation. I. Non-targeted and delayed effects of exposure to ionizing radiation: I. Radiation-induced genomic instability and by-stander effects *in vivo*. *Radiat. Res.*, 159:567–580.
565. Morse, D. E. and Mickley, G. A. (1988): Interaction of the endogenous opioid system and radiation in the suppression of appetite behavior. *Soc. Neurosci. Abstr.*, 14:1106.
566. Moskalev, Y. I. (1991): *The Remote Consequences of Exposure to Ionizing Radiation*. Medicina, Moscow.
567. Mothersill, C. and Seymour, C. (2001): Radiation-induced bystander effects: past history and future directions. *Radiat. Res.*, 155:759–767.
568. Mothersill, C. and Seymour, C. B. (2002): Bystander and delayed effects after fractionated radiation exposure. *Radiat. Res.*, 158:626–633.
569. Moulder, J. E., Fish, B. L., Holcenberg, J. S., and Sun, G. X. (1990): Hepatic function and drug pharmacokinetics after total body irradiation plus bone marrow transplant. *Int. J. Radiat. Oncol. Biol. Phys.*, 19: 1389–1396.
570. Mtskhvetadze, A. V. and Kucherenko, T. M. (1968): Direct and indirect effect of irradiation on the transmission of the stimulus in the upper neck sympathetic ganglion of cats. *Radiobiologiia*, 8:624–627.
571. Mukherjee, S. K., Goel, H. C., Pant, K., and Jain, V. (1997): Prevention of radiation induced taste aversion in rats. *Indian J. Exp. Biol.*, 35:232–235.
572. Mullenix, P. J., Kernan, W. J., Schunior, A. et al. (1994): Interactions of steroid, methotrexate, and radiation determine neurotoxicity in an animal model to study therapy for childhood leukemia. *Pediatr. Res.*, 35:171–178.
573. Mullenix, P. J. (1998): Radiation protection in the developing central nervous system: investigation of a biological approach. In: *Radiprotectors: Chemical, Biological, and Clinical Perspectives*, edited by E. Bump, pp. 349–371. CRC Press, Boca Raton, FL.
574. Mullin, M. J., Hunt, W. A., and Harris, R. A. (1986): Ionizing radiation alters the properties of sodium channels in rat brain synaptosomes. *J. Neurochem.*, 47:489–495.
575. Munro, T. R. (1970): The relative radiosensitivity of the nucleus and cytoplasm of Chinese hamster fibroblasts. *Radiat. Res.*, 42:451–470.
576. Murbeth, S., Rousarova, M., Scherb, H., and Lengfelder, E. (2004): Thyroid cancer has increased in the adult populations of countries moderately affected by Chernobyl fallout. *Med. Sci. Monit.*, 10(7):CR300–CR306.
577. Murphy, B. J., Laderoute, K. R., Short, S. M., and Sutherland, R. M. (1991): The identification of heme oxygenase as a major hypoxic stress protein in Chinese hamster ovary cells. *Br. J. Cancer*, 64:69–73
578. Muscel, R. J., Zhang, H. B., and McKenna, W. G. (1993): Differential effect of ionizing radiation on the expression of cyclin A and cyclin B in Hela cells. *Cancer Res.*, 53:1128–1135.

579. Myers, J. H., Blackwell, L. H., and Overman, R. R. (1972): Early functional hemodynamic impairment in baboons after 1000 R or less of gamma radiation as revealed by hemorrhagic stress. *Radiat. Res.*, 52:564–578.
580. Nagasawa, H. and Little, B. (1992): Induction of sister chromatid exchanges by extremely low doses of alpha particles. *Cancer Res.*, 52:6394–6396.
581. Nagasawa, H., Cremesti, Kolesnick, Fuks, Z., and Little, J. B. (2002): Involvement of membrane signaling in the bystander effect in irradiated cells. *Cancer Res.*, 62:2531–2534.
582. Nagata, S. and Golstein, P. (1995): The fas death factor. *Science*, 267:1449–1456.
583. Nakata, E., Mason, K. A., Hunter, N. et al. (2004): Potentiation of tumor response to radiation or chemoradiation by selective cyclooxygenase inhibitors. *Int. J. Radiat. Oncol. Biol. Phys.*, 58:369–375.
584. Nam, S. Y., Kim, J. H., Cho, C. K. et al. (1997): Enhancement of radiation-induced hepatic microsomal epoxide hydrolase gene expression by oltipraz in rats. *Radiat. Res.*, 147:613–620.
585. Nandchahal, K. (1990): Mitotic figures and pyknotic nuclei and necrotic cells in the mouse jejunum during injury and repair after whole-body gamma irradiation. *Radiobiol. Radiother.*, (Berlin), 31:333–336.
586. Narayanan, P. K., LaRue, K. E., Goodwin, E. H., and Lehnert, B. E. (1999): Alpha particles induce the production of interleukin-8 by human cells. *Radiat. Res.*, 152:57–63.
587. NCRP. (1987): *Ionizing Radiation Exposures of the Population of the United States*, NCRP Report No. 93. National Council on Radiation Protection and Measurements, Washington, D.C.
588. NCRP. (1990): *The Relative Biological Effectiveness of Radiations of Different Quality*, NCRP Report No. 104. National Council on Radiation Protection and Measurements, Bethesda, MD.
589. Neta, R. and Okunieff, P. (1996): Cytokine-induced radiation protection and sensitization. *Semin. Radiat. Oncol.*, 6:306–320
590. Newcombe, H. B. and McGregor, J. F. (1972): Increased embryo production following low doses of radiation to trout spermatogonia. *Radiat. Res.*, 51:402–409.
591. Nicklas, J. A., O'Neill, J. P., and Albertini, R. J. (1986): Use of T-cell receptor gene probes to quantify the *in vivo hprt* mutations in human T-lymphocytes. *Mutat. Res.*, 173:67–72.
592. Nicolaj, C. N., Grau, C., and Lindegaard, J. C. (2003): Chemical radioprotection: a critical review of amifostine as a cytoprotector in radiotherapy. *Semin. Radiat. Oncol.*, 13:62–72.
593. Nikiforov, Y. E., Nikiforova, M., and Fagin, A. (1998): Radiation-induced post-Chernobyl pediatric thyroid carcinomas. *Oncogene*, 17(15):1983–1988.
594. Nokta, M., Belli, J., and Pollard, R. (1992): X-irradiation enhances *in vitro* human immunodeficiency virus replication: correlation with cellular levels of cAMP. *Proc. Soc. Exp. Biol. Med.*, 200:402–408.
595. Norton, S. and Kimler, B. F. (1988): Comparison of functional and morphological deficits in the rat after gestational exposure to ionizing radiation. *Neurotoxicol. Teratol.*, 10:363–371.
596. Norton, S., Mullenix, P., and Culver, B. (1976): Comparison of the structure of hyperactive behavior in rats after brain damage from x-irradiation, carbon monoxide and pallidal lesions. *Brain Res.*, 116:49–67.
597. Norton, S., Kimler, BF., and Mullenix, P. J. (1991): Progressive behavioral changes in rats after exposure to low levels of ionizing radiation *in utero*. *Neurotoxicol. Teratol.*, 13:181–188.
598. Nozue, M. and Ogata, T. (1989): Correlation among lung damage after radiation, amount of lipid peroxides, and antioxidant enzyme activities. *Exp. Mol. Pathol.*, 50:239–252.
599. Obe, G., Johannes, I., Johannes, C. et al. (1997): Chromosomal aberrations in blood lymphocytes of astronauts after long-term space flights. *Int. J. Radiat. Biol.*, 72:727–734.
600. O'Connel, J. F. A. and Brunschwig, A. (1937): Observations on the Roentgen treatment of intracranial gliomata with special reference to the effects of irradiation upon the surrounding brain. *Brain*, 60:230–258.
601. Odum, E. P. (1971): *Fundamentals of Ecology.* 3rd ed W.B. Saunders, Philadelphia, PA.
602. Ogilvy-Stuart, A. L., Shalet, S. M., and Gattamaneni, H R. (1991): Thyroid function after treatment of brain tumors in children. *J. Pediatr.*, 119:733–737.
603. Okada, S., Okeda, R., Matsushita, S., and Kawano, A. (1998): Histopathological and morphometric study of the late effects of heavy-ion irradiation on the spinal cord of the rat. *Radiat. Res.*, 150:304–315.
604. Oleinick, N. L. and Rustad, M. (1976): Interrelationships between ionizing radiation, protein synthesis, and the physiological expressions of radiation damage. *Adv Radiat. Biol.*, 6:107–160.
605. Olinski, R., Nackerdien, Z., and Dizdaroglu, M. (1992). DNA-protein cross-linking between thymine and tyrosine in chromatin of gamma-irradiated or $H_2O_2$-treated cultured human cells. *Arch. Biochem. Biophys.*, 297:139–143.
606. Olschowka, J. A., Kyrkanides, S., Harvey, B. K. et al (1997): ICAM-1 induction in the mouse CNS following irradiation. *Brain Behav. Immun.*, 11:273–285.
607. Onoda, J. M., Piechocki, M. P., and Honn, K. V. (1992) Radiation-induced increase in expression of the alpha IIb beta 3 integrin in melanoma cells: effects on metastatic potential. *Radiat. Res.*, 130:281–288.
608. Ord, M. G. and Stocken, LA. (1968): Variations in the phosphate content and thiol/disulfide ratio of histones during the cell cycle. *Biochem. J.*, 107:403–410.
609. Ortin, T. T., Shostak, C. A., and Donaldson, S. S. (1990) Gonadal status and reproductive function following treatment for Hodgkin's disease in childhood: the Stanford experience. *Int. J. Radiat. Oncol. Biol. Phys.*, 19:873–880.
610. Osterreicher, J., Skopek, J., Jahns, J. et al. (2003): Beta-1 integrin and IL-1alpha expression as bystander effect of medium from irradiated cells: the pilot study. *Acta Histochem.*, 105:223–230.

611. Otake, M. and Schull, W. J. (1984): *In utero* exposure to A-bomb radiation and mental retardation: a reassessment. *Br. J. Radiol.*, 57:409–414.
612. Otake, M. and Schull, W. J. (1998): Radiation-related brain damage and growth retardation among prenatally exposed atomic bomb survivors. *Int. J. Radiat. Biol.*, 74:159–171.
613. Overgaard, J. and Matsui M. (1990): Effect of radiation on glucose absorption in the mouse jejunum *in vivo*. *Radiother. Oncol.*, 18:71–77.
614. Pallis, CA., Louis, S., and Morgan, R. L. (1961): Brain myelopathy. *Brain*, 84:460–479.
615. Parshkov, E. M., Chebotareva, I. V., Sokolov, V. A., and Dallas, C. E. (1998): Additional thyroid dose factor from transportation sources in Russia following the Chernobyl disaster. *Environ. Health Perspect.*, 105(6):1491–1496.
616. Parsons, P. A. (2002): Radiation hormesis: challenging LNT theory via ecological and evolutionary considerations. *Health Physics.*, 82:513–516.
617. Pastel, R. H. (2002): Radiophobia: long-term psychological consequences of Chernobyl. *Mil. Med.*, 167:134–136.
618. Patchen, M. L., MacVittie, T. J., Solberg, B. D., and Souza, L. M. (1990): Therapeutic administration of recombinant human granulocyte colony-stimulating factor accelerates hemopoietic regeneration and enhances survival in a murine model of radiation-induced myelosuppression. *Int. J. Cell Cloning*, 8:107–122.
619. Patchen, M. L., MacVittie, T. J., Solberg, B. D., and Souza, L. M. (1990): Survival enhancement and hemopoietic regeneration following radiation exposure: therapeutic approach using glucan and granulocyte colony-stimulating factor. *Exp. Hematol.*, 18:1042–1048.
620. Patchen, M. L., MacVittie, T. J., and Souza, L. M. (1992): Postirradiation treatment with granulocyte colony-stimulating factor and preirradiation WR-2721 administration synergize to enhance hemopoietic reconstitution and increase survival. *Int. J. Radiat. Oncol. Biol. Phys.*, 22:773–779.
621. Patchen, M. L., MacVittie. T. J., Williams, J. L. et al. (1991): Administration of interleukin-6 stimulates multilineage hematopoiesis and accelerates recovery from radiation-induced hematopoietic depression. *Blood*, 77: 472–480.
622. Pawlik, T. M. and Keyomarsi, K. (2004): Role of cell cycle mediating sensitivity to radiotherapy. *Int. J. Radiat. Oncol. Biol. Phys.*, 59:928–942.
623. Pearce, L. L., Kanai, A. J., Birder, L. A. et al. (2002): The catabolic fate of nitric oxide oxidase and peroxynitrite reductase activities of cytochrome oxidase. *J. Biol. Chem.*, 277:13556–13562.
624. Peimer, S. I., Dudkin, A. O., and Swerdlov, A. G. (1986): Response of hippocampal pacemaker-like neurons to low doses of ionizing radiation. *Int. J. Radiat. Biol.*, 49:597–600.
625. Pellmar, T. C. and Lepinski, D. L. (1993): Gamma radiation (5–10 Gy) impairs neuronal function in the guinea pig hippocampus. *Radiat. Res.*, 136:255–261.
626. Pellmar, T. C., Schauer, D. A., and Zeman, G. H. (1990): Time- and dose-dependent changes in neuronal activity produced by x radiation in brain slices. *Radiat. Res.*, 122:209–214.
627. Pellmar, T. C. and Rockwell, S. (2005): Priority list of research areas for radiological nuclear threat countermeasures. *Radiat. Res.*, 163:115–123.
628. Peter, R. U., Braun-Falco, O., Birioukov, A. et al. (1994): Chronic cutaneous damage after accidental exposure to ionizing radiation: the Chernobyl experience. *J. Am. Acad. Dermatol.*, 30(5, Part 1):719–723.
629. Peterson, L. M., Evens, M. L., Graham, M. M. et al. (1992): Vascular response to radiation injury in the rat lung. *Radiat. Res.*, 129:139–148.
630. Petridou, E., Trichopoulos, D., Dessypris, N. et al. (1996): Infant leukaemia after *in utero* exposure to radiation from Chernobyl. *Nature (Lond.)*, 382(8589):352–353.
631. Philpott, D. E., Sapp, W., Miquel, J. et al. (1985): The effect of high energy (HZE) particle radiation (40 Ar) on aging parameters of mouse hippocampus and retina. In: *Scanning Electron Microspy*, Vol. III, edited by A. M. F. O'Hare, pp. 1177–1182. SEM, Chicago, IL.
632. Pineau, C., Velez de la Calle, J. F., Pinon-Lataillade, G., and Jegou, B. (1989): Assessment of testicular function after acute and chronic irradiation: further evidence for an influence of late spermatids on Sertoli cell function in the adult rat. *Endocrinology*, 124:2720–2728.
633. Pinon-Lataillade, G., Viguier-Martinez, M. C., Touzalin, A. M. et al. (1991): Effect of an acute exposure of rat testes to gamma rays on germ cells and on Sertoli and Leydig cell functions. *Reprod. Nutr. Dev.*, 31:617–629.
634. Pollack, A. and Zagars, G. K. (1998): Androgen ablation in addition to radiation therapy for prostate cancer. Is there a true benefit? *Semin. Radiat. Oncol.*, 8:95–106.
635. Pollack, M. and Timiras, P. S. (1964): X-ray dose and electroconvulsive responses in adult rats. *Radiat. Res.*, 21:111–119.
636. Pourquier, H., Baker, J. R., Giaux, G., and Benirschke, K. (1958): Localized roentgen-ray beam irradiation of the hypophysohypothalamic region of the guinea pig with a 2 million volt van de Graaf generator. *Am. J. Roentgenol. Radium Ther. Nucl. Med.*, 80:840–850.
637. Powers, D. A., Kress, T. S., and Jankowski, M. W. (1987): The Chernobyl source term. *Nuclear Safety*, 28:10.
638. Prasad, A. V., Mohan, N., Chandrasekar, B., and Meltz, M. (1995): Induction of transcription of 'immediate early genes' by low-dose ionizing radiation. *Radiat. Res.*, 143:263–272.
639. Priestman, T., Challoner, T., Butcher, M., and Priestman, S. (1988): Control of radiation induced emesis with GR38032F (GR). *Proc. Am. Soc. Clin. Oncol.*, 7:281.
640. Qin, J. and Li, L. (2003): Molecular anatomy of the DNA damage and replication checkpoints. *Radiat. Res.*, 159:139–148.
641. Quastel, M. R., Goldsmith, J. R., Mirkin, L. et al. (1997): Thyroid-stimulating hormone levels in children from Chernobyl. *Environ. Health Perspect.*, 105(6):1497–1498.
642. Quastel, M. R., Goldsmith, J. R., Cwikel, J. et al. (1997): Lessons learned from the study of immigrants to Israel from areas of Russia, Belarus, and Ukraine contaminated by the Chernobyl accident [commentary]. *Environ. Health Perspect.*, 105(6):1523–1528.

643. Raber, J., Rola, R., LeFevour, A. et al. (2004): Radiation-induced cognitive impairments are associated with changes in indicators of hippocampal neurogenesis. *Radiat. Res.*, 162:39–47.
644. Rabin, B. M. (1996): Free radicals and taste aversion learning in the rat: nitric oxide, radiation and dopamine. *Prog. Neuropsychopharmacol. Biol. Psychiatry*, 20:691–707.
645. Rabin, B. M., Joseph, J. A., and Erat, S. (1998): Effects of exposure to different types of radiation on behaviors mediated by peripheral or central systems. *Adv. Space Res.*, 22:217–225.
646. Rabin, B. M., Joseph, J. A., and Shukitt-Hale, B. (2003): Long-term changes in amphetamine-induced reinforcement and aversion in rats following exposure to $^{56}$Fe particle. *Adv. Space Res.*, 31:127–133.
647. Rabin, B. M., Hunt, W. A., Joseph, J. A. et al. (1991): Relationship between linear energy transfer and behavioral toxicity in rats following exposure to protons and heavy particles. *Radiat. Res.*, 128:216–221.
648. Rades, D., Fehlauer, F., Bajrovic, A. et al. (2004): Serious adverse effects of amifostine during radiotherapy in head and neck cancer patients. *Radiother. Oncol.*, 70:261–264.
649. Rakic, P. (1986): Normal and abnormal neuronal migration during brain development. In: *Radiation Risks to the Developing Nervous System*, edited by H. Kriegel, W. Schmahl, G. B. Gerber, and F.-E. Stieve, pp. 35–44. Gustav Fischer Verlag, Stuttgart.
650. Raleigh, J. A. (1987): Radiation peroxidation in model membranes. In: *Prostaglandin and Lipid Metabolism in Radiation Injury*, edited by T. L. Walden, Jr., and H. N. Hughes, pp. 1–27. Plenum Press, New York.
651. Randolph, M. L. and Brewen, J. G. (1980): Estimation of whole-body doses by means of chromosome aberrations observed in survivors of the Hiroshima A-bomb. *Radiat. Res.*, 82:393–407.
652. Reed, J. C. (2004): Apoptosis mechanisms: implications for cancer drug discovery. *Oncology*, 18(Suppl. 10): 11–20.
653. Reinhold, H. S., Fajardo, L. F., and Hopewell, J. W. (1990): The vascular system. In: *Advances in Radiation Biology*. Vol. 14. *Relative Radiation Sensitivities of Human Organ Systems, Part II*, edited by K. I. Altman and J. T. Lett, pp. 177–226. Academic Press, San Diego, CA.
654. Reiter, R., Tang, L., Garcia, J. J., and Munoz-Hoyos, A. (1997): Pharmacological actions of melatonin in oxygen radical pathophysiology. *Life Sci.*, 60:2255–2271.
655. Reyners, H., Gianfelici de Reyners, E., and Maisin, J. R. (1982): The beta-astrocyte: a newly recognized radiosensitive glial cell type in the cerebral cortex. *J. Neurocytol.*, 11:967–983.
656. Reyners, H., Gianfelici de Reyners, E., and Maisin, J. R. (1986): Early cell regeneration processes after split-dose x-irradiation of the cerebral cortex of the rat. *Br. J. Cancer*, 7(Suppl.):53:218–220.
657. Reyners, H., Gianfelici de Reyners, E., and Maisin, J. R. (1986): The role of the glia in late damage after prenatal irradiation. In: *Radiation Risks to the Developing Nervous System*, edited by H. Kriegel, W. Schmahl, G. B. Gerber, and F.-E. Stieve, pp. 117–131. Gustav Fischer Verlag, Stuttgart.
658. Riches, A. C. (1995): Experimental radiation leukaemogenesis. In: *Radiation Toxicology: Bone Marrow and Leukaemia*, edited by J. H. Hendry and B. I. Lord, pp. 311–334. Taylor & Francis, Bristol, PA.
659. Riopelle, A. J. (1962): Some behavioral effects of ionizing radiation on primates. In: *Response of the Nervous System to Ionizing Radiation*, edited by T. J. Haley and R. S. Snider, pp. 719–728. Academic Press, New York.
660. Roach M., (2003): Hormonal therapy and radiotherapy for localized prostate cancer: who, where and how long? *J. Urology*, 170 (6 Pt. 2):S35–40.
661. Robbins, M. E., Campling, D., Rezvani, M. et al. (1989): Nephropathy in the mature pig after the irradiation of a single kidney: a comparison with the immature pig. *Int. J. Radiat. Oncol. Biol. Phys.*, 16:1519–1528.
662. Robbins, M. E., Campling, D., Rezvani, M. et al. (1989): Radiation nephropathy in mature pigs following the irradiation of both kidneys. *Int. J. Radiat. Biol.*, 56:83–98.
663. Robbins M. E., Bywaters, T., Rezvani, M. et al. (1991): Residual radiation-induced damage to the kidney of the pig as assayed by retreatment. *Int. J. Radiat. Biol.*, 60:917–928.
664. Robertson, J. B. (1989): Toxicology of ionizing radiation. In: *A Guide to General Toxicology*, 2nd ed., edited by J. K. Marquis, pp. 141–156. Karger, New York.
665. Rodvall, Y., Pershagen, G., Hrubec, Z. et al. (1990): Prenatal x-ray exposure and childhood cancer in Swedish twins. *Int. J. Cancer*, 46:362–365.
666. Roizin, L., Akai, K., Carsten, A. et al. (1976): Post-x-ray myelinopathy (pathogenic mechanisms). In: *Proc. of Int. Symp. on the Aetiology and Pathogenesis of Demyelinating Diseases*, edited by T. Yonawa, pp. 29–57. Japan Press, Neiho-Sha.
667. Rola, R., Raber, J., Rizk, A. et al. (2004): Radiation-induced impairment of hippocampal neurogenesis is associated with cognitive deficits in young mice. *Exp. Neurol.*, 188:316–330.
668. Romanova, L. K. and Zhorova, E. S. (1994): The effect of irradiation at small doses on human embryos and fetuses (Russia). *Ontogenez*, 25(3):55–65.
669. Ron, E., Lubin, J. H., Shore, R. E. et al. (1995): Thyroid cancer after exposure to external radiation: a pooled analysis of seven studies. *Radiat. Res.*, 141(3):259–277.
670. Rosenthal, F. and Timiras, P. S. (1961): Changes in brain excitability after whole-body x-irradiation in the rat. *Radiat. Res.*, 18:648–657.
671. Rosenthal, F. and Timiras, P. S. (1961): Threshold and pattern of electroshock seizures after 250 R whole-body x-irradiation in rats. *Proc. Soc. Exp. Biol. Med.* 208:267–270.
672. Ross, J. A. T., Levitt, S. R., Holst, E. A., and Clemente C. D. (1954): Neurological and electroencephalographic effects of x irradiation of the head in monkeys. *AMA Arch Neurol. Psychiat.*, 71:238–249.
673. Rothenberg, M. A. (1950): Studies on permeability in relation to nerve function, II. Ionic movements across axonal membranes. *Biochim. Biophys. Acta*, 4:96–114.

674. Rowland, R. E. and Lucas, Jr., H. F. (1984): Radium-dial workers. In: *Radiation Carcinogenesis: Epidemiology and Biological Significance*, edited by J. D. Boice, Jr., and J. F. Faumeni, Jr., pp. 231–240. Raven Press, New York.
675. Rubin, P., Johnston, C. J., Williams, J. P. et al. (1995): A perpetual cascade of cytokines postirradiation leads to pulmonary fibrosis. *Int. J. Radiat. Biol. Phys.*, 33:99–110.
676. Rudoltz, M. S., Kao, G., Blank, K. R. et al. (1996): Molecular biology of the cell cycle: potential for therapeutic applications in radiation oncology. *Semin. Radiat. Oncol.*, 6:284–294.
677. Rugo, R. E. and Schiestl, R. H. (2004): Increases in oxidative stress in the progeny of x-irradiated cells. *Radiat. Res.*, 162:416–425.
678. Russel, D. S., Wilson, C. W., and Tansley, K. (1949): Experimental radionecrosis in the brains of rabbits. *J. Neurol. Neurosurg. Psychiatry*, 12:187–195.
679. Russo, I. H. and Russo, J. (1996): Mammary gland neoplasia in long-term rodent studies. *Environ. Health Perspec.*, 105(9):938–967.
680. Rytomaa, T. (1996): Ten years after Chernobyl. *Ann. Med.*, 28(2):83–87.
681. Sabine, J. C. (1956): Inactivation of cholinesterases by gamma radiation. *Am. J. Physiol.*, 187:280–282.
682. Sadekova, S., Lehnert, S., and Chow, T. Y. (1997): Induction of PBP74/mortalin/Grp75, a member of the hsp 70 family, by low doses of ionizing radiation: a possible role in induced radioresistance. *Int. J. Radiat. Biol.*, 72:653–660.
683. Saenger, E. L. and Hinnefeld, J. (1991): Perception of radiation injury vs. radiogenic effect. In: *The Medical Basis for Radiation-Accident Preparedness*. Vol. III. *The Psychological Perspective*, edited by R. C. Ricks, M. E. Berger, and F. M. O'Hara, pp. 39–50. Elsevier, New York.
684. Sakuma, S., Saya, H., Ijichi, A., and Tofilon, P. (1995): Radiation induction of the receptor tyrosine kinase gene Ptk-3 in normal rat astrocytes. *Radiat. Res.*, 143:1–7.
685. Salter, C. A. (2001): Psychological effects of nuclear and radiological warfare. *Mil. Med.*, 166:17–18.
686. Samaan, N. A. (1990): Hypothalamic-pituitary failure after radiotherapy for tumors of the head and neck. In: *Advances in Radiation Biology*. Vol. 14. *Relative Radiation Sensitivities of Human Organ Systems, Part II*, edited by K. I. Altman and J. T. Lett, pp. 111–117. Academic Press, San Diego, CA.
687. Sambur, M. B., Mel'nikov, O. F., and Indyk, V. M. (2000): Condition of the immune system of various generations of mice exposed to chronic action of low dose ionizing irradiation. *Immun. Allergol.*, 1:81–85.
688. Sanders, C. L. (1986): *Toxicological Aspects of Energy Production*, pp. 253–284. Battelle Press, Columbus, OH.
689. Sarri, Y., Conill, C., Verger, E. et al. (1991): Effects of single dose irradiation on pancreatic beta-cell function. *Radiother. Oncol.*, 22:143–144.
690. Sato, M. (1978): Electrophysiological studies on radiation-induced changes in the adult nervous system. In: *Advances in Radiation Biology*. Vol. 7. *Relative Radiation Sensitivities of Human Organ Systems*, edited by J. T. Lett and H. Adler, pp. 181–21. Academic Press, New York.
691. Sato, M. and Austin, G. (1964): Acute radiation effects on mammalian synaptic activities. In: *Response of the Nervous System to Ionizing Radiation*, edited by T. J. Haley and R. S. Snider, pp. 279–289. Little, Brown & Co., Boston, MA.
692. Sato, M., Austin, G. M., and Stahl, W. (1962): The effects of ionizing radiation on spinal cord neurons. In: *Response of the Nervous System to Ionizing Radiation*, edited by T. J. Haley and R. S. Snider, pp. 561–671. Academic Press, New York.
693. Sato, M., Austin, G. M., and Stahl, W. (1962): Delayed radiation effects on neuronal activity in the spinal cord of the cat. In: *Effects of Ionizing Radiation on the Nervous System: Proceedings*, pp. 93–110. International Atomic Energy Agency, Vienna.
694. Sato, M., Stahl, W., and Austin, G. M. (1963): Acute radiation effects on synaptic activity in the mammalian spinal cord. *Radiat. Res.*, 18:307–320.
695. Savtsova, Z. D. et al. (2000): Changes in immune system of experimental animals resulting from constant irradiation of several generations in exclusion zone of Chernobyl Atomic Power Plant. *Ukrain. J. Radiol.*, 8:71–76.
696. Scheid, W., Weber, J., Petrenko, S., and Traut, H. (1992): Chromosome aberrations in human lymphocytes apparently induced by Chernobyl fallout. *Health Phy.*, 64:531–534.
697. Scherb, H. and Weigelt, E. (2000): Congenital malformation and stillbirth in Germany and Europe before and after the Chernobyl nuclear power plant accident. *Environ. Sci. Pollut. Res. Int.*, Special Issue, (1):117–125.
698. Schmidt-Ullrich, R. K., Valerie, K. C., Chan, W., and McWilliams, D. (1994): Altered expression of epidermal growth factor receptor and estrogen receptor in MCF-7 cells after single and repeated radiation exposure. *Int. J. Radiat. Oncol. Biol. Phys.*, 29:813–819.
699. Schmidt-Ullrich, R. K., Mikkelsen, R. B., Dent, P. et al. (1997): Radiation-induced proliferation of the human A431 squamous carcinoma cells is dependent on EGRF tyrosine phosphorylation. *Oncogene.*, 15:1191–1197.
700. Schmidt-Ullrich, R. K. Dent, P., Grant, S. et al. (2000): Signal transduction and cellular radiation responses. *Radiat. Res.*, 153:245–257.
701. Schmitt, G. and Zamboglou, N. (1990): Radiation effects on bone and cartilage. In: *Advances in Radiation Biology*. Vol. 14. *Relative Radiation Sensitivities of Human Organ Systems, Part II*, edited by K. I. Altman and J. T. Lett, pp. 157–176. Academic Press, San Diego, CA.
702. Scholz, W., Ducho, E. G., and Breit, A. (1959): Experimentelle Roentgenspatschaden am ruchenmark des erwachsenen kaninchens. Ein weiterer beitrag zur wirkungsweise ionisierender strahlen auf das zentralnervose gewebe. *Psychiat. Neurol. Jpn.*, 61:417–442.
703. Scholz, W., Schlote, W., and Hirschberger, W. (1962): Morphological effect of repeated low dosage and single high dosage of x-irradiation to the central nervous system. In: *Response of the Nervous System to Ionizing Radiation*, edited by T. J. Haley and R. S. Snider, pp. 211–232. Academic Press, New York.

704. Schrag, S. D. and Dixson, R. L. (1985): Occupational exposures associated with male reproductive dysfunction. *Ann. Rev. Pharmacol. Toxicol.*, 25:567–592.
705. Schull, W. J. (1995): *Effects of Atomic Radiation: A Half Century of Studies from Hiroshima and Nagasaki*, Wiley-Liss, New York.
706. Schull, W. J. (2003): The children of atomic bomb survivors: a synopsis. *J. Radiol. Prot.*, 23:369–384.
707. Schull, WJ. and Otake, M. (1986): Neurological deficit among the survivors exposed *in utero* to the atomic bombing of Hiroshima and Nagasaki: a reassessment and new directions. In: *Radiation Risks to the Developing Nervous System*, edited by H. Kriegel, W. Schmahl, G. B. Gerber, and F-E. Stieve, pp. 399–419. Gustav Fischer Verlag, Stuttgart.
708. Schultz-Hector, S., Bhm, M., Blchel, A. et al. (1992): Radiation-induced heart disease: morphology, changes in catecholamine synthesis and content, β-adrenoceptor density, and hemodynamic function in an experimental model. *Radiat. Res.*, 129:281–289.
709. Schwartz, G. N., Neta, R., Vigneulle, R. M. et al. (1988): Recovery of hematopoietic colony-forming cells in irradiated mice pretreated with interleukin 1 (IL-1). *Exp. Hematol.*, 16:752–757.
710. Scott, B. R. and Dillehay, L. E. (1990): A model for hematopoietic death in man from irradiation of bone marrow during radioimmunotherapy. *Br. J. Radiol.*, 63:862–870.
711. Segerstahl, B. (1991): The costs. In: *Chernobyl: A Policy Response Study*, edited by B. Segerstahl, p. 59. Springer-Verlag, New York.
712. Seigneur, L. J. and Brennan, J. T. (1966): *Incapacitation in the Monkey (Macaca mulatta) Following Exposure to a Pulse of Reactor Radiation*, AFRRI Scientific Rep. SR 66-2, Armed Forces Radiobiology Research Institute, Bethesda, MD.
713. Serkiz, Y. I., Indyk, V. M., Pinchook, N. K. et al. (2000): Short-term and long-term effects of radiation on laboratory animals and their progeny living in the Chernobyl nuclear power plant region. *Environ. Sci. Pollut. Res. Int.*, Special Issue, (1):107–116.
714. Severa, J. and B·r J. (1991): *Handbook of Radioactive Contamination and Decontamination*. Elsevier, New York.
715. Shalet, S. M., Tsatsoulis, A., Whitehead, E., and Read, G. (1989): Vulnerability of the human Leydig cell to radiation damage is dependent upon age. *J. Endocrinol.*, 120: 161–165.
716. Shao, C., Folkard, M., Michael, and Prise, K. M. (2004): Targeted cytoplasmic irradiation induces bystander responses. *Proc. Natl. Acad. Sci. U.S.A.*, 101:13495–13500.
717. Sharp, J. C., Kelly, D. D., and Brady, J. V. (1986): The radio-attenuating effects of *n*-decylaminoethanethiosulfuric acid in the rhesus monkey. In: *Use of Nonhuman Primates in Drug Evaluation*, edited by H. Vagtborg, pp. 338–346. Southwest Foundation for Research and Education, San Antonio, TX.
718. Sharplin, J. and Franko, A. J. (1989): A quantitative histological study of strain-dependent differences in the effects of irradiation on mouse lung during the intermediate and late phases. *Radiat. Res.*, 119:15–31.
719. Shaw, E. G., Gunderson, L. L., Martin, J. K. et al. (1990): Peripheral nerve and ureteral tolerance to intraoperative radiation therapy: clinical and dose–response analysis. *Radiother. Oncol.*, 18:247–255.
720. Sheline, G. E., Wara, W. M., and Smith, V. (1980): Therapeutic irradiation and brain injury. *Int. J. Radiat. Oncol. Biol. Phys.*, 6:1215–1228.
721. Shenkin, H. A. (1951): Effects of various drugs upon cerebral circulation and metabolism in man. *J. Appl. Physiol.*, 3:465–471.
722. Shevchenko, V. A., Pomerantseva, M. D., Ramaiya, L. K. et al. (1992): Genetic disorders in mice exposed to radiation in the vicinity of the Chernobyl nuclear power station. *Sci. Tot. Environ.*, 112:45–56.
723. Shigematzu, I. (1991): *The International Chernobyl Project: An Overview. Assessment of Radiological Consequences and Evaluation of Protective Measures*, Report by an International Advisory Committee. International Atomic Energy Agency, Vienna.
724. Shimada, Y., Yasukawa-Barnes, J., Kim, R. Y. et al. (1994): Age and radiation sensitivity of rat mammary clonogenic cells. *Radiat. Res.*, 137(1):118–123.
725. Shipman, T. L., Lushbaugh, C. C., Peterson, D. F. et al. (1961): Acute radiation death resulting from an accidental nuclear critical excursion. *J. Occup. Med.* 3:146–192.
726. Shubik, V. M., Zaitseva, M. B., and Kositskaya, L. S. (1996): Role of immune deviations in some diseases observed in areas contaminated with radionuclides after the accident in Chernobyl NPP (Russia). *Radiatsionnaya Biologiya Radioekologiya*, 36(3):332–337.
727. Shukitt-Hale, B., Casadesus, G., McEwen, J. J. et al. (2000): Spatial learning and memory deficits induced by exposure to iron-56–particle radiation. *Radiat. Res.*, 154:28–33.
728. Shukitt-Hale, B., Casadesus, G., Cantuti-Castelvetri, I. et al. (2003): Cognitive deficits induced by $^{56}$Fe radiation exposure. *Adv. Space Res.*, 31:119–126.
729. Sich, A. R. (1994): Chernobyl accident management actions: implications for source term estimates. *Nuclear Safety*, 35:1–24.
730. Siemann, D. W. and Shi, W. (2003): Chemical radioprotection: a critical review of Amifostine as a cytoprotector in radiotherapy. *Semin. Radiat. Oncol*, 13:62–72.
731. Sienkiewicz, Z. J. et al. (1994): Prenatal irradiation and spatial memory in mice: investigation of dose–response relationship. *Int. J. Radiat. Biol.*, 65:611–618.
732. Silasi, G., Diaz-Heijtz, R., Besplug, J. et al. (2004): Selective brain responses to acute and chronic low-dose x-ray irradiation in males and females. *Biochem. Biophys. Res Commun.*, 325:1223–1235.
733. Sine, R. C., Levine, I. H., Jackson, W. E. et al. (2001): Biodosimetery assessment tool: a post-exposure software application for management of radiation accidents. *Military Medicine*, 166(Suppl. 12):85–87.
734. Sklar, C. A., Kim, T. H., and Ramsay, N. K. C. (1982): Thyroid dysfunction among long-term survivors of bone marrow transplantation. *Am. J. Med.*, 73:668–694.
735. Sklar, C. A. et al. (1990): Effects of radiation on testicular function in long-term survivors of childhood acute lymphoblastic leukemia: a report from the Children Cancer Study Group. *J. Clin. Oncol.*, 8:1981–1987.

736. Sloan, S. R., Newconb, E. W., and Pellicer, A. (1990): Neutron radiation can activate K-ras via a point mutation in codon 146 and induce a different spectrum of *ras* mutations than does gamma irradiation. *Mol. Cell. Biol.*, 10:405–408.
737. Smith, R. A., Ingles, J., Lochemes, J. J. et al. (2001): Gamma irradiation of HIV-1. *J. Orthop. Res.*, 19:815–819.
738. Sokolov, V. E., Rjabov, I. N., Ryabtsev, I. A. et al. (1993): Ecological and genetic consequences of the Chernobyl atomic power plant accident. *Vegetatio*, 109:91–99.
739. Southwood, R. (1987): Opening remarks. In: *Radiation and Health. The Biological Effects of Low-Level Exposure to Ionizing Radiation*, edited by R. Jones and R. Southwood, pp. 3–6. John Wiley & Sons, New York.
740. Spiegler, B. J., Bouffet, E., Greenberg, M. L. et al. (2004): Change in neurocognitive functioning after treatment with cranial radiation in childhood. *J. Clin. Oncol.*, 22:706–713.
741. Spotheim-Maurizot, M., Gardier, F., Sabattier, R., and Charlier, M. (1992): Metal ions protect DNA against strand blockage induced by fast neutrons. *Int. J. Radiat. Biol.*, 62:659–666.
742. Stapleton, G. E. and Curtis, H. J. (1946): *The Effects of Fast Neutrons on the Ability of Mice to Take Forced Exercise*, U.S. Atomic Energy Rep. No. 9, MDDC-696. Oak Ridge National Laboratory, Oak Ridge, TN.
743. Stather, J. W., Dionian, J., Brown, J. et al. (1987): Assessing risks of childhood leukaemia in Seascale. In: *Radiation and Health: The Biological Effects of Low-Level Exposure to Ionizing Radiation*, edited by R. Jones and R. Southwood, pp. 65–80. John Wiley & Sons, New York.
744. Steel, L. K., Hughes, H. N., and Walden, Jr., T. L. (1988): Quantitative, functional, and biochemical alterations in the peritoneal cells of mice exposed to whole-body gamma-irradiation, I. Changes in cellular protein, adherence properties and enzymatic activities associated with platelet-activating factor formation and inactivation, and arachidonate metabolism. *Int. J. Radiat. Biol. Relat. Stud. Phys. Chem. Med.*, 53:943–64.
745. Stephan, G. and Oestreicher, U. (1989): An increased frequency of structural chromosome aberrations in persons present in the vicinity of Chernobyl during and after the reactor accident: is this effect caused by radiation exposure? *Mutation Res.*, 223:7–12.
746. Stevenson, M. A., Pollock, S. S., Coleman, N. C., and Calderwood, S. K. (1994): X-irradiation, phorbolesters, and $H_2O_2$ stimulate mitogen activated protein kinase activity in NIH-3T3 cells through the formation of reactive oxygen internediates. *Cancer Res.*, 54:12–15.
747. Stewart, F. A., Luts, A., and Lebesque, J. V. (1989): The lack of long-term recovery and reirradiation tolerance in the mouse kidney. *Int. J. Radiat. Biol.*, 56:449–462.
748. Stewart, F. A., Lundbeck, F., Oussoren, Y., and Luts, A. (1991): Acute and late radiation damage in mouse bladder: a comparison of urination frequency and cystometry. *Int. J. Radiat. Oncol. Biol. Phys.*, 21:1211–1219.
749. Stewart, F. A., Oussoren, Y., and Luts, A. (1990): Long-term recovery and reirradiation tolerance of mouse bladder. *Int. J. Radiat. Oncol. Biol. Phys.*, 18:1399–1406.
750. Stewart, F. A., Soranson, J. A., Alpen, E. L. et al. (1984): Radiation-induced renal damage: the effects of hyperfractionation. *Radiat. Res.*, 98(2):407–420.
751. Stewart, J. (2002, April 23): The mechanics of a 'dirty bomb.' *CBS News*, Washington, D.C.
752. Stewart, J. (2002, April 24): The fear of radiation. *CBS News*, Washington, D.C.
753. Stieve, F-E. (1986): Experiences with accidents and consequences for treatment. *Br. J. Radiol.*, 59(Suppl. 19): 18–22.
754. Stiehm, E. R. (1992): The psychologic fallout from Chernobyl. *Am. J. Dis. Child.*, 146:761–762.
755. Stryker, J. A., Robins, D. B., and Velkley, D. E. (1990): Relative radiosensitivity of the urinary bladder in cancer therapy. In: *Advances in Radiation Biology*. Vol. 14. *Relative Radiation Sensitivities of Human Organ Systems, Part II*, edited by K. I. Altman and J. T. Lett, pp. 1–21. Academic Press, San Diego, CA.
756. Stuart, B. O., Palmer, R. F., Filipy, R. E. et al. (1977): Respiratory tract carcinogenesis in large and small experimental animals following daily inhalation of radon daughters and uranium ore dust. In: *Proc. of the IVth Congress of the International Radiation Protection Association*, April 24–30, Paris, pp. 104–117.
757. Sugg, D. W., Bickham, J. W., Brooks, J. A. et al. (1996): DNA damage and radiocesium in channel catfish from Chernobyl. *Environ. Toxicol. Chem.*, 15:1057–1063.
758. Suh, J. J. (2004): Efaproxiral: a novel radiation sensitizer. *Exper. Opin. Invest. Drugs*, 13:543–550.
759. Suzuki, K., Takahashi, M., Ishii-Ohba, H. et al. (1990): Steroidogenesis in the testes and the adrenals of adult male rats after gamma-irradiation *in utero* at late pregnancy. *J. Steroid Biochem.*, 35:301–305.
760. Suzuki, R., Yamaguchi, T., Kirno, T. et al. (1983): The effects of 5-minute ischemia in mongolian gerbils. I. Blood–brain barrier, cerebral blood flow, and local cerebral glucose utilization changes. *Acta Neuropathol. (Berlin)*, 60:207–216.
761. Switzer, R. C., Bogo, V., and Mickley, G. A. (1991): High energy electron and proton irradiation of rat brain induces degeneration detectable with the cupric-silver stain. *Soc. Neurosci. Abstr.*, 17(2):1460.
762. Szekely, J. E., Ronai, A. Z., Duna-Kovacs, Z. et al. (1977): Cross tolerance between morphine and beta-endorphin *in vivo*. *Life Sci.*, 20:1259–1264.
763. Tamaki, Y. and Inouye, M. (1988): Go/no-go discriminated avoidance learning in prenatally x-irradiated rats. *Neurotoxicol. Teratol.*, 10:35–38.
764. Tanimura, H. (1957): Changes of the neurosecretory granules in hypothalamo-hypophysical system of rats by irradiating their heads with x-rays. *Acta Anat. Nippon*, 32: 529–533.
765. Tannehill, S. P. and Mehta, M. P. (1996): Amifostine and radiation therapy: past, present, and future. *Semin. Oncol.*, 23:69–77.
766. Tannehill, S. P., Mehta, M. P., Larson, M. et al. (1997): Effect of amifostine on toxicities associated with sequential chemotherapy and radiation therapy for unresectable non-small-cell lung cancer: results of a phase II trial. *J. Clin. Oncol.*, 15:2850–2857.
767. Tateno, H. and Mikamo K. (1989): Effects of neonatal ovarian x-irradiation in the Chinese hamster. I. Correlation between the age of irradiation and the fertility span. *J. Radiat. Res.* (Tokyo), 30:185–190.

768. Taylor, J. A., Watson, M. A., Devereux, T. R. et al. (1994): p53 mutation hotspot in radon-associtated lung cancer. *Lancet*, 343:86–87.
769. Taylor, G. M. (1995): Genetic effects of ionising radiation with respect to leukaemia. In: *Radiation Toxicology: Bone Marrow and Leukaemia*, edited by J. H. Hendry and B. I. Lord, pp. 275–310. Taylor & Francis, Bristol, PA.
770. Teicher, B. A., Lazo, J. S., and Sartorelli, A. C. (1981): Classification of antineoplastic agents by their selective toxicities toward oxygenated and hypoxic tumor cells. *Cancer Res.*, 41:73–81.
771. Teicher, B. A., Holden, S. A., Al-Achi, A., and Herman, T. S. (1990): Classification of antineoplastic treatments by their differential toxicity toward putative oxygenated and hypoxic tumor subpopulations *in vivo* in the FSaIIC murine fibrosarcoma. *Cancer Res.*, 50:3339–3344.
772. Teskey, G. C. and Kavaliers, M. (1984): Ionizing radiation induces opioid-mediated analgesia in male mice. *Life Sci.*, 35:1547–1552.
773. Thacker, J. and Stretch, A. (1985): Responses of four x-ray-sensitive CHO cell mutants to different radiations and to irradiation conditions promoting cellular recovery. *Mutat. Res.*, 146:99–108.
774. Thomas, D. B., Rosenblatt, K., Jimenez, L. M. et al. (1994): Ionizing radiation and breast cancer in men (United States). *Cancer Causes Control*, 5(1):9–14.
775. Thorne, M. C., Ed. (1986): *Developmental Effects of Irradiation on the Brain of the Embryo and Fetus*, ICRP Publication 49. Pergamon Press, Oxford.
776. Tillman, B. F., Loyd, J. E., Malcolm, A. W. et al. (1989): Unilateral radiation pneumonitis in sheep: physiological changes and bronchoalveolar lavage. *J. Appl. Physiol.*, 66:1273–1279.
777. Timmermans, R. and Gerber, G. B. (1984): The effect of x-irradiation on cardiac β-adrenergic receptors following local heart irradiation. *Radiat. Res.*, 100:510–518.
778. Tolliver, J. M. and Pellmar, T. C. (1987): Ionizing radiation alters neuronal excitability in hippocampal slices of the guinea pig. *Radiat. Res.*, 112:555–563.
779. Torii, Y., Shikita, M., Saito, H., and Matsuki, N. (1993): X-irradiation-induced emesis in *Suncus murinus*. *J. Radiat. Res. (Tokyo)*, 34:164–170.
780. Torubarov, F. S. (1991): Psychological consequences of the Chernobyl accident from the radiation neurology point of view. In: *The Medical Basis for Radiation-Accident Preparedness*. Vol. III. *The Psychological Perspective*, edited by R. C. Ricks, M. E. Berger, and F. M. O'Hara, pp. 81–91. Elsevier, New York.
781. Trabalka, J. R., Eyman, L. D., and Auerbach, S. I. (1980): Analysis of the 1957–1958 Soviet Nuclear disaster. *Science*, 209:345–353.
782. Trakhtenbrot, L., Kelman, Z., Rotter, V., and Haaran-Ghera, N. (1990): Chromosomal mapping of the murine c-abl proto-oncogene by *in situ* hybridization. *Leukemia*, 4:136–137.
783. Travis, E. L. (1987): Relative radiosensitivity of the human lung. In: *Advances in Radiation Biology*. Vol. 12. *Relative Radiation Sensitivities of Human Organ Systems*, edited by J. T. Lett and K. I. Altman, pp. 205–238. Academic Press, San Diego, CA.
784. Trnovec, T., Kallay, Z., and Bezek, S. (1990): Effects of ionizing radiation on the blood–brain barrier permeability to pharmacologically active substances. *Int. J. Radiat. Oncol. Biol. Phys.*, 19:1581–1587.
785. Trnovec, T., Volenec, K., Bezek, S. et al. (1991): The effect of high energy electron irradiation on blood-brain barrier permeability to haloperidol and stobadin in rats. *Radiat. Environ. Biophys.*, 30:277–287.
786. Trotti. A. (1998): Toxicity antagonists in head and neck cancer. *Semin. Radiat. Oncol.*, 8:282–291.
787. Turbyfill, C. L., Roudon, R. M., Young, R. W., and Kieffer, V. A. (1972): *Alteration of Radiation Effects by 2-(n-Decylamino) Ethanethiolsulfuric Acid (WR-1607) in the Monkey*, AFRRI Scientific Report SR72-3. Armed Forces Radiobiology Research Institute, Bethesda, MD.
788. Turbyfill, C. L., Roudon, R. M., and Kieffer, V. A. (1972): Behavior and physiology of the monkey (*Macaca mulatta*) following 2500 rads of pulse mixed gamma-neutron radiation. *Aerosp. Med.*, 7:41–45.
789. Turns, J. E., Doyle, T. F., and Curran, C. R. (1971): *Norepinephrine Effects on Early Post-Irradiation Performance Decrement in the Monkey*, AFRRI Scientific Report SR71-16. Armed Forces Radiobiology Research Institute, Bethesda, MD.
790. Uchiyama, M. et al. (1989): Radiocesium body burden of Japanese who returned from European countries following the Chernobyl accident. *J. Radiat. Res.*, 30:51.
791. United Nations Scientific Committee on the Effects of Atomic Radiation. (2000): *Report to the General Assembly*. Vol. II. *Effects*. United Nations, New York.
792. Upton, A. C. (2001): Radiation hormesis: data and interpretations. *Crit. Rev. Toxicol.*, 31:681–695.
793. Urmer, A. H. and Brown, W. L. (1960): The effect of gamma radiation on the reorganization of a complex maze habit. *J. Gen. Psychol.*, 97:67–76.
794. USEPA. (1986): *Citizen's Guide to Radon*. U.S. Environmental Protection Agency, Washington, D.C.
795. U.S.S.R. State Committee on the Utilization of Atomic Energy. (1986): *The Accident at Chernoyl Nuclear Power Plant and Its Consequences*. Information compiled for the IAEA Experts Meeting, August 24–29, Vienna. Working Document for the Post-Accident Review Meeting, Draft Part I: General Material, Part II: Annexes, August 1–7 1986 (hereafter referred to as *The Accident … Soviet IAEA Report)*. Part I was subsequently published in Russian in *Atomnaya Energiya*, 61(5).
796. Utley, J. F. (1987): Relative radiosensitivities of the oral cavity, larynx, pharynx, and esophagus. In: *Advances in Radiation Biology*. Vol. 12. *Relative Radiation Sensitivities of Human Organ Systems*, edited by J. T. Lett and K. I. Altman, pp. 129–146. Academic Press, San Diego, CA.
797. Vahakangas, K. H., Samet, J. M., Metcalf, R. A. et al. (1992): Mutations of p53 and *ras* genes in radon-associated lung cancer from uranium miners. *Lancet*, 339:576–580.
798. Valk, P. E. and Dillon, W. P. (1991): Radiation injury of the brain. *Am. J. Neuroradiol.*, 12:45–62.
799. van Bekkum, D. W., Jongeiper, H. J., Nieuwerkerk, H. T. M., and Cohenm, J. A. (1954): The oxidative phosphorylation by mitochondria isolated from the spleen of rats after total body exposure to x-rays. *Br. J. Radiol.*, 27:127–130.

800. van Buul, P. P., van Duyn-Goedhart, A., and Sankaranarayanan, K. (1999): *In vivo* and *in vitro* radioprotective effects of prostaglandin E1 analogue misoprostol in DNA repair-proficient and -deficient rodent cell systems. *Radiat. Res.*, 152:398–403.
801. van der Kogel, A. J. (1986): Radiation-induced damage in the central nervous system: an interpretation of target cell responses. *Br. J. Cancer*, 53(Suppl. 7):53:207–217.
802. van der Meer, Y. et al. (1992): The sensitivity of quiescent and proliferating mouse spermatogonial stem cells to x irradiation. *Radiat. Res.*, 130:289–295.
803. van der Meer, Y. et al. (1992): The sensitivity to x rays of mouse spermatogonia that are committed to differentiate and of differentiating spermatogonia. *Radiat. Res.*, 130:296–302.
804. van der Meeren, A. and Lebaron-Jacobs, L. (2001): Behavioural consequences of an 8 Gy total body irradiation in mice: regulation by interleukin-4. *Can. J. Physiol. Pharmacol.*, 79:140–143.
805. van Dongen-Melman, J. E., de Groot, A., van Dongen, J. J. et al. (1997): Cranial irradiation is the major cause of learning problems in children treated for leukemia and lymphoma: a comparative study. *Leukemia*, 11:1197–1200.
806. Varagic, V. et al. (1967): The effect of x-irradiation on the amount of catecholamines in heart atria and hypothalamus of the rabbit and in brain and heart of the rat. *Int. J. Radiat. Biol.*, 12:113–119.
807. Verger, P. (1997): Down syndrome and ionizing radiation. *Health Physics*, 73(6):882–893.
808. Vidal-Pergola, G. M., Kimler, B. F., and Norton, S. (1993): Effect of *in utero* irradiation on the postnatal development, behavior, and brain structure of rats: dose fractionation with a 6-h interval. *Radiat. Res.*, 134:369–374.
809. Viel, J. F., Curbakova, E., Dzerve, B. et al. (1997): Risk factors for long-term mental and psychosomatic distress in Latvian Chernobyl liquidators. *Environ. Health Perspect.*, 105(6):1539–1544.
810. Viinamäki, H., Kumpusalo, E., Myllykangas, M. et al. (1995): The Chernobyl accident and mental well being: a population study. *Acta Psychiat. Scand.*, 91:396–401.
811. Vijayalaxmi, Reiter, R. J., Sewerynek, E. et al. (1995): Marked reduction of radiation-induced micronuclei in human blood lymphocytes pretreated with melatonin. *Radiat. Res.*, 143:102–106.
812. Vijayalaxmi, Reiter, R. J., Tan, D. X., Herman, T. S., and Thomas, C. R. (2004): Melatonin as a radioprotective agent: a review. *Int. J. Radiat. Oncol. Biol. Phys.*, 59:639–653.
813. Voevodskaya, N. A. and Vanin, A. F. (1992): Gamma-irradiation potentiates L-arginine-dependent nitric oxide formation in mice. *Biochem. Biophys. Res. Commun.*, 186:1423–1428.
814. von Sonntag, C. (1987): *The Chemical Basis of Radiation Biology*. Taylor & Francis, New York.
815. von Sonntag, C. (1991): The chemistry of free-radical-mediated DNA damage. *Basic Life Sci.*, 58:287–317.
816. Vyatleva, O. A., Katargina, T. A., Puchinskaya, L. M., and Yurkin, M. M. (1997): Electrophysiological characterization of the functional state of the brain in mental disturbances in workers involved in the clean-up following the Chernobyl atomic energy station accident. *Neurosci. Behav. Physiol.*, 27:166–172.
817. Wagner, W., Prott, F., and Schonekas, K. (1998): Amifostine: a radioprotector in locally advanced head and neck tumors. *Oncol. Rep.*, 5:1255–1257.
818. Wakeford, R. (1995): The risk of childhood cancer from intrauterine and preconceptional exposure to ionizing radiation. *Environ. Health Perspec.*, 103(11):1018–1025.
819. Walden, Jr., T. L. (1989): Long-term and low level effects of ionizing radiation. In: *Medical Consequences of Nuclear Warfare*, edited by R. I. Walker and T. J. Cerveny, pp. 171–226. TMM Publications, Falls Church, VA.
820. Walden, Jr., T. L. and Farzaneh, N. K. (1990): *Biochemistry of Ionizing Radiation*. Raven Press, New York.
821. Walden, Jr., T. L. and Farzaneh, N. K. (1991): Biochemical response of normal tissues to ionizing radiation. In: *Radiation Injury to the Nervous System*, edited by P. H. Gutin et al., pp. 17–36. Raven Press, New York.
822. Walden, Jr., T. L., Farzaneh, N. K., and Richards, L. (1989): Lipoxygenase products in radiation injury and protection. *New Trends Lipid Mediators*, 3:154–160.
823. Walker, D. I. and Eisen, V. (1979): Effect of ionizing radiation on 15-hydroxy prostaglandin dehydrogenase (PGDH) activity in tissue. *Int. J. Radiat. Biol.*, 36:399–407.
824. Wallace, W. H., Shalet, S. M., Crowne, E. C. et al. (1989): Ovarian failure following abdominal irradiation in childhood: natural history and prognosis. *Clin. Oncol. (R. Coll. Radiol.)*, 1:75–79.
825. Wallace, W. H., Shalet, S. M., Hendry, J. H. et al. (1989): Ovarian failure following abdominal irradiation in childhood: the radiosensitivity of the human oocyte. *Br. J. Radiol.*, 62:995–998.
826. Ward, J. F. (1990): The yield of DNA double-strand breaks produced intracellularly by ionizing radiation: a review. *Int. J. Radiat. Biol.*, 57:1141–1150.
827. Waselenko, J. K., MacVittie, T. J., Blakely, W. F. et al. (2004): Medical management of the acute radiation syndrome: recommendations of the Strategic National Stockpile Radiation Working Group. *Ann. Intern. Med.*, 140:1037–1051.
828. Watanabe, H., Okamoto, T., Takahashi, T. et al. (1992): The effects of sodium chloride, miso or ethanol on development intestinal metaplasia after x-irradiation of the rat glandular stomach. *Jpn. J. Cancer Res.*, 83:1267–1272.
829. Weinstein, N. D. (1991): Public response to home radon exposure. In: *The Medical Basis for Radiation-Accident Preparedness*. Vol. III. *The Psychological Perspective*, edited by R. C. Ricks, M. E. Berger, and F. M. O'Hara, pp. 173–178. Elsevier, New York.
830. Weisaeth, L. and Tonnessen, A. (2003): Responses of individuals and groups to consequences of technological disasters and radiation exposure., In: *Terrorism and Disaster*, edited by R. J. Ursano, C. S. Fullerton, and A. E. Norwood. Cambridge University Press, New York.
831. Weiss, J. F. (1997): Pharmacologic approaches to protection against radiation-induced lethality and other damage. *Environ. Health Perspect.*, 105(Suppl. 6):1473–1478.
832. Weiss, J. F., Kumar, K. S., Walden, T. L. et al. (1990): Advances in radioprotection through the use of combined agent regimens. *Int. J. Radiat. Biol.*, 57:709–722.
833. Weiss, J. F. and Landauer, M. R. (2003): Protection against ionizing radiation by antioxidant nutrients and phytochemicals. *Toxicology*, 189:1–20.

834. Weshler, Z., Ligumsky, M., Brufman, G. et al. (1987): Functional and morphological alterations following isolated rat stomach irradiation: a model for estimation of radiation injury. *In Vivo*, 1:357–361.
835. Wheeler, T. G. and Hardy, K. A. (1985): Retrograde amnesia produced by electron beam exposure: causal parameters and duration of memory loss. *Radiat. Res.*, 101:74–80.
836. Wheeler, T. G. and Tilton, B. M. (1983): *Duration of Memory Loss Due to Electron Beam Exposure*. U.S.A.F. SAM Tech. Rep. TR 83-33. U.S.A.F. School of Aerospace Medicine, Brooks Air Force Base, TX.
837. Whicker, F. W. (1989): Impact on plant and animal populations. In: *Health Impacts of Large Releases of Radionuclides*, Ciba Foundation Symposium No. 203, pp. 74–93. The Ciba Foundation, London.
838. Whicker, F. W. and Schultz, V. (1982): *Radioecology: Nuclear Energy and the Environment*, Vols. I and II. CRC Press, Boca Raton, FL.
839. Whicker, F. W., Pinder III, J. E., Bowling, J. W. et al. (1990): Distribution of long-lived radionuclides in an abandoned reactor cooling reservoir. *Ecol. Monogr.*, 60:471–496.
840. Wientroub, S., Weiss, J. F., Catravas, G. N., and Reddi, A. H. (1990): Influence of whole body irradiation and local shielding on matrix-induced endochondral bone differentiation. *Calcif. Tissue Int.*, 46:38–45.
841. Williams, D. (1994): Chernobyl, eight years on. *Nature*, 371:556.
842. Winkler, H. (1957): Untersuchungen uber die Wirkung von Roentgensstrahlen auf die bluthirschranke mit hilfe von P32 Zbl allg. *Pathol. Anat.*, 97:301–307.
843. Winsauer, P. J. and Mele, P. C. (1993): Effects of sublethal doses of ionizing radiation on repeated acquisition in rats. *Pharmacol. Biochem. Behav.*, 44:809–814.
844. Winsauer, P. J., Bixler, M. A., and Mele, P. C. (1995): Differential effects of ionizing radiation on the acquisition and performance of response sequences in rats. *Neurotoxicology*, 16:257–269.
845. Wishkerman, Y. V., Quastal, M. R., Douvdevani, A., and Goldsmith, J. R. (1997): Somatic mutations at the glycophorin A (GPA) locus measured in red cells of Chernobyl liquidators who immigrated to Israel. *Environ. Health Perspect.*, 105(Suppl. 6):1451–1454.
846. Withers, R. H. (1975): The four R's of radiotherapy. *Adv. Radiat. Biol.*, 5:241–271.
847. Wixon, H. N. and Hunt, W. A. (1983): Ionizing radiation decreases veratridine stimulated uptake of sodium in rat brain synaptosomes. *Science*, 220:1073–1074.
848. WHO. (1995): *Health Consequences of the Chernobyl Accident. Results of the IPHECA Pilot Projects and Related National Programmes*. World Health Organization, Geneva.
849. Wolfle, G., Bleyer, H., Muller, D., and Klinger, W. (1991): The influence of the radiation syndrome on cytochrome P450–dependent monooxygenation in rat liver. *Exp. Pathol.*, 43:89–95.
850. Wondergem, J., van der Laarse, A., van Ravels, F. J. et al. (1991): *In vitro* assessment of cardiac performance after irradiation using an isolated working rat heart preparation. *Int. J. Radiat. Biol.*, 59:1053–1068.
851. Wright, E. G. (1995): The pathogenesis of leukaemia. In: *Radiation Toxicology: Bone Marrow and Leukaemia*, edited by J. H. Hendry and B. I. Lord, pp. 245–274. Taylor & Francis, Bristol, PA.
852. Wu, L-J., Randers-Pehrson, G., Xu, A. et al. (1999): Targeted cytoplasmic irradiation with alpha particles induces mutations in mammalian cells. *Proc. Natl. Acad. Sci. U.S.A.*, 96:4959–4964.
853. Yacoub, A., Park, J. S., Qiao, L. et al. (2001): MAPK Dependence of DNA damage repair: ionizing radiation and the induction of expression of the DNA repair genes XRCC1 and ERCC1 in DU145 human prostate carcinoma cells in a MEK1/2 dependent fashion. *Int. J. Radiat. Biol.*, 77:1067–1078.
854. Yaes, R. J. (1992): Radiation damage to the kidney. In: *Advances in Radiation Biology*. Vol. 15. *Relative Radiation Sensitivities of Human Organ Systems, Part III*, edited by K. I. Altman and J. T. Lett, pp. 1–35. Academic Press, San Diego, CA.
855. Yamada, T. and Ohyama, H. (1988): Radiation-induced interphase cell death of rat thymocytes is internally programmed (apoptosis). *Int. J. Radiat. Biol.*, 53:65–75.
856. Yamamoto, M., Ueno, K., Igarashi, Y. et al. (1990): Determination of low-level Ra-226 in human bone by α-spectrometry. *J. Radiat. Res.*, 31:85.
857. Yamashita, H. and Miyasaka, T. (1952): Effects of beta rays upon a single nerve fiber. *Proc. Soc. Exp. Biol. Med.*, 80:375–377.
858. Yamashita, S., Namba, H., and Nagataki, S. (1993): Thyroid and radiation (Japan). *Folia Endocrinologica Japonica*, 69(10):1035–1043.
859. Yeung, T. K., Lauk, S., Simmonds, R. H. et al. (1989): Morphological and functional changes in the rat heart after x-irradiation: strain differences. *Radiat. Res.*, 119: 489–499.
860. Yin, E., Nelson, D. O., Coleman, M. A. et al. (2003): Gene expression changes in mouse brain after exposure to low-dose ionizing radiation. *Int. J. Radiat. Biol.*, 79: 759–775.
861. Yoshii, Y., Maki, Y., Tsunemoto, H. et al. (1981): The effect of total-head irradiation C3H/He of x-irradiation of the head in monkeys. *Radiat. Res.*, 86:152–170.
862. Yoshimura, T., Matsuno, K., Miyazaki, T. et al. (1993): Electron spin resonance studies of free radicals in gamma-irradiated golden hamster embryo cells: radical formation at 77 and 295 K and radioprotective effects of vitamin C at 295 K. *Radiat. Res.*, 136:361–365.
863. Young, R. W. (1986): Mechanisms and treatment of radiation-induced nausea and vomiting. In: *Nausea and Vomiting: Mechanisms and Treatment*, edited by C. J. Davis G. V. Lake-Bakaar, and G. V. Grahame-Smith, pp. 94–109 Springer-Verlag, New York.
864. Young, R. W. (1987): Acute radiation syndrome. In: *Military Radiobiology*, edited by J. J. Conklin and R. I. Walker pp. 165–190. Academic Press, New York.
865. Young, R. W. and Myers, P. H. (1986): The human response to nuclear radiation. *Med. Bull.*, 43:20–23.
866. Young, R. W. and Landauer, M. R. (2002): Psychological consequences of military operations in low-level radiation environments. *Mil. Med.*, 167:139–140.

867. Yuhas, J. M. (1970): Biological factors affecting the radioprotective efficiency of S-2–(3–aminopropylamino) ethylphosphorothioic acid (WR-2721): LD 50/30 doses. *Radiat. Res.*, 44:621–628.
868. Yuhas, J. M. (1980): Active versus passive absorption kinetics as the basis for selective normal tissue protection by S-2-(3-aminopropylamino) ethylphosphorothioic acid. *Cancer Res.*, 40:1519–1524.
869. Zaman, M. S., Lancaster, F. E., and Hupp, E. W. (1997): Physical and motor development in male and female rat offspring prenatally exposed to gamma radiation. *J. Environ. Sci. Health B.*, 32(2):313–325.
870. Zamora, M. L., Tracy, B. L., Zielinski, D. et al. (1998): Chronic ingestion of uranium in drinking water: a study of kidney bioeffects in humans. *Toxicol. Sci.*, 43:68–77.
871. Zaridze, D. G. (1997): Epidemiology of leukemias in children. *Arkhiv. Patologii.*, 59(5):65–70.
872. Zeman, W. (1963): Disturbances of nuclei acid metabolism preceding delayed radionecrosis of nervous tissue. *Proc. Natl. Acad. Sci. U.S.A.*, 50:626–630.
873. Zhavoronkova, L. A., Kholodova, N. B., Zubovskii, G. A. et al. (1995): Electroencephalographic correlates of neurological disturbances at remote periods of the effect of ionizing radiation (sequelae of the Chernobyl NPP accident). *Neurosci. Behav. Physiol.*, 25:142–149.
874. Ziboh, V. A., Mallia, C., Mohart, E., and Taylor, L. (1982): Induced biosynthesis of cutaneous prostaglandins by ionizing radiation. *Proc. Soc. Exp. Biol. Med.*, 169:386–391.
875. Zmudzka, B. Z. and Beer, J. Z. (1990): Activation of human immunodeficiency virus by ultraviolet radiation. *Photochem. Photobiol.*, 52:1153-1162.

# Notes

# 19 Plant and Animal Toxins

*Frederick W. Oehme and Daniel E. Keyler*

## CONTENTS

## TOXINS IN NATURE

Nature is beautiful, but is it always safe? Nature and things natural can be hazardous and risky if individuals are not informed and constantly aware of the dangers lurking in the green foliage, the crystal-clear waters, and the bushes and rocks! Naturally occurring toxins may be grouped into those originating in plants, algae, fungi, bacteria, and various members of the animal kingdom. This chapter presents the toxic hazards from plants, algae, mushrooms, and toadstools, as well as animal-origin toxins.

Toxins from widely diverse, naturally occurring organisms can be found throughout our natural environment. Foliage may include a variety of toxic plants, waters can carry algae, and, wherever moisture is, various hazardous fungi may grow. The dry land may harbor reptiles, amphibians, spiders, insects, and even mammals capable of inflicting injury. Animals found in waters include various invertebrates and vertebrates found in the oceans, shellfish and fish carrying toxins from other sources, and such unsuspected bearers of poisons as sea turtles, polar bears, seals, and walruses. In short, naturally occurring toxins are everywhere and have continuing potential for intoxication.

Natural toxins are unique in that they exist in the organism's structure, yet are not harmful to that host. They are also special in that individuals potentially affected by these natural toxins may become exposed by consuming the toxin-bearing material, such as when consuming poisonous plants or toxic marine animals, or by coming into contact with them and experiencing durable irritation or traumatic injury, such as occurs from poison ivy or blistering agents in plants or from cactus spines penetrating body surfaces. Most commonly, such harmful effects result from animals carrying venoms and producing envenomations (injections) through stings, bites, or delivery systems. In many cases, these delivery systems are specialized to cause reactions on contact, embed materials that are then released by the victim's biological tissues, or result in direct administration into the victim's body tissues and fluids. As in all areas of toxicology, the amount of exposure or dose delivered to the victim largely determines the resulting clinical effect.

Exposure to the toxins present in nature are in many cases accidental, such as persons collecting green plants who mistake a toxic variety for a similarly appearing innocuous plant. Individuals may elect to chew and swallow plants because of ignorance. In some cases, the use of native herbal medications presents a hazard when inappropriate natural materials are selected or an incorrect amount is included in various preparations. In other instances, natural toxins may be used maliciously for homicidal or retaliatory purposes. However they are introduced into the human body, natural materials are capable of producing mild, moderate, or even lethal effects.

Naturally occurring toxins are complex. Unlike synthetic compounds used for therapeutic or specific chemical needs, toxins from plants or animals are mixtures of organic, proteinacious, enzymatic, and even mineral materials that are present in various proportions that vary with the individual plant or animal maturity, location, species, and various life changes that growing and maturing organisms go through. The relative proportion of the numerous components in these toxic packages varies sufficiently that their impact on biological systems is not only affected by the quantity administered but also by the relative proportion of each of the components in the mixture. At any one time, the dose received from a specific animal envenomation or plant will vary significantly, such that an exposure to the same amount of a specific plant will not always produce the same biological effect. This creates significant problems when attempting to identify the toxic components of naturally occurring materials. Fractionation of the mixed bag of components in any toxic mixture thus becomes problematic when trying to address the active component responsible for a specific harmful effect. It also impacts the resulting toxicity, as the clinical impact varies with these various factors, and it is often necessary to evaluate the toxicity based on potential effects or by observing the direct impact made by the exposure. A single bite from a known poisonous snake does not always induce toxicity or the expected amount of damage. The gestation of a known amount of recognized poisonous plant may not induce toxicity or may indeed cause more life-threatening damage than previously documented.

So, why are these toxic and often lethal components present in these organisms? One could make the point that they are there "just because" and are part of the organism's normal body component of compounds necessary for appropriate functioning. Their introduction into human systems and their adverse effect are just an unfortunate coincidence! Others might argue that the presence of a compound is a defensive mechanism to avoid consumption or destruction by other organisms or predators, essentially offering a distasteful menu. Others see the types of chemicals present as weapons to be used by the organism (e.g., a rattlesnake) to defend itself against predators or other individuals, to discourage attack or consumption by other individuals. Certainly, plant toxins have that effect by keeping browsing animals from consuming plants that produce illness in their species. Finally, the use of venoms or poisons to immobilize animals or fight off other animals provides insight into the real world of nature, as securing an appropriate diet involves capturing a victim and consuming it as needed. Interestingly, when animals consume prey that their bites have made available for consumption, the toxin has little impact on the consuming animal that has just envenomated the victim. Being able to consume one's own poison seems the ultimate path of self-preservation!

The vast amount of information to be gained from the study of these naturally occurring poisons has evolved into a highly specialized and unique aspect of science. The variety of fractions found in venoms and toxic plant materials has required elaborate instrumentation, knowledge, and expertise to deal with the understanding of how and why these toxic materials exist in nature. This study, known as *toxinology*, has a broad scope of interest and impact. The science varies geographically and with the wide range of toxic plant and animal species. The science has a widely respected journal devoted to it (*Toxicon* [1]) and has wide applicability to understanding our own environment as well as the components of nature and their potentially harmful or beneficial contributions to the human race.

Toxinology seeks to understand the mechanism by which these widely diverse and selective natural chemicals produce their effect. Isolation of various components and demonstrating their structure and biological effects are exciting activities that keep individual scientists busy with one species of organism over a lifetime. The information gathered is used to determine potential applications of individual venom or toxin components to biomedical research or the treatment of human or animal diseases. The recent explosion in the use of botox for cosmetic surgery and in the treatment of various neurological disorders is a special example of how applying natural toxins may evolve in future years. Understanding venoms and their chemical structures and functions allows the production of antivenins through the manipulation of animal and biological systems. These antivenins are essential for the successful treatment of human animal bites and serious illnesses. As these biological processes are better understood and the individual venin components become identified, their application and use in human medicine and in therapy for natural intoxications will be of significant health benefit. Understanding the toxins from plants, algae, fungi, and animals is the substance of the remainder of this chapter.

## TOXINS FROM PLANTS

Ever since the frequency of exposures of humans to toxins has been tallied, plants have ranked among the most frequent causes of exposures to toxic substances. In 2005, they were the 11th most common cause of human exposures, with 74,811 instances (3.1% of the total exposures) reported in the American Association of Poison Control Centers Toxic Exposure Surveillance System Annual Report [2]. At the same time, plants and their chemical composition have been investigated for the potential of providing new medical agents and derivatives to treat illnesses and have been a common component of herbal drugs and folk medicines. From previous times and cultures, herbalists and healers have gathered leaves, roots, blossoms, and bark to brew remedies, potions, and elixirs with mixed results. Now, alert clinicians have observed that, while some of these herbals seem to work well and others fail, many at best exert a positive placebo effect, while others induce illness. Plants are a true example of "balm or poison" and the care required to receive beneficial effects vs. life-threatening doses from these tricky poisons.

Only very small amounts of some plant parts (flowers, fruits, berries, leaves, stems, bark, or roots) may be toxic and produce illness or even rather dramatic effects. Some plants are poisonous if they are chewed or swallowed. Others cause poisoning by initiating allergies, dermatitis, or mechanical injury due to spines or needles. Some plants are harmful only if eaten or chewed at certain stages of their growth; others are toxic at all stages of development. Thus, both the species of the plant and its stage of growth or the season of the year in which exposure occurs is important in determining the potential toxicity.

Of the many plant varieties, only a relatively small number are truly poisonous and capable of producing life-threatening effects. A few of those pose a serious threat to life or health, and among those are some widely cultivated flowers, vegetables, and ornamental plants. Some of them are household plants in which specimens may come from Maine or California. Together with the toxic flora of field and forest, they represent an abnormal potential for poisoning, particularly in small children who are apt to appreciate beauty by chewing on it.

Species of the plant, its stage of growth, and the season of the year are all important factors in evaluating potential plant poisonings. In some cases, the part of the plant that is eaten also matters. For example, all portions of the potato plant—except the actual potato—are poisonous; the leaves of a rhubarb plant are highly dangerous, despite the fact that the stalks are used to make delicious pies. The only reliable rule is not to eat anything that you are not absolutely sure is safe [3].

Animals that are outdoors and those that consume plants as part of their natural dietary intake are at severe risk for consuming a salad bar of potential hazards. Of the approximately 1000 recognized poisonous plants that grow in North America, all are toxic to animals, particularly livestock sent to pastures and ranges for their dietary needs. Each year, significant economic losses due to consumption of these poisonous plants have incurred. Determining the extent of these losses is difficult, but estimates of the losses caused by poisonous plants in only the 17 Western states are dramatic: 1% of cattle death losses can be attributed to toxic foliage; 1% of the loss of calves due to abortions or illnesses surrounding births; a 3.5% death loss has been estimated in sheep; and a 1% loss of pregnancy potential and newborn lambs is estimated to occur in sheep from plant toxins. A total economic loss of almost $250 million is the best estimate available for poisonous plant damage in the 17 Western states [4]. The losses estimated may reasonably be expected to be considerably greater when one considers the fact that poisonous plants do not always produce death. Depending on the amount ingested, no clinically observable damage may occur but reduced weight gain and lowered economic productivity may result.

What types of chemicals are involved in potential poisonous plant chemistry and toxicity, and what are the plants that produce poisoning if they are ingested or come into contact with the human body? Although it is unusual to have a naturally occurring chemical present in only one plant, it is equally unusual for only one toxic substance to be found in a plant species. Indeed, it is the common finding that most plants have several toxic components present. It is the distribution of these materials and their concentrations in various parts of the plant and at various stages of growth that present the hazard. Berries consumed at an early stage of development may be highly toxic, but once the berries are ripe and are in full bloom their toxicity may be reduced. The greatest concentration of the toxin may have moved from the unripe berry to the stems or to the roots. The leaves of a plant may be highly toxic when the plant is young but as the plant matures they may become innocuous; however, when the seeds of the mature plant are digested, acute toxicity may occur. In each of these examples, the concentration and the type of poison varies with each environmental circumstance; hence when considering the various chemicals present in plants, one must recognize the potential for more than one toxic factor to be present and that toxicity is dependent on which toxin is present in the highest concentration and which portion of the plant is consumed by the potential victim.

## Toxic Chemical Principles in Plants

Some of the toxic principles in poisonous plants exist due to the plant biochemistry, but others are present in the soil or water nourishing the plant and have accumulated in the growing foliage. With such a heterogeneous potential for effects, it is understandable that numerous categories of chemical principles and components have been attributed to plant poisonings. Most plants contain numerous derivatives and related compounds. A general listing includes alcohols, various alkaloids, amines, anticholinergics, glycosides and glucosides, mechanical injury vectors and associated chemicals, minerals associated with plant growth and incorporated into plants, oxalates, various compounds capable of producing photodynamic pigments and resulting photosynthesis, polypeptides, various resins and resinoids, saponins, and toxalbumens. Some potentially toxic plants have enzymatic properties or contain hormones, phenolics, and unusual nitrogens.

### Alcohols

Alcohols are organic compounds formed from hydrocarbons and found in species of golden rod and snake root. Tremetol is a specific alcohol that is passed in milk and induces neurological effects in humans and animals consuming the toxin-containing milk.

### Alkaloids

The alkaloids are a widely disseminated group of various chemicals found in numerous plants. Their concentration in plants is mildly influenced by the climate and availability of water but varies greatly with the species and variety of plants studied. They are widely distributed throughout the plant's structure and produce strong physiological reactions in consuming individuals. The toxicity induced varies from neurological through liver to kidney damage. The veratrum alkaloids are found in *Aconitum* species, nicotine, and coniine and are also present in *Crotalaria*, *Senecio*, potatoes, tomatoes, lilies, and the highly toxic Japanese yew (*Taxus*). As a result of continuing studies on the chemistry of plants, new alkaloid compounds are being discovered and added to this list.

### Amines

Amines account for only a small number of the toxins found in plants, but many foods contain pressor amines, which constrict blood vessels and increase blood pressure. Clinical problems are induced in individuals medicated with tranquilizers or antidepressants, which inhibit monoamine oxidases. Generally, the toxic amines affect skeletal or nervous tissue; they are found in sweet peas (*Lathyrus*) and mistletoe (*Phoradendron*). A number of plants exhibit anticholinergic properties due to their atropine or scopolamine content. These principles are commonly found in deadly nightshade (*Atropa belladonna*), sacred datura (*Datura metaloides*), jimson weed (*D. stramonium*), trumpet lily (*D. arborea*), angel trumpet (*D. candida*, *D. suaveolens*), other *Datura* spp., henbane (*Hyoscyamus niger*), matrimony vine (*Lycium barbarum*), and mandrake (*Mandragora officinarum*). The entire portions of these plants are toxic, although the flowers, fruits, and seeds are especially high in the atropine-like principles. Agitation, hallucinations, flushed skin, dry mucous membranes, elevated heart rates, and dilated pupils are common effects. These plants account for many admissions to critical care units and are often associated with patients who have ingested seeds or brewed tea made from the seeds. Although death is rare, serious injuries result from the impaired judgment and functionality observed in the victims.

### Glycosides

Naturally occurring glycosides are widely distributed in the plant kingdom and are the most commonly found and largest group of toxic compounds in plants. Although many are nontoxic, some varieties do have toxic effects upon ingestion or other exposure. The amount of a particular glycoside present again depends on the plant genetics, the portion of the plant being sampled, and the plant's age, as well as factors such as climate, soil fertility, and available moisture. Toxicity from this group is a function of the aglycone component or a part of it. Numerous groups of toxic glycosides have been identified:

- Coumarin and vanillin glycosides—sweet clover, laurel, horse chestnut
- Cyanogenetic glycosides, which contain or liberate hydrocyanic acid—velvet grass (*Holcus lanatus*), hydrangea (*Hydrangea* spp.), flax (*Linum* spp.), cassava (*Manihot esculenta*), lima bean (*Phasecolus lunatus*), cherries (*Prunus* spp.), apple (*Pyrus malus*), sudan grass and Johnsongrass (*Sorghum* spp.), poison suckleya (*Suckleya suckleyana*), white clover (*Trifolium repens*), arrowgrass (*Triglochin* spp.), vetch seed (*Vicia sativa*), corn (*Zea mays*)
- Gloiterogenic glycosides—chard (*Beta vulgaris*), rape seed or meal (*Brassica napus*), black mustard seed (*Brassica nigra*), kale (*Brassica oleracca* var. *acephala*), Chinese cabbage (*Brassica pekinensis*), turnip root (*Brassica rapa*), soybean (*Glycine max*)

- Irritant oils derived from mustard oils—horseradish (*Armoracia*), white mustard (*Brassica hirta*), Indian mustard (*Brassica juncea*), wild radish (*Raphanus raphanistrum*), fanweed (*Thlaspi arvense*)
- Volatile, unstable protoanemonin oil released from the glycoside ranunculin—marsh marigold (*Caltha palustris*), buttercups (*Ranunculus* spp.)
- Steroid glycosides, which contain cardiac toxins capable of stimulating or disrupting cardiac function—dogbane (*Apocynum* spp.), lily-of-the-valley (*Convallaria majalis*), foxglove (*Digitalis purpurea*), oleander (*Nerium oleander*), and squill (*Urginea maritima*).

Plants containing grayanotoxins (rhododendrons, azaleas [*Rhododendron*], sheep laurel, mountain laurel [*Kalmia*], and andromeda [*Pieris*]) are also potent cardiotoxic plants but not due to steroid glycosides. Noncardioactive saponin steroid glycosides are not easily absorbed but have additional toxins that injure the digestive tract and allow entrance of the toxin into the bloodstream; examples include corn cockle (*Agrostemma githago*), English ivy (*Hedera helix*), alfalfa (*Medicago sativa*), pokeweed (*Phytolacca americana*), bouncing bet, cow cockle (*Saponaria* spp.), coffeeweed, and rattlebox (*Sesbania* spp.).

## Mechanical Injury

Some plants may inflict injury through mechanical processes that then allow additional adverse effects to occur. The nettle (*Urtica chamaedryoides*) is a common example, as it bears stinging hairs that contain significant amounts of acetylcholine and histamine that induce serious reactions. Other plants having such mechanisms include anemone (*Anemone patens*), poverty grasses, awns (*Aristida* spp.), squirreltail barley, barbed awns (*Hordeum jabatum*), foxtail grasses (*Setaria lutescens*), needle grasses (*Stipa* spp.), cocklebur (*Xanthium* spp.), and burdock (*Arctium lappa*).

## Mineral Toxicities

Plants may become toxic secondarily through the accumulation of minerals found in soil at high concentrations. Chief among those elements are nitrogen, which is accumulated in those plants as nitrate, and selenium, which has varying amounts in soils, depending on geologic formation and rainfall. Plants that frequently contain toxic concentrations of nitrates include pigweeds (*Amaranthus* spp.), bishop's weed (*Ammi majus*), tarweed (*Amsinckia* sp.), pigweed, lamb's quarters (*Chenopodium* spp.), Canada thistle (*Cirsium arvense*), poison hemlock (*Conium maculatum*), bindweed (*Covolvulus* spp.), Jimson weed (*Datura* spp.), fireball (*Kochia scoparia*), sweetclover (*Melilotus officinalis*), smartweeds (*Polygonum* spp.), dock (*Rumex* spp.), Russian thistle (*Salsola pestifier*), annual sage (*Salvia reflexa*), elder (*Sambucus pubens*), nightshades (*Solanum* spp.), goldenrods (*Solidago* spp.), and Johnsongrass (*Sorghum halepense*). In addition, certain crops will also accumulate exceedingly high levels of nitrate depending on the environmental conditions and the availability of nitrogen. Some of these are oat hay (*Avena sativa*), beet and mangold (*Beta vulgaris*), broccoli and kale (*Brassica oleracea*), turnip (*Brassica rapa*), soybean (*Glycine max*), barley (*Hordeum vulgare*), sweet potato vines (*Ipomoea batatas*), alfalfa (*Medicago sativa*), radish (*Raphanus sativus*), sudan grass (*Sorghum vulgare*), wheat (*Triticum aestivum*), and corn (*Zea mays*).

Plants taking up selenium are of significant concern to livestock and also to humans who might utilize these high selenium-containing forages. Crops that may do so include poisonvetches (*Astragalus*), prince's plume (*Stanleya*), goldenweeds (*Oonopsis*), woody asters (*Xylorrhiza*), aster (*Aster* spp.), saltbushes (*Atriplex* spp.), gumweeds (*Grindelia* spp.), snakeweed (*Gutierrezia* spp.) tansy aster (*Machaeranthera* spp.), and beard tongue (*Penstemon* spp.).

Other elements found in soil or the environment may contaminate plants or produce toxicity. Such elements include cadmium taken up from fertilizers in which the metal impurity is present; copper from copper-rich soils or copper-containing pesticide use; fluoride from fluoride-containing soils, rock formations, or industrial influence contaminating the soils; lead from surface contaminatio from smelters or lead-containing mine debris; and molybdenum found in soils abnormally high in this element.

## Oxalates

The organic acid oxalates found in plants occur in th form of soluble sodium and potassium salts or insolubl calcium oxalates or acid oxalate. Although small amount of oxalates are found in many plants, several of them wil have increased oxalate content with the season of the yea and geographic location. Highest oxalate concentration are in late summer and fall in the leafy stage of plan growth. The resulting toxicity may occur from the solubl oxalate salts that enter the bloodstream and later are pre cipitated in the kidney or as firm calcium oxalates or aci oxalates that irritate digestive tract oral membranes o contact, producing burning and irritation upon chewing The common plants containing variable amounts of solu ble oxalates include beet, mangold (*Beta vulgans*), lamb' quarters (*Chenopodium album*), halogeton (*Halogeto glomeratus*), sorrel, sour sob (*Oxlais* spp.), pokewee (*Phytolacca americana*), rhubarb (*Rheum rhaponticum* sorrel, dock (*Rumex* spp.), greasewood (*Sarcobatus ver miculatus*), and spinach (*Spinecia oleracea*). Commo plants containing the insoluble oxalates that produce irr tation upon contact with the mucous membranes include jack-in-the-pulpit (*Arisaema* spp.), caladium (*Caladium* spp.), dumbcane (*Dieffenbachia* spp.), philodendron (*Philodendron* spp.), skunk cabbage (*Symplocarpus foetidus*), and caladium (*Xanthosoma* spp.).

### Photosensitivity-Inducing Plants

Photosensitivity from plant substances is an interesting phenomenon in which individuals become hypersensitive to sunlight due to the presence of plant-originating material in the peripheral circulation. This may be produced by: (1) plants that carry specific photosynthesizers in themselves that directly produce the hypersensitivity; (2) plants that contain additional compounds that by themselves produce liver dysfunction with no photodynamic activity, but the resulting damage retards elimination of plant materials and allows them to reach the peripheral circulation, thus inducing photosynthesis; or (3) congenital metabolic defects that allow the accumulation of photodynamic pigments more materials within the individual circulation. Plants producing these types of photosynthesis risks include buckwheat (*Fagopyrum sagittatum*), St. John's wort or Klamath weed (*Hypericum perforatum*), lechuguilla (*Agava lechuguilla*), cultivated rape (*Brassica napus*), lantanta (*Lantana* spp.), horsebrush (*Tetradymia* spp.), water bloom (species of blue–green algae), oats (*Avena sativa*), milk purslane (*Euphorbia maculata*), alfalfa (*Medicago sativa*), smartweeds (*Polygonum* spp.), summer cypress (*Kolchia scoparia*), sudan grass (*Sorghum vulgare* var. *sudanese*), clovers (*Trifolium* spp.), and vetches (*Vicia* spp.).

### Phytotoxins

Phytotoxins are large, complex molecules similar to bacterial toxins in structure and reactions. They are antigenic with protein-like characteristics that have resulted in the terminology *toxalbumins* also being used as a designation for this group of plant toxins. Some of these plants have the highest concentration of the poison present in the seeds, such as the castor bean (*Ricinus communis*), from which commercial castor oil is produced. Such plants also include precatory bean, rosary pea (*Abrus precatorius*), and black locust (*Robinia pseudoacacia*). The toxin ricin has been specifically isolated as the toxin in castor bean. The highly potent nature of these phytotoxins has attracted attention from defense groups dealing with potential use by terrorists.

### Polypeptides

Only a small number of plants containing poisonous polypeptides are currently known. These include mushrooms, blue–green algae, soybeans, potatoes, lima beans, kidney beans, unripe bananas, mangos, and some legumes. Significant consumption for several days is usually required to induce toxicity.

### Resins and Resinoids

This heterogeneous group of compounds has the physical characteristic of being solid or semisolid at room temperature and easily melted or burned. They are soluble in organic filaments and do not contain nitrogen. They tend to be oily or greasy upon extraction and induce variable degrees of toxicity in both potency and clinical effects. Such compounds are found in milkweed (*Asclepias* spp.), hemp, marijuana (*Cannabis sativa*), water hemlock (*Cicuta* spp.), iris (*Iris versicolor*), laurel (*Kalmia* spp.), Japanese pieris (*Pieris japonica*), pine (*Pinus* spp.), laurel, rhododendron (*Rhododendron* spp.), and Japanese wisteria (*Wisteria floribunda*).

### Others

Various other individual poisonous principles produce toxicity, often in unique ways specific to that individual plant. Some contain enzymes that induce neurological deficits or cancer, such as bracken fern (*Pteridium aquilidum*), horsetails (*Equisetum* spp.), and male fern (*Dryopteris felix-mas*). The polyphenolic gossypol, found in the seed of cotton (*Gossypium* spp.), has commercial applications in human medicine, despite having toxicities in all species. Unusual nitrogen compounds found in some of the legume family foliages, such as creeping indigo (*Indigofera spicata*), sweet pea (*Lathyrus*), and koa hade (*Leucaena leucocephela*), produce specific effects on pregnancy, the nervous system, hair follicles, and general health. Estrogenic factors are found in subterranean clover (*Trofolium subterraneum*). The toxic factor found in the cotyledon stage of cocklebur (*Xanthium strumarium*) produces rapid weakness and death in consuming individuals. As clinical observations become more extensive and as analytical chemistry become more specific, no doubt additional toxic components will be identified and related to the varying illnesses and conditions associated with plant exposure and consumption.

## Organ Systems Affected

The complexity and variability of the various compounds present in plants make most plant varieties capable of inducing more than a single effect; in fact, they can produce several system-specific adverse changes. Depending on the toxins present and their individual concentrations, such harmful effects may influence every organ system in the body, individually or as complex combinations of various systemic effects.

### Gastroenteritis-Inducing Plants

The majority of plant intoxications occur from the ingestion of plants containing gastroenteric irritants. Symptoms range from burning in the mouth and throat when chewing the leaves of such common household plants, such as those from a vine growing on a windowsill (*Philodendron* spp.) or from a dumbcane (*Dieffenbachia* spp.) growing in a big pot by the front door, to severe vomiting, intestinal cramping, and purging diarrhea caused by consumption of the fresh roots and stems of pokeweed (*Phytolacca americana*), wisteria seeds, berries of the spurge laurel (*Daphne* spp.), or leaves of buttercups (*Ranunculus* spp.).

The onset of the response to the irritant is variable, as it depends on the nature of the toxin, a possible requirement for activation of the irritant, and various mechanisms that may illicit emesis or gastroenteritis, thus modifying the clinical response. Plants that cause minor abdominal discomfort in an adult may provoke profound emesis and diarrhea in a small child. Because dehydration and electrolyte imbalance can develop rapidly, it may be necessary to institute replacement therapy promptly to prevent shock.

Serious intoxications are produced by plants containing toxalbumins, such as the rosary pea (*Abrus precatorius*) and the castor bean (*Ricinus communis*). Following ingestion of the chewed seeds of either of these plants, there may be considerable delay prior to the onset of effects; this is related to the quantity of material present in the chewed seeds. Typically, the latent period is about 2 hours, but it may be a day or more. The clinical picture is one of severe hemorrhagic gastroenteritis with persistent nausea, emesis, colic, and profuse diarrhea. Dehydration may result in oliguria and cardiovascular collapse. In nonfatal cases, the period of illness varies from 2 to 10 days. Other examples of intoxications include the following:

- Rosary pea seed and precatory bean (*Abrus precatorius*) contain a phytotoxin related to the botchulinum toxin. It produces nausea, colic, diarrhea, weakness, and trembling with a hemolytic anemia and ultimately uremia in severe cases.
- Poinsettia (*Euphorbia pulchenima*) has irritating saps in the leaves and stems that lead to vomiting and delirium in patients, who are ill for several days.
- Philodendron (*Philodendron* spp.) has calcium oxalate crystals that severely affect kidney function and induce listlessness and potentially uremia in severely intoxicated patients.
- Dumbcane (*Dieffenbachia* spp.) contains insoluble oxalate crystals that are not absorbed but are a severe contact irritant. Swollen mucous membranes of the mouth and throat result that may induce suffocation due to respiratory tract swelling that occludes the airway.
- Ivy (*Hedera helix*) contains saponic glycosides in the berries. This plant then induces severe diarrhea and may contribute to excitement and excessive nervousness in poisoned individuals.
- Lantana (*Lantana camara*) has an alkaloid in the unripe berries and clippings. It is a commonly available plant that in low doses of intake produces mainly digestive tract irritation with some photosensitization leading to sunburn and dermatosis. If large amounts are consumed, circulatory collapse and muscle weakness develop that may be life threatening.
- Daffodil (*Narcissus*) is a common plant with a prominent bulb-like root. It has a potent alkaloid in high concentrations in the bulb. Ingestion or chewing on this bulb induces repetitive and prominent vomiting and digestive tract irritation.
- Rhubarb (*Rheum rhaponticum*) is well known for containing high levels of oxalic acid in the leaf blades. This acid is highly irritating and produces severe abdominal pain that proceeds to vomiting, convulsions, anuria, coma, and death as oxalate crystals occlude the tubules, ultimately producing uremia.
- Pokeweed (*Phytolacca americana*) has a prominent alkaloid (phytolaccatoxin) in all plant parts. Although the roots are most toxic, the prominent deep purple berries fortunately contain lower concentrations of the alkaloid; however, because of their attractiveness, the berries are most frequently eaten. The alkaloid produces gastrointestinal signs that often extend to inducing spasms and convulsions with respiratory paralysis occasionally developing into anoxia.
- Death camas (*Zigadenus* spp.) is another plant with a bulb-like root that resembles an onion. It is frequently collected as a "wild onion" and then added to cooked stews and soups. The alkaloid is not significantly changed by the heating process, and prominent gastrointestinal effects result from consumption. The vomiting and diarrhea it produces are often life saving because they rapidly remove the alkaloid from the intestinal tract.
- Mayapple (*Podophyllum peltatum*) contains a resinoid in high concentrations in the root. The rest of the plant is nontoxic and is the most commonly utilized portion. If the root is used as a food material, violent purgative effects result from the resinoid properties.
- Castor bean (*Ricinus communis*) has the prominent phytotoxin ricin in the seeds. This is a plant protein that induces all the effects that protein materials induce; gastrointestinal

effects are prominent, and fever, red-blood cell hemolysis, and, with high doses, severe convulsions are also induced.

- Elderberry (*Sambucus americana*) is prominently known for inducing gastrointestinal involvement, yet the toxic principle has not been specifically identified. Based on exposures, it is apparently present in all parts of the plant but remains an unknown. Elderberry also contains some cyanide that fortunately is volatilized by heating or through the fermentation process.
- Daphnea (*Daphne* spp.) contains irritating glycosides, particularly dihydroxycoumarin, in the berries and bark of the plant. These compounds produce severe gastrointestinal irritation, and bloody diarrhea is a common effect from significant ingestion.
- Wisteria (*Wisteria sinensis*) is another plant with an unidentified toxin in its seed. Gastrointestinal irritations result from its ingestion, and neuromuscular collapse occasionally follows the initial consumption or frequent repeated exposures.
- Black locust (*Robinia seudoccacia*) has a heat-labile phytotoxin and glycoside in all plant parts. The seeds and the bark are the most commonly involved source of poisoning, although other less-available plant portions also induce risk. The gastrointestinal signs are complicated by additional incidences of depression and neurological effects.

The ability of several different plant chemicals to induce similar digestive tract involvement and various systemic effects highlights the numerous various types of plant toxins that are capable of inducing similar yet complex toxic syndromes. The clinical effects of plant toxins are not simple, and accurate diagnosis and therapy are largely based on the history of exposure so appropriate management can be pursued. In the absence of a history of specific plant ingestion, general management and nursing care are required to alleviate the clinical effects.

## Plants Containing Digitalis

An important clinical group of plants are those with cardiotoxic properties. Many of these cardiotoxin-containing plants induce fatal toxicity when individuals eat the berries or chew on leaves or flowers. Occasionally, the consumption of water from vases containing the flowers of digitalis-containing plants has been implicated in poisonings. The initial syndrome from these ingestions is usually local irritation to the mouth followed by emesis. In contrast to toxicity from pure cardiac glycosides, poisoning from these plants will also be associated with diarrhea and abdominal pain due to the presence of other irritants and saponins. Examples of such plants include:

- Mistletoe (*Phoradendron serotinu*) has toxin amines in its berries. Sudden gastrointestinal effects are followed by cardiovascular collapse similar to that seen from the ingestion of an overdose of digitalis.
- Lily-of-the-valley (*Convallaria majalis*) has cardioactive glycosides in its leaves and flowers that induce irregular heart rates and vomiting.
- Foxglove (*Digitalis purpurea*) has significant concentrations of the digitalis glycosides in its leaves. This is the plant from which digitalis was originally isolated and serves as the template for cardioactive plant poisons. Following initial gastrointestinal irritation and its effects, poisoned individuals develop headaches, irregular heart rates, and cardiac arrhythmias that often lead to tremors and neurological disturbances.
- Oleander (*Nerium oleander*) has a cardioactive glycoside very similar to digitalis in all its plant parts. It is a potent toxin vehicle that has produced poisoning from minimal ingestion. One leaf has been accused of inducing life-threatening toxicity. The initial effects of nausea, abdominal pain, and bloody diarrhea lead rapidly to circulatory irregularities, unconsciousness, and coma, with death due to respiratory failure.

The cardioactive plant toxins have served as models for potent forensic situations used for suicide or for murder. The complexity of the plant chemistry has for many years confused identification of the toxic vehicle; however, current instrumentation and available chemical knowledge has now produced the ability to recognize, characterize and identify the presence of these cardioactive glycosides in suspected cases of intoxication.

## Plants Containing Nicotine, Cytisine, and Coniine

Certain neurological alkaloids exert similar actions. Numerous fatal poisonings have occurred because wild flora often contain complexes of these alkaloids. Tobacco leaves (*Nicotiniana* spp.) are commonly toxic to individuals harvesting tobacco and those who consume the wild leaves. Cytisine poisoning most commonly results from the consumption of seeds from the pea-like pods of the golden chain tree (*Laburnum anagyroides*). Intoxications from coniine usually result from nibbling the parsley-like leaves of poison hemlock (*Conium maculatum*) or from eating the seeds from that same plant. This alkaloid is related to nicotine, and all parts of the poison hemlock plant are toxic. It produces an ascending paralysis of the nervous system. The top of the hemlock is famous in Greek history for its contribution to the fatal poisoning of Socrates. The toxicity from these plants usually initiates vomiting within 15 minutes of ingestion which is accompanied by profuse salivation. Abdominal cramping and diarrhea are rare, yet the gastrointestinal syndrome is always present within an hour of ingestion. This is followed by confusion, hyperpyrexia, incoordination, occasional mydriasis, and then tachycardia that leads to fatalities from respiratory failure. Serious intoxications result from this group of plants.

## Plants Containing Atropine

Although a number of plants contain atropine and its related alkaloids, by far the most common and important poisoning is due to the consumption of jimson weed (*Datura stramonium*). An abundantly found wild weed, this plant affects persons who eat the seeds or suck the flowers, often when they are attempting to use the plant as a hallucinogen. The early symptoms are mydriasis and dryness of the mouth, and the poisoned individual will complain of excessive thirst. The skin becomes hot and dry with redness and rash-like dermatosis occurring around the head and neck of the victim. Severe intoxications feature pronounced hyperpyrexia, delirium, and hallucinations. These hallucinations may lead to dangerous behavior and activities that become life threatening. Convulsions may appear in severe overdoses, and ultimately coma can develop.

Atropine poisoning mimics a variety of pathologic states such as encephalitis, meningitis, and uremia. Atropine poisoning produces consistently equal and bilateral dilation of the pupil; hot, dry skin; and significantly increased heart rates. Young children are more sensitive to atropine poisoning because of their lower tolerance to elevated body temperatures. Fatalities due to atropine-containing plant toxins are uncommon, and recovery is usually complete within 24 hours. Aside from the much more persistent papillary dilation, the other clinical effects resolve within 12 to 24 hours. Atropine-containing plants include the following:

- Jimson weed (*Datura stramonium*) contains at least three alkaloids (hyoscyanine, atropine, scopolamine) in all its plant parts. There are considerably more poisoning in humans than in animals from this plant due to the frequent and often habitual use of jimson weed seeds and flowers to induce various hallucinogenic states. Ingestion of the alkaloids induces prominent thirst, delirium, and convulsion and can lead to coma.
- Belladonna (*Atropa belladonna*) contains most prominently the atropine alkaloid. Potatoes and tomatoes (*Solanum* spp.) contain the solanine alkaloid in the leaves, in green tomatoes, and in the sprouts of potatoes. In addition to the atropine-like effects, gastrointestinal irritation is usually seen, and significant neurological involvement occurs.
- The nightshades—black nightshade, Jerusalem cherry, bull nettle (*Solanum* spp.)—also contain the solanine alkaloid in the berries and fruits. The unripe berries of this group of nightshade plants are deadly and responsible for numerous hospitalizations and near fatalities. In addition to the gastrointestinal signs, neurological effects with hot skin and pupil dilation are prominent.

This entire group of atropine-like plants induces similar signs that make their clinical recognition of important medical value.

## Convulsion-Producing Plants

The principle type of plant responsible for producing convulsions as primary toxic manifestations is the water hemlock (*Cicuta* spp.). The water hemlock is found only in wet, swampy areas. Individuals are attracted to its carrot-like appearance, but usually within 15 minutes to 1 hour after ingestion the affected individual experiences nausea, salivation, emesis, and tremors. This is quickly followed by multiple grand mal seizures, with death occurring secondarily due to prolonged anoxia encountered during the severe tonic muscular contractions.

- Water hemlock (*Cicuta maculata*) is the most common of the *Cicuta* species and the most often involved in toxicity. The plant contains an unsaturated alcohol resinoid that is rapidly absorbed, quickly enters the nervous systems, and induces acute nervous signs, violent spastic convulsions, and death from respiratory depression. This plant produces such violent convulsions that fractures of jaw bones or limbs are commonly found postmortem in individuals who have died from this poisoning.
- Members of the heath family, such as laurel (*Laurus nobilis*), rhododendron (*Rhododendron* spp.), azaleas (*Rhododendron* spp.), and pieris (*Pieris japonica*), all contain a hydrocarbon resinoid (andromedotoxin) in all their plant parts. This makes them extremely hazardous, as they produce the characteristic gastrointestinal irritation, mental and muscular depression, paralysis, lowered blood pressure, and eventual coma.
- Crocus (*Colchicum autumnale*) has alkaloids in all parts of the plant. The compound is heat stable and is excreted in milk, but slowly.

Vomiting is characteristic as an initial effect followed by nervous signs of tremors and seizures.

- White snakeroot (*Eupatorium rugosum*) contains a unique 16-carbon alcohol (tremetol) that posed a serious risk to cattle and humans in the early days of Midwest settlements in the United States. Cattle grazing pastures that had this plant typically would develop ketosis with neurological signs characterized by muscular trembling. This became a serious human problem when the milk from these animals was consumed by families depending on the nourishment of this product. The tremetol was excreted in the milk in concentrations sufficient to induce "trembles" in children and adults drinking the milk.
- Yew (*Taxus* spp.) is an attractive plant that unfortunately induces acute cardiac and gastrointestinal problems if the leaves are chewed. The alkaloid is at its highest concentration in the foliage, with minimal amounts in the attractive red berries. This reduced concentration in the berries is often a life-saving feature, as children are attracted to the colorful fruit and consumption is not uncommon. Following the typical gastrointestinal response of vomiting, individuals rapidly become weak and have cardiac rhythm changes that induce serious life risks; these are followed by convulsions that include depression of the respiratory center. Death within a few hours of consumption of this plant is typical.
- Locoweed (*Astragalus* spp.) has been used for centuries by Native American tribes during their ceremonial rituals. The plant has several compounds present, all of which produce mind-altering hallucinogenic states. While minimal amounts are used in the ceremonies, larger consumption may lead to violent behavioral changes that can be life threatening.

## lants Containing Cyanide

A number of plants and their fruits contain cyanide or yanide precursors that release true cyanide upon enzyatic digestion either in the digestive tract or under approriate environmental conditions. These cyanide comounds are usually in highest concentration in the pits or eaves of the respective plants; the cyanide release from eeds or pits requires that the seed coat be broken by hewing, thus allowing the enzymatic juices of the intesnal tract to complete the conversion to the highly toxic yanide. Classical onset is extremely acute, resulting in e rapid onset of rapid respirations, collapse due to cellar anoxia, and the development of bright-red blood and rminal seizures leading to death usually within 10 to 30 minutes of ingesting the freely available active toxin. Fortunately, a commercially available cyanide kit is available and present in every emergency room. The treatment with intravenous sodium nitrite and sodium thiosulfate is highly effective but must be initiated before cardiac function ceases. Cyanide-containing plants include the following:

- Apples (*Malus* spp.) have the cyanogenic glycoside in the seeds; cherries (*Prunus* spp.) have the glycoside in their pits and leaves; other *Prunus* species (peaches and apricots) have the cyanogenic glycoside in their pits and smaller concentrations in the leaves.
- Hydrangea (*Hydrangea macrophylla*) is typical of a toxic plant in that it has a potent cyanogenic glycoside present in combination with irritating chemical principles that aggravate the clinical toxicity.

## Plants Producing Dermal Contact Effects

Depending on the chemical principle or structure involved, adverse effects from the consumption of these plants are rare but may occur under unique conditions. In general, skin irritation is due to blistering or mechanical injury from hairs or spines that penetrate the skin. Such plants include the following:

- Poison ivy (*Toxicodendron radicans*) and poison sumac (*Toxicodendron vernix*) are well recognized as having an allergenic sap containing 3-*n*-pentadecylcatechol, which is famous for the blistering it produces. The fact that some individuals do not respond to contact from this plant while others show violent reactions attests to the allergic nature of this compound.
- Snow-on-the-mountain (*Euphoria marginata*) contains an irritating vesicant sap that is highly irritating to the conjunctiva of the eye and locally irritating to the skin and to the gastrointestinal tract mucosa if consumed.
- Nettles (*Urtica* spp.) are clearly irritating due to the prickly hairs on the surface of the plant that produce an irritant reaction upon skin contact. The hairs also provide mechanical irritations that further accentuate the irritating nature of the toxin by allowing some skin penetration.

All of these irritant-containing plants are contact toxicants that allow removal of the irritant if abundant soap and water are used within 5 to 10 minutes of contact to wash off the toxicant. The immediate use of such cleansing will also help prevent the spread of the toxicant to other areas of the body due to itching and rubbing. Medications are available that will relieve the irritation and discomfort

**TABLE 19.1**
**Poisonous Plant Glossary**

| Common Name | Scientific Name | Toxic Effect |
|---|---|---|
| Apple | *Malus* sp. | Cyanide |
| Apricot | *Prunus* sp. | Cyanide |
| Azalea | *Rhododendron* sp. | Neurological |
| Belladonna | *Atropa belladonna* | Atropine-like |
| Black locust | *Robinia pseudoacacia* | Gastrointestinal |
| Black nightshade | *Solanum nigrum* | Atropine-like |
| Bull nettle | *Solanum* sp. | Atropine-like |
| Buttercup | *Ranunculus* sp. | Gastrointestinal |
| Cactus | *Pedilanthus* sp. | Dermatological |
| Castor bean | *Ricinus communis* | Gastrointestinal, convulsive, tremors |
| Cherry | *Prunus* sp. | Cyanide |
| Crocus | *Colchicum autumnale* | Neurological |
| Daffodil | *Narcissus* sp. | Gastrointestinal |
| Daphnea | *Daphne* sp. | Gastrointestinal |
| Death camas | *Zigadenus* sp. | Gastrointestinal |
| Dumbcane | *Dieffenbachia* sp. | Oxalate |
| Elderberry | *Sambucus americana* | Gastrointestinal, cyanide |
| Fescue | *Festuca prateusis* | Gangrene, hormonal, reproductive |
| Fireweed | *Kochia scoparia* | Liver |
| Foxglove | *Digitalis purpurea* | Cardiac, digitalis |
| Greasewood | *Sarcobatus* | Oxalate |
| Halogeton | *Halogeton glomeratus* | Oxalate |
| Hydrangea | *Hydrangea macrophylla* | Cyanide |
| Ivy | *Hedera helix* | Gastrointestinal |
| Jerusalem cherry | *Solanum pseudocapiscum* | Atropine-like |
| Jimson weed | *Datura stramonium* | Atropine-like |
| Lamb's quarters | *Chenopodium album* | Oxalate |
| Lantana | *Lantana camara* | Gastrointestinal, photosensitization, neurological |
| Laurel | *Laurus nobilis* | Neurological |
| Lily-of-the-valley | *Convallaria majalis* | Cardiac, digitalis |
| Locoweed | *Astragalus* sp. | Convulsive, tremors |
| Lupine | *Lupinus argenteus* | Liver |
| Marijuana | *Cannabis sativa* | Neurological |
| Mayapple | *Podophyllum peltatum* | Gastrointestinal |
| Milkweed | *Asclepias* sp. | Gastrointestinal, neurological |
| Mistletoe | *Phoradendron serotinum* | Cardiac, digitalis |

and may be used to prevent further mechanical spread of the compound. Unfortunately, however, no neutralizing material is available for use after the compound has come into contact with the skin; thus, strong and abundant washing is the only available relief to reduce the effect of these irritating compounds.

Cactus (*Pedilanthus* spp.) has strong spines capable of penetrating skin. The penetration is worsened by any force that drives the thorn-like spines further into the flesh. Rapid removal of the spines is important to avoid subsequent granulomas from developing. These body reactions represent attempts to wall off the irritation from the spine, but they often produce ulcers and highly inflamed lesions in several parts of the body that had contact with the cactus spines.

## Treatment for Plant Ingestion

The signs of most plant ingestions are gastrointestina involvement, with neurological effects, depression, sei zures, and coma if severe intoxication occurs. A histor of consumption of these plants or proximity to plar materials is most helpful in diagnosis and is an importar phase of the problem-solving aspects of dealing wit plant poisonings. Very few direct antidotes are availabl for treatment, although the cyanide kit is an exceptio and has historically been life-saving in numerous cya nide-containing plant exposures. In general, treatmer seeks to remove as rapidly as possible the plant materia from the digestive tract by inducing emesis or hastenin a laxative effect, as well as by the use of intestinal decor

**TABLE 19.1 (cont.)**
**Poisonous Plant Glossary**

| Common Name | Scientific Name | Toxic Effect |
|---|---|---|
| Mother-in-law's tongue | *Sanseivera* sp. | Oxalate |
| Mountain laurel | *Kalmia latifolia* | Neurological |
| Mustards, crucifers | *Brassica kaber* | Gastrointestinal |
| Nettles | *Urtica* sp. | Dermatological |
| Oak | *Quercus* sp. | Kidney |
| Oleander | *Nerium oleander* | Cardiac, digitalis |
| Peach | *Prunus* sp. | Cyanide |
| Philodendron | *Philodendron* sp. | Oxalate, kidney |
| Pieris | *Pieris japonica* | Cyanide |
| Pigweed | *Ameranthus retroflexus* | Kidney |
| Poinsetta | *Euphoribia pulchenima* | Gastrointestinal |
| Poison hemlock | *Conium maculatum* | Neurological |
| Poison ivy | *Toxicodendron radicans* | Dermatological |
| Poison sumac | *Toxicodedron vernix* | Dermatological |
| Pokeweed | *Phytolacca americana* | Gastrointestinal |
| Potato | *Solanum tuberosum* | Atropine-like |
| Precatory bean | *Abrus percatorius* | Gastrointestinal |
| Ragwort | *Senecio* sp. | Cardiac, digitalis |
| Red maple | *Acer rubrum* | Hemolytic |
| Rhododendron | *Rhododendron* sp. | Neurological |
| Rhubarb | *Rheum rhaponticum* | Oxalate |
| Rosary pea seed | *Abrus precatorius* | Gastrointestinal |
| Senna | *Cassia fasciculata* | Gastrointestinal |
| Snow-on-the-mountain | *Euphorbia marginata* | Dermatological |
| Sorghum | *Sorghum* sp. | Cyanide |
| St Johnswort | *Hypericum perforatum* | Photosensitization |
| Tobacco | *Nicotiana* sp. | Nicotine-like |
| Tomato | *Lycopersicon lycopersicum* | Atropine-like |
| Veratrum | *Veratrum californicum* | Teratology |
| Water hemlock | *Cicuta maculata* | Convulsive, tremors |
| White snakeroot | *Eupatorium rugosum* | Neurological |
| Wild indigo | *Baptisia* sp. | Gastrointestinal |
| Wisteria | *Wisteria sinensis* | Gastrointestinal |
| Yew | *Taxus* sp. | Convulsive, tremors |

amination through binding agents such as activated charcoal to minimize absorption of the plant's toxins. In severe cases, where prolonged vomiting and diarrhea have been present, the use of fluids and electrolyte infusions are helpful to maintain hydration and often are life saving. General symptomatic care to relieve pain and discomfort are also employed and found valuable by the affected victims.

Table 19.1 provides the common names, scientific names, and general toxic effect of each of the toxic plants of concern discussed above. This table should be useful for identifying, among all the many plants in the environment, those that are significant hazards and the type of toxicity or adverse effects that each can produce. Note that many of those listed affect more than one organ system.

## Laboratory Identification of Plant Toxins

The classification and identification of plant compounds associated with cases of poisoning is often subjective; indeed, in many cases plant chemicals are deemed toxic by association with poisoning episodes. They are often the major compound found in a plant known to be poisonous and associated with related compounds that have been shown to produce similar symptoms of poisoning. Years of analysis of plant toxins and the more recent development of specific and sensitive instrumentation have revealed that plants contain many chemicals that are harmful, often several within the same plant. Their ultimate effect depends on consumption and environmental factors that modify and vary the individual ratios of the plant's toxic components.

Some plant constituents are not particularly toxic in the form in which they exist in the plant, but upon ingestion they undergo transformations to more toxic principles; thus, when investigating the effect of plant toxins in mammalian systems, metabolites of the toxin have often been the only detectable compounds. For some groups, such as alkaloids, considerable variability can exist within a plant species with respect to the level and composition of the toxic fractions, which depend on location, stage of growth, and other environmental conditions [5].

Analytical resources available to detect plant toxins vary significantly from laboratory to laboratory; yet, many analytical procedures are available to identify the numerous categories and wide spectrum of potential plant toxins. In addition to the classic analytical techniques used (e.g., thin-layer chromatography, high-performance liquid chromatography, gas chromatography), other methodologies are also employed. Solvent extraction techniques, supercritical fluid extraction, lead-acetate precipitation, droplet counter current chromatography, centrifugal thin-layer chromatography, and reverse-phase low-pressure chromatography columns are also employed. Even more recently, the wide application of mass spectrometry and its various combinations with chromatography methods has allowed for identification of alkaloids and other plant toxins with minimal plant sample available. Even more recent advances have suggested that capillary electrophoresis and further combinations with mass spectrometry and specific detector systems will further improve the sensitivity and detection of illusive plant toxins.

The range of specific plant toxic chemicals now detectable include the following [5]: pyrrolizidine alkaloids; piperidine alkaloids; pyridine alkaloids; indole, tryptamine, β-carboline, and related alkaloids; quinolizidine alkaloids; steroid alkaloids; diterpene alkaloids; indolizidine alkaloids; tropane alkaloids; isoquinoline alkaloids; other alkaloids, such as colchicine, sesbanimide, and dioscorine; cyanogenic glycosides; glucosinolates; phenolic glycosides; saponins and cardiac glycosides; nitropropanol glycosides; other glycosides, such as azoxyglycosides, naphthalenes, anthracenones, and anthraquinones; lectins or hemagglutinins, including abrin, jatrophin, momordin, phoratoxin, and ricin; enzymes such as thiaminase; amino acid analogs and nonprotein amino acids, such as hypoglycin, β-cyano-L-alanine, and *S*-methylcysteine sulfoxide; phenylethylamines, including tyramine, *N*-methyl-β-phenethylamine, hordenine, β-phenylethylamine, and galegine; selenium compounds usually stored as amino acids (selenocystine, methylselenocysteine, γ-L-glutamyl-Se-methylseleno-L-cysteine; sesquiterpene lactones; diterpenes, such as croton oil, ingenol and tigliane esters, simplexin, and grayanotins I, II, III, IV, and XIV; terpenes (tremetol) and hydrocarbons (tanacetine, malvalic acid, cicutoxin, and crepenynic acid); and the numerous oxalates, nitrates, sulfides, and organofluorine compounds that exist seemingly throughout the plant kingdom as soluble acid salts, as insoluble salts of calcium and magnesium, and in numerous disulfide and similar inorganic forms [5]. Such a wide spectrum of organic and inorganic chemicals challenges the analytical chemist and provides continuing excitement for toxicologists.

## Plant Toxins Present in Milk

Mammary gland secretions are one of the frequent ways in which chemicals are eliminated from a consuming individual's body [6]. Compounds from plants are particularly suited for excretion via this route, as many are organic in nature and have considerable lipid partitioning into the mammary gland secretions. Exposures may occur directly from mothers having recently consumed plant materials and suckling infants receiving the milk. All species may be affected by this route of exposure. In commercial dairy herds, animals pasturing on plants that contain compounds likely to appear in milk pose a hazard for the general public purchasing such processed milk and milk products. Fortunately, when the milk from one herd exposed to toxins excreted in milk is mixed with a large volume of milk from nonexposed cows, the resulting dilution provides considerable protection; however, the potential risk is still there. This is especially true if home-produced dairy products are consumed from individual herds in which dilution of the plant toxins is unlikely. Although excretion of natural toxins via the milk is not considered a major route of excretion, the milk emulsion of lipids in an aqueous solution of proteins may contain mixtures of virtually any toxin or compound that is in solution in the mother's or cow's body. The toxins may be bound to blood proteins, be in solution in the circulating lipids, or be freely circulating in the plasma. All of these forms can cross mammary cell membranes, generally by simple diffusion.

Some poisonous principles can be excreted very readily in milk. These principles usually have a high fat solubility and are concentrated in the lipid portion of the milk; they have long biological half-lives and are frequently present in milk at high concentrations and for long periods of time. Suckling offspring or consumers of milk from cows excreting such toxins are especially vulnerable for toxicity. White snake root (*Eupatorium rugosum*) and rayless goldenrod (*Haplopappus heterophyllus*) contain tremetol, which is probably the best example of a natural plant toxicant transferred via milk. It induces a condition in cattle called "trembles" and in humans causes a serious debilitating disease referred to as "milk sickness." Two plants notorious for having that toxin present and available for milk contamination are white snake root (*Eupatorium rugosum*) and rayless goldenrod (*Haplopappus heterophyllus*). Both are widely disseminated in the northeast and Midwest, offering ample opportunities for inclusion of the toxin in milk products.

Pyrrolizidine alkaloids have been shown to be transferred in the milk of cattle grazing pastures infested with the numerous plants containing these natural toxins. The important species of toxic plants containing pyrrolizidine alkaloids are *Senecio*, *Crotalaria*, *Heliotropium*, *Trichodesma*, *Amsinckia*, and *Echium*. Of these, the *Senecio* species are the most potentially hazardous, particularly tansy ragwort (*Senecio jacobaea*) and threadleaf groundsel (*Senecio douglasii*). In addition to their contamination of milk, plant toxins from this group also present potent hepatotoxic risks.

Glucosinolates occur in the Cruciferae family and also may appear in milk. The genus *Brassica* includes cabbage, broccoli, kale, rape, mustard, turnips, and other plants. Other glucosinolate-containing plants include meadowfoam (*Limnanthes*), watercress (*Nasturtium*), radish (*Raphanus*), horseradish (*Amoracia*), and stinkweed (*Thlaspi*). Piperidine alkaloids are widely spread and are found in poison hemlock (*Conium maculatum*) and in many plants of the genera *Nicotiana*, *Conium*, *Lobelia*, *Pinus*, *Duboisia*, *Sedum*, *Withania*, *Carica*, *Hydrangea*, *Dichroa*, *Cassia*, *Prosopis*, *Genista*, *Amondendron*, *Lupinus*, *Liparia*, and *Collidium*. Lupines (*Lupinus* spp.) contain quinolizidine. Other members of this group include scotch broom (*Cytisus*), golden chain (*Laburnum*), and mountain thermopsis (*Thermopsis*).

Locoweed (*Astragalus lentiginosus*), used in ritualistic ceremonies, contains the indolizidine alkaloid swainsonine, which is transferred in milk. *Oxytropis serecia* is another swainsonine-containing plant in the locoweed category. Grasses of the lush green pastures and other plants that have high chlorophyll content are metabolized to indole and 3-methylindole in the rumens of cattle consuming these forages. These metabolites of L-tryptophan are also transferred in milk to the individuals consuming it. Colchicine is a poisonous alkaloid in autumn crocus (*Colchicum autumnale*) that is stable after drying, heating, or storage; it is also excreted primarily in lactating animals via the milk. Because of its slow excretion, a cumulative effect is likely from even small doses received daily over time.

Numerous other plants, particularly those on the western ranges of the United States, are capable of accumulating high levels of selenocompounds that not only are hazardous to consuming animals but are also are excreted in cow's milk at concentrations in direct proportion to the selenium intake. Certain plant species of *Astragalus* and *Stanleya* have this selenium-accumulating capability.

The presence of other compounds in milk may cause a disagreeable taste or odor. Plants of the genus *Allium* include species of onion and garlic, which can taint the milk. The sesquiterpene lactones (*Tenulen*) in sneezeweed and bitterweed impart a bitter flavor to the milk of lactating animals grazing these plants. Bracken fern (*Pteridium aquilinum*) produces milk from cows fed this plant that induces urinary carcinomas.

With increasing concern for food safety, the recognition and appreciation of the potential for plant toxins to contaminate human milk supplies are timely. The fact that plant alkaloids are frequent in these situations and are quite basic allows them to accumulate in milk. The combination of the chemical's basicity and fat solubility allows accumulation of the plant toxins in milk and reduces excretion by other natural processes. Factors to be considered in evaluating the human health risk from natural plant toxins with the heavy consumption of milk by infants and young children, who are already more susceptible to plant toxins, include consumption of milk produced by lactating mothers who use herbal remedies with potential toxicity, the availability of toxic plants to lactating dairy animals, and the consumption of milk from point sources where free-grazing milking animals have access to toxic plants. The fact that few laboratory methods have been developed for monitoring natural toxicants of plant origin in milk and milk products heightens the need to assess and monitor their potential danger to human health and safety.

## Some Interesting Plant Toxin Observations

Plants are capable of adversely affecting many specific organ system targets. One of the most common is skin, resulting in dermatitis as a result of hypersensitivity to plant toxins or direct contact with irritant saps or spines. Mucous membrane irritation is a close second due to contact or consumption of plants with irritant saps or those containing oxalate crystals or enzymes. Gastrointestinal reactions are by far the most common adverse effect of ingesting a toxic plant part. Indeed, it is almost a rarity to observe a true plant poisoning without some form of gastrointestinal irritation, which may be severe and may require aggressive fluid therapy. Several plant groups and individual plant species are capable of causing liver damage; although it usually results from the long-term use of plants for various purposes, some plant products produce a relatively acute hepatotoxicity. Renal impairment after plant ingestion is rare and is usually confined to associations with dehydration or profound shock. Some plants, such as rhubarb leaves, are noted to cause kidney failure due to the systemic absorption of soluble oxalate salts. Involvement of the hematopoietic system is not uncommon and often involves intravascular hemolysis, methemoglobinemia formation, and bleeding tendencies caused by coumarins in herbal teas or from penny royal oil. Parts of many plants, including pears, peaches, apples, apricots, cherries, certain lima beans, and hydrangea contain the amygdalin glycoside, which can release cyanide. These can cause acute episodes that in some cases have been related to ingestion of health-food supplements.

Common plants that contain cardiac glycosides, which can be responsible for effects and in some cases deaths, include fox glove (*Digitalis purpurea*), oleander (*Oleander nurium*), lily-of-the-valley (*Convallaria majalis*), or red squill (*Urginea maritima*). Other less-studied plants are reported to depress the heart and slow the pulse. These effects may be secondary to other organ effects. The ergot alkaloids used for migraine headache control and other purposes may cause severe arterial vasospasm. The autonomic nervous system effects are well described for anticholinergic toxicity from plant products such as jimson weed (*Datura* spp.) along with members of the Solanaceae family such as deadly nightshade (*Atropa belladonna*) and henbane (*Hyoscyamus niger*). The nicotine present in various members of the *Nicotiana* plant family causes peripheral ganglionic stimulation with seizures and death possible.

Central nervous system changes, including depression and stimulation, may also be frequent plant effects. Many plants, such as tobacco, coffee, tea, marijuana, and cocaine, are actively harvested to achieve such central nervous system alterations. More severely neurotoxic plants and their derivatives are water hemlock (*Cicuta maculata*) and poison hemlock (*Conium maculatum*). The former is well recognized to cause immediate seizures after ingestion, and poison hemlock causes death through paralysis [7].

Despite the widely recognized concern regarding problems arising from plant ingestions, some studies have concluded that most plant and herb exposures are mild and do not require aggressive treatment, despite the fact that complications have been detected in these studies that resulted especially from intentional or chronic cases of plant ingestion [8].

An example of the circumstances and resulting acute toxicity due to the ingestion of water hemlock (*Cicuta birosa*) is a report of two individuals who viewed the plants growing in a marshy area and believed them to be an edible species, such as wild parsnips, artichokes, celery, sweet potatoes, or sweet anise. They ate several bites of the root and 20 minutes later became nauseous, experienced salivation and stomach cramps, and vomited repeatedly before developing grand mal seizures and convulsions leading to coma. Prompt attention at an emergency center resulted in recovery by the 11th day [9].

The toxicity and significant allergenic and anaphylactic response to castor beans (*Ricinus communis*) and their dust present similar risks to exposed individuals. The acute gastroenteritis, gastrointestinal bleeding, hemolysis, hypoglycemia, and fluid and electrolyte depletion are common effects from exposure to these plant materials. Mortality rates of almost 10% have been reported [10].

Acute atropine poisoning from the ingestion of jimson weed (*Datura* spp.) is a significant risk to children who might ingest it. An average of almost 80 admissions per year occur in hospitals in New Zealand due to poisonings from these plants related to their belladonna, hyoscine, and atropine content. Although clinical poisoning results, fatalities are rare but do occur [11].

Juniper tar (cade oil) is distilled from the branches and wood from *Juniperus oxycedrus* containing etheric oils, triterpene, and phenols. It is a frequent constituent of many folk medicines and is used for various antipruritic, keratolytic, and antimicrobial benefits in human and veterinary dermatology. Significant antiinflammatory activity is also noted. The tar is ingested as a liquid extract of the tree by individuals in the Mediterranean region; yet, it has caused fever, severe hypotension, renal failure, hepatotoxicity, and cutaneous burns when applied to the face. A patient who received supportive and symptomatic care improved but required 11 days to return to good health [12].

Similarly, the inclusion of plant materials in a Chinese vegetable entrée at an international buffet was associated with multiple cases of burning and facial edema after lunch in an office cafeteria. Pain and stinging or burning of the oral mucosa and potential airway obstruction due to the severe edema resulted from the presence of plant material containing raphides. Plants of the genera *Dieffenbachia*, *Philodendron*, and *Brassica* contain needle-shaped crystals of calcium oxalate in their specialized cells. These raphides are excreted in response to mechanical pressure and cause irritation and subsequent injury that may be life threatening if the respiratory passageways become swollen [13].

Not all such toxicities result from direct plant ingestion. The rhododendron (*Rhododendron ponticum*) grows extensively on the eastern flat sea area of Turkey. This plant species contain toxic diterpenes (*Grayanotoxins*) in their nectar which is used by bees to generate honey from which commercial products are developed. Ingestion of honey derived from this plant (known locally as "mad honey") causes profound hypotension and bradycardia [14]. Although predominantly a problem in Turkey and the Mediterranean-bordering countries, honey intoxications from grayanotoxin contamination were reported as early as 1794 in the Great Smoky Mountain region of the United States and northward, probably associated with mountain laurel (*Kalmia latifolia*) and sheep laurel (*Kalmia augustifolia*) pollination. What may seem to be a regional concern is actually a global safety issue, and the availability of imported honey further suggests the need for continuing vigilance [15].

The oleanders—pink oleander (*Nerium oleander*) and yellow oleander (*Thebetia peruviana*)—contain cardenolides that exert positive ionotropic effects on the hearts of consuming animals and humans. Toxic exposures by humans, domestic animals, and wildlife from these oleander toxins occur with regularity throughout the geographi region where these plants grow. Although the human mortality associated with oleander ingestion is generally low due to the purgative action that occurs rapidly following ingestion, small children and domestic livestock consuming reasonable amounts of the plant are at increased risk of oleander poisoning [16]. The wide distribution and fondness of individuals to decorate their surroundings with this attractive yet toxic plant provide an interesting example of why the "one medicine" concept even applies to plant toxicities.

## TOXINS FROM ALGAE

Algae produce toxins; their presence in water should alert one to potential hazards. Toxic blue–green algae (cyanobacteria) are commonly found growing in fresh and salt water in temperate areas worldwide. They are commonly found in freshwater lakes, livestock ponds, rivers, streams, canals, and ditches. Under the appro-ɔriate warm environmental temperatures, calm weather, ınd stagnate nutritionally rich conditions, very rapid growth of blue–green algae results in blooms that accumulate on the surface of the water and provide oppor-unities for neurotoxins or hepatic toxins to be pro-luced. These blooms of blue–green algae often occur n the northern United States in the late summer or early vinter and are associated with hot, calm weather; lecreased rainfall; and increased nutrients in the water rom fertilization or animal wastes. Winds that blow in concentrate the algae along the shorelines and provide ample opportunity for exposure. In the southern states, ılgal blooms occur all year long when the optimal environmental conditions are met. The blue–green ılgae *Microcystis* and *Nodularia* produce the cyclic peptide hepatotoxins microcystin and nodularin, and he algae *Anabaena*, *Aphanizomenon*, and *Oscillatoria* produce the potent neurotoxins anatoxin-a and ana-ɔxin-a(s) [17].

The neurotoxin anatoxin-a is a bicyclical secondary mine that causes deep polarization of nicotinic mem-ranes and acts as a potent postsynaptic neuromuscular locking agent. The deep polarization of neurononico-inic membranes is rapid and persistent and leads to espiratory paralysis. The related neurotoxin ana-ɔxin-a(s) inhibits acetylcholinesterase in the peripheral ervous system. Fortunately, this toxin does not seem ɔ cross the blood–brain barrier, but the neurological ffects on skeletal muscle dysfunction are obvious. Cyanobacteria ingested with water are rapidly broken own in the gastrointestinal tract environment. In the tomach acids, the organisms are lysed and release tox-ıs that, when free, are rapidly absorbed further from ıe small intestine. The hepatotoxins microcystin and odularin are transported to the liver hepatocytes ırough a specific carrier-mediated uptake mechanism. There they inhibit protein phosphatasis which initiates a cascade of events leading to rapid massive hepatocyte necrosis, intrahepatic hemorrhage, and shock.

### THE EFFECTS OF ALGAE TOXINS

When the blue–green algae grow in polluted, nutrient-enriched waters or ponds, they can appear and disappear in a matter of days with varying temperatures and weather conditions. The rapid development of various toxins depends on the alga growth present, and the clinical effects may similarly be varied depending on the toxins present. Generally, all the toxic effects may be grouped into those that relate to nervous effects, producing relatively rapid death, or those that cause slower clinical damage requiring several days to weeks to resolve. The underlying feature of the latter group is that liver damage is the target effect, and if the victim survives the early dramatic hepatic injury or if small to moderate exposures occur over a relatively few days, the individual may develop subtle or subclinical hepatotoxicosis that may affect the victim's health for several weeks.

### NEUROTOXINS

The two primary groups of alganeurotoxins are the anatoxins produced by filamentous *Anabaena flosaquae* and the aphantoxins from *Aphanizomenon flosaquae*. The anatoxins are potent postsynaptic, depolarizing, neuromuscular blocking agents that affect the acetylcholine receptors and produce staggering, muscle fasciculations, gasping for breath, convulsions, and even opisthotonus. Death is by respiratory arrest, which may occur within minutes or a few hours, depending on the dosage and the species of animal involved. Victims need only ingest a few milliliters of the toxic, surface-bloom-contaminated water to receive a lethal dose. Some victims also demonstrate salivation, laceration, urinary incontinence, and defecation prior to death by respiratory failure. The aphantoxins neurotoxins inhibit nerve conduction by blocking sodium channels, yet induce clinical damage very similar to that of the anatoxins.

### HEPATOTOXINS

Microcystin is the fast death factor produced by *Microcystis aeruginosa*. Identified as a cyclic heptapeptide, this cyclic peptide structure is transferred to its morphologic appearance in waters where it characteristically shows up as small packets of glistening spheres under the microscope. Its unique microscopic appearance has been valuable for identification by public health officials and biologists in pollution situations. Another low-molecular-weight heptapeptide hepatotoxin with similar microscopic appearances is cyanoginosin (BE-4 toxin), a product of a unique strain of *Microcystis aeruginosa* WR-70 [18]. Characteristic liver injury is produced either acutely or more subtly with limited dose exposure. Other hepatotoxic toxins are generated by *Nodularia spumigena*, *Cylindrospermopsis raciborskii* (*Anabaena raciborskii*), and varieties of *Oscillatoria agardhii*.

All the microcystin hepatotoxins cause death by hypobolemic shock resulting from interstitial hemorrhage into the liver. The rapid and extensive centrilobular necrosis of the liver is a result of extensive fragmentation and vesiculation of hepatocyte cell membranes. It has been suggested that, instead of destroying the hepatocyte cell membrane directly, the hepatoxins affect a cytoskeletal component, which, depending on dosage exposure, may result in a sudden loss of total organ function or more prolonged effective damage [18].

## Other Algatoxins

Scytophycins produced by certain strains of *Scytonema pseudohofmanni* are strongly cytotoxic in cell cultures. *S. hofmanni* produces cyanobacterin that has strong anti-cyanobacterial activity and is considered to be an algaecide for cyanobacteria. A cytotoxic alkaloid has been isolated from *Hapalosiphon fontinalis* strain V-3-1 that has broad antialga and antimycotic activity. All of these toxins have relatively low lethal toxicity in laboratory studies, but, when combined with other cyanobacteria observed in almost all freshwater alga-poisoning reports, potent neurotoxic alkaloids, hepatotoxic peptides, and these several less lethal but more selective cytotoxic compounds demonstrate a unique combination of hazardous effects [18].

The individual cyanobacteria with their associated toxins may produce a spectrum of toxicity determined by the specific population of algae present and the concentration of their respective toxins. Acute death may occur with few clinical signs. Other victims may present with muscle tremors, rigidity, lethargy, convulsions, and death from respiratory paralysis within minutes to hours following the onset of effects. Another cascade of signs consistent with inhibition of cholinesterase includes salivation, urination, laceration, and defecation, together with tremors, convulsions, and respiratory difficulty; death from respiratory arrest may occur again within a few hours. Finally, a myriad of signs generally related to liver damage may present a cascade of depression, vomiting, diarrhea, gastrointestinal slow down, weakness, and anemia. Although death often occurs with this syndrome within several hours, it may be delayed for several days. Individuals who survive the initial toxicosis may develop secondary photosensitization due to the compromised liver function and presence of photodynamic sensitive metabolites being retained.

## Public Health Significance

The potential for mass populations of toxin-producing cyanobacteria in natural and controlled water bodies requires public health awareness to avoid such entities or, if circumstances permit their development, to respond appropriately with clinical management, chemical detection, and control of human and animal exposure. Toxic cyanobacterial populations have been reported in fresh waters in more than 45 countries and in numerous brackish, coastal, and marine environments. The variations in biology and circumstances that are characteristic of hazards of natural origin clearly suggest that uncertainties and numerous gaps in knowledge are present; however, the importance of identifying the exposure media of potable and recreational waters, as well as animal and plant foods that may contribute to the human population exposure, is paramount. Steps to develop and implement risk management strategies for cyanobacterial toxins in water bodies are recalled with each outbreak of toxicity, and the gradual recognition by public health communities of the collective experience and wisdom is slowly building [19].

Paramount to dealing with outbreaks of alga intoxication is detection of the cyanobacterial agents present. Their wide variety and structure have led to the common practice of reporting the presence of "toxins" as microcystin-LR equivalents, regardless of which hepatotoxic variant is present [20]. The increased availability of high-performance liquid chromatography with online mass spectral analysis has facilitated more accurate detection of toxic variance, but because several microcystins share the same molecular mass definitive identification may be difficult. A further difficulty is the requirement for sample processing before analysis. Recent new technologies employing recombinant antibodies and molecularly imprinted polymers are now being used to develop assays and biosensors for these algatoxins that are highly sensitive, do not require sample processing, and offer simple and less expensive alternatives to existing analytical techniques [20].

An additional hurdle is identifying some of the diverse groups of cyanobacteria capable of producing a wide range of toxic secondary metabolites that have been classified as neurotoxins, hepatotoxins, cytotoxins, dermatotoxins, and irritant toxins. Cyanobacterial blooms are particularly hazardous due to this production of secondary metabolites and their endotoxins, which may be broadly toxic to humans, animals, and plants. These compounds differ in mechanisms of uptake, affected organs, and molecular mode of action. Investigating and understanding the effects of these toxins on the organisms receiving them, as well as their basic toxic mechanisms, add to the challenges raised by the presence of the cyanobacteria organisms in water supplies [21].

### Prevention and Control

The wide extremes in adverse health effects resulting from cyanobacteria exposure in humans highlight the importance of public health measures to prevent and control the often unexpected development of toxicity scenarios. Measures to prevent and control future episodes induced by these organisms are extremely important due to the neurotoxic effects of anatoxins that mimic the effects of organophosphate insecticides; the dramatic hepatic and digestive tract injuries caused by microcystins that lead to acute death or lingering hepatic failure; the diverse health-ravaging protein synthesis inhibition of cytotoxins; the skin and mucous membrane irritation and tissue inflammation caused by irritant algatoxins; and the confusing effect of cyanobacteria lipopolysaccharides in inducing a toxic-shock-like syndrome.

To halt nutrification of ponds and lakes in natural wildernesses and agricultural and industrial areas is difficult, and merely placing cautionary notices and signs to prevent human exposure to blooming waters is inadequate. Education programs to alert rural inhabitants and campers seeking to enjoy the pleasures of nature are a start but still do not address the problem of human oversight. Continued observation and recognition of circumstances and environmental conditions that spawn the rapid development of the cyanobacteria toxins are currently the most commonly used preventive measure by public health and regulatory agencies.

Individuals who have become intoxicated from organisms through the public water supply or through contaminated waters used in preparing various commercial or medicinal formulations are particularly vulnerable. Recently, three human probiotics (*Lactobacillus rhamnosus* strains GG, *Lactobacillus rhamnosus* strain LC-705 and *Bifidobacterium lactis* strain Bb12) were found to bind the microcystin cyanobacteria peptide found in water solutions. As much as 46% removal occurred and offered the possibility of utilizing incubation of sensitive water materials with these probiotics to markedly reduce potential toxicity [22].

The increasing recognition of cyanobacterial toxins in drinking water from water treatment plants [23] and the dramatic documentation of a variety of cyanobacteria and microcystin concentrations in public water supply reservoirs [24] send up red flags regarding the importance of intensifying control measures currently in place to prevent and avoid contamination of these human water sources. The finding of six toxic genera of cyanobacteria with a more than 40% frequency of hepatotoxic strains provides sufficient evidence that the development and application of detection and control measures for avoiding widespread consumption of these contaminated waters are necessary. The wide distribution of water supplies from contaminated reservoirs offers good justification for resources directed to avoiding such catastrophes [24].

## TOXINS FROM MUSHROOMS AND TOADSTOOLS

The rapid and reasonably accurate identification of poisonous plants is difficult, but identifying toxic mushrooms through merely visible examination is virtually impossible. Untrue are the old wives' tales suggesting that if the skin of a mushroom is easily removed over the cap or if the mushroom will not blacken when cooked or if the mushroom's gills are gray then the mushrooms are nontoxic. Ultimately, there are no hard-and-fast rules to distinguish delicacy from deadly.

Mushrooms are the fruits of fungi that are made up of multiple threads called hyphae. Mushrooms grow on several different types of materials, such as wood, dirt, plant or animal material, or dung. The spores of mushrooms are a valuable identification aide. Spores are used to form "prints" as part of the identification process. Many different shapes have been catalogued for the various parts of mushrooms, and when put together they further assist in identification. Putting all these items together requires experience, knowledge, and skill—all of which are necessary for the proper identification of genus and species and the determination of whether the fungi is safe or toxic.

Although *Aminita* (e.g., *A. muscaria*, *A. phalloides*) is the most toxic genus of mushrooms, the specific toxic principle varies significantly with the specific species of mushroom collected and ultimately consumed. These toxic principles produce clinical signs that include excitement and hallucinations, severe gastrointestinal involvement with few fatalities, jaundice and explosive liver damage, circulatory failure, or coma and death within hours to days. Clinicians recognize the importance of not relaxing treatment measures until the patient is fully recovered from any one of the varied and complex clinical syndromes produced by ingestion of toxic mushrooms.

### Clinical Syndromes

The mushroom so frequently pictured in storybooks with the red cap covered in white polka dots is *Amanita muscaria*. Mushrooms in this group produce maniac excitement and hallucinations within 2 to 3 hours after ingestion. The more excitable phase is usually preceded by an initial period of drowsiness. The duration of central nervous system excitement usually does not exceed more than 3 to 4 hours. A specific group of muscarine-containing mushrooms will produce sweating, salivation, and colic that begins within 15 minutes of ingestion. Pulmonary edema and moist rales are frequently present. Fortunately, this toxin does not cross the blood–brain barrier in sufficient concentration to produce central nervous system effects.

Another subgroup of toxic mushrooms produces severe gastroenteritis that develops with a rapid onset of action. Whereas wild mushrooms eaten by adults often invariably have been cooked—a process that markedly reduces or inactivates several irritants and even some major toxins—this very often is not the case for children, which may account for some of the pediatric fatalities following ingestion of uncooked "nontoxic" species.

The most serious mushroom poisoning is that in which there is a latent period of about 12 hours prior to the onset of symptoms. These intoxications have a grave prognosis and consume considerable therapeutic resources. The mushrooms involved in this syndrome belong to the *Amanita phalloides* group. They contain two classes of thermostable cyclic polypeptides, the phallotoxins and amatoxins. Each of these has distinct pharmacological properties. The phallotoxins act initially at the end of a brief latent period by producing nausea, vomiting, painful colic, and a watery, profuse, and sometimes bloody diarrhea. Following an apparent recovery and a symptom-free period of 3 to 5 days, during which symptoms are relieved, the amatoxins exert their hepatorenal toxicity. Jaundice develops, abdominal pain and weight loss become obvious, and a severe extension of the initial poisoning occurs. If only a small amount has been ingested, only renal tubular necrosis may occur. Large mushroom ingestions often produce fatal hepatic necrosis even before the kidney damage becomes obvious. The hepatorenal effects terminating in liver failure that are produced by intermediate dosages of mushrooms are the most commonly accounted phenomena. An especially high mortality is associated with mushroom intoxication in children.

## Specific Mushroom Toxins

What are the various chemicals specifically toxic in mushrooms? There are various categories and types of mushroom toxins, most of which have a characteristic set of signs. Animals as well as humans are poisoned by mushrooms, but the safe ingestion of a fungal species by an animal does not make it safe for human consumption.

### Cyclopeptides

*Amanita* and *Galerina* species of mushrooms are most commonly involved in producing anatoxins [25]. These toxins inhibit mammalian nuclear RNA polymerase B forms, which are involved in the transcription of DNA to messenger RNA, thus causing inhibition of protein synthesis at the ribosome level. Their primary target organs are the gastrointestinal tract, kidney, and liver. After ingestion of these toxins, a latent period occurs during which little happens for several hours, but then vomiting, diarrhea, and intense abdominal pain occur for several hours. These then subside, and the patient may appear almost normal for a few days. The toxins gradually produce hepatic and kidney failure, which may occur within a few days if the dose received is high enough. Hepatic liver enzymes are elevated; the patient is jaundiced and anuric. Tachycardia, acidosis, hypotension, dehydration, electrolyte and blood coagulation abnormalities, hypoglycemia, hepatic coma, and shock usually occur. If *Amanita smithiana* is the ingested mushroom, primary renal tubulonecrosis may occur with minimal hepatic signs.

Treatment of cyclopeptide toxic ingestion is one of the few mushroom poisonings in which extensive therapies have been tried. Silymarin, derived from the milk thistle *Silybum marianum*, has offered some hope as a free-radical scavenger and inhibitor of the penetration by amatoxins into the liver cells. Therapy is required for 4 to 5 days at 20 to 50 mg/kg of body weight per day intravenously or 1.4 to 4.2 g/day orally. Often silymarin is used together with penicillin G. Penicillin G at 300,000 to 40,000,000 units per day seems to be of some benefit in treating this mushroom intoxication if given within a few hours of ingestion. The antibiotic seems to reduce the hepatic uptake of the toxin, but unfortunately such early therapy following mushroom ingestion is often not possible because of the several-hour latent period following ingestion. Thioctic acid has been recommended in the past, but it is difficult to obtain and has had mixed success.

Fairly classical additional treatments have met with moderate success in mushroom toxicities from cyclopeptides. Decontamination of the gastrointestinal tract within a few hours of ingestion using emesis or gastric lavage along with activated charcoal administration is classic. Multiple doses of charcoal given several hours apart are considered to be useful due to enterohepatic circulation of the toxin. Crucial is the supportive care required to maintain appropriate organ functioning and to avoid renal failure, hepatic coma, blood coagulation problems, acid-base abnormalities, decreased blood glucose, and dehydration. Because of the prominent liver damage, liver transplants have been required in humans suffering from this specific mushroom intoxication.

### Orelline and Orellanine

Species of *Cortinarius* mushrooms contain the orrellanine neurotoxin. Because clinical signs may not appear for several days following ingestion, diagnosis is a problem, although with larger doses the clinical signs appear earlier. Anorexia, gastrointestinal disturbances, oliguria, and renal failure are the usual sequence of events. Although no specific antidote exists, decontamination methods are useful if ingestion is recognized within the first few hours. Affected kidneys generally recover slowly if the mushroom dose is insufficient to produce death. Dialysis is very useful during the period of renal failure.

### Muscimol and Ibotenic Acid

Dramatic and mind-boggling clinical effects occur within a few hours after ingestion of *Amanita muscaria*, *Amanita pantherina*, *Amanita strobiliformus*, or *Tricholoma muscarium* mushrooms. The effects begin with drowsiness followed by giddiness, maniac behavior, hallucinations, and derangement of senses, as well as signs that resemble those seen in alcohol intoxication. These neurologic effects are complicated by nausea, vomiting, salivation, and diarrhea, which may further complicate the clinical picture. The patient becomes confused, disoriented, and delirious and may experience visual hallucinations, muscle spasms and twitching, and occasional convulsions. The periods of drowsiness and agitation often alternate, and enthusiastic, exaggerated physical activity may be seen. The effects usually peak within 3 hours after ingestion, and mortality is minimal, ranging only up to only 12%.

These variable effects suggest that toxin concentrations in the mushrooms are quite variable, which has been confirmed by chemical analyses. Muscimol and ibotenic are derivatives of gamma-aminobutyric acid and exert their effects on the central nervous system. They both appear to cross the blood–brain barrier, probably by means of an active transport system. Although clinical effects are usually still present more than 5 hours after peak excretion in urine, existing blood levels do not correlate with the clinical effects on the central nervous system. With no specific antidote available, care is generally symptomatic and supportive to avoid self-inflicted injuries due to abnormal behavior. The extreme drowsiness phase frequently requires observation to avoid lack of attention to dangerous behavior.

### Monomethylhydrazine

Several mushroom groups carry this toxin. *Gyromitra gigas*, *Gyromitra esculenta*, *Helvella elastica*, *Verpa bohemica*, *Piziza badia*, and *Gyromitra infula* contain a monomethylhydrazine precursor that becomes toxic when converted to a metabolite. The clinical signs (mostly gastrointestinal) occur several hours after the mushrooms are ingested but may not show up for 2 or more days. Once developed, the clinical syndrome is sudden, characterized by a bloated sensation followed by vomiting and prominent watery diarrhea. Abdominal cramping, abdominal pain, headaches, and general depression follow and may last for about 2 days. Some individuals may develop jaundice, methemoglobinemia, and hemolysis. In the later stages, the clinical effects may include dizziness, loss of muscle coordination, seizures, and coma. The prominent gastrointestinal effects produced require that affected patients be monitored for dehydration, electrolyte imbalances, red blood cell destruction, and methemoglobinemia. If ingestion is recognized prior to clinical signs developing, gastric decontamination is highly recommended. Although no directive antidote is available, the range of clinical effects (e.g., central nervous system signs, red blood cell alterations, blood glucose depression) and the mushrooms' effects on several organ systems suggest that hepatic and renal function should be monitored for several days.

### Muscarine

Prominent cholinergic effects result from consumption of *Clitocybe dealbata*, *Clitocybe truncicola*, *Inocybe lacera*, *Inocybe pudica*, and *Entoloma rhodopoilum* mushroom exposure. These effects occur within 30 minutes to 2 hours after ingestion, although when high concentrations of muscarine are present in the consumed mushrooms signs may appear in as little as 15 to 20 minutes. Affected individuals experience blurred vision; excessive perspiration, salivation, and lacrimation; decreased heart rates; lowered blood pressure; increased peristalsis; mild hypotension; urinary urgency; nasal discharge; pulmonary congestion; and difficult respirations. The excessive perspiration, salivation, and lacrimation are generally only seen with this type of mushroom poisoning. Patients may also exhibit flushing of the skin, vomiting, abdominal pain, and watery diarrhea. The toxin is readily absorbed and binds effectively to acetylcholine receptors, initiating what are well-recognized acetylcholine effects at muscarinic sites: the cells of glands, cardiac muscles, and smooth muscle. The effects are peripheral, as muscarine has an ionic character that prevents it from crossing the blood–brain barrier. Emesis should always be initiated if ingestion is observed. Activated charcoal may be of some value, but the threat of life-threatening cholinergic signs requires that atropine be administered. The atropine dose should be repeated until atropinization is present, as the endpoint for appropriate reversal is the cessation of secretions such as salivation.

### Psilocybin and Psilocin

*Psilocybe*, *Panaeolus*, *Gymnopilus*, *Conocybe*, and some *Stropharia* species of mushrooms contain these toxins. Effects are initiated within minutes after ingestion but may be delayed for a few hours. The effects last up to 4 hours, with peak activity occurring about an hour after consumption. The psychoactive effects from these compounds may last up to 15 hours, with other nonpsychological effects such as mydriasis, hypertension, and drowsiness persisting even longer. These potent neuroactive toxins alternate mood swings from euphoric to apprehensive, and visual hallucinations, sweating, yawning, flushing, tremulous speech, incoordination, paresthesias, intensified hearing, and unmotivated compulsive body movements typically occur. Nausea and vomiting are present in only about 20% of ingestions. Up to 20 mushroom caps are required to produce the psychological effects, including flashbacks. Psilocybin and psilocin are four-substituted tryptamine derivatives that are active when they reach the brain. The effects of psilocin and LSD on brain serotonin are similar. Both psilocin and psilocybin have serotonergic interactions in the peripheral and central nervous system. Most cases are effectively managed when the victim is placed in a low-stimulus environment and appropriate supportive care is provided. If panic or hyper reactions develop, diazepam therapy is beneficial. Phenothiazine tranquilizers are counter indicated.

### Irritants of the Gastrointestinal Tract

Although few fatalities have been reported from mushrooms that produce gastrointestinal irritation, the nausea, vomiting, abdominal pain, and diarrhea observed soon after ingestion are dramatic effects. Additionally, weakness, dizziness, paresthesias, sweating, and varying degrees of headaches have also been associated with the gastrointestinal syndrome. *Agaricus xanthodermus*, *Amanita volvata*, *Boleus satanus*, *Chlorophyllum molybdites*, *Entoloma lividum*, *Gomphus floccosus*, *Lamptero-myces japonica*, *Rhodophyllus rhodopolius*, and *Rhodophyllus sinatus* are the various mushroom varieties associated with gastrointestinal disturbances. In total, the onset of clinical signs may be variable, being as rapid as within 15 minutes of ingestion or delayed for a few hours. The actual chemicals involved in this digestive syndrome are diverse, and it is unknown what specific contributions each of these compounds makes to the gastroenteritis. Fortunately, these effects usually resolve spontaneously within a few hours although some may last 24 hours or more. There are no serious after-effects, and general symptomatic care is adequate to ensure recovery.

### Hypersensitivity Reactions

Various types of hypersensitizations may occur after exposure to mushrooms. These include gastrointestinal, dermatological, and respiratory effects that may be associated with the actual mushroom, smuts, rusks, sly moles, or dry rotten fungi. Species of *Agaricus* produce an upset stomach for some individuals. Even *Agaricus bisporus*, the mushroom most commonly found in U.S. grocery stores, has produced temporary gastrointestinal upset in some people who are particularly sensitive to this mushroom. Respiratory and dermal reactions to fungi and fungal spores are fairly common. Because these are usually short-term effects, no specific treatment except prevention by avoiding future exposure is applied. In rare cases where clinical signs are particularly significant, brief corticosteroid administration may be applied. In instances of liver failure, recent developments in liver transplantation have contributed to restoring health in liver-intoxicated mushroom victims.

#### Treatments

The diverse nature of these mushrooms often precludes the rapid application of therapy. No specific antidotes are available, but if treatment is sought within a few hours after ingestion the use of emetics to empty the stomach followed by the administration of activated charcoal is a common practice. The use of atropine is generally avoided. Additional treatments are symptomatic to relieve the uncomfortable effects of the mushroom ingestions and to support the patient during the recovery or healing process. If neurological effects are present, appropriate seizure or nervous control is applied, and diuretic fluid administration and appropriate electrolyte infusions are used to assist with excretion of the toxins and correction of dehydration and electrolyte imbalances. Careful patient monitoring and awareness of potential liver or kidney after-effects are necessary steps to ensure satisfactory prognosis and recovery.

An exceptional effort has been made to identify useful therapy for the ingestion of the cyclopeptide antitoxins commonly found in *Amanita* and *Galerina* species. The application of silymarin, derived from the milk thistle *Silybum marianum*, has been periodically favored as a specific treatment for these cyclopeptides intoxications. The flavonoligans in silymarin have been found to be protective in animals; they are thought to be free-radical scavengers and inhibitors of amnitoxin penetration into liver cells. When used for 4 to 5 days at 20 to 50 mg/kg of body weight per day intravenously or 1.4 to 4.2 g/day orally, often in combination of penicillin G, this combination comes as close as possible to being a specific antidotal regiment for mushroom toxicity.

A common-sense guideline is "Don't pick mushrooms yourself. If you are unsure of a mushroom's toxicity, then don't eat it!" Paramount in the basic toxicology syndrome is the recognition that dose is still a factor. A little taste might satisfy the yearning, while an abundant appetite may lead to serious consequences.

## ANIMAL TOXINS

Natural toxins are represented by an impressively wide spectrum of compounds in numerous botanical species; however, nature's evolutionary chemistry is equally impressive in its complex synthesis of toxins and their diversity and distribution in the world's animal phyla and taxa. From the oceans and continental land masses of all geographic regions and climates toxins have been discovered in an array of aquatic and terrestrial animals. The prominent animal kingdom phyla associated with toxins are the Cnidarins, Platyhelminthes, Echinodermata, Mollusca, Annelida, Arthropoda, and Chordata [26].

## What Is an Animal Toxin?

An animal toxin can be a protein, enzymatic protein, or a small peptide that is homogeneous in nature and is produced by living cells or organisms. When a toxin is introduced into the body tissues of another living organism an adverse physiological event results from the inhibition or promotion of another substance or biochemical reaction. In certain circumstances, once a toxin has been introduced into another organism's tissues an immune response is elicited, triggering the induction of neutralizing antibodies or antitoxins. Toxins may derive their chemistry from the environment or the host animal's diet, but the production of some toxins requires the codependence of associated microorganisms and the host animal. Additionally, the chemical make-up of specific animal toxins can be influenced by geographic variation and seasonal climatic factors and the influence of these on an animal's habitat. Toxins, biologically and chemically, range from quite simple to complex compounds. Certain animal species may possess only single toxins within specific anatomical sites in their bodies. The effect of a single toxin can be like a biochemical bullet that precisely finds its specific target; however, mixtures of toxins in animals are not uncommon, and each single toxin in the mixture may have an individual specific target.

## Venoms and Poisons

Some animals, such as amphibians and reptiles, are known to possess a mixture of different toxins usually associated with, and harbored in, highly evolved toxin delivery systems or glands. These toxin cocktails, collectively, are referred to as venoms or poisons, depending on their origin in the animal species. Venoms and poisons are usually mixtures of toxins that can elicit a cascade of adverse pharmacological events in the target victim or organ system. Venoms, being a composite mixture of toxins, are biologically manufactured or synthesized *de novo* in a well-developed secretory gland or organ associated with a venom delivery anatomical structure. A poison is generally considered to be the terminal toxin product of a metabolic process, a toxin product of an organism within the animal due to a symbiotic relationship, or a toxin product resulting from the biodegradation product of a substance. Toxins that are not directly synthesized by an animal but are present in the animal's tissues render the animal poisonous. The differentiation between venomous and poisonous animals is most applicable to creatures in marine ecosystems. Given these simplistic definitions, venomous animals are poisonous, but many poisonous animals are not venomous [27].

## Toxin Functions

Toxins, venoms, and poisons act as single agents or work in concert to repel other organisms or more effectively achieve a desired adverse end effect on another organism. Commonly, toxin functions are loosely related to their effects on a given physiological system, resulting in the use of such terms as neurotoxins, myotoxins, cytotoxins, and hemorrhagic toxins. Regardless of the physiological system affected, animal toxins appear to primarily exist for two important reasons. First, many toxins are useful in subduing prey for food, and second, toxins often provide a form of self-defense for the animals in which they are present. Both of these functions, independently or together, provide for the self-preservation and sustained survival of animal species; however, in some circumstances a toxin is present in an animal as a result of being passively taken into the body, causing no apparent adverse effects to the animal and having no apparent function. It is difficult to know if a mollusk harbors a toxic alga for the purpose of protection from predators or for predation of food vs. its simply being present due to the alga being present in the mollusk's environment. A toxin can enter the food chain without having specific function for the animal initially harboring the toxin.

## Bioaccumulation and Bioamplification of Animal Toxins

Animals can accumulate toxins as they are transferred up through the food chain. This can be an active or passive bioamplification process as animals ingest, filter in, or absorb certain toxins or toxin-producing organisms, with the end result being a passively acquired toxicity for the host animal. In turn, these animals may or may not become poisonous to larger animals that prey on them [28]. This particular scenario is prominent in marine biology, as many species of fish are the top consumers or predators of various toxic marine organisms; for example, a mollusk can feed on toxin-producing algae and then itself be eaten by a fish that survives the ingestion without complication, but a human might then consume this fish with its burden of bioaccumulated toxin. The mollusk or fish not being affected could be the result of tolerance to the algal toxin or a natural resistance to the toxin's effects, depending on the species. When humans or other mammals ingest these toxin-laden fish, severe toxicological medical consequences can occur. Animals may also acquire their toxins from a wide variety of toxin-containing plants on which they feed as part of their diet. A more thorough discussion of toxins in the food chain is discussed in Chapter 14.

**TABLE 19.2**
**Interrelationships of Animal Phyla, Environment, Toxins, and Toxin Targets**

| Phylum | Environment | Animal (*Genus species*) | Toxin | Toxin Target |
|---|---|---|---|---|
| Mollusca | Marine | Blue-ringed octopus (*Hapalochlaena lunulata*) | Tetrodotoxin | $Na^+$ ion channel |
| Chordata | Marine | Pufferfish (*Fugu* sp.) | Tetrodotoxin | $Na^+$ ion channel |
| Chordata | Terrestrial | Rough-skinned newt (*Taricha granulosa*) | Tetrodotoxin | $Na^+$ ion channel |
| Chordata | Terrestrial | Neotropical toad (*Atelopus carbonensis*) | Tetrodotoxin | $Na^+$ ion channel |
| Cnidaria | Marine | Sea anemone (*Aiptasia pallida*) | Phospholipase $A_2$ | Cell membranes |
| Arthropoda | Terrestrial | Honeybee (*Apis mellifera*) | Phospholipase $A_2$ | Cell membranes |
| Mollusca | Marine | Cone snail (*Conus geographicus*) | Omega-conotoxin | $Ca^{++}$ ion channel |
| Arthropoda | Terrestrial | American funnel web spider (*Agelenopsis aperta*) | Omega-agatoxin | $Ca^{++}$ ion channel |
| Mollusca | Marine | Cone snail (*Conus purpurascens*) | *K*-conotoxin | $K^+$ ion channel |
| Cnidaria | Marine | Sea anemone (*Bundosoma granulifera*) | BgK | $K^+$ ion channel |
| Chordata | Terrestrial | E. African green mamba (*Dendroaspis angusticeps*) | Dendrotoxin | $K^+$ ion channel |
| Arthropoda | Terrestrial | Yellow scorpion (*Leiurus quinquestriatus*) | Charbytoxin | $K^+$ ion channel |
| Chordata | Marine | Broad-banded sea snake (*Laticauda semifasciata*) | Erabutoxin | ACh receptor |
| Chordata | Terrestrial | Black-necked cobra (*Naja nigricollis*) | Neurotoxin-alpha | ACh receptor |
| Chordata | Terrestrial | E. African green mamba (*Dendroaspis angusticeps*) | Fasciculin 1 | ACh-esterase |

## Toxin Delivery Systems

Toxin delivery to a specific target can only occur following its introduction onto or into another living organism, and in this respect the animals of the world have proven to be quite accomplished, demonstrating considerable creativity. The introduction of a toxin, venom, or poison can occur via absorption onto or across a barrier and injection into or through a barrier. Usually, this barrier is dermal in nature and not readily penetrated by most nocuous external environmental substances. Toxin application and penetration processes are achieved by specific mechanical strategies such as superficial mucous secretions, bristles and hairs, glandular skin secretions, barbs, spines, and teeth (see Table 19.2). Pharmacological or toxicological actions of toxins following their topical deposition on body surfaces can cause superficial or severe localized reactions. This topical delivery process can be enhanced, depending on the chemical properties and pharmacological actions of the toxins, to result in transdermal absorption of the toxin and systemic toxicity. Toxin delivery into the systemic processes of another living organism can cause effects ranging from limited physiologic responses to severe systemic physiological alterations, resulting in the death of the unfortunate toxin recipient. Airborne toxin delivery, a method frequently employed by some insects, is achieved by the organism spraying a toxin topically onto dermal or ocular surfaces or even spraying it into the oral cavity of predators. More advanced and direct methods of toxin delivery are achieved by injection via a bite or string. These may be associated with an anatomically sophisticated delivery system such as that present in the fang-venom–duct-venom gland complex present in solenoglyphic venomous snakes or the pulsating stinging apparatus of the common honeybee.

## Complementation and Diversification of Animal Toxins

The sharing and interrelationships of phyla, toxins, toxin structures, and toxin functions exist across marine and terrestrial environments. Distinct animal species from different phyla from widely separated geographic regions and habitats have developed toxins with a variety of chemical structural, and functional relationships. Combinations of animal phyla, animal toxins, and the associated toxin structure–function relationships have yielded four basic schemes in nature: (1) interphylogenetic species producing a common toxin that acts on a common target, (2) interphylogenetic species producing different toxins with chemically related structures that act on a common target, (3) interphylogenetic species producing toxins with unrelated chemical structures that act on a common target, and (4) intraphylogenetic species producing toxins with similar chemical structures or toxins with unrelated chemical structures that act on the same target or different targets (Table 19.2). The interrelationships of these factors provide evidence of divergent evolutionary functions in toxin-producing animals, varying between phyla or within a phylum as well as a certain degree of convergent evolutionary function with respect to toxin properties and function.

The k-conotoxin from the cone snail (*Conus purpurascens*, phylum Mollusca) is quite different in chemical configuration from charbytoxin from the yellow scorpion (*Leiurus quinquestriatus*, phylum Arthropoda), yet both of these toxins target the potassium-ion channel [26]. In contrast, toxins of similar chemical structure may act on different targets; for example, neurotoxin-alpha in the venom of the black-necked cobra (*Naja nigricollis*) and fasciculin 1 from the venom of the east African gree mamba are structurally similar, but neurotoxin-alpha acts on the acetylcholine receptor while fasciculin 1 acts on acetylcholinesterase. Also, some toxins, such as tetrodotoxin, that are widespread in the animal kingdom (having been found in the skin of amphibians, fish organs, and blue-ringed octopus) have a common chemical structure and common pharmacological site of action [30]. Omega-conotoxin from the cone snail (*Conus geographicus*, phylum Mollusca) is similar in chemical configuration and structure to omega-agatoxin from the American funnel web spider (*Agelenopsis aperta*, phylum Arthropoda), and both act on calcium-ion channels [29].

## Mechanisms of Animal Toxin Actions

The biodiversity of organisms produces a wide spectrum of ways in which animal toxins act pharmacologically to help ensure the continued survival of their host animals, whether the toxins are necessary for acquiring or digesting food or providing defense. In certain circumstances, these toxic effects are exerted in animals that have ingested a toxin-containing organism and inadvertently become victims of poisoning. Whether the toxin effects are neurotoxic, myotoxic, cytotoxic, or coagulopathic, they are usually due to toxin proteins, enzymatic proteins, or peptides acting either singly or collectively, as in the case of some venoms, to create a single reaction or a cascade of reactions resulting in toxicological consequences to a living organism.

Toxicity resulting from the pharmacologic actions of different toxins can reflect both qualitative and quantitative differences. The ability of a toxin to elicit a toxic effect by acting at a single target may depend on the dose of toxin reaching the target or subtargets within a target. Receptors may possess only a single binding site for a particular toxin or more than one binding site for that toxin. The degree of toxin binding may be limited by conformational characteristics of the binding site, but binding to a single site or more than a single site may still depend on the toxin dose. If the toxin dose is low, minimal binding at only one site on the receptor may not result in any measurable toxicity. If the toxin dose is large, increased binding may occur at a single site or multiple binding sites on the receptor, resulting in significant toxicity. Thus, for different animal toxins and their quest for different receptor targets or subtargets, there potentially exists a flexible stoichiometric relationship. All of these factors, in addition to the toxin's own chemical configuration, can affect what binding site a toxin is specific for. Subtle differences in how these factors are related are especially important, as slight chemical and structural differences in toxins can not only allow the binding of a specific toxin to the same receptor as another toxin but also allow the toxin to bind at different sites on the same receptor. Additionally, some binding sites within a given target, such as a cell membrane, allow the binding of different toxins. This multiple-toxin–multiple-target relationship is commonly observed with snake venoms. The multiple toxins found in venoms may also function sequentially via synchronized pharmacologic actions when injected into another organism, with one toxin leading to the improved bioavailability of another more potent toxin to its target.

Although toxin dose is an important component in the development of toxicological effects, the inherent affinity of toxins for given targets relating to toxin specificity is also a factor that influences the potency of a given dose of toxin. Animal toxins can compete as agonists or antagonists with endogenous neurotransmitters in synaptisomal clefts and may either act presynaptically or postsynaptically. Toxins that target the acetylcholine receptor, or neuromuscular junction, are well established among the different animal phyla. A number of toxins also exist that target voltage-gated ion channels, primarily sodium-, potassium-, and calcium-ion channels. Again, nature has provided some unique examples of the selective pharmacological effectiveness of toxin specificity, such as the selective binding of toxins such as beta-scorpion toxin vs. tetrodotoxin from newts, which bind to different sites on the sodium-ion channel.

Toxins may also digest or directly inactivate neurotransmitters. Phospholipase $A_2$ is an enzymatic protein with wide representation in the animal kingdom; it is known to exist in more than 100 isoforms. Not surprisingly, the various forms are capable of exhibiting diverse pharmacologic activities and physiological properties ranging from the destruction of cell membranes to acting as presynaptic neurotoxins [31,32]. Other animal toxins can act on a variety of cell membranes, altering cellular membrane integrity and inducing secondary toxic effects from a toxin's cytolytic, myotoxic, and coagulopathic effects.

Collectively, toxin effects—whether myotoxic, agonistic, or antagonistic to receptors; ion channel altering; or neurotoxic based on their properties such as affinity and specificity, conformational structure and specificity, or length of amino acid sequences—contribute to the varying levels of toxicity, the diversity of pharmacological function, and the extent of pharmacological function leading to a toxicological end effect. A need to understand the role of animal toxins in nature has arisen from the increasing interface between humans and animals in all ecosystems. The number of ways in which animals use their toxins has been expanded due humans pursuing activities crossing all ecosystem barriers; consequently, animals have been forced to use their toxins and venoms in defensive circumstances far beyond their intended protective use against natural predators. The dynamics of these animal–natural predator and animal–human relationships will be further discussed with respect to toxinology: specifically, animal phyla, their specific toxins or venoms, toxin functions in the natural world, and their impact in biomedical research and clinical toxicology.

**TABLE 19.3**
**Marine and Freshwater Invertebrate Animals: Venomous or Toxin Producing**

| Phylum | Class | Common Names | Genera (Toxin Producing) |
|---|---|---|---|
| Annelida | Polychaeta | Bristle worms, fire worms | *Chloeia*, *Glycera*, *Eunice*, *Eurythoe*, *Hermodice* |
| | Hirudinea | Medicinal leech | *Hirudo* |
| Cnidaria | Anthozoa | Anemones | *Actinia*, *Aiptasia*, *Antheopsis*, *Condylactis*, *Entacmaea*, *Halcuriasp*, *Heteractis*, *Stomolphus*, *Tealia*, *Urticina* |
| | Anthozoa | Corals (soft, stoney, hard) | *Alcyonaria*, *Gorgonaria*, *Heliopora* |
| | Cubozoa | Box jellyfish | *Chironex*, *Chiropsalmusa*, *Carukia*, *Carybdea*, *Tamoya*, *Tripedalia* |
| | Hydrozoa | Fire corals | *Millepora* |
| | | Portuguese man-o'-war | *Physalia* |
| | | Sea fans | *Aglaophenia*, *Halecium*, *Lytocarpus*, *Macrorhynchia*, *Nemalecium*, *Thecocarpus* |
| | Scyphozoa | True jellyfish | *Cassiopea*, *Cyanea*, *Chrysaora*, *Pelagia*, *Rhizostoma* |
| Echinodermata | Echinoidea | Sea urchins | *Asthenosoma*, *Aerosoma*, *Diadema*, *Phormosoma*, *Tripneustes*, *Toxopneustes* |
| | Asteroidea | Starfish (sea stars) | *Acanthaster* |
| | Holothuroidea | Sea cucumbers | *Bohadschia*, *Thelonota* |
| Mollusca | Gastropoda | Cone snails | *Conus* sp. |
| | Cephalopoda | Blue-ringed octopus | *Hapalochlaena* |
| Porifera | Demospongiae | Sponge | *Fibula*, *Lissodendoryx*, *Microiona*, *Neofibularia*, *Tedania* |

## Marine and Freshwater Ecosystem Animal Toxins

More than two thirds of Earth is covered in ocean waters that are home to diverse and highly integrated marine communities. The remaining one third of the planet's land mass is dissected by rivers and scattered with lakes representing freshwater aquatic ecosystems that also contain integrated animal communities of great diversity. Similar to the relationship for venomous vs. poisonous animals, all marine animals are aquatic, but not all aquatic animals are marine. In organisms from marine environments alone, the chemical structures of more than 5000 natural products, as single chemical classes or mixed biosynthetic compounds, have been elucidated during the past two decades [33–35]. This number is probably far less for compounds derived from freshwater species because of the considerably smaller freshwater habitat on Earth. These organism–taxonomic and toxin–chemistry relationships are complex, the biogenetic diversity of animal toxins in these habitats is great, and the quest to identify animal communities and specific sources of animal toxins is a continual and ongoing one for scientists.

## Invertebrate Marine and Freshwater Animal Toxins and Venoms

Multiple genera representing 12 different classes derived from five different phyla are the phylogenetic roots for thousands of invertebrate species in marine ecosystems (Table 19.3). Although marine invertebrates are rather simplistic in form, ranging from simple hydrozoan corals to more complex forms such as gastropod cone snails, many possess some form of toxin or venom as a means for acquiring a meal or as a deterrent against predators. In addition to their toxin effects on other organisms, a remarkable number of marine invertebrate toxins are also harmful or even lethal to humans.

### Annelida

Bristle worms and fire worms (class Polychaeta) are reported to possess toxins (Table 19.3) [36]. Known by a variety of common names, they share the common feature of having parapodia, leg-like appendages on each of their body segments, with hollow monofilament-like bristles (setae) attached. In many cases, the bristles may have a feathery appearance. Despite anatomical likenesses, the different species can display a range of appearances reflected in their common names, such as palolo worms, featherduster worms, sea mouse, or lug worms. Some, such as *Eunice gigantea*, have been reported to grow up to 3 m in length [36].

*Eurythoe complanata*, a species indigenous to the tropical Pacific Ocean and Gulf of Mexico, has spiny barbed bristles that it promptly flares when threatened. If a predatory animal or human touches the bristles, the barbed glassy spines are readily released into the tissue of the offender. This action results in a mechanical or physical insult that produces a localized dermal inflammatory response and paresthesia that may last for weeks but is not believed to be associated with delivery of a toxin. Embedded bristles may be partially removed by the application of adhesive tape over the affected area. Topical use of a steroid cream and oral or topical antihistamines may be a beneficial symptomatic treatment; severe cases of secondary infection or even gangrene are possible [37,38].

Importantly, certain species such as *Eunice* and *Glyceria* have jaws associated with a defined venom gland and a fang-like proboscis that delivers a large-molecular-weight, ion-channel-forming alpha-glycerotoxin [39]. Clinical symptoms from the bite indicate that it can be quite painful; blanching observed around the punctures is followed after several days with persistent itching. Uneventful recovery generally follows.

The medicinal leech (class Hirudinea) is a freshwater annelid with a quite interesting history, as it has been used by the lay public for self remedies and by physicians since 200 B.C. for a variety of medical purposes, including its use as a postoperative aid in microvascular surgery patients [40,41]. These creatures have two suckers that lie at the anterior and posterior ends of the head, with the mouth being in the anterior sucker where three jaws of teeth are located. Once attached to the often unknowing victim, the leech sucks blood out using rhythmic contractions of its pharynx to deliver the blood to its crop for storage until digestion takes place. The leech can ingest a blood volume nearly ten times its weight, and as much as 15 mL of blood may be removed [42]. The anticoagulant effect of leech venom stems from the toxin hirudin, an antithrombin substance, but the venom also contains hyaluronidase, proteolytic enzymes, fibrinase, collagenenase, and salivary apyrases [43,44].

The most common effects of leech bites are the depletion of blood volume in small animals from which the leech draws its nutrition and continued oozing of blood from the bite site following detachment. It has been reported that oozing may continue for up to 24 hours after leech removal [45]. Leeches have caused hematuria due to urinary tract invasion and hemoptysis from upper airway attachment [46]. Treatment of leech bite is not standardized, and myriad means of leech removal have been proposed. Cocaine has been used to paralyze leeches, resulting in their release from their victim, but would certainly not be recommended in cases of a leech-lodged airway. The use of forceps to hold the leech until it releases has been effective; one patient required anesthesia with direct visualization for leech retrieval [47]. Regardless of the leech removal method, the wound should then be irrigated and pressure applied; a gelatin sponge can be used if bleeding persists, and follow-up evaluation for wound infection is recommended [45].

## Cnidaria

The Cnidaria, also known as Coelentrates, include anemones, corals, jellyfish, and hydrozoans (Table 19.3). This phylum is unique for being the only entirely venomous phylum in the entire animal kingdom. The creatures in these classes appear to be quite different visually, yet they frequently share a commonality in their toxin-delivery mechanism, a process that involves stinging organelles known as nematocysts, sometimes referred to as cnidocysts [48,49]. These microscopic structures are organelles contained within specialized stinging cells (or cnidocytes), which, in the resting stage, look like well-inflated balloons, each with a small spike protruding. The spike continues to the interior as a barbed luminous thread that spirals like a spring within a venom capsule. If a significant osmolar or pressure change occurs, the nematocyst explosively discharges its inverted, barbed filament into the tissue of its victim with a simultaneous injection of venom [50]. The discharged nematocyst then appears, grossly, as a slightly deflated balloon attached to a string [36]. Nematocysts are primarily located on tentacles or along folding anatomical ridges; thousands of these organelles are present, allowing for the delivery of considerable quantities of toxins. Interestingly, some creatures have extended the use of nematocysts and their toxins via an interesting phenomenon in which non-cnidarian marine creatures such as the octopus (*Tremoctopus violaceus*) attach fragments of Portuguese man-o'-war tentacles to their own tentacles as both offensive and defensive armaments [51].

Toxins that have been identified in cnidarians stem from a variety of peptides and proteins that possess neurotoxic, hemolytic, or cytolytic properties [52]. Although the number of species of Cnidaria has been estimated at 10,000 or greater, the number of species potentially dangerous to humans is not known but is believed to be relatively low [53]. Regardless of the unknown number of species dangerous to human health, the medical problems associated with cnidarian envenoming are global in nature [54].

### Anthozoa

Anemones are a beautiful and graceful form of sea life, yet these mostly sedentary cnidarians can readily deploy their paralyzing toxins from nematocysts into potential predators that brush against their tentacles or into unknowing animals seeking shelter among their tentacles. Not all sea anemone toxins have been confirmed to be actually located in tentacle nematocysts, as the contracting, intact body of some anemones will secrete cytolytic toxins with mechanical stimulation [55,56]. Anemones can also retract their tentacles under stress conditions, resulting in the secretion of large quantities of cytolytic or hemolytic toxins in mucus, rather than by nematocyst release, that can induce osmotic cell lysis via pore-forming ion channels in membranes. Interestingly, toxins that serve to acquire prey may also aid in the self-preservation of an individual that possesses resistance to its own cytolysins within an anemone colony by allowing intraspecies spatial competition [57].

An additional complexity concerning this marine life form is the interrelationship between some anemone genera (*Entacmaea* and *Heteractis*) and species of fish in the genera *Amphiprion* and *Premnas* that demonstrates a symbiotic relationship in which anemone toxins provide protection to fish hiding among the nematocyst-laden anemone tentacles [58]. The reason why fish can survive this symbiotic relationship has not been confirmed to be a universal one, as some fish species are protected by a mucus coating of their skin while others may have acquired immunity to certain sea anemone toxins [59].

An array of toxins has been discovered in anemones; lethal cytolytic proteins and peptides have been reported in more than 32 species of sea anemones in the order Actinaria alone [60]. Roughly 30 different pore-forming toxins, known as actinoporins, have been characterized in more than 20 sea anemone species [61]. Several toxins have been extensively characterized, such as sodium- and potassium-ion channel blockers [62] and cytolysins that are lethal to crustaceans or vertebrates. Also, a cardiostimulatory protein lacking cytolytic activity has been isolated from *Urticina piscivora* [63]. Some toxins, such as tealiatoxin produced by *Tealia felina*, exhibit both hemolytic and histaminolytic activities [64].

Although a multitude of toxins have been identified, the medical consequences of most anemone stings to humans are fairly similar. The common symptom is a painful, stinging cutaneous lesion, which is localized to the area of tentacle contact [36]. Severity is proportional to the number of discharged nematocysts in the skin. Nematocysts can be inactivated by an application of 5% acetic acid (vinegar) or a 70% isopropyl alcohol solution. Removal of remnant tentacles and nematocysts can be accomplished by unidirectional scraping of the affected area with a knife blade or the edge of a plastic card. The lesion should then be washed with antibacterial soap and generously rinsed with water. Topical steroid cream may be useful in reducing local inflammation [65]. Rare acute envenomations have resulted in symptoms of severe pulmonary edema and cardiac arrest; eight deaths following envenomation by *Stomolphus nomuri* have been reported [66]. A species of *Condylactis* is believed to have been responsible for the death of a young male due to rapid hepatocellular failure [67].

## Cubozoa

The box jellyfish are probably the most well known of the jellyfishes because of their unique anatomy and the potentially severe toxicity that can result from their painful stings. Although closely related, their distinct anatomy separates them from true jellyfish. Box jellyfish are four-sided, cow-bell-shaped, translucent, gelatinous creatures that have muscular pedalia at each corner, where tentacles ten or more times the height of the bell (some reported to be in excess of 60 m) stream in the underwater currents [68]. They inhabit most subtropical and tropical waters, and the coasts of Australia are notoriously famous for the presence of dangerous jellyfish. *Carybdea* species having been occasionally found in temperate waters [69]. Cubozoans possess four photosensory structures known as rhopalia, which each contain four simple eyes and two complex eyes. This vision system makes cubozoans exceptionally phototactic as evidenced by their light-sensitive behavior of descending to deeper water during bright sunlight and moving toward the surface in early morning, late afternoon, and evening.

In contrast to the majority of jellyfish, box jellies do not drift with the underwater currents and are not planktonic in nature. They are strong swimmers, and some species, such as *Chironex fleckeri*, the dangerous Australian species commonly referred to as the sea wasp, have been reported to move 3 to 6 m/min [70]. Cuboidal jellyfish from Australian waters are associated with interesting and unique local nomenclature such as "Irukandji" (*Carukia barnsei*), "jimble" (*Carybdea rastoni*), and "morbakka" (*Tamoya* spp.) [71–73].

Cubozoan tentacles are laden with nematocysts of several different types that collectively form a complement of nematocysts referred to as a cnidome [68]. The differing severity of toxicity among the various genera may correlate with the nematocyst concentration present in the tentacles of a given species, which is also related to cubozoan specie size [74]. Recent studies using antibodies against box jellyfish have shown a lack of cross-reactivity with other jellyfish venoms, clearly suggesting that the toxicity difference is also a result of different toxins in venoms from different genera [75]. The venom of *Chironex fleckeri* is proteinaceous and possesses dermonecrotic, hemolytic and myotoxic toxins that can work synergistically to cause lethal cardiorespiratory effects [76,77]. Myotoxins of 150 kDa and 600 kDa have been isolated from *C. fleckeri* and shown to be lethal in animal studies, but the characterization of other venom components has been illusive [78]. Commonly, the result of envenomation is a hypercatecholaminemia symptom profile, but this does not hold for all species, as the effects of venom from *Chriopsalmus*, species responsible for many fatalities in the Philippines does not respond to receptor antagonists or antibodies that are effective in reversing *C. fleckeri* venom-induced symptoms [79,80]. In cases of severe envenomation, death has occurred within minutes, and at least 67 deaths have been documented [76].

In general, cubozoan envenomations do not result in severe complications, but those inflicted by *Chironex fleckeri* can cause death, as evidenced by the numerous fatalities reported in Australian waters [81]. Treatment of cubozoan envenomations without severe symptoms is accomplished with the use of acetic acid to retard nematocyst firing and symptomatic support. More severe envenomations have required support with mechanical ventilation, which can be ineffective in cases of profound cardiotoxicity [82]. Antivenom for *C. fleckeri* envenomation treatment has been developed, but only one report in the literature documents its effectiveness in humans with severe cardiorespiratory symptoms [83].

Irukandji syndrome is a severe form of toxicity resulting from the sting of the small *Carybdea barnesi* jellyfish. It is a species of undefined geographic origin that has primarily been reported off the coasts in the northern-half of Australia, although an Irukandji-like syndrome has been reported in Hawaii [84]. Cardiopulmonary complications are predominant, but the exact underlying mechanism is unclear. Victims exhibit distress, generalized pain, nausea, frequently severe hypertension, and pulmonary edema, and the clinical profile suggests a catecholamine excess [85]. Treatment is generally supportive, requiring analgesics, antihypertensive drugs, ventilation, and pressor drugs, with recovery of cardiac compromise in several (3 to 4) days [76].

### Hydrozoa

Hydrozoa are primarily colonial benthic cnidarians, with the exception of floating *Physalia*, which are simple marine life forms that, like all other members of Cnidaria, possess nematocysts. Four different morphological forms of nematocysts are known to be present in Hydrozoa: atrichous isorhiza, microbasic euryteles, microbasic mastigophores, and stenoles [53]. They differ in the arrangement of spines on the shaft, which may result in varying degrees of anchoring to tissue and venom delivery into the victim [86]; thus, contact with these organisms produces mechanical as well as toxicological injury. Although the number of species in the class has been estimated at more than 10,000, the number of species known to be potentially dangerous to humans is small [76].

Fire coral (*Millepora*) is a hydrocoral rather than a true coral. It is familiar to many reef divers as it is frequently encountered among colonies of true corals. Antler-like in appearance, the velvety tines are covered with nematocyst-containing hydranths that look like small, segmented hairs with umbrellas at the tip. The venom contains thermolabile, proteinaceous components possessing both dermonecrotic and hemolytic properties. Limited toxicity studies in small rodents have shown that it is capable of causing death [87]. When contact with fire coral occurs, there is an immediate burning and stinging sensation that progresses rapidly to a painful pruritus. Wheals develop, and the area of contact becomes edematous and erythematous. Pain symptoms generally resolve within an hour or two, while dermatological symptoms may persist for up to a week, with pigmentation lesions persisting for 1 to 2 months [88]. Amelioration of symptoms usually requires only the application of a skin cream, but in more severe cases it may require deactivation and removal of the nematocysts, followed by the application of hydrocortisone cream. It has been reported that papain-containing meat tenderizers may be useful in retrieving ingrained nematocysts [65].

Portuguese man-o'-war (*Physalia physalis*) is a non-diving, oceanic, jellyfish-like hydrozoan with a global geographic distribution. Their gas-filled bladders, actually modified medusae, allow them to drift with wind and surface-water currents. Their numerous tentacles may be long, trailing 13 m or more, and they can deliver potent venom from their associated nematocysts [89]. A small, morphological form of this genus, *Physalia utriculus* (commonly called the blue bottle) is indigenous to the Micronesian waters of the Pacific, but lesions resulting from its stings are less extensive than those of *P. physalis* [36]. Venom from *Physalia* has not been fully characterized, but a glycoprotein-structured toxin, physaliatoxin, has been isolated and described as possessing strong cytolytic and hemolytic properties [90]. Additionally, enzymatic proteins with cardiotoxic effects have been isolated, and whole venom can elicit rapid histamine release from mast cells [91,92]. Complications to humans from the stings of *Physalia* can range from acute pain in tentacle contact areas to migratory joint pain to, in rare circumstances, death from rapid respiratory failure followed by cardiovascular collapse [93]. Unlike the inhibitory response to firing nematocysts of *Chironex* with acetic acid, this therapy has been reported to induce further nematocyst discharge in cases of *Physalia* sting [94]. Cold packs may reduce the superficial inflammatory and pain response of stung tissues, and papain solution has been useful for inactivating nematocysts [95]. The use of topical lidocaine preparations may also be beneficial in reducing localized inflammation and pain [96]. Antivenom for the treatment of *Chironex* stings provides no paraspecific coverage or protection to victims of *Physalia* envenomation, as it is not cross-reactive with *Physalia* venom [97]. Thus, no definitive treatment in cases of life-threatening envenomation is available, and basic cardiopulmonary resuscitation and support measures are essential if any chance of recovery is possible.

Sea fans are some of the more elegant appearing hydrozoans with their feathery, fan-leafed fern look. These benthic hydrozoans are found in shallow Atlantic, Indo-Pacific, and Australian waters. Despite their harmless appearance, their nematocysts can deliver painful stings associated with erythematous and edematous whealing when contact occurs [54]. The best-known genera for delivering venomous stings are the *Aglaophenia* and *Lytocarpus*, yet other genera may also cause dermal-contact-related problems (Table 19.3). Little is known of sea fan venoms, yet in recent years the discovery of new venomous species has occurred by accident as case reports of stings find their way into the literature. One such species, *Nemalcium lighti*, was found to be capable of delivering pruritus- and burn-sensation-inducing stings to an unfortunate snorkel victim that persisted for days [52]. Sea fan envenomations are not life threatening, but they produce an aggravating discomfort, and their superficial symptoms respond to deactivation of nematocysts still in contact with the skin, and the application of antiinflammatory creams [65].

Scyphozoa is the class representing the true jellyfish, which are probably the most frequently encountered jellyfish by beach-goers. They tend to be radially symmetrical, and their gelatinous medusal forms swim in currents from the surface waters down to the abyss of the oceans, depending on the species [36]. They have a variety of interesting and descriptive names: upside-down jellyfish (*Cassiopea xamachana*); sea nettle (*Chrysaora quinquecirrha*); hair jelly (*Cyanea* sp.), which was described by Halstead as a "mop hiding under a dinner plate"; and mauve stinger (*Pelagia noctiluca*) [36,98]. Although scyphozoan nematocysts are capable of delivering irritating stings that may rarely cause muscle cramps and unconsciousness, their venoms are not as dangerous as their cubozoan relatives, as no cases of death are found reported in the literature [99].

*Chrysaora quinquecirrha* crude venom (CQV) possesses a bioactive hyaluronidase, an enzyme thought to aid in the spread of venom, that is similar to the hyaluronidase found in the venom of the five pace snake (*Agkistodon acutus*), again demonstrating the similarity of toxins found across phyla in the animal kingdom. Studies have shown that CQV exhibits hemolytic activity [100]. Animal studies have predominantly supported observations that fatal reactions to jellyfish envenomation have been the result of cardiac, respiratory, or renal toxicity, but CQV studies also suggest that the venom is capable of inducing hepatic lesions, subsequent to lethal cytotoxicity, that may explain fatal reactions as a result of delayed hepatic failure [101].

*Cassiopea xamachana*, from Puerto Rican coasts, can cause painful stings, and recent research showed that their venom (CxTX) contains a low-molecular-weight (<10 kDa) fraction that binds to acetylcholine muscarinic receptors and demonstrated strong hemolytic activity in an animal model. Additionally, a phospholipase $A_2$ ($PLA_2$) fraction contributed significantly to paralytic activity and lethality [102]. It is likely that CxTX toxicity is the result of venom toxins acting synergistically. The degree to which scyphozoans share similarities in their toxins remains to be elucidated, but toxin effects appear to cause a similar clinical profile in humans who have had an unfortunate contact. The varying severity of envenomation apparently is dependent on the extent of the tentacle contact area and the duration of contact [36]. Treatment of scyphozoan envenomation is similar to that recommended for cubozoan stings, and with good supportive care a favorable outcome is to be expected.

## Echinodermata

The phylum Echinodermata is perhaps most familiar to people of all ages because of the amazing starfish (class Asteroidea). For many years, these creatures have been taken from the world's oceans, dried, and sold as curios to be placed on desktops, library shelves, and used in unthinkable numbers for school "show and tells." Sea urchins (class Echinodea) are also widely recognized due to the fear of a painful encounter with one of their spines. Many children and snorkelers have paid the price for that tempting moment of curiosity when an extended finger has tried to gently touch one of these creatures and been rapidly withdrawn in profound pain, as have unwary waders planting their feet on the needle-like spines. Unlike the better known echinoderms, the sea cucumbers (class Holothuroidea) have a more fragile, larger, slug-appearing form that lacks the mechanical-injury-inducing abilities of other echinoderms. Although these creatures are reported to number approximately 6000 different species, only about 1% is believed to be venomous [103].

Sea urchins represent the majority of venomous echinoderms and are painfully familiar to people who have spent any time swimming, beach walking, surfing, or diving and snorkeling in subtropical or tropical coastal waters. They have a pincushion appearance with stiff bristle-like spines that are structurally a calcium-carbonate matrix and may be hollow or solid, blunt or sharp; sea urchins may possess a venom-producing gland at the spinal apex [104]. They also may introduce venom via pedicellariae, which are jaw-like structures used to grasp unsuspecting prey. Urchin venoms are comprised of cholinergic substances, hemolysins, proteases, serotonin, and sterol glycosides [105].

Most urchin species are incapable of effectively injecting venom into humans, and the resultant sting is primarily a mechanical injury; however, two Indo-Pacific ranging species—the needle-spined sea urchin (*Diadema setosum*) and the flower sea urchin (*Toxopneustes pileolus*)—are known to be capable of inflicting serious envenomation symptoms [89]. The severity of symptoms varies depending on the number of penetrating spines, and the associated pain is greater than that of mechanical injury alone. In rare cases of severe envenomation, symptoms of bronchospasm, cranial nerve dysfunction, hepatitis, hypotension, palpitations, and syncope have occurred [104,105]. *Toxopneustes* envenomation has even been reported to have caused deaths in Japan [106]. Treatment of urchin envenomation symptoms is not well defined and is complicated by the fragile nature of the spines, which are frequently broken off in the wound. Methods of extricating spines have been relatively ineffective, and appropriate wound care may involve warm water immersion, analgesic drugs, and breaking large embedded spines into smaller fragments so they may be more readily absorbed or extruded [37]. Thus, the mainstay of treatment is symptomatic and supportive as needed.

Despite the many harmless species of starfish, one major species is responsible for envenomations to humans: he crown-of-thorns starfish (*Acanthaster planci*) of the ndo-Pacific region [36]. The crown-of-thorns starfish, as ts name suggests, is an ornately pigmented and large starfish found among the colorful coral reefs, where their eeding on coral polyps can destroy the fragile coral reef structure [107]. Like urchins, the spines of this starfish nflict a painful wound. This starfish can grow to 60 cm n diameter and possess 13 to 16 rays with spines up to 6 m in length [36]. Venom is believed to lie in the epidermal ell layer of the spines, where acidophilic glandular cells roduce venom that has been shown to have hemolytic nd lethal activities in rodents [108]. Crude venom has lso been shown to exhibit edema-forming and capillary ermeability-inducing properties [109]. More recent studes have revealed that the venom contains a 15.5-kDa, hospholipase $A_2$ that can induce skeletal muscle myonerosis in animals [110]. Their venom has also been eported to contain a 20- to 25-kDa basic glycoprotein vith lethal properties when injected into mice, and hepatic nd hypotensive effects have also been observed in animal xperiments [111,112].

The symptoms of crown-of-thorns starfish stings are uite similar to those observed for urchins, but the pain ubsequent to venom release from embedded spines has een described as extremely intense with throbbing and urning sensations that may last for a month [37]. Treatment is also similar to that for urchins, with hot-water ubmersion being of benefit in relieving pain initially, ccompanied by prudent follow-up wound care, observaon for secondary infection, and use of steroid creams nd antibiotics as indicated by symptoms [37].

Sea cucumbers, in stark contrast to their spiny-bodied rchin and starfish relatives, are soft-bodied, blob-like eatures with an amazing array of color patterns and a cretive nature. They have no true appendages other than nall tentacles around the mouth; internally, they harbor gans referred to as organs of Cuvier, also known as vierian tubules [89]. When divers and beachgoers hane sea cucumbers they release a glycoside known as olothurin [113]. This toxin is a saponin produced in the vierian tubules and readily deployed from the cloaca hen contact is made with the offending organism; it has glue-like consistency and can be most inflammatory to in and cause conjunctival complications to the eye 14]. To relieve symptoms, victims should decontaminate the areas of contact with gentle and voluminous rinsing and washing with mild detergent; ocular exposures require normal saline irrigation [113].

## Mollusca

The cone snails (class Gastropoda, superfamily Conidea) are dangerously beautiful, toxin-producing representatives of the phylum Mollusca. As a genus, *Conus* represents the largest genus of living marine invertebrates, and it encompasses more venomous species than any other single genus in the entire animal kingdom, with more than 500 species displaying an array of patterns of brilliant colors [115]. Their venom components have primarily been associated with modestly sized disulfide- and cysteine-rich peptides, collectively known as conotoxins [116]. Currently, the number of different conotoxins is believed to be more than 50,000, and within this number are many peptides and proteins with diverse structures and toxicological activities that target a wide range of ion channels and receptors [117,118]. Not surprisingly, given their venom diversity, cone snail toxins and their derived peptides have a potential role in the development of therapies for a variety of medical disorders.

Cone snails inhabit tropical waters of the world and are quite predatory; they possess a hollow, harpoon-like radular tooth within an extensible rostrum that can be rapidly deployed for envenomating prey [103]. Prey is usually engulfed and simultaneously envenomated. Cone snails have been roughly divided into three groups, depending on their preferred prey: molluscivorous (feeding on gastropods, making them somewhat cannibalistic), piscivorous (feeding on fish), and vermivorous (feeding on polychaete worms). They have also been known to prey on bivalve mollusks, octopus, and hemichordate worms [119].

Cone snail toxins have been classified according to their structure, principally their cysteine framework, which is based on the distribution and number of cysteines in the primary sequence [120]. Two major groups of toxins that have been intensely studied are the conotoxins with multiple disulfide bonds and the much less prevalent *Conus* peptides that possess only a single or no disulfide linkage. Superfamilies of these peptide toxins that have been most extensively researched are the A- (alpha-conotoxins, alphaA-conotoxins, and kappaA-conotoxins), M- (mu-conotoxins and psi-conotoxins), and O- (omega-conotoxins, delta-conotoxins, muO-conotoxins, and kappa-conotoxins) superfamilies. Those in the A-superfamily act as nicotinic acetylcholine receptor antagonists and voltage-gated potassium channel blockers, while the M-superfamily toxins impair voltage-gated sodium channels and act as noncompetitive nicotinic acetylcholine receptor antagonists. The O-superfamily conotoxins possess the greatest range of function pharmacologically, having actions to block voltage-sensitive calcium channels and the ability to block both sodium and potassium voltage-gated channels. The nondisulfide *Conus* peptides conantokin, contalakin, and conopressin exhibit *N*-methyl-D-aspartate receptor antagonism, neurotensin-like peptides, and vasopressin antagonism, respectively, but the pharmacological role of contryphans remains to be determined [117].

Cone-snail-related human fatalities were first reported by Rumphius in 1705, and the fatality rate in more recent times has been reported to be 70% [121]. Of all the cone snails, the geography cone (*Conus geographicus*) most frequently causes envenomation to humans and is responsible for most fatalities; however, *C. aulicus*, *C. gloriamaris*, *C. marmoreus*, *C. omaria*, *C. striatus*, and *C. tulipa* have also been reported to cause death in humans [121,122]. Victims of *C. geographicus* and *C. obscurus* envenomation experience a radiating numbness; in contrast, victims of *C. aulicus*, *C. textile*, and *C. tulipa* report intense pain. Acute envenomation symptoms involve muscular paralysis, respiratory distress, loss of gag reflex and ability to swallow, and blurred or double vision. Timely treatment involving cardiopulmonary support is essential, as no antivenom currently exists [117].

The blue-ringed octopus (class Cephalopoda) is a small octopus species native to the shallow coastal and intertidal waters of Australia. Two similar-appearing species, *Hapalochlaena maculosa* and *H. lunulata*, as their common name implies, have yellowish-gold bodies with dark mottled bands and blue rings, which appear to light up a brilliant fluorescent blue color when the animal is disturbed or frightened [89]. They have eight arms, are carnivorous, and use their two beak-like jaws to introduce venom via the salivary duct with the inflicted bite, but they have also been observed to subdue prey by releasing their saliva into water surrounding the prey [123].

*Hapalochlaena* venom contains two main toxins: hapalotoxin and tetrodotoxin, with tetrodotoxin playing a defensive role against predatory fish. Additional venom components such as histamine, serotonin, and tyramine have been found in the salivary glands [124]. Originally, the tetrodotoxin was referred to as maculotoxin, but careful spectral analysis using nuclear magnetic resonance imagery showed that the venom was identical to tetrodotoxin [125]. The bite of the blue-ringed octopus may not elicit a significant pain response, and it may go unnoticed until symptoms of paresthesia around the face and neck are felt. Envenomation to humans has proven fatal, and death has been reported to occur within 30 minutes of having been bitten. Because no antivenom is available, the rapid initiation of standard life support measures is critical, with an emphasis on maintaining respirator-supported breathing until spontaneous breathing returns [126]. Interestingly, recent documentation indicates that saxitoxin is present in *Abdophus* species of octopus from the northern coastline of Western Australia. The source of this curare-like neurotoxin, normally produced by dinoflagellates, in the diet of this octopus is uncertain, but the tentacle toxin concentration is believed to be capable of causing profound paralytic shellfish poisoning [127].

## Porifera

Sponges (class Demospongiae) are the simplest multicellular marine life forms, as they have no defined organ systems. They are sessile creatures with amorphous and radially symmetrical shapes that grow attached to nearly any substrate; they are found in deep ocean bottoms and shallow coastal waters. There are four classes of Porifera but the class Demospongiae has been most studied. The exact number of sponge species is still a major question but at least 7000 species are reported in the literature (far fewer are freshwater species), and it has been speculated that the number may be more than 15,000. [128].

Some have sharp spicules, but these are not believed to deliver toxin, and it may be that potential toxins are associated with the solid body rather than spicules [129]. Although detailed knowledge of sponge chemistry and pharmacology is in its infancy, the number of pharmacologically active compounds being discovered, such as the antitumor macrolides (spongistatins), is rapidly increasing [130]. Toxins specific to sponges have not been identified but it is clear that adverse dermal reactions can result when dermal contact occurs.

Skin contact with the Southern Australian toxic sponge *Neofibularia mordens* can cause severe dermatitis that may last for months and does not respond well to symptomatic medication. *Tedania* and *Lissodendoryx* species have also been reported to cause an irritating dermatitis, but o shorter symptom duration [89]. Ice packs and cooling solutions containing camphor and menthol have been reported to ameliorate the irritated skin; hot water and alcohol ar to be avoided as they intensify the agitation [37].

Freshwater sponges, such as the "pipera-coora (*Ephydatia* sp.) from Australia and *Spongilla lacustri* from Europe, may cause irritation upon handling, an undefined substances from both were found to cause diarrhea, respiratory distress, and death in animals [89].

## Vertebrate Marine and Freshwater Animal Toxins and Venoms

Marine and freshwater vertebrate animal toxins ar present in some of the most familiar creatures to human (Table 19.4), such as a stingray gliding through sand beach waters, a catfish caught on a fishing line, or a se snake or sea turtle swimming in a coral reef. Thes particular marine vertebrates are responsible for th most significant envenomations to humans. Fish are ve tebrates represented by 22,000 species, accounting fo half the world's vertebrate population [131]; more tha 200 species of fish are venomous [132]. It has bee

**TABLE 19.4**
**Marine and Freshwater Vertebrate Animals: Venomous or Toxin Producing**

| Phylum | Class | Common Names | Genera (Toxin Producing) |
|---|---|---|---|
| Chordata | Chondrichthyes (cartilaginous fish) | Hornsharks, spiny dogfish | *Heterodontus, Squalus* |
| | | Stingray, bat rays, eagle rays, butterfly rays, river rays, whip rays, skate rays, stingaree | *Aetobatus, Aetomylaeus, Dasyatis, Gymnura, Mobula, Myliobatis, Potamotrygon, Pteromylaeus, Taeniura, Rhinoptera, Urogymnus, Urolophus* |
| | Osteichthyes (boney fish) | Boxfish, trunkfish | *Lactophrys, Ostracion* |
| | | Catfish | *Aurius, Centrochir, Clarias, Heteropneustes, Ictalurus, Liobagrus, Noturus, Pimelodus, Plotosus, Pterodoras, Tandanus* |
| | | Scropionfishes | *Centropogon, Dendrochirus, Gymnapistes, Helicolenus, Hypodytes, Inimicus, Neosebastes, Notesthes, Pterois, Scorpaena, Sebastes, Sebastodes* |
| | | Stonefish | *Synanceia (Synanceja)* |
| | | Toadfish | *Batrachoides, Thalassophryne* |
| | | Blennies | *Meiacanthus* |
| | | Rabbitfish | *Siganus* |
| | | Stargazer | *Uranoscopus* |
| | | Weaverfish | *Echiichthys, Trachinus* |
| | Reptilia | Sea kraits | *Laticauda* |
| | | True sea snakes | *Acalyptophis, Aipysurus, Astrotia, Disteria, Emydocephalus, Enhydrina, Ephalophis, Hyderlaps, Hydrophis, Kerilia, Kolpophis, Lapemis, Parahydrophis, Pelamis, Thalassophina, Thalassophis* |

speculated that the evolution of the venom apparatus in a large number of these fish is a result of their nonmigratory, slow-moving habits and their preference for sheltered, shallow-water environments [133]. It is these various vertebrate marine forms that presented the need for a more distinct differentiation between the terms *poisonous* and *venomous*. Marine animals that are considered poisonous generally possess an acquired toxicity as a result of their having ingested other toxin-producing organisms, while venomous animals manufacture their venoms in a distinctive venom gland primarily from their own endogenously produced toxins and deliver their venom via barbs, fangs, spines, or teeth that cause trauma in addition to the toxicity associated with the venom. The collective toxins that make up of venoms tend to be heat-labile, large-molecular-weight proteins. Poisons, in contrast, are primarily heat-stable, low-molecular-weight substances, and the poisonous animals involved lack a distinct anatomical venom delivery system [134]. Fish of the genera *Batrachoides*, *Echiichthys*, and *Syanchceia* are commonly considered scorpionfishes but are singled out in Table 19.4 because they are prominent venomous piscine genera. Sea snakes of the *Laticauda* genus have periods in their reproductive life cycles that require brief excursions to coastal beaches for egg laying. All other sea snake genera spend their entire lives in water, including giving birth to live young [135]; thus, sea snakes are truly marine animals and are included along with the boney and cartilaginous fishes for presentation and discussion in this vertebrate marine animal section.

## Chondrichthyes

Sharks from only two genera are known to be venomous (Table 19.4), and they are sometimes described as shark-like fish of the subclass Elasmobranchii [136]. Although it is likely that envenomations by these sharks may be more frequent than reported, literature that documents the consequences of envenomation by either genus is lacking. Envenoming by these sharks usually occurs to fishermen as a result of a dermal penetrating puncture by the sharks' venomous spines. They have two sharp spines with luminal venom ducts and no associated integumentary sheath enclosure that contain a whitish mass of venom-producing vacuolated cells. One spine is located immediately in front of the anterior dorsal fin and the other immediately in front of the posterior dorsal fin [68]. The Port Jackson shark (*Heterodontus portusjacksoni*) is native to Australian waters, and their spines inflict rough-edged lesions with venom delivery resulting in muscle weakness of the envenomated appendage [137]. The spiny dogfish (*Squalus megalops/cubensis*) from Brazil is known in the local region as "cacao-bagre." It caused significant injury to a fisherman who handled it. Immediate and intense pain occurred at the site of spine puncture, followed by erythema, edema, and localized kerotosis that persisted for 14 days [138]. Shark spine injuries have the obvious trauma component that accompanies venom effects and have been treated with wound irrigation, hot-water submersion, and antibiotics [36,134].

Stingrays are elasmobranchs, which are represented by at least one genus in nearly all of the oceans as well as the fresh waters of Central and South America. Their dorsoventrally compressed bodies are the perfect design to provide cryptic cover while the rays rest in shallow water. The tail extends from the flattened, tetrahedral-shaped body, making them appear like a kite. The tail may possess a single spine or multiple spines with highly serrated edges. Their body size may range from centimeters to more than 3 m, and their barb-like spines have highly serrated edges [134]. Spine morphology varies considerably between different genera and species. The venom is contained within specialized secretory cells inside an integumental sheath that surrounds the caudal spine.

Little is known about the specific chemical make-up of freshwater or marine stingray venoms, but they are thermolabile [139]. Venoms of the stingray genera *Patamotrygon* and *Urobatis* have been reported to cause cardiocirculatory toxicity as a result of direct effects on the myocardium [139,140], and neurotoxic effects have also been reported [141]. Recent studies of freshwater stingray venom from *Potamotrygon falkneri* using acrylamide gel separation revealed 18 different venom components, with molecular weights ranging from 12 to 130 kDa and exhibiting caseinolytic, gelatinolytic, and hyaluronidase enzymatic activities [142].

Envenomation occurs when the tail whips the spine forward and upward into a predator or other unfortunate victim when the stingray is alarmed, frightened, or stepped on, causing simultaneous rupturing of the sheath and the release of venom. Considerable mechanical tissue trauma may occur at the site of the laceration or puncture, resulting in severe bleeding, and the venom can add further insult by intensifying severe localized pain, erythema, edema, skin necrosis and ulceration. Systemic symptoms of dizziness, fever, migraine, myalgia, and vomiting have occurred from freshwater stingray envenomations [142]. Victims of marine stingray envenomation have died of cardiovascular failure [143], and in one case the reported cause of death was the result of cardiac tamponade and myocardial necrosis following puncture of the thoracic cavity by the stingray spine [144].

Stingray envenomation routinely is inflicted on the lower extremities of humans, and treatment may include immediate attempts to reduce pain by immersion of the affected area in hot water. The wound area may also be anesthetized with 1% lidocaine and irrigated with povidone iodine, followed by prophylaxis for infection and tetanus. The wound should be carefully examined to confirm that no residual spine fragments or other debris remain, as the removal of these in a timely fashion greatly reduces pain and secondary complications [142]. Severe cases of envenomation, usually those involving puncture of the upper torso, may require careful radiographic evaluation of the puncture site and advanced life-support measures [144].

## Osteichthyes

Venomous boney fishes are primarily inhabitants of tropical oceans but also exist in some temperate oceans and seas. Additionally, many species have become quite popular and are maintained by fish hobbyists and in private collections. The venoms of the boney fishes are produced, maintained, and delivered by anatomical structures similar to all species. This involves venom production by specific secretory glands within an integumental sheath, associated with spines that are anatomically located in conjunction with fins of the anal, caudal, dorsal, opercular, pectoral, pelvic, or shoulder areas of a given species. When a spine pierces tissue, the integrity of the integumental sheath is disrupted, and the venom is delivered into the wound [145]. It is notable that the inflicted spine injury from many species may in itself be more traumatic and damaging than the venom.

Fish venoms of nearly all species are used almost exclusively for self-defense; as such, there is a great degree of similarity in their chemical and pharmacological properties, and their differences are quantitative [146]. Like the venom of their cartilaginous relatives, boney fish venoms also exhibit labile sensitivities, their stability and activity being easily abolished by fluctuations in temperature, heat, and pH [147]. Venom toxins identified for piscine fish range from 15 kDa [148] to 800 kDa [147]. The pharmacological effects of piscine venoms, collectively, are represented by three major observed toxic effects *in vivo*: (1) their ability to induce serious cardiovascular toxicity, (2) their neuromuscular activity, and (3) their cytolytic activity. Mechanistically, it is thought that both the cardiovascular and neuromuscular effects are the result of the potent cytolytic property and that a single toxin in the venom is responsible for expression of all these activities [146].

Not surprisingly, given the pharmacological and toxicity similarities for most piscine fish venoms, the treatment of their envenomations is also similar. Early initiation of treatment is important and involves immediate irrigation of the wound, removal of any easily retrieved material, and immersion of the wound in hot water (not so hot as to cause burns). Analgesic drugs and the use of local anesthetic at the wound site should be instituted. If specific antivenom is available it should be administered. Timely follow-up with wound care is important to rule out the possibility of retained solid materials in the wound and to evaluate for secondary infection. Marine wound infections, when they occur, are commonly polymicrobial, and appropriate antibiotic administration should be initiated if symptoms warrant [149]. A few of the more well known groups and genera of boney piscine fish are now discussed in more detail.

### *Catfish*

Catfish are an easily recognizable whiskered fish, and approximately 1000 different species inhabit fresh and salt waters worldwide. They have dorsal and pectoral fin spines that may or may not be edged with sharp serrations and are associated with a venom-containing integumental sheath. These spinal fins are usually kept flush with the body; however, when the fish is distressed, the spines become locked in an extended position to inflict their wounds [68]. Although multiple genera are associated with toxic reactions (Table 19.3), the main adverse consequence of their inflicted stings is related to the trauma associated with the puncture of tissue.

In North America, catfish of the genera *Ictalurus* and *Noturus* are responsible for the most stings [150]. Venom of the white catfish (*Ictalurus catus*) has been reported to contain two to eight lethal factors and two dermonecrotic factors that are moderately heat stable [151]. In the Indo-Pacific region, the Arabian Gulf catfish (*Arius thalassinus*) and the striped catfish (*Plotosus lineatus*) are notorious for their ability to inflict extremely painful wounds and deliver several proteinaceous toxins, known as crinotoxins, in the stinging process. Both species have also been shown to contain these toxins in their skin secretions as well. Investigation of the activities of *A. thalassinus* toxins revealed that they contain vasoconstrictor components that can induce vasoconstriction in vascular smooth muscle [152] and may be related to muscarinic receptor activity [153]. The skin toxin of *P. lineatus* has yielded a hemolysin, two edema-forming factors, and two lethal factors [154]. Venoms of the Indian catfish species *Heteropneustes fossili* and *Plotosus canius* contain or cause the release of prostaglandins that induce smooth muscle contractions in experimental models [155] and have caused severe neurotoxic symptoms when injected into mice [156]. A lethal component of *P. canius* venom, toxin-PC, acts presynaptically to block neurotransmitter release resulting in neuromuscular blockade [157]. Given the array of catfish venom toxinological effects it is unfortunate that there is little reporting of envenomation cases that would provide a basis to develop optimal clinical management.

Catfish stings and envenomation most often occur to fishermen who are actively pursuing the capture of fish. Symptoms of bleeding, stinging, throbbing sensation, intense pain, and even peripheral neuropathy may result [158]. Treatment is comprised of judicious wound care and removal of any residual spine fragments in the wound. Secondary infections are a frequent complication with catfish injuries, and antibiotic therapy may be necessary [159,160].

### *Scorpionfishes*

Scorpionfish is a somewhat vague name that is frequently used to encompass a broad general taxonomical classification when referring to any one of a number of genera of anatomically similar fish (Table 19.3). They are commonly known throughout their wide range, as they are present in all oceans of the Earth, but venomous species are most prevalent in tropical waters. Out of this diverse group, approximately 80 species have been reported to possess venom toxins or to have caused envenomation to humans [161]. They are easily recognized due to their unique and often ornate appearance, an attribute that also makes them desirable to collectors. Some species, such as the lionfish (*Pterois volitans*) and zebrafish (*Dendrochirus zebra*) of the Indo-West Pacific oceans and Red Sea, are highly sought after by aquarists [162].

Scorpionfish venom glands are associated with spines in the configuration of 3 anal spines, 10 to 15 dorsal spines, and 2 pelvic spines. Venom extracts from lionfish (*Pterois volitans*) and soldier fish (*Gymnapistes marmoratus*) have been found to increase intracellular $Ca^{2+}$ in cultured neurons possibly as a result of pore formation in cell membranes. The increased $Ca^{2+}$ was associated with increased cell death due to cellular swelling and neurite loss [163]. In support of the calcium-ion role in toxicity is the observation of both adrenergic and cholinergic cardiotoxic effects from *P. volitans* [164] and *G. marmoratus* [165] crude venoms in animal tissue and cell culture models. Cardiovascular effects of *Scorpaena guttata* venom in rat hearts are biphasic, with an early reduction in chronotropic and inotropic effects being followed later by an increase in these effects. Thus, the venom acts on muscarinic and adrenoreceptors, respectively, and this may be the result of endogenous autocoid release [166]. The venom of the bullrout (*Notesthes robusta*), a species native to eastern Australia, contains a 169.8- to 174.5-kDa protein (nocitoxin) that is capable of pain induction in humans via polymodal nociceptor stimulation [167]. Collectively, the cardiotoxic and neurotoxic effects suggest that scorpionfish venoms act pre- and postsynaptically to induce cell membrane depolarization, and that this action is dependent on extracellular divalent cations [146]. Future research with scorpionfish venoms should provide a number of pharmacologic toxins useful as research tools and possible medical uses.

Envenomation by scorpionfishes nearly always results in profound pain and the possibility of significant swelling; also, systemic effects such as nausea, vomiting, diaphoresis, dyspnea, hypotension, weakness, syncope and collapse have been reported. Systemic complications are possibly related to the intensity of pain rather than true toxic effects of venom [168,169]. Treatment of envenomation is supportive; however, stonefish antivenom has been implicated as potential therapy for lionfish and soldierfish envenomations [170].

### Stonefish

*Synanceia* species, also taxonomically referred to as *Synanceja*, are some of the most researched venomous fish genera, and their venoms have yielded unique toxins with a variety of pharmacological activities [171]. The different *Synanceia* species are native to waters throughout the Indo-Pacific oceans and are known by a variety of local names, such as dornorn or wary-phoul in Australia, devilfish in Java, gofu in Polynesia, and sherova in East Africa; in the United States, use of the name stonefish is almost universal [134]. These slightly grotesque but quite interesting looking fish are morphologically reminiscent of an algae-, moss-, or barnacle-covered rock, and as such their looks complement their cryptic behavior. They lay on the bottom waiting to ambush unsuspecting prey and may also be accidentally stepped on by unobservant humans wading in the shallow waters they frequent.

Stonefish venom glands are closely associated with the thirteen dorsal spines, which appear to be covered with rough, thick skin. Two lateral grooves run the length of the spine, and the venom glands reside in the middle third of the spine shaft. The glands are sheathed in a fibrous, multilayered capsule that is vascular and has a nerve network. Stonefish will release a milky white dermal secretion, independent of their spine venom, when handled or agitated. Crude, fresh stonefish venom has been characterized as being opalescent in appearance and heat labile, with a pH of 6.0 [172].

The venoms of three species, *Synanceia horrida*, *S. trachynis*, and *S. verricosa*, have all been found to contain catecholamines [173], exhibit various enzymatic activities [174], and increase vascular permeability [172]. A lethal substance called stonustoxin (SNTX) that is comprised of two large subunits (alpha, 79.4 kD; beta, 79.3 kDa) has been isolated from Indian stonefish (*S. horrida*) venom [175]. SNTX induces cardiovascular hypotensive responses in rats that are thought to be nitric oxide mediated [176]. SNTX also inhibits muscle cell contraction by acting directly on the myocyte rather than neurotransmission blockade [177]. Estuarine stonefish (*S. trachynis*) venom contains the 158-kDa toxin trachynilysin (TLY), which elicits atrial muscle cell membrane hyperpolarizing actions [178], causes massive neurotransmitter discharge at low concentrations, and both muscle and nerve injury at higher concentrations in tissue preparations [179]. Interestingly, it has been reported that TLY effect mechanisms are quite similar to those produced by alpha-latrotoxin from the black widow spider (*Latrodectus* spp.) [180]. The venom of the reef stonefish (*S. verrucosa*) also has its own special toxin: verrucotoxin (Vtx), a four-subunit toxin comprised of two alpha (83-kDa) and two beta (78-kDa) subunits [181]. This toxin can also elicit hypotensive [181] and neurotoxic effects in animals [182]. These different stonefish toxins represent the most bioactive components of their respective crude venoms and account for all of the suspected lethal activity, suggesting that their venoms are comprised of a relatively small number of toxins [164]. Their pharmacological and toxicological effects in various cell and animal models may be useful in biomedical research, but it is difficult to clearly define specific functions that impact human health.

Clinical features of stonefish envenoming in humans involve local tissue trauma with gradually intensifying pain at the spine penetration site. Swelling and erythema may progress for up to 24 hours. Systemically, the key serious complications are cardiovascular related, with hypotension, bradycardia, pulmonary edema, arrhythmia and cardiovascular collapse associated with neurologic symptoms of muscle weakness, paralysis, and convulsions, potentially leading to death [183]. Although death has been reported [184], there may be possible inaccuracies with the documentation, and stonefish sting should not be a terminal life event.

Treatment of stonefish-envenomated victims is supportive and symptomatic, and specific antivenom produced in Australia (CSL, Ltd.; Parkville, Victoria, Australia) is effective in reversing many of the severe effects of venom toxicity [185]. This antivenom has been considered to possibly provide paraspecific coverage against the toxic effects of *Gymnapistes marmoratus* and *Pterois volitans* venoms based on animal studies, but this has not been tested in humans [170].

### Toadfish

*Thalassophryne* species are similar to stonefish in behavior and appearance. They inhabit the bottoms of coastal waters worldwide, and some species can use their swim bladders to sing a watery trill. They have two dorsal and two opercular spines connected to venom glands, and their venom is delivered with rupture of the integumentary spinal sheath upon stinging [134]. Venom is composed of heat-labile, bioactive toxins ranging in molecular weight from 18 to 97 kDa, that alter arteriolar and venous microvascular hemodynamics and cause myonecrosis [186]. Envenomation to humans results in severe pain, dizziness, and edema with a rapidly ensuing tissue necrosis [187]. Treatment consists of immersion of the wound in hot water, analgesics or local anesthetics, and supportive wound care [188].

### Weeverfish

*Echiichthys* or *Trachinus* species (weeverfish) predominantly inhabit the flat, sandy bottoms of European coastal waters, the Black and Mediterranean seas, Chile's Pacific coast, and the South African coast [68]. Taxonomical confusion may come from the fact that four species of *Ech*-*ichthys* were previously listed in the genus *Trachinus* [30 Although multiple species have been identified, two European species are well known for their toxic stings: the greater weeverfish (*Echiichthys draco*) and the lesser wee verfish (*Echiichthys vipera*) [68]. As its name implies, the greater weeverfish is fairly large and may reach a half meter in length. Weeverfish venom delivery is accomplished by any of the five dorsal or two opercular spines, which have a thin-walled integumentary sheath near their base covering the venom gland [134].

Weeverfish toxins are quite heat labile but have been isolated intact. Dracotoxin, a 105-kDa polypeptide with hemolytic properties, is produced by *Trachinus draco* [189], and trachinine, a 324-kDa compound composed of four subunits, has been isolated from *T. vipera* [190]. The crude venom of *T. draco* contains histamine and catecholamines, exhibits cholinesterase activity, and produces hypertension with subsequent sustained hypotension in cats, demonstrating a biphasic effect [191].

Weeverfish stings vary widely in their severity, from the usual symptoms of intense pain with local inflammation to extremely rare cases of secondary infections associated with fatality [192]. Generally, symptoms resolve in about a week, and treatment is the same as that used for other Osteichthyes envenomations [149].

### Sabre-Tooth Blenny

*Meiacanthus nigrolineatus* is briefly mentioned here because it is the only piscine fish known to deliver its venom by grooved canine teeth, in contrast to other piscine fish that deliver their venom via spines. Their venom gland is located in the lower jaw, and there is no information as to the chemical or pharmacological properties of their venom [193]. Their bite to humans is said to be painful [194].

### Rabbitfish

*Siganus* species are native to the Indo-Pacific oceans and seas. They have 13 venom-delivering dorsal spines, and the associated venom glands are located within anterolateral grooves that run the length of the spine. The four pelvic spines are similar in structural anatomy. This venom delivery and venom gland arrangement is quite similar to that of the lionfish. Like the sabre-tooth blenny, little is known about their venom, but stings are said to be comparable to those of other similar scorpionfish [134].

### Stargazer

*Uranoscopus* species earned the descriptive name of stargazers due to their small eyes which are located on the flat dorsal aspect of their cuboidal-shaped head, making them appear to be gazing upward toward the surface of the water. The stargazer has a short but sharp stubby spine and a poorly defined venom gland that surrounds the base of the spine on each side of its body in the region immediately above the pectoral fins. Knowledge concerning the chemical composition or pharmacological activity of their venom is lacking, and it is only the reporting of human fatalities caused by the Mediterranean species *Uranoscopus scaber* that supports their being considered venomous [195]. Investigators have tested for toxic effects in animals by injecting crude venom extracts from stargazer venom glands but were unsuccessful in confirming any toxicity [196]. This lack of toxicity confirmation may have resulted from the chemical and toxin instability of the venom following its removal from the venom gland. Treatment of stargazer envenomation should be consistent with that of other piscine envenomations [197].

### Boxfish and Trunkfish

*Lactophrys* and *Ostracion* are two genera to be discussed briefly because they produce toxins in their skin and secrete them rather than delivering them with a specialized apparatus. They are native to the Indo-Pacific. Their toxins are used as a defense against other fish predators and therefore are ichthyotoxic. By releasing these natural fish-killing toxins into their immediate environment, they effectively deter their predatory counterparts and survive to live another day [134]. The primary toxin produced by both these genera is pahutoxin (PHN), a choline chloride ester of 3-acetoxypalmitc acid that has phospholipid membrane permeability disruptor and surfactant properties [198]; thus, the dual pharmacologic actions of the lipophilic and hydrophilic components appear to function synergistically. The regulatory mechanism of this action stems from two distinct fractions in trunkfish skin secretion and has been proposed to be receptor mediated [199]. The consequences to humans who make contact with these skin secretions are dermal irritation, and persistent complications would not be expected to follow [30].

## Reptilia

Sea snakes navigate coral reefs and oceans in the world's subtropical and tropical regions but have been observed thousands of miles away from these regions, even in the absence of oceanic currents or foul weather; thus, they are the most widely dispersed venomous reptile on Earth [200]. Their morphological appearance and physical body structure, although similar to terrestrial snakes, is usually distinguished by the presence of laterally compressed tails with a paddle-like look. Their physiology allows them to dive to depths of 100 m and stay submerged for more than 3 hours, but they rely on atmospheric air to survive. There are 17 different genera (Table 19.4), and they average approximately 1 m in length, but some species (e.g., *Laticauda colubrina*) approach 3 to 4 m in length. Sea snakes usually have short (<4 mm), conical, fixed-position fangs, and the fang length varies according to the prey eaten. The fangs are luminal and interface with ducts that connect the fangs to elongated venom glands lying in the upper jaw on each side of the head. Venom glands are fibrous encapsulated and attach to adductor muscles that contract to express venom through the venom duct network and into prey [135].

Sea snake venoms are complex toxin blends with varying combinations of protein toxin types and are extremely toxic. Their venoms are the most potent of any vertebrate animal on Earth. The venom yield delivered from most sea snake species ranges from 1 to 10 mg, but species such as *Astrotia stokesii* can deliver more than 20 mg. As little as 0.04 mg/kg of *Aipysurus duboisii* venom is fatal to mice. Venom dose–response studies suggest that the beaked sea snake (*Enhydrina schistosa*) delivers enough venom in its maximum yield to kill approximately 53 humans [135]. Whole sea snake venom possesses several enzymes and is comprised of acetylcholinesterase, hyaluronidase, leucine aminopeptidase, 5′-nucleotidase, phosphomonoesterase, phosphodiesterase, and phospholipase A [201]. Toxins isolated from venom usually contain 60 to 62 amino acid residues, and 29 different toxins have been isolated from 9 separate species of sea snakes. Purified toxins maintain their integrity, as they have been demonstrated to be heat stable, as well as acid and base pH tolerant, and they are more potent than crude venom based on animal $LD_{50}$ experimental data [202]. The potency of sea snake venoms is linked to both pre- and postsynaptic neurotoxins. An isolated postsynaptic protein toxin of 6- to 8-kDa molecular weight is highly stabilized by multiple disulfide bridges and is a primary pharmacologically active toxin [202]. The postsynaptic toxin binds to the acetylcholine receptor. Venom from the beaked sea snake contains a phospholipase, and the venom of *Laticauda semifasciata* contains the curaremimetic toxin erabutoxin [203], which has a demonstrated presynaptic toxic effect plus phospholipase activity [203]. However, the net effect of venom on living animals envenomated by sea snakes is paralysis resulting from irreversible binding of the postsynaptic toxin at the acetylcholine receptor in the neuromuscular junction [204].

Envenomation by sea snakes to humans is a serious medical concern. Sea snake fangs, being small, do not make deep punctures, and fang marks (which may appear as scratches) may not be easily visualized. Additionally, the bite is frequently asymptomatic, and local symptoms are usually absent. Envenomation is manifested by diplopia, dysarthria, dysphagia, muscle fasciculations, muscle pain, shock, trismus, and paralysis [205]. Myoglobinuria, myonecrosis, and nephropathy have also been observed [206]. Neurotoxic symptoms may be deferred for up to 3 hours, often causing the victim to delay seeking medical attention. Fatalities usually result from paralysis of the diaphragm, progressing to respiratory arrest due to a lack of timely treatment. A pressure immobilization method of first aid has been found to be beneficial and may be of major importance when a victim is not near a medical facility [205]. Cardiopulmonary support is effective therapy, and sea snake antivenom is a definitive therapy. Apparently, Australian sea snake antivenom appears to be effective for envenomation by all sea snake species [207].

## Terrestrial Animal Toxins

Land-dwelling animals of the world are found on all the continental land masses of the Earth, although the diversity and numbers of different species are greatest in tropical regions. As the distance from the equator becomes greater, the diversity and number of animal species decreases; thus, an inverse relationship exists between the number of species and distance from tropical climates, a relationship that is paralleled with respect to the diversity and number of animal toxins. In recent centuries, a confounding element in these relationships has been humans. As humans have increasingly traveled the Earth, there has been an associated increase in the dispersal of species to regions beyond their natural ranges. Additionally, as the world's population increases and expands into natural environments there is an ever-increasing interface between humans and terrestrial animal species. This leads to greater numbers of human encounters with animals and consequent envenomations from insects, centipedes, scorpions, spiders, amphibians, lizards, and snakes. The frequency of encounters with these toxin-producing species is probably greatest in tropical third-world countries, where the availability of medical treatment is nonexistent or limited. Unfortunately, the consequences of humans living in close proximity to potentially dangerous creatures may also lead to the disappearance of animal species over time and the loss of knowledge related to their natural toxins. The toxins from stinging, biting, and skin-secreting terrestrial species, both invertebrate and vertebrate, have received considerable attention in biomedical research for scientific and medical reasons.

## Terrestrial Invertebrate Animal Toxins and Venoms

The terrestrial invertebrates capable of delivering toxins via bites, stings, skin secretions, or bristle and hair contact are members of the largest phylum in the animal kingdom, Arthropoda. These relatively small, often thought of as creepy, creatures are probably responsible for envenomations to humans in numbers exceeding the sum of envenomations from all other animal phyla [132]. Arthropods all have legs and walk, run, or fly. There are three main classes of toxin- or venom-producing members (Table 19.5).

### Arachnida

Arachnids differ from centipedes and insects by possessing eight legs, in contrast to the large number of centipedes and the six on insects. They do not have antennae or wings but can live in a variety of ecosystem environments. Their toxins represent an amazing pharmacological arsenal capable of subduing a spectrum of

**TABLE 19.5**
**Terrestrial Invertebrate Animals: Venomous or Toxin Producing**

| Phylum | Class | Common Names | Genera (Toxin Producing) |
|---|---|---|---|
| Arthropoda | Arachnida | Scorpions | *Androctonus, Botothus, Buthus, Centuroides, Heterometrus, Leiurus, Mesobuthus, Opistophthalmus, Orothochirus, Palamnieus, Scorpio, Tityus* |
| | | Spiders | *Achaearanea, Aganippe, Agelenopsis, Araneus, Arbantis, Argiope, Artosa, Atrax, Badumna, Bothriocyrtum, Breda, Chiracanthium, Cupiennius, Diallomus, Dolomedes, Drassoides, Dysdrea, Eriophora, Fillistata, Hadronnyche, Harpactirella, Hermeas, Heteropoda, Holoplatys, Isopeda, Ixeuticus, Lampona, Lycosa, Latrodectus, Liocranoides, Loxosceles, Missulena, Misumenoides, Neoscona, Nephila, Olios, Opisthoncus, Peucetia, Phidippus, Selenocosima, Steatoda, Tegenaria, Thiodina, Trechona, Ummidia* |
| | | Bird/goliath spiders | *Ornithoctonus, Theraphosa* |
| | | Tarantulas | *Acanthoscurria, Avicularia, Brachypelma, Megaphobema, Phoeneutria, Phrixotrichus, Poecilotheria, Stromatopelma* |
| | Chilopoda | Centipedes | *Cormocephalus, Ethmostigmus, Otostigmus, Rhysida, Scolopendra, Scutigera* |
| | Insecta | Ants | *Mrymecia, Paraponera, Solenopsis* |
| | | Caterpillars | *Doratifera, Euproctis, Lagoa, Lonomia, Thaumetopoea* |
| | | Bees, bumblebees | *Apis, Bombus* |
| | | Hornets, wasps | *Dolichovespula, Dasymutilla, Vespa* |
| | | Yellowjackets | *Paravespula, Vespula* |
| | | Sawflies | *Arge, Lophyrotoma, Perreyia* |

prey, and multiple arachnid forms pose significant danger to human health in some regions of the earth.

### *Scorpions*

Scorpions are represented by at least 1500 individual species from multiple genera (Table 19.5) that live in arid and wet ecosystems. They are the largest of the arachnids in size and may possibly be the oldest living creatures on Earth (400 million years) [208]. Scorpions look like a small, earth-tone-colored lobster carrying a thin tail, with an inverted hook, arched over its back. The scorpion's body is comprised of three distinct sections, but the head and thorax are joined to form a cephalothorax, which has pincers, or pedipalps, which are used for restraining prey and feeding themselves. The pincer-like claws are actually a modification of the first set of legs, and the remaining three pairs of legs are used for walking. The body part most humans and prey are familiar with is the stinger, which is attached to the telson, a terminal distal segment that houses venom glands connecting to two orifices in the stinger. Scorpions sting with this apparatus and deliver their venom into a variety of prey organisms, including other arthropods and humans. Scorpions are fossorial and reclusive in nature, tending to be nocturnally active.

Scorpion venoms have been the focus of ion channel researchers ever since the isolation of noxiustoxin, a short-chain venom peptide, in 1928 [209]. Scorpion venoms are combinations of approximately 50 to 100 low-molecular-weight polypeptide toxins. It has been estimated that more than 100,000 different peptides exist among all of the scorpion species [208]. Of this large number of species and toxins, only 30 scorpion species and an associated 200 distinct polypeptides have been investigated and their functional properties reported [210]. Scorpion venoms contain various polypeptides with an array of pharmacological and physiological actions highly specific for crustaceans, insects, and mammals that have been of interest in bioinsecticidal research and development [211].

All scorpion toxins are constructed with an alpha-helical segment and two or three anti-parallel beta-sheet folds tightly held by three or four disulfide bridges that maintain the tertiary conformational structure. Four main families of scorpion ion channel toxins have been proposed, with their separation being based on their effects on membrane ion channels: $Na^+$ channels [212], $K^+$ channels [212], $Cl^-$ channels [213], and $Ca^{2+}$ channels [214]. Of these four ion channel families, two main groups, based on their pharmacological activity and molecular size, have been identified. The short-chain toxins are comprised of 23 to 47 amino acids that primarily affect $K^+$ channels [215]; the long-chain peptides are comprised of 59 to 76 amino acids that target their effects on voltage-gated $Na^+$ channels [216]. It is also evident that scorpions of nearly all of the toxin-producing genera possess varying combinations of ion channel toxins [217].

Scorpion potassium channel toxins are diverse and have been divided into alpha-, beta-, and gamma- scorpion toxin subfamilies based on their biophysical properties

[215]. These subfamilies have been further split based on their voltage-gated and ligand-gated $K^+$ channel actions [218]. Scorpion $K^+$ channel toxins have been identified from representatives of several different genera, including *Androctonus mauretanicus* (kalitoxin), *Buthus martensi* Karch (BmTX3), *Centuroides noxius* (noxiustoxin), *Leiurus quinquestriatus* (charybdotoxin), *Hterometrus spinnifer* (HsTX), and *Scorpio maurus* (maurotoxin) [219].

The scorpion sodium-ion-channel, long-chain toxins are important in their toxic effects, as 90% lethality is attributed to these toxins binding voltage-sensitive $Na^+$ channels in muscle and nerve cells of mammals [220]. Scorpion sodium-ion-channel toxins are designated as alpha- or beta-toxins depending on their binding site of action in the sodium-ion channel [216]. Some of these toxins have been found to be arthropod selective with regard to their toxicity in crickets, crabs, squids, and triatomides (kissing bugs) while lacking apparent toxicity in mammals. This is the case for the sodium-ion-channel toxin ardiscretin, isolated from *Tityus discrepans* venom [217]. In contrast, alpha-toxins from *Buthus* venoms have significant toxic effects in mammals [221]. Similarly, the sodium-ion-channel toxin Lqh-alpha-IT, isolated from *Leirurus quinquestriatus hebraeus* venom, contains an insect-specific alpha-toxin, while an alpha-toxin from *Androctonus australis* affects primarily mammals. Thus, despite alpha-toxin similarities between genera, the diverse effects on target organisms suggest that relatively minute differences in structure are responsible for the expression of different functions and that alpha insect and alpha mammal toxins may have been derived from a common genetic ancestor [222].

Scorpion envenomation is a serious public health problem in India and Mexico, the Mediterranean, the Middle East, northern and southern Africa, and South America, but relatively few species, despite the large number of toxin-producing species, are worthy of medical concern. In areas such as Tunisia, however, where up to 40,000 scorpion stings occur annually, resulting in 20 to 40 deaths, the medical treatment of scorpion envenomation is frequent [223].

Scorpions will sting when they are laid upon, sat on, stepped on, or agitated by any squeeze contact with clothing and shoes, and they can deliver a painful sting when their arched telson directs their stinger into the victim. Stings frequently occur in the evening hours when people are walking around in their bare feet. Although adult humans may suffer severe complications with scorpion stings, children suffer the greatest mortality [224,225]. Given the pharmacological properties of scorpion venom, it is not surprising that stings, depending on the genus and species involved, can cause local pain, autonomic and central nervous system complications, seizures, and cardiopulmonary dysfunction [225], with shock and pulmonary edema being responsible for most fatalities [226]. It is also important to note that hemolysis frequently develops in severe envenomations, and secondary nephrotoxic effects can occur [206].

Medical management strategies vary with different global geographic regions but have revolved around treatment being administered in intensive-care facilities, the use of various pharmacotherapies, and the administration of specific antivenom [225]. The use of antivenoms and their effectiveness have been of great controversy historically, with treatment involving large antivenom doses that result in anaphylaxis and delayed serum sickness [226]. Efforts are currently underway to produce more effective antivenoms with reduced adverse-effect profiles using recombinant technology [223].

### *Spiders*

More than 38,000 spider species, comprising 3526 genera are known [228]. They inhabit nearly all ecosystems, ranging from the arid hot deserts to the rain forests and from the peaking slopes of Mount Everest [229] to underwater worlds [230]. Spiders are masters of deception, existing in a wide variety of shapes and colors and exhibiting equally diverse behaviors. They have been documented to mimic different vegetation forms, insects, bird dung, twigs, and gastropods, and some, such as the crab spiders, can even change color to camouflage themselves against their preferred foliage [231]. They range in size from the size of a pin head (*Patu marplesi*, Samoan moss spider) to the 25-cm leg span of the bird spider (*Theraphosa blondi*) [232]. Like scorpions, their head and thorax are combined to form a cephalothorax, but their abdomen is differentiated from this anatomical structure by a narrow restricting band, or pedicel, making the abdomen pronounced. Spiders usually have eight eyes, but some, such as the brown spider (*Loxosceles* species), have only six and some cave-dwelling species are blind. Eyesight is limited in most species, but some hunting spiders such as the jumping spider (*Portia fimbriata*) have markedly acute vision [233].

Almost all species of spiders have venom glands and produce toxins, but their ability to be of harm to humans or large prey is usually limited by the size of their paired fangs, which reside with the chelicerae at the anterior of their mouth. The chelicerae or fangs are quite small in the majority of spider species, and as such are of no danger to humans. Spiders are predatory by nature, and their venom delivery is quite effective for prey immobilization. They do not swallow their prey following their paralytic envenomation but rather extract prey contents using their siphoning stomachs. The digestive enzymes used in the liquefaction of their prey are not venom components. Two infraorders represent most of the spiders with which humans are familiar. Mygalomorphae include the tarantulas and venomous funnel-web spiders of the *Atrax* and *Hadronyche* genera, which possess large paraxial fangs that move in a vertical up and down fashion. Fangs of these large spiders can puncture a fingernail [232]. The smaller Araneomorphae represent approximately 93% of the Earth's spiders; they have smaller diaxial fangs that move in a horizontal fashion side to side [228]. Although Araneomorphae spiders are not as large as the Mygalomorphae, the araneomorph species are responsible for most bites to humans and are truly dangerous to humans; thus, an inverse relationship exists with regard to size and toxicity, as the smaller spiders are generally more toxic. The spider genera known to produce toxic bites to humans are listed in Table 19.5.

Spider venoms are heterogeneous toxin mixtures composed of 3- to 8-kDa polypeptides, with a few being greater than 10 kDa in mass. The black widow (*Latrodectus* sp.) possesses a 100-kDa polypeptide toxin [234]. Spider venoms share many pharmacological properties similar to scorpion venoms with respect to ion channel actions. It has been estimated that the total number of diverse spider venom polypeptides may be in excess of 1.9 million [235]. Spider venom components have been generally identified as neurotoxins that affect glutamatergic transmission and $Ca^{2+}$, $Na^{+}$, $K^{+}$, and $Cl^{-}$ ion channels; they stimulate neurotransmitter release and block postsynaptic cholinergic receptors [236].

The chemical configurations of several glutamate receptor toxins have been elucidated and determined to be primarily acylpolyamines and polyamine amides [237]. A few of the identified toxins include joro spider toxin (JSTX) from *Nephila calvata*, nephila spider toxin (NSTX) from *Nephila maculata*, argiopine from *Argiope lobata*, curtatotoxins from *Hololena curta*, and PhTX-4 toxins from *Phoneutria nigriventer* [236].

Ion channel spider toxins have been extensively studied, and they have significant neuropharmacological properties that affect the release of neurotransmitters. $Ca^{2+}$ ion channel venom studies led to the discovery of omega-agatoxins from *Agelenopsis aperta* and paved the way for the discovery of omega-grammatoxin (*Grammostola spatulata*), atracotoxin (*Atrax robusta*), and many other omega-toxins. $Na^{+}$ ion channel toxins impair excitable cell membrane function by altering ligand-gated sodium channels such as nicotinic cholinoreceptors and voltage-gated sodium channels that impact neuromuscular function, making them effective paralytic toxins. Spiders in the *Atrax* and *Hadronchye* genera (funnel-web spiders) and the genus *Phoneutria* (wandering spiders) have venoms possessing important $Na^{+}$ ion channel toxins. $K^{+}$ ion channel toxin actions are similar to those reported for scorpions. $K^{+}$ ion channel venom toxins such as hannatoxin from the rose tarantula (*Phrixotrichus spatulata*), phrixotoxins from the Chilean fire tarantula (*Phrixotrichus auratus*), and heteropodatoxins from *Heteropoda venatoria* are just a few of the many pharmacologically active potassium channel toxins. $Cl^{-}$ ion channel spider toxins, although not critical in action potential mechanisms, are involved in transepithelial transport [238] and cellular excitability as receptor components for inhibitory neurotransmitters such as GABA [239]. Spider toxins and venom fractions that affect cholinergic transmission, in both the autonomic and central nervous system, have been isolated from several species of orb-weaving spiders in the genus *Argiope* [240].

Enzymes in spider venoms are important components responsible for a number of different pharmacological functions. Collagenase activity in venom fractions from Australian spiders of the genera *Eriophora* and *Nephila* (orb weavers) and *Isopeda* (huntsman spiders) has been identified [241]. Hyaluronidase, believed to enhance the spread of venom, has been found in venoms from the genera *Cupiennius*, *Lycosa*, *Loxosceles*, and *Phoeneutria* [236], and metalloproteinases with proteolytic activity have been isolated from *Loxosceles* venom [242]. Phosphodiesterases that cleave cyclic nucleotides have been found in the venoms from *Atrax robusta*, *Aphonopelma cratus*, and *Latrodectus mactans* [243], and sphingomyelinase D-type, a powerful necrotizing toxin, has been characterized from *Loxosceles reclusa* venom [244].

The effects of spider venoms and toxins in cases of spider bite to humans are surprisingly limited, and although many spider genera contain species that have bitten humans the majority of bites are relatively innocuous, with only a few species being responsible for significant envenomation profiles (Table 19.5) [245]. *Badumna* species (black house spiders), *Lycosa carolinensis* (wolf spider), and *Chiracanthium mildei* (sac spider) can cause pain and erythema, and *Tegenaria agrestis* (hobo spider) can cause necrotic lesions, but all of these symptoms generally resolve without long-term consequence [246]. Spider bites, in contrast to many insect bites and stings, are not associated with inducing allergic reactions [245].

Spiders of medical importance (capable of causing significant morbidity or mortality) include the Australian genera *Atrax* and *Hadronyche* (funnel web spiders), Brazilian *Phoeneutrea* species (wandering spiders), *Loxosceles* (recluse spiders), and *Latrodectus* and *Steotoda* (widow and comb-footed spiders) [246]. Funnel web spiders of Australia (*Atrax* spp.) are believed to be the most venomous and dangerous spiders. Although bites are uncommon and severe envenomation is infrequent, they can cause life-threatening symptoms, and deaths have occurred [247,248]. Symptoms associated with severe envenomation include both parasympathetic and sympathetic effects on neuromuscular and cardiopulmonary systems, with paralysis, coma, and profound hypotension resulting in general system failure [248,249]. Fortunately, a funnel web spider (*Atrax robustus*) venom-specific IgG antivenom is available that is effective in reversing venom-induced toxicity. This antivenom is poly-specific and is also effective in treating envenomations caused by other *Atrax* and *Hadronyche* species [246].

Armed, wandering, or banana spider (*Phoeneutria* spp.) envenomations in South America can result in significant medical complications. They have caused at least one pediatric death and are responsible for 42% of the 2850 spider bites reported each year in Brazil [250]. These spiders occasionally appear in other parts of the world after they have been inadvertently imported on agricultural produce. Less than 1% of victims experience severe complications (pulmonary edema). The common symptoms of envenomation are pain and edema. Local anesthetics and oral analgesics are of benefit, and a $Fab'_2$ antiarachnid antivenom has been used in moderate to severe envenomations, but its true efficiency is unknown [250].

Recluse, or brown, spiders (*Loxosceles* spp.) are capable of inflicting severe ulcerating and tissue-necrotizing bites and are considered a serious hazard in Central and South America, as well as south–central and southeastern parts of North America. They have also appeared in Europe, Asia, South Africa, and Australia, most likely having been transported into these countries. Little is known of their bites in these settings [246]. Their venoms are known for the 23-kDa enzyme shingomyelinase D, which has four isoforms and is the main pharmacologically active venom component responsible for most toxic effects [251]. Many spider bites are attributed to *Loxosceles* species, but rarely are bites confirmed to be true recluse bites, and misdiagnosis follows [252]. Bites frequently go unnoticed, as they are relatively painless, and victims delay in seeking medical attention.

Envenomation by *Loxosceles* may demonstrate a dermonecrotic profile or may appear as a viscerocutaneous form [246]. A necrotic lesion may take up to a week to express itself. The resulting ulceration is insidious, and may persist for months. Fever and fatigue may also be present. The viscerocutaneous symptoms of *Loxosceles* envenomation may be fulminate, with hemolysis, coagulopathy, shock, renal failure, and multiple system failure leading to death if untreated [253].

The persistent nature of recluse spider bite lesions has led to considerable controversy with regard to the most effective treatment, but in general good, thorough wound care, following an early diagnosis, leads to a favorable outcome [252]. Therapies ranging from hyperbaric oxygen to nitroglycerin patches have been used with varying effectiveness [254]. Life-support measures and good intensive care are best in cases of severe systemic envenoming [254]. Although antivenom is available in South America, its benefit is questionable [253].

Widow spiders (*Latrodectus* spp.) are found throughout the world, and as many as 40 species are known [246]. Envenomation may cause significant morbidity and has resulted in death [255,256]. *Latrodectus* venoms, and clinical features following envenomation, are similar and have a common clinical syndrome throughout all geographic regions [246]. Widow venoms contain latrotoxins, which are >100-kDa proteins. The best known is alpha-latrotoxin. They stimulate massive neurotransmitter release from a variety of nerves [236]. The tissue around a bite does not react to envenomation, but local pain may be present. Severe skeletal muscle pain in the abdomen, back, and chest; diaphoresis; hypertension, and cramping are symptoms requiring medial treatment [255]. Spiders in the genus *Steatoda* have been reported to have symptoms similar to widow spiders, but less severe in effect. Treatment may include parenteral benzodiazepines or opioids, and antivenoms are available in Australia and the United States. Antivenom, given intravenously, is beneficial in cases of severe envenomation [255,246]. The Australian redback spider antivenom has also been used in severe cases of *Steotoda* envenomation [245].

Tarantulas, due to their size and reputation, deserve a final word with respect to spiders. There are approximately 860 known species of tarantulas comprising 107 genera. They are found in semitropical, semitemperate, and tropical regions around the world. They are predators that inhabit many biodiverse ecosystems and are capable of physically overpowering and mechanically injuring invertebrate and vertebrate prey. Their uniqueness and notoriety have resulted in their becoming an amazingly significant portion of the world's pet trade, and as a result multiple species of several genera have been afforded legal protection. Their large fangs deliver rapid-acting venoms that have irreversible effects on central and peripheral nervous systems, leading to the induction of rapid paralysis, even in higher vertebrates [257]. Despite the magnitude of their notoriety and their anatomical and venom pharmacological traits, tarantulas are not particularly dangerous to humans.

Few details are available concerning tarantula venoms. Crude venom studies involving mice and 55 different tarantula species showed that certain species' venoms can induce massive neurotoxicity leading to death in 3 to 5 minutes, while other venoms have similar, but reduced effects, resulting in death after more than 2 hours. The most toxic venoms come from the arboreal genera *Heteroscodra*, *Poecilotheria*, and *Stomatopelma*; their venoms cause death to mice within seconds [257]. Venom of the Chinese bird spider (*Ornithoctonus huwena*), known as *dilaohu* (the "earth tiger") in China, contains a series of toxins known as huwentoxins [258]. These are small (3000- to 5000-Da) neurotoxin peptides with voltage-gated $Ca^{2+}$ and $Na^{+}$ ion channel activities. They represent a significant portion of crude venom and are capable of causing lethality in mice and birds in less than 2 minutes [259].

Bites to humans usually cause mild to severe pain, burning sensation, edema, erythema, itching, and joint stiffness. Severe envenomations, as seen with arboreal Asian (*Haplopelma* spp.), Sri Lankan and Indian (*Poecilotheria* spp.), and South African (*Stromatopelma* and *Pterinochilus* spp.) tarantulas have been associated with more severe clinical problems, such as intense pain and muscle cramps of several weeks' duration, temporary paralysis, and coma [260]. South American tarantulas (*Theraphosa* spp.) have body hairs (setae) capable of eliciting dermal urticaria [261] and ocular injury [262].

Tarantula bite treatment has not been specifically defined, but symptomatic treatment with analgesics, antihistamines, topical steroids, and supportive care as needed has been used with success, even in cases of severe envenomation such as those involving Sri Lankan species [263].

*Chilopoda*

Centipedes are arthropods familiar to most people, and it is usually an unpleasant familiarity. Perhaps the words of arthropod scientist J.L. Cloudsley-Thompson best characterize centipedes: "Centipedes seem to exert a weird fascination on the morbid appetites of the hysterical and insane" [264]. Centipedes are globally distributed, being native to the Americas, Africa, Australia, Asia, and Europe. They are nocturnally active, fast moving, and aggressive. Their segmented bodies have one pair of legs per segment, and their compound eyes have up to 200 optical units. Centipedes range in length from 3 to 250 mm [264]. They are carnivorous and use their powerful poison claws, which are actually modifications of their first pair of legs, to deliver venom from an internal sac into their prey, causing rapid paralysis [265]. Their prey includes other small arthropods, including other centipedes, and they have been known to feed on toads, snakes, and other small vertebrates [266]. There are approximately 3000 known species, but those of the Scolopendromorpha and Scutigeromorpha orders are most important clinically [267].

Centipede venoms, like snake venoms, are complex protein mixtures. They have been found to contain esterases, histamine, 5-hydroxytryptamine, lipids, polysaccharides, proteinases, and other various enzymes. An acidic, heat-labile, 60-kDa protein cardiotoxin (toxin S) has been isolated from *Scolopendra subspinipes* and could induce hypertension in cats [268]. In contrast, extract from *Scolopendra moristans* venom caused cardiac asystole in the toad heart [269]. Studies of centipede venom on other arthropod nerve systems, using *Scolpendra* venom fraction (SC1), revealed that the fraction possessed muscarinic agonist properties that acted directly on an insect muscarinic receptor subtype closely homologous to the M1–M3 (muscarin) receptors found in mammals [270].

Envenomation by *Scolopendra* centipedes frequently results in pain and occasionally local necrosis but generally does not cause serious complications [271]; however, rhabdomyolysis [272] and fatality have been reported [268]. A study of confirmed Australian centipede envenomations involving *Cormocephalus*, *Ethmostigmus*, and *Scolopendra* species showed that bites from all three genera could cause significant pain, erythema, swelling, and itching, with *Ethmostigmus* and *Scolopendra* symptoms being more severe [267]. Ingestion of *Scutigera morpha* by an infant (confirmed by an intact centipede found in the feces) resulted in systemic effects, evidenced by paleness, muscle hypotonia, and vomiting, followed by spontaneous recovery over the 48-hour period following the ingestion [273]. Treatment with ice packs, analgesics, and hot-water submersion has been an effective intervention [267].

## Insecta

Insects are the most ubiquitous of all terrestrial animal forms; they inhabit every ecological niche on the planet. They have six legs (or many more, in the case of larval forms such as caterpillars), which allows them to navigate a multitude of substrates. Many possess wings, they have three hinged body parts (head, thorax, abdomen), and they all are covered with a chitinous exoskeleton. Additionally, they possess a variety of biting mouth parts and venom-delivery devices. Although many insect species are vectors for a large number of infectious diseases, few have truly toxic venoms of concern. Those that do possess toxins in their venoms are incapable, as a single entity, of causing fatality to humans; however, single assaults can cause toxic reactions and elicit immune responses in humans, and multiple concurrent insults by some species (e.g., swarms of Africanized bees) can result in human fatality. In fact, the venoms of most insects in the order Hymenoptera, including fire ants, honeybees, and vespids, are highly immunogenic due to venom proteins [274]. Venom doses range from <1 μg for fire ants, 10 μg for wasps, and up to 50 μg for bees [274]. The allergic reactions resulting from insect stings are a more clinically prevalent problem than complications associated with direct venom toxicity.

*Ants*

Only a few species of ants capable of inflicting significant bites and stings are of any significance with reference to venom toxinology. Formic acid is a common component of venom sprayed for either defensive or prey-acquiring purposes by many genera (*Camponotus*, *Formica*, *Lasius*, *Poyergus*), but it is a chemical toxin of little consequence to humans, other than causing mild dermal or ocular irritation. In contrast, South American fire ants (*Solenopsis* spp.) [275], Australian/Tasmanian bulldog and "jack jumper" ants (*Myrmecia* spp.), and Costa Rican and South American tropical ants (*Pseudomymex triplarins* and *Pachycondyla goeldii*) can potentially inflict severe stings [276]. Unfortunately, these ants have been relocated to geographic regions far removed from their native lands, and fire ants have been present in the United States since the early 1920s [277].

Fire ant (*Solenopsis* spp.) venom contains the alkaloids, 2-methyl-6-alkyl and alkenyl disubstituted piperidines, commonly referred to as solenopsins. These toxins possess cytotoxic activity that alters the functional integrity of mast cell membranes, causing intracellular degranulation and the subsequent release of histamine. The allergenic component of venom is due to low-molecular-weight proteins, four of which are present in fire ant venom [278]. Fire ants have an aggressive stinging behavior with envenomation resulting in a wide range of effects including dermal necrosis, edema, sterile pustules (occurring several weeks after envenomation), urticaria, and, in extremely rare circumstances, anaphylactic shock and fatality [279]. The quantity of venom delivered by fire ants ranges from 10 to 100 ng [280], and the severity of envenomation appears to be dose dependent [281]. Management of fire ant stings is not consistent, but antihistamines, topical corticosteroids, cold compresses, and antibacterials have been used [277].

Jack jumper ant (*Myrmecia* spp.) venom has enzymatic activities due to its phospholipase $A_2$, phospholipase B, hyaluronidase, acid phosphatase, and alkaline phosphatase fractions [282]. Venom also contains two basic peptides: pilosulin 1, which elicits cytotoxic effects [283], and pilosulin 2, which has antihypertensive properties [276]. Envenomation by the jack jumper ant demonstrates that its venom is a strong insect allergen, as it is responsible for more than 90% of ant venom anaphylaxis cases in Australia as well as deaths in Tasmania [284]. The allergic problems associated with "re-stings" in victims became so serious that a major effort was made to develop effective immunotherapy to prevent sting anaphylaxis [284].

One mimicking insect often confused with an ant is the solitary female wasp in the genus *Dasymutilla* (Table 19.5). These wingless female wasps have the morphological appearance of a large ant and are commonly referred to as velvet ants (*Dasymutilla magnifica*). As their name implies, their bodies are covered in dense, brightly colored hair, giving them a velvety appearance. Venom of *Dasymutilla* species contains phospholipase and hyaluronidase but is low in other enzymatic activities [285]. Their sting is one of the most painful of all insects, and for one species (*Dasymutilla occidentalis*) the pain is so severe they have been dubbed "cow killers," [286] although in reality the sting is not strong enough to kill a cow.

#### *Caterpillars*

Caterpillars are the insect larval form of butterflies and moths (Lepidoptera). Familiar, captivating, and wondrous to children all over the nonarctic regions of the Earth, caterpillars are not necessarily innocent, fuzzy, colorful, crawly creatures. In many countries, they are considered a major economic agricultural pest because of their defoliating appetites. Many species probably acquire their toxins from the foliage they feed on, such as *Lagoa* spp. that feed on oak leaves and have high tannin content in their spines [287]. Though most species are harmless, a few should be left untouched. The caterpillars of concern have hairs (setae) on their slug-like or tubular-shaped bodies, sometimes giving them the appearance of a brush-like, fuzzy, or glass-haired cucumber. Multiple caterpillar species are known to have hairs that readily break off when handled; these are generally considered stinging hair caterpillars because of the needle-prick pain, burning, and itching caused by contact with the hairs. In some species (*Doratifera* spp.), stinging groups of hairs are paired and mounted on body protuberances attached to venom glands [89], and American and European species may have hairs containing complex peptides and proteins [288]. Examples of such caterpillar species include the saddleback caterpillar (*Sibine stimulea*), io moth (*Auomeris io*), hag-moth (*Phobetron pithecium*), buck moth (*Hemileuca maia*), hickory tussock moth (*Lophocampa caryae*), silver-spotted tiger moth (*Halisidota argentata*), and the stinging rose caterpillar (*Parasa indetermina*) [289–291].

Australian caterpillars, especially the mistletoe browntail moth (*Euproctis edwarsi*), have been confirmed to be of medical significance as their venom contains kallikrein-like serine proteases [292]. Those causing lesser effects are represented by the genera *Doratifera* (cup moths), *Theosa* (the genus of the billygoat plum stinging caterpillar), *Theretra* (hawk moths), and many more [89]. Symptoms vary in severity from intense pain (rarely) to mild or moderate pain and from slight redness to wheals; the symptoms may last for up to 48 hours [289].

Puss caterpillars found in the southern United States (*Lagoa cripata*) and Brazil (*Megalopyge opercularis*) have caused local pain, regional lymphadenopathy, headache, and inflammatory dermatitis. Despite analytical attempts to elucidate venom components from these species, they appear to be lacking in enzymes or toxins, as only a few amino acids have been found; however, nongastrointestinal body extracts reveal a high tannin content derived from the caterpillars' diet of oak leaves [287].

Saturniidae moth (*Lonomia oblique*), a caterpillar primarily found in southern Brazil, is a species of caterpillar second to none when it comes to envenomation. In addition to the local symptoms that result from hair contact, *Lonomia* caterpillar envenomation produces profound burning pain and can induce major hematological complications. Their sets of hard-bristled hairs contain venom with a prothrombin activator (*Lonomia oblique* prothrombin activator protease, or LOPAP), factor X activator, phospholipase $A_2$-like activity (lonomiatoxin) [293], and both nociceptive and edematogenic properties [294]. Severe envenomation from bristle contact with *Lonomia* results in a major hemorrhagic syndrome [295] with a decline in prothrombin and accelerated thrombin generation, manifested clinically by ecchymosis and hematuria, with bleeding of the nose, skin, gastrointestinal tract, lungs, and vagina as well as pulmonary and intracerebral hemorrhage leading to death [296,297]. This syndrome usually occurs from [illegible] to 72 hours after envenomation.

Caterpillar sting treatment from most species is symptomatic and involves removal of hairs using tape, washing the skin, ice packs, and antihistamines, leading to resolution of symptoms within hours [289]. *Lonomia* envenomation treatment is supportive; specific antivenom has been developed [298], but antivenom experience in human victims of *Lonomia* envenomation is lacking.

### *Bees, Bumblebees, Hornets, and Wasps*

Bees, bumblebees, hornets, and wasps are part of nature's air force, and human beings from all parts of the world are familiar with the consequences of their attack. Like the ants, primary complications from envenomation by these stinging hymenopterans stem not only from their limited venom toxicity but also from venom toxins that are hyperallergenic to humans [299]. Massive numbers of stings to a victim can result in a venom body burden capable of producing severe venom toxicity and even death [300]; thus, victims may die whether or not they are allergic to the venom.

Bumblebees, honeybees, and vespid wasps can all deliver a painful sting. Bumblebees tend to be less aggressive, Africanized honeybees ("killer bees"; *Apis mellifera scutellata*) and hornets will aggressively swarm victims, and wasps are aggressive solitary attackers. The bees use their venomized stingers for self-defense or the defense of the colony, while wasps also use their venom to acquire prey [301]. There are approximately 200 species of bumblebees in Canada, Europe, and the United States [302], as well as multiple honeybee (*Apis*) and hornet/wasp (*Vespa*) subspecies. The venom delivery system for all of these stinging hymenopterans is actually an anatomical adaptation of the ovipositor in females. A lengthy venom gland is connected to a channel that runs through the barbed stinger, and as such only females can sting. Following a sting, a wasp can usually remove its stinger and repeatedly sting its victim. In contrast, the honeybee's stinger has a more barbed anatomy and remains in the victim, along with the muscular venom gland reservoir, which continues to pulse venom into the victim until it is depleted or removed [30].

The venoms of bumblebees and honeybees are cross reactive, demonstrating similarities in their composition [303]. The injected dose of venom for a single bee is about 0 μg [274]. Bee (*Apis* and *Bombus*) venom components have been isolated and identified as a melittin peptide and tetramers (a major constituent of bee venom), 39-kDa acid phosphatase, 38-kDa hyaluronidase, 28-kDa protease, 14-kDa phospholipase, apamin, and mast-cell degranulating peptide [302,304]. The anaphylactic effects of bee venoms associated with IgE antibodies are due to phospholipase, hyaluronidase, and, to a lesser degree, melittin [304].

Hornets, wasps, and yellowjacket venom immunochemical studies have shown that hyaluronidase, phospholipase, and a substance of unknown function (called antigen 5) are the major venom components [305]. Studies of venom from the Taiwan yellow-legged hornet (*Vespa verutina*) identified three toxins designated as verutoxins (V1, V2a, and V2b). All three verutoxins possess phospholipase $A_1$, and their hemolytic activity can kill mice at doses as low as 0.87 mg/kg [306]. Given that an average vespid delivers approximately 10 μg per sting [274], this hornet species harbors potently toxic venom.

Human envenomation from either bees or vespids can trigger rapid acute immune reactions leading to vomiting, diarrhea, hypotension, coma, hemaglobinuria and myoglobinuria progressing to renal failure, and rhabdomyolysis [307,308]. Moderate venom reactions are more common and involve pain, dermal rashes, urticaria, and edema [309]. Toxicity, independent of immune system responses or in victims without allergic reactions, is due primarily to phospholipase activity.

Treatment of victims stung by bees or vespids is largely symptomatic and supportive but requires a timely response in the case of anaphylaxis, with epinephrine, antihistamine, steroids, and life support. Effective antivenoms are not available, but this therapeutic strategy has been investigated [299].

The spread of Africanized bees from South America to North America and their aggressive behavior have led to a large number of fatalities in the process of their northward migration [310]. The severity of envenomation by *Apis mellifera scutellata* relates to the venom dose injected with each individual sting and the potentially large number of stings a victim sustains. A killer bee, in contrast to other bees, delivers approximately 150 μg of venom per sting, a quantity that is threefold greater than that delivered by other bees. This characteristic, amplified by the large number of stings resulting from their strong swarming behavior [311], makes killer bee encounters extremely dangerous. The cumulative dose of venom, rather than anaphylaxis, can lead to irreversible organ system damage and death [312]; thus, it is important to remove the victim from the swarm as rapidly and safely as possible and seek medical attention.

### *Sawflies*

Sawflies (*Arge*, *Lophyrotoma*, and *Perreyia* spp.) cause significant problems for the livestock industry and the veterinary profession. Sawfly larvae have been documented as being responsible for large numbers of cattle and sheep deaths in Australia (*Lophyrotoma interrupta*), cattle, dogs and sheep in Denmark (*Arge pullata*), and cattle, pigs, and sheep in Uruguay (*Perreyia flavipes*) [313]. Sawflies infest certain species of trees and grasses for the larval phase of their metamorphosis, and with a change of seasons many are found on the ground where livestock graze. When they are consumed by livestock, only a brief period of time lapses between ingestion, symptoms, and death [314]. Necrotizing hepatotoxicity is the primary toxic insult. Symptomatic cattle have been described as being hyperexcitable and aggressive to humans [314]. The toxins—the octapeptide lophyrotomin and heptadecapeptide pergidin—are unique because these larvae synthesize the toxins rather than bioaccumulating them from their food sources, which appears to be a possible species survival strategy [315]. In feeding trial experiments, it was found that even dead larvae were toxic to cattle and sheep and that oven-drying larvae at 110°C for 24 hours did not abolish their toxicological activity. Of interest is the fact that pergidin appears to be the first phosphorylated peptide isolated from an animal species [313]. Sawfly larvae being found on three different continents suggests that no common toxin source from animal food roughage exists and that each larval species can synthesize its own toxic peptides. Further conjecture rests with the possible migration of these peptides to nonvisceral tissues in produce animals, as they are water soluble and nondegradable, which would likely result in their having long resident times in animal tissues.

## Terrestrial Vertebrate Animal Toxins and Venoms

Terrestrial venomous or poisonous vertebrates have caused the human mind to conjure up the most dark-sided images of nature. Snakes, lizards, frogs, salamanders, toads, and newts have all been used in witches' brews and are associated with the evil that exists in the natural world. The unusual appearances of these fascinating creatures and the fact that many possess potent poisons have resulted in their biological and chemical traits being associated historically with great mystical powers. Importantly, in today's world, the presence of some of these creatures is a growing medical concern in many countries where human and natural communities are converging. Equally, intriguing are the terrestrial creatures, such as certain species of birds and mammals, that are not usually thought of as likely sources of toxins and venoms. Modern analytical and technological capabilities allow more comprehensive analysis of animal toxins and a better understanding of animal systematics. As a result of the interface between these disciplines, our toxinological knowledge and its value in the biological sciences, conservation, and medicine continue to grow.

### Reptilia

An historian in ancient Greece, Diodorus Siculus, thought that the sands of the Theban desert in Egypt could spontaneously generate serpents [316]. For millennia since ancient times, terrestrial poisonous snakes, such as cobras and rattlesnakes, have consistently maintained a prominent place in the hierarchy of many cultures and religions and have fascinated us, even those who fear them most. Less well known, from a global perspective, are the venomous lizards (beaded lizard and Gila monster), as they are only distributed throughout small, confined arid geographic regions of the southwestern United States, Mexico, and Guatemala [317]. Whereas arachnids account for a large number of envenomations to humans, and a considerable number of marine animals inflict envenomations as well, the majority of consequences from these envenomations are associated most frequently with morbidity. In contrast, snake bite has been reported to be the most significant toxin-related disease because of its quantitative contribution to morbidity *and* mortality, particularly in the Americas, Africa, and Asia [318].

Historical studies of snake venom poisoning mortality are limited, and the only global comprehensive review revealed an average of 50,000 deaths annually, based on hospital records from 1945 to 1949 [319]. Current estimates of venomous snake bite envenomations are 2.5 million annually, with 125,000 deaths [320]. Thus, venomous snake bite is a medical concern of appreciable significance in the world today.

Although the problem of snake bites is important, it should not overshadow the role that venomous snakes play in the balance of the natural world. Venomous snakes using their venom for its primary biological function readily consume millions of pests that invade human food supplies in many countries of the world. They reduce the loss of food supplies by keeping pest numbers in check and by keeping populations from exploding they may reduce the spread of disease carried by their prey. The impact of their elimination from world ecosystems would definitely be noticed. Furthermore, venomous snakes synthesize an array of toxins that possess a spectrum of pharmacological properties and act on the physiology of numerous animal species. Their venoms provide the natural templates from which biochemical and biomedical research can generate useful therapeutic agents.

#### Snakes

There are approximately 2900 recognized species o snakes, comprising approximately 420 genera [321]. Venomous snakes are found in arid, subtropical, tropical, an temperate regions, as well as polar ecosystems, on six o the continents (they are not found on Antarctica) [318 They inhabit arboreal, aquatic, desert, mountain, subterranean, and diverse terrestrial microhabitats. Snakes ar the most common predator of other vertebrates, and mos venomous snake species feed on other reptiles, amphibians, birds, and mammals, with their venoms being highl selective for the preferred prey of a given species [321]

The taxonomical process of classifying and dividin venomous snakes into various groups has been constantl evolving. Generally, the four main groups of venomou snakes are the Atractaspididae; the advanced snakes of th super family Colubroidea; the Viperidae, which is a famil divided into vipers and pit vipers based on the presenc

**TABLE 19.6**
**Terrestrial Vertebrate Animals: Venomous Snakes**

| Phylum | Class | Family | | Common Names | Genera |
|---|---|---|---|---|---|
| Chordata | Reptilia | Atractaspididae | African | Burrowing, mole, stiletto vipers | *Atractaspis, Macrelaps* |
| | | Colubridae | Americas | Bimini racer, parrot, vine snakes | *Alsophis, Leptophis, Oxybelis* |
| | | | | Hognose, wandering garter snakes | *Heterodon, Thamnophis, Philodryas* |
| | | | African | Boom slang, twig snakes | *Disphlodidus, Thelotornis* |
| | | | Asian | Keelback, tree snakes | *Rabdophis, Ahaetulla* |
| | | Elapidae | African | Cobras, mambas | *Naja, Hemachatus, Dendroaspis* |
| | | | Americas | Coral snakes | *Micrurus, Micuroides* |
| | | | Asian | Cobras, coral snakes | *Naja, Calliophis, Maticora* |
| | | | | King cobras, kraits | *Ophiophagus, Bungarus* |
| | | | Australian | Black/brown snakes | *Pseudechis, Pseudonaja* |
| | | | | Copperheads, death adders | *Austrelaps, Acanthophis* |
| | | | | Taipans, tiger snakes | *Oxyuranus, Notechis* |
| | | Viperidae | | Crotalinae (pit vipers) | |
| | | | Americas | Bushmasters | *Lachesis* |
| | | | | Cantils, copperheads, cottonmouths | *Agkistrodon* |
| | | | | Lanceheads, fer-de-lances, urutu | *Porthidium, Bothrops* |
| | | | | Jumping vipers, montane vipers | *Atropoides, Cerrophidion* |
| | | | | Eyelash vipers, palm pit vipers | *Bothriechis, Bothriopsos* |
| | | | | Rattlesnakes | *Crotalus, Sistrurus* |
| | | | Asian | Bamboo, green tree vipers, habu | *Trimerersurus, Protobothrops* |
| | | | | Hundred pace snakes | *Deinagkistrodon* |
| | | | | Malayan pit vipers, mamushi | *Calloselasma, Agkistrodon* |
| | | | | Temple pit vipers | *Tropidolaemus* |
| | | | | Viperinae (true vipers) | |
| | | | African | Bush, saw scale, montane vipers | *Atheris, Echis, Montatheris, Proatheris* |
| | | | | Desert vipers, horned vipers, night adders | *Macrovipera, Cerastes* |
| | | | | Puff adders, Gaboon vipers, rhinoceros vipers | *Bitis* |
| | | | Asian | Fea's, leaf-nosed, Russell's vipers | *Azemiops, Eristicophis, Daboia* |
| | | | | Desert, false horned vipers | *Macrovipera, Pseudocerastes* |
| | | | European | Asps, adders, mountain vipers | *Vipera* |

of an infrared heat-sensing pit between the eye and nasal opening in some vipers; and the Elapidae (Table 19.6). Sea snakes, unique members of the family Elapidae (subfamily Hydrophiinae) in that they spend the majority of their lives in oceanic waters, are discussed with the marine vertebrates; however, they are true snakes [317].

Venomous snakes have evolved several anatomical features with a variety of functions important to their survival. The structures of importance in venomous snakes are those relating to their venom and its production and delivery. Biomanufacturing of venom is primarily accomplished via ophidian oral glands of two types, although a variety of subtle modifications exist. A salivary system modification, frequently referred to as the Duvernoy's gland, is the toxin-secreting (venom) structure of colubrid snakes. In some species, a seromucus cell, a secretory structure with a weakly defined lumen-like duct, produces toxic proteinaceous substances [322]. A quite similar structure is found in elapid snakes [321]. The exception to this similarity is the elongated, compressor-muscle-enclosed venom gland of the Asian coral snakes [316] and some Burrowing and Night adders, which have a venom gland approaching half their body length [321]. Venom glands in viperid snakes are slightly more sophisticated as the musculature, derived from the jaws, encompasses a well-defined duct that extends to the base of the accessory gland that interfaces with the fang sheath and terminates at the proximal fang–tissue juncture [27].

Fangs of venomous snakes generally are one of four basic dentition types: (1) aglyphous (conical fangs with no groove, typical of some colubrids); (2) opisthoglyphous (enlarged teeth with a groove in the back, commonly found in colubrid rear-fanged snakes and atractaspids); (3) proteroglyphous (large anterior, fixed-position fangs with deep frontal grooves, such as cobras); and (4) solenoglyphous (large tubular, canaliculated fangs with a hinge-like action due to their attachment to the highly flexible maxillary bone, such as observed in Gaboon vipers) [321].

Both the production of toxic saliva (or venom) and the delivery systems for these different dentitions are key to the acquisition of prey food, such as centipedes, fish, frogs, birds, mammals, and other snakes, and serve as a defense against predators.

Snake venoms, nature's true biochemical cocktails, are composed of a greater number of toxins than the venoms of other venomous or toxin-producing species of animals, making them the most complex of all poisons [323]. The venoms of the four major groups of terrestrial snakes (Table 19.6) possess activities that affect several physiologic systems in animals. Venoms primarily affect the central or peripheral nervous systems, blood coagulation system, cardiovascular system, musculoskeletal system, and renal system [318]. Some venom components, such as the purine and pyrimidine nucleosides, adenosine, guanosine, and hypoxanthine, seem to be present in nearly all elapid and viperid venoms and act as multifunctional toxins believed to be important in prey acquisition [324]. The net toxic effect, or effects, may be a result of direct toxin toxicity on a target within a single system but usually results from a combination of effects caused by different toxins acting directly or secondarily, or collectively.

### Atractaspididae

The snakes referred to as mole vipers, burrowing adders, or sometimes stiletto snakes comprise 14 genera and 65 species of the fossorial and subterranean-inhabiting family Atractaspididae. They are not actually vipers [321]. They are small, dark, and glossy scaled, with virtually no distinction between their head and neck; they have small bead-like eyes. Native to Africa, the Arabian Peninsula, and Israel, they have the appearance of a nonthreatening little snake; however, local names in Africa ("father of blackness" and "shroud bearer") reflect its potential danger to humans [321]. When scared, the snake will form a buried-head coil and release like an unwinding spring to deliver a slashing bite. Strangely, each of the four different dentition patterns is represented by at least one genus in the family. Most species in the family are rear fanged, but a notable exception is the front-fanged assembly of snakes in the genus *Atractaspis*. Fangs are usually grooved and connected to an elongated venom gland. Biting is accomplished with only a slight opening of the mouth, as a single fang can extend laterally outward from the side of the mouth to deliver venom in a slashing-stab motion. *Atractaspis* venom possesses a high percentage of low-molecular-weight toxins but is devoid of cholinesterase and hydrolase activity [325]. The most significant toxins isolated from the venom of this family are the sarafotoxins, which are a cysteine-rich group of acidic polypeptides capable of inducing cardiotoxic effects rapidly by the vasoconstriction of coronary vasculature [326]. As little as 2.5 μg of sarafotoxin (from *A. engaddensis* venom) in mice caused 100% fatality [325].

Envenomation from *Atractaspis* can be a serious event, and deaths, although not common, have occurred from bites by *A. irregularis* and *A. microlepidota* [327]. Symptoms of envenomation usually involve pain and swelling, with vomiting, abdominal pain, dyspnea, profuse salivation, and necrosis [328]. No specific antivenom is known to provide paraspecific protection from the effects of envenomation, and treatment is supportive [325].

### Colubridae

The colubrid snakes are highly diverse, comprised of 290 genera representing 1700 snake species. Those species in the family that produce some form of venom or toxic secretions account for more than 50% of all colubrid species [321]. The bites of a large number of colubrid snake species can cause varying degrees of toxicity, yet documentation of colubrid envenomation in humans has only been reported for approximately 50 species representing 30 genera [329] and, depending on how toxicity is defined, there may be many more [317,330]. Colubrids that lack a specific fang structure (aglyphous) and those with rear-fanged structures (opisthoglyphs) have also been documented to cause severe envenomation, and human fatalities have resulted from the bites of the rear-fanged South African boomslang (*Dispholidus typus*) [331] and twig snake (*Thelotornis capensis*) [332]; the East Asian yamakagashi (*Rhabdophis tigrinis*), which lacks a venom-injecting mechanism [333], and the Argentine racer (*Phylodryas* sp.) [334].

Given the obvious powerful effects of these colubrid venoms, it is unfortunate that very little is known about their toxins or composition. Early studies of salivary secretions or venoms from colubrids, with or without enlarged postmaxillary teeth, revealed the presence of 7 to 10 proteins [335]. More recent electrophoretic venom analyses have identified 10 to 20 constituents (4 to 200 kDa) with species-specific profiles, suggesting that the oral secretions of some colubrids are equally as complex as many front-fanged species of venomous snakes [336]. Enzymatic properties had not been associated with colubrid venom properties until recent investigations characterized the venoms and oral secretions of several colubrids, such as the keeled water snake (*Amphiesma stolata*), brown tree snake (*Boiga irregularis*), and false water cobra (*Hydrodynastes gigas*) and indicated the presence of proteolytic activity. Phosphodiesterase activity, although slight, exists in the western hognose snake (*Heterodon nasicus nasicus*) and wandering garter snake (*Thamnophis elegans vagrans*); acetylcholinesterase is present in the venom of the brown tree snake (*Boiga irregularis*); and moderate to high levels of phospholipase $A_2$ activity are present in mangrove snakes (*Boiga dendrophila*) and lyresnakes (*Trimorphodon biscutatus lambda*), respectively [336].

The neuromuscular activity effects of colubrid venom on avian skeletal nerve muscle are exhibited by the inhibition of postsynaptic actions of venoms from the dog toothed catsnake (*Boiga cynodon*), mangrove snake, and other *Boiga* species, as well as the Egyptian catsnake (*Telescopus dhara*). Inhibition by *B. cynodon* venom is a reversible process, but the venom of *Trimorophodon biscutatus* demonstrates irreversible presynaptic neurotoxic activity [337]. Studies of the venom of the Rufous beaked snake (*Rhamphiophis oxyrhyncus*), a relatively small South African colubrid of the subfamily Psammophiine, revealed a high protein fraction with phospholipase $A_2$ activity and in animal nerve muscle preparations showed postsynaptic neurotoxicity. The venom of this species can also induce hypotension followed by cardiovascular collapse in rodents, which would be effective in subduing such prey as the naked mole rat [338]. Collectively, these findings regarding the pharmacological and toxicological activities of colubrid venoms show their diversity and potentially high bioactive potency and only begin to reveal he potential for their use in biomedical research. The omposition of colubrid venoms not only confirms their imilarity in make-up to more venomous non-colubrid pecies, which undoubtedly relates to their effectiveness n acquisition of a preferred prey, but also raises concern s to their potential for severe toxicity to humans.

The bites of colubrid snakes demonstrate the extremes of toxic effects, with most bites resulting in signs and ymptoms confined to the bitten area and pronounced, lifehreatening, coagulopathic effects in only a few [339]. 3rief details of human envenomations by many species of olubrids are frequently all that is available, and it is mpossible to know if the symptoms observed represent a evere bite or those associated with a small amount venom neffectively delivered [317]. Many colubrid species may arbor quite toxic venoms, but in cases of envenomation, s with viper bites, the degree of envenomation is not lways fully expressed, and the incidence of bites by some pecies is so rare that the true significance of colubrid enom toxicity in humans is unknown. It is believed, owever, that at least one species in nearly every colubroid amily is capable of envenomation, leading to severe medcal complications in humans, including the potential for atality [340].

Envenomations by the boomslang, twig snake, and ed-necked keelback can be most severe and are known ) cause severe coagualpathic complications resulting in delayed onset of extensive and prolonged superficial emorrhaging that progresses to death [339]. Their bites o not induce local symptoms of envenomation. Fang unctures do not bleed at the time of the bite; their venoms re defibrinating and activate prothrombin, making them rocoagulant in effect, with anticoagulated blood resulting ue to consumption [341–343]. Unfortunately, only a spe-ies-specific antivenom exists for treating boomslang nvenomations, and, despite similarities in blood coagulpathy effects observed for envenomations by other simar species such as *Thelotornis capensis*, there is no antivenom cross-reactivity [341]. As such, supportive care and replacement of blood components may be the only potentially effective therapies available [343].

Not as potentially deadly as the genera previously mentioned but of interest because of its wide range across the North American continent and frequent exposure to humans, is the genus *Thamnophis*, the aglyphic colubrid snakes commonly known as garter snakes [344]. In addition to their possessing enlarged, grooved, postmaxillary teeth, the salivary secretions of the common garter snake (*Thamnophis sirtalis*) are elaborated from mandible glands and serve as evidence of a venom delivery system [345]. Several cases of envenomation have resulted from bites by the garter snake species *Thamnophis sirtalis* [346] and *Thamnophis elegans vagrans* [347]. The victims of these bites suffered local swelling, edema. hemorrhagic vesicles, and ecchymosis, to the extent that they required hospital admission [347,346]. Systemic symptoms did not occur, but the clinical presentation was likened to that observed with pit viper envenoming [347]. Thus, even colubrid species considered to be totally harmless are capable of causing some degree of toxinological insult. Colubroid species will continue to be redefined toxinomically and be of significant importance at the forefront of venom research, as the science of their venoms will play an important role in our understanding of venom evolution in the animal world and in the laboratories of biomedical and pharmacotherapeutic research [334].

### *Elapidae*

Elapid snakes rank high among animals in the toxin world and are predominantly found on four continents (Table 19.6). This family is represented by 63 genera that contain 272 species. Cobra, coral snake, mamba, and taipan are names of familiarity and notoriety [321]. Ranging in size from inches to meters, they are present in an abundance of brilliant colors and color patterns, such as the coral snakes of the *Micrurus* and *Micruroides* genera, and exhibit functional threatening anatomical adaptations, such as hooding in cobras (*Naja* spp.), an image respected by humans for thousands of years. The most noted of the elapids are the cobras, coral snakes, kraits, mambas, and Australian elapids, where the diversity of the family is greatest (Table 19.6) [348]. All elapids are venomous and possess proteroglyphous dentition [317].

The elapid venom delivery system consists of tubular or grooved front teeth that are usually located on the anterior of the maxilla, followed by conical solid teeth. In general, their fangs are shorter than vipers, and the associated venom glands have a lesser storage capacity. Elapid venoms are considered to be primarily neurotoxic in effect, as their venoms contain nonenzymatic proteins, some enzymatic proteins, and cytotoxic components [349]. The presynaptic neurotoxins are usually phospholipase $A_2$ isoforms that either inhibit or induce transmitter release from the presynaptic myoneural junction [350]. The postsynaptic neurotoxins are either short chain (60 to 62 amino acids) or long chain (66 to 74 amino acids), and their tertiary structure is maintained via four to five disulfide cross-linkages. They competitively block acetylcholine from binding to the receptor, forming a nondepolarizing block [26]. Elapid phospholipase $A_2$ toxins may induce pharmacological effects independent of their enzymatic activity and independent of other venom components, such as is the case with notexin from *Notechis* species (tiger snake) venom [350]. Other types of phospholipase $A_2$ require complexing with other protein factors to form subunits that express their pharmacologic actions, as in the case of taipoxin from *Oxyuranus* species (taipan) venom [351]. The phospholipase $A_2$ enzymes exist in a wide variety of forms, and each enzyme may be highly specific in its action, as is the case with beta-bungarotoxin from the venom of the many-banded krait (*Bungarus multicinctus*), which acts via presynaptic effects on nicotinic acetylcholine transmission but not on adrenergic transmission [352]. To add to the complexity of phospholipase $A_2$ enzyme functions is the fact that any number of the phospholipase $A_2$ isoforms may express identical pharmacological effects, but by different mechanisms; thus, generalizing phospholipase $A_2$ venom activities for all elapid species would be erroneous [31].

In contrast to neurotoxins, some elapid venoms induce hemotoxic effects by possessing procoagulant properties that enhance the conversion of prothrombin to thrombin. Australian elapid venoms possess toxins with this pharmacologic action; for example, notecarin from the tiger snake (*Notechis scutatus scutatus*), oscutarin from the taipan (*Oxyuranus scutellatus*), and pseutarin from the brown snake (*Pseudonaja textilis*) are all potent procoagulant toxins [343]. These varied venom pharmacological properties make snakes of the family Elapidae some of the most dangerous snakes in the world. The more prominent elapid genera will be briefly discussed here.

Cobras (*Naja*, *Boulengerina*, *Hemachatus*, and *Pseudohaje* spp.) are elapids that can attain a relatively large size (3 m), making their ability to inject large quantities of venom a concern to humans. Their highly toxic venom is frequently used to paralyze other venomous snakes they seek for food [321]. In addition to the traditional envenoming by biting, some species of cobras (*N. mossambica*, *N. nigricollis*, and *N. sputatrix*) can spit or spray venom from an orifice in the front of their luminal fangs for a distance of 2 to 3 m. All cobras can dorsoventrally compress their necks to form a hood. The king cobra (*Ophiophagus hanna*) is not a true cobra but shares similar anatomy and venom properties. A cytotoxic L-amino acid oxidase (hannatoxin) has been isolated from their venom [354], and the king cobra is perhaps the world's largest venomous snake, attaining a length of up to 5.7 m [355].

Cobra venoms contain cytotoxins and numerous phospholipase $A_2$ isoforms. The cytotoxins are basic proteins that can account for up to 40% of the total venom protein. They are membrane-active toxins that alter membrane permeability [349]. The cobra phospholipase $A_2$ toxins are of the group I type based on their structural similarity to pancreatic-secreted phospholipase $A_2$, and they exhibit myotoxic effects [356]. Many of these toxins have been referred to as cardiotoxins, and they are the most abundant of cobra venom constituents. They are basic proteins that are quite similar in conformation to the neurotoxic components of cobra venom but produce primarily cardiotoxic effects [357].

The combined effects of cobra venom toxins following envenomation yield a painful bite with swelling and neurological symptoms of respiratory distress and paralysis. Despite the common thought that cobra venoms are not tissue damaging, their cell membrane toxins frequently cause necrosis of the skin and subcutaneous tissues. In particular, the venom of African spitting cobras (*Naja mossambica* and *N. nigricollis*) can cause significant tissue damage in addition to neurotoxicity [348]; however, bites by cobra species (*Naja naja philippinensis*) of the Philippines cause minimal tissue damage but exhibit severe and rapidly developing neurotoxicity [358]. Spitting cobras also defensively eject their venom into the eyes of their predators, including humans, causing profound conjunctivitis, pain, photophobia, blepharitis, and corneal opacification. These oculotoxic effects are believed to be related to the free, unbound cardiotoxins in the venom [359]. Death following cobra envenomation can occur rapidly as a result of respiratory paralysis, which can develop in as little as 15 minutes; however, neurotoxic symptoms may also be delayed for several hours, causing a delay in seeking medical attention.

Antivenoms are available for treating envenomation by African and Asian species, and adequate dosing of the appropriate antivenom usually neutralizes the medically dangerous symptoms that threaten life [360]. Additionally, the anticholinesterase drugs edrophonium and neostigmine have been successfully used in treating Philippine cobra (*Naja naja philippinensis*) [358] and monocellate cobra (*Naja kaouthia*) [361] envenomations, respectively. The treatment of ocular exposure from the venom of spitting cobras has involved irrigation and the topical application of heparin–tetracycline solution [362].

Coral snakes (*Micrurus* and *Micuroides* spp.) are native to the southern United States, Mexico, and Central and South America, reaching their greatest diversity near the equatorial belt. They are relatively abundant compared with their Asian relatives (*Calliophis* and *Maticora* spp.). The genus *Micrurus* is represented by nearly 70 species, with a wide range of red-, black-, white-, or yellow-ringed color patterns. Many mimic nonvenomous species [317]. These elapids are consid erably smaller than cobras (<50 cm to 1.5 m), and their fangs are correspondingly smaller.

Coral snake venoms have pharmacological activities that are capable of inducing cardiotoxicity, hemorrhagic toxicity, myototoxicity, and neurotoxicity. The majority of *Micrurus* species possess phospholipase $A_2$ enzymatic activity capable of eliciting anticholinesterase and anticoagulant actions. [363]. Neurophysiological effects can result from irreversible neuromuscular blockade, reducing evoked acetylcholine release but increasing spontaneous acetylcholine release [364].

Coral snake envenomations to humans are characterized by a bite with minimal pain, followed by symptoms of nausea and vomiting, dizziness, generalized weakness, drowsiness, euphoria, muscle fasciculations, diaphoresis, bulbar paralysis, diplopia, and slurred speech. Complete paralysis with loss of respiratory function can last for several days. The neurological symptoms may not be immediately evident and are often delayed in their presentation for or up to 12 hours [365].

Treatment of corals snake envenomation is most effectively accomplished with vigilance of respiratory function, and the use of antivenoms. In the absence of appropriate antivenom it is possible to maintain life by providing artificial respiration for sustained periods until spontaneous breathing returns [366]. Antivenoms for coral snake species in Central and South America are available and are effective treatment, but many victims in these countries suffer symptoms without relief as they do not have timely access to a medical facility [360].

Mambas (*Dendroaspis*) are large (up to 4 m), agile, African snakes with extremely toxic venoms. They come in grayish and green color morphologies, including the terrestrial-inhabiting black mamba (*D. polylepis*) and the arboreal-inhabiting green mamba (*D. angusticeps*), of which there are three subspecies. The black mamba is named not for its body color but for the charcoal-colored interior of its mouth [367]. They are proteroglyphous with enlarged maxillary front fangs that are moderately movable; they have no teeth posterior to the fangs, which are associated with a well-developed venom gland.

Mamba venoms contain dendrotoxins, toxins of 57 to 60 amino acid residues cross-linked with disulfide bridges. These are neurotoxic in effect and stimulate acetylcholine release from neuromuscular junctions. They have minimal antiprotease activity and selectively block certain subtypes of voltage-dependent $K^+$ channels in neurons. Green (alpha-dendrotoxin) and black (toxin K) mamba venoms differ in the subtype of $K^+$ channels pharmacologically affected [368]. Via their blocking action on $K^+$ channels, they facilitate transmitter release at peripheral synapses, thereby inducing nerve action potentials repetitively. In the central nervous system, dendrotoxin induces seizures when injected into animals [369]. The mamba toxin fasciculin is an anticholinesterase toxin that is a selective potent inhibitor of acetylcholinesterase. Not surprisingly, mamba venoms also contain cardiotoxins [370] and muscarinic toxins of several types that selectively act on different muscarinic acetylcholine receptor subtypes [371]. Their venom is also high in hyaluronidase activity which enhances venom spread. These different toxins, with their respective pharmacologic actions, work synergistically and are likely the explanation for the pronounced cardiovascular problems in victims of mamba envenomation.

Mambas can store from 4 to 8 mL venom per gland, and they will bite repeatedly, injecting a significant quantity of venom. There may be only a tingling sensation at the site of the bite, and symptoms can appear rapidly (within 10 minutes) or can be delayed. Symptoms are progressive and commonly consist of unconsciousness, diaphoresis, flaccid paralysis, respiratory paralysis, sustained hypotension, and direct cardiotoxicity leading to death [372].

Treatment of mamba envenomation victims with polyvalent antivenom (South African Vaccine Producers [Pty.], Ltd.; Sandspringham, South Africa) is effective and is the treatment of choice. Symptomatic and supportive treatment is important and should complement antivenom therapy but not substitute for antivenom use [372].

Taipans, the quintessential venomous snakes of the elapid family, are native to Australia and Papua New Guinea. These are sizeable snakes (2 to 3.5 m) and are similar to mambas in their alert, nervous behavior. They are very aware of their environment and any intruder. The two types of taipans are the coastal taipan (*Oxyuarnus scutellatus*) and the inland taipan (*Oxyuranus microlepidotus*), sometimes referred to as the fierce snake and believed to be the most venomous terrestrial snake in the world. The inland taipan's average bite delivers enough of its potent venom to kill 62 humans (70-kg body weight); the venom of the coastal taipan is sufficient to kill 27 humans [367]. Taipan venom contains neurotoxins, a $Ca^{2+}$ complex-specific channel blocker, and a prothrombin activator [373]. The neurotoxin taipoxin is a trimeric phospholipase $A_2$ that inhibits the release of acetylcholine from the presynaptic cholinergic nerve terminal, and is perhaps the most clinically important toxin. Post-synaptic phospholipase neurotoxins have also been found to be present in taipan venom [374].

Human envenomation by taipans results in progressive neurotoxicity exhibited by ptosis, ophthalmoplegia, trismus, and paralysis of bulbar, peripheral, and respiratory muscles [376]. A pronounced consumptive coagulopathy may also occur [376]. Australian antivenom is available and effective if given early after the bite but is less efficacious in ameliorating neurotoxic symptoms after they are present, and respiratory support may be required [377]. The single most important measure to ensure survival of taipan envenomation is rapid presentation to a medical facility with life-support equipment.

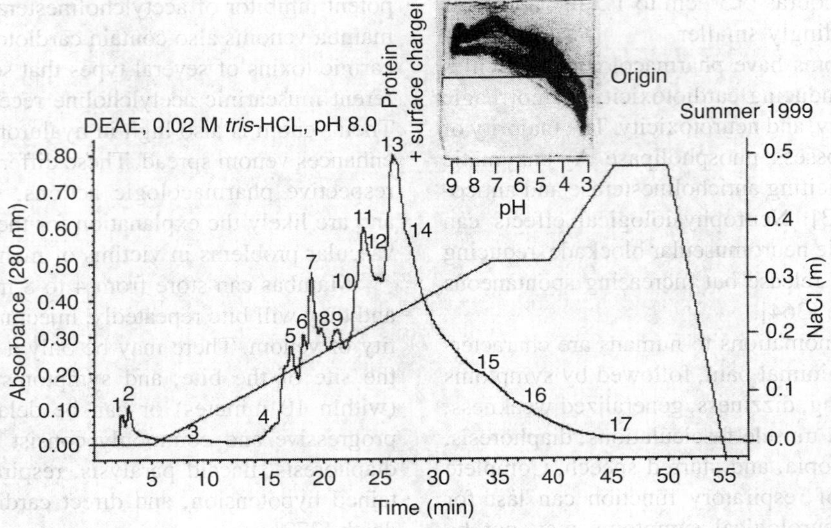


**FIGURE 19.1** Profile of *Crotalus horridus* (timber rattlesnake) crude venom obtained by cation-exchange high-performance liquid chromatography that illustrates the complexity of snake venom; each peak represents a different venom fraction possessing its own pharmacologic activity. The inset is an electrophoretic profile of venom showing the separation of venom components based on pH, acidic, or basic venom protein characteristics. (Used with permission of John Perez, Natural Toxins Research Center, Kingsville, TX.)

### *Viperidae*

Vipers and pit vipers, like the elapids, are indigenous to four continents and represent two subfamilies, comprising 30 genera and 230 species (Table 19.6). Their various unique anatomical structures and behaviors have given them a reputation as creatures to be reckoned with. The Viperidae do have a prominent and well-developed venom delivery system, as they possess the advanced solenoglyphic dentition interfaced with a highly flexible maxillary bone that is associated with a sizeable venom gland. As such, regardless of venom toxicity, they are capable of effectively delivering a large quantity of venom deep into the tissues of their victims. The Viperidae are subdivided to two subfamilies based on the presence or absence of infrared sensory facial pits that are used in sensing and sizing up their prey. Those without heat-sensing pits (Viperinae) are Afro-Asian and Eurasian genera that are not found in the Americas or Australia. Viperids with the specialized heat-sensing capability, pit vipers (Crotalinae), are found in both the Americas and Eurasia but are absent in Africa and Australia (Table 19.6). Given their multicontinental and latitudinal distribution (69°N to 47°S), they inhabit arboreal, semiaquatic, and terrestrial ecosystems from tropical to temperate climates [321]. The venoms of the Crotalinae and Viperinae subfamilies and their toxicological effects are probably the most extensively studied of all snakes.

Viperidae venoms, although they possess some toxins similar to elapid toxins such as phospholipase $A_2$, differ from those of elapids in their clinical pharmacologic effects. Viperidae venoms are complex, with a variety o constituents that are chemically different, and each con stituent may exhibit a different pharmacological activit (Figure 19.1). In contrast to neurological complication observed in most cases of elapid envenomation, Viperida venoms exhibit a wide variety of pharmacological mech anisms that primarily lead to venom-induced hemorrhagi effects. Exceptions to this basic concept are the neurolog ical effects caused by the South American rattlesnak (*Crotalus durissus*) toxin crotoxin and the Mojave rattle snake (*Crotalus scutulatus*) toxin mojavetoxin in certai regions of the southwestern United States. Asian Viperida venoms, such as that of the Malayan pit viper, contain serine protease glycoprotein that cleaves fibrinogen-A from fibrinogen; venom of the saw scale viper (*Echi carrinatus*) contains the prothrombin-activating zinc met alloproteinase ecarin; and venom of the Russell's vipe (*Daboia russelli*) has protease constituents that activat factor X in the clotting cascade [323].

Viperidae (Viperinae and Crotalinae) venoms alte blood coagulation and cause myonecrosis. They posses numerous enzymatic proteins that can impact nearl every organ system [378]. Prothrombotic venom toxin of the viper *Bitis arietans* (bitiscetin) and the pit viper *Bothrops jararaca* (botrocetin) and *Trimeresurus albo labris* (alboaggregin) contain, C-type lectin and metal loproteinase-disintegrins that promote platelet aggrega tion. In contrast, similar toxins within these same C type lectin and metalloproteinase-disintegrin families–

from *Trimeresurus flavoviridus* (flavocetin), *Bitis arietans* (bitistatin), *Agkistrodon halys blomhoffi* (mamushigin), and *Crotalus atrox* (catrocollistatin)—have been found to inhibit platelet function and demonstrate the commonality of some venom components across genera and across different continents [379]. Some Crotalinae and Viperinae (*Crotalus adamanteus*, *Crotalus ruber*, and *Agkistrodon halys blomhoffii*) venoms also contain phosphodiesterases [380]. The timber rattlesnake of the United States, one of the most widely distributed species of Crotalinae, has a venom component (crotalocytin) that causes rapid platelet aggregation [381].

Other venom toxins within the Viperidae family are myotoxins, which are expressed as phospholipase $A_2$ isoforms with and without enzymatic activities. In some pit vipers (*Bothrops asper*), these have been shown to change in expression depending on the age of the snake, with juveniles not expressing any phospholipase $A_2$ isoforms. These are quantitatively important venom constituents that contribute to muscle necrosis, and are triggered at the plasma membrane as a result of membrane destabilization with a loss of $Ca^{2+}$ permeability [382]. Within the pit vipers, zinc metalloproteinases have been isolated from venom of the western diamond-backed rattlesnake (*Crotalus atrox*; atrolysins) and from the eastern diamond-backed rattlesnake (*Crotalus adamanteus*; adamalysins). These differ from most venom proteases in that they have hemorrhagic activity and are known as the snake venom metalloproteinases (SVMPs) [383]. L-Amino acid oxidases are also common to most Viperidae venoms and serve as an aid in prey digestion [323].

Envenomation by pit vipers or vipers usually results in complications with blood coagulation, tissue damage, and a redistribution of body fluids due to alterations in vascular permeability. The severity of envenomation can vary considerably depending on how much venom was injected with the bite and where the bite was sustained. Large quantities delivered intravenously represent a worst-case scenario. Fortunately, most pit viper or viper bites are fairly superficial, and large quantities of venom are not injected; at times, no venom is injected [384]. Pronounced pain is almost always present following bites, and edema and ecchymosis are common.

Envenomations by pit vipers and vipers can result in profound hypotension and incoaguable blood due to venom anticoagulant effects or as a result of consumption of blood coagulation components. Red cell membranes may rupture, leading to hemolysis and delayed kidney damage. In general, victims may exhibit a disseminated intravascular coagulopathy (DIC)-like syndrome. A few species of Viperidae (*Crotalus scutulatus*, *C. durissus terrificus*, and *Vipera aspis*) can also cause significant neurotoxicity from envenomation [385,386].

Treatment of Viperidae envenomation usually involves the use of monovalent or polyvalent antivenoms, which are effective if administered within a reasonable timeframe [329]. Antivenoms are available for most prominent species of pit vipers and vipers [360]. Supportive care and wound care are essential when treating Viperidae envenomation, and necrotic lesions may be slow to heal. Morbidity from Viperidae envenomation frequently occurs when antivenom therapy is unavailable or delayed or an inadequate dose of antivenom has been administered [384].

### Lizards

*Heloderma*, the only genus of lizards known to be dangerously venomous in the world, is comprised of only two species: the beaded lizard (*H. horridum*) and the Gila monster (*H. suspectum*). These two species further divide into five subspecies: *H. suspectum suspectum*, *H. s. cinctum*, *H. horridum horridum*, *H. h. exasperatum, and H. h. alvarezi* [387]. The Gila monster is native to the extreme southwestern United States and northwestern Mexico. The range of the beaded lizard interfaces with that of *H. suspectum* until the southern extreme in Mexico but continues on down along the Pacific coast to Guatemala [317]. The beaded lizard is stout and solidly thick in the body, legs, and tail, and it has a beady covering that is thoroughly, and irregularly blotched with a pink and yellow pattern on a black and brown background. Interestingly, the forked tongue of *H. suspectum* is black, in contrast to the pink forked tongue of the *H. horridum*. The Gila monster prefers an arid, rocky terrain and will even dig burrows or use the burrows of another animal; the beaded lizard inhabits forested areas and is frequently observed in trees during the rainy season [388]. These lizards range in size from 35 to 70 cm, with the beaded lizard being the lengthier of the two.

*Heloderma* have short, slightly recurved, sharp, peg-like teeth that are grooved on both the anterior and posterior surfaces; grooved teeth are anchored to the maxillary and mandible jaw bones. These lizards appear sluggish and somewhat uncoordinated when they walk, as they seem to waddle, but they can rapidly flex their body and clamp their jaws around a victim. They will frequently begin chewing once they are attached to a victim and will not let go, making it difficult to remove them.

The delivery of venom by *Heloderma* stems from multilobed glands, lacking in musculature, that are located in the anterior portion of the lower jaw, on both sides, with each gland lobe independently having a duct that opens at the base of a tooth. When biting a victim, the muscle contraction of the jaw forces the venom from the ductal openings and channels it along tooth grooves, both upward and downward, with venom conduction occurring via capillary action.

Venom chemistry studies of *Heloderma* have not been comprehensive, but limited studies show the

**TABLE 19.7**
**Terrestrial Vertebrate Animals: Amphibians and Reptilia (Lizards)**

| Phylum | Class | Common Name | Genera (Toxin Containing) |
|---|---|---|---|
| Chordata | Amphibia | Frogs, poison dart frogs | *Aetlopus, Allobates, Brachycephalus, Colosthetus, Dendrobates, Epipedobates, Gastrophryne, Kaloula, Limnonectes, Mantella, Phylobates, Polypedates, Rana* |
| | | Newts | *Cynops, Notophthalmus, Paramesotrition, Taricha, Triturus* |
| | | Salamanders | *Salamandra* |
| | | Toads | *Ansonia, Bombina, Bufo, Leptophryne, Pedostypes* |
| | Reptilia | Lizards (beaded lizard, Gila monster) | *Heloderma* |

venom of both species to be similar [389]. The venoms are known to possess enzymatic properties, as demonstrated by the presence of hyaluronidase, arginine hydrolase, phospholipase $A_2$ [390], and kallikrein-like enzymes [389,391]. One of the kallikrein-like enzymes capable of inducing hypotensive effects has been dubbed helodermatine. Additionally, an acidic protein neurotoxin (gilatoxin), a hydrolase toxin (horridum toxin), and a hypothermic toxin (helothermine) from a *Heloderma* sp. with a hypothermic effect capable of significantly lowering body temperature in animals have all been identified in beaded lizard venom [392,393]. *Heloderma suspectum* venom appears to lack hemolytic, hemorrhagic, and proteolytic properties [394], but vasoactive peptides and a vasodilating venom component have been isolated [392].

*Heloderma* have few enemies other than humans who suffer the consequences of their powerful bite. Although reports of death from *Heloderma* envenomation exist, their legitimacy has been greatly questioned [387]. Envenomation from *Heloderma* is quite painful, but much of the pain is a result of pure mechanical trauma injury. These lizards will hang on for extended periods of time without lessening their grip. Teeth may even break off when human victims try to remove the tightly attached lizard. Prominent symptoms, aside from intense pain, are edema, erythema, nausea and vomiting, dizziness, hypotension, leukocytosis, weakness, tachycardia, diaphoresis, and lymphangitis [395]. Arteriospasm, with severe pain, of an envenomated digit may also occur [396]. Hypotension is a major concern and in some cases may be related to venom-induced anaphylaxis involving swelling of the tongue [397]. Symptoms may persist for more than 72 hours, and wound care is critical, as broken teeth may have to be removed and secondary infection may ensue. Treatment of envenomation consists of good supportive care and may involve fluids, antihistamines, and corticosteroids, with careful cleaning of the wound and antibiotics [395].

## Amphibia

In addition to the variety of freshwater and toxic marine animal life forms, such as sedentary anemones, swimming catfish, or drifting jellyfish, and the numerous semiaquatic and terrestrial toxic animal life forms such serpentine-swimming sea snakes, walking lizards, and crawling land snakes, nature has filled another ecological niche with amphibians that not only walk awkwardly but also hop. They may live in water or on the land, or utilize both as adults, but usually some stage of their life cycle combines these two habitats. Amphibians possessing toxic properties in their body or skin inhabit subtropical and tropical regions, but a few are also found in temperate regions. A few of the approximately 2600 Amphibia are known in the realm of toxinology for their toxic skin secretions (Table 19.7) [398]. As a group, amphibians commonly are the prey food items of a variety of animal predators, such as birds, fish, snakes, and humans. Given their limited modes of locomotion, escaping from hungry predators is difficult; antipredation strategies that ensure their continued survival often depend on their ability to fend off predators in toxinological ways, making them quite distasteful or even poisonous. The ability of some amphibians to secrete skin toxins has been a phenomenon for which toxinological explanations have been sought and these secretions are of great interest to medicine and science.

Interest in amphibian skin secretions began over 2 millennia ago in China, when powdered frog skins were used as heart stimulants and diuretics [89]. Today, the quest for knowledge concerning amphibian toxins and related biologically active compounds continues, as evidenced by the continuing studies of one of the most widely distributed and potent natural toxins, tetrodotoxin (TTX). Present in a spectrum of unrelated taxa, such as bacteria, a flatworm, goby fish, octopus, and pufferfish, TTX is also prevalent in species and subspecies from several amphibian genera [399], although, of the more than 5000 amphibian species categorized into 44 families [400], only appear to contain TTX [401]. First isolated from the pufferfish of oceanic waters and then from a terrestrial amphibian, the California newt (*Taricha torsa*), TTX continues to be discovered in new species [402].

Amphibian toxins are passively delivered, transdermally, when the creatures are harassed or are actually being engulfed by predators. The skin of amphibians is a complex organ of interrelated biochemical, morphological, and physiological functions utilized for hydration, reproduction, respiration, thermoregulation, antifungal and antimicrobial protection, and antipredator defense and is essential to amphibian survival [403]. Toxins are released from specialized mucus-secreting, or granular, dermal glands distributed throughout the dorsal skin, but in some cases they are also present in specialized anatomical structures, such as the bulbous parotid gland located in the postmaxillary region of some salamanders and toads [89]. These are sites of synthesis for alkaloids, biogenic amines, peptides, proteins, and steroids [403]. The consequence of toxin release is basically a distasteful adverse reaction to predators, but in some instances death to another creature may occur following ingestion of the toxin containing amphibian.

Frog, newt, and toad skin secretions, from several species representing several major amphibian genera, contain the potent, nonprotein neurotoxin TTX [404]. Frogs of the genera *Brachycephalus*, *Colosthetus*, and *Polypedates* [402]; toads of the genus *Atelopus* [404]; and newts of the genera *Cynops* and *Taricha* [405] all represent various amphibian genera with at least one TTX-bearing species. The origin of TTX in amphibians and all other animals has been a topic of great interest, and considerable evidence suggests that TTX may be of dietary origin and associated with a symbiotic relationship with TTX-producing bacteria; as such, it would be bioaccumulated [406]. However, recent studies in rough-skin newts (*Taricha granulosa*) suggest that TTX has been produced by granular skin glands in high concentrations, independent of bacteria symbiosis, alluding to the possibility of different TTX origins in different amphibian species [407]. A recently discovered TTX, 11-oxotetrodotoxin (11-oxoTTX), has been isolated from the skin extract of the Brazilian frog (*Brachycephalus ephippium*) and was found to be four to five times more toxic than TTX itself, demonstrating possible evolutionary changes in TTX [408]. Further research with two other species of *Brachycephalus* (*B. nodoterga* and *B. pernix*) has confirmed the presence of several deoxy-TTX and nor-TTX analogs [402].

The toxicity of TTX results from its ability to specifically and irreversibly block voltage-gated $Na^+$ ion channels, abolishing the action potential at very low concentrations (μg/kg) [409]. The result of this action is varying degrees of paralysis in animals, but some predators, such as the common garter snake (*Thamnophis sirtalis*), appear to be resistant to TTX toxicity [410]. TTX toxicity to humans results from ingestion of significant quantities of TTX. A human fatality has been reported that occurred within 24 hours of ingesting a rough-skin newt, which caused cardiopulmonary arrest [411]. Although TTX is relatively prevalent in amphibians, other significant amphibian toxins are also of interest.

### *Frogs and Toads*

Frogs and toads of several genera (Table 19.7) from particular bufonid, dendrobatid, mantellid, and myobatrachid anuran lineages have yielded over 500 lipophilic alkaloid compounds, representing 22 different structural classes [412]. The alkaloids represent several truly unique toxins that appear to be present in dermal secretions, not because of *de novo* synthesis or genetic factors but as a result of the arthropod diet of certain frogs. A diet preference for myrmicine ants appears to lead to alkaloid sequestration in several brilliantly colored Central and South American dendrobatid poison dart frogs, and these ants have comprised up to 73% of the stomach contents of *Dendrobates* species [413]. The frogs (poison dart frogs) best known for the more potent alkaloid toxins are those in the genera *Allobates*, *Dendrobates*, *Epidobates*, and *Phyllobates*. Toxic alkaloids present in these genera include the voltage-gated $Na^+$-channel-activating batrachotoxins; noncompetitive, nicotinic receptor-gated channel-blocking histrionicotoxins; cardiotoxic and myotoxic pumiliotoxins and allopumiliotoxins; and the nicotinic receptor agonist epibatidine, which has antinociceptive properties [414]. Additional skin alkaloid toxins include coccinellines, cyclopentaquinolizidines, decahydroquinolines, tricyclic gephrytoxins, indolizidines, piperidines, pseudophrynamines, and pyrrolidines [415]. Across the ocean from these dendrobatid frogs are several varieties of frogs and toads from Thailand that have toxic skin secretions. The ranid frogs (*Rana raniceps*, *R. signata*, and *R. hosei*), along with the rhacophorid frog *Polypedates leucomystax*, have skin extracts that cause locomotor difficulties and prostration when injected into mice. A skin extract from *Limnonectes kuhli* contains pumiliotoxin alkaloids. Virtually nothing is known about the potential toxic effects of these extracts on humans [416].

Toads of the Americas and Asia are also noted for their noxious secretions. The widely distributed marine toad (*Bufo marinus*), Amazonian (*B. aqua*), Asian (*B. gargrizans*), European (*B. vulgaris*), and North American Colorado river toad (*B. alvarius*) are a few of over 200 toad species that have parotid glands capable of producing biologically active compounds [417]. The biologically active substance serotonin is found in both humans and toads, and the 5-hydroxydimethyltryptamine form present in all toad species is known as bufotenine. When this compound is methylated it becomes bufotenidine, dehydrobufotenidine, and 5-methoxydimethyltryptamine (5-MeO-DMT), a potent bufotenine only known to exist in *B. alvarius* [418].

**TABLE 19.8**
**Marine Animals: Poisonous and Toxin Containing**

| Phylum | Class | Common Names | Genera (Potentially Toxic) | Toxins |
|---|---|---|---|---|
| Arthropoda | Crustacea | Crabs | *Atergatis*, *Atergatopsis*, *Birgus*, *Cancer*, *Carpilius*, *Demania*, *Eriphia*, *Lophozozymus*, *Platypodia*, *Zozymus* | Gonyautoxin, palytoxin, saxitoxin, tetrodotoxin |
| Chordata | Chondrichthyes | Sharks | *Carcharinus* | Carchatoxin |
| | Osteichthyes | Fugu (pufferfish) | *Arothon*, *Tetraodon* | Tetrodotoxin |
| | | Scombroid fish | | |
| | | Anchovies | *Engraulis* | Ciguatoxin, histamine |
| | | Barracuda | *Sphyraena* | Maitotoxin, histamine |
| | | Bluefish, bonito | *Pomatomus*, *Sarda* | Scaritoxin, histamine |
| | | Grouper | *Epinephelus*, *Mycteroperca* | Histamine |
| | | Herring, mackerel | *Clupea*, *Decapterus* | Histamine |
| | | Mahi mahi | *Coryphaena* | Histamine |
| | | Parrot fish | *Scarus*, *Ypscarus* | Histamine |
| | | Red snapper | *Lutjanus* | Histamine |
| | | Sardines | *Esulosa*, *Sardina*, *Sardinella*, *Sardinops* | Histamine |
| | | Skipjacks | *Euthynnus* | Histamine |
| | | Tuna | *Thunnis* | Histamine |
| Mollusca | Bivalvia | Clams, cockles, mussels, oysters, scallops, shellfish | *Mytilus*, *Saxidomus*, *Seliqua* | Brevitoxin, gonyautoxin, pectenotoxin, saxitoxin, yessotoxin, dinophysistoxin, domoic acid, okadaic acid |
| | Gastropoda | Snails | *Babylonia*, *Charonia*, *Nassarius*, *Natica*, *Niotha*, *Oliva*, *Polinices*, *Tectus*, *Tutufa*, *Turbo*, *Zeuxis* | Surugatoxin, neosurugatoxin, saxitoxin, neosaxitoxin, gonyautoxin, tetrodotoxin, tetramine |

It is the 5-MeO-DMT that is believed to induce hallucinations in humans following the licking of toads or smoking their venom [419]. In addition to bioactive toad venom components are the bufodienolides (bufogenins and bufotoxins), which are biosynthesized from cholesterol and are highly toxic cardioactive steroids that most likely work in concert with epinephrine and norepinephrine, which are also present in toad venom [420].

Toxicity associated with toad venom is similar to that of digoxin [421], with inhibition of the sodium–potassium ATPase pump [422]. Small companion animals have died as a result of toad ingestion [423]. Chan Su poisoning occurs in humans who ingest dried toad venom as an aphrodisiac, and people who have licked or mouthed toads to get high have suffered profound cardiotoxicity that has resulted in several fatalities [424]. Successful treatment of the cardiotoxic effects has been accomplished with the use of digoxin-specific antibody Fab fragments [424]. Still further demonstrating the diversity of toad toxins is the isolation of morphine-like toxins from *Bufo marinus* skin extract [425] and a sleep-inducing factor from extracts of the Indian toad (*Bufo melanostictus*) [426]; only males of this latter species possess a Bidder's organ, located rostroventrally to the kidney, that produces a cardiotoxin capable of inducing cardioneuromuscular blockade in animals [427].

### *Newts and Salamanders*

Newts are similar to salamanders in appearance excep[t] that their skin has a dryer, less clammy or slimy feel to it. Toxic species are native to the eastern and western coasts of the United States and Japan. Their primary toxin is tetrodotoxin (TTX), which has been discussed earlier. Salamanders of European terrestrial forested habitats may be bright orange and black bicolored or solid black, and their bodies have a wet sheen. The European fire salamander (*Salamandra salamandra terrestis*) and the Alpine salamander (*Salamandra atra atra*), like their dendrobatic amphibian relatives, secrete toxic alkaloids from thei[r] skin. Samamandarine and samamanderone are the steroidal alkaloids of prominence, but cycloneosamandione, samamandinine, samanine, samamandaridine, and samandenone may also be present [428]. These are possibly cholesterol-derived steroidal alkaloids, and they are synthesized by the salamanders rather than being taken up i[n] the diet or some other form. Detailed pharmacologi[c] actions of these salamander alkaloids are lacking, but the[y] have been shown to have nerve-blocking activity bot[h] peripherally and centrally in the nervous system [429]. Although these toxins may serve as deterrents to predators, their effects in humans are unknown.

**TABLE 19.9**
**Animal Toxins and Poisons from Unusual Sources**

| Phylum | Class | Common Names | Genus | Toxin/Poison |
|---|---|---|---|---|
| Chordata | Aves | Rubbish birds: | | |
| | | Blue-capped ifrita | *Ifria* | Batrachotoxin |
| | | Hooded pitohui | *Pitohui* | Homobatrachotoxin |
| | | Quail | *Coturnix* | Coniine, hyoscyamine, solanins |
| | Mammalia | Platypus | *Ornithorhyncus* | Enzymatic proteins, hyaluronidase |
| | | Polar bear | *Thalarctos* | Retinol (vitamin A) |
| | | Shrew | *Blarina, Neomys, Sorex, Solenodon* | Kallikrein |
| | | Vampire bat | *Desmodus* | Draculin |
| | Reptilia | Sea turtles: | | |
| | | Green | *Chelonia* | Chelonitoxin |
| | | Hawksbill | *Eretmochelys* | |
| | | Loggerhead | *Caretta* | |

## Marine Animals: Poisonous, Toxin Bioaccumulated

Poisonous marine animals do not synthesize the toxins they carry in their bodies and organ systems, but rather accumulate the toxins from their diet or environment; thus, the toxins are actually produced by other organisms that live in the surrounding marine environment. A variety of marine animals found in the human diet may potentially become toxin bearing and cause severe risk to human health and life following their consumption (Table 19.8); for example, a well-known seafood poisoning caused by eating pufferfish can produce neurotoxic symptoms and cardiopulmonary toxicity, occasionally leading to death in humans. Pufferfish is offered in gourmet restaurants as a delicacy ("fugu"), but the consumption of improperly prepared pufferfish can lead to severe tetrodotoxin poisoning in the innocent diner [430].

Marine animals most commonly acquire these toxins from dinoflagellate (toxic microalgae) species such as *Alexandrim* species, *Gymnodinium catenatum*, and *Pyrodinium bahamense* (all associated with paralytic shellfish poisoning and red tides) [431]; *Dinophysis* species (diarrhetic shellfish poisoning) [432]; *Nitzschia* species (amnesic shellfish poisoning); and *Gymnodinium brevis* (neurotoxic shellfish poisoning) [433]. Still other toxins associated with different poisoning symptoms also come from dinoflagellates, such as *Gabierdiscus toxicus* and *Ostreopsis lenticularis* (ciguatera poisoning), that contain toxin-producing bacteria [434]. *Pfiesteria* species have caused considerable human sickness yet no toxin has been identified [435]. Not surprisingly, the presence of dinoflagellate toxins in marine animals will fluctuate in some species with seasonal climate changes that trigger algal blooms. Additionally, some toxins form and accumulate in food fish as a result of spoilage, such as is the case for scombroid poisoning [436].

There are no simple antidotes for treating victims of marine food poisoning. Symptomatic and supportive care, including life support, are essential for successful outcomes.

## Animal Toxins and Poisons from Unusual Sources

The amazing thing about animal toxins is that, in addition to their being present in snakes, poison dart frogs, jellyfish, scorpions, and spiders, among other creatures considered poisonous, toxins and poisons also show up in the most unlikely of creatures (Table 19.9). No one would ever think of poisonous birds, such as the ones from Papua New Guinea that have toxins in their feathers to deter predators; these toxins are similar to those found in Central and South American poison dart frogs [437]. Or, would anyone have imagined that the powerful polar bear could be poisonous and cause serious vitamin A toxicity from ingestion of its liver as a human dietary food [30]? The gracefully swimming sea turtles, like many poisonous fish, may accumulate algal toxins from the oceans that, when their meat is consumed, cause death [89]. The littlest of mammals, the shrews, have salivary toxins capable of inducing hypotension and paralysis [439]. Even the most unusual of mammals, the egg-laying platypus, has a toxin-secreting, crural venom gland connected by a duct to an erectile keratinous spur just above the webbed foot on each hind leg. Only the males possess this venom apparatus, and if skin penetration to a predator occurs it can cause severe pain, edema, and hyperalgesia [440]. Of course, one might expect a vampire bat to have a toxin in its saliva that would alter blood coagulation, and such is the case with this flying mammal, as the anticoagulant draculin allows the free flow of blood from its dietary victims [441]. Toxins in the animal kingdom, whether used for defense against predation or for prey acquisition, have been an evolutionary key to the long-term survival of toxin-containing, toxin-producing, or poisonous creatures. The animal phyla, classes, genera, and species that make up nature's living synthesis laboratories will continue to be the true Nobel chemists of the world despite human intellect.

## BENEFICIAL USES OF NATURALLY OCCURRING TOXINS

So what does this massive amount of information on the diverse effects of multiple natural toxins come down to? Is there continuity to the available information? Can generalizations be made and extrapolated across species? What can be learned for the betterment of societal health and preventive and therapeutic medicine? The various potencies and variety of clinical damage produced by the plants, algae, mushrooms, and diverse representatives of the animal kingdom clearly reflect the wide-ranging toxicities resulting from these naturally occurring intoxications. Correlating the presence of specific toxic compounds with their physiological effects allows researchers to consider developing new chemicals derived from these natural products to influence physiological functioning. Such modifications could improve health, could block disease mechanisms, or could offer valuable insight into therapies and new modalities for therapeutic intervention. The abundant reports of individuals being bitten by snakes, the course of the toxicity, and the effectiveness of treatment-specific antivenoms offer information on future improved therapies. The report of a young male bitten by a king cobra who was treated with massive doses of Thai king cobra antivenom within an hour of envenomation and experienced rapid, uneventful recovery with no after effects, provides good clinical experience and insights into the value of such early and massive therapy [442]. This case and others document the efficacy of an efficient emergency medical system and the prompt utilization of intravenous dosing with specific volumes of antivenom.

Components of natural toxins may also be used to attenuate or reverse illnesses in humans, initially on an empirical basis but with later applications in the treatment of human functional disturbances. Snake-venom-derived arginine-glycine-aspartic acid containing disintegrins (rhodostomin) has been shown to inhibit cell adhesion and particularly the adhesion of breast and prostate carcinoma cells to bone extracellular matrices. Rhodostomin also inhibited the migration and invasion of breast and prostate carcinoma, markedly inhibiting tumor growth and bone destruction. These disintegrins from snake venom strongly inhibit the adhesion, migration, and invasion of tumor cells as well as tumor growth of human breast cell cancer in bone; thus, they have the potential to be developed as alternative therapies for bone metastasis of cancer cells [443].

Snake venoms have also been used to provide more effective therapy of human and animal ailments. Coagulopathy is a significant cause of morbidity and mortality in snake-bitten humans; yet, properly prepared fractions of snake antivenoms can be as effective as heparin, or more so [444]. Understanding the mechanisms of action for such potent benefits serves to advance medical therapeutics and prevent human suffering. The disintegrins and other snake venom proteins have shown extreme promise in the treatment of a range of hemostatic disorders; for example, ancrod from the Malayan pit viper (*Calloselasma rhodostoma*) has been effectively used as an anticoagulant for therapeutic defibrination [445]. These and other fractions from naturally occurring toxins offer the potential to benefit human and animal health.

The diverse animal and plant sources for these potentially useful products reflect the numerous opportunities for identifying beneficial fractions from plants and animals that could have wide use in medicine. As an example, extra virgin olive oil has been found to contain a compound that provides the same pain-relieving effect as the antiinflammatory compound ibuprofen. Because inflammation is believed to form the basis for a variety of chronic diseases and current antiinflammatory compounds have undesirable side effects, this discovery is just another example of the possibilities that chemicals in natural products may offer.

The opportunities are all around us in the plant and animal kingdoms. Much knowledge is to be gained from observing nonhuman animals and their continuing exposure to plants and other animals while living in the wild. The comparative value of studying animal responses to naturally occurring toxins emphasizes the importance of applying such information to human medicine and to toxicology. It is true that there is one medicine … and also one toxicology!

## QUESTIONS

1. What are the differences between the chemical characteristics of a venom and those of a synthetic chemical?
2. What conditions affect the concentration and type of toxin in a poisonous plant? What conditions affect the concentration and type of toxin in the venom of a poisonous animal?
3. What body organs are commonly affected by poisonous plants? Name two plants that primarily affect each such organ.
4. Discuss the laboratory procedures used to detect the presence of toxins in plants.
5. What plant toxins are secreted in breast milk or the milk from lactating animals?
6. What are the common toxins generated by blue–green algae? What body systems do they each affect?

7. Why are blue–green algae of public health significance?
8. What are the clinical effects produced by mushroom consumption?
9. Through what mechanisms do animal toxins exert their harmful effects?
10. What are five categories (classes) of toxic marine invertebrates?
11. Give three toxic marine animals in the Cnidaria group.
12. Discuss the potentially beneficial uses of the toxins found in plants and animals.
13. What are five boney fish known to be venomous and hazardous to humans?
14. What are the effects of envenomation by sea snakes?
15. How might terrestrial invertebrate animals deliver their toxins?
16. What are the largest arachnoids? What are the oldest living toxic creatures?
17. In what parts of the world are scorpion bites serious health problems?
18. Give the common names of five different venomous spiders.
19. What is the chemical composition of centipede venom?
20. What are the common names of five venomous members of the Insecta group of animals?
21. What is unique about Australian caterpillars?
22. What dose of venom is injected by the sting of one bumblebee? What is the dose of venom injected by one hornet sting? What is the dose of venom injected by the sting of one killer bee?
23. Name and characterize five venomous reptiles.
24. What various types of dentition do venomous snakes have?
25. Describe the differences in venom activity among colubrid, elapid, and pit viper snakes.
26. Discuss the various treatments available for the bites of venomous snakes.
27. Name the two venomous lizards in the world and describe how they deliver their venoms.
28. How are amphibian toxins delivered to victims?
29. What is the potent nonprotein neurotoxin found in skin secretions from several amphibian genera?
30. Discuss the unique toxins in the dermal secretions of frogs and toads.
31. What commonly dispensed drug induces toxicity similar to that associated with toad toxicity?
32. What is the primary toxin elaborated by newts?
33. What are five marine animals that commonly acquire toxins from dinoflagellate species?
34. Discuss the animals that are not usually considered to carry toxins but are capable of inducing poisoning in unsuspecting humans.

## REFERENCES


1. Harvey, A. L., Twenty years of dendrotoxins, *Toxicon*, 39, 15–26, 2001.
2. Watson, W. A. et al., Annual report of the American Association of Poison Control Centers toxic exposure surveillance system. *Am. J. Emerg. Med.*, 23, 589, 2005.
3. Oehme, F. W., The hazard of plant toxicities to the human population, in *Effects of Poisonous Plants on Livestock*, Keeler, R. F., Van Kampen, K. R., and James, L. F., Eds., Academic Press, New York, 1978, pp. 67–80.
4. Nielsen, D. B., Economic impact of poisonous plants on the rangeland livestock industry. *J. Anim. Sci.*, 66(9), 2330, 1988.
5. Muir, A. D. et al., Toxic plants: laboratory methods, in *Foodborne Disease Handbook*, Vol. 3, Hui, Y. H. et al., Eds., Marcel Dekker, New York, 1994, chap. 6.
6. Panter, K. E. and James, L. F., Natural plant toxicants in milk: a review, *J. Anim. Sci.*, 68, 893, 1990.
7. Kunkel, D. B., Prelude to plant poisonings, *Emerg. Med.*, 17, 93, 1985.
8. Ramon, M. F. et al., A survey of the Spanish Poison Control Centre of plant and herb exposures during one year, *J. Toxicol.*, 42, 526, 2004.
9. Constanza, D. J. and Hoversten, V. W., Accidental ingestion of water hemlock, *Calif. Med.*, 119, 78, 1973.
10. Challoner K. R. and McCarron, M. M., Castor bean intoxication, *Ann. Emerg. Med.*, 19, 1177, 1990.
11. Hudson, M. J., Acute atropine poisoning from ingesting of *Datura rosei*, *N. Z. Med. J.*, 77, 245, 1973.
12. Koruk, S. T. et al., Juniper tar poisoning, *Clin. Toxicol.*, 1, 47, 2005.
13. Watson, J. T. et al., Outbreak of food-borne illness associated with plant material containing raphides, *Clin. Toxicol.*, 1, 17, 2005.
14. Biberoglu, S. et al., Mad honey, *JAMA*, 259, 1943, 1988.
15. Lampe, K. F., Rhododendrons, mountain laurel, and mad honey, *JAMA*, 259, 2009, 1988.
16. Langford, S. D. and Boor, P. J. Oleander toxicity: an examination of human and animal toxic exposures, *J. Toxicol.*, 109, 1, 1996.
17. Hooser, S. B. and Talcott, P. A. Blue–green algae, in *Small Animal Toxicology*, Peterson, M. E. and Talcott, P. A., Eds., W.B. Saunders, Philadelphia, PA, 2001, chap. 25.
18. Carmichael, W. W., Toxins of freshwater algae, in *Handbook of Natural Toxins*, Tu, A. T., Ed., Marcel Dekker, New York, 1988, chap. 6.
19. Codd, G. A., Morrison, L. F., and Metcalf, J. S. Cyanobacterial toxins: risk management for health protection, *Toxicol. Appl. Pharmacol.*, 203, 264, 2005.
20. McElhiney, J. and Lawton, L. A., Detection of the cyanobacterial hepatotoxins microcystins, *Toxicol. Appl. Pharmacol.*, 203, 219, 2005.
21. Wiegand, C. and Pflugmacher, S., Ecotoxicological effects of selected cyanobacterial secondary metabolites a short review, *Toxicol. Appl. Pharmacol.*, 203, 201, 2005.
22. Meriluoto, J. et al., Removal of the cyanobacterial toxin microcystin-LR by human products, *Toxicon*, 46, 111, 2005.
23. Hoeger, S. J., Hitzfeld, B. C. and Dietrich, D. R. Occurrence and elimination of cyanobacterial toxins in drinking water treatment plants, *Toxicol. Appl. Pharmacol.*, 203, 231, 2005.

24. Vieriea, J. M. et al., Toxic cyanobacteria and microcystin concentrations in a public water supply reservoir in the Brazilian Amazonia region, *Toxicon*, 45, 901, 2005.
25. Spoerke, D., Mushroom exposure, in *Small Animal Toxicology*, Peterson, M. E. and Talcott, P. A., Eds., W.B. Saunders, Philadelphia, PA, 2001, chap. 42.
26. Menez, A., Functional architectures of animal toxins: a clue to drug design, *Toxicon*, 36, 1557, 1998.
27. Russell, F. E., *Snake Venom Poisoning*, 2nd ed., Scholium International, New York, 1983, p. 562.
28. Mebs, D., Occurrence and sequestration of toxins in food chains, *Toxicon*, 36, 1519, 1998.
29. Uchitel, O. D., Toxins affecting calcium ion channels in neurons, *Toxicon*, 35, 1161. 1997.
30. Mebs, D., *Venomous and Poisonous Animals: A Handbook for Biologists, Toxicologists and Toxinologists, Physicians and Pharmacists*, CRC Press, Boca Raton, FL, 2002.
31. Kini, R. M., Excitement ahead: structure, function and mechanism of snake venom phospholipase $A_2$ enzymes, *Toxicon*, 42, 827, 2003.
32. Rapuano, B. E., Yang, C. C., and Rosenburg, P., The relationship between high-affinity noncatalytic binding of snake venom phospholipases $A_2$ to brain synaptic plasma membranes and their central lethal potencies, *Biochim. Biophys. Acta*, 856, 457, 1986.
33. Faulkner, D. J., Marine natural products: metabolites of marine invertebrates, *Nat. Prod. Rep.*, 1, 251, 1984.
34. Faulkner, D. J., Marine natural products chemistry, *Chem. Rev.*, 93, 1671, 1993.
35. Wright, A. E., Isolation of marine natural products, in *Natural Products Isolation*, Cannel, R. J. P., Ed., Humana Press, Totowa, NJ, 1998, chap. 13.
36. Halstead, B. W., *Poisonous and Venomous Marine Animals*, Vol. 1, U.S. Government Printing Office, Washington, D.C., 1965.
37. Harrison, L. J., Dangerous marine life, *J. Fla. Med. Assoc.*, 79, 633, 1992.
38. Eckert, G. J., Absence of toxin-producing parapodial glands in amphinomid polychaetes (fireworms), *Toxicon*, 23, 350, 1985.
39. Bon, C. et al., Partial purification of alpha-glycerotoxin, a presynaptic neurotoxin from the venom gland of the polychaete annelid *Glycera convoluta*, *Neurochem. Int.*, 7, 63, 1985.
40. Bunker, T. D., The contemporary use of the medicinal leech, *Injury*, 12, 430, 1981.
41. Henderson, H. P. et al., Avulsion of the scalp treated by microvascular repair: the use of leeches for postoperative decongeston, *Br. J. Plast. Surg.*, 36, 235, 1983.
42. Stille, A. and Maisch, J. M., Hirudo, in *The National Dispensary*, 3rd ed., Carson, J., Ed., Lea's Son & Co., Philadelphia, PA, 1884, p. 766.
43. Lent, C., New medical and scientific uses of the leech, *Nature*, 323, 494, 1986.
44. Baskova, I. P., Khalil, S., and Nikonov, G. I., Effect of salivary gland secretion of *Hirudo medicinalis* on the extrinsic and intrinsic mechanisms of blood clotting, *Bull. Exp. Biol. Med.*, 98, 1016, 1985.
45. Adams, S. L., The emergency management of a medicinal leech bite, *Ann. Emerg. Med.*, 18, 139, 1989.
46. Almallah, Z., Internal hirudiniasis as an unusual cause of haemoptysis, *Br. J. Dis. Chest*, 62, 215, 1968.
47. Coghlan, C. J., Leeches and anaesthesia, *Anaesthesia*, 35, 520, 1980.
48. Kem, W. R., Sea anemone toxins: structure and action, in *The Biology of Nematocysts*, Hessinger, D. A. and Linhoff, H. M., Eds., Academic Press, San Diego, CA, 1988, p. 375.
49. Schuchert, P., Phylogenetic analysis of the *Cnidaria*, *Z. Zool. Syst. Evolutionforsch*, 31, 161, 1993.
50. Watson, G. M. and Hessinger, D. A., Cnidocyte mechanoreceptors are tuned to the movements of swimming prey by chemoreceptors, *Science*, 243, 1589, 1989.
51. Jones, E. C., *Tremoctopus violaceus* uses Physalia tentacles as weapons, *Science*, 139, 1963.
52. Marques, A. C., Haddad, Jr., V., and Migotto, A. E., Envenomation by a benthic Hydrozoa (Cnidaria): the case of *Nemalecium lighti*, *Haleciidae*, *Toxicon*, 40, 213, 2002.
53. Bouillon, J., Classe des hydrozaires, in *Trait de Zoologie*, *Cnidaires: Hydrozoaires, Scyphozoaires, Cubozoaires*, Grasse, P. P., Ed., Masson, Paris, 1994, p. 29.
54. Rifkin, J. F., Williamson, J. A., and Fenner, P. J., Anthozoans, hydrozoans, and scyphozoans, in *Venomous and Poisonous Marine Animals: A Medical and Biological Handbook*, Williamson, J. A. et al., Eds., University of New South Wales Press, Sydney, 1996, p. 180.
55. Grotendorst, G. R. and Hessinger, D. A., Purification and partial characterization of the phospholipase $A_2$ and colytic factor from sea anemone, *Aiptasia pallida*, nematocyst venom, *Toxicon*, 37, 1779, 1999.
56. Sencic, L. and Macek, P., New method for isolation of venom from the sea anemone *Actinia cari*: purification and characterization of cytolytic toxins, *Comp. Biochem. Physiol.*, 97B, 687, 1990.
57. Macek, P., Polypeptide cytolytic toxins from sea anemones Actinaria, *FEMS Microbiol. Immunol.*, 105, 121, 1992.
58. Mebs, D., Anemonefish symbiosis: vulnerability and resistance of fish to the toxin of the sea anemone, *Toxicon*, 32 1059, 1994.
59. Fautin, D. G., The anemonefish symbiosis: what is known and what is not, *Symbiosis*, 10, 23, 1991.
60. Anderluh, G. and Macek, P., Cytolytic peptide and protein toxins from sea anemones (Anthozoa: Actinaria), *Toxicon* 40, 111, 2002.
61. Monastyrnaya, M. M. et al., Biologically active polypeptides from the tropical sea anemone *Radianthus macrodactylus*, *Toxicon*, 40, 1197, 2002.
62. Norton, R. S., Structure and function of peptide and protein toxins from marine organisms, *J. Toxicol. Toxin Rev.*, 17 99, 1998.
63. Cline, E. I., Separation and characterization of cardiac stimulatory protein variants from the sea anemone *Urticina piscivora* from the west coast of Canada, *Int. J. Bio-Chromatog.*, 1, 249, 1996.
64. Elliott, R. C., Konya, R. S., and Vickneshawara, K., The isolation of a toxin from the Dahlia sea anemone *Teilia felina*, *Toxicon*, 24, 117, 1986.
65. Hillman, J. V., Marine animal exposures in Florida, *J. Fla Med. Assoc.*, 83, 187, 1996.
66. Zhang, M. and Qin, S., The marine stinger in the South China Sea, *Chin. J. Mar. Drugs*, 4, 36, 1991.

67. Burnett, J. W. et al., Coelentrate venom research 1991–1995: clinical, chemical and immunological aspects, *Toxicon*, 34, 1377, 1996.
68. Williamson, J. A. et al., *Venomous and Poisonous Marine Animals: A Medical and Biological Handbook*, University of New South Wales Press, Sydney, 1996.
69. Hartwick, R. F., Observations on the anatomy, behaviour, reproduction and life cycle of the cubozoan *Carybdea sivickisi*, *Hydrobiologica*, 216–217, 171, 1991.
70. Hammer, W. M., Jones, M. S., and Hammer, P. P., Swimming feeding, circulation, and vision in the Australian box jellyfish, *Mar. Freshwater Res.*, 46, 985, 1995.
71. Fenner, P. J. et al., The 'Irukandji syndrome' and acute pulmonary oedema, *Med. J. Aust.*, 149, 150, 1988.
72. Fenner P. J. and Williamson, J. A. H., Experiments with the nematocysts of *Carybdea rastoni* ('Jimble'), *Med. J. Aust.*, 147, 258, 1987.
73. Fenner, P. J. et al., 'Morbakka,' another cubomedusan, *Med. J. Aust.*, 143, 550, 1985.
74. Carette, T., Aldersade, P., and Seymour, J., Nematocyst ratio and prey in two Australian cubomedusans, *Chironex fleckeri* and *Chiropsalmus* sp., *Toxicon*, 40, 1547, 2002.
75. Ramasamy, S. et al., The *in vivo* cardiovascular effects of an Australian box jellyfish *Chiropsalmus* sp. venom in rats, *Toxicon*, 45, 321, 2005.
76. Bailey, P. M. et al., Jellyfish envenoming syndromes: unknown toxic mechanisms and unproven therapies, *Med. J. Aust.*, 178, 34, 2003.
77. Bailey, P. M. et al., A functional comparison of the venom of three Australian jellyfish—*Chironex fleckeri*, *Chirospsalmus* sp., and *Carybdea xaymacana*—on cytosolic $Ca^{2+}$, haemolysis, and *Artemia* sp. lethality, *Toxicon*, 45, 233, 2005.
78. Endean, R., Monks, S. A., and Cameron, A. M., Toxins from the box-jellyfish *Chrionex fleckeri*, *Toxicon*, 31, 397, 1993.
79. Currie, B., Clinical implications of research on the box-jellyfish *Chrionex fleckeri*, *Toxicon*, 32, 1305, 1994.
80. Ramasamy, S. et al., The *in vivo* cardiovascular effects of box jellyfish *Chironex fleckeri* venom in rats: efficacy of pre-treament with antivenom, verapamil, and magnesium sulfate, *Toxicon*, 43, 685, 2004.
81. O'reilly, G. M. et al., Prospective study of jellyfish stings from tropical Australia, including the major box jellyfish *Chironex fleckeri*, *Med. J. Aust.*, 175, 652, 2001.
82. Lumley, J. et al., Serious envenomation by the Northern Australian box-jellyfish *Chironex fleckeri*, *Med. J. Aust.*, 1, 13, 1988.
83. Maguire, E. J., *Chironex fleckeri* ('sea wasp') sting, *Med. J. Aust.*, 2, 1137, 1968.
84. Yoshimoto, C. M. and Yanagihara, A. A., Cnidarian (coelenterate) envenomations in Hawaii improve following heat application, *Trans. R. Soc. Trop. Med. Hyg.*, 96, 300, 2002.
85. Little, M. et al., Severe Irukandji syndrome, the epidemiology, management, and name change?, in *Proc. of the Int. Society on the Study of Toxinology*, 6th Asia–Pacific Congress on Animal, Plant, and Microbial Toxins, Cairns, Queensland, 2002, p. 41.
86. Mariscal, R. N., Nematocysts, in *Coelentrate Biology*, Muscatine, L. and Lenhoff, H. M., Eds., Academic Press, New York, 1974, p. 129.
87. Middlebrook, R. E. et al., Calcium-dependent smooth muscle excitatory effect elicited by the venom of the hydrocoral *Millepora complanata*, *Toxicon*, 40, 777, 2002.
88. Bee, M., Fish stings and other marine envenomations, *W. Va. Med. J.*, 87, 301, 1991.
89. Covacevich, J., Davie, P., and Pearn, J., Eds., *Toxic Plants and Animals: A Guide for Australia*, Queensland Museum, South Brisbane, 1987, p. 87.
90. Tamkum, M. M. and Hessinger, D. A., Isolation and partial characterization of a hemolytic and toxic protein from the nematocyst venom of the Portuguese man-of-war, *Physalia physalis*, *Biochem. Biophys. Acta*, 667, 67, 1981.
91. Burnett, J. W. and Calton, G. J., Venomous pelagic coelentrates: chemistry, toxicology, immunology, and treatment of their stings, *Toxicon*, 25, 581, 1987.
92. Flowers, A. L. and Hessinger, D. A., Mast cell histamine release induced by Portuguese man-of-war (*Physalia*) venom, *Biochem. Biophys, Res. Commun.*, 103, 1083, 1981.
93. Stein, M. R. et al., Fatal Portuguese Man-o'-war (*Physalia physalis*) envenomation, *Ann. Emerg. Med.*, 312, 131, 1989.
94. Exton, D. R., Treatment of *Physalia physalis* envenomation, *Med. J. Aust.*, 149, 54, 1988.
95. Exton, D. R., Fenner, P. J., and Williamson, J. A., Cold packs: effective topical analgesia in the treatment of painful stings by *Physalia* and other jellyfish, *Med. J. Aust.*, 151, 625, 1989.
96. Burnett, J. W. and Calton, G. J., Jellyfish envenomation syndromes updated, *Ann. Emerg. Med.*, 16, 1000, 1987.
97. Lumley, J. et al., Fatal envenomation by *Chironex fleckeri*, the Northern Australian box-jellyfish: the continuing search for lethal mechanisms, *Med. J. Aust.*, 128, 527, 1988.
98. Mebs, D., Marine animals, in *Venomous and Poisonous Animals: A Handbook for Biologists, Toxicologists and Toxinologists, Physicians and Pharmacists*, CRC Press, Boca Raton, FL, 2002.
99. Brewer, R. H., Morphological differences between, and reproductive isolation of, two populations of the jellyfish *Cyanea* in Long Island Sound, U.S.A., *Hydrobiologia*, 216–217, 471, 1991.
100. Long-Rowe, K. O. and Burnett, J. W., Characteristics of hyaluronidase and hemolytic activity in the fishing tentacle nematocyst venom of *Chrysaora quinquecirrha*, *Toxicon*, 32, 165, 1994.
101. Houck, H. E. et al., Toxicity of sea nettle (*Chrysaora quinquecirrha*) fishing tentacle nematocyst venom in cultured rat hepatocytes, *Toxicon*, 34, 771, 1996.
102. Radwan, F. F. Y. et al., Toxicity and mAChRs binding activity of *Cassiopea xamachana* venom from Pureto Rican coasts, *Toxicon*, 45, 107, 2005.
103. Kizer, K. W., Marine envenomations, *J. Toxicol. Clin. Toxicol.*, 21, 527, 1983.
104. Baden, H. P., Injuries from sea urchins, *Clin. Dermatol.*, 5, 112, 1987.
105. Wu, M. et al., Sea-uchin envenomation, *Vet. Hum. Toxicol.*, 45, 307, 2003.
106. Smith, M. M., *Sea and Shore Dangers: Their Recognition, Avoidance, and Treatment*, Smith J. L. B., Ed., Smith Institute of Ichthyology, Rhodes University, Grahamstown, 1977.

107. Shiomi, K. et al., Purification and characterization of a lethal factor in venom from the crown-of-thorns starfish (*Acanthaster planci*), *Toxicon*, 26, 1077, 1988.
108. Taira, E., Tanahara, N., and Funatsu, M., Studies on the toxin in the spines of starfish *Acanthaster planci*. I. Isolation and some properties of the toxin found in spines, *Ryukyu Daigaku Nogakubu Gakujutsu Hokoku*, 22, 203, 1975.
109. Shiomi, K. et al., Biological activity of crude venom from the crown-of-thorns starfish *Acanthaster planci*, *Nippon Suisan Gakkaishi*, 51, 1151, 1985.
110. Mebs, D., A myotoxic phospholipase $A_2$ from the crown-of-thorns starfish *Acanthaster planci*, *Toxicon*, 29, 289, 1991.
111. Shiomi, K. et al., Liver damage by the crown-of-thorns starfish (*Acanthaster planci*) lethal factor, *Toxicon*, 28, 469, 1990.
112. Yara, A. et al., Cardiovascular effects of *Acanthaster planci* venom in the rat: possible involvement of PAF in its hypotensive effect, *Toxicon*, 30, 1281, 1992.
113. Soppe, G. S., Marine envenomations and aquatic dermatology, *Alert Diver*, July/August, 3, 1990.
114. Exton, D. et al., Phylum Echinodermata, in *Venomous and Poisonous Marine Animals*, Williamson, J. A. et al., Eds., University of New South Wales, Sydney, 1999, p. 312.
115. Kohn, A. J., Tempo and mode of evolution in Conidae, *Malacologia*, 32, 55, 1990.
116. Oliver, B. M., *Conus* venom peptides, receptor and ion channel targets and drug design: 50 million years of neuropharmacology, *Mol. Biol. Cell*, 8, 2101, 1997.
117. McIntosh, J. M. and Jones, R. M., Cone venom: from accidental stings to deliberate injection, *Toxicon*, 39, 1447, 2001.
118. Conticello, S. G., Mechanisms for evoking hypervariability: the case of conopeptides, *Mol. Biol. Evol.*, 18, 120, 2001.
119. Joy, D. D. et al., Venomous cone snails: molecular phylogeny and the generation of toxin diversity, *Toxicon*, 39, 1889, 2001.
120. McIntosh, J. M., Oliver, B. M., and Cruz, L. J., *Conus* peptides as probes for ion channels, *Meth. Enzymol.*, 294, 605, 1999.
121. Cruz, L. J. and White, J., Clinical toxicology of *Conus* snail stings, in *Clinical Toxicology of Animal Venoms*, Meier, J. and White, J., Eds., CRC Press, Boca Raton, FL, 1995, p. 117.
122. Fegan, D. and Andersen, D., *Conus geographicus* envenomation, *Lancet*, 349, 1672, 1997.
123. Sutherland, S. K. and Lane, W. R., Toxins and mode of envenomation of the common ringed or blue banded octopus, *Med. J. Aust.*, 1, 893, 1969.
124. Howden, M. E. H. and Williams, P. A., Occurrence of amines in the posterior salivary glands of the octopus *Hapalochlaena maculosa* (Cephalopoda), *Toxicon*, 12, 317, 1974.
125. Sheumack, D. D. et al., Maculotoxin: a neurotoxin from the venom glands of the octopus *Hapalochlaena maculosa* identified as tetrodotoxin, *Science*, 199, 188, 1978.
126. Sutherland, S. K., *Australian Animal Toxins*, Oxford University Press, Melbourne, 1983, p. 353.
127. Robertson, A. et al., First report of saxitoxin in octopi, *Toxicon*, 44, 765, 2004.
128. Hooper, J. H., Memoirs of the Queensland Museum, in *Proc. of the 5th Int. Sponge Symp*, Hooper, J. H., Ed., Brisbane, 1999, p. 44.
129. Southcott, R. V. and Coulter, J. R., The effects of the southern Australian marine stinging sponges *Neofibularia mordens* and *Lissodendoryx* sp., *Med. J. Aust.*, 2, 895, 1971.
130. Pettit, G. R. et al., Isolation and structure of spongiostatin, part 1, *J. Org. Chem.*, 58, 1302, 1993.
131. Nelson, J. S., *Fishes of the World*, John Wiley & Sons, New York, 1984.
132. Russell, F. E., Toxic effects of animal toxins, in *Casarett & Doull's Toxicology: The Basic Science of Poisons*, Klaassen, C. et al., Eds., McGraw-Hill, New York, 1996, chap. 26.
133. Maretic, Z., Fish venoms, in *Handbook of Natural Toxins: Marine Toxins and Venoms*, Tu, A. T., Ed., Marcel Dekker, New York, 1988, p. 445.
134. Halstead, B. W., *Poisonous and Venomous Marine Animals*, Vol. 1, U.S. Government Printing Office, Washington, D.C., 1970.
135. Heatwole, H., *Sea Snakes*, Krieger, Malabar, 1999, p. 3.
136. Halstead, B. W., *Poisonous and Venomous Marine Animals of the World*, 2nd ed., Darwin Press, Princeton, NJ, 1988, p. 701.
137. Southcott, R. V., Notes on stings of some venomous Australian fishes, *Med. J. Aust.*, 2, 722, 1970.
138. Haddad, Jr., V. and Gadig, O. B. F., The spiny dogfish (cacao-bagre): description of an envenoming in a fisherman, with taxonomic and toxinologic comments on the *Squalus* gender, *Toxicon*, 46, 108, 2005.
139. Russell, F. E. and Van Harreveld, A., Cardiovascular effect of the venom of the round stingray *Urobatis halleri*, *Arch Intern. Physiol.*, 62, 332, 1954.
140. Rodrigues, R. J., Pharmacology of South American fresh water stingray venom (*Potamotrygon motoro*), *Trans. N.Y Acad. Sci.*, 34, 677, 1972.
141. Vellard, J., Mission scientifique au Goyaz et au Ria Ara guaya, *Mem. Soc. Zool. France*, 29, 513, 1932.
142. Haddad, Jr., V., Freshwater stingrays: study of epidemio logic, clinic, and therapeutic aspects based on 84 envenom ings in humans and some enzymatic activities of th venom, *Toxicon*, 43, 287, 2004.
143. Russell, F. E. et al., Studies of the mechanism of deat from stingray venom: a report of two fatal cases, *Am. J Med. Sci.*, 226, 611, 1958.
144. Fenner, P. J. et al., Fatal and non-fatal stingray envenom ation, *Med. J. Aust.*, 151, 621, 1989.
145. Williamson, J. A., Clinical toxicology of venomous Scor paenidae and other selected fish stings, in *Clinical Toxicology of Animal Venoms and Poisons*, Meier, J. and White J., Eds., CRC Press, Boca Raton, FL, 1995, p. 142.
146. Church, J. E. and Hodgson, W. C., The pharmacologic activity of fish venoms, *Toxicon*, 40, 1083, 2002.
147. Schaeffer, Jr., R. C., Carlson, R. W., and Russell, F. E Some chemical properties of the venom of the scorpionfis *Scorpaena guttata*, *Toxicon*, 9, 69, 1971.

148. Auddy, B. and Gomes, A., Indian catfish (*Plotosus canius*, Hamilton) venom: occurrence of lethal protein toxin (toxin-PC), *Exp. Med. Biol*., 391, 225, 1996.
149. Auerbach, P. S., Marine envenomation, in *Wilderness Medicine: Management of Wilderness and Environmental Emergencies*, 3rd ed., Mosby, St. Louis, MO, 1995, p. 1327.
150. Mann, J. W. and Werntz, J. R., Catfish stings to the hand, *J. Hand. Surg*., 16A, 318, 1991.
151. Calton, G. J. and Burnett, J. W., Catfish (*Ictalurus catus*) fin venom, *Toxicon*, 13, 399, 1975.
152. Al-Hassan, J. M. et al., Vasoconstrictor components in the Arabian Gulf catfish (*Arius thalasinus*) proteinaceous skin secretion, *Toxicon*, 24, 1009, 1986.
153. Thulesius, O. et al., Vascular responses elicited by venom of the Arabian catfish (*Arius thallasinus*), *Gen. Pharmacol*., 14, 129, 1983.
154. Shiomi, K. et al., Toxins in the skin secretion of the oriental catfish (*Plotosus lineatus*): immunological properties and immunocytochemical identification of producing cells, *Toxicon*, 26, 353, 1988.
155. Datta, A. et al., Pharmacodynamic actions of crude venom of the Indian catfish *Heteropneustes fossilis*, *Indian J. Med. Res*., 76, 892, 1982.
156. Fahim, F. A. et al., Biochemical studies on the effect of *Plutosus lineatus* crude venom (*in vivo*) and its effect on EAC cells (*in vitro*), *Adv. Exp. Med. Biol*., 391, 343, 1996.
157. Auddy, B., Alam, M. I., and Gomes, A., Pharmacological actions of the venom of the Indian catfish (*Plotosus canius* Hamilton), *Indian J. Med. Res*., 99, 47, 1994.
158. Das, S. K., Johnson, M. B, and Cohly, H. H., Catfish stings in Mississippi, *South. J. Med*., 88, 809, 1995.
159. Zeman, M. G., Catfish stings: a report of three cases, *Ann. Emerg. Med*., 18, 211, 1989.
160. Scoggin, C. H., Catfish stings, *JAMA*, 13, 176, 1975.
161. Russell, F. E., Pharmacology of toxins of marine origin, in *International Encyclopedia of Pharmacology and Therapeutics*, Raskova, H., Ed., Pergamon Press, Oxford, 1971, chap. 71.
162. Aldred, B., Erickson, T., and Lipscomb, J., Lionfish envenomations in an urban wilderness, *Wilderness Environ. Med*., 7, 291, 1996.
163. Church, J. E. et al., Modulation of intracellular $Ca^{++}$ levels by Scorpaenidae venoms, *Toxicon*, 41, 679, 2003.
164. Church, J. E. and Hodgson, W. C., Adrenergic and cholinergic activity contributes to the cardiovascular effects of lionfish (*Pterois volitans*) venom, *Toxicon*, 40, 787, 2002.
165. Hopkins, B. J. and Hodgson, W. C., Cardiovascular studies on venom from the soldierfish (*Gymnapistes marmoratus*), *Toxicon*, 32, 973, 1998.
166. Carlson, R. W. et al., Some pharmacological properties of the venom of the scorpionfish *Scorpaena guttata*, part II, *Toxicon*, 11, 167, 1973.
167. Hahn, S. T. and O'Connor, J. M., An investigation of the biological activity of bullrout (*Notesthes robusta*) venom, *Toxicon*, 38, 79, 2000.
168. Kizer, K. W., McKinney, H. E., and Auerbach, P. S., Scorpaenidae envenomation: a five-year poison center experience, *JAMA*, 253, 807, 1985.
169. Watkins, A. B. K., Bullrout stings, *Med. J. Aust*., 2, 212, 1969.
170. Church, J. E. and Hodgson, W. C., Stonefish (*Syanceia* spp.) antivenom neutralizes the *in vitro* and *in vivo* cardiovascular activity of soldierfish (*Gymnapistes marmoratus*) venom, *Toxicon*, 39, 319, 2001.
171. Gwee, M. C. E. et al., A review of stonefish venoms and toxins, *Pharmac. Ther*., 64, 509, 1994.
172. Weiner, S., The production and assay of stonefish antivenine, *Med. J. Aust*., 2, 715, 1959.
173. Garnier, P. et al., Presence of norepinephrine and other biogenic amines in stonefish venom, *J. Chromat. B: Biomed. Appl*., 685, 364, 1996.
174. Khoo, H. E. et al., Biological activities of *Synanceja horrida* (stonefish) venom, *Nat. Toxins*, 1, 54, 1992.
175. Poh, C. H. et al., Purification and partial characterization of Stonustoxin, (lethal factor) from Synanceja horrida venom, *Comp. Biochem. Physiol*., 99, 793, 1991.
176. Low, K. S. Y. et al., Stonustoxin: a highly potent endothelium-dependent vasorelaxant in the rat, *Toxicon*, 31, 1471, 1993.
177. Low, K. S. Y. et al., Stonustoxin: effects on neuromuscular function in vitro and in vivo, *Toxicon*, 31, 1471, 1994.
178. Sauviat, M. P., Effects of trachynilysin, a protein isolated from stonefish (*Synanceia trachynis*) venom, on frog atrial muscle, *Toxicon*, 38, 945, 2000.
179. Kreger, A. S. et al., Effects of stonefish (*Synanceia trachyns*) venom on murine and frog neuromuscular junctions, *Toxicon*, 31, 307, 1993.
180. Juzans, P. et al., Increase in spontaneous quantal ACh release and alterations of motor nerve terminals induced by an isolated peptide of stonefish (*Synanceia trachynis*) venom, *Toxicon*, 33, 1125, 1995.
181. Garnier, P. et al., Enzymatic properties of the stonefish (*Synanceia verrucosa* Block and Schneider, 1801) venom and purification of a lethal, hypotensive and cytolytic factor, *Toxicon*, 33, 143, 1995.
182. Breton, P. et al., Verrucotoxin and neurotoxic effects of stonefish (*Synanceia verrucosa*) venom, *Toxicon*, 37, 1213, 1999.
183. Lehmann, D. F., and Hardy, J. C., Stonefish envenomation, *New Engl. J. Med*., 329, 510, 1993.
184. Smith, J. L. B., Two rapid fatalities from the stonefish stabs, *Copeia*, 3, 249, 1957.
185. Currie, B. J., Marine antivenoms, *J. Toxicol. Clin. Toxicol*., 41, 301, 2003.
186. Sosa-Rosales, J. I. et al., Important biological activities induced by *Thalassophryne maculosa* fish venom, *Toxicon*, 45, 155, 2005.
187. Lopes-Ferreira, M. et al., Hemostatic effects induced by *Thalassophryne natteri* fish venom: a model of endothelium-mediated blood flow impairment, *Toxicon*, 40, 1141, 2002.
188. Sutherland, S. K., Antivenom use in Australia: premedication, adverse reactions and the use of venom detection kits, *Med. J. Aust*., 157, 734, 1992.
189. Chhatwal, I. and Dreyer, F., Isolation and characterization of dracotoxin from the venom of the greater weeverfish *Trachinus draco*, *Toxicon*, 30, 87, 1992.

190. Perriere, C. et al., Storage influence on stonefish venom components activity, *Toxicon*, 36, 1313, 1998.
191. Russell, F. E. and Emery, J. A., Venom of the weevers *Trachinus draco* and *Trachinus vipera*, *Ann. N.Y. Acad. Sci.*, 90, 805, 1960.
192. Russell, F. E., Marine toxins and venomous and poisonous marine animals, in *Advances in Marine Biology*, Vol. III, Academic Press, London, 1965, pp. 255–384.
193. Fishelson, L., Histology and ultrastructure of the recently found buccal toxic gland in the fish *Meiacanthus nigrolineatus* (Blenniidae), *Copeia*, 2, 386–392, 1974.
194. Keyler, D. E., personal communication, 2005.
195. Halstead, B. W. and Dalgleish, A. E., The venom apparatus of the European stargazer *Uranoscopus scaber* Linnaeus, in *Animal Toxins*, Russell, F. E. and Saunders, P. R., Eds., Pergamon Press, New York, 1967, p. 177.
196. Maretic, Z., Fish venoms, in *Handbook of Natural Toxins*, Vol. 3, Tu, A. T., Ed., Marcel Dekker, New York, 1988, p. 445.
197. Halstead, B. W. and Vinci, J. M., Venomous fish stings (Ichthyoacanthotoxicoses), *Clin. Derm.*, 5, 29, 1987.
198. Kalmanzon, E. et al., Receptor-mediated toxicity of pahutoxin, a marine trunkfish surfactant, *Toxicon*, 42, 63, 2003.
199. Kalmanzon, E. et al., Endogenous regulation of the functional duality of pahutoxin, a marine trunkfish surfactant, *Toxicon*, 44, 939, 2004.
200. Reid, H. A., Epidemiology of sea-snake bites, *J. Trop. Med. Hyg.*, 78, 106, 1975.
201. Tu, A. T., Biotoxicology of sea snake venoms, *Ann. Emerg. Med.*, 16, 149, 1987.
202. Tu, A. t., Hong, B. S. and Solie, T. N., Characterization and chemical modifications of toxins isolated from the venoms of the sea snake *Laticauda semifasciata* from Philippines, *Biochemistry*, 10, 1295, 1971.
203. Dowdall, M. J., Fohlman, J. P. and Eaker, D., Inhibition of high-affinity choline transport in peripheral cholinergic endings by presynaptic snake venom neurotoxins, *Nature*, 269, 700, 1977.
204. Ishizaki, H, Allen, M. and Tu, A. T., Effect of sulfhydral group modification on the neurotoxic action of sea snake toxin, *J. Pharm. Pharmacol.*, 36, 36, 1984.
205. Tu, A. T. and Fulde, M. B., Sea snake bites, *Clin. Derm.*, 5, 118, 1987.
206. Abuelo, J. G., Renal falilure caused by chemicals, foods, plants, animal venoms, and misuse of drugs, *Arch. Intern. Med.*, 150, 505, 1990.
207. Tu, A. T. and Salafranca, E. S., Immunological properties and neutralization of sea snake venoms (II), *Am. J. Trop. Med. Hyg.*, 23, 135, 1974.
208. Lourenco, W. R., Diversity and endemism in tropical versus temperate scorpion communities, *Biogeographica*, 70, 155, 1994.
209. Possani, L. D., Martin, B., and Svendsen, I., The primary structure of noxiustoxin: a $K^+$ channel blocking peptide from the venom of the scorpion *Centuroides noxius* Hoffman, *Carlsberg Res. Commun.*, 47, 285, 1982.
210. Srinivasan, K. N. et al., SCORPION, a molecular database of scorpion toxins, *Toxicon*, 40, 23, 2001.
211. Zlotkin, E. et al., Insect sodium channel as the target for insect-selective neurotoxins from scorpion venom, in *Molecular Action of Insecticides on Ion Channels*, Dark, J. M., Ed., American Chemical Society, Washington, D.C., 1995, p. 56.
212. Miller, C. et al., Charbytoxin, a protein inhibitor of single $Ca^{2+}$-activated $K^+$ channels from mammalian skeletal muscle, *Nature*, 313, 316, 1985.
213. DeBin, J. A., Maggio, J. E., and Strichartz, G. R., Purification and characterization of chlorotoxin, a chloride channel ligand from venom of the scorpion, *Am. J. Physiol. (Cell Physiol.)*, 264, C361–C369, 1993.
214. Valdivia, H. H. et al., Scorpion toxins targeted against the sarcoplasmic reticulum $Ca^{2+}$-release channel of skeleta and cardiac muscle, *Proc. Natl. Acad. Sci. U.S.A.*, 89 12185, 1992.
215. Tytgat, A unified nomenclature for short chain peptide isolated from scorpion venoms; alpha-Ktx molecular subfamilies, *Trends Pharmacol. Sci.*, 20, 445, 1999.
216. Possani, L. D. et al., Peptides and genes coding for scorpion toxins that affect ion channels, *Biochimie*, 82, 861 2000.
217. D'Suze, G. et al., Ardiscretin a novel arthropod-selectiv toxin from *Tityus discrepans* scorpion venom, *Toxicon*, 43 263, 2004.
218. Garcia, M. L. et al., Scorpion toxins: tools for studying K channels, *Toxicon*, 36, 1641, 1998.
219. Vacher, H. and Martin-Eauclaire, M.-F., Antigenic poly morphism of the 'short' scorpion toxins able to block K channels, *Toxicon*, 43, 447, 2004.
220. Martin-Eauclaire, M. F. and Couraud, F., Scorpion neuro toxins: effects and mechanisms, in *Handbook of Neurotox icology*, Chang, L. W. and Dyer, E. S., Eds., Marcel Dekker New York, 1995, p. 683.
221. Rochat, H., Bernard, P., and Couraud, F., Scorpion toxins chemistry and mode of action, in *Advances in Cytophar macology*, Vol. 3, Ceccarelli, B. and Clementi, F., Eds Raven Press, New York, 1979, p. 325.
222. Gurevitz, M. et al., Nucleotide sequence and structure anal ysis of a cDNA encoding an alpha insect toxin from th scorpion *Leiurus quinquestriatus hebraeus*, *Toxicon*, 29 1270, 1991.
223. Aubrey, N. et al., Engineering of a recombinant Fab fro a neutralizing IgG directed against scorpion neurotoxi AahI and functional evaluation versus other antibody frag ments, *Toxicon*, 43, 233, 2004.
224. Bucaretchi, F. et al., A comparitive study of severe scorpio envenomation in children caused by *Tityus bahiensis* an *Tityus serrulatus*, *Rev. Inst. Med. Trop. Sao Paulo*, 37, 33 1995.
225. Sofer, S., Shahak, E., and Gueron, M., Scorpion envenc mation and antivenom therapy, *J. Pediatr.*, 124, 973, 199
226. El-amin., E. O. et al., Scorpion sting: a management prob lem, *Ann. Trop. Ped.*, 11, 143, 1991.
227. Belgith, M. et al., Efficacy of serotherapy in scorpion stin a matched pair study, *J. Toxicol. Clin. Toxicol.*, 37, 51, 199
228. Platnick, N. I., *Advances in Spider Taxonomy, 1992–199 With Redescriptions 1940–1980*, New York Entomologic Society and The American Museum of Natural Histor New York, 1997.

229. Hillyard, P., *The Book of the Spider: From Arachniphobia to the Love of Spiders*, Random House, New York, 1994.
230. Seldon, P. A. et al., Missing links between *Argyroneta* and Cybaeidae revealed by fossil spiders, *J. Arachnol.*, 30, 189, 2002.
231. Thery, M. and Casas, J., Predator and prey views of spider camouflage, *Nature*, 415, 133, 2002.
232. King, G., The wonderful world of spiders: preface to the special *Toxicon* issue on spider venoms, *Toxicon*, 43, 471, 2004.
233. Jackson, R. R. and Pollard, S. D., Predatory behavior of jumping spiders, *Annu. Rev. Entomol.*, 41, 287. 1996.
234. Rash, L. D. and Hodgson, W. C., Pharmacology and biochemistry of spider venoms, *Toxicon*, 40, 225, 2002.
235. Tedford, H. W. et al., Australian funnel-web spiders: master insecticide chemists, *Toxicon*, 43, 601, 2004.
236. Rash, L. D. and Hodgson, W. C., Pharmacology and biochemistry of spider venoms, *Toxicon*, 40, 225, 2004.
237. McCormick, K. D. and Meinwald, J., Neurotoxic acylpolyamines from spider venoms, *J. Ecol. Chem.*, 19, 2411, 1993.
238. Jentsch, T. J. et al., The CIC chloride channel family, *Pjlugers Arch. Eur. J. Physiol.*, 437, 783, 1999.
239. Bornmann, J., Electrophysiology of $GABA_A$ and $GABA_B$ receptor subtypes, *Trends Neuro. Sci.*, 11, 112, 1988.
240. Usmanov, P. B. et al., Postsynaptic blocking of glutamatergic and cholinergic synapes as a common property of Araneidae spider venoms, *Toxicon*, 23, 528, 1985.
241. Atkinson, R. K. and Wright, L. G., The involvement of collagenase in the necrosis induced by the bites of some spiders, *Comp. Biochem. Physiol.*, 102C, 125, 1992.
242. Feitsosa, L. et al., Detection and characterization of metalloproteinases with gelatinolytic, fibrinolytic and fibrinogenolytic activities in brown (*Loxosceles intermedia*) venom, *Toxicon*, 36, 1039, 1998.
243. Russell, F. E., Phosphodiesterase of some snake and arthropod venoms, *Toxicon*, 4, 153, 1966.
244. Babcock, J. L., Civello, D. J., and Geren, C. R., Purification and characterization of a toxin from brown recluse spider (*Loxosceles reclusa*) venom gland extracts, *Toxicon*, 19, 677, 1981.
245. Ibister, G. K. and Gray, M. R., A prospective study of 750 definite spider bites, with expert identification, *Q. J. Med.*, 95, 723, 2002.
246. Ibister, G. K. and White, J., Clinical consequences of spider bites: recent advances in our understanding, *Toxicon*, 43, 477, 2004.
247. Beazley, R. N., Deaths from the bite of a trapdoor spider, *Med. J. Aust.*, 255, 1903.
248. Torda, T. A., Loong, E. and Greaves, I., Severe lung oedema and fatal consumption coagulopathy after funnel-web bite, *Med. J. Aust.*, 2, 442, 1980.
249. Browne, G. J., Near fatal envenomation from the funnel-web spider in an infant, *Ped. Emerg. Care*, 13, 271, 1997.
250. Bucaretchi, F. et al., Accidents caused by *Phoneutria* (armed spider), EAPCCT XIX International Congress, *J. Toxicol Clin. Toxicol.*, 37, 413, 1999.
251. Futrell, J., Loxoscelism, *Am. J. Sci.*, 304, 261, 1992.
252. Swanson, D. L. and Vetter, R. S., Bites of brown recluse spiders and suspected necrotic arachnidism, *N. Engl. J. Med.*, 352, 700, 2005.
253. White, J., Cardoso, J. L., and Hui, W. F., Clinical toxicology of spider bites, in *Handbook of Clinical Toxicology of Animal Venoms and Poisons*, Meier, J., Ed., CRC Press, Boca Raton, FL, 1995, p. 259.
254. da Silva, P. H. et al., Brown spiders and loxoscelism, *Toxicon*, 44, 693, 2004.
255. Clark, R. F. et al., Clinical presentation and treatment of black widow spider envenomation: a review of 163 case, *Ann. Emerg. Med.*, 21, 782, 1992.
256. Ibister, G. K. and Gray, M. R., Latordectism: a prospective cohort study of bites by formally identified redback spiders, *Med. J. Aust.*, 179, 88, 2003.
257. Escoubas, P. and Rash, L., Tarantulas: eight-legged pharmacists and combinatorial chemists, *Toxicon*, 43, 555, 2004.
258. Liang, S. P. et al., Properties and amino acid sequence of huwentoxin-I, a neurotoxin purified from the venom of the Chinese bird spider, *Selenocosmia huwena*, *Toxicon*, 31, 969, 1993.
259. Liang, S., An overview of peptide toxins from the venom of the Chinese bird spider *Selenocosima huwena* Wang (= *Ornithoctonus huwena* [Wang]), *Toxicon*, 43, 575, 2004.
260. Schmidt, G., Efficacy of bites from Asiatic and African tarantulas, *Trop. Med. Parasitol.*, 40, 114, 1989.
261. Castro, F. F., Antilla, M. A., and Croce, J., Occupational allergy caused by urticating hair of a Brazilian spider, *J. Allergy Clin. Immunol.*, 95, 1282, 1995.
262. Blaikie, A. J. et al., Eye disease associated with handling pet tarantulas: three case reports, *Br. Med. J.*, 314, 1524, 1997.
263. Blackman, J. R., Spider bites, *J. Am. Board Fam. Pract.*, 8, 288, 1995.
264. Cloudsley-Thompson, J. L., *Spiders, Scorpions, Centipedes, and Mites*, Pergamon Press, Oxford, 1968, 278.
265. Menez, A. et al., Venom apparatus and toxicity of the centipede *Ethmostigmus rubripes* (Chilopoda Scolopendridae), *J. Morphol.*, 206, 303, 1990.
266. Quistad, G. B., Dennis, P. A. and Skinner, W. S., Insecticidal activity of spider (Araneae), centipede (*Chilopoda*), scorpion (Scorpiondae) and snake (Serpentes) venoms, *J. Econ. Entomol.*, 85, 33, 1992.
267. Balit, C. R. et al., Prospective study of centipede bites in Australia, *J. Toxicol. Clin. Toxicol.*, 42, 41, 2004.
268. Gomes, A. et al., Isolation, purification and pharmacodynamics of a toxin from the venom of the centipede *Scolopendra subspinipes dehanni* Brandt, *Indian J. Exp. Biol.*, 21, 203, 1983.
269. Mohamed, A. H., Effects of an extract from centipede *Scolopendra moristans* on intestine, uterus and heart contractions and on blood glucose and liver and muscle glycogen levels, *Toxicon*, 18, 581, 1980.
270. Stankiewicz, M. et al., Effects of a centipede venom fraction on insect nervous system, a native *Xenopus* oocyte receptor and on an expressed *Drosophila* muscarinic receptor, *Toxicon*, 37, 1431, 1999.

271. Gomes, A. et al., Occurrence of histamines and histamine release by centipede venom, *Indian J. Med. Res.*, 76, 888, 1982.
272. Logan, J. L. and Ogden, D. A., Rhabdomyolysis and acute renal failure following the bite of the giant desert centipede *Scolopendra heros*, *West. J. Med.*, 4, 549, 1985.
273. Barnett, P. L. J., Centipede ingestion by a six-month-old infant: toxic side effects, *Ped. Emerg. Care*, 7, 229, 1991.
274. King, T. P., Immunochemical studies of stinging insect venom allergies, *Toxicon*, 34, 1455, 1996.
275. Haight, K. L. and Tschinkel, W. R., Patterns of venom synthesis and use in the fire ant, *Solenopsis invicta*, *Toxicon*, 42, 673, 2003.
276. Davies, N. W., Wiese, M. D., and Brown, S. A., Characterization of major peptides in "jack jumper" ant venom by mass spectrometry, *Toxicon*, 43, 173, 2004.
277. Ginsburg, C. M., Fire ant envenomation in children, *Pediatrics*, 73, 689, 1984.
278. Hoffman, D. R., Allergens in Hymenoptera venom. XXIV. The amino acid sequences of imported fire ant venom allergens Sol i II, Sol i III, and Sol i IV, *J. Allergy Clin. Immunol.*, 91, 71, 1993.
279. deShazo, R. D., Dermal hypersensitivity reactions to imported fire ants, *J. Allergy Clin. Immunol.*, 74, 841, 1984.
280. Freeman, T. M. et al., Imported fire ant immunotherapy: effectiveness of whole body extracts, *J. Allergy Clin. Immunol.*, 90, 210, 1992.
281. Read, G. W., Lind, N. K. and Oda, C. S., Histamine release by fire ant (*Solenopsis*) venom, *Toxicon*, 16, 361, 1978.
282. Matuszek, M. A. et al., Some enzymatic activities of two Australian ant venoms: a jumper ant *Myrmecia pilosula* and a bulldog and *Myrmecia pyriformis*, *Toxicon*, 32, 1543, 1994.
283. Wu, Q. X. et al., Cytotoxicity of pilosulin 1, a peptide from the venom of the jumper ant *Myrmecia pilosula*, *Biochim. Biophys. Acta*, 1425, 74, 1998.
284. Brown, S. G. A. et al., Prevalence, severity and natural history of jack jumper ant venom allergy in Tasmania, *J. Allergy Clin. Immunol.*, 111, 187, 2003.
285. Schmidt, J. O., Blum, M. S. and Overal, W. L., Comparative enzymology of venoms from stinging Hymenoptera, *Toxicon*, 24, 907, 1986.
286. Carpenter, J. M. and Wheeler, W. C., Towards simultaneous analysis of morphological and molecular data in Hymenoptera, *Zool. Scripta*, 28, 1, 1999.
287. Lamdin, J. M. et al., The venomous hair structure, venom and life cycle of *Lagoa crispate*, a puss caterpillar of Oklahoma, *Toxicon*, 38, 1163, 2000.
288. Degado, Q. A., Venoms of Lepidoptera, in *Arthropod Venoms*, Bettini, S., Ed., Springer-Verlag, Berlin, 1978, p. 20.
289. Balit, C. R. et al., Prospective study of definite caterpillar exposures, *Toxicon*, 42, 657, 2003.
290. Kawamoto, F. and Kumada, N., Biology and venoms of Lepidoptera, in *Insect Poisons, Allergens, and Other Invertebrate Venoms*, Tu, A. T., Ed., Marcel Dekker, New York, 1984.
291. Green, V. A. and Siegal, C. J., Bites and stings of Hymenoptera, caterpillar and beetles, *J. Toxicol. Clin. Toxicol.*, 21, 491., 1983.
292. Bleumink, E. et al., Protease acivities in the spicule venom of *Euproctis* caterpillars, *Toxicon*, 20, 607, 1982.
293. Arocha-Pinango, C. L., Marval, E., and Guerrero, B., *Lonomia* genus caterpillar toxin: biochemical aspects, *Biochimie*, 82, 937, 2000.
294. Bastos, L. deC. et al., Nociceptive and edematogenic responses elicited by a crude bristle extract of *Lonomia obiqua* caterpillars, *Toxicon*, 43, 273, 2004.
295. Kelen, E. M., Picarelli, Z. P., and Duarte, A., Hemorrhagic syndrome induced by contact with a caterpillar of genus *Lonomia* (Saturniidae, Hmeleucinae), *J. Toxicol. Toxin Rev.*, 14, 283, 1995.
296. Fan, H. W. et al., Hemorrhagic syndrome and acute renal failure in a pregnant woman after contact with Lonomia caterpillars: a case report, *Revista do Instituto de Medicina Tropical de Sao Paulo*, 40, 119, 1998.
297. Duarte, A. C. et al., Intracerebral haemorrhage after contact with *Lonomia* caterpillars, *Lancet*, 348, 1033, 1996.
298. daSilva, W. D. et al., Development of an antivenom against toxins of *Lonomia oblique* caterpillars, *Toxicon*, 34, 1045, 1996.
299. Schumacher, M. J., Egen, N. B., and Tanner, D., Neutralization of bee venom lethality by immune serum antibodies, *Am. J. Trop. Med. Hyg.*, 55, 197, 1996.
300. Barss, P., Renal failure and death after multiple stings in Papua New Guinea, Ecology, prevention and management of attacks by vespid wasps, *Med. J. Aust.*, 151, 659, 1989.
301. Free, J. B., The defence of bumblebee colonies, *Behaviour*, 12, 233, 1958.
302. Hoffman, D. R. and Jacobson, R. S., Allergens to hymenoptera venom. XXVII. Bumblebee venom allergy and allergens, *J. Allergy Clin. Immunol.*, 97, 812, 1996.
303. Kochuyt, A. M., Vanhoeyveld, E., and Stevens, E. A. M., Occupational allergy to bumble bee venom, *Clin. Exp. Allergy*, 23, 190, 1993.
304. Kemeny, D. M. et al., Antibodies to purified venom proteins and peptides. I. Development of a highly specific RAST for bee venom antigens and its application to bee sting allergy, *Allergy Clin. Immunol.*, 71, 505, 1983.
305. King, T. P. et al., Protein allergens of white-faced hornet, yellow hornet and yellow jacket venoms, *Biochemistry*, 17 5165, 1978.
306. Ho, C. H., Lin, Y. L., and Li, S. F., Three toxins with phospholipase activity isolated from the yellow-legged hornet (*Vespa verutina*) venom, *Toxicon*, 37, 1015, 1999.
307. Mejia, G. et al., Acute renal failure due to multiple stings by Africanized bees, *Ann. Intern. Med.*, 104, 210, 1986.
308. Sakuja, V. et al., Acute renal failure following multiple hornet sings, *Nephron*, 49, 319, 1988.
309. Lin, C. C., Chang, M. Y. and Lin, J. L., Hornet sting induced systemic allergic reaction and large local reaction with bulle formation and rhabdomyolosis, *J. Toxicol. Clin Toxicol.*, 41, 1009, 2003.
310. Guzman-Novoa, E. and Page, R. E., Impact of Africanized bees on Mexican beekeeping, *Am. Bee J.*, 134, 101, 1994
311. Rinderer, T. E., Oldroyd, B. P., and Sheppard, W. S., Africanized bees in the U.S., *Sci. Am.*, 269, 84, 1993.
312. Kolecki, P., Africanized bee attacks in Arizona: morbidity and mortality, *J. Toxicol. Clin. Toxicol*, 34, 590, 1996.

313. Oelrichs, P. B., Unique toxic peptides isolated from sawfly larvae in three continents, *Toxicon*, 37, 537, 1999.
314. Dutra, F. et al., Poisoning of cattle and sheep in Uraguay by sawfly, (*Perreyia flavipes*) larvae, *Vet. Human Toxicol.*, 39, 281, 1997.
315. Oelrichs, P. B. et al., Isolation and identification of the toxic peptides from *Lophyrotoma zonalis* (Pergidae) sawfly larvae, *Toxicon*, 39, 1933, 2001.
316. Minton, S. A. and Minton, M. R., *Venomous Reptiles*, Charles Scribner's Sons, New York, 1980, p. 130.
317. Campbell, J. A. and Lamar, W. W., *The Venomous Reptiles of the Western Hemisphere*, Vol. I, Comstock Publishing, Ithaca, NY, 2004.
318. White, J. et al., Clinical toxinology: where are we now?, *J. Toxicol. Clin. Toxicol.*, 41, 263, 2003.
319. Swaroop, S. and Grab, B., Snakebite mortality in the world, *Bull. WHO*, 10, 35, 1954.
320. Chippaux, J. P., Snake-bites: appraisal of the global situation, *Bull. WHO*, 76, 515, 1998.
321. Greene, H. W., *Snakes: The Evolutionary Mystery of Nature*, University of California Press, Los Angeles, 1997.
322. Rosenberg, H. I., Bdolah, A., and Kochva, E., Lethal factors and enzymes in the secretion from Duvernoy's gland of three colubrid snakes, *J. Exp. Zool.*, 233, 5, 1985.
323. Warrell, D. A., Animal poisons, in *Manson's Tropical Diseases*, 19th ed., Manson-Bahr, P. E. C. and Bell, D. R., Eds., Bailliere Tindall, London, 1987, chap. 48.
324. Aird, S. D., Taxonomic distribution and quantitative analysis of free purine and pyrimidine nucleosides in snake venoms, *Comp. Biochem. Physiol.*, 140, 109, 2005.
325. Kochva, E., Viljoen, C. C. and Botes, D. P., A new type of toxin in the venom of snakes of the genus *Atractaspis* (Atractaspidinae), *Toxicon*, 20, 581, 1982.
326. Stocker, K. F., *Medical Use of Snake Venom Proteins*, CRC Press, Boca Raton, FL, 1990, p. 44.
327. Chajek, T., Anaphylactoid reaction and tissue damage following bite by *Atractaspis engaddendsis*, *Trans. R. Soc. Trop Med. Hyg.*, 68, 333, 1974.
328. Christensen, P. A., Snakebite and the use of antivenom in South Africa, *S. Afr. Med. J.*, 59, 934, 1981.
329. Norris, R. L. and Minton, S. A., Non-North American venomous reptile bites, in *Wilderness Medicine*, 4th ed., Fletcher, J., Ed., Mosby, St. Louis, MO, 2001, p. 936.
330. McKinstry, D. M., Morphologic evidence of toxic saliva in colubrid snakes: a checklist of world genera, *Herp. Rev.*, 14, 12, 1983.
331. Pope, C. H., Fatal bite of captive African rear-fanged snake (*Dispholidus*), *Copeia*, 4, 280, 1958.
332. FitzSimons, D. C. and Smith, H. M., Another rear-fanged South African snake lethal to humans, *Herpetologica*, 14, 198, 1958.
333. Ogawa, H. and Sawai, Y., Fatal bite of the yamakagashi (*Rhabdophis tigrinis*), *Snake*, 18, 53, 1986.
334. Fry, B. G. et al., Analysis of Colubroidea snake venoms by liquid chromatography with mass spectrometry: evolutionary and toxinological implications, *Rapid Commun. Mass Spectrom.*, 17, 2047, 2003.
335. Minton, S. A. and Weinstein, S. A., Colubrid snake venoms: immunologic relationships, electrophoretic patterns, *Copeia*, 4, 993, 1987.
336. Hill, R. E. and Mackessy, S. P., Chracterization of venom (Duvernoy's secretion) from twelve species of colubrid snakes and partial sequence of four venom proteins, *Toxicon*, 38, 1663, 2000.
337. Lumsden, N. G. et al., *In vitro* neuromuscular activity of 'colubrid' venoms: clinical and evolutionary implications, *Toxicon*, 43, 819, 2004.
338. Lumsden, N. G. et al., A biochemical and pharmacological examination of *Ramphiophis oxyrhyncus* (Rufous beaded snake) venom, *Toxicon*, 45, 219, 2005.
339. Minton, S. A., Venomous bites by nonvenomous snakes: an annotated bibliography of colubrid envenomation, *J. Wild. Med.*, 1, 119, 1990.
340. Fry, B. G., personal communication, 2005.
341. Aitchison, J. M., Boomslang bite: diagnosis and management, a report of 2 cases, *S. Afr. Med. J.*, 78, 39, 1990.
342. Kornalik, F., Taborska, E., and Mebs, D., Pharmacological and biochemical properties of a venom gland extract from the snake *Thelotornis kirtlandi*, *Toxicon*, 16, 535, 1978.
343. Hoffmann, J. J. M. L. et al., Haemostatic effects *in vivo* after snakebite by the red-necked keelback (*Rhabdophis subminiatus*), *Blood Coag. Fibrinolysis*, 3, 461, 1992.
344. Vest, D. K., Envenomation following the bite of a wandering garter snake (*Thamnophis elegans vagrans*), *Clin. Toxicol.*, 18, 573, 1981.
345. Jansen, D. W. and Foehring, R. C., The mechanism of venom secretion from the Duvernoy's gland of the snake *Thamnophis sirtalis*, *J. Morphol.*, 175, 271, 1983.
346. Hayes, W. K. and Hayes, F. E., Human envenomation from the bite of the eastern garter snake, *Thamnophis sirtalis* (Serpentes: Colubridae), *Toxicon*, 23, 719, 1985.
347. Gomez, H. F. et al., Human envenomation from a wandering garter snake, *Ann. Emerg. Med.*, 23, 1119, 1994.
348. Minton, S. A., Neurotoxic snake envenoming, *Semin. Neurol.*, 10, 52, 1990.
349. Gasanov, S. E. et al., Cobra venom cytotoxin free of phospholipase $A_2$ and its effect on model membranes and T leukemia cells, *J. Membrane Biol.*, 155, 133, 1997.
350. Gubesek, F., Krizaj, I., and Pungercar, J., Monomeric phospholipase $A_2$ neurottoxins, in *Venom Phospholipase $A_2$ Enzymes: Structure, Function and Mechanism*, Kini, R. M., Ed., Wiley, Chichester, U.K., 1997, p. 245.
351. Bon, C., Multicomponent neurotoxic phospholipases $A_2$, in *Venom Phospholipase $A_2$ Enzymes: Structure, Function and Mechanism*, Kini, R. M., Ed., Wiley, Chichester, U.K., 1997, p. 269.
352. Abe, T., Alema, S., and Miledi, R., Isolation and characterization of presynaptically acting neurotoxins from the venom of *Bungarus* snakes, *Eur. J. Biochem.*, 80, 1, 1977.
353. Joseph, J. S. et al., Effect of snake venom procoagulants on snake plasma: implications for the coagulation cascade of snakes, *Toxicon*, 40, 175, 2002.
354. Ahn, M. Y., Lee, B. M., and Kim, Y. S., Characterization and cytotoxicity of L-amino acid oxidase from the venom of king cobra (*Ophiophagus hanna*), *Int. J. Biochem. Cell Biol.*, 29, 911, 1997.

355. Tun-Pe et al., Bites by the king cobra (*Ophiophagus hanna*) in Myanmar: successful treatment of severe neurotoxic envenoming, *Q. J. Med.*, 80, 751, 1991.
356. Lizano, S., Domont, G., and Perales, J., Natural phospholipase $A_2$ myotoxin inhibitor proteins from snakes, mammals and plants, *Toxicon*, 42, 963, 2003.
357. Agbaji, A. S., Conformation of cardiotoxins isolated from *Naja naja siamensis*, *Indian J. Biochem. Biophys.*, 23, 52, 1986.
358. Watt, G. et al., Bites by the Philippine cobra (*Naja naja philippinensis*): with prominent neurotoxicity with minimal local signs, *Am. J. Trop. Med. Hyg.*, 39, 306, 1988.
359. Ismail, M. et al., The ocular effects of spitting cobras: I. The Ringhals cobra (*Hemachatus haemachatus*) venom-induced corneal opacification syndrome, *J. Toxicol. Clin. Toxicol.*, 31, 31, 1993.
360. Lalloo, D. G. and Theakston, D. G., Snake antivenoms, *J. Toxicol Clin. Toxicol*, 41, 277, 2003.
361. Gold, B. S., Neostigmine for the treatment of neurotoxicity following envenomation by the Asiatic cobra, *Ann. Emerg. Med.*, 28, 87, 1996.
362. Pan, C. G. et al., Topical heparin with tetracycline versus heparin or tetracycline alone, in preventing ocular scarring due to the venom of the black spitting cobra (Naja sumatrana), *J. Toxicol. Clin. Toxicol.*, 43, 775, 2005.
363. Cecchini, A. L. et al., Biological and enzymatic activities of *Micrurus* sp. (Coral) snake venoms, *Comp. Biochem. Physiol.*, 140, 125, 2005.
364. Vita-Brazil, O., Coral snake venoms: mode of action and pathophysiology of experimental envenomation, *Rev. Inst. Med. Trop. Sao Paulo*, 29, 119, 1987.
365. Kitchens, C. S. and Van Mierop, L. H. S., Envenomation by the eastern coral snake (*Micrurus fulvius fulvius*), *JAMA*, 258, 1615, 1987.
366. Gaar, G. G., Assessment and management of coral snake and other exotic snake envenomations, *J. Fla. Med. Assoc.*, 83, 178, 1996.
367. O'Shea, M., *Venomous Snakes of the World*, Princeton University Press, Princeton, NJ, 2005, p. 78.
368. Harvey, A. L., Twenty years of dendrotoxins, *Toxicon*, 39, 15, 2001.
369. DeSarro, G. et al., Anticonvulsant activity of 5,7DCKA, NBQX, and felbamate against some chemoconvulsants in DBA/2 mice, *Pharmacol. Biochem. Behav.*, 58, 281, 1996.
370. Mbugua, P. M., Welder, A. A., and Acosta, E., Cardiotoxicity of Kenyan green mamba (*Dendroaspis angusticeps*) venom and its fractionated components in primary cultures of rat myocardial cells, *Toxicology*, 52, 187, 1988.
371. Jolkkonen, M. et al., Muscarinic toxins from the black mamba *Dendroaspis polylepis*, *Eur. J. Biochem.*, 234, 579, 1995.
372. Hodgson, P. S. and Davidson, T. M., Biology and treatment of mamba snakebite, *Wild. Environ. Med.*, 2, 133, 1996.
373. Arthur, C. K. et al., Effects of taipan (*Oxyuranus scutellatus*) venom on erythrocyte morphology and blood viscosity in a human victim *in vivo* and *in vitro*, *Trans. R. Soc. Trop. Med. Hyg.*, 85, 401, 1991.
374. Trevett, A. J. et al., Electrophysiological findings in patients envenomed following the bite of a Papuan taipan (*Oxyuranus scutellatus canni*), *Trans. R. Soc. Trop. Med. Hyg.*, 89, 415, 1995.
375. Trevett, A. J. et al., Failure of 3,4-diaminopyridine and edrophonium to produce significant clinical benefit in neurotoxicity following the bite of Papuan taipan (*Oxyuranus scutellatus canni*), *Trans. R. Soc. Trop. Med. Hyg.*, 89, 444, 1995.
376. Southern, D. A., Callanan, V. I., and Gordon, G. S., Severe envenomation by the taipan (*Oxyuranus scutellatus*), *Med. J. Aust.*, 165, 662, 1996.
377. Connolly, S. et al., Neuromuscular effects of Papuan taipan snake venom, *Ann. Neurol.*, 38, 916, 1995.
378. Gold, B. S., Dart, R. C. and Barish, R. A., Bites of venomous snakes, *N. Engl J. Med.*, 347, 347, 2002.
379. Andrews, R. K. and Berndt, M. C., Snake venom modulators of platelet adhesion receptors and their ligands, *Toxicon*, 38, 775, 2000.
380. Mori, N, Nikai, T and Sugihara, H., Phosphodiesterase from the venom of *Croatlus ruber ruber*, *Int. J. Biochem.* 19, 115, 1987.
381. Smith, S. V. and Brinkhous, K. M., Inventory of exogenous platelet-aggregating agents derived from venoms, *J Thromb. Haemos.*, 66, 259, 1991.
382. Gutierrez, J. M. and Lomonte, B., Phospholipase $A_2$ myotoxins from *Bothrops* snake venoms, *Toxicon*, 33, 1405 1995.
383. Bjarnason, J. B. and Fox, J. W., Snake venom metallo endopeptidases, *in Methods in Enzymology: Proteolytic Enzymes*, Vol. 248, Part E, Barrett, A. J., Ed., Academic Press, New York, 1995, p. 345.
384. Russell, F. E., Snake venom poisoning, *Vet. Hum. Toxicol.* 33, 584, 1991.
385. Dos-Santos, M. C. et al., Immunization of equines with phospholipase $A_2$ potects against the lethal effects of *Crotalus durissus terrificus* venom, *Braz. J. Med. Res.*, 22 509, 1989.
386. Antonini, G. et al., Neuromuscular paralysis in *Viper aspi* envenomation: pathogenic mechanism, *J. Neurol. Neurosurg. Psychiatry*, 54, 187, 1991.
387. Russell, F. E. and Bogart, C. M., Gila monster: its biology venom and bite—a review, *Toxicon*, 19, 341, 1981.
388. Beck, D. D. and Lowe, C. H., Ecology of the beaded lizard *Heloderma horridum*, in a tropical dry forest in Jalisco Mexico, *J. Herpetol.*, 25, 395, 1991.
389. Alagon, A. C. et al., Venom from two subspecies of *Heloderma horridum* (Mexican beaded lizard): general characterization and purification of *N*-benzoyl-L-arginine ethyl ester hydrolase, *Toxicon*, 20, 463, 1982.
390. Mebs, D. and Raudonat, H. W., Biochemical investigation on *Heloderma* venom, *Mem. Inst. Butantan Simp. Int.*, 33 907, 1966.
391. Alagon, A. C. et al., Helodermatine, a kallikrein-like hypotensive enzyme from the venom of *Heloderma horridum horridum* (Mexican beaded lizard), *J. Exp. Med* 164, 1835, 1986.
392. Hendon, R. R. and Tu, A. T., Biochemical characterization of the lizard gilatoxin, *Biochemistry*, 20, 3517 1981.
393. Mochca-Morales, J., Martin, B. M., and Possani, L. D. Isolation and characterization of helothermine, a novel toxin from *Heloderma horridum horridum* (Mexican beaded lizard), *Toxicon*, 28, 299, 1990.

394. Robberecht, R. et al., Evidence that helodermin, a newly extracted peptide from Gila monster venom, is a member of the secretin/VIP/PIH family of peptides with an original pattern of biological properties, *FEBS Lett.*, 166, 277, 1984.
395. Hooker, K. R. and Caravati, E. M., Gila monster envenomation, *Ann. Emerg. Med.*, 24, 731, 1994.
396. Stahnke, H. L., Heffron, W. A. and Lewis, D. L., Bite of the Gila monster, *Rocky Mt. Med. J.*, 67, 25, 1970.
397. Russell, F. E., Toxic effects of terrestrial animal venoms and poisons, in *Casarett & Doull's Toxicology*, 6th ed., Klaassen, C. D. et al., Eds., McGraw-Hill, New York, 2001, chap. 26.
398. Piacentine, J. et al., Life-threatening anaphylaxis following Gila monster bite, *Ann. Emerg. Med.*, 15, 9959, 1986.
399. Daly, J. W., Garaffo, H. M., and Spande, T. F., in *Amphibian Alkaloids*, Vol. 43, Cordell, G. A., Ed., Academic Press, San Diego, CA, 1993, p. 185.
400. Frost, D. R., Amphibian species of the world: an online reference. Vol. 3.0, American Museum of Natural History, New York, 2004 (http://research.amnh.org/herpetology/amphibia/index.html).
401. Daly, J. W. et al., First occurrence of tetrodotoxin in a dendrobatid frog (*Colostethus inguinalis*), with further reports fro, bufonid genus *Atelopus*, *Toxicon*, 32, 279, 1994.
402. Pires Jr., O. R. et al., Further report of the occurrence of tetrodotoxin and new analogues in the Anuran family Brachycephalidae, *Toxicon*, 45, 73, 2005.
403. Clarke, B. T., The natural history of amphibian skin secretions, their normal functioning and potential medical applications, *Biol. Rev.*, 72, 365, 1997.
404. Mebs, D. et al., Further report of the occurrence of tetrodotoxin in *Atelopus* species (Family: Bufonidae), *Toxicon*, 33, 246, 1995.
405. Tsuruda, K. et al., Secretory glands of tetrodotoxin in the skin of the Japanese newt *Cynops pyrrhogaster*, *Toxicon*, 40, 131, 2002.
406. Miyazawa, K. and Noguchi, T., Distribution and origin of tetrodotoxin, *J. Toxicol. Toxin Rev.*, 20, 11, 2001.
407. Cardall, B. L. et al., Secretion and regeneration of tetrodotoxin in the rough-skin newt (*Taricha grannulosa*), *Toxicon*, 44, 933, 2004.
408. Pires, O. R. et al., The occurrence of 11-oxotetrodotoxin, a rare tetrodotoxin analogue, in the brachycephalidae frog *Brachycephalus ephippium*, *Toxicon*, 42, 563, 2003.
409. Hardman, J. G. and Limbard, L. E., *Goodman & Gilman's*, 10th ed., McGraw-Hill, New York, 1996.
410. Brodie, E. D. and Brodie, Jr., E. E., Predator–prey arms races, *Bioscience*, 49, 557, 1999.
411. Bradley, S. G. et al., Fatal poisoning from the Oregon rough-skinned newt, *JAMA*, 246, 247, 1981.
412. Daly, J. W. et al., Bioactive alkaloids of frog skin: combinatorial bioprospecting reveals that pumilotoxins have an arthropod source, *Proc. Natl. Acad. Sci. U.S.A.*, 99, 11092, 2002.
413. Caldwell, J. P., The evolution of myrmecophagy and its correlates in poison frogs (family Dendrobatidade), *J. Zool. Soc. Lond.*, 240, 75, 1996.
414. Mortari, M. R., Main alkaloids from the Brazilian dendrobatidae frog *Epipedobates flavopictus*: pumiliotoxin 251D, historonicotoxin, and decahydroquinolones, *Toxicon*, 43, 303, 2004.
415. Daly, J. W., Thirty years of discovering arthropod alkaloids in amphibian skin, *J. Nat. Prod.*, 61, 162, 1998.
416. Daly, J. W. et al., Biologically active substances from amphibians: preliminary studies on anurans from twenty-one genera of Thailand, *Toxicon*, 44, 805, 2004.
417. Lyttle, T, Goldstein, D., and Gartz, J., Bufo toads and bufotenine: fact and fiction surrounding an alleged psychedelic, *J. Psychoactive Drugs*, 28, 267, 1996.
418. Davis, W. and Weil, A. T., Identity of a New World psychoactive toad, *Ancient Mesoamerica*, 3, 51, 1992.
419. Weil, A. T. and Davis, W., *Bufo alvarius*: a potent hallucinogen of animal origin, *J. Ethnopharmacol.*, 41, 1, 1995.
420. Lyttle, T., Misuse and legend in the 'toad licking' phenomenon, *Int. J. Addictions*, 28, 521, 1993.
421. Kwan, T. et al., Digitalis toxicity caused by toad venom, *Chest*, 102, 949, 1992.
422. Bagrov, A. Y. et al., Digitalis-like and vasconstrictor effects of endogenous digoxin-like factors from the venom of *Bufo marinus* toad, *Eur. J. Pharmacol.*, 234, 165, 1993.
423. Bedford, P. G. C., Toad venom toxicity and its clinical occurrence in small animals in the United Kingdom, *Vet. Rec.*, 94, 613, 1974.
424. Brubacher, J. R. et al., Efficacy of digoxin specific Fab fragments (Digibind®) in the treatment of toad venom poisoning, *Toxicon*, 37, 931, 1999.
425. Kazuhiro, O, Kantorwitz, J. D., and Spector, S., Isolation of morphine from toad skin, *Proc. Natl. Acad. Sci. U.S.A.*, 82, 1852, 1985.
426. Das, M. et al., A sleep inducing factor from common Indian toad (*Bufo melanostictus*) skin extract, *Toxicon*, 38, 1267, 2000.
427. Gomes, A. et al., A lethal cardiotoxic protein isolated from Bidder's organ of common Indian toad, *Bufo melanostictus* Schneider, *Indian J. Exp. Biol.*, 34, 211, 1996.
428. Mebs, D., Variability in alkaloids in the skin of the European fire salamander (*Salamandra salamandra terristris*), *Toxicon*, 45, 603, 2005.
429. Habermehl, G., Venoms of amphibia, in *Chemical Zoology*, Vol. 9, Academic Press, New York, 1974, p. 161.
430. Tsunenari, S., Uchimura, Y., and Kanda, M., Puffer poisoning in Japan: a case report, *J. Forens. Sci.*, 25, 240, 1980.
431. Lehane, L., Paralytic shellfish poisoning: a potential health problem, *Med. J. Aust.*, 175, 29, 2001.
432. James, K. J. et al., First evidence of an extensive northern European distribution of azaspiracid poisoning (AZP) toxins in shellfish, *Toxicon*, 40, 989, 2002.
433. Sierra-Beltran, A. P. et al., An overview of the marine food poisoning in Mexico, *Toxicon*, 36, 1493, 1998.
434. Gonzoles, I. et al., Role of associated bacteria in growth and toxicity of cultured benthic dinoflagellates, *Bull. Soc. Path. Exp.*, 85, 457, 1992.
435. Miller, T. R. and Belas, R. *Pfiesteria piscicida*, *P. shumwaye*, and other *Pfiesteria*-like dinoflagellates, *Res. Microbiol.*, 154, 85, 2003.
436. Chen, K. T. and Malison, M. D., Outbreak of scombroid fish poisoning, Taiwan, *Am. J. Public Health*, 77, 1335, 1987.
437. Diamond, J. M., Rubbish birds are poisonous, *Nature*, 360, 19, 1992.

438. Dumbacher, J. P. et al., Homobatrachotxoin in the genus *Pitohui*: chemical defense in birds, *Science*, 258, 799, 1992.
439. Pucek, M., Chemistry and pharmacology of insectivore venoms, in *Venomous Animals and Their Venoms*, Bucherl, W., Bucklley, E. A., and Deulofeu, V., Eds., Academic Press, New York, 1968, p. 43.
440. de Plater, G., Martin, R. L. and Milburn, P. J., A pharmacological investigation of the venom from the platypus (*Ornithorhynchus anatinus*), *Toxicon*, 33, 157, 1995.
441. Apitz-Castro, R. et al., Purification and partial characterization of draculin, the anticoagulant factor present in the saliva of vampire bats (*Desmodus rotundus*), *J. Thromb. Haemos*., 73, 94, 1995.
442. Wetzel, W. W. and Christy, N. P. A king cobra bite in New York City, *Toxicon*, 3, 393, 1989.
443. Yang, R. S. et al., Inhibition of tumor formation by snake venom disintegrin, *Toxicon*, 45, 661, 2005.
444. White, J., Snake venoms and coagulopathy, *Toxicon*, 45, 951, 2005.
445. Marsh, N. and Williams, V., Practical applications of snake venom toxins in haemostasis, *Toxicon*, 45, 1171, 2005.

# *Part 3*

## *Methods*

# 20 The Use of Laboratory Animals in Toxicology Research

*William J. White, C. Terrance Hawk, and Mary Ann Vasbinder*

## CONTENTS

This chapter provides an overview of laboratory animal science and medicine as they pertain to the use of laboratory animals in toxicologic research. The purpose of this chapter is to familiarize the toxicologist with a number of important issues involving the use of laboratory animals. It is not possible to explore all subjects that pertain to this topic nor to comprehensively review the subjects in this chapter. The reader is urged to seek more detail in the references provided and to consult specialists in the field for more information.

Veterinarians with specialized training in laboratory animal medicine have played important supporting roles in institutions conducting toxicologic research. Much of the information used by them, as well as by other laboratory animal professionals and toxicologists, has been developed as a direct result of studies initiated in the basic sciences and in toxicology. By their very nature, toxicologic studies have required an understanding of the biological characteristics and needs of laboratory animals as well as an understanding of those variables that impact the performance of laboratory animals in research studies. Our understanding of laboratory animals is far from complete and is further complicated by the wide variety of species from which the toxicologist may choose. Species such as mice and rats are extensively used for which vast amounts of background data are available and whose biological characteristics have been well explored. Less commonly used species, such as guinea pigs, hamsters, and gerbils, have proportionately less known about them and much less published background data available. Recent efforts by a number of organizations to collect, analyze, and publish background data on a continuing basis in a variety of disciplines are helping to extend the knowledge of certain laboratory animal species [48,191].

Like any field, the quality of published information can vary widely. Moreover, a lot of dogma and unsubstantiated opinion still remain in the literature. A number of recommended practices still cannot be well substantiated in the peer-reviewed literature. When designing experimental protocols, toxicologists should familiarize themselves with the important variables associated with the animals that will be used in their studies and put in place measures to control or account for these variables.

In today's society, the use of animals for research purposes has become a subject of much public discussion. Over the last century, economically developed countries in which most biomedical research takes place have moved further and further away from extensive public involvement with the agricultural use of animals and replaced that exposure with zoos, family pets, and stylized animal characters in the media. This has caused society to reexamine our relationships with animals, both in agriculture and in research, which has resulted in the development of laws, regulations, and guidelines that address responsible animal use. As with any controversial subject, there are those who find the measures taken to be insufficient and who advocate even further change through dialogue, protest, or terrorism. In many cases, radical views and actions are based on misinformation and a basic distrust of large organizations and governments; however, a large number of individuals concerned with animal welfare are open minded and seek to strike an appropriate balance in their use for biomedical research. They, like the members of the research community, are focused on responsible animal usage, including, where appropriate, the refinement, reduction, and replacement (the three Rs) of laboratory animals [278].

Responsible animal usage has stimulated interest in *in vitro* alternatives, computer-simulated models, and computer structure–activity analyses to screen for appropriate drug candidates. These efforts have had some limited success in the initial stages of the drug discovery process. They have not yet led to suitable *in vitro* replacements for most animal usage in toxicologic studies, especially product-registration studies, nor do they appear likely to do so in the near future. This failure to develop complete substitutes for intact living organisms used in research is likely due to the complex interactions that exist at organ, cellular, and subcellular levels. A suitable replacement for animals will have to reliably predict biological phenomena, including being at least as good and consistent a model for risk assessment in humans as animals. Such systems will have to be extensively validated and accepted by regulatory bodies as suitable substitutes.

The use of animals in toxicologic research cannot be taken for granted. With their use comes not only a responsibility to adhere to institutional, governmental, and scientific principles, policies, laws, regulations, and guidelines, but also an ethical and moral responsibility for the lives of the animals used in research or product manufacture. Each researcher is also responsible for the quality of the care that the animals receive, the appropriateness of their use, and minimization or relief of pain.

## REGULATIONS, LAWS, POLICIES, AND GUIDELINES

The use of animals in research, as well as the assurance that provisions for appropriate animal welfare and care have been made, is controlled by a number of mechanisms. These can be subdivided into two general categories: (1) guidelines and recommendations, and (2) laws and regulations. In addition, policies can be developed for either of these two categories; for example, guidelines or recommendations that are not regulated by law can be included in an overall policy that governs institutional activities or eligibility for receiving funding. Moreover, policies may also be created and used as accepted interpretations of regulations developed in response to laws.

Laws and regulations require mandatory compliance. Failure to meet the requirements imposed by laws or regulations usually is attended by legal actions that may culminate in fines, revocation of the ability to conduct animal-related activities, or imprisonment. In the case of laboratory animals in the United States, laws and regulations are administered by the U.S. Department of Agriculture (USDA). This regulatory body is charged with conducting regular inspections and registering all facilities using laboratory animals. Within the USDA, the Animal and Plant Health Inspection Service (APHIS) is charged with making such inspections. Within APHIS, the Animal Care (AC) unit is responsible for enforcing regulations developed in response to legislation governing the use of laboratory animals in research.

Guidelines and recommendations usually are developed by independent groups with expertise in one or more aspects of laboratory animal science and medicine. Compliance with guidelines or recommendations is voluntary; however, failure to do so may be accompanied by undesirable consequences, such as denial of funding by government institutions, the inability to have data accepted for publication, or the inability to use data in submissions filed in response to mandated regulatory processes [81,95]. Compliance with such guidelines or recommendations may have to be ensured through filing of legally binding statements, submission of regular reports to agencies tracking such activities, or by participation in a voluntary accreditation program, such as the one conducted by the Association for Assessment and Accreditation of Laboratory Animal Care International (AAALAC), who uses a combination of regular reports and periodic site visits to evaluate programs, facilities, and animal care.

## INTERNATIONAL ASSURANCE AND REGULATION OF LABORATORY ANIMAL CARE AND USE

The existence and complexity of laws, regulations, guidelines, and recommendations governing the care and use of laboratory animals vary significantly among countries. In general, nonindustrial countries commonly do not have laws governing the use of animals in research, teaching, or product production. Some international guidelines or recommendations may be followed, but only to the extent that they impact the suitability of work or products for registration in other countries.

Countries within the European Economic Union use certain minimal standards developed through the council of Europe and subsequently ratified by member states. These standards are used as the basis for individual country laws and regulations that meet or exceed these standards. Variations between countries can exist. Within an individual country, standards may be different for different aspects of animal production and use. These laws and standards may apply not only to research use of animals but also to transportation of animals, as well as their exhibition and sale for other purposes. A comprehensive review of all these items, including regulatory oversight and reporting requirements, is beyond the scope of this text; however, a few important considerations should be mentioned.

Protocol review at either a regional or national level is a common component of the regulations governing animal care and use in European countries. Protocols must fulfill certain guidelines and provide a detailed description of the proposed study as well as supporting rationale. The time from submission of a protocol until a decision is made by reviewers can be significant.

Licensing of researchers to perform specific procedures is often a component of the regulatory process. Guidelines for credentials and training are specified, and licensing is often done on an individual procedure basis.

The evaluation of programs and unannounced inspections of facilities are often done through a government agency. Activities can be suspended based on findings of inspections, and licensure can be revoked. Other penalties can be imposed, depending on the country.

Unlike in the United States, all species used in research conducted in Europe are covered by guidelines and regulations. Differences can exist between countries in terms of acceptable care and use practices, and no assumption should be made that all of these standards are similar to ones in the United States or elsewhere. The level of detail and the emphasis on various topics can differ substantially from United States standards.

The Federated European Laboratory Animal Science Associations (FELASA) is a European consortium of laboratory animal science associations. FELASA has developed a number of recommendations for health monitoring, accreditation of animal diagnostic laboratories, and other topics that can impact the quality of laboratory animals used in toxicologic research. Adherence to these recommendations is voluntary, and in some cases the recommendations may differ from accepted practices in other parts of the world.

In Canada, the Canadian Council on Laboratory Animal Care has produced guidelines for the care and use of laboratory animals. Inspection of research facilities is conducted on a voluntary basis by this organization [46,47]. In Japan, the Ministry of Education, Sports, Science, and Technology along with the Ministry of Health, Labor, and Welfare and the Ministry of Agriculture, Forestry, and Fisheries recently formulated animal experimentation guidelines. These guidelines cover programmatic issues (e.g., establish the use of institutional animal care and use committees) rather than experimental procedures. More detailed guidelines formulated by the Council of Japan were modeled after Institute for Laboratory Animal

Research (ILAR) guidelines. Japanese laws covering the use of laboratory animals include the Law for Humane Treatment and Management of Animals and the Standards Relating to the Care and Management of Laboratory Animals and Relief of Pain. These laws became enforceable on June 1, 2006.

## U.S. LAWS, REGULATIONS, GUIDELINES, RECOMMENDATIONS, AND POLICIES

### THE ANIMAL WELFARE ACT

The regulation of laboratory animals used in research is governed by the Laboratory Animal Welfare Act of 1966, Public Law (P.L.) 89-544 [308]. The USDA administers the Animal Welfare Act. The act establishes legal requirements for research facilities to provide certain minimum standards for the care of animals in research. In 1970, the Laboratory Animal Welfare Act was amended (P.L. 91-579), at which time its name was changed to the Animal Welfare Act because the term *animal* was changed to include all warm-blooded animals except those used for food or fiber [308]. The 1970 Act required registration of research facilities as well as annual reporting requirements. Animals used for other purposes, such as exhibition or as pets, were also extended coverage.

The Act was again amended in 1976 (P.L. 94-279) to include more rigorous standards for transportation of animals [308]. The Act was last amended in 1985 (P.L. 99-198) to include requirements for exercising dogs, the psychological well-being of nonhuman primates, the consideration of alternative procedures to ones causing pain or distress, training of personnel, and the establishment of an institutional animal care and use committee (IACUC) of specified composition [308].

In response to the Act and its amendments, a series of regulations and policies were developed. Significant additions to the regulations were added in August of 1989, April of 1990, and February of 1991 that refined definitions and terms; set standards for humane handling, care, treatment, and transportation; set space requirements for primary enclosures for guinea pigs, rabbits, and hamsters; and set standards for humane handling, care, treatment, and transportation of dogs, cats, and nonhuman primates [188,280]. The regulations promulgated by the USDA in response to the Animal Welfare Act are published in the Code of Federal Regulations (CFR, Title 9, Chapter A, Parts I, II, and III) [71].

Currently, the Animal Welfare Act covers warm-blooded animals including dogs, cats, nonhuman primates, hamsters, guinea pigs, rabbits, marine mammals, and any other domestic animal, as well as those animals normally found in the wild, used in research, testing, exhibition, or experimentation or kept as pets. The term *animal*, as defined in the Animal Welfare Act, excludes rats, mice, and birds that are bred for research, as well as horses and other farm animals used or intended for use as food or fiber or for use in improving animal nutrition, breeding, or production. Farm animals, including horses, are covered under the Animal Welfare Act if they are used for nonagricultural research or exhibition, which includes biomedical research [167].

Extension of the USDA's regulation of laboratory animals to include rats, mice, and birds bred for research is viewed by some as being inevitable. This exclusion was still in place as of 2000 and reflects a balancing of public and private resources with the perceived benefits to regulating these excluded species. The effects on toxicologic research in the United States of extending regulation to these species is difficult to predict, as these species represent the bulk of all animals used in such research; hence, the limited degree of experience with the current regulatory process may not be predictive of the consequences of extended coverage. Given the USDA's broad mandate for regulation, which includes transportation, the cost to the government and feasibility of providing the appropriate level of regulation of this activity may be difficult to accurately estimate.

### *THE GUIDE FOR THE CARE AND USE OF LABORATORY ANIMALS*

In the United States, the primary set of guidelines for the care and use of laboratory animals is *The Guide for the Care and Use of Laboratory Animals* ("*The Guide*") that was developed by the National Academy of Sciences [222]. *The Guide* is used as the primary reference for voluntary assurance and accrediting bodies, such as National Institutes of Health's Office for Laboratory Animal Welfare (OLAW) and AAALAC. This document was first prepared and published in 1963 and has been revised many times since. The 1996 revision made substantial changes in the overall approach to laboratory animal care and use.

*The Guide* cannot be appropriately used without an institutional animal care and use committee [222,243] which is appointed by the chief executive officer of the institution and is advisory to that person. The committee to comply with both the requirements of *The Guide* and USDA regulations, must include a doctor of veterinary medicine who is certified or has training or experience in laboratory animal science and medicine or in the use of the species used in research at the institution. The committee must also have at least one practicing scientist experienced in research involving animals as well as one nonscientist who may or may not be employed by the institution. The committee must also include at least one public member to represent community interest with respect to the care and use of laboratory animals. The public member must not be affiliated with the institution nust not be an immediate family member of anyone affil-ated with the institution, and must not be involved in the se of laboratory animals.

The IACUC is responsible for the evaluation and over-ight of the institution's animal care and use program and ll related issues set forth in *The Guide*. The IACUC must nspect all animal facilities used by the institution's esearch program as well as carry out a programmatic eview of the research program every six months. The ommittee subsequently analyzes the findings of their nspection and review and prepares a written report for ne responsible institutional official that details their find-ngs and any recommendations for action. The institu-onal official, in turn, much address any major deficien-ies detailed in the report by providing a reasonable plan or corrective action and a timetable. The committee must aintain written records of their meetings and all deci-ons taken. Meetings should be held as frequently as ecessary to accomplish their designated tasks, but it nould meet at a minimum of twice a year. A mechanism or documenting minority views must also be provided.

An essential component of the IACUC's activity is the eview and approval of animal use protocols prior to the itiation of research or other animal-related activities. rotocols submitted for committee review should be com-lete with respect to their description of the animal care nd use components. To prepare a protocol that covers all f the areas and issues posed both by the USDA regula-ons and *The Guide* recommendations, the IACUC often equires the researcher to place the protocol in a standard ormat that may take the form of a questionnaire. Topics ch as surgery, anesthesia, provision of analgesics during ainful procedures, determination of acceptable alterna-ves to painful procedures, methods of euthanasia, criteria r ending an animal's participation in a study or proce-ure, and assurance of appropriate housing and care prac-ces are some of the many issues that are considered in e review of any protocol.

*The Guide* charges the users of research animals with e responsibility of achieving specific outcomes with spect to the care and use of animals but, by using a erformance-based approach, provides latitude as to how ese outcomes are achieved [222]. Performance standards efine an outcome in detail and provide criteria for assess-g that outcome. They do not restrict the method by which is outcome is achieved. This is in contrast to engineering andards that do not provide for interpretation and modi-cation of prescribed methods or procedures in the event at acceptable alternative methods are available or unusual rcumstances occur. The IACUC is charged with devel-oing additional performance standards to evaluate alter-ative methods to achieve specified outcomes and is given e latitude to modify recommendations set forth in *The uide* based on performance data generated by researchers establish the adequacy of alternative methods.

### Public Health Service Policy

The administration and coordination of the Public Health Service's (PHS) policy on Humane Care and Use of Laboratory Animals ("The Policy") is performed by the OLAW [264]. This policy requires institutions to establish and maintain proper measures to ensure the appropriate care and use of all animals involved in research, research training, and biological testing activities conducted or supported by the PHS. The Policy also endorses a set of principles developed by the Interagency Research Animal Committee (IRAC) that covers the use and care of vertebrate animals used in testing, research, and training [140].

The Policy applies to all research supported by the PHS and places a heavy emphasis on training of all personnel involved in the care of use of animals in a research setting. The Policy requires the biannual preparation and evaluation of reports by the IACUC on the institution's programs and facilities that involve animal activities. The Policy requires the filing of an assurance statement with the OLAW binding the institution to The Policy as well as to the use of *The Guide* as a basis for developing and implementing institutional programs for activities involving animals. The OLAW also requires the filing of annual reports by the institution. Achieving and maintaining such assurance is a prerequisite for consideration for funding of activities supported by PHS and, by extension, may be adopted by agencies as essential criteria for consideration for funding.

## ASSOCIATION FOR ASSESSMENT AND ACCREDITATION OF LABORATORY ANIMAL CARE INTERNATIONAL

The Association for Assessment and Accreditation of Laboratory Animal Care International is a voluntary organization that was founded in 1965. AAALAC is a U.S.-based nonprofit corporation with national and international representation from scientific and educational organizations. Its goal is to promote responsible and high-quality animal care and use by research institutions through the use of a peer-review process. The organization assists institutions in developing and improving programs and facilities in developing and improving programs and facilities for animal care and use.

The accreditation process is designed to assess the institution's conformance with appropriate locally applicable guidelines, recommendations, laws, regulations, and policies, as well as evaluate the program's ability to ensure that animal health and well-being are safeguarded by institutional processes and operational procedures. In the United States, AAALAC uses *The Guide* as it primary standard; outside of the United States, applicable regulations, laws, and other standards are used that are relevant to that country and the institution's activities.

The accreditation process is confidential. It is initiated by the submission of a comprehensive application that is designed to provide a detailed description of the institution's animal programs and facilities. An initial site visit is made by two or more individuals representing AAALAC, one of whom may be an *ad hoc* consultant with special expertise in programmatic areas relevant to the institution applying for accreditation. Following the initial site visit, a written report is filed along with recommendations. This is then considered by AAALAC's Council on Accreditation to determine if the facility should be granted accreditation. Facilities that have obtained accreditation are revisited every 3 years to determine eligibility to maintain continued accreditation. Annual reports of activities that include programmatic changes and animal usage must be filed.

The achievement of AAALAC accreditation may reduce the level of some reporting requirements for organizations such as the OLAW but does not substitute for assurances filed with the OLAW or with other requirements set forth in laws or regulations. AAALAC does not publish separate guidelines or recommendations nor does it set other standards that are at variance with existing guidelines, laws, or regulations. Lists of institutions that have obtained full accreditation status are regularly published by AAALAC; however, details of an institution's accreditation history are not made available.

## DESIGN AND CONSTRUCTION OF ANIMAL HOUSING FACILITIES

The design of animal facilities for biomedical research has evolved over time, as have housing methods for research animals. Key design concepts that have survived the test of time relate to practical matters, such as waste handling, sanitation, investigator access, disease control, and size of the animal species being used. No universally accepted design for animal facilities exists, nor is it likely that any one given design for animal holding rooms will be appropriate for all species of animals that could be used in toxicologic research. Some reviews on the subject of animal facilities design have been prepared, as have design process considerations for specialized support facilities for research animals [9,17,45,61,64,79,113,128,130,139, 174,189,222,223,258,281,297,345].

The process of animal facility design requires, first and foremost, a clear documentation of the intended type and extent of use of the facility based on past and present toxicologic research programs, as well as a reasonable expectation of the growth of these programs over a period of 5 to 10 years [345]. In considering the intended facility use, the numbers and types of animals by species must be determined. While studies involving small rodent species often consume the bulk of toxicologic research that use animals, other larger species are frequently used and may have to be provided for within a facility. As an alternative, facilities may be designed to accommodate only a limited range of species, with housing and usage of other species being outsourced to other facilities that are better suited to their maintenance.

When the level of usage has been determined, specialized requirements—such as the use of biohazardous material requiring biocontainment facilities, a surgical program requiring facilities for aseptic procedures, specialized radiographic or imaging facilities, diet preparation facilities, specialized equipment or procedures facilities, diagnostic laboratory or necropsy facilities, or other specialized spaces requiring specific types of equipment or housing for animals used in toxicologic research—must be identified, as well as the level of activity that will occur in such facilities. Because these types of facilities often represent a significant cost to construct and operate and consume considerable amounts of floor space, it is important to carefully assess the realistic level of activity projected for such facilities.

The animal care program itself must also be carefully described, and details of the program should be well worked out before the design process is started. The reason for this is that the location of various components of the animal facility will depend greatly on the methodology and equipment to be used in the animal care program; for example, caging and racking within an animal facility must move to and from cage washing facilities. The amount of labor involved in performing these tasks is directly related to the distance of various components of the animal housing facility from the cage washing facilities, as is the opportunity for transmission of adventitious microorganisms potentially infecting one group of animals spreading to others through the cage sanitation process. The frequency of cage changing, as well as the location in which cages are changed and the equipment required for such cage changing, will also influence the amount of storage space and operational space within individual animal's rooms and storage space within the cage washing facility.

The type of caging and the level of bioexclusion of microorganisms to be achieved for certain groups of animals maintained within the animal research facility will also dictate many aspects of the design [344]. Facilities housing large numbers of rodents in microisolation cages or in flexible or semirigid isolators will require different designs and operational practices, as compared to facilities that use conventional open-cage housing.

Animal holding rooms can be classified into two general types. The first, *rodent housing*, is characterized by smaller rooms with few fixtures; the design is orientated toward removal of caging from the room for sanitation and perhaps sterilization [204,222,223,344]. Such rooms usually have a small sink for hand washing and, if microisolation caging is used, will often be outfitted with a laminar flow hood designed for cage changing using aseptic technique. When microisolation caging is used, such rooms commonly have only a single access door minimize traffic through the room. Such rooms generally do not contain floor drains or specialized equipment for spraying water under pressure during cleaning of the room or equipment. The second type of animal holding room is designed to handle larger caging or runs/pens to house animals the size of rabbits or larger [222]. Such rooms are equipped with floor drains or some form of trough-drain system to allow in-the-room cleaning of cage components or, in the case of rabbits, the frequent spillage of liquids onto the floor [222]. While some caging may be removed from the room and taken to the cage washing area for more complete sanitation, certain components of daily cleaning are carried out within the room itself.

With the exception of some aspects of cage washing, most routine animal care activities are conducted manually, making the animal facility labor intensive. Moreover, the use of the animal facility by researchers also involves regular and often complex manual manipulations of the animals for study purposes. For this reason, distances between support facilities and the animal holding areas should be minimized and provisions often have to be made within the design for the strategic location of procedural rooms and, where appropriate, critical laboratory facilities necessary to conduct determinations that cannot be adequately conducted at great distances from the animal facility. The final design of an animal facility must be analyzed for the efficient traffic flow of both personnel and equipment. It is important that both users of the animal facility, as well as the animal care staff, analyze the design with respect to day-to-day tasks that need to be accomplished and the distances that need to be traversed by both people and equipment to accomplish these tasks.

Over time, two common arrangements of rooms within animal facilities have developed [129]. These have evolved from considerations of material and personnel flow patterns and from the use of specialized equipment and practices to control the spread of adventitious organisms and contaminants within animal facilities [288]. The most common facility is one with a single corridor design. Often, these are a result of renovation of existing space originally designed for laboratory use in which a single corridor that provides both entrance and exit from individual animal rooms was necessitated by the available space and location or, in some cases, by the need to isolate specialized animal facilities, such as microbiological barriers, in which showering and clothing changes for people and decontamination of supplies are required. Such barriered animal facilities have become popular in support of some forms of toxicologic research, especially those programs using transgenic or immunologically deficient animals [344].

The other common animal facility arrangement is a dual-corridor system orientated toward cage-washing facilities. In such a system, one or more corridors (i.e., supply or clean) that exit the clean side of the cage wash connect with one of two entrance doors to each animal room. The second door in each room is used to remove caging and other material for return through a second corridor (i.e., return or dirty) to the cage wash by a separate entrance in which soiled caging is collected for subsequent sanitation. This design addresses the logical flow of personnel and equipment through animal rooms so as not to have crossing of equipment within common corridors. It is unlikely that such a design provides any realistic measure of disease control by preventing cross-contamination, given the many other opportunities for movement of microorganisms through a facility [344]. Commonly, an air-pressure gradient is maintained across the two-corridor system such that the supply corridor (clean) is at the highest pressure, the next highest air pressure is maintained in the animal room, and the lowest pressure is in the return (dirty) corridor [222,223].

Because toxicologic research programs are dynamic and requirements can change dramatically over time, most animal facility designs try to maintain some degree of flexibility [113]. Animal rooms designed for large or small animals usually are designed with a minimum of fixed equipment or specialized mechanical systems. This allows easy conversion from the use of one species to another within the categories of large or small animal holding areas. Because most caging and racking are more efficiently located along the perimeter of the room, rectangular designs for animal holding rooms are common, as they maximize wall space while minimizing the relatively unused center of the room space. It is the usual practice to keep the center of the room clear to allow for the orderly movement of equipment, supplies, and personnel through the room and for the conduct of animal care and research functions. For this reason, common designs, whether for a one- or two-corridor system, have rooms with widths ranging from 12 to 14 feet and lengths of 18 to 20 feet.

Room sizes are also often dictated by building support structures, which, in the case of multiple-story buildings, result in support columns that must be maintained throughout the structure. The spacing of these columns is done at fixed intervals; hence, room dimensions must conform to these constraints to prevent columns from being placed within a room rather than in the walls.

Animal rooms, as well as support areas, are often grouped by function to minimize the distance traveled between related functions and to minimize the runs of piping and other mechanical services between areas that have higher use of mechanical services (e.g., cage washing rooms, operating rooms, holding rooms for dogs).

In considering the design of animal research facilities, provisions must be made for future expansion [113]. Certain items, such as cage washing facilities, specialized laboratories, and surgical facilities, are expensive components of new construction. During the initial construction of animal facilities, it may be possible to design them with sufficient capacity to support larger animal populations through logical expansion of the existing building. When developing a design for an animal facility, planners often include provisions for future expansion that must be considered in the animal facility design so architectural and mechanical barriers are not put in place.

An overriding concern in the construction of animal facilities is the need to frequently clean and disinfect surfaces within the facility. In some cases, specialized components of the animal facility, such as biocontainment facilities or bioexclusion facilities, must also have the capability of being totally disinfected or sterilized using aggressive agents that would be unsuitable for use in the rest of the animal facility. Many of the construction guidelines presented below are directed toward making such tasks and routine cleaning and sanitization easier to accomplish on a regular basis, as well as preventing the deterioration of the physical facilities by conducting such practices. By their very nature, these guidelines are quite general and are based on common practices and equipment.

## Floors

Opinion varies greatly on the best type of flooring to use for animal facilities. Each type of flooring has its advantages and disadvantages, and it is likely that no perfect flooring exists for animal facilities. Those materials showing the greatest chemical resistance often are not able to withstand the impact of sharp or heavy objects, whereas those showing the greatest resistance to mechanical damage may not be resistant to the actions of chemicals and urine. To withstand repeated sanitation and the spillage of urine and water, floors must be moisture resistant and nonabsorbent [222]. They should be resistant to the action of hot water, cleaning agents, disinfectants, urine, and other biological materials. They should be relatively smooth for ease of cleaning, although they may have to have textured surfaces, especially in high-moisture areas where injury to personnel could occur from slipping or to animals, such as hoof stock kept in pens that are regularly cleaned with water.

Floors should be impact resistant and capable of supporting equipment, cage racks, and other stored items without becoming cracked, pitted, or gouged. To facilitate cleaning, there should be a minimum number of joints in which debris can become entrapped. Materials that have proved particularly satisfactory in many applications are epoxy aggregates, sealed concrete, and other hardened synthetic-based aggregates. Correct surface preparation of subflooring and experienced installation of the flooring itself are essential to its satisfactory performance.

## Drains

Drains in animal facilities are a source of contention, but in some areas they are a necessity. Facilities housing large animals require drains. Floors in such areas should be gently sloped toward the drain, and in the case of pens and runs a dedicated trough for draining should be considered. In such applications, drainage should be designed to allow rapid removal of water and the drying of surfaces to minimize elevation of humidity. Where waste is to be flushed down drains along with water, larger diameter drains (4 inches or greater) should be considered [222]. To facilitate this process, rim-flush drains that sweep waste collected from trough drainage systems down into drain pipes through coarse basket strainers fitted in the drains minimize maintenance considerations. It is critical in sanitation programs for animal facilities that drain traps be kept filled with liquid and that drains that are not in use for long periods of time be capped and sealed to prevent the backflow of sewage, gases, and other contaminants into the animal holding areas.

Floor drains are not essential in the rooms of all animals. Rooms containing rodents, for example, can be sanitized better by using alternative cleaning methods, such as wet vacuuming or mopping with cleaning compounds and disinfectants.

## Walls and Ceilings

Walls within animal facilities should be frequently sanitized, so they must be moisture resistant and nonabsorbent. They also should be free of cracks or holes, such as those created by unsealed utility penetrations. Junctions of the walls with the ceiling and floors, as well as with themselves, should be adequately caulked or otherwise sealed to prevent the build-up of debris and eliminate potential harborage for insects. The construction of the walls can take a variety of forms, with masonry being the most durable; however, other wall construction materials may be more appropriate in certain situations. The surface coating of walls should be resistant to the chemical actions of disinfectants and cleaning agents and should have significant resistance to abrasion and impact. Common wall materials acceptable in other applications, such as painted gypsum board, may not prove satisfactory in some animal facility applications given their low impact resistance and significant damage when exposed to moisture. Some of the difficulties associated with certain wall construction materials can be overcome by protecting the wall surface through the use of strategically placed guard rails or bull-nosed curbing on the wall-to-floor junction to minimize the impact of equipment on the walls [222].

Like walls, ceilings should be smooth and moisture resistant. Their surfaces should be smooth and moisture resistant. Their surfaces should be capable of withstanding detergents and disinfectants and be free of imperfect junctions that would allow the build-up of debris and harborage of insects. Generally, suspended ceilings are undesirable in animal facilities unless they are constructed in such a way as to be readily sanitizable and free of imperfect junctions or penetrations. Gypsum board that has been appropriately sealed and finished with a durable coating or concrete that has been smoothed and sealed or painted are the most appropriate and long-lasting choices for ceilings.

### Doors to Animal Rooms

Doors to animal rooms should be constructed of and coated with materials that resist corrosion and are amenable to disinfection. They should be appropriately sealed or gasketted and fitted with door sweeps to ensure adequate sealing with the floor and the door frames. This is critical in preventing the incursion of pests and the uncontrolled movement of air and particulates beneath and around the door. Doors should be self-closing and should open into animal rooms to prevent injury to personnel or damage to equipment moving in hallways. For safety reasons, viewing windows often are appropriate, but the ability to cover the viewing windows in certain instances where close control of lighting is required must be considered.

Doors should be equipped with recessed or shielded handles (to prevent damage from equipment) and equipped with metal or plastic strike plates to prevent equipment damage to the surfaces. Generally, doors that are 42 in. × 84 in. accommodate most animal equipment; however, larger doors may be required in certain portions of the animal facilities. Consideration of room-level security is important to limit access to the animals or in areas where bioexclusion or biocontainment is required. This is best accomplished with mechanical or electrical locks; however, it is important that doors be designed to be opened from the inside without a key for emergency egress purposes.

### Corridors

Corridors within the animal facility are the principal means of access to animal rooms and support areas. For this reason, they should be wide enough to accommodate the movement of equipment and personnel and be amenable to cleaning and disinfection. Generally, a width of 6 to 8 feet is sufficient for most purposes [45,130,222]. All of the considerations reviewed for floors, walls, and ceilings generally apply to corridors; however, suspended ceilings are more commonly used in corridors due to the need to access mechanical fixtures, such as electrical conduits and plumbing. It is good practice to ensure that access to water lines, drain pipes, and other utilities supporting animal rooms is available through panels or chases in the corridor outside of the animal rooms. Wall-mounted equipment in the corridors, such as telephones, fire extinguishers, fire alarms, and record-keeping stations, should be recessed or installed high enough to prevent damage from movement of equipment and personnel through the corridors.

### Support Areas

In addition to the basic animal rooms, animal facilities must incorporate certain facilities and equipment to support the basic day-to-day investigational and animal care needs, including locker rooms often coupled with shower facilities and restrooms, break areas, cage washing facilities, feed and bedding storage, waste handling and storage areas, supply rooms, general storage areas, administrative offices, and procedural rooms for use by researchers. Although a comprehensive treatment of these areas is not addressed in this chapter, information regarding their essential features and construction can be found elsewhere [222]. A few essential points, however, bear emphasizing.

When designing animal facilities, it is a common error to undersize these facilities in comparison to the intended animal populations or types of research to be conducted. This is especially true of support areas that experience cyclic levels of activity during their normal function. For example, locker rooms where clothing changes occur are intensively used at the beginning and end of work days and, depending on the type of facility, may also experience significant activity at lunch time or during designated break periods. Commonly, these areas are designed to allow only a few individuals to use them at a time, thus requiring adaptation of schedules and making it difficult to access the facility during certain times throughout the day.

Such areas must be large enough to accommodate reasonable numbers of employees and to provide sufficient locker space for clothing and the storage of uniforms and other supplies. If mandatory showering is required, as is the case in some barriered facilities, sufficient capacity for showers or other clothing change functions must be provided. From a programmatic standpoint, the necessity for certain types of clothing changes or the use of certain items of disposable clothing, such as face masks or shoe covers, must be carefully considered. The more complex the clothing change, the more complex the support area must be. Similarly, if reusable garments are used, adequate storage space and provisions for laundering and disinfection must be made. If clothing changes are to be made at the level of the animal holding room, such changes must be provided for in the design of the facility and the procedure must be logical if it is to be routinely followed. Human use support areas, such as locker rooms and break areas or administrative offices, should be separated from animal areas. They are often located at the periphery of the animal facility and are used as control points for entrance into the animal facility.

The location of support areas is dictated by the location of utilities and certain architectural features in the animal facility; for example, cage washing facilities are usually located at the end of corridors servicing groups of animal rooms. These facilities usually are located near storage areas for bedding, feed, and building utility runs to minimize distances traveled and construction costs. Waste-handling facilities are commonly located near cage washing facilities, and such facilities usually are located within a convenient distance of loading docks or other exterior entrances.

Cage washing facilities are designed to allow the use of chemical detergents, disinfectants, and large amounts of hot water. Regular cleaning of room surfaces is an essential part of their operation, and the facility must be designed to allow this to be easily done. Large amounts of equipment are moved in and out of these facilities, and they must be big enough to accommodate such equipment both in process and for short periods of storage. Substantial quantities of heat and moist air are generated within the cage-washing facility, so heating, ventilation, and air conditioning (HVAC) systems must be designed to accommodate these loads.

Often the locations of support areas are selected to decrease the amount of unwanted noise and traffic passing by animal rooms [222]. Noise-producing areas, such as mechanical rooms, waste processing areas, and cage washers, are often grouped together and separated from animal holding areas by noise traps, such as air locks or sound-proofing surfaces. Commonly, species of animals such as dogs, nonhuman primates, and swine that produce loud vocalizations as part of their normal behavior are also separated from other holding areas that contain more noise-sensitive animals, such as rodents, by using similar building concepts. Housing for these noise-producing species is commonly located near support areas that produce noise, although the two functions usually are separated by air locks, doors, or other means. Areas used for the storage of feed are kept at lower environmental temperatures than other areas to inhibit decomposition of liable ingredients and the development of immature forms of insect pests that may contaminate the packaging or the feed.

## Specialized Components of Animal Facilities

Within animal facilities, there is often a need for specialized structures and equipment to support toxicologic research. Such facilities include, but are not limited to, barrier facilities for maintenance of animals in defined microbiological states, biocontainment facilities for the housing and manipulation of animals exposed to biohazardous materials, surgical facilities, imaging facilities, necropsy facilities, and facilities for conducting diagnostic tests on animals.

### Barrier Facilities/Barrier Rooms

Barrier facilities or individual barrier rooms may be used in toxicologic research for either bioexclusion or biocontainment purposes [344]. Most commonly, they are used for bioexclusion. In this role, the room or series of rooms (facility) is separated from other components of the animal facility using a variety of construction methodologies in a manner to prevent undesirable microorganisms from entering the room whether carried on personnel equipment or materials. Barrier rooms are composed of durable materials that are impervious to liquids and free of unsealed penetrations. Personnel entry is usually through a lock system requiring progressive decontamination of personnel by showering and/or clothing change. Air to the room is independently supplied and high-efficiency particulate air (HEPA)-filtered. Water supplied to the area is independently decontaminated or sterilized, and provisions are available to provide sterilized feed, bedding, and other supplies for use in the room. Waste is removed using processes designed to prevent the incursion of unwanted microorganisms during the removal process.

Most commonly, multiple methods of introducing supplies and equipment are built into the barrier room. Generally, a through-the-wall autoclave is provided with sufficient capacity to handle reasonably large loads of materials. For maintenance purposes, it should be possible to service the autoclave from the outside of the barrier. It is essential that the autoclave be calibrated and validated based on material types and load configurations. This calibration and validation should be done on a regular basis, usually yearly. The use of cumulative heat-sensitive or biological indicators placed at one or two points in a load to be autoclaved is not sufficient to confirm adequate disinfection or sterilization.

In addition to autoclaves for processing large volumes of materials that are unaffected or marginally affected by heat, alternative means of disinfection must be provided for the introduction of heat-sensitive materials into the barrier room. This is accomplished by using a double-door chamber into which materials that have been immersed in disinfectants can be placed. Additional disinfectants are sprayed into the port before it is closed, and the materials are held for an appropriate contact time. Once this chemical disinfectant process has taken place, materials can be removed from the chamber using the inside door for subsequent use in the barrier. As with autoclaving procedures, the use of chemical disinfectants must be calibrated so adequate application procedures, concentrations, and contact times are maintained. Materials processed by this method must be wrapped in coverings that can withstand the application of the disinfectants, and the materials inside of the packaging must be already disinfected by alternative means. For example, vacuum-packed containers of paper products that have been previously gamma-irradiated could be introduced using this spray-port method.

The introduction of personnel into barrier rooms poses the greatest risk to the microbiological integrity of the barrier room. Because it is impossible to adequately disinfect the external surfaces of people, it is necessary to place a barrier between them and the animals maintained within the room. This is accomplished via an entry-lock system that allows the orderly removal of clothing and regowning in attire that covers the body surfaces using materials that are known to be free of organisms of concern [344].

The entry-lock system usually is composed of four compartments. The first compartment, often referred to as an *insect lock* or *air pressure lock*, is used for static pressure control between other areas of the animal facility and the lock system. This first lock also serves to inhibit the passage of insects into the other more critical areas of the lock system. It is also used in some instances for the storage of external clothing, such as jackets, or for depositing shoes worn elsewhere in the animal facility. The next component of the lock system is an undress lock, where all clothing is removed and placed in lockers or other suitable receptacles. Presumably, most, if not all, of the contamination that might be associated with personnel resides on such clothing.

At this point in the lock system, there are two possible configurations. The first is termed the *wet entry system* and involves personnel taking a water shower. Depending on the rigors of the barrier facility, an electromechanical system may be placed in this portion of the lock system to ensure that a wet shower is actually taken. This consists of interlocking doors with timers requiring a wait period within the shower area and an electrical interface with the water supply to the showers such that the release timers are only activated if water has run for a certain period of time. In the end, even such rigorous measures do not necessarily guarantee that the shower taken is adequate or consistent [344].

Although some decontamination of external surfaces of a person's body may occur to varying degrees with a shower, the principal rationale for the use of the water shower is to ensure that street clothing has been removed. A number of disadvantages are associated with a shower in a lock entry system, including the time constraints associated with the entry procedure and the considerable requirements for expendable supplies used in the showering procedure. Following the shower, a separate room is provided for drying off and putting on a sterile or disinfected uniform, including components such as a cap, mask, gloves, and dedicated footwear. When these procedures have been conducted, entrance into the barrier room can be made. Leaving the barrier room, the employee reverses the process; however, often the showering step is omitted.

Because the wet entry system is quite cumbersome and limits the number of employees that can enter the barrier facility at any given time, a new methodology, the dry entry system, has gained increasing favor in many research settings. This latter method involves elimination of the wet shower and a direct change in a separate room from the degowning portion of the lock into clean, disinfected clothing as occurs in the last step of the wet entry system. Once clothed in appropriate attire, the employee enters an air shower that blows high-velocity, HEPA-filtered air across all surfaces, removing particulates that may contain any contaminants imparted to the external surfaces of the gown during the changing process. Doors to the air shower are electrically interlocked, requiring that a full cycle be completed before the employee can enter the barrier room. Upon exiting the barrier room, all aspects of the process, including the air shower, are again used by the employee in reverse order.

In addition to the time savings, the air shower also provides a very positive interlock between the barrier room and the lock system, preventing the escape of any animals that have left their cages and any potential incursion of insects or other pests into the barrier room through the lock system. The air shower also greatly decreases the amount of particulates that are brought out of the barrier room serving as a safeguard for the general animal facility should a contamination occur within the barrier [344].

As a general rule, cage washing and other sanitation procedures are done within the barrier facilities rather than bringing caging materials out of the barrier. This is because each transport of materials across the barrier entails some degree of risk of introduction of unwanted microorganisms. Movement of caging from the barrier through common animal facility hallways and cage washing facilities also poses a risk of contamination that may or may not be eliminated by the disinfection processes used to reenter such materials into the barrier. Similarly, rigidly operated barrier rooms generally do not provide for personnel break areas or other administrative support facilities within the barrier. Such areas tend to encourage removal of protective clothing and inappropriate activities given the microbiological status of the barrier.

Barrier rooms are expensive to operate and are best used to house large numbers of animals or to house large animals under defined microbiological conditions. Due to their large size and the large amount of materials that must be passed across the barrier to maintain it, failure of the barrier, as measured by changes in the microbiological status of the animals maintained within the barrier, is common. Once an unwanted microorganism gains entrance, all animals within the barrier room are at risk of becoming contaminated unless some form of secondary containment or bioexclusion system (e.g., microisolation cages) is also used within the barrier [344]. This practice is seldom done, given the high cost of such secondary bioexclusion practices.

## Biohazard Containment Facilities

Toxicologic research often requires the use of biohazardous materials in conjunction with animals. Such materials, whether they be toxic, carcinogenic, radioactive, or infectious, must be contained within a very defined space and

**TABLE 20.1**
**Comparison of Conventional Animal Care Strategies with Strategies for Containing Biohazards**

| | Recommended Animal Care Practice Strategies[a] | Biocontainment Strategies[b] | Rationale[c] |
|---|---|---|---|
| Air flow | High | Low | A |
| Cage cleaning | Frequent | Infrequent | A |
| Bedding change | Frequent | Infrequent | A |
| Equipment/supplies | Reusable | Disposable | B |
| Animal handling | Frequent | Infrequent | C |

[a] Practices or approaches commonly used to provide a clean, healthful environment for animals.
[b] Practices or approaches designed to minimize the release of hazardous materials from animals or their environment.
[c] A, the more potentially contaminated material generated, the more that has to be decontaminated and the greater chance for error; B, processing materials for reuse while assuming proper decontamination is costly and logistically difficult and has the potential for error; C, the more handling, the greater the risk for personnel injury and contamination.

handled appropriately so as not to contaminate other animals or personnel [287,288]. It is a popular misconception that barrier facilities or certain other bioexclusion systems, such as microisolation cages, provide, in normal operation, the necessary level of biocontainment to meet most research needs. In general, good biocontainment practices are at odds with recommended animal husbandry and bioexclusion techniques (Table 20.1).

The type of biocontainment facilities required when using biohazardous materials in animals depends on the nature of the hazard and the level of risk posed by it. In the case of infectious materials, the Centers for Disease Control and Prevention (CDC) has published classification systems for microorganisms and has described biological containment facilities of four increasingly secure types designated biosafety level (BL) I through BL IV [49,50]. These safety practices and the equipment used in conjunction with these increasing levels of hazards are designed to prevent inadvertent transmission or release of contaminants from the work area. These guidelines set standards for reducing aerosol generation, hazards associated with sharp objects, the use of protective clothing, and the operational standards for hoods and other devices used to manipulate infected materials, including animals.

Biocontainment facilities are resource intensive, and their complexity increases with the size of the animals that must be contained at the various biosafety levels. Laboratory animals the size of rabbits or larger require considerable housing space and generate large amounts of waste material that is potentially contaminated. Normal operations of cleaning and disinfection, as well as removal of biological materials, including animal carcasses, may require complex systems, making such facilities expensive to construct and to operate. For this reason, those institutions conducting limited work in larger species involving biohazardous materials may choose to subcontract these tasks to more specialized institutions in which the volume of such work is sufficient to justify the cost of constructing and maintaining these facilities.

Most institutions, however, use cage-level biocontainment practices for studies involving rodents and other small mammals. Studies up through BL III can be conducted in small, dedicated facilities using these systems along with appropriate safety practices.

## Surgical Facilities

The use of surgically altered animals is common in toxicologic research. Traditionally, the complexity of surgical procedures performed and the nature of the surgical facilities required were directly proportional to the body size of the animals being used. Small laboratory animals, such as rats and mice, were often used in studies in which relatively simple procedures involving organ removal were required. The facilities to support these operative procedures were not complex in nature, and often a portion of a multipurpose area could be used on a temporary basis for conducting them. Larger, more complex procedures were usually reserved for animals the size of dogs or domestic farm animals in which the operative field was much larger and techniques used in human surgery, as well as the necessary support equipment, such as specialized gas anesthesia machines, bypass equipment, and specialized imaging equipment used in human or traditional veterinary practice, could be employed. Such procedures require a much larger, dedicated surgical facility.

A few guidelines for the design and construction of surgical facilities for use in biomedical research have been published [222,345]. In general, such facilities should have certain basic components, including the following an operating room in which the surgical procedure is conducted; a postoperative care area separated from the rest of the surgical facility in which animals can recover from the surgical procedure; an animal preparation area in which the animal can be anesthetized and other preoperative procedures conducted; an instrument and supply preparation area in which materials can be assembled, cleaned, and sterilized; and a surgeon preparation area where personnel can decontaminate their hands and then be suitable attired for the surgical procedure. A number of support facilities may also be associated with the surgical facility, including a diagnostic laboratory, imaging facilities, and specialized procedure rooms to study instrumented animals.

Existing guidelines for surgical facilities endeavor to separate the various functions to be performed by physical barriers or distance or by separating the functions in time [222]. Surfaces and construction materials used in surgical suites are designed to be easily cleaned and disinfected [222]. Air provided to surgical facilities should be appropriately conditioned to minimize temperature loss or gain during the surgical procedure. Such facilities are utility intensive and often require access to medical gases and emergency power.

Over the last decade there has been a trend toward miniaturization, and complex surgical procedures in support of toxicologic research can now be performed on small laboratory rodents. The increasing commercial availability of vascular catheterized rodents, coupled with numerous types of telemetry devices for continuous recording of a variety of parameters from nonrestrained animals, has offered new possibilities in drug discovery research. If sufficient numbers of such procedures are to be done, a dedicated surgical facility for rodents may be appropriate. In general, many of the same components used in traditional large-animal surgical facilities are present in such a dedicated rodent surgical facility but may not be separated by physical barriers, such as walls.

Due to the small size of the operative field in small laboratory animals, the use of sterile instrument tip procedures can be coupled with chemical disinfectants and table-top dry-heat sterilizers to allow large numbers of rodents to be instrumented aseptically during a single surgical session [38,68,69]. To facilitate these procedures, horizontal laminar-flow hoods may be used to minimize the chance of cross-contamination between animals during surgery and to reduce bacteriologic contamination of the operative area.

Unlike traditional operative procedures on larger animals, most rodent surgery is done in a sitting position and may require the use of magnification and focused, point-source illumination. For this reason, provision should be made for adequate counter space and electrical power. Because multiple groups of animals may be used in the surgical facility over a short period of time, preoperative and postoperative housing is often done in cage-level bioexclusion systems to prevent the chance of cross-contamination between projects when animals are used from either different sources of supply or from different locations employing different housing methods. It is important to remember that the size of the animal or the species of the animal does not diminish the need to maintain appropriate aseptic procedures during the course of the surgical manipulation [20,32]; however, due to the small size of rodents and the relatively small operating field, the risk of postoperative infection and the need for maintaining large areas of a facility in an aseptic fashion are greatly reduced [68,69].

## CAGING AND HOUSING SYSTEMS

The environment of the laboratory animals used in toxicologic research is classified by levels of enclosure. A cage, pen, or stall is the immediate limit of an animal's environment in a research facility and is designated as its primary enclosure. The room or space in which the primary enclosure is located is termed the secondary enclosure [222]. The environmental conditions in the secondary enclosure influence, but do not control, the environment in the primary enclosure [26,354,355]. The animal's primary enclosure should allow the animal to remain clean and dry, securely contain the animal so it cannot escape and cannot injure itself while residing in the enclosure, and provide adequate ventilation to allow the animal a sufficient supply of fresh air appropriately conditioned to meet its needs. The primary enclosure should also provide for the normal physiologic and behavioral requirements of the animals, including reproduction, movement, urination, and defecation. It should allow for investigator access and observation of the animal with minimal disturbance. In addition to these criteria, primary enclosures may also be used to maintain the animal and its environment in a specified microbiological status. This bioexclusion function requires a heavy reliance on aseptic techniques and personnel training, as well as significant support facilities, including autoclaves and laminar flow work stations [222,344].

### Primary Enclosures

#### Microisolation Cages

A microisolation cage consists of a plastic cage bottom with a stainless steel wire top that is covered by a fitted plastic lid containing a filter. The filter usually is comprised of a polyester material that can be fabricated to different thickness and average pore sizes. In principle, when the lid remains on the cage, all ventilation occurs through the filter, which excludes particulates that may contain infectious organisms. As long as all manipulations are done in a laminar flow work station and all materials are adequately disinfected or sterilized, unwanted organisms should be excluded from the cage environment [72]. For this to occur, all manipulations must be done using aseptic technique. Such systems can be effective in maintaining a particular microbiological status of animals, but manipulation of the animals may be cumbersome for researchers and requires that all research equipment and materials undergo rigorous and frequent decontamination or sterilization to maintain the microbiological status of the animals.

Microisolation caging does have some disadvantages in terms of the environmental conditions surrounding the animals. Because static microisolation caging does not allow for significant air exchange through the filter, the concentration of water vapor, gases, and heat may build up to levels well above that in the secondary enclosure [161,298]. Another difficulty may be related to the adequacy of the fit of the filtered top to the cage itself. Studies have shown that, in some designs of microisolator caging, ventilation occurs under the lip of the cage, allowing unfiltered air to migrate across cage surfaces and other areas into the sterilized environment of the microisolation cage [202,203]. This makes practices such as the exterior disinfection of the cage before manipulation in the change station an important consideration. In addition, when manipulating such caging in a laminar flow change station, it is important to minimize the amount of materials in the change station, as laminar flow air, upon striking objects, will eddy for a distance of at least three times the diameter of the object struck, causing unwanted movement of air between materials in the station, resulting in airborne cross-contamination [349].

### Ventilated Microisolation Caging

To address some of the difficulties associated with poor ventilation of static microisolation caging, specialized racks designed to hold the microisolation caging have been developed. These racks contain a source of HEPA-filtered air that is discharged either through a nozzle directly into the cage or through small openings above the filter cage top such that the air moves directly into the cage. Ventilated racks usually have a mechanism for removing air from the cage, often through a port located directly above the filter top in another area of microisolation cage. This exhaust air is either HEPA filtered and discharged into the room or directly discharged into the room exhaust system. These racks usually are equipped with individual blowers to pass air through the HEPA filters and generate the necessary static pressure. Like static microisolation caging, ventilated microisolators must be handled and managed in the same manner as static microisolation caging using aseptic technique, sterilized materials, and laminar flow change stations [344].

### Conventional Caging

By far the most commonly used type of caging in toxicologic research is conventional open caging. In the case of rodents, this is of two types. The first is solid-bottom plastic or metal caging with contact bedding. The cage body is covered with a lid that often contains an integrated feeder for holding feed pellets and space for a water bottle. As an alternative, water may be provided by automatic watering systems, in which case a small hole is found in one end of the cage through which a small water valve is introduced.

The most common type of cage is constructed of plastic with a solid bottom. These can be opaque or transparent. Transparent materials have the advantage of allowing easy visualization of the animals, whereas opaque materials provide a more sheltered environment for the animals by decreasing the amount of light striking them. The type of plastic materials used in the construction of the caging is important, as procedures such as frequent autoclaving and washing with high-temperature water and chemical compounds that are designed to disinfect can rapidly deteriorate some forms of plastic [58,177]. In addition, animal urine, which contains significant quantities of minerals and proteinaceous material, can cling tenaciously to plastic, requiring the use of mineral acids or other compounds to remove it [222]. Over time, all plastic caging will deteriorate and have to be replaced. Cracks or other deficits in the plastic make sanitation difficult and may affect the structural integrity of the caging, requiring replacement.

A second type of caging is referred to under the general classification of *suspended caging*. Such caging usually is constructed of metal, most commonly stainless steel. The floors usually are of wire mesh, containing either two or four wires per inch. Some types of suspended caging have punched metal floors that have been electropolished or otherwise smoothed to prevent injury to feet. Punched metal flooring is commonly used for housing rabbits, as it provides greater support for their feet. As an alternative, some suspended caging is constructed of heavy-duty plastics. Floors may also be constructed of the same plastic using the same principle as the punched metal flooring. Suspended cages for larger animals, such as dogs, cats, or swine, commonly use coated wire or coated metal punched flooring to minimize the chance of damage to the animal's feet.

The mesh size of any suspended cage flooring is important. It must be sufficiently wide to allow the passage of feces while still providing sufficient support to the feet to prevent pressure injuries. In theory, suspended caging should allow feces and urine to pass into catch pans containing sheet or loose bedding or, in some cases, into catch pans that can be flushed into trough drain systems. Most animals, including rodents, are coprophagic, but if feces pass through the flooring the animals will have limited access to it. Many species may consume feces as it is being defecated and before it can drop through the cage floor. Coprophagy is critical for the normal health of some species, such as rabbits.

Larger suspended caging usually is cleaned in place, whereas caging for rodents and other small mammals usually is taken to a wash area for cleaning. Watering can be provided to suspended caging by means of automatic watering devices, water bottles, or, in the case of some large animals, water bowls. Feed usually is dispensed either in feeders that hang on the cages or within the cage or, for some larger animals, by use of feed bowls. Caging for some larger animals, such as nonhuman primates, cats, and dogs, often contains a resting board or perch that is suspended off of the floor. This allows the animals to express normal behaviors and provides them an elevated platform on which to rest.

Neither solid-bottom nor suspended conventional caging is designed to prevent the airborne or fomite transmission of adventitious microorganisms. Control of such organisms when these cages are used is at the secondary enclosure level.

### Pens and Runs

Domestic farm animals and dogs often are maintained in either runs or pens. A pen is a large indoor enclosure whose floor usually is the floor of the secondary enclosure. In the case of those animals that do not exhibit climbing behavior, indoor pens may be constructed such that the top of the pen is open. Most commonly, however, pens form a complete enclosure. Pens usually incorporate a drainage system, either by the use of a slanted floor underneath a raised pen floor that allows liquids to drain toward the back of the pen into a trough drain system or through the construction of a trough drain system in the floor of the pen itself. Pens are cleaned using water under pressure and detergent or disinfectants. Pens may also be bedded with loose bedding that is removed and replaced on a regular basis. Whenever in-place cleaning of pens is done, it is important that the design of the pens be such that floor drains are large enough to accommodate fecal material and other items, such as bedding, that may be flushed down the drainage system.

In theory, pens provide large animals the ability to move about normally and to acquire some exercise [131]. Studies have shown, however, that such exercise is rarely spontaneous and usually is associated with the presence of humans [42,43,134]. For this reason, components of some legislation, as well as certain guidelines, require regular periods of exercise for certain penned animals [308]. Special provision must be made to allow such exercise to occur.

The term *run* is often reserved for pens that have both an indoor and outdoor component. In such cases, the indoor component of the run often has a smaller floor area than the outdoor component. The indoor component usually is in a conditioned space or has other provisions for the thermal comfort of the animal, such as heated flooring. Access to the outdoor portion of the run may be limited through the use of animal-operated doors or doors that must be opened by animal facility personnel. Feeding and provision of water to animals housed in runs usually are done in the indoor portion of the run. Cleaning is done using water under pressure accompanied by detergents or disinfectants. Care must be taken during the process of cleaning such enclosures so as not to unduly stress or cause the animals to become wet during the process [222].

Housing of animals in pens or runs requires that consideration be given to handling and training of animals housed in such enclosures [256]. Toxicologic research often requires repeated sampling or other manipulations. Removing animals from such enclosures, as well as protection of implanted sampling ports or catheters while animals are in the enclosures, must be given adequate attention. Many problems can be overcome by conditioning periods during which the animals are handled and trained to respond in certain ways.

## Secondary Enclosures

The construction of typical animal holding rooms has been previously discussed. A number of specialized secondary enclosures other than the conventional animal holding rooms have been developed for the purpose of minimizing transmission of adventitious microorganisms between groups of animals housed in adjacent cubicles. The air supplied to the cubicle may come from a separate air supply within the cubicle or from the room in which the cubicle is located. In the latter instance, a negative pressure is created within the cubicle by an exhaust system, usually located in the ceiling of the cubicle, that pulls air into the cubicle through a space under the doors to the cubicle in a fashion similar to a fume hood. The cubicle is entered by opening the doors, at which point all directional control of air movement is lost [347]. Unlike cage-level containment systems, cubicles only address airborne cross-contamination between cubicles, not fomite transmission, as might occur on dust particles or by manipulation by investigators or animal-care personnel. Due to the low face velocity associated with the air movement underneath the door to the cubicle, absolute prevention of airborne cross-contamination cannot be ensured [344,347]. Cubicles have been shown to be useful in some applications where the separation of species is required in limited space, and they apparently can limit the spread of certain microorganisms under the proper conditions [171]. They are probably best used in conjunction with cage-level bioexclusion systems if reliable prevention of microbiological contamination is required.

### Ventilated Cabinets

Ventilated cabinets consist of a cabinet with one or more doors. The cabinet has a small fan and HEPA filter that extracts air from the room and releases it into the cabinet. Air is exhausted either between the doors to the cabinet or through an exhaust duct into the room. The exhaust air may or may not be filtered [185]. Like cubicles, when the cabinet door is opened, caging within the cabinets no longer receives the benefit of the HEPA-filtered air supply. Airborne cross-contamination between cages within the system is not controlled unless a cage-level bioexclusion system is also used. Fomite transmission is not prevented by ventilated cabinets. The use of ventilated cabinets in toxicologic studies is limited.

### Mass Air Displacement and Laminar Air Flow Rooms

These secondary enclosure bioexclusion systems are designed to decrease the airborne transfer of microorganisms between groups of animals. Both systems use a HEPA filter and blower to supply air to either diffusers or a plenum located in the ceiling of an animal room. In the case of mass air displacement rooms, large volumes of appropriately filtered air are released through a series of diffusers located in the ceiling in an attempt to wash any airborne contaminants to the floor and out through exhaust ducts located at floor level at several locations in the room [186,194]. By contrast, laminar air flow rooms pressurize a space above the ceiling (plenum) and discharge air out through the thousands of tiny holes in the ceiling, creating laminar movement of air [19]. In theory, particles released from individual cages will be caught up in the laminar air stream and then drop to the floor [65]. In practice, once the laminar air flow strikes caging or other objects within the room, it eddies for distances of three or more times the diameter of any object that it strikes, causing mixing and producing an effect similar to mass air displacement [349]. Neither of these room types effectively addresses airborne cross-contamination between adjacent cages, and they do not address fomite or other means of transmission of microorganisms. By themselves they are not a complete bioexclusion system [344]. Depending on the design of the ventilation equipment, these rooms can be quite noisy and may be prone to excessive heat build-up. Prior to the advent of cage-level bioexclusion systems, mass air displacement and laminar flow rooms were a common feature in toxicological research facilities using animals. They are less commonly used today.

## ILLUMINATION

Light is an important environmental factor that can affect biological and behavioral processes in animals and the conduct of routine animal care duties. Numerous studies have documented the effects of light on the morphology, physiology, and behavior of various animals [33,82,233, 316,338]. Not all of these effects have been adequately explored in all species or in sufficient detail to predict the magnitude and the extent of impact on various toxicologic studies.

The three factors that characterize lighting are (1) the lighting spectrum (wavelength), (2) the light intensity as measured in foot candles (English) or lux (metric), and (3) photoperiodicity (light/dark cycle) [22,300]. In addition, factors such as the light history of the animal, animal pigmentation, time of light exposure during the circadian cycle, species, sex, age, body temperature, hormonal status, and stock or strain of the animal can alter the influence of specific characteristics of the light on an animal [33,76, 240,285,294,331].

The one characteristic of light that has the most profound effect on animals is photoperiodicity. It regulates circadian (one day in length) and ultradian (greater than one day) biological rhythms in animals. These rhythms, in turn, can alter a variety of basic processes. Photoperiodicity is a critical regulator of reproduction and behavior [118,196,268,339] that has a behavioral influence on processes such as feed consumption and resulting body weight gain, nutrient intake, and hormone secretion [34,54,316]. It appears that, in most species, a period of 10 to 14 hours of light is required to maintain normal biological rhythms [200]. Some species such as rats and mice are more tolerant of light-cycle length than such species as hamsters, which require at least 14 hours of daylight for normal reproductive function [54,99]. Continuous daylight, which can occur with malfunctioning light timers or inadvertent overriding of light timing devices, can have serious consequences for some toxicologic studies [28]. For this reason, it is important to regularly assess the function of light timing devices and to build in safeguards to any light timer overriding mechanisms to ensure that a consistent photoperiodicity is maintained [222].

A consistent photoperiodicity with adequate dark cycles is also essential to allow for the regular daily renewal of rods and cones in the eye, especially in nocturnal and crepuscular species [60]. Without adequate dark periods, retinal degeneration cannot occur and can magnify the effects of retinal degeneration associated with high light intensity [236]. Every effort should be made to minimize disruption of the dark cycle of the photoperiod. When procedures must be conducted during the dark cycle, infrared illumination, low-intensity red lighting and the use of point-source, low-intensity lighting, such as hand-held flashlights, may allow the necessary procedures to be performed without significant disruption. Alternatively, light cycles can be adjusted to fit the work required (reverse light cycles), or the necessary procedure may have to be rescheduled into the light cycle.

Consistency in the photoperiod is important; for this eason, the use of windows in animal rooms must be arefully considered. Exterior windows allow seasonal mbient lighting conditions to alter photoperiodicity. This nay not be important in certain species, such as nonhuman rimates, dogs, and some agricultural or other large mamnals, and may even be considered a form of enrichment. n small rodents in which reproduction may be a study arameter, however, this lack of control may be an imporınt variable [222]. Other significant concerns posed by xterior windows include temperature regulation within ıe animal holding room because of heat loss or gain ırough the windows and difficulties in providing adeuate security for exterior windows [222].

Light intensity is another important characteristic of ght that can dramatically affect certain laboratory aninals. Many commonly used laboratory animals, including ıts and mice, are nocturnal. Moreover, many of these pecies have been developed on albino backgrounds, ıereby lacking pigment in their eyes, skin, and other ssues that provide some protection from the effects of ght intensity. The most damaging effect appears to be hototoxic retinal atrophy, which occurs in albino rats and ice [172,175,271].

Because light intensity is affected by the type, locaon, and number of lighting fixtures in a room, considerıle variation can occur between toxicologic research cilities. The age of the light-producing fixture (e.g., ılbs, tubes) is also important, as the output intensity will ıry with length of use. It is unclear whether the effects ported for light intensity are related to a single wavength or group of wavelengths or occurs equally over the hole spectrum. Light intensity is also affected by caging pe and location as well as the reflective ability of other om surfaces [234,339]. Studies have shown that the osition of mice in cages at various levels on the caging ck can influence the incidence of retinal atrophy, with incidence of affected animals as high as 30.2% being corded for animals housed in upper cages, as compared 0.7% incidence for animals housed in lower cages 8,115]. For this reason, randomization of caging as well the rotation of cages are important considerations if ch effects are to be equally distributed between test and ontrol groups [116].

When specifying light intensity, it is important that a iform point of measure be used. In recent studies, as ell as recommendations by various organizations, a mearing point of 1 m off of a floor in the center of the animal ılding room has been used and is considered to be the andard point of measure for intensity [22,59,222]. This used only to set an overall room intensity level for the rposes of comparing animal rooms and does not reflect e actual intensity experienced by the animals.

The light exposure history of individual animals can ter their sensitivity to phototoxicity. At least one study suggests that light intensities that are 130 to 270 lux above the intensity under which the animal was raised may be the point at which retinal damage may begin to occur [294]. Young albino and pigmented mice appear to have the ability to reverse some of the retinal damage associated with elevated light intensities [331,341]. Current recommendations suggest that a light level of 325 lux (30-foot candles) at 1 m (3.4 feet) above the floor is sufficient for routine sanitation while still avoiding significant retinal degeneration in albino animals [22]. Other recommendations exist that are based on cage-level lighting intensities, but they are difficult to assess and administer [201].

The special distribution of light as it relates to adverse effects on laboratory animals has not been extensively studied. No artificial light has exactly the same spectrum as sunlight. Some fluorescent bulbs have a much wider spectrum with more output in the ultraviolet and infrared ranges than others. This can be partially compensated for by mixing standard fluorescent fixtures produced by different manufacturers in the same lighting fixture. Because different fluorescent compounds may be used by different manufacturers, this procedure can produce greater spectral diversity than bulbs from a single manufacturer. Overall, no definitive evidence suggests that a wide spectral diversity in artificial lighting provides any direct enhancement of animal health or well-being.

Some species of new-world primates are capable of synthesizing certain forms of vitamin D from dietary constituents and sunlight. Because it is common practice to provide enriched diets for these species that contain the appropriate forms of already converted vitamin D, access to full-spectrum artificial sunlight would not appear to be necessary.

A number of practical issues should be considered with respect to lighting in animal facilities. Many of these are covered in recommendations produced by the Illuminating Engineering Society of North American (IESNA) handbook, which provides practical information on the selection and installation of lighting [153,154]. It is generally recommended that light bulbs or fixtures have protective covers to ensure the safety of both the animals and animal care staff. Cleaning operations and the use of water in animal husbandry procedures can pose hazards from breakage and electrical shock.

## NOISE

There is little doubt that extreme levels of acoustic energy (80 to 120 decibels of sound pressure) can, under the right circumstances, produce auditory and extra-auditory changes in laboratory animals [18,92,107,108,224,249, 250,251,358]. The assessment of sound-related interactions and injuries in laboratory animals is complex, as the potential effects of noise on animals must take into consideration not only the intensity of the sound but also the frequency pattern of sound presentation, including the rate of onset, duration, and vibration effects [12,59,241, 252,255]. The hearing range, noise-exposure history, and susceptibility to adverse effects of sound based on species, age, and strain further complicate any analyses [37,145, 197,199,266,284,328]. Many diverse effects have been reported in the literature that could result in unwanted variation in any toxicologic study. No comparative damage risk criteria exist for each of the common laboratory animal species, making it difficult to know what sound presentation is harmful [22].

In general, noise-producing animals and animal care activities should be separated from species and activities that do not generate noise [103,222,252,283]. This can be done by careful considerations of facility design and operational practices [11,245]. Physical barriers such as doors, air locks, and sound-absorbing materials can be useful in minimizing the effect of sound [245]. Excessive and unpredictable noise patterns can be minimized through personnel training and the use of alternative practices and equipment in routine animal care duties. Environmental noise can be created by technical devices such as video monitors, equipment and electronics. The noise may be difficult to recognize because it can exist in the ultrasonic range. The differences in hearing and the effect of noise between strains and species have been reviewed [364].

## VENTILATION

The purpose of ventilating an animal holding room is to dilute gaseous and particulate contaminants; supply adequate oxygen; remove thermal loads caused by animal respirations, lights, and equipment; adjust the moisture content of room air; and, where appropriate, create static pressure differentials between adjoining areas [58,222]. It is, unfortunately, a common mistake to place heavy reliance on a ventilation system as the primary mechanism to prevent the movement of undesirable microbiological organisms from one location within a room or a facility to another. Other factors, including movement of personnel between areas, movement of equipment, and transfer of animals within a facility, can quickly overcome any benefits associated with well-controlled ventilation. In theory, high ventilation rates should dilute out particulate and gaseous contaminants including undesirable microorganisms and allergens [104,301,307]. Because it is difficult to control the movement of air within secondary enclosures, viable particulates can easily escape any containment provided strictly on the basis of ventilation.

As a general rule, air supplied to any secondary or primary enclosure should be free of contaminants and be properly conditioned. HEPA filtration is recommended to provide some assurance that supplied air is free of particulates that could potentially harbor infectious organisms [189]. HEPA filters are given efficiency ratings that can exceed 99% for particulate exclusion within certain particle size ranges [5]. Average pore diameters of HEPA filters are designed to exclude particulates the size of bacteria or larger but may not be small enough to exclude individual viral particles. Because viruses and other organisms tend to travel on particulates due to their high electrostatic charge, HEPA filters are effective in excluding them. Unfortunately, large volumes of air are processed by HEPA filters, and because their efficiency rating is not 100%, a certain amount of infectious particles will pass HEPA filters over time. Given the dilution in the air stream, this small amount of passage may not pose a significant threat in most situations; however, because incoming air is seldom challenged with high concentrations of infectious particles, this deficiency usually is not a problem.

If HEPA filters become wet or excessively laden with particulates, bacteria associated with such particles may grow through the HEPA filters and be discharged on the clean side [77]. Clogging of HEPA filters usually is monitored by measuring the static pressure across the HEPA filter. Increases in static pressure indicate that the HEPA filter is becoming laden with particulates and, at some predetermined static pressure reading, should be changed.

Correct installation of HEPA filters is critical. If they are incorrectly mounted, untreated air can pass by the HEPA filters, producing the same effect as if they contained a large hole. For this reason, HEPA filters should be tested in place using particulate generation equipment (e.g., DOP testing) [77]. It is important that such testing be done on a regular basis and that it be thoroughly done when the HEPA filter is first mounted. It is critical that such testing be done by individuals experienced in the intricacies of the testing procedure, as inappropriate testing can fail to reveal problems in the mounting or construction of the HEPA filter. HEPA filters should be protected by pre-filters designed to catch the majority of particulates of large to intermediate size. This usually is accomplished by a series of prefilters of different efficiencies. Such prefilters should be changed regularly.

Air exhausted from animal holding rooms may or may not be filtered prior to discharge to the environment. In the case of toxicologic research involving hazardous agents, exhaust air usually is filtered using filters that are capable of excluding the material in question from the air stream. If the level of hazard is low enough, dilution with large volumes of air by discharge into combined exhaust or into the environment directly may be sufficient.

Facility design guidelines emphasize the need for regulating air-pressure differentials either within animal holding facilities as a whole or within specialized support facilities, such as surgical, procedural, service, housing, or quarantine areas. In general, areas maintained under negative pressure with respect to adjacent areas are designed to prevent escape into the air stream of unwante materials from the area maintained under negative pressure [99,287,288]. This would be most appropriate, for example, in an animal holding area being used for quarantining animals. Conversely, maintenance of an animal holding area under positive pressure with respect to the surrounding areas would be undertaken under those circumstances where animals within the positive pressure area are free of certain microorganisms and whose exposure to particulates from other adjacent areas might increase their risk of contamination. Such positive pressure housing might be used in areas maintaining specific pathogen-free animals or used for surgery. While maintaining static pressure differentials may pose some advantage with respect to contamination control, it should not be relied on for containment of chemical or infectious agents that could be transferred between areas. Most air handling systems have neither the capacity nor the necessary control mechanisms to maintain pressure differentials when doors, pass-throughs, or other structures are opened even for brief periods [222].

For years, the ventilation rate for animal holding rooms and primary enclosures has been an important focus of many recommendations for research animal facilities. The ventilation rate or air-exchange rate in the context of animal facilities refers to the number of times the total volume of air in the room is exhausted and resupplied with appropriately conditioned air per unit of time. This often is expressed in air exchanges per hour. Rules of thumb have been developed based on experience and limited data. These guidelines have suggested that air-exchange rates of 10 to 15 changes per hour are generally satisfactory for animal facilities [45,222]. Unfortunately, these estimates do not take into account the range of possible heat loads, the type of bedding or frequency of cage cleaning, the species and number and size of animals housed, differences in room dimensions, or the efficiency of air distribution linking the secondary and primary enclosures [192,193,198,222,355]. Given these variables, either under- or overventilation can occur, with resulting problems in heat and odor accumulation [58].

Instead of using general guidelines, such as 10 to 15 air changes per hour, more recent recommendations suggest that the heat generated by animals be calculated using an average total heat gain formula as developed by the American Society of Heating, Refrigerating, and Air Conditioning Engineering [6]. Using this formula, the minimum ventilation required can be calculated from the total cooling load necessary to control heat generated by the animals and other heat sources. This, in turn, will allow determination of ventilation rates. Additional ventilation capacity may be added as a margin of safety and to address any unusual odor generation [222].

The adequacy with which air is distributed within a room is an important consideration [193]. If it is inadequately distributed, heat and water vapor build-up may occur unevenly. The coupling of secondary enclosure to primary enclosure environments may also vary within the room [354]. The use of computational fluid dynamic modeling has revealed that supply and exhaust diffuser placement, as well as type, can have a profound effect on air distribution [136,137,269]. This has important implications not only for the environment of the animals with the resulting effect on studies but also for researchers and technicians working in the animal room.

The concept of drafts as they relate to human comfort has been explored [192,193]. Unfortunately, little work has been done with animals to demonstrate whether or not there are biological consequences or if discomfort is associated with the detection of movement of air. In the case of small laboratory rodents and rabbits, caging designs and cage placements are unlikely to allow significant air movement within these enclosures to affect the animals contained therein. The one exception may be individually ventilated cages where high air exchange rates can be produced. Even so, there is no evidence to suggest that harmful effects occur. Similarly, larger species kept in pens, runs, or open cages have not been reported to have any ill effects associated with the detection of air movement.

Current guidelines allow the use of recycled air under certain conditions [222]. In general, no less than 50% fresh air should be mixed with recycled air for reuse in animal holding areas. Recycled air must be appropriately conditioned and mixed with sufficient fresh air to control temperature and humidity, and other appropriate conditioning should be applied to remove particulates. Mixing recycled air from different animal holding areas is discouraged [222]. Recycling animal room air or the use of appropriately conditioned air from human use areas for animal rooms can provide energy cost savings and be an adjunct to other energy recovery efforts.

Ventilation should not be used as a substitute for good husbandry practices. Adjusting cage densities and bedding or cage cleaning frequencies may prove a better solution to odor-generation problems than trying to increase the capacity of air-handling systems. The conditions within animal holding rooms should be regularly monitored to ensure that the ventilation system is working correctly. Plans should be made for actions to be taken in the event of conditions that may affect the operation of the air handling system such as power failures, failure of mechanical parts, or failure of air filters.

## TEMPERATURE AND HUMIDITY

The maintenance of body temperature within certain critical ranges is essential for animals if they are to maintain normal metabolic and physiologic processes [162,263,337]. Thermal regulation has been extensively studied in some species [110,111]. Clinically observable effects can occur in unadapted animals that have been exposed to temperatures above 85°F (29.4°C) or below 40°F (4.4°C) that are denied access to shelter or other protective mechanisms [110,162,265,343]. Depending on the extent and length of time that the animals are exposed to temperatures outside of these ranges, the effects produced can vary from changes in reproductive ability to loss of life. In general, animals are quite adaptable to temperature changes and use behavioral, morphologic, and physiologic mechanisms to maintain body temperature [106,110,111,121,246,343]. Unfortunately, such adaptation takes time and can affect the animal's performance in toxicologic studies.

For the most common laboratory animals, a dry-bulb temperature range from 61°F (16°C) to 84°F (29°C) has been suggested as acceptable [222]. Within this range, depending on species, clinical effects are unlikely to occur, and experience has shown that there is little likelihood of toxicologic interaction. Of the common laboratory animals, rabbits are generally kept in a somewhat cooler environment because of their dense hair coat. The recommended temperature range for rabbits is 61°F (16°C) to 72°F (22°C) [222]. Close control of temperatures within this recommended range may reduce some variation in certain studies. Unfortunately, very close control of temperature is not always possible within animal facilities, and some fluctuation during the course of any 24-hour period is to be expected. It has also been demonstrated that temperature will stratify within an animal room, especially one containing many cages. Temperature fluctuations of 2 to 6°F within any given animal room as measured from the floor to the ceiling are not uncommon [26,27]. This is another reason why the rotation of cages within any given animal room may be valuable. Variables such as cage construction, population density, the use of filter tops, and animal activity can also cause variations in temperature and humidity in the cage and the room and between cages [4,27,234,295,298,354].

Some means of regularly assessing the temperature within individual animal holding rooms should be used. Most commonly, temperature sensors placed in the return air ducts or mounted approximately 5 feet off the floor near the door are used to monitor room temperature.

In addition to temperature, humidity may add additional stress to animals and can affect an animal's ability to efficiency lose heat. Many animals, such as rats, mice, dogs, and certain domestic farm animals, cannot sweat and as such cannot take advantage of evaporative cooling [26,337]. These animals must use insensible heat loss through the respiratory tract or alterations in blood supply to heat-radiating structures, such as the tail, soles of the feet, and tips of the ears, to dissipate excess heat or to protect against excessive heat loss [110,111]. Some animals, such as dogs, can increase the frequency of their respiration (panting) to further lose heat. If the humidity in the environment is high, evaporative heat loss through insensible means is decreased and internal temperatures build. At very low humidities, significant insensible water loss occurs that must be compensated for to maintain adequate fluid balance. Within recommended temperature ranges, however, the effects of relative humidity appear to be minimal. Some human epidemiological evidence relates low relative humidity to an increased susceptibility to certain respiratory diseases [16]. Unfortunately, convincing evidence relating humidity to respiratory disease in animals in not available.

In rodents, low relative humidity has been suggested to cause ringtail (an annular notching of the tail that can lead to necrosis) in young rats and mice. Studies conducted on this condition indicate that it occurs at a high environmental temperature (81°F) with a low relative humidity (less than 30°) [93,235,314]. Although possible these conditions are unlikely to occur under most laboratory animal housing conditions. Moreover, in those studies that have been reported, a variety of other factors, including diet and caging type, can effectively eliminate the condition [93,235]. When the condition is seen, generally very few animals are affected and often no correlation to relative humidity exists.

The concept of thermal neutrality as a desirable state in which to maintain animals being studied has been explored [35,336]. In theory, a temperature or temperature range exists for any given animal species in which oxygen consumption is minimal and no energy is expended to heat or cool the animal to maintain a constant body temperature. Ambient temperatures above this range result in increases in metabolism, physiologic alterations, and behavioral changes that favor heat loss. Conversely, temperatures below the range result in adjustments in the same systems designed to produce or conserve heat. Ambient temperatures outside of the thermal neutral zone can result in observable changes, such as alternations in feed consumption, activity, reproduction, and growth. When coupled with relative humidity, climatograms can be constructed in which it is presumed that animals are most comfortable [336]. In practice, such optimal conditions are difficult to determine for various species of animals and evidence suggests that climatograms may vary significantly with age, reproduction, and other factors [336].

## BEDDING

Although many toxicologic studies are conducted on animals housed on suspended wire flooring with no contact bedding, the use of contact bedding is increasing. This is due principally to the increase in bioexclusion housing that uses cage-level containment. Even in the case of animals housed on suspended wire flooring, bedding is often placed below the flooring to absorb the moisture from urine and feces. Bedding can influence experimental data and animal well-being [41,239,253,254,312].

Bedding is constructed of natural products, most commonly wood, paper, or corncobs. No bedding is ideal for all occasions and all species but must be chosen based on its characteristics. A number of descriptions of the desirable characteristics of different types of bedding are available, as well as methods for evaluating various bedding types based on these characteristics [27,149,169,261,309,335]. Of all the characteristics of bedding, absorbency would seem to be the most important. It is the purpose of bedding to absorb moisture to keep the environment dry and thereby minimize the growth of microorganisms that would otherwise flourish on the organic material and moisture within the cage. Moreover, by diluting feces and urine with bedding, the animals within the cage have less contact with excreta and hence have less chance of being soiled by it.

Wood products, although probably the most common form of bedding, are not the most absorbent. Their absorbency is directly related to their total surface area; the finer the wood product bedding, the more surface area and hence the more moisture it can retain. Paper product bedding, unlike wood product, tends to absorb large amounts of water directly into the material, which causes it to swell. The absorbency of paper product bedding can be much greater than wood product bedding. Unfortunately, some forms of paper product bedding lose their structural integrity when wet, making them difficult to manipulate, which can be a problem when used under suspended catch pans. To address this, some types of paper product bedding have a thin plastic backing, which makes them easier to remove from catch pans.

Another important factor with respect to bedding is its potential for introducing contaminants. Wood product bedding is obtained as a waste product from the milling of lumber. Wood-chip bedding is composed of the small wood chips produced when sawing lumber with coarse saws. Prior to processing, it may contain a range of chip sizes, including very coarse and very fine materials, such as sawdust. Wood from which bedding is produced may come from a variety of locations with different histories of exposure to compounds and microorganisms. Raw wood chips are processed by first screening out large debris and then heating them to a temperature of 140 to 150°F in an air-convection furnace for approximately 20 minutes. This dries the product, which reduces bacterial contamination and the potential for growth and also drives off volatile oils. The wood-chip bedding is then run through a series of screens to develop products of varying sizes. After being screened, it is placed in bags and made available for animal housing.

Wood shavings are the product of planing kiln-dried lumber. Lumber is dried in large ovens maintained at a temperature of 140°F for periods of up to 14 days. The lumber is then run through shaping planes that produce shavings. These shavings are then bagged and used for bedding.

Both hardwood and softwood bedding are available as processed products. Storage conditions for these products are critical to preventing contamination. Certain softwoods, such as cedar, are no longer used for bedding for laboratory animals; their use is primarily relegated to bedding for pets. Studies using softwood bedding have shown that volatile oils in them have the ability to induce liver microsomal enzymes, which can alter drug metabolism [56,86,320,321,322,324,334] and increase the incidence of neoplasia [146,323]. Softwood bedding that is kiln dried and hardwood bedding are both much less likely to produce hepatic microsomal effects, as essential oils associated with these inductive effects are driven off in the drying process. Hardwood beddings have not been shown to induce liver microsomal enzymes but may have other effects [309]. If autoclaving or other heat treatments are applied to the bedding after receipt, the chance of essential oils producing any inductive effect are even further reduced [70,260]. In general, any such induction takes 3 or more days to occur and is rapidly lost when animals are removed from bedding materials, such as unprocessed cedar.

Paper bedding can come in a variety of shapes and sizes. Flat cardboard-like sheets, as well as plastic-backed absorbent pads, are commonly used in bedding trays under suspended wire cages. Paper chips and shredded paper products that may be bleached or unbleached provide a highly absorbent, contaminant-free contact bedding. Recently, compressed pads of paper that are cut to the size of individual cages have been introduced. These materials can come presterilized and are inserted into the cage at changing. When placed in cages bedded with this material, animals scratch up the paper product into a loose bedding that is highly absorbent. This activity by the animals is considered by some to be a form of psychological enrichment. Similarly, the ability to burrow in contact bedding is also considered to be beneficial for some rodents.

Corn cob bedding is produced from coarsely ground corn cobs that have been heat treated. Their absorbency is similar to wood product bedding. The bedding itself is of irregular shape, is in not easily compressed, and retains its structure when wet. There has been little indication that this bedding can cause significant research interactions; however, it is not widely used [259].

Larger animals, such as swine, dogs, cats, and other species, may also benefit from the use of contact bedding. In general, the use of such bedding is designed primarily to aid in cleaning of the cages; however, in the case of swine, it may assist in providing for their natural routing behavior.

A number of problems have been associated with the various types of bedding. Nude mice and SKH-1 mice that do not have eyelashes have a tendency to collect debris under their eyelids. This effect is less noticeable with screened wood product bedding; however, paper product bedding appears to release many more fines that are capable of lodging under the eyelids, causing swelling, and, in the presence of the right microorganism, producing abscesses. Corncob bedding seems to be similar to wood product bedding in this regard.

The ingestion of bedding can be problematic for newborn animals. If the particles are too small or do not easily pass the gastrointestinal tract, newborn animals can ingest the bedding and develop fatal gastrointestinal tract obstructions.

It is important to periodically assess bedding materials for the presence of toxic and carcinogenic materials. Whereas some of these materials may be driven off during the processing of the bedding, others may be more stable.

Bedding, especially contact bedding, must be changed frequently based on a variety of factors, including the wetness of the bedding, the amount of feces present, the number of animals in the cage, and a variety of other factors. In the case of large animals or group-housed animals, bedding change often is required several times a week. Smaller animals or individually housed animals may require less frequent bedding change. Failure to change the bedding and replace it with new bedding not only can result in an unhealthful environment but can also increase the concentration of materials excreted in the feces and urine within the animal's environment, making ingestion or inhalation much more likely. Also, evidence suggests that a soiled environment may also affect liver metabolism, which could interfere with toxicological research [322].

## WATER

Access to a sufficient quality of clean, potable water is essential for normal hydration of the animal and for the maintenance of normal physiologic and metabolic processes. Water consumption and drinking patterns of laboratory animals vary with the species [311]. Water supplied to toxicologic laboratories may be either from municipal sources or from privately owned wells. In either case, the potential exists for contaminants to be introduced into animal holding facilities or for significant variation to occur in water quality over time.

To address this problem, animal facility drinking water, as well as water used for other husbandry tasks, is treated to adjust its chemical and biologic characteristics. Water treatments may include one or more of the following: filtration, ion exchange, ultraviolet (UV) disinfection, halogenation (chlorination or iodination), reverse osmosis, or ozonization. At the local or room level, water may be acidified using mineral acids. Acidification usually is reserved for the treatment of water used in water bottles, although it is occasionally used in automatic drinking water systems.

Water filtration can be done using either depth or membrane filters. Depth filters are used for coarse filtration and involve the passage of water through a coarse medium, such as sand. Finer filtration, down to 0.1 μm, can be obtained using a series of membrane filters often constructed out of paper or synthetic materials. These filters are only reliable for removing nonviable particulates, as many bacteria that collect on such filters can grow through them. If membrane filters are to be used for bacterial filtration, some regular means of disinfecting the filters must be employed. Some chemical components in water, such as suspended gases and certain organic compounds, can be removed by charcoal filters. These filters consist of canisters of activated charcoal that have varying degrees of retentivity for different compounds. These filters must be changed regularly and suffer from similar problems with respect to microbiological growth as membrane filters. Ion exchange resins can be used to remove inorganic compounds but offer no biological treatment.

Often, reverse osmosis is applied to animal water supplies. This treatment involves placing water under pressure on one side of a membrane that is designed to exclude all but a few types of molecules and collecting the water that passes through the membrane in an area of lower pressure on the opposite side of the membrane. To be effective in animal facilities, the capacity of the reverse osmosis equipment must be large enough to accommodate the necessary water usage. This entails either many such units or very large units. Water produced by reverse osmosis is devoid of all minerals and hence is chemically very aggressive to surfaces. Over time, water treated by reserve osmosis can recapture metal from stainless steel surfaces and other materials. Because the reverse osmosis process depends on an intact membrane, absolute bacteriologic or viral exclusion is only possible if there are no deficits in the membrane that might allow untreated water to pass through. Although water produced by reverse osmosis, in theory, should be free of microorganisms, in practice this is not always the case.

Ultraviolet disinfection is a common and rapid means of killing a variety of microorganisms that might be found in water supplies. Because the process involves exposing the water to UV light, the water must be reasonably free of minerals so deposits do not form inside the exposure chambers. Moreover, light intensities of UV-producing fixtures decrease over time and so, too, does the effectiveness of the process. For this reason, regular measurement of the irradiance level of the light source, as well as regular replacement of the light source, must be done.

Halogenation, most commonly chlorination, is a relatively simple but effective means of disinfecting water supplies [133]. Chlorine is effective against a wide range of microorganisms [21]. Depending on the chlorine species being measured, ranges of effective concentration have been established that will not adversely affect animals but still accomplish disinfection. It is important to realize that a certain minimum contact time is required or the chlorine or any chemical disinfectant to be effective. For this reason, chlorination usually is done centrally with the treated water being held in a contact tank for a period of several minutes to several hours to ensure adequate disinfection.

Depending on the form of chlorine being used, the chlorine reacts in variable fashion with organic materials present in the water, causing the generation of halogenated organic compounds [225]. The importance and the likelihood of such occurrence are difficult to assess. It should be noted, however, that chlorine-based compounds are the most commonly used disinfectants for water and for husbandry practices in animal facilities, as well as for the treatment of drinking water for humans. The use of chlorine to disinfect drinking water has an important advantage over filtration and reverse osmosis methods in that it provides residual disinfection within watering systems. This prevents the build-up of bacteria in the watering systems due to back flow into water bottles or automatic watering devices of saliva and feed debris during drinking.

Ozonization is another means of ensuring microbiological decontamination of water supplies and the destruction of dissolved organic compounds in the water. Such systems use an ozone generator to produce ozone gas and bubble it into a water contact tank. Contact periods of less than 90 seconds are required to adequately sterilize water. The resulting water is highly aggressive to pipes and surfaces, requiring that the ozone be broken down into less aggressive oxygen-containing compounds. This is done either by exposure to ultraviolet light or by the addition of chlorine-containing compounds (sometimes both). Water that has been ozonated is devoid of any residual disinfection capacity and hence must be treated with halogen-containing compounds (e.g., chlorine) to have residual disinfection capabilities.

Acidification of drinking water was first used to control the growth of *Pseudomonas* in the drinking water in water bottles used for mice that had been irradiated [4,184,187]. Water of a low pH (acidic) is bacteriostatic to certain vegetative forms of some bacteria. To be most effective, acidification should be at pH 2.3 or lower [127]. Usually this is done with mineral acids; however, such acids have little buffering capacity, so the addition of saliva or other alkaline-containing compounds may rapidly increase the pH of the drinking water.

The use of acidification, as well as other forms of water treatment, is not without potential research effects [1,119]. Any water treatment should be considered an experimental variable and should be well understood by the toxicologist [58,127]. Usually, it is necessary to group together a number of different water treatments to provide a water system that is capable of not only adjusting the chemical composition of the water but also ensuring that it is free from harmful microorganisms.

## FEED

Feed is considered by some to be one of the principal uncontrolled variables in toxicology today. Many textbooks and articles, as well as guidelines, have been written on nutrition for laboratory animals [209–218,220,221, 226,230]. Numerous examples are available of how diet composition or administration can affect the outcome of toxicological research either directly or indirectly [30,57, 144,227,232,237,292,340], as well as examples of individual nutrients either in excess or deficiency producing a wide range of metabolic, morphologic, or physiologic conditions that have influenced the outcome of studies [8,44,229,231,330]. It is not the purpose of this review to cover all of these. The reader is encouraged to pursue such information from available texts, reports, and journal articles, as well as seek the advice of those skilled in laboratory animal nutrition for specific recommendations. Subcommittees of the National Research Council's Committee on Animal Nutrition have published reviews of the nutrient requirements of laboratory animals that can be used as a guide in assessing the adequacy of specific formulations [209–218,220,221].

Feed provided to laboratory animals should be palatable, nutritionally complete, and free of contaminants unless the lowering or removal of an nutrient or the addition of other compounds is required by the study design. The feed should be provided in sufficient quantity and be of the necessary quality to ensure normal growth and provide for the overall health of the animal [46,222]. Increased demands of reproduction and lactation may require slightly different feed formulations. Many commercially available diets are caloric dense and protein rich and often are fed *ad libitum*. Such practices encourage feed wastage and may adversely affect some toxicology studies [159].

The physical and taste characteristics of laboratory animal feed can be important variables in study design. Some common laboratory animal species, such as nonhuman primates, guinea pigs, and rabbits, have well-developed taste and dietary consistency preferences. Guinea pigs, for example, do not easily accept changes in the texture of the diet or in its taste. These animals may have to be habituated to changes in their diet by the gradual introduction of the new diet during the course of feeding a previously accepted diet. This can prove challenging when artificial diets must be substituted for natural ingredient diets.

Laboratory animal feeds can be classified as natural ingredient diets, semipurified (or purified) diets, or chemically defined diets [273]. These classifications are based on the ingredients used to construct the feed. Natural ingredient diets are composed of unprocessed or crudely processed materials such as cereal grains, milk products, and animal product meals such as fishmeal or meatmeal.

Semipurified diets (also referred to as synthetic, semisynthetic, and purified) are constructed of processed ingredients that are refined so as to be uniform in content. Examples of such ingredients include casein, dextran, soy protein, hemicellulose, and fats such as corn oil. Chemically defined diets are composed of chemically pure compounds, including specific amino acids, fatty acids, inorganic salts, vitamins, and sugars such as glucose and fructose [351].

Natural ingredient diets can be pelletized (extruded), expanded (baked), or processed as a meal (powdered). They can also be suspended in agar or other types of hydrocolloids, providing both feed and a source of water. Semipurified and chemically defined diets are provided in either a powdered or liquid form. Experimental compounds can be incorporated in any of these physical forms, but as a rule they are usually fed as a gel or in liquid or powdered form.

The exact formulation of specific feeds may or may not be available to the toxicologist. Open-formula diets are diets that are manufactured to published specifications based on quantities of specific ingredients [165,166]. The quantitative and often qualitative ingredient compositions of open-formula diets are readily available from the manufacturer. Open-formula diets usually are made from natural ingredients and as such may easily suffer from variability due to differences in source of ingredient supply, harvest times, and storage conditions [234]. Detailed ingredient analysis prior to manufacture in both open- and closed-formula diets is seldom done, and when analysis is conducted it usually is done for only a few key constituents. For this reason, open-formula diets can still exhibit significant variability between suppliers and within any given supplier.

Closed-formula diets are manufactured by feed companies who hold the exact composition of the diet in terms of the quantities of specific ingredients as proprietary information. They will provide a list of the ingredients used and usually will provide a calculated analysis. Closed-formula diets are often less costly than open-formula diets, as the manufacturer can adjust the formula periodically, based on the availability and costs of various ingredients. It must be remembered that, like open-formula diets, closed-formula diets may have variable nutrient levels. The calculated analysis on such products is not the same as a laboratory analysis of an individual batch of closed-formula diet.

Recently, the concept of constant nutrition or variable-ingredient diets has been introduced. This is a natural extension of closed-formula diets in that the manufacturer is allowed to select the type and quantity of various natural product ingredients to be used in a diet formulation based on availability and cost. The maximum percentage that any given ingredient can constitute in the diet is limited by published specifications; however, the final product must have analyzed nutrient levels that fall within prescribed ranges. Generally, constant nutrition diets are coupled with regular analysis by the manufacturer of the product for an extensive list of nutrients to ensure that the product provides the specified levels of such nutrients as set forth in the diet specifications. Given the inherent variability of both open- and closed-formula diets, as well as the lack of regular laboratory analysis for specific nutrients in such diets, constant nutrition diets provide a means for standardizing diets between locations as well as suppliers.

Certain basic analyses are required for the interstate shipment of dietary ingredients in the United States. Analyses for up to five constituents, including protein, are made, and certain minimal levels based on the guarantee analysis stated on the feed container must be met [17]. It is important to realize that the actual level of these nutrients is not guaranteed, only that the level is not less than that listed. Similarly, diets can be certified with respect to levels of certain toxic and carcinogenic compounds. The analysis for that particular lot of feed is used to certify that the diet contains specific levels of these compounds that may interfere with some toxicologic research [80,100, 227,230,296,350]. Certification of feed does not imply laboratory analysis for specific nutrients, only that the material does not contain more than a stated level of certain compounds as determined by analyses. Hence, a certified diet could be deficient in nutrients and yet still be certified [57,117,228,267,352,353].

For convenience in feeding and to minimize waste, diets often are pelletized [97]. During this process, the diets are briefly exposed to temperatures exceeding 60°C. The process of pelletizing diets is one of compression and extrusion. The amount of time that the pellet is exposed to elevated temperatures is very brief; hence, there is no assurance that vegetative or nonnegative forms of bacteria or other microorganisms are consistently killed.

Expanded diets undergo a process of prolonged mixing and heating with added liquid in a manner similar to mixing a cake batter. This material is mixed with air and expanded through a die into pellets that are dried with heated air. This process forms a much harder and durable product. The process also exposes the mixture to temperatures in excess of 100°C for periods of 15 minutes or more. This causes some reduction in microorganisms; however, the product itself is not sterile. Contamination can occur in both pelletized and expanded diets after processing. Expanded diets have a lower specific gravity and hence a lower caloric density than pelletized or powdered diets of equal volume. To obtain the same caloric intake, animals will consume greater volumes of an expanded diet.

A number of applications in toxicological research require the decontamination or sterilization of feed [204]. If true reductions in bacteria and viruses are to be achieved in feed products, feed can be treated by the application of steam under pressure (autoclaving), dry heat, or irradiation. Heat applied in these processes can destroy certain labile ingredients within the diets [62,96,242]. Irradiation produces very little elevation in temperature, and the feed is less likely to require additional fortification if sterilization is contemplated [66,356].

Any process used to sterilize feed must be calibrated and carefully controlled. Minor items, such as variation in stacking patterns of bags of feed within an autoclave or the magnitude and number of vacuum pulses, can yield significant differences in the total reduction of microorganisms during the autoclaving process [222]. Equipment used for such feed treatment must be regularly calibrated. The use of indicator tapes or other sterilization indicators in a single spot does not ensure the adequacy of the process. Moreover, cycles designed for pelletized feed seldom are effective in processing powdered feed or other forms of diet.

When irradiation is used, it is important to specify dosages based on the minimum dosage received throughout the entire load or container. When an item has been treated, it should be labeled as to the date of treatment so judgments can be made as to the appropriate time by which the product should be used. Feed should be sealed in durable and preferably water-resistant packaging prior to irradiation.

With or without heat or radiation treatment, certain labile components within diets will degrade over time [102,242]. Diet manufacturers place milling dates on containers of feed using either sequential dating codes or Julian dating. As a general rule, natural ingredient, dry laboratory animal diets that have been stored properly can be used for 180 days after manufacture [222]. Some diets that contain supplemental vitamin C or other very labile ingredients should be used within approximately 3 months from milling unless special stabilized forms of these especially labile nutrients are added to extend the shelf life [222]. Supplemental vitamins and other labile ingredients can be provided separately through the use of drinking water supplements, sprayed on dietary additives, or, in the case of nonhuman primates and certain other species, by the feeding of fruits and vegetables.

Feeding can also be used as part of a psychological enrichment program for certain species of animals. In the case of nonhuman primates, specialized feeders, the use of a variety of different types of food items, and the presentation of feed in a way that foraging is required can all be used to stimulate these animals and reinforce natural behaviors. For the purposes of toxicologic research, maintaining constancy in a diet is important. Varying the diet with a wide selection of unbalanced foods, as might occur through nonstructured supplementation of diets, can cause health-related problems [195]. It is also important to minimize abrupt changes in diet because in many species these can lead to digestive and metabolic disturbances.

The storage of feed and other dietary ingredients should be done in areas that are regularly cleaned and enclosed to prevent entry of pests. In general, feed should be stored off of the floor, and care should be taken to prevent exposure of these items to temperatures above 70°F for prolonged periods of time [102,228,272]. Containers used to hold loose feed should be covered and constructed of materials that prevent the entry of vermin and other pests.

The presentation of feed to animals is another important variable. With few exceptions, feed should be administered to animals in feeders that allow easy access to the feed while minimizing the possibility of contamination with feces or urine. Feeders should also minimize waste, which is particularly important with powdered diets. The size and number of openings in feeders, as well as the location of the feeder, can affect an animal's consumption of feed. Some designs of feeders can actually restrict an animal's access to the feed, resulting in caloric restriction.

Dogs, cats, nonhuman primates, and certain other species are commonly meal fed, such that feed is provided one to three times a day in limited quantities [178]. Rodents are more commonly fed *ad libitum*. Feed consumption can be influenced by the light cycle, with some species preferring to consume more feed in the dark part of the light cycle than others. Rodents consume feed both in the light and dark portions of the light cycle, with more feed being consumed during the dark part of the cycle than the light. Poultry consume feed primarily during the light cycle, as do many other nonrodent species.

Hamsters characteristically horde feed and for that reason will tend to remove feed from feeders to form feed piles on the floor. To minimize injury associated with this behavior, hamsters are commonly fed off of the floor when pelletized diets are used [122]. The ready availability of a water source is also important in feeding behavior for many animals. Many species eat and drink at the same time.

Of all the variables associated with feed, perhaps the one that has the greatest impact on toxicologic studies, other than outright nutritional deficiency, is overfeeding [159]. Animals will consume feed until an internal caloric limit is reached. The imposition of mild to moderate caloric restriction in all species, including rodents, can increase longevity; postpone the development of spontaneously occurring neoplasms; decrease the incidence of background lesions in the kidneys, cardiovascular, and endocrine systems; and cause other changes that in aggregate can produce a more consistent model for toxicologic research [3,23,24,52,53, 183,319]. To adequately interpret data from long-term studies on animals using caloric restriction, other primary studies, including range-finding studies for dosages, need to be conducted in calorically restricted animals.

Caloric restriction or limited feeding is not a new concept [225–227,274,302–306,317]. The effects of this procedure have been recognized for decades [63,150,182].

Several publications have reviewed various segments of this relatively large body of literature [123,124, 156–158,357]. The phenomenon appears to be closely correlated with total calories and not with specific nutrient concentrations, such as protein levels [170]. The technique appears to be most easily applied to those studies in which compounds are administered by gavage; however, there is little reason to believe that it could not be applied to studies in which the compound was administered by other routes. Errors associated with administering compounds incorporated into the diet and administered under a caloric restriction regimen may not pose any greater errors than those associated with similar administration using *ad lib* feeding, which assumes a consistent intake on a daily basis and a consistent level of wastage.

The usefulness of caloric restriction in group-housed animals, however, still must be explored [55]. It has been assumed that any dominance hierarchy established within a group would lead to significant differences in diet consumption among members of the group with, perhaps, the introduction of unwanted variation; however, any such dominance hierarchy that already exists within the group will still result in variation even when feed is available *ad libitum*. This problem could be further magnified if one or more members of the group died during the course of a study. The interaction of group size, sex, and caloric restriction still requires additional investigation.

The decision of the toxicologist to use caloric restriction as a means of decreasing the variability encountered in product registration studies using rodents must be based on a careful examination of the literature and a dialogue with regulatory authorities. Caloric restriction in rodents is used by a growing number of pharmaceutical companies and has gained acceptance by regulatory bodies, provided that adequate control data are provided.

## ENVIRONMENTAL ENRICHMENT

Considerations for animal enrichment must be given in toxicologic studies. Regulations and guidelines such as the USDA Animal Welfare Act or the *Guide for the Care and Use of Laboratory Animals* [222] can be difficult to accommodate in these studies. The regulations require exercise for canines and providing an environment that promotes psychological well-being in nonhuman primates. The multiple goals for the use of animal enrichment include: (1) improving the quality of the captive environment, (2) increasing behavioral diversity, (3) reducing abnormal behaviors, (4) increasing the use of the environment, and (5) increasing the animal's ability to cope with challenges [359]. Environmental enrichment has the potential for unintended consequences [360]. Historically, environmental enrichment for animals used in toxicology studies has been avoided due to the fear that the enrichment will alter study outcomes. More recently, however, efforts to provide optimal animal environments are being considered in toxicologic studies.

Social interaction is an important component of animal well-being. Social contact can be auditory, olfactory, visual, or tactile and can involve pair housing and group housing. Such contact can be accomplished by either full- or part-day housing arrangements. Caging with removable panels or doors can facilitate this type of interaction. Social stabilization of group-housed animals must be established before initiation of the study [361]. Animal losses in studies can occur if socially housed animals are not appropriately paired and monitored; fighting can result in injury or death. Treatment of injured animals is often perceived as a study confounder, so typically these animals are removed from the study. Social housing can make procedures such as food consumption measurements difficult. Consumption of feces from other animals can raise questions as to the amount of metabolized compound ingested [362]. Justification for singly housing animals should be considered on a case-by-case basis [363]. Positive human interaction enhances the well-being of dogs. Socialization and acclimation of dogs to the research environment can have a lifelong impact on the animals. These interactions can be various, including play activities, interaction, and the manipulation of animals in a sling or table. Aggressive behaviors should be carefully evaluated and an action plan put in place for amending the behavior or eliminating the animal from the study.

Structural enrichment comes in many forms and depends on the species used in the study. The International Symposium on Regulatory Testing and Animal Welfare [363] made recommendations that include housing for rodents in solid-bottomed cages with bedding and enrichment. Common enrichment for rodents includes shreddable nesting and cage complexities, such as domes that allow shelter for the animal. Dogs used in toxicology studies may be singly or group housed. Dogs that are singly housed in cages that do not provide twice the required cage space must be provided exercise. Visual contact with other dogs is highly desirable. The cage size and shape can limit the types of enrichment devices available for use. Regulations require that the cage must allow for normal postural movement and species-specific behaviors. Additionally, the cage must be safe for both the animal and the animal handlers.

Perches and other non-nutritive types of enrichment are among the most used enrichment strategies [360]. Such devices may include hammocks, perches, and items for manipulation. These devices must be safe for the animals to use and not easily destroyed or ingested. Edible items must come with certificates of analyses to be certain that the device will not impact the study. Ideally, the items will not obstruct animal observation or husbandry. The devices must be changed frequently, as most animals stop using such items after exposure for some duration. It is ideal to have an assortment of devices that can be rotated in the animal environment to prevent boredom [361]. Foraging opportunities for primates is a method of promoting cognitive skills and fine-tuned motor skills [362].

## HEALTH AND HEALTH MONITORING

Animals used in toxicologic research come from several different categories of sources. A limited number of animals are occasionally obtained from either wild populations or open colonies in which there is little attempt to control the introduction of adventitious organisms. Such conventional or random-source animals have the potential for significant interindividual variability with respect to health status.

A second source of animals is a closed colony of animals in which the introduction of new animals is minimized or eliminated and certain control measures are put in place by means of vaccination, antimicrobial therapy, or anthelminthic treatment to control and, in some cases, eliminate a limited number of microorganisms. Interindividual variability in health profiles may still exist; however, the maintenance of these control measures at the toxicology laboratory can provide some assurance that certain organisms have been eliminated.

The last source classification of animals represents animals from colonies that have originated through procedures designed to totally eliminate unwanted microorganisms and parasites, usually via caesarean section or embryo transfer. These animals are maintained within bioexclusion systems, such as barrier production facilities that are designed to prevent contamination with species-specific pathogens and, in the case of some bioexclusion systems, even common opportunistic organisms that could cause unwanted research interference [315].

The availability of any species in any one of these source categories may be limited or even nonexistent. Larger domestic animals, such as dogs, cats, and farm animals, and nondomestic animals, such as nonhuman primates, generally are not available in a highly defined microbiological status. Health programs designed to control or improve the health status of these animals will not eliminate the possibility of species-specific or opportunistic disease that may interfere with toxicologic research using these animals.

Health monitoring programs for non-barrier-maintained or random-source animals depend heavily on a comprehensive initial quarantine and screening process that is supplemented by individual or colony health data available from the supplier [84,85]. Some organisms that are seldom found in barrier-reared animals are not uncommon in many research facilities [147]. Routine vaccination and anthelminthic therapy, as well as the application of antimicrobial agents, may all be used to control or eliminate undesirable organisms. The toxicologist must be aware that, during the course of a study, individual animals may develop clinical illness or may exhibit individual variation as a result of underlying infectious processes. The frequency of such occurrences cannot be predicted, and allowances must be made for this in the study design when using such animals.

To minimize the impact of health issues on toxicologic research, each institution must develop an institutional philosophy with respect to the impact of certain organisms in their research animals. Not all organisms have a significant impact on research, and those that do may not interfere with certain types of research. Simplistically, organisms can be considered in terms of their relative importance and the amount of peer-reviewed information available to support their impact on research (Figure 20.1).

A number of reviews of microorganisms affecting laboratory animals have been assembled [29,120,205, 206,247]. Unfortunately, there is no universally accepted list of organisms that should be excluded from all animals under all circumstances. Many adventitious organisms do not cause clinical disease; instead, they cause subclinical infections that do not produce histologic changes associated with their presence. In some cases, they may have only limited or subtle research effects that are self-limiting and cease after protective antibodies are formed [344]. In a few instances, more significant effects occur only when the organism is first introduced into a naïve population (epizootic phase of infection). After the organism becomes established and protective antibodies are present either through maternal transfer or the development of a controlled infection, clinical signs or research effects may disappear. Often, agents that are commonly recommended to be screened for have been selected on a historical basis rather than through an analysis of their potential to cause effects or the prevalence of the organism.

Suppliers of barrier-produced animals have developed exclusionary lists of organisms that define the status of the animals that they produce. These exclusionary lists set forth microorganisms that are to be screened for on a regular basis and whose presence is deemed to be a cause for the supplier to eliminate that colony of animals from production. In some cases, an additional group of opportunistic microorganisms may be screened for by suppliers to document the presence of these organisms in the event that they might interfere with certain types of research. These opportunistic organisms may have little potential to cause research interaction and are not a reason to eliminate colonies. These organisms may be common human or environmental commensals that can only be excluded by very specialized bioexclusion housing techniques that may be impractical in the toxicologic research environment.

Animals free of one or more specified microorganisms are termed *specific pathogen free* (SPF) [206]. No universal definition of SPF exists with respect to what organisms are to be excluded from each common laboratory species.

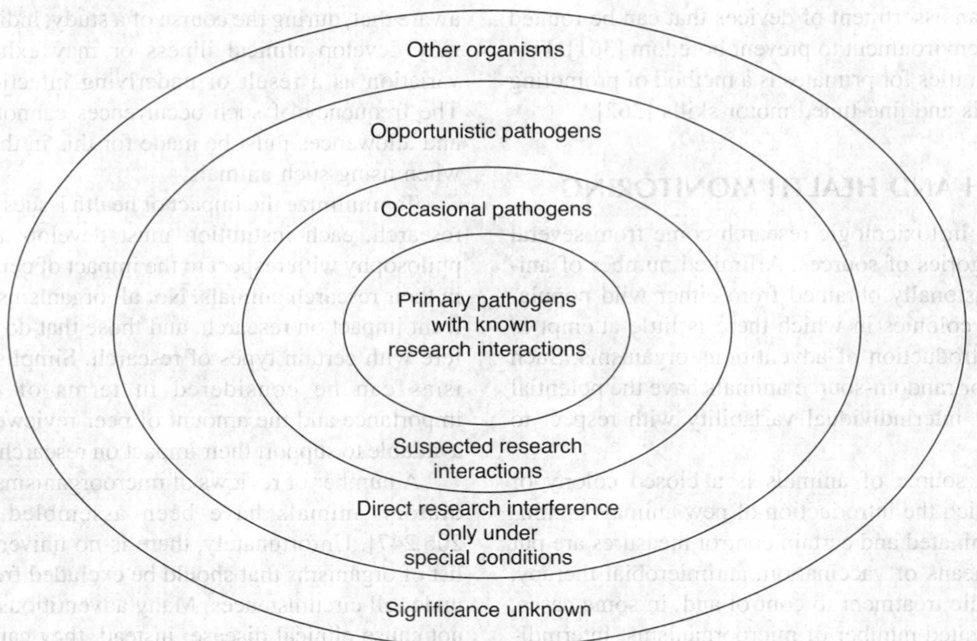


**FIGURE 20.1** Classification of organisms by their significance and impact on research.

Proprietary terms or abbreviations may be used to designate a particular supplier's or research institution's definition of SPF.

When developing an institutional health standard for laboratory animals to be used in toxicology research, the advice of a competent veterinary professional with experience in laboratory animal medicine science should be sought. Decisions on which organisms to exclude and to test for will have to be based on the nature of the research being conducted, species being utilized, a realistic assessment of the facilities and equipment available for maintaining animals, and a thorough review of peer-reviewed literature with respect to the disease-causing potential, host range, and research interactions posed by specific microorganisms. Table 20.2 lists a number of organisms that may infect common laboratory animal species. This table includes species-specific organisms and some common environmental and human commensals whose exclusion might require extraordinary efforts although their presence may have little research impact; however, such opportunistic organisms may be considered for inclusion in the screening program if their presence would have to be documented to help explain unexpected variation in certain specialized studies [84]. It is unrealistic for most toxicologic research facilities, however, to set as a goal the exclusion of all the organisms on this list, as the likelihood of long-term success, unless extraordinary measures are used, is quite low [344].

No matter what institutional exclusionary list is decided upon, each organism on the list should have documentation in the peer-reviewed literature to support its inclusion. Moreover, in developing an exclusionary list consideration should be given to the formulation of a pla to address the potential discovery of positive findings i research populations within the institution. Developmen of an exclusionary list presumes that some method i available to limit the spread of the organism if it gain entrance into the facility and that some course of action such as elimination of infected animals or the impositio of some form of control measures, such as drug therapy will be used to eliminate it [344]. Because toxicologist will be impacted by the health and health monitorin program at their institutions, it is important that they par ticipate in the formulation of such programs.

A number of microorganisms can be carried by ani mals that have health implications for humans. Zoonoti organisms that can be transmitted from animals to human are often included in the screening and control processe that are part of the health program for laboratory animals Many of these organisms, such as herpes B virus (nonhu man primates), *Salmonella*, *Shigella*, Hantaan virus, an lymphocytic choriomeningitis virus, may be included i screening programs not because of their possible researc effects but because of their ability to cause disease i people who work with the animals [101,138,289].

Health monitoring programs require consideration c sample size and frequency; the type, availability, sensitiv ity, and specificity of assays for various microorganism biological characteristics of the microorganisms; samplin procedures and preparation; and interpretation of th results. Discussions of these subjects are beyond the scop of this review but are available elsewhere [10,75,84,8

**TABLE 20.2**
**Organisms Capable of Infecting Laboratory Rodents[a]**

| Organism | Mice | Rats | Guinea Pigs | Hamsters |
|---|---|---|---|---|
| Sendai virus | X | X | X[b] | X |
| Pneumonia virus of mice | X | X | X | X |
| Mouse hepatitis virus | X | | | |
| Minute virus of mice | X | | | |
| GD-VIII[c] | X | | | |
| Reo-3 virus | X | X | X | X |
| Epizootic diarrhea of infant mice | X | | | |
| Lyphocytic choriomeningitis virus | X | X | X | X |
| Polyoma virus | X | | | |
| Mouse cytomegalovirus | X | | | |
| Ectromelia virus | X | | | |
| Mouse parvovirus | X | | | |
| K virus | X | | | |
| Mouse thymic virus | X | | | |
| Hanta viruses | X | X | | |
| Sialodacryoadenitis virus | | X | | |
| Rat coronavirus | | X | | |
| Rat parvovirus | | X | | |
| Kilham rat virus | | X | | |
| H-1 virus | | X | | |
| Mouse adenovirus | X | X | | |
| Guinea pig adenovirus | | | X | |
| Cilia-associated respiratory bacillus | X | X | | |
| *Bordetella bronchiseptica* | X | X | X | X |
| *Citrobacter freundii* 4248 | X | | | |
| *Corynebacterium kutscheri* | X | X | X | X |
| *Salmonella* spp. | X | X | X | X |
| *Mycoplasma pulmonis* | X | X | X | X |
| *Mycoplasma* spp. | | X | | |
| *Streptobacillus moniliformis* | X | X | | |
| *Helicobacter hepaticus* | X | X | | |
| *Helicobacter bilis* | X | X | | |
| *Helicobacter* spp. | X | X | X | X |
| *Klegsiella pneumoniae* | X | X | X | X |
| *Klebsiella pneumoniae* | X | X | X | X |
| *Pasteurella pneumotropica* | X | X | X | X |
| *Pasteurella multocida* | X | X | X | X |
| *Pasteurella* spp. | X | X | X | X |
| *Pseudomonas aeruginosa* | X | X | X | X |
| *Pseudomonas* spp. | X | X | X | X |
| *Staphylococcus aureus* | X | X | X | X |
| *Streptococcus pneumoniae* | X | X | X | X |
| β-Hemolytic *Streptococcus* spp. | X | X | X | X |
| *Streptococcus* spp. | X | X | X | X |
| *Streptococcus zooepidemicus* | | | X | |
| *Encephalitozooan cuniculi* | X | X | X | X |

[a] Infection does not imply disease or interference with research results.
[b] Also referred to as Theiler's murine encephalomyelitis virus.
[c] Probably is not really infected with this virus; can acquire other parainfluenza viruses that cross-react serologically.

76,293,325,342]. In light of all of these considerations, s well as the limitations of the health monitoring tech-iques currently available, it is imperative that no single set of results be used to make a determination of the health status of any group of animals used in toxicologic research. Any positive findings should be confirmed with alternative tests or additional samples. False-positive and false-negative results do, in fact, occur, and drastic actions taken without confirmatory data can cause irreparable damage to ongoing studies [342].

Special problems are imposed when immunocompromised animals are used in pharmacological research. The very property that makes these animals unique (i.e., defects in their immune system) makes them susceptible to many organisms that would not be of concern to animals with an intact immune system. Manipulation of these animals requires the use of aseptic technique and rigorous attention to detail to ensure that they are not exposed to even common opportunistic organisms that could set up life-threatening infections or cause research interactions. Health monitoring programs for these animals often require the use of immunocompetent sentinel animals that are housed either on soiled bedding obtained from the immunocompromised animals or in direct contact with the animals themselves [73,310,342]. Such animals must be obtained from suppliers in a microbiologically defined status that limits the microorganisms to which the animals are exposed to a small number of commensals that have no disease-causing capacity or research implications. Maintenance of these animals requires very special bioexclusion housing practices and modification of research techniques.

The collection of background data can be incorporated into the health monitoring programs. In addition to documenting the presence or absence of various microorganisms, samples may also be collected for histopathology, clinical chemistry, and hematology using representative animals obtained from the same populations as the animals participating in the study.

In some instances it may be necessary to eliminate one or more microorganisms from a group of animals to make it acceptable for introduction into a toxicologic research facility if alternatives to the use of these animals are not available. This process is referred to as *rederivation* and can be accomplished either by caesarean section with cross-fostering of young onto lactating females of the correct microbiological status or by hand rearing the young using the appropriate milk substitute. As an alternative, embryos collected from donor females can be transferred, after washing, to recipient mothers of the appropriate health status. Both of these techniques require considerable technical skill and extensive health monitoring after the procedures to ensure that the appropriate health status of the rederived animals has been obtained. Each technique has advantages and disadvantages. Both of these procedures are expensive and time consuming, especially if they are to be applied to outbred animals that require relatively large founding populations.

In the case of a few viral agents, cessation of breeding and the elimination of naïve individuals for a period of 6 weeks or more, coupled with environmental disinfection, have been shown to eliminate certain viruses from infected populations [342]. With some bacteria and parasites, medications can be used but usually are only practical when applied to small numbers of animals, as such therapies are seldom 100% effective.

## GENETICS AND GENETIC MONITORING

Laboratory animals must be appropriately identified in terms of nomenclature and genetics if studies are to be confirmed either at the same institution or at other institutions. Laboratory animals may be referred to by their common names as well as their scientific names for the purposes of description. In the case of laboratory rodents, specific rules of nomenclature exist for stock and strain designations that cover the genetic classification of the animal and identify its origins and its methods of derivation [87,90,142,143,207,208]. Listings of various stocks and strains, as well as other genetic classifications of animals, are periodically published [88,89]. The correct designation of the animals used should be included in any reports or publications as well as other information regarding the source of supply.

The majority of rodents used in toxicologic research fall into three general genetic classifications: outbred/random-bred, inbred, and F1 hybrid. Transgenic animals can be produced on any of these backgrounds and represent a rapidly growing segment of rodents used in toxicologic research.

Inbred animals are produced as the result of 20 generations or more of brother–sister mating and possess more than 98% genetic homozygosity. They are produced by most breeders by means of a colony structure that allows breeding of large numbers of these animals that are, at most, separated from a brother–sister mating by no more than one generation (Figure 20.2). A gnotobiotic foundation colony is maintained in isolators or other bioexlusion systems and is periodically used to replace the pedigreed nucleus colony of animals within the production facility. The nucleus colony animals are brother–sister mated with pedigrees being maintained. The progeny from these matings form nonpedigreed brother–sister matings in an expansion colony that produces breeders for either a pair mated or polygamous-mated production colony, the offspring of which are used in research.

Inbred animals are useful in toxicologic research in which genetic homogeneity is required. Inbred strains may differ significantly in their response to various compounds due to variations in their genotype and phenotype. They are particularly useful when studying pharmacological mechanisms. Conclusions drawn using a single inbred strain should be verified using other inbred strains or outbred stocks if the findings are to be applied to more genetically diverse populations, such as domestic animals or humans.

F1 hybrids are produced from the mating of two inbred strains. Offspring are genetically identical at all alleles a

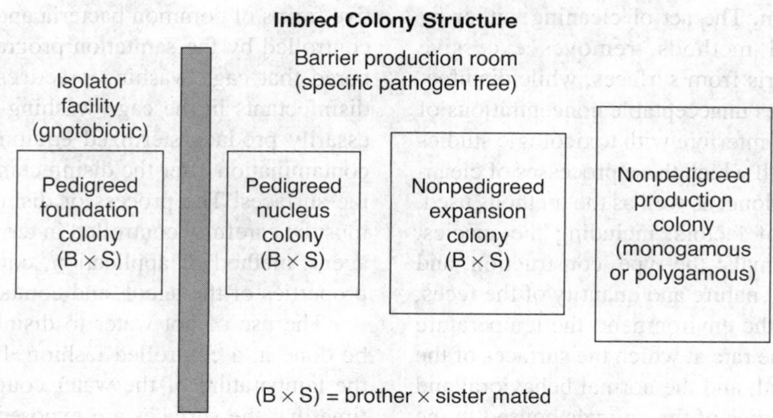


**FIGURE 20.2** Schematic representation of the segments of an inbred production colony.

which the parental strains differ. F1 hybrids are not self-perpetuating, in that breeding F1 hybrids to each other to yield F2 generations allows the various genes that were heterozygous in the F1 alleles to segregate. Commercial production of F1 hybrids requires maintenance of two inbred colonies. Use of F1 hybrids in toxicologic research is limited; however, they do exhibit greater genetic diversity than inbred strains.

Outbred or random-bred animals possess a high degree of individual genetic diversity. The term *random-bred* refers to a system of mating by which there is no conscious selection for specific traits or kinships. Unfortunately, random mating assumes that a population is of infinite size and that all reproductively fit members of the population participate in the mating programs. It also assumes that no conscious or unconscious selection process is used when new matings are set up. Even under the most careful circumstances, these assumptions cannot be fulfilled. Random mating programs, especially those that start with small numbers of individuals, will rapidly result in an increasing level of inbreeding or loss of heterozygosity. Because random-bred or outbred animals are presumed to be useful in toxicology and because of their great degree of individual diversity, loss of this diversity through random mating diminishes their usefulness.

Outbred animals are produced using a purposeful system of mating that seeks to retain the heterozygosity within a group of animals. A number of rotational systems for outbreeding have been devised, as have systems based on the coefficient of inbreeding of pedigreed groups of breeders [114,163,257]. The latter system provides the greatest assurance of continued genetic diversity, provided that the colony is started with a sufficiently large number of genetically unrelated individuals.

Because established colonies of outbred animals undergo independent genetic assortment of polymorphic alleles and will independently fix certain nondeleterious mutations, different colonies of outbred animals separated by time as well as by geographic location will undergo genetic divergence even if they were started from the same colony of animals. To minimize this genetic divergence, breed stock can be traded between colonies of the same stock within the same supplier's organization. Systems for linking colonies in this fashion have been described. These rely on well-established population genetic techniques [125].

The genetic authenticity of inbreds and F1 hybrids can be monitored by assaying a variety of polymorphic alleles located on multiple chromosomes [126,238]. If mismatings occur, variations in this allelic profile will be found, indicating a genetic contamination. Any polymorphic marker can be used, including biochemical, immunologic, and DNA assays.

The authenticity of outbred stocks or of individual outbred animals cannot be determined. Populations of outbred animals can be compared using markers similar to those used for inbreds and F1 hybrids, but large enough groups from a population must be sampled to characterize the allele frequency within the population. Using this allele frequency data, population genetic calculations can be made that will allow the similarity of different populations to be compared [48,125].

For transgenic animals, the presence of a transgene can be determined using polymerase chain reaction (PCR) or other DNA analysis techniques to verify its presence and zygosity. This information can be used to adjust the breeding program or to authenticate a transgenic animal. Finding the presence of a particular transgene, however, does not determine if it has any functional significance.

## SANITATION

Sanitation refers to the cleaning processes required for the maintenance of environmental conditions that are conducive to health. Sanitation of the animal's primary and secondary enclosure includes bedding change and cage cleaning and disinfection. The act of cleaning, either by manual or mechanical methods, removes excessive amounts of dirt and debris from surfaces, while disinfection reduces or eliminates unacceptable concentrations of microorganisms that can interfere with toxicologic studies [222]. The frequency with which these processes of cleaning and disinfection are done, as well as the methods used, depends on a number of factors, including the species, age, and size of the animals; the type, construction, and size of the enclosure; the nature and quantity of the feces, urine, or debris soiling the environment; the temperature and relative humidity; the rate at which the surfaces of the enclosure becomes soiled; and the normal behavioral and physiological characteristics of the animals housed in the enclosure [222,322].

To facilitate cleaning, the surfaces of both the secondary and primary enclosures should be impervious to liquids and smooth to allow easy application of cleaning materials and disinfectants. The surface should also be capable of withstanding scrubbing and should be free of cracks or other deficits that may hinder the cleaning process [177]. The use of agents designed to mask animal odors should not be used in animal facilities designed to support toxicologic research. Such agents expose the animals to chemical compounds that may interact with metabolic or physiologic processes involved in drug uptake and metabolism. Such agents simply obscure poor sanitation practices or inadequate ventilation. Conversely, characteristic animal odors released from the decomposition of urine or feces cannot be used as the sole criteria for the adequacy of sanitation practices or ventilation. Some studies suggest a relationship between certain decomposition products of urine and feces (principally ammonia) and potentiation of the effects of certain disease-causing organisms [36,104,179].

The sanitation of cages and assorted equipment, such as watering devices and feed containers, can be done either manually or with mechanical equipment. The frequency of cleaning is dictated by cage type and the husbandry program. The use of wire-bottom or perforated-bottom cages, the use of regularly changed contact or noncontact bedding, and regular flushing of catch pans suspended under certain types of caging, coupled with the number, size, and type of animals in the cage, will influence how often cages will need to be washed and sanitized. Hamsters, guinea pigs, and rabbits produce urine that contains high concentrations of minerals and proteins that can adhere to caging equipment. Regular cleaning will minimize the build-up of such materials on the caging; however, pretreatment with acid solutions prior to cleaning may be required in order to remove such deposits from cages housing these species of animals [223].

Disinfection of caging can be accomplished by applying hot water or chemicals either singly or in combination. The conditions under which these disinfection regimens are used must be adjusted so they adequately kill vegetative forms of common bacteria and other organisms to be controlled by the sanitation program. It must be remembered that cage washing practices, including the use of disinfectants in the cage washing regimen, will not necessarily produce sterilized equipment nor guard against contamination after the disinfectants have been rinsed off the surfaces. The process of disinfection with chemicals must be carefully controlled in terms of preparation of the agent, method of application, conditions of application, properties of the agent, and contact time [13,31,180].

The use of hot water to disinfect surfaces must also be done in a controlled fashion. It is the combination of the temperature of the water coupled with the length of time that the surfaces are exposed to a given water temperature (cumulative heat factor) that must be controlled. The same total exposure in terms of time and temperature can be obtained by washing at lower water temperatures for longer periods of time [327]. Water temperatures ranging from a 140 to 180°F have been used effectively in mechanical cage washers for disinfection with hot water alone. If detergents or disinfectants are combined with the hot-water washing process, it is important that these compounds be thoroughly rinsed from all surfaces, as residual materials may interfere with some types of studies. If manual washing of cages is undertaken, as might be necessary in some specialized housing facilities, special care must be given to standardize the washing process and to ensure that personnel are provided with the necessary personal protection equipment to minimize their exposure to hot water or chemical agents [222].

Although conventional methods for washing cages may be satisfactory for most toxicologic studies, the use of specialized bioexclusion systems, such as microisolation caging, may require a greater level of assurance of disinfection. In such cases, wrapping caging after cleaning in steam-permeable materials and autoclaving may be required to achieve the necessary level of decontamination.

Runs and pens may be sanitized by using high-pressure water coupled with detergents or disinfectants. Some organisms, such as gastrointestinal parasites, may not be eliminated by such treatments and may require periodic applications of chemical disinfectants for prolonged periods of time. In all cases, thorough rinsing of the enclosures is critical so carry-over of materials that might adversely affect the animals or research does not occur.

Secondary enclosures, such as animal holding rooms, hallways, and support areas, should also undergo regular cleaning and disinfection. These areas can easily become contaminated with infectious organisms that can reinfect clean and disinfected caging. Contamination of secondary enclosures can also allow the spread of organisms from one area of a facility to another. To break the cycle, it is necessary to regularly clean the floors, walls, and other surfaces in the room using agents similar to those used in the cage washing process.

The choice of disinfectants to use in an animal facility is difficult, may not be limited to a single agent, and will rely on a variety of factors. Discussion of disinfection principles is beyond the scope of this text, but useful information that will aid in selecting agents can be found in a number of references [13,31,180]. For general-purpose disinfection, agents should be active both by direct contact and in the vapor phase. The agent should have a broad spectrum of activity, including the ability to kill spores and nonenveloped viruses. Halogen-based compounds, most commonly those containing chlorine, appear to have the widest range of activity and are commonly selected. Care must be taken when using such agents, as regular use of any aggressive disinfectant can also damage surfaces.

## PEST CONTROL

Animal facilities by their very nature generate large amounts of waste and use significant amounts of feed and bedding. Their complex design and extensive mechanical systems provide excellent harborages for rodent and insect pests. Pests can serve as vectors for the introduction into research animal colonies of unwanted parasites and pathogenic microorganisms. Inappropriate application of materials designed to kill pests can pose a risk to toxicologic research either through direct toxic effects of the animals or by interaction on a molecular level with experimental protocols [98,155,329]. Pest control is best conducted by professionals who can design an integrated pest management system that puts in place control methods and monitoring systems to track the effectiveness of the program. The use of potentially toxic chemicals should be avoided whenever possible in favor of nonchemical control measures [73,105,109,164,223]. Application of pesticides should only be conducted with the knowledge of the animal care and research staff and done in compliance with federal, state, or local regulations [222].

## WASTE DISPOSAL

Animal facilities generate large amounts of waste in either solid or liquid form. These wastes can either be nonhazardous or hazardous in nature. Depending on the country, wastes from animal facilities may be regulated and may have to be tracked and disposed of in a prescribed manner [244]. In the United States, the Medical Waste Tracking Act (MWTA) of 1988 charged the U.S. Environmental Protection Agency with tracking certain potentially hazardous wastes. Additional state and local regulations may also be imposed, causing the complexity of research waste management to vary from state to state [282]. Although these regulations were designed to limit exposures to biohazardous wastes, all animal wastes may be subject to specific tracking and disposal procedures, depending on the interpretation of these regulations. For this reason, a local safety committee with expertise in these regulations should oversee the disposal process and ensure that all personnel handling waste are properly trained.

Some forms of waste may require special handling and may be incinerated on site. The operation of an on-site incinerator is heavily regulated in most countries and is not available at all research facilities. More commonly, contract waste handlers are used to dispose of research and medical waste using methods that comply with existing regulations. Waste water from cage washing, room cleaning, and flushing of pens and runs is flushed into drains to be subsequently treated on site using local sewage treatment facilities or sent to municipal treatment systems.

Bedding constitutes the largest waste component in animal facilities. This is seldom treated on site and usually is removed either as bagged waste or in large containers for off-site disposal. Final disposal of noninfectious waste bedding is often by composting, land application, or landfills. Noninfectious animal carcasses and carcasses exposed to infectious or other hazardous materials are commonly refrigerated or frozen in plastic bags for subsequent incineration [222]. Recently, cost-effective chemical digestion methods for carcasses and other solid biological wastes have become available, allowing on-site treatment of such items with minimal environmental impact. Wherever possible, waste that is known to contain infectious or heat-labile hazardous materials should be treated by autoclaving or other destructive methods prior to storage for final disposal. Care should be taken to ensure that biological materials with a high liquid content, such as carcasses, are placed in sturdy leak-proof containers that can be sealed to minimize the potential spread of contamination [222].

Areas in which wastes are stored should be adequately ventilated and appropriately labeled. Individual waste containers should also be labeled as to the contents and any hazardous materials that they might contain. In the case of biological materials that are being collected for disposal, refrigerated or freezer storage may be required; in such instances, a properly labeled, dedicated refrigerator or freezer should be available.

## ANESTHESIA, ANALGESIA, AND SURGERY

Some toxicologic research protocols involve the surgical modification of animals or may subject animals to procedures that are likely to cause pain. Such procedures have been an important focus of the animal use oversight processes in most countries that have legislation regulating the use of animals in research. The researcher bears ultimate responsibility for the use of appropriate procedures and administration of the necessary drugs in such circumstances to ensure the welfare of animals. Many references are available to assist toxicologists in developing an appropriate surgical, anesthetic, or analgesic regimen [39,68,69, 152,168]. Assistance in the process should be sought from veterinarians experienced in laboratory animal medicine. The conduct of surgery in a research setting is usually a team effort requiring advanced planning, training to acquire the necessary skills, appropriate facilities and equipment, and regular evaluation of outcomes and objective oversight of the entire surgical process [39,40,69].

Because both infection and lack of appropriate materials to complete the surgery successfully can result in the failure of a surgical procedure with the subsequent loss of animals, it is important that all aspects of the surgical procedure, from the selection of the animal through its preparation, anesthesia, surgical manipulation, and postoperative recovery, be carefully planned in detail and that all necessary supplies, equipment, and assistance are provided [14,15,222,291]. It is well established that animals are no less susceptible to infection than humans [20,32,68,332, 333]. For this reason, any materials or equipment used in the operative procedure must be appropriately disinfected or sterilized and techniques be used to maintain the critical components of surgical supplies and equipment in aseptic condition throughout the surgical procedure [25,151,270,279,290].

The complexity of this task varies with the size and number of animals that must undergo a surgical procedure. Large animals often require a much larger surgical field from which hair is removed and the skin is scrubbed, disinfected, and covered by sterile drapes [132,348]. Often the incision site is quite large, and the procedures undertaken may be more complex and require more instrumentation than those conducted on small laboratory rodents [69]. Personnel must be provided with appropriate operating attire that is decontaminated or sterile. The surgeon's hands and arms should be disinfected by scrubbing with an antibacterial soap and water, and the necessary clothing and gloves should be packaged in a manner that will allow them to be put on by the surgeon in a location and manner such that they do not become contaminated. The extent of surgical clothing required will vary with the species of animals being used and will be proportional to the size of the operative site and incision [51,248,348]. In the case of surgery involving rodents, the use of sterile gloves, a face mask, and a clean garment that covers the upper torso including the arms may be all that is required, whereas surgery on larger animals, such as rabbits, dogs, or nonhuman primates, may require much more extensive coverage of the surgeon with disinfected, sterile attire [38,67,69].

It is critical to review in detail each component of the procedure to determine if assistance is needed either from a person who is in surgical attire and capable of manipulating instruments or tissues in the operative field or by personnel who are not in full surgical attire and can handle materials that have not been disinfected or sterilized, administer anesthesia, or aseptically dispense surgical supplies.

To be qualified to conduct surgical procedures, toxicologists must undergo training. Most commonly, this is done using a combination of techniques, including assisting in similar surgical procedures, observing others performing the surgical procedure, practicing on cadavers, or, in the case of basic surgical technique, using training aids to acquire the necessary skills to conduct the surgery [1]. The surgeon should be prepared to deal with possible complications, including anesthesia difficulties, hemorrhage, difficulties in wound closure, and other similar, unwanted occurrences that could be associated with the procedure.

The postoperative recovery of the animal should be carefully planned, and clinical parameters that would signal a need for additional intervention should be prepared for; such intervention would include supportive therapy, which might consist of supplemental heat, administration of fluids, use of parenteral drugs (antibiotics, analgesics), nutritional support, or resuscitative procedures. The criteria for a successful surgical outcome should be determined, as should the criteria for terminating a process when the outcome is unsuccessful. Toxicologists should have planned for and be trained in appropriate methods of euthanasia when it might be required [7,160,181].

When new procedures are being developed, veterinary assistance and oversight may be necessary to further refine techniques and minimize failures. If animals die unexpectedly or unfavorable outcomes occur, the process should be analyzed and steps taken to improve it. Often critical to this is a comprehensive diagnostic examination of the animals that have undergone the procedure, either at the end of the study or when unexpected deaths or failure of the procedure occurs. Postmortem examinations, including histopathology and, where appropriate, other diagnostic measures, are essential to this process.

Surgical procedures can be classified as either recovery or nonrecovery and further subdivided as major or minor [222]. Each requires respect for the animals being used and provision of adequate anesthesia. Many complex issues arise when selecting the appropriate anesthetic regimen for any surgical procedure. Anesthetic techniques, equipment, and drugs are constantly being refined. With few exceptions, most techniques that can be applied to humans can be applied to animals. Most anesthetic regimens used in toxicologic research, however, tend to be relatively simple and effective. Commonly, only one or two drugs are administered, and agents are favored that are easy to administer and whose use is well characterized in the species selected.

Because most toxicologic research involves small rodent species, agents that can be given intraperitoneally or by inhalation are easiest to administer and are commonly used. Selecting a few anesthetic regimens and being skilled in their use in a particular species often proves to be the most successful approach for the toxicologist

Switching between new anesthetic regimens without adequate training can lead to undesirable outcomes. Conversely, if an anesthetic regimen does not work well in a particular species or for a specific procedure, the toxicologist should not be reticent to seek an alternative technique that is more successful.

If inhalation agents are used, care must be taken to minimize human exposure, especially to women of childbearing age. Simple systems have been devised to scavenge waste anesthetic gases, and devices exist to monitor personnel exposure. Many anesthetic agents and analgesics are categorized as controlled substances that require secure storage, recordkeeping, and licensure.

When administering anesthesia, it is necessary to be cognizant of the signs and reflexes used to gauge the depth of anesthesia and what dosages will be used to redose the animals should additional anesthesia be required. Similarly, a number of important factors can influence the course of anesthesia, including the state of hydration of the animal, decreases in the animal's core temperature due to loss of heat into its surroundings, the route and site of administration of the agent, and variability in the depth of anesthesia associated with failure to fast the animal. Techniques have been described to address all of these issues, as well as others [168,346].

When an animal has recovered from anesthesia, there may be instances in the postoperative period where analgesics may be required. Similarly, in some studies, pain may result from nonsurgical procedures or conditions that may have to be relieved for the study to continue. The diagnosis of pain in animals is difficult and does not rely on any single clinical sign [135,299]. Analgesics have not been well studied in laboratory animals. It is clear that animals can and do experience pain and that analgesics can provide relief from pain, as evidenced by alterations in parameters such as weight gain, feed consumption, and activity measurements [219]. Determination of the presence or absence of pain in small laboratory animals is particularly difficult, as is the assessment of the effectiveness of various analgesics administered prior to or following the onset of pain. Because the information available for choosing analgesics is limited and in some cases relatively subjective, toxicologists should seek to assess the effectiveness of any drug or regimen selected through the measurement of clinical parameters that are affected by pain.

## ACQUISITION, QUARANTINING, AND CONDITIONING ANIMALS

The choice of a particular species for use in toxicologic research may be based on past work using the species; evaluation of metabolic, physiologic, or morphologic characteristics of a given species; the incidence of development of certain spontaneous lesions; or an assessment of other biologic parameters, including lifespan, reproductive cycles, and behavioral characteristics that would make them suitable for a particular type of toxicological study. Laws and guidelines in many countries require consultation of literature or other sources of information to verify the appropriateness of a particular species to ensure unnecessary duplication of work and to determine if alternatives to painful procedures exist. Other factors, such as the availability of a particular species and its conservation status, may also impact the decision.

When ordering animals, the only truly verifiable specifications are weight and sex. With the exception of nonhuman primates, dogs, and cats, there are very few instance in which a complete clinical or historical record is kept on individual animals. In some cases, a vaccination history and records of preshipment conditioning programs, including dosing with anthelminthic and antimicrobial treatment, may also be available for cats, dogs, and nonhuman primates, as well as a limited colony health history, as previously discussed.

Commercially bred rodents and rabbits maintained under defined microbiological conditions by large commercial breeders are easily obtainable, but care must be taken when ordering these animals. Most commercial breeders produce rodents and rabbits in large numbers, often using polygamous or harem mating systems. Stock animals are maintained after weaning either in weight groups or in age groups by week of birth. Animals are seldom held by day of birth unless specific arrangements are made to set aside groups of animals for this purpose.

When ordering animals, they usually are specified by weight or age but seldom both. For outbred animals and, to a lesser extent, inbred and F1 hybrid animals, the weight range at any given age can be relatively broad and significantly overlap weight ranges for other age groups; hence, it is possible for animals of two different ages to have the same weight. These overlaps can span several weeks of growth. Suppliers construct growth charts that can be used to estimate age based on weight within specific weight ranges. These are useful when a specific weight range of animals is required for a particular study, but some assurance still must be given that this represents a certain age range. If both an exact age and a specific weight range are selected, only a small portion of an outbred population of animals (and, to a lesser extent, inbred and F1 hybrid animals) will be represented, resulting in unconscious selection for certain traits associated with animals within the population that fall within this specific age and weight range. Such overselection can adversely skew study populations and can lead to circumstances where findings cannot be repeated. For these reasons, it is best to specify animals by either age or weight.

When selecting animals for a study, it is important for toxicologists to carefully determine the size of the study group to be used. If very small groups of outbred animals are used, it is possible to select, purely through sampling error, a nonrepresentative group of animals from a much larger population [125]. Attempts to repeat the findings of the study with another small group or even with larger groups may yield different results purely due to inappropriate sampling. It is also a reason why care must be taken to adequately randomize the assignment of animals to both test and control groups.

Populations of animals are dynamic and constantly changing their profile of expressed phenotypes. This occurs to a much smaller degree with inbred animals and F1 hybrids but can occur, as previously described, through the process of fixation and assortment of natural mutations. The problem is magnified in outbred colonies due to random genetic drift [125]; hence, toxicologists should expect that historical controls will vary over time in an unpredictable fashion and that such variation will occur regardless of whether the same colony is used for a source of animals for subsequent studies or alternative colonies are selected. It is also important to appreciate that, while different suppliers may produce strains or stocks of similar designations, perhaps derived from a common source, the longer the time interval from the point of stocking until the present, the more the groups of animals have likely drifted apart genetically and the more likely they are to have different phenotypic expressions, either through continual genetic reassortment and fixation or through the development of new allelic polymorphisms. Switching sources of supply of animals can cause changes in historical controls as well as in standardized assays. These changes usually are not dramatic but can be a source of concern for those assuming that some consistency exists between different populations of animals.

Animals are transported from suppliers to toxicology laboratories in shipping containers that are designed to meet national and certain international standards [141]. Depending on the country, various regulations may be applicable to the control of animals in transport. Animals containing infectious agents or whose genetic material has been altered and animals that have been exposed to hazardous materials of a noninfectious nature must be shipped in conformance to regulations that can differ significantly between countries. Commercial animal suppliers are experienced in shipping animals to conform with these requirements; however, if a toxicologist intends to ship animals between institutions, it is important to work with experienced brokers or other shipping agents who can assist in making the necessary arrangements and acquire the necessary shipping containers to legally conduct this process.

For the most part, animals are shipped in new, disposable containers that often are filtered to prevent the incursion of microorganisms. Containers are sometimes sterilized or disinfected, as may be the bedding, feed, and water used in shipping, depending on the supplier and the animals being transported. Most transportation either entirely or in part is done by truck. Most large laboratory animal suppliers maintain independent trucking routes and dedicated vehicles that are disinfected between shipments that transport animals. Some animal suppliers have only limited trucking capabilities and rely on air shipments. All suppliers ship some portion of their animals by air to supply customers not served by truck routes or to accommodate special conditions or orders that cannot be handled by truck. Unless very special shipping containers are used, the shipment of animals by air allows the possibility of excessive stress and contamination during shipment [190,222,223]. Moreover, the supplier does not have control of the shipment while it is being handled by the air carrier; hence, the control of environmental conditions may vary depending on the circumstances.

It is generally good practice to process animals immediately upon receipt at the research institution. Disinfection of the outside of the containers with solutions of general-purpose disinfectants is often a prudent step, especially if the animals have been shipped by air. Animals should be removed from their containers and examined upon arrival to confirm their clinical condition and verify order specifications. They should be placed in appropriate caging that has been labeled to identify the animals, and they should be given access to feed and water as soon as possible after arrival. Most toxicologic research facilities provide a period of stabilization or quarantine for newly arrived animals. During quarantine periods, animals are observed for clinical signs of disease, samples are taken for health assessment, and, in some cases, vaccinations, treatment with antimicrobials, or treatment with anthelminthics may also be undertaken [222]. Other forms of diagnostic testing may also be conducted during the quarantine period. The length of quarantine can vary considerably, depending on the institutional exclusionary list or organisms for a particular species and health monitoring program.

A stabilization period differs from quarantine in that a stabilization period is designed to allow the animals time to recover from the stress of transportation, become rehydrated, and to gain back weight that may have been lost during transportation [2,74,78,148,326]. This period of acclimation also allows the animal to become accustomed to using the water and feed sources and to adapt to any changes in diet [112,173,262,313]. In the case of group-housed animals, it allows the establishment of both social hierarchies and other behavioral adaptations. Stabilization and quarantine should be done concurrently. Some institutions do not have a quarantine period in which health monitoring is conducted; instead, health monitoring information from the animal supplier is used to determine the fitness of the animals for incorporation into the research program. A stabilization period, however, may be instituted for the group to allow the necessary acclimation to occur prior to use. Typical stabilization periods range from 3 to 7 days for most species and are based on some limited work to suggest that periods of 48 or 72 hours are necessary to overcome the stress of transportation [173,218,286].

## QUESTIONS

1. What housing systems can be used to minimize the risk of introduction and spread to animals of unwanted microorganisms that could alter toxicologic research?
2. What microorganisms that infect animals have the ability to alter toxicologic research results?
3. If outbred (non-inbred) rodents are used in toxicologic research or product registration studies, what factors associated with their breeding methods and source colonies, as well as ordering specifications, can cause variation in research results?
4. What are the differences between closed formula, open formula, and constant nutrition natural ingredient diets used to feed research animals?
5. What laws, regulations, and guidelines affect the use of laboratory animals in toxicological and product registration research?
6. What constituents of the research animal's environment can cause variation in toxicologic research results?

## REFERENCES


1. Academy of Surgical Research (ASR). (1989): Guidelines for training in surgical research in animals. *J. Invest. Surg.*, 2:263–268.
2. Aguila, H. N., Pakes, S. P., Lai, W. C., and Lu, Y. S. (1988): The effect of transportation stress on splenic natural killer cell activity in C57BL/6J mice. *Lab. Anim. Sci.*, 38: 148–151.
3. Albanes, D. (1987): Total calories, body weight, and tumor incidence in mice. *Cancer Res.*, 47:1987–1992.
4. Allander, C. and Abel, E. (1973): Some aspects of the differences of air conditions inside a cage for small laboratory animals and its surroundings. *Z. Versuchstierkd. Bd.*, 15:20–34.
5. ASHRAE. (1992): Air cleaners for particulate contaminants. In: *1992 ASHRAE Handbook: HVAC Systems and Equipment, I-P Edition*, American Society of Heating, Refrigeration, and Air Conditioning Engineers, Atlanta, GA.
6. ASHRAE. (1993): Environmental control of animals and plants. In: *1993 ASHRAE Handbook: Fundamentals, I-P Edition*. American Society of Heating, Refrigeration, and Air Conditioning Engineers, Atlanta, GA.
7. American Veterinary Medical Association (AVMA). (1993): 1993 Report of the AVMA Panel on Euthanasia. *J. Am. Vet. Med. Assoc.*, 202:229–249.
8. Ames, B. N., Shigenaga, M. K., and Hagen, T. M. (1993): Review: oxidants, antioxidants and the degenerative disease of aging. *Proc. Natl. Acad. Sci. U.S.A.*, 90:7915–7922.
9. Animal Welfare Institute. (1979): *Comfortable Quarters for Laboratory Animals*. Animal Welfare Institute, Washington, D.C.
10. Anonymous (1976): Long-term holding of laboratory rodents. *ILAR News*, 19:L1–L25.
11. Anthony, A. (1962): Criteria for acoustics in animal housing. *Lab. Anim. Care*, 13:340–347.
12. Armario, A., Castellanos, J. M., and Balasch, J. (1985): Chronic noise stress and insulin secretion in male rats. *Physiol. Behav.*, 34:359–361.
13. Ascenzi, J. M. (1996): *Handbook of Disinfectants and Antiseptics*. Marcel Dekker, New York.
14. Association of Operating Room Nurses (AORN). (1982): Recommended practices for traffic patterns in the surgical suite. *Assoc. Oper. Room Nurs. J.*, 15:750–758.
15. Ayliffe, G. A. J. (1991): Role of the environment of the operating suit in surgical wound infection. *Rev. Infect. Dis.*, 13(Suppl. 10):S800–S804.
16. Baetjer, A. M. (1968): Role of environment temperature and humidity in susceptibility to disease. *Arch. Environ. Health.*, 16:565–570.
17. Barker, H. J., Lindsey, J. R., and Weisbroth, S. H. (1979): Housing to control research variables. In: *The Laboratory Rat*. Vol. 1. *Biology and Diseases*, edited by H. J. Baker, J. R. Lindsey, and S. H. Weisbroth, pp. 169–192. Academic Press, Orlando, FL.
18. Barrett, A. M. and Stockham, M. A. (1963): The effect of housing conditions and simple experimental procedures upon the corticosterone level in the plasma of rats. *J. Endocrinol.*, 26:97–105.
19. Beall, J. R., Torning, F. E., and Runkle, R. S. (1971): A laminar flow system for animal maintenance. *Lab. Anim. Sci.*, 21:206–212.
20. Beamer, T. C. (1972): Pathological changes associated with ovarian transplantation. In: *The 44th Annual Report of the Jackson Laboratory*, p. 104. Jackson Laboratory, Bar Harbor, ME.
21. Beck, R. W. (1963): The control of *Pseudomonas aeruginosa* in mouse breeding colony by the use of chlorine in the drinking water. *Lab. Anim. Care*, 13:41–45.
22. Belhorn, R. W. (1980): Lighting in the animal environment. *Lab. Anim. Sci.*, 30:440–450.
23. Berg, B. N., and Simms, H. S. (1960): Nutrition and longevity in the rat. II. Longevity and onset of disease with different levels of food intake. *J. Nutr.*, 71:255–263.
24. Berg, B. N. and Simms, H. S. (1961): Nutrition and longevity in the rat. III. Food restriction beyond 800 days. *J. Nutr.*, 74:23–32.
25. Berg, J. (1993): Sterilization. In: *Textbook of Small Animal Surgery*, 2nd ed., edited by D. Slatter, pp. 124–129. W.B. Saunders, Philadelphia, PA.
26. Besch, E. L. (1975): Animal cage-room dry-bulk and dew-point temperatures differential. *ASHRAE Trans.*, 88: 549–557.
27. Besch, E. L. (1980): Environmental quality within animal facilities. *Lab. Anim. Sci.*, 30:385–406.
28. Besch, E. L. (1990): Environmental variables and animal needs. In: *The Experimental Animal in Biomedical Research*. Vol. 1. *A Survey of Scientific and Ethical Issues for Investigators*, edited by B. E. Rollin and M. L. Kesel, pp. 113–131. CRC Press, Boca Raton, FL.
29. Bhatt, P. N., Jacoby, R. O., Morse III, H. C., and New, A. E. (1986): *Viral and Mycoplasmal Infections of Laboratory Rodents: Effects on Biomedical Research*. Academic Press, Orlando, FL.

30. Birt, D. F. and Conrad, R. D. (1981): Weight gain, reproduction, and survival of Syrian hamsters fed five natural ingredients diets. *Lab. Anim. Sci.*, 31:149–155.
31. Block, S. S. (1991): *Disinfection, Sterilization and Preservation*, 4th ed. Lea & Febiger, Philadelphia, PA.
32. Bradfield, J. F., Schachtman, T. R., McLaughlin, R. M., and Steffen, E. K. (1992): Behavioral and physiological effects of inapparent wound infection in rats. *Lab. Anim. Sci.*, 42:572–578.
33. Brainard, G. C. (1989): Illumination of laboratory animal quarters: participation of light irradiance and wavelength in the regulation of the neuroendocrine system. In: *Science and Animals: Addressing Contemporary Issues*, pp. 69–74. Scientists Center for Animal Welfare, Greebelt, MD.
34. Brainard, G. C., Vaughan, M. K., and Reitner, R. J. (1986): Effect of light irradiance and wavelength on the Syrian hamster reproductive system. *Endocrinology*, 119:648–654.
35. Brewer, N. R. (1964): Estimating heat produced by laboratory animals: new data on animal heat and vapor transmission account for activity and other factors to provide a more reliable basis for conditioning design calculations. *Heating Piping Air. Cond.*, 36:139–141.
36. Broderson, J. R., Lindsey, J. R., and Crawford, J. E. (1976): The role of environmental ammonia in respiratory mycoplasmosis of rats. *Am. J. Pathol.*, 85:115–130.
37. Brown, A. M. and Pye, J. D. (1975): Auditory sensitivity at high frequencies in mammals. *Adv. Comp. Physiol. Biochem.*, 6:1–73
38. Brown, M. J. (1994): Aseptic surgery for rodents. In: *Rodents and Rabbits: Current Research Issues*, edited by S. M. Niemi, J. S. Venable, and H. N. Guttman, pp. 67–72. Scientists Center for Animal Welfare, Bethesda, MD.
39. Brown, M. J., Pearson, P. T., and Tomson, F. N. (1993): Guidelines for animal surgery in research and teaching. *Am. J. Vet. Res.*, 54:1544–1559.
40. Brown, M. J. and Schofield, J. C. (1994): Perioperative care. In: *Essentials for Animal Research: A Primer for Research Personnel*, edited by B. T. Bennett, M. J. Brown, and J. C. Schofield, pp. 79–88. National Agricultural Library, Washington, D.C.
41. Burkhart, C. A. and Robinson, J. L. (1978): High rat pup mortality attributed to the use of cedar-wood shavings as bedding. *Lab. Anim.*, 12:221–222.
42. Campbell, S. A. (1990): Effects of exercise programs on serum biochemical stress indicators in purpose-bred beagle dogs. In: *Canine Research Environment*, edited by J. A. Mench and L. Krulisch, pp. 77–82. Scientists Center for Animal Welfare, Bethesda, MD.
43. Campbell, S. A., Hughes, H. C., Griffen, H. E., Landi, M. S., and Mallon, F. M. (1988): Some effects of limited exercise on purpose-bred beagle dogs. *Am. J. Vet. Res.*, 49:1298–1301.
44. Campbell, T. C. and Hayes, J. R. (1974): Role of nutrition in the drug-metabolizing enzyme system. *Pharmacol. Rev.*, 26:171–197.
45. Canadian Council of Animal Care. (1980): *Guide and Care and Use of Experimental Animals*, Vol. 1. Canadian Council on Animal Care, Ottawa, Canada.
46. Canadian Council on Animal Care. (1984): *Guide and Care and Use of Experimental Animals*, Vol. 2. Canadian Council on Animal Care, Ottawa, Canada.
47. Canadian Federation of Humane Societies. (1990): *Guidelines for Community Members of Animal Care Committees*. Experimental Animals Committee, Canadian Federation of Humane Societies, Nepean, Canada.
48. CD(SD)IGS Study Group. (1998): *Biological Reference Data on CD(SD)IGS Rats: 1998*. Best Printing, Yokohama, Japan.
49. CDC and NIH. (1993): *Biosafety in Microbiological and Biomedical Laboratories*, 3rd ed. HHS Publ. No. (CDC)93-8395, Centers for Disease Control and Prevention/National Institutes of Health, Washington, D.C.
50. CDC and NIH. (1995): *Primary Containment for Biohazards: Selection, Installation and Use of Biological Safety Cabinets*. Centers for Disease Control and Prevention/National Institutes of Health, Washington, D.C.
51. Chamberlain, G. V. and Houang, E. (1984): Trial of the use of masks in gynecological operating theatre. *Ann. R. Coll. Surg.*, 66:432–433.
52. Cheney, K. E., Liu, R. K., Smith, G. S., Leung, R. E., Mickey, M. R., and Walford, R. L. (1980): Survival and disease patterns in C57BL/6J mice subjected to undernutrition. *Exp. Gerontol.*, 15:237–258.
53. Cheney, K. E., Liu, R. K., Smith, G. S., Meredith, P. J., Mickey, M. R., and Walford, R. L. (1983): The effects of dietary restriction of varying duration on survival, tumor patterns, immune function, and body temperature in B10C3F$_1$ female mice. *J. Gerontol.*, 38:420–430.
54. Cherry, J. A. (1987): The effect of photoperiod on development of sexual behavior and fertility in golden hamsters. *Physiol. Behav.*, 39:521–526.
55. Chvedoff, M., Clarke, M. R., Irisarri, E., Faccini, J. M., and Monro, A. M. (1980): Effects of housing conditions on food intake, body weight and spontaneous lesions in mice: a review of the literature and results of an 18-month study. *Food Cosmetics Toxicol.*, 18:517–522.
56. Cinti, D. K., Lemelin, M. E., and Christian, J. (1976): Induction of liver microsomal mixed-function oxidases by volatile hydrocarbons. *Biochem. Pharmacol.*, 25:100–103
57. Clapp, M. J. L. (1980): The effect of diet on some parameters measured in toxicologic studies in the rat. *Lab. Anim.*, 14:253–261.
58. Clough, G. (1976): The immediate environment of the laboratory animal. In: *Control of the Animal House Environment*, edited by T. McSheehy, pp. 77–94. Trevor Laboratory Animals, London.
59. Clough, G. (1982): Environmental effects on animals used in biomedical research. *Biol. Rev.*, 57:487–523.
60. Clough, G. (1987): The animal: design, equipment and environmental control. In: *The UFAW Handbook on the Care and Management of Laboratory Animals*, 6th ed. edited by T. B. Poole, pp. 108–143. Longman, London.
61. Clough, G. and Gamble, M. R. (1976): *Laboratory Animal Houses: A Guide to the Design and Planning of Animal Facilities*, LAC Manual Series No. 4, Medical Research Council Laboratory Animals Council Laboratory Animal Centre, Abbey Press, Abingdon, Oxon.
62. Collins, T. F. X., Hinton, D. M., Welsh, J. J., and Black T. N. (1992): Evaluation of heat sterilization of commercial rat diet for use in FDA toxicological studies. *Toxicol. Indust. Health.*, 8:9–20.

63. Conybeare, G. (1979): Effect of quality and quantity of diet on survival of tumour incidence in outbred Swiss mice. *Food Cosmetics Toxicol.*, 18:65–75.
64. Cooper, E. C. (1989): Design considerations for research animal facilities. *Lab. Anim.*, 18:23–26.
65. Coriell, L. L. and McGarrity, G. J. (1973): Biomedical applications of laminar airflow. In: *Germ-Free Research Biological Effect of Gnotobiotic Environments*, edited by J. B. Henegham, p. 43. Academic Press, New York.
66. Cover, C. E. and Belcher, L. A. (1992): Effect of an irradiated rodent diet on growth and food consumption: a comparative study. *Contemp. Top. Lab. Anim. Sci.*, 31: 13–17.
67. Cunliffe-Beamer, T. L. (1983): Biomethodology and surgical techniques. In: *The Mouse in Biomedical Research*. Vol. III. *Normative Biology, Immunology and Husbandry*, edited by H. L. Foster, J. D. Small, and J. G. Fox, pp. 419–420. Academic Press, New York.
68. Cunliffe-Beamer, T. L. (1990): Surgical techniques. In: *Guidelines for the Well-Being of Rodents in Research*, edited by H. N. Guttman, pp. 80–85. Scientists Center for Animal Welfare, Bethesda, MD.
69. Cunliffe-Beamer, T. L. (1993): Applying principles of aseptic surgery to rodents. *AWIC Newsl.*, 4:3–6.
70. Cunliffe-Beamer, T. L. (1981): Barbiturate sleeptime in mice exposed in mice exposed to autoclaved or unautoclaved wood beddings. *Lab. Anim. Sci.*, 31:672–675.
71. Department of Agriculture. (1987): Animal and plant health inspection service: 9 CFR Parts 1 and 2; animal welfare; proposed rules. *Fed. Reg.*, 52:10292–10322.
72. Dillehay, D. L., Lehner, N. D. M., and Huerkamp, M. J. (1990): The effectiveness of a microisolator cage system and sentinel mice for controlling and detecting MHV and Sendai virus infections. *Lab. Anim. Sci.*, 40:367–370.
73. Donahue, W. A., VanGundy, D. N., Satterfield, W. C., and Coglan, L. G. (1989): Solving a tough problem. *Pest Control*, August:46–50.
74. Drozdowicz, C. K., Bowman, T. A., Webb, M. L., and Lang, C. M. (1990): Effect of in-house transport on murine plasma corticosterone concentration and blood lymphocyte populations. *Am. J. Vet. Res.*, 51:1841–1846.
75. Dublin, S. and Zietz, S. (1991): Sample size for animal health surveillance. *Lab. Anim.*, 20:29–33.
76. Duncan, T. E. and O'Steen, W. K. (1985): The diurnal susceptibility of rat retinal photoreceptors to light-induced damage. *Exp. Eye Res.*, 41:497–507.
77. Dyment, J. (1976): Air filtration. In: *Control of the Animal Housing Environment*, edited by T. McSheehy, pp. 209–246. Laboratory Animals, London.
78. Dymsza, H. A., Miller, S. A., Maloney, J. F., and Foster, H. L. (1963): Equilibrium of the laboratory rat following exposure to shipping stresses. *Lab. Anim. Care*, 13:60–65.
79. Eaton, P. (1987): Hygiene in the animal house. In: *The UFAW Handbook on the Care and Management of Laboratory Animals*, 6th ed., edited by T. B. Poole, pp. 144–148. Longman, London.
80. Edwards, G. S., Fox, J. G., Policastro, P., Goff, U., Wolf, M. H., and Fine, D. H. (1979): Volatile nitrosamine contamination of laboratory animal diets. *Cancer Res.*, 39: 1857–1858.
81. Environmental Protection Agency. (1978): Proposed guidelines for registering pesticides in the U.S. hazard evaluation: humans and domestic animals. *Fed. Reg.*, 43: 37336–37403.
82. Erkert, H. G. and Grober, J. (1986): Direct modulation of activity and body temperature of owl monkeys (*Aotus lemurinus griseimembra*) by low light intensities. *Folia Primatol.*, 47:171–188.
83. Everett, R. (1984): Factors affecting spontaneous tumor incidence rats in mice: a literature review. *CRC Crit. Rev. Toxicol.*, 13:235–251.
84. Federation of European Laboratory Animal Science Associations. (1994): Recommendations for the health monitoring of mouse, rat, hamster, guinea pig and rabbit breeding colonies. *Lab. Anim.*, 28:1–12.
85. Federation of European Laboratory Animal Science Associations. (1999): Health monitoring of non-human primate colonies, supplement on health monitoring. *Lab. Anim.*, 33(Suppl. 1):S1-3–S1-18.
86. Ferguson, H. C. (1966): Effect of red cedar chip bedding on hexobarbital and pentobarbital sleep time. *J. Pharm. Sci.*, 55:1142–1143.
87. Festing, M. and Staats, J. (1973): Standardized nomenclature for inbred strains of rats: fourth listing. *Tranplantation*, 16:221–245.
88. Festing, M. F. W. (1993): *International Index of Laboratory Animals*, 6th ed. University of Leicester, U.K. (available from M.F.W. Festing, P.O. Box 301, Leicester LEI 7RE, U.K.).
89. Festing, M. F. W. and Greenhouse, D. D. (1992): Abbreviated list of inbred strains of rats. *Rat News Lett.*, 26:10–22.
90. Festing, M. F. W., Kondo, K., Loosli, R., Poiley, S. M., and Spiegel, A. (1972): International standardized nomenclature for outbred stocks of laboratory animals. *ICLA Bull.*, 30:4–17.
91. Fidler, I. J. (1977): Depression of macrophages in mice drinking hyperchlorinated water. *Nature*, 270:735–736
92. Fletcher, J. L. (1976): Influence of noise on animals. In: *Control of Animal House Environment: Laboratory Animal Handbooks*, Vol. 7, edited by T. McSheehy, pp. 51–62. Trevor Laboratory Animals, London.
93. Flynn, R. J. (1959): Studies of the etiology of ringtail of rats. *Proc. Anim. Care Panel*, 9:155–160.
94. Flynn, R. J. (1963): *Pseudomonas aeruginosa* infection and radiobiological research at Argonne National Laboratory: effects, diagnosis, epizootiology, control. *Lab. Anim. Care*, 13:25–35.
95. Food and Drug Administration. (1978): Nonclinical laboratory studies, good laboratory practice recommendations. *Fed. Reg.*, 43:59986–60025.
96. Ford, D. J. (1977): Effect of autoclaving and physical structure of diet on their utilization by mice. *Lab. Anim.*, 11:235–239.
97. Ford, D. J. (1977): Influence of diet pellet hardness and particle size on food utilization by mice, rats and hamsters. *Lab. Anim.*, 11:241–246.
98. Fouts, J. R. (1970): Some effects of insecticides on hepatic microsomal enzymes in various animal species. *Rev. Can. Biol.*, 29:377–389.

99. Fox, J. G. (1986): Interrelationships of disease and environmental variables in laboratory animals. In: *Safety Evaluation of Drugs and Chemicals*, edited by W. E. Lloyd, pp. 91–114. Hemisphere Publishing, Washington, D.C.
100. Fox, J. G., Aldrich, F. D., and Boylen, Jr., G. W. (1976): Lead in animal foods. *J. Toxicol. Environ. Health*, 1:461–467.
101. Fox, J. G., Newcomer, C. E., and Rozmiarek, H. (1984): Selected zoonoses and other health hazards. In: *Laboratory Animal Medicine*, edited by J. G. Fox, B. J. Cohen, and F. M. Loew, pp. 614–648. Academic Press, New York.
102. Fullerton, F. R., Greenman, D. L., and Kendall, D. C. (1982): Effects of storage conditions on nutritional qualities of semipurified (AIN-76) and natural ingredient (NIH-07) diets. *J. Nutr.*, 112:567–573.
103. Gamble, M. R. (1979): Fire alarms and oestrus in rats. *Lab. Anim.*, 10:93–104.
104. Gamble, M. R. and Clough, G. (1976): Ammonia build-up in animal boxes and its effect on a rat tracheal epithelium. *Lab. Anim.*, 10:161–163.
105. Garg, R. C. and Donahue, W. A. (1989): Pharmacologic profile of methoprene, an insect growth regulator, in cattle, dogs and cats. *J. Am. Vet. Med. Assoc.*, 194:410–412.
106. Garrard, G., Harrison, G. A., and Weiner, J. S. (1974): Reproduction and survival of mice at 23°C. *J. Reprod. Fertil.*, 37:287–298.
107. Gerber, W. F. and Anderson, T. A. (1967): Cardiac hypertrophy due to chance audiogenic stress in the rat and rabbit. *Comp. Biochem. Physiol.*, 21:237.
108. Gerber, W. F., Anderson, T. A., and Van Dyne, B. (1966): Physiologic responses of the albino rat to chronic noise stress. *Arch. Environ. Health*, 12:751–754.
109. Gibson, S. V., Besch-Williford, C., Raisbeck, M. F., Wagner, J. E., and McLaughlin, R. M. (1987): Organophosphate toxicity in rats associated with contaminated bedding. *Lab. Anim.*, 37:789–791.
110. Gordon, C. J. (1990): Thermal biology of the laboratory rat. *Physiol. Behav.*, 47:963–991.
111. Gordon, C. J. (1993): *Temperature Regulation in Laboratory Animals*. Cambridge University Press, New York.
112. Grant, L., Hopkinson, P., Jennings, G., and Jenner, F. A. (1971): Period of adjustment of rats used for experimental studies. *Nature*, 232:135.
113. Graves, R. G., (1990): Animal facilities: planning for flexibility. *Lab. Anim.*, 19:29–50.
114. Green, E. L. (1981): Breeding systems. In: *The Mouse in Biomedical Research*. Vol. I. *History, Genetics and Wild Mice*, edited by H. L. Foster, J. D. Small, and J. G. Fox, pp. 91–104. Academic Press, New York.
115. Greenman, D. L., Bryant, P., Kodell, R. L., and Sheldon, W. (1982): Influence of cage shelf level on retinal atrophy in mice. *Lab. Anim. Sci.*, 32:353–356.
116. Greenman, D. L., Kodell, R. L., and Sheldon, W. G. (1981): Association between cage shelf level and spontaneous and induced neoplasms in mice. *J. Natl. Cancer Inst.*, 73: 107–113.
117. Greenman, D. L., Oller, W. L., Littlefield, N. A., and Nelson, C. J. (1980): Commercial laboratory animal diets: toxicant and nutrient variability. *J. Toxicol. Environ. Health*, 6:235–246.
118. Halberg, F., Halberg, E., Barnum, C. P., and Bittner, J. J. (1959): Physiologic 24-hour periodicity in human beings and mice, the lighting regimen and daily routine. In: *Photoperiodism and Related Phenomena in Plants and Animals: Proceedings of a Conference on Photoperiodism*, edited by R. G. Withrow, Publ. No. 55, pp. 803–879. American Association for Advancement of Science, Washington, D.C.
119. Hall, J. E., White, W. J., and Lang, C. M. (1980): Acidification of drinking water: Its effects on selected biologic phenomena in male mice. *Lab. Anim. Sci.*, 30:643–651.
120. Hamm, T. E. (1986): *Complications of Viral and Mycoplasmal Infections in Rodents to Toxicology Research and Testing*. Hemisphere Publishing, Washington, D.C.
121. Hardy, J. D. (1961): Physiology of temperature regulation. *Physiol. Rev.*, 41:521–606.
122. Harkness, J. E., Wagner, J. E., Kusewitt, D. F., and Frisk, C. S. (1977): Weight loss and impaired reproduction in the hamster attributable to an unsuitable feeding apparatus. *Lab. Anim. Sci.*, 27:117–118.
123. Hart, R. W., Keenan, K., Turturro, A., Abdo, K. M., Leakey, J., and Lyn-Cook, B. (1995): Symposium overview: caloric restriction and toxicity. *Fund. Apply. Toxicol.*, 25:184–195.
124. Hart, R. W., Leakey, J., Duffy, P. H., Feuers, R. J., and Turturro, A. (1996): The effects of dietary restriction on drug testing and toxicity. *Exp. Toxicol. Pathol.*, 48:24–35.
125. Hartl, D. L. (1988): *A Primer of Population Genetics*, 2nd ed. Sinauer Associates, Sunderland, MA.
126. Hedrich, H. J. and Adams, M. (1990): *Genetic Monitoring of Inbred Strains of Rats: A Manual on Colony Management, Basic Monitoring Techniques, and Genetic Variants of the Laboratory Rat*. Gustav Fischer Verlag, Stuttgart.
127. Hermann, L. M., White, W. J., and Lang, C. M. (1982): Prolonged exposure to acid, chlorine, or tetracycline in the drinking water: effects on delayed-type hypersensitivity hemagglutination titers and reticuloendothelial clearance rates in mice. *Lab. Anim. Sci.*, 32:603–608.
128. Hessler, J. R. (1991): Facilities to support research. In *Handbook of Facilities Planning*. Vol. 2. *Laboratory Animal Facilities*, edited by T. Ruys, pp. 34–54. Van Nostranc Reinhold, New York.
129. Hessler, J. R. (1991): Single versus dual-corridor systems advantages, disadvantages, limitations, and alternatives fo effective contamination control. In: *Handbook of Facilitie. Planning*. Vol. 2. *Laboratory Animal Facilities*, edited by T. Ruys, pp. 59–66. Van Nostrand Reinhold, New York.
130. Hessler, J. R., and Moreland, A. F. (1984): Design anc management of animal facilities. In: *Laboratory Anima Medicine*, edited by J. G. Fox, B. J. Cohen, and F. M Loew, pp. 505–526. Academic Press, Orlando, FL.
131. Hite, M., Hanson, H. M., Bohidar, N. R., Conti, P. A., anc Mattis, P. A. (1977): Effect of cage size on patterns of activit and health of beagle dogs. *Lab. Anim. Sci.*, 27:60–64.
132. Hoffman, L. S. (1979): Preoperative and operative patien management. In: *Small Animal Surgery: An Atlas of Operative Technique*, edited by W. E. Wingfield and C. A. Rawlings, pp. 14–23. W.B. Saunders, Philadelphia, PA.
133. Homberger, F. R., Pataki, Z., and Thomann, P. E. (1993) Control of *Pseudomonas aeruginosa* infection in mice b chlorine treatment of drinking water. *Lab. Anim. Sci* 43:635–637.

134. Hughes, H. C., Compbell, S., and Kenney, C. (1989): The effects of cage size and pair housing on exercise of beagle dogs. *Lab. Anim. Sci.*, 39:302–305.
135. Hughes, H. C., and Lang, C. M. (1983): Control of pain in dogs and cats. In: *Animal Pain: Perception and Alleviation*, edited by R. L. Kitchell and H. H. Erickson, pp. 207–216. American Physiological Society, Bethesda, MD.
136. Hughes, H. C. and Reynolds, S. (1995): The use of computational fluid dynamics for modeling air flow design in a kennel facility. *Contemp. Topics*, 34:49–53.
137. Hughes, H. C., Reynolds, S., and Rodriguez, R. (1996): Designing animal rooms to optimize air flow using computational fluid dynamics. *Pharm. Eng.*, 16(2):46–65.
138. Hugh-Jones, M. E., Hubbert, W. T., and Hagstad, H. V. (1995): *Zoonoses: Recognition, Control and Prevention*. Iowa State University Press, Ames, IA.
139. Institute of Laboratory Animal Resources (1978): *Laboratory Animal Housing*, proceedings of a symposium held at Hunt Valley, MD, September 22–23, 1976. National Academy of Sciences, Washington, D.C.
140. Interagency Research Animal Committee (IRAC). (1985): U.S. government principles for utilization and care of vertebrate animals used in testing, research, and training. *Fed. Reg.*, 50(97).
141. IATA. (1995): *IATA Live Animal Regulations*. International Air Transport Association, Montreal, Quebec (available from IATA, 2000 Peel Street, Montreal, Quebec H3A, Canada).
142. International Committee on Standardized Genetic Nomenclature for Mice. (1994): Rules for nomenclature of inbred strains. *Mouse Genome*, 92:xxviii–xxxii.
143. International Committee on Standardized Genetic Nomenclature for Mice. (1994): Rules and guidelines for gene nomenclature. *Mouse Genome*, 92:viii–xxiii.
144. International Life Sciences Institute. (1995): *Dietary Restriction: Implications for the Design and Interpretation of Toxicity and Carcinogenicity Studies*, edited by R. W. Hart, D. A. Neumann, and R. T. Robertson. ILSI Press, Washington, D.C.
145. Iturrian, W. B. (1971): Effect of noise in the animal house on experimental seizures and growth of weanling mice. In: *Defining the Laboratory Animal: Proc. of the IVth Int. Symp. on Laboratory Animals*, pp. 332–352. National Academy of Sciences, Washington, D.C.
146. Jacobs, B. B. and Dieter, D. K. (1978): Spontaneous hepatomas in mice inbred from Ha:ICR Swiss stock: effects of sex, cedar shavings in bedding, and immunization with fetal liver or hepatoma cells. *J. Natl. Cancer Inst.*, 61:1531–1534.
147. Jacoby, R. O. and Lindsey, J. R. (1997): Health care for research animals is essential and affordable. *FASEB J.*, 11:609–614.
148. Jelinek, V. (1971): The influence of the condition of the laboratory animals employed on the experimental results. In: *Defining the Laboratory Animal*, pp. 110–120. National Academy of Sciences, Washington, D.C.
149. Jones, D. M. (1977): The occurrence of dieldrin in sawdust used as bedding material. *Lab. Anim.*, 11:137.
150. Jose, D. G. and Good, R. A. (1973): Quantitative effects of nutritional protein and caloric deficiency on immune responses to tumors in mice. *Cancer Res.*, 33:807–812.
151. Kagan, K. G. (1992): Care and sterilization of surgical equipment. *Vet. Tech.*, 13:65–70.
152. Kagan, K. G. (1992): Aseptic technique. *Vet. Tech.*, 13: 205–210.
153. Kaufman, J. E. (1984): *IES Lighting Handbook Reference Volume*. Illuminating Engineering Society, New York.
154. Kaufman, J. E. (1987): *IES Lighting Handbook Application Volume*. Illuminating Engineering Society, New York.
155. Keast, D. and Coales, M. F. (1967): Lymphocytopenia induced in a strain of laboratory mice by agents commonly used in treatment of ectoparasites. *Aust. J. Exp. Biol. Med. Sci.*, 45:645–650.
156. Keenan, K. P. et al. (1997): The effects of diet, overfeeding and moderate dietary restriction on Sprague–Dawley rat survival, disease and toxicology. *J. Nutr.*, 127:8518–8568.
157. Keenan, K. P., Laroque, P., Ballam, G. C., Soper, K. A., Dixit, R., Mattson, B. A., Adams, S. P., and Coleman, J. B. (1996): The effects of diet, *ad libitum* overfeeding, and moderate dietary restriction on the rodent bioassay: the uncontrolled variable in safety assessment. *Toxicol. Pathol.*, 24:757–768.
158. Keenan, K. P., Laroque, P., and Dixit, R. (1998): Need for dietary control by caloric restriction in rodent toxicology and carcinogenicity studies. *J. Toxicol. Environ. Health (Part B)*, 1:135–148.
159. Keenan, K. P., Smith, P. F., and Soper, K. A. (1994): Effect of dietary (caloric) restriction on aging, survival, pathobiology and toxicology. In: *Pathobiology of the Aging Rat*, Vol. 2, edited by W. Notter, D. L. Dungworth, and C. C. Capen, pp. 609–628. ILSI Press, Washington, D.C.
160. Keller, G. L. (1982): Physical euthanasia methods. *Lab. Anim.*, 11(4):20–26.
161. Keller, L. S. F., White, W. J., Snider, M. T., and Lang, C. M. (1989): An evaluation of intra-cage ventilation in three animal caging systems. *Lab. Anim. Sci.*, 39:237–242.
162. Keplinger, M. L., Lanier, G. E., and Deichmann, W. B. (1959): Effects of environmental temperature on the acute toxicity of a number of compounds in rats. *Toxicol. Appl. Pharmacol.*, 1:156–161.
163. Kimura, M. and Crow, J. F. (1963): On maximum avoidance of inbreeding. *Genet. Res.*, 4:399–415.
164. King, J. E. and Bennett, G. W. (1989): Comparative activity of fenoxycarb and hydroprene in sterilizing the German cockroach (Dictyoptera: Blattellidae). *J. Econ. Entomol.*, 82:833–838.
165. Knapka, J. J. (1983): Nutrition. In: *The Mouse in Biomedical Research*, Vol. III, edited by H. L. Foster, J. D. Small, and J. G. Fox, pp. 51–67. Academic Press, New York.
166. Knapka, J. J., Smith, K. P., and Judge, F. J. (1974): Effect of open and closed formula rations on the performance of three strains of laboratory mice. *Lab. Anim. Sci.*, 24: 480–487.
167. Knauff, D. R. (1987): Revised laboratory animal policy, *Lab. Anim.*, 16:11.
168. Kohn, D. F., Wixson, S. K., White, W. J., and Benson, G. J. (1997): *Anesthesia and Analgesia In Laboratory Animals*. Academic Press, New York.
169. Kraft, L. M. (1980): The manufacture, shipping and receiving and quality control of rodent bedding materials. *Lab. Anim. Sci.*, 30:366–376.

170. Krichevsky, D., Weber, M. M., and Klurfeld, D. M. (1984): Dietary fat versus caloric content in initiation and promotion of 7. 12-dimethylbenz(*a*)anthracene induced mammary tumorigenesis in rats. *Cancer Res.*, 44:3174–3177.
171. Kuntz, M. J. (1989): Cubicles: rational approach to specialized laboratory animal housing. *Anim. Technol.*, 40:203–209.
172. Kupp, Jr., R. P., Pinto, C. A., Rubin, L. F., and Griffin, H. E. (1989): Effects of ambient lighting on the eyes of rats. *Lab. Anim.*, 18:32–35,37.
173. Landi, M. S., Kreider, J. W., Lang, C. M., and Bullock, L. P. (1982): Effects of shipping on the immune function in mice. *Am. J. Vet. Res.*, 43:1654–1657.
174. Lang, C. M. (1983): Design and management of research facilities for mice. In: *The Mouse in Biomedical Research*, Vol. III, edited by H. L. Foster, J. D. Small, and J. G. Fox, pp. 37–50. Academic Press, New York.
175. Lanum, J. (1979): The damaging effects of light on the retina: empirical findings, theoretical and practical implications. *Surv. Ophthalmol.*, 22:221–249.
176. LaRegina, M. C. and Lonigro, J. (1988): Serologic screening for murine pathogens: basic concepts and guidelines. *Lab. Anim.*, 17:40–47.
177. LeBlanc, D. A. and Danforth, D. D. (1992): Substrate compatibility of animal cage wash products. *Contemp. Top. Lab. Anim. Sci.*, 31:13–16.
178. Leveille, G. A. and Hanson, R. W. (1966): Adaptive changes in enzyme activity and metabolic pathways in adipose tissue from meal-fed rats. *J. Lipid Res.*, 7(1):7–46.
179. Lindsay, J. R. and Conner, M. W. (1978): Influences of cage sanitation frequency on intracage ammonia ($NH_3$) concentration and progression of murine respiratory mycoplasmosis in the rat. *Zentralbl. Bakteriol. Parasitenkd. Infektionskr. Hyg.*, 241:215–216.
180. Linton, A. H., Hugo, W. B., and Russell, A. D. (1987): *Disinfection in Veterinary and Farm Animal Practice*. Blackwell Scientific, Oxford.
181. Lumb, W. V. and Moreland, A. F. (1982): Chemical methods for euthanasia. *Lab. Anim.*, 11:29–35.
182. Maeda, H., Gleiser, C. A., Masoro, E. J., Murata, I., McMahan, C. A., and Yu, B. P. (1985): Nutritional influences on aging of Fischer 344 rats. II. Pathology. *J. Gerontol.*, 40:671–688.
183. Masaro, E. J., (1992): Aging and proliferative homeostasis: modulation by food restriction in rodents. *Lab. Anim. Sci.*, 42:132–137.
184. McDougall, P. T., Wolf, N. S., Stenback, W. A., and Trentin, J. J. (1967): Control of *Pseudomonas aeruginosa* in an experimental mouse colony. *Lab. Anim. Care*, 17:204–214.
185. McGarrity, G. J. and Coriell, L. L. (1973): Mass airflow cabinet for control of airborne infection of laboratory rodents. *Appl. Microbiol.*, 26:167–172.
186. McGarrity, G. J. and Coriell, L. L. (1976): Maintenance of axenic mice in open cages in mass air flow. *Lab. Anim. Sci.*, 26:746–750.
187. McPherson, C. W. (1963): Reaction of *Pseudomonas aeruginosa* and coliform bacteria in mouse drinking water following treatment with hydrochloric acid or chlorine. *Lab. Anim. Care*, 13:737–744.
188. McPherson, C. W. (1984): Laws, regulations, and policies affecting the use of laboratory animals. In: *Laboratory Animal Medicine*, edited by J. G. Fox, B. J. Cohen, and F. M. Loew, pp. 19–30. Academic Press, Orlando, FL.
189. Megna, V. A. (1984): Engineering needs and trends of a toxicology laboratory. *Concepts Toxicol.*, 1:118–137.
190. Meskin, L. H. and Shapiro, B. L. (1971): Teratogenic effect of air shipment on A/Jax mice. *J. Dent. Res.*, 50:169.
191. Middle Atlantic Reproduction and Teratology Association (MARTA) and Midwest Teratology Association (MTA). (1996): *Historical Control Data (1992–1994) for Developmental and Reproductive Toxicity Studies Using the CRL:CD®(SD)BR Rat*. Charles River Laboratories, Wilmington, MA.
192. Miller, P. L. and Nash, R. T. (1971): A further analysis of room air distribution performance. *ASHRAE Trans.*, 77: 205–215.
193. Miller, P. L. and Nash, R. T. (1979): Analysis, evaluation and comparison of room air distribution performance: a summary. *ASHRAE Trans.*, 78:235–242.
194. Miller, P. L. and Nevins, R. G. (1969): Room air distribution with an air distributing ceiling. Part II. *ASHRAE Trans.*, 75:118–131.
195. Moore, B. J. (1987): The California diet: an inappropriate tool for studies of thermogenesis. *J. Nutr.*, 117:227–231.
196. Mulder, J. B. (1971): Animal behavior and electromagnetic energy waves. *Lab. Anim. Sci.*, 21:389–393.
197. Mulligan, S. R. et al. (1993): Sound levels in rooms housing laboratory animals: an uncontrolled daily variable. *Physiol. Behav.*, 53:1067–1076.
198. Murakami, H. (1971): Differences between internal and external environments of the mouse cage. *Lab. Anim. Sci.*, 21:680–684.
199. Murata, M. and Takigawa, H. (1989): Teratogenic effects of noise in mice. *J. Sound Vibration*, 132:11–18.
200. Nair, V. and Casper, R. (1969): The influence of light on daily rhythm in hepatic drug metabolizing enzymes in rat. *Life Sci.*, 8(Part I):1291–1298.
201. National Aeronautics and Space Administration (NASA). (1988): Summary of conclusions reached in workshops and recommendations for lighting animal housing modules used in microgravity related projects. In: *Lighting Requirements in Microgravity: Rodent and Nonhuman Primates: NASA Technical Memorandum 101077*, edited by D. C. Holley, C. M. Winget, and H. A. Leon, pp. 5–8. Ames Research Center, Moffet Field, CA.
202. National Institutes of Health, Office of the Director, Division of Engineering Services, F. Memarzadeh Principal Investigator. (1998): *Ventilation Design Handbook on Animal Research Facilities Using Static Microisolator*, Vol. I. National Institutes of Health, Bethesda, MD.
203. National Institutes of Health, Office of the Director, Division of Engineering Services, F. Memarzadeh Principal Investigator (1998): *Ventilation Design Handbook on Animal Research Facilities Using Static Microisolator*, Vol. II. National Institutes of Health, Bethesda, MD.
204. National Research Council. (1989): *Immunodeficient Rodents: A Guide to their Immunobiology, Husbandry, and Use*. National Academy Press, Washington, D.C.

205. National Research Council. (1991): *Companion Guide to Infectious Diseases of Mice and Rats*. National Academy Press, Washington, D.C.
206. National Research Council. (1991): *Infectious Diseases of Mice and Rats*. National Academy Press, Washington, D.C.
207. National Research Council, Institute of Laboratory Animal Resources, Committee on Rat Nomenclature. (1992): Definition, nomenclature, and conservation of rats strains. *ILAR News*, 34:S1–S26.
208. National Research Council, Institute of Laboratory Animal Resources, Committee on Transgenic Nomenclature. (1992): Standardized nomenclature for transgenic animals. *ILAR News*, 34:45–52.
209. National Research Council. (1977): *Nutrient Requirements of Rabbits: A Report of the Committee on Animal Nutrition*. National Academy Press, Washington, D.C.
210. National Research Council. (1978): *Nutrient Requirements of Nonhuman Primates: A Report of the Committee on Animal Nutrition*. National Academy Press, Washington, D.C.
211. National Research Council. (1981): *Nutrient Requirements of Cold Water Fishes: A Report of the Committee on Animal Nutrition*. National Academy Press, Washington, D.C.
212. National Research Council. (1981): *Nutrient Requirements of Goats: A Report of the Committee on Animal Nutrition*. National Academy Press, Washington, D.C.
213. National Research Council. (1982): *Nutrient Requirements of Mink and Foxes: A Report of the Committee on Animal Nutrition*. National Academy Press, Washington, D.C.
214. National Research Council. (1983): *Nutrient Requirements of Warm Water Fishes and Shellfishes: A Report of the Committee on Animal Nutrition*. National Academy Press, Washington, D.C.
215. National Research Council. (1985): *Nutrient Requirements of Dogs: A Report of the Committee on Animal Nutrition*. National Academy Press, Washington, D.C.
216. National Research Council. (1985): *Nutrient Requirements of Sheep: A Report of the Committee on Animal Nutrition*. National Academy Press, Washington, D.C.
217. National Research Council. (1986): *Nutrient Requirements of Cats: A Report of the Committee on Animal Nutrition*. National Academy Press, Washington, D.C.
218. National Research Council. (1988): *Nutrient Requirements of Swine: A Report of the Committee on Animal Nutrition*. National Academy Press, Washington, D.C.
219. National Research Council. (1992): *Recognition and Alleviation of Pain and Distress in Laboratory Animals: A Report of the Institute of Laboratory Animals Resources Committee on Pain and Distress in Laboratory Animals*. National Academy Press, Washington, D.C.
220. National Research Council. (1994): *Nutrient Requirements of Poultry: A Report of the Committee on Animal Nutrition*. National Academy Press, Washington, D.C.
221. National Research Council. (1995): *Nutrient Requirements of Laboratory Animals: A Report of the Committee on Animal Nutrition*. National Academy Press, Washington, D.C.
222. National Research Council, Commission on Life Sciences, Institute of Laboratory Animal Resources. (1996): *Guide for the Care and Use of Laboratory Animals*. National Academy Press, Washington, D.C.
223. National Research Council, Commission of Life Sciences, Institute of Laboratory Animals Resources, Committee on Rodents. (1996): *Laboratory Animal Management: Rodents*. National Academy Press, Washington, D.C.
224. Nayfield, K. C. and Besch, E. L. (1981): Comparative responses of rabbits and cats to elevated noise. *Lab. Anim. Sci*., 31:386–390.
225. Newall, G. W. (1980): The quality, treatment and monitoring of water for laboratory rodents. *Lab. Anim. Sci*., 30: 377–384.
226. Newberne, P. M. (1975): Diet: the neglected experimental variable. *Lab. Anim*., 4:20–24.
227. Newberne, P. M. (1975): Influence on pharmacological experiments of chemicals and other factors in diets of laboratory animals. *Fed. Proc*., 34:209–218.
228. Newberne, P. M. and Fox, J. G. (1980): Nutritional adequacy and quality control of rodent diets. *Lab. Anim. Sci*., 30:352–365.
229. Newberne, P. M. and McConnell, R. G. (1979): Nutrition of the Syrian golden hamster. *Prog. Exp. Tumor Res*., 24: 127–138.
230. Newberne, P. M. and McConnell, R. G. (1980): Dietary nutrients and contaminants in laboratory animal experimentation. *J. Environ. Pathol. Toxicol*., 4:105–122.
231. Newberne, P. M. and Rogers, A. E., (1973): Rat colon carcinomas associated with aflatoxin in marginal vitamin A. *J. Natl. Cancer Inst*., 50:439–448.
232. Newberne, P. M., Roger, A. E., and Wogan, G. N. (1968): Hepatorenal lesions in rats fed a low lipotrope diet and exposed to aflatoxin. *J. Nutr*., 94:331–343.
233. Newbold, J. A., Chapin, L. T., Zinn, S. A., and Tucker, H. A. (1991): Effects of photoperiod on mammary development and concentration of hormones in serum of pregnant dairy heifers. *J. Dairy Sci*., 74:100–108.
234. Newton, W. M. (1978): Environmental impact on laboratory animals. *Adv. Vet. Sci. Comp. Med*., 22:1–28.
235. Njaa, L. R., Utne, F., and Braekkan, O. R. (1957): Effect of relative humidity on rat breeding and ringtail. *Nature*, 180:290–291.
236. Noell, W. K. and Albrecht, R. (1971): Irreversible effects of visible light on the retina: role of vitamin A. *Science*, 172:76.
237. Nolen, G. A. and Alexander, J. C. (1966): Effects of diet and type of nesting material on the reproduction and lactation of the rat. *Lab. Anim. Care*, 16:327–336.
238. Nomura, T., Esaki, K., and Tomita, T. (1984): *ICLAS Manual for Genetic Monitoring of Inbred Mice*. University of Tokyo Press, Tokyo.
239. Noris, M. L. and Adams, C. E. (1976): Incidence of pup mortality in the rat with particular reference to nesting material, maternal age and parity. *Lab. Anim*., 10:165–169.
240. O'Steen, W. K. (1980): Hormonal influences in retinal photodamage. In: *The Effects of Constant Light on Visual Processes*, edited by T. P. Williams and B. N. Baker, pp. 29–49. Plenum Press, New York.
241. Ogle, C. W. and Lockett, M. F. (1968): The urinary changes induced in rats by high pitched sound (20 kcyc/sec). *J. Endocrinol*., 42:253–260.
242. Oller, W. L., Greenman, D. L., and Suber, R. (1985): Quality changes in animal feed resulting from extended storage. *Lab. Anim. Sci*., 35:646–650.

243. Orlans, F. B., Simmonds, R. C., and Dodds, W. J. (1987): Consensus recommendations on effective institutional animal care and use committees. *Lab. Anim. Sci.*, 37(special issue):11–13.

244. Party, E. and Wilkerson, A. (1991): Implications of new medical waste regulations on laboratory animal research. *Lab. Anim.*, 20(8):28–36.

245. Pekrul, D. (1991): Noise control. In: *Handbook of Facilities Planning*. Vol. 2. *Laboratory Animal Facilities*, edited by T. Ruys, pp. 166–173. Van Nostrand Reinhold, New York.

246. Pennycuik, P. R. (1967): A comparison of the effects of a range of high environmental temperatures and of two different periods of acclimation on the reproductive performances on male and female mice. *Aust. J. Exp. Biol. Med. Sci.*, 45:527–532.

247. Percy, D. H. and Barthold, S. W. (1993): *Pathology of Laboratory Rodents and Rabbits*. Iowa State University Press, Ames, IA.

248. Pereira, L. J., Lee, G. M., and Wade, K. J. (1990): The effect of surgical handwashing routines on the microbial counts of operating room nurses. *Am. J. Infect. Control*, 18:354–364.

249. Peterson, E. A. (1980): Noise and laboratory animals. *Lab. Anim. Sci.*, 30:422–436.

250. Peterson, E. A., Augenstein, J. S., Tanis, D. C., and Augenstein, D. G. (1981): Noise raises blood pressure without impairing auditory sensitivity. *Science*, 211:1450–1452.

251. Pfaff, J. (1974): Noise as an environmental problem in the animal house. *Lab. Anim.*, 8:347–354.

252. Pfaff, J. and Stecker, M. (1976): Loudness levels and frequency content of noise in the animal house. *Lab. Anim.*, 10:111–117.

253. Pick, J. R. and Little, J. M. (1965): Effect of type of bedding material on thresholds of pentylenetetrazol convulsions in mice. *Lab. Anim. Care*, 15:29–33.

254. Plank, S. J. and Irwin, R. (1966): Infertility of guinea pigs on sawdust bedding. *Lab. Anim. Care*, 16:9–11.

255. Poche, Jr., L. B., Stockwell, C. W., and Ades, H. W. (1969): Cochlear hair cell damage in guinea pigs after exposure to impulse noise. *J. Accoust. Soc. Am.*, 46:947–951.

256. Podberscek, A. L., Blackshaw, J. K., and Beattie, A. W. (1991): The effects of repeated handling by familiar and unfamiliar people on rabbits in individual cages and group pens. *Appl. Anim. Behav. Sci.*, 28:365–373.

257. Poiley, S. M. (1960): A systematic method of breeder rotation for non-inbred laboratory animal colonies. *Proc. Anim. Care Panel*, 10:159–166.

258. Poiley, S. M. (1974): Housing requirements: general consideration. In: *Handbook of Laboratory Animal Science*, Vol. I, edited by E. C. Melby, Jr., and H. H. Altman. CRC Press, Cleveland, OH.

259. Port, C. D. and Kaltenbach, J. P. (1969): The effect of corncob bedding on reproductivity and leucine incorporation in mice. *Lab. Anim. Care*, 19:46–49.

260. Porter, G. and Lane-Petter, W. (1965): The provision of sterile bedding and nesting materials with their effects on breeding mice. *J. Anim. Technol. Assoc.*, 16:5–8.

261. Potgieter, F. J. and Wilke, P. I. (1991): Laboratory animal bedding: a review of specifications and requirements. *J. S. Afr. Vet. Assoc.*, 62:143–146.

262. Prasad, S., Gatmaitan, B. R., and O'Connell, R. C. (1978): Effect of a conditioning method on general safety test in guinea pigs. *Lab. Anim. Sci.*, 28:591–593.

263. Prychodko, H. (1958): Effect of aggregation of laboratory mice (*Mus cusculus*) on food intake at different temperatures. *Ecology*, 39:500.

264. Public Health Service. (1996): *Public Health Service Policy on Human Care and Use of Laboratory Animals*, PL99-158, Health Research Extension Act 1985. U.S. Department of Health and Human Services, Washington, D.C.

265. Pucak, G. J., Lee, C. S., and Zaino, A. S. (1977): Effects of prolonged high temperature on testicular development and fertility in the male rat. *Lab. Anim. Sci.*, 27:76–77.

266. Ralls, K. (1967): Auditory sensitivity in mice *Peromyscus and Mus musculus*. *Anim. Behav.*, 15:123–128.

267. Rao, G. N. and Knapka, J. J. (1987): Contaminant and nutrient concentrations of natural ingredient rat and mouse diet used in chemical toxicology studies. *Fundam. Appl. Toxicol.*, 9:329–338.

268. Reiter, R. J. (1973): Comparative effects of continual fighting and pinealectomy on the eyes, the Harderian glands and reproduction in pigmented and albino rats. *Comp. Biochem. Physiol.*, 44:503–509.

269. Reynolds, S. D. and Hughes, H. C. (1994): Design and optimization of air flow patterns. *Lab. Anim.*, 23:46–49.

270. Ritter, M. A. and Marmion, P. (1987): The exogenous sources and controls of microorganisms in the operating room. *Orthop. Nursing*, 7:23–28.

271. Robinson, Jr., W. G. and Kuwabara, T. (1976): Light-induced alterations of retinal pigment epithelium in black, albino, and beige mice. *Exp. Eye Res.*, 22:549–557.

272. Rogers, A. E. (1985): Factors influencing the results of animal experiments in toxicology. In: *Basic Toxicology: Fundamentals, Target Organs, and Risk Assessment*, edited by F. C. Lu, pp. 254–267. Hemisphere, Washington, D.C

273. Rose, R. J. (1990): Practical aspects of formulating research diets. *Lab. Anim.*, 19:47–49.

274. Ross, M. H. and Bras, G. (1971): Lasting influences of early caloric restriction on prevalence of neoplasma in the rat. *J. Natl. Cancer Inst.*, 47:1095–1113.

275. Ross, M. H. and Bras, G. (1973): Influence of protein under-and over-nutrition on spontaneous tumor prevalence in the rat. *J. Nutr.*, 103:944–963.

276. Ross, M. H. and Bras, G., and Ragbeer, N. S. (1970): Influence of protein and caloric intake upon spontaneous tumor incidence of the anterior pituitary gland and the rat *J. Nutr.*, 100:177–189.

277. Ross, M. H., Lustbader, E. D., and Bras, G. (1983): Body weight, dietary practices, and tumor susceptibility in the rat. *J. Natl. Cancer Inst.*, 71:1041–1046.

278. Russell, W. M. S. and Burch, R. L. (1959): *The Principles of Human Experimental Techniques*. Methuen & Co., London

279. Rutala, W. A. (1990): APIC guideline for selection and use of disinfectants. *Am. J. Infect. Control*, 18:99–117.

280. Ruys, T. (1991): Codes, regulations and standards. Appendix E. Comments on the federal animal welfare regulation dealing with dogs, cats and nonhuman primates (9CFR Part 3, Subpart A, Feb. 15, 1991). In: *Handbook of Facilities Planning*. Vol. 2. *Laboratory Animal Facilities*, edited by T. Ruys, pp. 398–405. Van Nostrand Reinhold, New York

281. Ruys, T. (1991): The effect of animal species and types of the design of animal facilities. In: *Handbook of Facilities Planning*. Vol. 2. *Laboratory Animal Facilities*, edited by T. Ruys, pp. 55–59. Van Nostrand Reinhold, New York.
282. Ruys, T. (1991): Waste. In: *Handbook of Facilities Planning*. Vol. 2. *Laboratory Animal Facilities*, edited by T. Ruys, pp. 241–244. Van Nostrand Reinhold, New York.
283. Sales, G. D. (1991): The effect of 22 kHz calls and artificial 38 kHz signals on activity in rats. *Behav. Processes*, 24:83–93.
284. Sales, G. D., Wilson, K. J., and Spencer, K. E. V. (1988): Environmental ultrasound in laboratories and animal houses: a possible cause for concern in the welfare and use of laboratory animals. *Lab. Anim.*, 22:369–375.
285. Saltarelli, D. G. and Coppola, C. P. (1979): Influenced of visible light on organ weights of mice. *Lab. Anim. Sci.*, 29:319–322.
286. Sanhouri, A. A., Jones, R. S., and Dobson, H. (1989): The effects of different types of transportation on plasma cortisol and testosterone concentrations in male goats. *Br. Vet. J.*, 145:446–450.
287. Sansone, E. G. and Losikoff, A. M. (1979): Potential contamination from feeding test chemicals in carcinogen bioassay research: evaluation of single- and double-corridor animal housing facilities. *Toxicol. Appl. Pharmacol.*, 50:115–121.
288. Sansone, E. G., Losikoff, A. M., and Pendleton, R. A. (1977): Potential hazards from feeding test chemicals in carcinogen bioassay research. *Toxicol. Appl. Pharmacol.*, 39:435–450.
289. Schnurrenberger, P. R. and Hubbert, P. R. (1981): *An Outline of Zoonoses*. Iowa State University Press, Ames, IA.
290. Schofield, J. C. (1994): Principles of aseptic technique. In: *Essentials for Animal Research: A Primer for Research Personnel*, edited by B. T. Bennett, M. J. Brown, and J. C. Schofield, pp. 59–77. National Agricultural Library, Washington, D.C.
291. Schonholtz, G. J. (1976): Maintenance of aseptic barriers in the conventional operating room. *J. Bone Joint Surg.*, 58A:439–445.
292. Schroeder, H. A., Balassa, J. J., and Vinton, Jr., W. H. (1965): Chromium, cadmium and lead in rats: effects on life span, tumors and tissue levels. *J. Nutr.*, 86:51–66.
293. Selwyn, M. R. and Shek, W. R. (1994): Sample sizes and frequency of testing for health monitoring in barrier rooms and isolators. *Contemp. Top.*, 33:56–60.
294. Semple-Rowland, S. L. and Dawson, W. W. (1987): Retinal cyclic light damage threshold for albino rats. *Lab. Anim. Sci.*, 37:289–298.
295. Serrano, L. J. (1971): Carbon dioxide and ammonia in mouse cages: effect of cage covers, population and activity. *Lab. Anim. Sci.*, 21:75–85.
296. Silverman, J. and Adams, J. D. (1983): *N*-Nitrosamines in laboratory animal feed and bedding. *Lab. Anim. Sci.*, 33:161–164.
297. Simmonds, R. C. (1991): Characteristics of laboratory animal facilities. In: *Handbook of Facilities Planning*. Vol. 2. *Laboratory Animal Facilities*, edited by T. Ruys, pp. 1–33. Van Nostrand Reinhold, New York.
298. Simmons, M. L., Robie, D. M., Jones, J. B., and Serrano, L. J. (1968): Effect of a filter cover on temperature and humidity in a mouse cage. *Lab. Anim.*, 2:113–120.
299. Soma, L. R. (1987): Assessment of animal pain in experimental animals. *Lab. Anim. Sci.*, 37:71–74.
300. Stoskopf, M. K. (1983): The physiological effects of psychological stress. *Zoo Biol.*, 2:179–190.
301. Swanson, M. C., Campbell, A. R., O'Hollaren, M. T., and Reed, C. E. (1990): Rate of ventilation, air filtration, and allergen product rate in determining concentrations of rat allergies in the air of animal quarters. *Am. Rev. Respir. Dis.*, 141:1578–1581.
302. Tannenbaum, A. (1942): The genesis and growth of tumors. II. Effect of caloric restrictions per se. *Cancer Res.*, 2:460–467.
303. Tannenbaum, A. (1942): The genesis and growth of tumors. III. Effects of a high-fat diet. *Cancer Res.*, 2: 468–475.
304. Tannenbaum, A. (1945): The dependence of tumor formation on the degree of caloric restriction. *Cancer Res.*, 5: 609–615.
305. Tannenbaum, A. (1945): The dependence of tumor formation on the composition of the calorie-restricted diet as well as on the degree of restriction. *Cancer Res.*, 5: 616–625.
306. Tannenbaum, A. (1959): Nutrition and cancer. In: *The Physiopathology of Cancer*, 2nd ed., edited by F. Homberger, pp. 517–562. Hoeber-Harper, New York.
307. Teelman, K. and Weihe, W. H. (1974): Microorganism counts and distribution patterns in air-conditioned animal laboratories. *Lab. Anim.*, 8:109.
308. The Animal Welfare Act of 1966, PL 89–544, as amended by Animal Welfare Act of 1970, PL 91–579, by the 1976 Amendments to the Animal Welfare Act, PL 94–297, and by the 1985 Food Security Act, PL 99–198.
309. Thigpen, J. E., Lebetkin, E. H., Dawes, M. L., Clark, J. L., Langely, C. L., Amyx, H. L., and Crawford, D. (1989): A standard procedure for measuring rodent bedding particle size and dust content. *Lab. Anim. Sci.*, 39:60–62.
310. Thigpen, J. E., Lebetkin, E. H., Dawes, M. L. et al. (1989): The use of dirty bedding for detection of murine pathogens in sentinel mice. *Lab. Anim. Sci.*, 39:324–327.
311. Thompson, R. (1971): The water consumption and drinking habits of a few species and strains of laboratory animals. *J. Inst. Anim. Technol.*, 22:29–36.
312. Torronen, R., Pelkonen, K., and Karenlampi, S. (1989): Enzyme-inducing and cytotoxic effects of wood-based materials used as bedding for laboratory animals: comparison by a cell culture study. *Life Sci.*, 45:559–565.
313. Toth, L. A. and January, B. (1990): Physiological stabilization of rabbits after shipping. *Lab. Anim. Sci.*, 40: 384–387.
314. Totton, M. (1958): Ringtail in new-born Norway rats: a study of the effect of environmental temperature and humidity on incidence. *J. Hyg.*, 56:190–196.
315. Trexler, P. C. (1987): Animals of defined microbiological status: animal production and breeding methods. In: *The UFAW Handbook on the Care and Management of Laboratory Animals*, 6th ed., edited by T. B. Poole, pp. 85–98. Longman, London.

316. Tucker, H. A., Peticlere, D., and Zinn, S. A. (1984): The influence of photoperiod on body weight gain, body composition, nutrient intake and hormone secretion. *J. Anim. Sci.*, 59:1610–1620.
317. Tucker, M. J. (1979): The effect of long-term food restriction on tumours in rodents. *Int. J. Cancer*, 23:803–807.
318. Tuli, J. S., Smith, J. A., and Morton, D. B. (1995): Stress measurements in mice after transportation. *Lab. Anim.*, 29:132–138.
319. Turnbull, G. J., Lee, P. N., and Roe, F. J. C. (1985): Relationship of body-weight gain to longevity and to risk of development of nephropathy and neoplasia in Sprague–Dawley rats. *Food Chem. Toxicol.*, 23:355–361.
320. Vesell, E. S. (1967): Induction of drug-metabolizing enzymes in liver microsomes of mice and rats by softwood bedding. *Science*, 157:1057–1058.
321. Vesell, E. S., Lang, C. M., White, W. J., Passananti, G. T., Hill, R. N., Clemens, T. L., Liu, D. K., and Johnson, W. D. (1976): Environmental and genetic factors affecting the response of laboratory animals to drugs. *Fed. Proc.*, 35:1125–1132.
322. Vesell, E. S., Lang, C. M., White, W. J., Passananti, G. T., and Tripp, S. L. (1973): Hepatic drug metabolism in rats: impairment in a dirty environment. *Science*, 179:896–897.
323. Vlahakis, G. (1977): Possible carcinogenic effects of cedar shavings in bedding of C3H-A$^{vyf}$ B mice. *J. Natl. Cancer Inst.*, 58:149–150.
324. Wade, A. E., Holl, J. E., Hilliard, C. C., Molton, E., and Greene, F. E. (1968): Alteration of drug metabolism in rats and mice by an environment of cedarwood. *Pharmacology*, 1:317–328.
325. Waggie, K., Kagiyama, N., Allen, A. M., and Nomura, T. (1994): *Manual of Microbiologic Monitoring of Laboratory Animals*, 2nd ed., NIH Publ. No. 94-2498. U.S. Department of Health and Human Services, Washington, D.C.
326. Wallace, M. E. (1976): Effect of stress due to deprivation and transport in different genotypes of house mouse. *Lab. Anim.*, 10:335–347.
327. Wardrip, C. L., Artwohl, J. E., and Bennett, B. T. (1994): A review of the role of temperature versus time in effective cage sanitation program. *Contemp. Topics*, 33:66–68.
328. Warfield, D. (1973): The study of hearing in animals. In: *Methods of Animal Experimentation*, Vol. IV, edited by W. Gay, pp. 43–143. Academic Press, London.
329. Wassermann, M., Wassermann, D., Gershon, Z., and Zellermayer, L. (1969): Effects of organochlorine insecticides on body defense systems. *Ann. N.Y. Acad. Sci.*, 160:393–401.
330. Wattenberg, L. W. (1975): Effects of dietary constituents on the metabolism of chemical carcinogens. *Cancer Res.*, 35:3326–3331.
331. Wax, T. M. (1977): Effects of age, strain, and illumination intensity on activity and self-selection of light-dark schedules in mice. *J. Comp. Physiol. Psychol.*, 91:51–62.
332. Waynforth, H. B. (1980): *Experimental and Surgical Techniques in the Rat*. Academic Press, London.
333. Waynforth, H. B. (1987): Standards of surgery for experimental animals. In: *Laboratory Animals: An Introduction for New Experiments*, edited by A. A. Tuffery, pp. 311–312. Wiley-Interscience, Chichester.
334. Weichbrod, R. H., Cisar, C. F., Miller, J. G., Simmonds, R. C., Alvares, A. P., and Ueng, T. H. (1988): Effects of cage beddings on microsomal oxidative enzymes in rat liver. *Lab. Anim. Sci.*, 38:296–298.
335. Weichbrod, R. H., Hall, J. E., Simmonds, R. C., and Cisar, C. F. (1986): Selecting bedding material. *Lab. Anim.*, 15(6):25–29.
336. Weibe, W. H. (1965): Temperature and humidity climatograms for rats and mice. *Lab. Anim. Care*, 15:18–28.
337. Weibe, W. H. (1973): The effect of temperature on the action of drugs. *Annu. Rev. Pharmacol.*, 13:409–425.
338. Weibe, W. H. (1976): The effect of light on animals. In: *Control of the Animal House Environment: Laboratory Animal Handbooks*, Vol. 7, edited by T. McSheehy, pp. 63–76. Trevor Laboratory Animals, London.
339. Weibe, W. H., Schidlow, J., and Strittmatter, J. (1969): The effect of light intensity on the breeding and development of rats and golden hamsters. *Int. J. Biometerol.*, 13:69–79.
340. Weindruch, R. and Walfor, R. L. (1988): *The Retardation of Aging and Disease by Dietary Restriction*. Charles C Thomas, Springfield, IL.
341. Weis, I., Stotzer, H., and Seitz, R. (1974): Age- and light-dependent changes in the rat eye. *Vichows Arch. A Pathol. Anat. Histopathol.*, 362:145–156.
342. Wesibroth, S. H., Peters, R., Riley, L. K., and Shek, W (1998): Microbiological assessment of laboratory rats and mice. *ILAR J.*, 39:272–290.
343. White, W. J. (1990): The effect of cage space and environmental factors. In: *Guidelines for the Well-being of Rodents in Research*, edited by H. N. Guttman, pp. 29–45. Scientists Center for Animal Welfare, Bethesda, MD.
344. White, W. J., Anderson, L. C., Geistfeld, J., and Martin D. C. (1998): Current strategies for controlling/eliminating opportunistic microorganisms. *ILAR J.*, 39:291–305.
345. White, W. J. and Blum, J. R. (1997): Design of surgica suites and postsurgical care units. In: *Anesthesia and Anal gesia in Laboratory Animals*, edited by D. F. Kohn, S. K Wixson, W. J. White, and G. J. Benson, pp. 149–163 Academic Press, New York.
346. White, W. J. and Field, K. J. (1987): Anesthesia and sur gery of laboratory animals. *Vet. Clin. North Am.*, 17 989–1017.
347. White, W. J., Hughes, H. C., Singh, S. B., and Lang, C M. (1983): Evaluation of a cubicle containment system i preventing gaseous and particulate airborne cross-contam ination. *Lab. Anim. Sci.*, 33:571–576.
348. Whyte, W. (1988): The role of clothing and drapes in th operating room. *J. Hosp. Infect.*, 11(Suppl. C):2–17.
349. Whyte, W. and Shaw, B. H. (1974): The effect of obstruc tions and thermals in laminar-flow systems. *J. Hygiene* 72: 415–423.
350. Williams, G. M. (1984): The significance of environmenta chemicals as modifying factors in toxicity studies. In: *Con cepts in Toxicology*, Vol. I. *Toxicology Laboratory Desig and Management for the 80s and Beyond*, edited by A. S Tegeris, pp. 14–19. S. Kargar, Basel.
351. Wise, A. (1982): Interaction of diet and toxicity: the futur role of purified diet in toxicological research. *Arch. Tox col.*, 50:287–299.

352. Wise, A. (1980): The variability of dietary fiber in laboratory animal diets and its relevance to the control of experimental conditions. *Cosmet. Toxicol.*, 18:643–648.
353. Wise, A. and Gilbert, D. J. (1981): Variation of minerals and trace elements in laboratory animal diets. *Lab. Anim.*, 15:299–303.
354. Woods, J. E. (1975): Influence of room air distribution on animal cage environments. *ASHRAE Trans.*, 81:559–571.
355. Woods, J. E., Nevins, R. G., and Besch, E. L. (1975): Experimental evaluation of heat and moisture in metal dog cage environments. *Lab. Anim. Sci.*, 25:425–433.
356. Wostman, B. S. (1975): Nutrition and metabolism of the germfree mammal. *World Rev. Nutr. Diet*, 22:40–92.
357. Yu, B. P. (1990): Food restriction research: past and present status. *Rev. Biol. Res. Aging*, 4:349–371.
358. Zondek, B. and Tamari, I. (1964): Effect of audiogenic stimulation on genital function and reproduction. III. Infertility induced by auditory stimuli prior to mating. *Acta Endocrinol.*, 45(Suppl. 90):227–234.
359. Bauman, V. (2005): Environmental enrichment for laboratory rodents and rabbits: requirement of rodents, rabbits, and research. *ILAR J.*, 46(2):162–170.
360. Bayne, K. A. L. (2005): Potential for unintended consequences of environmental enrichment for laboratory animals and research results. *ILAR J.*, 46(2):129–139.
361. Fillman-Holliday, D. and Landi, M. S. (2002): Animal care best practices for regulatory testing. *ILAR J.*, 43(Suppl.): 49–58.
362. Bayne, K. A. L. (2003): Environmental enrichment of nonhuman primates, dogs and rabbits used in toxicology studies. *Toxicol. Pathol.*, 31(Suppl.):132–137.
363. Morris T., Goulet, S., and Morton, D. (2002): The International Symposium on Regulatory Testing and Animal Welfare: recommendations on scientific practices for animal care in regulatory toxicology. *ILAR J.*, 43(Suppl.): S123–S125.
364. Turner, J. G., Parrish, J. L., Hughes, L. F., Tooth, L. A., and Caspary, D. M. (2005): Hearing in laboratory animals: strain differences and nonauditory effects of noise. *Comp. Med.*, 55(1):12–23.

# Notes

# 21 Validation and Regulatory Acceptance of New, Revised, and Alternative Toxicological Methods

*William S. Stokes and Leonard M. Schechtman*

## CONTENTS

## INTRODUCTION

Toxicological test methods are necessary to assess the hazard and safety of various substances such as medicines, consumer products, and industrial chemicals. Many of these methods have traditionally used animals as the test system; however, in recent years, there has been increasing interest in developing alternative methods that reduce or replace animal use and that refine animal use to lessen or eliminate pain and distress. For any new or revised test method to be used to meet regulatory testing requirements, including alternative methods, the method must first undergo adequate validation and then be determined to be acceptable by regulatory authorities. This chapter discusses the criteria and processes for validation and regulatory acceptance of new, revised, and alternative methods. In addition, examples of new alternative test methods that have been accepted by national and international authorities are reviewed.

## THE CONCEPT OF ANIMAL USE ALTERNATIVES IN TOXICOLOGY

The concept of animal use alternatives was first described in 1959 by Rex Burch and William Russell in their book *The Principles of Humane Experimental Technique* [1]. Commonly referred to as the "3Rs of alternatives," this concept involves *reducing* the number of animals needed for a specific study, *replacing* animals with nonanimal systems and approaches, and *refining* animal use to lessen or avoid pain and distress. In the 1980s, animal protection groups began to emphasize the need to identify and use alternative methods for animal testing. Industry responded with various initiatives that included support to establish the Center for Alternatives to Animal Testing (CAAT) at The Johns Hopkins University in 1981 [2]. Public concern and increased awareness about animal use contributed to the passage of new laws requiring consideration of alternative methods prior to the use of animals in the United States in 1985 [3,4] and in Europe in 1986 [5]. Additional laws in 1993 and 2000 directed the National Institutes of Health to conduct research on alternative methods, to develop and validate alternative methods for testing, and to establish a formal process for consideration of proposed alternative testing methods [6,7].

## Regulatory Requirements for Consideration of Alternative Methods

In the United States, Animal Welfare Act regulations implemented in 1989 require investigators to consider alternative methods prior to the use of animals whenever proposed procedures involve more than slight or momentary pain or distress [8]. Before animals can be used, the investigator must provide evidence of the sources used to determine if alternative methods to procedures that cause more than slight or momentary pain or distress are available. The investigator must document this search, and both the search for alternatives and the proposed animal use must be reviewed and approved by an institutional animal care and use committee (IACUC). Institutions using animals for research and testing must register with the U.S. Department of Agriculture (USDA) and are subject to periodic compliance inspections by the Animal Care Unit of the USDA Animal and Plant Health Inspection Service (APHIS).

Investigators subject to the provisions of the Public Health Service (PHS) Policy on the Humane Care and Use of Laboratory Animals must also consider refinement, reduction, and replacement alternatives prior to the use of animals [9]. These include organizations that receive funding from PHS agencies (e.g., NIH, FDA, CDC, ATSDR), as well as organizations that participate in the voluntary animal facility accreditation program of the Association for Assessment and Accreditation of Laboratory Animal Care International (AAALAC). The PHS Policy implements relevant provisions of the Health Research Extension Act of 1985 and requires that studies using animals comply with the U.S. Government Principles for the Utilization and Care of Vertebrate Animals Used in Testing, Research, and Training (Table 21.1) [9]. These principles effectively require incorporation of refinement, reduction, and replacement alternatives into animal studies to the extent that they are consistent with obtaining testing and research objectives.

## Refinement Alternatives

Toxicity testing often involves pain and distress as a result of direct or indirect tissue damage from the test article. Additional pain and distress may occur as severe toxicity, progresses to a lethal outcome as a result of the significant disruption of normal homeostatic mechanisms. The goals of refinement alternatives are to minimize or eliminate pain and distress and to enhance the well-being of animals used in testing and research. Refinements not only provide for improved animal welfare but also enhance the quality of experiments by reducing or eliminating pain and distress as an experimental variable [10,11].

Death has been used historically as an experimental endpoint in toxicity testing; however, considerable pain and distress may precede death. With recent changes to national and international testing guidelines, death is no longer a required endpoint for toxicity studies conducted for regulatory safety purposes. Toxicity testing regulations and guidelines now allow for humane euthanasia of moribund animals and animals that show evidence of severe pain and distress [12]. These include national and international test guidelines for acute oral toxicity conducted to provide an estimate of the oral $LD_{50}$ [13–16]. International guidance has also been developed for selecting appropriate endpoints for toxicity studies [17].

Refinement can be achieved in toxicity studies by identifying earlier, more humane endpoints that are predictive of traditional study endpoints that involve pain and distress [18]. Clinical signs, physiologic parameters, biochemical measurements, and other parameters can serve as potential earlier biomarkers of humane endpoints. Detailed data should be collected to confirm the validity of the earlier biomarker. When it has been determined that the earlier biomarker provides the same or better accuracy as the traditional biomarker, it can be proposed for acceptance by regulatory authorities.

The local lymph node assay (LLNA) is an example of an alternative test method where the use of an earlier mechanistic endpoint completely eliminates the pain and distress previously involved in determination of allergic contact dermatitis potential of chemicals [18–21]. In the traditional test method using the Buehler or guinea pig maximization test, the test requires observation for actual elicitation of allergic dermatitis manifested by redness, swelling, and pruritis. In contrast, the LLNA uses an earlier, more sensitive biomarker that avoids the need to evoke the potentially painful elicitation phase. This method is discussed in greater detail later in this chapter.

**TABLE 21.1**
**U.S. Government Principles for the Utilization and Care of Vertebrate Animals Used in Testing, Research, and Training**

The development of knowledge necessary for the improvement of the health and well-being of humans as well as other animals requires *in vivo* experimentation with a wide variety of animal species. Whenever U.S. government agencies develop requirements for testing, research, or training procedures involving the use of vertebrate animals, the following principles shall be considered, and whenever these agencies actually perform or sponsor such procedures the responsible institutional official shall ensure that these principles are adhered to:

I. The transportation, care, and use of animals should be in accordance with the Animal Welfare Act (7 U.S.C. 2131 et seq.) and other applicable federal laws, guidelines, and policies.

II. Procedures involving animals should be designed and performed with due consideration of their relevance to human or animal health, the advancement of knowledge, or the good of society.

III. The animals selected for a procedure should be of an appropriate species and quality and the minimum number required to obtain valid results. Methods such as mathematical models, computer simulation, and *in vitro* biological systems should be considered.

IV. Proper use of animals, including the avoidance or minimization of discomfort, distress, and pain when consistent with sound scientific practices, is imperative. Unless the contrary is established, investigators should consider that procedures that cause pain or distress in human beings might cause pain or distress in other animals.

V. Procedures with animals that may cause more than momentary or slight pain or distress should be performed with appropriate sedation, analgesia, or anesthesia. Surgical or other painful procedures should not be performed on anaesthetized animals paralyzed by chemical agents. (For guidance throughout these principles, the reader is referred to *The Guide for the Care and Use of Laboratory Animals* prepared by the Institute for Laboratory Animal Research, National Academy of Sciences.)

VI. Animals that would otherwise suffer severe or chronic pain or distress that cannot be relieved should be painlessly killed at the end of the procedure or, if appropriate, during the procedure.

VII. The living conditions of animals should be appropriate for their species and contribute to their health and comfort. Normally, the housing, feeding, and care of all animals used for biomedical purposes must be directed by a veterinarian or other scientist trained and experienced in the proper care, handling, and use of the species being maintained or studied. In any case, veterinary care shall be provided as indicated.

VIII. Investigators and other personnel shall be appropriately qualified and experienced for conducting procedures on living animals. Adequate arrangements shall be made for their in-service training, including the proper and humane care and use of laboratory animals.

IX. Where exceptions are required in relation to the provisions of these principles, the decisions should not rest with the investigators directly concerned but should be made, with due regard to principle II, by an appropriate review group such as an institutional animal care and use committee. Such exceptions should not be made solely for the purposes of teaching or demonstration.

### Reduction Alternatives

Reduction alternatives are approaches and methods that result in attainment of study objectives with fewer animals; for example, minimizing one or more experimental variables can often improve statistical power, allowing for fewer animals per group. Using inbred rodent strains is one way to reduce experimental variation associated with genetic differences found in outbred stocks. Optimal statistical designs of studies will also contribute to ensuring the use of the most appropriate number of animals. One approach to reduction is to periodically conduct a retrospective review of testing results to determine if the number of animals can be reduced without significantly affecting the outcome of the study. For example, six rabbits were routinely used to conduct an ocular irritation assay; however, a retrospective statistical evaluation determined that the number could be reduced to three in most situations [22]. Further reductions can be accomplished by testing one animal at a time sequentially and stopping i evidence indicates severe irritation or ocular corrosion i one animal [23]. The up-and-down procedure is an exam ple of how animal use for assessing acute oral toxicity ha been reduced by up to 80% by use of an innovative sta tistical approach and sequential animal testing [13,16,24 The acute toxic class method and the fixed dose procedu also provide for reduced animal use for acute oral toxicit studies [14,15,24].

### Replacement Alternatives

Replacement alternatives are those that use nonanim methods, such as cell, tissue, and organ cultures or no sentient phylogenetically lower species such as insect Nonanimal test methods for assessing dermal corrosivit potential have been approved [25–29]. When these tes are positive, no further animal testing is required whe the substance is classified and labeled as a corrosive.

*vitro* test methods are increasingly being used as components in an integrated approach to assess the safety or potential toxicity of various chemicals, medicines, and products [11,31]. Their development has been stimulated by advances in new technologies and an enhanced understanding of the molecular and cellular mechanisms of toxicity. Advances in cell and tissue culture methods and the development of genetically modified stable cell lines have contributed to improved *in vitro* model systems. New scientific tools such as toxicogenomics, proteomics, and metabonomics are facilitating the identification of more sensitive and earlier biomarkers of toxicity that will likely be incorporated into future *in vitro* and animal safety testing methods. Efforts are underway toward investigating the validation strategies that would be required to adopt such new technologies for regulatory decision-making purposes [32,33]. The number and diversity of *in vitro* test systems incorporating these sensitive biomarkers will undoubtedly expand greatly in the coming years.

### Tiered Testing Strategies

Tiered testing strategies utilize a step-wise assessment, after the addition of relevant information at each tier, to determine whether the available information is adequate to make a decision regarding a specific toxicity. If, based on a weight-of-evidence evaluation, the information is not sufficient, then the testing progresses to the next tier, where additional information is generated and considered. The testing progresses until sufficient information is available to make a decision. Utilization of tiered testing strategies can sometimes allow for toxicity hazard classifications to be made with fewer or no animals, or with less pain and distress when animals are required. Proposals for tiered testing strategies for ocular and dermal irritation and corrosion testing have been incorporated into the Globally Harmonized System for chemical hazard classification and labeling and as supplemental guidance for international test guidelines [34–35].

## THE CONCEPT OF VALIDATION FOR NEW AND ALTERNATIVE SAFETY EVALUATION METHODS

Prior to using data from new and alternative methods for regulatory safety assessment decisions, the test methods used to generate such data must be determined to be scientifically valid and acceptable for their proposed use [36,38]. Adequate validation is therefore a prerequisite for test methods to be considered for regulatory acceptance. Demonstration of scientific validity requires evidence of the relevance and reliability of a test method and is necessary to determine the usefulness and limitations of a test method for a specific intended purpose. Regulatory acceptance involves reviewing the results of validation studies to determine the extent to which a test method can be used to fulfill specific regulatory needs and requirements. This section reviews established criteria for validation.

### Test Method Validation Criteria

Validation is the scientific process by which the relevance and reliability of a test method are determined for a specific purpose [36,39]. *Relevance* is defined as the extent to which a test method correctly measures or predicts a biological or toxic effect of interest. Relevance incorporates consideration of the accuracy of a test method for a specific purpose and consideration of mechanistic and cross-species or other test system relationships. *Reliability* is an objective measure of the degree to which a test method can be performed reproducibly within and among laboratories over time. A test method is considered adequately validated when its performance characteristics have been adequately determined for a specific purpose.

Criteria that should be met for a new or revised test method to be considered adequately validated for regulatory risk assessment purposes have been developed by national and international authorities (Table 21.2) [36–38]. These criteria serve as principles that should be followed in the validation of new test methods and provide clarity as to the critical information that should be collected and provided to substantiate the validity of test methods to regulatory authorities. The extent to which these criteria are met will vary with the test method and its proposed use. Accordingly, there must be flexibility in assessing a test method given its intended purpose and the supporting database. Test methods can be designed and used for different purposes by different organizations and for different categories of substances. Accordingly, the determination by regulatory authorities as to whether a specific test method is adequately validated and useful for a specific purpose will be on a case-by-case basis. Regulatory acceptance of new test methods generally requires a determination that use of the information from the test method will provide for equivalent or improved protection of human health, animal health, or the environment, as appropriate for the proposed use. Further guidance on adequately addressing established validation criteria is provided in this section.

### Test Method Purpose and Regulatory Rationale

New proposed test methods should have a clearly stated regulatory rationale and a clearly defined specific proposed use. The proposed use should describe how a test method is to be used in the context of current or anticipated regulatory requirements, regulations, and guidelines. Regulatory authorities have developed guidance and numerous standardized test guidelines that can be used to meet regulatory safety and hazard assessment requirements for various

**TABLE 21.2**
**Validation Criteria**

For a new or revised test method to be considered validated for regulatory risk assessment purposes, it should generally meet the following criteria (the extent to which these criteria are met will vary with the method and its proposed use); however, there must be flexibility in assessing a method given its purpose and the supporting database:

The scientific and regulatory rationale for the test method, including a clear statement of its proposed use, should be available.

The relationship of the test method's endpoints to the biologic effect of interest must be described. Although the relationship may be mechanistic or correlative, tests with biologic relevance to the toxic process being evaluated are preferred.

A detailed protocol for the test method must be available and should include a description of the materials required, a description of what is measured and how it is measured, acceptable test performance criteria (e.g., positive and negative control responses), a description of how data will be analyzed, a list of the species for which the test results are applicable, and a description of the known limitations of the test, including a description of the classes of materials that the test can and cannot accurately assess.

The extent of within-test variability and the reproducibility of the test within and among laboratories must have been demonstrated. Data must be provided describing the level of intra- and interlaboratory reproducibility and how it varies over time. The degree to which biological variability affects this test reproducibility should be addressed.

The performance of the test method must have been demonstrated using reference chemicals or test agents representative of the types of substances to which the test method will be applied and should include both known positive and known negative agents. Unless it is hazardous to do so, chemicals or test agents should be tested under code to exclude bias.

Sufficient data should be provided to permit a comparison of the performance of a proposed substitute test with that of the test it is designed to replace. Performance should be evaluated in relation to existing relevant toxicity testing data and relevant toxicity information from the species of concern. Reference data from the comparable traditional test method should be available and of acceptable quality.

The limitations of the method must be described; for example, *in vitro* or other nonanimal test methods may not replicate all of the metabolic processes relevant to chemical toxicity that occur *in vivo*.

Ideally, all data supporting the validity of a test method should be obtained and reported in accordance with Good Laboratory Practices (GLPs). Aspects of data collection not performed according to GLPs must be fully described, along with their potential impact.

All data supporting the assessment of the validity of the test method must be available for review:

- Detailed protocols should be readily available and in the public domain.
- The methods and results should be published or submitted for publication in an independent, peer-reviewed publication.
- The methodology and results should have been subjected to independent scientific review.

Because tests can be designed and used for different purposes by different organizations and for different categories of substances, the determination of whether a specific test method is considered by an agency to be useful for a specific purpose must be made on a case-by-case basis. Validation of a test method is a prerequisite for it to be considered for regulatory acceptance.

toxicity endpoints [36,40,41]. Data generated for substances using these standardized test method protocols can serve as the basis for comparing the performance of a new test method proposed to evaluate the same toxicity endpoint.

The specific purpose of test methods currently included or proposed for inclusion in regulations and guidelines can vary widely. For example, many methods serve as definitive test methods that provide sufficient information for regulatory hazard classification and labeling, while others may serve as screening tests, mechanistic adjunct tests, or components of a testing battery. A new test method may be proposed as a complete replacement for an existing test method or may be proposed to substitute for an existing test method in certain testing situations, such as for the evaluation of test articles in specific well-defined product or chemical classes or those with specific physical or chemical properties.

*Definitive test methods* are those that provide sufficient data to characterize a specific hazard potential of a substance for hazard classification and labeling purpose without further testing. Examples include specific anima tests for skin irritation, eye irritation, allergic contact dermatitis, acute oral toxicity, multigenerational reproductiv toxicity, and the rodent carcinogenicity bioassay.

*Screening test methods* are those that may in som situations allow for hazard decisions in a tiered testing strategy or that may provide information helpful in making decisions on prioritizing chemicals for more definitive testing. As an example, a test method could be proposed as a screening test in a tiered testing strategy wher positive results can be used to classify and label th hazard of a substance, and negative results woul undergo testing using the currently accepted testing procedure. Whenever a test method is proposed to be use as a screening test, the specific decisions that will b made with each possible test result must be clearl defined.

Several *in vitro* screening tests have been accepted for determining if a substance has the potential to cause dermal corrosion [25–29]. Positive results can be used to classify and label substances as corrosives, while negative results would undergo additional testing in animals to identify any false-negative corrosive substances and to determine the dermal irritation potential. The use of information from screening tests to meet regulatory requirements must take into consideration the precautionary principle and the need to avoid potential underclassification of the hazard.

*Mechanistic adjunct test methods* are those that provide data that add to or help interpret the results of other assays or that otherwise provide information useful for the hazard assessment process. An example is the estrogen receptor-binding assay [53]. A positive result in this assay indicates that a substance has the potential to bind to the estrogen receptor in an *in vitro* system; however, it does not definitively indicate that the substance will be active *in vivo* because it does not take into account absorption, distribution, metabolism, and excretion (ADME) factors. When considered in conjunction with other testing information, such as a positive rodent uterotrophic bioassay, a positive result in this *in vitro* assay contributes mechanistic information for a weight-of-evidence decision supporting the likelihood that the *in vivo* bioassay response resulted from an estrogen-active substance. When an adjunct test method is proposed to generate data for use in a weight-of-evidence decision, it is important to provide data that substantiate and quantitatively characterize the weight, or likelihood, that a toxic effect will be associated with the outcome from the mechanistic test.

A *testing battery* is a series of test methods that are generally performed at the same time or in close proximity to reach a decision on hazard potential. In such cases, the component test methods of the proposed battery must undergo validation as individual test methods. For the individual test methods proposed for inclusion in a test battery, it is essential that each individual test method validation study use the same reference substances or at least a sufficient number of the same substances to adequately evaluate the usefulness and limitations of the proposed test battery. This is necessary to allow calculation of the accuracy of each possible combination of component test methods and to identify the most accurate combination for a given classes of substances.

Test methods proposed to replace an existing definitive test method will require evidence from validation studies that the use of the proposed method will provide for a comparable or better level of protection than the currently used test method or approach. In some cases, there may be limitations of a new test method with regard to certain types of physical or chemical properties (e.g., solubility in an *in vitro* system) that do not allow for it to completely replace an existing test. In this case, it may be determined to be an adequate *substitute* for the existing test method for many but not all test substances or testing circumstances.

## Test Method Scientific Rationale and Relationship of the Test Method Endpoint to the Biological Effect of Interest

The scientific rationale for a new test method should always be provided. This should include the mechanistic basis and relationship of the biological model used in the test system compared to that for the species of interest for which the testing is being performed (e.g., humans for health-related testing). The extent to which the mechanisms and modes of action for the toxicity endpoint of interest are similar or different in the proposed test system compared to that in the species of interest must be considered. Other uncertainties regarding mechanisms and modes of action and their potential impact on the relevance of the test method must be discussed. The potential role and impact of *in vivo* ADME on the toxicity of interest must also be considered, as well as the extent to which each of these parameters is or is not addressed by the proposed test method. For an *in vitro* test system, the impact of any ADME limitations of the *in vitro* test system must be discussed. It is also important to consider what is known or not known about similarities and differences in responses between the target tissues in the species of interest, the surrogate species used in the currently accepted test method, and the cells or tissues of the proposed *in vitro* test system.

## Detailed Test Method Protocol

### Protocol

The outcome of a validation study should be a detailed standardized and transferable test method protocol that has been adequately evaluated to characterize its performance. The test method protocol should be sufficiently detailed that it can be reproduced in other appropriately equipped laboratories with trained personnel. Because most testing conducted for regulatory purposes must be conducted in accordance with national or international good laboratory practice (GLP) regulations [42–44], the test method protocol should be prepared so it can be used as the basis for a GLP-compliant study protocol in specific laboratories.

The test method protocol should provide a detailed description for all aspects of the proposed test method (Table 21.3) [38], including a description of all materials, equipment, and supplies. Detailed procedures for dose selection for animal studies or concentration selection for *in vitro* studies should be provided. Where

**TABLE 21.3**
**Selected *In Vitro* Test Method Protocol Components**

| |
|---|
| Biological systems, materials, equipment, reagents, and supplies |
| Concentration selection procedures: for example, defined limit concentration, range-finding studies, procedures for determining limit of solubility, highest noncytotoxic concentration |
| Test system endpoints measured |
| Duration of test article exposure, postexposure incubation |
| Positive, vehicle, negative, and benchmark control substances; basis for their selection |
| Acceptable response ranges for positive, vehicle, and negative control substances, including historical control data and basis for acceptable ranges |
| Decision criteria for interpreting the outcome of a test result, basis for the decision criteria for classifying a chemical, accuracy characteristics of the selected decision criteria |
| Information and data to be included in the study report |
| Standard data collection and submission forms |

appropriate, this should include procedures for dose-range finding studies and solubility testing to select appropriate solvents, as appropriate. Criteria should be provided for the highest concentration or dose that should be used. For *in vivo* studies, this may be a maximum tolerated dose, with carefully defined criteria, or a defined upper-limit dose. For *in vitro* methods, this may be a defined limit concentration (e.g., 1 millimolar), the highest noncytotoxic concentration, or the highest soluble concentration. The duration and basis for test substance exposure and postexposure incubation for *in vitro* systems should be provided. The nature of data to be collected and the methods and procedures for data collection must be specified.

## Positive, Vehicle, and Negative Controls

Nearly every toxicological test will have untreated controls that serve as the basis for detecting whether the test article produces an increased response above the control response. When a vehicle or solvent is used with the test article, a vehicle or solvent control should also be used. In addition, it may also be desirable or necessary to have concurrent positive and negative controls. For *in vitro* test systems, vehicle and positive controls should be designated and used for every test. For some *in vivo* tests, it may also be necessary to use positive controls. These are necessary to ensure that the test system is operating properly and capable of providing appropriate positive and negative responses. A positive control substance should normally be selected that is intermediate in the potential dynamic response range of the test system. For *in vitro* tests, an acceptable positive control response range should be developed for each laboratory. Test results are not normally considered acceptable if the positive control is outside of the established (historical) acceptable positive control range.

## Benchmark Controls

In some cases it may be desirable to include substances for which potential toxicity has previously been established in human, animal, and/or *in vitro* test systems. These substances, commonly referred to as benchmark controls, could include substances that are in the expected response range of the test articles or that have similar chemical structure or physical-chemical properties as the test articles [32]. Benchmark controls can be helpful in providing information about the relative toxicity of a test article compared to other well-characterized similar substances and can also be used to ensure that the test system is functioning properly in specific areas of the response range.

## Decision Criteria

For test methods that determine the hazard classification category of a test substance, the test method protocol will have to specify the decision criteria used to determine the classification category based on results from the test system. For methods that provide qualitative assessments of toxicity, these may be the criteria used to determine if a substance is positive, negative, or equivocal; for example for Corrositex®, a test method for determining the corrosivity category of substances, the corrosivity hazard category is based on the time that it takes for the substance to penetrate a biobarrier membrane [29]. A formula or algorithm that incorporates the decision criteria for a test outcome is often used to convert test method results into a prediction of the toxic effect [36,38]. Accordingly, decision criteria are sometimes referred to as a "prediction model." Test method decision criteria, or prediction models, should contain four elements: (1) a definition of the specific purpose of the test method, (2) specifications of all possible results that may be obtained when using th test method, (3) an algorithm that converts each study result into a prediction of the toxic effect of interest, and (4) specification of the accuracy associated with the selected decision criteria (i.e., sensitivity, specificity, false-positive and false-negative rates) [36,38]. Decision criteria should always be specified in the test method protocol. It is important to note that decision criteria and prediction models for the final proposed test method protocol may have to be revised following a validation study to obtain a sensitivity and specificity appropriate for the intended regulatory use. Such modifications should seek to minimize false-negative and false-positive rates appropriate for the toxicity endpoint being assessed.

## Test System

The basis for selection of the test system should be described in the test method protocol and should include a detailed description and specifications for animals, cells, tissues, or other critical components used. Procedures for ensuring the correct identity and critical parameters of animal stocks and strains, cells, and tissues should be provided in the test method protocol, including the basis for determining that the components are acceptable [36,38].

## Evaluation of Test Method Reliability

Test method reliability involves determining the intralaboratory repeatability and intra- and interlaboratory reproducibility of a test method [36]. *Intralaboratory repeatability* of a test method is the closeness of agreement between test results obtained in a single laboratory when the test method is used to evaluate the same substance under identical conditions at the same time. These data provide an estimate of the variation that is inherent in the biological responses of a test system and the study conditions in a single laboratory. *Intralaboratory reproducibility* is the determination of the extent that qualified personnel within the same laboratory can successfully replicate results using a specific test method protocol at different times. Acceptable intralaboratory reproducibility should be achieved before evaluating interlaboratory reproducibility.

*Interlaboratory reproducibility* is a measure of the extent to which different qualified laboratories using the same test method protocol and the same substances can produce qualitatively and quantitatively similar results. This assessment is necessary to determine if the test method protocol contains sufficient procedural detail that will result in qualified laboratories obtaining similar and consistent results and indicates the extent to which a test method can be transferred successfully among laboratories. Interlaboratory reproducibility should be assessed using the same or a subset of the reference substances used to assess test method accuracy. Most importantly, reference substances representing the full range of possible test outcomes, chemical and physical properties, and mechanisms of toxicity should be evaluated. This can sometimes be accomplished with a smaller number of reference chemicals than used to characterize accuracy; however, there should be a compelling scientific and statistical rationale for using a reduced number of substances for this determination. Interlaboratory reproducibility has typically been assessed using three qualified laboratories. The impact of the results of test method reliability assessments on laboratory transferability and erroneous results should always be considered. Situations where evidence indicates poor reproducibility, such as for certain chemical classes, physicochemical properties, or specific areas of response, should be identified as potential limitations of the proposed test method.

## Reference Substances

Reference substances are those for which the response of the substance is known in the existing reference test method; they are used to characterize the accuracy and reproducibility of the proposed test method [36,38]. Test method reliability and accuracy must be evaluated using reference substances representative of the types of substances to which the test method will be applied and should include both known positive and negative substances. The selection of appropriate reference substances is a critical aspect of validation studies. The ideal reference chemicals are those for which high-quality testing data are available from both the reference test method and from the species of interest (e.g., humans); however, adequate human testing data are rarely available for ethical reasons. Exceptions are for substances and endpoints that do not result in severe or irreversible effects, such as allergic contact dermatitis and mild to moderate dermal irritation. These studies are usually limited to the premarketing assessment of products that are intended for human contact, such as cosmetics and some mild consumer products. For test methods proposed for predicting human health effects, reference substances for which there are accidental human exposures and toxic effects should be considered.

The number and types of reference substances selected must adequately characterize the accuracy and reproducibility of a test method for its specific proposed use [36,38]. Reference chemicals should represent the range of chemical classes, product classes, and physical and chemical properties (e.g., pH, solubility, color, solids, liquids) for which the test method is expected or proposed to be applicable. Reference chemicals should also represent the range of expected responses proposed for the test method, including negatives and weak-to-strong positives. Reference chemicals and formulations should ideally be of known purity and composition and should be readily available from commercial sources. Formulations should provide detailed information on the type, purity, and percentage of

**TABLE 21.4**
**Two-by-Two (2×2) Table**

| | | New Test Outcome | | |
|---|---|---|---|---|
| | | **Positive** | **Negative** | **Total** |
| **Reference Test Classification** | **Positive** | a | c | a + c |
| | **Negative** | b | d | b + d |
| | **Total** | a + b | c + d | a + b + c + d |

The 2 × 2 table can be used for calculating accuracy (concordance) (a + d/a + b + c + d); negative predictivity (d/c + d); positive predictivity (a/a + b); prevalence (a + c/a + b + c + d); sensitivity (a/a + c); specificity (d/b + d); false positive rate (b/b + d); and false negative rate (c/a + c).

each ingredient. Unless justified, chemicals should not normally pose an extreme environmental or human health hazard, should not be prohibitively expensive, and should not involve exorbitant disposal costs.

## Evaluation of Test Method Accuracy

*Accuracy* reflects the closeness of agreement between results from a new proposed test method and reference values from a currently accepted test method. A two-by-two table can be used to calculate accuracy and the associated parameters (Table 21.4) [38]. *Sensitivity* is the proportion of all positive substances that are correctly classified as positive by the new test method. *Specificity* is the proportion of all negative substances that are correctly identified as negative substances in the new test method. The *false-positive rate* is the proportion of all negative (inactive) substances that are falsely identified as positive, and the *false-negative rate* is the proportion of all positive (active) substances that are falsely identified as negative.

Ideally, test methods should be highly accurate, have a high level of sensitivity and specificity, and have negligible false-positive and false-negative rates; however, in toxicity testing, this is rarely achievable. Accordingly, decision criteria for interpreting the outcome of a test method must be adjusted depending on the desired performance characteristics and the impact of an erroneous result; for example, from the perspective of protecting public health, it is most desirable to use decision criteria for a test method that provide for a high level of sensitivity and have no or minimal false negatives. This is because a false-negative result incorrectly indicates a lack of hazard or lower hazard than actually exists for a substance. The real hazard of the substance will subsequently not be indicated on packaging or in worker safety information. Products without a proper warning of their real hazard, such as skin and eye corrosives, could then result in human injury or disease to exposed persons. Conversely, a low level of specificity and high false-positive rate can result in overlabeling the true hazard of a substance. This has economic implications in that some hazards require more expensive packaging and shipping precautions. It is also desirable not to overlabel the hazard of a substance because this could lead to complacency in consumer and worker compliance with recommended exposure precautions if accidental exposures frequently do not result in adverse effects.

## Test Method Limitations

The limitations of a test method and test system must be described; for example, the extent to which an *in vitro* test system does not replicate all of the metabolic processes relevant to the *in vivo* toxicity for a specific toxicity endpoint should be discussed. Furthermore, limitations may be identified with regard to the ability of the test method to reliably and accurately detect the toxicity or biological activity of specific chemical classes and specific physical or chemical properties.

## Quality of Validation Data

Ideally, all data supporting the validity of a test method should be obtained and reported in accordance with national or international GLP regulations and guidelines [42–44]. Because nearly all safety testing for regulatory purposes must be accomplished in accordance with GLP requirements, it is logical that validation studies for a test method proposed for safety testing be carried out in accordance with those same GLP requirements. This will provide increased confidence in data quality and documentation as to the extent of laboratory adherence to the test method protocols under evaluation. GLPs provide a formal quality assurance system for data collected in the study. If validation studies are not conducted in accordance with GLPs, then aspects of data collection or auditing not performed according to GLPs should be documented. International guidance for the application of

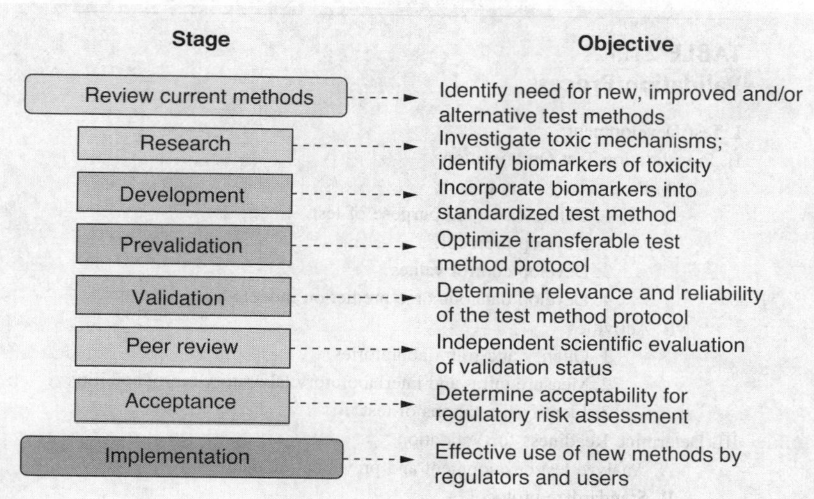


**FIGURE 21.1** Test method evolution process. (Adapted from Interagency Coordinating Committee on the Validation of Alternative Methods, *Guidelines for the Nomination and Submission of New, Revised, and Alternative Test Methods*, NIH Publ. No. 03-4508, National Institute of Environmental Health Sciences, Research Triangle Park, NC, 2003.)

GLPs to *in vitro* testing is available [30]. In any case, all laboratory notebooks, raw and transformed data, and all other relevant test-related information should be retained and available for audit if requested by the reviewing authorities.

### Availability of Validation Data

All data supporting the assessment of the validity of a proposed test method should be made available for review. This includes raw data collected from the test system, as well as transformed data that are derived from the raw data. All test method protocols used to generate data must also be available. Ideally, the test method protocols and results from validation studies should be subjected to independent scientific peer review and published or submitted for publication in an independent peer-reviewed publication. Additional information on scientific peer review is provided below in the section on the Interagency Coordinating Committee on the Validation of Alternative Methods (ICCVAM) [36,38].

## EVOLUTION AND VALIDATION OF A TEST METHOD

Validation of a test method is one of many stages involved in the evolution of a test method from concept to regulatory acceptance (Figure 21.1) [36]. These stages may begin with determination of the need for a new test method. New test methods are often sought that provide for improved prediction of adverse effects, that are more humane or do not use animals, or that involve less expense and time to conduct. Additional research may be needed to understand critical mechanisms and critical modes of action for a toxicity endpoint of interest and to identify potential biomarkers that can be included in a test method. The test method development stage involves incorporation and evaluation of one or more promising predictive biomarkers in a test system. This usually involves testing a limited number of substances with well-known toxicity to determine if the critical biomarker is capable of detecting the toxic effect. If so, then a decision may be made to initiate validation of the test system. Validation of a proposed test method is an iterative process that typically evolves through several phases (Table 21.5) [36].

### Prevalidation and Test Method Optimization

The early phases of validation are often referred to as prevalidation. The objectives are to first develop a standardized test method protocol and then to optimize the protocol to maximize accuracy and reliability. Careful planning is essential prior to the initiation of any validation study. The validation study plan must adequately address established validation criteria. The objectives of each phase of the validation study should be clearly defined and a validation study design selected that will adequately address the defined objectives. Modifications are often necessary during the prevalidation phases to reduce sources of intra- and interlaboratory variation and to optimize the accuracy of the test method to measure or predict the toxicity or biological activity of interest. The version of the test method protocol determined to be sufficiently accurate and reproducible during the last phase of prevalidation should be finalized for the formal validation phase. Because the objective of the formal validation phase is to

**TABLE 21.5**
**Validation Process**

I. Test Development
II. Prevalidation/Test Optimization
  A. Preliminary planning:
    1. Define basis and purpose of test.
    2. Develop protocol.
    3. Develop control values.
    4. Develop data/outcome prediction model.
  B. Activities:
    1. Qualify and train laboratories.
    2. Measure intra- and interlaboratory laboratory reproducibility.
    3. Identify limitations of test.
III. Determine Readiness for Validation
  A. Analyze test development and prevalidation data.
  B. Standardize protocol.
IV. Test Validation
  A. Form steering committee/ management team:
    1. Define purpose of validation study.
    2. Design study.
    3. Select participating laboratories.
    4. Establish management evaluation and oversight procedures.
  B. Pretest procedures:
    1. Implement data recordkeeping procedures.
    2. Select reference chemicals.
    3. Code and distribute reference chemicals.
  C. Test coded chemicals:
    1. Measure interlaboratory performance.
    2. Compile and evaluate data.
  D. Evaluate test:
    1. Analyze and summarize test results.
    2. Challenge data with prediction model.
    3. Conduct peer review of protocol and data.
    4. Accept, revise, or reject model.
V. Submission of Test for Regulatory Approval
  A. Prepare report.
  B. Make supporting data available.
  C. Prepare results for publication.

determine the reproducibility and accuracy of this optimized and standardized test method protocol, no changes should be made to the protocol during the final phase of validation.

## Coding and Distribution of Test Substance

Test substances should normally be coded during both the prevalidation and formal validation phases to exclude bias. This can be accomplished by the use of a chemical distribution facility not directly associated with the participating laboratories. Each substance should be uniquely coded for each different laboratory so the identity is not readily available to laboratory personnel; however, provisions must be made to ensure that the designated safety officer in each laboratory has the Safety Data Sheets available for each coded substance in case the need arises t[illegible] access the information. One approach is to provide par[illegible]ticipating laboratory testing staff with sealed package[illegible] containing all relevant health and safety data, includin[illegible] instructions for accidental exposures or other laborator[illegible] accidents. The envelopes can then be returned to the stud[illegible] sponsor at the end of study, with an explanation for an[illegible] opened envelopes. Laboratories will need to ensure tha[illegible] all environmental, safety, handling, and disposal proce[illegible]dures are in compliance with regulatory requirements.

## Selection of Laboratories for Validation Studies

Laboratories selected for validation studies should be ade[illegible]quately equipped and have personnel with appropriat[illegible] training; for example, validation of an *in vitro* test metho[illegible]

**TABLE 21.6**
**Regulatory Acceptance Criteria**

Validated methods are not automatically accepted by regulatory agencies; they need to fit into the regulatory structure. Flexibility is essential in determining the acceptability of methods to ensure that appropriate scientific information is considered in regulatory risk assessment. A test method proposed for regulatory acceptance generally should be supported by the following attributes:

The method should have undergone independent scientific peer review by disinterested persons who are experts in the field, knowledgeable in the method, and financially unencumbered by the outcome of the evaluation.

There should be a detailed protocol with standard operating procedures (SOPs), a list of operating characteristics, and criteria for judging test performance and results.

Data generated by the method should adequately measure or predict the endpoint of interest and demonstrate a linkage between either the new test and an existing test or the new test and effects in the target species.

There should be adequate test data for chemicals and products representative of those administered by the regulatory program or agency and for which the test is proposed.

The method should generate data useful for risk assessment purposes (i.e., for hazard identification, dose–response assessment, and exposure assessment). Such methods may be useful alone or as part of a battery or tiered approach.

The specific strengths and limitations of the test must be clearly identified and described.

The test method must be robust (relatively insensitive to minor changes in protocol) and transferable among properly equipped and staffed laboratories.

The method should be time and cost effective.

The method should be one that can be harmonized with similar testing requirements of other agencies and international groups.

The method should be suitable for international acceptance.

The method must provide adequate consideration for the reduction, refinement, and replacement of animal use.

hat involves aseptic tissue culture should utilize laboraories that have demonstrated proficiency in successfully conducting tissue culture experiments or testing. The use of three laboratories has generally been found to be adequate for assessing the interlaboratory reproducibility of est methods during validation studies. It is helpful to designate the laboratory most experienced with the test method as the *lead laboratory* during prevalidation studies o serve as a resource for technical issues that develop during the studies.

### Phased Validation Studies

n a recent *in vitro* validation study managed by the National Toxicology Program (NTP) Interagency Center or the Evaluation of Alternative Toxicological Methods NICEATM), dividing the prevalidation study into three phases was found to aid in efficiently optimizing the test method protocol [45]. The first phase involved a series of multiple testing with the positive control, with cycles of modifications and additions to the protocol until all laboratories were able to obtain reproducible results. This phase also was used to establish acceptance criteria for the test system, including positive control acceptance values for each laboratory. The second phase tested three coded substances representing three different areas of the response range (low, moderate, and high toxicity) and was again followed by minor protocol revisions to minimize variation within and among the participating laboratories. The third phase tested nine coded substances, again representing the range of responses as well as range of solubility. Additional minor protocol revisions were made after this phase, and an optimized test method protocol was finalized for testing the 60 remaining reference chemicals in the formal validation phase.

## REGULATORY ACCEPTANCE CRITERIA FOR NEW SAFETY EVALUATION METHODS

After a new test method has been evaluated in a validation study, regulatory authorities do not automatically accept it; rather, the regulatory authorities must determine if there is sufficient evidence that use of test method will provide for equivalent or improved protection of human health, animal health, or the environment, as defined by the intended purpose of the test method. Furthermore, although a validated test method can be found to be technically acceptable, its utility by different regulatory bodies is often dictated by the products they regulate. Regulatory acceptance criteria that should be adequately addressed by a test method proposed for regulatory applications have been developed for the United States (Table 21.6) [36], as well as internationally [37]. This section discusses the regulatory acceptance criteria for new test methods.

### Independent Scientific Peer Review

The test method should have undergone independent scientific peer review by a group of persons that includes experts in the respective field and who are knowledgeable in the method. These individuals should not have financial or otherwise influential conflicts of interest such that they, any family members, or their organizations stand to gain financially or otherwise from either a positive or negative outcome of the peer review process. Ideally, all materials substantiating the scientific validity of the test method should be made available to the public, the public should have the opportunity to comment on the test method, and the peer review panel should conduct its deliberations in public session. ICCVAM has implemented a peer review process that incorporates these features [38].

### Detailed Test Method Protocol

A detailed test method protocol must be provided that includes all relevant standard operating procedures, operating characteristics, and decision criteria (e.g., assay acceptance criteria, response criteria). The performance of the test method protocol should be substantiated by appropriate validation studies, and any changes in the protocol from the version used in the validation study should be scientifically justified. The test method protocol should provide sufficient information to allow for development of laboratory specific GLP-compliant test protocols.

### Adequate Measurement or Prediction of the Endpoint of Interest

Data generated by the test method should adequately measure or predict the endpoint of interest and demonstrate a linkage between either the new test and an existing test or the new test and effects in the target species of interest. Such a determination should ideally include an objective assessment of the accuracy of the existing test for measuring or predicting the toxic effect of interest; however, much of the data generated for an existing test method may be proprietary and therefore not readily available for such an assessment. In such instances, regulatory authorities may not be able to provide the actual data to the public that they use for making regulatory acceptance decisions.

### Adequate Test Data

Adequate testing data should be available for the chemicals and products for which the test method is proposed for use; however, in some cases these data may not adequately represent the complete spectrum of chemicals and products regulated by a specific regulatory agency or program. This lack of data might serve as the basis for nonacceptance by one or more regulatory authorities. It is possible for test methods to be found valid and acceptable for some defined chemical classes or physical and chemical properties that may not encompass the entire range of substances regulated by a specific agency. In such cases, acceptance may include specific restrictions on the substances for which the test method may be used.

### Usefulness for Risk Assessment

A test method should generate data useful for risk assessment purposes to be accepted. This will most commonly be for hazard identification purposes but could also include dose–response or exposure assessment purposes. The specific use for which the test method is proposed should be provided, such as whether the test method is proposed as a substitute or complete replacement for an existing test method, whether the test method is proposed as a screening test in a tiered testing strategy, or if the test is proposed as part of a battery of tests using a weight-of-evidence approach.

### Identification of Strengths and Limitations

The specific strengths and limitations of the test method must be clearly identified and described. This description may be in terms of demonstrated usefulness and limitations based on high-quality data, or it may be based on the fact that certain types of substances have not yet been evaluated in the test system. The limitations associated with a test method will necessarily influence its regulatory utility and could restrict its applicability in certain regulatory domains.

### Robustness and Transferability

The validation study must have demonstrated that the test method is sufficiently robust and that it is transferable among properly equipped and staffed laboratories. There must be sufficient evidence to assure regulatory authorities that similar results will be obtained with the same substance regardless of the geographic area and laboratory where the test method is conducted or the laboratory personnel performing the test.

### Time and Cost Effectiveness

The test method should be time and cost effective compared to the test method that it is proposed to substitute or replace. Obviously, a test method that takes considerably more time and expense to conduct than an existing test would have to have sufficient other advantages to warrant acceptance by regulatory authorities.

### Harmonized for Use by Other Agencies and International Groups

The test method should be capable of being harmonized with similar testing requirements of other agencies and international groups. This criterion is especially important in light of the impending implementation of a Globally Harmonized System (GHS) of Classification and Labeling of Chemicals [34]. Accordingly, new test methods should include assessment of the test method accuracy for the GHS hazard classification scheme where appropriate.

### Suitability for International Acceptance

New test methods should be suitable for acceptance by international authorities, such as the United Nations, the Organization for Economic Cooperation and Development (OECD), and the International Organization for Standardization (ISO). This is to ensure that the method can be accepted for use internationally and thereby avoid the need for duplicative testing.

### Adequate Consideration of the 3Rs

New test methods must provide for adequate consideration of the reduction, replacement, and refinement (3Rs) of animal use if they involve animals or test system components derived from animals. Such consideration is necessary to comply with U.S. and European animal welfare regulations, policies, and guidelines [8,9,46].

## TEST METHOD PERFORMANCE STANDARDS

Many new and alternative test methods are proprietary in nature and are protected by intellectual property laws such as patents, trademarks, and copyrights. Such intellectual protections stimulate innovation by providing financial incentives for companies to develop and market new products, such as *in vitro* testing methods that may reduce, refine, or replace animal use. U.S. laws, however, require that government regulatory authorities cannot simply endorse or approve proprietary methods until they first convey the basis by which the proprietary methods have been determined to be acceptable for use [38,47]. This issue was recently addressed by ICCVAM, which further developed the concept of performance standards for application to toxicological test methods [38,47,49,50].

### Defining Test Method Performance Standards

Performance standards are defined as the basis by which a proprietary or nonproprietary test method has been determined to have sufficient accuracy and reliability for a specific testing purpose [38]. Performance standards are based on an adequately validated test method and provide a basis for evaluating the comparability of mechanistically and functionally similar test methods. This process involves first determining that a test method has sufficient accuracy and reliability for a defined specific testing purpose. This information is then used to develop performance standards that can be used as the basis for evaluating the acceptability of proposed test methods based on similar scientific principles and that measure or predict the same biological or toxic effect. If a similar test method adequately addresses and meets these standards, then it would be considered to be comparable, in terms of performance, to the test method used to establish the performance standards. These performance standards can then be used by regulatory authorities to communicate the basis by which they find the original reference test method to be acceptable for specific regulatory testing purposes, as well as to judge the acceptability of subsequent similar methods. ICCVAM now routinely develops and proposes performance standards during test method evaluations for both proprietary and nonproprietary methods that have undergone adequate validation [38,47]. The availability of test method performance standards should streamline the validation and regulatory acceptance of structurally and functionally similar methods.

### Components of Performance Standards

Performance standards consist of three elements: (1) essential test method components, (2) a minimum list of reference chemicals, and (3) accuracy and reliability values [38,47]. *Essential test method* components are the requisite structural, functional, and procedural elements of a validated test method that should be included in the protocol of a proposed mechanistically and functionally similar test method. These components include unique characteristics of the test method, critical procedural details, and quality control measures. If there are deviations from the recommended essential test method components, then a scientific rationale must be provided and any potential impact of the deviations discussed. Incorporation of and adherence to essential test method components will help ensure that a proposed test method is based on the same concepts as the corresponding validated test method.

The *minimum list of reference chemicals* is used to assess the accuracy and reliability of a mechanistically and functionally similar test method that incorporates all of the essential test method components. These chemicals are a representative subset of those used to demonstrate the reliability and accuracy of the validated reference test method on which the performance standards are based. To the extent possible, these reference chemicals should:

- Represent the range of responses that the validated test method is capable of measuring or predicting (e.g., negative, and weak-to-moderate-to-strong positives).
- Produce consistent results in the validated test method and in the *in vivo* reference test method or target species of interest.
- Reflect the accuracy of the validated test method.
- Have well-defined chemical structures.
- Be readily available (i.e., can be purchased from commercial sources).
- Not be associated with excessive hazard or prohibitive disposal costs.
- Represent the range of known or suspected mechanisms or modes of action for the toxicity measured or predicted by the test method.
- Represent the range of physical and chemical properties for which the test method is proposed to be capable of testing (e.g., solubility, pH, volatility).

These reference chemicals are the minimum number that should be used to evaluate the performance of a proposed mechanistically and functionally similar test method. These chemicals should not be used to develop the decision criteria or prediction model for the proposed test method. If any of the recommended chemicals are unavailable, other chemicals for which adequate reference data are available could be substituted with adequate scientific justification. To the extent possible, any substituted chemicals should be of the same chemical class and potency as the original chemicals. If desired, additional chemicals representing other chemical or product classes and for which adequate reference data are available can be used to more comprehensively evaluate the accuracy of the proposed test method; however, these additional chemicals should not include those used to develop the proposed test method.

*Accuracy* and *reliability* values are the comparable performance that should be achieved by the proposed test method when evaluated using the minimum list of reference chemicals. Reference chemicals should be designated for performance standards that will result in accuracy and reliability values similar to the overall values determined from the entire validation database for the reference test method.

## Process for Developing Performance Standards

The Interagency Coordinating Committee on the Validation of Alternative Methods has developed a process for establishing performance standards during the evaluation of proposed new test methods [38]. The process is designed to ensure rigorous scientific review and to provide the opportunity for broad stakeholder and public comment. The ICCVAM process for developing performance standards for new test methods is as follows:

- The National Toxicology Program Interagency Center for the Evaluation of Alternative Toxicological Methods (NICEATM) and the appropriate ICCVAM working group develop proposed performance standards for consideration during the ICCVAM evaluation process. If a sponsor proposes performance standards, these are considered by ICCVAM at this stage. Generally, the proposed performance standards will be based on the information and data provided in the test method submission or on other available applicable data.
- The ICCVAM/NICEATM peer review panel evaluates the proposed performance standards for completeness and appropriateness during its evaluation of the validation status of the proposed test method. The proposed performance standards are made available with the test method submission to the public for comment prior to and during the peer review panel meeting.
- The appropriate ICCVAM working group, with the assistance of NICEATM, prepares the final performance standards for ICCVAM approval, taking into consideration the recommendations of the peer review panel and public comments.
- Performance standards recommended by ICCVAM are incorporated into ICCVAM test method evaluation reports, which are published, provided to federal agencies, and made available to the public. Availability of ICCVAM test method evaluation reports are announced routinely in the *Federal Register*, NTP newsletters, and ICCVAM/NICEATM e-mail listserv groups.
- Regulatory authorities can then reference the performance standards in the ICCVAM report when they communicate their acceptance of a new test method. In addition, performance standards adopted by regulatory authorities can be provided in guidelines issued for new test methods.

## Performance Standards for Dermal Corrosivity Test Methods

Performance standards are available for three propr etary—Corrositex®, EPISKIN™, and EpiDerm™—an one nonproprietary—rat skin transcutaneous electrica resistance (TER)—*in vitro* dermal corrosivity test meth ods [49]. Due to the structural and functional difference of the four methods, three different sets of performanc standards were developed [49]. EPISKIN™ and Ep Derm™ are structurally and functionally similar; there fore, one set of performance standards was developed f these two methods. The standards were based o EPISKIN™, as this method had a larger validation dat base than EpiDerm™ (60 vs. 24). In addition to the essential test method components, a minimum list of 24 reference chemicals was selected from the 60 chemicals used for the validation of EPISKIN™. This list included 12 corrosives and 12 noncorrosives. All of the selected reference chemicals are commercially available. Accuracy and reliability values for the 24 minimum reference chemicals closely matched the overall performance for the 60 chemicals in the validation database. For the rat skin TER, a minimum list of 24 reference chemicals was selected, which also provided accuracy and reliability values similar to those for the total validation database of 60 chemicals.

Performance standards based on Corrositex® were developed for a generic *in vitro* membrane barrier test system for skin corrosion [28]. This test method is capable of identifying the three subcategories of corrosivity described by the United Nations Packing Group (PG) classification system. Accordingly, the validation database contained a larger number of substances (129). The selected minimum list of reference chemicals contained a total of 40 chemicals, including 12 noncorrosive methods and 28 corrosive chemicals [49]. As with the other *in vitro* methods, the accuracy and reliability values for the minimum list of reference chemicals were similar to those for the total validation database. These performance standards were subsequently included in the proposed OECD Test Guideline 435 for an *in vitro* membrane barrier test system for skin corrosion [28].

### Using Performance Standards for Validation Studies

The availability of performance standards can significantly expedite the validation and acceptance of new test methods that are structurally and functionally similar to previously accepted methods for which there has been adequate validation [38,47,49]. For example, using performance standards, validation studies on a generic version of Corrositex® could potentially be accomplished with 40 substances compared to the over 129 chemicals used for the original validation [29]. Performance standards should also facilitate the validation of improved versions of existing tests. The concept and definition of test method performance standards has recently been included in international guidance on validation [37].

## ICCVAM ROLE IN VALIDATION AND REGULATORY ACCEPTANCE

### History

The Interagency Coordinating Committee on the Validation of Alternative Methods was first established as an *ad hoc* interagency committee in 1994 [11,36]. It consisted of representatives from 15 federal agencies and programs that require, generate, use, or disseminate toxicological testing information (Table 21.7) [36]. This committee was charged with developing validation and regulatory acceptance criteria and recommending a process for achieving the regulatory acceptance of scientifically valid alternative test methods [36]. The principles embodied in the validation and regulatory acceptance criteria are based on good science and the need to ensure that the use of new test methods will provide for equivalent or better protection of human health and the environment than previous testing methods or strategies. ICCVAM issued its report in 1997 [36].

To implement a process for achieving regulatory acceptance of proposed new, revised, and alternative test methods with regulatory applicability, a standing ICCVAM was established to evaluate the scientific validity of these test methods. The National Institute of Environmental Health Sciences also established the National Toxicology Program (NTP) Interagency Center for the Evaluation of Alternative Toxicological Methods (NICEATM) to administer ICCVAM and to provide scientific and operational support for the committee and its activities. NICEATM collaborates with ICCVAM to carry out scientific peer review and interagency consideration of new test methods of multiagency interest. The Center also performs other functions necessary to ensure compliance with provisions of the ICCVAM Authorization Act of 2000 and conducts independent validation studies on promising new test methods.

**TABLE 21.7**
**ICCVAM Committee**

The Interagency Coordinating Committee on the Validation of Alternative Methods (ICCVAM) was established by the NIEHS to develop a report recommending criteria and processes for validation and regulatory acceptance of toxicological testing methods. Fifteen federal regulatory and research agencies have participated in this effort, including:

Agency for Toxic Substances and Disease Registry (ATSDR)
Consumer Product Safety Commission (CPSC)
Department of Agriculture (USDA)
Department of Defense (DOD)
Department of Energy (DOE)
Department of the Interior (DOI)
Department of Transportation (DOT)
Environmental Protection Agency (EPA)
Food and Drug Administration (FDA)
National Institute for Occupational Safety and Health (NIOSH)
National Institutes of Health (NIH)
National Cancer Institute (NCI)
National Institute of Environmental Health Sciences (NIEHS)
National Library of Medicine (NLM)
Occupational Safety and Health Administration (OSHA)

**TABLE 21.8**
**Specific Purposes of the ICCVAM (P.L. 106-545, Section 3(b))**

(1) Increase the efficiency and effectiveness of federal agency test method review,
(2) Eliminate unnecessary duplicative efforts and share experiences between federal regulatory agencies,
(3) Optimize utilization of scientific expertise outside the federal government,
(4) Ensure that new and revised test methods are validated to meet the needs of federal agencies, and
(5) Reduce, refine, and replace the use of animals in testing, where feasible.

**TABLE 21.9**
**Duties of the ICCVAM (P.L. 106-545, Section 3(e))**

(1) Review and evaluate new or revised or alternative test methods, including batteries of tests and test screens, that may be acceptable for specific regulatory uses, including the coordination of technical reviews of proposed new or revised or alternative test methods of interagency interest.
(2) Facilitate appropriate interagency and international harmonization of acute or chronic toxicological test protocols that encourage the reduction, refinement, or replacement of animal test methods.
(3) Facilitate and provide guidance on the development of validation criteria, validation studies, and processes for new or revised or alternative test methods and help facilitate the acceptance of such scientifically valid test methods and awareness of accepted test methods by federal agencies and other stakeholders.
(4) Submit ICCVAM test recommendations for the test method reviewed by the ICCVAM, through expeditious transmittal by the Secretary of Health and Human Services (or the designee of the Secretary), to each appropriate federal agency, along with the identification of specific agency guidelines, recommendations, or regulations for a test method, including batteries of tests and test screens, for chemicals or class of chemicals within a regulatory framework that may be appropriate for scientific improvement, while seeking to reduce, refine, or replace animal test methods.
(5) Consider for review and evaluation petitions received from the public that:
 (A) identify a specific regulation, recommendation, or guideline regarding a regulatory mandate; and
 (B) recommend new or revised or alternative test methods and provide valid scientific evidence of the potential of the test method.
(6) Make available to the public final ICCVAM test recommendations to appropriate federal agencies and the responses from the agencies regarding such recommendations.
(7) Prepare reports to be made available to the public on its progress under this Act. The first report shall be completed not later than 12 months after the date of the enactment of this Act, and subsequent reports shall be completed biennially thereafter.

## Purposes and Duties

The ICCVAM Authorization Act of 2000 formally established ICCVAM as a permanent interagency committee under NICEATM [7]. The Act mandates specific purposes and duties of the ICCVAM (Table 21.8 and Table 21.9) [7]. ICCVAM also continues to coordinate interagency issues on test method development, validation, regulatory acceptance, and national and international harmonization. The public health goal of NICEATM and ICCVAM is to promote the scientific validation and regulatory acceptance of new toxicity testing methods that are more predictive of human health, animal health, and ecological effects than currently available methods. Methods are emphasized that provide for improved toxicity characterization and savings in time and costs and that provide the refinement, reduction, and replacement of animal use whenever feasible.

## Test Method Nomination and Submission Process

Any organization or individual can submit a test metho for which adequate validation studies have been completed to ICCVAM for evaluation. ICCVAM has published guidelines for the information that should be submitted and has developed an outline to organize th information and data supporting the scientific validit of a proposed test method [38]. Any organization o individual can also nominate test methods for whic adequate validation studies have not been completed t the ICCVAM for further study. Nominations are prioritized based on established ICCVAM prioritization criteria (Table 21.10) [38]. Specific activities, such a workshops and validation studies, are then conducte for those test methods with the highest priority and fo which resources are available.

**TABLE 21.10**
**ICCVAM Prioritization Criteria**

Preliminary evaluations summarize the extent to which proposed test method submissions or nominations address the following ICCVAM prioritization criteria:

The extent to which the proposed test method is:
- Applicable to regulatory testing needs
- Applicable to multiple agencies/programs
- Warranted, based on the extent of expected use or application and impact on human, animal, or ecological health

The potential for the proposed test method, compared to current test methods accepted by regulatory agencies, to:
- Refine animal use (decreases or eliminates pain and distress)
- Reduce animal use
- Replace animal use

The potential for the proposed test method to provide improved prediction of adverse health or environmental effects, compared to current test methods accepted by regulatory agencies

The extent to which the test method provides other advantages (e.g., reduced cost and time to perform) compared to current methods

The completeness of the nomination or submission with regard to ICCVAM test method submission guidelines

### ICCVAM Test Method Recommendations

Several *in vitro* and *in vivo* alternative test methods that have been reviewed and recommended by ICCVAM have now been accepted by national and international authorities [25,29,51,52]. These methods have resulted in significant refinement, reduction and partial replacement of animal use. These include four *in vitro* methods for identifying dermal corrosives [25,29], the local lymph node assay for assessing allergic contact dermatitis [51], and the revised up-and-down procedure for determining acute oral toxicity [52]. Additional *in vitro* methods have been evaluated or are currently undergoing evaluation, while still others are being developed for a wide range of human health and ecological testing purposes [53–55].

## OTHER ORGANIZATIONS INVOLVED IN VALIDATION

In addition to the United States, several other countries have also established government centers to conduct independent validation studies of new alternative methods. These include, among others, the European Centre for the Evaluation of Alternative Methods (ECVAM) in the European Union, the Japanese Center for the Validation of Alternative Methods (JaCVAM) in Japan, and the National Centre for Documentation and Evaluation of Alternative Methods to Animal Experiments (ZEBET) in Germany.

### ECVAM

The European Centre for the Evaluation of Alternative Methods was established by the European Union in 1992 as a component within the European Commission [56]. ECVAM is administratively located as a unit within the Institute for Health and Consumer Protection within the European Commission's Joint Research Centre in Ispra, Italy. The purpose of ECVAM is to promote the development and validation of methods that can reduce, refine, and replace the use of animals for research and testing. The Centre has intramural research laboratories in which it conducts relevant *in vitro* research. ECVAM also conducts extramural validation studies. ECVAM has received significant support to assist with the development and validation of *in vitro* methods to meet EU Parliament deadlines that will largely prohibit the use of animals for the testing of cosmetic ingredients in 2009 [57]. EU legislation adopted in 2004 already bans the use of animals for testing cosmetic products [58]. ICCVAM and ECVAM work together on projects and validation studies of common interest to facilitate harmonized methods and to leverage resources [59,60].

### JaCVAM

The Japanese Center for the Validation of Alternative Methods was officially established in November 2005. The Center is a component of the National Institute of Health Sciences, which is part of the Ministry of Health and Welfare, and is located in Tokyo, Japan. The Center was established to develop, validate, and review alternative test methods in Japan, and it interacts closely with the Japanese Society for Alternatives to Animal Experimentation (JSAAE) [61]. The JSAAE has provided national leadership in the development and validation of alternative methods in Japan and provides an important scientific network for related *in vitro* studies in Japan.

## ZEBET

The German Centre for the Documentation and Validation of Alternative Methods was established in Germany in 1989 [62]. It is a component of the Federal Institute for Health Protection of Consumers and Veterinary Medicine (BgVV) located in Berlin, Germany. ZEBET has been instrumental in the development and validation of several important *in vitro* alternative methods, including methods for dermal irritation and corrosivity, embryotoxicity, ocular irritation, and photosensitization.

# THE LLNA: AN ALTERNATIVE METHOD FOR ALLERGIC CONTACT DERMATITIS

The murine local lymph node assay (LLNA) is an alternative *in vivo* method for assessing the allergic contact dermatitis (ACD) potential of chemicals [51,63–67]. The LLNA is a mechanism-based test method that provides dose–response information, uses fewer animals, and eliminates pain and distress compared to the standard testing methods for which it can be substituted [51]. The LLNA was the first alternative test method evaluated by ICCVAM in 1998 [65–67]. Based on the results of a comprehensive scientific peer review and technical evaluation, ICCVAM determined that the LLNA was a valid substitute to currently accepted test methods that use guinea pigs and concluded that the LLNA provides for the refinement and reduction of animal use [51,65]. ICCVAM forwarded test recommendations to agencies for their consideration, and the LLNA was subsequently accepted in 1999 by the U.S. Environmental Protection Agency, U.S. Food and Drug Administration, Consumer Product Safety Commission, and Occupational Safety and Health Administration. A new internationally harmonized test guideline (Test Guideline 429) on skin sensitization using the LLNA was adopted in 2002 by the Test Guidelines Programme at the Organization for Economic Cooperation and Development (OECD) [68].

## Regulatory Rationale for the LLNA

Allergic contact dermatitis (ACD) is associated with chemical exposure in the workplace and at home. An assessment of the potential for chemicals to cause ACD is an important component of routine safety testing. Traditionally, guinea pigs have been used to assess the ACD potential of chemicals, pharmaceuticals, and consumer products [51,69]. Although these test methods vary, the guinea pig maximization test (GPMT) and the Buehler assay (BA) have been the most commonly used methods for ACD testing. Both of these tests rely on the induction and elicitation phases of ACD and require about a month to perform the assay. The GPMT also may involve the use of complete Freund's adjuvant, which can be highly irritating to animals. The endpoint measured in the guinea pig methods is a visual assessment of erythema and edema at the challenge location and requires substantial technical expertise [70].

## Mechanistic Basis of Allergic Contact Dermatitis

Allergic contact dermatitis develops in two phases: an initial induction phase followed by an elicitation phase. The induction phase begins with initial skin contact with a sensitizing agent. At the site of chemical contact with the skin, an immediate release of signaling factors and activation of the skin dendritic cells occur. Dendritic cells process the chemical and subsequently mature and migrate to the draining lymph node where they serve as antigen-presenting cells. Lymphocytes within the nodes upon antigen presentation, undergo cellular proliferation. Lymphocyte proliferation is the mechanistic endpoint assessed in the LLNA and indicates that the induction phase of ACD has taken place [71]. Following proliferation, T-lymphocytes are considered primed, as they have a specific recall for the sensitizing agent. Upon subsequent exposure to the agent, an antigen-specific response occurs which is referred to as the *elicitation phase*. This second phase occurs only if there is elicitation of specific mediators that cause an inflammatory cell influx to the dermal site. Elicitation is a systemic response that can occur at locations other than the original site of sensitization. The elicitation phase is characterized by erythema and edema and occurs 24 to 72 hours after the challenge exposure. This response is the endpoint assessed in traditional guinea pig tests [71].

## The LLNA Procedure

The basic principle underlying the LLNA is that sensitizers induce proliferation of lymphocytes in the lymph node draining the site of chemical application. Generally, this proliferation is proportional to the dose applied and potency of the sensitizer. The test measures cellular proliferations as a function of *in vivo* radioisotope incorporation into the DNA of dividing lymphocytes in the draining lymph nodes proximal to the application site (Figure 21.2) [72]. The proliferation in the test groups is compared to that in concurrent vehicle-treated controls. A positive control is added to each assay to provide an indication of appropriate assay performance.

### Animals

Young adult female mice (nulliparous and not pregnant) of the CBA/Ca or CBA/J strain at ages 8 to 12 weeks of age are used for the assay. Females are used because the existing database is predominantly based on this gender

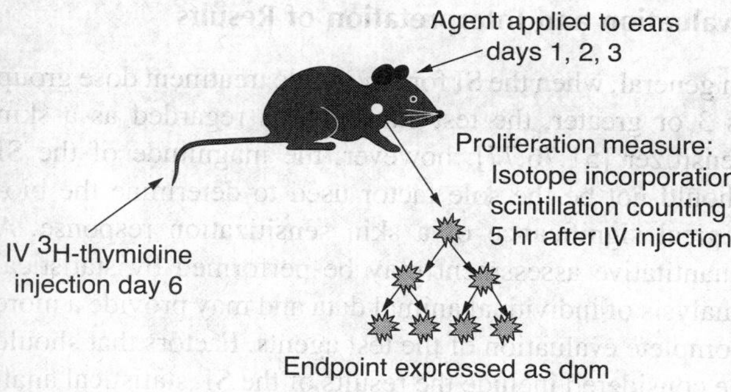


**FIGURE 21.2** Local lymph node assay.

Other strains and males can be used if they have been evaluated and shown to produce equivalent results to the validated method. Mice should be carefully observed for any clinical signs, including local irritation at the application site and signs indicative of systemic toxicity. Weighing mice prior to treatment and at the time of necropsy will aid in assessing systemic toxicity. All observations are systematically recorded, with records being maintained for each individual mouse.

## Test Articles

Solid test substances should be dissolved in appropriate solvents or vehicles and diluted, if appropriate, prior to dosing of the animals. Liquid test substances may be dosed directly or diluted prior to dosing. Fresh preparations of the test substance should be prepared daily unless stability data demonstrate the acceptability of storage. The solvent or vehicle should be selected on the basis of maximizing the test concentrations while producing a solution or suspension suitable for application of the test substance. In order of preference, recommended solvents and vehicles are acetone/olive oil (4:1 v/v), *N,N*-dimethylformamide (DMF), methyl ethyl ketone (MEK), propylene glycol (PG), and dimethylsulfoxide (DMSO), but others may be used [73]. Particular care should be taken to ensure that hydrophilic materials are incorporated into a vehicle system that wets the skin and does not immediately run off. Thus, wholly aqueous vehicles are to be avoided. It may be necessary for regulatory purposes to test the chemical in the clinically relevant solvent or product formulation.

Five successfully treated animals are used per dose group, with a minimum of three consecutive concentrations of the test substance plus a solvent or vehicle control and a positive control group. Test substance treatment doses should be based on the recommendations given in Kimber and Basketter [73] and in the ICCVAM Peer Review Panel Report [51]. Doses are selected from the concentration series 100%, 50%, 25%, 10%, 5%, 2.5%, 1%, 0.5%, etc. The maximum concentration tested should be the highest achievable level that does not result in local irritation or overt systemic toxicity. To identify the appropriate maximum test substance dose, an initial toxicity test, conducted under identical experimental conditions except for an assessment of lymph node proliferative activity, may be necessary. To support an ability to identify a dose–response relationship, data must be collected on at least three test substance treatment doses, in addition to the concurrent solvent or vehicle control group. For negative LLNA studies, the concurrent positive control must induce a stimulation index (SI) greater than 3 relative to its vehicle-treated control.

## Controls

Concurrent solvent/vehicle and positive controls should be included in each test. In some circumstances, it may be useful to include a naïve control. Except for treatment with the test substance, animals in the control groups should be handled in a manner identical to that for animals of the treatment groups. Positive controls are used to ensure the appropriate performance of the assay. The positive control should produce a positive LLNA response at an exposure level expected to give an increase in the SI greater than 3 over the negative control group. The positive control dose should be chosen such that the induction is clear but not excessive. Preferred positive control substances are hexyl cinnamic aldehyde (HCA) or mercaptobenzothiazole.

## Protocol Schedule

Day 1—Individually identify and record the weight of each mouse prior to dermal applications. Apply 25 μL/ear of the appropriate dilution of the test substance, the positive control, or the vehicle alone to the dorsum of both ears.

Days 2 and 3—Repeat the application procedure as carried out on day 1.

Days 4 and 5—No treatment.

Day 6—Record the weight of each mouse. Inject 250 μL of sterile phosphate-buffered saline (PBS) containing 20 μCi of $^{3}$H-methyl thymidine ($^{3}$H-TdR) or 250 μL PBS containing 2 μCi of $^{125}$I-iododeoxyuridine ($^{125}$IU) and $10^{-5}$ *M* fluorodeoxyuride into each experimental mouse via the tail vein [74,75]. Five hours later, the draining (auricular) lymph node of each ear is excised and pooled in PBS for each animal [51]. Both bilateral draining lymph nodes must be collected.

## Lymphocyte Measurements

A single-cell suspension of lymph node cells (LNCs) is prepared for each mouse. The single-cell suspension is prepared in PBS either by gentle mechanical separation through 200-mesh stainless steel gauze or by another acceptable technique for generating a single-cell suspension. LNCs are washed twice with an excess of PBS and the DNA precipitated with 5% trichloroacetic acid (TCA) at 4°C for approximately 18 hours. For the $^{3}$H-TdR method, pellets are resuspended in 1 mL TCA and transferred to 10 mL of scintillation fluid. Incorporation of tritiated thymidine is measured by β-scintillation counting as disintegrations per minute (dpm) for each mouse and expressed as dpm/mouse. For the $^{125}$IU method, the 1-mL TCA pellet is transferred directly into gamma counting tubes. Incorporation of $^{125}$IU is determined by gamma counting and also expressed as dpm/mouse.

## Calculation of the Stimulation Index

The LLNA measures lymphocyte proliferation in the draining lymph nodes of mice topically exposed to the test article using the incorporation of radioactive thymidine or iododeoxyuridine into DNA. The results are expressed as a ratio, the stimulation index (SI), of the mean number of disintegrations per minute for treated mice as compared to controls. Chemicals with a SI of 3.0 or more are considered positive, and those with a SI less than 3.0 are considered negative. This scoring differs from the scoring in guinea pig assays, where a test substance is classified as positive based on the percentage of animals in a group that are responders (at least 15% in a nonadjuvant assay and at least 8% in an adjuvant test) [70].

Results for each treatment group are expressed as the mean SI. The SI is the ratio of the mean dpm/mouse within each test substance treatment group and the positive control treated group against the mean dpm/mouse for the solvent- or vehicle-treated control group. The investigator should be alert to possible outlier responses for individual animals within a group that may necessitate the use of an alternative measure of response (e.g., median rather than mean) or elimination of the outlier. Each SI should include an appropriate measure of variability that takes into account the interanimal variability in both the dosed and control groups [51].

In addition to an assessment of the magnitude of the SI, a statistical analysis should be conducted. This assessment should include an assessment of the dose–response relationship, as well as pair-wise dosed group vs. concurrent solvent or vehicle control comparisons (e.g., linear regression analysis to assess dose–response trends; Dunnett's test for pair-wise comparisons). When choosing an appropriate method of statistical analysis, the investigator should be aware of the possible inequality of variances and other related problems that may necessitate a data transformation or a nonparametric statistical analysis.

Individual mouse disintegrations per minute data should be presented in tabular form, along with the group mean dpm/mouse, its associated error term, and the SI (and associated error term) for each dose group compared against the concurrent solvent or vehicle control group.

## Evaluation and Interpretation of Results

In general, when the SI for any single treatment dose group is 3 or greater, the test substance is regarded as a skin sensitizer [51,76,77]; however, the magnitude of the SI should not be the sole factor used to determine the biological significance of a skin sensitization response. A quantitative assessment may be performed by statistical analysis of individual animal data and may provide a more complete evaluation of the test agents. Factors that should be considered include the results of the SI, statistical analyses, the strength of the dose–response relationship, chemical toxicity, solubility, and the consistency of the vehicle and positive control responses. Equivocal results should be clarified by considering statistical analysis, structural relationships, available toxicity information, and dose selection. A test substance not meeting the above criteria is considered a nonsensitizer in this test.

## Training and Preparation for Node Identification

There are several methods that can be used to provide colo identification of the draining nodes. These techniques may be helpful to provide training for initial identification anc should be performed to ensure proper isolation of th appropriate node. Examples of such treatments that can b used for training are listed below:

- *Evan's blue dye treatment*—Inject approximately 0.1 mL of 2% Evan's blue dye (prepared in sterile saline) intradermally into the pinnae of an ear. Humanely kill the mouse after several minutes and then proceed with the dissection to remove the auricular lymph nodes.
- *Colloidal carbon and other dye treatments*—Colloidal carbon and India ink are examples of other dye treatments that may be used [78]. For the purpose of node identification during training, a strong sensitizer is recommended. This agent should be applied in the standard acetone:olive oil vehicle (4:1). Suggested sensitizers used for this training exercise include 0.1% oxazolone, 0.1% (w/v) 2,4-dinitrochlorobenzene, and 0.1% (v/v) dinitrofluorobenzene. After treating the ear with a strong sensitizer, the draining node will dramatically increase in size, thus aiding in the identification and location of the node. Due to the exacerbated response, the suggested sensitizers for node identification are not recommended as controls for the assay performance and should only be used for training.

The node draining the ear (auricular) is located distal t the masseter muscle, away from the midline, and near th bifurcation of the jugular vein. Nodes can be distinguishe from the glandular and connective tissue in the area b the uniformity of the nodal surface and a shiny translucent appearance. The application of sensitizing agents will cause an enlargement of the node.

## ALTERNATIVE METHODS FOR SKIN CORROSION

Four *in vitro* test methods are available for assessing the dermal corrosivity hazard potentials of chemicals: EPISKIN™, EpiDerm™ (EPI-200), the rat skin transcutaneous electrical resistance (TER) assay, and Corrositex® [25–29]. Based on the available data and performance for each test method, these methods are recommended for use as screening assays for the identification of corrosive substances in a tiered testing strategy, such as that proposed by the OECD (Table 21.11) [35]. In this strategy, positive results can be used to classify and label a substance as a dermal corrosive. Negative results require additional testing to properly identify any corrosive substances that may have been falsely identified as noncorrosives and to determine whether the substance is a dermal irritant.

### *In Vitro* Membrane Barrier Test Systems for Skin Corrosion

Validation studies have been completed for an *in vitro* membrane barrier test system commercially available as Corrositex® [29]. Based on its scientific validity, this test method has been recommended for use as part of a tiered testing strategy for assessing the dermal corrosion hazard potential of chemicals, whereby any substance that qualifies for testing can be evaluated [28,29]. In addition, this test method may be used to make decisions on the corrosivity and noncorrosivity of specific classes of chemicals (e.g., organic and inorganic acids, acid derivatives,* and bases) for certain transport testing circumstances. The basis of this test system is that it detects membrane damage caused by corrosive test substances.

The test substance is first evaluated to determine if it is compatible with the test procedure (i.e., if it qualifies for testing). If compatible, the substance is evaluated for the category of acid or base (strong or weak) to determine the appropriate time scale for classifying the potential corrosivity of the test substance. Finally, a compatible substance is applied to the surface of the artificial membrane barrier. The time it takes for the test substance to penetrate through the membrane barrier to an underlying indicator solution determines the corrosivity classification of that test substance. Penetration of the membrane barrier (or breakthrough) is indicated by a color change in a pH indicator dye in the solution below the barrier.

Performance standards have been established for *in vitro* membrane barrier test systems for corrosivity, as discussed in an earlier part of this chapter [49]. Essential test method components include a description of physical components of the test method (e.g., membrane barrier, categorization solutions, indicator solution); the test substance categorization system; the processes for determining test substance compatibility and test substance categorization; assembly of the physical components of the test method; application of a test substance; the appropriate control substances (solvent controls, positive [corrosive] controls, negative [noncorrosive] controls, benchmark controls); measurement of membrane barrier penetration; interpretation of results; and classification of test substances with regard to corrosivity potential. The test report should include the following information, if relevant to the conduct of the study: test and control substances, justification of the test method and protocol used, test method integrity, criteria for an acceptable test, test conditions, results, description of other effects observed, discussion of the results, and conclusions.

### *In Vitro* Human Skin Cell Culture Systems for Skin Corrosion

The European Centre for the Validation of Alternative Methods conducted validation studies on two *in vitro* test methods that use cultured human skin cell test systems for assessing skin corrosivity: EpiDerm™ (MatTek; Ashland, MA) and EPISKIN™ (EPISKIN SNC; Lyon, France) [79]. These two methods utilize a three-dimensional tissue culture model of human skin comprised of a reconstructed epidermis and a functional stratum corneum composed of human keratinocytes. These test methods have been recommended for the testing of all classes of chemicals and for inclusion in tiered testing strategies as part of a tiered or weight-of-evidence evaluation. Neither test method has been validated for categorizing the corrosive properties of chemicals across the three U.N. Packing Group subcategories of corrosivity, although data are available for differentiating between Packing Group 1 and the two less severe Packing Groups. The methods have been accepted by the European Commission [80,81] and have been recommended by the ICCVAM for use as screening assays in a tiered testing strategy [25]. An international test guideline is also now available [27].

The test material is applied topically to a three-dimensional human keratinocyte culture model, comprised of at least a reconstructed epidermis with a functional stratum corneum. Corrosive substances are identified by their ability to induce a decrease in cell viability below defined threshold levels at specified exposure periods. The principle of the human skin model assay is based on the

"Acid derivative" is a non-specific class designation and is broadly defined as an acid produced from a chemical substance either directly or by modification or partial substitution. This class includes anhydrides, haloacids, salts, and other types of chemicals.

**TABLE 21.11**
**Testing and Evaluation Strategy for Dermal Irritation/Corrosion**

| Activity | Finding | Conclusion |
|---|---|---|
| 1 Existing human and/or animal data showing effects on skin or mucous membranes | Corrosive | Apical endpoint—considered corrosive; no testing is necessary. |
| | Irritating | Apical endpoint—considered an irritant; no testing is necessary. |
| | Not corrosive/not irritating | Apical endpoint—considered not corrosive or irritating; no testing is necessary. |
| ↓ | | |
| *No information is available, or available information is not conclusive.* | | |
| ↓ | | |
| 2 Perform SAR evaluation for skin corrosion/irritation | Predict severe damage to skin | Considered corrosive; no testing is necessary. |
| | Predict irritation to skin | Considered an irritant; no testing is necessary. |
| ↓ | | |
| *No predictions can be made, or predictions are not conclusive or negative.* | | |
| ↓ | | |
| 3 Measure pH (consider buffering capacity, if relevant). | pH ≤ 2 or ≥ 11.5 (with high buffering capacity, if relevant) | Assume corrositivity; no testing is necessary. |
| ↓ | | |
| *2 < pH < 11.5, or pH ≤ 2.0 or ≥ 11.5 with low or no buffering capacity, if relevant.* | | |
| ↓ | | |
| 4 Evaluate systemic toxicity data via dermal route[a] | Highly toxic | No further testing is necessary. |
| | Not corrosive or irritating when tested to limit dose of 2000 mg/kg body weight or higher, using rabbits | Assumed not to be corrosive or irritating; no further testing is necessary. |
| ↓ | | |
| *Such information is not available or is not conclusive.* | | |
| ↓ | | |
| 5 Perform validated and accepted *in vitro* or *ex vivo* test for skin corrosion. | Corrosive response | Assume corrosivity *in vivo;* no further testing is necessary. |
| ↓ | | |
| *Substance is not corrosive, or internationally validated in vitro and ex vivo testing methods for skin corrosion are not yet available.* | | |
| ↓ | | |
| 6 Perform validated and accepted *in vitro* or *ex vivo* test for skin irritation. | Irritant response | Assume irritancy *in vivo*; no further testing is necessary. |
| ↓ | | |
| *Substance is not an irritant, or internationally validated in vitro and ex vivo testing methods for skin irritation are not yet available.* | | |
| ↓ | | |
| 7 Perform initial *in vivo* rabbit test using one animal. | Severe damage to skin | Considered corrosive; no further testing is necessary. |
| ↓ | | |
| *No severe damage.* | | |
| ↓ | | |
| 8 Perform confirmatory test using one or two additional animals. | Corrosive or irritating | Considered corrosive or irritating; no further testing is necessary. |
| | Not corrosive or irritating | Considered not corrosive or irritating; no further testing is necessary. |

[a] Can be considered before steps 2 and 3.

premise that corrosive chemicals are able to penetrate the stratum corneum by diffusion or erosion and are cytotoxic to the keratinocytes in the underlying layers. The use of test systems that include human-derived cells or tissue should be in accordance with applicable national and international laws, regulations, and policies.

Investigators using a similar *in vitro* human skin cell culture model system for skin corrosion must be able to demonstrate that the assay is valid for its intended use. This includes demonstrating that different preparations are consistent in barrier properties (i.e., capable of maintaining a barrier to noncorrosive substances, able to respond appropriately to weak and strong corrosive substances) and that any modification to the existing validated reference test method does not adversely affect its performance characteristics. Performance standards are now available that can be used for this purpose, as described earlier in this chapter [49]. Essential test method components are also available for *in vitro* human skin model test methods for skin corrosivity [49]. The components are essentially the same as those described for Corrositex® with the addition of components unique to *in vitro* human skin model systems. Human skin models can be obtained commercially (e.g., EPISKIN™, EpiDerm™ [EPI-200]) or they can be developed or constructed in the testing laboratory.

### *In Vitro* Skin Transcutaneous Electrical Resistance Tests for Skin Corrosion

The European Centre for the Validation of Alternative Methods also conducted validation studies on the rat skin transcutaneous electrical resistance (TER) assay, another *in vitro* test method for assessing skin corrosivity [79]. The rat skin TER assay measures the extent to which a chemical alters the transcutaneous electrical resistance of a skin disc during a defined exposure period. Based on its scientific validity, this test method has been recommended for the testing of all classes of chemicals and for inclusion in tiered testing strategies as part of a tiered or weight-of-evidence evaluation [25,26].

The test substance is applied for up to 24 hours to the epidermal surface of skin discs in a two-compartment test system in which the skin discs function as the separation between the compartments. The skin discs are prepared from humanely killed 28- to 30-day-old rats. Corrosive substances are identified by their ability to produce a loss of normal stratum corneum integrity and barrier function, which is measured as a reduction in the TER below a specified level. For rat skin TER, a cutoff value of 5 kΩ has been selected based on extensive data for a wide range of substances where the majority of values were either clearly well above or well below this value. Generally, substances that are noncorrosive in animals but are irritating do not reduce the TER below this cutoff value; however, the use of other skin preparations or other equipment to measure resistance may require the use of a different cutoff value. In such situations, more extensive validation would be required. A dye-binding step is incorporated into the test procedure to confirm positive results. The dye-binding step determines if the increase in ionic permeability is due to physical destruction of the stratum corneum.

Investigators using an *in vitro* skin TER corrosivity test that is different from the validated test method protocol must be able to demonstrate that the assay is valid for its intended use. This includes demonstrating that different preparations are consistent in barrier properties (i.e., capable of maintaining a barrier to noncorrosive substances, able to respond appropriately to weak and strong corrosive substances) and that any modification to the existing validated reference test method does not adversely affect its performance characteristics. Performance standards are now available to accomplish the necessary validation of test method modifications of the TER [49].

## FUTURE PROGRESS

Significant progress has been made in recent years regarding the scientific principles and processes for adequate validation of *in vitro* and other test methods proposed for regulatory applications. Regulatory authorities have committed resources and personnel to this effort to identify new and alternative test methods with potential regulatory applicability that are as good as or better than traditionally employed test methods. Regulatory authorities have also communicated the criteria that they will use as the basis for making decisions on the regulatory acceptability of new, revised, and alternative methods. The ICCVAM provides an efficient process for the interagency evaluation of new, revised, and alternative methods of multiagency interest, thereby limiting or eliminating duplicative efforts by independent agencies. These established criteria and processes will facilitate the validation and regulatory acceptance of proposed test methods that incorporate new science and technology. Future progress in the development, validation, and adoption of improved testing methods can be expected to support enhanced protection of public health, animal health, and the environment. Adoption of scientifically valid alternative methods will also support improved animal welfare by reducing, replacing, and providing for more humane use of laboratory animals.

## ACKNOWLEDGMENTS

The authors acknowledge the expert editorial assistance of Debbie McCarley in the preparation of this chapter, and Drs. Frank Johnson and Rajendra Chhabra for their careful review and suggestions. We also acknowledge the many federal scientists who have served as members of the ICCVAM to develop the criteria and processes described in this chapter and those who served as members of the Interagency Regulatory Alternatives Group (IRAG), a predecessor to the ICCVAM. The contributions of the many stakeholders who have worked to develop, validate, and review new alternative test methods that have now been adopted are also gratefully acknowledged. The content of this manuscript was supported by the Intramural Research Program of the National Institutes of Health.

## QUESTIONS

1. What are the basic requirements for adequate validation of a proposed new test method?
2. What are the considerations used by regulatory authorities in determining the acceptability of a new test method?

## REFERENCES

1. Russell, W. M. and Burch, R. L., *The Principles of Humane Experimental Technique*, Methuen & Co., London, 1959 (reprinted and available from the Universities Federation for Animal Welfare, 8 Hamilton Close, South Mimms, Potters Bar, Herts EN6 3QD England).
2. Zurlo J., Rudacille D., and Goldberg A.M., *Animals and Alternatives in Testing: History, Science, and Ethics*, Mary Ann Liebert, New York, 1994.
3. 7 U.S.C. §2131 et seq. (P.L. 89-544), as amended (P.L. 91-579, 94-279, 99-198), 1966.
4. 42 U.S.C., Health Research Extension Act of 1985 (P.L. 99-158, Section 495), 1985.
5. Balls, M. et al., The three R's: the way forward, *ATLA*, 23(6), 838–867, 1995.
6. National Institutes of Health Revitalization Act of 1993, P.L. 103–43, U.S. Government Printing Office, Washington, DC, 1993.
7. 42 U.S.C. §2851-2, ICCVAM Authorization Act of 2000 (P.L. 106-545), 2000.
8. USDA, Animal Welfare, Final Rules: Code of Federal Regulations, Title 9, Chapter 1, Subchapter A, Parts 1, 2, and 3, U.S. Department of Agriculture, Washington, D.C., 1989.
9. PHS, *Public Health Service policy on Humane Care and Use of Laboratory Animals*. U.S. Department of Health and Human Services, Washington, D. C., 1986.
10. Stokes, W. S., Jensen, D. J., and Latg, B. S., Guidelines for institutional animal care and use committees: consideration of alternatives, *Contemp. Top.*, 34(3), 51–60, 1995.
11. Stokes, W. S., Animal use alternatives in research and testing: obligation and opportunity, *Lab. Anim.*, 26(3), 28–32, 1997.
12. Stokes, W. S., Humane endpoints for laboratory animals used in regulatory testing, *ILAR J.*, 43, S31–S38, 2002.
13. USEPA, *Health Effects Test Guidelines, OPPTS 870.1100: Acute Oral Toxicity*, Office of Pesticides, Prevention, and Toxic Substances, U.S. Environmental Protection Agency, Washington, D.C., 1999 (http://www.epa.gov/OPPTS_Harmonized/870_Health_Effects_Test_Guidelines/Series/870-1100.pdf).
14. OECD, *Test Guideline 420, Guideline for Testing of Chemicals: Acute Oral Toxicity–Fixed Dose Method*, Paris, France: Organization for Economic Cooperation and Development, Paris, 2001.
15. OECD, *Test Guideline 423, Acute Oral Toxicity–Acute Toxic Class Method*, Organization for Economic Cooperation and Development, Paris, 2001.
16. OECD, *Test Guideline 425, Acute Oral Toxicity–Up and Down Procedure*, Organization for Economic Cooperation and Development, Paris, 2001.
17. OECD, *Guidance Document on the Recognition, Assessment, and Use of Clinical Signs as Humane Endpoints for Experimental Animals Used in Safety Evaluations*, Organization for Economic Cooperation and Development, Paris, 2000.
18. Stokes, W. S., Humane endpoints for laboratory animals used in toxicity testing, in *Progress in the Reduction, Refinement, and Replacement of Animal Experimentation*, Balls M. et al., Eds., Elsevier, Amsterdam, 2000, pp. 897–906.
19. Stokes, W. S., Reducing unrelieved pain and distress ir laboratory animals using humane endpoints, *ILAR J.*, 41 59–61, 2000.
20. Griffin, G., Stokes, W. S., Paks, S. P., and Gauthier, C. The ICLAS/CCAC International Symposium on Regulatory Testing and Animal Welfare: introduction and overview, *ILAR J.*, 43, S1–S4, 2002.
21. Dean, J. H. et al., ICCVAM evaluation of the murine loca lymph node assay. II. Conclusions and recommendation of an independent scientific peer review panel, *Reg. Toxicol. Pharmacol.*, 34, 258–273, 2001.
22. Springer, J. A. et al., Number of animals for sequentia testing, *Food Chem. Toxicol.*, 31, 105–109, 1993.
23. OECD, *Test Guideline 405, Guideline for the Testing o Chemicals: Acute Eye Irritation/Corrosion*, Organizatio for Economic Cooperation and Development, Paris, 2002
24. Botham, P. A., Acute systemic toxicity, *ILAR J.*, 43(S) S7–S30, 2002.
25. Interagency Coordinating Committee on the Validation c Alternative Methods (ICCVAM), *ICCVAM Evaluation o EPISKIN™, EpiDerm™ (EPI-200), and Rat Skin Transcu taneous Electrical Resistance (TER): In Vitro Test Method for Assessing Dermal Corrosivity Potential of Chemical* NIH Publ. No. 02-4502, National Institute of Environmen tal Health Sciences, Research Triangle Park, NC, 200 (http://iccvam.niehs.nih.gov/methods/epioderm.htm).
26. OECD, *Test Guideline 430, In Vitro Skin Corrosion Transcutaneous Electrical Resistance Test (TER)*, Orga nization for Economic Cooperation and Developmen Paris, 2004.
27. OECD, *Test Guideline 431, In Vitro Skin Corrosior Human Skin Model Test*, Organization for Economic Coop eration and Development, Paris, 2004
28. OECD, *OECD Test Guideline 435, In Vitro Membran Barrier Test Method for Skin Corrosion*, Organization fo Economic Cooperation and Development, Paris, 2006.
29. Interagency Coordinating Committee on the Validation c Alternative Methods (ICCVAM), *Corrositex®: An In Vit Test Method for Assessing Dermal Corrosivity Potential Chemicals*, NIH Publ. No. 99-4495, National Institute Environmental Health Sciences, Research Triangle Park, NC 1999 (http://iccvam.niehs.nih.gov/doc/reports/corprrep.htm

30. OECD, *OECD Series on Principles of Good Laboratory Practice and Compliance Monitoring, No. 14: The Application of the Principles of GLP to In Vitro Studies*, Organization for Economic Cooperation and Development, Paris, 2004.
31. Stokes, W. S. and Marafante, E., Alternative testing methodologies: the 13th meeting of the scientific group on methodologies for the safety evaluation of chemicals—introduction and summary, *Environ. Health Perspect.*, 106(2), 405–412, 1998.
32. Stokes, W. S., Validation of *in vitro* methods for toxicology studies, in *Biological Concepts and Techniques in Toxicology*, Riviere, J. E., Ed., Taylor & Francis, New York, 2006.
33. Corvi, R. et al., Validation of toxicogenomics-based test systems: ECVAM–ICCVAM/NICEATM considerations for regulatory use, *Environ. Health Perspect.*, 44, 111–117, 2006.
34. United Nations, *Globally Harmonized System of Classification and Labelling of Chemicals (GHS)*, United Nations, Geneva, 2003.
35. OECD, *Test Guideline 404, Acute Dermal Irritation/Corrosion*, Organization for Economic Cooperation and Development, Paris, 2002.
36. Interagency Coordinating Committee on the Validation of Alternative Methods (ICCVAM), *Validation and Regulatory Acceptance of Toxicological Test Methods: A Report of the Ad Hoc Interagency Coordinating Committee on the Validation of Alternative Methods*, NIH Publ. No. 97-3981, National Institute of Environmental Health Sciences, Research Triangle Park, NC, 1997 (http://iccvam.niehs.nih.gov/docs/docs.htm#general).
37. OECD, *Guidance Document No. 34, Guidance Document on the Validation and International Acceptance of New or Updated Test Methods for Hazard Assessment*, Organization for Economic Cooperation and Development, Paris, 2005.
38. Interagency Coordinating Committee on the Validation of Alternative Methods (ICCVAM), *Guidelines for the Nomination and Submission of New, Revised, and Alternative Test Methods*, NIH Publ. No. 03-4508, National Institute of Environmental Health Sciences, Research Triangle Park, NC, 2003 (http://iccvam.niehs.nih.gov/docs/guidelines/subguide.htm).
39. Balls, M. et al., Practical aspects of the validation of toxicity test procedures, the report and recommendations of ECVAM Workshop 5, *ATLA*, 23, 129–147, 1995.
40. OECD, *Guidance Document for the Development of OECD Guidelines for Testing of Chemicals*, Organization for Economic Cooperation and Development, Paris, 1995,
41. ISO, *Biological Evaluation of Medical Devices, Part 11. Test for Systemic Toxicity*, Ref. No. ISO 10993-11:1993(E), International Organization for Standardization, Geneva, Switzerland, 1993.
42. OECD, *OECD Principles of Good Laboratory Practice*, Organization for Economic Cooperation and Development, Paris, 1998.
43. EPA, *EPA–FIFRA Good Laboratory Practice Standards (GLPs) Enforcement Policy*, U.S. Environmental Protection Agency, Washington, D.C. (http://www.epa.gov/compliance/resources/policies/civil/fifra/fifraglp).
44. FDA, *Good Laboratory Practices*, U.S. Food and Drug Administration, Washington, D.C. (http://www.fda.gov/ora/compliance_ref/bimo/glp/default.htm).
45. Paris, M. W. et al., Reproducibility analyses for *in vitro* neutral red uptake methods from a validation study to evaluate *in vitro* cytotoxicity assays for estimating rodent acute systemic toxicity, *Toxicologist*, 90(S-1), 403, 2006.
46. Statutory Instrument 2002 No. 473, The Animals (Scientific Procedures) Act 1986, as amended (http://www.opsi.gov.uk/si/si2002/20020473.htm.)
47. Stokes, W. S. et al., The use of test method performance standards to streamline the validation process, in *Proc. of the 5th World Congress on Alternatives and Animal Use in the Life Sciences*, August 21–25, 2005, Berlin, Germany.
48. Blaauboeur, B. J., Barratt, M. D., and Houston, J. B., The integrated use of alternative methods in toxicological risk evaluations: ECVAM integrated testing strategies task force report I, *ATLA*, 27, 229–237, 1999.
49. Interagency Coordinating Committee on the Validation of Alternative Methods (ICCVAM), *Recommended Performance Standards for In Vitro Test Methods for Skin Corrosion*, NIH Publ. No. 04-4510, National Institute of Environmental Health Sciences, Research Triangle Park, NC, 2004 (http://iccvam.niehs.nih.gov/docs/docs.htm#invitro).
50. Balls, M., Defined structural and performance criteria would facilitate the validation and acceptance of alternative test procedures [editorial], *ATLA*, 25, 483–484, 1997.
51. Interagency Coordinating Committee on the Validation of Alternative Methods (ICCVAM), *The Murine Local Lymph Node Assay: A Test Method for Assessing the Allergic Contact Dermatitis Potential of Chemicals/Compounds*, NIH Publ. No. 99-4494, National Institute of Environmental Health Sciences, Research Triangle Park, NC, 1999 (http://iccvam.niehs.nih.gov/methods.llna.htm).
52. Interagency Coordinating Committee on the Validation of Alternative Methods (ICCVAM), *The Revised Up-and-Down Procedure: A Test Method for Determining the Acute Oral Toxicity of Chemicals*, NIH Publ. No. 02-4501, Research Triangle Park, NC, 2001 (http://iccvam.niehs.nih.gov/docs/docs.htm#udp).
53. Interagency Coordinating Committee on the Validation of Alternative Methods (ICCVAM), *ICCVAM Evaluation of In Vitro Test Methods for Detecting Potential Endocrine Disruptors: Estrogen and Androgen Receptor Binding and Transcriptional Activation Assays*, NIH Publ. No. 03-4503, National Institute of Environmental Health Sciences, Research Triangle Park, NC, 2003 (http://iccvam.niehs.nih.gov/methods/endodocs/edfinrpt/edfinrpt.pdf).
54. Interagency Coordinating Committee on the Validation of Alternative Methods (ICCVAM), *ICCVAM Biennial Progress Report: 2004-2005*, NIH Publ. No. 06-4516, National Institute of Environmental Health Sciences, Research Triangle Park, NC, 2006.
55. Interagency Coordinating Committee on the Validation of Alternative Methods (ICCVAM), *Report of the International Workshop on In Vitro Methods for Assessing Acute Systemic Toxicity: Results of an International Workshop Organized by the Interagency Coordinating Committee on the Validation of Alternative Methods and the National Toxicology Program Interagency Center for the Evaluation*

*of Alternative Toxicological Methods (NICEATM)*, NIH Publ. No. 01-4499, National Institute of Environmental Health Sciences, Research Triangle Park, NC, 2001 (http://iccvam.niehs.nih.gov/methods/invidocs/finalall.pdf.)

56. Balls, M., The establishment of ECVAM and its progress since 1993, *ATLA*, 30(Suppl. 2), 5–11, 2002.
57. Hartung, T. et al., ECVAM's response to the changing political environment for alternatives: consequences of the European Union Chemicals and Cosmetics Policies, *ATLA*, 31, 473–481, 2003.
58. Eskes, C. and Zuang, V., Eds., Alternative (non-animal) methods for cosmetics testing: current status and future prospects, a report prepared in the context of the 7th amendment to the cosmetics directive for establishing the timetable for phasing out animal testing, *ATLA*, 33, S1, 2005.
59. Stokes, W. S. and Schechtman, L. M., The Interagency Coordinating Committee on the Validation of Alternative Methods (ICCVAM): a review of the ICCVAM test method evaluation process and current international collaborations with the European centre for the validation of alternative methods (ECVAM), *ATLA*, 30(Suppl. 2), 23–32, 2002.
60. Schechtman, L. M. and Stokes, W. S., ECVAM–ICCVAM: prospects for future collaboration, *ATLA*, 30(Suppl. 2), 227–236, 2002.
61. Ohno, Y., ICH Guidelines, implementation of the 3Rs (refinement, reduction, and replacement): incorporating best scientific practices into the regulatory process, *ATLA*, 43(Suppl.), S95–S98, 2002.
62. Spielmann, H. and Liebsch, M., Validation successes: chemicals, *ATLA*, 30(Suppl. 2), 33–40, 2002.
63. Basketter, D. A. et al., Dinitrohalobenzenes: evaluation of relative skin sensitization potential using the local lymph node assay, *Cont. Dermat.*, 36, 97–100, 1997.
64. Basketter, D. A., Gerberick, G. F., and Kimber, I., Applying immunology to allergen identification: the local lymph node assay, *Trends Pharmacol. Sci.*, 22, 264–265, 2001.
65. Sailstad, D., Hattan, D., Hill, R., and Stokes, W., Evaluation of the murine local lymph node assay (LLNA). I. The ICCVAM review process, *Regul. Toxicol. Pharmacol.*, 34(3), 249–257, 2001.
66. Dean, J., Twerdok, L., Tice, R., Sailstad, D., Hattan, D., and Stokes, W., Evaluation of the murine local lymph node assay (LLNA). II. Conclusions and recommendations of an independent scientific peer review panel, *Regul. Toxicol. Pharmacol.*, 34(3), 258–273, 2001.
67. Haneke, K. Tice, R., Carson, B., Margolin, B., and Stokes, W. S., Evaluation of the murine local lymph node assay (LLNA). III. Data analyses completed by the National Toxicology Program (NTP) Interagency Center for the Evaluation of Alternative Toxicological Methods (NICEATM), *Regul. Toxicol. Pharmacol.*, 34(3), 274–286, 2001.
68. OECD, *Test Guideline 429, Skin Sensitization: Loca Lymph Node Assay*, Organization for Economic Coopera tion and Development, Paris, 2002.
69. Landsteiner, K. and Jacobs, J., Studies on sensitization c animals with simple chemical compounds, *J. Exp. Med* 61, 643–656, 1935.
70. Marzulli, F. N. and Maibach, H. I., Eds., *Dermatotoxico ogy*, 5th ed., Taylor & Francis, New York, 1996.
71. Selgrade, M. J. K. et al., Immunotoxicity. In: *Introductio to Biochemical Toxicology*, 3rd ed., Hodgson, E. an Smart, R. C., Eds., John Wiley & Sons, New York, 200 pp. 597–561.
72. Sailstad, D. M., The murine local lymph node assay: a alternative test method for skin hypersensitivity testin *Lab. Anim.*, 31, 1–6, 2002.
73. Kimber, I. and Basketter, D. A., The murine local lymp node assay: a commentary on collaborative studies an new direction, *Food Chem. Toxicol.*, 30, 165–196, 1992
74. Loveless, S. E. et al., Further evaluation of the local lymp node assay in the final phase of an international collabo rative trial, *Toxicology*, 108, 141–152, 1996
75. Kimber, L. et al., An international evaluation of the muri local lymph node assay and comparison of modified pr cedures, *Toxicology*, 103, 63–73, 1995.
76. Basketter, D. A. et al., The local lymph node assay: a viab alternative to currently accepted skin sensitization test *Food Chem. Toxicol.*, 34, 985–997, 1996.
77. Kimber, I. et al., The local lymph node assay: develo ments and applications, *Toxicology*, 93, 13–31, 1994.
78. Tilney, N. L., Patterns of lymphatic drainage in the adu laboratory rat, *J. Anat.*, 109, 369–383, 1971.
79. Fentem, J. H. et al., The ECVAM international validati study on *in vitro* tests for skin corrosivity. 2. Results a evaluation by the management team, *Toxicol. In Vitro*, 1 483–524, 1998.
80. Balls, M. and Corcelle, G., Statement on the scienti validity of the rat skin transcutaneous electrical resistan (TER) test (an *in vitro* test for skin corrosivity), *ATLA*, 2 275–277, 1998.
81. Balls, M. and Hellsten, E., Statement on the application the EpiDerm™ human skin model for corrosivity testin *ATLA*, 28, 365–367, 2000.
82. Balls, M. and Corcelle, G., Statement on the scienti validity of the EPISKIN™ test (an *in vitro* test for sk corrosivity), *ATLA*, 26, 278–280, 1998.

# 22 Acute Toxicity and Eye Irritancy

*Mark E. Blazka and A. Wallace Hayes*

**CONTENTS**

The methods and principles of evaluating two categories of hazards—acute systemic toxicity and eye irritation, both resulting from a single or very short-term exposure—are described in this chapter. In recent years, economics and concerns over animal welfare have raised many issues in animal testing. Alternative methods for determining acute toxicity and eye irritation are being developed and, in some cases, accepted by regulatory agencies for hazard assessment purposes. This chapter describes the classical and currently accepted methods for evaluating the potential of a test article for acute systemic toxicity and eye irritation and gives an overview of the current regulatory testing requirements in the United States, European Union, and Japan.

## PRINCIPLES OF ACUTE TOXICOLOGY

Acute toxicity testing began nearly a century ago when physicians and pharmacologists were concerned with potent poisons and drugs. In 1927, Trevan [196] introduced the concept of a median lethal dose ($LD_{50}$) for the standardization of digitalis extracts, insulin, and diphtheria toxin. He recognized that the precision of the $LD_{50}$ value was dependent on many factors, such as seasonal variation and the number of animals used in a test. High-precision $LD_{50}$ values can only be established with a large number of animals.

The list of extraneous factors that affect the precision of the $LD_{50}$ includes, among other factors, sex, species, strain, age, diet, nutritional status, general health conditions, animal husbandry, experimental procedures, route of administration, stress, dosage formulation (vehicle) and intra- and interlaboratory variations. In spite of the many variables affecting the $LD_{50}$ determination, many governmental agencies still regard the $LD_{50}$ as the sole measurement of the acute toxicity of all materials; however, recent research has resulted in the development and gradual acceptance of alternative methods to assess the acute oral toxicity potential of test materials.

It is important to accurately determine the killing power of highly toxic substances, as a small difference in exposure can distinguish a safe from a lethal exposure; however, a precise $LD_{50}$ is not necessary when evaluating less toxic materials, such as pesticides and consumer products. Although evaluating the safety of products with low to moderate toxic potential is still desirable, an approximate measurement of their toxic potential is sufficient for the purposes of risk and exposure assessment. Furthermore, it may not be scientifically sound to determine a precise $LD_{50}$ value for substances with a low to moderate potential for toxicity. This is because extraneous variables, some of which cannot be controlled by the experimenter, and errors inherent in the determination of $LD_{50}$ values can have a significant impact on the stud results. Many methods have been developed over the years to calculate the $LD_{50}$ and evaluate the acute toxicity potential of chemicals with a small number of animals. Some of these methods are discussed later in this chapter.

Many scientists have advocated changes in the emphasis of acute toxicity testing. To date, there is a general consensus among toxicologists in academia, industry, and government that the emphasis of acute toxicity testing must change [6,16,58,71,110,117,140,184,188]. The value of a precise $LD_{50}$, except for highly toxic substances, should be deemphasized, and the focus should be on obtaining as much information as possible on the toxic manifestation and mechanism using the fewest number of animals. Alternative methods for assessing the oral [159–161], dermal [148], and inhalation [149,152] toxic potential of test material have been proposed or adopted by various regulatory agencies. Undoubtedly, such information will be more useful than the $LD_{50}$ to physicians in treating overexposure.

Even though the emphasis of acute toxicity testing is changing, the principles of dose–response and development of signs of toxicity remain the basis of the science of toxicology. It is the objective of this section to refresh the experienced and introduce the novice to these general concepts.

## Definition of Acute Toxicity

Toxicity is defined as the harmful effect of a chemical or a drug on a living organism. Various expert groups have defined acute and subchronic toxicities. The Organization for Economic Cooperation and Development (OECD) [157] defines acute toxicity as "the adverse effects occurring within a short time of (oral) administration of a single dose of a substance or multiple doses given within 24 hours." In terms of human exposure, this definition of acute toxicity refers to life-threatening crises such as accidental catastrophes, overdoses, and suicide attempts.

## Dose–Response Relationship

Toxicologists often obtain two types of data: quantal and graded. The quantal response is an "all-or-none" response; it either happens or it does not happen. On the other hand, the graded response can be determined quantitatively, and it is continuous. Mortality and incidences of pharmacotoxic signs are examples of quantal data, whereas enzyme activity, protein concentration, body weight, feed consumption, and electrolyte concentration are quantitative parameters. Many apparently quantal responses are quantitative. If technical measurements permit, they may be graded; for example, the severity of a pharmacotoxic sign can be graded if detection methods are available.

At the molecular level, the graded dose–response relationship often can be explained by the receptor, a relatively old concept but still a valid one. Let $S$ be a particular substance that produces a specific response by interacting with a target protein molecule, the receptor ($R$), in the body to form a substance–receptor complex ($SR$). Assuming that the reaction is reversible and there is only one binding site on every target receptor molecule, this process can be described by the following expression:

$$S + R \underset{k_2}{\overset{k_1}{\rightleftarrows}} SR$$

and the mass equation for this reversible process is:

$$\frac{k_2}{k_1} = K_d = \frac{[S][R]}{[SR]} \tag{22.1}$$

where $[S]$, $[R]$, and $[SR]$ are the concentrations of the substance, the receptor, and the substance–receptor complex, respectively, at any particular time, and $K_d$ is the dissociation constant of the process. Let $[R]_0$ be the initial concentration of the receptor, which is usually very small and constant in number when compared with the concentration of the substance. Then,

$$[R]_0 = [R] = [SR]$$

thus,

$$[R] = [R]_0 - [SR]$$

Substituting the above into the mass equation (Equation 22.1) and rearranging:

$$[SR]K_d = [S]([R]_0 - [SR])$$

or

$$[SR](K_d + [S]) = [R]_0[S]$$

which can be rearranged to:

$$\frac{[SR]}{[R]_0} = \frac{[S]}{K_d + [S]} \tag{22.2}$$

where $[SR]/[R]_0$ is the fraction of receptor that has reacted with the substance to form the substance–receptor complex. If we assume that the response ($E$) resulting from the interaction of the substance with the receptor is dependent on the fraction of total receptor concentration that has reacted with the substance, then

$$E = \frac{[S]}{K_d + [S]} \tag{22.3}$$

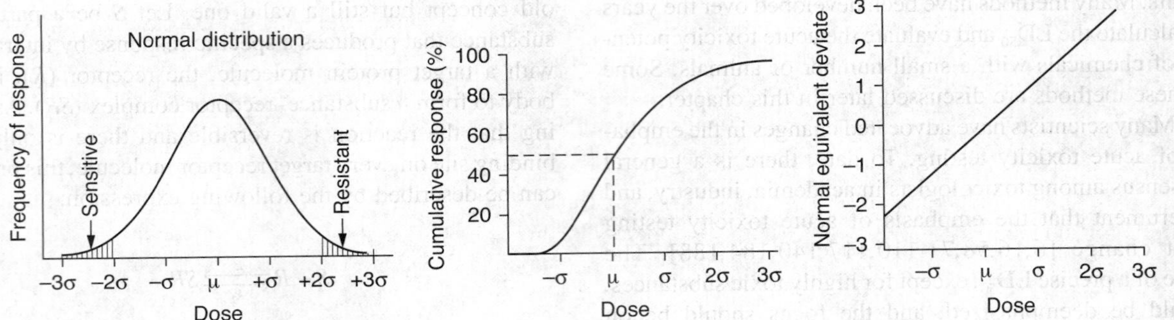


**FIGURE 22.1** Normal distribution of dose–response relationships: frequency of response, cumulative response, and cumulative response in terms of normal equivalent deviate.

Equation 22.3 is a hyperbolic function; therefore, the response ($E$) is related to the concentration of the substance in a hyperbolic function relationship. If the concentration of the substance at the receptor site is dependent on the dose, then the response is dependent on the dose administered. This phenomenon is perhaps the simplest version of the receptor kinetic concept relating the dose of the chemical to a biological response. The kinetics of the substance–receptor interaction may be more complicated, and different dose–response relationships could be drawn based on these complicated kinetics. Readers who are interested in different substance–receptor kinetics are referred to Ferdinard [72].

The quantal dose–response relationship often is difficult to conceptualize based on the receptor theory; however, a quantal response can also be viewed as a graded response if the whole population is considered as an individual. This relationship can best be explained in terms of a probability distribution. For a particular response, members of a population (e.g., all the rats in the world) respond differently to a particular stimulus, such as exposure to a chemical. Some rats will be highly sensitive whereas others will be very resistant. If these different responses are distributed normally within the population (i.e., with most members of the population being neither extremely sensitive nor resistant), the well-known bell-shaped population distribution curve results. If the probability of dose–response is expressed in terms of cumulative response, a sigmoidal curve can be obtained as shown in Figure 22.1. Most biological response distributions, however, are not exactly normal and tend to be skewed to the higher dose; that is, extreme resistors have a larger *range of dose* to response than the extremely sensitive portion of the population. In general, a logarithmic dose transformation can normalize the distribution (i.e., convert the skewed distribution to a normal distribution) (Figure 22.2). After this logarithmic dose transformation, if the probability of the log dose–response is expressed cumulatively, the sigmoidal response curve is obtained (Figure 22.2). How is this lognormal transformation related to a regular dose–response curve? Is there justification or basis for a log dose transformation? To answer these questions let us again look at Equation 22.3. This equation can be arranged to:

$$E = \frac{[S]}{k_2/k_1 + [S]}$$

which can be rearranged to:

$$E = \frac{k_1[S]}{k_2 + k_1[S]} \quad (22.4$$

Over a certain concentration range, Equation 22.4 will produce a curve very similar to the logarithmic function $E = K_1 \log(k_2[S] + 1)$ [34]; therefore, there may be justification for the log transformation besides simply a mathematical convenience.

Because a sigmoidal curve is more difficult to analyze than a straight line, many experts feel that further transformation of the log dose–response hyperbolic function is necessary to obtain a straight-line function curve. Perhaps the most widely used transformation is the normal equivalent deviate (NED) or the similar probit transformation [12,35,38,44,50,75,130]. This technique involves the log dose transformation and the transformation of the cumulative response probability to the NED or probit. After both the probability and the dose are transformed, their transformed values are directly related to each other. A brief derivation of the straight-line direct function relationship between the log dose and NED or probit will be presented later in this chapter.

## $LD_{50}$ and Its Determination

### Definition

The $LD_{50}$ in its simplest form is the dose of a compound that causes 50% mortality in a population. A more precise definition has been provided by the OECD panel of exper

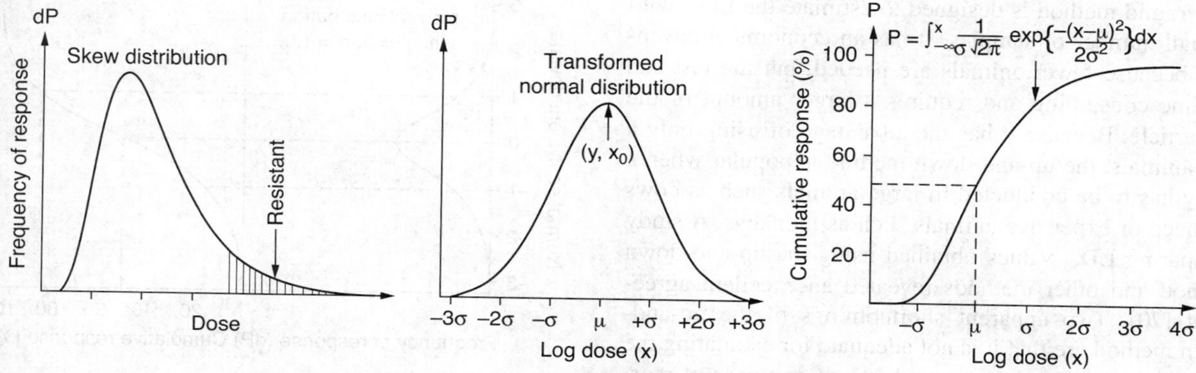


**FIGURE 22.2** Skew of dose–response can be normalized by log dose transformation.

as the "statistically derived single dose of a substance that can be expected to cause death in 50% of the animals" [157]. In other words, an $LD_{50}$ of a compound is not a constant, as it has been treated by many; rather, it is a statistical term designed to describe the lethal response of a compound in a particular population under a discrete set of experimental conditions.

The significance of the $LD_{50}$ has been examined by many scientists [6,58,71,110,117,140,184,188] who have concluded that it is an imprecise value and not a biological constant; furthermore, its importance should be deemphasized for most types of materials. For most materials, an approximate $LD_{50}$ value is sufficient and more emphasis should be placed on characterizing the signs of toxicity, identifying the target organs, and elucidating the mechanism of action of the material.

The numeric value of the $LD_{50}$ has been widely used to classify and compare the toxic potential of chemicals; the importance placed on the $LD_{50}$ and how it is used in a safety evaluation has almost reached a level of abuse. Although determining the $LD_{50}$ under a set of experimental conditions can provide valuable information about the toxicity of a compound, the numeric $LD_{50}$ *per se* is not equivalent to acute toxicity. It should always be remembered that lethality is only one of many reference points used to characterize acute toxicity. When evaluating the acute toxic potential of a material, the slope (response/dose) of the dose–response curve, time to death, pharmacotoxic signs, and pathological findings all must be considered, as they are critical endpoint; therefore, defining acute toxicity based only on the numeric value of an $LD_{50}$ is inappropriate.

As pointed out in a previous paragraph, lethality is a quantal response, and the probability of a cumulative response is related to dose in a hyperbolic (sigmoidal) function. The cumulative probability of response is directly related to the standard deviates of a log dose population (Figure 22.1); therefore, the slope of the log dose–response curve will indicate the relationship between the range of dose and the lethal response. This relationship is more important in risk assessment than the numeric value of the $LD_{50}$ because more insight is available about the intrinsic toxic characteristics of a compound. Sometimes the slope can give a clue to the mechanism of toxicity; for example, a steep slope may indicate rapid onset of action or faster absorption. A large margin of safety is predicted when a compound has a flat slope (i.e., only a small increase in response with a large increase in dose). With the slope, it often is possible to extrapolate the response to a low dose (e.g., $LD_0$, $LD_1$) or even to a no-observed-effect level (NOEL). It is especially important to know the slope when comparing a set of compounds. Two compounds may have identical $LD_{50}$ values but different slopes and thus quite different toxicological characteristics, depending on the range of doses. Parallel dose–response curves may indicate a similar mechanism of toxicity, kinetic pattern, and probably similar prognosis. Neither the $LD_{50}$ nor the slope can reveal absolutely a specific mechanism, but with pharmacokinetic and other biochemical studies elucidation of the mechanism of toxicity may be possible.

## Determination of $LD_{50}$

Many methods are available for the determination of the $LD_{50}$. They can be grouped into two categories: the *normal population assumption* and the *normal population assumption-free* methods. The former usually can be analyzed by graphic procedures.

The *normal population assumption-free* methods are represented by Thompson's moving average interpolation [194] and the up-and-down method [19–22,34,46]. The former method is widely accepted, and convenient tables [50,202] are available for estimation of the value of the $LD_{50}$ with confidence limits when either 0 or 100% mortality incidences are observed. There are some restrictions on the use of the Thompson method, such as the four doses must be at equal log dose intervals and the number of animals per dose level must be equal. The up-and-down or pyramid method is designed to estimate the $LD_{50}$ with a small number of samples. It has an economical advantage because fewer animals are needed, but the test can be time consuming and requires a larger amount of the test article. Because it has the advantage of using only a few animals, the up-and-down method is popular when a study has to be conducted in large animals such as cows or sheep or expensive animals such as monkeys. A study comparing $LD_{50}$ values obtained using the up-and-down method and other methods revealed an excellent agreement [20]. Two apparent shortcomings of the up-and-down method are that it is not adequate for estimating the incidence of delayed deaths and a dose–response of mortality or signs of toxicity cannot easily be obtained; however, Weil [203] has adapted the up-and-down method to calculate the slope of acute toxicity response.

The *normal population assumption* method is represented by the probit analysis approach, which can be performed either by graphic means [75] or by mathematical calculation [76]. Because the probit analysis is used widely in evaluating acute toxicity data, the principles in performing this analysis will be discussed briefly. This method involves the transformation of both the cumulative response probability and dosage data.

When the dose is transformed into a log dose ($x$), the frequency of response vs. log doses follows a normal distribution (Figure 22.2), which can be expressed mathematically as

$$dP = \frac{1}{\sigma\sqrt{2\pi}}\exp\left(\frac{-(x-u)^2}{2\sigma^2}\right) \quad (22.5)$$

where $\sigma^2$ and $u$ are the variance and the mean of the population, respectively, and $P$ is the probability corresponding to each value of $x$ (Figure 22.2). The $LD_{50}$ is defined as the log dose that can produce 50% mortality in a population (i.e., $P = 0.5$ or 50% cumulative response). Let $x$ be the log $LD_{50}$. Then, $P = 0.5$ will correspond to the area under the log normal distribution curve from $-\infty$ to $x_0$, or $P = 0.5$ will correspond to the integration of the Equation 22.5 from $-\infty$ to $x_0$: That is,

$$P = 0.5 = \int_{-\infty}^{x_0} \frac{1}{\sigma\sqrt{2\pi}}\exp\left(\frac{-(x-u)^2}{2\sigma^2}\right)dx \quad (22.6)$$

The solution of Equation 22.6 is $x = u$, the true mean or the median of the log-normal distribution. One way to solve this equation is by a graphic method. The integration of Equation 22.5 from $x = -\infty$ to $+\infty$ can be represented graphically by a sigmoidal curve as illustrated in Figure 22.2. Analysis of the sigmoidal curve is more difficult than analysis of a straight line. One way to transform the sigmoidal curve to a straight line is by NED analysis or similarly by probit analysis. For a detailed description of

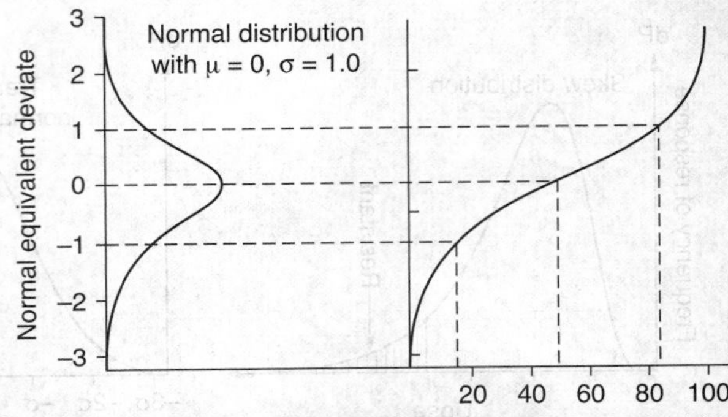


**FIGURE 22.3** Probability of response can be expressed in term of percentage of the population or the NED of a normal distr bution with mean = 0 and standard deviation = 1.

this analysis, the reader should consult Finney's text [75 A brief derivation of the straight-line function betwee log dose and the transformed probability of response i described below.

Probability ($P$) is normally expressed in terms of pe centage or with values between 0 and 1, but Gaddum [8 proposed measuring the probability of response on a trans formed scale, the normal equivalent deviate, or the sta dard deviation of a normal distribution, which can b described mathematically by Equation 22.5. In a particul case, the normal distribution of response with mean equ to 0 and the standard deviation equal to 1, Equation 22 can be written as:

$$dP = \frac{1}{\sqrt{2\pi}}\exp\left(\frac{-x^2}{2}\right)dx$$

Similarly, if this distribution of response is plotted on th $y$-axis (Figure 22.3), then:

$$dP = \frac{1}{\sqrt{2\pi}}\exp\left(\frac{-y^2}{2}\right)dx \quad (22.7)$$

The probability in such a case is defined by a value c the $y$-axis of Figure 22.3 (i.e., the integration of Equatic 22.7 from $-\infty$ to $y$):

$$P = \frac{1}{\sqrt{2\pi}}\int_{-\infty}^{y}\exp\frac{-y}{2}dy \quad (22.$$

In other words, for each value of $y$ (from $-\infty$ to $+\infty$ expressed in terms of the standard deviation of a norm distribution with the mean equal to 0 and the standa deviation equal to 1, there is a corresponding value probability ($P$) expressed in terms of percentage or havi a value ranging from 0 to 1. Thus, equivalent values the $y$-axis can be used to define the value of $P$ or *vice versa*; $y$ and $P$ define each other. This relationship is illustrated in Figure 22.3.

The particular probability of response to a particular log dose value $x$, as described in Equation 22.6, will be:

$$P = \int_{-y}^{-x} \frac{1}{\sigma\sqrt{2\pi}} \exp\left(\frac{-(x-u)^2}{2\sigma^2}\right) dx \qquad (22.9)$$

where $u$ and $\sigma$ are the mean and standard deviation of the log dose, respectively.

If $P$ is expressed by a value of $y$ on the $y$-axis (standard deviations), then:

$$P = \frac{1}{\sqrt{2\pi}} \int_{-\infty}^{y} \exp\left(\frac{-y^2}{2}\right) dy$$
$$= \int_{-\infty}^{x} \frac{1}{\sigma\sqrt{2\pi}} \exp\left(\frac{-(x-u)^2}{2\sigma^2}\right) dx$$

The solution of this equation is $x = u + \sigma y$, or:

$$y = \frac{(x-u)}{\sigma} = \frac{1}{\sigma}x - \frac{u}{\sigma} \qquad (22.10)$$

Therefore, the probability when expressed in terms of $y$ (the NED scale) is related linearly to $x$, the log dose. If $x$ is plotted against the corresponding $y$, a straight line with slope = $1/\sigma$ will be obtained. To further facilitate calculation, Bliss [11] suggested a slightly different NED unit called the *probit*, such that the new $y$ value is equal to $[(x - u)/\sigma] + 5$. This procedure eliminates the negative values of NED when $P$ has a value of less than 50%; therefore, the probit is equal to the NED plus 5. The linear relationship between probits and log dose is similar to the relationship between NED and log dose; thus, when $y = 5$, from Equation 22.10,

$$5 = \frac{(x-u)}{\sigma} + 5$$

and $x = \mu$ (i.e., the median log dose which has a probability of response of 50%).

## ESTIMATION OF $LD_{50}$ BY PROBIT ANALYSIS

The basic linear equation for the probit analysis as described in the previous section is:

$$y = 5 + \frac{1}{\sigma}(x - u)$$

where $y$ is the probit, $\sigma$ is the standard deviation of a log-normal distribution with mean $u$, and $x$ is the log dose. This equation is linear with respect to $y$ and $x$ and often can be expressed as a linear equation; for example, $y = \alpha + \beta x$, where $\beta = 1/\sigma$ = slope, and $\alpha = 5 - (u/\sigma)$. When $y = 5$, $(x - u)/\sigma = 0$; thus, $x = \mu$ (the median log dose). Further, $y$ is related to $P$ (the probability of response that has a value of 0 to 1) by the following equation:

$$P = \frac{1}{\sqrt{2\pi}} \int_{-x}^{y} \exp\left(\frac{-y}{2}\right) dy$$

The reader should bear in mind that both the $u$ and $x$ are in log dose scale.

The following steps should be taken for graphic estimation of $LD_{50}$ by probit analysis:

1. Convert response probabilities to probit units by a probit transformation table (see Diem and Lentner [45], pp. 54–55).
2. Convert all doses into log dose units (e.g., $\log_{10}$ dose = $x$). (Steps 1 and 2 may be eliminated if probit–log graphic paper is available.)
3. Using the probit as the abscissa and $\log_{10}$ dose as the ordinate, plot the response probit units against the $\log_{10}$ dose.
4. Draw a straight line such that the vertical deviations of points (the probits) at each $x$ value are as small as possible. Extreme probits (e.g., those outside the range of probit 7 and 1) carry little weight in the fitting of the probit–log dose response line and thus should be excluded.
5. From the regression of the probit–log dose line, extrapolate the log dose corresponding to probit units of 5, which also correspond to the $P = 0.5$. This extrapolated dose should be the median lethal *log dose*, and the $LD_{50}$ value would be the *antilog* of this log dose value.
6. Calculate the slope of the probit–log dose line. This slope, $\beta = 1/\sigma$, is defined as the number of increases in probit units for a unit increase in log dose. The slope defined by Litchfield and Wilcoxon [116] is equal to:

$$\frac{1}{2}\left(\frac{LD_{84}}{LD_{50}} + \frac{LD_{50}}{LD_{16}}\right) = \sigma$$

   This slope is different but related to the slope described here; thus, the larger the slope value, the steeper the probit–log dose response. The opposite is true in the Litchfield and Wilcoxon definition.
7. A $\chi^2$ test should be conducted to determine if the fitted line is adequate. A small value of the $\chi^2$ statistic (within the limits of random variation) may indicate satisfactory agreement between the

theoretically expected line and the fitted line. A significantly large $\chi^2$ statistic may indicate either that the animals do not respond independently or that the fitted line (probit–log dose) does not adequately describe the dose–response relationship of the test substance. If the latter is true, forms of the dose–response curve other than the probit–log dose linearity may exist, and further transformation may be needed [75]. If the former is the case, then precision of the line is reduced.

8. Determination of precision is by weighting the coefficient. The standard deviation of a binomial distribution is $\sqrt{PQ/n}$, where $P$ and $Q$ are the mean probabilities, $P$ equals $(1 - Q)$, and $n$ is the number of test subjects. Thus, the variance is $PQ/n$, the square of the standard deviation. It is obvious that the variance (i.e., the spread of a distribution) is inversely related to $n$. This relationship means that the larger the number of test subjects, the smaller the variance and the better the precision. The reciprocal of the variance is invariance, which measures the weight ($nW$). Here, $W$ (weighting coefficient) = $Z^2/PQ$, where:

$$Z = \left(1/\sqrt{2\pi}\right)\exp\left(-y^2/2\right)$$

and is related to the normal frequency function corresponding to the NED. A table of weighting coefficients (see Diem and Lentner [45], p. 55) corresponding to probits ($y$) is available [76]. The standard error for the log $LD_{50}$ is given by:

$$\sigma/\sqrt{\sum nW}$$

if the estimated log $LD_{50}$ does not greatly differ from the true mean log $LD_{50}$, because this estimation does not take into consideration the error in the estimation for the probit–log dose–response line. A better equation for the estimation of the variance of the estimated log $LD_{50}$ is given by:

$$V(m) = \sigma^2\left(\frac{1}{\sum nW} + \frac{(m-\bar{x})^2}{\sum nW(x-\bar{x})^2}\right)$$

where $V(m)$ is the variance of $LD_{50}$, $\bar{x}$ is the weighted mean log dose, $m$ *is* the median log dose, $x$ is the log dose, and $1/\sigma = 1/\Sigma nW(x - \bar{x})$. If the $\chi^2$ is large, indicating that the test subjects do not respond independently to the dose, the estimation of variance of log $LD_{50}$ may not apply, and adjustment due to the sampling variation of the slope $(1/\sigma)$ of the probit–log dose line may have to be made [75]. For a quick estimation of the $LD_{50}$, this adjustment may be dropped, and the standard error would be the square root of the variance (i.e., $\sqrt{V(m)}$). One must remember that the dose is expressed in log dose; therefore, estimation of the standard error (SE) for the $LD_{50}$ in the original dose unit (e.g., mg/kg) is impossible. However, an approximation is given by:

$$SE(LD_{50}) = (10^m) \times \left[(\log_e(10) \times (S_m)\right]$$

where $S_m$, which equals:

$$\sigma/\sqrt{\sum W} \text{ or } \sqrt{V/(m)}$$

is the estimated standard error for the median log dose $m$ (i.e., $m = \log LD_{50}$ or $10^m = LD_{50}$). A more rapid approximation of the standard error of log $LD_{50}$ was given by Litchfield and Wilcoxon [116] as:

$$S_m = \frac{S}{N'/2}$$

where $S$ is the difference between two log doses of expected effects (as indicated by the probiting dose line) that differ by one unit of probit, and $N'$ is the total number of animals between the log dose limits, corresponding to the expected probit 4.0 to 6.0 (i.e., the 16% and 80% responses).

9. The concept of a *fiducial limit* is similar to the confidence limit. The value of the two may be the same, but they are not always identical. The fiducial probability $F$ (e.g., 95%) can be defined as the situation when the true value of a parameter lies between the calculated upper and lower limits, which would not be contradicted by a significance test at the $1/2(1 - F)$ probability level. These higher and lower limits are referred to as the fiducial limits. For rapid analysis, the fiducial limits at the $F = 95\%$ level can be estimated by log $LD_{50} \pm 1.96$ $(S_m)$. A more detailed estimation can be obtained by the maximum likelihood estimation [75]. Another simple approximation of the fiducial limits is given by Litchfield and Wilcoxon [116] as $LD_{50}/fLD_{50}$ and $LD_{50} \times fLD_{50}$ for the lower and upper limits, respectively, where $LD_{50}$ is defined as the $LD_{50}$ factor equal to:

$$(s)\left(2.77/\sqrt{N'}\right)$$

Here, $s$ is the slope, which is defined as:

$$\frac{1}{2}\left(\frac{LD_{54}}{LD_{50}} + \frac{LD_{50}}{LD_{16}}\right) = \frac{1}{2}(3.55 + 3.55) = 3.55$$

in this example, and $N'$ is the total number of animals used between response probabilities 16% and 84% (i.e., probit 4 and 6, equal to 30 in this example). Then, $fLD_{50}$ equals 1.896; therefore, the lower fiducial limit is equal to 8.91/1.896 = 4.70, and the upper fiducial limit is equal to 8.96 × 1.896 = 16.90.

## Logistic Transformation

Waud [201] suggested a logistic approach to calculate the $LD_{50}$. Thus,

$$P = \frac{D^E}{(D^E + K^E)}$$

where $P$ is the probability of response; $D$ is the dose; $E$ and $K$ are scale and location parameters, respectively; and $K$ corresponds to the $LD_{50}$. With the procedure of iteration, $K$ and $E$ can be estimated with a range of confidence. The derivation of this equation is beyond the scope of this chapter, and interested readers are referred to the original article by Waud [201].

## Nonlethal Parameters

Although the $LD_{50}$ and the slope of the dose–response curve can provide valuable information on the toxicity of a compound, the $LD_{50}$ *is not equivalent to toxicity*. Chemicals can induce damage to the physiological, biochemical, immunological, neurological, or anatomical systems not characterized by the $LD_{50}$. Depending on the severity and extent of the disturbance of the normal biological functions, the animal may survive the toxic response even though some irreversible tissue damage may have occurred. Nonlethal, adverse effects are as undesirable as lethality and certainly should be taken into consideration during the risk assessment of a chemical.

A major problem in analyzing nonlethal responses is that in many cases the data are not quantal; for example, dermal toxicity ranges from slight to severe. These polyhotomous data may be handled by RIDIT analysis, which was designed to analyze quantal responses with more than two outcomes [1,18,92].

Although toxic effects may contribute to lethality, any attempt to correlate a particular nonlethal response to mortality may be irrational [189] unless that response is the only one responsible for the eventual death of the animal. Identification of the response or responses related to mortality is not often a straightforward matter. Nonlethal responses that affect the general well-being of an animal should be considered in the risk assessment of a compound.

If nonlethal responses can be viewed as true quantal data, the *median effective dose* ($ED_{50}$) and the corresponding dose–response curve may apply. The $ED_{50}$, which often is used in the standardization of biologically active compounds such as a drug, has a meaning similar to the $LD_{50}$ except that it is designated to examine nonlethal parameters such as pharmacological responses and other nonlethal adverse effects. The $ED_{50}$ is defined as a statistically derived single dose of a substance that can be expected to cause a particular effect to occur in 50% of the animal population. The therapeutic index (TI), defined by the ratio of $LD_{50}/ED_{50}$ or $LD_1/ED_{99}$, has been applied to establish the safety margin of some biologically active drugs. The higher the index, the greater the margin of safety with the drug; that is, a large difference exists between the amount of compound predicted to kill 50% of the animals and the amount of compound predicted to elicit a particular response in 50% of the animals. The TI gives an even greater estimate of safety when the $LD_1$ is compared with the $ED_{99}$.

## Reversibility of Nonlethal Parameters

In general, reversible responses are those that diminish with elimination of the chemical from the body. A true reversible response will cause no residual effects when the chemical is completely eliminated from the body. Such responses are commonly seen in drugs used at therapeutic dose levels. As the amount of drug in the body increases, the magnitude of the effect also increases. If it is truly reversible, the effect will wear off when the drug is completely eliminated.

The reversibility of a particular response is dependent on the organ or system involved, intrinsic toxicity of the chemical, length of exposure, total amount of the chemical in the body at a specific time, and the age and general health of the animal. If the amount of chemical in the body is high enough, the intensity of the response may overwhelm a particular organ. Effects indicated through hormonal imbalance such as thyroid effects generally are reversible unless the threshold is surpassed. Damage in rapidly regenerating organs such as liver is usually more likely to be reversible than damage in nonregenerating tissues such as nerves. A good example is the delayed onset of neuropathy caused by many organophosphate insecticides. The chemical may be completely eliminated from the body before the effect manifests itself. Animals with renal or liver diseases are often more susceptible to damage (reversible or irreversible) by a chemical insult because of decreased ability to eliminate the chemical. Exposure to a chemical at an early age may induce irreversible damage more easily than at an older age because of the limited development of the kidneys and functional capacity of other organs such as the liver. In risk assessment, it is important to know whether a toxic effect is reversible. Irreversible effects seen in animals obviously are weighted more heavily in reaching a conclusion on the toxicity and hazard a chemical may pose for humans.

## ACUTE TOXICITY TESTING

The objectives of acute toxicity testing are to define the intrinsic toxicity of the chemical, predict hazard to nontarget species or toxicity to target species, determine the most susceptible species, identify target organs, provide information for risk assessment of acute exposure to the chemical, provide information for the design and selection of dose levels for prolonged studies, and, most important and practical of all, provide valuable information for clinicians in the prediction, diagnosis, and treatment for acute overexposure (poisoning) to chemicals. Acute studies often are considered the first line of defense in the absence of data from long-term studies. These data help industrial, governmental, and academic institutions formulate safety measures for their researchers and for limited segments of their worker population during the early stage of the development of a chemical. From a regulatory standpoint, acute toxicity data are essential in the classification, labeling, and transportation of a chemical. From an academic standpoint, a carefully designed acute toxicity study can often provide important clues on the mechanism of toxicity and the structure–activity relationship for a particular class of chemicals.

Many acute toxicity studies have been conducted solely for the purpose of determining the $LD_{50}$ of a chemical; however, the reader is reminded that acute toxicity is not equivalent to the $LD_{50}$ and that the $LD_{50}$ is not an absolute biological constant to be equated, as many investigators have, with such chemical constants as pH, $pK_a$, melting point, and solubility. The $LD_{50}$ is only one of many indices used in defining acute toxicity. A well-designed acute toxicity study should include consideration of the dose–response relationship of both lethal and nonlethal parameters, as discussed above. Sometimes biochemical measurements in an acute test can aid in elucidating the mechanism of toxic actions. Histopathology of organs may be helpful in determining the cause of death and identifying the target organs.

The use of animals in acute toxicity studies has been widely debated for many years. Aside from valid scientific concerns regarding the usefulness of classic $LD_{50}$ values (e.g., uncertainty in species extrapolation, seldom needed for potent drug standardization) are broader issues on animal testing, some political in nature and others economically based. The cost of animal testing has been increasing at a skyrocketing rate over the last decades, and even without animal rights activism the scientific community will require less costly alternatives to cope with the increasing demand for safety evaluations of a vast number of existing and new chemicals. Reduction in the number of animals used and refinement of existing testing methods to minimize pain and suffering of animals represent the short-term objective, but replacement of animal testing with nonanimal-based methods is the ultimate goal. Currently, genuine, validated, and regulatory accepted nonanimal alternative methods to replace whole animal acute toxicity testing are still more of a goal than a reality, even though the concept has been widely accepted by scientists from industry, professional societies, and certain regulatory bodies [4,128,187,188].

### Types of Acute Testing

Because acute toxicity data may provide the first line of defense, a battery of tests under different conditions and exposure routes should be considered. In general, these tests should include oral, dermal, and inhalation toxicities as well as skin and eye irritation studies. Other tests such as acute preneonatal and neonatal exposure, dermal contact sensitization and phototoxicity should also be considered. Depending on sound scientific factors, which may vary from one chemical to another, the number and kind of acute tests required to establish the initial toxicity database may not be the same; for example, inhalation testing may not be conducted when inhalation exposure is not expected to occur because of the physical properties of the chemical (e.g., respirable particles cannot be generated). Generally, oral, dermal, and inhalation toxicity along with eye irritation tests should be considered as part of an initial acute investigation. These tests are often used for the regulatory purposes of labeling and classifying the hazard potential of a chemical or formulation, although increasing concerns also are placed on skin sensitization studies. This chapter is concerned only with acute oral, dermal, and inhalation toxicity and eye irritation testing.

### Acute Oral Toxicity

Classical oral toxicity tests, which use a large number of animals and precisely determine the $LD_{50}$, continue to be required by some regulatory bodies for the purpose of classification and labeling of chemicals; however, many studies have shown that adequate acute toxicity and lethality information can be obtained by using fewer animals than classical $LD_{50}$ studies require. DePass [42], Lipnick et al. [115], and Gribaldo et al. [87] have reviewed several modified $LD_{50}$ tests. Although the main endpoint remains lethality, these tests generally fulfill the goal of reducing the number and suffering of animals and, in some cases, provide adequate information for hazard classification and labeling.

One example of a modified $LD_{50}$ test is the *approximate lethal dose method*; this method involves sequential dosing until the lowest lethal dose is obtained. Initially, an arbitrary dose is given to an animal. If the animal survives, a second animal is given 1.5 times the initial dose, and sequentially several animals are given increasing doses in the same manner until a lethal dose is achieved. The lethal dose is the approximate lethal dose (ALD). In general, only 6 to 10 animals are required to achieve the

**TABLE 22.1**
**Comparison Of Acute Oral Toxicity In The Rat Using The Approximate Lethal Dose (ALD) Vs. Classical (Conventional) Method**

| | Classical Method | | Approximate Method | |
|---|---|---|---|---|
| **Chemical** | *n* | $LD_{50}$ (mg/kg) | *n* | ALD (mg/kg)[a] |
| Tetraethyl lead | 36 | 20 | 5 | 26 |
| Methomyl | 53 | 40 | 5 | 26 |
| Hexachlorophene | 46 | 165 | 11 | 90 |
| Adiponitrile | 65 | 301 | 7 | 300 |
| Caffeine | 40 | 483 | 8 | 450 |
| *N*-Butylhexamethylene diamine | 35 | 536 | 7 | 1000 |
| Hexamethylene diamine | 92 | 1127 | 5 | 1500 |
| Bromobenzene | 35 | 3591 | 8 | 3400 |
| Carbon tetrachloride | 105 | 10,054 | 5 | 7500 |

[a] Lowest dose at which death was produced.

*Source:* Kennedy, G.L. et al., *J. Appl. Toxicol.*, 6, 145, 1986. With permission.

ALD. Comparison of classical $LD_{50}$ values and the ALD ndicates that the ALD can be used to closely predict the $LD_{50}$ (Table 22.1).

Recently developed alternative methods such as the imit test [80], British Society of Toxicology method [17], p-and-down method [20–22], fixed dose procedure [197], nd acute toxic class method [177] reduce the number of nimals needed to assess the oral toxic potential of chemcals. A number of these alternative methods have been ndorsed by scientific researchers, regulatory authorities, nd animal advocates [4,128,187,188], and some have een adopted as regulatory guidelines [159–161]. As use nd general acceptance of these alternative methods ncrease, regulatory bodies will likely need to redefine heir approach to how chemicals are classified and labeled.

## Classical Method

Due to regulatory acceptance of the fixed dose, acute toxic lass, and up-and-down methods [153], use of the classical ethod to assess the toxic potential of chemicals is no onger accepted in the European Union [147].

### Principle

he test material, undiluted or diluted with the appropriate olvent or suspending vehicle, is given to several groups f animals by gavage with a feeding needle or by gastric tubation. A vehicle control group is included if needed, ut generally this group is not necessary if the toxicity of e vehicle is known. Clinical signs, morbidity, and morlity are observed at specific intervals. Animals that die become extremely moribund during the study are subcted to necropsies. At the conclusion of the study's bservation period, surviving animals are sacrificed and necropsied. Tissues may be saved for histopathological examination to facilitate the understanding of the acute toxicity of the compound. To increase the reproducibility of the study, all experimental conditions and procedures should be standardized, and the study should be conducted according to the generally recognized good laboratory practices (GLPs) outlined by the Environmental Protection Agency (EPA) and the OECD [56,57,163].

### Animals

Responses elicited by a compound often vary greatly among species. Ideally, toxicity tests should be conducted with an animal that will elicit compound-related toxic responses similar to those that occur in humans; that is, the animal metabolizes the compound identically to humans and has the same susceptible organ systems. Under such conditions, the animal data may be extrapolated to humans. Unfortunately, finding such an ideal animal is a difficult if not impossible task.

A less ideal but more manageable approach is to conduct acute toxicity studies in a variety of animal species under the assumption that, if the toxicity of a compound is consistent in all the species tested, then a greater chance exists that such a response may also occur in humans. Even though the response in different species is not consistent, it generally is considered better to err on the safe side with the risk assessment being based on the most sensitive species, unless there is justification that such responses are less likely to occur in humans. An example of when a sensitive species would not be considered would be when it is known there is a dissimilarity in test article metabolism between a more sensitive animal species and humans. Although these are logical assumptions and generally quite reliable, it is possible that the results in the animals may underestimate or overestimate the response in humans; therefore, no absolute criterion exists for selecting a particular animal species. Priority should be given to species with metabolism or other physiological and biochemical parameters similar to humans. Animal species also should be selected on the basis of convenience, economical factors, and the existing database for the animal. Rats, mice, rabbits, and guinea pigs are most commonly chosen for acute toxicity studies.

Acute toxicity, even within a particular species, can vary with health conditions; age; sex; genetic makeup; body weight; differences in absorption, distribution, metabolism, and excretion of the compound; and the influence of hormones [48]. Immature animals, for example, may lack an effective drug-metabolizing enzyme system; this may contribute to higher toxicity of the compound in an immature animal if the enzyme is responsible for detoxification of the compound. On the other hand, if the enzyme responsible for generating a toxic metabolite is inactive, then it is possible that the toxic potential of the material will be underestimated. Obesity may affect the distribution and storage of a compound, especially if it is highly lipophilic. Sex hormones may be the target, or sex hormones may modify a particular toxic response, which then may account for different toxic responses between sexes. Liver and renal diseases associated with old age may contribute to higher toxicity. Variations in genetic makeup among different strains may alter metabolism or other parameters, which may affect the toxicity of a particular compound. It is therefore important to document all data on animals: age, sex, body weight, strain, general health condition, and source. In general, healthy, young adults should be used.

### *Number and Sex*

The precision of the acute test is dependent to a large extent on the number of animals employed per dose level. Historically, ten rats (five per sex) have been recommended in most regulatory guidelines [51,61,66,157,186], although more recently modified protocols are acceptable using as few as three animals per dose level [160,161]. The degree of precision required and in turn the number of animals per dose group required depend on the purpose of the study. In screening tests or tests designed to define the range of toxicity, fewer animals per dose level or fewer numbers of dose levels may be considered. In rare situations where a fairly precise $LD_{50}$ is needed, the number of dose levels (at least three dose levels) and animals per dose group may have to be increased. Literature surveys have shown that, when sensitivity differences between the sexes exist, females, in general, are more sensitive [115].

### *Grouping, Preparation of Animals, and Randomization*

Animals not previously treated with test substances in other studies should be identified individually by coded marks, metal ear tags, or tattoos. The animals then should be quarantined for at least a week prior to dosing to acclimatize them to the conditions of the animal room. The animals should be fasted prior to administration of the test substance if the route of administration is oral. The purpose of fasting the animal is to eliminate feed in the gastrointestinal tract, which may complicate absorption of the test substance. Rats usually are fasted overnight. Because mice have a higher metabolic rate, withholding feed for 3 to 4 hours may be adequate. Overfasting small animals with a high metabolic rate may induce undesirable effects. The animals should be randomly assigned to dose groups. Randomization ensures a homogeneous population and can minimize errors due to sampling bias. All animals with body weights and health conditions out of the normal range should be eliminated prior to the randomization procedure.

### *Dose Levels*

In general, the dose levels should be sufficient in numbe to allow a clear demonstration of a dose–response rela tionship and to permit an acceptable determination of th $LD_{50}$, if required. Three dose levels generally are suffi cient. The selected dose levels should bracket the expecte $LD_{50}$ value with at least one dose level higher than th expected $LD_{50}$ but not causing 100% mortality, and on dose level below the expected $LD_{50}$ value but not causin 0% mortality, when the probit analysis method is applie to estimate the $LD_{50}$. However, with a method such as th moving average under some specific conditions (at leas four dose levels with equal logarithmic intervals betwee each dose level and with equal numbers of animals in eac dose group), the $LD_{50}$ can be estimated even with 0% mortality at the lower dose levels and 100% mortality a the two higher dose levels. In any event, three or mor dose levels with a wide range of toxicity responses ar recommended if no other toxicity data are available.

### *Dosages*

If necessary, the test substance should be dissolved o suspended in a suitable vehicle, preferably in water, salin or an aqueous suspension such as 0.5% methylcellulos in water. If a test substance cannot be dissolved or su pended in an aqueous medium to form a homogeneou dosage preparation, corn oil or another solvent can b used. If the toxicity of the vehicle is not known, a vehic control group should be included in the test. The anima in the vehicle control group should receive the same vo ume of vehicle given to animals in the highest dose grou The test substance can be administered to animals at constant concentration across all dose levels (i.e., varyi the dose volume) or at a constant dose volume (i.e., var ing the dose concentration); however, the investigat should be aware that the toxicity observed by administr tion in a constant concentration may be different from th observed when given in a constant dose volume. The maximum dose volume in rodents should not exceed 10 mL/kg body weight for nonaqueous vehicles or 20 mL/kg body weight for aqueous solution or suspension. In any event, for scientific and humane reasons the dose volume should be as small as possible.

*Observations*

The emphasis in acute toxicity studies is on determination of the dose–response and the onset of toxic signs. The observation period should be flexible depending on the purpose of the study. This period should be based on the onset of signs, the nature of the toxicity, time to death, and the rate of recovery. For most highly toxic substances, the onset of toxic signs and the time to death may be very short, and prolonged observation may not be necessary. The slope of the dose–response curve for such test substances is usually very steep, and the treated animals either die or survive within a very short time. The observation period also should be long enough for the determination of reversibility or the recovery of an adverse effect. Under specific circumstances, the observation period might be longer, but it normally does not exceed 14 days.

Clinical examination, observation, and mortality checks should be made shortly after dosing, at frequent intervals over the next 4 hours and at least once daily thereafter. The intervals and frequency of observation should be flexible enough to determine the onset of signs, onset of recovery, and the time to death. The mortality checks should be frequent enough to minimize unnecessary loss of animals due to autolysis or cannibalism. Cage-side observations should include any changes in the skin, fur, eyes, mucus membranes, circulatory system, autonomic and central nervous systems, somatomotor activities, behavior, etc. Any pharmacotoxic signs such as tremor, convulsions, salivation, diarrhea, lethargy, sleepiness, morbidity, fasciculation, mydriasis, miosis, droppings, discharges, or hypotonia should be recorded. The most common pharmacotoxic signs are listed in Table 22.2 through Table 22.4. Individual body weights should be determined just prior to dosing, once weekly, and at death or at termination. Necropsies should be performed on animals that are moribund, found dead, and sacrificed at the conclusion of the study. All changes in the size, color, or texture of any organ should be recorded. Any gross change observed at necropsy should be described according to the size, color, and position of the lesion. While a complete microscopic examination of tissues and organs is ideal and would be helpful in defining acute toxicity, economic and time factors may preclude such a study. If the investigator feels that microscopic examination of a lesion is essential, tissues from these lesions should be preserved in an appropriate fixative such as 10% buffered formalin.

### Fixed Dose Procedure (Test Limit)

The traditional test limit for acute oral toxicity was considered to be 5.0 g/kg body weight, but more recently accepted protocols for acute toxicity [159–161] have a test limit of 2.0 g/kg body weight. The protocol used is a modification of the protocol developed by the British Society of Toxicology (BST) as described by Van den Heuvel et al. [198]. Basically, this procedure calls for dosing animals in a stepwise fashion using fixed doses of 5, 50, 500, and 2000 mg/kg. The initial dose level selected (discriminating dose) is a dose expected to produce some signs of toxicity and should be nonlethal, nonpainful, and nonstressful. The dose could be selected by using available information or by conducting a sighting study using three or four animals. If no mortality is observed at the highest dose level, a higher dose level is generally not necessary. The focus of the test should not be limited to mortality (found dead or killed for humane reasons), but should include other toxicity endpoints such as time course of signs of toxicity and necropsy findings. These data and the discriminating dose should provide adequate data for hazard assessment, comparative reference, and labeling classification (Table 22.5).

A multinational validation study in 33 laboratories with 20 materials using the fixed dose approach produced consistent results on the time course of signs of toxicity which was adequate for acute toxicity risk assessment and acute toxicity classification based on the EEC criteria. Compared to the classical method, fewer animals were used and less stress occurred [198]. This test has been adopted by the OECD as an alternative acute oral toxicity method [159].

### Acute Toxic Class Method

This method, described by Roll et al. [177], is based on the assumption that using a minimum number of animals in a stepwise procedure will provide enough information on the acute toxicity of a substance to allow classification according to the most commonly used classification schemes. Three animals of one sex are used for each step; normally, females are used (considered to be generally slightly more sensitive [115]), but either sex can be used. The initial dose is selected from one of four fixed dose levels—5, 50, 300, or 2000 mg/kg body weight—and should be chosen to produce some mortality. If existing information suggests that mortality is unlikely at the 2000-mg/kg dose, then a limit test at that level may be conducted with three animals of each sex. If deaths occur, further testing at the lower dose levels may be necessary. This method was evaluated in national and international validation studies [182,183] and has been adopted by the OECD as an alternative acute oral toxicity method [160]. Like the fixed dose method (see above), this method

**TABLE 22.2**
**Common Signs and Observations in Acute Toxicity Tests**

| Clinical Observation | Observed Signs | Organs, Tissues, or Systems Most Likely To Be Involved |
|---|---|---|
| I. Respiratory blockage in the nostril, changes in rate and depth of breathing, changes in color of body surface | A. Dyspnea (difficult or labored breathing, essentially gasping for air, respiration rate usually slow) | |
| | 1. Abdominal breathing (breathing by diaphragm, greater deflection of abdomen upon inspiration) | CNS respiratory center, paralysis of costal muscles, cholinergic |
| | 2. Gasping (deep labored inspiration, accompanied by a wheezing sound) | CNS respiratory center, pulmonary edema, secretion accumulation in airways, increased cholinergic |
| | B. Apnea (a transient cessation of breathing following a forced respiration) | CNS respiratory center, pulmonary cardiac insufficiency |
| | C. Cyanosis (bluish appearance of tail, mouth, foot pads) | Pulmonary–cardiac insufficiency, pulmonary edema |
| | D. Tachypnea (quick and usually shallow respiration) | Stimulation of respiratory center, pulmonary–cardiac insufficiency |
| | E. Nostril discharges (red or colorless) | Pulmonary edema, hemorrhage |
| II. Motor activities: changes in frequency and nature of movements | A. Decrease or increase in spontaneous motor activities, curiosity, preening, or locomotions | Somatomotor, CNS |
| | B. Somnolence (animal appears drowsy but can be aroused by prodding and resumes normal activities) | CNS sleep center |
| | C. Loss of righting reflex (loss of reflex to maintain normal upright posture when placed on the back) | CNS, sensory, neuromuscular |
| | D. Anesthesia (loss of righting reflex and pain response—animal will not respond to tail and toe pinch) | CNS, sensory |
| | E. Catalepsy (animal tends to remain in any position in which it is placed) | CNS, sensory, neuromuscular autonomic |
| | F. Ataxia (animal is unable to control and coordinate movement while walking with no spasticity, epraxia, paresis, or rigidity) | CNS, sensory, autonomic |
| | G. Unusual locomotion (spastic, toe walking, pedaling, hopping, and low body posture) | CNS, sensory, neuromuscular |
| | H. Prostration (immobile and rests on belly) | CNS, sensory, neuromuscular |
| | I. Tremors (involving trembling and quivering of the limbs or entire body) | Neuromuscular, CNS |
| | J. Fasciculation (involving movements of muscles, seen on the back, shoulders, hindlimbs, and digits of the paws) | Neuromuscular, CNS, autonomic |
| III. Convulsion (seizure): marked involuntary contraction or seizures of contraction of voluntary muscle | A. Clonic convulsion (convulsive alternating contraction and relaxation of muscles) | CNS, respiratory failure, neuromuscular, autonomic |
| | B. Tonic convulsion (persistent contraction of muscles, attended by rigid extension of hindlimbs) | CNS, respiratory failure, neuromuscular, autonomic |
| | C. Tonic–clonic convulsions (both types may appear consecutively) | CNS, respiratory failure, neuromuscular, autonomic |
| | D. Asphyxial convulsion (usually of clonic type but accompanied by gasping and cyanosis) | CNS, respiratory failure, neuromuscular, autonomic |
| | E. Opisthotonos (tetanic spasm in which the back is arched and the head is pulled toward the dorsal position) | CNS, respiratory failure, neuromuscular, autonomic |
| IV. Reflexes | A. Corneal eyelid closure (touching of the cornea causes eyelids to close) | Sensory, neuromuscular |
| | B. Primal (twitch of external ear elicited by light stroking of inside surface of ear) | Sensory, neuromuscular |
| | C. Righting (ability of animal to recover when placed dorsal side down) | CNS, sensory, neuromuscular |
| | D. Myotact (ability of animal to retract its hindlimb when limb is pulled down over the edge of a surface) | Sensory, neuromuscular |
| | E. Light (pupillary; constriction of pupil in presence of light) | Sensory, neuromuscular, autonomic |
| | F. Startle reflex (response to external stimuli such as touch, noise) | Sensory, neuromuscular |

| | | |
|---|---|---|
| V. Ocular signs | A. Lacrimation (excessive tearing, clear or colored) | Autonomic |
| | B. Miosis (constriction of pupil regardless of the presence or absence of light) | Autonomic |
| | C. Mydriasis (dilation of pupils regardless of the presence or absence of light) | Autonomic |
| | D. Exophthalmos (abnormal protrusion of eye in orbit) | Autonomic |
| | E. Ptosis (dropping of upper eyelids, not reversed by prodding animal) | Autonomic |
| | F. Chromodacryorrhea (red lacrimation) | Autonomic, hemorrhage, infection |
| | G. Relaxation of nictitating membrane | Autonomic |
| | H. Corneal opacity, iritis, conjunctivitis | Irritation of the eye |
| VI. Cardiovascular signs | A. Bradycardia (decreased heart rate) | Autonomic, pulmonary-cardiac insufficiency |
| | B. Tachycardia (increased heart rate) | Autonomic, pulmonary-cardiac insufficiency |
| | C. Vasodilation (redness of skin, tail, tongue, ear, foot pad, conjunctivae, sac, and warm body) | Autonomic, CNS, increased cardiac output, hot environment |
| | D. Vasoconstriction (blanching or whitening of skin, cold body) | Autonomic, CNS, decreased cardiac output, cold environment |
| | E. Arrhythmia (abnormal cardiac rhythm) | CNS, autonomic, pulmonary-cardiac insufficiency, myocardiac infarction |
| VII. Salivation | A. Excessive secretion of saliva (hair around mouth becomes wet) | Autonomic |
| VIII. Piloerection | A. Contraction of erectile tissue of hair follicles resulting in rough hair | Autonomic |
| IX. Analgesia | A. Decrease in reaction to induce pain (e.g., hot plate) | Sensory, CNS |
| X. Muscle tone | A. Hypotonia (generalized decrease in muscle tone) | Autonomic |
| | B. Hypertonia (generalized increase in muscle tension) | Autonomic |
| XI. Gastrointestinal signs: | | |
| Droppings (feces) | A. Solid, dried, and scant | Autonomic, constipation, GI motility |
| | B. Loss of fluid, watery stool | Autonomic, diarrhea, GI motility |
| Emesis | A. Vomiting and retching | Sensory, CNS, autonomic (in rat, emesis is absent) |
| Diuresis | A. Red urine | Damage in kidney |
| | B. Involuntary urination | Autonomic sensory |
| XII. Skin | A. Edema (swelling of tissue filled with fluid | Irritation, renal failure, tissue damage, long-term immobility |
| | B. Erythema (redness of skin | Irritation, inflammation, sensitization |

**TABLE 22.3**
**Autonomic Signs**

| | |
|---|---|
| Sympathomimetic | Piloerection<br>Partial mydriasis |
| Sympathetic block | Ptosis<br>Diagnostic if associated with sedation |
| Parasympathomimetic | Salivation<br>(examined by holding blotting paper)<br>Miosis<br>Diarrhea<br>Chromodacryorrhea in rats |
| Parasympathomimetic block | Mydriasis (maximal)<br>Excessive dryness of mouth<br>(detect with blotting paper) |

**TABLE 22.4**
**Toxic Signs of Acetylcholinesterase Inhibition**

| Muscarinic Effects[a] | Nicotinic Effects[b] | CNS Effects[c] |
|---|---|---|
| Bronchoconstriction | Muscular twitching | Giddiness |
| Increased broncho-constriction | Fasciculation | Anxiety |
| | Cramping | Insomnia |
| Nausea and vomiting (absent in rats) | Muscular weakness | Nightmares |
| | | Headache |
| Diarrhea | | Apathy |
| Bradycardia | | Depression |
| Hypotension | | Drowsiness |
| Miosis | | Confusion |
| Urinary incontinence | | Ataxia |
| | | Coma |
| | | Depressed reflex |
| | | Seizure |
| | | Respiratory depressi |

[a] Blocked by atropine.
[b] Not blocked by atropine.
[c] Atropine might block early signs.

allows a judgment with respect to classifying the test material in one of a series of toxicity classes in accordance with the Globally Harmonized System of Classification and Labeling of Chemicals (GHS) [154].

## Up-and-Down Procedure

The up-and-down procedure [20–22,161], one of the more modern methods for estimating the $LD_{50}$, is based on the maximum likelihood method. Like the acute toxic class method, animals are dosed following a stepwise proc dure; however, animals are dosed one at a time at a mi imum of 48-hour intervals, with the first animal receivi a dose just below the estimated $LD_{50}$. If the first anim

**TABLE 22.5**
**Investigation of Acute Oral Toxicity Using the Fixed-Dose Method To Interpret Results**

| Fixed Dose | Results | Interpretation |
|---|---|---|
| 5 mg/kg[a] | Less than 100% survival[b] | Compounds that may be very toxic if swallowed |
| | 100% survival but evident toxicity | Compounds that may be toxic if swallowed |
| | 100% survival; no evident toxicity | Retested at 50 mg/kg if not already tested at that level |
| 50 mg/kg | Less than 100% survival[b] | Compounds that may be toxic or very toxic if swallowed; retested at 5 mg/kg if not already tested at that level |
| | 100% survival but evident toxicity | Compounds that may be harmful if swallowed |
| | 100% survival; no evident toxicity | Retested at 500 mg/kg if not already tested at that level |
| 500 mg/kg | Less than 100% survival[b] | Compounds that may be toxic or harmful if swallowed; retested at 50 mg/kg if not already tested at that level |
| | 100% survival but evident toxicity | Compounds that do not present a significant acute toxic risk if swallowed |
| | 100% survival; no evident toxicity | Retested at 2000 mg/kg if not already tested at that level |
| 2000 mg/kg[c] | Less than 100% survival[b] | Compounds that may be harmful if swallowed; retested at 500 mg/kg if not already tested at that level |
| | 100% survival with or without evident toxicity | Compounds that do not present a significant acute toxic risk if swallowed |

[a] Where a dose of 5 mg/kg produces significant mortality or where a sighting study suggests that mortality will result at that dose level, the substance should be investigated at a lower dose level. The level chosen should be one that is likely to produce evident toxicity but no mortality.
[b] Includes compound-related mortality and humane kills but not accidental deaths.
[c] It should be noted that testing mortality at this dose level is carried out primarily for risk assessment purposes; however, where no evident toxicity is seen at 500 mg/kg its results are relevant to classification if there is greater than 50% mortality (including humane kills).

*Source:* Adapted from Van den Heuvel, M. J. et al., *Food Chem. Toxicol.*, 28, 469, 1990.

**TABLE 22.6**
**Comparison of Rat Oral $LD_{50}$ Using the Up-and-Down Method vs. Classical (Conventional) Method**

| Chemical No. | Classical Method | | Up-and-Down Method | |
|---|---|---|---|---|
| | *n* | $LD_{50}$ (g/kg) | *n* | $LD_{50}$ (g/kg) |
| 1 | 50 | 0.273 | 6 | 0.388 |
| 2 | 40 | 0.344 | 9 | 0.421 |
| 3 | 40 | 3.490 | 8 | 4.120 |
| 4 | 40 | 3.520 | 6 | 4.020 |
| 5 | 40 | 4.040 | 6 | 3.520 |
| 6 | 40 | 5.560 | 6 | 5.700 |
| 7 | 40 | 9.280 | 6 | 8.770 |
| 8 | 20 | >10.00 | 3 | >10.10 |
| 9 | 50 | 10.11 | 7 | 11.09 |
| 10 | 10 | >20.00 | 8 | 22.40 |

*Source:* Bruce, R.D., *Fund. Appl. Toxicol.*, 5, 151, 1985. With permission.

survives, the next one receives a higher dose. If the first animal dies, the next one receives a lower dose. The spacing of doses generally is adjusted up or down by a factor of 3.2 (default factor corresponding to a dose progression of one half log unit), depending on the outcome of the previous animal. Comparison of classical $LD_{50}$ values to the up-and-down-derived $LD_{50}$ shows close agreement (Table 22.6). This test has been adopted by the OECD as an alternative to the more traditional methods of $LD_{50}$ determination [161]. The OECD test guideline also contains the provision for a limit test that uses a maximum of five animals. This test is used when information suggests that the test material has a low potential to be toxic; in this study, a test dose of 2000 mg/kg or, as required, 000 mg/kg is used. Animals are dosed in a sequential manner with the second animal receiving the dose only if the first animal survives the limit dose.

Data from this method can be analyzed using the SAS 181] or BMDP [47] computer program packages, which are available to many toxicology laboratories. Other examples of programming for the estimation of the $LD_{50}$ with small computer have been reported [113,178].

## ACUTE DERMAL TOXICITY

Dermal exposure is an important route of exposure. The objective of conducting an acute dermal toxicity study is the same as an acute oral toxicity study: to assess the adverse effects resulting from a single dermal application of a test substance. The acute dermal test also provides the initial toxicity data for regulatory purposes, labeling, classification, transportation, and subsequent subchronic and chronic dermal toxicity studies. In addition, results from this type of test could provide information on dermal absorption and the mode of toxic action of the test material. Comparison of acute toxicity by the oral and dermal routes may provide evidence of the relative penetration of a test material. Although the general experimental design and principles of acute dermal toxicity testing are similar to those of acute oral testing, there are differences. These differences include selection of the animal species, number of animals per dose level, preparation of animals, dosage, and administration of the test substance. Only differences in the acute dermal test are described in this section.

Recently, an OECD test method similar in principle to the acute oral toxicity fixed dose procedure [197] has been developed and is currently under review [148]; when approved, this method would reduce the number of animals required to assess the dermal toxic potential of chemicals. Because this method has not been finalized and formally approved it will not be discussed here.

### Principle

The test material is applied dermally, undiluted or diluted with the appropriate solvent, in graduated doses to several groups of animals. A vehicle control group is included if needed, but generally this group is not necessary if the toxicity of the vehicle is known. Clinical signs, morbidity, and mortality are observed at specific time intervals. Animals that die or become extremely moribund during the study are necropsied. At the conclusion of the study's observation period, surviving animals are sacrificed and necropsied. Tissues may be saved for histopathological examination to facilitate the understanding of the acute toxicity of the compound. To increase the reproducibility of the study, all experimental conditions and procedures should be standardized, and the study should be conducted according to generally recognized GLPs outlined by the EPA and the OECD [56,57,163].

### Animals

The three most commonly used animal species for this type of test are rabbits, rats, and guinea pigs; however, other species can be used for this type of test. At the start of the study, healthy animals from 8 to 12 weeks old with a range of weight variation not exceeding ±20% of the appropriate mean value should be used. Variables such as species used, age and health of the animal, body weight, sex, and housing environment can affect the outcome of an acute dermal toxicity test. The animals should be housed individually in a controlled environment. Quarantine, acclimatization, and randomization are as described above for acute oral toxicity studies. The back of the animal or a band around the trunk should be clipped free of hair. When clipping the hair, care must be taken not to abrade the skin. If abraded skin is called for, a needle may be used, but care must be taken not to damage the dermis. Increasingly, investigators have come to question the value of conducting tests on abraded skin, and many consider such tests to be irrelevant. To date, almost all testing guidelines call for conducting the dermal test only on intact skin [51,62,67,151,157,186]. In contrast to the acute oral toxicity test method, fasting animals overnight is not necessary. Generally, ten animals per dose level (five per sex) are sufficient to allow for an acceptable estimation of the dermal $LD_{50}$; however, depending on the nature of the test substance and available safety information, smaller numbers of animals can be used. Females used in the study should be nulliparous and not pregnant.

## Dose Levels

Dose selection is similar to the acute oral toxicity test. Higher doses do not need to be tested when a test substance at 2000 mg/kg (considered the limit dose) has not produced test-substance-related mortality. This is because administration of additional test substance would only be applied on top of the test substance layer already present. This layering may form a physical barrier to prevent further absorption of the test substance from the application site. Although a control group generally is not needed, a vehicle control group should be included in the study if the toxicity of the vehicle is not known. Its influence on dermal penetration of the test substance should be fully established prior to the study.

## Preparation of Dosage and Dosing Procedure

The test substance should be applied uniformly to approximately 10% of the body surface of the animal (e.g., 4 cm × 5 cm for rats; 12 cm × 14 cm for rabbits; 7 cm × 10 cm for guinea pigs). Under certain conditions, the area of application may vary; for example, the area of application for highly toxic substances may be small because a lower volume that is applied.

Liquid test substances generally are applied undiluted. If the test substance is a solid, it should be pulverized, weighed, placed on a plastic sheet or porous gauze dressing, moistened so as to form a paste with normal saline (one part test substance for one part saline) or an appropriate solvent, and then spread evenly on the closely clipped skin to ensure uniform contact with the skin. Grinding of the solid test substances may not be needed under some conditions; for example, when a granular formulation is tested, it may be more relevant to test the substance in its formulation state than to destroy the formulation by grinding.

The test substance can be applied under semiocclusive, occlusive, or nonocclusive (open) conditions; choice of the application method depends on what the most likely exposure pattern is in humans. The application method with the highest potential for skin irritation is occlusiv followed by semiocclusive and nonocclusive exposure. should be noted that skin irritation not only may cau stress to the animal but can also increase dermal penetr tion of the test substance.

For nonocclusive application, the application si remains uncovered but the volume of liquid test mater that can be applied to the skin may be limited dependi on the volatility of the liquid. It may be necessary immobilize the animal or use a device such as an Eliz bethan collar so as to prevent the animal from ingesti the test material as a result of licking the application si For occlusive application, the application site is cover with an impervious material such as a plastic sheet. F semiocclusive application, the application site is cover with a porous gauze dressing as described in the followi paragraph. The volume that can be applied with the occl sive or semiocclusive patch generally is larger than th of the nonocclusive method.

## Dosing Procedures for Liquid Test Substances

The dosing procedure for the rabbit is detailed becau rabbits are the most widely used species for this type testing. Rabbits are clipped free of hair with an electr animal hair clipper. The rabbit may have to be restrain by tying the hind legs to a secured post and holding th nape of the neck during clipping. When using the occl sive method, a plastic cuff in a cylindrical shape (appro imately 12 to 15 inches long and 10 inches in diamete and open at both ends can be used. The cuff is put on the trunk of the rabbits, covering the application sit With the help of another investigator, the plastic cuff folded around the trunk and secured at the thorax an flank of the rabbit with surgical adhesive tapes. Ca should be exercised so the cuff is sufficiently secured b not too tight to affect breathing. Using a long feedi needle, the correct amount of the liquid test substance drawn into a syringe of appropriate size. The needle the is placed under the cuff and half of the dose is deliver evenly on each side of the vertebral column. After wit drawal of the needle, the test substance is evenly distri uted over the application site by gently rubbing the to of the plastic cuff. A piece of cloth of appropriate si is then wrapped around the plastic cuff and taped in pla to absorb any test substance that may spill off the cu After dosing, the investigator should observe the anim for a moment to see if breathing is affected, prior putting the animal back into the cage. In the semioccl sive method, a porous gauze dressing replaces the plast cuff. In nonocclusive exposure, the test substance applied uniformly over the skin; care must be taken minimize runoff from the skin, especially for aqueo dosing solutions. Applying the test substance in sma amounts at a time may help.

## Dosing Procedure for Solid Test Substances

If the test substance is a solid, it should be ground with a mortar and pestle unless there is justification not to pulverize. The correct dose of the ground solid is weighed, placed in the center of a plastic sheet of appropriate size, and moistened with sufficient normal saline or another appropriate vehicle. If a vehicle other than saline or water is used, the effect of the vehicle on the skin penetration of the test substance should be considered, and its toxicity should be known. The type of vehicle selected should be based on the expected mode of exposure of the test substance and should be mixed into a paste. The paste then is spread evenly around the center of the plastic sheet. With one person holding the rabbit by grasping it at the back, another person moistens its belly and its back with paper towels soaked with saline, then the rabbit is placed with its belly on the test substance paste on the plastic sheet, and another investigator wraps the sheet around the trunk of the rabbit. The plastic cuff is secured in place with surgical tape at the thorax and the flank. A piece of cloth of appropriate size then is wrapped around the plastic cuff and secured in place in the same manner. In the semiocclusive method, a porous gauze dressing replaces the plastic sheet.

## Dosing Procedures for Rats and Guinea Pigs

The method for dosing rats and guinea pigs is similar to that of the rabbit. Liquid samples should be placed on the back instead of the belly or on the lateral trunk. If nonocclusive exposure is called for in rats, the test substance should be applied to the skin as near to the head as possible to prevent ingestion by preening of the application site. A plastic collar may be used to further limit access to the treatment site. Generally, the plastic collar produces more stress in the rat, as indicated by chromodacryorrhea (red stain around the eyes), than in the rabbit. To minimize stress in rats, small collars can be handmade from light cardboard. The collar is lined around the neck area with cut rubber tubing that is stapled in place. The cardboard collar is lighter and easier to place on small animals. It can readily be replaced if needed (the collar placed on the neck usually will last about 3 days), and it is more economical than the commercially available plastic collars.

## Exposure Period and Removal of Cuff

Almost all testing guidelines [51,61,66,151,157,186] call for 24-hour continuous exposure. Upon completion of the exposure, the cuff is removed and the application site is gently wiped with a paper towel soaked with saline, water, or any appropriate solvent to remove residual test substance remaining on the application site.

## Observation Period

As in the acute oral toxicity test, the recommended minimum observation period is 14 days; however, the duration and intervals of observation should be flexible enough to establish the onset of signs, time to death, and time to recovery but should be frequent enough such that the loss of animals due to autolysis and cannibalizing is minimal. In addition, skin irritation should be assessed according to a scoring system such as the one described by Draize et al. [49].

# Acute Inhalation Toxicity

Inhalation exposure is an important route of exposure. The objective of conducting an acute inhalation toxicity study is to evaluate the toxic potential of a test material that may be inhaled, such as a gas, volatile substance, or aerosol. Such testing may provide information on the adverse effects resulting from exposure to inhalation application of a single dose of a test substance. The acute inhalation test provides the initial toxicity data for regulatory purposes, labeling, classification, transportation, and subsequent subchronic and chronic dermal toxicity studies. Comparison of acute toxicity by the oral and inhalation routes may provide evidence of the relative penetration and bioavailability of a test material. Although the general experimental design and principles of acute inhalation toxicity testing are similar to those of acute oral testing, there are differences. These differences include selection of the animal species, number of animals per dose level, preparation of animals, dosage, and administration of the test substance.

Recently, OECD test methods similar in principle to the acute oral toxicity fixed dose [197] and acute toxic class [160] methods have been developed and are currently under review [149,152]; when approved, these methods would reduce the number of animals necessary to assess the inhalation toxic potential of chemicals. Because these methods have not been finalized and formally approved, they will not be discussed here.

## Principle

Several groups of animals are exposed to a fixed concentration of test material by inhalation for a short period of time, one concentration per group. Although whole body exposure inhalation data are accepted by regulatory agencies, it is recommended that nose-only or head-only exposure be used, as this minimizes oral exposure from animals licking the compound off their fur. When a vehicle is used to help attain an appropriate concentration of test material, a vehicle control group should be included in the study. Clinical signs, morbidity, and mortality are observed at specific time intervals. Animals that die or become extremely moribund during the study are subjected to necropsies. At the conclusion of the study's observation period, surviving animals are sacrificed and necropsied. Tissues may be saved for histopathological examination to facilitate the understanding of the acute toxicity of the compound.

### Animals

Although several mammalian species have been used, the preferred species is the rat. When selecting a test species, priority should be given to the species with metabolism or other physiological and biochemical parameters similar to humans. At the start of the study, healthy animals from 8 to 12 weeks old with a range of weight variation within and between test groups should not exceed ±20% of the mean weight. The animals should be housed individually in a controlled environment. Quarantine, acclimatization, and randomization are as described above for acute oral toxicity studies.

At least ten animals (five per sex) for each concentration level are recommended in most regulatory guidelines [63,64,68,150,186], as this number is sufficient to allow for an acceptable estimation of the inhalation $LC_{50}$. Depending on the nature of the test substance and available safety information, however, smaller numbers of animals can be used. Females used in the study should be nulliparous and not pregnant.

### Environmental Conditions

Inhalation equipment used should be able to sustain a dynamic airflow of 12 to 15 air changes per hour, ensure adequate oxygen content of 19%, and provide an evenly distributed exposure atmosphere [63,150]. If a whole body chamber is used, individual housing must be used, and the total volume of test animals should not exceed 5% of the volume of the test chamber. Temperature and relative humidity must be monitored continuously and should be maintained at 22 ± 2°C and 30 to 70%, respectively.

### Dose Levels

In general, three concentration levels should be used and spaced to produce a concentration–response curve that permits acceptable determination of the $LC_{50}$. Animals are usually exposed for a period of 4 hours. Dose range finding studies using single animals may help determine the doses to be used in the main study when the toxic potential of the test material is unknown. When selecting the dose levels to be used, particle size analysis should be performed to determine the consistency of particle size distribution. Although not specifically required by the OECD, EPA guidelines recommend that the mass median aerodynamic diameter (MMAD) particle size range should be between 1 and 4 μm (a particle size distribution that permits deposition throughout the respiratory tract). When a vehicle is used to attain the desired test material concentration, a vehicle control group should be included in the test.

### Observations

The recommended minimum observation period is 14 days; however, the duration and intervals of observation should be flexible enough to establish the onset of signs, time to death, and time to recovery but should be frequent enough such that the loss of animals due to autolysis and cannibalizing is minimal. Clinical examination, observation, and mortality checks should be made shortly after dosing, at frequent intervals over the next 4 hours and at least once daily thereafter. Cage-side observations should include any changes in the skin, fur, eyes, mucus membranes, circulatory system, autonomic and central nervous systems, somatomotor activities, behavior, etc. Individual body weights should be determined just prior to dosing, once weekly, and at death or at termination. Gross necropsies should be performed on animals that are moribund, found dead, and sacrificed at the conclusion of the study, with particular attention paid to any changes in the respiratory tract.

### Test Limit

If no test-substance-related mortality is observed at an exposure of 5 mg/L or the maximum attainable concentration for 4 hours, then it is not necessary to conduct a full study [64,150].

## ASSESSMENT OF EYE IRRITATION INDUCED BY CHEMICALS

The eye captures visible energy and converts the energy to neurosignals, which are transmitted to the intricate central nervous system in which they form neuroimage (vision). The importance of having this ability to perceive the external environment through vision is a giant step i the evolution process. In humans, vision and hearing ar vital for the development of speech, learning, and intelligence. Loss of vision can greatly curtail normal living.

The three basic components of vision are optics, photoreceptors, and conducting nerves. All three component must function properly to form a clear and sharp neuroimage in the visual cortex. The optics of the eye (cornea aqueous humor, iris, lens, and vitreous humor) mu remain transparent and be able to refract and focus ligh at the correct position on the photoreceptors. The photoreceptors (the cones and rods) of the retina must be abl to undergo photolysis and convert light energy to neuro potential impulses. The optic nerves must be able to carr these neuroimpulses to the visual cortex.

Because the eye is constantly exposed to the external environment, the cornea must be protected from drying, dust, and microorganisms. The eyelids, the lacrimal system, and the somatosensory response of the cornea all work together to protect this outermost structure of the eye. Like other organs, the major portion of the eye is nourished by blood vessels. The retinal, circumcorneal, and uveal vessels also nourish and help maintain the eye. These vessels are so arranged and constructed that they normally do not alter the transparency of the ocular optics. Nutrients reach the transparent tissues of the eye via tears, the aqueous humor, and vitreous fluids.

Normal ocular functions are in delicate balance and are interdependent. Any traumatic insult, chemical or physical, can upset one or many of these ocular functions, thus creating a disturbance in vision. Depending on the extent of the traumatic injury (ranging from drying of the tear film to corneal ulceration or optic nerve damage), partial or complete loss of vision can result. Ocular injury can result from not only accidental physical trauma but also radiation and chemicals.

Chemicals can cause ocular damage locally by accidental exposure to the eye or systemically by ingestion of chemicals such as food contaminants and drugs. Because many chemicals can produce ocular damage either locally or systemically [85,95,129,176], it is important to test products for ocular effects before exposing workers during manufacturing and, ultimately, before subjecting consumers to products on the market. Ocular effects resulting from systemic exposure are beyond the scope of this chapter. This section focuses on eye irritation resulting from direct ocular contact.

Conducting ocular tests in humans is not only impractical but also unethical; consequently, many methods and techniques have been developed over the years for testing ocular effects in animals. This section describes the animal methods for detecting potential eye irritants and discusses their limitations. In recent years, *in vitro* methods intended to replace eye irritancy tests in animals have evolved; an overview of some of these methods is discussed in another chapter in this book.

Testing for potential eye irritancy is required for labeling and classification of chemicals by most regulatory agencies worldwide. The test protocol, interpretation of results, and classification scheme vary among countries. The differences among major industrial countries are also discussed here.

## Definition of Chemically Induced Eye Irritation and Corrosion

Irritation can be defined as reversible inflammatory changes in the eye and its surrounding mucus membranes following direct exposure to a material on the surface of anterior portion of the eye. Corrosion is irreversible ocular tissue damage following exposure to a material. From a practical point of view, the distinction between reversible and irreversible changes sometimes is limited by the length of the observation period; therefore, the term "eye corrosion" should be reserved for gross tissue destruction of the eye which generally occurs rapidly following exposure. When interpreting results from an eye irritation study, one must take into consideration the biological nature and significance of the ocular changes; for example, conjunctival redness is considered a mild ocular effect.

## Normal Physiology and Anatomy of the Eye

A brief description of the normal physiology and anatomy of the eye is essential for understanding the development of eye irritation. Details can be found in a variety of textbooks and reviews [74,126,172]. Functionally, the eye can be divided into three basic parts (Figure 22.4). From posterior to anterior, they are as follows:

- *Photoreceptors (retina)*—The part of the eye that connects to the central nervous system via the optic nerve.
- *Optics*—Structure that focuses visible light (image) onto the retina; it includes (from anterior to posterior) the cornea, iris, aqueous humor in the anterior chamber, the lens and its related organelles such as the zonules and ciliary body (muscles), and the vitreous in the posterior chamber.
- *Protective, lubricating, and nutritional structures*—These include the anterior eyelids and conjunctiva and associated secretory glands, the sclera and its outside layer (the fibrous tunic) and inside layer (uvea-vascular), and the ciliary body (secretory).

For chemically induced eye irritation, the main concern is generally on the directly exposed organelles such as the cornea, conjunctiva, and the iris. Effects on these structures can easily be detected by gross observation. If the chemical can penetrate deeper into the eye, other organelles also can be affected. Detection of the effects on these deeper structures requires special aids.

### Cornea

The cornea is composed of, from anterior to posterior, the epithelium, Bowman's membrane, stroma, Descemet's membrane, and endothelium. The epithelium is about five cells deep in the transitional zones at the periphery. The basal cells are columnar, the other cells are squamous, and the cells between the two layers are polygonal (wing cells). The Bowman's membrane (12 μm) is an acellular layer of collagen and ground substance that provides a functional

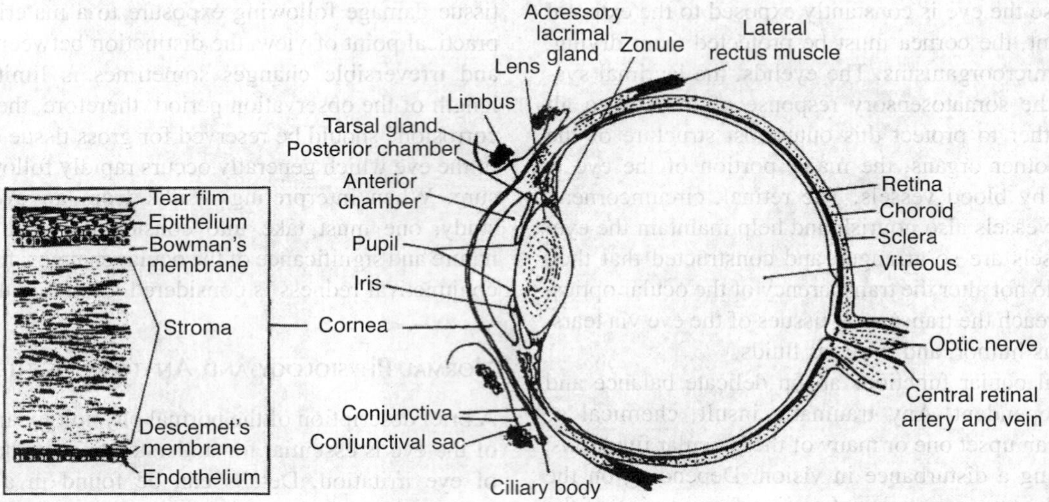


**FIGURE 22.4** Schematic illustration of the eye.

interface between the stroma and epithelium. An intact Bowman's membrane and the epithelial basal cell layer are vital to the regeneration of damaged epithelium. Damage to the Bowman's membrane may predispose the cornea to fibrosis. The stroma consists of lamellae of collagen fibrils and fibroblasts supported by ground substances. The stroma forms most (nine tenths) of the cornea and is limited on its inner surface by Descemet's membrane.

In addition to the organization of sheets of fibrils, other unique features such as proper hydration also contribute to corneal transparency. The Descemet's membrane (5 to 10 µm), like the Bowman's membrane, is an acellular layer and is the basement membrane of the endothelium. The endothelium is a single layer of cells that completely covers the posterior surface of the cornea. The cells are hexagonal with large nuclei. This layer of the cornea is particularly rich in the active transport enzyme adenosine triphosphatase (ATPase). The maintenance of proper hydration of the cornea has been attributed to the activity of this enzyme, which catalyzes an active sodium–potassium pump [15,108,118]. The limbus is a transitional region between the cornea and the sclera. This region, rich in vascularization, is the source of fluid and infiltration cells during corneal injury.

The epithelium and the overlying tear film provide the intrinsic protection for the cornea. Other layers have almost no intrinsic resistance to injury. Penetration into deeper layers of the cornea and other structures of the eye is limited by the solubility and lipophilicity of the chemical. Chemicals that are lipophilic and water soluble penetrate more rapidly and probably deeper into the eye than other chemicals.

The cornea is always covered with a film of tears consisting of several oily and aqueous layers. Proper tear formation and drainage as well as the stability of the precorneal tear film are important for a normal precorne optical surface, proper lubrication, nutrition for the corne removal of bacteria and debris from the cornea, and acti ity on the cornea. Reduction of tear formation can lead a dry eye, mechanical friction, irritation, or infection. discussion of the assessment of tear film formation, st bility, and drainage is available [31].

The cornea is a powerful refractive biological opti Its refractive power is dependent on its being transpare and on proper hydration. Maintenance of proper transpa ency and hydration is dependent on many mechanism such as proper tear flow, absence of deposits and bloc vessels, proper arrangement of collagen fibrils, unin paired nutritional supply for the metabolic active pun ($Na^+$–$K^+$), and proper intraocular pressure. Decrease transparency or hydration can be a result of corneal sca (decreased corneal thickness) or corneal edema (increase corneal thickness). Corneal edema can be caused by ep thelial damage, endothelial damage, increased intraocul pressure, lack of oxygen, or inhibition of the electroly balance pump ($Na^+$-$K^+$-activated ATPase), which located mainly in the endothelial membrane but also found in the epithelium. Methods for measuring corne curvature, corneal thickness, intraocular pressure, th blood/aqueous humor barrier, and corneal endotheliu damages have been reviewed [31].

## Conjunctiva

The conjunctiva is part of the eyelid. It is the delica membrane that lines the eyelid (palpebral conjunctiva) an covers the exposed surface of the eyeball (bulbar conjun tiva). Histologically, the conjunctiva is an aqueous, no keratinized epithelium with numerous mucus-secretin cells. Accessory lacrimal glands are present in the co junctiva, which contribute to the aqueous layer of a precorneal tear film. The Meibomian gland, a specialized sebaceous gland in the eyelid, secretes the outer oily layer of the tear film.

The main function of the eyelid is to protect the eye, especially the cornea, from external trauma through proper blinking reflexes and secretion of tears. Normal secretory and excretory functions of the tear are also important for normal optical function of the eye. The precorneal tear film can form an optically uniform layer over the microscopically irregular surface of the corneal epithelial cells. The tear flow continuously flushes cellular debris or foreign bodies from the eye, lubricates and protects the corneal surface from the mechanical friction caused by blinking, provides nutrients to the cornea, and induces antibacterial activities by proteolytic enzymes and immunoglobulin. All of these functions are important to maintain an optically intact corneal surface. Substances that affect the stability of the precorneal tear film by interfering with the secretory and excretory functions or with the blinking mechanism can cause serious damage to the cornea and may even cause corneal ulceration.

The nictitating membrane, or the third eyelid, is an mportant and prominent structure in many species of animals, including the rabbit, but it is not as important in numans and nonhuman primates. It aids in protecting the conjunctiva and the cornea when the eyeball is retracted. The nictitating membrane, like the conjunctiva, also contains acrimal glands and its secretion contributes to the aqueous ayer of the precorneal tear film. In addition, the nictitating nembrane helps to support the position of low eyelids and orms the lacrimal lake in the medial canthus. Vascularization in the conjunctiva generally consists of superficial and leep groups, mainly in the bulbar conjunctiva.

Three endpoints generally are associated with irritation in the conjunctiva: redness, chemosis, and discharge. n response to an irritant, the eyelids blink, the tear secreion increases, and the conjunctiva vessels dilate. Blinking nd tearing (discharge) aid in removing the irritant from he eye, and tear flow also may reduce the acidity or lkalinity of the irritant. Vessel dilation may be triggered y histamine, prostaglandins, or other inflammatory meditors, resulting in an apparent increase in vascularity (redess) in the conjunctiva. If irritation is severe, the dilation f the vessel increases and vascular fluid and proteins leak ito the conjunctiva, resulting in edema (chemosis). If the dema is severe, bulging may hinder normal functioning f the eyelids.

### is

he iris forms the pupil and functions in regulating the nount of light that may reach the retina. High-intensity ght causes constriction of the diameter of the pupil, hereas low-intensity light dilates it. It does so by two sets of muscles acting opposite each other to control the diameter of the pupil. These muscles, circulatory and radiating, are innervated by both the autonomic and sympathetic nervous systems. The set of muscles forms the distinct characteristic of iridic furrows of the iris.

The iris is located posterior to the cornea and is a very vascular structure made of loose connective tissues, muscle, and pigmented cells. The amount of pigment in the iris varies. Heavily pigmented cells are found in most species except albinos. Only a small amount of pigment is found in the albino rabbit eye. This is an advantage in ocular studies because it allows easier and better examination of the iridal vessels, lens, and retina.

The observation endpoints of local iridic injury are increased vascularity, edema (increased thickness of the stroma/swelling), reaction to light, aqueous flare, and gross destruction of tissue. These are the manifestations of an inflammatory process (iritis) responding to an irritant. Like the conjunctival vessels, the iridic vessels dilate and leak vascular fluid in response to irritants. Dilation of vessels and leakage cause edema and apparent changes in vascularity such as injection of iridic vessels (hyperemia). Aqueous flare is a result of protein leaking from the iridic vessels into the aqueous humor of the anterior chamber. Protein leakage into the anterior chamber alters the refractive index of the aqueous humors. Light beams entering the anterior chamber are scattered, giving the anterior chamber a cloudy appearance that contrasts with a clear appearance in normal eyes as a light beam passes through the pupil and the anterior chamber (e.g., during examination with a slit lamp). This is referred to as the aqueous flare or Tyndall phenomenon and is usually not noted during routine gross examination of the eye. In a more severe form of iritis, tissue destruction may result and nerve innervation may be disrupted, causing the pupil to be unresponsive to light. Failure to react to light, from a practical standpoint, is the most reliable observation of a severe iridic reaction, as severe iritis is usually accompanied by severe opacity in the cornea, which may obscure the visible detection of changes on the iris.

## THE DRAIZE TEST

The Draize test was developed in 1944 by Draize et al. [49] to study eye irritation. The test was based on the original work of Freidenwald et al. [79]. For years, the Draize test has been used as the animal test to identify human eye irritants. It is a simple and generalized test. It is easy to conduct and requires no special instruments. Whereas simplicity is probably the main reason for its popularity, it is also a limitation of the test. Undeniably, the Draize test can adequately identify most of the moderate to severe human eye irritants, but the test may fail to detect mild or subtle ocular irritation even with proper modification.

In the original Draize test, a standard 0.1 mL or 0.1 g of test substance is applied to the conjunctival sac of an albino rabbit's eye. The eyelid is held together for a few seconds and then released. The degree or extent of opacity on the cornea, the redness on the iris, and the chemosis and discharge on the conjunctiva are scored subjectively according to an arbitrary scale at preselected intervals (1, 24, 48, and 96 hours) after exposure. Scoring is based on the degree of effects caused by the testing substance. More emphasis is placed on the opacity of the cornea, which has a maximum score of 80, whereas emphasis is progressively less with other effects: conjunctival changes (maximum score of 20) and iritis (maximum score of 10) [24,84,121]. The Draize test has been a subject of controversy among animal rights groups [93,179] and even in the scientific community [9,24,84,88,96,121,174,204]. This test has been criticized on the dose volume, use of animals as models, methods of exposure, irrigation, number of animals, observation and scoring including laboratory procedure variability, and interpretation of results, all of which are discussed below.

## Dose Volume

The 0.1-mL dose volume used in the original test was based on the volume used earlier by Friedenwald et al. [79] to study the mechanism of acid- and base-induced ocular damage. This dose volume was selected arbitrarily as a standard volume for intraocular injection. Draize et al. [49] adopted it solely for convenience, which unfortunately has set a seemingly unchangeable doctrine for years even though the 0.1-mL dose volume lacks a scientific basis and, in conjunction with the conjunctival dosing method, often overpredicts the eye irritancy of a chemical.

Proponents of the 0.1-mL dose volume argue that this dose is a maximized test for the worst case and that it can better predict human eye irritants. Although the purpose of the Draize test is to predict what would happen to human eyes within the expected range of exposure, the 0.1-mL dose is out of the range of human exposure. Even though the intent is to maximize the dose that is applied to the rabbit eye, use of 0.1 mL is excessive when you consider the fact that the maximal volume that the cul-de-sac of a rabbit's eye can hold is only 30 to 50 μL [134]. Because the cul-de-sac will not retain more than 50 μL, the remaining amount of test material will simply fall from the eye. Furthermore, the worst case is not necessarily the best case. Constantly estimating the eye irritation potential of test materials will have a desensitizing effect on consumers' and workers' awareness of potential eye irritation, thereby defeating the purpose of testing for eye irritancy to protect consumers and workers.

No data exist to substantiate the argument that the 0.1-mL dose can better predict human eye irritants. On the contrary, in at least one survey little correlation was found between human accidental exposure experience and data generated by the traditional 0.1-mL maximal dose. The survey did not support the general presumption that rabbit eyes are more sensitive than human eyes [28]. Simply reducing the dose volume has produced data closer to eye irritation experienced in humans. Comparison of human eye irritation resulting from accidental exposure to many consumer products has revealed that a lower dose volume (0.01 mL) predicts the eye irritancy potential much better than the 0.1-mL dose volume [78,88]. In one of the studies, the time required for recovery from eye irritation in humans was compared with animal tests in monkeys and rabbits [78]. Results of this study clearly demonstrated that the modified Draize test (Federal Hazardous Substances Act [FHSA] protocol) with a dose volume of 0.1 mL was the poorest predicting test. Although all three animal tests overpredicted the eye irritancy experienced in humans, the low dose volume and monkey tests were better than the standard Draize test.

In 1977, a panel on eye irritancy test of the National Academy of Sciences (NAS), formed at the request of the Consumer Product Safety Commission (CPSC), recommended lowering the dose volume [144]. Subsequently, dose volumes ranging from 0.003 to 0.03 mL were proposed because they appeared to predict human eye irritants accurately, to cause less pain to animals, and to be able to discriminate slight to moderate eye irritants [88,90,209]. Williams et al. [209] showed that direct corneal application in a dose volume of 0.01 mL increased the response on the cornea when compared with the standard 0.1-mL dose but did not change the response on the conjunctiva. These results in the absence of compounding effects of a high-dose volume suggest that the lower dose volume is just as sensitive a method for eye irritancy testing as the higher dose volume.

## Animal Models

As with other toxicological tests in animals, the primary reason for assessing the ocular irritation potential of test articles in animals is their predictability for humans. Recognizing that there are anatomical, physiological, and biochemical differences between human and animal eyes, researchers are confronted with the difficult task of selecting the appropriate animal model and suitable test conditions to identify potential human eye irritants. The cornea thickness of dogs and rhesus monkeys is similar to that of humans (approximately 0.5 mm) [121,125,135], whereas rabbit corneal thickness is somewhat thinner (0.37 mm) [121]. There is a lack of a recognizable Bowman's membrane in rabbits, but they have a well-developed nictitating membrane (an additional target tissue). Rabbits have thick fur around the eyes, loose eyelids susceptible to mild irritants, an ineffective tear drainage system, and a poorly developed blinking mechanism [144

There are also species differences in biochemistry (e.g., variation in enzyme content [112] and different penetration rates of various substances [120]).

Despite these shortcomings and exceptions in predictability, the rabbit has been the preferred species for eye irritancy studies. Advantages of using the rabbit include a large established database, the fact that they are a relatively inexpensive animal to use, their availability, their ease of handling, and their large, unpigmented eyes, which are suitable for various ophthalmological examinations. With some exceptions [86,174], the rabbit eye is generally more sensitive to irritating materials than human or monkey eyes [7,25]; thus, there are built-in safety factors for making extrapolation and assessment of hazard to humans.

In addition to rabbits, dogs and primates sometimes are used for ocular testing. Eye irritancy in primates generally is more closely correlated with the exposure experience in humans, although dogs also have been shown to be suitable under certain circumstances [10]. Because they are more expensive and less available, dogs and primates are only used occasionally to assess eye irritancy.

Regardless of which animal is used, the investigator should always have a good understanding of the animal eye being observed. Background ocular findings, if not observed prior to exposure, can be recorded falsely as chemically induced damage.

## Methods of Exposure

Basically, there are two ways of administering a test article to the eye: (1) instilling the test material into the cul-de-sac of the conjunctiva, or (2) applying it directly onto the cornea. Of these methods, the conjunctival exposure procedure has been more frequently used historically because of the ease of application and has been perceived as an accurate method of dosing; however, some studies [10,88] have shown that conjunctival instillation of the test article is inappropriate under many circumstances, especially when the test article is a solid powder. The possibility exists that a solid test article can become trapped in the conjunctival sac, producing some undesirable mechanical effects that makes it difficult to interpret the observed ocular irritation results. It is also known that a considerable amount of the standard 0.1-mL or 0.1-g dose (especially as a solid powder) either falls or is blinked from the eye once the animal's eyelids are released. Based on this evidence, the claim that conjunctival dosing is more accurate than direct application to the cornea may not be valid.

The corneal exposure method, on the other hand, mimics more closely the actual accidental exposure experience in humans. When assessing the hazard of most chemical accidents, this method should be considered except when the chemical is intended for pharmaceutical use [144]. When applicators that had been developed for the corneal exposure method [7,25] were used, a more uniform corneal lesion was observed, resulting in less observation variability [7]. For a study as specific as corneal wound healing, it is recommended that a corneal applicator be used [139]; however, for hazard assessment, it is desirable to apply the test substance directly onto the cornea while the lids of the test eye are gently held open. Immediately afterwards, the eyelids are closed for a second and then released to allow blinking; this action more closely mimics actual human exposure [88].

## Irrigation

Washing the eye is a typical emergency remedy after accidental exposure to chemical substances. In experimental studies, the treated eye usually is irrigated 20 to 30 seconds after exposure to the test substance. Water is rapidly but gently squeezed from a plastic bottle to produce a constant gentle stream of water irrigating the entire treated eye. Irrigation should last for at least 1 minute. The effect of irrigation on the interpretation of test results has been the subject of many studies [7–9,14,40,77,86,89,165,185]. Although irrigation of the treated eye right after exposure can prevent or minimize eye irritation in rabbits, the effectiveness of irrigation is dependent on the chemical, the concentration, the time lag between exposure and initiation of the irrigation, and the volume of irrigation. Early washing (less than 1 minute after test article application) generally is recommended to reduce irritation [40,77, 89,185]; however, in some cases, irrigation has been shown to increase ocular irritation [82,322]. In other cases, ocular damage was almost instantaneous if irrigation was not initiated within a few seconds [40].

## Number of Animals

As a rule the precision of a study increases with the number of animals used. Sometimes, the desired precision may be offset by animal-to-animal variabilities. Economic considerations are also important in determining the number of animals used in a test group. A balance between economic considerations and reliability of test results should determine the number of animals tested in a study.

For eye irritation studies, a group size of nine rabbits was recommended in the original Draize test, and group sizes of at least six, three, three, and four rabbits have been recommended by the Federal Hazardous Substances Act (FHSA) [73], Interagency Regulatory Liaison Group (IRLG) [101], OECD [158], and NAS [144], respectively. The relationship of variability, classification, and group size is addressed in the literature [8,89,205]. With larger group size, smaller variability has been noted [205], whereas with a decreased group size lesser differentiation of irritancy has been suggested [8]. Recognizing these facts, Guillot et al. [89] suggested that with three rabbits in an initial study, there was a 96% chance that a positive or negative eye irritation result would be obtained. A similar conclusion was reached in another study in which the ocular irritation potential of 67 petroleum products was evaluated using six rabbits per product [43]. The eye irritation scores for the petroleum products based on all six rabbits were compared statistically with the scores using two, three, four, or five animals. The comparison showed that a subsample size of two, three, four, or five rabbits correctly classified (compared with the original six rabbits per test classification) the chemicals at 88, 93, 95, and 96% accuracy, respectively.

## Observations and Scoring

Reversibility and severity are the two major criteria used to measure eye irritancy in the Draize test. Reversibility refers to the time required for the ocular effects to disappear and for the eye to return to its normal state. To determine this reference time, treated eyes are examined periodically at 24-hour intervals, on day 7 after exposure, or at longer intervals as necessary to establish reversibility [49]. The observation period varies for different guidelines: The FHSA uses 24-, 48-, and 72-hour time spans [73]; the OECD uses 1-, 24-, 48-, and 72-hour time spans and, if needed, extended observations [158]; and the NAS recommends time spans of 1, 3, 7, 14, and 21 days [144]. The observation period should be flexible so one can confidently assess the persistence of ocular effects and fully characterize the degree of involvement, as the onset and healing of ocular effects often are unpredictable [86].

Assessing the severity of different ocular effects is subjective. This subjective evaluation is the major source of error for intra- and interlaboratory variation [204]. To minimize at least the intralaboratory variability in scoring, uniformity in scoring techniques must exist among investigators regardless of which scoring system is followed. Pictorial references such as those prepared by the FDA [70] and the CPSC [39] can be extremely helpful in the standardization of scoring eye irritation.

The types of ocular effects observed in the Draize test involve the cornea, iris, nictitating membrane, and conjunctiva. A system for grading ocular responses (Table 22.7) was originally proposed by Draize et al. [49]; subsequently, a number of modifications were proposed [39,70,144]. In the Draize system, the intensity and area of involvement on the cornea are graded separately on a scale of 0 to 4. The product of the two scores is multiplied by 5 to obtain a weighted corneal score. The congestion, swelling, circumcorneal injection, hemorrhage, and iridic failure of reactions to light are graded collectively on a scale of 0 to 2, and this score is multiplied by 5 to obtain a weighted iridic score. The redness, chemosis, and discharge of the conjunctivae are graded on scales of 0 to 3, 0 to 4, and 0 to 3, respectively; the sum of the conjunctival scores is then multiplied by 2 to obtain a weighted conjunctival score. Other lesions also are recorded, such as pannus (corneal neovascularization), phylyctena, and rupture of the eyeball.

In the guidelines set forth by the EPA, CPSC, FHSA, OECD, European Economic Community (EEC), and Japan's Ministry of Agriculture, Forestry, and Food (MAFF) [39,51,61,73,186], only the degree (intensity of cornea damage, iritis, and redness and chemosis [swelling]) of the conjunctivitis is scored (Table 22.8). The area involved on the cornea as well as the discharge of the conjunctiva are not taken into consideration in scoring. Various aids are used at times to facilitate or increase the resolution power of these observations. These aids include fluorescein staining and ophthalmoscopic or slit-lamp microscopic examinations. A scoring system has been developed for the slit-lamp and fluorescein staining examination [144] (Table 22.9). Other scoring systems have been proposed for lacrimation, blepharitis, chemosis, injection of conjunctival blood vessels, iritis, kerectasis, and corneal neovascularization [3].

## Interpretation of Results

Essentially, four categories of data are generated by the Draize test that are considered when interpreting the results of ocular testing: (1) type of ocular effects, (2) severity, (3) reversibility, and (4) rate of incidence. Weighting the scores in the original Draize test has to some extent taken the first category into consideration, yet it biases toward the cornea, one of the most critical ocular tissues. Severity is measured according to a graded scoring system, whereas reversibility is expressed as the time required for the affected ocular tissue to return to the normal state. Incidence is the number of animals that show some kind of ocular effect during the study. Interpretation of the data is a multiple and factorial undertaking. All four categories of data are somewhat interrelated; the individual scores do not represent an absolute standard for the irritancy of a material [156].

In one study, how the eye irritation was interpreted was not considered to be the major factor contributing to interlaboratory variability [204]. This finding is not surprising, if one assumes that everyone adheres to the same interpretation criteria; however, the question is what are the appropriate criteria for interpreting eye irritation results that would have an impact on placing eye irritants into different categories? The individual tissue scores do not represent an absolute standard for the irritancy of a material [157].

Many classification systems for eye irritants have been proposed. Some have been published in the literature [86,89,107,144] and in various testing guidelines [51,73,186], yet many others are used in individual laboratories. There is general agreement among investigators on how to classify test substances when no irritation

**TABLE 22.7**
**Scale of Weighted Scores for Grading the Severity of Ocular Lesions**

| Lesion | Score[a] |
|---|---|
| I. Cornea | |
| A. Opacity—degree of density (area which is most dense is taken for reading): | |
| Scattered or diffuse area—details of iris clearly visible | 1 |
| Easily discernible translucent areas, details of iris clearly visible | 2 |
| Opalescent areas; no details of iris visible; size of pupil barely discernible | 3 |
| Opaque; iris invisible | 4 |
| B. Area of cornea involved: | |
| One quarter (or less) but not zero | 1 |
| Greater than one quarter but less than one-half | 2 |
| Greater than one half but less than three quarters | 3 |
| Greater than three quarters, up to entire area | 4 |
| Score = A × B × 5; total maximum = 80. | |
| II. Iris | |
| A. Values: | |
| Folds above normal, congestion, swelling, circumcorneal injection (any one or all of these or combination of any thereof), iris still reacting to light (sluggish reaction is positive | 1 |
| No reaction to light; hemorrhage, gross destruction (any one or all of these) | 2 |
| Score = A × 5; total maximum = 10. | |
| III. Conjunctivae | |
| A. Redness (refers to palpebral conjunctivae only): | |
| Vessels definitely injected above normal | 1 |
| More diffuse, deeper crimson red; individual vessels not easily discernible | 2 |
| Diffuse beefy red | 3 |
| B. Chemosis | |
| Any swelling above normal (includes nictitating membrane) | 1 |
| Obvious swelling with partial eversion of the lids | 2 |
| Swelling with lids about half closed | 3 |
| Swelling with lids about half closed to completely closed | 4 |
| C. Discharge | |
| Any amount different from normal (does not include small amounts observed in inner canthus of normal animals) | 1 |
| Discharge with moistening of the lids and hairs just adjacent to the lids | 2 |
| Discharge with moistening of the lids and considerable area around the eye | 3 |
| Score = (A + B + C) × 2; total maximum = 20. | |

[a] The maximum total score is the sum of all the scores obtained for the cornea, iris, and conjunctivae.

*Source:* Buehler, E.V., in *Toxicology Annual*, Winek, C. L., Ed., Marcel Dekker, New York, 1974, p. 53. With permission.

bserved or when severe irritation or corrosion is seen, ut there is little agreement on how to classify irritancy hat falls between these two extremes. The manner in hich data are evaluated directly affects the conclusions eached.

Because of the complexity of eye irritancy data and eir interdependence, some investigators have chosen to mplify the interpretation to a pass-or-fail approach. In e FHSA guideline [73], for example, if four or more of e six test rabbits show ocular effects within 72 hours ter a conjunctival sac exposure (0.1 mL or 0.1 g of the st material), the test material is considered to be a pos- ive eye irritant. The ocular effects in consideration are lceration of the cornea (other than a fine stippling), corneal opacity (other than a slight deepening of the normal luster), inflammation of the iris (other than deepening of folds), an obvious swelling with partial eversion of the lids, or a diffuse crimson red with individual vessels but not easily discernible." If only one of the six animals tested shows ocular effects within 72 hours, the test is considered negative. If two or three of the six animals tested shows ocular effects, the test is repeated. The test substance is considered to be a positive irritant if three or more animals show ocular effects in the repeated test; otherwise, the test is repeated. Any positive ocular effect observed in the third test automatically classifies the test substance as an irritant. A similar approach has been adopted in the IRLG guideline [101], but an option is given that declares a test

**TABLE 22.8**
**Grades for Ocular Lesions**

| Lesion | Grades | Lesion | Grades |
|---|---|---|---|
| *Cornea* | | *Conjunctivae* | |
| No ulceration or opacity | 0 | **Redness** (refers to palpebral and bulbar conjunctivae excluding cornea and iris) | |
| Scattered or diffuse areas of opacity (other than slight dulling of normal luster); details of iris clearly visible | 1[a] | Vessels normal | 0 |
| Easily discernible translucent areas; details of iris slightly obscured | 2 | Some vessels definitely injected | 1 |
| Nacreous areas; no details of iris visible; size of pupil barely discernible | 3 | Diffuse, crimson red, individual vessels not easily discernible | 2[a] |
| Complete corneal opacity; iris not discernible | 4 | Diffuse, beefy red | 3 |
| *Iris* | | **Chemosis** | |
| Normal | 0 | No swelling | 0 |
| Markedly deepened folds, congestion, swelling, moderate circumcorneal injection (any of these separately or combined); iris still reacting to light (sluggish reaction is positive) | 1[a] | Any swelling above normal (includes nictitating membrane) | 1 |
| No reaction to light; hemorrhage; gross destruction (any or all of these) | 2 | Obvious swelling with partial eversion of lids | 2[a] |
| | | Swelling of lids about half closed | 3 |
| | | Swelling of lids more than half closed | 4 |

[a] Lowest grade considered positive.

positive when two or three of six rabbits tested show a positive ocular effect, and the test is not repeated. The pass-or-fail interpretation is too simplistic, however, and it does not separate eye irritants, especially those that fall between the two extreme irritancy categories (from nonirritating to severely irritating). Gradation of potential eye irritation is important to denote an anticipated hazard and to convey to consumers or workers that a specific degree of precaution should be exercised whenever a potential exposure to the substance exists.

Green et al. [86] used a different approach in which eye irritancy was classified into four easily recognizable categories based on the most severe responder in a group:

- *Nonirritation*—Exposure of the eye to the test article under the specified conditions causes no significant ocular changes. No tissue staining with fluorescein was observed. Any changes that did occur cleared within 24 hours and were no greater than those caused by normal saline under the same conditions.
- *Irritation*—Exposure of the eye to the test article under the specified conditions causes minor, superficial, and transient changes of the cornea, iris, or conjunctiva as determined by external or slit-lamp examination with fluorescein staining. The appearance at any grading interval of any of the following changes was sufficient to characterize a response as an irritation: opacity of the cornea (other than a slight dulling of the normal luster), hyperemia of the iris, or swelling of the conjunctiva. Any changes cleared within 7 days.
- *Harmfulness*—Exposure of the eye to the test article under specified conditions causes significant injury to the eye, such as loss of the corneal epithelium, corneal opacity, iritis (other than a slight infection), conjunctivitis, pannus, or bullae. The effect healed or cleared within 21 days.
- *Corrosion*—Exposure of the eye to the test article under specified conditions results in the types of injury described in the previous category and also results in significant tissue destruction (necrosis) or injuries that adversely affect the visual process. Injuries persisted for 21 days or more.

This classification system took into consideration th nature of ocular effects, reversibility of those effects, an to a certain extent, the qualitative severity, but not th incidence. The NAS committee that revised NAS publ cation 1138 [144] proposed a system of classification sim ilar to that of Green et al. [86] even though the categori were named differently: inconsequential or complete lac of irritation, moderate irritation, substantial irritation, ar severe or corrosive irritation. The NAS classification w also based on the most severe responder, and inciden

**TABLE 22.9**
**Scoring Criteria for Ocular Effects Observed in Slit-Lamp Microscopy**

| Location of Observations | Grades |
|---|---|
| *Corneal observations* | |
| Intensity | |
| Only epithelial edema (with only slight stromal edema or without stromal edema) | 1 |
| Corneal thickness 1.5× normal | 2 |
| Corneal thickness 2× normal | 3 |
| Cornea entirely opaque so corneal thickness cannot be determined | 4 |
| Area involved | |
| ≤25% of total corneal surface | 1 |
| >25% but ≤50% | 2 |
| >50% but ≤75% | 3 |
| >75% | 4 |
| Fluorescein staining | |
| ≤25% of total corneal surface | 1 |
| >25% but ≤50% | 2 |
| >50% but ≤75% | 3 |
| >75% | 4 |
| Neovascularization and pigment migration | |
| ≤25% of total corneal surface | 1 |
| >25% but ≤50% | 2 |
| >50% but ≤75% | 3 |
| >75% | 4 |
| Perforation | 4 |
| Maximal corneal score | 20 |
| *Iridal observations* | |
| Cells in aqueous chamber | |
| A few | 1 |
| A moderate number | 2 |
| Many | 3 |
| *Iridal observations (cont.)* | |
| Aqueous flare (Tyndall effect) | |
| Slight | 1 |
| Moderate | 2 |
| Marked | 3 |
| Iris hyperemia | |
| Slight | 1 |
| Moderate | 2 |
| Marked | 3 |
| Pupillary reflex | |
| Sluggish | 1 |
| Absent | 2 |
| Maximal iridal score | 11 |
| *Conjunctival observations* | |
| Hyperemia | |
| Slight | 1 |
| Moderate | 2 |
| Marked | 3 |
| Chemosis | |
| Slight | 1 |
| Moderate | 2 |
| Marked | 3 |
| Fluorescein staining | |
| Slight | 1 |
| Moderate | 2 |
| Marked | 3 |
| Ulceration | |
| Slight | 1 |
| Moderate | 2 |
| Marked | 3 |
| Maximal conjunctival score | 12 |

vas not considered. A provision for repeating the test was ;iven as an option to increase the confidence level in naking a judgment in some borderline cases. This eye rritancy classification system has been widely adopted. One shortcoming of the NAS system was that too wide a pectrum was created for moderate irritancy, which may ead to overuse of the cautionary term *moderate*. Many nvestigators have experienced problems in interpreting esults from fluorescein staining of the cornea when the NAS gradation system is used. The confusion arises nainly from the occasional artifacts inherent in fluorescein staining. Experience and sound scientific judgment re needed to properly interpret the fluorescein staining esults (see the discussion on ophthalmological techiques).

Griffith et al. [88] disagreed with using the most severe esponder for classification of eye irritancy, claiming that o epidemiological evidence suggested that the most evere rabbit responder would correlate with the worst possible case of human exposure. Instead, these investigators used the median time for recovery for classification according to the same temporal criteria as in the NAS system. The underlying logic is that the incidence of responders is being considered indirectly.

The classification systems of Green et al. [86], Griffith et al. [88], and NAS [144] have not taken into account the severity of irritancy, although there is a perception of a direct relationship between severity and reversibility. Examining the data of Griffith et al. [88] supports the conclusion that a direct correlation exists between median time to recovery and the severity of irritancy.

Kay and Calandra [107] proposed yet another rating system based on the Draize scores, taking into account the extent and persistence of irritation and the overall consistency of the data. The Kay and Calandra system has not been verified for correlation to human exposure experience nor has it been compared with other classification systems.

Guillot et al. [89] proposed a scoring system in which the greatest mean irritation score within an observation period is identified. On the basis of this score, the test substance is classified into six categories, ranging from nonirritating to maximum or extremely irritating. To maintain this initial rating, the data also must meet the arbitrary criteria for reversibility and frequency of occurrence; otherwise, the rating is upgraded one category. Guillot et al. [89] did attempt to compare their rating with the OECD protocol and claimed that one third of the 56 materials tested could be classified into a lower category by the OECD protocol.

The most current modification of the OECD protocol [158] is an effort to minimize the number of animals used to produce data suitable for hazard classification. In this simplified scheme, a Draize eye test is conducted using one animal if severe effects are expected, or three animals if no severe eye irritation is anticipated. Scoring is based on ocular lesions that occur within 72 hours of exposure, and results are expressed in terms of the lesions and their reversibility (eye irritation) or irreversibility (eye corrosion). The EPA has revised its health effects test guidelines for acute eye irritation [65] to be more consistent with the OECD protocol. A revised EEC directive, based on the OECD approach, provides hazard classification corresponding to risk phrases (e.g., R36, irritating to eyes; R41, risk of serious damage to eyes). These risk phrases are assigned to the label of a chemical when two or more of the three animals exhibit scores within certain arbitrary numbers [52].

A summary of the current international classification systems and major features for eye irritancy testing is provided in Table 22.10. Despite such a range of classification schemes, few differences exist in the actual scoring systems (all basically adhere to the original Draize) [49].

## SPECIAL OPHTHALMOLOGICAL TECHNIQUES

The Draize test is a generalized test concentrating on the effects of the material on the cornea, iris, and conjunctiva. Examination usually is performed using a hand light. Accurate observations are limited by the experience and training of the investigator, and it is possible for subtle ocular changes to be missed. If subtle ocular changes are to be detected and ambiguous gross observations resolved or if internal tissues (e.g., the lens and the retina) are to be examined, the investigator must rely on special techniques. Many such techniques have been developed over the years, most of which are more objective than the gross examination itself. A few comments on the fluorescein staining technique and several of the more objective methods are presented.

### Fluorescein Staining for Corneal Damage

Fluorescein is a weak organic acid (Figure 22.5) that is only slightly soluble in water, but its sodium salt is moderately soluble in water. It is very efficient in absorbing ultraviolet light and emitting fluorescent light. The maximum absorption of fluorescein is 490 nm (excitation) in the violet region, and its maximum emission is 520 nm in the green region of the spectrum. Its nonionized form is less fluorescent than its ionized form. At pH 7.4, fluorescein does not appear to bind to tissue and is nontoxic in animals, making it an ideal marker for an ocular fluid dynamics study. Because fluorescein is a deeply colored and highly fluorescent chemical, it can be detected at very low concentrations in biological tissues or fluid; however its detection sensitivity often is limited by the background fluorescence of biological tissues.

Because sodium fluorescein is a polar molecule, it can easily diffuse into aqueous medium and does not readily traverse lipophilic membranes; for example, if ulceration occurs on the cornea, the lipophilic membrane barrier is compromised, and the fluorescein diffuses freely through the ulcerated area of the cornea and either is dissolved or suspended in the aqueous medium of the stroma. More detailed information on the chemical and biological properties of fluorescein is provided in two excellent reviews [124,136].

Since its first use in studying the origin of aqueous humor secretion more than a century ago [55], fluorescein has become an important aid in ophthalmology. It has been used as a marker in detecting obstructions in the nasolacrimal drainage systems, for studying changes in the flow dynamics of different ocular fluids, for demonstrating leakage of retinal vessels in angiography, for estimating permeability of the cornea and lens, and for identifying ulcerations on the cornea [124]. Among these uses, the ability of fluorescein to detect subtle changes on the corneal epithelium [36,98] has made this a routine procedure in animal eye irritation studies.

An intact corneal epithelium is a lipophilic barrier to sodium fluorescein, but when the barrier is damaged and an ulceration or change in membrane structure occurs some of the fluorescein is able to penetrate into the intercellular aqueous spaces of the stroma. When light is cast on the cornea, fluorescence is detected in the damaged area of the epithelium. When fluorescein enters the stroma it eventually passes through Descemet's membrane and endothelium into the aqueous humor.

Staining is usually performed using either a prepared solution of fluorescein or fluorescein-impregnated paper strips. Commercially available solutions can contain 0.25, 1.0, or 2.0% fluorescein sodium salt. These solutions will contain preservatives that act to minimize bacterial contamination [29]. A drop of the solution is instilled onto the eye, with excessive fluorescein bein

**TABLE 22.10**
**Major Features of Eye Irritation Tests and International Classification Schemes**

| Methodology | FHSA (CPSC FDA OSHA) | OECD | EPA (Modified OECD) | Canada (Modified OECD) | EU (EEC) |
|---|---|---|---|---|---|
| nitial considerations | | | | | |
| Screen for pH (<2 or >11.5) | NS | Yes | Same as OECD | Same as OECD | Same as OECD |
| Results from skin irritation | NS | Yes | Same as OECD | Same as OECD | Same as OECD |
| Number of animals | | | | | |
| Screen for severe effects | NS | 1 | Same as OECD | Same as OECD | Same as OECD |
| Main test | ≥6 | ≥3 | Same as OECD | Same as OECD | 3 |
| Volume administered | 0.1 mL or 100 mg | 0.1 mL or ≤ 100 mg | Same as OECD | Same as OECD | Same as OECD |
| Scoring times | 1, 2, 3 d | 1 hr; 1, 2, 3 d (may be extended to assess reversibility) | 1 hr; 1, 2, 3 d (may be extended to assess reversibility ≤21 d) | 1 hr; 1, 2, 3 d (may be extended to assess reversibility) | 1 hr; 1, 2, 3 d |
| Minimal positive response | | | | | |
| Corneal opacity | 1 | NS[a] | 1 | 2.0[b] | ≥2.0, < 3.0[c] |
| Iritis | 1 | NS[a] | 1 | 1.0[b] | ≥1.0, <1.5[c] |
| Conjuctival | | | | | |
| Redness | 2 | NS[a] | 2 | 2.5[b] | ≥2.5[c] |
| Chemosis | 2 | NS[a] | 2 | 2.5[b] | ≥2.0[c] |
| Positive test | ≥4 positive of 6 animals | NS | NS[a] | | ≥2 positive of 3 animals |
| Label categories | | | | | |
| Irritant | Reversible inflammatory effect | Same as FHSA | Same as FHSA | Positive response requires labeling as a poisonous and infectious material | R 36 |
| Severe irritant | NS | NS | NS | NS | R 41 |
| Corrosive | Visible destruction or irreversible alterations | Same as FHSA | Same as FHSA | NS | NS |

Individual scores do not represent an absolute standard for the irritant properties of a material.
Mean of at least three animals.
Mean of three scoring intervals and scores representing two or more animals.
ote: NS, not specified.

ushed immediately with a sufficient amount of water. he eye is then examined under a cobalt-filtered ultra-olet light for epithelial defects. Fluorescein-impreg-ated paper strips [111] are free of contamination and asy to use. Moistened with collyria (medicated eye tion), a strip is touched lightly to the dorsal bulbar onjunctiva. The small amount of fluorescein should stribute uniformly on the cornea by either diffusion or inking. Flushing is not usually necessary if the strips e applied properly; nonetheless, if the strip touches e cornea, it becomes necessary for the cornea to be ushed with water before examination. Better results e obtained with the fluorescein-impregnated strip hen examination is by slit-lamp microscopy.

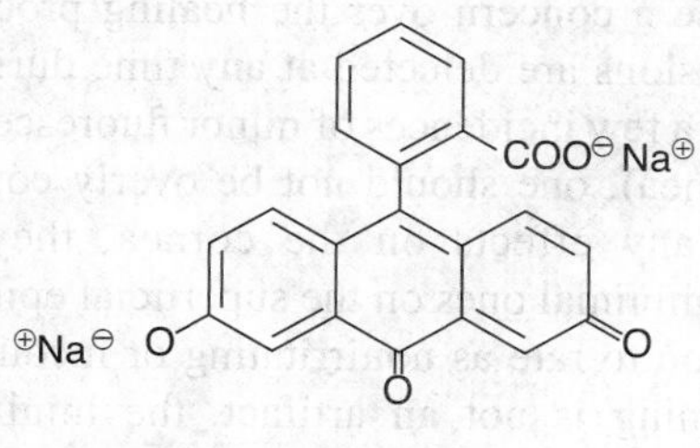

**FIGURE 22.5** Structure of fluorescein.

Fluorescein staining has two valuable applications in a routine eye irritation test: It can be used to screen eyes prior to the study to ensure that only healthy eyes are used, and it can be used to evaluate the cornea's recovery

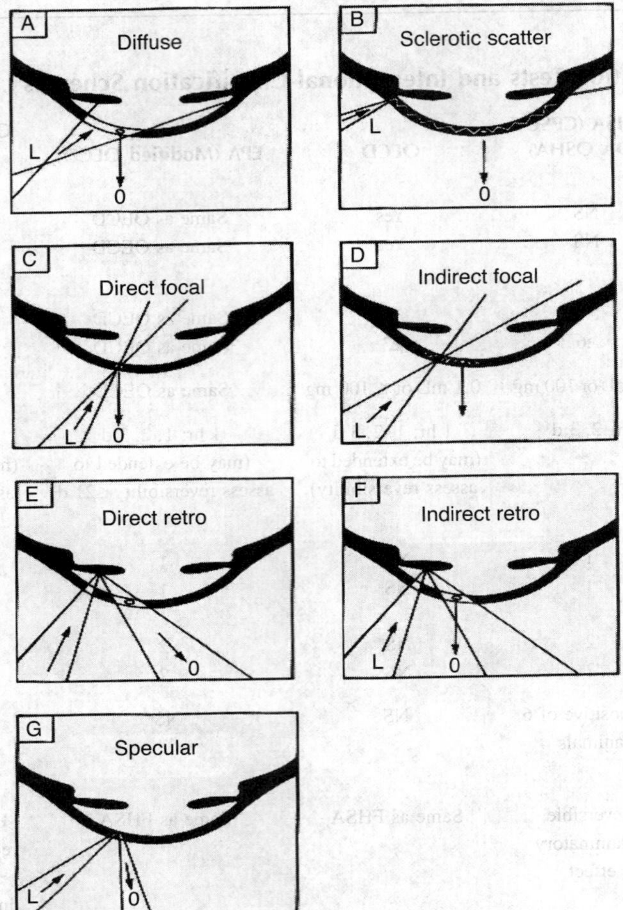


**FIGURE 22.6** Seven basic methods of illumination in slit-lamp microscopy: (A) diffuse, (B) sclerotic scatter, (C) direct focal, (D indirect focal, (E) direct retroillumination, (F) indirect retro-illumination, and (G) specular reflection. O, observer; L, illuminatc light. (Adapted from McDonald, T.O. et al., *J. Soc. Cosmet. Chem.*, 24, 163, 1973. With permission.)

from grossly observed damage. Slight epithelial effects still can be detected by fluorescein even though they are not visible during gross observation. Although most of these subtle effects on the cornea will disappear in a relatively short period of time, prolonged effects detected by fluorescein staining, but not by gross examination, should raise a concern over the healing process. When no gross lesions are detected at any time during a study (except for a few incidences of minor fluorescein staining on the cornea), one should not be overly concerned. If there are any effects on the cornea, they must be extremely minimal ones on the superficial epithelium for eye irritation to rate as nonirritating or inconsequential. If the staining is not an artifact, the minimal ocular effects detected under such circumstances should be readily reversible.

Although fluorescein staining can detect very subtle corneal epithelial changes, significant background staining can alter the interpretation of the actual amount of damage present because of an increase in the number of artifacts present. Apparent staining of the cornea can als result from incomplete flushing of excessive fluorescei with water or even from reflected light. A strong jet c water during irrigation can cause mild damage to th cornea. Damage also can occur if the eye is not handle properly during gross examination. These changes ar not related to the test article but may be detected wit fluorescein staining. Sometimes fluorescein staining ca cause haziness on the cornea even though a clear corne is seen prior to fluorescein staining. Whether the haz appearance of the cornea is a reflection of mild chang or artifact depends on several factors. If the hazy appea ance also is visible under a cobalt filter and is precede by grossly visible lesions, it generally is considered be a residual effect of mild severity that will disappe within a short time; however, if the hazy appearance seen intermittently or is not preceded by ocular effect it is likely an artifact. Proper training and experience a necessary to recognize artifacts and to obtain reliabl reproducible, and consistent results from fluoresce staining. In general, it is not necessary to stain lesions that are obvious and grossly evident. It is when lesions would otherwise go undetected by gross examination that fluorescein staining is of value.

## SLIT-LAMP MICROSCOPY

The slit-lamp biomicroscope is an important instrument for studying ocular tissues, especially the cornea. As its name suggests, a slit lamp consists of a microscope that views optical sections of different layers of the cornea made by an intense light beam acting as a surgical knife or microtome cutting through different layers of the eye. Many lesions can be observed with the slit-lamp biomicroscope that would remain undetected by gross examination. Using recent models of slit-lamp microscopes, one can not only observe the different layers of the cornea but also examine other transparent parts of the eye such as the aqueous humor, lens, and vitreous body.

The slit-lamp biomicroscope consists of an illuminating light source and a microscope. Both components are movable and adjustable, allowing the eye to be illuminated and observed from different angles and with different width and height adjustments of the slit light beam. An area of the cornea can simultaneously be illuminated and magnified by aligning the incidence of the light beam and the focus of the microscope. The light beam also can be directed at the area from different angles, providing several views of the same area.

Two types of slit images are used for illumination: parallelepiped and optical section [127]. For the parallelepiped slit image, a rectangular light beam (approximately 1 to 2 mm wide and 5 to 10 mm high) is projected onto the cornea. The shape of the illuminated area is similar to a parallelepiped prism where the outer and inner surfaces are bent because of the shape of the cornea. For the optical section slit image, the width (20 μm) of the light beam is narrowed to its minimum and is projected onto the cornea, providing a sagittal view that is similar to a thin histological section. Several basic illumination techniques are available (Figure 22.6): diffuse illumination, sclerotic scatter illumination, direct and indirect focal illumination, direct and indirect retroillumination, and specular reflection [135,190].

### Diffuse Illumination

In diffuse illumination, a slightly out-of-focus wide beam is used to scan and localize any gross lesions of a large area of the eye. Usually, the first step in examining the eye under a microscope is to detect gross lesions and their extent of change. This technique is similar to observing the eye with a hand light, except that the observation is made under a microscope (Figure 22.6a).

### Sclerotic Scatter Illumination

In sclerotic scatter illumination (Figure 22.6b), a narrow light beam is directed at the temporal limbus, and the microscope is focused centrally on the area of the cornea to be examined. The light reflected from the sclera will transmit within the cornea by total reflection. Under normal conditions, nothing will be seen, but if even minor changes are present the reflected light will be obstructed and the damaged area (e.g., mild corneal edema) will be illuminated. This technique is useful for detecting minimal changes in the cornea.

### Direct Focal Illumination

In direct focal illumination, the light beam and the microscope are focused sharply at the same point of interest in the same plane (Figure 22.6c). If a rectangular slit image is used for illumination and focused on the cornea, three general areas are seen when the parallelepiped is formed on the cornea: the epithelium (anterior bright line), the stroma (central clear marble-like area), and the endothelium (posterior thin bright line). If an optical section slit image is used for illumination, the corneal layers seen from anterior to posterior are a thin bridge layer, a thin dark layer, a granular layer, and another thin bright layer. These correspond to the tear film, the epithelium, the stroma, and the endothelium, respectively. Altering the angle of incidence of the light beam decreases or increases the reflection. This allows for detection of the depth of the lesion. Opacities on the different layers can be detected easily as obstructions of the incident light beam.

### Indirect Focal Illumination

Indirect focal illumination (Figure 22.6d) is accomplished by a narrow beam of light directed at an opaque area of the cornea; for example, changes in blood vessels at the cornea adjacent to the opaque area are illuminated and can be detected by focusing the microscope at these areas.

### Direct and Indirect Retroillumination

In direct (Figure 22.6e) and indirect (Figure 22.6f) retroillumination, the light beam is directed at tissues behind the cornea (e.g., the iris or the fundus). The reflected light illuminates the area of interest of the corneal tissue and can be focused under the microscope. The microscope can be located directly on the path of the reflected light (direct retroillumination), thus permitting subtle changes to be observed against a contrasting background. Any optical obstruction by lesions such as scars, pigment, or vessels located along the reflection light path will appear as darker areas on a brighter background. Lesions such as corneal edema and precipitates that can scatter the reflection light will show up as a brighter area against a darker background. When the microscope is located off the reflection light path (indirect retroillumination), the corneal structure is observed against a dark background such as the pupil or iris. Indirect retroillumination is better for observing opaque structures, whereas direct illumination often is used to detect corneal edema and precipitates.

### Specular Reflection Illumination

Specular reflection (Figure 22.6g) is most useful in studying the endothelium and precorneal tear film. This technique makes use of the difference in refractive properties between the corneal surface and the adjacent medium of the posterior and anterior surfaces of the cornea. The microscope is focused on the cornea adjacent to the path of the incident slit light beam. By alternating the angle of incidence, a point can be reached such that a total reflection is obtained on the junction between the aqueous medium and the most posterior corneal surface, thus illuminating endothelial cell patterns and the Descemet's membrane. Similar techniques can be performed on the anterior corneal surface to visualize precorneal tear film.

### Scoring System for Slit-Lamp Examinations

By using slit-lamp microscopic techniques, many subtle changes can be observed that would not otherwise be evident from the Draize test. A different scoring system must be developed to reflect such subtle changes. Baldwin et al. [2] proposed a scoring system for the cornea, anterior chamber, iris, and lens. Subsequently, the NAS [144] developed a scoring system for slit-lamp examinations that is similar to the Draize system in that an emphasis was placed on changes occurring in the cornea, iris, and conjunctiva. In the NAS system, the intensity and area involved are the two main criteria for scoring. Using this scoring system, the investigator must have a good understanding of the physiology of the normal eye. Like the Draize score, the NAS system is based on corneal effects; total maximal corneal score is 20 as compared with 11 and 15 for iridic and conjunctival scores, respectively. A detailed scoring scale and criteria are listed in Table 22.9.

## Corneal Pachymetry

Because corneal transparency is so important to vision (over 70% of the Draize score is derived from assessment of damage to the cornea), objective procedures to quantify corneal effects are an important element in eliminating intra- and interlaboratory variability in assessing the effects of ocular exposure to exogenous agents. Corneal transparency, thickness, and hydration are related in a linear fashion [94]; therefore, changes in corneal thickness can be used as an indicator of irritant affects, which may impair normal corneal hydration and transparency. When corneal thickness is measured appropriately, it can be used to objectively quantify swelling of the corneal stroma, which is a typical early irritant response. The measurement of corneal thickness is called "pachymetry," which comes from the Greek words *pachys* (thick) and *metry* (the process of measuring).

One method for measuring corneal thickness uses an optical pachometer attached to a slit-lamp microscope. Optical pachometers provide essentially indirect measurements of apparent corneal thickness based on displacement of light beams bouncing off the endothelial and epithelial surfaces of the cornea. The principles of this method have been previously described [31].

Various investigators [27,37,109,138] have reported that corneal thickness is significantly correlated with the Draize corneal score using a variety of substances from different chemical classes. Moreover, Kennah et al. [109 clearly demonstrated a substantial reduction in the coefficient of variation when comparing corneal swelling to Draize scores for various surfactants, alcohols, ketones acetates, and aromatic chemicals.

Recent advances in human ophthalmological procedures to correct visual acuity (e.g., radial keratotomy eximer-laser photorefractive keratectomy) have resulted in the development of improved devices to measure corneal thickness, which guide the practitioner before and after the procedure and provide a means to measure the effectiveness of the treatment. The ultrasonic pachymete is such a device [13,142,193,199], and it may have useful application to *in vivo* ocular irritation testing.

The ultrasonic pachymeter is an instrument with a hand-held probe that emits an ultrasonic signal of fixe velocity. The probe is placed directly on the anterior sur face of the cornea, and, after signal emission, a senso directly measures the time difference between echoes o signal pulses reflected from the front and back surface of the cornea. This time differential is directly proportiona to the thickness of the cornea via a function that is com puted as the product of the time delay between the tw echoes (in seconds) and the velocity of sound in the cor neal tissue (in meters/second). Whereas the optica pachometer indirectly equates displacement of inciden light to corneal thickness, the ultrasonic pachymeter pro vides a direct measurement.

Comparative evaluations of the sources of variabilit in human corneal thickness measurements via optical an ultrasonic [83,97,114,166,180] devices and variabilit occurring between different ultrasonic devices [206] hav been reported and discussed. Salz et al. [180] found tha sources of variation include intra- and intersession varia tion, interobserver variation, left and right eye variatio and variations due to alternative settings of ultrason sound frequency. They reported that the optical pachyme ter had significant intersession variation, significant inte observer variation, and significant differences in left and right eye thickness measurements, whereas the ultrasonic pachymeter demonstrated high reproducibility, no inter-observer variation, and no left and right eye variation.

The ultrasonic pachymeter has many desirable features such as relatively low cost, portability, and ease of operation, and it requires less operator skill and training than the optical pachometer. When used in humans, a topical anesthetic is employed because the tip of the measuring probe must be in contact with the corneal surface before a measurement can be taken; however, it has been reported that because of a lower corneal sensitivity in rabbits [33,131] an anesthetic is not necessary before taking corneal thickness measurements.

Because the velocity of sound can vary in different tissue, accurate readings for absolute corneal thickness require that the ultrasonic sound frequency of the instrument be matched to the tissue of interest. The velocity of sound in human corneal tissue has been variously reported as 1502 m/sec [143], 1586 m/sec [175], and 1610 m/sec [146]. Salz et al. [180], in their human cornea comparison of optical to ultrasonic pachymetry, used an approximate velocity of 1590 m/sec and found good agreement between the two measurement methods. The velocities of sound in cat [114], rabbit [32], and bovine [164] corneal tissue were found to be 1590 m/sec, 1580 m/sec, and 1550 m/sec, respectively. Empirical methods to determine the velocity of sound in corneal tissue have been described [114,164].

The utility of ultrasonic pachymetry in measuring corneal thickness changes in rabbits [119,141] and rats [119] after treatment with ocular irritants has been reported. The findings, albeit limited to a small number of chemicals, support the continued pursuit of this method as a relatively inexpensive, objective way to measure corneal irritant effects.

## Confocal Microscopy

The confocal microscope is another instrument that can be used to measure corneal thickness, as well as provide high-resolution microscopic images to study the cellular structure within corneal tissue. The first confocal microscope was described by Minsky [132,133] in a 1957 patent application. This device had a pinhole and a lens (objective and condenser) located on either side of the specimen to be viewed. The intent of the design was to eliminate any scattered light that might pass through the specimen, thus concentrating all light at a point source that was the focal point. The term "confocal" originated because the objective lens and the condenser lens were focused on the same specimen point.

Whereas the image seen in a conventional light microscope includes the in-focus image in the $x,y$ (horizontal) plane and the out-of-focus image above and below in the $z$ (vertical) plane, the confocal microscope only focuses in the $x,y$ plane. Indeed, defocusing a confocal microscope makes the image totally disappear rather than appear blurred. Reducing the out-of-focus signal above and below the focal plane results in enhanced resolution. In contrast to the light microscope, which is focused by moving the objective, moving the specimen focuses the confocal microscope. This feature provides an optical sectioning capability that allows thick tissue sections such as the cornea to be viewed *in vivo* or *in vitro* in both the horizontal and vertical planes. Because of the point-source light illumination, however, scanning the specimen is necessary to produce a full field of view with the confocal microscope. Scanned images can be viewed through a video monitor on a real-time basis, imported into a videocassette recorder, or stored as a digital image [168] for later viewing and analysis. For a complete review of the principles and applications of the scanning confocal microscope, see Petroll et al. [169].

By successively scanning the cornea and capturing a series of optical sections, it is possible to reconstruct a three-dimensional image of the tissue. Methods for three-dimensional imaging of rabbit cornea *in vitro* [104,122,170] and *in vivo* [69,104,137,167] have been described. These methods have been used to characterize the changes in area and depth of corneal injury of surfactant-induced eye irritation in the rabbit [123] and to examine the relationship between area and depth of injury to corneal cell death [103].

Mauer et al. [123] used *in vivo* scanning confocal microscopy to qualitatively and quantitatively characterize the initial changes occurring after treatment with surfactants known to produce slight, mild, moderate, and severe corneal irritation. Materials were applied directly to the corneas of six rabbits per group at a dose of 10 μL with macroscopic (Draize) and microscopic evaluations beginning at 3 hours after treatment and continuing periodically through day 35. Microscopic three-dimensional images were obtained from the surface epithelium to the endothelium and measurements made for surface epithelial cell size, epithelial layer thickness, total corneal thickness, and depth of keratocyte necrosis. The average Draize scores at 3 hours for the slight, mild, moderate, and severe irritants were 6.0, 39.3, 48.5, and 68.7, respectively. Confocal microscopic images at 3 hours showed that corneal injury with the slight irritant was limited to the epithelium (cell size and thickness 59 and 82% of control). The mild irritant had removed the surface epithelium, increased the corneal thickness to 158% of control, and produced keratocyte necrosis to a depth of 4.3 μm. With the moderate irritant, the epithelium was markedly attenuated, the corneal thickness was 156% of controls, and keratocyte necrosis extended to a depth of 19 μm. For the severe irritant, the epithelium was significantly thinned, the corneal thickness was 166% of controls, and keratocyte necrosis extended to a depth of 391 μm.

The use of confocal microscopy in studies designed to provide semiquantitative information on the nature and depth of injury to the cornea after chemical treatment has the potential to serve as an important link to the development of physiologically relevant and mechanistically based *in vitro* alternatives to the Draize eye test [123].

## LOCAL ANESTHETICS

For humane and scientific reasons, guidelines such as those established by the IRLG [101] and the OECD [158] provide options for using local anesthetics in eye irritation studies. Tetracaine, lidocaine, butacaine, proparacaine, and cocaine have all been tested for their eye irritation, with the results being mixed and inconclusive. Although most of these anesthetics can alleviate pain, they also can inhibit or reduce the somatosensory area of the eye and the blinking reflex. Tear flow is reduced, causing the test substance to be trapped and remain undiluted on the cornea instead of being blinked from the eye or diluted and flushed away by the tear flow. The blinking and tearing reflexes are important defense mechanisms, especially among higher primates, against accidental exposure to any substance [96]. Some local anesthetics can cause delay in corneal epithelial regeneration and loss of surface cells from the cornea [90]. Some local anesthetics such as procaine, lignocaine, piperocaine, amylocaine, amethocaine, and cinchocaine are cytotoxic to cultured human cells, including conjunctival cells [41]; however, at least one study has shown that a 0.5% tetracaine solution apparently had no effect on corneal healing [171]. Further research is needed to reveal the interaction of local anesthetics and chemically induced ocular effects. Local anesthesia is sometimes useful to induce akinesia of the eyelid during eye examination. Local anesthetics are desirable to alleviate pain, but one must be aware of the potential physical, chemical, physiological, and toxicological incompatibilities before considering the use of local anesthetics.

## HISTOLOGICAL APPROACHES

Histological examination of the eyes has been included routinely in subchronic and chronic toxicity studies, but because it is time consuming and costly it is performed only occasionally in eye irritation studies. Results may be no more informative than those from observations and measurements by other techniques; however, histological examination of ocular tissue can reveal the type of damage, tissues involved, and certain subtle changes in ocular tissue.

Both electron and light microscopic examinations have been used to evaluate local ocular injury [86,95,105, 106,176,192,195,200]. Although these methods sometimes can reveal morphological changes of different parts of the cornea, conjunctiva, lens, and retina, as well as visual nerve degeneration, there are shortcomings with electron and light microscopy. Issues with using histological techniques to examine damage include being able to section the precise lesion, problems in slide preparation, and subjective interpretation of observations. Another problem is that histological examination generally is made on dehydrated tissue [26], which makes some lesions, such as corneal edema, difficult to detect; however, histological examination of ocular tissues in local eye irritation studies has been considered an objective method because of its high sensitivity in detecting very mild ocular effects [96].

### Protocol Refinement

Because it is generally agreed that *in vitro* techniques will not replace animal testing immediately, efforts should be made to reduce the number of animals used and to refine the study design to minimize pain. Because precision of an eye irritancy test is a function of the number of animals used, the question arises as to whether or not it is justified to use a large number of animals to increase precision. The answer is no, as there is seldom an advantage to testing eye irritancy with more than three to six animals. The largest variable in an eye irritancy test is among animals, and the test itself is designed to be a bioassay; therefore, to use a large number of animals in hope of achieving a higher level of precision is neither realistic nor scientifically sound. A statistical analysis of 15[illegible] Draize irritancy studies with six-rabbit scores has shown that reducing the number of animals to five, four, three, or two retains a 98, 96, 94, or 91% agreement, respectively, with an irritant classification of these chemicals based on the six-rabbit scores [191]. The correlation coefficients for randomly selected subsets of five, four, three, or two scores were 0.998, 0.996, 0.992, and 0.984, respectively. The results of this study show that sufficient accuracy can be obtained by reducing the number of animals used in the Draize test. A combination of lower test substance dose volume (one tenth the Draize test dose volume) and fewer animals (three) also has yielded good correlation with the standard Draize test [23].

Another proposal is to test only for skin irritation. If the material causes severe skin irritation, it is presumed to be severely irritating to the eye as well; thus, the argument concludes, an eye irritation test is not needed. Extrapolation from skin to eye is not always valid. In at least one study of 60 severe skin irritants, only 39 also caused severe eye irritation, 15 caused mild or no ocular effects, and the other 6 caused moderate eye irritation [207,208]. Nonetheless, this approach has been proposed as one element of a tier system to prevent conducting an eye irritancy test when other potentially relevant information is available [102].

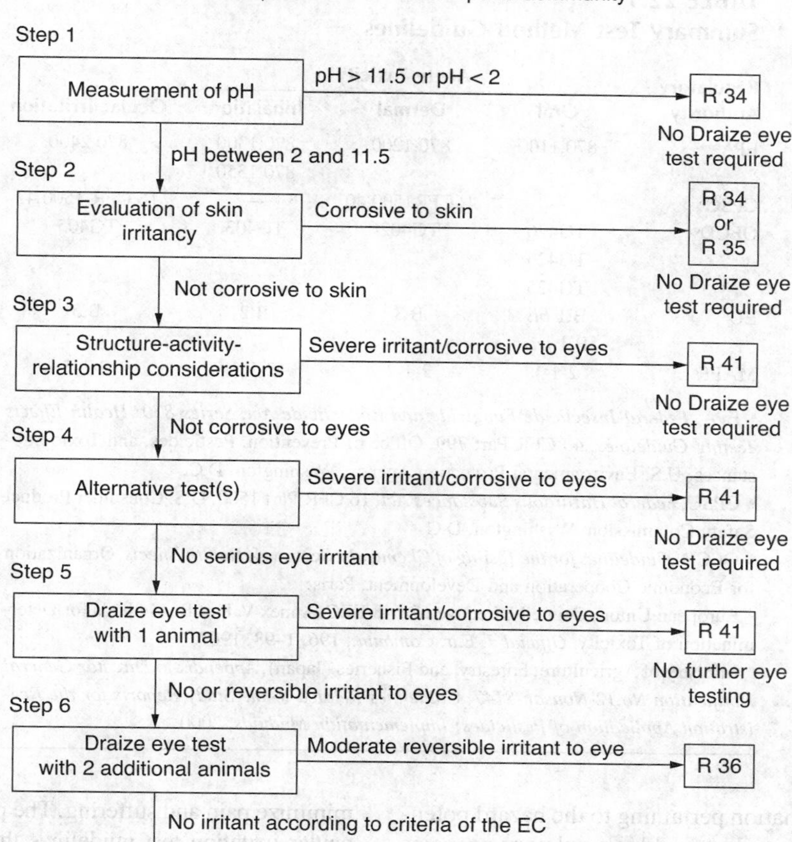


**IGURE 22.7** Tier scheme for eye irritation testing. (From OECD, *OECD Series on Testing and Assessment. Detailed Review on lassification Systems for Eye Irritation/Corrosion in OECD Member Countries,* ENV/JM/MONO(99)4, Office of Economic and ommunity Development, Paris, 1999. With permission.)

Many company guidelines specify that materials with xtremely high or low pH values do not have to be tested or eye irritancy. This approach is fully justified, especially or highly basic compounds. Alkali compounds generally ave a higher potential of causing severe eye irritation an acidic compounds.

## ier Testing Strategies

variety of tier testing strategies have been proposed to duce the number of animals in eye irritation testing ,91,99,102]. These strategies usually begin with a weight- -evidence approach in an effort to review existing infor- ation that would allow classification and labeling a mate- al as a severe ocular irritant without animal testing, or to onduct testing with a reduced number of animals. An exam- e of this approach is shown in Figure 22.7, a tier testing heme proposed by the OECD to support the harmonization eye irritation testing and classification [162]. Stages 1 to 3 involve information on the physicochemical characterization of the test material and use decision points to preclude animal testing, such as if the test material has a high or low pH (<2 or >11.5), if the agent is a known corrosive or severe dermal irritant, or if relevant information from structure–activity relationships (SARs) is available. If the weight of evidence suggests that the test material is a severe irritant, it should be so labeled. If information suggests that the test material is not a severe eye irritant, conducting an alternative test is the next step. If the results of the alternative test are indicative of a severe response, the material is classified as a severe irritant. If not, then testing in one or two animals is necessary before a final evaluation can be made.

## REGULATORY STATUS

The purpose of conducting safety testing is to obtain information that enables the toxicologist to evaluate the hazard potential of a test material to determine if and how it can

**TABLE 22.11**
**Summary Test Method Guidelines**

| Regulatory Authority | Acute Toxicity | | | |
|---|---|---|---|---|
| | **Oral** | **Dermal** | **Inhalation** | **Ocular Irritation** |
| EPA[a] | 870.1100 | 870.1200 | 870.1300<br>870.1350 | 870.2400 |
| CPSC[b] | — | 16CFR1500.40 | — | 16 CFR 1500.42 |
| OECD[c] | TG420<br>TG423<br>TG425 | TG402 | TG403 | TG405 |
| EU[d] | B.1 *bis*<br>B.1 *tris* | B.3 | B.2 | B.5 |
| MAFF[e] | 2-1-1 | 2-1-2 | 2-1-3 | 2-1-5 |

[a] EPA, *Federal Insecticide Fungicide and Rodenticide Act, Series 870: Health Effects Testing Guidelines*, 40 CFR Part 799, Office of Prevention, Pesticides, and Toxic Substances, U.S. Environmental Protection Agency, Washington, D.C.

[b] CPSC, *Federal Hazardous Substances Act*, 16 CFR Part 1500. U.S. Consumer Product Safety Commission, Washington, D.C.

[c] OECD, *Guidelines for the Testing of Chemicals*. Section 4. *Health Effects*, Organization for Economic Cooperation and Development, Paris.

[d] European Union, Council Directive 67/548/EEC, Annex V, Part B: Methods for Determination of Toxicity. *Official J. Eur. Commun.*, 196, 1–98, 1967.

[e] Ministry of Agriculture, Forestry, and Fisheries (Japan), *Appendix to Director General Notification No.12-Nousan-8147, Guidelines Related to the Study Reports for the Registration Application of Pesticides: Implementation Methods*, 2000.

be used safely. Information pertaining to the hazard potential of a test material is also used by regulatory agencies for the purposes of classification and labeling when a "new" or "existing" chemical is being registered or "notified." Ultimately, the goal of safety testing is to be able to predict the probability of risk to human health under conditions of intended or accidental exposure. Understanding the potential for toxicity of a chemical or product allows one to manage the risk associated with use of the material through communication (e.g., labeling, use instructions, Material Safety Data Sheets), package design (e.g., use of a child-resistant closure), availability (e.g., institutional vs. consumer product), and medical management following an accidental exposure. The following sections summarize the primary test guidelines, international chemical inventories that require the conduct of acute toxicity and ocular irritation testing, and classification schemes used by regulators in the United States and European Union for labeling purposes.

## Test Guidelines

The regulatory status of test methods used to evaluate the acute toxic and ocular irritation potential of test materials is in a state of flux. Most international regulatory agencies are attempting to reduce the number of animals necessary to assess the acute toxicity potential and ocular hazard and to minimize pain and suffering. The primary acute toxicity an ocular irritation test guidelines that have been adopted b various regulatory agencies are summarized in Table 22.1 The International Conference on Harmonization of Techni cal Requirements for Registration of Pharmaceuticals fo Human Use (ICH) is an ongoing effort to bring togethe regulatory authorities of the United States (FDA), Europ (EMEA), and Japan and representatives from the pharma ceutical industry with the goal of harmonizing the technic aspects of test method design, data interpretation, and proc uct registration [100]. Although acute toxicity testing is critical first step in the evaluation of a new pharmaceutic compound, the only recommendation pertaining to acu toxicity testing made by ICH is that the $LD_{50}$ determinatio for pharmaceuticals should be abandoned [100].

## Chemical Inventories

Table 22.12 lists some of the chemical inventories of exis ing and new chemicals which require data from acu toxicity and irritation studies. In general, *existing chem cals* are defined as the chemicals that were in use at t time legislation creating the inventory went into effec *New chemicals* are those chemicals that are not includ in an inventory and must go through a formal notificati and registration process before they can be manufactur or imported into a country for commercial purpose

**TABLE 22.12**
**Chemical Inventories with an Acute Toxicity Data Requirement**

| Country | Inventory | Regulatory Authority |
|---|---|---|
| Australia | Australian Inventory of Chemical Substances (AICS) | National Occupational Health and Safety Commission (Worksafe Australia) |
| Canada | Domestic Substances List (DSL) | Environment Canada |
| | Non-Domestic Substances List (NDSL) | |
| | New Substances Notification Regulations (NSNR) | |
| China | Inventory of Existing Chemical Substances in China (IECSC) | State Environmental Protection Administration |
| European Union | European Inventory of Existing Commercial Chemical Substances (EINECS) | European Chemicals Bureau |
| | European List of Notified Chemical Substances (ELINCS) | European Chemicals Bureau |
| | Screening Information Data Set (SIDS) | Organization for Economic Cooperation and Development |
| Korea | Korean Existing Chemicals Inventory (KECI) | Ministry of Environment[a] |
| New Zealand | ERMA New Zealand's Register | Environmental Risk Management Authority |
| Philippines | Philippines Inventory of Chemicals and Chemical Substances (PICCS) | Department of Environment and Natural Resources |
| United States | Toxic Substances Control Act (TSCA) | Environmental Protection Agency |
| | High Production Volume (HPV) | |

[a] Since 1995, authority for conducting toxicity reviews has been delegated to the National Institute of Environmental Research (NIER).

Whereas registration requirements vary from country to country, most do require, at a minimum, that data from some acute toxicity testing be provided. In some cases, irritation testing (ocular, dermal) is also required. The chemical inventories listed in Table 22.12 include chemicals and chemical mixtures but do not include agrochemicals, cosmetics, food additives, pesticides, pharmaceuticals, and radionucleotides, which are regulated by separate legislation not discussed here.

Within the past 10 years, legislation has been or is being implemented to expand and update existing inventories [30,145], address an existing data gap [60,155] or unifying multiple databases [173]. When it is approved by the European Commission, the REACH (Registration, Evaluation, and Authorization of Chemicals) program will replace the current databases for existing chemical inventories (EINECS, European Inventory of Existing Commercial Chemical Substances) and new chemical inventories (ELINCS, European List of Notified Chemical Substances) in the European Union with a single system managed by the European Chemicals Agency. The REACH regulation had not been entered into force as of 2006, but it is planned that the European Chemicals Agency will become fully operational by 2008.

**TABLE 22.13**
**Dangerous Substances Directive Acute Toxicity Classification Scheme**

| | Criteria[a] | | | |
|---|---|---|---|---|
| | | | Inhalation (mg/L/4 hr) | |
| Classification (R Phrase)[b] | Oral (mg/kg) | Dermal (mg/kg) | Gas/Vapor | Aerosol/Particulate |
| Harmful (R20/R21/R22) | $200 < X \le 2000$ | $400 < X \le 2000$ | $2 < X \le 20$ | $1 < X \le 5$ |
| Toxic (R23/R24/R25) | $25 < X \le 200$ | $50 < X \le 400$ | $0.5 < X \le 2$ | $0.25 < X \le 1$ |
| Very toxic (R26/R27/R28) | $X \le 25$ | $X \le 50$ | $X \le 0.5$ | $X \le 0.25$ |

[a] *Official J. Eur. Commun.*, L225, 263, 2001.
[b] *Official J. Eur. Commun.*, L225, 85, 2001.

*Note:* $X = LD_{50}$ (oral, dermal) or $LC_{50}$ (inhalation).

**TABLE 22.14**
**FIFRA Classification Scheme**

| | Toxicity Categories | | | |
|---|---|---|---|---|
| **Toxicological Endpoint** | **I** | **II** | **III** | **IV** |
| Acute oral toxicity | $X \leq 50$ mg/kg | $50 < X \leq 500$ mg/kg | $500 < X \leq 5000$ mg/kg | $X > 5000$ mg/kg |
| Acute dermal toxicity | $X \leq 200$ mg/kg | $200 < X \leq 2000$ mg/kg | $2000 < X \leq 5000$ mg/kg | $X > 5000$ mg/kg |
| Acute inhalation toxicity | $X \leq 0.05$ mg/l/4 hr | $0.05 < X \leq 0.5$ mg/l/4 hr | $0.5 < X \leq 2$ mg/l/4 hr | $X > 2$ mg/l/4 hr |
| Primary eye irritation | Corrosive (irreversible destruction of ocular tissue) or corneal involvement or irritation persisting for more than 21 d | Corneal involvement or other eye irritation clearing in 8–21 d | Corneal involvement or other eye irritation clearing in 7 d or less | Minimal effects clearing in within 24 hr |

*Note:* $X = LD_{50}$ (oral, dermal) or $LC_{50}$ (inhalation).

*Source:* Adapted from EPA, *Label Review Manual*, 3rd ed., EPA document 735-B-03-001, Office of Prevention, Pesticides, and Toxic Substances, U.S. Environmental Protection Agency, Washington, D.C., 2003, chap. 7.

## Classification Schemes

Two examples of how acute toxicity and ocular irritation data are used for the purposes of classifying and labeling chemicals and products are those detailed in European Council Directive 67/548/EEC Annex VI [53] and U.S. EPA FIFRA guidelines [59]. The types of chemicals and products evaluated in the European and U.S. schemes are different, but both schemes (as shown in Tables 22.10, 22.13, and 22.14) highlight the manner in which acute toxicity and ocular irritation data are used to define the hazard potential of individual chemicals and products. Chemicals and products evaluated in accordance with criteria defined in Council Directive 67/548/EEC Annex VI [53] are assigned risk, or R, phrases [54]. The choice of R phrases is made on the basis of the classification to ensure that the potential danger identified in classification is expressed on the label.

## QUESTIONS

1. **Q.** What is the importance of acute toxicity testing, and is a precise $LD_{50}$ value necessary to adequately define acute toxicity?

   **A.** Acute toxicity testing is the way in which we define the intrinsic toxicity of a chemical, identify target organs, provide information for risk assessment of acute exposure, provide information for the design and selection of dose levels for more prolonged studies (i.e., subchronic, chronic), and, most importantly, provide information to clinicians for use in the treatment of acute chemical poisoning. Information from acute toxicity testing is also used to provide insight into the mechanism of action of a chemical, to formulate safety measures during the early stages in the development of a new chemical, and for categorization and labeling purposes when handling and shipping chemicals. One should not confuse the concept of *acute toxicity* with the term $LD_{50}$. The $LD_{50}$ is a statistically defined measure of acute toxicity but is only one of many ways to define acute toxicity. Indeed, a precise $LD_{50}$ is seldom required in acute toxicity testing, and its use is being deemphasized to reduce the total number of animals and pain and suffering involved in their use. The $LD_{50}$ is being replaced by more modern methods. The up-and-down procedure and the acute toxic class method are alternatives to the $LD_{50}$ test that can be used to estimate the medial lethal dose and to provide hazard classification for labeling, respectively.

2. **Q.** Name some of the factors that can influence the results of an acute toxicity study.

   **A.** Physiochemical properties of the test article (lipophylicity, molecular weight, solubility), species used, age of the animals, route of exposure, and rate of test article metabolism.

3. **Q.** The following mortality data were obtained from an acute oral toxicity study:

| Dose (mg/kg) | 1 | 2 | 4 | 8 | 16 | 32 |
|---|---|---|---|---|---|---|
| Mortality | 0/10 | 1/10 | 3/10 | 4/10 | 7/10 | 10/10 |

   Calculate the $LD_{50}$, the SE of the $LD_{50}$, the fiducial limits, and the slope of the dose–response curve:

| Log Dose (x) | n | Probits Observed | Probits Expected | Probabilities Expected (P) | Responses Observed | Responses Expected | $\chi^2$ |
|---|---|---|---|---|---|---|---|
| 0.30 | 10 | 3.72 | 3.82 | 11.9 | 1 | 1.19 | 0.0344 |
| 0.60 | 10 | 4.48 | 4.36 | 26.1 | 3 | 2.61 | 0.0344 |
| 0.90 | 10 | 4.75 | 4.75 | 46.4 | 4 | 4.64 | 0.0789 |
| 1.20 | 10 | 5.52 | 4.45 | 67.4 | 7 | 6.74 | 0.1646 |
| 1.50 | 10 | — | — | — | 10 | — | 0.0307 |

$\Sigma\chi^2 = 0.386$, df = 2

A. Procedure

a. Determine the log dose and probits:

| Log dose | 0.0 | 0.3 | 0.6 | 0.9 | 1.2 | 1.5 |
|---|---|---|---|---|---|---|
| Probits | — | 3.72 | 4.48 | 4.75 | 5.52 | — |

b. Plot log dose vs. probits (Figure 22.8) and fit the best points to a straight line (see Figure 22.8).

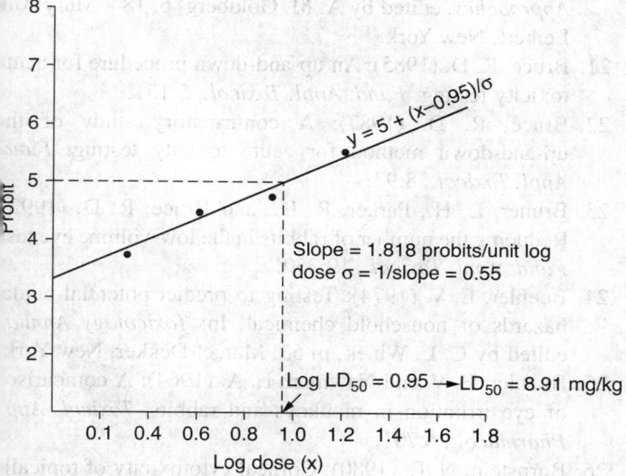


FIGURE 22.8 Example of probit vs. log–dose plot.

c. From the log dose probits line, extrapolate the log $LD_{50} = 0.95$; then $LD_{50}$ = antilog 0.95 = 8.91 mg/kg body weight.

d. From the same line, calculate the slope as (numbers of probit units)/unit log dose = 2/11 = 1.818. Thus (Figure 22.8),

$$\sigma = \frac{1}{\text{Slope}} = 0.55$$

e. *$\chi^2$ test of goodness-of-fit*—Expected probability is converted from the expected probits. The test is conducted by converting each expected probit (*y*) back to the expected probability (*P*) and then to the number of expected responses (*E*); that is, multiply the expected probability *P* by *n*. The difference between expected and observed number of responses will be used to calculate the $\chi^2$ statistic, but instead of using $\Sigma[(E-0)^2/E]$, the weighted value will be used: $\Sigma[(E-0)^2/E(1-P)]$. The degree of freedom (df) is $N-2$, where $N$ is the number of dose levels used in the calculation of $\chi^2$. The critical $\chi^2$ for $(4-2) = 2$ degrees of freedom is 6.0 at $p = 0.05$, and the calculated $\chi^2 = 0.386$, which is less than the critical value, indicating that the fitted line is adequate.

f. *Determination of precision of $LD_{50}$ by weighting*—The SE of log $LD_{50}$ is equal to:

$$S_m = \sigma\Big/\sqrt{\sum mW} = 0.55/\sqrt{18.5} = 0.129$$

The approximation of SE ($LD_{50}$) = $(10^m) \times (S_m)$ = $8.91 \times 2.302 \times 0129 = 2.646$. The precision of $LD_{50} = 8.91 \pm 2.646$ mg/kg.

| | Dose (mg/kg) 2 | 4 | 6 | 16 |
|---|---|---|---|---|
| *W* | 0.277 | 0.423 | 0.541 | 0.564 |
| *nW* | 2.77 | 4.23 | 5.41 | 5.64 |
| *nW* | — | — | 18.05 | — |

g. *Fiducial limits*—Using the approximation formula, the fiducial limit calculated at the $F = 95\%$ level is given by log $LD_{50} \pm 1.96\ (S_m)$. Thus, the lower log $LD_{50}$ limit $= 0.5 - 1.96 \times 0.129 = 0.697$, and the antilog $0.697 = 4.977$. The upper log $LD_{50}$ limit $= 0.95 + 1.96 \times 0.129 = 1.20 = 15.849$. Antilogs of 0.697 and 1.2 give the fiducial $LD_{50}$ limit 4.98 to 15.85 mg/kg.

4. **Q.** What type of distribution provides a precise description of a lethality response to a toxic test article?
**A.** Log-normal distribution.

5. **Q.** Compare and contrast eye irritation and eye corrosion, including in the discussion a description of the observation endpoints usually associated with irritancy in the three major tissues of the eye.

**A.** Eye irritation can be defined as reversible inflammatory changes in the eye and its surrounding mucus membranes following exposure to a material on the surface of the anterior portion of the eye. To contrast, corrosion represents irreversible tissue damage to the eye following exposure to a material. The amount of damage to each of the three major eye tissues—the cornea, the conjunctiva, and the iris—is what differentiates irritancy from corrosion. Gross tissue destruction that follows rapidly after exposure and persists for an extended period in any or all of these tissues is usually an indication of eye corrosion. Irritancy, however, can occur to various degrees. Assessment of injury is based on the presence and severity of cloudiness (opacity) and swelling of the cornea; redness, edema (chemosis), and discharge in the conjunctiva; and increased vascularity, edema, absence of reaction to light, and cloudiness (aqueous flair) in the iris.

## REFERENCES


1. Ashford, J. R. (1959): An approach to the analysis of data for semiquantal responses in biological assay. *Biometrics*, 156:573.
2. Baldwin, H. Q., McDonald, T. D., and Beasley, C. H. (1973): Slit-lamp examination of experimental animal eyes. II. Grading scales and photographic evaluation of induced pathological conditions. *J. Soc. Cosmet. Chem.*, 24:181.
3. Ballantyne, B. and Swanston, D. W. (1974): The irritant effects of dilute solutions of dibenzoxyazepine (CR) on the eye and tongue. *Acta Pharmacol. Toxicol.*, 35:412.
4. Balls, M. (1991): Why modification of the $LD_{50}$ test will not be enough. *Lab. Anim.*, 25:198.
5. Barratt, M. D. et al. (1995): The integrated use of alternative approaches for predicting toxic hazard. *ATLA*, 23:410.
6. Bass, R. et al. (1982): $LD_{50}$ versus acute toxicity. *Arch. Toxicol.*, 51:183.
7. Battista, S. P. and McSweeney, E. S. (1965): Approaches to a quantitative method for testing eye irritation. *J. Soc. Cosmet. Chem.*, 16:199.
8. Bayard, S. and Hehir, R. M. (1976): Evaluation of proposed changes in the modified Draize rabbit irritation test. *Toxicol. Appl. Pharmacol.*, 37:186.
9. Beckley, J. H. (1965): Comparative eye testing: man vs. animal. *Toxicol. Appl. Pharmacol.*, 7:93.
10. Beckley, J. H. (1965): Critique of the Draize eye test, now and then: eighteen, nine or six rabbits. *Am. Perf. Cosmet.*, 80:5.
11. Bliss, C. I. (1934): The method of probits: a correction. *Science*, 79:409.
12. Bliss, C. I. (1964): Insecticide assays. In: *Statistics and Mathematics in Biology*, edited by O. Kempthorne, T. A. Bancroft, J. W. Gowen, and J. L. Lush, p. 345. Hofner, New York.
13. Bohnke, M. et al. (1996): High-precision, high-speed mea surement of eximer laser keratectomies with a new optica pachymeter. *Ger. J. Ophthalmol.*, 5:338.
14. Bonifield, C. T. and Scala, R. A. (1965): The paradox i testing for eye irritation: a report on thirteen shampoos *Proc. Sci. Sect. Toilet Goods Assoc.*, 43:34.
15. Bonting, S. L., Simon, K. A., and Hawkins, N. M. (1961) Studies on sodium-potassium-activated adenosine triphos phatase. 1. Quantitative distribution in several tissues o the rat. *Arch. Biochem.*, 95:416.
16. Botham, P. (2002): Acute systemic toxicity. *ILAR J* 43(Suppl.):S27.
17. British Toxicology Society. (1984): A new approval t classification of substances and preparations on the basi of their acute toxicology. *Human Toxicol.*, 3:85.
18. Bross, I. D. J. (1958): How to use RIDIT analysis. *Bio metrics*, 14:18.
19. Brownlee, K. A., Hodges, J. L., and Rosenblatt, M. (1953) The up-and-down method with small samples. *J. Am. Sta Assoc.*, 48:262.
20. Bruce, R. D. (1984): An up-and-down procedure for acut toxicity testing. In: *Acute Toxicity Testing: Alternativ Approaches*, edited by A. M. Goldberg, p. 184. Mary An Leibert, New York.
21. Bruce, R. D. (1985): An up-and-down procedure for acut toxicity testing. *Fund. Appl. Toxicol.*, 5:151.
22. Bruce, R. D. (1987): A confirmatory study of th up-and-down method for acute toxicity testing. *Fun Appl. Toxicol.*, 8:97.
23. Bruner, L. H., Parker, R. D., and Bruce, R. D. (1992 Reducing the number of rabbits in the low-volume eye tes *Fund. Appl. Toxicol.*, 19:330.
24. Buehler, E. V. (1974): Testing to predict potential ocul hazards of household chemical. In: *Toxicology Annua* edited by C. L. Winek, p. 53. Marcel Dekker, New York
25. Buehler, E. V. and Newman, E. A. (1964): A compariso of eye irritation in monkeys and rabbits. *Toxicol. App Pharmacol.*, 6:701.
26. Burnstein, N. L. (1980): Corneal cytotoxicity of topicall applied drugs, vehicles, and preservatives. *Surv. Ophtha mol.*, 25:15.
27. Burton, A. B. G. (1972): A method for the objective asses ment of eye irritation. *Food Cosmet. Toxicol.*, 10: 209.
28. Calabrese, E. J. (1983): Ocular toxicity. In: *Principles Animal Extrapolation*, p. 400. John Wiley & Sons, Ne York.
29. Cello, R. M. and Lasmanis, J. (1958): *Pseudomonas* infe tion of the eye of the dog resulting from the use of co taminated fluorescein solution. *J. Am. Vet. Med. Asso* 132:297.
30. CEPA. (1999): Canadian Environmental Protection Ac new substances notification regulations. *Canada Gazet* Part III, 22(3):1.
31. Chan, P. K. and Hayes, A. W. (1985): Assessment of chem ically induced ocular toxicity: a survey of methods. I *Toxicology of the Eye, Ear, and Other Special Sense* edited by A. W. Hayes. Raven Press, New York.
32. Chan, T., Payor, S., and Holden, B. A. (1983): Corne thickness profiles in rabbits using an ultrasonic pachom ter. *Invest. Ophthalmol. Vis. Sci.*, 24:1408.

33. Chan-Ling, T. et al. (1987): Long-term neural degeneration in the rabbit following 180° limbal incision. *Invest. Ophthalmol. Vis. Sci.*, 28:2083.
34. Choi, S. C. (1971): An investigation of Wetherill's method of estimation for the up-and-down experiment, *Biometrics*, 27:961.
35. Clark, A. J. (1933): *Mode of Action of Drugs on Cells*. Williams & Wilkins, Baltimore, MD.
36. Cohen, I. J. (1983): Use of fluorescein in eye injuries. *J. Occup. Med.*, 5:540.
37. Conquet, P. H. et al. (1977): Evaluation of ocular irritation in the rabbit: objective versus subjective assessment. *Toxicol. Appl. Pharmacol.*, 39:129.
38. Cornfield, J. (1964): Measurement and composition of toxicities: the quantal response. In: *Statistics and Mathematics in Biology*, edited by O. Kempthorne, T. A. Bancroft, J. W. Gowen, and J. L. Lush, p. 327. Hofner, New York.
39. CPSC. (1976): *Illustrated Guide for Grading Eye Irritation by Hazardous Substances*. Directorate for Engineering and Science, Consumer Product Safety Commission, Washington, D.C.
40. Davies, R. G., Kynoch, S. R., and Liggett, M. P. (1976): Eye irritation tests: an assessment of the maximum delay time for remedial irrigation. *J. Soc. Cosmet. Chem.*, 27:301.
41. Dawson, M. and Mustafa, A. F. (1985): Use of cultured human conjunctival and other cells to assess the relative toxicity of six local anesthetics. *Food Chem. Toxicol.*, 23:305.
42. DePass, L. R. (1989): Alternative approaches in median lethality ($LD_{50}$) and acute toxicity testing. *Toxicol. Lett.*, 49:159.
43. DeSousa, D. J., Rosue, A. A., and Smolon, W. J. (1984): Statistical consequences of reducing the number of rabbits utilized in eye irritation testing: data on 67 petrochemicals. *Toxicol. Appl. Pharmacol.*, 76:234.
44. Dews, P. B. and Berkson, J. (1964): On the error of bioassay with quantal response. In: *Statistics and Mathematics in Biology*, edited by O. Kempthorne, T. A. Bancroft, J. W. Gowen, and J. L. Lush, p. 361. Hofner, New York.
45. Diem, K. and Lentner, C. (1970): *Documenta Geigy Scientic Tables*, 7th ed. Geigy Pharmaceuticals, Ciba-Geigy Corp., Ardsley, NY.
46. Dixon, W. J. (1965): The up-and-down method for small samples. *J. Am. Stat. Assoc.*, 60:967.
47. Dixon, W. J., Ed. (1983): *BMDP Statistics Software*. University of California Press, Berkeley, CA.
48. Doull, J. (1980): Factors influencing toxicology. In: *Casarett & Doull's Toxicology: The Basic Science of Poisons*, edited by J. Doull, C. D. Klaassen, and M. O. Amdur, p. 70. Macmillan, New York.
49. Draize, J. H., Woodward, G., and Calvery, H. O. (1944): Methods for the study of irritation and toxicity of substances applied topically to the skin and mucous membranes. *J. Pharmacol. Exp. Ther.*, 82:377.
50. Dunnett, C. W. (1968): Biostatistics in pharmacological testing. In: *Selected Pharmacological Testing Methods*, edited by A. Burger, p. 7. Edward Arnold, London.
51. EEC. (1996): *Methods for the Determination of Toxicity and Other Health Effects*, Council Directive 67:548, EEC Annex V (Part B), European Economic Commission, Brussels.
52. EEC. (2004): Acute toxicity: eye irritation/corrosion, EEC guideline for testing of chemicals no. B5. *Official J.*, L152:1.
53. EEC. (2001): General classification and labeling requirements for dangerous substances and preparations, Commission Directive 59, EC 28th Adaption to Technical Progress of Council Directive 67:548, EEC, Annex VI. *Official J.*, L225:263.
54. EEC. (2001): Nature of special risks attributed to dangerous substances and preparations. Commission Directive 59, EC 28th Adaption to Technical Progress of Council Directive 67:548, EEC, Annex III. *Official J.*, L225:85.
55. Ehrlich, P. (1882): Uber provocirte flurescenzer-Scheinungen am Auge. *Dtsch. Med. Wochenschr.*, 2: 21.
56. EPA. (1989): Good Laboratory Practice Standards; Federal Insecticide, Fungicide and Rodenticide Act (FIFRA); Final Rule. *Fed. Reg.*, 54:34052.
57. EPA. (1989): Good Laboratory Practice Standards; Toxic Substances Control Act (TSCA); Final Rule. *Fed. Reg.*, 54:34034.
58. EPA. (1984): *EPA Fact Sheet: Background on Acute Toxicity Testing for Chemical Safety*, U.S. Environmental Protection Agency, Washington, D.C.
59. EPA. (2003): *Label Review Manual*, 3rd ed., EPA document 735-B-03-001, chap. 7. Office of Prevention, Pesticides, and Toxic Substances, U.S. Environmental Protection Agency, Washington, D.C.
60. EPA. (2005): *High Production Volume (HPV) Challenge Program*. U.S. Environmental Protection Agency, Washington, D.C. (www. epa.gov/chemrtk/volchall.htm).
61. EPA/OPPTS. (1996): *OPPTS Harmonized Test Guidelines. Health Effects Test Guideline 870. 1100: Acute Oral Toxicity*, U.S. EPA document 712-C-96-190. Office of Prevention, Pesticides, and Toxic Substances, U.S. Environmental Protection Agency, Washington, D.C.
62. EPA/OPPTS (1996): *OPPTS Harmonized Test Guidelines: Health Effects Test Guideline 870. 1200: Acute Dermal Toxicity*, U.S. EPA document 712-C-96-192. Office of Prevention, Pesticides, and Toxic Substances, U.S. Environmental Protection Agency, Washington, D.C.
63. EPA/OPPTS (1996): *OPPTS Harmonized Test Guidelines. Health Effects Test Guideline 870. 1300: Acute Inhalation Toxicity*. U.S. EPA document 712-C-96-193. Office of Prevention, Pesticides, and Toxic Substances, U.S. Environmental Protection Agency, Washington, D.C.
64. EPA/OPPTS. (1996): *OPPTS Harmonized Test Guidelines. Health Effects Test Guideline 870. 1350: Acute Inhalation Toxicity With Histopathology*. U.S.EPA document 712-C-96-191. Office of Prevention, Pesticides, and Toxic Substances, U.S. Environmental Protection Agency, Washington, D.C.
65. EPA/OPPTS. (1996): *OPPTS Harmonized Test Guidelines, Health Effects Test Guideline 870. 2400: Acute Eye Irritation*. U.S. EPA document 712-C-96-195. Office of Prevention, Pesticides, and Toxic Substances, U.S. Environmental Protection Agency, Washington, D. C.
66. EPA/TSCA. (2005): *Health Effects Test Guidelines: TSCA Acute Oral Toxicity*. Title 40, Code of Federal Regulations, Part 799.9110. U.S. Environmental Protection Agency, Washington, D.C.

67. EPA/TSCA (2005): *Health Effects Test Guidelines: TSCA Acute Dermal Toxicity*. Title 40, Code of Federal Regulations, Part 799.9120. U.S. Environmental Protection Agency, Washington, D.C.
68. EPA/TSCA. (2005): *Health Effects Test Guidelines: TSCA Acute Inhalation Toxicity*. Title 40, Code of Federal Regulations, Part 799.9130. U.S. Environmental Protection Agency, Washington, D.C.
69. Essepian, J. P. et al. (1994): The use of confocal microscopy in evaluating corneal wound healing after eximer laser surgery. *Scanning*, 16:300.
70. FDA. (1976): *Illustrated Guide for Grading Eye Irritation by Hazardous Substances*. U.S. Food and Drug Administration, Washington, D.C.
71. FDA. (1983): *Final Report on Acute Studies Workshop*. U.S. Food and Drug Administration, Washington, D.C.
72. Ferdinard, W. (1976): *The Enzyme Molecule*. John Wiley & Sons, New York.
73. FHSA. (1979): Regulations under the Federal Hazardous Substance Act. Title 16, Code of Federal Regulations, Part 1500. U.S. Consumer Product Safety Commission, Washington, D.C.
74. Fine, B. S. and Yanoff, M. (1972): *Ocular Histology: A Text and Atlas*. Harper & Row, New York.
75. Finney, D. J. (1971): *Probit Analysis*, 3rd ed., chaps. 3 and 4. Cambridge University Press, Cambridge, U.K.
76. Fisher, R. A. and Yates, F. (1963): *Statistical Tables for Biological, Agricultural and Medical Research*, 6th ed. Oliver and Boyd, Edinburgh, Scotland.
77. Floyd, E. P. and Stockinger, H. G. (1958): Toxicity studies of certain organic peroxides and hydroperoxides. *Am. Indust. Hyg. Assoc. J.*, 19:205.
78. Freeberg, F. E. et al. (1984): Correlation of animal test methods with human experience for household products. *J. Toxicol. Cutan. Ocul. Toxicol.*, 1(3):53.
79. Friedenwald, J. S., Hughes, W. F., and Hermann, H. (1944): Acid–base tolerance of the cornea. *Arch. Ophthalmol.*, 31:279.
80. Gad, S. C. et al. (1984): Innovative designs and practices for acute systemic toxicity studies. *Drug Chem. Toxicol.*, 7:423.
81. Gaddum, J. H. (1983): *Reports on Biological Standards*. III. *Methods of Biological Assay Depending on a Quantal Response*, Special Report Series of the Medical Research Council, No. 813. Medical Research Council, London.
82. Gaunt, I. F. and Harper, K. H. (1964): The potential irritancy to rabbit eye mucosa of certain commercially available shampoos. *J. Soc. Cosmet. Chem.*, 15:209.
83. Giasson, C. and Forthomme, D. (1992): Comparison of central corneal thickness measurements between optical and ultrasound pachometers. *Optom. Vis. Sci.*, 69:236.
84. Giovacchini, R. P. (1972): Old and new issues in the safety evaluation of cosmetics and toiletries. *CRC Crit. Rev. Toxicol.*, 1:361.
85. Grant, W. M. (1974): *Toxicology of the Eye*, 2nd ed. Charles C Thomas, Springfield, IL.
86. Green, W. R. et al. (1978): *A Chemically Induced Eye Injury in the Albino Rabbit and Rhesus Monkey*. Soap and Detergent Association, New York.
87. Gribaldo, L. et al. (2005): Acute toxicity. *Altern. Lab Animals*, 33(Suppl. 1):27.
88. Griffith, J. F. et al. (1980): Dose–response studies wit chemical irritants in the albino rabbit eye as a basis fo selecting optimum testing conditions for predicting hazar to human eye. *Toxicol. Appl. Pharmacol*, 55:501.
89. Guillot, J., Gonnet, J. F., and Clement, C. (1982): Evalu ation of the ocular irritation potential of 56 compounds *Food Chem. Toxicol.*, 20:573.
90. Gunderson T. and Liebman, S. D. (1944): Effect of loca anesthetics on regeneration of corneal epithelium. *Arch Ophthalmol.*, 31:29.
91. Gupta, K. C. et al. (1993): An eye irritation test protoc and an evaluation and classification system. *Food Chem Toxicol.*, 31:117.
92. Gurland, J., Lee, L., and Dahm, P. A. (1960): Polychoto mous quantal response in biological assay. *Biometrics* 16:382.
93. Harriton, L. (1981): Conversation with Henry Spira Draize test activist. *Lab. Anim.*, 10:16.
94. Hedbys, R. O. and Mishima, S. (1966): The thickness hydration relationship of the cornea. *Exp. Eye Res* 5:221.
95. Henkes, H. and Canta, L. R. (1973): Drug-induced diso ders of the eye. In: *Proc. of the European Society for th Study of Drug Toxicity*, edited by W. A. M. Duncan, p. 14 Elsevier, New York.
96. Heywood, R. and James, R. W. (1978): Towards objectivit in the assessment of eye irritation. *J. Soc. Cosmet. Chem* 29:25.
97. Hitzenberger, C. K., Drexler, W., and Fercher, A. F. (1992 Measurement of corneal thickness by laser Doppler inte ferometry. *Invest. Ophthalmol. Vis. Sci.*, 33:98.
98. Holland, M. C. (1964): Fluorescein staining of the corne *JAMA*, 188:81.
99. Hurley, P. M. et al. (1993): Screening procedures for ey irritation. *Food Chem. Toxicol.*, 31:87.
100. ICH. (2005): International Conference on Harmonisati of Technical Requirements for Registration of Pharmace ticals for Human Use, www. ICH. org.
101. IRLG. (1981): *Recommended Guidelines for Acute Ey Irritation Test*. Interagency Regulatory Liaison Grou Washington, D.C.
102. Jackson, J. and Rutty, D. A. (1985): Ocular toleran assessment: integrated tier policy. *Food Chem. Toxico* 23:309.
103. Jester, J. V. et al. (1998): Area and depth of surfacta induced corneal injury correlates with cell death. *Inve Ophthalmol. Vis. Sci.*, 39:922.
104. Jester, J. V. et al. (1992): Comparison of *in vivo* and *vivo* cellular structure in rabbit eyes detected by scanni confocal microscopy. *J. Microsc.*, 165:169.
105. Jumblatt, M. M., Fogle, J. A., and Neufeld, A. H. (198 Cholera toxin stimulates adenosine 3′5′-monophospha synthesis and epithelial wound closure in the rabbit corn *Invest. Ophthalmol. Vis. Sci.*, 19:1321.
106. Jumblatt, M. M. and Neufeld, A. H. (1981): Characteri tion of cyclic AMP-mediated wound closure of the rab corneal epithelium. *Curr. Eye Res.*, 1:189.

107. Kay, J. H. and Calandra, J. C. (1962): Interpretation of eye irritation tests. *J. Soc. Cosmet. Chem.*, 13:281.
108. Kaye, G. I. and Tice, L. W. (1966): Studies on the cornea. V. Electron microscopic localization of adenosine triphosphatase activity in the rabbit cornea in relation to transport. *Invest. Ophthalmol.*, 5:22.
109. Kennah, H. E. et al. (1989): An objective procedure for quantitating eye irritation based on changes in corneal thickness. *Fundam. Appl. Toxicol.*, 12:258.
110. Kennedy, Jr., G. L., Ferenz, R., and Burgess, B. A. (1986): Estimation of acute oral toxicity in rats by determination of the approximate lethal dose rather than the $LD_{50}$. *J. Appl. Toxicol.*, 6:145.
111. Kimura, S. J. (1951): Fluorescein paper: simple means of insuring use of sterile fluorescein. *Am. J. Ophthalmol.*, 34:466.
112. Kuhlman, R. E. (1959): Species variation in the enzyme content of corneal epithelium. *J. Cell. Comp. Physiol.*, 53:313.
113. Lieberman, H. R. (1983): Estimating $LD_{50}$ using the probit technique: a basic computer program. *Drug Chem. Toxicol.*, 6:111.
114. Ling, T., Ho, A., and Holden, B. A. (1986): Method of evaluating ultrasonic pachometers. *Am. J. Optom. Physiol. Optics.*, 63:462.
115. Lipnick, R. L. et al. (1995): Comparison of the up-and-down, conventional $LD_{50}$, and fixed dose acute toxicity procedures. *Food Chem. Toxicol.*, 33:223.
116. Litchfield, J. T. and Wilcoxon, F. (1949): A simplified method of evaluating dose–effect experiments. *J. Pharmacol. Exp. Ther.*, 96:99.
117. Lorke. D. (1983): A new approach to practical toxicity testing. *Arch. Toxicol.*, 54:275.
118. Maeda, K. and Sakagudin, K. (1965): Studies on sodium-potassium-activated adenosine triphosphatase in the cornea: electron-microscopic observations on the rat cornea. *Jpn. J. Ophthalmol.*, 9:195.
119. Martins, T., Pauluhn, J., and Machemer, L. (1992): Analysis of alternative methods for determining ocular irritation. *Food Chem. Toxicol.*, 30:1061.
120. Marzulli, F. N. (1965): New data on eye and skin tests. *Toxicol. Appl. Pharmacol.*, 7:79.
121. Marzulli, F. N. and Simmon, M. E. (1971): Eye irritation from topically applied drugs and cosmetics: preclinical studies. *Am. J. Optom.*, 48:61.
122. Masters, B. and Paddock, S. (1990): *In vitro* confocal imaging of the rabbit cornea. *J. Micros.*, 158:267.
123. Mauer, J. K. et al. (1997): Confocal microscopic characterization of initial corneal changes of surfactant-induced eye irritation in the rabbit. *Toxicol. Appl. Pharmacol.*, 143:291.
124. Maurice, D. M. (1967): The use of fluorescein in ophthalmological research. *Invest. Ophthalmol.*, 6:465.
125. Maurice, D. M. and Giardini, A. A. (1951): A simple optical apparatus for measuring the corneal thickness and the average thickness of the human cornea. *Br. J. Ophthalmol.*, 35:169.
126. McCaa, C. S. (1985): Anatomy, physiology and toxicology of the eye. In: *Toxicology of the Eye, Ear, and Other Special Senses*, edited by A. W. Hayes, p. 1. Raven Press, New York.
127. McDonald, T. O., Baldwin, H. A., and Beasley, C. H. (1973): Slit-lamp examination of experimental animal eyes, I. Techniques of illumination and the normal eye. *J. Soc. Cosmet. Chem.*, 24:163.
128. Mehlman, M. A., Pfitzer, E. A., and Scala, R. A. (1989): A report on methods to reduce, refine, and replace animal testing in industrial toxicology laboratories. *Cell Biol. Toxicol.*, 5:349.
129. Meier-Ruge, W. (1973): Eye toxicity. In: *Proceedings of the European Society for the Study of Drug Toxicity*, Vol. 14, edited by W. A. M. Duncan, p. 133. Elsevier, New York.
130. Miller, L. C. (1964): The quantal response in toxicity tests. In: *Statistics and Mathematics in Biology*, edited by O. Kempthorne, T. A. Bancroft, J. W. Gowen, and J. L. Lush, p. 315. Hofner, New York.
131. Millodot, M., Lim, C. H., and Ruskell, G. L. (1978): A comparison of corneal sensitivity and nerve density in albino and pigmented rabbits. *Ophthalmol. Res.*, 10:307.
132. Minsky, M. (1988): Memoir on inventing the confocal scanning microscope. *Scanning*, 10:128.
133. Minsky, M. (1961): Microscopy Apparatus, U.S. Patent No. 30313467.
134. Mishima, S. (1981): Clinical pharmacokinetics of the eye: Proctor lecture. *Invest. Ophthalmol. Vis. Sci.*, 21:504.
135. Mishima, S. and Hedbys, B. O. (1968): Measurement of corneal thickness with the Haag–Streit pachometer. *Arch. Ophthalmol.*, 80:710.
136. Mishima, S. and Maurice, D. M. (1971): *In vivo* determination of the endothelial permeability to fluorescein. *Acta Soc. Ophthalmol. (Jpn.)*, 765:236.
137. Moller-Pedersen, T. et al. (1998): Confocal microscopic characterization of wound repair after photorefractive keratectomy. *Invest. Ophthalmol. Vis. Sci.*, 39:487.
138. Morgan, R. L., Sorenson, S. S., and Castles, T. R. (1987): Prediction of ocular irritation by corneal pachymetry. *Food Chem. Toxicol.*, 25:609.
139. Muir, C. K. (1983): The toxic effect of some industrial chemicals on rabbit ileum *in vitro* compared with eye irritancy *in vivo*. *Toxicol. Lett.*, 19:309.
140. Muller, H. and Kley, H. P. (1982): Retrospective study on the reliability of an 'approximate $LD_{50}$' determined with a small number of animals. *Arch. Toxicol.*, 51:189.
141. Myers, R. C. et al. (1998): Comparative evaluation of several methods and conditions for the *in vivo* measurement of corneal thickness in rabbits and rats. *Toxicol. Meth.*, 8:219.
142. Nagy, Z. Z., Suveges, I., and Nemeth, J. (1995–96): Interoperative pachymetry during eximer photorefractive keratectomy. *Acta Chir. Hung.*, 35:217.
143. Nakajima, A., Kimura, T., and Yamazaki, M. (1967): Applications of ultrasound in biometry of the eye. In: *Ultrasonics in Ophthalmology Diagnostic and Therapeutic Applications*, edited by R. E. Goldberg and L. K. Sarin, p. 124. W.B. Saunders, Philadelphia, PA.
144. National Research Council, Committee for Revision of NAS Publication 1138. (1977): Dermal and eye toxicity tests. In: *Principles and Procedures for Evaluating the Toxicity of Household Substances*, p. 41. National Academy of Sciences, Washington, D.C.

145. NIER. (1999): *Korean Existing Chemicals Inventory: Data Requirements and Hazard Evaluation*. NIER Public Notice, No. 1999-39. National Institute of Environmental Research, Incheon, Korea.
146. Nover, A. and Glanschneider, D. (1965): Untersuchungen uber die fortpflanzungsgeschwindigkeit und absorptiondes ultraschalls im Gewebe. Experimentelle beitrage zur ultraschalldiagnostik intraocular tumoren, *Albtecht von Graefes Arch. Klin. Exp. Ophthalmol.*, 168:304.
147. OECD. (2002): *OECD Test Guideline 401 Will Be Deleted. A Major Step in Animal Welfare: OECD Reaches Agreement on the Abolishment of the $LD_{50}$ Acute Toxicity Test*. Office of Economic and Community Development, Paris.
148. OECD. (2004): *OECD Guideline for Testing of Chemicals. Proposal for a New Draft Guideline 434: Acute Dermal Toxicity: Fixed Dose Procedure*. Office of Economic and Community Development, Paris.
149. OECD. (2004): *OECD Guideline for Testing of Chemicals. Draft Proposal for a New Guideline 433: Acute Inhalation Toxicity: Fixed Concentration Procedure*. Office of Economic and Community Development, Paris.
150. OECD. (1981): *OECD Guideline for Testing of Chemicals. Guideline 403: Acute Inhalation Toxicity*. Office of Economic and Community Development, Paris.
151. OECD. (1987): *OECD Guideline for Testing of Chemicals. Guideline 402: Acute Dermal Toxicity*. Office of Economic and Community Development, Paris.
152. OECD. (2004): *OECD Guideline for the Testing of Chemicals. Draft Proposal for a New Guideline 436: Acute Inhalation Toxicity: Acute Toxic Class (ATC) Method*. Office of Economic and Community Development, Paris.
153. OECD. (2001): *OECD Series on Testing and Assessment No. 24: Guidance Document on Acute Oral Toxicity Testing*. Office of Economic and Community Development, Paris.
154. OECD. (1998): *Harmonized Integrated Hazard Classification System for Human Health and Environmental Effects of Chemical Substances as Endorsed by the 28th Joint Meeting of the Chemicals Committee and the Working Party on Chemicals*. Office of Economic and Community Development, Paris.
155. OECD. (2005): *Cooperation on the Investigation of Existing Chemicals: Description of OECD Work on Investigation of High Production Volume Chemicals (SIDS Program)*. Office of Economic and Community Development, Paris (www.oecd.org/document/21/0,2340.en_2649_34379_ 1939669_1_1_1_1,00.html).
156. OECD. (1984): *Data Interpretation Guides for Initial Hazard Assessment of Chemicals*. Office of Economic and Community Development, Paris.
157. OECD. (1981): *OECD Guidelines for Testing of Chemicals*. Office of Economic and Community Development, Paris.
158. OECD. (2002): *OECD Guidelines for Testing of Chemicals. Guideline 405: Acute Eye Irritation/Corrosion*. Office of Economic and Community Development, Paris.
159. OECD. (2001): *OECD Guideline for Testing of Chemicals. Guideline 420: Acute Oral Toxicity: Fixed Dose Method*. Office of Economic and Community Development, Paris.
160. OECD. (2001): *OECD Guideline for Testing of Chemicals. Guideline 423: Acute Oral Toxicity: Acute Toxic Class Method*. Office of Economic and Community Development, Paris.
161. OECD. (2001): *OECD Guideline for Testing of Chemicals. Guideline 425: Acute Oral Toxicity: Up and Down Procedure*. Office of Economic and Community Development, Paris.
162. OECD. (1999): *OECD Series on Testing and Assessment. Detailed Review on Classification Systems for Eye Irritation/Corrosion in OECD Member Countries*. Office of Economic and Community Development, Paris.
163. OECD. (1981): *Test Guidelines: Decision of the Council Concerning Mutual Acceptance of Data in the Assessment of Chemicals*. Annex 2. *OECD Principles of Good Laboratory Practices*. Office of Economic and Community Development, Paris.
164. Oksala, A. and Lehtinen, A. (1958): Measurement of the velocity of sound in some parts of the eye. *Acta Ophthalmol.*, 36:633.
165. Olson, K. J. et al. (1962): Toxicological properties of several commercially available surfactants. *J. Soc. Cosmet Chem.*, 13:469.
166. Patel, S. and Stevenson, R. W. W. (1994): Clinical evaluation of a portable ultrasonic and a standard optical pachometer. *Optom. Vis. Sci.*, 71:43.
167. Petroll, W. M., Cavanagh, H. D., and Jester, J. V. (1993): Three dimensional imaging of corneal cells using *in vivo* confocal microscopy. *J. Micros.*, 170:213.
168. Petroll, W. M. et al. (1992): Digital image acquisition in *in vivo* confocal microscopy. *J. Micros.*, 165:61.
169. Petroll, W. M., Jester, J. V., and Cavanagh, H. D. (1994): *In vivo* confocal imaging: general principles and applications. *Scanning*, 16:131.
170. Petroll, W. M., Jester, J. V., and Cavanagh, H. D. (1996): Quantitative three-dimensional confocal imaging of the cornea *in situ* and *in vivo*: system design and calibration. *Scanning*, 18:45.
171. Pfister, R. R. and Burstein, N. (1976): The effects of ophthalmic drugs, vehicles, and preservatives on corneal epithelium: a scanning electron microscope study. *Invest Ophthalmol.*, 15: 246.
172. Prince, J. H. et al. (1960): *Anatomy and Histology of the Eye and Orbit in Domestic Animals*. Charles C Thomas, Springfield, IL,
173. REACH. (2003): Proposal for a regulation of the European Parliament and of the Council concerning the Registration, Evaluation, Authorisation and Restriction of Chemicals (REACH), establishing a European Chemicals Agency and amending Directives 1999/45/EC and 67/548/EEC. COM (2003) 0644 final.
174. Reiger, M. M. and Battista, G. W. (1964): Some experiences in the safety testing of cosmetics. *J. Soc. Cosmet Chem.*, 15:161.
175. Rivera, A. and Sanna, G. (1962): Determiniazione della velocita degli ultrasuoni nei tessuti oculari di uomo et di maiale, *Annali di Ottalmologia e Clinica Oculistica*, 88:675.

176. Roeig, D. L. et al. (1980): Occurrence of corneal opacities in rats after acute administration of 1-alpha-acetylmethadol. *Toxicol. Appl. Pharmacol.*, 56:155.
177. Roll, R., Hoffer-Bosse, T., and Kayser, D. (1986): New perspectives in acute toxicity testing of chemicals. *Toxicol. Lett.*, 31(Suppl.):86.
178. Rosiello, A. P., Essigmann, J. M., and Wogan, G. N. (1977): Rapid and accurate determination of the median lethal dose ($LD_{50}$) and its error with a small computer. *J. Toxicol. Environ. Health*, 3:797.
179. Rowan, A. (1981): The Draize test: political and scientific issues. *Cosmet. Tech.*, 3:32.
180. Salz, J. J. et al. (1983): Evaluation and sources of variability in the measurement of corneal thickness with ultrasonic and optical pachymeters. *Ophthalmol. Surg.*, 14:750.
181. SAS. (1991): *SAS Users' Guide: Statistics*. SAS Institute, Cary, NC.
182. Schlede, E. et al. (1994): The international validation study of the acute toxic class method (oral). *Arch. Toxicol.*, 69:659.
183. Schlede, E. et al. (1992): A national validation study of the acute toxic class method: an alternative to the $LD_{50}$ test. *Arch. Toxicol.*, 66:455.
184. Schutz, E. and Fuchs, H. (1982): A new approach to minimizing the number of animals used in acute toxicity testing and optimizing the information of test results. *Arch. Toxicol.*, 51:197.
185. Seabaugh, V. M. et al. (1976): A comparative study of rabbit ocular reactions of various exposure times to chemicals. In: *Proc. of the Fifteenth Annual Meeting*, Society of Toxicology, Atlanta, GA.
186. Society of Agricultural Chemical Industry. (1985): *Agricultural Chemicals Laws and Regulations* [English transl.]. Society of Agricultural Chemical Industry, Japan.
187. Society of Toxicology, Animals in Research Committee. (1989): SOT position paper: comments on $LD_{50}$ and acute eye and skin irritation tests. *Fund. Appl. Toxicol.*, 13:621.
188. Society of Toxicology of Canada. (1985): Position paper on the $LD_{50}$, adopted at the STC Annual meeting.
189. Sperling, F. (1976): Nonlethal parameters as indices of acute toxicity: inadequacy of the acute $LD_{50}$. In: *New Concepts in Safety Evaluation*, edited by M. A. Mehlman, R. E. Shapiro, and H. Blumenthal, p. 177. Hemisphere, Washington, D.C.
190. Sugar, J. (1980): Corneal examination. In: *Principles and Practice of Ophthalmology*, Vol. 1, edited by G. A. Peyman, D. R. Sanders, and M. F. Goldberg, p. 393. W.B. Saunders, Philadelphia, PA.
191. Talsma, D. M. et al. (1988): Reducing the number of rabbits in the Draize eye irritancy test: a statistical analysis of 155 studies conducted over 6 years. *Fundam. Appl. Toxicol.*, 10:146.
192. Tanaka, N. et al. (1982): Evaluation of ocular toxicity of two beta blocking drugs, cereteolol and practolol, in beagle dogs. *J. Pharmacol. Exp. Ther.*, 224:424.
193. Terry, M. A. and Ousley, P. J. (1996): Variability in corneal thickness before, during, and after radial keratotomy. *J. Refract. Surg.*, 12:700.
194. Thompson, W. R. (1947): Use of moving averages and interpolation to estimate median effective dose. *Bacterial. Rev.*, 11:115.
195. Tonjum, A. M. (1975): Effects of benzalkonium chloride upon the corneal epithelium: studies with scanning electron microscopy. *Acta Ophthalmol.*, 53:358.
196. Trevan, J. W. (1927): The error of determination of toxicity. *Proc. R. Soc. Lond.*, 101B:483,
197. van den Heuvel, M. J. et al. (1990): The international validation of a fixed dose procedure as an alternative to the classical $LD_{50}$ test, *Food Chem. Toxicol.*, 28:469.
198. van den Heuvel, M. J., Dayan, A. D., and Shillaker, R. O. (1987): Evaluation of the BTS approach to the testing of substances and preparations for their acute toxicity. *Human Toxicol.*, 6:279.
199. Villasenor, R. A. et al. (1986): Comparison of ultrasonic corneal thickness measurements before and during surgery in the prospective evaluation of radial keratotomy (PERK) study. *Ophthalmology*, 93:327.
200. Waltman, S. R. and Kaufman, H. E. (1970): *In vivo* studies of human corneal and endothelial permeability. *Am. J. Ophthalmol.*, 70:45.
201. Waud, D. R. (1972): On biological assays involving quantal responses. *J. Pharm. Exp. Ther.*, 183:577.
202. Weil C. S. (1952): Tables for convenient calculation of median effective dose ($LD_{50}$ or $ED_{50}$) and instruction in their use. *Biometrics*, 8:249.
203. Weil, C. S. (1983): Economical $LD_{50}$ and slope determinations. *Drug Chem. Toxicol.*, 6:595.
204. Weil, C. S. and Scala, R. A. (1971): Study of intra- and interlaboratory variability in the results of rabbit eye and skin irritation tests. *Toxicol. Appl. Pharmacol.*, 19:276.
205. Weltman, A. S., Sharber, S. B., and Jurtshuk, T. (1968): Comparative evaluation and influence of various factors on eye irritation tests. *Toxicol. Appl. Pharmacol.*, 7:308.
206. Wheeler, N. C. et al. (1992): Reliability coefficients of three corneal pachymeters. *Am. J. Ophthalmol.*, 113:645.
207. Williams, S. J. (1984): Prediction of ocular irritancy potential from dermal irritation test results. *Food Chem. Toxicol.*, 22:157.
208. Williams, S. J. (1985): Changing concepts of ocular irritation evaluation: pitfalls and progress. *Food Chem. Toxicol.*, 23:189.
209. Williams, S. J., Grapel, G. J., and Kennedy, G. I. (1982): Evaluation of ocular irritancy: potential intralaboratory variability and effect of dosage volume. *Toxicol. Lett.*, 12:235.

# Notes

# 23 Genetic Toxicology

*David J. Brusick, Wanda R. Fields, Brian C. Myhr, and David J. Doolittle*

## CONTENTS

Genetic toxicology, a specialized field of toxicology, involves the identification and analysis of agents with toxicity directed toward the hereditary components of living organisms. A large proportion of human disease is either directly or indirectly associated with genetic damage. Although many agents are capable of indirectly producing genetic damage at very high exposure concentrations by altering cellular homeostasis, the primary objective of genetic toxicologists is to detect and assess genetic hazard from agents that are highly specific for nucleic acids and are capable of producing genetic damage at subtoxic concentrations. Such agents are classified as *genotoxic*.

The term *genotoxic* is a general descriptor used to distinguish chemicals that have an intrinsic affinity for DNA from those that do not. Genotoxic substances have several common chemical or physical properties (e.g., electrophilicity) that facilitate their interaction with nucleic acids. A report of the International Commission for Protection Against Environmental Mutagens and Carcinogens (ICPEMC) provides a more detailed definition of genotoxicity and emphasizes that categorization of a chemical as genotoxic is not an *a priori* indication of a health hazard [90].

Genotoxicants are characterized by their ability to induce specific classes of stable changes in (1) the nucleotide sequence of genes, (2) the chromosome structure, or (3) the chromosome number. Changes in nucleotide sequence are described as *mutations*, chromosomal damage is referred to as *clastogenicity*, and changes in chromosome number are called *aneuploidy*. These major classes of genetic damage are responsible for a large proportion of the array of human genetic diseases and contribute significantly to the production of congenital malformations. Table 23.1 provides examples of the role of these three mechanisms in human germ and somatic cell diseases.

Because the original goal of genetic toxicology was to protect the human gene pool from environmentally induced increases in new mutations, the discipline of genetic toxicology initially focused on transmissible damage. Test methods employed during this early period were primarily *in vivo* and focused on detecting damage to mammalian germ cells. Later, reports from Bruce Ames as well as several other investigators were published showing a strong relationship between genotoxicity and rodent carcinogenicity [2,25,169,170].

Thus, genetic toxicology has evolved to play a dual role in safety evaluation programs. One role is the implementation of testing and risk assessment methods to determine the impact of genotoxic agents found in the environment on the integrity of the human gene pool. The second role is the application of genetic toxicology data to achieve a better understanding of carcinogenic mechanisms.

## BASIC GENETIC CONCEPTS

### GENE STRUCTURE

The molecules responsible for the hereditary characteristics of all living systems, with the exception of some viruses that use RNA, are composed of DNA. Even those organisms that store their hereditary information in RNA go through a DNA intermediate during replication. The structure and biochemical characteristics of human DNA have been summarized by Baltimore [15]. Some common characteristic features of DNA molecules are listed in Table 23.2. The functions of information storage and gene expression are similar for all DNA-based organisms.

**TABLE 23.1**
**Examples of Mutagenic Mechanisms for Some Human Diseases**

| Mutation Type | Examples of Inherited Effects | Examples of Somatic Effects |
|---|---|---|
| Single base changes | Sickle cell disease | Epithelial cancers, activation of *ras* oncogenes |
| Small deletions and/or translocations | Phenylketonuria | Lymphomas, leukemias, enhanced activation of *myc*, *abl* oncogenes |
| Whole chromosome losses or gains | Hemophilias | |
| | Duchenne muscular dystrophy | Loss of tumor suppressor genes, retinoblastoma, Wilms' tumor, breast cancer |
| | Down's syndrome | |
| | Turner's syndrome | |

**TABLE 23.2**
**Basic Biochemical Characteristics of All Double-Stranded DNA**

1. DNA consists of two purines (guanine, adenine) and two pyrimidines (thymine and cytosine).
2. A nucleotide pair consists of one purine and one pyrimidine (adenine/thymine [AT] or guanine/cytosine [GC]).
3. Nucleotide pairs are connected to a double helix molecule by sugar-phosphate backbone linkages and hydrogen bonding.
4. The AT base pair is held by two hydrogen bonds, and the GC is held by three (see Figure 23.2).
5. The distance between each base pair in a molecule is 3.4 Å, producing 10 nucleotide pairs per turn of the DNA helix.
6. The number of adenine molecules must equal the number of thymine molecules in a DNA molecule. The same relation exists for guanine and cytosine molecules. The ratio of AT to GC base pairs, however, may vary in DNA from species to species.
7. The two strands of the double helix are complementary and antiparallel with respect to the polarity of the two sugar-phosphate backbones, one strand being 3′,5′ and the other 5′,3′ with respect to the terminal OH group on the ribose sugar.
8. DNA replicates by a semiconservative method in which the two strands separate and each is used as a template for the synthesis of a new complementary strand.
9. The rate of DNA nucleotide polymerization during replication is approximately 600 nucleotides per second. The helix must unwind to form templates at a rate of 3600 rpm to accommodate this replication rate.
10. The DNA content of cells is variable ($1.8 \times 10^9$ Da for *Escherichia coli* to $1.9 \times 10^{11}$ Da for human cells).

The basic complete functional unit in a DNA molecule is termed a *gene*. Most of the early knowledge concerning structure and operation of genes was acquired from studies with bacteria or bacteriophages. Advances in understanding the molecular biology of mammalian cells has resulted in equivalent information in eukaryotic cells. The DNA of prokaryotic (bacteria) and eukaryotic (plant and animal cells) organisms differs in a number of ways (Table 23.3). In prokaryotic cells, there is a single *chromosome* with little or no substructure along the DNA molecule. DNA in eukaryotic cells, on the other hand, is highly differentiated, containing nonfunctional, repeated sequences and regions of noncoding DNA called *introns* that are inserted between informational coding sequences called *exons*. The exact functions of repeated DNA sequences and the intron regions are not known.

The nucleotide composition and the mechanisms by which information encoded in a gene is transformed into gene products are universal. Universality was confirmed by recombinant DNA genetic engineering studies in which genes continue to function properly after having been transplanted from human cells to bacterial cells or from bacterial cells to plant cells [26,105].

In eukaryotic cells, the process of gene expression (Figure 23.1) is controlled by regulatory genes. During *transcription* (reading of the DNA code into mRNA), RNA polymerase transcribes both exons and introns into pre-mRNA. Enzymes located in the nucleus of the cell excise intron regions and splice the coding sequence back together. The resulting mature mRNA is then transported to *ribosomes* outside the nucleus for *translation* (reading of the mRNA into polypeptides). In some instances, the

**TABLE 23.3**
**Characteristics of DNA in Prokaryotic and Eukaryotic Cell Types**

| Prokaryotic Cells | Eukaryotic Cells |
|---|---|
| Cells are primarily haploid. | Cells are primary diploid. |
| DNA is uncomplexed. | DNA is complexed with proteins forming chromosomes. |
| DNA is not localized in the cell cytoplasm. | DNA is localized primarily within the nucleus of the cell. |
| No morphologic stages occur in DNA replication. | DNA replication is described by mitotic cycle consisting of specific cytologic stages. |
| DNA is often found as a closed circle. | DNA is found in linear chromosomes. |
| Replication is not associated with cellular organelles. | Replication and separation of chromosomes are associated with cellular organelles called *centrioles*. |
| All genes encoded in the DNA are functional. | Repetitive, nonfunctional gene sequences are common. |
| Spacer sequences have not been identified. | Noncoding spacer sequences identified as introns occur along the DNA molecule. |

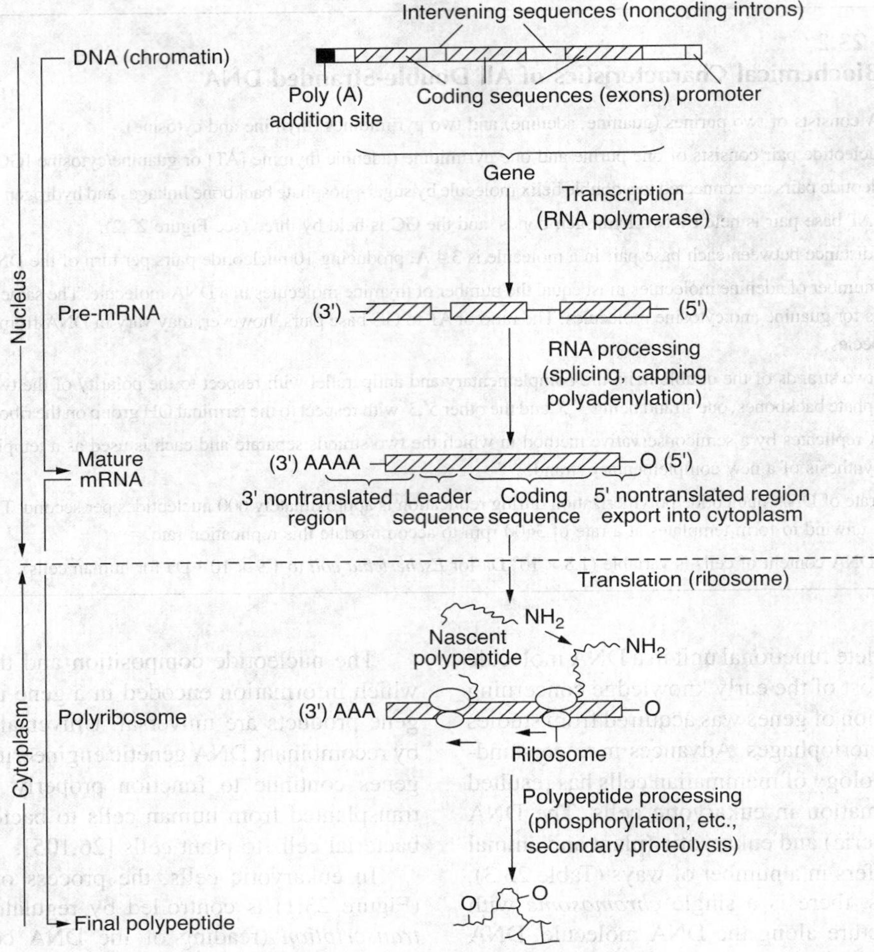


**FIGURE 23.1** Transcription and translocation of mammalian genes.

resultant polypeptides undergo posttranslational modification (glycosylation, hydroxylation, proteolytic cleavage) depending on their cellular functions.

## Somatic and Germ Cell Characteristics

From a genetic perspective, multicellular eukaryotic organisms are composed of two cell types: somatic cells and germ cells. Somatic cells constitute the major portion of the mammalian organism. The *genomes* of most somatic cells are *diploid* (two complete sets of chromosomes), and genetic alterations in somatic cells are not transmissible to subsequent generations. Virtually all *in vitro* mammalian cell assays used in genetic toxicology employ somatic cell types. Germ cells (sperm and eggs) are a special cell population in multicellular organisms. Their function is to form the next generation. Mutations carried in these cells produce a broad array of heritable genetic diseases, congenital malformations, and other disorders. Germ cells are derived from diploid stem cells in gonadal tissues, and following meiosis they carry a *haploid* set of chromosomes. *Mutations* carried in germ cells are classified as recessive or dominant depending on their expression in the diploid state. A large proportion of human genetic diseases are associated with recessive mutant genes, which are expressed only when two mutant alleles (one contributed by each parent) are present in the homozygous condition. Recessive mutations are maintained in the gene pool in the heterozygous state by individuals who are phenotypically normal.

# GENOTOXICITY

## Background Genetic Alterations

DNA synthesis and replication are not flawless processes and in rare instances genetic alterations occur spontaneously during normal cell division. In addition, normal aerobic metabolism produces reactive species capable of damaging DNA. The endogenous level of DNA damage (Table 23.4) forms the basis of background or spontaneous mutation frequencies. Background DNA alterations may als

**TABLE 23.4**
**Estimates of Human Endogenous DNA Damage and Repair Processes**

| Type of Damage | Estimated Occurrences of Damage per Hour per Cell[a] | Maximal Repair Rate, Base Pairs per Hour per Cell |
|---|---|---|
| Depurination | 1000 | —[b] |
| Depyrimidination | 55 | —[b] |
| Cytosine deamination | 15 | —[b] |
| Single-stranded breaks | 5000 | $2 \times 10^5$ |
| $N^7$-methylguanine | 3500 | Not reported |
| $O^6$-methylguanine | 130 | $10^4$ |
| Oxidation products | 120 | $10^5$ |

[a] Might be higher or lower by a factor of 12.

[b] Not reported, but the rates are at least $10^4$ to judge from the concentration of repair activities in cell extracts.

*Source:* Modified from data in Setlow [161].

arise from exposure to cosmic radiation, sunlight, and food. Although DNA damage due to these and other natural sources can be minimized, it cannot be totally eliminated.

## Classification Scheme for Genotoxic Effects

DNA damage may be classified into several broad categories based on the nature (presumed mechanism) of the DNA change. The following is one such classification:

1. *DNA macro-lesions* involve the breakage or interchange of DNA segments between chromosomal structures. This type of damage may be visible through cytologic analysis of condensed chromosomes. Although many genotoxicants cause DNA macro-lesions, this type of DNA damage may also be caused by secondary mechanisms (i.e., processes that result in altered cellular homeostasis such as high temperature) which do not target nucleic acids specifically [67,68,84].
2. *DNA micro-lesions* are nonvisible alterations occurring at the nucleotide level. Nucleotide damage generally produces point mutations through base-pair substitution or insertion/deletion, or it may induce recombination between sister chromatids. Micro-lesions are generally induced by agents that specifically target nucleic acids (e.g., electrophilic agents).

In addition, some genotoxic effects may be induced by other mechanisms that do not fall into either of the two categories defined above. Examples of such genotoxic damage are aneuploidy caused by damage to proteins responsible for chromosome segregation and altered gene expression produced by DNA methylation.

# EFFECTS OF GENETIC DAMAGE

## Processes Leading to Mutation

The normal DNA base pairings are shown in Figure 23.2. Adenine and thymine form two hydrogen bonds, and guanine and cytosine form three. Hydrogen bonds are weak electrostatic forces involving oxygen and nitrogen atoms at specific sites on the purine and pyrimidine molecules. When electrophilic chemical species covalently bind to portions of the DNA bases involved in the formation of hydrogen bonds, the resultant (*adduct*) structures (Figure 23.3) can produce electron shifts from the H-bonding sites to areas within the molecules giving rise to opportunities for short-lived mispaired bases (e.g., A:C or G:T). If this mispairing occurs before or during a DNA replication cycle, the result may be the substitution of an incorrect base pair for the original pair. A cycle of DNA replication is needed to create base-pair substitution mutations and is referred to as the *expression period* in mutation assays.

Base-pair addition/deletion mutations, also called *frameshift mutations*, result from the addition or deletion of one or a few nucleotide pairs from the nucleotide

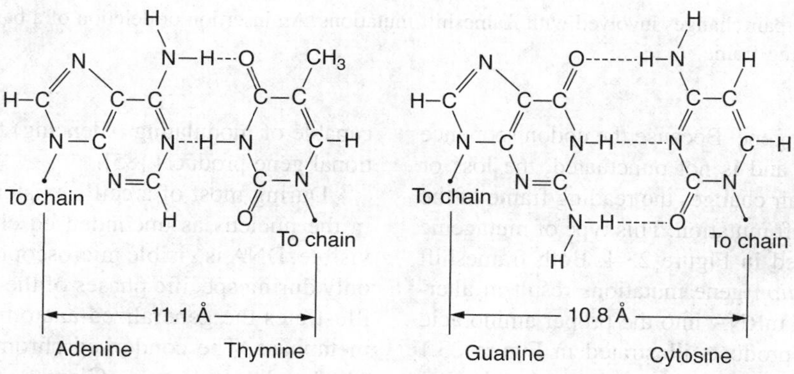

FIGURE 23.2 Hydrogen bonding of nucleotides normally found in DNA.

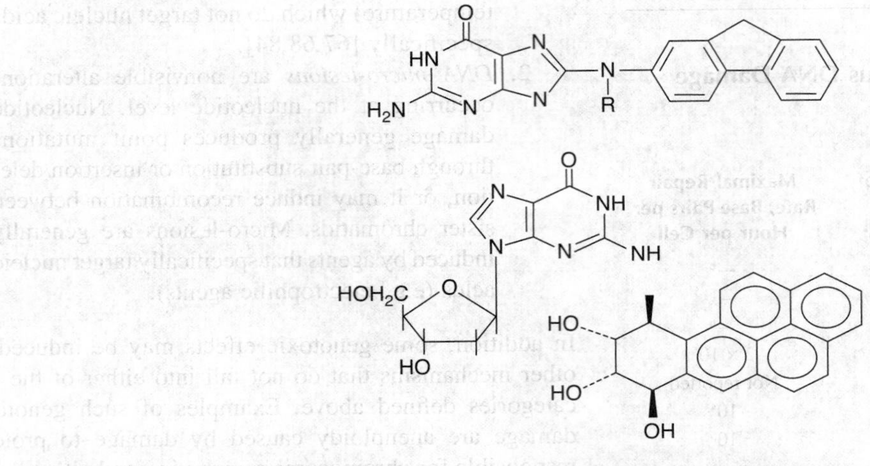

**FIGURE 23.3** Two examples of DNA chemical adducts. (Top) An aromatic amine adduct to guanine. (Bottom) A benzo(*a*)pyrene metabolite adduct to guanine.

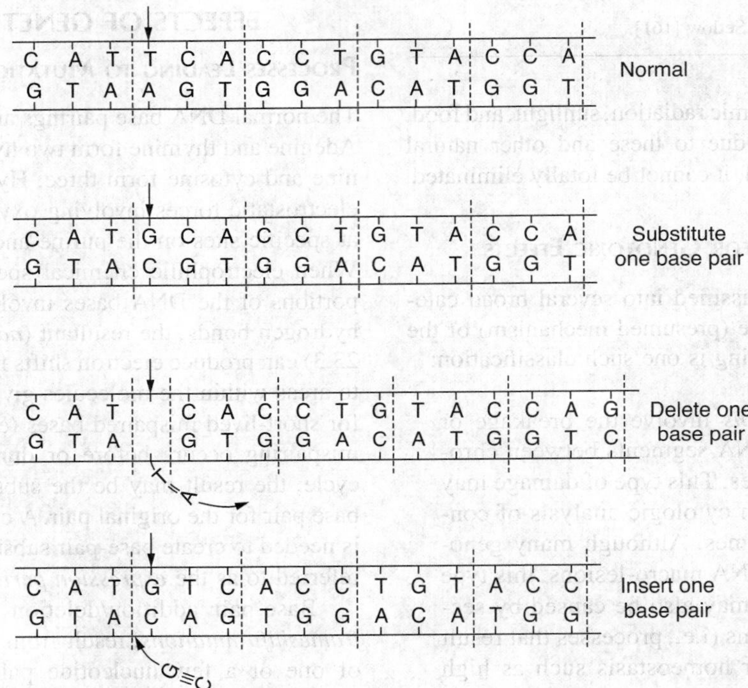


**FIGURE 23.4** The basic pair changes involved with frameshift mutations. An insertion or deletion of a base pair results in the shift of the gene or exon reading frame.

sequence in an exon or gene. Because the codon sequence reads in one direction and is not punctuated, the loss or gain of a single base pair changes the reading frame of the gene—hence, frameshift mutation. This type of mutagenic mechanism is illustrated in Figure 23.4. Both frameshift and *base-pair substitution* gene mutations result in alterations in translation of mRNA into the proper amino acid sequence in the gene products illustrated in Figure 23.1 that may produce a mutant cell or organism. Mutations or methylation of DNA (e.g., 5-methyl cytosine) in genes are capable of modulating (silencing) the production of functional gene products [85].

During most of a cell's existence, DNA is packaged in the nucleus as uncondensed chromosome and is not visible. DNA is visible microscopically as a chromosome only during specific phases of the cell cycle. Figure 23. illustrates the generalized anatomy of a chromosome a metaphase. The condensed chromosome shows distinctive banding following Giemsa staining and contains centrally located constriction known as the *centromere*

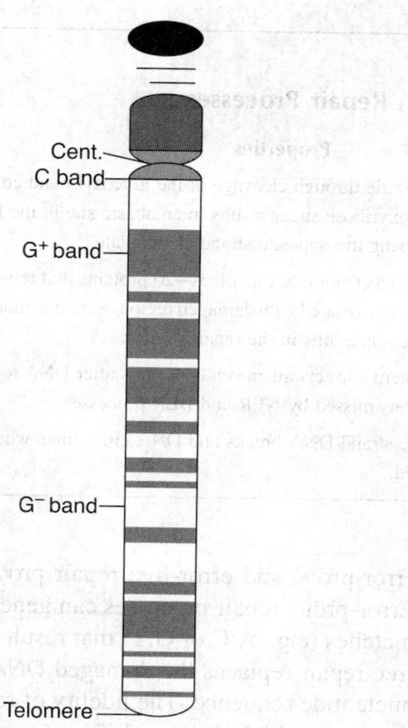


**FIGURE 23.5** The anatomy of a typical chromosome to show active (unstained) and inactive (stained) regions.

To either side of the centromere are the chromosome arms that terminate in unique sequences called *telom-eres*. The darker bands represent highly condensed DNA believed to represent areas of little or no gene expression (transcription). The lighter bands represent relaxed DNA sequences believed to be active genomic areas where gene expression (transcription) is occurring. The telom-eric regions at the ends of the chromosome structures are important for chromosome stability [150]. Telomeres consist of repeat units of a small number of nucleotides hat stabilize the linear DNA molecule. At each cell livision, chromosomes lose small amounts of telomeric DNA. The ability of the cell to replace the lost DNA vith telomerase determines whether the cell continues o proliferate or, if the telomeric loss reaches a critical evel, stops dividing [41,47].

Damage to chromosomal DNA is visible at metaphase nd can be subdivided into changes in chromosome num-er (gain or loss single chromosomes or sets of chromo-omes) and changes in chromosome structure (breaks, eletions, rearrangements). Each type of chromosome hange has a characteristic physical description, so a rea-onably high degree of uniformity can be maintained when coring them microscopically. Variations in chromosome umber can result from incomplete dissociation of a single et or entire sets of chromosomes at metaphase, resulting n aneuploidy or polyploidy [55]. A wide range of *in vitro* nd *in vivo* genetic test methods is available to assess numerical changes and structural breaks. Chromosome breaks can be induced by several mechanisms, including disruption of DNA synthesis or DNA cross-linking. The toxicological relevance of chromosome damage is associ-ated with the ability of the damage to persist in a viable cell where the damage can be expressed. Balanced chro-mosome translocations, stable deletions, and numerical changes such as aneuploidy are all associated with human genetic diseases.

In addition to the tests that specifically measure nucle-otide substitutions and chromosome alterations, a third group of tests to measure other mechanisms of genotoxic-ity has been identified. Included in this group are tests for DNA binding and repair, DNA strand breakage, sister chro-matid exchange (SCE), and mitotic recombination. These methods are categorized as tests for measuring primary DNA damage. Few regulatory testing guidelines specify primary DNA damage tests any longer, as these tests have been found to yield results redundant with the tests for mutagenicity and clastogenicity and in some cases (e.g., DNA strand breakage) may yield false-positive results due to non-DNA-specific cytotoxicity. Although many studies suggest a strong association between somatic cell death (cytotoxicity) and the production of DNA damage, the relationship is not well-understood [13].

A significant amount of research has focused on the induction and measurement of DNA binding. The initial hope of this technology was to use adduct formation as a method to prove exposure to a specific agent (adducted to DNA) and to use the data to predict cancer risk. To date, that hope has not been fulfilled. Adduct formation has proven to be a sensitive method to demonstrate exposure; however, not enough is known about adduct processing and repair to use this information to define genetic or cancer risk except for a small number of agents. Binding varies between tissues of the same animal, and different adducts have different biological consequences influenced by complex factors that are not yet understood. What is known is that organs and tissues in mammals have differ-ent capacities for repairing adducts, and that some adducts, such as the *O*-alkyl adducts, are repaired ineffi-ciently and probably produce most of the mutagenic dam-age responsible for initiation [147].

## Repair of DNA Damage

The fidelity and integrity of genetic information in organ-isms are maintained by several types of enzymatic DNA repair. The characteristic of self-repair is unique to DNA and emphasizes how important the integrity of this mole-cule is to the survival of an organism. Table 23.5 summa-rizes the general types of repair processes that exist in organisms. Components of DNA repair operate constitu-tively in organisms, and other types are adaptive or induc-ible following cellular exposure to genotoxicants [159,190].

**TABLE 23.5**
**Classification and Properties of the Major DNA Repair Processes**

| Class | Properties |
|---|---|
| Base excision repair (BER) | Elimination of single nucleotide through cleavage of the glycosyl bond connecting the altered base to the deoxyribose sugar results in an abasic site in the DNA followed by resynthesis using the opposite strand as template. |
| Nucleotide excision repair (NER) | Removal of bulky DNA adducts from DNA involves ~20 proteins that remove up to several hundred nucleotides associated with damaged region. Repair synthesis using the opposite strand as a template fills in the repair patch. |
| Mismatch repair (MMR) | A second-chance repair system corrects mismatch base pairs after DNA replication. This process catches lesions missed by NER and BER processes. |
| Recombinational repair (RR) | This process acts on double-strand DNA breaks and DNA cross-links where both DNA strands are damaged. |

DNA damage is induced by both environmental agents as well as normal intracellular processes of DNA replication and the production of free radicals and oxidation byproducts (oxidative stress). Some estimates of endogenous damage and repair capacities for human cells are given in Table 23.4. Constitutive cellular repair capacities are generally adequate to compensate for background damage; however, recurrent exposures to excessive concentrations of exogenous genotoxic agents may saturate the DNA repair capacity, leading to mutation and associated chronic diseases such as cancer and aging.

The common feature of repair processes is their ability to detect, remove, and replace damaged segments of DNA [66]. If a DNA lesion (i.e., DNA strand break or adduct) can be repaired prior to mutation fixation, the net effect of the DNA damage to an organism may be nil. This is especially true following intermittent, low-level exposures to genotoxicants where repair enzymes are not fully saturated by excessive numbers of damaged DNA sites. DNA adducts are not all recognized or repaired equally by repair. The effect of an adduct on the conformation of DNA often determines how readily it is detected by the repair enzymes and removed. As a consequence of this situation, it is not advisable to use the number of adducts per cell as a predictor of damage or genetic hazard unless definitive information is available about the elimination kinetics for the specific adducts [110,183].

Test systems measuring some parameter of the DNA repair process have been used as screens for detection of primary DNA damage. Normal organisms are capable of some type of DNA repair activity following chemical insult; thus, stimulation or induction of repair activity following chemical treatment at sublethal concentrations is a good indicator that the target organism has experienced DNA-directed toxicity. These tests are generally identified as measures of unscheduled DNA synthesis (UDS).

Studies of DNA repair kinetics indicate that once premutational lesions have been induced in the DNA, both error-prone and error-free repair processes are activated. Error-prone repair processes can generate nucleotide mismatches (e.g., A:C or G:T) that result in mutations. Error-free repair replaces the damaged DNA site with a correct nucleotide sequence. The fidelity of repair depends on the degree to which the two different processes are involved. Factors that determine whether error-prone or error-free pathways predominate include (1) the target species, (2) the cell type involved, (3) the chemical mutagen, and (4) the specific DNA lesion induced. Some data suggest that the error-free repair pathways predominate at low exposure levels and error-prone pathways come into play only following saturation of the error-free enzymes.

The basic processes of nucleotide excision repair (NER) and base excision repair (BER), the primary repair mechanisms for chemical damage, are shown in Figure 23.6. The enzyme complex responsible for NER of bulky adducts requires more than 20 proteins and consists of several steps [129]. An endonuclease cleaves the DNA at the site of the damage, an exonuclease cuts out the damaged region including nucleotides to either side, and new bases are replaced by a DNA polymerase using an editing function to ensure that the correct bases are incorporated into the repair patch. Finally, DNA ligase seals the repair patch. Occasionally, even in error-free repair, incorrect bases are incorporated by the polymerase, resulting in mismatched bases that do not properly hydrogen bond. In BER, a number of different glycosylases participating in mismatch repair catalyze release of the inappropriate base followed by replacement and ligation [71]. NER serves the genome on a global basis (scanning the entire length of the DNA for damage) in addition to repair localized around DNA sites actively engaged in transcription (TCR or transcription-coupled repair).

Mismatch repair enzymes recognize non-hydrogen bonding (aberrant) base pairs. A short segment of the DNA duplex is excised and filled by the repair polymerase [66]. This is a second-chance repair process that occurs afte

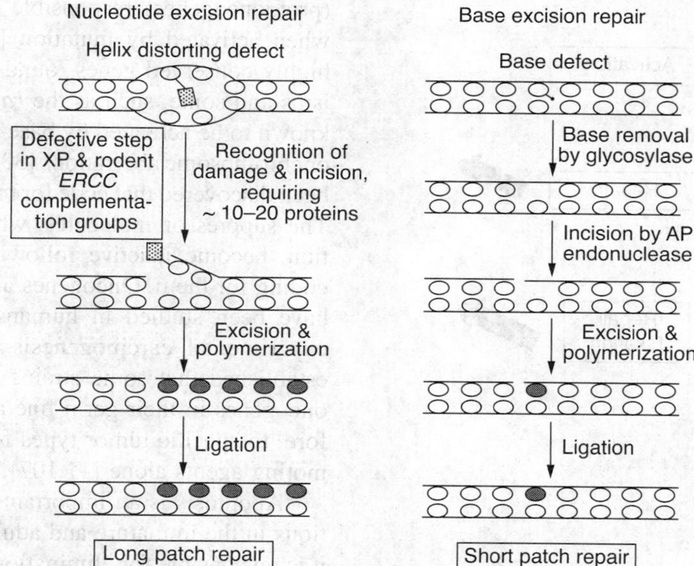


**FIGURE 23.6** Long and short patch excision repair processes.

BER and NER and improves the accuracy of those processes. Finally, recombinational repair acts on double-strand breaks and interstrand cross-links, resulting in damage to both strands of the DNA. Collectively, these processes maintain the integrity of DNA against endogenous and exogenous sources of damage.

Repair of mutational damage has been shown to be inducible by low-level exposures to DNA-damaging agents. The inducibility of repair processes above constitutional levels will increase the magnitude of exposure required to exceed the intrinsic capacity, producing a threshold for mutation and a protective effect for subsequent exposures. Theoretical assumptions and data from studies of repair support the belief that at background or low exposure levels an error-free removal of alkyl groups from DNA can be virtually 100% effective. Thus, one observes survival shoulders and nonlinear kinetics for mutation induction in repair-proficient cells and the loss of apparent no-effect regions in repair-deficient cells.

Because of the influence that the species- and cell-specific genetic background has on repair capacity, an appreciation of the variability of DNA repair in somatic and germ cells of humans as well as the animal models is essential in assessing genetic risk [60,70]. Several general characteristics of mammalian DNA repair are summarized in Table 23.6.

## Relationship of Genetic Damage to Disease

The proposed sequence of events from DNA interaction to the expression of human toxicity is represented in Figure 23.7. The biological consequences of DNA damage is dependent on whether it is induced in somatic or germ cells. Different techniques and strategies are used in genetic toxicology to detect somatic and germ cell hazards. As a consequence of research in medical genetics and new information flowing from the human genome sequence, the mutational basis for many human disorders and anomalies is well documented [95]. The genetic burden in humans has been estimated to be approximately 5% (Table 23.7). Diseases caused by this genetic burden contribute significantly to the healthcare costs of most developed countries and the high mortality rates in less well-developed regions. The origin of preexisting DNA

**TABLE 23.6**
**Some Characteristics of Global DNA Repair**

DNA repair is tissue and species specific (e.g., human capacity is approximately tenfold greater than mouse).

DNA repair is increased in genes involved in transcription (e.g., chromosome loops); the process is identified as transcription-coupled repair (TCR).

DNA adducts are repaired with different efficiency (e.g., bulky > small).

Effects of DNA repair are tied closely to cell stringency and apoptosis (e.g., cell lines such as CHO, V79, and CHL that have low cycling stringency have higher error rates).

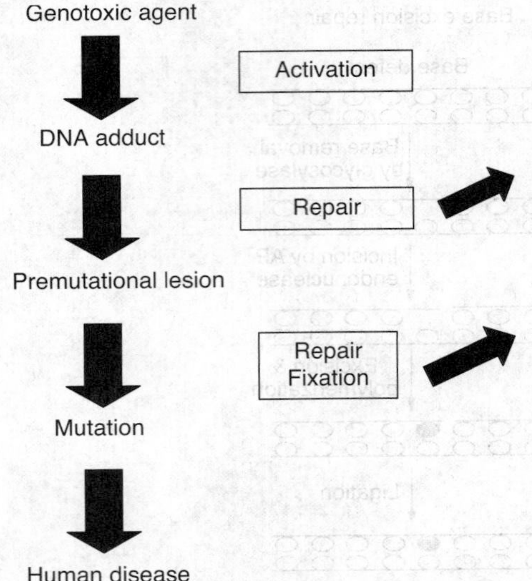


**FIGURE 23.7** The hypothetical process from exposure to genotoxic to the induction of genetic disease in humans. Repair processes can eliminate the formation of mutations that produce disease.

alterations found in the human gene pool is unknown, but Table 23.8 provides some possible sources for new mutations. Animal experiments offer convincing evidence that environmental agents are capable of inducing permanent, transmissible mutations in either somatic or germ cells and support the presumption that humans are susceptible to environmental mutagens. Damage to somatic cells may produce a range of dysfunctions, including inherited and induced tumors [33], teratogenesis [98], reproductive failure [34], and atherosclerosis [91], and may be involved in aging [4].

## Carcinogen–Mutagen Relationship

The introduction of a *Salmonella* assay for detecting mutation, combined with an *in vitro* metabolic activation system, appeared to offer a rapid, inexpensive solution to the identification of chemical carcinogens [6]. The Ames test was the forerunner for an array of submammalian and mammalian cell assays proposed as rapid screens for carcinogens [148]. The rationale for the use of these methods consisted of investigations demonstrating that properties associated with the transformed (malignant) cell phenotype were encoded in the DNA. In the late 1970s, studies were published showing that DNA isolated from transformed mouse cells and purified of other contaminants, cut into discrete fragments by bacterial restriction enzymes, and transfected into "normal" cells caused the conversion of the "normal" cells into a transformed phenotype [188]. Investigators conducting these experiments ultimately identified a series of genes in the restriction fragments (protooncogenes) responsible for transforming the cells when activated by mutation [152]. Protooncogenes are highly conserved genes found in most eukaryotic organisms and some, such as the *ras*, *myc*, and *neu* genes, are known to be activated by base-pair substitution mutations or chromosome aberrations [33,55]. Additional genes have been discovered that code for tumor suppressor molecules. The suppressor molecules, which prevent cell proliferation, become inactive following mutation in the genes coding for them. Oncogenes and tumor suppressor genes have been studied in humans and in rodents used for experimental carcinogenesis studies [33]. It has been established that some strains of animals carry activated oncogenes in their germ line and are predisposed, therefore, to specific tumor types following exposures to promoting agents alone [11,107].

Apoptosis is an important cellular process that functions in the immature and adult animal. In adult animals, it is responsible for elimination of potentially preneoplastic cells by nuclear degradation. The p53 tumor suppressor gene codes for a major apoptosis-inducing protein. Elimination or reduction in the expression of p53 by mutation is believed to be a critical step in the process of neoplasia. Apoptosis proceeds through the actions of a family of proteases called *caspases*. The internal elements of the cells disintegrate, and they are ultimately phagocytized [40]. Some genotoxicity attributed to environmental treatment conductions (e.g., hypo- and hyperthermia, pH alterations, osmotic shock) may be due to an induction of apoptosis [124].

Initial optimism that genotoxicity tests might be a short-cut means of carcinogen detection was generated by studies correlating the results of specific genotoxicity tests (or batteries of tests) with rodent bioassay responses for the same chemicals. During the early 1970s, concordance between rodent carcinogenicity and results from the Ames test appeared to be as high as 90 to 95%. By 1984, however the concordance between genotoxicity and rodent bioassay results dropped to just over 60%. There are several reasons for this reduction in concordance [12]; however, the factor that had the most influence was an expanded set of chemical classes used in the comparisons. Reports in the early to mid-1970s showing high concordance employed groups of chemicals highly biased toward electrophilic carcinogens. Studies suggesting lower concordance employed groups of carcinogens that were not electrophilic and had a greater proportion of epigenetic-carcinogens [194]. These new analyses also emphasized a perilous lack of specificity (noncarcinogens scoring as negative) across most genetic toxicology tests. In fact, most chemicals classified as noncarcinogens by agencies such as the National Toxicology Program (NTP) and the International Agency for Research on Cancer (IARC) produced positive responses in one or more of a battery of *in vitro* genetic assays (Table 23.9), suggesting that the specificity of i

**TABLE 23.7**
**Examples of Genetic Disorders in Humans**

| Category of Genetic Alteration | Estimated Frequency/ $10^3$ Population | Typical Examples[a] |
|---|---|---|
| Chromosome abnormalities | 6.86 | Down's syndrome (trisomy) |
| | | Klinefelter's syndrome (XXY) |
| | | Turner's syndrome (XO) |
| | | Cri du chat (deletion of chromosome) |
| | | Numerous other trisomies (XYY) |
| Dominant mutations | 1.85–2.64 | Familial polysposis (AD) |
| | | Neurofibromatosis (AD) |
| | | Huntington's chorea (AD) |
| | | Crouzon's craniofacial dystosis (AD) |
| | | Retinoblastoma (AD) |
| | | Anitidia (AD) |
| | | Chrondodystrophy (AD) |
| Recessive mutations | 2.23–2.54 (AR) | Xeroderma pigmentosa (AR) |
| | 0.78–1.99 (XR) | Duchenne muscular dystrophy (XR) |
| | | Hemophilia (XR) |
| | | Lesch–Nyhan syndrome (XR) |
| | | Sickle cell disease (AR) |
| | | Galactosemia (AR) |
| | | Phenylketonuria (AR) |
| | | Diabetes mellitus (AR) |
| | | Fanconi's syndrome (AR) |
| | | Albinism (AR) |
| | | Cystic fibrosis (AR) |
| Polygenic (complex inheritance) | 26.00–32.00 | Cleft lip |
| | | Anenchephaly |
| | | Spina bifida |
| | | Clubfoot |
| | | Idiopathic epilepsy |
| | | Congenital heart defects |

[a] AD = autosomal dominant; AR = autosomal recessive; XR = X-linked recessive.

*Source:* Committee on Mutagenicity, *A Consultative Document on Guidelines for the Testing of Chemicals for Mutagenicity*, Committee on Mutagenicity of Chemicals in Food, Consumer Products, and the Environment, Department of Health, Social Services, and Public Safety, U.K., 1979.

*itro* tests is too poor to rely upon individual tests or test batteries to predict carcinogenic profiles for new molecules 104]. Subsequent addition of *in vivo* test results (e.g., mouse micronucleus) to the battery database failed to mprove its predictive performance [194].

A more recent application of genetic toxicology results to carcinogenicity has been the use of genotoxicity tests to help interpret the mode of action for rodent carcinogens [10,32]. Carcinogens with positive effects in genetic tests are considered to act through direct effects

**TABLE 23.8**
**Possible Origin of New Mutations in the Human Gene Pool**

A small number arising from spontaneous events which occur during normal DNA replication and repair
Reactive oxidative products formed during normal metabolism
Unavoidable environmental exposures (e.g., food, radiation, products of combustion, mycotoxin, pesticides, manufacturing emissions)
Therapeutic treatments that are directly mutagenic
Therapeutic treatments that are indirectly mutagenic by successful treatment of formerly lethal genetic diseases thereby elevating their frequency in the reproducing population

**TABLE 23.9**
**A Comparison of the Sensitivity Specificity and Concordance of Three *In Vitro* Genotoxicity Tests for Rodent Carcinogenicity Responses**

| | Test Systems Battery | | | |
|---|---|---|---|---|
| **Characteristic** | **Ames** | **Mouse Lymphoma** | **Chromosome Aberrations** | **All Tests Combined** |
| Sensitivity | 58.8% | 73.1% | 65.6% | 84.7%[a] |
| Specificity | 73.9% | 39.0% | 30.8% | 22.9%[b] |
| Concordance | 62.5% | 62.9% | 67.8% | — |

[a] At least one positive assay when all three are conducted as a battery.
[b] All three tests clearly negative for all carcinogens when combined into a battery.

*Note:* Sensitivity is the proportion of positive results for carcinogens. Specificity is the proportion of negative results for noncarcinogens. Concordance is the proportion of both positive and negative results that are correct for a test.

on DNA. Carcinogens that test negative for genotoxicity are presumed to produce tumors through one of several epigenetic modes of action. Thus, a classification scheme of genotoxic and nongenotoxic carcinogens has been proposed to aid in the selection of the most appropriate data extrapolation model for cancer risk assessment. A genotoxic mechanism implies a no-threshold mechanism calling for conservative risk estimations (e.g., linear extrapolation), whereas a threshold may exist for epigenetic modes of action. Although conceptually this scheme appears reasonable, a better understanding of carcinogenic processes is necessary to support its general application to risk assessment. With the concordance between rodent carcinogens and mutagens averaging about 60 to 70 percent, the appropriate integration of genetic toxicology results into toxicological assessments is not a straightforward process.

Mutations and chromosome aberrations are found in the DNA of tumor tissues and, as such, the use of genetic tests as part of an assessment for carcinogenic potential may be justified by mechanistic considerations alone and does not have to depend on high correlations with rodent bioassay data. New mutation models such as the *in vivo* transgenic models developed for measuring mutation may begin to offer a much more relevant, mechanistic-based assessment of carcinogenicity than *in vitro* assays. Two transgenic models for mutation detection employing transgenes from the bacterial *lac* operon have been developed and partially validated [73]. Transgenic mouse models altering gene expression are also being used [172]. Models with altered tumor suppressor genes ($p53^{+/-}$) or activated protooncogenes (Tg.AC and *ras*H2 mouse) are under validation in short-term bioassays for chemical carcinogens [181]. The $p53^{+/-}$ and *ras*H2 models are useful for studying potential genotoxic agents, as the mode of action for enhanced tumorigenesis is inactivation of a normal allele by chemical-induced mutation, deletion, or rearrangement. In the $p53^{+/-}$ model, the normal allele undergoes mutation, resulting in a homozygous mutant lacking the p53 protein responsible for programmed cell death. In the *ras*H2 model, mutation in the normal hemizygous allele results in elevation of the Ras protein predisposing the mouse to tumor development.

## Effects of Mutagens on the Human Gene Pool

Induction of damage to the germline DNA of plant and animal species has the potential to create serious adverse consequences for the health and survival of those organisms. In humans, genetic damage is a cause of hereditary diseases, cancer, congenital anomalies, and even reduced life expectancy. The genes imbedded in the current human population were all acquired from the previous generation's *gene pool*. The gene pool is the sum total of genes in gametes available, in the reproductively active population of a species, for transmission to the next generation. Deleterious genes are present in the current gene pool at a set frequency, as evidenced by predictable rates of recurring genetic diseases in the human species. The origin of this genetic burden, or genetic load, is not known, but it is imperative that the current caretakers of the gene pool use all precautions to transmit it to the next generation in no worse shape than it was received.

Genetic disease in humans appears to be produced by the same types of mutations identified in animal models: (1) chromosome abnormalities resulting in stable changes in chromosome number or structure, (2) dominant gene mutations in which only a single mutant allele (of the normal gene pair) is required to produce the disease, (3) recessive gene mutations in which both alleles of the pair must be mutant for expression of the trait, or (4) polygenic mutations in which the mutant trait is determined by the interaction of several genes.

**TABLE 23.10**
**Examples of Human Diseases and Conditions Caused by Mutations in Germs Cells**

| Genetic Disease or Condition | Estimated No. of Cases in the United States |
|---|---|
| Dyslexia | 15,000,000 |
| Manic depression | 2,000,000 |
| Schizophrenia | 1,500,000 |
| Juvenile diabetes | 1,000,000 |
| Adult polycystic kidney disease | 500,000 |
| Familial Alzheimer's disease | 250,000 |
| Multiple sclerosis | 250,000 |
| AAT deficiency (emphysema) | 120,000 |
| Myotonic muscular dystrophy | 100,000 |
| Fragile X chromosome syndrome | 100,000 |
| Sickle-cell anemia | 65,000 |
| Duchenne muscular dystrophy | 32,000 |
| Cystic fibrosis | 30,000 |
| Huntington's disease | 25,000 |
| Hemophilia | 20,000 |
| Phenylketonuria | 16,000 |
| Retinoblastoma (childhood eye cancer) | 10,000 |

*Source:* McKusick, V.A., *Mendelian Inheritance in Man: Catalogs of Autosomal Dominant, Autosomal Recessive, and X-Linked Phenotypes*, 5th ed., The Johns Hopkins University Press, Baltimore, MD, 1978.

Table 23.10 gives some examples of human genetic diseases as well as the estimated number of these disorders in the U.S. population. The examples given in the table represent only a small portion of the total human diseases and defects known to be of genetic origin [121]. It is estimated that the human genome consists of approximately 25,000 genes controlling all aspects of an organism's biology and behavior; of that number, we currently know the relationship with disease for only about 1000. Although the human genome has been sequenced, it is expected to take years before all genes are associated with their functional roles.

The deleterious consequences of germ-cell alterations to the human gene pool (listed in Table 23.7) are not equivalent; for example, other than balanced translocations, most chromosome abnormalities produce cell lethality. The most common outcome of chromosome damage induced in either the ova or sperm is dominant lethal effects not transmitted to the next generation. Conversely, dominant viable mutations are expressed in the first generation after induction but may contribute only moderately to the genetic burden. The impact of dominant mutations on the gene pool is limited because the affected individuals are affected by the disease; they are aware that they are carrying the mutant form of the gene and that they have 50% probability of transmitting the dominant gene to their children. Thus, depending on the severity of the effect, parents can decide, prior to reproducing, if the risk associated with possible transmission is acceptable.

Unlike dominant mutations, recessive mutations are not expressed unless both alleles of the affected gene are defective. Recessive mutations located on the X-chromosome are recessive in females but will be expressed in male offspring because they carry only a single X-chromosome (sex-linked mutations). Two phenotypically normal heterozygous carriers for a recessive mutant allele (on autosomal chromosomes) will theoretically produce offspring that have a 25% incidence of exhibiting a recessive disease. Additionally, 50% of their offspring will be the same phenotypically normal heterozygous carriers as the parents. Consequently, increases in the incidence of new recessive mutations pose the most serious threat to the integrity of the gene pool because they tend to accumulate over time in the heterozygous configuration. Due to the generational latent period, the ultimate expression of new recessive mutations in the current population's gene pool will have no apparent association with the environmental exposure that induced them until generations later. This situation severely limits the opportunity to use human epidemiological studies for proof of human genetic risk [14,60].

## Intrinsic Differences Affecting Susceptibility to Genotoxic Effects

Gene polymorphisms affecting traits such as metabolic and DNA repair pathways are most likely to influence human susceptibility to genetic damage. Individuals carrying different polymorphic forms of the same genes can vary significantly in their response to the same toxic exposures. Unlike the genetic uniformity intrinsic in the animals used as experimental models in toxicology testing, there is a broad range of metabolic and DNA repair capacity in human populations, and that variability affects the risk from exposure to genotoxicants [50,77,129]. Studies comparing the DNA repair capacity of humans with that of experimental animals also suggest some significant differences in repair capacity [76], limiting the direct extrapolation of results from mice to humans.

As described previously, DNA repair in humans is controlled by a complex genetic system of structural and regulatory genes involving almost 50 different proteins. Human *polymorphisms* exist for many of these genes. Individual DNA repair capacity may range to as low as 65% of the normal average level [129], placing such individuals at a higher risk to mutagens than most of the population.

Additional information about repair genes comes from studies of human genetic diseases such as xeroderma pigmentosum (XP). Diseases such as XP are caused by mutations in NER genes. Individuals with XP lack one of the enzymes in the repair process, and affected individuals, with as little as 1 to 2% of the normal repair capacity, usually experience high levels of intrinsic DNA damage [161]. Cancer susceptibility among XP cases is believed to be a consequence of the genetic damage and can be as much as 1000 times greater than for non-XP individuals. Other human syndromes associated with reduced repair capacity (e.g., ataxia telangiectasia, Bloom's syndrome, Fanconi's anemia) are inherited traits that also exhibit increased cancer risk [66,161]. Thus, it is unlikely that genetic damage induced in genetically homogenous animal models with uniform repair will be easily extrapolated to the human populations because factors such as DNA repair exhibit such broad response ranges. It is also clear from this illustration that quantitative estimates of hazardous or safe exposures to genotoxic chemicals based on results from animal data are likely to have large margins of error.

### Pharmacogenetics

In addition to DNA repair polymorphisms, other allele variations confer important differences that will influence the impact of genotoxic agents. Some of these differences affect metabolic conversion kinetics. Individuals genetically predisposed to be either a fast or slow metabolizer of specific chemicals might be at higher risk from the parent compound or metabolites. Determination of an individual's genotype can be made with only a small sample of their DNA.

Identification of gene polymorphisms that influence the susceptibility of an individual to the pharmacologic and toxic effects of therapeutic agents has developed into a major field of medical genetics. Institutions have assembled alleles from hundreds of polymorphic genes, and large human populations can be screened for single gene polymorphism with automated high-throughput screening systems employing techniques such as DNA binding and transcription activation. The integration of the information into a genetic profile may eventually be used to direct an individual's lifestyle optimization. Studies in humans exposed to polycyclic aromatic hydrocarbons (PAHs) documented a genotypic influence on the level of adducts. Polymorphism in GSTMI 2 (glutathione transferase), CYP1A1, and CYP2D6 alleles all affect the formation of DNA adducts found in white blood cells of exposed individuals [31].

### Life Style and Exposure to Genotoxicants

The numerous examples of heritable cancer, where acquisition of specific genes confers a high risk of cancer, include retinoblastoma, colon cancer, and breast cancer. These alleles may contribute to as much as 5% of the total cancer incidence; the other 95% is a product of lifestyle. It is well documented that occupation and life-style can influence genetic hazard [64,92,164]. Tobacco smoke contains a broad range of mutagenic agents [100]. Consumption of alcoholic beverages has been associated with genetic alterations in humans [141]. Diet is also a direct and indirect source of mutagens. The average human consumes about 10 tons (dry weight) of food by the age of 50 [169], and genetic toxicology studies have demonstrated that a number of common foods contain substances that are mutagenic [59,136,137]. Other ubiquitous materials reported to produce positive effects in mutagenicity tests include cosmetic ingredients, drugs, food additives, and pesticide residues. At present, it is not possible to assess, in quantitative terms, the relevance, if any, of these agents to a mutational risk to human somatic or germ cells.

In addition to tobacco smoke, alcoholic beverages, and consumption of mutagens in or on food products, other life-style factors that may be important are less subject to individual control. Exposure to ultraviolet light and ionizing radiation are common. Ambient air contains carbonaceous particles coated with agents producing mutagenic responses in a range of assays [113]. The level of particle loading is location dependent, and exposures encountered in urban areas or in certain occupations (e.g., coke oven workers) are considerably higher than exposures in more rural areas [110].

Ames hypothesized that endogenously formed mutagens and intermediates found in various plants are significant contributors to genotoxic risk [3]. He and others have reported the formation of numerous mutagens resulting from lipid peroxidation of fatty acids and as byproducts of the inflammatory process [7,29]. The agents include aldehydes, peroxides, and other free radicals. Considerable efforts are currently underway to evaluate the relative contributions of dietary, environmental, and endogenously formed mutagens to the human disease burden, as well as the modulating influence of antimutagenic agents (e.g., antioxidants) on mutation expressions and cancer [64].

## METHODS FOR ASSESSING GENETIC HAZARD AND RISK

Observations of new dominant mutations arising in human populations and extensive experimental data demonstrating chemically induced mutations in somatic cells as well as germlines of mice document the assumption that all mammalian species, including humans, are susceptible to environmentally generated mutation. The ability to demonstrate genetic toxicity to human populations through epidemiological methods has been limited by:

- The small number of instances in which sufficient induced sentinel mutations were induced to allow detection in an epidemiology study
- The small number of genetic diseases or marker genes identified with specific genetic diseases
- The difficulty in identifying reproductively active populations exposed to biologically significant levels of mutagenic agents

**TABLE 23.11**
**Generalized Testing Requirements for Environmental Chemicals**

| Level (Tier) | Test Types | Assessment Function |
|---|---|---|
| 1 | Bacteria reverse mutation (Ames)<br>*In vitro* gene mutation in mammalian cells (e.g., mouse lymphoma)<br>*In vitro* chromosome or micronucleus | Rapid, broad-based screening for genotoxic activity. For chemicals with low human exposure, the testing may be reduced to the Ames and *in vitro* chromosome testing. Uniform negative results in these tests will be sufficient for most chemicals. |
| 2 | One or more *in vivo* somatic cell endpoints (e.g., bone marrow micronucleus, comet assay, DNA binding, transgenic mice, UDS) | The results of levels 1 and 2 constitute hazard identification; the test results from level 2 will be added to those for level 1 for chemicals with high human exposure potential. Negative results in levels 1 and 2 will be sufficient testing for chemicals with high exposure potential. Level 2 testing may also be required to clarify positive results from a test in level 1. |
| 3 | *In vivo* tests in gonadal cells (e.g., dominant lethal, micronucleus, DNA binding, UDS) | Used to assess the ability of chemicals to access germ cells; it is not a direct measure of heritable damage but increases the level of concern for human risk. |
| 4 | *In vivo* tests to provide quantitative risk data for gonadal cells (e.g., specific locus tests, visible or biochemical; heritable translocation tests in mice) | This level measures heritable mutation damage, and these tests can be used to estimate the risk for induction of new mutations into the human gene pool. |

In 1992, the United Nations Environment Program reviewed the status of and methods available for genetic risk assessment [177]. A more recent classification scheme was proposed by the U.S. Environmental Protection Agency [46]. The system is built on a four-level testing scheme that starts with *in vitro* methods and progresses through *in vivo* somatic tests in level 2 and into germ cells assays in levels 3 and 4 (Table 23.11). Based on results, chemicals will fall into one of the classes shown in Table 23.12. When only a general characterization of potential risk is required, a qualitative assessment may be sufficient.

When essential, quantitative risk analysis for genetic damage can be performed using dose–response results from animal models such as the mouse specific locus test or the mouse heritable translocation assay. From these data, the population incidence of mutation can be calculated for anticipated exposure levels and expressed as the probability of new disease occurrence in the population. The same problems that confound the production of cancer risk estimates from animal studies affect the ability to generate accurate genetic risks. These include understanding the dose–response kinetics, species differences in susceptibility, and factors that influence the phenotypic expression of new mutations. A few quantitative genetic risk assessments have been made [49] using data from mouse models for mutation and clastogenesis (Table 23.13). The major deficiencies in current risk assessment models are the lack of accurate methods to extrapolate animal mutation data to estimates of anticipated new disease incidences in exposed human populations (Table 23.14). The difficulties in human genetic risk estimation are summarized in several publications [59,153].

**TABLE 23.12**
**Proposed Classification from the Results of Genetic Testing**

| Classification | Somatic Cells | Germ Cells |
|---|---|---|
| Possible human mutagen | Positive response in one level 1 or 2 tests | Positive results from one or more level 1 or 2 tests with evidence of germ cell interaction from a level 3 test |
| Probable human mutagen | Positive *in vivo* results with supporting positive results in level 1 tests | Sufficient evidence of germ interaction supported by valid positive results in heritable (level 4) events in animal germ cells |
| Human mutagen | Positive findings in human somatic cells following an *in vivo* exposure | Positive findings in human germ cells following an *in vivo* exposure |
| Not mutagenic | Uniformly negative results across sufficient level 1 and 2 tests | |

*Note:* See Table 23.11 for descriptions of testing levels 1 to 4.

*Source:* Adapted from Dearfield, K.L. et al., *Mutation Res.*, 521, 121–135, 2002.

**TABLE 23.13**
**Risk Estimates for Somatic and Germ Cell Mutations for Selected Mammalian Mutagens**

| Compound | Germ Cell Stage | Dose (mg/kg) Required To Double the Spontaneous Background | | |
|---|---|---|---|---|
| | | Gene | Chromosomal | Somatic Cell |
| Cyclophosphamide | pg | 8 | 69 | nt |
| | g | 320 | nt | |
| Methylmethane sulfonate | pg | 2 | 4 | 7 |
| | g | 17 | nt | |
| Procarbazine | pg | 43 | 33 | 11 |
| | g | 110 | nt | |
| Mitomycin C | pg | nt | 4 | 0.3 |

*Note:* pg, postgonial sperm stages exposed to mutagen; g, gonial sperm stages exposed to mutagen; nt, not tested.

*Source:* Ehling, U.H., *Risk Analysis*, 8, 45–56, 1988. With permission.

**TABLE 23.14**
**Information Necessary To Conduct Quantitative Genetic Risk Analysis**

| | |
|---|---|
| Endogenous DNA repair | Capacities and mechanisms for repairing damaged DNA differ among organisms. |
| Metabolic specificity | Many agents are not mutagenic themselves but require conversion to a chemical form that can react with DNA and cause mutation. This conversion is called *metabolic activation* and is accomplished by enzymes that vary in specificity and levels among (a) species, (b) individual organisms within a species, and (c) different tissues in an organism. |
| Background mutation rate | All organisms and genes have an inherent background rate of mutations. Many factors, including mutations, may alter that background rate. |
| Age | Somatic mutations may accumulate during the lifetime of an organism. An older organism may thus be more vulnerable to disease, due to a greater body burden of background and induced mutations, than a younger organism. |
| Diet | Deficiencies in the levels of some vitamins, such as folate, may increase susceptibility to chromosomal mutations. Many foods contain both mutagens and antimutagens. Consumption of large quantities of mutagenic foods may account for the occurrence of certain types of cancer. |
| Economic and societal factors | Poor diet, inadequate health care, prevalence of infectious diseases, and excess exposure to known environmental mutagens such as cigarette smoke and sunlight could interact to increase susceptibility to mutagens. |
| Duration of exposure | Duration of exposure to a mutagenic substance may affect the resulting genetic risk, depending on the form of the dose–response curve and the specificity of the agent for particular stages of germ-cell development. |
| Germ cell specificity | Acute exposure to chemicals that induce mutation in late stages of germ-cell growth will result in a transitory genetic risk, confined to conceptions resulting from the gametes exposed during the sensitive stage. Acute exposure to chemicals that induce mutation in early stages of germ-cell growth will result in a permanent genetic risk. |

## GENETIC RISK AND HUMAN POPULATIONS

Assessing human germ cell risk to mutagenic substances represents a formidable task, and no chemicals have been proven to induce new mutations in offspring of exposed individuals [59]. Animal mutagens have been detected in rodent germ cells, and quantitative estimates of induced mutation rates per gene locus or the dose required to double a specific mutation rate have to be calculated from results of the *in vivo* specific locus or heritable translocation assays [153]. These estimates may be of limited value in calculating human risk or in setting safe exposure levels because they are based only on male gametes and, in the case of specific locus assay, generally on premeiotic stem cells (spermatogonia). The data do not reflect the risk to later cell stages in spermatogenesis or in female germ cells. Estimates of mutation in postmeiotic sperm and from female gametes will become available, but, even so other important biological variables would interfere with reliable risk estimates and extrapolation across species boundaries. Factors such as differences in endocrine profiles, gene structure, mutation specificity, DNA repair, mutation expression, and disease homology between rodents and humans will make extrapolation tenuous. In addition, postzygotic repair of damaged sperm has been demonstrated in mice and is a factor that may affect genetic risk. Exposure assessment is a critical component of risk assessment. Exposure may occur by different routes and duration; however, most *in vivo* mutagenicity studies used in hazard assessment are dosed acutely by oral or intraperitoneal routes. Information derived from the physical/chemical properties of the agent, its concentration in environmental matrices, and exposure modeling are important factors that must be included in the development of quantitative genetic risk assessments.

## GENETIC TOXICOLOGY TESTING STRATEGIES AND DATA EVALUATION

Genetic toxicology assessments should not consist of a single assay. Due to the multiplicity of mechanisms involved in genetic damage, evaluations should be made using a battery of tests with different endpoints. Test batteries may consist of screening tests (*in vitro*), hazard/risk assessment tests (*in vivo*), or both. It is important at the outset of testing to carefully define the objectives desired in a testing program. Screening to prioritize agents for further testing will require different types of tests than would be used to quantify somatic cell hazard to humans.

After conducting the tests, the more complex task is that of interpreting the results generated from the test battery. A genotoxic compound may be defined as *an agent that produces a positive response in a bioassay measuring any genetic endpoint* (e.g., mutation, unscheduled DNA synthesis, chromosome breakage). Although this definition considers virtually all forms of damage to DNA to classify an agent as genotoxic, genotoxicity should not be interpreted *a priori* as an indication of hazard or risk. The label "genotoxic" is only a convenient method of classifying chemicals according to their DNA reactivity. Additional experimental information beyond this initial classification is necessary to resolve concerns of genetic hazard to somatic or germ cells.

Test results are interpreted on an individual test basis (i.e., positive or negative in the specific assay) and on the results of the overall test battery. Most decisions made for regulatory purposes are based on the response profile from a battery of tests specified by regulations or guidelines. The international scientific community has proposed a number of genetic test batteries for the evaluation of new chemicals, pesticides, food additives, and pharmaceutical products.

### REGULATORY REQUIREMENTS FOR ENVIRONMENTAL CHEMICALS

Standard study designs for the routine genetic toxicology assays have been published by the EPA [54], Organization for Economic Cooperation and Development (OECD) [142,143], Canadian Health and Welfare [36], and the European Economic Community [48]. Over the past several years, attempts have been made to harmonize genetic testing requirements and the protocols used to conduct the tests. For general environmental chemicals, a tiered approach is common to quickly and inexpensively screen for chemicals that need additional testing. The testing process summarized in Table 23.11 covers most of the recommended testing requirements for North America, Europe, and Japan regulatory agencies. Most test batteries include, at a minimum, (1) the Ames test, (2) a test for *in vitro* or *in vivo* cytogenetic analysis, and, in some batteries, (3) an *in vitro* test for gene mutation in mammalian cells. Other tiers and tests may be included to expand the profile on the agent if positive results were obtained in these tests. It is also possible to add special tests that may be particularly informative for special chemical classes. The new EPA guidelines for interpretation of data provides a framework for qualitative assessment of test results using a weight-of-evidence approach [46].

### REGULATORY REQUIREMENTS FOR PHARMACEUTICAL PRODUCTS

A standard core battery of tests for evaluating pharmaceuticals was developed under the International Conference on Harmonization of the Technical Requirements for Registration of Pharmaceuticals for Human Use (ICH). The ICH process resulted in two guidelines: (1) ICH S2A, which provides guidance and recommendations for the conduct of genetic tests, and (2) ICH S2B, which establishes a standard genotoxicity test battery. The ICH core battery consists of *in vitro* bacterial mutagenesis assays, the rodent *in vivo* micronucleus test, and either the mouse lymphoma test or an *in vitro* test for chromosome aberrations. This approach has been adopted for pharmaceutical registration in the United States, Europe, and Japan.

The use of the mouse lymphoma assay was initially preferred because it could serve as an *in vitro* measure of both gene mutation and chromosome aberrations, as it is believed that induction of mutation at the target gene (thymidine kinase) can be induced by either base-pair substitution mutation at the mutant site on the normal chromosome or by deletions of the allele through chromosome breakage [127]. The former mechanism results in large colony mutants, and the latter mechanism produces small mutant colonies; consequently, use of the mouse lymphoma assay for ICH purposes requires colony sizing [193]. Chromosome analysis using cultured cell lines or human lymphocytes is an acceptable alternative to the mouse lymphoma assay, and some have proposed use of the *in vitro* micronucleus test. Table 23.15 summarizes the basic elements of the ICH core battery. In 2004, the U.S. Food and Drug Administration published a guid-

**TABLE 23.15**
**ICH Test Battery with General Requirements**

| Test[a] | OECD Guideline Design | Testing Expectations |
|---|---|---|
| Bacteria reverse mutation | 471/472 | *Salmonella* and *Escherichia* tester strains<br>Maximum concentration of 5 mg/plate or 5 μL/plate<br>Equivocal response requires further testing<br>Negative response may require confirmation |
| *In vitro* chromosome aberrations | 473 | Cell may be continuous or primary culture<br>Maximum concentration of 5 mg/mL or 10 m*M*<br>Test into the insoluble range<br>Toxicity should equal or exceed the $LC_{50}$<br>One harvest time at 1.5× normal cell cycle length<br>If negative without S9, use continuous treatment for 1.5 cycles<br>Record polyploidy and endo reduplication<br>Equivocal responses require further testing<br>Negative test without S9 may require confirmation |
| Mouse lymphoma | 476 | Maximum concentration of 5 mg/mL or 10 m*M*<br>Test into the insoluble range<br>Toxicity should achieve 80 to 90%<br>Duplicate treatments or eight concentrations<br>Sizing of mutant colonies required<br>Negative studies must be confirmed<br>Equivocal results require further testing |
| Rodent micronucleus (mouse or rat) | 474 | Maximum dose of 2 g/kg b.w.<br>Single sex sufficient if toxicity similar in both sexes<br>May use acute or multiple dosing regimens<br>Three dose levels<br>For acute dosing, harvest high dose at 48 hr<br>Other doses harvested at 24 hr<br>Must have five scorable animals/group<br>Score 2000 PCE per animal<br>Equivocal results require further testing |

[a] Either an *in vitro* cytogenetics (473) test or the mouse lymphoma (476) test may be used in the basic battery.

ance document recommending approaches for the integration of genetic toxicology study results [61]. This is a first step toward building a uniform evaluation process for mixed test results incorporating mechanistic data and a weight-of-evidence approach.

## Limitations of Current Testing Strategies

By 1980, over 150 different tests for genotoxicity had contributed data to scientific journals. Many of the tests were redundant (e.g., measured the same endpoint in a different organism or cell type), but each method had a champion who attempted to define the unique value of the particular technique or target organism. The following decade saw extensive efforts to validate and evaluate the best test or set of tests for the purposes of detecting genotoxicants. In the process, several valuable lessons were learned [27]:

- Many test methods appear to be hypersensitive and respond positively to most chemicals whether or not they directly cause DNA damage. Examples of such tests are alkaline elution and other measures of single- and double-strand breakage. The use of these tests in genetic toxicology evaluations has dropped dramatically.
- Several test methods that worked extremely well in the laboratory of the developer did not reliably transfer to other testing laboratories. Even when two testing laboratories used identical methods, test results in one could not be duplicated in another. The most notorious tests in this category were the *in vitro* cell transformation assays.
- The array of mechanisms detected by routine tests does not include all of the toxic endpoints that are relevant to genetic hazard in mammals.

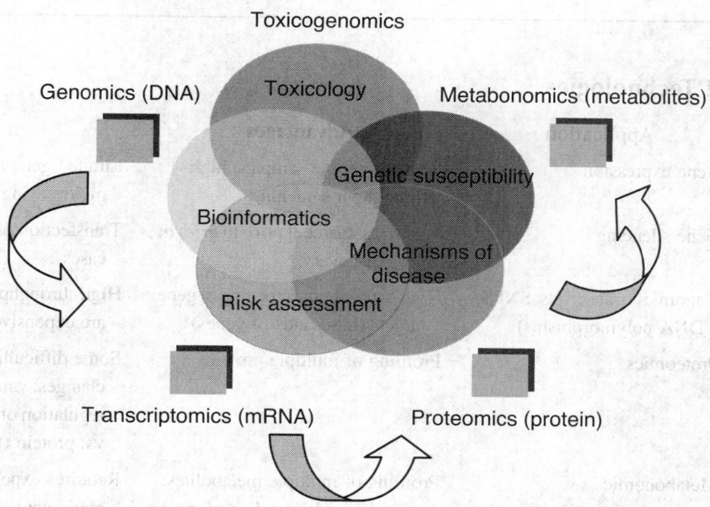


**FIGURE 23.8** Interdisciplinary aspects of toxicological analyses.

The ability to reliably detect aneuploidy, induction of tandem repeat errors, transposons, or effects from DNA methylation is absent from most testing batteries.

- Some tests are susceptible to false-positive responses generated by (1) nonphysiological treatment conditions such as high osmolarity, (2) excessive cytoxicity, and (3) the production of reactive oxidation species generated by S9 mix chemistry.

Several actions were taken in an attempt to resolve these ssues. Genetic toxicology testing schemes were simplied, and most agencies now expect to see the results from a limited test battery consisting of tests for gene mutation and chromosome aberrations before asking for further ests. Alternative tests, such as *in vitro* cell transformation, DNA breakage and repair, UDS, mitotic recombination, and DNA adduct formation, are now being used in basic esearch or as supplemental methods to support or explain indings from core battery results. Finally, treatment conditions have been spelled out in more detail and modified o minimize false-positive responses.

Weight-of-evidence methods have been developed to evaluate complex response patterns from large test batteries. Some of the methods, such as the mutagenic activity profiles created by ICPEMC, have been validated with large datasets and encoded into computer software [125].

### JSE OF OTHER TECHNIQUES TO UPPLEMENT CORE BATTERY TEST DATA

Several technologies may be applicable to filling the gaps dentified in the existing methods used in genetic toxicology. All of these methods can be performed *in vivo* which adds a dimension of relevance that can put *in vitro* findings into better perspective. The *in vivo/in vitro* UDS assay is the most common follow-up test for positive findings in the ICH core battery. The *in vivo* transgenic rodent mutation models are just about the only *in vivo* follow-up tests available to confirm *in vitro* test results for gene mutation. The single-cell gel electrophoresis (comet) assay responds to a wide range of genotoxic mechanisms [9]. The use of alternative methods for carcinogenicity testing such as the p53, H-*ras*, and Tg.AC transgenic mouse models and the Syrian hamster embryo (SHE) cell transformation assay are both faster and more information rich than conventional rodent cancer bioassays [112]. However, there will continue to be a need to assess the hazard to somatic and germ cell DNA from exposure to new products.

## TOXICOGENOMICS: AN EMERGING TECHNOLOGY

Testing for genetic alterations has broadened over the past decade [196] with the emergence of -omics technologies. Genomics, transcriptomics, proteomics, and metabonomics have been applied to assess the genetically mediated responses of chemicals (Figure 23.8) [17,30,57,58,134]. The Human Genome Project (which involved sequencing the entire human genome) fostered the expansion of toxicological testing to include a new field termed *toxicogenomics*. Toxicogenomics combines genomics and bioinformatics to characterize and identify mechanisms of toxicity induced by chemicals, including drugs. Governmental research agencies such as the National Institute of Environmental Health Sciences (NIEHS) are actively studying toxicogenomic applications to toxicology. The Toxicogenomics Research Consortium (TRC) of the National Center for Toxicogenomics (NCT), a division of the NIEHS, is comprised of the

**TABLE 23.16**
**Comparison of Current Technologies**

| Technique | Application | Advantages | Limitations |
|---|---|---|---|
| Quantitative RT/PCR | Gene expression | Screen multiple samples; high-throughput screening | Limited gene targets per assay, even with multiplexing and gene cards |
| SiRNA/RNAi | Gene silencing | *In vitro* assessment prior to *in vivo*; drug discovery | Transfection efficiencies; toxicity in certain cases |
| Microarray (filter, chip) | Genomics (transcripts, SNPS, DNA polymorphisms) | High-throughput screening of gene targets (thousands of genes) | High-throughput analyses (i.e., many samples) are expensive |
| Protein array (filter, chip, solution) | Proteomics | Profiling of multiple proteins | Some difficulties in correlating with mRNA changes; variations in time requirements for regulation of mRNA (transcription/stability) vs. protein (translation, modifications) |
| Metabolic profiling (NMR, mass spectrometry) | Metabonomics | Profiling of multiple metabolites; provides understanding of gene functions | Requires expensive/extremely specialized equipment |
| Methylation-specific PCR | Epigenetic DNA modification (e.g., methylation) | Provides gene-specific analysis | Multistep process leading to potential for sample lost; limited to gene-specific characterization; global methylation patterns must be assessed by other measures |

NCT Microarray Group and five institutions (University of North Carolina at Chapel Hill, Fred Hutchinson Cancer Research Center, Oregon Health and Science University, Massachusetts Institute of Technology, and Duke University). The TRC plans to: (1) evaluate toxicant-specific patterns of gene expression; (2) integrate gene expression profiling with proteomics, metabonomics, and phenotypic anchoring; (3) study the toxicological effects of chemical mixtures; and (4) contribute gene expression and proteomics data to the Chemical Effects in Biological Systems Database (CEBS; www.niehs.nih.gov).

This section summarizes the current roles and methodology of the -omics in toxicology. Because the -omics fields are rapidly advancing, readers are encouraged to consult current scientific literature to most effectively apply the latest techniques to their research [30,44,93,108, 123,128,146,185].

## OVERVIEW OF CURRENT TECHNOLOGIES

The four major -omics are (1) genomics, the study of gene sequences and genetic variability; (2) transcriptomics, the quantitative measurement of gene expression (mRNA) in a cell or tissue by various measures; (3) proteomics, the measurement of all cellular protein production and levels in a cell or tissue; and (4) metabonomics, the multiparametric measurement of metabolites. A number of new techniques have been developed that enable high-throughput screening, enhanced biological analyses, and quantitative assessments (Table 23.16).

### Polymerase Chain Reaction and Quantitative RT-PCR

By 1997, polymerase chain reaction (PCR) was emerging as an integral part of *in vitro* toxicology laboratories as applied to: (1) gene expression analysis from *in vitro* and *in vivo* samples, (2) *in situ* genetic analysis of paraffin embedded tissue, (3) differential display, (4) gene cloning, and (5) genotyping [179]. The application of PCR technology has led to the development of new methods for mutation analysis in mammalian systems from mouse to human, and the types of damage detected are relatively unrestricted to specific mechanisms of toxicity. Mutation targets selected by survival of chemical resistance (e.g., HPRT, 6-thioguanine) or by genetic polymorphism detected by single-strand conformational polymorphism (SSCP), restriction-fragment-length polymorphism (RFLP), or single nucleotide polymorphisms (SNP) technologies [111] are amenable to analysis by reverse transcriptase–polymerase chain reaction (RT-PCR). Polymorphisms represent variations in the sequence of specific genes or gene products. Polymorphisms probably occur in many types of genes. Polymorphisms in phase I and II metabolic enzymes, oncogenes, and tumor suppressor genes have been observed to affect drug metabolism, toxic responses, cellular growth rates, and disease progression; therefore, investigating and understanding the roles of polymorphisms in toxicology are essential.

Analysis of gene expression initially used non-PCR requiring assays such as subtractive hybridization (SH) and serial analysis of gene expression (SAGE), as well as differential display (DD) polymerase chain reaction [114,179,184]. These assays illustrated the significance and value of comparing gene expression differences between control and chemically-treated samples or normal and diseased tissue. The limitations of these assays were the number of known genes that were identified by the analysis, limitations on sample throughput, and labor intensity of the assay. The combination of reverse transcription of RNA into complementary DNA (cDNA) and subsequent amplification via PCR in the presence of gene specific primers led to the advent of RT-PCR. Since the late 1990s, RT-PCR assays have been continually improved by advances in primers, probes, and reaction materials which have resulted in advancements in the quantitative and qualitative assessment of gene expression (Figure 23.9) [65,81].

Improvements in quantitative RT-PCR facilitated the adoption of gene expression technology into toxicological testing; for example, differential gene expression profiles following chemical treatments have been characterized in *in vitro* and *in vivo* models (Table 23.17). Primary lung cell cultures of human bronchial epithelial cells have been used to characterize *c-myc* expression following benzo(*a*)pyrene and benzo(*a*)pyrene diol epoxide treatment. Differential mRNA responses were observed which were also represented by distinctions in DNA adduct accumulation and cell cycle regulation [62]. Normal, immortalized, and cancerous cell lines were also screened via RT-PCR to characterize gene expression changes in tumor progression, drug responses and tissue specificity of environmental toxins (Table 23.17) [189].

## Microarrays

Microarray technology includes the use of gene chips that allow for the simultaneous assessment of a large number of genes involved in signal transduction, stress response, cell-cycle regulation, DNA synthesis and repair, inflammatory responses, metabolism, and many other gene categories. The genes on microarrays can be designed for different species (e.g., human, mouse, rat) and for specific disease and stress states (e.g., cancer, inflammation). Arrays can also be designed to separate gene targets by functions, pathways, and gene families. Array designs are developed by genome sequences from public databases and by proprietary gene sequences. Photolithographic techniques (adapted from the semiconductor industry) are used in the manufacturing process [109]. As a result of the progress made in the Human Genome Project and the application of expressed sequence tags (ESTs), gene chips have been designed to encompass the entire genome for human, rat, and mouse. With the aid of various data analysis methods, including cluster and hierarchical analyses, the researcher can generate complex details on interactions within gene pathways.

Advances in commercially available arrays have been facilitated by shifting detection methods from radioactivity to chemiluminescence. Aspects of microarray research that are still under discussion center on validation and standardization. Some of these issues have been addressed in a recent article by Rockett and Hellmann [155] and include measures for validation (quantitative RT-PCR, western or northern blots, *in silico* analyses), uniform formats for RNA amplification, data analysis and representation, journal requirements for publication of array data and assay design, and standardizations for across-platform comparisons. A document that provides a reference framework for microarray investigations is the Minimum Information About a Microarray Experiment (MIAME) checklist (www.mged.org). While adhering to MIAME, working groups such as the Microarray Gene Expression Data (MGED) Society and the NCT Microarray Group have facilitated the development of assay designs that have been instrumental in moving microarray analyses to applications in toxicological investigation (www.niehs.nih.gov).

Microarray technology has revolutionized the early concepts of differential gene expression analysis by offering the researcher the option of screening very large numbers of genes, even entire genomes, simultaneously. For example, samples isolated following exposure to pure chemicals and complex mixtures of chemicals, as well as various disease states (e.g., lung cancer, breast cancer) have been characterized with this technology (Table 23.17) [108,126,140,156].

## DNA Methylation

DNA methylation is an epigenetic mechanism that regulates gene expression and is mediated by the covalent addition of a methyl group to the 5-carbon of cytosine, typically located in a CpG dinucleotide, a process catalyzed by DNA methyltransferases and the methyl donor *S*-adenosylmethionine [21,97]. Appropriate methylation in GC-rich regions of DNA are required for normal maintenance of gene expression. Chemically induced hyper- or hypomethylation of genomic DNA can lead to phenotypic changes that may be causally involved in disease pathogenesis (i.e., carcinogenesis) [94,96,154,186,187]. The concept that DNA methylation plays a multifaceted role in carcinogenesis is compatible with key features of this disease process, including progressive clonal selection of abnormal cells, the reversibility of tumor promotion, and the multiplicity of tumor phenotypes [72,197].

Comparative studies on gene methylation status during disease progression are very useful in assessing the relevance of disease models in animals to human risk assessment; for example, the mouse skin multistage carcinogenesis model reflects the aberrant DNA methylation patterns of human tumors. Furthermore, Counts et al. [198] found that cell proliferation and global methylation

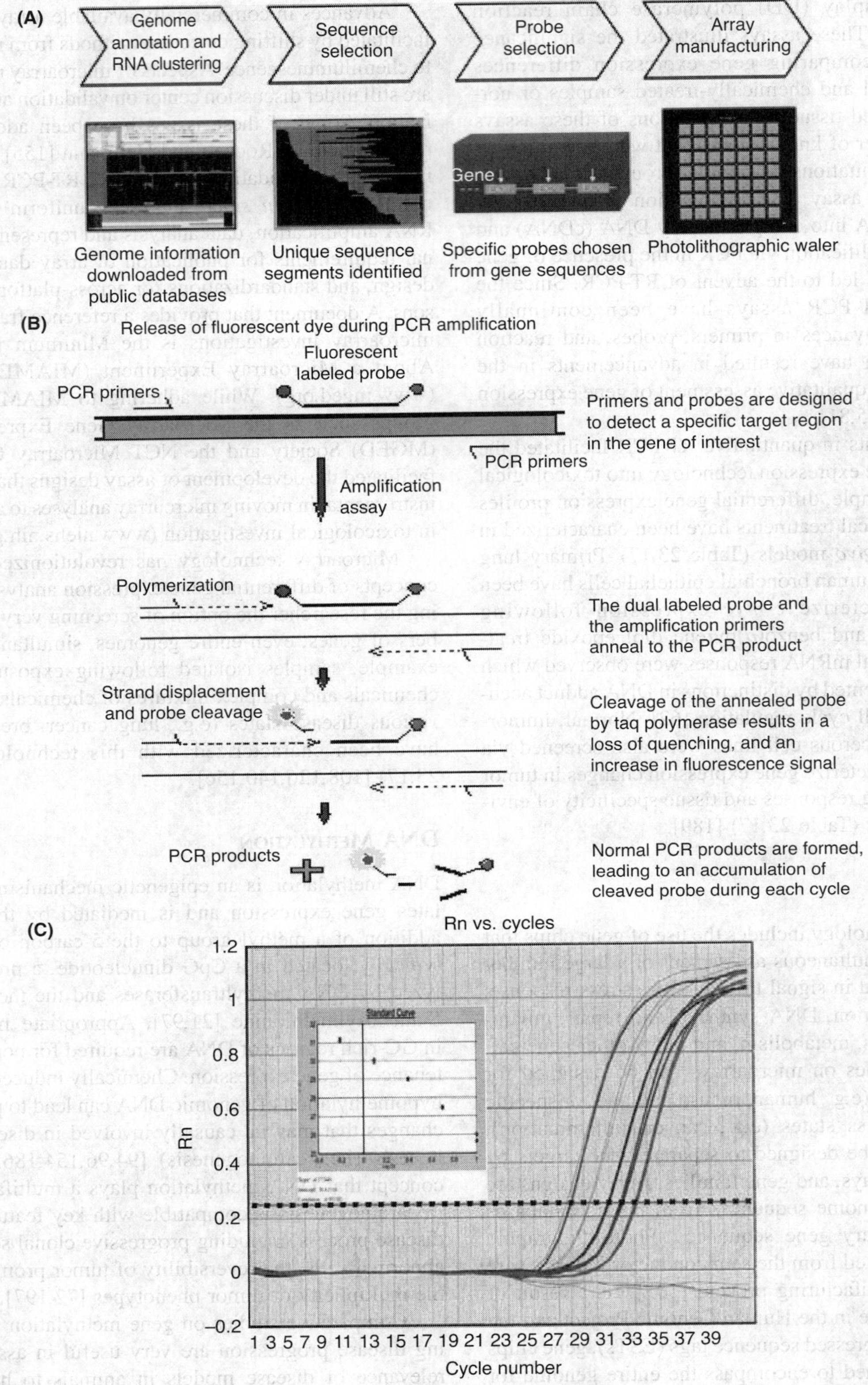


**FIGURE 23.9** Advances in gene expression technologies. (A) Designing arrays. (From Kreiner, T. and Buck, K.T., *Am. J. Healt Syst. Pharm.*, 62, 296–305, 2005. With permission.) (B) Real-time PCR aided by fluorescence technology. (Courtesy of Applie Biosystems; Foster, CA.) (C) Representative data from real-time quantitative RT/PCR: quantitation of *c-myc* mRNA facilitated b amplification of a relative standard curve prepared with cDNA from a small-cell lung carcinoma cell line (NCI-H82).

**TABLE 23.17**
**Gene Expression Profiles Characterized by *In Vitro* and *In Vivo* Models**

| Agent/Disease | Model | Method | Refs. |
|---|---|---|---|
| Diesel exhaust | *In vitro* (human airway cells, rat alveolar macrophages) | Semiquantitative RT/PCR | Baulig et al. [18], Koike et al. [106] |
| Hepatotoxins | *In vitro* (human hepatocytes and HepG2 cells | cDNA filter array; Real-time PCR | Harris et al. [78] |
| Tricoethane mycotoxin deoxynivalenol | *In vivo* | Macroarray, QRT/PCR | Kinser et al. [103] |
| Dichloroacetic acid | *In vivo* (mice) | cDNA microarray | Thai et al. [174] |
| Bromobenzene | *In vivo* (rats) | Transcriptomics, proteomics | Heijne et al. [82] |
| Cytotoxic antiinflammatory drugs, DNA-damaging agents | *In vitro* (HepG2) | cDNA microarray | Burcynski et al. [30] |
| Arsenic | *In vivo* (Tg.AC mice) | DNA methylation | Xie Y et al. [191] |
| Benzo(*a*)pyrene | *In vitro* | QRT/PCR | Fields et al. [62] |
| Lung cancer | *In vitro* | cDNA microarray | Hellmann et al. [83] |
| Cigarette smoke | *In vivo* | cDNA microarray; Gene Chip | Hackett et al. [75], Shah et al. [162], Gebel et al. [69] |
| Cigarette smoke condensate and extracts | *In vitro* (human bronchial cells, Swiss 3T3 cells) | Real-time QRT/PCR | Fields et al. [63], Bosio et al. [24] |
| Asbestos | *In vitro* | cDNA microarray, real-time PCR | Ramos-Nino et al. [151] |
| Nephrotoxins | — | cDNA microarray | Kramer et al. [108] |
| Oxidative stress | *In vitro* (Hep-G2) | cDNA microarray, real-time PCR | Morgan et al. [135] |

tatus changes in mouse liver after phenobarbital and/or holine-devoid, methionine-deficient diet administration, ndicating that the high propensity of the B6C3F$_1$ mouse o develop liver tumors is due, in part, to an altered ability o maintain normal methylation status. They concluded hat mouse liver tumor response may not be an appropriate ndpoint for human risk assessment.

Methods for global and gene-specific DNA methy-ation analysis can be divided into separate categories. 'or global methylation analysis, methyl-accepting apacity (SssI DNA methyltransferase) assays and chro-natographic assays (high-performance liquid chroma-ography, thin-layer chromatography, or liquid chroma-ography/mass spectrometry) measure the overall level f methylated cytosines in a genome. The latter assay equires digestion of the DNA into single nucleotides. Numerous methods have been developed for analyzing ene-specific methylation, including bisulfite-based nethods. Bisulfite can selectively deaminate cytosine, ut not 5-methylcytosine, to uracil which leads to a equence change in the DNA via PCR amplification that nables discrimination between cytosine and 5-methyl-ytosine. The DNA sequence differences that result etween a methylated and unmethylated cytosine can be ssessed by several methods, including direct sequenc-g, restriction enzyme digestion, nucleotide extension ssays (MS-SnuPE), and methylation-specific PCR MSP). Genome-wide techniques such as restriction landmark genomic scanning for methylation (RLGS-M) and CpG island microarray have been developed for assessing unknown methylation hot spots or methylated CpG islands in the genome.

DNA methylation changes may be either global or gene specific (e.g., p16, GST HOXA5). An important question currently under active investigation is the level of DNA methylation required to alter cellular and organ functions. This is addressed by studying the combination of methylation status, gene expression profiles, and organ function, including carcinogenesis. *In vivo* investigations have studied the role of methylation in controlling cell proliferation and cell cycling and in disease states such as lung cancer and skin tumorigenesis. Subchronic toxic effects of oral arsenite, arsenate, monomethylarsonic acid, and dimethylarsinic acid were assessed in v-Ha-*ras* transgenic (Tg.AC) mice [191]. These chemicals led to global DNA hypomethylation and altered gene expression patterns in the liver. Gene methylation status has also been used in developmental toxicology efforts to investigate the effect of synthetic estrogen and diethylstilbestrol [122].

## Proteomics

The field of proteomics seeks to quantify the expression profile of cellular proteins as an indicator of disease risk. Tissue profiling by proteomic techniques has led to the identification of biomarkers of effect following chemical exposure [20,35]. Early proteomic technologies used two-dimensional gel electrophoresis followed by LC/MS and western blot or enzyme-linked immunosorbent assay (ELISA). The field of proteomics now uses primarily matrix-assisted laser desorption ionization mass spectrometry (MALDI-MS) and surface-enhanced laser desorption/ionization (SELDI)-based protein chip technologies to determine protein expression patterns. MALDI-MS and SELDI-based protein chip technologies provide time savings, higher throughput, and increased reproducibility. On a smaller scale, protein arrays (filter and suspension) facilitate screening protein profiles of tissue samples and biological fluids.

Thousands of peptides and proteins can be rapidly detected from a variety of tissues with MALDI-MS, which produces protein profiles of normal cells as well as various stages of disease. Data from tissue protein profiles might be used in the future to predict tumor behavior, diagnosis, and prognosis and reveal mechanisms of disease. MALDI-MS is also amenable to generating protein profiles from small sample amounts when used in conjunction with laser capture microdissection (LCM), a microanalytical technique used to procure homogenous tissue samples via micro- and single-cell tissue extractions. When applied in tandem with MALDI-MS, LCM enables the investigator to assess the spatial distribution of proteins in the tissues.

Commercially available protein arrays are used to assess biomarkers for various diseases and pathogenic states (e.g., alterations in cytokine expression) and transcription factors in cell lysates, conditioned media, and serum samples. Applications of protein filter arrays are similar to those of gene microarrays, except that the screening and detection process is performed with antibody probes.

## Metabonomics

As defined by Nicholson et al. [138], metabonomics is "the quantitative measurement of the dynamic multiparametric metabolic response of living systems to pathophysiological stimuli or genetic modification." There is a great need to advance our understanding of the biological relevance of genomic data. Metabonomics holds the promise of being an important tool in understanding this relationship as it offers the possibility of relating gene expression patterns to biological effects. Just as the genome represents all of the gene sequences of an organism, the metabolome represents all of the low-molecular-weight molecules in a cell at a given developmental or physiological state.

The leading techniques used in metabonomics are mass spectrometry (MS) and nuclear magnetic resonance (NMR). MS analysis allows for positive and negative ion detection primarily through electrospray ionization. This may be preceded with an initial sample separation via liquid chromatography. In concert with chemometric and bioinformatics tools, metabonomic data can generate biochemically based fingerprints of the effects of drugs and toxicants in biological systems and hence tie biologically relevant changes to data obtained from various -omics technologies.

Metabonomics offers the promise of obtaining rea drug-effect and disease endpoints to support drug devel opment, risk, and toxicological assessments in living sys tems via analysis of biofluids (urine, serum) as well a tissue and cell lysates [138,139]. The Consortium o Metabonomics in Toxicology (COMET) project, a collab orative project hosted by Biological Chemistry, Imperia College (London), and several major pharmaceutical com panies, has the goal of evaluating the utility of metabo nomics as a new technology in providing information o early-stage toxicity and safety issues in the pharmaceutica industry [115]. Metabonomic technology has been applie in the ecotoxicological assessment of environmental con taminants [28], physiological influences on biofluids [22] systems biology in pharmaceutical research [116], physi ological monitoring, drug safety assessments, diseas diagnosis [117], toxicological assessments (e.g., hydra zine, carbon tetrachloride, bromobenzene) [86,99], an hepatic and renal toxicity [39,87].

## Genetically Modified Systems

In cell culture models, gene reporter assays can serve a measures to detect toxicological affects of chemicals o promoters of genes involved in cellular regulation (i.e transcription factors). These gene reporter assays are facil itated by the introduction of gene reporter constructs (i.e linking the gene of interest promoter to coding sequence for luciferase or green fluorescence protein) into a variet of cell types and subsequent quantization of signal c protein generation in response to chemical treatmen Most recently, transgenic applications have contributed t the application of gene silencing (RNAi) to toxicologic testing. RNAi is a recent technology and has only bee used in mammalian cells since 2001, while prior invest gations were in *Drosophila* and *Caenorhabditis elegan* RNAi or siRNA provides a means of knocking out gen activity (i.e., gene silencing) by the introduction of sho (21 to 23 nucleotides), sequence-specific RNA duplexe into the host cell, which initiates posttranscription ger knockdown. Because this process may lead to toxi effects, Stealth™ RNA oligonucleotide duplexes (Invitr gen), for example, have been designed to eliminate suc toxic effects.

In eukaryotes, the RNAi process occurs by the clea age of long double-stranded RNA (dsRNA) into 21- 23-nucleotide short (or small) interfering RNA (siRN duplexes. The cleavage is facilitated by the dicer enzym

**TABLE 23.18**
**Cancer Databases and Tools for Analyzing DNA, RNA, and Protein Interactions**

| Agency | Database Link |
|---|---|
| NIH Center for Bioinformatics | www.discover.nci.nih.gov/tools |
| NIEHS: National Center for Toxicogenomics | www.niehs.nih.gov/nct/home.htm |
| FDA's National Center for Toxicological Research (NCTR) | www.fda.gov/nctr/science/centers/toxicoinformatics/arraytrack |
| EMBL-EBI The European Bioinformatics Institute | www.ebi.ac.uk/arrayexpress/ |
| Environmental Genome Project | www.genome.utah.edu/genesnps/ |
| Biocarta | www.biocarta.com |
| MatchMiner, GoMiner, CIMMiner, MedMiner, AbMiner, LeadScope/LeadMiner, MIMmimer | www.discover.nci.nih.gov/matchminer |

A silencing complex is subsequently formed as the siRNA associates with an intracellular multiprotein RNA-induced silencing complex (RISC). This complex recognizes and cleaves complementary cellular mRNA. The cleaved mRNA is targeted for degradation, resulting in a reduction of posttranscriptional gene expression in the cell. A detailed review of this methodology and its application in defining gene functions in various biological systems (whole organism, cell culture, and cell-free systems) has been provided by Sohail et al. [168].

Transgenic animals for mutation detection were developed in 1990 using shuttle vector technology [74,163]. A few models commercially available in the 1990s, such as the MutaMouse and BigBlue Mouse, measured *lac I* and *lac Z* gene activation [178,181]. These models laid the foundation for the development of genetically modified animals with altered gene expression patterns via either gene knockout or enhancement. Knockout and transgenically enhanced models are also being investigated to advance knowledge on the response of specific gene modifications on toxic responses and disease pathogenesis. The Tg.AC (transgene expression of oncogenic v-Ha-*ras* within the skin) and UL53-3 × A/J (harbors a dominant-negative p53 mutation) have been used to assess the multistage aspects of tumorigenesis via exposures to 7,12-dimethylbenz(*a*)anthracene (DMBA), 12-*O*-tetradecanoylphorbol-13-acetate (TPA), or cigarette smoke [45,144,145].

Further efforts to assess the role of genotype variations in toxicology have been undertaken by the Environmental Genome Project (EGP). The EGP is a program initiated by the NIEHS, which has the mission to "improve understanding of human genetic susceptibility to environmental exposures." To address this issue, EGP has undertaken several major research initiatives, including the Comparative Mouse Genomics Centers Consortium (CMGCC), Human DNA Polymorphisms Discovery and Characterization, and the GeneSNPs Database of Human and Mouse Environmental Responsive Genes. To improve the biological significance of human DNA polymorphisms, the CMGCC is charged with developing transgenic and knockout mouse models based on human DNA sequence variants in environmentally responsive genes. The candidate target genes include genes involved in DNA repair, cell cycle control, cell signaling, cell structure, gene expression, apoptosis, and metabolism. Commercially available animal models from the Jackson Laboratory, Taconic, and Mouse Models of Human Cancers Consortium (MMHCC) repositories offer transgenic models for research in apoptosis, cancer, diabetes and obesity, immunology and inflammation, cardiovascular diseases, and metabolism, for example. Pertinent publications associated with NIEH's investigations of SNPs may be found at www.niehs.nih.gov/cmgcc/htm. Additionally, bioinformatics tools have been designed to allow for functional assessment of human gene variants in the mouse; hence, data obtained within and outside of the consortium on mouse models of human genetic variations are maintained in the CMGCC Mouse Federated Database.

## Bioinformatics

A scientific discipline known as bioinformatics has been born out of the need to maintain, correlate, and statistically evaluate volumes of biological data arising from the -omics revolution. Bioinformatics encompasses biological sciences, computer science, and statistics. Integration of toxicogenomic data into risk assessment, mechanistic and predictive toxicology, and pattern recognitions in disease progression requires the compilation of large sets of data from toxicogenomic studies, cellular assays, molecular assays, and animal studies. Such efforts have been initiated by various agencies and research teams (Table 23.18) and are critical to an understanding of toxicological responses at the molecular and mechanistic levels; however, some of the challenges with database design and maintenance involve data integration (standardized data storage and exchange), uniform nomenclature, and standardized assay design, for example [119].

In addition to the repositories of data, the Biocarta system, a commercially developed web-based database, provides information on gene function, displays specific gene pathways coordinated from gene expression and toxicological data, provides links to the scientific citations (PubMed) that contributed to the pathway designs, and provides reagent and assay resources. Additionally, in efforts to link toxicological and pharmacological data to disease mechanisms, the National Cancer Institute has developed databases that display mechanisms for cancer and various tools for analyzing DNA, RNA, and protein interactions in disease (Table 23.18).

## EMERGING TECHNOLOGIES

New techniques are still emerging for characterizing gene expression and mechanisms of disease, such as micro RNA (miRNA) expression analysis, applications with activity-based probes (ABPs), enhanced cell-based assays for high-throughput screening, and micro-/nanobiotechnologies. The application of these methods to toxicology is briefly introduced here.

### Micro RNA

Micro RNAs (miRNAs) are small noncoding RNAs consisting of approximately 22 nucleotides that have been recently identified as regulators of gene expressions; however, specific details of the target genes and expression levels of the miRNAs themselves are not currently known. *In vitro* investigations combined with oligonucleotide filter arrays have been employed to investigate high-throughput analysis of miRNA expression from *O*-tetradecanoylphorbol-13-acetate treatment in HL-60 cells [167]. miRNAs have been implicated in hematopoietic lineage differentiation in mammals and control of cell death and proliferation in flies.

### Activity-Based Probes

Genomic data alone will not lead to complete discernment of mechanisms of toxicity and disease; hence, as discussed earlier, applications of proteomic techniques have been incorporated in recent years to expand such investigations. Some standard proteomic technologies have afforded advancement of the knowledge supplied by the genomic era; however, limitations in the assay resolutions and sample complexities have limited some of the advancement. In an attempt to focus proteomic efforts on certain proteins that serve physiological relevance, ABPs, alternatively known as chemical probes, have been developed. Activity-based probes are small molecules that serve as a method to tag, enrich, and isolate distinct sets of proteins based on their enzymatic activity. ABPs allow for direct assessment of enzymatic activity within complex proteomes.

### Cellular-Based Technologies: Enhanced *In Vitro* Assays and Models

Improvements in detection methods for enzyme-linked immunosorbent assays (ELISAs) have led to the use of in-cell-based assays for high-throughput screening of chemical affects on posttranslational protein modifications. Functional and phenotypic assays for various signal transduction pathways (e.g., NF-κB, AKT, MAPK) have been developed by companies such as ActiveMotif, Inc. for effective screening of chemical responses without isolating cellular material. To address the needs of analyzing biological and toxicological changes within an organ-type environment without the technical and financial issues associated with animal models, researchers at the National Cancer Institute and various corporations (e.g., HmREL Corp.) are developing organotypic and *in vivo* surrogate models for certain organ sites (e.g., breast, lung, prostate), disease states (e.g., cancer), and pseudo organ/tissue interactions [182]. These models are amenable to toxicological and pharmacological assessment for toxic responses and drug efficacy through measuring physiological changes and are expected to help bridge *in vitro* and *in vivo* toxicological investigations.

### Micro- and Nanobiotechnologies

Although assessment of toxicological effects is necessary in the whole organism, interactions and alterations at the molecular level must be characterized as well; for example, cellular functions and structures can be disrupted by modifications of the nanometer structure of critical molecules, hence micro- and nanobiotechnologies (devices) can be used to assess biological and toxicological processes at the nanoscale. To facilitate detection, separation, manipulation, and analysis of cells and biological samples at this scale, nanofabricated tools and technologies such as laser tweezers, surface plasmon resonance (SPR), laser capture microdissection (LCM), atomic force microscopy (AFM), and multi-photon microscopes have been developed [195]. These miniaturized devices with enhanced sensors such as the micro total analysis system (micro-TAS), or "lab-on-a-chip," can be used for toxicological assessment of cellular structures and functions, such as cellular adhesion, signal transduction, motility, membrane elasticity, chemotaxis, phagocytosis, protein expression, and metabolism.

Because many of the micro- and nanobiotechnologies described by Zieziulewicz et al. [195] offer promise for biological and toxicological analysis with nano-scale analysis and for the purpose of brevity, LCM will be covered here as it relates to tissue-specific analysis as well as cellular and molecular interactions, as it is currently being employed in toxicogenomic applications. LCM provides a homogenous cell population and allows gene expression analysis of pooled single cells, cell subpopulations, and cell populations, thereby eliminating the confounding of gene expression data that may occur as whole tissue analyses are performed [19,51, 149,176]. LCM was developed through an initiative of the National Institutes of Health and Arcturus Engineering, Inc., to aid in macromolecule (DNA, RNA, and protein) analyses from various tissues [43,52]. The LCM system consists of an inverted microscope that is equipped with a low-power, near-infrared laser. Following the preparation of tissue sections on glass slides, which includes placement of a thermal plastic (ethylene–vinyl acetate) film, specific cell populations or single cells are extractable from the tissue as energy liberated from the laser melts the thermoplastic film at designated locations, which is aided by visualization of the desired cell locations with the inverted microscope. The ability to dissect single cells or clusters of cells is achieved by adjusting the laser diameter from 7.5 to 30 μm. The thermal plastic film provides a means for accruing the cells with either no or minimal detectable damage to the underlying macromolecules. The dissected cells are amenable to transfer to a transparent cap that fits into a microcentrifuge tube, and they are subsequently subjected to designated extraction methods for various molecular analyses [16,23,166].

Due to the advancement of technologies in the postgenomic era, LCM has been used across multiple toxicogenomic applications. As a result, detection mechanisms and sensitivity limits of technologies such as quantitative RT/PCR, microarrays, and protein assays are being enhanced to handle limited quantities of macromolecules. LCM has aided in the proteomic analyses of human breast cancer tissue, in the identification of protein biomarkers [42,165], in microarray analysis [126], and in gaining an understanding of the signal transduction cascades induced by environmental toxic agents, such as oxidants and asbestos fibers [171].

## SELECTED STUDY DESIGNS AND GENETIC TESTING

This section contains a set of study designs that will meet the OECD and ICH guidelines for the United States, European Community, and Japan. Although not fully described in every protocol, it is recommended that each test article be examined for several chemical, physical, and biological parameters. Included in this preliminary assessment is an examination of the solubility, volatility, and light sensitivity of the test article, as well as its stability in the vehicle chosen for the study. A preliminary toxicity study is performed to establish a maximum concentration or dose level. Careful selection of concentrations is important for obtaining definitive evaluations of test articles. Because of the potential for artifacts produced by excessive ionic levels or low pH levels in mammalian cell cultures [27], these parameters should be monitored and kept within acceptable ranges during the exposure periods for these protocols. The protocols all contain specific criteria related to the interpretation of toxicity, but it is often possible to save time and materials by knowing the approximate concentration range required to achieve a desired toxicity range. This group of protocols does not represent the entire repertoire of methods available to genetic toxicologists; however, it does provide the essential tests required for routine screening for most new compounds under development for use in commerce. Attached to all genotoxicity study reports should be the results of analytical chemical analyses for stability and concentration verification of the dosing solutions or suspensions. The laboratory historical positive and vehicle control response ranges, means, and standard deviations for each protocol should also be included in the study reports to meet OECD guidelines.

### Tests Measuring Gene Mutation

#### Protocol 1: *Salmonella–Escherichia coli*/ Mammalian–Microsome Reverse Mutation Assay

The objective of the bacteria/microsome assay is to evaluate a test article for mutagenic activity in a bacterial reverse mutation system with and without a mammalian S9 activation component. The assay design is based on OECD Guideline 471 (1997 revision) and the ICH S2A Guideline.

##### Materials

*Indicator Cells*. The strains of *Salmonella typhimurium* and *Escherichia coli* that are used routinely in the plate assay are described in Table 23.19. Frozen stocks are maintained at –60 to –80°C and are used to prepare master plates for assays. The bacterial strains from the master plates are checked for the appropriate genetic markers and the characteristic number of spontaneous revertants before or during each assay.

**TABLE 23.19**
**Strains Used in the Ames Plate Assay**

| Tester Strains | Genotypes | | | Additional Mutations |
|---|---|---|---|---|
| *Salmonella typhimurium* | | | | |
| TA1535 | *his* G46 | *rfa* | *uvr*B | — |
| TA1537 | *his* C3076 | *rfa* | *uvr*B | — |
| TA98 | *his* D3052 | *rfa* | *uvr*B | pKM101 |
| TA100 | *his* G46 | *rfa* | *uvr*B | pKM101 |
| *Escherichia coli* | | | | |
| WP2 | *trp* | | *uvr*A | — |

**TABLE 23.20**
**Positive Control Chemicals for the Standard Plate Incorporation Test**

| Assay | Chemical | Vehicle | Concentrations per Plate (µg) | Responding Strains |
|---|---|---|---|---|
| Nonactivation | Sodium azide | Water | 2.0 | TA1535, TA100 |
| | 2-Nitrofluorene | DMSO | 1.0 | TA98 |
| | ICR-191 | DMSO | 2.0 | TA1537 |
| | 4-Nitroquinoline-*N*-oxide | DMSO | 1.0 | WP2*uvr*A |
| S9 Activation | 2-Aminoanthracene | DMSO | 2.5 | All TA strains except TA98 |
| | 2-Aminoanthracene | DMSO | 25 | WP2*uvr*A |
| | Benzo(*a*)pyrene | DMSO | 2.5 | TA98 |

### *Media*

Vogel–Bonner salt solution supplemented with 2.5% Oxoid Nutrient Broth #2 powder is used for culture growth. Minimal bottom agar plates are prepared with Vogel–Bonner minimal medium E supplemented with 1.5% agar and 0.2% glucose. Top (overlay) agar is prepared with 0.7% agar and 0.5% NaCl supplemented with 0.05-m*M* histidine/biotin solution for selection of histidine revertants or 0.05-m*M* tryptophan for the selection of tryptophan revertants.

### *S9 Metabolic Activation System*

S9 homogenate prepared from the livers of Aroclor 1254-induced Sprague–Dawley male rats [5] is obtained commercially. S9 mix is prepared immediately prior to experimental use and contains 100-µL/mL S9, 4-m*M* NADP, 5-m*M* glucose-6-phosphate, and salts [8].

### *Control Articles*

Negative control plates exposed in triplicate to the vehicle chosen for the test article are included for all 5 bacterial strains in the presence and absence of the rat liver S9 metabolic activation system. Common vehicles include deionized water, dimethylsulfoxide, ethanol or dimethylformamide. The vehicle concentration is the same as that used with the test article treatments. Known mutagens are used as concurrent positive control cultures for each assay and establish the historical database for the laboratory. The positive controls selected for each strain and for demonstration of appropriate metabolic activation by the S9 mix are given in Table 23.20.

### *Experimental Design*

*Dose Selection*. A dose range-finding study is conducted with tester strains TA100 and WP2*uvr*A with and without S9 metabolic activation. Usually 10 doses are tested in half-log dilution steps using only one plate per dose. The test article is weighed and diluted in an appropriate vehicle to set up the dosing solutions. The test article and other components are prepared fresh and added to tubes of molten overlay agar. The contents are then mixed and poured onto the surfaces of the Vogel–Bonner plates. For activation studies, 0.5 mL of the S9 mix is included in each overlay tube. For highly reactive chemicals that have short half-lives in aqueous environments, a liquid preincubation method has been developed in which the bacteria and test article (and S9 mix for the activation portion) are incubated for 20 minutes prior to adding the molten overlay agar and pouring onto the plates [192]. Vehicle control plates are included for comparison with the test article treatments. The plates (once the overlay has solidified) are incubated at 37°C for 48 to 56 hr and scored for numbers of revertants per plate. Cytotoxicity is assessed as a decrease in the number of revertant colonies or by a thinning or disappearance of the bacterial background lawn.

*Mutagenicity Testing*. For the standard plate incorporation test, at least 5 to 8 dose levels of the test article are selected, based on the dose range-finding results. When no toxicity is observed, concentrations up to 5000 nL or 5000 µg per plate should be employed. Triplicate plates per dose group for all five bacterial strains, with and without S9 mix, are used in standard assays. The appropriate vehicle and positive controls are included in triplicate plates for all strains (Table 23.20). Plate test data consist of direct revertant colony counts obtained from a set of selective agar plates seeded with populations of mutant cells suspended in a semisolid overlay. Because the test article and the cells are incubated in the overlay for at least 48 hr, and a few cell divisions occur during the incubation period, the test is semiquantitative in nature. Although these features of the assay reduce the quantitation of results, they provide certain advantages not contained in a quantitative suspension test, including:

- The small number of cell divisions permits potential mutagens to act on replicating DNA, which is often more sensitive than nonreplicating DNA.
- The combined incubation of the test article and the cells in the overlay permits constant exposure of the indicator cells for at least 48 hr.

*Assay Acceptance Criteria*. Before assay data can be considered valid for evaluation, the following criteria must be met:

- The *Salmonella* tester strains must be shown to be sensitive to crystal violet, which serves as evidence for the presence of the *rfa* cell wall mutation. Strains TA98 and TA100 must be shown to be sensitive to ampicillin, as evidence of the presence of the pKM101 plasmid.
- The spontaneous reversion rate for each strain must be within the characteristic range for the testing laboratory. These ranges are typically as follows:

| Strain | Average Revertants/Plate |
|---|---|
| TA98 | 8–60 |
| TA100 | 60–240 |
| TA1535 | 4–45 |
| TA1537 | 2–25 |
| WP2*uvrA* | 5–40 |

- The bacteria suspensions used for inoculation of the treatment tubes must be at least $0.5 \times 10^9$ bacterial/mL.
- The positive control treatments must induce at least a threefold increase over the vehicle control plates for each strain for test condition (nonactivation or with S9 mix). If a strain or test condition does not meet this criterion, a repeat assay must be performed for that strain and test condition.

*Assay Evaluation Criteria*. Several methods for statistical analysis of the Ames test are suggested in the EPA Gene-Tox work group report for the *Salmonella typhimurium* mutation assay [101]. Because the procedures used to evaluate the mutagenicity of the test article are semiquantitative, the criteria used to determine positive effects are inherently subjective and based primarily on historical databases. Most datasets are evaluated using the following considerations and criteria:

- *Surviving populations*—Plate test procedures do not permit exact quantitation of the number of cells surviving chemical treatment. At low concentrations of the test article, the surviving populations on the treated plates are essentially the same as on the vehicle negative control plates. At high concentrations, the surviving population is usually reduced by some unknown fraction. The plate incorporation protocol normally employs a 2- or 3-log range of doses; the highest of these doses is selected to show toxicity as determined by subjective criteria such as background clearing or a reduction in the number of spontaneous colonies on treated plates compared with the vehicle control.
- *Dose–response phenomena*—The demonstration of a dose-related increase in revertant (mutant) counts is an important criterion for establishing mutagenicity. A factor that might modify dose–response results for a mutagen is the selection of doses that are too low (usually mutagenicity and toxicity are related). If the highest dose is far lower than a toxic concentration, no increase may be observed over the dose range selected. Conversely, if the lowest dose employed is highly cytotoxic, the test article may kill any mutants that are induced and, thus, does not appear to be mutagenic. Occasionally, high levels of toxicity produce microcolonies (which do not represent genetic reversions), which can be confused for revertants by inexperienced investigators.
- *Strains TA-1535 and TA-1537*—A test article that produces a positive dose response over three concentrations, with the highest increase in revertants equal to three times the vehicle control, is considered mutagenic.
- *Strains TA-98, TA-100, and WP2uvrA*—A test article that produces a positive dose response over three concentrations, with the highest increase in revertants equal to twice the vehicle control, is considered mutagenic. Occasionally, a doubling is not necessary for TA-100 if a clear dose-related pattern is observed over several concentrations.
- *Pattern*—Because TA-1535 and TA-100 are derived from the same G-46 parental strain, to some extent the microbial assay has a built-in redundancy. In general, these two strains should respond to the same mutagen, and such a pattern is sought. Also, if a strain responds to a mutagen under nonactivation conditions, it generally also responds in the presence of S9 mix.
- *Replication*—OECD Guideline 471 recommends that equivocal results be clarified by further testing, preferably with some modification of the treatment conditions. Negative results should also be confirmed in an independent trial or justification given for accepting the results of a single negative trial.

The preceding criteria are not absolute, and other extenuating factors may enter into a final evaluation decision; however, these criteria can be applied to most situations and are presented to aid individuals not familiar with this procedure. It must be emphasized that modifications of the procedure involving preincubation conditions or source of S9 (animal/tissue) are necessary for evaluation of specific chemicals or classes of chemicals.

## Protocol 2: Forward Mutation at the TK Locus in L5178Y $tk^{+/-}$ Mouse Lymphoma Cells

The objective of the mouse lymphoma assay is to evaluate a test article for its ability to induce forward mutation in the L5178Y $tk^{+/-}$ mouse lymphoma cell line, as assessed by colony growth in the presence of a selective agent, 5-trifluorothymidine (TFT). The assay design is based on OECD Guideline 476 (1997 revision) and the ICH S2B guidance document. Thymidine kinase (TK) is a cellular enzyme that allows cells to salvage thymidine from the surrounding medium for use in DNA synthesis. If a thymidine analog such as TFT is included in the growth medium, the analog will be phosphorylated via the TK pathway and will cause cell death by inhibiting the *de novo* synthesis of thymidine monophosphate, thereby inhibiting DNA synthesis and resulting in cellular death. Cells that are heterozygous at the TK locus ($tk^{+/-}$) may undergo a single-step forward mutation to the $tk^{-/-}$ genotype, in which little or no TK activity remains. Such mutants are as viable as the heterozygotes in normal media because DNA synthesis proceeds by a *de novo* synthetic pathway that does not involve thymidine as an intermediate. The basis for selection of the $tk^{-/-}$ mutants is the lack of any ability to utilize toxic analogs of thymidine, which enable only $tk^{-/-}$ mutants to grow in the presence of TFT. Cells that grow to form colonies in the presence of TFT are therefore assumed to have mutated to the $tk^{-/-}$ genotype either spontaneously or by the action of a test article.

### Materials

*Indicator Cells*. The cell line used in this assay—L5178Y $tk^{+/-}$, clone 3.7.2 C—was derived from the L5178Y mouse lymphoma cell line by Donald Clive. Stocks are maintained in liquid nitrogen and laboratory cultures are periodically checked for the absence of *Mycoplasma* contamination and karyotype stability. To keep the negative control frequency (spontaneous frequency) of $TK^-$ mutants at an acceptably low level, cell cultures are grown in medium containing methotrexate to inhibit *de novo* synthesis of thymidylate and select against the $TK^-$ phenotype, then returned to normal growth medium for 3 or more days before use.

*Media*. The cells are maintained as suspension cultures in RPMI 1640 or Fischer's medium for leukemia cells of mice supplemented with L-glutamine, sodium pyruvate, Pluronic F68, and 10% horse serum. Fischer's medium, with the horse serum reduced to 5%, is used for the test article treatment periods. Cloning medium consists of RPMI 1640 growth medium (minus Pluronic F68) containing 20% horse serum and the addition of agar to achieve a semisolid state. Selection medium is cloning medium containing 3 µg/mL of TFT [37,38].

*S9 Metabolic Activation System*. S9 homogenate prepared from the livers of Aroclor 1254-induced Sprague–Dawley male rats is obtained commercially. The *in vitro* activation system is obtained by mixing the S9 with NADP and isocitrate just before addition to the cell cultures. The final concentrations in the cultures are 3 m*M* NADP (sodium salt), 15 m*M* isocitrate, and typically about 15 µL/mL of S9 (determined from a preliminary titration experiment for 3-methylcholanthrene mutagenicity).

*Control Articles*. A negative control consisting of assay procedures performed on cells exposed to vehicle in the medium is assayed to determine any effects on survival or mutation caused by the vehicle used for the test article. For test articles assayed with S9 activation, the vehicle negative control articles include the activation mixture. Reference substances for use as positive control articles establish historical databases for the laboratory performing the assay. The positive control articles listed below are chosen because they are well established mutagens and are known to induce both small and large mutant colonies [130,131]. Methyl methanesulfonate (MMS) is generally used as a positive control for nonactivation assays at concentrations of 1? µg/mL for 4-hr treatments and 6.5 µg/mL for 24-hr treatments. This mutagen induces a preponderance of small TK mutant colonies. 3-Methylcholanthrene requires metabolic activation by microsomal enzymes to become mutagenic and is generally used at 2 to 4 µg/mL as a positive control article for mutation assays performed with S9 activation.

### Experimental Design

*Dose Selection*. The solubility of the test article in water or an organic vehicle (such as DMSO or ethanol) is determined first, then a wide range of test article concentrations are tested for cytotoxicity, starting with a maximum applied concentration of 5 mg/mL or 10 m*M* (whichever is lower), solubility or homogeneity of suspension permitting, and using twofold dilution steps. After an exposure time of 4 hr with and without S9 activation (and 24 hr without S9 for ICH S2B compliance), the cells are washed and resuspended in culture medium. A viable cell count is obtained 24 hr after treatment initiation (and in another 24 hr for the 24-hr treatments). Relative cytotoxicities expressed as the reduction in cell number/mL compared to the negative control cells are used to select six or more doses that cover the range from 0 to about 80% reduction in growth. These selected doses subsequently are applied to cell cultures prepared for mutagenicity testing, but fewer dose groups (at least four) may be carried through the mutant selection process. This procedure compensates for trial-to-trial variations in cellular cytotoxicity and helps to ensure the choice of at least four doses appropriately stepped over a cytotoxicity range corresponding to high cytotoxicity (or maximum concentration) to lower doses having little or no apparent effect on cell growth.

*Mutagenicity Testing*. The procedure used is based on published methods [1,37]. Single or duplicate cultures are exposed to test article for 4 or 24 hr at the preselected doses. The test article is washed away, and the cells are resuspended in growth medium for 2 days after the treatment period to allow recovery, growth, and expression of the induced $TK^-$ phenotype. Cell counts are determined daily and appropriate dilutions made to allow optimal growth rates. At the end of the 2-day expression period, 3 × $10^6$ cells from each culture selected for analysis are seeded into soft agar plates with selection medium, and the resistant (mutant) colonies are counted after 12 to 13 days of incubation. To determine the number of seeded cells capable of forming colonies, 600 cells from each culture selected for analysis are also cloned in soft agar plates in normal, nonselective medium. The mutant frequency is calculated from the ratio of resistant colonies o colonies in the nonselective plates, adjusted for the number of cells seeded under both conditions. The S9 activation and nonactivation assays are independent and can be run separately or concurrently. The only difference s the addition of the S9 activation system during the 4-ır treatment period and the use of different positive control articles.

The mutant colonies in this assay develop into a bimodal distribution of sizes. The large mutant colonies are considered to have few genetic changes beyond a small *tk* mutation, whereas the small mutant colonies may generally represent more extensive genetic alterations. An ncreased proportion of small mutant colonies correlates with clastogenic activity by the test article [131]. To meet he ICH S2B guideline, the mutant colony size distribution must be determined and the percent small and large mutant olonies determined. The colony counts and size analysis re accomplished simultaneously with a high-resolution ounting system obtained from Loats Associates, Inc. Westminster, MD).

The mouse lymphoma assay may also be performed y seeding the cells for selection and colony-forming abil-y under nonselective conditions in 96-well plates, rather ıan cloning in soft agar. In this method, an average of .6 cells is seeded per well for cloning efficiency deterıination, and 2000 cells/well are seeded for mutant selecon. The mutant frequency is calculated from the number f empty wells by assuming a Poisson distribution. The mall and large mutant frequencies are similarly calcuıted by counting wells with no small mutant colonies or o large mutant colonies, respectively. This method genrally yields higher mutant frequencies in both control and eated cultures than the soft agar method because small ıutant colonies often do not grow well in soft agar and ıay be too small to be detected by automated counters. he soft agar and microwell methods are equally acceptɔle to regulatory agencies.

An assay is usually performed with an initial trial using 4-hr treatments with and without S9 activation. If the results are negative or equivocal, a repeat trial is performed using a 24-hr treatment period under nonactivation conditions and a different range or stepping of doses with S9 activation. Alternatively, an assay may be performed using all three treatment conditions in one trial, and a second trial performed for any treatment condition that yielded equivocal results in the first trial.

*Assay Acceptance Criteria*

An assay is considered acceptable for evaluation of test results only if all of the criteria in the following list are satisfied. The S9 activation and nonactivation portions of assays are evaluated independently, and each portion is repeated, as necessary, to satisfy the general acceptance and evaluation criteria:

- The average absolute cloning efficiency of the vehicle controls (from two or more cultures) must be in the range of 65 to 120%. This range was established by the International Workshop on Genotoxicity Tests (IWGT) mouse lymphoma cell mutation working group as a reasonable acceptance criterion for experienced laboratories to achieve in performing either the soft agar or microwell versions of the assay [133].
- The average fold increase in vehicle control cell number (suspension growth) should be in the range of 8 to 32 over the 2-day expression period. This criterion was established by the IWGT working group as providing evidence that sufficient control of cell growth conditions was established to ensure that mutagenesis by test chemicals would be adequately measured [133].
- The average background mutant frequency of the vehicle control cultures is calculated separately for concurrent activation and nonactivation treatments, even though the same population of cells is used for both portions of a trial. For both conditions, the acceptable ranges of background frequencies were established by IWGT as 35 to 140 × $10^{-6}$ for soft agar assays and 50 to 170 × $10^{-6}$ for microwell assays [132]. Data collected from experienced laboratories were used to establish these ranges to derive a global method of evaluating test articles for mutagenic activity. Assays with backgrounds outside this range are not necessarily invalid but should not be used as evidence for weak or no mutagenic activity by a test chemical.
- A positive control is included with each assay to provide confidence in the procedures used to detect mutagenic activity. The induced mutant

frequencies must be consistent with each laboratory's positive control historical database and show significant induction of small mutant colonies. An assay is considered acceptable in the absence of a positive control (loss due to contamination or technical error) only if the test article clearly shows mutagenic activity as described in the evaluation criteria.

- An assay must include applied concentrations that (1) reach 5 mg/mL or 10 m*M* (whichever is lower) for test articles causing little or no cytotoxicity, (2) reduce the relative total growth (RTG) to approximately 20%, or (3) reach a concentration that exceeds the solubility limit. RTG is the measure of cytotoxicity in the mouse lymphoma assay and is obtained by multiplying the relative suspension growth by the relative cloning efficiency.

### *Assay Evaluation Criteria*

A common condition considered necessary to demonstrate mutagenesis for any given treatment is a mutant frequency that is at least two times the concurrent background mutant frequency, defined as the average mutant frequency of the vehicle control cultures; however, the induction of mutations is an additive process relative to the existing background of mutants and not a multiple of the background mutant frequency. In recognition of this, and as an attempt to derive a global method for assay evaluation, the IWGT working group established a new criterion based on multilaboratory mutant frequency distributions for acceptable vehicle controls obtained for both the agar and microwell versions of the assay [132]. A global evaluation factor (GEF) was established as the mean of each distribution plus one standard deviation. The GEF values are 90 for agar assays and 126 for microwell assays. Thus, a mutant frequency in a culture exposed to a test article is considered significantly elevated only if the induced mutant frequency (mutant frequency minus the concurrent background mutant frequency) meets or exceeds the GEF. In addition, a statistical trend test should be applied to determine if the induced mutant frequency response is dose related.

A test article is evaluated as mutagenic if both the induced mutant frequency for any treatment meets or exceeds the GEF and a positive trend test is obtained. A test article is evaluated as nonmutagenic if the GEF is not achieved and the trend analysis is negative. If only one of the criteria is met, additional trials may be necessary to resolve the evaluation. If a repeat trial does not confirm an earlier GEF or trend response, the test article is evaluated as nonmutagenic in this assay system.

Finally, it should be noted that for some test articles the correlation between toxicity and applied concentration is poor. The proportion of the applied article that effectively interacts with the cells to cause genetic alterations is not always repeatable or under control. Conversely, measurable changes in the frequency of induced mutants may occur with concentration changes that cause only small changes in observable toxicity. Therefore, either parameter—applied concentration or cytotoxicity (RTG)—can be used to establish whether or not the observed mutagenic activity is dose related.

## Tests Detecting Chromosome-Breaking (Clastogenic) Agents *In Vitro*

### Protocol 3: Chromosomal Aberrations in Chinese Hamster Ovary Cells

The objective of the Chinese hamster ovary (CHO) cel *in vitro* assay is to evaluate the ability of a test article tc induce structural chromosome aberrations in mammaliar cells. Structural aberrations are a consequence of failur or mistakes in repair processes such that breaks either dc not rejoin or rejoin in abnormal configurations [56] Numerical aberrations (a change in the modal number o chromosomes) are not determined by this protocol. Th protocol design parallels the human peripheral blood lym phocyte chromosomal assay (Protocol 4) and meet OECD Guideline 473 (1997 revision).

#### *Materials*

*Indicator Cells*. Cells to be used in this assay can b obtained from the American Type Culture Collectio (Manassas, VA; Repository No. CCL61). The original cell were obtained from an ovarian biopsy of a Chinese ham ster, but the cells evolved to a permanent, fibroblastic cel line with an average cycle time of approximately 12 h The CHO–WBL subclone has a modal chromosomal num ber of 21 and a low frequency of spontaneous aberration

*Medium*. The cells are grown in McCoy's 5a mediur supplemented with 10% heat-inactivated fetal bovin serum, 2-m*M* L-glutamine, and antibiotics. The cells ar maintained as monolayer cultures in a humidified 37° incubator with 5% $CO_2$.

*Control Articles*. Vehicle control cultures are prepared t contain the vehicle at the highest concentration as used fc the test-article-treated cultures. Known clastogens are use for concurrent positive control articles. Mitomycin C doe not require metabolic activation and is used at concentra tions in the range of 0.02 to 5 μg/mL for nonactivatio assays. Cyclophosphamide does require metabolic activa tion to become clastogenic and is used at concentration of 1 to 25 μg/mL for S9 metabolic activation assays.

*S9 Metabolic Activation System*. An *in vitro* metaboli activation system [118] consists of a rat liver postmitc chondrial fraction (S9) and an energy-producing reactic (NADP plus isocitrate). S9 from the livers of Aroclor 125 induced Sprague–Dawley male rats can be purchased con mercially. Just prior to use, the S9 is thawed and mixed with NADP (sodium salt) and isocitrate to obtain the following concentrations after addition to cell cultures: 1.8 m*M* NADP, 2.7 m*M* isocitrate, and 15 μL/mL S9.

*Experimental Design*

*Solubility and Cytotoxicity Determination*. Prior to setting up the assay, the solubility of the test article is determined in appropriate vehicles and after dosing into culture medium. The vehicles of choice are water or culture medium, followed by DMSO, ethanol, or special vehicles such as emulsifiers in water. DMSO or ethanol are diluted 1:100 into cultures during dosing to avoid vehicle cytotoxicity. The vehicle selected for an assay is one that maintains solubility or an evenly dispersed suspension. For most test articles, dose range finding is conveniently conducted as an integral part of the chromosomal aberrations assays. Approximately 15 doses are applied to duplicate cultures, using dilution steps of 70% of the preceding dose and starting at a high concentration of 5 mg/mL or 10 m*M* (whichever is lower) for soluble test articles or a high concentration of about twice the solubility limit in culture medium. Thus, a series of treated cultures over a wide concentration range covering 4 orders of magnitude for each treatment condition is available from which cultures with the appropriate range of test-article-induced cytotoxicity can be selected to analyze for chromosomal aberrations.

*Chromosomal Aberrations Assay*. The initial assay is performed with 3- or 4-hr treatments with the test article absent and present in the S9 metabolic activation system. The series of cultures is seeded at about $0.9 \times 10^6$ cells per 75 cm$^2$ culture flask. Duplicate cultures are treated the next day with the doses of test article, vehicle control, and positive control chemicals. The activation cultures also receive the S9 activation mixture. After the treatment period, the cultures are washed with saline and reincubated in fresh medium until 18 hr after the initiation of the treatments, at which point 0.1 μg/mL Colcemid is added to arrest cell division. Incubation of the cultures is continued for another 2 hr. Before harvesting the cells, the cultures are examined by microscope to estimate the cytotoxicity. Observations of the degree of confluency and apparent availability of mitotic cells (large, rounded cells on the monolayer surface or floating in the medium) are recorded. Cells are then harvested by trypsinizing the cultures treated with the highest concentration of test article expected to yield any metaphases and cultures treated with at least four lower concentrations. The harvested cells are swollen in 75-m*M* KCl hypotonic solution, fixed in methanol:glacial acetic acid (3:1), dropped onto glass slides, and air dried. After drying, the slides are stained with 5% Giemsa, dried, and mounted permanently. At least 1000 cells, if available, are analyzed for mitotic index (MI; percentage of cells in mitosis) in all of the control cultures and the selected range of test-article-treated cultures. The dose groups selected for chromosomal analysis are those that have a sufficient number of metaphases and suppression of the MI in a range of approximately 0% to 50%. Treatments causing greater than 60% MI reduction are considered excessively cytotoxic and are not analyzed.

*Scoring for Aberrations*. Cells are selected for scoring on the basis of good chromosome morphology, and only cells with the number of centromeres equal to the modal number for the CHO cell line are analyzed (except when chromosomal damage is so extensive that an exact count is not possible). Normally, 200 metaphases will be analyzed per dose group (100 from each of the duplicate cultures). At least 25 metaphases are analyzed from the cultures having greater than 25% of cells with one or more aberrations. Standard forms are used to record breaks, fragments, and reunion figures, as well as percent cells with gaps and polyploidy or endoreduplication. Gaps, polyploidy, and endoreduplication are reported but are not included in the evaluation of percent cells containing aberrations. All slides are coded prior to scoring to control bias. The complete list of chromosomal alterations that are recorded is as follows:

- *Simple breaks*—Chromatid and chromosome breaks, acentric fragments
- *Complex exchanges*—Interstitial deletions, triradials, triradial fragments, quadriradials, chromosome intrachanges, sister unions, distal nonunions, proximal nonunions, complex rearrangements, dicentrics, dicentric fragments, rings, ring fragments, abnormal monocentric chromosomes, and multiple aberrations (cells with more than four aberrations)
- *Gaps*—Chromatid and chromosome gaps
- Polyploid cells
- Endoreduplicated cells

*Replication*. If the initial assays with 3- to 4-hr treatments are negative or equivocal, a second assay trial is usually performed. The treatment period for the nonactivation treatments is increased to about 20 hr, and the doses for the 3- to 4-hr S9 activation assay may be altered to refine the cytotoxicity range in which the analysis will be conducted to confirm the negative results.

*Assay acceptance criteria*. The nonactivation and activation portions of assays are independent units, and each portion is repeated as necessary to achieve the following assay acceptance criteria:

- The vehicle negative control cultures must contain less than about 5% cells with aberrations. The positive control cultures must contain a statistically significant ($p \leq 0.01$) increase in percent cells with aberrations relative to the vehicle control cultures.

- To accept a negative evaluation for a test article, the chromosomal assays must include the highest applicable concentration. This concentration is one that exceeds the solubility limit in culture medium or achieves a reduction in the MI of approximately 50% or reaches the limit concentration of 5 mg/mL or 10 mm (whichever is lower).
- For each of the three treatment conditions, cultures exposed to at least three concentrations of test article should be analyzed for aberrations.

*Assay Evaluation Criteria*. A test article is considered clastogenic in this assay if a Fisher's exact test [175] for pairwise comparisons of percent cells containing one or more aberrations in treated cultures compared to the vehicle control cultures is significant ($p \leq 0.01$) for one or more dose groups. In addition, a Cochran–Armitage test for linear trend of a dose-related response should also be obtained. Test articles that do not induce statistically significant increases under all three treatment conditions used in the assay are considered negative for clastogenic activity.

A number of general guidelines have been established as an aid to determining the meaning of CHO chromosome aberrations. It should be borne in mind that the detected chromosomal changes are of a type that can survive more than one mitotic cycle of the cell and that these aberrations all result from breaks in the chromatin that either fail to repair or repair in atypical combinations. It is anticipated that many of the cells bearing breaks or reunion figures would be eliminated (i.e., fail to divide again) after their first mitotic division, and, as a corollary, those cells that survive the first division would primarily bear balanced lesions. The detection of these lesions, and hence a complete risk evaluation, must usually rely on additional testing. In general, a cell bearing configurations such as small deletions or reciprocal translocations may be perpetuated and therefore constitutes a greater risk to an individual than one with large deletions or complex rearrangements.

Gaps are not counted as significant aberrations unless they are present at a much higher than usual frequency. Open breaks are considered indicators of genetic damage, as are configurations resulting from the repair of breaks. The latter includes, for example, translocations, multiradials, rings, and multicenters. Reunion figures such as these, which may lead to stable chromosome configurations, are weighed as more significant in safety analysis than simple breaks. The number of aberrations per cell is also considered significant. Cells with more than one aberration indicate more (undetected) genetic damage than those containing evidence of single events.

Comparison of a treated group response with a concurrent vehicle negative control having an unusually low frequency of aberrations can cause undue statistical significance; therefore, treatment data are considered against historical control data. The type of aberration, its frequency, and its correlation to dose trends within a given exposure period are all considered when evaluating a test article as being positive or negative in this assay.

## Protocol 4: Chromosomal Aberrations in Human Peripheral Blood Lymphocytes

The objective of the human peripheral blood lymphocyte *in vitro* assay is to evaluate the ability of a test article to induce structural chromosome aberrations in normal human cells. Structural aberrations are a consequence of failure or mistakes in repair processes such that breaks either do not rejoin or rejoin in abnormal configurations [56]. Numerical aberrations (a change in the modal number of chromosomes) are not determined by this protocol. The protocol design parallels the CHO cell chromosomal assay (Protocol 3) and meets OECD Guideline 473 (1997 revision).

### Materials

*Indicator Cells*. Human venous blood is collected in heparinized "vacutainers" from one or more healthy donors and used to establish whole blood cultures in 15-mL centrifuge tubes. Usually 0.6 mL of blood is diluted with sufficient culture medium to obtain 10 mL cultures during treatment with the test article.

*Medium*. The medium for the blood cultures is RPMI 1640 with 25-m*M* HEPES buffer supplemented with 20% heat-inactivated fetal bovine serum, 2-mM L-glutamine, 2% phytohemaglutinin M, and antibiotics. The cells are maintained as suspension cultures in a humidified 37°C incubator with 5% $CO_2$.

*Control Articles*. Vehicle control cultures are prepared to contain the vehicle at the highest concentration as used for the test-article-treated cultures. Known clastogens are used for concurrent positive control articles. Mitomycin C does not require metabolic activation and is used at concentrations in the range of 0.025 to 3.0 μg/mL for nonactivation assays. Cyclophosphamide does require metabolic activation to become clastogenic and is used at concentrations of 0 to 300 μg/mL for S9 metabolic activation assays.

*S9 Metabolic Activation System*. See Protocol 3.

### Experimental Design

*Solubility and Cytotoxicity Determination*. See Protocol 3.

*Chromosomal Aberrations Assay*. The initial assay is performed with 3- or 4-hr treatments with the test article absent and present in the S9 metabolic activation system. Two days after initiation of the blood cultures, duplicate cultures are treated with the selected range of test article concentrations, vehicle control, and positive control chemicals. The activation cultures also receive the S9 activation mixture. After the treatment period, the cultures are washed with saline and reincubated in fresh medium until 20 hr after the initiation of the treatments, at which point 0.1 µg/mL Colcemid is added to arrest cell division. Incubation of the cultures is continued for another 2 hr. The assay is terminated by centrifuging the cultures and resuspending the cells in 75-m*M* KCl hypotonic solution, which causes the lymphocytes to swell and the red blood cells to lyse. The lymphocytes are fixed in methanol:glacial acetic acid (3:1), dropped onto glass slides, and air dried. After drying, the slides are stained with 5% Giemsa, dried, and mounted permanently. At least 1000 cells, if available, are analyzed for mitotic index (MI; percentage of cells in mitosis) in all of the control cultures and the selected range of test-article-treated cultures. The dose groups selected for chromosomal analysis are those that have a sufficient number of metaphases and suppression of the MI in a range of approximately 0 to 50%. Treatments causing greater than 60% MI reduction are considered excessively cytotoxic and are not analyzed.

*Scoring for Aberrations*. Cells are selected for scoring on the basis of good chromosome morphology, and only cells with the number of centromeres equal to the modal number of 46 ± 2 for human cells are analyzed (except when chromosomal damage is so extensive that an exact count is not possible). Normally, 200 metaphases will be analyzed per dose group (100 from each of the duplicate cultures). At least 25 metaphases are analyzed from the cultures having greater than 25% of cells with one or more aberrations. As for Protocol 3, standard forms are used to record breaks, fragments, and reunion figures, as well as percent cells with gaps and polyploidy or endoreduplication. Gaps, polyploidy, and endoreduplication are reported but are not included in the evaluation of percent cells containing aberrations. All slides are coded prior to scoring to control bias. The complete list of chromosomal alterations that are recorded is the same as given in Protocol 3.

*Replication*. If the initial assays with 3- to 4-hr treatments are negative or equivocal, a second assay trial is usually performed. The treatment period for the nonactivation treatments is increased to about 22 hr, and the doses for the 3- to 4-hr S9 activation assay may be altered to refine the cytotoxicity range in which the analysis will be conducted to confirm the negative results.

*Assay Acceptance Criteria*. Same as given in Protocol 3.

*Assay Evaluation Criteria*. Same as given in Protocol 3. The use of human lymphocytes for *in vitro* chromosomal aberrations studies has increased in recent years with the realization that CHO cells often indicate genomic damage not detected in other genotoxicity assays or in human lymphocytes. This may be a consequence of chromosome fragility in mammalian cells lines as opposed to the normal chromatin structure in primary cells. DNA repair capacity is also lost in mammalian cell lines; therefore, human lymphocytes may eliminate some responses obtained with cell lines that are irrelevant to normal human cells. It is not uncommon to obtain low-level clastogenic activity in CHO cells, particularly for very cytotoxic treatments, that are not induced by the test article in lymphocyte cultures. Caution is in order, however, when using lymphocytes in the testing of nucleoside analogs because intracellular nucleoside pools in lymphocytes are much lower than for other cell types and imbalances in this pool can cause abnormalities in chromosomal structure [102].

### Protocol 5: Rodent Bone Marrow Micronucleus Assay

This objective of this assay is to detect *in vivo* clastogenic activity by detecting the formation of micronuclei in polychromatic erythrocyte cells in rodent bone marrow after dosing the animals with a test article. The assay design is based on OECD Guideline 474 (1997 revision) and the ICH S2A Guideline. The micronucleus test can serve as a rapid screen for clastogenic agents and test articles that interfere with spindle fiber formation or function in normal mitotic cell division [79,80,160]. Micronuclei are small chromatin bodies consisting of chromosome fragments or even an entire chromosome, which lag behind at mitotic anaphase. At telophase, these fragments and chromosomes are not segregated to either daughter nucleus and form single or multiple micronuclei in the cytoplasm. During maturation of hematopoietic cells from erythroblasts to immature erythrocytes, the nucleus is extruded, but the micronuclei, if present, persist in the cytoplasm of these enucleated cells. The micronuclei in the enucleated immature blood cells (polychromatic erythrocytes [PCEs]) are detected and counted in this protocol. The detection of micronuclei in enucleated cells eliminates the time required to search for metaphase spreads in treated cell populations.

#### Materials

*Indicator animals*. Adult male and/or female mice or rats are normally used for this study. If mice are used, the CD-1® (ICR) BR outbred strain is often used to maximize genetic heterogeneity, which tends to minimize strain-specific responses. Similarly, if rats are used, the Sprague–Dawley CD® (SD)IGS BR strain is often used. The animals are usually 8 to 10 weeks old at the time of dosing.

*Control Articles*. Vehicle control animals are dosed concurrently with the same vehicle and by the same route as used for the test-article-treated animals. The administered volume of vehicle is equal to the maximum dosing volume of the test article preparation. The known clastogen, cyclophosphamide, is used as a concurrent positive control article. Cyclophosphamide requires metabolic activation to become clastogenic. Water solutions (about 8 mg/mL) are dosed by a single oral gavage at about 80 mg/kg.

### Experimental Design

*Animal Husbandry*. Animals are isolated by sex. Mice are group housed at five or less per polycarbonate cage, and rats are housed at one or two per stainless steel hanging wire cage. Environmental lighting, temperature, and humidity are controlled and monitored. A commercial diet and tap water are provided *ad libitum* unless contraindicated by the particular experimental design. Sanitary cages and bedding are used. Personnel handling animals or working within the animal facilities are required to wear suitable protective garments. Animals are randomly assigned to dose groups and are uniquely identified by tag. Prior to study initiation, animals are weighed to calculate dose levels. Mice are selected in the 20- to 40-g range, and rats are selected in the weight range of 150 to 400 g. Also, the weight variation within each dose group is controlled to ±20% of the group mean weight for each sex.

*Dose Selection*. Prior to any treatment of animals, the solubility characteristics of the test article are determined from any available information and additional solubility testing. The vehicles generally used in this protocol are deionized water, normal saline, corn oil, or 0.5% carboxymethyl cellulose solution. The vehicle providing the best solubility or homogeneity of dispersion is selected. If reliable acute toxicity information (e.g., $LD_{50}$) is available for the same animal strains, sex, and route of administration, appropriate dose levels can be selected. Usually this information is insufficient, and an up-and-down approach to a dose range-finding study is performed with both sexes and the route of administration and dosing regimen selected for the micronucleus assay. The limit dose for test articles having little or no expected toxicity is 2 g/kg. To use the minimum number of animals in locating an appropriate high dose, one to three animals of each sex are exposed to the limit dose or a dose expected to cause toxic signs. Depending on the toxic signs obtained, other groups of one to three animals of each will be dosed with lower or higher doses, as appropriate. This procedure is continued until a dose causing toxic signs without lethality can be estimated. If both sexes exhibit essentially the same toxicity profile, only male animals are used in the micronucleus assay. When significant differences in toxicity are found between the sexes, both sexes are included in micronucleus assay, as recommended by the ICH S2A guideline. Only female animals are used if a test article is intended specifically for use in human females. For toxic test articles, the high dose selected is based on some evidence of toxicity (e.g., clinical signs, death, depression in bone marrow cell maturation). Two lower doses at one half and one fourth of the high dose are also included for the micronucleus assay.

*Dosing Schedule and Route of Administration*. The treatment regimen is most often a single administration of the test article. Sometimes more than one administration is given several hours apart on the same day, if there are issues with excessive dosing volumes or a short half-life *in vivo*. Another common procedure is to administer a test article in two or more daily doses given 24 hr apart. The most common method of administration is by oral gavage; however, other acceptable routes include intraperitoneal, intravenous, subcutaneous, or intramuscular injections. If the test article has sufficient aqueous solubility, intravenous injections have the advantage of ensuring bone marrow exposure and eliminating the need to demonstrate this exposure by an analytical method.

*Micronucleus Assay*. A minimum of five animals of one or both sexes, as appropriate, is used for each of the three dose groups and the concurrent vehicle and positive control groups. The bone marrow from these animals is extracted 24 hr after treatment (or the last treatment afte multiple dosing). If a single administration of test article is used, two additional groups of at least five animals of each appropriate sex are included for bone marrow extrac tion 48 hr after treatment. These two groups are the high dose animals and a vehicle control. If some lethality may occur in the high dose group, a secondary group of animal exposed to the high dose may be included and substitute as needed for animals lost in the high-dose group for eithe bone marrow sampling time.

*Extraction of Bone Marrow*. Animals are euthanized by CO inhalation followed by incision of the diaphragm. The hin limbs (tibias or femurs) are removed. The marrow is flushe from the bone and transferred to centrifuge tubes containin 3 to 5 mL fetal bovine serum (one tube per animal).

*Preparation of Slides*. After centrifugation to pellet th marrow, the supernatant fluid is aspirated and portions c the pellet spread on slides, air dried, and fixed in methano Mouse marrow slides are then stained in May–Grünwal solution and Giemsa and analyzed by microscope. Ra marrow slides are stained in acridine orange and analyze by fluorescence microscopy. The use of acridine orang is necessary for rat marrow analysis because heparin par ticles with morphologies essentially identical to micronu clei are released from rat blood mast cells during slid processing. The micronuclei are differentiated by yello fluorescence as opposed to red fluorescence from the hep arin particles.

*Slide Analysis*. All slides are coded prior to analysis t control bias. At least 2000 PCEs per animal are score when possible. The frequency of micronucleated cel is expressed as the percent micronucleated cells based on the total PCEs present in the scored optical fields. The frequency of PCEs vs. mature erythrocytes (normochromatic erythrocytes [NCEs]) is also determined by scoring the number of PCEs and NCEs observed in the optical fields while scoring at least the first 500 erythrocytes on each slide. The unit of scoring is the micronucleated cell, not the micronucleus, so the occasional cell with more than one micronucleus is counted as one micronucleated PCE. The staining procedures permit the differentiation by color of the polychromatic and normochromatic erythrocytes.

*Assay Acceptance Criteria*. The assay is repeated as necessary to achieve the following assay acceptance criteria:

- The percent micronucleated PCEs in the vehicle control group must lie within the historical range for the laboratory. For both the mouse and rat strains given in this protocol, this range should be less than 0.4% [158].
- There must be a statistically significant increase in the percent micronucleated PCEs relative to the concurrent vehicle control group, and the positive control response should be consistent with the historical data for the laboratory.
- The high dose should reach the limit dose of 2 g/kg for relatively nontoxic test articles or produce some indication of toxicity, such as toxic signs or mortality or a reduction in the ratio of PCEs to NCEs. A reduction in the PCE:NCE ratio, when it occurs in a dose-related manner, is an indication of cytotoxic action by the test article in the bone marrow.

*ssay Evaluation Criteria*. The individual animal is onsidered the unit of variation. An analysis of variance n the proportions of PCEs with micronuclei per animal s performed when the variances are homogeneous. anked proportions are used for heterogeneous varinces. If the analysis of variance is significant ($p \leq$ .05), a Dunnett's t-test is used to determine which dose roups have significantly elevated micronucleated PCEs elative to the vehicle control group. This analysis is erformed separately for each sex (if both sexes are ncluded in the study) and each bone marrow sampling me. Additionally, a statistical test for trend (e.g., ochran–Armitage) may be employed to identify any ose-related response. A test article is evaluated as posive in this assay if there is a statistically significant esponse for at least one dose group or a statistically gnificant dose-related response. Statistical anomalies an occur, however, so the biological relevance of the esults must be considered or a repeat study performed evaluate equivocal results.

## QUESTIONS

1. Would transcription-coupled DNA repair (TCR) be more active in an exon or an intron?
2. Which type of gene mutation would be corrected by mismatch repair: a base-pair substitution or a frameshift mutation?
3. Which part of a chromosome is believed to influence cell mortality: the centromere, the telomere, or the centriole?
4. Which short-term test shows the best overall concordance with animal carcinogenicity: SCEs, the Ames test, or the mouse micronucleus test?
5. The p53 gene is a member of which class of genes: an oncogene, a dominant gene, or a tumor suppressor gene?
6. A disruption in the nucleotide sequence of which of the following is likely to result in a mutation: a codon, a gene, or a chromosome?
7. Which of the following technologies was critical to the development of the transgenic mouse models for mutation: shuttle vector, PCR, or gel electrophoresis?
8. Chemically induced heritable mutations have not been documented in which of the following species: fruit fly, mouse, or human?

## REFERENCES

1. Amacher, D. E., Paillet, S. C., Turner, G. N., Ray, V. A., and Salsburg, D. S. (1980): Point mutations at the thymidine kinase locus in L5178Y mouse lymphoma cells. II. Test validation and interpretation. *Mutat. Res.*, 72: 447–474.
2. Ames, B. N. (1979): Identifying environmental chemicals causing mutations and cancer. *Science*, 204:587–593.
3. Ames, B. N. (1984): Dietary carcinogens and anti-carcinogens. *Clin. Toxicol*., A22:291–301.
4. Ames, B. N. (1989): Endogenous DNA damage as related to cancer and aging. *Mutat. Res*., 214:41–46.
5. Ames, B. N., Durston, W. E., Yamasaki, E., and Lee, F. D. (1973): Carcinogens are mutagens: a simple test system combining liver homogenates for activation and bacteria for detection. *Proc. Natl. Acad. Sci. USA*, 70:2281.
6. Ames, B. N., Durston, W. E., Yamasaki, E., and Lee, F. D. (1973): Carcinogens are mutagens: a simple test system combining liver homogenates for activation and bacteria for detection. *Proc. Natl. Acad. Sci. USA*, 70:2281.
7. Ames, B. N., Gold, L. S., and Willett, W. C. (1995): The causes and prevention of cancer. *Proc. Natl. Acad. Sci. USA*, 92:5258–5265.
8. Ames, B. N., McCann, J., and Yamasaki, E. (1975): Methods for detecting carcinogens and mutagens with the *Salmonella* mammalian microsome mutagenicity test. *Mutat. Res*., 31:347–364.
9. Anderson, D. and Plewa, M. J. (1998): The international comet assay workshop. *Mutagenesis*, 13:67–73.

10. Anderson, M. E., Meek, M. E., Boorman, G. A., Brusick, D. J., Cohen, S. M. et al. (2000): Lessons learned in applying the USEPA proposed cancer guidelines to specific compounds. *Toxicol. Sci.*, 53:159–172.
11. Anderson, M. W., Maronpot, R. R., and Reynolds, S. H. (1988): Role of oncogenes in chemical carcinogenesis: extrapolation from rodents to humans. In: *Methods for Detecting DNA Damaging Agents in Humans*, edited by H. Bartsch, K. Hemminki, and I. K. O'Neill, Publ. No. 89, pp. 477–485. International Agency for Research on Cancer, Lyon.
12. Ashby, J. and Purchase, I. F. H. (1988): Reflections on the declining ability of the *Salmonella* assay to detect rodent carcinogens as positive. *Mutat. Res.*, 205:51–58.
13. Ashby, J. and Richardson, C. R. (1985): Tabulation and assessment of 113 human surveillance cytogenetics studies conducted between 1965 and 1984. *Mutat. Res.*, 154: 111–133.
14. Au, W. W., Cajas-Salazar, N., and Salama, S. (1998): Factors contributing to discrepancies in population monitoring studies. *Mutat. Res.*, 400:467–478.
15. Baltimore, D. (2001): Our genome unveiled. *Nature*, 409:814–816.
16. Banks, R. E. et al. (1999): The potential use of laser capture microdissection to selectively obtain distinct populations of cells for proteomic analysis: preliminary findings. *Electrophoresis*, 20:689–700.
17. Barrios, R. A. and Boelsterli, U. (2003): *Mechanistic Toxicology: The Molecular Basis of How Chemicals Disrupt Biological Targets*. Taylor & Francis, Boca Raton, FL.
18. Baulig, A., Garlatti, M., Bonvallot, V., Marchand, A, Barouki, R. et al. (2003): Involvement of reactive oxygene species in the metabolic pathways triggered by diesel exhaust particles in human airway epithelial cells. *Am. J. Physiol Lung Cell. Mol. Physiol.*, 285:671–679.
19. Best, C. J. and Emmert-Buck, M. R. (2001): Molecular profiling of tissue samples using laser capture microdissection. *Expert Rev. Mol. Diagn.*, 1:53–60.
20. Bichsel, V. E., Liotta, L. A., and Petricoin III, E. F. (2001): Cancer proteomics: from biomarker discovery to signal pathway profiling. *Cancer J.*, 7(1):69–78.
21. Bird, A. (1992): The essentials of DNA methylation. *Cell*, 70:5–8.
22. Bollard, M. E., Stanley, E. G., Lindon, J. C., Nicholson, J. K., and Holmes, E. (2004): NMR-based metabonomic approaches for evaluating physiological influences on biofluid composition. *NMR Biomed.*, 18:143–162.
23. Bonner, R. F. et al. (1997): Laser capture microdissection: molecular analysis of tissue. *Science*, 278:1481–1483.
24. Bosio, A., Knorr, C., Janssen, U., Gebel, S., Haussmann H., and Muller, T. (2002): Kinetics of gene expression profiling in Swiss 3T3 cells exposed to aqueous extracts of cigarette smoke. *Carcinogenesis*, 23:741–748.
25. Bridges, B. A. (1976): Short-term screening tests for carcinogens. *Nature*, 261:195–200.
26. Brousseau, R., Scarpulla, R., Sung, W., Hsing, H. M., Narang, S. A., and Ulu, R. (1982): Synthesis of a human insulin gene. V. Enzymatic assembly, cloning and characterization of the human proinsulin DNA. *Gene*, 17:279–289.
27. Brusick, D. J. (1987): Implications of treatment-condition-induced genotoxicity for chemical screening and data interpretation. *Mutat. Res.*, 189:1–6.
28. Bundy, J. G., Spurgeon, D. J., Svendsen, C., Hankard, P K., Weeks, J. M. et al. (2004): Environmental metabonomics: applying combination biomarker analysis in earthworms at a metal contaminated site. *Ecotoxicology* 13:797–806.
29. Burcham, P. C. (1999): Internal hazards: baseline DNA damage by endogenous products of normal metabolism *Mutat. Res.*, 443:11–36.
30. Burcynski, M. E., McMillian, M. Ciervo, J., Li, L, Parker J. B. et al. (2000): Toxicogenomic-based discrimination o toxic mechanism in HepG2 human hepatoma cells. *Toxicol. Sci.*, 58:399–415.
31. Butkiewicz, D., Grzybowska, E., Hemminki, K., Ovrebo S., Haugen, A. et al. (1998): Modulation of DNA adduc levels in human mononuclear white blood cells and granulocytes by CYP1A1, CYP2D6, and GSTM1 genetic polymorphisms. *Mutat. Res.*, 415:97–108.
32. Butterworth, B. E., Conolly, R. B., and Morgan, K. T (1995): A strategy for establishing mode of action of chemical carcinogens as a guide for approaches to risk assessments. *Cancer Lett.*, 93:129–146.
33. Buzard, G. S. (1996): Studies of oncogene activation an tumor suppressor gene inactivation in normal and neoplastic rodent tissue. *Mutat. Res.*, 365:43–58.
34. Cacheiro, N. L. A., Russell, L. B., and Swartout, M. S (1974): Translocations, the predominant cause of total sterility in sons of mice treated with mutagens. *Genetic* 75:73–91.
35. Caldwell, R. L. and Caprioli, R. M. (2005): Tissue profilin by mass spectrometry: a review of methodology and appl cations. *Mol. Cell. Proteomics*, 4:394–401.
36. Canadian Health and Welfare. (1986): Guidelines on th use of mutagenicity tests in the toxicological evaluation c chemicals. *Environ. Mol. Mutagen.*, 11:261–304.
37. Clive, D., Caspary, W., Kirby, P. E., Krehl, R., Moore, M et al. (1987): Guide for performing the mouse lymphom assay for mammalian cell mutagenicity. *Mutat. Res* 189:143–156.
38. Clive, D., Johnson, K. O., Spector, J. F. S., Batson, A. G and Brown, M. M. M. (1979): Validation and characteriza tion of the L5178Y TK$^{+/-}$ mouse lymphoma mutagen assa system. *Mutat. Res.*, 59:61–108.
39. Coen, M., Lenz, E. M., Nicholson, J. K., Wilson, I. D Pognan, F., and Lindon, J. C. (2003): An integrated metabonomic investigation of acetaminophen toxicity in th mouse using NMR spectroscopy. *Chem. Res. Toxico* 16:295–303.
40. Corey, S. and Adams, J. M. (2002): The *bcl*2 family: regulators of the cellular life-or-death switch. *Nat. Rev. Cancer*, 2:647–656.
41. Counter, C. (1996): The roles of telomeres and telomera in cell life span. *Mutat. Res.*, 366:45–63.
42. Cowherd, S. M., Espina, V. A., Petricoin III, E. F., ar Liotta, L. A. (2004): Proteomic analysis of human brea cancer tissue with laser-capture microdissection a reverse-phase protein microarrays. *Clin Breast Cance* 5:385–92.

43. Curran, S. et al. (2000): Laser capture microscopy. *Mol. Pathol.*, 53:64–68.
44. Darva, F., Dorman, G. Krajcsi, P., Puskas, L. G. Kovari, Z. et al. (2004): Recent advances in chemical genomics. *Curr. Med. Chem.*, 11:3119–3145.
45. De Flora, S., Balansky, R. M., D'Agostini, F., Izzotti, A., Camoirano, A. et al. (2003): Molecular alterations and lung tumors in p53 mutant mice exposed to cigarette smoke. *Cancer Res.*, 63:793–800.
46. Dearfield, K. L., Cimino, M. C., McCarrol, N. E., Mauer, I., and Valcovic, L. R. (2002): Genotoxicity risk assessment: a proposed classification strategy. *Mutat. Res.*, 521:121–135.
47. deLange, T. (1994): Activation of telomerase in a human tumor. *Proc. Natl. Acad. Sci. U.S.A.*, 91:2882–2885
48. European Economic Community. (1983): 6th Amendment to Directive 67/548/EEC, Annex VII, 15.10.79, and Annex V, EEC Directive 79-831, Part B, Toxicological Methods of Annex VIII, Draft.
49. Ehling, U. H. (1988): Quantification of the genetic risk environmental mutagens. *Risk Anal.*, 8:45–56.
50. Ehling, U. H. et al. (1983): Review of the evidence for the presence or absence of thresholds in the induction of genetic effects by genotoxic chemicals. *Mutat. Res.*, 123:281–341.
51. Elkahloun, A. G., Gaudet, J., Robinson, G. S., and Sgroi, D. C. (2002): *In situ* gene expression analysis of cancer using laser capture microdissection, microarrays and real-time quantitative PCR. *Cancer Biol. Ther.*, 1:354–358.
52. Emmert-Buck, M. R. et al. (1996): Laser capture microdissection. *Science*, 274:998–1001.
53. EPA (1986): EPA guidelines for mutagenicity risk assessment. *Fed. Reg.*, 51:34006–34012.
54. EPA. (1982): *Health Effects Test Guidelines*, EPA Publ. No. 560/682–001. Office of Pesticides and Toxic Substances, U.S. Environmental Protection Agency, National Technical Information Service, Springfield, VA.
55. Evans, H. J. (1990): Cytogenetics: Overview. In: *Mutation and the Environment*, Part B, edited by M. L. Mendelsohn and R. J. Albertini, pp. 301–323. Wiley-Liss, New York.
56. Evans, H. J. (1976): Cytological methods for detecting chemical mutagens. In: *Chemical Mutagens: Principles and Methods for their Detection*, Vol. 4, edited by A. Hollandaer, pp. 1–29. Plenum Press, New York.
57. Ezendam, J., Staedtler, F., Pennings, J., Vandebriel, R. J., Pieters, R. et al. (2004): Toxicogenomics of subchronic hexachlorobenzene exposure in Brown Norway rats. *Environ. Health Perspect.*, 112:782–791.
58. Farr, S. and Dunn, R. T. (1999): Concise review: gene expression applied to toxicology. *Toxicol. Sci.*, 50:1–9.
59. Favor, J. and Shelby, M. D. (2005): Transmitted mutational events induced in mouse germ cells, following acrylamide or glycimide exposure. *Mutat. Res.*, 580:21–30.
60. Favor, J., Layton, D., Sega, G., Wassom, J., Burkart, J. et al. (1995): Genetic risk extrapolation from animal data to human disease. *Mutat. Res.*, 330:23–34.
61. FDA (2004): *Guidance for Industry: Recommended Approaches to Integration of Genetic Toxicology Study Results*. Center for Drug Evaluation and Research, Food and Drug Administration, U.S. Department of Health and Human Services, Washington, D.C.
62. Fields, W. R., Desiderio, J. G., Leonard, R. M., Burger, E. E., Brown, B. G., and Doolittle, D. J. (2004): Differential *c-myc* expression profiles in normal human bronchial epithelial cells following treatment with benzo[*a*]pyrene, benzo[*a*]pyrene-4,5 epoxide, and benzo[*a*]pyrene-7,8,9,10 diol expoxide. *Mol. Carcinog.*, 40:79–89.
63. Fields, W. R., Leonard, R. M., Odom, P. S., Nordskog, B. K., Ogden, M. W., and Doolittle, D. J. (2005): Gene expression in normal human bronchial epithelial (NHBE) cells following *in vitro* exposure to cigarette smoke condensate. *Toxicol. Sci.*, 86:84–91.
64. Ferguson, L. (1999): Natural and man-made mutagens and carcinogens in the human diet. *Mutat. Res.*, 443:1–10.
65. Ferre, F., Marchese, A., Pexxoli, P., Griffin, S., Buxton, E., and Boyer, V. (1994): Quantitative PCR: an overview. In: *The Polymerase Chain Reaction*, edited by K. B. Mullis and R. A. Gibbs, pp. 67–88. Birkhauser, Boston.
66. Friedberg, E. C. (2003): DNA damage and repair. *Nature*, 421:436–440.
67. Galloway, S. M., Deasy, D. A., Bean, C. L., Kraynak, A. R., Armstrong, M. J., and Bradley, M. O. (1987): Effects of high osmotic strength on chromosome aberrations, sister chromatid exchanges, and DNA strand breaks, and the relation to toxicity. *Mutat. Res.*, 189:15–25.
68. Galloway, S. M., Miller, J. E., Armstrong, M. J., Bean, C. L., Skopek, T. R., and Nichols, W. W. (1998): DNA synthesis inhibition as an indirect mechanism of chromosome aberrations: comparison of DNA-reactive and non-DNA-reactive clastogens. *Mutat. Res.*, 400:169–186.
69. Gebel, S., Gerstmayer, B., Bosio, A., Haussmann H., Miert, E. V., and Muller, T. (2004): Gene expression profiling in respiratory tissue from rats exposed to mainstream cigarette smoke. *Carcinogenesis*, 25:169–178.
70. Generoso, W. M., Cain, K. I., and Banby, A. J. (1983): Some factors affecting the mutagenic response of mouse germ cells to chemicals. In: *Utilization of Mammalian Specific Locus Studies in Hazard Evaluation and Estimation of Genetic Risk*, edited by F. J. de Serres and W. Sheridan, pp. 227–239. Plenum Press, New York.
71. Glassner, B. J., Posnick, M., and Samson, L. D. (1998): The influence of DNA glycosylases on spontaneous mutation. *Mutat. Res.*, 400:33–44.
72. Goodman, J. I. (1998): The traditional toxicologic paradigm is correct: dose influences mechanism. *Environ. Health Perspect.*, 106:285–288.
73. Gorelick, N. J. (1995): Overview of mutation assays in transgenic mice for routine testing. *Environ. Mol. Mutagen.*, 25:218–230.
74. Gossen, J., de Leeuw, W., Tan, C., Zwartgoff, E., Berends, F. et al. (1989): Efficient rescue of integrated shuttle vectors from transgenic mice: a model for studying mutations *in vivo*. *Proc. Nat. Acad. Sci. U.S.A.*, 86:7071–7975.
75. Hackett, N. R., Heguy, A., Harvey, B. G., O'Conner, T. P., Luettich, K. et al. (2003): Variability of antioxidant-related gene expression in the airway epithelium of cigarette smokers. *Am. J. Respir. Cell Mol. Biol.*, 29:331–343.
76. Hanawalt, P. (1998): Genomic instability: environmental invasion and the enemies within. *Mutat. Res.*, 400: 117–125.

77. Harris, C. C. (1998): Interindividual variation among humans in carcinogen metabolism, DNA adduct formation and DNA repair. *Carcinogenesis*, 10:1563–1566.

78. Harris, A. J., Dial, S. L., and Casciano, D. A. (2004): Comparison of basal gene expression and effects if hepatocarcinogenes in gene expression in cultured primary human hepatocytes and HepG2 cells. *Mutat. Res.*, 549: 79–99.

79. Heddle, J. A., Cimino, M. C., Hayashi, M., Romagna, F., Shelby, M. D. et al. (1991): Micronuclei as an index of cytogenetic damage: past, present, and future. *Environ. Mol. Mutagen.*, 18:277–291.

80. Heddle, J. A., Hite, M., Kirkhart, B., Mavournin, K. H., MacGregor, J. T. et al. (1983): The induction of micronuclei as a measure of genotoxicity. *Mutat. Res.*, 123:61–118.

81. Heid, C. A., Stevens, J., Livak, K. J., and Williams, P. M. (1996): Real-time quantitative PCR. *Genome Meth.*, 6:986–994.

82. Heijne, W. H., Stierum, R. H., Slijper, M., van Bladeren, P. J., and van Ommen, B. (2003): Toxicogenomics of bromobenzene hepatotoxicity: a combined transcriptomics and proteomics approach. *Biochem. Pharmacol.*, 65: 857–875.

83. Hellmann, G. M., Fields, W. R., and Doolittle, D. J. (2001): Gene expression profiling of cultured human bronchial epithelial and lung carcinoma cells. *Toxicol. Sci.*, 61: 154–163.

84. Hilliard, C., Armstrong, M., Bradt, C., Hill, R., Greenwood, S., and Galloway, S. (1998): Chromosome aberrations *in vitro* related to cytotoxicity of nonmutagenic chemicals and metabolic poisons. *Environ. Mol. Mutagen.*, 31:316–326.

85. Holliday, R. and Ho, T. (1988): Gene silencing in mammalian cells by uptake of 5-methyl deoxycytidine 5′ phosphate. *Somatic Cell Mol. Genet.*, 17:537–542.

86. Holmes, E., Nicholls, A. W., Lindon, J. C., Conner, S. C., Connelly, J. C. et al. (2000): Chemometric models for toxicity classification based on NMR spectra of biofluids. *Chem. Res. Toxicol.*, 13:471–478.

87. Holmes, E., Nicholson, J. K., and Tranter, G. (2001): Metabonomic characterization of genetic variations in toxicological and metabolic responses using probabilistic neural networks. *Chem. Res. Toxicol.*, 14:182–191.

88. ICH S2A Guideline. (1996): Specific aspects of regulatory genotoxicity tests for pharmaceuticals. *Fed. Reg.*, 61: 18199.

89. ICH S2B Guideline. (1997): Genotoxicity: a standard battery for genotoxicity testing of pharmaceuticals. *Fed. Reg.*, 62:62472.

90. ICPEMC. (1988): Testing for mutagens and carcinogens: the role of short-term genotoxicity assays. *Mutat. Res.*, 205:3–12.

91. ICPEMC. (1990): The possible involvement of somatic mutations in the development of atherosclerotic plaques. *Mutat. Res.*, 239:143–148.

92. Inger-Lise, H. (1990): Occupational and lifestyle factors and chromosomal aberrations of spontaneous abortions. In: *Mutation and the Environment*, Part B, edited by M. L. Mendelsohn and R. J. Albertini, pp. 467–475. Wiley-Liss, New York.

93. Inoue, T. and Pennie, W. D. (2003): *Toxicogenomics*. Springer, New York.

94. Issa, J. P. (2000): CpG-island methylation in aging and cancer. *Curr. Top. Microbiol. Immunol.*, 249:101–118.

95. Jimenez-Sanchez, G., Childs, B., and Valle, D. (2001): Human disease genes. *Nature*, 409:853–855.

96. Jones, P. A. and Baylin, S. B. (2002): The fundamental role of epigenetic events in cancer. *Nat. Rev. Genet.*, 3: 415–428.

97. Jones, P. A. and Takai, D. (2001): The role of DNA methylation in mammalian epigenetics. *Science*, 293:1068–1070.

98. Kalter, H. (1977): Correlation between teratogenic and mutagenic effects of chemicals in mammals. In: *Chemical Mutagens: Principles and Methods for Their Detection*, Vol. 6, edited by A. Hollandaer. Plenum Press, New York.

99. Keun, H. C., Ebbels, T. M., Bollard, M. E., Beckonert, O., Antti, H. et al. (2004): Geometric trajectory analysis of metabolic responses to toxicity can define treatment specific profiles. *Chem. Res. Toxicol.*, 17:579–587.

100. Kier, L. D., Yamasaki, E., and Ames, B. N. (1974): Detection of mutagenic activity in cigarette smoke condensates. *Proc. Natl. Acad. Sci. U.S.A.*, 71:4159–4163.

101. Kier, L. D., Brusick, D. J., Auletta, A. E., Von Halle, E. S., Brown, M. M. et al. (1986): The *Salmonella typhimurium*/mammalian microsomal assay: a report of the U.S. EPA GeneTox program. *Mutat. Res.*, 168:69–240.

102. Kihlman, B. A., Nichols, W. W., and Levan, A. (1963): Effects of deoxyadenosine and cytosine arabinoside on the chromosomes of human leucocytes *in vitro*. *Hereditas*, 50:139–143.

103. Kinser, S., Jia, Q., Li, M., Laughter, A., Cornwell, P. et al. (2004): Gene expression profiling in spleens of doxynivalenol-exposed mice: immediate early genes as primary targets. *J. Toxicol Environ Health A*, 67:1423–1441.

104. Kirkland, D., Aardema, M., Henderson, L., and Muller, L. (2005): Evaluation of the ability of a battery of three *in vitro* genotoxicity tests to discriminate rodent carcinogens and non-carcinogens. I. Sensitivity, specificity, and relative predictivity. *Mutat. Res.*, 548:1–256.

105. Kleinhofs, A. and Behki, R. (1977): Prospects for plant genome modification by nonconventional methods. *Annu. Rev. Genet.*, 11:79–101.

106. Koike, E., Hirano, S., Shimojo, N., and Kobayashi, T. (2002): cDNA microarray analysis of gene expression in rat alveolar macrophages in response to organic extract of diesel exhaust particles. *Toxicol Sci.*, 67:241–246.

107. Kopelovich, L., Bias, N. E., and Helson, L. (1979): Tumor promotor alone induces neoplastic transformation of fibroblasts from humans genetically predisposed to cancer. *Nature*, 282:619–621.

108. Kramer, J. A. (2004): Overview of the application of transcription profiling using selected nephrotoxicants for toxicology assessment. *Environ. Health Perspect.*, 112:460–46

109. Kreiner, T. and Buck, K. T. (2005): Moving toward whole genome analysis: a technology perspective. *Am. J. Health Syst. Pharm.*, 62:296–305.

110. Kriek, E., Rojas, M., Alexandrov, K., and Bartsch, H. (1998): Polycyclic aromatic hydrocarbon–DNA adducts in humans: relevance as biomarkers for exposure and cancer risk. *Mutat. Res.*, 400:215–231.

111. Kristensen, V. N., Kelefioti, D., Kristensen, T., and Børresen-Dale, A. (2001): Genotyping techniques: high-throughput methods for detection of genetic variation. *BioTechniques*, 30:318–332.
112. LeBoeuf, R. A., Kerckaert, G., Aardema, M. J., Gibson, D. P., Brauninger, R., and Isfort, R. J. (1996): The pH 6.7 Syrian hamster embryo cell transformation assay for assessing the carcinogenic potential of chemicals. *Mutat. Res.*, 356:85–127.
113. Lewtas, J. (1986): A quantitative cancer risk assessment methodology using short-term genetic bioassays: the comparative potency method. In: *Risk and Reason: Risk Assessment in Relation to Environmental Mutagens and Carcinogens*, edited by P. Oftedal and A. Bragger, pp. 107–120. Alan R. Liss, New York.
114. Liang, P. and Pardee, A. B. (1998): Differential display: a general protocol. *Mol. Biotechnol.*, 10:261–267.
115. Lindon, J. C., Nicholson, J. K., Holmes, E., Antti, H., Bollard, M. E. et al. (2003): Contemporary issues in toxicology: the role of metabonomics in toxicology and its evaluation by the COMET project. *Toxicol. Appl. Pharmacol.*, 187:137–146.
116. Lindon, J. C., Holmes, E., and Nicholson, J. K. (2004): Metabonomics: systems biology in pharmaceutical research and development. *Curr. Opin. Mol. Ther.*, 6: 265–272.
117. Lindon, J. C., Holmes, E., Bollard, M. E., Stanley, E. G., and Nicholson, J. K. (2004): Metabonomics technologies and their applications in physiological monitoring, drug safety assessment and disease diagnosis. *Biomarkers*, 9:1–31.
118. Maron, D. M. and Ames, B. N. (1983): Revised methods for the *Salmonella* mutagenicity test. *Mutat. Res.*, 113: 173–215.
119. Mattes, W. B., Pettit, S. D., Sansone, S. A., Bushel, P. R., and Waters, M. D. (2004): Database development in toxicogenomics: issues and efforts. *Environ. Health Perspect.*, 112:495–505.
120. McCann, J., Choi, E., Yamasaki, E., and Ames, B. N. (1975): Detection of carcinogens as mutagens in the *Salmonella*/microsome test: assay of 300 chemicals. *Proc. Natl. Acad. Sci. U.S.A.*, 72:5135–5139.
121. McKusick, V. A. (1978): *Mendelian Inheritance in Man: Catalogs of Autosomal Dominant, Autosomal Recessive and X-linked Phenotypes*, 5th ed. The Johns Hopkins University Press, Baltimore, MD.
122. McLachlan, J. A., Burow, M., Chiang, T. C., and Li, S. F. (2001): Gene imprinting in developmental toxicology: a possible interface between physiology and pathology. *Toxicol. Lett.*, 120:161–164.
123. Medlin, J. (2002): Two committees tackle toxicogenomics. *Environ. Health Perspect.*, 110:A746–747.
124. Meintieres, S. and Marzin, D. (2004): Apoptosis may contribute to false-positive results in the *in vitro* micronucleus test performed in extreme osmolality, ionic strength and pH conditions. *Mutat. Res.*, 560:101–118.
125. Mendelsohn, M. L., Moore II, D. H., and Lohman, P. H. M. (1992): A method for comparing and combining short-term genotoxicity test data: results and interpretation. *Mutat. Res.*, 266:43–60.
126. Mills, J. C., Roth, K. A., Cagan, R. L., and Gordon, J. I. (2001): DNA microarrays and beyond: completing the journey from tissue to cell. *Nat. Cell Biol.*, 3:E175–E178.
127. Mitchell, A. D., Auletta, A. E., Clive, D., Kirby, P. E., Moore, M. M., and Myhr, B. C. (1997): The L5178Y/$tk^{+/-}$ mouse lymphoma specific gene and chromosomal mutation assay: a phase III report of the U.S. Environmental Protection Agency GeneTox Program. *Mutat. Res.*, 394:177–303.
128. Moggs, J. G. (2004): Phenotypic anchoring of gene expression changes during estrogen-induced uterine growth. *Environ. Health Perspect.*, 112:1589.
129. Mohrenweiser, H. W. and Jones, I. M. (1998): Variation in DNA repair is a factor in cancer susceptibility: a paradigm for the promises and perils of individual and population risk estimation. *Mutat. Res.*, 400:15–24.
130. Moore, M. M. et al. (1985): Analysis of trifluorothymidine ($TFT^r$) mutants of L5178Y $TK^{+/-}$ mouse lymphoma cells. *Mutat. Res.*, 151:61–174.
131. Moore, M. M. and Doerr, C. L. (1990): Comparison of chromosome aberration frequency and small-colony TK-deficient mutant frequency in L5178Y/$TK^{+/-}$-3.7.2 C mouse lymphoma cells. *Mutagenesis*, 5:609–614.
132. Moore, M. M., Honma, M., Clements, J., Ballantyne, M., Bolcsfoldi, G. et al. (2005): *Mutat. Res.* (submitted).
133. Moore, M. M., Honma, M., Clements, J., Bolcsfoldi, G., Cifone, M. et al. (2003): Mouse lymphoma thymidine kinase locus gene mutation assay: international workshop (Plymouth, England) on genotoxicity test procedures workgroup report. *Mutat. Res.*, 540:127–140.
134. Mori, C. (2003): Application of toxicogenomic analysis to risk assessment of delayed long-term effects of multiple chemicals, including endocrine disruptors in human fetuses. *Environ. Health Perspect.*, 111:803–809.
135. Morgan, K. T., Ni, H., Brown, H. R., Yoon, L., Qualls, C. W. et al. (2002): Application of cDNA microarray technology to *in vitro* toxicology and the selection of genes for a real-time RT-PCR-based screen for oxidative stress in HepG2 cells. *Toxicol. Pathol.*, 30:435–451.
136. Nagao, M., Takahashi, Y., Yamanaka, H., and Sugimura, T. (1979): Mutagens in coffee and tea. *Mutat. Res.*, 68:101–106.
137. Nagao, M., Yahagi, T., Kawachi, T., Seino, Y., Honda, M. et al. (1977): Mutagens in foods, and especially pyrolysis products of protein. In: *Progress in Genetic Toxicology*, edited by D. Scott, B. A. Bridges, and F. H. Sobels, pp. 259–264. Elsevier/North-Holland, New York.
138. Nicholson, J. K., Lindon, J. C., and Holmes, E. (1999): 'Metabonomics': understanding the metabolic responses of living systems to pathophysiological stimuli via multivariate statistical analysis of biological NMR spectroscopic data. *Xenobiotica*, 29:1181–1189.
139. Nicholson, J. K., Connelly, J., Lindon, J. C., and Holmes, E. (2002): Metabonomics: a platform for studying drug toxicity and gene function. *Nat. Rev. Drug Discov.*, 1:153–161.
140. Nie, A. Y., McMillian, M., Parker, J. B., Leone, A., Bryant, S., Yie, H. L., Bittner, A., Nelson, J., Carmen, A., Wan, J., and Lord, P. G. (2006): Predictive toxicogenomics approaches reveal underlying molecular mechanisms of non-genotoxic carcinogenicity. *Mol. Carcinog.*, 45:914–933.

141. Obe, G. and Anderson, D. (1987): Genetic effects of ethanol. *Mutat. Res.*, 186:177–200.
142. OECD. (1997): *Updated OECD Guidelines for Testing of Chemicals*. Section 4. *Health Effects*, *Ninth Addendum*. Organization for Economic Cooperation and Development, Brussels.
143. OECD. (1997): *Guideline 471: Bacterial Reverse Mutation Test; Guideline 473: In Vitro Mammalian Chromosome Aberration Test; Guideline 476: In Vitro Mammalian Cell Gene Mutation Test; Guideline 474: Mammalian Erythrocyte Micronucleus Test*, OECD Guidelines for the Testing of Chemicals, Organization for Economic Cooperation and Development, Brussels.
144. Owens, D. M., Spalding, J. W., Tennant, R. W., and Smart, R. C. (1995): Genetic alterations cooperate with v-Ha-*ras* to accelerate multistage carcinogenesis in Tg.AC transgenic mouse skin. *Cancer Res.*, 55:3171–3178.
145. Owens, D. M., Wei, S.-J.C., and Smart, R. C. (1999): A multihit, multistage model of chemical carcinogenesis. *Carcinogenesis*, 20:1837–1844.
146. Pennie, W., Pettit, S. D., and Lord, P. G. (2004): Toxicogenomics in risk assessment: an overview of an HESI collaborative research program. *Environ. Health Perspect.*, 2:417–419.
147. Phillips, D., Farmer, P., Beland, F., Nath, R., Poirier, M. et al. (2000): Methods of DNA adduct determination and their application to testing compounds for genotoxicity. *Environ. Mol. Mutagen.*, 35:222–233.
148. Pienta, R. J., Kuschner, L. M., and Russell, L. S. (1984): The use of short-term tests and limited bioassays in carcinogenicity testing. *Regul. Toxicol. Pharmacol.*, 4: 249–260.
149. Player, A., Barrett, J. C., and Kawasaki, E. S. (2004): Laser capture microdissection, microarrays and the precise definition of a cancer cell. *Expert Rev. Mol. Diagn.*, 4: 831–840.
150. Preston, J. R. (1997): Telomeres, telomerase and chromosome stability. *Radiat. Res.*, 147:529–534.
151. Ramos-Nino, M. E., Heintz, N., Scappoli, L., Martinelli, M., Land, S. et al. (2003): Gene profiling and kinase screening in asbestos-exposed epithelial cells and lungs. *Am. J. Respir. Cell Mol. Biol.*, 29:S51–S58.
152. Reedy, E. P., Reynolds, R. K., Santos, E., and Barbacid, M. (1982): A point mutation is responsible for the acquisition of transforming properties by the T24 human bladder carcinoma oncogene. *Nature*, 300:145–152.
153. Rhomberg, L., Dellarco, V. L., Siegel-Scott, C., Dearfield, K. L., and Jacobson-Kram, D. (1990): Quantitative estimation of the genetic risk associated with the induction of heritable translocations at low dose exposure: ethylene oxide as an example. *Environ. Mol. Mutagen.*, 16:104–125.
154. Richardson, B. (2003): Impact of aging on DNA methylation. *Ageing Res. Rev.*, 2:245–261.
155. Rockett, J. C. and Hellmann, G. M. (2004): Confirming microarray data: is it really necessary? *Genomics*, 83: 541–549.
156. Rom, W. N. and TchouWong, K.-M. (2003): Functional genomics in lung cancer and biomarker detection. *Am. J. Respir. Cell Mol. Biol.*, 29:153–156.
157. Russell, L. B., Selby, P. B., van Halle, E., Sheridan, W., and Valcovic, L. (1981): The mouse specific locus test with agents other than radiation: interpretation of data and recommendations for future work. *Mutat. Res.*, 86:329–354.
158. Salamone, M. F. and Mavournin, K. H. (1994): Bone marrow micronucleus assay: a review of the mouse stocks used and their published mean spontaneous micronucleus frequencies. *Environ. Mol. Mutagen.*, 23:239–273.
159. Samson, L. and Schwartz, J. L. (1980): Evidence for an adaptive DNA repair pathway in CHO and human skin fibroblast cell lines. *Nature*, 287:861–863.
160. Schmid, W. (1975): The micronucleus test. *Mutat. Res.* 31:9–15.
161. Setlow, R. B. (1978): Repair deficient human disorders and cancer. *Nature*, 271:713–717.
162. Shah, V., Sridhar, S., Beane, J., Brody, J. S., and Spira, A (2005): SIEGE: smoking induced epithelial gene expression database. *Nucleic Acids Res.*, 33:573–579.
163. Short, J. M., Kohler, S. W., Provost, G. T. S., Feick, A. and Kretz, P. L. (1990), The use of lambda phage shuttle vectors in transgenic mice for development of a short-term mutagenicity assay. In: *Mutation and the Environment*, Part A, edited by M. Mendelsohn and R. Albertini, pp 355–367. Wiley-Liss, New York.
164. Simic, M. C. and Bergtold, D. S. (1991): Dietary modulation of DNA damage in human, *Mutat. Res.*, 250:17–24.
165. Simone, N. L., Paweletz, C. P., Charboneau, L., Petricoin III, E. F., and Liotta L. A. (2000): Laser capture microdissection: beyond functional genomics to proteomics. *Mol Diagn.*, 5:301–307.
166. Simone, N. L. et al. (1998): Laser-capture microdissection opening the microscopic frontier to molecular analysis *Trends Genet.*, 4:272–276.
167. Sioud, M. and Rosok, O. (2004): Profiling microRNA expression using sensitive cDNA probes and filter arrays *Biotechniques*, 37:574–576.
168. Sohail, M., Doran, G., Riedemann, J., Macaulay, V., and Southern, E. D. (2003): A simple and cost-effective method for producing small interfering RNAs with high efficiency *Nucleic Acids Res.*, 31:e38.
169. Sugimura, T. (1978): Let's be scientific about the problem of mutagens in cooked food. *Mutat. Res.*, 55:149–152.
170. Sugimura, T., Sato, S., Nagao, M., Yahagi, T., Matsushima T. et al. (1976): Overlapping of carcinogens and mutagens In: *Fundamentals in Cancer Prevention*, edited by P. N Magee, T. Matsushima, T. Sugimura, and S. Takayama, pp 191–215. University of Tokyo Press, Tokyo; University Park Press, Baltimore, MD.
171. Taatjes, D. J., Palmer, C. J., Pantano, C., Hoffmann, S. B Cummins, A., and Mossman, B. T. (2001): Laser-based microscopic approaches: application to cell signaling in environmental lung disease. *Biotechniques*, 31:880–894.
172. Tennant, R. W., Spalding, J., and French, J. E. (1996 Evaluation of transgenic mouse bioassays for identifying carcinogens and noncarcinogens. *Mutat. Res.*, 36 119–127.
173. Tennant, R. W. (2002): The National Center for Toxicogenomics: using new technologies to inform mechanistic toxicology. *Environ. Health Perspect.*, 110:A8–A10.

174. Thai. S. F., Allen, J. W., DeAngelo, A. B., Georger, M. H., and Fuscoe, J. C. (2003): Altered gene expression in mouse livers after dichloroacetic acid exposure. *Mutat. Res.*, 543: 167–180.
175. Thakur, A. J., Berry, K. J., and Mielke, Jr., P. W. (1985): A FORTRAN program for testing trend and homogeneity in proportions. *Comput. Prog. Biomed.*, 19:229–233.
176. Todd, R., Lingen, M. W., and Kuo, W. P. (2002): Gene expression profiling using laser capture microdissection. *Expert Rev. Mol. Diagn.*, 2:497–507.
177. UNEP–ICPEMC; Brusick, D. J., Gopalon, W. B., Hesletine, E., Huismans, J. W., and Lohman, P. H. M., Eds. (1992): *Assessing the Risk of Genetic Damage*. Hodder & Stoughton, London.
178. Van Delft, J. et al. (1998): Gene-mutation assays in lacZ transgenic mice: comparison of lacZ with endogenous genes in splenocytes and small intestinal epithelium. *Mutat. Res.*, 415:85–96.
179. Vanden Heuval, J. P. (1998): *PCR Protocols in Molecular Toxicology*. CRC Press, Boca Raton, FL.
180. Venkatachalam, S. and Donehower, L. A. (1998): Murine tumor suppressor models. *Mutat. Res.*, 400:391–407.
181. Vijg, J. and van Steeg, H. (1998): Transgenicassays for mutations and cancer: current status and future perspectives. *Mutat. Res.*, 400:337–354.
182. Viravaidya, K. and Shuler, M. L. (2004): Incorporation of 3T3-L1 cells to mimic bioaccumulation in a microscale cell culture analog device for toxicity studies. *Biotechnol. Prog.*, 20:590–597.
183. Vrieling, H., van Zeeland, A., and Mullenders, L. H. F. (1998): Transcription coupled repair and its impact on mutagenesis. *Mutat. Res.*, 400:135–142.
184. Wan, J. S., Sharp, S. J., Poirer, G. M., Wagaman, P. C., Chambers, J. et al. (1996): Cloning differentially expressed mRNAs. *Nat. Biotechnol.*, 14:1685–1691.
185. Waters, M. D., Olden, K., and Tennant, R. W. (2003): Toxicogenomic approach for assessing toxicant-related disease. *Mutat. Res.*, 544:415–424.
186. Watson R. E., McKim, J. M., Cockerel, G. L., and Goodman J. I. (2004): The value of DNA methylation analysis in basic, initial toxicity assessments. *Toxicol. Sci.*, 79:178–188.
187. Watson, R. E. and Goodman, J. I. (2002): Epigenetics and DNA methylation come of age in toxicology. *Toxicol. Sci.*, 67:11–6.
188. Weinberg, R. A. (1981): Use of transfection to analyze genetic information and malignant transformation. *Biochem. Biophys. Acta*, 651:25–35.
189. Willey, J. C., Crawford, E. L., Jackson, C. M. et al. (1998): Expression measurement of many genes simultaneously by quantitative RT-PCR using standardized mixtures of competitive templates. *Am. J. Respir. Cell Mol. Biol.*, 19: 6–17.
190. Wolff, S., Afzal, V., and Olivieri, G. (1990): Inducible repair of cytogenetic damage to human lymphocytes: adaption to low-level exposures to DNA-damaging agents. In: *Mutation and the Environment*, Part B, edited by M. L. Mendelsohn and R. J. Albertini, pp. 397–405. Wiley-Liss, New York.
191. Xie, Y., Trouba, K. J., Liu J., Waalkes, M. P., and Germolec, D. R. (2004): Biokinetics and subchronic toxic effects of oral arsenite, arsenate, monomethylarsonic acid, and dimethylarsinic acid in v-Ha-*ras* transgenic (Tg.AC) mice. *Environ. Health Perspect.*, 112:1255–1263.
192. Yahagi, T., Nagao, M., Seino, Y., Matsushima, T., Sugimura, T., and Oleada, M. (1977): Mutagenicities of *N*-nitrosamines on *Salmonella*. *Mutat. Res.*, 48:121–130.
193. Young, R., Oveisistork, F., Harrington-Brock, K., Schalkowsky, S., Moore, M., and Myhr, B. (1991): Quantitative size analysis of L5178Y $TK^{+/-}$ mutant colonies in soft agar; an interlaboratory comparison. *Environ. Mol. Mutagen.*, 17(Suppl. 19):79.
194. Zeiger, E. (1998): Identification of rodent carcinogens and noncarcinogens using genetic toxicology tests: premises, promises and performance. *Regul. Toxicol. Pharmacol.*, 28:85–95.
195. Zieziulewicz, T. J., Unfricht, D. W., Hadjout, N., Lynes, M. A., and Lawrence D. A. (2003): Shrinking the biologic world: nanobiotechnologies for toxicology. *Toxicol. Sci.*, 74:235–244.
196. Zucco, F., De Angelis, I., Testai, E., and Stammati, A. (2004): Toxicology investigations with cell culture systems: 20 years after. *Toxicol. In Vitro*, 18:153–163.
197. Goodman, J. I. and Watson, R. E. (2002): Altered DNA methylation: a secondary mechanism involved in carcinogenesis. *Ann. Rev. Pharmacol. Toxicol.*, 42:501–525.
198. Counts, J. L., Sarmiento, J. I., Harbison, M. L., Downing, J. C., McClain, R. M., and Goodman, J. I. (1996): Cell proliferation and global methylation status changes in mouse liver after phenobarbital and/or choline-devoid, methionine-deficient diet administration. *Carcinogenesis*, 17(6):1251–1257.

# Notes

# 24 Short-Term, Subchronic, and Chronic Toxicology Studies

*Nelson H. Wilson, Jerry F. Hardisty, and Johnnie R. Hayes*

## CONTENTS

## INTRODUCTION

Repeated-dose toxicity studies are conducted to screen for potential adverse effects of a chemical using laboratory animals as surrogates for a target species, most often the human. Repeated-dose studies may be of varying duration, generally 1 to 4 weeks for short-term studies, 3 months for *subchronic* studies, and 6 to 12 months for *chronic* studies. Many variables associated with the health of the test species are monitored in short-term, subchronic, and chronic toxicity studies, resulting in the ability to detect a variety of adverse effects. At some time during their careers, most toxicologists are involved in the design, performance, or review of data from these toxicology studies. This results from the central role played by these studies in the safety assessment of pharmaceuticals, pesticides, food additives, and other chemicals. It has been suggested that subchronic data alone may be sufficient to predict the hazard of long-term low-dose exposure to a compound [51]. Although this ma[y] be true for compounds where adequate structure–activit[y] relationships exist and with no indication of genetic tox[-]icity, it generally is not true when compounds have com[-]pletely unknown toxicity or when structure–activity rela[-]tionships predict a potential adverse effect. For certai[n] chemicals or mixtures, results from a short-term or a sub[-]chronic toxicity study may represent the most sophist[i-]cated toxicology data available. With many chemicals, [a] subchronic study is critical to the design of longer ter[m] hazard assessment studies.

It is essential that toxicologists become familiar wi[th] the scientific principles upon which repeated-dose tox[i-]cology studies are based and understand the methodolog[y] used to perform these studies. It is the purpose of th[e] authors to provide an introduction to these studies an[d] certain of the principles upon which they are based.

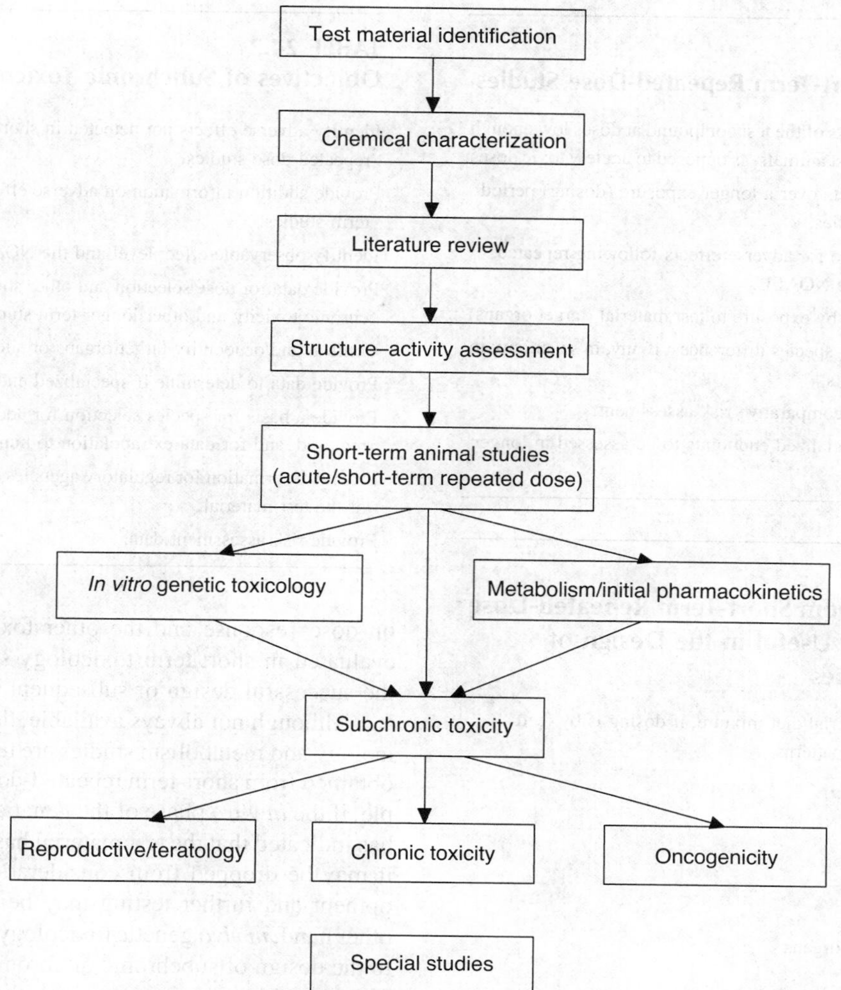


**FIGURE 24.1** Schematic representation of a typical hazard assessment program illustrating a step-by-step, tiered approach and the interactions between the various elements. It should be noted that the approach presented here is not the only approach that could be used.

## REPEATED-DOSE STUDIES AND HAZARD (SAFETY) ASSESSMENT

A typical hazard assessment program is illustrated in Figure 24.1; however, in the practice of toxicology, there is no such thing as a "typical" hazard assessment program. Each program is unique based on the individual material being tested and its intended use.

The first step in any hazard assessment is determination of the material to be studied and the purpose of the assessment. Both of these factors will influence the final design of the program. At the time of selection of the material to be tested, its intended use in the marketplace will have been defined. In most cases, this dictates factors such as exposure route and the types of studies required.

The next step is the chemical and physical characterization of the material or mixture to be investigated. Knowledge of the major components and any associated contaminants is required for almost every phase of the hazard assessment program. The more detailed the chemical and physical characterization, the greater the likelihood that the entire program will be successful.

When the toxicologist has information concerning the chemical nature of the *test material*, the scientific literature is searched to determine what is known, if anything, about its biological activity. If no information is available, then the potential biological activity of chemicals with similar chemical structure should be ascertained. Based on the chemical characterization data and any toxicology information available in the scientific literature, the toxicologist can then perform a structure–activity assessment. This will aid in developing the hazard assessment program and the specific designs for each of the toxicology studies.

**TABLE 24.1**
**Objectives of Short-Term Repeated-Dose Studies**

Determine adverse effects of the test compound at doses low enough to allow survival of most animals, as opposed to acutely toxic doses.
Determine adverse effects over a longer exposure (dosing) period than used in acute studies.
Determine dose response for adverse effects following repeated dosing and identify the NOAEL.
Identify organs affected by exposure to test material (target organs).
Provide data concerning species differences, if any, in sensitivity to potential adverse effects.
Provide initial data for comparative risk assessment.
Determine need for specialized endpoints to be assessed in longer term studies.

**TABLE 24.2**
**Data Obtained from Short-Term Repeated-Dose Studies That Are Useful in the Design of Subsequent Studies**

Palatability of test material/diet mixture, if dosing is by feed
Body weight response patterns
Physical observations
Behavioral changes
Clinical pathology
Toxicokinetics
Gross necropsy
Histopathology
Identification of target organs
Dose responses

In most cases, the first series of studies will be short-term toxicity studies. The initial short-term toxicity study is generally a single dose, *acute* toxicity study. Results from acute toxicity studies are used to estimate dosages to be used in short-term repeated-dose studies and obtain initial data useful in assessing relative toxicity. A number of regulatory agencies have made recommendations concerning designs for acute studies, as discussed in Chapter 22 of this text. If warranted, additional variables may be added to these recommended designs based on the structure–activity assessment and literature review.

Data obtained from short-term repeated-dose studies are generally required for the successful design of subchronic toxicity studies. Similarly, data from subchronic studies are essential for the design of chronic studies. Table 24.1 lists objectives for short-term repeated-dose toxicity studies. Table 24.2 lists examples of data obtained from these studies that are useful in the design of subsequent toxicity studies. As will be discussed later, one of the difficult decisions for a toxicologist designing a toxicology study is selecting the dose range to be used. Data on dose–response and the other toxicological endpoints evaluated in short-term toxicology studies are critical to the successful design of subsequent toxicity studies.

**TABLE 24.3**
**Objectives of Subchronic Toxicology Studies**

Identify adverse effects not detected in shorter-term acute or repeated-dose studies.
Provide addition information on adverse effects identified in short-term studies.
Identify observable effect level and the NOAEL.
Provide data for dose selection and other study design features for chronic toxicity and other longer term studies.
Confirm and/or identify target organs or sites of action.
Provide data to determine if specialized endpoints are required.
Provide a basis for species selection for additional studies, if required, and for data extrapolation to humans.
Provide information for regulatory agencies in support of the safety of the test material.
Provide risk assessment data.

Although not always available, data from genetic toxicology and metabolism studies are useful adjuncts to data obtained from short-term repeated-dose studies; for example, if the *in vitro* phase of the genetic toxicology program has indicated that the test material has genotoxic potential, it may be dropped from consideration for further development and further testing may be unnecessary. On the other hand, *in vivo* genetic toxicology studies can be added to the design of subchronic or chronic toxicology studies. This can, in many cases, reduce the number of animals used in the overall safety assessment and conserve other resources. If information concerning the metabolism and pharmacokinetics of the test material is available, it should be used in the design of repeated-dose toxicity studies; for example, if pharmacokinetic or toxicokinetic studies indicate a potential for accumulation of the test material from multiple dosing, this is an important consideration in deciding the most appropriate dose range. Also, because different species may metabolize the test material differently, this information is important in the selection of appropriate animal species and strains for subsequent testing. The major objectives of subchronic toxicity studies are presented in Table 24.3.

In some cases, subchronic toxicity testing may complete the data required for a hazard assessment. In other cases, as illustrated in Figure 24.1, data from subchronic studies are used in the design of additional studies, including chronic toxicity studies. Data from subchronic toxicity studies are useful in the design of oncogenicity and reproduction studies as well as any "special" studies that may be either warranted by the results of other studies or requested by a regulatory agency.

As noted above, data from a short-term repeated-dose study should be available before initiating subchronic toxicity testing, and, generally, data should be available from a subchronic study before proceeding with chronic testing. The major utility in progression from shorter term to longer term studies is to ensure, to the extent possible, that proper dose ranges are selected. Requirements for a scientifically valid short-term repeated-dose study are similar to those of subchronic toxicity studies; therefore, the general aspects of short-term studies will not be discussed separately. Three aspects of short-term repeated-dose studies are especially important, however, and should be mentioned here.

First, as implied by the description "short-term repeated-dose study," these studies are of shorter duration than subchronic toxicity studies. Generally, the duration of these studies is either 14 or 28 days with the compound administered daily. A 28-day study can produce more valuable information than a 14-day study which, in turn, will be more valuable than a 7-day study because the longer term exposure increases the probability of detecting adverse effects.

Second, dose selection is somewhat dependent on the purpose of the study. If the short-term dosing is designed to produce information to be used in the design of a subchronic study, then dosing should ensure that any potential adverse effect is observed. Higher doses are generally used in short-term studies to determine target organs and it may not be as critical to ascertain a no-observed-adverse-effect level (NOAEL); however, it is always useful to have a NOAEL in a toxicology study. If no further studies are anticipated with the compound, a dose range that includes a NOAEL becomes very important. Because little information other than acute toxicity data is generally available during early compound development, it may be necessary to run a more comprehensive range of dose levels in short-term repeated-dose studies. Four, five, or more dose groups are generally used in short-term studies. This increases the chances of defining the dose response and should increase the confidence of dose range selection for subsequent studies.

A third aspect of short-term studies that is different from subchronic studies is the number of animals per group. Although fewer animals may be used, ten animals of each gender in each dose group is recommended for short-term rodent studies. Additional animals may be required for nonroutine endpoints, such as a group added to ascertain the reversibility of an adverse effect upon cessation of dosing. Also, additional animals may be placed in the high-dose group if potential compound related mortality is expected.

Many other factors influence the design of all repeated-dose toxicity studies. Before initiation of a study, it must be decided if the specific chemical or chemical mixture (i.e., the test material) to be used in the study is appropriate. The appropriate animal model must be chosen, the correct route of exposure must be selected, and the study duration must be decided. Control and treatment groups and their doses have to be selected. Variables to be evaluated must be selected to maximize the probability of detecting potential adverse effects.

## Regulatory Requirements

National and international regulatory bodies have issued guidelines for the design and conduct of repeated-dose toxicity studies. Even though study designs proscribed by these various guidelines have similar characteristics, some differences between the requirements have existed historically within and between regulatory agencies. This has resulted, in some cases, in the duplication of studies to ensure that the guidelines of different agencies are satisfied. The agencies themselves and the regulated community have considered this duplication of the use of animals and other resources undesirable. As a result, regulatory authorities worldwide have harmonized many of their guidelines to ensure toxicity studies conducted under a single set of regulations will be universally acceptable. In the United States, the Environmental Protection Agency (EPA) issued a single set of harmonized toxicity testing guidelines in 1998 [17] to blend the requirements of the separate guidelines previously promulgated through the Federal Insecticide, Fungicide and Rodenticide Act (FIFRA), the Toxic Substances Control Act (TSCA), and the Organization for Economic Cooperation and Development (OECD). The OECD, an international organization of 30 countries including the United States, has issued its own toxicity testing guidelines [55]. Each member country has agreed to accept studies conducted according to these guidelines. In 1991, regulatory authorities and trade organizations from the United States, Japan, and the European Union initiated a cooperative effort through the International Conference on Harmonization (ICH) to produce guidance documents concerning requirements for safety studies with pharmaceutical products. ICH guidance documents generally address specific issues related to toxicity testing (e.g., the duration of chronic toxicity testing [43]) and are not detailed guidelines for protocol design.

Each regulatory harmonization effort discussed above has been useful; however, a well-planned, scientifically valid, adequately conducted repeated-dose study should satisfy the requirements of any regulatory agency. The scientific foundation for toxicological hazard assessment is continually expanding, and to compromise this foundation by pursuing standard checklist protocols is irrational, wasteful, and unscientific. Today, regulations generally recognize the importance of scientific judgment and encourage discussions between the regulated community and the regulator about alternative study designs. Investigator and regulator each share in the obligation to do a scientifically sound study.

This chapter focuses on hazard assessment for chemicals that are small molecules. Today, biotechnology is a rapidly developing field. Hazard assessments for biologicals (e.g., recombinant proteins) are a special case that will not be discussed in detail in this chapter. Suffice it to say that guidelines for testing such materials continue to be developed. The hazard assessment for a biological is tailored specifically to the compound and may be more or less comprehensive than a classical assessment with a small molecule compound; however, many of the principles and methods of individual studies with small molecules are applicable to testing with biotechnology products. The toxicologist responsible for investigating the safety of these materials should become familiar with the current guidelines and suggestions for their testing.

The balance of this chapter relates to the typical design and conduct of subchronic and chronic repeated-dose toxicity studies. The information provided should suffice to satisfy most regulatory guidelines. Variables determined in subchronic and chronic toxicity studies are essentially identical, the major difference being the schedule for data collection. Table 24.4 compares the minimum requirements for subchronic and chronic toxicity studies using rodents as described in the EPA [14,15], U.S. Food and Drug Administration (FDA) [25,27], and OECD [54,58] testing guidelines. For repeated-dose study designs intended to satisfy guidelines of a particular regulatory authority, it is recommended that the reader: (1) use the testing guidelines of that authority as a starting point during the design of a study, (2) refine the design based on the compound of interest, (3) document the design in a detailed written protocol, and (4) discuss the protocol with a representative of the regulatory authority prior to initiating a study.

### Good Laboratory Practice Regulations

In addition to regulations and guidelines concerning the design of repeated-dose studies, U.S. and international regulatory authorities have issued good laboratory practice (GLP) regulations or principles concerning the manner in which all non-clinical hazard assessment studies (e.g., animals studies) are to be conducted, documented, and reported [11,19,36,46–48,57,62]. These GLPs are designed to ensure the quality of the study data and report. Requirements from each of the regulatory agencies are similar. GLPs set standards for the *test system*, laboratory organization, personnel, facilities, equipment, operations, and recordkeeping. They require that studies be conducted according to written protocols (see Table 24.5) and validated, written standard operating procedures (SOPs). Chemical analyses are required to characterize the *test article* and control article administered during the studies. Study procedures and data must be clearly and completely documented and reported. Automated data collection systems must be qualified and validated prior to use. Regulatory agencies inspect testing laboratories to ensure GLP compliance. Safety studies not conducted according to GLPs will not be accepted by regulatory agencies. In addition, some versions of GLPs, including those of the U.S. regulatory agencies, provide for disqualification of laboratories that are in major noncompliance with GLP requirements. For these reasons, laboratories conducting studies for regulatory submission generally take great care to comply with GLPs.

## STUDY DESIGN

### Chemical and Physical Characterization of the Test Material

It may seem that selection of the test material for repeated-dose toxicity testing should require little involvement by the toxicologist because it is usually provided by a chemist or product manager; however, several important factors must be considered by the toxicologist to ensure that a study will adequately assess the potential toxicity of the chemical and be accepted by regulatory agencies. Obviously, one of these factors is that the batch or lot of test material is representative of the chemical intended to be tested. Adequate chemical and physical characterization is essential for this determination. Several regulatory agencies and organizations have issued guidelines to describe the information needed and some of the methodology to be used to characterize a chemical [16,24,42,56].

A safety assessment should include a partnership between the toxicologist and a chemist that understands the chemical characteristics of the test material. An adequate chemical characterization of the test material should be provided to the toxicologist. Prior to initiation of a repeated-dose study, the toxicologist should carefully review the chemical and physical characterization data. This characterization should include methods of synthesis precursors used in the synthesis, and any solvents and manufacturing aids used in the manufacturing process. It should also include a quantitative assessment of the major components, with associated accuracy and precision information, and at least a qualitative analysis of minor components. Ideally, the toxicologist should be provided with a mass balance for the test material. Also, the methods of purification of the test material should be provided. This allows the toxicologist to determine the potential for occurrence of residues and impurities that could produce adverse effects. It also provides the toxicologist with the information necessary to determine if any specific residues or impurities should be specifically targeted for additional analytical determinations.

The specific characterization information required by the toxicologist may vary with each test material. The required data may be different from that used to establish manufacturing and quality control specifications. It i essential that the manufacturing and quality control specifications and procedures meet the specific regulatory requirements for toxicity testing; for example, good manufacturing practices (GMPs), which are somewhat similar to GLPs, are required for some chemicals [21,30,32, 33,40]. The chemist should ensure that adequate GMP records exist, if required.

Before initiation of a repeated-dose toxicity study, chemical analyses should be conducted to ensure that the test material is stable over the range of anticipated concentrations and for the maximum period of use during the study. According to GLPs, this evaluation can be conducted concurrently with the study, but detection of instability during the study could invalidate the study results. If the material is stable only under certain conditions (e.g., frozen storage), special arrangements for its storage should be made to eliminate instability problems. Instability of the test material may result in lower than expected doses and exposure of the animals to degradation products. The products of degradation may either have their own unique toxicity or alter the toxicity of the test material. This may make it impossible to correctly interpret data from the study.

The toxicologist will have assessed the chemical characterization of the test material at the initial stages of the hazard assessment; however, more complete information may become available before initiation of repeated-dose toxicity studies. Sometimes the chemical synthesis or other production methods may change between the initial assessment and the start of subchronic or chronic studies; therefore, it may be necessary to reassess the chemical characterization before initiation of these studies. Unanticipated changes in the chemistry of the test material or mixture may necessitate changes in the toxicity study design.

Natural products of plant origin (e.g., some of the herbal remedies that are popular today) are often complex mixtures that present challenges to the chemist and toxicologist. The quantitative chemical composition can be quite variable based on the environmental growing conditions of the plant, such as soil composition, rainfall, or temperature, and the particular cultivar. This results in a range of quantitative values for the components of the mixture and this must be considered by the toxicologist.

The test material should be as similar as possible to the chemical to which humans will be exposed; therefore, every attempt should be made to ensure that the test material is either identical to the final commercial product or representative of the anticipated final product. This may not always be possible with commercial products because large-scale production facilities are usually not available during this phase of a safety assessment. When such facilities become available, it may be possible to bridge between the final commercial product and the test material by chemical analysis. "Bridge chemistry" should identify any differences between the chemical that was tested and the commercial product. This will allow a determination of the toxicological bioequivalence of the test material and final commercial product.

If a "new" impurity of unknown toxicity is detected in a drug product after chronic animal studies have been conducted, the second International Conference on Harmonization [42] has proposed the following:

- If the intended human use is short term:
  - Single-dose toxicology comparison of old and new test substance
  - Repeated-dose, 4-week comparison of old and new test substance
  - Mutagenicity tests
- If intended use is long term, with a high dose of the active compound:
  - Single-dose toxicology comparison of old and new test substance
  - Repeated-dose, 3-month subchronic comparison of old and new test substance
  - Mutagenicity tests (if results of the mutagenicity test are positive, carcinogenicity testing of the impurity may be considered)
- If the product is to be given to women of child-bearing potential:
  - Consider the need for a comparative embryotoxicity and teratogenicity (segment II) study in one suitable species

The test material should not only have the same chemical characteristics as the material of commerce but should also have the same physical characteristics. If the test material is a solid intended to be used as a powder, it should be administered to the test animals in powder form. Particle size of the powder should be similar for the test material and the material to which humans will be exposed. If humans will be exposed to the material in solution, a solution of the material should be used in the repeated-dose study. Analyses should be conducted to ensure that the composition of the test material falls within the limits of anticipated or known product specifications. Physical specifications will allow the toxicologist, often times in consultation with the chemist, to determine the most appropriate method of adding the test material to the dosing matrix, such as the diet or drinking water (also see the route-of-exposure discussion, below).

It is important for the toxicologist to ensure that an adequate supply of test material is available before initiating a toxicity study. A single lot of the test material should be used throughout the subchronic or chronic study whenever possible. Furthermore, if the repeated-dose study is part of a series of studies in a safety assessment program, it is desirable to use a single lot of test material

**TABLE 24.4**
**Subchronic and Chronic Rodent Oral Toxicity Studies Based on Various Regulatory Guidelines**

| | EPA OPPTS Guidelines | | FDA Redbook | | OECD Guidelines | |
|---|---|---|---|---|---|---|
| | Subchronic (14) | Chronic (15) | Subchronic (25) | Chronic (27) | Subchronic (58) | Chronic (54) |
| Study duration | ≥90 days | ≥12 mo | ≥90 days | ≥12 mo | ≥90 days | ≥12 mo |
| Number of treated groups | | | | | | |
| Standard study | ≥3 | ≥3 | ≥3 | ≥3 | ≥3 | ≥3 |
| Limit test[a] | 1 | 1 | NA | NA | 1 | NA |
| No. of negative control groups[b] | | | | | | |
| Untreated control | 1 | 1 | 1 | 1 | 1 | 1 |
| Vehicle control | 1 | 1 | 1 | 1 | 1 | 1 |
| No. of animals/gender/group[c] | ≥10 | ≥20 | ≥20 | ≥10 | ≥10 | ≥20 |
| Age of animals (at start of study) | ≤8-9 wk | ≤8 wk | <6–8 wk | ~6 wk | <9 wk | ASAP[f] |
| Body weight measurement | | | | | | |
| Frequency through 13 weeks | Weekly | Weekly | Weekly | Weekly | Weekly | Weekly |
| Frequency after 13 weeks | NA | Every 4th wk | NA | Weekly | NA | Every 4th wk |
| Feed consumption measurement | | | | | | |
| Frequency through 13 weeks | Weekly | Weekly | Weekly | Weekly | Weekly | Weekly |
| Frequency after 13 weeks | NA | Monthly | NA | Weekly | NA | Every 3rd mo |
| Observations | | | | | | |
| For mortality and morbidity (times/day) | 2 | 2 | 2 | 2 | 2 | 2 |
| For general condition (times/day) | 1 | 1 | 1 | 1 | 1 | 1 |
| For detailed clinical findings (frequency) | Weekly | Weekly | Weekly | Weekly | Weekly | Daily |
| Neurotoxicity evaluation (at term)[d] | Y | Y | Y | N | Y | N |
| Ophthalmology | | | | | | |
| No. of animals pretest | AA | AA | AA | AA | AA | N |
| No. of animals/gender/high dose and control every 3 months[e] | NA | n | NA | AS | NA | N |
| No. of animals/gender/high dose and control at term[e] | AS | 10 | AS | AS | AS | N |

| | | | | | | |
|---|---|---|---|---|---|---|
| Hematology and clinical chemistry (no./gender/group) | AS | 10 | 10 | 10 | AS | 10 |
| Intermediate time(s) | N | 6 mo | 14 and 45 d | Every 3rd mo | N | 3 and 6 mo[g] |
| Term | Y | Y | Y | Y | Y | Y |
| Urinalysis (no./gender/group) | AS | 10 | 10 | 10 | AS | 10 |
| Intermediate time(s) | N | N | Y | Every 3rd mo | N | 3 and 6 mo |
| Term | O | Y | Y | Y | O | Y |
| Gross necropsy and tissue collection | AA | AA | AA | AA | AA | AA |
| Organ weights (no./gender/group at term) | AS | AS | AS | AS | AS | 10 |
| Adrenals, kidneys, liver | Y | Y | Y | Y | Y | Y |
| Brain | Y | Y | Y | Y | Y | Y |
| Testes/ovaries | Y/Y | Y/Y | Y | Y | Y/Y | Y/Y |
| Epididymides, heart, uterus | Y | Y | Y | N | Y | N |
| Spleen | Y | Y | Y | Y | Y | N |
| Thymus | Y | N | Y | Y | Y | N |
| Histopathology | | | | | | |
| All tissues (all high dose and control animals) | Y | Y | Y | Y | Y | Y |
| All tissues (all animals killed or died on study) | Y | Y | Y | Y | N | Y |
| Target tissues and gross lesions (all animals) | Y | Y | Y | Y | Y | Y |
| Selected tissues (all intermediate dose animals) | N | N | Y | N | N | N |

[a] If a test at one dose level of at least 1000 mg/kg body weight/day produces no observed adverse effects and if toxicity would not be expected based on data from structurally related compounds, then a full study using three dose levels may not be considered necessary.

[b] Contingent upon the route of exposure in the test article treated groups.

[c] Extra animals must be added for interim sacrifices and special determinations (e.g., toxicokinetics and reversibility of effects).

[d] Not required if similar data are available from other studies or if other clinical signs are noted to an extent that would interfere with evaluation.

[e] Animals in intermediate groups are to be evaluated if treatment-related findings are noted in the high-dose group.

[f] As soon as possible after weaning and acclimation (e.g. ≤6 to 8 weeks of age).

[g] No clinical chemistry at 3 months.

*Note:* NA, not applicable; Y, yes/required; N, no/not required; O, optional; AA, all animals; AS, all survivors.

**TABLE 24.5**
**Protocol Contents Required by FDA Good Laboratory Practices Regulations**

| |
|---|
| Title and study objective |
| Identification of test and control articles |
| Identification of sponsor and testing facility |
| Justification of test system (animal model) |
| Test system information (e.g., number, body weight, gender, source, species/strain, age, method of identification) |
| Description of study design and methods for control of bias (e.g., random assignment of animals to treatment groups, processing of clinical pathology samples in replicates) |
| Animal husbandry information |
| Dosing information, including dose form preparation and route of administration |
| Methods by which degree of absorption of the test and control articles by the test system will be determined, if necessary |
| Types and frequencies of assays, analyses, and measurements to be made |
| Description of statistical methods |
| Records to be maintained |

for the entire program. This reduces the probability of encountering inconsistent results in different studies with the same test material, resulting from interlot differences in chemical or physical characteristics. If a single lot of test material is not available in sufficient quantity to complete a study or studies, multiple lots may be used. Chemical characterization of each new lot is required to ensure that it meets all specifications and reasonably duplicates previous lots.

The U.S. FDA has issued guidelines addressing the chemistry data requirements for direct food additive petitions [24]. The ICH has issued guidelines for chemistry requirements for drug candidates during preclinical hazard assessment [42]. Although the toxicologist may not have as complete a data package as described above before the initiation of a repeated-dose study, sufficient data must be available to meet GLP regulations. FDA GLPs state that, "The identity, strength, purity, and composition or other characteristics which will appropriately define the test or control article shall be determined for each batch and shall be documented. Methods of synthesis, fabrication, or derivation of the test and control articles shall be documented" [35]. U.S. EPA GLP statements concerning the requirements for chemical characterization of the test material are essentially identical to those of the FDA [12]. The OECD GLP guidelines for chemical characterization state: "For each study, the identity, including batch number, purity, composition, concentrations, or other characterizations to appropriately define each batch of the test or reference items should be known" [57]. Further information on the chemical characterization of the test material is provided by the FDA in its guidelines for toxicity studies, "The composition of the test substance should be known: Information should include the name and quantities of all major components, known contaminants and impurities, and the percentage of unidentifiable materials. The test substance in toxicology studies should be the same substance that the petitioner intends to market" [26].

## Route of Exposure and Method of Test Material Administration

The anticipated human route of exposure to the test material dictates the route of exposure for most subchronic and chronic toxicity studies. Nonintended routes of human exposure also should be considered during selection of exposure route. The most common routes of exposure in these studies are dietary, oral (gavage or capsule), dermal and inhalation. *Parenteral* administration by infusion using implanted or external pumps or intravenous, subcutaneous, intraperitoneal, or other types of injection is becoming increasingly common. Less frequently, test materials are administered in the drinking water. Frequently, several potential routes of human exposure to a single chemical exist; for example, consumer exposure to a pesticide may occur by dietary consumption of food crops containing residues of the chemical. Farm worker exposure to the same pesticide may occur by either inhalation or dermal routes during application and harvesting. In such cases, subchronic testing may be required to assess the effects of exposure by all three routes. Emphasis is generally placed on the route by which the most widespread human exposure would occur. Chronic testing is usually only conducted using that route of exposure. Several of the more common routes of exposure are discussed in detail below.

### Capsule Administration

If expected human exposure is by the oral route, a solid test material (e.g., a dry powder) can be administered by capsule. Unformulated bulk test material can be given to some large animals, such as dogs and cats, in gelatin capsules inserted into the esophagus manually or with the aid of a mechanical device designed to prevent the animal from biting the individual administering the capsule. Capsules are available in a wide range of sizes and can hold, depending on the density of the test material, as much as 1.5 g. The amount of test material that can be administered is limited by the capacity of each capsule and the practical number of capsules that can be administered at one time. For smaller species, such as rodents, this method is generally impractical, and the test material must be administered orally as a solution or suspension in an appropriate *vehicle*.

## Oral Gavage

Another common route of oral administration is oral gavage by *intubation*. This technique may be used for rodent and nonrodent species. For oral gavage, a solution or suspension of the test material is deposited into the stomach via the esophagus using an intubation tube attached to a graduated syringe or other device. A test material is added to an appropriate vehicle, usually aqueous. If the material is not readily soluble, a suspension may be prepared. For suspensions, a material such as methylcellulose, carboxymethylcellulose, or gum tragacanth is added to water to increase the viscosity, and the test material is homogeneously suspended in the vehicle. A wetting agent such as Tween® 80 can be used to increase the suspendability of the material. Although aqueous vehicles are preferred, it is possible to use an oil vehicle for lipid-soluble materials. Food oils, such as corn oil, may be used at appropriate volumes but mineral oils must be avoided. It should be remembered that food oils add to normal dietary caloric intake and, at high volumes, oils may interfere with the absorption of fat-soluble nutrients, such as fat-soluble vitamins. Also, absorption of the test material may be quite different from an oil vehicle compared with an aqueous vehicle and this may be highly volume dependent.

The volume of test solution or suspension administered can influence gastric emptying time. Because gastric emptying time may affect gastrointestinal uptake and bioavailability, it generally is preferred that the dose levels of the test substance be administered on a constant-volume (mL/kg), variable-concentration (mg/mL) basis. For dosing using a constant volume, 10 mL/kg body weight is most commonly used as the upper limit. At 10 mL/kg, a 260-g rat would receive a dose volume of 2.6 mL, a 35-g mouse would receive 0.35 mL, and a 10-kg dog would receive 100 mL. Dosing usually is conducted once daily but may be performed more frequently in order to mimic the intended human dosing regimen or if concentration and/or pharmacokinetics considerations limit the achievable single-dose exposure to less than the amount desired.

Daily exposure by oral gavage results in a *bolus dose*. Administration via the diet or drinking water results in a more constant exposure throughout the duration of the study that is more dependent on feeding patterns than dosing schedule. Because rodents are nocturnal, the peak periods for activity and feeding behavior are the end of the light and dark periods (i.e., just before lights-on or just after lights-off). If high volumes of dosing solution (e.g., 20 to 50 mL/kg) are to be administered to rodents or if the presence of feed may interfere with the absorption of the test substance, consideration should be given to dosing animals during late morning or in the afternoon. For nonrodents, timing of dosing relative to feeding (i.e., before or at a certain time after daily feeding) accomplishes the same purpose. In any case, the timing of gavage dose in relation to feeding should be stated clearly in the study protocol.

Administration of a test chemical as an undiluted liquid, as a diluted liquid, as a dissolved solid, or as a solid in suspension generally results in accurate delivery of the intended dose to each of the animals in a repeated-dose study when exposure is by oral gavage. To deliver an accurate dose of test chemical in solution or suspension, the chemical must be mixed with solvent or suspending agent at the proper concentration. The appropriate concentration of test solution or suspension can be calculated using the intended dose level and the dose volume of the solution or suspension to be administered to the animals. This calculation is illustrated in Figure 24.2. Obviously, this same calculation applies to solutions or suspensions for other exposure routes (e.g., administration by capsule, dermally, or parenterally).

Use of a vehicle to carry the test material into the animal has been mentioned several times in the preceding discussion. Choice of vehicle is a critical decision for toxicology studies. Obviously, the vehicle should be nontoxic at the dose (volume) administered. It should not act in an additive or synergistic manner to enhance the toxicity of the test material nor should it interfere with the expression of any potential toxicity. Because such interactions are not always predictable, it is imperative that vehicle controls be used in all oral gavage toxicology studies; however, even use of a vehicle control may not always compensate for the administration of vehicle in the groups receiving the test material. Interactions between vehicle and test materials can be encountered and must be considered during the design of studies. Vehicles may produce physiological or nutritional alterations that may affect the toxicity of the test material, especially in longer term studies; for example, an oil vehicle may result in poor absorption of fat-soluble nutrients. Food oil vehicles add to the caloric intake of the animal and may produce effects in longer term studies. Absorption of fat-soluble materials from oil vehicles can be slower than when the same material is administered in an aqueous matrix. The resulting change in the toxicokinetics of the test material can produce profound differences in toxicity. Because toxicity testing generally uses exaggerated doses compared to human exposure, it is sometimes difficult to obtain solutions. In such cases, suspensions are sometimes used. This again can affect toxicokinetics and influence toxicity. For these reasons, among others, great care must be utilized in selecting an appropriate vehicle for administration of test materials to animals.

## Dermal Application

For a drug, cosmetic, industrial chemical, or pesticide with dermal exposure potential for humans, dermal application is the most appropriate exposure route for repeated-dose

**Formula:**

$$\textit{Concentration}\text{ (mg test material/mL solution or suspension)} = \frac{\textit{Intended dose level}\text{ (mg test material/kg body weight/interval)}}{\textit{Dose volume}\text{ (mL solution or suspension/kg body weight/interval)}}$$

**Example:**

The concentration of a solution intended to deliver a daily 1500-mg/kg dose level of test material A to rats at a dose volume of 5 mL/kg body weight (b.w.) is calculated as follows:

$$\textit{Concentration}\text{ (mg A/mL)} = \frac{\text{1500 mg A/kg b.w./day}}{\text{5 mL solution/kg b.w./day}}$$

$$= \text{300 mg A/mL solution}$$

As indicated in the following, a 300-g rat would receive 1.5 mL of this solution each day and, therefore, would receive the intended daily dose level. First, calculate the volume of solution administered to the rat:

$$\text{0.3 kg b.w.} \times \text{5 mL solution/kg b.w./day} = \text{1.5 mL/day}$$

Then, calculate the amount of test material A in that volume:

$$\text{1.5 mL solution/day} \times \text{300 mg A/mL solution} = \text{450 mg A/day}$$

Finally, calculate the dose level resulting from administration of that amount of test material:

$$\text{450 mg A/day} \div \text{0.3 kg b.w.} = \text{1500 mg A/kg b.w./day}$$

**FIGURE 24.2** Calculation of test material concentration in a solution or suspension.

studies. The test material is applied to a defined area of skin from which the hair has been removed. Removal of hair by both shaving and depilatories can alter skin permeability to applied materials. Because the test animal can ingest some of a dermal dose during grooming, it is common practice to cover the test material application site by wrapping the trunk of the animal with a semi-occlusive material (e.g., gauze strips) during the exposure period. Occasionally, if exaggerated dermal absorption is desired, the application site is covered with an occlusive material (e.g., rubber dam). Either method of wrapping (occlusion or semi-occlusion) maintains the test material in contact with the skin. Wrapping also increases the accuracy of the administered dose because, on unoccluded animals, dry test materials may fall off the application site and liquid materials may run off or evaporate from the site. Restraint or Elizabethan collars are also sometimes used to prevent the animal from tampering with the wrap or to prevent oral ingestion if the application site is unoccluded; however, the use of restraints or collars stresses the animals and, therefore, may alter the outcome of a study. To avoid unnecessary stress, the animals should be trained to accept such procedures before actual dosing is initiated.

## Parenteral Administration

A pharmaceutical intended for administration to humans by injection should be administered to the test animals by the same parenteral route. A common type of parenteral exposure is subcutaneous injection. To avoid irritation and other potential effects, such as fibrotic reactions, the specific site of injection should be changed daily. It is recommended that the dosing vehicle be aqueous, as oil vehicles may track back along the needle path and be deposited in the lipophilic skin and hair. As with oral gavage exposure, absorption from an oil vehicle may differ compared with an aqueous vehicle. For subcutaneous administration, care must be taken to ensure that the material is deposited subcutaneously rather than intradermally or intramuscularly. This can be accomplished by lifting the skin of the animal to form a pocket and injecting the dosage into the subcutaneous pocket.

Other common routes of parenteral exposure include intravenous (IV), intramuscular (IM), and intraperitoneal (IP) injection. Special considerations are required for each of these routes. For IV administration, the test material must be soluble in an aqueous vehicle, as physiologic saline is the usual vehicle. Air bubbles must be cleared from the delivery system (i.e., needle, catheter, or syringe) before injecting the material intravenously. The dosing solution for IV administration must be at a physiologic pH and must not be extremely irritating or corrosive. Repeated use of the same injection site must be avoided. To prevent inadvertent delivery of the test material into the muscle or other surrounding tissue, care must be taken to ensure that the needle tip is in the lumen of the vein before injecting the test material. Proper placement of the needle can be assessed by drawing a small amount of blood back into the delivery device (e.g., catheter or syringe). If no blood is observed, the needle is not located in the lumen of the vein. For IM and IP routes of exposure, proper placement of the needle tip also must be assessed before administering the dose material. In contrast to IV exposure, drawing back on the delivery device without obtaining blood is required prior to administering material by the IM route. Similarly, drawing back on the device without obtaining blood or intestinal contents prior to injecting test material is necessary for correct IP administration. Most other considerations for IM and IP exposure are similar to those for IV administration (e.g., the material must not be corrosive or irritating, repeated injections at the same site should be avoided).

Continuous infusion may be used to simulate a constant human exposure or, for intravenous infusion of poorly soluble test substances, to deliver a sufficiently large daily dose for safety assessment. Infusions may be administered using either implanted or external pumps. Implanted pumps, may be battery-operated mechanical pumps (nonrodents) or osmotic pumps (rodents and nonrodents). These pumps permit unrestricted movement of the animal. Practical considerations of volume and weight of implanted pumps restrict their use to very soluble or potent test substances that can be administered slowly over a long period. External pumps may be connected to rodents and nonrodents using tethers and swivels or affixed to nonrodents as backpacks. External pumps allow for infusion of larger volumes of test substance but also require more waste substance because of the larger dead space of the tether catheter. Use of external pumps may be problematic for subchronic or chronic studies because of their relatively long duration. The patency of the catheter for implanted and external pumps must be confirmed. The dose received by each animal should be based on measurements of the actual amount of test solution infused during the dosing period. This is determined by measuring the weight of the pump and solution prior to initiation and following completion of exposure.

### Implantation

Subcutaneous or intramuscular implantation is often used for evaluating biopolymers for medical devices or prostheses. In addition, test materials have been embedded in special matrices that allow continuous, sustained release for weeks or months after subcutaneous implantation. Demonstration of the stability of the test substance in the matrix, both during preparation of the pellet and after implantation, is required when using this technology. Determination of plasma concentrations of the test material and its metabolites can be used to confirm proper dosing. Because of size limitations, these pellets are only useful for delivery of very potent substances, such as hormones and biological proteins.

### Dietary Administration

Dietary administration would be appropriate for a food additive or a pesticide that has potential to become a residue in or on food crops. This type of administration, although frequently used in subchronic and chronic toxicity studies, is less accurate than most other routes of dose delivery. This is primarily because of differences between body weight and feed consumption of individual animals, which results in variable compound consumption. A further complication in rodent studies is that these animals tend to add body fat and not lean body mass as they age; therefore, lipophilic compounds have a larger mass of fat into which they can partition. This may actually decrease the effective dose. In addition, fat does not have significant detoxication enzymes; therefore, as dose is increased to account for increased body weight in older rodents, detoxication capacity may be stressed, resulting in unanticipated toxicity. This complication is not unique to dietary exposure studies and will affect the effective dose of a test material in rodents exposed by most other routes. Although potential variability in effective dose is important to interpretation of the results of a study, no attempt is generally made to compensate for this variability during preparation of the test diet or calculation of the administered dose in repeated-dose toxicity studies. Spillage or soiling of feed, which occurs fairly frequently with some test animals, is a factor that contributes variability to the calculation of dietary dose. Significant spillage or soiling must be considered when measuring feed consumption to be used in calculation of dietary concentration or compound consumption. Dietary administration can be conducted by either of two methods: (1) adjusting the dietary concentration of the test material to account for changing body weight and feed consumption, or (2) feeding a constant concentration in the diet.

Routinely adjusting the dietary concentration of a test material based on the changing body weight and feed consumption of the animal provides reasonably good control over the delivered dose during a repeated-dose study. Dietary concentration is usually adjusted weekly during a subchronic study. In a chronic study, diet concentration is usually adjusted weekly during the first 13 to 14 weeks (especially for rodents because of the rapid growth of the animals during this period) and biweekly or sometimes monthly thereafter. Using this method, the mean feed consumption and body weight of a dose group during a given study interval are used to calculate the dietary concentration to be fed to that group during the following interval. There are several ways to calculate the concentration; one formula for this calculation and an example of its use are presented in Figure 24.3.

Feeding a constant concentration of the test material in the diet throughout the study is the second form of dietary administration. This method provides less control

**Formula:**

$$\textit{Concentration}\ (\text{mg test material/kg diet}) = \frac{\textit{Intended dose level}\ (\text{mg test material/kg body weight/day})}{\textit{Projected feed consumption}\ (\text{kg feed/kg body weight/day})}$$

where projected feed consumption (PFC) for a study week is based on body weight (b.w.) and absolute feed consumption (AFC) data from the previous week and is calculated as follows:

$$\text{PFC (kg/kg/day) for week}_n = \frac{\text{AFC week}_{n-1}\ (\text{kg})}{7\ \text{days}} \div \left[\text{b.w. end of week}_{n-1}\ (\text{kg}) + \frac{\text{b.w. gain during week}_{n-1}\ (\text{kg})}{2}\right]$$

**Example:**

In a subchronic rat study, the mean body weight of the males in the 15-mg/kg/day dose group at the beginning of week 11 is 520 g. At the end of week 11, the mean weight for these males is 540 g. Mean feed consumption of these animals is 154 g during the 7 days of week 11. The dietary concentration intended to deliver a 15-mg/kg/day dose level of compound A to this group of rats during week 12 of the study is calculated as follows:

$$\text{PFC (week 12)} = \frac{0.154\ \text{kg feed}}{7\ \text{days}} \div \left[0.540\ \text{kg b.w.} + \frac{0.020\ \text{kg b.w. gain}}{2}\right]$$

$$= 0.04\ \text{kg feed/kg b.w./day}$$

Therefore:

$$\textit{Concentration}\ (\text{week 12}) = \frac{15\ \text{mg A/kg b.w./day}}{0.04\ \text{kg feed/kg b.w./day}}$$

$$= 375\ \text{mg A/kg feed}$$

Note that, because their body weight and feed consumption differ, the diet concentrations for males and females in the same dose group will generally differ throughout the study.

**FIGURE 24.3** Calculation of adjusted diet concentration to yield constant dose level.

over the administered dose level because it is a function of the amount of feed (and, therefore, test material) consumed and body weight gained by each animal. The consumed dose, or compound consumption, of each individual animal can be calculated using its feed consumption and body weight. The mean of the individual animal compound consumptions represents the compound consumption for each dose group. A commonly used formula for the calculation of individual compound consumption is illustrated in Figure 24.4.

When feeding a constant concentration in a rodent study, the toxicologist must be aware that the compound consumption may vary significantly for an individual animal or group of animals during the course of the study. For example, compound consumption during the first week of a 13-week rat study in which the test compound is fed at a constant dietary concentration is frequently more than twice the compound consumption during the last week of the study. This variation results from the rapid decrease in feed consumption relative to body weight (g feed per kg body weight per day) by young rats during the first several months of life. This variability in compound consumption is illustrated in the example given in Figure 24.5.

## Drinking Water

Water-soluble test substances can be offered as a mixture in the drinking water, providing adequate dosage can be achieved. This route may be preferred when it mimics human exposure conditions of constant exposure, compared with event-oriented exposure such as pill taking. Spillage can be a significant problem when administering material in drinking water, and recovery of spilled water is usually not feasible. Another complication with drinking water administration is that evaporation of water and volatilization of the test material can occur from the tip of the drinking (sipper) tube, resulting in alteration of the concentration of the material in the water. Use of a sipper tube containing a ball bearing tip minimizes this problem.

## Assessment of the Adequacy of Test Material Preparations

Whatever the route of exposure, it is critical to determine if the test material is delivered to the animals at the intended doses. For dietary studies, test diets should be prepared before initiation of a repeated-dose study using the intended diet preparation method. It must be shown

**Formula:**

$$\textit{Compound consumption}\text{ (mg test material/kg b.w./day)} = \underset{\text{(mg test material/kg feed)}}{\textit{Concentration}} \times \underset{\text{(kg feed/kg body weight/day)}}{\textit{Relative feed consumption}}$$

where relative feed consumption (RFC) for a study week is based on body weight (b.w.) and absolute food consumption (AFC) during that week and is calculated as follows:

$$\text{RFC (kg/kg/day)} = \frac{\text{AFC (kg)}}{\text{7 days}} \div \left[\text{b.w. start of week (kg)} + \frac{\text{b.w. gain during week}_{n-1}\text{ (kg)}}{2}\right]$$

**Example:**

In a subchronic rat study, a group of males is fed test material A at a constant dietary concentration of 2% (w/w). The body weight of one animal in this group is 175 g at the start of week 1 and 225 g at the end of week 1. Its feed consumption during the 7 days of week 1 is 168 g. Subsequently, that rat weighs 490 g at the start and 510 g at the end of week 13, and its feed consumption during that week is 196 g. The compound consumption of this rat for each week is calculated as follows;

$$\text{RFC (week 1)} = \frac{\text{0.168 kg feed}}{\text{7 days}} \div \left[\text{0.175 kg b.w.} + \frac{\text{0.050 kg b.w. gain}}{2}\right]$$

$$= \text{0.120 kg feed/kg b.w./day}$$

$$\text{RFC (week 13)} = \frac{\text{0.196 kg feed}}{\text{7 days}} \div \left[\text{0.490 kg b.w.} + \frac{\text{0.020 kg b.w. gain}}{2}\right]$$

$$= \text{0.056 kg feed/kg b.w./day}$$

Therefore:

$$\begin{aligned}\textit{Compound consumption}\text{ (week 1)} &= \text{2\% test material A} \times \text{RFC (week 1)}\\ &= \text{2 g A/100 g feed} \times \text{0.120 kg feed/kg b.w./day}\\ &= \text{20,000 mg A/kg feed} \times \text{0.120 kg feed/kg b.w./day}\\ &= \text{2400 mg A/kg b.w./day}\end{aligned}$$

$$\begin{aligned}\textit{Compound consumption}\text{ (week 13)} &= \text{2\% test material A} \times \text{RFC (week 13)}\\ &= \text{20,000 mg A/kg feed} \times \text{0.056 kg feed/kg b.w./day}\\ &= \text{1120 mg A/kg b.w./day}\end{aligned}$$

**FIGURE 24.4** Calculation of compound consumption resulting from constant diet concentration.

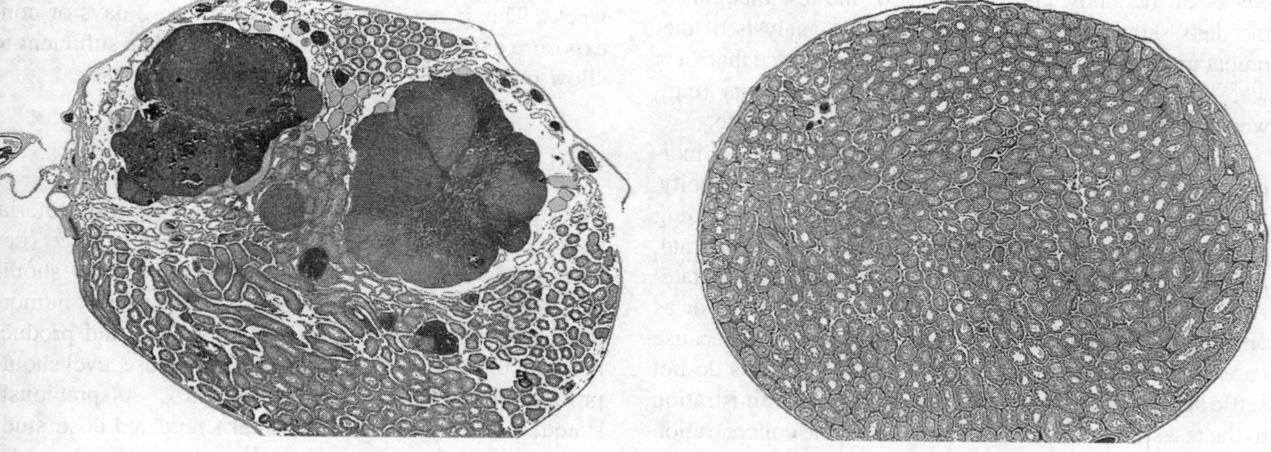

**FIGURE 24.5** Photomicrographs of hematoxylin- and eosin-stained testes from Fischer 344 rats (original magnification, 2.5×). The testis on the left exhibits an interstitial cell tumor. The testis on the right is without pathologic changes.

that this preparation method yields diets containing the appropriate amounts of homogeneously mixed test material. Chemical analysis of samples taken from several locations in each test diet preparation should be conducted to determine if the proper concentrations of test material have been achieved and to assess the homogeneity of the dietary admixtures. If the results of these analyses indicate that the anticipated concentrations were not achieved or the distribution of test material in the diet was not homogenous, the diet preparation method should be revised and retested. Diet preparation must be validated before the study can be initiated.

During the prestudy homogeneity determinations, additional diet samples should be collected and analyzed to show that, within the range of concentrations to be used in the study, the test material is stable in the diet. These samples should be stored under animal room conditions and under frozen conditions for the maximum period of time during which the diet will be used or stored. For a study in which dietary admixtures will be prepared and fed once per week, stability of the test material in the diet would commonly be assessed for samples stored under animal room conditions for at least seven and fourteen days. This allows estimation of the degradation rate at room temperature. Analysis of frozen diet samples stored for several intervals is also advisable. Demonstration of stability under frozen storage conditions makes chemical analysis of diet samples immediately after collection during the toxicity study unnecessary. It also validates the possibility of confirming analytical results by reanalysis of stored frozen samples, if needed. Adequate stability of the test material in the diet should be demonstrated prior to initiation of the study.

Even when the adequacy of the diet preparation method and stability of the test material in the diet have been demonstrated, it is important to monitor diet preparation during the study. For each diet preparation during the first several weeks of the study, concentrations of the test material in the diets should be assessed. Subsequent analysis of diet preparations every 2 to 4 weeks will add assurance that diets were prepared properly. More frequent analysis (e.g., weekly throughout the study) is even more desirable.

For routes of exposure other than dietary, the principles cited above also apply. Concentration, homogeneity, and stability of the test material in solvents or suspending agents must be determined for studies using oral, dermal, inhalation, or other routes of exposure when the test material is to be administered in solution or suspension or as an aerosol. Suspensions represent a special case because care must be taken to ensure that the suspensions do not settle and become nonhomogeneous during administration to the test species. For inhalation studies, the concentration of gas, aerosol, or particulates to which the animals are exposed should also be assessed using appropriate analytical methods.

## Duration of Exposure

As stated previously, the duration of subchronic toxicity studies involves exposure of the test species to a chemical during a significant portion of its lifetime. Classically, these studies are conducted for 90 consecutive days or approximately 13 weeks. Chronic toxicity studies in rodents most commonly involve exposure for a major portion of their life span, generally 12 months although studies of shorter duration (e.g., 6 to 9 months) are considered acceptable by some groups [43]. Rodent chronic toxicity studies are sometimes combined with lifetime oncogenicity studies to achieve efficiencies during some of the study procedures (e.g. diet preparation). During these combination studies, the animals in the chronic toxicity segment are generally studied during the first 6 to 12 months and then are terminated. Those in the oncogenicity segment continue on study generally for at least 24 months. In nonrodents (e.g., dogs, nonhuman primates) that are longer lived than rodents, 6 to 12 months represents a significantly smaller portion of their life span but is currently considered an adequate duration of exposure to detect chronic effects.

Daily test material administration during a repeated-dose toxicity study can be continuous, intermittent, or repeated. In most dietary and drinking water studies, the animals have free access to diets or water containing the test material throughout the study, and exposure is essentially continuous, although influenced by diurnal patterns of consumption. In dermal or inhalation studies, exposure to the test material is intermittent, generally 4 to 6 hours per day. When the route of administration is intravenous infusion, exposure may be either continuous or intermittent. With bolus dose parenteral administration or oral gavage, test material administration is generally once or at most, a few times each day. Labor-intensive methods of administration, such as oral gavage, are sometimes done only during the standard work week (i.e., 5 days per week). This is not recommended because 2 days of nonexposure each week during the study may be sufficient to allow modification or reversal of toxic responses.

## Dose Groups

The minimum number of groups receiving test material in a repeated-dose toxicity study is generally three (i.e. low-, mid-, and high-dose). The high-dose level should produce evidence of toxicity but should not result in more than 10% mortality. The mid-dose level should produce no more than slight toxicity, and the low dose level should produce no toxicity yielding a NOAEL. As previously stated, a short-term (2- to 4-week) repeated-dose study should be conducted to aid in the selection of doses for subchronic testing. For test materials where a dose response has not been well defined during a short-term repeated-dose study, additional dose groups may be required in the subchronic study to ensure that the range of desired responses (i.e., no toxicity to significant toxicity) is achieved; however, it is sometimes difficult to completely satisfy these criteria. Before selecting doses for a chronic toxicity study, a subchronic study that defines no-effect and effect levels should be completed.

### Limit Studies

For test materials that possess very low potential for toxicity, the inclusion of only one test material dose group in a repeated-dose study is sometimes acceptable. A study with this design is termed a *limit study*. Limit testing is inappropriate for materials with anticipated high human exposure. The dose level for the test material group in dietary and dermal limit studies is normally at least 1000 mg/kg/day. Another type of limit study involves utilization of the maximal exposure level under the conditions of the study. For example, suppose the majority of toxicology data concerning a lipophilic drinking water contaminant has been collected using oral administration in a corn oil vehicle and additional data are desired using a water vehicle. The limited water solubility of the test material may result in the maximal dose being significantly lower than the dose used in the corn oil gavage studies; however, the test material can be tested as a saturated water solution. Such a study may reveal the test material to be either more or less toxic in water than in the oil vehicle. The data are relevant to the assessment of hazard associated with exposure in the drinking water because the maximal possible exposure by this route of administration was tested; thus, although a limit study may not define the "complete" toxicology of a test material, it can define the "practical" toxicology of the material.

## Control Groups

Adequate controls are essential to successful toxicity studies of all types, including repeated-dose studies. Studies should contain at least one control group for comparison with the groups receiving the test material. The control group should be treated identically to the treated groups except the control group should receive no test material. Control groups can be either negative or positive controls.

Negative control groups are intended to demonstrate the normal state of the animal for comparison to data from the groups treated with the test material. They also provide an opportunity to compare baseline data for the current study to baseline data from previous studies. There are several types of negative controls. If the test material is dissolved or suspended in a vehicle for administration, a vehicle control group should receive, by the same route of exposure, the maximum amount of solvent or suspending agent administered to any of the test material groups. If the test material is administered in the diet, an untreated control group should receive the same diet without test material. For test materials administered undiluted, a sham control group should receive the same physical treatment as the treated groups (e.g., insertion of an intubation tube with or without delivery of an innocuous substance such as water, administration of empty capsules, injection of physiological saline).

Positive control groups are intended either to demonstrate susceptibility of the animal to a specific toxicity or to compare the response of test material treated animals with that of animals treated with a chemical that produces a known toxicity similar to the test material. If a positive control group is included in a study design, at least one negative control group should also be included. Positive control groups are infrequently used in repeated-dose toxicity testing; however, if the chemical structure of a test material suggests that it may possess a specific toxicity (e.g., neurotoxicity), then it may be important to demonstrate that the species and strain selected for testing are susceptible to that toxicity.

A positive control that is sometimes useful in repeated-dose studies is the reference control. This control consists of a material that is chemically or physically similar to the test material but has either a comprehensive toxicology database associated with it or a history of use without adverse effects. Inclusion of a reference control group allows a comparison between reference and test material within the same study. This can assist in identifying any effects related to the general characteristics of the reference material; for example, oral administration of a poorly absorbed oil can decrease the absorption of fat-soluble vitamins. If the test material is known or suspected to produce this effect, use of a reference material, such as mineral oil, can be useful. This would distinguish effects related to vitamin depletion from effects produced directly by the test material. Additionally, if the test material were to add substantially to the caloric intake, a reference control diet isocaloric to the test diet would be useful, especially in longer term studies. A reference control group may also be useful to compare the degree of anticipated toxicity of the test material to a reference material of known toxicity; for example, it could be important to demonstrate that the hepatotoxicity of a test material intended for use as an anesthetic is significantly less severe than that produced by an anesthetic already in use.

## Animal Models

To increase the probability of testing in a species that may respond to the test material in a manner similar to humans, two species are generally used. Routinely, one rodent species and one nonrodent species are utilized. Rats and dogs are the generally preferred species for most routes of exposure. The rabbit is preferred for dermal exposure. Mice,

**TABLE 24.6**
**Selection Criteria for Species and Strain in Repeated-Dose Studies**

| |
|---|
| Requirements by regulatory agencies |
| Metabolism of test material in a manner similar to humans |
| Availability of historical control data |
| Most sensitive species and strain |
| Responsiveness of particular organs and tissues to specific toxicities |
| Availability of the species and strain |
| Availability of appropriate animal housing and husbandry |
| Experience of the laboratory in the use of the species and strain |

hamsters, miniature swine, guinea pigs, nonhuman primates, and a few other species are used on occasion in these studies. Many factors should be carefully considered during the selection of the most appropriate species and strain for testing with a specific chemical. Some of these factors are summarized in Table 24.6.

## Toxicokinetics

Ideally, selection of an animal model for repeated-dose toxicity studies should be based on the similarity between the toxicokinetics or metabolism of the test chemical in that species and strain to its toxicokinetics or metabolism in humans. This selection criterion assumes that these factors are known in potential test animals and in humans. Often, these data are unavailable during the initial phase of a hazard assessment. Although the metabolism of a chemical may be understood in one strain of one species of laboratory animal before initiation of repeated-dose testing, it is seldom known in several species and strains. With the exception of pharmaceuticals, the metabolism of a chemical in humans is almost never known before initiation of a subchronic or chronic study; consequently, similarity in metabolism between humans and animal models is seldom the initial basis for selection of test species and strain. This may change, however. Currently, human microsomes and systems that express specific human detoxification enzymes are commercially available. This opens the possibility of having *in vitro* data concerning human metabolism before initiating a hazard assessment [49].

## Sensitivity to Test Material

Another commonly used criterion for selection of the animal species and strain for repeated-dose testing is sensitivity to the test material. As a conservative approach to the extrapolation of toxic effects seen in animals to humans, the animal model selected should be the most sensitive to the effects of the chemical. Data required for this decision are often not available until a significant portion of the total hazard assessment program for the chemical has been completed. Acute and short-term repeated-dose studies may reveal information concerning species sensitivity; however, the relative sensitivity of different species and strains frequently only can be determined following completion of longer term studies with their more comprehensive endpoints. Nevertheless, sensitivity to the chemical should be considered during selection of the test animal; for example, differences in sensitivity of particular organs and tissues to toxic compounds among different species should be considered. Strains that have aberrant metabolic pathways, especially those associated with detoxification, should not be used except in special cases. The Gunn rat does not produce certain glucuronides [65] and would not be an appropriate animal model for a hazard assessment. Cats are deficient in their ability to produce glucuronides but can produce sulfate conjugates.

Availability of historical control data for the variables evaluated during repeated-dose toxicity testing is an important consideration in selecting the test species and strain. These data are frequently useful in determining the significance of a finding when comparison of data from treated and concurrent control groups suggests a potential treatment-related effect. Historical data concerning growth, feed consumption, clinical pathology, and other variables are often useful in interpreting findings from a subchronic or chronic study. Historical histopathology data are of particular importance due to the subjective nature of these data. Although published data can be useful, historical data from the laboratory at which the study is being conducted are more applicable. Most laboratories have historical databases for commonly used species and strains. If less common species are being considered, the availability of historical data should be assessed before final selection.

## Other Animal Model Considerations Involved in Study Design

After consideration of the above criteria, pragmatic considerations are necessary during selection of a species and strain. The animals should be obtained from a reputable reliable supplier who will guarantee their health and will arrange expeditious and controlled shipment of the animals to the laboratory. The supplier must maintain careful records concerning the animal colony and maintain a healthy colony. The supplier should provide disease-free animals, as it is often not possible to treat for disease once a study has initiated. The quantity of available test material may influence the selection of the animal model. For example, it may be necessary to select a rodent species if the amount of test material available is insufficient for long-term administration to a larger species such as the dog. Capabilities of the testing laboratory should be con

**TABLE 24.7**
**Serum Antibody Analyses in Rodents**

| Rat | Mouse |
|---|---|
| Sendai virus | Sendai virus |
| Pneumonia virus of mice | Pneumonia virus of mice |
| Reovirus type III | Reovirus type III |
| *Mycoplasma pulmonis* | *Mycoplasma pulmonis* |
| Lymphocytic choriomeningitis virus | Lymphocytic choriomeningitis virus |
| Mouse adenovirus FL/K87 | Mouse adenovirus FL/K87 |
| Mouse polio virus | Mouse polio virus |
| Hantaan virus | Hantaan virus |
| Encephalitozoon cuniculi | Encephalitozoon cuniculi |
| Cilia associated respiratory bacillus | Cilia associated respiratory bacillus |
| Rat parvovirus-IFA | Mouse parvovirus-IFA |
| Rat caronavirus/sialodacryoadenitis virus | Murine hepatitis virus |
| Kilham rat virus | Minute virus of mice |
| Toolan H-1 virus | Ectomelia virus |
| | Mouse pneumonitis virus |
| | Polyomavirus |
| | Mouse thymic virus |
| | Epizootic diarrhea of infant mice virus |
| | Mouse cytomegalovirus |

sidered during test animal selection. The laboratory must have appropriate caging and other equipment and must be able to maintain the proper environmental conditions in the animal room. In addition, the laboratory conducting the study should have experience with use of the chosen species in toxicology studies. This can avoid problems associated with species specific physiology and anatomy.

### Age of Animals

The age of animals used in subchronic and chronic toxicity studies is relatively standard. For rodents, initiation of test material administration at 6 weeks of age will satisfy virtually all guidelines for testing. Dogs approximately 4 to 6 months of age at initiation of exposure to the chemical are usually acceptable, but, if the test material is expected to produce toxicity in reproductive organs, older dogs should be used to avoid confusing results due to the variability in the degree of sexual maturity noted in 4- to 6-month-old dogs. The precise age of nonhuman primates is frequently not known; however, age can be approximated by experienced suppliers. For nonhuman primates and other less commonly used species, young animals should be used.

### Prestudy Health Assessment

To the extent possible, each animal included in a repeated-dose study must be in good health. The animals must not have been previously used for any other type of experimental procedures. An exception is sometimes made for nonhuman primates, which may occasionally be used for more than one study, with a reasonable period between studies to ensure that any residual test material is absent. These animals should undergo extensive health screening, including clinical pathology, between studies.

For rodent studies, enough animals of each gender should be obtained to allow culling of those with conditions that could either interfere with completing the study or be interpreted as treatment related at completion of the study. It is good practice to obtain at least 10% more animals than will be required to fill the study groups. Minimally, a pretest physical examination and body weight measurement should be conducted to assess the health of each animal before study initiation. Pretest ophthalmologic examination and clinical pathology evaluations are advisable. Animals in poor health or exhibiting ocular or other defects should be eliminated from consideration for the study.

To ensure that the toxicologist is aware of any infection that the animals may be exposed to during the study, a sentinel group is often maintained in the room with the study animals. For rodents, this group normally contains 5 to 10 animals of each gender. Serum antibody titers are assessed in these sentinels at the initiation of the study and at the termination of the in-life phase of the study. If necessary, antibody titers and other evidence of infection can also be obtained from these animals during the study without disturbing the animals on test. A relatively complete list of antibody analyses used in rodent species is presented in Table 24.7.

Health assessment of nonrodent species by the supplier is generally more comprehensive than rodents. This reduces the need to obtain many extra non-rodents. It is good practice, however, to conduct procedures after receipt of nonrodents to make sure that their health has not changed before use in a subchronic or chronic toxicity study.

### Number of Animals

To satisfy most regulatory guidelines, a minimum of 10 to 20 rodents of each gender should be included in each control and test-material-dosed group in a repeated-dose study. For nonrodents, the minimum number of animals of each gender in each group is 4; however, the minimum number of animals is frequently exceeded in an attempt to compensate for unexpected mortality or to increase the sensitivity of the study. Twenty rodents or 5 or 6 nonrodents of each gender per group are often used as the base number of animals for the study. Some study designs include an interim necropsy at one or more intervals for detection and evaluation of the progression of potential effects during the study. Other designs may contain treated animals that will be maintained without exposure after the termination of the main study groups to determine the reversibility of any adverse effects. Still other designs include satellite groups for special purposes (e.g., toxicokinetic determinations or untreated sentinel animals used to monitor the health of the study animals). The base number of animals placed on study at its initiation should be increased by the number of animals to be used for these enhanced study designs.

### Individual Animal Identification

Before assignment to the repeated-dose toxicity study, each animal must be assigned a unique identification number. This number will be associated with the animal throughout the study and will be used to identify specimens, tissues, and data from the animal after the in-life portion of the study is completed. This number must stay with the animal continuously during the study so there is no chance of misidentification. It is not adequate to simply attach the animal identification to its cage. Animals may escape from their cages or may be placed in the wrong cage during cage-changing operations. Unique identification numbers can be placed on the animals by a number of methods. Whatever the identification method, it should remain permanent and readable for the duration of the study. Older methods used for rodents included toe clipping, where a small portion of the toe was removed in a specified coded manner, and ear punching, where holes were punched through the ears in a specified coded manner. These methods are less acceptable today as more precise and humane methods have become widely available. Currently, the use of a numbered tag attached to an ear or tattoos placed on the tail or ear are commonly used methods for large and small animal identification. A newer method that has gained considerable acceptance in recent years involves subcutaneous implantation of a miniature electronic device that can be read by a hand-held scanner.

### Randomization of Animals

After culling all animals that do not meet the study criteria, such as animals that do not pass physical examination or are not within specified body weight boundaries, and assigning unique identification numbers to the remaining animals, the next step is to randomize them into the various study groups. This is a critical step in the study to ensure the greatest ability to detect statistical differences between the groups in the study without bias. Although a number of randomization methods have been devised, some more appropriate than others, one of the most popular methods is the utilization of random number tables. After randomization into the various study groups, some method should be employed to determine if the animals are truly randomized based on a variable critical to the study. The most commonly used variable is body weight. Statistical analysis of mean body weight data is conducted to show that there are no statistically significant intergroup differences in mean body weight, or other variables, at the initiation of the study. It is not uncommon to find that the mean body weight of the animals in one of the study groups is significantly different from one or more of the other groups. In cases where there may be a significant difference between the mean body weights of any of the study groups, the animals are again randomized into study groups and the process repeated. Randomization must be conducted independently for each gender because of body weight differences between males and females.

## Animal Husbandry

Proper care and maintenance of animals in a repeated-dose toxicity study are essential not only for ethical reasons but also to minimize mistakenly attributing adverse findings to the test material. The Animal Welfare Act (AWA), enforced by the Animal and Plant Health Inspection Service (APHIS) of the U.S. Department of Agriculture, mandates standards for acceptable handling, care, treatment, and transportation of many species, including most laboratory species except rodents [63]. In its *Guide for the Care and Use of Laboratory Animals*, the Institute for Laboratory Animal Research (ILAR) of the National Research Council has published guidelines that are widely accepted as standards for laboratory animal husbandry [44]. From their arrival at the laboratory, animals must be maintained in an appropriately controlled environment. They must be provided an adequate quantity and quality of feed and water and housed in clean cages of appropriate design. The animals should be acclimated to the study room conditions for at least one week before study initiation.

### Environmental Factors

Temperature and humidity should be controlled within limits specified in the documents referenced above. Table 24.8 is taken from the ILAR document and contains the

**TABLE 24.8**
**ILAR Recommended Dry-Bulb Temperatures for Common Laboratory Species**

| Species | Dry-Bulb Temperature °C | °F |
|---|---|---|
| Mouse, rat, hamster, gerbil, guinea pig | 18–26 | 64–79 |
| Rabbit | 16–22 | 61–72 |
| Cat, dog, non-human primate | 18–29 | 64–84 |
| Farm animals and poultry | 16–27 | 61–81 |

recommended temperature ranges for various laboratory animal species [44]. Low humidity can result in drying of the mucous membranes and eyes of laboratory animals. High humidity can result in growth of bacterial and fungal populations that result in respiratory distress and dermal involvement such as ringworm. In addition, urine and excreta may not dry as readily, thereby increasing room odor. Relative humidity of 30 to 70% is considered acceptable by ILAR for most laboratory species [44].

Adequate ventilation is a key factor in maintaining good animal health during a toxicology study. Establishing a positive room air pressure reduces possible exposure of animals to test materials being used in other animal rooms. When more air is forced into a room than can be completely cleared by exhaust systems, air flows through the cracks around the door, and the partial pressure of air in the room becomes positive with respect to the hallway or area outside the room. Ventilation should be homogeneous throughout the room; this generally is controlled by adjustable diffusers, and the ventilation of all rooms in a facility must be balanced periodically to provide the same relative air flow and positive pressure with regard to hallways. In general, 10 to 15 fresh air changes per hour is considered acceptable, but this range is highly dependent on a number of factors, such as the number of animals in the room.

Common lighting schedules used are 12 hours of continuous light and 12 hours of darkness for rats, mice, dogs, and monkeys and 14 hours of light and 10 hours of dark for hamsters. This schedule allows the animals to become acclimated to a light cycle. This stimulates a constant pattern of secretion of thyroid hormones, ACTH, and growth hormone. Regulated lighting cycles are necessary in reproduction studies because rodents enter continuous estrus under conditions of constant light phases without darkness. Because high-intensity fluorescent light can cause blindness in albino rodents, current practice is to limit their exposure to high-intensity light to times when observations are collected by providing dual-intensity (high-low) lighting systems.

## Animal Caging

In the United States, rodents are commonly housed one per cage during subchronic and chronic toxicity studies. In other countries, rodents are frequently multiply housed during these studies because it is believed that multiple housing increases survival and decreases background pathology. Multiple caging of animals can produce problems associated with unique identification and trauma to the animals from fighting, etc. Multiply housed rodents are more susceptible to transmitted disease and other health concerns. Furthermore, a multiply caged rodent that dies on study may sustain tissue destruction from cannibalism. It is also not possible to determine individual feed consumption when multiple animals are housed in a single cage. This results in the loss of important data because it is not possible to correlate body weight with individual feed intake. Additionally, if the test material is fed as part of the diet, it is not possible to calculate actual exposure doses for individual rodents in the absence of individual feed consumption data. Although some of the same problems exist for nonrodents, multiple housing of some species (e.g., nonhuman primates and dogs) on a regular or continuous basis during the study is accepted practice to permit the social interaction and exercise considered necessary for these species.

Rodents generally are housed in metal (stainless or galvanized steel) or plastic (polyethylene, polypropylene, or polycarbonate) cages. Metal caging or floor pens are used for dogs. Minimum cage sizes for all species are stipulated in the ILAR publication [44] and, for nonrodents, in the AWA [63]. Compliance is monitored by federal and state health agencies. Because minimum sizes for cages are stipulated, only caging type remains to be decided. The two major types of caging are solid floor cages or pens and suspended-floor cages.

Solid floor caging requires bedding to be added to the cage to absorb and contain waste materials and may introduce dust. Sawdust and chips of some conifers induce hepatic cytochrome P450 monooxygenase activity, which may affect the outcome of the study. In addition, this type of caging allows animals to have access to their waste. Shoebox-style cages used for rodents may clear the atmosphere at a slower rate than suspended wire cages.

Cages with suspended wire floors also have disadvantages. Traumatic foot and leg injuries can occur, particularly with smaller animals. Plantar foot pad lesions are common in long-term studies of rats housed in wire-bottom cages. This type of cage also exposes the animal to room drafts.

Most gradients in light, temperature, or airborne products in an animal room will occur vertically. Animals within groups should be distributed in cage racks so members of each study group are present equally at all vertical caging levels. This practice, and the practice of periodically changing the relative position of each cage rack within the room, avoids confounding treatment group with cage position. Documentation of environmental conditions and of cage and rack rotation is essential.

Cleaning of cages at frequent intervals is essential. Poor husbandry may result in skin lesions, alopecia, or the appearance of signs and behavior that may be interpreted as a possible effect of the test article.

## Diets

The influence of diet and nutrition on the toxicity of xenobiotics is another important aspect of the design of toxicology studies. The diet fed the animals during toxicology studies can influence the results; therefore, the decision made by the toxicologist concerning what diet to feed the animals may have a profound impact on the outcome of the studies. The diet fed during a study should have been designed for the study species. Although diets can be custom made, they are generally obtained from commercial suppliers. The supplier chosen should be reputable and capable of supplying information concerning basic diet composition and nutritional information. Although not feasible for a long-term study (e.g., 12 months), the same lot of diet should be used for the entire study whenever possible. Diets should be used before their expiration dates. They should be stored under appropriate conditions to maintain their nutritional value, prevent insect and rodent infestations, and ensure that they are not contaminated by environmental chemicals.

Commercial diets are available in either ground powder or pelleted form. When the test material is to be incorporated into the feed, a powdered diet is generally used. Use of powdered diet also facilitates the determination of feed consumption. Pelleted diets are most frequently used when test material administration is by routes other than dietary. A powdered diet can be pelleted after a test material has been added, which reduces the dust from the diets. Pelleted diets also decrease the potential exposure of animal room personnel to the test material; however, the heat and pressure involved during the pelleting process may cause degradation of test materials sensitive to these conditions.

Diets used in toxicology studies are of two basic compositions. Currently, diets made from natural ingredients are most commonly used; however, semipurified diets made from refined macronutrients, such as protein and carbohydrate, and micronutrients, such as vitamin and mineral mixes, are sometimes used. Each of these diets has particular advantages and disadvantages that must be carefully considered by the toxicologist.

Natural diets are formulated from unrefined plant and animal products to meet the nutritional requirements of a particular species. In closed-formula diets, the manufacturer does not provide the exact proportions of the constituents. The diets are formulated based on nutritional specifications without emphasis on consistency of specific ingredients between lots. Plant materials contain a number of non-nutritive components that can affect various physiological and biochemical functions, including detoxification and metabolic activation, in the test animal. These components may vary with plant species, strain, growing conditions, and site; therefore, individual lots of closed-formula diets may differ in these constituents. Open-formula diets are formulated with constant quantities of specified ingredients. This results in a more consistent composition than closed-formula diets. An example of an open-formula diet is the NIH-07 rodent diet, which has been relatively well characterized [60]. Open-formula diets have advantages for long-term studies because of their consistent formulation. An additional consideration concerning both open- and closed-formula natural diets is their potential to contain contaminants, such as pesticides, heavy metals, and mycotoxins. To overcome this problem, some manufacturers provide diets that have been assayed for certain potential contaminants to ensure that they are below stated specifications. It is highly recommended that these certified diets be used in toxicology studies. A disadvantage of natural commercial diets is that their nutritional composition cannot be readily altered. It is possible to supplement these diets but not possible to remove constituents. An important advantage of natural diets is their long history of use and the resulting large quantities of historical control data.

As noted above, semipurified diets are made from refined macroconstituents such as protein, carbohydrate fiber, and micronutrient mixes containing individual minerals and vitamins and a defined fat source, such as corn oil. Their constituents can be varied to design diets for specific nutritional purposes and allow for the inclusion of test materials that may provide nutrient activity or result in nutritional deficits. Nutrient composition can be reproduced exactly from lot to lot of semipurified diet. As opposed to natural ingredient diets, semipurified diets do not contain pesticides, mycotoxins, and other constituents that may alter the animal's response to the test material; however, a major problem with semipurified diets is a lack of historical data from their use in long-term studies. A large number of different dietary compositions are currently in use, and data obtained from one semipurified diet may not extrapolate to another semipurified diet. Even with commonly used semipurified diets, such as the AIN-76A [1] and the AIN-93 [61] diets, the data are insufficient to determine their impact on long-term toxicology studies, especially carcinogenicity studies [23]. Although these diets can be utilized in subchronic toxicity studies, the data obtained may not be as useful as that from studies with natural diets, especially when these data are used to design longer term studies with natural diets. Although it may be necessary to use semipurified diets with specific test materials, care must be taken if they are to be used in a safety assessment.

## Drinking Water

Drinking water free of contaminants that could interfere with the objectives of the study should be available to the animals during repeated-dose toxicity studies. Water is frequently provided to the rodent and nonrodent animals through automatic watering systems. In these systems, a common water supply is piped to the animal cages and each cage contains a valve that allows the animal *ad libitum* access to the water. Water bottles are another, more labor-intensive method of providing water to the rodents and some nonrodents. For this method, each bottle is fitted with a stopper containing a sipper tube through which the animal can drink; the bottle is suspended on the cage. A third method, generally only used with nonrodents, is to provide the animals with water in a drinking bowl. Any of these methods is acceptable as long as procedures are in place to ensure that the animals are provided an adequate supply of potable water.

# In-Life Evaluations

## Physical Examination

Several variables are routinely evaluated during the treatment phase of subchronic toxicity studies. Each animal should be observed twice daily at least 4 hours apart (a.m. and p.m.) for overt signs of toxicity, moribundity, and mortality. During these a.m. and p.m. observations, the cage of each animal should be opened to permit unobstructed observation. In addition, each animal should be removed from its cage for a complete physical examination at least once per week. These examinations should include detailed observations for approximate time of onset of any changes, as well as the degree and duration of changes involving the skin, fur, eyes, mucous membranes, respiratory function, circulatory system, autonomic and central nervous system, somatomotor function, and general behavior. A study-specific or SOP-specific glossary of clinical terms and descriptive criteria for each finding is recommended. The terms should be simple and descriptive, using a minimum of medical or diagnostic terminology.

## Body Weight Measurement

It is recommended that body weight should be measured at least once per week, even though biweekly or monthly measurement after the first three months of a chronic toxicity study is acceptable to most regulatory agencies. Weekly measurement is recommended because body weight is one of the most sensitive indicators of the condition of an animal if it is monitored frequently and carefully during a study. Rapid or marked body weight loss is usually a harbinger of ill health or death. Rapid body weight loss can be due to either decreased feed consumption, water consumption, disease, or specific toxic effects.

## Feed Consumption

In rodents, feed consumption generally is measured once per week during subchronic studies and the first 3 months of chronic studies. After the third month of a rodent chronic study, feed consumption may be measured less frequently (e.g., biweekly or monthly). For nonrodents, in which the quantity of feed required usually does not allow weekly feeding, feed consumption is evaluated for shorter intervals, often once or twice per day. Accurate measurement of feed consumption is essential for studies in which the test material is administered in the diet. As discussed earlier in this chapter, feed consumption and the dietary concentration of the test material are used to calculate the dose of test material consumed by the animals in such studies. Some species, especially the mouse and the nonhuman primate, frequently soil or waste feed. This makes accurate measurement of consumption difficult. In these species, feed consumption measurement can be attempted using either a feed container designed to minimize wastage or by attempting to estimate feed wastage. Limitations of such data should be considered in evaluating test material consumption and the significance of any apparent differences between feed consumption in test-material-treated and control animals. Feed consumption measurement is another means of monitoring animal well-being. Animals that are ill or suffering adverse effects from exposure to the test material frequently will exhibit significantly decreased feed intake.

## Ophthalmologic Examination

Ophthalmological examination of all test animals should be conducted before initiation and at the completion of the test material administration period. Slit lamp examination and use of an indirect ophthalmoscope are two common methods of ophthalmologic examination during toxicity testing. These evaluations should be conducted by a veterinary ophthalmologist experienced in the observation of the species used for the study.

## Clinical Pathology

Clinical pathology variables such as hematology, clinical chemistry, and urinalysis are important indicators of general health and toxicity and are assessed at termination of a subchronic or chronic study. In addition, pretest and interim (typically at 4 weeks in subchronic studies and at 13 weeks in chronic studies) clinical pathology may be conducted to allow evaluation of progression of any treatment-related effects noted at termination of the study. In rodents, clinical pathology determinations are usually conducted for ten animals of each gender in each group. For nonrodents, clinical pathology should be done for all animals.

## Sample Collection

Proper sample collection and handling are critical to completion of a meaningful clinical pathology evaluation. Whenever possible, the method of sample collection should be the same throughout the study and should be one that distributes variance (such as run-to-run variation in an enzyme assay) equally across groups. Samples generally are collected according to either a totally random design or a stratified random design. A stratified random design ensures that approximately the same numbers of animals of each gender and from each group are sampled within any block of time or during any set of assays.

Repeated blood sampling of rodents can be accomplished by serial collection from the same animals by nonterminal procedures (such as puncture of the orbital sinus or jugular vein) or by collection from the abdominal aorta or vena cava at termination of subgroups of animals. Because of practical restraints on the frequency and volume of blood collection in rodents, pretest studies often are not performed. Repeated collection of adequate volumes of blood is usually not a problem for nonrodents, and pretest clinical pathology is often included in studies using these animals.

The effect of repeated blood sampling on the animals and on the sample volume that can be reliably obtained at each sampling interval should be considered. Sample volumes should be sufficient both to conduct the assays indicated in the protocol and, if possible, to provide a reserve sample for any necessary repeat test; however, significant reduction of total blood volume (more than 10%) by blood collection should be avoided.

Plasma or serum to be used for clinical evaluation should be clear and straw colored. Red or pink plasma suggests that some hemolysis has occurred, either as a result of pathology or, more commonly, as an artifact of the sample collection or preparation procedures. Severe artifactual hemolysis may alter the results for some of the clinical variables to be evaluated. Slight hemolysis, commonly observed in serum and plasma collected from rodents by orbital sinus or jugular venipuncture, generally is acceptable for clinical pathology studies as long as historical laboratory ranges have been established for blood collected by these methods. If unusual or unexpected results are obtained, aliquots of serum or plasma can be spiked with ascending amounts of test material to determine if it interferes with the assay.

## Clinical Chemistry, Hematology, and Urinalysis

Clinical pathology should include determination of a number of serum or plasma chemistry and hematology variables. The variables included in the clinical chemistry evaluation should assess electrolyte balance, protein and carbohydrate metabolism, and organ function. An acceptable list of clinical chemistry variables is shown in Table 24.9. Additional variables should be assessed, as appropriate, to address other anticipated effects of the test material (e.g., serum cholinesterase levels in the case of carbamate or organophosphate insecticides). It goes without saying that assays designed for assessment of clinical chemistry in humans must be validated for use with the species used in the toxicology studies. Typical hematologic variables assessed during repeated-dose testing are shown in Table 24.10. The reader is referred to Chapter 26 for a more detailed discussion of these clinical chemistry and hematology determinations. In addition, two excellent veterinary texts are available for further reference [45,50].

**TABLE 24.9**
**Clinical Chemistry Variables Normally Obtained In Repeated-Dose Studies**

| | |
|---|---|
| Glucose* | Potassium* |
| Urea nitrogen* | Chloride* |
| Creatinine* | Bilirubin (total) |
| Total protein* | Cholesterol |
| Albumin | Triglycerides |
| Globulin | Alkaline phosphatase* |
| Albumin/globulin ratio | Aspartate aminotransferase |
| Inorganic phosphorus | Alanine aminotransferase* |
| Calcium | Gamma glutamyl transferase* |
| Sodium* | Ornithine carbamyl transferase* |

*Note:* Table is based, in part, on the recommendations of the U.S. FDA [26]. This list does not include all clinical chemistry variables that could be obtained; additional variables could be added dependent on the test material. When the blood volume obtained for analysis is small, the FDA recommends that priority be given to those assays marked with an asterisk (*).

**TABLE 24.10**
**Hematology and Urinalysis Variables Generally Determined in a Repeated-Dose Study**

| Hematology | Urinalysis |
|---|---|
| Hematocrit | Appearance |
| Hemoglobin | Urine volume |
| Erythrocyte count | Specific gravity |
| Mean corpuscular volume | pH |
| Mean corpuscular hemoglobin | Glucose |
| Mean corpuscular hemoglobin concentration | Protein |
| Total leukocyte count | Microscopic evaluation of urinary sediment |
| Differential leukocyte count | |
| Reticulocyte count | |
| Platelet count | |
| Prothrombin time | |

**TABLE 24.11**
**Tissues Collected for Histopathology in Repeated-Dose Studies**

| | | |
|---|---|---|
| Adrenals | Kidneys | Rectum |
| Aorta | Lacrimal gland | Salivary gland (submandibular) |
| Bone marrow smear | Larynx | Sciatic nerve |
| Brain | Lesions | Seminal vesicles |
| Cecum | Liver | Skin |
| Colon | Lungs | Spinal cord (cervical, thoracic, lumbar) |
| Duodenum | Lymph nodes (mandibular and mesenteric) | Spleen |
| Epididymides | Mammary gland (females) | Sternum and bone marrow |
| Esophagus | Muscle (thigh) | Stomach |
| Eyes | Nose | Testes |
| Femur and bone marrow | Ovaries | Thymus |
| Gall bladder (when present) | Pancreas | Thyroid with parathyroid |
| Heart | Pharynx | Trachea |
| Ileum | Pituitary | Urinary bladder |
| Jejunum | Prostate | Uterus |

Urinalysis is often included in the clinical pathology evaluation and may be important, especially for test materials that are nephrotoxins. Urinalysis variables typically evaluated are listed in Table 24.10. Urinalysis is frequently of limited value because the collection of satisfactory urine samples is fraught with technical difficulties. Urine generally is collected in containers or tubes from troughs or trays placed below the cages in which the animals are housed and, therefore, steps must be taken to minimize fecal contamination. Because urine is frequently collected during an extended period (e.g., overnight), bacterial growth in the sample is a concern. Collection of the sample on ice can reduce bacterial growth but presents technical challenges in and of itself. Care should be taken that water is either freely available to the animals throughout the urine collection period or is withdrawn at the appropriate time before collection. Care also must be taken to ensure that the sample is not inadvertently contaminated by feed or drinking water spilled by the animal. Because of these difficulties, the utility of urinalysis should be discussed with an experienced veterinary clinical pathologist prior to its inclusion in the design of a repeated-dose toxicity study. If urinalysis is conducted, its limitations must be kept in mind when the data are reviewed.

## Postmortem Evaluations

One of the more definitive assessments of toxicological effects conducted during a repeated-dose study is the macroscopic and microscopic examination of tissues and organs from treated and control animals. In typical subchronic and chronic studies, samples of approximately 50 tissues and organs are collected during the necropsy of each animal. Table 24.11 presents a list of tissues that are commonly collected for potential histopathological examination.

### Necropsy

*Necropsy* of an animal is conducted when it dies during the study, when it is killed during the study for humane reasons (e.g., in cases of moribundity), or when it is killed at a scheduled interval (interim sacrifice or termination of the study). Necropsy should be completed as quickly as possible after the death of an animal to avoid *autolysis*, which can interfere with the subsequent microscopic examination of its tissues. Autolysis is defined as the enzymatic self-digestion of cells or tissues that occurs after death. Autolysis is an especially important consideration for animals that die during the study, as their death may not be discovered for a significant period of time. Animals found to be moribund (i.e., about to die) during the study should be terminated for humane reasons and to avoid tissue autolysis.

During necropsy, tissues and organs are systematically removed, and macroscopically visible abnormalities are noted. These abnormalities include changes in color, shape, size, or consistency of a tissue. The documentation of an abnormality in the necropsy records should include its location and a clear description of the change, using nondiagnostic terminology. Completeness of the examination during necropsy and the quality of the description of abnormalities are critical to the determination of pathologic effects. An abnormality in a tissue can only be prepared for microscopic examination if it was collected and accurately described during necropsy. Because of the central role that the necropsy plays in detecting effects in a repeated-dose toxicity study, it is extremely important that necropsy technicians are highly trained and experienced in the necessary techniques.

After collection, tissue samples are usually preserved by immersion in an appropriate fixative, commonly 10% neutral buffered formalin. In some cases, particularly for organs such as testes or eyes, special fixatives may be used [64]. To ensure adequate *fixation*, the volume of fixative should be at least ten times the volume of tissues. Certain organs (e.g., the lung and urinary bladder) are frequently filled with fixative prior to immersion to improve fixation. During collection of large numbers of tissues from many animals, it is possible to inadvertently miss a tissue; therefore, it is highly advisable to inventory and document the samples as they are placed into the fixative containers. This inventory will be invaluable during subsequent preparation of the tissues for microscopic examination and in reconstructing the study during post-study auditing of the data. In addition, the identity of each tissue must be clearly maintained while in fixative. This is not a problem for tissues that are large or have distinctive *morphology*. To ensure subsequent identification of extremely small tissues and those with indistinct morphology, they are frequently placed in labeled plastic cassettes or cloth bags prior to being placed in the fixative container.

## Organ Weights

Collection of terminal body weight and organ weights for all animals during necropsy is normal practice in repeated-dose toxicity studies. Minimally, weights should be recorded for the brain, liver, kidneys, testes, and adrenal glands. Frequently, other organs such as the thyroid, parathyroid, ovaries, spleen, thymus, uterus, epididymis, heart, and lungs are also weighed. Consideration should be given to the residual blood that may remain in organs such as the spleen, heart, and lungs. Residual blood may be variable between animals due to the method of sacrifice and blood collection. Organs should be weighed as soon as possible after removal from the animal. They should be trimmed free of fat and connective tissue prior to weighing and placed into fixative immediately thereafter.

It is common practice to normalize organ weights by expressing them relative to body weight and brain weight. Relative organ weights are used to eliminate the influence of normal variation in animal growth on the interpretation of organ weight data; however, normalized organ weight data should be reviewed with the knowledge that they have some limitations. Expressing organ weights relative to body weight can yield apparent, but artificial, treatment-related effects on organ weights in studies where the test material affects body weight gain. Organ weights normalized to brain weights help overcome this problem because test materials that alter body weight generally do not alter brain weight. The best practice is to consider all three types of data (i.e., actual organ weight and organ weight relative to body and brain weight). Histopathological data are often used to help assess the significance of apparent differences between organ weights of test-material-treated and control animals.

## Microscopic Pathology

Microscopic examination of the tissues and organs of treated and control animals is one of the most time-consuming laboratory functions in toxicity studies. In nonrodent studies, sections of all tissues and organs from all animals should be prepared for microscopic examination. Generally, in rodent studies, only tissues and organs from the controls and high-dose group animals and animals that were killed or died during the study are examined microscopically. For the other treatment groups, only a few major organs (e.g., liver, kidney, and any other organs in which macroscopic abnormalities were noted at necropsy or in which test-material related effects are detected in high-dose animals) are examined. Although initial histopathological examination of control and high-dose tissues followed by examination of other doses is typical in rodent studies, simultaneous histopathological examination of all tissues from all animals is not uncommon. This practice yields the most expeditious completion of the histopathological evaluation phase of a study because sequentia examinations are not required. Simultaneous examination also reduces the intergroup variability in diagnoses tha might occur when tissues from intermediate groups are examined considerably after completion of the evaluatio of the control and high dose groups. Such variability ca lead to incorrect conclusions concerning treatment relationship of lesions noted in the intermediate dose groups In their guidance for safety studies, the U.S. FDA ha suggested that all tissues from all control and high-dose animals and any animal that does not survive the study duration should be subjected to histopathologic examina tion [26]. Tissues found to be affected in high-dose ani mals should also be examined for lower dose animals.

Routinely, tissues are prepared for light microscopi examination by embedding in paraffin, sectioning at 5 t 7 microns, and staining with hematoxylin and eosi (H&E) stain. Special stains, such as stains for the presenc of fat (as Oil Red O) or connective tissue (trichrome stain may be used for some tissues. Use of a protein-specifi stain (Mallory–Heidenhain) is illustrated in Figure 24.6 If desired, representative samples of selected tissues ma be frozen at necropsy and stored for biochemical or immu nohistochemical analyses or specially prepared for elec tron microscopy. In the histology laboratory, it is impor tant that tissues be prepared according to standardize procedures especially with respect to type of section (e.g cross, longitudinal), location, and orientation on th microscope slide. The histology technician must review the observations recorded during necropsy of the animal to ensure that all grossly observed lesions are properl sectioned and mounted on the slide for subsequent micro scopic examination. Whenever, possible, samples o lesions prepared for examination should include the lesio and portions of surrounding "normal" tissue.

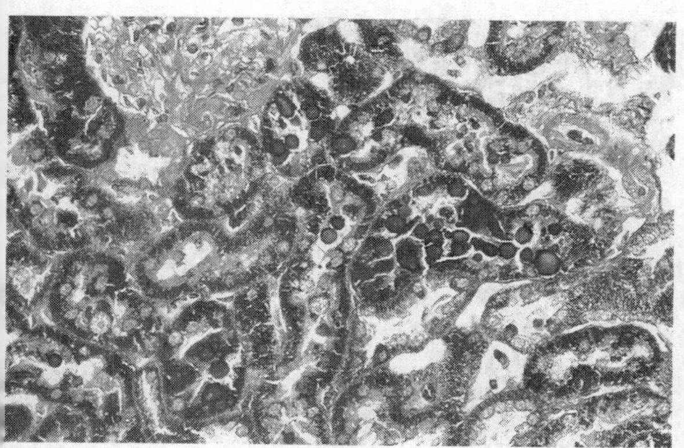

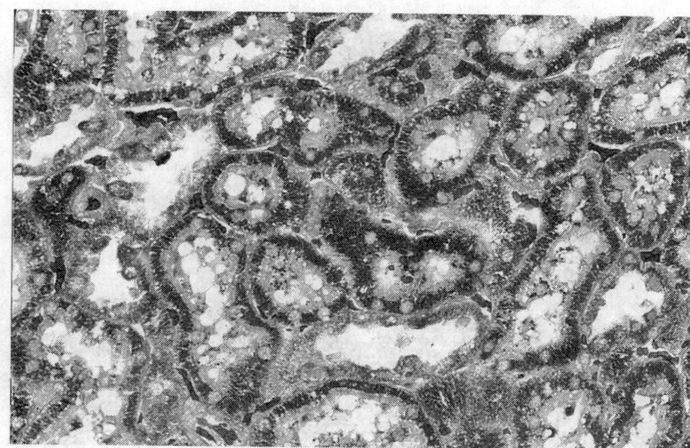

**FIGURE 24.6** Photomicrographs of kidneys from Fischer 344 male rats stained for protein using Mallory–Heidenhain stain (original magnification, 100×). The kidney tubules on the left are stained heavily and contain many hyalin droplets (arrow). The kidney tubules on the right exhibit normal levels of protein staining (arrow).

Histopathologic examination of the tissues requires specialized training and is performed by a pathologist trained and experienced in the evaluation of toxicologic pathology. Such training and experience provide the pathologist with knowledge of the normal features and naturally occurring lesions that can be observed microscopically in tissues from laboratory animals. However, the pathologist's responsibility is more than evaluation of the tissues and accounting for all the lesions reported at necropsy. The pathologist should be an integral part of the protocol design team to provide input into many factors, such as the selection of the species and strain to be used and the clinical pathology variables to be evaluated. The pathologist should also provide guidance concerning the list of tissues to be collected to ensure that they are processed, stained, and evaluated in a manner that satisfies the study objective.

It is critical that the pathologist review the data generated during the in-life and necropsy phases of the study before proceeding with the histopathologic evaluation. Results of clinical observations, clinical chemistry and hematology determinations, organ weight measurements, and necropsy examinations can lead the pathologist to focus on particular organs as potential targets for toxicity during the microscopic examination. For example, increased liver weight should lead to a more careful examination for hepatocellular *hyperplasia* or *hypertrophy*, while elevated serum creatinine along with a necropsy description of the surface of the kidneys as "rough" should result in a more thorough examination for nephropathy.

Some have suggested that knowledge of in-life and necropsy findings will bias the pathologist, causing a more stringent examination of the potentially affected tissue. Similarly, some are concerned that knowledge of the dose level administered to the animal during the study will bias the pathologist, resulting in a more thorough examination of the tissues from animals that received the test material. The second situation in particular could result in a higher incidence or severity of microscopic findings in treated animals compared with controls, a higher incidence that is simply an artifact of the thoroughness of the microscopic examination. One way to prevent potential bias is to keep the pathologist ignorant of other study findings and of the identity of the animal until histopathologic examination of the tissues has been completed. This so-called blinded reading does prevent bias but may also prevent the pathologist from identifying certain subtle, dose-related changes; therefore, blinded reading is not recommended for routine histopathologic evaluations and should only be conducted in special situations.

Pathologic changes in cellular or subcellular structure can occur either spontaneously, such as with aging, or as a result of exposure to a chemical. Deciding which changes are significant and what severity should be assigned to a change during histopathologic examination is quite subjective. Because of this, it is possible that different pathologists looking at the same tissues will produce different diagnoses. It is even possible that, during histopathologic evaluation of a large number of tissues that extends over a number of months, the criteria for diagnosis of the same finding by a single pathologist will change somewhat. The variability that results from the subjective nature of histopathologic evaluation is unavoidable. To minimize its effect on the results of toxicity studies, several procedures can be useful. First, during each reading period, the pathologist should examine tissues from a small subset of the study animals. Each subset should contain approximately equal numbers from each control and treated group to be examined. Second, if a potential target organ is identified over an extended period of time, the pathologist should reexamine that organ from all animals during a compressed reading period of a few days or less to assess to what extent, if any, diagnostic drift occurred over time. Third, informal peer consultation concerning unusual

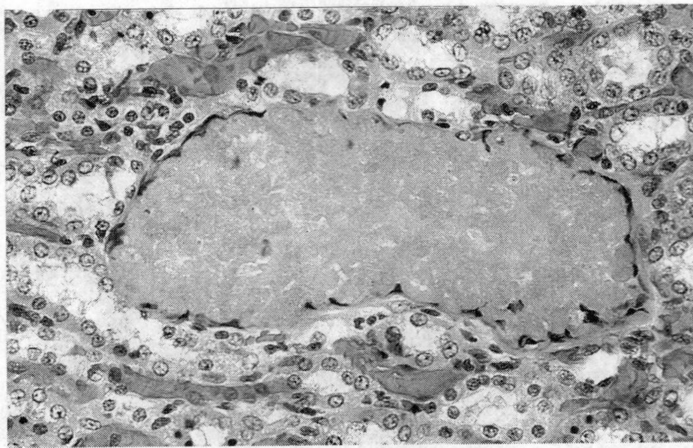
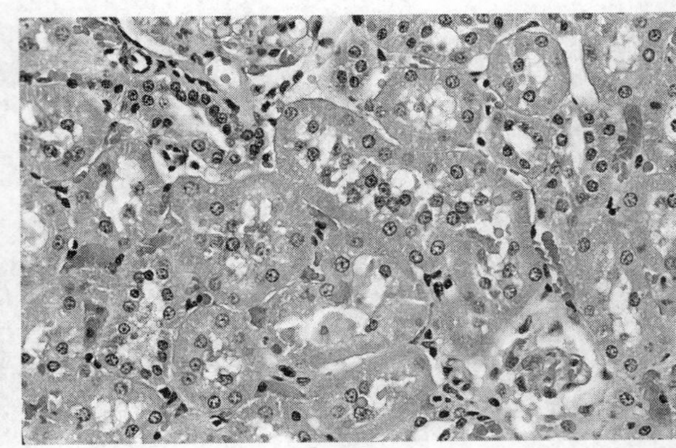

**FIGURE 24.7** Photomicrographs of hematoxylin- and eosin-stained kidneys from Fischer 344 male rats (original magnification 100×). The lumen of one tubule in the kidney on the left is obstructed by a large accumulation of granular casts. The kidney on the right is without pathologic changes.

or subtle tissue changes observed during examination of the tissues can be conducted to arrive at a consensus diagnosis. Additionally, a formal peer review process involving (1) reexamination of target organs, (2) reexamination of a representative percentage of other tissues from the study, and (3) review of interpretation of the pathology findings can be conducted by a second pathologist to ensure consistency in diagnosis and grading of tissue changes and accuracy of the pathology conclusions [37].

The objective of the histopathologic evaluation is the same as the objective of all other determinations during a repeated-dose toxicity study (i.e., to detect adverse effects that could be relevant to humans or any other target species exposed to the test compound). Because of certain idiosyncrasies, some animals and strains are not useful for this purpose. In some cases, the animal exhibits a high spontaneous rate of pathology in a particular organ that prevents detection of any compound-induced increase in that rate. As an example, severe testicular pathology occurs spontaneously in a very high percentage of old male Fischer 344 rats (see Figure 24.5); therefore, this strain of rat is generally not useful for the detection of testicular effects. In other cases, a pathologic change occurs in the laboratory animal only in response to test-compound exposure, but that pathology would not be expected to occur in the target species. An example of this is light hydrocarbon nephropathy that occurs only in male rats. Many hydrocarbons produce this nephropathy (e.g., d-limonene, unleaded gasoline). The pathology is caused by accumulation of a male-rat-specific protein, α-2-microglobulin, in the renal tubule following exposure to these chemicals. This nephropathy is characterized by hyalin droplets in the cytoplasm (Figure 24.6), granular casts in the lumen (Figure 24.7), and cellular regeneration in the tubules. Because humans do not produce α-2-microglobulin, this gender- and species-specific pathologic finding has no relevance to human hazard assessment and the male rat is not a suitable model for human renal effects related to light hydrocarbon exposure. It should be noted, however, that for both examples cited the test animal is perfectly acceptable for use as a human surrogate in toxicity testing as long as the limitations imposed by their idiosyncrasies are taken into account.

Even though the idiosyncratic situations discussed above do occur occasionally, the pathology that occurs in most laboratory animals and most tissues is considered relevant to the assessment of hazard in humans. The histopathologic examination of tissues in a subchronic or chronic toxicity study can yield a vast array of diagnoses. A detailed discussion of possible chemical-related pathologic findings is beyond the scope of this chapter. For detailed descriptions of methods for and diagnosis of veterinary toxicologic pathology, the reader is referred to a review by Hardisty [37] and two comprehensive texts on the subject [2,38].

## Additional Endpoints for Repeated-Dose Toxicology Studies

Evaluation of special endpoints can be added to repeated-dose toxicology studies to maximize the utilization of animal resources, minimize the time and cost of a hazard assessment, and obtain additional data. Care should be used in selecting these endpoints to ensure that valid methodology is used and the data will be accepted by regulatory agencies. Guidelines for safety assessment of direct food and color additives suggest that data concerning the immunotoxic and neurotoxic potential of the test material be generated during subchronic toxicity studies [26]. To be sure that the methodology and data presentation are acceptable to the regulatory agencies, meetings should be held with the appropriate agency during design of the study.

It is possible, through the addition of special determinations, to make a subchronic or chronic toxicity study so complicated that the main objectives are jeopardized. All the ramifications of addition of special endpoints to a study design, including practical considerations such as daily workload, should be considered to ensure that basic study endpoints are not compromised. Rather than overwhelming the capabilities of the testing laboratory, conduct of a separate study designed to evaluate the special endpoints may be preferable. This is not to say that special evaluations should never be added to subchronic or chronic studies. With appropriate consideration of the possible complications, they can be and often are added to standard study designs.

If a question still remains whether additional endpoints should be added to short-term repeated-dose studies or to longer term studies, the following should be considered. If these endpoints were added to short-term studies, then the data would be available to aid in design of longer term study; however, if the particular variable to be evaluated appears only after longer exposure periods, any change might not occur in short-term studies. The most conservative approach is to add the endpoints to each study type. The following paragraphs provide examples of how data concerning nonstandard toxicological endpoints may be obtained from classical repeated-dose toxicity study designs.

## Genetic Toxicology

Certain *in vivo* and *in vivo*/*in vitro* genetic toxicology data may be obtained during or at the termination of a repeated-dose toxicity study. Addition of these endpoints to a study could decrease the number of animals used in a hazard assessment and shorten its duration. *In vivo* genetic toxicology studies increase the value of a hazard assessment. They use the route of administration by which humans are exposed to the test material, and all processes of absorption, distribution, metabolism, excretion, and DNA repair are intact; therefore, these *in vivo* studies provide important information that cannot be obtained from *in vitro* studies using cellular systems.

A number of *in vivo* genetic toxicology assays are currently in use, some of which are suitable for incorporation into repeated-dose toxicity assays. For example, the authors have incorporated the bone marrow micronucleus assay into a classic subchronic study design. During many repeated-dose toxicity studies, bone marrow smears are made (see Table 24.11). These bone marrow smears are made essentially in the same manner as those used for the *in vivo* micronucleus assay [39]. The slides are stained with acridine orange and the polychromatic erythrocytes analyzed for micronuclei. Use of bone marrow slides from the repeated-dose study has several advantages in that it: (1) incurs no additional time and cost for collecting the samples, (2) allows the assay to be conducted on animals exposed to the test material for long periods of time, (3) eliminates the need for resources to conduct an independent study, and (4) does not interfere with the histopathological assessment of the tissues.

It may be possible to use bone marrow slides from repeated-dose toxicity studies for the bone marrow chromosomal aberration assay [4]. This is another *in vivo* genetic toxicology assay that could provide additional data without conducting an independent study. Another such genetic toxicology assay is the *in vivo*/*in vitro* unscheduled DNA synthesis assay [52]. In this assay, freshly isolated hepatocytes are used to determine unscheduled DNA synthesis. Although this assay is not compatible with histopathological use of the liver, as few as two or three extra animals per group is all that is required. One problem associated with incorporation of additional endpoints, such as genetic toxicology, into a repeated-dose study design is that the laboratory conducting the study must have valid assay methodology. This is not always the case.

Not all *in vivo* genetic toxicology assays are completely suitable for incorporation into repeated-dose studies; for example, the *in vivo* sister chromatid exchange assay requires the administration of deoxybromouridine to the animals, and this compound can compromise the classical endpoints used in toxicity studies, including histopathology. It is possible, however, to incorporate extra animals to assess specific endpoints while not compromising the main study animals.

## Neurotoxicity

It is possible to incorporate neurotoxicity screening into repeated-dose toxicity study designs. In fact, in its food additive toxicology testing recommendations, the FDA has suggested that neurotoxicity screening be incorporated into these study designs [26]. The EPA has similar recommendations [14,15]. Neurotoxicity screening is designed to determine if the test material has the potential to produce adverse effects on the nervous system. Screening is conducted to determine if additional, more sophisticated, neurotoxicity testing is required. The first indication of a requirement for neurotoxicity testing may come from the structure–activity assessments; however, the neurotoxicity database is not as extensive as those for certain other types of toxicity and may not provide useful insight into the need for neurotoxicity testing. Drugs and pesticides that target the nervous system are a well-known exception; for example, there is no question that a new organophosphate insecticide will require neurotoxicity testing. For most compounds, it will be necessary to develop data through empirical testing.

Most classical repeated-dose study designs contain elements that may provide some information on neurotoxicity potential. These include cage-side observations of

**TABLE 24.12**
**FDA Draft Criteria for a Neurotoxicity Screen as a Component of Short-Term and Subchronic Studies**

Histopathological examination of tissues representative of the nervous system, including the brain, spinal cord, and peripheral nervous system

Quantitative observations and manipulative tests to detect neurological, behavioral, and physiological dysfunctions, including:

- General appearance
- Body posture
- Incidence and severity of seizure
- Incidence and severity of tremor, paralysis or other dysfunction
- Level of motor activity and arousal
- Level of reactivity to stimuli
- Motor coordination
- Strength
- Gait
- Sensorimotor response to primary sensory stimuli
- Excessive lacrimation or salivation
- Piloerection
- Diarrhea
- Polyuria
- Ptosis
- Other signs of neurotoxicity deemed appropriate

the animals, physical examinations, and measurement of variables, such as food consumption, that may relate to behavior modifications during the in-life phase of the study. Additional information is obtained during histopathological examination of the structures of the nervous system collected at necropsy, such as the brain and spinal cord. The FDA believes that these procedures, as well as others, should be specifically included into the design of repeated-dose toxicity protocols [26]. Table 24.12 lists the design elements recommended by the FDA for a neurotoxicity screen. Specific behavioral and neurotoxicity tests exist to provide most of the requested data. The particular test designs should be chosen based on their validity, history of use, lack of undue stress to the animal, and the experience of the laboratory with the specific test. Care should be taken to perform these procedures on all the treatment groups and controls in the study. A concern is that some of these tests may stress the animals and produce changes in the traditional variables measured during a toxicity study. If this were to happen, it is assumed that the control group would also demonstrate these changes. This may or may not be true; therefore, a conservative approach would be the addition of extra animals to the study that would be subjected to the manipulative procedures for neurotoxicity screening and not be used in the traditional phases of the study. For a more complete discussion of the FDA recommendations, the reader is referred to the draft guidelines for safety of direct food additives [27].

### Immunotoxicity

Rapid advances have been made during the last 20 to 30 years with respect to the detection of immunotoxicity. The two major forms of immunotoxicity are immunosuppression and hyperactivity of the immune system. Immunosuppression results in a reduction in the animal's resistance to infection and potential increase in susceptibility to tumorigenesis by a suppression of critical immunological responses. Hyperactivity of the immune system can result in autoimmune diseases and increased sensitivity to allergic disorders. Determination of the mechanisms associated with these disorders can be extremely complex because of the large number of biochemical, cellular, and physiological factors that can be affected as well as the cellular interactions required to mount an immunological defense. The detection of potential immunological changes is less complex, and a number of tests exist that can provide warning of a potentially immunotoxic compound.

The FDA has published recommendations for the inclusion of immunotoxicity evaluations in repeated-dose toxicity studies [23]. It suggests that such evaluations be conducted in rodents. Immunotoxicology testing procedures were divided into two broad categories. Type I tests are those assays that do not require the study animals to be treated with an agent that presents an immunological challenge. Type II tests are assays that require the study animals to be challenged with an agent that elicits an immune response, such as antigens, vaccines, infectious agents, or tumor cells. Because type I tests do not require manipulation of animals, they can be included in the routine assays done during a repeated-dose toxicity study. Because type II tests require treatment of the animals with an immunological challenge, these animals are not suitable for evaluations conducted during toxicity studies; therefore, additional animals must be included in the study design.

Table 24.13 lists the immunotoxicology evaluations the FDA suggests be included in type I tests. Generally, these evaluations are those currently recommended for repeated-dose toxicity studies with the exception of a more comprehensive histopathology of lymphoid tissues. Inclusion of these evaluations into current study designs should have no impact on the validity of the study. Inclusion of the expanded type I tests listed in Table 24.14 would require more planning during the study design phase and assurance that the laboratory performing the study would be capable of performing the assays. Generally, the expanded type I test would only be conducted after consultation with the FDA. Type II immunotoxicity tests recommended by the FDA could be done as a component of a repeated-dose toxicity study, but they appear to be more appropriately conducted as independent studies and are beyond the scope of this discussion.

**TABLE 24.13**
**FDA Draft Recommendation for Type I Immunotoxicity Tests That Can Be Included in Repeated-Dose Toxicity Studies**

Hematology
- White blood cell counts
- Differential white blood cell counts
  - Lymphocytosis
  - Lymphopenia
  - Eosinophilia

Histopathology
- Lymphoid tissues
  - Spleen
  - Lymph nodes
  - Thymus
  - Peyer's patches in gut
  - Bone marrow
- Cytology (if needed)[a]
  - Prevalence of activated macrophages
  - Issue prevalence and location of lymphocytes
  - Evidence of B-cell germinal centers
  - Evidence of T-cell germinal centers
  - Necrotic or proliferative changes in lymphoid tissues

Clinical Chemistry
- Total serum protein
- Albumin
- Albumin-to-globulin ratio
- Serum transaminases

[a] More comprehensive cytological evaluation of the tissues would not be done unless there is evidence of potential immunotoxicity from the preceding evaluations

**TABLE 24.14**
**FDA Draft Recommendation for Expanded Type I Immunotoxicity Tests That Can Be Included in Repeated-Dose Toxicity Studies**

Hematology
- Flow cytometric analysis
  - B-lymphocytes
  - T-lymphocytes
  - T-lymphocyte subsets (TH and TS)
- Immunostaining of blood or spleen fraction
  - B-lymphocytes
  - T-lymphocytes

Histopathology
- Immunostaining of B-lymphocytes in spleen and lymph nodes with polyclonal antibodies to IgG
- Immunostaining of T-lymphocytes and subsets with monoclonal or polyclonal antibodies
- Micrometric measurements of germinal centers and periarteriolar lymphocyte sheath of the spleen and follicles and germinal centers of lymph nodes

*In vitro* analysis of functional capacity of specific immune cells
- Activity of natural killer cells
- Mitogenic stimulation of B- and T-lymphocytes
- Macrophage phagocytic index
- Stem cell assays

Serum chemistry
- Serum protein electrophoresis
  - Albumin
  - α-Globulin
  - β-Globulin
  - γ-Globulin
- Quantification of γ-globulin fraction
  - IgG, IgM, IgA, IgE
- Complement
  - Cytokines (IL-2, IL-1, γ-interferon)
  - Auto-antibodies
  - Antiparietal cell antibodies

## Toxicokinetics

Advances in modern analytical chemistry, especially high-performance liquid chromatography/mass spectroscopy, has provided the toxicologist with the capability to obtain toxicokinetic data in the early stages of a hazard assessment. These data can be important in the design of subsequent studies and the demonstration of systemic exposure, and they can provide insight into the factors that may influence the toxicity of the test material. For these reasons, toxicokinetic assessments are increasingly being incorporated in more protocols for short-term and long-term repeated dosing studies.

It is currently possible to obtain *in vitro* metabolism and preliminary *in vivo* metabolism data before initiation of short-term repeated dosing studies. These data can be useful in the design of 14-day repeated dosing studies and aid in the integration of toxicokinetic procedures into the short-term protocols. Because of the small quantities of blood required with the current analytical procedures, blood collection from studies with larger animals can be done using the study animals, and they eliminate the need for animals devoted to toxicokinetics alone. With rodent species, however, it still may be necessary to have an extra set of animals in the study for assessment of toxicokinetics. Using a separate set of animals in rodent toxicology studies to obtain toxicokinetic data still has advantages over doing an independent toxicokinetic study. It ensures that the animals are similar to those in the toxicology study in respect to age, environment factors, and dosing parameters.

Whereas the initial *in vivo* studies may be single-dose studies, repeated dosing studies provide the opportunity to obtain data from repeated dosing. They can, therefore, provide information concerning induction and inhibition of the test material's metabolism associated with the previous exposures. Toxicokinetic data obtained during different time periods in a subchronic study provide information concerning the effects of prolonged dosing on the absorption, distribution, metabolism, and excretion of the test article and also information concerning the disposition of metabolites and may provide explanations for unexpected results in the toxicology study. Data obtained during the initiation, midpoint, and near study termination may indicate important changes in toxicokinetics based on biochemical or toxicological alterations of biodisposition of the test article and it metabolites; for example, hepatotoxicity that is manifested after several weeks of dosing may significantly alter the metabolism of the test article. These data can also provide information useful in the design of longer term studies. Gender differences in the toxicokinetics, for example, may indicate a need for differential dosing of males and females. Species differences in the toxicokinetics may indicate the most suitable animal model for longer term studies and aid in the extrapolation of the animal data to humans.

### Miscellaneous Other Endpoints

A wide variety of additional endpoints can be evaluated in repeated-dose studies. In part, any additional endpoints would depend on the questions to be addressed and the creativity of the toxicologist designing the studies; for example, the increased availability of electron microscopy makes this endpoint a more viable option today than in previous decades. Development of an increased number of histochemical assays, especially those employing specific antibodies, makes the preserved wet tissues, embedded tissues, and the histopathology slides obtained from these studies valuable for future use. In many cases, the need for additional endpoints is unknown until the initial results of the study are known. The preserved materials then become a valuable resource. If it is found that the liver from a study contains vacuoles and the toxicologist suspects these to be fat vacuoles, it is possible to use special lipid stains to determine if the vacuoles are lipid. Other special stains exist for a variety of purposes. An example of a protein-specific stain (Mallory–Heidenhain) is illustrated in Figure 24.6. If the toxicologist believes that the histopathological data indicate that a particular compound is producing cellular proliferation, there are a number of approaches to investigate this hypothesis. A standard method to determine cellular proliferation is by injecting the animal with $^3$H-thymidine and measuring the incorporation of the radiolabel into cellular DNA by methods such as radioautography; however, if the study has been terminated, this is not possible. Also, the toxicologist may not want to administer radiolabel to the animals for a number of reasons, even if the in-life portion of the study has not been terminated. If the evidence of increased cellular proliferation occurs in tissues with a relatively high rate of normal proliferation, such as the gut mucosa, it is possible to determine the mitotic index by counting mitotic figures in cells from slides previously prepared for histopathological analysis. Alternatively, it is now possible to immunostain for specific proteins associated with cellular proliferation in preserved tissues. Again, the particular additional endpoints added to the toxicology study should address a specific issue.

## DATA ANALYSIS AND INTERPRETATION

### Compilation and Summarization of Study Data

When the data have been collected from a repeated-dose toxicity study, the next steps involve data analysis, interpretation, and reporting. These data are derived either from the measurement of a variable associated with the test animal (e.g., body weight, feed consumption, serum enzyme activity) or from observation of the animal (e.g., physical examination findings, macroscopic and microscopic pathology). The frequency, number, and variety of these measurements and observations in repeated-dose toxicity studies yield an extremely large volume of data that must be organized and summarized prior to analysis. Historically, individual animal data were recorded manually and then were compiled either manually or following entry into a computer. Some specialized data are still handled manually today; however, most routine data are collected, compiled and statistically analyzed electronically using custom or commercially available computer programs.

For quantitative data such as body weights and feed consumption, individual animal data are used to calculate the mean values with a measure of statistical variation for each treated or control group. For other types of quantitative data, such as the number of animals exhibiting a behavioral effect or the number of animals in which a particular finding is determined histopathologically, the incidence of the observation in each treated and control group (i.e., the number of animals affected as a fraction of the total number of animals observed) is presented. Some data, such as the results of microscopic examination of urinary sediment, cannot be effectively summarized using a group mean or incidence value. Summarization of these types of data involves listing the individual animal data in their appropriate groups.

### Determination of Treatment-Related Effects

Data from the treatment groups are compared with data from the control group to determine if any treatment-related effects have occurred. In virtually all cases, the data from one group of animals will differ somewhat from the data for any other group; therefore, differences between the sets of data that are potentially related to treatment must be differentiated from spurious occurrences and from normal biological variation. This is accomplished by two methods. The first is simple examination of the data and detection of differences worthy of

**TABLE 24.15**
**Factors That Determine the Significance of Differences Between Treated and Control Groups**

| |
|---|
| Dose-related trends |
| Reproducibility |
| Related findings |
| Magnitude and type of difference |
| Occurrence in both sexes |

further consideration based on the experience of the toxicologist and comparison with historical data. The second method uses statistical tests to detect differences for which the probability that the difference occurred by chance is low. These methods should always be used in combination. Although it is an extremely powerful and useful tool, statistical analysis alone should not be used to detect treatment-related effects because, as stated by the FDA, "Statistical outliers are not always biological outliers and a 'significant' statistical test ($p \leq 0.05$) does not always indicate biological significance" [26].

Differences between the data from the control and treated animals that are detected using the methods cited above may indicate an adverse effect associated with the test material, physiological adaptation to the test material, or normal biological variation unrelated to the test material. Determination of the significance of differences between treated and control groups is based on a number of factors that are frequently considered in combination with each other. These factors are listed in Table 24.15 and discussed in the following text.

## Dose-Related Trend

One of the best indicators of an effect related to treatment is the presence of a dose-related trend in the data; that is, the magnitude of the effect varies directly with the dose level. Such an effect is reflective of the basic principle of toxicology stated by Paracelsus in the sixteenth century and often paraphrased as: "The dose makes the poison." If a difference from controls is noted in two or more dose groups and the severity or incidence of the difference increases as the dose level of the test material increases, it is probably a treatment-related effect. When a difference from controls is noted only in animals receiving the highest dose level of the test material, it may or may not be treatment related, and other factors must be considered in determining its significance. Differences from control data in test-material-treated animals are probably not associated with treatment if a dose-related trend is not observed. Because of this, as stated previously, selection of an appropriate range of dose levels is extremely important and facilitates data interpretation.

## Reproducibility of Effect

Another reliable indicator of a treatment-related effect is its reproducibility. If a difference from controls is noted in the treated animals at multiple intervals during a study, the difference is likely related to treatment. Further weight is given to the determination of the treatment relationship if the same difference is noted in other, independent studies in which the test material was administered to the same species, and even more weight is given if the difference is observed in a second species. The absence of reproducibility, especially in the same species, is an indication that the difference may have occurred by chance.

## Correlated Findings

Another consideration in the assessment of significance of an intergroup difference is the presence of related findings; for example, an elevation in the activity of serum alanine aminotransferase in treated animals when compared with the control group is probably related to treatment if there is an elevation in serum aspartate aminotransferase with concomitant hepatic necrosis. If no correlation with other findings is observed, the elevation may be of no significance, or at least its significance must be determined considering other factors.

## Magnitude and Type of Intergroup Difference

The magnitude and type of difference observed between treated and control data may also give an indication of its potential association with test material administration; for example, a doubling of an organ weight in treated animals compared with controls should be considered more likely to be treatment related than a 10% increase, even if the smaller increase is statistically significant. Furthermore, a fairly large decrease in the activity of serum alanine aminotransferase in treated animals is generally assigned limited clinical significance, whereas an increase of the same magnitude in the activity of this enzyme may be considered indicative of a toxic effect. It is obvious that the assessment of the significance of a change on the basis of its magnitude or type requires knowledge of normal trends and ranges for the data.

## Gender Differences

Determination of the significance of an apparently treatment-related effect is also influenced by whether or not the difference occurs in animals of both genders. Because treatment-related effects often occur in both genders, a difference from controls that is noted only in treated animals of one gender may not be associated with the test material. It must be remembered, however, that in some cases one gender or the other is more sensitive to the toxicity of a chemical, so only the sensitive gender will exhibit the effect at a given dose. For this reason, a difference should not be considered insignificant solely on the basis of its absence in one gender. Male rats are well known for their greater capacity to detoxify certain compounds because of their higher activity of P450-dependent monooxygenases; therefore, they may demonstrate less toxicity to these compounds than females. If the metabolic product is more toxic than the parent compound, however, they may demonstrate higher sensitivity than females. This gender difference is not seen in a number of other species, including humans [53].

## STUDY REPORT

After the data from a study have been analyzed, a report is prepared. Depending on the intended use of the report, it may contain various levels of detail. A report prepared for submission to a regulatory agency in support of the safety of a chemical will generally contain much more detail (including all individual animal data) than a research report intended to be used by an organization or individual only to give guidance for future testing or a manuscript prepared for publication in the scientific literature. Whatever the purpose of the report, it should be written in sufficient detail to permit peer review of the conduct and conclusions of the study and to allow the study to be reproduced exactly, if required.

### Report Content

All reports, regardless of their purpose, should contain certain common elements that are essential to adequately describe the conduct and results of a study. The report should clearly state the objectives of the study. It should precisely define the test material, indicate the test species used, and describe the methods and equipment employed to collect and analyze the data. Protocol deviations and an assessment of their impact on the study should be presented. The report should present the data pertinent to the study objectives in a form that facilitates its review and should discuss these data in the depth required to support the conclusions of the study. The discussion should describe any treatment-related effects observed and should explain how the various factors described above were used to determine the significance of any differences between treated and control animals in the study. Finally, the conclusions drawn from the results of the study should be clearly and concisely stated.

### Retrospective Report Audits

After the study has been completed and reported, retrospective audits of the study are frequently conducted. The manufacturer of the test material may audit the study prior to submitting it to a regulatory agency in support of registration or approval of the test material for its intended use. The regulatory agency to which the study report was submitted may also conduct an audit. Regulators also audit study reports to assess compliance of the testing laboratory with GLP regulations. Whatever the reason for retrospective auditing of the study, the process is essentially the same. The *raw data* are inventoried to ensure that they have been properly maintained. The data are reviewed and compared with the study report to ensure that the report accurately and completely presents the methods used and the data collected. Any deviations from laboratory standard operating procedures, the study protocol, or GLP regulations that occurred during conduct and documentation of the study should have been clearly explained in the report. Individual animal data should support summary tables, and discussion of the results should be consistent with the individual and summary data. Retrospective auditing is of great value to all parties involved. The manufacturer of the test material can feel comfortable that they will receive no surprises during a subsequent audit by the regulatory agency. Regulators will be more confident about the quality of the study if the data have been audited. When regulators are confident about the data, their reviews will proceed more smoothly and regulatory decisions will be expedited.

## Data and Specimen Collection and Retention

The preceding sections of this chapter describe the design of repeated-dose studies and the variables to be determined during those studies. Knowledge of this information is critical to the study conduct. Just as important to the registration or acceptance of the data from these studies is proper documentation of the procedures and results. All GLP regulations require complete and accurate documentation of the study, documentation that would allow an appropriately trained and experienced individual to repeat the study with little variation from the original design.

Good laboratory practices require that documentation of study procedures and the resulting data must be recorded in indelible ink that is suitable for photocopying. The individual responsible for the records must be documented along with the date the record was created. Any change or correction to the documentation must be made as soon as possible after the original entry and must clearly explain the reason for the change, identify the individual making the change, and indicate the date the change was made. If portions of the data are to be recorded electronically, the accuracy and reliability of any system used to collect and manipulate the data must be validated prior to its use. Electronic data collection systems are subject to the same basic requirements as hand-recorded data; that is, the individual responsible for entry or revision of the data must be identified, the date of these entries must be detailed, and an explanation of the reason for any change to the original entry must be documented.

While the study is in progress, records and specimens must be stored in an orderly and secure manner that will facilitate report preparation when the study has been completed. At completion of the study, GLP regulations require that study documentation and specimens must be placed into an archive facility that will ensure their integrity and security. The study protocol, all study records, test article samples, fixed tissues, paraffin blocks, microscopic slides, and the study reports must be archived following completion of the data. Specimens with a limited useful life-span such as blood samples for clinical pathology determinations and samples from mutagenicity assays do not need to be retained.

Access to the archive facility should be controlled by a designated archivist and limited to authorized individuals. The conditions of storage must prevent deterioration of the data. Archives must be temperature and humidity controlled and should have provisions for fire suppression and vermin control. Removal of any data or specimens from the archives should require documentation of the transfer, and procedures should be in place for retrieval of the items if they have not been returned to the archives in a timely manner.

The duration of retention for archived materials is dependent on the regulatory agency to which the data will be submitted and the reason for the submission. Archival requirements for several regulatory agencies and types of submissions are shown in Table 24.16. After the materials have been archived for the required period, the data and specimens may be discarded without compromising the regulatory status of the study; however, many companies effectively archive data from pivotal studies indefinitely even though there is no regulatory requirement to do so. This extra precaution may prove useful in addressing future liability issues related to the test material or in designing additional studies if new safety questions for the test material arise at some time in the future.

**TABLE 24.16**
**GLP Requirements for Duration of Data and Specimen Archiving**

| Regulatory Agency | Retention Requirement |
|---|---|
| FDA | The shorter of:<br>At least 2 years following date of approval of application for research or marketing permit<br>At least 5 years following date of nonclinical laboratory study submission supporting application for research or marketing permit<br>At least 2 years following date of study submission supporting application for research or marketing permit if submission is not approved or in other similar situations |
| EPA FIFRA | The longer of:<br>For the life of the approval for any study submitted in support of an approved research or marketing permit<br>At least 5 years following the date study is submitted in support of a research or marketing permit<br>At least 2 years following the date of study submission if application for research or marketing permit is not approved or after the decision has been reached not to seek a research or marketing permit |
| EPA TSCA | At least 10 years following the effective date of the final test rule or publication of the acceptance of a negotiated test agreement |
| OECD | For the period specified by the appropriate authorities |

## REGULATIONS CONCERNING GENERATION AND USE OF DATA FROM REPEATED-DOSE TOXICITY STUDIES

Almost every industrialized country in the world has regulations governing the introduction, transportation, and use of new pesticides, food additives, human and animal drugs, and other chemicals. Many of these countries also have regulations governing medical devices, workplace exposure to chemicals, the introduction of industrial chemicals into commerce, and the disposition of chemical waste. There is general uniformity in the objectives of these laws (e.g., not to impede the beneficial use of chemicals but at the same time to ensure their safety in use). Even though the regulatory agencies can agree, in broad terms, on one framework of toxicity testing guidelines, toxicologists must become familiar with the details of particular guidelines to fulfill their role as a bridge between scientific and regulatory concerns. Several regulations governing repeated-dose toxicity and other types of toxicity testing are briefly described below.

### U.S. Laws and Regulatory Guidelines

#### Federal Food, Drug and Cosmetic Act

The Federal Food, Drug, and Cosmetic Act (FFDCA) [29] as amended by the Food and Drug Administration Modernization Act of 1997 [28] controls the introduction of human and animal drugs, direct food additives, indirect food additives (such as packaging materials), and components of cosmetics. In the case of new human or animal drugs, safety and efficacy must be established for a particular therapeutic application before the FDA grants approval for marketing. The approval process is comprehensive and involves two sequential phases. For the investigational new drug (IND) phase, industry is required to file preclinical toxicity data with the FDA before investigation of the safety and potential therapeutic value of a drug in limited numbers of humans. When the efficacy and safety of the drug in the treatment of a particular disease have been established through extensive clinical trials, these data together with additional animal toxicity data are provided to the agency as part of a New Drug Application (NDA) or new animal drug application. The NDA is reviewed by the FDA with respect to the safety and efficacy of the drug. As a consequence of the FDA review, the NDA is approved or disapproved or deficiencies in the data are cited. The summary of a NDA must address benefit and risk relationships, clinical data, nonclinical pharmacology and toxicology, human pharmacokinetics, bioavailability, and microbiology. It must also contain information on pharmacologic class, scientific rationale, and clinical benefits, as well as chemistry and manufacturing [34].

With respect to food additives, industry must show that a material either intended for direct addition to food, such as a preservative or flavoring agent, or having indirect contact with food, such as a packaging or can-coating material, is safe for its intended use. Results from a hazard assessment are submitted to the FDA for review as part of a Food Additive Petition. If the data demonstrate the safety of the chemical, a regulation is published allowing the chemical to be used for a particular purpose in food or in contact with food. In 1982, the FDA published *Toxicological Principles for the Safety Assessment of Direct Food Additives and Color Additives Used in Food* [22]. Revised versions of this document have been issued by the FDA [23,26]. This so-called "Redbook" delineates the nature of evaluations necessary to determine food additive safety. It provides a basic scheme for scientifically sound decisions for the development of the safety assessment. These guidelines include a priority-setting system that increases the efficiency of the food additive safety assessment.

The FFDCA also provides an alternative method by which materials can be approved for use in or on food: the generally recognized as safe (GRAS) process [31]. For a material to be considered GRAS, its safety evaluation must satisfy three basic requirements. First, a group of experts qualified by scientific training and experience must conclude that the material, when used as intended, is safe for human consumption. The sponsor of the potential additive convenes this expert group, and the group operates without regulatory oversight. Second, the information considered by the expert group during the GRAS review must be common knowledge (i.e., available in the scientific literature). Third, there must be general agreement within the scientific community with the conclusion of the expert group. Final GRAS approval does not require regulatory review; however, many producers of such materials petition the FDA to affirm their GRAS status prior to marketing them.

At the present time, there are no requirements for the FDA to review cosmetic formulations for safety prior to marketing. Although the FFDCA only requires that the cosmetics be free of any "poisonous and deleterious" substances, responsible suppliers of ingredients for use in cosmetics and manufacturers of final products conduct toxicity studies relevant to the specific cosmetic.

The FFDCA also addresses the concentrations of pesticides permitted in foods in the United States. Under FFDCA Section 408, the EPA establishes maximum allowable concentrations for pesticide residues in raw agricultural commodities (i.e., food crops, eggs, raw milk, and meat). Under FFDCA Section 409, the EPA also establishes a maximum allowable concentration for a pesticide in processed food, if processing concentrates the residue of the pesticide in the raw agricultural commodity. These maximum concentrations are referred to as *tolerance levels* or *tolerances*. Human exposure resulting from the consumption of foods containing tolerance levels of a pesticide must not exceed the maximum permissible intake of the pesticide established by the EPA under the Federal Insecticide, Fungicide, and Rodenticide Act. Tolerances for most foods are enforced by the FDA, while those for meat, poultry, and some egg products are enforced by the Food Safety and Inspection Service within the U.S. Department of Agriculture (USDA).

## Federal Insecticide, Fungicide and Rodenticide Act

The Federal Insecticide, Fungicide and Rodenticide Act (FIFRA) was administered initially by the USDA and is now administered by the EPA [10]. Under the FIFRA, the EPA is responsible for the registration of pesticides for use in the United States. This act requires extensive toxicity testing to be conducted in mammalian, avian, and aquatic species to support the safety of a pesticide. Detailed guidelines for toxicity study design and reporting have been issued by the EPA [17]. Toxicity data submitted in an application for registration of a pesticide are reviewed by toxicologists in the Office of Pesticide Programs at the EPA. Other data specifically required by the FIFRA as a condition of pesticide registration include product chemistry, residue chemistry, environmental fate, reentry protection, spray drift, plant protection, nontarget insects, and product performance.

## Food Quality Protection Act

The Food Quality Protection Act (FQPA) was signed into law in 1996 and amends sections of both the FFDCA and FIFRA [13]. It is intended to establish more consistent regulation of pesticides and to protect human health using an approach that places increased emphasis on the scientific evaluation of pesticide safety data. Among other provisions affecting pesticide registration and tolerance setting, it (1) mandates special considerations for protection of infants and children, (2) requires determination of aggregate pesticide exposure from all sources (e.g., food, home use, drinking water), (3) requires summation of exposures from multiple chemicals that exhibit a common mechanism of toxicity, (4) expedites approval of pesticides considered to be most safe, and (5) requires periodic reevaluation of tolerances for registered pesticides to be certain that the data supporting registration remain acceptable and complete by current standards.

### Toxic Substances Control Act

The Toxic Substances Control Act (TSCA) is administered by the Office of Pollution Prevention and Toxics within the EPA; it is a complex and far-reaching law that affects industrial chemicals existing in commerce as well as new chemicals in the United States [20]. One of the first requirements of the TSCA was the compilation of an inventory of chemicals that were active in commerce in the years 1977 to 1979 and determination of the need for toxicity testing of these existing chemicals. Manufacturers or importers of industrial chemicals that are considered new under the TSCA definition are required to notify the EPA at least 90 days before the manufacture or import of a new chemical. The act requires that certain information regarding the new chemical be submitted to the EPA for review in a premanufacture notification (PMN). Although the act does not require toxicity testing to be conducted on a new chemical prior to manufacture, it does require submission of all existing health and safety data for the new chemical so its risk to health or the environment can be assessed. If the EPA determines in its review that the new chemical does present an unreasonable risk, one of several actions it may take is to require that the chemical be tested for specific toxic effects. The EPA may also issue a testing rule requiring that a specified chemical or chemical mixture be tested for certain toxic effects. The EPA has issued test standards for the conduct of toxicity studies on chemicals or chemical mixtures for which testing will be required under the TSCA. These standards are the same guidelines used for pesticide testing under the FIFRA [17].

### Transportation Act

Regulations promulgated by the Department of Transportation (DOT) require that materials shipped in interstate commerce be labeled and contained in a manner consistent with the degree of hazard they present [5]. The DOT requires that acute toxicity data for chemicals be used to place them into packing groups. Labeling requirements for chemicals are based on this packing group.

### The Coast Guard

Prior to importing a chemical into the United States, the Coast Guard requires a set of acute mammalian toxicity data for the chemical [6]. This acute toxicity profile should minimally include the following: acute oral toxicity, acute dermal toxicity, and skin and eye irritation studies.

### Consumer Product Safety Act)

Prior to passage of the Consumer Product Safety Act (CPSA), the classification and testing for acute toxicity of household products were conducted under regulations promulgated by the FDA, which administered the Federal Hazardous Substances Act (FHSA). The function of administering FHSA now resides with the Consumer Product Safety Commission. If results obtained in acute oral or dermal toxicity tests conducted using methods outlined in the Code of Federal Regulations (CFR) for hazardous substances meet prescribed criteria of toxicity, labeling and packaging as prescribed in the regulation must be used [3].

### Occupational Safety and Health Act

This law, administered by the Occupational Safety and Health Administration (OSHA) of the Department of Labor, is designed to ensure safety in the workplace [59]. No requirements exist in this law for manufacturers to test substances for toxicity prior to their use in the workplace. The impact of the TSCA has an overlapping effect, in that occupational exposure to new chemicals is considered in premanufacture notices. As indicated previously, specific test requirements under the TSCA affect new or existing chemicals that are manufactured or processed in the United States.

### Resource Conservation and Recovery Act

The Resource Conservation and Recovery Act (RCRA) authorizes the EPA to institute a national program to control hazardous waste defined as "solid, liquid, semisolid or gaseous waste that may cause increased mortality or serious illness, or may cause substantial hazard to the health or the environment when improperly managed" [18]. The main purpose of these regulations is to control the generation, storage, treatment, transportation, disposal, recordkeeping, and reporting of hazardous waste. The RCRA places the primary responsibility of identifying and managing hazardous waste on the waste generators. Other persons or institutions involved in waste disposal and management also have an obligation to know if the waste is hazardous. The degree of toxicity, concentration, migration to the environment, persistence or degradation in the environment, bioaccumulation in the ecosystem, types of improper management, quantities of waste, past human and environmental damage records, and other factors are all taken into consideration.

## International Laws and Regulations Concerning Hazard Assessment and Toxicity Testing Guidelines

The United States has not been alone in developing laws to protect humans and their environment from possibly dangerous effects of new industrial chemicals in the marketplace. In the European Union, the Council of Ministers of the European Economic Community (EEC) has issued a directive concerning laws, regulations, and administrative procedures that relate to the classification, packaging, and labeling of dangerous substances [7]. An amendment to this directive was adopted later to protect humans and their environment from potential risks that might arise through the marketing of new chemicals [8]. It requires that a new chemical be subjected to a base set of tests to define its physical and chemical properties, mammalian toxicity, and ecotoxicologic effects. The base set of mammalian studies consists of the following: acute oral $LD_{50}$, acute dermal $LD_{50}$, acute inhalation $LC_{50}$ (if applicable), skin and eye irritation, and dermal sensitization. A manufacturer or importer is required to furnish the appropriate authorities in his EEC member state with a notification containing, in part, the results of these tests with the new chemical. Such notification must be filed 45 days before marketing in the member state in which it is to occur.

The amendment to the EEC directive is the counterpart of the U.S. TSCA. There are, however, some differences, not the least of which is in the approach to toxicity testing. The EEC requires only notification of a new chemical prior to marketing, whereas TSCA demands that the notification, in the case of a domestic manufacturer, be given to the EPA at least 90 days before manufacture.

The EEC has also provided guidance on the evaluation of safety and efficacy of drugs. The EEC has now adopted guidance notes for efficacy, testing of pharmacokinetics in humans, and bioavailability of drugs for long-term use, as well as a number of more specific activity groups including cardiac glycosides, oral contraceptives, topical corticosteroids, nonsteroidal antiinflammatories, antimicrobials, anticonvulsants, antianginals, and chronic peripheral arterial disease agents. Safety guidance notes also were adopted for single-dose and repeated-dose toxicity, pharmacokinetic metabolism, mutagenic potential, carcinogenic potential, and reproduction studies [9].

As discussed earlier in this chapter, with many countries establishing their own regulations for safety assessment, it became more likely that manufacturers would have to perform several different versions of the same tests to satisfy the requirements of different countries. In an attempt to avoid unnecessary and wasteful duplication of work, the OECD, a group comprised of experts from a number of nations throughout the world, produced a set of toxicity testing guidelines that enable tests to be carried out in a similar manner in different countries [55]. The OECD package, which has been revised 18 times to implement improvements to individual study designs, includes guidelines on acute oral, dermal, and inhalation studies; eye and skin irritation and skin sensitization studies; subchronic oral, dermal, and inhalation studies; and teratogenicity, carcinogenicity, and chronic and combined chronic/carcinogenicity studies. Results from studies conducted according to OECD guidelines are generally fully acceptable to the various regulatory bodies throughout the world.

To facilitate more universal acceptance of data generated to support approval and use of pharmaceutical products, the International Conference on Harmonization has developed guidance documents concerning the efficacy, quality, and safety of these drugs. The ICH is made up of regulatory authorities and trade organizations from the United States, Japan, and the European Union. Instead of providing detailed instructions for study design or conduct, the ICH guidances address specific issues related to testing, such as the duration of chronic toxicity testing, definition of an acceptable battery of genotoxicity studies, and required elements of chemical stability testing of new drug products [41,43]. Because of ICH efforts, many hindrances to the approval and use of valuable pharmaceuticals around the world have been removed.

## Regulatory Internet Sites

Many of the regulatory agencies and organizations worldwide that have responsibility for protection of humans from the hazardous effects of chemicals maintain Internet websites. These websites usually include information concerning the history, structure, and specific responsibilities of these organizations. The sites also generally contain or reference the statutes and guidelines under which the organizations operate. Internet addresses for a number of informative regulatory sites are listed in Table 24.17.

# QUESTIONS

1. Five related compounds are under consideration for development by a pharmaceutical company named Nomohats, Ltd. The compounds are hair regrowth agents. The intended treatment population includes males and females from 24 to 90+ years of age. All of the compounds have been thoroughly characterized chemically, and each has shown adequate efficacy following oral exposure during preliminary testing. The effects of oral exposure in single-dose acute studies were similar for each of the compounds, and none of the compounds was genotoxic in preliminary screens. Nomo-

**TABLE 24.17**
**Regulation Websites**

| Regulatory Subject | Site Address (http://) |
|---|---|
| **International** | |
| European Union | www.eurunion.org |
| EU General Product Safety Directive | ec.europa.eu/consumers/cons_safe/gpsd/index_en.htm |
| Organization for Economic Cooperation and Development | www.oecd.org |
| OECD Testing Guidelines | www.oecd.org/document/22/0,2340,en_2649_34377_1916054_1_1_1_1,00.html |
| International Conference on Harmonization | www.ich.org |
| ICH Guidance Documents | www.ich.org/cache/compo/276-254-1.html |
| **United States** | |
| Food and Drug Administration | www.fda.gov |
| FDA Center for Food Safety and Applied Nutrition | www.cfsan.fda.gov/list.html |
| FDA Center for Drug Evaluation and Research | www.fda.gov/cder/index.html |
| FDA Center for Devices and Radiological Health | www.fda.gov/cdrh/index.html |
| Environmental Protection Agency | www.epa.gov |
| EPA Office of Pollution Prevention and Toxic Substances | www.epa.gov/oppts |
| EPA Office of Pesticide Programs | www.epa.gov/pesticides |
| Toxic Substances Control Act | www.epa.gov/region5/defs/html/tsca.html/tsca.htm |
| Federal Insecticide, Fungicide and Rodenticide Act | www.epa.gov/region5/defs/html/tsca.html/fifra.htm |
| Federal Food, Drug and Cosmetic Act | www.fda.gov/opacom/laws/fdcact/fdctoc.htm |
| Food Quality Protection Act | www.epa.gov/oppfead1/fqpa |
| Food and Drug Administration Modernization Act of 1997 | www.fda.gov/cder/guidance/105-115/htm, |
| EPA Testing Guidelines | www.epa.gov/opptsfrs/home/guidelin.htm |
| FDA Guidance Documents | www.fda.gov/cder/guidance/index.htm |

hats has budgeted funding to allow development of one of the compounds. You are the toxicologist responsible for generating additional toxicity data to assist in selecting the compound to be carried forward in the development program. You have 2 months because a meeting with the FDA has been scheduled by Nomohats for 10 weeks from today. The purpose of the meeting is to review the chemical characterization and preliminary toxicity data for the compound and to suggest a toxicology program to demonstrate that there is no unreasonable risk associated with human exposure to the compound. Describe the studies that you would conduct to aid in selection of the compound for further development, remembering that these studies will also generate most of the new toxicity data to be reviewed during the meeting with FDA.

2. After consulting with the FDA, you have determined that the hazard assessment program for your compound, *Fuzzypate,* should include subchronic toxicity testing. In a 14-day repeated-dose study with this compound in Göettingen minipigs at 5 mg/kg/day, dermal administration resulted in renal toxicity in 40% of the animals. These animals also exhibited mildly decreased body weight gain, urinary incontinence, and slightly enlarged and discolored kidneys. In the mid-dose group, 2.5 mg/kg/day, no renal toxicity was reported, and the body weight of the animals was slightly lower than the untreated control group. At 1 mg/kg/day, only a statistically significant increase in serum levels of liver enzymes was observed. Efficacy testing revealed that *Fuzzypate* is effective for its intended use at 0.2 mg/kg/day. Pharmacokinetic determinations indicate that the material is excreted almost entirely in the urine and its serum half-life is 120 minutes. What dose levels would you select for the subchronic toxicity study? Why? In addition to the standard variables, what special endpoints would you include for investigation during the subchronic study?

3. The results in the table on the following page were obtained during the subchronic study with *Fuzzypate*. In what dose groups is there a treatment-related effect on body weight? Are liver and kidney target organs for *Fuzzypate* toxicity? If so, in which dose groups is there a treatment-related effect? Based on the data in the table, what is the NOAEL for this study?

**Selected Results from the Subchronic Toxicity Study of *Fuzzypate* in Minipigs**

| Variable/Finding/Interval | Treatment Group (Males) | | | | Treatment Group (Females) | | | |
|---|---|---|---|---|---|---|---|---|
| | Control | Low | Mid | High | Control | Low | Mid | High |
| Body weight (kg) | 9.3 ± 0.7 | 8.6 ± 1.5 | 10.2 ± 0.6 | 7.2 ± 1.1[a] | 9.9 ± 1.2 | 10.1 ± 0.9 | 9.2 ± 0.7 | 8.1 ± 0.6[a] |
| Hepatocellular hypertrophy (number with finding/total number) | | | | | | | | |
| Week 4 | 0/4 | 1/4 | 0/4 | 2/4 | 1/4 | 0/4 | 1/4 | 1/4 |
| Week 13 | 1/6 | 1/6 | 2/6 | 3/6 | 0/6 | 2/6 | 1/6 | 1/6 |
| Renal tubule hyperplasia (number with finding/total number) | | | | | | | | |
| Week 4 | 1/4 | 0/4 | 4/4 | 4/4 | 0/4 | 1/4 | 2/4 | 2/4 |
| Week 13 | 1/6 | 1/6 | 2/6 | 6/6 | 0/6 | 0/6 | 3/6 | 4/6 |

[a] Statistically significantly different from controls ($p < 0.05$).

## REFERENCES


1. American Institute of Nutrition. (1977): Report of the American Institute of Nutrition Ad Hoc Committee on Standards for Nutritional Studies. *J. Nutr.*, 107:1340–1348.
2. Boorman, G. A., Eustis, S. L., Elwell, M. R., Montgomery, Jr., C. A., and MacKenzie, W. F. (1990): *Pathology of the Fischer Rat: Reference and Atlas*. Academic Press, San Diego, CA.
3. Consumer Product Safety Commission. (2006): 16 CFR Parts 1015–1402, *Consumer Product Safety Act*. Office of the Federal Register, National Archives and Records Administration, U.S. Government Printing Office, Washington, D.C.
4. Datta, P. K., Friger, H., and Schleiermacher, E. (1970): The effect of chemical mutagens on the mitotic chromosomes of the mouse *in vivo*. In: *Chemical Mutagenesis in Mammals and Man*, edited by F. Vogel and G. Rohrborn, pp. 194-213. Springer-Verlag, New York.
5. U.S. Department of Transportation. (2005): 49 CFR Part 173, *Shippers: General Requirements for Shipments and Packaging*, Office of the Federal Register, National Archives and Records Administration, U.S. Government Printing Office, Washington, D.C.
6. U.S. Department of Transportation. (2005): 49 CFR Part 176, *Carriage by Vessel*, Office of the Federal Register, National Archives and Records Administration, U.S. Department of Transportation, Washington, D.C.
7. European Economic Community. (1967): Council Directive 67/548/EEC, *OJEC*, L196:1–98.
8. European Economic Community. (1979): Sixth Amendment to Council Directive 79/831/EEC, *OJEC*, L259/79:11.
9. European Economic Community. (1987): Council Recommendation 87/176/EEC, *OJEC*, L073:1–46.
10. EPA. (2006): 40 CFR Parts 152–186, *Federal Insecticide, Fungicide and Rodenticide Act*, U.S. Environmental Protection Agency, Washington, D.C.
11. EPA. (2006): 40 CFR Part 160, *Good Laboratory Practice Standards*, U.S. Environmental Protection Agency, Washington, D.C.
12. EPA. (2006): 40 CFR Part 160, *Good Laboratory Practice Standards*, Sect. 160.105. U.S. Environmental Protection Agency, Washington, D.C.
13. EPA. (1996): Public Law 104-170, *Food Quality Protection Act*, U.S. Environmental Protection Agency, Washington, D.C.
14. EPA. (1998): *90-Day Toxicity in Rodents*, OPPTS Test Guideline 870-3100. Office of Prevention, Pesticides, and Toxic Substances, U.S. Environmental Protection Agency Washington, D.C.
15. EPA. (1998): *Chronic Toxicity*, OPPTS Test Guideline 870-4100. Office of Prevention, Pesticides, and Toxic Substances, U.S. Environmental Protection Agency, Washington, D.C.
16. EPA. (1998): *OPPTS Test Guidelines*, Series 830, *Physica Chemical Properties*. Office of Prevention, Pesticides, and Toxic Substances, U.S. Environmental Protection Agency Washington, D.C.
17. EPA. (1998): *OPPTS Test Guidelines*, Series 870, *Healtl Effects*. Office of Prevention, Pesticides, and Toxic Substances, U.S. Environmental Protection Agency, Washing ton, D.C.
18. EPA. (2006): 40 CFR Parts 240–271, *Resource Conserva tion and Recovery Act*, U.S. Environmental Protectio Agency, Washington, D.C.
19. EPA/TSCA. (2006): 40 CFR Part 792, *TSCA Good Labo ratory Practice Standards*, U.S. Environmental Protectio Agency, Washington, D.C.
20. EPA. (2006): 40 CFR Parts 700–799, *Toxic Substance Control Act*, U.S. Environmental Protection Agency, Wash ington, D.C.
21. European Union. (2003): *Medicinal Products for Humar and Veterinary Use: Good Manufacturing Practice*, Direc tive 2003/94/EC. Commission of the European Communi ties, Brussels, Belgium.
22. FDA. (1982): *Toxicological Principles for the Safet Assessment of Direct Food Additives and Color Additive Used in Food*, U.S. Food and Drug Administration, Rock ville, MD.
23. FDA. (1993): *Toxicological Principles for the Safet Assessment of Direct Food Additives and Color Addititive Used in Food, Redbook II, Draft*. Center for Food Safet and Applied Nutrition, U.S. Food and Drug Administra tion, Rockville, MD.
24. FDA. (2006): *Recommendations for Submission of Chen ical and Technological Data for Direct Food Additive Pet tions*. U.S. Food and Drug Administration, Rockville, ML

25. FDA. (1993): *Toxicological Principles for the Safety Assessment of Direct Food Additives and Color Additives Used in Food "Redbook II,"* Chapter IV.C.7. Center for Food Safety and Applied Nutrition, U.S. Food and Drug Administration, Rockville, MD.
26. FDA. (2003) *Redbook 2000: Toxicological Principles for the Safety Assessment of Food Ingredients*. Center for Food Safety and Applied Nutrition, U.S. Food and Drug Administration, Rockville, MD.
27. FDA. (2003) *Redbook 2000: Toxicological Principles for the Safety Assessment of Food Ingredients*, Chapter IV.C.3.a. Center for Food Safety and Applied Nutrition, U.S. Food and Drug Administration, Rockville, MD.
28. FDA. (1997): Public Law 105-115, *Food and Drug Administration Modernization Act of 1997*. U.S. Food and Drug Administration, Rockville, MD.
29. FDA. (2002): 21 USC 301 *et seq., Federal Food, Drug, and Cosmetic Act, as Amended, and Related Laws*. U.S. Food and Drug Administration, Rockville, MD.
30. FDA. (2006): 21 CFR 110, *Current Good Manufacturing Practice in Manufacturing, Packaging or Holding Human Food*. U.S. Food and Drug Administration, Rockville, MD.
31. FDA. (2006): 21 CFR 170, *Food Additives*. U.S. Food and Drug Administration, Rockville, MD.
32. FDA. (2006): 21 CFR 210, *Current Good Manufacturing Practice in Manufacturing, Processing, Packaging or Holding of Drugs*. U.S. Food and Drug Administration, Rockville, MD.
33. FDA. (2006): 21 CFR 211, *Current Good Manufacturing Practice for Finished Pharmaceuticals*. U.S. Food and Drug Administration, Rockville, MD.
34. FDA. 2006): 21 CFR 314, *Applications for FDA Approval To Market a New Drug or an Antibiotic Drug*. U.S. Food and Drug Administration, Rockville, MD.
35. FDA. (2006): 21 CFR 58, *Good Laboratory Practice for Nonclinical Laboratory Studies*, Sect. 58.105. U.S. Food and Drug Administration, Rockville, MD.
36. FDA. (2006): 21 CFR 58, *Good Laboratory Practice for Nonclinical Laboratory Studies*. U.S. Food and Drug Administration, Rockville, MD.
37. Hardisty, J. F. and Eustis, S. L. (1990): Toxicological pathology: a critical stage in study interpretation. In: *Progress in Predictive Toxicology*, edited by D. B. Clayson, I. C. Nunro, P. Shubik, and J. A. Swenberg. Elsevier, New York.
38. Haschek, W. M. and Rousseaux, C. G. (1991): *Handbook of Toxicologic Pathology*. Academic Press, San Diego, CA.
39. Heddle, J. A., Hite, M., Kirkhart, B., Larson, K., MacGregor, J. T., Newell, G. W., and Salamone, M. F. (1983): The induction of micronuclei as a measure of genotoxicity, *Mutat. Res.*, 123:61–118.
40. HPB. (2002): *Good Manufacturing Practices (GMP) Guidelines*. Health Protection Branch, Ottawa, Canada.
41. ICH. (1997,2003): *Guidelines: Quality, Guidance Q1A(R2) and Safety*, Guidance S2A. ICH Secretariat, Geneva, Switzerland.
42. ICH. (1999): *Guidelines: Quality*, Guidance Q6A. ICH Secretariat, Geneva, Switzerland.
43. ICH. (1998): *Harmonized Tripartite Guidelines: Duration of Chronic Toxicity Testing in Animals (Rodent and Non-Rodent Toxicity Testing)*, Guidance 54. ICH Secretariat, Geneva, Switzerland.
44. Institute for Laboratory Animal Research. (1996): *Guide for the Care and Use of Laboratory Animals*, National Academy Press, Washington, D.C.
45. Jain, N. C. (1986): *Schalm's Veterinary Hematology*, 4th ed. Lea & Febiger, Philadelphia, PA.
46. MAFF. (1984): *Good Laboratory Practice Standards for Toxicological Studies in Agricultural Chemicals*. Agricultural Production Bureau, Ministry of Agriculture, Forestry, and Fisheries, Japan.
47. MITI. (1984): *Good Laboratory Practice Standards Applied to Industrial Chemicals*. Basic Industries Bureau, Ministry of International Trade and Industry, Japan.
48. MOHW. (1982): *Good Laboratory Practice Standards for Safety Studies on Drugs*, Pharmaceutical Affairs Bureau, Ministry of Health and Welfare, Japan.
49. Lee, M.-Y., Park, C. B., Dordick, J. S., and Clark, D. S. (2005): Metabolizing enzyme toxicology assay chip (MetaChip) for high-throughput microscale toxicity analyses. *PNAS*, 102:983–987.
50. Loeb, W. F. and Quimby, F. W. (1989): *The Clinical Chemistry of Laboratory Animals*. Pergamon Press, New York.
51. McNamara, B. P. (1976): Concepts in health evaluation of commercial and industrial chemicals. In: *New Concepts in Safety Evaluation*, edited by M. A. Mehlman, R. E. Shapiro, and H. Blumenthal, pp. 61–140. Hemisphere, Washington, D.C.
52. Mirsalis, J. C. and Butterworth, B. E. (1980): Detection of unscheduled DNA synthesis in hepatocytes isolated from rats treated with genotoxic agents: an *in vivo-in vitro* assay for potential carcinogens and mutagens. *Carcinogenesis*, 1: 621–625.
53. Mugford, C. A. and Kedderis, G. L. (1998) Sex-dependent metabolism of xenobiotics, *Drug Metab. Rev.*, 30: 441–498.
54. OECD. (1981): *Chronic Toxicity Studies: OECD Guidelines for Testing of Chemicals*, Test Guideline 452. Organization for Economic Cooperation and Development, Paris.
55. OECD. (1981–2007): *OECD Guidelines for Testing of Chemicals*. Section 4. *Health Effects*. Organization for Economic Cooperation and Development, Paris.
56. OECD. (1981–2007): *OECD Guidelines for Testing of Chemicals*. Section 1. *Physical Chemical Properties*. Organization for Economic Cooperation and Development, Paris.
57. OECD. (1997): *OECD Principles of Good Laboratory Practice*. Organization for Economic Cooperation and Development, Paris.
58. OECD. (1998): *Repeated Dose 90-Day Oral Toxicity Study in Rodent: OECD Guidelines for Testing of Chemicals*, Test Guideline 408. Organization for Economic Cooperation and Development Paris.
59. OSHA. (2006): 29 CFR Parts 1910, 1915, 1918, and 1926, *Occupational Safety and Health Act*, U.S. Occupational Safety and Health Administration, Washington, D.C.

60. Rao, G. N. and Knapka, J. J. (1987): Contaminant and nutrient concentrations of natural ingredient rat and mouse diet used in chemical toxicology studies. *Fundam. Appl. Toxicol.*, 9:324-238.
61. Reeves, P. G., (1997): Components of the AIN-93 diets as improvements in the AIN-76A diet. *J. Nutr.*, 127: 838S–841S.
62. U.K. DHSS. (1999): *The Good Laboratory Practice Regulations 1999*. Department of Health and Social Security, London.
63. USDA. (2007): 9 CFR Parts 1–3, *Animal Welfare Act*. Animal and Plant Health Inspection Service, U.S. Department of Agriculture, Washington, D.C.
64. Latendresse, J. R., Warbrittion, A. R., Jonassen, H., and Creasy, D. M. (2002): Fixation of testes and eyes using a modified Davidson's fluid: comparison with Bouin's fluid and conventional Davidson's fluid. *Toxicol. Pathol.*, 30(4): 524–533.
65. Zakim, D., Hochman, Y., and Vessey, D. A. (1985): Methods for characterizing the function of UDP-glucuronyltransferases. In: *Biochemical Pharmacology and Toxicology*, Vol. 1, edited by D. Zakim and D. A. Vessey, p. 189. John Wiley & Sons, New York.

# 25 Principles of Testing for Carcinogenic Activity

*Gary M. Williams, Michael J. Iatropoulos, and Harald G. Enzmann*

## CONTENTS

**TABLE 25.1**
**Regulations or Agreements under Which Carcinogenicity Testing May Be Required**

| Legislation/Guidance | Agency | Agents of Concern |
|---|---|---|
| Commission Directive 414/EEC (1991) | EU | Plant protection products |
| Commission Directive 67/EEC (1993) | EU | New notified substances |
| Commission Regulation 1488/EEC (1994) | EU | Existing substances |
| Dangerous Substances Directive (1967; amended 1992) | EU | Industrial chemicals |
| Food, Drug, and Cosmetics Act (1906, 1938, amended 1992)/ FDA Redbook II | U.S. FDA | Food, medicines, cosmetics, food additives, color additives, new drugs, animal and feed additives, medical devices |
| Federal Hazardous Substances Act (1960; amended 1988) | U.S. CPSC | Household products |
| Federal Insecticide, Fungicide, and Rodenticide Act (FIFRA) (1948; amended 1978) | U.S. EPA | Pesticides |
| Guidance on Toxicology Study Data for Application of Agriculture Chemical Registration (1985) | MAFF (Japan) | Agricultural chemicals |
| Guideline for Toxicity Testing of Chemicals (1990) | MHW (Japan) | Chemicals |
| Pharmaceutical Affairs Law (1980)/ Guidelines for Toxicity Studies of Drugs Manual (1990) | MHW (Japan) | Medicines |
| Guidelines for Toxicity Studies of New Animal Drugs (1988) | MHW (Japan) | Animal medicines |
| Pesticide Registration Directive (1991) | EU | Pesticides |
| Technical Requirements for the Registration of Pharmaceuticals for Human Use (1995) | ICH | Medicines |
| Toxic Substances Control Act (TCSA) (1976; amended 1992) | U.S. EPA | Hazardous chemicals not covered by other laws, includes premarket review |

*Note:* EEC, European Economic Communities; EU, European Union; ICH, International Conference on Harmonization; MAFF, Ministry of Agriculture, Forestry, and Fisheries; MHW, Ministry of Health and Welfare; U.S. CPSC, United States Consumer Product Safety Commission; U.S. EPA, United States Environmental Protection Agency; U.S. FDA, United States Food and Drug Administration.

## CHEMICALS WITH CARCINOGENIC ACTIVITY

Cancer is a leading cause of death [8,162], and it can result from exposure to exogenous chemicals [359]; thus, in the toxicological assessment of chemicals, testing for carcinogenicity constitutes one of the most important evaluations and is required under numerous regulations or agreements (Table 25.1). A large database on the carcinogenic or oncogenic activities of chemicals in rodents has accrued [114,115,236] as a consequence of over 80 years of basic research and the output from national testing programs in several countries, particularly the United States and Japan. In the U.S. National Toxicology Program (NTP), routine rodent cancer bioassays (RCBs), mainly in mice and rats, have been reported for over 530 chemicals [244], and further testing is ongoing.

The definitive bioassay for animal carcinogenic activity is the RCB, which in its various forms is detailed here and elsewhere [152,224,280,339,340,358]. The current method for the RCB is exemplified by that currently in use by the NTP [23]. This method was refined from basic procedures developed beginning in the 1960s largely by the pharmaceutical industry under guidance from the U.S. Food and Drug Administration (FDA) and by the U.S. National Cancer Institute (NCI) Bioassay Program [49,50, 63,356]. Findings from a wide variety of chemicals subjected to RCBs have led to the recognition that results cannot be unquestioningly extrapolated to humans [229,358], for whom, of course, carcinogenic activity is established by epidemiological studies. Available mechanistic procedures to be discussed assist in assessment of RCB results for potential human hazard.

### Carcinogenic Activity

A committee of the International Federation of Societies of Toxicologic Pathologists (IFSTP) has adopted the definition of a chemical carcinogen as a "substance that causes a cell or group of normal cells, which would not otherwise have shown this property, to change its biological behavior and demonstrate progressive growth of a malignant character" [89]. The carcinogenic or oncogenic activity of a test substance (TS) is best documented by the finding of unequivocal evidence in either gender of the experimental animal of induction of a type of malignant neoplasm not seen in controls [376,388], either as preneoplasia or overt neoplasia, thereby indicating *ab initio* development of neoplasia. A marginal increase of a very rare malignancy of a certain type and site under some circumstances may incriminate a TS. Another generally

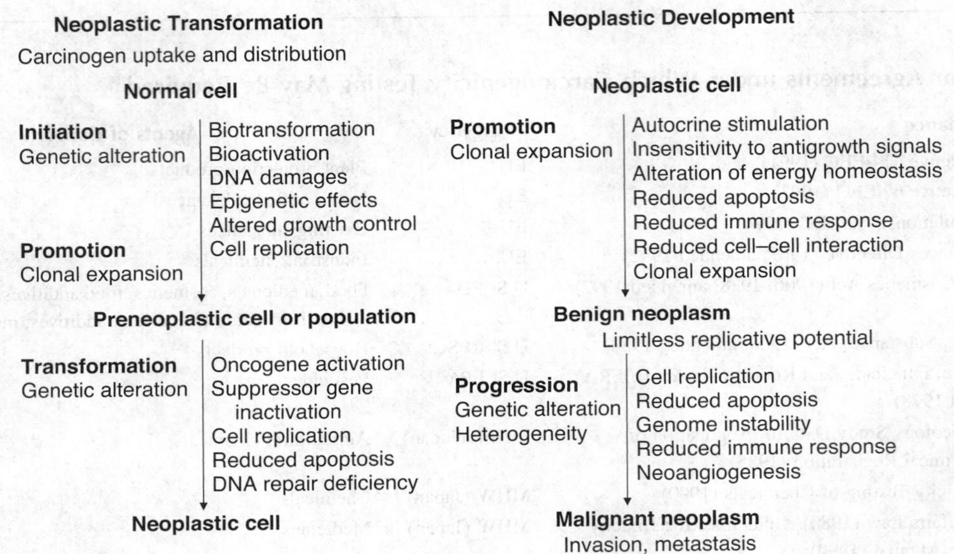


**FIGURE 25.1** Sequence of oncogenesis.

used criterion is an increase in the incidence of the types of malignant neoplasms that occur in controls. The malignant neoplasms can be of any histological type, epithelial or mesenchymal, and, although a clear increase in malignant neoplasms is most persuasive, an increase of the combined number of benign and malignant neoplasms of the same cell type of origin, where the latter are not significantly increased, is generally accepted as reflecting carcinogenic activity [151,218]. The evidence of malignancy is best established by the presence of invasion or metastasis. For most rodent neoplasms, however, the diagnosis of malignancy is made histologically on evidence of cellular atypia; thus, some diagnoses are controversial, as discussed in the later sections on anatomic pathology and cancer hazard evaluation. The finding of an increase in only benign neoplasms, especially if the type of neoplasm is not established to be premalignant, does not constitute sufficient evidence for carcinogenicity but does provide limited evidence. In addition to these criteria, an increase in the multiplicity of neoplasms above that in controls or a reduction in the latency period for development of neoplasms have also been considered [388]. Although these latter findings may indicate an influence of the chemical on neoplastic development, they are less definitive evidence of carcinogenic activity. The criteria discussed apply to findings in any species, strain, or gender. A strong effect in one gender or species, in more than one dose group or in independent experiments, strengthens the evidence.

Of course, each expert body applies criteria as they deem appropriate, often influenced by a high level of concern for hazard identification in the interests of public health protection. It must be kept in mind that carcinogenicity in a rodent study is not necessarily conclusive proof of a cancer hazard to humans, as discussed below. Ultimately, in assessing human risk (which is the reason wh animal tests are done), scientific judgment must be use in evaluating all the information available, including importantly, mechanism (or mode) of action.

## Types of Carcinogens

A wide variety of organic chemical structures has exhibite carcinogenic activity in rodents [114,115,153, 158,159]. Thi reflects the fact that the multistep process of oncogenes (Figure 25.1) can be influenced by chemicals in variou ways, mainly involving either chemical reactivity in produc ing neoplastic transformation (initiation) or epigenetic effec leading to neoplastic transformation or enhancement of ce growth facilitating neoplastic development. Thus, chemica can give rise to increases in neoplasms through a variety modes of action (MOAs) that have been broadly characte ized as DNA reactive or epigenetic (Table 25.2 and Tab 25.3) [368,373]. DNA-reactive carcinogens are defined compounds that react covalently with DNA in the target ce of carcinogenicity to produce procarcinogenic mutatio which lead to neoplastic transformation. Epigenetic carci ogens are defined as compounds that under conditions carcinogenicity produce a non DNA-reactive cellular effe in the target tissue of carcinogenicity, which either indirect results in mutations or facilitates development of preexisti neoplastic cells into neoplasms. The types of chemicals th can be assigned to these two categories, as well as inorgan compounds, are given in Table 25.2 and have been discuss in detail elsewhere [359,373]. Carcinogens are, of cours both naturally occurring and synthetic.

Of the many chemicals with carcinogenic activity RCBs, few are associated with cancer in humans, assessed by the World Health Organization (WHO) Inte

**TABLE 25.2**
**Classification of Substances with Experimental Carcinogenic Activity**

| Class | Examples |
|---|---|
| A. DNA-reactive chemicals | |
| 1. Activation independent | Alkylating agents: nitrogen mustards, chlorambucil, cyclophosphamide |
| | Epoxides: ethylene oxide |
| 2. Activation dependent | Aliphatic halides: vinyl chloride |
| | Aromatic amines, aminoazo dyes and nitro-aromatic compounds: *o*-toluidine, 2-amino-1-methyl-6-phenyl-imidazo[4,5-*b*]pyridine (PhIP), polycyclic 4 aminobiphenyl, benzidine, dimethylaminoazo benzene, 1-nitropropane |
| | Polycyclic aromatic hydrocarbons: benzo(*a*)pyrene |
| | *N*-nitroso compounds: dimethylnitrosamine, *N*-nitrosonornicotine |
| | Hydrazine derivatives: 1,2-dimethylhydrazine, azoxymethane, methyl-azoxymethanol |
| | Mycotoxins: aflatoxin $B_1$, pyrrolizidine alkaloids |
| | Pharmaceuticals: phenacetin, tamoxifen (rodent liver) |
| | Triazine (diazoamino compounds): 3,3-dimethyl-1-phenyltriazene |
| B. Epigenetic chemicals | |
| 1. Promoter | Liver enzyme-inducer type hepatocarcinogens: chlordane, DDT, pentachlorophenol, phenobarbital, polybrominated biphenyls, polychlorinated biphenyls |
| | Urothelial cell proliferation enhancers: saccharin |
| 2. Endocrine-modifier | Hormones: estrogens, atrazine, diethylstilbestrol, chloro-*S*-triazines |
| | Antiandrogens: finasteride, vinclozolin |
| | Antithyroid thyroid tumor enhancers: thyroperoxidase inhibitors ( amitrole, sulfamethazine); thyroid hormone conjugation enhancers (phenobarbital, spironolactone) |
| | Gastrin-elevating inducers of gastric neuroendocrine tumors: omeprazole, lansoprazole, pantoprazole |
| 3. Immunomodulator | Purine analogs |
| | Cyclosporine |
| 4. Cytotoxin | Mouse forestomach toxicants: butylated hydroxyanisole, propionic acid, diallyl phthalate, ethyl acrylate |
| | Rat nasal toxicants: chloracetanilide herbicides |
| | Rat renal toxicants: potassium bromate, nitrilotriacetic acid |
| | Male rat $\alpha_{2\mu}$-globulin nephropathy inducers: d-limonene, *p*-dichlorobenzene |
| 5. Peroxisome proliferators | Hypolipidemic fibrates: ciprofibrate, clofibrate, gemfibrozil |
| | Phthalates: di(2-ethylhexyl) phthalate (DEHP), di(isononyl) phthalate (DINP) |
| | Lactofen |
| 6. Nucleoside analogs | Zidoriudine, zalcitabine |
| C. Inorganic compounds[a] | |
| 1. Metals | Beryllium, cadmium, chromium, nickel, silica |
| 2. Fibers | Asbestos |
| D. Unclassified | Acrylamide, acrylonitrile, dioxane, furfural, methapyrilene, sugar alcohols |

[a] Some are categorized as genotoxic because of evidence for damage of DNA; others may operate through epigenetic mechanisms such as alterations in fidelity of DNA polymerases.

ıational Agency for Research on Cancer (IARC) (Table 25.3). Most of these are of the DNA-reactive type [376], ndicating the importance of this MOA in human hazard. n support of this, several DNA-reactive carcinogens have been active in primates [313], including transplacentally 276]. Apart from cigarette smoke and mycotoxins, human contacts with virtually all DNA-reactive carcinogens occur in the workplace or as therapeutic interventions, and he external exposures are substantial compared to those esulting from environmental chemicals. A variety of neoplasms is induced by these agents, including notably in he lung, urinary bladder, and liver. The few human epigenetic carcinogens [376] are mainly pharmaceuticals, nd these are associated with cancer increases only at therapeutic exposures that produce the cellular effect that underlies their carcinogenicity in rodents, mainly immunosuppression or hormonal effects. With immunosuppression, liver risk is greatly increased, while hormonal agents target responsive tissues such as uterus; thus, a primary objective of cancer hazard assessment is to identify chemicals with DNA reactivity [381] and those with epigenetic effects known to be relevant to human risk.

Because of the significance of DNA reactivity and the reliability with which it can be identified, few agents of this type are proposed for uses in which there is any intentional human contact, apart from chemotherapeutic alkylating agents. In contrast, agents that elicit rodent neoplasm increases by epigenetic MOAs are widely used;

**TABLE 25.3**
**Classification of Chemicals and Mixtures Considered Carcinogenic to Humans by the International Agency for Research on Cancer[a]**

| | |
|---|---|
| **DNA-Reactive** | |
| Aflatoxins | Melphalan |
| 4-Aminobiphenyl | MOPP (nitrogen mustard, vincristine, procarbazine, and prednisone) |
| 5-Azacytidine | 2-Napththyamine |
| Benzidine | Nickel and nickel compounds |
| Betel quid with tobacco | Phenacetin containing analgesic mixtures |
| *bis*(Chloromethyl)ether | Soot |
| 1,4-Butanediol dimethanesulfonate (myleran) | Sulfur mustard |
| Chlorambucil | Triethylethiophosphoramide (thiotepa) |
| 1-(2-chloroethyl)-3-(4-methylcyclohexyl)-1-nitrosourea (methyl-CCNU) | Tobacco smoke and products |
| Chromium compounds, hexavalent | Vinyl chloride |
| Coal tars | |
| Cyclophosphamide | |
| Ethylene oxide[b] | |
| **Epigenetic** | |
| Azathioprine | Oral contraceptives |
| Cyclosporine | Tamoxifen |
| Diethylstilbestrol | 2,3,7,8-Tetrachloro-dibenzo-*p*-dioxin (TCDD)[b,c] |
| Estrogens, steroidal, conjugated | |
| **Unclassified** | |
| Alcoholic beverages | Mineral oils |
| Arsenic and arsenic compounds | Shale oils |
| Benzene | |

[a] IARC (1997).
[b] Based on evidence for a relevant mechanism in humans.
[c] Based on overall increase in neoplasia.

for example, more than 80 medicines in current use are oncogenic in the RCB [61]. Generally, epigenetic carcinogens have not been carcinogenic in primates [298,313]. Accordingly, an important aspect of safety assessment is to elucidate any MOA that might lead to an increase in neoplasia in an RCB to guide mechanistic research for informed hazard evaluation [381].

## Potency

The magnitude of the carcinogenic activity in rodents of chemicals with respect to dose varies more than 10 million-fold [114]. The most extensive system for expressing numerical indices of carcinogenic potency is the Carcinogenic Potency Database [114,115], which uses $TD_{50}$ values, defined as the daily dose rate required to halve the probability of an experimental animal remaining tumor free at the end of its standard lifespan [264]. A simplified method proposed for use in the regulatory setting is the $TD_{25}$, defined as the chronic dose rate in mg per kg body weight per day that will cause neoplasia in 25% of the animals at a specific site, after correction for the spontaneous incidence, within the standard lifetime of that species [73]. Also, for DNA-reactive carcinogens, a general but not exact, relationship exists between carcinogenicity and DNA binding, which can be expressed as the chemical binding index (CBI) [209].

## REQUIREMENTS FOR TESTING

Testing of chemicals in experimental animals for carcinogenic activity is done to assess potential human cancer hazard under conditions to which humans might be exposed to the chemical. The U.S. federal government has enacted numerous laws covering requirements for carcinogenicity testing of substances for which human or environmental exposure occurs (Table 25.1). Other countries have similar provisions (Table 25.1). The requirements for testing of industrial chemicals in the United States and the European Union (EU) have been undergoing revision.

For pharmaceuticals, circumstances requiring carcinogenicity studies have been agreed upon by the International Conference on Harmonization (ICH) and are available on the Center for Drug Evaluation and Research (CDER) website (http://www.fda.gov/cder); they apply in the United States, the European Union, and Japan.

Cosmetics, as with foods, usually do not require carcinogenicity testing. In fact, under the seventh amendment to the EU cosmetics directive [85], after 2009 the use of animals in the safety assessment of cosmetics will not be permitted.

The requirements for the testing of biopharmaceuticals present specific issues [295] for which the FDA has provided guidance (http://www.fda.gov/cber). Several nucleoside analogs have been found to be carcinogenic in rodents [389], possibly through perturbation of nucleotide pools. Growth factors and immunosuppressive antibodies are noted as raising concern for carcinogenic potential. Normal hormones or growth factors that are intended to correct deficiency states (e.g., insulin) may not need to be tested, unless they are administered at doses exceeding the physiological levels or by a route that results in substantially increased exposure in some tissues. Modified proteins with new biologic properties will require testing, because, for example, it is known that a modified insulin [305] with an increased affinity for the insulin-like growth factor 1 (IGF-1) receptor produces mammary tumors in rats [68]. Also, a fragment of parathyroid hormone with bone trophic activity produces bone neoplasms in rats [266]. Certain modifications of proteins to improve bioavailability, such as with conjugation to polyethylene glycol, have not raised carcinogenicity issues.

An emerging aspect of regulatory concern is photochemical carcinogenicity [172]. Indications for the possible need for photochemical carcinogenicity testing include: (1) long-term exposure of the skin to chemicals that can undergo photoactivation, (2) alteration of the structure of the epidermis, (3) sensitization of the skin to ultraviolet radiation (UVR), and (4) exacerbation of suspected UVR-induced carcinogenesis [98]. Regulations requiring specific data on toxicity of nanosized particle (NSP) materials (<100 nm) before introduction into commerce are currently under discussion [16].

For the use of many types of chemicals, proscribed RCBs are required, usually in both rats and mice. Nevertheless, expedited approaches to assessment of potential carcinogenic activity are available to obtain data quickly and economically, with minimal use of experimental animals, to assess potential hazard before extensive development or exposures of humans. Moreover, application of the threshold of toxicological concern (TTC) (see Risk Assessment section) can obviate the need for testing of materials with negligible human contact [18,104,198,235, 343].

## SYSTEMATIC APPROACH TO TESTING

The goal of a systematic approach to testing is to obtain reliable data for hazard assessment and evaluation of the TS at the earliest possible stage. An approach that incorporates the mechanistic concepts described above is the decision point approach (DPA) [358], which is presented here as a general framework for the concept (Table 25.4). An expedited approach to testing can assist in prioritizing from the large reservoir of existing untested chemicals and in selecting desirable new molecular (or chemical) entities (NMEs), both small and large (e.g., protein) molecules, which are being synthesized, particularly in the pharmaceutical industry.

The endpoints detailed below provide guidance in identifying a potentially carcinogenic TS, but negative results do not preclude carcinogenicity, and in any event carcinogenicity testing may be required. In some instances, data from short-term toxicity studies may provide an indication of potential carcinogenicity as with liver enlargement [7]. A review of the FDA/CDER database of pharmaceuticals, however, demonstrates that short-term toxicity studies, including in transgenic mice, do not accurately and reliably predict neoplastic findings in long-term assays [171].

### IMPLICATIONS OF CHEMICAL STRUCTURE

From the classes of DNA-reactive carcinogens given in Table 25.2, the types of electrophiles that are involved in chemical reactivity and hence DNA binding are well known (Figure 25.2). Such molecular features also have been referred to as structural alerts [10]. In general, a relationship exists between DNA binding and carcinogenicity [209], although not all DNA binding is necessarily mutagenic [349] or carcinogenic [246]. Among both DNA-reactive and epigenetic carcinogens, numerous classes have structural features in common within the class. The presence of one of these features in an NME of unknown carcinogenicity suggests potential activity. The FDA Center for Food Safety and Nutrition (CFSAN) has grouped food additives into classes by chemical structure, estimating their potential toxicity [339,342]. These structural classes are used for assignment to levels of concern. Additives with functional groups of high probable toxicity are assigned to Category C. Additives of intermediate or unknown probably toxicity are assigned to Category B, and compounds of low probable toxicity are assigned to Category A. Artificial intelligence systems (*in silico*) for assessing potential toxicities related to structures have been developed [51,192].

### SHORT-TERM ASSAYS

#### *In Vitro*

A large number of short-term assays for various genetic endpoints is available [365], and all regulatory agencies have specific recommendations or requirements that may extend beyond the intent to predict potential carcinogenicity. For pharmaceuticals, a core battery including both *in vitro* and *in vivo* assays has been agreed upon by the ICH

**TABLE 25.4**
**Decision Point Approach in Carcinogen Testing**

Stage A. Structure of chemical
- 1. Possible electrophiles
- 2. Relation to known carcinogens

Stage B. Short-term genotoxicity assays
- 1. Bacterial mutagenesis; hepatocyte DNA repair
- 2. Other

*Decision Point 1: Evaluation of findings in stages A and B.*

Stage C. Assays for epigenetic effects
- 1. Cultured cells
  - Mitogenesis
  - Induction of cytochrome P450
  - Peroxisome proliferation
  - Gap junction protein downregulation
  - Inhibition of cell-cell communication
  - Hormone agonist effect
  - Altered gene expression
- 2. *In vivo*
  - Increased cell proliferation
  - Reduced cell apoptosis
  - Induction of cytochrome P450
  - Peroxisome proliferation
  - Hormone perturbation
  - Gap junction protein downregulation
  - Enhancement of preneoplastic lesions
  - Immunosuppression
  - Altered gene expression

*Decision Point 2: Evaluation of results from stages A through C.*

Stage D. *In vivo* assays
- 1. DNA reactivity
  - DNA damage assays
- 2. Limited bioassays
  - Preneoplastic lesions (rat liver, mouse skin, mouse lung, rat breast)
  - Transgenic mice

*Decision Point 3: Evaluation of results from stages A to C and selected tests in stage D*

Stage E. Carcinogenicity bioassays
- 1. Accelerated bioassays
- 2. Long-term bioassays

*Decision Point 4: Final evaluation of all results and cancer hazard assessment*

[59]. For cosmetics, more extensive testing has been proposed in the EU [292], because animal testing for cosmetics will be eliminated in 2009 [85]; however, an expedited approach may still be possible [189]. These batteries generally include a bacterial mutagenicity assay, a mammalian cell mutagenicity/chromosome aberration assay, and an *in vivo* mutagenicity/chromosome aberration assay.

The results of short-term assays are usually interpreted in terms of potential carcinogenicity of the TS; however, the predictivity of most assays (i.e., the percentage of positives that prove to be carcinogens) is limited to DNA-reactive carcinogens as a consequence of the fact that DNA alteration is a MOA of carcinogens. Moreover, some assays such as mammalian cell transformation appear to respond to both DNA-reactive and epigenetic agents [202]. A bacterial mutagenicity assay [114] is required in all testing batteries and has reasonably high predictivity for carcinogenicity [391]. Nevertheless,

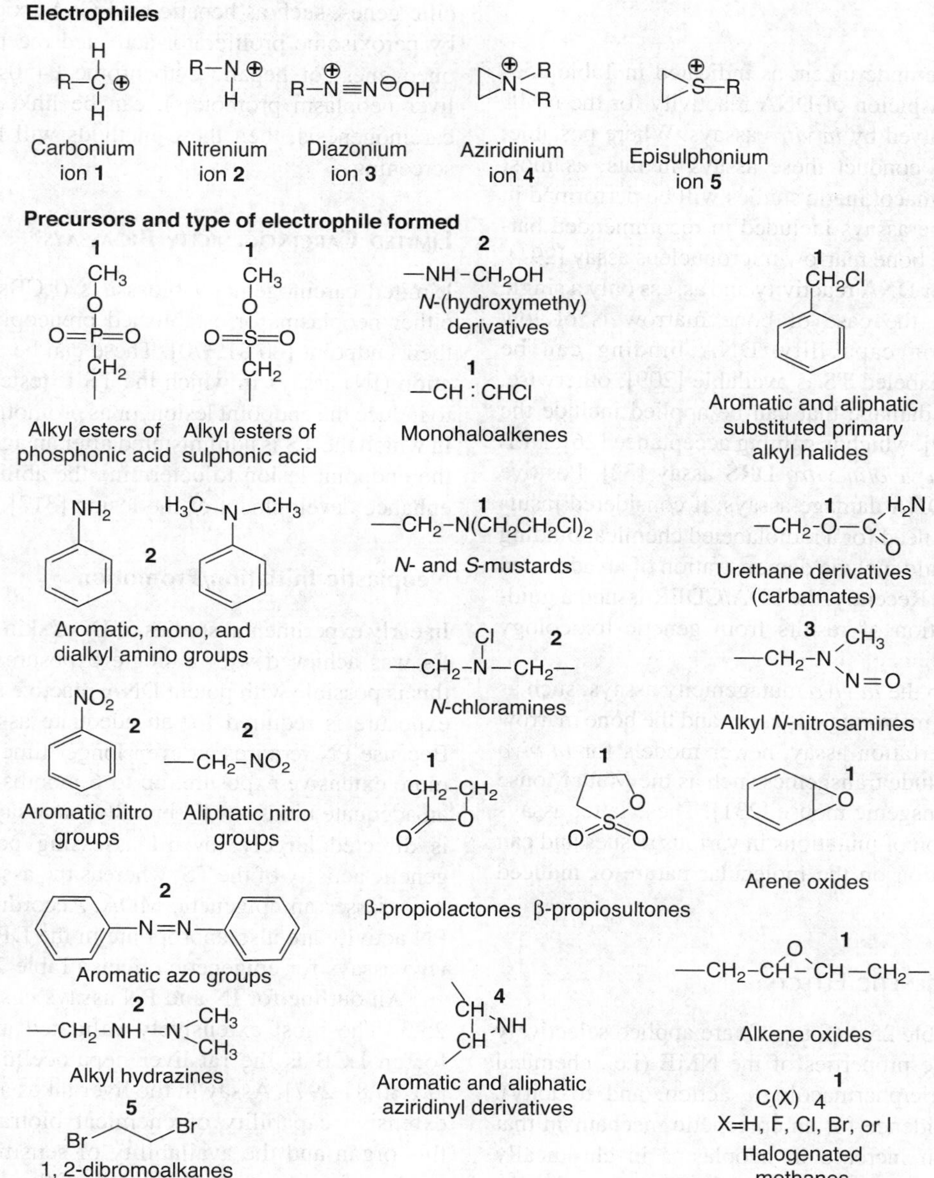


FIGURE 25.2 Structure of reactive electrophiles and precursors.

substantial number of bacterial mutagens, such as quercetin, is noncarcinogenic. Moreover, recent experience from the NTP reveals that certain *in vivo* effects are better predictors than bacterial mutagenicity [7]. The ability of *Salmonella* mutagenicity to differentiate carcinogens from noncarcinogens is not increased by certain other standard *in vitro* assays, such as mammalian cell mutagenicity and chromosome aberration assays [105,391]. To reinforce assurance of the predictiveness of a positive *Salmonella* mutagenicity finding in the DPA (Table 25.4, Stage B), another well-established assay with a defined protocol, the hepatocyte/DNA repair synthesis (DRS) assay [380], can be used. Positive results in the two assays provide essentially perfect predictivity for the detection of DNA-reactive carcinogens [379]. Other DNA damage assays include binding of radiolabeled TS, such as the alkaline single-cell gel electrophoresis assay (comet) [90] and the nucleotide postlabeling (NPL) or $^{32}P$-postlabeling assay [271]. The utility of hepatocytes is that they provide intrinsic bioactivation. Also, hepatocytes from various species [222], including humans, can be used [223]. Clear positive results in both of these assays, therefore, raise serious concerns for many uses of such a chemical. If testing at this level, however, yields equivocal findings, *in vivo* assays for DNA reactivity are available.

## In Vivo

*In vivo* assays are undertaken, as indicated in Table 25.4 (Stage D), if a suspicion of DNA reactivity for the NME has not been resolved by *in vitro* assays. Where possible, it is desirable to conduct these assays in rats, as most toxicity and pharmacokinetic studies will be performed in this species. Some assays included in recommended batteries, such as the bone marrow micronucleus assay [291], are not specific for DNA reactivity and assess only a single tissue, which in the case of bone marrow is of low biotransformation capability. DNA binding can be assessed if radiolabeled TS is available [209]; otherwise, assays for DNA damage that can be applied include the comet assay [289], which is gaining acceptance [26], NPL [265], and the *in vivo/in vitro* DRS assay [32]. Positive results in these DNA damage assays, if considered insufficient, direct the need for a radiolabeled chemical binding assay, which should include demonstration of an adducted DNA base [289]. Recently, the FDA/CDER issued a guidance for integration of results from genetic toxicology studies [348].

In addition to the *in vivo* mutagenicity assays, such as the bone marrow micronucleus assay and the bone marrow chromosome aberration assay, newer models for *in vivo* mutagenicity include transgenics such as the Muta Mouse and Big Blue transgenic mouse [231]. These latter assays allow for detection of mutations in various tissues and can provide information on the molecular nature of induced mutations.

## Assays for Epigenetic Effects

The assays in Table 25.4 (Stage C) are applied selectively depending on the properties of the NME (i.e., chemical structure, biologic/pharmacologic action, and toxicity). They provide evidence for an epigenetic mechanism that may result in an increase in neoplasms in chronically exposed rodents. Many can be conducted in cultured cells, particularly hepatocytes, to obtain intrinsic bioactivation. Even when applied *in vivo*, the experiments are of short duration, except for assays for promoting activity, the detection of which is described further in the Limited Carcinogenicity Bioassays section. Positive results for the TS indicate potential oncogenicity, in which case the potency of established oncogenes with a similar mechanism and organotropism provides a guide to whether a favorable risk assessment can be made for the NME. A particularly valuable effect to monitor is cell proliferation, because many nonspecific enhancements can lead to enhancement of neoplasia [33,45].

Technologies are available for screening for effects on gene expression and function, including microarray hybridization methods for RNA expression and proteomics for protein levels [1,67,186]. If the expression of specific genes, such as hepatic acyl-CoA oxidase (increased by peroxisome proliferator-activated receptor alpha liver oncogenes) or hepatic cytochrome P450s (increased by liver neoplasm promoters), can be linked to epigenetic carcinogenesis, then these methods will have utility for screening.

## Limited Carcinogenicity Bioassays

Limited carcinogenicity bioassays (LCBs) are based on either neoplasms or established preneoplastic lesions as their endpoint [80,81,370]. These can be applied as initiation (IN) assays in which the TS is tested for its ability to induce the endpoint lesion, or as promotion (PN) assays, in which the TS is administered after an agent that induces the endpoint lesion to determine the ability of the TS to enhance development of the lesion [317].

### Neoplastic Initiation/Promotion

In early experimental studies of IN of skin carcinogenesis, IN was achieved with a single exposure [14]. Although this is possible with potent DNA-reactive agents, repeated exposure is required for an adequate assessment of IN. Because PN requires an even longer time for expression, more extensive exposure, up to 6 months, is required for an adequate bioassay. Essentially, an assay for IN activity is directed largely toward assessing potential *in vivo* genetic activity of the TS, whereas the assay for PN activity assesses an epigenetic MOA. Accordingly, assays for PN activity are also appropriate in the DPA at Stage C *in vivo* assays for epigenetic effects (Table 25.4).

An outline for IN and PN assays is shown in Figure 25.3. The most extensively validated and used model for an LCB is the rat liver hepatocellular focus assay [71,80,81,297]. Assay in the liver takes advantage of the extensive capability of chemical biotransformation in this organ and the availability of sensitive and reliable markers for preneoplastic lesions. Other commonly used LCBs are the mouse skin papilloma/carcinoma, the mouse lung adenoma/carcinoma, and the rat mammary gland adenoma/carcinoma assays [80,81,358]. These have advantages for specific types of agents; for example, mouse skin is very responsive to polycyclic aromatic hydrocarbons and is appropriate for assessment of topical products.

Positive findings for IN are highly indicative of potential carcinogenic activity [160]. In fact, it has been calculated that an IN/PN assay (of 40-week duration) can be as sensitive as the 2-year chronic bioassay in detecting carcinogenic activity [386]. PN activity also suggests a likelihood of carcinogenic activity [160]. In either case, it is possible to establish dose–response data and no-effect levels for design of chronic bioassays or risk assessment.

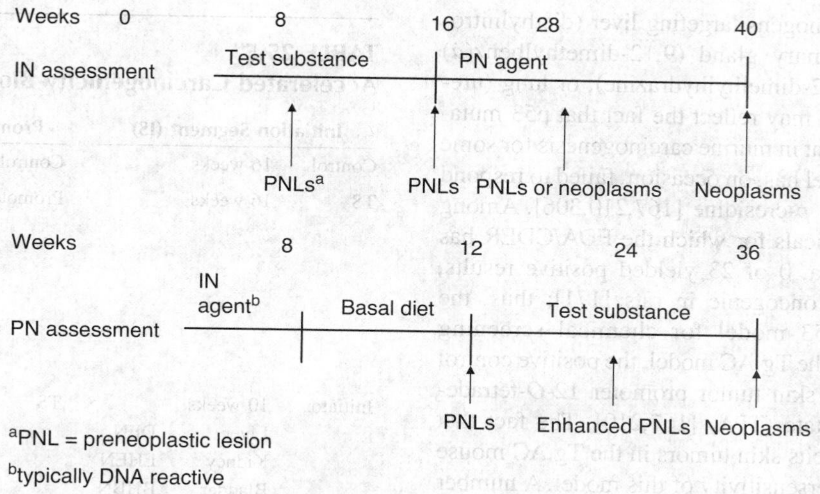


**FIGURE 25.3** Limited bioassays for initiation and promotion in rodent liver.

## Transgenic Mice

Another type of LCB, increasingly in use, is one that employs transgenic mice [86,210] in which the principal genetic targets in the transgene are specific oncogenes (H-ras model), tumor suppressor genes (p53 model), or the entire genome in DNA-repair deficient animals (XPA-deficient model). Four models that have received the greatest attention are the p53 heterozygous mouse ($p53^{+/-}$), the Tg.AC mouse, the CB6F1-Tg-H-ras2 mouse, and the $XPA^{-/-}$ mouse. Other models, such as Min, p53/XPC double mutant, Eμ-pim-1, and ARF-deficient, have shown responsiveness to carcinogens but have not been adopted for regulatory studies [86]. The studied transgenic models have responded appropriately to a number of carcinogenic and noncarcinogenic agents [167]; they have been introduced as alternatives to the RCB in mice and accepted as providing evidence of carcinogenicity [160,210].

Three of the commonly used models are based on alterations in genes that are relevant to gene changes in many human and rodent neoplasms (e.g., the p53 tumor suppressor gene and the H-ras oncogene), while the $XPA^{-/-}$ model provides an enhanced response to DNA damage as a consequence of the elimination of nucleotide excision repair. Each model has specific features. Mice heterozygous for p53 differ in response depending on the strains used as parents; for example, the C57BL mouse, which is the most widely used [70], has a low incidence of background liver cell neoplasms, whereas liver neoplasms in unexposed controls are more frequent in C3H mice. The Tg.AC model carries a v-Ha-ras oncogene fused to the promoter of ζ-globin gene in the FVB/N mouse strain [203], a strain not commonly used in toxicology and which is susceptible to audiogenic seizure. The Tg-H-ras2 mouse carries five to six copies of human c-Ha-ras gene integrated in tandem array in the genome of F1 mice of transgenic male C57BL/6J mice and normal female BALB/ cByJ mice [283]. The C57BL/6J is not very susceptible to hepatocarcinogenesis or lung carcinogenesis, whereas the BALB/c is susceptible to lung carcinogenesis [258]. Currently, the $p53^{+/-}$ and Tg.AC mice are widely available, but the Tg-H-ras2 is less accessible outside Japan, where it was developed.

Although any route of exposure can be used with these models, most data have been obtained by the oral route for the $p53^{+/-}$ and Tg-H-ras2 assays, whereas topical application is the preferred route for the Tg.AC. In these models, neoplasms can be elicited within 6 months with few or no neoplasms in controls. Beyond 6 months, these animals begin to develop a high incidence of genetically determined neoplasms as follows: the $p53^{+/-}$ with a C57BL parent develops lymphomas and sarcomas [213]; the Tg.AC develops odontogenic tumors, erythrocytic leukemia, and salivary gland and ovarian neoplasms [212,213]; and the H-ras2 develops benign and malignant lung neoplasms, splenic hemangiosarcomas, and forestomach papillomas [283]. The evaluation of increases in any of these neoplasm types must consider whether the increase is attributable to induction of neoplasia as opposed to acceleration of development. The XPA-deficient mice do not show appreciably increased incidences of background neoplasms (only some liver neoplasms in the C3H-derived strain) [66,351]. All of these models respond primarily to DNA-reactive carcinogens, although the Tg.AC model has the potential to identify epigenetic skin carcinogens [69,210,304], and the Tg-H-ras2 has also responded to epigenetic carcinogens [210,390].

The $p53^{+/-}$ mouse, in the studies available so far, clearly responds with accelerated development of thymic lymphomas (e.g., phenolphthalein, cyclophosphamide, and melphalan [167]), but compared to the wild-type background it has not exhibited an accelerated response to DNA-reactive carcinogens targeting liver (diethylnitrosamine) [184], mammary gland (9,12-dimethylbenz(*a*) anthracene), colon (1,2-dimethylhydrazine), or lung (urethane) [184,258]. This may reflect the fact that p53 mutation is not an early event in murine carcinogenesis for some tissues. Alas, this model has, on occasion, failed to respond to the positive control *p*-cresidine [167,210,306]. Among candidate pharmaceuticals for which the FDA/CDER has received bioassay data, 0 of 23 yielded positive results, although some were oncogenic in rats [171]; thus, the usefulness of the p53 model for chemical screening remains uncertain. In the Tg.AC model, the positive control generally used is the skin tumor promoter 12-*O*-tetradecanoylphorbol-13-acetate (TPA) [167,210]. The fact that this agent by itself elicits skin tumors in the Tg.AC mouse demonstrates the hypersensitivity of this model. A number of innocuous materials have been positive in the Tg.AC, strongly indicating that it should be avoided where possible. For all of these models, evaluation may be enhanced by measurement of cell proliferation in critical target organs. In the case of Tg.AC, stimulated cell proliferation can be the basis for skin neoplasm enhancement.

### Other Models

The newborn mouse is also used in LCBs [93]. In this model, newborn mice of any strain are administered the TS by intraperitoneal injection or oral intubation at days 8 and 15 after birth and then held for observation for up to 1 year of age. The model exhibits high sensitivity to DNA-reactive carcinogens but is unlikely to respond to epigenetic agents because of the limited exposure that is provided. An interesting new approach is the use of avian eggs for an *in ovo* carcinogenicity assay (IOCA) [28,81]. TS is injected into the egg white of fertilized turkey or quail eggs prior to incubation. The embryonic liver is removed 3 to 4 days before hatching for the evaluation of preneoplastic lesions. This assay is very effective in monitoring the processes of proliferation, differentiation, and apoptosis, and DNA damage can be assessed [262]. It also has the advantage of being defined as a nonanimal method for carcinogenicity testing. Currently, IOCA is undergoing interlaboratory validation [28].

## Accelerated Cancer Bioassay

The accelerated cancer bioassay (ACB) model can be used to develop data on carcinogenicity when there is not a requirement for a RCB or there is an urgent need to obtain data, or it can be used as an alternative RCB for one species. The ACB is essentially a composite of six or more IN/PN LCBs for rodent organs in which carcinogenicity has been found for known human carcinogens (i.e., liver, lung, kidney, urinary bladder, stomach, and mammary gland). The protocol involves two segments: one in which

**TABLE 25.5**
**Accelerated Carcinogenicity Bioassay**

| Initiation Segment (IS) | | | Promotion Segment (PS) | | |
|---|---|---|---|---|---|
| Control | 16 weeks | | Control | 24 weeks | |
| TS | 16 weeks | | Promoter | 24 weeks | |
| | | | | Liver | PB |
| | | | | Kidney | NTA |
| | | | | Bladder | NTA |
| | | | | Stomach | BHA |
| | | | | Lung | BHT |
| | | | | Breast | DES |
| Initiator | 10 weeks | | TS | 24 weeks | |
| | Liver | DEN | | | |
| | Kidney | EHEN | | | |
| | Bladder | BHBN | | | |
| | Stomach | MNU | | | |
| | Lung | DMN | | | |
| | Breast | DMBA | | | |
| TS | 16 weeks | | TS | 24 weeks | |

*Note:* BHA, butylated hydroxyanisole; BHBN, *N*-butyl-*N*(4-hydroxybutyl)nitrosamine; BHT, butylated hydroxytoluene; DEN, diethylnitrosamine; DES, diethylstilbestrol; DMBA, 7,12-dimethylbenz(*a*)anthracene; DMN, dimethylnitrosamine; EHEN, *N*-ethylhydroxy-ethylnitrosamine; MNU, methylnitrosourea; NTA, nitrilotriacetic acid; PB, phenobarbital; TS, test substance.

the TS is administered at an appropriate high dose (see Dose Selection Studies for Bioassay section) for 16 weeks in an IN segment followed by promoters for the target organs, and a second part in which the TS is administered in a PN segment for 24 weeks after administration of initiating agents for the target organs (Table 25.5). The TS is also given alone for 40 weeks to assess carcinogenicity.

The ACB has a number of advantages: (1) It takes less time than the RCB, as the name implies; (2) it provides mechanistic data on IN/PN; and (3) the animals exhibit much less age-related pathology at termination because they are less than 1 year of age at termination and background rodent neoplasms occur predominately after 50 weeks (Table 25.6) [302]. Of course, the chief limitation is that the ACB is not as comprehensive as the RCB although it has been calculated that, because of the PN stimulus, the IN is as sensitive as a chronic bioassay [386]

## RODENT CANCER BIOASSAY

Usually, RCBs are required in rats and mice, although other species may be used (as discussed in the Animals and Their Environment section). For pharmaceuticals, the value of a mouse RCB has been questioned [350]; nevertheless, a mouse RCB would be indicated when (1) the biological effect of the chemical is not expressed in rats

**TABLE 25.6**
**Profiles of Percent Incidences of Common Spontaneous Neoplasms Expressed by Time of Death in Rats and Mice**

| | Males | | | Females | | |
|---|---|---|---|---|---|---|
| Species, Sites/Neoplasms | Rank | <50 Wk[a] (%) | >50 Wk[a] (%) | Rank | <50 Wk[a] (%) | >50 Wk[a] (%) |
| SD rats | | | | | | |
| Pituitary gland adenoma | 1 | 0 | 100 | 1 | 8 | 92 |
| Mammary gland fibroadenoma | 2 | 9 | 91 | 2 | 7 | 93 |
| Skin fibroma | 3 | 18 | 82 | NA[b] | NA[b] | NA[b] |
| Lymphoma, multicentric | 4 | 60 | 40 | 5 | 40 | 60 |
| Skin fibrosarcoma | 5 | 11 | 89 | NA[b] | NA[b] | NA[b] |
| Mammary gland adenocarcinoma | NA[b] | NA[b] | NA[b] | 3 | 27 | 73 |
| Pituitary gland carcinoma | NA[b] | NA[b] | NA[b] | 4 | 0 | 100 |
| CD-1 mice | | | | | | |
| Lymphoma, multicentric | 1 | 39 | 61 | 1 | 24 | 76 |
| Bronchiolo-alveolar adenoma | 2 | 14 | 86 | 2 | 7 | 93 |
| Hepatocellular adenoma | 3 | 13 | 87 | NA[b] | NA[b] | NA[b] |
| Bronchiolo-alveolar carcinoma | 4 | 5 | 95 | NA[b] | NA[b] | NA[b] |
| Hepatocellular carcinoma | 5 | 14 | 86 | NA[b] | NA[b] | NA[b] |
| Histiocytic sarcoma | NA[b] | NA[b] | NA[b] | 3 | 14 | 86 |
| Mammary gland adenocarcinoma | NA[b] | NA[b] | NA[b] | 4 | 12 | 88 |
| Myeloid leukemia | NA[b] | NA[b] | NA[b] | 5 | 25 | 75 |

[a] Values reflect percent of total at the two arbitrary intervals (i.e., before and after 50 weeks).
[b] NA, not analyzed either because inappropriate or not common neoplasms.

*Source:* Adapted from Son, W. C. and Gopinath, C., *Toxicol. Pathol.*, 32, 371–374, 2004.

(2) the chemical has a pharmacologic effect on the gall bladder (absent in rats), or (3) mice are more representative of human biotransformation [171]. Alternatives to the mouse RCB are discussed in the Limited Carcinogenicity Bioassays section.

The RCB was developed for testing of small molecules but is also being used to test large biological molecules. Such biomolecules were introduced into medicine beginning with vaccines and now also include hormones, oligodeoxy nucleotides, genes, recombinant human proteins, humanized monoclonal antibodies, blood products, and cellular therapies. Each of these presents specific issues in the conduct of an RCB. The ICH has produced a framework for preclinical safety evaluation of biotechnology-derived pharmaceuticals [295,344]. In the systematic approach discussed here, genetic toxicology studies unfortunately are not particularly informative for biopharmaceuticals because the large molecules may not enter the cells used in assays, particularly bacteria. Moreover, it is exceedingly unlikely that such macromolecules would produce genetic effects. An issue with impurities seems remote because chemical synthesis is not involved in most cases.

Customized approaches are usually required for chronic preclinical safety assessment of biotechnology-derived pharmaceuticals and genetically engineered food products [344]. Some of these products are intended for intravenous administration, which is a difficult route of administration for an RCB. Immunogenicity presents an additional complication with proteins, because with the development of neutralizing antibodies the biological activity of the protein can be abolished, and the chronic antigenic stimulation can compromise carcinogenicity testing. Guidances concerning the design of carcinogenicity studies are generally based on the clinical indication or in-use exposure. Exceptionally, with these products, the possibility arises of using relevant but nontraditional species or using an animal model of the target disease [295]. Any testing should, of course, be done in a species in which the molecule has biological activity. For an immunogenic protein, one solution to this problem may be to test a homologous rodent molecule in the corresponding species.

## Design

Most RCBs are performed to meet regulatory requirements, as listed in Table 25.1; otherwise, there are more efficient approaches to carcinogen identification (as discussed in the earlier Systematic Approach to Testing section). An RCB performed for regulatory requirements must follow guidelines [83,89,340,344] and, in particular, the regulations for good laboratory practices (GLPs; described below). Detailed descriptions of standard procedures for an RCB with chemicals have been published [224,280,358]. Aspects of the design, conduct, analysis, evaluation, reporting, and interpretation are given below.

## Feeding Procedures

During the past 20 years, increased variability in body weights, survival, and incidences of background neoplasms in Sprague–Dawley, Wistar, and Fischer strain rats and CD-1 and B6C3F$_1$ mice have been noted [3,20,42,75,182,267, 274,309,310,323,354]. The changes in these parameters can confound and even jeopardize the interpretation of an RCB [20,75,180,182,242,272,294,309,310,314].

Thus far, most RCBs have been conducted using *ad libitum* (AL) feeding. To overcome the problem of overeating with AL feeding, two solutions have been proposed [42,182,242,314]. One is referred to as the caloric optimization diet (COD), which consists of limiting caloric intakes to 50 to 80% of AL consumption [42,124]. The other is the diet-restricted (DR) model in which animals are given diets limited in the offered quantity of feed sufficient to produce a 15% reduction in body weight compared to the AL controls [2,125,242]. In addition to these procedures, the use of weight-matched groups that are fed in such a way that their mean body weight is matched with that of the high-exposure AL group has been evaluated [2,242] but is not currently being used. The COD and DR procedures clearly improve the health and survival of animals, as they reduce the occurrence of age-related pathology, such as chronic progressive nephropathy and myocardiopathy in rats [134,150], and of certain background neoplasms [14,134], such as uterine polyps, pituitary gland neoplasms, and mononuclear cell leukemia (see Species and Strain section) [27,134].

In the past, dietary control was routinely used in testing of oral contraceptives [176]. Recently, an emphasis has been placed on the use of dietary control for all medicinal and chemical products [3]. Differences in the incidences of neoplasias between AL-fed and DR rodents are available for the two commonly used rat strains, Sprague–Dawley and Fischer 344, and the B6C3F$_1$ hybrid mouse [24,27, 42,133,134,221,242,307,323]. In both rats and mice, especially the latter, hepatocellular neoplasia in both genders is reduced in DR groups. Similarly, reductions in pancreatic (both acinar and islet) neoplasia (especially in male rats), pituitary neoplasia (especially in female rats), and adrenomedullary neoplasia (especially in male Sprague–Dawley rats) were achieved in both species. In general, these decreases in DR groups are due to changes in metabolism and hormonal homeostasis induced by DR. In addition, certain decreases in the DR groups are only present in one species (skin fibroma in rats), one strain (thyroid C-cell, mammary gland, ovarian, and hematopoietic neoplasms in F344 rats), or one gender (pulmonary neoplasia in male mice). Also in female F344 rats, uterine polyps and pituitary gland neoplasia have been significantly reduced in feed studies with NTP-2000 diet [134], which was developed to reduce the background burden of pathology in rats. The mechanism of these decreases cannot be explained currently. Available information, however, on the impact of either COD or DR on response to well-studied carcinogens generally reveals a reduced response. Accordingly, detection of carcinogenic effects may be masked at sites sensitive to tumor reduction by body weight gain inhibition. Analyzing tumor incidence within body weight strata can reduce bias introduced by weight differences (see Statistical Analyses section).

## Groups and Identification

After acclimatization, animals should be assigned to groups using randomization procedures immediately prior to going on study. Randomization eliminates bias, but if there is another source of variation, such as gender, cage position, or order of euthanasia at termination, then a stratified randomization is more appropriate. This involves separate randomization within each level of stratifying variable, such as body weight or cage position [204].

The use of two independent control groups helps to assess biological variability in incidences of commonly occurring background neoplasms [11,129]. The minimum group size is 50, which permits detection of neoplasms with incidences in the range of 5 to 10%. Larger groups can be used to allow for an interim sacrifice (e.g., 12 months) of control and exposed animals and is useful in that it may provide some information on time to tumor. Such information is not otherwise available when the neoplasia is not life threatening and no unscheduled deaths occur due to neoplasms [103,204], as recently reported [302] (Table 25.6). To monitor for intercurrent diseases (see Animals and their Environment section), a satellite group of six or nine is used.

The groups should consist of a high-dose group (see Dose Selection Studies for Bioassay section) and at least two lower doses. Many testing laboratories, including the NTP, favor a second group at half the high dose to provide sufficiently exposed animals in the event that the high dose impairs adequate long-term survival. Others space doses by 1/3 or 1/4. The factor between doses should not exceed 5. Because mainly test substances that are not DNA reactive advance to an RCB, a valuable third group is one at the no-effect level (NEL) for any epigenetic effect identified at higher doses that may lead to oncogenesis. This NEL should yield a cancer NEL, which is valuable for risk assessment. If quantitative risk assessment is envisioned, four or more dose levels may be needed.

Cages must be identified with study information including study number and animal numbers. Animals must be individually identified. This can be done by tattoo or by implantation of an electronic transponder

The latter is not advisable for transgenic mice, as subcutaneous sarcomas have been induced [19].

### Duration

The anticipated lifespan for commonly used strains of rats is 24 to 30 months, and it is 18 to 24 months for mice. The usual duration for both a rat and a mouse RCB is 24 months, although for a mouse RCB, 18 months may be acceptable if the strain will not allow a 24-month period with sufficiently high survival [49]. Administration should be daily and should start shortly after weaning at about 4 to 6 weeks of age. Test groups are not allowed to live longer than control groups; otherwise, background neoplasms could appear to be induced. If a high-dose group experiences high mortality (greater than 50%) due to TS administration, it should be terminated; the other test groups and controls should continue until 24 months are completed. In general, survival should not be less than 50% for mice at 15 months and rats at 18 months or 25% for mice at 18 months and rats at 24 months [49].

## GOOD LABORATORY PRACTICE

As regards the RCB, the main intent of good laboratory practices (GLPs) is maintaining the integrity of a complex system of data management. This presupposes that the study is designed, conducted, evaluated, and reported according to standard operating procedures (SOPs) and that records are maintained in a manner that ensures a comprehensive and independent review. GLPs must be conducted in such a way that all data can be validated. GLP is a global process that has been implemented by several national bodies [165,257,330,338].

In a GLP study, the study director plays the critical roles of moderator, catalyst, and gatekeeper and is responsible for the integrity of study data. As with any process, the management of the GLP process has three components: planning (organizing, goal setting, prioritizing, and scheduling), operating (implementing, conducting), and controlling (monitoring, evaluating, and taking of remedial action). The important role of the test facility manager (TFM) is crucial, as the TFM has the ultimate responsibility for ensuring that all work from all studies is carried out in accordance with GLP principles. The specific responsibilities of the TFM include appointment and replacement of the study director, committing adequate and trained human resources to each study, providing appropriate housing for experimental animals, and utilizing scientists and equipment dedicated to the conduct of a study and assurance of the integrity of the TS.

At the core of the GLP process are information transfer and data acquisition systems and programs. Documented and validated performances (using a standard dataset) of these systems and programs are requirements in GLP. Manual systems are subject to SOP review and approval. Computer-based systems (both hardware and software) are subject to validation. The validation methods must address common features relating to system definition, documentation, and management. These methods must be clear, specific, operational, and periodically reviewed. Validation is the responsibility of the end user. The validation control system includes: (1) system definition, (2) test protocol, (3) validation testing, (4) performance evaluation, (5) operational procedures, and (6) validation report. The basic objectives of the test protocol should include determination of accuracy, reliability, performance, and reporting of activities and errors. Some software, such as for statistical analysis, vendor-supplied graphics, word processors, spreadsheet applications, and calculator programs, ordinarily have a validation control system provided by the vendor. For these, documentation of the source of codes and formulas is required. For other programs requiring a control system, validation of the hardware, manufacturer's name, model, serial number, configuration options, peripheral devices, memory board sensors, interface boards, controlling devices, communication links, and references to system documentation is necessary. For other software, development resources (e.g., compilers), function libraries, source code management tools, and debugging utilities are necessary.

All raw data must be properly identified and stored, eventually in study notebooks that ensure the integrity of the data. All data must be entered in a permanent and legible manner. Any changes to data must be dated and identified as changes, with reasons noted for any change, as well as identification of the scientist making the change. Further, the professionals responsible for data entry, verification, and review should also be identified. Finally, documentation of the data so described must be maintained at all times, and secure audit trails must be created for authorized changes in the database and also in the study notebook [330]. In histopathology, the most important study materials are the tissues on glass slides and their respective blocks, which require a specific trail leading from sampling records during necropsy and trimming records after necropsy [22,57,338].

To obtain proper material, tissues are prepared according to standardized procedures with respect to location, type of section (e.g., cross), and orientation on the slide. In addition, slides of lesions should include both lesion and surrounding normal tissue. Any failure to adhere to this regimen is a form of censorship and results in noncompliance. It is of paramount importance to integrate clinical (cage-side observation), structural (gross and microscopic), and functional (cellular and biochemical) data. Failure to integrate these three types of data will result in compromise of data and loss of data integrity. The pathologist must keep the study director apprised of events as they occur. The responsibilities of the study pathologist include keeping account of all the lesions reported at necropsy and performing microscopic evaluation of the normal and abnormal tissue changes. In this way, the pathologist ensures that appropriate tissues are collected, processed, and evaluated in a manner that satisfies the objective of the study. The pathologist also functions in a quality control capacity for the morphologic aspects of a carcinogenicity study; for example, if a compound discolors adipose tissue, the pathologist ascertains whether all adipose tissues of exposed animals are discolored or whether the intensity is more pronounced in the high dose group. Finally, the pathologist ensures consistency in diagnosis, integration of data, and grading of pertinent lesions, avoiding diagnostic drifts and censorship. For a facility to be GLP compliant, specified environmental conditions (see the Animals and Their Environment section) must be maintained and monitored by a program applied at appropriate intervals.

The final regulatory endproduct for every study is the Compliance Statement, which is provided in the final report and signed by the study director. In this section of the report, all modifications, deviations, and amendments to the protocol should be listed. A second page with a Quality Assurance Statement is signed by the quality assurance auditor. During the last 25 years of GLP implementation, several common findings of importance (e.g., listing deficiencies in form FD-483) have been compiled by various agencies. The deficiencies noted were in all subparts of the FDA regulations (subparts A–J, 58.10–58.190). Those pertaining to pathology are detailed in the Anatomic Pathology section.

Implementation of GLP has benefited the conduct of chronic toxicity and carcinogenicity studies in many ways, including (1) improved documentation, (2) refinement in bioavailability and bioequivalance information, (3) crystallization of the thinking process in safety assessment, and (4) strong, science-based regulated professionalism.

## HEALTH AND SAFETY PROCEDURES

A comprehensive, rigorously followed health and safety plan is necessary for the proper conduct of an RCB. The fact that a substance is being tested for carcinogenic activity makes it, in effect, a suspect carcinogen, although information in the Material Safety Data Sheet (MSDS), such as genotoxicity or reproductive toxicity, influences the stringency of the handling procedures instituted, from receipt of the chemical through disposal of animal waste and processing of tissues for histopathological examination. All measures are subject to quality control.

The safety plan [237,253] must address the responsibility within management for development and adherence of the plan; medical surveillance for employees; employee training; safe handling practices for the chemical; animal handling; general laboratory safety; safe personnel practices; safe work area practices (e.g., spill control and decontamination); handling of air, liquid, and solid wastes; monitoring of workers and physical equipment; emergency control; recordkeeping; design of facilities; and pollution potential. Applicable regulations of the U.S. Occupational Safety and Health Administration (OSHA) provide only a minimum structure from which to work, and lessening of the hazard within the particular facility must be addressed individually and with ingenuity. Laboratory directors must appreciate that chemicals may penetrate protective clothing and travel a considerable distance from their point of use [252,285–287,333,335]. Indeed, the finding of TS in the blood of control animals [249] testifies to migration.

No safety measure is unique to an RCB; it is the degree of adherence to such procedures that distinguishes the conduct of these studies from all others. It is beyond the scope of this section to address each individually. A few examples of aspects that are often inadequately addressed include the following: (1) use of a properly ventilated cage dumping area or an enclosed animal bedding disposal cabinet to prevent inhalation of contaminated dust and aerosols by employees; (2) an air-handling system that provides decreasing gradations of air pressure from clean corridor to the animal rooms to the dirty corridor and that is periodically tested under such stress as several doors being opened at one time or with all possible chemical hoods in operation; (3) maintenance personnel, as well as scientific supervisors, following the same rules as technicians for personal protection; (4) storage facilities that protect the integrity of the TS over extended periods of time during which unused material may be held and the immediate containers checked for deterioration; (5) a breathable air line available for use with an air-supplied respirator in the TS preparation areas; and (6) workers, including weekend staff, who are familiar with emergency safety instructions within the laboratory and know whom to notify in the event of various types of potential emergency situations.

## ANIMALS AND THEIR ENVIRONMENT

The use of animals in research is subject to national regulations. In the United States, the use of rats and mice is not regulated, but usually follows the guidelines for other animal use that have been provided by the U.S. Department of Health and Welfare for Care and Use of Laboratory Animals [237,238,328], the latest amendment of the U.S. Congress Animal Welfare Act [324], the U.S. Public Health Service Policy of Humane Care and Use of Laboratory Animals [238], the U.S. Department of Agriculture Animal Welfare Rules [327], the Animal Welfare Act of the National Research Council [241], and the U.S. EPA Health Effects Test Guidelines [336]. Institutional responsibilities include making available all protocols for review by a committee and providing veterinary care. More extensive coverage is provided in Chapter 20.

## Species and Strain (Genotype)

For the rat RCB, several strains have been widely used. The NTP generally uses the inbred Fischer F344, while the pharmaceutical and chemical industries have favored the outbred Sprague–Dawley or Wistar. For the mouse RCB, industry generally favors the CD-1 strain, whereas the NTP uses the B6C3F$_1$ hybrid, which is a first-generation cross between male C3H and female C57BL/6 strains.

Both rat and mouse strains, including hybrids, differ substantially in their background of neoplasms and their susceptibility to induction of tumors [43]; for example, the F344 has very high incidence of several neoplasms, notably testicular (Table 25.7). Among mouse strains, those derived from the C3H (i.e., B6C3F$_1$) have high incidences of liver tumors. The A/J mouse has a high lung tumor incidence, and this is taken advantage of in limited cancer bioassays. The comparative percent incidences of the principal spontaneous neoplasms from five different strains of rats and mice are given in Table 25.7.

Several factors that influence response to carcinogens differ among genotypes. Notably, biotransformation activities for certain chemicals differ considerably. In addition, gender-dependent differences in xenobiotic biotransformation are most pronounced in rats. The differences involve mainly cytochromes P450 (CYPs), sulfotransferases, glutathione transferases, and glucuronyltransferases [232,233,245].

A troublesome feature of the F344 rat and B6C3F$_1$ mouse is that their average survival has progressively decreased, and increases have occurred in the incidences of liver neoplasms in female and male mice; pituitary neoplasms in female mice; thyroid neoplasms, adrenal pheochromocytomas, and leukemias in male rats; and mammary neoplasms in female rats [273,321]. Some of these neoplasm increases are positively correlated with excessive body weight [321] due to overeating (a detailed discussion of this issue was presented earlier in the Feeding Procedures section). Likewise, in Charles River Sprague–Dawley rats, decreases in survival have been reported [181,182]. These problems may be due to breeding practices. Currently, the Wistar strain does not appear to present this problem [267].

## Feed and Water

Four major types of diets are available: (1) natural-product, unrefined, largely cereal-based formulations usually referred to as "chow-type" diets, such as NIH-07 (Purina 5018 or Purina 5001); (2) semipurified diets, formulated from refined nutrient ingredients, such as AIN76A (with sucrose) or modified AIN76A (with dextrose); (3) open-formula diets, such as NIH-31 (Purina 7017), which are formulated to contain researcher-specified quantities of nonproprietary ingredients; and (4) chemical-defined diets, which are individually specialized [193,241]. Industry generally favors the open-formula diet. In 2000, the NTP introduced the NTP-2000 diet, which was developed to reduce the background tumor burden [134]. Interestingly, the natural, unrefined diet is reported to protect against the effects of several chemical carcinogens [248]. Details about various regimens of feed availability were given earlier in the Feeding Procedures section.

## Caging and Stratification

Following randomization, animals can be caged individually or with more than one animal in each cage; each approach has advantages and disadvantages. Individually caged animals tend to overeat. Multiple animal caging leads to a conflict for hierarchy in the cage and consequent cage differences. Group-caged female mice develop pseudopregnancy which results in uterine decidual reactions, whereas wounding as a result of fighting is common in group-housed male mice. Also, group caging of mice produces almost a doubling of the lymphoma incidence in both males and females [312]. In general, group-caged rodents demonstrate higher survival rates and lower background pathology [279]. For inhalation and dermal administration studies, single housing is required. All cages used should be either metallic (stainless or galvanized steel) or plastic (polycarbonate, polyethylene, or polypropylene), with a minimum stipulated cage size [328]. The floor of the cages can either be solid or suspended. Solid-floor cages require bedding that does not have the enzyme-induction properties of, for example, hardwood bedding [352]. A problem with wire-bottom cages is foot injury. Cages and racks should be rotated periodically to balance known confounding sources of variability such as proximity to fluorescent lights [200,275]. Consequently, rodents within groups should be distributed in cage racks in such a way as to be present equally at all vertical levels of caging. A thorough documentation of such cage and rack rotation is mandated.

## Environment and Emergency Power

Environmental stress experienced by test animals must be minimized, particularly with mice, which are easily stressed even when maintained under conventional housing conditions and handled in the usual manner. The incidence of tumors in mice infected with viruses can be increased by chronic stress [278,293]. Standards for care are detailed in the *Guide for the Care and Use of Laboratory Animals* [241]. These include: (1) 10 to 15 fresh air changes per hour occur in each animal room; (2) air pressure is adjusted so the animal rooms are slightly positive to the "dirty" corridor and negative to the "clean" one, with minimal crossovers between the corridors; (3) all air is adequately filtered before it enters the animal facility and is diluted or filtered after it leaves to prevent possibly toxic concentrations of

**TABLE 25.7**
**Comparative Percent Incidence of Pertinent Neoplasia in Different Strains of Rats and Mice (104 Weeks Old)**

| Types of Neoplasia | Sprague–Dawley Rats[a] (%) | | Wistar Rats[b] (%) | | Fischer 344 Rats[c] (%) | | 1CRCr:CD-1 Mice[d] (%) | | B6C3F$_1$ Mice[e,f] (%) | |
|---|---|---|---|---|---|---|---|---|---|---|
| | Males | Females | Males | Females | Males | Females | Males | Females | Males | Females |
| Hepatocellular neoplasia | 5 | 3 | 1 | 2 | 3 | 1 | 18 | 3 | 22 | 17 |
| Pancreas islet neoplasia | 7 | 3 | 4 | 2 | 6 | 1 | 1 | 1 | 1 | 0 |
| Pancreas acinar neoplasia | 1 | <1 | 13 | 1 | 6 | 0 | <1 | 0 | 2 | 0 |
| Pheochromocytoma | 11 | 4 | 8 | 2 | 21 | 4 | <1 | <1 | 0 | 2 |
| Adrenocortical neoplasia | 2 | 3 | 6 | 6 | 1 | 2 | 1 | <1 | <1 | 0 |
| Pituitary adenoma | 46 | 78 | 16 | 29 | 25 | 42 | 0 | 5 | 2 | 8 |
| Thyroid C-cell neoplasia | 5 | 5 | 6 | 8 | 12 | 8 | 0 | 0 | 0 | 0 |
| Thyroid follicular neoplasia | 3 | 2 | 2 | 1 | 2 | 1 | 1 | <1 | 2 | 6 |
| Mammary gland fibroadenoma | 1 | 18 | 3 | 9 | 4 | 57 | <1 | 1 | 0 | 0 |
| Mammary gland carcinoma | <1 | 40 | 1 | 27 | 0 | 25 | 0 | 6 | 0 | 0 |
| Skin fibroma | 5 | <1 | 5 | 1 | 10 | 2 | <1 | <1 | 1 | 2 |
| Skin papilloma | 2 | 1 | 2 | <1 | 5 | <1 | <1 | 0 | 0 | 0 |
| Pulmonary neoplasia | <1 | <1 | <1 | 0 | 4 | 4 | 15 | 15 | 22 | 6 |
| Preputial gland neoplasia | 1 | NA | <1 | NA | 10 | NA | <1 | NA | <1 | NA |
| Leydig cell neoplasia | 4 | NA | 6 | NA | 84 | NA | 1 | NA | 0 | NA |
| Clitoral gland neoplasia | NA | <1 | NA | <1 | NA | 14 | NA | 0 | NA | <1 |
| Uterine polyps | NA | 4 | NA | 12 | NA | 14 | NA | <1 | NA | 1 |
| Ovarian neoplasia | NA | 1 | NA | 5 | NA | 6 | NA | 1 | NA | 6 |
| Mononuclear cell leukemia | <1 | <1 | <1 | <1 | 47 | 22 | 2 | 2 | 0 | 0 |
| Lymphoma | 2 | 2 | 3 | 2 | <1 | <1 | 8 | 22 | 14 | 24 |
| Forestomach neoplasia | <1 | <1 | 0 | <1 | 0 | 2 | <1 | <1 | 5 | 2 |
| Scrotal mesothelioma | 1 | NA | 2 | NA | 5 | NA | 0 | NA | 0 | NA |

[a] Data from Brix et al. [27], Christian et al. [42], Giknis and Clifford [113], and McMartin et al. [221].
[b] Data from Bomhard and Rinke [20], Eiben and Bomhard [75], Giknis and Clifford [112], Poteracki and Walsh [267], Tennekes et al. [309,310], and Walsh and Poteracki [354].
[c] Data from Christian et al. [42], Haseman et al. [133], and NTP [242].
[d] Data from Giknis and Clifford [111] and Maita et al. [214].
[e] B6C3F$_1$ mice = (C57BL/6N + C3H/HeN)F1.
[f] Data from NTP [242], Tamano et al. [307], and Turusov et al. [323].

*Note:* NA, not applicable.

the test chemical from entering the outside air (a process that is particularly important with inhalation studies because of the large amounts of chemical used); (4) temperature and humidity are maintained within those ranges reported to be optimal (i.e., 23.3 ± 1.1°C or 74 ± 2°F) and a relative humidity of 40 ± 5% in rat and mouse rooms; and (5) automatic control systems record both temperature and humidity at least three or four times per day. Control of lighting is essential. Usually, a 12-hr continuous light interval per day is used for both rats and mice (14-hr light day for hamsters). Moreover, 4 hr of high-intensity light (only during cage-side observation) and 20 hr of low-intensity fluorescent light are recommended to avoid blindness caused by high-intensity light. In addition, because the light and temperature gradients occur vertically, a cage-rack rotation is mandated [241]. Consideration should be given to a reversed light cycle. With animals awake and feeding during usual working hours, the activities of staff are not as stressful to the animals.

An emergency power source is essential to maintain operation of storage freezers and refrigerators, lighting, autotechnicons, and some degree of air conditioning, as well as air handling, during power failure or when personnel are unable to reach the facility. The emergency power and alarm systems should be tested on a regular schedule. In a GLP-compliant facility, all these conditions must be monitored.

Also important is control of pests by adequate facility design and sanitary procedures. Pesticides must not be allowed to contaminate the animal rooms, feed rooms, or cage washing areas and accordingly should be dispensed only in closed traps in limited areas. Detergents and cleaning agents for use on floors, cage washers, and other equipment must be nonvolatile and must not leave a residue. The environment and welfare of animals can be enriched by providing wooden chew blocks.

## DOSE SELECTION STUDIES FOR BIOASSAY

### Dose Selection: The Maximum Tolerated Dose

Dose setting can be based on a number of endpoints, including toxicity, toxicokinetics, saturation of absorption, and maximum feasible dose [96,165]. The various methods for selection of the high-dose level have been reviewed by a working group on dose selection convened by the International Life Sciences Institute [96]. Generally, it is expected that the high-dose level in an RCB should be a toxicity-based dose—the maximum tolerated dose (MTD) [303]—which is also referred to as the minimum toxic dose (or the minimally toxic dose by the NTP) [131]. Testing at the MTD ensures that potential carcinogenicity has been fully evaluated; nevertheless, it creates issues in the interpretation of positive findings, because of high dose effects, as discussed later.

The first widely used definition of the MTD was formulated by the NCI [303] as follows: "The MTD is defined as the highest dose of the test agent during the chronic study that can be predicted not to alter the animals' normal longevity from effects other than carcinogenicity." This definition does not stipulate that any toxicity must be produced; hence, a slight, but significant, reformulation was introduced by the U.S. Interagency Staff Group on Carcinogens [240,341]: The MTD is "the highest dose which when given for the duration of the chronic study is just high enough to elicit signs of minimal toxicity without significantly altering animals' normal lifespan due to effects other than carcinogenicity." The MTD so defined is a dose used in the chronic study [9], which, of necessity, is selected from subchronic studies (normally 90-day studies). The ICH defines the MTD as the dose predicted from a range-finding study to produce minimum toxicity over the course of the carcinogenicity study [341]. Such an effect may be predicted from alterations in physiological function that would be predicted to alter survival, target organ toxicity, significant alterations in clinical pathological parameters, or no more than 10% suppression of weight gain relative to controls calculated as the difference between the starting weights and those at the end of the study [240,341]. A fortuitously selected target MTD in the RCB should suppress weight gain by 10% or slightly greater and produce only minimal other toxicities. The ideal dose selection infrequently happens, and issues arise when the high dose is either below or above what is considered by a regulatory agency to be an MTD.

Toxicokinetic endpoints have gained acceptance for dose setting for pharmaceuticals [99]. A dose that produces saturation of absorption is considered an MTD, and a dose that produces a plasma concentration that is 25 times greater than human exposure is considered pragmatic for test substances with no genotoxicity signal [163]. When using this parameter, it must be determined whether among humans there are poor metabolizers who would achieve higher achieved blood levels than average individuals.

Toxicodynamic endpoints can be used to establish a high dose that will produce a cellular effect, beyond which the validity of the study would be compromised. An example is renal toxicity, which would be expected to become more severe with aging in rats as they develop chronic progressive nephropathy.

To identify the target MTD or high dose, subchronic studies are performed, as described below. If the target MTD (or the high dose) cannot be established using the criteria discussed, it is usually recommended that the high dose level or limit dose not exceed 5% in the feed [152,225,256] which translates into approximately 3 to 4 g/kg/day for rats and 7 to 8 g/kg/day for mice. This limit appears to be based on concern for the nutritional impact of high proportions of TS in the diet. Certainly, some TSs at high exposure interfere with nutritional elements, such as impairment of vitamin K

**TABLE 25.8**
**Species Physiologic Differences Important in Carcinogenicity[a]**

| Parameter | Species | | |
|---|---|---|---|
| | Rat | Mouse | Human |
| Life span (years) | 2.5 | 2 | 70 |
| Body weight (kg) | 0.40 | 0.02 | 60 |
| Body surface (cm²) | 325 | 46 | 1600 |
| Food consumption (g/kg/day) | 50 | 150 | 2.5 |
| Human equivalents in months[b] | 34 | 38 | 1 |
| Basal metabolic rate (kcal/kg/day) | 109 | 188 | 26 |
| Heart rate (beats/min) | 350 | 600 | 75 |
| Respiratory rate (beats/min) | 97 | 120 | 12 |
| Stomach pH | 4 | 3 | 2 |
| Bacterial flora | Numerous | Numerous | Few |
| Reproductive cycle | Estrus | Estrus | Menstrual |
| Placental barrier | HE | HE | HC |
| Tissue volume ratio direction compared to humans (increased, decreased, or same)[c] | | | |
| Liver | 2.4× decreased | 1.3× increased | NA |
| Kidney | 1.5× increased | Same | NA |
| Lung | 6× increased | 204× increased | NA |
| Adipose tissue | 5× increased | 2× decreased | NA |
| Intestine | 4.3× increased | 3.3× increased | NA |
| Heart | 2.8× increased | 6× increased | NA |
| Muscle | 9× increased | 5× increased | NA |

[a] Data from Iatropoulos [144], Iatropoulos et al. [147], Monro [230], and Williams and Iatropoulos [377].
[b] Human xenobiotic exposure of 1 month is 4.1% of the lifespan of rats and 5.5% of the lifespan of mice.
[c] Calculated ratio based on organ volume (mL) and organ plasma flow rate (mL/min).

*Note:* HC, hemochorial consisting of fetal trophoblast, fetal connective tissue and fetal endothelium; HE, hemoendothelial consisting of fetal endothelium; NA, not applicable.

function by butylated hydroxytoluene [53,168], but such problems can be overcome by appropriate nutritional supplements. Nevertheless, reduced food intake and, hence, caloric intake will affect the outcome of the RCB, usually reducing neoplasm incidence [2,42,190,242].

In the United States, the dose selection for pharmaceuticals should be submitted to the FDA/CDER and its Carcinogenicity Assessment Committee (CAC), which strongly favors a toxicity-based endpoint. Such consultation is not yet practiced in Europe, but the European Medicine Agency (EMEA) as well as several national authorities offers various forms of scientific advice procedures. The selection of other dose levels is discussed in the earlier Rodent Cancer Bioassay section.

## Subchronic Study

The MTD is usually identified from results of a 90-day study using the route of administration to be used in the RCB. It is unusual that a tested dose would qualify as the predicted MTD, so interpolation is usually made. When the two genders show different MTDs, different sets of doses are administered.

## Chemical Disposition Studies

Before starting an RCB, it is helpful and, in the case of medicinal products, a requirement [99,163] to determine the time course of internal exposure and eventually the relationship between exposure and observed effects. For this purpose, multiple-dose toxicokinetic (TK) studies are conducted [35,163]. Plasma protein binding of the TS in rodents and humans should be determined before initiation of the carcinogenicity studies. Should binding exceed 80%, it is advisable to express exposure (plasma TS concentration) in terms of the free fraction [39,259]. In addition, tissue distribution and accumulation data can be valuable [38,269], especially because species differ in their composition of certain tissues. For example, the liver represents 4.8% of the body weight of a rat, whereas it is only 2.8% of the body weight of a human [144]. Conversely, adipose tissue is only 11% of the body weight of a nonobese rat [110], whereas it can be about 21% of the body weight of a human [144]. Furthermore, appropriate and useful extrapolation from rodents to humans involves understanding of physiologic (Table 25.8), histokinetic, and xenodynamic considerations [40]; for example, smaller species

**TABLE 25.9**
**Comparative Adult Tissue Volume Ratios**

| Tissue | Volume Ratios | | |
|---|---|---|---|
| | Mouse | Rat | Human |
| Muscle | 0.05 | 0.10 | 0.01 |
| Adipose | 0.01 | 0.10 | 0.02 |
| Heart | 3.00 | 1.40 | 0.50 |
| Lung | 36.70 | 1.10 | 0.18 |
| Liver[a] | 0.85 | 0.24 | 0.59 |
| Kidney | 2.35 | 3.50 | 2.50 |
| Intestine | 0.60 | 1.30 | 0.33 |
| Plasma | 4.40 | 4.36 | 1.19 |

[a] 77.8% of the volume comes from the portal and 22.8% from the arterial circulation.

*Note:* Volume ratio is calculated by dividing plasma flow rate of the tissue by its volume.

*Source:* Data from Iatropoulos [144] and Iatropoulos et al. [147].

**TABLE 25.10**
**Comparative Adult Tissue Relative Weights[a]**

| Tissue | Percent Relative Weights | | |
|---|---|---|---|
| | Mouse | Rat | Human |
| Muscle | 43.5 | 42.5 | 41.4 |
| Adipose | 10.5 | 11.0 | 21.4 |
| Heart | 0.3 | 0.4 | 0.5 |
| Lung[b] | 0.5 | 0.5 | 1.7 |
| Liver | 4.4 | 4.8 | 2.8 |
| Kidney | 0.8 | 0.8 | 0.4 |
| Total of six tissues | 60.0 | 60.0 | 68.2 |

[a] In percent of body weight.
[b] Dry weight.

*Source:* Adapted from Iatropoulos, M.J., in *Drug Toxicokinetics*, Welling, D.G.P. and de la Iglesia, F.A., Eds., Marcel Dekker, New York, 1993, pp. 245–266.

(e.g., rodents) have higher rates of xenobiotic biotransformation per kilogram of body weight, faster rates of tissue distribution, and shorter tissue half-lives [144,147]. Tissue volume directly affects the volume of distribution of xenobiotics, bioavailability, half-lives, and systemic clearance [17,173,187]. In general, with increasing age, as occurs over the course of a carcinogenicity study, body weight increases as does the proportion of adipose tissue, which has a lower metabolic rate than skeletal muscle mass (Table 25.8 through Table 25.10) [106]; thus, the volume of distribution decreases with age, resulting in an increased concentration of hydrophilic TS during distribution. In addition, the increased adipose tissue in rats, but not in mice (Table 25.9 and Table 25.10), and the decreased water content in tissues can disrupt the energy homeostasis, resulting in a deficiency in energy availability in rats compared to humans. Both influence the outcome of kinetic studies. In dose-proportional linear kinetics, half-life and clearance are independent of xenobiotic concentration. In contrast, in nonlinear kinetics, these parameters are dependent, because the various processes (absorption, distribution, metabolism, and excretion) can become saturated (Table 25.8 and Table 25.9) [206].

The TK component of a carcinogenicity study often entails satellite groups run in parallel and consists of at least three rodents per group, gender, and time interval, with interim sacrifices at least every 6 months. The time-course factor is very important in explaining how the various concentrations accumulate and relate to different exposures [99]. The animals are maintained and exposed under identical conditions with the main study. Approximately four blood samples are taken from each animal over the duration of the RCB, with the blood removed not exceeding 15% of the total blood volume. If the route of administration is feed admixture, it is essential that sampling be done during the feeding phase; otherwise, unrepresentatively low values will be obtained. The parameters examined may include the maximum achieved concentration ($C_{max}$), the minimum concentration ($C_{min}$), the time to $C_{max}$ ($T_{max}$), and the area under the plasma concentration time curve (AUC) [144], although only the AUC is required to establish exposure [163]. In an RCB, together with measuring the systemic concentration of exposure (AUC), determination of the total (cumulative) amount of exposure over time is also essential so exposure can be related to induced effects [163,269]. Target site compound levels potentially connect the target site of both chronic toxicity and carcinogenicity; for example, essentially 100% of a compound absorbed from the intestine passes through the liver (portal circulation), and most biotransformation takes place in this organ (first pass). To accomplish accurate extrapolation of exposure over time, human time equivalent (HTE) values are employed (Table 25.8). At 12 months of exposure, these values amount to an HTE of 38 years for the mouse and 34 years for the rat; at 24 months, to HTEs of 76 and 64 years for the mouse and rat, respectively [144,147].

Toxicokinetic data are critical not only for delineating blood levels but also for understanding the effects of exposure, such as if the parent or a metabolite is the main active moiety, if the plasma concentration reflects the cellular site of action, and what is the interspecies concentration (blood)–response (site) relationship [144,228,260]. Furthermore, biotransformation can also distort concentration–response homeostasis. Here, at high exposures, together with absorption and elimination being saturable (i.e., capacity limited), biotransformation can also significantly alter what is bioavailable at tissue target sites, including through the process of bioactivation, which is especially important in interspecies extrapolation [40]. Moreover, saturation of biotransformation and activation of secondary routes of biotransformation should also be taken into consideration [38,62,109,123]. Knowledge of disposition (including species differences in xenobiotic biotransforming enzymes) of the TS is essential to interpreting the basis for an increase of neoplasia and for extrapolating to humans.

An important aspect of TK studies is whether only the TS (parent compound) is measured or if all the TS-derived metabolites are. Also, in the conduct of TK studies, it is imperative to be aware that controls can be contaminated, and precautions to prevent such contamination have been developed [249].

## QUALITY CONTROL OF THE TEST SUBSTANCE

The TS should be of a high quality and stability and should be manufactured in the same way and contain the same concentrations of impurities as the final product. Impurities in excess of 0.1% should be individually identified. For pharmaceuticals, the preclinical and clinical final product tested should be preferably the same. The product should have a well-defined and described scale-up process [164]. Most medicines will be formulated with excipients and, accordingly, the noncarcinogenicity of these must be established. In other situations, as with agricultural chemicals and cosmetics, the technical-grade product or a representative technical grade of active ingredients is tested. In some instances, complex mixtures are tested (e.g., polychlorinated biphenyls). This aspect is discussed further in the Complex Mixtures section.

### Test Substance

#### Chemicals or Small Molecules

In some situations, it is desirable or necessary to test pure chemicals, as with candidate medicines or food additives. With isomeric compounds, if the two enantiomers exhibit inversion *in vivo* or have the same biological activity, then they are considered one entity, and carcinogenicity testing of the racemate is appropriate. If the biological activity is not the same, then carcinogenicity testing of the active enantiomer is indicated.

#### Biopharmaceuticals

Biopharmaceuticals include recombinant human proteins, humanized monoclonal antibodies, oligo-deoxynucleotides, and genes. Recombinant human proteins are being produced by a variety of techniques, often in bacteria, where glycosylation does not occur. Quality control of biopharmaceuticals, as with small molecules, is necessary to establish that they are actually what they are expected to be at all times during the study. For this purpose, all relevant information, such as synonyms, trade names, structural and molecular formulas, and weights, as well as methods of analysis and chemical and physical properties of the pure substance, should be available.

#### Complex Mixtures

Most chemical exposures involve mixtures rather than single agents; yet, the scientific data for some mixtures is generated mainly from studies of individual agents. Mixtures are comprised of chemicals with several isomers, chemicals with major contaminants, hazardous waste in solid or liquid form, and air pollutants. The mixtures can either have a common source (e.g., tobacco smoke, aluminum production, coal tars) or be formulated deliberately (coal gasification, footware processing work exposure). Several such mixtures are recognized as human carcinogens (e.g., tobacco smoke and coal tars). A daunting challenge has been to ascertain the role of individual components in the carcinogenicity of such mixtures [154]. To accomplish this, complex mixtures can be fractionated and characterized into chemically defined entities that can be individually tested. Nevertheless, in the mixture, individual chemicals can enhance or inhibit the activity of others, as discussed in the Interactive Carcinogenesis section.

#### Nanoparticles

Nanosized particles (NSPs; <100nm) have recently been applied in the diagnosis, monitoring, drug delivery, and even treatment (gene therapy) of some types of cancer [227]; consequently, information about safety and potential hazards is emerging as the NSPs have inflammatory prooxidant, and antioxidant activities [251].

### Impurities or Contaminants

It is highly recommended that all TSs should be pure chemicals of analytical grade, as even traces of impurities of <1% can produce confounding effects. An example of this is *o*-toluenesulfonamide (OTS), an impurity of saccharin present in the early carcinogenicity studies [234]. Impurities can occur in the starting materials to be used in the formulation or in materials used in the manufacturing process of the TS [92,174,345]. Impurities in excess of 0.1% should be thoroughly identified. If the TS with the identified impurity is intended for long-term use, such as with a high dose of active compound, then mutagenicity and multiple-dose toxicity studies (up to 3 months) should be performed. If the results of the mutagenicity assays are positive, carcinogenicity testing of the impurity may be considered [164].

### Preparation of Dose

The most essential aspects in preparing a TS for dosing are the identity of all ingredients, homogeneity of the product, particle size, stability of all active ingredients, and vehicle (carrier) to be used [92,174]. All batches of dosage preparations must be analyzed for concentration to confirm accurate preparation prior to use. Homogeneity of the TS–diet mixture and the stability of the TS in diet under the intended in-study conditions should be established prior to the start of an RCB. During the conduct of the RCB, samples of the TS–diet mixture are taken approximately every 6 months for analysis and correction if necessary. It is recommended that the same batch be used for the entire RCB. This presupposes that there is a validated method of analysis for the active ingredients at hand. Quality control data as described above should be provided and monitoring should continue throughout the duration of the study.

## ROUTE OF ADMINISTRATION

The route of administration should be appropriate to potential human contact and reflect knowledge about comparative bioavailability; thus, chemical disposition data are essential and should be used in designing the 2-year RCB [144,147]. The most common route is the oral, followed by inhalation or parenteral exposure. The selection of a delivery system is crucial for all routes as it has the potential to significantly affect bioavailability. Second, producing uniform and homogenous TS exposure throughout the bioassay is equally important. This requires the availability of a validated analytical methodology [163,296]. For some agents, a rationale exists for prenatal exposure, but generally exposure is begun after weaning at 4 to 6 weeks of age.

### Oral

The oral route of administration is generally used for medicines and food additives for which ingestion is the usual route of contact of humans. For oral administration, the TS can be admixed in the feed or given by intragastric instillation (gavage). Comparable systemic exposures can be achieved with either, but with the intragastric route, the bolus dose results in a higher blood $C_{max}$ [194]. With feed additives, attention must be paid to possible inappetence due to impalatibility.

The concentration of TS in the diet is adjusted to compensate for changes in body weight. During the rapid growth phase, the concentrations are adjusted biweekly first, weekly, and later monthly after the growth plateau has been reached. Here, TS stability and homogeneity data are essential. Periodically during the study, samples of TS–diet mixtures should be analyzed to confirm intended concentrations and enable corrective action to be implemented. In general, this is the most cost-efficient mode of exposure with most test substances.

Intragastric administration affords the most precise oral delivery of the TS. For this route of administration, a vehicle is required, such as carboxymethylcellulose (CMC). Edible oils (corn, olive) have been used, but these present problems [35], including effects on body weight [130]. For a TS that cannot be dissolved in water or CMC, polyethylene glycol 400 or a uniform suspension in CMC can be used. For a TS that is unstable or volatile in diet, microencapsulation can be used. For more details, see the Feeding Procedures section. In general, the TS volume in rats is 3 to 5 mL/kg body weight. In mice, a volume of up to 10 mL/kg has been shown to be desirable as this enables more accurate measurement of the dose. In addition, dilution enhances absorption and decreases local irritation, especially with the weak bases or acids that comprise many TSs [25]. Unscheduled deaths should be closely investigated at necropsy by inspection of the lungs and trachea for the TS. If deaths are due to gavage accidents, then a monitoring program should be implemented. In general, the baseline rate for accidents per technician should not exceed 1 per 10,000 gavages.

### Dermal

Under certain circumstances of dermal contact by humans, such as with cosmetics, test substances are delivered via the dermal route. Bioactivation of activation-dependent carcinogens does occur in the skin, although not to the degree as in the liver or intestinal tract [35]. This route is also used for photochemical carcinogen testing. Dermal application is usually to the superior dorsal area of the back (interscapular), where the skin is clipped at least weekly, 24 hours prior to application (unless hairless mice are used) [201]. In dermal studies, animals are routinely housed singly to minimize ingestion by other cage mates. The TS (0.25 to 1 mL) is applied topically over the clipped area at intervals (e.g., 2 or 3 times weekly) to allow for recovery (especially when the TS is irritating). Skin penetration varies with species, chemistry of the TS, and vehicle [290]; thus, one should consider in dermal dosing by weight or surface area that rats are large enough to vary significantly in size, whereas mice are not. Any dermal study should be preceded by or accompanied by measurement of cell proliferation. This is because skin neoplasms can be elicited by nonspecific enhancement of cell proliferation.

### Inhalation (Intratracheal)

With some test substances, inhalation administration is indicated. The TS can be delivered to animals in chambers for 8 hr/day [138] or by nose-only devices for several hours [301]. After delivery, the animals can either remain in the chamber or be taken to an adjacent room. Either way, constant airflow through the chamber during and after exposure prevents the build-up of ammonia. Single cage occupancy is recommended to avoid ingestion due to grooming and licking. The cages should be rotated within the chamber periodically. Inhalation administration is an expensive, labor-intensive route of delivery that requires frequent monitoring of achieved concentrations. Sampling should be done from several fixed locations in the chamber after documenting the homogeneity of the test atmosphere [216]. With nose-only or head-only delivery units [301], major advantages include minimization of external contamination and effective monitoring of respiratory parameters. The major disadvantages include restraining the animals, which leads to alteration of many physiological parameters and entails substantial manpower requirements. Intratracheal administration can be conducted under anesthesia to achieve chronic delivery (up to 2 years) directly into the bronchial passages [300]. Here, a second control group receiving vehicle should be added and exposure should be limited to once or twice weekly.

### Parenteral Injection

In special cases, such as when the TS is destroyed in the gastrointestinal tract, intraperitoneal, intramuscular, or subcutaneous injections are employed. Here, new factors such as molecular size and pH have the potential to affect absorption and cause irritation [12]. Again, if a carrier is used, then a second control group is needed. The delivery regimen here is limited to 2 or 3 times per week with rotation of injection sites. When using these routes, the potential for local tumor formation due to physical factors and sustained initiation must be recognized [280]. With dermal injection, subcutaneous sarcomas have been induced by the irritant or physical properties of the TS [118]. An insulin analog, insulin glargine, which produces injection site initiation and inflammation, was found to be associated with increases in malignant fibrous histocytomas found at the injection site in both rats and mice [266,305].

### Multigenerational, Transplacental, and Perinatal

Experimental transplacental carcinogenesis has been extensively studied, mainly in rats and mice [315] but also primates [276,313]. The design of experiments ranges from multigenerational, involving exposure of germ cells of one or both parents and subsequently the progeny, to exposure of embryonal or fetal cells [4,315]. It has frequently been discussed whether such exposures should be included in an RCB, but so far it is accepted that a conventional RCB can identify carcinogens that might have activity in developmental stages. Specifically, efforts to map the approximate start of the RCB to equivalent ages in humans yielded a range of from 10.6 to 15.1 years [135].

## CLINICAL AND PATHOLOGICAL EXAMINATION

The RCB is not an extended chronic toxicity study; standard chronic toxicity assays (6 and 9 or 12 months, depending on requirements) involve more clinical observations than are necessary or appropriate for the RCB. In the RCB, animals should be observed at the beginning and at the end of each work day. Attention should be paid to signs of dermal irritation, as this may lead to skin neoplasia. Unwell animals should be euthanized before they become moribund or are lost to autolysis (or cannibalization, if multiple caging is used). Real-time automated programs for carcinogenicity studies have been developed [250] that allow for monitoring, at least biweekly, the appearance, location, and growth of palpable cutaneous or subcutaneous masses. Standard parameters measured in the RCB include body weight, food consumption, temperature, health status, and mortality.

### Body Weight and Survival

In RCBs using an MTD, controls frequently show greater weight gain and often poorer survival than the high-dose group in which body weight is reduced both by toxicity as well as secondary effects resulting in reduced food consumption or energy utilization [208].

### Intercurrent Diseases

Laboratory strains of rats and mice are susceptible to a variety of diseases, both genetic and acquired. The genetically determined conditions importantly include predisposition to the development of neoplasms (as discussed in the Species and Strain section). Also, pathologies such as amyloidosis in mice and chronic progressive nephropathy in rats are common. These conditions increase with age and complicate long-term studies. Acquired diseases such as sialoadenitis and murine hepatitis, can be minimized by proper animal husbandry. Many strains of mice harbor *Helicobacter hepaticus* in the gastrointestinal tract. Infection with this organism can lead to an increased incidence of liver neoplasms, particularly in male mice [121]. In hamsters, the chronic form of the wet-tail disease (of unknown etiology, perhaps *Escherichia coli*) enhances the susceptibility of the small intestine to carcinogenicity resulting even in adenocarcinomas [175,378]. Satellite sentinel animals are included in all chronic studies to effectively monitor intercurrent acquired diseases.

### Clinical Pathology

Several regulatory agencies have suggested the monitoring of continuous variables including hematology, clinical chemistry, urinalysis, and organ weights [170,339]. It must be realized that, in spite of initial randomization, aged rodents are no longer homogeneous because of non random attrition and development of diseases in the second half of the in-life phase. Although such measures may be helpful at times, they need not be implemented without a specific reason (e.g., hematology smears to assist in the diagnosis of leukemia).

## Anatomic Pathology

Pathology, an integral part of the protocol design, plays a pivotal role throughout the conduct, evaluation, and interpretation of carcinogenicity studies (as discussed in the section Good Laboratory Practice section) [23,57]. In general, emphasis is placed on routine methods, but special pathology methods may be required depending on the target organs. Whatever can be anticipated from previous subchronic and chronic studies should be addressed in the protocol for pathology. Furthermore, microscopic changes must be correlated with clinical observations, body weight effects, survival patterns, clinical pathology data, and other information. The general and specific GLP and statistical considerations are discussed in separate sections elsewhere in this chapter.

All animals euthanized for humane reasons and those found dead should be submitted to a complete necropsy. At scheduled necropsies, the pathology team should be prepared for potential outcomes by participating in a prenecropsy briefing, where all known clinicopathological correlations are discussed. At necropsy, body weights are obtained. Examination of all recorded palpable masses constitutes an initial procedure with examination of all body orifices and skin. A ventral midline incision, with reflection of the skin so subcutaneous tissues are exposed, initiates the opening of the abdominal cavity followed by the thoracic and finally the cranial cavities. All gross lesions are described as to their location, size, shape, consistency, and color. In general, organs should be examined *in situ* as well as after removal from the animal. Artifactual tissue damage (e.g., crushing or tearing of tissues) must be avoided, but if this is not possible, then such damage should be minimized and noted. Tissues should be cleaned by rinsing in physiologic saline solution (tapwater is not acceptable because the low osmolarity damages cells). Alimentary tract hollow organs are opened to avoid tissue autolysis and are examined after removal of their contents (their anatomic integrity, if possible, should be maintained). Individual lungs are instilled (with 10% neutral buffered formalin: ~ 4 mL in rats and ~ 2 mL in mice), taking care not to overinflate and, if indicated, the urinary bladder also (with formalin ~0.2 to 0.5 mL), after which the trachea and urethra are ligated to maintain the inflated state. Lesions and neoplasms are dissected to include regional lymph nodes, if possible, and a small portion of surrounding (normal) tissue. At least 55 standard tissues and lesions are sampled, trimmed, and processed for histopathological examination, including blood and bone marrow smears [41,102,217,280] and brown adipose tissue (either mediastinal or periadrenal), which is emerging as an important neglected tissue [149]. Any significant deviation from these procedures amounts to censoring, which potentially compromises the integrity of the study (see the Good Laboratory Practice and Statistical Analyses sections). Organ weights are usually not taken in carcinogenicity studies because of the variability in weights caused by disease, neoplasms, or body weight fluctuations. In some cases, selected organs are weighed [339,340], usually including the adrenals, brain, heart, kidneys, liver, lungs, spleen, testes (with epididymides), and uterus (including horns). Where organ weights are taken, organ to body weight percentage values are recommended [102].

The preparation of routine microscopic slides should be in accordance with a standard operating procedures. Each slide should be matched with blocks and routinely taken tissues or grossly observed findings. Special methods, such as immunostaining for proliferating cell nuclear antigen (PCNA), compatible with formalin fixation, should be either described in detail in the protocol or be part of an SOP. Recently, these methods have shown utility, as they are capable of identifying cell proliferation and preneoplasia and early neoplasia [148,386].

During microscopic examination, an open slide evaluation is recommended initially [142,143,247]. It consists of evaluation of the concurrent control groups first and then the high-exposure groups. After this, the rest of the exposure groups are evaluated and the presence or absence of a treatment-response pattern established. Finally, the findings can be compared to in-house historical or published control data to minimize subjectivity and diagnostic drift. Open evaluation is preferably conducted by one pathologist. If the study entails more than 1000 animals, then one pathologist may read the males and a second the females to reduce the length of time to complete the histopathologic evaluation. Open evaluation is also performed when quantitation is performed (e.g., PCNA). Under certain circumstances, when a reevaluation of certain tissue-specific lesions is necessary, a blinded microscopic examination of selected target tissues may be performed. In this type of examination, all slides are reevaluated in a blinded manner in a random sequence by a second pathologist, preferably one not previously familiar with the slides. Also, the valuable practice of peer review can be utilized. This consists of an independent examination by a second pathologist of all tissues from a representative sampling of randomly selected animals of both genders from the control and high-dose groups and a representative sampling of proliferative lesions to substantiate the data from the initial evaluation for neoplasms and other proliferative lesions. Here, it is also recommended that the evaluation be open and that all changes (not only microscopic) are taken into consideration.

**TABLE 25.11**
**Format for Showing Summary Data Incidences and *p*-Values of Neoplastic Lesions in Male and Female Rats and Mice**

| Organ/Tissue and Neoplasm | Control | Low | Medium | High |
|---|---|---|---|---|
| Number of animals at the beginning | 50 | 50 | 50 | 50 |
| Liver | (40) | (45) | (50) | (43) |
| Hepatocellular adenoma | 4 | 5 | 7 | 10 |
| Context of observation of the neoplasm[a] | | | | |
| Unadjusted *p*-values:[b] | | | | |
| Exact test | $p =$ | $p =$ | $p =$ | $p =$ |
| Asymptotic test | $p =$ | $p =$ | $p =$ | $p =$ |
| Hepatocellular carcinoma | 2 | 2 | 5 | 3 |
| Context of observation of the neoplasm[a] | | | | |
| Unadjusted *p*-values:[b] | | | | |
| Exact test | $p =$ | $p =$ | $p =$ | $p =$ |
| Asymptotic test | $p =$ | $p =$ | $p =$ | $p =$ |
| Lung | (45) | (47) | (49) | (45) |
| Bronchiolar/alveolar adenoma | 2 | 1 | 4 | 8 |
| Context of observation of the neoplasm[a] | | | | |
| Unadjusted *p*-values:[b] | | | | |
| Exact test | $p =$ | $p =$ | $p =$ | $p =$ |
| Asymptotic test | $p =$ | $p =$ | $p =$ | $p =$ |
| Bronchiolar/alveolar carcinoma | 2 | 2 | 5 | 4 |
| Context of observation of the neoplasm[a] | | | | |
| Unadjusted *p*-values:[b] | | | | |
| Exact test | $p =$ | $p =$ | $p =$ | $p =$ |
| Asymptotic test | $p =$ | $p =$ | $p =$ | $p =$ |

[a] Contexts of observation of the neoplasm, if information is available, should be one of the four possibilities: fatal, incidental, mortality independent, and nature of fatal and incidental. Use "NA" to indicate that the information is not available.

[b] Unadjusted *p*-values are *p*-values unadjusted for the effect of multiple tests.

*Note:* Numbers in parentheses are the numbers of animals with the tissues examined microscopically. The *p*-values under the control group are from trend tests; the *p*-values under each dosed group are from pairwise comparisons between that dosed group and the control group.

It is vital to record all pathology data in a consistent manner and to depict them as individual data in appendices. A summary of all treatment-related data should be in tables (see Table 25.11 and Table 25.12). An appendix depicting all missing tissues is highly desirable. Neoplastic diagnoses should adhere to an accepted nomenclature of histopathologic terms, according to recommendations made by the Society of Toxicologic Pathologists (*Guides for Toxicologic Pathology*, STP/ARP/AFIP, Washington, D.C.), with particular care to distinguish among proliferative non neoplastic lesions, benign neoplasms, and malignant neoplasms [88,89,116,119,218]. The pathologist must integrate all clinical, structural (macroscopic and microscopic), and functional (cellular and biochemical) data. Moreover, the pathologist ensures proper accounting of gross and microscopic lesions and changes (as detailed in the Good Laboratory Practice and Statistical Analyses sections).

In pathology, the most common deficiencies in GLP are as follows: (1) Gross observations are not fully provided and the exposure related ones are not compared side by side with microscopic findings; (2) gross and microscopic pathology data do not match and no explanation is provided; (3) organ weights in the notebook and report do not match; (4) sacrifice dates before completion of the study are not in the final report and no explanation is given; (5) differences between forms of data recording by the pathologist and those in the SOPs are not explained; (6) individual animal data in the notebook and data in tabular form in the report on the corresponding animal are not the same and no explanation is provided; (7) there is a lack of uniformity in pathology nomenclature and no explanation is provided; (8) there is lack of lesion accountability and important tissues are missing without any explanation being given; (9) the method of slide evaluation

**TABLE 25.12**
**In-House Historical Control Data Based on Carcinogenicity Studies Conducted at XYZ Laboratory from 1995 to 2005**

Species: mouse; sex: male; strain: Crl:CD-1

| | Historical Control Incidences | | |
|---|---|---|---|
| **Studies** | **Neoplasm Type 1** | **Neoplasm Type 2** | **Neoplasm Type 3** |
| Study 1 (1995) | 1/49 | 4/49 | 8/50 |
| Study 2 (1998) | 1/50 | 3/50 | 4/50 |
| ⋮ | ⋮ | ⋮ | ⋮ |
| Study 12 (2005) | 0/50 | 2/50 | 5/50 |
| Total number | 20/347 | 23/417 | 34/417 |
| Standard deviation | 1.0% | 3.2% | 4.0% |
| Range | 0–2% | 0–10% | 3–17% |

is not stated (e.g., open, peer-reviewed); and (10) there is a lack of initialing and dating in various records such as gross data, tissue trimming, microscope evaluation, tissue recuts, etc. Each of these is important, and no explanation being provided can compromise the RCB.

## RODENT CANCER BIOASSAY EVALUATION

### Tumor Increases or Decreases

If the RCB has been conducted properly, it should provide adequate evidence to assess whether exposure has led to changes in the incidences of specific neoplasms. A conclusion of lack of carcinogenic activity requires survival of adequate numbers of animals given sufficient relevant exposures, usually at the MTD, with no evidence of neoplasm increases according to the criteria discussed earlier in the Chemicals with Carcinogenic Activity section. A conclusion for a positive outcome is generally less rigorous; that is, in the interest of conservative hazard identification, a statistical increase in neoplasms is often accepted as valid in spite of artifacts in the study such as excessively reduced body weight gain or poor survival. Nevertheless, the possibility must be considered that a neoplasm increase can be a consequence of the toxicity of the treatment conditions [118] and not a chemical action of the TS. Moreover, it is possible that the differences occurred by chance alone. This can be evaluated to some extent by the use of two control groups to document random variation [11,129]. A statistically significant greater incidence in neoplasms in a treatment group that does not exceed that of one of the control groups can be regarded as a "numerical imbalance," not a true increase.

A critical aspect in the evaluation is the interpretation of pathological diagnoses. Findings of increases in malignant neoplasms are universally accepted as evidence of carcinogenic activity. Combining malignant and benign neoplasms of the same cell type of origin is also widely accepted, and guidelines have been published by the NTP [218]. It should be noted that in some tissues certain benign tumors are not related to malignancies (e.g., benign mammary fibroadenomas and malignant adenocarcinomas). When evaluating increases in neoplasms, attention must be paid to pathology in the tissue that is the site of neoplasms because cell injury and compensatory cell proliferation can facilitate neoplastic development (see the Cancer Hazard and Risk Assessment section).

When assessing the incidences of neoplasms in groups, both the incidence in animals surviving to termination and the combined incidences of decedent and terminal sacrifice animals are considered. Evaluation should be made with regard to the ranges of commonly occurring neoplasms (Table 25.13 and Table 25.14), as well as the percent incidence profiles expressed by time of death (i.e., the neoplasm distribution) (Table 25.6) [11,20,27,42, 128,134,214,221,267,307,309,310,323,354]. Among the commonly occurring neoplasms in rats, pituitary, mammary gland, and skin neoplasms occur predominantly after 50 weeks, whereas lymphoma occurs almost equally before 50 weeks (Table 25.6) [302]. In mice, lung, mammary gland, and liver neoplasms occur after 50 weeks, whereas lymphoma or leukemia occurs almost equally before 50 weeks (Table 25.6) [302].

The most appropriate comparison of an experimental group is with its matched control or controls, where two groups are used [11,107]. Nevertheless, in arriving at a conclusion of neoplasm increase, historical control data [128] from the strain used may be helpful, preferably if animals from the same breeder have been kept under identical conditions in the same facilities more or less at the same time as the study under evaluation. Differences between controls and exposed groups can be analyzed by a variety of statistical methods, as discussed below. Whether any true increase in neoplasia has relevance for human cancer hazard is discussed in the Cancer Hazard and Risk Assessment section.

**TABLE 25.13**
**Percent Incidence Ranges of Common Spontaneous Neoplasms in 104-Week-Old Rats**

| | Males | | Females | |
|---|---|---|---|---|
| Sites/Neoplasms | SD[a] (%) | Wi[b] (%) | SD[a] (%) | Wi[b] (%) |
| Pituitary gland adenoma | 1–70 | 22–51 | 26–93 | 2–61 |
| Pituitary gland carcinoma | 1–36 | 2–20 | 1–58 | 1–2 |
| Pancreas islet adenoma | 2–26 | 2–20 | 1–14 | <1–2 |
| Pancreas islet carcinoma | 1–14 | 1–2 | 1–6 | <1–2 |
| Pancreas acinar adenoma | 1–11 | 2–7 | 1–3 | 2–5 |
| Pheochromocytoma | 1–29 | 2–13 | 1–18 | 2–4 |
| Thyroid C cell neoplasia | 2–30 | 6–23 | 3–28 | 5–24 |
| Thyroid follicular adenoma | 2–12 | 2–13 | 1–6 | 2–10 |
| Skin fibroma | 1–11 | 2–11 | 1–4 | 2–5 |
| Skin keratoacanthoma | 1–10 | 2–15 | 1–3 | 1–4 |
| Mammary gland fibroadenoma | 1–6 | 1–4 | 13–62 | 11–34 |
| Mammary gland adenocarcinoma | 1–4 | <1–1 | 9–58 | 2–13 |
| Mammary gland adenoma | 1–2 | <1–1 | 1–32 | <1–1 |
| Adrenocortical adenoma | 1–8 | 2–7 | 1–34 | 2–4 |
| Endometrial sarcoma | NA | NA | 1–18 | <1–1 |
| Endometrial hemangioma | NA | NA | 1–15 | 17–37 |
| Endometrial polyp | NA | NA | 1–10 | 2–17 |
| Hepatocellular adenoma | 1–8 | <1–1 | 2–13 | <1–1 |
| Lymph node hemangiosarcoma | <1–1 | 2–15 | <1–<1 | 1–2 |
| Lymph node hemangioma | <1–1 | 2–13 | <1–<1 | 2–11 |
| Leydig cell testicular neoplasia | 1–9 | 2–11 | NA | NA |
| Lymphosarcoma, multicentric | 1–6 | 2–10 | 1–10 | 2–6 |
| Thymoma, benign | <1–1 | 2–5 | <1–<1 | 2–10 |
| Vaginal adenoma | NA | NA | <1–<1 | 2–10 |

[a] Data from Christian et al. [42], Brix et al. [27], Giknis and Clifford [113], and McMartin et al. [221].
[b] Data from Bomhard and Rinke [20], Eiben and Bomhard [75], Giknis and Clifford [112], Poteracki and Walsh [267], Tennekes et al. [309,310], Walsh and Poteracki [354].

*Note:* SD, Sprague–Dawley rats; Wi, Wistar rats; NA, not applicable.

**TABLE 25.14**
**Percent Incidence Ranges of Common Spontaneous Neoplasms in 100-Week-Old CD-1 Mice**

| Sites/Neoplasms | Males | Females |
|---|---|---|
| Alveolar/bronchiolar adenoma | 2–42 | 2–27 |
| Alveolar/bronchiolar carcinoma | 1–26 | 1–18 |
| Hepatocellular adenoma | 3–28 | 1–8 |
| Hepatocellular carcinoma | 2–16 | 1–4 |
| Thymoma, malignant | 2–15 | 1–2 |
| Lymphoma, multicentric | 4–28 | 2–50 |
| Histocytic sarcoma | 1–8 | 2–18 |
| Hemangiosarcoma | 2–12 | 2–12 |
| Pituitary adenoma | 2–3 | 1–14 |
| Harderian gland adenoma | 2–14 | 1–8 |
| Skin neurilemmoma, malignant | 1–2 | 2–14 |
| Endometrial polyp | — | 2–17 |
| Uterine leiomyoma/sarcoma | — | 3–10 |

*Source:* Data from Giknis and Clifford [111] and Maita et al. [214].

A controversial issue is the interpretation of neoplasm decreases in the RCB [60], which are quite frequent [132]. One consideration is that the reduction should not be attributable to nonspecific weight gain suppression (as discussed in the Feeding Procedures section). A variety of mechanisms for specific anticarcinogenesis have been delineated (refer to the Anticarcinogenesis section).

## Statistical Analyses

An expert biostatistician should be involved in the RCB. The methods ultimately employed can be diverse, and only some general guidance is provided here. More extensive coverage is provided in Chapter 8, and authorities have provided guidance on statistical aspects [337,346]. Statistics are employed in all aspects of the study (i.e., design, conduct, evaluation, analysis, and interpretation). The object of statistics is to substantiate whether exposure has elicited a neoplastic response. Other than exposure, two other factors can underlie an increase in neoplasia. One is bias (a systematic difference other than what is caused by exposure), and the other is chance (a random difference). It is indeed highly desirable to avoid bias and minimize (and control) chance, although it cannot be completely excluded, as identically nonexposed or exposed animals do not respond identically (the biological reality). This is an important reason for using two control groups. Differences between these controls provide a measure of chance variation between small groups. The probability of chance can be statistically measured; the smaller the probability, the higher the confidence [204]. Of the two mathematical models used in statistics, the stochastic one that contains uncertainties or random variables is mainly used to analyze RCBs. The deterministic model, which does not involve any uncertainties, is not as useful.

A one-sided (one-tailed) *p*-value is the probability of getting by chance an exposure effect in a specified direction as great as or greater than that observed [91]. A two-sided (two-tailed) *p*-value is the probability of getting by chance an exposure difference (effect) in either direction which is as great as or greater than that observed. The calculation of a *p*-value of $<0.05$ signifies that the effect could have occurred by chance in less than 1 time in 20. Theoretically, at least, randomization eliminates nonsystematic bias, but, if there is another major source of variation (e.g., gender of the same strain or batch of the same strain), then a stratified randomization is more appropriate. To achieve that, separate randomization within each level of the stratifying variable (e.g., cage position, order of weighting, order of killing at termination) must be performed [204]. The power of any statistical measurement refers to the probability that the subject test will reject the hypothesis that there is a difference when it is not true [103].

There are three aspects that determine the nature of collected data: the biological system, the study design instrumentation, and the methodologies applied. Censoring any of these must be minimized. In general, exposure variables are independent, and effect variables are dependent [103,204]. Representative samples should be appropriately collected and of sufficient size. In addition, samples should be accurate (of high quality) and precise (reproducible). The pattern of distribution of data in the sample is very important, because it shows their central tendency and dispersion. In the Gaussian distribution (the most common), two thirds of all values are within one standard deviation. Others include binomial and Chi-square patterns of distribution. To successfully combine data for analysis, stratification is applied (e.g., early, late, or total deaths).

To enhance the capture of an exposure effect, the exposure-related trend is employed [308]. Furthermore, if there are differences in survival, age adjustment to avoid bias is necessary. Even if there are no survival differences, age adjustment increases the power of detecting differences between groups, as well as sharpening their contrast by avoiding dilution of pivotal (key) data with less pivotal data. Age adjustment is effective when the context of observation is taken into account; that is, the condition (neoplasm) is assumed to have caused death, or death was incidental, or even the condition is visible [263]. If the study pathologist feels that it is not possible to determine the context of observation, then it is difficult to apply this method. Finally, when multiple comparisons are to be made with unequal numbers in the groups, a number adjustment should be employed [101].

In the past, the single most important statistical consideration in the analysis of bioassays was a simple quantal response; that is, either neoplasia did occur or it did not occur. Currently, the mechanisms underlying neoplasia induced by chemicals are more fully understood and must be given individual consideration. These mechanisms include effects on survival rate, body weight gain, age at first tumor, time-to-tumor, patterns (trends) of tumor incidence, tumor multiplicity, rates of proliferation at target sites, presence of markers of preneoplasia or early neoplasia, and exposure response. As noted earlier, at least one interim sacrifice should be included and possibly also a short (1- to 3-month) recovery segment before final termination. The first statistical test used should be one that makes a general assessment, not a series of pairwise comparisons. The nature of the data should be established (i.e., whether it is continuous or discontinuous, scalar, rank, or quantal), and the distribution of these data should be examined [204].

In an RCB, the time course of adverse effects is of importance; consequently, life-table methods are employed to determine duration of survival and time until neoplasms develop, as well as the probability of survival or time course of neoplastic development [56,58,263]. The presentation of life-tables should depict gender and exposure level and include the time of selected interval, as well as the number of animals that were alive during the interval, excluding the animals that were either withdrawn (e.g., interim sacrifices) or died during the same interval. For continuous outcome measures, which can be either scalar or ranked, group means are compared against the control mean at each termination point using a one-way analysis of variance (ANOVA) followed by Dunnett's method for multiple comparisons, which is a powerful *post hoc* test [72]. Moreover, a square-root transformation can be added to stabilize the variance. In addition, the continuous mean outcome levels are compared among termination times at each exposure level using one-way ANOVA followed by Tukey's multiple comparison procedure [95].

To assess the course of exposure response (its linearity), ordinary least-squares regression analysis can be used fitting the outcome level vs. exposure and squared exposure terms. For the incidence of specific site neoplasia comparing all test groups, the Pearson chi-square test followed by pairwise comparisons of each exposure group with control, adjusted for multiple comparisons, can be applied [100]. For incidence trend analysis, the Cochran–Armitage test, partitioning the chi-square statistic into the overall trend and departure from linearity (*p* nonlinear) can be tested [94]. Furthermore, survival data can be used by applying log-rank test for both homogeneity and exposure-related trend [126,127]. Neoplastic data can be analyzed using a survival-adjusted trend test that discriminates among fatal, incidental, and palpable neoplasms [263]. If one or more tumor types in a valid RCB show a significant positive trend in neoplasm incidence rates, the significance level (*p*-value) for rare (≤1%) neoplasms would be 0.025 and for common neoplasms 0.05 [346]. For pairwise comparisons (control vs. high dose), the significance of rare neoplasms would be 0.05 and of common neoplasms 0.01 [346].

When animal weights differ across dose groups, as is usually the case with high-dose testing, such differences can contribute to differences in neoplasm incidence (see the Feeding Procedures section). Analyzing tumor incidence within body weight strata can reduce the bias resulting from weight differences [108]. Sustained body weight gain suppression can lead to stress responses resulting in general protein catabolism and concomitant upregulation of heat shock protein (Hsp) 70 in the liver [149]. Over time, this results in systemic antigluconeogenesis, fat anabolism, and reduction of the intermediary metabolism [145]. Here, there is a significant species difference between rodents and humans—namely, subtle sustained hypoglycemia resulting from systemic antigluconeogenesis does not trigger an increase in food consumption in rodents [140,316]. Thus, the increased protein catabolism and body temperature, with concomitant slow decrease in tissue hydration, cannot be compensated for, leading to a chronic disruption (deficit) in energy homeostasis and metabolism.

The interpretation of analyzed data is the final critical step of the whole process. Of importance here is the existence of extensive historical control data, both published and unpublished, for the specific species and strains used [65,120,127,195,252]. The operational concepts here involve the biological and statistical data differences between control and exposed groups. Important here is the nature of these differences and the main reason for the differences. The final interpretation should be based on both biological and statistical consideration [89,344]. The outcome can indicate concurrence that is either positive or negative; these two outcomes simplify the interpretation. On the other hand, when there is no biological but only statistical significance, a false-positive outcome is the result. This constitutes a type I error. Conversely, few but rare neoplasms, constituting a biological significance in the absence of any statistical one, may render a false negative, representing a type II error. In both cases, by providing an explanation for the differences (causality) and demonstrating the proof for the underlying mechanism, both types of error are minimized.

In conclusion, eight considerations have proven to be very helpful in evaluation and interpretation when applying statistical methodology: (1) exposure–effect relationship, (2) incidence of proliferation and preneoplastic and early neoplastic markers at the target site of neoplasia, (3) presence of gender and species similarities or differences at the target sites, (4) convergence in target sites of nonproliferative chronic toxicity and neoplasia, (5) combined neoplasia increases in tissues affected by chronic toxicity, (6) neoplasms of similar histogenetic target sites in other genders or species, (7) concurrent and historical control data, and (8) relative survival of control and exposed groups. Statistical data from all these considerations help with the final interpretation.

## BIOASSAY REPORTING

The final critical step of the RCB is the study final report. The final report consists first of an introductory section containing the compliance statement signed by the study director, followed by the quality assurance and study identification statements. The study identification statement provides the study title and number, the test substance, the testing facility, the test facility manager, the sponsor, the study director, the principal investigator of all study aspects, the exact specific study timetable, and approved signatures from all final report authors, including the study director and investigators of all aspects of the study (e.g., analytical, toxicokinetics, during-life, pathologist). The first part of the final report itself is the summary, which is an abstract of the entire study. It contains, in this order, an introduction, a listing of the materials and methods, the results, and the conclusions The summary is followed by a summary table that depicts all pertinent findings in tabular form.

The summary section is followed by an extensive introduction explaining the origin and purpose of the RCB, followed by listings of test animals, test materials methods, results, discussion, conclusions, and references All sections should fully describe all methods used and all data obtained. All individual data (e.g., analytical body weight, necropsy, microscopic) should be in appropriate appendices. All relevant summary data (e.g., analytical, body weight) should be in tables. Numerical incidences should precede percent incidences. Two examples of specific formats for critical tables are given in Table 25.11 and Table 25.12. Tables and graphs presenting special issues and arguments should be included in the text (text tables or text graphs). Appropriate statistical analysis of correlation of survival patterns, clinical observations, body weight gain pattern, and toxicokinetic data with gross and microscopic findings should be conducted. All of these considerations have been discussed in other sections in detail.

An effective way to summarize the findings is a format used in Europe known as the tabulated study report. All relevant data are presented in standardized tabular form without narrative. This corresponds to the summary tables of the final report described above.

## CLASSIFICATION OF EVIDENCE OF CARCINOGENICITY

Completed RCBs must be reported to the regulatory agency under whose purview they were performed [49,63,82,83,334]. The results are then subject to evaluation and classification. In the United States, the results of RCBs on pharmaceuticals tested under investigational new drugs (INDs) approved by the FDA/CDER are submitted to the Reviewing Division which then evaluates them, often through the CDER Carcinogenicity Assessment Committee (CAC). The final interpretation of the results will appear in the labeling of the medicine, if approved. The FDA normally describes the RCB data without comment on human relevance, except to note multiples of exposure in the rodents compared to humans. The CAC consists of a chair, an executive secretary, and members from several divisions, the Office of Epidemiology and Biostatistics, the Office of Testing and Research, and the Office of Pharmaceutical Sciences. The Reviewing Division (of the FDA) notifies the sponsor when a CAC meeting is scheduled after all RCB studies are submitted, and the sponsor may attend.

The U.S. Environmental Protection Agency formerly used an alphabetical/numerical classification ranging from Group A (human carcinogen, based on animal data) to Group E (noncarcinogen) (Table 25.15) [334,336], but it has converted to a narrative classification [335,336], which allows incorporation of mechanistic data, similar to the IFSTP classification described below. At the EPA, an *ad hoc* CAC of the EPA Science Advisory Committee evaluates the submitted dossier.

Rodent cancer bioassays conducted by the NTP are reviewed by a peer review panel and published as technical reports. They are available to the EPA, FDA, and OSHA for regulatory action [334]. The NTP uses a classification system of no, limited (some), or clear evidence of carcinogenicity [242]. The NTP also publishes biennial reports on carcinogenesis [243].

In the European Union and Japan, similar classification schemes are used by various health boards and the Commission for Human Medicinal Products of the European Medicine Agency [63,73,84]. The International Agency for Research on Cancer convenes external working groups several times each year to evaluate groups of chemicals with published carcinogenicity data. They publish the IARC monographs and IARC biennial reports with evaluations of experimental and human data. The grouping ranges from Group 1 (carcinogenic to humans) to Group 4 (evidence suggesting lack of carcinogenicity) (Table 25.15). A deficiency in this classification is that Group 3 contains chemicals with data that are considered not likely to be a human cancer hazard.

The IFSTP [89] has proposed a classification as follows: (1) carcinogens for humans based on epidemiological data, (2) genotoxic carcinogens for animals based on experimental data, (3) epigenetic carcinogens for animals based on experimental data, and (4) suspected carcinogens insufficiently tested. This is the only classification that explicitly incorporates mechanistic distinction (Table 25.15).

## CANCER HAZARD AND RISK ASSESSMENT

In cancer risk assessment, the first step is hazard identification [169,239], which involves an RCB to identify exposure-related neoplasms (see Rodent Cancer Bioassay Evaluation section). Using dose–response data from the bioassay and potential human contact, a cancer risk is assessed [239], often involving allomorphic scaling [332]. To identify a potential human cancer hazard, such as, for example, when following IFSTP recommendations, the RCB results must be interpreted together with other mechanistic data [89]. An important aspect to be considered in the interpretation is that of comparing percent incidences of the common human and rodent neoplasms; for example, the magnitude of the percent incidence of background neoplasms in rodents ranges from 4 times greater to 4500 times greater than the incidences in humans (Table 25.16), demonstrating the proclivity of rodents to certain neoplasms. Indeed, in various regulatory databases, the incidence of positive findings in either mouse or rat or both RCBs is 41 to 54% [211].

As discussed below, if the agent is clearly DNA reactive, then a neoplasm increase strongly implies a potential hazard to humans [361,364,368]. On the other hand, it is now recognized that epigenetic carcinogens may exert the effects underlying tumorigenicity only in particular rodent species (e.g., $\alpha_{2\mu}$-globulin nephropathy inducers in male rats) or only at high toxic doses (e.g., nitrilotriacetic-acid-induced nephropathy). Such effects either are considered to be irrelevant to human hazard [331] or can be subjected to a margin of exposure (MOE) risk assessment [336]. The MOAs of carcinogenesis for epigenetic (nongenotoxic) agents are complex, involving a variety of secondary organ and tissue target sites, with indirect interference with the organ or tissue homeostasis. Sustained adaptive effects [377] and disruption of endocrine, paracrine, nervous, and immune systems can be involved in the pathogenesis of neoplasia induced by such agents. Accordingly, the carcinogenetic effects of these agents are typically species, gender, and tissue specific.

Most of these secondary mechanisms involve effects such as enhancement of cell proliferation, disruption of hormonal feedback pathways, inhibition of trophic activ-

**TABLE 25.15**
**Classification of Carcinogens**

| IARC | HWC | EPA | IFSTP |
|---|---|---|---|
| Group 1 | Group I | Group A | Group 1 |
| Group 2A | Group II | Group $B_1$ | Group 2a |
| Group 2B | Group III | Group $B_2$ | Group 2b |
| — | — | Group C | — |
| Group 3 | Group IV | Group D | Group 3a |
| — | — | — | Group 3b |
| — | — | — | Group 3c |
| Group 4 | Group V | Group E | Group 4 |

*Notes:* IARC, International Agency for Research on Cancer [153], HWC, Health and Welfare Canada [136], EPA, U.S. Environmental Protection Agency [329]; IFSTP, International Federation of Societies of Toxicologic Pathologists [89]; —, no correspondence.

IARC Group 1, HWC Group I, EPA Group A, IFSTP Group 1: The agent is *carcinogenic* to humans. There is sufficient evidence in humans of a positive relationship between cancer and human exposure such that chance, bias, and confounding variables can be reasonably ruled out.

IARC Group 2A, HWC Group II, EPA Group $B_1$, IFSTP Group 2a: The agent is *probably carcinogenic* to humans. There is limited evidence in humans of a positive relationship between cancer and human exposure, but chance, bias, and confounding variables cannot be ruled out. There is sufficient evidence of carcinogenicity in animals based on an increased incidence of benign and malignant neoplasms in at least two species or in two independent studies.

IARC Group 2B, HWC Group III, EPA Group $B_2$, IFSTP Group 2b: The agent is *possibly carcinogenic* to humans. There is either limited evidence or absence of data in humans; there is either sufficient or limited and weak evidence of carcinogenicity in animals (e.g., presence of other relevant data, genotoxic agents cause only benign tumors or increases in certain spontaneous neoplasms).

EPA Group C: The agent is possibly carcinogenic to humans. There is either an absence of data in humans or limited evidence of carcinogenicity in animals (e.g., agents that cause only benign tumors, or neoplasm incidence increases are marginal and not consistent).

IARC Group 3, HWC Group IV, EPA Group D, IFSTP Group 3a: The agent is not classifiable as to its carcinogenicity in humans.

HWC Group IV: The data are inadequate for evaluation, or these agents cannot be classified in other groups.

IFSTP Group 3a: The experimental data of epigenetic carcinogens show a threshold level within the range of human exposure.

IFSTP Group 3b: The experimental data of epigenetic carcinogens show a threshold level beyond the range of human exposure.

IFSTEP Group 3c: The experimental data of epigenetic carcinogens show that their mechanism of action is not applicable in humans.

IARC Group 4, HWC, Group V, EPA Group E, IFSTP Group 4: The agent is *probably not carcinogenic* to humans. There is evidence suggesting a lack of carcinogenicity in humans (even if inadequate) and in animals (negative animals studies). In IFSTP Group 4, the suspected carcinogens have not been sufficiently tested.

ity in tissues (including long-standing tissue ischemia), immune surveillance dysfunction, inhibition of enzymatic reaction/activation in cells, sustained exaggerated pharmacological effect, and sustained accumulation of normally low levels of endogenous products. All of these effects result in sustained cellular toxicity, leading to compensatory proliferation, which is a common pathway through which agents with diverse cellular effects ultimately induce or enhance neoplasia [45,54,148]. The effects that lead to compensatory cellular proliferation usually require high levels of exposure and exhibit thresholds. It is probably for this reason that, of the NCI/NTP rodent carcinogens tested, 6% had increased incidences of neoplasms, which were limited to the top dose for all sites of carcinogenicity [131]. Examples of neoplastic effects with no significance or limited significance for human hazard are given in Table 25.17, and some mechanisms and neoplastic findings are described in more detail in the following.

## Modes of Action Not Indicative of Cancer Hazard to Humans

For several neoplastic responses in rodents, sufficient mechanistic information has accrued to support the generally accepted conclusion that the underlying MOAs for certain agents that are not DNA reactive in the target tissue are species-specific and do not operate in humans (Table 25.17).

**TABLE 25.16**
**Comparison of Percent Incidences of Selected Common Human and Rodent Neoplasms That Occur in All**

| | Humans[a] | | | | SD Rats[b] | | | | CD-1 Mice[c] | | | |
|---|---|---|---|---|---|---|---|---|---|---|---|---|
| | Males | | Females | | Males | | Females | | Males | | Females | |
| Site | Percent | Rank | Percent | Rank | Percent | Rank | Percent | Rank | Percent | Rank | Percent | Rank |
| Lung | 0.032 | 2 | 0.027 | 2 | 0.1 | 23 | 0.1 | 23 | 15 | 2 | 15 | 2 |
| HLR[d] | 0.011 | 6 | 0.010 | 5 | 2 | 10 | 2 | 11 | 8 | 3 | 22 | 1 |
| Breast | NA | NA | 0.071 | 1 | NA | NA | 58 | 2 | NA | NA | 6 | 3 |
| Liver | 0.004 | 9 | 0.002 | 13 | 5 | 6 | 3 | 6 | 18 | 1 | 3 | 5 |

[a] Data from American Cancer Society [8]; in humans, the percent figure reflects new cancer cases estimated from 2004 U.S. population.
[b] Data from Christian et al. [42], Giknis and Clifford [113], and McMartin et al. [221].
[c] Data from Giknis and Clifford [111]; in rats and mice, the percent figure reflects spontaneous neoplasms observed in excess of 1000 control animals.
[d] Hematolymphoreticular tissue neoplasms correspond to the diagnosis of lymphoma in humans.

*Note:* NA, not applicable.

**TABLE 25.17**
**Examples of Neoplastic Effects in Rodents with No or Limited Significance for Human Safety**

| Neoplastic Effect | Pathogenesis (Agents) |
|---|---|
| *No significance* | |
| Gastric neuroendocrine cell neoplasia mainly in rats | Gastric secretory suppression, gastric atrophy induction (cimetidine, omeprazole, butachlor) |
| Hepatocellular neoplasia in rats and mice | Peroxisome proliferation (clofibrate, phthalate esters, phenoxy agents) |
| | Phenobarbital-like promotion |
| Renal tubular neoplasia in male rats | $\alpha_{2\mu}$-Globulin nephropathy/hydrocarbons (d-limonene, trimethylpentane) |
| Subcutaneous sarcomas in rodents | Local irritant effect (iron dextran) |
| Thyroid follicular cell neoplasia in rats | Hepatic enzyme induction, thyroid enzyme inhibition (oxazepam, amobarbital, sulfonamides, thioureas) |
| Urinary bladder neoplasia in rats | Crystalluria, carbonic anhydrase inhibition, urine pH extremes (melamine, saccharin carbonic anhydrase inhibitors, dietary phosphates) |
| *Limited significance* | |
| Adenohypophysis neoplasia in rats | Feedback interference/neuroleptics (dopamine inhibitors) |
| Adrenal medullary neoplasia in rats | Feedback interference (lactose, sugar alcohols) |
| Forestomach neoplasia in rats and mice | Stimulation of proliferation (butylated hydroxyanisole, phthalate esters, proprionic acid) |
| Harderian gland neoplasia in mice | Feedback interference (misoprostol [$PGE_1$], nalidixic acid, aniline dyes) |
| Pancreatic islet cell neoplasia in rats | Feedback interference (neuroleptics) |
| Endometrial neoplasia in rats | Feedback interference (proestrogens, dopamine agonists) |
| Leydig cell testicular neoplasia in mice | Feedback interference (proestrogens, diethylstilbestrol, tamoxifen) |
| Leydig cell testicular neoplasia in rats | Feedback interference (lactose, sugar alcohols, $H_2$ antagonists, carbamazepine, vidarabine, isradipine, dopaminergics, finasteride) |
| Mesovarial leiomyoma in rats (occasionally in mice) | Feedback interference ($\beta_2$ agonists) |
| Mononuclear cell leukemia in rats (mainly F344) | Immunosuppression (in part unknown) (furan, iodinated glycerol) |
| Osteomas in mice | Feedback interference (calcineurin, in part unknown) (cyclosporine, misoprostol, proestrogen) |
| Ovarian tubulostromal neoplasia in mice | Feedback interference (cytotoxic agents, nitrofurantoin) |
| Skin fibromas in rats | Nonspecific physicochemical effects locally (in part unknown) (recombinant human insulin) |
| Skin papillomas or carcinomas in rodents | Chronic local irritation |
| Splenic sarcomas in rats | Methemoglobinemia (in part unknown) (dapsone) |
| Uterine leiomyoma in mice | Feedback interference ($\beta_1$ antagonists) |

### Rat Gastric Neuroendocrine Neoplasm (Carcinoid) Elicited by Suppression of Gastric Acid Secretion

Hyperplasia and neoplasia of gastric neuroendocrine (NE) cells (enterochromaffin-like cells) are stimulated by gastrin in rats and to a lesser degree in mice. Elevations of gastrin are elicited by reduced hydrochloric acid production, which can be caused by either gastric antisecretory medicines such as proton pump inhibitors (e.g., omeprazole) or $H_2$ antagonists (e.g., cimetidine) [29,226]. Agents that cause gastric atrophy (e.g., alachlor) have also elicited this neoplasm [311]. Rats have a high density of gastric NE cells, achieve high levels of gastrin (over 1000 pg/mL), and are very responsive to elevation of gastrin [318]. For most hypoacidity-producing agents, female rats are more susceptible than males to development of NE cell neoplasms. NE cell neoplasms have been observed in patients with multiple endocrine neoplasia syndrome (MEN-1), associated with elevated gastrin but not with antiulcer therapy [226]. Significant NE cell proliferation in humans is seen only with gastrin levels above 400 pg/mL, and this can be controlled in the clinical setting.

### Rat Kidney Neoplasm Resulting from $\alpha_{2\mu}$-Globulin Nephropathy

A variety of xenobiotics induce kidney neoplasms in male rats, mainly F344, which excrete $\alpha_{2\mu}$-globulin in the urine. This protein is associated with hyaline droplet formation, atypical hyperplasia of the epithelium of the $P_2$ segment of the proximal tubules and neoplasia. Male rats (especially the F344) are very proteinuric compared to humans, and no human renal protein is similar to $\alpha_{2\mu}$-globulin [331]. Accordingly, it has been accepted that renal tubule neoplasms produced as a result of the $\alpha_{2\mu}$-globulin accumulation mechanism are not an appropriate endpoint for human hazard identification [277]. Likewise, an IARC working group concluded that an agent that acts solely through $\alpha_{2\mu}$-globulin nephropathy in the production of renal cell neoplasms alone in male rats is not a cancer hazard to humans [277].

### Rodent Liver Neoplasm Elicited by Hepatic Peroxisome Proliferators

A wide variety of xenobiotics elicit increases in rodent liver tumors associated with proliferation of peroxisomes [157]. Rodents are more susceptible to induction of hepatic peroxisome proliferation by peroxisome proliferator-activated receptor $\alpha$ (PPAR$\alpha$) agonists than primates or humans [37,157,374], apparently because of high expression of the PPAR$\alpha$ in rodent liver [320]. Perhaps related to this, it has been reported that, in cultured rat hepatocytes, peroxisome proliferators enhance DNA synthesis and suppress apoptosis, whereas in human hepatocytes DNA synthesis is suppressed and apoptosis enhanced [261]. Whereas the mechanism of carcinogenicity of these agents is not fully understood but appears to involve sustained hepatocellular proliferation [191], none is associated with cancer in humans, and an IARC group has recommended that a liver neoplasm response in mice or rats secondary only to peroxisome proliferation could modify the evaluation of carcinogenicity [157].

### Rodent Subcutaneous Sarcoma Resulting from Implantation of Solid Materials or Injection of Irritant Materials

Rodents, especially males, are highly susceptible to the development, both as background and induced, of subcutaneous neoplasms variously diagnosed as fibrous histiocytomas, fibrous sarcomas, liposarcomas, and leiomyosarcomas. These can arise from a solid-state effect known as the Oppenheimer effect [255] with implantation of solid materials. They are also induced by injection of irritant materials subcutaneously in rodents elicits mesenchymal sarcomas [117]. None of these materials (e.g., recombinant human insulin [266]) has been associated with cancer in humans.

### Rodent Thyroid Neoplasm Resulting from Thyroid–Pituitary Disruption

Few DNA-reactive carcinogens elicit thyroid neoplasms probably because bioactivation does not occur to a significant extent in this gland. On the other hand, thyroid-pituitary disruption is a common mechanism of thyroid carcinogenesis in rodents, particularly rats [312]. Reduced thyroid hormone levels, either through inhibition of synthesis by antithyroid agents (e.g., amitrole) or increased clearance as a result of enhanced conjugation (e.g., phenobarbital) can lead to a feedback increase of thyroid-stimulating hormone (TSH) levels which produces thyroid follicular cell hypertrophy, hyperplasia, and eventually neoplasia. Species differ in their susceptibility to disruption of thyroid economy, with the rat being particularly sensitive [78]. Several inducers of hepatic thyroid hormone conjugation in rats (which often is also associated with increased liver neoplasms) do not affect mice [353]. No nonradioactive chemical exposure is known to cause thyroid follicular neoplasms in humans [277]. Accordingly, it has been concluded that chemical-specific data on thyroid effects in rodents can be applied to risk assessment [139], and that agents that cause thyroid neoplasia through an adaptive hormonal mechanism belong to a different category from those acting through genotoxic effects or involving pathological response to tissue injury [277].

### Rat Urinary Bladder Transitional Cell Neoplasm Resulting from Luminal Milieu Modification

Many studies have used rat models for urothelial neoplasia. In early investigations, it was recognized that placement of inert pellets in the bladder lumen would elicit increases in urothelial neoplasia. Rats have been shown to be more sensitive than mice to urothelial damages, apparently because the rat bladder lacks tight junctions, thus rendering the superficial urothelial layer ineffective as an intraluminal barrier and leaving the underlying layers vulnerable to chronic stimulation [146,199]. Moreover, rats, particularly males, have more intraluminal proteins, silicate precipitation, crystal formation, and urolithiasis than other species [44,46]. In particular rats, unlike humans, develop calcium phosphate urinary precipitates [277]. The consequences of this are exacerbated by the fact that the bladder is horizontal in rodents and does not empty as effectively as the vertical human bladder, despite the fact that rats urinate frequently. Thus, rats are more prone to chronic cell damage to the bladder urothelium, which results in cell proliferation and neoplasia [45,146]. This effect does not occur in humans [45,76,382]. An IARC working group has concluded that production of bladder cancer in rats under conditions of formation of calcium-phosphate-containing urinary precipitates is not predictive of cancer hazard to humans [277].

## Modes of Action Probably Not Indicative of Cancer Hazard to Humans

Several modes of action of epigenetic oncogenesis in rodents appear not to be relevant to human cancer hazard (Table 25.17) [6]. If only neoplasms resulting through one of these MOAs is elicited by the TS, that is not sufficient evidence of carcinogenicity.

### Rodent Liver Neoplasm Induced by Phenobarbital-Like Enzyme Inducers

A variety of compounds elicit an increase in rodent liver neoplasms through a MOA involving liver hyperplasia [369]. The prototype compound is phenobarbital [369]. Consideration of kinetic and dynamic factors led to the determination that the MOA was unlikely in humans [141]. Indeed, extensive epidemiologic study revealed no increased risk of cancer [254].

### Rodent Testicular Leydig Cell Neoplasms Resulting from Hormone Disruption

Leydig or interstitial cell neoplasms occur spontaneously in high incidence (>80%) in aged F344 rats [133]. These tumors are invariably benign. In rats, they result from increases in luteinizing hormone (LH) [87] and in mice from increases in estrogen [270] or administration of estrogen agonists such as diethylstilbestrol (DES) or tamoxifen [215,319]. The human counterpart is extremely rare, and no agent that produces increases in rat testicular neoplasms (i.e., cimetidine, hydralazine, gemfibrozil, carbamazepine, vidarabine, isradipine, exogenous gonadotropins, LHRH analogs, flutamide, ergolines, and finasteride) has been associated with induction of this or any other neoplasm in humans. The data, therefore, suggest that nongenotoxic compounds that induce Leydig cell neoplasms in rodents do not represent a human cancer hazard [52].

## Tumors of Questionable Significance to Human Cancer Hazard

Tumors of questionable significance for hazard assessment [6] are those whose pathogenesis may be unique to rodents, in some cases due to their cell type of origin, and for which no association with human cancer hazard has been established. If one of these is the only neoplasm increased in the RCB without any further evidence suggesting a possible carcinogenic or genotoxic effect of the TS, such a finding is questionable evidence of human cancer hazard.

### Mouse Bladder Mesenchymal Lesion

This recognized lesion occurs in the posterior portion of the urinary bladder close to the trigone area and has been known under various names for some time [177]. Recently, the lesion has been referred to as a "mesenchymal tumor" [31,122]. The lesion has been found in mice given agents that bind to progesterone (mainly) and estrogen receptors (e.g., many oral contraceptives) [219]. Persuasive evidence has been provided [177,178] that the lesion represents a decidual reaction of mesenchymal cells carrying or developing progesterone receptors. No known counterpart of this lesion has ever been described in humans; therefore, its significance is questionable.

### Mouse Hemangioma/Hemangiosarcoma

This vascular neoplasia, usually multifocal, affects many organs almost equally in male and female CD-1 mice. The range of incidence is 6 to 33% in males and 7 to 32% in females [111]. Organs involved include spleen (12%), liver (9%), lymph nodes (4%), testes (4%), skin (3%), and pancreas (2%) in males and uterus (9%), spleen (6%), liver (5%), ovary (5%), lymph nodes (4%), and heart (2%) in females [111]. In humans, no comparable lesion has been reported, and no inducing agent has been established [322].

### Mouse Histiocytic Sarcoma

This neoplasm of histiocytes found in older mice affects mainly the liver and uterus. No agent established to produce an increase in only this neoplasm is associated with cancer development in humans [88,322].

## Mouse Ovary Tubular Adenoma

This is a benign neoplasm with tubular, stromal, or mixed components that occurs mainly in mice [5,36]. Tubulostromal adenomas have been observed with cytotoxic agents but are not seen in other laboratory animals or humans and are considered questionable for human hazard assessment [5,36].

## Rat Clitoral Gland Neoplasm

These rare neoplasms occur sporadically and if found in a treatment group can be a source of concern. No chemical has been shown to reproducibly induce these, and no underlying mechanism has been identified.

## Rat Granular Cell Neoplasm

A proliferative lesion of "granular" cells with granular eosinophilic cytoplasms occurs in the vaginal–cervical regions of female Sprague–Dawley, Donryu, and Wistar rats [55,288]. This lesion is probably under hormonal influence, mainly estrogen. Granular cell aggregates occur rarely in the vulva of women, but no evidence suggests that the pathogenesis is similar to that of rat granular cell tumors. Thus, this lesion is considered probably not relevant for humans.

## Rat Hibernoma

Neoplasms of brown adipose tissue (BAT) occur in the mediastinum or periadrenal area of rats. Recently, these have been reported in a flat low incidence pattern, mainly in male rats, with the administration of several agents. Chemically, these agents include some thiazolidinediones (TZDs; antihyperglycemics), imidazopyridine hypnotics [266], phentolamine (PHEN; an alpha-antagonist) [149,268], and triazolone herbicides. Nevertheless, to date, no chemical has been conclusively shown to induce BAT preneoplastic lesions or early (small *in situ*) neoplasms. Some TZDs, which have conclusively induced dose-related BAT hyperplasia in both rats and mice, do not stimulate progression of this chronic hyperplasia into preneoplasia and neoplasia after 2 years of continuous exposure [137]. Moreover, in a study with chronic (12 months) exposure to PHEN, focusing on the structure, function, and pathology of BAT in younger (6 weeks of age) and older (36 weeks of age) rats, no increased BAT hyperplasia, decreased BAT apoptosis, or even preneoplastic and neoplastic lesions were present [149]. In addition, PHEN doses after 12 months of exposure increased rectal temperature, decreased epinephrine (EPIN) and norepinephrine (NEPI) levels, and did not induce BAT hyperplasia. Temperature, EPIN, and NEPI changes were more pronounced in younger rats but were not present after 4 weeks of recovery. Hypothermia and sustained increases in EPIN and NEPI are considered important conditions for induction of BAT hyperplasia. BAT neoplasms are very rare benign neoplasms in both rodents and humans [149]; however, this tissue has been excluded from the list of tissues routinely sampled in RCBs, so BAT neoplasm data are present only when, at the end of the 2 years, a BAT neoplasm is evident. To date, malignant hibernomas have never been conclusively described in humans [79,185,360]. This is further supported by the TZD [137] results in rodents; consequently, these tumors are of questionable significance to humans.

## Rat Mammary Gland Fibroadenoma

This is a benign neoplasm with a minor glandular epithelial component and a predominant pericanalicular type of proliferation of connective tissue. It bears no resemblance to the common intracanalicular type of fibroadenoma seen in women [281,282], which is hormonally responsive [13]. Fibroadenoma is the most common breast neoplasm in all the major rat strains (Table 25.7 and 25.13) and does not progress to malignancy; thus, combining fibroadenomas and carcinomas is inappropriate, and fibroadenomas by themselves are of questionable relevance.

## Rat Mesovarial Leiomyoma

Smooth muscle tumors of the ovarian suspensory ligament have developed in female rats after long exposures to $\beta_2$-adrenoceptor stimulant medicines [183]. This neoplasm is rare in humans and the agents that have induced it in rats (soterenol, mesuprine, zinterol, terbutaline, reproterol, and salbutamol) are not associated with cancer in humans.

## Rat Mononuclear Cell Leukemia

Mononuclear cell leukemia (MCL) occurs in high incidence (62% in males and 42% in females) in F344 rats [133]. MCL, also referred to as large granular lymphocyte leukemia (LGL), is a spontaneously occurring lethal neoplasm that first develops in the spleen and then in the liver, lungs, lymph nodes, and bone marrow. It occurs at over 18 months of age. No agent has been demonstrated to reproducibly induce this neoplasm, so nothing is known about its pathogenesis. A number of chemicals (furan, C.I. Direct Blue 15, iodinated glycerol, diisononyl phthalate, and dimethyl-morpholinophosphoramidate) are known to be associated with increased incidences of MCL [77,207]. These are not implicated in human cancer.

## Rat Scrotal Vaginal Tunic Mesothelioma

This mesenchymal lesion, which includes hyperplasia and neoplasia [220], arises from the serous membranes of the scrotal tunica vaginalis testes. It is common (about 3%) in F344 rats [133]. Because the scrotal lesion often arise in association with testicular neoplasms, especially the commonly occurring Leydig cell neoplasms, which assume a large size, an element of physical initiation may be involved in the pathogenesis. Chemicals that produce increases in this tumor (e.g., acrylamide, potassium bromate, pentachlorophenol) are chemically diverse, and no mode of action has been established. None of these is associated with mesothelioma in humans or any other cancer.

### Rodent Forestomach Squamous Cell Carcinoma

The rodent forestomach is a portion of the stomach between the esophagus and glandular stomach that is lined by squamous epithelium and does not exist in humans. A number of DNA-reactive agents have induced neoplasms of the forestomach in rodents [197], usually through a direct effect. Nongenotoxic agents such as butylated hydroxyanisole, propionic acid, and HMG CoA-reductase inhibitors also have produced increases in this neoplasm. The epigenetic mechanism appears to involve chronic irritation leading to a promoting action, which requires high exposure, as shown for butylated hydroxyanisole [161,382,384]. None of these epigenetic agents has been associated with cancer in humans.

## Types of Cancer Hazards

Formerly, all rodent carcinogens were considered to be potential human cancer hazards. This concept was embodied in the Delaney Clause to the 1958 Federal Food, Drug, and Cosmetic Act (FFDCA), which provided that no chemical determined to be carcinogenic in animals could be allowed as a food or color additive, regardless of concentration. Subsequently, expanded understanding of the mechanisms of carcinogenesis has led to refinements of hazard assessment. Beginning in 1992, the IARC accepted data on mechanisms as being relevant to evaluation of the carcinogenic risk of an agent to humans [155], and this has been further developed [161,277]. A critical aspect is the distinction between DNA-reactive and epigenetic carcinogens (see Types of Carcinogens section). This has been applied by the EPA in concluding that the rat kidney neoplasm response to $\alpha_{2\mu}$-globulin nephropathy inducers is not relevant to humans [331]. The potential hazards of a variety of DNA-reactive and epigenetic carcinogens has been explored in detail by an international group of experts, drawing on comprehensive reviews of modes of action of ten prototype carcinogens both epigenetic and DNA reactive [166].

Currently, when assessing potential human cancer hazard, regulatory agencies often refer to findings implicating an agent as a genotoxic carcinogen, usually with only an operational definition [48]—that the chemical produced positive results in some short-term assays [188]. All chemicals that are reliably positive in a variety of short-term assays, especially including a DNA damage assay, are in fact DNA reactive and thus belong to that category of carcinogen. Many chemicals, however, are positive in only some assays (e.g., phenobarbital and DES [153,160]), sometimes because of intrinsic spurious positive results [30]; thus, the issue becomes a question of which assay results are to be accepted as evidence of genotoxicity. A scientifically sound approach is to define genotoxicity as a mechanism of carcinogenesis; that is, a genotoxic carcinogen is one that forms molecular lesions (such as DNA adducts) that lead to mutations in the cells that are the precursors of neoplasms induced by the agent. Under this definition, genotoxic carcinogens would likely be confined to DNA-reactive agents fulfilling the criterion for carcinogenic activity of an agent that induces malignant neoplasms not seen in controls. Such chemicals are generally multispecies carcinogens, inducing a high yield of neoplasms with short latent periods and often in several organs. For such carcinogens, assumption of human hazard is well founded [364], although evidence is accruing that even DNA-reactive carcinogens have thresholds at low levels of administration [383,386]. Moreover, some of the underlying mechanisms of these agents are species specific and do not operate in humans because of differences either in disposition (such as tamoxifen, which is poorly bioactivated and readily detoxified in humans) or in toxicodynamics [21,74]. Epigenetic carcinogens, in contrast, generally do not present human hazards, as evidenced by the few associated with human cancer (Table 25.3), in spite of the large number of such agents to which humans are regularly exposed [159,371]. The lack of relevance stems from the fact that their effects are either rodent specific, as with the $\alpha_{2\mu}$-globulin nephropathy inducers, or require high and long duration exposure in rodents to elicit the cellular effect leading to carcinogenicity [362,363]. Epigenetic mechanisms that either are not indicative of a cancer hazard to humans or are probably not indicative have been discussed earlier. These considerations should be taken into account in formulating hazard assessment.

## Cancer Risk Assessment

In the United States, under the provision of the Delaney Clause to the 1958 FFDCA, a zero tolerance in food was established for chemicals determined to be carcinogenic in either humans or animals. This applies to residues of veterinary drugs used in food-producing animals [30,347]. Likewise, the EPA Cancer Principles of 1970 stated that no level of exposure to a chemical carcinogen should be considered toxicologically insignificant for humans. These are all risk-management approaches. Notably, in Europe, governmental agencies have not been required by legislatures to impose standards of no exposure for carcinogenic agents and have used more flexible approaches than those imposed in the United States [380], although often this entails restricting external contact to that as low as reasonably achievable (ALARA), which again is a risk management procedure.

The prototype chemical safety assessment, introduced by Lehman and Fitzhugh [205] of the FDA, is the acceptable daily intake (ADI), calculated as the no-observed-adverse-effect level (NOAEL) divided by uncertainty factors (UFs), typically 100 to 1000. It is becoming increasingly accepted that for epigenetic carcinogens, even if a potential cancer hazard to humans is presumed, such a safety margin can be deployed [299,336]. For pharmaceuticals with animal neoplasm findings, the FDA/CDER calculates a safety factor that is the ratio of the AUC in rodents for the highest noncarcinogenic dose (nontumorigenic effect level, or NTEL) to the maximum AUC at the therapeutic dose [15,48], allowing for individuals who are poor metabolizers.

In 1996, the FFDCA was amended by removing the zero-risk provision of the Delaney Clause and replacing it with a new standard of "a reasonable certainty of no harm" [49]. The new standard applies to pesticide residues in both raw and processed foods, allowing the presence of some residues that have been shown to cause cancer in humans or animals [325,326]. In the new EPA draft cancer assessment guidelines, the safety margin is referred to as a margin of exposure using the NTEL [336]. In the European Union, it has been proposed to incorporate potency considerations, including $TD_{25}$ values and to classify carcinogens into three potency groups [73]. The FDA CFSAN has published a "threshold of regulation" procedure for indirect food additives [343]. Based on a large database of chemical carcinogens, it was determined that an external dose of 1.5 µg/person/day is unlikely to represent a cancer risk. This concept has been developed into a threshold of toxicologic concern (TTC) [198]. The Joint World Health Organization (WHO)/Food and Agriculture Organization (FAO) Joint Expert Committee on Food Additives (JECFA) reviews safety studies on food additives and contaminants and the Joint FAO/WHO Meeting on Pesticide Residues (JMPR) reviews those on pesticides to set global exposure limits. These groups develop ADIs and tolerable daily intakes (TDIs), respectively.

Another approach proposes a carcinogen safe exposure level (SEL) as the no-effect level for the molecular or cellular effect that is the basis for carcinogenicity divided by a safety factor similar to UFs [367,371]. This actually is more conservative than using the NTEL because the NEL for molecular and cellular effects leading to cancer is lower than the dose required to elicit cancer, as illustrated in reviews of MOAs for phenobarbital [363] and butylated hydroxyanisole [362].

For genotoxic or DNA-reactive carcinogens, authorities regulate such agents either by prohibiting human exposure (risk management) or by using a linear no threshold model for quantitative risk assessment [329,334,387]. Health Canada in its Human Health Risk Assessment for Priority Substances under the Canadian Environmental Protection Act applies an exposure potency index (EPI) calculated as an average exposure in the population divided by the dose in experimental animals that produces a 5% incidence of neoplasms. Currently, the concept of TTC is being considered for contaminants of medicines [179] and food substances [198]. This recognizes the accruing evidence that even DNA-reactive carcinogens have thresholds [383,386]. The SEL concept described earlier can also be applied to DNA-reactive carcinogens using a NEL for DNA binding [371].

## INTERACTIVE CARCINOGENESIS

Interactive carcinogenesis comprises the enhancement or inhibition of carcinogenesis by combined or sequential exposures to more than one carcinogen or to a carcinogen and a noncarcinogen [366]. Various types of interaction can occur between chemicals [34], including syncarcinogenesis, neoplasm promotion, cocarcinogenesis, and anticarcinogenesis, and between chemical and physical agents, as with photochemical carcinogenesis. Studies of interactive effects for regulatory purposes are being increasingly applied. Also, in the testing of complex mixtures, these types of interactions can influence the outcome and must be taken into consideration.

### SYNCARCINOGENESIS

Syncarcinogenesis is the enhancement of carcinogenesis produced by concurrent or sequential administration of two carcinogens, usually of the DNA-reactive type. This interaction in the case of DNA-reactive carcinogens represents a summation of the genetic effects of each of the agents. Generally, the enhancement occurs only in a target organ where both carcinogens produce a neoplastic effect [372,378].

### PROMOTION

Neoplasm promotion is the enhancement of neoplastic development by a second epigenetic agent given after an initiating, typically DNA-reactive, carcinogen [14], when a sufficient interval has been allowed for acute molecular effects of the initiating carcinogen (e.g., DNA adducts) to be processed. If the second agent is administered when molecular lesions are still present, the enhancement may be due to cocarcinogenesis (see below). Promoting agents essentially facilitate clonal expansion of initiated preneoplastic cells and their evolution into neoplasms. The growth advantage of preneoplastic populations can be achieved either by an enhanced rate of cell proliferation or by a decreased rate of apoptosis in either the tissue or in incipient neoplasms. Promoting agents are usually assumed to be noncarcinogens, but in fact most are weak carcinogens under some circumstances, probably because they facilitate neoplasm development from cryptogenically initiated cells that are the source of background neoplasms. Thus, most, if not all, promoters are epigenetic carcinogens (Table 25.2), albeit weak. An essential characteristic of a promoting agent is that it is not DNA reactive and is not an initiating agent; otherwise, its enhancement of the effect of an initiating agent is likely to be due to syncarcinogenesis. For experimental demonstration that promotion is the mode of action of a test substance that has been shown to induce neoplasms, the exclusion of both genotoxicity and initiating activity is paramount.

### Cocarcinogenesis

Cocarcinogenesis is the enhancement by a noncarcinogen of the carcinogenicity of a carcinogen when the chemical is administered prior to or concurrently with the carcinogen or when given shortly after a carcinogen at a time when molecular damage is still present. Cocarcinogens can enhance the uptake of the carcinogen, enhance its tissue localization, increase the proportion that is bioactivated, or enhance the induced neoplastic transformation, usually by transiently increasing cell proliferation. Cocarcinogens do not act as promoters, although most promoters have cocarcinogenic activity, often due to enhancement of cell proliferation, when administered concurrently with the initiating agent.

### Anticarcinogenesis

Anticarcinogenesis is the reduction of occurrence of neoplasia and may involve inhibition of the carcinogenicity of a concurrently or previously administered agent. Anticarcinogens are typically noncarcinogenic, although certain epigenetic carcinogens, such as phenobarbital [47] and butylated hydroxyanisole [385], are effective anticarcinogens. Three operational pathways that have been recognized are preventing the formation of carcinogens, blocking the effects of carcinogens, and suppressing neoplastic development [355].

### Photochemical Carcinogenesis

A specific type of interactive carcinogenesis is photochemical carcinogenesis, which is the combined skin carcinogenicity of a chemical and ultraviolet light radiation (UVR) [98,156]. Photochemical carcinogenicity can result from several types of interactions among the chemical, the UVR, and the skin. Some chemicals, such as psoralens, can be photoactivated to DNA-reactive chemical species. Others, such as fluoroquinolones, can undergo photoactivation to generate secondary reactive molecular species such as reactive oxygen [375]. Also some chemicals can affect the structure of skin, causing, for example, thinning of the epidermis in the case of retinoids which results in sensitization of the skin to effects of UVR. Finally, immunosuppression can enhance skin carcinogenesis [196]. Any photochemical carcinogenicity study should be preceded by or accompanied by measurement of proliferation to monitor for nonspecific enhancement that can lead to enhancement of skin carcinogenesis.

Photochemical carcinogenicity studies can be required for topically applied medicines and even for some oral medicines, as well as for topically applied cosmetics and consumer products [341,346]. The test species is usually the SKH1 albino hairless mouse, which has the advantage that it does not require hair clipping and allows easy detection of UVR-induced squamous cell papillomas and carcinomas. In a typical protocol [64,97,284], the TS is applied before UVR (at 290 to 400nm by a UV solar simulator) on Monday, Wednesday, and Friday and after UVR on Tuesday and Thursday for 40 weeks, followed by a 12-week observation period without exposure. This pattern of exposure allows detection of photoactivated chemicals, as well as those that may modulate photocarcinogenesis. Typically, separate groups are administered low and high UVR in addition to the TS. The endpoints of evaluation include neoplasm incidence (prevalence), multiplicity (yield), and latency (time to tumor).

## CONCLUDING REMARKS

It has been evident for 25 years or more that the carcinogenicity of a chemical could be predicted from data other than an RCB [357]. Nevertheless, for a variety of reasons, the RCB remains an essential component in the safety assessment of chemicals. The RCB appears to be highly effective in as much as, to our knowledge, no chemical that was negative in properly conducted RCBs has subsequently proven to be carcinogenic in humans. The basic methodology of the RCB has changed little since its inception, but both the analysis and the interpretation are evolving. Various types of neoplastic responses in rodents are now recognized as not being relevant to human safety. Moreover, responses considered to be relevant are increasingly accepted as having thresholds; hence, safe levels of contact can be established.

## ACKNOWLEDGMENTS

We thank Barbara Iatropoulos and Klaus Brunnemann for preparation of the manuscript. Also, we gratefully acknowledge the thoughtful comments by many colleagues in the chemical and pharmaceutical industries. The preparation of this chapter was made possible by support provided to the Medicine, Food, and Chemical Safety Program at New York Medical College.

## QUESTIONS

1. What are the pertinent mechanisms of carcinogenesis operational in rodents and humans?
2. What are the mechanisms of carcinogenesis operational only for rodents? Explain why.
3. Which neoplasms are of questionable significance to human cancer hazards? Explain why.
4. What are the types of cancer hazards?
5. What constitutes adequate exposure in a rodent cancer bioassay?

## REFERENCES


1. Aardema, M. J. and MacGregor, J. T. (2002): Toxicology and genetic toxicology in the new era of 'toxicogenomics': impact of '-omics' technologies. *Mutat. Res.*, 499: 13–25.
2. Abdo, K. M. and Kari, F. W. (1996): The sensitivity of the NTP bioassay for carcinogen hazard evaluation can be modulated by dietary restriction. *Exp. Toxicol. Pathol.*, 48: 129–137.
3. Allaben, W. T., Turturro, A., Leakey, J. E. A., Seng, J. E., and Hart, R. W. (1996): FDA points-to-consider documents: the need for dietary control for the reduction of experimental variability within animal assays and the use of dietary restriction to achieve dietary control. *Toxicol. Pathol.*, 24:776–781.
4. Alexandrov, V. A., Popovich, I. G., Anisimov, V. N., and Napalkov, N. P. (1989): Influence of hormonal disturbances on transplacental and multigeneration carcinogenesis in rats. In: *Perinatal and Multigeneration Carcinogenesis*, edited by N. P. Napalkov, J. M. Rice, L. Tomatis, and H. Yamasaki, IARC Publ. No. 96, pp. 35–49. Lyon, France.
5. Alison, R. H. and Morgan, K. T. (1987): Ovarian neoplasms in F344 rats and B6C3F$_1$ mice. *Environ. Health Perspect.*, 73:91–106.
6. Alison, R. H., Capen, C. C., and Prentice, D. E. (1994): Neoplastic lesions of questionable significance to humans. *Toxicol. Pathol.*, 22:179–186.
7. Allen, D. G., Pearse, G., Haseman, J. K., and Maronpot, R. R. (2004): Prediction of rodent carcinogenesis: an evaluation of prechronic liver lesions as forecasters of liver tumors in NTP carcinogenicity studies. *Toxicol. Pathol.*, 32:393–401.
8. Jemal, A., Siegel, R., Ward, E., Murray, T., Xu, J., and Thun, M., Eds. (2005): *Cancer Statistics 2007*, pp. 43–66. American Cancer Society, Atlanta, GA.
9. Apostolou, A. (1990): Relevance of maximum tolerated dose to human carcinogenic risk. *Regul. Toxicol. Pharmacol.*, 11:68–80.
10. Ashby, J. and Tennant, R. W. (1991): Definitive relationships among chemical structure, carcinogenicity and mutagenicity for 301 chemicals tested by the U.S. NTP. *Mutat. Res.*, 257:229–306.
11. Baldrick, P. (2005): Carcinogenicity evaluation: comparison of tumor data from dual control groups in the Sprague–Dawley rat. *Toxicol. Pathol.*, 33:283–291.
12. Ballard, B. E. (1968): Biopharmaceutical consideration in subcutaneous and intramuscular drug administration. *J. Pharmacol. Sci.*, 57:357–378.
13. Bartow, S. A. (1994): The breast. In: *Pathology*, edited by E. Rubin and J. L. Farber, pp. 73–992. J.B. Lippincott, Philadelphia, PA.
14. Berenblum, I. (1974): *Frontiers of Biology: Carcinogenesis as a Biological Problem*, edited by A. Neuberger and E. L. Tatum, North Holland, Amsterdam.
15. Bergman, K., Olofsson, I.-M., and Sjoeberg, P. (1998): Dose selection for carcinogenicity studies of pharmaceuticals: systemic exposure to phenacetin at carcinogenic dosage in the rat. *Regul. Toxicol. Pharmacol.*, 28: 226–229.
16. Bergeson, L. L. and Auerbach, B. (2004): *The Environmental Regulatory Implications of Nanotechnology*, Daily Environmental Report No. 71. Bureau of National Affairs, Washington, D.C. (http://www.lawbc.com/other_pdfs).
17. Bischoff, K. B. (1975): Some fundamental considerations of the applications of pharmacokinetics to cancer chemotherapy. *Cancer Chemother. Rep.*, 59:777–793.
18. Blackburn, K., Stickney, J. A., Carlson-Lynch, H. L. McGinnis, P. M., Chappell, L., and Felter, S. P. (2005): Application of the threshold of toxicological concern approach to ingredients in personal and household care products. *Regul. Toxicol. Pharmacol.*, 43:249–259.
19. Blanchard, K. T., Barthel, C., French, J. E., Holden, H. E. Moretz, R., Pack, F. D., Tennant, R. W., and Stoll, R. E (1999): Transponder-induced sarcoma in the heterozygous p53$^{+/-}$ mouse. *Toxicol. Pathol.*, 27(5):519–527.
20. Bomhard, E. and Rinke, M. (1994): Frequency of spontaneous tumours in Wistar rats in 2-year studies. *Exp. Toxicol. Pathol.*, 46:17–29.
21. Boocock, D. J., Maggs, J. L., Brown, K., White, I. N. and Park, B. K. (2000): Major inter-species differences in the rates of *O*-sulphonation and *O*-glucuronylation of alpha-hydroxytamoxifen *in vitro*: a metabolic disparity protecting human liver from the formation of tamoxifen DNA adducts. *Carcinogenesis*, 21(10):1851–1858.
22. Boorman, G. A., Montgomery, Jr., C. A., Eustis, S. L. Wolfe, M. J. McConnell, E. E., and Hardisty, J. F. (1985): Quality assurance in pathology for rodent carcinogenicity studies. In: *Handbook of Carcinogen Testing*, edited by H Milman and E. Weisburger, pp. 345–357, Noyes, Park Ridge, NJ.
23. Boorman, G. A., Maronpot, R. R., and Eustis, S. L. (1994): Rodent carcinogenicity bioassay: past, present and future *Toxicol. Pathol.*, 22:105–111.
24. Boorman, G. A., Haseman, J. K., Waters, M. D., Hardisty J. F., and Sills, R. C. (2002): Quality review procedures necessary for rodent pathology databases and toxicogenomic studies: the National Toxicology Program experience. *Toxicol. Pathol.* 30(1):88–92.
25. Borowitz, J. L., Moore, P. F., Yim, G. K. W., and Miya, T. S. (1971): Mechanism of enhanced drug effects produced by dilution of the oral dose. *Toxicol. Appl. Pharmacol* 19:164–168.
26. Brendler-Schwaab, S., Hartmann, A., Pfuhler, S., and Speit, G. (2005): The *in vivo* comet assay: use and status in genotoxicity testing. *Mutagenesis*, 20(4):245–254.

27. Brix, A. E., Nyska, A., Haseman, J. K., Sells, D. M., Jokinen, M. P., and Walker, N. J. (2004): Incidences of selected lesions in control female Harlan Sprague–Dawley rats from two-year studies performed by the National Toxicology Program. *Toxicol. Pathol.*, 33:477–483.
28. Brunnemann, K. D., Enzmann, H. G., Perrone, C. E., Iatropoulos, M. J., and Williams, G. M. (2002): *In ovo* carcinogenicity assay (IOCA): evaluation of mannitol, caprolactam and nitrosoproline. *Arch. Toxicol.*, 76:606–612.
29. Brunner, G. H. G., Lamberts, R., and Creutzfeldt, W. (1990): Efficacy and safety of omeprazole in the long-term treatment of peptic ulcer and reflux oesophagitis resistant to ranitidine. *Digestion*, 47:64–68.
30. Brynes, S. D. (2005): Demystification of 21 CRF Part 556: tolerances for residues of new animal drugs in food. *Regulatory Toxicol. Pharmacol.*, 42:324–327.
31. Butler, W. H., Cohen, S. H., and Squire, R. A. (1997): Mesenchymal tumors of the mouse urinary bladder with vascular and smooth muscle differentiation. *Toxicol. Pathol.*, 25:268–274.
32. Butterworth, B. E., Ashby, J., Bermudez, E., Casciano, D., Mirsalis, J., Probst, G., and Williams, G. (1987): A protocol and guide for the *in vivo* rat hepatocyte DNA repair assay. *Mutat. Res.*, 189:123–133.
33. Butterworth, B. E. and Goldsworthy, T. L. (1991): The role of cell proliferation in multistage carcinogenesis. *Proc. Soc. Exp. Biol. Med.*, 198:683–687.
34. Calabrese, E. J. (1991): *Multiple Chemical Interactions*, Lewis, Chelsea, MI.
35. Caldwell, J., Gardner, I., and Swales, N. (1995): An introduction to drug disposition: the basic principles of absorption, distribution, metabolism and excretion. *Toxicol. Pathol.*, 23:148–157.
36. Capen, C. C., Beamer, W. G., Tennant, B. J., and Stitzel, K. A. (1995): Mechanisms of hormone-mediated carcinogenesis of the ovary in mice. *Mutat. Res.*, 333:143–151.
37. Cariello, N. F., Romach, E. H., Colton, H. M., Ni, H., Yoon, L. et al. (2005): Gene expression profiling of the PPAR-alpha agonist ciprofibrate in the cynomolgus monkey liver. *Toxicol. Sci.*, 88(1):250–264.
38. Cayen, M. N. (1995): Considerations in the design of toxicokinetic programs. *Toxicol. Pathol.*, 23:148–157.
39. Cayen, M. N. and Black, H. E. (1993): Role of toxicokinetics in dose selection for carcinogenicity studies. In: *Drug Toxicokinetics*, edited by P. G. Welling and F. A. de la Iglesia, pp. 69–83. Marcel Dekker, New York.
40. Chappell, W. R. and Mordenti, J. (1991): Extrapolation of toxicological and pharmacological data from animals to humans. *Adv. Drug Res.*, 20:2–116.
41. Chengelis, C. P., Gad, S. C., and Holston, J. (1995): *Regulatory Toxicology*, Raven Press, New York.
42. Christian, M. S., Hoberman, A. M., Johnson, MD., Brown, W. R., and Bucci, T. J. (1998): Effect of dietary optimization on growth, survival, tumor incidences and clinical pathology parameters in CD Sprague–Dawley and Fischer-344 rats: a 104-week study. *Drug Chem. Toxicol.*, 21: 97–117.
43. Clayson, D. B. and Kitchin, K. T. (1999): Interspecies differences in response to chemical carcinogens. In: *Carcinogenicity*, edited by K. T. Kitchin, pp. 837–880. Marcel Dekker, New York.
44. Cohen, S. M. (1998): Urinary bladder carcinogenesis. *Toxicol. Pathol.*, 26:121–127.
45. Cohen, S. M. and Ellwein, L. B. (1991): Genetic errors, cell proliferation, and carcinogenesis. *Cancer Res.*, 51: 6493–6505.
46. Cohen, S. M., Klaunig, J., Meek, M. F., Hill, R. N., Pastoor, T. et al. (2004): Evaluating the human relevance of chemically induced animal tumors. *Toxicol. Sci.*, 78:181–186.
47. Conney, A. H. (2003); Enzyme induction and dietary chemicals as approaches to cancer chemoprevention: the seventh DeWitt S. Goodman lecture. *Cancer Res.*, 63(21):7005–7031.
48. Contrera, J. F., Jacobs, A. C., Prasanna, H. R., Mehta, M., Schmidt, W. J., and DeGeorge, J. J. (1995): A systemic exposure-based alternative to the maximum tolerated dose for carcinogenicity studies of human therapeutics. *J. Am. Coll. Toxicol.*, 14:1–10.
49. Contrera, J. F., Jacobs, A. C., DeGeorge, J. J., Chen, C., Choudary, J. et al. (1996): *Carcinogenicity Testing and the Evaluation of Regulatory Requirements for Pharmaceuticals*, Docket No. 96D-0235. U.S. Department of Health and Human Services, Public Health Service, Washington, D.C.
50. Contrera, J. F., Jacobs, A. C., and DeGeorge, J. J. (1997): Carcinogenicity testing and the evaluation of regulatory requirements for pharmaceuticals. *Regul. Toxicol. Pharmacol.*, 25:130–145.
51. Contrera, J. F., Matthews, E. J. and Benz, R. D. (2003): Predicting the carcinogenic potential of pharmaceuticals in rodents using molecular structural similarity and E-state indices. *Regul. Toxicol. Pharmacol.*, 38:243–259.
52. Cook, J. C., Klinefelter, G. R., Hardisty, J. F., Sharpe, R. M., and Foster, P. M. D. (1999): Rodent Leydig cell tumorigenesis: a review of the physiology, pathology, mechanisms, and relevance to humans. *Crit. Rev. Toxicol.*, 29: 169–261.
53. Cottrell, S., Andrews, C. M., Clayton, D., and Powell, C. J. (1994): The dose-dependent effect of BHT (butylated hydroxytoluene) on vitamin K-dependent blood coagulation in rats. *Food Chem. Toxicol.*, 32(7):589–594.
54. Counts, J. L. and Goodman, J. I. (1995): Principles underlying dose selection for, and extrapolation from, the carcinogen bioassay: dose influences mechanism. *Regul. Toxicol. Pharmacol.*, 21:418–421.
55. Courtney, C. L., Hawkins, K. L., Meierhenry, E. F., and Graziano, M. J. (1992): Immunohistochemical and ultrastructural characterization of granular cell tumors of the female reproductive tract in two aged Wistar rats. *Vet. Pathol.*, 29:86–89
56. Cox, D. R. (1972): Regression models and life tables. *J. R. Stat. Soc.*, 13:187–220.
57. Crissman, J. W., Goodman, D. G., Hildebrandt, P. K., Maronpot, R. R., Prater, D. A., Riley, J. H., Seaman, W. J., and Thake, D. C. (2004): Best practices guideline: toxicologic histopathology. *Toxicol. Pathol.*, 32:126–131.
58. Cutler, S. J. and Ederer, F. (1958): Maximum utilization of life table method in analyzing survival. *J. Chron. Dis.*, 8:699–712.
59. D'Arcy, P. F. and Harron, D. W. G. (1996): *Proceedings, Third International Conference on Harmonization*, Yokohama, Japan, 1995.

60. Davies, T. S. and Monro, A. (1994): The rodent carcinogenicity bioassay produces a similar frequency of tumor increases and decreases: implications for risk assessment. *Regul. Toxicol. Pharmacol.*, 20:281–301.
61. Davies, T. S. and Monro, A. (1995): Marketed human pharmaceuticals reported to be tumorigenic in rodents. *J. Am. Coll. of Toxicol.*, 14:90–107.
62. Dedrick, R. L. (1986): Interspecies scaling of regional drug delivery. *J. Pharmaceut. Sci.*, 175:1047–1052.
63. DeGeorge, J. J. and Contrera, J. F. (1996): A regulatory perspective of the guidance on the utility of two rodent species. In: *Proc. of the Third Int. Conf. on Harmonization, Yokohama, Japan, 1995*, edited by P. F. D'Arcy and D. W. G. Harron, pp. 274–277. Greystone Books, Antrim, North Ireland.
64. De Gruijl, F. R. and Forbes, P. D. (1995): UV-induced skin cancer in a hairless mouse model. *BioEssays*, 17:651–660.
65. Deschl, U., Kittel, B., Rittinghausen, S., Morawietz, G., Kohler, M., Mohr, U., and Keenan, C. (2002): The value of historical control data: scientific advantages for pathologists, industry and agencies. *Toxicol. Pathol.*, 30(1): 80–87.
66. De Vries, A., von Oostrom, C. Th. M., Dortant, P. M., Beems, R. B., van Kriejl, C. F., Capel, P. J. A., and van Steeg, H. (1997): Spontaneous liver tumours and benzo(*a*)pyrene-induced lymphomas in XPA-deficient mice. *Mol. Carcinog.*, 19:46–53.
67. Dickinson, D. A., Warnes, G. R., Quievzyn, G., Messer, J., Zhitkovich, A., Rubitski, E., and Aubrecht, J. (2004): Differentiation of DNA reactive and non-reactive genotoxic mechanisms using gene expression profile analysis. *Mutat. Res.*, 549:29–41.
68. Dideriksen, L. H., Jorgensen, L. N., and Drejer, K. (1992): Carcinogenic effect on female rats after 12 month administration of the insulin analogue B10Asp. *Diabetes*, 41(Suppl. 1):143A (abstract no. 507).
69. Doi, A. M., Hailey, J. R., Hejtmancik, M., Toft, I. J. D., Vallant, M., and Chhabra, R. S. (2005): Topical application of representative multifunctional acrylates produced proliferative and inflammatory lesions in F344/N rats and B6C3F(1) mice, and squamous cell neoplasm in Tg.AC mice. *Toxicol. Pathol.*, 33(6):631–640.
70. Donehower, I., Harvey, M., Slagle, B. L., McArthur, M. J., Montgomery, Jr., C. A., Butel, J. S., and Bradley, A. (1992): Mice deficient for p53 are developmentally normal but susceptible to spontaneous tumors. *Nature*, 356: 215–221.
71. Dragan, Y. P., Rizvi, T., Xu, Y.-H., Hully, J. R., Bawa, N., Campbell, H. A., Maronpot, R. R., and Pitot, H. C. (1991): An initiation-promotion assay in rat liver as a potential complement to the 2-year carcinogenesis bioassay. *Fundam. Appl. Toxicol.*, 16:525–547.
72. Dunnett, C. W. (1955): A multiple comparison procedure for comparing several treatments with a control. *J. Am. Stat. Assoc.*, 50:1096–1122.
73. Dybing, E., Sanner, T., Roelfzema, H., Kroese, D., and Tennant, R. W. (1997): T25: A simplified carcinogenic potency index: description of the system and study of correlations between carcinogenic potency and species/site specificity and mutagenicity. *Pharmacol. Toxicol.*, 30:272–279.
74. Edwards, R. J., Murray, B. P., Murray, S., Schulz, T., Neubert, D., Gant, T. W., Thorgeirsson, S. S., Boobis, A. R., and Davis, D. S. (1994): Contribution of CYP1A1 and CYP1A2 to the activation of heterocyclic amines in monkeys and humans. *Carcinogenesis*, 15:829–836.
75. Eiben, R. and Bomhard, E. M. (1999): Trends in mortality, body weights and tumor incidences of Wistar rats over 20 years. *Exp. Toxicol. Pathol.*, 51:523–536.
76. Ellwein, L. G. and Cohen, S. (1990): The health risk of saccharin revisited. *CRC Crit. Rev. Toxicol.*, 20:311–326.
77. Elwell, M. R., Dunnick, J. K., Hailey, J. R., and Haseman, J. K. (1996): Chemicals associated with decreases in the incidence of mononuclear cell leukemia in the Fischer rat. *Toxicol. Pathol.*, 24:238–245.
78. Emerson, C. H., Cohen III, J. H., Young, R. A., Alex, S., and Fan, S.-L. (1990): Gender-related differences of serum thyroxine-binding proteins in the rat. *Acta Endocrinol.*, 123:72–78.
79. Enterline, H. T., Lowry, L. D., and Richman, A. V. (1979): Does malignant hibernoma exist? *Am. J. Surg. Pathol.*, 3:265–271.
80. Enzmann, H., Bomhard, E., Iatropoulos, M. J., Ahr, H. J., Schlueter, G., and Williams, G. M. (1998): Short- and intermediate-term carcinogenicity testing: a review. Part 1. The prototypes mouse skin tumour assay and rat liver focus. *Food Chem. Toxicol.*, 36:979–995.
81. Enzmann, H., Iatropoulos, M. J., Brunnemann, K. D. Bomhard, E., Ahr, H. J., Schlueter, G., and Williams, G M. (1998): Short- and intermediate-term carcinogenicity testing: a review. Part 2. Available experimental models *Food Chem. Toxicol.*, 36:997–1013.
82. EEC. (1967): Directive 67/548/EEC with Amendments and Adaptions. Annex VI: Criteria for Classification of Carcinogenic Substances. European Economic Communities Brussels, Belgium.
83. EEC. (1983): Note for Guidance Concerning the Application of Chapter I(E) of Part 2 of the Annex to Directive 75/398/EEC, with a View to the Granting of a Marketing Authorization of a New Drug. European Economic Communities, Brussels, Belgium.
84. EEC. (1988): Directive 88/379/EEC with Amendments Annex I: Criteria for Classification of Carcinogenic Substances. European Economic Communities, Brussels, Belgium.
85. EMEA. (2004): *CHMPSWP Conclusions and Recommendations on the Use of Genetically Modified Animal Model for Carcinogenicity Assessment*. European Medicines Agency, London.
86. EU. (2003): Directive 2003/15/EC of the European Parliament and of the Council 2/27/03 Amending Directive 76/768/EEC on the Approximation of the Laws of the Member States Relating to Cosmetic Products. European Union, Brussels, Belgium (http://europa.eu.int/eur-fex/en dat/2003/1_D66/1_06620030311en00260035.pdf).
87. Ewing, L. L. (1989): The trophic effect of luteinizing hormone on the rat Leydig cell. *J. Am. Coll. Toxicol.*, 8 473–485.
88. Faccini, J. M., Abbott, D. P., and Paulus, G. J. J. (1990) *Mouse Histopathology: A Glossary for Use in Toxicity and Carcinogenicity Studies*, Elsevier, Amsterdam.

89. Faccini, J. M., Butler, W. R., Friedmann, J.-C., Hess, R., Reznik, G. K., Ito, N., Hayashi, Y., and Williams, G. M. (1992): IFSTP guidelines for the design and interpretation of the chronic rodent carcinogenicity bioassay. *Exp. Toxicol. Pathol.*, 44:443–456.
90. Fairbairn, D. W., Olive, P. L., and O'Neill, K. L. (1995): The comet assay: a comprehensive review. *Mutat. Res.*, 339:37–59.
91. Feinstein, A. R. (1975): Clinical biostatistics. *Clin. Pharmacol. Ther.*, 17:499–513.
92. Fitzgerald, J. M., Boy, V. F., and Manus, A. G. (1984): Formulation of insoluble and immiscible test agents in liquid vehicles for toxicity testing. In: *Chemistry for Toxicity Testing*, edited by C. W. Jameson and D. B. Walters. Butterworth, Stoneham, MA.
93. Flammang, T. J., von Tungeln, L. S., Kadlubar, F. F., and Fu, P. P. (1997): Neonatal mouse assay for tumorigenicity: alternative to the chronic rodent bioassay. *Regul. Toxicol. Pharmacol.*, 26:230–240.
94. Fleiss, J. L. (1981): *Statistical Methods for Rates and Proportions*, 2nd ed., pp. 145–146. John Wiley & Sons, New York.
95. Fleiss, J. L. (1986): *The Design and Analysis of Clinical Experiments*, pp. 58–59. John Wiley & Sons, New York.
96. Foran, J. A., Editor. (1997): *Principles for the Selection of Doses in Chronic Rodent Bioassays*. ILSI Press, Washington, D.C.
97. Forbes, P. D., Sambuco, C. P., and Davies, R. E. (1993): Photocarcinogenesis safety testing. *J. Am. Coll. Toxicol.*, 12:417–424.
98. Forbes P. D. and Sambuco C. P. (1998): Assays for photocarcinogenesis: relevance of animal models. *Int. J. Toxicol.*, 17:577–588.
99. Frantz, S. W., Beatty, P. W., English, J. C., Hundley, S. G., and Wilson, A. G. E. (1994): The use of pharmacokinetics as an interpretive and predictive tool in chemical toxicology testing and risk assessment: a position paper on the appropriate use of pharmacokinetics in chemical toxicology. *Reg. Toxicol. Pharmacol.*, 19:317–337.
100. Gabriel, K. (1966): Simultaneous test procedures for multiple comparisons on categorical data. *J. Am. Stat. Assoc.*, 61:1081–1096.
101. Gabriel, K. R. (1978): A simple method of multiple comparison of means. *J. Am. Stat. Assoc.*, 73:724–729.
102. Gad, S. C. (1996): Histologic and clinical pathology in the safety assessment and development of new therapeutic agents. *Scand. J. Lab. Anim. Sci.*, 13:325–334.
103. Gad, S. C. and Weil, C. S. (1986): *Statistics and Experimental Design for Toxicologists*, pp. 1–17. Telford Press, Caldwell, NJ.
104. Galer, D. M. and Monro, A. M. (1998): Veterinary drugs no longer need testing for carcinogenicity in rodent bioassays. *Regul. Toxicol. Pharmacol.*, 28:115–123.
105. Galloway, S. M. et al. (1994): Report from working group on *in vitro* tests for chromosomal aberrations. *Mutat. Res.*, 312:241–261.
106. Garby, L., Garrow, J. S., Jorgensen, B., Lammert, O., Madsen, K., Sorensen, P., and Webster, J. (1988): Relationship between energy expenditure and body composition in man: specific energy expenditure *in vivo* of fat and fat-free tissue. *Eur. J. Clin. Nutr.*, 42:301–305.
107. Gart, J. J., Chu, K. C., and Tarone, R. E. (1979): Statistical issues in the interpretation of chronic bioassay tests for carcinogenicity. *J. Natl. Cancer Inst.*, 62:957–974.
108. Gaylor, D. W. and Kodell, R. I. (1999): Dose–response trend tests for tumorigenesis, adjusted for body weight. *Toxicol. Sci.*, 49:318–323.
109. Gehring, P. J., Watanabe, P. G., and Park, C. N. (1978): Resolution of dose response toxicity data for chemicals requiring metabolic activation—example: vinyl chloride. *Toxicol. Appl. Pharmacol.*, 44:581–591.
110. Geyer, H. J., Scheuntert, I., Rapp, K., Kettrup, A., Korte, F., Greim, H., and Rozman, K. (1990): Correlation between acute toxicity of 2,3,7,8-tetrachlorodibenzo-*p*-dioxin (TCDD) and total body fat content in mammals. *Toxicology*, 65(1–2):97–107.
111. Giknis, M. L. A. and Clifford, C. B. (2000): *Spontaneous Neoplastic Lesions in the Crl:CD-1® (1CR)BR Mouse*. Charles River Laboratories, Worcester, MA.
112. Giknis, M. L. A. and Clifford, C. B. (2003): *Spontaneous Neoplasms and Survival in Wistar Han Rats: Compilation of Control Group Data*. Charles River Laboratories, Worcester, MA.
113. Giknis, M. L. A. and Clifford, C. B. (2004): *Compilation of Spontaneous Neoplastic Lesions and Survival in Crl:CD® (SD) Rats from Control Groups*. Charles River Laboratories, Worcester, MA.
114. Gold, L. S. and Zeiger, E. (1996): *Handbook of Carcinogenic Potency and Genotoxicity Data Bases*. CRC Press, Boca Raton, FL.
115. Gold, L. S. et al. (2005): Supplement to the Carcinogenic Potency Database (CPDB): results of animal bioassays published in the general literature through 1997 and by the National Toxicology Program in 1997–1998. *Toxicol. Sci.*, 85:745–808.
116. Gopinath, C., Prentice, D. E., and Lewis, D. J. (1987): *Atlas of Experimental Toxicological Pathology*. MTP Press, Lancaster, U.K.
117. Grasso, P. and Golberg, L. (1966): Subcutaneous sarcoma as an index of carcinogenic potency. *Food Cosmet. Toxicol.*, 4:297–320.
118. Grasso, P., Sharratt, M., and Cohen, A. J. (1991): Role of persistent, non-genotoxic tissue damage in rodent cancer and relevance to humans. *Annu. Rev. Pharmacol. Toxicol.*, 31:253–287.
119. Greaves, P. C. (2000): *Histopathology of Preclinical Toxicity Studies*, 2nd ed. Elsevier, Amsterdam.
120. Greim, H., Gelbke, H. P., Reuter, U., Thielmann, H. W., and Edler, L. (2003): Evaluation of historical control data in carcinogenicity studies. *Hum. Exp. Toxicol.*, 22(10): 541–549.
121. Hailey, J. R., Haseman, J. K., Bucher, J. R., Raadovsky, A. E., Malarkey, D. E. et al. (1998): Impact of *Helicobacter hepaticus* infection in $B6C3F_1$ mice from twelve National Toxicology Program two-year carcinogenesis studies. *Toxicol. Pathol.*, 26:602–611.
122. Halliwell, W. H. (1998): Submucosal mesenchymal tumors of the mouse urinary bladder. *Toxicol. Pathol.*, 26:128–136.
123. Halpert, J. R., Guengerich, F. P., Bend, J. R., and Correia, M. A. (1994): Selective inhibitors of cytochrome P450. *Toxicol. Appl. Pharmacol.*, 124:163–175.

124. Hart, R. W., Keenan, K. P., Turturro, A., Abdo, K. M., Leakey, J., and Lyn-Cook, L. (1995): Caloric restriction and toxicology. *Fundam. Appl. Toxicol.*, 25:184–195.
125. Hart, R. W., Neumann, D. A., and Robertson, R. T. (1995): *Dietary Restriction: Implications for the Design and Interpretation of Toxicity and Carcinogenicity Studies*. ILSI Press, Washington, D.C.
126. Haseman, J. K. (1984): Statistical issues in the design analysis and interpretation of animal carcinogenicity studies. *Environ. Health Persp.*, 58:385–392.
127. Haseman, J. K. (1990): Use of statistical decision rules for evaluating laboratory animal carcinogenicity studies. *Fundam. Appl. Toxicol.*, 14:637–648.
128. Haseman, J. K., Huff, J., and Boorman, G. A. (1984): Use of historical control data in carcinogenicity studies in rodents. *Toxicol. Pathol.*, 2:126–135.
129. Haseman, J. K., Winbush, J. S., and O'Donnell, Jr., M. W. (1986): Use of dual control groups to estimate false positive rates in laboratory animal carcinogenicity studies. *Fundam. Appl. Toxicol.*, 7:573–584.
130. Haseman, J. K. and Rao, G. N. (1992): Effects of corn oil, time-related changes, and inter-laboratory variability on tumor occurrence in control Fischer 344 (F344/N) rats. *Toxicol. Pathol.*, 20:52–60.
131. Haseman, J. K. and Lockhart, A. (1994): The relationship between use of the maximum tolerated dose and study sensitivity for detecting rodent carcinogenicity. *Fundam. Appl. Toxicol.*, 22:382–391.
132. Haseman, J. K. and Johnson, L. (1996): Analysis of National Toxicology Program rodent bioassay data for anticarcinogenic effects. *Mutat. Res.*, 350:131–141.
133. Haseman, J. K., Hailey, J. R., and Morris, R. W. (1998): Spontaneous neoplasm incidences in Fischer 344 rats and B6C3F$_1$ mice in two-year carcinogenicity studies: a National Toxicology Program update. *Toxicol. Pathol.*, 26: 428–441.
134. Haseman, J. K., Ney, E., Nyska, A., and Rao, G. N. (2003): Effect of diet and animal care/housing protocols on body weight, survival, tumor incidences, and nephropathy severity of F344 rats in chronic studies. *Toxicol. Pathol.*, 31: 674–681.
135. Hattis, D., Goble, R., and Chu, M. (2005): Age-related differences in susceptibility to carcinogenesis. II. Approaches for application and uncertainty analyses for individual genetically acting carcinogens. *Environ. Health Perspect.*, 113:509–516.
136. Health and Welfare Canada. (1989): *Guidelines for Canadian Drinking Water Quality: Supporting Documentation*, Part 1, pp. 1–5. Health and Welfare, Ottawa, Canada.
137. Herman, J. R., Dethloff, L. A., McGuire, E. J., Parker, R. F., Walsh, K. M., Gough, A. W., Masuda, H., and de la Iglesia, F. A. (2002): Rodent carcinogenicity with the thiazolidinedione antidiabetic agent troglitazone. *Toxicol. Sci.* 68:195–235.
138. Hesseltine, G. R., Wolff, R. K., Hanson, R. L., Mauderly, J. L., and McClellan, R. O. (1984): Effect of day vs. night inhalation exposure on lung burdens of galliumoxide in rats. In: *Inhalation Toxicology Research Institute Annual Report*. Inhalation Toxicology Research Institute, Albuquerque, NM.
139. Hill, R. N., Crisp, T. M., Hurley, P. M., Rosenthal, S. L., and Singh, D. V. (1998): Risk assessment of thyroid follicular cell tumors. *Environ. Health Perspect.*, 106: 447–457.
140. Himms-Hagen, J. (1995): Role of brown adipose tissue thermogenesis in control of thermoregulatory feeding in rats: a new hypothesis that links thermostatic and glucostatic hypotheses for control of food intake. *Proc. Soc. Exp. Biol. Med.*, 208:159–169.
141. Holsapple, M. P., Pitot, H. C., Cohen, S. H., Boobis, A. R., Klaunig, J. E., Pastoor, T., Dellarco, V. L., and Dragan, Y. P. (2006): Mode of action in relevance of rodent liver tumors to human cancer risk. *Toxicol. Sci.*, 89(1):51–56.
142. Iatropoulos, M. J. (1984): Appropriateness of methods for slide evaluation in the practice of toxicologic pathology. *Toxicol. Pathol.*, 12:4–5.
143. Iatropoulos, M. J. (1988): Society of Toxicologic Pathologists position paper: 'blinded' microscopic examination of tissues from toxicologic or oncogenic studies. In: *Carcinogenicity*, edited by H. C. Grice and J. L. Ciminera, pp. 133–135. Springer-Verlag, New York.
144. Iatropoulos, M. J. (1993): Comparative histokinetic and xenodynamic considerations in toxicity. In: *Drug Toxicokinetics*, edited by D. G. P. Welling and F. A. de la Iglesia. pp. 245–266. Marcel Dekker, New York.
145. Iatropoulos, M. J. (1994): Endocrine considerations in toxicologic pathology. *Exp. Toxicol. Pathol.*, 45:391–410.
146. Iatropoulos, M. J., Newman, A. J., Dayan, A. D., Brughera M., Scampini, G., and Mazue, G. (1994): Urinary bladder hyperplasia in the rat: non-specific pathogenetic considerations using a beta-lactam antibiotic. *Exp. Toxicol. Pathol.* 46:265–274.
147. Iatropoulos, M. J., Williams, G. M., Wang, C.-X., and Karlsson, S. H. (1996): New histopathologic and histokinetic methods in preclinical safety studies. *Scand. J. Lab Anim. Sci.*, 13:339–343.
148. Iatropoulos, M. J. and Williams, G. M. (1996): Proliferation markers. *Exp. Toxicol. Pathol.*, 48:175–181.
149. Iatropoulos, M. and Williams, G. (2004): The function and pathology of brown adipose tissue in animals and humans *J. Toxicol. Pathol.*, 17:147–153.
150. Imai, K., Yoshimura, S., Yamaguchi, K., Matsui, E., Isaka H., and Hashimoto, K. (1990): Effects of dietary restriction on age-associated pathological changes in F-344 rats. *J Toxicol. Pathol.*, 3:209–221.
151. Interdisciplinary Panel on Carcinogenicity (IPC/AIHC) (1984): Criteria for evidence of chemical carcinogenicity *Science*, 225:682–687.
152. IARC. (1980): *Long-Term and Short-Term Screening Assays for Carcinogens: A Critical Appraisal*, IARC Monographs Suppl. 2. International Agency for Research on Cancer, Lyon, France.
153. IARC. (1987): *IARC Monographs on the Evaluation of Carcinogenic Risks to Humans: Preamble*, IARC Technical Report No. 87/001. International Agency for Research on Cancer, Lyon, France.
154. IARC. (1990): *Complex Mixtures and Cancer Risk*, IARC Publ. No. 104. International Agency for Research on Cancer, Lyon, France.

155. IARC. (1992): *Mechanisms of Carcinogenesis in Risk Identification*, IARC Publ. No. 116. International Agency for Research on Cancer, Lyon, France.

156. IARC. (1992): *Solar and Ultraviolet Radiation*, IARC Tech. Rep. No. 55. International Agency for Research on Cancer, Lyon, France.

157. IARC. (1995): *Peroxisome Proliferation and Its Role in Carcinogenesis: Views and Expert Opinions of an IARC Working Group*, IARC Tech. Rep. No. 24. International Agency for Research on Cancer, Lyon, France.

158. IARC. (1996): *Directory of Agents Being Tested for Carcinogenicity*, IARC Publ. No. 134. International Agency for Research on Cancer, Lyon, France.

159. IARC. (1997): *IARC Monographs on the Evaluation of Carcinogenic Risks to Humans*, Vols. 1–69. International Agency for Research on Cancer, Lyon, France.

160. IARC. (1999): *The Use of Short- and Medium-Term Tests for Carcinogens and Data on Genetic Effects in Carcinogenic Hazard Evaluations*, Consensus Report, IARC Tech. Rep. No. 146, pp. 1–18. International Agency for Research on Cancer, Lyon, France.

161. IARC. (2003): Benzofuran, pp. 27–30; Butylated hydroxyanisole, pp. 31–40; Dichlorovos, pp. 49–56. In: *IARC Monographs in the Predictive Value of Rodent Forestomach and Gastric Neuroendocrine Tumours in Evaluating Carcinogenic Risks to Humans*, IARC Tech. Publ. No. 39. International Agency for Research on Cancer, Lyon, France.

162. IARC. (2003): *World Cancer Report*. International Agency for Research on Cancer, Lyon, France.

163. ICH. (1994): *Guideline III/5081 on Toxicokinetics*, Commission of the European Communities, Directorate General III: Industry: Industrial Affairs III: Consumer Goods Industries: Pharmaceuticals. International Conference on Harmonization, Brussels, Belgium.

164. ICH. (1996): *ICH Harmonized Tripartite Guideline: Impurities in New Drug Products*. International Conference on Harmonization, Brussels, Belgium.

165. ICH. (1997): *ICH Harmonized Tripartite Guideline: Dose Selection for Carcinogenicity Studies of Pharmaceuticals*, Recommended for Adoption of Step 4 of the ICH Process on October 27, 1994, by the ICH Steering Committee and Addendum on the Limit Dose Related to Dose Selection for Carcinogenicity Studies of Pharmaceuticals, Step 4 Consensus Guideline. International Conference on Harmonization, Brussels, Belgium.

166. International Expert Panel on Carcinogen Risk Assessment (IEPCRA). (1996): The use of mechanistic data in the risk assessments of ten chemicals: an introduction to the chemical-specific reviews. *Pharmacol. Ther.*, 71:1–5.

167. International Life Sciences Institute (ILSI); Health and Environmental Sciences Institute (HESI). (2001): Alternatives to carcinogenicity testing. *Toxicol. Pathol.*, 29(Suppl.): 1–351.

168. IPCS. (1996): Butylated hydroxytoluene (BHT). In: *Toxicological Evaluation of Certain Food Additives and Contaminants*, prepared by the Expert Committee on Food Additives (JECFA) for the 4th Joint Meeting of FAO/WHO, pp. 3–86. International Programme on Chemical Safety, World Health Organization, Geneva.

169. IPCS. (2004): *IPCS Risk Assessment Terminology*. International Programme on Chemical Safety, World Health Organization, Geneva.

170. International Workshop. (1992): Clinical pathology testing in preclinical safety assessment. *Toxicol. Pathol.*, 20: 469–543.

171. Jacobs, A. (2005): Prediction of 2-year carcinogenicity study results for pharmaceutical products: how are we doing? *Toxicol. Sci.*, 88(1):18–23.

172. Jacobs, A., Avalos, J., Brown, P., and Wilkin, J. (1999): Does photosensitivity predict photococarcinogenicity? *Int. J. Toxicol.*, 18:191–198.

173. Jain, R. K., Gerlowski, L. E., Weissbrod, J. M., Wang, J., and Pierson, Jr., R. N. (1981): Kinetics of intake, distribution and excretion of zinc in rats. *Ann. Biomed. Eng.*, 9: 345–361.

174. Jameson, C. W. (1984): Analytical chemistry requirements for toxicity testing of environmental chemicals. In: *Chemistry for Toxicity Testing*, edited by C. W. Jameson and D. B. Walters, Butterworth, Stoneham, MA.

175. Jonas, A. M., Tomita, Y., and Wyand, D. S. (1965): Enzootic intestinal adenocarcinoma in hamsters. *J. Am. Vet. Med. Assoc.*, 147:1102–1108.

176. Jordan, A. (1992): FDA requirements for nonclinical testing of contraceptive steroids. *Contraception.*, 46:499–509.

177. Karbe, E. (1999): 'Mesenchymal tumor' or 'decidual-like reaction'? *Toxicol. Pathol.*, 27:354–362.

178. Karbe, E., Hartmann, E., George, C., Wadsworth, P., Harleman, J. and Geiss, V. (1998): Similarities between the uterine decidual reaction and the 'mesenchymal lesion' of the urinary bladder in aging mice. *Exp. Toxicol. Pathol.*, 50:4–6.

179. Kasper, P. (2004): Assessment and acceptance of thresholds of genotoxic impurities in new drug substances: a regulatory perspective. *Int. J. Pharmaceut. Med.*, 18: 209–214.

180. Keenan, K. P. (1996): The uncontrolled variable in risk assessment: *ad libitum* overfed rodents—fat, facts and fiction. *Toxicol. Pathol.*, 24:376–383.

181. Keenan, K., Smith, P., Hertzog, P., Soper, K., Ballam, G., and Clark, R. (1994): The effects of overfeeding and dietary restriction on Sprague–Dawley rat survival and early pathology biomarkers of aging. *Toxicol. Pathol.*, 22:300–331.

182. Keenan, K. P. Laroque, P., Ballam, G. C., Soper, K. A., Dixit, R., Mattson, B. A., Adams, S. P., and Coleman, J. B. (1996): The effects of diet, *ad libitum* overfeeding, and moderate dietary restriction on the rodent bioassay: the uncontrolled variable in safety assessment. *Toxicol. Pathol.*, 24:757–768.

183. Kelly, W. A., Marler, R. J., and Weikel, J. H. (1993): Drug-induced mesovarial leiomyomas in the rat: a review and additional data. *J. Am. Coll. Toxicol.*, 12: 13–22.

184. Kemp, C. J. (1995): Hepatocarcinogenesis in p53-deficient mice. *Mol. Carcinog.*, 12:132–136.

185. Kempson, R. L., Fletcher, C. D. M., Evans, H. L., Hendrickson, M. R., and Sibley, R. K. (2001): *Tumours of the Soft Tissues: Atlas of Tumour Pathology*, pp. 195–235. AFIP, Washington, D.C.

186. Kier, L. D., Neft, R., Tang, L., Suizu, R., Cook, T., Onsurez, K., Tiegler, K., Sakai, Y., Ortiz, M., Nolan, T., Sankar, U., and Li, A. P. (2004): Applications of microarrays with toxicologically relevant genes (tox genes) for evaluation of chemical toxicans in Sprague–Dawley rats *in vivo* and human hepatocytes *in vitro*. *Mutat. Res.*, 549:101–113.
187. King, F. G., Dedrick, R. L., and Farris, F. F. (1986): Physiological pharmacokinetic modeling of cisdichlorodiamine platinum (II) (DDP) in several species. *J. Pharmacokin. Biopharmaceut.*, 14:131–155.
188. Kirkland, D. J., Aardema, M., Henderson, L., and Müller, L. (2005): Evaluation of the ability of a battery of 3 *in vitro* genotoxicity tests to discriminate rodent carcinogens and noncarcinogens. I. Sensitivity, specificity and relative predictivity. *Mutat. Res.*, 584:1–256.
189. Kirkland, D. J., Henderson, L., Marzin, D., Müller, L., Parry, J. M., Speit, G., Tweats, D. J., and Williams, G. M. (2005): Testing strategies in mutagenicity and genetic toxicology: an appraisal of the guidelines of the European Scientific Committee for Cosmetics and Non-Food Products for the evaluation of hair dyes. *Mutat. Res.*, 588: 88–105.
190. Klaassen, C. D. (1999): The role of diet and caloric intake in aging, obesity and cancer. *Toxicol. Sci.*, 2(Suppl.): 1–146.
191. Klaunig, J. E., Babich, M. A., Baetcke, K. P., Cook, J. C., Corton, J. C. et al. (2003): PPAR-alpha agonist-induced rodent tumors: modes of action and human relevance. *Crit. Rev. Toxicol.*, 33(6):655–780.
192. Klopman, G. and Rosenkranz, H. S. (1994): Approaches to SAR in carcinogenesis and mutagenesis: prediction of carcinogenicity/mutagenicity using MULTICASE. *Mutat. Res.*, 305:33–46.
193. Knapka, J. J. (1979): Laboratory animal feed. *Science*, 204: 1367–1368.
194. Komulainen, H. (1996): Pharmacokinetic experiments in animals: needs and application of data. *Scand. J. Lab. Anim. Sci.*, 23:315–316.
195. Krewski, D., Smythe, R. T., Dewanji, A., and Colin, D. (1988): Statistical tests with historical controls. In: *Carcinogenicity*, edited by H. C. Grice and J. L. Ciminera, pp. 23–38. Springer-Verlag, New York.
196. Kripke, M. L. (1994): Ultraviolet radiation and immunology: something new under the sun (Presidential address). *Cancer Res.*, 54:6102–6105.
197. Kroes, R. and Wester, P. W. (1986): Forestomach carcinogens: possible mechanisms of action. *Food Chem. Toxicol.*, 24:1083–1089.
198. Kroes, R., Renwick, A. G., Cheeseman, M., Kleiner, J., Mangelsdorf, I., Piersma, A., Schilter, B., Schlatter, J., van Schothorst, F., Vos, J. G., and Wurtzen, G. (2004): Structure-based thresholds of toxicological concern (TTC): guidance for application to substances present at low levels in the diet. *Food Chem. Toxicol.*, 42(1):65–83.
199. Kunze, E. and Chowaniec, J. (1990): Tumors of the urinary bladder. In: *Pathology of Tumours in Laboratory Animals*. Vol. I. *Tumors of the Rat*, IARC Publ. No. 99, pp. 345–373. International Agency for Research on Cancer, Lyon, France.
200. Lai, Y. L., Jacoby, R., and Jonas, A. (1978): Age related and light associated retinal changes in Fischer rats. *Invest. Ophthalmol. Vis. Sci.*, 17:634–638.
201. Lavbelin, G., Roba, J., Roncucci, R., and Parmentier, R. (1975): Carcinogenicity of 6-aminochrysene in mice. *Eur. J. Cancer*, 11:327–334.
202. Le Boeuf, R. A., Kerckaert, G. Aardema, M., Gibson, D., Brauninger, R., and Isfort, R. (1996): The pH 6.7 Syrian hamster embryo cell transformation assay for assessing the carcinogenic potential of chemicals. *Mutat. Res.*, 356: 85–127.
203. Leder, A., Kuo, A., Cardiff, R. D., Sinn, E., and Leder, P. (1990): v-Ha-ras transgene abrogates the initiation step in mouse skin tumorigenesis: effects of phorbol esters and retinoic acid. *Proc. Natl. Acad. Sci. U.S.A.*, 87:9178–9182.
204. Lee, P. (1988): Assumptions in analyses of the bioassay: a statistician's view. In: *Carcinogenicity*, edited by H. C. Grice and J. L Ciminera, pp. 1–10. Springer-Verlag, New York.
205. Lehman, A. J. and Fitshugh, O. G. (1954): Ten-fold safety factor studies. *U.S. Q. Bull.*, XVIII (1):33–35.
206. Leung, H. W. and Paustenbach, D. J. (1988): Application of pharmacokinetic to derive biological exposure indexes from threshold limit values. *Am. Indust. Hyg. Assoc.*, 49: 445–450.
207. Lington, A. W., Bird, M. G., Plutnick, R. T., Stubblefield W. A., and Scala, R. A. (1997): Chronic toxicity and carcinogenic evaluation of diisonoyl phthalate in rats. *Fundam. Appl. Toxicol.*, 36:79–89.
208. Long, G. G., Symanowski, J. T., and Roback, K. (1998). Precision in data acquisition and reporting of organ weights in rats and mice. *Toxicol. Pathol.*, 26:316–318.
209. Lutz, W. K. (1986): Quantitative evaluation of DNA binding data for risk estimation and for classification of direct and indirect carcinogens. *J. Cancer Res. Clin. Oncol.*, 112 85–91.
210. MacDonald, J., French, J. E., Gerson, R. J., Goodman, J. Inoue, T. et al. (2004): The utility of genetically modified mouse assays for identifying human carcinogens: a basic understanding and path forward. *Toxicol. Sci.* 77:188–194.
211. MacDonald, J. S. (2004): Human carcinogenic risk evaluation. Part IV. Assessment of human risk of cancer from chemical exposure using a global weight-of-evidence approach. *Toxicol. Sci.*, 82:3–8.
212. Mahler, J. F., Stokes, W., Mann, P. C., Takaoka, M., and Maronpot, R. R. (1996): Spontaneous lesions in aging FVB/N mice. *Toxicol. Pathol.*, 24:710–716.
213. Mahler, J. P., Flagler, N. D., Malarkey, D. E., Mann, P. C. Haseman, J. K., and Eastin, W. (1998): Spontaneous and chemically-induced proliferative lesions in Tg.AC transgenic and p53-heterozygous mice. *Toxicol. Pathol.*, 26 501–511.
214. Maita, K., Hirano, M., Harada, T., Mitsumori., K., Yoshida A., Takahashi, K., Nakashima, N., Kitazawa, T., Enomoto A., Inui, K., and Shirasu, Y. (1988): Mortality, major cause of moribundity, and spontaneous tumors in CD-1 mice *Toxicol. Pathol.*, 16:340–349.

215. McAnulty, P. A. and Skydsgaard, M. (2005): Diethylstilbestrol (DES): carcinogenic potential in *Xpa*$^{-/-}$, *Xpa*$^{-/-}$/*p53*$^{+/-}$, and wild-type mice during 9 months' dietary exposure. *Toxicol. Pathol.*, 33:609–620.
216. McClellan, R. O. and Hobbs, C. H. (1986): Generation, characterization and exposure systems for test atmospheres. In: *Safety Evaluation of Chemicals*, edited by W. E. Lloyd. Hemisphere, Washington, D.C.
217. McConnell, E. E. (1983): Pathology requirements for rodent two year studies. I. A review of current procedures. *Toxicol. Pathol.*, 11:60–64.
218. McConnell, E. E., Solleveld, H. A., Swenberg, J. A., and Boorman, G. A. (1986): Guidelines of combining neoplasms for evaluation of rodent carcinogenesis studies. *J. Natl. Cancer Res.*, 76:283–289.
219. McConnell, R. F. (1989): General observations on the effects of sex steroids in rodents with emphasis on long-term oral contraceptive studies. In: *Safety Requirements for Contraceptive Steroids*, edited by F. Michael, pp. 211–229. Cambridge University Press, New York.
220. McConnell, R. F., Westen, H. H., Ulland, B. M., Bosland, M. C., and Ward, J. M. (1992): Proliferative lesions of the testes in rats with selected examples from mice. In: *Guides for Toxicologic Pathology.* STP/ARP/AFIP, Washington, D.C.
221. McMartin, D. N. Sahota, P. S., Gunson, D. E., Hsu, H. H., and Spaet, R. H. (1992): Neoplasms and related proliferative lesions in control Sprague–Dawley rats from carcinogenicity studies: historical data and diagnostic considerations. *Toxicol. Pathol.*, 20:212–225.
222. McQueen, C. A. and Williams, G. M. (1987): The hepatocytes primary culture/DNA repair test using hepatocytes from several species. *Cell Biol. Toxicol.*, 3:209–218.
223. McQueen, C. A. and Williams, G. M. (1988): Genotoxicity of carcinogens in human hepatocytes: application in hazard assessment. *Toxicol. Appl. Pharmacol.*, 96:360–366.
224. Milman, H. A. and Weisburger, E. K. (1985): *Handbook of Carcinogen Testing.* Noyes, Park Ridge, NJ.
225. MHW. (1989): *Guidelines for Toxicity Studies Required for Applications for Approved To Manufacture (Import) Drugs: Carcinogenicity Study.* Ministry of Health and Welfare, Tokyo, Japan.
226. Modlin, I. M. and Sachs, G. (1998): *Age-Related Diseases: Biology and Treatment*, pp 242–245. Schnetztor-Verlag GmbH, Konstanz, Germany.
227. Moghimi, S. M., Hunter, A. C., and Murray, J. C. (2005): Nanomedicine: current status and future prospects. *FASEB J.*, 19:311–330.
228. Monro, A. (1992): What is an appropriate measure of exposure when testing drugs for carcinogenicity in rodents? *Toxicol. Appl. Pharmacol.*, 112:171–181.
229. Monro, A. (1993): How useful are chronic (life-span) toxicology studies in rodents in identifying pharmaceuticals that pose a carcinogenic risk to humans? *Adv. Drug React. Toxicol. Rev.*, 12:5–34.
230. Monro, A. (1996): Are lifespan rodent carcinogenicity studies defensible for pharmaceutical agents? *Exp. Toxicol. Pathol.*, 48:155–166.
231. Morrison, V. and Ashby, J. (1994): A preliminary evaluation of the performance of the Muta Mouse (bac Z) and Big Blue (bac I) transgenic mouse mutation assays. *Mutagenesis*, 9:367–376.
232. Mugford, C. A. and Kedderis, G. L. (1998): Sex-dependent metabolism of xenobiotics. *Drug Metab. Rev.*, 30(3): 441–498.
233. Mulder, G. J. (1986): Sex differences in drug conjugation and their consequences for drug toxicity: sulfation, glucuronidation and glutathione conjugation. *Chem. Biol. Interact.*, 57:1–15.
234. Munro, I. C. (1977): Considerations in chronic testing: the chemical, the dose, the design. *J. Environ. Pathol. Toxicol.*, 1:183–197.
235. Munro, I. C., Ford, R. A., Kennepohl, E., and Sprenger, J. G. (1996): Thresholds of toxicological concern based on structure–activity relationships. *Drug Metab. Rev.*, 28: 209–217.
236. NCI. (1994): *Survey of Compounds Which Have Been Tested for Carcinogenic Activity*, NIH Publ. 94-3765. National Cancer Institute Washington, D.C.
237. NIH. (1981): *NIH Guidelines for the Laboratory Use of Chemical Carcinogens:* NIH Publ. 81-2385. National Institutes of Health, Washington, D.C.
238. NIH. (1986): *Humane Care and Use of Laboratory Animals*, NIH Publ. 86-23. National Institutes of Health, Washington, D.C.
239. National Research Council. (1983): *Risk Assessment in the Federal Government: Managing the Process.* National Academy Press, Washington, D.C.
240. National Research Council. (1993): *Use of Maximum Tolerated Dose in Animal Bioassays for Carcinogenicity.* National Academy Press, Washington, D.C.
241. National Research Council. (1996): *Guide for the Care and Use of Laboratory Animals*, National Academy Press, Washington, D.C.
242. NTP. (1997): *Effect of Dietary Restriction on Toxicology and Carcinogenesis Studies in F344/N Rats and B6C3F$_1$ Mice*, NTP Tech. Rep. 460, NIH Publ. No. 97-3376, pp. 1–411. National Toxicology Program, Research Triangle Park, NC.
243. NTP. (2000): *Report on Carcinogens, Ninth Edition: Carcinogen Profiles 2000.* Prepared by Technology Planning and Management Corporation, Durham, NC.
244. NTP. (2005): *NTP Technical Reports Index.* National Toxicology Program, Research Triangle Park, NC.
245. Nelson, D. R., Koymans, L., Kamatski, T., Stegeman, J. J., Feyereisen, R. et al. (1996): P450 superfamily: update on new sequences, gene mapping accession numbers and nomenclature. *Pharmacogenetics*, 6:1–42.
246. Neumann, H.-G. (1986); The role of DNA damage in chemical carcinogenesis of aromatic amines. *J. Cancer Res. Clin. Oncol.*, 112:100–106.
247. Newberne, P. M. and de la Iglesia, F. A. (1985): Philosophy of blind slide reading in toxicologic pathology. *Toxicol. Pathol.*, 13:225.
248. Newberne, P. M. and Sotnikov, A. V. (1996): Diet: the neglected variable in chemical safety evaluations. *Toxicol. Pathol.*, 24:746–756.

249. Nicholls, I., Kolopp, M., Pommier, F., and Scheiwiller, M. (2005): The presence of drug in control samples during toxicokinetic investigations: a Novartis perspective. *Reg. Toxicol. Pharmacol.*, 42:172–178.
250. Noble, J. F. (1984): Automated data acquisition systems in the 80s and beyond. II. Operation. In: *Toxicology Laboratory Design and Management for the 80s and Beyond*, edited by A. Tegeris, pp. 143–158. Karger, New York.
251. Oberdörster, G., Oberdörster, E., and Oberdörster, J. (2005): Nanotoxicology: an emerging discipline evolving from studies of ultrafine particles. *Environ. Health Perspect.*, 113:823–839.
252. Office of Science and Technology Policy. (1985): Chemical carcinogens: a review of the science and its associated principles. *Fed. Reg.*, 50(184):10371–10442.
253. Office of Technology Assessment. (1987): *Identifying and Regulating Carcinogens: A Background Paper*, pp. 1–251. U.S. Congress, Washington, D.C.
254. Olsen, J. H., Schulgen, G., Boice, J. D., Whysner, J., Travis, L. B., Williams, G. M., Johnson, F. B., and O'McGee, J. (1995): Antiepileptic treatment and risk for hepatobiliary cancer and malignant lymphoma. *Cancer Res.*, 55:294–297.
255. Oppenheimer, E. T., Willhite, M., Stout, A. F., Damishofsky, I., and Fishman, M. M. (1964): A comparative study of the effects of embedding cellophane and polystyrene films in rats. *Cancer Res.*, 24:379–386.
256. OECD. (1981): *Adopted Guidelines for Testing of Chemicals*. Section 4. *Health Effects*, No. 451, Carcinogenicity Studies. Organization of Economic Cooperation and Development, Paris, France.
257. OECD. (OECD) (1998): *Principles of Good Laboratory Practices*, ENV/mc/CHEM(98)17. Organization of Economic Cooperation and Development, Paris, France.
258. Ozaki, M., Ozaki, K., Watanabe, T., Uwagawa, S., Okuno, Y., and Shirai, T. (2005): Susceptibilities of p53 knockout and rasH2 transgenic mice to urethane-induced lung carcinogenesis are inherited from their original strains. *Toxicol. Pathol.*, 33:267–271.
259. Parkinson, A. (2001): Biotransformation of Xenobiotics. In: *Casarett & Doull's Toxicology*, 6th ed., edited by C. D. Klaassen, pp. 133–224. McGraw-Hill, New York.
260. Peck, C. C., Barr, W. H., and Benet, L. Z. (1992): Opportunities for integration of pharmacokinetics, pharmacodynamics and toxicokinetics in rational drug design. *J. Pharmaceut. Sci.*, 81:605–610.
261. Perrone, C. E., Shao, L., and Williams, G. M. (1998): Effect of rodent hepatocarcinogenic peroxisome proliferators on fatty acyl-CoA oxidase, DNA synthesis, and apoptosis in cultured human and rat hepatocytes. *Toxicol. Appl. Pharmacol.*, 150:277–286.
262. Perrone, C., Ahr, H.-J., Duan, J. D., Jeffrey, A. M., Schmidt, U., Williams, G. M., and Enzmann, H. G. (2004): Embryonic turkey liver: activities of biotransformation enzymes and cultivation of DNA-reactive carcinogens. *Arch. Toxicol.*, 78:589–598.
263. Peto, R., Pike, M. C., Day, N. E., Gray, R. G., Lee, P. N., Parish, S., Peto, J., and Wahrendorf, J. (1980): Guidelines for simple, sensitive significance tests for carcinogenic effects in long-term animal experiments. In: *Long-Term and Short-Term Screening Assays for Carcinogens: A Critical Appraisal*, edited by R. Montesano, H. Bartsch, and L. Tomatis, IARC Monographs on the Evaluation of the Carcinogenic Risk of Chemicals to Humans, Suppl. 2, pp. 311–426. International Agency for Research on Cancer, Lyon, France.
264. Peto, R., Pike, M. C., Bernstein, L., Gold, L. S., and Ames, B. N. (1984): A proposed general convention for the numerical description of carcinogenic potency of chemicals in chronic-exposures animal experiments. *Environ. Health Perspect.*, 58:1–8.
265. Phillips, D. H. (1997): Detection of DNA modifications by the $^{32}$P-postlabeling assay. *Mutat. Res.*, 378(102):1–12.
266. PDR. (2005): Bone marrow metabolism, pp. 1840–1844; Recombinant human insulin (Lantus), pp. 715–719; Tamoxifen, pp. 622, 2966. In: *Physician's Desk Reference*, edited by L. Murray. Thompson PDR, Montvale, NJ.
267. Poteracki, J. and Walsh, K. M. (1998): Spontaneous neoplasms in control Wistar rats: a comparison of reviews. *Toxicol. Sci.*, 45:1–8.
268. Poulet, F. M., Berardi, M. R., Halliwell, W., Hartman, B., Auletta, C., and Bolte, H. (2004): Development of hibernomas in rats dosed with phentolamine mesylate during the 24-month carcinogenicity study. *Toxicol. Pathol.*, 32(5):558–566.
269. Powles, P. (1996): Interpretation of data from toxicokinetic studies. *Scand. J. Lab. Anim. Sci.*, 23:317–323.
270. Prahalada, S., Majka, J. A., Soper, K. A., Nett, T. M., Bagdon, W. J., Peter, C. P., Burek, J. D., MacDonald, J. S., and van Zwieten, M. J. (1994): Leydig cell hyperplasia and adenomas in mice treated with finasteride, a 5α-reductase inhibitor: possible mechanism. *Fundam. Appl. Toxicol.*, 22: 211–219.
271. Randerath, K., Reddy, M. V., and Gupta, R. C. (1981): $^{32}$P-postlabeling test for DNA damage. *Proc. Natl. Acad. Sci U.S.A.*, 78:626–6129.
272. Rao, G. N., Piegorsch, W. W., and Haseman, J. K. (1987): Influence of body weight on the incidence of spontaneous tumors in rats and mice of long-term studies. *Am. J. Clin Nutr.*, 45:252–260.
273. Rao, G., Haseman, J., Grumbein, S., Crawford, S., and Eustis, S. (1990): Growth, body weight, survival and tumor trends in (C57B1/6xC3H/NeN)F1 (B6C3F$_1$) mice during a nine year period. *Toxicol. Pathol.*, 18:71–77.
274. Registry of Industrial Toxicology Animal Data (RITA) (1999): Optimization of carcinogenicity bioassays. *Exp Toxicol. Pathol.*, 51:461–475.
275. Reuter, J. and Hobbelen, H. (1977): The effect of continuous light exposure on the retina in albino and pigmented rats. *Physiol. Behav.*, 18:939–944.
276. Rice, J. M., Williams, G. M., Palmer, A. E., London, W. T., and Sly, D. L. (1981): Pathology of gestational choriocarcinoma induced in patas monkeys by ethylnitrosourea given during pregnancy. *Placenta*, 3(Suppl.):223–230.

277. Rice, J. M., Baan, R. A., Blettner, M., Genevois-Charneau, C., Grosse, Y., McGregor, D. B., Partensky, C., and Wilbourn, J. D. (1999): Rodent tumors of urinary bladder, renal cortex, and thyroid gland in IARC monographs evaluation of carcinogenic risk to humans. *Toxicol. Sci.*, 49:166–171.
278. Riley, V. (1975): Mouse mammary tumors: alteration of incidence as apparent function of stress. *Science*, 189: 465–467.
279. Riley, V. (1981): Psychoneuroendocrine influences on immune-competence and neoplasia. *Science*, 212: 1100–1109.
280. Robens, J. F., Calabrese, E. J., Piegorsch, W. W., Schueler, R. L., and Hayes, A. W. (1994): Principles of testing for carcinogenicity. In: *Principles and Methods of Toxicology*, 3rd ed., edited by A. W. Hayes, pp. 697–728. Raven Press, New York.
281. Rosai, J. (1981): Breast. In: *Ackerman's Surgical Pathology*, pp. 1098–1149. Mosby, St. Louis, MO.
282. Russo, J., Russo, I. H., Rogers, A. E., Van Zwieten, M. J., and Gusterson, B. (1990): Tumours of the mammary gland. In: *Pathology of Tumours in Laboratory Animals*. Vol. I. *Tumors of the Rat*, edited by V. S. Turusov and U. Mohr, IARC Publ. No. 99, pp. 47–78. International Agency for Research on Cancer, Lyon, France.
283. Saitoh, A., Kimura, M., Takahashi, R., Yokoyama, M., Nomura, T., Izawa, M., Sekiya, T., Nishimura, S., and Katsuki, M. (1990): Most tumors in transgenic mice with human c-Ha-ras gene contained somatically activated transgenes. *Oncogene*, 5:1195–1200.
284. Sambuco, C. P., Davies, R. E., Forbes, P. D., and Hoberman, A. M. (1991): Photocarcinogenesis and consumer product testing: technical aspects. *Toxicol. Methods*, 1: 75–83.
285. Sansone, E. B. and Losikoff, A. M. (1978): Contamination from feeding volatile test chemicals. *Toxicol. Appl. Pharmacol.*, 46:703–708.
286. Sansone, E. B. and Tewari, Y. B. (1978): Penetration of protective clothing materials by 1,2-dibromo-3-chloropropane, ethylene dibromide, and acrylonitrile. *J. Am. Indust. Hyg. Assoc.*, 39:921–922.
287. Sansone, E. B. and Tewari, Y. B. (1978): The permeability of laboratory gloves to selected solvents. *J. Am. Indust. Hyg. Assoc.*, 39:169–174.
288. Sasahara, K., Ando-Lu, J., Nishiyama, K., Takahashi, M., Yoshida, M., and Maekawa, A. (1998): Granular cell foci of the uterus in Donryu rats. *J. Comp. Pathol.*, 119: 195–199.
289. Sasaki, Y. F., Izumiyama, F., Nishidate, E., Matsusaka, N., and Tsuda, S. (1997): Detection of rodent liver carcinogen genotoxicity by the alkaline single-cell gel electrophoresis (comet) assay in multiple mouse organs (liver, lung, spleen, kidney, and bone marrow). *Mutat. Res.*, 391:201–214.
290. Scheuplein, R. J. and Blank, I. H. (1971): Permeability of the skin. *Physiol. Rev.*, 51:702–743.
291. Schmid, W. (1975): The micronucleus test. *Mutat. Res.*, 31:9–15.
292. SCCNFP. (2003): *Updated Recommended Strategy for Testing Hair Dyes for their Potential Genotoxicity/Mutagenicity/Carcinogenicity*, SCCNFP/0720/03. Scientific Committee on Cosmetic and Non-Food Products Intended for Consumers, Brussels, Belgium.
293. Seifter, E., Rettura, G., Zisblatt, M., Levenson, S M., Levine, N., Davidson, A., and Seigter, J. (1973): Enhancement of tumor development of physically-stressed mice inoculated with an oncogenic virus. *Experientia*, 29: 1379–1882.
294. Seilkop, S. K. (1995): The effect of body weight on tumor incidence and carcinogenicity testing in $B6C3F_1$ mice and F344 rats. *Fundam. Appl. Toxicol.*, 24:247–259.
295. Serabian, M. A. and Pilaro, A. M. (1999): Safety assessment of biotechnology-derived pharmaceuticals: ICH and beyond. *Toxicol. Pathol.*, 27:27–31
296. Shah, V. P., Midha, K. K., Dighe, S., McGilveray, I. J., Skelly, J. P. et al. (1992): Analytical methods validation: bioavailability, bioequivalence and pharmacokinetic studies. *J. Pharmaceut. Res.*, 81:309–312.
297. Shirai, T., Hirose, M., and Ito, N. (1999): Medium-term bioassays in rats for rapid detection of the carcinogenic potential of chemicals. In: *The Use of Short- and Medium-Term Tests for Carcinogenic Hazard Evaluation*, edited by D. B. McGregor, J. M. Rice, and S. Venitt, IARC Publ. No. 146, pp. 251–271. International Agency for Research on Cancer, Lyon, France.
298. Sieber, S. M. and Adamson, R. H. (1978): Long-term studies on the potential carcinogenicity of artificial sweeteners in non-human primates. In: *Health and Sugar Substitutes*, edited by B. Guggenheim, pp. 266–271. Karger, Basel, Switzerland.
299. Silva, L. B. and van der Laan, J. W. (2000): Mechanisms of nongenotoxic carcinogenesis and assessment of the human hazard. *Regul. Toxicol. Pharmacol.*, 32:135–143.
300. Smith, D. M., Rogers, A. E., and Newberne, P. M. (1975): Vitamin A and benzo(*a*)pyrene carcinogenesis in the respiratory tract of hamsters fed a synthetic diet. *Cancer Res.*, 35:1485–1488.
301. Smith, D. M., Ortiz, L. W., Archuleta, R. F., Spalding, J. F., Tillery, M. I., Ettinger, H. J., and Thomas, R. G. (1981): A method of chronic nose-only exposures of laboratory animals to inhaled fibrous aerosols. In: *Inhalation Toxicology and Technology*, edited by H. P. Leong, pp. 89–105. Ann Arbor Science, Ann Arbor, MI.
302. Son, W. C. and Gopinath, C. (2004): Early occurrence of spontaneous tumors in CD-1 mice and Sprague–Dawley rats. *Toxicol. Pathol.*, 32:371–374.
303. Sontag, J. R., Page, N. P., and Safiotti, U. (1976): *Guidelines for Carcinogen Bioassay in Small Rodents*. DHHS Publ. (NIH)76-801. National Cancer Institute, Bethesda, MD.
304. Spalding, J. W., French, J. E., Tice, R. R., Furedi-Machek, M., Haseman, J. K., and Tennant, R. W. (1999): Development of a transgenic mouse model for carcinogenesis bioassays: evaluation of chemically induced skin tumors in Tg.AC mice. *Toxicol. Sci.*, 49:241–254.

305. Stammberger, I., Bube, A., Durchfeld-Meyer, B., Donaubauer, H., and Troschau, G. (2002): Evaluation of the carcinogenic potential of insulin glargine (Lantus) in rats and mice. *Int. J. Toxicol.*, 21:171–179.
306. Storer, R. D., French, J. E., Haseman, J., Hajian, G., LeGrand, E. K., Long, G. G., Mixon, L. A., Ochoa, R., Sagartz, J. E., and Soper, K. A. (2001): $p53^{+/-}$ Hemizygous knockout mouse: overview of the available data. *Toxicol. Pathol.*, 29(Suppl):30–50.
307. Tamano, S., Hagiwara, A., Shibata, M., Kurata, Y., Fukushima, S., and Ito, N. (1988): Spontaneous tumors in aging $B6C3F_1$ mice. *Toxicol. Pathol.*, 16:321–326.
308. Tarone, R. E. (1975): Tests for trend in life table analysis. *Biometrika*, 62:679–682.
309. Tennekes, H., Gembardt, C., Dammann, M., and van Ravenzwaay, B. (2004): The stability of historical control data for common neoplasms in laboratory rats: adrenal gland (medulla), mammary gland, liver, endocrine pancreas, and pituitary gland. *Regul. Toxicol. Pharmacol.*, 40(1):18–27.
310. Tennekes, H., Kaufmann, W., Dammann, M., and van Ravenzwaay, B. (2004): The stability of historical control data for common neoplasms in laboratory rats and the implications for carcinogenic risk assessment. *Regul. Toxicol. Pharmacol.*, 40(3):293–304.
311. Thake, D. C., Iatropoulos, M. J., Hard, G. C., Hotz, K. J., Wang, C.-X., Williams, G. M., and Wilson, A. G. E. (1995): A study of the mechanism of butachlor-associated gastric neoplasms in Sprague–Dawley rats. *Exp. Toxicol. Pathol.*, 47:107–116.
312. Thomas, G. A. and Williams, E. D. (1991): Evidence for and possible mechanisms of non-genotoxic carcinogenesis in the rodent thyroid. *Mutat. Res.*, 248:357–370.
313. Thorgiersson, U., Dalgard, D., Reeves, J., and Adamson, R. (1994): Tumor incidence in a chemical carcinogenesis study of nonhuman primates. *Regul. Toxicol. Pharmacol.*, 19:130–151.
314. Thurman, J. D., Bucci, T. J., Hart, R. W., and Torturro, A. (1994): Survival, body weight, and spontaneous neoplasms in *ad libitum*-fed and food-restricted Fischer-344 rats. *Toxicol. Pathol.*, 22:1–9.
315. Tomatis, L. (1989): Overview of perinatal and multigeneration carcinogenesis. In: *Perinatal and Multigeneration Carcinogenesis*, edited by N. P. Napalkov, J. M. Rice, L. Tomatis, and H. Yamasaki, IARC Publ. No. 96, pp. 1–15. International Agency for Research on Cancer, Lyon, France.
316. Trayhurn, P. (1993): Brown adipose tissue: from thermal physiology to bioenergetics. *J. Biosci.*, 18:161–173.
317. Tsuda, H., Park, C. B., and Moore, M. A. (1999): Short- and medium-term carcinogenicity tests. In: *The Use of Short- and Medium-Term Tests for Carcinogenic Hazard Evaluation*, edited by D. B. McGregor, J. M. Rice, and S. Venitt, IARC Publ. No. 146, pp. 203–249. International Agency for Research on Cancer, Lyon, France.
318. Tuch, K., Ockert, D., Hauschke, D., and Christ, B. (1992): Comparison of the ECL-cell frequency in the stomachs of 3 different rat strains. *Pathol. Res. Pract.*, 188:672–675.
319. Tucker, R. W. and Barrett, J. C. (1986): Decreased number of spindle and cytoplasmic microtubules in hamster embryo cells treated with a carcinogen, diethylstilbestrol. *Cancer Res.*, 46:2088–2095.
320. Tugwood, J. D. and Elcombe, C. R. (1999): Predicting carcinogenicity: peroxisome proliferators. In: *Carcinogenicity*, edited by K. T. Kitchin, pp. 337–360. Marcel Dekker, New York.
321. Turturro, A., Duffy, P., Hart, R., and Allaben, W. T. (1996): Rationale for the use of dietary control in toxicity studies: $B6C3F_1$ mouse. *Toxicol. Pathol.*, 24:769–775.
322. Turusov, V. S. (1994): Histiocytic sarcoma. In: *Pathology of Tumours in Laboratory Animals*. Vol. II. *Tumors of the Mouse*, edited by V. S. Turusov and U. Mohr, IARC Publ. No. 111, pp. 671–680. International Agency for Research on Cancer, Lyon, France.
323. Turusov, V. S., Torii, M., Sills, R. C., Willson, G. A., Herbert, R. A., Hailey, J., Haseman, J. K., and Boorman, G. A. (2002): Hepatoblastomas in mice in the U.S. National Toxicology Program (NTP studies). *Toxicol. Pathol.*, 30:580–591.
324. U.S. Congress. (1985): Animal Welfare Act, CFR 9, Parts 1, 2, 3.
325. U.S. Congress. (1996): Food Quality Protection Act, Public Law 104-170.
326. U.S. Congress. (1998): Food Quality Protection Act Amendment, Public Law 105-324.
327. U.S. Department of Agriculture. (1989): Animal Welfare Rules, CFR 9, Parts 1, 2.
328. U.S. DHW. (1977): *Guide for Care and Use of Laboratory Animals*, Publ. No. NIH 77-23. U.S. Department of Health and Welfare, Washington, D.C.
329. U.S. EPA. (1986): Guidelines for carcinogen risk assessment. *Fed. Reg.*, 51:33992–34005.
330. U.S. EPA (1990): *Good Automated Practices*, Draft 12-28-90. Scientific Systems Staff, Office of Information Resources Management U.S. Environmental Protection Agency, Washington, D.C.
331. U.S. EPA. (1991): *Alpha 2μ-Globulin: Association with Chemically Induced Renal Toxicity and Neoplasia in the Male Rat, Risk Assessment Forum*, EPA/625/3–91/019F U.S. Environmental Protection Agency, Washington, D.C
332. U.S. EPA. (1992): A cross-species scaling factor for carcinogen risk assessment based on equivalence of mg/kg/day: notice. *Fed. Reg.*, 57:24152–24173.
333. U.S. EPA. (1992): Guidelines for exposure assessment *Fed. Reg.*, 57:22888–22938.
334. U.S. EPA. (1996): Proposed guidelines for carcinogen risk assessment. *Fed. Reg.*, 61:17960–18011.
335. U.S. EPA. (1997): *Exposure Factors Handbook* EPA/600/P-95–002A. Office of Research and Development, U.S. Environmental Protection Agency, Washington D.C.
336. U.S. EPA. (2003): *Final Draft Guidelines for Carcinogenic Risk Assessment*, EPA/630/P-03/001A. U.S. Environmental Protection Agency, Washington, D.C. (www.epa.gov ncea/raf/cancer2003.htm).

337. U.S. EPA. (2005): *Supplemental Guidance for Assessing Susceptibility from Early-Life Exposure to Carcinogens*, pp. 1–42. U.S. Environmental Protection Agency, Washington, D.C.

338. U.S. FDA. (1978): Good Laboratory Practices for nonclinical laboratory studies, Code of Federal Regulations, Title 21, Part 58. *Fed. Reg.*, 43:59986–60025.

339. U.S. FDA. (1993): *Toxicological Principles for the Safety Assessment of Direct Food Additives and Color Additives Used in Food (Redbook II)*. U.S. Food and Drug Administration, Washington, D.C.

340. U.S. FDA. (1993): Advisory committee for protocols for safety evaluation, panel on carcinogenesis: report on cancer testing in the safety of food additives and pesticides. *Toxicol. Appl. Pharmacol.*, 20:419–438.

341. U.S. FDA. (1994): International Conference on Harmonization, draft guideline on dose selection for carcinogenicity studies of pharmaceuticals. *Fed. Reg.*, 59:9752–9760 (http://www.ifpma.org/ich5s.html).

342. U.S. FDA. (1994): *General Principles for Evaluating the Safety of Compounds Used in Food-Producing Animals*. Center for Veterinary Medicines, U.S. Food and Drug Administration, Washington, D. C.

343. U.S. FDA. (1995): Food additives: threshold of regulation for substances used in food contact articles. *Fed. Reg.*, 60:36582–36596.

344. U.S. FDA. (1997): International Conference on Harmonization, guidance on preclinical safety evaluation of biotechnology-derived pharmaceuticals. *Fed. Reg.*, 62:61515–61519 (http://www.iben.gov).

345. U.S. FDA. (1997): International Conference on Harmonization, guidelines on impurities in new drug products. *Fed. Reg.*, 62:27454–27461.

346. U.S. FDA. (2001): *Guidance for Industry: Statistical Aspects of the Design, Analysis, and Interpretation of Chronic Rodent Carcinogenicity Studies of Pharmaceuticals*. U.S. Food and Drug Administration, Washington, D.C.

347. U.S. FDA. (2002): Revision of the definition of the term 'no residue' in the new animal drug regulations. *Fed. Reg.*, 67:78172–78174.

348. U.S. FDA. (2004): *Guidance for Industry: Recommended Approaches to Integration of Genetic Toxicology Study Results*, pp. 1–6. U.S. Food and Drug Administration, Washington, D.C.

349. Utzat, C. D., Clement, C. C., Ramos, L. A., Das, A., Tomasz, M., and Basu, A. K. (2005): DNA adduct of the mitomycin C metabolite 2,7-diaminonitrosene is a nontoxic and nonmutagenic DNA lesion *in vitro* and *in vivo*. *Chem. Res. Toxicol.*, 18:213–223.

350. van Oosterhout, J. P. J., van der Laan, J. W., de Waal, E. J., Olejniczak, K., Hilgenfeld, M., Schmidt, V., and Bass, R. (1997): The utility of two rodent species in carcinogenic risk assessment of pharmaceuticals in Europe. *Regul. Toxicol. Pharmacol.*, 25:6–17.

351. van Steeg, H., Klein, H., Beems, R. B., and van Kreijl, C. F. (1998): Use of DNA repair-deficient XPA transgenic mice in short-term carcinogenicity testing. *Toxicol. Pathol.*, 26:742–749.

352. Vessel, E. S. (1967): Induction of drug metabolizing enzymes in liver microsomes of mice and rats by softwood bedding. *Science*, 157:1057–1058.

353. Viollon-Abadie, C., Lassere, D., Debruyne, E., Nicod, L., Carmichael, N., and Richert, L. (1999): Phenobarbital, β-naphthoflavone, clofibrate, and pregnenolone-16α-carbonitrile do not affect hepatic thyroid hormone UDP–glucuronosyl transferase activity, and thyroid gland function in mice. *Toxicol. Appl. Pharmacol.*, 155:1–12.

354. Walsh, K. M. and Poteracki, J. (1994): Spontaneous neoplasms in control Wistar rats. *Fundam. Appl. Toxicol.*, 23: 65–72.

355. Wattenberg, L. W. (1985): Chemoprevention of cancer. *Cancer Res.*, 45:1–8.

356. Weisburger, E. K. (1983): History of the bioassay program of the National Cancer Institute. *Prog. Exp. Tumor Res.*, 26:187–201.

357. Weisburger, J. H. and Williams, G. M. (1981): Carcinogen testing: current problems and new approaches. *Science*, 214:401–407, 1981.

358. Weisburger, J. H. and Williams, G. M. (1984): Bioassay of carcinogens: *in vitro* and *in vivo* tests. In: *Chemical Carcinogenesis*, Vol. 2, 2nd ed., ACS Monograph 182, pp. 1323–1373. American Chemical Society, Washington, D.C.

359. Weisburger, J. H. and Williams, G. M. (1995): Causes of cancer. In: *American Cancer Society Textbook of Clinical Oncology*, edited by G. P. Murphy, W. Lawrence, Jr., and R. E. Lenhard, Jr., pp. 10–39. American Cancer Society, Atlanta, GA.

360. Weiss, S. W. and Goldblum, J. R. (2001): Benign lipomatous tumors and liposarcoma. In: *Soft Tissue Tumors*, pp. 571–695. Mosby, St. Louis, MO.

361. Whysner, J. and Williams, G. M. (1992): International cancer risk assessment: the impact of biologic mechanisms, *Regul. Toxicol. Pharmacol.*, 15:41–50.

362. Whysner, J. and Williams, G. M. (1996): Butylated hydroxyanisole mechanistic data and risk assessment: conditional species-specific cytotoxicity, enhanced cell proliferation, and tumor promotion. *Pharmacol. Ther.*, 71(1/2): 137–151.

363. Whysner, J., Ross, P. M., and Williams, G. M. (1996): Phenobarbital mechanistic data and risk assessment: enzyme induction, enhanced cell proliferation, and tumor promotion. *Pharmacol. Ther.*, 71(1/2):153–191.

364. Williams, G. M. (1987): Definition of a human cancer hazard. In: *Nongenotoxic Mechanisms in Carcinogenesis*, Banbury Report 25, pp. 367–380. Cold Spring Harbor Laboratory, Cold Spring Harbor, NY.

365. Williams, G. M. (1989): Methods for evaluating chemical genotoxicity. *Ann. Rev. Pharmacol. Toxicol.*, 29:189–211.

366. Williams, G. M. (1989): Interactive carcinogenesis in the liver. In: *Liver Cell Carcinoma*, edited by P. Bannasch, D. Keppler, and G. Weber, Falk Symposium 51, pp. 197–216. Kluwer, Boston, MA.

367. Williams, G. M. (1990): Screening procedures for evaluating the potential carcinogenicity of food-packaging chemicals. *Regul. Toxicol. Pharmacol.*, 12:30–40.

368. Williams, G. M. (1992): DNA reactive and epigenetic carcinogens. *Exp. Toxicol. Pathol.*, 44:457–464.
369. Williams, G. M. (1997): Chemicals with carcinogenic activity in the rodent liver; a mechanistic evaluation of human risk. *Cancer Lett.*, 118:1–14.
370. Williams, G. M. (1999): Chemical-induced preneoplastic lesions in rodents as indicators of carcinogenic activity. In: *The Use* of *Short- and Medium-Term Tests for Carcinogens and Data on Genetic Effects in Carcinogenic Hazard Evaluations*, edited by D. B. McGregor, J. M. Rice, and S. Venitt, IARC Publ. No. 146, pp. 185–202. International Agency for Research on Cancer, Lyon, France.
371. Williams, G. M. (1999): Mechanistic considerations in cancer risk assessment. *Inhal. Toxicol.*, 11:549–554.
372. Williams, G. M. and Furuya, K. (1984): Distinction between liver neoplasm promoting and syncarcinogenic effects demonstrated by exposure to phenobarbital or diethylnitrosamine either before or after *N*-2-fluorenylacetamide. *Carcinogenesis*, 5:171–174.
373. Williams, G. M. and Weisburger, J. H. (1991): Chemical carcinogenesis. In: *Toxicology: The Basic Science of Poisons*, edited by M. O. Amdur, J. Doull, and C. D. Klaassen, pp. 127–200. Pergamon Press, New York.
374. Williams, G. M. and Perone, C. (1996): Mechanism-based risk assessment of peroxisome proliferating rodent hepatocarcinogens. In: *Peroxisomes: Biology and Role in Toxicology and Disease*, Vol. 804, edited by J. K. Reddy, T. Suga, G. P. Mannaerts, P. B. Lazarow, and S. Subramani, pp. 554–572. The New York Academy of Sciences, New York.
375. Williams, G. M. and Jeffrey, A. M. (2000): Oxidative DNA damage: endogenous and chemically induced. *Regul. Toxicol. Pharmacol.*, 32:283–292.
376. Williams, G. M. and Iatropoulos, M. J. (2001): Principles of testing for carcinogenic activity. In: *Principles and Methods of Toxicology*, 4th ed., edited by A. Wallace Hayes. Taylor & Francis, Philadelphia, PA, pp. 959–1000.
377. Williams, G. M. and Iatropoulos, M. J. (2002): Alteration of liver cell function and proliferation: differentiation between adaptation and toxicity. *Toxicol. Pathol.*, 30: 41–53.
378. Williams, G. M., Chandrasekaran, V., Katayama, S., and Weisburger, J. H. (1981): Carcinogenicity of 3-methyl-2-naphthylamine and 3,2′-dimethyl-4-aminobiphenyl to the bladder and gastrointestinal tract of the Syrian golden hamster with atypical proliferative enteritis. *J. Natl. Cancer Inst.*, 67:481–488.
379. Williams, G. M., Laspia, M. F., and Dunkel, V. C. (1982): Reliability of the hepatocyte primary culture/DNA repair test in testing of coded carcinogens and non carcinogens. *Mutat. Res.*, 97:359–370.
380. Williams, G. M., Mori, H., and McQueen, C. A. (1989): Structure–activity relationships in the rat hepatocyte DNA-repair test for 300 chemicals. *Mutat. Res.*, 221:263–286.
381. Williams, G M., Iatropoulos, M. J., and Weisburger, J. H. (1996): Chemical carcinogen mechanisms of action and implications for testing methodology. *Exp. Toxicol. Pathol.*, 48:101–111.
382. Williams, G. M., Karbe, E., Fenner-Crisp, P., Iatropoulos, M. J., and Weisburger, J. H. (1996): Risk assessment of carcinogens in food with special consideration of non-genotoxic carcinogens. *Exp. Toxicol. Pathol.*, 48:209–215.
383. Williams, G. M., Iatropoulos, M. J., Jeffrey, A. M., Luo, F. Q., Wang, C.-X., and Pittman, B. (1999): Diethylnitrosamine exposure-response for DNA ethylation, hepatocellular proliferation and initiation of carcinogenesis in rat liver display non-linearities and thresholds. *Arch. Toxicol.*, 73:394–402.
384. Williams. G. M., Iatropoulos, M. J., and Whysner, J. (1999): Safety assessment of butylated hydroxyanisole and butylated hydroxytoluene as antioxidant in food additives. *Food Chem. Toxicol.*, 37:1027–1038.
385. Williams, G. M., Iatropoulos, M. J., and Jeffrey, A. M. (2002): Anticarcinogenicity of monocyclic phenolic compounds. *Eur. J. Cancer Prev.*, 11(Suppl. 2):S101–S107.
386. Williams, G. M, Iatropoulos, M. J., and Jeffrey, A. M. (2005): Thresholds for DNA-reactive (genotoxic) organic carcinogens. *J. Toxicol. Path.*, 18:69–77.
387. Wiltse, J. and Dellarco, V. L. (1996): U.S. Environmental Protection Agency guidelines for carcinogen risk assessment: past and future. *Mutat. Res.*, 365:3–15.
388. WHO. (1969): *Principles for the Testing and Evaluation of Drugs for Carcinogenicity*, WHO Technical Report Series No. 426. World Heath Organization, Geneva.
389. Wutzler, P. and Thust, R. (2001): Genetic risks of antiviral nucleoside analogues: a survey. *Antiviral Res.*, 49:55–74.
390. Yamamoto, S., Urano, K., Koizumi, H., Wakana, S., Hioki K., Mitsumori, K., Kurokawa, Y., Hayashi, Y., and Nomura T. (1998): Validation of transgenic mice in carrying the human prototype c-Ha-ras gene as a bioassay model for rapid carcinogenicity testing. *Environ. Health Perspect.*, 106:57–69.
391. Zeiger, R. (1998): Identification of rodent carcinogens and non-carcinogens using genetic toxicity tests: premises promises, and performance. *Regul. Toxicol. Pharmacol.* 28:85–95.

# 26 Principles of Clinical Pathology for Toxicology Studies

*Robert L. Hall and Nancy E. Everds*

## CONTENTS

Clinical pathology is an integral component of preclinical safety assessment and toxicity studies designed to identify target organs and establish dose–response relationships. In the context of these studies, clinical pathology usually consists of relatively routine hematology, clinical chemistry, and urinalysis tests. The majority of parameters evaluated are identical to those used in human and veterinary medicine because the fundamental physiology and pathophysiology of blood and major organ systems are similar in most mammalian species. There are, of course, species differences for reference intervals, some methodologies, the value or appropriateness of individual tests, and interpretation of findings. Selection of tests for a toxicology study is dependent on several factors, including study objectives, test species, regulatory requirements, and test article characteristics.

Clinical pathology tests are best characterized as screening tools to identify general metabolic or pathologic processes and target tissues. Although specific diagnoses and precise mechanisms for a toxic effect are infrequently identified, test results narrow the possibilities and help direct further studies. Clinical pathology tests also provide one measure for determining the biological importance of effects associated with test article administration. Alterations in clinical pathology test results are typically not the only evidence of adverse or pathologically significant toxicologic effects. In-life clinical observations and anatomic pathology findings usually corroborate pathologically meaningful laboratory findings.

Interpretation of clinical pathology data from a toxicology study is considerably different from the assessment of data from an individual patient suffering from an unknown illness. The most obvious difference is that data from groups of treated subjects receiving increasing dose levels of a test article are compared with data from a group of age-, weight-, and sex-matched control subjects that are concurrently exposed to the same environmental and experimental conditions. For larger laboratory animals (e.g., rabbits, dogs, monkeys), pretreatment clinical pathology data for each individual are also available for comparison with posttreatment results. Finally, clinical pathology results from a toxicology study can be correlated with carefully recorded in-life observations and necropsy observations, organ weight data, and histopathologic findings for an extensive tissue list. Given the uniformity of the animals studied and the analytical precision of modern clinical pathology instrumentation, identification of subtle changes that would not be apparent for an individual patient is the norm. One of the most challenging aspects of clinical pathology data interpretation for a toxicology study is differentiating potentially harmful toxic effects from changes representing homeostatic or metabolic responses to benign test article effects or study-related procedures. Proper interpretation of clinical pathology results from a toxicology study requires not only an understanding of the tests but also knowledge of species differences, study design, unique study-related procedures, clinical observations, anatomic pathology findings, and the test article. Interpretation of one test result is frequently dependent upon the results of another test, and pattern recognition is essential.

This chapter addresses (1) experimental design considerations, including test selection, timing and frequency of testing, sources of variability in clinical pathology test results with emphasis on preanalytical factors, and quality control; (2) basic principles of clinical pathology data interpretation, including the use and misuse of reference intervals; and (3) the characteristics and interpretation of routine hematology, clinical chemistry, and urinalysis tests used in toxicology studies. For in-depth descriptions of clinical pathology tests, including methods, the reader is referred to References 1 through 10.

## EXPERIMENTAL DESIGN CONSIDERATIONS

The value of clinical pathology in toxicology studies is heavily influenced by the experimental design and technical skill of the laboratory. Selection and timing of appropriate tests, consideration of unique study procedures, reduction of sources of variation, proper sample collection and handling, and controlled analytical technique are all factors that determine the value of clinical pathology test results.

**TABLE 26.1**
**Examples of Basic Tests Applicable to Most Rat, Dog, and Monkey Studies**

| Hematology and Coagulation | Clinical Chemistry | Urinalysis |
|---|---|---|
| RBC count | Glucose | Color and clarity |
| Hemoglobin | Urea nitrogen (or urea) | Overnight volume (e.g., 16-hour) |
| Hematocrit | Creatinine | Urine specific gravity |
| Mean corpuscular volume | Total protein | Reagent strip tests: pH, protein, glucose, ketones, bilirubin, urobilinogen, blood |
| Mean corpuscular hemoglobin | Albumin | Microscopic examination of sediment: cells, casts, crystals, bacteria, sperm |
| Mean corpuscular hemoglobin concentration | Globulin (calculated) | |
| RBC morphology | A/G ratio (calculated) | |
| WBC count | Cholesterol | |
| WBC differential count | Total bilirubin | |
| Platelet count | Alanine aminotransferase | |
| Blood and bone marrow smears | Aspartate aminotransferase | |
| Prothrombin time | Alkaline phosphatase | |
| Activated partial thromboplastin time | Gamma-glutamyltransferase | |
| | Creatine kinase | |
| | Calcium | |
| | Inorganic phosphorus | |
| | Sodium | |
| | Potassium | |
| | Chloride | |

## TEST SELECTION

The selection of appropriate clinical pathology tests for a toxicology study is ultimately dependent on the study objective or purpose (Table 26.1). If the objective is simply to screen a number of similar chemical entities for potential hepatocellular toxicity, the laboratory evaluation might be limited to a few targeted tests such as alanine aminotransferase, glutamate dehydrogenase, and sorbitol dehydrogenase activities. Conversely, if the study is part of the package of studies required to support government approval of a new drug or other chemical entity with potential human exposure (e.g., food additives, pesticides, chemicals used in manufacturing), several tests are required or recommended in study guidelines published by the presiding governmental regulatory agencies (e.g., U.S. Food and Drug Administration [FDA], U.S. Environmental Protection Agency [EPA], Japanese Ministry of Health, Labor, and Welfare [MHLW]) or professional standards organization (e.g., the European Organization for Economic Cooperation and Development [OECD]) [11]. Although the various published guidelines are relatively similar with respect to recommended clinical pathology tests, several differences and instances of ambiguous or inappropriate testing requirements exist.

In an effort to encourage global harmonization of regulatory guidelines, a Joint Scientific Committee for International Harmonization of Clinical Pathology Testing was formed in 1992 to provide recommendations for clinical pathology testing of laboratory animals used in regulated safety assessment and toxicity studies. The committee was comprised of representatives from ten professional organizations located throughout the world, with scientific expertise in animal clinical pathology, and was independent of the International Conference on Harmonization of Technical Requirements for Registration of Pharmaceuticals for Human Use (ICH). The committee prepared a document listing minimum recommendations for clinical pathology testing in regulated safety assessment and toxicity studies [12]. These recommendations are described in the following paragraphs, along with comments from the authors reflecting updated approaches and new methodologies.

With respect to hematology, the core recommended tests are total white blood cell (WBC) count, absolute differential WBC count, red blood cell (RBC) count, hemoglobin concentration, hematocrit (or packed cell volume), mean corpuscular volume (MCV), mean corpuscular hemoglobin (MCH), mean corpuscular hemoglobin concentration (MCHC), evaluation of RBC morphology, and platelet count. The importance of calculating absolute WBC differential counts from the total WBC count and the relative (percent) WBC differential counts is stressed. The method for evaluation of RBC morphology is not defined, but many laboratories prepare blood smears and examine the RBCs microscopically for morphologic characteristics such as size variation (anisocytosis, microcytosis, or macrocytosis), color (polychromasia), shape (poikilocytosis), and hemoglobin content (hypochromasia). Other laboratories may choose to evaluate RBC morphology with the use of automated measurements such as MCV, MCH, MCHC, red cell distribution width (RDW), and hemoglobin distribution width (HDW). If automated methods are used for determination of hematology parameters, it is prudent to routinely prepare blood smears in the event that the data indicate a need to examine the cells microscopically.

At the time the paper was published, the Joint Scientific Committee did not recommend routine determination of reticulocyte count because these counts were typically done manually and were, therefore, labor intensive and relatively imprecise. At present, however, most commonly used hematology analyzers provide accurate automated reticulocyte counts for multiple species, and absolute reticulocyte count should be included as a standard hematology test. If the hematology analyzer is unable to perform an automated reticulocyte count, peripheral blood smears stained with a supravital dye (e.g., new methylene blue) should be prepared for possible manual enumeration of reticulocytes. Manual counts would then be indicated whenever a significant unexplained decrease in red cell mass existed.

The Joint Scientific Committee recommends that bone marrow smears be prepared at study termination for possible cytologic examination. Unexplained nonregenerative anemia, leukopenia, thrombocytopenia, and pancytopenia are possible reasons for performing bone marrow cytologic examinations. Routine performance of cytologic examinations, myeloid-to-erythroid (M:E) ratios, or bone marrow differential cell counts was not recommended.

Prothrombin time (PT) and activated partial thromboplastin time (APTT), or appropriate alternatives such as the Thrombotest [13], and platelet count are the core recommended tests for hemostasis assessment. If blood volume limitations or sample quality are a concern (e.g., multiple blood collections for a rat study; retroorbital plexus/sinus collections), it may be necessary to perform PT and APTT only at study termination.

Current regulatory guidelines for carcinogenicity and oncogenicity studies recommend performing hematology tests (e.g., WBC differential counts) on some or all animals at set intervals (e.g., weeks 26, 52, 78, and 104). In contrast, the Joint Scientific Committee recommends preparation of blood smears for all animals at unscheduled sacrifices (e.g., moribund animals) and at terminal sacrifice for possible examination as an adjunct to histopathology for the identification and differentiation of hematopoietic neoplasia. For example, if the histopathologist is unsure whether leukocytic infiltrates in multiple tissues represent leukemia or a leukemoid response (i.e., marked leukocytosis secondary to an inflammatory stimulus), the blood smear from that animal can be examined to help differentiate between the two conditions. Because rodent hematopoietic neoplasias often become leukemic only late in the disease process [14], after histologic evidence is sufficient for diagnosis, WBC differentials at set intervals are not useful.

With respect to clinical chemistry, the core recommended tests are glucose, urea nitrogen, creatinine, total protein, albumin, globulin (calculated from total protein and albumin), cholesterol, calcium, sodium, potassium, and selected tests of hepatocellular and hepatobiliary health and function. Measurement of at least two scientifically appropriate tests for hepatocellular evaluation (e.g., alanine aminotransferase, sorbitol dehydrogenase, glutamate dehydrogenase, aspartate aminotransferase, or total bile acids) and at least two scientifically appropriate tests for hepatobiliary evaluation (e.g., alkaline phosphatase, gamma-glutamyltransferase, 5′-nucleotidase, total bilirubin, or total bile acids) are recommended. Because several acceptable tests are used to evaluate hepatic health and function, the Joint Scientific Committee recommends that each laboratory should have the freedom to choose those tests that best meet their individual needs and with which they have the most experience. For example, glutamate dehydrogenase is commonly evaluated in Europe, but the availability of commercial kits for this enzyme assay is limited in the United States. A recent publication from the American Society for Veterinary Clinical Pathology provides an additional description of clinical pathology indicators of hepatic injury in preclinical studies [15].

Core recommended urinalysis tests performed on an overnight sample (i.e., approximately 16-hour collection) are an assessment of urine appearance (color and turbidity), volume, specific gravity or osmolality, pH, and either the quantitative or semiquantitative determination of protein and glucose.

The Joint Scientific Committee lists several tests that are specifically not recommended for routine use in animal toxicity and safety studies. These tests include ornithine decarboxylase, ornithine carbamoyltransferase, lactate dehydrogenase, creatine kinase, serum or plasma protein electrophoresis, microscopic examination of urine sediment, and urinary mineral and electrolyte excretion (e.g., urine sodium, potassium, chloride, calcium, or inorganic phosphorus excretion). Although ornithine decarboxylase has appeared in the test lists of several regulatory guidelines, it has no value as a diagnostic clinical chemistry test [16]. This enzyme may have been included in the original FDA guidelines by mistake, and the error was repeated by other organizations. The FDA may have intended to include ornithine carbamoyltransferase, a liver-specific enzyme involved in the urea cycle that enjoyed limited popularity as a diagnostic test in the late 1970s. Ornithine carbamoyltransferase has never demonstrated a clear diagnostic advantage over other more common liver enzymes (e.g., alanine aminotransferase) and is rarely measured. Lactate dehydrogenase is similar to aspartate aminotransferase (i.e., nonspecific) but generally more variable and is not considered beneficial. Creatine kinase may be helpful for evaluating test articles that caus muscle injury but is not considered necessary for most test articles. Blood collection techniques that potentially damage tissue and contaminate the blood sample (e.g., cardiac puncture and retroorbital plexus/sinus collection) can diminish the value of measuring muscle enzyme activities by increasing variability.

As a diagnostic test for patients, serum or plasma protein electrophoresis is used to evaluate large, unexplained increases or decreases in globulin concentration. With respect to increased globulin concentration, protein electrophoresis is used to rule out a monoclonal gammopathy caused by some cancers of lymphoid origin (e.g., plasma cell myeloma). Monoclonal gammopathies and large, unexplained decreases in globulin concentration are rare in toxicity and safety studies; therefore, the routine use of protein electrophoresis is inappropriate. Analysis of stored serum samples can easily be accomplished if electrophoresis appeared indicated. Monoclonal protein spikes that represent the test article are occasionally observed following intravenous administration of some biotechnology products (e.g., a monoclonal antibody) if the blood sample is collected relatively soon after dosing.

Microscopic examination of urine sediment may be helpful for screening test articles that are known to cause severe renal or bladder toxicity, but histopathology is a more sensitive tool for detecting lesions of the kidney and bladder. In part, this is because the collection of high-quality urine specimens from many animals at one time is very difficult. On rare occasions, examination of urine sediment may be valuable for detecting the presence of crystals specific for a test article. Measurement of urinary mineral or electrolyte excretion may be appropriate for test articles that are known to affect renal function (e.g., diuretics) or bone metabolism (e.g., parathyroid hormone), but as routine screening tests these are excessive. If serum/plasma mineral and electrolyte concentrations are greatly affected by a test article and other causes for these findings are ruled out (e.g., vomiting, diarrhea, renal failure), an assessment of the renal handling of the mineral or electrolyte in question may then be valuable.

The recommendations of the Joint Scientific Committee and the requirements listed in the various regulatory guidelines should be viewed as a minimum database for screening the potential toxicity of a test article that will likely undergo extensive evaluation in studies of varying duration with multiple species before regulatory approval. Many additional tests may be appropriate to perform, depending on the study objectives and known characteristics of the test article. Platelet function tests (e.g., platelet aggregation and bleeding times) may be appropriate for evaluating drugs that target platelets. Analysis of methemoglobin concentration or the enumeration of Heinz bodies in erythrocytes can be valuable for assessing oxidative injury caused by a compound. Serum troponin I and T can be useful markers of cardiac muscle damage. Activity of plasma or RBC cholinesterase is a measure of exposure to organophosphates or carbamates. Activity of brain cholinesterase is a measure of the toxic effect of these compounds. Determination of urinary enzyme activities and characterization of urinary protein excretion can be used for screening related compounds known or suspected to cause renal toxicity [17–20]. Various hormones may be measured when endocrine dysfunction is suspected, and cholesterol fractions may help define effects on lipid metabolism. Serum inorganic phosphorus and chloride concentrations are commonly measured as part of standard chemistry profiles in the United States and can be helpful when assessing changes in calcium and sodium concentrations, respectively.

With the renewed industry and regulatory interest in biomarkers for both efficacy and toxicity, the list of potentially valuable clinical pathology tests will grow. All routine laboratory tests are biomarkers, but novel biomarkers are being sought to provide more specific information about the status of individual organs and overall health. These new tests should be used judiciously. That is, a nonstandard or novel test should be used for a specific purpose and only after the test has been proven useful for the species in question.

### Frequency and Timing of Testing

Frequency and timing of clinical pathology testing are dependent on study objectives and duration, the biological activity of the test article, and the species tested. The Joint Scientific Committee made minimum recommendations [12] that may be modified because of these factors. With respect to regulated acute or single-dose toxicity studies, it is interesting to note that the regulatory guidelines have no clinical pathology requirements. Traditionally, in these studies, animals are dosed once and sacrificed following a 2-week observation period. Although the purpose of many of these studies is to determine appropriate dose levels for future repeat-dose studies, a great deal can be learned about potential test article toxicities because the dose levels are frequently higher than those administered in repeat-dose studies. If clinical pathology tests are utilized, it is a mistake to wait until the end of the observation period to obtain samples. Two weeks following the administration of near-lethal doses of carbon tetrachloride or mercuric chloride to rats, for example, standard clinical pathology tests will fail to recognize the effects of these compounds on the liver or kidney, respectively, because of the regenerative capacity and large functional reserve of these tissues. Although no single optimal time exists for all test articles, a general guideline for clinical pathology testing following an acute dose is to obtain samples approximately 48 to 72 hours postdose. Occasionally, specific clinical pathology tests may be most appropriate within hours after dosing because the test article has a very short half-life or a limited duration of action (e.g., some cytokines, peptides, and organophosphates). Following administration of acute renal toxins, urinary enzyme activities tend to be greatest in the first 24 to 48 hours postdose [20]. Alternatively, the effects of some test articles (e.g., cytotoxic chemotherapeutic agents) may take longer to reach their peak. For these test articles, it may be best to wait several days before performing clinical pathology tests or to take samples at multiple times to determine the time of greatest effect and pace of recovery (e.g., days 4, 7, 10, and 14).

Limited blood volume in mice dictates that blood sample collection for clinical pathology is practical only when the animals are sacrificed. Because the blood volume of a 30-g mouse is less than 2 mL, it is difficult to acquire a full milliliter of blood from a single mouse, regardless of the blood collection technique. If several tests are required or desired (e.g., full hematology and clinical chemistry profiles), it may be necessary to specify certain animals for hematology tests and others for clinical chemistry tests. A preferred alternative is to limit the number of clinical chemistry tests to those that are the most relevant as screening tools for major organ toxicity (e.g., urea nitrogen or creatinine, alanine aminotransferase or another liver-specific enzyme, total protein, albumin). Pooling of blood samples is inappropriate for clinical pathology testing.

Prestudy clinical pathology testing is not recommended for rats because of the relatively large number of animals per group, homogeneity of the population, irrelevance of prestudy data due to rapid age-related changes in values for young rodents, and risk of adversely affecting the health of young animals due to blood loss or the blood collection procedure. For repeat-dose studies in rats, testing should at least be done at study termination. Interim testing may be unnecessary for long-duration studies (e.g., a 13-week study) if testing was done in short-duration studies (e.g., a 4-week study) that used dose levels not substantially lower than those of the long-duration studies. Conversely, interim testing (e.g., a week 6 testing point during a 13-week study) can be beneficial for interpretation of subtle effects. Clinical pathology testing is not routinely recommended for rodents after 52 weeks because naturally occurring geriatric disease conditions (e.g., ulceration and infection of mammary gland tumors, chronic progressive nephropathy, and pituitary tumors disrupting normal endocrine function) obscure meaningful interpretation of laboratory data.

For repeat-dose studies in large animals (e.g., rabbits, dogs, and monkeys), testing should be done before initiation (i.e., prestudy or baseline), at least once during the study, and at study termination. Animals shipped from a supplier should have several days to acclimate to the new environment before baseline testing is performed. The baseline data are important for screening out animals with potential health problems or results that notably differ from those of their peers. Findings that may signal possible subclinical health problems and eliminate animals from study consideration include low values for erythrocyte parameters (e.g., RBC count, hemoglobin concentration, hematocrit, MCV), low or high WBC and neutrophil counts, low total protein and albumin concentrations, high globulin concentration, and high liver enzyme activities. There may be a specific need to eliminate otherwise healthy animals because they have findings that might complicate interpretation of test-article-related effects; for example, beagles with factor VII deficiency have slightly prolonged PTs. Even though these beagles are clinically normal, it may be inappropriate to include one in a study evaluating a product that targets coagulation. One might choose to exclude an animal with a low-normal neutrophil count from a study of a chemotherapeutic drug because it may be more difficult to determine if, or how much, the drug is impacting neutrophil production. By the same token, if an animal has an unusually high neutrophil count at a baseline interval simply because of excitement or fear during blood collection (i.e., physiological leukocytosis), then a subsequent postdose decrease in the neutrophil count, when the animal is less fearful of the procedure, could be overinterpreted as caused by the test article.

The advantage of using larger test animals with greater blood volume, enabling more frequent clinical pathology evaluations, is offset by the low number of animals in each treatment group (often four or fewer per sex per group) and the increased variability between animals for many tests. Data interpretation is enhanced by having more than one baseline interval for studies using monkeys and dogs, regardless of the number of animals per sex per group. An additional benefit to multiple baseline intervals is that the animals become more accustomed to the blood collection procedures, and variability caused by excitement or fear is generally reduced. Some range-finding studies only use one animal per sex per group, and each animal serves as its own control. Using at least two baseline intervals affords an appreciation for normal day-to-day, intra-animal variability. Finally, two baseline intervals are generally desirable for animals that have been surgically manipulated (e.g., placement of indwelling intravenous catheters) to avoid using animals with iatrogenic complications.

The nature of the test article often dictates the timing and number of clinical pathology intervals for repeat-dose studies in dogs and monkeys. For studies of 6 weeks' duration or less, an interim testing interval within 7 days of initiation of dosing may be useful. The primary purpose for this early interval is to detect transient changes, such as increases in serum enzyme activities, that may be absent at later intervals [21]. This information can be very important for clinical trials. For studies with cytotoxic chemotherapeutic agents, the number and frequency of hematology intervals are often considerable because common objectives are to identify the nadir for circulating leukocyte

**TABLE 26.2**
**Examples of Sources of Preanalytical Variation**

| Physiological | Procedural | Artifact |
|---|---|---|
| Age | Blood collection site and technique | Poor-quality sample (e.g., hemolyzed or clotted) |
| Sex | Order of sample collection and analysis | Improper anticoagulant |
| Strain | Anesthesia | Too much anticoagulant |
| Diet | Other study design factors (e.g., continuous infusion, vehicle, iatrogenic blood loss) | Improper sample storage |
| Fasted/nonfasted | | |
| Time of sample collection | | |
| Excitement/fear | | |
| Stress | | |

counts and the timing of hematopoietic recovery. A single-dose study of a chemotherapeutic agent in dogs might include hematology tests twice before initiation of treatment and at days 4, 7, 10, 14, 21, and 28 and clinical chemistry tests twice before initiation of treatment and at days 7 and 28.

Urinalysis testing should be conducted at least once during a repeat-dose study. It is best to conduct the urinalysis testing at the same time as other clinical pathology tests. Although not stated in the Joint Scientific Committee's document, urinalysis testing for mice is impractical and not recommended as a routine test.

## Sources and Control of Preanalytical Variation

Although most toxicology studies are relatively well controlled, many study design and procedural factors affect variability of the data and impact clinical pathology evaluation and interpretation. To identify subtle (and, in some cases, not so subtle) effects on clinical pathology test results, preanalytical sources of variation should be reduced whenever possible within the limitations of the study. Sources of variation can be loosely categorized as physiological, procedural, and artifactual (Table 26.2). Physiological sources of variation include differences associated with age, strain, sex, diet, fasted condition, excitement or fear, stress, and time of blood collection. Procedural sources of variation include order of sample collection (i.e., group order vs. randomized), blood collection site and technique, and anesthesia. Causes of artifactual or spurious results include poor-quality specimens (e.g., partially clotted hematology samples, hemolyzed serum or plasma samples), inappropriate use of an anticoagulant, improper sample storage, and iatrogenic blood loss.

Initiation of treatment for most regulated toxicology studies occurs when the animals are relatively young and still in a growth phase (e.g., rats 6 to 8 weeks old and beagles 4 to 6 months old). As the animals mature, the results of several clinical pathology parameters change. Typical changes in most species include increasing RBC count, hemoglobin concentration, hematocrit, absolute neutrophil count, total protein, and globulin concentration and decreasing reticulocyte count, MCV, absolute lymphocyte count, alkaline phosphatase activity, and inorganic phosphorus concentration. Although these and other age-related changes may be subtle, they are sufficient to evoke false conclusions if interpretation of posttreatment data were based solely on comparisons with pretreatment or baseline data collected more than 1 to 2 weeks earlier. As previously mentioned, the variability of a number of parameters becomes much greater for aging rodents, especially those over 1 year of age. This variability significantly reduces the likelihood of drawing meaningful conclusions from clinical pathology data collected in the latter half of chronic rodent studies.

Strain differences, especially for rodents, are important to consider when evaluating clinical pathology data. Differences for hematology parameters tend to be the most obvious; for example, Fischer 344 rats tend to have lower leukocyte counts than those of Sprague–Dawley rats but are also more predisposed to developing large, granular lymphocytic leukemia [22,23]. An important difference has been identified for RBCs of cynomolgus monkeys (*Macaca fascicularis*) based on geographic origin. Cynomolgus monkeys from China and contiguous areas such as Vietnam have much larger, but fewer, RBCs than cynomolgus monkeys from Indonesia, the Philippines, or Mauritius [24]. The differences are so great that reference intervals for these animals may not overlap for parameters such as RBC count and MCV. Interpretation of hematologic effects could easily be compromised if monkeys from these different geographic origins were used in the same toxicology study or perhaps different studies of the same test article. Although not yet reported, it is possible that other differences for clinical pathology test results exist between these populations of cynomolgus monkeys.

Diet clearly affects many clinical pathology parameters, and standard diets are necessary to avoid small differences that might be misinterpreted. Comparison of data from animals fed purified or unusual diets with data from animals fed standard diets (e.g., historical reference ranges) should be done with caution. Some species, such as the rabbit and hamster, are prone to the effects of atherogenic diets and exhibit very high cholesterol concentrations when fed these diets. The amount of protein in the diet is known to affect urea nitrogen concentration but likely has subtle effects on other parameters over time. Because some diets, especially for dogs, are more prone than others to cause false-positive results for fecal occult blood, diet can be an important factor when assessing the potential of a test article to cause gastrointestinal ulceration or hemorrhage.

Much has been published on the effects of fasting animals prior to blood collection for clinical pathology [25–29], and differences of opinion exist concerning this practice, especially with respect to rodents. Although most laboratories routinely fast dogs and monkeys overnight prior to blood collection, procedures for rodents differ among laboratories. Historically, fasting has been encouraged in clinical practice as a means of reducing the variability of certain parameters, most conspicuously glucose concentration, so the physician or veterinarian can more readily compare the results from a single patient with reference intervals. In toxicology studies, because the concurrent control group is much more relevant for comparison purposes than are historical reference intervals, the key principle is to treat all of the groups the same with respect to conditions prior to blood collection.

Fasting of mice for longer than a few hours is discouraged because mice reduce their water consumption when fasted and rapidly become dehydrated. Not only does dehydration affect many clinical pathology parameters (e.g., RBC count, serum protein concentrations, urea nitrogen concentration), it results in more difficult blood collection and can reduce sample quality and volume. Mice are sometimes fasted for a limited time before blood collection (e.g., 4 hours) because blood collection from mice is usually a terminal procedure and fasting reduces hepatocyte glycogen stores and may improve histopathologic detection of hepatocellular injury or change. Because the necropsy of many animals may take several hours to complete, care must be taken to keep the period of fasting similar for all mice.

Most laboratories in the United States prefer to fast rats overnight. Although one frequently cited reason for this is to reduce variability for certain parameters such as glucose, perhaps the most important reason is to standardize the conditions for all animals. If animals in the high-dose group are eating poorly, providing all animals with access to feed before blood collection can have the effect of comparing fed animals (i.e., the control animals) to fasted animals (i.e., the high-dose animals). Because several differences for clinical pathology parameters exist between fed and fasted animals, it is more difficult to determine if differences between the control and high-dose groups are due to a test article effect or simply to differences in overnight feed consumption. When compared with rats having access to feed, fasted rats tend to have lower WBC counts; lower urea nitrogen, cholesterol, triglyceride, calcium, and bilirubin concentrations; and lower alanine aminotransferase and alkaline phosphatase activities [26,28].

Excitement, fear, and stress can have pronounced effects on clinical pathology test results. Excitement and fear are associated with acute endogenous catecholamine release ("fight or flight" phenomenon), and stress is associated with endogenous corticosteroid release. Effects of catecholamines are immediate but short lasting (e.g., less than 30 minutes). Effects of corticosteroids tend to be more gradual and long lasting. The most obvious changes observed in very excited or frightened animals are increased erythrocyte and leukocyte counts and glucose concentrations. These changes are observed occasionally with overexcited beagles and unanesthetized monkeys that are not used to handling for blood collection. Monkeys may also react to the presence of several people in the animal room performing additional study-related procedures at the same time as blood collections. Endogenous corticosteroids affect leukocyte counts, but somewhat differently than catecholamines. Whenever possible, clinical pathology testing should be delayed for at least 1 week following shipping or surgical procedures to avoid stress-related changes.

Blood samples should be collected in a manner that minimizes the possibility of temporal biases. Examples of time-related biases include differences between morning and afternoon results (circadian effects), differences caused by delayed separation of serum from clotted blood, and day-to-day differences (may be preanalytical or analytical). These biases can be eliminated or at least minimized by randomizing the animals for blood collection and using procedures that enable blood collection and sample processing over the shortest reasonable amount of time. An alternative to true randomization is a structured pattern of bleeding such that one animal from each group is bled in succession (round-robin). When samples have been collected, they should be analyzed in the same order as the blood collection. Rearranging the samples back to group order has the potential of causing false-positive findings that are actually due to analytical drift. For example, a small drift in the analysis of chloride by ion-selective electrode (e.g., an increase of 2 mmol/L over 60 samples) may be sufficient to produce a statistically higher mean chloride concentration for the high-dose animals if the groups are sufficiently large (e.g., 15 animals per sex per group), the control animals are analyzed first, and the high-dose animals are analyzed last. If animals must be bled over 2 days because of laboratory or necropsy capacity issues, the problem of day-to-day variability can be reduced by collecting and analyzing samples from males on 1 day and females the next.

Randomization of animals for blood collection is occasionally impractical (e.g., if timed collections must follow intravenous administration of the test article). When animals must be sampled by group, it is better to sample the high-dose group immediately before or after the control group than to sample the animals in consecutive group order (i.e., control, low-dose, mid-dose, high-dose). If it is necessary to sample the animals in consecutive group order, then procedures must be in place to analyze or process the samples in a similar time frame. When control animals are bled 1 hour or more before the high-dose animals, analysis of the hematology samples and separation of the chemistry samples should not be delayed until all groups are bled. Such a delay can result in differences between the control and high-dose groups that are due solely to time-related, *in vitro* changes. This is most likely to occur for rodents when blood collection is one of the terminal procedures and several hours are necessary to bleed, sacrifice, and necropsy all of the animals. If clotted blood from the control group is allowed to stand 1 or 2 hours longer than that from the high-dose group before separation of serum, the high-dose group will often have statistically significant differences for several chemistry parameters, including higher glucose concentration, lower potassium and inorganic phosphorus concentrations, and lower aspartate aminotransferase and lactate dehydrogenase activities. These differences result from *in vitro* changes in the control animal samples (i.e., the consumption of glucose by erythrocytes and the release of cell constituents by erythrocytes, leukocytes, and platelets). Circadian effects are another potential source of variation when blood collection is protracted; for example, rodents bled early in the morning can have slightly higher leukocyte counts than those bled in the afternoon during their normal period of inactivity.

Collection site and technique and use of anesthesia are perhaps the most commonly cited procedural influences on clinical pathology test results. Many investigators have analyzed differences in data resulting from these variables [22,26,30–42], especially for the rat. Although differences in the results do exist (e.g., total leukocyte counts in samples from the retroorbital venous plexus or sinus are higher than those from larger vessels such as the abdominal aorta or posterior vena cava; glucose concentrations in samples from the abdominal aorta are higher than those from other sources), laboratories should use the technique with which they have had good success obtaining high-quality samples and are most comfortable. Use of an unfamiliar or unpracticed blood collection technique introduces unnecessary variability and spurious results. With appropriate instruction and practice, any of the commonly used techniques can generate adequate results. To optimize the value of the data, however, a single method of collection should be used throughout a study or series of comparable studies. For example, anesthesia tends to decrease interanimal variability for clinical pathology results from monkeys, especially their cell counts and electrolyte concentrations. If pretreatment or baseline blood collection is performed on animals anesthetized for physical examinations or other baseline procedures, the relatively heterogeneous data obtained from unanesthetized animals during treatment may cause confusion or be misinterpreted. Similar problems can occur when interim blood samples from rodents are collected from the tail or retroorbital plexus or sinus and terminal blood samples collected from the heart or posterior vena cava. Variability may also be related to differences in bleeding technique proficiency rather than the method itself [33].

Blood collection from dogs, monkeys, and rabbits is facilitated by relatively easy access to large vessels (e.g., jugular, cephalic, and saphenous veins for dogs; femoral, cubital, and saphenous veins for monkeys; auricular artery and jugular vein for rabbits). Blood collection from mice is complicated by size and volume limitations, and terminal procedures are often used (e.g., cardiac puncture or sampling of the posterior vena cava or abdominal aorta at necropsy). Many acceptable methods are used for blood collection from rats, and the choice of technique depends on a number of factors, not the least of which is frequency of opportunities to collect blood from rats. If a laboratory performs enough studies on rats to require blood collection every 1 to 2 days, then it is probably worthwhile for the technical staff to become proficient at collecting blood from the jugular vein [43]. Although every bleeding technique has procedural advantages and disadvantages, this technique offers the greatest number of advantages if mastered. Several high-quality blood samples can be collected from one animal, directly from a large vessel with needle and syringe, without anesthesia or expensive equipment or risk of damaging important structures such as the eye or heart. Because no time-consuming ancillary procedures are necessary (e.g., warming the tail to dilate the tail vein or anesthesia), samples can be collected quickly, and it is possible to accurately collect timed samples (e.g., at 1, 5, 10, and 15 minutes postdose) from a single animal. More than other techniques, blood collection from the jugular vein of rats requires regular practice to remain proficient.

Inappropriately collected or prepared samples increase variability of the data by introducing spurious or artifactual results. Fibrin or clot formation in a hematology sample always results in a spuriously low platelet count but may also cause low erythrocyte and leukocyte counts. Small clots may form because of an insufficient amount of anticoagulant (potassium ethylenediamine tetraacetic acid [potassium EDTA] for hematology samples), inadequate mixing of blood with anticoagulant, or poor blood collection technique with exposure of blood to substances from traumatized tissue. Excess anticoagulant can cause dilutional errors for cell counts. In addition to the dilutional error, manual packed cell volumes are further lowered because of erythrocyte shrinkage of the erythrocytes. The use of an inappropriate anticoagulant can cause spurious results and must be avoided. Intentional or accidental exposure of clinical chemistry samples to potassium EDTA results in very high potassium concentrations, very low calcium and magnesium concentrations due to chelation, and very low activities of enzymes such as alkaline phosphatase and creatine kinase that use magnesium as a cofactor [44]. Trisodium citrate, the anticoagulant of choice for the coagulation assays, also chelates divalent cations and would additionally increase sodium concentration if incorrectly used for the clinical chemistry sample. Excess anticoagulant in samples for coagulation assays can cause prolonged coagulation times.

Although hemolysis can be caused by test articles, it can also be caused by poor technique for sample collection, sample transport, or serum/plasma separation. Hemolysis can result in spuriously increased or decreased test results by two principal mechanisms: release of erythrocyte constituents and interference with test methodology [44–48]. Techniques and procedures should be chosen that eliminate *in vitro* hemolysis to the greatest extent possible.

As previously indicated, prolonged contact between serum and clotted blood causes spurious changes that can be controlled by prompt separation of the serum [49]. Although most analytes are relatively stable for a reasonable amount of time [6,50,51], unnecessary delay or storage of samples before analysis should be avoided. Ideally, hematology samples should be analyzed within a few hours of collection and no later than 24 hours after collection. If samples for coagulation tests cannot be run on the day of collection, the plasma should be frozen at –20°C and thawed only once before analysis [52]. If samples for clinical chemistry tests cannot be run on the day of collection, the serum or plasma (lithium heparin is the recommended anticoagulant when plasma is used for clinical chemistry) should be refrigerated or frozen overnight. If there will be a long delay before analysis or the desired analyte is relatively labile, samples should be frozen at –70°C. Samples from all animals for any given testing interval (and all intervals, if practical) should be handled the same way.

Other sources of variation for clinical pathology data are possible, but the impact of each on data interpretation in toxicology studies is generally minimized by the inclusion of age- and sex-matched control groups exposed to the same environmental conditions and undergoing the same experimental procedures. Occasionally overlooked are the effects of procedures that may differ between the control and treated groups. Because toxicology studies frequently include blood collections for analyses other than clinical pathology (e.g., toxicokinetic measurements or detection of antibody directed at a peptide test article), it is imperative that control animals be bled in a manner similar to that for the treated animals. The collection procedures and volume of blood taken for these tests can have a significant effect on hematology and clinical chemistry results [53,54], and data interpretation is seriously compromised if control animals do not undergo the same procedures. Even when control animals are bled in a like manner, iatrogenic blood loss can complicate data interpretation [55]. This is especially true for rats and small monkeys.

## Analytical Variation and Quality Control

In addition to sources of variation that occur before sample analysis, the analytical procedure itself is a source of variation. Analytical variation is minimized by a robust quality control system within the clinical pathology laboratory. Detailed discussions of quality control systems are available in many textbooks [2,56,57]. At a minimum, the quality control system should include initial verification that a new method satisfies the goals of the laboratory for accuracy and precision for the analyte being measured; standard operating procedures for all laboratory functions necessary for analysis and reporting of test results; documentation of routine instrument calibration procedures; documentation of routine and nonroutine instrument maintenance; documentation of appropriate personnel training; proper labeling of all reagents, controls, standards, calibrators, and other chemicals in the laboratory; routine analysis of quality control specimens and review of quality control data for detection of systematic errors; routine review of subject data for detection of random errors; standard procedures for responding to and documenting out-of-control situations; and participation in some form of external quality control or proficiency testing with survey samples for analysis.

Analytical variation is affected by the accuracy and precision of a test procedure. Accuracy is a measure of the extent to which the mean estimate of a quantity approaches its true value, and precision is a measure of the agreement among replicate measurements (i.e., the reproducibility of a test result). Accuracy is generally determined in a prescribed way when a new method is introduced into the laboratory, and it is continually reassessed, albeit somewhat crudely, by means of proficiency testing. Most analytical procedures exhibit a small amount of systematic bias or inaccuracy that is either constant or proportional. In other words, the mean estimate of the quantity of analyte is always in error in the same direction, either higher or lower, than the true quantity. Precision of a test can be assessed within a single run (e.g., 20 consecutive analyses of the same sample for within-run precision) and from day to day (e.g., the same sample analyzed several days in a row for between-run precision) and is reflected by the coefficient of variation (CV) of a test. The CV expresses the error or variability of replicate test results as a percentage of the mean value: (standard deviation/mean) × 100; the lower the CV, the greater the precision or repeatability of the test.

Accuracy and precision are desirable independent qualities for a clinical pathology test. Tests can be accurate but imprecise or inaccurate but precise. In the context of most toxicology studies, where results from groups of treated animals are compared with results from concurrent control groups and their own pretreatment results, precision is more valuable than accuracy. This is in contrast to the clinical setting, where the physician or veterinarian evaluates individual patients under less controlled conditions and uses broad historical reference intervals for making decisions. In a toxicology study, an imprecise test, regardless of accuracy, is less useful in detecting small differences between the control and treated groups. If the true mean glucose concentrations for 4 control dogs and 4 treated dogs are 100 and 115 mg/dL, respectively, but the standard deviations for the groups are large because of imprecision, then it is unlikely that the observed difference between the means (i.e., 15 mg/dL) will be considered a real difference. A precise test, regardless of a systematic bias or inaccuracy, is better able to identify small test-article-related differences between the groups. In the same example, if the glucose test had a positive bias but was more precise, the means of the two groups might be 110 and 125 mg/dL, respectively, with smaller standard deviations. The improved precision permits a more accurate interpretation of the same 15-mg/dL difference between group means.

Clinical pathology laboratories that analyze animal samples frequently use tests that are optimized for accuracy using human samples. It is likely that many of these tests have small systematic biases when used for animal specimens. In most cases, however, the excellent precision afforded by using standardized commercial reagent kits, standards, and calibrators is preferable to time-consuming, costly efforts to optimize the accuracy of a test for different animal species by using home-brew materials that undermine precision. Occasionally, standard human chemistry tests do not work well for certain laboratory animal species (e.g., albumin for rabbits), and these tests require specialized methods. Immunoassays utilizing monoclonal or polyclonal antibodies raised against human substances (such as hormones) are frequently inappropriate for use on animal specimens. Hematology analyzers that enumerate and differentiate blood cells must be validated as appropriate for laboratory animal species because of differences between animal and human cell morphology.

For human hospital laboratories, federal regulations allow each laboratory to set its own policies for assaying control materials as long as at least two control samples of different concentrations (i.e., normal and abnormal levels) are assayed every 24 hours. The data from each quality-control analysis are used to make decisions about the validity of the patients' data. In the setting of preclinical toxicology studies, it may be beneficial to assay at least two control samples with each study run for documentation purposes. In other words, if a laboratory is scheduled to run samples from two or more regulated toxicology studies (e.g., studies needed to make submissions to regulatory agencies and using good laboratory practices), then quality control samples may be assayed with each of the studies and the results maintained with their respective study data. If results from a control sample are found to exceed the allowable limits for one or more analytes, then steps must be taken to resolve the problem. Actions that may be taken include, but are not limited to, the following: Check for obvious problems such as reagent levels, clots, or mechanical fault; repeat the assays on control samples using fresh aliquots; repeat the assays using newly reconstituted control samples; recalibrate the instrument for the analytes in question and then reassay the control samples; change the reagents, recalibrate, and reassay the control samples; and perform maintenance, recalibrate, and reassay the control samples.

## PRINCIPLES OF DATA INTERPRETATION

Interpretation of data from toxicology studies begins with the identification of differences between control and treated groups and ends with an assessment of toxicological or biological relevance. Or, in simple terms, do real differences exist between the groups, and, if so, are those differences bad? Interpretation of clinical pathology data requires an understanding of each test's characteristics, species differences, and principles of internal medicine. Factors that influence the interpretation of a potential effect include study design and conditions, clinical observations, other clinical pathology results, anatomic pathology findings, and the test article itself. Interpretations of many clinical pathology findings are interdependent, and pattern recognition is critical. After range-finding studies have been conducted, appropriate dose selection usually precludes large, dramatic effects on clinical pathology test results. The most common effects are relatively mild and often appear secondary to small alterations in metabolic or homeostatic mechanisms; however, test articles causing significant damage to liver, kidney, or hematopoietic tissue can produce marked changes in clinical pathology results. Effects on clinical pathology results are rarely the only evidence of biologically important or adverse toxic effects. Clinical observations or anatomic pathology findings usually corroborate biologically important laboratory findings.

### STATISTICAL COMPARISONS

Statistical analysis of clinical pathology data is commonly performed in toxicology studies and often results in identification of several statistically significant differences between control and treated groups; however, all test article

**TABLE 26.3**
**Factors To Consider for Determining Whether Differences Between Control and Treated Groups Are Due to the Test Article**

| |
|---|
| Magnitude of the difference |
| Dose dependency |
| Consistency over time |
| Consistency between sexes |
| Correlative findings |
| Number of animals tested (e.g., 2/sex/group vs. 15/sex/group) |
| Number of animals affected (i.e., group or individual differences) |
| Statistical significance |
| Timing of the difference with respect to dosing |
| Known characteristics of the test article/vehicle |
| Potential sources of preanalytical variability (e.g., expected variability for species, age, and tests in question; effects of study-related procedures such as route of administration, toxicokinetic sampling, and anesthesia) |

effects caused by a test article need not be statistically significant, and all statistically significant differences do not necessarily represent test-article-related effects. If used, statistical tests should be viewed simply as a tool to help identify differences between groups and not as the principal justification for decisions concerning potential test article effects [58,59]. It is important to remember that the number of animals per group affects the power of a statistical test. Using fewer test subjects increases the likelihood that statistical tests will fail to identify a true effect. Because the number of animals per group is usually quite small for studies with dogs or monkeys (e.g., four or fewer per sex per group), it is imperative that the data for each animal at the different test intervals be examined to look for patterns of change over time among the treated animals that are absent among the control animals. As the number of animals per group increases, the frequency of identifying statistically significant differences of very small magnitude increases. In rat studies with 15 or more animals per sex per group, it is common to observe statistically significant differences that are unrelated to treatment.

## Is an Apparent Difference Real?

When faced with an apparent difference between control and treated groups, the first question the investigator must answer is whether or not that difference represents a true test-article effect or an incidental finding. Many factors can influence the answer (Table 26.3). A large difference is obviously more likely to be real than a small one. Additional factors that suggest that a difference is test article related include dose dependency, consistency over time, consistency between sexes, correlation with clinical observations (e.g., low chloride concentration and emesis), correlation with other clinical pathology findings (e.g., low hematocrit and high reticulocyte count), correlation with anatomic pathology findings (e.g., high globulin concentration and lymph node hyperplasia), presence in a large number of animals (e.g., 15 per sex per group vs. 3 per sex per group), and consistency with previously identified effects of the test article or related compounds. With large animals, it may be possible to determine whether an apparent difference was present before treatment was initiated and is, therefore, incidental. The chronology of an apparent difference is also important for interpretation; for example, following a single administration of most test articles, it is more likely that a real difference will occur within a few days rather than only after 2 weeks. Following chronic repeated administration of most test articles, it is more likely that a real difference will be apparent after a few months of treatment rather than only after a longer treatment period. In other words, it would be unusual for a real difference at 6 months to be completely absent at 3 months.

With regard to specific tests and test species, the amount of expected analytical, interanimal, and intra-animal variability can influence the interpretation of apparent differences. For example, because alanine aminotransferase activity has much greater interanimal and intra-animal variability for monkeys than for dogs, a relatively modest difference for this enzyme between control and treated groups is less likely to be a true effect for monkeys. Interanimal variability increases dramatically for older rodents (e.g., >52 weeks of age) because of naturally occurring disease conditions, so small differences between groups of older rodents are less likely to be true effects.

Procedural factors such as the route of test article administration, blood collection technique, animal handling, and randomization for blood collection also affect variability and must be considered. For example, continuous intravenous infusion increases interanimal variability for several tests (e.g., WBC count, hematocrit, and serum proteins), and interpretation of small differences in results of these tests is difficult. Blood collection by cardiac puncture affects variability of muscle enzymes such as creatine kinase and aspartate aminotransferase. The potential for spurious findings associated with lack of randomization for blood collection was discussed previously.

## Is a Real Difference Bad?

If a difference between control and treated groups is clearly real, the next question the investigator must answer is whether or not that difference represents a bad or adverse effect. Does the finding itself (e.g., low hemoglobin concentration) or the condition that caused the finding (e.g., blood loss from gastrointestinal erosions or ulcerations) represent a toxic effect that compromises the animal's health? The answer is often ambiguous and quite subjective. Many of the factors previously considered are important for this decision, and the size of the difference is clearly a focal point. Although a large difference for a given parameter is more likely to be adverse than a small difference, it is impossible to define set limits for each test that represent an adverse effect. Many other factors must be considered. The same magnitude of change for a given test can have completely different connotations depending on the mechanism for that change, correlative findings, the test species, the animals' age, the study design and procedures, and the test article itself (e.g., a clinical pathology parameter may be the target for the pharmacological activity of a drug). Urea nitrogen may be markedly increased because of dehydration (e.g., mice that have been fasted too long before blood collection or animals that refuse to drink water containing test article) but only mildly increased in the early stages of renal toxicity. Increased alanine aminotransferase activity associated with histopathological evidence of hepatocellular degeneration and necrosis is likely more important toxicologically than the same level of increase for which no correlative findings exist. Because of species differences for interanimal variability, a threefold increase for alanine aminotransferase activity for dogs (i.e., treated vs. control group means) is more likely to represent an adverse condition than the same increase for monkeys. A 10% decrease in hemoglobin concentration is less likely to represent an adverse condition in animals bled repeatedly during a study or that received the test article by continuous intravenous infusion than in animals that were not bled repeatedly or were treated by oral gavage. Very high neutrophil counts would normally reflect an adverse condition, but if the test article was a granulocyte colony-stimulating factor, high neutrophil counts would be a desirable effect.

Some tests measure analytes that are critical to good health (e.g., neutrophil count, hemoglobin concentration, glucose concentration, calcium concentration, potassium concentration), and correlative in-life observations (e.g., bacterial infections, lethargy, weakness, weight loss) may help determine whether an observed effect on such a test has impacted the animals' health. Other tests measure analytes that are markers for effects best evaluated by histopathologic examination (e.g., alkaline phosphatase, urea nitrogen). All clinical pathology tests can be altered by more than one process or mechanism, and some mechanisms have worse implications than others; for example, a 10% decrease in hemoglobin concentration associated with poor food consumption and reduced body weight gain is of less concern than the same decrease caused by Heinz body hemolysis (oxidative injury). When determining the biologic or toxicologic significance of an effect on a clinical pathology test result, consideration must be given to the analyte's normal function for maintaining health, correlative findings that may better define the overall impact of the test article on health, and the mechanism that brought about the change. All study data must be considered.

## Reference Intervals

Reference intervals (often incorrectly referred to as reference or normal ranges) are constructed with values obtained from reference individuals. In most clinical pathology laboratories used for toxicology studies, the reference individuals are control animals from previous studies and clinically healthy animals that have not received treatment (e.g., monkeys or dogs that have clinical pathology tests performed before initiation of dosing). In human medicine, the National Committee for Clinical Laboratory Standards recommends that reference intervals be estimated by the nonparametric method and that a minimum of 120 values from reference individuals be used [60]. Test results from the reference individuals are subjected to statistical treatment such that rare values at both ends of the distribution are eliminated; for example, if the lowest and highest 2.5% of the values are eliminated, the resulting reference interval represents the central 95% of the distribution of values. If the distribution is Gaussian, the interval corresponds to the mean $\pm$ 1.96 SD. By definition, when the central 95% is used as the reference interval, 5% of the results from ostensibly normal individuals (1 of 20) are outside of the reference interval for any given test. Clearly, a value outside of the reference interval does not necessarily indicate an abnormality [61].

Like statistical comparisons, historical reference intervals can be used as a tool when assessing apparent differences between control and treated groups. Unfortunately, the value of reference intervals for data interpretation is sometimes overestimated, and the potential for their improper use is great [62–64]. It is tempting to invoke historical reference intervals when trying to decide whether apparent differences for clinical pathology results are true or adverse effects. While historical reference intervals can be helpful for establishing some perspective concerning what is typical or expected, the conditions of every toxicology study are unique, and it is inappropriate to use a reference interval as the primary reason for dismissing an apparent difference between control and treated animals as being incidental or biologically insignificant. It can be equally inappropriate to use a reference interval as the primary reason for determining that an apparent difference is real or adverse. As detailed in the previous sections, many factors must be considered when evaluating the nature of an apparent test article effect.

The suitability of a historical reference interval for data interpretation for a given study is a function of the parameters or *partitioning factors* that define the reference population used to construct the interval. Many potential partitioning factors exist with respect to toxicology studies

**TABLE 26.4**
**Examples of Partitioning Factors for Constructing Reference Intervals**

| General Factors | Factors for Control Animals | Laboratory Factors |
|---|---|---|
| Species | Route of administration | Instrument |
| Strain | Vehicle/control article | Reagents/methodology |
| Age | Study-related procedures prior to clinical pathology (e.g., blood collections, anesthesia, therapeutic treatments) | Sample storage (length and temperature) |
| Sex | | |
| Supplier | | |
| Site of blood collection | | |
| Use and type of anesthesia | | |
| Diet | | |
| Fasted/Nonfasted | | |
| Time of sample collection | | |
| Matrix (i.e., serum or plasma) | | |

(Table 26.4), and their importance is sometimes overlooked. The most commonly used partitioning factors are species, strain, sex, and age; in other words, a typical reference population might be defined as male Fischer 344 rats from 8 to 10 weeks of age. Many other partitioning factors, however, can influence the reference interval and make it broader, narrower, higher, or lower. These factors include animal supplier, site of blood collection, use of anesthetic, type of anesthetic, diet, fasting status, time of sample collection, and sample matrix (e.g., serum or plasma). If control animals are used for the reference population, additional partitioning factors include route of administration (e.g., dietary, oral gavage, or intravenous infusion), vehicle or control article administered (e.g., sterile water or corn oil in a gavage study; sterile saline or 5% dextrose in an intravenous infusion study), and whether or not the animals were bled repeatedly for toxicokinetic analyses or multiple clinical pathology intervals. Finally, laboratory considerations include instrumentation or techniques used to analyze the specimen and sample storage conditions (e.g., storage temperature and time interval between collection and analysis). Ideally, whenever a new instrument or reagent system is introduced, a new reference interval is constructed. In practice, laboratories often rely on evidence that demonstrates relative consistency between the old and new analytical methods to avoid complete replacement of old intervals.

The number of partitioning factors used obviously has a direct impact on the number of reference individuals available for each reference interval. With the exceptions of reference intervals for dogs and monkeys tested before treatment is initiated (e.g., male beagles, 4 to 6 months old, XYZ supplier, jugular vein, no anesthetic, fasted), it can be difficult to obtain enough data for meaningful reference intervals. Because of this problem, laboratories often ignore many of the partitioning factors and lump together the data from control animals of dissimilar types of studies or from samples that have been handled differently. The result is broader reference intervals with less relevance to data interpretation.

Even when reference intervals are appropriate for a given study (i.e., the partitioning factors match the study animals, conditions, and procedures), it is wrong to assume that a difference between control and treated groups represents a true or adverse effect simply because values for the treated group fall outside of the reference interval. Concurrent control animal data for a given study are never a perfect reflection of the historical reference data; that is, they do not exhibit the same means and distribution. The mean for the control group may be higher or lower than the mean for the reference interval, and the distribution of results for the control group is almost always narrower than that of the reference interval. Depending on the location of the control group data within the reference interval, a very small, incidental difference for the treated group may fall outside of the reference interval, and a relatively large, adverse test-article-induced difference may fall within the reference interval. Many situations exist when a small difference, within the reference interval, represents a very significant adverse effect; for example, dehydration can result in normal RBC counts and protein concentrations in anemic, hypoproteinemic animals. Loss of an entire subpopulation of lymphocytes is not necessarily enough to cause the absolute total lymphocyte count to fall below the reference interval. When reference intervals are broad, as occurs when few partitioning factors are used or normal interanimal variability is great (e.g., alkaline phosphatase activity for monkeys), significant toxicity can occur without individual test results exceeding the limits of the intervals. Investigators must understand the limitations of reference intervals with respect to data interpretation. By themselves, reference intervals do not determine whether or not an apparent difference is real or adverse. They are simply an adjunct to sound scientific judgment.

Regardless of the many pitfalls affecting their use for data interpretation, historical reference intervals do have other important functions. They serve as a nonspecific measure of quality control that can detect changes over time in assays, study conditions, or animal characteristics. For example, it may be noticed that liver enzyme activity in mice has increased over time, and the cause of this finding might be traced to changes in handling practices or animal supplier. Reference intervals also serve as a nonspecific measure of analyte variability. The cause of seemingly excessive variability might be inadequate assay precision, nonstandardized preanalytical procedures, or true interanimal variability. Finally, reference intervals can be valuable when only treated animals are tested at an unscheduled interval because of signs of toxicity or when very few animals are used for small investigational studies with no concurrent control group. Although pretreatment baseline data are more relevant to interpretation of potential effects, it may be discovered that the few animals acquired for a small study have preexisting problems or are atypical with respect to historical data.

## HEMATOLOGY TESTS AND INTERPRETATION

The hematology tests most often performed for toxicology studies evaluate erythrocytes, leukocytes, platelets, and coagulation. Technological advances by manufacturers of hematology and coagulation analyzers have increased the availability of highly sophisticated instruments to perform tests on animal blood with precision and accuracy suitable for preclinical safety assessment studies. Automated hematology methods are a necessity in toxicology because manual methods are unacceptably imprecise, labor intensive, and slow. Regardless of the method of analysis, however, absolute WBC and reticulocyte counts (cells/unit of volume) should always be reported rather than relative counts (percent of total). Relative counts are irrelevant for interpretation of effects.

### HEMATOLOGY AND COAGULATION ANALYZERS

Many currently marketed hematology instruments are capable of measuring or calculating all of the routinely recommended tests from less than 300 μL of anticoagulated whole blood. The instruments typically use a combination of two or more principles including electrical aperture impedance, laser light scatter, and differential staining characteristics to determine cell number (RBC, WBC, and platelet counts), cell type (WBC differential count and reticulocyte count), and cell size and size distribution (MCV, mean platelet volume [MPV], and RDW). Hemoglobin concentration is measured directly. From the above parameters, hematocrit, MCH, and MCHC are calculated, and a few less common parameters may also be determined. Because the blood cells of laboratory animal species differ morphologically from human blood cells, only instruments with software validated for animal samples should be used [65].

Coagulation analyzers measure the time required for fibrin clot formation in a plasma sample to which reagent has been added. Clot detection can be accomplished by evaluating changes in conductivity, physical resistance, light scatter, or optical density. Because most instruments can now be programmed to permit detection of clot formation at times much faster (e.g., PT for dogs) or slower (e.g., PT for guinea pigs) than are typically observed for human samples, they can more easily be adapted for use on laboratory animal samples.

### ERYTHROCYTES

Effects on erythrocyte parameters typically reflect a change in the balance between RBC production and RBC loss or destruction. Changes in plasma volume (e.g., dehydration or volume expansion) can also indirectly affect erythrocyte parameters. Although mechanisms for effects on erythrocyte parameters may be obvious, such as hemorrhage from gastric ulceration, they are often relatively obscure. Even when the exact mechanism is unclear, the effects can usually be described in terms of broad mechanistic categories and impact on health. Red cell mass is evaluated by RBC count, hemoglobin concentration, and hematocrit. These parameters may or may not change proportionally, depending on the cause of the decreased red cell mass and whether or not cell size and hemoglobin content are affected. Red cell mass may be decreased or increased as a result of experimental treatment.

#### Decreased Red Cell Mass

Decreases in red cell mass must be differentiated from anemia. Anemia is defined as a condition characterized by a hemoglobin concentration below the lower reference limit but can also be defined functionally as a decrease in red cell mass sufficient to cause decreased oxygen delivery to peripheral tissues. Decreased oxygen delivery may result in clinical signs such as pallor, weakness, exercise intolerance, tachycardia, or tachypnea. Humans generally have no clinical signs of anemia until hemoglobin concentration decreases below 8 g/dL [66].

In many toxicology studies, animals in a treated group have lower RBC counts, hemoglobin concentrations, or hematocrits than those for the control animals, but the differences are less than those necessary to cause clinical signs or affect tissue oxygenation and are, therefore, not indicative of anemia. For example, reductions from control values for these parameters up to approximately 10% are relatively mild and probably do not have an adverse effect on the health of the animal. Reductions from approximately 10 to 30% may be considered a moderate effect and may or may not be clinically adverse. Reductions of more than 30% are generally marked and clearly represent a clinically adverse anemic condition. It is important to understand that, although a 5% reduction in hemoglobin concentration may be insufficient to adversely affect health, the cause of the reduction (e.g., liver toxicity, gastric ulceration, or immune-mediated red cell destruction) may be very adverse. Unless the differences for RBC count, hemoglobin concentration, and hematocrit are quite large, it is preferable to simply discuss the magnitude of the differences between the control and treated groups and avoid using the term *anemia*.

An animal's response to decreased red cell mass can be broadly categorized as regenerative or nonregenerative. A regenerative response is characterized by an appropriate increase in erythropoiesis, and a nonregenerative response is characterized by the absence of an appropriate increase. The most important parameter for determining whether the response is regenerative or nonregenerative is the absolute reticulocyte count. Reticulocytes can be counted automatically with most of the newer hematology analyzers or manually by staining blood with a supravital stain such as new methylene blue before making a blood film. For proper interpretation, the absolute reticulocyte count (i.e., reticulocytes/μL) should always be determined by multiplying the relative reticulocyte count (i.e., percent of erythrocytes) by the RBC count. If an animal is severely anemic, its relative reticulocyte count may be increased, yet its absolute reticulocyte count (most relevant because it measures the erythropoietic response) may be decreased.

### Conditions Characterized by a Regenerative Response

Regenerative responses result from two general causes of reduced red cell mass: blood loss (hemorrhage) and accelerated erythrocyte destruction (hemolysis). Following acute blood loss or hemolysis, it takes approximately 3 to 4 days for new erythrocytes (i.e., reticulocytes) to increase in number in peripheral blood. Reticulocytes are larger and contain less hemoglobin per volume of red cells than do mature RBCs. During a strong regenerative response, the increased number of reticulocytes will usually increase MCV and decrease MCHC. In practice, increased MCV is more commonly observed than decreased MCHC, especially for dogs and monkeys. In addition, the presence of variable cell size (anisocytosis) often results in increased RDW. Reticulocytes stain slightly more basophilic than mature erythrocytes with Romanowsky-type stains typically used for assessing RBC morphology and doing manual WBC differential counts (e.g., Wright or Giemsa stain). During a strong regenerative response, erythrocyte morphology is described by the terms *anisocytosis* (variable size) and *polychromasia* (variable color) because of the increased numbers of reticulocytes. Greater numbers of nucleated RBCs, also called *metarubricytes*, and Howell–Jolly bodies (small pieces of nuclear material not cleared from the erythrocyte) may also be observed during a regenerative response.

During a strong regenerative response, erythrocyte production can increase as much as sixfold to eightfold over normal levels. As long as the accelerated erythropoiesis is able to match erythrocyte loss or destruction, red cell mass parameters will remain relatively stable. If the cause of loss or destruction is removed or if erythropoiesis exceeds loss or destruction, the red cell mass will return to normal. When recovery periods are included in toxicity studies with compounds causing regenerative responses, affected animals occasionally exhibit transiently increased red cell mass parameters, during or following recovery, compared with control animals. Although the impetus for increased erythropoiesis is removed at the beginning of the recovery period, the committed erythrocyte precursors continue to mature into RBCs, resulting in this rebound effect.

#### *Blood Loss*

Blood loss occurs secondary to a variety of conditions or study-related procedures and should always be considered when a decrease in red cell mass is accompanied by a similar decrease in serum protein concentration. The source of the blood loss may be identified by clinical observations (e.g., dermal ulceration, melena, epistaxis), necropsy findings (e.g., gastrointestinal ulceration or cystitis), or other laboratory tests (e.g., fecal or urine occult blood tests). Blood loss associated with serial blood collection for pharmacokinetic investigations, clinical pathology tests, or other study-specific requirements must always be considered when interpreting changes in erythrocyte parameters. Blood loss typically results in increased reticulocyte count, MCV, and polychromasia unless the condition is very acute (i.e., less than the 3 to 4 days necessary for the regenerative response to build up) or complicating factors affect erythropoiesis (see indirect causes of nonregenerative conditions).

#### *Hemolysis*

Hemolysis is characterized by premature elimination of RBCs from circulation and is routinely classified as either intravascular or extravascular. Hemolysis is considered intravascular when the RBCs are destroyed directly within circulation and extravascular when RBCs are phagocytized, most commonly in the spleen and liver, by cells of the mononuclear phagocyte system (i.e., macrophages). Of the two, extravascular hemolysis is more commonly encountered in toxicity studies and usually results from oxidative or immune-mediated injury to the RBCs [67]. Regardless of the underlying mechanism extravascular hemolysis is characterized by splenomegaly bone marrow hypercellularity, and/or extramedullary hematopoiesis (more pronounced in rodents). Histologically, increased pigment is generally observed in the spleen and sometimes other organs, indicating increased red cell turnover and processing of hemoglobin. Increased serum bilirubin (bilirubinemia) and urine bilirubin (bilirubinuria) are possible, but less common, correlative findings. Intravascular hemolysis is characterized by the presence of free hemoglobin in plasma or urine. Processes resulting in intravascular hemolysis generally occur rapidly and may cause rapid and profound changes in red cell mass. Some hemolytic processes have both extravascular and intravascular components.

Extravascular hemolysis secondary to oxidative injury is sometimes identified by the presence of Heinz bodies. Heinz bodies are particles of irreversibly denatured hemoglobin attached to the inner surface of the RBC membrane. They result when test articles with oxidative properties cause disulfide bonds to form from the sulfhydryl groups of hemoglobin. RBCs containing Heinz bodies may be removed from circulation by macrophages or become morphologically distinct (e.g., ghost and blister cells) following selective removal of the Heinz bodies. Although Heinz bodies can be difficult to detect with the Romanowsky-type stains, they stain prominently with supravital stains (e.g., methylene blue, crystal violet, or brilliant cresyl blue) used for manual reticulocyte counts. The number of cells containing Heinz bodies can be determined in the same manner as that for manual reticulocyte counts. The size and number of Heinz bodies are dependent on the causative agent, dose, and time after exposure. Acute exposure to a potent oxidative agent typically causes an acute anemia characterized by the observation of many RBCs containing a single, large Heinz body or, less frequently, multiple, small Heinz bodies. Ghost cells, blister cells, and other morphologic abnormalities may be present. Increased reticulocyte counts (reticulocytosis) develop after 3 to 4 days. Chronic, low-level exposure to an oxidizing agent usually does not cause a significant reduction in red cell mass. Although absolute reticulocyte count and MCV may be slightly increased, hematocrit and hemoglobin concentration are relatively unchanged or only slightly reduced. The number of RBCs containing Heinz bodies is low, and they may go unrecognized if not looked for specifically. Identification of test articles that cause Heinz body formation is particularly important clinically because a deficiency of the erythrocyte enzyme glucose-6-phosphate dehydrogenase is relatively common in humans and results in greater susceptibility to oxidant-induced hemolysis [68].

Oxidant test articles that cause Heinz body formation may also cause methemoglobinemia and *vice versa* [69]. Methemoglobin is hemoglobin in which the iron has been reversibly oxidized from the ferrous state ($Fe^{2+}$) to the ferric state ($Fe^{3+}$). Cells containing increased methemoglobin concentrations are not more susceptible to hemolysis but transport less oxygen than normal. The clinical signs of significant methemoglobinemia, therefore, are those of hypoxia [70]. Mucous membranes become cyanotic at methemoglobin concentrations above 10%. Lethargy and weakness occur at concentrations of approximately 30% or more. At greater than 80%, methemoglobinemia may be fatal. Blood containing a high concentration of methemoglobin appears brown. Accurate and efficient measurement of methemoglobin concentration can be accomplished with instruments called *hemoximeters* or *cooximeters*. These multiwavelength, microprocessor-controlled photometers are designed to measure hemoglobin pigments such as carboxyhemoglobin and methemoglobin as well as the percentage of hemoglobin oxygenation. Samples collected for methemoglobin determination should be analyzed quickly (e.g., within 30 to 60 minutes) because methemoglobin is normally reduced by the erythrocyte enzyme methemoglobin reductase. The combination of a sensitive, precise instrument and a well-executed toxicology study allows detection of test-article-induced methemoglobin formation at concentrations well below those necessary to cause clinical signs [71]. Among laboratory animals, the mouse is a poor model for studying the potential of test articles to cause methemoglobinemia because their erythrocytes have very high methemoglobin reductase activity [72]. Test-article-induced methemoglobin in mice is quickly reduced to hemoglobin and, therefore, is more difficult to detect.

Immune-mediated RBC destruction has been associated with many drugs [73] but is largely an idiosyncratic phenomenon and is, therefore, detected infrequently in preclinical toxicology studies. When it is observed, only one or two animals are typically affected, and they may or may not be in the high-dose group. In contrast to hemolytic conditions that occur soon after exposure to the test article, immune-mediated hemolysis is typically not observed until the test article has been administered for at least 1 week. Three general mechanisms exist for immune-mediated hemolysis: (1) the test article acts as a hapten bound to the RBC membrane; (2) the test article elicits an antibody response, and the resulting antigen-antibody complex binds to the RBC; and (3) the test article causes the immune system to mistakenly recognize normal RBC membrane antigens as foreign, and true autoantibodies are produced. On rare occasion, immune-mediated hemolysis is complement mediated and occurs intravascularly. More commonly, cells of the mononuclear phagocyte system, especially in the spleen, recognize antibody-coated RBCs and either phagocytize the entire cell or remove a portion of its membrane, creating morphologically distinct cells called *spherocytes*. In a blood film, spherocytes appear smaller and denser than other RBCs; they are round and lack central pallor. Although small numbers of spherocytes can be observed with other conditions, they are the predominant morphologic feature of immune-mediated hemolysis. RBC autoagglutination is occasionally observed in the blood film, especially if the antibody response is primarily immunoglobulin M (IgM). Alternatively, autoagglutination may be detected in a wet mount of fresh blood diluted with saline. Direct antiglobulin tests may be used to confirm the presence of antibody or complement on RBCs, but species-specific antiimmunoglobulin or anticomplement must be used [74]. Animals with test-article-induced immune-mediated hemolytic anemia become severely anemic with repeated test article administration; however, they usually exhibit a strong regenerative response and will nearly always recover if test article administration is stopped. Confirmation that the anemia was test article induced can be accomplished by rechallenging the animal following recovery. Upon rechallenge, hemolysis should be evident within 1 to 2 days. Immune-mediated hemolytic anemia, unrelated to test article administration, is a common sequela of the large granular lymphocyte leukemia observed in a high percentage of Fischer 344 rats older than 1 year of age [23].

Additional mechanisms associated with extravascular hemolysis include any that result in RBC structural changes recognized as abnormal by the mononuclear phagocyte system. These include effects on RBC membrane lipids and proteins, intercalation of test article into the cell membrane bilayer, and effects on RBC metabolic processes.

Direct damage to the RBC membrane is most commonly caused by test articles administered intravenously. The lipid bilayer of the cell membrane is sensitive to test articles with detergent-like properties, and intravascular hemolysis can occur rapidly when RBCs are exposed to high concentrations of the test article at the site of intravenous infusion (e.g., the tip of the catheter or needle). Administration of a very hypotonic solution, as might occur if sterile water is inappropriately chosen as the vehicle for a low-concentration solution, is another potential cause of acute intravascular hemolysis. Water passively enters erythrocytes because of the ionic concentration gradient, and the cells swell and rupture. Intravascular administration of hypertonic solutions typically does not cause hemolysis. If the amount of released hemoglobin exceeds the carrying capacity of circulating haptoglobin, unbound hemoglobin passes through the glomerulus and is excreted in the urine. Visible hemoglobinuria (red-tinged urine) may be observed within a few hours of treatment, and a regenerative response is detectable within 3 or 4 days of treatment. Intravenously administered test articles that cause extensive intravascular hemolysis are usually associated with local damage to endothelium. The effect of the vascular damage may be observed grossly (e.g., tail lesions in rodents treated via the tail vein) or histologically. Additional histopathologic evidence of intravascular hemolysis is the presence of hemoglobin pigment within renal tubular epithelial cells and hemoglobinuric nephrosis. Compounds administered by other routes rarely cause significant intravascular hemolysis.

Whenever regenerative anemia is identified in a nonhuman primate study, consideration must be given to the possibility of hemolysis associated with the hemotropic parasite *Plasmodium*. Many imported monkeys, even though they are captive bred, harbor subclinical infections with this organism that causes malaria [75,76]. The readily identifiable intracellular organism is frequently observed in blood films from healthy animals that have no signs of anemia. Parasitemia, however, is inconsistent or cyclical, and multiple blood samples may be examined from an animal before the organism is identified. On rare occasions, test article administration or the stress of shipment and study-related procedures precipitates a hemolytic crisis. When this occurs, the parasitemia is usually quite obvious in the blood film.

Another infrequent cause of hemolysis in toxicology studies is mechanical fragmentation or microangiopathic hemolysis. This form of hemolysis typically occurs when RBCs are forced to pass through small, fibrin-obstructed vessels in highly vascular tissue. The resulting fragmented RBCs are called *schizocytes* (helmet cells) and are easily identified microscopically. Disseminated intravascular coagulation (DIC) is the best example of a condition causing microangiopathic hemolysis, but widespread vascular injury in tissues such as lung, liver, or intestine also causes some degree of fragmentation. These conditions, especially DIC, are generally so severe that an animal may fail to mount a significant regenerative response before it dies or is humanely euthanized.

## Nonregenerative Conditions

Nonregenerative conditions are characterized by the absence of appropriately increased erythropoiesis in the presence of declining red cell mass. This can result in progressively decreased red cell mass over time because red cells that are normally destroyed or lost are no longer replenished. Erythrocyte morphology in nonregenerative conditions is characterized by the absence of changes observed with regeneration (e.g., anisocytosis, polychromasia, increased MCV, decreased MCHC). Most often erythrocytes appear normocytic (normal size) and normochromic (normal color), and the MCV is unchanged or decreased. Absolute reticulocyte count is unchanged or decreased, depending on whether the problem is simply an inability to adequately respond to increased needs or more significantly, failure of erythropoiesis. Nonregenerative conditions result from indirect or direct causes.

### *Indirect Causes*

A common finding in preclinical toxicology studies is mildly decreased red cell mass (i.e., RBC count, hemoglobin concentration, and hematocrit) without a regenerative response and without an obvious mechanism for the effect. Red cell mass is usually no more than about 10% lower than that of the respective control group (e.g., mean control group hematocrit = 44%; mean high-dose group hematocrit = 40%). Slightly lower MCV is sometimes observed, especially in rodents. The animals may or may not exhibit clinical signs of poor health (e.g., poor grooming habits or decreased activity) or reductions in body weight, body weight gain, or, less frequently, feed consumption. In some cases, no clinical signs are observed. Potential concurrent effects on other clinical pathology test results include mildly decreased serum total protein and albumin. These relatively mild, nonspecific findings for red cell mass are identified most frequently in rat studies where the number of animals per group is high, the dose levels used are typically higher than those for dog or monkey studies, and the normal interanimal variability of hematology data is relatively low. In addition, because the circulating life spans of mouse and rat erythrocytes (approximately 25 to 40 and 45 to 65 days, respectively) are shorter than those for dogs (approximately 100 to 120 days) and nonhuman primates (varies with species; for rhesus monkeys, approximately 85 to 100 days) [4], similar reductions in RBC production will become apparent more quickly for rodents. In many cases, specific mechanisms for these changes are not identified; however, they suggest a generalized reduction of anabolic processes, including those of hematopoietic tissues. Anything that affects the normally brisk pace of RBC production (in humans, approximately 100 billion new cells per day) will ultimately be reflected in the test results. It is also possible that decreased physical activity and correspondingly decreased tissue oxygen demand may contribute to reduced erythropoiesis.

Chronic inflammatory diseases [77–79] and significant kidney [80], liver [81], and endocrine dysfunction (e.g., hypothyroidism and hypoadrenocorticism) [82] negatively affect erythropoiesis and erythrocyte survival, and all of these conditions can be associated with this type of mild to moderate reduction in red cell mass. With chronic inflammatory conditions, the causes of reduced erythropoiesis are thought to be decreased availability or transfer of iron to developing erythrocytes, decreased responsiveness of the bone marrow to erythropoietin, and decreased red cell life span. With renal disease, reduced erythropoiesis is attributed to decreased renal production of erythropoietin and the effects of uremic toxins. Liver disease is sometimes associated with acanthocytosis, a morphologic abnormality of erythrocytes characterized by several blunt cytoplasmic projections resembling pseudopodia. The acanthocytes are thought to result from the accumulation of free, nonesterified cholesterol in the RBC membrane. Acanthocytes are relatively inflexible and removed prematurely from circulation by cells of the mononuclear phagocyte system. Small reductions in circulating red cell mass are observed with hypothyroidism and may result from reduced basal metabolic rate and cellular requirements for oxygen. Changes in thyroid metabolism may play a role in the mild erythrocyte effects observed in some animals with reduced food consumption because caloric malnutrition can result in decreased $T_3$ and decreased responsiveness to $T_3$, which may, in turn, lead to reduced production of erythropoietin. These indirect effects on erythropoiesis are relatively common in toxicology studies and result from toxic effects on other tissues or organ systems. Collectively, they are sometimes referred to as *anemia of chronic disease*. As stated above, indirect effects on the erythron tend to be relatively mild and are not as toxicologically important as the direct effects on the primary target tissue.

Chronic iron deficiency, most commonly associated with chronic blood loss or inadequate dietary iron, is relatively rare in toxicology studies and, in contrast to the anemia of chronic disease, is characterized by microcytic (low MCV), hypochromic (low MCHC) erythrocytes.

*Direct Causes*

Sustained direct toxic effects on the development and maturation of erythrocytes result in severe anemia because senescent erythrocytes are not replaced. Rodents become anemic faster than dogs or monkeys because of the short circulating lifespan of their erythrocytes. In contrast to most toxicities that indirectly affect erythropoiesis, direct toxic effects markedly reduce absolute reticulocyte count.

Direct injury to pluripotent hematopoietic stem cells or their stromal microenvironment causes failure of blood cell production, resulting in the condition called *aplastic anemia* [83,84]. Aplastic anemia is characterized by varying degrees of pancytopenia—decreased erythrocytes, leukocytes (primarily neutrophils), and platelets—and hypocellular bone marrow. Because leukocytes are affected, decreased resistance to bacterial infections usually causes severe illness or death before the anemia becomes life-threatening. Irradiation is a classic model of stem cell injury and is used therapeutically as part of the process for bone marrow transplantation. In addition to irradiation, several chemicals and drugs are known to cause aplastic anemia in humans. These include benzene, toluene, lindane, pentachlorophenol, chloramphenicol, phenylbutazone, penicillamine, gold salts, and acetazolamide. Chemotherapeutic drugs such as alkylating agents (e.g., busulfan and cyclophosphamide), antimetabolites (e.g., fluorouracil and methotrexate), and cytotoxic antibiotics (e.g., doxorubicin and daunorubicin) have the potential to cause aplastic anemia because of their pharmacologic activity. Normally, however, their effects are reversible following completion of each treatment cycle.

In toxicology studies with test articles such as these, decreased WBC and platelet counts are typically recognized earlier in the study (generally 5 to 10 days after initiation of treatment) than decreased RBC counts because the circulating lifespans for neutrophils (12 to 24 hours) and platelets (7 to 10 days) are much shorter than that for erythrocytes. Conversely, absolute reticulocyte count can be a sensitive indicator of hematopoietic injury and may be better than WBC or platelet counts for identifying the onset of toxic effect and subsequent recovery. This is most apparent in rodent studies because the normally high absolute reticulocyte counts for young rodents facilitate identification of decreases, while their normally low neutrophil counts make neutropenia more difficult to recognize. Furthermore, reticulocyte counts are not compromised by poor blood collection technique that may cause increased interanimal variability for platelet count.

The timing of sample collection following administration of a test article that temporarily interrupts hematopoiesis will dictate the findings in peripheral blood and can impact interpretation. During the peak effect, reticulocytes, neutrophils, monocytes, eosinophils, platelets, and often lymphocytes are decreased in number, and histopathologic examination of the bone marrow reveals hypocellularity. Once the effect ends, hematopoietic tissue usually mounts a strong recovery, often called a *rebound effect*, that is characterized peripherally by increased counts for reticulocytes, platelets, and sometimes neutrophils. Increased extramedullary hematopoiesis in the spleen, especially for rodents, is often apparent before the bone marrow repopulates. Because the time to peak effect and the duration of effect vary for different test articles, it is necessary to perform hematology tests at multiple intervals for proper interpretation.

Pure red cell aplasia, a condition in which only the production of red cells fails, is rarely observed in toxicology studies, even though several drugs are known to cause the disorder in humans [85]. Because drug-induced pure red cell aplasia is usually an idiosyncratic condition and may be immune mediated, recognition of this toxic effect is low in animal studies using a limited number of test subjects. Furthermore, it would be very difficult to prove that unexplained nonregenerative anemia in a single animal was a direct test article effect.

Megaloblastic anemia is a nonregenerative reduction in red cell mass characterized by macrocytic erythrocytes (increased MCV), asynchronous maturation of cytoplasm and nucleus in hematopoietic precursors, and hypersegmented or giant neutrophils [86]. In humans, it is associated with a variety of disorders that cause folate or vitamin $B_{12}$ deficiency (e.g., sprue, alcoholic cirrhosis, and pernicious anemia) and drugs (e.g., methotrexate) that impair DNA synthesis. Macrocytosis results because developing erythrocytes undergo fewer divisions before maturation. Megaloblastic conditions are rarely identified in laboratory animals, perhaps due to differences in uptake and metabolism of folate and vitamin $B_{12}$. On the other hand, nonhuman primates have been used frequently as animal models for folate and vitamin $B_{12}$ deficiency [87]. Some antiviral drugs impair cellular DNA synthesis and cause mildly increased MCV that may or may not be associated with decreased red cell mass.

Finally, nonregenerative anemia is a feature of some hematopoietic tumors and leukemias. Nonregenerative anemias caused by neoplastic processes are due to a combination of factors, including blunted response to erythropoietin, excessive cytokine release (IL-1, IL-6, and TNFα) suppressing erythropoiesis, hemolysis, nutritional deficiencies, and competition for bone marrow space [88,89]. It is not unusual in carcinogenicity studies to observe severe anemia secondary to naturally occurring leukemia in a few animals.

## Leukocytes

Leukocytes are evaluated by counting the concentration of WBCs in peripheral blood and differentiating them into specific cell types: neutrophils, lymphocytes, monocytes, eosinophils, and basophils. Hematology analyzers capable of automated differentials provide more accurate data and enable the detection of changes in even minor leukocyte types (e.g., decreased eosinophils). When interpreting and reporting differential WBC count results, it is essential to evaluate the absolute cell counts (i.e., cells/μL). Relative or percent counts are simply a means for determining the absolute counts and have little or no inherent value for assessing an animal's condition. A 50% neutrophil count in a dog could be normal, increased, or decreased, depending on the total WBC count. As with changes in red cell mass, small changes in leukocyte counts should generally not be described using diagnostic terminology. When appropriate, however, commonly used terms for increases and decreases include *leukocytosis* and *leukopenia*, *neutrophilia* and *neutropenia*, *lymphocytosis* and *lymphopenia*, *monocytosis* and *monocytopenia*, *eosinophilia* and *eosinopenia*, and *basophilia*. Neutrophils and lymphocytes are the most numerous WBC types in peripheral blood and are usually the cells affected when peripheral blood leukocyte counts change due to toxicity. Indirect effects on these two cell lines are often observed in response to study-related procedures or test article effects on other tissues. Direct test article effects on these two cell lines can occur but are less common [90].

### Physiological Leukocytosis

Physiological leukocytosis occurs in excited or frightened animals secondary to endogenous catecholamine release. Increased heart rate, blood pressure, and muscular activity shift leukocytes from the marginal pool (i.e., cells that adhere to the endothelium of small vessels or are sequestered in vascular beds of tissues such as the spleen) to the circulating leukocyte pool. The total WBC count may double in number. The specific cell type responsible for the majority of the increase varies because of species-dependent differences in the normal distribution of leukocyte types. Increased neutrophil count is the most obvious change for dogs, while increased lymphocyte count is most conspicuous for rats. The physiological leukocytosis observed for primates is fairly evenly distributed between neutrophils and lymphocytes. Because physiological leukocytosis is most frequently observed in animals not accustomed to handling or blood collection, it is common for a few animals to have notably high WBC counts only at the first blood collection interval for a study. Recognition of this phenomenon is critical in studies using few animals (usually dogs or monkeys) and only one pretreatment interval. Overinterpretation of the data might lead to the false conclusion that a test article has a myelosuppressive effect because posttreatment WBC counts, determined when the animal is more accustomed to being handled, may be much lower than pretreatment counts. By the same token, if the test article is an antineoplastic drug and myelosuppression is a critical endpoint, strong consideration should be given to acquiring at least two pretreatment hematology samples to facilitate proper data interpretation.

## Steroid- or Stress-Induced Leukocyte Response

This leukocyte pattern occurs following exogenous corticosteroid administration or when stressful conditions result in increased production of endogenous corticosteroids. It is characterized primarily by increased neutrophil count (immature neutrophils, such as bands, are absent) and decreased lymphocyte and eosinophil counts. Increased monocyte counts may or may not be present. It is relatively unusual to observe this pattern for an entire group of study animals, even though study-related procedures or test article administration appear to create conditions typically considered stressful; however, individual animals, especially those in a moribund condition resulting from toxicity, often exhibit this pattern.

## Increased and Decreased Neutrophil Counts

Although the classically ascribed function of neutrophils is to protect the host from bacterial infection, they are a component of many nonseptic inflammatory lesions as well. It is, therefore, common to observe increased neutrophil counts secondary to a variety of inflammatory conditions resulting from test article toxicity (e.g., intravascular hemolysis or tissue necrosis) or study-related procedures (e.g., chronic catheterization or repeated injections), with or without the involvement of an infectious agent. Various cytokines and hematopoietic growth factors have been developed as therapeutic agents, and many of these directly affect neutrophil kinetics, resulting in mild to marked neutrophilic leukocytosis. The term *left shift* refers to an increased number of immature neutrophils (e.g., band neutrophils) in circulation, usually in response to an inflammatory lesion with a significant demand for neutrophils or administration of inflammatory cytokines. Inflammatory lesions resulting in left shifts or marked neutrophilia are usually easily identified, either by physical examination or at necropsy. The term *degenerative left shift* describes the combination of a normal or decreased absolute neutrophil count with more immature than mature neutrophils. This pattern indicates that the supply of neutrophils is insufficient to meet the peripheral need for neutrophils. It generally reflects a very severe infection as might occur with aspiration pneumonia, gastrointestinal perforation, or septicemia associated with bacterial contamination of an indwelling intravenous catheter. In conditions such as these, *toxic neutrophils* may be observed. These neutrophils have distinct morphological characteristics including cytoplasmic basophilia, vacuolation, and granulation and the presence of Döhle bodies (small, bluish-gray cytoplasmic inclusions made of aggregated rough endoplasmic reticulum). The term *toxic neutrophils* is a misnomer in that these cells only indicate greatly accelerated neutrophil production, regardless of cause; for example, toxic neutrophils may be observed following the therapeutic administration of hematopoietic growth factors.

Unless the test article is a potent cytotoxic agent (e.g., a conventional chemotherapeutic), the observation of severe neutropenia, with or without a left shift or toxic neutrophils, is generally limited to one or two individuals that have severe complications secondary to test article toxicity or study-related procedures. Leukocyte effects of cytotoxic agents, on the other hand, are typically observed in most or all of the animals in groups receiving toxic dose levels. Because mice and young rats normally have very low neutrophil counts (e.g., less than 1000/μL), recognition of neutropenia is more difficult for these species than for others. In addition, although a neutrophil count below 500/μL in a clinical patient is generally considered an indicator of great risk for bacterial infection in humans, monkeys, and dogs, the same is not true for rodents.

Single or short-term administration of a potent cytotoxic agent is usually characterized by neutropenia within a few days to a week of treatment, followed by a recovery that may include the presence of immature neutrophils for a short time and may result in a rebound neutrophilia. Detection of these changes is dependent on the timing and frequency of hematology testing. Similarly, the appearance of the bone marrow, whether hypocellular or hypercellular, is also dependent on the timing of sample collection. Hematology findings for test articles that directly injure pluripotent stem cells or rapidly dividing committed precursor cells are usually characterized by decreased numbers of reticulocytes, platelets, neutrophils, monocytes, eosinophils, and, possibly, lymphocytes. Selective damage of granulocyte precursors without affecting erythropoiesis or thrombopoiesis is unusual.

Immune-mediated reductions in neutrophil counts are relatively rare but can occur [91–93]. Like immune-mediated hemolytic conditions, however, they are often idiosyncratic and difficult to identify in preclinical safety studies using limited numbers of test subjects.

## Increased and Decreased Lymphocyte Counts

Lymphocytes are responsible for a wide variety of immune system functions. Several subpopulations of lymphocytes exist, but they are indistinguishable by light microscopy. Lymphocytes are relatively long-lived compared with other leukocytes and have the ability to leave the vascular system through venules in lymph nodes and eventually reenter the blood via the thoracic duct. Physiological leukocytosis should be considered whenever increased lymphocyte count is present in only a few animals. Cynomolgus monkeys occasionally exhibit markedly increased lymphocyte counts (e.g., >60,000/μL) of unknown etiology; these are generally transient but can obfuscate test-article-related findings. Increased lymphocyte count is an uncommon test-article-related effect, although it may be observed in conjunction with compounds that cause chronic inflammatory lesions, especially in rodents, or with administration of test articles that are antigenic and elicit an immune response by the test animals.

Decreased lymphocyte count is most frequently observed as part of the stress- or steroid-induced leukocyte response. Marked lymphopenia is commonly observed in moribund animals. Cytotoxic test articles often cause decreased absolute lymphocyte counts, but the magnitude of the decrease is usually less prominent than that for neutrophils in nonrodent species. In rodents, it is generally easier to detect effects on lymphocytes because of the normally high number of lymphocytes compared with neutrophils. During recovery from effects of cytotoxic test articles, lymphocyte counts may remain decreased longer than neutrophils and typically do not exhibit a prominent rebound response. Because of the many subpopulations of lymphocytes, it is difficult to gauge the biological importance of small decreases in absolute lymphocyte count. Selective reduction or elimination of a subpopulation may occur (e.g., the human immunodeficiency virus effect on $CD^{4+}$ lymphocytes) without greatly affecting the total lymphocyte count. Evaluation of corollary information (e.g., overall health and findings that might suggest immunosuppression) is essential to putting small changes into context.

## Monocytes, Eosinophils, Basophils, and Large Unstained Cells

Although absolute counts for these cell types are generally quite low (e.g., <1000/μL), test-article-induced effects can often be detected using analyzers that do automated differential counts. As with neutrophils and lymphocytes, increases in circulating numbers of these cells are generally secondary phenomena unless the test article is a hematopoietic growth factor or other cytokine that directly stimulates cell production (e.g., interleukin-5 increases eosinophil count).

Monocytes phagocytize and digest particulate matter such as senescent cells and necrotic cell debris. They also serve as precursors to cells of the mononuclear phagocyte system and as a source of macrophages in inflamed tissues. Monocytes participate in modulation of the inflammatory response through cytokine production and antigen processing and presentation to lymphocytes. Increased monocyte counts may occur secondarily to any condition with substantial tissue destruction, such as widespread inflammation, tumor-associated necrosis, or hemolytic anemia. Decreased monocyte counts may be identified following administration of a cytotoxic chemotherapeutic agent.

Eosinophils are part of the body's defense against helminthic parasite infections, and eosinophilia is occasionally observed in nonhuman primates with a heavy parasite load. Increased eosinophil count is also observed secondarily to some hypersensitivity reactions. Decreased eosinophil count is sometimes identified as part of the stress-induced leukocyte pattern or following administration of cytotoxic chemotherapeutic agents.

Effects on basophil counts are extremely rare, but basophils may play a role in some hypersensitivity reactions. Large unstained cells are counted by automated hematology analyzers manufactured by Bayer Corporation. These cells cannot be classified by the instrument into one of the five major cell types but generally are thought to represent lymphocytes, monocytes, or some immature cells. Changes in large unstained cell counts are typically observed with changes in the other cell types.

## Leukemia

In most 2-year carcinogenicity studies using rodents, some percentage of the animals will develop leukemia, a neoplastic proliferation of a hematopoietic cell line. The diagnosis of leukemia is best made by histopathologic examination of tissues infiltrated with the neoplastic cells. The odds of correctly identifying animals with leukemia are much greater by doing routine histopathology than by doing periodic hematologic examinations at regularly scheduled intervals. Although leukemias are sometimes characterized by markedly elevated WBC counts and the presence of neoplastic cells (e.g., blasts) in circulation, many animals with leukemia exhibit neither of these characteristics. Even when neoplastic cells are present in peripheral blood, it is often difficult to determine from which cell line they were derived (e.g., granulocytic, lymphocytic, myelomonocytic) because immature, anaplastic, or blast-stage cells from different cell lines can be indistinguishable by light microscopy using standard staining techniques.

Lymphocytic leukemias are the most commonly observed leukemias in laboratory rats and are occasionally observed as an incidental finding in 90-day toxicity studies [14]. Fischer 344 rats may develop large granular lymphocyte leukemia (also called *mononuclear cell leukemia*) at incidences of 30 to 40% in the second year of a carcinogenicity study [23]. This neoplasm appears to arise in the spleen and commonly infiltrates other tissues, particularly the liver. Affected rats develop an immune-mediated hemolytic anemia and often exhibit hyperbilirubinemia and elevated liver enzyme activities in the serum. Neoplastic cells in peripheral blood appear as large, immature lymphocytes and may contain prominent azurophilic granules. Erythrophagocytosis by the neoplastic cells is occasionally observed.

## Platelets

Platelets play a key role in primary hemostasis. When blood vessels are damaged, platelets quickly adhere to the subendothelium, undergo a shape change, and begin to aggregate, forming a primary platelet plug that is sufficient to control bleeding from minor injuries to small vessels. These activated platelets secrete a variety of substances that stimulate vasoconstriction and promote fibrin formation. Fibrin serves to cement the aggregated platelets into a stable hemostatic plug. Healthy endothelial cells in close proximity to the damaged vessel release inhibitors of platelet aggregation and fibrin formation in order to limit the clot size.

### Increased Platelet Count

Extremely high platelet counts have the potential to increase the risk of thrombosis in humans, but increases in platelet count are generally asymptomatic in laboratory animals. Increased platelet counts in toxicology studies are usually indirect or secondary responses, but some occur as a direct test article effect.

Increased platelet counts occurring as an indirect effect are generally of small magnitude and not likely to have any biological significance. The terms *reactive* or *secondary thrombocytosis* have been used to describe the increased counts observed in conjunction with generalized bone marrow stimulation as observed with hemolysis, blood loss, and many types of acute and chronic inflammation. Release of hematopoietic growth factors such as erythropoietin and cytokines such as interleukin-6 and interleukin-11 may be at least partially responsible for the increased platelet production in some of these conditions [94,95]. Acute but transient increases in platelets may occur in association with physiological leukocytosis because catecholamine-induced splenic contraction releases platelets sequestered within the sinusoids of the spleen. A rebound thrombocytosis often follows recovery from significant thrombocytopenia caused by test articles such as chemotherapeutic agents that reversibly inhibit platelet production or injure megakaryocytes.

A few test articles directly stimulate platelet production and the release of platelets from megakaryocytes. These test articles include hematopoietic growth factors such as thrombopoietin and erythropoietin and small molecules such as vincristine.

### Decreased Platelet Count

Signs of decreased platelet count include petechial and ecchymotic hemorrhages (most commonly observed in mucous membranes or at mucocutaneous sites), epistaxis, melena, menorrhagia, and prolonged bleeding from small wounds such as venipuncture sites. These signs typically do not occur spontaneously unless the platelet count is very low (e.g., less than 20,000/μL) or there is some type of hemostatic challenge (e.g., surgery) [96]. Because of their protected environment, animals in toxicity studies are less prone to exhibit signs of hemorrhage when their platelet counts are extremely low.

Decreased platelet counts are relatively common spurious findings associated with difficult venipuncture or inadequate anticoagulation of blood samples and subsequent *in vitro* platelet aggregation. *In vitro* aggregation can generally be confirmed by the observation of platelet clumps at the feathered edge of the blood film or by the characteristic scattergram produced by some automated hematology analyzers. Although typically observed for individual mice and rats, platelet clumps and spuriously low platelet counts can sometimes appear group related because animals receiving the test article may be more difficult to bleed as a result of poor health, dehydration, or small size relative to the control animals.

When not a spurious finding, decreased platelet counts result from decreased production or increased consumption of platelets. Test articles that reduce erythroid and myeloid cell production, such as chemotherapeutic agents, frequently inhibit production of platelets by megakaryocytes as well. Moderately to markedly reduced platelet counts tend to occur a few days after obvious reductions in neutrophil and reticulocyte counts because the circulating lifespan of platelets is about 5 to 10 days.

Decreased platelet counts due to increased consumption of platelets can occur secondarily to acute lesions of highly vascular tissues (e.g., the gastrointestinal tract) or result from extensive hemorrhage, especially from multiple sites. If lesions affecting blood vessels are severe and widespread, disseminated intravascular coagulation may develop, and platelet counts will be markedly decreased. Test articles may also stimulate immune reactions against platelets. Immune-mediated thrombocytopenia, like immune-mediated hemolysis, has been associated with many drugs [97] but is largely an idiosyncratic phenomenon and is, therefore, detected infrequently in preclinical toxicology studies. When observed, only one or two animals are typically affected, and these animals may or may not be in the high-dose group. Immune-mediated thrombocytopenia and immune-mediated hemolysis may occur together. Anti-platelet antibody can be detected using flow cytometry; however, the best evidence that thrombocytopenia is immune mediated may come from a rechallenge exposure with the test article following cessation of treatment and recovery. Upon rechallenge, platelet count should drop acutely if the mechanism is immune-mediated destruction. Compounds may also result in activation of platelets, which leads to their premature removal from circulation. Activated platelets can be detected in some species by flow cytometric evaluation of activation markers.

Like immature red cells, young platelets have residual RNA that stains with nucleic acid dyes. Using methods analogous to those used for erythrocytes, these reticulated platelets can be counted using flow cytometry. Decreased platelet production is associated with inappropriately decreased concentrations of reticulated platelets, while platelet destruction or consumption is associated with markedly increased reticulated platelet counts.

### Platelet Function

Test articles that affect platelet function have the potential to cause the same clinical signs as marked thrombocytopenia, but the tendency to do so is much less because of the complexity of platelet function and the presence of alternative or redundant pathways *in vivo* for the various platelet functions. Platelet function tests such as bleeding time and platelet aggregation may be beneficial when evaluating safety of therapeutic agents related to coagulation and platelet function, but it is important to recognize the considerable analytical and interanimal variability for these tests. Group results may be less meaningful than assessment of results from individual animals before and after test article administration.

## Bone Marrow Smear Evaluation

The most important aspect of bone marrow smear evaluation is understanding when it is indicated. Although preparation of smears is advisable for most repeat-dose toxicology studies, evaluation of those smears is usually unnecessary. The standard hematology tests provide considerable information concerning bone marrow function, and if results from these tests are unaffected, it is very unlikely that bone marrow evaluation will provide any additional knowledge concerning potential significant test article effects on the hematopoietic system. Likewise, if the effects observed in peripheral blood are relatively small, it is unlikely that bone marrow smear evaluation will be beneficial.

Even when results of hematology tests are affected by test article administration, bone marrow evaluation has no benefit if mechanisms for the hematology findings are clear from the peripheral blood data; for example, decreased red cell mass parameters associated with increased absolute reticulocyte counts indicate a normal regenerative erythropoietic response to hemorrhage or hemolysis. Bone marrow evaluation would simply confirm the presence of erythroid hyperplasia and provide no new information. Likewise, increased absolute neutrophil count in response to an inflammatory condition is normal, and bone marrow evaluation would only confirm granulocytic hyperplasia.

The main indications for bone marrow smear evaluation in toxicology studies are moderate to marked nonregenerative anemia, leukopenia, or thrombocytopenia (or any combination of the three) with no apparent etiology or significant morphological abnormalities of peripheral blood cells. The primary objective of the bone marrow smear evaluation is to assess the relative numbers and the maturation of precursor cells to explain peripheral blood changes. Miscellaneous observations, including increased iron stores, plasma cell hyperplasia, and excessive cytophagia may also be recognized. Because bone marrow smears are relatively poor indicators of the actual cellularity of the bone marrow, it is necessary to consider the histopathologic findings for sections of sternum, rib, or femur. Although histologic sections generally are inadequate for evaluating individual cell types and abnormal cell morphology, they provide a good assessment of overall cellularity and are useful in providing estimations of cell density (e.g., for megakaryocytes or mast cells).

There are three major approaches to bone marrow smear evaluation. Regardless of the approach taken, the results of the bone marrow evaluation (both bone marrow smear and histologic evaluation) must be interpreted in conjunction with peripheral blood test results. The most simplistic and least informative bone marrow evaluation is to determine the myeloid-to-erythroid (M:E) ratio by counting a certain number of nucleated cells (usually between 200 and 500 cells) and classifying cells as either granulocytic or erythroid. The results generally are not useful in interpretation of peripheral blood changes and generally do not help to understand the underlying problem. For example, a test article might cause increased M:E ratios. This finding is consistent with granulocytic hyperplasia and erythroid hypoplasia. If the peripheral blood data indicate that the animal has a high neutrophil count and its hematocrit is normal, then an increased M:E ratio likely indicates granulocytic hyperplasia. If the animal has a normal neutrophil count and a nonregenerative anemia, then an increased M:E ratio likely indicates erythroid hypoplasia. In both cases the outcome of the M:E ratio could have been predicted from the peripheral blood results, and the bone marrow evaluation provided no additional information. If the animal is neutropenic and has a nonregenerative anemia, a increased M:E ratio only indicates that there are relatively more granulocytic cells than erythroid cells and has not increased our understanding of the process.

The most time-consuming and labor-intensive bone marrow evaluation is to perform a bone marrow cell differential count by differentiating the cell type and stage of development of at least 500 cells. When completed, an M:E ratio can be calculated from the results. This thorough evaluation yields more information but at a very high cost. Differential counts provide numeric information concerning the relative numbers of different precursor cells and whether a cell line is maturing normally. Unusual or abnormal morphologic characteristics of the cells must be described separately. Techniques for conducting bone marrow differentials using flow cytometry have been described and have the potential to rapidly provide much more accurate information than manual 500-cell differential counts [98].

The most cost-effective and informative approach to bone marrow evaluation is the subjective cytological examination. In this approach, the bone marrow smear is examined in much the same manner as a morphologic pathologist examines a histologic section of liver, and a diagnosis or interpretation is recorded. The person performing the examination, usually a veterinary anatomic or clinical pathologist, assesses the quality and cellularity of the smear, the presence and relative number of precursors for each of the three major cell lines (erythrocytes, granulocytes, and platelets), and the maturation of each of the cell lines. Abnormal morphology is noted, as well as unusual numbers or characteristics of other cell types such as lymphocytes, plasma cells, monocytes, macrophages, and mast cells. A diagnosis or interpretation is rendered based on the examination of the smear and the peripheral blood tests results.

Regardless of the method used for evaluation of bone marrow smears, the end goal of the evaluation is to assess any negative impact of the test article on the number or maturation of hematopoietic cell precursors [99,100]. Conclusions from the bone marrow evaluation should address the relationship between the bone marrow findings (from both smears and histologic sections) and peripheral blood changes. For example, if a test article causes decreased platelet count, the absence of megakaryocytes in bone marrow smears and sections indicates that the decrease is due to failure of platelet production rather than increased platelet consumption peripherally. Likewise, if bone marrow evaluations find increased numbers of morphologically normal megakaryocytes, the results indicate that thrombocytopenia is due to a consumptive process.

## Coagulation

The clotting mechanism or cascade has traditionally been divided into two pathways. *In vivo*, the intrinsic pathway begins with the activation of zymogen factor XII following exposure to negatively charged subendothelial components such as collagen. Factors XI, IX, and VIII are also part of the intrinsic pathway. The extrinsic pathway begins with the activation of zymogen factor VII following exposure to tissue factor (also called *tissue thromboplastin*) expressed by cells deep in the vessel wall. Both pathways share the same terminal sequence of events including the activation of factor X, conversion of prothrombin to thrombin, and conversion of fibrinogen to fibrin. It is thought that the extrinsic pathway is the primary initiator of coagulation *in vivo* [101]. The extrinsic and intrinsic pathways are routinely evaluated by the one-stage prothrombin time (OSPT) and activated partial thromboplastin time (APTT), respectively. An alternative test, the activated coagulation time (ACT) test, is a simple, rapid measure of the intrinsic pathway that does not require a coagulation analyzer [102–104]. These three coagulation assays are relatively insensitive and nonspecific. Activity of a single clotting factor must be reduced to approximately 30% of normal before noticeably prolonged times are detected for an individual animal. When the results from groups of animals are compared, statistically significant differences are occasionally observed that are smaller than what would generally be considered an important change for an individual animal (e.g., less than 2 seconds of difference between the means for the control and high-dose groups). The toxicologic significance of differences such as these is sometimes difficult to determine. Although they clearly do not represent an effect likely associated with a bleeding diathesis for individual animals, they may be an indication of an important change in coagulation homeostasis. It may be valuable to design a longer study or increase the dose level to see if the effect is repeatable, dose related, and associated with clinical signs.

Under the conditions of most toxicology studies, where animals are exposed to high concentrations of a test article for a prolonged period of time, any major effect on the production of a clotting factor will likely result in a clinically obvious, bleeding diathesis. The administration of vitamin K antagonists such as dicumarol or the ingestion of synthetic or poorly absorbed fat substitutes is associated with bleeding and prolonged PT and APTT because the fat-soluble vitamin K is required by the liver for production of functional forms of factors II, VII, IX, and X. In theory, PT will be affected before APTT because factor VII has the shortest half-life of the clotting factors. Although the liver synthesizes nearly all of the clotting factors, PT and APTT are insensitive measures of liver function. Because of the liver's large functional reserve, liver injury must be relatively severe before coagulation times are noticeably affected. Disseminated intravascular coagulation is characterized by depletion of all clotting factors, including fibrinogen, and moderately to markedly prolonged coagulation times. In many cases of disseminated intravascular coagulation, the plasma samples fail to clot during the coagulation assays. These changes are observed in conjunction with decreased platelets (discussed previously).

Similarly to platelet count, coagulation times can be spuriously prolonged by difficult blood collection or poor collection technique. The combination of low platelet count and prolonged coagulation times, in an otherwise healthy animal, is an indication of poor sample quality. Inherited factor VII deficiency affects a small number of laboratory beagles [52]. These animals can usually be distinguished during pretreatment screening by PTs that are 2 or 3 seconds longer than those of the other animals acquired for the study. Although the deficiency rarely causes a clinical problem, it would be inappropriate to use these animals if the test article is known or suspected to affect coagulation. PT and APTT can both be artifactually prolonged because of excessive sodium citrate anticoagulant in the plasma sample [105,106]. This can occur if insufficient blood volume is added to standardized collection tubes or if the animal's hematocrit is elevated because of hemoconcentration (e.g., dehydration) or drug-induced polycythemia. Normal coagulation times vary from one laboratory animal species to another. Among the notable differences are the relatively fast PTs for dogs (e.g., 6 to 8 seconds) and slow PTs for guinea pigs (e.g., 30 to 40 seconds).

Although fibrinogen is occasionally measured along with PT and APTT as a coagulation assay, its primary value is as an acute phase protein produced in response to inflammation. Increasing fibrinogen concentration almost always indicates an inflammatory process.

## CLINICAL CHEMISTRY TESTS AND INTERPRETATION

Routinely performed clinical chemistry tests provide information concerning hepatocellular and biliary integrity and function, renal function, carbohydrate, lipid and protein metabolism, and mineral and electrolyte balance. Modern clinical chemistry analyzers require very small sample volumes; less than 250 μL of serum is needed to perform as many as 20 tests. It is possible, therefore, to obtain complete biochemical profiles from rats at multiple time points during a study without excessive blood collection. Most of the common clinical chemistry assays developed for human testing are applicable, without modification, to animal clinical chemistry testing.

### Hepatocellular and Hepatobiliary Integrity and Function

Many routine clinical chemistry tests can be affected by liver toxicity because of the critical metabolic, synthetic, and excretory functions of the liver and the abundant enzymatic machinery required to perform these functions [107,108]. Conversely, a significant loss of liver tissue with little or no detectable change in routine tests is possible because of the liver's large functional reserve. No single test is superior for detecting all of the various types of liver toxicity, but the pattern of abnormal findings in a battery of tests may help characterize the location and severity of liver lesions [109].

#### Liver Enzymes

Serum activities of liver enzymes are used primarily to identify hepatocellular injury and cholestasis, with or without hepatobiliary injury. Although sometimes referred to as liver function tests, they do not provide specific information about liver function. The liver can be severely dysfunctional in the absence of effects on serum liver enzyme activities. Animals with end-stage liver cirrhosis, for example, can exhibit normal serum enzyme activities. Conversely, focal lesions causing marked elevation of certain liver enzyme activities may have no appreciable effect on overall hepatic function because of the large functional reserve of the liver.

The specificity, sensitivity, and predictive value of liver enzyme tests are largely dependent on the models of hepatotoxicity used to make those determinations. This fact is at least partially responsible for the practice of including multiple liver enzymes in the clinical chemistry test panels for toxicology studies. The apparent absence of changes in liver enzyme activities does not necessarily rule out the possibility of hepatotoxicity. Possible reasons for this include suboptimal timing of clinical pathology testing and excessive variability of control animal results, especially for mice and monkeys.

Serum activities of many enzymes normally present within hepatocytes are increased following hepatocellular injury (i.e., degeneration or necrosis). The utility of a particular enzyme for the identification of hepatocellular injury depends on factors such as relative specificity to liver, intrahepatic location, intracellular location, the concentration gradient between the hepatocyte and serum, serum half-life, *in vitro* stability, and economy of measurement [110–112]. The most frequently used enzymes to assess hepatocellular injury are alanine aminotransferase (ALT), formerly serum glutamic pyruvic transaminase (SGPT), and aspartate aminotransferase (AST), formerly serum glutamic oxaloacetic transaminase (SGOT) [15]. Sorbitol dehydrogenase (SDH) glutamate dehydrogenase (GDH), and lactate dehydrogenase (LDH) are measured less frequently.

In general, ALT is the most useful enzyme for detection of hepatocellular injury in the majority of laboratory animal species. Although the enzyme is present in many tissues, its greatest concentration in most species is within hepatocytes, and, in general, significant elevations of serum ALT activity indicate release of ALT by hepato cytes. Species for which ALT is less useful because of relatively low hepatocyte concentrations include the guinea pig [113] and large domestic animals such as pigs (including minipigs), goats, sheep, cows, and horses [30,111,115]. Because the enzyme is primarily cytosolic, and its concentration within the cell is up to 10,000 times greater than that in the serum, ALT may enter the serum in any condition that sufficiently alters cell membrane integrity. In addition to simple leakage from degenerating or necrotic cells, there may be other mechanisms for movement of the enzyme across the cell membrane [116] because high serum activities of ALT are occasionally observed with no apparent cell death. The magnitude of serum activity elevation is proportional to the number of affected hepatocytes and is not necessarily indicative of the reversibility of the lesion; however, the greatest elevations result from severe lesions affecting a large portion of the liver. As a general guideline, test-article-related ALT increases in excess of 200 IU/L are usually accompanied by histopathologic evidence of hepatocellular injury, while activities below this level may or may not have correlative findings.

Following an acute but reversible hepatotoxic event, serum ALT activity increases relatively rapidly, peaks within 1 or 2 days, and then declines over the next few days. Significant hepatotoxicity can go undetected if clinical pathology tests are delayed for 1 or 2 weeks following a single administration of the test article. Prolonged elevations following a single insult may reflect increased concentrations of ALT in regenerative liver tissue or continued loss of ALT from cells in close proximity to the primary lesion that undergo degenerative changes as a result of the altered microenvironment.

Increased serum ALT activity does not always indicate primary hepatocellular injury. Biliary disease or toxicity and bile duct obstruction may cause increased serum ALT activity at least in part due to the effect of retained bile salts on the cell membranes of neighboring hepatocytes. Muscle damage, when severe and extensive, can increase serum ALT activity in the absence of hepatic injury [117,118]. Increased intracellular activity of ALT will cause serum ALT activity to increase proportionately. Some studies have shown an association between toxins or drugs (e.g., corticosteroids and anticonvulsants) and increased serum ALT activity that may be due, at least in part, to increased intracellular activity [116]. Because the compounds may also have pathologic effects on hepatocytes, it may be difficult to determine whether a serum activity elevation is due to increased intracellular activity or cell injury.

Assessment of hepatotoxicity using serum ALT activity in monkeys may be complicated by the presence of subclinical, enzootic hepatitis A infection [119]. Transiently increased serum ALT activity occurs concurrently with seroconversion to the virus and periportal inflammation. Because animals entering a facility may not have been exposed to the virus previously, it is possible to observe sporadic, increased ALT activities (e.g., up to approximately 300 IU/L) for a few individual monkeys during the course of a toxicology study. Some facilities choose to bank serum collected from monkeys before a study is initiated for possible serologic testing to help clarify ambiguous ALT results. Interpretation of serum ALT activities for mice is complicated by considerable interanimal variability. In toxicology studies using mice, it is relatively common for a few animals, including the control animals, to have much higher serum ALT activities than those of the majority (e.g., 200 IU/L vs. 40 IU/L). The cause of these high activities is thought to be physical damage to the liver, especially when mice are handled by grasping the body [120]. Unfortunately, if the only animals affected happen to belong to the high-dose group, it may be difficult to rule out a test article effect.

Serum SDH and GDH activities have been recommended as good indicators of hepatic toxicity in laboratory animal species [12,109,122] because increased serum activities are liver specific and relatively sensitive. SDH is a cytosolic enzyme, and GDH is located in mitochondria. Elevations in serum SDH activity generally return to baseline levels faster than for other liver enzymes because of a short serum half-life. The addition of either of these tests to a standard clinical chemistry profile is a good choice if potential liver toxicity is of particular interest. SDH is routinely used in large animal veterinary practices, and GDH enjoys popularity in Europe. Because neither enzyme is used in human medicine in the United States, assay test kits are less available and may be difficult to obtain.

Serum AST and LDH activities tend to parallel serum ALT activity with respect to liver damage, but these enzymes are much less liver specific because of high concentrations in muscle and other tissues. Compared with larger animals, rodents tend to exhibit more interanimal variability for these enzymes. The variability may be due to contamination with muscle tissue during blood collection procedures such as cardiac puncture and rupture of the retroorbital plexus or sinus. There is little advantage to determining both AST and LDH. Generally, only AST is determined, as LDH tends to have greater interanimal variability. Elevations in serum AST activity caused by hepatotoxicity are usually less pronounced than concurrent elevations in serum ALT activity. Because a portion of intracellular AST is located in mitochondria, a more severe injury may be necessary for the release of like quantities of AST. As with ALT, drugs such as corticosteroids and anticonvulsants may increase intracellular activity of AST.

Decreased serum activities of ALT and AST are occasionally observed in toxicology studies. Among the potential causes for these findings are decreased hepatocellular synthesis or release of the enzymes, inhibition or reduction of enzyme activity, and assay interference. The most widely recognized of these causes involves an effect on pyridoxal 5′-phosphate (vitamin $B_6$), a coenzyme cofactor required for full catalytic activity of the aminotransferases. If a test article negatively affects this cofactor, directly or indirectly, serum aminotransferase activities may decrease [122–124]. This phenomenon is perhaps most commonly observed in monkeys that develop chronic watery diarrhea and may result simply from loss of the water-soluble cofactor. Because the aminotransferase assays can be run with or without additional pyridoxal 5′-phosphate, a test-article-related effect on pyridoxal 5′-phosphate can be identified by analyzing for enzyme activity with and without excess cofactor. Regardless of the mechanism involved, decreased serum activities of the aminotransferases are generally not associated with toxicologically significant effects on the liver.

Several enzymes that originate from hepatocytes and biliary epithelial cells are increased in serum as a result of increased synthesis or release following intrahepatic or extrahepatic cholestasis or in conjunction with biliary hyperplasia. These include serum alkaline phosphatase (ALP), gamma-glutamyltransferase (GGT), leucine aminopeptidase (LAP), and 5′-nucleotidase (5′-N). Of these, the most commonly measured are ALP and GGT.

A variety of related enzymes contribute to total serum ALP activity. In humans, at least four genes have been identified that code for different ALP isoenzymes: tissue nonspecific (found in liver, bone, and kidney), intestinal, placental, and germ cell. In most laboratory animals, only two ALP genes have been identified; these code for the tissue nonspecific and intestinal isoenzymes. Tissue nonspecific ALP enzymes originating in liver, bone, and kidney are the product of the same gene and are, therefore, isoforms rather than isoenzymes. The isoforms can be distinguished because of differences in degree of post-translational glycosylation and tissue of origin [125,126].

The contribution of each isoenzyme and isoform to total serum ALP activity is dictated by tissue production and serum half-life. Because bone ALP is produced by osteoblasts, bone ALP activity is highest in young, growing animals, regardless of species, and decreases as animals mature. In puppies, the bone isoform may be responsible for up to 95% of total serum ALP activity. In adult dogs, the liver isoform is most prevalent. Bone ALP is responsible for approximately 60% of serum ALP activity in 6-week-old rats [125].

In rats, serum activity of the intestinal isoenzyme increases after feeding and is reduced with fasting [127]. Dog intestinal ALP is rarely identified in serum due to its short half-life, and diseases of the intestine are not typically associated with increased serum ALP. Kidney ALP may be found in urine but is not released into blood to any significant extent and has a short serum half-life. In response to corticosteroids (either administered or endogenous), dogs can produce a unique hyperglycosylated form of the intestinal ALP isoenzyme (corticosteroid-induced ALP isoenzyme) from the liver. The relative amount of corticosteroid-induced ALP activity is small compared with the others, and it is completely absent from the serum of most dogs.

In spite of the different ALP isoenzymes and isoforms, serum ALP activity is most frequently considered a measure of cholestasis. It is a sensitive indicator of cholestasis in the dog and usually increases well before other markers such as GGT and total bilirubin. Because of cell swelling and pressure obstruction of small bile ductules, primary hepatocellular toxicities often cause enough intrahepatic cholestasis to elevate serum ALP activity. Periportal lesions result in higher activities than do centrilobular lesions. Extrahepatic cholestasis, as might occur with pancreatitis, biliary calculi, or complications of bile duct cannulation, stimulates higher serum ALP activity than intrahepatic cholestasis. The degree of elevation, however, is rarely sufficient for differentiating primary hepatocellular toxicity from primary biliary toxicity. In contrast to the dog, the value of serum ALP activity for distinguishing cholestatic lesions in monkeys is reduced because of marked interanimal variability.

Elevations of serum ALP activity can be the first indication of a toxic effect on bone formation. Elevations of serum ALP activity due to increased osteoblast activity tend to be less pronounced than those due to cholestasis (e.g., no greater than twofold to threefold higher than control animals). If the test article is administered for sufficient duration, the effect on ALP activity is usually accompanied by clinical or histopathologic evidence of bone changes.

Drugs such as anticonvulsants and corticosteroids can induce synthesis of liver ALP, with or without evidence of hepatobiliary disease. Following corticosteroid administration to dogs, serum activity of liver ALP increases within a few days, while the corticosteroid-induced ALP activity does not increase noticeably in serum for about 10 days [128]. Although an increase in corticosteroid-induced ALP activity has been observed in dogs with a variety of chronic disease conditions [65] and may be related to increased endogenous corticosteroid release, this isoenzyme has not been closely evaluated in toxicology studies using dogs.

The measurement of serum GGT activity has gained popularity because it is more specific than ALP and was shown to be effective in certain models of biliary toxicity in the rat [129]. Although the highest tissue concentrations of this membrane-localized enzyme are in the kidney and pancreas, serum elevations have been reported only with hepatobiliary toxicity and following induction by drugs that stimulate microsomal enzyme production [107,130]. Unlike ALP, GGT is unaffected by bone growth or disease; furthermore, its serum activity is less likely to increase secondary to primary hepatocellular toxicity or intrahepatic cholestasis due to hepatocellular swelling. In rodents, serum GGT activity is often undetectable, and even small increases may be significant. As in dogs, rat serum GGT activity can increase substantially with enzyme induction due to xenobiotic administration.

Serum LAP and 5′-N activities have been investigated as alternatives to ALP and GGT but have not found general acceptance. In some models of liver toxicity, 5′-N appears more sensitive than ALP or GGT [109,131].

### Bilirubin

In contrast to serum liver enzyme activities, serum total bilirubin concentration is primarily a liver function test. In the absence of hemolysis, hyperbilirubinemia indicates liver dysfunction. Bilirubin results from the breakdown of heme by cells of the mononuclear phagocyte system. Hemoglobin from senescent erythrocytes accounts for approximately 85% of all serum bilirubin. When macrophages release bilirubin into circulation, it is known as free, unconjugated, prehepatic, or indirect bilirubin. It is water insoluble and circulates bound to albumin. Hepatocytes efficiently remove unconjugated bilirubin from plasma and prepare it for removal from the body by a four-step process that includes uptake, conjugation, secretion, and excretion. Secretion of conjugated bilirubin across the canalicular membrane is the rate-limiting step in the process, and small amounts of conjugated or direct bilirubin escape into plasma. Conjugated bilirubin is not bound to albumin and is freely filtered through the glomerulus. In most species, conjugated bilirubin is completely reabsorbed by renal tubular epithelial cells unless the amount of filtered bilirubin is excessive. In the dog, the renal threshold is low and traces of bilirubin are normal in concentrated urine.

Even though the liver is a frequent target organ of toxicity, hepatotoxicity infrequently results in increased total bilirubin concentration, whether due to conjugated bilirubin, unconjugated bilirubin, or both, because of the large functional reserve of the liver. In the dog, a 70% hepatectomy will not increase total bilirubin concentration.

Conjugated hyperbilirubinemia occurs as a result of impaired secretion of bilirubin, cholestasis, or both. Because bilirubin secretion is the rate-limiting step, any disease that damages enough hepatocytes can potentially increase serum conjugated bilirubin concentration. Periportal lesions cause higher serum bilirubin concentrations than do centrilobular lesions, and extrahepatic cholestasis causes higher serum bilirubin concentration than does intrahepatic cholestasis. When increased bilirubin concentration results from a cholestasis process, serum ALP activity is generally elevated, particularly in the dog.

Unconjugated hyperbilirubinemia occurs almost exclusively as a result of relatively severe, acute hemolysis. If hepatocytes cannot process the large amount of unconjugated bilirubin produced by macrophages during a hemolytic episode, serum bilirubin concentration increases. A hemolytic event sufficient to overload a normal liver always produces other findings indicative of hemolysis. It is possible, however, for relatively modest hemolysis to cause unconjugated hyperbilirubinemia if hepatic function is already compromised. Although a number of nonhemolytic, unconjugated hyperbilirubinemia syndromes are known, these syndromes are usually due to hereditary defects in the uptake and conjugation of free bilirubin.

Unconjugated (or indirect) bilirubin can be differentiated from conjugated (or direct) bilirubin by the Van den Bergh test. The test is used clinically to help distinguish prehepatic causes of hyperbilirubinemia, such as hemolysis, from hepatic or posthepatic causes such as hepatitis or biliary obstruction. In well-designed toxicology studies, the combination of clinical observations, other laboratory data (e.g., hematocrit or liver enzyme activities), and anatomic pathology findings (e.g., hemosiderin accumulations in splenic macrophages or periportal hepatocellular necrosis) is usually more than sufficient to determine the primary mechanism for any observed hyperbilirubinemia. Laboratory determination of direct and indirect bilirubin is rarely necessary and not recommended as part of the routine panel of tests performed in toxicology studies.

Decreased serum bilirubin concentration is occasionally associated with administration of test articles that induce microsomal enzyme production [130]. Human patients receiving phenobarbital therapy have lower serum bilirubin levels than the general population as a whole [132]. Enzyme induction apparently enhances the metabolism and excretion of bilirubin and could potentially mask an otherwise elevated bilirubin level.

### Other Liver-Related Parameters

Like serum bilirubin concentration, total serum bile acid concentration is a measure of hepatic function. Bile acids are synthesized from cholesterol by hepatocytes, conjugated to an amino acid, secreted into the biliary system, and excreted into the intestine, where they facilitate fat absorption. Bile acids undergo efficient enterohepatic circulation, with most reabsorption occurring at the level of the ileum. Portal blood conveys the bile acids to the liver for uptake, reconjugation, and resecretion. Liver toxicity has the potential to alter bile acid metabolism at any of multiple steps and cause increased serum bile acid concentration. Unfortunately, although it is considered a sensitive and specific test for hepatobiliary disease in clinical veterinary medicine [133], measurement of total serum bile acids has not proven effective at increasing the identification of hepatotoxicity beyond that of the standard battery of tests routinely performed in toxicology studies. Like total bilirubin, increased serum bile acid concentration does not discriminate between different types of hepatic lesions.

The liver is wholly or partially responsible for the synthesis of many substances including glucose, cholesterol, urea, and a variety of proteins. Severe hepatocellular dysfunction can cause decreased serum urea nitrogen concentration, hypoglycemia, hypocholesterolemia, hypoproteinemia (especially hypoalbuminemia), and prolonged coagulation times. On the other hand, liver disease can also result in hypercholesterolemia and hyperglobulinemia. The patterns of change in clinical pathology tests caused by different types of liver toxicity, whether primary (e.g., chloroform-induced hepatic necrosis) or secondary (e.g., hypoxia-induced centrilobular necrosis), are often overlapping. Examination of the entire biochemical profile, along with other clinical pathology and anatomic pathology findings, is necessary to properly evaluate data for potential liver toxicity.

## Renal Function

Serum urea nitrogen (or urea) and creatinine concentrations, in conjunction with measures of urine concentration (urine specific gravity or osmolality) and volume, are the most common tests used to evaluate renal function [134–138]. These tests are easy and inexpensive to perform but are relatively insensitive to small effects on the kidney and can be affected by nonrenal factors.

Urea is synthesized by the liver from ammonia that is absorbed from the intestine or produced by endogenous protein catabolism. Urea is freely filtered through the glomerulus and excreted in urine. Some urea is reabsorbed passively with water in the proximal tubule; the amount reabsorbed is inversely related to the rate of urine flow through the tubule. Serum urea nitrogen concentration is, therefore, affected by rate of urea production, glomerular filtration rate (GFR), and flow rate of urine through the renal tubule. Mechanisms for increased urea nitrogen are categorized as prerenal, renal, or postrenal.

Prerenal causes for increased urea nitrogen are increased urea synthesis and decreased renal blood flow. Increased urea synthesis results from consumption of high-protein diets or conditions that increase protein catabolism such as starvation, fever, infection, tissue necrosis, or high gastrointestinal hemorrhage. Decreased renal blood flow decreases GFR and may be caused by conditions such as dehydration (the most common cause of increased urea nitrogen in toxicology studies), cardiovascular disease, or shock. Changes in urea nitrogen concentration caused by increased urea synthesis are typically small. Changes caused by decreased renal perfusion may also be small, but if GFR is severely affected, the increase in urea nitrogen is indistinguishable from that which would occur due to primary renal failure. When increased urea nitrogen is due to prerenal causes, renal concentrating ability is typically maintained. If the prerenal condition is dehydration, urine volume will be reduced and urine concentration will be increased as the kidneys attempt to conserve water.

Increased urea nitrogen due to renal causes results from diseases or toxicity of the renal parenchyma. Like the liver, kidneys have a large functional reserve capacity. In clinical practice, it is commonly said that serum urea nitrogen concentration does not increase notably until approximately 75% of the kidneys' nephrons are nonfunctional. In preclinical toxicity studies, however, it is likely that differences between control and treated animals can be detected prior to that degree of impairment. When the cause of increased urea nitrogen is primary renal disease, renal concentrating ability may be impaired, and urine specific gravity may be isosthenuric (i.e., the same as the glomerular filtrate; approximately 1.008 to 1.012). Increased urea nitrogen due to renal causes is generally accompanied by histopathologic evidence of renal damage (e.g., proximal tubular nephrosis or chronic progressive nephropathy), and the animals may exhibit signs of poor health such as inappetence, weight loss, or inactivity.

Postrenal causes of increased urea nitrogen reduce GFR by obstructing the outflow of urine. Obstruction by naturally occurring urinary calculi is occasionally observed as an incidental finding in rodent studies, but test articles that promote urinary calculi formation can also be responsible for this condition.

Creatinine is a nonprotein nitrogenous waste product formed at a relatively constant rate by the nonenzymatic breakdown of creatine. Creatine is a breakdown product of phosphocreatine, a molecule that stores energy in muscle. Serum creatinine concentration is therefore, influenced by muscle mass and conditioning but it is relatively independent of dietary influences and protein catabolism. Creatinine is freely filtered by the glomerulus, but unlike urea it is not reabsorbed by the tubules. Following alterations in renal blood flow, renal function, or urine outflow, the changes in serum creatinine concentration tend to parallel those for serum urea nitrogen concentration. The timing and magnitude of the changes for serum creatinine may lag behind those for serum urea nitrogen. This usually occurs as a result of the tubular reabsorption of urea, especially when urine flow is slow or when there is increased formation of urea. Serum creatinine is usually a better reflection of glomerular filtration than serum urea nitrogen because it is influenced by fewer secondary factors. Unfortunately, the most commonly used method for determining serum creatinine concentration, the Jaffe reaction, is nonspecific, and interfering compounds called *noncreatinine chromagens* affect its accuracy

Noncreatinine chromagens are typically of insufficient quantity to complicate data interpretation; however, if serum creatinine concentrations are increased in the absence of correlative effects on serum urea nitrogen or renal histopathology, other analytical methods for creatinine (e.g., enzymatic) can be used to investigate the possibility of analytical interferences. Endogenous creatinine clearance is sometimes used as a noninvasive measure of GFR because blood levels of creatinine are relatively stable over short intervals, creatinine is freely filtered, and creatinine is not significantly secreted or reabsorbed [134,139].

Other clinical chemistry findings sometimes observed when renal function is significantly impaired include increased serum inorganic phosphorus concentration and decreased serum sodium and chloride concentrations. Whereas increased inorganic phosphorus is primarily due to reduced filtration, decreased sodium and chloride result from loss of tubular function and normal reabsorption.

## Proteins, Carbohydrates, and Lipids

### Serum Proteins

Serum total protein concentration is a measure of all plasma proteins with the exception of those consumed in clot formation such as fibrinogen and other clotting factors. Serum total protein concentration is, therefore, about 0.3 to 0.5 g/dL lower than plasma total protein concentration. Albumin is the most abundant serum protein and is largely responsible for maintaining intravascular osmotic pressure. Albumin serves as a storage reservoir of amino acids and as a transport protein for plasma constituents that do not have a specific transport protein. Many test articles bind to and are transported by albumin. Globulins are a heterogeneous population of proteins that include specific transport proteins (e.g., transferrin for iron, lipoproteins for lipids, haptoglobin for hemoglobin, and thyroxine-binding globulin for thyroxine), mediators of inflammation (e.g., complement), acute-phase proteins, clotting factors, enzymes, and immunoglobulins. Globulins are loosely categorized by their electrophoretic migration pattern as $\alpha$, $\beta$, and $\gamma$ globulins. Several different globulin proteins are present in each category or region of the electrophoretic pattern [140]. The regions can be further subdivided (e.g., most species have two $\alpha$ regions), but serum protein electrophoresis cannot distinguish specific globulin proteins. Immunoglobulins are generally thought of as $\gamma$ globulins, but some, particularly IgM, extend into the $\beta$ regions of the electrophoretogram.

The liver synthesizes albumin and most of the globulins, while lymphocytes and plasma cells synthesize immunoglobulins. Serum total protein and albumin concentrations are measured directly, and serum globulin concentration is calculated by subtracting albumin from total protein. Hydration status must be considered for proper interpretation of changes in serum protein concentrations. Low protein concentrations, like low red cell mass, can be masked by dehydration.

In toxicology studies, the most frequent reason for increased serum total protein concentration is reduced hydration of the treated animals relative to the control animals. Serum albumin and globulin concentrations increase proportionately when simple dehydration occurs. The effect on hydration status of the treated animals may or may not be detectable as clinical dehydration. Possible correlative clinical observations include gastrointestinal fluid losses (e.g., vomiting, diarrhea, excessive salivation), polyuria, and reduced water consumption. Because water consumption in rodents is closely associated with food consumption, any cause of decreased food consumption in rodents has the potential to cause dehydration. In short-term dietary studies, for example, palatability problems may result in higher serum protein and urea nitrogen concentrations because of differences in hydration; however, if decreased feed consumption is protracted and body weights or body weight gains are affected, serum protein concentrations will typically decrease over time.

The other common causes for increased serum total protein concentration in toxicology studies are inflammatory conditions that stimulate the production of globulins called *acute-phase proteins* (e.g., fibrinogen, haptoglobin, $\alpha_2$-macroglobulin, C-reactive protein) and *immunoglobulins* [141–143]. Often with inflammatory conditions, however, there is a concurrent decrease in serum albumin concentration because albumin is a negative acute-phase protein. The opposite direction of changes in albumin and globulin concentration may result in the absence of a change in total protein concentration. The albumin-to-globulin ratio will be reduced. This serum protein pattern is commonly observed in animals affected by complications of long-term intravenous catheterization.

Decreased serum protein concentrations result from either decreased protein synthesis or increased protein loss. Protein synthesis can be negatively affected by decreased food consumption, maldigestion or malabsorption, and hepatic dysfunction. Although the functional reserve capacity of the liver is quite substantial and hepatic injury must be relatively severe before protein synthesis is notably diminished for individual patients, small differences between control and treated groups in large studies may be apparent with relatively modest hepatotoxicity. Loss of both albumin and globulin occurs with hemorrhage, exudative lesions such as burns or severe dermal toxicity, and, occasionally, severe diarrhea. Albumin may be the principal protein lost as a result of protein-losing enteropathies and glomerulopathies because of its relatively small size. Globulin concentrations may increase secondarily to enteropathies because of inflammation and increased systemic exposure to gastrointestinal toxins and bacterial flora. Decreased globulin concentrations, without concurrent or proportional decreases in albumin concentration, may be indicative of decreased synthesis of immunoglobulins. Histopathologic evidence of effects on lymphoid tissue strengthens this interpretation. Reduced serum globulin concentrations have been observed following prolonged administration of antibiotics (e.g., 4 weeks) to young animals, perhaps as a result of inhibition of normal bacterial flora and subsequently reduced antigenic stimulation.

A small decrease in serum albumin concentration is one of the most frequent findings in toxicology studies for animals given poorly tolerated test articles. Like the small decreases for other parameters (e.g., hematocrit, glucose concentration, cholesterol concentration, body weight, and body weight gain) that may occur concurrently, decreased albumin due to nonspecific causes is usually considered an indication of the animals' overall poor condition. Decreases in serum albumin concentration are more commonly observed for rodents than for the larger species. Although this may be due simply to the increased numbers of animals per group, it may also be a function of differences among species for the circulating half-life of albumin. Smaller species tend to have faster turnover of albumin [140]; for example, the half-life of albumin is approximately 2 days for mice, 8 days for dogs, and 16 days for baboons.

## Serum Glucose

Serum glucose concentration is a reflection of intestinal glucose absorption, hepatic glucose production, and tissue uptake of glucose. The balance between hepatic production and tissue uptake is influenced by many hormones, including insulin, glucagon, corticosteroids, adrenocorticotropic hormone, growth hormone, and catecholamines. In oversimplified terms, insulin promotes uptake of glucose by tissues, glucocorticoids and glucagon stimulate hepatic gluconeogenesis, and catecholamines and glucagon stimulate glycogenolysis.

The most frequently encountered causes of increased serum glucose concentration in toxicology studies are failure to fast an animal and catecholamine release secondary to excitement or fear. Moribund animals occasionally exhibit marked hyperglycemia, probably as a result of both corticosteroid and catecholamine release. If blood collection is performed in group order (starting with the control group) rather than random order and the samples are not processed promptly, the high-dose group may appear to have higher serum glucose concentrations than those of the control group because of greater glucose consumption by erythrocytes in the control group's samples. Glucose concentration decreases at a rate of approximately 7 to 10 mg/dL/hr when serum is not separated from the blood clot. Infrequent causes of increased serum glucose concentration in toxicology studies include some relatively common clinical conditions such as diabetes mellitus, pancreatitis, hyperadrenocorticism, and steroid therapy.

In toxicology studies, test-article-related decreases in serum glucose concentration are most commonly observed in animals that fail to thrive and gain body weight, with or without a concurrent decrease in food consumption. When this occurs, the difference for serum glucose concentration between the control and treated animals is usually no more than 10 or 15 mg/dL and likely does not adversely affect the animals. Although a precise mechanism for decreased glucose is typically undetermined, the difference may reflect the overall process that has caused the animals to do poorly and is frequently accompanied by small decreases in circulating red cell mass and serum protein concentrations. Clinical conditions that cause decreased serum glucose include intestinal disease with malabsorption, severe hepatic disease, endotoxemia, and some tumors, in particular insulinomas and hepatomas.

## Serum Lipids

The two major serum lipids are cholesterol and triglycerides. Cholesterol is utilized for the synthesis of cellular membranes, bile acids, corticosteroids, and some sex steroid hormones, and triglycerides are an important source of energy. Serum cholesterol and triglycerides are derived from dietary intake and endogenous synthesis, primarily by the liver. Cholesterol and triglycerides circulate as components of chylomicrons and lipoprotein particles: high-density lipoprotein (HDL), low-density lipoprotein (LDL), and very-low-density lipoprotein (VLDL) [144]. Chylomicrons are produced by intestinal cells after ingestion of a fatty meal and are rich in triglycerides. Hepatocytes synthesize VLDLs, particles with less triglyceride but more cholesterol than chylomicrons. The triglycerides in chylomicrons and VLDLs are broken down to free fatty acids and monoglycerides by lipoprotein lipase attached to the surface of endothelial cells, especially in the capillaries of adipose tissue and muscle. Adipocytes tend to re-esterify fatty acids for storage as triglycerides. Myocytes tend to oxidize fatty acids for energy. The loss of triglyceride transforms VLDL to LDL. In humans, LDLs transport about two thirds of serum cholesterol. In contrast HDL is responsible for a significant majority of cholesterol transport in most laboratory animal species. Species differences in lipid metabolism cause difficulties in correlating lipid effects in animal models with potential effects in humans [145]. Male hamsters, some rabbit strains, and genetically engineered mice have been used as animal models for lipid research.

Small changes in serum cholesterol and triglyceride concentrations, both increases and decreases, are relatively frequent findings in toxicology studies. The change are generally believed to represent minor alterations in lipid metabolism that do not adversely affect the animals' health. Unfortunately, exact mechanisms for the changes are rarely identified. Factors to consider include alterations in food consumption and assimilation, body weight and composition, liver function, and hormones.

Serum triglyceride concentration is elevated postprandially, while serum cholesterol concentration is relatively stable. When fat is mobilized to meet energy requirements because of significant anorexia, starvation, malabsorption, or maldigestion, serum triglycerides are usually increased, sometimes markedly. Cholesterol levels, however, are variable. Clinical conditions that often increase both serum cholesterol and triglycerides include hypothyroidism and diabetes mellitus. Lipoprotein lipase activity is reduced in both conditions. In most species, hypercholesterolemia is more prominent in hypothyroidism, and hypertriglyceridemia is more prominent in diabetes mellitus. Cholestasis and other forms of liver disease can increase serum cholesterol concentration because the liver is the major excretory pathway for cholesterol. Conversely, liver disease may also be associated with hypocholesterolemia. Hypercholesterolemia, increased urinary protein excretion (due to glomerular disease), and hypoalbuminemia are characteristics of nephrotic syndrome. Young rats (e.g., 7 to 20 weeks of age) usually exhibit less interanimal variability than dogs and monkeys. It is, therefore, more common to detect subtle effects on serum lipid concentrations in short-term rat studies, rather than dog or monkey studies, especially because the number of animals per group is typically much higher for rat studies. Serum cholesterol and triglyceride concentrations are extremely variable for older rats because of a number of naturally occurring diseases.

## Minerals and Electrolytes

### Serum Calcium and Inorganic Phosphorus

Serum calcium concentration is controlled primarily by parathyroid hormone, calcitonin, and vitamin D and reflects a balance among intestinal absorption, bone formation and reabsorption, and urinary excretion [146]. Serum inorganic phosphorus concentration is affected by the same hormones but is more sensitive to changes in dietary intake and urinary excretion. Approximately 50% of serum calcium is in its biologically active, ionized form. Ionized calcium is critical for neuromuscular activity, bone formation, coagulation, and other biochemical processes. Approximately 40% of serum calcium is bound to albumin, and the remainder is complexed to anions such as phosphate and citrate.

Although many disease conditions are associated with hypercalcemia [146], increased serum calcium concentration is relatively uncommon in toxicology studies unless the test article specifically targets calcium metabolism or has properties of either parathyroid hormone or vitamin D. Because approximately 40% of serum calcium is bound to albumin, mildly increased serum calcium concentration is occasionally observed when serum albumin concentration is increased. This change is physiologically appropriate and should not be considered adverse. Rare causes of increased serum calcium in toxicology studies include primary hyperparathyroidism, pseudohyperparathyroidism (a paraneoplastic syndrome), hypervitaminosis D, and renal disease.

Mildly decreased serum calcium concentration, as a result of decreased serum albumin concentration, is a frequent finding in toxicology studies. Signs of toxicity secondary to decreased calcium, such as neurological and neuromuscular abnormalities, are absent because ionized calcium is relatively unaffected. Rare causes of decreased serum calcium in toxicology studies include hypoparathyroidism, nutritional hyperparathyroidism, acute pancreatitis, and renal disease.

Young, rapidly growing animals have high serum inorganic phosphorus concentrations (e.g., greater than or equal to serum calcium concentration) that decrease with maturity. Serum inorganic phosphorus concentration is affected by GFR, and increased concentrations parallel changes in serum urea nitrogen and creatinine and should be interpreted in conjunction with those parameters. Rare causes of increased serum inorganic phosphorus in toxicology studies include hypoparathyroidism, nutritional hyperparathyroidism due to excess dietary phosphorus, and hypervitaminosis D. Decreased serum inorganic phosphorus concentration observed in toxicology studies is most commonly associated with significantly reduced food consumption.

### Serum Sodium, Potassium, and Chloride

Sodium, the major extracellular cation in plasma, is the principal determinant of extracellular fluid volume. Chloride, the major extracellular anion in plasma, supports fluid homeostasis and balances cation secretion. Potassium is the major intracellular cation and has a critical role in neuromuscular and cardiac excitation. Clinically determined reference intervals for electrolyte concentrations tend to be much wider than the range of values obtained from animals in a well-controlled toxicity study. It is common to observe very small, but statistically significant, differences between control and treated groups with no apparent mechanism or effect on animal health. Many of these differences are likely incidental, but others probably represent subtle homeostatic effects or uncontrolled preanalytical influences. Changes in serum sodium and chloride concentrations tend to parallel each other when they are associated with relative water content, but serum chloride concentrations are disproportionately affected in disorders affecting acid–base balance.

Significant increases in serum sodium are rare in toxicology studies. Increases in serum chloride are occasionally observed secondarily to metabolic acidosis resulting from diarrhea. In this condition, renal tubular reabsorption of chloride is increased because of decreased availability of bicarbonate. Approximately proportional decreases in serum sodium and chloride concentrations can occur with gastrointestinal losses (e.g., vomiting or diarrhea), renal losses (e.g., renal failure), diuretic effects, and hypoadrenocorticism (rare in toxicology studies). Vomiting may cause decreased serum chloride concentration without affecting sodium because stomach secretions are rich in chloride.

Increased serum potassium concentration may be observed with a variety of conditions causing acidosis because extracellular hydrogen ions are exchanged for intracellular potassium ions. Severe tissue necrosis and anuric or oligouric renal failure are infrequent causes of increased serum potassium. Serum potassium may be falsely elevated because of hemolysis (either technique or test-article-related) in species that have high intraerythrocytic potassium (e.g., nonhuman primates). Marked thrombocytosis and thrombocytopenia can be associated with increased and decreased serum potassium, respectively, because potassium is released from platelets during clot formation. Serum potassium is very sensitive to potassium intake, and decreased concentrations are often associated with anorexia. Similar to effects on sodium and chloride, decreased serum potassium is sometimes associated with gastrointestinal losses and polyuric renal losses. Disorders resulting in alkalosis (e.g., persistent vomiting) may cause decreased serum potassium because intracellular hydrogen ions are exchanged for extracellular potassium ions.

### Miscellaneous Serum Chemistry Tests

Serum creatine kinase (CK) activity is measured primarily as a marker for skeletal muscle toxicity. Test-article-related increases in CK activity can be detected only if complicating factors or study-related procedures do not obscure results. For example, intramuscular injections, surgical procedures, and poor venipuncture technique can all give rise to marked elevations in CK that preclude identification of meaningful differences between control and treated animals. Serum aldolase activity is a less frequently used marker for skeletal muscle injury.

Several tests have been used with varying degrees of success in human medicine as markers of acute myocardial injury [147], and a few of these tests, such as CK-MB (one of the isoenzymes of creatine kinase), troponin I, and troponin T, may have applicability in toxicology studies designed specifically to evaluate myocardial injury [148–152]. The troponins may be sensitive enough to detect subclinical myocardial toxicity in long-term studies. Troponin I analysis is more readily available than that for troponin T because troponin T requires specific instrumentation.

Amylase and lipase activities are measured clinically to diagnose diseases causing acute pancreatic necrosis. These enzymes have limited value in toxicology studies, especially repeat-dose studies, because toxin-induced pancreatitis will cause severe illness and have prominent, unmistakable morphologic consequences. Amylase and lipase are not sufficiently sensitive to be markers for test articles with specific action against the islet cells of the endocrine pancreas.

## URINALYSIS AND URINE CHEMISTRY TESTS AND INTERPRETATION

### Urinalysis

Urinalysis has traditionally been considered part of the minimum laboratory database for evaluating patients. Although urinalysis provides specific information about the urogenital tract and general information regarding some systemic conditions, it is not particularly well suited for most toxicology studies. The cost-to-benefit ratio is poor because sample collection and analysis are labor intensive and relatively few toxicities produce detectable effects on urinalysis parameters. In addition, technical difficulties associated with collecting a large number of urine specimens from small laboratory animals or uncooperative large animals can impact the accuracy of test results. If a test article is known or suspected to affect the urinary system, then measures can be taken to provide appropriate specimens for urinalysis (e.g., by catheterization, cystocentesis, or carefully collected fresh voided samples). Usually, however, when a large number of animals are being administered a test article of unknown toxic potential, the most efficient method of urine collection (in a collection vessel at the bottom of a metabolic cage) produces many artifacts that affect test results and complicate interpretation. Because voided urine traverses the urethra, vagina or prepuce, and perineal or preputial hairs it can acquire cells and bacteria that are not indicative of toxicity. Drinking water can dilute the urine. When urine incubates overnight at the bottom of the metabolic cage with contaminants (e.g., feces, bacteria, food, hair, and cleaning chemicals), it is understandable that urinalysis data can produce questionable results. Preservatives and methods for keeping the urine chilled may help limit artifactual changes, but these procedures are not without cost and other issues.

The standard urinalysis includes measurement of physicochemical properties and microscopic evaluation of urine sediment. The physicochemical properties include volume (for timed collections), color, clarity, specific gravity or osmolality, and reagent strip tests (pH, protein glucose, ketones, bilirubin, urobilinogen, and blood). Some reagent strips have additional tests for nitrite (indicates presence of nitrite-producing bacteria) and leukocyte esterase, but these are not particularly valuable for animal specimens, especially those collected overnight [153]. Urinary sediment evaluation is a semiquantitative microscopic measure of the presence of cells (urogenital cells, WBCs, and RBCs), casts, bacteria, crystals, and other formed elements. This is the most expensive component of the urinalysis and, in most instances, the least informative or necessary [154]. In general, disorders that increase the number of cells or casts in urine sediment will be better characterized by histopathologic examination of the kidneys and bladder. Occasionally, however, test-article-specific crystals that might otherwise go undetected are identified by urine sediment examination.

## Urine Volume and Concentration

Timed urine volume and a measure of urine concentration (urine specific gravity or osmolality) can be valuable in assessing renal function because these parameters demonstrate the concentrating ability of the kidneys. Loss of urine-concentrating ability usually precedes development of azotemia as a consequence of chronic renal disease. Timed urine volume (e.g., 16 or 24 hours) and a measure of urine concentration are probably the most beneficial of the routine urinalysis tests. Urine concentration generally varies inversely with urine volume and is most commonly assessed clinically and in toxicity studies by measuring urine specific gravity. Urine specific gravity (usually determined by refractometry) is an approximation of urine solute concentration. Although urine osmolality (usually determined by freezing-point depression osmometry) is a more accurate estimation of urine solute concentration, it is measured infrequently. Reagent strip specific gravity measurements are not sufficiently accurate for use in preclinical toxicity studies and should not be used. Animals with impaired ability to concentrate urine due to renal disease have decreased urine specific gravity or osmolality and increased urine volume. With advanced renal disease, urine specific gravity may become fixed in a range from approximately 1.008 to 1.012 (termed *isothenuria*). Isosthenuria and hyposthenuria (i.e., urine specific gravity below 1.008) are particularly meaningful when serum urea nitrogen concentration is elevated. This combination usually indicates primary renal dysfunction. Test articles with diuretic activity also produce dilute urine and increased urine volume, but serum urea nitrogen is usually unaffected. Increased urine concentration (e.g., urine specific gravity greater than 1.030 in dogs or 1.050 in rats) indicates that the kidneys have functional concentrating ability. Very high urine concentrations may be observed in dehydrated animals because the kidneys attempt to conserve water. In toxicology studies, urine specific gravity is sometimes high in treated animals that are anorectic and consequently not drinking normally. This is particularly true of rodents. Problems with watering systems can create interpretive issues because faulty sipper tubes or animals that habitually play with their water source may cause water contamination of urine, while dysfunctional systems that limit water availability may result in dehydrated animals.

## Reagent Strip Tests

Reagent strip tests provide a semiquantitative biochemical evaluation of urine. When one of the following tests indicates a test article effect, a more quantitative determination of that parameter may prove valuable. Because urine reagent strip reactions are usually measured by reagent strip readers that use a grayscale system, highly concentrated (dark) or abnormally colored urine may interfere with test results. In these cases, the reagent strips can be manually evaluated, or the parameters can be measured by alternate methods.

Urine pH can be affected by diet, test article pH, and sample handling. Animals consuming high-protein meat diets tend to produce urine of lower pH than do animals consuming cereal or vegetable diets. If administered in large enough quantities, an acid or alkaline test article can affect urine pH. Urine pH is often artifactually elevated in samples collected overnight for two reasons: urease-producing bacteria cause ammonia formation, and carbon dioxide is lost from open containers. Urine pH is generally not a good indicator of acid–base balance.

A low concentration of protein is a normal finding in the urine of most animals, especially if the urine is concentrated. A high concentration of protein is abnormal, especially in dilute urine. Increased urine protein concentration may result from glomerular injury, defective tubular reabsorption, hemorrhage, inflammation, or proteinaceous secretions from the lower urogenital tract in voided specimens. Sediment examination may help to identify the cause of proteinuria. Older rats, particularly males, may develop marked proteinuria as a consequence of chronic progressive nephropathy, a common, naturally occurring disease of rats. Highly alkaline urine and quaternary ammonium disinfectants can cause false-positive findings for urine protein measured by reagent strips. Urine protein can be determined quantitatively by automated clinical chemistry analyzers.

The finding of glucose in urine is abnormal. Under normal conditions, renal tubules reabsorb all glucose filtered through the glomerulus. If the glucose load is excessive as a result of markedly increased serum glucose (e.g., >180 mg/dL for the dog), glucose spills into urine. Urine glucose may be observed with renal toxins that target proximal tubular epithelial cells because of failure to adequately reabsorb filtered glucose. If urine glucose is considered an important endpoint for a study, precautions should be taken to avoid false-negative findings resulting from a proliferation of bacteria that consume glucose. While diabetes mellitus is the most frequent clinical disease associated with glucosuria, it is rarely observed in toxicology studies.

Ketones are not normally present in the urine of most species. Urine of fasted male rats, however, is often positive for the presence of ketones. For other species and female rats, increased urine ketones are occasionally observed in anorectic animals and animals that have been fasted for a prolonged period of time. Ketonuria indicates that energy metabolism has shifted from gluconeogenesis to incomplete oxidation of fatty acids. False-negative findings for urine ketones occur as a result of bacterial degradation and the loss of volatile ketones from open containers.

Bilirubin is normally absent in the urine of most laboratory animals. A small amount, however, can often be measured in the urine of male dogs, especially in concentrated urine. Increased urine bilirubin results from the same conditions that cause increased serum bilirubin, and it may precede the change in serum. False-negative findings for urine bilirubin occur from prolonged exposure of urine to light, causing oxidation of bilirubin to biliverdin.

Urobilinogen is normally present in the urine of animals. Theoretically, urine urobilinogen tests for patency of the bile duct. Intestinal bacteria convert conjugated bilirubin to urobilinogen, a portion of which is reabsorbed by the intestine. Because a small amount of the reabsorbed urobilinogen is normally excreted in the urine, a negative urine urobilinogen is purported to indicate obstruction of the bile duct, an extremely rare occurrence in toxicology studies. The test has little value and is generally determined only because it exists on the same reagent strip as the other tests.

Blood is occasionally present in the urine of animals. Although the origin of the blood is usually unknown, estrus is a common source in female dogs and monkeys. Reagent strips do not effectively discriminate among erythrocytes, hemoglobin, and myoglobin, but examination of the urine sediment may help to differentiate hematuria from hemoglobinuria. Hematuria may result from bleeding disorders or inflammation, trauma, or neoplasia of the urogenital tract.

## Urine Sediment Evaluation

Cells, casts, and crystals are poorly preserved in urine during overnight or 24-hour collections; for example, it is common to find no erythrocytes in the urine sediment even though the reagent strip blood test is positive and hemolysis has been ruled out. If sediment detail is deemed an important endpoint for a study, other means of urine collection (e.g., cystocentesis or free catch) should be considered.

Small numbers of erythrocytes, leukocytes, and epithelial cells are normal findings in urine sediment obtained from voided urine specimens. Large numbers of these cells may or may not be abnormal, and gross or histopathologic correlates are necessary to determine their origin. Increased numbers of large epithelial cells (i.e., squamous and transitional cells) generally do not indicate a significant abnormality, but increased numbers of small epithelial cells (i.e., renal tubular cells), especially in conjunction with granular or cellular casts, are indicative of kidney disease.

Urinary casts are infrequently observed in the urine of normal animals. A cast is the cylindrical mold of a segment of renal tubule formed by protein alone or protein and cells. Casts are generally classified as hyaline, cellular, waxy, or broad. Hyaline casts are made of protein alone, and increased numbers are sometimes observed with glomerular disorders that cause excessive proteinuria. An occasional hyaline cast is normal. Cellular casts (erythrocyte, leukocyte, or epithelial) usually indicate renal tubular lesions but are rarely observed in animal urine. If not moved rapidly into the urine, cellular casts become granular casts as the cells degenerate and take on a granular appearance. Waxy casts represent the final stage of degeneration of the cellular cast and indicate prolonged intrarenal urine stasis. Granular and waxy casts, therefore, may also be an indication of renal tubular disease. Broad casts are identified by their width and represent casts formed in collecting ducts or pathologically dilated portions of the nephron. Broad casts also indicate intrarenal urine stasis. Although the origin of individual cells in urine cannot always be determined, increased numbers of casts (cylindriuria) indicate that renal tubular injury has occurred proximal to the renal pelvis.

Bacteria are a consistent finding in the urine of normal animals, given the routine methods used to collect urine during toxicology studies. If the test article is an antibiotic a test-article-related decrease in the number of bacteria may be observed.

Crystals are common in the urine of laboratory animals. The type of crystals observed is dependent on urine pH. Crystals observed frequently in alkaline urine include triple phosphate, amorphous phosphate, calcium carbonate, and ammonium urate crystals. Urate and oxalate crystals are associated with acid urine. Although rarely observed, ammonium biurate crystals are associated with liver failure. Test-article-specific crystals will occasionally form when a test article or metabolite is highly concentrated in the urine. These crystals may be pathologically significant if they obstruct renal tubules or lead to the development of calculi. Urine crystals may be important in the establishment of rodent-specific mechanisms of bladder carcinogenesis. Mechanistic studies with specific urine collection techniques may be designed to evaluate urine crystals for this purpose [155].

## Quantitative Urine Chemistry Tests

Because serum urea nitrogen and creatinine and standard urinalysis tests are relatively insensitive markers of renal injury, several urine chemistry tests have been proposed as better methods for identifying and quantifying early renal injury or dysfunction. For the most part, these tests are impractical as part of the general screen in routine, regulated toxicology studies; however, they may be valuable as tools for assessing early renal toxicity at dose levels below those that cause histopathologic lesions and for determining the intrarenal location of the earliest toxic insult. Some of these tests are also well suited for acute screening studies to determine the relative nephrotoxicity of different analogues of a compound with known nephrotoxic action for lead candidate selection.

### Urinary Enzyme Activity

Many urinary enzymes have been evaluated for use as early markers of nephrotoxicity [17,20], and several have been proven effective in specific models of nephrotoxicity [19,156–160]. Perhaps the two most frequently measured urinary enzymes are GGT and *N*-acetyl-β-D-glucosaminidase (NAG) because they are relatively stable at room temperature and have somewhat different cellular locations; GGT is located in the brush border of proximal tubular epithelial cells, and NAG is a lysosomal enzyme with apparently greater distribution along the nephron. Other enzymes that have been evaluated as indicators of nephrotoxicity include ALP (another brush-border enzyme), LDH, ALT, and AST (all primarily cytosolic enzymes). Urinary enzyme activities should be corrected for variations in urine concentration by calculating the total activity excreted per unit time or the ratio of urinary enzyme activity to urinary creatinine concentration. Urinary enzyme activities are most effective for assessing acute renal injury. They appear to have much less utility for assessing chronic conditions (e.g., repeat-dose studies of several weeks' duration) and, like liver enzymes, do not provide information concerning renal function.

### Urinary Proteins

Quantitative measurement of urinary protein excretion has historically been used to evaluate protein-losing nephropathies. A variety of chronic renal diseases can result in significant proteinuria [161–164]. Nephrotoxins that produce readily identifiable proteinuria typically exhibit correlative histopathologic findings for glomeruli, tubules, or both. As with urinary enzyme activities, urine protein excretion should be corrected for urine concentration by calculating the total amount excreted/unit time (e.g., mg/16 hour) or the ratio of urinary protein concentration to urinary creatinine concentration. Although glomerular disorders (increased protein filtration) tend to exhibit greater proteinuria than tubular disorders (decreased protein reabsorption), the exact source of the protein loss cannot be determined by simply quantifying total urinary protein.

$\beta_2$-microglobulin is a low-molecular-weight plasma protein that is freely filtered through glomeruli and almost completely reabsorbed (>99%) by proximal tubular epithelial cells. Immunoassays for urinary $\beta_2$-microglobulin have been developed and used to differentiate glomerular from tubular protein loss and assess tubular function [165–167]. Unfortunately, antibodies specific for animal $\beta_2$-microglobulin are not commercially available, and the structure of the protein appears to be highly species specific [136]. The use of sodium dodecyl sulfate–polyacrylamide gel electrophoresis has been proposed as a means of classifying renal injury by the molecular weight pattern of excreted urinary proteins [18]. An increase in high-molecular-mass proteins (e.g., >69,000 Da) is associated with glomerular injury, and an increase in low-molecular-mass proteins (e.g., 12,000 to 60,000 Da) is associated with tubular injury.

### Urinary Electrolytes

Urinary electrolyte concentrations (sodium, potassium, and chloride) from timed urine collections are the most commonly performed quantitative urine chemistry tests for regulated toxicology studies because they have been listed in the study guidelines for preclinical evaluation of new pharmaceutical products by Japan's Ministry of Health, Labor, and Welfare [11]. Unfortunately, these tests have a very poor cost-to-benefit ratio as screening tools for nephrotoxicity. If only concentrations are assessed, they offer little advantage or information beyond that obtained from urine specific gravity or osmolality. Additional information can be obtained by calculating the total amount of each electrolyte excreted per unit time (e.g., mmol/16 hour) or the fractional clearance of each electrolyte. In contrast to effects on urinary enzyme activities or protein concentrations, effects on urinary electrolyte excretion are most often a reflection of the normal homeostatic mechanisms required to maintain electrolyte and fluid balance in the face of changes in intake (e.g., anorexia) and output (e.g., gastrointestinal losses from diarrhea) rather than specific measures of renal injury or dysfunction. Of course, increased excretion of urinary electrolytes can occur secondarily to administration of many nephrotoxins and compounds with diuretic activity, but routine tests are typically more sensitive and cost effective for detecting test-article-related changes. One reason for the relative insensitivity of urinary electrolyte measurements is that they tend to exhibit considerable interanimal variability.

With respect to the effect of diuretic agents on urinary electrolyte excretion, the period of time for which the urine is collected has a major impact on the results obtained. For example, if a diuretic is administered in the morning, fluid and electrolyte excretion may be quite high during the first several hours postdose. But, if the urine sample collection is performed overnight, after the effect of the diuretic has subsided, electrolyte excretion may appear decreased because of compensatory electrolyte reabsorption to counteract the loss of fluid during the day.

## QUESTIONS

1. Time-related biases are sources of preanalytical variation. Give at least three examples of time-related bias, and describe how these biases might affect interpretation of several different clinical pathology test results.
2. Describe how reference intervals are constructed, and compare their uses with those of concurrent control groups and correlative findings for a given toxicology study.
3. Differentiate regenerative from nonregenerative conditions as they relate to decreased red cell mass and give examples of each.
4. Describe how the common tests of hepatocellular and hepatobiliary integrity and function are used to characterize hepatic toxicity.

## REFERENCES


1. Beutler, E. et al., *Williams Hematology*, 6th ed., McGraw-Hill, New York, 2001.
2. Burtis, C. A., Ashwood, E. R., and Bruns, D. E., *Tietz Textbook of Clinical Chemistry and Molecular Diagnosis*, 4th ed., W.B. Saunders, Philadelphia, PA, 2005.
3. Hasegawa, A. and Furuhama, K, *Atlas* of *the Hematology of the Laboratory Rat*, Elsevier, Amsterdam, 1998.
4. Feldman, B. F., Zinkl, J. G., and Jain, N. C., *Schalm's Veterinary Hematology*, 5th ed., Lippincott Williams & Wilkins, Baltimore, MD, 2000.
5. Kaneko, J. J., Harvey, J. W., and Bruss, M. L., *Clinical Biochemistry of Domestic Animals*, 5th ed., Academic Press, San Diego, CA, 1997.
6. Kaplan, L. A., Pesce, A., and Kazmierczak, S., *Clinical Chemistry: Theory, Analysis, Correlation*, 4th ed., Mosby, St. Louis, MO, 2003.
7. Loeb, W. F. and Quimby, F. W., *The Clinical Chemistry of Laboratory Animals*, 2nd ed., Taylor & Francis, Philadelphia, PA, 1999.
8. Latimer, K. S. et al., *Duncan & Prasse's Veterinary Laboratory Medicine: Clinical Pathology*, 4th ed., Iowa State Press, Ames, IA, 2003.
9. Greer, J. P., Foerster, J., and Lukens, J. N., *Wintrobe's Clinical Hematology*, 11th ed., Lippincott Williams & Wilkins, Baltimore, MD, 2003.
10. Thrall, M. A. et al., *Veterinary Hematology & Clinical Chemistry*, Lippincott Williams & Wilkins, Baltimore, MD, 2004.
11. Hall, R. L., Clinical pathology for preclinical safety assessment: current global guidelines, *Toxicol. Pathol*., 20, 472, 1992.
12. Weingand, K. et al., Harmonization of animal clinical pathology testing in toxicity and safety studies, *Fundam. Appl. Toxicol*., 29, 198, 1996.
13. Godsafe, P. A. and Singleton, B. K., The use of the whole blood Thrombotest (1/51) as a routine monitor of vitamin K-dependent blood coagulation factor levels in the rat, *Comp. Haematol. Int*., 2, 51, 1992.
14. Frith, C. H., Ward, J. M., and Chandra, M., The morphology, immunohistochemistry, and incidence of hematopoietic neoplasms in mice and rats, *Toxicol. Pathol*., 21, 206, 1993.
15. Boone, L. et al., Selection and interpretation of clinical pathology indicators of hepatic injury in preclinical studies, *Vet. Clin. Pathol*., 34, 182, 2005.
16. Carakostas, M. C., What is serum ornithine decarboxylase?, *Clin. Chem*., 34, 2606, 1988.
17. Clemo, F. A. S., Urinary enzyme evaluation of nephrotoxicity in the dog, *Toxicol. Pathol*., 26, 29, 1998.
18. Kolaja, G. J. et al., The use of sodium dodecyl sulfate-polyacrylamide gel electrophoresis to detect renal damage in Sprague–Dawley rats treated with gentamicin sulfate. *Toxicol. Pathol*., 20, 603, 1992.
19. Stonard, M. D. et al., Urinary enzymes and protein patterns as indicators of injury to different regions of the kidney, *Fundam. Appl. Toxicol*., 9, 339, 1987.
20. Price, R. G., Urinary enzymes, nephrotoxicity, and renal disease, *Toxicologist*, 23, 99, 1982.
21. Davies, D. T., Enzymology in preclinical safety evaluation, *Toxicol. Pathol*., 20, 501, 1992.
22. Matsuzawa, T., Nomura, M., and Unno, T., Clinical pathology reference ranges of laboratory animals, *J. Vet. Med. Sci*., 55, 351, 1993.
23. Stromberg, P. C., Large granular lymphocyte leukemia in F344 rats, *Am. J. Pathol*., 119, 517, 1985.
24. Butterfield, L., Are there just 'races' (subspecies) of cynomolgus monkeys or should the name of *Macaca irus* be revived?, *Lab. Primate Newslett*., 36, 19, 1997.
25. Apostolou, A., Saidt, L., and Brown, W. R., Effect of overnight fasting of young rats on water consumption body weight, blood sampling, and blood composition, *Lab Anim. Sci*., 26, 959, 1976.
26. Kimball, J. P. et al., Short-term carbon dioxide/oxygen anesthesia for laboratory rats and mice, *Clin. Chem*., 41 S163, 1995.
27. Maejima, K. and Nagase, S., Effect of starvation and refeeding on the circadian rhythms of hematological and clinico-biochemical values and water intake of rats, *Exp Anim. Jpn*., 40, 389, 1991.
28. Matsuzawa, T. and Sakazume, M., Effects of fasting on haematology and clinical chemistry values in the rat and dog, *Comp. Haematol. Int*., 4, 152, 1994.
29. Thompson, M. B., Avoiding pitfalls in clinical chemistry quality control is not quality assurance, in *Managing Conduct and Data Quality of Toxicology Studies: Sharing Perspectives, Expanding Horizons*, Conference Proceedings Princeton Scientific, Princeton, NJ, 1986, p. 199.

30. Bennett, J. S. et al., Effects of ketamine hydrochloride on serum biochemical and hematological variables in rhesus monkeys (*Macaca mulatto*), *Vet. Clin. Pathol.*, 21, 15, 1992.
31. Dameron, G. W. et al., Effect of bleeding site on clinical laboratory testing of rats: orbital venous plexus versus posterior vena cava, *Lab. Anim. Sci.*, 42, 299, 1992.
32. Khan, K. N., Effect of bleeding site on clinical pathologic parameters in Sprague–Dawley rats: retro-orbital venous plexus versus abdominal aorta, *Contemp. Top.*, 35, 63, 1996.
33. Matsuzawa, T. et al., A comparison of the effect of bleeding site on haematological and plasma chemistry values of F344 rats: the inferior vena cava, abdominal aorta, and orbital venous plexus, *Comp. Haematol. Int.*, 4, 207, 1994.
34. Millis, D. L. et al., Comparison of coagulation test results for blood samples obtained by means of direct venipuncture and through a jugular vein catheter in clinically normal dogs, *J. Am. Vet. Med. Assoc.*, 207, 1311, 1995.
35. Neptun, D. A., Smith, C. N., and Irons, R. D., Effect of sampling site and collection methods on variations in baseline clinical pathology parameters in Fischer-344 rats. I. Clinical chemistry, *Fundam. Appl. Toxicol.*, 5, 1180, 1985.
36. Roncaglioni, M. C., de Gaetano, G., and Donati, M. B., Some aspects of hematological toxicity in animals, in *Animals in Toxicological Research.*, Bartosek, I., Ed., Raven Press, New York, 1982, p. 77.
37. Smith, C. N., Neptun, D. A., and Irons, R. D., Effect of sampling site and collection methods on variations in baseline clinical pathology parameters in Fischer-344 rats. II. Clinical hematology, *Fundam. Appl. Toxicol.*, 7, 658, 1986.
38. Stringer, S. K. and Seligmann, B. E., Effects of two injectable anesthetic agents on coagulation assays in the rat, *Lab. Anim. Sci.*, 46, 430, 1996.
39. Suber, R. L. and Kodell, R. L., The effect of three phlebotomy techniques on hematological and clinical chemical evaluation in Sprague–Dawley rats, *Vet. Clin. Pathol.*, 14, 23, 1985.
40. Upton, P. K. and Morgan, D. J., The effect of sampling technique on some blood parameters in the rat, *Lab. Anim.*, 9, 85, 1975.
41. Nemzek, J. A. et al., Differences in normal values for murine white blood cell counts and other hematological parameters based on sampling site, *Inflamm. Res.*, 50, 523, 2001.
42. Schnell, M. A. et al., Effect of blood collection technique in mice on clinical pathology parameters, *Hum. Gene Ther.*, 13, 155, 2002.
43. Sawyer, M. L., Douglas, S. E., and Mielke, P. M., Blood collection via the jugular vein in rats, *Contemp. Top.*, 36, 64, 1997.
44. Dufour, D. R., Sources and control of analytical variation, in *Clinical Chemistry: Theory, Analysis, Correlation*, 3rd ed., Kaplan, L. A. and Pesce, A. J., Eds., Mosby, St. Louis, MO, 1996, p. 65.
45. Feldman, B. F. and O'Neil, S., Hemolysis as a factor in clinical chemistry and hematology of the dog, *Vet. Clin. Pathol.*, 17, 20, 1988.
46. Frank, J. J. et al., Effect of *in vitro* hemolysis on chemical values for serum, *Clin. Chem.*, 24, 1966, 1978.
47. Laessig, R. H. et al., The effect of 0.1 and 1.0% erythrocytes and hemolysis on serum chemistry values, *Am. J. Clin. Pathol.*, 66, 639, 1976.
48. Leard, B. L. et al., The effect of haemolysis on certain canine serum chemistry parameters, *Lab. Anim.*, 24, 32, 1990.
49. Zhang, D. J. et al., Effect of serum-clot contact time on clinical chemistry laboratory results, *Clin. Chem.*, 44, 1325, 1998.
50. Ono, T. et al., Serum-constituents analysis: effect of duration and temperature of storage of clotted blood, *Clin. Chem.*, 27, 35, 1981.
51. Thoreson, S. I. et al., Effects of storage time on chemistry: results from canine whole blood, heparinized whole blood, serum, and heparinized plasma, *Vet. Clin. Pathol.*, 21, 88, 1992.
52. Dodds, W. J., Hemostasis, in *Clinical Biochemistry of Domestic Animals*, 5th ed., Kaneko, J. J., Harvey, J. W., and Bruss, M. L., Eds., Academic Press, San Diego, CA, 1997, p. 241.
53. Scipioni, R. L. et al., Clinical and clinicopathological assessment of serial phlebotomy in the Sprague–Dawley rat, *Lab. Anim. Sci.*, 47, 293, 1997.
54. Nahas, K. and Provost, J.-P., Blood sampling in the rat: current practices and limitations, *Comp. Clin. Pathol.*, 11, 14, 2002.
55. Hulse, M., Feldman, S., and Bruckner, J. V., Effect of blood sampling schedules on protein drug binding in the rat, *J. Pharmacol. Exp. Ther.*, 218, 416, 1981.
56. Passey, R. B., Quality control for the clinical chemistry laboratory, in *Clinical Chemistry: Theory, Analysis, Correlation*, 3rd ed., Kaplan, L. A. and Pesce, A. J., Eds., Mosby, St. Louis, MO, 1996, p. 382.
57. Westgard, J. O. and Klee, G. G., Quality assurance, in *Fundamentals of Clinical Chemistry*, 3rd ed., Tietz, N. W., Ed., W.B. Saunders, Philadelphia, PA, 1987, p. 238.
58. Carakostas, M. C. and Banerjee, A. K., Interpreting rodent clinical laboratory data in safety assessment studies: biological and analytical components of variation, *Fundam. Appl. Toxicol.*, 15, 744, 1990.
59. Chanter, D. O., Tuck, M. G., and Coombs, D. W., The chances of false negative results in conventional toxicology studies with rats, *Toxicologist*, 43, 65, 1987.
60. NCCLS, *How To Define and Determine Reference Intervals in the Clinical Laboratory: Approved Guideline.* NCCLS document C28-A, National Committee for Clinical Laboratory Standards, Villanova, PA, 1995.
61. Desbiens, N. A., Turney, S. L., and Gani, K. S., Multichannel 18-test panels: Are 60% of the panels abnormal by chance? *J. Lab. Clin. Med.*, 115, 292, 1990.
62. Hall, R. L., Lies, damn lies, and reference intervals (or hysterical control values) for clinical pathology data, *Toxicol. Pathol.*, 25, 647, 1997.
63. Waner, T., Nyska, A., and Chen, R., Population distribution profiles of the activities of blood alanine and aspartate aminotransferase in the normal F344 inbred rat by age and sex, *Lab. Anim. Sci.*, 25, 263, 1991.
64. Weil, C. S. and Carpenter, C. P., Abnormal values in control groups during repeated-dose toxicologic studies, *Toxicol. Appl. Pharmacol.*, 14, 335, 1969.

65. Knoll, J. S., Clinical automated hematology systems, in *Schalm's Veterinary Hematology*, 5th ed., Feldman, B. F., Zinkl, J. G., and Jain, N. C., Eds., Lippincott Williams & Wilkins, Boston, MA, 2000.
66. Szaflarski, N. L., Physiologic effects of normovolemic anemia: implications for clinical monitoring, *AACN Clin. Issues*, 7, 198, 1996.
67. McGrath, J. P., Assessment of hemolytic and hemorrhagic anemias in preclinical safety assessment studies, *Toxicol. Pathol.*, 21, 158, 1993.
68. Beutler, E., Glucose-6-phosphate dehydrogenase deficiency and other enzyme abnormalities, in *Williams Hematology*, 5th ed., Beutler, E. et al., Eds., McGraw-Hill, New York, 1995, p. 564.
69. McGrath, J. P. et al., Oxidative erythrocytic injury in preclinical toxicity testing, *Vet. Pathol.*, 30, 429, 1993.
70. Mansouri, A. and Luri, A. A., Concise review: methemoglobinemia, *Am. J. Hematol.*, 42, 7, 1993.
71. Davis, J. A., Greenfield, R. E., and Brewer, T. G., Benzocaine-induced methemoglobinemia attributed to topical application of the anesthetic in several laboratory animal species, *Am. J. Vet. Res.*, 54, 1322, 1993.
72. Stolk, J. M. and Smith, R. P., Species differences in methemoglobin reductase activity, *Biochem. Pharmacol.*, 15, 343, 1966.
73. Packman, C. H. and Leddy, J. P., Drug-related immune hemolytic anemia, in *Williams Hematology*, 5th ed., Beutler, E. et al., Eds., McGraw-Hill, New York, 1995, p. 691.
74. Wardrop, K. J., The Coombs' test in veterinary medicine: past, present, future, *Vet. Clin. Pathol.*, 34, 325, 2005.
75. Donovan, J. C. et al., Hematologic characterization of naturally occurring malaria (*Plasmodium inui*) in cynomolgus monkeys (*Macaca fascicularis*), *Lab. Anim. Sci.*, 33, 86, 1983.
76. Riley, J. H., Safety testing of immunomodulatory drugs in primates: difficulties in differentiating test article effects from occult diseases—malaria, *Toxicol. Pathol.*, 33, 802, 2005.
77. Erslev, A. J., Anemia of chronic disease, in *Williams Hematology*, 5th ed., Beutler, E. et al., Eds., McGraw-Hill, New York, 1995, p. 518.
78. Feldman, B. F., Kaneko, J. J., and Farver, T. B., Anemia of inflammatory disease in the dog: clinical characterization, *Am. J. Vet. Res.*, 42, 1109, 1981.
79. Feldman, B. F., Kaneko, J. J., and Farver, T. B., Anemia of inflammatory disease in the dog: ferrokinetics of adjuvant-induced anemia, *Am. J. Vet. Res.*, 42, 583, 1981.
80. Caro, J. and Erslev, A. J., Anemia of chronic renal failure, in *Williams Hematology*, 5th ed., Beutler, E. et al., Eds., McGraw-Hill, New York, 1995, p. 456.
81. Palek, J., Acanthocytosis, stomatocytosis, and related disorders, in *Williams Hematology*, 5th ed., Beutler, E. et al., Eds., McGraw-Hill, New York, 1995, p. 557.
82. Erslev, A J., Anemia of endocrine disorders, in *Williams Hematology*, 5th ed., Beutler, E. et al., Eds., McGraw-Hill, New York, 1995, p. 462.
83. Shadduck, R. K., Aplastic anemia, in *Williams Hematology*, 5th ed., Beutler, E. et al., Eds., McGraw-Hill, New York, 1995, p. 238.
84. Weiss, D. J. and Klausner, J. S., Drug-associated aplastic anemia in dogs: eight cases (1984–1988), *J. Am. Vet. Med. Assoc.*, 196, 472, 1990.
85. Erslev, A. J., Pure red cell aplasia, in *Williams Hematology*, 5th ed., Beutler, E. et al., Eds., McGraw-Hill, New York, 1995, p. 448.
86. Babior, B. M., The megaloblastic anemias, in *Williams Hematology*, 5th ed., Beutler, E. et al., Eds., McGraw-Hill, New York, 1995, p. 471.
87. Wixson, S. K. and Griffith, J. W., Nutritional deficiency anemias in nonhuman primates, *Lab. Anim. Sci.*, 36, 231, 1986.
88. Cazzola, M., Mechanisms of anaemia in patients with malignancy: implications for the clinical use of recombinant human erythropoietin, *Med. Oncol.*, 17, 11, 2000.
89. Kurzrock, R., The role of cytokines in cancer-related fatigue, *Cancer*, 92, 1684, 2001.
90. Weiss, D. J., Leukocyte response to toxic injury, *Toxicol. Pathol.*, 21, 135, 1993.
91. Bloom, J. C. et al., Cephalosporin-induced immune cytopenia in the dog: demonstration of erythrocyte-, neutrophil-, and platelet-associated IgG following treatment with cefazedone, *Am. J. Hematol.*, 28, 71, 1988.
92. Lorenz, M. et al., Atypical antipsychotic-induced neutropenia in dogs, *Toxicol. Appl. Pharmacol.*, 155, 227, 1999.
93. Iverson, S., Zahid, N., and Uetrecht, J. P., Predicting drug-induced agranulocytosis: Characterizing neutrophil-generated metabolites of a model compound, DMP 406, and assessing the relevance of an *in vitro* apoptosis assay for identifying drugs that may cause agranulocytosis, *Chem. Biol. Interact.*, 142, 175, 1002.
94. Williams, W. J., Secondary thrombocytosis, in *Williams Hematology*, 5th ed., Beutler, E. et al., Eds., McGraw-Hill, New York, 1995, p. 1361.
95. Beguin, Y., Erythropoietin and platelet production, *Haematologica*, 84, 541, 1999.
96. Boon, G. D., An overview of hemostasis, *Toxicol. Pathol.*, 21, 170, 1993.
97. George, J. N., El-Harake, M., and Aster, R. H., Thrombocytopenia due to enhanced platelet destruction by immunologic mechanisms, in *Williams Hematology*, 5th ed., Beutler, E. et al., Eds., McGraw-Hill, New York, 1995, p. 1315.
98. Martin, R. A. et al., Differential analysis of animal bone marrow by flow cytometry, *Cytometry*, 13, 638, 1992.
99. Rebar, A. H., General responses of the bone marrow to injury, *Toxicol. Pathol.*, 21, 118, 1993.
100. Bollinger, A. P., Cytologic evaluation of bone marrow in rats: indications, methods, and normal morphology, *Vet Clin. Pathol.*, 33, 58, 2004.
101. Jesty, J. and Nemerson, Y., The pathways of blood coagulation, in *Williams Hematology*, 5th ed., Beutler, E. et al. Eds., McGraw-Hill, New York, 1995, p. 1227.
102. Byars, T. D. et al., Activated coagulation time (ACT) of whole blood in normal dogs, *Am. J. Vet. Res.*, 37, 1359, 1976.
103. Schiffer, S. P., Gillett, C. S., and Ringler, D. H., Activated coagulation time for rhesus monkeys (*Macaca mulatta*), *Lab. Anim. Sci.*, 34, 191, 1984.

104. Wilkerson, R. D., Conran, P. B., and Greene, S. L., Activated coagulation time test: a convenient monitor of heparinization for dogs used in cardiovascular research, *Lab. Anim. Sci*., 34, 62, 1984.
105. Kurata, M. et al., Prolongation of PT and APTT under excessive anticoagulant in plasma from rats and dogs, *J. Toxicol. Sci*., 23, 149, 1998.
106. O'Brien, S. R., Sellers, T. S., and Meyer, D. J., Artifactual prolongation of the activated partial thromboplastin time associated with hemoconcentration in dogs, *J. Vet. Int. Med*., 3, 163, 1995.
107. Sherwin, J. E. and Sobenes, J. R., Liver function, in *Clinical Chemistry: Theory, Analysis, Correlation*, 3rd ed., Kaplan, L. A. and Pesce, A. J., Eds., Mosby, St. Louis, MO, 1996, p. 505.
108. Sturgill, M. G. and Lambert, G. H., Xenobiotic-induced hepatotoxicity: mechanisms of liver injury and methods of monitoring liver function, *Clin. Chem*., 43, 1512, 1997.
109. Carakostas, M. C. et al., Evaluating toxin-induced hepatic injury in rats by laboratory results and discriminant analysis, *Vet. Pathol*., 23, 264, 1986.
110. Boyd, J. W., The mechanisms relating to increases in plasma enzymes and isoenzymes in diseases of animals, *Vet. Clin. Pathol*., 12, 9, 1983.
111. Boyd, J. W., Serum enzymes in the diagnosis of diseases in man and animals, *J. Comp. Pathol*., 98, 381, 1988.
112. Keller, P., Enzyme activities in the dog: tissue analyses, plasma values, and intracellular distribution, *Am. J. Vet. Res*., 42, 575, 1981.
113. Clampitt, R. B. and Hart, R. J., The tissue activities of some diagnostic enzymes in ten mammalian species, *J. Comp. Pathol*., 88, 607, 1978.
114. Tenant, B. C., Hepatic function, in *Clinical Biochemistry of Domestic Animals*, 5th ed., Kaneko, J. J., Harvey, J. W., and Bruss, M. L., Eds., Academic Press, San Diego, CA, 1997, p. 327.
115. Kramer, J. W. and Hoffman, W. E., Clinical enzymology, in *Clinical Biochemistry of Domestic Animals*, 5th ed., Kaneko, J. J., Harvey, J. W., and Bruss, M. L., Eds., Academic Press, San Diego, CA, 1997, p. 303.
116. Solter, P. F., Clinical pathology approaches to hepatic injury, *Toxicol. Pathol*., 33, 9, 2005.
117. Swenson, C. L. and Graves, T. K., Absence of liver specificity for canine alanine aminotransferase, *Vet. Clin. Pathol*., 26, 26, 1997.
118. Watkins, J. R., Gough, A. W., and McGuire, E. J., Drug-induced myopathy in beagle dogs, *Toxicol. Pathol*., 17, 545, 1989.
119. Slighter, R. G. et al., Enzootic hepatitis A infection in cynomolgus monkeys (*Macaca fascicularis*), *Am. J. Primatol*., 14, 73, 1988.
120. Swaim, L. D., Taylor, H. W., and Jersey, G. C., The effect of handling techniques on serum ALT activity in mice, *J. Appl. Toxicol*., 5, 160, 1985.
121. Dooley, J. F., Sorbitol dehydrogenase and its use in toxicology testing in lab animals, *Lab. Anim*., 13, 20, 1984.
122. Cornish, H. H., The role of vitamin $B_6$ in the toxicity of hydrazines, *Ann. N.Y. Acad. Sci*., 166, 136, 1969.
123. Dhami, M. S. I. et al., Decreases in aminotransferase activity of serum and various tissues in the rat after cefazolin treatment, *Clin. Chem*., 25, 1263, 1979.
124. Waner, T. et al., Gingival hyperplasia in dogs induced by oxodipine, a calcium channel blocker, *Toxicol. Pathol*., 16, 327, 1988.
125. Hoffman, W. E. et al, Automated and semiautomated analysis of rat alkaline phosphatase isoenzymes, *Toxicol. Pathol*., 22, 633, 1994.
126. Hoffman, W. E. and Solter, P. F., Alkaline phosphatase isoenzymes: biochemistry and clinical evaluation in domestic and laboratory animals, *Curr. Top. Vet. Res*., 1, 171, 1994.
127. Waner, T. and Nyska, A., The influence of fasting on blood glucose, triglycerides, cholesterol, and alkaline phosphatase in rats, *Vet. Clin. Pathol*., 23, 78, 1994.
128. Solter, P. F. et al., Hepatic total 3α-hydroxy bile acids concentration and enzyme activities in prednisone-treated dogs, *Am. J. Vet. Res*., 55, 1086, 1994.
129. Leonard, T. B., Neptun, D. A., and Popp, J. A., Serum gamma glutamyl transferase as a specific indicator of bile duct lesions in the rat liver, *Am. J. Pathol*., 116, 262, 1984.
130. Goldberg, D. M., The expanding role of microsomal enzyme induction and its implications for clinical chemistry, *Clin. Chem*., 26, 691, 1980.
131. Carakostas, M. C., Power, R. J., and Banerjee, A. K., Serum 5′-nucleotidase activity in rats: a method for automated analysis and criteria for interpretation, *Vet. Clin. Pathol*., 19, 109, 1990.
132. Jaynes, P. K., Antiepileptic drug therapy: the laboratory effects on enzyme induction, *Lab. Manage*., March, 40, 1984.
133. Center, S. A. et al., Bile acid concentrations in the diagnosis of hepatobiliary disease in the dog, *J. Am. Vet. Med. Assoc*., 187, 935, 1985.
134. Bovee, K. C., Renal function and laboratory evaluation, *Toxicol. Pathol*., 14, 26, 1986.
135. Finco, D. R. and Duncan, J. R., Evaluation of blood urea nitrogen and serum creatinine concentrations as indicators of renal dysfunction: a study of 111 cases and a review of related literature, *J. Am. Vet. Med. Assoc*., 168, 593, 1976.
136. Loeb, W. F., The measurement of renal injury, *Toxicol. Pathol*., 26, 26, 1998.
137. Perrone, R. D., Madias, N. E., and Levey, A. S., Serum creatinine as an index of renal function: new insights into old concepts, *Clin. Chem*., 38, 1933, 1992.
138. Stonard, M. D., Assessment of renal function and damage in animal species, *J. Appl. Toxicol*., 10, 267, 1990.
139. Bovee, K. C. and Joyce, T., Clinical evaluation of glomerular function: 24-hour creatinine clearance in dogs, *J. Am. Vet. Med. Assoc*., 174, 488, 1979.
140. Kaneko, J. J., Serum proteins and the dysproteinemias, in *Clinical Biochemistry of Domestic Animals*, 5th ed., Kaneko, J. J., Harvey, J. W., and Bruss, M. L., Eds., Academic Press, San Diego, CA, 1997, p. 117.
141. Burton, S. A. et al., C-reactive protein concentration in dogs with inflammatory leukograms, *Am. J. Vet. Res*., 55, 613, 1994.

142. Solter, P. F. et al., Haptoglobin and ceruloplasmin as determinants of inflammation in dogs, *Am. J. Vet. Res.*, 52, 1738, 1991.
143. Ceron, J. J., Eckersall, P. D., and Martinez-Subiela, S., Acute phase proteins in dogs and cats: current knowledge and future perspectives, *Vet. Clin. Pathol.*, 34, 85, 2005.
144. Bruss, M. L., Lipids and ketones, in *Clinical Biochemistry of Domestic Animals*, 5th ed., Kaneko, J. J., Harvey, J. W., and Bruss, M. L., Eds., Academic Press, San Diego, CA, 1997, p. 83.
145. Bauer, J. E., Comparative lipid and lipoprotein metabolism, *Vet. Clin. Pathol.*, 25, 49, 1996.
146. Rosol, T. J. and Capen, C. C., Calcium-regulating hormones and diseases of abnormal mineral (calcium, phosphorus, magnesium) metabolism, in *Clinical Biochemistry of Domestic Animals*, 5th ed., Kaneko, J. J., Harvey, J. W., and Bruss, M. L., Eds., Academic Press, San Diego, CA, 1997, p. 619.
147. Christenson, R. H. and Azzazy, H. M. E., Biochemical markers of the acute coronary syndromes, *Clin. Chem.*, 44, 1855, 1998.
148. Beck, M. L. et al., Cardiac troponin T is a sensitive and specific biomarker of cardiac injury in laboratory animals, *Clin. Chem.*, 43, S192, 1997.
149. Dameron, G. W. et al., Tissue and species specificity of two generations of cardiac troponin-T immunoassays, *Clin. Chem.*, 43, 5192, 1997.
150. Evans, G. O., Biochemical assessment of cardiac function and damage in animal species, *J. Appl. Toxicol.*, 11, 15, 1990.
151. Hossein-Nia, M. et al., Creatine kinase MB isoforms and troponins T and I: sensitive markers of myocardial damage in pre-clinical studies. *Clin. Chem.*, 42, 5241, 1996.
152. O'Brien, P. J., Landt, Y., and Ladenson, J. H., Differential reactivity of cardiac and skeletal muscle from various species in a cardiac troponin I immunoassay, *Clin. Chem.*, 43, 2333, 1997.
153. Vail, D. M., Allen, T. A., and Weiser, G., Applicability of leukocyte esterase test strip in detection of canine pyuria, *J. Am. Vet. Med. Assoc.*, 189, 1451, 1986.
154. Fettman, M. J., Evaluation of the usefulness of routine microscopy in canine urinalysis, *J. Am. Vet. Med. Assoc.*, 190, 892, 1987.
155. Cohen, S. M., Comparative pathology of proliferative lesions of the urinary bladder, *Toxicol. Pathol.*, 30, 663, 2002.
156. Ellis, B. G., Price, R. G., and Topham, J. G., The effect of papillary damage by ethyleneimine on kidney function and some urinary enzymes in the dog, *Chem. Biol. Interact.*, 7, 132, 1973.
157. Ellis, B. G., Price, R. G., and Topham, J. C., The effect of tubular damage by mercuric chloride on kidney function and some urinary enzymes in the dog, *Chem. Biol. Interact.*, 7, 101, 1973.
158. Grauer, G. F. et al., Estimation of quantitative enzymuria in dogs with gentamicin-induced nephrotoxicosis using urine enzyme/creatinine ratios from spot urine samples, *J. Vet. Intern. Med.*, 9, 324, 1985.
159. Greco, D. S. et al., Urinary gamma-glutamyl transpeptidase activity in dogs with gentamicin-induced nephrotoxicity, *Am. J. Vet. Res.*, 46, 2332, 1985.
160. McAuley, F. T. et al., The predictive value of enzymuria in cyclosporin A-induced renal toxicity in the rat, *Toxicol. Lett.*, 32, 163, 1986.
161. Center, S. A. et al., 24-Hour urine protein/creatinine ratio in dogs with protein-losing nephropathies, *J. Am. Vet. Med. Assoc.*, 187, 820, 1985.
162. Grauer, G. F., Thomas, C. B., and Eicker, S. W., Estimation of quantitative proteinuria in the dog, using the protein-to-creatinine ratio from a random, voided sample, *Am. J. Vet. Res.*, 46, 2116, 1985.
163. Hall, R. L., Wilke, W. L., and Fettman, M. J., The progression of adriamycin-induced nephrotic syndrome in rats and the effect of captopril, *Toxicol. Appl. Pharmacol.*, 82, 164, 1986.
164. White, J. V. et al., Use of protein-to-creatinine ratio in a single urine specimen for quantitative estimation of canine proteinuria, *J. Am. Vet. Med. Assoc.*, 185, 882, 1984.
165. Schardijn, G. H. C. and van Eps, L. W. S., $\beta_2$-microglobulin: its significance in the evaluation of renal function. *Kidney Int.*, 32, 635, 1987.
166. Viau, C., Bernard, A., and Lauwerys, R., Determination of rat $\beta_2$-microglobulin in urine and in serum. I. Development of an immunoassay based on latex particles agglutination *J. Appl.* Toxicol., 6, 185, 1986.
167. Viau, C. et al., Determination of rat $\beta_2$-microglobulin in urine and in serum. II. Application of its urinary measurement to selected nephrotoxicity models, *J. Appl. Toxicol.* 6, 191, 1986.

# 27 Dermatotoxicology

*Benjamin B. Hayes, Esther Patrick, and Howard J. Maibach*

## CONTENTS



Human skin is a dynamic, multilayered organ comprising approximately 10% of the normal adult body. It has numerous functions that are essential for terrestrial life, including thermoregulation, sensory perception, nutrient storage, vitamin synthesis, and barrier protection [256,257,307,347]. Resiliency and tensile strength protect against physical injury, pigmentation protects against ultraviolet light, barrier properties protect against the entry

**TABLE 27.1**
**Regional Variation of Skin Thickness in Humans**

| Region | Thickness (μm)[a] | Ref. |
|---|---|---|
| Stratum corneum of abdomen | 8.2 | Holbrook and Odland [156] |
| Abdomen | 46.6 | Whitton and Ewell [379] |
| Stratum corneum of back | 9.4 | Holbrook and Odland [156] |
| Back | 43.2 | Bergstressor et al. [23] |
| Stratum corneum of thigh | 10.9 | Holbrook and Odland [156] |
| Thigh | 54.3 | Bergstressor et al. [23] |
| Stratum corneum of forearm | 15.0 | Holbrook and Odland [156] |
| Forearm | 60.9 | Bergstressor et al. [23] |
| Cheek | 38.8 | Whitton and Ewell [379] |
| Forehead | 50.3 | Whitton and Ewell [379] |
| Back of hand | 84.5 | Whitton and Ewell [379] |
| Fingertip | 369.0 | Whitton and Ewell [379] |

[a] Values are for full thickness skin unless stratum corneum is specified.

of environmental chemicals into the body, and the growth pattern and surface characteristics protect against microbial colonization and invasion. The regenerative capacity following wounding and the number of processes by which skin can deal with environmental insults provide strong evidence of the importance of healthy skin to the organism. The psychological value of healthy skin has led in part to the development of multibillion-dollar cosmetics and personal care industries.

The policies of agencies such as the Occupational Safety and Health Administration (OSHA), Department of Transportation (DOT), Consumer Product Safety Commission (CPSC), and Food and Drug Administration (FDA) in the United States and the Organization for Economic Cooperation and Development (OECD) and European Economic Community (EEC) internationally indicate that the identification of chemicals hazardous to the skin and the protection of society from exposure to those chemicals should be given high priority. These agencies mandate specific assays to evaluate the effects of skin exposure prior to registration, transport, and marketing of chemicals of formulated products. The adverse skin responses associated with repetitive, low-dose exposure to industrial chemicals and consumer products all too often are not accurately predicted by the required assays. The need to market products with low risk of producing dermal and systemic injury to increase consumer satisfaction has led to the development of numerous assays to rank chemicals for their ability to injure the skin. Although these assays are not mandated by regulatory agencies, the frequency with which they are conducted and their utility warrants attention.

The discipline of dermatotoxicology evaluates the toxic effects of chemicals on the skin and includes both the pharmacokinetics of epicutaneously applied chemicals and assays that evaluate the development of neoplasms, trigger an immune response, directly destroy the skin (corrosion), irritate the skin, produce urticaria (hives), and produce noninflammatory painful sensations. The inflammatory responses of skin are the most common chemically induced diseases in humans.

## SKIN STRUCTURE AND FUNCTION

To understand the variety of adverse responses to skin and the basis for the predictive assays for skin injury, some understanding of skin anatomy and physiology is necessary. Approximately 20,000 to 23,000 cm$^2$ of skin cover the body of an adult human. Skin is heterogeneous. Its characteristics (e.g., sweat glands, hair follicles, sebaceous glands, the thickness of skin) vary by body region [156,379]; for example, the thickness of the skin of the eyelid is approximately 0.51 mm, whereas that of the palm and sole are approximately 4.1 mm (see Table 27.1).

A film composed of triglycerides, phospholipids, esterified cholesterol, and other materials released by holocrine sebaceous glands, as well as salts and water released by eccrine sweat glands, normally covers the outer surface [264]. This surface film has been referred to as an *acid mantle*; the pH of the skin normally varies between 4.2 and 5.6. Micrococciae and *Corynebacterium* species normally colonize the skin surface [217,256]. Changes in surface film composition (e.g., changes caused by inflammatory conditions of occlusion) may result in a 1000-fold increase in the absolute number of microorganisms colonizing the area and a shift in flora present [33,34]. The surface film penetrates the outermost cellular layers of the skin.

Based on structure and embryonic origin, the cellular layers of the skin are divided into two distinct regions (Figure 27.1): the epidermis and the dermis. The outer region, the epidermis, develops from embryonic ectoderm

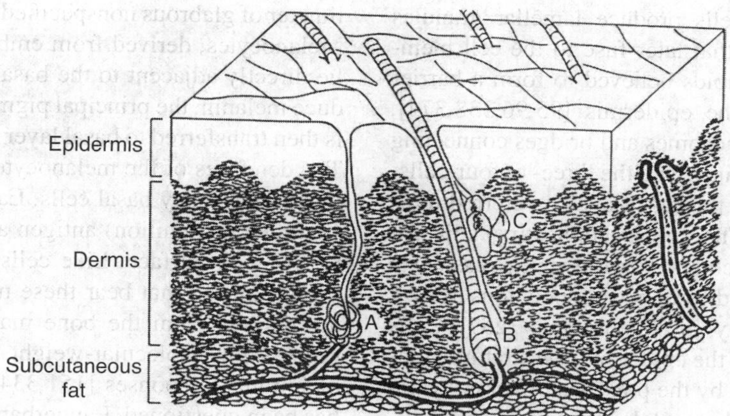


FIGURE 27.1 Schematic of human skin. (A) Eccrine sweat glands are located in the dermis; a duct transports sweat through the epidermis to the surface. (B) Hair follicles are located deep in the dermis; each hair extends through the skin via an epithelized channel. (C) Contents of sebaceous glands are released into the follicular channel as the sebocytes die. Each skin appendage has its own blood supply. Plexuses formed in the upper dermis (shown in the drawing to the far right) supply nutrients to the upper epidermis and upper dermis.

and covers the connective tissue; the dermis is derived from the mesoderm [155,157,248]. The epidermis accounts for approximately 5% of full-thickness skin [216,257]. For descriptive purposes, the epidermis is subdivided into five to six layers based on cellular characteristics (Figure 27.2). The layers of keratinocytes are formed by the ordered differentiation of cells from one layer of mitotic basal cells. The number of distinguishable layers varies by anatomical site.

Basal layer keratinocytes are metabolically active cells with the capacity to divide. Some daughter cells of the basal layer move upward and differentiate. Cells adjacent to the basal layer contain large mitochondria; the Golgi apparatus and rough endoplasmic reticulum (RER) are

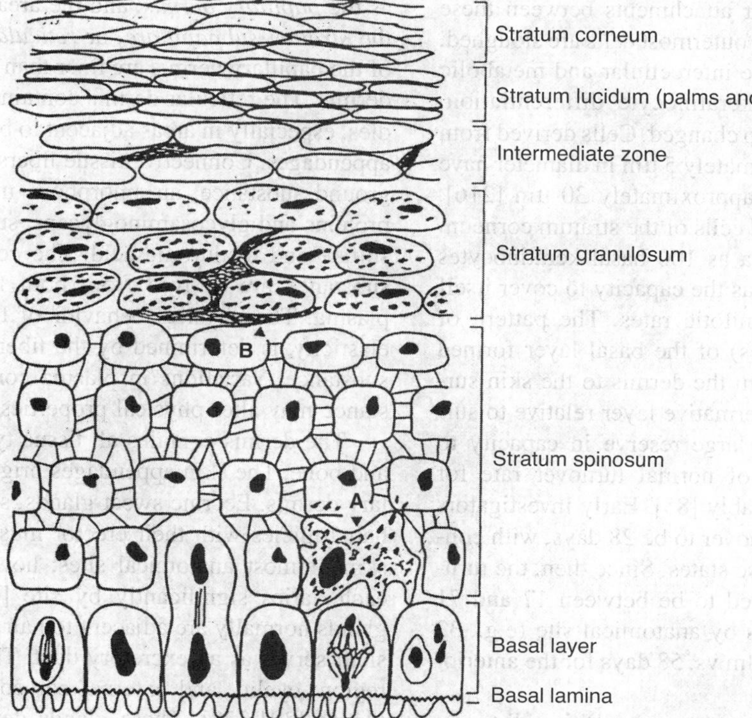


FIGURE 27.2 The epidermis, showing all possible cell layers and locations of the two dendritic cell types: (A) melanocytes, and (B) Langerhans cells.

well developed. These cells produce lamellar granules (intracellular organelles) that later fuse to the cell membrane to release neutral lipids believed to form a barrier to penetration through the epidermis [95,96,338,374]. Microscopically, the desmosomes and bridges connecting adjacent cells resemble spines, and the three- to four-cells-thick layer of cells above the basal layer is referred to as the *stratum spinosum*. The spines connecting adjacent cells are temporary structures; keratinocytes dissociate from neighboring cells and form new associations as they move upward, individually, in the epidermis [257]. Cells of the third subdivision of the epidermis, the stratum granulosum, are characterized by the presence of keratohyalin granules, polyribosomes, large Golgi apparatus, and RER. Cells of the granular layer are the uppermost viable cells of the epidermis. Here, the lamellar granules are released at the cell surface. An intermediate zone of cells separates the cornified layers of the outer epidermis from the viable granulosum. In the palms and soles, the stratum lucidum, or clear cell layer, lies above the stratum granulosum. This layer is indistinguishable in skin sections from other areas. Cells of the intermediate zone may contain enzymes capable of metabolizing exogenous chemicals but have lost the ability to synthesize proteins. The outermost cornified layer, the stratum corneum, consists of cells that have lost their nucleus and all capacity for metabolic activity. The dominant constituent of these cells is keratin, a scleroprotein with chains linked by both disulfide and hydrogen bonds that are synthesized and stored in the deep epidermal layers. The intracellular attachments between these cells gradually break and the outermost cells are sloughed.

In addition to the visible intercellular and metabolic changes observed during keratinocyte differentiation, their size and shape have also changed. Cells derived from basal cuboidal cells approximately 5 μm in diameter have elongated and flattened to approximately 30 μm [216]. Four differentiated cornified cells of the stratum corneum (2 × 2) cover the same area as 100 basal keratinocytes (10 × 10). Each basal cell has the capacity to cover itself many times with modest mitotic rates. The pattern of papillae (ridges and grooves) of the basal layer formed by accessory structured from the dermis to the skin surface increases the area of germative layer relative to surface area. This provides a large reserve in capacity to cover the area. Estimates of normal turnover rate for keratinocytes vary considerably [85]. Early investigators [292] estimated normal turnover to be 28 days, with considerable increases in disease states. Since then, the turnover rate has been calculated to be between 17 and 71 days [103]. Turnover varies by anatomical site (e.g., 32 to 36 days for the human palm vs. 58 days for the anterior surface of the forearm).

The epidermis also contains two dendritic cell types: melanocytes and Langerhans cells. Between 460 and 1000 melanocytes and Langerhans cells per square millimeter of glabrous nonspecified skin is normal [216,267]. Melanocytes, derived from embryonic neural crest cells, lie directly adjacent to the basal layer. Melanocytes produce melanin, the principal pigment of human skin, which is then transferred to basal layer keratinocytes in granules. The dendrites of the melanocyte allow one cell to supply melanin to many basal cells. Langerhans cells express Ia (immune recognition) antigen and receptors for IgG and C3 on their surface. Like cells of the monocyte/macrophage lineage that bear these markers, Langerhans cells are derived from the bone marrow mesenchyme. They process low-molecular-weight haptens during induction of immune responses [151,334]. Although this function has been questioned, Langerhans cells take up small molecules (nonlipid) and increase in number in areas that have developed allergic reactions [333]. Note that Langerhans cells lie in epidermal layers containing enzymes that can metabolize exogenous chemicals. In some cases, metabolites of the agent applied to the skin cause allergic contact dermatitis.

The dermis and epidermis are separated by a basal lamina. The dermis is attached to this membrane by fine fibers of connective tissue. Cells of the basal layer are anchored to the lamina by radicles. This area of attachment, called the *marginal layer*, is identified histologically by periodic acid Schiff reaction. There are occasional breaks in the attachments. Large breaks are observed in exfoliative skin conditions [43,216]. The dermal connective tissue enclosed by the epidermal papilla is referred to as the *papillary dermis*, and the area below the papilla is the *stratum subpapillare*, or *reticular dermis*. The fibers of the papillary dermis are finer than fibers of the reticular dermis. The reticular dermis contains thick collagen bundles, especially in areas adjacent to blood vessels and skin appendages. Connective tissue fibers are separated by the ground substance, an amorphous material consisting of proteins and glycosaminoglycans, such as chrondroitin A sulfate and hyaluronic acid. The constituents of ground substance are derived from both fibrocytes and blood plasma. The physical behavior of the dermis, including elasticity, is determined by the fiber bundles and ground substance. Variations in plasma content of ground substance may alter physical properties substantially.

The dermis contains all tissue types, except cartilage and bone. The skin appendages originate in the subpapillary dermis. Eccrine sweat glands, sebaceous glands, and hair follicles with their erector muscles are found in the skin of most anatomical sites; however, the number of each varies significantly by site [309,336]. Sebaceous glands normally are adjacent to hair follicles, and the hair shaft serves as an excretory duct. The axillae, anogenital region, eyelid, and external ear contain apocrine sweat glands [301,328]. These glands develop at puberty and form odorless secretions that are decomposed by bacteria to produce characteristic odors. The dermis also contains nerve cells with highly specialized sensory endings in some areas, fat lobules, migratory white blood cells, and mast cells. Mast cells are indistinguishable from fibroblasts in size or appearance; however, they contain granules that stain metachromically with agents of a thiazine group. Mast cells are most numerous in areas adjacent to blood vessels, skin appendages, and nerves. The precise function of mast cells is unknown; however, they appear to be involved in the pathogenesis of some inflammatory conditions [342,368]. Their granules contain histamine, heparin, and other vasoactive agents that may be released upon stimulation of the cell surface by IgE cross-linking 48/80, activated serum compounds, and some enzymes. Release of these mediators is accompanied by formation of other agents, such as the metabolites of arachidonic acid [342], which are inflammatory mediators in some conditions.

The dermis and fascia of muscles are separated by the subcutis, a layer of fatty tissue. The extent and development of subcutis depends on sex, age, diet, and body region. Blood vessels supplying the skin arise from the subcutis. Vascular plexuses are formed in the transition zone of the subcutis and dermis adjacent to coils of the eccrine sweat gland. Arteries extend upward to mid-dermis, forming anastomoses there. Similar, but independent, plexuses form at the base of the hair shaft and sebaceous glands. A third vascular network is formed from arteries branching off from vessels at the level of eccrine sweat glands that branch into finer vessels that form plexuses in the papillary dermis. Plexuses of the papillary dermis supply the upper dermis, including the upper hair shaft, and the epidermis with nutrients. The adjacent but separate vascular units in the dermis sometimes react differently in pathological processes (i.e., follicular rash).

A simple visual comparison allows one to conclude that the skin of humans and animals varies considerably. The most obvious difference is hair coat covering the skin. In lower mammals, each hair shaft may contain several follicles, a large follicle arising from the subpapillary dermis and several accessory follicles arising from the papillary dermis. In humans, sebaceous gland density varies from 100 to 900 glands per square centimeter; in other mammals, sebaceous glands are more evenly distributed [328]. Human sweat is produced by eccrine sweat glands. Apocrine sweat glands are the dominant sweat gland of animals. Eccrine sweat glands open directly to the skin surface, whereas apocrine glands empty into the hair shaft. Apocrine sweat is less acidic than eccrine sweat, and the pH of the skin surface of animals usually is somewhat higher than that of humans [203]. The thickness of skin also varies extensively by species and body site. Differences in content of granules of mast cells from different species have been reported [368], as have differences in sensitivity to various inflammatory mediators applied to skin. These differences undoubtedly contribute to the lack of correlation between the results of some animal and human predictive assays [285]. The lack of correlation justifies predictive skin testing in humans after preliminary screening in animals if the risks to subjects are minimal.

**TABLE 27.2**
**Factors Determining Percutaneous Absorption**

- Release from vehicle
    - Varies with solubility in vehicle
    - Varies with concentration
    - Varies with pH
- Kinetics of skin penetration
    - Influenced by anatomical site
    - Influenced by degree of occlusion
    - Influenced by intrinsic skin condition
    - Influenced by animal age
    - Influenced by concentration of dosing solution
    - Influenced by surface area dosed
    - Influenced by frequency of dosing
- Tissue distribution
- Excretion kinetics
- Substantivity to the skin
- Volatility
- Wash and rub resistance
- Binding to skin components
- Cutaneous metabolism
- Anatomic pathways

*Source:* Adapted from Wester, R.C. and Maibach, H.I., *Drug Metab. Rev.*, 14, 169–205, 1983.

## PHARMACOKINETICS FOLLOWING APPLICATION OF CHEMICALS TO THE SKIN

Until the beginning of the twentieth century, skin was considered a relatively inert barrier to chemicals that might enter the body [312]. We now know that this view is incorrect. Although the barrier properties of the skin are impressive, many chemicals penetrate the skin, and the skin can metabolize exogenous compounds. Because of its large surface area, skin may be a major route of entry into the body for some exposure situations. Delivery of drugs through the skin to treat systemic conditions has become almost commonplace. Interest in cutaneous pharmacokinetics has increased as the skin has been reconsidered to be a route for systemic administration of drugs and chemicals, as well as a route of entry for toxins. A variety of assays, both *in vivo* and *in vitro*, for measuring absorption through the skin have been developed [11,12, 376,377], and many factors that govern absorption through the skin have been determined (Table 27.2).

The stratum corneum is a major diffusion barrier of the skin [146,308]. Removal of the stratum corneum by tape stripping increases the rate of absorption of some chemicals [34]. Absorption of chemicals through shunts, openings of skin appendages, and gaps in the stratum corneum associated with these structures have been considered [137,311,346,380]. Because of the relative surface area of these shunts (0.1% to 1.0% of the total area), they do not play a decisive role in absorption [311]; however, they may be important initially after application of the penetrant [313], and sebaceous glands may act as a drug reservoir for some materials [146,159,308]. The stratum corneum is not viable and has no capacity for active transport processes; therefore, absorption can be described as passive diffusion across this membrane: $J = K_m \times C_v \times D_m \times y$ [89], where $J$ is the rate of absorption, $K_m$ is the partition coefficient between the vehicle and stratum corneum, $C_v$ is the concentration, $D_m$ is the diffusion constant of the penetrant in the stratum corneum, and $y$ is the thickness of the stratum corneum. It is obvious that skin from different animals or sites of different thickness from the same animal will vary in barrier properties to absorption.

The concentration term is concentration at the skin surface. Application of suspensions of penetrant with slow dissolution rates, emulsions, or penetrants in vehicles in which the diffusion rate is slow will alter the surface concentration and may control the fate of penetration [48,56,65,154,393,394]. This principle has been used in designing slow-release transdermal delivery devices. Other factors that affect thermodynamic activity of the solution at the skin surface, such as pH and temperature, may alter the absorption rate [312,313]. The influence of the vehicle cannot be overstated for a specific concentration of drug; thermodynamic activity may vary by 1000-fold from one vehicle to another. Some vehicles may promote penetration by altering the characteristics of the stratum corneum [196]. Other factors that affect percutaneous absorption include the condition of the skin [110]; age [149,230,304]; surface area to which the material is applied [377]; penetrant volatility; temperature and humidity [119]; substantivity; wash and rub resistance to removal from the skin; and binding to the skin [277]. Skin may become saturated by a penetrant and thus resist penetration from subsequent applications.

In an intuitive sense, it has been assumed that multiple applications would provide significantly greater mass transfer; however, dermatopharmacokinetics (DPK) studies demonstrate that, in any one given day, multiple applications may add little transfer. This may explain why many dermatologic drugs can be successfully dosed on a once-daily basis. Over many days, the above phenomenon does not appear to occur [80,375].

When a chemical has gained access to the viable epidermal layers, it may initiate a local effect, may be absorbed into the circulation and produce an effect, or may produce no local or systemic effects. The viable epidermis contains many enzymes capable of metabolizing exogenous chemical [24–27,239,266,401], including cytochrome P450 isoenzymes, mixed-function oxidases, and glucuronyl transferases. Enzymes have been identified in three compartments [10,71]. Skin enzymes are inducible by systemic phenobarbital, rifampicin, and 3-methyl cholanthrene [360] and by topical 2,3,7,8-tetrachlorodibenzo-*p*-dioxin (TCDD) [293], 3-methyl cholanthrene, and Aroclor® 1254 [79].

Early studies indicated that enzymatic activity in skin was only a fraction of the activity of the liver. Those studies were conducted *in vitro* using whole skin; the enzymatic activity is in the epidermis, which makes up less than 5% of whole skin [266]. When enzymatic activities of the epidermis were calculated, activities ranged from 80 to 240% of those in liver. In some cases, different metabolites are formed in liver and skin from the same parent compound.

The list of enzymes isolated from skin continues to grow, but the skin does not have the capacity to metabolize all chemicals; for example, topically applied hexachlorophene does not appear to be metabolized. At present, it is not possible to predict metabolic pathways or rates following topical application; these must be determined experimentally. Comprehensive reviews of the metabolic capability of skin have been published [10,177].

## *In Vivo* Percutaneous Absorption Assays

Percutaneous absorption can be determined by applying a known amount of chemical to a specified surface area and then measuring the level of the chemical in the urine or feces. To correct for excretion of the material through the lungs, sweat levels (of retention in the body) measured following topical administration usually are expressed as a percentage of levels following parenteral administration of the chemical [46,376]. Because the analytical techniques to measure the chemical are not always available and because some chemicals may be metabolized, radioactive-labeled chemicals, usually carbon-14 or tritium, are customarily used in these assays. Although studies with radiolabeled compounds accurately reflect absorption they may not provide accurate estimates of bioavailability for example, comparison of bioavailability from the nitroglycerin (unmetabolized drug) level and the level of radioactive tracer indicates that use of the tracer overestimates available drug by as much as 20%. This corresponds to the metabolism of the drug to an inactive form.

*In vivo* studies have been conducted in humans and in a number of species [11,12]. Comparison of the absorption fates of a number of compounds shows that absorption rates in the rat and rabbit tend to be higher than in humans and that the skin permeability of monkeys and swine more closely resembles that of humans (Table 27.3). Although these differences are not predicted by any single factor, such as epidermal thickness, they are not unexpected in light of differences in skin characteristics. There are interspecies

**TABLE 27.3**
**Species Differences in *In Vivo* Absorption (% Dose Absorbed)**

| Compound | Rat | Rabbit | Pig | Squirrel Monkey | Human | Ref. |
|---|---|---|---|---|---|---|
| Haloprogin | 95.8 | 113.0 | 19.7 | — | 11.0 | Bartek et al. [11] |
| Acetylcysteine | 3.5 | 2.0 | 6.0 | — | 2.4 | Bartek et al. [11] |
| Cortisone | 24.7 | 30.3 | 4.1 | — | 3.4 | Bartek et al. [11] |
| Caffeine | 53.1 | 69.2 | 32.4 | — | 47.6 | Bartek et al. [11] |
| Butter yellow | 48.2 | 100.0 | 41.9 | — | 21.6 | Bartek et al. [11] |
| Testosterone | 47.4 | 69.6 | 29.4 | — | 13.2 | Bartek et al. [11] |
| DDT | 46.3 | — | 43.4 | 1.5 | 10.4 | Bartek and LaBudde [12] |
| Lindane | 51.2 | — | 37.6 | 16.0 | 9.3 | Bartek and LaBudde [12] |
| Parathion | 97.5 | — | 14.5 | 30.3 | 9.7 | Bartek and LaBudde [12] |
| Malathion | 64.6 | — | 15.5 | 19.3 | 8.2 | Bartek and LaBudde [12] |

differences in routes of excretion of some chemicals as well. This may be due in part to metabolism of the chemical, and the metabolic capabilities of the species should be considered when selecting an animal model and designing the experiment. Ingestion of the test material by the animal must be prevented, and this may require restraint of the animal or design of specialized protective apparatus for the site of application. Because urine and feces are collected for analysis, specialized cages are also required.

The difficulties in conducting these types of pharmacokinetic assays, such as collecting excrement for relatively long periods of time (24 hours), requirements for specialized cages and protective apparatus, and the increased space requirements for housing animals individually, have led to the use of other *in vivo* assays and to the development of *in vitro* models. Loss of radioactive material from the skin surface has been used to estimate *in vivo* percutaneous absorption [234]. The difference in applied dose and residue on the skin is assumed to be absorbed. The characteristics of the radioisotope, penetrant, and vehicle may limit the usefulness of this procedure. Volatile materials leave the surface without penetrating, and it is difficult to recover all material from the skin surface. In addition, skin may retain a reservoir of the penetrant that has not entered the circulation.

A clear relationship exists between the mass of a chemical residing in the stratum corneum that has been washed 30 minutes after application and the eventual penetration that can be measured in urine or blood. This principle has led to a facile method for estimating percutaneous absorption in animals and humans. The approach of determining the stratum corneum content via cellophane tape sampling is also used for bioequivalence determination [46].

*In vivo* biological responses, such as vasoconstriction assays to estimate the absorption of corticosteroids [245] and changes in blood flow being used to study penetration through various types of skin and under diverse conditions [144,145], have been used to estimate rates of penetration. These endpoints are complicated biological processes and may vary with the ability of the tissue to produce the response. For example, application of histamine produces increased blood flow; however, the degree of change would depend not only on the rate of penetration but also on the reactivity of receptors at that time. Most exogenous chemicals produce their vascular effects by triggering the formation and release of endogenous mediators; thus, the usefulness of penetration studies using biological endpoints is limited to comparisons between closely related chemical structures that can be assumed to trigger the same process.

## *In Vitro* Percutaneous Absorption Assays

The excised skin of humans or animals can be used to measure penetration of chemicals. *In vitro* assays utilizing excised skin use specially designed diffusion cells [12,47,117]. The skin is stretched over the opening of a collecting receptacle, epidermal side up. The chemical to be studied is applied to the epidermis, and fluid from the receptacle is assayed to measure the penetration of the chemical. Chemicals usually are radioactively labeled. Some investigators have used diffusion cells in which the epidermis was covered with fluid containing the chemical; however, the preferred method for toxicological relevance is a one-chambered cell in which the stratum corneum is exposed to the air and the underside of the skin is bathed in saline or other receptacle fluid. Because diffusion through a membrane depends on relative concentrations on each side, some chambers have been designed to allow periodic replacement of the receptacle fluid. Fluid in the receptacle base usually is constantly stirred and maintained at a physiological temperature. Either full-thickness skin or epidermis alone may be used in *in vitro* assays. With relatively hairless skin, epidermis can be separated from the dermis by heat treatment.

This type of *in vitro* assay offers advantages over *in vivo* assays. Highly toxic compounds can be studied in human skin. Large numbers of cells can be run simultaneously. Diffusion through the membrane, eliminating other pharmacokinetic factors, can be studied. In addition, these assays may be less expensive and easier to conduct; however, these assays do not mimic human exposure in some important areas. Because excised skin must often be stored prior to use, it cannot be assumed that the skin will retain full enzymatic activity. This may alter the metabolic profile of compounds entering the receptacle. In intact skin, chemical penetrating the epidermis would enter the circulation through vessels and lymphatics located just below the epidermis. In excised full-thickness skin, the dermis is also involved in the absorptive process. The influence of the dermis can be minimized by using heat-separated epidermis or by removal of the skin with a dermatome at the level of the upper dermis. In the intact animal, the chemical enters the peripheral circulation in plasma; the collecting fluid of diffusion chambers is usually saline or water. The relative solubility of hydrophobic and hydrophilic chemicals in these collecting fluids may alter the rate at which they leave the skin. Surface conditions of excised skin may vary from normal skin; changes in the surface emulsion occurring during storage have not been studied. Storage conditions and procedures for preparing the tissue may affect skin absorption and metabolism. It has been proposed that the suitability of each specimen of excised skin be verified by measurement of penetration of a standard, tritiated water through the tissue prior to its use to study penetration of other chemicals. Comparisons of penetration rates obtained from *in vitro* and *in vivo* assays have been made [12,47]. Often a good correlation between the two methods has been obtained; however, with some compounds, correlation of the methods is poor. Differences between *in vivo* and *in vitro* results for some compounds can be explained because of solubility in the receptacle fluid and blood. Differences observed for other compounds cannot be explained.

*In vitro* penetration rates through skin of various species have also been compared (Table 27.3). Skin of the weanling pig and miniature swine appears to serve as good *in vitro* models for most compounds [11]. The skin of monkeys seems to be a good model, as well [376]. For most compounds, mouse and hairless mouse skin appears more permeable than skin of other species. Rat skin appears to be a good model for some compounds; however, when differences have been noted, they have been large.

A few investigators have estimated percutaneous absorption using "model" membranes, including excised stratum corneum, and physiochemical data have been used to predict absorption. Lipid/water partition coefficients have been correlated with skin permeability. Smaller molecules (molecular weight ~400) are more readily absorbed than large molecules. Molecules with polar groups, in general, do not penetrate as well as nonpolar molecules [307]. The addition of hydroxyl groups also lowers the permeability. Substitutions that increase lipid solubility may increase penetration, depending on the vehicle in which the chemicals are applied [312]. Electrolytes do not penetrate the skin well [313]; shunt diffusion through skin appendages for these molecules may be important.

The presumed simplicity of *in vitro* penetration assays has led to their universal acceptance for preclinical and other screening purposes. This acceptance and wide-scale use have resulted in many variations in how studies have been conducted. It is not surprising that confusion regarding the interpretation of data generated by variations of this method occurs. Bronaugh [47] collated the experimental variables leading to discrepancies, and his text provides a catalog of variables that, properly considered, can lead to experimental designs that may have *in vivo* relevance, especially for hydrophilic materials.

## NEOPLASTIC RESPONSE OF SKIN

The skin is the most common site of cancer in humans. Both benign and malignant tumors may be derived from viable keratinocytes and melanocytes of the epidermis and rarely from skin appendages, blood vessels, peripheral nerves, and lymphoid tissue of the dermis [216,390]. Historically, basal cell and squamous cell carcinomas, which develop from keratinocytes, account for 60% and 30% respectively, of all skin cancer. The remainder includes malignant melanoma and the rare tumors developing from other cell types; however, the incidence of malignant melanomas appears to be increasing. Melanomas often metastasize, and the prognosis for patients with this disease is poor. Only 4 to 5% of squamous cell carcinomas are metastatic, and basal cell tumors rarely metastasize. The relatively noninvasive nature of the common forms of skin cancer accounts for a cure rate of over 95%. Skin cancer accounts for less than 0.3% of all cancer deaths.

The association between exposure to environmental carcinogens and the development of basal cell and squamous cell carcinomas is strong [326]. Epidemiological studies have demonstrated a strong correlation between exposure to ultraviolet radiation and development of skin cancer [99]. Clinical experience leaves little doubt that radiographs can also produce cancer of the skin. Both forms of radiation have induced tumors in experimental animals [44]. The association between environmental chemicals and skin cancer was first demonstrated by Sir Percival Potts in 1775. Following Potts' association between skin cancer of the scrotum and soot exposure experimental studies in animals revealed that polycyclic aromatic hydrocarbons such as benzo(*a*)pyrene are the carcinogens in soot, coal tar, pitch, and various cutting oil [107]. The same types of experiments demonstrated that skin cancer development is a multistage process [43,44].

Despite abundant experimental evidence that chemicals can produce skin cancer, few chemicals have been associated with increased incidences of skin cancer in humans. Epidemiological studies have demonstrated associations between polycyclic aromatic hydrocarbons and arsenic and the increased incidence of benign, precancerous lesions and basal cell and squamous cell carcinomas [166]. The ability to establish relationships between chemical exposure and the development of skin cancer by epidemiology is minimized by many confounding factors, including exposure to ultraviolet light, high background incidence rates, long latency periods for the development of cancer [107], and incomplete reporting due to the nonlethal nature of the disease. Even without strong epidemiological evidence, the experimental evidence that exposure of the skin can lead to tumor development and the degree of dermal exposure to chemicals in the workplace justifies the practice of evaluating carcinogenic potential by dermal exposure. Furthermore, it is likely that certain internal tumors (bladder cancer from occupational aniline exposure) result from chemicals absorbed through the skin.

*In vivo* skin carcinogenesis studies are generally conducted using Sprague–Dawley rats, but mice have also been used frequently. Differences in species sensitivity to various agents have been demonstrated [44,280], and a review of the data suggests that when differences in species were noted rats tended to be more sensitive. The design of carcinogenesis assays has been reviewed in other texts [115,116,164,220,276] and is described in Chapter 25. The method, in brief, is daily application of the test material to the clipped skin of up to six groups of animals (untreated, vehicle control, and up to four dose groups) for 104 weeks. The dose is generally administered at a constant volume of 1.0 mL/kg body weight to 5 to 10% of the animal's body surface. Typically, 70 male and 70 female animals are included in each group, with additional groups dosed on the same schedule used for periodic blood analysis and for evaluating the toxicokinetics of the test material. Two variations of standardized skin carcinogenicity studies have also been reported. In the first, the skin is treated with the chemical of interest. Then, a promoting agent, such as tetradecanoyl phorbol acetate (TPA), may be applied to reduce the latency period. In the second, the skin is treated with a noncarcinogenic dose of a carcinogen, such as dimethyl benzanthracene (DMBA), followed by repeated doses of the agent under study. These approaches may be helpful in elucidating the mechanisms of action of the chemicals.

Numerous factors may influence the outcome of dermal carcinogenic assays, and the choice of what to test is crucial in such assays. Although it is tempting to evaluate pure chemicals, it should be remembered that other agents in mixtures could act as promoting agents; for example, coal tar and pitch may contain catechol and pyrogallol, which are promoters of the carcinogen benzo(*a*)pyrene found in these mixtures. Wounding increases the number of tumors that spontaneously develop, and severe inflammatory responses may cause tissue destruction. Care should be taken in selecting an appropriate nonirritant dose.

Epidemiology and experimental studies have shown that sunlight (ultraviolet light) is a skin carcinogen [101,102]. Various investigators [352] have shown that exposure to some chemicals in combination with a dose of ultraviolet light increases the number of tumors or decreases the latency period for tumor development following exposure to the chemical or ultraviolet (UV) light alone. No standard protocol has yet been put forth by any regulatory agency, and typically regulatory agencies request to review specific protocols prior to study conduct [115,116,164]. In general, the studies have involved repeated intercurrent exposures to simulated sunlight and the test article. The endpoint is the time required for UV radiation (UVR) to produce skin cancer. Two doses of UVR are used as a control: a high dose that will result in a short latency period for tumor development and a weaker dose that would result in a lower tumor yield and longer latency period. The interaction of the test article and the weaker dose of UVR is compared to each control and to a vehicle control group irradiated with the lower dose of UVR. Complete details of a typical protocol have been published by Forbes et al. [115,116].

The FDA has expressed concern that some topical drugs and cosmetic ingredients may promote neoplastic changes in the skin induced by other agents, such as sunlight. Although to date there is no epidemiological evidence for such an occurrence, the exploration of this possibility will require developing new approaches to conventional assays for carcinogenicity.

A number of *in vitro* assays for studying chemical carcinogenesis have been developed [165]. Of particular interest for dermal carcinogenesis is the ability to cultivate epidermal keratinocytes of rats, mice, and humans [204]. Cultured human keratinocytes can metabolize polycyclic aromatic hydrocarbons, and chemical transformation using human fibroblasts has been achieved [204]. The establishment of human epidermal lines for *in vitro* carcinogenesis testing will provide an important new predictive tool. Chapter 23 reviews the relationship of mutagenicity and carcinogenicity.

## SKIN ALLERGY (DELAYED-TYPE HYPERSENSITIVITY)

Since the turn of the twentieth century, certain forms of eczema have been recognized as allergic in nature. Jadassohn [167,168] demonstrated that in some patients, dermatitis was due to increased sensitivity following repeated contact with a substance, not the toxic (irritant) properties of the material [166,167]. By 1930, a procedure for producing this hypersensitivity to chemicals in guinea pigs had been developed [35]. The pioneering work of Landsteiner and his colleagues demonstrated that low-molecular-weight chemicals conjugate with proteins to form an antigen that stimulates the immune system to form a hyperreactive state [214]. They demonstrated that immunogenicity is related to chemical structure [213] and that two types of immunologic response exist, one transferable by serum and another transferred by suspensions of white blood cells [212].

It is now known that most cases of allergic contact dermatitis are of the cell-mediated type, transferable by lymphocytes. This type of skin response is often referred to as *delayed-contact hypersensitivity* because of the relatively long period (approximately 24 hours) required for the development of the inflammation following exposure.

Some understanding of the processes by which this hypersensitivity develops is helpful in selecting and interpreting results of predictive sensitization tests. During ontogenesis, stem cells from the yolk sac, fetal liver, and bone marrow migrate to the central lymphoid organs, the thymus, and bone marrow in mammals. After birth, stem cells derive from bone marrow. In the central lymphoid organs, stem cells differentiate into immunocompetent lymphocytes. This results in two classes of lymphocytes: thymus-processed T lymphocytes and B lymphocytes processed in bone marrow. B lymphocytes are precursors of antibody-producing cells responsible for immune responses transferable by serum. T lymphocytes are responsible for producing delayed-type hypersensitivity (DTH) and for regulation of the immune system. This regulation is accomplished by subsets of T cells (i.e., T helper and T suppressor cells). Lymphocytes leaving the lymphoid organs are programmed to recognize a specific chemical structure via receptor molecules. If, during circulation through body tissues, a cell encounters the structure it is programmed to recognize, an immune response may be induced. The ability to develop and express a hypersensitivity response is determined by the relative activities of the T helper and T suppressor cell types [294].

To stimulate an immune response, a chemical must be presented to lymphocytes in an appropriate form [213,214]. Chemicals usually are haptens that must conjugate with proteins in the skin or other tissues to be recognized by the immune system. Haptens conjugate with multiple proteins to form a number of different antigens that may stimulate an allergic response by stimulating T lymphocytes with different recognition capabilities [295].

Hapten–protein conjugates are processed by macrophages, Langerhans cells, or other cells expressing immune-response Ia proteins on their surface. Although the exact nature of this process is not completely understood, it is known that physical contact between macrophage and T cells is required [70,350], suggesting that receptor interactions are necessary. Physical interaction is accompanied by the release of interleukins, a family of soluble, regulatory proteins that stimulate cell division, act as growth factors, and increase expression of immune proteins on the surface of some cells [109,161,253,258].

Following antigen stimulation in the skin, lymphocytes enter the lymphatic system and migrate to the draining lymph nodes. Disruption of lymphatic drainage prevents sensitization of an animal [118]. Stimulated cells settle in the paracortical regions of the lymph nodes, and T lymphocytes differentiate into immunoblasts. This differentiation involves interaction with other cell types. Immunoblasts eventually give rise to T effector cells, which enter the systemic circulation and, upon encountering the antigen that they are programmed to recognize, release lymphokines that initiate a local inflammatory response. Immunoblasts also give rise to memory cells that enter the systemic circulation. These memory cells are capable of similar activities as the processed T lymphocytes; they recognize antigen and can be stimulated to divide. Memory cell production is essentially an expansion of the number of cells capable of recognizing a given antigen. The lymphokines released by primed effector cells that encounter their stimulating antigen directly and, indirectly, by stimulation of other white blood cells produce a local inflammatory response. Actions of lymphokines include direct tissue damage, chemotactic factors, stimulation of mitosis, increased phagocytic activity of macrophages, and factors that inhibit migration of some cell types from the area [77]. Only a small percentage of lymphocytes in an area of skin exhibiting a delayed hypersensitivity response are specifically stimulated by antigen [261]. Most cells in the lesion are recruited by lymphokines. Histologically, the response has been described as a hyperproliferative epidermis with intracellular edema, spongiosis, intraepidermal vesiculation, and mononuclear cell infiltrate by 24 hours. The dermis shows perivenous accumulation of lymphocytes, monocytes, and edema. No reaction occurs if the local vascular supply is interrupted, and the appearance of epidermal changes follows the invasion of monocytes. Vascular changes (i.e., increased blood flow) occur early (2 to 6 hours) in the response.

The histology of the response varies somewhat by species; for example, a higher proportion of polymorphonuclear cells in the cellular infiltrate has been observed in DTH reaction sites of mice than in guinea pigs or humans [169]. These differences may be due, in part, to mixed immune responses. Mice develop both antibody and DTH responses to haptens applied to the skin [6]. Exposure via the skin is believed to preferentially lead to DTH in guinea pigs and humans.

The biological processes necessary for producing hypersensitivity in predictive tests are often grouped into two phases: induction of the capability to respond and elicitation of a response. Induction has been referred to as the *afferent phase*, the initial exposures through clona expansion and the release of memory cells that enter the system circulation. Elicitation, referred to as the *efferent phase*, consists of local recognition of the antigen by the memory cells, their release of lymphokines, and the activity of inflammatory mediators that are generated locally which produce the dermatitis. All standardized predictive tests in guinea pigs and some early tests in mice [6,128,233] use the efferent phase response as an indication of immune reactivity to the chemical. The local lymph node assay (LLNA) in mice uses stimulation of lymphocytes [185–187] in local draining lymph nodes during the afferent phase as the endpoint.

Modulation of development of DTH in experimental animals and in humans is complex. The intrinsic biological variables controlling sensitization can be influenced by the selection of animals likely to be capable of mounting an immune response to the hapten. The extrinsic variables of dose, vehicle, route of exposure, adjuvant, etc., can be manipulated to develop sensitive predictive assays. The method of skin exposure is important. Keratinocytes produce interleukins [77], important regulatory proteins for induction of DTH. Langerhans cells express Ia antigen and may act as antigen-presenting cells [70,303,350]. Intradermal injection in animals bypasses the processes but ensures entry of the chemical into the skin.

Vehicle plays an important role in percutaneous penetration and hence presumably in sensitization. The theory, at least with regard to flux, should be simple: Maximum solubility leads to maximum thermodynamic activity and enhanced flux. The experimental literature only partially documents this presumed truth. Analysis of the relevant literature provides a partial interpretative key to this fundamental area [229]. Increasing the dose per unit area increases the sensitization rate. Upadhye and Maibach [350] reviewed in detail the influence of area application on the development of human sensitization. When early publications viewed as dogma were examined with statistical methods, the findings failed to be significant, yet dose per unit area and occlusion [224,225] appear to be highly important variables.

Application of haptens to damaged skin (i.e., irritated, tape-stripped) increases the sensitization fate. Although effects of vehicle, dose level, and damaged tissue have been studied more extensively in guinea pigs and humans than in mice, until conclusive studies are reported it is prudent to consider their influence in the design of all studies. Repeated applications to the same site are more effective for inducing sensitization than applications to new sites each time [104,224]. The incidence of sensitization increases with increased number of exposures [224], and an interval between exposures of 2 to 6 days increases the sensitization rate [6,187,224]. This may be due to the booster effect of memory cells. Materials such as Freund's complete adjuvant nonspecifically enhance development of immune responses but may selectively trigger humoral immunity in some species [54]. Treatment of animals with adjuvant, either simultaneously or shortly after hapten exposure, increases sensitization rates [225–227]. The development of DTH is under genetic control; within the human and guinea pig populations, all individuals do not have the capability to respond to a given hapten [224]. The status of the immune system will determine if an immune response can be induced; for example, animals may become tolerant to a hapten, and pregnancy may suppress expression of the allergy [200].

Appropriate planning and execution of predictive sensitization assays are critical. All too often, techniques are discredited when, in fact, the performance of the tests was inferior or the study design (e.g., choice of vehicle or dose) was inappropriate. The first priority is to choose an appropriate experimental design. Often the assay to be used is chosen on a *pro forma* basis without realizing the inherent weaknesses and strengths of the method. A common error in choosing an animal assay is using Freund's complete adjuvant when one wishes to determine dose–response relationships. The adjuvant provides such sensitivity that dose–effect relationships are muted.

Choice of dose and vehicle appropriate to the assay and the study question is the second priority. Although dose must be sufficient to ensure penetration, it must be below the irritation threshold at challenge to avoid misinterpretation of irritant inflammation as allergic. For example, quaternary ammonium compounds, such as benzalkonium chloride, rarely sensitize, but they have been identified as allergens in some guinea pig assays. Knowing the irritation potential of compounds allows the investigator to appropriately design and execute these studies. Vehicle choice determines, in part, the absorption of the test material and can influence sensitization rate, ability to elicit response at challenge, and the irritation threshold. Inappropriate selection of vehicle and dose effectively invalidates studies.

Sensitization assays often are assigned to novices when they should be performed and read by persons experienced with the method being used. Experienced investigators will recognize marginal reactions that should be further investigated and positives that may be irritant in nature, and they will be able to assist in the estimation of risk associated with the proposed use of the material. Working with laboratories and personnel with extensive experience greatly decreases errors and increases the reliability of all of the standard assays described below [158,180,181,191,300].

Data from various sensitization assays have been broadly used to determine the likelihood of induction of clinical allergic contact dermatitis in populations to be exposed. Interpretation of these assays requires experience, judgment, and sophistication because each assay has its own strengths, weaknesses, and limitations. Central to the issue of interpretation is definition of allergic contact dermatitis in humans and animals; for example, in the mouse, the endpoint is only documented reliably by measuring ear swelling. In humans, the patch test, albeit a highly valuable bioassay, has often been misinterpreted. Guidance for proper interpretation is found in the original references for many of the assays discussed below; for human repeat insult patch tests, guidance is found in a review article by Stotts [235]. Several authors and groups are refining criteria for allergic contact dermatitis that are operational rather than mechanistic. Fundamentally, each of the systems acknowledges the complexity of the biological process. Each parameter is qualitative or quantitative and is multifunctional; for example, in humans, it is necessary to have a pertinent history, carefully performed patch test results with appropriate virgin controls, and sufficient follow-up to define that removal of the allergen improved clinical status [175,365].

## Guinea Pig Sensitization Tests

Predictive animal tests to determine the potential of substances to induce delayed hypersensitivity in humans are conducted most often in guinea pigs. Several tests have been described. Each offers its own advantages and disadvantages; most have many features in common. All use young (1 to 3 months old or 250 to 550 g), randomly bred albino guinea pigs. To reduce the possibility of seasonal variability in reactivity, animals are maintained in facilities with a temperature of approximately 20 ± 1°C, a relative humidity of 40 to 50%, a 12-hour automatic light cycle, a standard vitamin C-supplemented chow, and water available at all times. Test sites are clipped free of hair with electric clippers; some assays specify chemical depilation as well. Almost all evaluate the response as production of visible dermatitis, using descriptive scales for erythema and edema. Because of genetic influences, sensitivity to a common chemical, such as dinitrochlorobenzene, is usually confirmed periodically for animals from each vendor. There is disagreement as to which sex, if either, is more susceptible to sensitization. Males are more aggressive and may damage the skin of cage mates. Some assays specify use of one sex or half of each sex. The tests differ significantly in route of exposure, use of adjuvants, induction interval, and number of exposures. Table 27.4 summarizes the principal features of the most commonly used assays to predict sensitization and assays acceptable to regulatory agencies [67,100,105,170,262,300].

### The Draize Test

The Draize sensitization test (DT) [87,171,188] was the first predictive sensitization test and is still widely used. One flank of 20 guinea pigs is shaved and 0.05 mL of a 1% solution of test material in saline, paraffin oil, or polyethylene glycol is injected into the anterior flank on day 0. The next day, and every other day through day 20, 1.1 mL of the test solution is injected into a new site on the same flank. Challenge follows a 2-week rest period. The opposite untreated flank is shaved, and 0.05 mL of test solution is injected into each animal. Twenty previously untreated controls are injected at the same time. The test site is visually evaluated 24 and 46 hours after injection. The intensity of the responses of test animals is compared with that of controls. A larger more erythematous response than that of controls is considered a positive response. Results are expressed as the percentage of animals positive or as the ratio of positive animals tested.

### Open Epicutaneous Test

The open epicutaneous test (OET) [180,188,189) simulates the conditions of human use by using topical application of the test material. The procedure determines the dose required to induce sensitization and to elicit a response in sensitized animals. The irritancy profile is evaluated by testing various concentrations (typically, undiluted, 30%, 10%, 3%, and 1%) in ethanol, acetone, water, polyethylene glycol, or petrolatum. In six to eight guinea pigs, 0.025 mL of the dosing solutions is applied to 2-cm$^2$ areas of the shaved flanks. Vehicle solubility and use conditions (e.g., direct application to skin or dilution during normal use) is considered when selecting the concentrations to be tested. Test sites are visually evaluated 24 hours after application of test solutions for the presence or absence of erythema. The dose not causing a reaction in any animal (maximum nonirritant concentration) and the dose causing a reaction in 25% of the animals (minimal irritant concentration) are determined. During induction, 0.10 mL of test solution is applied to an 8-cm$^2$ area of flank skin of 6 to 8 guinea pigs for 3 weeks, or 5 times a week for 4 weeks. As many as six groups of animals are treated with different doses; a control group is treated with vehicle only. The highest dose tested usually is the minimal irritant concentration; lower doses are based on usage concentration or a stepwise reduction (e.g., 30%, 10%, 3%, 1%). Solutions are applied to the same site each day unless a moderate inflammatory response develops. Each animal is challenged on the previously untreated flank 24 to 72 hours after the last induction treatment. The minimal irritant concentration, the maximum nonirritant concentration (from the irritancy screen), and five solutions of lower concentrations are applied (0.025 mL to a 2-cm$^2$ area). Skin reactions are read on an all-or-none basis at 24, 48, and 72 hours after applications of the solutions. The maximum nonirritating concentration in the vehicle-treated group is calculated. Animals in test groups that develop inflammatory response to lower concentrations are considered sensitized. The dose required to sensitize is determined by comparing the number of positive animals in the test groups. The minimal concentration necessary to elicit a positive response in a sensitized animal is apparent from the challenge responses.

**TABLE 27.4**
**Principal Features of Guinea Pig Sensitization Assays**

| Feature of Test | Draize | Open Epicutaneous Test | Buehler Assay | Freund's Complete Adjuvant Test | Optimization Test | Split Adjuvant | Guinea Pig Maximization |
|---|---|---|---|---|---|---|---|
| Number in test group | 20 | 6–8 | 10–20 | 8–10 | 20 | 10–20 | 20–25 |
| Number in control group | 20 | 6–8 | 10–20 | 8–10 | 20 | 10–20 | 20–25 |
| *Induction* | | | | | | | |
| Exposure route | ID | Open epicutaneous | Patch | ID | ID | Patch | ID and patch |
| Number of exposures | 10 | 20–21 | 3 | 3 | 9 | 4 | 1 ID; 1 topical |
| Duration of each patch | No patch | Continuous (no patch) | 6 hr | 5–50% | 0.1% | 0.1–1.2 mL | Maximum tolerated |
| Concentration | 0.2 | Nonirritating | Slightly irritating | 5–50% | 0.1% | 0.1–0.2 mL | Maximum tolerated |
| Test group(s) | TS | TS | TS | TS in FCA | TS in FCA | TS, FCA | TS, TS + FCA, FCA |
| Control group | None | Vehicle only | Vehicle only | FCA only | — | — | FCA, FCA+V, V |
| Site for dosing | Left flank | Right flank | Left flank | Shoulder | Back (flank first injection) | Mid-back | Shoulder |
| Frequency of exposure | Every second day | Daily | Every 5–7 days | Every 4 days | Every other day | Days 0, 2, 4, 7 | Day 0 (ID); day 7 patch |
| Duration (days) | 1–18 | 0–20 | 0–14 | 0–9 | 0–21 | 0–9 | 0–9 |
| Misc. | — | — | 9 exposure version | — | — | Dry ice treatment day 0; FCA (ID) day 4 | Irritant dose or SLS pretreatment |
| Rest period (days) | 19–34 | 21–34 | 15–27 | 9–21; 22–34 | 22–34 | 10–21 | 9–20 |
| *Challenge* | | | | | | | |
| Exposure route | ID | Open | Patch | ID; patch | ID | Patch | Patch |
| Number of exposures | 1 | 2 | 1 | 2 | 2 | 1 | 1 |
| Duration of exposure | — | — | 24 hr | — | — | 24 hr | 24 hr |
| Exposure day(s) | 35 | 21 and 35 | 28 | 22; 35 | 14–28 | 22 | 21 |

*Note*: FCA, Freund's complete adjuvant; ID, indeterminable; SLS, sodium lauryl sulfate; TS, test substance; V, vehicle.

## Buehler Test

The Buehler test [52–54,121,188] also employs topical application of the test material. An absorbent patch —20 × 20 mm Webril®, backed by Blenderm™ tape and saturated with 0.4 mL of the test material—is placed on the shaved flanks of 10 to 20 guinea pigs. The test concentration varies from undiluted to usage levels. An optimum concentration that produces slight erythema is selected based on an irritancy screen conducted in other animals. The patch is held in place by wrapping the animal with an occlusive wrapping. The animal is then placed in a special restrainer fitted with a rubber dam to maintain even pressure over the patch for a 6-hour exposure period. This procedure is repeated 7 and 14 days after the initial exposure. A control group of 10 to 20 animals is patched with vehicle only. Two weeks after the last induction patch, animals are challenged with patches saturated with a nonirritating concentration of test material applied to both flanks and with the vehicle (if other than water or acetone). Wrapping and restraint are as during induction. After 6 hours the, the patch is removed and the area is depilated. Test sites are visually evaluated 24 and 48 hours after removal. Animals developing erythematous responses are considered sensitized (if irritant control animals do not respond). The incidence of positive reactions and the average intensity of the response are calculated.

## Freund's Complete Adjuvant Test

Freund's complete adjuvant test (FCAT) is an intradermal technique incorporating test materials in a 50/50 mixture of FCA and distilled water. The test has been significantly modified since originally described [188]. The latest published description [189] is summarized here. A 6 × 2-cm area across the shoulders of two groups of 10 to 20 guinea pigs is shaved and used as the injection site. Animals of one group are injected with a 5% solution of the test material in FCA/water; injection volume is 0.1 mL. Control animals are injected with FCA/water. Injections are repeated every 4 days until three injections are given. The minimal irritating and maximum nonirritating concentrations following topical application of 0.025-mL solutions to a 2-cm$^2$ area of skin are determined in a minimum of four naïve guinea pigs (see OET procedure). Twenty-one days after the first induction injection, 0.025 mL of the minimal irritant concentration, the maximum nonirritant concentration, and two lower concentrations is applied to 2-cm$^2$ areas of the shaved flank. Test sites are not covered and are evaluated for the presence of erythema at 24, 46, and 72 hours after application. The minimum nonirritating concentration in FCA/water-treated controls is determined. Animals injected with the test material during induction that respond to lower doses are considered sensitized. The incidence of sensitization and the threshold of concentration for elicitation of the response in these animals are calculated.

## Optimization Test

The optimization test resembles the DT but incorporates the use of adjuvant for some induction injections and both intradermal and topical challenges [188,189,244]. Injections during induction are 0.1 mL of 0.1% concentration of test material in 0.9% saline or in 50/50 FCA/saline. In total, ten injections are given. On day 1 of the first week, one injection into the shaved flank and one into a shaved area of dorsal skin are given. Two and 4 days later, one injection into a new dorsal site is given. The test material is administered in saline during the first week. During the second and third weeks, test material is administered in FCA/saline every other day to a shaved area over the shoulders. Twenty test animals are treated; 20 controls are injected with saline during week 1 and FCA/saline during weeks 2 and 3. The intensity of the 24-hour responses during week 1 is calculated as reaction volume. Thickness of a skin fold over the injection site is measured with a caliper (mm), and the two largest diameters of the erythematous reaction are recorded (mm). The reaction volume is calculated by multiplying fold thickness times both diameters and is expressed as microliters. The mean reaction volume of each animal to the intradermal injections using saline as a vehicle (week 1) is calculated. Animals are challenged with 0.1 mL of 0.1% test material in saline 35 days after the first injection. The challenge reaction volume for each animal is calculated and compared to the mean reaction volume for that animal. Any animal developing a reaction volume at challenge greater than the mean plus one standard deviation during induction is considered sensitized. Vehicle control animals are injected with saline at challenge. A second challenge is conducted 45 days after the first injection. A nonirritating concentration of the test material in a suitable vehicle is applied to the flank skin, away from injection sites; 0.05 mL is applied to an area of approximately 1 cm$^2$.

The area is covered with a 2 × 2-cm piece of filter paper backed by an occlusive dressing, which remains in place for 24 hours. Reactions are visually evaluated using the four-point erythema scale of the Draize primary irritancy scale (see Table 27.9). The control animals are patched with vehicle alone. The number of positive animals in the test group is statistically compared with the number of pseudo-positive animals in the control group using the exact Fisher test. Separate comparisons of intradermal and epicutaneous challenges are made. A *p*-value of ≤0.01 is considered significant. To classify materials as strong, moderate, weak, or nonsensitizers, a classification scheme has been devised using results of the exact Fisher test and number of positives detected (Table 27.5).

**TABLE 27.5**
**Classification Scheme for the Optimization Test**

| Interdermal Positive Animals (%) | Epidermal Positive Animals (%) | Classification |
|---|---|---|
| NS, 0–30 | NS, 0 | Nonsensitizer |
| S, 30–50 | NS, 0–30 | Weak sensitizer |
| S, 50–75 | S, 30–50 | Moderate sensitizer |
| S, >50 | S, >75 | Strong sensitizer |

*Note:* S, statistically significant; NS, not statistically significant by exact Fisher test.

## The Split Adjuvant Test

The split adjuvant test [188,226,227] uses skin damage and FCA as adjuvants; application of the test material is topical. An area of back skin just behind the scapulas of 10 to 20 guinea pigs is clipped, shaved to glistening, then treated with dry ice for 5 to 10 seconds. A dressing of a layer of loose mesh gauze and stretch adhesive with a 2 × 2-cm opening over the shaved area is placed around the animal and secured by adhesive tape. This dressing remains in place throughout induction. Approximately 0.2 mL of creams or solid test material or 0.1 mL or liquid is spread over the test site and covered with two layers of #2 filter paper backed by occlusive tape and attached to the dressing by adhesive tape. The concentration tested varies by irritancy potential, use conditions, etc. Two days later, the filter paper is lifted from the test site, the test material is reapplied, and the filter paper covering is replaced. On day 4, the filter paper cover is removed; two injections of 0.075 cc FCA are given into the edges of the test site, the test material is reapplied, and the site is resealed. On day 7, the test material is reapplied, and on day 9 the dressing is removed. Twenty-two days after the initial treatment animals are challenged by topical application of 0.5 mL of test material to a 2 × 2-cm area of the shaved mid-back. The test site is covered by filter paper backed with adhesive tape, held in place by wrapping the animal with an elastic adhesive bandage secured with adhesive tape. A group of naïve controls (10 to 20 animals) is treated by the same procedure at challenge. The dressing is removed 24 hours after application, and the test site is evaluated visually at 24, 48, and 72 hours, using a seven-point descriptive visual scale. Sensitization of individual animals is indicated by significantly stronger reactions than reactions of controls.

## Guinea Pig Maximization Test

The guinea pig maximization test (GPMT) [188,224, 225,364] combines FCA, irritancy, intradermal injection, and occlusive topical application during the induction period. The shoulder regions of two groups of 20 to 25 guinea pigs are shaved. Two identical sets of intradermal injections of 0.1 mL 50/50 FCA/water; test material in water, paraffin oil, or propylene glycol; and the same dose of test material in FCA/vehicle are placed in a 2 × 4-cm area. Seven days later, the test article is placed on filter paper over the injection site. The filter paper is covered with approximately 4 × 8 cm occlusive surgical tape and secured in place with an elastic bandage wrapped around the animal. If the test material is nonirritating, the test site is pretreated with 10% sodium lauryl sulfate in petrolatum on day 6 to provoke an irritant reaction. If a vehicle other than petrolatum is used for topical application of the test material, the filter is saturated with the solution. Control animals are patched with the vehicle alone. The dressing is removed from the animals 48 hours after application. Test and control animals are challenged on the shaved flank with the highest nonirritating concentration, with approximately one half of the highest nonirritating concentration, and with the vehicle. Solutions are applied to 1 × 1-cm pieces of filter paper secured in place as during induction; patches are removed 24 hours later. The challenge area is shaved, if needed, 21 hours after patch removal. Reactions are evaluated visually 24 and 48 hours after patch removal. The intensity of responses to test material and vehicle in the test group is compared to the responses in controls. Reactions are considered positive when they are more intense than the response to vehicle and the responses to the test material in controls. Based on the incidence of positives in the test group, test materials are rated as weak to extreme sensitizers (Table 27.6).

**TABLE 27.6**
**Classification of Materials by Maximum Test**

| Sensitization Rate (% Responding at Challenge) | Grade | Classification |
|---|---|---|
| 0–8 | I | Weak |
| 9–28 | II | Mild |
| 29–64 | III | Moderate |
| 65–80 | IV | Strong |
| 81–100 | V | Extreme |

## Sensitization Tests in Mice

Although guinea pigs have been the animal of choice for predictive sensitization assays for 40 years, interest and activity in developing and validating standardized predictive assays in mice have been intensive during the last 10 years. Classical guinea pig sensitization assays are relatively costly and time consuming. All use subjective endpoints, and data interpretation is prone to difficulties. Manipulations of the animals are sufficiently stressful in some assays to alter normal physiological parameters. With the development of new techniques for studying

DTH [41] and evaluating cellular response [130], it has become possible to study the response in other laboratory animals [6,74,75] in shorter time frames. Numerous less subjective techniques for evaluating the allergic response have been proposed, including changes in the water content of challenged ears [41], measurement of ear thickness with an engineer's micrometer [8,354], and responses of lymphocytes [8,251,252,288]. Although numerous approaches to developing a predicative sensitization assay in mice have been proposed, only two methods have been sufficiently developed to warrant consideration as standardized assays. The test site for each is the mouse ear, yet these methods employ distinctively different approaches. The mouse ear swelling test (MEST) uses both topical and injection exposures for induction and a topical challenge of the pinnae in which visual evaluation is replaced by measuring ear thickness with an engineer's micrometer. The local lymph node assay (LLNA) consists of a topical induction followed by measurement of the mitotic activity of the draining lymph node. LLNA is unique out of all predicative assays in evaluating the response of the efferent phase of the response. A third mouse sensitization assay, the vitamin A enhancement test (VAET), has been used for the evaluation of ingredients of consumer products but has not been used by a sufficient number of laboratories to be considered standard. In the VAET, the reactivity of the immune system is heightened by maintaining animals on a diet with high doses of vitamin A for a preconditioning period and throughout induction and challenge. Challenge is topical and is assessed by measuring ear thickness. These assays are contrasted in Table 27.7.

### Local Lymph Node Assay

Kimber and colleagues [182,185,186,269] investigated measuring lymphocyte proliferation as an alternative approach to visual evaluations or measurement of edema of the mouse ear. They found that exposure of the dorsum of the ear pinnae to sensitizers produced hyperplasia of T cells in auricular lymph nodes. A dose divided into three exposures over three consecutive days produced a more intense response than the same dose delivered in a single exposure [187]. In a limited trial, Balb/c mice were shown to be more sensitive than other strains [186]. Initial studies used 24-hour cultures of excised lymph node cells labeled with $^{3}$H-thymidine, with or without exogenous IL-2 [185,186]; however; to simplify the assay, the method was modified to expose proliferating cells to $^{3}$H-thymidine *in situ* via intravenous injection. The method described is the final method used in intralaboratory validation studies that have been reported [15,314]. A complete list of references was made available in 1999 by an independent review evaluation of the method sponsored by the Interagency Coordinating Committee on the Validation of Alternative Methods (ICCVAM) [263]. Several new reviews have recently been published [395–397].

The local lymph node assay employs multiple topical *in vivo* doses of the material of interest to the mouse ear. This is followed by *in vitro* evaluation of mitotic activity of cells from draining lymph nodes. At least three dose levels are evaluated in separate groups of four CBA/ca mice. Experimental animals between 8 and 12 weeks of age are used. Either males or females may be used, but each assay should use only a single sex.

Vehicle selection and test concentration are based primarily on the solubility and viscosity of the solution or suspension. Investigators should be sure that the doses selected are nontoxic to the animals. Vehicles shown to be acceptable include 4:1 acetone/olive, methyl ethyl ketone, dimethylformamide, propylene glycol, dimethyl sulfoxide, and 2.5% hydroxypropyl cellulose in methanol. Investigators have proposed testing three consecutive concentrations from the series 100%, 50%, 25%, 10%, 5%, 2.5%, 1.0%, 0.5%, 0.25%, 1.0%, 0.05%, and 0.001%.

For three consecutive days, 25 μL of the appropriate test solution or the vehicle alone is applied to the dorsal surface of the pinnae each day. Five days after the first exposure, 250 mL phosphate-buffered saline (PBS) containing 20 μCi methyl thymidine is injected via the tail vein of each animal. Five hours after injection, animals are euthanized by carbon dioxide asphyxiation. Draining auricular lymph nodes are excised and pooled with nodes from other animals in the same group. A single-cell suspension is prepared by gently passing the nodes through stainless steel, 200-mesh gauze with the plunger of a syringe. The cell suspension is centrifuged at 190 × g for 10 minutes, and the pellet is then washed twice with 10 mL PBS. Cells are resuspended in 3 mL of 5% trichloroacetic acid (TCA) and incubated overnight at 4°C. The precipitated macromolecules are recovered by the centrifugation, the supernate is removed, and the precipitate is resuspended in 1 mL 5% TCA. The suspension is transferred to 10 mL scintillation fluid, and disintegrations/min are counted using a β scintillation counter. Disintegrations per lymph node are calculated for each experimental group and the control group. The ratio of $^{3}$H-thymidine incorporation into the test group and the control is calculated for each dose. Some investigators prefer to pool the lymph nodes from all animals in the dosage group. If the ratio is 2 to 3 for any dose, the material is considered a sensitizer.

Several groups of workers have reported comparisons of the LLNA with various guinea pig assays and have suggested variations of the method [13,14,181,182 162,314]. It is clear that LLNA is not as stringent as some guinea pig assays; however, it is expected to retain utility as a rapid screening assay for materials with strong sensitization potential, as it offers advantages in the low number of animals used, lower cost, and less time required for conducting the assay. A validation meeting to review the

**TABLE 27.7**
**Principal Features of Human Sensitization Assays**

| Feature | Complete Schwartz–Peck | Shelanski–Shelanski | Repeat Insult Patch Tests Draize | Griffith–Voss–Stotts | Modified Draize | Human Maximization |
|---|---|---|---|---|---|---|
| Number of subjects | 200 | 200 | 200 | 200 | 100–200 | 25 |
| *Induction* | | | | | | |
| Exposure site | Upper arm | Upper arm, same site | Upper arm or back; naïve site for each exposure | Upper arm, same site | Upper arm or back, same site | Upper arm or lower back, same site |
| Number of exposure | 1 | 15 | 10 | 9 | 10 | 5 |
| Duration of exposures | 24–72 hr | 24 hr | 24 hr | 24 hr | 48–72 hr | 48 hr |
| Frequency of exposure | — | 3 per week | 3 per week | 3 per week | 3 per week | 24 hr between patches |
| Evaluation schedule | At removal, 24 hr, 48 hr | At removal | At removal | 48–72 hr | 30 minutes after removal | Before each application |
| Miscellaneous | 4-week usage period | Fatiguing index | Pilot group | Continuous exposure | SLS/irritation adjuvant | Rest period duration |
| *Challenge* | | | | | | |
| Exposure site | Upper arm | Upper arm | Upper arm or back | Upper arm | Upper arm or back | Lower back, upper arm |
| Duration of exposure | 24–72 hr | 48 hr | 48 hr | 24 hr | 72 hr | SLS 1 hr; 48 hr |
| Evaluation schedule | At removal, 24 hr, 48 hr | At removal | At removal | 48 hr, 96 hr | At removal, 24 hr | At removal, 24 hr, 48 hr |
| Miscellaneous | — | — | Naïve test site | Original and naïve test sites | Naïve test site; may use two 48-hr exposures | Sensitization index |

*Note:* SLS, sodium lauryl sulfate 5% for induction adjuvancy and 10% at challenge.

strengths and weaknesses of the assay has been reported [263]. Certain materials, such as metals, have not been reliably identified as allergens. The validation attempt matched LLNA results with guinea pig Buehler and maximization assays. Both have false positives and false negatives. No attempt was made to determine the clinical relevance of the LLNA data nor the benchmarks used. Key to the clinical relevance of this assay will be a more precise series of databases that permit clinical collaboration between the LLNA and human experience; that is, how does one relate the LLNA data with the frequency of allergic contact dermatitis in humans?

## Mouse Ear Swelling Test

Gad and coworkers [128,129] used ear thickness measured with a caliper-type engineer's micrometer to evaluate the response of mice after challenge with potential sensitizers. They optimized a protocol in which tape-stripped skin sites that had been injected with Freund's complete adjuvant were topically exposed to the test material each day for 4 days. Seven days after the last topical exposure, animals were challenged by topical application of the test material to one ear. Early work also showed that Balb/c, CF-1, or SW mice could be used in the assay. Females were selected because their less aggressive behavior allows group housing with minimally damaged ears. Responses of animals less than 5 weeks old or more than 13 weeks old were weaker than animals 6 to 10 weeks of age. Administration of induction doses to the stomach region yielded a higher rate of sensitization than application to the back of the animals. The efficacy of the induction method was increased somewhat by both tape stripping and preinjection of the test site with Freund's complete adjuvant. Preliminary work showed that exposure via occlusive patches during induction did not increase the sensitivity of the assay, and in some cases the response was diminished when the patching system was employed. Their final protocol incorporated these findings.

For the MEST, 6- to 8-week-old CF-1, Balb/c, or SW female mice are gang housed in direct bedding cages. Following a 5- to 7-day quarantine period, the fur of the abdomen is shaved by electric clippers from 10 to 15 test animals and 5 controls. Animals with damaged pinnae are excluded from testing at this time. After the area is shaved, it is tape stripped with Dermaclear®, a surgical adhesive tape, until the skin appears glossy. Then a divided dose of 0.05 mL Freund's complete adjuvant is injected intradermally with a tuberculin syringe fitted with a 30-gauge needle. The Freund's complete adjuvant is injected into two sites within the shaved area but along the borders. Following the injection of adjuvant, 100 μL of vehicle containing the test material or vehicle alone (controls) is applied to the center of the shaved area. The abdomen is allowed to dry, and the mouse is then returned to its cage. The process of tape stripping and application of appropriate material to the abdomen is repeated each day for the next 3 days.

Vehicle is chosen based on solubility and chemical compatibility with the test substance. Acetone, methyl ethyl ketone, or 70%, 80%, and 95% ethanol in water have been shown to be acceptable vehicles. Mixed ethanol/olive oil systems are not satisfactory. The dose selection is based on dermal irritation toxicity range finding studies prior to testing each compound. Four groups of two mice each are subjected to the induction procedure, including shaving, tape stripping, and application of the test material, and the ears are then exposed as during challenge. At least four concentrations are evaluated in the range study. Minimally irritating or nonirritating concentrations are selected for the induction application. The highest nonirritating concentration is used at challenge.

Challenge is performed 7 days after the last topical induction application by applying 20 μL of the test material in vehicle to one ear of each animal (test and control) and 10 μL of the vehicle to the opposite ear. Ear thickness is measured before application of the challenge dose and 24 and 48 hours after the challenge application. Animals are lightly anesthetized with ether, and the thickness of both pinnae is measured with an engineer's micrometer. Positive respondents are defined as animals in which the ear dosed with the test material shows at least a two- to threefold greater increase in thickness than the vehicle-treated control ear. The control group should not show greater than 10% increase in ear thickness for the test to be considered valid. If the control groups show more than a 10% increase, the study should be repeated using lower doses. The percentage of respondents is calculated. In addition, the degree of ear swelling is calculated by dividing the thickness of the ear to which the test material was supplied by the thickness of the vehicle-treated control ear. Measurements from all animals in the test group are included. No additional explanations or examples of the use of the degree of ear swelling in interpreting results are available. A later paper proposed that the data generated by MEST (and classical guinea pig assays) could be used to calculate a potency index of sensitization [127].

The original paper of Gad et al. [128] included validation studies of 72 compounds. They reported a false-negative rate of only 2% and no false positives when MSET was compared to GPMT data on the same materials. The incidence of sensitization in MSET was consistently lower than that produced by GPMT. Similar findings were reported for comparisons between the Buehler assay and MEST in the same paper; however, published guinea pig data used for comparisons included data from other topical guinea pig techniques that did not employ the restraint procedure specified by Buehler. Intralaboratory validation of the method includes a comparison of test results for eight materials tested in five laboratories [129]

That report indicates that MSET did not identify weak sensitizers in two of three laboratories. Other studies have confirmed that the incidence of positive response in MEST is consistently lower than in GMPT, and weak and moderate sensitizers are not identified correctly [73,91]. The MEST has been accepted by the Environmental Protection Agency (EPA) for registration of chemicals under the Toxic Substances Control Act (TSCA) [68].

Although not reported in papers describing this method, the type of micrometer used may affect interpretation of the test results. Van Loveren [354] compared the use of spring-loaded, caliper-type instruments with screw and friction thimble micrometers. Spring-loaded instruments were best. Electronic instruments have been used in other types of immunological investigations. In our experience, the use of anesthetic can be eliminated by operator training prior to handling animals.

### Vitamin A Enhancement Test

Miller and colleagues [233–236,250] decreased the dose of strong sensitizers required to induce sensitization by maintaining the animals on diet supplemented with high levels of vitamin A acetate. The mechanism of the increase in response was studied using radioisotopes, but, because of ease in performing the assay, ear thickness measurement was selected for use in the predicative assay. Principal features of the test included a preconditioning period for the diet of 28 days, six exposures to the shaved abdomen and thorax during the 12-day induction period, and challenge 4 days later. Results from the test group were compared statistically to those of the controls, and a 50% increase over the response to controls was observed that indicated sensitization. The minimally irritating concentration was used for induction, and the highest nonirritating dose was used at challenge (determined by dose response in separate groups of mice). The vehicle was selected based on nonirritancy and solubility of the test substance. An obvious difficulty with the method was the long conditioning period required prior to the study. General comments concerning choice of micrometer for MEST also apply to VAET. The test was never widely adopted or submitted to formal validation procedures.

## Human Sensitization Assays

Chemicals can be tested for their ability to induce contact hypersensitivity in panels of human volunteers from whom informed consent is obtained. Human studies should be undertaken only after the results of predictive tests in animals are available if the test material is a new compound or if it contains significantly increased levels of common ingredients. Testing higher doses in animals provides some margin of safety for potential human subjects. Generally, materials identified as sensitizers in animals are not tested on humans; however, if the potential benefit of the material warrants, a small group of human subject may be tested with materials inducing sensitization in animals. Such situations should be reviewed by an instructional review board. Test subjects should be informed of the risks, and the number of subjects should be limited (additional subjects can be exposed if members of a small group do not respond).

Subjects should be randomly selected; however, some precautions are indicated. Some investigators believe intact draining lymph nodes adjacent to the application site are necessary to induce sensitization, and patches should not be placed on areas adjacent to mastectomies. Persons with unilateral mastectomy who wish to participate may be tested by applying patches to the opposite sides of the body. Tests should not be performed on scar tissue. Recurrence of skin conditions in remission (e.g., psoriasis, eczema) has been associated with patch testing and other minor physical traumas. Subjects at risk should be informed of this possibility and encouraged to consult their dermatologist prior to testing. Subjects should not be tested with materials to which they are known to be allergic (demonstrated by diagnostic patch test or in previous predictive assays). It is prudent to routinely question potential subjects concerning their history of dermatologic disease and allergies. Allergic contact dermatitis to materials already in commerce is sometimes detected by early induction patches. This indicates that, under patch conditions, the material may elicit a response in presensitized individuals. Although the incidence of preexisting sensitization to the material should be considered in risk assessment decisions regarding marketing, detection of preexisting sensitization is not helpful in evaluating the ability of a material to induce sensitization. Records of previous responses of individuals participating in multiple predicative assays should be reviewed prior to testing to eliminate subjects presensitized to components of the product.

Although numerous variations have been reported, four basic predicative human sensitization tests are in current use: (1) single induction/single challenge patch test; (2) repeated insult patch test (RIPT); (3) RIPT with continuous exposure (modified Draize); and (4) maximization test. The principal features of human sensitization assays are summarized in Table 27.7. As originally described, all methods used customized patches, and patch selection was governed by available adhesive systems. Description of customized patches would be of historical interest only; as currently conducted, all human assays use similar patches. Occlusive patches, consisting of a nonwoven pad (usually Webril®) of four-ply gauze sponges backed by a good occlusive surgical tape, such as Blenderm™, are available commercially or may be custom made in strips of four or five pads. Acceptable alternatives include the Hill Top Chamber® [176], which contains a Webril® pad inside an occlusive plastic disc backed by porous tape; the Duhring chamber

[124], a stainless steel disc that contains a Webril® pad; and the large Finn Chamber®. Duhring and Finn chambers usually are secured in place by porous surgical tape. Occasionally, semiocclusive patches made of Webril® backed by porous tape may be used. Semiocclusive patches are decidedly inferior to occlusive patches in induction of sensitization.

For assays other than maximization, usually 150 to 200 subjects are tested. Henderson and Riley [153] showed statistically that, if no positive reactions are observed in 200 randomly selected subjects, as many as 15/1000 of the general population may react (95% confidence). As the sample size is reduced, the likelihood that the test will not correctly predict adverse reactions in the general population increases.

Results of the RIPT, modified Draize test, and human maximization tests have been accepted as valid by regulatory agencies; however, some sponsors routinely use one of the methods described and defines its use as the "standard of the industry." This is a simplistic view of the methods and their strengths and weaknesses. As for all toxicity endpoint, the method used should provide a reasonable exaggerated exposure over the anticipated exposure from use of the product. The device group of the FDA held several meetings to review details of sensitization procedures. These deliberations led to a guidance document [63] for evaluating skin sensitization to chemicals in natural rubber products. The modified Draize procedure is recommended. Tests for the transdermal products are currently being evaluated by the FDA.

## Schwartz–Peck Test (and Modifications)

A single application induction patch followed by a single application patch test was described by Schwartz [316,317] and Schwartz and Peck [318] with a usage test of 1 month after challenge to verify patch results. The test has been modified by some to eliminate the usage test [50], to eliminate patching altogether, and to place a usage period between induction and challenge patches [345]. The term *complete Schwartz–Peck method* refers to a single induction patch, usage period, and single challenge patch test. This may also be referred to as the *Traub–Tusing–Spoon method*. Incomplete Schwartz–Peck tests do not incorporate a usage period. A patch saturated with the test material, diluted if necessary, is applied to the outer upper arm of 200 test subjects and remains in place for 24 to 72 hours. The dose tested and duration of patch contact vary with intended use. Cosmetics may be tested without a covering (open application) or with semiocclusive patches. The test site is visually evaluated at patch removal and at 24 and 48 hours after removal for erythema and edema. A 4-week normal usage period follows the induction patch in the complete Schwartz–Peck test, with a challenge patch applied to the same site on the upper arm at the conclusion of the usage period. For the incomplete Schwartz–Peck test, a second patch procedure is performed 10 to 14 days after the induction patch. Duration of contact and evaluation of the site are similar to those during induction. The development of dermatitis at challenge that is much stronger than during induction signifies sensitization.

Schwartz originally described the incomplete Schwartz–Peck test. A usage test was to be conducted after the challenge patch using 1000 different subjects. Although Schwartz and Peck referred to their assay as a "prophetic patch test," experience has shown that only potent haptens will induce sensitization in this assay [191]. In fairness, it should be noted that the test was originally designed to evaluate the effect of nylon garments on the skin. It was intended to detect adverse effects, irritation, and secondary irritation (sensitization). The mechanism of skin allergy was not understood when the test was designed. Although the test was useful for its original purpose, unfortunately its use was expanded without considering new information generated by immunologists. Clearly, the assay is inferior to all other predictive human sensitization assays; however, a few groups continue to use the method.

## Repeat Insult Patch Test

Three major variations on the RIPT are in common use: (1) the Draize human sensitization test [86,87]; (2) the Shelanski–Shelanski test [320–322]; and (3) the Voss–Griffith test [52,138,139,362]. Although the Shelanskis first published a description of a RIPT, they based its development on a verbal description of a method Draize was revising [322]. Voss modified the Shelanski–Shelanski test [362], and his assay was later modified by Griffith [138]. As one would expect, the three assays have much in common. There are some significant differences in the assay as originally described, however. Several groups at the FDA evaluated various RIPT protocols, leading to guidances that are published on their websites (see www.FDA.gov, medical devices section)

In the Draize human sensitization test, an occlusive patch containing the test material is applied to the upper arm or upper back of 200 volunteers. The patch remains in place for 24 hours and is then removed. The test site is evaluated at patch removal for erythema and edema. A second patch test is applied to a new site 24 hours after the first patch is removed. This process is repeated until ten patches are applied. For convenience, the test may be run on a Monday-to-Friday schedule, with subjects removing their own patches on Saturday (72 hours between Friday and Monday applications). Ten to 14 days after application of the last induction patches, subjects are challenged via a patch applied to new site. Duration of contact is 24 hours, and sites are visually evaluated at removal or the patch. The response at challenge is compared to the responses to patches applied early in induction, and the incidence of sensitization is reported.

Like the Draize RIPT, the Shelanski–Shelanski test employs occlusive patches that remain in contact with the skin of the upper arm for 24 hours. The patching cycle is the same; however, patches are placed on the same test site each time and a total of 15 sets of patches is applied during induction. The test site is evaluated before application of a new patch to the site. If inflammation has developed, the patch is placed on an adjacent uninflamed site. Two to 3 weeks after the induction period, subjects are challenged by application of a patch that remains in place for 48 hours. Test sites are evaluated at patch removal for erythema and edema, and the incidence of positive response is reported. Patch responses during induction were considered by Shelanski and Shelanski to be evidence of "skin fatigue" (cumulative irritation); the time of development (i.e., number of patches) was reported as a fatigue index. Voss [362] reduced the number of 24-hour patch exposures to 9 over a 3-week period. At challenge 2 weeks after the last induction patch, duplicate patches applied to the original test site are worn for 24 hours. Patch sites are evaluated 48 and 96 hours after patch application. A pilot group of 10 to 12 subjects was tested prior to exposing the full panel of 60 to 70 subjects.

Griffith later published more details of the method [138,139]. A maximum of four dissimilar materials was tested simultaneously, and duplicate challenges were applied to the sites of induction and to the opposite arm, thus testing areas drained by different regional lymphatics. The concept of a rechallenge of subjects with reactions difficult to interpret was also introduced. The number of subjects was increased to 200 by conducting tests on multiple panels. Stotts [335] presented detailed examples of proper interpretation of human repeat insult patch tests. Sensitization is characterized by challenge reactions stronger than reactions early in the induction phase, by persistence of responses through delayed readings, by delayed appearance of response, or by weak responses in a few subjects when the material has not produced irritation in the panel. Examples of patterns of responses indicating presensitization and weak responses that warrant rechallenge are also presented in the paper.

As currently conducted, the differences in the Draize and Voss–Griffith RIPTs are minimal. Many investigators apply patches to the same site during induction and refer to the procedure as a Draize RIPT. The value of multiple grades at challenge is widely recognized. Multiple test materials are tested simultaneously in all RIPTs for reasons of efficiency and economy. Although the distinctions between the Draize and Voss–Griffith procedures have blurred with common usage, the Shelanski–Shelanski test, with five to six more induction applications, remains distinct.

## Human Maximization Test

Kligman [191] reviewed the common human predicative sensitization test methods in use in 1966 and found them to be unsatisfactory in inducing sensitization in to nine clinical allergens. In panels of 200 subjects, the Shelanski–Shelanski method induced sensitization to four materials, the original Draize test and the complete Schwartz–Peck test induced sensitization to two allergens each, and the incomplete Schwartz–Peck test failed to induce sensitization to any allergen. Kligman concluded [193]:

> Emphasis must shift from prophecy to the more practical objective of identifying potential allergens. Once the allergenic potential is known with reasonable certainty, a judgment of risk might be ventured after examining all the pertinent variables.

This represented a profound change in the intent of predicative sensitization assays. Based on his studies of factors affecting rates of sensitization in predicative assays [190,197], Kligman designed the human maximization test [197]. He later modified the procedures somewhat to reduce the difficulties in performing the test and in interpreting the test [197]. The maximization test uses irritancy as an adjuvant. During induction, irritating compounds are tested at a concentration that produces moderate erythema within 48 hours. If materials are nonirritating, the test site is pretested with a 24-hour patch of 5% sodium lauryl sulfate (SLS). Additional pretreatment SLS patches may be applied before each patch application until a brisk erythema is produced. Induction concentrations are at least five times higher than use levels; petrolatum is the preferred vehicle. Often, custom-made Webril®/Blenderm™ patches or Duhring chambers are used. Patches are applied to the outer aspect of the arm or lower back, and up to four dissimilar materials may be tested at one time. Wrapping with extra tape is often necessary to ensure occlusion. Bandage sprays may be used to ensure sealing of the test site. Five sets of patches are worn on the same site for 48 hours each, with a 24-hour rest period between removal and reapplication.

Following a 2-week rest period, an SLS provocative patch is applied to prepare the skin for challenge. A patch saturated with a 2.5 to 5.0% solution of SLS is applied to previously untreated sites on the lower back. SLS concentration is based on the season and on individual subject response. The SLS patch is removed after 1 hour, and a patch containing the test material is applied. A control site is patched with SLS (1 hour) and petrolatum (48 hours) to aid in interpretation of the results. The patch is removed 48 hours after application, and the test sites are evaluated. Test sites are reexamined 24 and 48 hours after patch removal. The number of subjects developing a positive response is reported, and a sensitization index based on percentages of subjects responding is assigned to the test material. In common practice, the human maximization procedure is performed on either the outer upper arm or the back. Although it is clear that the maximization test is a sensitive tool for detection of allergenicity, the skin damage produced is dramatic and unacceptable to many subjects.

### Modified Draize Human Sensitization Test

The RIPT procedure was modified to provide for continuous patch exposure. Materials are applied to the outer upper arm each Monday, Wednesday, and Friday until ten patches have been applied during a 3-week period [240,241]. Patches remain in place until approximately 30 minutes before application of a fresh patch. This brief rest period allows some clearing of responses to tape and facilitates grading. Fresh patches are applied to the same site unless moderate inflammation has developed; the patches are placed on adjacent, noninflamed skin, should inflammation become pronounced. This produces a continuous exposure of 504 to 552 hours (some investigators apply only nine patches) compared to a total exposure period of 216 to 240 hr for RIPT of comparable induction applications. In addition, induction concentration was increased to levels above usage exposure. Two weeks after induction, subjects are challenged by exposure of a new site to a patch of 72 hours' duration at a nonirritating concentration. Test sites are evaluated at patch removal and 24 hours after removal. Jordan and King [173] proposed modifying the challenge procedure to two consecutive 48-hour patch periods. The modified Draize test has been selected as the test of choice for chemicals in natural rubber products [63].

## *In Vitro* Assays for Allergic Contact Dermatitis

As our understanding of cell biology of delayed contact hypersensitivity has increased, *in vitro* assays to replace diagnostic patch testing and as early-screening predicative assays have been proposed. Much of the work toward diagnostic procedures has centered on measuring the effect of cytokines released by sensitized lymphocytes on target cells as a marker for allergy. Proposals for *in vitro* predicative assays have focused on the afferent phase of antigen binding and stimulation of target cells. To date, neither approach has proved satisfactory, but some limited experimental successes have been reported.

Migration inhibition of peritoneal exudate cells from capillary tubes has been reported to be inhibited by the antigen to which donor guinea pigs were sensitized [131]. Inhibition of macrophage migration is medicated by soluble factors produced by sensitized lymphocytes in the presence of sensitizing antigen [88]. This factor, identified as migration inhibition factor (MIF), has been shown to aggregate macrophages, increase macrophage adherence to glass, and decrease macrophage mobility [81]. Rocklin et al. [303] reported that MIF was produced by highly purified populations of proliferating T cells. Moorehead et al. [258] later demonstrated that MIF production is dependent on Ia/T cell subsets. Mitogenic factors, such as phytohemaglutinin, initiate mitotic activity and transform lymphocytes in their blast forms. Pearmain et al. [289] demonstrated that tuberculin produced the same effect in peripheral blood leukocyte cultures from sensitive patients but not from unsensitized patients. Similar work experiments were conducted in nickel-sensitive subjects [7]. Lymphocytes from unsensitized controls also underwent blast transformation when exposed to nickel. The nonspecificity of nickel and mercury transformation was confirmed by numerous other investigators [281,288,294]. Using $^{14}$C-thymidine uptake to measure blast transformation, MacLeod et al. [218] demonstrated transformation in over half of the nickel-sensitive subjects and no unsensitized controls.

Experimental studies of blast transformation in which animal or human subjects were intentionally sensitized have been somewhat more successful. At least two investigators [130,251] have demonstrated that lymphocytes from guinea pigs sensitized with dinitrochlorobenzene (DNCB) and incubated with dinitroflurobenzene (DNFB) were transformed to a greater degree than cells from unsensitized controls (note that a dinitrophenyl group would attach to protein using either material). Tritiated thymidine was used to measure blast formation when exposed to DNFB conjugated to epidermal protein. Similar responses were demonstrated using human volunteers and epidermal extracts for conjugation [252]. Miller and Levis [249] produced the same effect using leukocyte and erythrocyte membranes for conjugation.

Cytotoxicity consistent with that produced by lymphotoxin has been demonstrated in experimentally induced allergic contact dermatitis [84]. Peripheral lymphocytes from DNFB-sensitized guinea pigs were incubated with DNFB-coated radiolabeled chick erythrocytes. Increased radioscope leakage was produced by lymphocytes from sensitized animals than produced by controls. Similar effects were demonstrated using epidermal cells [341].

The use of exclusively *in vitro* assays for predicting sensitization potential of schemes has been proposed. In on hypothetical system [41], binding to Langerhans cells would be measured. If no binding occurred, the activation state of the Langerhans cell would be evaluated. If no activation was detected the material would not be considered to induce sensitization. If Langerhans cells were activated an autologous lymphocyte blastogenesis assay would be preformed. Although this approach has not been supported experimentally, current commercial activity in producing skin recombinants may make this type of approach feasible Blomberg et al. [36] reviewed the use of *in vitro* assays to study mechanisms of contact dermatitis. There is grea enthusiasm for developing such assays to serve as substitutes for testing on animals and humans [398,399].

## SKIN IRRITATION AND CORROSION

Historically, skin irritation has been described by exclusion as localized inflammation not medicated by sensitized lymphocytes or antibodies (e.g., that which develops by a process not involving the immune system). The application of some chemicals destroys tissues directly, producing skin damage (including necrosis) at the site of application. Chemicals producing necrosis that result in the formation of scar tissue are described as *corrosives*. Chemicals producing inflammation after a single exposure are termed *active irritants*. Some chemicals do not produce acute irritation from a single exposure but may produce inflammation following repeated application to the same area of skin. The cumulative irritation from repeated exposures was originally called *skin fatigue* [320]. Because of the possibility of skin contact during transport and the wide use of many chemicals, regulatory agencies have mandated that chemicals be screened for their ability to produce skin corrosion and acute irritation. These studies have been conducted in animals using standardized protocols; however, recent efforts to replace animal studies with *in vitro* or human assays have had some success. It is not appropriate to conduct screening studies for corrosion in humans. Acute irritation can be evaluated in humans after animal studies have shown that the risk of systemic toxicity is low and if the material is known to be noncorrosive. Tests for cumulative irritation in both animals and humans have been reported. In general, cumulative irritation is evaluated in humans unless the toxicity of the material necessitates testing in animals (pesticides, industrial chemicals).

The processes that result in any form of skin irritation are not well understood. In addition to destroying tissue directly, chemicals can disrupt cell functions or trigger the release, formation, or activation of autocoids. Autocoids that are generated following exposure of tissue to some chemicals produce low increases in blood flow, increase vascular permeability, attract white blood cells in the area, and damage cells indirectly. The additive effects of the autocoid mediators would result in local skin inflammation. No agent has yet met all the criteria to establish it as a mediator of skin irritation [297]; however, histamine, 5-hydrotryptamine, prostaglandins, leukotrienes, kinins, complement, reactive oxygen species, and products of white blood cells [278] have been strongly implicated as mediators of some irritant reactions.

We have studied the process by which chemicals produce acute skin irritation following open topical applications to the ear of the mouse [258,283,285]. The time course of the inflammatory responses varied from compound to compound and was independent of vehicle and applied dose. Because the differences in time course could not be attributed to differing rates of penetration, this suggested that the materials tested triggered different inflammatory processes. Differences in the irritation process triggered by three chemicals were confirmed by histology, albumin leakage, and changes in rate of blood flow. Using a series of pharmacological antagonists of putative mediators of irritation, enzyme inhibitors to prevent formation of suspected mediators and agents that deplete the body of serum mediators, we confirmed that different pathways of mediator involvement existed for skin irritation. The implications for this finding are clear. A battery of *in vitro* assays would be required to screen materials for skin irritation; no single assay would be effective.

Many factors may modulate the development of skin irritation (Table 27.8). As with delayed-contact hypersensitivity (DCH), these factors have been classified as extrinsic or intrinsic [210,232,243,381]. Some extrinsic factors have been shown experimentally to be important considerations in designing predicative tests for skin irritation. A few investigators have also considered intrinsic variables when designing studies.

Like other toxic responses, skin irritation is related to dose. If the duration of contact and the dosing procedure are held constant, the intensity of the response increases as the concentration of the solution increases. Under patch test conditions, the rate of increase in intensity decreases as the concentration increases [383]. The type of appliance, chamber, and tape used to secure patches in place have been shown to influence the intensity of irritant responses [67,191,202]. More intense inflammatory responses are produced as the degree of occlusion is increased. The search for good techniques of occlusion ultimately led to the development of the Finn, Duhring, and Hill Top chambers now routinely used in patch testing.

Increases in occlusion usually are accompanied by local increases in surface temperature, and increased temperature is believed to predispose subjects to irritation. The temperature of solutions used in immersion assays is usually around 105°F [140,174,273]. Although systematic studies demonstrating that those temperatures were necessary were not presented, increased temperature has been shown to be necessary to reproduce irritant dermatitis in some instances [305].

The influence of vehicle in diagnostic patch testing for allergy is well recognized. Similar effects are seen when irritation is studied. These effects are demonstrated convincingly in open systems; for example, a dose of croton oil that produces no measurable edema in the mouse ear when applied in olive oil produces the maximum response when applied in acetone [285]. Patch occlusivity and interactions between vehicles and adhesives used in patch systems also influence the intensity of response and make it more difficult to demonstrate vehicle effects under patch conditions. In most predicative irritation tests, the choice of solvent is related to use conditions, and water is the solvent often used.

**TABLE 27.8**
**Factors Influencing Sensitivity of Skin to Development of Irritation**

| Variables | Refs. |
|---|---|
| *Extrinsic* | |
| Degree of occlusion | Magnusson and Hersle [221,222] |
| Choice of vehicle | Patrick et al. [287] |
| Frequency of dosing | Kligman and Wooding [199], Frosch and Kligman [124] |
| Duration of exposure | Wooding and Opdyke [383] |
| Dose (concentration) | Patrick et al. [287], Kligman and Wooding [199] |
| Temperature | Rothenberg et al. [306] |
| Environmental conditions | Hannuksela et al. [148], Carter and Griffith [62] |
| Altered barrier function (including abrasion) | Frosch and Kligman [121] |
| Chemical damage | Finkelstein et al. [111,112]; Patrick and Maibach [284] |
| Tape stripping | Kligman [194] |
| *Intrinsic* | |
| Anatomical site | Magnusson and Hersle [221,222] |
| Concomitant disease | Skog [325]; Bjornberg [30] |
| Species differences | Davies et al. [82] |
| Age | Rockl et al. [302] |
| Gender (effect disputed) | Wagner and Purschel [363], Bjornberg [31], Frosch and Kligman [122] |
| Race | Weigand et al. [371], Weigand and Gaylor [370] |

Dosing schedules have been developed that maximize the development of the responses of interest. In general, the longer the duration of contact to the same dose of a given chemical, the greater the intensity of the response. Multiple exposures at frequent intervals are the basis for most cumulative irritation assays, although there is some disagreement on the optimal time between exposures. In developing the soap chamber test (discussed later), Frosch and Kligman [125] varied both the frequency and duration of exposure in order to produce a more sensitive test.

The seasonal variability in human response to normal exposure to irritants is well documented [148]. Conducting usage studies in late fall and winter increases the discriminating ability of the test [62]. Some investigators have demonstrated similar effects using patch test procedures [190,383] and small-scale exaggerated exposure tests [174]. Although the basis for this variability is not well understood, it is believed to be due in part to changes in the barrier properties of the skin. Many investigators have experimentally altered the barrier properties of skin to develop assays that are more sensitive. Alterations vary from the abrasion of Draize-type tests [88] and the chamber scarification test [121] to tape stripping to remove the outer surface of the epidermis [194]. Pretreating the test sites with damaging agents has been shown to increase the reactivity of the skin to other chemicals [111,112,284]. Although these extrinsic factors modify the barrier function, intrinsic factors governing barrier properties are also important [238]. Barrier function would be expected to contribute to responses observed in screening tests used to identify sensitive subjects [122,126]. The demonstration that persons with some skin disease develop more intense response to irritants [30,325] and have diminished barrier function [356] was not unexpected. Susceptibility to the development of irritant responses is believed to be under genetic control. The prevalence of irritant responses in atopic individuals supports this theory. The response to identical patches applied to different sites is convincing evidence of regional variation in susceptibility to irritants. The reactivity of the various sites appears to correlate to the ability of chemicals to penetrate the skin in that area. Regional differences in skin response is not limited to humans [358].

Susceptibility to irritation is believed to vary by age, sex, and race. Children have been shown to develop inflammatory responses to lower levels of a variety of chemicals than adults [302]. The inflammatory response to some materials is decreased in the elderly [195,340]. Some investigators have suggested that the skin of women is more sensitive to irritants than men [362]; however, sex differences in reactivity were not confirmed by other investigators [31]. Investigators have reported that higher doses of irritants are required to produce inflammation of black skin [370,371]. This difference in reactivity disappeared when black skin was tape stripped, leading some investigators to hypothesize that black skin forms a more efficient barrier. Berardesca and Maibach [21] questioned whether these differences are more real than apparent.

**TABLE 27.9**
**Draize Scoring System in Albino Rabbits[a]**

| Description | Score Assigned[b] |
|---|---|
| Erythema and eschar formation | |
| No erythema | 0 |
| Very slight erythema (barely visible ) | 1 |
| Well-defined erythema | 2 |
| Moderate to severe erythema | 3 |
| Severe erythema (beet redness) to slight formation (injuries in depth) | 4 |
| Edema formation | |
| No edema | 0 |
| Very slight edema (barely perceptible) | 1 |
| Slight edema (edges of area well defined by definite raising) | 2 |
| Moderate edema (raised approximately 1 mm) | 3 |
| Severe edema (raised more than 1 mm and extending beyond the area of exposure) | 4 |

[a] The scale as defined by Draize and adopted by various regulatory agencies.
[b] The primary irritation index (PPI) is calculated by averaging the erythema values and averaging the edema values from all animals and then combining the averages (maximum PII = 8).

The principles of general toxicology should be remembered when designing and conducting any animal (and human) assay for skin irritation. One should consider dose–response relationships. Draize scores require careful interpretation; they are best interpreted by comparison to related compounds of formulations with a history of human exposure. Knowledge of intended human use (and foreseeable misuse) permits a more rational interpretation. With occlusive application techniques, one should remember that occlusion increases the permeability of some but not all moieties. Although there is a consistent, reasonably good correlation between responses in rabbits and humans, occasional inconsistencies have occurred [133]. Investigators [179,285] have found that different species exhibit widely varying reactivity under identical test conditions, especially in substances with only minor irritant potential; thus, the accuracy of the Draize test and other animal testing as it relates to humans has been called into question [152]. In addition, results from the animal methods currently in use differ due to the subjective visual test scoring. These differences occur most frequently in the assessment of the toxicity of mild irritants and colored material [272]. Wise investigators conduct carefully planned and executed tests in humans when rabbit tests indicate that materials are possible irritants. This is particularly true when one wishes to compare the irritancy potential of mild irritants. One should follow the guidelines of the National Academy of Science (NAS) committee [262] when this course of action is taken. The clinical and basic knowledge of what is now called the *irritant dermatitis syndrome* continues to advance, and interest in irritant dermatitis has led to an international symposium being held approximately every 3 years. Several textbooks summarize current advances [98,355,400].

## Irritation Tests in Animals

### Draize-Type Tests

Primary irritation and corrosion are most often evaluated by modifications described by Draize and colleagues in 1944 [88]. The Federal Hazardous Substances Act (FHSA) adopted one modification as a standard procedure [66]. The backs of six albino rabbits are clipped free of hair.

Each material is tested on two 1-inch-square sites on the same animal; one site is intact and one is abraded in such a way that the stratum corneum is opened but no bleeding produced. Abrasion can be performed using the tip of a hypodermic needle drawn across the skin repeatedly or commercial instruments such as the Berkeley Scarifier [147].

Materials are tested undiluted, and 0.5 mL liquid or 0.5 g solid or semisolid material is applied. In some cases, the skin may be moistened to help solids adhere to the site, or an equal volume of solvent may be used to moisten the material.

Each test site is covered with two layers of 1-inch-square surgical gauze secured in place with tape. The entire trunk of the animal is the wrapped with rubberized cloth or other occlusive impervious material to retard evaporation of the substances and hold the patches in one position. The wrappings are removed 24 hours after application, and the test sites are evaluated for erythema and edema using a prescribed scale (Table 27.9). Evaluations of abraded and intact sites are recorded separately. Test sites are evaluated again 48 hours later (72 hours after application) using the same scale.

**TABLE 27.10**
**Comparison of Skin Irritation Tests Based on the Draize Method**

| Feature | Draize | FHSA | DOT | FIFRA | OECD[a] |
|---|---|---|---|---|---|
| Number of animals | 3[b] | 6 | 6 | 6 | 3 |
| Abrasion | Abraded and intact | Abraded and intact | Intact | 2 abraded and 2 intact | Intact |
| Dose liquids | 0.5 mL undiluted | 0.5 mL undiluted | 0.5 mL | 0.5 mL undiluted | 0.5 mL |
| Dose solids | 0.5 g | 0.5 g in solvent | 0.5 g | 0.5 g moistened | 0.5 g moistened |
| Wrapping materials | Gauze and rubberized cloth | Impervious material | — | — | Semiocclusive |
| Length of exposure | 24 hr | 24 hr | 4 hr | 4 hr | 4 hr |
| Evaluated at[c] | 24 hr, 72 hr | 24 hr, 72 hr | 4 hr, 48 hr | 0.5 hr, 1 hr, 24 hr, 48 hr, 72 hr | 0.5 hr, 1 hr, 24 hr, 48 hr, 72 hr |
| Treatment at removal | Not specified | Not specified | Skin washed | Skin wiped, not washed | Skin washed |
| Excluded from testing | — | — | — | Materials with pH < 2 or > 11.5 | Materials with pH < 2 or > 11.5 |

[a] Although other species are acceptable, the albino rabbit is the preferred species.
[b] Draize tested four materials on six rabbits. Three abraded and three intact sites were tested with each material.
[c] Times listed are after removal for FIFRA and OECD. Times listed for Draize, FHSA, and DOT are after application of the test material.

*Note:* DOT, Department of Transportation; FHSA, Federal Hazardous Substances Act; FIFRA, Federal Insecticide, Fungicide, and Rodenticide Act; OECD, Organization for Economic Cooperation and Development.

The reproducibility of the FHSA procedure [358,372] and the relevance of test results to human experience have been questioned [82,158,243,296,331]. Numerous modifications to the Draize procedure have been proposed to improve its prediction of human experience. Modifications that have been proposed include changing the species tested [260], reduction of the exposure period, the use of fewer animals, and testing on intact skin only [93,142,256]. A few investigators have supplemented visual evaluation with other techniques [247,248], but these additions have not been considered for the standard method. Several governmental bodies have used their own modifications of the Drake procedure for regulatory decisions. The FHSA, DOT, Federal Insecticide, Fungicide, and Rodenticide Act (FIFRA), and OECD guidelines are contrasted to the original Draize methods in Table 27.10. Cruzan et al. [76] proposed a composite test that meets requirements of major agencies.

Summaries and evaluations of the scores vary somewhat. Draize reported values for individual animals at each time point, combined the erythema and edema values at each time point, and then averaged the 24- and 72-hour evaluations for intact and abraded sites separately. He also calculated a primary irritation index (PII) that was the average of the intact and abraded sites. Agents producing a PII of 2 were considered only mildly irritating, 2 to 5 moderately irritating, and more than 5 severely irritating. The primary irritation calculated for the FHSA is essentially the PII of Draize. A minimum PII of 5 defines an irritant by CPSC standards. The method of the National Institute of Occupational Safety and Health (NIOSH) does not combine responses of abraded sites and includes probable effects on normal and damaged skin in their evaluation.

Although vesiculation, ulceration, and severe eschar formations are not included in the Draize scoring scales, all Draize-type tests are used to evaluate corrosion as well as irritation. When severe reactions, which may not be reversible, are noted, test sites are observed for a longer period. Delayed evaluations usually are made on days 7 and 14; however, evaluations have been made as late as 35 days after application. The EPA bases interpretations on 7-day observations. The basic exposure procedures of the OECD guidelines for skin irritation or corrosion have been further modified to test for corrosion during shorter periods [275]. Under a directive of the European Economic Community, a shorter 3-minute exposure was added (with no wrapping procedure). The United Nations' recommendations for the transport of dangerous goods is based on exposure times of 4 hours, 1 hour, and 3 minutes with the recommendation that the 1-hour exposure be conducted first. Evaluations are made 1, 24, 48, and 72 hours and 7 days after dosing.

It should be noted that the Draize method has generally erred on the side of safety in that it overpredicts the severity of skin damage produced by chemicals, thus providing a safety factor for those exposed. One criticism that is often repeated is that the test is not sensitive enough to separate mild from moderate irritants. The purpose when designing the Draize test was to identify chemicals that posed a severe hazard to the public, not to compare products. Criticisms of the Draize test have been embraced by groups supporting the elimination of animal testing as they serve to demonstrate that use of the method is unwarranted. These criticisms overlook the tremendous value the test has provided in warning consumers, workers, and manufacturers of potential dangers associated with specific chemicals so appropriate precautions can be taken. Although Draize-type tests will be replaced by *in vitro* assays at some time in the future, we have no validated *in vitro* substitutes at present [297].

## Non-Draize Animal Studies

Animal assays to evaluate the ability of chemicals to produce cumulative irritation have been developed [291]; however, they are not required by any regulatory agency. The impetus for their usage is largely the development of products that are better tolerated by consumers and industrial workers. Although many such tests have been described, only a few are used extensively enough to summarize here. Even those used more often are not as well standardized as Draize-type tests, and many variables have been introduced by multiple investigators.

Repeat application patch tests using several species where diluted materials are applied to the same site each day for 15 to 21 days have been reported [291]; the guinea pig or rabbit is most commonly used. Patches used vary considerably, with gauze-type dressings and metal chambers being the extremes. Some authors recommended testing the materials with no covering, presumably with a restraining collar to prevent grooming of the area and ingestion of the material. A material of similar use or that produces a known effect in humans is included in almost all of the repeat application procedures as a control. The degrees of inflammation produced by the materials in a single assay are compared. Test sites are evaluated for erythema and edema using the scales of the Draize-type tests or more descriptive scales developed by the investigator. Although interpretation ratings such as "slight," "moderate," or "severe" irritant are not usually made, the data from cumulative irritancy assays in rabbits have been used to predict reactions in humans. Other investigators have used multiple application with shorter periods of time to evaluate materials [174].

A 5-day dermal irritation test in rabbits was used to compare consumer products of various types [219]. After the animals' backs were shaved, 0.5 mL of test materials was spread over a 5 × 4.5-cm area of skin. The test sites were protected from grooming by placing the animals in a leather harness or Elizabethan collar. After 4 hours, sites were cleaned and graded using the Draize scoring system. This procedure was repeated each day for 5 days. The authors showed good agreement between this assay and 21-day human patch tests of liquid detergents, after-bath colognes, and hair preparations; the technique was less satisfactory for other types of materials.

The guinea pig immersion assay has been used to evaluate the irritancy of aqueous detergent solutions [48,273,374]. Ten guinea pigs are placed in restraining devices that are immersed in a 40°C test solution for 4 hours. The apparatus is designed to maintain the head of the guinea pig above the solution. Immersion is repeated daily for three treatments. The flank is shaved 24 hours after the final immersion, and the skin is evaluated for erythema, edema, and fissures. A photographic grading scale for this assay was presented in MacMillan et al. [219]. Only materials of limited toxic potential are suitable for this assay because systematic absorption of a lethal dose is possible. Concentrations of test materials varies somewhat but is usually below 10% to limit systematic toxicity of the agents. A second group of animals is usually tested with a reference material as a control for the material of interest.

An open application procedure in guinea pigs uses microscopic examination of skin biopsies of sites treated with weak irritants to rank materials [5]. Biopsies are taken after three daily applications of 10% of solvent or 5% aqueous test solutions to 1-cm areas of the shaved flank. Sites are evaluated visually for erythema and edema, and 3-m histological sections are stained with May–Grunward–Giemsa under oil immersion to evaluate microscopically epidermal thickness and dermal infiltration. A composite score reflecting the macroscopic evaluation, number of applications before development of visible response, the epidermal thickness, and the cellular response is used to rank chemicals. Although this method provides information on pathogenesis of the response to each chemical, the extensive processing may limit its application to special studies.

Uttley and Van Abbe [353] developed a mouse ear test in which undiluted shampoos were applied to one ear daily for 4 days. The degree of inflammation was quantified visually as vessel dilation, erythema, and edema. The degree of inflammation produced by materials of interest was compared to that produced by a reference material tested on another group of mice. One confounding factor with this assay may be the use of anesthetics to facilitate performance of the procedure which may alter the development of inflammation.

To distinguish between mild and moderate irritants in an acute exposure test, Finkelstein and colleagues [111,112] used pretreatment of test sites with an irritant and enhanced visualization of the response by injection of Trypan blue to increase test sensitivity. The technique was performed in anesthetized rabbits, rats, or guinea pigs. A circular area of the shaved abdomen was painted with a 20% solution of formaldehyde and then was allowed to dry for 5 minutes. This was repeated three times and then 1-inch cotton flannel pads saturated with test material were applied to each site. A control substance of known irritancy was tested in each study. Pads were secured in place and the entire trunk was then wrapped in polyethylene. A solution of Trypan blue was injected into subcutaneous tissue away from the dosage sites. The dye was absorbed and served as a marker for plasma leakage because it spontaneously binds to albumin. After 16 hours, patches were removed and the degree of bluing at each site was evaluated on a 0 to 100% scale. In light of work comparing the reactivity of dorsal and abdominal animal skin [358], one wonders if the enhanced sensitivity was due in part to choice of test site. Another study reported quantitating the amount of Evans blue dye recovered from rat skin after exposing the skin to inflammatory agents [160].

A few tests in which material is not applied topically have been developed that claim to evaluate the intrinsic irritancy of test materials. The persistence of edema in the skin of depilated juvenile white mice following intracutaneous injection of solutions has been used to assess local irritation [51,367]. The number of wrinkles observed when reefing the skin with thin pincers is counted before and at selected time points through 6 hours after injection of 0.01-mL test solution. Although the number of test animals has varied between 8 and 25, the developers considered 20 to 25 optimal. An obvious limitation of this method is that materials must be administered as isoosmotic solutions, which requires substantial pretest formulation. Although the developers claim this procedure has good predicative power for eye, skin, and mucosal irritation, it has not been adopted extensively, and no validation studies comparing this method to the standard assays have been published.

Justice et al. [174] described a repeat animal patch test (RAPT) test for comparing the irritation potential of surfactants. Solutions were applied to the clipped backs of immobilized albino mice with saturated cotton-tipped applicators. The test sites were covered with rubber dams to prevent evaporation. This process was repeated seven times at intervals of 10 minutes. The skin was then evaluated microscopically for epidermal erosion.

Brown [49] used both open and closed exposures to rank surfactants for skin irritation potential. Tests ranged from 6-hour patch exposures each day for 3 consecutive days in rabbits to daily open application to the skin of rabbits, guinea pigs, or hairless mice for up to 4.5 weeks. Good agreement among the test methods was not obtained, and none of the methods gained wide acceptance, although they are similar to techniques developed by others later.

We have used an assay in which dilute solutions of surfactants and other chemicals are applied to one ear of five or six mice each day for 4 days [284]. Ear thickness measurements at various time points after each treatment was used to quantify the degree of inflammation. Multiple groups (at least four) were tested with different doses of the material. The dose producing a 50% maximum response following a single treatment, and the slopes of the dose–response lines for the chemicals were compared. Pretreating the ear with croton oil or TPA 72 hours before application of the material of interest increased the sensitivity of the assay. Although the procedure was useful for most surfactant-based products, it is not suitable for oily and highly perfumed materials because animals attempt to remove the materials by grooming. Moloney and Teal [255] also used ear thickness to quantify inflammatory changes produced by *n*-alkanes applied to the ears of mice. They dosed animals twice per day for 4 days to produce inflammation. Dithranol-induced skin irritation and the modulating effects of different pharmacological agents, such as the corticosteroid and the lipoxygenase and cyclooxygenase inhibitor studies, were studied using the mouse ear model [357].

## Human Irritation Tests

Because only a small area of skin must be tested, it is possible to conduct predicative irritation assays in humans, provided that systematic toxicity (from absorption) is low and informed consent is obtained. Although regulatory agencies do not routinely require testing in humans, human tests are preferred to animal tests in some cases because of the uncertainties of interspecies extrapolation. New materials, those of unknown or unfamiliar composition, should be tested on animal skin first to determine if application to humans is warranted [263]. Patch test responses generally heal rapidly, within a week or so. More severe reactions should be evaluated periodically over a longer period of time to determine how the inflammatory response is resolved. Some subjects may develop changes in pigmentation level at the test site following severe responses. It is prudent to arrange for medical consultation whenever human clinical tests are undertaken.

### Single-Application Irritation Patch Tests

Many forms of single-application patch tests have been published. The duration of patch exposure has varied between 1 and 72 hours. Custom-made apparatus to hold the test material has been designed [221,321,323], and a variety of adhesives that are no longer commercially available have been used [223]. Although the individual assays provided important information to the investigators of the period, they were never standardized or gained widespread acceptance.

**TABLE 27.11**
**Human Patch Test Grading Scales**

| Score | Description |
|---|---|
| A. Simple Patch Test Grading Scale | |
| 0 | Negative, normal skin |
| ± | Questionable erythema not covering entire area |
| 1 | Definite erythema |
| 2 | Erythema and induration |
| 3 | Vesiculation |
| 4 | Bullous reaction |
| B. Detailed Human Patch Test Grading Scale | |
| 0 | No apparent cutaneous involvement |
| 1/2 | Faint, barely perceptible erythema or slight dryness (glazed appearance) |
| 1 | Faint but definite erythema, no eruptions or broken skin *or*<br>No erythema but definite dryness; may have epidermal fissuring |
| 1-1/2 | Well-defined erythema or faint erythema with definite dryness; may have epidural fissuring |
| 2 | Moderate erythema, may have a *few* papules or deep fissures, moderate-to-severe erythema in the cracks |
| 2-1/2 | Moderate erythema with barely perceptible edema *or*<br>Severe erythema not involving a significant portion of the patch (halo effect around the edges), may have a few papules *or*<br>Moderate to severe erythema |
| 3 | Severe erythema (beet red ). May have generalized papules *or*<br>Moderate to severe erythema with slight edema (edges well defined by raising) |
| 3-1/2 | Moderate to severe erythema with moderate edema (confined to patch area) *or*<br>Moderate to severe erythema with isolated eschar formations or vesicles |
| 4 | Generalized vesicles *or*<br>Eschar formation *or*<br>Moderate to severe erythema and/or edema extending beyond patch area |

The single-application patch procedure outlined by the NAS [262] incorporates important aspects of assays used by many investigators. The procedure is similar to FHSA tests in rabbits. Commercial patches, chambers, gauze squares, or cotton bandage material, such as Webril®, applied to either the intrascapular region of the back or to the dorsal surface of the upper arms are expected to produce equivalent reactions [196]. Patches are secured in place with surgical tape without wrapping the trunk of the arm. For new volatile materials, a relatively nonocclusive tape, such as Micropore®, Dermicel™, or Scanpore®, should be used. Increasing the degree of occlusion with occlusive tapes such as Blenderm™ or chamber devices such as the Duhring and Hill Top chambers generally increases the severity of responses. A 4-hour exposure period was suggested by the NAS panel; however, it is desirable to test new materials and volatiles for shorter periods (30 minutes to 1 hour), and many investigators apply materials intended for skin contact for 24- to 48-hour periods. Subjects should routinely be instructed to remove patches immediately if any unusual discomfort develops. After the period of exposure, the patches should be removed, the area cleaned with water to remove any residue, and the test site marked by study personnel.

Responses are evaluated 30 minutes to 1 hour after each patch removal (to allow hydration and pressure effects to subside) and again 24 hours after the patch is removed. Persistent reactions may be evaluated for 3 to 4 days. The Draize scale for erythema and edema (Table 27.9) has been used for grading human skin responses; however, the scale has no provision for scoring papular, vesicular, or bullous responses. Integrated scales ranging from 4 to 16 points have been published (see Table 27.11) and are generally preferred to the Draize scale. Up to ten materials can be tested simultaneously on each subject. Skin reactivity differs by body region, and some patch sites may receive more pressure (e.g., from chairs or clothing); therefore, the location where the materials are placed on the skin (e.g., upper right back, lower left back) should be systematically varied within each study. Each study should include at least one reference material. Scores from all subjects are averaged for each material, and comparisons are made between standards and other test materials.

Some investigators have accepted an average difference of one unit on the grading scale as meaningful. Other investigators analyze the data by standard nonparametric statistical tests. It is also possible to test multiple doses and calculate the $ID_{50}$ [199].

Wooding and Opdyke [383] investigated the effects of modifying some test parameters on intensity of the response, and the intensity of inflammation has been shown to increase after removal of patches in some cases [77,299]. Kooyman and Snyder [202] used 6-hour exposures to 8% solutions of bar soaps and evaluated test sites 24 hours after patch application. Griffith et al. [140] reported using single-application patch tests with exposures of less than 24 hours to evaluate laundry detergents containing enzymes. Justice et al. [174] varied exposure time between 18 and 24 hours to test bar soaps, liquid detergents, and laundry detergents. Others have suggested that 48-hour patch exposures are more suitable for some products [305].

A standardized procedure for evaluating the irritation potential of new chemicals in humans as a replacement for the Draize rabbit test has been proposed [16,17,92,141, 385,386]. The method has been tested in several different laboratories, and results seem to be reasonably reproducible. The classification of irritancy is based on a comparison of the length of exposure to the chemical being tested vs. the length of exposure producing irritation following application of a 20% solution of sodium laurel sulfate. Chemicals producing irritation after shorter exposures than used for SLS are considered irritants and would be labeled as such. Each test subject is exposed to the undiluted test material in occlusive chambers (e.g., Hill Top Chamber®) and to a 20% solution of SLS. Length of exposure begins with a 15-minute exposure with evaluation at removal and at 24 and 48 hours. If no response is observed, then another set of patches is applied and worn for 30 minutes. This process of patching, evaluation, and patching for a longer exposure interval is repeated until the subject responds to the SLS exposure or until a 4-hour exposure has been completed. The test was developed for chemicals but may have use for evaluating consumer products as well. For most household chemicals and cosmetics, this cycle can probably be shortened considerably. Nixon et al. [265] have used a 4-hour FHSA-type procedure (including abrasion) to evaluate a range of household products.

## Repeat-Application Irritation Patch Tests

The term *skin fatigue* was used to explain the development of inflammation late in the induction phase of sensitization tests without positive responses at challenge [320]. The phenomenon was also referred to as *secondary irritation* and later as *cumulative irritation*. The human repeated insult patch test for skin allergy was modified to evaluate skin irritation. As with single-application patch test, many investigators developed their own version of the repeat application patch test. Most were patterned after human sensitization studies with 24-hour exposures with or without a rest period between patches. Kligman and Wooding [199] applied the Litchfield and Wilcocon probit analysis to cumulative irritation testing with calculation of $IT_{50}$ and $ID_{50}$ values, and statistical comparison of those values for different materials. Their early work forms the basis for the 21-day cumulative irritation assay, which is currently widely used.

The cumulative irritation assay, as described by Lanman et al. [215], was developed to compare antiperspirants, deodorants, and bath oils to provide guidance for product development. A 1-inch square of Webril® was saturated with liquid or up to 0.5 g of viscous substances and applied to the skin. The patch was applied to the upper back and sealed into place with occlusive tape. After 24 hours, the patch was removed, the area evaluated, and a fresh patch applied. The procedure was repeated for up to 21 days. The sensitivity of the assay was increased by increasing the number of test subjects from 10 to 24. The $IT_{50}$, as described by Kligman and Wooding [199], was used to evaluate and compare test materials.

Modifications of the cumulative irritation assay have been reported. Intensity of response has been evaluated using other evaluation schemes, the interval between application of fresh patches has been varied, and other methods of data evaluation have been proposed [22,61,298]. The newer chamber devices have replaced Webril® with occlusive tape in some laboratories. Some investigators currently use cumulative scores to compare test materials and do not calculate an $IT_{50}$. The necessity of 20 applications has been questioned [22]. Although the procedure came to be known as the 21-day cumulative irritation assay, the number of applications used was varied by Lanman, depending on the types of materials to be tested; 21 days was the maximum period of testing. Kligman and Wooding performed their studies on surfactants in 10 days. Lanman found that 21 applications were necessary to discriminate between baby lotions. The number of applications used to rank materials should be chosen based on the class of material being studied.

Numerous other human repeated application schedules have been used for comparing commercial products. Finkelstein et al. [111,112] described tests using either a 5- to 6-hour or a 17- to 18-hour exposure each day for 14 days. Test sites were evaluated 1 hour after patch removal. Modifications of this procedure have also been used to evaluate shaving creams and toilet soaps [327].

Repeated application patch tests on intact skin fail to predict some adverse reactions due to repeated application to "damaged" skin (e.g., acne, shaved underarms, or sensitive areas such as the face) [18]. The chamber scarification test [121,123,125] was developed to evaluate materials that would normally be applied to damaged skin. Light-skinned Caucasians who develop severe erythema with edema and vesicles following a 24-hour exposure to 5%

**TABLE 27.12**
**Grading Scale for the Chamber Scarification Tests**

| Grade | Description |
|---|---|
| 0 | Scratch marks barely perceptible |
| 1 | Erythema confined to scratches |
| 2 | Broader bands of erythema with or without rows of vesicles, pustules, or erosions |
| 3 | Severe erythema with partial confluence with or without other lesions |
| 4 | Confluent, severe erythema sometimes with edema, necrosis, or bulla |

SLS in Duhring chambers applied to the inner forearm are preselected subjects. Six to eight 10-mm² areas on the mid-volar forearm are scarified with eight crisscross scratches made with a 30-gauge needle. Four scratches are parallel, with another four at right angles. In scarifying the tissue, the bevel of the needle is to the side and is drawn across the tissue at a 45% angle with enough pressure to scratch the epidermis without drawing blood. Duhring chambers containing the test material (0.1 g for ointments, creams, or powders or Webril® saturated with 0.1 mL for liquids) are placed over the scarified areas and are secured in place with nonocclusive tape wrapped around the forearm. Fresh chambers containing the same materials are applied daily for 3 days. The test sites are evaluated on a 0 to 4 scale (Table 27.12) 30 minutes after removal of the last set of chambers. The responses are averaged, and materials are classified as low (0 to 0.41), slight (0.5 to 1.4), moderate (1.5 to 2.4), or severe irritants. A scarification index (SI) may be calculated by dividing the score on intact skin by the score from the scarified site. The SI is used to estimate the relative risk for damaged and normal tissue; it is not used to rank test materials.

Although bar soaps produce erythema when tested by conventional patch test techniques, the typical clinical response is dryness and flaking with occasional erythema and fissuring. Frosch and Kligman [125] developed the soap test chamber test to compare the "chapping" potential of bar soaps. Sensitive subjects are preselected as described for the chamber scarification test or by ammonium hydroxide blistering time [102]. Duhring chambers fitted with Webril® pads are used to apply 0.1 mL of an 8% solution of soap to the forearm. Chambers are secured in place by encircling the arm with porous tape. Patch contact time is 24 hours on day 1 (Monday) and 6 hours each day for the next 4 days (Tuesday through Friday). Test sites are monitored each day before application of fresh solutions. If severe erythema is noted, dosing is discontinued. Unless treatment was discontinued before the fifth exposure, skin reactions are evaluated on day 8 (Monday). This test has shown good agreement with skin-washing procedures but has overpredicted irritant responses to some materials [120].

### Exaggerated Exposure Irritant Tests

Although patch tests have been useful in detecting differences in the irritation potential of some materials, in some cases predicted differences were not apparent when materials were used by consumers. Exaggerated exposure tests have been developed to bridge the gap between responses occurring during product use and patch testing. Perhaps the oldest nonpatch irritancy test still in use is the arm immersion technique [202], in which the relative irritancy of two soap or detergent products is compared. As was originally described, soap solutions of up to 8% were prepared in troughs. Temperature was maintained at 105°F while subjects immersed one hand and arm to just above the elbow in one test solution and the other arm in a solution containing a second product. The period of exposure varied between 10 and 15 minutes, 3 times a day, for 5 days or until observable irritation was produced on both arms.

In most volunteers, the first sign of irritation was erythema of the anticubital surface of the arm [174,202]. Later, the hands developed dryness and cracking. These observations led to the development of separate assays on the anticubital area and the hands. Numerous versions of the anticubital washing test (also known as the flex washing test and elbow crease washing test) have been used. Published methods compare two products; however, dosing regimes differ somewhat. Investigators have used two [120] or three [140,337] washing procedures per day, and some specify that lather is allowed to remain on the skin for a brief period. Erythema and edema are evaluated as endpoints for all studies. Frosch [120] used a similar procedure on the cheeks to evaluate toilet soaps. A simple 1 to 4 (i.e., slight, moderate, severe, very severe) grading scale is used to evaluate the severity of the response. Products can be compared in terms of the average grades or the number of washes producing an effect. Some investigators have tested up to four samples per forearm by washing in glass cylinders and then rinsing the area [162].

At least two types of hand immersion procedures have been used. In small studies (i.e., 10 subjects), relatively concentrated solutions (up to 2%) of two materials are tested. Up to four hand-dishwashing products have been compared at near-use concentrations in studies on 64 subjects using a Latin-square dosing pattern [9]. Exposure conditions have varied from two to three 10- to 15-minute immersions each day [140] to a single 30-minute exposure each day [9]. Grading scales for this type of assay focus on scaling and cracking as well as erythema.

Evaluation of skin conditions before and after use in the home has also been used to compare the irritation potential of various products. These tests represent skin tolerance studies, as either irritation or allergy could be detected. The clinical method published by Johnson et al. [172] has been varied to include tests of bar soaps, laundry soaps and detergents, and dishwashing detergents. Essentially, the method is a double-blind crossover study with usage periods of 2 weeks [62]. Skin condition is evaluated by a dermatologist before the study and after the use of each product. Magnification of the area is used to facilitate grading using a 0 to 10 scale. Tests are conducted using large panels (more than 300 subjects per product), and up to eight materials can be evaluated simultaneously using a Latin-square design. The principles used in conducting this type of large-scale usage study have been applied to laundry powders for diapers and bar soaps used in infants [97,140] and to fabric softeners in adult populations [369]. Special emphasis should be placed on the statistical design of clinical trials of this type to ensure validity of the study [1].

### Use of Bioengineering Devices in Irritancy Evaluations

Measurements of biophysical parameters of skin function have been proposed as adjuncts to visual elevations of the inflammatory response [237,239,343,344,382]. In many instances, investigators constructed their own instruments to perform these measurements. As the availability of commercial instruments increases, assessment of the biophysical changes in skin has become widely used to supplement visual evaluations. The benefits of the methods vary and include more objective measurements of erythema (as change in blood flow), more precise determination of the color change, and measurement of parameters of damage that cannot be evaluated visually. When combined with various techniques of exposure, it is possible to vastly decrease the number of subjects and probable clinical relevance of the studies.

Laser Doppler velocimetry has been used to quantify the increased blood flow to inflamed tissue. This device is an optical technique for estimation of microcirculation, based on the Doppler principle. When the laser beam, a 632-nm helium–neon (He–Ne) laser source, is directed toward the tissue, reflection, transmission, absorption, and scattering occur. Laser light backscattered from moving particles, such as red cells, is shifted in frequency according to the Doppler principle, whereas radiation that is backscattered from nonmoving structures remains at the same frequency. Detailed guidelines for using this device are provided by the Standardization Group of the European Society of Contact Dermatitis [28].

Skin reflectance spectrophotometry has been used to measure the color change in skin associated with inflammation. Polychromatic light is directed into the skin. The reflected light is collected in an integrating sphere and guided to a monochromator, where the light is split into five bands in the spectral range of 355 to 700 nm. Melanin content and oxygenized and deoxygenized hemoglobin are analyzed at different spectra, and the relative changes are expressed as a percentage of chromophore content in control skin [3,29].

Changes in transepidermal water loss (TEWL) and electrical resistance have been detected before inflammatory changes are apparent. Early investigations clearly established that chemicals that provoked inflammation increased TEWL. TEWL reflects the integrity of the water-barrier function of stratum corneum as determined by the use of an evaporimeter (open chamber method, where the skin capsule is open to the atmosphere). TEWL is calculated from the slope provided by two hygrosensors precisely oriented in the chamber. Air movement and humidity are the drawbacks of this method. A report reviewing interindividual, environmental, and instrument-related variables with respect to the Evaporimeter EPI (ServoMed), its use, and good practice guidelines can be obtained from Pinnagoda et al. [292]. Malten and Thiele [238] showed that increase in TEWL occurred before visible inflammation when ionic, polar, water-soluble substances (e.g., sodium hydroxide, soaps, detergents) were used as irritants. Unionized, polar irritants, such as dimethyl sulfoxide, and unionized, nonpolar, water-soluble irritants, such as hexanediol diacrylate and butanediol diacrylate, did not induce increased water loss until visible inflammation occurred. This method thus seems to be possibly valuable for detecting irritancy with no perceivable irritancy for ionizable, polar, water-soluble substances. These techniques have been used to compare the irritancy potential of soaps tested at near-use levels [150].

Water content of the skin can be estimated by electrical measurements of skin resistance and capacitance. Subtle degrees of skin damage can be measured using skin resistance before skin inflammation occurs [343]. The corneometer is used to register the electrical capacitance of the skin surface. The principle of this instrument is to decipher distinctly different dielectric constants of water and other constitutional materials (fewer than 7) with a probe applied to the skin at constant pressure. Another instrument working on similar principle is the skin hygrometer, which measures conductance of the skin using high-frequency electric current of 3.5 MHz with the help of a sensitive probe [341]. Several other devices using different frequencies and technologies have been developed to measure water content including instruments that measure impedance, resistance, phase angle, microwave transmission, and photoacoustic methods. Noninvasive bioengineering techniques, such as colorimetry, remittance spectroscopy measurement of surface pH, and skin surface topography have been reviewed by Berardesca and Maibach [20,21]. One unique new way of evaluating irritancy uses *in vivo* measurements of the water-binding capacity of the stratum corneum after occlusion [20]. Bioinstrumentation techniques have now also been used in *in vitro* studies. These techniques permit more clinically realistic bioassays that more closely mimic use experience.

### *In Vitro* Assays of Skin Irritation and Corrosion

Since 1980, much effort has been expended toward developing *in vitro* alternatives to Draize-type tests for both eye and skin irritation. Approaches to development of an *in vitro* model include cell toxicity [37,39], measurement of inflammatory mediators, effect on cell recovery and survival [94,201,259], effect on cellular physiology [282], cell morphology [37–39], biochemical endpoints [254], and effect on membranes [32] and artificial membranes [134–136] constructed to release dye indicator. Some investigators have included metabolic activators in their systems [40]. Nonmammalian cells [55] have also been used.

*In vitro* assays for skin irritation are being explored as tools in toxicologic research and as aids in formulating mild products, as well as for replacement of the Draize-type tests. This mirrors the overall use of *in vitro* assays during the last 15 years [17]. Numerous proposals for validation of the methods have been put forth; however, tests in intact animals or humans are currently the only means of assessing the potential irritation hazard from skin exposure [329]. *In vitro* assays for skin corrosion are near acceptance and validation.

Two commercial systems, Skin2® and EpiDerm™, have been used extensively for skin irritation testing. Skin2® [114,348,349] is a three-dimensional coculture of human fibroblasts and keratinocytes. In Skin2®, human neonatal fibroblasts are seeded on a nylon mesh to which they attach and lay down collagen. When the proper degree of confluence is reached, human keratinocytes are seeded onto the fibroblast culture. The epidermis is exposed to air and a partially differentiated stratum corneum develops. Several cytotoxicity assays and assays for release of inflammatory mediators are available for use with these systems. EpiDerm™ [59] consists of a multilayered cornified epithelium with no dermal element. It appears to have a well-differentiated stratum corneum. Perkins et al. [290] used these systems to evaluate skin corrosion *in vitro* using the dimethyltriazoldiphenyl tetrazolium-formazan (MTT) [60] cell viability assay with some success.

SKINTEX™ and Corrositex® [116–119] are described as membrane barrier/protein matrices. They are two-component systems consisting of a barrier matrix that contains an indicator dye. Dye release is expected to correlate with protein disruption and denaturation. The second compartment is a reagent system that increases in turbidity when exposed to irritants. Corrositex® has been accepted by the DOT as an alternative to the Draize test for skin corrosion [163] and has been validated for limited purposes by ICCVAM [391].

Several European investigators have reported success in using excised rat or human skin to assess skin corrosivity [16,270,271,366,378] based on changes in transcutaneous electrical resistance (TER). A rubber O-ring is used to secure full-thickness skin (including dermis), epidermal surface uppermost, onto the top of small tubes. These tubes are then suspended in a larger tube containing an electrolyte solution in distilled water. Applied to the surfaces of at least three discs is 150 μg of the test material. After 24 hours the skin is rinsed, the surface is treated with ethanol, and electrolyte solution is added to the skin surface. The TER is then measured using a commercial instrument. Values above 11 to 12.5 kΩ/disc are indicative of corrosivity (varies by investigator and source of skin). With the judicial use of reference materials to set the threshold values for classifying materials as corrosive, this method appears to be reproducible. Full validation of the method has not yet been completed.

## CONTACT URTICARIA AND URTICARIA-LIKE SYNDROMES

Circumscribed, erythematous, evanescent areas of edema involving the epidermis and superficial portions of the dermis are referred to as *urticaria* [361]. Classically, the reaction has been described as a wheal-and-flare response that develops within 30 to 60 minutes after exposure of the skin to certain agents. Symptoms of immediate contact reactions can be classified according to their morphology and severity; itching, tingling, and burning with erythema is the weakest type of immediate contact reaction. Local superficial wheal-and-flare with tingling and itching represents the prototype reaction of contact urticaria. Generalized urticaria after local contact is rare but can occur from strong urticants. Signs in other organs appear with the skin symptoms in cases of immunologic contact urticaria syndrome. Urticaria and angioedema (edematous processes in the deep dermis, subcutaneous, and submucosal tissue) may persist for up to 72 hours [72]. The strength of the reactions may vary greatly, and often the entire range of local signs—from slight erythema to strong edema and erythema—can be seen from the same substance if different concentrations are used in skin tests [205]. Not only the concentration but also the site of the skin contact can affect the reaction. A certain concentration of contact urticant may produce strong edema and erythema reactions on the skin of the upper back and face, but only erythema on the volar surfaces of the lower arms or legs. In some cases, contact urticaria can be demonstrated only on slightly or previously eczematous skin, and it can be part of the mechanism responsible for maintenance of chronic eczemas [2,228,268]. Some agents, such as formaldehyde, produce urticaria on healthy skin following repeated but not single applications to the skin. Diagnosis of immediate-contact urticaria is based on a thorough history and skin testing with suspected substances [72]. Skin tests for human diagnostic testing have been summarized [361]. Because of the risk of

**TABLE 27.13**
**Agents Reported To Cause Urticaria in Humans**

| |
|---|
| Immunologic mechanism |
| Grains |
| Nuts |
| Bacitracin |
| Parabens |
| Seafood (protein extracts) |
| Penicillin |
| Butylated hydroxy toluene |
| Nonimmunologic mechanism |
| Aspirin |
| Balsam of Peru |
| Benzoic acid |
| Cayenne pepper |
| Cinnamic aldehyde |
| Codeine |
| Dimethyl sulfoxide |
| Unknown mechanisms |
| Lettuce/endive |
| Cassia oil |
| Formaldehyde |
| Ammonium persulfate |
| Neomycin |

*Note:* More comprehensive lists and the original references can be found in Amin, S. et al., in *Dermatotoxicology Methods: The Laboratory Worker's Vade Mecum*, Marzulli, F.N. and Maibach, H.I., Eds., Taylor & Francis, New York, 1998, pp. 161–176.

systemic reactions, such as anaphylaxis, human diagnostic tests should be performed only by experienced personnel with facilities for resuscitation on hand.

Contact urticaria has been divided into two main types: immunologic and nonimmunologic [207]. Several reviews list agents suspected to cause each type of urticarial response [208,209,361]. A few common urticants are listed in Table 27.13. Nonimmunologic contact urticaria occurs without previous exposure in most individuals and is the most common type. The reaction remains localized and does not cause systemic symptoms or spread to become generalized urticaria. Typically, the strength of this type of contact urticaria reaction varies from erythema to a generalized urticarial response, depending on the concentration, skin site, and substance. The mechanism of nonimmunologic contact urticaria has not been delineated, but a direct influence upon dermal vessel walls and nonimmunologic release of mast cell mediators are possible mechanisms [319]. Several reports suggest that nonimmunologic urticaria produced by different agents may involve different combinations of mediators [207].

The most potent and best studied substances producing nonimmunologic contact urticaria are benzoic acid, cinnamic aldehyde, and nicotinic esters. Under optimal conditions, more than half of a random sample of individuals shows local edema and erythema reactions within 45 minutes of application of these substances if the concentration is high enough. Benzoic acid and sodium benzoate are used as preservatives for cosmetics and other topical preparations at concentrations from 0.1% to 0.2% and are capable of producing immediate contact reactions at the same concentrations. Cinnamic aldehyde at a concentration of 0.01% may elicit an erythematous response associated with a burning or stinging feeling in the skin [242]. Mouthwashes and chewing gums contain cinnamic aldehyde at concentrations high enough to produce a pleasant tingling or lively sensation in the mouth that enhances the sale of the product. Higher concentrations produce lip swelling of typical contact urticaria in normal skin. Eugenol in the mixture inhibits contact sensitization to cinnamic aldehyde and inhibits nonimmunologic contact urticaria from this same substance. The mechanism of the quenching effect is not certain, but competitive inhibition at the receptor level may be the explanation [124].

Immunologic contact urticaria is an immediate type 1 allergic reaction in people previously sensitized to the causative agent [361]. The molecules of a contact urticant react with specific IgE molecules attached to mast cell membranes. The cutaneous signs are elicited by vasoactive substances, including histamine, released from mast cells. The role of histamine is conspicuous, but other mediators of inflammation (i.e., prostaglandins, leukotrienes, kinins) may influence the degree of response. Immunologic contact urticaria reactions can extend beyond the contact site, and generalized urticaria may be accompanied by other symptoms, such as rhinitis, conjunctivitis, asthma, and even anaphylactic shock. The term *contact urticaria syndrome* was therefore suggested by Maibach and Johnson [231]. The name has been accepted for a symptom complex where local urticaria occurs at the contact site with symptoms in other parts of the skin or in target organs, such as the nose and throat, lung, and gastrointestinal and cardiovascular systems. Fortunately, the appearance of systemic signs is rare, but it may be seen in cases of strong hypersensitivity or in a widespread exposure and abundant percutaneous absorption of an allergen.

Foodstuffs are the most common causes of immunologic contact urticaria. The otolaryngeal area is a site where immediate contact reactions frequently are provoked by food allergens, most often in atopic individuals. The actual antigens are proteins or protein complexes. As a proof of immediate hypersensitivity, specific IgE antibodies against the causative agent typically can be found in the patient's serum using the RAST technique and skin test for immediate allergy. The passive transfer test (Prausnitz–Kustner test) also often gives a positive result.

Predictive assays for evaluating the ability of materials to produce nonimmunologic contact urticaria have been developed. No predictive assays for immunologic contact urticaria have been published. Lahti and Maibach [206] developed an assay in guinea pigs using materials known to produce urticaria in humans. One tenth of a milliliter of the material is applied to one ear of the animal; it is also applied to the opposite ear as a control. Ear thickness is measured prior to application and then every 15 minutes for 1 or 2 hours after application. The swelling response is dependent on the concentration of the eliciting substance. The maximum response is about a 100% increase in ear thickness, and it appears within 50 minutes after application of a contact urticant. In histologic sections, marked dermal edema and intra- and perivascular infiltrate of heterophilic granulocytes appear 40 minutes after application of test substances. This assay is the predictive test of choice for nonimmunologic contact urticaria if animals are to be tested. Guinea pig body skin reacts with quick-appearing erythema to cinnamic aldehyde, methyl nicotinate, and dimethyl sulfoxide but not benzoic acid, sorbic acid, or cinnamic acid. Analogous reactions can be elicited in the earlobes of another animal species. Cinnamic aldehyde and dimethyl sulfoxide produce a swelling reaction in the guinea pig, rat, and mouse. Benzoic acid, sorbic acid, cinnamic acid, diethyl fumafate, and methyl nicotinate produce no response in the rat or mouse, but the guinea pig ear reacts to all of them [207]. This suggests that either there are several mechanisms of nonimmunologic contact urticaria or that differences are due to the relative sensitivity of the species to the mediators.

Materials can also be screened for nonimmunologic contact urticaria in humans. A small amount of the test material is applied to a marked site on the forehead, and the vehicle is applied to a parallel site. The areas are evaluated at approximately 20 to 30 minutes after application for erythema or edema [361]. Differentiating between nonspecific irritant reactions and contact urticaria may be difficult. Strong irritants, such as hydrochloric acid, lactic acid, cobalt chloride, formaldehyde, and phenol, can cause clear-cut immediate whealing if the concentration is high enough, but the reactions usually do not fade away within a few hours. Instead, they are followed by signs of irritation; erythema and scaling or crusting are seen 24 hours later. Some substances have only urticant properties (e.g., benzoic acid, nicotinic acid esters), some are pure irritants (e.g., sodium lauryl sulfate), and some have both of these features (formaldehyde, dimethyl sulfoxide). Contact urticaria reactions are much less frequently encountered than either skin irritation or skin allergy [209]; however, increasing awareness of contact urticaria has expanded the list of etiologic agents and perhaps will lead to the development of adequate predictive assays for detecting causative agents of other forms of urticaria.

## SUBJECTIVE IRRITATION AND PARAESTHESIA

Cutaneous application of some chemicals elicits sensory discomfort, tingling, and burning without visible inflammation. This noninflammatory painful response has been termed *subjective irritation* [122,126]. Materials reported to produce subjective irritation include dimethyl sulfoxide, some benzoyl peroxide preparations, and the chemicals salicylic acid, propylene glycol, amyl-dimethyl-paramino benzoic acid, and 2-ethoxyethyl-*p*-methoxy cinnamate, which are ingredients of cosmetics and over-the-counter (OTC) drugs. Pyrethroids, a group of broad-spectrum insecticides, produce a similar condition that may lead to temporary numbness, or paraesthesia [11,57,200]. As in subjective irritation, the nasolabial folds, cheeks, and periorbital areas are frequently involved. The ear is also sensitive to the pyrethroids. Herbst et al. investigated an assay to quantify subjective irritation using a pyrethoid chemical model [401].

Only a portion of the human population seems to develop nonpyrethroid subjective irritation. Frosh and Kligman [122] found they needed to prescreen subjects to identify "stingers" for conducting predictive assays. Only 20% of subjects exposed to 5% aqueous lactic acid in a hot humid environment developed stinging responses [122]. All stingers in their series reported a history of adverse reactions to facial cosmetics, soaps, etc. A similar screening procedure by Lammintausta et al. [211] identified 18% of their subjects as stingers. Prior skin damage, such as sunburn, pretreatment with surfactants, and tape stripping, increases the intensity of responses in stingers. Persons not normally experiencing a response report pain upon exposure to lactic acid or other agents that produce subjective irritation [122]. Attempts to identify reactive subjects by association with other skin descriptions (e.g., atopy, skin type, degree of skin dryness) have not yet been fruitful; however, data show that stingers develop stronger reactions to materials, causing nonimmunologic contact urticaria and some increase in transepidermal water loss and blood flow following application of irritants via patches than those of non-stingers [211].

The mechanisms by which materials produce subjective irritation have not been investigated extensively. Pyrethroids act directly on the axon by interfering with the channel-gating mechanism and impulse firing [359]. It has been suggested that agents causing subjective irritation act via a similar mechanism because no visible inflammation is present. An animal model was developed to rate paraesthesia to pyrethroids and may be useful for other agents [57,113]. The test site is the flank of 300- to 450-g guinea pigs. Both flanks are shaved, and animals are housed individually in observation cages. A volume of 100 μL of the test material is spread over approximately 30 $mm^2$ on one flank. The animal's behavior is monitored by an unmanned video camera for 5 minutes and at 0.5, 1, 2, 4, and 6 hours after application of the materials. Subsequently, the film is analyzed for the number of full turns of the head made to the control and pyrethroid-treated flanks. Head turns were usually accompanied by attempted licking and biting of the application sites. It was possible to rank pyrethroids for their ability to produce paraesthesia using this technique. The ranking corresponded to the ranking available from human exposure.

As originally published, the human subjective irritation assay required the use of a 110°F environmental chamber with 80% relative humidity [122]. Volunteers were seated in the chamber until a profuse facial sweating was observed. Sweat was removed from the nasolabial fold and cheek. A 5% aqueous solution of lactic acid was then rubbed briskly over the area. Those who reported stinging for 3 to 5 minutes within the first 15 minutes were designated as stingers and used for subsequent tests. Subjects were asked to evaluate the degree of stinging as 0 = no stinging, 1 = slight stinging, 2 = moderate stinging, and 3 = severe stinging. Stinging was evaluated 10 seconds and 2.5, 5, and 8 minutes after application of the test material. Other investigators [211] used a 15-minute treatment with a commercial facial sauna to produce facial sweating. The subjects turn away from the sauna for application of the test materials, then turn back to face the sauna for the observation period. The facial sauna technique is less stressful to both subjects and investigators and produces similar results.

Advances in understanding the somatosensory processes in humans is leading to the development of more objective methods for evaluating skin sensory effects [384,387]. Although these approaches are not yet sensitive enough to warrant use in predictive assays, they have been used to study pain and itch responses to histamine [389] and to some solvents [388]. In time, these techniques will undoubtedly be applied to predictive testing.

## QUESTIONS

1. A chemically exposed occupational worker developed generalized urticaria and shortness of breath at the work site. Differential diagnosis includes:
   A. Photoirritation
   B. Allergic contact dermatitis
   C. Contact urticaria syndrome (answer)
2. On introduction of a revised topical formulation, complaints of acute burn, sting, and itch increase beyond the expected level. Likely diagnostic possibilities include:
   A. Photoallergic contact dermatitis
   B. Sensory irritation (answer)
   C. Photoirritation
3. Topical formulations may be contraindicated during pregnancy because of concern regarding:
   A. Irritant dermatitis
   B. Allergic contact dermatitis
   C. Percutaneous penetration (answer)

## REFERENCES

1. Allen, A. M. (1978): Clinical trial design in dermatology: experimental design, I. *Int. J. Dermatol.*, 17:42–51.
2. Amin, S., Lahti, A., and Maibach, H. I. (1998): Contact urticaria syndrome. In: *Dermatotoxicology Methods: The Laboratory Worker's Vade Mecum*, edited by F. N. Marzulli and H. I. Maibach, pp. 161–176. Taylor & Francis, New York.
3. Anderson, P. Y. and Bjerring, P. (1990): Noninvasive computerized analysis of skin chromophores *in vivo* by reflectance spectroscopy. *Photodermatol. Photoimmunol. Photomed.*, 7:247–257.
4. Andersen, K. E. and Maibach, H. I. (1983): Multiple-application delayed-onset contact urticaria: possible relation to certain unusual formalin and textile reactions. *Contact Dermatitis*, 10:227–234.
5. Anderson, C., Sundberg, K., and Groth, O. (1986): Animal model for assessment of skin irritancy. *Contact Dermatitis* 15:143–151.
6. Asherson, G. L. and Ptak, W. (1968): Contact and delayed hypersensitivity in the mouse 1: active sensitization and passive transfer. *Immunology*, 15:405–416.
7. Aspergren, N. and Rorsman, H. (1962): Short-term culture of leucocytes in nickel hypersensitivity. *Acta Derm. Venereol. (Stockholm)*, 42:412–417.
8. Back, O. and Larsen, A. (1982): Contact sensitivity in mice evaluated by means of ear swelling and a radiometric test. *J. Invest. Dermatol.*, 78:309–312.
9. Bannan, E. A. (1975): Personal communication.
10. Baron, J., Voigt, J. M., Whitter, T. B., Kawabata, T., Knipp S. A., Gruengerich, F. P., and Jacoby, W. B. (1986): Identification of intratissue sites for xenobiotic activation and detoxication. *Adv. Exp. Biol. Med.*, 197:119–144.
11. Bartek, M., LaBudde, J. A., and Maibach, H. I. (1972) Skin permeability *in vivo* comparison in rat, rabbit, pig and man. *J. Invest. Dermatol.*, 58:114–123.
12. Bartek, M. J. and LaBudde, J. A. (1975): Percutaneous absorption *in vitro*. In: *Animal Models in Dermatology* edited by H. I. Maibach, pp. 103–120. Churchill Livingstone, New York.
13. Basketter, D. A., Robertts, D. W., Cronin, M., and Scholes E. W. (1992): The value of the local lymph node assay in quantitative structure–activity investigation. *Contact Dermatitis*, 26:137–142.
14. Basketter, D. A. and Scholes, E. W. (1992): Comparison of the local lymph node assay with the guinea pig maximization test for the detection of a range of contact allergens. *Food Chem. Toxicol.*, 30:65–69.
15. Basketter, D. A., Scholes, E. W., and Kimber, I. et al (1991): Interlaboratory evaluation of the local lymph node assay with 25 chemicals and comparison with guinea pig test data. *Toxicol. Methods*, 1:30–43.

16. Basketter, D. A., Whittle, E., and Chamberlain, M. (1994): Identification of irritation and corrosion hazard in skin: an alternative strategy to animal testing. *Food Chem. Toxicol.*, 32:539–542.
17. Basketter, D. A., Whittle, E., Griffiths, H. A., and York, M. (1994): The identification and classification of skin irritation hazard by a human patch test. *Food Chem. Toxicol.*, 32:769–775.
18. Battista, C. W. and Rieger, M. M. (1971): Some problems of predictive testing. *J. Soc. Cosmet. Chem.*, 22:349–359.
19. Benvenuto, A. J. and Cohen, A. (1990): A realistic role for non-animal tests. *Pharmaceut. Exec.*, June.
20. Berardesca, E. and Maibach, H. I. (1989): Physical anthropology of skin. In: *Models in Dermatology*, Vol. 4, edited by H. I. Maibach and N. Lowe, pp. 202–206. Karger, New York.
21. Berardesca, E. and Maibach, H. I. (1989): Effect of non-visible damage on the water-holding capacity of the stratum corneum, utilizing the plastic occlusion stress test (POST). In: *Current Topics in Contact Dermatitis*, edited by P. Frosch et al., pp. 554–559. Springer-Verlag, New York.
22. Berger, R. W. and Bowman, J. P. (1962): A reappraisal of the 21-day consecutive irritation test in man. *J. Toxicol. Cutan. Ocul. Toxicol.*, 1:109–115.
23. Bergstressor, P. R., Paniser, R. J., and Taylor, J. R. (1978): Counting and sizing of epidermal cells in human skin. *J. Invest. Derm.*, 70:280–284.
24. Bickers, D. R. (1991): Xenobiotic metabolism in skin. In: *Physiology, Biochemistry, and Molecular Biology of the Skin*, 2nd ed., edited by L. A. Goldsmith, pp. 205–236. Oxford University Press, New York.
25. Bickers, D. R., Dutta-Choudhury, T., and Mukhtar, H. (1982): Epidermis—a site of drug metabolism in neonatal rat skin: studies on cytochrome P450 content and mixed-function oxidase and epoxide hydrolase activity. *Mol. Pharmacol.*, 21:241–249.
26. Bickers, D. R., Eiseman, J., Kappas, A., and Alvares, A. P. (1975): Microscope immersion oils: effects of skin application on cutaneous and hepatic drug-metabolizing enzymes. *Biochem. Pharmacol.*, 24:779–783.
27. Bickers, D. R., Kappas, A., and Alvares, A. P. (1974): Differences in inducibility of cutaneous and hepatic drug metabolizing enzymes and cytochrome P450 by polychlorinated biphenyls and 1,1,1-trichloro-2,2-*bis*(*p*-chlorophenyl) ethane (DDT). *J. Pharmacol. Exp. Ther.*, 188:300–309.
28. Bircher, A., DeBoer, E. M., Agner, T., Wahlberg, J. E., and Serup, J. (1994): Guidelines for measurement of cutaneous blood flow by laser Doppler flowmetry: a report from the standardization group of the European Society of Contact Dermatitis. *Contact Dermatitis*, 30:65–72.
29. Bjerring, P. and Anderson, P. H. (1990): Skin reflectance spectrophotometry. *Photodermatol. Photoimmunol. Photomed.*, 4:167–171.
30. Bjornberg, A. (1974): Skin reactions to primary irritants and predisposition to eczema. *Br. J. Dermatol.*, 91:425.
31. Bjornberg, A. (1975): Skin reactions to primary irritants in men and women. *Acta. Derm. Venereol.* (*Stockholm*), 55:191.
32. Blake-Haskins, J. C., Scala, D., Rhein, L. D. et al. (1986): Predicting surfactant irritation from the swelling response of a collagen film. *J. Soc. Cosmet. Chem.*, 317:199–210.
33. Blank, H. I. (1952): Water content of the stratum corneum. *J. Invest. Dermatol.*, 18:433–440.
34. Blank, H. I. (1953): Further observations on factors which influence the water content of the stratum corneum. *J. Invest. Dermatol.*, 21:259–269.
35. Bloch, B. and Steiner-Wourlisch, A. (1930): Die Sensibilisierung des Meerschweinchens gegen primeln. *Arch. Dermatol. Syph.*, 162:349–378.
36. Blomberg, B. M. E., Bruynzeel, D. P., and Scheper, R. J. (1991): Advances in mechanisms of allergic contact dermatitis *in vitro* and *in vivo* research. In: *Dermatotoxicity*, 4th ed., edited by F. N. Marzulli and H. I. Maibach, pp. 255–362. Hemisphere, New York.
37. Borenfreund, E., Babich, H., and Martin-Alguacil, N. (1988): Comparison of two *in vitro* cytotoxicity assays: the neutral red (NR) and tetrazolium MTT tests. *Toxicol. In Vitro*, 2:1–6.
38. Borenfreund, E. and Puerner, J. A. (1984): A simple quantitative procedure using monolayer cultures for cytotoxicity assays. *J. Tissue Culture Methods*, 9:7–9.
39. Borenfreund, E. and Puerner, J. A. (1985): Toxicity determined *in vitro* by morphological alterations and neutral red absorption. *Toxicol. Lett.*, 24:119–124.
40. Borenfreund, E. and Puerner, J. A. (1987): Short-term quantitative *in vitro* cytotoxicity assay involving an S-9 activation system. *Cancer Lett.*, 34:243–248.
41. Bos, J. D. (1984): A new approach to contact allergenicity screening. *Med. Hypoth.*, 15:103–108.
42. Botham, P. A., Hall, T. J., Dennett, R., McCall, J. C., Basketter, D. A., Whittle, E., Cheeseman, M., Esdaile, D. J., and Gardner, J. (1992): The *in vitro* skin corrosivity tests: results of a interlaboratory trial. *Toxicol. In Vitro*, 6: 191–194.
43. Boutwell, R. K. (1981): Chemical carcinogenesis. Part a. Biochemical role. In: *Biology of Skin Cancer (Excluding Melanomas)*, edited by D. D. Laerum and O. H. Iverson, pp. 134–150. International Union Against Cancer, Geneva, Switzerland.
44. Boutwell, R. K., Urbach, F., and Carpenter, G. (1981): Chemical carcinogenesis. Part b. Experimental models. In: *Biology of Skin Cancer (Excluding Melanomas)*, edited by D. D. Laerum and O. H. Iverson, pp. 109–123. International Union Against Cancer, Geneva.
45. Briggaman, R. A., Toshliki, R., and Cronce, D. J. (1991): The epidermal–dermal junction and genetic disorder of this area. In: *Physiology, Biochemistry, and Molecular Biology of the Skin*, 2nd ed., edited by L. A. Goldsmith, pp. 1243–1265. Oxford University Press, New York.
46. Bronaugh, R. L. and Maibach, H.I. (2005): *Percutaneous Absorption: Drugs, Cosmetics, Mechanisms, Methodology.* Taylor & Francis, Boca Rotan, FL.
47. Bronaugh, R. and Maibach, H. I. (1991): *In Vitro Percutaneous Absorption*. CRC Press, Boca Raton, FL.
48. Bronaugh, R. L., Congolon, E. R., and Scheuplein, R. J. (1981): The effect of cosmetic vehicles on the penetration of *N*-nitrodiethanolamine through excised skin. *J. Invest. Dermatol.*, 76:94–96.

49. Brown, V. K. H. (1971): A comparison of predictive irritation tests with surfactants on human and animal skin. *J. Soc. Cosmet. Chem.*, 22:411–420.
50. Brunner, M. J. and Smiljanic, A. (1952): Procedure for evaluation of skin sensitizing power of new materials. *Arch. Derm. (Chicago)*, 66:703–705.
51. Bucher, K., Bucher, K. B., and Walz, D. (1981): The topically irritant substance: essentials, biotests, predictions. *Agents Actions*, 11:515–519.
52. Buehler, E. V. (1964): A new method for detecting potential sensitizers using the guinea pig. *Toxicol. Appl. Pharmacol.*, 6:341.
53. Buehler, E. V. (1965): Experimental skin sensitization in the guinea pig and man. *Arch. Dermatol.*, 91:171.
54. Buehler, E. V. (1985): A rationale for the selection of occlusion to induce and elicit delayed contact hypersensitivity in the guinea pig: a prospective test. In: *Contact Allergy Predictive Tests in Guinea Pigs*, edited by K. E. Anderson and H. I. Maibach, pp. 38–58. Karger, Basel.
55. Bulich, A. A., Greene, M. W., and Isenberg, D. L. (1981): Reliability of bacterial luminescence assay for determination of the toxicity of pure compounds and complex effluents. In: *Aquatic Toxicology and Hazard Assessment, 4th Conference*, edited by D. R. Branson and K. L. Dickson, pp. 338–347. American Society for Testing and Materials, Washington, D.C.
56. Busse, M. J., Hunt, P., Lees, K. A., Maggs, P. N. D., and McCarthy, T. M. (1969). Release of betamethasone derivatives from ointments: *in vivo* and *in vitro* studies. *Br. J. Dermatol.*, 81:103.
57. Cagen, S. Z., Malloy, L. A., Parker, C. M., Gardiner, T. H., Celder, C. A., and Jud, V. A. (1964): Pyrethroid mediated sensory stimulation characterized by a new behavioral paradigm. *Toxicol. Appl. Pharmacol.*, 6:270–279.
58. Calandra, J. (1971): Comments on the guinea pig immersion *CTFA Cosmet. J.*, 3:47.
59. Cannon, C. L., Neal, P. J., Kubilus, J., Klausner, M., Swartzendruber, D. C., Squier, C. A., and Wertz, P. W. (1994): Lipid ultrastructure and barrier function characterization of a new *in vitro* epidermal model. *J. Invest. Dermatol.*, 102:600.
60. Carmichael, J., Degraff, W. G., Gazdar, A. F., Minna, J. D., and Mitchell, J. B. (1987): Evaluation of a tetrazolium-based semiautomated colorimetric assay: assessment of chemo-sensitivity testing. *Cancer Res.*, 47:936–942.
61. Carabello, F. B. (1985): The design and interpretation of human skin irritation studies. *J. Toxicol. Cutan. Ocul. Toxicol.*, 4:61–71.
62. Carter, R. O. and Griffith, J. F. (1965): Experimental basis for the realistic assessment of safety of topical agents. *Toxicol. Appl. Pharmacol.*, 7:60–73.
63. CDRH. (1999): *Guidance for Industry and FDA Reviewers/Staff: Premarket Notification (510(K)) Submissions for Testing for Skin Sensitization to Chemicals in Natural Rubber Products*. Center for Devices and Radiological Health, U.S. Department of Health and Human Services, Washington, D.C.
64. Choman, B. R. (1963): Determination of the response of skin to chemical agents by an *in vitro* procedure. *J. Invest. Dermatol.*, 44:177–182.
65. Christie, O. A. and Moore-Robinson, M. (1970): Vehicle assessment: methodology and results. *Br. J. Dermatol.*, 82:93.
66. Code of Federal Regulations. (1997): Title 16, Parts 1500.40, 1500.41, 1500.42. Office of the Federal Registrar, National Archives of Records Service, General Services Administration, Washington, D.C.
67. Code of Federal Regulations. (1998): Title 40, Parts 162.10, 163.31, 771. Office of the Federal Registrar, National Archives of Records Service, General Services Administration, Washington, D.C.
68. Code of Federal Regulations. (1998): Title 49 Part 173, Appendix A. Office of the Federal Registrar, National Archives of Records Service, General Services Administration, Washington, D.C.
69. Code of Federal Regulations. (1998): Title 49, Part 173.240. Office of the Federal Registrar, National Archives of Records Service, General Services Administration, Washington, D.C.
70. Cohen, P. J. and Katz, S. I. (1992): Cultured human Langerhans cells process and present intact protein antigens. *J. Invest. Dermatol.*, 99:331–336.
71. Coomes, M. W., Norling, A. M., Pohl, R. J., Muller, D., and Fouts, J. R. (1983): Foreign compound metabolism by isolated skin cells from the hairless mouse. *J. Pharmacol. Exp. Ther.*, 225:770–777.
72. Cooper, K. D. (1991): Urticaria and angioedema: diagnosis and evaluation. *J. Am. Acad. Dermatol.*, 25:166–175.
73. Cornacoff, J. B., House, R. V., and Dean, J. H. (1988): Comparison of a radioisotopic incorporation method and the mouse ear swelling test (MEST) for contact sensitivity to weak sensitizers. *Fundam. Appl. Toxicol.*, 10:40–44.
74. Crowle, A. J. (1975): Delayed hypersensitivity in the mouse. *Adv. Immunol.*, 20:197–264.
75. Crowle, A. J. and Crowle, C. M. (1961): Contact hypersensitivity in mice. *J. Immunol.*, 32:302–320.
76. Cruzan, G., Dalbey, W. E., D'Aleo, C. J., and Singer, E J. (1986): A composite model for multiple assays of skir irritation. *Toxicol. Indust. Health*, 2:309–320.
77. Cunningham-Rundles, S. (1981): Cell-mediated immunity In: *Immunodermatology*, edited by B. Safai and R. A Good pp. 1–33. Plenum Press, New York.
78. Dahl, M. V. and Trancik, R. J. (1977): Sodium lauryl sul fate irritant patch test: degree of inflammation at variou times. *Contact Dermatitis*, 3:263–266.
79. Das, M., Bickers, D. R., and Mukhtar, H. (1986): Epidermis the major site of cutaneous benzo(*a*)pyrene 7,8-diol metabo lism in neonatal Balb/c mice. *Drug Metab. Disp.*, 14:637–642
80. Wester, R. C. and Maibach, H. I. (1999): Effect of singl versus multiple dosing in percutaneous absorption. In: *Per cutaneous Absorption: Drugs—Cosmetics—Mechanism—Methodology*, 3rd ed., edited by R. Bronaugh and H. Maibach, pp. 463–474. Marcel Dekker, New York.
81. David, J. R. and Remold, H. G. (1976): Macrophage acti vation by lymphocyte mediators and studies on the inte action of macrophage inhibitory factor (MIF) and its targe cell. In: *Immunology of the Macrophage*, edited by D. S Nelson, pp. 401–427. Academic Press, New York.
82. Davies, R. E., Harper, K. H., and Kymoch, S. R. (1972) Interspecies variation in dermal reactivity. *J. Soc. Cosme Chem.*, 23:371–381.

83. DeLeo, V., Harber, L. C., Kong, B. M., and DeSalva, S. J. (1987): Surfactant induced alteration of archadonic acid metabolism of mammalian cells in culture. *Proc. Soc. Exp. Biol. Med.*, 184:477–482.
84. Delescluse, J. and Turk, J. L. (1970): Lymphocyte cytotoxicity: a possible *in vitro* test for contact dermatitis. *Lancet*, 2:75–77.
85. Dover, R. and Wright, N. A. (1991): The cell proliferation kinetics of the epidermis. In: *Physiology, Biochemistry, and Molecular Biology of the Skin*, 2nd ed., edited by L. A. Goldsmith, pp. 1480–1501. Oxford University Press, New York.
86. Draize, J. H. (1955): Procedures for the appraisal of the toxicity of chemicals in foods, drugs, and cosmetics. VIII. Dermal toxicity. *Food Drug Cosmet. Law J.*, 10:722–731.
87. Draize, J. H. (1959): Dermal toxicity. In: *Association of Food and Drug Officials, U.S. Appraisal of the Safety of Chemicals in Food, Drugs, and Cosmetics*, pp. 46–59. Texas State Department of Health, Austin.
88. Draize, J. H., Woodard, G., and Calvery, H. O. (1944): Methods for the study of irritation and toxicity of substances applied topically to the skin and mucous membrane. *J. Pharmacol. Exp. Ther.*, 82:377–390.
89. Dugard, P. J. (1983): Skin permeability theory in relation to measurements of percutaneous absorption. In: *Dermatoxicology*, 2nd ed., edited by F. N. Marzulli and H. I. Maibach, pp. 91–115. Hemisphere, Washington, D.C.
90. Dumonde, D. C., Wolstencroft, R. A., Panayi, G. S. et al. (1969): Lymphokines: non-antibody mediators of cellular immunity generated by lymphocyte activation. *Nature (Lond.)*, 224:38–43.
91. Dunn, B. J., Rusch, G. M., Siglin, J. C., and Blaszcak, D. L. (1990): Variability of a mouse ear swelling test (MEST) in prediction of weak and moderate contact sensitizers. *Fundam. Appl. Toxicol.*, 15:242–248.
92. Dykes, P. J., Black, D. R., York, M., Dickens, A. D., and Marks, R. (1995): A stepwise procedure for evaluating irritant materials in normal volunteer subjects. *Human Exp. Toxicol.*, 14:204–211.
93. Edwards, C. C. (1972): Hazardous substances: proposed revision of test for primary skin irritants. *Fed. Reg.*, 37: 37635–27636.
94. Ekwal, B. (1963): Screening of toxic compounds in mammalian cell cultures. *Ann. N.Y. Acad. Sci.*, 407:64–77.
95. Elias, P. M. (1987): Lipids and the epidermal permeability barriers. *Arch. Derm. Res.*, 270:95–117.
96. Elias, P. M., Cooper, E. R., Korc, A., and Brown, B. E. (1981): Percutaneous transport in relation to stratum corneum structure and lipid composition. *J. Invest. Dermatol.*, 76:297–301.
97. Ellickson, B. E. and Jungermann, E. (1987): Comparative soap mildness test on infants. *Curr. Ther. Res.*, 9:441–446.
98. Elsner, P. and Maibach, H. I. (1995): Irritant dermatitis: new clinical and experimental aspects. In: *Current Problems in Dermatology*, Basel, Karger.
99. Emmett, E. A. (1975): Occupational skin cancer: a review. *J. Occup. Med.*, 17:44–49.
100. EPA. (1982): Pesticides registrations: proposed data requirements, Section 158.135, toxicology data requirements, *Fed. Reg.*, 47:53192.
101. Epstein, J. H. (1965). Comparison of the carcinogenic and cocarcinogenic effects of ultraviolet light on hairless mice. *J. Nat. Cancer Inst.*, 34:741–745.
102. Epstein, J. H. (1985). Animal models for studying photocarcinogenesis. In: *Models in Dermatology*, Vol. 2, edited by H. I. Maibach and N. Lowe, pp. 303–312. Karger, Basel.
103. Epstein, W. and Maibach, H. I. (1965): Cell renewal in human epidermis. *Arch. Dermatol.*, 92:462–468.
104. Epstein, W. L., Kligman, A. M., and Senecal, L. P. (1963): Role of regional lymph nodes in contact sensitization. *Arch. Dermatol.*, 88:789.
105. EEC. (1963): Sixth amendment to the council directive on the classification and labeling of dangerous substances, Annex VI. *Off. J. Eur. Commun.*, L257:13–33.
106. Everall, J. D. (1981): Chemical carcinogenesis. A. Environmental carcinogens. In: *Biology of Skin Cancer (Excluding Melanomas)*, edited by D. D. Laenum and O. H. Iverson, pp. 105–108. International Union Against Cancer, Geneva.
107. Everall, J. D. and Dowd, P. M. (1978): Influence of environmental factors excluding ultraviolet radiation on the incidence of skin cancer. *Bull. Cancer*, 65:241–248.
108. Fare, G. (1966): Rat skin carcinogenesis by topical applications of some azo dyes. *Cancer Res.*, 26:2466–2408.
109. Farrar, J. J., Banjamin, W. R., Hilficker, M. L., Howard, M., Farrar, W. V., and Fuller-Farrar, J. F. (1982): The biochemistry, biology, and role of interleukin in the induction of cytotoxic T-cell and antibody-forming B-cell responses. *Immunol. Rev.*, 63:129–166.
110. Feldman, R. J. and Maibach, H. I. (1967): Regional variation in percutaneous penetration of $C^{14}$ cortisone in man. *J. Invest. Dermatol.*, 48:151–183.
111. Finkelstein, P., Laden, K., and Meichowski, W. (1963): New methods for evaluating cosmetic irritancy. *J. Invest. Dermatol.*, 40:11–14.
112. Finkelstein, P., Laden, K., and Meichowski, W. (1965): Laboratory methods for evaluating skin irritancy. *Toxicol. Appl. Pharmacol.*, 7:74–48.
113. Flannigan, S. A. and Tucker, S. B. (1986): Variation in cutaneous sensation between synthetic pyrethroic insecticides. *Contact Dermatitis*, 13:140–147.
114. Fleishmajer, R., Contard, P., Schwartz, E. et al. (1991): Elastin-associated microfibrils in a three-dimensional fibroblast culture. *J. Invest. Dermatol.*, 97:638–643.
115. Forbes, P. D. (1997): Carcinogenesis and photocarcinogenesis test methods. In: *Dermatotoxicology*, 5th ed., edited by F. N. Marzulli and H. I. Maibach, pp. 535–544. Taylor & Francis, New York.
116. Forbes, P. D., Sambuco, C. P. Dearlove, G. E., Parker, R. M., Kiorpes, A. L., and Wedig, J. H. (1997): Sample protocols for carcinogenesis and photocarcinogenesis. In: *Dermatotoxicology Methods: The Laboratory Worker's Vade Mecum*, edited by F. N. Marzulli and H. I. Maibach, pp. 281–302. Taylor & Francis, New York.
117. Franz, T. J. (1975): Percutaneous absorption: on the relevance of *in vitro* data. *J. Invest. Dermatol.*, 64:190–195.
118. Frey, J. R. and Wenk, P. (1957): Experimental studies on the pathogenesis of contact eczema in the guinea pig. *Int. Arch. Allergy*, 11:81–100.

119. Fritsch, W. C. and Stoughton, R. B. (1963): The effect of temperature and humidity on the penetration of [$^{14}$C] acetylsalicylic acid in excised human skin. *J. Invest. Dermatol.*, 41:307.
120. Frosch, P. J. (1982): Irritancy of soap and detergent bars. In: *Principles of Cosmetics for the Dermatologist*, edited by P. Frost and S. N. Howitz, pp. 5–12. Mosby, St. Louis, MO.
121. Frosch, P. J. and Kligman, A. M. (1976): The chamber scarification test for irritancy. *Contact Dermatitis*, 2: 314–324.
122. Frosch, P. J. and Kligman, A. M. (1977): A method for appraising the stinging capacity of topically applied substances. *J. Soc. Cosmet. Chem.*, 25:197–207.
123. Frosch, P. J. and Kligman, A. M. (1977): The chamber scarification test for assessing irritancy of topically applied substances. In: *Cutaneous Toxicity*, edited by V. A. Drill and P. Lazar, pp. 127–144. Academic Press, New York.
124. Frosch, P. J. and Kligman, A. M. (1979): The Duhring chamber: an improved technique for epicutaneous testing of irritant and allergic reactions. *Contact Dermatitis*, 5:73.
125. Frosch, P. J. and Kligman, A. M. (1979): The soap chamber test: a new method for assessing the irritancy of soaps. *J. Am. Acad. Dermatol.*, 1:35–41.
126. Frosch, P. J. and Kligman, A. M. (1982): Recognition of chemically vulnerable and delicate skin. In: *Principles of Cosmetics for the Dermatologist*, edited by P. Frost and S. N. Howitz, pp. 287–296. Mosby, St. Louis, MO.
127. Gad, S. C. (1988): A scheme for the prediction and ranking of relative potencies of dermal sensitizers based on data from several systems. *J. Appl. Toxicol.*, 8:361–368.
128. Gad, S. C., Dunn, B. J., Dobbs, D. N. et al. (1986): Development and validation of an alternative dermal sensitization test: the mouse ear swelling test (MEST). *Toxicol. Appl. Pharmacol.*, 84:93–114.
129. Gad, S. C., Dunn, B. J., Gavigan, F. A., Reilly, C., and Walsh, R. D. (1987): Development, validation, and transfer of a new test system technology in toxicology. In: *New Test System in Toxicology*, edited by A. M. Goldberg, pp. 275–292. Mary Ann Liebert, New York.
130. Geczy, A. F. and Baumgarten, A. (1970): Lymphocyte transformation in contact sensitivity. *Immunology*, 19: 189–203.
131. George, M., and Vaughan, J. H. (1962): *In vitro* cell migration as a model for delayed hypersensitivity. *Proc. Soc. Exp. Biol. Med.*, 111:514–521.
132. Gibson, W. T. and Teall, M. R. (1983): Interactions of $C^{12}$ surfactants with the skin: changes in enzymes and visible and histological features of rat skin treated with sodium laurel sulfate. *Food Chem. Toxicol.*, 21:587–593.
133. Gilman, M. E., Evans, R. A., and DeSalva, S. J. (1978): The influence of concentration, exposure duration, and patch occlusivity upon rabbit primary dermal irritation indices. *Drug Chem. Toxicol.*, 1:391–400.
134. Gordon, V. C. (1990): An *in vitro* dermal safety test. *Drug Cosmet. Indust.*, pp. 32.
135. Gordon, V. C., Kelly, C. D., and Bergman, H. C. (1989): Skintex™: An *In Vitro* Method for Determining Dermal Irritation, presented at the Fifth International Congress of Toxicology.
136. Gordon, V. C., Kelly, C. D., and Bergman, H. C. (1990): Evaluation of Skintex™: an *in vitro* method for determining dermal irritation. *Toxicologist*, 10:75.
137. Grasso, P. (1971): Some aspects of the role of skin appendages in percutaneous absorption. *J. Soc. Cosmet. Chem.*, 22:523–534.
138. Griffith, J. F. (1969): Predictive and diagnostic test for contact sensitization. *Toxicol. Appl. Pharmacol.*, S3:90–102.
139. Griffith, J. F. and Buehler, E. (1976): Prediction of skin irritancy and sensitization potential by testing with animals and man. In: *Cutaneous Toxicity*, edited by V. Drill and P. Lazer, pp. 155–173. Academic Press, New York.
140. Griffith, J. F., Weaver, J. E., Whitehouse, H. S., Poole, R. L., Newman, E. A., and Nixon, C. A. (1969): Safety evaluation of enzyme detergents: oral and cutaneous toxicity, irritancy and skin sensitization studies. *Food Cosmet. Toxicol.*, 7:501–573.
141. Griffiths, H. A., Wilhelm, K. P., Robinson, M. K., Wang, S. M., McFadden, J., York, M., and Basketter, D. A. (1997): Interlaboratory evaluation of a human patch test for the identification of skin irritation potential/hazard. *Food Chem. Toxicol.*, 35:255–260.
142. Guillot, J. P., Gonnet, J. F., Clement, C., Caillard, L., and Truhaut, R. (1982): Evaluation of the cutaneous-irritation potential of 56 compounds. *Food Chem. Toxicol.*, 20: 563–572.
143. Guin, J. D., Meyer, B. N., Drake, R. D., and Haffley, P. (1984): The effect of quenching agents on contact urticaria caused by cinnamic aldehyde. *J. Am. Acad. Dermatol.*, 10: 45–51.
144. Guy, R. H., Tur, E., Bugatto, B., Gaebel, C., Sheiner, L. and Maibach, H. I. (1984): Pharmacodynamic measurements of methyl nicotinate percutaneous absorption *Pharm. Res.*, 1:76–51.
145. Guy, R. H., Wester, R. C., Tur, E., and Maibach, H. I (1983): Noninvasive assessments of the percutaneous absorption of methyl nicotinate in humans. *J. Pharm. Sci.* 72:1077–1079.
146. Hadgraft, J. (1979): The epidermal reservoir: a theoretica approach. *Int. J. Pharm.*, 2:265–274.
147. Haley, T. and Hunziger, J. (1974): Instrument for produc ing standardized skin abrasions. *J. Pharm. Sci.*, 63:106.
148. Hannuksela, M., Prilia, V., and Salo, O. P. (1975): Ski reactions to propylene glycol. *Contact Dermatitis*, 1:112
149. Harpin, V. A. and Rutter, N. (1983): Barrier properties o the newborn infant's skin. *J. Pediatr.*, 102:419–425.
150. Hassing, J. H., Nater, J. P., and Bleumink, E. (1982): Irri tancy of low concentrations of soap and synthetic deter gents as measured by skin water loss. *Dermatologica*, 164 312–314.
151. Hauser, C., Elbe, A., and Stingl, G. (1991): The Langer hans cell. In: *Physiology, Biochemistry, and Molecula Biology of the Skin*, 2nd ed., edited by L. A. Goldsmith pp. 1144–1164. Oxford University Press, New York.
152. Helman, R. G., Hall, J. W., and Kao, J. Y. (1986): Acu dermal toxicity: *in vivo* and *in vitro* comparisons in mic *Fundam. Appl. Toxicol.*, 7:94–100.
153. Henderson, C. R. and Riley, B. C. (1945): Certain statis tical considerations in patch testing. *J. Invest. Dermatol* 6:227–230.

154. Higuchi, T. (1960): Physical chemical analysis of percutaneous absorption process from creams and ointments. *J. Soc. Cosmet. Chem.*, 11:85–97.
155. Holbrook, K. A. (1991): Structure and function of the developing human skin. In: *Physiology, Biochemistry, and Molecular Biology of the Skin*, 2nd ed., edited by L. A. Goldsmith, pp. 64–111. Oxford University Press, New York.
156. Holbrook, K. A. and Odland, G. F. (1974): Regional differences in the thickness (cell layers) of the human stratum corneum: an ultrastructural analysis. *J. Invest. Dermatol.*, 62:415–422.
157. Holbrook, K. A. and Smith, L. T. (1981): Ultrastructural aspects of human skin during the embryonic, fetal, premature, neonatal, and adult periods of life. In: *Morphogenesis and Malforming of the Skin*, edited by R. J. Blandau, pp. 9–38. Alan R. Liss, New York.
158. Hood, D. B., Neher, R. J., Reinke, R. E., and Zapp. J. A. (1965): Experience with the guinea pig in screening primary irritants and sensitizers. *Toxicol. Appl. Pharmacol.*, 7:455–456.
159. Hueber, F., Wepierre, J., and Schaefer, H. (1992): Role of transepidermal and transfollicular routes in percutaneous absorption of hydrocortisone and testosterone: *in vivo* study in the hairless rat. *Skin Pharmacol.*, 5:99–107.
160. Humphrey, D. M. (1993). Measurement of cutaneous microvascular exudates using Evans blue. *Biotechnic. Histochem.*, 68:342–349.
161. Ihle, J. N., Rebar, K., Keller, J., Lee, J. C., and Hapel, A. J. (1982): Interleukin 3: possible roles in the regulation of lymphocyte differentiation and visual assessment. *Br. J. Dermatol.*, 92:131–142.
162. Imokowa, G., Sumura, K., and Katsumi, M. (1975): Study on skin roughness caused by surfactants. I. A new method *in vivo* for evaluation of skin roughness. *J. Am. Oil Chem. Soc.*, 52:479–483.
163. In Vitro International (1993): Application for Exemption (49 CFR 173.136 and 173.137) for the Corrositex™ Test to the U.S. Department of Transportation, Exemption E-11082m, April 28.
164. IARC. (1992): *Solar and Ultraviolet Radiation*, IARC Monographs on the Evaluation of Carcinogenic Risks to Humans, Vol. 55. International Agency for Research on Cancer, Lyon, France.
165. Iverson, O. H. (1981): Chemical carcinogenesis. Part e. Short term tests for carcinogens. In: *Biology of Skin Cancer (Excluding Melanomas)*, edited by D. D. Laerum and O. H. Iverson, pp. 151–163. International Union Against Cancer, Geneva, Switzerland.
166. Jackson, R. and Grainge, J. W. (1975): Arsenic and cancer. *Can. Med. Assoc. J.*, 113:396–401.
167. Jadassohn, J. (1896): Zur Kenntniss der medicamentosen Dermatosen. *Verhdlg. Deutsch. Derm. Gesellsch. 5.* Congress, pp. 103–129.
168. Jadassohn, J. (1896): A contribution to the study of dermatoses produced by drugs. In: *Selected Essays and Monographs* (transl. L. Elking, 1900), pp. 207–229. New Syndenham Society, London.
169. Jaffee, B. D. and Maguire, Jr., H. C. (1981): Delayed-type hypersensitivity and immunological tolerance to contact allergens in the rate. *Fed. Proc.*, 40:991 (abstract 4312).
170. Japan/MAFF. (1985): *Testing Guidelines for Evaluation of Safety of Agriculture Chemicals*. The Ministry of Agriculture, Forestry, and Fisheries, Tokyo, Japan.
171. Johnson, A. W. and Goodwin, B. F. J. (1985): The Draize test and modifications. In: *Contact Allergy Predictive Tests in Guinea Pigs*, edited by K. E. Andersen and H. I. Maibach, pp. 31–38. Karger, Basel.
172. Johnson, S. A. M., Kile, R. L., Kooyman, D. J., Whitehouse, H. S., and Brod, J. S. (1953): Comparison of effects of soaps and detergents on the hands of housewives. *Arch. Dermatol. Syph.*, 68:643–650.
173. Jordan, W. P. and King, S. E. (1977): Delayed hypersensitivity in females during the comparison of two predictive patch tests. *Contact Dermatitis*, 3:19–26.
174. Justice, J. D. et al. (1961): The correlation between animal tests and human tests in assessing product mildness. *Proc. Sci. Sect. Toilet Goods Assoc.*, 35:12–17.
175. Kanerva, L. (2000): *Handbook of Occupational Dermatology*. Springer-Verlag, Berlin.
176. Kaminsky, M. Szivos, M. M., and Brown, K. R. (1986): Application of the hill top patch test chamber to dermal irritancy testing in the albino rabbit. *J. Toxicol. Cutan. Ocul. Toxicol.*, 5:81–87.
177. Kao, J. and Carver, M. P. (1991): Skin metabolism. In: *Dermatotoxicity*, 4th ed., edited by F. N. Marzulli and H. I. Maibach, pp. 143–200. Hemisphere, New York.
178. Kao, J., Hall, T., and Holland, T. M. (1983): Quantitation of cutaneous toxicity: an *in vitro* approach using skin organ culture. *Toxicol. Appl. Pharmacol.*, 65:206–217.
179. Kastner, D. (1977): Irritancy potential of cosmetic ingredients. *J. Soc. Cosmet. Chem.*, 28:741–754.
180. Kero, M. and Hannuksela, M. (1980): Guinea pig maximization test, open epicutaneous test and chamber test in induction of delayed contact hypersensitivity. *Contact Dermatitis*, 6:341–344.
181. Kimber, I. and Basketter, D. A. (1992): The murine local lymph node assay: a commentary on collaborative studies and new directions. *Food Chem. Toxicol.*, 30:165–169.
182. Kimber, I., Hilton, J., and Botham, P. A. (1990): Identification of contact allergens using the murine local lymph node assay: comparisons with the Buehler occluded patch test in guinea pigs. *J. Appl. Toxicol.*, 10:173–180.
183. Kimber, I., Hilton, J., Botham, P. A. et al. (1991): The murine local lymph node assay: results of an inter-laboratory trial. *Toxicol. Lett.*, 55:203–213.
184. Kimber, I., Hilton, J., and Weisenberger, C. (1989): The murine local lymph node assay for identification of contact allergens: a preliminary evaluation in *in situ* measurement of lymphocyte proliferation. *Contact Dermatitis*, 21:215–220.
185. Kimber, I., Mitchell, J. A., and Griffin, A. C. (1986): Development of a murine local lymph node assay for the determination of sensitizing potential. *Food Chem. Toxicol.*, 24: 481–494.
186. Kimber, I. and Weisenberger, C. (1989): A murine local lymph node assay for the identification of contact allergens: assay development and results of an initial validation study. *Arch. Toxicol.*, 63:274–282.
187. Kimber, I., and Weisenberger, C. (1991): Anamnestic responses to contact allergens: application in the murine local lymph node assay. *J. Appl. Toxicol.*, 11:129–133.

188. Klecak, G. (1983): Identification of contact allergens: predictive tests in animals. In: *Dermatotoxicology*, edited by F. N. Marzulli and H. I. Maibach, pp. 193–236. Hemisphere, New York.
189. Klecak, G. (1985): The Freund's complete adjuvant test and the open epicutaneous test. In: *Contact Allergy Predictive Tests in Guinea Pigs*, edited by K. E. Andersen and H. I. Maibach, pp. 152–171. Karger, Basel.
190. Kligman, A. M. (1964): Quantitative testing of chemical irritants. In: *Evaluation of Therapeutic Agents and Cosmetics*, edited by M. Steinberg et al., pp. 186–192. McGraw-Hill, New York.
191. Kligman, A. M. (1966): The identification of contact allergens. *J. Invest. Dermatol.*, 47:369–374.
192. Kligman, A. M. (1966): The identification of contact allergens by human assay. II. Factors influencing the induction and measurement of allergic contact dermatitis. *J. Invest. Dermatol.*, 47:375–392.
193. Kligman, A. M. (1966): The identification of contact allergens by human assay. III: The maximization test: a procedure for screening and rating contact sensitizers. *J. Invest. Dermatol.*, 47:393–409.
194. Kligman, A. M. (1969): Evaluation of cosmetics for irritancy. *Toxicol. Appl. Pharmacol.*, 53:30–44.
195. Kligman, A. M. (1976): Perspectives and problems in cutaneous gerontology. *J. Invest. Dermatol.*, 73:39–46.
196. Kligman, A. M. (1983): A biological brief on percutaneous absorption. *Drug Dev. Indust. Pharmacol.*, 521–560.
197. Kligman, A. M. and Epstein, W. (1959): Some factors affecting contact sensitization in man. In: *Mechanism of Hypersensitivity*, edited by H. Shaffer et al., pp. 713–722. Little Brown, Boston.
198. Kligman, A. M. and Epstein, W. (1975): Updating the maximization test for identifying contact allergens. *Contact Dermatitis*, 1:231–239.
199. Kligman, A. M. and Wooding, W. M. (1967): A method for the measurement and evaluation of irritants on human skin. *J. Invest. Dermatol.*, 49:75–94.
200. Knox, J. M., Tucker, S. B., and Flannigan, S. A. (1984): Paraesthesia from cutaneous exposure to synthetic pyrethroid insecticide. *Arch. Dermatol.*, 120:744–746.
201. Knox, P., Uphill, P. F., Fry, J. R., Benford, J. et al. (1986): The FRAME multicentre project on *in vitro* cytotoxicology. *Food Chem. Toxicol.*, 24;457–463.
202. Kooyman. D.J. and Snyder, F. M. (1942): Tests for the mildness of soaps. *Arch. Dermatol. Syph.*, 46:846–855.
203. Kral, F. and Schwartzman, R. M. (1964): *Veterinary and Comparative Dermatology*, Lippincott, Philadelphia, PA.
204. Kuroki, T., Nemoto, N., and Kitano, Y. (1980): Use of human epidermal keratinocytes in studies on chemical carcinogenesis. In: *Carcinogenesis: Fundamental Mechanisms and Environmental Effects*, edited by B. Pullman, P. O. P. Tso, and H. Gelboin, pp. 417–426. Reidel, Boston, MA.
205. Lahti, A. (1980): Non-immunologic contact urticaria. *Acta. Derm. Venereol. (Stockholm)*, 605:1–49.
206. Lahti, A. and Maibach, H. I. (1984): An animal model for non-immunologic contact urticaria. *Toxicol. Appl. Pharmacol.*, 76:219–224.
207. Lahti, A. and Maibach, H. I. (1985): Species specificity of non-immunologic urticaria: guinea pig, rat and mouse. *J. Am. Acad. Dermatol.*, 13:66–69.
208. Lahti, A. and Maibach, H. I. (1991): Immediate contact reactions: contact urticaria and the contact urticaria syndrome. In: *Dermatotoxicity*, 4th ed., edited by F. N. Marzulli and H. I. Maibach, pp. 473–495. Hemisphere Publishing, New York.
209. Lahti, A., von Krogh, G., and Maibach, H. I. (1985): Contact urticaria syndrome: an expanding phenomenon. In: *Dermatologic Immunology and Allergy*, edited by J. Stone, pp. 379–390. Mosby, St. Louis, MO.
210. Lammintausta, K. and Maibach, H. I. (1988): Exogenous and endogenous factors in skin irritation. *Int. J. Dermatol.*, 27:213–222.
211. Lammintausta, K., Maibach, H. I., and Wilson, D. (1988): Mechanisms of subjective (sensory) irritation: propensity of non-immunologic contact urticaria and objective irritation in stingers. *Dermatosen Beruf Umwelt*, 36:45–49.
212. Landsteiner, K. and Chase, M. W. (1937): Studies on the sensitization of animals with simple chemical compounds. IV. Anaphylaxis induced by picryl chloride and 2:4 dinitrochloro-benzene. *J. Exp. Med.*, 66:337–351.
213. Landsteiner, K. and Jacobs, J. (1935): Studies on the sensitization of animals with simple chemical compounds. *J. Exp. Med.*, 61:643–648.
214. Landsteiner, K. and Jacobs, J. (1936): Studies on the sensitization of animals with simple chemical compounds. II. *J. Exp. Med.*, 64:625–629.
215. Lanman, B. M., Elvers, W. B., and Howard, C. S. (1968): The role of human patch testing in a product development program. In: *Proceedings of the Joint Conference on Cosmetic Sciences*, pp. 135–145. The Toilet Goods Association, Washington, D.C.
216. Lever, W. F. and Schaumburg-Hevor, C. (1983): *Histopathology of the Skin*, 6th ed. Lippincott, Philadelphia, PA.
217. Leyden, J. J., Nordstrom, K. M., and McGinley, K. J. (1991): Cutaneous microbiology. In: *Physiology, Biochemistry, and Molecular Biology of the Skin*, 2nd ed., edited by L. A. Goldsmith, pp. 1403–1423. Oxford University Press, New York.
218. MacLeod, T. M., Hutchinson, F., and Raffle, E. F. (1970): The uptake of labeled thymidine by leucocytes of nickel sensitive patients. *Br. J. Dermatol.*, 82:487–492.
219. MacMillan, F. S. K., Rafft, R. R., and Elvers, W. B. (1975): A comparison of the skin irritation produced by cosmetic ingredients and formulations in the rabbit, guinea pig, beagle dog to that observed in the human. In: *Animal Models in Dermatology*, edited by H. I. Maibach, pp. 12–22 Churchill Livingstone, Edinburgh.
220. Magee, P. N. (1970): Tests for carcinogenic potential. In *Methods in Toxicology*, edited by G. E. Paget, pp. 158–196 Davis, Philadelphia, PA.
221. Magnusson, B. and Hersle, K. (1965): Patch test methods I. A comparative study of six different types of patch tests *Acta. Dermatol.*, 45:123–128.
222. Magnusson, B. and Hersle, K. (1965): Patch test methods II. Regional variations of patch test responses. *Acta. Dermatol.*, 45:257–261.

223. Magnusson, B. and Hersle, K. (1966): Patch test methods. III. Influence of adhesive tape on test response. *Acta. Dermatol.*, 46:275–278.
224. Magnusson, B. and Kligman, A. M. (1969): The identification of contact allergens by animals assay: the guinea pig maximization test. *J. Invest. Dermatol.*, 52:268–276.
225. Magnusson, B. and Kligman, A. M. (1970): *Allergic Contact Dermatitis in the Guinea Pig*. Charles C Thomas, Springfield, IL.
226. Maguire, H. C. (1973): Mechanism of intensification by Freund's complete adjuvant of the acquisition of delayed hypersensitivity in the guinea pig. *Immunol. Commun.*, 1:239–246.
227. Maguire, H. C. (1974): Alteration in the acquisition of delayed hypersensitivity with adjuvant in the guinea pig. *Monogr. Allergy*, 8:13–26.
228. Maibach, H. I. (1976): Immediate hypersensitivity in hand dermatitis: role of food contact dermatitis. *Arch. Dermatol.*, 112:1289–1291.
229. Lee, E. E. and Maibach, H. I. (2002): The role of vehicles in diagnosing contact allergy: an update. *Exogenous Dermatol.*, 1:107–111.
230. Maibach, H. I. and Boisits, E. (1982): *Neonatal Skin*. Marcel Dekker, New York.
231. Maibach, H. I. and Johnson, H. L. (1975): Contact urticaria syndrome: contact urticaria to diethyltoluamide (immediate-type hypersensitivity). *Arch. Dermatol.*, 111:726–720.
232. Maibach, H. I., Lamminatusta, K., Berardesca, E., and Freeman S. (1989): Tendency to irritation: sensitive skin. *J. Am. Acad. Dermatol.*, 21:833–835.
233. Maisey, J. and Miller, K. (1986): Assessment of the ability of mice fed on vitamin A supplemented diet to respond to a variety of potential contact sensitizers. *Contact Dermatitis*, 15:17–23.
234. Malkinson, F. D. (1958): Studies on the percutaneous absorption of $C^{14}$-labeled steroids by use of the gas-flow cell. *J. Invest. Dermatol.*, 31:19.
235. Malkovsky, M. et al. (1983): Enhancement of specific antitumor immunity in mice fed a diet enriched in vitamin A acetate. *Proc. Nat. Acad. Sci. U.S.A.*, 80:6322–6326.
236. Malkovsky, M. et al. (1983): T-cell mediated enhancement of host-versus-graft reactivity in mice fed a diet enriched in vitamin A acetate. *Nature*, 302:338–340.
237. Malten, K. E. and Thiele, F. A. J. (1973): Evaluation of skin damage. II. Water loss and carbon dioxide release measurements related to skin resistance measurements. *Br. J. Dermatol.*, 89:565–569.
238. Malten, K. E. and Thiele, F. A. J. (1973): Some theoretical aspects of orthoergic (= irritant) dermatitis. *Arch. Belges Dermatol.*, 28:9–22.
239. Martin, R., Denyer, S., and Hadgraft, J. (1987): Skin metabolism of topically applied compounds. *Int. J. Pharm.*, 39:23–32.
240. Marzulli, F. N. and Maibach, H. I. (1973): Antimicrobials: experimental contact sensitization in man. *J. Soc. Cosmet. Chem.*, 24:399–421.
241. Marzulli, F. N. and Maibach, H. I. (1974): The use of graded concentration in studying skin sensitizers: experimental contact sensitization in man. *Food Cosmet. Toxicol.*, 12:219–227.
242. Mathias, C. G. T., Chappler, R. R., and Maibach, H. I. (1980) Contact urticaria from cinnamic aldehyde. *Arch. Dermatol.*, 116:74–76.
243. Mathias, C. G. T. and Maibach, H. I. (1978): Dermatoxicology monographs. I. Cutaneous irritation: factors influencing the response to irritants. *Clin. Toxicol.*, 13:333–346.
244. Maurer, T., Thomann, P., Weirich, E. G., and Hess, R. (1975): The optimization test in the guinea pig: a method for the predictive evaluation of the contact allergenicity of chemicals. *Agents Actions*, 5:174–179.
245. McKenzie, A. W. and Stoughton, R. M. (1962): Method for comparing the percutaneous absorption of steroids. *Arch. Dermatol.*, 86:608–610.
246. McKillop, C. M., Brock, J. A. C., Oliver, C. J. A., and Rhodes, C. (1987): A quantitative assessment of pyrethroid-induced paraesthesia in the guinea pig flank model. *Toxicol. Lett.*, 36:1–7.
247. Mezei, M. (1970): Dermatitic effect of nonionic surfactants. V. The effect of nonionic surfactants on rabbit skin as evaluated by radioactive tracer techniques *in vivo*. *J. Invest. Dermatol.*, 54:510–516.
248. Mezei, M., Sager, R. W., Stewart, W. D., and DeRuyter, A. L. (1966): Dermatitic effect on nonionic surfactants. I. Gross, microscopic, and metabolic changes in rabbit skin treated with nonionic surface-active agents. *J. Pharm. Sci.*, 55:584–590.
249. Miller, A. E. and Levis, W. R. (1973): Studies on the contact sensitization of man with simple chemicals. I. Specific lymphocyte transformation in response to dinitrochlorobenzene sensitization. *J. Invest. Dermatol.*, 61: 261–269.
250. Miller, K., Maisey, J., and Malkovsky, M. (1984): Enhancement of contact sensitization in mice fed a diet enriched in vitamin A acetate. *Int. Arch. Allergy Appl. Immunol.*, 75:120–125.
251. Milner, J. E. (1970): *In vitro* lymphocyte response in contact hypersensitivity. *J. Invest. Dermatol.*, 55:34–38.
252. Milner, J. E. (1971): *In vitro* lymphocyte response in contact hypersensitivity, II. *J. Invest. Dermatol.*, 56:349–352.
253. Mizel, S. B. (1982): Interleukin 1 and T cell activation. *Immunol. Rev.*, 63:51–72.
254. Mol, M. A. E., Van de Ruit, A. B. C., and Kluivrs, A. W. (1989): NAD+ levels and glucose uptake of cultured human epidermal cells exposed to sulfur mustard. *Toxicol. Appl. Pharmacol.*, 98:159–165.
255. Moloney, S. J., and Teal, J. J. (1988): Alkane-induced edema formation and cutaneous barrier dysfunction. *Arch. Dermatol. Res.*, 280:375–379.
256. Montagna, W. (1962): *The Structure and Function of Skin*. Academic Press, New York.
257. Montagna, W. and Lobitz, W. C. (1964): *The Epidermis*. Academic Press, New York.
258. Moorehead, J. W., Murphy, J. W., Harvey, R. P. et al. (1962): Soluble factors in tolerance and contact sensitivity to 2,4-dinitrofulurobenzene in mice, IV. *Eur. J. Immunol.*, 12:431–436.
259. Mosman, T. (1983): Rapid colorimetric assay for cellular growth and survival: application to proliferation and cytotoxic assays. *J. Immunol. Methods*, 65:55–63.

260. Motoyoshi, K., Toyoshima, Y., Sato, M., and Yoshimura, M. (1979): Comparative studies on the irritancy of oils and synthetic perfumes to the skin of rabbit, guinea pig, rat, miniature swine and man. *Cosmet. Toiletries*, 94:41–42.
261. Najarian, J. S. and Feldman, J. D. (1963): Specificity of passive transfer or delayed hypersensitivity. *J. Exp. Med.*, 118:341–352.
262. National Academy of Sciences, Committee for the Revision of NAS Publication 1138. (1977): *Principles and Procedures for Evaluating the Toxicity of Household Substances*, pp. 23–59. National Academy of Sciences, Washington, D.C.
263. National Toxicology Program. (1999): *The Murine Local Lymph Node Assay: A Test Method for Assessing the Allergic Contact Dermatitis Potential of Chemicals/Compounds*, NIH Publ. No. 99–4494. National Institutes of Health, Research Triangle Park, NC.
264. Nicolaides, N. (1963): Human skin surface lipids: origins, composition and possible function. In: *Advances in Biology of Skin, Vol. IV, The Sebaceous Glands*, edited by W. Montagna, R. A. Ellis, and A. F. Silver, pp. 167–187. Pergamon Press, Oxford.
265. Nixon, G. A., Tyson, C. A., and Wertz, W. C. (1975): Interspecies comparisons of skin irritancy. *Toxicol. Appl. Pharmacol.*, 31:481–490.
266. Noonan, P. K., and Wester, R. C. (1983): Cutaneous biotransformations and some pharmacological and toxicological implications. In: *Dermatotoxicology*, 2nd ed., edited by F. N. Marzulli and H. I. Maibach, pp. 71–90. Hemisphere, New York.
267. Odland, G. F. (1991): Structure of the skin. In: *Physiology, Biochemistry, and Molecular Biology of the Skin*, 2nd ed., edited by L. A. Goldsmith, pp. 3–62. Oxford University Press, New York.
268. Odom, R. B., and Maibach, H. I. (1976): Contact urticaria: A different contact dermatitis. *Cutis*, 18:672–676.
269. Oliver, G. J. A., Botham, P. A., and Kimber, I. (1986): Models for contact sensitization: Novel approaches and future developments. *Br. J. Dermatol.*, 115:53–62.
270. Oliver, G. J. A., Pemberton, M. A., and Rhodes, C. (1986): An *in vitro* skin corrosivity test-modification and validation. *Food Chem. Toxicol.*, 24:507–512.
271. Oliver, G. J. A., Pemberton, M. A., and Rhodes, C. (1986): The identification of corrosive agents for human skin *in vitro*. *Food Chem. Toxicol.*, 24:513–515.
272. Oliver, G. J. A., Pemberton, M. A., and Rhodes, C. (1988): An *in vitro* model for identifying skin-corrosive chemicals. I. Initial validation. *Toxicol. In Vitro*, 2:7–17.
273. Opdyke, D. (1971): The guinea pig immersion test: a 20-year appraisal. *CTFA Cosmetic J.*, 3:46–47.
274. Opdyke, D. L. and Burnett, C. M. (1965): Practical problems in the evaluation of the safety of cosmetics. *Proc. Sci. Sect. Toilet Goods Assoc.*, 44:3–4.
275. OECD (1993): *Guidelines for Testing of Chemicals: Acute Dermal Irritation/Corrosion* (404). Organization for Economic Cooperation and Development, Paris, France.
276. OECD (1981): *Guidelines for Testing of Chemicals: Carcinogenicity Studies* (451) and *Combined Chronic Toxicity/Carcinogenicity Studies* (453). Organization for Economic Cooperation and Development, Paris, France.
277. Ostrenga, J., Steinmetz, C., Poulsen, B., and Yett, S. (1971): Significance of vehicle composition. II. Prediction of optimal vehicle composition. *J. Pharm. Sci.*, 60:1180–1183.
278. Page, A. R. and Good, R. A. (1958): A clinical and experimental study of the function of neutrophils in the inflammatory response. *Am. J. Pathol.*, 34:645–656.
279. Page, N. P. (1977): Concepts of a bioassay program in the environmental carcinogenesis. In: *Environmental Cancer*, Vol. 3. *Advances in Modern Toxicology*, edited by H. F. Kraybill and M. A. Mehlman, pp. 87–171. Hemisphere, Washington, D.C.
280. Palotay, J. L., Adachi, K., Dobson, R. L., and Pinto, J. S. (1986): Carcinogen-induced cutaneous neoplasms in non-human primates. *J. Nat. Cancer Inst.*, 97:1269–1272.
281. Pappas, A., Orfanos, C. E., and Bertram, R. (1970): Non-specific lymphocyte transformation *in vitro* by nickel acetate. *J. Invest. Dermatol.*, 55:198–200.
282. Parce, J. W., Owicki, J. C., Kercso, K. M., Sigal, G. B. et al. (1989): Detection of cell-affecting agents with a silicon biosensor. *Science*, 246:243–247.
283. Patrick, E., Burkhalter, A., and Maibach, H. I. (1987): Recent investigations of mechanisms of chemically induced skin irritation in laboratory mice. *J. Invest. Dermatol.*, 88:24s–31s.
284. Patrick, E. and Maibach, H. I. (1987): A novel predictive irritation assay in mice. *Toxicologist*, 7:84.
285. Patrick, E. and Maibach, H. I. (1989): Comparison of the time course, dose response, and mediators of chemically induced skin irritation in three species. In: *Current Topics in Contact Dermatitis*, edited by P. Frosch, A. Dooms-Goossens, R. Lachapelle, R. Rycroft, and R. Scheper, pp. 399–404. Springer-Verlag, Berlin.
286. Patrick, E. and Maibach, H. I. (1991): Predictive skin irritation tests in animals and humans. In: *Dermatotoxicity*, 4th ed., edited by F. N. Marzulli and H. I. Maibach, pp. 201–222. Hemisphere, New York.
287. Patrick, E., Maibach, H. I., and Burkhalter, A. (1985): Mechanisms of chemically induced skin irritation. I. Studies of time course, dose response, and components of inflammation in the laboratory mouse. *Toxicol. Appl. Pharmacol.*, 81:476–490.
288. Pauly, J. L., Caron, G. A., and Suskind, R. R. (1969): Blast transformation of lymphocytes from guinea pigs, rats, and rabbits induced by mercuric chloride *in vitro*. *J. Cell Biol.*, 40:847–850.
289. Pearmain, G. E., Lycette, R. R., and Fitzgerald, P. H (1963): Tuberculin induced mitoses in peripheral blood lymphocytes. *Lancet*, 1:637–638.
290. Perkins, M. A., Osborne, R., and Johnson, G. R. (1996) Development of an *in vitro* method for skin corrosion testing. *Fundam. Appl. Toxicol.*, 31:9–18.
291. Phillips, L., Steinberg, M., Maibach, H. I., and Akers, W A. (1972): A comparison of rabbit and human skin response to certain irritants. *Toxicol. Appl. Pharmacol.*, 21 369–382.
292. Pinnagoda, J., Tupker, R. A., Agner, T., and Serup, J (1990): Guidelines for transepidermal water loss (TEWL measurements: a report from the standardization group of the European Society of Contact Dermatitis. *Contact Dermatitis*, 22:164–178.

293. 293.Pohl, R. J., Philpot, R. M., and Fouts, J. R. (1976): Cytochrome P450 content and mixed function oxidase activity in microsomes isolated from mouse skin. *Drug Metab. Dis.*, 4:442–450.
294. Polak, L. (1977): Immunological aspects of contact sensitivity. In: *Dermatotoxicology and Pharmacology*, edited by F. N. Marzulli and H. I. Maibach, pp. 225–288. Hemisphere, Washington, D.C.
295. Polak, L., Polak, A., and Frey, J. R. (1974): The development of contact sensitivity to DNFB in guinea pigs genetically differing in their response to DNP-skin protein conjugate. *Int. Arch. Allergy*, 46:417–426.
296. Potokar, M. (1985): Studies on the design of animal tests for the corrosiveness of industrial chemicals. *Food Chem. Toxicol.*, 23:615–617.
297. Prottey, C. (1978): The molecular basis of skin irritation. In: *Cosmetics*, Vol. 1, edited by M. M. Breuer, pp. 275–349. Academic Press, London.
298. Rapaport, M., Anderson, D., and Pierce, U. (1978): Performance of the 21-day patch test in civilian populations. *J. Toxicol. Cutan. Ocul. Toxicol.*, 1:109–115.
299. Rietschell, R. L. (1982): Advances and pitfalls in irritant and allergen testing. *J. Soc. Cosmet. Chem.*, 33:309–313.
300. Ritz, H. L. and Buehler, E. V. (1980): Planning conduct and interpretation of guinea pig sensitization patch tests. In: *Current Concepts in Cutaneous Toxicity*, edited by V. A. Drill and P. Lazar, p. 25. Academic Press, New York.
301. Robertsaw, D. (1991): Apocrine sweat glands. In: *Physiology, Biochemistry, and Molecular Biology of the Skin*, 2nd ed., edited by L. A. Goldsmith, pp. 763–775. Oxford University Press, New York.
302. Rockl, H., Muller, E., and Haltermann, W. (1966): Zum aussagewert positiver epicutantest bei sauglingen and kindern. *Arch. Klin. Exp. Dermatol.*, 226:407.
303. Rocklin, R. E., MacDermott, R. P., Chess, L. et al. (1974): Studies on mediator production by highly purified human T and B lymphocytes. *J. Exp. Med.*, 140:1303–1316.
304. Roskos, K. and Maibach, H. (1992): Percutaneous absorption and age: implications for therapy. *Drugs Aging*, 2: 432–449.
305. Rostenberg, A. (1961): Methods for the appraisal of the safety of cosmetics. *Drug Cosmet. Indust.*, 88:592.
306. Rothenborg, H. W., Menne, T., and Sjolin, K. E. (1977): Temperature dependent primary irritant dermatitis from lemon perfume. *Contact Dermatitis*, 3:37.
307. Rothman, S. (1954): *Physiology and Biochemistry of the Skin*. The University of Chicago Press, Chicago, IL.
308. Rouigier, A., Dupuis, D., Lotte, C., Roguet, R., and Schafer, H. (1983): *In vivo* correlation between stratum corneum reservoir function and percutaneous absorption. *J. Invest. Dermatol.*, 81:275–278.
309. Sato, K., Kang, W. H., and Sato, F. (1991): Eccrine sweat glands. In: *Physiology, Biochemistry, and Molecular Biology of the Skin*, 2nd ed., edited by L. A. Goldsmith, pp. 741–762. Oxford University Press, New York.
310. Sauder, D. N. (1991): Interleukens. In: *Physiology, Biochemistry, and Molecular Biology of the Skin*, 2nd ed., edited by L. A. Goldsmith, pp. 1188–1198. Oxford University Press, New York.
311. Scheuplein, R. J. (1967): Mechanism of percutaneous absorption. II. Transient diffusion and the relative importance of various routes of skin penetration. *J. Invest. Dermatol.*, 48:79–88.
312. Scheuplein, R. J. (1978): Permeability of skin: a review of major concepts. *Curr. Probl. Dermatol.*, 7:58–68.
313. Scheuplein, R. and Bronough, R. L. (1983): Percutaneous absorption. In: *Biochemistry and Physiology of the Skin*, edited by L. A. Goldsmith, pp. 1255–1295. Oxford University Press, New York.
314. Scholes, E. W., Basketter, D. A., Saril, A. E. et al. (1992): The local lymph node assay: results of a final inter-laboratory validation under field conditions. *J. Appl. Toxicol.*, 12:217–222.
315. Schopf, E., Schulz, K. H., and Isensee, I. (1969): Untersuchungen uber den lymphocyten transformations test be: Quecksilberallerge. *Arch. Klin. Exp. Dermatol.*, 234:420.
316. Schwartz, L. (1951): The skin testing of new cosmetics. *J. Soc. Cosmet. Chem.*, 2:321–324.
317. Schwartz, L. (1969): Twenty-two years' experience in the performance of 200,000 prophetic patch tests. *South. Med. J.*, 53:478–484.
318. Schwartz, L. and Peck, S. M. (1944): The patch test in contact dermatitis. *Public Health Rep.*, 59:546–557.
319. Schwartz, L. B. (1991): Mast cells and their role in urticaria. *J. Am. Acad. Dermatol.*, 25:190–204.
320. Shelanski, H. A. (1951): Experience with and considerations of the human patch test method. *J. Soc. Cosmet. Chem.*, 2:324–331.
321. Shelanski, H. A. and Shelanski, M. V. (1953): A new technique of human patch tests. *Proc. Sci. Sect. Toilet Goods Assoc.*, 19:46–49.
322. Shelanski, H. A. and Shelanski, M. V. (1953): New technique of patch tests. *Drug Cosmet. Ind.*, 73:186.
323. Shellow, W. V. R. and Rapaport, M. J. (1981): Comparison testing of soap irritancy using aluminum chamber and standard patch methods. *Contact Dermatitis*, 7:77–79.
324. Simpson, W. L. and Cramer, W. (1943): Fluorescence studies of carcinogens in skin. *Cancer Res.*, 3:362–369.
325. Skog, E. (1960): Primary irritant and allergic eczematous reactions in patients with different dermatoses. *Acta. Derm. Venerol.*, 40:307–312.
326. Slaga, T. J. et al. (1989): *Skin Carcinogenesis: Mechanisms and Human Relevance*. Alan R. Liss, New York.
327. Smiles, K. A. and Pollack, M. E. (1977): A quantitative human patch testing procedure for low level skin irritants. *J. Soc. Cosmet. Chem.*, 26:755–764.
328. Sokolov, U. E. (1962): *Mammal Skin*. University of California Press, Berkeley.
329. SOT Position Paper. (1989): Comments on the $LD_{50}$ and acute eye and skin irritation tests. *Fundam. Appl. Toxicol.*, 13:621–623.
330. Soter, N. A. (1991): Acute and chronic urticaria and angioedema. *J. Am. Acad. Dermatol.*, 25:146–154.
331. Steinberg, M., Akers, W. A., Weeks, M., McCreesh, A. H., and Maibach, H. I. (1975): A comparison of test techniques based on rabbit and human skin responses to irritants with recommendations regarding the evaluation of mildly or moderately irritating compounds. In: *Animal Models in Dermatology*, edited by H. I. Maibach, pp. 1–11. Churchill Livingstone, Edinburgh.

332. Stephens, T. J., Silber, P. M., Recce, B. et al. (1990): Testskin™ : an *in vitro* model for detecting cytotoxicity and inflammation. *Toxicologist*, 10:78.

333. Stingl, G. and Abever, W. (1983): The Langerhans cell. In: *Biochemistry and Physiology of the Skin*, edited by L. A. Goldsmith, pp. 907–921. Oxford University Press, New York.

334. Stingl, G., Katz, S. I., Clement, L., Green, I., and Shevach, E. (1978): Immunologic functions of Ia-bearing epidermal Langerhans cells. *J. Immunol.*, 121:2005–2013.

335. Stotts, J. (1980): Planning, conduct, and interpretation: human predictive sensitization patch tests. In: *Current Concepts in Cutaneous Toxicity*, edited by V. A. Drill and P. Lazar, pp. 41–53. Academic Press, New York.

336. Strauss, J. S., Downing, D. T., Ebling, F. J., and Steward, M. E. (1991): Sebaceous glands. In: *Physiology, Biochemistry, and Molecular Biology of the Skin*, 2nd ed., edited by L. A. Goldsmith, pp. 712–740. Oxford University Press, New York.

337. Strube, D. D., Koontz, S. W., Murahata, R. I., and Theiler, R. F. (1989): The flex wash test: a method for evaluating the mildness of personal washing products. *J. Soc. Cosmet. Chem.*, 40:297–306.

338. Sweeney, T. M. and Downing, D. T. (1970): The role of lipids in the epidermal barrier to water diffusion. *J. Invest. Dermatol.*, 55:135–140.

339. Swisher, D. A., Johnson, J., and Ledger, P. W. (1987): A method for screening *in vitro* cytotoxicity of agents toward human keratinocytes. *J. Invest. Dermatol.*, 88:520.

340. Tagami, H. (1971): Functional characteristics of aged skin. *Acta Dermatol. Venereol. (Stockholm)*, 66:19–21.

341. Tagami, H., Masatoshi, O., and Iwatsuki, K. (1986): Evaluation of the skin surface hydration *in vivo* by electrical measurements. *J. Invest. Dermatol.*, 75:500–597.

342. Tharp, M. D. (1991): The mast cell and its mediators. In: *Physiology, Biochemistry, and Molecular Biology of the Skin*, 2nd ed., edited by L. A. Goldsmith, pp. 1019–1083. Oxford University Press, New York.

343. Thiele, F. A. J. and Malten, K. E. (1973): Evaluation of skin damage. I. Skin resistance measurements with alternating current (impedance measurements). *Br. J. Dermatol.*, 89:373–382.

344. Thiele, F. A. J. and Malten, K. E. (1973): Some measuring methods for the evaluation of orthoergic contact dermatitis. Arch. Belges Dermatol., 28:23–46.

345. Traub, E. F., Tusing, T. W., and Spoor, H. J. (1954): Evaluation of dermal sensitivity: animal and human tests compared. *Arch. Derm. (Chicago)*, 69:399–409.

346. Tregear, R. T. (1964): Relative penetrability of hair follicles and epidermis. *J. Physiol. (Lond.)*, 156:303–313.

347. Tregear, R. T. (1966): *Physical Function of Skin*. Academic Press, New York.

348. Triglia, D., Braa, S. S., Donnelly, T., Kidd, I., and Naughton, G. K. (1991): A three-dimensional human dermal model substrate for *in vitro* toxicological studies. In: *In Vitro Toxicology: Mechanisms and New Technology*, edited by A. M. Goldberg, pp. 351–362. Mary Ann Lieber, New York.

349. Triglia, D., Braa, S. S., Yonan, C., and Naughton, G. K. (1991): *In vitro* toxicity of various classes of test agents using the neutral red assay on a human three-dimensional physiologic skin mode. *In Vitro Cell Dev. Biol.*, 27A:239–244.

350. Unanue, E. R. (1984): Antigen-presenting function of the macrophage. *Annu. Rev. Immunol.*, 2:395–428.

351. Upadhye, M. and Maibach, H. (1992): Influence of area of application of allergens in contact dermatitis. *Contact Dermatitis*, 27:186.

352. Urbach, F., Davies, R. E., and Forbes, P. D. (1988): Chemical modifiers of photocarcinogenesis. *Arch. Toxicol.*, 12(Suppl.):47–51.

353. Uttley, M. and Van Abbe, N. J. (1973): Primary irritation of the skin: mouse ear test and human patch procedures. *J. Soc. Cosmet. Chem.*, 24:217–227.

354. Van Loveren, H., Kato, K., Ratzlaff, R. E., Meade, R., Ptak, W., and Askenase, P. W. (1984): Use of micrometers and calipers to measure various components of delayed-type hypersensitivity ear swelling reactions in mice. *J. Immunol. Methods*, 67:311–319.

355. Van der Valk, P. G. M. and Maibach, H. I. (1996): *The Irritant Contact Dermatitis Syndrome*. CRC Press, Boca Raton, FL.

356. Van der Valk, P. O. M., Nater, J. P. K., and Bleumink, E. (1985): Vulnerability of the skin to surfactants in different groups of eczema patients and controls as measured by water vapor loss. *Clin. Exp. Dermatol.*, 101:98.

357. Viluksela, M. (1991): Characteristics and modulation of dithranol (anthralin)-induced skin irritation in the mouse ear model. *Arch. Dermatol. Res.*, 283:262–268.

358. Vinegar, M. B. (1979): Regional variation in primary skin irritation and corrosivity potentials in rabbits. *Toxicol. Appl. Pharmacol.*, 49:63–69.

359. Vivjeberg, H. P. and Vandenbercken, J. (1979): Frequency dependent effects of the pyrethroid insecticide decamethrin in frog myelinated nerve fibers. *Eur. J. Pharmacol.*, 58:501–504.

360. Vizethum, W., Ruzicka, R., and Goetz, G. (1980): Inductibility of drug-metabolizing enzymes in the rat skin. *Chem Biol. Interact.*, 31:215–219.

361. Von Krogh, G. and Maibach, H. I. (1982): The contac urticaria syndrome. *Semin. Dermatol.*, 1:59–66.

362. Voss, J. G. (1958): Skin sensitization by mercaptans of low molecular weight. *J. Invest. Dermatol.*, 31:273–279.

363. Wagner, G. and Purschel, W. (1962): Klinisch-analytische stu die die zum neuroderm-itisproblem. *Dermatologica*, 125:1.

364. Wahlberg, J. E. and Boman, A. (1985): Guinea pig maxi mization test. In: *Contact Allergy Predictive Tests in Guinea Pigs*, edited by K. E. Andersen and H. I. Maibach pp. 9–106. Karger, Basel.

365. Menne, T. and Wahlberg, J. E. (2002): Risk assessmen failures of chemicals commonly used in consumer prod ucts. *Contact Dermatitis*, 46(4):189–190.

366. Walker, A. P., Basketter, D. A., Baveral, M., Diembeck W., Matthies, W., Mougin, D., Paye, M., Rothlisberg, R and Dupuis, J. (1996): Test guidelines for assessment o skin compatibility of cosmetic finished products in mar *Food Chem. Toxicol.*, 34:651–660.

367. Walz, D. (1985): Quantitative assessment of irritation i the mouse skin test. *Food Chem. Toxicol.*, 23:199–203.

368. Wassrman, S. J. (1983): The mast cell and its mediator In: *Physiology, Biochemistry, and Molecular Biology o the Skin*, 2nd ed., edited by L. A. Goldsmith, pp. 878–89 Oxford University Press, New York.

369. Weaver, J. E. (1976): Dermatologic testing of household laundry products: a novel fabric softener. *Int. J. Dermatol.*, 15:297–300.
370. Weigand, D. A. and Gaylor, J. R. (1976): Irritant reaction in Negro and Caucasian skin. *South. Med. J.*, 67:548–551.
371. Weigand, D. A., Haygood, C., and Gaylor, J. R. (1974): Cell layer and density of Negro and Caucasian stratum corneum. *J. Invest. Dermatol.*, 62:563–568.
372. Weil, C. S. and Scala, R. A. (1971): Study of intra- and interlaboratory variability in the results of rabbit eye and skin irritation tests. *Toxicol. Appl. Pharmacol.*, 19:276–360.
373. Werner, Y., Lindberg, M., and Forslind, B. (1982): The water binding capacity of stratum corneum in dry non-eczematous skin of atopic eczema. *Acta. Derm. Venereol. (Stockholm)*, 62:334–336.
374. Wertz, P. W. and Downing, D. T. (1991): Epidermal lipids. In: *Physiology, Biochemistry, and Molecular Biology of the Skin*, 2nd ed., edited by L. A. Goldsmith, pp. 205–236. Oxford University Press, New York.
375. Wester, R. C. and Maibach, H. I. (1999): *In vivo* methods for percutaneous absorption measurements. In: *Percutaneous Absorption: Drugs, Cosmetics, Mechanisms, Methodology*, 3rd ed., edited by H. I. Maibach and R. L. Bronaugh, Marcel Dekker, New York.
376. Wester, R. C. and Maibach, H. I. (1975): Rhesus monkey as a animal model for percutaneous absorption. In: *Animal Models in Dermatology*, edited by H. I. Maibach, pp. 133–137. Churchill Livingstone, New York.
377. Wester, R. C., and Maibach, H. I. (1983): Cutaneous pharmacokinetics: 10 steps to percutaneous absorption. *Drug Metab. Rev.*, 14:169–205.
378. Whittle, E. and Basketter, D. A. (1993): The *in vitro* skin corrosivity test: development of a method using human skin. *Toxicol. In Vitro*, 7:265–268.
379. Whitton, J. T. and Ewell, J. D. (1973): The thickness of epidermis. *Br. J. Dermatol.*, 89:467–478.
380. Wiechers, J. (1989): The barrier function of the skin in relation to percutaneous absorption of drugs. *Pharm. Week. (Sci.)*, 11:185–198.
381. Wilhelm, K. P. and Maibach, H. I. (1990): Factors predisposing to cutaneous irritation. *Contact Dermatitis*, 8:17–22.
382. Wilhelm, K. P., Surber, C., and Maibach, H. I. (1989): Quantification of sodium lauryl sulfate irritant dermatitis in man: comparison of four techniques: Skin color reflectance, transepidermal water loss, laser Doppler flow measurement and visual scores. *Arch. Dermatol. Res.*, 281:293–295.
383. Wooding, W. H. and Opdyke, D. L. (1967): A statistical approach to the evaluation of cutaneous responses to irritants. *J. Soc. Cosmet. Chem.*, 16:809–829.
384. Yarnitsky, D. and Fowler, C. J. (1994): Quantitative sensory testing. In: *Manual of Clinical Neurophysiology*, edited by J. W. Osselton, pp. 253–320. Butterworths, London.
385. York, M., Basketter, D. A., Cuthbert, J. A., and Neilson, L. (1995): Skin irritation testing in man for hazard assessment-evaluation of four patch systems. *Human Exp. Toxicol.*, 14:729–734.
386. York, M., Griffiths, H. A., Whittle, E., and Basketter, D. A. (1996): Evaluation of a human patch test for the identification and classification of skin irritation potential. *Contact Dermatitis*, 34(3):204–212.
387. Yosipovitch, G. and Yarnitsky, D. (1997): Quantitative sensory testing. In: *Dermatoxicology Methods: The Laboratory Worker's Vade Mecum*, pp. 313–318. Taylor & Francis, New York.
388. Yosipovitch, G., Szolar, C., Hui, X. Y., and Maibach, H. I. (1996): Effect of topically applied menthol on thermal pain and itch sensations and biophysical properties of the skin. *Arch. Dermatol. Res.*, 288:245–248.
389. Yosipovitch, G., Szolar, C., Hui, X. Y., and Maibach, H. I. (1996): High potency corticosteroid rapidly decreases histamine induced itch but not thermal sensation and pain in man. *J. Am. Acad. Dermatol.*, 55:118–120.
390. Yuspa, S. H. and Blugosz, A. A., (1991): Cutaneous carcinogenesis. In: *Physiology, Biochemistry, and Molecular Biology of the Skin*, 2nd ed., edited by L. A. Goldsmith, pp. 1365–1401. Oxford University Press, New York.
391. National Toxicology Program (1999): Corrositex®: an *in vitro* test method for assessing dermal corrosivity potential of chemicals, NIH Publ. No. 99-4495. National Institutes of Health, Research Triangle Park, NC.
392. Smith, E. W. and Maibach, H. I. (2006): *Percutaneous Penetration Enhancers*, 2nd ed. Taylor & Francis, Boca Raton, FL.
393. Shah, V. P. and Maibach, H. I. (1993): *Topical Drug Bioavailability, Bioequivalence, and Penetration*. Plenum Press, New York.
394. Cockshott A., Evans P., Ryan C. A., Gerberick G. F., Betts C. J., Dearman R. J., Kimber I., and Basketter D. A. (2006) The local lymph node assay in practice: a current regulatory perspective. *Human Exp. Toxicol.*, 25(7):387–394.
395. Gerberick, G. F., Ryan, C. A., Kern, P. S., Schlatter, H., Dearman, R. J., Kimber, I., Patlewicz, G. Y., and Basketter, D. A. (2005): Compilation of historical local lymph node data for evaluation of skin sensitization alternative methods. *Dermatitis*, 16(4):157–202.
396. Kimber, I., Dearman, R. J., Basketter, D. A., Ryan, C. A., and Gerberick, G. F. (2002): The local lymph node assay: past, present and future. *Contact Dermatitis*, 47(6):315–328.
397. Divkovic, M., Pease, C. K., Gerberick, G. F., and Basketter, D. A. (2005): Hapten-protein binding: from theory to practical application in the *in vitro* prediction of skin sensitization. *Contact Dermatitis*, 53(4):189–200.
398. Gerberick, G. F., Vassallo, J. D., Bailey, R. E., Chaney, J. G., Morrall, S. W., and Lepoittevin, J. P. (2004). Development of a peptide reactivity assay for screening contact allergens. *Toxicol. Sci.*, 81(2):332–343.
399. Chew, A. L. and Maibach, H. I. (2005): *Irritant Dermatitis*. Springer, New York.
400. Herbst, R. A., Strimling, R. B., and Maibach, H. I. (2001). Assay to quantify subjective irritation caused by the pyrethroid insecticide alpha-cypermethrin. In: *Toxicology of the Skin*, edited by H. I. Maibach, pp. 105–114. Taylor & Francis, New York.
401. Wang, R. G., Knaak, J. B., and Maibach, H. I. (1993): *Health Risk Assessment: Dermal and Inhalation Exposure and Absorption of Toxicants*. CRC Press, Boca Raton, FL.

# Notes

# 28 Inhalation Toxicology

*Rudolph Valentine and Gerald L. Kennedy*

## CONTENTS



## INTRODUCTION

Inhalation of substances through the nose or mouth into the lungs represents a major route of contact for environmental substances. Along the way, substances or their metabolites may be available for interaction with tissues lining the respiratory tract. Cells and tissues at the site of entry can respond or the substances can be transported systemically where interaction with other internal organs and tissues can occur. As with the routes of exposure (dermal or oral) the ultimate response of the organism to inhaled substances will depend on the integrated amount and form of the substance in the body.

Because the respiratory tract forms a critical link between the blood supply of the body and the external environment, an understanding of the factors that can modify the integrity or function of the nose, upper airways, or lungs is of great importance to the inhalation toxicologist. By virtue of the rich vasculature of the lungs, inhaled substances gain ready access to the internal milieu of the body. External influences that interfere with the respiratory system may have profound effects on the organism, which in turn can be reflected by a broad spectrum of untoward local responses (ranging from upper respiratory tract irritation to pulmonary edema and systemic effects ranging from minor organ or tissue damage to death).

This chapter is intended to give the reader an overview of the principles and methods used by the inhalation toxicologist. It will focus on the methods used in assessing the effects of chemical substances on the respiratory system following inhalation. The emphasis is not on describing the many types of physical or toxicological changes that can be produced by substances but on providing selected examples of these interactions that demonstrate applications of the experimental methodologies.

## GENERAL ANATOMY AND FUNCTION OF THE RESPIRATORY TRACT

An overview of the anatomy, function, and physiology of the mammalian respiratory system is presented to clarify some of the issues that face the inhalation toxicologist. The respiratory system can be simplified into three major components: (1) nasopharyngeal, (2) tracheobronchial, and (3) pulmonary. These components represent distinctly different functions, clearance processes, and susceptibilities to inhaled substances.

### Nasopharyngeal Region

The nasopharyngeal region includes the nasal turbinates, epiglottis, glottis, pharynx, and larynx. As the entry port for inspired air, the nares and nasal cavity serve to remove the larger inhaled particles (through impaction in the turbinates and filtration by nasal hairs) and to condition (by moderating the temperature and raising the humidity) the incoming air. Given its location, the nose is exposed to the highest concentrations of inhaled substances within the respiratory tract; accordingly, the nasopharyngeal region is the subject of considerable investigation. The nasopharyngeal region also plays an important role in the physiological responses to inhaled irritants. There are three basic types of irritants [2]: (1) sensory irritants that act on the trigeminal nerve, (2) pulmonary irritants that act on irritant receptors of the airways, and (3) mixed sensory and pulmonary irritants.

The initial response to an inhaled irritant may involve an immediate burning or stinging sensation in the eyes, nose, or throat. Such stimuli may range from unpleasant to extreme pain and are mediated by interaction of irritants with chemoreceptive nerve endings from the trigeminal nerve in the cornea, nose, tongue, oral cavity, and upper respiratory tract. Once stimulated, these nerve endings can cause systemic responses, resulting not only in a burning sensation of the nose and eyes but also a reflex reduction in respiratory rate. This response occurs once a threshold concentration at the irritant receptor is exceeded and develops immediately or within a few minutes of exposure and is both time and concentration dependent. Irritant responses are generally protective in nature, limiting further exposure of the offending substance, especially to the lower respiratory tract. Although the effects of sensory irritants *per se* are not usually life threatening, irritants have the capacity to cause changes in the respiratory tract ranging from minor epithelial injury to fatal lung edema hours or days after exposure.

In our environment, many different types of respiratory tract irritants exist. Chlorine, for example, has been used as a war gas, and an accidental release of methyl isocyanate caused massive human exposures in India. Chemical irritants act on specific targets; for example, formaldehyde acts primarily on respiratory epithelial cells, chlorine acts on the nasal cilia and bronchial functions, dimethylamine acts on olfactory sensory cells, and cigarette smoke affects the laryngeal epithelium. Furthermore, a number of disease states (asthma, asphyxiation, chronic bronchitis, emphysema, pulmonary fibrosis, and pneumoconiosis) can be induced or exacerbated by exposure to irritants. Some of the effects of inhaled materials on the nose are shown in Table 28.1.

#### Nose

The nasal airway is divided into two passages by the nasal septum. Each passage extends from the nostrils to the nasopharynx. The nasopharynx is the airway posterior to the termination of the nasal septum and proximal to the soft palate. Air moves through the nostril openings (nares) into the nasal cavity, which contains turbinates (bony structures lined by well-vascularized respiratory or olfactory mucosal tissue). The neuroepithelium of the olfactory epithelium has the primary role of odor detection. Olfactory cells process odors to help identify predators, prey, and hazardous substances or conditions. Olfactory cells are rich with hydrolytic enzymes (especially carboxylesterases) necessary to detect inhaled substances (such as aromatic esters). This epithelium also contains mucus-secreting goblet cells, accessory (sustentacular) cells, and basal cells, the progenitor cells for olfactory epithelium. Exposure to inhaled substances may injure the epithelium, leading to either temporary or permanent loss of smell (anosmia).

**TABLE 28.1**
**Effects of Inhaled Materials on the Nose**

| Effect | Examples |
|---|---|
| Restrict airflow | Temperature changes (cold air), irritants (acids, bases) |
| Mucociliary flow | Slowing by sulfur dioxide, formaldehyde, methyl amines |
| Cellular | |
| Olfactory degeneration | Chloroform, aliphatic esters, methyl bromide |
| Respiratory/olfactory irritation | Chlorine, sulfur dioxide, formaldehyde |
| Nasal tumors | Hexamethyl phosphoramide, acetaldehyde, formaldehyde, dimethyl sulfate |

Importantly, the olfactory cells differ from other nerve tissue cells in that they have the capability to regenerate following injury if the olfactory epithelium and progenitor cells are not completely destroyed [18,134].

As a portal of entry, the nasal passages are a target site for a wide range of inhaled substances. For the epithelial mucosa in the nose, squamous tissue is the target of glutaraldehyde toxicity [103], transitional epithelial tissue may be damaged by ozone [190], respiratory mucosal tissue by formaldehyde [191], and the olfactory region by β-β-iminodipropionitrile [94]. The effects produced at these locations may be attributable to the local dose of the substances reaching the site, site-specific tissue susceptibility, or a combination of these factors.

Importantly, the location and proper characterization of nasal lesions play a role in assessing the mode of action and for interspecies comparisons. Young [335] proposed that a set of four standard transverse sections from the rat nasal cavity should be prepared for histopathological examination. These included sections taken at the following landmarks: (1) immediately posterior to the upper incisor teeth, (2) at the incisive papilla, (3) at the second palatal ridge, and (4) at the level of the first upper molar teeth. This method applies to both short-term and chronic toxicity studies and allows detailed examination of the nasal cavity without destructive invasion at the time of necropsy. Other sampling strategies include that of the National Toxicology Program [301,302], which uses three sites for sectioning. Regardless, section level selection and description are essential for comparing the extent, severity, and relevance of the lesions.

A series of diagrams designed for mapping nasal lesions of both the Fischer 344 rat and the B6C3F$_1$ mouse has been published by Mery et al. [183]. The diagrams present each of the major cross-sectional airway profiles and provide space for recording nasal mucosal lesions. Sagittal diagrams are also provided to permit transfer of data from transverse sections onto the long axis of the nose. The use and application of these diagrams help unify and make the collection and presentation of data more consistent.

It should be emphasized that recording the lesion distribution is only one step in the process of identifying patterns of nasal toxicity. Because the distribution of lesions results from the interplay of local dose and tissue susceptibility, the researcher needs to keep in mind their relative influence and the relevance of such responses to humans. For example, the airflow-driven local dose of highly reactive water-soluble substances such as formaldehyde can dictate the lesion distribution, but airflow is not likely to be a major factor in the lesions produced by less water-soluble substances such as ozone, chloroform, and methyl bromide [146].

It has been shown that nasal enzymes acting at localized sites are responsible for the conversion of chemicals such as dibasic acid esters and hexamethylphosphoramide to reactive intermediates that cause injury at enzyme-rich cellular locations [25,58]. The role of local tissue metabolism has important implications in the risk assessment of inhaled materials that cause injury solely in the nose; for example, Bogdanffy and Valentine [26] described the role of olfactory cell carboxylesterases in the metabolism of vinyl acetate in the development of nasal tumors in rats. Intracellular acidification resulting from the liberation of acetic acid and protons from the metabolism of vinyl acetate and the attendant cytotoxicity and regenerative hyperplasia of olfactory epithelium at high doses are thought to be threshold-related phenomena in the mode of action for vinyl acetate and perhaps other related esters. For most substances however, further study is needed to elucidate the dosimetry, susceptibility, and mechanisms of action for inhaled substances on nasal tissue, particularly for use in extrapolating animal data to humans.

## Larynx

The laryngeal region, located between the upper respiratory tract and the trachea, is also a potential target tissue of inhaled toxicants, as a variety of substances have produced epithelial injury to this area. The ventral laryngeal epithelium of rats and mice is highly responsive to inhaled materials such as cobalt sulfate [41], tobacco smoke [252], and a wide variety of industrial chemicals, pharmaceuticals, and aerosol propellants [101]. Interspecies differences also exist among laboratory rodents with regard to the anatomy of these sensitive areas of the laryngeal mucosa. In Sprague–Dawley rats, the mucosa covering the epiglottis differs from that of the Syrian golden hamsters in that it is thinner and composed of a mixture of cell types. In contrast, the cartilage of the hamster is much more prominent and forms a distinct protrusion into the lumen at the base of the epiglottis. These anatomic differences can play a role in the use of this information in extrapolating these findings to humans [244].

Laryngeal lesions commonly involve degeneration of the epithelium with subsequent regeneration, hyperplasia, and squamous metaplasia. In more severe reactions, the larynx may have epithelial ulceration with exudation. Recovery or regression from induced changes is variable and dependent on the time scale involved and severity of the type of initial lesion [164]. Specific areas of the rodent laryngeal mucosa appear to be sensitive to inhaled materials. These sites include the epithelium covering the base of the epiglottis, ventral pouches, and the medial surfaces of the vocal processes of the arytenoid cartilage; therefore, the detection of induced changes requires a consistent, thorough, and detailed histological examination.

### Tracheobronchial Airways

The tracheobronchial airways conduct inspired air through a series of branching ducts ending at the terminal bronchioles. Many differing epithelial cell types line this airway and are distinct on the basis of ultrastructural morphology, functionality, and sensitivity to inhaled materials. Considerable variation exists in the abundance of each of these cell types at defined airway levels among species; however, tracheobronchial epithelium contains predominantly basal, ciliated mucous, Clara, and neuroendocrine cells [131]. Mucociliary clearance is a protective mechanism by which contaminants of inhaled air are trapped or dissolved in the mucous layer then removed by ciliary transport. The two critical components of the system are the mucous blanket and ciliated cells. Methods to define the mucous composition include carbohydrate histochemistry and cytochemistry, autoradiography, immunohistochemistry, and biochemistry [156,247,276,278]. Assessment of the mucociliary "escalator" transport function has been useful in measuring the biological response to inhaled gases and cigarette smoke in particular.

Marked species differences exist in the geometric structure of the tracheobronchial region. Typically, the branching patterns of the tracheobronchial tree are either monopodal or regular dichotomous. The monopodal pattern is characterized by long tapering airways with asymmetric, smaller lateral branches that leave the main tube at a shallow (<60°) angle as seen in cats, dogs, hamsters, horses, mice, monkeys, pigs, rabbits, and rats. In contrast, regular dichotomous branching, typical of humans, involves division of a tube into nearly identical, equal-diameter smaller tubes with approximately equal branching angles. Variations in the symmetry of branching, in airway diameters and lengths, and in the number of airway generations can affect the deposition of particles in the respiratory tract and can account for observed differences. Mathematical airway models suggest that the uptake kinetics of inhaled gases (oxygen, nitrogen dioxide, and sulfur dioxide) vary substantially among rats, dogs, and humans. Such variations are at least partially the result of anatomical and physiological differences in their airways [298].

### Pulmonary Region

The pulmonary region is composed of respiratory bronchioles, alveolar ducts, and alveoli. The alveolus is the functional gas exchange unit of the lung between the air and blood compartments. The alveolus itself has walls formed by epithelial cells. Major cell types of alveolar epithelium are known as type I and type II cells. The type I epithelial cell is very thin, has a smooth surface, and covers nearly the entire alveolar surface. The type II cell contains numerous microvilli, is metabolically active, and manufactures and secretes surfactants, which, by reducing surface tension at the air–liquid interface of the alveolus reduce the tendency for alveoli to collapse.

Another important cell type in the lung is the alveola macrophage. This is a large, mononucleated cell that functions in part to engulf foreign particulate materials depositing in this region. These cells are scavengers of inhale particulate material by the process of phagocytosis, an their location on the alveolar surfaces preserves the ste rility of the deep lung by keeping inhaled bacteria an particles from accumulating.

The function of macrophages can be affected by cyto toxic airborne substances, such as asbestos, quartz, an heavy metals. This response is not limited to cytotoxi particles; for example, the instillation of 4 mg of carbo particles (considered a nuisance dust due to its relativ lack of cytotoxicity) into the lung of a mouse induces a acute macrophage response, with the number of macro phages increasing tenfold [32]. A spectrum of responses resulting in alterations in macrophage numbers, viability morphology, and changes in the phagocytic capacity, ca be produced by chemical insult [91]. The effects of inhale gases and particles on macrophage functions are importar because the function of these cells in homeostasis. Som of the studies of macrophage function describe the inju rious effects of gases on macrophage bactericidal activity In these models, animals preexposed to ozone had highe mortality when subsequently exposed to pathogenic bac teria, ostensibly by impairing macrophage phagocytosi and lysosomal enzyme activity [53,91,98].

In addition to cellular and functional adaptations, spe cialized cells have evolved to protect the respiratory trac The anterior portions of the nose, larynx, trachea, bronch and bronchioles are lined by ciliated mucosa. Intersperse between these cells are columnar goblet cells that manu

**TABLE 28.2**
**General Breathing Characteristics of Common Species**

| Species | Mass (g) | Frequency (breaths/min) | | Tidal Volume (mL) | | Minute Ventilation (mL) | |
|---|---|---|---|---|---|---|---|
| Human | 70,000 | | | | | | |
| Rest | | 12 | | 750 | | 9000 | |
| Light exercise | | 17 | | 1700 | | 28,900 | |
| Dog | 10,000 | 20 | | 200 | | 3600 | |
| Monkey | 3000 | 40 | | 21 | | 840 | |
| Guinea pig | 500 | 90 | | 2.0 | | 180 | |
| Rat | 350 | 160 | | 1.4 | | 240 | |
| Mouse | 30 | 180 | | 0.25 | | 45 | |
| Allometric scaling factors | | *k* | *x* | *k* | *x* | *k* | *x* |
| ($Y = kM^x$; *M* in kg) | | 53.5 | −.26 | 7.69 | 1.04 | 411 | .78 |

*Source:* Adapted from Boggs, D.F., in *Treatise on Pulmonary Toxicology*. Vol. 1. *Comparative Biology of the Normal Lung*, Parent, R., Ed, CRC Press, Boca Raton, FL, 1992.

facture and secrete mucus. The cilia and mucus work together in a coordinated fashion to propel particulate material entrained in the mucus away from the lungs.

Measurement of breathing patterns, lung flows and volumes, lung capacities, forced expirograms, and blood gas and pH not only is important in determining the functional performance of the lung but is also used to approximate the dose of substance received under experimental conditions. Obtaining reliable values for these parameters is difficult because of the need for cooperation in maximal effort respiratory maneuvers. The use of masks, mouthpieces, or anesthesia to obtain this information may alter normal breathing patterns. Species such as dog, horse, and sheep, as well as some rodents, have been trained to wear masks or, alternatively, appropriate correction factors have been applied to describe pulmonary function under normal conditions. A comparison of typical breathing characteristics of a number of commonly studied species is provided in Table 28.2.

## PHYSICAL AND TOXICOLOGICAL PROPERTIES OF INHALED SUBSTANCES

### Gases and Vapors

Gases and vapors share a common physical property of interest to the inhalation toxicologist; both can be present in the gas phase under ambient testing conditions. Gases, by definition, exist in the gaseous state at standard temperature and pressure, while vapors represent the gas phase components from substances that are solid or liquid at standard temperature and pressure. As the temperature increases, the amount of gas phase components (represented by its vapor pressure) from a liquid will increase; thus, liquids with higher vapor pressure will have a greater amount (concentration) of molecules in the gas phase relative to those with lower vapor pressure. For the inhalation toxicologist, liquid- or solid-phase test substances present a special challenge because test atmospheres may exist in both gas and solid–liquid forms, necessitating consideration of their different depositional, clearance, analytical, and biological response characteristics.

Inhaled gases and vapors can be absorbed throughout the respiratory tract to produce either local injury or systemic toxicity. Absorption depends largely on the physicochemical characteristics (e.g., concentration, water solubility, partition coefficient) of the inhaled substance and physiologic characteristics of the respiratory tract (e.g., airflow, respiratory rate, tissue perfusion, local metabolism).

The net driving force for absorption of inhaled gases or vapors is diffusion, the net movement of molecules down a concentration gradient from regions of high to low concentrations. Substance absorption into tissues continues until the vapor concentrations in the airspace and tissue are the same. At this point, the system is at equilibrium and no further transport of substance occurs; however, substance absorption into tissues can continue if the absorbed vapor is removed by transport into the bloodstream, metabolism to a different chemical, or reaction with tissue macromolecules. These processes serve to lower the substance concentrations in the tissue so further diffusion into the tissue may again proceed. Conversely, when the vapor concentration is higher in tissue than the inspired air, the net diffusive force is external—that is, from tissue to air. Passive diffusion is the mode by which carbon dioxide, a byproduct of aerobic metabolism, is conducted from the tissue, to the bloodstream, and to the alveolar airspace and ultimately is removed in the expired breath.

Inhaled gases or vapors may be absorbed directly within the nasal cavity. Although this spares the lung from direct exposure, the nose is consequently exposed to high, potentially injurious concentrations of inhaled substances. Vapors may be either physically absorbed within nasal tissues, a reversible process driven in part by the relative solubility of the vapor in air or tissue, or irreversibly bound to tissue components. Vapors highly soluble in aqueous solutions, such as the mucus covering the epithelial surfaces of the upper respiratory tract, generally tend to be absorbed within these regions. Highly water-soluble substances such as acetone, methanol, hydrogen fluoride, and propylene glycol monomethyl ether are essentially completely absorbed in the nose [193,194,287]. In contrast, gases or vapors with low water solubility, such as nitrogen dioxide or methylene chloride, are not well absorbed in the upper respiratory tract but are absorbed deeper in the respiratory tract [48,195]. These generalizations oversimplify nasal absorption of vapors, but improvements in regional air flow and physiologic parameter measurement can allow deposition to be extrapolated from animal to human. Physicochemical and physiological parameters (diffusivity, tissue:air partition coefficient, chemical reactivity, metabolism) have been incorporated into computational fluid dynamic models of the rodent and human nose to predict species dependent nasal deposition of vapor [196]; the results suggest that good agreement between rats and humans is attained with either one or multiple-tissue compartment models. Additional determinants of vapor absorption are the upper respiratory tract vascular blood flow and tissue metabolism or binding [195,287]. Irreversible binding to nasal tissues may occur if the inhaled vapor is reactive or is metabolized within nasal epithelium, with subsequent formation of covalent bonds with tissue macromolecules. Such reactivity reduces the substance concentration in tissues and facilitates further diffusion and absorption.

Vapors with low chemical reactivity and water solubility generally bypass the nasal cavity and are conducted into the conductive airways. Limited absorption occurs in the conducting airways, and because gas exchange does not occur here the conducting airways represent a respiratory dead space, amounting to approximately 30% of the resting tidal volume (i.e., about 150 mL in humans). Beyond the conducting airways, inhaled substances enter the alveolar (gas exchange) region and diffuse across very thin epithelial and endothelial membranes (typically 0.1 µm thick) into the tissue or blood.

## Aerosols

Aerosols are simply particle suspensions, either solid or liquid, of a substance in a gaseous medium. From the hydrocarbon fume emitted from a burning candle to the salt spray emitted from ocean waves by wind action, myriad sources of natural and manmade aerosols exist in everyday life. Aerosols comprise a diverse spectrum of particle physicochemical characteristics with different origins, composition, density, size, shape, and aerodynamic properties. Moreover, aerosols are quasi-stable, as aerosol size, composition, and concentration can vary with time from external forces, including collision with other particles, gravitational settling, volatilization, and gas absorption. Depending on the aerosol and external forces, aerosols may be stable (time frame of hours to days) or unstable (seconds to minutes).

Given the ubiquity of aerosols, it is important for the inhalation toxicologist to understand the basic concentration, size, and physicochemical properties of aerosols to appreciate the potential harm they may cause in the respiratory tract. Before discussing the relationships between aerosol properties and deposition, a brief description of terminology is appropriate:

- *Particle*—A small, discrete object
- *Dusts*—Solid particles formed from mechanical processes (e.g., crushing, grinding, milling), generally >0.5 µm ($10^{-6}$ m) in size
- *Fume*—Particles arising from high temperature or combustion processes with subsequent condensation of vapors; fumes exist initially as very small (<100 nm; $10^{-9}$ m) particles but form larger (~1 µm) particle clusters through agglomerative growth
- *Mists*—Similar to fogs, a suspension of liquid particles in gas that can be formed from condensation of vapors or from atomization of liquids (e.g., sprayers); mists are characterized by particles larger than about 40 µm, fogs by their smaller particle size
- *Nanoparticles*—Particles, either of manmade or biological origin, with at least one dimension less than 100 nm; also referred to as ultrafine particles, which have all dimensions in the nanometer range
- *Particulate*—An adjective describing particle-related properties
- *Smokes*—A complex mixture of solid or liquid particles, such as soot, liquid droplets, or mineral ash particles from incomplete combustion of organic materials; smoke particles are generally ~0.5 µm

Atmospheric aerosols exist in three size regimes, as represented in Figure 28.1 [325,326], reflecting the origin, growth, and loss of particles spanning a size range of almost five orders of magnitude. The smallest size regime is known as the *nuclei mode* and the largest is known as the *coarse particle mode*. The nuclei mode consists of particles formed via condensation of vapors forming primary particles o

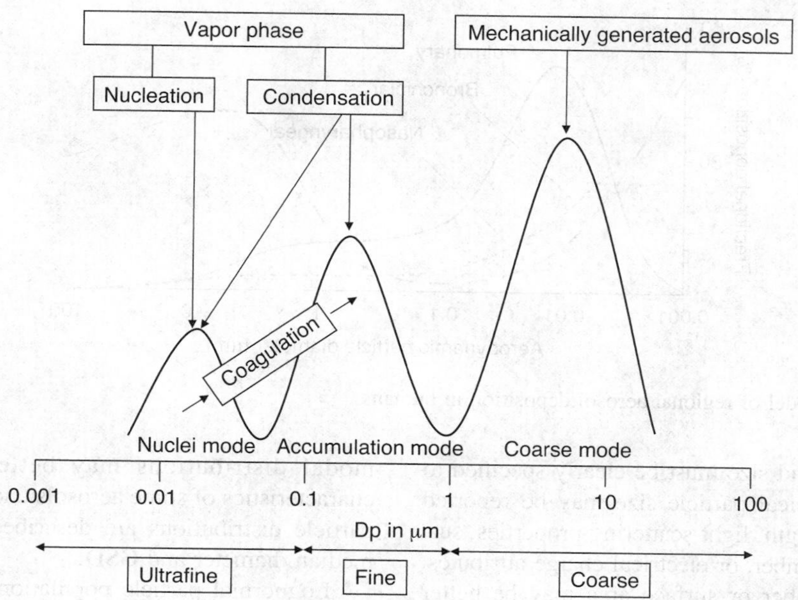


**FIGURE 28.1** Particle size regimes and aerosol formation mechanisms. (Adapted from Whitby, K.T. et al., in *The General Motors/Environmental Protection Agency Sulfate Dispersion Experiment*, Stevens, R.K. et al., Eds., EPA-600/3-76-035, U.S. Environmental Protection Agency, Research Triangle Park, NC, 1976, pp. 29–80; Whitby, K.T. and Cantrell, B.K., in *Proc. of Int. Conf. on Environmental Sensing and Assessment (ICESA)*, Institute of Electrical and Electronic Engineers, Las Vegas, NV, September 14–19, 1975.)

nanometer size. Further condensation can increase particle size, but more generally these primary particles coagulate by interparticle collisions. Particles in nuclei mode are less than approximately 0.1 mm and have short lifetimes in the atmosphere because these particles tend to coagulate with other nuclei mode particles. The particle size regime intermediate between nuclei and coarse modes is known as the *accumulation mode* and consists of particles derived from chemical reaction condensation products and coagulation with nuclei mode particles. Particle growth rate is slower in the accumulation mode as these larger particles do not diffuse as rapidly as the smaller nuclei mode particles; hence, the particle lifetime for accumulation mode particles is comparatively longer (approximately days to weeks), enabling them to undergo long-range transport. Coarse mode particles are formed from mechanical processes such as grinding, spraying, etc., thus their composition differs from nuclei and accumulation modes. Coarse mode particles are typically between approximately 2 and 10 μm in size and have relatively short lifetimes in the environment due to losses from gravitational settling or rain washout. Depositional losses with particles larger than 10 μm favor very short atmospheric lifetimes, as these particles settle rapidly out of the air by gravity and their high settling velocity precludes significant respiratory tract inhalation or deposition.

Until comparatively recently, the possible occupational health hazards of particles has focused on the coarse mode particles, in part because they are commonly encountered by humans through mechanical processing and in part because their size allows them to penetrate and deposit in the human respiratory tract (Figure 28.2). As will be discussed later, attention is turning now toward the finer particle modes in response to a growing body of evidence indicating greater health hazards for these smaller particles. Regardless of origin or physical properties, an understanding of particle size as it relates to particle respirability is key to evaluating possible health hazards of ambient or occupational aerosols and to ensure that aerosols are inhalable when generating test atmospheres for inhalation toxicity studies with laboratory animals.

The interaction between the individual's breathing pattern and respiratory tract anatomy, coupled with the size (reflected in the inherent aerodynamic properties of the aerosol), dictates the amounts and sites of aerosol deposition. The individual's breathing pattern affects deposition by a changing breathing rate (increases or decreases air velocity and the velocity of the entrained aerosol through airways), tidal volume (allows more or less penetration into alveoli), and breath holding (allows more time for particle and tissue interactions). In addition to concentration, the single most important determinant of toxicological effect to respiratory tract tissue is particle size. Particle size drives aerosol deposition. Because size is a relative term, it may be expressed as some physical attribute (e.g., generally some measure of physical diameter) or as the aerodynamic size (a measure of how the particle behaves in an airstream relative to other particles of defined size,

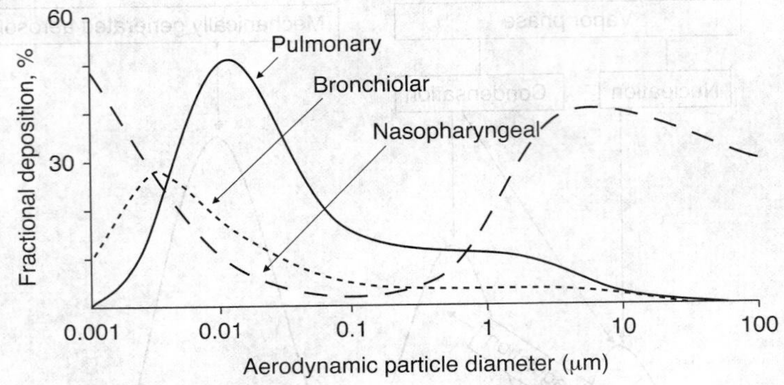


**FIGURE 28.2** ICRP model of regional aerosol deposition in humans.

shape, and density), and size must be clearly specified to avoid confusion. Physical particle size may be reported based on diameter, length, light-scattering properties, surface area, volume, number, or electrical charge attributes. Whereas particle number or surface area may be better correlated in some cases with toxicity, it is aerosol mass that is most often associated with particle toxicity and thus represents the most important physical attribute of the particle size population. It may seem obvious, but achieving a respirable aerosol for the species of interest is absolutely critical to the inhalation toxicologist. Generating test atmospheres that are not respirable may serve to demonstrate a lack of toxicity but has no bearing on the intrinsic toxicological properties of a substance if the substance cannot gain entrance to potentially susceptible regions of the respiratory tract.

As particles are rarely of uniform size (or mass), a description of the particle population's average size and uniformity is required. By convention, particles are defined by their physical or aerodynamic size (count or mass median aerodynamic diameters, respectively) and a measure of the variability of the particle size distribution (the geometric standard deviation [GSD]). Empirically, the lognormal distribution most commonly describes particle size data from test atmospheres generated for inhalation studies, although Gaussian and bimodal or multimodal distributions may better describe the size characteristics of some aerosols and mixtures. Lognormal particle distributions are described completely by their median diameter and GSD.

Lognormal particle populations can be visualized if the frequency or mass of particles with a given size (measured either with size selective atmospheric samplers such as the impactor cut point or physical diameters) is plotted against the measured particle size (mm) using logarithmic axes; aerosol populations that are lognormal appear as a straight line. An example size population (data from a six-stage cascade impactor) is shown in Table 28.3 and used to describe the data transformations that describe a lognormal particle population. Table 28.3 shows the cut points (the median aerodynamic particle size that collects 50% of the aerosol mass larger than the cut point) for each impactor stage, expressed as a percentage of the total mass collected from all stages. The amount of aerosol mass collected on each stage is plotted against the cut point for each stage using linear axes (Figure 28.3A); the resulting distribution curve is typically skewed. Replotting the data using a logarithmic scale for particle size (impactor cut point), the curve is sharpened and is more symmetric resembling a normal distribution (Figure 28.3B). By replotting the same data but expressing the frequency as the cumulative percentage of particles less than a given

**TABLE 28.3**
**Example Cascade Impactor Data Used in Figure 28.3**

| Impactor Stage | Impactor Cut Point (Dpc) (μm) | Aerosol Mass per Stage (mg) | Mass per Stage (%) | Cumulative Mass < Dpc |
|---|---|---|---|---|
| 1 | 9 | 0.08 | 0.6 | 99.4 |
| 2 | 5.1 | 0.90 | 7.3 | 92.1 |
| 3 | 3 | 2.96 | 24.0 | 68.1 |
| 4 | 1.6 | 4.50 | 36.5 | 31.6 |
| 5 | 0.9 | 3.18 | 25.8 | 5.8 |
| 6 | 0.4 | 0.70 | 5.7 | 0.2 |
| Final | 0 | 0.02 | 0.2 | 0.0 |

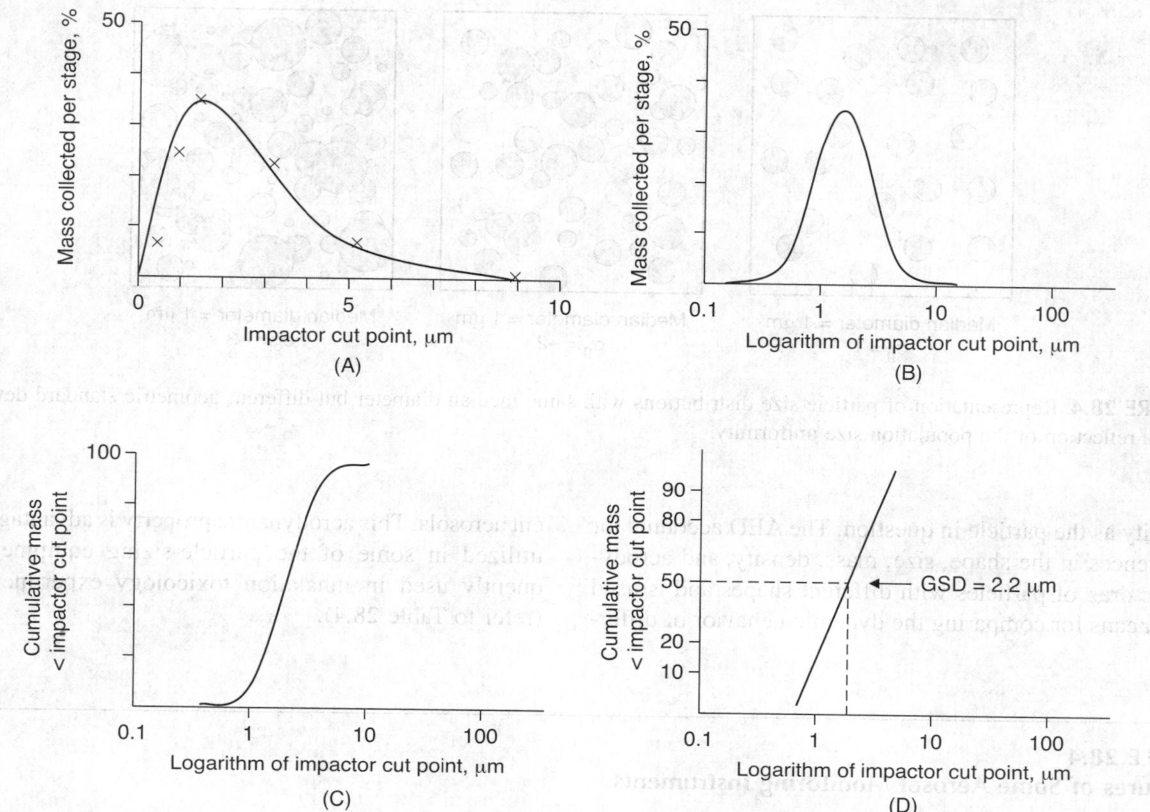


**FIGURE 28.3** (A) Cascade impactor data from Table 28.3, plotted on arithmetic axes. (B) Size distribution data from A, replotted ısing logarithmic scale for impactor cut points. (C) Size distribution data from B, replotted using cumulative mass less than cut point liameters. (D) Size distribution data from C, replotted using probability scale for cumulative mass. Geometric standard deviation is 2.2 $\mu$m.

ize (such as the impactor cut point; see Table 28.3), again ısing a logarithmic scale for particle size, a relatively inear central region with curvilinear tails is obtained (Figıre 28.3C). A straight line results if the cumulative mass lata on the abscissa are transformed and graphed using a probability scale for the cumulative frequency (Figure 28.3D). The mass median diameter (MMD) is read lirectly from this plot; it represents the size where 50% of the particles are larger (or smaller) than the median ize. The GSD ($\sigma_g$) is derived from the lognormal data as ollows:

$$\sigma_g = \frac{\text{Size at 84.1\%}}{\text{Size at 50\% mass}} \quad \text{or} \quad \frac{\text{Size at 50\% mass}}{\text{Size at 15.9\% mass}}$$

particle population with uniform size (i.e., all the articles comprising the aerosol are relatively uniform ı size) is referred to as *monodisperse* and has $\sigma_g$ values ess than 1.2. In contrast, a polydisperse particle poputation encompasses a broader range of particle sizes and as $\sigma_g$ values greater than approximately 1.8. This is escribed in Figure 28.4.

The simplest aerosols have spherical shapes and may be characterized by their physical diameter; generally, this attribute is measured by either optical or scanning electron microscopy techniques. Using the same graphic method of presentation as described above, using the number of particles as a function of a given size, the count median diameter and GSD of a particle population may be determined. For a given lognormal distribution, the mean mass-, surface area-, or volume-equivalent diameters of these spheres may be calculated according to the equations developed by Hatch and Choate [108] when one knows one of these values and the geometric standard deviation.

Most aerosols are not simple spheres, but they exist in various and often complex shapes. Indeed, aerosols may be fibrous (defined as particles with length to width ratios greater than three), globular, flake, crystal-like, or fractal (an irregularly shaped particle agglomerate). The size and shape of these aerosols are more difficult to characterize based solely on physical diameter or length; however, the size of these nonspherical aerosols may be normalized by expressing size as the aerodynamic equivalent diameter (AED), defined as the diameter of a spherical particle of unit density (1 g/mL) that has the same terminal settling

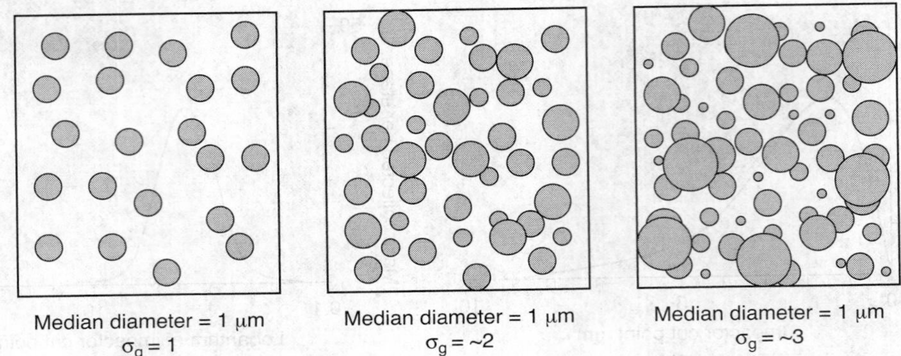


**FIGURE 28.4** Representation of particle size distributions with same median diameter but different geometric standard deviations (μg), a reflection of the population size uniformity.

velocity as the particle in question. The AED accounts for differences in the shape, size, mass, density, and aerodynamic drag of particles with different shapes and is used as a means for comparing the dynamic behavior of different aerosols. This aerodynamic property is advantageously utilized in some of the particle-sizing equipment frequently used in inhalation toxicology experimentation (refer to Table 28.4).

**TABLE 28.4**
**Features of Some Aerosol Monitoring Instruments**

| Instrument | Aerosol size (μm) | Response Time | Measured Parameter | Factors Affecting Response | Advantages | Disadvantages |
|---|---|---|---|---|---|---|
| Photometer | 0.3–15 | Continuous | Total light scattering | Density, size distribution, refractive index | Continuous readout | Response changes with dust type |
| β-Attenuation monitor | 1–15 | 1–30 min | Absorption of γ-radiation | Atomic number | Direct mass measurement | Low sensitivity |
| Optical particle counter | 0.3–15 | 0.1–10 min | Light scattering, size, and count | Density, refractive index | Indication of size, high sensitivity | Low resolution and accuracy |
| Piezobalance | 0.02–10 | 0.5–2 min | Mass | Particle size | Direct mass measurement | Sensor cleaning |
| Piezobalance cascade impactor | 0.05–25 | 1–60 min | Mass | Particle size | Direct mass measurement, size, distribution | Sensor cleaning, internal losses |
| Condensation nucleus counter | 0.01–1 | 0.5–30 sec | Particle count | Particle count | Small particle sensitivity | Alcohol emission |
| Diffusional mobility particle size | 0.003–1 | <1–5 min | Size based on electrical mobility | Electrical charge | Real-time display of ultrafine particle size | Measures only fine particles |
| Fibrous aerosol monitor | Optically visible fibers | 1–1000 min | Light scattering size of fibers | Fiber length, size | Specific for fibers | Nonfibrous interferences |
| Aerodynamic particle sizer | 0.8–15 | 2–10 min | Aerodynamic size | Density | Direct measure of aerodynamic diameter | Coincidence and density effects |
| Tapered element oscillating microbalance | 0.002–15 | 0.01–30 min | Mass | Absorbed water | Direct mass measurement | Frequent filter replacement |

*Source:* Adapted from Baron, P.A., *Appl. Indust. Hyg.*, 3, 97–103, 1988.

## Fibers

A special type of aerosol, and one of particular significance to human health, is fibers, which are particles with an elongated shape. Because of the association of naturally occurring (e.g., asbestos and erionite) and manmade synthetic fibers with the development of pulmonary fibrosis and carcinogenesis in humans [126], interest in the hazards of inhaled fibers is heightened. Major differences in the nature and persistence of lung injury have been identified based on the fiber composition, fiber durability (a factor determining biologic persistence in the lung), fiber size (another factor governing fiber penetration and reactivity into the lung), and fiber exposure concentration.

For the purposes of describing the physical dimensions of fibers of toxicological concern, general characteristics of fibers relating size with respiratory tract toxicity have emerged from both animal and occupational exposure studies. In animal studies, a variety of fiber types and dimensions have been instilled or injected into the pleura or abdominal cavities [62,236,279,312]. Although these methods of fiber introduction into the body are not representative of normal exposure pathways, in terms of both the rate and location of fiber deposition and quantitative aspects of the cellular response, this early research provided fundamental insights into the physical basis for the biological reactivity of fibers. Based on this work, a general relationship between the fiber dimension and lung response emerged. Notably, Stanton and Wrench [279] found that intrapleural sarcomas were likely, especially if the implanted fibers were >8 mm in length and <1.5 mm diameter. In contrast, nonfibrous preparations of these same materials had a much less severe pulmonary response [334]. Walton [313] suggested that asbestos fibers greater than 5 to 10 mm in length with diameters less than 1.5 to 2 mm and an aspect ratio (i.e., length-to-width ratio) greater than 5 are most hazardous. A considerable body of evidence has accumulated supporting the hypothesis that especially long and thin fibers present the greatest health hazard. Recognizing that clear distinctions between fiber dimension and pathogenicity do not necessarily exist, the U.S. Environmental Protection Agency has defined a fiber as a particle with a length of >5 mm with as aspect ratio of at least 3. In recognition of the higher hazard of the smaller diameter fibers, the World Health Organization has also stipulated a diameter of <3 mm in their fiber definition [328].

Fiber diameter, length, and shape all influence lung fiber deposition and, thus, potential tissue injury. Fiber diameter is recognized as the most important factor in determining fiber respirability; the thinner the fiber, the more respirable and penetrating it becomes for the lung. The probability of fiber impaction and sedimentation in the respiratory tract is governed by the aerodynamic diameters of the fibers, which are approximately three times their physical diameter [281,294]; thus, a fiber with a physical diameter of about 3.5 mm will behave aerodynamically as a spherical particle of about 10-mm diameter, a size considered the upper limit of alveolar deposition. Fibers that are very thin are more likely to reach alveoli than larger diameter fibers. In humans, the upper limit for aerosol deposition in the lung is considered to be around 10 mm in diameter, so fibers with actual diameters exceeding about 3 mm are considered to be nonrespirable. Conversely, fibers with diameters under 3 mm but which are very long (50 to 100 mm) are highly respirable. Timbrell and Skidmore [295] showed that finite limits exist for fiber respirability, so fibers with lengths exceeding about 200 mm and diameters greater than about 3 mm would not penetrate the airways to deposit in distal alveoli to a significant amount.

Empirical observations of fiber deposition are supported by modeled predictions of alveolar deposition in rats and humans [60]. Alveolar deposition curves were found to be similar in both species but were maximal at an aerodynamic diameter of 2 to 3 mm in rats (15% alveolar deposition) and ~4 mm (about 30% alveolar deposition) in humans. Essentially no alveolar deposition would occur in humans at ~10-mm aerodynamic diameter, but for rats it would occur at ~5 mm. As the fiber aspect ratio increases, the deposition maxima occur at smaller aerodynamic diameters in both species and the extent of alveolar fiber deposition decreases. Finally, fiber shape is another determinant of deposition. Timbrel and Skidmore [295] found that long, straight fibers tended to align their long axis in the direction of airflow, facilitating penetration into the lung. Fibers with any curvature, bends or twists can twist and rotate in the airstream, increasing the fiber's aerodynamic diameter and therefore reducing its respirability and deposition.

Fiber-containing test atmospheres generated for inhalation toxicity studies require assessment of both fibrous and nonfibrous components to fully characterize the test atmosphere. Large amounts of respirable nonfibrous dust may exacerbate the lung response by taxing the ability of alveolar macrophages to remove inhaled particles. To document exposure conditions, it is essential to minimally characterize fiber composition (including surface modifying agents, such as binding agents), nonfibrous and fibrous length and width distributions, and air sampling data describing fiber concentration (typically by count, such as fibers per cubic centimeter); supplemental measurements can include morphology and aerodynamic particle size and mass. Fiber sizing techniques involve fiber sample collection on an appropriate filter media and counting fiber numbers and/or length and width dimensions by using light, scanning, or transmission electron microscopy.

The three general fiber categories are based on their origin: naturally occurring fibers, manmade synthetic mineral fibers, and synthetic organic fibers (Figure 28.5). The

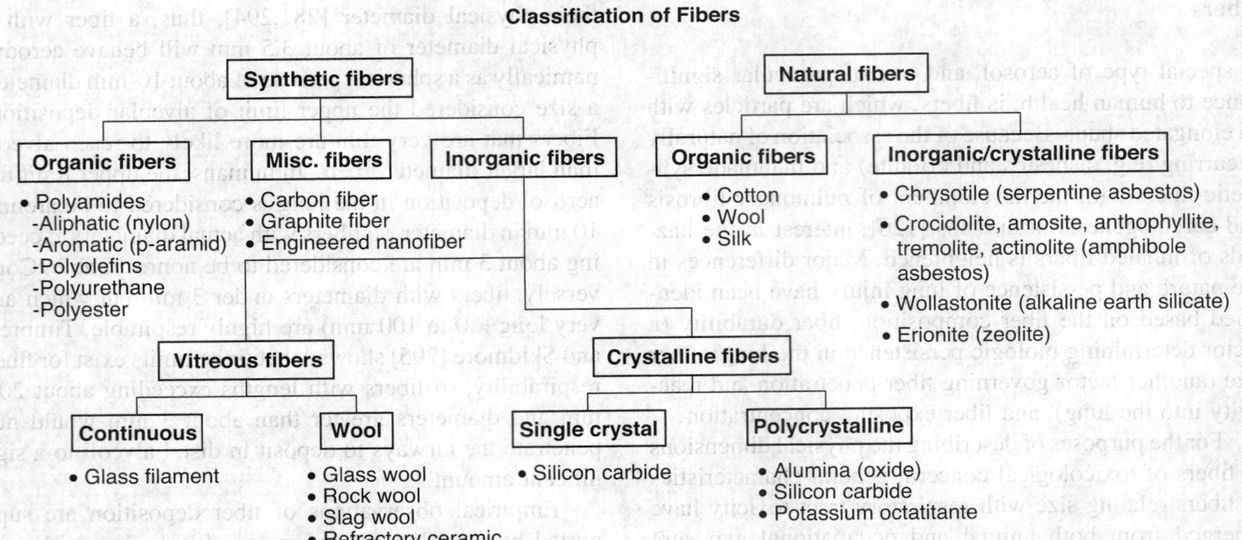


**FIGURE 28.5** Classification scheme for synthetic and natural fibers. (Adapted from ILSI, *Inhal. Toxicol.*, 17, 497–537, 2005.)

most well known and studied naturally occurring class of mineral fibers is asbestos, which is grouped into serpentine (e.g., chrysotile, a hydrated magnesium silicate) and amphibole (e.g., amosite, crocidolite, and tremolite, which are hydrated metal silicates) asbestos. The surface morphology of the two is very distinct; whereas chrysotile is a curly fiber comprised of parallel fibrillar subunits, amosite fibers have a straight, rod-like structure. These are important distinctions that relate to the durability and pathogenicity of these fibers in the lung. Amphibole asbestos has less water solubility than chrysotile; chrysotile is more likely to fragment axially, releasing long thin fibrils that are more reactive, but amphiboles are more likely to fragment longitudinally, releasing shorter fiber fragments. Of the manmade mineral fibers, fibrous glass is perhaps the most studied. Unlike the naturally occurring mineral fibers, fibrous glass is amorphous and does not possess a crystalline structure. Glass fibers are produced by drawing molten glass through a small orifice, producing very small diameter fibers, typically 3 to 10 mm but some as fine as 1 mm. The synthetic organic fibers are useful due to their high tensile strength and chemical or flame resistance. *para*-Aramid fibers, for example, have been used as asbestos replacements, in automotive and aerospace applications, and, more visibly, in ballistic armor. The *para*-aramid fibers exist as either a continuous filament or as pulped material. As produced, *para*-aramid fibers are too thick to be deposited in alveoli, but fiber processing operations can physically break interfibril bonds, literally peeling away respirably sized, curly fibrils [163].

Fibers possess broad differences in biological reactivity in lung tissue that can be related to three general properties: fiber dimension, fiber durability (or biopersistence), and fiber dose. The physical dimensions of fibers determine how and where fibers will deposit (just as with other aerosols, the deposition site will impact bioavailability of the fiber due to site-dependent processes responsible for facilitating fiber clearance). An early clue to the importance of fiber length in the pathogenicity of inhaled fibers was the observation by Stanton and Wrench [279] that, regardless of composition, fibers longer than about 8 mm were more potent in the development of pulmonary mesothelial tumors than shorter fibers. For fibers to have any significant lung deposition, fiber diameters must be less than about 3 mm, so the fibers most likely to be pathogenic will have diameters less than ~1 to 2 mm. Similarly, there is an upper boundary on length, as fibers with lengths greater than ~200 mm do not readily penetrate to peripheral alveoli in humans. In addition to the ability of very-fine-diameter, high-aspect-ratio fibers to penetrate deeply into the respiratory tract, long fiber pathogenicity may relate to whether alveolar macrophages are capable of phagocytizing the fibers. Fibers longer than about 10 to 20 mm can no longer be readily engulfed and affected by intracellular hydrolases. The longer fibers are thus less likely to be cleared by macrophages, thereby enhancing fiber retention, interaction with lung epithelium, and translocation to the pleura. Collectively, the results from various *in vivo* and *in vitro* studies of fiber toxicity indicate that long, thin fibers (i.e., fiber lengths longer than 20 mm and fiber diameters less than 1 mm) are the most pathogenic for lung tumors [186].

Fiber durability is another critical determinant of ultimate fiber toxicity as fibers that resist dissolution or degradation (i.e., are biopersistent) in the lung or other organs of the body increase the likelihood for adverse effects Fiber properties that impart chemical or mechanical resistance to fiber disintegration or fragmentation in the lung will prolong fiber presence in tissue. This property of fibers to disintegrate in tissues was first noted when natural and manmade fibers of different compositions were found to have different toxicological properties and lung retention characteristics [132,188,189]. Investigators found lower burdens of some fibers in exposed lungs than expected based on the fiber exposure concentrations, and in general this correlated with a less severe spectrum of pulmonary pathologic effects [116]. Indeed, a sizeable body of data has accumulated showing that the biological effects of inhaled fibers depended not so much on their composition but rather on their durability and resistance to disintegration in the lung [166]. Studies show that biopersistence of fibers longer than 20 mm are good predictors of pathological responses following inhalation exposure [22,116].

Fiber degradation may proceed by two general pathways. In the first, the fiber dissolves and is completely solubilized within the lung milieu. In the second, the physical dimensions of the fiber (either diameter or length) may be altered by the action of lung fluids on the fiber surface. If solubilization alters the mechanical integrity of the fiber such that the fiber breaks along its longitudinal axis, both the dimensions and number of subsequent fibers change. If a fiber breaks transversely, an additional fiber or particle is produced (depending on the dimensions of the resulting fragments) and the overall length of the fiber is reduced; this can result in facilitated clearance, especially if the fiber fragments are now of a size that may be readily engulfed and removed from the lung by macrophages. Conversely, some fiber types are especially prone to mechanical disruption, releasing more fibers. Chrysotile asbestos deposited in the lung, for example, may split longitudinally, releasing numerous fine fibrils and increasing the lung fiber burden as demonstrated by Bellmann et al. [16]. Similarly, *para*-aramid fibers can be disrupted transversely at fibril defect sites by enzymatic attack to release shorter fibers [137,321]. Dissolution processes that affect fiber length are more likely to reduce potential adverse lung effects by allowing phagocytosis of fiber fragments and alveolar clearance. The various factors influencing fiber durability has been reviewed by Searl [267] and experimental methods for the evaluation of fiber solubility *in vitro* and *in vivo* have been reviewed by Hesterberg et al. [115,116].

Experimental demonstration of fiber degradation *in vivo* has been reported by measuring dimensions of fibers recovered from the lungs of animals. Techniques vary, but common lung digestion methods utilize either thermal or microwave ashing techniques, chemical digestion with strong alkali, or oxidizing agents [136,317], followed by standard fiber size and count techniques. It is imperative to determine the effects of the lung digestion method on the fibers alone to ensure that artifactual reductions in fiber size or number do not occur. This consideration may be bypassed with *in situ* observations of fiber dimensions by using histological techniques and scanning or transmission electron microscopy [38].

## Fine, Ultrafine, and Nanoscale Particles

Besides the coarse particle mode, Figure 28.1 also shows two other aerosol size regimes: fine particles, which are 0.1 to 1 mm in diameter, and ultrafine particles (UFPs), which are less than 0.1 mm in diameter. Data are emerging that show increased pulmonary toxicity in rodents and acute respiratory and cardiovascular distress and increased morbidity and mortality in humans associated with these finer particle size regimes.

Fine and ultrafine particles are incidental to the processes that created them, such as condensation of combustion by-products or high-temperature manufacturing processes. Environmental UFPs may be formed through high-temperature oxidation processes that produce vapors. A common example of an UFP is the zinc oxide combustion fume produced during welding of galvanized steel. Once the vapor saturation point is exceeded for ambient conditions, the vapors can condense, forming high concentrations of ultrafine (smaller than ~30 nm) primary particles. Given their small size, individual UFPs have vanishingly low masses but have relatively high surface areas with respect to their mass. High number concentrations—$10^8$ to $10^9$ particles per cubic centimeter of air (p/cm$^3$)—can be attained but only transiently. At such particle concentrations, the distances between UFPs are very small so individual particles readily collide with and adhere to other UFPs, rapidly forming a particle cluster (chain agglomerate) of lower concentration but much larger size (often up to several microns in dimension); soot is a visible example of such an agglomerate. Supersaturated vapors can also condense on molecular clusters, forming nucleation sites (condensation nuclei) that contribute to further vapor condensation and particle growth. Because agglomeration is proportional to the square of the particle concentration, UFPs at a low number concentration can remain stable indefinitely, whereas UFPs at a high number concentration rapidly agglomerate (milliseconds to minutes) [117,121]. For example, the time required to reduce the initial particle concentration by half due to particle coagulation is estimated to be around 0.2 seconds for $10^{10}$ p/cm$^3$, 20 seconds for $10^8$ p/cm$^3$, and 33 minutes for $10^6$ p/cm$^3$. Particle growth through coagulation and condensation shifts the UFP size distribution to the larger, fine particle mode.

In addition to studies on incidentally formed UFPs, a new field of materials science is emerging with specifically engineered UFP, also referred to as *nanoscale particles* or *engineered nanoparticles*, referred to hereafter as *engineered nanomaterials* (ENs). As a class, the ENs hold great promise for a variety of emerging new applications

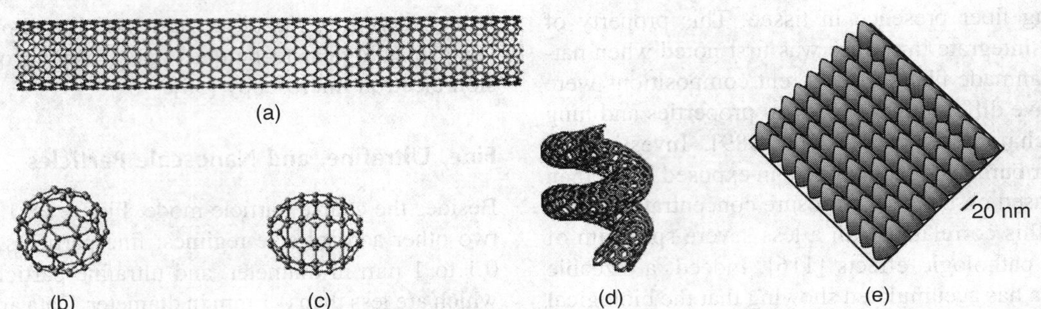


**FIGURE 28.6** Morphology of engineered nanomaterials: (a) carbon nanotube consisting of carbon atoms arranged in a pentagonal/hexagonal matrix coiled in a tubular form; (b) the basic building block of nanotubes is the carbon lattice structure known as a buckyball, a C60 fullerene; (c) an ellipsoid C70 buckyball; (d) a coiled nanotube; (e) a nanodot array.

based on the unique optical, structural, and electrical properties of substances that exist at the nanoscale. ENs are expected to transform materials technology by providing more environmentally sustainable products that utilize energy and resources effectively and minimize waste. For materials scientists, this class of substances is a target for intense technical innovation. Given their likely widespread commercial use and potential for human exposure, ENs have become the focus of strong industry, academic, and regulatory interest in attempting to understand their potential risks [221].

Engineered nanomaterials differ from ambient UFPs by being specifically made to achieve desired physical, chemical, optical, magnetic, catalytic, and morphological properties that are not evident in larger, bulk particles. ENs are fabricated in the laboratory using pure substances such as carbon or metals and are more homogeneous in composition than ambient UFP. Because ENs are intended for specific applications, they can be made to be highly uniform (monodisperse) or of varied size and shape. ENs have at least one dimension less than 100 nm, a size regime that approximates that of proteins. General nomenclature for the various shapes of EN is as follows: Engineered substances with one dimension in the nanoscale are planar structures known as *nanoplates* (e.g., nanoclays); particles with two dimensions in the nanoscale are known as *nanotubes* or *nanowires*; and particles with three dimensions in the nanoscale are known as *spherical nanoparticles*.

The size and shape distributions of ENs are important when attempting critical particle and biological receptor interactions (e.g., EN penetration through specific cell membrane pores). Regulating the size characteristics of EN is very important if the material properties are to be predictably exploited. Strict controls of the size and shape are important when developing fluorescent probes that emit light within a narrow range of wavelengths, a characteristic important when developing molecular probes (biomarkers) with selective but diverse spectral properties.

Given the ability to impart functionality to ENs through chemical modification of the ENs, various biomaterial applications are under investigation. These engineered biomaterials consist of a core EN particle of either inorganic or polymeric materials and an outer layer of a specific, functionalized molecule for use in protein/DNA identification, targeted delivery of drugs, genes, or antitumor therapeutics. Nanoscale $TiO_2$, for example, can adsorb ultraviolet light, creating an electron hole that can interact with water at the particle interface to produce a variety of reactive oxygen species [35,152]. Unless it can be modified to eliminate the generation of reactive oxygen species, nanoscale $TiO_2$ would not be useful in biomedical applications. The reactivity of nanoscale $TiO_2$ can be overcome by coating the $TiO_2$ surface with various metal oxides to preserve its electromagnetic properties while minimizing the generation of reactive oxygen species.

Carbon ENs (based on cage-like, hexagonal, carbon lattice structures known as *fullerenes*) are hollow structures that can be spherical (e.g., buckyballs), tubular (nanotubes with high aspect ratios and fullerene caps), or ellipsoid in shape (Figure 28.6). Carbon nanotubes are formed from vapor condensation around a metal catalyst and therefore are not pure carbon, but they can contain significant quantities of various metals (e.g., iron, yttrium, nickel). ENs find favor with materials scientists because of their potential applications in high-strength mechanical reinforcements, semiconductors, photocatalysis, pharmaceuticals for novel therapeutic delivery systems, protective coatings, and myriad other uses, many yet to be conceived.

The arrival of ENs into commerce will most likely involve some exposure to humans during their manufacture, use, or disposal. EN development will entail the manufacture of and their application to diverse products with foreseeable dermal and inhalation routes of exposure, the latter perhaps deserving most attention. There is insufficient data so far to say what metric of exposure, size, mass, count, particle surface area, etc., is the most significant feature dictating biological responses to ENs, but generally the smaller the particles the more toxic they are likely to be [222]. It is not certain what the physical form of the inhaled ENs will involve; for example, will exposures to ENs involve isolated, singlet particles or aggregates containing large numbers of singlet EN particles?

Recent studies by Maynard et al. [175] suggest that airflow over bulk, single-wall carbon nanotubes (SWNTs) can release low levels of submicron-sized particles and that agitation of bulk powder containing SWNTs with a fluidized bed releases small quantities of both ultrafine and fine particles as detected by electrical mobility and aerodynamic particle analyzers. Although the generation methods were not efficient in producing high levels of respirable SWNT aerosols and are unlikely to represent the types of processes occurring in manufacture, they nonetheless show that both nanosized and micrometer-sized (as aerodynamic diameter) aerosols could be generated. However, actual workplace air sampling during physical handling of SWNTs did not result in increased airborne mass or count concentrations of SWNTs.

Not only will new technologies be required to develop appropriate monitoring and detection procedures for the various processes employing ENs, but due consideration will also be needed to ensure that measurements are conducted using the most relevant concentration- or composition-related parameter of nanoparticle exposure. At present, the most appropriate size-related parameter is not known. Analysis of EN size distributions (i.e., particle count, surface area, or mass of ENs per unit volume of air) [337] or surface morphological or chemical properties may provide the ability to discriminate process-specific ENs from other ENs or ambient nanoparticles. Although nanoparticle surface area measurements may be a prognostic indicator of potential inflammation and attendant health sequelae [285,297], other studies suggest that surface area might not be, at least for $TiO_2$-based nanostructures [321]. Further complicating the analytical picture is the composition of the ENs themselves. Based on *in vitro* studies with human cells, Sayes et al. [254,255] reported that increasing the degree of functional derivitization (e.g., hydroxylated or oxidized) of SWNTs or fullerene structures reduces cytotoxicity, as measured by the generation of reactive oxygen species. Pure fullerene structures appear to have a high degree of rapid reactive oxygen species activity, which can be inhibited by acetylcysteine *in vitro*; in contrast, functionalized polyhydroxylated fullerenes can cause cytotoxicity and apoptosis by a delayed, non-reactive oxygen species pathway [127].

Despite the expected benefits from the development of novel EN materials, as of this writing there is a paucity of material characterization, toxicological, and exposure data on ENs to allow reasoned judgments regarding the potential health and environmental hazards these materials might pose in commerce. More insights into the toxicological hazards of these particles are discussed in the Fine, Ultrafine, and Nanoscale Particle Effects section.

### Aerosol Deposition Processes

From the nares to the lungs, inhaled air follows a tortuous pathway. Inspired air has a relatively high velocity in the nose and follows the convoluted pathways of the nasal cavity. In the conducting airways, air velocity moderates, given the increase in total cross-sectional airway area with each bifurcation, but the airstream changes direction with each inhalation/exhalation cycle. Finally, in the pulmonary region, air velocity slows greatly because the inspired volume of air is now distributed over the enormous total cross-sectional area of alveolar gas exchange region. Different deposition mechanisms exist within these regions, and each is influenced by the aerodynamic diameter of the aerosol and the individual's respiratory tract anatomy and breathing characteristics. These deposition mechanisms are shown schematically in Figure 28.7 and include impaction, sedimentation, interception, and diffusion:

- *Impaction*—When airflow changes direction, the inertia of any suspended particle will cause it to continue in its original direction for some finite time before changing direction. Thus, impaction occurs when the inertial characteristics of the particles are such that they cannot follow the airstream; the particles depart from the airstream and impact on the airway surface downstream from the directional change. Impaction is dependent on the velocity and angular change in the airstream direction and is also dependent on particle density and the aerosol diameter squared. Impaction is the predominant deposition mechanism for particles larger than 1 μm and regions of the respiratory tract where air velocities are high and airstream directional changes are abrupt.
- *Sedimentation*—Particles suspended in a gas will slowly settle under the influence of gravity. The speed at which particles settle is proportional to the density of the particle and its diameter squared. Sedimentation is also an important deposition mechanism for particles larger than 1 μm and regions of the respiratory tract where air velocity is low, particle residence times are high, and airway diameters are small.
- *Interception*—Interception occurs when the trajectory of the particles brings them sufficiently close to airway walls such that the particle contacts the airway wall. Interception is an important deposition mechanism for fibers, especially for high-aspect-ratio fibers as they will have a greater chance of airway contact. Fibers generally align themselves with the airflow; however, when airflow direction suddenly changes such as where airways branch, deposition can occur at

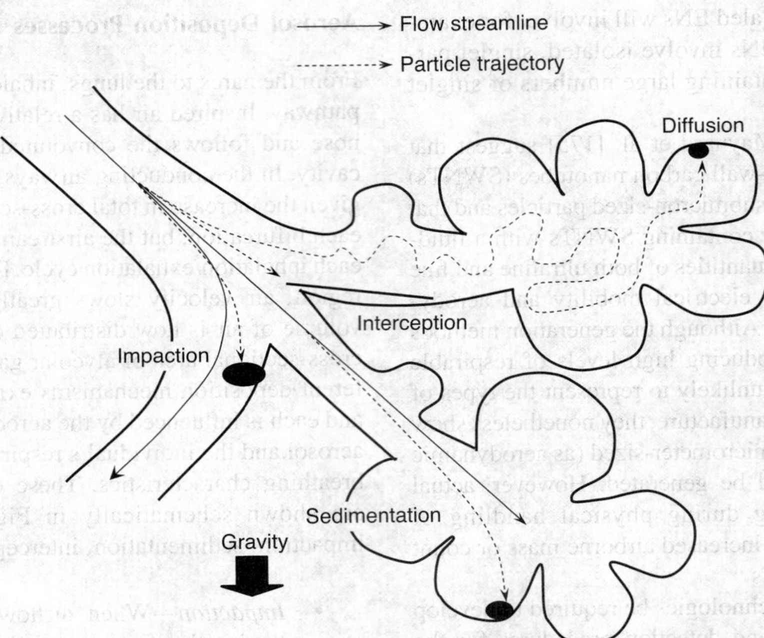


FIGURE 28.7 Mechanisms of particle deposition.

the bifurcation (i.e., the ridge where the airways diverge). Interception increases as airway turbulence increases and airway diameter decreases.

- *Brownian diffusion*—Gas molecules that surround aerosols continuously strike these submicrometer particles, inducing random particle movement. This process is inversely proportional to particle diameter but is independent of their density. Diffusion is not an important deposition mechanism for particles larger than about 0.5 μm but represents the major mode of particle deposition in the alveoli where airflow is low. Diffusion may occur throughout the respiratory tract.

Physiologic and anatomic factors also influence aerosol deposition. Air entering the nose must flow through a series of convoluted passages. Compared to humans, the nasal passages in rodents are more complex and tortuous [263,301], resulting in comparatively higher particle deposition in the nose of rodents. In contrast, mouth breathing effectively bypasses the particle deposition processes of the nose and results in a higher proportion of particles that may reach the lower respiratory tract.

The branching pattern of the tracheobronchial tree influences airflow dynamics and the amount and location of particle deposition in three ways. First, the diameter of the conducting airways becomes progressively smaller with each successive airway division. In humans, the major bronchi have diameters of about 1 cm, and bronchioles and alveolar ducts are about 1 mm in diameter; thus, the chances of particle deposition in a given period of time due to diffusion or sedimentation are much higher in smaller airways as inhaled particles will have shorter distance to travel before contacting airway walls in these areas. Second, the total number of airways increases with each airway branch. Although the airway branches become progressively smaller, the number of these branches also increases such that the total cross-sectional area increases with successive airway divisions. A given volume of air thus moves more and more slowly as it enters the deep lung. Third, variations in airway branching patterns affect the amount of particle deposition. In humans, for example, the airway branching pattern is dichotomous (symmetrical), whereas in rodents the branching pattern is monopodal (asymmetric). Airflow in dichotomous airways is highest toward the airway centerline and results in localized areas of particle deposition by impaction and local turbulence primarily at bifurcations. Indeed, bifurcations of the upper airways are hotspots for deposition of inhaled particles. In contrast, little change in either airstream direction or velocity occurs in animals with monopodal branching patterns because the branching angles of the daughter airways are much shallower compared to dichotomous branching. As a consequence, particle deposition is comparatively lower in animals with monopodal branching.

Quantitative particle deposition studies have been conducted in both humans and animals by numerous investigators. Due to the variability in the experimental techniques, airway anatomy, and respiratory physiology differences between individuals, however, aerosol deposition models for humans have not been defined precisely. Nevertheless, a widely cited model was developed by the International Commission on Radiologic Protection (ICRP) [124] for regional aerosol deposition in humans (Figure 28.2). This model has been found to be in general agreement with experimental deposition data for humans [182]. Although the latest ICRP aerosol deposition model includes several improvements over previous ICRP models, such as accounting for oral and nasal cavity deposition and deposition of ultrafine particles, the model still represents average deposition behavior but does not fully describe the range of regional particle deposition due to intersubject variability.

The collective deposition data for humans show that deposition in the pulmonary, bronchiolar, and nasopharyngeal regions is at a minimum with approximately 0.3- to 0.5-μm (aerodynamic equivalent diameter) aerosols. Particles in this size range are minimally affected by Brownian diffusion, sedimentation, and impaction processes and consequently are mostly expired; however, total deposition increases markedly with aerosols either larger or smaller than 0.3 to 0.5 μm. With smaller particles, diffusion dominates deposition so particles in the ultrafine size range behave aerodynamically like gas molecules. In addition, particle diffusion is not affected by particle density so aerodynamic diameters do not apply to UFPs. Although actual deposition data for UFPs in the various compartments of the respiratory tract are scarce, modeling studies indicate that UFPs have very high deposition efficiencies, as much as 50% for ~10-nm particles in alveoli [124].

The difficulty in obtaining regional deposition data is due to methodological and analytical limitations involving technical isolation of the target respiratory tract areas and analytical sensitivity for measuring low levels of UFPs. Some workers have measured total depositions of UFPs that corroborate model predictions. Wilson et al. [330] and Daigle et al. [61] measured differences in inhaled and exhaled UFP concentration to demonstrate an inverse relationship between particle size and total respiratory tract deposition of UFPs in resting, normal humans: ~80% for 8.7-nm particles, decreasing to ~71% deposition for 24-nm particles and 37% for 240-nm particles. Similar findings were obtained in groups of resting elderly and young adult humans, suggesting no significant differences in the dose of inhaled UFPs as a function of age [143]. Also, breath holding appears to increase the deposition of UFP, an observation suggesting a diffusion-mediated deposition mechanism [129]. Based on studies using nasal casts, Cheng et al. [51,52] and Swift et al. [289] showed that UFPs (<0.2 μm) do not preferentially deposit in the lower respiratory tract but have relatively high deposition efficiency in the nose. In this work, upper airway and nasal deposition were increased proportionately with decreasing particle size (<0.2 μm).

For very small UFPs (below 2 nm), deposition in the nose was found to predominate, with deposition approaching 80%, presumably via a turbulent diffusional mechanism. As described earlier, singlet UFPs have short atmospheric lifetimes as agglomeration rapidly increases particle size to the fine particle size regime (~1 μm diameter). Fine mode particles are not likely to be extensively deposited compared to singlet UFPs as the deposition processes of diffusion and sedimentation are not very efficient for 0.1- to 1-μm aerosols (see Figure 28.2).

Aerosols larger than ~0.5 μm deposit predominantly in the nasopharyngeal region by impaction or interception processes. Particles larger than approximately 30 to 80 μm are generally not inhalable through the nose [310]. Due to the higher collection efficiency of particles within the nose, the inhalable particle diameter increases and total deposition decreases with particles greater than 0.2 μm for mouth breathing. For the same size aerosol (including UFPs), the net effect of nose breathing is to reduce the total amount of aerosol available for delivery to the lung relative to mouth breathing.

Aerosol deposition in experimental animals has been investigated and reported [261,262]. As already noted, deposition in the upper respiratory tract of particles between approximately 0.5 and 1 μm is minimal in humans but this deposition minimum is also seen in most species. The relationship between the amount of nasal deposition with increasing particle size is similar among dogs and humans, although the amount deposited in the nose is lower in dogs compared to humans. The data also suggest that the human nose has higher deposition efficiency than rodents for particles greater than 1 μm. Given the anatomic complexity of the rat nose, this unexpected observation may result from the experimental conditions used (i.e., decreased inertial impaction in rodents from an anesthesia-related reduction in breathing rate and airflow velocity). In the tracheobronchial region, deposition among test animals is minimal but relatively constant for 0.1- to 5-μm particles; by comparison, deposition efficiency in humans is generally higher than in dogs or rodents. In the pulmonary region, aerosol deposition is similar for rats and hamsters but is considerably lower than for humans and dogs; alveolar deposition efficiencies for dogs and nasal-breathing humans are comparable. Although peak alveolar deposition efficiencies in animals occur with about 1- to 2-μm aerosols, most alveolar deposition in humans occurs with approximately 2- to 4-μm particles.

Overall, the relative particle deposition efficiencies within various portions of the human and nonhuman respiratory tracts are quite similar, with somewhat lower deposition efficiencies for rodents compared to humans. However, when differences in lung size, body size, and ventilation rates are considered, smaller animals will receive a greater dose (for a given exposure time) per unit of lung or body weight than humans. For particles approximately 1 μm in diameter, the rat can be expected to receive a dose roughly five to ten times and the dog about three times that of humans on a per-unit-lung-mass basis [231].

The practical implications of defining particle size in the interpretation of inhalation toxicology studies center on the need to generate an aerosol respirable for the species under investigation. The aerosol generation system, connecting ductwork, and exposure chamber conditions must be evaluated and optimized, if necessary, so particles of respirable size for the test species are produced and delivered. For acute studies in rodents, the goal is to expose animals to an aerosol that can deposit throughout the respiratory tract. The objective then is to ensure that every anatomic compartment of the respiratory tract has some exposure to the test substance without selective deposition in any one area. This topic was investigated by an *ad hoc* group with the recommendation that, for acute studies, test aerosols should be produced with a mass median aerodynamic diameter (MMAD) of 1 to 4 μm to allow deposition in both upper and lower regions of the rat respiratory tract [141]. This recommendation was based on the fact that, because aerosol toxicity may be expressed in specific anatomic regions (e.g., the nose), testing should not be conducted with aerosols having a size range that might preclude deposition at other, perhaps more sensitive sites. On the other hand, for subchronic or chronic inhalation studies, the emphasis should be on detecting functional or pathologic changes in the lower respiratory tract such as alterations in lung clearance, pulmonary function, or chronic disease (fibrosis or cancer). Thus, with consideration to detecting adverse lung effects in repeated exposure inhalation studies, a comparatively smaller sized aerosol (i.e., 1 to 3 mm) was recommended [165].

## Fine, Ultrafine, and Nanoscale Particle Effects

Toxicological interest in the mechanisms and potential health impacts of the fine particle mode inhalation continues but is being overshadowed by interest in ultrafine particle effects, both those associated with ambient UFPs as well as engineered nanoparticles. Interest in the former is based on two general but widely researched observations linking the fine and ultrafine particle fractions with an exacerbation of pulmonary injury compared to larger particles and epidemiological studies reporting associations between exposure to low concentrations of ambient fine particles with sickness and deaths in elderly populations. More recently, interest in the hazards of ENs is due to a paucity of data on the hazards of these novel materials and the observation that at least some types of nanomaterials (e.g., SWNTs) are more toxic to the lung than carbon black, essentially an ultrafine carbon particle soot.

In the early 1990s, a series of epidemiological studies described associations between the levels of ambient particulate matter and the incidence and severity of health effects in humans [64,65,81,234,235,264–266]. Although UFPs were not monitored, the association of adverse health outcomes with aerosol size was generally stronger for fine particles (<2.5 μm; $PM_{2.5}$) than coarser particles (<10 μm; $PM_{10}$) [64,65]. Specifically, these studies indicated that inhalation of $PM_{10}$ at levels below the federal ambient air quality standard of 150 μg/m$^3$ (24-hour average), was associated with increased daily mortality. Other workers have also found an association between $PM_{10}$ inhalation and increased hospital admissions and emergency room visits for respiratory distress [253], increased incidence of asthma attacks [327], increased use of asthma medications, and reduced pulmonary function [233]. Where deaths were reported, they were primarily in those individuals whose health status was compromised (i.e., people typically over 60 years old and with preexisting cardiovascular or chronic pulmonary diseases). Interestingly, the data do not suggest that ambient particle concentrations are a cause of death among *healthy* individuals. The severity of the health effects also differed among the cities evaluated, suggesting that the physicochemical properties of the aerosol at a given location impact the shape of the exposure response curve [277].

Although these data suggest a role for inhaled fine particles in adverse health effects, the mechanisms or roles of copollutants require further study [307]; for example, some current research is exploring the roles of immune system suppression [336]; blood coagulability [207]; particle- or particle-associated endotoxin-mediated inflammatory [259,290] and thrombogenic events following translocation into the blood; oxidant stress induced at extrapulmonary sites leading to adverse cardiovascular outcomes; and particle-induced airway inflammation as moderators of pulmonary function and heart rhythm variability [86,99]. By whatever mechanism, what makes these epidemiologic observations so intriguing is that acute health effects are reported to occur in susceptible individuals below current regulatory standards, sparking debate over whether and to what extent the national ambient particulate matter standard of the 1996 Clean Air Act should be reduced [305]. As of this writing, the U.S. Environmental Protection Agency has proposed revisions to the National Ambient Air Quality Standard for particles by recommending a limit of 35 μg/m$^3$ ($PM_{2.5}$, 24-hr) and 70 μg/m$^3$ ($PM_{2.5-10}$).

Experimentally, very fine particles have been observed to produce a more severe pulmonary response compared to larger particles of the same material. Oberdörster et al. [215] described differences in the inflammatory response between fine (0.25 μm) and ultrafine (0.02 μm) $TiO_2$ particles following instillation in rat lungs. Only the instillation of the ultrafine $TiO_2$ produced inflammatory changes described as an increase in total number of cells recovered primarily polymorphonuclear leukocytes, and increased protein in the bronchial lavage fluid. Similar studies showed that release of proinflammatory mediators might be exaggerated following exposure to UFP. Driscoll and Maurer [72], for example, showed that ultrafine (0.02 μm) $TiO_2$ recruited a greater number of alveolar macrophages and polymorphonuclear leukocytes, increased levels of inflammatory cytokines in bronchoalveolar lavage fluid, and increased interstitial inflammation and collagen compared to fine (0.3 μm) $TiO_2$. These authors have also reported that the pulmonary burden of ultrafine $TiO_2$ was greater than the equivalent mass of fine $TiO_2$. Taken together, these authors have suggested that UFPs, even those of inherently low cytotoxicity and solubility such as $TiO_2$, have greater pulmonary inflammatory activity and toxicity than larger particles.

The hypothesis that UFPs are comparatively more reactive and toxic to the lung has received some support from the work with UFPs from the combustion of the perfluoropolymer polytetrafluoroethylene. Studies begun by Waritz and Kwon [322] and later pursued by Seidel et al. [268], Warheit et al. [316], and Oberdörster et al. [217], indicated that exposure to freshly generated polytetrafluoroethylene UFPs (approximately 30 nm in diameter) was lethal to rats at aerosol concentrations of approximately 100 μg/m$^3$ (approximately 1 × 10$^6$ p/cc in 30 minutes); conversely, allowing the aerosol to age for 5 minutes significantly reduced toxicity, ostensibly by allowing the particles to agglomerate to a larger size [268]. Although there is no exposure to perfluoropolymer UFPs in ambient air, studies with particles of different compositions provide opportunities to explore the mechanistic bases for UFP-mediated lung injury. Overall, the various hypotheses proposed suggest that UFPs are comparatively more reactive, perhaps due to their comparatively greater surface area and higher particle concentration per unit of mass, and are more toxic either by translocating more rapidly through the epithelium and into the interstitium [84], thus causing greater impairment of alveolar macrophage clearance processes [216], or by enhancing activation of alveolar macrophages to release inflammatory mediators [73].

As noted earlier, ENs have prompted great interest among inhalation toxicologists and regulators given the paucity of data relating to the potential health hazards of these materials. The hazards may relate to the intrinsic physical properties that occur at the molecular scale. Importantly, when particles are made progressive smaller, the surface-to-volume ratio of the particles increases exponentially. Upon reaching nanoscale dimensions, the proportion of atoms on the particle surface available for interaction with other surfaces and particles also increases exponentially. One expects that, as a result of this increased availability of substances on particle surfaces, the chemistry at the surface of the nanoscale particle can also change; that is, the particle can become chemically or catalytically reactive as particle size is reduced.

Besides potential catalytic activity at the nanoscale, the intended size of ENs belies the fact that they also have a tendency to agglomerate via van der Waal's forces, forming larger particle aggregates. ENs may be coated with a variety of substances, both to impart functionality as well as to stabilize the particles against aggregation. Doing so, however, not only changes the particle composition but can also alter ease of atmospheric entrainment, the nature of particle–tissue interactions, and distribution within the body to further complicate EN hazard assessment.

Single-wall carbon nanotubes (see Figure 28.6) are a form of ENs that possess a fiber-like structure of nanometer diameter but up to several micrometers in length. SWNTs can be engineered to possess a range of desirable chemical, mechanical, electronic, fire-retardant, and magnetic properties [54], but to date they have not been extensively studied for their potential health effects. SWNTs contain the same types of hexagonal molecular structures as graphite, and, indeed, some manufacturers consider SWNTs to be essentially carbon from a health-hazard perspective. SWNTs, however, are fibrous nanoparticles that are essentially insoluble and highly nondegradable, factors that may predispose them to remain in the lung and increase their toxicity relative to graphite or more soluble materials.

The first studies addressing the pulmonary toxicity of SWNTs were conducted by Lam et al. [155] and Warheit et al. [320]. Because they are expensive to manufacture, limited quantities of ENs are available for traditional inhalation toxicity studies. For this reason, available studies on EN effects to the lung have used intratracheal (IT) instillation as the route of exposure. Although they permit the use of minute quantities of these costly materials, the physiological relevance of IT instillation techniques vs. inhalation exposure has been questioned, but in this case IT instillation is appropriate as it allows the generation of data related to the SWNT mechanisms of action.

Lam et al. [155] intratracheally instilled a single bolus of either 0.1 or 0.5 mg of carbon SWNT, carbon black (a negative particle control), and quartz (a reference inflammatory positive control particle) into mice. Three types of carbon SWNTs were tested, fabricated using different processes and containing different levels of residual metallic iron, nickel, or yttrium manufacturing catalysts. All three SWNTs produced dose-dependent granulomas and interstitial inflammation that became more severe after 90 days. Large aggregates of SWNT-containing macrophages were seen in alveoli among all test materials, with some SWNT aggregates migrating into the interstitium, causing interstitial inflammation and forming granulomas. By comparison, carbon-black-treated mice were normal, and animals treated with quartz had mild to moderate inflammation. The observation that both raw and pure SWNT (those containing high and low levels of residual metal catalysts, respectively) produced lung granulomas suggested that metal content alone was not responsible for the granulomas. Lam et al. concluded that, on a mass basis, SWNTs were more toxic than carbon black or quartz and that, regardless of the residual metal catalyst present in the EN, SWNTs produced more severe injury than other nanosized materials.

In contrast, Warheit et al. [320] reported somewhat different results with SWNTs. Warheit instilled proportionately more (1 or 5 mg/kg) of SWNTs (1.4-nm diameter; >1 μm length) and micrometer-sized quartz (positive control), carbonyl iron (negative control), or graphite particles into rat lungs and evaluated bronchoalveolar lavage biomarkers, cell proliferation, and lung histopathology up to 3 months after treatment. Significant inflammatory changes were reported with quartz but no significant effects were seen with either carbon black or carbonyl iron particles. Unlike Lam et al., Warheit's group observed that SWNTs produced only transient inflammatory changes; multifocal granulomas were observed from SWNT exposure, but these effects were not dose dependent, were nonuniformly distributed, and were not progressive beyond 1 month. Warheit et al. concluded that the transient inflammatory changes and granulomas produced by SWNTs did not occur with expected changes in pulmonary biomarkers or increases in cell proliferation and suggested that SWNTs may be producing these changes by a potentially new mechanism of toxicity. Collectively, the studies by Lam and Warheit serve to highlight the need for more extensive physicochemical characterization of test substances and more rigorous species, dose, and endpoint controls to allow better understanding of the mechanisms of SWNT-induced lung injury.

In addition to direct effects on the lung, recent studies reported that within hours inhaled nanoparticles may be translocated to the heart, brain, liver, and other parts of the body. Although the mechanisms by which nanoparticles penetrate tissues and distribute to different body compartments are not known, *in vitro* studies with alveolar macrophages or red blood cells exposed to radiolabeled polystyrene microspheres (0.08, 0.2, and 1 μm in diameter) indicated that particle uptake may be by diffusion; endocytic (membrane invagination of particles) processes did not appear to occur. Internalized particles were not membrane bound and could access intracellular organelles and DNA. Similarly, inhaled $TiO_2$ nanoparticles (22-nm count median diameter [CMD]) were found in airway and airspace lumens, within tissues (epithelial and endothelial cells and fibroblasts), and within intracellular (cytoplasm, nuclei) compartments within an hour of exposure; within 24 hours, approximately 80% of the particles remained on the epithelial side of the airspaces, suggesting that $TiO_2$ nanoparticles had strongly adhered to epithelial surfaces [92].

Some evidence suggests that tissue penetration and extrapulmonary translocation of ultrafine particles may be dependent on their size and composition. Kreyling et al. [154] conducted mass balance measurements on rats following intratracheal inhalation of $^{192}$Ir nano particles (15 and 80 nm CMD) for 1 hour. Particles were found to be cleared mainly via airways into the gastrointestinal tract with small (<1%) amounts translocating into secondary organs such as liver, spleen, heart, and brain; larger nanoparticles translocated about tenfold less than the smaller nanoparticles. Oberdörster et al. [219] studied the uptake of inhaled ultrafine carbon particles ($^{13}$C nanoparticles, ~20 to 29 nm CMD at 80 and 180 μg/m$^3$) from rat lung in various tissues (liver, heart, brain, olfactory bulb, and kidney). Although some carbon nanoparticles were found in the liver 0.5 hour after inhalation at 180 μg/m$^3$, by 24 hours, the amount in the liver was much greater, amounting to about a fivefold higher mass than the amount observed in the lung. Interestingly, no significant accumulation of carbon nanoparticles occurred in the other tissues.

These findings contrast with more recent studies where inhaled carbon nanoparticles ($^{13}$C, ~36 nm CMD at 160 μg/m$^3$ for 6 hours) were tracked from the lungs to the cerebellum, cerebrum, and olfactory bulb [220]. Although concentrations of nanoparticles were inconsistently increased in the cerebellum and cerebrum, particle concentrations in the olfactory bulb were significantly elevated and increased with time after exposure. The presence of nanoparticles in the olfactory bulb is thought to follow particle deposition on nasopharyngeal tissues with subsequent translocation along olfactory neurons axons into the brain [220,296]. The presence of this particle transport pathway may have special importance for ENs, as the nose is a site of major nanoparticle deposition, with the smallest nanoparticles having the highest nasal deposition (Figure 28.2).

## Biological Mechanisms of Particle Injury

As described in the previous section, a growing body of evidence suggests that fine particles are inherently more toxic than larger particles. Proposed hypotheses for fine-particle-induced injury include: (1) reductions in the integrity of the pulmonary epithelial and endothelial barriers, (2) impaired pulmonary host defense mechanisms, (3) release of inflammatory mediators to produce either local or systemic effects, (4) impaired particle clearance with airway hypersecretion of mucus, (5) aggravation of preexisting airway occlusion, and (6) direct or indirect effects on cardiovascular function [47]. Although all appear to be involved to varying degrees, particle-mediated inflammation may well be the predominant mechanism for respiratory tract injury and cardiovascular effects.

Inhaled particles can be directly cytotoxic by effecting transudation of serum proteins from alveolar capillaries. This may involve peroxidation of membrane lipids, resulting in defects in membrane integrity, function, and subsequently cell death; if injury is sufficiently severe, the airways and alveoli fill with fluid, causing pulmonary edema and ultimately death. The source of these radicals may be membrane-based oxidases [303].

Some metal-containing particles, especially those that either contain or can complex the ferric form of iron ($Fe^{+3}$), are particularly active in inducing membrane peroxidation through a reduction–oxidation pathway, producing hydroxyl radical (OH·) [95]; the inclusion of chelators markedly reduces the oxidant-generating capacity of the particles [96]. This pathway may be especially important for nanoscale materials, as these particles may have a disproportionately larger number of catalytic sites on the surface relative to larger particles [78,297]. By a related mechanism, particles, either by themselves [331] or by interaction with pulmonary alveolar macrophages or alveolar type II cells [133], pulmonary epithelium [269], or polymorphonuclear leukocytes [237], can stimulate generation of reactive oxidant species such as hydrogen peroxide ($H_2O_2$), superoxide anion ($O_2^-$), or hydroxyl radical. These radicals may cause either direct tissue injury or genetic damage. A variety of metal-containing dusts have been shown to stimulate production of reactive oxidant species from macrophages [17], with iron being among the most active metals [87,245,308]; other metal dusts such as titanium dioxide are either inactive or minimally active [23,100,213].

Oxidant production, whether caused by fine or ultrafine particles, may also oxidize and deplete glutathione, a cellular antioxidant [286]. The relative levels of reduced and oxidized glutathione appear to play a pivotal role in determining the lung's inflammatory response to oxidant stresses. Alterations in the redox levels of glutathione can affect the degree of acetylation or deacetylation of histone proteins on DNA and allow transcription to occur [242]. Conformational changes in DNA allow access of the nuclear transcription factors nuclear factor kappa B (NF-κB) and the associated activation protein [89,106], permitting gene transcription. Studies with SWNTs *in vitro*, for example, show cells undergoing oxidant stress with a decreased ratio of reduced/oxidized glutathione, thereby triggering activation of the AP-1 and NF-κB nuclear transcription factors [63,172]. Gene transcription results in the production of various proinflammatory cytokines, a class of chemical messengers that regulate cellular homeostasis and cell proliferation. Notably, the initiating or proinflammatory cytokines, such as tumor necrosis factor α (TNFα) and interleukin-1 (IL-1), can induce production of other cytokines that are responsible for the recruitment of inflammatory cells such as neutrophils to sites of particle deposition [76] with subsequent activation of those recruited cells to produce reactive oxygen species, providing further toxic insult to airway epithelium integrity.

Noncytotoxic, poorly soluble particles can lead to tissue injury from the prolonged inflammatory response induced by high particle concentrations (either from high exposure levels or inability to remove inhaled particles) in the lung [201]. This phenomenon, known as *lung overload*, can produce pathologic lesions in the lung similar to more cytotoxic dusts such as quartz [214] but requires much higher lung burdens than with quartz. Lung overload arises when the normal clearance processes become less and less effective with high particle exposures, leading to accumulation of particles within the lungs [200]. Pathologically, overload appears initially as an increased number of particles in the interstitium in macrophages in airspaces; with time, chronic inflammation may ensue, resulting in alveolar cell hyperplasia, granulomas, fibrosis, and, potentially, lung tumors [199]. Lung tumors have not been reported if particle clearance and pulmonary inflammation do not occur, suggesting a threshold-related phenomenon for overload-related lung tumors.

The functional decrements in clearance due to dust overloading are apparent if the clearance kinetics is measured; overloaded lungs exhibited a marked increase in particle retention compared to the normal lung. The mechanism for the overload phenomenon is believed to be due directly to a saturation of the capability of alveolar macrophages to ingest particles. The data suggest that alveolar macrophages become overloaded not by the total mass of ingested particles but by the cumulative volume of particles within macrophages. Morrow [199] estimated that overload may occur when the average volumetric load of particles exceeds 60 $\mu m^3$/macrophage and that macrophage-mediated clearance ceases when the volumetric load exceeds 600 $\mu m^3$/macrophage. Once the phagocytic activity of macrophages is reduced by large numbers of particles within alveoli, the mobility of macrophages also diminishes even if the particles are not inherently cytotoxic. The immobilization of alveolar macrophages within the lung results in two detrimental effects. First, subsequent clearance is greatly slowed, and, second, the macrophages may continue to elaborate bioactive substances (e.g., proteases, cytokines, growth factors, oxidants, immunomodulating agents) within the lungs. The overall effect is to increase the likelihood for a prolonged inflammatory response with attendant lung injury or tumor development.

The impetus for the overload theory lies, in part, from rodent chronic inhalation bioassays with diesel soot [169,174,176], carbon black [110,211], antimony trioxide [68], toner [204], titanium dioxide [162], and talc [212], particles that previously had been considered benign and do not ordinarily cause pathologic changes if lung clearance processes are not overwhelmed. As a class, these particles have low solubility and low cytotoxicity and, with the exception of diesel soot, are nongenotoxic, although genotoxicity *per se* is not a rigid requirement for tumor development with these particles [111]. Consistent with all, however, is the observation that high levels of dust exposure are associated with an increasing lung particle burden. At some burden above the capacity of alveolar macrophages to remove particles, lung overload occurs [319].

Several features of the lung overload phenomenon impact risk assessment. First, species differences are known in the levels at which overloading and the pathologic sequelae occur, complicating extrapolation of results in test animals to humans; for example, rats have developed lung tumors when mice or hamsters exposed to similar

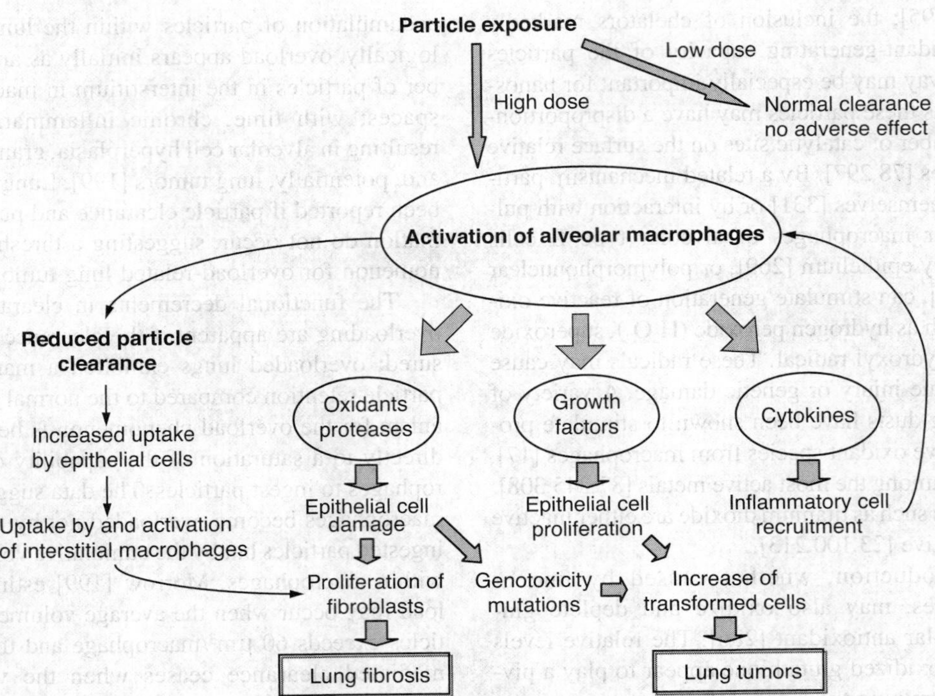


**FIGURE 28.8** Relationships between particle exposure, macrophage activation and development of lung fibrosis and cancer. (Adapted from Oberdörster, G. et al., *Reg. Toxicol. Pharmacol.*, 27, 123–135. 1995.)

particle concentrations and producing impaired alveolar macrophage clearance did not [36,111]. Rats had more severe inflammation after repeated inhalation of carbon black or ultrafine $TiO_2$ compared to hamsters or mice [19,79]. These studies concluded that hamsters had elevated rates of particle clearance relative to the other strains; while mice and rats experienced lung overload with similar but high $TiO_2$ lung burdens, mice did not exhibit the progressive fibrosis observed in rats. Elder [79] also noted that the lung injury and particle burdens with $TiO_2$ tended to increase as the surface area of the particles increased. So, a second consideration is that local tissue dose and subsequent injury from inhaled particles may also be related to particle surface area. Other metrics of exposure concentration may include particle mass, count, volume, or surface area, but for overload to occur surface area appears to be the particle characteristic most closely associated with pulmonary inflammation, fibrosis, and lung tumor development [40,47,216].

By whatever metric, the most important outcome of overload is the persistent inflammation produced by the particles. The consequences of this have an important bearing on the mechanism of tumor development. Marked impairment of particle clearance occurred in rats at 7.1 $mg/m^3$ or greater with carbon black and was associated with lung injury, chronic inflammation, epithelial hyperplasia, and fibrosis [77]; such changes were not seen in comparably exposed rats at 1.1 $mg/m^3$, suggesting a threshold phenomenon. In the overloaded rat lung, the investigators reported a significant, dose-dependent increase of *hprt* mutations in alveolar epithelial cells at inflammatory concentrations. Inflammatory cell products, including reactive oxygen species and cytokines, were thought to provide the environment, including induction of cellular mutations and cell proliferation that are necessary for tumor development from oxidant induced genotoxicity. The observation that particle-elicited rat inflammatory lung cells can be directly mutagenic suggests that there will be an exposure level (or lung burden) below which persistent inflammatory changes are not induced and therefore mutations should not occur for nongenotoxic particles. On the other hand, at inflammatory concentrations, both the inherent genotoxicity of the particle and the resultant genotoxicity ensuing from the particle-elicited proinflammatory milieu may drive tumor development. The proposed theory harmonizing these observations is shown in Figure 28.8. Overall, the overload theory provides a plausible mechanism with regard to the development of lung tumors from low-cytotoxicity, low-solubility, nongenotoxic dusts.

## Particle Clearance Mechanisms

Depending on the site of aerosol deposition within the respiratory tract, different particle clearance mechanisms exist that govern both the rate and route of particle removal.

In the anterior nose, for example, the coarse nasal hairs filter out larger particles, so any particles that deposit on the mucous layer overlying the nonciliated respiratory epithelium are generally removed by sneezing or nose wiping. Particles that deposit in the more distal regions of the nasal cavity are cleared on a mucous film overlying a ciliated epithelium. The coordinated movement of cilia facilitates the mucous flow in this region, which propels the mucus toward the nasopharynx; subsequent removal of insoluble particles is through swallowing or expectoration. In contrast, soluble particles may be absorbed into the bloodstream by diffusion into the mucus and through the nasal epithelium. As already described, nanoscale particles may penetrate the olfactory epithelium and migrate along olfactory nerve axons to the olfactory bulb [220,296].

The tracheobronchial region, like much of the nasopharyngeal region, is lined with a mucus-covered, ciliated epithelium. Insoluble particles that deposit in this area are cleared via mucociliary transport, a process where mucus-entrained particles are swept toward the oral cavity for removal by swallowing or expectoration. Particle clearance from this region is generally faster in the larger, rather than the smaller, airways [198]. Substances such as formaldehyde, which causes mucostasis and ciliastasis [192], or sulfur dioxide ($SO_2$), which produces an increase in mucus secretion, inhibit normal mucociliary activity and slow net ciliary movement, delaying particle clearance [260]. Also, particles may directly enter the peribronchiolar regions by crossing bronchial epithelium or be absorbed into the circulation after dissolution in the mucous layer.

Depending on the physicochemical properties and amount of particles deposited, several processes exist to clear particles entering the alveolar (gas exchange) region. These clearance processes include dissolution, macrophage phagocytosis, direct passage and uptake by the vascular system, or movement into the lymphatic system. Dissolution of particles is a process whose rate is generally inversely related to particle size; it may occur within the fluid lining the conducting airways or alveoli, the interstitium, the alveolar macrophages, or the lymphatic system, and the solute may subsequently be absorbed through the epithelium into tissue or blood.

Macrophages serve important roles in particle clearance from the alveoli because secretory mucous cells and a ciliated epithelium are absent in alveoli. As mobile phagocytic cells of alveolar surfaces, macrophages deliver ingested particles or fibers to the terminal bronchiole, the distal terminus for mucociliary transport. Alternatively, particle-laden macrophages may pass directly through alveolar epithelium to the interstitium for bronchiolar or lymphatic system removal. Of these two processes, particle clearance from the alveoli is believed to proceed primarily by mucociliary transport [33]. Free or macrophage-associated particles that enter the lymphatic system may be translocated to tracheobronchial lymph nodes, where

**TABLE 28.5**
**Particle Clearance**

| Anatomic Region | Clearance Mechanism | Approximate Clearance Half-Time |
|---|---|---|
| Nasopharynx | Mucociliary transport | Minutes |
| Tracheobronchial | Mucociliary transport | Minutes to hours |
| Alveolar | Macrophages | Days to weeks |
| Alveolar | Interstitial migration | Months |
| Alveolar | Dissolution | Months to years |

they may accumulate (if the particles are insoluble) or, alternatively, enter the blood stream. Particles that are not removed by the lymphatic or bronchiolar clearance routes may eventually migrate within the interstitium toward the pleura. Finally, particles may directly cross the capillary endothelium and enter the blood [123]. Failure to remove deposited particles by one of these processes may result in particle accumulation or sequestration in lung tissue, a condition known as *pneumoconiosis*.

The clearance rates of inhaled, insoluble particles from the respiratory tract vary depending on the anatomic location. Particle clearance kinetics may be described in five phases, each with a different clearance rate (Table 28.5). As considerable interindividual variations in the rapidity of clearance can exist within species, these phases represent idealized depictions of generally accepted clearance processes within the different regions of the respiratory tract. The first phase reflects the removal of particles deposited in the nasopharyngeal region. Here, the clearance rate is dependent on regional variations in mucus transport velocity, a process that can vary widely among individuals [240]. Typically, particle clearance from the nasopharynx occurs with a removal time of approximately 10 minutes [249]. A second phase of particle clearance involves removal of particles from the tracheobronchial region. Again, particle clearance via tracheobronchial mucociliary transport depends on mucus transport velocities. Regional differences in transport velocity are seen, with the highest clearance rates occurring in the trachea and lower rates in the terminal bronchioles; this regional dependency is due to the reduced thickness and velocity of the mucus blanket toward the peripheral airways [197]. Although particle transport from the tracheobronchial region is relatively rapid, with clearance of most particles occurring within ~24 hours, particle clearance is not always complete even when evaluated weeks later. Prolonged retention of insoluble particles has been shown with particles instilled in the upper respiratory tract [225], an observation that may reflect particle translocation into tissues. Particle clearance from the alveolar region represents the third, fourth, and fifth phases. The third phase is believed to represent particle transport by the alveolar macrophages to the ciliated epithelium or the lymphatic system; this phase proceeds at variable rates and has a clearance half-time of approximately 2 to 6 weeks. The fourth phase represents particle removal by interstitial pathways; this has been found to be especially true for UFPs, which readily migrate into the interstitium, escaping phagocytosis and elimination by alveolar macrophage. The fifth phase represents removal by particle dissolution. Clearance half-times for the last two processes are very slow, generally measured in months and years, respectively. The efficiency of alveolar macrophage mediated particle clearance may be greatly reduced if the particles are cytotoxic so macrophages do not have an opportunity to participate in particle transport.

## Alveolar Macrophages and Particles

Due to the absence of a mucociliary transport apparatus in alveoli, alveolar macrophages play a central role in particle clearance from alveoli. The lung has adapted to dusty environments by its ability to recruit macrophages to deposition sites to help defend the integrity and maintain the sterility of lung surfaces. The mechanisms whereby macrophages are attracted to sites of particle deposition are not fully understood but may involve several mechanisms. Complement activation and liberation of complement-derived chemotactic factors may be one mechanism for getting macrophages to particle deposition sites [314,315]; this pathway, however, does not account for the inflammatory response for all particle types, as complement depression has no effect on silica-induced inflammatory parameters [318]. Given that particles are more likely to initially contact lung tissue rather than macrophages, interest has centered around the possible role of lung epithelial cells as modulators of cell recruitment. Several studies indicate that type II lung epithelium may serve as the source of chemotaxins by the generation of reactive oxygen species with attendant oxidative stress that ultimately causes the release of proinflammatory mediators [273]. Data also suggest that some particle types may generate reactive oxygen species acellularly [12], but more likely they are liberated from lung epithelium. Within hours, elicited chemotactic factors recruit both polymorphonuclear leukocytes (from the pulmonary vasculature) and alveolar macrophages to deposition sites. Recruited macrophages and neutrophils ingest foreign materials and either migrate toward the terminal bronchioles or enter the interstitium.

Phagocytosis plays an important role in alveolar macrophage function. Once within the phagosome, the particle is subjected to a diverse array of hydrolases and other lysosomal enzymes, as well as activated oxygen species, including superoxide anions and hydroxyl radicals. These bioactive substances are involved in killing inhaled microorganisms but may also play a role in several particle-induced diseases, possibly through altering the biochemical microenvironment near the macrophage. Pulmonary emphysema, for example, is thought to be related to a protease and protease-inhibitor balance; the secretion of excessive amounts of protease by macrophages or neutrophils and an insufficient amount of protease inhibitor are factors leading to the hydrolysis of lung connective tissues. Alternatively, the phagocytosis of asbestos fibers may be incomplete so the phagolysosome does not fully encapsulate the fiber and the macrophage releases the lysosomal enzymes into the alveoli, causing local tissue injury. Macrophages may also become activated upon particle phagocytosis, releasing inflammatory mediators such as reactive oxygen species, macrophage and neutrophil chemotactic factors, prostaglandins and platelet-activating factors which play roles in smooth-muscle contraction and vascular permeability, growth factors for fibroblasts and collagen production, and various cytokines, such as TNF$\alpha$ and IL-1, which modulate the inflammatory process. Particles that are readily cleared or rapidly dissolved may only transiently activate macrophages; however, continued macrophage activation by certain cytotoxic dusts such as silica or asbestos is thought to be responsible for chronic lung diseases ostensibly by initiating a prolonged inflammatory response within the lung [75].

As cytokines are critical messengers for maintaining homeostasis, they play an important role in lung response to inhaled toxicants by modulating the recruitment and activation of inflammatory cells. Cytokines are low-molecular-weight proteins that govern intercellular communication by interaction with membrane receptors on target cells. Low levels are normally expressed in the body and are necessary for cell replication, differentiation, and tissue maintenance; however, as cytokines modulate these functions, substances that reduce or amplify cytokine levels can have major effects on the outcome of the body's response to the inhaled material.

For the purposes of describing effects of cytokines on lung homeostasis, Driscoll et al. [70] described three main categories of cytokines. One category is the *initiating cytokines*, generally acknowledged to be pivotal in moderating the initial host response to inhaled materials. Prototypic cytokines include TNF$\alpha$ and IL-1 and are involved in triggering a secondary cascade of responses through additional cytokine networks. Although peripheral blood monocytes and alveolar macrophages are primary sources of TNF$\alpha$ and IL-1, alveolar macrophages appear to contribute more TNF$\alpha$ than blood monocytes and relatively more TNF$\alpha$ than IL-1 [71]. The presence of pathogenic dusts, such as crystalline silica, serves as a chronic stimulus for TNF$\alpha$ and IL-1 expression because the cytotoxic silica cannot be readily cleared from the lung [77].

A second category is the *recruitment cytokines*, which govern the recruitment of certain inflammatory cell populations to sites of TNF$\alpha$ and IL-1 release. This class of cytokines also contains chemokines that regulate the influx of neutrophils and other immune system cells. The source of chemokines is not clear, but recent work suggests that pulmonary epithelial cells, alveolar type II cells, fibroblasts and alveolar macrophages can all release the chemokine macrophage inflammatory protein 2 (MIP-2) in response to cytotoxic dusts [74]. There appears to be marked specificity in the types of leukocytes responding to a given recruitment cytokine. Thus, although MIP-1$\alpha$ is chemotactic for monocytes, neutrophils, and lymphocytes, MIP-2 is chemotactic for neutrophils.

Finally, a third class of cytokines is the *resolution cytokines*. These cytokines can moderate fibroblast proliferation and the production of collagen and also downregulate expression of the initiating cytokines. In so doing, these resolution cytokines ultimately terminate or resolve the inflammatory response to inhaled substances by affecting repair of damaged tissues. Examples of resolution cytokines include transforming growth factor $\alpha$ (TGF$\alpha$), transforming growth factor $\beta$ (TGF$\beta$), and interleukins 4, 6, and 10 (IL-4, IL-6, and IL-10). Overall, a considerable body of evidence has accumulated implicating the central role of cytokines in the pathophysiology of fibrotic lung disease by regulating inflammatory cell recruitment, fibroblast and epithelial cell replication, and ultimately tissue repair.

## EXPERIMENTAL INHALATION TOXICOLOGY

The potential toxic effects that need to be appreciated following inhalation exposure include irritation of the respiratory tract, tissue permeability changes, behavioral changes, pathologic changes to vital organs or tissues distal to the respiratory tract, immune system responses, pulmonary function alterations, metabolic disturbances, carcinogenicity, or ultimately death. Studies to measure the effects of chemical and physical agents on the biological system after entering the respiratory tract must follow carefully designed protocols and be sufficiently detailed. The generation of aerosols requires particular skills and must be carefully described so others can replicate the experimental conditions to compare results from test to test. Indeed, the process of establishing constant and reproducible exposure conditions is considerably more complex than that required for other portals of entry such as oral or dermal administration. This is a direct result of the type of equipment required to generate, maintain, and measure experimentally produced atmospheres in a form that can be inhaled by the test species. The total dose received depends on the physical and chemical properties of the material, physiologic attributes of the test animal, and factors involved in deposition and clearance. One must also remember that inhalation exposures often result in simultaneous chemical exposures via the skin and the gastrointestinal tract from aerosol deposits on fur subsequent ingested from grooming.

### ANIMAL MODELS

The choice of an animal species selected for an inhalation study is an important one for practical, technical, and ethical reasons. It is apparent that there is no one ideal human surrogate but that each species presents its own advantages and disadvantages. Most commonly used are rodents, including the rat, mouse, guinea pig, and hamster (probably in that order). As a consequence of the extensive use of rodents in inhalation toxicology, investigators have generated a large body of information on their responses. Rodents have been used in toxicological tests as models for vapor and particle deposition and clearance, mechanisms of toxicant-induced lung and airway injury, and as models for infectivity and immunologic function. Due to their relatively short lifespan, the use of rodents represents a reasonable approach for studying chronic toxicity and carcinogenesis. When extrapolating from experimental animal data to humans, factors that must be considered include the comparative anatomy of the respiratory tract, presence or absence of concurrent diseases or infections, and similarities of the biochemical and physiological responses in the intended species.

The use of animals in product development and safety assessment must continually reflect the welfare and ethical treatment of animals. Intelligent strategies to minimize animal testing by using *in vitro* or tiered approaches to data evaluation and collection should be practiced where possible, recognizing that *in vitro* tests for inhalation-related endpoints have only recently begun to be explored. Before they can be developed for regulatory acceptance, harmonization and validation activities must occur. In practice, inhalation testing should be conducted at laboratories that possess the appropriate facilities, have a trained and competent staff capable of humane animal care, and have institutional review committees to review their testing protocols for compliance with animal welfare policies. All inhalation studies should be scientifically and ethically justified, and every study should be designed to avoid or minimize pain, discomfort, and stress. Care must be taken to avoid unnecessary or duplicative testing. Finally, the use of animal enrichment measures, wherever practical, should be encouraged.

In practice, selection of the test species is more often based on criteria such as the size and availability of the test animals, the number of animals required to differentiate chemically induced changes from background or incidental changes, and the expense involved in obtaining and maintaining the requisite number of animals for statistical validity in light of animal use and welfare considerations. Indeed, techniques to reduce the use of animals without compromising study integrity are an important consideration in study design.

In the selection of an appropriate animal species, one must consider the objective of the study; the available toxicological information; the unique functional or structural characteristics, if any, of the species that make it a good (or bad) animal model; what the anticipated response might be that would enable the investigator to most accurately determine the number of animals needed for the duration of the study; and, finally, the appropriate controls. Historically, the use of small rodents has predominated because their size allows the testing of larger groups, their relatively short lifetime allows testing over the entire lifespan (a large body of data already exists on these species), and, finally, the costs of their acquisition and upkeep are relatively low.

The guinea pig has been used in studies of immune function and respiratory sensitization and has been particularly useful in determining the relative potencies of isocyanates [324,333]. The guinea pig's unusually abundant bronchial smooth muscle makes this species a useful model to study airway hyperresponsiveness or bronchoconstriction in asthma models.

The hamster, having a relatively low spontaneous lung tumor rate and resistance to pulmonary infections, has been used in respiratory tract cancer studies. Some investigators feel that the hamster is the best animal model for the study of experimental lung cancer [251].

Using rodents to predict effects in humans has several disadvantages. Because the nasal and pharyngeal anatomy of rodents is unlike that of humans, the amounts and sites of particle deposition may be quite different. Rodents are also obligate nose breathers, with superior nasal filtering efficiency compared to humans. Assessment of pulmonary function in nonanesthetized rodents is difficult, although recent miniaturization of probes and detectors has allowed some success here [56]. The most serious problems with rodents, especially rats, are those involving spontaneous respiratory infections and their sequelae. Rats appear to be uniquely susceptible to chronic inflammation, pulmonary fibrosis, and cancer from insoluble, noncytotoxic particles by overwhelming lung clearance mechanisms (particle overload). Because rats appear most susceptible to particle overloading, their selection for some types of particle inhalation studies often complicates extrapolation of the results to humans [218]. In oncogenicity studies, rats appear to be less susceptible to fiber-induced mesothelioma, a malignant tumor of the pleural lung surface. Conversely, hamsters appear to be more susceptible to the development of fiber-induced mesothelioma but less sensitive to the induction of lung tumors than rats. Despite these considerations, rats remain the most favored animal model for both short- and long-term inhalation studies.

Inhalation toxicity studies have also been conducted on dogs, principally the beagle. The dog is a convenient size for a number of laboratory measurements including evaluation of pulmonary and cardiac function. Several natural disease states exist in the dog, making it a good model for evaluating the impact of the substance on conditions such as asthma. The costs associated with the facilities necessary to properly care for dogs are a disadvantage relative to rodents. The dog also may be relatively insensitive to certain inhaled gases such as ozone [284]. An advantage, though, is the extensive use of the dog as an experimental model to study the effects of radioactive materials following inhalation. A wealth of information is available concerning the long-term inhalation of radionuclides.

Because the nasal anatomy of monkeys is similar to humans, monkeys are sometimes used in inhalation experiments. Lung function can be determined; however, considerable costs are involved in obtaining and maintaining monkey colonies, and their lack of general availability is a problem limiting their use. Note that within the species identified as "monkey" intraspecies differences do exist. Also, the subgross pulmonary anatomy differs between types of monkeys and humans [180].

Other species such as the ferret, horse, donkey, sheep, cat, rabbit, and pig have been suggested for use in inhalation studies. Each of these species has been used for special applications, which may take advantage of a particular anatomical feature, chemical sensitivity, or research curiosity. In some instances, the endpoint of concern drives the animal model chosen. The Brown Norway rat has been used as an asthma model is a study design including analysis of cellular infiltrate in the lung, inflammatory factors in bronchoalveolar lavage, immunoglobulin E in the serum, and changes in delayed-onset respiratory reaction upon inhalation challenges (demonstrated with diphenylmethane diisocyanate) [229].

## Study Types

Evaluating the inhalation toxicity of substances can involve short- or long-term exposures. Acute studies generally define the amount of substance per volume air (ppm or mg/m$^3$) required to produce a given response (most often death) and any associated clinical signs suggestive of target organ toxicity. Another major goal of the acute study is to help define the exposure concentrations that should be used for further repeated-exposure studies. Longer term studies (up to and including lifespan studies) are conducted to determine the target organs for repeated exposures and carcinogenic potential.

Acute studies generally involve a single, 4- or 6-hour exposure of small groups of rodents to the test substance at a series of concentrations ranging from those producing little or no signs of response (e.g., clinical signs, body weight changes) to those producing death. Studies may also be conducted at shorter durations (e.g., 10, 30, or 60 minutes) to help assess time-dependent responses arising from very short-term exposure. The practice of using many groups and relatively many animals per group to define the $LC_{50}$ (i.e., the concentration calculated to kill half of the exposed animals under the particular set of experimental

**TABLE 28.6**
**Number of Rats Needed To Determine the Approximate Lethal Dose vs. the $LD_{50}$**

| Test Substance[a] | Number of Rats Used ALD | $LD_{50}$ |
|---|---|---|
| Adiponitrile | 7 | 65 |
| Bromobenzene | 8 | 35 |
| *n*-Butylhexamethylene diamine | 7 | 35 |
| Caffeine | 8 | 40 |
| Carbon tetrachloride | 5 | 105 |
| Hexachlorophene | 11 | 46 |
| Hexamethylenediamine | 5 | 92 |
| Methomyl | 5 | 53 |
| Tetraethyl lead | 5 | 36 |
| Average | 6.8 | 56.3 |

[a] Test substances were administered orally to rats, and the minimal number of animals necessary to derive an approximate or median lethal dosage is shown.

*Source:* Kennedy, Jr., G. L. et al., *J. Appl. Toxicol.* 6, 145–148, 1986. With permission.

conditions) has been replaced to some extent by the determination of the approximate lethal concentration (ALC), which is the lowest concentration at which the a single death is observed, or by acute toxicity classification schemes currently being reviewed. An ALC-based protocol (Table 28.6) can dramatically reduce the number of animals needed per study (from an average of 56 to 7) [139] and may find value in toxicity screening; however, regulatory agencies may still require formal $LC_{50}$ determinations with multiple groups of animals for use in risk assessment. To minimize animal use and improve the utility of inhalation screening assays, more detailed examinations of respiratory tract histopathology and analysis of bronchoalveolar lavage fluid for evaluation of cellular injury may be useful in mechanistic evaluations. Some testing guidelines, such as draft guideline OPPTS 870.1350, embrace this concept [306].

A relevant, nonlethal endpoint in acute inhalation testing of substances is the sensory irritation potential. Its usefulness is based on the fact that a large number of substances are sensory irritants (i.e., depression of respiratory rate through stimulation of irritant receptors in the upper respiratory tract) and that animal models can be used to predict both the degree of irritancy and their relative potency. A large portion of the substances for which occupational exposure limits (e.g., American Conference of Governmental Industrial Hygienists Threshold Limit Values) have been established are based on sensory irritation. In these cases, no clear target-organ toxicity other than irritation to the eyes, nose, or upper respiratory tract has been identified from either animal studies or reports of human experiences in the workplace. Efforts to define exposure limits based solely on the use of sensory irritation data from animal testing have been proposed by several investigators [3,256]. Although equating exposure limits to sensory irritation levels simplifies exposure limit setting in one regard, caution must be exercised in the use of sensory irritation data alone [31]. Some workers have found that pathological changes may occur in the respiratory tract below acutely irritating levels in rodents, limiting the usefulness of sensory irritating assays by themselves for exposure limit setting [42,338].

In sensory irritation studies, changes in the respiratory rate and depth are recorded in response to known concentrations of test substances [3,5]. Animals (usually mice) are placed in small cylindrical tubes with their heads protruding through a rubber dam. The neck fitting forms a relatively airtight seal, which, with appropriate attachments, allows the unit to be used as a body plethysmograph. The animal in the tube is then placed with the head extending into the exposure chamber. Following an acclimatization period, the animal is exposed to pre-equilibrated concentrations of either vapors or aerosols (Figure 28.9), and the changes in respiratory rate and pattern are continuously monitored. Each animal is exposed to a given concentration for approximately 10 to 30 minutes. This technique allows determination of the upper respiratory tract irritation potential in a simple and reproducible manner. The data are expressed as $RD_{50}$ values (i.e., the calculated concentration expected to reduce the normal respiratory rate by 50%).

Repeated-exposure (subchronic) studies generally last 14, 28, or 90 days. The subchronic study precedes the chronic study and is used to determine the target organs affected under defined exposure conditions. Satellite groups can be included to monitor the progression or regression of injury following a defined recovery period (typically, 2 to 4 weeks). This sequence also allows a preliminary estimation of the potential for cumulative toxicity. Although individual practices may vary, one approach is to perform initial testing of 2 weeks' duration. Groups of animals, generally 10 rats, are exposed to concentrations of 1/5, 1/15, and 1/50 of the approximate lethal concentration or $LC_{50}$ for 6 hours per day, 5 days a week, for 9 days over a 2-week period. During this test, exposure concentrations can be refined in either direction depending on the presence and severity of effects seen in the test rats. The selection of exposure concentrations can also be based on the clinical observations seen in the acute study; for example, if large body weight losses occur following a single 4-hour exposure at 1/5 the lethal dose, the range of exposure concentrations used in the study would be scaled downward. In this subchronic study design, animals are observed for signs of response and weighed daily. Clinical chemistry, hematology, and urine analyses are conducted at the end of the ninth exposure, and complete pathologic

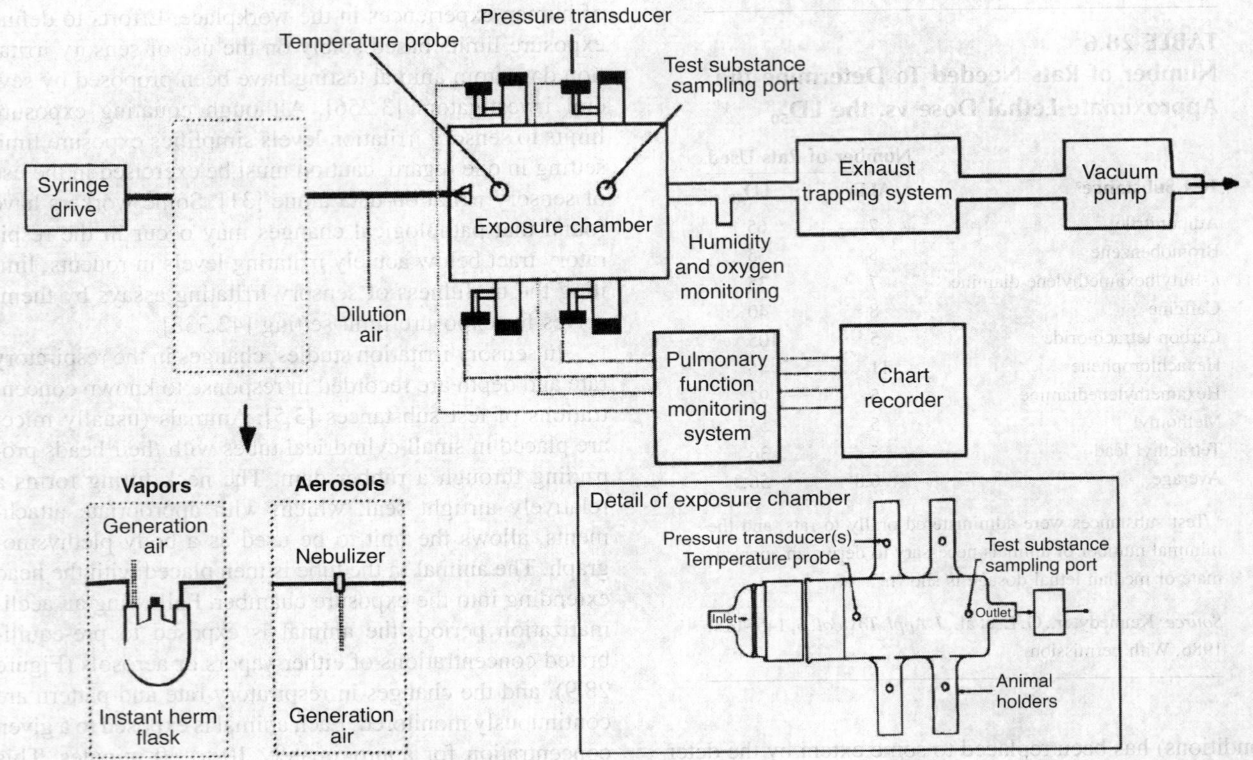


**FIGURE 28.9** Apparatus for determining sensory irritation properties of vapors or aerosols. Insert: Details of exposure chamber.

examinations, including weights of major internal organs, are performed. Another group of rats is allowed a 14-day recovery period, during which no additional exposures are conducted and the above parameters are again measured to determine the reversibility of any test substance-induced changes. This same general approach may be scaled appropriately for use in 28- or 90-day exposure protocols typically specified by regulatory agencies.

Chronic studies involve measurement of effects resulting from exposure over the animal's lifetime at concentrations where acutely toxic effects are not observed. Exposures are selected in anticipation of only slight body weight effects (~10% reduction from controls) and no change in survival. This corresponds to the definition of the maximum tolerated dose (MTD). Here, the investigator is trying to detect chronic pathological changes, usually cancer, produced by test substance exposure. Large numbers of animals are required and special care must be taken to distinguish the background incidence of effects from those produced by exposure to the test substance. In most studies, chronic exposure conditions usually mimic those encountered in the workplace with animals exposed 6 hours per day, 5 days a week, for their lifetime (typically 18 to 24 months for mice and 24 to 30 months for rats). For environmental substances, exposures may be conducted 24 hours per day. It is obvious that this latter type of exposure protocol is labor intensive, and in practice this limitation makes it rarely considered. In both situations, the need to relate cause and effect necessitates the use of fixed exposure concentrations, a situation that rarely exists in either the environment or the workplace.

In aerosol exposure studies, particle clearance (such as the lung particle burden or the clearance rate of a radiolabeled tracer particle) should be evaluated. Indeed, an important but often overlooked principle in designing inhalation toxicology experiments to evaluate potential biological sequelae of substance exposures is that tolerated doses (i.e., exposure concentrations) should be defined and employed as previously described. This is particularly important with particles that are relatively insoluble and have a low order of systemic and respiratory tract toxicity where particle overload is likely to occur. Once overloaded, biological effects may ensue from the unusually high lung burdens.

Fibers represent a particular type of particle that requires careful attention. A recent conference evaluated short-term assays for fiber toxicity that could assist in prioritizing fibers for chronic assays. The working group objectives were to summarize the current state of the science around the use of short-term assay systems in assessing potential fiber toxicity and carcinogenicity, to review strengths and limitations of these approaches, and to suggest a testing strategy to combine available methods to identify hazards that may require further evaluation [125].

The end use of the data helps determine the most appropriate exposure pattern; for example, investigators interested in the effects of substances in confined places such as a submarine may opt for continual 24-hour exposure conditions. Examples of 24-hour exposures of several species simultaneously for extended periods of time to determine dose–response data for application to continuous exposures include studies of carbon tetrachloride by Prendergast et al. [238] and ethanolamine by Weeks [323]. This type of exposure pattern also would be preferred by those interested in setting guidelines for ambient air levels to protect the health of the community; however, given the cost of these studies, temporal adjustments are commonly made assuming that the effects are linearly related to exposure duration. Those who need to provide guidance for workplace situations would best be served by intermittent daily exposure patterns.

In contrast to the extended exposure scenarios, emergency releases of large volumes of substances, particularly gases or those substances that vaporize readily at ambient conditions, may lead to relatively short exposures for community evacuation purposes. Indeed, incidents such as the accidental release of approximately 50 tons of methyl isocyanate in Bhopal, India, in 1984, which resulted in the death of approximately 3000 individuals, highlight the need for emergency planning measures to prevent the catastrophic release of toxic substances into the community. Inhalation toxicity data generated over various exposure durations are useful to define tolerable limits associated with the (1) onset of death, (2) production of irreversible damage, (3) production of reversible damage without impaired escape capability, and (4) production of no measurable damage. These concentration vs. exposure time profiles are designed to prevent both unwanted, nonreversible injury and impairment of self-escape. As one example of this type of work, short-term lethality responses of rats to perfluoroisobutylene, a highly toxic gas, have been studied at time intervals ranging from 15 seconds to 6 hours [275] and, from these data, both short- and long-term exposure control limits have been derived [138]. This type of data is used in the Emergency Response Planning Guides (ERPGs) from the American Industrial Hygiene Association [1] and the Acute Exposure Guideline Levels (AEGLs) for hazardous substances [304] of the U.S. Environmental Protection Agency. A comparison of ERPG-2 and TLV® values for selected industrial chemicals is provided in Table 28.7. These values provide guidance to emergency-response specialists for establishing short-term exposure limits appropriate for workplace or community emergency-escape planning [209].

In looking at the effect of varying concentrations and time with regard to endpoint effects, current work practice often involves other than traditional 40-hour weeks made up of 5 consecutive 8-hour workdays (e.g., novel-work schedules). Some repeated-exposure inhalation studies have examined the response to a given substance following other than 6- or 8-hour exposures per day, 5 days a week, to assist in setting airborne control levels for these novel work schedules. The responses of rats to carbon tetrachloride following 11.5-hour exposures per day were different from those following 8 hours per day [230]. Both concentration- and time-dependent responses to aniline were seen when daily scheduled exposure to rats went from 8 hours per day for 5 days to 12 hours per day for 4 days [144]. Burgess et al. [43] found concentration was the primary toxicity determinant when three aniline concentrations were tested using daily exposures of 3, 6, or 12 hours. Concentration was also the more significant factor when rats were exposed to dimethylacetamide for daily exposures of 3, 6, or 12 hours [149].

**TABLE 28.7**
**Comparison of Emergency vs. Workplace Airborne Control Limits for Selected Substances**

| Substance | ERPG-2[a] (ppm) | TLV®[b] (ppm; 8-hr TWA) |
|---|---|---|
| Ammonia | 150 | 25 |
| Benzene | 150 | 0.5 |
| Carbon tetrachloride | 100 | 5 |
| Chlorine | 3 | 0.5 |
| Ethylene oxide | 50 | 1 |
| Formaldehyde | 10 | 0.3 (ceiling) |
| Hydrogen chloride | 20 | 2 (ceiling) |
| Hydrogen cyanide | 10 | 4.7 (ceiling) |
| Methanol | 1000 | 200 |
| Phosgene | 0.2 | 0.1 |
| Trichloroethylene | 500 | 50 |
| Vinyl chloride | 5000 | 1 |

[a] Maximum airborne concentration below which nearly all individuals could be exposed for 1 hour without irreversible or serious health effects or symptoms that could impair the ability to take protective action.
[b] Airborne concentration of a chemical substance to which it is believed that nearly all workers may be repeatedly exposed, day after day over a working lifetime, without adverse health effects.

## INHALATION TOXICITY TESTING

National and international regulatory agencies have implemented testing guidelines regarding the conduct of inhalation toxicity studies. These guidelines attempt to standardize testing procedures so toxicity data can be obtained and evaluated using a defined protocol. Test data generated in this fashion may be used in a variety of ways: (1) to investigate the relationship between exposure concentrations and adverse effects, (2) to provide information on the mechanism of toxicity and permit a reasoned extrapolation of the experimental animal data to potential human health risks, (3) to form the basis for dose-level

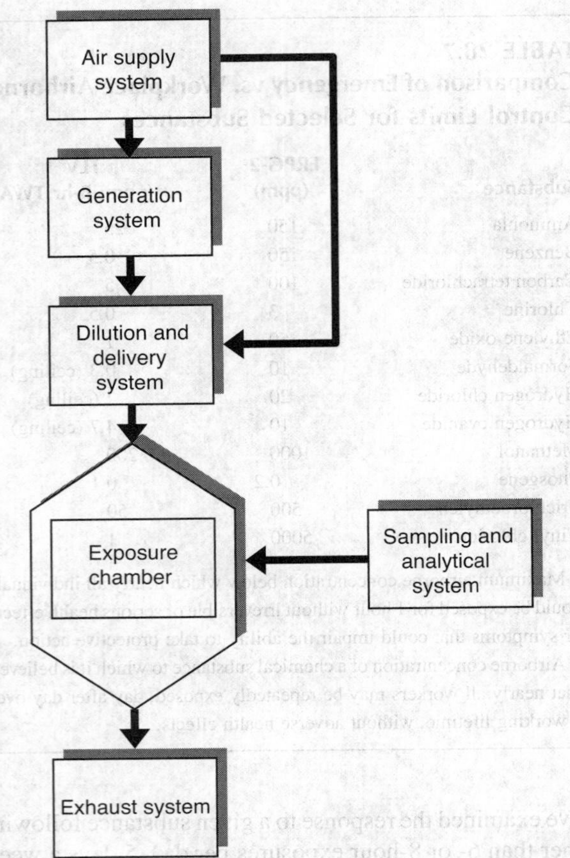


FIGURE 28.10 Components of an inhalation exposure system.

selection for repeated-exposure studies, or (4) to categorize the inhalation toxicity of a test material relative to other substances. Such data are the foundation for establishing defensible exposure limits for the protection of human health and appropriate toxicity classification and labeling of substances. To reduce redundant testing, regulatory agencies have adopted standardized protocols for both short-term and long-term inhalation testing and are harmonizing protocols internationally. Guidance on general study design, exposure conditions, measurement frequencies, and major endpoints of concern has been published for acute, subchronic, and chronic inhalation toxicity testing [223,306].

## Exposure Systems

Inhalation exposure systems involve the harmonious integration of several subsystems whose design, construction, and operation are critical to a safe, functional system allowing generation of meaningful results. The essential subsystems are shown schematically in Figure 28.10 and include a conditioned air-supply system, a suitable vapor or aerosol generator for the test substance, an atmosphere dilution and delivery system, one or more exposure chambers, an atmosphere sampling and analytical system, and an exhaust/scrubbing system. The two inhalation modes of exposure to test substances are nose only or whole body. Although the design, operational parameters, testing objectives, and technical requirements for these two approaches are similar, there are several important issues to consider when choosing one mode vs. another (Table 28.8). Specific components typically used for test substances generation and analysis and animals exposure are shown in Figure 28.11.

**TABLE 28.8**
**Considerations in Choosing Whole-Body or Nose-Only Exposure Models**

| | |
|---|---|
| Whole-body | Exposes entire body surface, not just respiratory tract |
| | Potential for secondary exposure (ingestion via grooming) |
| | Better simulation of real-world exposures |
| | Minimal animal handling and stress involved |
| | Allows free movement (and observation) during exposures |
| Nose-only | Exposure via respiratory tract only |
| | Less test material needed |
| | Reduced potential exposure of test substance to operator |
| | Animal restraint during exposure (observations are limited) |
| | Animal stress associated with restraint/handling |
| | Restrainer selection that minimizes temperature-induced stress |
| | Labor associated with loading and unloading animals |

### Conditioned Air Supply

A sufficient amount of clean, conditioned air for chamber operation must be supplied. The availability of high-quality, contaminant-free (e.g., particles or organics) air for chamber supply is essential, especially for long-term inhalation studies. Ordinarily, ambient air is dried, filtered, and purged of organic vapors prior to adjusting its temperature and humidity to the desired levels. Where this is not practical for very small, low-airflow chambers with limited numbers of animals, a commercial source of compressed, purified air may be utilized. The size of the air-conditioning equipment should be scaled to allow some excess flow capacity for the required number of chambers operating at the desired flow rates. Airflow rate, temperature, and humidity must be maintained within fairly tight ranges to ensure that environmental conditions are acceptable to the well-being of the test species. Excursions beyond the normal range could result in varying overall exposures and could lead to altered test species responses to any given concentrations. Typically, environmental conditions are targeted to a temperature from 20 to 24°C, humidity from

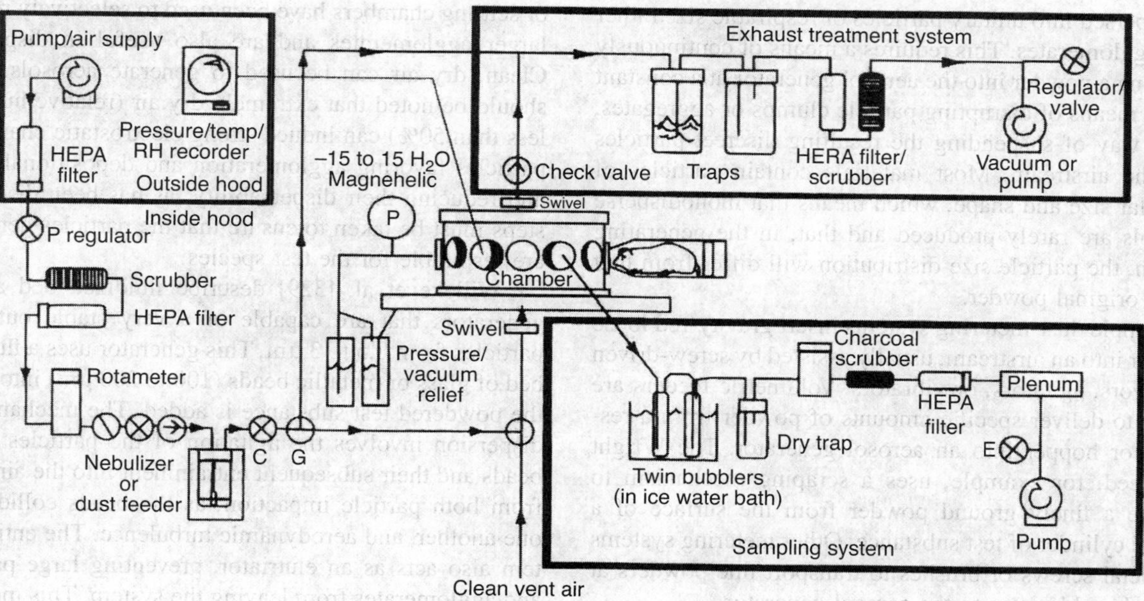


**FIGURE 28.11** Details of exposure system components. Shown are the air supply, test atmosphere generation and dilution systems, the sampling system, and the exhaust system used for nose-only inhalation exposures.

30 to 70%, and airflow greater than 10 air changes per hour. Modern inhalation systems have been automated for chamber monitoring and exposure control, thus eliminating the need for manual adjustments and data logging. It is important that these conditions remain relatively constant for the duration of the exposure (and from exposure to exposure).

## Generation of Test Atmospheres

Gases are the simplest atmospheres to generate. They can be metered by flowmeters, syringe drives, diffusion tubes, or other techniques into a dilution airstream; allowed to mix; and then introduced into the exposure chamber. A number of flow-dilution devices are available. Vapors of either liquid or solid compounds can be formed by heating within a temperature-controlled heating device (care must be taken to prevent chemical decomposition or reaction with air or water at the elevated temperatures used to generate vapor). Another technique, depending on physical properties such as boiling point, viscosity, or chemical purity, is to use an infusion pump to meter the liquid test substance onto a heated surface. A variation of this technique is the J-tube developed by Miller et al. [185]. This device uses a heated J-shaped tube filled with glass beads to increase the available surface area for test substance vaporization. This unit works well with substances that decompose at temperatures near their boiling point. Other liquid materials may be vaporized in fritted-glass bubblers. Frequently, nitrogen is used to vaporize the test substance to minimize chemical reaction or oxidation with air at elevated temperatures. The saturated stream can then be diluted with filtered air or oxygen to raise the oxygen concentration to 19 to 21% and to adjust the test substance level to the desired concentration. The application of an automatic proportional-integral-derivative (PID) control algorithm to an inhalation exposure system operating with a concentration monitoring system that samples on a 30-minute cycle was shown to produce exposures that accurately achieved and maintained the targeted concentration [332].

Special care must be taken when generating vapors from mixtures of test substances with low boiling points. If test atmospheres are generated from the liquid mixture by flash evaporation, the composition of the resulting vapor will be representative of the composition of the liquid mixture. In contrast, if the headspace of the liquid mixture is tapped, the resulting vapor composition in the mixture will reflect the partial pressures of the components with a disproportionate amount of the component with the highest vapor pressure relative to the other components. This is a problem in longer term inhalation studies, because the composition of the test atmosphere will vary from day to day as the higher vapor pressure components of the mixture are depleted from the headspace as the mixture is consumed.

The generation of particulate materials in a uniform manner is considerably more difficult than vapor generation. Particulate materials may be generated from dry powders or from liquids. The generation of aerosols using dry dispersion techniques presents problems that are unique to each dust being studied. In toxicology testing, particle concentration and size distribution must remain constant over long periods of time. The powder being tested must be dispersed into unitary particles of respirable size rather than agglomerates. This requires a means of continuously metering a powder into the aerosol generator at a constant rate, a means of disrupting particle clumps or aggregates, and a way of suspending the resulting discreet particles into the airstream. Most materials contain particles of irregular size and shape, which means that monodisperse aerosols are rarely produced and that, in the generating system, the particle size distribution will differ from that in the original powder.

Simple dust-metering systems insert gravity-fed loose powder into an airstream, usually assisted by screw-driven conveyors, agitators, or vibrators. Volumetric feeders are useful to deliver specific amounts of powder from a reservoir or hopper into an aerosol generator. The Wright dust feed, for example, uses a scraping mechanism to remove a finely ground powder from the surface of a packed cylinder of test substance. Other metering systems use metal screws or brushes to transport fine powders at user-selectable rates to the aerosol generator.

Aerosol generation involves the dispersion of a powder by supplying sufficient energy, such as a high-velocity airstream, to a relatively small volume of the bulk powder to separate the particles by overcoming their own attraction forces. Hydrophobic materials such as talc are more easily dispersed than hydrophilic materials such as limestone or quartz because particle agglomeration or clumping is minimal; thus, dry powders are considerably easier to disperse than humidified ones. The metered powder may be dispersed and agglomerates broken up directly by a turbulent air jet, or the dust-laden airstream can be passed through an impactor or fluidized bed. Elutriators or settling chambers have been used to selectively remove larger agglomerates and are also useful in dispersion. Clean, dry air can be used to generate aerosols, but it should be noted that extremely dry air (relative humidity less than 50%) can induce strong electrostatic charges on particles favoring agglomeration and depositional losses and reducing their dispersability; as has been discussed, steps must be taken to ensure that the particles generated are respirable for the test species.

Willeke et al. [329] describe fluidized bed aerosol generators that are capable of a very stable output of particles from 0.5 to 3 µm. This generator uses a fluidized bed of glass or metallic beads (100 to 200 µm) into which the powdered test substance is added. The mechanism of dispersion involves the agitation of the particles by the beads and their subsequent entrainment into the airstream from both particle impaction, as the beads collide with one another, and aerodynamic turbulence. The entire system also acts as an elutriator, preventing large particles and agglomerates from leaving the system. This means of generation can only be used for dry, nonadhering powders and produces electrically charged aerosols that must be neutralized upon leaving the fluidized bed. Compared to some other dust generators, fluidized bed generators do not generally have high particle output, limiting their use in acute toxicity testing.

Lee et al. [161] used a high-pressure, air-impingement device (microjet) to separate the finer fibrils from a *para*-aramid fiber matrix (Figure 28.12). Once separated from the bulk fiber matrix, the smaller fibers pass through a cyclone into the exposure chamber. This system was used to generate fiber mass concentrations as high

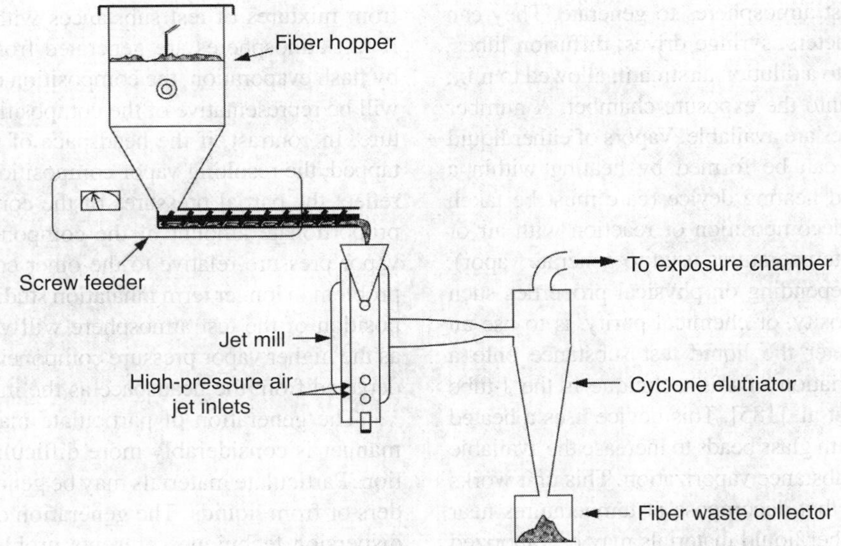


**FIGURE 28.12** Generation system used to produce airborne *para*-aramid fibrils using a controlled-flow powder dispenser coupled to a commercially available jet mill to disperse fiber aggregates. A cyclonic elutriator was used to remove nonrespirable fibers from the test atmosphere. A cross-section of the jet mill is shown in Figure 28.13.

as 18 mg/m$^3$ or fiber count concentrations as low as 2.5 fibers/cm$^3$ for periods ranging from 2 weeks to 2 years. The microjet is a versatile aerosol generator that has found practical applications for both fibrous and nonfibrous materials. By increasing airflow to the secondary high pressure air jets, the microjet can be adjusted to either disrupt particle aggregates or to effect some reduction in particle size through trituration in the oval inner chamber (Figure 28.13). Other examples of the use of this technology for particle dispersal are described by Cheng et al. [50] and Bernstein et al. [21]. Bernstein et al. [21] described a brush-feed, micronizing jet mill that produces a relatively wide concentration range of respirable particles. Test concentrations from 0.22 to 7.48 mg/L were attained and maintained with particles less than 3 μm. Only at higher aerosol concentrations were these small particle sizes not attainable.

There is a need for aerosols to reach the peripheral lung with dry powder medications as well as with collected ambient aerosols in environmental research projects. In a novel aerosol generator to create short-duration, high-concentration exposures, a fixed volume of compressed air was used to create a short burst of highly concentrated aerosol in a 300-mL holding chamber. Using this system, collected diesel soot was deagglomerated to a fine aerosol with a mass median aerodynamic diameter of 0.55 μm. Aå fine powder such as 3-μm silica particles was completely deagglomerated to an aerosol with a MMAD of 3.5 μm. More than 97% of the deposited soot was located distal to the extrapulmonary bronchi, indicating that the system delivers a highly respirable aerosol. The aerosol system is particularly useful for peripheral lung delivery of collected ambient aerosols or dry powder pharmaceuticals following a minimal effort in formulation of the powder [90].

Given the interest in the relationships between ambient aerosols and morbidity and mortality, researchers are attempting to study collected ambient particle samples in animals. One such high-volume, ultrafine particle concentrator has been described by Gupta et al. [104]. The concentrator consists of several units, including a size-selective inlet, a condensation aerosol growth unit, two virtual impactors (concentrators), a thermal size restoration device, an air cooler, and a size-selective outlet. This system was evaluated using single-component artificial aerosols with a variety of physicochemical properties as well as ambient air. All UFPs grew and were concentrated by about the same enrichment factor, regardless of their composition and surface properties. A workshop concerning experimental work with concentrated ambient air particles (CAPs) was convened. Efforts were made to harmonize experimental and analytical protocols to permit standardized outcome measures based on spirometry and response markers to lung lavage markers relevant to cardiac responses. The need

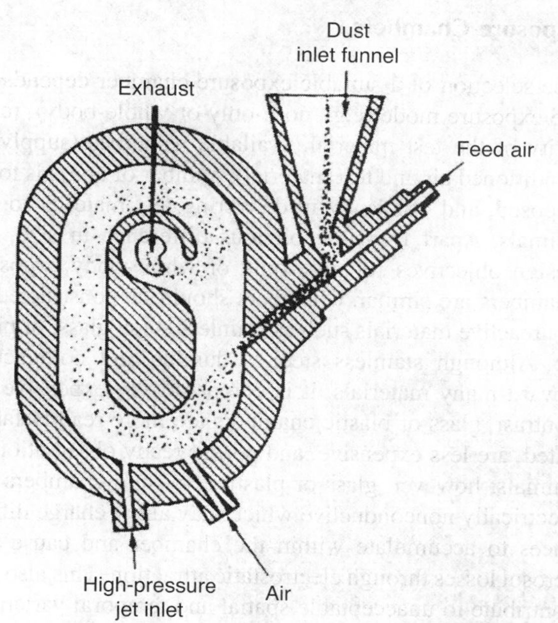


**FIGURE 28.13** Cross-section of jet mill used for fiber and particle generation.

for more complete analyses of these samples was emphasized, as was obtaining a consistent set of parameters characterizing exposure atmospheres and the ambient particulates from which the CAPs are sampled [167].

## Dilution and Delivery Systems

When a suitable atmosphere has been generated, it must be delivered to the exposure chambers at an appropriate concentration. Dilution is effected by mixing the test atmosphere with conditioned, filtered air before its introduction to the exposure chamber. It is important that the delivery system be fabricated with materials that minimize wall losses, either through absorptive or reactive processes, and designed to minimize physical losses through aerosol deposition within ductwork. The former situation may be avoided by using nonreactive materials and the latter by minimizing the number of bends and maintaining laminar flow in the ductwork. In general, the delivery system should be nonreactive with tubing as wide, short, and straight as practical to minimize test material losses and pressure drop. It is especially important to use conductive materials in the choice of tubing materials. Substantial aerosol deposition can occur in areas of high electrostatic charge, leading to excessive and variable aerosol losses along transfer lines. A broad selection of mechanical and electrical valves and flow measurement devices is available to measure and control the flow of test material and dilution air.

## Exposure Chambers

The selection of a suitable exposure chamber depends on the exposure mode (e.g., nose-only or whole-body), reactivity of the test material, available resources (supply of conditioned air and test material), number of animals to be exposed, and efficiency in delivering test material to the animals. Apart from the obvious difference in size, the design objectives for nose-only or whole-body exposure chambers are similar. Chambers should be constructed of nonreactive materials such as stainless steel, glass, or plastic. Although stainless steel is durable and nonreactive toward many materials, it is comparatively expensive. In contrast, glass or plastic chambers are more readily fabricated, are less expensive, and permit ready observation of animals; however, glass or plastic exposure chambers are electrically nonconductive which may allow charge differences to accumulate within the chamber and cause test aerosol losses through electrostatic attraction. This also can contribute to unacceptable spatial and temporal variations in chamber aerosol concentrations. Normally, chambers are operated under dynamic exposure conditions (i.e., a continuous supply of air is flowing through the chamber) with a slight negative pressure within the chamber to prevent leakage of test material into the exposure room.

## Atmosphere Analysis

In many cases, analysis of the test atmospheres is the most challenging part of conducting inhalation studies. The investigator must not only sample the chamber for proper environmental conditions (temperature, relative humidity, and oxygen content) but also select, validate, and conduct analyses for the test substances in terms of particle size (if a solid–liquid aerosol), concentration, and solvent concentration (if present in a mixture); the problem magnifies when testing materials that are not stable in air or in the presence of moisture and when multiple component substances are tested. All of these issues must be addressed and resolved for the study to be considered well conducted. This section attempts to identify some of the available analytical approaches and methods that may be useful.

Analysis of test atmospheres begins with the sampling system. An appropriately designed sampling system is central to accurate analysis of test substance concentrations in the exposure chamber. First of all, the sampler probe should be located in the immediate proximity of the breathing zone of the animals, so regional distribution differences within the chamber do not skew the actual concentrations inhaled by the animals. The sampling train should be designed to transmit a representative air sample from the animal's breathing zone to the sensor or collection medium without significant losses. It is not necessary for the collection efficiency to be 100%, but if sample losses cannot be avoided at least the collection efficiency of the sampling train should be measured with the test substance under the experimental conditions to be used. Above all, the investigator should strive for a constant collection efficiency. For aerosols, the sampling train should be as straight and short as possible and so designed to minimize impaction sites (i.e., those locations with bends, reducers, or turns). Rarely are aerosols monodisperse and, because most particle entrapment mechanisms are size dependent, the collection characteristics of a given sampler will change with aerodynamic particle size and upon sampler air flow. Aerosol collection efficiency will also change in response to sampler particle loading. A filter will be more efficient and trap finer particles as dust collects on the surface, but efficiency will gradually decrease as the pressure drop across the filter increases, reducing airflow. In contrast, the efficiency of an electrostatic precipitator will drop as a resistive layer of particles accumulates on the collection electrode.

Sampling errors generally reflect the contribution of many small errors in the system rather than any single large source of error. The items that must be carefully considered and evaluated in any such system include sampling train leakage and losses, variations in flow rate and sample volumes, sampler aerosol or vapor collection efficiency, sample stability in the collection media, extraction efficiency from the sampling substrate, and analytical background or interferences introduced by the sampling substrate. Additionally, the air sampling rate should be known with a high degree of precision, and this can be achieved through the use of previously calibrated rotameters, critical orifices, displacement monitors (dry or wet test meters), or electronic mass flow sensors. Much of this technology is described in classical texts on industrial hygiene.

For gases or vapors, sampling of the test atmosphere may be either instantaneous (grab sampling) or continuous in nature. If grab sampling is conducted, samples should be taken with sufficient frequency to identify fluctuations in concentration. Samples may be collected in evacuated glass or metal containers, in inflatable flexible polymer bags, or by gas-tight syringes. Alternatively, gas or vapor samples may be taken from remote locations by directing the sample through nonreactive sample lines by using a vacuum pump and introducing aliquots of the sample stream directly into the analytical instrument. Continuous samples can be taken in this manner by pumping (or drawing) a constant stream of gas directly through a dedicated detector or by having the analytical device continuously cycle through sample streams from multiple exposure chambers.

Gases and vapors offer the least difficulty in sampling as they follow the laws of diffusion, mix freely with the general atmosphere, and equilibrate rapidly. Sampling typically involves removal of the substance from air by scrubbing through a solvent or reagent, adsorption to a collection surface, or collection in some type of container (e.g., an impermeable, nonreactive gas bag).

An extractive sampling train for gases, vapors, and aerosols includes sample lines from the animal's breathing zone, a scrubbing device, an airflow or volume meter to determine sample volume, and a vacuum pump. Commonly, solvent-filled impingers are used for extractive sampling. Impingers consist of a sampling tube extending through a reservoir of collecting media; the end of the tube at the bottom of the reservoir may be open, narrowed, or fritted; the latter two modifications provide comparatively higher collection efficiency. The increased collection efficiency is due to the smaller bubbles produced by the narrow or fritted impinger tips and which allow more complete diffusion of vapor from the bubbles into the surrounding solvent. The solvent used in impingers can quickly evaporate, so sampling cannot be conducted indefinitely.

Solid adsorbent sampling devices are used for vapors and some gases. Solid adsorbent samplers consist of a small tube (about 100 mm long by 6 mm diameter) containing a sorbent (typically charcoal), which serves as the collection medium, and a vacuum source to draw a known rate of air over the sorbent. Charcoal is an effective sorbent for many compounds as its surface structure is microporous, thereby providing a large surface area for chemical adsorption. Importantly, charcoal possesses a large number of pores that exist in a range of dimensions to allow molecules of different sizes to enter the pores; thus, charcoal affords nonspecificity and can trap a wide variety of organic compounds. Other trapping agents include polymers engineered to trap certain chemicals, silica gel for inorganic substances, or molecular sieves. Once trapped, however, the sorbent must also allow the vapor or gas to be desorbed with high efficiency into a solvent for subsequent quantitation. Carbon disulfide is broadly used as a desorption solvent for charcoal tubes as it has good solvent properties and a low background with the flame ionization detector generally used in analysis. Sampling is done at relatively low airflows to allow time for chemical adsorption, yet must be of brief duration so as not to exceed the adsorbent capacity of the sorbent. Although relatively simple and inexpensive, the use of gas sampling tubes in most toxicology studies is uncommon due to the labor involved in sample collection and preparation compared to continuous, direct-reading analyzers.

Gas sampling using containment systems is also relatively uncommon in toxicology studies. They serve as temporary storage vessels to allow transport of gas or vapor samples from the exposure system to the analytical device. Examples include evacuated glass or stainless steel containers or Tedlar® or Mylar® plastic bags. Whereas some substances are stable in such containers, numerous logistical and practical limitations exist with regard to their routine use in inhalation toxicology studies. Gas sampling bags are physically cumbersome (their bulk depends on the sample volume necessary), they require a logistical commitment to having a sufficient quantity of sample vessels available for the duration of testing and vessel transport to the analytical laboratory, they require demonstration of chemical inertness and storage stability of the substance between the interval of sample collection and analysis, they must be decontaminated if vessels are reused for other substances, and they may confound results from undetected leaks or sample losses during shipment.

Instruments for gas or vapor component identification and quantitation have brief response times (seconds or minutes) and are used after air sampling. Sensors in these instruments typically utilize absorbance of infrared or ultraviolet radiation, flame- or photoionization, electron capture, or chemiluminescence detection.

As with gas and vapor sampling, instantaneous (grab) or continuous (integrated) sampling techniques may be also used to collect particulate matter. Instantaneous sampling involves removing a small volume of the atmosphere and separating the particle phase by either filtration or impaction of the aerosol against a solid substrate on which the deposited particles are later characterized (counted, sized, or subjected to chemical analysis). For continuous sampling, a constant stream of the test atmosphere is directed to the analytical instrument. Analytical techniques that rely on physical forces such as gravity, impaction, electrophoresis, thermophoresis, inertia, or diffusion techniques are used for particle collection; collected particles are typically measured using mass or optical properties. An overview of the common instrumentation useful for determining particle mass and size for use in inhalation studies is provided in Table 28.4.

Grab sampling for aerosols usually involves either impactors or filters. Fluid-filled impingers (as used for gases and vapors) have been used for aerosols but because collection efficiency is variable they are not commonly used. Instead, grab sampling for aerosols generally involves the use of filters for determining aerosol mass concentration in the chamber. For nonhydroscopic aerosols, the accuracy of chamber aerosol concentrations is limited by the sensitivity of the volume-measuring device, as the weight of collected aerosol can be determined with great precision using modern analytical scales. Hydroscopic aerosols pick up water vapor from the air and consequently can change weight during the brief interval required for filter weighing. To more accurately measure the weight of hydroscopic aerosols, standardized weighing techniques are necessary, such as desiccating (especially glass or cellulose acetate fiber filters) or equilibrating filters in a constant humidity chamber prior to weighing.

The choice of filter media available is broad and includes glass fiber and various polymer membranes (e.g., cellulose acetate, polycarbonate, polyvinyl chloride, polyvinylidene fluoride, Teflon®). Membrane filters are perforated with holes of a defined size. Unlike fiber filters that trap particles within the filter matrix, membrane filters collect particles on the surface of the membrane. The choice of filter media is dictated by the need for high collection efficiency for particles of the size of interest, chemical inertness (to allow good recovery of the test substance for subsequent chemical analysis), and minimal pressure drop across the filter (to allow high air-sampling rates). Pressure drop is especially important for membrane filters because smaller holes create a higher pressure drop, limiting airflow through the filter. To some degree, smaller size holes in the filter membrane ensure a higher sample collection efficiency, but even larger pore membrane filters collect many particles smaller than the pore size due to deposition by interception or diffusion. Membrane filters are widely used when analyzing the morphology of mineral dusts by either optical or electron microscopy.

In addition to determining aerosol chamber concentrations via measurement of collected mass on filters, the filters may be extracted and analyzed to determine test substance content. This is particularly valuable for mixed aerosol atmospheres where the content of each component must be evaluated, but this will require a specific analytical method for each component of interest. Clearly, in cases where test substances are to be extracted from the filter medium, the filter itself must be chemically unreactive with the test substance and should allow rapid and quantitative desorption of test substances for chemical analysis. Filters made from superfine glass fibers with diameters below 1 μm are available for collecting virtually all suspended particles; they have high particle loading levels and low airflow resistance. While glass fiber filters interfere negligibly with most subsequent chemical analyses, it should be noted that glass surfaces have imbedded hydroxyl groups that may bind metals or react with some substances. Regardless of composition, filter spiking and recovery studies should be evaluated before use to determine whether the particular filter substrate will interfere with analysis.

Complementary to aerosol mass monitoring by filtration, real-time aerosol monitors can be used to provide analytical data for concentration adjustments. Light-based particle monitoring systems rely on light scattering through refraction, reflection, or diffraction, all of which will vary with the optical size of the particles. The theory and explanation behind light-scattering devices has been well described by others [28,142,309]. Although light-scattering techniques for measuring aerosol concentrations provide instantaneous output and have found general acceptance for chamber monitoring, it should be noted that instrument response is a function of aerosol size, shape, color, concentration, and chemical composition. Thus, light scattering aerosol monitors must be calibrated against other analytical methods with the specific aerosol of interest to allow quantitative use; even then, they are subject to coincidence errors and are useful only at relatively low aerosol concentrations. These particle monitors do, however, provide an immediate display of changes in concentrations and are useful to detect deviations in aerosol generator output or chamber airflow dynamics.

For particle sizing, cascade impactors are frequently used. Impactors are flow-through devices that withdraw aerosol from the test atmosphere. The impactor contains a series of impingement stages, each stage with progressively smaller nozzle dimensions. Because airflow through the impactor is constant, each stage induces progressively increasing airstream velocities coupled with sequentially reduced impaction distances to effect a progressive separation of smaller and smaller particles as the aerosol progresses through the unit. Particles deposited on each stage can be weighed and examined microscopically. The aerodynamic median size of an aerosol may be determined with the cascade impactor by calculating the percentage of aerosol by weight on each stage using weight, radioactivity, or chemical analyses to determine the amount of material deposited on each stage. This process is shown in Figure 28.3. The same caveats with respect to selection and use of filter media and filter weighing techniques apply with cascade impactors. A major liability when operating cascade impactors is the potential for overloading impactor stages with excessive particle mass. This may lead to the erroneous conclusion that aerodynamic sizes are smaller than actual due to particles bouncing off of the target stage and subsequent capture on lower stages, indicating finer particle size.

Advances in electronics permit bypassing the manual weighing of individual impactor stages and allow the direct, nearly instantaneous readout of aerosol size. An example of a single-particle aerodynamic particle sizer is the Aerodynamic Particle Sizer (TSI; St. Paul, MN). This device displays aerodynamic diameters by measuring the particle velocity in a laser light-scatter sensing zone. Particles of different size are accelerated to different speeds when directed through a nozzle. The airstream is directed through dual focused laser beams, one downstream of the other. The system's circuitry measures the speed of the particle, represented by the time required to traverse the distance between the two lasers, and converts each time-of-flight measurement into an aerodynamic diameter. This device provides high-resolution, real-time aerodynamic measurements of particles between 0.5 to 20 μm but can only process particles at a concentration of about 1000 p/cm$^3$; otherwise, coincidence errors occur. Higher concentrations require dilution before measurement. Unlike the TSI unit, the piezobalance cascade impactor—for example, the Quartz Crystal Microbalance (QCM) by California Measurements (Sierra Madre, CA)—measures mass concentrations (and aerodynamic diameters) directly. The QCM is a self-contained unit with a 6- to 10-stage cascade impactor that uses a quartz sensing crystal in lieu of a filter on each stage. The crystal has an inherent oscillating frequency that changes in proportion to the mass of aerosol deposited on the crystal. This allow very small amounts of mass to be detected and thus is ideal for very low (such as ambient aerosol) concentrations; size analysis of high concentrations can also be measured by very short sampling durations to avoid stage overloading.

### Exhaust and Scrubbing Systems

Exhaust systems are included in the chamber design not only to assist in the flow of air through the exposure chambers but also to minimize discharge of test material into the environment. Federal, state, and local regulations on air emissions can be stringent and require that adequate abatement systems must be incorporated. Depending on the physical properties of the test material, liquid or solid absorbents (e.g., water for water-soluble solvents and activated charcoal for low-level organics) may be employed to remove gases or vapors, whereas porous filtering material may be appropriate for aerosols. Because liquid or solid absorbents become saturated with test material, they must be monitored with sufficient frequency to ensure that the adsorbent capacity has not been exceeded. In contrast, particle filters generally become more efficient as particle loading increases, but because airflow may be reduced concurrently when filters become loaded, the exposure chamber aerosol concentration may inadvertently increase. Thermal oxidation or incineration is another means of treating chamber effluent and can be applied to scrubbing combustible test material.

## Exposure Modes

A variety of exposure models have been used to evaluate the effect of substances following inhalation. The whole-body exposure mode is encountered frequently and allows the exposed test species to move freely in the atmosphere containing the test substance. Whole-body inhalation exposures will result in a portion of the substance being ingested regardless of physical form. Gases or vapors can dissolve in the mucus fluid lining the respiratory tract and, via the mucociliary escalator, reach the pharynx, where they are swallowed. Droplets or solid particles also reach the gastrointestinal tract via this mechanism. Further, contributions to total absorbed dose can occur by dermal absorption of the test substance. The normal grooming and preening activities of rodents both during and after inhalation exposures can deliver the substance to the gastrointestinal tract. For example, rats were exposed to respirable zinc chromate dust either in conventional wire-mesh cages or nose-only in fiberglass tubes (so that the only exposure was through the external nares). The caged rats excreted 8.4 times as much fecal and 5.5 times as much urinary chromium as rats in the tubes, indicating that a significant amount of dust could be ingested or absorbed following whole-body exposure [158].

Conducting whole-body aerosol exposures in group-caged rodents might appear to reduce the actual amount of substance to which each animal is exposed due to the filtering action of fur in groups of rodents huddled together; however, Ulrich and Marold [300] tested 3-μm aerosols of dodecyl alcohol and measured lung concentrations in rats housed singly or in groups of three or seven. They found similar amounts of substance in the lungs of each group.

A mobile whole-body exposure system was developed for exposing mice to concentrated ambient particulate matter smaller than 2.5-μm MMAD. Each 20-L exposure cage was designed to hold 9 mice within individual compartments. The actual performance of the exposure system was determined for 0.5- to 2.0-μm diameter aerosols by measuring the uniformity of aerosol distribution and particle deposition in the tracheobronchial and pulmonary regions of mice exposed in the system. The measured deposition efficiencies in this mobile exposure system were similar to those reported for mice exposed in a nose-only exposure system [224].

Dermal absorption may contribute to the total body burden from whole-body vapor exposures. The skin is a partial or incomplete barrier to inward or outward migration of most vapors. Because the surface area of the skin is large, a small flux of vapor across the entire skin surface can result in absorption of sufficient amounts of vapor to cause adverse health effects. McDougal and colleagues [177] described a whole-body chamber in which rats were fitted with face masks and received positive-pressure, charcoal-filtered fresh air while being exposed to vapors of either dibromomethane (DBM) or bromochloromethane (BCM), both volatile organic chemicals whose major metabolites can be easily measured in blood. In rats exposed to the vapor by inhalation for 4 hours, the percent absorption through the skin for DBM was 5.8%; for BCM, 4.2%. These uptakes were independent of inhalation exposure concentration. Determination of the relative rates of absorption of vapor through the skin and the respiratory tract can be determined by comparing dermal absorption in rats without inhalation contact and inhalation by the usual methods. When inhalation is the route of exposure, the rate-limiting process in the absorption of substances with large blood:air partition coefficients (e.g., bromomethane) is the rate at which the vapor is transferred to the blood by alveolar ventilation [6]. The rate-limiting process of absorption for dermal exposure is passage through the stratum cornea which can be represented by the permeability constant. Blood concentration measurements allow the input of both vapor inhalation and dermal vapor absorption integration when relating total delivered dose to biologic effects [178].

Head-only exposures are useful for single or repeated brief exposures. Because nose-only systems limit exposure of the test substances to the respiratory tract and not the

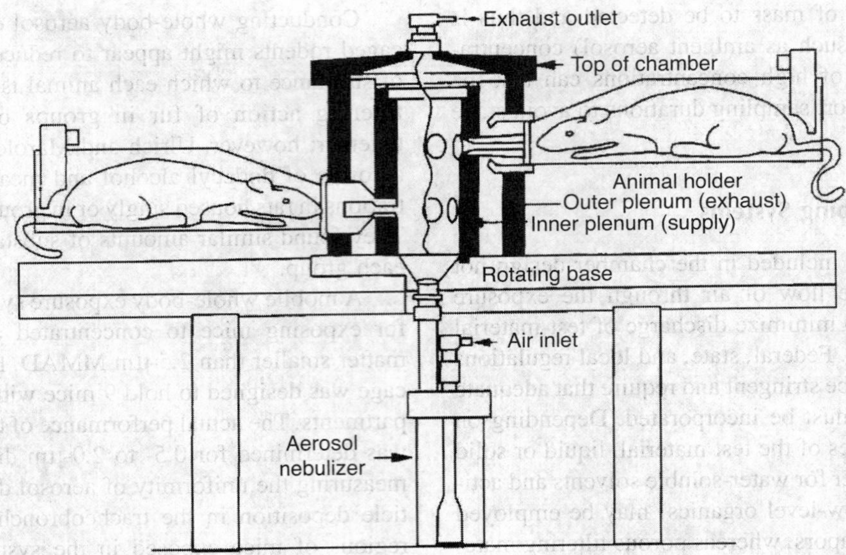


**FIGURE 28.14** Exposure chamber and rodent holder designed for nose-only exposures to aerosol or vapors.

entire body, the amount of test material required is considerably reduced. This feature facilitates testing highly toxic or limited-availability test substances. In addition, nose-only exposure units simplify test substance containment and subsequent clean-up and allow the chamber concentrations to be more readily altered. Disadvantages of this system may include loss of material to the fur of the head, difficulty in obtaining a proper seal at the neck without impairing circulation, the possibility of restraint-induced stress and handling of the animals, and the relatively limited number of animals that can be tested simultaneously.

All of the above disadvantages can be dealt with in a manner that makes this type of testing useful in safety evaluation programs. To ensure uniformity of exposure conditions on an animal-to-animal basis, a relatively large air flow is needed. For larger animals such as the dog and monkey, helmet exposures can be conducted. Pressure fluctuations in these systems may be great, and records of breathing patterns during exposure may be needed to accurately measure the exposure doses. In addition, the fitting at the neck must be comfortable and easy to adjust. Designs including both inflatable collars and thin rubber membranes have been used successfully.

Kirk [150] described a system for exposing guinea pigs to radioactive gases. The animals were placed in plethysmographs to record breathing patterns, and a rubber seal isolated the head inside the exposure area: the animals also could be exposed body-only while breathing fresh air if desired. Thomas and Lie [292] describe a similar system for aerosol testing in rats and mice. Dogs [288] and monkeys [257] have also been tested using head-only techniques.

A typical nose-only exposure unit begins employs an animal exposure cylinder, generally constructed from rigid plastic (or glass). This tube is connected to an enclosure or exposure chamber that allows introduction and evacuation of test atmospheres and contains sampling probes to determine test concentrations, particle-size distributions, temperature, and humidity. Rodent holders are usually fabricated of plastic (polymethylmethacrylate, polycarbonate), aluminum, or stainless steel cylinders to conform to the general shape of the animals, and they are fitted with conical headpieces projecting into the plenum of the exposure chamber. Some nose-only systems minimize rebreathing of test atmospheres by attempting to maintain a constant flow of fresh test atmosphere directed onto the face of the animal while having a local exhaust near the nose as shown in Figure 28.14. The animal holders must be of various sizes to accommodate rodents of differing sizes to minimize stress. They should be supported carefully to minimize discomfort to the animal and adequately ventilated to minimize heat retention within the tubes.

Both physical restraint and inadequate temperature control within the chamber can subject the test animals to additional stress. This seems to be especially true for inhalation developmental and reproductive studies. Pathological effects on the male reproductive system, independent of the inhaled test substance, have been reported in restrained male rats where inadequate temperature control occurred leading to excessive body temperature [37]. Inhalation developmental toxicity studies conducted with restrained pregnant rodents have produced slight increases in maternal toxicity and either slight [184] or no [299] increases in fetal anomalies when dams were restrained during the period of organogenesis. To minimize confinement-induced stress in inhalation developmental toxicity studies in mice, Dorman et al. [66] utilized a rat restrainer tube to expose an unrestrained pregnant mouse, making essentially a whole-body exposure chamber for individual mice. Adaptation to the stress induced in rats and mice in nose-only inhalation tubes was examined. In naïve animals, during the first hour of restraint, heart rate increased by 58 beats per minute in rats (18.6%) and by 174 beats per minute in mice (32.3%) as compared to cage controls. Temperature increased by 2°C in mice and was unchanged in rats compared to cage controls. Heart rate and temperature values remained within normal physiologic values during restraint. In rats, the response to restraint in nose-only holders was the same after 4 days regardless of whether the duration of restraint was increased gradually to 4 hr/day or kept constant at 4 hr/day. In mice, the group that was gradually adapted had a statistically significant higher heart rate and temperature after 4 days than the fixed-duration adapted group. Rats and mice restrained for 4 hr/day every day showed a gradual decrease in heart rate and temperature over time. Full adaptation to restraint required 14 days of fixed-duration daily restraint [208].

Overall, the selection of the nose-only exposure mode can affect the outcome of some inhalation toxicity studies and represents a decision that investigators should properly control for in experimental design. Nonetheless, high-capacity, nose-only exposure systems, such as those described by Cannon et al. [45], have been used for various acute and subchronic inhalation studies. A thorough description of the components used in nose-only inhalation systems, with details on automated collection methods for airflow regulation and test atmosphere analysis, was presented by Pauluhn [226]. In a study by Moss et al. [203], which used a directed-flow, nose-only exposure system at flow rates close to the minute ventilation of the animal, it was possible to maintain desired concentrations, thereby greatly minimizing the amount of test substance needed. At an airflow rate of 2.5 times the animal's minute ventilation, the concentration of test substance of inhaled air was reduced less than 10%.

Design considerations of head-only units are identical to those of nose-only units. Mask designs represent a unique kind of nose-only exposure and usually are limited to relatively large animals. Masks for dogs [9], monkeys [102], ponies [173], donkeys [4], chickens [14], and rats [280] have been used successfully.

It may be necessary to conduct brief or instantaneous exposures to airborne materials. This might be required when developing data to deal with setting emergency exposure limits or when only small quantities of materials are available. Approaches include airlock and very low internal volume nose-only exposure chambers. To evaluate the lethal doses of highly toxic perfluoroisobutylene in rats with exposure times from 0.25 to 10 minutes, a low-volume chamber was designed [275]. A 1.5-L cylindrical Plexiglas® chamber (75 cm × 5 cm in diameter) with ten staggered animal ports was developed using three equidistant sampling ports to monitor gas uniformity. With a 10-L/min airflow rate, chamber equilibrium was rapidly achieved in less than 1 minute.

Testing using head-only exposures has been conducted [274] in 0.5-m$^3$ chambers that accommodated up to 60 rats each. Rats were exposed to fibrous aerosols for 6 hr/day, 5 days a week, for up to 24 months with relatively little test-condition-related stress as measured by growth, body temperature, and plasma corticosteroid levels. The number of personnel necessary to accomplish these exposures was not considered excessive.

The use of intratracheal instillation as an alternative to exposure of animals by inhalation deserves some comment. Conditions exist in which the pulmonary effects of a substance cannot be easily evaluated by inhalation. Although not entirely valid reasons, space, time, or economic considerations can result in the decision not to use inhalation as the route of test material exposure but rather to use intratracheal instillation techniques to get the material directly into the respiratory tract. The reason for choosing intratracheal instillation over inhalation can also rest on a lack of sufficient quantities of test material or on safety issues (e.g., extreme toxicity, flammability, explosivity). Using this method, the actual dose delivered to the lung of the experimental animal can be directly and precisely measured. This technique is inexpensive in that very small amounts of substance are required (and expensive chambers, generating apparatus, and support personnel are avoided). Also, because the technique is contained and uses relatively little material, exposure hazards to laboratory workers are greatly reduced compared to those of an inhalation study. Finally, materials that are not readily respirable in rodents can be introduced into the lungs with this technique; notably, long fibers that can be inhaled by humans but not by rodents can be tested via this route. The problem that limits the usefulness of intratracheal instillation as an exposure technique relative to inhalation is that the dose to the respiratory tissues can be variable, local, and highly artificial and does not accurately reflect the lung distribution of substance following inhalation exposure. Additionally, intratracheal instillation techniques focus on the lower respiratory tract and, as such, cannot produce responses that would occur in the nasopharyngeal region by inhalation exposure.

Intratracheal instillation involves a suspension of particles in a carrier liquid that is injected directly into the lumen of the trachea or nebulized as very fine droplets into the airway. The fluid and particles flow into the gravity-dependant areas of the lung. The carrier liquid is then rapidly absorbed into the pulmonary circulation, leaving the particles on the internal surfaces of the lung. This technique permits the introduction of a wide range of doses and substances to the lung in a short period of time. In larger animals, localized exposures to specific areas or lobes of the lung are possible, often allowing the contralateral lung to serve as a control (for nonsystemically acting agents). The technique was first applied by Kimura [148], using rabbits and guinea pigs for responses to various coal tars. Early works have described in detail the methodologies found to be useful in the mouse [153,210], in the rat [24], and in the hamster [250]. The procedures used in these studies are quite similar, with the main difference being found in the maximal amount of total liquid that can be used to deliver the test substance to the lung without harming the animal.

In small rodents, intratracheal instillation is accomplished by inserting a catheter or needle transorally through the mouth and epiglottis into the tracheal lumen. In larger species (including humans), a fiberoptic bronchoscope can be used to more precisely visualize the instillation site. Because the animal must not move during the procedure, the choice of anesthetic is important, with short-acting materials that suppress reflexes for a minimal period of time being preferred. Physiologic saline is the vehicle most frequently used to suspend or solubilize the test substance, although even this may evoke a mild transient inflammatory response. Surfactants can be used to improve the suspension properties for dusts, but the surfactant effects on the lung must be evaluated. In addition, dosage volumes must be adjusted for the body weight of the animal.

Although intratracheal instillation enables the administration of large amounts and nonrespirable sizes of particulate matter that would otherwise not be able to gain access to the lung, highly localized deposition of particle results. Indeed, the major obstacle for the routine use of intratracheal instillation as a replacement for inhalation bioassays lies in the fact that the patterns of particle distribution in the lung following instillation are uneven and are unlike those resulting from inhalation. Particle deposition by inhalation is focal; that is, inhaled particles deposit at selected sites in the lung depending on their size. Brain and coworkers [34] showed that intratracheal instillation of particles produced nonuniform deposition patterns, largely dependent on gravitational settling. These investigators studied the distribution of particles labeled with $^{99}$TC in both rats and hamsters following either intratracheal injection or aerosol inhalation. Particle distribution patterns in the lung following inhalation were distributed evenly within a given lobe; among lung lobes, most of the dust deposited in the apical lobes [39]. Pritchard et al. [239] found that variability in the deposition of cerium oxide particles in rats within a specific lobe was considerably greater following instillation than inhalation with little penetration of instilled particles to the peripheral lung. Similarly, greater peripheral lung loading was seen following inhalation of ferric oxide particles than following instillation [67]. In contrast, the distribution of both short and long glass fibers in rats was reportedly similar using either inhalation or instillation [114]. Drew et al. [69] and Muller et al. [205] found that both routes produced the same relative lobular distribution of uranium oxide particles.

**TABLE 28.9**
**Advantages and Disadvantages of the Intratracheal Instillation Technique**

*Advantages*
It delivers an exact amount of material to the lung (high local exposures are possible).
The procedure is simple (compared to inhalation) and requires little equipment.
It requires only a small amount of test material.
Skin absorption confounders are avoided.
Safety concerns are minimal.
Issues of extreme toxicity, flammability, and explosivity are minimized.
It can evaluate test substances not readily respirable.

*Disadvantages*
Doses to various portions of the respiratory tract can vary.
Distribution and deposition of the test substance within the respiratory tract do not reflect distribution following inhalation.
Repeated dosing is possible but limited.
High local doses of administered substances can occur.

In summary, the pulmonary effects of test substances, for reasons of space, time, and economics, along with non-availability of sufficient sample quantities, cannot always be evaluated following inhalation exposures. Uses of intratracheal administration involves methodology that is simple and uses relatively little material so the risk to personnel is much reduced (from an inhalation study). One can deliver relatively large amounts of material to the lung in a short period of time.

Against these advantages, however, is the primary fact that amounts and sizes of particulate that would otherwise not be able to gain access to the lung, do so. The patterns of particle distribution in the lung following instillation are uneven and are unlike those resulting from inhalation The nonuniformity is partially random but represents systematic and reproducible regional differences. Another serious problem is that the instillation technique totally bypasses the upper respiratory tract. Problems can result from altering dosing rates and the use of differing suspending agents. Despite the advantages stated earlier these latter concerns should alert the inhalation toxicologist to the limitations of results obtaining using this technique (Table 28.9).

## Chamber Selection and Operation

Conventional whole-body chambers are constructed of stainless steel and have transparent observation ports of plastic or glass. Both stainless steel and glass are well suited as chamber construction materials due to their excellent chemical resistance and their low adsorption characteristics. The Rochester chamber design described by Leach et al. [159] consists of a hexagonal chamber fitted at the base apex end with hexagonal pyramids. This design was shown to have a good airflow pattern (e.g., no stagnant areas) and uniform distribution of test substances within the chamber. As originally designed, the chamber also allows for good visibility of the test animals and could simultaneously house 8 dogs, 4 monkeys, and 40 rats. Hinners et al. [122] found the hexagonal construction to be unnecessary and obtained satisfactory performance using chambers with a square cross-section, thereby simplifying the design and reducing manufacturing costs. This design provides a large access door for efficient animal loading, large windows for animal observations, appropriate sampling ports, control of air and test substance flows, signal devices for equipment failure, and temperature and humidity controls, and it has interior surfaces of stainless steel to prevent corrosion and to facilitate cleaning. A novel exposure chamber design described by Moss et al. [202] uses a vertical flow chamber of approximately 2 $m^3$ with an offset pyramidal inlet, three levels of caging, and solid-sheet catch pans below the cages. By incorporating offset catch pans, this design creates multiple areas of turbulence and promotes more uniform aerosol distribution within the chamber.

Horizontal flow chambers seem to be comparable to vertical feed chambers in concentration gradient characteristics but require higher flow rates to maintain uniform test atmospheres. A system described by Hemenway and MacAskill [112] utilizes an inlet and outlet diffuser/plenum configuration, which, along with premixing of the test substance, allows uniform exposure concentrations. These chambers also permit a higher animal packing density due to the lack of large inlet and outlet cones, and they can fit into most conventional animal rooms without major modification. Ferin and Leach [83] described a horizontal flow unit comprised of separate models for air supply, contaminant addition, animal exposure, and exhaust. The use of high airflow rates may cause laminar flow entering the module to become turbulent as it passes through the animal cages.

Most inhalation experiments utilize dynamic exposure conditions; that is, the test atmosphere, comprised of test substance diluted with fresh air, is continually replenished. Thus, the test atmosphere flows through the exposure chamber and is then exhausted. Once the generation apparatus is turned on, the concentration in the chamber rises to a theoretical equilibrium value, which is the ratio of the flow of the substance to the total flow in the chamber (Figure 28.15). The equation describing the concentration–time curve is:

$$C_t = f/F\left[1 - e^{-(F/V)t}\right]$$

where $C_t$ is concentration after $t$ minutes, $f$ is the flow of substance, $F$ is total flow through the chamber, and $V$ is the chamber volume. Usually, this equation is converted to an expression that defines the time required to attain a given percentage of the equilibrium concentration ultimately attained:

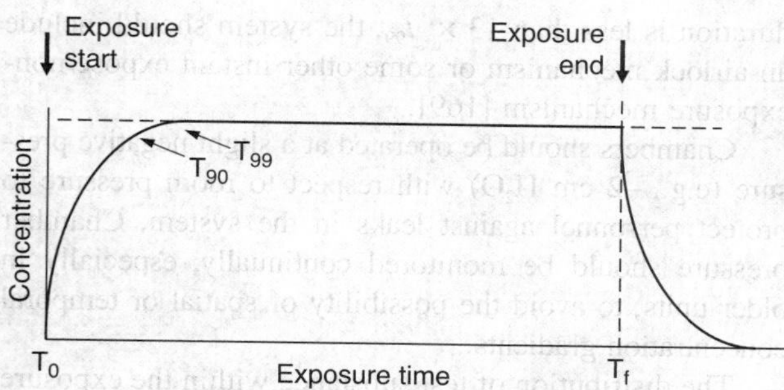


**FIGURE 28.15** Conceptual concentration and time relationships in a dynamic exposure chamber.

$$t_x = K \times V/F$$

where $t_x$ is the time required to attain $x$% of the equilibrium concentration, and $K$ is a constant whose value depends on $x$. Most frequently, the concentration–time relationship is described by $t_{99}$, the time required to attain 99% of the theoretical equilibrium concentration. For determination of $t_{99}$, $K = 4.605$. At this time point, the concentration in the chamber may be considered constant, assuming no changes in air or test substance supply rates. The concentration–time characteristics of a given chamber are described by the values of $V$ and $F$, which are preferred, and more descriptive than the outdated practice of giving air changes per hour [170,272].

Airflow to a chamber can vary from approximately 10 to 60 changes per hour. This is understood as the addition and withdrawal of a volume of air equal to the volume of the chamber. Because added air mixes with that already present, a complete change of air has not occurred. The term "air change" is misleading, and the dynamics of mixing in inhalation chambers has been described by the statistical considerations first put forward by Silver [272]. As already discussed, a more descriptive term for chamber airflow is the $t_{99}$.

The duration of exposure is defined as the interval from the start of flow of test substance to the point where delivery is discontinued. The exposure is terminated by stopping the flow of substance to the atmosphere generator, which leads to the decline in the chamber concentration on an exponential curve that is the inverse of the rising curve (Figure 28.15). Animals are not removed from the chamber, nor are the chamber doors opened to observe the animals, until at least $t_{99}$ minutes, to allow sufficient time for the chamber contents to be fully eliminated. For longer term exposures, the rising and falling section of the curves can be neglected and the exposure profile becomes a square-wave form. For short exposures, where exposure duration is less than 13 × $t_{99}$, the system should include an airlock mechanism or some other instant expose/non-exposure mechanism [169].

Chambers should be operated at a slight negative pressure (e.g., –2 cm $H_2O$) with respect to room pressure to protect personnel against leaks in the system. Chamber pressure should be monitored continually, especially in older units, to avoid the possibility of spatial or temporal concentration gradients.

The distribution of test substance within the exposure chamber is determined by taking repeated samples from different areas of the chamber in relation to a reference sampling location, using statistically valid sampling strategies. After homogeneous distribution of test substance within the chamber has been established, the actual concentration should be measured several times during the exposure. For exposures of 4- to 6-hour duration, a common practice is to sample every 0.5 hour (or at least hourly), although for regulatory purposes much less frequent sampling (e.g., 2 to 4 times per exposure) may be acceptable; the frequency of sampling should be dictated by the temporal stability of the test atmosphere generation and airflow supply systems. Continuous readout monitoring and control instruments are especially useful in long-term studies, because they are capable of preventing concentration excursions when appropriate alarms are integrated into the system. Some type of alarm instrumentation can prevent many weeks or months of effort from being destroyed by sudden elevations in test substance concentration.

Inhalation chambers operated in a dynamic mode should have a reasonably uniform distribution of test substance to avoid differential animal exposures. Due to their comparatively higher diffusion rates, it is generally acknowledged that gas and vapor concentrations show less intrachamber variation than aerosols. Aerosols are also subject to size-dependent impaction and sedimentation losses, compounding the difficulty in maintaining uniform chamber concentrations. MacFarland [171], however, calculated the percentage differences in concentrations between aerosols and gases or vapors and found little difference in the chamber uniformity between aerosols and gases or vapors. Examples from the literature support the variability from mean gas or aerosol concentrations ranging from 0.7 to 18.0%. Because variations exist in all chamber types and airflow control operating systems, investigators are well advised to carefully characterize the exposure chamber spatial and temporal distributions of test substances under actual generation conditions to prevent concentration-gradient errors from complicating the study.

Control of the relative humidity and, more importantly, the internal temperature of inhalation chambers is crucial to the integrity and interpretation of animal inhalation studies. As already noted, elevated temperatures can alter animal physiology, affect metabolic rates, and increase the rate and type of chemical interactions. Ideally, the total volume of animals in a chamber should not exceed 5% of the chamber volume to avoid heat-induced artifacts among test animals [85]. Coincidentally, chamber loading also affects the test substance equilibrium concentration in the exposure chamber. Silver [272] reported, for example, that, when the animal loading (total animal weight to chamber volume) exceeded 5%, excessive losses of test substance occurred, presumably by absorption to body surfaces. In some instances, such as when chamber loadings exceed this recommendation, chamber walls or inlet air may be cooled to maintain normal interior temperatures. Heat-balance studies with rats in stainless steel or glass chambers of equal size and using a room air intake of 100 L/min showed that the chamber walls were effective at removing approximately 90% of the animal body heat as compared with the heat loss through the airstream. If conducting studies in chambers with low airflow rates, heat transfer to the surrounding environment can be increased by painting the chambers, by attaching cooling coils to the chambers walls, or by directing air conditioning ducts directly onto the chamber [20]. In addition, control of relative humidity is essential for proper heat balance in rodents. Although it is generally recommended that relative humidity should be controlled within relatively narrow limits (e.g., 40 to 60%) to minimize adverse effects on feed consumption and behavior on study outcome [243], recent work showed that rats tolerated humidity levels of 3, 40, or 80% in repeated-exposure studies without effects on body weights, feed or water consumption, or respiratory tract histopathology [228].

## Administered Dose

The fundamental concept of dose (i.e., the amount of substance introduced into the test system) is not excluded from the domain of inhalation toxicology. Yet, due to the difficulty in measuring individual respiratory parameters and the test substance's physicochemical properties (two primary characteristics that affect the respiratory tract deposition of inhaled materials), the determination of the internalized dose is much more complex by inhalation compared to other exposure routes. Of the total amount of test material inhaled, a variable fraction may actually be deposited, absorbed, and reach target tissues. This absorbed fraction is, in turn, dependent on the local absorptive, metabolic, and clearance processes that may modify the chemical composition or concentration of the inhaled substance along deposition sites in the respiratory tract. Although these processes are not easily determined and have restricted the ability of inhalation toxicologist to provide quantitative measures of inhaled dose, progress in the area of dose determination has evolved in the related areas of biomarkers and mathematical modeling. Increasingly sophisticated models are being used to estimate toxicant concentrations in the blood and organs by treating the uptake, distribution, metabolism, and excretion of toxicants as a series of independent processes occurring at experimentally determined rates [57].

The administered dose in inhalation studies is most often characterized in terms of the exposure conditions (i.e., the exposure concentration and duration according to certain conventions); for example, "rats were exposed to *x* ppm (or *y* mg/m$^3$) of substance *y* for 6 hr/day, 5 days per week, for 13 weeks." For gases and vapors, concentration may be expressed on a volume basis, such as mole *x* per total moles, parts per million (ppm), or parts per billion (ppb), or on a weight basis, such as milligrams per liter (mg/L) or milligrams per cubic meter (mg/m$^3$). For solid or liquid aerosols, concentrations generally are expressed on a weight basis, such as milligrams per liter (mg/L), or, for fibrous materials, on a number basis, such as the number of fibers per cubic centimeter (fibers/cm$^3$). Furthermore, for fibrous aerosols, consideration should be given to expressing concentration on both a respirable mass (mg fibers/m$^3$) and a particle count (number of fibers/cm$^3$) basis; with some low-density fibers, the number count may be extremely high but the mass concentration may be very low.

An important note regarding the description of exposure concentration is needed. There is a fundamental distinction between the measured concentration and nominal concentration of a test substance. Nominal concentration refers to the amount of substance introduced into the exposure chamber in a given volume of air. Nominal concentrations are a holdover from the earliest days of inhalation testing when describing exposure conditions and were used when analytical methods were not as developed as today. As an example, if 200 g of a powdered substance were delivered to the aerosol generator over a 4-hour period and 40 L/min of dilution air was consumed, the nominal chamber concentration of test substance would be 200 g/(40 L/min × 240 minutes) or 0.02 g/L (or 20 mg/L or 20,000 mg/m$^3$). For a dust, 20 mg/L is a very high exposure concentration and, in practice, is very difficult to achieve or maintain while ensuring particle respirability. In reality, the nominal concentration of powdered substances greatly overestimates the actual or measured concentration as unpredictable internal losses (through impaction, interception, and settling) of aerosol inevitably occur in the generator, along conducting passages to the exposure chamber, and within the exposure chamber itself. In contrast, nominal concentrations for vapors or gases are generally much closer in agreement with the analytically determined chamber concentration as long as the substance does not react or decompose within the generation or exposure systems. For gases or vapors, nominal concentrations can be useful as analytical adjuncts to corroborate results from the chamber analysis. In any case, nominal concentrations are not as accurate or descriptive of the actual or mass-derived concentration of test substance in the animal's breathing zone, which is the proper measure of exposure conditions.

A fundamental interrelationship between exposure concentration and time exists for many inhaled substances. Haber [105], for example, determined that the response of an animal to a gas could be related to the product of the concentration and the exposure time. This relationship, known as Haber's law, states that the product of the concentration ($C$) and exposure time ($T$) required to produce a specific physiologic or toxicologic effect is equal to a constant ($K$): $C \times T = K$. The specific effect can be something other than death, but death is a commonly applied endpoint. Haber's law has been used to predict response for exposure conditions that have not been described experimentally—for example, predicting the expected 1-hour $LC_{50}$ for a chemical substance when only 4-hour data exist. There are short but finite time periods during which some endpoint may never be attained; that is, death may not occur at practically attainable concentrations in short time periods (e.g., 0.1, 1, or 10 minutes). At the other extreme, some exposure levels exist where the substance does not produce measurably adverse effects despite continuous, prolonged exposures. For these reasons, Haber's law does not apply to all substances but generally is applicable when extrapolated to $C \times T$ values that differ by a factor of about 3 to 4 for a given $C \times T$ product. Rinehart and Hatch [246] and Kelly et al. [135] have shown, for example, that Haber's law applies for phosgene gas and titanium tetrachloride aerosol when evaluated for the respiratory rate and lethality endpoints, respectively. Gelzleichter et al. [93] also showed that Haber's law applied for nitrogen dioxide or ozone in rats exposed for 3 days over a fourfold range of exposure concentrations; however, if rats were exposed to mixtures of ozone and nitrogen dioxide, lung damage was found to be related to the peak concentration rather than the cumulative dose of the gases. This observation was related to the synergistic, adverse effects of the mixture of these two gases. A quantitative assessment of the temporal and concentration relationship for lethality with several irritant and systematically active substances was evaluated by ten Berge et al. [291], who reported that, for the substances evaluated, the $C \times T$ relationship does not always represent the response data over the exposure periods reported. They reported that $C^n \times T$, where the exponent $n$ may vary from 0.8 to 3.5, is often a better predictor of mortality. Work by Pauluhn et al. [227] to explain $C \times T$ relationships from acute inhalation testing for product classification purposes suggests that 4-hour exposure data provide essentially the same projected acute lethality value as would 1-hour testing. When only 4-hour data were available, a default value of 4 should be used for conversion of 4-hour $LC_{50}$ values to 1-hour $LC_{50}$ values, independent of the physical state of the test substance. With sensory irritation as the endpoint, varying the exposure concentration has a proportionately greater effect than does changing exposure duration; thus, with the exception of calcium oxide, four test substances (chlorine, ammonia, formaldehyde, and 1-octane) were found not to follow Haber's law [270].

The internalized dose is related to the atmospheric concentration of the test substance, the duration of exposure, the individual's respiratory volume and frequency, and the deposition efficiency within the individual. These parameters are related according to the following general equation:

$$D = E_d V_m C T$$

where $D$ is the deposited dose (mg), $E_d$ is the deposition efficiency for the substance within the respiratory tract, $V_m$ is the minute volume (L/min), $C$ is the concentration of the test substance (mg/L), and $T$ is the time of exposure (minutes).

Use of this equation presupposes that the above parameters are well characterized for the species and test substance under investigation and a given set of experimental conditions. In practice, such data are not generally available, especially for gases or vapors and when other methods of dose determination have been used. For a series of vapors, the internalized dose has been determined by measuring the net decrease in test substance concentration in the exhaled air compared to the concentration in the inhaled air [287]; similar methodology has been used to measure the disposition of ultrafine aerosols in the human respiratory tract. In a system that is designed to validate a rat-based, physiologically based pharmacokinetic (PBPK) model of nasal extraction and metabolism of vinyl acetate (a water-soluble, reactive vapor), Hinderliter et al. [119] exposed human volunteers to radiolabeled vinyl acetate via nasal inhalation. Measurements of the concentration of vinyl acetate and its metabolite acetaldehyde in nasal airways compared favorably with the model predictions, thus providing support for human risk assessments conducted for workplace exposures.

Alternative methods to estimate the deposited dose have included chemical assays for the total amount of test substance or its metabolite in excreta, a useful technique for radioactively labeled materials. Measurement of metabolite levels and DNA or protein adducts in biologic fluids or tissues represents another means of assessing substance exposure and internalized dose [29,59,168]. Related studies have measured the amount and formation rate of DNA and protein cross-links in nasal mucosa following exposure to formaldehyde [46], the formation of DNA adducts from polycyclic aromatic hydrocarbon exposure [29], butadiene monoepoxide levels in tissues [293], complex mixtures such as tobacco smoke [258], and the dose-dependent differences in benzene metabolic formation [113,181]. An advantage of measuring DNA adducts in evaluating dose–response relationships is the comparatively higher sensitivity associated with the measurement of such biomarkers. In contrast to the comparatively high exposure concentrations employed in cancer bioassay studies, changes in the amount of metabolite–DNA adducts may be measured at much lower exposure concentrations compared to those associated with changes in tumor incidences. This method provides a better estimation of the shape of the dose–response curve at low concentrations, providing information is available to address the stability, extraction efficiency, and accuracy of the methods used to measure DNA adducts.

The need for quantifying the internalized dose, particularly as it relates to human risk assessment, has encouraged research in the area of pharmacokinetic (PK) modeling [7]. Use of PK modeling is finding wide acceptance among risk assessors and regulatory agencies for predicting the dose of toxicant to a given tissue across species and under various exposure conditions. This approach describes the kinetic relationships between physiologic factors (such as organ and body weights, respiratory rate, and blood flow), biochemical factors (such as substrate affinity for an enzyme and reaction velocity), and physicochemical factors (such as the extent of substance partitioning into air, blood, or tissue) on the disposition of that substance within the body. Pharmacokinetic modeling is based on developing a thorough understanding of these processes in an experimental animal and then extrapolating these parameters, with appropriate validation in experimental animals, to predict target tissue doses and toxic effects in humans. Validated PBPK models, using accepted physiological values, may then be useful in predicting expected tissue levels of a substance (and thus organ toxicity) under various exposure regimens. Indeed, PBPK models have been developed to describe the expected tissues doses of substances in rodents or humans for vapors such as butadiene [118], benzene [248], halogenated hydrocarbons [311], and butoxyethanol [160], as well as particulate matter such as powdered fire suppressants [147] or diesel soot, titanium dioxide [282], quartz dust [283], formaldehyde [145], vinyl acetate [232], and carbon tetrachloride [80]. When PK modeling is employed, due caution must be exercised with respect to the applicability of the animal model and experimental techniques used and their relationship to human values for the data to be useful in human health risk assessment.

A practical example of the utility of PK modeling was presented in a study evaluating the hazard of formaldehyde to humans [55]. Anatomically accurate, computational fluid dynamics models of the nasal airways of F344 rats, Rhesus monkeys, and humans were used to predict the regional flux of formaldehyde to the respiratory and olfactory mucosa. Statistical optimization was used to identify parameter values, and good simulations of the data were obtained. The parameter estimates for rats and monkeys were used to guide allometric scale-up to the human case. The relative levels of nasal mucosal DNA cross-links (DPX) in rats, Rhesus monkeys, and humans for a given inhaled concentration of formaldehyde were predicted by the model to vary with concentration. This modeling approach reduces uncertainty in the prediction of human nasal mucosal DPX resulting from formaldehyde inhalation. Use of such PK models may reduce the uncertainty for human cancer risk assessment by decreasing the uncertainty in interspecies extrapolations.

## *In Vitro* Toxicity Test Methods

The focus of toxicology has shifted somewhat, since the mid 1980s, from whole-animal toxicity tests to alternative *in vitro* toxicity methods [10,88,271]. *In vitro* toxicology describes a field of study that applies to technology using isolated organs, tissues, and cell culture to study the toxic effects of substances [109]. In addition to scientific advances, the development of *in vitro* toxicity test methods has been influenced by a variety of socioeconomical factors. The ethical and technical bases for continued use of animals in research balances a reduction in the number of test animals, refinement of test protocols to minimize suffering, and replacement of current animal tests with appropriate *in vitro* tests with the need for valid toxicological data. The necessity of determination of the potential toxic effects of a large number of substances and formulations has sparked the need for rapid, sensitive, and specific test methods. At present, *in vitro* methods cover a broad range of techniques and models, and a standardized battery of *in vitro* tests can be used for assessing acute local and systemic toxicity.

A practical strategy for incorporating the concepts of *in vitro* approaches for inhalation toxicity testing has been proposed by the European Centre for the Evaluation of Alternative Methods (ECVAM) [157]. This strategy uses existing literature, evaluates the physicochemical characteristics of test substances, and predicts potential toxic effects based on structure–activity relationships (SARs). Physicochemical characteristics of substances such as molecular structure, solubility, vapor pressure, pH, electrophilicity, and chemical reactivity are important properties that may provide critical information for hazard identification and toxicity prediction [82,88].

*In vitro* exposure techniques include direct exposure of cells to the test substance itself, cell exposure to collected air samples containing the substances of interest, or submerged cells undergoing intermittent or continuous direct exposure. *In vitro* screening tests may identify cellular responses allowing estimation of test substance toxic potency. Based on the obtained result, *in vitro* tests may be followed by a second phase using isolated cells specific to the target organs (e.g., nasal olfactory cells, fibroblasts, or mesothelial cells) [157]. Freshly isolated cells maintained in suitable culture media that can mimic biotransformation activities and cellular functions comparable to the *in vivo* environment are preferred to long-term cultures or cell lines that may differentiate, lose their organ-specific functions, and lack the enzyme systems required for biotransformation [11,157].

The effects of air contaminants have been studied using several *in vitro* exposure techniques. Most techniques have involved studies of isolated particulate substances added to culture medium [15,206]. Although this exposure condition may be adequate for soluble test materials, this may not simulate *in vivo* exposure conditions of dose and dose rate from airborne aerosols. Moreover, such techniques may often ignore the effects of particle size which are crucial in *in vivo* toxicity testing. Some *in vitro* research has investigated the effects of suspended and extracted particles from atmospheric aerosols [97,107] or cigarette smoke condensate [179,241].

The toxic effects of sulfuric acid aerosols on human airway epithelial cells [49] and the effects of ozone on cultures of primate bronchial epithelial cells [128] or human bronchial cells [187] have been studied using cells grown on collagen-coated membranes. Toxic effects of individual airborne substances, such as ozone and nitrogen dioxide, and complex mixtures, such as diesel motor exhaust [151] and cigarette smoke [8], were studied using cultured human lung cells on porous membranes permeable to culture media. The direct exposure technique at the air–liquid interface offers a reproducible contact between chemically and physically unmodified airborne contaminants and target cells and technically may reflect more closely inhalation exposure *in vivo* [8,151].

Regardless of the techniques employed, before *in vitro* methods can be formally used in lieu of traditional inhalation testing protocols for use in hazard classification, labeling, or risk assessment, validation efforts must be continued to ensure that the approaches, results, and conclusions reached from such techniques are applicable to and consistent with the objectives of international testing guidelines.

## Risk Assessment Considerations

Direct calculation of delivered dose in the species of interest potentially affects the magnitude of an uncertainty factor needed to address extrapolation of laboratory animal data to equivalent human exposure scenarios to improve human health risk estimates. The development of an inhalation reference concentration (RfC) typically involves extrapolation of an effect level observed in a laboratory animal exposure study to a level of exposure in humans that is not expected to result in an appreciable health risk. The average deposited dose normalized by regional surface area is the default dose metric often used. The most relevant dose metric is generally one that is most closely associated with the mode of action leading to the response. Critical factors in determining the best dose metric to characterize the dose–response relationship include the nature of the biological response being examined—that is, the magnitude, duration, and frequency of the intended exposure scenario and the mechanisms by which the toxicants exert their effects [30,130].

Route-specific risk assessments are performed to establish on a substance-by-substance basis reference doses for daily intake in humans (RfD, RfC). These values generally are derived from repeated exposure and dose studies in animals where effects are titrated from no (NOAEL), through minimal (LOAEL), to greater (EL) responses. This requires a rather complete database so the no-effect levels can be identified with relative confidence. Butenhoff et al. [44] presented such a risk assessment for the organofluoride surfactant, ammonium perfluorooctanoate. In this assessment, the authors had human exposure data (serum levels of the surfactant) and animal effects data relating effects to serum levels. By comparing serum chemical levels occurring in humans to those producing effects in animal studies (varying endpoints and exposure conditions, such as route and length of exposure), it was possible to establish a margin of exposure difference (e.g., animals begin to adversely respond at serum levels 1, 10, 100,… times greater than serum levels seen in humans).

To establish an inhalation reference dose for this surfactant, however, the database is limited to a single study [139]. Although the test established both the target tissue and exposure at which the effect can be produced, the experiment is limited in that the design included only 2 weeks of exposure and the exposure concentrations were separated by an order of magnitude. To allow extrapolation from the broader database of repeated-dose oral studies, Hinderliter et al. [120] identified serum levels of the chemical in the rat that correlated to both single and repeated inhalation exposures. These serum levels then were compared to serum levels derived from both single and multiple oral dose studies in this species. Similar serum levels were found to produce similar effects following either oral or inhalation exposure; hence, it became possible to read across exposure routes and establish internal doses that corresponded to external doses from either route.

## CONCLUSION

The study of the toxic effects of inhaled chemicals is typically more challenging compared to dermal or oral routes of exposure due to the high level of technical expertise necessary for the safe generation, control, and characterization of test atmospheres. This chapter has provided insights into the toxicological properties of inhaled materials and respiratory tract responses alongside descriptions of the basic technology needed to properly conduct inhalation exposures. Where more detail was appropriate, we have noted some of the emerging aspects of the field. Notwithstanding the need for a strong background in the technical aspects of study design and exposure system operation, we acknowledge that textbook guidance (such as provided by this chapter) is no substitute for hands-on experience; inhalation studies are almost as much art as they are science, and these skills are not acquired via the written word. For this reason, the experimenter is urged to gain practical experience wherever possible and to refer to some of the references cited to assist in gaining further appreciation of developing technologies. Despite the introduction of such novel materials as embodied in nanotechnology, a foundation in basic inhalation toxicology concepts is still necessary to evaluate the hazards that these and all other materials pose. In the end, however, the mark of a well-conducted inhalation study is one of clarity. Details of the rationale for test atmosphere generation, composition, concentration, and uniformity will be systemically addressed so the investigator can focus on the significance of the experimental results. We hope we have made the task a little less daunting for the reader beginning or continuing to work on inhalation toxicology-related questions.

## QUESTIONS

1. What are the main determinants affecting dose in inhalation studies?
2. In inhalation toxicity studies, what are the differences between nominal concentration, analytical (measured) concentration, internal concentration, and tissue concentration? How do these measures relate to dose in humans?
3. What are the special challenges presented by nanoparticles? (Consider study design, technical feasibility, generation and analytical needs, personnel safety, and waste treatment issues.)
4. In characterizing test atmospheres, should the analysis of very fine particles be based on mass, size, surface area, surface morphology, surface chemistry, or other parameters? Describe the merits and limitations of each.
5. What are some key considerations when sampling and analyzing test atmospheres of a solid particle aerosol? Of a low-volatility liquid mixture with both liquid aerosol and vapor components?
6. How might the toxicity differ between the same test substance generated primarily as an aerosol vs. primarily as a vapor?
7. You are asked to evaluate the inhalation toxicity of a pesticide formulation consisting of a granular powder. What considerations must be made to generate and analyze test atmospheres?
8. Under what circumstances would you *not* conduct an inhalation study on a test substance?
9. What considerations should be given to assess the inhalation hazard of a fibrous material intended to replace asbestos?

## REFERENCES


1. AIHA. (1998): *Emergency Response Planning Guidelines and Workplace Environmental Exposure Level Guides*. American Industrial Hygiene Association, Fairfax, VA.
2. Alarie, Y. (1973): Sensory irritation by airborne chemicals. *CRC Crit. Rev. Toxicol.*, 2:299–363.
3. Alarie, Y. (1981): Bioassay for evaluating the potency of airborne sensory irritants and predicting acceptable levels of exposure in man. *Food Cosmet. Toxicol.*, 19:623–626.
4. Albert, R. E., Berger, J., Sanburn, K., and Lippmann, M. (1974): Effects of cigarette smoke components on bronchial clearance in the donkey. *Arch. Environ. Health*. 29: 96–101.
5. Amdur, M. O. (1957): The influence of aerosols upon the respiratory response of guinea pigs to sulfur dioxide. *Am. Indust. Hyg. Assoc. Q.*, 18:149–155.
6. Andersen, M. E. (1981): Pharmacokinetics of inhaled gases and vapors. *Neurobehav. Toxicol. Teratol.*, 3:383–389.
7. Andersen, M. E. (1987): Tissue dosimetry in risk assessment or what's the problem here anyway? In: *Pharmacokinetics and Risk Assessment, Drinking Water and Health*, Vol. 8, pp. 8–26. National Academy Press, Washington, D.C.
8. Aufderheide, M., Knebel, J. W., and Ritter, D. (2003): An improved *in vitro* model for testing the pulmonary toxicity of complex mixtures such as cigarette smoke. *Exp. Toxicol. Pathol.*, 55:51–57.
9. Bair, W. J., Porter, N. S., Brown, D. P., and Wehner, A. P. (1969): Apparatus for direct inhalation of cigarette smoke by dogs. *J. Appl. Physiol.*, 26:847–850.
10. Bakand, C. et al. (2005): Toxicity assessment of industrial chemicals and airborne contaminants: transition from *in vivo* to *in vitro* test methods: a review. *Inhal. Toxicol.*, 17: 775–787.
11. Barile, F. A. 1994. *Introduction to In Vitro Cytotoxicity Mechanisms and Methods*. CRC Press, Boca Raton, FL.
12. Barlow, P. G., Donaldson, K., MacCallum, J., Clouter, A., and Stone, V. (2005): Serum exposed to nanoparticle carbon black displays increased potential to induce macrophage migration. *Toxicol. Lett.*, 155:397–401.
13. Baron, P. A. (1988): Modern real-time aerosol monitors. *Appl. Indust. Hyg.*, 3:97–103.
14. Batista, S. P., Guerin, M. R., Gori, B. G., and Kensler, C. J. (1973): A new system for quantitatively exposing laboratory animals by direct inhalation. *Arch. Environ. Health*, 27:376–382.
15. Becker, S., Soukup, J. M., and Gallagher, J. E. (2002): Differential particulate air pollution induced oxidant stress in human granulocytes, monocytes and alveolar macrophages. *Toxicol. In Vitro*, 16:209–218.
16. Bellmann, B., Konig, H., Muhle, H., and Pott, F. (1986): Chemical durability of asbestos and of man-made mineral fibres *in vivo*. *J. Aerosol Sci.*, 17:341–345.
17. Berg, I., Schluter, T., and Gercken, G. (1993): Increase of bovine alveolar macrophage superoxide anion and hydrogen peroxide release by dusts of different origin. *J. Toxicol. Environ. Health*, 39:341–354.
18. Bergman, U., Östergren, A., Gustafson, A. L., and Brittebo, E. B. (2002) Differential effects of olfactory toxicants on olfactory regeneration. *Arch. Toxicol.*, 76:104–112.
19. Bermudez, E., Mangum, J. B., Wong, B. A., Asgharian, B., Hext, P. M., Warheit, D. B., and Everitt, J. I. (2004): Pulmonary responses of mice, rats, and hamsters to subchronic inhalation of ultrafine titanium dioxide particles. *Toxicol. Sci.*, 77:347–357.
20. Bernstein, D. M. and Drew, R. T. (1980): The major parameters affecting temperature inside inhalation chambers. *Am. Indust. Hyg. Assoc. J.*, 41:420–426.
21. Bernstein, D. M., Moss, O. R., Fleissner, H., and Bretz, R. (1984): A brush feed micronising jet mill powder aerosol generator for producing a wide range of concentrations of respirable particles. In: *Aerosols: Science, Technology, and Industrial Application of Airborne Particles*, edited by B. Y. H. Liu, D. Y. H. Pui, and H. Fissan, pp 721–724. Elsevier, New York.
22. Bernstein, D. M., Riego Sintes, J. M., Ersboell, B. K., and Kunert, J. (2001): Biopersistence of synthetic mineral fibers as a predictor of chronic inhalation toxicity in rats. *Inhal. Toxicol.*, 13:823–849.
23. Blackford, J. A., Antonini, J. M., Castronova, V., and Dey, R. D. (1994): Intratracheal instillation of silica upregulates inducible nitric oxide synthase gene expression and increases nitric oxide production in alveolar macrophages and neutrophils. *Am. J. Respir. Cell Mol. Biol*, 11:426–431.
24. Blair, W. H. (1974): Chemical induction of lung carcinomas in rats. In *Experimental Lung Cancer*, edited by E. Karbe and J. F. Park, pp. 199–206. Springer-Verlag, New York.
25. Bogdanffy, M. S. (1990): Biotransformation enzymes in the rodent nasal mucosa: the value of a histochemical approach. *Environ. Health Perspect.*, 85:177–186.
26. Bogdanffy, M. S. and Valentine, R. (2003): Differentiating between local cytotoxicity, mitogenesis and genotoxicity in carcinogenic risk assessments: the case of vinyl acetate. *Toxicol. Lett.*, 140–141:83–98.
27. Boggs, D. F. (1992): In *Treatise on Pulmonary Toxicology*. Vol. 1. *Comparative Biology of the Normal Lung*, edited by R. Parent. CRC Press, Boca Raton, FL.
28. Bohren, C. F. and Huffman, D. R. (1998): *Absorption and Scattering of Light by Small Particles*. John Wiley & Sons, New York.
29. Bond, J. A., Harkema, J. R., Henderson, R. F., Mauderly, J. L., McClellan, R. O., and Wolff, R. K. (1989): Molecular dosimetry of inhaled diesel exhaust. In: *Assessment of Inhalation Hazards*, edited by D. V. Bates, D. L. Dungworth, P. N. Lee, R. O. McClellan and F. J. C. Roe, pp. 315–324, Springer-Verlag, New York.
30. Bond, J. A., Wallace, L. A., Osterman-Golkar, S., Lucier, G. W., Buckpitt, A., and Henderson, R. F. (1992): Assessment of exposure to pulmonary toxicants: use of biologic markers, *Fundam. Appl. Toxicol.*, 18:161–174.
31. Bos, P. M. J., Zwart, A., Ruezel, P. G. J., and Bragt, P. C. (1992): Evaluation of the sensory irritation test for the assessment of occupational health risk. *CRC Crit. Rev. Toxicol.*, 21:423–450.
32. Bowden, D. H. and Adamson, I. Y. R. (1978): Adaptive responses of the pulmonary macrophagic system to carbon. I. Kinetic studies, *Lab. Invest.*, 42:422–429.
33. Bowden, D. H. and Adamson, I. Y. R. (1984): Pathways of cellular efflux and particulate clearance after carbon instillation to the lung. *J. Pathol.*, 143:117–125.

34. Brain, J. D., Knudson, D. E., Sorokin, S. P., and Davis, M. A. (1976): Pulmonary distribution of particles given by intratracheal instillation or by aerosol inhalation. *Environ. Res.*, 11:13–33.
35. Brezova, V., Gabcova, S., Dvoranova, D., and Stasko, A. (2005): Reactive oxygen species produced upon photoexcitation of sunscreens containing titanium dioxide (an EPR study). *J. Photochem. Photobiol. B*, 79:121–134.
36. Brightwell, J., Fouillet, X., Cassano-Zoppi, A. L., Bernstein, D., Crawley, F., Duchosal, F., Gatz, R. S., and Pfeifer, H. (1989): Tumours of the respiratory tract in rats and hamsters following chronic inhalation of engine exhaust emissions. *J. Appl. Toxicol.*, 9:25–31.
37. Brock, W. J., Trochimowicz, H. J., Farr, C. H., Millischer, R. J., and Rusch, G. M. (1996): Acute, subchronic, and developmental toxicity and genotoxicity of 1,1,1-trifluoroethane (HFC-143a): *Fundam. Appl. Toxicol.*, 31:200–209.
38. Brody, A. R., Hill, L. H., Adkins, B., and O'Connor, R. W. (1981): Chrysotile asbestos inhalation in rats: deposition patterns and reaction of alveolar epithelium and pulmonary macrophages. *Am. Rev. Respir. Dis.*, 123:670–679.
39. Brody, A. R. and Roe, M. W. (1983): Deposition pattern of inorganic particles at the alveolar level in the lungs of rats and mice. *Am. Rev. Respir. Dis.*, 128:724–729.
40. Brown, D. M., Wilson, M. R., MacNee, W., Stone, V., and Donaldson, K. (2001): Size dependent proinflammatory effects of ultrafine polystyrene particles: a role for surface area and oxidative stress in the enhanced activity of ultrafines. *Toxicol. Appl. Pharmacol.*, 175:191–199.
41. Bucher, J. R., Elwell, M. R., Thompson, M. B., Chou, B. J., Renne, R., and Ragan, H. A. (1990): Inhalation toxicity studies of cobalt sulfate in F344/N rats and B6C3F$_1$ mice. *Fundam. Appl. Toxicol.*, 15:357–372.
42. Buckley, L. A., Jiang, X. Z., James, R. A., Morgan, K. T., and Barrow, C. S. (1984): Respiratory tract lesions induced by sensory irritants at the RD50 concentration. *Toxicol. Appl Pharmacol*, 74:417–429.
43. Burgess, B. A., Pastoor, T. P., and Kennedy, G. L., Jr. (1984): Aniline-induced methemoglobinemia and hemolysis as a function of exposure concentration and duration. *Toxicologist*, 4:64.
44. Butenhoff, J. L., Gaylor, D. W., Moore, J. A., Olsen, G. W., Rodricks, J., Mandel, J. H., and Zobel, L. R. (2004): Characterization of risk for general population exposure to perfluorooctanoate. *Reg. Toxicol. Pharmacol.*, 39: 363–380.
45. Cannon, W. C., Blanton, E. F., and McDonald, K. E. (1983): The flow-past chamber: an improved nose-only exposure system for rodents. *Am. Indust. Hyg. Assoc. J.*, 44:923–928.
46. Casanova, M., Deyo, D. F., and Heck, H. D'A. (1989): Covalent binding of inhaled formaldehyde to DNA in the nasal mucosa of Fischer 344 rats: analysis of formaldehyde and DNA by high-performance liquid chromatography and provisional pharmacokinetic interpretation. *Fundam. Appl. Toxicol.*, 12:397–417.
47. Castranova, V. (1998): Particulates and the airways: Basic biological mechanisms of pulmonary pathogenicity. *Appl. Occup. Environ. Hyg.*, 13:613–616.
48. Chang, L, Graham, J. A., Miller, F. J., Ospital, J. J., and Crapo, J. D. (1986): Effects of subchronic inhalation of low concentrations of nitrogen dioxide. I. The proximal alveolar region of juvenile and adult rats. *Toxicol. Appl. Pharmacol.*, 83:46–61.
49. Chen, L. C., Fang, C. P., Quo, Q. S., Fine, J. M., and Schlesinger, R. B. (1993): A novel system for the *in vitro* exposure of pulmonary cells to acid sulfate aerosols. *Fundam. Appl. Toxicol.*, 20:179–176.
50. Cheng, Y. S., Marshall, T. C., Henderson, R. F., and Newton, G. J. (1985): Use of a jet mill dispersing dry powder for inhalation studies. *Am. Indust. Hyg. Assoc. J.*, 46: 449–454.
51. Cheng, Y. S., Yamada, Y., Yeh, H. C., and Swift, D. L. (1988): Diffusional deposition of ultrafine aerosol in a human nasal cast. *J. Aerosol Sci.*, 19:741–751.
52. Cheng, Y. S., Hansen, G. K., Su, Y. F., Yeh, H. C., and Morgan, K. T. (1990): Deposition of ultrafine aerosols in rat nasal molds. *Toxicol. Appl. Pharmacol.*, 106:222–233.
53. Coffin, D. L., Gardner. D. E., Holzman, R. S., and Walock, F. S. (1968): Influence of ozone on pulmonary cells. *Arch. Environ. Health*. 16:633–636.
54. Collins, P. G. and Avouris, P. (2000): Nanotubes for electronics. *Sci Am.*, 283:62–69.
55. Connolly, R. B., Lilly, P. D., and Kimbell, J. S. (2000): Simulation modeling of the tissue disposition of formaldehyde to predict nasal DNA–protein cross-links in Fischer 344 rats, Rhesus monkeys, and human. *Environ. Health Perspect.*, 108(Suppl. 5):919–924.
56. Costa, D. L. (1985): Interpretation of new techniques used in the determination of pulmonary function in rodents. *Fundam. Appl. Toxicol.*, 5:423–434.
57. Crapo, J. D., Smolko, E. D., Miller, F. J., Graham, J. A. and Hayes, A. W. (1989): *Extrapolation of Dosimetric Relationships for Inhaled Particles and Gases*. Academic Press, San Diego, CA.
58. Dahl, A. R. and Hadley, W. H. (1991): Nasal cavity enzymes involved in xenobiotic metabolism: effects on the toxicity of inhalants, *CRC Crit. Rev. Toxicol.*, 21:345–372
59. Dahl, A. R., Schlesinger, R. B., Heck, H. D., Medinski M. A., and Lucier, G. W. (1991): Comparative dosimetry of inhaled materials: differences among animal species and extrapolation to man. *Fundam. Appl. Toxicol.*, 16:1–13.
60. Dai, Y. T. and Yu, C. P. (1998): Alveolar deposition of fibers in rodents and humans. *J. Aerosol Med.*, 11:247–258.
61. Daigle C. C., Chalupa, D. C., Gibb, F. R., Morrow, P. E., Oberdörster, G., Utell, M. J., and Frampton, M. W. (2003): Ultrafine particle deposition in humans during rest and exercise. *Inhal. Toxicol.*, 15:539–552.
62. Davis, J. M. G. (1974): Pathological aspects of the injections of glass fibers into the pleural and peritoneal cavities of rats and mice. In: *Occupational Exposure to Fibrous Glass*, DHEW Publ. No. 76-151, pp. 141–149. U.S. Department of Health, Education, and Welfare, Washington, D.C.
63. Dick, C. A., Brown, D. M., Donaldson, K., and Stone, V. (2003): The role of free radicals in the toxic and inflammatory effects of four different ultrafine particle types. *Inhal. Toxicol.*, 15:39–52.

64. Dockery, D. W., Schwartz, J., and Spengler, J. D. (1992): Air pollution and daily mortality: associations with particulates and acid aerosols. *Environ. Res.*, 59:362–273.
65. Dockery, D. W., Pope, C. A., Xiping, X., Spengler, J. D., Ware, J. H., Fay, M. E., Ferris, B. G., and Speizer, F. E. (1993): An association between air pollution and mortality in six U.S. cities, *New Engl. J. Med.*, 329:1753–1759.
66. Dorman, D. C. et al. (1996): Development of a mouse whole body exposure systems from a directed flow, rat nose only system. *Inhal. Toxicol.*, 8:107–120.
67. Dorries, A. M. and Valberg, P. A. (1992): Heterogeneity of phagocytosis for inhaled versus instilled material. *Am. Rev. Respir. Dis.*, 146:831–837.
68. Drew, R. T., Terrill, J. B., Daly, I. W., and Sheldon. A. (1986): Dose-dependent clearance of antimony from rat lungs. *Toxicologist*, 6:141.
69. Drew, R. T., Kuschner, M., and Bernstein, D. M. (1987): The chronic effects of exposure of rats to sized glass fibres. *Ann. Occup. Hyg.*, 31:711–729.
70. Driscoll, K. E., Higgins, J. M., Leytart, M. J., and Crosby, L. L. (1990): Differential effects of mineral dusts on the *in vitro* activation of alveolar macrophages eicosanoid and cytokine release. *Toxicol. In Vitro*, 4:284–288.
71. Driscoll, K. E., Lindenschmidt, R. C., Maurer, J. K., Higgins, J. M., and Ridder, G. (1990): Pulmonary responses to silica or titanium dioxide: inflammatory cells, alveolar macrophage-derived cytokines and histopathology. *Am. J. Resp. Cell Mol. Biol.*, 2:381–390.
72. Driscoll, K. E., and Maurer, J. K. (1991): Cytokine and growth factor release by alveolar macrophages: potential biomarkers of pulmonary toxicity. *Toxicol. Pathol.*, 19: 398–405.
73. Driscoll, K. E. (1994): Macrophage inflammatory proteins: biology and role in pulmonary inflammation. *Exp. Lung Res.*, 20:473–490.
74. Driscoll, K. E., Strzelecki, J., Hassenbein, D., Janssen, Y. M. W., Marsh, J., Oberdörster, G., and Mossman, BT. (1994): Tumor necrosis factor (TNF): evidence for the role of TNF in increased expression of manganese superoxide dismutase after inhalation of mineral dusts. *Ann. Occup. Hyg.*, 38(Suppl. 1):375–382.
75. Driscoll, K. E. (1995): Role of cytokines in pulmonary inflammation and fibrosis. In: *Concepts in Inhalation Toxicology*, edited by R. O. McClellan and R. F. Henderson, pp. 471–503. Taylor & Francis, New York.
76. Driscoll, K. E. (1996): The role of interleukin-1 and tumor necrosis factor α in the lung's response to silica. In: *Silica and Silica-Induced Lung Disease*, edited by V. Castranova et al. CRC Press, Boca Raton, FL.
77. Driscoll, K. E., Howard, B. W., Carter, J. M., Asquith, T., Johnston, C., Detilleux, P., Kunkel, S. L., and Isfort, R. J. (1996): Alpha-quartz-induced chemokine expression by rat lung epithelial cells: effects of *in vivo* and *in vitro* particle exposure. *Am. J. Pathol*, 149:1627–1637.
78. Duffin, R., Clouter, A., Brown, D. M., Tran, C. L., MacNee, W., Stone, V., and Donaldson, K. (2002): The importance of surface area and specific reactivity in the acute pulmonary inflammatory response to particles. *Ann. Occup. Hyg.*, 46:242–245.
79. Elder, A., Gelein, R., Finkelstein, J. N., Driscoll, K. E., Harkema, J., and Oberdörster, G. (2005): Effects of subchronically inhaled carbon black in three species. I. Retention kinetics, lung inflammation, and histopathology. *Toxicol. Sci.*, 88:614–629.
80. El-Masri, H. A., Thomas, R. S., Sabados, G. R., Phillips, J. K., Constan, A. A., Benjamin, S. A., Andersen, M. E., Mehendale, H. M., and Yang, R. S. H. (1996): Physiologically based pharmacokinetic–pharmacodynamic modeling of the toxicological interaction between carbon tetrachloride and kepone. *Arch. Toxicol.*, 71:704–713.
81. Fairly, D. (1990): The relationship of daily mortality to suspended particulates in Santa Clara County, 1980–1986. *Environ. Health Perspect.*, 89:159–168.
82. Faustman, E. M. and Omenn, G. S. (2001): Risk assessment. In: *Casaret & Doull's Toxicology: The Basic Science of Poisons*, 6th ed., edited by C. D. Klaassen, pp. 83–104. McGraw-Hill, New York.
83. Ferin, J. and Leach, L. J. (1980): Horizontal air flow inhalation exposure chambers. In: *Generation of Aerosols and Facilities for Exposure Experiments*, edited by K. Willeke, pp. 517–523. Ann Arbor Science Publishers, Ann Arbor, MI.
84. Ferin, J., Oberdörster, G., and Penny, D. P. (1992): Pulmonary retention of ultrafine and fine particles in rats. *Am. J. Respir. Cell Mol. Biol.*, 6:535–542.
85. Fraser, D. A., Bales, R. E., Lippmann, M., and Stockinger, H. E. (1959): *Exposure Chambers for Research in Animal Inhalation*, Public Health Monograph 357, U.S. Government Printing Office, Washington, D.C.
86. Frampton, M. W. et al. (2005): Effects of exposure to ultrafine carbon particles in healthy subjects and subjects with asthma. *Res. Rep. Health Eff. Inst.*, 126:1–47.
87. Fubini, B., Mollo, L., and Giamello, E. (1995): Free radical generation at the solid/liquid interface in iron containing minerals. *Free Rad. Res.*, 23:593–614.
88. Gad, S. C. (2000): *In Vitro Toxicology*. Taylor & Francis, New York.
89. Galter, D., Mihm, S., and Droge, W. (1994): Distinct effects of glutathione disulphide on the nuclear transcription factor kappa B and the activator protein-1. *Eur. J. Biochem.*, 221:639–648.
90. Garde, P., Ewing, P., Lastbom, L., and Ryrfeldt, A. (2004): A novel method to aerosolize powder for short inhalation exposures at high concentrations: isolated rat lungs exposed to respirable diesel soot. *Inhal. Toxicol.*, 16:45–52.
91. Gardner, D. E. (1984): Alterations in macrophage functions by environmental chemicals, *Environ. Health Perspect.*, 55:343–358.
92. Geiser, M., Rothen-Rutishauser, B., Kapp, N., Schürch, S., Kreyling, W., Schulz, H., Semmler, M., Im Hof, V., Heyder, J., and Gehr, P. (2005): Ultrafine particles cross cellular membranes by nonphagocytic mechanisms in lungs and in cultured cells. *Environ. Health Perspect.*, 113:1555–1560.
93. Gelzleichter, T. R., Witschi, H., and Last, J. A. (1992): Concentration-response relationships of rat lungs to exposure to oxidant air pollutants: a critical test of Haber's law for ozone and nitrogen dioxide. *Toxicol. Appl. Pharmacol.*, 112:73–80.

94. Genter, M. B., Llorens, J., O'Callaghan, J. P., Peele, D. B., Morgan, K. T., and Crofton, K. M. (1992): Olfactory toxicity of β-β-iminodiproprionitrile (IDPN) in the rat. *J. Pharmacol. Exp. Ther.*, 263:1432–1439.
95. Ghio, A. J., Kennedy, T. P., Whorton, A. R., Crumbliss, A. L., Hatch, G. E., and Hoidal, J. R. (1992): Role of surface complexed iron in oxidant generation and lung inflammation induced by silicates. *Am. J. Physiol.*, 263:L511–L518.
96. Ghio, A. J., Zhang, J., and Piantadosi, C. A. (1992): Generation of hydroxyl radical by crocidolite asbestos is proportional to surface [$Fe^{3+}$]. *Arch. Biochem. Biophys.*, 298:646–650.
97. Glowala, M., Mazurek, A., Piddubnyak, V., Fiszer-Kierzkowska, A., Michalska, J., and Krawczyk, Z. (2002): HSP70 overexpression increases resistance of V79 cells to cytotoxicity of airborne pollutants, but does not protect the mitotic spindle against damage caused by airborne toxins. *Toxicology*, 170:211–219.
98. Goldstein, E., Bartlema, H. C., Van der Ploeg, M., van Duijn, P., Van der Stap, J. G. M. M., and Lippert, W. (1978): Effect of ozone on lysosomal enzymes of alveolar macrophages engaged in phagocytosis and killing of inhaled *Staphylococcus aureus*, *J. Infect. Dis.*, 138:299–311.
99. Gong, Jr., H., Linn, W. S., Terrell, S. L., Clark, K. W., Geller, M. D., Anderson, K. R., Cascio, W. E., and Sioutas, C. (2004): Altered heart-rate variability in asthmatic and healthy volunteers exposed to concentrated ambient coarse particles. *Inhal. Toxicol.*, 16:335–343.
100. Goodglick, L. A. and Kane, A. B. (1986): Role of reactive oxygen metabolites in crocidolite asbestos toxicity to mouse macrophages. *Cancer Res.*, 46:5558–5566.
101. Gopinath, C., Prentice, D. E., and Lewis, D. J. (1987): The respiratory system. In: *Atlas of Experimental Toxicologic Pathology Current Histopathology*, Vol. 13, edited by G. A. Gresham, pp. 22–42. MTP Press, Norwell, MA.
102. Greenberg, H. L., Avol, E. L., Bailey, R. M., and Bell, K. A (1977): *Effects of Sulfate Aerosols upon Cardiopulmonary Function in Squirrel Monkeys*, pp. 279–393. National Technical Information Service, Springfield, VA.
103. Gross, E. A., Mellick, P. W., Kari, F. W., Miller, F. J., and Morgan, K. T. (1994): Histopathology and cell replication responses in the respiratory tract of rats and mice exposed by inhalation to glutaraldehyde for up to thirteen weeks. *Fundam. Appl. Toxicol.*, 23:348–362.
104. Gupta, T., Demokritou, P., and Koutrakis, P. (2004): Development and performance evaluation of a high-volume ultrafine particle concentrator for inhalation toxicological studies. *Inhal. Toxicol.*, 16:851–862.
105. Haber, F. (1924): *Funf vortrage aus den jahren* 1920–1923. Springer-Verlag, Berlin.
106. Haddad, J. J., Olver, R. E., and Land, S. C. (2000): Antioxidant/pro-oxidant equilibrium regulates HIF-1alpha and NF-kappa B redox sensitivity: evidence for inhibition by glutathione oxidation in alveolar epithelial cells. *J. Biol. Chem.*, 275:21130–21139.
107. Hamers, T., Van Schaardenburg, M. D., Felzell, E. C., Murk, A. J., and Koeman, J. H. (2000): The application of reporter gene assay for the determination of the toxic potency of diffuse air pollution. *Sci. Total Environ.*, 262: 159–174.
108. Hatch, T. and Choate, S. P. (1929): Statistical description of the size properties non-uniform particulate substances. *J. Franklin Inst.*, 29:66–78.
109. Hayes, A. J. and Markovic, B. (1999): Alternatives to animal testing for determining the safety of cosmetics. *Cosmet. Aerosols Toiletries Aust.*, 12:24–30.
110. Heinrich, U., Dungworth, D. L., Pott, F., Peters, L., Dasenbrock, C., Levsen, K., Kock, W., Creutzenberg, O., and Schulte, A. (1994): The carcinogenic effects of carbon black particles in rats and tar-pitch condensation aerosol after inhalation exposure in rats. *Occup. Hyg.*, 38:351–356.
111. Heinrich, U., Fuhst, R., Rittinghausen, S., Creutzenberg, O., Bellmann, B., Koch, W., and Levsen, K. (1995): Chronic inhalation exposure of Wistar rats and two different strains of mice to diesel engine exhaust, carbon black, and titanium dioxide. *Inhal. Toxicol.*, 7:533–556.
112. Hemenway, D. R. and MacAskill, S. (1982): Design, development and test results of a horizontal inhalation toxicology facility *Am. Indust. Hyg. Assoc. J.*, 43:874–879.
113. Henderson, R. F., Sabourin, P. J., Bechtold, W. E., Griffith, W. C., Medinsky, M. A., Birnbaum, L. S., and Lucier, G. W. (1989): The effect of dose, dose rate, route of administration and species on tissue and blood levels of benzene metabolites. *Environ. Health Perspect.*, 82:9–18.
114. Henderson, R. F., Driscoll, K. E., Harkema, J. R., Lindenschmidt, R. C., Chang, I. Y., Maples, K. R., and Barr, E. B. (1995): A comparison of the inflammatory response of the lung to inhaled versus instilled particles in F344 rats. *Fundam. Appl. Toxicol.*, 24:183–197.
115. Hesterberg, T. W., Miller, W. C., Musselman, R. P., Kamstrup, O., Hamilton, R. D., and Thevenaz, P. (1996): Biopersistence of man-made vitreous fibers and crocidolite asbestos in the rat lung following inhalation. *Fundam. Appl. Toxicol.*, 29:267–279.
116. Hesterberg, T. W., Chase, G., Axten, C., Miller, W. C., Musselman, R. P., Kamstrup, O., Hadley, J., Morscheidt, C., Bernstein, D. M., and Thevenaz, P. (1998): Biopersistence of synthetic vitreous fibers and amosite asbestos in the rat lung following inhalation. *Toxicol. Appl. Pharmacol.*, 151:262–275.
117. Hidy, G. M. (1984): *Aerosols: An Industrial and Environmental Science*. Academic Press, Orlando, FL.
118. Himmelstein, M. W., Acquavella, J. F., Recio, L., Medinsky M. A., and Bond, J. A. (1997): Toxicology and epidemiology of 1,3-butadiene. *CRC Crit. Rev. Toxicol.*, 27:1–108.
119. Hinderliter, P. M., Thrall, K. D., Corley, R. A., Bloemen L. J., and Bogdanffy, M. S. (2005): Validation of human physiologically based pharmacokinetic model for vinyl acetate against human nasal dosimetry data. *Toxicol. Sci.* 85:460–467.
120. Hinderliter, P. M., DeLorme, M. P., and Kennedy, G. L (2006): Perfluorooctanoic acid: relationship between repeated inhalation exposures and plasma PFOA concentration in the rat. *Toxicology*, 222:80–85.
121. Hinds, W. C. (1982): *Aerosol Technology: Properties Behavior, and Measurement of Airborne Particles*. John Wiley & Sons, New York.
122. Hinners, R. G., Burkart, J. K., and Punte, C. L. (1968) Animal inhalation exposure chambers. *Arch. Environ Health*, 16:194–204.

123. Holt, P. F. (1981): Transport of inhaled dust to extrapulmonary sites. *J. Pathol.*, 133:123–129.
124. ICRP. (1994): *Human Respiratory Tract Model for Radiological Protection, A Report of Committee 2 of the ICRP.* Pergamon Press, Oxford.
125. ILSI. (2005): Testing of fibrous particles: short-term assays and strategies. *Inhal. Toxicol.*, 17:497–537.
126. IPCS. (1986): *IPCS Environmental Health Criteria, EHC 53: Asbestos and Other Natural Mineral Fibers.* International Programme on Chemical Safety, World Health Organization, Geneva.
127. Isakovic, A., Markovic, Z., Todorovic-Markovic, B., Nikolic, N., Vranjes-Djuric, S., Mirkovic, M., Dramicanin, M., Harhaji, L., Raicevic, N., Nikolic, Z., and Trajkovic, V. (2006): Distinct cytotoxic mechanisms of pristine versus hydroxylated fullerene. *Toxicol. Sci.*, 65: 166–176.
128. Jabbour, A. J., Altman, L. C., Wight, T. N., and Luchtel, D. L. (1998): Ozone alters the distribution of betal integrins in cultured primate bronchial epithelial cells. *Am. J. Respir. Cell Mol.*, 19:357–365.
129. Jaques, P. A. and Kim, C. S. (2000): Measurement of total lung deposition of inhaled ultrafine particles in healthy men and women. *Inhal. Toxicol.*, 12:715–731.
130. Jarabek, A. M., Asgharian, B., and Miller, F. J. (2005): Dosimetric adjustments for interspecies extrapolation of inhaled poorly soluble particles *Inhal. Toxicol.*, 17: 317–334.
131. Jeffrey, P. K. (1983): Morphologic features of airway surface epithelial cells and glands. *Am. Rev. Respir. Dis.*, 128:S14–S20.
132. Johnson, N. F., Griffiths, D. M., and Hill, R. J. (1984): Size distributions following long-term inhalation of MMMF. In: *Biological Effects of Man-Made Mineral Fibres: Proceedings of a WHO/IARC Conference*, Vol. 2, pp. 102–125. World Health Organization, Geneva.
133. Kanj, R. S., Kang, J. L., and Castranova, V. (2005): Measurement of the release of inflammatory mediators from rat alveolar macrophages and alveolar type II cells following lipopolysaccharide or silica exposure: a comparative study. *J. Toxicol. Environ. Health A*, 68:185–207.
134. Keenan, C. M., Kelly, D. P., and Bogdanffy, M. S. (1990): Degeneration and recovery of rat olfactory epithelium following inhalation of dibasic esters. *Fundam. Appl. Toxicol.*, 15:381–393.
135. Kelly, D. P., Lee, K. P., and Burgess, B. A. (1981): Inhalation toxicity of titanium tetrachloride atmospheric hydrolysis products. *Toxicologist*, 1:76–77.
136. Kelly, D. P., Williams, S. J., Kennedy, G. L., and Lee, K. P. (1985): Recovery and characterization of lung-deposited Kevlar aramid fibers in rats. *Toxicologist*, 5:129.
137. Kelly, D. P., Merriman, E. A., Kennedy, G. L., and Lee, K. P. (1993): Deposition, clearance and shortening of Kevlar *para*-aramid fibrils in acute, subchronic and chronic inhalation studies in rats. *Fundam. Appl. Toxicol.*, 21: 345–354.
138. Kennedy, Jr., G. L. and Geisen, R. J. (1985): Setting occupational exposure limits for perfluoroisobutylene, a highly toxic chemical following acute exposure. *J. Occup. Med.*, 27:675.
139. Kennedy, Jr., G. L., Ferenz, R. L., and Burgess. B. A. (1986): Estimation of acute oral toxicity in rats by determination of the approximate lethal dose rather than the $LD_{50}$. *J. Appl. Toxicol.*, 6:145–148.
140. Kennedy, Jr., G. L., Barnes, J. R., and Chen, H. C. (1986): Inhalation toxicity of ammonium perfluorooctanoate, *Food Chem. Toxicol.*, 24:1325–1329.
141. Kennedy, Jr., G. L., Morris, J. B., Roloff, M. V., Salem, H., Ulrich, C. E., Valentine, R., and Wolff, R. K. (1992): Recommendations for the conduct of acute inhalation limit tests. *Fundam. Appl. Toxicol.*, 18:321–327.
142. Kerker, M. (1969): *The Scattering of Light and Other Electromagnetic Radiation*. Academic Press, New York.
143. Kim, C. S. and Jaques, P. A. (2005): Total lung deposition of ultrafine particles in elderly subjects during controlled breathing. *Inhal. Toxicol.*, 17:387–399.
144. Kim, Y. C. and Carlson, G. P. (1986): The effect of an unusual workshift on chemical toxicity. II. Studies on the exposure of rats to aniline. *Fundam. Appl. Toxicol.*, 7: 144–152.
145. Kimbell, J. S. (1994): Issues in modeling dosimetry in rats and primates. *Inhal. Toxicol.*, 6:S73–83.
146. Kimbell, J. S., Subramaniam, R. P., Gross, E. A., Schlosser, P. M., and Morgan, K. T. (2001): Dosimetry modeling of inhaled formaldehyde: comparisons of local flux predictions in the rat, monkey, and human nasal passages. *Toxicol. Sci.*, 64:100–110.
147. Kimmel, E. C., Carpenter, R. L., Smith, E. A., Reboulet, J. E., and Black, B. H. (1998): Physiologic models for comparison of inhalation dose between laboratory and field-generated atmospheres of a dry powder fire suppressant. *Inhal. Toxicol.*, 10:905–922.
148. Kimura, T. (1923): Artificial production of a cancer in the lungs following intrabronchial insufflation of coal-tar. *Gann*, 7:15–21.
149. Kinney, L. A., Burgess, B. A., Stula, E. F., and Kennedy, Jr., G. L. (1993): Inhalation studies in rats exposed to dimethylacetamide (DMAc) from 3 to 12 hours per day. *Drug Chem. Toxicol.*, 16:175–194.
150. Kirk, W. P., Rennberg, B. F., and Morken, D. A. (1975): Acute lethality in guinea pigs following respiratory exposure to $^{85}$Kr. *Health Phys.*, 28:275–284.
151. Knebel, J. W., Ritter, D., and Aufderheide, M. (2002): Exposure of human lung cells to native diesel motor exhaust: development of an optimized *in vitro* test strategy. *Toxicol. In Vitro*, 16:185–192.
152. Konaka, R., Kasahara, E., Dunlap, W. C., Yamamoto, Y., Chien, K. C., and Inoue, M. (1999): Irradiation of titanium dioxide generates both singlet oxygen and superoxide anion. *Free Rad. Biol. Med.*, 27:294–300.
153. Kouri, R. E., Rude, T., Thomas, P. E., and Whitmire, C. J. (1976): Studies on pulmonary aryl hydrocarbon hydroxylase activity in inbred strains of mice. *Chem. Biol. Interact.*, 13:317–331.
154. Kreyling, W.G., Semmler, M., Erbe, F., Mayer, P., Takenaka, S., Schulz, H., Oberdörster, G., and Ziesenis, A. (2002): Translocation of ultrafine insoluble iridium particles from lung epithelium to extrapulmonary organs is size dependent but very low. *J. Toxicol. Environ. Health A.*, 65: 1513–1530.

155. Lam, C. W., James, J. T., McCluskey, R., and Hunter, R. L. (2004): Pulmonary toxicity of single wall carbon nanotube in mice 7 and 90 days after intratracheal instillation. *Toxicol. Sci.*, 77:126–134.
156. Lamb, D. and Reid, L. (1969): Histochemical types of acidic glycoprotein produced by mucous cells of the tracheobronchial glands in man. *J. Pathol.*, 98:213–229.
157. Lambre, C. R., Auftherheide, M., Bolton, R. E., Fubini, B., Haagsman, H. P., Hext, P. M., Jorissen, M., Landry, Y., Morin, J. P., Nemery, B., Nettesheim, P., Pauluhn, J., Richards, R. J., Vickers, A. E. M., and Wu, R. (1996): *In vitro* tests for respiratory toxicity: the report and recommendations of ECVAM workshop 18. *Altern. Lab. Anim. J.*, 24:671–681.
158. Langard, S. and Nordhagen, A. L. (1980): Small animal inhalation chambers and the significance of dust ingestion from the contaminated coat when exposing rats to zinc chromate. *Acta Pharmacol. Toxicol.*, 46:43–46.
159. Leach, L. J., Spiegl, C. J., and Wilson, R. H. (1959): A multiple chamber exposure unit designed for chronic inhalation studies. *Am. Indust. Hyg. Assoc. J.*, 20:13–22.
160. Lee, K., Dill, J. A., Chou, B. J., and Roycroft, J. H. (1998): Physiologically based pharmacokinetic model for chronic inhalation of 2-butoxyethanol. *Toxicol. Appl. Pharmacol.*, 153:211–226.
161. Lee, K. P., Kelly, D. P., and Kennedy, Jr., G. L. (1983): Pulmonary response to inhaled Kevlar aramid synthetics fibers in rats. *Toxicol. Appl. Pharmacol.*, 71:242–253.
162. Lee, K. P., Trochimowicz, H. J., and Reinhardt, C. F. (1985): Pulmonary response of rats exposed to titanium dioxide ($TiO_2$) by inhalation for two years. *Toxicol. Appl. Pharmacol.*, 79:179–192.
163. Lee, K. P., Kelly, D. P., O'Neal, F. O., Stadler, J. C., and Kennedy, G. L. (1988): Lung response to ultrafine Kevlar aramid synthetic fibrils following 2-year inhalation exposure in rats. *Fundam. Appl. Toxicol.*, 11:1–20.
164. Lewis, D. J. (1991): Morphological assessment of pathological changes within the rat larynx. *Toxicol. Pathol.*, 19: 352–357.
165. Lewis, T. R., Morrow, P. E., McClellan, R. O., Raabe, O. G., Kennedy, G. L., Jr., Schwetz, B. A., Goehl, T. J., Roycroft, J. H., and Chabra, R. S. (1989): Establishing aerosol exposure concentration for inhalation toxicity studies. *Toxicol. Appl. Pharmacol.*, 99:377–383.
166. Lippmann, M. (1990): Man-made mineral fibers MMMF: human exposures and health risk assessment. *Toxicol. Indust. Health*, 6:225–246.
167. Lippmann, M., Cassee, F. R., Costa, D. L., Costantini, M., van Erp, A. M., and Gordon, T. (2005): International workshop on the design and analysis of experimental studies using PM concentrator technologies, Boston, May 5, 2004. *Inhal. Toxicol.*, 17:839–850.
168. Lucier, G. W., Belinsky, S., and Thompson, C. (1989): Molecular dosimetry of chemical carcinogens: implications for epidemiology and risk assessment. In: *Assessment of Inhalation Hazards*, edited by D. V. Bates, D. L. Dungworth, P. N. Lee, R. O. McClellan, and F. J. C. Roe, pp. 85–101. Springer-Verlag, New York.
169. MacFarland, H. N. (1976): Respiratory toxicology, In: *Essays in Toxicology*, Vol 7, edited by W. J. Hayes, pp. 121–154. Academic Press, New York.
170. MacFarland, H. N. (1981): A problem and a non-problem in chamber inhalation studies. In: *Inhalation Toxicology and Technology*, edited by B. K. J. Leong, pp. 11–18. Ann Arbor Science Publishers, Ann Arbor, MI.
171. MacFarland, H. N. (1983): Designs and operational characteristics of inhalation exposure equipment: a review. *Fundam. Appl. Toxicol.*, 3:603–613.
172. Manna, S. K., Sarkar, S., Barr, J., Wise, K., Barrera, E. V., Jejelowo, O., Rice-Ficht, A. C., and Ramesh, G. T. (2005): Single-walled carbon nanotube induces oxidative stress and activates nuclear transcription factor-κB in human keratinocytes. *Nano Lett.*, 5; 1676–1684.
173. Mauderly, J. L. (1974): Evaluation of the female pony as a pulmonary function model. *Am. J. Vet. Res.*, 35:1025–1029.
174. Mauderly, J. L., Jones, R. K., Griffith, W. C., Henderson, R. F., and McClellan, R. O. (1987): Diesel exhaust is a pulmonary carcinogen in rats exposed chronically by inhalation. *Fundam. Appl. Toxicol.*, 9:208–221.
175. Maynard, A. D., Baron, P. A., Foley, M., Shvedova, A. A., Kisin, E. R., and Castranova, V. (2004): Exposure to carbon nanotube material: aerosol release during the handling of unrefined single-walled carbon nanotube material. *J. Toxicol. Environ. Health*, 67:87–107.
176. McClellan, R. O. (1985): Health effects of diesel exhaust: a case study in risk assessment. *Am. Gov. Indust. Hyg.*, 12: 3–12.
177. McDougal, J. N., Jepson, G. W. Clewell III, H. J., and Andersen, M. E. (1985): Dermal absorption of dihalomethane vapors. *Toxicol. Appl. Pharmacol.*, 79:150–158.
178. McDougal, J. N., Jepson, G. W., Clewell III, H. J., MacNaughton, M. G., and Andersen, M. E. (1986): A physiological pharmacokinetic model for dermal absorption of vapors in the rat. *Toxicol. Appl. Pharmacol.*, 85:286–294.
179. McKarns, S. C., Bombic, D. W., Morton, M. J., and Doolittle, D. J. (2000): Gap junction intercellular communication and cytotoxicity in normal human cells after exposure to smoke condensates from cigarettes that burn or primarily heat tobacco. *Toxicol. In Vitro*, 14:41–51.
180. McLaughlin, Jr., R. F., Tyler, W. S., and Canada, R. O. (1961): A study of the subgross pulmonary anatomy in various mammals. *Am. J. Anat.*, 108:149–166.
181. Medeiros, A. M., Bird, M. G., and Witz, G. (1997): Potential biomarkers of benzene exposure. *J. Toxicol. Environ Health*, 51,519–539.
182. Mercer, T. T. (1975): The deposition model of the task group on lung dynamics: a comparison with recent experimental data. *Health Phys.*, 29:673–680.
183. Mery, S., Larson, J. L., Butterworth, B. E., Wolf, D. C. Harden, R. and Morgan, K. T. (1994): Nasal toxicity of chloroform in male F344 rats and female B6C3F$_1$ mice following one week inhalation exposure. *Toxicol. Appl Pharmacol.*, 125:214–227.
184. Miller, D. B. and Chernoff, N. (1995): Restraint-induced stress in pregnant mice: degree of immobilization affects maternal indices of stress and developmental outcome in offspring. *Toxicology*, 98:177–186.
185. Miller, R. R., Leto, R. L., Potts, W. J., and McKenna, M. T. (1980): Improved methodology for generating controlled test atmospheres. *Am. Indust. Hyg. Assoc. J.*, 41 844–846.

186. Miller, B. G., Jones, A. D., Searl, A., Buchannan, D., Cullen R. T., Soutar, C. A., Davis, J. M. G., and Donaldson, K. (1999): Influence of characteristics of inhaled fibers on development of tumours in the rat lung. *Ann. Occup. Hyg.*, 43:167–179.
187. Mogel, M., Kruger, E., and Seidel, A. (1998): A new coculture system of bronchial epithelial and endothelial cells as a model for studying ozone effects on airway tissue. *Toxicol. Lett.*, 96:25–32.
188. Morgan, A., Holmes, A., and Davison, W. (1982): Clearance of sized glass fibers from the rat lung and their solubility *in vivo*. *Ann. Occup. Hyg.*, 25:317–331.
189. Morgan, A. and Holmes, A. (1984): The deposition of MMMF in the respiratory tract of the rat, their subsequent clearance, solubility *in vivo* and protein coating. In: *Biological Effects of Man-Made Mineral Fibres: Proceedings of a WHO/IARC Conference*, Vol 2, pp. 1–17. World Health Organization, Geneva.
190. Morgan, K. T. (1991): Approaches to the identification and recording of nasal lesions in toxicology studies. *Toxicol. Pathol.*, 19:337–351.
191. Morgan, K. T., Kimbell, J. S., Monticello, T. M., Patra, A. L., and Fleishman, A. (1991): Studies of inspiratory airflow patterns in the nasal passages of the F344 rat and rhesus monkey using nasal molds: relevance to formaldehyde toxicity. *Toxicol. Appl. Pharmacol.*, 110:223–240.
192. Morgan, K. T., Patterson, D. L., and Gross, E. A. (1986): Responses of the nasal mucociliary apparatus of F-344 rats to formaldehyde gas. *Toxicol. Appl. Pharmacol.*, 82:1–13.
193. Morris, J. B. and Smith, F. A. (1982): Regional deposition and absorption of inhaled hydrogen fluoride in the rat. *Toxicol. Appl. Pharmacol.*, 62:81–89.
194. Morris, J. B. and Cavanagh, D. G. (1986): Deposition of ethanol and acetone vapors in the upper respiratory tract of the rat. *Fundam. Appl. Toxicol.*, 6: 78–88.
195. Morris, J. B. (1990): First-pass metabolism of inspired ethyl acetate in the upper respiratory tracts of the F344 rat and Syrian hamster. *Toxicol. Appl. Pharmacol.*, 102:331–345.
196. Morris, J. B., Frederick, B., Ultman, J. S., Gentry, P. R., Bush, M. L, Lomax, L. G., Black, K. A., Finch, L., Kimbell, J. S., Morgan, K. T., and Subramaniam, R. P. (2001): A hybrid computational fluid dynamics and physiologically based pharmacokinetic model for comparison of predicted tissue concentrations of acrylic acid and other vapors in the rat and human nasal cavities following inhalation exposure. *Inhal. Toxicol.*, 13:359–376.
197. Morrow, P. E., Gibb, F. R., and Gazioglu, K. M. (1967): A study of particulate clearance from the human lungs. *Am. Rev. Respir. Dis.*, 96:1209–1221.
198. Morrow, P. E. (1970): Models for the study of particle retention and elimination in the lung. In: *Inhalation Carcinogenesis*, edited by M. G. Hanna, Jr., P. Nettesheim, and J. R. Gilbert, CONF-691001, pp. 103–115. U.S. Atomic Energy Commission, Division of Technical Information, Oak Ridge, TN.
199. Morrow, P. E. and Mermelstein, R. (1988): Chronic inhalation toxicity studies: protocols and pitfalls. In: *Inhalation Toxicology: The Design and Interpretation of the Inhalation Studies and Their Use in Risk Assessment*, pp. 103–117. Springer-Verlag, New York.
200. Morrow, P. E. (1992): Dust overloading of the lungs: update and appraisal. *Toxicol. Appl. Pharmacol.*, 113:1–12.
201. Morrow, P. E., Haseman, J. K., Hobbs, C. H., Driscoll, K. E., Vu, V., and Oberdörster, G. (1996): The maximum tolerated dose for inhalation bioassays: toxicity vs. overload. *Fundam. Appl. Toxicol.*, 29:155–167.
202. Moss, O. R., Decker, J. R., and Cannon, W. C. (1982): Aerosol mixing in an animal exposure chamber having three levels of caging and excreta pans. *Am. Indust. Hyg. Assoc. J.*, 25:28–36.
203. Moss, O. R., James, R. A., and Asgharian, B. (2006): Influence of exhaled air on inhalation exposure delivered through a directed-flow nose-only exposure system. *Inhal. Toxicol.*, 18:45–51.
204. Muhle, H., Bellmann, B., Creutzenberg, O., Dasenbrock, C., Ernst, H., Klipper, R., MacKenzie, J. C., Morrow, P., Mohn, U., Takenaka, S., and Mermelstein, R. (1991): Pulmonary response to toner upon chronic inhalation exposure in rats. *Fundam. Appl. Toxicol.*, 17:280–299.
205. Muller, H. L., Drosselmeyer, E., Hotz, G., Seidel, A., Thiele, H., and Pickering, S. (1989): Behaviour of spherical and irregular (U,Pu)$O_2$ particles after inhalation or intratracheal instillation in rat lung during *in vitro* culture with bovine alveolar macrophages, *Int. J. Radiat. Biol.*, 55:829–842.
206. Nadeau, D., Vincent, R., Kumarathasan, P., Brook, J., and Dufresne, A. (1995): Cytotoxicity of ambient air particles to rat lung macrophages: comparison of cellular and functional assays. *Toxicol. In Vitro*, 10:161–172.
207. Nadziejko, C., Fang, K., Chen, L. C., Cohen, B., Karpatkin, M., and Nadas, A. (2002): Effect of concentrated ambient particulate matter on blood coagulation parameters in rats. *Res. Rep. Health Eff. Inst.*, 111:7–29.
208. Narciso, S. P., Nadziejko, E., Chen, Lung Chi; Gordon, T., and Nadziejko, C. (2003): Adaption to stress induced by restraining rats and mice in nose-only inhalation holders. *Inhal. Toxicol.*, 15:1133–1143.
209. National Research Council, Committee on Toxicology. (1986): *Criteria and methods for preparing Emergency Exposure Guidance Level (EEGL), Short-Term Public Emergency Guidance Level (SPEGL), and Continuous Exposure Guidance Level (CEGL) Documents*. National Academy Press, Washington, D.C.
210. Nettesheim, P. and Hammonds, A. S. (1971): Induction of squamous cell carcinoma in the respiratory tract of mice. *J. Nat Cancer Inst.*, 47:697–701.
211. Nikula, K. J., Snipes, M. B., Barr, E. B., Griffith, W. C., Henderson, R. F., and Mauderly, J. L. (1995): Comparative pulmonary toxicities and carcinogenicities of chronically inhaled diesel exhaust and carbon black in F344 rats. *Fundam. Appl. Toxicol.*, 25:80–94.
212. NTP. (1993): *Toxicology and Carcinogenesis Studies of Talc in F344/N Rats and B6C3F Mice*, Tech. Rep. Ser. No. 421, NIH Publ. No. 93-315, National Toxicology Program, Washington, D.C.
213. Nyberg, P. and Klockars, M. (1990): Measurement of reactive oxygen metabolites produced by human monocyte-derived macrophages exposed to mineral dusts. *Int. J. Exp. Pathol.*, 71:537–544.
214. Oberdörster, G. (1988): Lung clearance of inhaled insoluble and soluble particles. *J. Aerosol Med.*, 1:289–330.

215. Oberdörster, G., Ferin, J., Gelein, R., Soderholm, S. C., and Finkelstein, J. (1992): Role of the alveolar macrophage in lung injury: studies with ultrafine particles. *Environ. Health Perspect.*, 97:193–199.
216. Oberdörster, G., Ferin, J., and Lehnert, B. E. (1994): Correlation between particle size, *in vivo* particle persistence and lung injury. *Environ. Health Perspect.*, 102(Suppl. 5):173–179.
217. Oberdörster, G., Gelein, R. M., Ferin, J., and Weiss, B. (1995): Association of particulate air pollution and acute mortality: involvement of ultrafine particles? *Inhal. Toxicol.*, 7:111–124.
218. Oberdörster, G. (1995): Lung particle overload: Implications for occupational exposure to particles. *Reg. Toxicol. Pharmacol.*, 27:123–135.
219. Oberdörster, G, Sharp, Z., Atudorei, V., Elder, A., Gelein, R., Lunts, A., Kreyling, W., and Cox, C. (2002): Extrapulmonary translocation of ultrafine carbon particles following whole-body inhalation exposure of rats. *J. Toxicol. Environ. Heath A*, 65:1531–1543.
220. Oberdörster, G, Sharp, Z., Atudorei, V., Elder, A., Gelein, R., Lunts, A., Kreyling, W., and Cox, C. (2004): Translocation of inhaled ultrafine particles to the brain. *Inhal. Toxicol.*, 16:437–445.
221. Oberdörster, G., Maynard, A., Donaldson, K., Castranova, V., Fitzpatrick, J., Ausman, K., Carter, J., Karn, B., Kreyling, W., Lai, D., Olin, S., Monteiro-Riviere, N., Warheit, D., and Yang, H. (2005): Principles for characterizing the potential human health effects from exposure to nanomaterials; elements of a screening strategy. *Particle Fibre Toxicol.*, 2:1–35.
222. Oberdörster, G., Oberdörster, E., and Oberdörster, J. (2005): Nanotoxicology: an emerging discipline evolving from studies of ultrafine particles. *Environ. Health Perspect.*, 113:823–839.
223. OECD. (2006): *Section 4 Health Effects Test Guidelines*. Organization for Economic Cooperation and Development, Brussels (http://www.oecd.org/document/55/0,2340,en_2649_34377_2349687_1_1_1_1,00.html).
224. Oldham, M. J., Phalen, R. F., Robinson, R. J., and Kleinman, M. T. (2004): Performance of a portable whole-body mouse exposure system. *Inhal. Toxicol.*, 16:657–662.
225. Patrick, G. and Stirling, C. (1977): The retention of particles in large airways of the respiratory tract. *Proc. R. Soc. Lond. Serv. V*, 198:455–462.
226. Pauluhn, J. (1994): Validation of an improved nose-only exposure system for rodents. *J. Appl. Toxicol.*, 14:55–62.
227. Pauluhn, J., Bury, D., Fost, U., Gamer, A., Hoernicke, E., Hofmann, T., Kunde, M., Neustadt, T., Schlede, E., Schnierle, H., Wettig, K., and Westphal, D. (1996): Acute inhalation toxicity testing: considerations of technical and regulatory aspects. *Arch. Toxicol.*, 71:1–10.
228. Pauluhn, J. and Mohr, U. (1999): Repeated 4-week inhalation exposure of rats: Effect of low-, intermediate-, and high-humidity chamber atmospheres. *Exp. Toxicol. Pathol.*, 51:178–187.
229. Pauluhn, J. (2005): Brown Norway rat asthma model of diphenylmethane 4,4′-diisocyanate. *Inhal. Toxicol.*, 17: 729–739.
230. Paustenbach, D. J., Christina, J. E., Carlson, G. P., and Born, G. S. (1986): The effect of an 11.5 hr/day exposure schedule on the distribution and toxicity of inhaled carbon tetrachloride in the rat. *Fundam. Appl. Toxicol.*, 6:472–483.
231. Phalen, R., Kenoyer, J., and Davis, J. (1977): Deposition and clearance of inhaled particles: comparison of mammalian species. In: *Proceedings of the Annual Conference on Environmental Toxicology*, Vol. 7, pp. 159–170. National Technical Information Services, Springfield, VA.
232. Plowchalk, D. R., Andersen, M. E., and Bogdanffy, M. S. (1997): Physiologically based modeling of vinyl acetate uptake, metabolism, and intracellular pH changes in the rat nasal cavity. *Toxicol. Appl. Pharmacol.*, 142:386–400.
233. Pope, C. A., Dockery, D. W., Spengler, J. D., and Raizenne, M. E. (1991): Respiratory health and PM-10 pollution: a daily time series analysis. *Am. Rev. Respir. Dis.*, 144: 668–674.
234. Pope, C. A., Schwartz, J., and Ransom, M. R. (1992): Daily mortality and PM-10 pollution in Utah Valley. *Arch. Environ. Health*, 47:211–217.
235. Pope, C. A., Burnett, R. T., Thun, M. J., Calle, E. E., Krewski, E., Ito, K., and Thurston, G. D. (2002): Lung cancer, cardiopulmonary mortality and long term exposure to fine particulate air pollution. *J. Am. Med. Assoc.*, 287:1132–1141.
236. Pott, F. (1978): Some aspects of the dosimetry on the carcinogenic potency of asbestos and other fibrous dusts. *Staub-Reinhalt Luft*, 38:486–490.
237. Prahalad, A. K., Soukup, J. M., Inmon, J., Willis, R., Ghio, A. J., Becker, S., and Gallagher, J. E. (1999): Ambient air particles: effects on cellular oxidant radical generation in relation to particulate elemental chemistry. *Toxicol. Appl. Pharmacol.*, 158:81–91.
238. Prendergast, J. A., Jones, R. A., Jenkins, L. J., and Siegel J. (1967): Effects on experimental animals of long-term inhalation of trichloroethylene, carbon tetrachloride, 1,1,1-trichloroethane, dichlorodifluoromethane, and 1,1-dichloroethylene. *Toxicol. Appl. Pharmacol.*, 10:270–289.
239. Pritchard, J. N., Holmes, A., Evans, J. C., Evans, N., Evans R. J., and Morgan, A. (1985): The distribution of dust in the rat lung following administration by inhalation and by single intratracheal instillation. *Environ. Res.*, 36:268–297
240. Proctor, D. F. (1980): The upper respiratory tract. In: *Pulmonary Diseases and Disorders*, edited by A. P. Fishman pp. 209–223. McGraw-Hill, New York.
241. Putnam, K. P., Bombic, D. W., and Doolittle, D. J. (2002) Evaluation of eight *in vitro* assays for assessing the cytotoxicity of cigarette smoke condensate. *Toxicol. In Vitro* 16:599–607.
242. Rahman, I., Gilmour, P. S., Jimenez, L. A., and MacNee W. (2002): Oxidative stress and TNF-alpha induce histone acetylation and NF-kappaB/AP-1 activation in alveolar epithelial cells: potential mechanism in gene transcription in lung inflammation. *Mol. Cell. Biochem.*, 234–235 239–248.
243. Rao, G. N. (1986): Significance of environmental factor on the test system. In: *Managing Conduct and Data Quality of Toxicology Studies*, edited by B. K. Hoover, pp 173–185. Princeton Scientific, Princeton, NJ.

244. Renne, R. A., Wehner, A. P., Greenspan, B. J., DeFord, H. S., Ragan, H. A., Westerburg, R. B., Buschhom, R. L., Burger, G. T., Hayes, A. W., Suber, R. L., and Mosberg, A. T. (1993): 2-week and 13-week inhalation studies of aerosolized glycerol in rats. *Inhal. Toxicol.*, 4:95–111.
245. Rice, T. M, Clarke, R. W., Godeleski, J. J., Al Autairi, E., Jiang, N. F., Hauser, R., and Paulauskis, J. D. (2001): Differential ability of transition metals to induce pulmonary inflammation. *Toxicol. Appl. Pharmacol.*, 177:46–53.
246. Rinehart, W. E. and Hatch, T. (1964): Concentration product (Ct) as an expression of dose in sublethal exposures to phosgene, *Am. Indust. Hyg. Assoc. J.*, 25:545–553.
247. Rose, M. C., Lynn, W. S., and Kaufman, B. (1979): Resolution of the major components of human lung mucosal gel and their capabilities for reaggregation and gel formation. *Biochemistry*, 18:4030–4037.
248. Roy, A. and Georgopoulos, P. G. (1998): Reconstructing week-long exposures to volatile organic compounds using physiologically based pharmacokinetic models. *J. Expo Anal. Environ. Epidemiol.*, 8:407–422.
249. Rutland, J. and Cole, P. J. (1981): Nasal mucociliary clearance and ciliary beat frequency in cystic fibrosis compared with sinusitis and bronchiectasis. *Thorax*, 36:654–658.
250. Saffiotti, U., Cefis, F., and Kolb, L. A. (1968): A method of the experimental induction of bronchogenic carcinoma. *Cancer Res.*, 28:104–124.
251. Saffiotti, U. (1970): *Morphology of Experimental Respiratory Carcinogenesis*, A.E.C. Symp. Series 21, pp. 2, 45–250. Division of Technical Information, U.S. Army Environmental Center, Washington, D.C.
252. Sagartz, J. W., Madarasz, A. J., Forsell, M. A., Burger, G. T., Ayres, P. H., and Coggins, C. E. (1992): Histological sectioning of the rodent larynx for toxicity testing. *Toxicol. Pathol.*, 20:118–121.
253. Samet, J. M., Bishop, Y., Speizer, F. E., Spengler, J. D., and Ferris, B. G. (1981): The relationship between air pollution and emergency room visits in an industrial community. *J. Air Pollut. Control Assoc.*, 31:236–240.
254. Sayes, C. M., Fortner, J. D., Guo, W., Lyon, D., Boyd, A. M., Ausman, K. D., Tao, Y. J., Sitharaman, B., Wilson, L. J., Hughes, J. B., West, J. L., and Colvin, V. (2004): The differential cytotoxicity of water-soluble fullerenes. *Nano Lett.*, 4:1881–1887.
255. Sayes, C. M., Liang, F., Hudson, J. L., Mendez, J., Guo, W., Beach, J. M., Moore, V. C., Doyle, C. D., West, J. L., Billups, W. E., Ausman, K. D., and Colvin, V. L. (2006): Functionalization density dependence of single-walled carbon nanotubes cytotoxicity *in vitro*. *Toxicol. Lett.*, 161: 135–142.
256. Schaper, M. (1993): Development of a database for sensory irritants and its use in establishing occupational exposure limits. *Am. Indust. Hyg. Assoc. J.*, 54:488–544.
257. Scheimberg, J., McShane, O. P., and Carson, S. (1973): Inhalation of a powdered aerosol medication by nonhuman primates in individual space-type exposure helmets. *Toxicol. Appl. Pharmacol.*, 25:478–479.
258. Scherer, G. and Richter, E. (1997): Biomonitoring exposure to environmental tobacco smoke: a critical reappraisal. *Hum. Exp. Toxicol.*, 16:449–459.
259. Schins, R. P., Lightbody, J. H., Borm, P. J., Shi, T., Donaldson, K., and Stone, V. (2004): Inflammatory effects of coarse and fine particulate matter in relation to chemical and biological constituents. *Toxicol. Appl. Pharmacol.*, 195:1–11.
260. Schlesinger, R. B., Chen, L. C., and Driscoll, K. E. (1984): Exposure-response relationship of bronchial mucociliary clearance in rabbits following acute inhalations of sulfuric acid mist. *Toxicol. Lett.*, 22:249–254.
261. Schlesinger, R. B. (1985): Comparative deposition of inhaled aerosols in experimental animals and humans: a review, *J. Toxicol. Environ. Health*, 15:197–214.
262. Schlesinger, R. B. (1995): Deposition and clearance of inhaled particles. In: *Concepts in Inhalation Toxicology*, 2nd ed., edited by R. O. McClellan and R. F. Henderson, pp. 191–224. Taylor & Francis, New York.
263. Schreider, J. P. (1986): Comparative anatomy and function of the nasal passages. In: *Toxicology of the Nasal Passages*, edited by C. S. Barrow, pp. 1–25. Hemisphere, Washington, D.C.
264. Schwartz, J. (1991): Particulate air pollution and daily mortality in Detroit. *Environ. Res*, 56:204–213.
265. Schwartz, J. and Dockery, D. W. (1992): Particulate air pollution and daily mortality in Steubenville, Ohio. *Am. J. Epidemiol.*, 135:12–19.
266. Schwartz, J. and Dockery, D. W. (1992): Increased mortality in Philadelphia associated with daily air pollution concentrations. *Am. Rev Respir. Dis*, 145:600–604.
267. Searl, A. (1994): A review of the durability of inhaled fibres and options for the design of safer fibers. *Ann. Occup. Hyg.*, 38:839–855.
268. Seidel, W. C., Scherer, K. V., Cline, D. T., Olson A. H., Bonesteel, J. K., Church D. F., Nuggehalli, S., and Pyror, W. A. (1991): Chemical, physical, and toxicological characterization of fumes produced by heating polytetrafluoroethylene homopolymer and its copolymers with hexafluoropropylene and perfluoro(propyl vinyl ether). *Chem. Res. Toxicol.*, 4:229–236.
269. Shukla, A., Timblin, C., BeruBe, K., Gordon, T., McKinney, W., Driscoll, K., Vacek, P. and Mossman, B. T. (2000): Inhaled particulate matter causes expression of nuclear factor (NF)-κB-related genes and oxidant-dependent NF-κB activation *in vitro*. *Am. J. Respir. Cell Mol. Biol.*, 23:182–187.
270. Shusterman, D., Matovinovic, E., and Salmon, A. (2006): Does Haber's law apply to human sensory irritation? *Inhal. Toxicol.*, 18:457–471.
271. Silbergeld, E. K. (1998): Toxicology. In: *Encyclopedia of Occupational Health and Safety*, 4th ed., edited by J. M. Stellman, p. 33. International Labor Organization, Geneva.
272. Silver, S. D. (1946): Constant flow gassing chambers: principles influencing design and operation. *J. Lab. Clin. Med.*, 31:1153–1161.
273. Simon, R. H. and Paine, R. III. (1995): Participation of pulmonary alveolar epithelial cells in lung inflammation. *J. Lab. Clin. Med.*, 126:108–118.
274. Smith, D. M., Ortiz, L. W., Archuleta, R. et al. (1981): A method for chronic nose-only exposures of laboratory animals to inhaled fibrous aerosols. In: *Inhalation Toxicology and Technology*, edited by B. K. J. Leong, pp. 89–105. Ann Arbor Science Publishers, Ann Arbor, MI.

275. Smith, L. W., Gardner, R. J., and Kennedy, Jr., G. L. (1982): Inhalation toxicity of perfluoroisobutylene. *Drug Chem. Toxicol.*, 5:295–303.
276. Spicer, S. S., Schulte, B. A., and Thomopoulos, G. N. (1983): Histochemical properties of the respiratory tract in different species. *Am. Rev. Respir. Dis.*, 128:S20–S26.
277. Spurney, K. R. (1993): Atmospheric, anthropogenic aerosol and its toxic and carcinogenic components. *Wiss Unwelt*, 2:139–151.
278. St. George, J. A., Cranz, D. L., Zicker, S., Etchison, J. R., Dungworth, D. L., and Plopper, C. G. (1985): An immunohistochemical characterization of rhesus monkey respiratory secretions using monoclonal antibodies, *Am. Rev. Respir. Dis.*, 132:556–563.
279. Stanton, M. F. and Wrench, C. (1972): Mechanisms of mesothelioma induction with asbestos and fibrous glass. *J. Natl. Cancer Inst.*, 48:797–821.
280. Stavert, D. M., Archuleta, D. C., Behr, M. J., and Lehnert, B. E. (1991): Relative acute toxicities of hydrogen fluoride, hydrogen chloride and hydrogen bromide in nose- and pseudo-mouth-breathing rats. *Fundam. Appl. Toxicol.*, 16: 636–655.
281. Stöber, W., Flachsbart, H., and Hochrainer, D. (1970): Der aerodynamische durchmesser von latexaggregaten and asbestifasern. *Staub-Reinhalt Luft*, 30:277–285.
282. Stöber, W., Morrow, P. E., and Morawietz, G. (1990): Alveolar retention and clearance of insoluble particles in rats simulated by a new physiology-oriented compartmental kinetics model. *Fundam. Appl. Toxicol.*, 15:329–349.
283. Stöber, W. (1999): Pock model simulation of pulmonary quartz dust retention data in extended inhalation exposures of rats. *Inhal. Toxicol.*, 11:269–292.
284. Stockinger, H. E. (1957): Evaluation of the hazards of ozone and oxides of nitrogen-factors modifying toxicity. *Arch. Indust. Health*, 15:181–190.
285. Stoeger, T., Reinhard, C., Takenaka, S., Schroeppel, A., Karg, E., Ritter, B., Heyder, J., and Schulz, H. (2006): Instillation of six different ultrafine carbon particles indicates a surface area threshold dose for acute lung inflammation in mice. *Environ. Health Perspect.*, 114:328–333.
286. Stone, V., Shaw, J., Brown, D. M., MacNee, W., Faux, S. P., and Donaldson, K. (1998): The role of oxidative stress in the prolonged inhibitory effect of ultrafine carbon black on epithelial cell function. *Toxicol. In Vitro*, 12:649–659.
287. Stott, W. T. and McKenna, M. J. (1984): The comparative absorption and excretion of chemical vapors by the upper, lower, and intact respiratory tract of rats. *Fundam. Appl. Toxicol.*, 4:594–604.
288. Stuart, B. O, Willard, D. H., and Howard, E. B. (1971): Studies of inhaled radon daughters, uranium ore dust, diesel exhaust and cigarette smoke in dogs and hamsters. In: *Inhaled Particles III*, edited by W. H. Walton, pp. 543–553. Unwin, Surrey, U.K.
289. Swift, D. L., Montassier, N., Hopke, P. K., Karpen-Haves, K., Cheng, Y. S., Su, Y. F., Yeh, H. C., and Strong, J. C. (1992): Inspiratory deposition of ultrafine particles in human nasal replicate cast. J. *Aerosol Sci.*, 23:65–72.
290. Tao, F., Gonzalez-Flecha, B., and Kobzik, L. (2003): Reactive oxygen species in pulmonary inflammation by ambient particulates. *Free Radicals Biol. Med.*, 35:327–340.
291. ten Berge, W. F., Zwart, A., and Appelman, L. M. (1986): Concentration-time mortality response relationship and systemically acting vapors and gases. *J. Hazard. Mater.*, 13:301–309.
292. Thomas, R. G. and Lie, R. (1963): *Procedures and Equipment Used in Inhalation Studies on Small Animals*, Lovelace Foundation Report No. LF-11. U.S. Atomic Energy Commission, Albuquerque, NM.
293. Thorton-Manning, J. R., Dahl, A. R., Allen, M. L., Bechtold, W. E., Griffith, W. C., and Henderson, R. F. (1998): Disposition of butadiene epoxides in Sprague–Dawley rats following exposures to 8000 ppm 1,3-butadiene: comparisons with tissue epoxide concentrations following low-level exposures. *Toxicol. Sci.*, 41:167–173.
294. Timbrell, V. (1965): The inhalation of fibrous dusts. *Ann. N.Y. Acad. Sci.*, 132:255–273.
295. Timbrell, V. and Skidmore, J. W. (1970): The effect of shape on particle penetration and retention in animal lungs. In: *Inhaled Particles*, Vol. III, edited by W. H. Walton, pp 49–57. Unwin, Surrey, U.K.
296. Tjalve, H., Henriksson, J., Tallkvist, J., Larsson, B., and Lindquist, N. (1996): Uptake of manganese and cadmium from the nasal mucosa into the central nervous system via olfactory pathways in rats. *Pharmacol. Toxicol.*, 79: 347–356.
297. Tran, C. L., Buchanan, D., Cullen, R. T., Searl, A., Jones, A. D., and Donaldson, K. (2000): Inhalation of poorly soluble particles. II. Influence of particle surface area on inflammation and clearance. *Inhal. Toxicol*, 12: 1113–1126.
298. Tsujino, I., Kawakami, Y., and Kaneko, A. (2005): Comparative simulation of gas transport in airway models of rat, dog, and human. *Inhal. Toxicol.*, 17:475–485.
299. Tyl, R. W., Ballantyne, B., Fisher, L. C., Fait, D. L., Savine, T. A., Pritts, I. M., and Dodd, D. E. (1994): Evaluation of exposure to water aerosol or air by nose-only or whole-body inhalation procedures for CD-1 mice in developmental toxicity studies. *Fundam. Appl. Toxicol.*, 23:251–260.
300. Ulrich, C. E. and Marold, B. W. (1979): Pulmonary deposition of aerosols in individual and group-caged rats. *Am Indust. Hyg. Assoc. J.*, 40:633–636.
301. Uriah, L. C. and Maronpot, R. R. (1990): Normal histology of the nasal cavity and application of special techniques *Environ. Health Perspect.*, 85:187–208.
302. Uriah, L. C., Morgan, K. T., and Maronpot, R. R. (1990) Proc. of Symposium on Toxicologic Pathology of the Upper Respiratory System, National Toxicology Program NIEHS, Durham, NC, Sept. 14–16. *Environ. Health Perspect.*, 85:163–352.
303. Vallyathan, V. and Wallace, W. E., Eds. (1986): *Airborne Asbestos Health Assessment Update*, EPA/600/8–84/003F pp. 163–184. CRC Press, Boca Raton, FL.
304. USEPA. (1997): National Advisory Committee for Acute Exposure Guideline Levels for Hazardous Substances, *Fed Reg.*, 62:58840–58851.
305. USEPA. (2004): *Air Quality Criteria for Particulate Matter*, National Center for Environmental Assessment, U.S Environmental Protection Agency, Washington, D.C.

306. USEPA. (2006): *Series 870: Health Effects Test Guidelines*. Office of Prevention, Pesticides, and Toxics, U.S. Environmental Protection Agency, Washington, D.C. (http://www.epa.gov/opptsfrs/publications/OPPTS_Harmonized/870_ Health_Effects_Test_Guidelines/index.html).
307. Valberg, P. A. and Watson, A. Y. (1998): Alternative hypotheses linking outdoor particulate matter with daily morbidity and mortality. *Inhal. Toxicol.*, 10:641–662.
308. Vallyathan, V, Mega, J. F., Shi, X., and Dalal, N. S. (1992): Enhanced generation of free radicals from phagocytes induced by mineral dusts. *Am. J. Respir. Cell Mol. Biol.*, 6:404–413.
309. Van de Hulst, H. C. (1957): *Light Scattering by Small Particles*. John Wiley & Sons, New York.
310. Vincent, J. H. and Armbruster, L. (1981): On the quantitative definition of the inhalability of airborne dust. *Ann. Occup. Hyg.*, 24:245–248.
311. Vinegar, A., Jepson, G. W., and Overton, J. H. (1998): PBPK modeling of short term (0–5 min) human inhalation exposures to halogenated hydrocarbons. *Inhal. Toxicol.*, 10:411–429.
312. Wagner, J. C., Berry, G., and Timbrell, V. (1973): Mesothelioma in rats after inoculation with asbestos and other materials. *Br. J. Cancer*, 28:175–185.
313. Walton, W. H. (1982): The nature, hazards, and assessment of occupational exposure to airborne asbestos dust: a review. *Ann. Occup. Hyg.*, 25:115–247.
314. Warheit, D. B., Hill, L. H., George, G., and Brody, A. R. (1986): Time course of chemotactic factor generation and the corresponding macrophage response to asbestos inhalation. *Am. Rev. Respir. Dis.*, 134:128–133.
315. Warheit, D. B., Overby, L. H., George, G., and Brody, A. R. (1986): Pulmonary macrophages are attracted to inhaled particles through complement activation. *Exp. Lung Res.*, 14:51–66.
316. Warheit, D. B., Seidel, W. C., Carakostas, M. C., and Hartsky, M. A. (1990): Attenuation of perfluoropolymer fume pulmonary toxicity: effects of filters, combustion method and aerosol age. *Exp. Mol. Pathol.*, 52:309–329.
317. Warheit, D. B., Hwang, H. C., and Achinko, L. (1991): Assessments of lung digestion methods for recovery of fibers. *Environ. Res.*, 54:183–193.
318. Warheit, D. B., Carakostas, M. C., Bamberger, J. R., and Hartsky, M. A. (1991): Complement facilitates macrophage phagocytosis of inhaled iron particles but has little effect in mediating silica-induced lung inflammatory and clearance responses. *Environ. Res.*, 56:186–203.
319. Warheit, D. B., Hansen, J. F., Yuen, I. S., Kelly, D. P., Snajdr, S., and Hartsky, M. A. (1997): Inhalation of high concentrations of low toxicity dusts in rats results in impaired pulmonary clearance mechanisms and persistent inflammation. *Toxic Appl. Pharmacol.*, 145 :10–22.
320. Warheit, D. B., Laurence, B. R., Reed, K. L., Roach, D. H., Reynolds, G. A. M., and Webb, T. R. (2004): Comparative pulmonary toxicity assessment of single-wall carbon nanotubes in rats. *Toxicol. Sci.*, 77:117–125.
321. Warheit, D. B., Reed, K. L., Stonehuerner, J. D., Ghio, A. J., and Webb, T. R. (2006): Biodegradability of *para*-aramid respirable-sized fiber shaped particulates (RFP) in human lung cells. *Toxicol. Sci.*, 89:296–303.
322. Waritz, R. S. and Kwon, B. K. (1968): The Inhalation toxicity of pyrolysis products of polytetrafluoroethylene heated below 500 degrees Centigrade. *Am. Indust. Hyg. Assoc. J.*, 68:19–26.
323. Weeks, W. M., Downing, T. O., Musselman, N. P., Carson, B. S., and Groff, W. A. (1960): The effects of continuous exposure of animals to ethanolamine vapor. *Am. Indust. Hyg. Assoc. J.*, 21:374–381.
324. Weyel, D. A. and Schaeffer, R. B. (1985): Pulmonary and sensory irritation of diphenylmethane-4,4′- and dicyclohexylmethane-4,4′-diisocyanate. *Toxicol. Appl. Pharmacol.*, 77:427–433.
325. Whitby, K. T., Killelson, D. B., Cantrell, B. K., Barsic, N. J., Dolon, D. F., Tarvestad, L. D., Nieken, D. J., Wolf, J. L., and Wood, J. R. (1976): Aerosol size distributions and concentrations measured during the General Motors Proving Grounds sulfate study. In: *The General Motors/Environmental Protection Agency Sulfate Dispersion Experiment*, EPA-600/3-76-035, edited by R. K. Stevens, P. J. Lamothe, W. E. Wilson, J. L. Durham, and T. G. Dzubay, pp 29–80. U.S. Environmental Protection Agency, Research Triangle Park, NC.
326. Whitby, K. T. and Cantrell, B. K. (1976): Atmospheric aerosols: characteristics and measurement. In: *Proc. of Int. Conf. on Environmental Sensing and Assessment (ICESA)*, Institute of Electrical and Electronic Engineers (IEEE), Las Vegas, NV, September 14–19, 1975.
327. Whittemore, A. and Korn, E. (1980): Asthma and air pollution in the Los Angeles area. *Am. J. Public Health*, 70: 687–696.
328. WHO. (1985): *World Health Organization Reference Methods for Measuring Man-Made Mineral Fibers (MMMF)*. World Health Organization, Geneva.
329. Willeke, K., Lo, C. S. K., and Whitby, K. J. (1974): Dispersion characteristics of a fluidized bed. *J. Aerosol Sci.*, 5:449–455.
330. Wilson, Jr., F. J., Hiller, F. C., Wilson, J. D., and Bone, R. C. (1985): Quantitative deposition of ultrafine stable particles in the human respiratory tract. *J. Appl. Physiol.*, 58: 223–229.
331. Wilson, M. R., Lightbody, J. H., Donaldson, K., Sales, J., and Stone, V. (2002): Interactions between ultrafine particles and transition metals *in vivo* and *in vitro*. *Toxicol. Appl. Pharmacol.*, 184:172–179.
332. Wong, B. A. (2003): Automated feedback control of an inhalation exposure system with discrete sampling intervals: testing, performance, and modeling. *Inhal. Toxicol.*, 15:729–743.
333. Wong, K. L. and Alarie, Y. (1982): A method for repeated evaluation of pulmonary performance in unanesthetized, unrestrained guinea pigs and its application to detect effects of sulfuric acid mist inhalation. *Toxicol. Appl. Pharmacol.*, 63:72–90.
334. Wright, G. W. and Kuschner, M. (1977): The influence of varying lengths of glass and asbestos fibers on tissue response in guinea pigs. In: *Inhaled Particles*, Vol. IV, edited by W. H, Walton, pp 455–474. Pergamon Press, New York.
335. Young, J. T. (1981): Histopathologic examination of the rat nasal cavity. *Fundam. Appl. Toxicol.*, 1:309–312.

336. Zelikoff, J. T., Chen, L., Cohen, M. D., Fang, K., Gordon, T., Li., Y., Nadziejko, C., and Schlesinger, R. B. (2003): Effects of inhaled ambient particulate matter on pulmonary antimicrobial immune defense. *Inhal. Toxicol.*, 15: 131–150.
337. Zimmer, A. T. and Maynard, A. D. (2002): Investigation of the aerosols produced by a high-speed, hand-held grinder using various substrates. *Ann. Occup. Hyg.*, 46:663–672.
338. Zissu, D. (1995): Histopathological changes in the respiratory tract of mice exposed to ten families of airborne chemicals. *J. Appl Toxicol.*, 15:207–213.

# 29 Detection and Evaluation of Chemically Induced Liver Injury

*Gabriel L. Plaa and Michel Charbonneau*

## CONTENTS



Liver injury induced by chemicals has been recognized as a toxicological problem for more than 100 years. During the late 1800s, scientists were concerned about the mechanisms involved in the hepatic deposition of lipids following exposure to yellow phosphorus. Also, hepatic lesions produced by arsphenamine, carbon tetrachloride, and chloroform were studied in laboratory animals during the first 40 years of the twentieth century. During this same period, the correlation between hepatic cirrhosis and excessive ethanol consumption was recognized.

"Liver injury" is not a single entity; the lesion observed depends not only on the chemical agent involved but also on the period of exposure. After acute exposure, one usually finds lipid accumulation in the hepatocytes, cellular necrosis, or hepatobiliary dysfunction, whereas cirrhotic or neoplastic changes are usually considered to be the result of chronic exposures. Different biochemical alterations may lead to the same endpoint; no single mechanism governs the appearance of hepatocellular degenerative changes or alterations in function. Some forms of liver injury are reversible, whereas others result in a permanently deranged organ. The mortality associated with various forms of liver injury is variable. The incidence of injury can differ among species, and the presence of a dose-dependent relation may not always be apparent.

The marked vulnerability of the liver to chemically induced damage is a function of: (1) its anatomical proximity to the blood supply from the digestive tract, (2) its ability to concentrate and biotransform foreign chemicals, and (3) its role in the excretion of xenobiotics or their metabolites into the bile. The diverse nature of the functional activity of the liver and its varied response to injury make the selection of appropriate testing procedures a difficult task. This chapter discusses the major tests that are useful in the detection and evaluation of liver injury in laboratory animals.

## CLASSIFICATION OF CHEMICALLY INDUCED LIVER INJURY

The morphological changes observed following hepatic injury produced by chemical and biological agents can be classified according to two parameters: location and type of lesion produced.

### Location Within the Hepatic Parenchyma

An early system of describing pathological lesions of the liver originated from the concept of the hexagonal lobule introduced by F. Kiernan in 1833 (Figure 29.1). This configuration, the classical manner of presenting the relations between the hepatic cell, its vascular supply, and the biliary system, was considered to represent the functional unit of the liver. The terminal hepatic venule (central vein) is found in the center of the lobule, and the portal space, containing a branch of the portal vein, a hepatic arteriole, and a bile duct, is located at the periphery of the lobule. Based on this configuration, lesions of the hepatic parenchyma have been classified as centrilobular, midzonal, or periportal.

The hexagonal lobule configuration does not correspond to the functional unit of the liver. The hexagonal lobule is not conspicuous under microscopic examination. Injection of colored gelatin mixtures in the portal vein or hepatic artery shows that terminal afferent vessels supply

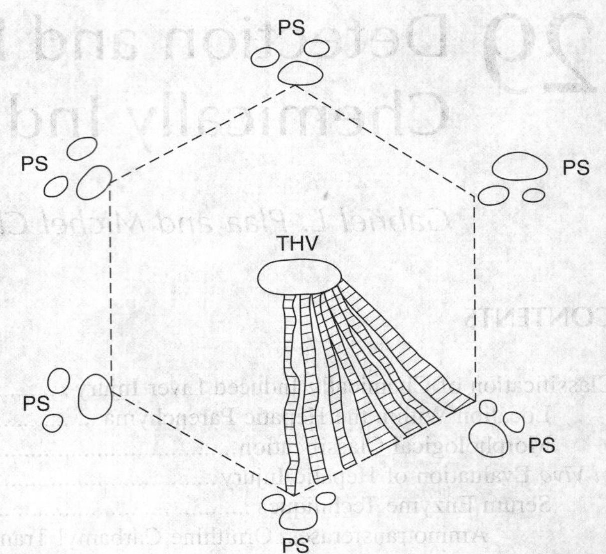


**FIGURE 29.1** Schematic representation of the traditional hexagonal lobule. (PS) portal space, consisting of a branch of the portal vein, hepatic arteriole, and a bile duct; (THV) terminal hepatic venule (central vein). (From Plaa, G.L., in *Casarett & Doull's Toxicology: The Basic Science of Poisons*, 4th ed., Amdur, M.O. et al., Eds., Pergamon Press, New York, 1991, pp. 334–353. With permission.)

blood only to sectors of adjacent hepatic lobules. These sectors are situated around terminal portal branches and extend from the terminal hepatic venule of one hexagon to the terminal hepatic venule of an adjacent hexagon. Rappaport [295] defined the parenchymal mass in terms of functional units called the *liver acini*. A simple liver acinus consists of a small parenchymal mass that is irregular in size and shape and is arranged around an axis consisting of a terminal portal venule, a hepatic arteriole, a bile ductule, lymph vessels, and nerves (Figure 29.2). This acinus lies between two or more terminal hepatic venules with which its vascular and biliary axis interdigitates. There is no physical separation between two liver acini. The hepatic cells of the simple acini are in cellular and sinusoidal contact with the cells of adjacent or overlapping acini. Even with this extensive communication, the hepatic cells of one particular acinus are preferentially supplied by their parent vessels. Three relatively discrete circulatory zones appear within each acinus (Figure 29.2). Hepatocytes in close juxtaposition to the terminal afferent vessel constitute zone 1; these cells are the first to be supplied with fresh blood, rich in oxygen and nutrients. The higher order of zones 2 and 3 is indicative of the greater distance between the cells comprising these zones and the supply of fresh blood.

One of the interesting correlates of the concept of zonal acinar circulation is the growing realization that not all hepatic parenchymal cells within the liver lobule have the same kind of functional specificity. Rappaport's acinar

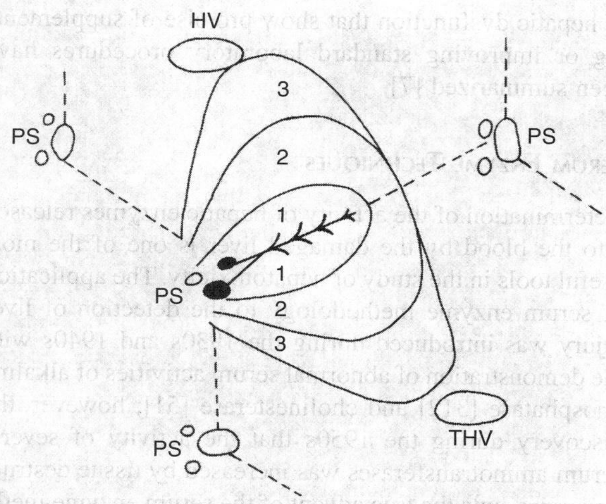


**FIGURE 29.2** Schematic representation of a simple hepatic acinus, according to A.M. Rappaport. (PS) portal space, consisting of a branch of the portal vein, hepatic arteriole, and a bile duct; (THV) terminal hepatic venule (central vein); (1, 2, 3) zones draining off the terminal afferent vessel. (From Plaa, G.L., in *Casarett & Doull's Toxicology: The Basic Science of Poisons*, 4th ed., Amdur, M.O. et al., Eds., Pergamon Press, New York, 1991, pp. 334–353. With permission.)

concept has been modulated by others [136,216] to account for differences in enzyme distribution and redox state. Areas of differing metabolic activity exist within the liver [186]. Respiratory enzyme activity is particularly high in the zone closest to the terminal afferent vessel (zone 1) (Figure 29.2), whereas the most distant zone (zone 3) is particularly rich in cytochrome P450-dependent enzyme systems. The perivenous (zone 3) cells are relatively rich in some NADPH-dependent enzymes, and periportal cells (zone 1) are relatively poor [136]. The concept of *metabolic zonation* is based on differences observed between enzyme activities in periportal and perivenous regions [186]. Thurman and Kauffmann [345] reported on the lobular distribution of maximal enzyme activities measured by immunohistochemical or microchemical techniques; these parameters do not always correlate with metabolic flux rates as measured by microfluorometry and miniature $O_2$ electrodes. Periportal–perivenous gradients are described for cytochrome P450s, sulfatation, glucuronidation, and glutathione *S*-transferases [136]. Functional gradients are also reported for hepatobiliary activity [1]. The implications of such findings are not well understood; however, zonal and cellular enzymatic specificity and metabolic heterogeneity may permit the rationalization of differing mechanisms of action in the development of hepatic lesions associated with hepatotoxic agents.

The classical hexagonal descriptions of focal, midzonal, periportal, and centrilobular lesions, although functionally incorrect, are compatible with Rappaport's zonal acinar configuration. Centrilobular necrosis, for example, occurs in cells located in the distal acinar zone (zone 3) (Figure 29.2). When several such zones are affected, a concentric lesion can be visualized. Regeneration is said to occur from cells located in the midzonal region of the hexagonal representation, which corresponds to the acinar zone closest to the terminal afferent vessel (zone 1), a zone shown to be particularly high in cytogenic activity; therefore it appears that the acinal circulatory concept of the hepatic lobule does not come into serious conflict with the earlier descriptions of pathologic lesions.

## Morphological Classification

Morphologically, liver injury can manifest itself in different ways [281]. The acute effects can consist of an accumulation of lipids (steatosis) and the appearance of degenerative processes, leading to cell death (necrosis). The necrotic process can affect small groups of isolated parenchymal cells (focal necrosis), groups of cells located in zones (centrilobular, midzonal, or periportal necrosis), or virtually all of the cells within a hepatic lobule (massive necrosis). The accumulation of lipids can also be zonal or more diffuse in nature. Although acute injury may consist in both necrosis and fat accumulation, it is not necessary that both features be present. The cholestatic type of lesion, resulting in diminution or cessation of bile flow with retention of bile salts and bilirubin, is also an important form of liver injury [277,287]; this lesion leads to the appearance of jaundice. A type of massive necrosis that resembles a viral infection [374] is produced by certain chemicals. A number of drugs are also associated with a mixed type of lesion (e.g., one that possesses both cholestatic and viral-like hepatic components) [374]. Chemically induced liver injury resulting from chronic exposure can produce marked alteration of the entire liver structure, with degenerative and proliferative changes observed in the various forms of cirrhosis. Neoplastic changes may be another endpoint of chemical liver injury.

Through the years, a number of classification systems have evolved to describe the chemicals involved. The schemes are beyond the scope of this chapter. In brief, however, some are based on morphological changes [281] and others deal with the postulated mechanisms of action or the circumstances of exposure [374].

In the morphological classifications, one finds chemicals, such as carbon tetrachloride ($CCl_4$), chloroform ($CHCl_3$), phosphorus, tannic acid, ethionine, and ethanol, that produce zonal hepatocellular alterations, such as necrosis or steatosis. Intrahepatic cholestasis is a lesion produced by a number of drugs (e.g., phenothiazine derivatives, antimicrobial agents, anabolic steroids, oral hypoglycemics) and is characterized by biliary dysfunction. In addition, massive hepatocellular necrosis is produced by other drugs (e.g., iproniazid, monoamine oxidase inhibitors, halothane).

With regard to the mechanisms of action involved [189,337,374], one finds a variety of possible effects. At least six different sites of action have been described [221], including bleb formation on the cell membrane, transport pumps in the canaliculus, enzyme–drug adduct formation in the endoplasmic reticulum, vesicle movement to the cell surface, apoptotic processes, and inhibition of β-oxidation or respiration in mitochondria. Predictability of the appearance of the injury and the production of lesions in laboratory animals are important considerations. Some forms of drug-induced liver injury in humans are due to hypersensitivity (allergic reactions) or an expression of individual susceptibility (idiosyncratic reactions) [189,337,374]. Reactive metabolites and immunological mechanisms appear to be important components in the elucidation of such lesions [263].

From these classifications, one sees that a variety of pathological processes are involved in what is called, in general terms, *liver injury*. Furthermore, many different substances can cause injury. Although classification schemes assist in conceptualizing what is occurring, it should be understood that, with additional knowledge of the events actually involved in the elaboration of the biochemical lesion, changes in the classifications certainly occur. Regardless of this fact, the pathological types of injury produced by hepatotoxicants largely determine the biochemical and functional manifestation of injury and thus the battery of toxicological tests required to detect and evaluate liver injury.

## *IN VIVO* EVALUATION OF HEPATIC INJURY

Many laboratory procedures have been performed to diagnose and monitor liver diseases in human clinical medicine. The performance characteristics of common tests used in humans were evaluated in 1999, and guidelines were published by the National Academy of Clinical Biochemistry [105,106]. Over the years, many techniques used in humans [130], as well as other procedures, have been applied to laboratory animals. The major tests that have proved useful for evaluation of experimental hepatic injury in laboratory animals can be placed in four primary categories: (1) serum enzyme tests, (2) hepatic excretory tests, (3) alterations in the chemical constituents of the liver, and (4) histological analysis of liver injury. Two important processes (i.e., repair and recovery of liver parenchyma and apoptosis of hepatocytes) are also the bases for tests to study the effects of chemicals on the liver. This chapter covers methods used to detect hepatic dysfunction or injury but does not address biochemical approaches employed to define the different mechanisms involved. The reader is referred to Kodavanti and Mehendale [204] for additional biochemical techniques not covered in this chapter. Also, several biomarkers indicative of hepatic dysfunction that show promise of supplementing or improving standard laboratory procedures have been summarized [7].

### SERUM ENZYME TECHNIQUES

Determination of the activity of hepatic enzymes released into the blood by the damaged liver is one of the most useful tools in the study of hepatotoxicity. The application of serum enzyme methodology to the detection of liver injury was introduced during the 1930s and 1940s with the demonstration of abnormal serum activities of alkaline phosphatase [312] and cholinesterase [51]; however, the discovery during the 1950s that the activity of several serum aminotransferases was increased by tissue destruction represents the true advent of the serum enzyme methods. Subsequently, a number of other enzymes were identified in blood, several of which demonstrate abnormal activity in the presence of liver injury.

Zimmerman [374] identified four major categories of serum enzymes based on their specificity for and sensitivity to different types of liver injury. The first group contains enzymes such as alkaline phosphatase (AP), 5′-nucleotidase (5′-NT), and γ-glutamyltranspeptidase (γ-GT). Elevated serum activities of these enzymes appear to reflect cholestatic injury more effectively than necrogenic injury. In contrast, the second group of enzymes includes those that are more sensitive to cytotoxic hepatic injury; this group has been further subdivided into:

1. Enzymes that are somewhat nonspecific and can reflect injury to extrahepatic tissue, such as aspartate aminotransferase (AST), lactate dehydrogenase (LDH), malic dehydrogenase (MDH), and aldolase (ALD)
2. Enzymes found mainly in the liver, such as alanine aminotransferase (ALT), isocitrate dehydrogenase (ICDH), and glutamate dehydrogenase (GDH)
3. Enzymes that are almost exclusively located in the liver, such as ornithine carbamyl transferase (OCT), sorbitol dehydrogenase (SDH), LDH, guanase, and fructose-1-phosphate aldolase

Assay of the more hepatospecific subgroup of enzymes may be particularly useful when studying agents with unknown hepatotoxic potential. Although elevated serum activity of the aminotransferases may reflect injury to extrahepatic organs such as the heart, skeletal muscle, or kidney, elevated activities of OCT and SDH are reliable reflections of hepatic injury. The third and fourth serum enzyme categories contain, respectively, enzymes that are relatively insensitive to hepatic injury but are elevated with extrahepatic diseases, such as creatine phosphokinase (CPK), and enzymes that demonstrate a depressed serum activity in liver disease, such as cholinesterase (ChE).

## Aminotransferases, Ornithine Carbamyl Transferase, and Sorbitol Dehydrogenase

The selection of a battery of enzymes for evaluating the hepatotoxic potential of an unknown chemical in laboratory animals is complicated by the varying sensitivity of the enzymes to different types of lesion. In an early series of experiments, Molander et al. [252] found that the measurement of serum AST provided a more sensitive index of hepatocellular injury in rats treated with $CCl_4$ than did the measurement of either ChE or AP. A number of experimentally induced necrotic states are also detectable by an elevation in the serum activity of ALT, a liver cytoplasmic enzyme. Balazs et al. [24] assessed serum ALT as a liver test in rats after treatment with ethionine, $CCl_4$, thioacetamide, dimethylnitrosamine, or allyl alcohol. Serum ALT elevation occurred following the acute administration of all of these agents, but with ethionine the elevation was not pronounced. This finding is understandable in that ethionine does not produce extensive centrilobular necrosis but usually results in fatty infiltration.

Other investigators also found that ALT is an insensitive measure of hepatic steatosis [14,138]. On the other hand, those agents that are associated with severe necrotic lesions produce pronounced elevation of serum ALT. Balazs et al. [23,24] found a good correlation between the elevation in serum activity and the severity of the lesion when the gross pathological changes or the severity of the histopathology were compared to the elevation in ALT. Others have reported an excellent correlation between the severity of quantified histologic damage produced by $CHCl_3$ and the elevation in serum ALT activity in rats [285]. Therefore, it seems that, with ALT, not only is it possible to detect the presence of liver injury but under some circumstances the severity of the lesion can be estimated by the elevation in serum enzyme activity. In chemical interaction studies, where the severity of the hepatic lesion of interest may be enhanced or diminished because of the presence of a second or third chemical [282], measurement of serum ALT activity is a useful investigative tool.

Aminotransferase activity in different tissues varies and distinct species differences occur [368]. In most instances, AST activity is greater than ALT. High AST activity occurs in skeletal muscle, diaphragm, heart muscle, and liver tissue. ALT is not as widely distributed; in humans, the greatest activity is found in the liver. Cornelius [82] studied the hepatic distribution of ALT in various animals and found that a relation exists between body size and the amount of hepatic ALT. The smaller the animal is within the weight range studied, the greater the hepatic ALT activity. More than 90% of the ALT was found to be located within the liver of all mammals of small body size; this group includes the common laboratory animals used in toxicity studies. Cornelius et al. [83] showed that, whereas AST was present in almost all tissues of pigs, cattle, dogs, and horses, low activity of ALT was found in horses, cattle, and pigs. When these species were subjected to $CCl_4$ intoxication, significant elevation of serum ALT occurred only in the dog. On the other hand, when serum AST activity was used for measuring hepatotoxicity, all species exhibited an increase in serum enzyme activity after $CCl_4$. In the rat, both serum AST and ALT activities are markedly elevated after experimental injury; either enzyme could probably be used for detecting injury in this species. Hemolysis, however, has a marked effect on serum AST in the rat, whereas its effect is practically negligible in the case of serum ALT [278]. This fact should be kept in mind when one is using rats for assessing liver function. In addition, because the hepatic specificity of ALT is greater than AST, measurement of serum ALT activity, rather than AST, might be preferable for determining the status of the liver.

Ornithine carbamyl transferase is found predominantly in the mitochondrial fraction of liver cells [91,302,303] and normally occurs only in minute amounts in serum. The mucosa of the small intestine contains a small amount (1 to 2% that of the liver), and tissue such as brain and kidney contains only trace amounts [303–305]. OCT serum activity is markedly elevated in both acute and chronic liver disease in humans [256,305]. With experimentally induced hepatotoxicity, Reichard [302] found that serum activity of OCT increased considerably more than those of the aminotransferases but it followed a similar temporal phase. OCT was also as sensitive an index of liver injury as GDH and AST in cattle and sheep poisoned with $CCl_4$, dimidium bromide, or sporidesmin [122]. Tegeris et al. [342] found that OCT activity was markedly elevated in dogs and swine poisoned with $CCl_4$, whereas the serum activity remained within normal limits in animals treated with uranyl nitrate, a nephrotoxicant. In addition, the peak serum activities of OCT (expressed as multiples of control values) in $CCl_4$-challenged animals were markedly greater that those of ALT, AST, LDH, or ICDH; the temporal pattern of OCT response was similar to that of the aminotransaminases. Serum OCT activity is a useful monitor of liver injury in rats treated with various hepatotoxicants [91,103,104,208]. Indeed, Drotman [103] described a dose-dependent relationship between the amount of $CCl_4$ administered to rats and serum OCT activity. In addition, a sixfold increase in OCT activity was found at a $CCl_4$ dose that did not produce distinctive liver damage upon light microscopic examination of the tissue, suggesting that OCT may be as sensitive an index of liver injury as histopathological examination [104]. The correlation between elevation in serum OCT activity and quantified histological changes following $CHCl_3$ administration is good [285].

Sorbitol dehydrogenase, a cytoplasmic enzyme, is also relatively specific for liver [15,91], and an increase in the serum activity of this enzyme is a relatively sensitive index of hepatocellular damage. In an elegant series of experiments, Korsrud et al. [207] determined the serum activity of nine hepatic enzymes in $CCl_4$-poisoned rats in an attempt to identify those enzymes that would respond quantitatively to varying $CCl_4$ doses and would indicate minimal liver damage. They placed the enzymes in three groups based on the lowest dose of $CCl_4$ required to elevate serum activity and concluded that SDH was the most sensitive enzymic index of liver injury. Four enzymes—ICDH, fructose-1,6-aldolase (F-1,6-ALD), ALT, and AST—were less sensitive to $CCl_4$ than SDH but were more responsive to liver injury than were alcohol dehydrogenase (ADH), 6-phosphodigluconase (6-PDG), LDH, and MDH; however, histological alterations were observed at $CCl_4$ doses that did not elevate serum enzyme activity. Subsequently, Korsrud et al. [208] studied thioacetamide and dimethylnitrosamine. As before, serum SDH was the most sensitive enzymic index of liver necrosis.

Sorbitol dehydrogenase, however, was not a preferentially sensitive index of liver injury in diethanolamine-treated rats. Six enzymes (SDH, ICDH, F-1,6-ALD, ALT, AST, and MDH) were equally responsive to diethanolamine hepatotoxicity, whereas a higher dose of this compound was required to elevate the serum activity of OCT, GDH, and LDH. Based on these observations Korsrud et al. [208] suggested that SDH was the best enzymic index of liver injury when minimal damage or minimal changes are being assessed; however, histological changes characteristic of each hepatotoxicant were noted at doses that did not result in an elevation of serum activity. Thus, the serum enzyme assays were less sensitive than histopathological examination for detecting liver damage.

Later work by Travlos et al. [349] compared the relative sensitivities of SDH, ALT, and histopathology as indices of liver injury in rats. When increases in both enzymes occurred simultaneously, terminal histopathological changes were very highly predictable (75 to 100%). They concluded that clinical chemistry evaluations could be useful for detecting potential treatment effects throughout a study, although histopathological evaluation can only be performed on termination.

Serum ALT activity is probably the most frequently used enzymic parameter to assess hepatic injury in laboratory animals. Because of the high sensitivity of OCT and SDH, however, it appears reasonable that one of these two enzymes could be used in conjunction with ALT when examining the hepatotoxic potential of an unknown chemical. In this manner, the battery of tests might better reflect the range of sensitivity encompassed by light microscopy. When the effects of several hepatotoxicants on plasma ALT and GDH values were compared in rats, the latter enzyme was reported to be a more effective marker than ALT, based on plasma elevations following injury, persistence following injury, and sensitivity [260]. Also, plasma ICDH/ALT ratios appear useful for differentiating mild-to-moderate degrees of centrilobular hepatic necrosis from periportal necrosis in rats [76]. These avenues should be pursued in future studies.

### Lactate Dehydrogenase Isoenzymes

In addition to serum enzyme activities, serum isoenzyme patterns have been utilized for the detection of organ damage in humans and laboratory animals [84,85,150,365]. Isoenzymes are enzymatically active proteins that catalyze the same reactions and occur in the same species but differ in their physicochemical properties. The isoenzymes of LDH are used as diagnostic agents in clinical medicine [85] and in some instances have been evaluated for use in experimentally induced organ damage in laboratory animals. Cornish et al. [85] utilized LDH isoenzymes to detect specific organ damage in rats; they found that the serum isoenzyme patterns resulting from liver or kidney damage differed markedly and concluded that these differences could be utilized to distinguish the damaged organ. Liver damage resulted primarily in an increase in serum LDH-5 isoenzyme activity, whereas the activity of the LDH-1 and LDH-2 isoenzymes was elevated in rats with kidney injury. Grice et al. [150] treated rats with $CCl_4$, mercuric chloride, thioacetamide, or diethanolamine at doses that would produce either minimal or pronounced tissue damage. Although AST activity was a more sensitive indicator of organ damage than LDH, it did not provide isoenzyme patterns that could identify the specific target organ. In contrast, LDH isoenzyme patterns were capable of identifying the specific target organ: The LDH-5 bands indicative of liver injury were increased in rats poisoned with $CCl_4$, diethanolamine, and thioacetamide, whereas mercuric chloride, a potent nephrotoxicant, increased the activity of LDH-1 and LDH-2. Morphologic damage generally occurred at dosage levels considerably below those producing detectable serum enzyme alterations. Thus, these authors concluded that serum enzyme activities and isoenzyme patterns are an important supplement to, but not a substitute for, histopathological examination of tissues.

A number of other enzymes of clinical interest occur in multiple forms, among which are AST and alkaline and acid phosphatase. Although these enzymes may have well-established roles in experimental toxicology, the use of their isoenzyme patterns does not yet have a definitive role.

### Enzymes Useful for Detecting Obstructive Disorders

Most of the preceding discussion concerns the use of serum enzymes to detect necrotic or degenerative processes following the administration of toxicants. In general, these enzymes are not as useful for detecting those types of hepatic alteration that are associated with diminution or cessation of bile flow. The degree of change in

**TABLE 29.1**
**Effect of Various Hepatotoxic Procedures on Four Liver Function Tests in Mice[a]**

| Hepatotoxic Procedure[b] | BSP Retention (mg/dL) | Alkaline Phosphatase (units) | Bilirubin Concentration (mg/dL) | ALT Activity (units/mL) |
|---|---|---|---|---|
| Control (no treatment) | 0.3 ± 0.3 | 3.0 ± 0.5 | 0.2 ± 0.1 | 25 ± 5 |
| ANIT (150 mg/kg p.o.) | 45.0 ± 23 | 5.6 ± 2.6 | 1.1 ± 0.4 | 282 ± 126 |
| $CCl_4$ (1 mL/kg p.o.) | 13.0 ± 7 | 5.3 ± 1.3 | 0.4 ± 0.2 | 8510 ± 1930 |
| Bile duct ligation | 26.0 ± 3 | 19.0 ± 10 | 3.8 ± 0.8 | 655 ± 132 |

[a] Values are expressed as means ± SE; each group contained 10 mice.
[b] Hepatotoxic procedure was performed 24 hours before assessing function.

*Source:* Data from Plaa, G.L., in *Selected Pharmacological Testing Methods*, Burger, A., Ed., Marcel Dekker, New York, 1968, pp. 255–288.

serum enzyme activities that one can obtain by the induction of experimental hepatotoxicity in mice is demonstrated in Table 29.1. Three hepatotoxic procedures were employed in this study [278]. One group of animals received α-naphthylisothiocyanate (ANIT), another received $CCl_4$, and the third group had their bile ducts ligated. Serum enzyme activities (ALT and AP) were determined 24 hours later. For comparative purposes, sulfobromophthalein (BSP) retention and serum bilirubin concentrations were also measured in these animals. It is evident (Table 29.1) that a necrotizing agent such as $CCl_4$ produces sufficient parenchymal injury to cause a large increase in serum ALT activity, whereas those experimental procedures that markedly impair biliary excretion (ANIT treatment, bile duct ligation) cause only a mild increase in ALT activity. The reciprocal relation is obtained when serum AP activity is assessed; obstruction of biliary flow (bile duct ligation) markedly increases serum AP activity, whereas the necrotizing challenge ($CCl_4$) produces only a mild elevation.

Alkaline phosphatase is the prototype of those enzymes (Zimmerman's group 1) that reflect pathological reductions in biliary flow. In the rat, this enzyme is found in the liver and the intestine. Alkaline phosphatase exerts a role in downregulation of the secretory activities of the intrahepatic biliary epithelium [6]. After bile duct ligation, the activity in the liver increases due to *de novo* synthesis of the membranous form of the enzyme [321]. The use of this enzyme in chemically induced liver dysfunction has been fairly extensively investigated. In the dog, the enzyme is useful for detecting biliary dysfunction; however, in the cat, ligation of the common bile duct results in only a slight increase in serum AP activity. The normal level of serum AP in the rat is exceptionally high, independent of growth, and unusually susceptible to variations in diet [157]. Increases in serum AP activity were not remarkable in a comparative study of clinical chemistry and liver histopathology performed in a subchronic study in rats, whereas increases in serum total bile acid concentration were [349]. Thus, serum AP activity may not be useful for detecting cholestatic changes in the rat.

In addition to AP, other enzymes, such as 5′NT, γ-GT, and leucine aminopeptidase (LAP), may be of use in assessing obstructive liver injury. The serum activity of these enzymes, which are localized in the membranes of hepatocytes and bile duct cells, is increased during extrahepatic cholestasis in humans [29,364]. Kryszewski et al. [211] found a significant elevation in serum activity of AP, 5′NT, and γ-GT 12 hours after bile duct ligation in the rat: AP and 5′NT peaked at 24 hours and then gradually decreased, and γ-GT peaked at 48 hours and remained elevated even 192 hours after bile duct ligation. Fujii [130] found serum AP and γ-GT to be useful indicators of cholestasis in dogs. Thus, changes in the serum activities of these enzymes are useful for detecting toxicant-induced cholestatic changes in laboratory animals. A simplified electrophoretic method was developed to separate and quantify multiple forms of human serum 5′NT [270]; three isoforms were identified in normal subjects and hepatobiliary dysfunction resulted in the increased activity of only one form of serum 5′NT. Comparable studies have not been performed in animals.

The increase in serum γ-GT activity during ANIT-induced cholestasis in rats appears to be of biliary cell origin and not from hepatocytes [56]. In this respect, elevated serum γ-GT differs from serum AP increases in activity also observed during cholestasis; AP appears to originate from the canalicular pole of the hepatocyte [56].

Of interest is the observation of Moritz and Snodgrass [254] that acute obstruction of the bile duct in the rat produced a rapid rise in the serum activity of SDH and OCT. From 1 to 24 hours following bile duct ligation, the activities of these two enzymes increased in serum to levels approximating those found after a single dose of $CCl_4$, even though the histological degree of hepatic necrosis was substantially less with obstruction than with

$CCl_4$ poisoning. This finding confirmed and extended the observation of Hallberg et al. [160] that bile duct obstruction in dogs resulted in increased serum OCT activity. These observations, if confirmed in other models of experimentally induced obstructive disorders, suggest that OCT and SDH could serve to identify hepatic alterations associated with a diminution or cessation of bile flow.

### Analytical Determination of Aminotransferase Activity

Essentially, two major techniques are employed for the measurement of serum aminotransferase activity. For AST, one measures the conversion of aspartic acid and α-ketoglutaric acid to glutamic acid and oxaloacetic acid; for ALT, one measures the conversion of alanine and α-ketoglutaric acid to glutamic acid and pyruvic acid. With the ultraviolet method of analysis, the enzyme processes are coupled with ones in which nicotinamide adenine dinucleotide (NAD) is converted from its reduced form (NADH) to the oxidized form (NAD). The course of the reaction is followed by the decrease in absorbance at 340 nm produced by the oxidation of NADH.

The colorimetric procedure involves the reaction of the product (oxaloacetic or pyruvic acid) with dinitrophenylhydrazine to form a colored hydrazone. This product can be determined by its absorbance in the visible range. The principal advantages of the colorimetric method are that an ultraviolet spectrophotometer is not required and temperature control of the enzymic reaction is more easily attained.

A certain amount of controversy exists in the literature over the relative accuracies of both of these procedures. Most of the argument, however, concerns the use of AST for the detection of coronary occlusion [8]. For experimentally induced hepatic injury, these objections do not seem to be as pertinent as they might be for the diagnosis of coronary occlusion. If one is primarily interested in following the kinetics of the enzyme reaction, the ultraviolet method is probably preferred. With this procedure, the product of the enzyme reaction does not accumulate because it is converted to another product through the use of either MDH or LDH. It is also true that when AST is measured by the colorimetric procedure one of the substrates (α-ketoglutaric acid) does interfere with the final colorimetric analysis; however, serum aminotransferase activity in laboratory animals is generally used to detect the presence or absence of liver injury. In this situation, one is not primarily interested in the absolute value of activity but rather the degree of change; therefore, the colorimetric procedure of Reitman and Frankel [307] is of sufficient accuracy. The relative ease of this procedure compared to the spectrophotometric method seems to make it more advantageous for use with laboratory animals when large numbers of samples are to be analyzed. Wells and To [360] developed a microanalytical technique that allows repetitive plasma ALT measurements on tail vein blood from individual mice using the Reitman and Frankel procedure.

### Analytical Determination of Ornithine Carbamyl Transferase Activity

Mammalian OCT catalyzes the transfer of the carbamyl group from carbamyl phosphate to ornithine and results in the formation of citrulline. OCT activity may be measured directly by following the appearance of citrulline or indirectly by arsenolysis, in which the enzyme catalyzes the reverse reaction of citrulline to ammonia, carbon dioxide, and ornithine. In the forward reaction, citrulline is determined colorimetrically with diacetyl monoxime after destruction of serum urea with urease [65,66,353]. In the reverse reaction, OCT activity is determined by production of $^{14}CO_2$ from ($^{14}$C-ureido)-L-citrulline [103,208,306] or by production of ammonia. The formation of ammonia can be analyzed by conversion to indophenol [206] or by a microdiffusion procedure in a Conway cell [301].

With the isotopic OCT method described by Korsud et al. [208], $^{14}CO_2$ produced from ($^{14}$C-ureido)-L-citrulline after an 18-hour incubation is trapped and the activity counted by scintillation spectrophotometry. The enzyme activity is expressed as nanomoles citrulline converted per minute per milliliter of plasma. Drotman [103] used a modification of this method with success.

With the colorimetric method of Konttinen [206], citrulline is decomposed by arsenolysis catalyzed by OCT. The liberated ammonia is determined as indophenol after reaction with phenol-nitroprusside and alkaline hypochlorite. The results are expressed as international units (IU) per liter of plasma; 1 IU of OCT catalyzes the transformation of 1 micromole of citrulline to ornithine per minute. The colorimetric determination of OCT activity has the advantage of not requiring a liquid scintillation spectrophotometer, and it eliminates the use of radiolabeled chemicals and scintillation mixtures.

## Hepatic Excretory Function

Chemicals entering the systemic circulation may be excreted by the liver unchanged or after modification within the hepatocyte. Compounds that undergo biliary excretion have been divided arbitrarily into three classes (A through C) based on the bile/plasma concentration ratios obtained during their excretion [203]. Examples of class A substances include sodium, potassium, and chloride ions, as well as glucose; these compounds have a bile/plasma ratio of about 1.0. Class B substances (e.g., bile salts, bilirubin, BSP, many xenobiotics) achieve a bile/plasma ratio of greater than 1.0, usually between 10 and 1000. Among class C substances, which have a bile/plasma ratio of less than 1.0, are macromolecules such as inulin, phospholipids, mucoproteins, and albumin.

Transport systems in the sinusoidal and canalicular membranes of hepatocytes have been characterized [182,193,274,335]: sodium-taurocholate cotransporting polypeptide (NTCP), organic anion transporting polypeptides (OATPs), organic cation transporters (OCTs), organic anion transporters (OATs), and the ATP-dependent transporters multidrug resistance P-glycoproteins (MRP1, MRP2, MRP3, and MRP6), bile salt export pump (BSEP), and multidrug-resistance-associated proteins (MDR1, MDR2, and MDR3). NTCP is associated with the sinusoidal sodium-dependent uptake of conjugated bile salts, while OATP is a multispecific carrier for sodium-independent uptake of bile salts, organic anions, and other amphipathic organic solutes. Several MRP efflux pumps are present in the basolateral membrane. They are said to function as reverse transporters [182] and may reduce intracellular concentrations (monovalent glucuronides and glutathione-*S*-conjugates) when secretion via the canalicular route is impaired.

Canalicular transport systems are important for the biliary excretion of bile salts (BSEP), bilirubin glucuronide (MRP2), non-bile-salt organic anions (MRP2), glutathione (MRP2), and phospholipids (MDR2 and MDR3). Microsomal enzyme inducers can increase bile flow and biliary excretion of glutathione-derived sulfhydryls by affecting MRP2 in rats [185]; they also induce the MRP family in mice [237] via distinct transcription factors. Hereditary defects of intrahepatic transporters have been described (e.g., progressive familial intrahepatic cholestasis, Dubin–Johnson syndrome, Wilson's disease) [235,274]. Perturbations associated with extrahepatic or chemical-induced cholestasis [87,88,137,182,335,348] are known, although the mechanisms involved and causality are still unclear. MRP2 and BSEP can be affected by taurolithocholate-, troglitazone-, or estradiol glucuronide-induced acute decreases in bile flow in rats [87,88,132,137,329], but MDR2 does not appear to be involved in the bile flow deficiency following manganese–bilirubin combined treatment in mice [2]. Differential expression of hepatic transporter genes has been demonstrated in mice [3] following treatment with the hepatotoxicants acetaminophen and $CCl_4$.

## Sulfobromophthalein Retention

The most common class B chemical used for the detection of liver injury is BSP. In 1925, this anionic phthalein dye was used in clinical medicine by S. Rosenthal and E. White after preliminary tests with other phthalein dyes proved less satisfactory. Since its introduction, this substance has been used extensively for the assessment of liver function in humans and laboratory animals. After intravenous injection, BSP is present in the cardiovascular compartment. Its disappearance from the circulatory system depends on its uptake by the liver. The use of BSP to assess liver function is based on the observation that dye removal from blood is delayed by hepatic dysfunction. Commonly, BSP concentration in plasma is determined at a specific time after a standard dose of dye (per unit of body weight) is administered intravenously. Selection of the optimal dose of BSP is essential for correct interpretation of functional impairment.

The removal of BSP from the plasma is dependent on the simultaneous operation of a number of hepatic processes, such as active transport across the plasma membrane into a storage compartment, metabolic transformation, and ATP-dependent transport across the canalicular membrane [372]. The most critical step in this process is thought to be the transfer of BSP from liver to bile. Most important in terms of selecting an optimal BSP dose is that its biliary excretion can be saturated and a transport maximum ($T_m$) exists; the clearance of BSP by laboratory animals is dose dependent [198]. Usually one observes that small doses are rapidly removed from the circulation; this rate of removal continues as one increases the dose, until a dosage level of BSP is reached where the rate of disappearance becomes longer. For example, for isolated perfused rat liver [286], 5, 10, and 20 mg of BSP are cleared at the same exponential rate; however, the capacity of the liver to extract BSP from the perfusate becomes saturated when 30 or 40 mg is injected. With the latter dose, the rate of disappearance becomes zero order, which indicates that the maximal capacity of the transport system is reached. This same type of phenomenon occurs in mice *in vivo* [108].

A marked species difference exists in the ability of the rat, rabbit, and dog to remove BSP from the plasma; this difference can be readily discerned by administering varying BSP doses to these laboratory animals (Figure 29.3). Both the rat and the rabbit have a remarkable ability to clear BSP from the plasma, whereas the dog has a relatively poor capacity [198]. If the overall BSP $T_m$ for biliary excretion is measured, large differences are observed [198]. The rat and rabbit excrete BSP at a rate of about 1 mg/min/kg, whereas the dog excretes it at a rate of about 0.2 mg/min/kg.

The significance of these findings is that the optimal dose of BSP for measuring liver function depends on the species employed. The dose should be one that is relatively close to the one that indicates BSP clearance capacity is being exceeded. For the rabbit, dog, and rat, these dosages are about 75, 15, and 50 mg/kg, respectively [198]. In mice the optimal dosage depends on the strain employed but is somewhere between 75 and 100 mg/kg [278]. The dose selected should result in about 2 to 3% retention at 30 min in normal animals.

Determination of the BSP $T_m$ has been used as an index of hepatic function in humans. Wheeler et al. [363] devised a procedure that can be employed in conscious

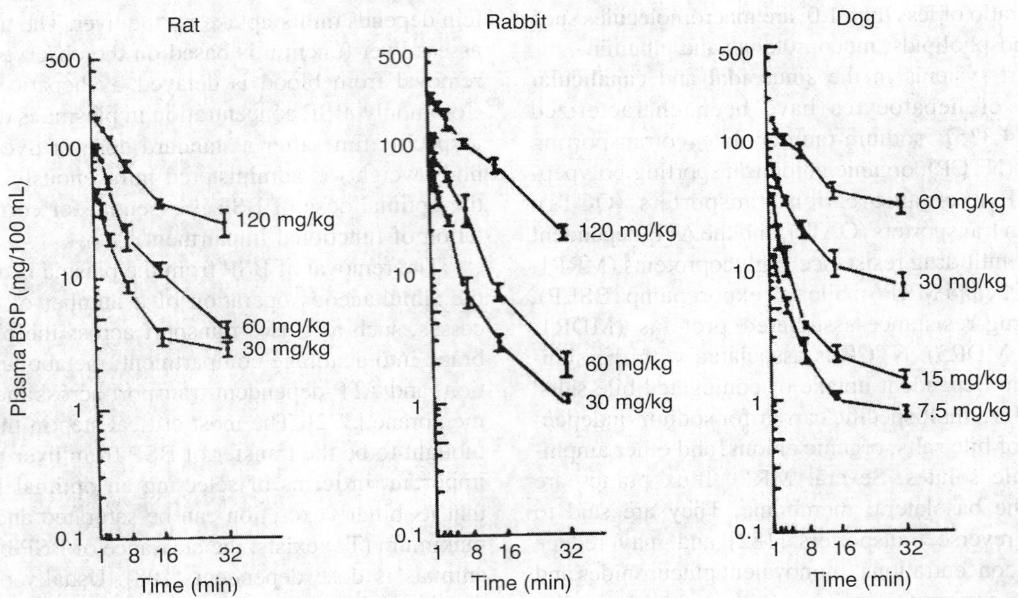


**FIGURE 29.3** Plasma disappearance curves for sulfobromophthalein (BSP) administered in varying doses in the rat, rabbit, and dog. (From Klaassen, C.D. and Plaa, G L., *Am. J. Physiol.*, 213, 1322–1326, 1967. With permission.)

dogs. With this technique, one infuses BSP at three different rates and measures the serum concentration at varying times. From the data, one can calculate the $T_m$ and the relative storage capacity (S) for BSP. In addition, methods have been devised for making similar measurements in the rabbit and rat [197]. Use of these techniques has not been widespread in laboratory animals. They are useful for assessing excretory capacity, however, and are employed in mechanistic studies to determine specific functional lesions involved in the reduction of BSP clearance; for example, defects in the transfer of BSP from plasma to liver, the storage of BSP within the hepatocyte, the conjugation of BSP with glutathione, or the transfer of BSP from the liver cell into the bile could participate in the $CCl_4$-induced depression of BSP clearance.

When these possibilities were evaluated, it appeared that the major effect of $CCl_4$ was to decrease the transfer of BSP from the hepatocyte to the bile [199,294]. Klaassen and Plaa [199] found that 24 hours after a single intraperitoneal dose of $CCl_4$ both the BSP $T_m$ and hepatic BSP conjugating activity were depressed; no change in hepatic BSP storage was detected. Because $CCl_4$ reduced plasma disappearance and the $T_m$ of phenol-3,6-dibromophthalein disulfonate (DBSP), a nonmetabolized analog of BSP, and depressed excretion of both BSP and DBSP under submaximal conditions, it was concluded that the excretory parameter was probably the prime event altered by $CCl_4$. Subsequently, Priestly and Plaa [294] demonstrated that impaired BSP excretion, bile flow rate, relative hepatic storage, and BSP retention were observed as early as 3 hours after $CCl_4$ administration. Impaired BSP conjugation, however, was not unequivocally demonstrated until 12 hours after $CCl_4$. Thus, although impairment of both conjugation and excretion contributes to BSP retention, the effect on excretion appears to be more important.

Although BSP was introduced in 1925, it was not until 1950 that it was realized that this dye is excreted in a conjugated form into the bile [49,50]. Up to that time, it was assumed that this material did not undergo biotransformation prior to its excretion. BSP is conjugated with glutathione in humans, rats, and dogs. A number of other conjugates, including BSP–cysteinylglycine and BSP–cysteine, are also formed, presumably by cleavage of glutamic acid and glycine from the glutathione moiety. BSP conjugation, catalyzed by a glutathione *S*-transferase, is a cytoplasmic process. Under certain conditions, impairment of BSP conjugation with hepatic glutathione can lead to depression of BSP excretion in the bile without impairment of general excretory function [48,79,292,293].

Although BSP is a useful and sensitive test of liver function, a variety of events can cause BSP retention. Diffuse and severe hepatocellular damage is associated with an increase in dye retention [161]. Liver injury of the cholestatic type, however, usually decreases biliary excretion to a greater extent than does parenchymal cell injury [369]. For example, using ANIT, Becker and Plaa [30] showed that the amount of BSP retained following such treatment is much greater than that observed after the necrotic effects of $CCl_4$. In this instance, ANIT affects BSP retention by decreasing the biliary excretion of BSP [283]. In rabbits, treatment with anabolic steroids can result in BSP retention due to a decrease in excretory capacity [223]. In contrast, bunamiodyl (sodium 3-(3-butyramide-2,4,6-triiodophenyl)-2-ethylacrylate) seems to diminish uptake of BSP by the hepatocyte [40]. Finally, decreased hepatic blood flow can also cause BSP retention [108].

Phenobarbital markedly enhances the excretion of BSP [200,279]. This effect is apparently not related to an increase in BSP conjugation, as phenobarbital pretreatment also enhances the biliary excretion of DBSP [279]. Klaassen showed [195] that when seven microsomal enzyme inducers were examined only phenobarbital produced a significant increase in bile flow and a significant increase in anion excretion; benzo(*a*)pyrene and 3-methylcholanthrene did not increase bile flow. Chlordane, nikethamide, phenylbutazone, and chlorcyclizine treatments tended to increase bile flow, but the increases were not statistically significant. These substances also failed to enhance BSP or DBSP biliary excretion [195]. Subsequently, two other microsomal enzyme inducers, spironolactone and pregnenolone-16α-carbonitrile, were shown to increase bile flow and the biliary excretion of BSP and DBSP in rats [375]. The canalicular transporter MRP2 is involved in the biliary elimination of BSP and DBSP [183,333]. A number of microsomal enzyme inducers (including TCDD, phenobarbital, and pregnenolone-16α-carbonitrile) were shown to induce the MRP family in rodents [184,237].

## Indocyanine Green Retention

Several other compounds have been introduced into clinical medicine for the purpose of measuring liver function by the dye-clearance principle. The rose bengal test appeared in 1931 [98], and a third useful dye, indocyanine green (ICG), was introduced in 1959 [222]. Indocyanine green was originally used to measure cardiac output by the indicator–dilution technique. It was subsequently found [362] in the dog, however, that 97% of the administered dose was eventually recovered from the bile in an unaltered form. No dye was found in the urine. ICG has about the same spectrum of sensitivity and specificity as BSP, but it has a number of properties that make it more desirable to employ under certain circumstances. Cherrick et al. [71] found the following: (1) ICG is rapidly and completely bound to plasma protein, of which albumin is the principal carrier; (2) the dye is excreted in bile in an unconjugated form; (3) there seems to be no extrahepatic mechanism for removing the material; (4) ICG is nonirritating when inadvertently introduced subcutaneously, and it produces no untoward reactions upon single or repeated intravenous injections; and (5) the plasma disappearance of ICG is similar to that of BSP.

In laboratory animals, ICG is usually employed to supplement BSP tests. In the dog, Hunton et al. [175] found that: (1) the plasma disappearance rate of ICG is exponential for at least 15 minutes and usually for 30 to 60 minutes; (2) the amount removed per minute seems to be inversely related to the dose administered; (3) the maximal rate of excretion of ICG into the bile is about 0.4 mg/min/kg; and (4) substances such as bilirubin, rose bengal, and BSP interfere with ICG excretion. Klaassen and Plaa [202] found that over a 32-minute period the rate of disappearance of ICG was exponential in the rat, rabbit, and dog (Table 29.2). The rabbit exhibited a greater capacity to remove ICG from plasma than did the rat, and the dog had the lowest capacity. It appears that the optimal dosage for ICG clearance in the dog is about 1.5 to 2.0 mg/kg; in the rat, approximately 16 mg/kg; and in the rabbit, 25 to 30 mg/kg.

**TABLE 29.2**
**ICG Plasma Disappearance Rates in Rats, Rabbits, and Dogs**

| ICG Dose (mg/kg) | $T_{1/2}$ (min) | $K$ (% removed/min) |
|---|---|---|
| *Rat* | | |
| 4 | 2.5 | 28 |
| 8 | 4.0 | 17 |
| 16 | 6.5 | 11 |
| 32 | 8.5 | 8 |
| 64 | 18.0 | 4 |
| *Rabbit* | | |
| 8 | 1.5 | 46 |
| 16 | 3.5 | 20 |
| 32 | 7.0 | 10 |
| *Dog* | | |
| 1 | 7.0 | 10 |
| 2 | 17.0 | 4 |
| 4 | 30.0 | 2 |

Biliary excretory maximum and hepatic storage values for ICG could not be determined [202], as infusion rates sufficient to produce a biliary excretory $T_m$ produced a marked decrease in bile flow. Decreased bile flow was observed in all three species but was most pronounced in the rat and least in the dog. Rapid administration of ICG, as used in plasma clearance experiments, does not produce marked alterations in bile flow. Other investigators [162] reported maximal biliary excretion rates for rats given ICG that were comparable to the peak excretion rates obtained by Klaassen and Plaa [202]. In MDR2-deficient mice, the biliary excretion of ICG was reduced by 90%, while the excretion of total glutathione was decreased by 65%, relative to wild mice [174].

The major advantage of ICG in the detection and evaluation of hepatic function is that the material is not biotransformed prior to excretion. In addition, ICG is directly determined in plasma, without chemical treatment. In practice, it simply involves diluting an aliquot of plasma (0.1 to 1.0 mL) with water and determining the absorbance at 805 nm, the wavelength for peak ICG absorption. ICG, however, is unstable in aqueous solutions, but it can be made more stable by mixing it directly with serum or with an albumin solution. The dye is unstable when mixed with heparin solutions containing bisulfite [77], indicating that preservatives of the same type may also have an effect on ICG.

### Other Anionic Chemicals

Although rose bengal was introduced before ICG, it has not been extensively employed in laboratory animals; in humans, the dye is used to diagnose hepatic disorder, especially in children. Rose bengal, like ICG, has the advantage that it is apparently not biotransformed before excretion into the bile [196,212]. It is available commercially in a radioactive form, so its concentration can be quantified in small blood samples. Its uptake in isolated hepatocytes is similar to that of BSP [372]. Klaassen [196] examined the pharmacokinetics of rose bengal in four species (rat, rabbit, dog, guinea pig) and found that, with the exception of biotransformation, the dye appears to be handled by the liver in a manner similar to BSP; it is a class B anion that is actively excreted into the bile. A marked species variation in the rate of biliary excretion of rose bengal exists [196]. The rat and rabbit excrete rose bengal into the bile at comparable rates, whereas the guinea pig is much more efficient and the dog much less efficient.

Like BSP, the removal of rose bengal from the blood is altered by changes in hepatic excretory function [196,248,341], and Klaassen [196] indicated that the dye could be used as a measure of hepatic excretory function in laboratory animals. Because it is not biotransformed, alterations in its clearance would reflect changes in its uptake into the liver or its excretion into the bile; however, if it is used as a hepatic function test by measuring the concentration of rose bengal in the blood at only one time after its administration, the selection of this time interval is critical. A blood sample at 15 to 20 minutes after administration appears optimal [196]. Additional studies are needed to determine if rose bengal is as sensitive an index of hepatic dysfunction in laboratory animals as is BSP.

Other agents exist for monitoring hepatic excretory function. Mehendale and coworkers [245,247] used phenolphthalein glucuronide (PG), imipramine (IMP), and the polar metabolites of imipramine (PMIMP) as model compounds to characterize the hepatobiliary dysfunction produced by mirex. This pesticide did not suppress the hepatic uptake and metabolism of IMP but inhibited the movement of PMIMP from the hepatocyte to bile; it also inhibited the biliary excretion of PG, a model anionic substrate that does not undergo biotransformation. These model substrates allowed the investigators to localize the site of mirex-induced dysfunction. In another report [90], they suggested that PG may be a more sensitive model compound than BSP for detection of hepatobiliary dysfunction. $CCl_4$ at 100/μL/kg depressed biliary excretion of PG in rats; hepatobiliary dysfunction was undetectable with BSP at this dosage of $CCl_4$ [50,199]. The plasma disappearance and biliary excretion kinetics of PG in the rat have been described [245]. The biliary excretion of phenolphthalein sulfate is reported to be markedly delayed in MRP2-deficient rats, suggesting that this substance is handled by MRP2 [340].

The effects of some known potential cholestatic agents were investigated *in vitro* using taurocholate in regular and collagen-sandwich cultured human hepatocytes [210]. Cyclosporin A, bosentan, glyburide, erythromycin estolate, and troleandomycin all inhibited bile-acid efflux. Glyburide administered to rats resulted in elevated serum total bile acids in rats. These preliminary findings are quite encouraging and warrant more extensive studies.

### Endogenous Cholephiles

At least one endogenous substance, bilirubin, has been used to evaluate chemically induced hepatic injury. Normally, bilirubin is excreted into the bile. Elevation of serum bilirubin concentration accompanies sufficiently severe parenchymal injury, but it is a relatively insensitive measure of chemically induced hepatic injury. The degree of change in serum bilirubin that one obtains with experimental hepatotoxicity in mice is summarized in Table 29.1. $CCl_4$, although producing sufficient parenchymal injury to cause a large increase in serum ALT activity, does not affect bilirubin concentrations greatly. On the other hand, bile duct occlusion does elevate serum bilirubin considerably. ANIT also elevates serum bilirubin but not to the extent that biliary occlusion does. The BSP retention values indicate that those experimental procedures that markedly impair biliary excretion also affect BSP retention, whereas $CCl_4$ causes a lesser degree of retention. However, if one assumes that BSP retention and bilirubin concentrations measure relatively the same type of liver function, it is evident that the changes occurring with BSP are considerably greater than those occurring with bilirubin. A likely explanation is that the measurement of endogenous concentrations of bilirubin may not assess the total capacity of the liver to clear bilirubin, as does a load of BSP selected to be a near-capacity dose for the particular species of animal employed. Indeed, if one does administer exogenous amounts of bilirubin and follows its plasma disappearance, as with BSP, the sensitivity of the bilirubin clearance procedure can be increased. Nevertheless, BSP clearance is simpler and more sensitive than bilirubin and is therefore preferred for measurement of hepatocellular injury.

Bile acids, a second group of endogenous chemicals that are normally excreted into bile, have been used to assess some hepatotoxicants. Unlike serum bilirubin retention, elevation of serum bile acid concentrations, presumably because of decreased biliary secretion, appears to be a highly responsive index of hepatobiliary dysfunction [21,124]. At least following $CHCl_3$ and $CCl_4$ treatments in rats, elevation of serum bile acids occurred at dosages that exerted no effect on serum enzyme activity or bilirubin concentration [21]. Furthermore, elevations of individual serum bile acids occurred at dosages of $CCl_4$ that produced no consistent histological change. Neghab and Stacey [258] recently demonstrated that toluene, a nonhalogenated aromatic hydrocarbon, also results in increases of serum bile acids in rats in a dose-dependent manner; the elevations occur in the absence of other abnormal liver enzyme findings. Xylene also results in toluene-like interference of hepatocellular uptake of bile acids is isolated hepatocytes [258]. Although these findings are consistent with toluene and xylene actions on hepatocellular bile acid transport, they are not necessarily indicative of liver injury. The specificity of serum bile acid elevations and the role of such events in the evaluation of hepatotoxicity are yet to be determined.

### Biliary Secretory Function

Techniques designed to assess bile secretory function are also available; however, these methods lend themselves more to specific research problems than to overall toxicological assessment. Fujimoto [131] reviewed older methods that can be applied *in vivo*. The so-called bile-acid-dependent and bile-acid-independent fractions of total bile flow [25,46,203] have been studied extensively in several species; secretin-sensitive bile flow is thought to be small in the rat but more important in the rabbit and dog.

Fujimoto [131] developed a number of new techniques, where marker substances are injected retrogradely into the biliary tree to assess the permeability characteristics of the biliary system. ANIT, which produces intrahepatic cholestasis in rats, and bile duct ligation increase the distended capacity of the biliary tree, whereas another cholestatic agent, taurolithocholate, decreases the distended capacity; $CCl_4$ seems to exert no effect [129,268,276].

The retrograde technique has been modified to become the segmented retrograde intrabiliary injection (SRII) procedure [267]. With the use of radioactive marker substances of varying molecular weights (D-glucose, mannitol, sucrose, inulin, or dextran), one can assess the membrane characteristics of the biliary tree (canalicular and tight-junction complexes). This procedure was used to study the hepatobiliary dysfunction produced by *Amanita phalloides* [129], taurolithocholate [131], colchicine [22], S,S,S-tributylphosphorotrithioate [22], manganese and manganese-bilirubin combinations [18], and sequential treatments with ketones and chloroform [165]. These studies have been useful for discerning the site and possible mechanisms of action involved in their hepatobiliary effects.

### Determination of Biliary Function

#### Sulfobromophthalein Clearance

The BSP test dose is first injected intravenously. After 30 minutes, a suitable amount of blood is withdrawn and plasma is prepared by centrifugation. Aliquots of plasma are placed into tubes containing alkalinized or acidified saline. The BSP plasma content is determined by the difference between the absorbance in alkalinized and acidified saline. The BSP dose administered depends on the species of animal being employed. It should be one that normal animals can readily clear in a 30-minute collection period; thus, the amount of BSP retained at 30 minutes should be about 1% of the dose administered. For the dog, rat, rabbit, and mouse, these dosages are about 15, 75, 50, and 100 mg/kg, respectively [198,278].

#### Biliary Sulfobromophthalein Transport Maximum

The procedure can be performed in anesthetized rats or rabbits [197]. With the rat, the bile duct is cannulated as well as the femoral vein; the rectal temperature is maintained at 37°C to prevent hypothermic alterations in $T_m$ [311]. The BSP solution is infused (2.5 mg/kg/min BSP) at a rate of 0.03 mL/min for 60 minutes, and bile is collected at 15-minute intervals. The amount of BSP excreted is calculated, and the maximum value attained is the $T_m$. In the Sprague–Dawley rat, the BSP $T_m$ is about 1.0 mg/kg/min.

#### SRII Method

The procedure can be performed in anesthetized rats [131], and the rectal temperature should be maintained at 37°C to prevent hypothermic alterations in bile flow [311]. The bile duct is cannulated with PE-20 tubing just distal to the bifurcation of the biliary tree; this tubing is attached to a longer length of PE-20 tubing capable of containing a calibrated (40 mL) amount of solution. An exact amount of radioactive marker substance (D-glucose, sucrose, mannitol, inulin, or dextran) is infused into the bile duct and washed through with saline. Bile flow is then reestablished and bile drops collected serially. The bile flow rate is calculated by determining the time required to form each drop; the content of each drop is determined by liquid scintillation spectrometry, and the recovery is expressed as a percentage of total marker recovered. The volume of the distended biliary tree can also be evaluated by the SRII method [131,276].

## Mitochondrial Function

Mitochondrial function was one of the first biochemical processes investigated in chemical-induced hepatotoxicity, and many studies were performed to establish its role in various experimental models [374]. With carbon tetrachloride, unfortunately, it fell in disrepute as a possible initiating event, because the temporal appearance of the biochemical lesions in this organelle seemed to follow, rather than precede, histological changes [282]. More recent studies with acetaminophen [189] and 1,1-dichloroethylene (1,1-DCE) [240] have demonstrated the importance of mitochondrial dysfunction in liver injury. With 1,1-DCE, dose-dependent and time-dependent mitochondrial parameters (oxygen consumption, state 3/state 4 oxygen consumption ratio, ADP phosphorylation per oxygen consumed) in isolated liver fractions obtained from treated mice showed that state 3 (ADP-stimulated) respiration rates for glutamate (complex I)- and succinate (complex II)-supported respiration were decreased 20 and 90 minutes, respectively, after treatment, whereas state 4 (resting) respiration was unaffected. Serum ALT was significantly elevated at 2 hours, and centrilobular necrosis was observed 24 hours after treatment. Thus, with 1,1-DCE mitochondrial dysfunction was one of the early events observed in the progression of the lesion.

Mitochondria play a vital cellular role in fat oxidation and energy production [275]; the β-oxidation and tricarboxylic acid cycles are key elements; cytochrome c oxidase and ATP synthase are critical enzymes. Microvesicular steatosis results from acute impairment of fatty acid oxidation. This lesion has been observed in humans and in laboratory animals following exposure to valproic acid, pirprofen, tetracyclines, and amiodarone [189,374]. Several dideoxynucleosides and the experimental agent fialuridine have resulted in inhibition of mtDNA replication in humans with mitochondrial cytopathy [275]. The treatment of rats with the cholestatic agent ANIT for 16 weeks resulted in impaired mitochondrial bioenergetics [269].

Mitochondria are the main source of reactive oxygen species (ROS), which, in turn, can lead to oxidative damage of mtDNA and cell death. ROS are important cytotoxic and signaling mediators in inflammatory liver diseases [180]. Ethanol exposure can promote oxidative stress by increasing ROS formation and decreasing cellular defense mechanisms [169]. Oxidative stress also appears involved in acetaminophen hepatotoxicity [181]. The mitochondrial permeability transition (MPT) pore, located in the inner membrane, allows cell survival when closed but leads to cell death when open [275]. If all mitochondria have open pores, ATP synthesis is decreased, and ATP depletion and necrosis occur. If only some pores are open, unaffected organelles still synthesize ATP, whereas the open ones cease to form ATP and release cytochrome *c*, thus activating caspases which leads to apoptosis. Overall, the importance of mitochondria in the development of hepatic dysfunction seems well established. Pessayre and coworkers [275] have even suggested that the possible action of new pharmaceutical agents on mitochondria should be evaluated before marketing.

## Alterations in Chemical Constituents of the Liver

In addition to producing elevations in serum enzyme activities and altering hepatocyte transport processes, chemical hepatotoxicants can produce changes in structural and functional hepatic constituents that have been found useful for detecting and quantifying the degree of liver injury produced, as well as elucidating the mechanisms involved in producing the lesions. Alterations in the pharmacological effects of drugs can be used to detect and in some instances quantify liver dysfunction. Plaa et al. [284] demonstrated that prolongation of pentobarbital sleeping time could be used to quantify the relative hepatotoxicity of seven haloalkanes (Figure 29.4). Pentobarbital sleeping time is directly dependent on the ability of the liver to biotransform the barbiturate. Hepatocellular injury can lead to decreased activity of hepatic drug metabolizing enzymes and, therefore, a prolongation of pentobarbital hypnosis. In these experiments, the upper limit of nor-

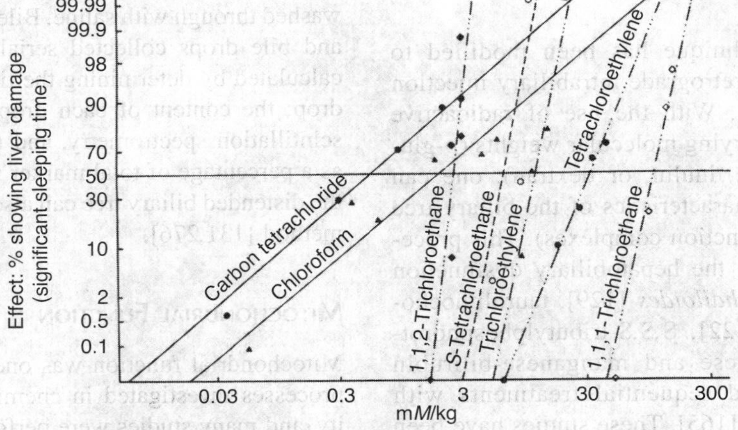


**FIGURE 29.4** Dose–response curves for the effect of seven halogenated hydrocarbons on the prolongation of pentobarbital sleeping time in mice. (From Plaa, G.L. et al., *J. Pharmacol. Exp. Ther.*, 123:224–229, 1958. With permission.)

malcy (mean sleeping time + 2 SD) for pentobarbital sleeping time was established in a large number of control mice. Subsequently, mice were administered various doses of one of the seven haloalkanes, and the pentobarbital sleeping time was determined 24 hours later. The frequency of abnormal sleeping times was then plotted against the haloalkane dose as one normally plots lethality data. These data permit comparison of dose–response curves, tests for parallelism, and statistical analyses of potency differences; thus, the relative hepatotoxic potential of these agents could be assessed. Kutob and Plaa [214] also used pentobarbital sleeping time in conjunction with BSP retention and liver succinic dehydrogenase activity to study the ability of ethanol to potentiate $CHCl_3$-induced liver damage in mice. Dingell and Heimburg [101], Lal et al. [215], Jaeger et al. [179], and Anderson et al. [11] made use of barbiturate sleeping time to assess chemically induced hepatotoxicity.

## Hepatic Lipid Content

A number of agents that produce liver injury also cause the accumulation of abnormal amounts of fat, predominantly triglycerides, in the parenchymal cells. In general, triglyceride accumulation can be thought of as resulting from an imbalance between the rate of synthesis and the rate of release of triglyceride by the parenchymal cells into the system circulation. Nonesterified fatty acids (NEFAs) removed from the circulation or synthesized endogenously are processed through two major pathways in the liver: (1) mitochondrial β-oxidation for production of metabolic energy, and (2) incorporation into complex lipids, especially triglycerides, phospholipids, cholesteryl esters, and glycolipids. Once synthesized, the complex lipids may be used for production of cellular membranes (structural lipids) or be continuously secreted from the liver into the blood. The latter pathway appears to be of greatest interest in the triglyceride accumulation observed in steatosis.

Blockage of the secretion of hepatic triglyceride into the plasma is the basic mechanism underlying the fatty liver induced in the rat by $CCl_4$, ethionine, phosphorus, puromycin, or tetracycline; by feeding a choline-deficient diet; or by feeding orotic acid [172,232]. When hepatic triglyceride is released into the plasma, it is not released as such but as a lipoprotein. The very-low-density fraction of the lipoproteins (VLDL) is the major transport vehicle for endogenously synthesized triglyceride; there is some evidence indicating that $CCl_4$ and ethionine cause a fall in the level of circulating lipoprotein, principally VLDL. The composition of VLDL by weight is 8 to 10% protein and 90 to 92% lipid. Of the lipids, triglyceride is the most abundant component (56%); the average content of phospholipid is 19 to 21% and cholesterol 17% [178].

Elevated triglyceride could result because of an increase in the rate of synthesis of this substance. Evidence suggests that the rate of synthesis is directly proportional to the concentration of the substrates present (NEFAs and glycerophosphate), so it is theoretically possible that increased hepatic triglyceride synthesis could occur because of increased NEFAs or increased glycerophosphate. Increased NEFAs could result from decreased oxidation, increased synthesis, or increased mobilization from peripheral stores. In the case of ethanol-induced fatty liver, impaired mitochondrial oxidation of NEFAs appears to be the primary abnormality seen in humans [172] due to a shift in redox potential (increased NADH/NAD ratio). It may be accompanied, however, by other abnormalities [177]. There is little evidence to support the idea that fatty acid synthesis is involved in the development of fatty liver.

In humans, two types of steatosis, macrovacuolar and microvacuolar, have been characterized as hepatic lesions. With macrovacuolar steatosis, the hepatocyte contains a single, large vacuole of fat and the nucleus is displaced to the periphery of the cell. In microvacuolar steatosis, the lipid exists in the hepatocyte as numerous small lipid vesicles and the nucleus remains in the center of the cell. Impaired mitochondrial β-oxidation of NEFAs has received more attention as a possible explanation for the presence of microvacuolar steatosis [128]. Among the drugs reported to be associated with such a lesion are salicylates, hypoglycin, valproic acid, amineptine, 2-arylpropionic acids, tetracyclines, zidovudine, and fialuridine.

Determination of hepatic fat content remains a reliable technique for demonstrating alterations by agents that produce steatosis with little or no necrosis (ethionine, phosphorus, tetracycline) and that are poorly reflected by serum enzyme measurement [374]. Alterations in hepatic triglyceride content have also been used as one of a battery of tests to determine the relative hepatotoxic potential of various halogenated hydrocarbons in rats [201] and to determine the ability of various alcohols to potentiate the hepatotoxic actions of $CCl_4$ (Table 29.3) [347].

## Lipid Peroxidation and Oxidative Stress

It is generally accepted that the toxicity of $CCl_4$ depends on the cleavage of a carbon–chlorine bond to generate a trichlor-methyl free radical ($\cdot CCl_3$); this free radical reacts rapidly with oxygen to form a trichlormethyl peroxy radical ($\cdot CCl_3O_2$), which may contribute to the toxicity [357]. The work of a number of investigators [80,282,298,357] demonstrates that: (1) the cleavage occurs in the endoplasmic reticulum and is mediated by the cytochrome P450 mixed-function oxidase system, (2) the product of the cleavage can bind irreversibly to hepatic proteins and lipids, and (3) the $CCl_4$-derived free radicals can initiate a process of autocatalytic lipid peroxidation by attacking the methylene bridges of unsaturated fatty acid side chains of microsomal lipids.

**TABLE 29.3**
**Effect of Alcohol Pretreatment on $CCl_4$-Induced Hepatotoxicity in Rats[a]**

| Treatment | ALT Activity (units/mL) | Triglycerides (mg/g liver) | G-6-Pase Activity (mg Pi per g liver per 20 min) |
|---|---|---|---|
| Ethanol (5.0 mL/kg) | 50 | 8 | 6.7 |
| Isopropanol (2.5 mL/kg) | 50 | 6 | 6.2 |
| $CCl_4$ (0.1 mL/kg) | 100 | 9 | 6.0 |
| $CCl_4$ (1.0 mL/kg) | 500 | 17 | 3.8 |
| Isopropanol + $CCl_4$ (0.1 mL/kg) | 2250[b] | 22[b] | 2.8[b] |
| Ethanol + $CCl_4$ (0.1 mL/kg) | 500[b] | 13[b] | 4.8[b] |

[a] Alcohol given p.o. 18 hours before i.p. $CCl_4$; tests run 24 hours after $CCl_4$ in 10 rats.
[b] Significantly different from group given alcohol alone ($p < 0.05$).

*Source:* Data from Traiger, G.J. and Plaa, G.L., *Toxicol. Appl. Pharmacol.*, 20, 105–112, 1971.

The peroxidative process initiated by the $\cdot CCl_3$ radical is thought to result in early morphologic alteration of the endoplasmic reticulum, loss of activity of the cytochrome P450 xenobiotic metabolizing system, loss of glucose-6-phosphatase activity, loss of protein synthesis, loss of the capacity of the liver to form and excrete VLDL, and eventually, through as yet unidentified pathways, cell death [80,282,298,357]. Alterations in these parameters have been used to monitor the course and extent of $CCl_4$-induced hepatic damage and have been applied to the evaluation of other hepatotoxicants.

Normal cellular metabolism can result in the production of reactive oxygen species (superoxide, hydrogen peroxide, singlet oxygen, and hydroxyl radical) and all cells contain defense systems to prevent or limit damage; glutathione is the major component of this system, but α-tocopherol and ascorbic acid play important roles [226]. The imbalance between prooxidants and antioxidants is known as *oxidative stress* [300].

Three groups of agents have been described to characterize the toxicants associated with the induction of lipid peroxidation or oxidative stress in liver cells [80,300]. One group consists of agents biotransformed to reactive free radicals that promote membrane lipid peroxidation directly; $CCl_4$ and $CBrCl_3$ are examples of this group. A second group consists of chemicals that are biotransformed to electrophilic intermediates, which then conjugate with glutathione and result in glutathione depletion; bromobenzene and acetaminophen are examples. The third group consists of substances that are converted to nonalkylating intermediates and generate reactive oxygen species by redox cycling; diquat and menadione serve as examples.

It is now established that nonparenchymal cells can be involved in oxidative stress responses leading to hepatotoxicity [218]. Reactive oxygen intermediates are generated by macrophages, endothelial cells, and stellate cells (Ito cells), but under physiologic conditions cellular antioxidants normally present prevent the intermediates from producing cytotoxicity. Enhanced formation of oxygen intermediates was demonstrated with $CCl_4$, galactosamine, and 1,2-dichlorobenzene. With regard to 1,2-dichlorobenzene, recent evidence indicates that Kupffer-cell-derived oxygen species are largely responsible for the lipid peroxidation [170]; in the case of ANIT, neutrophils appear to be involved [205].

Several procedures for the detection and quantification of lipid peroxidation in tissue samples or whole animals have been developed [288,298]. The reaction of malonaldehyde, a degradation product of peroxidized lipids, with thiobarbituric acid (TBA) to produce a TBA–malonaldehyde chromophore has been taken as an index of lipid peroxidation and is the most widely used method for detecting lipid peroxidation *in vitro*. Because malonaldehyde is rapidly metabolized in whole animals [288] and in whole liver homogenates [297], the failure to detect TBA-reacting material is not an indication of the absence of lipid peroxidation.

The determination of conjugated dienes in lipid extracts of hepatic subcellular fractions is a second approach for detecting lipid peroxidation [298]. The ultraviolet difference spectra of peroxidized lipids show an absorption maximum at 233 nm with a secondary absorption maximum between 260 and 280 nm due to the presence of ketone dienes. The appearance of conjugated dienes after treatment *in vivo* with toxicants is an unmistakable indication that lipid peroxidation has taken place.

Several other methods for the measurement of lipid peroxidation have been described; for example, the iodometric procedure of Bunyan et al. [57] has been employed. A variety of molecules that occur commonly in tissue may react with malonaldehyde and yield characteristic fluorescent chromophores [74]. Malonaldehyde undergoes decomposition, and the decomposition products may also lead to fluorescent products when they react with proteins

[325]. Measurement of these fluorescent products seems to offer a workable way for detecting lipid peroxidation in biological systems and tissues [298]. A second approach is the measurement of hydrocarbon gases. These gases appear early in the course of autoxidation of edible fats [125,171]. Two gases, ethane and pentane, are useful for measuring the peroxidative process *in vitro* and *in vivo*. Ethane is the predominant gas produced during autoxidation of linolenic acid [225], and pentane is the major gaseous hydrocarbon arising during thermal decomposition [112,113] and iron-catalyzed decomposition of linoleic and arachidonic acid hydroperoxides [107]. Riely et al. [309] initiated the use of ethane analysis in biological systems. They observed that ethane production was a characteristic of spontaneously peroxidizing mouse tissue (liver and brain) and was found in mice injected with $CCl_4$. In addition, they found that $CCl_4$-induced ethane evolution *in vivo* was potentiated by prior administration of phenobarbital and diminished by α-tocopherol, an antioxidant. Several other groups of investigators used ethane production to monitor the course of lipid peroxidation *in vivo* [58,159, 209,228]. Dillard et al. [100] suggested that pentane expiration was a more sensitive index of lipid peroxidation than ethane in rats fed a vitamin-E-deficient diet containing a high content of linoleic acid. Pentane production has also achieved considerable use as an index of the lipoperoxidative process [229,317,318]. A method has been devised for monitoring lipid peroxidation in humans by quantifying the excretion of ethane and pentane in exhaled breath during a 2-hour period [355]. Although the measurement of hydrocarbon gas production is an alternative procedure for the determination of lipid peroxidation, this technique cannot identify the tissue or subcellular organelle from which these substances arise. In addition, precautions must be taken to prevent or estimate the evolution of hydrocarbon gases by microorganisms in the gastrointestinal tract when *in vivo* studies are undertaken.

Prostaglandin $F_2$-like compounds ($F_2$-isoprostanes) are produced by peroxidation of arachidonic acid in phospholipids and released into the circulation [80,324]. The presence of lipid peroxidation in rats exposed to halothane has been demonstrated by the quantification of $F_2$-isoprostane in plasma and liver [16]. The measurement of $F_2$-isoprostanes *in vivo* is said to represent a promising method for detection of lipid peroxidation because of its reliability and sensitivity [17,80], but its utility in various forms of chemical-induced lipid peroxidation remains to be investigated.

## Hepatic Glucose-6-Phosphatase Activity

Regardless of the mechanism by which a chemical exerts its hepatotoxic effect, the biochemical sequelae may be useful in detecting and quantifying the damage produced. Glucose-6-phosphatase (G-6-Pase) is associated with the endoplasmic reticulum, and depression of its activity appears to specifically reflect injury to this organelle. The functional integrity of the enzyme is dependent on the presence of phospholipid, and peroxidative decomposition of microsomal lipids, as occurs with $CCl_4$, results in a significant loss of G-6-Pase activity [299].

Feuer et al. [116] administered a series of 19 compounds, including 10 known hepatotoxicants, to rats and monitored their effects on 8 hepatic enzymes. The enzymes selected represented the mitochondrial, microsomal (G-6-Pase), lysosomal, and cytoplasmic fractions. These investigators found that each of the ten known hepatotoxicants decreased hepatic G-6-Pase activity. Most of the other compounds, not known to affect the liver, did not alter G-6-Pase. Two compounds not considered to be hepatotoxicants, however, also reduced the activity of this enzyme. Based on these and other data, Grice [149] suggested that alterations in G-6-Pase activity might serve as an indicator of incipient liver damage and might occur in advance of histologically detectable organ damage; however, other data suggest that reduction of G-6-Pase is not the most sensitive test for detecting minimal hepatic damage. Klaassen and Plaa [201] found a significant depression of G-6-Pase in rat liver only at $CCl_4$ dosages of 0.3 mg/kg or greater, whereas hepatic triglyceride accumulation occurred at dosages of $CCl_4$ below 0.3 mL/kg. BSP retention occurs in rats after a $CCl_4$ dosage of 0.1 mL/kg [199], and, with light microscopy, morphologic changes are observed in rats treated with 0.13 mL $CCl_4$/kg [217].

Regardless of the relative sensitivity of this enzyme, it can be used as a diagnostic tool to quantify the effects of experimental maneuvers designed to increase or reduce the toxicity of known hepatotoxicants. An example is found in Table 29.3. This study [347] was designed to determine the ability of ethanol and isopropanol to potentiate $CCl_4$-induced hepatotoxicity. It is evident that the two alcohols themselves had no significant effect on ALT activity, triglyceride accumulation, or G-6-Pase activity; however, when pretreatment with either of these alcohols occurred 18 hours before the administration of the challenge dosage of $CCl_4$ (0.1 mL/kg), the response was greatly increased and was evident using all three parameters of hepatotoxicity. The challenge dose of $CCl_4$ merely caused a slight hepatotoxic response when it was given alone. Increasing the challenge dose of $CCl_4$ tenfold (1.0 mL/kg) resulted in a response that was less than that produced with isopropanol plus $CCl_4$ (0.1 mL/kg). Ethanol plus $CCl_4$ resulted in a response that was about equal to the response exerted by the tenfold dose of $CCl_4$ given alone.

Glucose-6-phosphatase has also been used to study mechanisms of $CCl_4$ and $CHCl_3$ liver injury. $CHCl_3$, which causes pathologic changes similar to those produced by $CCl_4$, does not alter hepatic G-6-Pase activity or the formation of conjugated dienes in hepatic microsomal lipids

[201]. These observations, along with others [288], suggest that $CHCl_3$ does not exert its hepatotoxic action by initiation of lipid peroxidation. Subsequently, $CHCl_3$ was found to produce conjugated dienes [53] and depress G-6-Pase activity [219] in phenobarbital-pretreated rats. Because $CHCl_3$-induced liver injury is more severe in phenobarbital-pretreated rats, the possibility exists that peroxidation is not the primary pathway by which $CHCl_3$ produces injury; rather, the putative initial lesion induced by $CHCl_3$ in these animals is only aggravated by the appearance of lipid peroxidation. Similarly, Jaeger et al. [179], using differences in the temporal sequences of serum (ALT) and hepatic (G-6-Pase) enzyme changes, found that $CCl_4$ and 1,1-DCE have different modes of action. In conjunction with an analysis of the ability of 1,1-DCE to initiate lipid peroxidation, the enzyme data suggest that: (1) the initial site of injury differs for $CCl_4$ and 1,1-DCE, and (2) 1,1-DCE does not act through a lipoperoxidative mechanism. Thus, G-6Pase activity aided in distinguishing separate mechanisms of action for these hepatotoxicants.

Depression of G-6-Pase activity has been used to document microsomal damage produced by the addition of agents to *in vitro* incubation systems. Glende et al. [143] and Benedetti et al. [34] used this enzyme (among other parameters) to demonstrate that the key event in $CCl_4$-induced alteration of microsomal enzyme activity is lipid peroxidation and not covalent binding of $CCl_4$-derived free radicals to microsomal lipids. Similarly, reduction of activity of this enzyme *in vitro* was used to document the destructive properties of degradation products of the lipoperoxidative process [42,173].

## Formation and Binding of Reactive Metabolites

Although the concept of lipid peroxidation is one of the truly important concepts of experimental pathology and toxicology, it does not appear to serve as a universal mechanism of liver injury. A number of drugs and chemicals (e.g., acetaminophen [140,251], furosemide [149,251], 1,1-DCE [179,241,242], trichloroethylene [4], bromobenzene [140,308], and dimethylnitrosamine [140]) produce hepatic damage but do not appear to promote lipid peroxidation. Rather, these agents are converted to highly reactive, electrophilic metabolites by the hepatic mixed-function oxidase (MFO) system. Following formation, the metabolites, which are considered to be the ultimate toxicants, can interact with hepatic constituents (e.g., protein, lipid, RNA, DNA) to form alkylated or arylated derivatives. Even with $CCl_4$, the effects on lipids and lipoprotein metabolism (leading to steatosis) appear likely due to covalent binding [357]. Various investigators postulate that the binding of reactive metabolites to hepatic macromolecules can initiate cellular damage through processes as yet unidentified. This can result in intrinsic or idiosyncratic liver injury.

A detailed discussion of the formation and detoxification of reactive metabolites and their interaction with hepatic constituents is beyond the scope of this chapter, and the reader is referred to several excellent reviews [140,141,249,251,327]. However, the experimental work carried out to unravel the mechanisms by which several toxicants produce liver injury has led to some important observations that should be discussed. One is that hepatotoxicity need not be correlated with the pharmacokinetics of the parent substance or even its major metabolites but may be correlated with the formation of quantitatively minor, highly reactive intermediates. Assuming that a relation exists between the severity of the lesion and the amount of covalently bound metabolite, the covalent binding of the metabolite can be used as an index of its formation. Indeed, this parameter might well be the most reliable estimate of the availability of the metabolite for production of damage at the target site, as most of the metabolite may undergo decomposition or further metabolism before it can be isolated in body fluids or urine [249]. Thus, one widely used maneuver to assess the contribution of formation and binding of metabolites in chemically induced hepatotoxicity is to determine if radiolabeled chemicals administered to animals over a wide dosage range are covalently bound to macromolecular constituents in tissues that subsequently become necrotic [249].

A second concept is that a threshold tissue concentration of the metabolite must be attained before liver injury is elicited; if it is not attained, injury does not occur. Endogenous substances such as glutathione play an essential role in protecting hepatocytes from injury by chemically reactive metabolites. The mechanism that establishes a dose threshold, however, may vary from compound to compound. Finally, other enzymic pathways (e.g., glutathione *S*-transferase and epoxide hydrolase) also play a role in protecting the hepatocyte by catalyzing the further degradation of the toxic reactive intermediates.

The studies mentioned above have provided relatively straightforward biochemical strategies for uncovering the possible existence of potentially toxic chemically reactive metabolites in new compounds [249]. In general, a dose–response study employing the radiolabeled compound over a wide dosage range is perhaps the single most important facet of the overall study in that it can provide information relating to a dose threshold for toxicity, possible mechanisms for the threshold response (e.g., glutathione depletion), and the degree of covalent binding of metabolites to target organs or constituents with the target organ. This latter information, in conjunction with dose–response studies documenting the dosages required to produce necrosis (or other endpoint), provides strong presumptive evidence favoring toxicity mediated by a reactive metabolite. Subsequent efforts should include the use of inducers (e.g., phenobarbital, 3-methylcholanthrene) and inhibitors (e.g., SKF-525A, $CoCl_2$, piperonyl butoxide) of the MFO system. Enhancement of *in vivo* or *in vitro* covalent binding of the radiolabeled compound, as well as toxicity by an inducing agent, can provide support for the contention that a reactive metabolite mediates toxicity. A similar conclusion can be drawn if an inhibitor of the MFO system depresses covalent binding and toxicity; however, the observation that inhibitors increase the response or that inducers decrease the response does not preclude the possibility that the toxicant exerts its effect through the formation of chemically reactive metabolite [249]. For example, an inducing agent may stimulate the activity of a detoxifying pathway to a greater extent than a toxifying pathway. In addition, manipulation of the concentration of hepatoprotective substances such as glutathione (e.g., depression of hepatic glutathione concentration by diethylmaleate administration) can alter covalent binding or toxicity of a compound and support the likelihood that the toxic effects are mediated by a metabolite. Correlation of the data from several of these studies with a pharmacokinetic analysis can delineate the participation of a chemically reactive metabolite in the production of toxicity. For a more detailed discussion of these concepts, the reader is referred to the reviews cited above.

## Liver Fibrosis

Fibrosis (the accumulation of collagen) represents a key phenomenon in chronic liver disease [314]. Septal fibrosis is the principal feature of experimentally induced liver cirrhosis. $CCl_4$ and ethanol have been the toxicants most frequently used to induce liver fibrosis and cirrhosis. In the rat, twice weekly administrations of $CCl_4$ for 7 to 12 weeks have been shown to produce cirrhosis [63,70,164]. Ito cells (also called lipocytes, fat-storing cells, or stellate cells) acquire characteristics of myofibroblasts [313] and play an important role in the formation of collagen in liver fibrosis [339]. Ito cells isolated from fibrotic livers have significant increases in mRNA levels of type I, III, and IV procollagen compared to normal cells [358]. Serum concentration of the amino-terminal propeptide of procollagen type III (PIIIP) can be used as a fibrogenic marker for the period progressing to cirrhosis, but its use in cirrhosis seems to be limited by factors other than liver fibrogenesis [164].

High concentrations of the amino acid 4-hydroxy-L-proline are found only in collagen. Determination of hepatic hydroxyproline content represents a valuable marker to evaluate total collagen content and thus fibrosis in liver tissue [314]. The hepatic levels of hydroxyproline are well correlated with the degree of liver fibrosis measured histologically [70,164,191,265]. Figure 29.5 illustrates the effect of repeated $CCl_4$ gavage on hepatic hydroxyproline concentrations in corn oil- and acetone-pretreated rats; after 10 weeks of treatment, acetone + $CCl_4$-treated animals showed a fully developed cirrhosis, whereas a much less severe lesion was observed in corn oil + $CCl_4$-treated rats (37 ± 2% and 16 ± 3% of the liver occupied by fibrous connective tissue, respectively). Finally, the serum activity of immunoreactive prolyl hydroxylase, the enzyme responsible for proline hydroxylation, is elevated in rats with fibrotic livers, but Okuno et al. [265] suggested that these elevations should be carefully evaluated when being used as a parameter to estimate the activity of fibrogenesis in the liver.

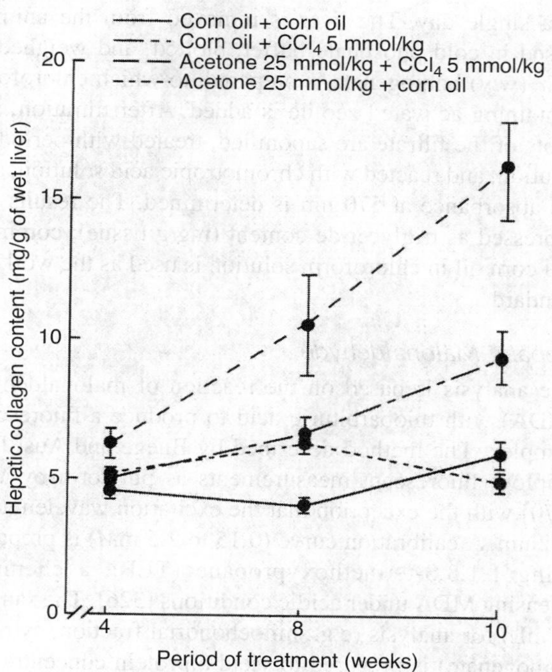


**FIGURE 29.5** Temporal progression of hepatic collagen content in rats treated with corn oil or acetone twice weekly (Tuesday and Thursday) and challenged with corn oil or $CCl_4$ 18 hours later. Values represent the mean ± SE determined in eight rats. (Data from Charbonneau, M. et al., *Hepatology*, 6, 694–700, 1986.)

## Analytical Determination of Hepatic Triglyceride, Hepatic Malonaldehyde, and Hepatic Collagen Contents

### *Hepatic Triglycerides*

The hepatic triglyceride assay described by Butler et al. [61] is an adaptation of the method of Van Handel and Zilversmit [352], originally developed for the direct determination of serum triglycerides. The procedure consists in five steps: (1) homogenization of tissue; (2) adsorption of phospholipids onto zeolite, followed by extraction of triglycerides into chloroform; (3) hydrolysis of triglycerides to fatty acids and glycerol; (4) oxidation of glycerol with $NaIO_4$ to formic acid and formaldehyde; and (5) formation of a colored complex of formaldehyde and chromotropic acid [61]. The procedure is relatively simple and can be used to analyze a large number of tissue samples in a single day. The liver is removed from the animal, rinsed in cold phosphate buffer, blotted, and weighed; a 10% (w/v) homogenate is prepared, to which chloroform containing activated zeolite is added. After filtration, aliquots of the filtrate are saponified, treated with periodate solution, and reacted with chromotropic acid solution, and the absorbance at 570 nm is determined. The results are expressed as triglyceride content (mg/g tissue); commercial corn oil in chloroform solution is used as the working standard.

### *Hepatic Malonaldehyde*

The analysis is based on the reaction of malonaldehyde (MDA) with thiobarbituric acid to produce a fluorescent complex. The method described by Buege and Aust [55] employs fluorescent measurements as put forth by Yagi [370] with the exception that the excitation wavelength is 532 nm. A calibration curve (0.15 to 2.5 m*M*) is prepared using 1,1,3,3-tetraethoxypropane (TEP), a chemical releasing MDA under acidic conditions [326]. The sample (1 mL) for analysis (e.g., mitochondrial fraction, cytosol, homogenate) is prepared to yield a protein concentration of 4 to 6 mg/mL in phosphate buffer, is mixed with 2 mL of TBA solution, and is incubated at 100°C. After adding butanol and centrifuging, the organic phase is analyzed for fluorescent content (532 and 553 nm for excitation and emission, respectively). The MDS content is expressed as pmol MDA per mg protein.

### *Hepatic Collagen*

The liver content of hydroxyproline is measured by the colorimetric method of Edwards and O'Brien [110], and total collagen content is calculated assuming 12.5% of collagen is constituted of hydroxyproline residues. Hydroxyproline is liberated from collagen by acid hydrolysis, oxidized to pyrrole with chloramine-T, then transformed into a red chromogen using a *p*-dimethyl-aminobenzaldehyde solution, commonly called Ehrlich reagent. The liver (about 1 g) is prepared as a 1% liver homogenate in 6-*N* HCl. Aliquots are placed in conical Pyrex® tubes and autoclaved (125°C, 240 KPa, 3 hours). Dried hydrolysates are prepared and resuspended to yield solutions containing from 0.1 to 4.0 µg hydroxyproline per mL. Chloramine-T is added as an oxidant, as well as Ehrlich reagent to form a red chromogen after incubation at 60°C. The hydroxyproline is calculated from the absorbance at 500 nm; assuming 12.5% of collagen is constituted of hydroxyproline residues, the total collagen content is derived.

## Repair and Recovery

The liver is a remarkable organ in its ability to regulate its growth and gain or loss of optimal mass. The biology of proliferative properties of the organ, the mechanisms initiating regeneration, the relationship between hepatocyte proliferation and apoptosis, and the involvement of nonparenchymal cells (Kupffer cells and stellate cells) have been well described [115,357]. Liver regeneration is a multistep process involving priming and progression. Although various forms of chemical-induced liver injury have been studied extensively in terms of biochemical events for over 50 years [282], interest generally has focused on the early initiating effects leading to the appearance of hepatocellular dysfunction, rather than on the later recovery phase of the lesion. In the rat, organ recovery after acute hepatotoxic insult is largely dose dependent; recovery time is longer when the lesion is more extensive. As the lesion progresses, hepatocellular regeneration appears; it appears within 6 hours after administration of a low dose of $CCl_4$ in the rat, even though the centrilobular necrosis is just becoming evident [230,231]. When the hepatic lesion is enhanced by the introduction of another agent, however, recovery time can be an important consideration for assessing possible mechanisms of action involved in the potentiations [280]. With the combination of $CCl_4$ and several potentiating agents (*n*-hexane, 2-hexanone, 2,5-hexanedione, isopropanol, and acetone), this relationship was assessed using both biochemical indices (serum ALT and OCT activities) and morphological patterns (quantitative histology) of liver injury; appropriate dose–response curves were established from the percentage of animals affected [69]. Time of recovery was shown to be due to the maximal severity of the lesion, regardless of the potentiation. Although pretreatment with the potentiator resulted in an enhanced hepatotoxic response from a low dose of $CCl_4$, the dose– response curve for the enhanced response was no different than that produced by a higher but equitoxic dose of $CCl_4$ administered alone. These data were interpreted as an indication that the five potentiators did not alter the temporal progression of $CCl_4$-induced liver injury.

Mehendale and his collaborators have performed an extensive series of experiments to assess the role of tissue repair in potentiated liver injury. The studies originated from the observation that chlordecone-potentiated $CCl_4$ hepatotoxicity in rats was quantitatively quite remarkable and resulted in enhanced lethality. Two tissue repair responses were observed after exposure to a low dose of $CCl_4$ [62]; the early phase regeneration (EPR) response (arrested $G_2$ hepatocytes activated to proceed through mitosis) occurs quickly (peaks at about 6 hours) and is followed (at about 24 hours) by the secondary phase regeneration (SPR) response (hepatocytes mobilized from $G_0/G_1$ to proceed through mitosis). During chlordecone potentiation of $CCl_4$ hepatotoxicity, the EPR phase appears to be eliminated and the SPR phase decreased; thus, the progression of the severe injury is facilitated and leads to lethality. Evidence suggests that induction of EPR may accelerate SPR. Interestingly enough, large doses of $CCl_4$ given alone also result in regeneration responses similar to those obtained with chlordecone and a small dose of $CCl_4$. Experiments performed with colchicine, partial hepatectomy, $CCl_4$ autoprotection, nutritional factors, and different animal species have provided data consistent with the purported roles attributed to EPR, SPR, and liver injury [62, 332].

The role of tissue repair has been assessed with other hepatotoxicants [246,332]. Dose–response studies indicate that thioacetamide when given alone affects hepatic tissue regeneration (measured with $^3$H-thymidine incorporation into DNA; proliferating nuclear cell antigen) in a manner not unlike that observed with $CCl_4$. Comparable observations were obtained with *o*-dichlorobenzene and trichloroethylene, although the dose-dependent relationships were not as evident with these agents. Increased lethality was not observed with isopropanol-potentiated $CCl_4$ liver injury. When the expression of stimulators of promitogenic signaling interleukin-6, inducible nitric oxide synthase, hepatocyte growth factor, transforming growth factor $\alpha$, and epidermal growth factor receptor was studied after thioacetamide hepatotoxicity in rats, it was modified by moderate caloric restriction [246]. With a binary mixture of chloroform and allyl alcohol, tissue repair was shown [10] to exert a key role in the final outcome of the liver injury in rats. Mehendale and his collaborators [227,332] have proposed a two-stage model for chemical-induced hepatotoxicity. Stage one would involve initiation and infliction of injury; stage two would lead to recovery or progression to massive injury, depending on the effects of the toxicant on cellular regeneration (enhancement would lead to recovery; inhibition would lead to massive injury). They hypothesized that progression of the liver injury might be mediated by cytoplasmic and lysosomal degradative enzymes leaking from dying cells [227]. The experimental results carried out following $CCl_4$- and acetaminophen-induced hepatotoxicity suggest that calpains may be involved. Although various aspects of the repair–recovery process are still hypothetical or speculative, the concept as such is thought provoking and certainly an important contribution to the understanding of chemical-induced liver injury. It will be interesting to see how it evolves.

## Apoptosis

Previous sections highlighted the fact that necrosis of hepatocytes is a common type of cell death occurring after chemical exposure. Another type of cell death, called *apoptosis* (from the Greek "apo" meaning away from and "ptosis" meaning falling—cells falling away from a tissue) [81], is recognized as an important process in chemically induced effects in the liver and liver disease [316]. In a landmark article, Kerr et al. [192] defined a form of cell death morphologically distinct from necrosis, which they named apoptosis. Apoptotic cells exhibited nuclear and cytoplasmic condensation followed by dissociation of the cell into membrane-bound fragments similarly to the events observed in a phenomenon previously described by embryologists as programmed cell death; for this reason, apoptosis has been inappropriately used as a synonym for programmed cell death. Apoptosis is defined as an active mode of cell death, as it requires RNA and protein synthesis, is controlled by pro- and antiapoptotic genes, and is induced by physiological stimuli in addition to the typical pathological stimuli associated with necrosis [78]. On the other hand, necrosis is considered to be a form of passive cell death. Recently, however, necrosis has also emerged as an alternate form of programmed cell death, whose activation might have biological consequences, including the induction of an inflammatory response [109].

Apoptosis is morphologically defined by a progressive condensation of the chromatin to the inner face of the nuclear membrane (DNA hyperchromicity and cresenteric caps), convoluted cell shape (blebbing or budding), dilatation of the endoplasmic reticulum, cell shrinkage with consequent loss of membrane contact with neighboring cells, and fragmentation of the cell with formation of membrane-bound acidophilic globules (apoptotic bodies) often containing nuclear material. The latter are frequently found within the cytoplasm of intact cells, indicating that they are phagocytized by adjacent cells [93]. In addition, apoptosis is not commonly associated with the inflammatory response that accompanies necrosis [192]. Other morphological criteria specific for necrotic cells are cell and nuclear swelling, patchy chromatin condensation, swelling of mitochondria, vacuolization in cytoplasm, plasma membrane rupture ("ghost-like" cells), and dissolution of DNA (karyolysis). Histological criteria of cell death, such as pyknosis and karyorrhexis, can be applied to both apoptosis and necrosis at certain stages and hence cannot distinguish between these two modes of cell death [273]. Finally, from the above, it is apparent that apoptosis involves only scattered single cells, whereas necrosis affects large areas of liver lobules. In the normal liver, apoptosis is predominant in acinar zone 3 and is thought to subserve the elimination of senescent cells [33].

Apoptosis and mitosis play complementary contrasting roles in tissue homeostasis, as the former leads to cell removal and tissue hypoplasia whereas the latter causes tissue hyperplasia; for example, the induction of altered hepatic foci appears to be related to compensatory cell proliferation in PCB-77-treated rats, whereas the inhibition of apoptosis appears to be important in PCB-153-treated rats [343]. Because apoptosis plays a critical role in deleting cells from tissues, it is not surprising that failure of apoptosis leads to imbalanced cell proliferation and is now recognized as a phenomenon associated with carcinogenesis. Bursch et al. [59], however, recently suggested that, in contrast to rat liver, inhibition of apoptosis in mice appears to be a minor determinant of tumor promotion.

Tumor promoters and nongenotoxic carcinogens inhibit active cell death, thereby increasing the accumulation of (pre)neoplastic cells and accelerating the development of cancer [320]. Sustained activation of the aryl hydrocarbon receptor (AhR) is postulated to generate a strong selective pressure in liver treated with 2,3,7,8-tetrachlorodibenzo-dioxin (TCDD), leading to selection and expansion of clones evading growth arrest and apoptosis [43]. Suppression of apoptosis with TCDD coincided with a marked hyperphosphorylation of p53 mediated by the AhR [319]. Using the inhibition of UV-induced apoptosis in rat hepatocytes in primary culture as an *in vitro* model for mechanistic studies on the inhibition of apoptosis, Bohnenberger et al. [44] showed that non-dioxin-like polychlorinated biphenyls (PCBs) are likely to promote liver carcinogenesis via the suppression of apoptosis. Heptachlor strongly inhibited transforming growth factor (TGF)-induced apoptosis and cytochrome *c* release into the cytosol in rat hepatocytes [264]. Peroxisome proliferators, a large and chemically diverse family of nongenotoxic rodent hepatocarcinogens that activate peroxisome proliferator-activated receptor $\alpha$ (PPAR$\alpha$), have been shown to suppress apoptosis induced by both spontaneous and TGF$\beta_1$, the physiological negative regulator of liver growth [310]. Experimental evidence supports a mechanism involving diminished apoptosis in the age-related difference in sensitivity where PPAR agonists produce a five- to sevenfold higher yield of grossly visible hepatic tumors in old relative to young animals [373].

### *Examples of Chemically Induced Liver Apoptosis*

In addition to being present in various viral, immunological, malignant, or drug-induced human liver diseases, hepatocyte apoptosis in animals can be triggered either *in vivo* or *in vitro* by many toxic agents [117]. Histopathologic features of apoptosis are frequently observed in chronic cholestatic disorders as a result of accumulation of toxic bile salts within hepatocytes [273]. Bile-salt-induced apoptosis occurs in a concentration-dependent manner at concentrations that are far smaller than those critical for micelle formation and do not cause cell necrosis [272]. Hydrophobic bile salts, such as glycodeoxycholate, glycochenodeoxycholate, or taurochenodeoxycholate cause apoptosis in isolated hepatocytes [73,273]. The hydrophilic bile acid tauroursodeoxycholic acid significantly reduced glycochenodeoxycholate-induced hepatocyte apoptosis [36]. Death receptor activation is an important endpoint in bile acid-induced apoptosis. Higuchi et al. [168] reported that deoxycholate is more potent than glycodeoxycholate, chenodeoxycholate, and glycochenodeoxycholate in upregulating death receptor 5 (DR5)/TRAIL receptor 2 sensitizing hepatocytes to tumor necrosis factor-related apoptosis-inducing ligand (TRAIL)-mediated apoptosis. Glycochenodeoxycholate-induced hepatocyte apoptosis is modulated by caspase cascade activation [356]. Hepatobiliary diseases are characterized by elevated caspase activation and apoptosis, which can be specifically detected *in situ* by a cleavage site-specific antibody against cytokeratin-18 [27].

Increases in the number as well as changes in the distribution of apoptotic bodies within the liver were observed in ethanol-fed rats and mice [35,145]. When compared with normal livers, the increased apoptosis was more pronounced as the duration of ethanol exposure increased and was reversed by ethanol withdrawal. All the putative mechanisms for alcohol-induced hepatocellular damage, such as oxidative stress, toxic acetaldehyde adduct formation, hypoxia, or immunologically mediated destruction, can induce apoptosis such that it may represent a final common mechanism mediating hepatocellular damage by ethanol [273]. Alcohol increases liver apoptosis predominantly through an intrinsic signaling pathway [97].

Metals cause changes in liver apoptosis. Mice injected intraperitoneally with 5 to 60 µmol/kg of cadmium showed both a time- and dose-dependent increase in liver apoptosis, which peaked at 9 to 14 hours after cadmium administration then decreased [158]. The time course of the apoptotic DNA fragmentation index, monitored by quantification of oligonucleosomal DNA fragments, correlated with the results obtained by histopathological analysis and a commercial *in situ* apoptotic DNA detection kit. An interesting conclusion of this work is that apoptosis is a major mode of elimination of critically damaged cells in acute cadmium hepatotoxicity in the mouse, and it precedes necrosis. Cadmium also causes a biphasic elevation in parenchymal cell apoptosis in the rat [351]; apoptosis of nonparenchymal cells is the basis of the pathogenesis of peliosis hepatitis. An organic metal, tributyltin, which is a highly toxic water contaminant, also induces apoptosis in rainbow trout hepatocytes through a step involving $Ca^{2+}$ efflux from the endoplasmic reticulum or other intracellular pools and by mechanisms involving cysteine proteases, such as calpains, as well as the phosphorylation of apoptotic proteins, such as Bcl-2 homologs [296]. Denizeau and coworkers [187] reported the involvement of initiator and effector caspases, cleavage of their intracellular substrates, and activation of both death receptor and mitochondrial pathways in tributyltin-induced apoptosis in rat hepatocytes. Long–Evans Cinnamon (LEC) rats, an animal model for inherited copper toxicosis, show liver apoptosis with increasing age [121]. Finally, chronic arsenic administration induces a specific pattern of apoptosis called postmitotic apoptosis [28].

### *Assessment of Apoptosis*

Precision-cut liver slices provide a valuable *in vitro* system for studying drug-induced liver apoptosis [255]. Using a model of apoptosis-driven regression of rat liver hyperplasia induced by cyproterone acetate, the mean duration of the histological stages of apoptosis was found to be approximately 3 hours [60]. Because apoptosis occurs in isolated single cells and does not stimulate persisting tissue changes, such as inflammation and scarring that characterize necrosis, procedures based on biochemical endpoints, such as DNA cleavage patterns, have been developed as complementary tools for the analysis of cell morphology. Internucleosomal DNA cleavage caused by $Ca^{2+}/Mg^{2+}$-dependent endonucleases is a prominent feature of apoptosis [13]. During apoptosis, DNA is cleaved in a nonrandom manner into 50 to 300 kilobases, then into 180 base-pair fragments. Several techniques are based on the analysis of DNA fragments. Separation of nuclear DNA from apoptotic cells on agarose gel yields the typical ladder appearance of fragments, whereas DNA from necrotic cells produces a smear on electrophoresis. The use of DNA electrophoresis as the sole criterion for apoptosis is not recommended, as DNA ladders can be observed in cells with necrotic-type morphology, and cells with ultrastructural characteristics of apoptosis can fail to produce DNA ladders.

Individual cell electrophoresis can be performed using the so-called *comet assay* [93]; the apoptotic cells appear as comets having characteristic heads, which represent the remains of the cells containing high-molecular-weight DNA, and tails, which represent a fraction of the degraded DNA. Commonly used techniques involve *in situ* labeling of DNA strand breaks such as the TUNEL (TdT-mediated dUTP-digoxigenin nick end labeling) assay; deoxynucleotides can be either fluorochrome or enzyme (phosphatase or peroxidase) tagged. For a description of this procedure used in the context of liver toxicity assessment, see Habeebu et al. [158]. Wheeldon et al. [361] suggested that further validation is required before *in situ* end labeling can be used confidently alone, because they observed variation in apoptotic body indices after cyproterone acetate withdrawal between *in situ* end labeling and hematoxylin and eosin staining techniques. Similarly, Grasl-Kraupp et al. [148] concluded that DNA fragmentation is common to different kinds of rat liver cell death and that its detection *in situ* should not be considered a specific marker of apoptosis.

Several techniques for measuring apoptosis rely on flow or laser scanning cytometry (for a review, see pp. 49–61 in Darzynkiewicz and Traganos [93]). Dive et al. [102] proposed a rapid multiparameter flow cytometric assay that discriminates and quantifies viable, apoptotic, and necrotic cells via measurement of forward- and side-light scatter (proportional to cell diameter and internal granularity, respectively) and the DNA-binding fluorophores Hoechst 33342 and propidium. For a comparison of approaches for quantifying hepatocyte apoptosis, see Goldsworthy et al. [146], and for a discussion of the selection of methods and the inappropriate uses of methodology, see pp. 61–69 in Darzynkiewicz and Traganos [93]. Annexin V is a 36-kDa protein that binds with high affinity to phosphatidylserine lipids in the cell membrane, a component that goes from the inner membrane leaflet to the outer cell surface in apoptotic cells; therefore, Annexin V has been shown useful for detecting the earliest stages of apoptosis. Recent advanced positron emission tomography (PET) techniques using either $^{18}$F-Annexin V [371] or $^{124}$I-Annexin V [190] have opened the way for *in vivo* measurement of chemically induced apoptosis in rats.

Modulation of molecular endpoints based on proteins involved in signaling cascades controlling apoptosis can be used to demonstrate chemically induced apoptosis; for example, the caspase (cysteine aspartyl proteases) enzymes play a central role in the cell death machinery as indicated in chemically induced apoptosis, discussed above. They are subdivided into initiator (caspase-2, -8, -9, and -10) and effector (caspase-3, -6, and -7) caspases. Two main pathways are well established in apoptosis: the death receptor pathway and the mitochondrial pathway. The former requires membrane receptors such as Fas, which is involved in a signaling complex activating caspase-8, ensuring activation of caspase-3; the latter involves the release of a small apoptogenic molecule, cytochrome *c*, leading to formation of apoptosome, which also leads to caspase-3 activation. The two pathways can be linked through Bid, a pro-apoptotic member of the Bcl-2 family of proteins [187]. At least 15 different Bcl-2 proteins have been identified and can be subdivided into pro- and anti-apoptotic members: Bcl-2, Bcl-xL, Bcl-w, Mcl-1, Bfl-1, Brag-1, and A1 inhibit apoptosis, whereas Bax, Bak, Bcl-xS, Bag, Bid, Bik, Hrk, and Bad promote apoptosis in mammalians [316].

A simple cytological method that is routinely used in different laboratories for evaluating apoptosis will now be described. The assay can be applied to tissue slices, primary cells, cell lines, and freshly isolated cells and can be performed with cells of virtually any origin. The method has been also used for evaluating the apoptotic rate of human neutrophils, as these cells are known to spontaneously undergo apoptosis without any stimulation and are recognized as an excellent model for studying this biological process [134]. Furthermore, observations made using this technique correlate well with the results obtained by conventional Hoechst 33342 and propidium iodide cytofluorimetric analyses [142]. This is a simple and low-cost histologically based procedure. The assay is performed at room temperature. Other techniques to measure apoptosis can be used in parallel. For new applications using isolated or cultured cells, it is strongly recommended that cell viability be verified by trypan blue exclusion before performing the assay; apoptotic, but not necrotic, cells exclude trypan blue. The level of nonviable cell loss can also be monitored by such an approach. Isolated cells (suspension of $10 \times 10^6$ cells/mL) are loaded into a cytospin chamber and gently spun with a cytocentrifuge to adhere the cells onto a microscope slide, whereas adhering cells can be directly grown on glass coverslips and tissues slices are used. A rapid coloration procedure is performed using the Diff-Quick® staining kit (Baxter; Miami, FL) according to the manufacturer's instructions. The slides are rinsed and sealed with warm parafilm or nail varnish; the sealed slides can be kept for several days. Cells are observed with a light microscope at a magnification of 400× or 1000×; nuclei appear as dark purple structures. Apoptotic cells are easily distinguished from normal cells by examining morphological changes: cell shrinkage, nuclear collapse (crescent shape or small dot profiles), dense white inner cell vacuoles, and apoptotic bodies (plasma membrane blebs/buds or granules). It is recommended that a minimum of 100 cells from different fields be counted in two replicate slides. Results are expressed as the percentage of cells in apoptosis: (100 × number of apoptotic cells)/total number of cells.

## *IN VITRO* EVALUATION OF HEPATIC INJURY

*In vitro* systems offer the possibility of assessing liver injury in the absence of extrahepatic factors, such as the absorption, distribution, and extrahepatic metabolism of the chemical, humoral factors, and toxic effects caused at other sites. For this reason, they are especially valuable for studying specific mechanisms involved in chemically induced liver injury. In this chapter, however, we focus on the use of these systems to detect and evaluate hepatotoxic properties of chemical agents.

The *in vitro* approaches available to assess hepatotoxic properties of chemical agents encompass different levels of organization ranging from isolated hepatocytes and cell cultures to precision-cut liver slices and isolated livers. Liver cell organization is not disrupted in these four *in vitro* systems, such that the injury caused by a chemical results from the overall effects of phase I and II biotransformation reactions, defense and repair systems, and cellular processes. Zonal architecture of the liver, normal polarity of hepatocytes (biliary and plasmatic poles), and presence of all liver cell types are observable only with the liver slice system and the isolated perfused liver. Isolated hepatocytes, however, offer the following advantages: They are relatively easy to prepare without sophisticated equipment; a large number of experiments with the liver of one animal, serving as its own control, can be performed; and sampling throughout the experiment is feasible. Selection of a particular *in vitro* approach depends on specific research needs; xenobiotic biotransformation is generally comparable from one system to another as exemplified by similar metabolic profiles for caffeine biotransformation in liver slices, hepatocyte cultures, and microsomes, with a rate of metabolite formation close to that calculated from *in vivo* caffeine elimination [41].

Technical aspects and applications of the isolated perfused liver system are well covered in Chapter 38 and will not be discussed here. For additional reading on the use of the isolated perfused liver as a model for studying drug- and chemical-induced hepatotoxicity, the reader can consult a monograph edited by Ballet and Thurman [26] and a review by Sweeny and Diasio [338].

### ISOLATED HEPATOCYTES

The liver is constituted of six different cell populations—the hepatocytes, biliary epithelial cells, endothelial cells, Kupffer cells, Ito cells (lipocytes, fat-storing cells), and pit cells. The cells other than the hepatocytes are designated as nonparenchymal cells. The hepatocytes (or parenchymal cells) constitute 60% of the liver cell population and occupy 80% of the total liver volume. Much of the effort in isolation of liver cells has focused on viable hepatocytes, as they are the main metabolic unit of the liver. Procedures, however, are also available for isolating nonparenchymal cells [127].

The literature on the preparation, properties, and application of isolated hepatocytes is abundant, and excellent monographs and reviews have been published [37,153,243, 244,338]. The *in situ* two-step procedure is the most commonly used technique to isolate hepatocytes. It is derived from the pioneering work of Berry and Friend [38], who introduced *in situ* liver perfusion with digestive enzymes, and from the subsequent work of Seglen [322,323], who enhanced the recovery of isolated viable cells by initially perfusing the liver with a calcium-free buffer and then placing the digestive enzymes in a calcium-supplemented buffer. In current techniques, buffers containing a chelating agent and collagenase as the digestive enzyme are used for the first and second steps; suspensions exhibiting greater than 90% viability are usually obtained. Nonviable hepatocytes can be separated from viable ones using a dibutyl phthalate separation technique [114].

Hepatocytes are most commonly isolated from rat liver; the two-step procedure, however, has been successfully adapted to other species such as the mouse, rabbit, and guinea pig [243,244]. Specialized techniques are also available for preparing suspensions enriched with periportal or perivenous hepatocytes [39]. Increasing interest in toxicological risk assessment has prompted the development of procedures such as the biopsy perfusion methods [5] to isolate hepatocytes from human liver samples. Successful approaches have also been achieved to optimize cryopreservation procedures for human hepatocytes [72,220,234].

Isolated hepatocytes have been extensively employed for studies on chemical (including drugs) biotransformation and toxicity. Over the years, the bulk of the reports published on chemical metabolism in isolated hepatocytes has defined the capabilities and limits of this system. There is no doubt that isolated cells have proved valuable for investigating biotransformation [37,39,155]. The activity of phase I or phase II metabolizing enzymes in viable cells is maintained for a few hours [153]; however, in standard metabolic/toxicological studies, the incubation period of the cells with the chemical is relatively short (less than 3 hours), such that the losses in activity do not generally invalidate the experiments. The substrate metabolic rates are often similar or slightly slower in isolated hepatocytes than in corresponding 9000 *g* supernatant or microsomes [153].

Cytotoxicity of chemicals in freshly isolated hepatocyte suspensions is a valuable tool when screening xenobiotics for hepatotoxic properties. As with the *in vivo* tests, the major cytotoxicity parameters studied are based on the structural integrity of the cell membrane. The uptake of normally nonpermeable dyes, such as trypan blue and neutral red, is one of the most common tests; the percentage of nonviable colored cells requires a cell-counting analysis under a light microscope. Leakage into the medium (separated from the cells) of the cytosolic enzyme LDH is a biochemical test that is used as frequently as the dye exclusion assay. The former, however, is more convenient to perform than the dye assay, and it also offers the advantage of summing up the release of all damaged cells, including disintegrated hepatocytes. For some chemicals, decreases in intracellular concentrations of LDH have been reported at dosages where LDH leakage into the medium could not be detected [238]. Measurement of intracellular potassium, sodium, or calcium content is a more sensitive marker of cell membrane integrity. Fariss et al. [114] observed a decrease in potassium levels 3 hours prior to LDH leakage in cells treated with adriamycin in combination with 1,3-*bis*(2-chloroethyl)-1-nitrosourea or ethyl methanesulfonate. Indices of cellular metabolic competence are also sensitive markers; modifications of glycogen deposits and protein synthesis have been shown to detect early changes in isolated cells treated with chlorpromazine, promethazine, bromobenzene, acetaminophen, and isoniazid [144]. Frazier [126] reported that cellular potassium concentration was the most sensitive toxicity index (with inhibition of protein synthesis second and trypan blue staining the least sensitive) to evaluate cadmium, copper, and zinc toxicity. Finally, morphological changes in isolated cells can be assessed by electron microscopy.

The isolation procedure inevitably introduces changes and renders cells more susceptible to the effect of chemicals compared to the intact liver. When studying the hepatotoxic properties of organic solvents, care should be given to the selection of concentrations used, as these agents are thought to exert direct solvent effects [37]. In addition, it is preferable to use equilibration techniques in stoppered flasks to study volatile chemicals with low water solubility.

Studies have shown that known *in vivo* hepatotoxicants tested in isolated cell suspensions are cytotoxic *in vitro*; individual chemicals such as acetaminophen, ethanol, methotrexate, fentanyl, bromobenzene, and chlorinated aliphatic compounds [155,350,354], as well as mixtures such as $CCl_4/CHCl_3$ [261] and combinations of trichloroethylene, tetrachloroethylene, and 1,1,1-trichloroethane [334] are hepatotoxicants *in vivo* and *in vitro*. There are also examples of species and strain differences observed *in vivo* that also occur *in vitro* [253,366]. Differences between *in vivo* and *in vitro* hepatotoxic potency, however, have been observed on several occasions. Dimethylnitrosamide and thioacetamide are not as potent in the isolated cell system as expected from their *in vivo* hepatotoxicity [336]. Rankings of relative toxicity for different haloalkanes or bile salts in isolated hepatocytes differ from those observed *in vivo* [92,147,334]. Finally, the ability of isolated cell suspensions to accurately detect the hepatotoxic properties for chemicals with unknown *in vivo* effects remains to be fully demonstrated, in particular for chemicals found to be weak or moderate hepatotoxicants.

Cholestatic responses *per se* cannot be seen in isolated hepatocytes. It is possible, however, to evaluate chemical interference with processes involved in the transport of endogenous substances into and out of hepatocytes. Such approaches have been successful for thioacetamides and cyclosporin A [54,213]. Hepatocyte couplets (two adjacent hepatocytes surrounding a lumen or vacuole) represent the primary bile secretory unit and are thought to be the equivalent of the bile canaliculus [47]. Secretory polarity is retained in hepatocyte couplets, and methods that permit the isolation of these couplets, as well as their utility for assessing canalicular function, have been perfected. The functional aspects include the excretion of fluid, organic anions and cations, lipids, and proteins; the regulation of cellular and canalicular pH; transcytosis and protein transport; cytoskeletal function and canalicular contractility; signal transduction, calcium signaling, and paracellular permeability; and electrophysiological events [47]. The effects of cholestatic agents on tight junctional permeability in isolated rat hepatocyte couplets has been assessed by monitoring the retention of a fluorescent bile acid analogue (cholyl-lysyl-fluoresein) or the penetration of horseradish peroxidase [315]. Incubation of couplets in the presence of taurolithocholic acid, cyclosporin A, estradiol 17β-glucuronide, menadione, or *t*-butyl hydroperoxide (substances known to affect hepatobiliary function *in vivo*) resulted in a quantifiable dose-dependent decrease in canalicular retention.

### Hepatocyte Cultures

Hepatocyte cell cultures represent an extension of the isolated hepatocyte assay that permits longer test periods by attaching cells to a matrix placed in a periodically refreshed medium to keep them viable. The comprehensive review prepared by Grisham [151] in 1979 is one indicator of the long-standing popularity of hepatic cell cultures to detect and evaluate the mechanisms of actions of toxic chemicals. Also, Berry et al. reviewed the subject in their excellent monograph on the use of isolated hepatocytes for *in vitro* studies [37; pp. 265–354]. Technical aspects of culturing hepatocytes are presented in a recent monograph: methods for preparing primary monolayer cultures of postnatal rat liver cells to assess xenobiotic hepatotoxicity [94], the preparation of human hepatocyte cultures [152], and the culturing of hepatocytes from various laboratory species [243]. Factors such as matrix used for cell attachment, culture medium, addition of hormones, oxygen tension, cell density, and the presence of other cell types determine enzyme and gene expression and affect cell viability in monolayer cultures of hepatocytes [243].

Biochemical parameters used to measure liver injury in isolated cells also apply to hepatocyte cultures. $CCl_4$-induced biochemical changes in cultured hepatocytes followed nearly the same continuum as observed *in vivo*, although the progression was much more rapid *in vitro* [233]; leakage of ALT was increased and G-6-Pase activity was decreased in intact liver and cultured cells, whereas 5′-nucleotidase activity was unaffected in either preparation. In hepatocyte cultures incubated with tetracycline or norethindrone, Anuforo et al. [12] found that the leakage into the medium of arginosuccinate lyase, ALT, or LDH was more pronounced than that for AP and AST. Chao et al. [68] reported that intracellular LDH content is a better indicator of the number of viable hepatocytes in contrast to LDH released into the medium.

Cultures of viable hepatocytes can be maintained for several days. Xenobiotic biotransformation enzyme activities decrease, however, after one or two days in culture [153,244,328]. For this reason, the assessment of cytotoxicity in hepatocyte cultures is routinely performed during the first two days of culturing; however, hepatocytes cultured in hormone-supplemented, serum-free medium maintain high biotransformation enzyme activities for several days [99,156,244]. Long-term cultures of functional hepatocytes can be achieved by coculturing hepatocytes with another rat liver cell type and represent a promising tool for investigating chronic liver toxicity [156]. Human hepatocytes, particularly when mixed with rat liver epithelial cells, may provide a valuable tool for predicting the hepatotoxicity of new drugs in humans [154]. Finally, evidence for the reconstruction of functionally intact biliary polarity in hepatocytes in culture indicates the possible utility of this system for studies on intrahepatic cholestasis [135].

Toxicity studies involving cultured hepatocytes usually assess three types of effects: cytotoxicity, genotoxicity, and enzyme induction [224]. For cytotoxicity, one can follow morphology, dye permeation (trypan blue), and the release of cytoplasmic enzymes (ALT, AST, LDH). Promutagen activation and induction of unscheduled DNA synthesis can be used to assess genotoxicity. Finally, enzyme induction effects can be discerned by following peroxisomal induction or cytochrome P450 induction. Cultured hepatocytes are considered useful for the study of peroxisome proliferation, as the cells can be used for screening new chemicals, for evaluating the sequence of events involved in their mechanism of action, and for assessing the potential for producing such effects in humans [123].

## Liver Slices

Liver slices maintained for short periods of time have been used with success in biochemical and pharmacological research for over 50 years. With regard to hepatotoxicity, Weldon et al. [359] reported in 1965 a clear dose-dependent reduction in the metabolism of L-leucine by liver slices incubated with $CCl_4$. Smuckler [331] subsequently observed a marked reduction in amino acid incorporation into proteins by liver slices from animals that had received $CCl_4$, as well as in control slices incubated with $CCl_4$. Marsh and Bizzi [239] concluded that the rat liver slice was a valid and useful system for the study of the action of drugs on lipid and lipoprotein metabolism. In a discussion of techniques for studying drug metabolism *in vitro*, Gillette [139] presented the liver slice system in the following terms:

> Since the circulation of nutrients and oxygen through the capillaries is lost, the transfer of these substances from the medium to the innermost cells must depend on passive diffusion. It is therefore imperative that the slices be thin because the rate of diffusion is inversely related to the square of the distance the substances must traverse. Even with slices as thin as 0.5 mm, the concentrations of the nutrients and other substrates which are rapidly utilized by cells in the slice may not be identical throughout the slice.

In recent years, technological improvements in the preparation of precision-cut rat liver slices have allowed this system to assess liver injury in a reproducible fashion. Precision-cut liver slice technology was pioneered at the University of Arizona. Brendel et al. [52] reviewed the tissue slice technology. Their dynamic organ culture system allows both the upper and lower surfaces of the cultured slice to be exposed to the gas phase during the course of incubation to overcome disintegration of the slice-medium interface in long-term incubation of tissue slices. One advantage the liver slice system offers over isolated or cultured hepatocytes is the fact that the tissue architecture is preserved. The liver slice procedure also appears to be more efficient for the collection of serial data than the perfused liver preparation.

As with isolated hepatocytes, the parameters measured to evaluate liver injury include LDH leakage, decreased protein synthesis, decreased intracellular potassium content, and histological alterations. Pig and human liver slices have also been cryopreserved with success and used for future toxicological and metabolic studies [120]. High viability of rat liver slices has been observed when tissues were rapidly frozen after preincubation with 18% $Me_2SO$ or VS4 (a 7.5-*M* mixture of $Me_2SO$, 1,2-propanediol, and formanide with a weight ratio of 21.5:15:2.4) [95]. Fisher et al. [118] observed that three dichlorobenzene isomers were more toxic to human liver slices when the incubation medium was Krebs–Henseleit buffer compared to Waymouth's medium.

Two different tissue slicers (Krumdieck/Alabama Research and Development Tissue Slicer; Brendel/Vitron Slicer) have been described [133,271]. Both slicers appear to readily produce liver slices that are adequate for biotransformation or toxicity studies [290]. Thickness of the liver slice is very critical for slice viability, the optimum being 200 to 250 μm. The slicing buffer medium should be similar in inorganic composition to the culture medium used in liver cultures; HEPES is recommended. The slices are pooled and transferred to culture vessels (stainless steel screens located in cylindrical Teflon® cradles, loaded horizontally in glass scintillation vials containing culture medium to wet the screen from the underside). Waymouth's medium is recommended as the culture medium, and a viability of 120 hours is claimed for liver slices. A roller culture system is claimed to be most effective for the extended viability of precision-cut tissue slices. The parameters for assessing viability of liver slices originally described by the Arizona group include: enzyme (LDH) leakage, intracellular potassium content, nonprotein sulfhydril content, lipid peroxidation, protein synthesis, gluconeogenesis, phase I and II biotransformation, and histological evaluation. Olinga et al. [266] compared different methods (magnetically stirred 24-well, fully immersed system; shaker-stirred, six-well, fully immersed system; rocker platform, six-well culture plate system; roller incubation-vial, partially immersed system) for incubation of liver slices for periods of 1.5 to 24.5 hours; in terms of cell viability, the stirred systems generally fared the worst, whereas the rocker and roller methods appeared to be superior. In another study [346], where precision-cut liver slices were incubated in either a shaking platform or a roller incubation system for 72 hours, the roller system was reported as the better one for preserving histological and functional integrity of the slices.

Behrsing et al. [32] took advantage of refinements in procedure that prevent necrosis at the contact point with the support mesh to demonstrate that rat liver slices can retain their integrity and viability for at least 10 days in culture. Refinements included: (1) flushing the liver and slicing the cores in cold UW (University of Wisconsin) solution, (2) loading the slices onto cellulose-ester filters placed on the mesh inside the inserts in a dynamic roller culture system rather than directly in contact with the titanium mesh or Teflon® screen, and (3) culturing in a defined medium based on Waymouth's MB 752/1 under a high $O_2$ content atmosphere. Geldanamycin, a known hepatobiliary toxicant, produced biochemical and morphological deficits in both hepatocytes and biliary cells in this model system, consistent with its target *in vivo*. Subsequent studies showed that in rat liver slices it is possible to detect compound-, concentration-, and cell-specific toxicities in response to geldanamycin and 17-allylaminogeldanamycin [31] and that species-related differences in toxicity seen *in vivo* are replicated using dog liver slices [9].

Assessing the relative toxicities of bromobenzene, *o*-bromobenzonitrile, *o*-bromobenzene, *o*-bromoanisole, *o*-bromotoluene, and *o*-bromo-benzomethyltrifluoride, Fisher et al. [119] concluded that results obtained with precision-cut liver slices appear to be more representative of those observed *in vivo* than do results obtained from isolated hepatocytes. Wormser et al. [367] observed species and age differences in toxicity and reported that the relative toxicities of various chemicals in the liver slice system were similar to those reported *in vivo*. The hepatotoxic properties of $CCl_4$ and $CHCl_3$ in precision-cut liver slices have suggested that this system is a useful tool for the investigation of site-specific toxicants [19,20]. In rats, bromobenzene produces centrilobular necrosis *in vivo*, whereas allyl alcohol produces periportal lesions. In liver slices, centrilobular lesions following bromobenzene exposure were observed, but the expected periportal lesion with allyl alcohol was not observed [133]. The lack of the expected site-specific response with allyl alcohol is attributed to the absence of circulatory influences in the tissue slice system. Indications of oxidative stress (GSH/GSSG ratio), as a mechanism of action to explain the hepatotoxic properties of atractyloside, were reported in studies using precision-cut liver slices obtained from rats and domestic pigs [262]. The hepatotoxic properties of coumarin, menadione, and allyl alcohol on protein synthesis, potassium content, and mitochondrial dimethylthiazoldiphenyl tetrazolium-formazan production (MTT assay) were assessed in liver slices prepared from different species (rats, guinea pigs, Cynomolgus monkeys, and humans) and incubated for 24-hour periods [291]. Menadione toxicity was evident in all four species, while allyl alcohol toxicity was less evident in rats; coumarin concentration-dependent toxicity was evident in guinea pigs and rats but less effective in monkeys or humans. These interesting species-dependent observations demonstrate the utility of the liver slice technique for evaluating species differences in chemical-induced hepatotoxicity.

The application of the precision-cut slice technique to the biotransformation of xenobiotics in liver, including the coupling of phase I and phase II metabolic pathways, has been extensive [89,133,235,271,344]. The utility of cultured human liver slices for assessing the effects of chemicals on P450 enzymes has been demonstrated [111]. Liver slices from several species (rats, dogs, guinea pigs, and humans) have been employed with success. Liver slices also were shown to respond to several known different metabolic situations: peroxisome proliferation, hormone-regulated glucose metabolism, unscheduled DNA synthesis, and formation of neoantigens after halothane exposure. Other applications of the technique have been demonstrated. The live-time evaluation of cell toxicity by following production of specific fluorescent dyes in liver slices by confocal microscopy was studied after exposing precision-cut slices to 1,1-dichloroethylene and then examining them microscopically; very few dead cells were observed at 2 hours, but progressively more were seen after 3 to 7 hours [86].

Apoptosis (detected by DNA fragmentation, amplification of the Bax gene, and histology) in liver slices was assessed after incubating slices in the presence of staurosporine, a protein kinase C inhibitor [75]; features of apoptosis were successfully detected even in the presence of necrosis. Histopathological changes, DNA fragmentation, caspase-3 activation, and the release of cytochrome *c* induced by thioacetamide in both rat liver slices and *in vivo* treated rats indicated that precision-cut liver slices provide a valuable system for studying chemical-induced apoptosis [255]. Chemical–DNA adduct formation has been observed in benzo(*a*)pyrene-exposed rat liver slices [163]. Precision-cut liver slices from two transgenic mouse strains, one of which couples the promoting region of CYP 1A1 to β-galactosidase and another that couples two forward and two backward 12-*O*-tetradecanoyl phorbol-13-acetate (TPA) repeat elements (TREs) to luciferase (termed AP-1/luciferase), showed promotional effects in the presence of β-naphthoflavone and TPA, respectively, indicating that precision-cut livers slices from transgenic mice offer a novel *in vitro* method for toxicity evaluation while maintaining normal cell heterogeneity [64]. Precision-cut liver slices from neonatal rats showed fewer morphological alterations and increased viability compared to adult rats and excellent induction of cytochrome P450 isoforms [236]. Finally, precision-cut liver slices have been reported to test specificity and efficiency of gene transfer as well as of viral replication using human tissue [194].

The diverse utility of the precision-cut liver slice technique for evaluating liver function and injury has become well established since its original introduction in 1985 by Smith et al. [330]. Furthermore, it is readily adaptable to the evaluation of human tissue. One advantage the system offers over isolated or cultured hepatocytes is the fact that tissue architecture is preserved and multiple analyses coupled with morphological examination are possible. Thus, detailed mechanistic toxicological investigations seem achievable *in vitro* with the technique.

## HISTOLOGICAL ANALYSIS OF LIVER INJURY

Analysis of the hepatotoxic potential of a chemical agent is incomplete without a histological description of the lesion produced. The characteristic hepatic lesions defined by light microscopy are mentioned above. The reader is referred to Zimmerman [374], Newberne [259], and Kanel [188] for a more detailed discussion of the various expressions of hepatotoxicity as observed by light microscopy. Quantification of the degree of injury observed by light microscopy can be achieved using the method of Chalkley [67] essentially as described by Mitchell et al. [250]. One ocular of a microscope is fitted with a micrometer eyepiece containing a grid on which 16 points of reference are chosen. A section of suitably stained liver tissue is selected from each animal and examined at 400× magnification. In a study of acetaminophen hepatotoxicity in mice, Mitchell et al. [250] found that a single section of liver could be considered to be representative of the entire organ. Each section is evaluated by scanning a series of 25 microscopic fields chosen at random. In each field, the tissue element immediately underneath each of the 16 points of reference is termed a "hit"; thus, 16 hits are examined per field, and 400 hits are examined in each section. The hits are categorized as: (1) normal parenchymal hepatocyte, (2) degenerated parenchymal hepatocyte, (3) necrotic parenchymal hepatocyte, and (4) other cellular structure [167]. Hits on each of the first three categories are recorded and expressed as a percentage of the total number of hepatocytes examined in that section. Accumulation of the data from three to five animals in each treatment group provides a base of sufficient size for statistical analysis.

This type of quantitative histological analysis was useful for determining the relative ability of various solvents to potentiate $CHCl_3$-induced hepatotoxicity (Table 29.4). In this study [167], rats were pretreated with an equimolar dose of *n*-hexane (H), acetone (A), 2,5-hexanedione (2,5-HD), or methyl *n*-butyl ketone (MBK); 18 hours later the rats received a small challenging dose of $CHCl_3$, calculated to produce minimal signs of liver injury. Hepatic damage was assessed 24 hours after $CHCl_3$ administration.

Although each of the solvents was capable of potentiating the hepatotoxic effects of $CHCl_3$, it is clear (Table 29.4) that a marked difference exists in the severity of the potentiation produced. Normal hepatocytes accounted for approximately 88, 75, 52, and 45% of the total of $CHCl_3$-challenged rats pretreated with H, A, 2,5-HD, or MBK, respectively. Degenerated hepatocytes accounted for approximately 6% of the total in rats receiving the combination of H + $CHCl_3$ and rose to approximately 18% in rats treated with MBK + $CHCl_3$. The percentage of necrotic hepatocytes was greatest in rats treated with MBK + $CHCl_3$ (37%) and decreased in the following order: 2,5-HD + $CHCl_3$ (35%), A + $CHCl_3$ (15%), and H + $CHCl_3$ (6%).

**TABLE 29.4**
**Histological Evaluation of the Effects of Pretreatment with Various Agents on $CHCl_3$-Induced Hepatotoxicity in Male Rats[a]**

| Treatment | Normal Hepatocytes (%) | Degenerated Hepatocytes (%) | Necrotic Hepatocytes (%) | ALT Activity (units/mL) | Total Bilirubin (mg/dL) |
|---|---|---|---|---|---|
| $CHCl_3$ | 99.7 ± 0.1 | 0.3 ± 0.1 | 0 | 37 ± 3 | 0.18 ± 0.01 |
| *n*-Hexane + $CHCl_3$ | 97.9 ± 2.3 | 5.8 ± 0.9 | 6.3 ± 2.2 | 347 ± 100 | 0.24 ± 0.01 |
| Acetone + $CHCl_3$ | 76.2 ± 4.7 | 9.2 ± 2.1 | 14.6 ± 3.0 | 1177 ± 534 | 0.26 ± 0.02 |
| 2,5-Hexandedione + $CHCl_3$ | 51.9 ± 5.0 | 13.5 ± 2.6 | 34.7 ± 3.1 | 2228 ± 477 | 0.82 ± 0.32 |
| Methyl *n*-butyl ketone + $CHCl_3$ | 45.2 ± 2.0 | 17.5 ± 2.0 | 27.2 ± 1.8 | 4910 ± 631 | 1.35 ± 0.17 |

[a] $CHCl_3$ (0.5 mL/kg i.p.) given 18 hours after agent (15 mmol/kg p.o.); rats killed 24 hours later. Values are mean ± SE of 4 to 15 rats.

*Source:* Data from Hewitt, W.R. et al., *Toxicol. Appl. Pharmacol.*, 53, 230–248, 1980.

These results indicate that this method for quantifying histological alterations provides an index of toxicity sensitive enough to discern the varying potentiating capacities of the four solvents tested. It is also noteworthy that the use of histological criteria to rank these solvents in order of increasing potentiating ability (H < A < 2,5-HD MBK) provided results similar to those obtained via determination of serum ALT (Table 29.4) and serum OCT [167] activity. Furthermore, the quantitative histological analysis provided a greater degree of discrimination between the solvents tested than did determination of total plasma bilirubin content (Table 29.4). In general, this procedure for quantifying histological abnormalities correlates well with other indices of liver injury [285]. Examples of the correlation between histopathological alterations and changes in functional indices of liver injury are given in Table 29.5. The severity of the hepatic lesion, expressed as the percentage of degenerated hepatocytes, the percent-

**TABLE 29.5**
**Linear Regression Analysis of the Relation Between Histological Evaluation of Severity of Liver Injury and Alterations in Various Parameters of Hepatic Damage**

| Correlation Coefficient (*r*) | *x*-Axis (% Hepatocytes) | *y*-Axis (Parameter) | Regression Line ($y = m(x) + (b)$) | Points/Line |
|---|---|---|---|---|
| 0.959 | Abnormal | log(ALT activity) | $y = 0.0397(x) + (1.5538)$ | 49 |
| 0.950 | Necrotic | log(ALT activity) | $y = 0.0566(x) + (1.5881)$ | 49 |
| 0.926 | Degenerated | log(ALT activity) | $y = 0.1186(x) + (1.5386)$ | 49 |
| 0.922 | Necrotic | ALT activity | $y = 99(x) + (-74)$ | 49 |
| 0.903 | Abnormal | ALT activity | $y = 67(x) + (-106)$ | 49 |
| 0.885 | Abnormal | log(OCT activity + 1) | $y = 0.0550(x) + (0.5728)$ | 47 |
| 0.879 | Degenerated | log(OCT activity + 1) | $y = 0.1691(x) + (0.5281)$ | 47 |
| 0.865 | Necrotic | log(OCT activity + 1) | $y = 0.0773(x) + (0.6331)$ | 47 |
| 0.830 | Necrotic | Bilirubin concentration | $y = 0.0025(x) + (0.15)$ | 49 |
| 0.820 | Abnormal | Bilirubin concentration | $y = 0.017(x) + (0.14)$ | 49 |
| 0.816 | Degenerated | OCT activity | $y = 116(x) + (-64)$ | 47 |
| 0.813 | Degenerated | ALT activity | $y = 188(x) + (-69)$ | 49 |
| 0.799 | Abnormal | OCT activity | $y = 37(x) + (19)$ | 47 |
| 0.770 | Necrotic | OCT activity | $y = 51(x) + (27)$ | 47 |
| 0.755 | Degenerated | Bilirubin concentration | $y = 0.0.49(x) + (0.14)$ | 49 |
| 0.650 | Abnormal | Relative liver weight[a] | $y = 0.016(x) + (4.22)$ | 49 |
| 0.645 | Necrotic | Relative liver weight | $y = 0.023(x) + (4.23)$ | 49 |
| 0.628 | Degenerated | Relative liver weight | $y = 0.049(x) + (4.21)$ | 49 |

[a] Relative liver weight = liver weight/body weight × 100.

*Source:* Data from Plaa, G.L. and Hewitt, W.R., in *Toxicology of the Liver*, Plaa, G.L. and Hewitt, W.R., Eds., Raven Press, New York, 1982, pp. 103–120.

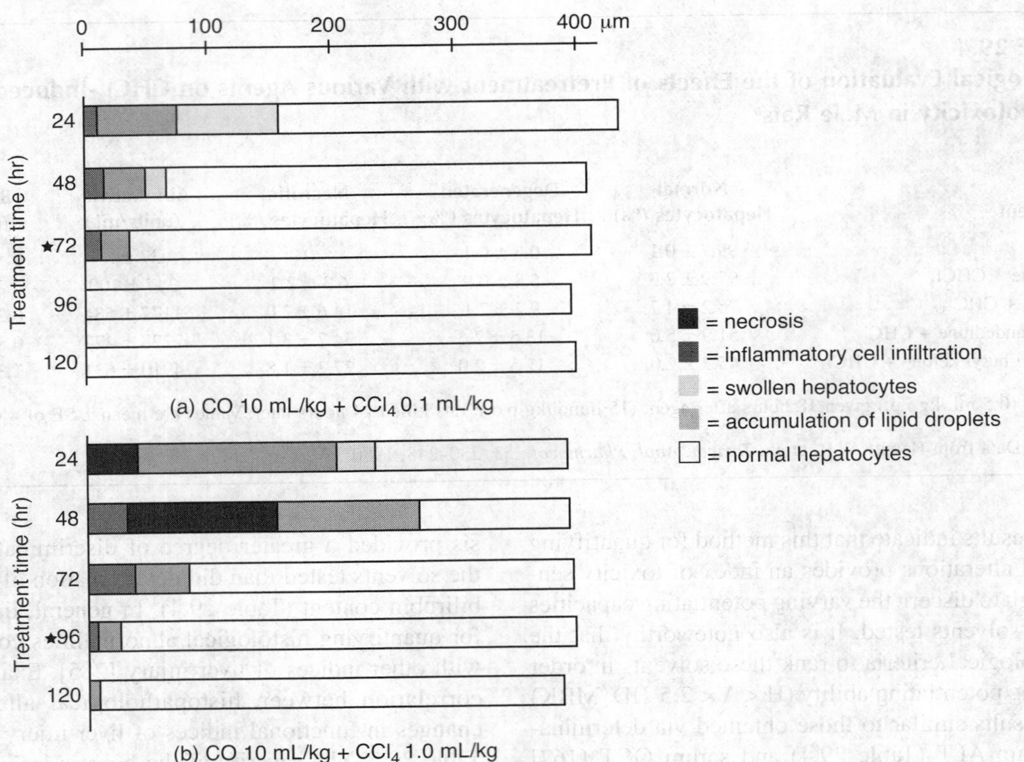


**FIGURE 29.6** Lobular morphologic patterns of $CCl_4$-induced liver injury for a 120-hour posttreatment period, according to the technique of Iijima et al. [176]. The left side (0 μm) of the figure corresponds to the centrilobular region, whereas the right side (400 μm) corresponds to the periportal space of the hepatic lobule. (From Charbonneau, M. et al., *Toxicology*, 35, 95–112, 1985. With permission.)

age of necrotic hepatocytes, or the percentage of abnormal hepatocytes (necrotic plus degenerated) is compared to alterations in ALT and OCT or to the plasma content of bilirubin. Regardless of the parameters assessed, a linear correlation is observed between the extent of the lesion as quantified by light microscopy and the severity of the biochemical alteration. Marked differences, however, can be observed in the strength of correlation between the different combinations of parameters examined. In general, elevations in the serum activity of ALT are most strongly correlated with the histopathological alterations. The correlations between the severity of the lesion and alterations in relative liver weight, however, are not strong.

Minor modifications of this method have been used. Mitchell et al. [250] used an eyepiece containing 8 points of reference and examined 50 random microscopic fields to collect 400 hits. It appears that the arrangement and number of reference points examined per field are of little consequence as long as a sufficient number of hits are collected for analysis. In one study [166], 30 points of reference per field and 50 fields were examined for a total of 1500 hits per section. However, H. Miyajima (unpublished observations) found no statistical difference between the results obtained by examining 1500 vs. 400 hits per section. Thus, for routine usage, it appears that the collection of 400 hits per section is satisfactory.

The quantitative method described above does not allow one to visualize the lobular distribution relative to the zonal configuration of the hepatic lobule. Iijima et al [176] devised a semiquantitative morphologic method that permits such visualization. To evaluate the morphologic patterns, ten hexagonal lobules are chosen randomly for each liver section. The distance of the injured area from the hepatocytes adjacent the terminal hepatic venule (THV; central vein) to the portal area is measured in one fixed direction per lobule using a micrometer ocular disc (5 mm divided into 100 parts). This distance is measured at a magnification of 100×. The sections are then examined at a magnification of 400× to classify the cellular changes observed. The damage is classified using six categories: (1) necrosis; (2) ballooning of hepatocytes; (3) swelling of hepatocytes; (4) inflammatory cell infiltration; (5) presence of lipid droplets; and (6) normal hepatocytes. The results are expressed in absolute mean distances (micrometers) from the THV. The mean distance for each category observed in a treatment group is calculated for four to six animals (total of 40 to 60 hexagonal lobules). The graphic representation of such an analysis is depicted in Figure 29.6, where the lesions produced by two dosage of $CCl_4$ (0.1 and 1.0 mL/kg, i.p.) were monitored over a 120-hour period [69]. It is evident that the severity of the lesion and recovery were dose dependent. The major advantage of this semiquantitative morphologic procedure is that it permits the investigator to prepare a graphic representation of what is visualized after examination of many microscopic sections. It can be particularly useful for the preparation of toxicological reports.

Electron microscopy is also of value in toxicological studies, as it permits a correlation between the ultrastructural and functional changes induced by foreign chemicals. Grice [149] delineated several of the advantages and disadvantages encountered in the application of electron microscopic techniques to the study of chemically induced liver injury. In general, electron microscopy provides a much earlier demonstration of hepatocyte injury and is of value for detecting minimal and often reversible pathologic changes that may be evident before they are detectable by light microscopy. The ability to detect subtle ultrastructural defects early in the course of poisoning often permits identification of the initial site of the lesion and thus can provide clues to possible biochemical mechanisms involved in the pathogenesis of liver injury. In addition, the power of these techniques can be enhanced through a quantitative, morphometric analysis of chemical effects [45,96]. Serious restrictions involving proper fixation techniques, sampling procedures, and the complexity of sample preparation, however, argue against the routine use of electron microscopy in the initial evaluation of the hepatotoxic potential of a new chemical; rather, this technique is probably of greatest use for confirming a suspected alteration or defining a pathologic event [149].

## CONCLUDING REMARKS

It can be seen that, although techniques for the determination of chemically induced liver injury in laboratory animals are readily available, no single technique is satisfactory for the detection and quantification of all forms of injury. Instead, a battery of procedures consisting of one or more of the biochemical and functional techniques coupled with a histological analysis of the liver is essential for the correct evaluation of the hepatotoxic potential of a chemical agent.

## QUESTIONS

1. If you are interested in a general screen to assess the possible hepatotoxic properties of a new medicinal agent currently in development, what battery of tests would you perform? Defend your choices.
2. What specific tests would you use to assess if a new therapeutic agent possesses cholestatic properties? Defend your choices.
3. A new industrial chemical is suspected of potentially interacting with other solvents in the workplace and resulting in hepatotoxic responses in exposed workers. Design an exploratory experimental study in laboratory animals to assess this possibility.
4. How best can liver histological studies be integrated with biochemical evaluation of hepatic dysfunction?
5. A chemical agent is suspected of resulting in hepatic oxidative stress in rats. Design an experimental study in rats to assess this possibility.

## REFERENCES

1. Aiso, M., Takikawa, H., and Yamanaka, M. (2000): Biliary excretion of bile acids and organic anions in zone 1- and zone 3-injured rats. *Liver*, 20:38–44.
2. Akoume, M.-Y., Tuchweber, B., Plaa, G. L. and Yousef, I. M. (2004): The role of *mdr2* in manganese–bilirubin-induced cholestasis in mice. *Toxicol. Lett.*, 148:41–51.
3. Aleksunes, L. M., Slitt, A. M., Cherrington, N. J., Thibodeau M. S., Klaassen, C. D., and Manautou, J. E. (2004): Differential expression of mouse hepatic transporter genes in response to acetaminophen and carbon tetrachloride. *Toxicol. Sci.*, 83:44–52.
4. Allemand, H., Pessayre, D., Descatoire, V., Degott, C., Feldman, G., and Benhamou, J.-P. (1978): Metabolic activation of trichloroethylene into a chemically reactive metabolite toxic to the liver. *J. Pharmacol. Exp. Ther.*, 204: 714–723.
5. Allen, K. L. and Green, C. E. (1993): Isolation of human hepatocytes by biopsy perfusion methods. In: *In Vitro Biological Systems*, edited by C. A. Tyson and G. N. Frasier, pp. 262–270. Academic Press, New York.
6. Alvaro, D., Benedetti, A., Marucci, L., Delle Monache, M., Monterubbianesi, R., Di Cosimo, E., Perego, L., Maccarri, G., Glaser, S., Le Sage, G., and Alpini, G. (2000): The function of alkaline phosphatase in the liver: regulation of intrahepatic biliary epithelium secretory activities in the rat. *Hepatology*, 32:174–184.
7. Amacher, D. E. (2002): A toxicologist's guide to biomarkers of hepatic response. *Hum. Exp. Toxicol.*, 21:243–262.
8. Amador, E., Frany, R. J., and Massod, M. F. (1966): Serum glutamic-oxaloacetic transaminase activity: diagnostic accuracy of the revised spectrophotometric and the dinitrophenylhydrazine methods. *Clin. Chem.*, 12: 475–481.
9. Amin, K., Ip, C., Jimenez, L., Tyson, C., and Behrsing, H. (2005). *In vitro* detection of differential and cell-specific hepatobiliary toxicity induced by geldanamycin and 17-allylaminogeldanamycin using dog liver slices. *Toxicol. Sci.*, 87, 442–450.
10. Anand, S. S., Murthy, S. N., Vaidya, V. S., Mumtaz, M. M., and Mehendale, H. M. (2003): Tissue repair plays pivotal role in final outcome of liver injury following chloroform and allyl alcohol binary mixture. *Food Chem. Toxicol.*, 41:1123–1132.

11. Anderson, M. E., French, J. E., Gargas, M. L., Jones, R. A., and Jenkins, L. J. J. (1979): Saturable metabolism and the acute toxicity of 1,1-dichloroethylene. *Toxicol. Appl. Pharmacol.*, 47:385–393.
12. Anuforo, D. C., Acosta, D., and Smith, R. V. (1978): Hepatotoxicity studies with primary cultures of rat liver cells. *In Vitro*, 14:981–988.
13. Arends, M. J., Morris, R. G., and Wyllie, A. H. (1990): Apoptosis: the role of the endonuclease. *Am. J. Pathol.*, 136:593–608.
14. Asada, M. (1958): Transaminase activity in liver damage. 1. Study on experimental liver damage. *Med. J. Osaka Univ.*, 9:45–51.
15. Asada, M. and Galambos, R. J. (1963): Sorbitol dehydrogenase and hepatocelluar injury: an experimental and clinical study. *Gastroenterology*, 44:578–587.
16. Awad, J. A., Horn, J. L., Roberts, L. J., and Franks, J. J. (1996): Demonstration of halothane-induced hepatic lipid peroxidation in rats by quantification of F2-isoprotanes. *Anesthesiology*, 84:910–916.
17. Awad, J. A., Roberts, L. J., Burk, R. F., and Morrow, J. D. (1998): Isoprotanes, prostaglandin-like compounds formed *in vivo* independently of cyclooxygenase: use as clinical indicators of oxidant damage. *Gastroenterol. Clin. North Am.*, 25:409–427.
18. Ayotte, P. and Plaa, G. L. (1986): Modification of biliary tree permeability in rats treated with a manganese–bilirubin combination. *Toxicol. Appl. Pharmacol.*, 84:205–303.
19. Azri, S., Mata, H. P., Reid, L. L., Gandolfi, A. J., and Brendel, K. (1992): Further examination of the selective toxicity of $CCl_4$ in rat liver slices. *Toxicol. Appl. Pharmacol.*, 112:81–86.
20. Azri-Meehan, S., Mata, H. P., Gandolfi, A. J., and Brendel, K. (1992): The hepatotoxicity of chloroform in precision-cut rat liver slices. *Toxicology*, 73:239–250.
21. Bai, C. L., Canfield, P. J., and Stacey, N. H. (1992): Individual serum bile acids as early indicators of carbon tetrachloride- and chloroform-induced liver injury. *Toxicology*, 75:221–224.
22. Bajwa, R. S., and Fujimoto, J. M. (1983): Effect of colchicine and S,S,S-tributyl phosphorotrithioate (DEF) on the biliary excretion of sucrose, mannitol and horseradish peroxidase in the rat. *Biochem. Pharmacol.*, 32:85–90.
23. Balazs, T., Airth, J. M., and Grice, H. C. (1962): The use of serum glutamic pyruvic transaminase test for the evaluation of hepatic necrotropic compounds in rats. *Can. J. Biochem. Physiol.*, 40:1–6.
24. Balazs, T., Murray, R. K., McLaughlan, J. M., and Grice, H. C. (1961): Hepatic tests in toxicity studies on rats. *Toxicol. Appl. Pharmacol.*, 3:71–79.
25. Ballatori, N. and Truong, A. T. (1992): Glutathione as a primary osmotic driving force in hepatic bile formation. *Am. J. Physiol.*, 263:G617–G624.
26. Ballet, F. and Thurman, R. G. (1991). *Research in Perfused Liver.* INSERM/John Libbey, Paris.
27. Bantel, H., Ruck, P., Gregor, M., and Schulze-Osthoff, K. (2001): Detection of elevated caspase activation and early apoptosis in liver diseases. *Eur. J. Cell Biol.*, 80: 230–239.
28. Bashir, S., Sharma, Y., Irshad, M., Nag, T. C., Tiwari, M., Kabra, M., and Dogra, T. D. (2006): Arsenic induced apoptosis in rat liver following repeated 60 days exposure. *Toxicology*, 217:63–70.
29. Batsakis, J. G., Kremers, B. J., Thiessen, M. M., and Shilling, J. M. (1968): Biliary tract enzymology: a clinical comparison of serum alkaline phosphatase, leucine aminopeptidase, and 5′-nucleotidase. *Am. J. Clin. Pathol.*, 50:485–490.
30. Becker, B. A. and Plaa, G. L. (1965): Quantitative and temporal delineation of various parameters of liver dysfunction due to α-naphthylisothiocyanate. *Toxicol. Appl. Pharmacol.*, 7:708–718.
31. Behrsing, H. P., Amin, K., Ip, C., Jimenez, L., and Tyson, C. A. (2005): *In vitro* detection of differential and cell-specific hepatobiliary toxicity induced by geldanamycin and 17-allylaminogeldanamycin in rats. *Toxicol. In Vitro*, 19:1079–88.
32. Behrsing, H. P., Vickers, A. E., and Tyson, C. A. (2003): Extended rat liver slice survival and stability monitored using clinical biomarkers. *Biochem. Biophys. Res. Commun.*, 31:209–213.
33. Benedetti, A., Brunelli, E., Risicate, R., Cilluffo, T., Jezequel, A.-M., and Orlandi, F. (1988): Subcellular changes and apoptosis induced by ethanol in rat liver. *J. Hepatol.*, 6:137–143.
34. Benedetti, A., Casini, A. F., Feralli, M., and Comporti, M. (1977): Studies on the relationships between carbon tetrachloride-induced alterations of liver microsomal lipids and impairment of glucose-6-phosphatase activity. *Exp. Mol. Pathol.*, 27:309–323.
35. Benedetti, A. A., Jezequel, A.-M., and Orlandi, F. (1988): Preferential distribution of apoptotic bodies in acinar zone 3 or normal human and rat liver. *J. Hepatol.*, 7:319–24.
36. Benz, C., Angermüller, S., Töx, U., Klöters-Plachky, P., Riedel, H. D., Sauer, P., Stremmel, W., and Stiehl, A. (1998): Effect of tauroursodeoxycholic acid on bile-acid-induced apoptosis and cytolysis in rat hepatocytes. *J. Hepatol.*, 28:99–106.
37. Berry, M. N., Edwards, A. M., and Barritt, G. J. (1991): *Isolated Hepatocytes: Preparation, Properties, and Applications*. Elsevier, New York.
38. Berry, M. N. and Friend, D. S. (1969): High-yield preparation of isolated liver parenchymal cells: a biochemical and fine structural study. *J. Cell Biol.*, 43:506–520.
39. Berry, M. N., Halls, H. J., and Grivell, M. B. (1992): Techniques for pharmacological and toxicological studies with isolated hepatocyte suspensions. *Life Sci.*, 51:1–16.
40. Berthelot, P. and Billing, B. H. (1966): Effect of bunamiodyl on hepatic uptake of sulfobromophthalein in the rat. *Am. J. Physiol.*, 211:395–399.
41. Berthou, F., Ratanasavanh, D., Riche, C., Picart, D., Voirin T., and Guillouzo, A. (1989): Comparison of caffeine metabolism by slices, microsomes and hepatocyte cultures from adult human liver. *Xenobiotica*, 19:401–417.
42. Bertone, G. and Dianzani, M. U. (1977): Inhibition by aldehydes as a possible further mechanism for glucose-6 phosphatase inactivation during $CCl_4$ poisoning. *Chem Biol. Interact.*, 19:91–100.

43. Bock, K. W. and Kohle, C. (2005): Ah receptor- and TCDD-mediated liver tumor promotion: clonal selection and expansion of cells evading growth arrest and apoptosis. *Biochem. Pharmacol.*, 69:1403–1408.
44. Bohnenberger, S., Wagner, B., Schmitz, H. J., and Schrenk, D. (2001): Inhibition of apoptosis in rat hepatocytes treated with 'non-dioxin-like' polychlorinated biphenyls. *Carcinogenesis*, 22:1601–1606.
45. Bolender, R. P. (1978): Morphometric analysis in the assessment of the response of the liver to drugs. *Pharmocol. Rev.*, 30:429–443.
46. Bouchard, G., Tuchweber, B., and Yousef, I. M. (2000): Bile salt independent flow during bile salt-induced choleresis and cholestasis in the rat: role of biliary thiol secretion. *Liver*, 20:27–37.
47. Boyer, J. L. (1997): Isolated hepatocyte couplets and bile duct units: novel preparations for the *in vitro* study of bile secretory function. *Cell Biol. Toxicol.*, 13:289–300.
48. Boyland, E. and Grover, P. L. (1967): The relationship between hepatic glutathione conjugation and BSP excretion and the effect of therapeutic agents. *Clin. Chim. Acta*, 16:205–213.
49. Brauer, R. W., Krebs, J. S., and Pessotti, R. L. (1950): Bromosulfophthalein as a tool for study of liver physiology. *Fed. Proc.*, 9:259.
50. Brauer, R. W., Pessotti, R. L., and Krebs, J. S. (1955): The distribution and excretion of $S^{35}$-labeled sulfobromophthalein sodium administered to dogs by continuous infusion. *J. Clin. Invest.*, 34:35–43.
51. Brauer, R. W. and Root, M. A. (1946): The effect of carbon tetrachloride induced liver injury upon the acetylcholine hydrolyzing activity of blood plasma of the rat. *J. Pharmacol. Exp. Ther.*, 88:109–118.
52. Brendel, K., Fisher, R. L., Krumdieck, C. L., and Gandolfi, A. J. (1993): Precision-cut rat liver slices in dynamic organ culture for structure-toxicity studies. In: *In Vitro Biological Systems*, edited by C. A. Tyson and G. N. Frasier, pp. 222–230. Academic Press, New York.
53. Brown, B. R., Sipes, I. G., and Sagalyn, A. M. (1974): Mechanisms of acute hepatic toxicity: chloroform, halothane and glutathione. *Anesthesiology*, 41:554–561.
54. Brown, D. J. and Hunter, A. (1984): The effect of thioacetamide on sulfobromophthalein and ouabain transport in isolated rat hepatocytes. *Toxicology*, 32:165–176.
55. Buege, J. A. and Aust, S. D. (1978): Microsomal lipid peroxidation. *Methods Enzymol.*, 53:302–310.
56. Bulle, F., Mavier, P., Zafrani, E. S., Preaux, A.-M., Lescs, M.-C., Siegrist, S., Dhumeaux, D., and Guellaen, G. (1990): Mechanism of gamma-glutamyl transpeptidase release in serum during intrahepatic and extrahepatic cholestasis in the rat: a histochemical, biochemical and molecular approach. *Hepatology*, 11:545–550.
57. Bunyan, J., Murrell, E. A., Green, J., and Diplock, A. T. (1969): On the existence and significance of lipid peroxides in vitamin E-deficient animals. *Br. J. Nutr.*, 21: 475–495.
58. Burk, R. F. and Lane, J. M. (1979): Ethane production and liver necrosis in rats after administration of drugs and other chemicals. *Toxicol. Appl. Pharmacol.*, 50:467–478.
59. Bursch, W., Chabicovsky, M., Wastl, U., Grasl-Kraupp, B., Bukowska, K., Taper, H., and Schulte-Hermann, R. (2005): Apoptosis in stages of mouse hepatocarcinogenesis: failure to counterbalance cell proliferation and to account for strain differences in tumor susceptibility. *Toxicol. Sci.*, 85: 515–529.
60. Bursch, W., Paffe, S., Putz, B., Barthel, G., and Schulte-Hermann, R. (1990): Determination of the length of the histological stages of apoptosis in normal liver and in altered hepatic foci of rats. *Carcinogenesis*, 11:847–853.
61. Butler, W. N., Maling, H. M., Horning, M. G., and Brodie, B. B. (1961): The direct determination of liver triglycerides. *J. Lipid Res.*, 2:95–96.
62. Calabrese, E. J. and Mehendale, H. M. (1996): A review of the role of tissue repair as an adaptive strategy: why low doses are often non-toxic and why high doses can be fatal. *Food Chem. Toxicol.*, 34:301–311.
63. Cameron, G. R. and Karunaratne, W. A. E. (1936): Carbon tetrachloride cirrhosis in relation to liver regeneration. *J. Pathol. Bacteriol.*, 92:1–21.
64. Catania, J. M., Parrish, A. R., Kirkpatrick, D. S., Chitkara, M., Bowden, G. T., Henderson, C. J., Wolf, C. R., Clark, A. J., Brendel, K., Fisher, R. L., and Gandolfi, A. J. (2003): Precision-cut tissue slices from transgenic mice as an *in vitro* toxicology system. *Toxicol. In Vitro*, 17:201–205.
65. Ceriotti, G. and Gazzaniga, A. (1966): A sensitive method for serum ornithine carbamyltransferase determination. *Clin. Chim. Acta*, 14:57–62.
66. Ceriotti, G. and Gazzaniga, A. (1967): Accelerated micro and ultramicro procedure for ornithine carbamyltransferase (OCT) determination. *Clin. Chim. Acta*, 16: 436–439.
67. Chalkley, H. W. (1943): Method for the quantitative morphologic analysis of tissues. *J. Natl. Cancer Inst.*, 4:47–53.
68. Chao, E. S., Dunbar, D., and Kaminsky, L. S. (1988): Intracellular lactate dehydrogenase concentration as an index of cytotoxicity in rat hepatocyte primary culture. *Cell Biol. Toxicol.*, 4:1–11.
69. Charbonneau, M., Iijima, M., Côté, M. G., and Plaa, G. L. (1985): Temporal analysis of rat liver injury following potentiation of carbon tetrachloride hepatotoxicity with ketonic or ketogenic compounds. *Toxicology*, 35: 95–112.
70. Charbonneau, M., Tuchweber, B., and Plaa, G. L. (1986): Acetone potentiation of chronic liver injury induced by repetitive administration of carbon tetrachloride. *Hepatology*, 6:694–700.
71. Cherrick, G. R., Stein, S. W., Leevy, C. M., and Davidson, C. S. (1960): Indocyanine green: observations on its physical properties, plasma decay and hepatic extraction. *J. Clin. Invest.*, 39:592–600.
72. Chesne, C., Guyomard, C., Grislain, L., Cleerc, C., Fautrel, A., and Guillouzo, A. (1991): Use of cryopreserved animal and human hepatocytes for cytotoxicity studies. *Toxicol. In Vitro*, 5:479–482.
73. Chieco, P., Romagnoli, E., Aicardi, G., Suozzi, A., Forti, G. C., and Roda, A. (1997): Apoptosis induced in rat hepatocytes by *in vivo* exposure to taurochenodeoxycholate. *Histochem. J.*, 29:875–883.

74. Chio, K. S. and Tappel, A. L. (1969): Synthesis and characterization of the fluorescent products derived from malonaldehyde and amino acids. *Biochemistry*, 8:2821–2827.
75. Chitkara, M. K., Petrick, J. S., Parrish, A. R., and Gandolfi, A. J. (1999): Generation and detection of apoptosis in rat liver slices. *Toxicologist*, 48(1S):90.
76. Chung, Y. H., Kim, J. A., Song, B. C., Song, I. H., Koh, M. S., Lee, H. C., Yu, E., Lee, Y. S., and Suh, D. J. (2001): Isocitrate dehydrogenase as a marker of centrilobular hepatic necrosis in the experimental model of rats. *J. Gastroenterol. Hepatol.*, 16:328–332.
77. Cobb, L. A. (1965): Effects of reducing agents on indocyanine green dye. *Am. Heart J.*, 70:145–146.
78. Columbano, A. (1995): Cell death: current difficulties in discriminating apoptosis from necrosis in the context of pathological processes *in vivo*. *J. Cell. Biochem.*, 58:181–190.
79. Combes, B. (1965): The importance of conjugation with glutathione for sulfobromophthalein sodium (BSP) transfer from blood to bile. *J. Clin. Invest.*, 44:1214–1224.
80. Comporti, M. (1998): Lipid peroxidation as a mediator of chemical-induced hepatocyte death. In: *Toxicology of the Liver*, 2nd ed., edited by G. L. Plaa and W. R. Hewitt, pp. 221–257. Taylor & Francis, New York.
81. Corcoran, G. B., Fix, L., Jones, D. P., Treinen Moslen, M., Nicotera, P., Oberhammer, F. A., and Buttyan, R. (1994): Apoptosis: molecular control point in toxicity. *Toxicol. Appl. Pharmacol.*, 128:169–181.
82. Cornelius, C. E. (1963): Relation of body weight to hepatic glutamic pyruvic transaminase activity. *Nature*, 200: 580–581.
83. Cornelius, C. E., Bishop, J., Switzer, J., and Rhode, E. A. (1959): Serum and tissue transaminase activities in domestic animals. *Cornell Vet.*, 49:116–126.
84. Cornish, H. H. (1971): Problems posed by observations of serum enzyme changes in toxicology. *CRC Crit. Rev. Toxicol.*, 1:1–32.
85. Cornish, H. H., Barth, M. L., and Dodson, V. N. (1970): Isoenzyme profiles and protein patterns in specific organ damage. *Toxicol. Appl. Pharmacol.*, 16:411–423.
86. Cromey, D., Lantz, C., Parrish, A. R., and Gandolfi, A. J. (1999): Live-time evaluation of cell toxicity in precision-cut tissue slices using confocal microscopy. *Toxicologist*, 48(1S): 71.
87. Crocenzi, F. A., Mottino, A. D., Cao, J., Veggi, L. M., Sanchez Pozzi, E. J., Vore, M., Coleman, R., and Roma, M. G. (2003): Estradiol-17β-D-glucuronide induces endocytic internalization of Bsep in rats. *Am. J. Physiol. Gastrointest. Liver Physiol.*, 285:G449–G459.
88. Crocenzi, F. A., Mottino, A. D., Sánchez Pozzi, E. J., Pellegrino, J. M., Rodriquez Garay, E. A., Milkiewicz, P., Vore, M., Coleman, R., and Roma, M. G. (2003): Impaired localization and transport function of canalicular Bsep in taurolithcholate induced cholestasis in the rat. *Gut*, 52:1170–1177.
89. Cui, X., Thomas, A., Han, Y., Palamanda, J., Montgomery, D., White, R. E., Morrison, R. A., and Cheng, K. C. (2005): Quantitative PCR assay for cytochromes P450 2B and 3A induction in rat precision-cut liver slices: correlation study with induction *in vivo*. *J. Pharmacol. Toxicol. Methods*, 52:234–243.
90. Curtis, L. R., Williams, W. L., and Mehendale, H. M. (1979): Potentiation of the hepatotoxicity of carbon tetrachloride following preexposure to chlordecone (Kepone) in the male rat. *Toxicol. Appl. Pharmacol.*, 51:283–293.
91. Curtis, S. J., Moritz, M., and Snodgrass, P. J. (1972): Serum enzymes derived from liver cell fractions. I. The response to carbon tetrachloride intoxication in rats. *Gastroenterology*, 62:84–92.
92. Dahlström-King, L., Couture, J., Lamoureux, C., Vaillancourt, T., and Plaa, G. L. (1990): Dose-dependent cytotoxicity of chlorinated hydrocarbons in isolated rat hepatocytes. *Fundam. Appl. Pharmacol.*, 14:833–841.
93. Darzynkiewicz, Z. and Traganos, F. (1998): Measurement of apoptosis. In: *Apoptosis*, edited by M. Al-Rubeai, pp. 33–73. Springer-Verlag, New York.
94. Davilla, J. C. and Acosta, D. (1993): Preparation of primary monolayer cultures of postnatal rat liver cells for hepatotoxic assessment of xenobiotics. In: *In Vitro Biological Systems*, edited by C. A. Tyson and G. N. Frasier, pp. 244–254. Academic Press, New York.
95. de Graaf, I. A. and Koster, H. J. (2001): Water crystallization within rat precision-cut liver slices in relation to their viability. *Cryobiology*, 43:224–237.
96. de la Iglesia, F. A., Sturgess, J. M., and Feuer, G. (1982): New approaches for the assessment of hepatotoxicity by means of quantitative functional-morphological interrelationships. In: *Toxicology of the Liver*, edited by G. L. Plaa and W. R. Hewitt, pp. 47–102. Raven Press, New York.
97. Deaciuc, I. V., D'Souza, N. B., Burikhanov, R., Nasser, M. S., Voskresensky, I. V., De Villiers, W. J., and McClain, C. J. (2004): Alcohol, but not lipopolysaccharide-induced liver apoptosis, involves changes in intracellular compartmentalization of apoptotic regulators. *Alcohol. Clin. Exp. Res.*, 28:160–172.
98. Delprat, G. D. and Stowe, W. P. (1931): The rose bengal test for liver function. *J. Lab. Clin. Med.*, 16:923–925.
99. Dich, J., Vind, C., and Grunnet, N. (1988): Long-term culture of hepatocytes: effect of hormones on enzyme activities and metabolic capacity. *Hepatology*, 8:39–45.
100. Dillard, C. J., Dumelin, E. E., and Tappel, A. L. (1977): Effect of dietary vitamin E on expiration of pentane and ethane by the rat. *Lipids*, 12:109–114.
101. Dingell, J. F. and Heimburg, M. (1968): The effects of aliphatic halogenated hydrocarbons on hepatic drug metabolism. *Biochem. Pharmacol.*, 17:1269–1278.
102. Dive, C., Gregory, C. D., Phipps, D. J., Evans, D. L. Milner, A. E., and Wyllie, A., H. (1992): Analysis and discrimination of necrosis and apoptosis (programmed cell death) by multiparameter flow cytometry. *Biochim. Biosphys. Acta*, 1133:275–285.
103. Drotman, R. B. (1975): A study of kinetic parameters for the use of serum ornithine carbamoyltransferase as an index of liver damage. *Food Cosmet. Toxicol.*, 13:649–651
104. Drotman, R. B. and Lawhorn, G. T. (1978): Serum enzymes as indicators of chemically induced liver damage *Drug Chem. Toxicol.*, 1:163–171.
105. Dufour. D. R., Lott, J. A., Nolte, F. S., Gretch, D. R., Koff R. S., and Seeff, L. B. (2000): Diagnosis and monitoring of hepatic injury. I. Performance characteristics of laboratory tests. *Clin. Chem.*, 46:2027–2049.

106. Dufour. D. R., Lott, J. A., Nolte, F. S., Gretch, D. R., Koff, R. S., and Seeff, L. B. (2000): Diagnosis and monitoring of hepatic injury. II. Recommendations for use of laboratory tests in screening, diagnosis, and monitoring. *Clin. Chem.*, 46:2050–2068.
107. Dumelin, E. E. and Tappel, A. L. (1977): Hydrocarbon gases produced during *in vitro* peroxidation of polyunsaturated fatty acids and decomposition of preformed hydroperoxides. *Lipids*, 12:894–900.
108. Eckhardt, E. T. and Plaa, G. L. (1963): Role of biotransformation, biliary excretion and circulatory changes in chlorpromazine-induced sulfobromophthalein retention. *J. Pharmacol. Exp. Ther.*, 139:383–389.
109. Edinger, A. L. and Thompson, C. B. (2004): Death by design: apoptosis, necrosis and autophagy. *Curr. Opin. Cell Biol.*, 16:663–669.
110. Edwards, C. A. and O'Brien, W. D. J. (1980): Modified assay for the determination of hydroxyproline in tissue hydrolyzate. *Clin. Chim. Acta*, 104:161–167.
111. Edwards, R. J., Price, R. J., Watts, P. S., Renwick, A. B., Tredger, J. M., Boobis, A. R., and Lake, B. G. (2003): Induction of cytochrome P450 enzymes in cultured precision-cut human liver slices. *Drug Metab. Dispos.*, 31: 282–288.
112. Evans, C. D., List, G. R., Doles, A., McConnell, D. G., and Hoffman, R. L. (1967): Pentane from thermal decomposition of lipoxidase-derived products. *Lipids*, 2: 432–434.
113. Evans, C. D., List, G. R., Hoffman, R. L., and Moser, H. H. (1969): Edible oil quality as measured by thermal release of pentane. *J. Am. Oil Chem. Soc.*, 46:501–504.
114. Fariss, M. W., Brown, M. K., Schmitz, J. A., and Reed, D. J. (1985): Mechanism of chemical-induced toxicity. I. Use of a rapid centrifugation technique for the separation of viable and nonviable hepatocytes. *Toxicol. Appl. Pharmacol.*, 79:283–295.
115. Fausto, N. (2000): Liver regeneration. *J. Hepatol.*, 32(Suppl. 1):19–31.
116. Feuer, G., Golberg, L., and LePelley, J. R. (1965): Liver response tests. I. Exploratory studies on glucose-6-phosphatase and other liver enzymes. *Food Cosmet. Toxicol.*, 3:235–249.
117. Feldmann, G. (1997): Liver apoptosis. *J. Hepatol.*, 26: 1–11.
118. Fisher, R., Barr, J., Zukoski, C. F., Putnam, C. W., Sipes, I. G., Gandolfi, A. J., and Brendel, K. (1991): *In vitro* hepatotoxicity of three dichlorobenzene isomers in human liver slices. *Hum. Exp. Toxicol.*, 10:357–363.
119. Fisher, R., Hanzlik, R. P., Gandolfi, A. J., and Brendel, K. (1991): Toxicity of *ortho*-substituted bromobenzenes in rat liver slices: a comparison to isolated hepatocytes and the whole animal. *In Vitro Toxicol.*, 4:173–186.
120. Fisher, R., Putman, C. W., Koep, L. J., Sipes, I. G., Gandolfi, A. J., and Brendel, K. (1991): Cryopreservation of pig and human liver slices. *Cryobiology*, 28:131–142.
121. Fong, R. N., Gonzalez, B. P., Fuentealba, I. C., and Cherian, M. G. (2004): Role of tumor necrosis factor-alpha in the development of spontaneous hepatic toxicity in Long–Evans Cinnamon rats. *Toxicol. Appl. Pharmacol.*, 200:121–130.
122. Ford, E. J. H. (1965): Changes in the activity of ornithine carbamyltransferase (OCT) in the serum of cattle and sheep with hepatic lesions. *J. Comp. Pathol.*, 75: 299–308.
123. Foxworthy, P. S. and Eacho, P. I. (1994): Cultured hepatocytes for studies of peroxisome proliferation: methods and applications. *J. Pharmacol. Toxicol. Methods*, 31: 21–30.
124. Franco, G. (1991): New perspectives in biomonitoring liver function by means of serum bile acids: experimental and hypothetical biochemical basis. *Br. J. Indust. Med.*, 48:557–561.
125. Frankel, E. N., Nowakowska, J., and Evans, C. D. (1961): Formation of methyl azelaaldehydate on autoxidation of lipids. *J. Am. Oil Chem. Soc.*, 318:161–162.
126. Frazier, J. M. (1990): Multiple endpoints measurements to evaluate the intrinsic cellular toxicity of chemicals. *In Vitro Toxicol.*, 3:349–357.
127. Friedman, S. L. (1993): Isolation and culture of hepatic non parenchymal cells. In: *In Vitro Biological Systems*, edited by C. A. Tyson and G. N. Frasier, pp. 292–310. Academic Press, New York.
128. Fromenty, B. and Pessayre, D. (1995): Inhibition of mitochondrial beta-oxidation as a mechanism of hepatotoxicity. *Pharmacol. Ther.*, 67:101–154.
129. Fuhrman-Lane, C. L., Erwin, C. P., Fujimoto, J. M., and Dibben, M. J. (1981): Altered hepatobiliary permeability induced by *Amanita phalloides* in the rat and the protective role of bile duct ligation. *Toxicol. Appl. Pharmacol.*, 58:370–378.
130. Fujii, T. (1997): Toxicological correlation between changes in blood biochemical parameters and liver histopathological findings. *J. Toxicol. Sci.*, 22:161–183.
131. Fujimoto, J. M. (1982): Some *in vivo* methods for studying sites of toxicant action in relation to bile formation. In: *Toxicology of the Liver*, edited by G. L. Plaa and W. R. Hewitt, pp. 121–145. Raven Press, New York.
132. Funk, C., Pantze, M., Jehle, L., Ponelle, C., Scheuermann, G., Lazendic, M., and Gasser, R. (2001): Troglitazone-induced intrahepatic cholestasis by an interference with the hepatobiliary export of bile acids in male and female rats. Correlation with the gender difference in troglitazone sulfate formation and the inhibition of the canalicular bile salt export pump (BSEP) by troglitazone and troglitazone sulfate. *Toxicology*, 167: 83–98.
133. Gandolfi, A. J., Wijeweera, J., and Brendel, K. (1996): Use of precision-cut liver slices as an *in vitro* tool for evaluating liver function. *Toxicol. Pathol.*, 24:58–61.
134. Gauthier, M. and Girard, D. (2001): Activation of human neutrophils by chlordane: induction of superoxide production and phagocytosis but not chemotaxis or apoptosis. *Hum. Exp. Toxicol.*, 20:229–235.
135. Gebhardt, R. (1983): Primary cultures of rat hepatocytes as a model system of canalicular development, biliary secretion, and intrahepatic cholestasis. *Gastroenterology*, 84:1462–1470.
136. Gebhardt, R. (1992): Metabolic zonation of the liver: regulation and implications for liver function. *Pharmacol. Ther.*, 53:275–354.

137. Gerk, P. M., and Vore, M. (2002): Regulation of expression of the multidrug resistance-associated protein 2 (MRP2) and its role in drug disposition. *J. Pharmacol. Exp. Ther.*, 302:407–415.
138. Ghoshal, A. K., Porta, E. A., and Hartroft, W. S. (1969): The role of lipoperoxidation in the pathogenesis of fatty livers induced by phosphorous poisoning in rats. *Am. J. Pathol.*, 54:275–291.
139. Gillette, J. R. (1971): Techniques for studying drug metabolism *in vitro*. In: *Fundamentals of Drug Metabolism and Drug Disposition*, edited by B. N. La Du, H. G. Mandel, and E. L. Way, pp. 400–418. Williams & Wilkins, Baltimore, MD.
140. Gillette, J. R. (1975): Mechanisms of hepatic necrosis induced by halogenated aromatic hydrocarbons. In: *The Pathogenesis and Mechanisms of Liver Cell Necrosis*, edited by D. Keppler, pp. 239–254. MTP Press, London.
141. Gillette, J. R. (1977): Kinetics of reactive metabolites and covalent binding *in vivo* and *in vitro*. In: *Biological Reactive Intermediates*, edited by D. J. Jollow, J. J. Kocsis, R. Snyder, and H. Vanio, pp. 25–41. Raven Press, New York.
142. Girard, D., Paquet, M. E., Paquin, R., and Beaulieu, A. D. (1996): Differential effects of interleukin-15 (IL-15) and IL-2 on human neutrophils: modulation of phagocytosis, cytoskeletal rearrangement, gene expression, and apoptosis by IL-15. *Blood*, 88:3176–3184.
143. Glende, Jr., E. A., Hruszkewycz, A. M., and Recknagel, R. O. (1976): Critical role of lipid peroxidation in carbon tetrachloride-induced loss of aminopyrine demethylase, cytochrome P-450 and glucose-6-phosphatase. *Biochem. Pharmacol.*, 25:2163–2170.
144. Goethals, F., Krack, G., Deboyser, D., Vossen, P., and Roberfroid, M. (1984): Critical biochemical functions of isolated hepatocytes as sensitive indicators of chemical toxicity. *Fundam. Appl. Toxicol.*, 4:441–450.
145. Goldin, R. D., Hunt, N. C., Clark, J., and Wickramasinghe, S. N. (1993): Apoptotic bodies in a murine model of alcoholic liver disease: reversibility of ethanol-induced changes. *J. Pathol.*, 171:73–76.
146. Goldsworthy, T. L., Fransson-Steen, R., and Maronpot, R. R. (1996): Importance of and approaches to quantitation of hepatocyte apoptosis. *Toxicol. Pathol.*, 24:24–35.
147. Gottschall, D. W., Wiley, R. A., and Hanzlik, R. P. (1983): Toxicity of *ortho*-substituted bromobenzenes to isolated hepatocytes: comparison to *in vivo* results. *Toxicol. Appl. Pharmacol.*, 69:55–65.
148. Grasl-Kraupp, B., Ruttkay-Nedecky, B., Koudelka, H., Bukowska, K., Bursch, W., and Schulte-Hermann, R. (1995): *In situ* detection of fragmented DNA (TUNEL assay) fails to discriminate among apoptosis, necrosis, and autolytic cell death: a cautionary role. *Hepatology*, 21:1465–1468.
149. Grice, H. C. (1972): The changing role of pathology in modern safety evaluation. *CRC Crit. Rev. Toxicol.*, 1:119–152.
150. Grice, H. C., Barth, M. L., Cornish, H. H., Foster, G. V., and Gray, R. H. (1971): Correlation between serum enzymes, isoenzyme patterns and histologically detectable organ damage. *Food. Cosmet. Toxicol.*, 9:847–855.
151. Grisham, J. W. (1979): Use of hepatic cell cultures to detect and evaluate the mechanisms of action of toxic chemicals. *Int. Rev. Exp. Pathol.*, 20:123–210.
152. Guguen-Guillouzo, C. and Guillozo, A. (1993): Human hepatocyte cultures. In: *In Vitro Biological Systems*, edited by C. A. Tyson and G. N. Frasier, pp. 271–278. Academic Press, New York.
153. Guillouzo, A. (1986): Use of isolated and cultured hepatocytes for xenobiotic metabolism and cytotoxicity studies. In: *Research in Isolated and Cultured Hepatocytes*, edited by A. Guillouzo and C. Gugen-Guillouzo, pp. 314–331. John Libbey/INSERM, Paris.
154. Guillouzo, A., Begue, J.-M., Campion, J. P., Gascoin, M.-N., and Gugen-Guillouzo, C. (1985): Human hepatocyte culture: a model of pharmaco-toxicological studies. *Xenobiotica*, 15:635–641.
155. Guillouzo, A. and Gugen-Guillouzo, C. (1986): *Research in Isolated and Cultured Hepatocytes*. John Libbey/INSERM, Paris.
156. Guillouzo, A., Morel, F., Ratanasavanh, D., Chesne, C., and Guguen-Guillouzo, C. (1990): Long-term culture of functional hepatocytes. *Toxicol. In vitro*, 4:415–427.
157. Gutman, A. D. (1959): Serum alkaline phosphatase activity in diseases of the skeletal and hepatobiliary systems: a consideration of the current status. *Am. J. Med.*, 27: 875–901.
158. Habeebu, S. S. M., Liu, J., and Klaassen, C. D. (1998): Cadmium-induced apoptosis in mouse liver. *Toxicol. Appl. Pharmacol.*, 149:203–209.
159. Hafeman, D. G. and Koekstra, W. G. (1977): Protection against carbon tetrachloride-induced lipid peroxidation in the rat by dietary vitamin E, selenium and methionine as measured by ethane evolution. *J. Nutr.*, 107:656–665.
160. Hallberg, D., Jonson, G., and Reichard, H. (1960): Serum alkaline phosphatases, transaminases and ornithine carbamyl transferase in biliary obstruction. *Acta Chir. Scand.* 120:251–257.
161. Hallesy, D. and Benitz, K. F. (1963): Sulfobromophthalein sodium retention and morphological liver damage in dogs *Toxicol. Appl. Pharmacol.*, 5:650–660.
162. Hargreaves, T. (1966): Bilirubin, bromosulfophthalein and indocyanine green excretion in bile. *Q. J. Exp. Physiol.* 51:184–195.
163. Harrigan, J. A., Vezina, C. M., McGarrigle, B. P., Ersing N., Box, H. C., Maccubbin, A. E., and Olson, J. R. (2004) DNA adduct formation in precision-cut rat liver and lung slices exposed to benzo(*a*)pyrene. *Toxicol. Sci.*, 77 307–314.
164. Hayasaka, A., Koch, J., Schuppan, D., Maddrey, W. C. and Hahn, E. G. (1991): The serum concentrations of the aminoterminal propeptide of procollagen type III and the hepatic content of mRNA for the alpha chain of procollagen type III in carbon tetrachloride-induced rat liver fibrogenesis. *J. Hepatol.*, 13:328–338.
165. Hewitt, L. A., Ayotte, P., and Plaa, G. L. (1986): Modifications in rat hepatobiliary function following treatment with acetone, 2-butanone, 2-hexanone, mirex, or chlordecone and subsequently exposed to chloroform. *Toxicol. Appl. Pharmacol.*, 83:465–473.

166. Hewitt, W. R., Miyajima, H., Côté, M. G., and Plaa, G. L. (1979): Acute alteration of chloroform-induced hepato- and nephrotoxicity by mirex and Kepone. *Toxicol. Appl. Pharmacol.*, 48:509–527.
167. Hewitt, W. R., Miyajima, H., Côté, M. G., and Plaa, G. L. (1980): Acute alteration of chloroform-induced hepato- and nephrotoxicity by acetone, *n*-hexane, methyl *n*-butyl ketone, and 2,5-hexanedione. *Toxicol. Appl. Pharmacol.*, 53:230–248.
168. Higuchi, H., Grambihler, A., Canbay, A., Bronk, S. F., and Gores, G. J. (2004): Bile acids up-regulate death receptor 5/TRAIL-receptor 2 expression via a c-Jun N-terminal kinase-dependent pathway involving Sp1. *J. Biol. Chem.*, 279:51–60.
169. Hoek, J. B., Cahill, A., and Pastorino, J. G. (2002): Alcohol and mitochondria: a dysfunctional relationship. *Gastroenterology*, 122:2049–2063.
170. Hoglen, N. C., Younis, H. S., Hartley, D. P., Gunawardhana, L., Lantz, R. C., and Sipes, I. G. (1998): 1,2-Dichlorobenzene-induced lipid peroxidation in male Fischer 344 rats is Kupffer cell dependent. *Toxicol. Sci.*, 45: 376–385.
171. Horvat, R. J., Lane, W. G., Ng, H., and Shepherd, A. D. (1964): Saturated hydrocarbons from autooxidizing metyl linoleate. *Nature*, 203:523–524.
172. Hoyumpa, A. M., Greene, H. L., Dunn, D. D., and Schenker, S. (1975): Fatty liver: biochemical and clinical considerations. *Dig. Dis.*, 20:1142–1170.
173. Hruszkewycz, A. M., Glende, Jr., E. A., and Recknagel, R. O. (1978): Destruction of microsomal cytochrome P-450 and glucose-6-phosphatase by lipids extracted from peroxidized microsomes. *Toxicol. Appl. Pharmacol.*, 46:695–702.
174. Huang, L. and Vore, M. (2001): Multidrug resistance to *p*-glycoprotein 2 is essential for the biliary excretion of indocyanine green. *Drug Metab. Dispos.*, 29:634–637.
175. Hunton, D. B., Bollman, J. L., and Hoffman II, H. N. (1961): The plasma removal of indocyanine green and sulfobromophthalein: effect of dosage and blocking agents. *J. Clin. Invest.*, 40:1648–1655.
176. Iijima, M., Côté, M. G., and Plaa, G. L. (1983): A semi-quantitative morphologic assessment of chlordecone-potentiated chloroform hepatotoxicity. *Toxicol. Lett.*, 17: 307–314.
177. Isselbacher, K. J. (1977): Metabolic and hepatic effects of alcohol. *N. Engl. Med.*, 296:612–626.
178. Jackson, R. L., Morissett, J. D., and Gotto, Jr., A. M. (1976): Lipoprotein structure and metabolism. *Physiol. Rev.*, 56:259–316.
179. Jaeger, R. J., Trabulus, M. J., and Murphy, S. D. (1973): Biochemical effects of 1,1-dichloroethylene in rats: dissociation of its hepatotoxicity from a lipoperoxidative mechanism. *Toxicol. Appl. Pharmacol.*, 24:457–467.
180. Jaeschke, H. (2000): Reactive oxygen and mechanisms of inflammatory liver injury. *J. Gastroenterol. Hepatol.*, 15:718–724.
181. Jaeschke, H., Knight, T. R., and Bajt, M. L. (2003): The role of oxidant stress and reactive nitrogen species in acetaminophen hepatotoxicity. *Toxicol. Lett.*, 144: 279–288.
182. Jansen, P. L. M. and Müller, M. (2003): The role of membrane transport in drug-induced hepatotoxicity and cholestasis. In: *Drug-Induced Liver Disease*, edited by N. Kaplowitz and L. D. DeLeve, pp. 97–124. Marcel Dekker, New York.
183. Johnson, D. R. and Klaassen, C. D. (2002): Role of rat multidrug resistance protein 2 in plasma and biliary disposition of dibromosulfophthalein after microsomal enzyme induction. *Toxicol. Appl. Pharmacol.*, 180:56–63.
184. Johnson, D. R. and Klaassen, C. D. (2002): Regulation of rat multidrug resistance protein 2 by classes of prototypical microsomal enzyme inhibitors that activate distinct transcription pathways. *Toxicol. Sci.*, 67:182–189.
185. Johnson, D. R., Habeebu, S. S., and Klaassen, C. D. (2002): Increase in bile flow and biliary excretion of glutathione-derived sulfhydryls in rats by drug-metabolizing enzyme inducers is mediated by multidrug resistance protein 2. *Toxicol. Sci.*, 66:16–26.
186. Jungermann, K. and Katz, N. (1989): Functional specialization of different hepatocyte populations. *Physiol. Rev.*, 69:708–764.
187. Jurkiewicz, M., Averill-Bates, D. A., Marion, M., and Denizeau, F. (2004): Involvement of mitochondrial and death receptor pathways in tributyltin-induced apoptosis in rat hepatocytes. *Biochim. Biophys. Acta*, 1693:15–27.
188. Kanel, G. C. (2003): Histopathology of drug-induced liver disease. In: *Drug-Induced Liver Disease*, edited by N. Kaplowitz and L. D. DeLeve, pp. 243–286. Marcel Dekker, New York.
189. Kaplowitz, N. and DeLeve, L. D. (2003): *Drug-Induced Liver Disease*. Marcel Dekker, New York.
190. Keen, H. G., Dekker, B. A., Disley, L., Hastings, D., Lyons, S., Reader, A. J., Ottewell, P., Watson, A., and Zweit, J. (2005): Imaging apoptosis *in vivo* using $^{124}$I-annexin V and PET. *Nucl. Med. Biol.*, 32:395–402.
191. Kent, G., Fels, G., Dubin, A., and Popper, H. (1959): Collagen content based on hydroxyproline determinations in humans and rat livers. *Lab. Invest.*, 8:48–56.
192. Kerr, J. F. R., Wyllie, A. H., and Currie, A. R. (1972): Apoptosis: a basic biological phenomenon with wide ranging implications in tissue kinetics. *Br. J. Cancer*, 26:239–257.
193. Kipp, H. and Arias, I. M. (2002): Trafficking of canalicular ABC transporters in hepatocytes. *Annu. Rev. Physiol.*, 64: 595–608.
194. Kirby, T. O., Rivera, A., Rein, D., Wang, M., Ulasov, I., Breidenbach, M., Kataram, M., Contreras, J. L., Krumdieck, C., Yamamoto, M., Rots, M. G., Haisma, H. J., Alvarez, R. D., Mahasreshti, P. J., and Curiel, D. T. (2004): A novel *ex vivo* model system for evaluation of conditionally replicative adenoviruses therapeutic efficacy and toxicity. *Clin. Cancer Res.*, 10:8697–8703.
195. Klaassen, C. D. (1971): Studies on the increased biliary flow produced by phenobarbital in rats. *J. Pharmacol. Exp. Ther.*, 176:743–751.
196. Klaassen, C. D. (1976): Pharmacokinetics of rose bengal in the rat, rabbit, dog, and guinea pig. *Toxicol. Appl. Pharmacol.*, 38:85–100.
197. Klaassen, C. D. and Plaa, G. L. (1967): Determination of sulfobromophthalein storage and excretory rate in small animals. *J. Appl. Physiol.*, 22:1151–1155.

198. Klaassen, C. D. and Plaa, G. L. (1967): Species variation in metabolism, storage, and excretion of sulfobromophthalein. *Am. J. Physiol.*, 213:1322–1326.
199. Klaassen, C. D. and Plaa, G. L. (1968): Effect of carbon tetrachloride on the metabolism, storage and excretion of sulfobromophthalein. *Toxicol. Appl. Pharmacol.*, 12: 132–139.
200. Klaassen, C. D. and Plaa, G. L. (1968): Studies on the mechanism of phenobarbital-enhanced sulfobromophthalein disappearance. *J. Pharmacol. Exp. Ther.*, 161:361–366.
201. Klaassen, C. D. and Plaa, G. L. (1969): Comparison of the biochemical alterations elicited in livers from rats treated with carbon tetrachloride, chloroform, 1,1,2-trichloroethane and 1,1,1-trichloroethane. *Biochem. Pharmacol.*, 18:2019–2027.
202. Klaassen, C. D. and Plaa, G. L. (1969): Plasma disappearance and biliary excretion of indocyanine green in rats, rabbits and dogs. *Toxicol. Appl. Pharmacol.*, 15:374–384.
203. Klaassen, C. D. and Watkins, J. B. (1984): Mechanisms of bile formation, hepatic uptake, and biliary excretion. *Pharmacol. Rev.*, 36:1–67.
204. Kodavanti, P. R. S. and Mehendale, H. M. (1991): Biochemical methods of studying hepatotoxicity. In: *Hepatotoxicology*, edited by R. G. Meeks, S. D. Harrison and R. J. Bull, pp. 241–325. CRC Press, Boca Raton, FL.
205. Kongo, M., Ohta, Y., Nishida, K., Sasaki, E., Harada, N., and Ishiguro, I. (1999): An association between lipid peroxidation and alpha-naphthylisothiocyanate-induced liver injury in rats. *Toxicol. Lett.*, 105:103–110.
206. Konttinen, A. (1968): A further simplified method of ornithine carbamoyltransferase measurement. *Clin. Chim. Acta*, 21:29–32.
207. Korsrud, G. O., Grice, H. C., and McLaughlan, J. M. (1972): Sensitivity of several serum enzymes in detecting carbon tetrachloride-induced liver damage in rats. *Toxicol. Appl. Pharmacol.*, 22:474–483.
208. Korsrud, G. O., Grice, H. G., Goodman, R. K., Knipfel, J. E., and McLaughlan, J. M. (1973): Sensitivity of several serum enzymes for the detection of thioacetamide-, dimethylnitrosamine-, and diethanolamine-induced liver damage in rats. *Toxicol. Appl. Pharmacol.*, 26:299–313.
209. Köster, U., Albrecht, D., and Kappus, H. (1977): Evidence for carbon tetrachloride-and ethanol-induced lipid peroxidation demonstrated by ethane production in mice and rats. *Toxicol. Appl. Pharmacol.*, 42:639–648.
210. Kostrubsky, V. E., Strom, S. C., Hanson, J., Urda, E., Rose, K., Burliegh, J., Zocharski, P., Cai, H., Sinclair, J. F., and Sahi, J. (2003): Evaluation of hepatotoxic potential of drugs by inhibition of bile-acid transport in cultured primary human hepatocytes and intact rats. *Toxicol. Sci.*, 76: 220–228.
211. Kryszewski, A. J., Neale, G., Whitfield, J. F., and Moss, D. W. (1973): Enzyme changes in experimental biliary obstruction. *Clin. Chim. Acta*, 47:175–182.
212. Kubin, R. H., Grodsky, G. M., and Carbone, J. V. (1960): Investigation of rose bengal conjugation. *Proc. Soc. Exp. Biol. Med.*, 104:650–653.
213. Kukongviriyapan, V. and Stacey, N. H. (1991): Chemical-induced interference with hepatocellular transport: role in cholestasis. *Chem. Biol. Interact.*, 77:245–261.
214. Kutob, S. D. and Plaa, G. L. (1962): The effect of acute ethanol intoxication on chloroform-induced liver damage. *J. Pharmacol. Exp. Ther.*, 135:245–251.
215. Lal, H., Puri, S. K., and Fuller, G. C. (1970): Impairment of hepatic drug metabolism by carbon tetrachloride inhalation. *Toxicol. Appl. Pharmacol.*, 16:35–39.
216. Lamers, W. H., Hilberts, A., Furt, E., Smith, J., Jonges, G. N., van Noorden, C. J. F., Janzen, J. W. G., Charles, R., and Moorman, A. F. N. (1989): Hepatic enzymic zonation: a reevaluation of the concept of the liver acinus. *Hepatology*, 10:72–76.
217. Larson, R. E., Plaa, G. L., and Crew, L. M. (1964): The effect of spinal cord transection on carbon tetrachloride hepatotoxicity. *Toxicol. Appl. Pharmacol.*, 6:154–162.
218. Laskin, D. L. and Gardner, C. R. (1998): The role of nonparenchymal cells and inflammatory macrophages in hepatotoxicity. In: *Toxicology of the Liver*, 2nd ed., edited by G. L. Plaa and W. R. Hewitt, pp. 297–320. Taylor & Francis, New York.
219. Lavigne, J. G. and Marchand, C. (1974): The role of metabolism in chloroform hepatotoxicity. *Toxicol. Appl. Pharmacol.*, 29:312–326.
220. Lawrence, J. N. and Benford, D. J. (1991): Development of an optimal method for the cryopreservation of hepatocytes and their subsequent monolayer culture. *Toxicol. In Vitro*, 5:39–50.
221. Lee, W. M. (2003): Drug-induced hepatotoxicity. *N. Engl. J. Med.*, 349:474–485.
222. Leevy, C. M., Stein, S. W., Cherrick, G. R., and Davidson, C. S. (1959): Indocyanine green clearance: a test of liver excretory function. *Clin. Res.*, 7:290.
223. Lennon, H. D. (1966): Relative effects of 17a-alkylated anabolic steroids on sulfobromophthalein (BSP) retention in rabbits. *J. Pharmacol. Exp. Ther.*, 151:143–150.
224. Li, A. P. (1994): Primary hepatocyte culture as an *in vitro* toxicological system of the liver. In: *In Vitro Toxicology* edited by S. C. Gad, pp. 195–220. Raven Press, New York.
225. Lieberman, M. and Mapson, L. W. (1964): Genesis and biogenesis of ethylene. *Nature*, 204:343–345.
226. Liebler, D. C. and Reed, D. J. (1997): Free-radical defense and repair mechanisms. In: *Free Radical Toxicology*, edited by K. B. Wallace, pp. 141–171. Taylor & Francis, New York.
227. Limaye, P. B., Apte, U. M., Shankar, K., Bucci, T. J. Warbritton, A., and Mehendale, H. M. (2003): Calpain released from dying hepatocytes mediates progression of acute liver injury induced by model hepatotoxicants. *Toxicol. Appl. Pharmacol.*, 191:211–226.
228. Lindstrom, R. D. and Anders, M. W. (1978): Effect of agents known to alter carbon tetrachloride hepatotoxicity and cytochrome P-450 levels on carbon tetrachloride-stimulated lipid peroxidation and ethane production in the intact rat. *Biochem. Pharmacol.*, 27:563–567.
229. Litov, R. E. et al. (1978): Lipid peroxidation: a mechanism involved in acute ethanol toxicity as demonstrated by *in vivo* pentane production in the rat. *Lipids*, 13:305–307.
230. Lockard, V. G., Mehendale, H. M., and O'Neal, R. M. (1983): Chlordecone-induced potentiation of carbon tetrachloride hepatotoxicity: a light and electron microscopic study. *Exp. Molec. Pathol.*, 39:230–245.

231. Lockard, V. G., Mehendale, H. M., and O'Neal, R. M. (1983): Chlordecone-induced potentiation of carbon tetrachloride hepatotoxicity: a morphometric and biochemical study. *Exp. Mol. Pathol.*, 39:246–256.
232. Lombardi, B. (1966): Considerations on the pathogenesis of fatty liver. *Lab. Invest.*, 15:1–20.
233. Long, R. M. and Moore, L. (1988): Biochemical evaluation of rat hepatocyte primary cultures as a model for carbon tetrachloride hepatotoxicity: comparative studies *in vivo* and *in vitro*. *Toxicol. Appl. Pharmacol.*, 92:295–306.
234. Loretz, L. J., Li, A. P., Flye, M. W., and Wilson, A. G. E. (1989): Optimization of cryopreservation procedures for rat and human hepatocytes. *Xenobiotica*, 15:489–498.
235. Lupp, A., Danz, M., and Muller, D. (2001): Morphology and cytochrome P450 isoforms expression in precision-cut rat liver slices. *Toxicology*, 161:53–66.
236. Lupp, A., Danz, M., and Muller, D. (2005): Histomorphological changes and cytochrome P450 isoforms expression and activities in precision-cut liver slices from neonatal rats. *Toxicology*, 206:427–438.
237. Maher, J. M., Cheng, X., Slitt, A. L., Dieter, M. Z., and Klaassen, C. D. (2005): Induction of the MRP family of transporters by chemical activators of receptor-mediated pathways in mouse liver. *Drug Metab. Dispos.*, 33: 956–962.
238. Malledant, Y., Siproudhis, L., Tanguy, M., Clerc, C., Chesne, C., and Saint-Marc, C. (1990): Effects of halothane on human and rat hepatocyte cultures. *Anesthesiology*, 72:526–534.
239. Marsh, J. B. and Bizzi, A. (1972): Effects of amphetamine and fenfluramine on the net release of triglycerides of very low density lipoproteins by slices of rat liver. *Biochem. Pharmacol.*, 21:1143–1150.
240. Martin, E. J., Racz, W. J., and Forkert, P-G. (2003): Mitochondrial dysfunction is an early manifestation of 1,1-dichloroethylene-induced hepatotoxicity in mice. *J. Pharmacol. Exp. Ther.*, 304:121–129.
241. McKenna, M. J., Zempel, J. A., Madrid, E. O., Braun, W. H., and Gehring, P. J. (1978): Metabolism and pharmacokinetic profile of vinylidene chloride in rats following oral administration. *Toxicol. Appl. Pharmacol.*, 45:821–835.
242. McKenna, M. J., Zempel, J. A., Madrid, E. O., and Gehring, P. J. (1978): The pharmacokinetics of [14C]vinylidene chloride in rats following inhalation exposure. *Toxicol. Appl. Pharmacol.*, 45:599–610.
243. McQueen, C. A. (1993): Isolation and culture of hepatocytes from different laboratory species. In: *In Vitro Biological Systems*, edited by C. A. Tyson and G. N. Frasier, pp. 255–270. Academic Press, New York.
244. McQueen, C. A. and Williams, G. M. (1987): Toxicology studies in cultured hepatocytes from various species. In: *The Isolated Hepatocyte: Use in Toxicology and Xenobiotic Biotransformations*, edited by E. J. Rauckman and G. M. Padilla, pp. 51–67. Academic Press, New York.
245. Mehendale, H. M. (1990): Assessment of hepatobiliary function with phenolphthalein and phenolphthalein glucuronide. *Clin. Chem. Enzyme Comm.*, 2:195–204.
246. Mehendale, H. M. (2005): Tissue repair: an important determinant of final outcome of toxicant-induced toxicity. *Toxicol. Pathol.*, 33:41–51.
247. Mehendale, H. M., Ho, I. K., and Desaiah, D. (1979): Possible molecular mechanisms of mirex-induced hepatobiliary dysfunction. *Drug Metab. Disp.*, 7:28–33.
248. Meurman, L. (1960): On the distribution and kinetics of injected $^{131}$I-rose bengal. *Acta Med. Scand.*, 167(Suppl. 354):7–85.
249. Mitchell, J. R. and Boyd, M. R. (1978): Dose thresholds, host susceptibility, and pharmacokinetic considerations in the evaluation of toxicity from chemically reactive metabolites. In: *Proceedings of the First International Congress on Toxicology*, edited by G. L. Plaa and W. A. Duncan, pp. 169–175. Academic Press, New York.
250. Mitchell, J. R. et al. (1973): Acetaminophen-induced hepatic necrosis. I. Role of drug metabolism. *J. Pharmacol. Exp. Ther.*, 187:185–194.
251. Mitchell, J. R., Nelson, S. D., Thorgeirsson, S. S., McMurty, R. J., and Dybing, E. (1976): Metabolic activation: biochemical basis for many drug-induced liver injuries. *Prog. Liver Dis.*, 5:259–279.
252. Molander, D. W., Wroblewski, F., and LaDue, J. S. (1955): Serum glutamic oxalacetic transaminase as an index of hepatocellular injury. *J. Lab. Clin. Med.*, 46:831–839.
253. Moldeus, P. (1978): Paracetamol metabolism and toxicity in isolated hepatocytes from rat and mouse. *Biochem. Pharmacol.*, 27:2859–2863.
254. Moritz, M. and Snodgrass, P. J. (1972): Serum enzymes derived from liver cell fractions. II. Responses to bile duct ligation in rats. *Gastroenterology*, 62:93–100.
255. Moronvalle-Halley, V., Sacre-Salem, B., Sallez, V., Labbe, G., and Gautier, J. C. (2005): Evaluation of cultured, precision-cut rat liver slices as a model to study drug-induced liver apoptosis. *Toxicology*, 207:203–214.
256. Musser, A. W., Ortigoza, C., Vazquez, M., and Riddick, J. (1966): Correlation of serum enzymes and morphologic alterations of the liver; with special reference to serum guanase and ornithine carbamyl transferase. *Am. J. Clin. Pathol.*, 46:82–88.
257. Neghab, M. and Stacey, N. H. (1997): *In vitro* interference with hepatocellular uptake of bile acids by xylene. *Toxicology*, 120:1–10.
258. Neghab, M. and Stacey, N. H. (1997): Toluene-induced elevation of serum bile acids: relationship to bile acid transport. *J. Toxicol. Environ. Health*, 52:249–268.
259. Newberne, P. (1982): Assessment of the hepatocarcinogenic potential of chemicals: response of the liver. In: *Toxicology of the Liver*, edited by G. L. Plaa and W. R. Hewitt, pp. 243–290. Raven Press, New York.
260. O'Brien, P. J., Slaughter, M. R., Polley, S. R., and Kramer, K. (2002): Advantages of glutamate dehydrogenase as a blood biomarker of acute hepatic injury in rats. *Lab. Anim.*, 36:313–321.
261. O'Hara, T. M., Sheppard, M. A., Clarke, E. C., Borzelleca, J. F., Gennings, C., and Condie, L. W. J. (1991): A $CCl_4/CHCl_3$ interaction study in isolated hepatocytes: non-induced and phenobarbital-pretreated cells. *J. Appl. Toxicol.*, 11:147–154.
262. Obatomi, D. K., Brant, S., Anthonypillai, V., and Bach, P. H. (1998): Toxicity of atractyloside in precision-cut rat and porcine renal and hepatic tissue slices. *Toxicol. Appl. Pharmacol.*, 148:35–45.

263. Obermayer-Straub, P. and Manns, M. P. (2003): Immunological mechanisms in liver injury. In: *Drug-Induced Liver Disease*, edited by N. Kaplowitz and L. D. DeLeve, pp. 125–149. Marcel Dekker, New York.
264. Okoumassoun, L. E., Averill-Bates, D., Marion, M., and Denizeau, F. (2003): Possible mechanisms underlying the mitogenic action of heptachlor in rat hepatocytes. *Toxicol. Appl. Pharmacol.*, 193:356–369.
265. Okuno, M., Muto, Y., Kato, M., Moriwaki, H., Noma, A., Tagaya, O., and Tanabe, Y. (1991): Changes in serum and hepatic levels of immunoreactive prolyl hydroxyalse in two models of hepatic fibrosis in rats. *J. Gastroenterol. Hepatol.*, 6:271–277.
266. Olinga, P., Groen, K., Hof, I. H., De Kanter, R., Koster, H. J., Leeman, W. R., Rutten, A. A. J. J. L., Van Twillert, K., and Groothuis, G. M. M. (1997): Comparison of five incubations systems for rat liver slices using functional and viability parameters. *J. Pharmacol. Toxicol. Methods*, 38: 59–69.
267. Olson, J. R. and Fujimoto, J. M. (1980): Evaluation of hepatobiliary function in the rat by the segmented retrograde intrabiliary injection technique. *Biochem. Pharmacol.*, 29:205–211.
268. Olson, J. R., Fujimoto, J. M., and Peterson, R. E. (1977): Three methods for measuring the increase in the capacity of the distended biliary tree in the rat produced by α-naphthylisothiocyanate treatment. *Toxicol. Appl. Pharmacol.*, 42:33–43.
269. Palmeira, C. M., Ferreira, F. M., Rolo, A. P., Oliveira, P. J., Santos, M. S., Moreno, A. J., Cipriano, M. A., Martins, M. I., and Seica, R. (2003): Histological changes and impairment of liver mitochondrial bioenergetics after long-term treatment with α-naphthylisothiocyanate (ANIT). *Toxicology*, 190:185–196.
270. Panteghini, M. (1994): Electrophoretic fractionation of 5′-nucleotidase. *Clin. Chem.*, 40:190–196.
271. Parrish, A. R., Gandolfi, A. J., and Brendel, K. (1995): Precision-cut tissue slices: applications in pharmacology and toxicology. *Life Sci.*, 57:1887–1901.
272. Patel, T., Bronk, S., and Gores, G. (1994): Increases of intracellular magnesium promote glycodeoxycholate-induced apoptosis in rat hepatocytes. *J. Clin. Invest.*, 94: 2183–2192.
273. Patel, T. and Gores, G. J. (1995): Apoptosis and hepatobiliary disease. *Hepatology*, 21:1725–1741.
274. Pauli-Magnus, C. and Meier, P. J. (2003): Pharmacogenetics of hepatocellular transporters. *Pharmacogenetics*, 13:189–198.
275. Pessayre, D., Fromenty, B., Mansouri, A., and Berson, A. (2003): Hepatotoxicity due to mitochondrial injury. In: *Drug-Induced Liver Disease*, edited by N. Kaplowitz and L. D. DeLeve, pp. 55–83. Marcel Dekker, New York.
276. Peterson, R. E., Olson, J. R., and Fujimoto, J. M. (1976): Measurement and alteration of the capacity of the distended biliary tree in the rat. *Toxicol. Appl. Pharmacol.*, 36:353–368.
277. Phillips, M. J., Poucell, S., and Oda, M. (1986): Mechanisms of cholestasis. *Lab. Invest.*, 54:593–608.
278. Plaa, G. L. (1968): Evaluation of liver function methodology. In: *Selected Pharmacological Testing Methods, Medical Research Series*, edited by A. Burger, pp. 255–288. Marcel Dekker, New York.
279. Plaa, G. L. (1977): Factors influencing biliary excretion and apparent $T_m$ for bilirubin and related anions. In: *Chemistry and Physiology of Bile Pigments*, edited by P. D. Berk and N. I. Berlin, pp. 396–403. National Institutes of Health, Bethesda.
280. Plaa, G. L. (1988): Experimental evaluation of haloalkanes and liver injury. *Fundam. Appl. Toxicol.*, 10:563–570.
281. Plaa, G. L. (1991): Toxic responses of the liver. In: *Casarett & Doull's Toxicology: The Basic Science of Poisons*, 4th ed., edited by M. O. Amdur, C. D. Klaassen, and J. Doull, pp. 334–353. Pergamon Press, New York.
282. Plaa, G. L. (2000): Chlorinated methanes and liver injury: highlights of the past 50 years. *Annu. Rev. Pharmacol. Toxicol.*, 40:43–65.
283. Plaa, G. L. and Becker, B. A. (1965): Demonstration of bile stasis in the mouse by a direct and an indirect method. *J. Appl. Physiol.*, 20:534–537.
284. Plaa, G. L., Evans, E. A., and Hine, C. H. (1958): Relative hepatotoxicity of seven halogenated hydrocarbons. *J. Pharmacol. Exp. Ther.*, 123:224–229.
285. Plaa, G. L. and Hewitt, W. R. (1982): Quantitative evaluation of indices of hepatotoxicity. In: *Toxicology of the Liver*, edited by G. L. Plaa and W. R. Hewitt, pp. 103–120. Raven Press, New York.
286. Plaa, G. L. and Hine, C. H. (1960): The effect of carbon tetrachloride on isolated perfused rat liver function. *Arch. Indust. Health*, 21:114–123.
287. Plaa, G. L. and Priestly, B. G. (1976): Intrahepatic cholestasis induced by drugs and chemicals. *Pharmacol Rev.*, 28:207–273.
288. Plaa, G. L. and Witschi, H. (1976): Chemicals, drugs and lipid peroxidation. *Annu. Rev. Pharmacol. Toxicol.*, 16 125–141.
289. Popper, H. and Udenfriend, S. (1970): Hepatic fibrosis: correlation of biochemical and morphological investigations. *Am. J. Med.*, 49:707–721.
290. Price, R. J., Ball, S. E., Renwick, A. B., Barton, P. T. Beamand, J. A., and Lake, B.G. (1998): Use of precision-cut rat liver slices for studies of xenobiotic metabolism and toxicity: comparison of the Krumdieck and Brendel tissue slicers. *Xenobiotica*, 28:361–371.
291. Price, R. J., Mistry, H., Wield, P. T., Renwick, A. B. Beamand, J. A., and Lake, B. G. (1996): Comparison of the toxicity of allyl alcohol, coumarin and menadione in precision-cut rat, guinea-pig, Cynomolgus monkey and human liver slices. *Arch. Toxicol.*, 71:107–111.
292. Priestly, B. G. and Plaa, G. L. (1969): Effects of benziodarone on the metabolism and biliary excretion of sulfobromophthalein and related dyes. *Proc. Soc. Exp. Biol Med.*, 132:881–885.
293. Priestly, B. G. and Plaa, G. L. (1970): Sulfobromophthalein metabolism and excretion in rats with iodomethane induced depletion of hepatic glutathione. *J. Pharmacol Exp. Ther.*, 174:221–231.

294. Priestly, B. G. and Plaa, G. L. (1970): Temporal aspects of carbon tetrachloride-induced alteration of sulfobromophthalein excretion and metabolism. *Toxicol. Appl. Pharmacol.*, 17:786–794.
295. Rappaport, A. M. (1979): Physioanatomical basis of toxic liver injury. In: *Toxic Injury of the Liver*, edited by E. Farber and M. M. Fisher, pp. 1–57. Marcel Dekker, New York.
296. Reader, S., Moutardier, V., and Denizeau, F. (1999): Tributyltin triggers apoptosis in trout hepatocytes: the role of $Ca^{2+}$, protein kinase C and proteases. *Biochim. Biophys. Acta*, 1448:473–485.
297. Recknagel, R. O. and Ghoshal, A. K. (1966): New data on the question of lipoperoxidation in carbon tetrachloride poisoning. *Exp. Mol. Pathol.*, 5:108–117.
298. Recknagel, R. O., Glende, Jr., E. A., Waller, R. L., and Lowrey, K. (1982): Lipid peroxidation: biochemistry, measurement, and significance in liver cell injury. In: *Toxicology of the Liver*, edited by G. L. Plaa and W. R. Hewitt, pp. 213–241. Raven Press, New York.
299. Recknagel, R. O. and Lombardi, B. (1961): Studies of biochemical changes in subcellular particles of rat liver and their relationship to a new hypothesis regarding the pathogenesis of carbon tetrachloride fat accumulation. *J. Biol. Chem.*, 236:564–569.
300. Reed, D. J. (1988): Evaluation of chemical-induced oxidative stress as a mechanism of hepatocyte death. In: *Toxicology of the Liver*, 2nd ed., edited by G. L. Plaa and W. R. Hewitt, pp. 187–220. Taylor & Francis, New York.
301. Reichard, H. (1957): Determination of ornithine carbamyl transferase with microdiffusion technique. *Scand. J. Clin. Invest.*, 9:311–312.
302. Reichard, H. (1959): Ornithine carbamoyl transferase in dog serum on intravenous injection of enzyme, choledochus ligation and carbon tetrachloride poisoning. *J. Lab. Clin. Med.*, 53:417–425.
303. Reichard, H. (1960): Ornithine carbamoyl-transferase activity in human tissue homogenates. *J. Lab. Clin. Med.*, 56:218–221.
304. Reichard, H. (1961): Ornithine carbamoyl transferase activity in human serum in diseases of the liver and biliary system. *J. Lab. Clin. Med.*, 57:78–87.
305. Reichard, H. (1962): Studies on ornithine carbamoyl transferase activity in blood and serum. *Acta Med. Scand.*, 172(Suppl. 390):1–8.
306. Reichard, H. (1964): Determination of ornithine carbamoyl transferase in serum: a rapid method. *J. Lab. Clin. Med.*, 63:1061–1064.
307. Reitman, S. and Frankel, S. (1957): A colorimetric method for the determination of serum oxaloacetic and glutamic pyruvic transaminases. *Am. J. Clin. Pathol.*, 28:56–63.
308. Reynolds, E. S. (1972): Comparison of early injury to liver endoplasmic reticulum by halomethanes, hexachloroethane, benzene, toluene, bromobenzene, ethionine, thioacetamide and dimethylnitrosamine. *Biochem. Pharmacol.*, 21:2555–2561.
309. Riely, C. A., Cohen, G., and Lieberman, M. (1974): Ethane evolution: a new index of lipid peroxidation. *Science*, 183: 208–210.
310. Roberts, R. A., Michel, C., Coyle, B., Freathy, C., Cain, K., and Boitier, E. (2004): Regulation of apoptosis by peroxisome proliferators. *Toxicol. Lett.*, 149:37–41.
311. Roberts, R. J., Klaassen, C. D., and Plaa, G. L. (1967): Maximum biliary excretion of bilirubin and sulfobromophthalein during anesthesia-induced alteration of rectal temperature. *Proc. Soc. Exp. Biol. Med.*, 125:313–316.
312. Roberts, W. M. (1933): Blood phosphatase and the Van Den Bergh reaction in the differentiation of the several types of jaundice. *Br. Med. J.*, 1:734–738.
313. Rockey, D. C., Boyles, J. K., Gabbiani, G., and Friedman, S. L. (1992): Rat hepatic lipocytes express smooth muscle actin upon activation *in vivo* and in culture. *J. Submicrosc. Cytol. Pathol.*, 24:193–203.
314. Rojkind, M. and Dunn, M. A. (1979): Hepatic fibrosis. *Gastroenterology*, 76:849–863.
315. Roma, M. G., Orsler, D. J., and Coleman, R. (1997): Canalicular retention as an *in vitro* assay of tight junctional permeability in isolated hepatocyte couplets: effects of protein kinase modulation and cholestatic agents. *Fundam. Appl. Toxicol.*, 37:71–81.
316. Rust, C. and Gores, G. J. (2000): Apoptosis and liver disease. *Am. J. Med.*, 108:567–74.
317. Sagai, M. and Tappel, A. L. (1978): Effect of vitamin E on carbon tetrachloride-induced lipid peroxidation as demonstrated by *in vivo* pentane production. *Toxicol. Lett.*, 2: 149–155.
318. Sagai, M. and Tappel, A. L. (1979): Lipid peroxidation induced by some halomethanes as measured by *in vivo* pentane production in the rat. *Toxicol. Appl. Pharmacol.*, 49:283–291.
319. Schrenk, D., Schmitz, H. J., Bohnenberger, S., Wagner, B., and Worner, W. (2004): Tumor promoters as inhibitors of apoptosis in rat hepatocytes. *Toxicol. Lett.*, 149:43–50.
320. Schulte-Hermann, R., Bursch, W., Grasl-Kraupp, B., Török, L., Ellinger, A., and Müllauer, L. (1995): Role of active cell death (apoptosis) in multi-stage carcinogenesis. *Toxicol. Lett.*, 82–83:143–148.
321. Seetharam, S., Sussman, N. L., Komoda, T., and Alpers, D. H. (1986): The mechanism of elevated alkaline phosphatase activity after bile duct ligation in the rat. *Hepatology*, 6:374–380.
322. Seglen, P. O. (1972): I. Effect of calcium on enzymatic dispersion of isolated, perfused liver. *Exp. Cell Res.*, 74: 450–454.
323. Seglen, P. O. (1976): Preparation of isolated rat liver cells. In: *Methods in Cell Biology*, edited by D. M. Prescott, pp. 29–83. Academic Press, New York.
324. Sevanian, A. and McLeon, L. (1997): Formation and biological reactivity of lipid peroxidation products. In: *Free Radical Toxicology*, edited by K. B. Wallace, pp. 47–70. Taylor & Francis, New York.
325. Shin, B. C., Huggins, J. W., and Caraway, K. L. (1972): Effects of pH, concentration and aging on the malonaldehyde reaction with proteins. *Lipids*, 7:229–233.
326. Sinnhuber, R. O. and Yu, T. C. (1958): 2-Thiobarbituric acid method for the measurement of rancidity in fishery products. *Food Technol.*, 12:9–12.

327. Sipes, I. G., and Gandolfi, A. J. (1982): Bioactivation of aliphatic organohalogens: formation, detection, and relevance. In: *Toxicology of the Liver*, edited by G. L. Plaa and W. R. Hewitt, pp. 181–212. Raven Press, New York.
328. Sirica, A. E. and Pitot, H. C. (1980): Drug metabolism and effects of carcinogens in cultured hepatic cells. *Pharmacol. Rev.*, 31:205–228.
329. Smith, M. T. (2003): Mechanisms of troglitazone hepatotoxicity. *Chem. Res. Toxicol.*, 16:679–687.
330. Smith, P. F., Gandolfi, A. J., Krumdieck, C. L., Putnam, C. W., Zukoski, C. F. I., Davis, W. M., and Brendel, K. (1985): Dynamic organ culture of precision liver slices for *in vitro* toxicology. *Life Sci.*, 36:1367–1375.
331. Smuckler, E. A. (1966): Studies on carbon tetrachloride intoxication IV. Effect of carbon tetrachloride on liver slices and isolated organelles *in vitro*. *Lab. Invest.*, 15: 157–166.
332. Soni, M. G. and Mehendale, H. M. (1998): Role of tissue repair in toxicologic interactions among hepatotoxic organics. *Environ. Health Persp.*, 106(Suppl, 6): 1307–1317.
333. Soto, A., Foy, B. D., and Frazier, J. M. (2002): Effect of cadmium on bromosulfophthalein kinetics in the isolated perfused rat liver system. *Toxicol. Sci.*, 69:460–469.
334. Stacey, N. H. (1989): Toxicity of mixtures of trichloroethylene, tetrachloroethylene and 1,1,1-trichloroethane: similarity of *in vitro* and *in vivo* responses. *Toxicol. Indust. Health*, 5:441–450.
335. Stanca, C., Jung, D., Meier, P. J., and Kullak-Ublick, G. (2001): Hepatocellular transport proteins and their role in liver disease. *World J. Gastroenterol.*, 7:157–169.
336. Story, D. L., Gee, S. J., Tyson, C. A., and Gould, D. H. (1983): Response of isolated hepatocytes to organic and inorganic cytotoxins. *J. Toxicol. Environ. Health*, 11:483–501.
337. Sturgill, M. G. and Lambert, G. H. (1997): Xenobiotic-induced hepatotoxicity: mechanisms of liver injury and methods of monitoring hepatic function. *Clin. Chem.*, 43: 1512–1426.
338. Sweeny, D. J. and Diasio, R. B. (1991): The isolated hepatocyte and isolated perfused liver as models for studying drug- and chemical-induced hepatotoxicity. In: *Hepatotoxicology*, edited by R. G. Meeks, S. D. Harrison, and R. J. Bull, pp. 215–239. CRC Press, Boca Raton, FL.
339. Szende, B., Lapis, K., Kovalszky, I., and Timar, F. (1992): Role of the modified (glycosaminoglycan producing) perisinusoidal fribroblasts in the $CCl_4$-induced fibrosis of the rat liver. *In Vivo*, 6:355–361.
340. Tanaka, H., Sano, N., and Takikawa, H. (2003): Biliary excretion of phenolphthalein sulfate in rats. *Pharmacology*, 68:177–182.
341. Taplin, G. V., O.M., M., and Kade, H. (1955): The radioactive ($I^{131}$-tagged) rose bengal uptake–excretion test for liver function using external gamma ray scintillation counting techniques. *J. Lab. Clin. Med.*, 45:655–678.
342. Tegeris, A. S., Smalley, Jr., H. E., Earl, F. L., and Curtis, J. L. (1969): Ornithine carbamyl transferase as a liver function test: comparative studies in dog, swine and man. *Toxicol. Appl. Pharmacol.*, 14:54–66.
343. Tharappel, J. C., Lee, E. Y., Robertson, L. W., Spear, B. T., and Glauert, H. P. (2002): Regulation of cell proliferation, apoptosis, and transcription factor activities during the promotion of liver carcinogenesis by polychlorinated biphenyls. *Toxicol. Appl. Pharmacol.*, 179:172–184.
344. Thohan, S., Zurich, M. C., Chung, H., Weiner, M., Kane, A. S., and Rosen, G. M. (2001): Tissue slices revisited: evaluation and development of a short-term incubation for integrated drug metabolism. *Drug Metab. Dispos.*, 29:1337–1342.
345. Thurman, R. G. and Kauffmann, F. C. (1985): Sublobular compartmentation of pharmacologic events (SCOPE): metabolic fluxes in periportal and pericentral regions of the liver lobule. *Hepatology*, 5:144–151.
346. Toutain, H. J., Moronvalle-Halley, V., Sarsat, J. P., Chelin, C., Hoet, D., and Leroy, D. (1998): Morphological and functional integrity of precision-cut rat liver slices in rotating organ culture and multiwell plate culture: effects of oxygen tension. *Cell Biol. Toxicol.*, 14:175–190.
347. Traiger, G. J. and Plaa, G. L. (1971): Differences in the potentiation of carbon tetrachloride in rats by ethanol and isopropanol pretreatment. *Toxicol. Appl. Pharmacol.*, 20: 105–112.
348. Trauner, M., Fickert, P., and Stauber, R. E. (1999): Inflammation-induced cholestasis. *J. Gastroenterol. Hepatol.*, 14: 946–959.
349. Travlos, G. S., Morris, R. W., Elwell, M. R., Duke, R., Rosenblum, S., and Thompson, M. B. (1996): Frequency and relationships of clinical chemistry and liver and kidney histopathology findings in 13-week toxicity studies in rats. *Toxicology*, 107:17–29.
350. Tyson, C. A., Gee, S. J., Hawk-Prather, K., Story, D. L. and Milman, H. A. (1989): Correlation between *in vivo* and *in vitro* toxicity of some chlorinated aliphatics. *Toxicol In Vitro*, 3:145–150.
351. Tzirogiannis, K. N., Panoutsopoulos, G. I., Demonakou M. D., Hereti, R. I., Alexandropoulou, K. N., Basayannis A. C., and Mykoniatis, M. G. (2003): Time-course of cadmium-induced acute hepatotoxicity in the rat liver: the role of apoptosis. *Arch. Toxicol.*, 77:694–701.
352. Van Handel, E. and Zilversmit, D. B. (1957): Micromethod for the direct determination of serum triglycerides. *J. Lab Clin. Med.*, 50:152–157.
353. Vassef, A. A. (1978): Direct micromethod for colorimetry of serum ornithine carbamoyltransferase activity, with use of a linear standard curve. *Clin. Chem.*, 24:101–107.
354. Vonen, B. and Mørland, J. (1984): Isolated rat hepatocytes in suspension: potential hepatotoxic effects of six different drugs. *Arch. Toxicol.*, 56:33–37.
355. Wade, C. R. and van Rij, A. M. (1985): *In vivo* lipid peroxidation in man as measured by the respiratory excretion of ethane, pentane, and other low-molecular-weight hydrocarbons. *Anal. Biochem.*, 150:1–7.
356. Wang, K., Brems, J. J., Gamelli, R. L., and Ding, J. (2005): Reversibility of caspase activation and its role during glycochenodeoxycholate-induced hepatocyte apoptosis. *J. Biol. Chem.*, 280:23490–23495.

357. Weber, L. W. D., Boll, M., and Stampfl, A. (2003): Hepatotoxicity and mechanism of action of haloalkanes: carbon tetrachloride as a toxicological model. *Crit. Rev. Toxicol.*, 33:105–136.
358. Weiner, F. R., Shah, A., Biempica, L., Zern, M. A., and Czaja, M. J. (1992): The effects of hepatic fibrosis on Ito cell gene expression. *Matrix*, 12:36–43.
359. Weldon, P. R., Rubenstein, B., and Rubenstein, D. (1965): The direct action of $CCl_4$ on the metabolism of liver slices. *Can. J. Biochem.*, 43:647–659.
360. Wells, P. G. and To, E. C. A. (1986): Murine acetaminophen hepatotoxicity: temporal interanimal variability in plasma glutamic–pyruvic transaminase profiles and relation to *in vivo* chemical covalent binding. *Fundam. Appl. Toxicol.*, 7:17–25.
361. Wheeldon, E. B., Williams, S. M., Soames, A. R., James, N. H., and Roberts, R. A. (1995): Quantitation of apoptotic bodies in rat liver by *in situ* end labeling (ISEL): correlation with morphology. *Toxicol. Pathol.*, 23:410–415.
362. Wheeler, H. O., Cranston, W. I., and Meltzer, J. I. (1958): Hepatic uptake and biliary excretion of indocyanine green in the dog. *Proc. Soc. Exp. Biol. Med.*, 99:11–14.
363. Wheeler, H. O., Meltzer, J. I., and Bradley, S. E. (1960): Biliary transport and hepatic storage of sulfobromophthalein sodium in the unanesthetized dog, in normal man, and in patient with hepatic disease. *J. Clin. Invest.*, 39:1131–1144.
364. Whitfield, J. B., Pounder, R. E., Neale, G., and Moss, D. W. (1972): Serum γ-glutamyl transpeptidase activity in liver disease. *Gut*, 13:702–708.
365. Wilkinson, J. H. (1970): Clinical application of isoenzymes. *Clin. Chem.*, 16:733–739.
366. Willson, R. A., Hart, J., and Hall, T. (1991): The concentration and temporal relationships of acetaminophen-induced changes in intracellular and extracellular total glutathione in freshly isolated hepatocytes from untreated and 3-methylcholanthrene pretreated Sprague–Dawley and Fischer rats. *Pharmacol. Toxicol.*, 69:205–212.
367. Wormser, U., Ben Zakine, S., Stivelband, E., and Eizen, O. (1990): The liver slice system: a rapid *in vitro* acute toxicity test for primary screening of hepatotoxic agents. *Toxicol. In Vitro*, 4:783–789.
368. Wroblewski, F. (1959): The clinical significance of transaminase activities of serum. *Am. J. Med.*, 27:911–923.
369. Yaari, A., Sikuler, E., Keyman, A., and Ben-Zvi, Z. (1992): Bromosulfophthalein disposition in chronically bile duct obstructed rats. *Hepatology*, 15:67–72.
370. Yagi, K. (1984): Assay for blood plasma or serum. *Methods Enzymol.*, 105:328–331.
371. Yagle, K. J., Eary, J. F., Tait, J. F., Grierson, J. R., Link, J. M., Lewellen, B., Gibson, D. F., and Krohn, K. A. (2005): Evaluation of $^{18}$F-annexin V as a PET imaging agent in an animal model of apoptosis. *J. Nucl. Med.*, 46: 658–66.
372. Yamazaki, M., Suzuki, H., Sugiyama, Y., Iga, T., and Hanano, M. (1992): Uptake of organic anions by isolated rat hepatocytes: a classification in terms of ATP-dependency. *J. Hepatology*, 14:41–47.
373. Youssef, J. A. and Badr, M. Z. (2005): Aging and enhanced hepatocarcinogenicity by peroxisome proliferator-activated receptor alpha agonists. *Ageing Res. Rev.*, 4: 103–118.
374. Zimmerman, H. J. (1999): *Hepatotoxicity*, 2nd ed., Lippincott Williams & Wilkins, Philadelphia, PA.
375. Zsigmond, G. and Solymoss, B. (1972): Effect of spironolactone, pregnenolone-16α carbonitrile and cortisol on the metabolism and biliary excretion of sulfobromophthalein and phenol-3,6-dibromophthalein disulfonate in rats. *J. Pharmacol. Exp. Ther.*, 183:499–507.

# Notes

# 30 Principles and Methods for Renal Toxicology

*Lawrence H. Lash*

## CONTENTS



## INTRODUCTION

This chapter focuses on guiding principles and methods for assessing renal function and toxicity on both the organ and cellular levels. The first section briefly summarizes some key aspects of renal physiology, anatomy, and biochemistry as they relate to the kidneys as a target organ for chemical induced toxicity. Material presented addresses the central issue of what it is that is distinctive about renal physiology and anatomy that makes the kidneys especially susceptible to various forms of injury. The second section presents a discussion of experimental models and approaches that are used to study renal function and toxicity. Models range from *in vivo* to *in vitro*, including those with intact tubular structure to subcellular and molecular approaches. Key considerations in the use of these models are discussed, focusing on advantages and limitations to their use in renal toxicology. The third section considers several selected examples of assays used to quantify renal cellular function. These are presented to illustrate examples of both some classic assays as well as newer approaches that can provide insight into cellular metabolism and molecular regulation. It is important to note that this chapter is meant to provide a summary and overview of factors that determine or contribute to nephrotoxicity and the experimental models and assay methods used in studies of renal function and toxicity. Hence, more detailed information on renal structure and function can be found in several texts, books, and book chapters [1,2]. Presentation of methods and assays is not in the form of detailed, step-by-step procedures; rather, presentation of the highlights, advantages, and limitations is the major emphases.

## RENAL PHYSIOLOGY, ANATOMY, AND BIOCHEMISTRY

The kidneys are uniquely sensitive to toxicants because they receive and filter a large quantity of blood relative to its weight. On average, the two kidneys receive 25% of the cardiac output, while comprising only about 1% of total body weight. This fact as well as the presence of a myriad of plasma membrane transport proteins, which can result in a high degree of accumulation and concentration of chemicals within the tubular epithelial cells, and drug metabolism enzymes, which can result in bioactivation of inert or nontoxic chemicals, contribute to the susceptibility of the kidneys. The major function of the kidneys is to excrete waste products while maintaining total body salt, water, and acid–base balance; thus, the kidneys are the primary organs responsible for maintaining the constancy of the internal environment. They accomplish this task by three general mechanisms: (1) glomerular filtration, (2) tubular reabsorption, and (3) tubular secretion. Each of these mechanisms is discussed later. The next section briefly discusses some of the key structural and functional features of the mammalian kidney. A critical point to understand is that the kidneys exemplify form following function. It is also important to understand how these functions, in a toxicological or pathological rather than a physiological setting, make the kidneys a unique target organ.

### Overall Structural Organization of Kidneys

The major tissue types in the kidneys are vascular tissue and tubular epithelia. Kidney structure can be considered at multiple levels of organization. At the highest level, the overall structure of the kidney is subdivided into the cortex, outer stripe of the outer medulla, inner stripe of the outer medulla, and inner medulla or papilla. Humans and larger mammals have multipapillate kidneys, whereas small mammals, such as rat and rabbit, are unipapillate. Each papilla descends into a renal fornix. In multipapillate kidneys, fornices merge to form the renal pelvis, which is an expanded upward extension of the ureter. There is a close association, both physically and functionally, between the renal vasculature and the epithelial tissue. The renal artery enters the kidney alongside the ureter, branching to become progressively the interlobular arterioles that lead to the glomerular capillary network. The venous system has subdivisions with similar designations, terminating in the renal vein, which also flows besides the ureter.

At the closed end, the nephron is extended to form the cup-shaped Bowman's capsule. The lumen of the capsule is continuous with the narrow lumen that extends through the renal tubule. Associated with the capsule is the tuft of capillaries that forms the glomerulus inside Bowman's capsule. This structure is responsible for the initial step in urine formation. A filtrate of plasma passes through the single-cell layer of capillary walls, through the basement membrane, and finally through another single-cell layer of epithelium that forms the wall of Bowman's capsule. The filtrate then passes into the lumen of the tubules to begin its passage through the various segments of the nephron, eventually passing into the collecting duct and the renal pelvis. The wall of the renal tubule is one cell layer thick and this epithelium functions to separate the plasma/urinary filtrate in the lumen from the interstitial fluid.

The functional unit of the mammalian kidney, the nephron, consists of a glomerular capillary network surrounded by Bowman's capsule, a proximal tubule (convoluted and pars recta or straight segment), a loop of Henle (thin descending limb, thick ascending limb), a distal tubule (convoluted tubule and collecting tubule), and a collecting duct (Figure 30.1). The nephron is an epithelial tubule that is closed on one end and opens into the renal pelvis and collecting ducts at the other end. Nephrons are divided into three classes, based on the location of their glomerulus: superficial, midcortical, and juxtamedullary. Glomeruli located near the surface of the kidney give rise to superficial nephrons, which generally only descend into the outer medulla. These nephrons are also called *short looped nephrons* and lack thin ascending limbs. Glomeruli located deep within the cortex, near the corticomedullary border, give rise to juxtamedullary nephrons. These nephrons are called *long-looped nephrons* and enter the inner medulla and have thin descending limbs. In general, mammals that produce highly concentrated urine have longer papillae and a higher proportion of long-looped vs. short looped nephrons. Glomeruli located between the superficial and juxtamedullary glomeruli give rise to midcortica

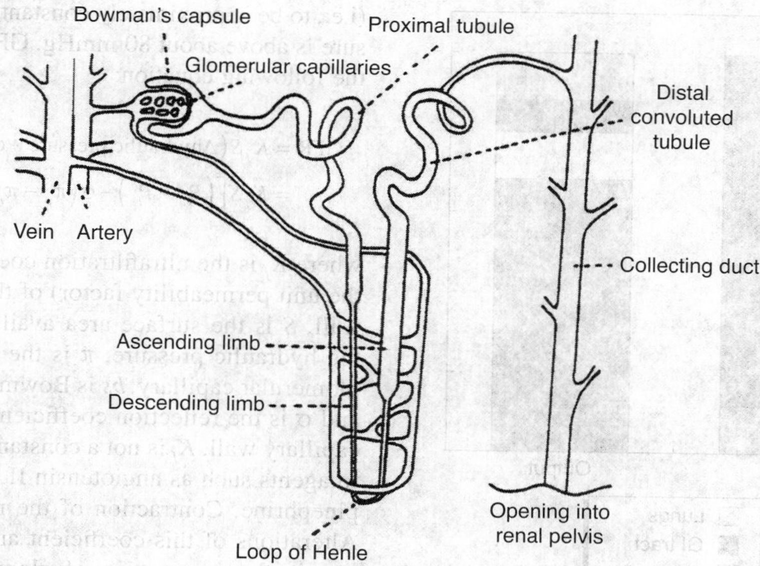


**FIGURE 30.1** Diagram of basic nephron structure and the associated renal vasculature.

nephrons. These may be either long-looped or short-looped, although the deeper glomeruli are more likely to be associated with long-looped nephrons.

Basic nephron structure can also be considered on the basis of localization near either the vascular or urinary pole. The urinary pole starts at the proximal convoluted tubule in the cortex. The tubule enters the medulla where it forms the loop of Henle. The distal tubule returns to the glomerulus of origin and then enters the cortex where it is somewhat convoluted. The distal tubule then leads into the collecting duct, which goes into the medullary pyramid and then into the papillae. The nephron ends at the junction of the distal tubule and the collecting duct. Several nephrons feed into a single collecting duct. The vascular pole is the juxtaglomerular region, which contains mesangial cells and the returning distal portion of the distal tubule. This region has many closely packed nuclei and is called the *macula densa*. The afferent arteriole breaks up into 8 to 10 capillary loops in the glomerulus. On the opposite side of the glomerulus is the urinary pole where the proximal tubule begins.

An important region of the nephron that is critically involved in regulation of renal blood flow and urine formation is the juxtaglomerular apparatus (JGA), which consists of specialized epithelial cells, the macula densa cells, and specialized secretory or granular cells at the vascular pole where the afferent and efferent arterioles enter and leave the glomerulus. The JGA is thus a combination of specialized tubular and vascular (i.e., smooth muscle) cells. A key function of the JGA is secretion of renin, which is involved in formation of angiotensin and, ultimately, in secretion of the mineralocorticoid aldosterone. This process is part of the feedback system that regulates glomerular filtration and renal blood flow.

## Glomerular Filtration and Urine Formation

Each kidney receives its blood supply from a single renal artery and renal vein. The renal arteries originate from the aorta, and the renal veins drain into the interior vena cava. The urine formed in each kidney is drained via a single ureter into the bladder. The initial stage of urine formation is filtration of plasma and accumulation of the ultrafiltrate in the lumen of Bowman's capsule. The glomerular filtrate contains nearly all of the constituents of the blood except for the blood cells and proteins. Approximately 15 to 25% of the water and solutes are removed from the plasma during a single pass through the renal circulation by filtration alone. In humans, the glomerular filtrate is typically produced at a rate of 125 mL/min or about 200 L/day. Consideration of normal fluid intake implies that most of the glomerular filtrate must be reabsorbed into the bloodstream to maintain water balance and prevent dehydration. The critical importance of the kidneys in this function is illustrated in a bar graph comparing major sources of fluid input and output under normal conditions (Figure 30.2). Input of fluid comes from metabolism and the gastrointestinal (GI) tract. Both of these are highly adaptable, although water production from metabolism typically is geared toward energy production rather than volume regulation. Input of water from the GI tract can be modulated by activation of the thirst mechanism in dehydration. The prominence of the kidneys is clearly seen in a comparison of the primary sources of water output: Although the lungs, gastrointestinal tract, and skin are significant sources, the kidneys provide 68% of the output under basal conditions.

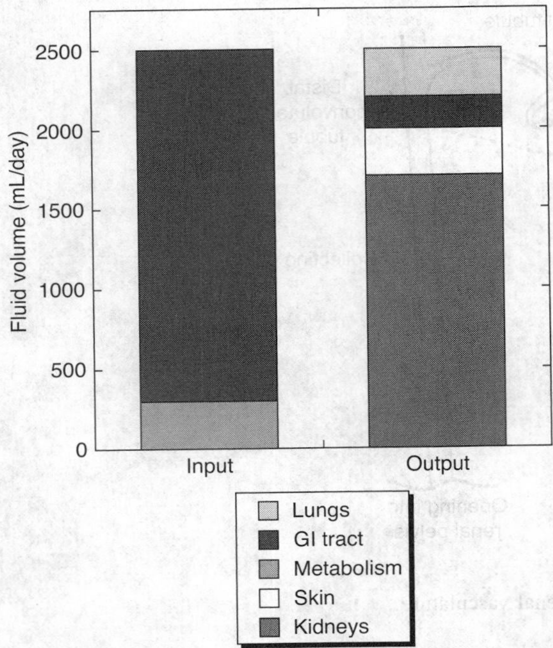


**FIGURE 30.2** Summary histogram showing the function of different organ systems in fluid volume homeostasis.

Glomerular filtration refers to the movement of water and solutes across the glomerular membrane to form an ultrafiltrate of plasma in the tubular lumen and is the first step in urine formation. It depends on two factors: (1) a pressure gradient from the lumen of the capillary to the lumen of Bowman's capsule, and (2) the sieve-like properties of the three-layered tissue separating the two compartments. Water and nonprotein solutes (e.g., inorganic anions and cations, intermediary metabolites such as glucose and urea) are found in the filtrate at approximately the same concentrations at which they occur in plasma. The exclusion of most proteins above a threshold molecular weight from the glomerular filtrate is critical for maintenance of the osmotic pressure difference across the glomerular membrane. As described below, the appearance of significant amounts of high-molecular-weight proteins in the urine is an obvious indicator of glomerular damage. Thus, filtration occurs because forces moving fluid out of the capillary (i.e., hydrostatic pressure) exceed forces preventing both movement out of the capillary (i.e., oncotic pressure due to plasma proteins) and into Bowman's space (i.e., hydraulic pressure within Bowman's capsule).

The glomerular filtration rate (GFR) is affected by changes in blood pressure, plasma protein concentration, and the ultrafiltration coefficient ($K_f$). GFR varies directly with blood pressure and inversely with protein concentration in the afferent arteriole. Although GFR is highly blood-flow dependent, renal blood flow is normally independent of systemic blood pressure and is autoregulated (i.e., to be approximately constant) as long as blood pressure is above about 80 mmHg. GFR can be calculated by the following equation:

$$\begin{aligned} \text{GFR} &= K_f S(\Delta\text{hydraulic pressure} - \text{oncotic pressure}) \\ &= K_f S\left[\left(P_{gc} - P_{bs}\right) - \sigma\left(\pi_p - \pi_{bs}\right)\right] \end{aligned} \tag{30.1}$$

where $K_f$ is the ultrafiltration coefficient (also known as the unit permeability factor) of the glomerular capillary wall, $S$ is the surface area available for filtration, $P$ is the hydraulic pressure, $\pi$ is the oncotic pressure, $gc$ is glomerular capillary, $bs$ is Bowman's space, $p$ is plasma, and $\sigma$ is the reflection coefficient of proteins across the capillary wall. $K_f$ is not a constant but varies in response to agents such as angiotensin II, vasopressin, and norepinephrine. Contraction of the mesangium reduces $K_f$. Alterations of this coefficient are thought to be important in the progression of glomerular damage but not tubular damage.

The two other processes, besides glomerular filtration, that account for the ability of the kidneys to regulate electrolyte, organic metabolite, and water balance are tubular reabsorption and tubular secretion. Tubular reabsorption refers to the movement of substances from the tubular fluid, through the epithelial cell, back into plasma. In this manner, substances that would otherwise have been excreted in urine are reabsorbed. The functional significance of tubular reabsorption is evident from an evaluation of the renal handling of $Na^+$ ions and glucose. The kidneys actively reabsorb >99% of the $Na^+$ ions and glucose that emerge in the tubular fluid after glomerular filtration. Although reabsorption of the filtered NaCl load occurs along all segments of the nephron, the proximal tubules account for approximately 60% of the total reabsorption under normal conditions. The remainder is reabsorbed in the thick ascending limb (~30%), distal convoluted tubule (~7%), and collecting duct (~2 to 3%). Depending on the filtered NaCl load, however, these proportions can be greatly modified, as $Na^+$ ion transport in the more distal portions of the nephron and the collecting duct is under hormonal control. Tubular secretion, in contrast, refers to the movement of substances from plasma, through the epithelial cell and into the tubular fluid and ultimately the urine. Tubular secretion is a critical component of how the body handles many drugs and xenobiotics.

The processes described above lead to a glomerular filtrate in the proximal tubular lumen that is isosmotic with plasma (i.e., ~300 mOsm). As the filtrate progresses along the nephron, however, there is a countercurrent mechanism that generates an extremely high osmolarity in the medullary interstitium; thus, the medullary interstitium at the base of the loop of Henle in long-looped nephrons can reach as high as 1200 mOsm

As fluid progresses through the thick ascending limb to the distal and cortical collecting tubules, osmolarity decreases to ~100 mOsm. This behavior is dependent primarily on three factors: (1) active chloride or sodium reabsorption in the ascending limb, (2) the anatomical relationship between the descending and ascending limbs (i.e., both limbs are close together and flow in each is in the opposite direction), and (3) the parallel vascular structure to the nephron structure that forms a countercurrent exchange of blood and prevents solutes from coming out of the descending limb and being washed away.

## Nephron Heterogeneity: Physiological and Biochemical Implications

The mammalian kidney is a complex organ whose basic structural unit, the nephron, is composed of several cell populations, each exhibiting diverse morphological, biochemical, and physiological properties [3,4]. As briefly described above, specific properties of each segment are critical to overall kidney function in reabsorption and secretion. As presented below, each segment is both morphologically and biochemically designed to perform their specific functions in urine formation and maintenance of electrolyte, fluid, and metabolite homeostasis. Additionally, these features also make each nephron segment possess a characteristic susceptibility to either chemical toxicants or to pathological states, such as hypoxia or ischemia and reperfusion. As is briefly described below in the following subsections, these susceptibility differences are due in part to the distinctive composition of membrane transporters and enzymes in each segment but also to differences in cellular energetics and oxygenation. An important feature of all the nephron cell types that is characteristic of transporting epithelia is that they are polarized cells, meaning that the plasma membrane is physically separated into distinct regions (luminal or brush-border membrane and basolateral membrane). A critical feature of this separation is that each region is comprised of distinct enzymes, carriers, channels, and other proteins. This separation is essential for function and disturbances in the processing of key membrane proteins have been identified as a response to nephrotoxicants and disease states.

### Nephron Cell Types: Structure and Function

The subsections below summarize some key morphological and functional features of several major nephron segments. Additional details and discussion of other nephron segments can be found in any number of texts or book chapters [1,2]. An overview of the major nephron segments, highlighting some of their key morphological features and physiological roles, is presented in Table 30.1.

#### *Proximal Tubules*

The first nephron segment to which the tubular filtrate comes in contact after glomerular filtration is the proximal tubule. The proximal tubule is subdivided into three segments: S1, S2, and S3. The first segment (S1) is located within the pars convoluta or proximal convoluted tubule and is found in the cortex; the S2 segment is localized at the end of the convoluted and beginning of the straight section (pars recta or proximal straight tubule); and the S3 segment is localized entirely within the proximal straight tubule, which is in the outer stripe of the outer medulla.

The epithelial cells of the S1 segment are rather leaky, allowing movement of solutes between cells by a paracellular route. Approximately 60 to 70% of the initial tubular filtrate is reabsorbed in the S1 segment. Reabsorption occurs via an array of active and passive transport processes, with water flowing passively down the osmotic gradient. Hence, the tubular fluid remains isosmotic with respect to the plasma in spite of the large reduction in fluid volume to approximately 25% of its original volume. Cells of the S1 segment are tall and have a prominent brush border, extensive infoldings of the basal membrane, and interdigitations of the lateral borders of adjacent cells. The extensive brush border, or microvilli, and the infoldings of the basal membrane illustrate the correlation between form and function, as these structures provide a large surface area for reabsorption. The cells possess a high content of mitochondria near the basal pole, a characteristic feature of $Na^+$-transporting epithelia. The vast majority of filtered glucose and amino acids are reabsorbed in this segment as well.

Cells of the S2 segment are structurally less complex than those of the S1 segment; S2 cells have a shorter, less dense brush border, less basolateral interdigitation, and a lower content of mitochondria. Although the rate of volume reabsorption is thought to decrease *in vivo* going from S1 to S2 segment, *in vitro* studies show a similar capacity for the two segments. The results *in vivo* are, therefore, likely due to the decreased concentration of electrolytes in the S2 tubular fluid caused by the large reabsorption that occurred in the S1 segment. Certain carriers, however, are present at higher activities in the S2 segment, such as many of those for organic anions.

The transition between S2 and S3 segments is relatively subtle, and species-dependent differences exist in S3 ultrastructure; for example, S3 cells are shorter and less complex than S2 cells in rabbits, whereas S3 cells in rats have a well-developed brush border that is longer than that in other proximal segments. Although glucose reabsorptive capacity is much lower in S3 segments than in either S1 or S2 segments, organic anion and organic cation transport is high in S3. Consistent with this distribution, several enzymes involved in drug metabolism are present at the highest levels in the S3 segment among the three proximal segments.

**TABLE 30.1**
**Summary of Morphology and Physiology of Selected, Key Nephron Segments from Mammalian Kidney**

| Nephron Cell Type | Morphology | Physiology |
|---|---|---|
| Proximal tubule | Tall and prominent microvilli on luminal membrane (especially S1)<br>Cuboidal shape<br>Extensive basolateral infoldings<br>Leaky epithelium<br>High density of mitochondria | 75% of filtered $Na^+$ removed by active transport<br>Passive water and $Cl^-$ reabsorption<br>Organic anion secretion (S2, S3)<br>Most glucose, amino acid reabsorption<br>Tubular fluid remains isosmotic to plasma |
| Thin descending limb | Squamous epithelium<br>Little cellular interdigitation (type I)<br>Better developed lateral interdigitations (type II, III)<br>Deep tight junctions (type I)<br>Shallow tight junctions (type II, III)<br>Low density of mostly spherical mitochondria | Extremely high osmotic water permeability<br>Aquaporin 1 on luminal and basolateral membranes<br>Low permeability to $Na^+$<br>Variable permeability to urea (low to moderate)<br>Tubular urine concentration |
| Thin ascending limb | Flat, moderately interdigitated cells<br>Shallow tight junctions<br>Low density of mostly spherical mitochondria | Water impermeable; no aquaporin<br>Little, if any, active $Na^+$ transport<br>Highly permeable to NaCl<br>Vasopressin-sensitive C1C–K1 $Cl^-$ channel<br>Variable permeability to urea (low to moderate) |
| Thick ascending limb | Extensive interdigitation<br>Shallow tight junctions<br>Large number of elongated, rod-shaped mitochondria<br>Cortical cells are thinner and have less mitochondria than medullary cells | Cytochrome P450-dependent arachidonic acid metabolism<br>Water impermeable; no aquaporin<br>$Na^+$–$K^+$–$2Cl^-$ cotransport<br>Active $Na^+$ transport<br>Active $Ca^{2+}$ and $Mg^{2+}$ transport<br>Dilution of hyperosmotic tubular urine |
| Distal tubule (distal convoluted and connecting tubule) | DCT: appears bright under microscope; short; deep tight junctions; numerous long mitochondria<br>CCT: appears granular under light microscope; wider than DCT | High rates of $Na^+$ reabsorption<br>Thiazide-inhibitable $Na^+$–$Cl^-$ cotransporter<br>$K^+$-$Cl^-$ cotransporter<br>$Ca^{2+}$ reabsorption<br>DCT: water impermeable<br>CCT: vasopressin-regulated aquaporin 2 water channel |
| Cortical collecting duct | Two cell types: principal cells (PCs) and intercalated cells (ICs) (2:1)<br>PC: few cellular organelles including mitochondria; sparse, short microprojections on apical membrane<br>IC: abundant cytoplasmic vesicles; higher mitochondrial content; densely packed microvilli on apical membrane | IC: abundant carbonic anhydrase<br>PC: no carbonic anhydrase<br>Vasopressin-dependent water permeability<br>Low urea permeability<br>Active and passive transport of $Cl^-$ and $K^+$ ions<br>Acid–base regulation |

### *Thin Descending and Thin Ascending Limbs*

Thin descending limbs are present in both short-looped and long-looped nephrons. They are subdivided into three subtypes in most mammalian species: types I, II, and III. Short-looped nephrons contain only type I cells, which form a squamous epithelium with little cellular interdigitation and deep tight junctions. Long-looped nephrons of rat and hamster, in contrast, contain two cell types: type II cells in the more proximal portion of the thin limb and type III cells as the thin limb enters the inner medulla. Type II cells contain well-developed lateral interdigitation and shallow tight junctions. A key physiological feature of type I, II, and III thin descending limb cells is their extremely high osmotic water permeability, which is mediated by aquaporin 1 water channels in both luminal and basolateral plasma membranes. Urea and sodium permeability exhibits a good deal of heterogeneity among both the different cell types of thin descending limbs and across species that have been studied, ranging from very high to low. Thin ascending limbs are only present in long-looped nephrons and begin slightly before the bend of the loop of Henle in most species.

There is an abrupt conversion of cell type to the medullary thick ascending limb as the tubule progresses past the inner-outer medullary border. The thin ascending limb contains flat, moderately interdigitated cells that are connected by shallow tight junctions. Unlike the thin descending limb, cells of the thin ascending limb are water impermeable in all species studied and do not express any aquaporin water channels. Thin ascending limb cells have very low $Na^+$ pump activity, suggesting little if any active $Na^+$ transport. In contrast, there is a high passive NaCl permeability, which is thought to involve paracellular flux. Vasopressin-sensitive, transcellular Cl transport occurs and is mediated by the C1C-K1 Cl channel that is present on both the luminal and basolateral plasma membranes. Cells from both thin descending and thin ascending limbs contain a low density of mostly spherical mitochondria, which is consistent with the low activities of active transport processes.

### *Thick Ascending Limb*

The thick ascending limb is divided into two segments: the medullary thick ascending limb, which is located in the outer medulla, and the cortical thick ascending limb, which is located in the medullary rays within the cortex. The medullary thick ascending limb originates at the inner–outer medullary border, either from the thin ascending limb of long-looped nephrons or from the thin descending limb of short-looped nephrons. The cell type becomes the cortical thick ascending limb at the outer medulla–cortex junction, although the transition between medullary and cortical cell types is not abrupt.

Medullary thick ascending limbs are longer in long-looped as compared with short-looped nephrons. There are extensive interdigitation and shallow tight junctions between adjacent medullary thick ascending limb cells. They contain a large number of elongated, rod-shaped mitochondria and one or two cilia on their luminal surface. Medullary thick ascending limbs are subdivided into two types, based on the smoothness or roughness of their luminal surfaces. Although the transition is fairly subtle from medullary to cortical thick ascending limb, the cortical thick ascending limbs are thinner than those of the medullary thick ascending limb, and cells from the cortical region contain a lower density of mitochondria. The cortical thick ascending limb extends past the macula densa, varying in length with species.

Although most of the renal cytochrome P450 activity is found in the proximal tubules (predominantly the S3 segment), both medullary and cortical thick ascending limbs contain cytochrome P450 activity that metabolizes arachidonic acid to several biologically active products, indicating the importance of these nephron segments in regulation of renal hemodynamics. Like the thin ascending limb, both medullary and cortical thick ascending limbs are water impermeable and do not express aquaporin water channels. Unlike the thin ascending limb, medullary and cortical thick ascending limbs have active transport processes. Active transporters include the bumetanide-sensitive $Na^+K^+2Cl^-$ cotransporter (NKCC2, BSC1) in the luminal plasma membrane, the ($Na^+$ + $K^+$)-stimulated ATPase on the basolateral plasma membrane, and the membrane-potential-dependent, divalent cation transport of $Ca^{2+}$ and $Mg^{2+}$. These active solute reabsorption processes provide the primary mechanism for dilution of the luminal fluid as it moves up the nephron through the thick ascending limbs.

### *Distal Tubule*

The term *distal tubule* is generally used to refer to the portion of the nephron that lies between the macula densa and the first junction of two tubules, although it is actually composed of several structurally and functionally distinct cell types. The most proximal portion begins just beyond the macula densa and contains cortical thick ascending limb cells. These cells abruptly transition to distal convoluted tubules. The distal convoluted tubules appear bright under the microscope and are short. The cells contain deep tight junctions, short blunt apical microvilli, deep basolateral invaginations, and numerous long mitochondria oriented perpendicular to the basolateral membrane. The third portion of the distal tubule, the connecting segment or connecting tubule, appears granular under light microscopy and is wider than the distal convoluted tubule. This segment is also sometimes referred to as the *granular distal tubule*. Functionally, both the distal convoluted tubule and connecting tubule exhibit high rates of $Na^+$ reabsorption that are mediated by the thiazide-inhibitable $Na^+Cl^-$ cotransporter (NCC1, TSC). Potassium is transported by the $K^+Cl^-$ cotransporter (KCC1) on the luminal plasma membrane and a conductive $K^+$ channel in rabbit distal convoluted tubules. Calcium ion reabsorption is also a prominent function of the distal convoluted tubule, but it occurs at a lower rate than in the cortical thick ascending limb. Finally, the distal convoluted tubule is water impermeable but the connecting tubule in rat has been shown to possess the vasopressin-regulated water channel, aquaporin-2, suggesting that this segment is water permeable.

### *Cortical Collecting Duct*

The cortical collecting duct originates at the convergence of two collecting tubules and extends down to the corticomedullary border. This segment contains two cell types, the principal (or light) cells and the intercalated (or dark) cells, at a ratio of approximately 2:1. In comparison with the intercalated cells, principal cells have fewer cellular organelles, including mitochondria; a less electron-dense cytoplasm; and sparse, short microprojections on their apical membranes in contrast with the abundant cytoplasmic vesicles and more densely packed microvilli on apical membranes of intercalated cells. Another important difference between the two subtypes of cortical collecting duct cells is that carbonic anhydrase is abundant in intercalated cells but is either absent or present at relatively low levels, depending on species, in principal cells. This has functional implication for pH regulation in the urine.

Physiologically, the collecting duct regulates the final composition of the urine. Although the collecting duct has a low capacity for reabsorption, it is important in regulating

**TABLE 30.2**
**Substrate Preferences and Predominant Pathways for Energy Metabolism Among Nephron Cell Types**

| Nephron Cell Type | Substrate Preference | Mitochondrial Density | Pathway Preference for ATP Generation |
|---|---|---|---|
| Proximal tubule | Fatty acids, ketone bodies, lactate, glutamine, glutamate, pyruvate, citrate, acetate | High | Oxidative phosphorylation, citric acid cycle, gluconeogenesis |
| Thick ascending limb (medullary and cortical) | Lactate, glucose, ketone bodies, fatty acids, acetate | Moderate to high | Oxidative phosphorylation and glycolysis |
| Distal convoluted tubule | Glucose, lactate, β-hydroxybutyrate | High | Glycolysis |
| Cortical collecting duct | Glucose, lactate, β-hydroxybutyrate, fatty acids | Low | Glycolysis |

the reabsorption of water, solutes, electrolytes, bicarbonate, and protons. Water permeability of the cortical collecting duct is almost completely dependent on vasopressin, which stimulates baseline osmotic water permeability by a factor of 10 to 100 in rats and rabbits. The cortical collecting duct has a low permeability for urea that is unaffected by vasopressin. In the presence of vasopressin, therefore, this property enables the kidneys to increase the concentration of urea within the lumen of the cortical collecting duct. This nephron segment also contains both active and passive transport processes for $Cl^-$ and $K^+$ ions and plays an important role in acid–base regulation.

## Cellular Energetics

Patterns of cellular energy metabolism differ widely among the cell types of the nephron. These patterns are based on a number of factors, including cellular energy (i.e., ATP) requirements, mitochondrial density, and oxygenation [5]. The first factor, cellular energy requirements, is largely dependent on active transport capacity. As described earlier, each nephron cell population exhibits markedly different levels of various active transport pathways. Thus, while the proximal tubules have high activities of active $Na^+$, organic anion, glucose, and amino acid transport pathways and have correspondingly high needs for ATP and high densities of mitochondria, the thin descending and ascending limb cells have low activities of active transport pathways and correspondingly low needs for ATP and low densities of mitochondria. Thick ascending limb cells, particularly those in the medullary region, have high activities of active $Na^+$ and divalent cation transport and high densities of mitochondria. Distal tubular cells also have high rates of active $Na^+$ transport and high amounts of mitochondria.

In terms of substrate utilization, each segment has a characteristic preference for energy needs and exhibits distinct patterns of activity for pathways of carbohydrate utilization and synthesis (Table 30.2); for example, although many segments contain a high density of mitochondria, ATP generation primarily comes from glycolysis in the more distal segments of the nephron. Although fatty acids and ketone bodies can be used by several nephron segments, they are the preferred source of reducing equivalents for ATP generation in the proximal tubules. Because many toxicants target mitochondria as an early step in their mechanism of action, cells such as those in the proximal tubule are particularly susceptible to injury, provided the chemical is transported into the cell.

An example of two nephron cell populations exhibiting a markedly different susceptibility to inhibition of mitochondrial function and ATP depletion is illustrated in our studies of the susceptibility of suspensions of freshly isolated proximal tubular and distal tubular cells from the rat to hypoxia or chemically induced ATP depletion [6,7]. In the hypoxia study [6], cells were incubated under either hyperoxic (95% $O_2$/5% $CO_2$), normoxic (21% $O_2$/74% $N_2$/5% $CO_2$), or hypoxic (95% $N_2$/5% $CO_2$) conditions to determine the extent of cell death over time. Distal tubular cells were markedly more sensitive to cytotoxicity from $O_2$ deprivation than those from the proximal tubule; moreover, this difference was not correlated with the extent of ATP depletion, as proximal tubular cells could exhibit ~75% ATP depletion yet little or no cytotoxicity. Essentially similar results were obtained with the chemical model of cellular ATP depletion: Proximal tubular cells exposed to iodoacetate + KCN exhibited little cytotoxicity but significant ATP depletion, whereas distal tubular cells exhibited a similar degree of ATP depletion but markedly higher release of lactate dehydrogenase (LDH). This suggests that nephron cell populations have differing abilities to withstand marked (>80%) ATP depletion.

Another factor that relates to differences in cellular energetics is *in vivo* oxygenation. As described above and in Lash et al. [6], proximal and distal tubular cells respond differently *in vitro* to $O_2$ deprivation. In the intact animal or in humans, however, the situation is more complex. On the one hand, each nephron cell population has different ATP consumption rates and differing mitochondrial densities. Beyond that, however, it is known that an $O_2$ gradient exists in the kidneys as one goes from cortex to medulla such that the inner medulla has been described

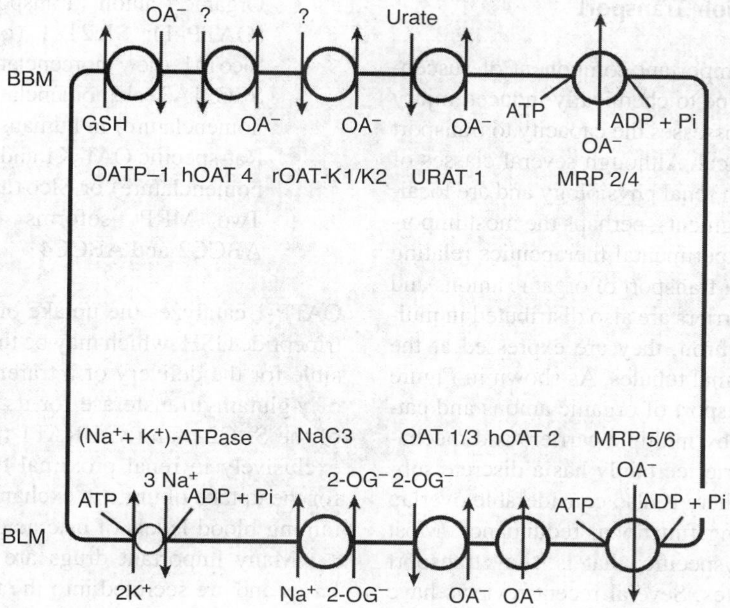

OA⁻ ?
?
Urate
BBM
GSH
OA⁻
OA⁻
OA⁻
ATP
ADP + Pi
OA⁻
OATP–1 hOAT 4 rOAT-K1/K2 URAT 1
MRP 2/4
(Na⁺+ K⁺)-ATPase
NaC3
OAT 1/3 hOAT 2
MRP 5/6
3 Na⁺
2-OG⁻ 2-OG⁻
OA⁻
ATP
ADP + Pi
ATP
ADP + Pi
BLM
2K⁺
Na⁺ 2-OG⁻
OA⁻ OA⁻

**FIGURE 30.3** Scheme for organic anion (OA⁻) transport in renal proximal tubule. *Abbreviations:* BBM, brush-border membrane; BLM, basolateral membrane; MRP, multidrug-resistance-associated protein; NaC3, sodium-dicarboxylate 3 cotransporter; OAT, organic anion transporter; OATP, organic anion transporting polypeptide; 2-OG⁻, 2-oxoglutarate. The "h" in front of some of the carriers indicates that the carrier has only been found in humans.

Na⁺
H⁺
Carnitine Na⁺
BBM
H⁺
OC⁺
OC⁺
ATP
ATP + Pi
OC⁺
Na⁺/H⁺ Exchanger OCTN 1
OCTN 2
MDR1
rOCT 1
OCT 2/3
BLM
OC⁺
OC⁺

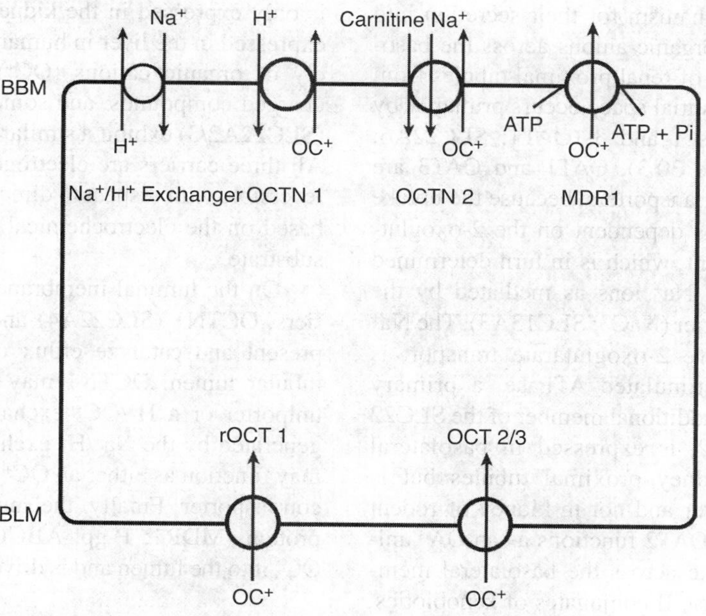

**FIGURE 30.4** Scheme for organic cation (OC⁺) transport in renal proximal tubule. *Abbreviations:* BBM, brush-border membrane; BLM, basolateral membrane; MRP, multidrug-resistance-associated protein; OCT, organic cation transporter. The "r" in front of some of the carriers indicates that the carrier has only been found in rats.

as being on the brink of anoxia. Accordingly, the pars recta (S3) segment of the proximal tubule and the medullary thick ascending limb in isolated, perfused rat kidneys are particularly susceptible to injury from hypoxia [8–12]. An additional series of studies from Epstein and colleagues [13–16] correlated the degree of cellular injury from hypoxia or ischemia with either ATP consumption (cellular work) or the inverse of mitochondrial activity.

## Organic Anion and Cation Transport

As suggested earlier, an important component of susceptibility of a specific cell type to chemically induced injury is whether that cell type possesses the capacity to transport and accumulate the chemical. Although several classes of transporters are involved in renal physiology and are localized in various nephron segments, perhaps the most important for toxicology and experimental therapeutics relating to the kidneys are those for transport of organic anions and cations. Although these carriers are also distributed in multiple segments of the nephron, they are expressed at the highest levels in the proximal tubules. As shown in Figure 30.3 and Figure 30.4, transport of organic anions and cations, respectively, occurs by multiple carriers and mechanisms. Although each carrier certainly has a discrete substrate specificity pattern, there is also considerable overlap between carriers, providing functional redundancy. Most of these carriers are polyspecific; that is, they transport multiple different substrates. Several recent reviews have been published on the properties and regulation of these carriers and their importance in renal function [17–23].

Organic anions, including many phase II metabolites of xenobiotics such as glucuronide, glutathione, and sulfate conjugates, are transported across the basolateral and brush-border membranes of renal proximal tubular cells which constitutes the mechanism for their secretion into tubular urine. Uptake of organic anions across the basolateral plasma membranes of renal proximal tubules from the renal plasma and interstitial space occurs primarily by organic anion transporters 1 and 3 (OAT1: SLC22A6; OAT3: SLC22A8) (Figure 30.3). OAT1 and OAT3 are considered tertiary active transporters, because the uptake of organic anions ($OA^-$) is dependent on the 2-oxoglutarate concentration gradient, which is in turn determined by counter-transport with $Na^+$ ions as mediated by the sodium–dicarboxylate carrier (NaC3: SLC13A3). The $Na^+$ ion gradient used to drive 2-oxoglutarate transport is achieved by ($Na^+$+$K^+$)-stimulated ATPase, a primary active ion transporter. An additional member of the SLC22 transporter family, OAT2, is expressed in basolateral membrane of human kidney proximal tubules but is expressed primarily in liver and not in kidney of rodent kidney proximal tubules. OAT2 functions as an $OA^-$ uniporter. Finally, $OA^-$ uptake across the basolateral membrane, particularly for phase II conjugates of xenobiotics, can occur by the multidrug resistance proteins 5 and 6 (MRP5/6: ABCC5/6), which are primary active transporters using the free energy of ATP hydrolysis to drive the uptake of $OA^-$.

In some respects, less is known about the mechanisms of the exit step by which $OA^-$ are transported across the luminal or brush-border membrane of renal proximal tubules into the tubular lumen (Figure 30.3). The major carriers involved in the efflux step include:

- Organic anion transporting polypeptide 1 (OATP-1): Slc21a1 (old nomenclature) or Slco1a1 (new nomenclature) in rats and mice; SLC21A3 (old nomenclature) or SLCO1A2 (new nomenclature) in humans
- Rat-specific OAT-K1 and OAT-K2: Slc21a4 (old nomenclature) or Slco1a3 (new nomenclature)
- Two MRP isoforms (MRP2 and MRP4): ABCC2 and ABCC4

OATP-1 catalyzes the uptake of $OA^-$ in exchange for the tripeptide GSH, which may be the principal carrier responsible for the delivery of intrarenal GSH to the active site of γ-glutamyltransferase for its turnover [24]. A member of the SLC22 family, URAT1 (SLC22A12), is expressed exclusively in renal proximal tubules and is responsible for the uptake of urate in exchange for $OA^-$, thereby maintaining blood levels of uric acid.

Many important drugs are organic cations or weak bases and are secreted into the tubular lumen by an array of organic cation transporters (OCTs) (Figure 30.4). These carriers, which are mostly members of the SLC22 carrier family, include two or three carriers, depending on species, on the basolateral plasma membrane and three carriers on the brush-border plasma membrane of the renal proximal tubule. On the basolateral membrane, rOCT1 (SLC22A1) is only expressed in the kidney in rodents and is mainly expressed in the liver in humans. rOCT1 transports a variety of organic cations ($OC^+$), some weak bases, noncharged compounds, and some anions. OCT2 and OCT3 (SLC22A2/3) exhibit a similar, broad substrate specificity. All three carriers are electrogenic, $Na^+$ independent, and reversible with respect to direction. Driving force is solely based on the electrochemical gradient of the transported substrate.

On the luminal membrane, two SLC22A family carriers, OCTN1 (SLC22A4) and OCTN2 (SLC22A5), are present and catalyze efflux of organic cations into the tubular lumen. OCTN1 may function as either an $OC^+$ uniporter or a $H^+/OC^+$ exchanger, with the $H^+$ gradient generated by the $Na^+/H^+$ exchanger. OCTN2, in contrast, may function as either an $OC^+$ uniporter or a $Na^+$/carnitine cotransporter. Finally, the multidrug resistance P-glycoprotein (MDR1; P-gp: ABCB1) catalyzes the efflux of $OC^+$ into the lumen and is driven by the hydrolysis of ATP.

## Drug Metabolism

Another key component of the intoxication process for most chemicals involves metabolism, as most chemicals are not toxic in their native form but must be metabolized to exert their effects. Thus, for polar or hydrophilic chemicals, when they have been transported into the renal cell by one of the various carriers discussed above, they can then be bioactivated by a large array of phase I and phase

II enzymes to yield potentially reactive metabolites. The traditional view of renal function is that the kidneys act as filters to remove toxic waste products from the blood via glomerular filtration or tubular secretion. Whereas the focus of most drug metabolism studies has been the liver, it became apparent that the kidneys have a significant capacity to carry out extensive oxidation, reduction, hydrolysis, and conjugation reactions [25,26]. Enzyme systems similar to those present in the liver and other extrarenal tissues are involved in renal drug metabolism. An important difference between the kidneys and other tissues is that many of the drug metabolism enzyme systems, such as cytochrome P450s, glutathione *S*-transferases, and others, are differentially distributed among the different nephron cell populations [3,4,27]. Thus, for example, whereas cytochrome P450s are exclusively found in the cortex (primarily the proximal tubules), prostaglandin synthetase, which catalyzes cooxidation of several drugs and xenobiotics, is localized in the medulla.

Knowledge of the nephron cell-type localization of drug metabolism enzymes and drug accumulation patterns is critical to understanding nephrotoxicity. A clinically important example of this is for acetaminophen [26]. Whereas the acute nephrotoxicity of acetaminophen is characterized by proximal tubular necrosis and is associated with metabolism by cytochrome P450, chronic nephrotoxicity of acetaminophen is characterized by papillary necrosis and interstitial fibrosis and is dependent on oxidation by prostaglandin synthetase in the inner medulla. This difference is largely due to the accumulation of acetaminophen in the cortex under acute exposure conditions and in the inner medulla under chronic exposure conditions. The differential distribution of acetaminophen thereby provides access to different drug metabolism enzymes.

Nonpolar or hydrophobic chemicals may also be bioactivated within the kidneys. Because of their chemical nature, these chemicals generally enter renal cells by passive diffusion rather than by specific membrane carrier proteins. Two related examples of this process are the halogenated solvents trichloroethylene [28] and perchloroethylene [29]. Each of these related chemicals may be metabolized by either cytochrome P450s or glutathione *S*-transferases, although trichloroethylene is a far better substrate than perchloroethylene for cytochrome P450s. Both chemicals, besides sharing metabolism by multiple enzymatic pathways, also share having multiple target organs. For the renal-specific, toxic effects of tri- and perchloroethylene, metabolism by the glutathione *S*-transferases is the relevant pathway. Both chemicals are converted, through a series of metabolic steps involving both intracellular and plasma membrane-bound enzymes, to the corresponding cysteine *S*-conjugates. These metabolites are then substrates for either the cysteine conjugate β-lyase or the flavin-containing monooxygenase, which convert them to chemically reactive species that are directly associated with alkylation of DNA and protein, oxidative stress, and various signal transduction mechanisms that lead to altered cellular function [28–30].

## Regulation of Renal Function

Glomerular filtration rate is primarily regulated by three factors: (1) The balance of pressures acting across the capillary wall (glomerular capillary hydraulic pressure and Bowman's space oncotic pressure tend to favor filtration, whereas glomerular capillary oncotic pressure and Bowman's space hydraulic pressure tend to retard filtration); (2) the rate of plasma flow through the glomeruli; and (3) the permeability and total surface area of the filtering capillaries. Each of these factors is, in turn, regulated by several physiologic factors, including local and systemic hormones. GFR is also regulated by a process called *tubuloglomerular feedback*, which involves transmission of a stimulus at the macula densa to the arterioles of the same nephron, based on the composition of the tubular fluid flowing past the macula densa. Tubuloglomerular feedback causes glomerular vasoconstriction when $Na^+$ delivery to the macula densa increases and glomerular vasodilatation when $Na^+$ delivery is decreased. The extent of the feedback response is also modulated by extracellular fluid volume. Thus, because $Na^+$ ions are the major solute available for renal excretion and $Na^+$ ions are the major cationic component of the extracellular fluid, changes in renal $Na^+$ excretion provide a primary driving force for these regulatory responses.

Renal effector mechanisms that modulate body fluid volume homeostasis include GFR, several peritubular and luminal factors (e.g., peritubular capillary Starling forces, luminal composition, medullary interstitial composition, and transtubular ion gradients), humoral mechanisms (e.g., the renin–angiotensin–aldosterone system, vasopressin, prostaglandins, atrial peptides, and endothelium-derived factors), and renal nerves. The reader is referred to textbooks of renal physiology [1] for more detailed information on these regulatory mechanisms.

Besides control of ionic composition and extracellular fluid volume, the kidneys play a central role in the handling of acids and bases. pH is controlled largely by the $HCO_3/pCO_2$ ratio. While the lungs regulate $pCO_2$, the kidneys control the $HCO_3$ concentration. Considering a plasma $HCO_3$ concentration of 24 mEq/L and the normally high GFR, the kidneys typically must reabsorb 5000 mEq $HCO_3$ each day. This occurs by two processes: (1) the carbonic anhydrase reaction (Equation 30.2), which is found in most cells but is highest on the brush-border membranes and cytoplasm of proximal tubular cells, and (2) transport across the luminal and basolateral plasma membranes.

$$CO_2 + H_2O \rightarrow H_2CO_3 \rightarrow HCO_3 + H^+ \qquad (30.2)$$

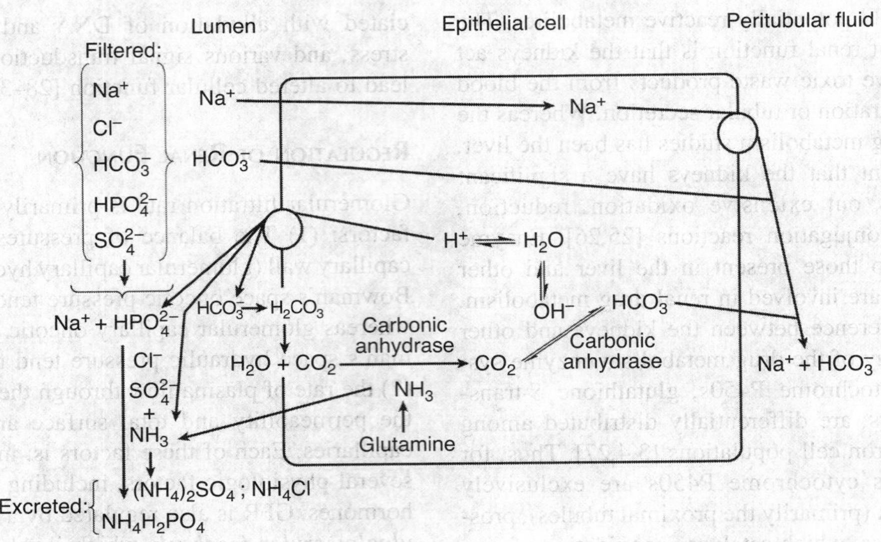


**FIGURE 30.5** Summary scheme of how renal epithelium regulates acid–base balance.

More than 95% of filtered $HCO_3$ is reabsorbed in the proximal tubules, with the remainder being reabsorbed in the loop of Henle. By the time the filtrate reaches the distal tubule or collecting duct, essentially all the $HCO_3$ has been reabsorbed. The pH in plasma and extracellular fluids is determined by the following equation:

$$pH = 6.1 + \log\{[HCO_3]/0.03\ pCO_2\} \qquad (30.3)$$

where $[HCO_3]$ is regulated by the kidneys and $pCO_2$ is determined by the lungs. At the physiological extracellular pH of 7.40, the ratio of $[HCO_3]/0.03\ pCO_2 = 20$. Normal arterial plasma concentrations of $HCO_3$ are 24 m*M* and normal arterial $pCO_2$ is 40 mmHg. Any primary change in either parameter causes the kidneys and lungs to respond to reattain the normal ratio.

Acid–base balance is also maintained by secretion of ammonia derived from amino acid catabolism. This occurs primarily in the distal nephron, where ammonia is secreted into the lumen and protonated. The protonated form of ammonia ($NH_4^+$) is impermeable to the luminal membrane; it becomes trapped in the lumen and is excreted. During metabolic acidosis, for example, renal ammonia production is increased by induction of amino acid deamination reactions, thereby increasing excretion of $H^+$ ions. A summary scheme of mechanisms by which kidneys maintain acid–base balance is shown in Figure 30.5. Factors that regulate or influence renal acidification and alkalinization are summarized in Table 30.3. These factors include renal hemodynamics, filtered load of ions, systemic acid–base conditions, and hormones, among others.

An important step in regulation of the volume, electrolyte, and acid–base homeostatic functions mediated by the kidneys, which influences susceptibility to toxicants, is organic anion transport. Modulation of organic anion secretion, which in turn influences $Na^+$ and fluid balance, is under hormonal control, with the subsequent receptor-mediated responses being signaled by protein kinases or other second messengers [17,23,31]. Protein kinase C activation reduces $OA^-$ uptake mediated by OAT1 and OAT3, reduces $OA^-$ efflux by MRP2, and reduces $OA^-$ transport by OATP1. Several potential biochemical and molecular mechanisms are involved in this downregulation. For OAT1, for example, regulation has been shown to occur by the modulation of: (1) gene transcription, (2) mRNA stability, (3) mRNA translation, (4) phosphorylation of the carrier protein, (5) internalization of membrane transporters, (6) recruitment of preformed transporters into the membrane, or (7) allosteric control by regulatory proteins. Additionally, the function of $OA^-$ transport may be indirectly modulated by regulation of $(Na^+ + K^+)$-stimulated ATPase activity.

In addition to protein kinase C, other protein kinases have been either shown or suggested to possibly regulate $OA^-$ transport, including protein kinase A (inhibition or no effect), mitogen-activated protein kinase/extracellular regulated kinase (MEK; inhibition), tyrosine kinase (inhibition or no effect), phosphatidylinositol 3-kinase (PI3K; inhibition), and calcium/calmodulin-dependent protein kinase II (inhibition). Hormones or hormone-like compounds acting through these various kinases and other second messengers such as cAMP, either inhibit (e.g., phenylephrine, dopamine, parathyroid hormone, bradykinin, endothelin, estradiol) or stimulate (e.g., catecholamines, epidermal growth factor, thyroid hormone, testosterone, dexamethasone) $OA^-$ transport and secretion. $OA^-$ secretion is faster in males than in females, illustrating the influence of the estrogen–androgen balance.

**TABLE 30.3**
**Factors Influencing or Regulating Renal Acid–Base Balance**

| Proximal Nephron | Distal Nephron |
|---|---|
| Filtered load: GFR, ultrafiltrate [$HCO_3^-$] | Aldosterone |
| Peritubular [$HCO_3^-$] and pH | Luminal buffering capacity |
| $pCO_2$ | Systemic pH, $pCO_2$ |
| Angiotensin II | $Na^+$ delivery and transport (cortical collecting tubule) |
| Extracellular volume | $K^+$ homeostasis |
| NaCl reabsorption | Anion transport |
| $K^+$ homeostasis | Parathyroid hormone, $Ca^{2+}$ homeostasis |
| Parathyroid hormone, $Ca^{2+}$ homeostasis | |
| Adrenergic nerve activity | |

## EXPERIMENTAL MODELS FOR ASSESSING RENAL FUNCTION AND TOXICITY

### General Considerations for Choosing a Model

A large variety of experimental models can be used to assess renal function and toxicity. Each model has its advantages and limitations. The choice of experimental model is often dictated by the questions being asked and the extent of knowledge of the process under investigation. Factors to be taken into consideration include the potential involvement of extrarenal processes and hormones in the renal response, knowledge of the nephron cell-type localization of any effects being studied, and the molecular mechanisms mediating the renal response. Some models are limited by an inability to distinguish nephron cell-type localization or by the lack of functional expression of certain biochemical processes. The various experimental models that can be used to study renal physiology and toxicological responses are summarized in Table 30.4, and each is discussed in more detail below. The reader is also referred to two reviews on this subject [32,33].

To gain a full understanding of how a chemical may cause nephrotoxicity, more than one experimental model is usually necessary, both to enable different types of questions to be addressed and to validate certain approaches (e.g., *in vivo* model to validate an *in vitro* model). An example of a situation in which the use of only one approach and experimental model could lead to incorrect conclusions can be found in studies on the cytotoxicity of inorganic mercury salts [34,35]. In these studies, Lash and colleagues found that inorganic mercury was a more potent cytotoxicant in suspensions of freshly isolated rat renal distal tubular cells than in rat renal proximal tubular cells. *In vivo*, however, the proximal tubule is the primary nephron segment target for inorganic mercury, and little cellular injury is seen in the distal nephron, except at very high doses [36]. The primary reason for the *in vivo* nephron-segment selectivity for mercury is that the proximal tubules are the first epithelial cells to be exposed to filtered or renal plasma mercury and possess sufficient membrane transporters and membrane and intracellular target molecules to bind and accumulate virtually all the mercury to which it is exposed. From the *in vitro* studies, therefore, one might erroneously conclude that it was the distal tubules and not the proximal tubules that were the primary target for mercury.

### Whole Animal and Clinical Assessments: Markers of Renal Function and Toxicity in Blood and Urine

Studies of renal function in whole animals or in humans typically involve measurement of parameters in blood or urine [37]. Such studies permit the collection of large amounts of data but are often limited in the ability to define biochemical or molecular mechanisms of injury. *In vivo* studies are often necessary or are one necessary component to establish target organ specificity or the involvement of extrarenal processes in nephrotoxicity or when more invasive procedures are not feasible, such as with humans. In experimental animals, however, *in vivo* studies have the advantage of allowing for the isolation of tissue for morphological analysis following chemical exposure or drug administration.

#### Urinalysis

Quantitative urine collection is used to more precisely determine treatment-related changes in renal excretory function and to define renal clearances. To enable quantitative and complete urine collection in animals, they are often housed in metabolism cages. A metabolism cage consists of an animal chamber mounted above an excrement collection system, which usually consists of a funnel and a urine and feces separator. Both urine and feces are collected for metabolism and metabolic balance studies. For assessment of renal function, urinalysis is what is performed. Major parameters determined in typical urinalysis are listed in Table 30.5. The various parameters

**TABLE 30.4**
**Summary of Selected Experimental Models Used To Study Renal Physiology and Toxicity**

| Model System | Uses/Advantages | Limitations |
|---|---|---|
| Clinical measurements and whole animals | Assess integrated physiological responses and regulatory mechanisms<br>Measure noninvasive markers of renal function<br>Applicable to humans | Minimal control of experimental exposure conditions<br>Limited ability to distinguish nephron cell-type involved<br>Limited ability to distinguish biochemical/molecular mechanisms |
| Isolated perfused kidney | Intact tissue structure<br>Distinguish intrarenal from extrarenal effects | Short-term viability (≤2 hours)<br>Interanimal variability<br>Incomplete definition of conditions<br>Expensive |
| Micropuncture and microperfusion | Determine tubular site of action<br>Quantitate transport processes in single nephrons | Sophisticated technology<br>Subject to technical errors and misinterpretation of results |
| Renal slices | Intact tissue structure<br>Ease of preparation<br>Drug screening<br>Measurement of drug metabolism | Short-term viability (≤2 hours)<br>Often limited access to luminal membrane<br>Poor oxygenation<br>Multiple nephron cell types present |
| Isolated perfused tubules | Determine tubular site of action<br>Quantitate transport processes in specific nephron segments | Short-term viability (≤2 hours)<br>Sophisticated technology |
| Isolated tubules/tubular fragments | Intact tubular structure<br>Precise definition of conditions<br>Ease of preparation<br>Several test samples with paired controls | Short-term viability (≤4 hours)<br>Often limited access to luminal membrane<br>Some mixture of cell types |
| Freshly isolated renal cells | Bidirectional exposure<br>Ease of preparation<br>Several test samples with paired controls<br>Cells from specific nephron segments<br>Drug screening | Short-term viability (≤4 hours)<br>Loss of plasma membrane polarization |
| Primary renal cell cultures | Similar to *in vivo* kidney<br>Longer-term viability<br>Maintenance of polarity | Difficult to maintain<br>Dedifferentiation<br>Limited lifetime relative to cell lines |
| Renal cell culture lines | Precise definition of incubation conditions<br>Immortalized<br>Reproducible<br>Easy to subculture<br>Transfect with cDNAs | Dedifferentiation<br>Often of ill-defined origin |

provide indications of the function of the kidneys in maintenance of fluid, pH, electrolyte, and solute balance. Additionally, the appearance of specific components in the urine at concentrations outside of the physiological range can provide information about the function of specific nephron segments. This application of urinalysis is particularly applicable in the clinical setting, where the search for biomarkers that can be early and sensitive indicators of exposure to toxic chemicals is a significant emphasis in research and was the subject of a National Research Council/National Academy of Sciences publication [38].

## Clearance Methods

Methods to assess renal clearance are fundamental to whole animal studies and have wide clinical applicability as well. These procedures involve the measurement of blood or plasma ($P_x$ in mg/mL) and urine ($U_x$ in mg/mL) concentrations of a substance. Urine flow rate ($V$ in mL/min) is also taken into account. Depending on the test substance used, either glomerular or tubular function can be assessed. Combining these parameters, gives the classical clearance equation:

$$C_x = U_x / P_x \quad (30.4)$$

The clearance ($C$) represents the quantity of blood or plasma cleared of the substance per unit time and has units of volume per unit time (usually mL/time).

The most common substances monitored for determination of renal clearance are inulin or creatinine. Inulin is

**TABLE 30.5**
**Parameters Often Determined in a Typical Urinalysis**

| Parameter | Purpose or Use of Measurement |
|---|---|
| Urine osmolarity or specific gravity | Overall fluid homeostasis |
| Urinary pH | Acid–base balance |
| Urinary volume | Overall fluid homeostasis |
| Urinary electrolyte and solute concentrations: $Na^+$, $K^+$, $Cl^-$, urea | Extracellular fluid balance |
| Creatinine excretion | Indicator of glomerular filtration |
| Glucose excretion | Proximal tubule function |
| Aminoaciduria | Proximal tubule function |
| Proteinuria: | |
| Low-molecular-weight proteinuria ($\beta_2$-microglobulin, $\alpha_1$-microglobin, retinol-binding protein) | Proximal tubule function |
| Albuminuria | Glomerular filtration |
| Enzymuria: | Specific enzymes derive from specific nephron segments |
| GSH *S*-transferase-alpha (proximal tubule) | |
| GSH *S*-transferase-pi (distal tubule) | |
| *N*-Acetyl-β-glucosaminidase (papillary collecting duct, proximal tubule > distal tubule) | |
| γ-Glutamyltransferase (proximal tubule) | |
| Tamm–Horsfall glycoprotein (thick ascending limb) | |
| Lactate dehydrogenase (distal tubule > proximal tubule) | |

the preferred indicator of GFR, as it is a fructose polysaccharide ($M_r$ = 5200 Da) that is freely filtered and then is neither reabsorbed nor secreted; hence, all of the inulin that appears in the urine arises from plasma via glomerular filtration. Inulin is considered the gold standard of exogenously administered markers for GFR. Its scarcity and high cost, however, have made its use less favored in the clinical setting. Inulin is readily measured in plasma or urine by a colorimetric assay, although glucose is also measured; separate measurement of glucose is necessary for correction to obtain the inulin concentration.

Creatinine is also used to monitor GFR and has the advantage of not requiring infusion of an exogenous substance. This makes creatinine useful for monitoring of GFR over a protracted period of time in the same individual. Because creatinine undergoes some reabsorption, however, measurement of its clearance tends to underestimate that of inulin by 10% or more, depending on species. Creatinine ($M_r$ = 113) is a metabolic product of creatine and phosphocreatine, which are both found almost exclusively in muscle. Creatinine production is proportional to muscle mass and varies little from day to day, although it can be affected by diet. Because of its convenience and low cost, creatinine is the most widely used indirect measure of GFR; however, correct interpretation of data in a clinical setting can be problematic.

One can also measure renal plasma flow (RPF) if a test substance is used that undergoes both filtration and active tubular secretion. *p*-Aminohippurate (PAH) is often used for that purpose because it is freely filtered by glomeruli and undergoes active secretion by several OATs. PAH extraction by the kidneys can vary from about 70% to 90%, depending on species, due to heterogeneity of the distribution of OATs and blood flow, thereby causing incomplete extraction. Renal blood flow (RBF) is converted from RPF by dividing RPF by the plasma fraction of whole blood, as estimated from the hematocrit (Hct), according to:

$$\text{RBF} = \text{RPF}/(1 - \text{Hct}) \tag{30.5}$$

The clearance ratio relative to a marker substance (usually inulin) can be calculated and used to determine whether a test substance is reabsorbed or secreted; thus, if a substance has a *C* value less than that of inulin, this indicates reabsorption. Glucose is a good example of a highly reabsorbed substance whose *C* value is always less than that of inulin. $C_{PAH}$, in contrast, is usually greater than $C_{inulin}$, indicating secretion.

Another method for measurement of GFR is to determine plasma clearance of an intravenous bolus injection of an indicator radionuclide-labeled marker. Renal clearance is measured as the plasma clearance, or the amount of indicator injected divided by the integrated area under the plasma concentration curve; models to estimate plasma clearance assume that the volume of distribution and renal excretion are constant over time. Two indicators that have been used and validated are $^{125}$I-iothalamate and $^{51}$Cr-labeled ethylenediaminetetraacetic acid (EDTA). The basic procedure involves injection of radiolabeled marker into arterial blood and sampling of venous blood. After correcting for noninstantaneous equilibration between arterial and venous circulations, plasma levels are measured and clearance is calculated from the slope and intercept of a line plotted on a logarithmic scale, using the formula:

$$C = V_o\left(\ln(2)\right)/t_{1/2} \tag{30.6}$$

where $V_o$ is the volume of distribution, and $t_{1/2}$ is the half-time for decay of marker levels in plasma.

An additional parameter to indicate GFR that is popularly used in whole animal, toxicity screening studies is measurement of blood urea nitrogen (BUN). As filtration slows, BUN rises and is generally paralleled by levels of creatinine; thus, in the absence of changes in protein intake or metabolism, BUN or plasma creatinine are useful indicators of GFR. An important point to realize is that, although BUN may be the most commonly used measurement for glomerular filtration, substantial loss of renal function (on the order of at least 50%) must occur before BUN rises above normal levels. BUN, like creatinine, is thus not a very sensitive indicator of renal injury. Additional urinary and plasma parameters, some of which are proposed to serve as sensitive biomarkers of injury in specific nephron cell populations, are discussed later.

## Experimental Models

### Isolated Perfused Kidney

As compared with *in vivo* studies in whole animals, the isolated perfused kidney shares the advantage of being a model that maintains intact renal structure. It also possesses the advantage of enabling separation of extrarenal from intrarenal factors that may be part of the mechanism of action. The methods used in animal preparation and kidney isolation and perfusion have been described in several reviews [39,40] and are only highlighted briefly here.

Because it is at the highest level of structural organization for *in vitro* kidney models, it is not surprising that the isolated perfused kidney preparation has been in use for nearly 150 years. The basic isolated perfused kidney preparation consists of the whole kidney isolated from the systemic vasculature and usually removed from the animal; it is then perfused through the renal artery with a synthetic plasma- or blood-like medium. Oxygenated perfusate is delivered into the renal vasculature with an adjustable pump so a constant, mean arterial pressure can be achieved and maintained during the course of experiments. Urine is collected through a cannula placed in the ureter. Several variations of the basic method have been implemented, depending on the purpose of the experiment; for example, various in-line monitoring devices have been included to measure perfusate temperature, pH, and oxygen content. The perfusate leaving the renal vein can either be recirculated or collected in a reservoir to achieve single-pass perfusion. The single-pass method has the advantage of ensuring a constant arterial perfusate composition, as it is not modified by renal metabolism or excretion, but has the disadvantage of being expensive because it consumes large amounts of perfusate, particularly if the perfusate contains physiologic colloids such as albumin. Although isolated perfused kidney preparations have been described for several species, including the dog, rabbit, guinea pig, pig, and toad, the most common species used in recent years has been the rat.

As summarized in Table 30.4, advantages of the isolated perfused kidney preparation include the fact that it retains kidney structure so it is particularly well suited for studies requiring intact renal morphology. It is the only *in vitro* preparation in which glomerular filtration and tubular functions are both retained in manners similar to the *in vivo* kidney. Control over many of the aspects of the perfusate composition is possible, including concentrations of dissolved and gaseous metabolites and electrolytes, colloids, oxygen carriers, drugs and their metabolites, and other chemicals added to modify function. This type of control enables establishment of reproducible exposures to chemicals or drugs under investigation. Also, because the isolated tissue is being used, higher concentrations of test chemicals are possible to achieve than could be used or tolerated *in vivo*.

Some of the limitations with the isolated perfused kidney model are those that are inherent with any *in vitro* model; for example, once the tissue is removed from the animal, function (e.g., GFR and tubular function) declines progressively over the course of experimentation, so the model can only be useful for a limited period of time, which is generally on the order of 2 hours. Because of this time limitation, studies are naturally restricted to acute effects as responses that occur over longer periods of time, such as changes in gene expression or enzyme induction, require longer exposure and study times. As explained by Diamond [40], the isolated perfused kidney exhibits several functional abnormalities that must be considered when evaluating results. These abnormalities include relatively high perfusion rates and low filtration fractions as compared with the *in vivo* kidney, loss of urinary concentrating ability, and glomerular proteinuria. Various structural abnormalities have also been observed in tubules of isolated perfused kidneys, including cellular necrosis of the medullary thick ascending limb and the S3 segments of the proximal tubule. The latter is particularly significant for drug metabolism and nephrotoxicity studies because the S3 segment of the proximal tubule is the site of a large proportion of the metabolism and toxic effects of exogenous compounds. Finally, other limitations exist in the ability of studies with the isolated perfused kidney to discriminate mechanisms of action and localization of processes in specific nephron cell types.

## Micropuncture and Microperfusion Techniques

*In vivo* microperfusion and micropuncture are two techniques that are used to study the function of individual nephron segments in the intact kidney [41]; for example, these methods can be used to directly determine the tubular handling of drugs and nephrotoxicants, characterize the function of specific carrier proteins, or determine the effect of toxicant administration on tubular function. A particular application of micropuncture is to determine the *in vivo* concentration and profile of luminal fluid along the nephron. Microperfusion, on the other hand, is a valuable approach because a drug or other chemical of interest can be applied directly to the tubule in the luminal perfusate and the effect on tubular function determined.

Although micropuncture and microperfusion are performed in several mammalian species, the rat is the most common experimental animal that is used. Moreover, specific strains of rats are often used in preference to others for particular purposes; for example, the Munich–Wistar rat is often used in studies of glomerular function because glomeruli are visible and accessible at the surface of the kidneys. Other strains, such as the spontaneously hypertensive or Brattleboro rat, are used to study nephron function under specific pathophysiologic conditions. Rats of 200- to 250-g body weight are usually used for micropuncture of superficial cortical nephron segments, whereas young rats weighing less than 150 g are used for micropuncture of collecting ducts, loops of Henle, or vasa recta because they are more readily accessible at this stage of development. Micropuncture or microperfusion are most often performed on the left kidney because of its accessibility and longer vascular pedicle as compared with the right kidney.

When the kidney has been isolated and placed in a holder, superficial nephron segments can be visualized with a stereomicroscope. The majority of the tubules visible on the surface are proximal convoluted (S1) tubules. As noted above, superficial glomeruli are visible in kidneys of strains such as the Munich–Wistar rat. Peritubular capillaries are visible as thin red lines between and around the tubules. A few distal tubular segments are also visible on the surface of the rat kidney and can be distinguished from proximal tubules by their thinner epithelial walls and smaller lumina. Typically, approximately 60% of the length of proximal tubules from dog or rat is accessible to micropuncture. Besides visual identification, proximal and distal tubules can also be distinguished by use of dyes such as lissamine green. After intravenous infusion of a small bolus of the dye, a distinct time course for appearance in the tubules is used to enable identification of specific tubular segments. Because of natriuretic effects and an increase in single-nephron GFR caused by lissamine green, micropuncture or microperfusion experiments should begin only after the dye has completely disappeared from the kidney and the effects of the dye on renal function have reverted to normal, which typically takes approximately 30 minutes.

Micropuncture has been instrumental in demonstrating tubular reabsorption and secretion of a large variety of endogenous and exogenous chemicals. It has the advantage that the tubule is studied in the intact, functioning kidney and can be studied as a physiologically stable preparation for several hours. One is limited, however, to using certain species. Because the method requires surgery, anesthesia is required, which necessitates consideration of the potential effects of anesthesia on tubule function. This concern can be taken into account by the use of paired control and experimental conditions in the same anesthetized animal. Another concern is that because of internephron heterogeneity, sampling from a single nephron may not be representative of all nephrons. Microperfusion is useful in determining and discriminating between luminal and peritubular actions of a drug or nephrotoxicant in the *in vivo* kidney, with the chemical being added specifically to either the tubular perfusion fluid or the systemic circulation to study luminal or peritubular action, respectively. Another advantage of microperfusion is the ability to distinguish net reabsorption from secretion, although other models can accomplish that as well (see below). Both micropuncture and microperfusion involve surgical procedures and careful placement of micropipets; thus, significant skill is required to conduct such microprocedures.

## Renal Slices

Renal slices have been a useful tool for many years to study renal physiology and biochemistry. Organic anion transport, for example, has been extensively studied with this technique. Most studies using renal slices have used slices from the cortex, in part because of the predominance of the proximal tubules in renal transport and toxic responses and because of its ready accessibility. The simplest method to prepare renal slices is freehand, using a razor blade and a glass microscope slide as a slicing guide [42]. Slices need to be uniform in thickness and be taken only from the section of kidney desired. To address the issue of uniformity of slice thickness and the attainment of slices that are as thin as possible, several automated tissue slicers have been developed. The first slicer developed to prepare so-called *precision-cut tissue slices* was the Krumdieck slicer [43]. The slicer is filled with an oxygenated, physiological buffer solution to preserve function, and it operates on the principle that a core of tissue is fed into an oscillating blade that is drawn across the core to make a slice. The slice is then swept away into a collection chamber. Slice thickness is adjustable, with optimum thickness considered to be around 250 μm [44]. The diameter of the core is preset and typically ranges from 4 to 8 mm. The Brendel/Vitron slicer, developed more recently [45], differs from the Krumdieck slicer in having a simpler design that is possibly more rugged and user friendly [44].

Once prepared, slices are typically incubated in a Dubnoff metabolic shaking bath, which permits control over temperature, shaking rate, and gas environment. Many assays are carried out in a bicarbonate-based buffer system, such as Krebs–Ringer, thus necessitating an atmosphere containing $CO_2$ (typically 95% $O_2$, 5% $CO_2$). There is some concern over oxygenation of the tissue slice and nephron heterogeneity; however, improvements in the procedure have made poor tissue slice oxygenation less of a problem. Incubations are typically done as for short-term cell cultures, using a serum-free, 1:1 mixture of Dulbecco's Modified Eagle's Medium (DMEM) and Ham's F-12. The medium is gassed with 95% $O_2$, 5% $CO_2$, and pH is adjusted to 7.4. Other media, such as Waymouth's culture medium or a Krebs–HEPES buffer, have also been used.

Several assays of renal function and toxicity have been used with renal slices. In most cases, parameters are normalized to slice weight rather than other parameters, such as cell number or protein, as this is the most convenient and appropriate factor to use. If the tissue slice becomes edematous or dehydrated, then tissue weight will be an inappropriate normalizing factor. Inasmuch as loss of protein may occur in nephrotoxicity, slice DNA content is recommended to normalize assays [44]. Tissue slice $K^+$ content is a good indicator of cell viability, as it reflects ($Na^+$ + $K^+$)-stimulated ATPase activity and membrane integrity. Losses in $K^+$ content are indicators of toxicity. As with other *in vitro* models, intracellular enzyme leakage (e.g., lactic dehydrogenase) is used as an indicator of membrane disruption and necrosis. Both cellular ATP content and respiration ($O_2$ consumption) are useful to assess energetic status of the tissue slice. Protein synthesis activity, as measured by incorporation of $^3$H-L-leucine, is also used to assess functional competence. Because active transport of organic anions and cations is a primary, energy-dependent function of proximal tubular cells, measurement of medium-to-slice ratios of transported substrates is often used as a measure of tubular viability.

Advantages of the renal slice technique include its versatility, correspondence with *in vivo* tissue, the general ease of preparation, its relative low cost, and the ease with which incubation conditions can be manipulated. Limitations to the use of renal slices include the possibility of collapsed lumens; poor oxygenation, particularly if slices are too thick; the presence of some damaged tubules, which is more problematic if slices are too thin; and the presence of multiple nephron cell types, thus making it difficult to ascribe observed effects to specific cell populations.

## Isolated Perfused Tubules

The techniques of isolating and perfusing individual nephron segments *in vitro* have been widely used in transport physiology studies since their development in the 1960s. Although nephron segments can and have been obtained from many species, tubules are the easiest to dissect from the rabbit kidney without the use of digestive enzymes. Detailed procedures for the harvesting of renal tissue, dissection and identification of nephron segments, and the design of perfusion and collection pipettes are described by Zalups and Barfuss [46]. S1, S2, and S3 segments of the proximal tubule, cortical collecting ducts, and medullary and cortical ascending thick limbs of Henle's loop have all been isolated from rabbit kidney and studied for their transport function and viability after exposure to various xenobiotics.

In most studies, perfusing and bathing solutions are simple electrolyte solutions supplemented with D-glucose and glutamine or glutamate, with pH adjusted to 7.4. When it is desirable to more closely mimic the *in vivo* state, isolated segments of the nephron can be perfused with an ultrafiltrate of rabbit plasma, made by passing the plasma through a 50-kDa cutoff filter. Thus, a great deal of control over the fluid environment can be achieved with this method. Another application of this model is to determine the transepithelial transport of solutes, including nephrotoxicants. Depending on whether the substrate is placed in the perfusate or bathing medium, lumen-to-bath or bath-to-lumen fluxes can be determined. Limitations in the application of this model for renal toxicology include the requirement for sophisticated and expensive equipment, the difficulty in preparing the nephron segments, the rather poor ability to isolate and perfuse nephron segments from species other than the rabbit, and a relatively limited lifetime for the model once nephron segments are isolated.

## Isolated Tubular Fragments and Isolated Cells: Freshly Isolated and Primary Culture

Tubular segments derived from specific nephron cell populations can be isolated by microdissection. These preparations maintain their normal tubular architecture, function, and polarity. Because the isolation method is rather tedious and the yield of tubules is typically low, other procedures, most notably treatment with collagenase, were developed; for example, perfusion of renal cortical tissue with buffers containing chelators such as EDTA and collagenase have been used to prepare tubule fragments from rats and rabbit. One problem was that these preparations were comprised of tubules from multiple nephron segments; hence, procedures, such as density-gradient centrifugation in Percoll or Ficoll or electrophoresis were then applied to obtain enriched preparations of tubules from either proximal tubule or distal tubule [47–53]. Although less commonly used than proximal tubular suspensions as an experimental model, tubule fragments from the medullary thick ascending limb have also been used [54–58].

In general, procedures for isolation of proximal tubules involve either enzymatic digestion and separation or mechanical separation [59]. In the enzymatic methods, cortical tissue can be digested either by perfusion with a collagenase-containing, balanced salt buffer or medium or can be dissected, minced, and incubated with the collagenase-containing buffer. Either of these methods provides a high yield of highly enriched proximal tubules that are largely S1 and S2 derived. Tubules can then be used for either short-term studies of metabolism, transport, and toxicity, or they can serve as seed material for primary culture [60–63]. Variations of culture and incubation conditions to increase oxygenation have also been used to improve the ability of rabbit proximal tubule primary cultures to maintain differentiated function [64]. Contamination with glomeruli can be minimized by including iron-oxide in the initial perfusate which collects in glomeruli and can be removed with magnets.

Individual epithelial cells, rather than tubule fragments, can also be isolated and used as either fresh suspensions for acute studies or as seed material for primary culture [32]. Isolation of single cells generally requires somewhat more vigorous enzymatic treatment than for nephron segments. As with methods to isolate tubules, once isolated, suspensions of cells require a secondary enrichment step to obtain highly enriched cells derived from a specific nephron segment. Hence, procedures such as Percoll density-gradient centrifugation [64,65] can be applied to enhance purity of a mixed cortical cell population (see below).

The decision to use either tubule segments or single cells as a model depends on several factors. First, the ease with which either tubule fragments or single cells can be prepared varies with species, so the choice of model is partly dictated by the species of interest. As noted above, tubule fragments are readily isolated from rabbit kidneys. Although tubule fragments can also be isolated from rat, it is easiest to obtain single cells [66]. Isolation procedures have also been adapted to prepare suspensions of proximal tubular cells from human kidneys [67,68]. This is a significant advance because use of cells derived from human kidneys eliminates the uncertainty involved in extrapolation of data from experimental animals to humans for risk assessment purposes. A second consideration is experimental design. One major difference between tubule fragments and isolated cells is that epithelial polarity is lost in the latter; thus, all membrane surfaces have equal access to substrates in the suspending medium. Additionally, collapse of the lumen in tubule fragments may occur, thus limiting access of substrate to the brush-border membrane.

Although single cells in suspension can be isolated from nephron cell types derived from multiple regions of the kidney, most studies have been performed with cells derived from the renal cortex and outer stripe of the outer medulla [32,33,65,66]. The basic procedure involves perfusion of the kidneys through the aorta below the renal arteries, first with a $Ca^{2+}$-free Hank's buffer containing ethylene glycol tetraacetic acid (EGTA) and then with a Hank's buffer containing $Ca^{2+}$ added back and collagenase. Because the inner medulla is poorly perfused, it can be removed intact with forceps, thus yielding a preparation that is predominantly of proximal tubular origin but also containing other cell types from the cortex and outer stripe of the outer medulla. Glomeruli, tubular fragments, and multicellular aggregates are removed by filtration through polypropylene mesh. The suspension of isolated renal cortical cells can then serve as a starting material for additional purification or enrichment steps to prepare multiple epithelial cell populations [32,33,65].

Advantages of the single-cell preparation include the high degree of homogeneity of the biological material, the equal access of all membrane surfaces to chemicals in the suspending medium (if membrane polarity is not important for the process being studied), the adequate amounts of material from a single animal to permit paired control and treated samples, and control of incubation conditions (e.g., temperature, pH, osmolarity, buffer and medium composition), thereby eliminating some of the uncertainty for mechanistic studies. Limitations of the procedure include potential damage to the cells during the isolation procedure, the limited life time of the cell suspensions (≤4 hours), and the loss of membrane polarity (if membrane polarity is important for the process being studied). Although potential damage can occur just by the collagenase perfusion step, any damage is typically modest and the versatility of the procedure allows paired controls and treated samples to be used for incubation, thereby helping to ensure that observed responses are due to treatments and not to cell isolation artifacts. Moreover, cell viability is routinely checked upon completion of the cell isolation by both trypan blue exclusion and LDH leakage assays. In the first case, cells are mixed with trypan blue in saline and are viewed under a light microscope on a hemacytometer. Only those cells whose plasma membranes are damaged will take up the dye and stain blue. Cells are thus counted for the proportion of staining, which is typically 5% to at most 15% after isolation. Similarly, for the LDH leakage assay, cells are mixed with pyruvate and NADH; oxidation of NADH is measured by the decrease in absorbance at 340 nm, and only those cells whose plasma membranes are damaged will either leak out LDH or allow NADH to enter and be oxidized. After recording the decrease in $A_{340}$ for 2 minutes, Triton X-100 is added to fully solubilize the membranes, thereby providing complete access of the cellular LDH to the added NADH; $A_{340}$

**TABLE 30.6**
**Composition of Medium Used in Primary Culture of Proximal Tubular Cells from Rat or Human Kidney**

Basal medium: Dulbecco's Modified Eagle's Medium/Ham's F12 (1:1)
HEPES (15 m*M*, pH 7.4)
$NaHCO_3$ (20 m*M*)
Antibiotics (only through day 3): penicillin (192 IU/mL), streptomycin sulfate (200 μg/mL), amphotericin B (2.5 μg/mL)
Insulin (5 μg/mL)
Transferrin (5 μg/mL)
Sodium selenite (30 n*M*)
Hydrocortisone (100 ng/mL)
Epidermal growth factor (100 ng/mL)
3,3′,5-Triiodo-DL-thyronine ($T_3$) (7.5 pg/mL)

is then recorded for another 2 minutes. Comparison of the slopes representing the rate of NADH oxidation (decrease $A_{340}$) gives a measure of cellular viability. Typically, the slope of the two rates for control cells is between 0.05 and 0.15, indicating intact cellular membranes and high viability. If measurements of trypan blue exclusion or LDH leakage fall outside of these typical ranges, the cells are not used for further studies.

As with the tubule fragments, specific nephron cell types in the cortical cell preparation can be enriched by a variety of methods. Although microdissection methods provide material of extremely high purity, the procedure is not particularly useful for most typical biochemical or toxicological studies, particularly those with cells from the rat or mouse, because of low yield of material. Additionally, such methods are fairly tedious and time consuming. The limited time frame for which the isolated cells are viable requires that they be used in studies as soon as possible after their isolation. Electrophoretic separation methods also suffer from the same limitations. Any procedure used to obtain enriched populations of renal epithelial cells must satisfy four primary criteria [32]: (1) the procedure must be relatively rapid, (2) any material used in the separation process must be biologically inert, (3) final yield must be high enough to enable performance of a sufficient number of paired control and test incubations, and (4) purity must be high enough to allow unambiguous conclusions to be made regarding the nephron cell type of origin.

Differences in cell density are one common means to separate different cell populations from one another. Although methods relying on cell density differences provide enrichment rather than absolute purification (because cell density is generally not a discrete property but exhibits a range of values for a given cell population), many of the procedures readily satisfy the four criteria noted above. Procedures for application of density-gradient centrifugation with Percoll to separate proximal tubular and distal tubular cells from cortical cell suspensions from the rat are described in detail elsewhere [32,65]. Percoll is a liquid consisting of microscopic, carbohydrate-coated beads that spontaneously forms a continuous density gradient when placed in a centrifugal field. Colored, density marker beads are commercially available to calibrate the density gradients. When distinct cell populations have been obtained, marker enzymes are used to confirm identity and to assess the degree of enrichment. Markers include parathyroid-hormone-sensitive adenylate cyclase, alkaline phosphatase, γ-glutamyltransferase, and glucose-6-phosphatase for proximal tubules and hexokinase and renal kallikrein for distal tubules.

A limitation with the use of suspensions of either tubule fragments or single cells is their short lifespan. To accommodate experiments that study processes that occur for longer time frames than the few hours that are possible with the freshly isolated tubules or cells and yet retain the use of a model that comprises material derived directly from the *in vivo* biological material, many investigators have established primary cultures of renal epithelial cells. The proximal tubule from rat, rabbit, or mouse has been the most common biological source for primary cell culture [32,60–64,69–79], although investigators have also developed culture methods for thick ascending limb and distal tubular cells [57,74,80]. Although tissue from rodents is more readily obtained, investigators have applied essentially identical methods to grow primary cultures of human proximal tubular cells [68,81–85]. Regardless of their species of origin, primary cultures of proximal tubular cells are generally grown in a serum-free, hormonally defined medium, one version of which is shown in Table 30.6. Serum is omitted because it encourages or enables the growth of fibroblasts, which comprise a very minor contamination of the cell preparation but can readily overtake the culture and prevent growth of the proximal tubular epithelial cells. The disadvantage is that the cells do grow more slowly in the absence of serum.

A photomicrograph showing human proximal tubular cells grown in monolayer culture and exposed to either control conditions or to the nephrotoxicant *S*-(1,2-dichlo-

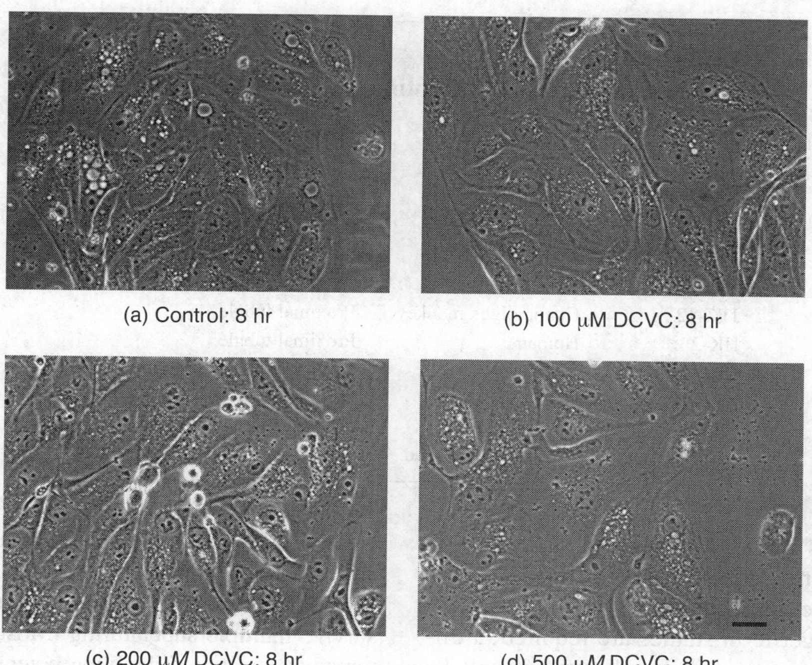


**FIGURE 30.6** Photomicrographs of confluent primary cultures of hPT cells treated for 8 hours with DCVC. hPT cells were cultured for 5 days in supplemented, serum-free, hormonally defined DMEM:F12 medium on 35-mm polystyrene dishes coated with vitrogen until they were confluent. Cells were then incubated for the indicated times up to 48 hours (results for 8 hours are shown) with either medium (= control) or various concentrations of DCVC. At the indicated times, photomicrographs were taken at 100× magnification on a Carl–Zeiss confocal laser microscope. Bar = 5 μm.

rovinyl)-l-cysteine (DCVC) illustrates the morphology of primary cell cultures under both physiological and toxicological conditions (Figure 30.6). Control cells exhibit normal epithelial morphology, with a prominent nucleus and a generally cuboidal shape with close cell-to-cell contacts. In contrast, cells incubated with increasing concentrations of DCVC exhibit an increased number of intracellular vesicles and many cells with elongated shape, indicating cellular injury. At the highest dose of DCVC used (i.e., 500 μ*M*), abnormal cellular shape is observed with numerous intracellular vesicles and vacuoles. We have shown that human proximal tubular cells incubated with a relatively low dose of DCVC (≤100 μ*M*) for a relatively short time period (≤8 hours) undergo both apoptosis and enhanced proliferation whereas cells incubated with a higher dose of DCVC (≥100 μ*M*) for a longer period of time (≥24 hours) exhibit primarily necrosis [86,87].

Use of primary cell cultures has many advantages. Inasmuch as cells grow in monolayers, plasma membrane polarity is maintained. Depending on the growth surface used, processes such as transepithelial transport can be studied. Many studies use renal cells grown on collagen-coated, plastic tissue culture flasks or dishes. Cells will grow with their luminal membrane face up; thus, although membrane polarity is maintained, media has direct access to only the luminal membrane. This problem has been solved by the use of semipermeable filters that are raised over the culture dish surface, thereby creating both a luminal and basolateral compartment to enable media to come in contact with both membrane surfaces. The only disadvantage of using such filters is that they are usually somewhat opaque, thus making it impossible to view the cells on the filter surface on a microscope. Another advantage is that the primary cultures are relatively simple to prepare. Because these cells remain viable in primary culture for up to 7 days (or longer if passaged), a large variety of assays and tests can be conducted with them, including all the assays one can perform with the freshly isolated tubules or cells, such as acute cellular necrosis as determined by trypan blue exclusion or LDH leakage, metabolism, transport, cellular oxygen consumption, and levels of key intracellular metabolites (e.g., ATP, glutathione), as well as assessments of apoptosis, repair and proliferation, gene expression, DNA damage, etc. Hence, the primary cell culture model is extremely versatile and, if cultured properly, should reflect very well the normal *in vivo* biochemistry and physiology of the proximal tubule.

To be sure, limitations also exist in the use of primary cultures. First, primary cultures are often difficult to maintain in a differentiated and functional state. This problem has been addressed by the use of serum-free, hormonally defined media and growth on appropriately coated sur-

**TABLE 30.7**
**Selected Immortalized, Continuous Renal Cell Lines**

| Cell Line | Species | Presumed Cell Type of Origin |
|---|---|---|
| $SGE_1$ | Wistar rat | Glomerulus |
| NRK-52E | Norway rat | Proximal tubule |
| $LLC\text{-}PK_1$ | Hampshire pig | Proximal tubule |
| OK | American opossum | Proximal tubule |
| MCT | Mouse | Proximal tubule |
| JTC-12 | Cynomolgus monkey | Proximal tubule |
| HK-2 | Human | Proximal tubule |
| GRB-MAL | Rabbit | Medullary thick ascending limb |
| M-m TAL-lc | Mouse | Medullary thick ascending limb |
| MDCK | Cocker spaniel | Distal tubule and collecting duct |
| A6 | African clawed toad | Distal tubule and collecting duct |

faces [32,33,60–64,74,75,78]. Another potential limitation is that primary cultures, if not passaged, remain viable only until they become confluent, which is typically within 7 days; hence, if longer term studies are required, a different experimental model must be chosen, as will be discussed in the following section.

In some cases, primary cell cultures can be passaged, thereby extending their usable lifetime to as much as several weeks [82]. In spite of the use of serum-free, hormonally defined media, such passaged cells often exhibit altered phenotypes with respect to drug metabolism, cellular energetics, and membrane transport properties [68]. This loss of differentiated function makes them a questionable model for use in assessing cell-type-specific toxicity, but they may be useful for study of specific processes whose expression or function are retained.

## Renal Cell Culture Lines

In contrast with primary cell cultures, cell lines are cells that have been transferred at least once to another culture surface (i.e., subcultured) and, more often, have been transferred multiple times. In some instances, the cellular genetics has been altered so that the cells become immortalized and can thus be passaged indefinitely. Such immortalized cells are referred to as *continuous cell lines*. The process by which cells can be immortalized can be spontaneous or can be induced by various agents (e.g., mutagenic chemicals, radiation, viruses). Some continuous renal cell lines that have been used for studying renal cell function and toxicology are listed in Table 30.7.

The development of immortalized cell lines of epithelial cells generally requires transfection of short-term cultures with plasmid or viral DNA containing an immortalizing gene (e.g., an oncogene). Taub [88] has reviewed in detail procedures for obtaining such continuous cultures. Others [89,90] have also successfully developed immortalized cell lines from proximal tubular and distal tubular primary cell cultures using the SV40 virus as a transforming agent. The need to use an immortalizing agent is necessitated by the generally poor ability of primary cultures to undergo and survive multiple subculturing. Most of the short-term cultures have a limited doubling potential and undergo senescence and eventually die. It is unclear whether the limited doubling potential of normal cells is an inherent property or is due to imperfections in the tissue culture environment; for example, primary cultures of human proximal tubular cells can be subcultured up to five or six times, but it is important that cell density is not decreased by more than 50% in each passage or cells will not grow [68,82].

Continuous epithelial cell lines offer several advantages in renal toxicology studies; for example, examination of processes that occur over relatively long time periods can be studied in these cultures, whereas the life time of primary cell cultures is insufficient for such studies. Other advantages are that these cell lines are relatively easy to use and produce highly reproducible results, without many of the difficulties inherent in the use of primary cultures. That said, however, the mere fact that the cell lines are immortalized indicates that they have undergone genotypic and phenotypic changes that make them different from the cell types from which they were derived, so it is critical to validate that the processes under study are expressed and regulated properly. Even when the process under investigation is expressed in the cell line, other processes that are normally found in the *in vivo* cell type or in the kidney as a whole may not be expressed or regulated in the same manner. This makes interpretations of physiological significance difficult and highlights a point made at the beginning of this chapter that multiple experimental models are required to fully understand processes of physiological, pathological, or toxicological importance. Nonetheless, immortalized cell lines have been invaluable for numerous studies of renal function and will continue to provide significant insight as long as the constraints and limitations to their use are kept in mind.

## ASSAYS OF RENAL CELLULAR FUNCTION

### URINARY ENZYMES AND OTHER PROTEINS

Measurements of urinary enzymes and other proteins were discussed earlier in the context of clinical and whole animal assessments of renal function and toxicity. In this section, the focus is on measurement of renal enzymes and other proteins that can be used as specific biomarkers of renal function or toxicant exposure. Biomarkers can be divided into three categories: biomarkers of susceptibility, biomarkers of exposure, and biomarkers of effect [38]. After a brief discussion of how each biomarker is defined, this section focuses on biomarkers of effect that can be used to provide specific, mechanistic insight into pathological or chemically induced renal injury.

A marker of susceptibility is an indicator of an organism's inherent or acquired sensitivity to toxicity from exposure to a specific xenobiotic. Some general factors that serve as markers of susceptibility include age, gender, and environmental exposures. Genetic polymorphisms may also render an individual more susceptible to the nephrotoxic effects of a given chemical. In terms of a urinary biomarker, for example, one may administer a test chemical that is metabolized by a specific enzyme that is known to exhibit genetic polymorphisms and then quantify metabolite levels in urine. Because determinants of susceptibility are multifactorial, validation of a marker of susceptibility must be carefully performed.

A marker of exposure is a chemical or a metabolite of the chemical or a product of an interaction between the chemical and some target molecule. One typically measures markers of exposure as concentrations of chemicals or their metabolites in blood, urine, or other body fluid or tissue. Difficulties in the use of such markers include that many drugs and other chemicals have the same or similar metabolites, making identification of the causative agent uncertain. Exposures to large doses of toxic chemicals may alter their rates of metabolism and excretion, thereby making it difficult to obtain a correlation between measured levels of the chemical or metabolite and the actual exposure. Moreover, without taking susceptibility into account, a marker of exposure in one individual may have markedly different functional implications than that in another individual.

As suggested by its name, a marker of effect should indicate the influence of a chemical exposure on some biological process. Markers of adverse effect can be indicators of altered biochemical processes in a target organ or cellular signals of tissue toxicity. The effects that occur may only be indirect in that they may only indicate a potential for toxicity but by themselves cannot be construed to directly produce toxicity. Examples of this type of marker are protein or DNA adducts. As discussed above in the section on whole animal and clinical assessments, parameters such as BUN and serum creatinine and urinary proteinuria can be indicators of renal injury. Their disadvantages, however, include a lack of sensitivity and an inability to indicate information about site or mechanism of action. As noted earlier, BUN and serum creatinine do not become significantly elevated until renal toxicity occurs to a significant extent. An ideal biomarker of effect, then, would be one that is highly sensitive, is measurable prior to a significant degree of renal toxicity has occurred, persists throughout the time course of the alterations in renal function, is specific to the kidneys, correlates in amount with the degree of injury, and is specific to a nephron cell type or specifically indicates a mechanism of action. Ideal markers should also be stable in urine over a long enough period of time and at a wide enough range of urinary pH values, and if they are enzymes they should not be readily inactivated in the urine. Some of the more sensitive and specific markers are shown in the bottom row of Table 30.5; examples include the urinary secretion of enzymes that are derived from various nephron segments. The ultimate goal of using biomarkers of effect for nephrotoxicity is to have an easily measured, sensitive, and stable indicator of renal injury at its very early stages so measures to counter the nephrotoxicity can be instituted prior to a significant decrease in renal function or to the advent of irreversible renal cell death.

A recently discovered marker for renal injury that has been described in a series of studies by Bonventre and colleagues [91–94] is kidney injury molecule-1 (Kim-1). Kim-1 is a type 1 transmembrane protein that is not detected in normal kidney tissue but is expressed at very high levels in dedifferentiated proximal tubular epithelial cells of human and rodent kidneys after either ischemic or chemically induced injury. It appears to satisfy several of the criteria for being an ideal biomarker of effect for renal injury: Kim-1 is stable in urine for prolonged periods of time and is specific to the kidneys. Its expression increases markedly from a baseline of essentially zero, and the increased expression occurs early in the pathologic or toxic event, thus indicating a high degree of sensitivity. Both protein and mRNA expression are increased in injured renal tissue. It also satisfies the criterion of being absent in normal kidney tissue and only being detectable in tissue that has been injured or is undergoing repair and proliferation.

Besides the biologic markers discussed above, a large number of other markers have been suggested to be linked to cell injury. These and the ones discussed above may be divided into several categories [38]: (1) immunologic factors (e.g., humoral factors such as antibodies and antibody fragments, components of the complement cascade, and coagulation factors); (2) lymphokines; (3) major histocompatibility antigens; (4) growth factors and cytokines (e.g., platelet-derived growth factor, transforming growth factor, tumor necrosis factor, interleukins); (5) lipid mediators (e.g., prostaglandins, thromboxanes, leukotrienes); (6) extracellular-matrix components (e.g., collagens, lami-

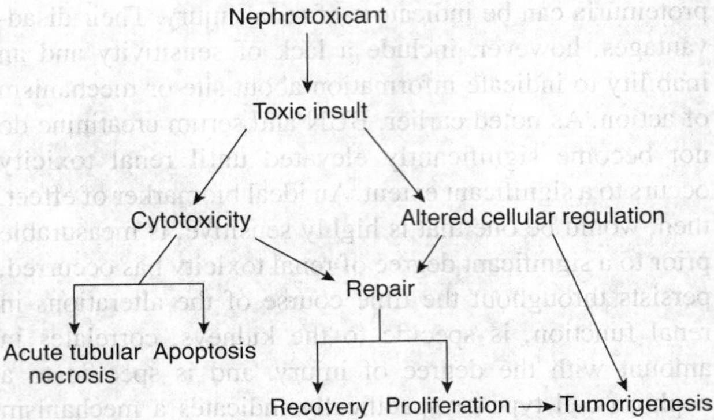


**FIGURE 30.7** General scheme of renal cellular responses to nephrotoxicants.

nin, fibronectin); (7) adhesion molecules; (8) reactive oxygen and nitrogen species; (9) transcription factors and protooncogenes (e.g., c-*myc*, c-*fos*, c-*jun*, c-Ha-*ras*, *Egr-1*); (10) tubule antigens (e.g., Tamm–Horsfall protein, brush-border enzymes, cystatin); (11) heat shock proteins; and (12) endothelin. Validation is required, however, for many of these markers.

## Assays of Cellular Necrosis and Apoptosis

Most of our knowledge about mechanisms of renal cell injury comes from studies that either focus on or use preparations of proximal tubular cells. This is because the proximal tubule is the easiest nephron cell type to obtain in high purity and is the primary target site for the majority of drugs and other toxic chemicals that target the kidneys, due to the large array of plasma membrane carriers and drug metabolism enzymes present. Although most of the mechanistic information about cell death has come from *in vitro* studies, some information, particularly about target cell identity and involvement of hormonal, neuronal, or other extrarenal factors, has come from *in vivo* studies. Renal cells that are intoxicated by drugs or other chemicals can undergo several potential responses, including cell death, growth arrest, or proliferation and repair. It has become clear over the past several years of research, that for a very large number of toxic chemicals, exposure can elicit all three responses. The interplay between these responses is illustrated in Figure 30.7; hence, as a response to a toxic insult, epithelial cells may either exhibit altered regulation of cellular function or cytotoxicity. The altered cells may either progress further to undergo neoplastic transformation or may be repaired. Cells experiencing cytotoxicity may either be repaired or, depending on exposure conditions, may undergo cell death by either apoptosis or necrosis. We have proposed such a scheme for trichloroethylene-induced nephrotoxicity [87]. Although the kidney can repair itself after sublethal injury, it has been hypothesized that repeated cycles of sublethal injury followed by repair may lead to uncontrolled cell proliferation and neoplastic transformation [30]. The factors that determine which response occurs and predominates are not completely understood but certainly include dose and time of exposure.

Cell death, both in general and for renal tubular epithelial cells in particular, can occur by two basic mechanisms, apoptosis or necrosis (oncosis) [95]. Some basic features of necrosis and apoptosis are compared in Table 30.8. The many differences between the two forms of cell death are very dramatic. Thus, whereas apoptosis is a highly regulated, energy-dependent process, necrosis can be viewed as an uncontrolled, pathological response; whereas apoptosis involves events in single cells, necrosis typically involves large numbers of cells in a wide area of tissue and is often mediated *in vivo* by infiltrating macrophages and similar inflammatory mechanisms.

Cell death by necrosis is typically measured by assays for plasma membrane permeability, such as LDH release or trypan blue uptake. The LDH release assay method is shown schematically in Figure 30.8, which illustrates typical results for a control incubation (left panel), which exhibits a small amount of LDH release, and an incubation

**TABLE 30.8**
**Comparison of Selected Features of Necrotic and Apoptotic Mechanisms of Cell Death**

| Parameter | Necrosis | Apoptosis |
|---|---|---|
| Cell shape | Swelling | Shrinkage |
| Membrane | Increased permeability | Intact |
| Energy dependence | No | Yes |
| Protein synthesis required | No | Yes |
| Toxicant dose | Relatively high | Relatively low |
| ATP depletion | Yes; early effect | Not until just before cell death |
| Second messengers | Not involved | Often involved |
| DNA damage | Random SS/DS breaks | Laddering |
| Nuclear damage | Chromatin damage, lysis | Chromatin condensation |

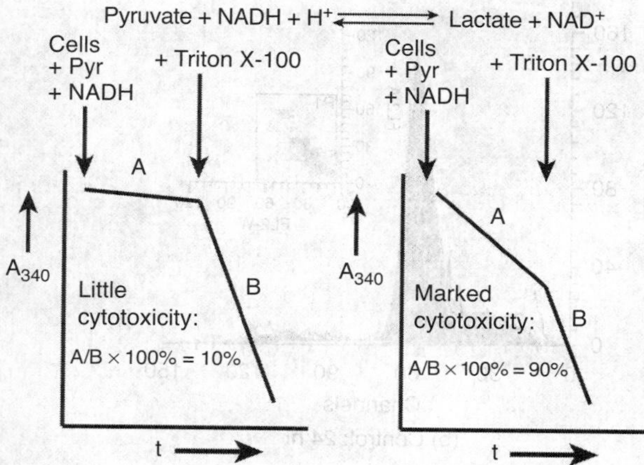


**FIGURE 30.8** Lactate dehydrogenase (LDH) release assay.

with a highly cytotoxic concentration of a chemical (right panel), which exhibits a large amount of LDH release. Results with trypan blue exclusion (or with another vital dye) would be similar to those with LDH release. Trypan blue exclusion is measured by mixing an aliquot of cells with the dye and applying it to a hemacytometer; this enables both assessment of viability by dye exclusion but also determination of cell number. For routine assessments of cell viability, LDH release is recommended because of its simplicity, its widespread use which has been validated in numerous studies with isolated cells from multiple tissues, and its lack of potential bias as can exist with the counting of stained cells under a microscope.

The LDH release assay was originally developed and used in suspensions of isolated cells [66]. With the advent of so many studies being conducted with either primary cell cultures or continuous cell lines, some variation to the basic method is needed, although the principles of the assay remain the same. For adherent cells attached to a stationery matrix, LDH activity is first measured in media as NADH oxidation at $A_{340}$. After removal of media and washing with PBS, cells are solubilized with 0.1% (v/v) Triton® X-100, and LDH activity is determined in the total cell extract. The fraction of LDH release is then used as an index of irreversible injury or necrosis, and is calculated according to:

$$\%\text{LDH release} = \left[\frac{(\text{LDH activity in media})}{(\text{LDH activity in media} + \text{LDH activity in total cells})}\right] \times 100\% \quad (30.7)$$

In some cases, however, LDH release cannot be used as a measure of cell viability because the toxicant directly inhibits LDH activity, such as occurs with inorganic mercury [34,35]. The likely mechanism of inhibition is direct binding to the critical cysteinyl sulfhydryl group of LDH; hence, any sulfhydryl-reactive chemical or a chemical that is metabolized to a sulfhydryl-reactive species will likely inhibit LDH, although potency will likely vary among compounds. In the situation in which LDH activity is inhibited, total LDH activity can be used as an estimate of cell viability, with activity calculated according to:

$$\text{Total LDH activity} = \text{LDH activity in media} + \text{LDH activity in total cells} \quad (30.8)$$

An additional assay that can be used to estimate cell viability is the MTT cell proliferation assay. MTT, or 3-(4,5-dimethylthiazol-2-yl)-2,5-diphenyltetrazolium bromide, is reduced by live cells to a strongly pigmented, purple formazan product. After solubilization, the absorbance of the formazan product can be measured in a microplate reader at 570 nm. A kit is available from Molecular Probes (http://probes.invitrogen.com) that is designed as a high-throughput assay of cell viability and proliferation. The manufacturer cautions that, because MTT can be reduced intracellularly by both various dehydrogenases and glutathione *S*-transferases, MTT may not always be a reliable probe of cell viability in cells exposed to chemicals that affect these enzymes. Nonetheless, it has become a commonly used and convenient assay.

Cells undergoing apoptosis, in contrast to necrosis, exhibit several unique morphological features (see Table 30.8). Detection of these features has been used in several assays to detect apoptosis. One common method involves *in situ* histochemical staining of cells and light microscopy [96]. This method, called the *TUNEL* method (terminal deoxynucleotidyltransferase [TdT]-mediated dUTP-biotin nick end labeling), relies on the *in situ* labeling of DNA breaks in individual nuclei in tissue sections processed by routine histopathology procedures. TdT specifically binds to 3′-OH ends of DNA that are exposed by proteolytic treatment. This is followed by synthesis of a labeled polydeoxynucleotide molecule, formed by the use of TdT to incorporate biotinylated deoxyuridine into the sites of DNA breaks. Signal is amplified by avidin-peroxidase, enabling detection by light microscopy.

Some forms of apoptosis can be detected by running agarose gels with ethidium bromide staining to detect DNA laddering that often, but not always, occurs in cells undergoing this form of cell death. The laddering is based on activation of endonucleases that cleave the DNA between nucleosomes, producing 200-bp multimers. While this is a useful method, it is not quantitative and is not necessarily associated with apoptosis.

Several flow cytometry methods have been developed to analyze the cell cycle or to label cellular components with, for example, fluorescent tags, and then sort cells according to their content of the fluorescently-labeled component. Advantages of methods using flow cytometry include their ability to identify and readily quantify multiple subpopulations of cells based on a large variety of parameters, and their ease of use. Flow cytometers can

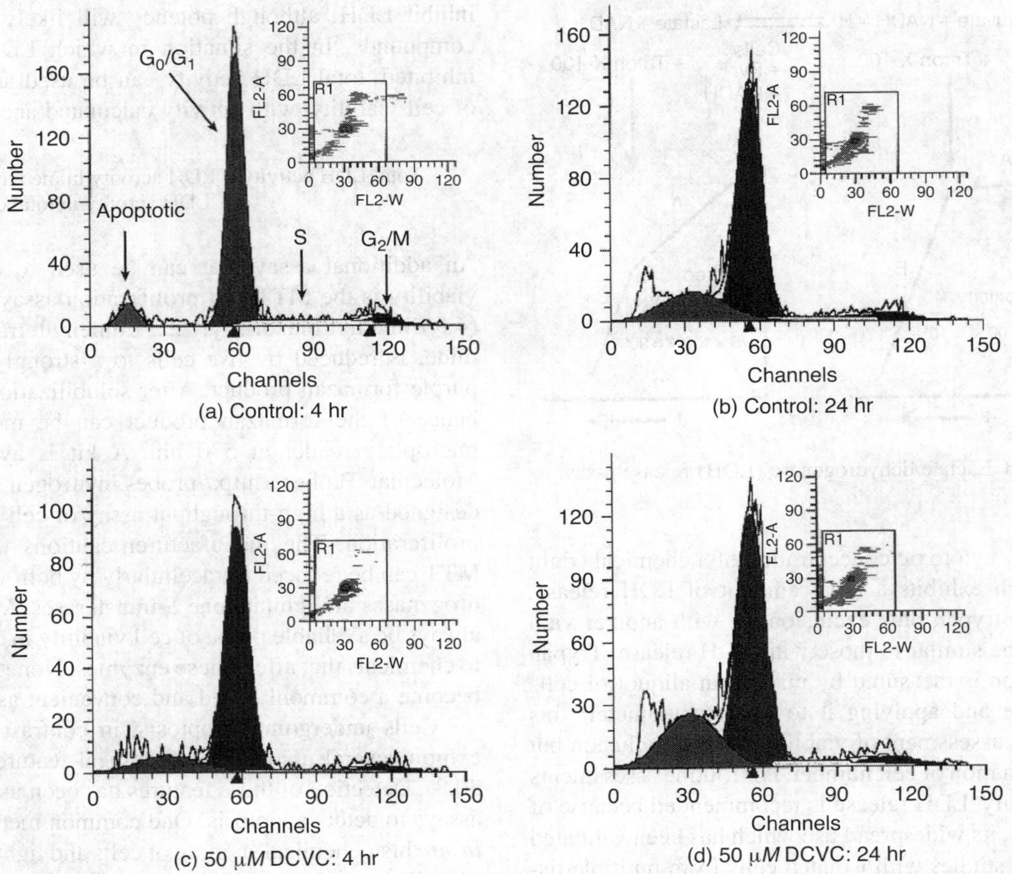


**FIGURE 30.9** Flow cytometry analysis of propidium iodide (PI)-stained hPT cells treated for 4 or 24 hours with 50 μ*M* DCVC. Confluent hPT cells were grown on 35-mm polystyrene culture dishes and were treated for the indicated times with PBS (= control) or 50 μ*M* DCVC. Cells were then harvested by trypsin–EDTA digestion, washed in sterile PBS, fixed overnight in ethanol, stained with propidium iodide, and then analyzed by flow cytometry with a Becton Dickinson FACSCalibur® flow cytometer. Peaks from left to right represent apoptotic cells, cells in $G_0/G_1$, cells in S phase, and cells in $G_2/M$. Results are from a single preparation and are typical of those from 6 separate experiments. Note that $G_0/G_1$ peaks comprise the majority of cells in most incubations and are somewhat broad because the cells are confluent. The coefficient of variation (CV) for these peaks ranged between 3 and 18%, with approximately 80% of the peaks in samples exhibiting CV values of 5 to 8%. *Insets:* Distribution of cells according to fluorescence intensity. Cells outside the box are those that were excluded from the analysis due to aggregation.

quantify the proportion of cells in $G_0/G_1$, S, and $G_2/M$ phase of the cell cycle by the use of fluorescent DNA dyes. For example, one can easily identify cells in $G_2/M$ phase as compared to those in $G_0/G_1$ phase because the former have twice the amount of DNA as the latter. Besides emitted light, the flow cytometer also analyzes scattered light from cells passing through the laser beam. Forward scatter and side scatter indicate cell size and granularity (density), respectively. Because apoptotic cells are typically condensed, the flow cytometer detects them as smaller, less fluorescent entities. Cells are stained with propidium iodide (PI), which is a DNA-intercalating agent like ethidium bromide. The analytical procedure used to detect the different cell populations is fluorescence-activated cell sorting (FACS) analysis. An example of such an analysis is shown in Figure 30.9, where primary cultures of human proximal tubular cells were incubated with either medium alone (= control) or with 50 μ*M* DCVC [86]. A marked increase in the fraction of apoptotic cells is evident after incubation with DCVC.

Another flow cytometric assay for apoptosis involves assessment of the changes in plasma membrane morphology, which is an early effect involved in cells that undergo apoptosis. In such cells, the membrane phospholipid phosphatidylserine is translocated to the outer face of the plasma membrane, thereby exposing it to the extracellular environment. Annexin V is a 35- to 36-kDa, $Ca^{2+}$-dependent, phospholipid-binding protein with a high affinity for phosphatidylserine. In measurement of apoptosis, Annexin V is conjugated to a fluorochrome such as fluorescein isothiocyanate (FITC), enabling binding of Annexin V–FITC to be a sensitive probe of early apoptosis. Cells are stained with both Annexin V–FITC and a vital dye such as propidium iodide (PI); they are then subjected to flow cytometry and FACS analysis and are identified as early apoptotic (Annexin V–FITC positive and PI negative), late apoptotic or necrotic (Annexin V–FITC positive and PI positive), or viable (Annexin V–FITC negative and PI negative).

Several other assays for apoptosis in renal epithelial cells are available, involving flow cytometry, western blotting, or enzymatic assays. Examples include measurements of apoptosis-associated proteins, such as Bcl-2 family members, cytochrome *c*, and cleavage products of poly(ADP-ribose)-polymerase (PARP). For more details on mechanisms of apoptosis and assays used to quantify apoptosis, the reader is referred to recent reviews [97]. As a general rule, it is advisable to assess cell death, and apoptosis in particular, by more than one method and to study cellular responses over a long enough time course and a broad enough concentration range of test chemical.

## Renal Cellular Function and Cytotoxicity

Although measurements of cell death by necrosis or apoptosis provide endpoints for screening potentially nephrotoxic chemicals, it is often important to assess effects of chemical exposures or pathological states that are more subtle or that provide an indication of potential toxic effects prior to irreversible cell death. These would include measurements of cellular respiration, active ion transport, such as ($Na^+ + K^+$)-stimulated ATPase activity; cellular redox status; and concentrations of key intracellular metabolites, such as ATP and GSH. Mitochondria are common and early targets for many nephrotoxic chemicals; hence, sensitive assessments of cellular energetics and mitochondrial function should provide good indicators of agents whose effects can be mediated through this organelle [98]. One reason the mitochondria are such a critical and prominent intracellular target for nephrotoxicants is that they contain a large number of protein thiol groups that can become inactivated by oxidation or alkylation and many toxic chemicals are bioactivated to oxidants and reactive electrophiles.

## Molecular Markers of Renal Cellular Repair, Regeneration, and Proliferation

As illustrated in Figure 30.7, renal epithelial cells can respond to toxicant exposures in a variety of ways, depending in part on dose and time of exposure. Other factors are surely important in determining whether the exposure leads to cell death, repair, or uncontrolled proliferation. There are a number of markers for repair and proliferation for the kidneys [99]; for example, the normal, differentiated proximal tubular cell expresses cytokeratins but not vimentin, which is a marker for endothelial cells. Renal proximal tubular cells that have either undergone neoplastic transformation or have dedifferentiated and are undergoing repair and proliferation do express vimentin. Morphologically, dedifferentiated proximal tubular cells exhibit differences from the normal, differentiated state, such as the lack of a brush border and an elongated rather than a cuboidal shape. Although the renal epithelium has regenerative capacity, the normal, terminally differentiated renal epithelium is nonproliferating, as evidenced by the flow cytometry FACS analysis shown in Figure 30.9, which shows that <10% of control cells are in S phase. Assays such as radiolabeled thymidine incorporation, 5-bromo-2′-deoxyuridine (BrdU) incorporation, and expression of proliferating cell nuclear antigen (PCNA) can be used to demonstrate repair and proliferation of the renal tubular epithelium.

## Applications of Genomics, Proteomics, and Metabolomics to Renal Physiology and Toxicology

Recent advances in genomics, proteomics, and metabolomics provide new opportunities for studying the molecular and cell biology of the renal cellular response to toxicant exposure. In the area of genomics, cDNA microarrays can be used to determine genes or classes of genes that are either upregulated or downregulated as a consequence of toxicant exposure. Patterns of gene expression changes can provide a signature for a chemical or class of chemicals that modulate renal function. Although microarrays are indicative of changes in mRNA expression, they may not translate into functional changes. The additional application of proteomics approaches can provide more detailed information on changes in both protein expression and posttranslational modifications of proteins (e.g., phosphorylation status), which can thus provide functional information. Proteomics can also be used to determine the identity of proteins that are specifically modified by reactive species generated by xenobiotic metabolism. In this manner, detailed molecular information about how a toxicant produces its effects can be gained.

The area of metabolomics can also be applied to provide insight into how nephrotoxicants produce their effects. The general approach relies on measurements of intermediary metabolites in urine, using such techniques as $^{13}C$-nuclear magnetic resonance spectroscopy. Just as cDNA microarrays can provide a signature for the effects of a toxicant on mRNA expression, metabolomics can provide a signature for the effects of a toxicant on pathways such as glycolysis, oxidative phosphorylation, and amino acid metabolism. Although none of these approaches can be readily performed in most laboratories, due to the requirement for specialized equipment, they are becoming more commonly used in toxicology and promise to open up new areas of investigation for the future.

## HUMAN DISEASE AND RISK ASSESSMENT

### Experimental Models of Renal Diseases

An important experimental approach to understanding and design of improved treatments for human diseases is the development of experimental models that mimic various aspects of the disease under study. Genetically modified animals, such as transgenic mice, have been used to mimic conditions that exist in a variety of human renal diseases. Renal diseases that have been studied include diabetic nephropathy, polycystic kidney disease, and various glomerular nephropathies, to name a few. Special rat strains are also available, such as the spontaneously hypertensive rat, to study the role of renal function in specific pathologies. Experimental approaches used to study renal function and toxicity will be similar with these animals and normal animals. Caution must be exercised, however, in the interpretation of findings because a genetic modification may alter multiple processes, so a complete characterization of these animals is necessary.

### Extrapolation of Animal Data to Humans

Although experimental animals such as rats and mice are frequently used to study nephrotoxicity and renal function, humans are the ultimate species of interest. Laboratory animals or *in vitro* preparations derived from such animals are used instead of humans or tissue from humans for several reasons. First, the design of clinical studies cannot include investigator-initiated study of mechanisms of nephrotoxicity. It is only when humans are inadvertently exposed to a nephrotoxic chemical, such as might occur in an occupational setting or due to environmental contamination or poisoning, can information about toxicity be gleaned. Several limitations exist, however, with such studies; for example, exposure information is often incomplete or limited, requiring the use of assumptions and approximations. The level of control of experimental design that exists with animal studies is not possible with clinical studies. Second, although *in vitro* studies with renal tissue from humans are possible and allow the investigator to exert the same degree of control over exposure conditions as are available with *in vitro* studies from animal tissues, human tissue is not always available and is significantly more expensive than studies with animal tissues. Nonetheless, mechanistic studies with renal tissue from humans are feasible and offer many advantages. The use of human tissue for studies of chemically induced nephrotoxicity and renal function removes the need for interspecies extrapolation. Although such extrapolations have been applied successfully for many chemicals, they have been much less successful for other chemicals. A prime example of a situation where rodent-to-human extrapolations have yielded equivocal results for human health risk assessments is that of trichloroethylene [30]. For halogenated solvents such as trichloroethylene, rodents appear to be poor surrogates for humans because of inherent differences that cannot be readily corrected, so, even though rodent data are used, the uncertainty factors applied to the human health risk assessments are rather large compared to those used for many other classes of chemicals.

## FUTURE CONSIDERATIONS

While many advances have been achieved in recent years in the study of nephrotoxicity, several areas of research require continued development or refinement and many new techniques and methods are only beginning to be exploited for the study of renal function and toxicity. Examples of the former include the various cell culture models and biomarkers of effect, exposure, and susceptibility. Continued application of molecular biology techniques to cell culture can help improve the ability of these models to mimic the *in vivo* renal cell. Application of approaches such as siRNA technology to and the use of transgenic animals in renal toxicology should also improve our ability to define exposure conditions, understand susceptibility to renal cellular injury, and better define mechanisms of action. Although numerous markers of effect are available for the kidneys, most do not satisfy enough of the criteria to be truly useful as indicators of early stages of damage. Continued development of biomarkers that can indicate mild alterations in some parameter of renal function at early times after exposure is needed. Some recently developed markers, such as Kim-1, appear to satisfy the criteria needed for an effective biomarker.

## SUMMARY

When considering the kidneys as a target organ for chemically induced toxicity, many unique features are apparent that determine susceptibility. Four major factors are critical: (1) high renal blood flow, (2) the ability of the kidneys to concentrate chemicals in the intraluminal fluid, (3) the ability to actively reabsorb or secrete chemicals through the tubular epithelial cells with high activity, and (4) the presence of enzymes for the bioactivation of a protoxicant to reactive intermediates.

In choosing an experimental model with which to study renal function and toxicity, several questions should be asked first: (1) What is known about tissue and nephron cell type specificity? (2) What is known about metabolism? (3) What is known about transport? (4) Are extrarenal factors or processes required for the expression of nephrotoxicity? (5) What is the time frame over which the nephrotoxic response occurs? Choices of models using experimental animals (e.g., rat, mouse, rabbit) range from the whole animal to a range of *in vitro* model systems, including the isolated perfused kidney, isolated perfused tubules, renal slices, isolated tubular fragments, isolated cells, primary cell culture, and immortalized renal cell lines. Each model has distinct advantages and limitations. Depending on answers to the questions above, one should choose multiple models to address renal function and toxicity.

When certain types of information about a putative nephrotoxicant are known, the model can then be chosen. The whole animal model is important for determining target organ specificity and the influence of extrarenal factors, including immunological and hormonal. Certain of the *in vitro* models have the advantage of maintaining tubular structure and epithelial membrane polarity. A critical issue in using many of the isolated tubule or cell models, including primary cell culture and immortalized cell lines, is how well differentiated function is maintained. The importance of this issue is dependent on the process being studied, as some models that only express some of the normal, *in vivo* phenotype will be good models if the process being studied is retained. Several of the *in vitro* models have the added advantage of being amenable to being used for screening of potentially nephrotoxic chemicals. High-throughput assays, such as those that can use a plate reader, are available for many of the assays of renal cellular function and toxicity. The screening approach is particularly useful to eliminate candidate drugs in the earliest phases of investigation.

## QUESTIONS

1. Describe procedures used to measure GFR.
2. Explain the biological significance of high-molecular-weight and low-molecular-weight proteinuria.
3. Explain how nephron heterogeneity determines the pattern of effects of nephrotoxicants.
4. Explain how tubuloglomerular feedback regulates GFR.
5. Describe the desired properties of a chemical that would be ideal for use in renal clearance measurements.
6. Compare and contrast whole animals and various *in vitro* models for their use in studies of nephrotoxicity and renal function in terms of their advantages and limitations.
7. Describe properties of an ideal biomarker of effect for use in human health studies.
8. Described differences between necrotic and apoptotic cell death at both cellular and organ levels.

## REFERENCES


1. Brenner, B. M., *The Kidney*, 5th ed., W.B. Saunders, Philadelphia, PA, 1996.
2. Sands, J. M. and Verlander, J. W., Anatomy and physiology of the kidneys, in *Toxicology of the Kidney*, 3rd ed., Tarloff, J. B. and Lash, L. H., Eds., CRC Press, Boca Raton, FL, 2005, chap. 1.
3. Walker, L. A. and Valtin, H., Biological importance of nephron heterogeneity. *Annu. Rev. Physiol*., 44, 203, 1982.
4. Guder, W. G. and Ross, B. D., Enzyme distribution along the nephron, *Kidney Int*., 26, 101, 1984.
5. Soltoff, S. P., ATP and the regulation of renal cell function. *Ann. Rev. Physiol*., 48, 9, 1986.
6. Lash, L. H., Tokarz, J. J., Woods, E. B., and Pedrosi, B. M., Hypoxia and oxygen dependence of cytotoxicity in renal proximal tubular and distal tubular cells. *Biochem. Pharmacol*., 45, 191, 1993.
7. Lash, L. H., Tokarz, J. J., Chen, Z., Pedrosi, B. M., and Woods, E. B., ATP depletion by iodoacetate and cyanide in renal distal tubular cells. *J. Pharmacol. Exp. Ther*., 276, 194, 1996.
8. Brezis, M., Rosen, S., Silva, P., and Epstein, F. H., Selective vulnerability of the medullary thick ascending limb of anoxia in the isolated perfused rat kidney. *J. Clin. Invest*., 73, 182, 1984.
9. Brezis, M., Shanley, P., Silva, K., Spokes, K., Lear, S., Epstein, F. H., and Rosen, S., Disparate mechanisms for hypoxic cell injury in different nephron segments: studies in the isolated perfused rat kidney. *J. Clin. Invest*., 76, 1796, 1985.
10. Ruegg, C. E. and Mandel, L. J., Bulk isolation of renal PCT and PST II: differential responses to anoxia or hypoxia. *Am. J. Physiol*., 259, F176, 1990.
11. Shanley, P. F., Brezis, M., Spokes, K., Silva, P., Epstein, F. H., and Rosen, S., Hypoxic injury in the proximal tubule of the isolated perfused rat kidney. *Kidney Int*., 29, 1021, 1986.
12. Shanley, P. F., Rosen, M. D., Brezis, M., Silva, P., Epstein, F. H., and Rosen, S., Topography of focal proximal tubular necrosis after ischemia with reflow in the rat kidney. *Am. J. Pathol*., 122, 462, 1986.
13. Brezis, M., Rosen, S., Spokes, K., Silva, P., and Epstein, F. H., Transport-dependent anoxic cell injury in the isolated perfused rat kidney. *Am. J. Pathol*., 116, 327, 1984.
14. Brezis, M., Rosen, S., Silva, P., Spokes, K., and Epstein, F. H., Mitochondrial activity: a possible determinant of anoxic injury in renal medulla. *Experientia*, 42, 570, 1986.
15. Shanley, P. F., Brezis, M., Spokes, K., Silva, P., Epstein, F. H., and Rosen, S., Transport-dependent cell injury in the S3 segment of the proximal tubule. *Kidney Int*., 29, 1033, 1986.
16. Epstein, F. H., Silva, P., Spokes, K., Brezis, M., and Rosen, S., Renal medullary Na-K-ATPase and hypoxic injury in perfused rat kidneys. *Kidney Int*., 36, 768, 1989.
17. Berkhin, E. B. and Humphreys, M. H., Regulation of renal tubular secretion of organic compounds. *Kidney Int*., 59, 17, 2001.
18. Hagenbuch, B. and Meier, P., Organic anion transporting polypeptides of the OATP/SLC21 family: phylogenetic classification as OATP/SLCO superfamily, new nomenclature and molecular/functional properties. *Pflügers Arch*., 447, 653, 2004.
19. Koepsell, H. and Endou, H., The SLC22 drug transporter family. *Pflügers Arch*., 447, 666, 2004.
20. Lee, W. and Kim, R. B., Transporters and renal drug elimination. *Annu. Rev. Pharmacol. Toxicol*., 44, 137, 2004.

21. Markovich, D. and Murer, H., The SLC13 gene family of sodium sulphate/carboxylate cotransporters. *Pflügers Arch.*, 447, 594, 2004.
22. van Aubel, R. A. M. H., Masereeuw, R. and Russel, F. G. M., Molecular pharmacology of renal organic anion transporters. *Am. J. Physiol.*, 279, F216, 2000.
23. Wright, S. H. and Dantzler, W. H., Molecular and cellular physiology of renal organic cation and anion transport. *Physiol. Rev.*, 84, 987, 2004.
24. Lash, L. H., Role of glutathione transport processes in renal function. *Toxicol. Appl. Pharmacol.*, 204, 329, 2005.
25. Lash, L. H., Role of renal metabolism in risk to toxic chemicals. *Environ. Health Perspect.*, 102(Suppl. 11), 75, 1994.
26. Lash, L. H., Role of metabolism in chemically induced nephrotoxicity, in *Mechanisms of Injury in Renal Disease and Toxicity*, Goldstein, R. S., Ed., CRC Press, Boca Raton, FL, 1994, chap. 9.
27. Mohandas, J. et al., Differential distribution of glutathione and glutathione-related enzymes in rabbit kidney: possible implications in analgesic nephropathy. *Biochem. Pharmacol.*, 33, 1801, 1984.
28. Lash, L. H., Fisher, J. W., Lipscomb, J. C., and Parker, J. C., Metabolism of trichloroethylene. *Environ. Health Perspect.*, 108(Suppl. 2), 177, 2000.
29. Lash, L. H. and Parker, J. C., Hepatic and renal toxicities associated with perchloroethylene, *Pharmacol. Rev.*, 53, 177, 2001.
30. Lash, L. H., Parker, J. C., and Scott, C. S., Modes of action of trichloroethylene for kidney tumorigenesis. *Environ. Health Perspect.*, 108(Suppl. 2), 225, 2000.
31. Terlouw, S. A., Masereeuw, R., and Russel, F. G. M., Modulatory effects of hormones, drugs, and toxic events on renal organic anion transport. *Biochem. Pharmacol.*, 65, 1393, 2003.
32. Lash, L. H., Use of freshly isolated and primary cultures of proximal tubular and distal tubular cells from rat kidney, in *Methods in Renal Toxicology*, Zalups, R. K. and Lash, L. H., Eds., CRC Press, Boca Raton, FL, 1996, chap. 11.
33. Lash, L. H., *In vitro* methods of assessing renal damage. *Toxicol. Pathol.*, 26, 33, 1998.
34. Lash, L. H. and Zalups, R. K., Mercuric chloride-induced cytotoxicity and compensatory hypertrophy in rat kidney proximal tubular cells. *J. Pharmacol. Exp. Ther.*, 261, 819, 1992.
35. Lash, L. H., Putt, D. A., and Zalups, R. K., Influence of exogenous thiols on mercury-induced cellular injury in isolated renal proximal tubular and distal tubular cells from normal and uninephrectomized rats. *J. Pharmacol. Exp. Ther.*, 291, 492, 1999.
36. Zalups, R. K. and Lash, L. H., Recent advances in understanding the renal transport and toxicity of mercury. *J. Toxicol. Environ. Health*, 42, 1, 1994.
37. Tarloff, J. B. and Kinter, L. B., *In vivo* methodologies used to assess renal function, in *Comprehensive Series in Toxicology*. Vol. 7. *Kidney Toxicology*, Goldstein, R. S., Ed., Elsevier, Oxford, 1997, chap. 7.06.
38. Commission on Life Sciences, *Biologic Markers in Urinary Toxicology*, National Academy Press, Washington, D.C., 1995.
39. Bekersky, I., Use of the isolated perfused kidney as a tool in drug disposition studies. *Drug Metab. Rev.*, 14, 931, 1983.
40. Diamond, G. L., The isolated perfused kidney, in *Methods in Renal Toxicology*, Zalups, R. K. and Lash, L. H., Eds., CRC Press, Boca Raton, FL, 1996, chap. 4.
41. Ramsey, C. R. and Knox, F. G., Micropuncture and microperfusion techniques, in *Methods in Renal Toxicology*, Zalups, R. K. and Lash, L. H., eds., CRC Press, Boca Raton, FL, 1996, chap. 5.
42. Ford, S. M., *In vitro* toxicity systems, *in vivo* methodologies used to assess renal function, in *Comprehensive Series in Toxicology*. Vol. 7. *Kidney Toxicology*, Goldstein, R. S., Ed., Elsevier, Oxford, 1997, chap. 7.07.
43. Krumdieck, C. L., Santos, J. E. D., and Ho, K. J., A new instrument for the rapid preparation of tissue slices. *Anal. Biochem.*, 104, 118, 1980.
44. Gandolfi, A. J., Brendel, K., and Fernanado, Q., Preparation and use of precision-cut renal cortical slices in renal toxicology, in *Methods in Renal Toxicology*, Zalups, R. K. and Lash, L. H., Eds., CRC Press, Boca Raton, FL, 1996, chap. 7.
45. Parrish, A. R., Gandolfi, A. J., and Brendel, K., Precision-cut tissue slices: applications in pharmacology and toxicology. *Life Sci.*, 57, 1887, 1995.
46. Zalups, R. K. and Barfuss, D. W., *In vitro* perfusion of isolated nephron segments: a method for renal toxicology, in *Methods in Renal Toxicology*, Zalups, R. K. and Lash, L. H., Eds., CRC Press, Boca Raton, FL, 1996, chap. 8.
47. Gesek, F. A., Wolff, D. W., and Strandhoy, J. W., Improved separation method for rat proximal and distal renal tubules. *Am. J. Physiol.*, 253, F358, 1987.
48. Heidrich, H. G. and Dew, M. E., Homogeneous cell populations from rabbit kidney cortex: proximal, distal tubule, and renin-active cell isolated by free-flow electrophoresis. *J. Cell Biol.*, 74, 780, 1977.
49. Kreisberg, J. I., Pitts, A. M., and Pretlow, T. G., II, Separation of proximal tubule cells from suspensions of rat kidney cells in density gradients of Ficoll in tissue culture medium. *Am. J. Pathol.*, 86, 591, 1977.
50. Kreisberg, J. I., Sachs, G., Pretlow II, T. G., and McGuire, R. A., Separation of proximal tubule cells from suspensions of rat kidney cells by free-flow electrophoresis. *J. Cell. Physiol.*, 93, 169, 1977.
51. Rodeheaver, D. P., Aleo, M. D., and Schnellmann, R. G., Differences in enzymatic and mechanical isolated rabbit renal proximal tubules: comparison in long-term incubation. *In Vitro Cell. Dev. Biol.*, 26, 898, 1990.
52. Scholer, D. W. and Edelman, I. S., Isolation of rat kidney cortical tubules enriched in proximal and distal segments. *Am. J. Physiol.*, 237, F350, 1979.
53. Vinay, P., Gougoux, A., and Lemieux, G., Isolation of a pure suspension of rat proximal tubules. *Am. J. Physiol.* 241, F403, 1981.
54. Allen, M. L., Nakao, A., Sonnenburg, W. K., Burnatowska-Hledin, M., Spielman, W. S., and Smith, W. L., Immunodissection of cortical and medullary thick ascending limb cells from rabbit kidney. *Am. J. Physiol.*, 255, F704, 1988
55. Chamberlin, M. E., LeFurgey, A., and Mandel, L. J., Suspension of medullary thick ascending limb tubules from the rabbit kidney. *Am. J. Physiol.*, 247, F955, 1984.

56. Eveloff, J., Haase, W., and Kinne, R., Separation of renal medullary cells: Isolation of cells from the thick ascending limb of Henle's loop. *J. Cell Biol.*, 87, 672, 1980.
57. Pizzonia, J. H., Gesek, F. A., Kennedy, S. M., Coutermarsh, B. A., Bacskal, B. J., and Friedman, P. A., Immunomagnetic separation, primary culture, and characterization of cortical thick ascending limb plus distal convoluted tubule cells from mouse kidney. *In Vitro Cell. Dev. Biol.*, 27A, 409, 1991.
58. Trinh-Trang-Tan, M.-M., Bouby, N., Coutaud, C., and Bankir, L., Quick isolation of rat medullary thick ascending limbs: enzymatic and metabolic characterization. *Pflügers Arch.*, 407, 228, 1986.
59. Groves, C. E. and Schnellmann, R. G., Suspensions of rabbit renal proximal tubules, in *Methods in Renal Toxicology*, Zalups, R. K. and Lash, L. H., Eds., CRC Press, Boca Raton, FL, 1996, chap. 9.
60. Chung, S. D., Alavi, N., Livingston, D., Hiller, S., and Taub, M., Characterization of primary rabbit kidney cultures that express proximal tubule functions in a hormonally defined medium. *J. Cell Biol.*, 95, 118, 1982.
61. Taub, M. L., Yang, I. S., and Wang, Y., Primary rabbit kidney proximal tubule cell cultures maintain differentiated functions when cultured in a hormonally defined serum-free medium. *In Vitro Cell. Dev. Biol.*, 25, 770, 1989.
62. Aleo, M. D., Taub, M. L., Nickerson, P. A., and Kostyniak, P. J., Primary cultures of rabbit renal proximal tubule cells. I. Growth and biochemical characteristics. *In Vitro Cell. Dev. Biol.*, 25, 776, 1989.
63. Nowak, G. and Schnellmann, R. G., Improved culture conditions stimulate gluconeogenesis in primary cultures of renal proximal tubule cells. *Am. J. Physiol.*, 268, C1053, 1995.
64. Aleo, M. D. and Kostyniak, P. J., Characterization and use of rabbit renal proximal tubular cells in primary culture for toxicology research, in *Methods in Renal Toxicology*, Zalups, R. K. and Lash, L. H., Eds., CRC Press, Boca Raton, FL, 1996, chap. 10.
65. Lash, L. H. and Tokarz, J. J., Isolation of two distinct populations of cells from rat kidney cortex and their use in the study of chemical-induced toxicity. *Anal. Biochem.*, 182, 271, 1989.
66. Jones, D. P., Sundby, G.-B., Ormstad, K., and Orrenius, S., Use of isolated kidney cells for study of drug metabolism. *Biochem. Pharmacol.*, 28, 929, 1979.
67. Cummings, B. S. and Lash, L. H., Metabolism and toxicity of trichloroethylene and *S*-(1,2-dichlorovinyl)-L-cysteine in freshly isolated human proximal tubular cells. *Toxicol. Sci.*, 53, 458, 2000.
68. Cummings, B. S., Lasker, J. M., and Lash, L. H., Expression of glutathione-dependent enzymes and cytochrome P450s in freshly isolated and primary cultures of proximal tubular cells from human kidney. *J. Pharmacol. Exp. Ther.*, 293, 677, 2000.
69. Blumenthal, S. S., Lewand, D. L., Buday, M. A., Mandel, N. S., Mandel, G. S., and Kleinman, J. G., Effect of pH on growth of mouse renal cortical tubule cells in primary culture. *Am. J. Physiol.*, 257, C419, 1989.
70. Boogaard, P. J., Zoeteweij, J. P., van Berkel, T. J. C., van't Noordende, J. M., Mulder, G. J., and Nagelkerke, J. F., Primary culture of proximal tubular cells from normal rat kidney as an *in vitro* model to study mechanisms of nephrotoxicity: toxicity of nephrotoxicants at low concentrations during prolonged exposure. *Biochem. Pharmacol.*, 39, 1335, 1990.
71. Chen, T. C., Curthoys, N. P., Lagenaur, C. F., and Puschett, J. B., Characterization of primary cell cultures derived from rat renal proximal tubules. *In Vitro Cell. Dev. Biol.*, 25, 714, 1989.
72. Elliget, K. A. and Trump, B. F., Primary cultures of normal rat kidney proximal tubule epithelial cells for studies of renal cell injury. *In Vitro Cell. Dev. Biol.*, 27A, 739, 1991.
73. Hatzinger, P. B. and Stevens, J. L., Rat kidney proximal tubule cells in defined medium: the roles of cholera toxin, extracellular calcium and serum in cell growth and expression of γ-glutamyltransferase. *In Vitro Cell. Dev. Biol.*, 25, 205, 1989.
74. Lash, L. H., Tokarz, J. J., and Pegouske, D. M., Susceptibility of primary cultures of proximal tubular and distal tubular cells from rat kidney to chemically induced toxicity. *Toxicology*, 103, 85, 1995.
75. Miller, J. H., Restricted growth of rat kidney proximal tubule cells cultured in serum-supplemented and defined media. *J. Cell. Physiol.*, 129, 264, 1986.
76. Rosenberg, M. R. and Michalopoulos, G., Kidney proximal tubular cells isolated by collagenase perfusion grow in defined media in the absence of growth factors. *J. Cell. Physiol.*, 131, 107, 1987.
77. Sakhrani, L. M., Badie-Dezfooly, B., Trizna, W., Mikhail, N., Lowe, A. G., Taub, M., and Fine, L. G., Transport and metabolism of glucose by renal proximal tubular cells in primary culture. *Am. J. Physiol.*, 246, F757, 1984.
78. Taub, M. L., Yang, I. S., and Wang, Y., Primary rabbit kidney proximal tubule cell cultures maintain differentiated functions when cultured in a hormonally defined serum-free medium. *In Vitro Cell. Dev. Biol.*, 25, 770, 1989.
79. Toutain, H., Vauclin-Jacques, N., Fillastre, J.-P., and Morin, J.-P., Biochemical, functional, and morphological characterization of a primary culture of rabbit proximal tubule cells. *Exp. Cell Res.*, 194, 9, 1991.
80. Scott, D. M., Zierold, K., and Kinne, R., Development of differentiated characteristics in cultured kidney (thick ascending loop of Henle) cells. *Exp. Cell Res.*, 162, 521, 1986.
81. Courjault-Gautier, F., Chevalier, J., Abbou, C. C., Chopin, D. K., and Toutain, H. J., Consecutive use of hormonally defined serum-free media to establish highly differentiated human renal proximal tubule cells in primary culture. *J. Am. Soc. Nephrol.*, 5, 1949, 1995.
82. Detrisac, C. J., Sens, M. A., Garvin, A. J., Spicer, S. S., and Sens, D. A., Tissue culture of human kidney epithelial cells of proximal tubule origin. *Kidney Int.*, 25, 383, 1984.
83. Rodilla, V., Miles, A. T., Jenner, W., and Hawksworth, G. M., Exposure of cultured human proximal tubular cells to cadmium, mercury, zinc and bismuth: toxicity and metallothionein induction. *Chem. Biol. Interact.*, 115, 71, 1998.

84. Trifillis, A. L., Regec, A. L., and Trump, B. F., Isolation, culture and characterization of human renal tubular cells. *J. Urol.*, 133, 324, 1985.
85. Van Der Biest, I., Nouwen, E. J., Van Dromme, S. A., and De Broe, M. E., Characterization of pure proximal and heterogeneous distal human tubular cells in culture. *Kidney Int.*, 45, 85, 1994.
86. Lash, L. H., Hueni, S. E., and Putt, D. A., Apoptosis, necrosis and cell proliferation induced by *S*-(1,2-dichlorovinyl)-L-cysteine in primary cultures of human proximal tubular cells. *Toxicol. Appl. Pharmacol.*, 177, 1, 2001.
87. Lash, L. H., Putt, D. A., Hueni, S. E., and Horwitz, B. P., Molecular markers of trichloroethylene-induced toxicity in human kidney cells. *Toxicol. Appl. Pharmacol.*, 206, 157, 2005.
88. Taub, M., Immortalized cell lines of renal cells, in *Methods in Renal Toxicology*, Zalups, R. K. and Lash, L. H., Eds., CRC Press, Boca Raton, FL, 1996, chap. 12.
89. Lacave, R., Bens, M., Cartier, N., Vallet, V., Robine, S., Pringault, E. Kahn, A., and Vanderwalle, A., Functional properties of proximal tubule cell lines derived from transgenic mice harboring L-pyruvate kinase-SV40 (T) antigen hybrid gene. *J. Cell Sci.*, 104, 705, 1993.
90. Vandewalle, A., Lelongt, B., Geniteau-Legendre, M., Baudouin, B., Antoine, M., Estrade, S., Chatelet, F., Verroust, P., Cassingena, R., and Ronco, P., Maintenance of proximal and distal cell functions in SV40-transformed tubular cell lines derived from rabbit kidney cortex. *J. Cell. Physiol.*, 141, 203, 1989.
91. Han, W. K., Bailly, V., Abichandani, R., Thadhani, R., and Bonventre, J. V., Kidney injury molecule-1 (KIM-1): a novel biomarker for human renal proximal tubule injury. *Kidney Int.*, 62, 237, 2002.
92. Ichimura, T., Bonventre, J. V., Bailly, V., Wei, H., Hession, C. A., Cate, R. L., and Sanicola, M., Kidney injury molecule-1 (KIM-1), a putative epithelial cell adhesion molecule containing a novel immunoglobulin domain, is upregulated in renal cells after injury. *J. Biol. Chem.*, 273, 4135, 1998.
93. Ichimura, T., Hung, C. C., Yang, S. A., Stevens, J. L., and Bonventre, J. V., Kidney injury molecule-1: a tissue and urinary biomarker for nephrotoxicant-induced renal injury. *Am. J. Physiol.*, 286, F552, 2004.
94. Vaidya, V. S., Ramirez, V., Ichimura, T., Bobadilla, N. A., and Bonventre, J. V., Urinary kidney injury molecule-1 (Kim-1): a sensitive quantitative biomarker for early detection of kidney tubular injury. *Am. J. Physiol. Renal Physiol.*, 290(2), F517, 2006.
95. Harriman, J. F. and Schnellmann, R. G., Mechanisms of renal cell death, in *Toxicology of the Kidney*, 3rd ed., Tarloff, J. B. and Lash, L. H., Eds., CRC Press, Boca Raton, FL, 2005, chap. 7.
96. Ben-Sasson, S. A., Sherman, Y., and Gavrieli, Y., Identification of dying cells: *in situ* staining. *Methods Cell Biol.*, 46, 29, 1995.
97. Danial, N. N. and Korsmeyer, S. J., Cell death: critical control points. *Cell*, 116, 205, 2005.
98. Lash, L. H. and Jones, D. P., Mitochondrial toxicity in renal injury, in *Methods in Renal Toxicology*, Zalups, R. K. and Lash, L. H., Eds., CRC Press, Boca Raton, FL, 1996, chap. 16.
99. Lash, L. H., Molecular and cell biology of normal and diseased or intoxicated kidney, in *Toxicology of the Kidney*, 3rd ed., Tarloff, J. B. and Lash, L. H., Eds., CRC Press, Boca Raton, FL, 2005, chap. 2.

# 31 Gastrointestinal Toxicology

*Robert W. Kapp, Jr.*

## CONTENTS

**TABLE 31.1**
**Summary of Common Macro- and Micronutrients in Humans**

| Nutrient Classification | Nutrient Types | End Product |
|---|---|---|
| Macronutrients | Carbohydrates | Monosaccharides |
| | Lipids | Fatty acids and cholesterol and monoglycerides |
| | Proteins and nucleic acids | Amino acids, bases |
| | Water | Water |
| Micronutrients | Minerals | Calcium, chromium, copper, fluorine, iodine, iron, magnesium, manganese, molybdenum, phosphorus, potassium, selenium, sodium, zinc |
| | Fat-soluble vitamins | A, D, E, K |
| | Water-soluble vitamins | $B_1$(thiamine), $B_2$ (riboflavin), $B_3$ (niacin), $B_5$ (pantothenic acid), $B_6$ (pyroxidine), $B_7$ (biotin), $B_9$ (folacin), $B_{12}$ (cobalamine), C |

## INTRODUCTION

The intensity of the effect of a toxicant on an organism is dependent primarily upon the relative effective concentration and the duration or persistence of the ultimate toxicant at a specific site of action. There are numerous portals of entry for toxicants. This chapter focuses on the normal structure and function of the gastrointestinal tract and how the process of digestion and absorption influences and is influenced by toxicants that enter the gastrointestinal tract. For those toxicants, the intensity of the ultimate toxicant is determined by a number of factors that are part of the normal functioning of the gastrointestinal tract including digestion, absorption, biotransformation, distribution, and elimination. The last part of this chapter describes the types of testing that can be performed to examine the effects of toxicants on various aspects of the normal functioning of the gastrointestinal tract.

## OVERVIEW OF GASTROINTESTINAL TRACT

All living cells require energy in the form of nutrients. Nutrients are substances that provide energy to basic body processes. In addition, nutrients are critical to the building of new or the replacement of old components for various body structures. Basically, nutrients are divided into two classes: macronutrients and micronutrients. In comparative terms, macronutrients are required in larger quantities than are micronutrients, which are require in much smaller amounts. The daily diet of most individuals consists of about 2 to 3 pounds of food and 1.5 liters of fluid containing perhaps 100,000 differences substances in amounts varying from 1 or 2 liters to as little as nanogram units. Of those 100,000 substances, only about 300 are classified as nutrients and of that number 45 are essential for the survival of the individual. Many of the consumed substances have value in preserving the food or add aroma or color to improve or enhance the taste or appearance of the food. Some of the substances are helpful in the digestion process, such as fibers and cellulose [1]. The vast majority of ingested food is broken down into more elemental units for use by the body. Table 31.1 provides a summary of the most common macronutrients and micronutrients consumed by humans.

The resulting breakdown products of these nutrient classes provide the organism with energy for the numerous life processes necessary to sustain the organism, such as active transport, protein and enzyme synthesis, and synthesis of structural molecules for cells. The minerals and vitamins can act as catalysts in thousands of body functions and are needed for growth and maintenance of body structures. Water is necessary as a means to regulate body temperature, as a solvent, and as a lubricant, plus it plays a role in hydrolysis reactions.

Although single-cell organisms can obtain nutrients directly from their surroundings, multicellular organisms have developed highly specialized structures to collect, digest, and absorb these nutrients. The gastrointestinal tract (GI tract) or alimentary canal is that specialized structure in multicellular organisms that essentially obtains nourishment by the ingestion of organic material, generally termed *holozoic nutrition*. This process of nutrition provides the trillions of body cells with nutrients in order to function as described above. Table 31.2 identifies the five basic stages of holozoic nutrition [2].

To complete these various stages of nutrition, the GI tract has specialized structures. The cell wall of the GI tract is composed of numerous tissue types that serve not only to break down foodstuff and fluid into absorbable components but also to create an environment for these components, as well as vitamins and minerals, to be absorbed into the bloodstream, assimilated, and then ultimately eliminated from the body. The GI tract is essentially a 25- to 30-foot series of hollow organs connected together to form a pathway that starts at the mouth and proceeds through the pharynx, esophagus, stomach, small

**TABLE 31.2**
**Stages of Holozoic Nutrition**

| Stage | Function |
|---|---|
| Ingestion | Physical act of engulfing foodstuff into the alimentary canal |
| Digestion | Mechanical and chemical reduction of foodstuff to elemental particles |
| Absorption | Passive and active mechanisms involved in transporting the nutrients across mucosal cell membranes |
| Assimilation | The body's ultimate use of the absorbed nutrients |
| Egestion | The elimination of the unused and undigested gastrointestinal tract contents |

*Source:* Data from Rothery, M., *A Level Biology*, Module 1: *Biochemistry, Cells, Cell Transport, Exchange, Enzymes, and Digestion*, www.mrothery.co.uk, 2005.

intestines, large intestines, rectum, and anus. Although the lumen is found within the body, the contents within the lumen of the GI tract are considered to be external to the body. The digestive system functions to modify the ingested luminal contents so they can be absorbed from the lumen (exterior of the body) by the blood and lymph circulatory system (interior of the body). Traditionally, the digestion system is organized into two divisions: (1) the GI tract or alimentary canal, which includes the mouth, pharynx, esophagus, stomach, small intestine, large intestine, rectum, and anus; and (2) accessory structures that include teeth, tongue, salivary glands, pharynx, liver, gallbladder, and pancreas. The major organs of the digestive tract and their relationship to one another are shown in Figure 31.1.

Each portion of the GI tract is highly specialized with respect to both structure and function. Accessory digestive organs are connected to the GI tract through a series of ducts, including the salivary glands, the pancreas, the liver, and the gall bladder. The teeth aid in the physical breakdown of foodstuffs that enter the mouth. The tongue assists in mastication and swallowing. The pharynx assists in sealing the respiratory tract during the swallowing reflex. The esophagus transports the ingested foodstuff to the stomach. The stomach serves to mix the foodstuffs with acidic gastric secretions and prepare the resulting mixture, which is now a semiliquid mass of partially digested food (termed *chyme*), for entrance into the small intestine. The

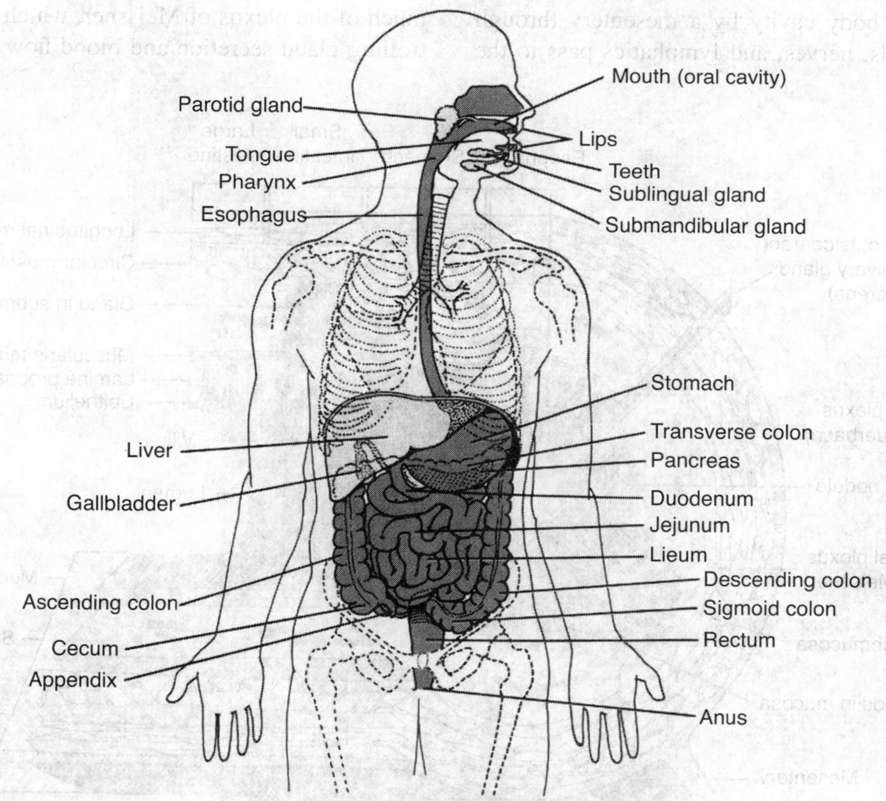


**FIGURE 31.1** Overview of digestive tract. (From Tortora, G.J. and Anagnostakos, N.P., in *Principles of Anatomy and Physiology*, 6th ed., Tortora, G.J., Ed., HarperCollins, New York, 1990, pp. 731–778. With permission.)

small intestine is a 20- to 25-foot tube in which the vast majority of the digestion and absorption of chyme occurs. The large intestine absorbs some electrolytes and water and prepares the contents for expulsion through the rectum and anus.

## HISTOLOGICAL ORGANIZATION OF GASTROINTESTINAL TRACT

To understand the GI tract, it is necessary to understand its microscopic as well as macroscopic structures. Throughout the length of the GI tract, a cross-section of the tissue consists of a series of four concentric layers or tunics. The intestinal wall varies in the various sections of the alimentary canal, reflecting the different roles in each locality; however, the basic structure remains the same (Figure 31.2).

### Serosa

The outermost layer is the adventitia or serosa, which includes loose supportive tissue proximally and is contained by a thin membrane distally that covers the entire GI tract. Those portions located within the abdominal cavity are covered by a layer of squamous cells and an underlying connective tissue. The entire GI tube is suspended within the body cavity by a mesentery through which blood vessels, nerves, and lymphatics pass to the organ systems. It is part of the peritoneum and facilitates movement of the structure within the body cavity.

### Muscularis

The next layer is the muscularis, which usually contains the inner circular layer of smooth muscle fibers and an outer longitudinal layer of smooth muscle fibers and the myenteric plexus. These muscle layers provide coordinated movements that allow for the mixing, digestion, and motility of ingested substances through the GI tract. The primary control of GI motility is centered in the myenteric plexus (plexus of Auerbach), which contains connections from both autonomic innervation connections, and is also found in the muscularis layer. The muscularis layer in the mouth, pharynx, and upper third of the esophagus has not only smooth muscles but also skeletal or striated muscles that produce the voluntary initial phase of swallowing.

### Submucosa

The next layer, moving toward the lumen, is the submucosa, which is comprised primarily of loose connective tissue, lymphatic and blood vessels, glands, and the submucosal plexus. It is a very vascular layer and contains much of the plexus of Meissner, which is critical in controlling gland secretion and blood flow to the GI tract.

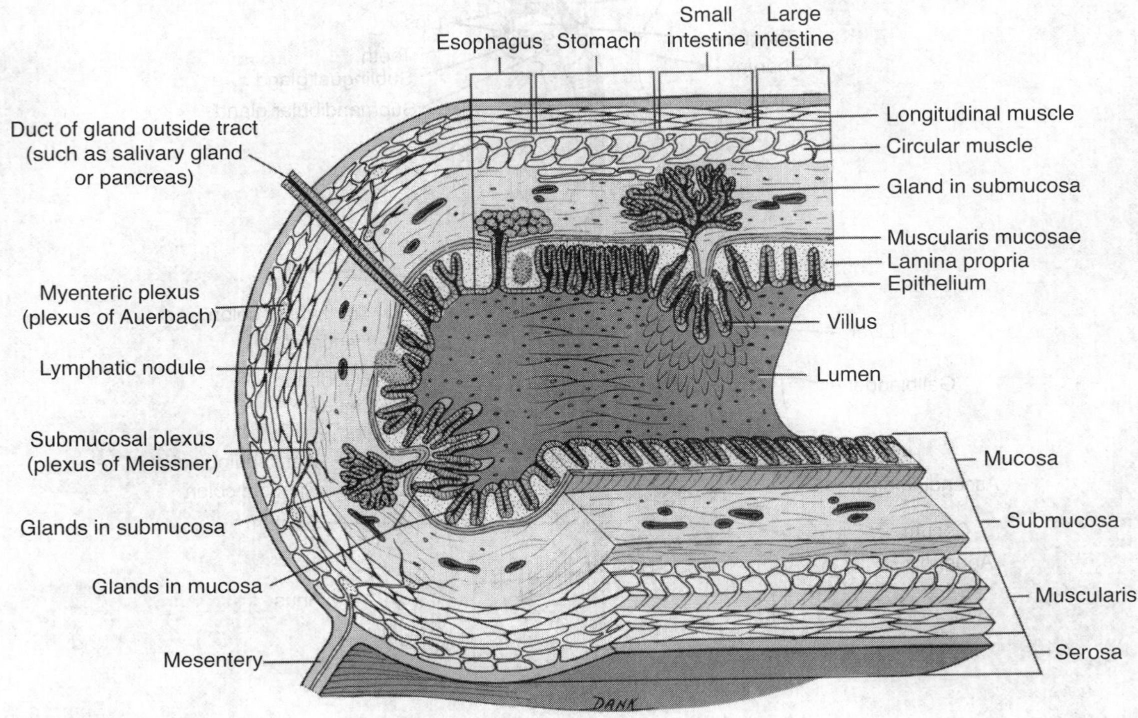


**FIGURE 31.2** Representative sections of the gastrointestinal tract. (From Tortora, G.J. and Anagnostakos, N.P., in *Principles of Anatomy and Physiology*, 6th ed., Tortora, G.J., Ed., HarperCollins, New York, 1990, pp. 731–778. With permission.)

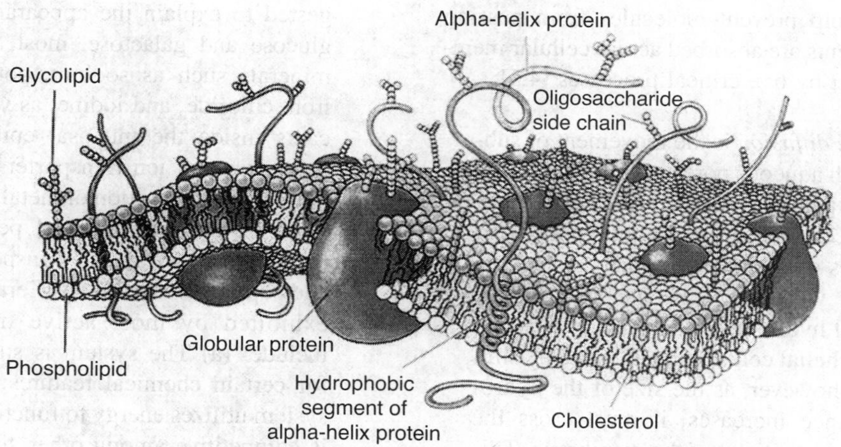


**FIGURE 31.3** Schematic representation of the Singer–Nicolson membrane model. (Courtesy of Materials Science and Engineering Laboratory, Center for Neutron Research, National Institute of Standards and Technology, Gaithersburg, MD.)

## Mucosa

The innermost layer of the GI tract is the mucosa, which is in contact with the external environment and actually allows for the absorption of the nutrients. It is comprised of three additional subdivisions. Starting from the layer proximate to the submucosa, the muscularis mucosa includes a smooth muscle and elastic connective tissue layer that creates the folds in the mucous membrane that enhance the area of absorption throughout the GI tract. The next distinctive layer of the mucosa is the lamina propria, which is composed of loose connective tissue and contains blood vessels, lymphatics, and perhaps some glands. This layer supports the epithelial cells, provides the blood and lymph supply, and binds the layer to the muscularis mucosa. The final layer of the muscosa that lines the lumen of the GI tract and is in direct contact with nutrients is the epithelial layer. The structure and function of these epithelial cells vary along the course of the tract. Depending on the location, the epithelial cells may be stratified squamous cells, such as in the esophagus, or secretory cells found in the stomach or absorptive cells in the small intestines. The functions of the stratified epithelium are secretion and protection. The functions of the simple epithelium, on the other hand, are secretion and absorption.

## Cellular Absorption

Cell membranes are plasma barriers between the environment and the cell cytoplasm. The membrane is specialized in that it contains specific proteins and lipid components that allow it to perform its unique roles for that particular cell. Because the lumen of the GI tract is considered external to the body itself, a critical factor in the nutrient absorption process is crossing this mucosal epithelial cell membrane. The membranes are essential for the integrity and function of the cell. The functions of GI tract cell membranes include:

- Protection
- Transport into and out of the cell
- Enzymatic activity
- Providing receptors for signal transduction
- Providing intercellular adhesion proteins
- Cell-to-cell recognition via glycocalyx
- Attachment to the cytoskeleton and extracellular matrix

According to the currently accepted Singer–Nicholson fluid mosaic model, the 7- to 9-nm-thick lipid bilayer is a neutral, two-dimensional structure with a mosaic of proteins embedded in it [3]. In this model, the phospholipids are amphipathic, meaning they are both hydrophilic and hydrophobic. Each phospholipid possesses a polar phosphate group at the head facing the outside of the membrane and two nonpolar fatty acids at the tail facing inward. Within model bilayers, lipids can exist in different phases: as gels or as liquid-ordered or liquid-disordered states (as described in the Singer–Nicholson model). See Figure 31.3 for the overall basic membrane structure. In the liquid-ordered state, saturated hydrocarbonic chains of phospholipids are tightly packed with cholesterol and glycolipids, thus horizontal mobility in the membrane is limited. The cells that line the GI tract are similar in thickness (7 to 9 nm) and structure (phospholipid bilayer), with polar head groups consisting of phosphatidylcholine and phosphatidylethanolamine. These polar groups are arranged perpendicularly to the cell membrane and are located facing the inner and outer membrane surfaces. The zonula occludens or tight junction is the outer layer of the junctional complex between other epithelial cells which line the luminal surface of the digestive tube. The tight junction is a series of integral membrane proteins that form a fused ring around intestinal cells to help prevent molecules from passing between cells. Nutrients are absorbed across cellular membranes in the GI tract by five critical processes [4,5]:

1. *Simple passive diffusion* is the movement of substances through aqueous pores in cell membranes without the expenditure of energy. Most hydrophilic substances (molecular weight < 600) cross the membranes of the intestinal mucosa by passive transport via these aqueous pores in the mucosa. Small hydrophobic substances can also access the epithelial cell's aqueous pores by simple diffusion; however, as the size of the hydrophobic substance increases, it must cross the through the lipid portions of the membrane. This movement is due primarily to the random motion that attempts to equalize the concentration of that substance on either side of the membrane. If higher concentrations of a substance exist on one side of a membrane, the law of mass tends to equilibrate the concentrations. The membrane is constructed of fat-soluble components that permit fat-soluble substances such as oxygen, nitrogen, carbon dioxide, and alcohols to easily diffuse across it.
2. *Filtration* occurs when water flows across membranes in large quantities because of hydrostatic or osmotic pressure. This water flow tends to pull some substances with molecular weights of 150 to 300 kDa through the same channels in the cell membrane. The substances do not cross the membrane because of random motion, but rather they are forced through the membrane by the movement of the water.
3. *Facilitated diffusion* is the common mode of transport postulated for sugars across both sides of the intestinal cell membrane. In this case, the facilitated carriers are proteins that reside in the cell membrane. The substance to be transported becomes lightly bound to the carrier protein, which, in turn, changes shape to open a pathway for transport across the cell. The substance moves as if it were actively transported, except no energy is expended and it is not moved against the concentration gradient. The carrier then returns to the outside of the membrane.
4. *Active transport* occurs when compounds require assistance in moving across the mucosal epithelial membrane. This is generally because the substance is large or insoluble, or it is moving against a concentration gradient or otherwise defying the logics of physics in its direction of movement. Because these substances nonetheless do successfully traverse the membrane, a system of transport mechanisms has been suggested to explain the appearance of the sugars glucose and galactose, most amino acids, and minerals such as sodium, potassium, calcium, iron, chloride, and iodine, as well as some toxicants inside the mucosal epithelial cells. The divalent metal ion transporter (DMT) assists in the GI tract absorption of metals, while the nucleotide transporter (nt) and peptide transporter (pept) are active in the transport of nucleotides and peptides, respectively. Certain characteristics exhibited by most active transport systems include: (a) The system is structurally specific for certain chemical features; (b) the transport system utilizes energy to function; (c) the system is competitive among other similarly structured chemicals that utilize the same transport system; (d) the system is transport-rate limiting, by either saturation or competition, and exhibits a transport maximum ($T_m$); and (e) the system moves substances against concentration and energy gradients. Generally, a substance being actively transported across the mucosal epithelial cell forms a membrane-bound complex on the low-concentration side of the membrane; subsequently, the substance then crosses the membrane through specific channels by the expenditure of energy, where it is released on the opposite, or high-concentration, side of the membrane.
5. *Vesicular transport*—Most substances either diffuse or are actively transported into the mucosal cell membrane; however, some are ingested by a process known as vesicular transport, which includes both endocytosis (the movement of substances from extracellular spaces into intracellular spaces) and exocytosis (the movement of substances from the cell interior to extracellular spaces). This process can be further broken down into *pinocytosis* (or fluid-phase endocytosis), where the cell membrane engulfs dissolved substances and fluids, and *phagocytosis* (solid-phase endocytosis), where the cell membrane engulfs microscopically visible particles. The exocytic vesicles are created from a number of sources, including endosomes traversing the cell, while others are derived from the endoplasmic reticulum and Golgi apparatus expelling various substances from the intracellular space.

## DIGESTIVE SYSTEM ORGANIZATION OVERVIEW

An average adult consumes about 1100 pounds of food and about 150 gallons of water each year. On average, approximately 10 liters of chyme flow through the alimentary canal per day; about 100 to 150 mL of fluid remain in the feces and are egested through the anus [6]. The major subdivisions of the gastrointestinal tract from the proximal to distal end include the mouth, where preparation of the nutrients occurs by mastication; the pharynx, which participates in swallowing; the esophagus, which functions as a conduit; the stomach, which serves as a reservoir for mixing and digestion and regulates content delivery; the small intestine, which serves as a digestive and absorptive organ; the large intestine, which absorbs water and electrolytes and processes the luminal contents for elimination; the rectum, which serves to accumulate and expel fecal material; and the anus, which controls defecation.

Digestion itself includes mechanical as well as chemical functions. The mechanical functions include chewing to reduce the size of the food to smaller particles, churning activities of the stomach that continue the reduction process, and finally the peristaltic action of the small and large intestines that propels food through the system. The chemical functions include a series of catabolic chemical reactions that ultimately break down protein into basic amino acids, convert carbohydrates into glucose, or convert fat into fatty acids and glycerol. As the food moves through the GI tract, it is mixed with various secretions which, combined with mechanical digestion, allow nutrients to pass through the walls of the epithelium cells into the body's blood and lymph capillaries.

## Control of Digestive System Function

Because the digestive system is critical for survival of the organism, the entire GI tract is under robust control by electrical and chemical regulatory processes. Some of the signals originate from within the digestive tract and some originate external to the GI tract [7]. Some signals coordinate various sections of the digestive system, while others link the digestive system and the conscious brain. The GI tract has its own nervous system—referred to as the *intrinsic* or *enteric nervous system* (ENS)—which is quite complex and is recognized as an integrative neuronal system separate from the central nervous system [8,10]. The GI tract is also affected by hormones produced in distant endocrine glands as well as many hormones produced within the GI tract itself [9]. The endocrine cells within the GI tract are collectively referred to as the *enteric endocrine system* (EES).

### Enteric Nervous System

The ENS can and does perform many of its tasks with little or no CNS control; however, normal digestive function requires coordination between the ENS and the CNS. These links are either parasympathetic connections (usually excitatory signals from the vagus and spinal nerves that stimulate motility and GI secretions) or sympathetic connections (usually inhibitory signals from the postganglionic fibers that reduce peristalsis and secretory activity) which permit signals from the CNS or the ENS to directly connect to the GI tract.

The entire length of the walls of the GI tract contains two complete networks of neurons. These neuronal networks include: (1) the myenteric plexus (also known as the plexus of Auerbach), which is located between the circular and longitudinal layers in the muscularis and is active in the control of motility in the GI tract, and (2) the submucosal plexus (also known as the plexus of Meissner), which is located in the submucosa and is active in controlling blood flow and cellular secretions. Three basic types of neurons are located throughout the enteric plexuses. Sensory neurons respond to mechanical, osmotic, thermal, and chemical stimulation and provide the enteric plexus with a continuous and comprehensive status report on the state of the GI tract wall. Interneurons process signals from the sensory neurons and relay them to the enteric motor neurons. Motor neurons exert control over GI tract motility and some gastric secretions. These neurons secrete many neurotransmitters, but the primary excitatory neurotransmitter is acetylcholine; the primary inhibitory neurotransmitter is norepinephrine.

### Enteric Endocrine System

The GI tract is the largest endocrine organ in the body, and it controls digestive function by producing hormones from many single-hormone-secreting cells that are located throughout the mucosal of the stomach and small intestine. Over 25 hormones have been identified throughout the length of the GI tract. In 1969, Pearse [11] coined the term *amine precursor uptake and decarboxylation* (APUD) cells to describe cells dispersed throughout the body having a common ability to add and decarboxylate monoamines to bioactive amines. Characteristic amines and peptides have since been found widely distributed in not only APUD cells but also neurons located in different organs, thus giving rise to the concept of a common regulatory system termed the *diffuse neuroendocrine system* (DNES) [12,13]. APUD cells have been identified in many organs and include over 60 cell types, such as hypothalamus, adenohypophysis, pineal, parathyroid, thyroid, adrenal, placental, pancreas, GI tract, lung, urogenital tract, skin, carotid body, and sympathetic ganglia cells. These cell types have the common ability to produce biogenic amines by absorbing 5-hydroxytryptophan and L-dihydroxyphenylalanine and subsequently decarboxylating them.

Regulation of the GI tract and most other biological systems is very complex. Complicated interactions occur among the endocrine system, the central nervous system, the peripheral nervous system, and cells located throughout the organism. These systems communicate with neurotransmitters and neurohormones both locally and dis-

**TABLE 31.3**
**Enteric Hormones**

| Hormone | Location | Function |
|---|---|---|
| Cholecystokinin | Duodenum (small intestine) | Control of pancreatic enzymes and bile secretion |
| Gastrin | Stomach | Control of gastric acid secretion |
| Gastric-inhibitory peptide | Duodenum and jejunum (small intestine) | Control of gastric acid secretion |
| Glucagon-like peptides | Ileum and the cecum, where proglucagon is cleaved into glucagon-like peptide 1 (GLP-1), GLP-2, and oxyntomodulin | GLP-1 lowers blood glucose levels and inhibits gastric and pancreatic secretion; GLP-2 may increase proliferation rate of GI tract cells; oxyntomodulin may inhibit gastric and pancreatic secretion and slow motility |
| Guanylin | Small intestine | Activates production of cGMP; inhibits sodium absorption |
| Ghrelin | Stomach | Control of growth hormone and energy balance |
| Histamine | Small intestine | Activates acid secretion |
| Motilin | Duodenum (small intestine) | Control of motility pattern in the small intestine |
| Neuropeptide Y (NPY) | Small intestine | Stimulates appetite and increases fat storage |
| Neurotensin | Endocrine cells and neurons in gastrointestinal tract | Initiates vasodilation |
| Peptide $YY_{3-36}$ | Small intestine | Inhibits appetite and stimulates pancreatic and gall bladder secretions |
| Secretin | Small intestine | Control of bicarbonate-rich secretions from the pancreas and liver |
| Serotonin (5-HT) | Small intestine | Activates intrinsic motor neurons |
| Somatostatin | Small intestine | Control of gastric secretion |
| Substance P | Endocrine cells and neurons in gastrointestinal tract | Activates pancreatic and bile secretion, increases motility, increases release of histamine |
| Uroguanylin | Small intestine | Binds to guanylyl cyclase receptors and activates synthesis of cGMP |
| Vasoactive intestinal peptide | Gastrointestinal tract | Control of musculature of gastrointestinal tract; control of water and gastric aid secretion |

*Source:* Data from Bowen [10], Sternber [16], and Ojeda and Griffin [18].

tally, thus they appear to have regulatory roles through neurocrine, endocrine, and paracrine mechanisms. Paracrine hormones are those released by endocrine cells within the GI tract which can affect other local endocrine cells that are normally not targets for these hormones. Several peptides are active endocrine modulators of GI function and can act in a paracrine fashion as well. Some of the paracrine hormones include guanylin, histamine, motilin, neuropeptide Y, serotonin, somatostatin, substance P, uroguanylin, and vasoactive intestinal peptide. Studies have shown that immune cells produce regulatory peptides as well as antibodies that can affect cells locally and at other mucosal sites, as well as within the brain itself. This finding has led researchers to utilize the phrase *common mucosal immune system* (CMIS) [14,15]. Hence, there is an intricate and critical commonality among the endocrine, nervous, and immune systems, which act in concert to regulate the GI tract [16]. APUD and DNES cells are polypeptide-secreting cells that are able to act upon contiguous or nearby cells (paracrine), distant cells (endocrine), or neuronal cells (neurocrine) to regulate numerous signals. It has recently been suggested, however, that the term *diffuse neuroimmunoendocrine system* (DNIES) be used instead of DNES to more correctly describe the total integration of signaling mechanisms found in GI tract regulation [17]. Table 31.3 summarizes some of the most common hormones found in the alimentary canal [10,16,18].

## Mouth and Tongue

Food enters the digestive system through the oral or buccal cavity, the outer opening of which is protected by the lips (or labia). The oral cavity is line with mucous membranes and has as its lateral walls the cheeks; the hard palate which forms the anterior roof; and the soft palate, which forms the posterior roof. The tongue constitutes the floor of the oral cavity and is a muscular organ that is attached underneath by the fenulum. The teeth (primarily the anterior incisors) begin the ingestion process by biting of suitable portions that can be retained in the mouth. When sufficient food is retained in the mouth, it is masticated by the molars and premolars and mechanically broken down into smaller particles by tearing, grinding, and chewing. The tongue is made up of six different muscles which allow for finite control during the mastication process. The pro cess is aided by saliva, which is secreted from three pairs of salivary glands: parotid, submaxillary, and sublingual. As much as 1.5 L of saliva is secreted in concert by these three glands over a 24-hour period [19]. Ordinarily, the mucous membranes produce only enough saliva to simply keep the mouth and pharynx moist by the parasympathetic nervous system; however, when food is introduced into the mouth, salivary secretions increase dramatically.

Food stimulates some of the 2000 taste bud receptors located primarily on the tongue. The taste buds not only provide a sense of taste to the brain but also send impulses to the salivary nuclei located in the medulla and pons. Impulses returning to the salivary glands from the salivary nuclei activate the increased secretion of saliva. Saliva contains two main types of protein secretions: (1) serous secretion by the parotid glands which contains α-amylase and digests starches, and (2) mucous secretion by the submandibular and sublingual glands which contains mucin for lubrication. Saliva itself is 99.5% water and 0.5% solutes [20]. The solutes include chlorides, bicarbonates, sodium phosphates, urea, uric acid, serum albumin, serum globulin, mucin, lysozyme and α-amylase [21].

The saliva contents help liquefy the food for ease of passage through the GI tract. Ptyalin or salivary amylase initiates the breakdown of starches into glucose which also begins the chemical digestion process. Only a small portion (<5%) of the starch is actually digested in the mouth because of the quick transit time from the mouth to the esophagus and beyond. Bicarbonate ions tend to neutralize acids, and lysozyme helps to inhibit the bacteria present in many foodstuffs. Saliva also helps to cleanse the mouth and protect the teeth, as they are constantly bathed in fresh saliva. In addition to saliva, lingual lipase is secreted at the base of the tongue to initiate the process of fat digestion. It appears to have a minor impact on fat digestion [22].

## Pharynx, Larynx, and Swallowing

The posterior exit of the mouth leads to the pharynx through a ring of palatine and lingual tonsils. The pharynx is the passageway for the foodstuff from the mouth to the esophagus and also serves as a conduit to the respiratory tract. Food movement must be coordinated to keep the respiratory tract clear of food as it passes on its way to the esophagus. The *swallowing reflex*, or *deglutition*, moves food into the esophagus by a combination of voluntary and involuntary muscle movements. The swallowing action moves food from the rear of the mouth through the pharynx and past the larynx into the esophagus and ultimately into the stomach. It can be divided into three separate functions: (1) the voluntary stage, where the bolus is directed to the rear of the oral cavity and into the pharynx; (2) the pharyngeal stage, an involuntary stage where the bolus moves through the pharynx, past the larynx, and into the top of the esophagus; and (3) the esophageal stage, also an involuntary stage, where the bolus moves through the esophagus into the stomach.

When an appropriate mass of food is chewed and moistened, a bolus is moved toward the pharynx primarily by the voluntary action of the tongue and lips, which direct the food toward the pharynx. As the bolus moves out of the mouth, it produces pressure on the rear of the pharynx, which sends an impulse to the deglutition center in the medulla and pons in the brain stem. The returning impulses trigger a complicated set of involuntary muscular contractions and relaxations in the pharynx that elevate the soft palate, close off the nasopharynx, and push the food downward toward the larynx. These contractions elevate the hyoid and larynx and move the epiglottis downward, closing the entrance to the trachea to prevent food from entering the respiratory tract. When the bolus has passed the larynx and entered the esophagus, the epiglottis reopens and breathing resumes. Figure 31.4 illustrates the swallowing process. A summary of the basic interactions and functions of the swallowing reflex is presented in Table 31.4.

## Esophagus

The primary function of the esophagus is to simply transport masticated foodstuffs from the oral cavity to the stomach. A bolus of food leaves the oral cavity and continues to move through the cricopharyngeal sphincter located at the upper end of the esophagus which subsequently closes, preventing upward movement of the bolus. The esophagus is the third and final organ involved in the swallowing reflex process. Normally, this sphincter is contracted, closing the mouth of the esophagus. The esophagus is a muscle-walled tube about 10 inches long which connects the pharynx to the stomach. It passes behind the heart through the diaphragm and into the abdominal cavity. In its resting stage, the esophagus is collapsed. The upper third of the esophagus contains striated muscle while the lower two thirds contains smooth muscle. When the bolus has entered the esophagus, its movement is controlled by the smooth muscles in the wall of the esophagus; they contract in a rhythmic sequence and propel the bolus downward toward the stomach. These waves of muscular contractions are called *peristaltic waves* and are the mode of transit throughout the remainder of the GI tract. In the esophagus, the transit time is quite brief and no meaningful digestion occurs in the esophagus.

The lower end of the esophagus just above the level of the diaphragm has a sphincter called the *lower esophageal* or *gastroesophageal sphincter*. This structure helps separate the lumen of the esophagus from the lumen of the stomach and prevents reflux of the stomach contents backward into the esophagus. Unlike the rest of the GI tract, the esophagus does not have an outer serosa covering, which makes it susceptible to the spread of tumor cells. Blood is supplied from the inferior thyroid artery to

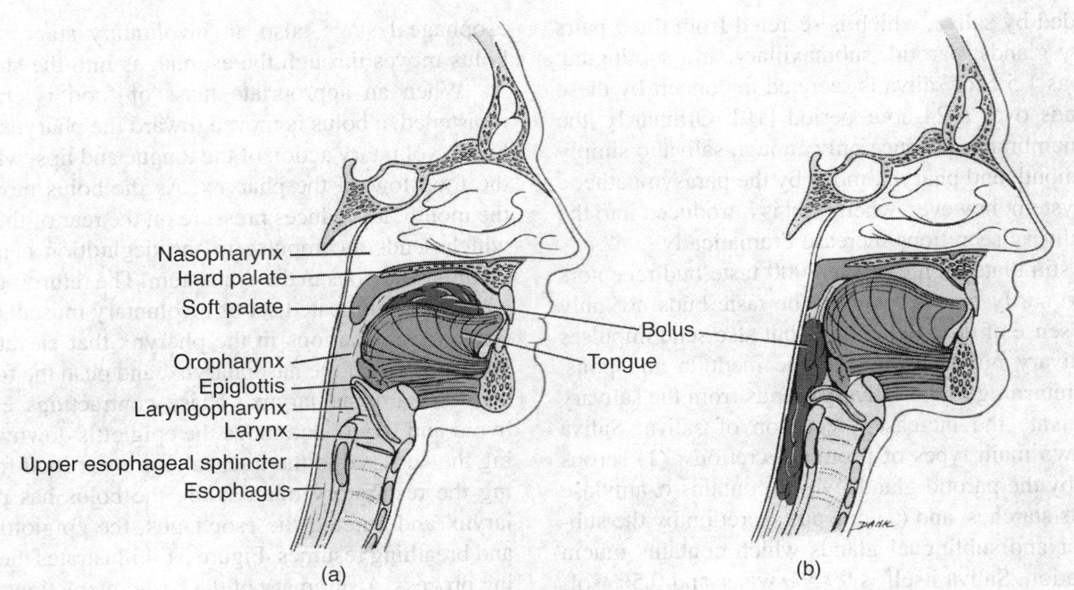


**FIGURE 31.4** Oral and pharyngeal movements during deglutition, or the swallowing reflex. (From Tortora, G.J. and Anagnostakos, N.P., in *Principles of Anatomy and Physiology*, 6th ed., Tortora, G.J., Ed., HarperCollins, New York, 1990, pp. 731–778. With permission.)

the upper esophagus, the tracheobronchial arteries to the middle esophagus, and the left gastric artery to the middle esophagus. The esophagus is also drained of blood by three venous pathways: the upper third into the superior vena cava; the middle third into the azygous system; and the lower third into the portal vein.

Sensory stimulation of the esophagus is transported through the vagus nerve to the tractus solitaries. The thoracic sympathetic nerves also carry afferent nerve impulses to the esophagus. The parasympathetic nervous system is primarily derived from the vagus nerve, with supplemental impulses from the upper portion by the glossopharyngeal and spinal accessory nerves [24,25]. The peristaltic action in the esophagus is initiated by the presence of the bolus, which expands the esophageal walls; subsequently, stretch receptors are activated to contract behind the bolus and relax in front of the bolus. This occurs as one continuous wave that propels the bolus past the sphincter and into the stomach. The esophageal lumen is lined with stratified squamous epithelia, which are continually being renewed. As the squamous cells migrate toward the lumen, they change physical characteristics and are eventually sloughed off [26,27]. The outer lumen is protected by a coat of keratin that can resist degradation by the various proteolytic enzymes. Mucus-secreting glands located in the submucosal layer lubricate the esophagus and assist with the passage of the bolus. There are species differences in the amount of keratin present (e.g., dogs have less keratin than rodents do) and in the density of the mucous glands (e.g., rodents have none).

**TABLE 31.4**
**Summary of the Swallowing Reflex**

| Structure | Activity | Function | Control Status |
|---|---|---|---|
| Mouth/teeth/tongue | Tearing, mastication and bolus manipulation | Preparation of appropriate-sized bolus; movement into the pharynx | Voluntary |
| | Secretion of mucus | Lubrication for passage into pharynx | Voluntary/involuntary |
| Pharynx | Triggers deglutination center in brainstem | Elevates the soft palate, closes off the nasopharynx | Involuntary |
| Larynx | Impulses received from brainstem | Elevates the hyoid and larynx and moves epiglottis over trachea | Involuntary |
| Esophagus | Receipt of bolus from pharynx | Opening of upper esophageal sphincter | Involuntary |
| | Initiation of peristalsis | Movement of bolus through esophagus | Involuntary |
| | Secretion of mucus | Lubrication for passage into stomach | Involuntary |

*Source:* Data from Tortora and Anagnostakos [21] and Tamir [23].

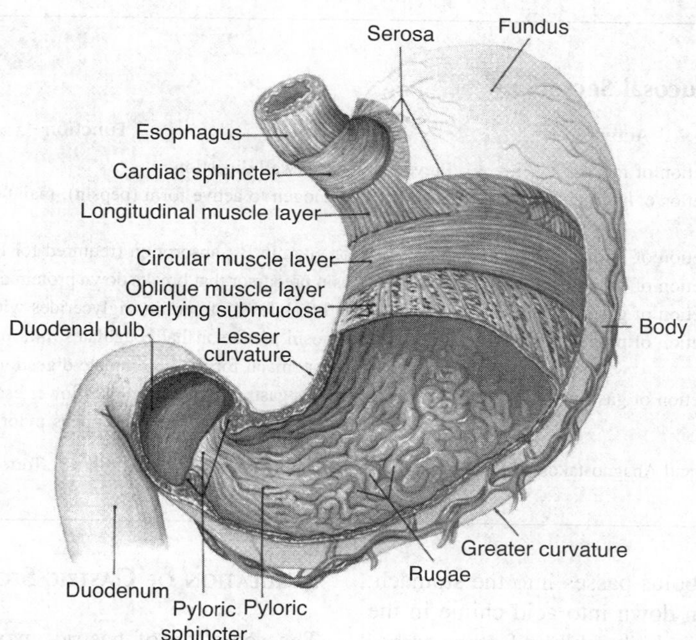


**FIGURE 31.5** Structure and general organization of the stomach. (From Yamaha, T. et al., Eds., *Atlas of Gastroenterology*, 4th ed., Lippincott Williams & Wilkins, Philadelphia, PA. With permission.)

## STOMACH

The bolus continues from the lower esophageal sphincter into the stomach through an upper opening termed the *cardia*. There is no circumferential muscle at the cardia; however, the anatomical structure of the area prevents the reflux of the stomach contents backward into the esophagus. The stomach is a J-shaped muscular structure with a capacity of about 2 liters. It is located in the upper left anterior quadrant of the abdominal cavity. Much of the surface of the stomach is behind the lower rib cage. The stomach serves as a reservoir and as a digestive organ. It is divided into four distinct areas: the cardiac region, the fundus, the corpus, and the pyloric region. The part of the corpus that is adjacent to the pyloric region is sometimes referred to as the *antrum* [25,28]. The structure and cellular composition of the tall columnar stomach epithelium remain constant throughout the different areas of the stomach. These cells contain mucus-producing apical vacuoles that produce a high-molecular-weight polymer consisting of a protein backbone with carbohydrate side-chains that are held together by disulfide bonds [29]. This polymer forms a 1-mm-thick secretory sheath that protects the mucosa from the acidic gastric fluids found in the stomach, most likely by retarding the normal diffusion of hydrogen ions. These surface epithelium are renewed every 72 hours by the isthmus of the gastric glands, where mitotic activity is increased, and the new cells migrate toward the surface.

The major biochemical feature of the stomach is its acid-secreting properties. The stomach mucosa contains many folds, called *rugae*, which further contain narrow openings, termed *gastric pits*, which are formed as a result of mucosal cells penetrating the connective tissue downward into the lamina propria. The wall of the stomach is lined with gastric glands that are interspersed in the surface or in these gastric pits. The gastric pits contain, besides the above-noted mucous cells, three types of cells: parietal cells, which are found in the middle of the pits and produce hydrochloric acid and intrinsic factor; chief cells, also known as zymogenic cells, which are found at the bottom of the pits and produce pepsinogen, prochymosin, and gastric lipase; and enteroendocrine or G cells, which are found in the glands in the lower fundus or antrum and produce gastrin. During a meal, the stomach gradually fills up to a capacity of approximately 1 to 2 liters. The secretions of the various cells types are controlled by impulses from the vagus nerve and by gastrointestinal reflexes mediated through neural, autacoid, and endocrine signals [30]. Figure 31.5 shows the general organization of the stomach.

Hydrochloric acid does not directly function in the digestion process; however, it inhibits microorganisms and lowers the stomach acidity which, in turn, inhibits emptying of the stomach. In addition, stomach pH is a critical determinant in the absorption kinetics of electrolytes and it activates pepsinogen. Carbohydrate digestion, which begins in the mouth upon exposure of the foodstuff to

**TABLE 31.5**
**Summary of Stomach Mucosal Secretions**

| Structure | Activity | Function |
|---|---|---|
| Mucous cells | Secretion of mucus | Prevents mucosal wall digestion |
| Parietal cells | Secretion of hydrochloric acid | Converts pepsinogen to active form (pepsin); maintains pH at ~2 for optimal pepsin functioning |
| | Secretion of intrinsic factor | Necessary for vitamin $B_{12}$ absorption (required for RBC formation) |
| Chief (zymogenic) cells | Secretion of pepsinogen | Provides pepsin precursor that breaks down protein chains into peptides for absorption |
| | Secretion of gastric lipase | Assists in the breakdown of certain triglycerides with short-chain fatty acids |
| | Secretion of prochymosin | Provides chymosin precursor that coagulates milk protein, allowing it to be retained longer in the stomach for more complete digestion |
| Enteroendocrine cells | Secretion of gastrin | Stimulates further gastric secretion, closes lower esophageal sphincter, increases muscular activity in the stomach, and opens pyloric sphincter |

*Source:* Data from Tortora, G.J. and Anagnostakos, N.P., in *Principles of Anatomy and Physiology,* 6th ed., Tortora, G.J., Ed., HarperCollins, New York, 1990, pp. 731–778.

amylase, continues as the bolus passes into the stomach. The bolus is further broken down into acid chime in the lower third of the stomach which inhibits further carbohydrate breakdown. At this point, the enzyme pepsinogen initiates protein digestion. Pepsinogen is activated by hydrochloric acid cleaving off a portion of the molecule, thus creating the enzyme pepsin, which breaks off peptides from proteins. Intrinsic factor is a protein that binds vitamin $B_{12}$ to enable absorption by the ileum of the small intestine. Table 31.5 presents a summary of mucosal secretions in humans [21].

When sufficient food enters the stomach, peristaltic movement, termed *mixing waves*, passes through the stomach every 15 to 25 seconds [21]. This movement combined with the stomach secretions reduces the foodstuff to a thin liquid termed *chyme*. The fundus section of the stomach shows little mixing movement and can function as a holding area for the chyme for up to an hour or more. During the time in the fundus, amylase digestion continues. The peristaltic wave action of the stomach walls propels the contents toward the pyloric sphincter. The pylorus is usually nearly closed and permits only very thin (and hence thoroughly digested) chyme into the duodenum of the small intestine. The larger contents are then turned back into the body of the stomach for further agitation and breakdown into smaller particles. Food remains in the stomach for variable periods of time depending on its size and composition. This period can be from 30 minutes to several hours or longer. The continuous peristaltic action of the stomach and the degree of contraction of the pyloric sphincter allow for the proper amount of chyme to enter the small intestine to allow proper absorption of nutrients. Little absorption occurs in the stomach because the stomach wall is impermeable to most materials. There are a few exceptions, however. The stomach does absorb some electrolytes, some water and alcohol, and acetosalicylic acid.

## Regulation of Gastric Secretion

The secretion of gastric enzymes is controlled in three separate phases: (1) cephalic phase, (2) gastric phase, and (3) intestinal reflex [21,31]. The cephalic phase is stimulated by chemoreceptors and mechanoreceptors in the nasal and buccal cavities; these receptors send impulses from the cerebral cortex in the hypothalamus through the vagus nerve that cause the release of acetylcholine, gastric lipase, and gastrin-releasing peptides from interneurons in the stomach. These impulses arising from the sight, smell, or even thought of food act on the parietal cells and G cells. In this way, the body prepares gastric secretion levels in the stomach to receive the anticipated food.

The gastric phase is both nerve and hormonally controlled. When food reaches the stomach, it causes distention and stimulates nervous receptors in the stomach wall. Distention activates the long vagovagal reflexes and the short local reflexes which, in turn, send impulse to the medulla and back to the stomach, which release acetylcholine [32]. The release of acetylcholine stimulates the flow of hydrochloric acid from parietal cells and gastrin from G cells. The presence of peptides will also stimulate the production of hydrochloric acid through gastrin release by direct action on the G cells. This phase controls about 80% of the secretions released from the stomach [31].

The intestinal phase accounts for a small amount of gastric secretion. The presence of chyme in the duodenum stimulates the mucosa to release enteric gastrin. This substance causes the circulating gastrin to increase acid production through activation of the G cells, but in limited amounts. A summary of the factors that influence the stimulation of gastric secretion is provided in Figure 31.6.

The inhibition of gastric secretion occurs when partially digested carbohydrates and lipids enter the duodenum and initiate the enterogastric reflex, which involves impulses sent from the walls of the duodenum to the

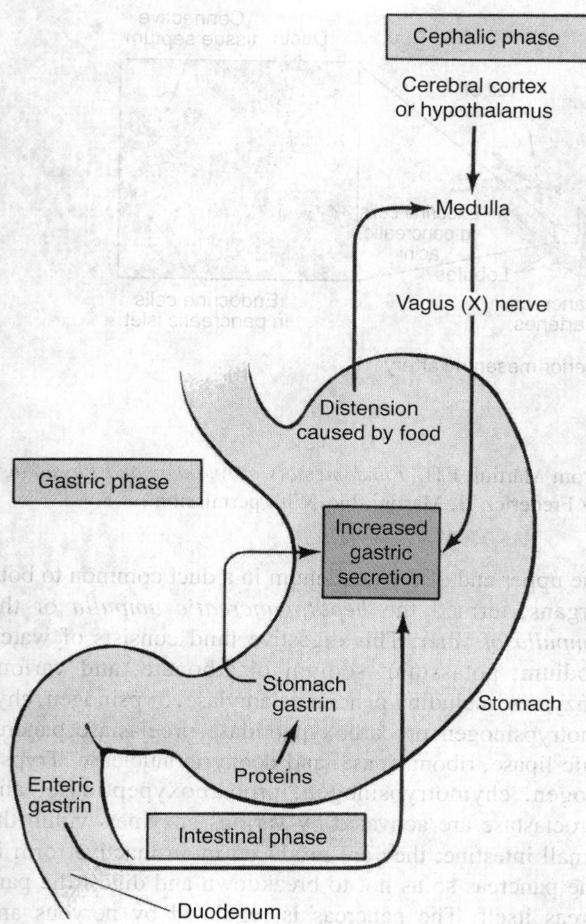


**FIGURE 31.6** Schematic of the phases of gastric secretory regulation. (From Tortora, G.J. and Anagnostakos, N.P., in *Principles of Anatomy and Physiology*, 6th ed., Tortora, G.J., Ed., HarperCollins, New York, 1990, pp. 731–778. With permission.)

medulla and back to the gastric glands in the stomach. Additional chemical regulatory actions that ensure that excess acid production does not accumulate in the stomach include the following [31]:

- *Lowered pH*—Gastric secretion is inhibited at pH < 3.
- *Cholecystokinin*—Triggered by the presence of chyme lipids and carbohydrates, this hormone is secreted by the intestinal mucosa; it inhibits gastric secretions, decreases GI tract motility, stimulates secretions from the pancreas while relaxing the sphincter at the hepatopancreatic ampulla, and increases the ejection of bile from the gall bladder.
- *Gastric inhibitory peptide*—In the presence of fatty acids in the duodenum, this hormone is secreted by the intestinal mucosa; it inhibits gastric secretions and decreases GI tract motility.
- *Secretin*—In the presence of acid, chyme, or fluids in the duodenum, this hormone is secreted by the intestinal mucosa; it inhibits gastric secretions, decreases GI tract motility, and stimulates secretions from the pancreas, liver, and small intestinal glands.

## Small Intestine

The small intestine is where the digestion of fat, protein, and almost all carbohydrates is completed and much of the absorption of nutrients occurs. The small intestine begins at the pylorus of the lower end of the stomach and ends at the juncture of the large intestine, a 15- to 20-foot distance [21,33]. The tube averages approximately 1 inch in diameter and is subdivided into three segments. The duodenum connects to the stomach and is the shortest segment at about 10 inches. The jejunum is about 8 to 10 feet in length and extends to the ileum. The ileum is about 10 to 12 feet in length and attaches to the large intestine at the ileocecal sphincter. Most digestion occurs in the duodenum, where digestive enzymes are added from the pancreas, gallbladder, and glands in the wall of the duodenum. The jejunum and ileum primarily absorb the digested nutrients from the duodenum.

The entire small intestine is comprised of the four basic layers that make up the vast majority of the GI tract. There are some variations that permit the small intestine to finish the digestion process and initiate the majority of the absorption process. The small intestine mucosa contains gastric pits which are lined with glandular epithelium glands called the *crypts of Lieberkühn*. These glands secrete "intestinal juice," which is a clear, yellow, alkaline fluid containing primarily water and mucus in amounts of up to 3 liters per day. Additional enzymes such as maltose, α-dextrinase, sucrase, lactase, aminopeptidase, dipeptidase, ribonuclease, and deoxyribonuclease are added to the chyme in the small intestines. Many of these enzyme-producing cells digest the chyme on the cell surface rather than in the lumen of the intestine. This intestinal juice is quickly reabsorbed by the small intestinal villi, and that physical process brings the chyme into the proximity of epithelial cells that secrete intestinal enzymes. The submucosa contains duodenal Brunner's glands, which secrete alkaline mucus. As in the stomach, the mucus protects the mucosa from breakdown by the hydrochloric acid and helps to neutralize acid present in the chyme.

The structure of the small intestine is especially suited for digestion and absorption. Absorptive cells with finger-like projections, called *microvilla*, greatly expand the surface area of the membrane. The mucosa is arranged in a series of villa that protrude into the intestinal lumen and further increase the functional surface area. Each villus has a core of lamina propria. Embedded in this connective tissue are an arteriole, a venule, a capillary network, and

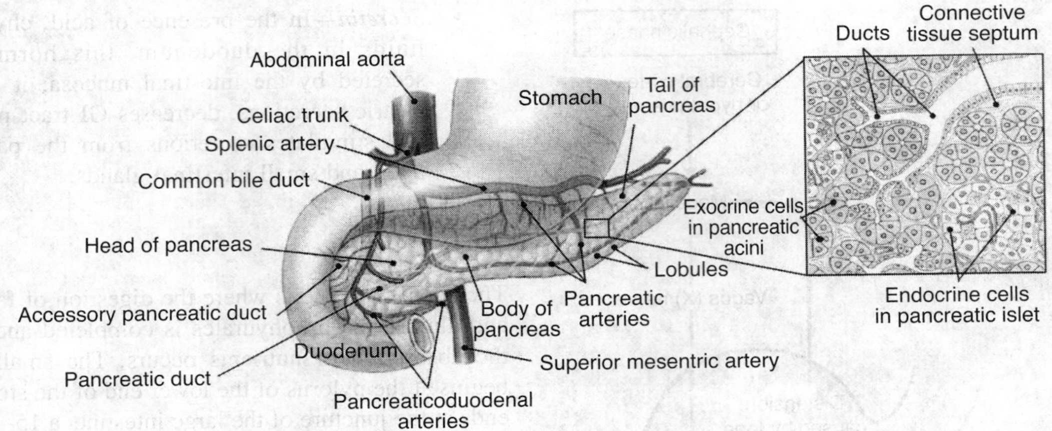


**FIGURE 31.7** Structure and general organization of the pancreas. (From Martini, F.H., *Fundamentals of Anatomy and Physiology*, 7th ed., Pearson Education, Glenview, IL, 2006. Copyright © 2006 by Frederick H. Martini, Inc. With permission.)

a lymphatic vessel (lacteal) which allow nutrients to pass through the capillary wall directly into the lymphatic and cardiovascular systems. Additional projections, called *plicae circulares*, are located in the proximal end of the duodenum and terminate at about the middle portion of the ileum. These plicae circulares not only increase the surface area but also cause the chyme to swirl as peristalsis moves it through the small intestine, thus increasing the absorption process. These adaptations in surface area increase the absorptive surface area many times and provide an estimated 250 to 500 square yards of absorptive surface. Finally, located primarily in the lower part of the small intestine (the ileum) are 30 to 40 lymphatic nodules called *Peyer's patches*, which are part of the autoimmune system; they detect microorganisms and produce antibodies designed to fight infection [25,33]. In addition to its own enzyme secretions, the small intestine utilizes the fluid from three additional structures: (1) the pancreas, (2) the gallbladder, and (3) the liver. The contributions of each of these accessory organs are briefly summarized below.

## Pancreas

The pancreas is a tubuloacinar gland that measures approximately 1 inch wide by 5 inches in length. The pancreas contains two types of glands. One type is hormonal in nature and consists of clusters of glandular epithelial cells called *pancreatic islets* (islets of Langerhans), which produce the hormones glucagon, insulin, and somatostatin. The remaining cells, called *asini*, comprise a vast majority of the organ; they represent the exocrine part of the pancreas and produce digestive enzymes or pancreatic juice. Up to 2 liters of clear, colorless pancreatic juice are produced each day. The asini communicates directly with the intralobular ducts, which eventually collect into the major pancreatic duct or the duct of Wirsung. Generally, this duct unites with the common bile duct, which enters the upper end of the duodenum in a duct common to both organs, termed the *hepatopancreatic ampulla* or the *ampulla of Vater*. This digestive fluid consists of water, sodium, potassium, sodium bicarbonate, and various enzymes, including pancreatic amylase, trypsinogen, chymotrypsinogen, procarboxypeptidase, proelastase, pancreatic lipase, ribonuclease, and deoxyribonuclease. Trypsinogen, chymotrypsinogen, procarboxypeptidase, and proelastase are activated by various enzymes within the small intestine; they are produced in an inactive form in the pancreas so as not to breakdown and digest the pancreas itself. The pancreas is regulated by nervous and hormonal mechanisms. Impulses are also sent along the vagus nerve concurrent with the cephalic and gastric phases of gastric regulation noted previously, which stimulates the production of pancreatic juice. Hormonal regulation includes responses from chyme entering the duodenum which can release secretin or cholecystokinin (CCK). Secretin, which is produced by S cells of the duodenum in response to hydrogen ions, stimulates the production of pancreatic juice rich in sodium bicarbonate ions; cholecystokinin, which is produced by the I cells of the duodenum in response to peptides and fatty acids stimulates the production of pancreatic juice rich in digestive enzymes[25,34]. The general structure and relative location of the pancreas are shown in Figure 31.7.

## Liver

The liver is a triangular/cone-shaped organ located under the diaphragm; it weighs about 3 pounds in the average adult. In addition to its digestive functions, the liver also participates in the detoxification of blood, synthesis of blood proteins, replenishment of erythrocytes, storage of glucose as glycogen, and the production of urea. The digestive functions include the production of bile. The liver consists of two main lobes—a smaller left lobe and

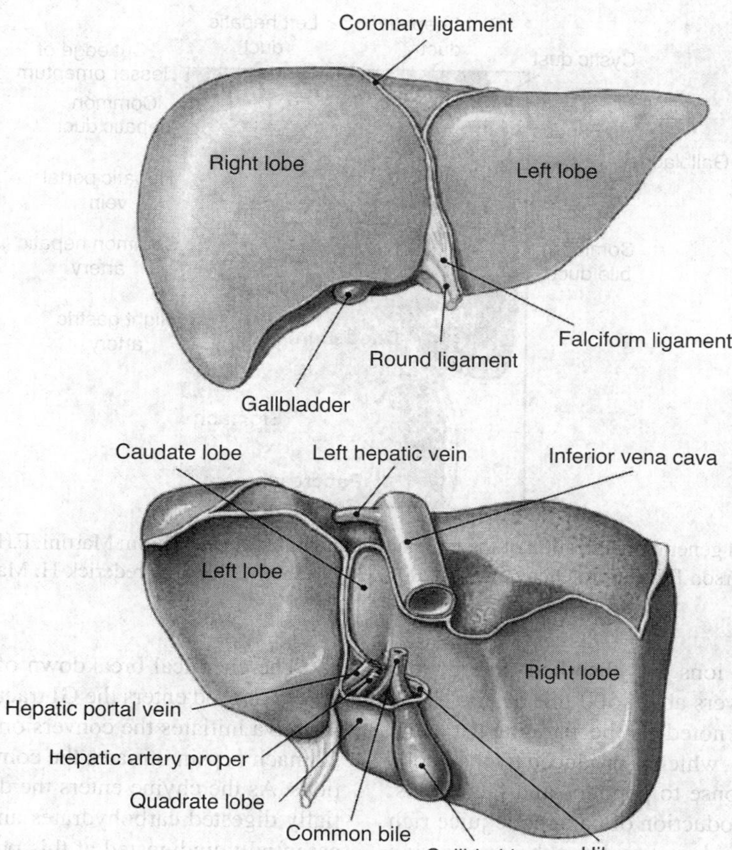


**FIGURE 31.8** Structure and general organization of the liver. (From Martini, F.H., *Fundamentals of Anatomy and Physiology*, 7th ed., Pearson Education, Glenview, IL, 2006. Copyright © 2006 by Frederick H. Martini, Inc. With permission.)

a larger right lobe—both of which are made up of thousands of lobules separated by the falciform ligament. Contained within each lobule is an arteriole, a venule, and a biliary duct. Arteriole blood and venous blood are mixed in the canals of these lobules. Hepatocytes are in direct contact with the blood flowing in these canals. The biliary ducts in the canals serve to drain the bile from hepatocytes into the lobules. These lobules are connected to small ducts that connect with larger ducts to ultimately form the hepatic duct. The liver produces and secretes as much as 1 liter of bile each day, and the hepatic duct transports the bile to the gallbladder and upper section of the duodenum. Bile is a watery green fluid with a pH of 7.6 to 8.6; it contains bile salts, bile pigments (primarily bilirubin), cholesterol, neutral fats, electrolytes, and a phospholipid called *lecithin*. Bile salts and lecithin emulsify fats, and the remaining fluid contents are excreted as waste. Bile salts are emulsifiers that break down fat globules into small (1-μm diameter) droplets that are more easily absorbed in the presence of pancreatic lipase. Absorption of cholesterol is enhanced by bile salts and lecithin. The liver is also regulated by nervous and hormonal mechanisms. Vagus nerve impulses concurrent with the cephalic and gastric phases of gastric regulation noted previously can double the production of bile. Hormonal regulation includes the release of secretin when chyme enters the duodenum. Secretin, which is produced by S cells of the duodenum in response to hydrogen ions, stimulates the production of bile. High blood concentrations of bile salts can also act to increase the rate of bile secretion [25,35]. The structure and general organization of the liver are shown in Figure 31.8.

## Gallbladder

The gallbladder is a hollow, muscular, pear-shaped shaped organ about 3 to 4 inches in length located immediately under the liver. The primary function of the gallbladder is to store, concentrate, and deliver bile from the liver to the small intestine. The gallbladder is normally relaxed and full of bile between meals. The gallbladder lumen mucosa contains simple columnar epithelial cells arranged in deep folds. The middle layer possesses smooth muscle instead of a submucosa which allows for the constriction process during the emptying of the organ. The outer layer consists of the visceral peritoneum. During the concentration of

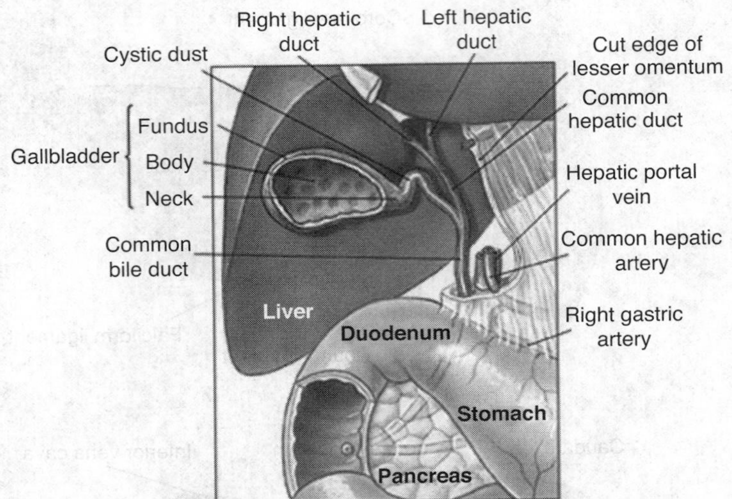


**FIGURE 31.9** Structure and general organization of the gallbladder and biliary ducts. (From Martini, F.H., *Fundamentals of Anatomy and Physiology*, 7th ed., Pearson Education, Glenview, IL, 2006. Copyright © 2006 by Frederick H. Martini, Inc. With permission.)

the liver bile, water and ions are absorbed through the muscosa. This organ delivers about 500 mL of bile to the duodenum each day. As noted in the paragraph on the pancreas, cholecystokinin, which is produced by the I cells of the duodenum in response to peptides and fatty acids, not only stimulates the production of pancreatic juice rich in digestive enzymes but also stimulates the contraction of the musculature of the walls of the gallbladder while simultaneously relaxing the sphincter at the mouth of the common bile duct, thus allowing the gallbladder to empty its contents into the duodenum [25,36]. The structure and general organization of the gallbladder and biliary ducts are shown in Figure 31.9.

## Digestion and Absorption in the Small Intestine

The digestion process in the small intestine can be divided into the mechanics of movement and the chemical breakdown of foodstuffs. Movements in the small intestine include segmentation and peristalsis. Segmentation is a localized contraction that is not involved in the forward motion of the chyme through the GI tract. Segmentation simply forces the chyme into close proximity with the surface of the small intestine by contractions of 12 to 16 times per minute [21]. Segmentation is primarily controlled by nerves in the intestinal walls that detect distention. Peristalsis, on the other hand, is a series of contractions and relaxations propelling the chyme through the GI tract. The peristalsis in the small intestine is very slow and deliberate, pushing the chyme through at a rate of about 1 cm/minute and leaving the chyme in the small intestine for approximately 3 to 5 hours. Peristalsis is similarly controlled by nerves in the intestinal walls that detect distention.

The chemical breakdown of the chyme begins as the article of food enters the GI tract in the mouth. The ptyalin in saliva initiates the conversion of starch to maltose. The stomach pepsin initiates the conversion of proteins to peptides. As the chyme enters the duodenum, it contains partially digested carbohydrates and proteins. The lipids are essentially undigested at this point in the process. As the chyme traverses the length of the small intestine, the digestive process is completed.

When the foodstuffs are prepared and digested as described above, the next critical phase is the absorption of the nutrients, which move across the epithelial cell lining of the GI tract and into the blood and lymphatic circulatory systems. It is estimated that about 90% of absorption occurs within the small intestine. The remainder of occurs in the stomach prior to the foodstuff entering the small intestine or after the foodstuff leaves the small intestine and enters the large intestine. Absorption occurs in the villi by one of the following mechanisms: passive diffusion, facilitated diffusion, osmosis, or active transport. Roughly 10 liters of fluid enter the small intestine and are absorbed each day. About 1 to 1.5 liters is derived from saliva, 2 to 3 liters are secreted by the stomach, about 2 liters are secreted by the glands of the small intestine, about 1 liter is contributed by the pancreas, and another liter enters as bile. Approximately 2 liters are ingested as food and fluid each day. Of the 10 liters, about 90% of the liquid is absorbed via osmosis by the blood capillaries of the villi in the small intestine [6,21]. The remaining 1 liter is what actually passes through the ileocecal sphincter into the large intestine. The normal rate of absorption is approximately 200 to 400 mL/hour.

It has been shown that the water moves in either direction across the intestinal mucosal cells to maintain an osmotic balance between the surface of the lumen, the cell interior, and the blood during uptake of elements of the chyme as well as electrolytes. Absorption in the small intestine is passive and requires the movement of solutes, so the rate of water uptake is a function of the solute absorption. Active transport of sugars and amino acids in the jejunum, in turn, causes passive movement of salt and water across the mucosal epithelial cells (cotransport). In the ileum, on the other hand, most of the water movement is by active transport via a specialized mechanism that pumps sodium into the lateral spaces within the cells. Water subsequently enters the lateral spaces from the cell (transcellular flux) or from the lumen (paracellular flux), resulting in an increased hydrostatic pressure and the transport of isotonic fluid into the extracellular space.

Gastrointestinal secretions and components of various ingested foods are sources of electrolytes for absorption in the small intestine. Intestinal transport of sodium has both active and passive mechanisms. Active transport can be independent or linked to the transport of other solutes such as sugar in a relationship referred to as *cotransport*. It is believed that cotransport accounts for most of the sodium absorption in the small intestine. The energy source for active sodium transport is the hydrolysis of adenosine triphosphate (ATP). Passive transport of sodium occurs primarily through the lateral spaces. Interestingly, in the presence of bicarbonate, sodium can be absorbed against an electrochemical gradient. It is suspected that a sodium–hydrogen exchange can account for this absorption. Chloride absorption is closely associated with sodium absorption by passive diffusion via the paracellular route, $Na^+Cl^-$ cotransport, and $Cl^-$–$HCO_3^-$ exchange. Iodine and nitrate ion absorption is also closely related to sodium absorption and usually passively follows the sodium ions. Evidence suggests that these electrolytes can also be actively transported. Potassium absorption across the intestinal mucosal cells is determined by differences between two opposing unidirectional fluxes down the concentration gradient; it is passive in nature.

Gastric acid solubilizes calcium salts which permits their being actively transported as the divalent cation. Parathyroid hormone in the presence of vitamin D (1,25-dihydroxycholecalciferol) increases the rate of $Ca^{2+}$ absorption. Iron can be absorbed as heme iron, which is iron bound to hemoglobin, or as free $Fe^{2+}$. In the cytoplasm of the mucosal epithelial cells, the heme iron is broken down to the free $Fe^{2+}$, which binds to apoferritin. As the free $Fe^{2+}$ circulates in the blood, it binds to transferrin, which transports the free $Fe^{2+}$ from the small intestine to the liver for storage.

Fat-soluble vitamins such as vitamins A, D, E, and K are incorporated into micelles and absorbed along with other lipids. Most water-soluble vitamins, including C and B complex (but not $B_{12}$, which requires intrinsic factor from the stomach), are absorbed by sodium-dependent cotransport or facilitated diffusion. Vitamin $B_{12}$ absorption depends on gastric acid, pepsin, and intrinsic factor secreted from gastric parietal cells and the pancreatic duct cells. Intrinsic factor binds to and protects vitamin $B_{12}$ from the digestion processes in the small intestine. Vitamin $B_{12}$ binds to specific receptor sites in the mucosal epithelial cells of the ileum, where it enters by vesicular transport. Fat-soluble vitamins such as A, D, E, and K appear to be absorbed in micelles along with the dietary fats and are released into the lymphatic circulatory system. Water-soluble vitamins, such as the B vitamins and vitamin C, are absorbed across the gastrointestinal tract by simple diffusion.

These digestion and absorption processes can also be categorized by type of substance entering the GI tract. The primary categories of foodstuffs include the following four major categories:

- Carbohydrates
- Proteins
- Lipids
- Nucleic acids

### *Carbohydrates*

Generally, the digestion of carbohydrates occurs through the catalytic hydrolysis of α-1,4-glycosidic bonds situated between monosaccharides. The reaction is initiated by ptyalin, which is inactivated in the stomach by the acidic environment; however, it is completed by pancreatic amylase in the small intestine. These enzymes break down starches to the disaccharide maltose and to glucose polymers between three and nine molecules in length. Pancreatic amylase and ptyalin do not hydrolyze α-1,6-glycosidic bonds. As these maltose and other polysaccharides approach the lumen, they are exposed to maltose, sucrase, and lactase, which break down the disaccharides and glucose polymers to various monosaccharides such as glucose, fructose, and galactose. The branched polysaccharides are hydrolyzed by the action of the enzyme α-dextrinase. Table 31.6 provides a summary of the critical digestive processes of carbohydrates. Polysaccharides are not absorbable into the small intestine; hence, they must be reduced to monosaccharides which are easily absorbed. Glucose and galactose are transported into cells by a sodium-dependent cotransport mechanism that transports glucose and sodium ions. The process uses energy generated by the sodium ion gradient to transport glucose and galactose into the epithelial cell. The transporter does not move sodium ions unless glucose is chemically bound to it. Once in the cell, fructose is converted from fructose to glucose. The resulting intracellular glucose from all sources is subsequently transported out of the cell into the extracellular space from that point by facilitated diffusion. Table 31.7 provides a summary of the critical absorption mechanisms for carbohydrates found in the GI tract.

**TABLE 31.6**
**Summary of Carbohydrate Digestion**

| Enzyme/ Digestive Fluid | Source | Process | Result |
|---|---|---|---|
| Ptyalin | Salivary glands | Converts polysaccharides (starch) to disaccharides; the process is initiated in the mouth and is incomplete when the chyme reaches the small intestine | Incomplete reaction to maltose (disaccharide); not absorbable in this form |
| Pancreatic amylase | Pancreas | Completes the process initiated by the pytalin | Maltose (disaccharide); not absorbable in this form |
| Maltase | Small intestine | Splits the disaccharide maltose into a monosaccharide | Glucose (monosaccharide) |
| α-Dextrinase | Small intestine | Splits the disaccharide trehalase and branched polysaccharides into glucose (monosaccharide) | Glucose (monosaccharide) |
| Sucrase | Small intestine | Splits the disaccharide sucrose into two monosaccharides | Glucose + fructose (monosaccharides) |
| Lactase | Small intestine | Splits the disaccharide lactose into two monosaccharides | Glucose + galactose (monosaccharides) |

**TABLE 31.7**
**Summary of Carbohydrate Absorption**

| Type of Nutrient | Type of Movement | Site of Transport | Destination |
|---|---|---|---|
| Maltose (disaccharide) | Not absorbable in this form | — | — |
| Glucose (monosaccharide) | Sodium-dependent active transport | Villi (small intestine) | Bloodstream to hepatic portal vein |
| Fructose (monosaccharides | Facilitated diffusion | Villi (small intestine) | Bloodstream to hepatic portal vein |
| Galactose (monosaccharides | Sodium-dependent active transport | Villi (small intestine) | Bloodstream to hepatic portal vein |

*Proteins*

Protein digestion is initiated in the stomach by the secretion of pepsinogen, which is activated to its active form—pepsin—in the presence of hydrochloric acid. Interestingly, pepsin is the only enzyme capable of also digesting collagen. Additional inactive proteases are released by the pancreas in response to cholecystokinin. Trypsinogen is activated to trypsin, its active form, by enterokinase or enteropeptidase in the duodenum. Trypsin subsequently activates the additional pancreatic enzymes chymotrypsinogen and procarboxypeptidase to their active forms—chymotrypsin and caroxypeptidase. This protease activity converts proteins to one of the following: amino acids, dipeptides, or tripeptides. This breakdown of proteins is completed by the enzymes aminopeptidase and dipeptidase, both of which are secreted by enterocytes of the small intestine located on the lumen. Table 31.8 provides a summary of the critical digestive processes of proteins. In contrast to carbohydrates, which must be monosaccharides to be absorbed, the digested protein products of proteins can be absorbed as amino acids, dipeptides, and tripeptides. Once the dipeptides and tripeptides are transported into the intestinal cells, however, cytoplasmic peptidase hydrolyzes these peptides to amino acids. These substances are absorbed into and out of the epithelial cells by specific sodium-dependent amino acid cotransporters: neutral, acidic, basic, and imino [31]. Table 31.9 provides a summary of the critical absorption mechanisms of proteins.

*Lipids*

Lipids in the diet include triglycerides, neutral fats, cholesterol, cholesterol compounds, and phospholipids. As fats are ingested, they do not dissolve in water; fats tend to congeal into large masses, and this separation of lipids and water limits the effectiveness of the lipase-fat-digesting enzymes. Lipid digestion occurs primarily in the small intestine and involves the emulsification of fats by bile. Bile acids are derivatives of cholesterol synthesized within the hepatocytes in the liver. Interestingly, they are also facial amphipathic, meaning they possess both hydrophilic (water soluble) and hydrophobic (fat soluble) components This amphipathic quality allows the bile acids to emulsify lipid aggregates and solubilize and transport lipids in an aqueous milieu. The process begins as triglycerides are broken down or emulsified in the duodenum by bile acids into monoglycerides and fatty acids [19,142]. Table 31.10 provides a summary of the critical digestive processes of

## TABLE 31.8
## Summary of Protein Digestion

| Enzyme/Digestive Fluid | Source | Process | Result |
|---|---|---|---|
| Pepsin (derivative of pepsinogen) | Stomach (chief cells) | Initial breakdown of collagen and proteins into peptides | Peptide fragments |
| *Endopeptidases (hydrolyze interior peptide bonds)* | | | |
| Trypsin (derivative of trypsinogen) | Pancreas | Continues breakdown of proteins into peptides | Peptide fragments |
| Chymotrypsin (derivative of chymotrypsinogen) | Pancreas | Continues breakdown of proteins into peptides | Peptide fragments |
| *Exopeptidases (hydrolyze exterior peptide bonds)* | | | |
| Carboxypeptidase (derivative of procarboxypeptidase) | Pancreas | Continues breakdown of proteins into peptides | Peptide fragments |
| Aminopeptidase | Small intestine | Cleaves terminal amino acids at amino ends of peptides | Amino acids (dipeptides, tripeptides) |
| Dipeptidase | Small intestine | Cleaves two amino acid molecules joined by a peptide bond (dipeptide) | Amino acids (dipeptides, tripeptides) |

## TABLE 31.9
## Summary of Protein Absorption

| Type of Nutrient | Type of Movement | Site of Transport | Destination |
|---|---|---|---|
| Polypeptide fragments | Not absorbable in this form | — | — |
| Dipeptide α amino acids (by cytoplasmic peptidase) | Sodium-dependent cotransport at lumen; facilitated diffusion to blood | Duodenum and jejunum | Bloodstream to hepatic portal vein |
| Tripeptide α amino acids (by cytoplasmic peptidase) | Sodium-dependent cotransport at lumen; facilitated diffusion to blood | Duodenum and jejunum | Bloodstream to hepatic portal vein |
| Amino acids | Sodium-dependent amino cotransport (four types of amino acid carriers: neutral, acidic, basic, imino) | Duodenum and jejunum | Bloodstream to hepatic portal vein |

## TABLE 31.10
## Summary of Lipid Digestion

| Enzyme/Digestive Fluid | Source | Process | Result |
|---|---|---|---|
| Lingual lipase | Salivary glands at the base of the tongue | Minor digestion of triglycerides to monoglycerides and fatty acids | Monoglycerides and fatty acids |
| Bile acids | Hepatocytes in liver lobules | Emulsification of lipid aggregates; solubilization and transport of lipids | Breakdown of triglycerides into 1-μm globules; hydrophobic products of lipid digestion are solubilized in micelles |
| Pancreatic lipases (pancreatic lipase, cholesterol ester hydrolase, phospholipase $A_2$) | Pancreas | Removes two of the three fatty acids from glycerol | Fatty acids, monoglycerides, cholesterol, and lysolecithin |
| Lipoprotein lipase | Capillary endothelial cells | Breaks down triglycerides into fatty acids and glycerol | Removal of chylomicrons from the circulation system |

lipids. The short-chain fatty acids (<12 carbon molecules) are absorbed by the small intestine by simple diffusion. The long-chain fatty acids (>12 carbon molecules) and monoglycerides are dissolved and absorbed into the center of structures called *micelles*, which consist of 20 to 50 molecules of bile salt. As the micelle comes into contact with the intestinal cells, the fatty acids, cholesterol, and monoglycerides cross the luminal membranes into the

**TABLE 31.11**
**Summary of Lipid Absorption**

| Type of Nutrient | Type of Movement | Site of Transport | Destination |
|---|---|---|---|
| Short-chain fatty acids | Facilitated diffusion | Small intestine | Bloodstream to hepatic portal vein |
| Long-chain fatty acids, monoglycerides, cholesterol, and lysolecithin in micelles | Facilitated diffusion | Ileum | Bloodstream to hepatic portal vein |
| Chylomicrons | Exocytosis | Small intestine | Lymph vessels through the thoracic duct |

**TABLE 31.12**
**Summary of Nucleic Acid Digestion**

| Enzyme/ Digestive Fluid | Source | Process | Result |
|---|---|---|---|
| Ribonuclease | Pancreas and small intestine | Converts ribonucleic acid into nucleotides | Nucleotides |
| Deoxyribonuclease | Pancreas and small intestine | Converts deoxyribonucleic acid into nucleotides | Nucleotides |
| Nucleosidase | Small intestine epithelial cells | Converts nucleosides into purines and pyrimidines | Nucleosides, purines, and pyrimidines |
| Phosphatase | Small intestine epithelial cells | Catalyzes hydrolysis of esters of phosphoric acid | Phosphate |
| Nucleotidase | Ileum epithelial cells | Converts nucleotides to sugars, pentose, phosphates, and nitrogenous bases | Sugars, pentose, phosphates, and nitrogenous bases |

intestinal cell wall, leaving the empty micelles in the chyme to repeat the process. Several transporter-mediated mechanisms have been postulated for the uptake of fatty acids such as CD36 [37] and FATP4 [38], among others [39,40]; however, some postulate that the transport is strictly passive diffusion [41]. It has been suggested that the mode of lipid uptake by the intestinal cells is dependent on the concentration of available fatty acids; that is, if the concentration of lipids is low and the body needs fatty acids, an active component appears to be a key factor in lipid transport. On the other hand, when the lipid concentration is high, most transport is postulated to be passive [42]. The bile acids are reabsorbed primarily in the ileum and returned to the liver in a cycle called *enterohepatic circulation*. Once the fatty acids and monoglycerides are in the epithelial cells, they are re-esterified to triglycerides, cholesterol ester, and phospholipids, which become coated with apoproteins. These newly formed structures, called *chylomicrons*, are transported out of the epithelial cells into the lymphatic system by exocytosis. Once in the circulatory system, the triglycerides in the chylomicrons are removed from the circulatory system as they pass through the liver. This removal is mediated by lipoprotein lipase, which is present in the capillary endothelial cells and breaks down the triglycerides into fatty acids and glycerol. To be mobile and functional in the blood, plasma lipids are combined with apoproteins. These structures are termed *lipoproteins*. There are several types of lipoproteins, depending on their specific composition of triglycerides, phospholipids, cholesterol, and protein. These lipoprotein include high-density lipoproteins (HDLs), low-density lipoproteins (LDLs), and very-low-density lipoproteins (VLDLs). Table 31.11 provides a summary of the critical absorption mechanisms of lipids.

### *Nucleic Acids*

A wide variety of plant and animal foods contain DNA, RNA, nucleotides, and free nucleic acids. Once in the GI tract, DNA and RNA are broken down into free nucleic acids and nucleotides by various nucleases found in both intestinal and pancreatic digestive juices. The pancreatic nucleases convert DNA and RNA into nucleotides along the length of the small intestines. DNA nucleotides include the following bases: adenine, thymine, guanine and cytosine. RNA nucleotides contain the following bases: adenine, uracil, guanine, and cytosine. These nucleotides are further digested by enzymes produced by the small intestinal epithelial cells, including nucleosidase and phosphatase. Nucleotidase enzymes found in the membrane of the ileum epithelial cells further break down the nucleotides into pentoses, phosphates, and nitrogenous bases. Table 31.12 is a summary of the critical digestive processes of nucleic acids. Nucleotides can be partially absorbed in the duodenum and the jejunum. The nucleosides are absorbed in greater amounts than the nucleotides because they lack the phosphoric acids. Transport of nucleosides into the mucosal cell wall occurs by facilitated diffusion and specific sodium-dependent carrier-mediated transport mechanisms [43]. When the nucleic acids have been broken down into their component

**TABLE 31.13**
**Summary of Nucleic Acid Absorption**

| Type of Nutrient | Type of Movement | Site of Transport | Destination |
|---|---|---|---|
| RNA nucleotides | Sodium-dependent active transport | Duodenum and jejunum | Bloodstream to hepatic portal vein |
| DNA nucleotides | Sodium-dependent active transport | Duodenum and jejunum | Bloodstream to hepatic portal vein |
| Sugars, pentose, phosphates and nitrogenous bases | Sodium-dependent active transport | Duodenum and jejunum | Bloodstream to hepatic portal vein |

sugars, phosphates, and nitrogenous bases, they are absorbed directly into the epithelial cells by sodium-dependent active transport. Table 31.13 provides a summary of the critical absorption mechanisms of nucleic acids. Nutrients are absorbed along the length of the small intestine as shown in Table 31.14 [44].

**TABLE 31.14**
**Summary of Small Intestine Nutrient Absorption**

| Region of Small Intestine | Nutrient Absorbed |
|---|---|
| Duodenum | Calcium |
| | Galactose |
| | Glucose |
| | Iron |
| | Magnesium |
| Jejunum | Amino acids |
| | Chloride |
| | Fatty acids |
| | Fat-soluble vitamins |
| | Fructose |
| | Galactose |
| | Glucose |
| | Potassium |
| | Sodium |
| | Sugars |
| | Water-soluble vitamins |
| Ileum | Alcohols[a] |
| | Amino acids |
| | Bile salts |
| | Chloride |
| | Fatty acids |
| | Potassium |
| | Sodium |
| | Some small peptides |
| | Vitamin $B_{12}$ |
| | Water |

[a] Also absorbed in the stomach and oral cavities.

*Source:* Adapted from Pearce, K., *Absorption*, University of Waterloo, Ontario, Canada, 2004 (http://www.student.math.uwaterloo.ca/~km pearce/ SummaryGroup 12.doc).

## Large Intestine

The large intestine, also referred to as the *colon*, is the final part of the digestive tract; it measures about 5 feet in length and 2 to 3 inches in diameter and extends from the ileocecal sphincter at the end of the termination of the small intestine through the rectum, which connects to the anus. The large intestine is divided into several segments. Immediately below the ileocecal sphincter is the cecum, a sac-like enlargement at the junction of the small and large intestines which measures about 2 to 3 inches in length. Attached to the cecum is a small tube called the *appendix*, which measures approximately a quarter of an inch wide and 2 to 3 inches in length; its function remains unclear [25]. Adjacent to the cecum on the right side of the abdomen is the ascending colon, which ascends to the bottom of the liver, where it turns left and forms the right colic flexure, which becomes the transverse colon. When the ascending colon reaches the end of the spleen, it turns downward and forms the left colic flexure, which becomes the descending colon. The next segment is the sigmoid colon, which descends into the pelvis and connects to the rectum. The last 8 inches of the large intestine comprise the rectum, which is a reservoir for waste. The final inch of the rectum is the anal canal, which ultimately becomes the anus. The anus has two sphincters, one that is involuntary and comprised of smooth muscle and a second external voluntary sphincter that is comprised of skeletal muscle. The anus remains closed except during the elimination of wastes [45]. The wall of the large intestine is comprised of the same layers as the rest of the GI tract; however, there are several differences. First, the muscularis layer consists of two layers, but the external layer of longitudinal muscles is thickened and forms three conspicuous longitudinal bands called *teniae coli*. See Figure 31.10 for the overall structure and general organization of the large intestine. Together, the three bands run the length of the large intestine. By virtue of their tonic contractions, the walls of the colon pucker into pocket-like sacs called *haustra*, which give the colon its segmented appearance. Epiploic appendages or fat-filled bags of visceral peritoneum are attached to the teniae coli. Their function remains largely unknown.

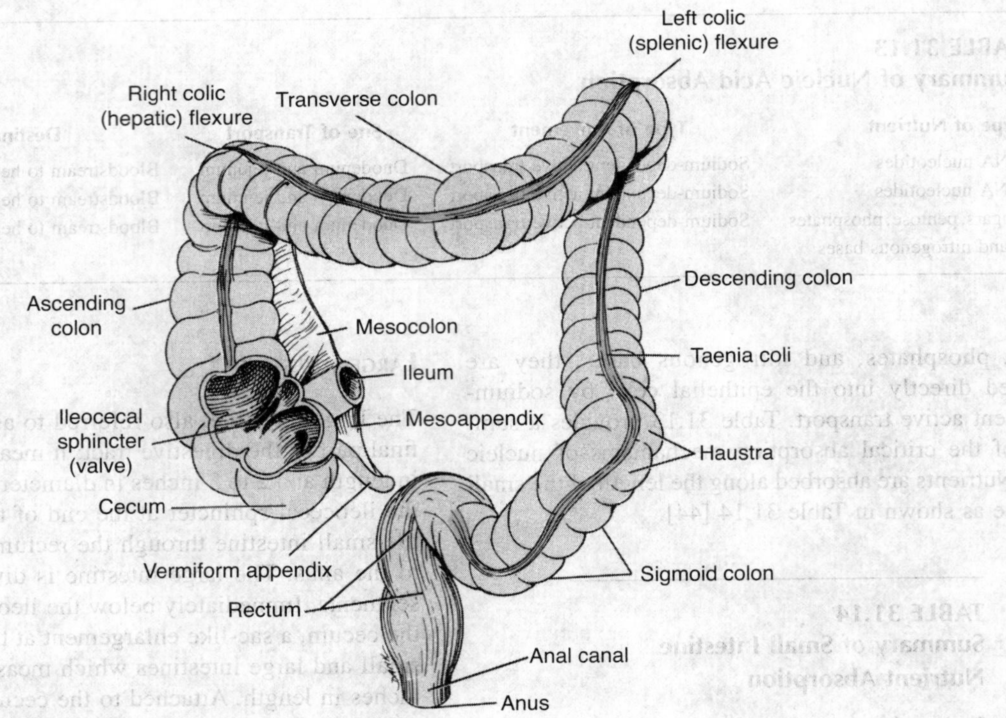


**FIGURE 31.10** Structure and general organization of the large intestine. (From Tortora, G.J. and Anagnostakos, N.P., in *Principles of Anatomy and Physiology*, 6th ed., Tortora, G.J., Ed., HarperCollins, New York, 1990, pp. 731–778. With permission.)

## Digestion

After the small intestine absorption is completed, the remaining fluid passes into the cecum of the large intestine. The ileocecal sphincter acts as a governor because it is normally slightly contracted and slows the passage of chyme from the small intestine to the cecum. Not only does this sphincter control the rate of chyme passage but it also prevents the reverse flow of chyme back into the ileum. Interestingly, when the chyme reaches the colon, the primary mechanism of digestion is by bacterial action. The secretion of glands in the mucosal epithelial cell wall is a protective measure against potential damage from bacterial activity. The bacteria ferment carbohydrates, further breakdown amino acids, decompose bilirubin, and synthesize some B complex vitamins as well as vitamin K.

## Bacteria

Bacteria are present throughout the length of the GI tract; however, a significant bacterial concentration is present in the large intestine. A critical step in the preparation of chyme for elimination is the fermenting action of bacteria normally present in the colon; therefore, in addition to the actions of the digestive juices and mixing action of the contents, most of the final phase of digestion in the large intestine depends on the symbiotic activity of bacteria in the system. Few bacteria are present in the gut at birth; however, rapid colonization occurs shortly after birth. It has been estimated that the average adult has $10^{14}$ intestinal flora, which equates to about 4 to 5 pounds that are critical to the digestive process. The large intestine contains as many as 400 different species of bacteria, and an individual's flora is immunologically unique to that person [46]. The stomach has the least amount of microflora (<$10^3$ CFU/mL), primarily because the low pH destroys many organisms; in contrast, almost the entire digestion process can be attributed to the large number (>$10^{14}$ CFU/mL) of anaerobic bacteria present in the large intestine [46]. Table 31.15 provides a summary of the various types of microflora found in the GI tract from the stomach through the large intestine [47–49].

Most of the final phase of digestion in the large intestine depends on the symbiotic activity of bacteria in the system. The normal microflora digest materials that enter through the ileocecal sphincter and maintain a barrier to infections that could develop from hostile organisms. This barrier consists of an overwhelming number of resident microflora that give invaders little space or food, as well as a by-product of metabolism that tends to reduce the reproductive capacity of foreign microflora. When foreign bacteria enters the GI tract, the resident bacteria are immunologically activated and help prevent colonization by the

**TABLE 31.15**
**Microflora Found in the Gastrointestinal Tract**

| Location | Concentration (CFU/mL) | Types of Bacteria |
|---|---|---|
| Stomach | <10$^3$ | Gram-positive aerobic microorganisms (e.g., *Streptococcus*, *Staphylococcus*, *Lactobacillus*) |
| Duodenum | 10$^3$ to 10$^4$ | Gram-positive microbes ↓↓↓ |
| | | Gram-negative microbes ↑ |
| Jejunum | 10$^4$ to 10$^5$ | Gram-positive microbes ↓↓ |
| | | Gram-negative microbes ↑↑ |
| Ileum | 10$^6$ to 10$^7$ | Gram-positive microbes ↓ |
| | | Gram-negative microbes ↑↑↑ |
| Large intestine | >10$^{14}$ | Gram-negative anaerobic microbes (e.g., *Bacteroides*, *Bifidobacterium*, *Fusobacterium*, *Clostridium*, *Eubacterium*) |

*Source:* Data from Wright et al. [141], Ringler and Daibach [142], and Tso [143].

invading organisms. Normal intestinal bacteria help to prevent infections because the normal bacteria occupy all of the available niches for bacteria, thus depriving the invading bacteria of suitable places to begin reproduction. Further, the robust occupying bacteria will utilize available food, which keeps the invading flora from multiplying. Finally, some of the normal bacteria produce antibacterial chemicals (called *bacteriocins*) as a result of their metabolism. This by-product can generate a local antibiotic effect that hinders the reproduction of invading flora. Secretions in the large intestine consist primarily of alkaline mucus that protects the epithelium and neutralizes acids produced by bacterial metabolism. The most common include *Enterobacteria*, *Bacteroides*, *Bifidobacterium*, *Streptococcus faecalis*, *Eubacterium*, *Peptococcus*, *Peptostreptococcus*, *Ruminococcus*, *Clostridia*, and *Lactobacilli*. The environment of the large intestine is primarily anaerobic, which favors obligate anaerobes. Table 31.16 provides a summary of bacterial action in the large intestine [47,49].

## Motility

When the chyme enters the cecum, the peristaltic action is triggered by distention of the wall in the colon; however, it occurs at a rate much slower than in the small intestine. The movement of fecal contents from the cecum to the rectum for elimination can take several days. The movements of the cecum and ascending colon are more vigorous and promote mixing to hasten the absorption process; however, the movements of the descending and sigmoid colon and rectum are slower and more deliberate to permit the formation of stool, which collects in the rectum. Although the origin of these coordinated contractions is not well understood, they appear to be dependent on a combination of internal rhythmic contractions and neural, paracrine, and humoral factors. Following a meal, colonic motility increases significantly, due to signals propagated throughout the enteric nervous system from the walls of the stomach, which triggers the gastrocolic reflex. Several

**TABLE 31.16**
**Bacterial Action in the Large Intestine**

| Category | Action | Result |
|---|---|---|
| Carbohydrates | Fermentation | Release of hydrogen, carbon dioxide, methane gas |
| Proteins | Breakdown remaining proteins | Release of amino acids |
| Amino acids | Breakdown of to basic components | Release of indole, hydrogen sulfide, fatty acids, and skatole |
| Bilirubin | Breakdown to basic components | Simple pigments |
| Soluble fiber | Fermentation | Lubricating gel to maintain stools in soft and flexible consistency to assist in elimination |
| Bacterial synthesis | Synthesis of vitamins | Creation of vitamin K and some B-complex vitamins, including riboflavin, nicotinic acid, folic acid, and biotin |
| Bacterial synthesis | Synthesis of nutrients | Bacterial self-supporting creation of nutrients for their own food supply |
| Immunological responses | Activation upon presence of foreign bacteria | Prevents colonization by invading pathogens |

*Source:* Data from Wright et al. [141] and Tso [143].

**TABLE 31.17**
**Muscular Action of the Large Intestine**

| Type of Motion | Description | Function |
|---|---|---|
| Haustral churning | Haustrum pouch relaxes and fills to capacity, then, as it becomes distended, contracts and moves the contents to the next haustrum. | This action provides a mixing motion to the contents which facilitates fluid extraction, creates larger masses, and helps advance the contents through the colon. |
| Peristalsis | A wave motion is produced by joint muscular coordination between the interior and exterior muscles of the muscularis layer. | This wave motion propels the contents forward through the colon; slower than similar action in the small intestine. |
| Mass peristalsis reflexes: | | |
| Gastrocolic reflex | As foodstuffs enter the stomach, a reflex is triggered at the mid-transverse colon that initiates a strong mass peristaltic action. | The strong wave action forces any latent contents in the lower large intestine into the rectum to make room for the new stomach contents. |
| Duodenocolic reflex | As fats enter the duodenum, a reflex is triggered at the mid-traverse colon that initiates a strong mass peristaltic action. | The strong wave action forces any latent contents in the lower large intestine into the rectum to make room for the new stomach contents |
| Defecation reflex | As colon contents move into the rectum, a signal is sent to the brain. | This signal tells the conscious brain that the rectum is full and defecation must occur. |
| Cooperative abdominal effort | As defecation occurs, voluntary and involuntary abdominal muscles are contracted. | This cooperative action of the abdominal muscles assists in pushing the waste from the rectum and anus. |

times a day, mass movements push feces into the rectum in this manner. In addition, the presence of fats in the duodenum appears to trigger another mass peristalsis reflex, termed the *duodenocolic reflex*, through the enteric nervous system. The general state of the large intestine musculature also depends upon the distention of the colon walls which can initiate contractions within the colon. The large intestine has six types of movement as summarized in Table 31.17.

## Absorption

Of the about 1 liter of fluid that enters the large intestine each day, approximately 90% is reabsorbed, leaving only about 100 mL to be excreted with the feces. When the chyme reaches the cecum of the large intestine, most of the digestion and secretion have been completed. The primary functions of the large intestine are the absorption of water, vitamins, and electrolytes; the formation of feces; and movement of the feces to the anus for elimination. The intestinal mucosal of the large intestine absorbs water, sodium, and chloride and secretes potassium and bicarbonate.

Water is absorbed in the large intestine in response to an osmotic gradient. Absorption of water in the large intestine is passive and requires the movement of solutes (primarily sodium); thus, the rate of water uptake is a function of the solute absorption. The mechanism responsible for generating this osmotic pressure involves sodium ions being transported from the lumen across the epithelium because of the sodium pumps and the ability of the luminal membrane to absorb sodium. In fact, the colonic epithelium is very effective at absorbing water and sodium (95% retention rate). Absorption in the colon is enhanced by the hormone aldosterone. This uptake of water serves to concentrate the contents of the large intestine to a semisolid form. Sodium and subsequently water are absorbed primarily in the cecum, the ascending colon, and, to a lesser extent, the transverse colon. Bicarbonate is secreted against a concentration gradient, and chloride absorption is linked to both bicarbonate secretion and sodium transport. The resulting secretion of bicarbonate ions into the lumen aids in neutralization of the acids generated by microbial fermentation in the large intestine. Potassium, on the other hand, is passively transported and can be absorbed if the potassium concentration is greater than 15 mEq/L or secreted if the potassium concentration is less than 15 mEq/L. Under normal circumstances, the luminal concentration of potassium is less than 15 mEq/L, which favors secretion.

Vitamins are important to the body as cofactors or coenzymes in numerous metabolic pathways. Some critical vitamins produced by intestinal bacteria are also absorbed in the large intestine. Approximately 50% of the daily requirement for vitamin K is produced by bacteria found in the large intestine. Pantothenic acid (vitamin $B_5$), which is necessary in the production of steroids and some neurotransmitters, is also produced by intestinal bacteria. Finally, biotin, a necessary cofactor in glucose metabolism, is also produced by intestinal bacteria. Although each of these is generated by various intestinal bacteria, the resulting vitamins are absorbed primarily in the cecum and ascending colon. Organic wastes from the breakdown of heme produce bilirubin, which is broken down into

**TABLE 31.18**
**Function of the Lining of the Large Intestine**

| Function | Location | Result |
|---|---|---|
| Protection | Nerve endings in wall of large intestine | Permits intact neural messages between the central nervous system and the colon |
| Lubrication | Entire large intestinal wall | Promotes stool transit |
| Facilitation | Movement of fluids through the wall of the large intestine | Optimizes nutritional uptake of vitamins and electrolytes |
| Bacterial growth | Large intestine wall | Promotes growth of healthy intestinal bacteria which breakdown waste, create additional lubrication, cleanse the colon, and protect the colon against infection |

urobilinogens and stercobilinogens. Ultimately, these are converted to urobilins and stercobilins, respectively. Bacterial activity on these latter substances in the large intestine generate:

- Ammonium ions ($NH^{4+}$)
- Hydrogen sulfide ($H_2S$), which produces a characteristic odor
- Indole, which produces characteristic odor
- Skatole, which produces a characteristic odor

Ammonia and some other organic wastes are absorbed by the large intestinal mucosal epithelial cells and enter the hepatic portal circulatory system, where they are processed by the liver and ultimately eliminated by the kidneys. Indigestible polysaccharides are also eventually deposited in the large intestine. The resident bacteria begin to break down these substances and utilize them as a source of nutrients.

The large intestine has a delicate lining that is moisturized by mucus and also by a gel that is a by-product of the bacterial fermentation process. Despite the many similarities in the cellular structure of the mucosa in the large and small intestine, clearly the most obvious difference is that the mucosa of the large intestine lacks absorptive villi. The mucosal epithelial walls of the large intestine have numerous crypts that extend deep into and open onto a flat luminal surface. The stem cells that support the process of renewal of the epithelium are located in the crypts. These cells divide to populate the crypt and surface epithelium. Mucus-secreting goblet cells are also much more abundant in the colonic epithelium than in the small intestine. These cells secret mucus in response to parasympathetic stimuli and the presence of lumen contents. The lining of the large intestine has numerous functions, as summarized in Table 31.18.

The chyme remains in the colon for 3 to 24 hours. During that time, numerous processes occur. In addition to the muscular activity and the bacterial actions noted above, several cellular exchanges take place, as summarized in Table 31.19. After 90% of the water is absorbed and completion of a myriad of additional actions in the large intestine, the chyme is transformed into a semisolid called *feces*, consisting of residual water, inorganic salts, bacteria, products of bacterial decomposition, mucus cells, cell fragments from the GI tract mucosa, and any undigested food. Under normal physiological conditions, normal feces consist of approximately 75% water and 25% solids. The characteristic brown color of feces is due to stercobilin and urobinin, both of which are produced by bacterial degradation of bilirubin. The characteristic fecal odor is a result of gases produced by bacterial metabolism, including skatole, mercaptans, and hydrogen sulfide.

### Defecation Reflex

As described previously, mass peristalsis moves the fecal material from the transverse, descending, and sigmoid colon into the rectum. When distention at 25% or more is

**TABLE 31.19**
**Cellular Exchanges of the Colon**

| Function | Process | Location | Result |
|---|---|---|---|
| Secretion | Secretion of mucus by goblet cells | Goblet cells located in tubular intestinal glands | Secreted mucus lubricates and protects the colon lining and nerve tissues. |
| Fluid absorption | Secondary to solute absorption, passive diffusion into epithelium cells | Primarily occurs in the columnar cells of the cecum and ascending colon | Approximately 80% of the water entering the colon is absorbed; maintains water balance. |
| Electrolyte absorption | Passive diffusion via sodium channels by columnar cells | Columnar cells located in tubular intestinal glands | $Na^+$, $Cl^-$, and $K^+$ are absorbed; maintains water balance |

**TABLE 31.20**
**Factors Affecting Degree of Toxicity**

| Factor | Location | Effect |
|---|---|---|
| Chemical characteristics of toxicant | Lumen of gastrointestinal tract | Lipophilicity, molecular weight, pH, and pKa; affect extent and rate of absorption |
| Dissolution rate of chemical | Cell wall of gastrointestinal tract | Absorption rate due to particulate size |
| Hydrochloric acid concentration | Stomach | Hydrolysis of toxicant into another compound |
| Dilution in gastric secretions | Lumen of gastrointestinal tract | Absorption rate due to competition and concentration |
| Dilution in food content | Lumen of gastrointestinal tract | Absorption rate due to competition and diluted concentration |
| Time of residence at absorption site | Lumen of various gastrointestinal tract locations | Absorption rate due to time at absorption site |
| Microflora | Lumen of gastrointestinal tract | Chemical changes to the toxicant that alter its absorption characteristics |
| Biotransformation | Lumen of gastrointestinal tract, liver | Clearance of toxicant |
| Fecal excretion | Large intestine | Clearance of toxicant |

*Source:* IPCS, *Principles and Methods for Evaluating the Toxicity of Chemicals, Part I*, International Programme on Chemical Safety, World Health Organization, Geneva, 1978 (http://www.inchem.org/documents/ehc/ehc/ehc006.htm).

detected in the rectal wall, a stimulus is sent from pressure-sensitive receptors that initiates the defecation reflex, which empties the rectum. When distention is detected, receptors send nerve impulses to the spinal cord and then back through the parasympathetic nervous system to the descending colon, sigmoid colon, rectum, and anus. This impulse contracts the rectum and increases the pressure along the diaphragm and abdominal muscles, forcing the internal sphincter open and ultimately expelling the feces through the rectum and out of the anus. The external sphincter is striate muscle and is voluntarily controlled. If it is relaxed, the defecation process is completed. If it is voluntarily constricted, defecation can be temporarily postponed. As described previously, the voluntary cooperative abdominal and diaphragm effort increases the pressure inside the abdomen, which creates a backup into the sigmoid colon until the next mass peristalsis reflex occurs once again, stimulating the pressure-sensitive receptors that signal the central nervous system of the need to defecate.

## EXPOSURE TO TOXIC SUBSTANCES

Because the GI tract is the normal portal of entry for foodstuff, exposure to any kind of ingestible toxic agent is possible, and the degree of exposure can be high. A toxicant can enter the body via water, foodstuff, pharmacological agents, or even inhaled substances which are subsequently swallowed via the resulting secretions from the lungs, trachea, or oral cavity. Toxicants enter the body by the same absorption pathways as necessary nutrients; they pass through the GI tract mucosa membranes and subsequently make their way into either the blood or lymphatic system. These substances may exert toxic events locally within the GI tract itself or at a distant site, depending on the nature of the ingested material, the rate of absorption, the biochemistry of the toxicant, and the physiology of the target organ. The degree of toxicity depends on the concentration and duration of exposure of a toxicant at the sensitive site. Several factors can alter the extent of potential toxicity when a toxicant has gained entrance into the lumen of the GI tract, including poor absorption from the GI tract, intestinal and hepatic first-pass effects as mediated by biotransformation, and the rate of incorporation of lipid-soluble substances into micelles. A toxicant can interact with the specific moieties at the site of action, or it can alter the microenvironment in such a way that the resulting action is toxic to the functioning of the specific site. The critical factors affecting the degree of toxicity are summarized in Table 31.20 [50].

### BIOTRANSFORMATION

The process by which substances are changed into different chemicals by enzyme reactions in the organism is called *biotransformation*, and it can be an important defense mechanism. This is possible because some toxicants and even some body wastes are changed into less toxic substances or are simply prevented from forming toxic derivatives. A chemical can be toxic in its original form as it enters the body, or toxic derivatives of a nontoxic parent compound can be produced by biotransformation. Biotransformation occurs in GI tract mucosal cells, in the liver, and, to a lesser extent at various other places throughout the circulatory system.

Biotransformation reactions can be divided into two basic types of reactions: phase I and phase II. Generally, in phase I reactions, the structure of the toxicant is changed in such a way that it is readied for additional changes during the next step in the process. Toxicants are neutralized in three ways during phase I: (1) the chemical

**TABLE 31.21**
**Phase I Reaction Summary**

| Type of Reaction | Specific Function |
|---|---|
| Oxidation | Aromatic and aliphatic hydroxylation; epoxidation; *N*-hydroxylation; *O*-, *N*-, and *S*-dealkylation; *S*-oxidation; dechlorination; oxidative desulfuration; amine oxidation; dehydrogenation |
| Reduction | Azoreduction, nitroreduction, carbonyl reduction |
| Hydrolyses | Esterases, amidases, proteases |
| Mixed function oxidase system | In many of the above reactions, the substrate binds to cytochrome P450, transferring electrons and oxygen in the following general reaction: $SH + NADPH + H^+ + O_2 \rightarrow SOH + NADP^+ + H_2O$ |

*Note:* Phase I reactions tend to make the substrates more polar and more readily excreted from the organism by exposing the functional group and readies it to be changed further during phase II. Some of these reactions result in bioactivation of the substance.

*Source:* Parkinson, A., in *Casarett & Doull's Toxicology: The Basic Science of Poisons*, 6th ed., Claassen, C.D., Ed., McGraw-Hill, New York, 2001, pp. 133–244. With permission.

**TABLE 31.22**
**Phase II Reaction Summary**

| Type of Reaction | Specific Function |
|---|---|
| Glucuronidation | Diphosphate glucuronosyl transferase forms glucuronides with activated glucuronic acid, which can catalyze the conjugation of a substance with a polar group. |
| Sulfation | Sulfotransferase activates sulfate, which can react with a polar compound becoming more water soluble. |
| Glutathione conjugation | Glutathione *S*-transferase catalyzes the formation of strong electrophiles such as epoxides, haloalkanes, nitroalkanes, alkenes, and aromatic halo and nitrocompounds. |
| Epoxide hydrolase | The reaction proceeds through the activation of water which detoxifies the epoxides of hydration. |
| Acetylation | Acetyl co-A is attached to aromatic amines and sulfonamides |
| Methylation | Several methyltransferases are utilized to increase lipophilicity. |
| Amino acid conjugation | Proceeds by two pathways: (1) the COOH group conjugated with $NH_2$ requires CoA activation; (2) aromatic $NH_2$ or NHOH conjugated with COOH requires ATP activation. |

*Note:* Phase II reactions are generally the result of the formation of compounds less biologically reactive and are conjugated with endogenous substances in order to make them more water soluble. The final product can be eliminated from the body either in the urine or the bile.

*Source:* Parkinson, A., in *Casarett & Doull's Toxicology: The Basic Science of Poisons*, 6th ed., Claassen, C.D., Ed., McGraw-Hill, New York, 2001, pp. 133–244. With permission.

structure is changed so it is hydrophilic; (2) the toxicant is broken down into two or more less toxic substances; or (3) the toxicant is transformed into an activated form that can be detoxified by other enzymes. Most of the phase I reactions are mediated by the cytochrome P450 enzymes, which primarily oxidize lipophilic toxicants and are responsible for adding polar groups to the molecule. Interestingly, the creation of these activated toxins can result in a more toxic derivative, so it is critical for the activated toxin to move forward to phase II reactions [51]. Seven major biochemical reactions occur in phase II: glutathione conjugation, amino acid conjugation, methylation, acetylation, sulfation, glucuronidation, and sulfoxidation. Each of these reactions transforms specific types of activated toxins by adding a molecule to the activated toxicant. A brief overview of the critical factors is summarized in Table 31.21 and Table 31.22 [52].

The metabolic activity of the small intestine is referred to as *intestinal first-pass effect*. The small intestine is considered an absorptive organ, but it also has the ability to metabolize drugs via many pathways involving preconjugation (phase I) and conjugation (phase II) reactions [53–55]. Virtually all of the drug-metabolizing enzymes present in the liver are also found in the small intestine; however, the enzyme levels are considerably lower in the small intestine than in the liver [56]. Another factor that plays a critical role in the amount of nutrient or toxicant absorbed is referred to as the *hepatic first-pass effect*. In this case, orally ingested nutrients or toxicants are absorbed by the intestinal mucosal cells, enter the capillaries and veins of the GI tract, and are transported by the portal vein directly to the liver before entering the general circulation of the body. The absorbed substance is exposed to the liver before its first pass through the body. Because most toxi-

**TABLE 31.23**
**Summary of Common Detoxication Pathways**

| Detoxication Pathway | Description | Result |
|---|---|---|
| Nucleophiles | Hydroxylated nucleophilic groups are conjugated by sulfation, glucuronidation, or methylation; thiols are conjugated by methylation or glucuronidation; amines and hydrazines are conjugated by acetylation or oxidation. | Reduction in the rate of nucleophile conversion to free radicals and/or electrophilic quinines and quinoneomines |
| Electrophiles | Electrophilic toxicants are conjugated by conjugation with thiol nucleophile glutathione. | Reduction in the rate of electrophile conversion to reactive free radicals |
| Free radicals | Superoxide anion radical is converted by superoxide dismutase (SOD), glutathione peroxidase (GPO), and catalase (CAT) to HOOH and $H_2O$. | Reduction in the amount of superoxide anion radical available for conversion to free radicals such as peroxynitrite, nitrosoperoxy carbonate, and carbonate anion radical |
| No functional groups | Functional groups such as hydroxyl or carboxyl are first added to the substance by the cytochrome P450 enzymes, then a transferase adds a glucuronic acid, a sulfuric acid, or an amino acid to the functional group. | Toxicants rendered much less active or more easily eliminated |
| Proteins | Intra and extracellular proteases interact with polypeptides. | Inactivation of toxic polypeptides |

*Source:* Data from Gregus and Klaassen [57] and Sreedharan and Mehta [58].

cants are lipophilic and nonpolar and have low molecular weights, they are easily absorbable through the GI tract mucosal cells and yet are difficult to eliminate and can accumulate throughout the body to toxic levels. Most lipophilic toxicants are difficult for the body to eliminate, and they accumulate to toxic levels. Many lipophilic toxicants are biotransformed into hydrophilic metabolites that do not easily enter the membranes of the target tissues. Endogenous materials such as bilirubin are also biotransformed into hydrophilic derivatives that are excreted into the bile and ultimately eliminated in the feces. This process is termed *detoxication*, and the general pathways are summarized in Table 31.23 [57,58]. When a toxicant is biotransformed to another compound whose derivative can be more toxic than the parent compound, this process is known as *bioactivation*, and it can be very toxic to the organism. This process, also referred to as *toxication*, has several pathways depending on the chemical (see Table 31.24) [57,58].

## Factors Affecting Absorption

The rate at which a toxicant is absorbed across the mucosal cell wall in the gastrointestinal tract is a function of a number of physical and chemical factors. Some toxicants may directly increase the permeability of the intestinal mucosal cell wall, such as ethylenediaminetetraacetic acid (EDTA). Also, the longer the exposure time to the mucosal membranes, the higher the rate of absorption, so another factor affecting the rate of toxicant absorption is the rate of peristalsis. Agents such as laxatives or sedatives work indirectly on the rate of absorption by speeding or slowing the rate of motility through the GI tract. If the substance is subject to a high hepatic clearance (i.e., it is rapidly metabolized by the liver), then a substantial fraction of the absorbed substance is extracted from the blood and metabolized and excreted as bile before it reaches the systemic circulation. The bile is then reintroduced into the GI tract and eventually eliminated through the feces. The consequence of this phenomenon is a significant reduction in bioavailability, which can be an excellent protective mechanism if the material is a toxicant and can make therapeutic treatment difficult if, in fact, the substance orally ingested is part of a drug therapy regimen [4].

With a few notable exceptions, most substances (toxicants and nutrients alike) are poorly absorbed from either the oral cavity or the esophagus. This is due primarily to the fact that the transit time in these two sections is relatively brief. In addition, these structures have no villi or similar such absorbing cells. Exceptions include nicotine from tobacco products, alcohol, and nitroglycerin, which is immediately absorbed when placed sublingually for treatment in certain heart conditions.

As the toxicant passes into the stomach and on through the small intestine, it can be absorbed by one and more of several mechanisms. Although no known mechanisms specifically transport only toxicants into the body, many of existing pathways for nutrient absorption in the GI tract are also used by foreign substances to gain entry into the body. Generally, if a toxicant is an organic acid or base, it most likely will be absorbed via simple diffusion. By virtue of the structure of these substances, one can assume that an organic acid exists primarily in lipid-soluble form

**TABLE 31.24**
**Summary of Common Toxication Pathways**

| Toxication Pathway | Description | Result |
|---|---|---|
| Nucleophile formation | Nucleophile activation is rare; some nucleophiles such as HCN and CO are reactive as nucleophiles, but many can be further activated to electrophiles. | These reactions increase the rate of nucleophile conversion to free radicals or electrophilic quinines and quinoneomines. |
| Electrophile formation | Electrophiles are often produced by adding an oxygen atom to a molecule which removes electrons and results in an electronic-deficient structure that is electrophilic. | These electrophiles are electron deficient and are reactive, creating polarized substances that are frequently catalyzed by cytochrome P450. |
| Free radical formation | Free radicals contain one or more unpaired electrons and are formed by the addition of an electron, loss of an electron, or hemolytic fission of a covalent bond. | Free radicals can increase the amount of superoxide anion radical available for conversion to free radicals such as peroxynitrite, nitrosoperoxy carbonate, and carbonate anion radical. |
| Redox-active reactants | Conversion of superoxide anion radicals to hydroxyl radicals by superoxide dismutase (SOD) and catalyzed by transition metal ions. | Reactive metabolites such as some electrophiles and neutral or cationic free radicals can be activated by conversion to electrophiles, while free radicals with extra electrons can produce hydroxyl radicals upon hemolytic cleavage. |

*Source:* Data from Gregus and Klaassen [57] and Sreedharan and Mehta [58].

in the stomach and in the ionized form in the small intestines. Based on that assumption, an organic acid would be absorbed more readily in the acid stomach than in the neutral small intestine. Organic bases, on the other hand, would be absorbed more readily in the neutral small intestine rather than the acidic stomach; however, because the pH is about neutral in the small intestine (pH 5.0 to 8.0), both weak acids and bases are nonionized and readily absorbed by passive diffusion. Significant exceptions to this situation do arise, primarily due to the large surface area of the small intestine, varying blood flow rate, and law of mass action maintaining a gradient. It should be noted, however, that small lipid-soluble substances generally enter the mucosal epithelial cells by passive diffusion.

In addition, a variety of active transport and facilitated mechanisms are responsible for transporting nutrients, including monosaccharides, amino acids, and minerals, into the mucosal epithelial cells. A xenobiotic transport mechanism that has been identified features multidrug resistance proteins (MDRs) or P-glycoprotein. Unfortunately, this transport system moves some chemotherapeutic agents out of some tumor cells, rendering these agents less effective, but it also transports some other potentially toxic chemicals out of cells, thus protecting the organism from toxic doses. This mechanism, then, functions to move certain xenobiotics out of the cell and reduce the net amount of GI absorption of some toxic chemicals. MDR-associated proteins (MRPs) have been shown to transport glucuronides and glutathione metabolites out of cells, as well [4]. The iron transport mechanism also absorbs toxins such as thallium, and manganese and cobalt both utilize and compete for access to the iron transport system. The calcium transport mechanism absorbs lead, and the pyrimidine transport system absorbs 5-fluorouracil. These toxicants are actively transported into the mucosal epithelial cells of the GI tract, but the vast majority of toxicants enter by simple passive diffusion. Low-molecular-weight, lipid-insoluble compounds are absorbed through aqueous membrane pores at the tight junctions in the membrane via passive diffusion. Another factor in the absorption of toxicants is that particulate toxicants can be absorbed via vesicular transport, either by pinocytosis or phagocytosis.

In addition, another phenomenon has been identified as a rate-limiting barrier to the absorption of toxicants in the small intestine: the unstirred water layer. Normally, the intestinal mucosal cell membranes are separated from the aqueous phase in the lumen by an unstirred water layer. This layer surrounds the villi and becomes a significant rate-limiting step with regard to the entry of all substances (nutrients and toxicants alike) into the mucosal cells. This water layer is not well mixed, and the solute molecules in the chyme must diffuse across the unstirred water layer to gain access to the membrane of the mucosal cells [59]. Micellar solubilization, on the other hand, greatly enhances the number of molecules that are available for uptake by the mucosal cells in the lumen. Tso et al. [42] suggest that micellar molecules are in equilibrium with molecules in the unstirred water, so the momeric fatty acids, monoglycerides, and amino acids cross the microvillar membrane. The large, polar bile salts are not absorbed well by intestinal mucosa, so the bile salts remain in the chyme to participate in forming more mixed micelles as they cycle within the intestine. The integration of segmentation and propulsion enhances absorption by mixing luminal contents with digestive juices and by increasing the contact time between chyme and absorptive cell surfaces. Intestinal motility also decreases the thickness of the unstirred layer of water so nutrients and electrolytes can diffuse readily to mucosal cell membranes.

## Factors Affecting Distribution

When a toxicant gains entry into either the lymphatic or blood circulatory system, it can move rapidly throughout the body. Its effect at the target tissue is based on the affinity of the toxicant to the target. The routes of entry into various target tissues include either active transport or simple passive diffusion. As previously noted, small, low-molecular-weight, water-soluble toxicants are able to pass through pores in the membrane via filtration and diffusion, while polar toxicants with larger molecular weights must be actively transported across the membrane. Lipid-soluble toxicants are able to easily penetrate the membrane. The rate of uptake is ultimately determined by either the diffusion rate, which is governed by the specific characteristics of the toxicant, or the perfusion rate, which is governed by the physical rate of delivery to the target tissue [4].

Generally, the concentration of a toxicant is dependent on the volume of distribution through the body compartments. Some toxicants are limited in their distribution by size, charge, or shape; others can pass through the various compartments and become distributed throughout the organism. The plasma concentration of a toxicant is critical, as it reflects the available concentration of the substance at the target tissue. Generally, if the concentration in plasma is high and the interstitial and intracellular concentrations are low, the amount of unbound toxicant available at the target tissue increases. Once a toxicant has entered the blood plasma, it may be excreted or biotransformed into more or less toxic substances, or it may be stored. Because of protein binding, active transport, or some other physiological factor, some toxicants accumulate in certain areas of the body. This can be a protective measure if the site of accumulation is other than a potential target tissue. In this particular case, the compartment in which a certain toxicant is accumulated is termed a *storage depot*. Because the toxicant is in equilibrium with its counterparts in the plasma and elsewhere, it can be released from the storage depot as the plasma level decreases over time. Interestingly, the initial distribution of various toxicants can change based on flow rates and toxicant affinities. The details and chemical characteristics of the common storage depots available to many organisms are described elsewhere in this book. For our discussion here, the reader should be aware of the following storage depots and how they influence the distribution of toxicants in the organism.

Albumin is the major plasma protein that functions as a storage depot. Although there are other plasma proteins, albumin can bind a large number of toxicants; at least six binding regions have been identified on the protein [60]. The binding of various toxicants as well as endogenous substances to plasma proteins such as albumin inhibits that substance from crossing the capillary beds and entering into extracellular space for distribution. The binding mechanism is reversible, so equilibrium can be maintained as various compartmental concentration levels change over time.

The liver and kidney both function as storage depots and can store a considerable number and amount of toxicants; both organs concentrate and remove toxicants as a primary function. Proteins found in these organs, such as metallothione and ligandin, have been shown to bind organic acids, cadmium, and zinc, which are concentrated in these organs. It is estimated that more toxicants are stored in these two organs than in all of the other organs combined [4].

Body fat functions as a storage depot. Because of the lipophilic nature of many toxicants, many are stored in neutral body fat in large amounts. Bone also functions as a storage depot. The hydroxyapatite crystals of bone provide a very large surface area for toxicants such as lead, strontium, and fluoride to easily penetrate. The toxicants in bone are reversibly bound, so these materials can be released back into the plasma by ionic exchange and dissolution of bone lattice by osteoclastic activity.

## Factors Affecting Elimination

Fecal elimination is another pathway for the GI tract to effectively remove toxicants. Three major components contribute to the process of toxicant excretion through the feces.

### Biliary Excretion

Because the liver is first in line for the newly absorbed nutrients or toxicants, it can and does extract and biotransform toxicants of all sorts so the metabolites can be excreted into the bile. The bile is then reintroduced into the GI tract and excreted in the feces unless it is reabsorbed in the enterohepatic circulation system. An increase in the amount of biliary excretion can lessen the ultimate toxicity of a toxicants unless some biological effect such as hydrolysis occurs that increases the lipophilic nature and absorbability of the toxicant. Bile components are divided into three classes of compounds based on their bile/plasma concentration ratios. Class A compounds have a bile/plasma ratio of ~1 and include substances such as sodium, potassium, glucose, mercury, and thallium. Class B compounds have a bile/plasma ratio of >1 but usually between 10 and 1000. Some Class B compounds would include bile acids, lead, arsenic, and many more common toxicants. Class C compounds have a bile/plasma ratio of <1 and include materials such as inulin, albumin, zinc, and chromium. The class B compounds are usually rapidly excreted by active transport directly into the bile. In gen-

**TABLE 31.25**
**Transport Systems in the Liver Hepatocyte**

| Transport System | Abbreviation | Function | Location | Refs. |
|---|---|---|---|---|
| Sodium-dependent taurocholate peptide | NTCP | Transports bile acid taurocholate into the liver | Sinusoidal side | [61,62] |
| Organic-anion polypeptides 1 and 2 | OATP1, OATP2 | Transport organic polypeptides into the liver | Sinusoidal side | [62–65] |
| Liver specific transporter | LST1 | Transports toxicants into the liver | Sinusoidal side | [66] |
| Organic cation transporter | OCT | Transports toxicants into the liver | Sinusoidal side | [67] |
| Bile salt excretory protein | BSEP | Transports bile acids out of the liver | Bile canaliculi side | [68] |
| Multidrug-resistant protein 1 | MDR1 | Transports toxicants into the bile | Bile canaliculi side | [69,70] |
| Multiresistant drug protein 2 | MRP2 | Transports toxicants into the bile | Bile canaliculi side | [71,72] |
| Multiresistant drug proteins 3 and 6 | MRP3, MRP6 | Transport toxicants back into blood circulation | Blood capillaries | [73,74] |

eral, conjugates with molecular weights greater than 325 and conjugates of both glutathione and glucuronide appear to have a higher concentration in the bile. Numerous transport systems in the hepatocytes have been identified whose function it is to move these various substances to and from the liver and back into the GI tract [4,10]. Table 31.25 provides a summary of the better-known transport systems in the liver hepatocytes [61–74].

## Intestinal Excretion

Studies have been performed in bile-duct-ligated animals in which the presence of a number of substances in the feces could only be attributed to direct transfer from the circulating blood to lumen of the small intestine [75,76]. Hence, direct excretion into the small intestine from the mucosal epithelial cells constitutes a portion of the source of toxicants excreted in the feces. In addition to the secretion of various chemical directly into the GI tract, the microflora have been shown to biotransform some toxicants to forms that become eliminated through the feces rather than reabsorbed. Because toxicants in the lumen of the intestine are also ingested by the resident microflora, it is estimated that some portion of the toxicant has been biotransformed by these intestinal microflora.

## Nonabsorbed Excretion

Some ingested nutrients and toxicants simply do not get absorbed during their transit through the GI tract; for example, because of their chemical characteristics, ingested sucrose polyester, cholestyramine, and ionized compounds such as quaternary ammonium are not well absorbed and are found in fecal excretions.

# TESTING THE GASTROINTESTINAL TRACT FOR TOXICANTS

A number of tests are available to detect any effects caused by the entry of toxic agents. Among them are tests that examine the structural integrity of the luminal cell well. Table 31.26 provides a summary of the types of testing utilized to examine the structural integrity and continuity of the gastrointestinal tract [77–92]. Toxicants can also effect changes in the rate of cell proliferation in the intestinal mucosal walls. Antineoplastic agents tend to slow the rate of cell division as part of their therapeutic mechanism, but some effects are indirect in that the toxicant can interfere with normal cell regulatory mechanisms, such as hormones, or cause tissue injury [93,94]. Several tests have been developed to examine the proliferation of cells of the mucosa. Table 31.27 is a summary of some of the available tests for mucosal cell proliferation [95–98].

The gastric secretion of hydrogen chloride from the parietal cells is a critical factor for the initiation and optimal functioning of pepsinogen and prochymosin within the stomach; therefore, toxicants that alter the gastric secretory activity can have a profound effect on the functioning of the stomach. The rate of acid secretion in the parietal cells is under both neurological and hormonal control, so the measurement of acid secretion provides an index of measurement for the status of digestion. Table 31.28 summarizes some of the procedures available for measuring gastric secretion [99–107].

Gastric emptying is a critical process in the GI tract because it controls the rate at which properly triturated chyme is released into the small intestine for optimal absorption [108,109]. The pylorus (from the Greek word for "keeper of the gate") determines what enters the small intestine from the stomach. Under normal circumstances, it allows only chyme of the proper texture and size to enter the small intestine. The control of gastric emptying is complex and can be altered by stress [110], meal composition and size [111], pathological processes, and pharmacological interactions [109]. Methods for the assessment of gastric emptying can be divided into three basic types: (1) tracer studies, (2) imaging studies, and (3) electrical resistance studies. Representative studies from each of these categories are summarized in Table 31.29 [109–123].

Understanding the factors affecting absorption of the GI tract is important in order to characterize the absorption, metabolism, and excretion of toxicants and

**TABLE 31.26**
**Testing for Structural Integrity**

| Test | Description | Examined Entity | Notes | Refs. |
|---|---|---|---|---|
| Direct observation of mucosal cell wall | Invasive visual inspection of gastrointestinal tract for lesions; quantitation of damage | Gastrointestinal walls | Duodenal and gastric ulcer indices have been created to quantitate findings. | [77–79] |
| | Invasive examination of the structure and organization of the gastrointestinal tract wall | Gastrointestinal walls | Slides are prepared to evaluate cellular organization and structure; endoscopic examination and biopsy methods. | [80–82] |
| | Invasive visual inspection of gastrointestinal tract pretreated with dyes for lesions; quantification of damage | Gastrointestinal walls | Sky Blue Dye 6, Evans blue dye, or Monastral Fast Blue B is injected intravenously. | [78,83,84] |
| Mucosal permeability | Invasive exam of blood to lumen clearance of $^{51}$Cr-EDTA | Assessment of leakage into lumen from blood stream | Blood and perfusate are collected to measure EDTA clearance. | [85,86] |
| Fecal blood loss | Noninvasive examination of feces for occult blood | Assessment of blood leakage into the gastrointestinal tract by colorimetric methods | Guaiac (hemoccult) is the least sensitive, *o*-toluidine the most sensitive; HemoQuant measures fluorescence of porphyrins; recent studies indicate these tests are insensitive. | [87–90] |
| | Intravenous injection of $^{59}$Fe sulfate for *in vivo* labeling of red blood cells | Assessment of blood leakage into the gastrointestinal tract by radioisotopic methods | Procedure is more sensitive than colorimetric methods; 24-hour fecal collection and assay for radioactivity. | [91] |
| Cell shedding | Monitoring rate of cell loss in the gastrointestinal tract | Measure DNA content of luminal fluid from surgically implanted gastrointestinal tubes | Fluid is quantified for displacement of $^{125}$Iododeoxyuridine DNA from DNA antibodies. | [92] |

**TABLE 31.27**
**Testing for Mucosal Cell Proliferation**

| Test | Description | Examined Entity | Notes | Refs. |
|---|---|---|---|---|
| Cell kinetic analysis | Invasive intravenous injection of $^{3}$H-thymidine to label crypt cell DNA | Assessment of the duration of the cell cycle and cell migration progress | Pulse exposure to $^{3}$H-thymidine with tissue harvest and fixed slides is a 2- to 4-week process. | [95] |
| Antibodies to cell cycle antigens | Use of proliferation cell nuclear antigen (PCNA) to label S-phase cells in crypts requiring incubation of the tissues | Assessment of proliferation of gastrointestinal cells | Method can be completed more quickly than autoradiography. | [96] |
| Flow cytometry | Incubation of intestinal crypt and villa cells and separation by DNA content | Assessment of proliferation of gastrointestinal cells | Method can be completed more quickly than autoradiography. | [95,97,98] |

what effect toxicants may have on the absorption and metabolism of necessary nutrients or pharmacological substances being used for therapeutic purposes. *In vitro* models have been developed that mimic the human gastrointestinal tract in an attempt to predict reactions of food and bioactive components in physiological conditions. These simulated digestion studies have been used for a number of years to investigate the digestion, behavior, and ultimate bioaccessibility (amount of a contaminant that reaches the systemic circulation and exerts a toxic effect) of animal proteins [124], plant proteins [125], and food additives [126]; to study the effects of the foodstuff matrix on the bioavailability and bioaccessibility of ingested contaminants [20]; to examine carotenoids in biological emulsions [127,128]; and to determine the effects of ingesting contaminants from soil [129]. *In vitro* digestion models reflecting the conditions of the gastrointestinal tract in the presence or absence of food have been used to study the bioaccessibility of compounds from soil. It is known that fasting can have a significant impact on the oral bioavailability of compounds, as the presence of food alters conditions in the gastrointestinal tract. To mimic intestinal absorption, the Caco-2 transport model was used [130].

**TABLE 31.28**
**Testing for Gastric Secretory Activity**

| Test | Description | Examined Entity | Notes | Refs. |
|---|---|---|---|---|
| *In vivo* direct gastric acid collection | In this invasive procedure, the pylorus is tied off in an anesthetized animal; the gastric content is sampled periodically via syringe. | The sample is titrated to pH 7.0 for hydrogen ions associated with specific endogenous compounds (i.e., mucoproteins) | Titration to pH 3.5 provides an estimate of total HCl present. | [99,100] |
| *In vivo* Azure A measurement | In this noninvasive procedure, Azure A–resin complex (azuresin, Diagnex Blue) is administered by gavage and the animal is housed to collect a 24-hour urine. | Allows the quantification of the breakdown of Azure A from azuresin, which is pH dependent. | The 24-hour urine is measured for the concentration of Azure A by spectrophotometry. | [101] |
| *In vivo* fundic pouch | In this invasive model, a dog is surgically fitted with a Heidenhain pouch to permit controlled exposures. | Permits repeated measurement of volume changes, electrolyte concentrations, and mucosal blood flow. | The denervated pouch decreases the basal acid secretory rate, which must be factored in to the findings. | [102,103] |
| *In vitro* tissue in a Ussing chamber | Segment of the stomach is surgically removed and used to separate two solutions. | Acid secretion and electrolyte flux can be quantified. | The muscle layer is removed prior to use in the Ussing chamber. | [104,105] |
| *In vitro* culturing of gastric gland cell types | Gastric tissue is prepared with pronase and collagenase and the cell types can be isolated and examined with no anatomical barriers between the mucosal and serosal surfaces. | With appropriate corrected measures, oxygen consumption can be measured with a polarographic electrode or a Gilson respirometer as an index of secretory activity. | A variation of this method is monitoring the uptake of aminopyrine; this substance moves into the intracellular space of the parietal cell and becomes trapped in the secretory vesicles. | [104,106,107] |

The general procedure for studies such as these is that the digestion process of the gastrointestinal tract is simplified by applying physiologically based conditions such as synthetic saliva, gastric juice, duodenal juice, and bile to samples with controlled pH and residence time periods that mimic each compartment of the GI tract. Two-stage digestion systems represent the stomach and the small intestine, and three-stage digestion systems represent either (1) the mouth, stomach, and small intestine, or (2) the stomach, small intestine, and large intestine [20]. Because practically no absorption occurs in the mouth or the stomach, bioaccessibility is usually determined from the chyme resulting from the small intestine compartment. Figure 31.11 is a flow diagram of a typical *in vitro* digestion model.

The Bioavailability Research Group Europe (BARGE; http://www.bgs.ac.uk/barge/home.html) was established in 1999 to promote cooperation among European countries interested in developing and comparing data and protocols for *in vitro* digestion systems designed to analyze data on oral bioavailability of soil contaminants. A comparison study of the findings of various European laboratories was recently published [131], and data collected by BARGE are summarized in Table 31.30 [132–136]. Table 31.31 summarizes tests that can be used to help characterize and quantify absorption within the GI tract [138–161].

Normal peristalsis is necessary for digestion, the delivery of chyme to various areas for absorption, and waste elimination. Factors that control peristalsis are numerous and complex. The motility characteristics change in pattern and periodicity from one segment of the GI tract to the next. The activity is mediated by a complex system of neurons in the wall of the gastrointestinal tract that coordinate contractions between the longitudinal and circular musculature. The contractions are both neurologically and hormonally controlled. The neurological control is dependent on the following enteric (intrinsic) neurotransmitters: vasoactive intestinal peptide, enkephalin, substance P, gastrin-releasing peptide, neuropeptide Y, and somatotropin. Extrinsic neurotransmitters from the vagus and pelvic nerves include choline, acetylcholine, and norepinephrine. The hormonal control is mediated by gastrin, secretin, and cholescystokinin, all of which are secreted by the walls of the GI tract in response to certain dietary components. Table 31.32 summarizes the more common methods used to analyze motility in the gastrointestinal tract [162–168].

As noted previously, the entire length of the GI tract contains microflora, many of which are located in the large intestine [169,170]. Pathogenic strains of some bacteria such as *Shigella* and *Helicobacter pylori* can

## TABLE 31.29
## Testing for Gastric Emptying Activity

| Test | Description | Examined Entity | Notes | Refs. |
|---|---|---|---|---|
| Tracer studies | *Breath tracers* are a noninvasive procedure involving ingestion of a $^{13}C$-labeled substrate that produces a detectable increases in $^{13}CO_2$. | After gastric emptying, the $^{13}C$-labeled substrate can be measured in the exhaled breath as $^{13}CO_2$. | Exhaled breath samples are examined by isotope-ratio mass spectrometry for the $^{13}C$:$^{12}C$ ratio over time. | [112,113] |
| | *Gastric tracers* are an invasive procedure involving gastric gavage of a known amount of nonabsorbable marker. | Serial aspiration samples are taken through a nasogastric tube or a catheter in the gastric wall. | The serial gastric samples are assumed to be homogeneous and are examined over time. | [109,114] |
| | *Plasma tracers* are an invasive procedure involving ingestion of paracetamol, which is slowly absorbed into the blood in the stomach and rapidly absorbed in the small intestine. | Serial blood samples are taken via an intravenous catheter. | The paracetamol concentration is measured in the blood L\plasma levels over time and is an indirect measure of gastric emptying. | [109,115] |
| Imaging studies | *Radiography* is a noninvasive procedure involving ingestion of radiopaque solids and liquids. | After ingestion, serial lateral and ventral radiographic images are taken over 4 to 12 hours | The movement of the radiopaque material is monitored qualitatively via the radiographic images. | [109,116] |
| | *Radioscintigraphy* is a noninvasive procedure involving ingestion of food containing radioisotopes. | Radiographic images are integrated with an integrated nuclear medicine computer system and the radioactive counts are recorded over time. | With the application of various correction factors to allow for radioactive decay, the movement of gastric contents can be recorded over time. | [117] |
| | *Ultrasonography* is a noninvasive procedure used to measure flow through areas of the antral portion of the stomach. | Ultrasonographic images are taken as materials move through the stomach in real time. | Movement is measured in real time; however, further validation of this method is necessary at this time. | [118] |
| | *Magnetic resonance imaging* is a noninvasive procedure used to examine liquid- and solid-phase gastric emptying and motility. | Magnetic resonance images provide a three-dimensional image of the stomach emptying. | Emptying and motility are measured simultaneously. | [119,120] |
| Electrical resistance studies | *Impedance epigastrography* is a noninvasive procedure that uses two pairs of electrodes to monitor electric current after the ingestion of food. | An electric current is applied to the epigastric region with a pair of electrodes while another pair of electrodes monitors the changes in the current. | Ingestion of food causes changes in the pattern of electrical resistance, which is an indirect measure of gastric emptying. | [109,121,122] |
| | *Applied potential tomography* is a noninvasive procedure that uses multiple pairs of electrodes to monitor electric current after the ingestion of food. | An electric current is applied to the epigastric region with multiple pairs of electrodes that monitor the changes in electric current over the entire upper abdomen. | Ingestion of food causes changes in the pattern of electrical resistance, which is an indirect measure of gastric emptying. | [121,123] |

mediate some toxic effects because they invade the gastrointestinal mucosa and produce physical changes in the cell wall that can lead to gastritis and peptic ulcers. Intestinal bacteria such as *Vibrio cholere*, *Clostridium difficile*, and *Escherichia coli* are also capable of generating toxins that change water and electrolyte flux rates in the cell wall and ultimately produce diarrhea. Other bacteria are able to activate certain foreign chemicals, such as cycasin, which is found in the nuts of cycad plants, to a form that is carcinogenic and can cause tumors in the liver, kidneys, and colon [171,172]. Table 31.33 provides a summary of some methods used to test for the potential toxic effects of bacteria in the gastrointestinal tract [Table 31.33].

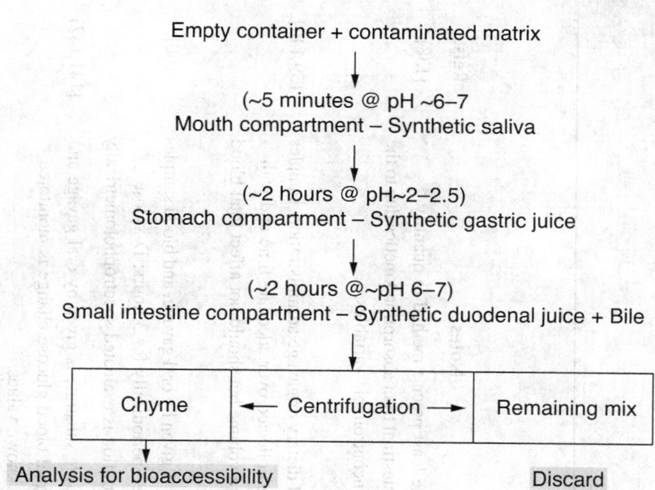


**FIGURE 31.11** Depiction of a typical *in vitro* digestion study. (Adapted from Versantvoort, C.H.M. et al., *Development and Applicability of an In Vitro Digestion Model in Assessing the Bioaccessibility of Contaminants in Food*, RIVM Rep. No. 320102002/2004, National Institute for Public Health and the Environment, Bilthoven, The Netherlands, 2002; http://www.rivm.nl/bibliotheek/rapporten/320102002.html.)

**TABLE 31.30**
**Various *In Vitro* Digestion Methods Assessed by BARGE**

| Test | Description | Stomach pH | Stomach Incubation Interval | Intestine pH | Intestine Incubation Time | Bile Concentration | Separation Method for Chyme | Refs. |
|---|---|---|---|---|---|---|---|---|
| SBET | Static gastric model | 1.5 | 1 hr | NA | NA | NA | 0.45 μm filter | [132] |
| RIVM | Static gastro-intestinal model | 1.1 | 2 hr | 5.5 | 2 hr | 0.9 g/L | Centrifugation @ 3000 *g* | [129,133] |
| SHIME | Static gastro-intestinal model | 4.0 | 3 hr | 6.5 | 5 hr | 1.5 g/L | Centrifugation @ 7000 *g* | [134] |
| DIN | Static gastro-intestinal model | 2.0 | 2 hr | 7.5 | 6 hr | 4.5 g/L | Centrifugation @ 7000 *g* | [133,135] |
| TIM | Dynamic gastro-intestinal model | 5.0→2.0 over 90 minutes | Gradual secretion @ 0.5 mL/min | 6.5–7.2 | Gradual secretion at 1 mL/min | Variable | Hollow-fiber membrane | [136] |

*Note:* SBET, Simplified Bioaccessibility Extraction Test (used in lead analysis in the United Kingdom); RIVM, Rijks Instituut voor Volksgezondeid and Milieu (National Institute for Public Health and the Environment) method (The Netherlands); SHIME, Simulator of Human Intestinal Microbial Ecosystems Method (Belgium); DIN, Deutsches Institut fur Normung method (Germany); TIM–TNO Nutrition and Food Research computer-controlled dynamic gastrointestinal model (The Netherlands).

*Source:* Data from Oomen [131] and Sips [137].

## CONCLUSION

The effect of a toxicant on an organism depends primarily on the relative effective concentration and the duration or persistence of the ultimate toxicant at a specific site of action. Numerous portals of entry exist for toxicants. This chapter has focused on the normal structure and function of the gastrointestinal tract and how the processes of digestion and absorption influence and are influenced by toxicants that enter the gastrointestinal tract. The intensity of effect of the toxicant is determined by a number of factors that are part of the normal functioning of the gastrointestinal tract, such as digestion, absorption, biotransformation, distribution, and elimination. Finally, the last part of this chapter described the types of testing that can be performed to examine the effects of toxicants on various aspects of the normal functioning of the GI tract.

**TABLE 31.31**
**Testing of Gastrointestinal Tract Absorptive Function**

| Test | Description | Examined Entity | Notes | Refs. |
|---|---|---|---|---|
| *In vivo* determination of gastrointestinal absorption | Follows a predetermined amount of a substance administered orally over time by determination of its concentration in systemic fluids | Test measures the bioavailability of a xenobiotic and the absorption kinetics on systemic exposure via a metabolic cage. | This is the least precise method to quantify gastrointestinal tract absorption kinetics due to the many uncontrollable variables. | [138] |
| | Quantification of absorption | Depending on the substance, measurements can be made in plasma, urine, feces, saliva, and breath. | After oral dietary or gavage administration, samples must be collected over adequate time and with adequate volume but should not affect total blood volume. | [139,140] |
| | Malabsorption test (D-xylose test) | Test measures the functional integrity of the proximal small intestine. | D-xylose is given by oral gavage, and blood samples are taken periodically for 3 hours; D-xylose concentration is evaluated spectrophotometrically. | [100] |
| *In vivo* determination of overall gastrointestinal absorption | Malabsorption test (disaccharides) | Test measures the functional integrity of the disaccharidases ability to cleave disaccharides into monosaccharides. | Disaccharide substrate is given by oral gavage and the normal blood glucose change is monitored against known values. | [141,142] |
| | Malabsorption test (fats) | Focus of fat test is to distinguish between defects in the digestion of fats to fatty acids from fatty acid resorption. | An examination of feces for increased fat content (steatorrhea) is followed by oral dosing with radiolabeled triglyceride assay; the reduction of triglycerides indicates digestive defect, and a reduction in fatty acid absorption suggests bile salt defect or other mucosal defects. | [143,144] |
| *In vivo* closed intestinal segment | Invasive procedure where an intact intestinal segment is evaluated for absorption kinetics | An intact segment of the intestine is closed off and injected with the test substance; after a period of time, the intestinal segment is removed and the rate of substance reduction is quantified. | If metabolism of the test substance occurs in the lumen, for the test results to be valid the fate of all metabolites must be known. | [145] |

| | | | | |
|---|---|---|---|---|
| *In vivo* intestinal perfusion | Invasive procedure for quantifying intestinal transport by measuring the amount of substance in the effluent against the amount infused into gastrointestinal tract over time | The net flux of water transport into the gastro-intestinal tract can be evaluated by monitoring changes in a nonabsorbable marker. | The proximal and distal ends of the intestine to be examined are cannulated, and the test substance is administered through the proximal cannula; after a period of equilibration, samples are drawn from the distal cannula and the volumes of the test substance are compared. | [146,147] |
| *In vitro* inverted sac technique | Characterizes the carrier-mediated processes by quantification of the test substance inside an inverted length of intestine | A small segment of intestine is everted and filled with a fluid and tied at both ends; absorption is quantified by monitoring the appearance of the test substance in the fluid on the inside of the everted sac. | Because of inadequate oxygen diffusion and tissue distention, the incubations must be no longer than 5 minutes with rat intestine. | [148,149] |
| *In vitro* isolated gastrointestinal tract mucosal cells | Isolation of various gastrointestinal tract mucosal cells | Gastrointestinal tract mucosal cells are isolated in cultures which permits analysis of membrane transport kinetics without contamination from extraneous substance metabolism. | Isolation of these cells has facilitated more detailed understanding of the biochemical and structural characteristics of cellular membranes. | [150–152] |
| *In vitro* electrical potential with a Ussing chamber | Short-term exposure of a segment of gastrointestinal tract suspended in a Ussing chamber to defined solutions with a voltmeter to measure electrical current | In addition to voltage, the flux of electrolytes can be measured by a pulse of isotope to one side of the chamber with subsequent quantification on the other side of the membrane. | In a variation of this technique, a unidirectional flux chamber is used to assess mucosal cell transport mechanisms. | [153,154] |
| *In vitro* organ cultures | Examination of mucosal explants for studies of cell proliferation and differentiation over longer periods of time | Test permits the study of colonic carcinogenesis and chemotherapy with measurement of toxicants and their effects. | Some human colonic cultures have been extended to up to 28 days with normal anatomical cell arrangement. | [155,156] |
| *In vitro* $CaCO_2$, $T^{84}$, and $HT^{29}$ cultures | Utilizes cultures of colonic carcinoma $T^{84}$ and subclone $HT^{29}$ which exhibit normal colon cell and goblet cell characteristics, respectively, in cell cultures | These cultures can be examined for electrolyte transport of nutrients and toxicants. | These cell lines exhibit normal cell characteristics, including tight junctions, apical microvilli, and vectorial electrolyte transport. | [157–161] |

**TABLE 31.32**
**Testing of Gastrointestinal Tract Motility**

| Test | Description | Examined Entity | Notes | Refs. |
|---|---|---|---|---|
| *In vivo* measurement of transit time | Gastric or duodenal administration of a suitable marker such as a $^{51}$Cr-labeled sodium chromate or $^{125}$I- or $^{131}$I-labeled polyvinyl pyrollodine or $^{99m}$technetium and subsequent recovery of sample | Using gamma-emitting radioisotopes allows for direct gastrointestinal segment assay without any processing. Small intestine movements must be administered to the duodenum via an indwelling catheter. | A rapid euthanization procedure is used to quickly extract the gastrointestinal tract; to stop movement within the tract, ligatures are placed every 3 to 5 cm. Counts are recorded within each segment | [162–165] |
| *In vitro* rabbit jejunum | Maintenance of rhythmic contractions *in vitro* | This technique is used to analyze substances that manifest inhibitory effects on gastrointestinal smooth muscle. | — | [166,168] |
| *In vitro* guinea pig ileum | Excision of a guinea pig ileum with the mesenteric plexus | This technique is used to examine substances that influence neuronal conduction and neurotransmitter release and to evaluate gastrointestinal smooth muscle contractions. | This method has been used extensively to assess changes in contractility in smooth muscle. | [167,168] |

**TABLE 31.33**
**Testing of Gastrointestinal Tract Microflora**

| Test | Description | Examined Entity | Notes | Refs. |
|---|---|---|---|---|
| Toxin study: rabbit ileum | Bacterial toxins are injected into closed segments of surgically ligated 3-inch sections of rabbit ileum; data have revealed surface receptors for enterotoxins. | The toxins cause fluid accumulation when injected into the intestine; quantification of the fluid can provide information about intestinal antiabsorptive or secretory activity. | Rabbit ileum provides more consistent results; the ligated segments are placed at 6-inch intervals, and samples are taken at 4- to 6-hour intervals for examination. | [173–175] |
| *In vivo* bacterial metabolism study in germ-free animals | Aseptically delivered animals maintained in germ-free conditions are treated against normal rats. | Various toxicants are administered to both types of animals, which are housed in metabolism cages where their urine and feces can be quantitated. | Differences in metabolites in urine and feces can indicate changes mediated by the intestinal microflora. | [176] |
| *In vitro* bacterial metabolism study | Defined bacterial cultures of luminal contents are cultured with various substances. | Demonstration of enzymatic activity in a bacterial strain *in vitro* can provide evidence of possible bacterial role in metabolite production. | Data can be flawed by synergistic action between an extract of *Bacteroides fragilis* and the microsomal enzyme activity of the mucosa of the large intestine. | [177] |

## QUESTIONS

1. List and describe the functions of each the various parts of the digestive system and accessory organs.
2. Compare and contrast the cellular structure of each section of the gastrointestinal tract.
3. Describe the various cellular absorption processes.
4. Several types of regulatory control of the alimentary canal exist. Describe them and how they interact to regulate the digestive system.
5. Describe the three phases of gastric secretory regulation.
6. What mechanisms are associated with lipid digestion and absorption?
7. List and describe the excretions of and digestive functions of the accessory organs.

8. Explain how substances are chemically altered when they are ingested.
9. Describe the intestinal first-pass effect and how it influences toxicity.
10. What is enterohepatic circulation?
11. Name and explain the critical factors influencing toxicity of ingested substances.
12. Name and explain the critical factors influencing distribution of ingested substances.
13. Describe three tests that examine the structural integrity of the gastrointestinal tract.
14. Describe three tests that examine gastric secretory activity.
15. Describe the three primary types of tests used to examine gastric emptying activity.
16. Describe the use of SHIME, SBET, and RIVM tests in food and soil contamination studies.
17. Describe and explain three *in vitro* tests used to measure intestinal absorption.
18. Describe and explain three *in vivo* tests used to measure intestinal absorption.
19. How do storage depots influence gastric toxicity?
20. Describe three factors that contribute to the process of toxicant excretion through the fecal elimination.

## ACKNOWLEDGMENT

The author wishes to acknowledge Carol T. Walsh, author of this chapter in the fourth edition of *Principles and Methods of Toxicology*, of which some of the text and material organization have been retained in this edition.

## REFERENCES

1. Beers, M. H. and Berkow, M. B., Eds. (1999–2005): *The Merck Manual of Diagnosis and Therapy*, 17th ed., John Wiley & Sons, New York (http://www.merck.com/mrk-shared/mmanual/home.jsp).
2. Rothery, M. (2005): *A Level Biology*. Module 1. *Digestion* (www.mrothery.co.uk).
3. Singer, S. J. and Nicolson, G. L. (1972): The fluid mosaic model of the structure of cell membranes. *Science*, 175(23):720–731.
4. Rozman, K. K. and Klaassen, C. D. (2001): Adsorption, distribution, and excretion of toxicants. In: *Casarett & Doull's Toxicology: The Basic Science of Poisons*, 6th ed., edited by C. D. Claassen, pp. 107–132. McGraw-Hill, New York.
5. Answers.com, GuruNet Corp., Wesley Hills, NY.
6. Ellert, M. (1998): *Nutrient Absorption*, Faculty Authored Resources, Southern Illinois University School of Medicine, Carbondale (http://www.siumed.edu/mrc/research/nutrient/gi42sg.html).
7. Sheng, H. P. (1999): *How Do We Digest and Absorb Food Intake: Gastrointestinal System and Functions*. Faculty of Medicine, Department of Physiology, The University of Hong Kong (http://web.hku.hk/~hpsheng/index.html).
8. Gershon, M. D. (1999): The enteric nervous system: a second brain, *Hosp. Pract. (Minneap.)*, 34(7):31–42.
9. Turner, B., Cowie, R., and Young, J. (1999): *The Functional Anatomy of Digestive and Urogenital Reflexes*. Howard University Medical School, Washington, D.C. (http://www.med.howard.edu/anatomy/gas/wk6/lect19gi.htm).
10. Bowen, R. (2004): *Pathophysiology of the Digestive System*. Colorado State University, Ft. Collins (http://arbl.cvmbs.colostate.edu/hbooks/index.html).
11. Pearse, A. G. (1969): The cytochemistry and ultrastructure of polypeptide hormone-producing cells of the APUD series and the embryologic, physiologic and pathologic implications of the concept. *J. Histochem. Cytochem.*, 17(5):303–313.
12. Waldum, H. L., Sandvik, A. K., Syversen, U., and Brenna, E. (1993): The enterochromaffinlike (ECL) cell: physiological and pathophysiological role. *Acta Oncol.*, 32:141–147.
13. Kvetnoy, I., Sandvik, A. K., and Waldum, H. L. (1997): The diffuse neuroendocrine system and extrapineal melatonin. *J. Mol. Endocrinol.*, 18:1–3.
14. Mayer, L. (2000): Mucosal immunity and gastrointestinal antigen processing. *J. Pediatr. Gastroenterol. Nutr.*, 30:S4.
15. Bondy, G. S. and Pestka, J. J. (2005): Gut mucosal immunotoxicology in rodents. In: *Investigative Immunotoxicology*, edited by H. Tryphonas et al., pp. 197–210. CRC Press, Boca Raton, FL.
16. Sternber, E. M. (2000): Interactions between the immune and neuroendocrine systems. In: *Progress in Brain Research*, Vol. 122, edited by E. A. Mayer and C. B. Saper, pp. 328–348. Elsevier, New York.
17. Kvetnoy, I. M. (2002): Neuroimmunoendocrinology: where is the field of study? *Neuroimmunoendocrinol. Lett.*, 23(2):119–120.
18. Ojeda, S. R. and Griffin, J. E. (2004): Organization of the endocrine system. In: *Textbook of Endocrine Physiology*, 5th ed., edited by J. E. Griffin and S. R. Ojeda, S. R., pp. 1–17. Oxford University Press, New York.
19. Smeltzer, S. C., Bare, B. G. (2003): Assessment of digestive and gastrointestinal function. In: *Brunner and Suddarth's Textbook of Medical–Surgical Nursing*, 10th ed., edited by L. S. Brunner and D. S. Suddarth, pp. 940–957. Lippincott Williams & Wilkins, Baltimore, MD.
20. Versantvoort, C. H. M., van de Kamp, E., and Rompelberg, C. J. M. (2002): *Development and Applicability of an In Vitro Digestion Model in Assessing the Bioaccessibility of Contaminants in Food*, RIVM Rep. No. 320102002/2004. National Institute for Public Health and the Environment, Bilthoven, The Netherlands (http://www.rivm.nl/bibliotheek/rapporten/320102002.html).
21. Tortora, G.J. and Anagnostakos, N.P. (1990): The digestive system. In: *Principles of Anatomy and Physiology*, 6th ed., edited by G. A. Thibodeau and K. T. Patton, pp. 731–778. Elsevier, New York.
22. Insel, P., Turner, R. E., and Ross, D. (2004): *Nutrition*, 2nd ed. Jones and Bartlett, Sudbury, MA, pp. 66–97.

23. Tamir, E. (2002): *The Human Body Made Simple*. Churchill Livingstone, Edinburgh, pp. 71–92.
24. Boyce, Jr., H. W., Boyce, G. A. (2003): Esophagus: anatomy and structural anomalies. In: *Atlas of Gastroenterology*, 4th ed., edited by T. Yamaha et al., pp. 163–170. Lippincott Williams & Wilkins, Philadelphia, PA.
25. Thompson, A. B. R. and Shaffer, E. A. (2002): *First Principles of Gastroenterology*, 3rd ed. Canadian Association of Gastroenterology, Oakville, Ontario (http://gastroresource.com/GITextbook/En/Default.htm).
26. Iatropoulos, M. J. (1986): Morphology of the gastrointestinal tract. In: *Gastrointestinal Toxicology*, edited by K. Rozman and O. Hänninen, pp. 246–266. Elsevier, New York.
27. Toner, P. G., Carr, K. E., and Wyburn, G. M. (1971): *The Digestive System: An Ultrastructural Atlas Review*. Appleton Century Crofts, New York.
28. Raufman, J.-P. (2003): Stomach: anatomy and structural anomalies. In: In: *Atlas of Gastroenterology*, 4th ed., edited by T. Yamaha et al., pp. 218–226. Lippincott Williams & Wilkins, Philadelphia, PA.
29. Forstner, J. F. and Forstner, G. G. (1994): Gastrointestinal mucus. In: *Physiology of the Gastrointestinal Tract*, Vol. 2, edited by L. R. Johnson, pp. 1255–1284. Raven Press, New York.
30. Soll, A. H. and Berglindh, T. (1987): Physiology of isolated gastric glands and parietal cells: receptors and effectors regulating function. In: *Physiology of the Gastrointestinal Tract*, edited by L. R. Johnson, pp. 883–909. Raven Press, New York.
31. Costanzo, L. S. (1998): *Physiology: Board Review Series*, 2nd ed. Lippincott Williams & Wilkins, Baltimore, MD, pp. 205–239.
32. Johnson, L. R. (2001): Gastric secretion. In: *Gastrointestinal Physiology*, 6th ed., edited by L. R. Johnson, pp. 75–94. Mosby, St. Louis, MO.
33. Rubin, D. C. (2003): Small intestine: anatomy and structural anomalies. In: *Atlas of Gastroenterology*, 4th ed., edited by T. Yamaha et al., pp. 1466–1485. Lippincott Williams & Wilkins, Philadelphia, PA.
34. Simeone, D. M. and Mulholland, M. W. (2003): Pancreas: anatomy and structural anomalies. In: *Atlas of Gastroenterology*, 4th ed., edited by T. Yamaha et al., pp. 2013–2025. Lippincott Williams & Wilkins, Philadelphia, PA.
35. Kanel, G. C. (2003): Anatomy, microscopic structure, and cell types of the liver. In: *Atlas of Gastroenterology*, 4th ed., edited by T. Yamaha et al., pp. 2263–2276. Lippincott Williams & Wilkins, Philadelphia, PA.
36. Simeone, D. M. (2003): Gallbladder and biliary tract. In: *Atlas of Gastroenterology*, 4th ed., edited by T. Yamaha et al., pp. 2166–2167. Lippincott Williams & Wilkins, Philadelphia, PA.
37. Abumrad, N. A., Sfeir, Z., Connelly, M. A., and Coburn, C. (2000): Lipids transporters: membrane transport systems for cholesterol and fatty acids. *Curr. Opin. Clin. Nutr. Metab. Care*, 3(4):255–262.
38. Stahl, A., Hirsch, D. J., Gimeno, R. E., Punreddy, S. G., P., Watson, N., Patel, S., Raimondi, A., Tartaglia, L. A., and Lodish, H. F. (1999): Identification of the major fatty acid transport protein. *Mol. Cell*, 4(3):299–308.
39. Chow, S. L. and Hollander, D. (1979): A dual, concentration-dependent absorption mechanism of linoleic acid by rate jejunum *in vitro*. *J. Lipid Res*., 20(3):349–356.
40. Stremmel, W., Lotz, G., Strohmeyer, G., and Berk, P. D. (1985): Identification, isolation, and partial characterization of a fatty acid binding protein from rat jejunal microvillous membranes. *J. Clin. Invest*., 75(3):1068–1076.
41. Hamilton, J. A. (2002): Mechanism of cellular uptake of long-chain fatty acids: do we need cellular proteins? *Mol. Cell. Biochem*., 239(1/2):17–23.
42. Tso P., Nauli A., and Lo C. M. (2004): Enterocyte fatty acid uptake and intestinal fatty acid-binding protein. *Biochem. Soc. Trans*., 32(1):75–78.
43. Bronk, J. R. and Hastwell, J. G. (1987): The transport of pyrimidines into tissue rings cut from rat small intestine. *J. Physiol*., 382:475–488.
44. Pearce, K. (2004): *Absorption*. University of Waterloo, Ontario, Canada (http://www.student.math.uwaterloo.ca/~kmpearce/SummaryGroup12.doc).
45. Cohn, S. M. and Birnbaum, E. H. (2003): Colon: anatomy and structural anomalies. In: *Atlas of Gastroenterology*, 4th ed., edited by T. Yamaha et al., pp. 1685–1698. Lippincott Williams & Wilkins, Philadelphia, PA.
46. Patel, J. S. et al. (2005): *Intestinal Bacteria: Normal Flora*. The University of Georgia, Athens (http://www.arches.uga.edu/~jspatel/normalflora.html).
47. Todar, K. (2005): *Todar's Online Textbook of Bacteriology*. Department of Bacteriology, University of Wisconsin–Madison (http://www.textbookofbacteriology.net/).
48. Rumney, C. J. and Rowland, I. R. (1992): *In vivo* and *in vitro* models of the human colonic flora. *CRC Crit. Rev. Food Sci. Nutr*., 31:299–331.
49. Van der Waaij, D. (1991): The microflora of the gut: recent findings and implications *Digest. Dis. Sci*., 9:36–48.
50. IPCS. (1978): *Principles and Methods for Evaluating the Toxicity of Chemicals*, Part I. International Programme on Chemical Safety, World Health Organization, Geneva (http://www.inchem.org/documents/ehc/ehc/ehc006.htm).
51. Coppoc, G. L. (2004): *Biotransformation*. Purdue Research Foundation, W. Lafayette, IN (http://www.vet.purdue.edu/bms/courses/bms513/pr_biot.htm#top).
52. Parkinson, A. (2001): Biotransformation of xenobiotics. In: *Casarett & Doull's Toxicology: The Basic Science of Poisons*, 6th ed., edited by C. D. Claassen, pp. 133–224. McGraw-Hill, New York
53. Renwick, A. G. and George, C. F. (1989): Metabolism of xenobiotics in the gastrointestinal tract. In: *Xenobiotic Metabolism in Animals: Methodology, Mechanisms, and Significance*, edited by D. H. Huston et al., pp. 13–40. Taylor & Francis, London.
54. Ilett, K. F., Tee, L. B. G., Reeves, P. T., and Minchin, R. F. (1990): Metabolism of drugs and other xenobiotics in the gut lumen wall. *Aliment. Pharmacol. Therap*., 46: 67–93.
55. Krishna, D. R. and Klotz, U. (1994): Extrahepatic metabolism of drugs in humans. *Clin. Pharmacokinet*., 26: 144–160.
56. Lin, J. H., Chiba, M., and Baillie, T. A. (1999): Is the role of the small intestine in first-pass metabolism overemphasized? *Pharmacol.l Rev*., 51:135–158.

57. Gregus, Z. and Klaassen, C. D. (2001): Mechanisms of toxicity. In: *Casarett & Doull's Toxicology: The Basic Science of Poisons*, 6th ed., edited by C. D. Claassen, pp. 35–81. McGraw-Hill, New York.
58. Sreedharan, R. and Mehta, D. L. (2004): Gastrointestinal tract. *Pediatrics*, 113(4):1044–1050.
59. Levitt, M. D. et al. (1990): Physiological measurements of luminal stirring in the dog and human small bowel. *J. Clin. Invest.*, 86:1540–1547.
60. Kragh-Hansen, U. (1981): Molecular aspects of ligand binding to serum albumin. *Pharmacol. Rev.*, 33:17–53.
61. Lücke, H. et al. (1978): Taurocholate–sodium co-transporter by brush-border membrane vesicles isolated from rat ileum. *Biochem. J.*, 174:951–958.
62. Kouzauki, H., Suzuki, H., Stieger, B., Meier, P. J., and Sugiyama. (2000): Characterization of the transport properties of organic anion transporting polypeptide 1 (oatp1) and $Na^+$/taurocholate contransporting polypeptide (ntcp): comparative studies on the inhibitory effect of their possible substrates in hepatocytes and cDNA-transfected COS-7 cells. *Pharmacol. Exp. Therap.*, 292(2):505–511.
63. Eckhardt, U., Schroeder, A., Stieger, B., Höchli, M., Landmann, L., Tynes, R. Meier, P. J., and Hagenbuch, B. (1999): Polyspecific substrate uptake by organic anion transporter oatp1 in stably transfected CHO cells. *Am. J. Physiol.*, 276:G1037–G1042.
64. Meier, P. J., Eckhardt, U., Schroeder, A., Hagenbuch, B., and Stieger, B. (1997): Substrate specificity of sinusoidal bile acid and organic uptake systems in rate and human liver. *Hepatology*, 26:1667–1677.
65. Meier, P. J. (1995): Molecular mechanisms of hepatic bile salt transport from sinusoidal blood into bile. *Am. J. Physiol. Gastrointest. Liver Physiol.*, 269:G801–G812.
66. Abe, T. et al. (1999): Identification of a novel gene family encoding human liver-specific organic anion transporter LST-1. *J. Biol. Chem.*, 274:17159–17163.
67. Koepsell, H., Schmitt, B. M., and Gorboulev, V. (2003): Organic cation transporters. *Rev. Physiol. Biochem. Pharmacol.*, 150:36–90.
68. Nathanson M. H. and Boyer J. L. (1991): Mechanisms and regulation of bile secretion. *Hepatology*, 14(3):551–566.
69. Shoemaker, R. H., Curt, G. A., and Carney, D. N. (1983): Evidence for multidrug-resistant cells in human tumor cell populations. *Cancer Treat. Rep.*, 67(10):883–888.
70. Kazumitsu, U., Pastan, I., and Gottesman, M. M. (1987): Isolation and sequence of the promoter region of the human multidrug-resistance (P-glycoprotein) gene. *J. Biol. Chem.*, 262(36):17432–17436.
71. Haimeur, A., Conseil, G., Deeley, R. G., and Cole, S. P. (2004): The MRP-related and BCRP/ABCG2 multidrug resistance proteins: biology, substrate specificity and regulation. *Curr. Drug Metab.*, 5(1):21–53.
72. Prime-Chapman, H. M., Fearn, R. A., Cooper, A. E., Moore, V., and Hirst, B. H. (2004): Differential multidrug resistance-associated protein 1 through 6 isoform expression and function in human intestinal epithelial Caco-2 cells. *J. Pharmacol. Exp. Therap.*, 311(2):476–484.
73. Borst, P., Evers, R., Kool, M., and Wijnholds, J. (2000): A family of drug transporters: the multidrug resistance-associated proteins. *J. Natl. Cancer Inst.*, 92(16):1295–1302.
74. Taipalensuu, J. et al. (2001): Correlation of gene expression of ten drug efflux proteins of the ATP-binding cassette transporter family in normal human jejunum and in human intestinal epithelial Caco-2 cell monolayers. *J. Pharmacol. Exp. Therap.*, 299(1):164–170.
75. Rozman, K. (1986): Fecal excretion of toxic substances. In: *Gastrointestinal Toxicology*, edited by K. Rozman and O. Hänninen, pp. 119–145. Elsevier, New York.
76. Rozman K. (1988): Disposition of xenobiotics: species differences. *Toxicol. Pathol.*, 16(2):123–129.
77. Brodie, D. A., Cook, P. G. M., Bauer, B. J., and Dagle, G. E. (1970): Indomethancin-induced intestinal lesions in the rat. *Toxicol. Appl. Pharmacol.*, 17:615–624.
78. Brodie, D. A. Tate, C. L., and Hooke, K. F. (1970): Aspirin: intestinal damage in rats. *Science*, 170:183–185.
79. Szabo, S. (1978): Duodenal ulcer disease. Animal model: cysteamine-induced acute and chronic duodenal ulcer in the rat. *Am. J. Pathol.*, 93:273–276.
80. Glavin, G. B. and Szabo, S. (1992): Experimental gastric mucosal injury: laboratory models reveal mechanisms of pathogenesis and new therapeutic strategies. *FASEB J.*, 6:825–831.
81. Kvietys, P. R., Perry, M. A., Gaginella, T. S., and Granger, D. N. (1990): Ethanol enhances leukocyte-endothelial cell interactions in mesenteric venules. *Am. J. Physiol.*, 259: G578–G583.
82. Iishi, H., Tatsuta, M., and Okuda, S. (1985): Endoscopic diagnosis of minute gastric cancer of less than 5 mm in diameter. *Cancer*, 56(3):655–659.
83. Satoh, H., Nada, I., Hirata, T., and Maki, Y. (1981): Indomethacin produces gastric antral ulcers in the refed rat. *Gastroenterology*, 81:719–725.
84. Morales, R. E., Johnson, B. R., and Szabo, S. (1992): Endothelin induces vascular and mucosal lesions, enhances the injury by HCl/ethanol, and the antibody exerts gastroprotection. *FASEB J.*, 6:2354–2360.
85. Lavo, B., Colombel, J. F., Knutsson, L., and Hallgren, R. (1992): Acute exposure of small intestine to ethanol induces mucosal leakage and prostaglandin E2 synthesis. *Gastroenterology*, 102:468–473.
86. Yamada, T., Specian, R. D., Granger, D. N., Gaginella, T. S. and Grisham, M. B. (1991): Misoprostol attenuates acetic acid-induced increases in mucosal permeability and inflammation: role of blood flow. *Am. J. Physiol.*, 261: G332–G339.
87. Simon, J. B. (1998): Fecal occult blood testing: clinical value and limitations. *Gastroenterologist*, 6:66–78.
88. Ahlquist, D. A., McGill, D. B., Schwartz, S., Taylor, W. F., and Owen, R. A. (1985): Fecal blood levels in health and disease: a study using HemoQuant. *New Engl. J. Med.*, 312:1422–1428.
89. Ahlquist, D. A., Wieand, H. S., Moertel, C. G., McGill, D. B., Loprinzi, C. L., O'Connell, M. J., Mailliard, J. A., Gerstner, J. B., Pandya, K., and Ellefson, R. D. (1993): Accuracy of fecal occult blood screening for colorectal neoplasia: a prospective study using hemoccult and HemoQuant tests. *J. Am. Med. Assoc.*, 269(10): 1262–1267.
90. Winawer, S. (1976): Fecal occult blood testing. *Am. J. Digest. Dis.*, 21:885–888.

91. Phillips, B. M. (1973): Aspirin-induced gastrointestinal microbleeding in dogs. *Toxicol. Appl. Pharmacol.*, 24: 182–189.
92. Hurst, B. C., Ress, W. D. W., and Garner, A. (1984): Cell shedding by the stomach and duodenum. In: *Mechanisms of Mucosal Protection in the Upper Gastrointestinal Tract*, edited by A. Allen et al., pp. 21–26. Raven Press, New York.
93. Lipkin, M. and Higgins, P. P. (1988): Biological markers of cell proliferation and differentiation in human gastrointestinal diseases. *Adv. Cancer Res.*, 50:1–24.
94. Lipkin M. (1992): Gastrointestinal cancer: pathogenesis, risk factors and the development of intermediate biomarkers for chemoprevention studies. *J. Cell. Biochem.*, 16G: 1–13.
95. Cheng, H., Bjerknes, M., and Amar, J. (1984): Methods for the determination of epithelial cell kinetic parameters of human colonic epithelium isolated from surgical and biopsy specimens. *Gastroenterology*, 86:78–85.
96. Bostick, R. M., Fosdick, L., Lillemoe, T. J., Overn, P., Wood, J. R., Grambsch, P., Elmer, P., Potter, J. D. (1997): Methodological findings and considerations in measuring colorectal epithelial cell proliferation in humans. *Cancer Epidemiol. Biomarkers Prevent.*, 6(11):931–942.
97. Cheng, H. and Bjerknes, M. (1982): Whole population cell kinetics of mouse duodenal, jejunal, ileal and colonic epithelia as determined by radioautography and flow cytometry. *Anatom. Rec.*, 203:251–164.
98. Cheng, H. and Bjerknes, M. (1990): Whole population cell kinetics of jejunal and colonic epithelium in lactating dams. *Anatom. Rec.*, 228(3):262–266.
99. Szabo, S., Reynolds, E. S., Lichtenberger, L. M., Haith, L. R., and Dzau, V. J. (1977): Pathogenesis of duodenal ulcer: gastric hyperacidity caused by propionitrile and cysteamine in rats. *Res. Commun. Chem. Pathol. Pharmacol.*, 16: 311–323.
100. Tietz, N. W. (1976): Gastric, pancreatic and intestinal function. In: *Fundamentals of Clinical Chemistry*, edited by N. W. Tietz, pp. 1063–1099. W.B. Saunders, Philadelphia, PA.
101. Kamada, T., Hiramatsu, K., Fusamoto, H., Masuzawa, M., and Abe, H. (1976): Endoscopic observation of the gastric mucus *in vivo* stained with Azure A. *Am. J. Gastroenterol.*, 65(6):532–538.
102. Okabe, S., Shimosako, K., and Harada, H. (1995): Antisecretory effect of leminoprazole on histamine-stimulated gastric acid secretion in dogs: potent local effect. *Jpn. J. Pharmacol.*, 69:91–100.
103. Warrick, M. W. and Lin, T. N. (1977): Action of glucagons and aspirin on ionic flux, mucosal blood flow and bleeding in the fundic pouch of dogs. *Res. Commun. Chem. Pathol. Pharmacol.*, 16:325–335.
104. Soll, A. H. and Berglindh, T. (1987): Physiology of isolated gastric glands and parietal cells: receptors and effectors regulating function. In: *Physiology of the Gastrointestinal Tract*, edited by L. R. Johnson, pp. 883–909. Raven Press, New York.
105. Larsen, R., Mertz-Nielsen, A., Hansen, M. B., Poulsen, S. S., and Bindslev, N. (2001): Novel modified Ussing chamber for the study of absorption and secretion in human endoscopic biopsies. *Acta Physiol. Scand.*, 173(2): 213–222.
106. Vigna, S. R., Mantyh, C. R., Soll, A. H., Maggio, J. E., and Mantyh, P. W. (1989): Substance P receptors on canine chief cells: localization, characterization and function. *J. Neurosci.*, 9:2878–2886.
107. Soll, A. H. (1980): Secretagogue stimulation of $^{14}$C-aminopyrine accumulation by isolated canine parietal cells. *Am. J. Physiol.*, 238:G366–G375.
108. Camilleri, M., Brown, M. L., and Malagelada, J. R. (1986): Relationship between impaired gastric emptying and abnormal gastrointestinal motility. *Gastroenterology*, 9l: 94–99.
109. Wyse, C. A., McLellan, J., Dickie, A. M., Sutton, D. G. E., Preston, T., and Yam, P. S. (2003): A review of methods for assessment of the rate of gastric emptying in the dog and cat: 1989–2002. *J. Vet. Intern. Med.*, 17: 609–621.
110. Gue, M., Peeters, T., Depoortere, I., Vantrappen, G., and Bueno, L. (1989): Stress-induced changes in gastric emptying, postprandial motility, and plasma gut hormone levels in dogs. *Gastroenterology*, 97(5):1101–1107.
111. Googin, J. M, Hoskinson, J. J., Butine, M. D., Foster, L. A., and Myers, N. C. (1998): Scintigraphic assessment of gastric emptying of canned and dry diets in healthy cats. *Am. J. Vet. Res.*, 59(4):388–392.
112. Sutton, D. G. et al. (2003): Validation of the $^{13}$C-octanoic acid breath test for measurement of equine gastric emptying rate of solids using radioscintigraphy. *Equine Vet. J.*, 35(1):27–33.
113. Wyse, C. A. et al. (2001): Use of the $^{13}$C-octanoic acid breath test for assessment of solid-phase gastric emptying in dogs. *Am. J. Vet. Res.*, 62(12):1939–1944.
114. Rennie, M. J. (1999): An introduction to the use of tracers in nutrition and metabolism. *Proc. Nutr. Soc.*, 58:935–944.
115. Heading, R. C., Nimmo, J., Prescott, L. F., and Tothill, P. P. (1973): The dependence of paracetamol absorption on the rate of gastric emptying, *Br. J. Pharmacol.*, 47:415–421.
116. Burns, J. and Fox, S. M. (1986): The use of a barium meal to evaluate total gastric emptying time in the dog. *Vet. Radiol.*, 27:169–172.
117. Donohoe, K. J., Maurer, A. H., Ziessman, H. A., Urbain, J. L., and Royal, H. D. (1999): Procedure guideline for gastric emptying and motility. *J. Nucl. Med.*, 40: 1236–1239.
118. Bolondi, L., Bortolotti, M., Santi, V., Calletti, T., Gaiani, S., and Labo, G. (1985): Measurement of gastric emptying time by real-time ultrasonography. *Gastroenterology*, 89(4):752–759.
119. Schwizer, W., Fraser, R., Borovicka, J., Crelier, G., Boesiger, P., and Fried, M. (1994): Measurement of gastric emptying and gastric motility by magnetic resonance imaging (MRI). *Digest. Dis. Sci.*, 39(12, Suppl.): 101S–103S.
120. Feinle, C., Kunz, P., Boesiger, P., Fried, M., and Schwizer, W. (1999): Scintigraphic validation of a magnetic resonance imaging method to study gastric emptying of a solid meal in humans. *Gut*, 44(1):106–111.
121. Spyrou, N. M. and Castillo, F. D. (1993): Electrical impedance measurement. In: *An Illustrated Guide to Gastrointestinal Motility*, edited by D. Kumar and D. Wingate, pp. 276–278. Churchill Livingstone, Edinburgh.

122. Sutton, J. A., Thompson, S., and Sobnack, R. (1985): Measurement of gastric emptying rates by radioactive isotope scanning and epigastric impedance. *Lancet*, 1:898–900.
123. Mangnall, Y. F. et al. (1988): Comparison of applied potential tomography and impedance epigastrography as methods of measuring gastric emptying. *Clin. Phys. Physiol. Meas.*, 9(3):249–254.
124. Zikakis, J. P., Rzucidlo, S. J., and Biasotto, N. O. (1977): Persistence of bovine milk xanthine oxidase activity after gastric digestion *in vivo* and *in vitro*. *J. Dairy Sci.*, 60: 533–541.
125. Marquez, U. M. L. and Lajolo, F. M. (1981): Composition and digestibility of albumin, globulins, and glutelins from *Phaseolus vulgaris*. *J. Agric. Food Chem.*, 29:1068–1074.
126. Tilch, C. and Elias, P. S. (1984): Investigation of the mutagenicity of ethylphenylglycidate. *Mutat. Res.*, 138:1–8.
127. Borel, P., Grolier, P., Armand, M., Partier, A., Lafont, H., Lairon, D., and Azais-Braesco, V. (1996): Carotenoids in biological emulsions: solubility, surface-to-core distribution, and release from lipid droplets. *J. Lipid Res.*, 37(2): 250–261.
128. Deming, D. M. and Erdman, Jr., J. W. (1999): Mammalian carotenoid absorption and metabolism. *Pure Appl. Chem.*, 71(12):2213–2223.
129. Oomen, A. G. et al. (2003): *Development and Suitability of in vitro Digestion Model in Assessing Bioaccessibility of Lead from Toy Matrices*. RIVM Rep. No. 320102001/2003, National Institute for Public Health and the Environment, Bilthoven, The Netherlands, 2002 (http://www.rivm.nl/bibliotheek/rapporten/320102001.html).
130. Oomen, A. G., Versantvoort, C., and Sips, A. (2003): Digestion Models Simulating Fasting and Fed Conditions, abstract presented at 19th Annual International Conference on Soils, Sediments, and Water, University of Massachusetts, Amherst, October 21, 2003 (http://www.umasssoils.com/posters2003/bioavailposter.htm#top).
131. Oomen, A. G., Hack, A., Minekus, M., Zeijder, E., Cornelis, C., Schoeters, G., Verstraete, W., Van de Wiele, T., Wragg, J., Rompelberg, C. J., Sips, A. J., and van Wijnen, J. H. (2002): Comparison of five *in vitro* digestion models to study the bioaccessibility of soil contaminants. *Environ. Sci. Technol.*, 36:3326–3334.
132. Grøn, C. and Anderson, L. (2003): *Human Bioaccessibility of Heavy Metals and PAH from Soil*. Danish Environmental Protection Agency, Copenhagen, Denmark.
133. Rotard, W., Christmann, W., Knoth, W., and Mailahn, W. (1995): Bestimmung der resorptionsverfüghbaren PCDD/PCDF aus Kieselrot. *UWSF-Z. Umweltchem. Ökotox.*, 7:3–9.
134. Molly, K. et al. (1994): Validation of the simulator of the human intestinal microbial ecosystems (SHIME) reactor using microorganism-associated activities. *Microb. Ecol. Health Dis.*, 7:191–200.
135. Hack, A. and Selenka, F. (1996): Mobilization of PAH and PCB from contaminated soil using a digestive tract model. *Toxicol. Lett.*, 88:199–210.
136. Minekus, M., Marteau, P., Havenaar, R., and Huis in't Veld, J. H. J. (1995): A multicompartmental dynamic computer-controlled model simulating the stomach and small intestine. *ATLA*, 23:197–209.
137. Sips, A. (2004): What Can *In Vitro* Digestion Models Add to Human Risk Assessment of Contaminated Soil? Presentation made at Bioavailability Workshop at the Petroleum Environmental Research Forum (PERF), Berkeley, CA, September 29–30, 2004.
138. Acra, S. A. and Ghishan, F. K. (1991): Methods of investigating intestinal transport. *J. Parenter. Enteral Nutr.*, 15(3):93S–98S.
139. Barr, W. H. and Reigelman, S. (1970): Intestinal drug absorption and metabolism. I. Comparison of methods and models to study physiological factors of *in vitro* and *in vivo* intestinal absorption. *J. Pharmacol. Sci.*, 59:154–163.
140. Peltzmann, K. S. and Havemeyer, R. N. (1971): Portal vein blood sampling in intestinal drug absorption studies. *J. Pharmaceut. Sci.*, 60:331.
141. Wright, E. M., Hirayama, B. A., Loo, D. D. F., Turk, E., and Hager, K. (1994): Intestinal sugar transport. In: *Physiology of the Gastrointestinal Tract*, Vol. 2, edited by L. R. Johnson, pp. 1751–1772. Raven Press, New York.
142. Ringler, D. H. and Daibich, L. (1979): Hematology and clinical biochemistry. In: *The Laboratory Rat*. Vol. 1. *Biology and Disease*, edited by H. J. Baker et al., pp. 105–121. Academic Press, New York.
143. Tso, P. P. (1994): Intestinal lipid absorption. In: *Physiology of the Gastrointestinal Tract*, Vol. 2, edited by L. R. Johnson, pp. 1867–1908. Raven Press, New York.
144. Ellefson, R. D. and Caraway, W. T. (1976): Lipids and lipoproteins. In: *Fundamentals of Clinical Chemistry*, edited by N. Tietz, pp. 474–541. W.B. Saunders, Philadelphia, PA.
145. Walsh, C. T. (1982): The influence of age on the gastrointestinal absorption of mercuric chloride and methyl mercury chloride in the rat. *Environ. Res.*, 27:412–420.
146. Lewis, L. D. and Fordtran, J. S. (1975): Effect of perfusion rate on absorption, surface area, unstirred water layer thickness, permeability and intraluminal pressure in the rat ileum *in vivo*. *Gastroenterology*, 68:1509–1516.
147. Mailman, D., Womack, W. A., Kvietys, P. R., and Granger, D. N. (1990): Villous motility and unstirred water layers in canine intestine. *Am. J. Physiol.*, 258:G238–G246.
148. Wilson, T. H. and Wiseman, G. (1954): The use of sacs of everted small intestine for the study of the transference of substances from the mucosal to the serosal surface. *J. Physiol.*, 123:116–125.
149. Chen, Y., Ping, Q., Guo, J., Lv, W., and Gao, J. (2003): The absorption behavior of cyclosporin A lecithin vesicles in rat intestinal tissue. *Int. J. Pharmaceut.*, 261(1–2): 21–26.
150. Aw, T. Y., Bai, C., and Jones, D. P. (1993): Small intestinal enterocytes. In: *Methods in Toxicology: In Vitro Biological Systems*, edited by C. A. Tyson and J. M. Frazier, pp. 193–201. Academic Press, Boston, MA.
151. Bell, G. I., Kayano, T., Buse, J. B., Burant, C. F., Takedo, J., Lin, D., Fukumoto, H., and Seino, S. (1990): Molecular biology of mammalian glucose transporters. *Diabetes Care*, 13:198–208.
152. Towler, C. M., Pugh-Humphreys, G. P., and Porteous, J. W. (1978): Characterization of columnar absorptive epithelial cells isolated from rat jejunum. *J. Cell Sci.*, 29: 53–75.

153. Binder, H. J. and Sandle, G. I. (1994): Electrolyte transport in the mammalian colon. In: *Physiology of the Gastrointestinal Tract*, Vol. 2, edited by L. R. Johnson, pp. 2133–2172. Raven Press, New York.
154. Stevens, B. R., Fernandez, A., Hirayama, B., Wright, E. M., and Kempner, E. S. (1990): Intestinal brush border membrane $Na^+$/glucose cotransporter functions *in situ* as a homotetramer. *Proc. Nat. Acad. Sci.*, 87:1456–1460.
155. Moorghen, M., Chapman, M., and Appleton, D. R. (1996): An organ-culture method for human colorectal mucosa using serum-free medium. *J. Pathol.*, 180:102–105.
156. Howdle, P. D. (1984): Organ culture of gastrointestinal mucosa. *Postgrad. Med.*, 60:645–652.
157. Dharmsathaphorn, K., McRoberts, J. A., Mandel, K. G., Tisdale, L. D., and Masui, H. (1984): A human colonic tumor cell line that maintains vectorial electrolytes transport. *Am. J. Physiol.*, 246: G204–G208.
158. Hecht, G., Koutsouris, A., Pothoulakis, C., LaMont, J. T., and Madara, J. L. (1992): *Clostridium difficile* toxin B disrupts the barrier function of T84 monolayers. *Gastroenterology*, 102:416–423.
159. Leroy, A., Lauwaet, T., De Bruyne, G., Cornelissen, M., and Mareel, M. (2000): *Entamoeba histolytica* disturbs the tight junction complex in human enteric T84 cell layers. *FASEB J.*, 14:1139–1146.
160. Lencer, W. I., Reinhard, F. D., and Neutra, M. R. (1990): Interaction of cholera toxin with cloned human goblet cells in monolayer culture. *Am. J. Physiol.*, 258:G96–G102.
161. Phillips, T. E., Huet, C., Biblio, P. R., Podolsky, D. K., Louvard, D., and Neutra, M. R. (1988): Human intestinal goblet cells in monolayer culture: characterization of a mucus-secreting subclone cleaved from the HT-29 colon adenocarcinoma cell line. *Gastroenterology*, 94(6): 1390–1403.
162. Gustavsson, S., Jung, B., and Nilsson, F. (1977): Simultaneous measurements of the propulsion and mixing of small bowel contents of the rat. *Acta Chir. Scand.*, 143(6): 359–364.
163. Summers, R. W. Kent, T. H., and Osborne, J. W. (1970): Effects of drugs, ileal obstruction and irradiation on rat gastrointestinal propulsion. *Gastroenterology*, 59: 731–739.
164. Walsh, C. T. and Ryden, E. B. (1984): The effect of chronic ingestion of lead on gastrointestinal transit in rats. *Toxicol. Appl. Pharmacol.*, 75:485–495.
165. Miller, M. S., Galligan, J. J., and Burks, T. F. (1981): Accurate measurement of intestinal transit in the rat. *J. Pharmacol. Meth.*, 6:211–217.
166. Dickson, E. W., Tubbs, R. J., Porcaro1, W. A., Lee, W. J., Blehar, D. J., Carraway, R. E., Darling, C. E., and Przyklenk, K. (2002): Myocardial preconditioning factors evoke mesenteric ischemic tolerance via opioid receptors and KATP channels. *Am. J. Physiol. Heart Circ. Physiol.*, 283(1):H22–H28.
167. Gilani, A. H., Bashir, S., Janbaz, K. H., and Khan, A. (2005): Pharmacological basis for the use of *Fumaria indica* in constipation and diarrhea. *J. Ethnopharmacol.*, 96(3):585–589.
168. Chidume, F. C., Kwanashie, H. O., Adekeye, J. O, Wambebe, C., and Gamaniel, K. S. (2002): Antinociceptive and smooth muscle contracting activities of the methanolic extract of *Cassia tora* leaf. *J. Ethnopharmacol.*, 81(2):205–209.
169. Rumney, C. J. and Rowland, I. R. (1992): *In vivo* and *in vitro* models of the human colonic flora. *CRC Crit. Rev. Food Sci. Nutr.*, 31:299–331.
170. Van der Waaij, D. (1991): The microflora of the gut: recent findings and implications. *Digest. Dis. Sci.*, 9:36–48.
171. Mickelsen, P. P. (1972): Introductory remarks, symposium on cycads. *Fed. Proc.*, 31:1465–1546.
172. Laqueur, G. L., McDaniel, E. G., and Matsumoto, H. (1967): Tumor induction in germfree rats with methylazooxymethanol (MAM) and synthetic MAM acetate. *J. Natl. Cancer Inst.*, 39:355–371.
173. Kandel, G., Donohue-Rolfe, A., Donowitz, M., and Keusch, G. T. (1989): Pathogenesis of *Shigella* diarrhea. VXI. Selective targeting of *Shiga* toxin to villus cells of rabbit jejunum explains the effect of the toxin on intestinal electrolyte transport. *J. Clin. Invest.*, 84:1509–1517.
174. Triadafilopoulos, G., Pothoulakis, C., O'Brien, M. J., and LaMont, J. T. (1987): Differential effects of *Clostridium difficile* toxins A and B on rabbit ileum. *Gastroenterology*, 93:273–279.
175. Triadafilopoulos, G., Pothoulakis, C., Weiss, R., Giampaolo, C., and La Mont, J. T. (1989): Comparative study of *Clostridium difficile* toxin A and cholera toxin in rabbit ileum. *Gastroenterology*, 97:1186–1192.
176. Foster, H. L. (1980): Gnotobiology. In: *The Laboratory Rat*. Vol. II. *Research Applications*, edited by H. J. et al., pp. 43–57. Academic Press, New York.
177. Tasich, M. and Piper, D. W. (1983) Effect of human colonic microsomes and cell-free extracts of *Bacteroides fragilis* on the mutagenicity of 2-aminoanthracene. *Gastroenterology*, 85:30–34.
178. Columbia Presbyterian Medical Center. (2007): Gastroenterology Resources, http://cpmcnet.columbia.edu/dept/gi/elsewhere.html.

# 32 Principles and Methods of Cardiac Toxicology

*Y. James Kang*

## CONTENTS

## INTRODUCTION AND DEFINITIONS

Cardiac toxicology is concerned with the adverse effects of exposure to extrinsic and intrinsic toxic substances in the heart. The measurement of cardiac toxic effects ranges from morphological changes to molecular alterations. These measurements are based on the anatomy, physiology, and biochemistry of the heart. Toxic substances can cause adverse effects but at the same time also cause myocardial protective responses. The cardiomyocytes in the heart respond individually to the same insult, and the same insult may have different effects on the population of cardiomyocytes, depending on the location of the cells in the heart; for example, cells localized in endocardium would respond differently from cells localized in the epicardium. The measurement of cardiac toxic effects requires taking the sum of all different responses of individual cardiomyocytes and noncardiac cells within the heart. Toxic exposures can result in alterations in biochemical pathways, energy metabolism, cellular structural and function, electrophysiology, and contractility leading to toxicologic cardiomyopathy. The defense mechanisms are also activated at the same time, but the balance between detrimental and protective actions determines the phenotype of the heart under the toxic exposure condition. The manifestations of toxicologic cardiomyopathy include cardiac arrhythmia, hypertrophy, and overt heart failure. The ultimate functional effect of these manifestations is decreased cardiac output and reduced peripheral tissue perfusion. The critical cellular event leading to toxicologic cardiomyopathy is myocardial cell death. Recognition during the last decade of the role of apoptosis in the development of heart failure significantly enhanced our knowledge of cardiac toxicology.

The experimental approaches employing direct manipulation of genes responsible for normal and abnormal cardiac function began in the mid-1990s. Thousands of embryonic stem cell lines and hundreds of genetic manipulated animal models have been produced [1]. The most important conclusion of these studies is that a sustained expression of any single mutated functional gene, either in the form of gain of function or loss of function, can lead to a significant phenotype, often in the form of cardiac hypertrophy and heart failure [1,2]. At the same time, however, it is difficult to apply this knowledge to patients for the following reasons: First, acquired cardiac disease, such as heart failure, is the result of interaction between environmental factors and genetic susceptibility, thus suggesting the role of polymorphism. Second, extrinsic and intrinsic stresses produce complicated lesions that cannot be explained by a single gene or a single pathway, which is a reflection of the complexity of the interactions between deleterious factors and the cardiac organ. Cardiac toxicity is the critical link between environmental factors and myocardial pathogenesis.

Figure 32.1 presents a new model of cardiac toxicity known as triangle analysis. This model recognizes the complexity of interactions between environmental stresses and the cardiac organ, as well as the balance between myocardial protection and deleterious dose and time effects. It is important to recognize that drugs or chemicals can lead to heart failure without heart hypertrophy. Also

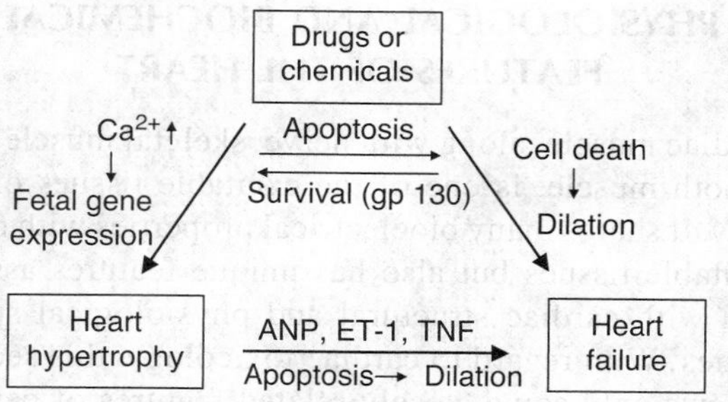


**FIGURE 32.1** Triangle analysis of cardiac toxic effects. Drugs and chemicals can directly cause both heart failure and heart hypertrophy. Under toxic insults, myocardial cell death becomes the predominant response leading to cardiac dilation and heart failure. In most cases, myocardial survival mechanisms can be activated so myocardial apoptosis is inhibited. The surviving cardiomyocytes often become hypertrophic through activation of calcium-mediated fetal gene expression and other hypertrophic programs. If toxic insult continues, the counter mechanisms against heart hypertrophy, such as activation of cytokine-medicated pathways, eventually lead to activation of the apoptotic program, cardiac dilation, and heart failure.

the same stimulus can lead to activation of both protective and destructive responses in the myocardium. Myocardial protection from toxic exposure often, if not always, expresses in the form of maladaptive hypertrophy, which primes the heart for malignant arrhythmia leading to sudden cardiac death or for a transition to heart failure.

In the study of cardiac toxicology, the manifestations of cardiac toxicity in human patients and animal models are critical parameters serving as indices of cardiac toxicity. These manifestations are expressed in the forms of cardiac arrhythmia, hypertrophy, and heart failure. These abnormal changes reflect myocardial functional alterations resulting from both acute and chronic cardiac toxicity. Although some changes such as cardiac hypertrophy were viewed as compensatory responses to hemodynamic changes, evidence accumulating from studies using more advanced technologies in human patients and animal models suggests that cardiac hypertrophy is a maladaptive process of the heart in response to intrinsic and extrinsic stresses [3–5]. Cardiac hypertrophy is a risk factor for sudden cardiac death and has a high potential to progress to overt heart failure; therefore, a distinction between a compensatory response and a maladaptive response is critical for the treatment of clinical patients with toxicologic cardiomyopathy.

## Cardiac Arrhythmia

Cardiac rhythms under physiological conditions are set by pacemaker cells that are normally capable of developing spontaneous depolarization and are responsible for generating the cardiac rhythm (the so-called *automatic rhythm*). Cardiac rhythm that deviates from the normal automatic rhythm is referred to as *cardiac arrhythmia*, often manifested in the form of tachycardia. The several classes of tachycardia include sinus tachycardia, atrial tachycardia, ventricular tachycardia, and torsade de pointes (a life-threatening ventricular tachycardia). In addition, subclasses such as atrial fibrillation, atrial flutter, accelerated idioventricular rhythm, and right ventricular outflow tract tachycardia provide further descriptions of the manifestations of arrhythmia. Mechanisms for the different classes of arrhythmia are discussed later in the section addressing QT prolongation and sudden cardiac death. Depending on the cause of the tachycardia, it is divided into abnormal automatic arrhythmia and triggered arrhythmia, which are discussed later.

## Cardiac Hypertrophy

There are two basic forms of cardiac hypertrophy. Concentric hypertrophy is often observed during pressure overload, when new contractile protein units are assembled in parallel, resulting in a relative increase in the width of individual cardiac myocytes [6]. By contrast, eccentric hypertrophy is characterized by the assembly of new contractile protein units in series, which represents a relatively greater increase in the length rather than width of individual myocytes [7]. Toxicologic cardiomyopathy is often manifested in the form of eccentric hypertrophy. The development of cardiac hypertrophy can be divided into three stages: (1) developing hypertrophy, when the cardiac workload exceeds cardiac output; (2) compensatory hypertrophy, where the workload-to-mass ratio is normalized and normal cardiac output is maintained; and (3) decompensatory hypertrophy, where ventricular dilation develops, cardiac output progressively declines, and overt heart failure occurs [8].

## Heart Failure

A traditional definition of heart failure is the inability of the heart to maintain cardiac output sufficient to meet the metabolic and oxygen demands of peripheral tissues. This definition has been modified recently to include changes in systolic and diastolic function that reflect specific alterations in ventricular function and abnormalities in a variety of subcellular processes [9]. A detailed analysis to distinguish right ventricular from left ventricular failure can provide a better understanding of the nature of the heart failure and predicting the prognosis.

## Acute Cardiac Toxicity

It is not difficult to define acute cardiac toxicity; however, it sometimes is technically difficult to measure acute cardiac toxicity; in particular, the impact of acute cardiac toxicity on the ultimate outcome of cardiac function is not often easily recognized. For example, a single high dose of arsenic can lead to sudden cardiac death, which is easily to measure [10]; however, a single oral dose of monensin (20 mg/kg) in calves leads to a diminished cardiac function progressing to heart failure that requires long-term observation (often a few months) for clinical signs of heart failure [11,12].

### Chronic Cardiac Toxicity

Environmental exposure to polluted air, such as air contaminated with particulate matter, can lead to cardiomyopathy, which has been recognized very recently [13,14]. An interesting clinical observation is that about 25% of human patients with cardiomyopathy have been identified as having idiopathic cardiomyopathy. It is expected that at least a portion of the idiopathic cardiomyopathy would be toxicologic cardiomyopathy. Recognition of the role of chronic cardiac toxicity in the pathogenesis of cardiomyopathy is of significantly clinical relevance and could help develop prevention and treatment measures for patients with toxicologic cardiomyopathy.

### Compensatory Response

The compensatory response involves cardiac functional adaptations to decreased contractility of the heart or increased metabolic and oxygen demands of peripheral tissues. Adaptations in myocardial structure include compensatory hypertrophy; in neurohormonal regulation, the angiotensin system has an increased impact on the heart. Under conditions of toxicological stress, the net effect of the compensatory response is to maintain cardiac output with a compromised cardiac system; therefore, the cardiac compensatory response is more related to functional maintenance of the heart under conditions of cardiac injury or cardiomyopathy.

### Maladaptive Response

Cardiac hypertrophy can be viewed as a compensatory response as well as a maladaptive response. Cardiac hypertrophy can normalize wall tension, but it is a risk factor for sudden cardiac death and has a high potential to progress to overt heart failure. It is the hypertrophy that triggers alterations in neurohormonal regulation that lead to the overproduction of inflammatory cytokines such as tumor necrosis factor α (TNFα) and atrial natriuretic peptide (ANP), which can lead to myocardial apoptosis and heart failure. A distinction between a compensatory response and a maladaptive response is more related to a functional definition than phenotype definition, such that the phenotype of cardiac hypertrophy can be viewed as either compensatory or maladaptive.

## PHYSIOLOGICAL AND BIOCHEMICAL FEATURES OF THE HEART

Cardiac muscle, along with nerve, skeletal muscle, and smooth muscle, is one of the excitable tissues of the body. It shares many bioelectrical properties with other excitable tissues but also has unique features associated with cardiac structural and physiological specificities. With regard to cardiac toxicology, this section reviews only some closely related features of cardiac physiology and biochemistry. Many textbooks on cardiac anatomy and physiology are available that provide extensive information on cardiac physiological and biochemical properties which will not be repeated in this section.

### Electrophysiology

The electrophysiology of the heart is concerned with bioelectricity and its related cardiac physiological function. Bioelectricity is the result of a charge generated from the movement of positively and negatively charged ions in tissues. In cardiac myocytes, three major positively charged ions make a significant contribution to the bioelectricity of the heart: calcium ($Ca^{2+}$), sodium ($Na^+$), and potassium ($K^+$). Each of the ions has specific channels and transporters (pumps) on the membrane of cardiac myocytes. Through the movement of these ions across the cell membrane, an action potential is generated and propagated from one cell to another, so electric conductance is produced in the heart.

#### Action Potential

Cardiac myocytes produce an action potential when activated. In the resting state, the resting potential of a myocyte is about –60 to –90 mV relative to the extracellular fluid potential. A sudden depolarization changes the membrane potential from negative inside to positive inside, followed by a repolarization to reset the resting potential. The process of an action potential from depolarization to the completion of repolarization is divided into five phases in cardiac Purkinje fibers, as shown in Figure 32.2. Phase 0 represents a rapid depolarization due to the inward current of $Na^+$. Phase 1 is associated with an immediate rapid repolarization, during which $Na^+$ inward current is inactivated and a transient $K^+$ outward current is activated, which is followed by an action potential plateau, or phase 2, which is characterized by a slowly decreasing inward $Ca^{2+}$ current and slow activation of an outward $K^+$ current. Phase 3 reflects a fast $K^+$ outward current and inactivation of the plateau $Ca^{2+}$ inward current, and phase 4 is the diastolic interval for the resetting of resting potential.

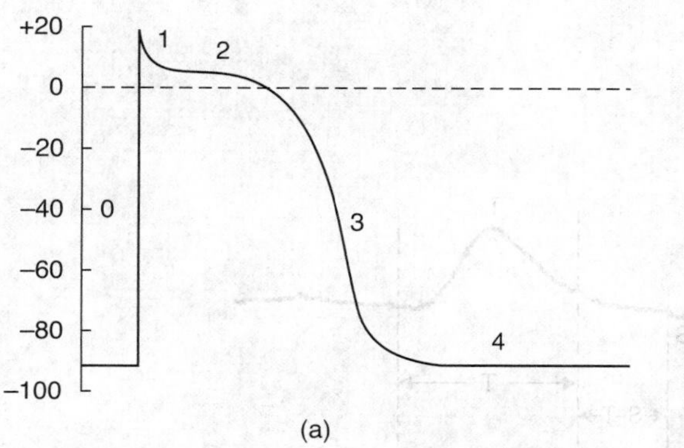


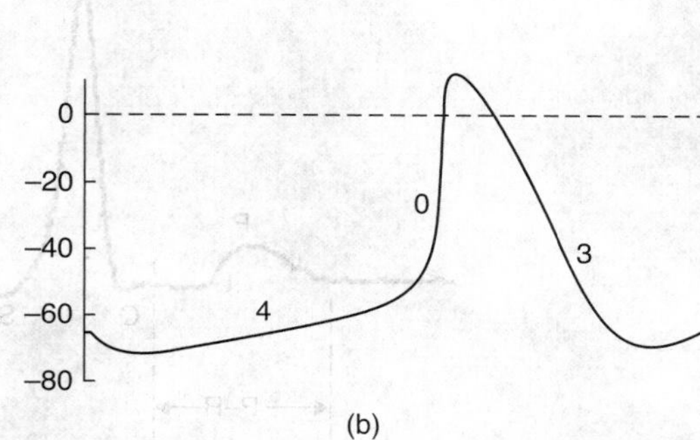


**FIGURE 32.2** Typical action potentials (in millivolts) recorded from cells in the (a) ventricle and (b) sinoatrial node. (Adapted from Hoffman, B.F. and Cranefield, P.E., *Electrophysiology of the Heart*, McGraw-Hill, New York, 1960.)

## Automaticity

Specialized cells in the heart capable of repetitively spontaneous self-excitation generate and distribute each impulse through the heart in a highly coordinated manner to control the normal heartbeat. These cells include the sinus node P cells and Purkinje fibers in the ventricles. Under normal conditions, other cells, such as atrial specialized fibers, do not have such automaticity but can become automatic under abnormal conditions. The sinus node P cells, or pacemaker cells, have only three distinct phases of action potential (Figure 32.2): phase 0, rapid depolarization; phase 3, plateau and repolarization; and phase 4, slow depolarization (often referred to as *pacemaker potential*). It is the pacemaker potential that brings the membrane potential to a level near the threshold for activation of the inward $Ca^{2+}$ current, which triggers the phase 0 rapid depolarization and makes the pacemaker cells automatic. In pacemaker cells, phase 0 is mediated almost entirely by increased conductance of $Ca^{2+}$ ions.

## Contractility

Cardiac myocytes, like other muscle cells, have the unique functional feature of contractility. Myocyte contraction occurs when an action potential causes the release of $Ca^{2+}$ from the sarcoplasmic reticulum (SR) as well as the entry of extracellular $Ca^{2+}$ into the cell. This action-potential-triggered $Ca^{2+}$ increase in the plasma and myocyte contraction is referred to as *excitation–contraction coupling*. The increase in $Ca^{2+}$ concentrations in the cell allows $Ca^{2+}$ to bind to troponin and tropomysin, leading to a conformational change in the contractile unit of the cardiac myocyte—the thin filament. This conformational change permits interaction between the actin and myosin filaments through the crossbridges (myosin heads). Adenosine triphosphate (ATP) is hydrolyzed by ATPase present in the crossbridges to release energy so the crossbridges can move in a ratchetlike fashion. This action increases the overlap of the actin and myosin filaments, resulting in shortening of the sarcomeres and contraction of the myocardium.

### Electrotonic Cell-to-Cell Coupling

Myocardium has to synchronize the contraction and relaxation of individual myocytes to perform its pump function. This is achieved by a special structural feature of cell-to-cell interaction—electrotonic cell-to-cell coupling, or the gap junction. Through the gap junction, major ionic fluxes between adjacent cardiomyocytes are spread, thus allowing electrical synchronization of contraction. Each single gap junction is composed of 12 connexin 43 (Cx43) units assembled in two hexametric connexons (hemichannels), which are contributed, one each, by the two participating cells. The connexins interact with other proteins within the cell, so connexons are not only important for cell-to-cell coupling but are also involved in cell signaling and volume regulation. An important feature of the connexon-controlled electrotonic cell-to-cell coupling is that the electrotonic current flow attenuates the differences in action potential duration of individual cardiac myocytes.

### Electrocardiograph

The electrophysiological features of cardiac myocytes and the electrotonic cell-to-cell coupling give rise the charge at any given locus in the heart with a magnitude and a direction; therefore, at any given moment in the cardiac cycle, there is a complex pattern of electrical charges across the membranes of myriad cells in the heart. The sum of all of the individual cells that exist at any given time within the heart is the *resultant cardiac vector*. Changes in the resultant cardiac vector throughout the cardiac cycle can be recorded as a *vector cardiograph*. Lead systems are used to record certain projections of the resultant cardiac vector. The potential difference between two recoding electrodes represents the projection of the vector on the line between the two leads. Components of vectors projected on such lines lack direction

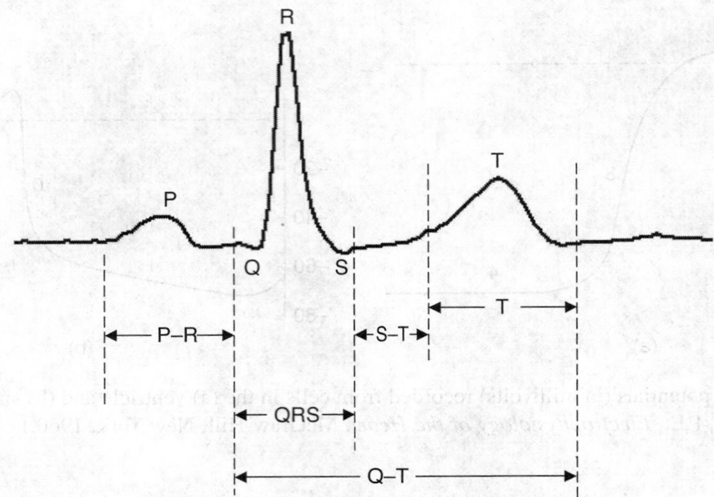


**FIGURE 32.3** Typical electrocardiogram with an illustration of important deflections and intervals.

and have a sum of magnitude, being scalar quantities; thus, a recording of changes over time in the potential differences between two points on the surface of the skin is so referred to as a *scalar electrocardiograph*.

In general, the pattern of the scalar electrocardiograph consists of P, QRS, and T waves, as shown in Figure 32.3. The PR interval is a measure of the time from the onset of atrial activation to the onset of ventricular activation. The QRS complex represents the conduction pathways through the ventricles. The ST segment is the interval during which the entire ventricular myocardium is depolarized and lies on the isoelectric line under normal conditions. The QT interval is sometimes referred to as the period of *electrical systole* of the ventricles and reflects the action potential duration. QT interval prolongation has recently been recognized as a major life-threatening factor of drug cardiac toxicity, which occurs due to a reduction of outward currents or enhanced inward currents during phase 2 and 3 of the action potential.

## Neurohormonal Regulation

Although the heartbeat is governed by the automaticity of the sinus node P cells, the neurohormonal regulation of cardiac electrophysiology and contraction is the crucial control of cardiac function under normal and abnormal conditions. Toxicants often exert their effects on the cardiac system by interfering with neurohormonal regulation, so this regulatory system is of significant relevance to cardiac toxicology. Many neurohormonal systems have significant impact on the heart. A detailed description of the regulatory system is provided in the following sections in association with specific discussion of cardiac functional regulation and compensatory and maladaptive responses to toxic exposures.

## ATP and the Heart

It is easy to understand that the energy needs of the heart are high. Chemical energy in the form of ATP is absolutely necessary to support the systolic and diastolic work of the heart [15]. In the heart, the primary ATP-utilizing reactions are catalyzed by actomyosin ATPase in the myofibril, the $Ca^{2+}$-ATPase in the sarcoplasmic reticulum, and the $Na^+$,$K^+$-ATPase in the sarcolemma [16]. ATP is also required for molecular synthesis and degradation in the heart, just as it is in other organ systems. ATP synthesis by oxidative phosphorylation in the mitochondria is usually sufficient to support the normal need for the work of the heart, even when the work output of the heart changes three- to fivefold [16]. In addition, the glycolytic pathway and the tricarboxylic acid cycle also make quantitatively small contributions to ATP synthesis. When the energetic state of the heart is analyzed, an important consideration that must be taken into account is the fact that the concentration of ATP does not define the energetic state of the heart. The amount of ATP manufactured and used at any given time is many times greater than what is found in the measurable ATP pool [16]. Thus, it is reasonable to assume that cardiac myocytes contain high concentrations of mitochondria that ensure that ATP levels remain constant through oxidation of a variety of carbon-based fuels for ATP synthesis under different conditions.

## Phosphocreatine and the Heart

A unique feature in energy metabolism of the heart is that the heart uses energy reserve systems such as phosphocreatine (PCr) to maintain a high phosphorylation potential to drive ATPase reactions under highly demanding conditions [17]. PCr is present in the heart in concentrations twice those of ATP [18]. The enzyme creatine kinase (CK) transfers the phosphoryl group between ATP and PCr at a rate about 10 times faster than the rate of ATP synthesis by oxidative phosphorylation. The reaction catalyzed by CK is PCr + ADP + $H^+$ ↔ creatine + ATP. When ATP demand exceeds ATP supply, the use of PCr is a major pathway for maintaining a constant supply of ATP. The CK reaction is also important to maintaining low adenosine diphosphate (ADP) and inorganic phosphate (Pi) concentrations, thereby retaining a high phosphorylation potential [19]. Creatine is not made in the heart but accumulates against a large concentration gradient facilitated by a creatine transporter. In the normal heart, about two thirds of the total creatine pool is phosphorylated through the CK reaction to form PCr [20,21].

### Metabolic Pathways

The continuous synthesis of ATP via mitochondrial oxidative phosphorylation is mandatory for the work of the heart [22]. Under normal conditions, the oxidation of fatty acids (FAs) is the major pathway, as it meets about 65% of the total energy demand. In contrast, the oxidation of glucose provides about 30% of the total energy demand [23,24]. In hypertrophic and failing hearts, a metabolic shift in the energy supply occurs, from its being FA dependent to glucose dependent; thus, decreased FA oxidation and increased glucose utilization in association with depressed FA deposition and increased glucose uptake are observed in hypertrophic and failing hearts [25]. This shift enhances the glycolytic pathway, thus increasing anaerobic metabolism; however, it remains a subject of debate as to whether this metabolic shift to the so-called *fetal phenotype* is adaptive or maladaptive. On the other hand, it is important to note that the shift often is only partial, and, even when the proportion of ATP synthesized from glucose increases many fold, aerobic metabolism still remains dominant [25]. With regard to cardiac toxicology, the metabolic shift is often observed with mitochondrial dysfunction. In response to toxic exposure, mitochondrial damage leads to impaired oxidative phosphorylation and a metabolic shift from aerobic to anaerobic and a reliance on glucose utilization.

## PATHOLOGIC ASSESSMENT OF CARDIAC TOXICITY

Myocardial toxic exposures lead to electrophysiological, biochemical, and morphological changes. Detection of these changes and assessment of the extent of the changes are the measurements of cardiac toxicity. The most direct examination of cardiac toxicity is the pathological assessment of the heart. These assessments include examining the gross anatomic changes of the whole heart, histopathological changes by light microscope, ultrastructural changes by electron microscope, and immunohistochemical alterations by fluorescent and confocal microscopes.

### Gross Anatomic Changes of the Heart

Anatomic examination of the whole heart is the direct and relatively easy way to observe cardiac morphological changes without the assistance of such instruments as microscopes. Changes that can be observed include those in the shape, color, and weight of the whole heart, as well as individual chamber size as measured by the weight and wall thickness. The heart weight after the heart has been removed and flushed with saline is a good index of cardiac hypertrophy. At the same time, the weights of different chambers can be obtained. Both absolute heart weight and the ratio of heart weight to body weight are used to estimate heart hypertrophy. An important note is that, at the time the heart weight is measured, the ventricular wall thickness and the chamber size must be measured. This will provide information regarding concentric or eccentric hypertrophy. A dilated heart (in particular, left-ventricle dilation) is readily observed by examining the chamber size (enlarged) and the wall thickness (thinner). In contrast, concentric hypertrophy is characterized by reduced chamber volume and increased wall thickness. The shape change of the heart can be observed when left ventricular dilation or hypertrophy occurs but the right ventricle remains normal, or *vice versa*. Atrial enlargement is also easily observed due to the shape change.

### Histopathological Changes by Light Microscopy

The most useful and simple tool for myocardial pathology study is the light microscope. In general, routine tissue section preparations and histopathological examination procedures used for other organ systems can be applied to the heart; however, the uniqueness of the heart dictates that some different procedures must also be used. For example, if a mouse model is used, a microphotograph of the whole heart can be obtained. If the observation of specific cellular components and proteins in the heart is required, the staining procedures are particularly important and are discussed later in the section on fluorescent and confocal microscopes. In brief, the procedures for histopathological observation of cardiac toxic effects involve tissue fixation, sectioning, staining, microscopic examination, microphotograph processing, and quantitatively morphometrical analysis.

For histopathological analysis, the dissected heart or a whole heart from a mouse is usually fixed in 4% paraformaldehyde in phosphate-buffered saline overnight and processed for paraffin embedding following the standard procedure for tissue embedding. The paraffin-embedded tissue block is then sectioned (about 5 μm), and the sections are stained with hematoxylin and

**TABLE 32.1**
**Criteria for Semiquantitative Analysis of Cardiomyopathy Based on Morphometric Data**

| Degree | Severity |
|---|---|
| 1 | Sarcoplasmic microvacuolizations and/or interstital or cellular edema |
| 2 | Same as 1 plus sarcoplasmic macrovacuolizations or atrophia, necrosis, fibrosis, endocardial lesions, and thrombi |
| | **Extension** |
| 0 | No lesions |
| 0.5 | Fewer than 10 single altered myocytes in the whole heart section |
| 1 | Scattered single altered myocytes |
| 2 | Scattered small groups of altered myocytes |
| 3 | Widely spread small groups of altered myocytes |
| 4 | Confluent groups of altered myocytes |
| 5 | Most cells damaged |

*Note:* The product of the severity and the extension gives the single scores of the cardiomyopathy (0 to 10) for statistical analysis.

eosin for overall structural examination. If some specific observation is required, such as myocardial fibrosis, then the staining procedure must be adjusted accordingly. Masson's trichrome stains collagen in the tissue section blue, and Sirius red stains the collagen red. Oil red O stains lipid droplets red. All of these staining procedures and others have been successfully used for cardiac toxicological studies. Under the light microscope, a histopathological examination focuses on intracellular organelle changes and nuclear and plasma membrane abnormalities. These changes can be graded according to their severity and extension. Based on the degree of severity and extension, a scoring system can be used as shown in Table 32.1. In this system, the degree of severity is divided into two categories, depending on the involvement of intracellular organelles (e.g., microvacuolization involves fewer organelles and macrovacuolization involves more organelles). Extension is concerned with how many cardiac myocytes are involved in the injury at different degrees of severity. The combination of severity and extension (severity × extension) produces a single score; for example, if the severity is 1 and the extension is 2 then the overall score is $1 \times 2 = 2$, and if the severity is 2 and the extension is 2 then the overall score is 4. This semiquantitative morphometrical analysis has been used for many cardiac toxicity studies [26–28].

For a better observation of the histopathological changes of the heart, a semi-ultrathin sectioning of the heart tissue should be used. The heart is fixed following the procedure for the ultrathin tissue sections for electron microscopic examination, as described below. The tissue block is then sliced by an ultratome into 0.5-μm sections and examined by light microscope. This procedure generates high-resolution images of the histopathological changes.

## Ultrastructural Changes by Electron Microscopy

To define detailed changes in intracellular organelles, particularly with regard to the cardiac toxicological concern of mitochondrial damage, electron microscopic examination is a gold standard procedure. Routine procedures for electron microscopic examination of subcellular ultrastructural changes can be found in standard textbooks of electron microscopy. With regard to the concern of cardiac toxicology, or cardiac ultrastructural examination in general, the tissue preservation and fixation procedures have to be undertaken with a special care. When a mouse or a rat model is used, an *in situ* perfusion and fixation procedure must be used to preserve the integrity of the ultrastructures of the heart, as described below.

Briefly, the hearts of anesthetized mice are perfused *in situ* as follows: A cannulated needle (25-gauge, 3/8-in., silastic medical grade, 0.02-in. ID, and 0.037-in. OD) is inserted about 3 mm into the cardiac apex. The heart is perfused with wash-out solution (6.6 g paraformaldehyde dissolved in 500 mL distilled water; 1.6 mL 50% glutaraldehyde; 360 mL 0.2-*N* sodium cacodylate buffer, pH 7.2 to 7.4; 0.5 g $CaCl_2$; 0.25 g $AlCl_3$; 1.0 g procaine hydrochloride; plus water to make a final volume of 1000 mL at room temperature) at a flow rate of 5 mL/min for 2 to 3 minutes until the wash-out fluid became clear. This is followed by Karnovsky's fixative (10 g paraformaldehyde dissolved in 500 mL distilled water; 50.0 mL 25% glutaraldehyde; 360 mL of 0.2-*N* sodium cacodylate buffer, pH 7.2 to 7.4; 0.5 g $CaCl_2$; 0.25 g $AlCl_3$; plus water to make final volume of 1000 mL) at the rate of 3 mL/min for 10 minutes at 4°C. The heart is then removed and immediately immersed in precooled (4°C) Karnovsky's fixative. To each tissue vial is added 2 mL of precooled Karnovsky's fixative (about 10-fold volumes of tissue blocks). A piece of dental wax is placed onto a Petri dish, and adequate fixative is added to the tissue. The left ventricle, right ventricle, and interventricular septum are dissected and cut into 1-mm$^3$ pieces with a razor blade. Each block is then placed into a vial filled with precooled Karnovsky's fixative at 4°C for 90 minutes and rinsed 3 times with 0.2-*N* sodium cacodylate buffer (pH 7.4) and postfixed with 1% ostium tetroxide at 4°C in the dark for 90 minutes (osmium solution is mixed: 1 part of 2% osmium tetroxide to 1 part of s-collidine buffer at pH 7.4). The block is rinsed with distilled $H_2O$, dehydrated with ethanol and propylene oxide, and processed following a routine procedure of ultrathin sectioning and examination. This *in situ* perfusion and fixation procedure is highly recommended for cardiac ultrastructural studies of mice and rats.

### Immunohistochemical Alterations by Fluorescent and Confocal Microscopy

The application of fluorescent and confocal microscopes in cardiac toxicity studies greatly enhances the capacity of experimental approaches and data acquisition. These procedures are often applied to determine the dynamic changes of cellular molecules in cardiac cells through a multiple staining procedure using specific antibodies against various proteins of interest. Molecular interaction and subcellular localization are readily observed using confocal microscope in association with different molecular fluorescent probes. Because the immunohistochemical assay involves the reaction of antibodies, cryostat heart sections are required. The cryostat tissue block can be sectioned into 5-μm slices, air dried, and fixed in acetone for 20 minutes at –20°C for specific antibody staining procedures. It is noteworthy that much of the immunohistochemical assay is associated with immunoperoxidase; therefore, endogenous peroxidase activity must often be quenched by incubating tissue sections in 3% $H_2O_2$. Nonspecific binding sites must be blocked by using normal goat serum. Sections are then incubated with polyclonal or monoclonal antibodies overnight at 4°C, followed by incubation with horseradish-peroxidase-conjugated anti-IgG antibody. The antibody binding sites are visualized by incubation with diaminobenzidine–$H_2O_2$ solution. If a double immunofluorescence staining is required, the tissue section is first stained with a specific antibody conjugated to a fluorescent probe. The same section is then stained by another antibody conjugated to a different fluorescent probe. The overlap of the labels with different fluorescents under the microscope will reveal the interaction or colocalization of different molecules in the cell.

## MYOCARDIAL CELL DEATH AND EVALUATION

### Apoptosis and Necrosis

Toxic insults trigger a series of reactions in cardiac cells leading to measurable changes. Mild injuries can be repaired; however, severe injuries will lead to cell death via apoptosis and necrosis. If the cell survives the insults, structural and functional adaptations will take place. *Apoptosis* was found to be involved in cardiomyopathy in 1994 [29]. The loss of cardiac myocytes is a fundamental part of myocardial injury that initiates or aggravates cardiomyopathy. Apoptosis is an important mode of myocardial cell loss, as has been demonstrated in heart failure patients [30]. Myocardial apoptosis has been shown to play an important role in the cardiac toxic effects induced by Adriamycin® [31,32], an important anticancer agent for which has limited clinical application due to its major side effect, cardiotoxicity. Exposure of primary cultures of cardiomyocytes to cadmium also induces apoptosis [33].

Many *in vivo* studies have shown that only a very small percentage of myocardial cell population undergo apoptosis under pathological conditions; for example, less than 0.5% of cells appeared apoptotic in myocardial tissue under the stress of dietary copper deficiency in mice [34]. At first glance, this number seems to be too insignificant to account for myocardial pathogenesis; however, this would be a false assessment. In a carefully designed time course study [35], it has been estimated that cardiomyocyte apoptosis may be completed in less than 20 hours in rats. Because the heart is a terminally differentiated organ, myocytes undergoing apoptosis will be lost; thus, the total cell loss can simply be accounted for by the rate of apoptosis plus necrosis. If apoptosis occurs at a constant rate of about 0.5% myocytes a day [34], the potential contribution of apoptosis to the overall loss of myocytes over a long period of time is significant.

*Necrosis* is a term that was widely used to describe myocardial cell death in the past. Myocardial infarction, in particular, had been considered as a consequence of necrosis [36]. Although it is now recognized that apoptosis contributes significantly to myocardial infarction [37], the significance of necrosis in myocardial pathogenesis cannot be underestimated. The contribution of necrosis to cardiomyopathy induced by environmental toxicants and pollutants is particularly important. A critical issue is how to distinguish apoptosis from necrosis.

Apoptosis and necrosis were originally described as two distinct forms of cell death that can be clearly distinguished [38]; however, these two modes of cell death can occur simultaneously in tissues and cultured cells. The intensity and duration of insults may decide the outcome; thus, triggering events can be common for both types of cell death. A downstream controller, however, may direct cells toward a programmed execution of apoptosis. If the apoptotic program is aborted before this control point and the initiating stimulus is severe, cell death may occur by necrosis [39]. Alternatively, in acute injury, apoptotic cells can progress along a continuum to eventual necrosis. To distinguish apoptosis from necrosis, more specific oligonucleotide probes have been developed to allow the recognition of different aspects of DNA damage [40]. They have been successfully applied, in combination with confocal microscopy, to identify apoptotic and necrotic cell death in the heart with different pathogenic challenges.

### Single-Strand DNA Breaks

A monoclonal mouse anti-ssDNA antibody has been developed that is specifically reactive with ssDNA but does not recognize dsDNA. An immunohistochemical assay for detection of ssDNA using this antibody in combination with a TUNEL (terminal-deoxynucleotidyl-transferase-mediated dUTP nick end labeling) assay can distinguish repairable ssDNA breaks from apoptotic DNA damage in the heart.

### Apoptotic DNA Damage

The endproducts of apoptotic DNA damage are fragments of double-strand DNA cleavage with a 3′ overhang [40] that can be specifically identified by *Tag*-polymerase-generated probes [40]. The specificity of this molecular probe to identify apoptosis has been confirmed by other methods such as dual labeling of terminal deoxynucleotidyltransferase (TdT) and caspase-3 [41]. In addition, this apoptotic-specific probe in combination with fluorescence labeling of different cellular components allows quantitative evaluation of apoptotic cells and the possibility of identifying the origin of the apoptotic cells, such as myocytes (stained with α-sarcomeric actin), endothelial cells (stained with factor VIII), and fibroblasts (stained with vimentin), in the heart [42].

### Necrotic DNA Damage

Double-strand DNA cleavage with a blunt end is characteristic of necrotic DNA damage. This is because, during necrosis, the release of lysosomal proteases degrades histones, resulting in the loss of DNA protection and exposure to endonucleases and exonucleases. Endonucleases produce double-strand DNA cleavage with 3′ overhangs, but exonucleases remove terminal nucleotides, giving the damaged DNA a blunt end. A probe generated by *pfu* polymerase can specifically recognize blunt-end DNA [42]. Its specific reaction with necrotic DNA has been confirmed by other methods, such as by observing the permeability of myosin antibody into necrotic cells [43] and disruption of the sarcolemma by vinculin staining, which can clearly define the continuity of the sarcolemmal surface [44].

### Quantitation of Apoptotic and Necrotic Cell Death

The proportion of apoptotic and necrotic cell death in the heart can be estimated by using a combination of the above procedures. First, a conventional TUNEL procedure can be used to identify the total TUNEL-positive cells. Second, the procedure to identify double-stranded, blunt-end DNA breaks can be used to quantify the proportion of necrotic cells in the total TUNEL-positive population. Finally, combining the procedure to identify double-stranded DNA breaks with 3′ overhangs with specific antibodies to identify total ssDNA breaks can distinguish the proportion of apoptotic cells from those with ssDNA breaks only.

### Myocyte vs. Nonmyocyte Apoptosis

Distinguishing apoptotic myocytes from nonmyocytes in the myocardium is another issue in the study of cardiac toxicology. An *in situ* TUNEL assay in combination with a dual immunohistochemical detection of α-sarcomeric actin has been used to distinguish apoptotic myocytes from nonmyocytes [18]. Apoptotic myocytes are dually stained by TUNEL and α-sarcomeric actin, and apoptotic nonmyocytes are stained only by TUNEL. Another procedure is immuno-gold TUNEL and electron microscopic examination of the apoptotic cells [31]. It is well known that the gold standard for identification of apoptotic cells is morphological examination by electron microscopy. The immuno-gold TUNEL and electron microscopic procedure defines cell types and the morphological characteristics of apoptotic cells.

## EVALUATION OF CARDIAC HYPERTROPHY AND HEART FAILURE

The pathological assessments discussed above provide information regarding cardiac hypertrophy by measuring the heart weight, dimensions, and cardiomyocyte enlargement and by microscopic examination. The most convenient method is to examine the ratio of heart weight to body weight or heart weight to tibia bone length in animal studies. In addition, molecular markers help define cardiac hypertrophy by examining the activation of fetal gene programs, such as increases in the expression of cardiac β-myosin heavy-chain protein and skeletal α-actin. The combination of these measurements can evaluate cardiac hypertrophy, but all of these measurements are invasive. The challenge is how to obtain the same information noninvasively. Echocardiography is one of the most important developments for cardiac disease diagnosis. Echocardiograms provide noninvasive and reliable information regarding cardiac mass change and functional alteration. With regard to heart failure, pathological assessments cannot make such determinations, and functional assessment is required.

*Myocardial adaptation* refers to the general process by which the ventricular myocardium changes in structure and function. This process is often referred to as *remodeling*. During maturation, myocardial remodeling is a normal feature that is a useful adaptation to increased demands; however, in response to pathological stimuli such as exposure to environmental toxicants, myocardial remodeling is adaptive in the short term but maladaptive in the long term and often results in further myocardial dysfunction. The central feature of myocardial remodeling is an increase in myocardial mass associated with a change in the shape of the ventricle [45].

At the cellular level, the increase in myocardial mass is reflected by cardiac myocyte hypertrophy, which is characterized by enhanced protein synthesis, heightened organization of the sarcomeres, and eventual increases in cell size. At the molecular level, the phenotype changes in cardiac myocytes are associated with reintroduction of the so-called *fetal gene program*, characterized by the patterns of gene expression mimicking those seen during embryonic development. These cellular and molecular changes are observed in both adaptive and maladaptive responses, making it difficult to distinguish between adaptive responses and maladaptive responses in cardiac toxicological studies.

## Physiological Hypertrophic Response

Hypertrophy of the heart can be physiological or pathological. Physiological hypertrophy is considered to be an adaptive response accompanied by an adjustment of cardiac function with an increased demand of cardiac output. Such an adaptive hypertrophy is observed in the increase in cardiac mass after birth and in response to exercise. A biochemical distinction of adaptive hypertrophy is that myocardial accumulation of collagen does not accompany the hypertrophy. Functionally, the increased mass is associated with enhanced contractility and cardiac output. In response to toxicological stresses, the heart also often increases its mass, which was viewed in the past as an adaptive response as well; however, evidence obtained from studies using more advanced technologies in human patients and animal models suggests that cardiac hypertrophy is a maladaptive process of the heart in response to intrinsic and extrinsic stresses.

## Pathological Hypertrophic Response

Although toxic-stress-induced hypertrophy can normalize wall tension, it is a risk factor for sudden cardiac death and has a high potential to progress to overt heart failure. A distinction between physiologic and pathologic hypertrophy is whether or not the hypertrophy is necessary for the compensatory function of the heart under physiologic and pathologic stress conditions. Many studies using genetically manipulated mouse models, either gain of function or loss of function, have supported the hypothesis that cardiac hypertrophy is neither required nor necessarily compensatory. For example, forced expression of a dominant negative calcineurin mutant confers protection against hypertrophy and fibrosis after abdominal aortic construction [46]. Also, the elimination of hypertrophy in animals by calcineurin suppression did not cause compromised hemodynamic changes for an observation period of several weeks [47]. In these experimental approaches, hypertrophic growth could be abolished in the presence of continuous pressure overload, but the compensatory response could not be compromised. An interesting observation is that an almost complete lack of cardiac hypertrophy in response to aortic banding in a transgenic mouse model was accompanied by a significant slower pace of deterioration of systolic function [48]. These observations indicate that cardiac hypertrophy in response to extrinsic and intrinsic stress is not a compensatory response, but hypertrophy puts the heart at high risk for malignant arrhythmia and heart failure so it is now viewed as a maladaptive response.

## Heart Failure

Pathological hypertrophy is a risk factor for malignant arrhythmia and heart failure. The link between heart hypertrophy and malignant arrhythmia is discussed in the next section. The transition from cardiac hypertrophy to heart failure is an important distinction between adaptive and maladaptive cardiac hypertrophy or between physiological and pathological hypertrophy. The critical cellular event of this transition is myocardial apoptosis triggered by inflammatory cytokines such as TNF$\alpha$ and neurohormonal factors such as ANP, leading to dilated cardiomyopathy and deterioration of cardiac function. Toxicological exposures may cause dilated cardiomyopathy or heart failure without an intermediate hypertrophic stage. Myocardial cell death also plays an essential role in the direct cardiac dilation pathogenesis. The manifestation of heart failure is cardiac dysfunction. Although many parameters can be clinically and experimentally employed to define heart failure, the overall definition of heart failure is the inability of the heart to maintain cardiac output sufficient to meet the metabolic and oxygen demands of peripheral tissues. To specifically define diastolic or systolic cardiac failure so clinical treatment and management can be more specific, measurements to define diastolic vs. systolic function must be implemented. In animal studies, cardiac failure is often defined by noninvasive echocardiography, along with invasive measurement of cardiac hemodynamic changes.

## Echocardiography

An important and widely used noninvasive technique in cardiology is ultrasound. Ultrasound (millions of cycles per second) has the ability to detect the position of both stationary and moving structures within the body. Ultrasound can determine the depth and position of echoes returned from the body and accurately record the motion of structures over a period of time. Because the movement of various cardiac components is related to function and dysfunction of the heart and the patterns of changes are consistent with specific cardiac diseases, the heart and great vessels are suitable for ultrasound examination. The transmission of ultrasound through the heart, with detection of the returning echoes detailing the position and movement of the cardiac acoustic interfaces, is referred to as *cardiac echocardiography*.

The basic ultrasonic equipment currently used in clinical and experimental studies is an echocardiograph, which consists of an oscilloscope, a transducer, and a video recorder. In studies using animal models such as the mouse or rat, anesthesia must be administered before acquisition of cardiac ultrasound recordings using the echocardiograph. Cardiac ultrasound imaging can be obtained by using a high-frequency, 15-MHz linear transducer at a maximum frame rate of 120 frames/second in the mouse model. Parasternal long- and short-axis views can be obtained after adjusting gain settings from optimal epicardial and endocardial wall visualizations. Left-ventricle, M-mode tracings can be obtained from the short-axis view. From an unconventional, more superior parasternal long-axis view, ascending aorta two-dimensionally targeted M-modes can be recorded. Ascending aortic pulse wave Doppler velocities can be obtained from the suprasternal window by means of a pediatric short focal length, 12-MHz phased array transducer. Echocardiographic loops of ≥20 cardiac cycles containing the above-mentioned two-dimensional data, M-mode tracings, and Doppler velocity panels can be stored digitally for offline analysis.

From short-axis views, epicardial and endocardial left-ventricle areas can be measured offline at the end systole and end diastole. In general, images are considered sufficient for measurement when >75% of the epicardial and endocardial contour can be adequately visualized. The American Society of Echocardiography recommends that the short-axis endocardial boarder be traced on the innermost endocardial edge, whereas the epicardial boarder should be traced along the first bright pixel immediately adjacent to the darker myocardium. The left-ventricle length, defined as the distance between the apex and the midmitral annulus, can be obtained from the parasternal long-axis views in which the mitreal annular plane and apex are well defined. The left-ventricle measurements must be made from at least three cardiac cycles at both the end systole and end diastole. Left-ventricle mass is calculated using the following formula: Left ventricle mass = $[1.05(5/6)A_1(L + t) - (5/6)A_2L]$, where 1.05 is the specific gravity of muscle; $A_1$ and $A_2$ are the epicardial and endocardial parasternal short-axis area, respectively; $L$ is the parasternal long-axis length; and $t$ is the wall thickness calculated from $A_1$ and $A_2$.

Two-dimensionally targeted M-mode echocardiographic images can be obtained at the level of the papillary muscles from the parasternal short-axis view and recorded at a speed of 150 cm/sec. Left-ventricle internal diameters and wall thickness can be obtained at the end systole and end diastole from cross-sectional short-axis views. The shortening fraction (SF), the echocardiographic equivalent of the ejection fraction, is calculated using the following formula: $\%SF = [(LVedd - LVeds)/(LVedd)] \times 100$, where LVeds and LVedd are the left ventricle internal diameter at end systole and end diastole, respectively. Measurement of the shortening fraction can indicate functional changes of the heart; however, more precise functional analysis of the heart can be obtained by invasive analysis.

## Cardiac Functional Assessment

Measurements of heart performance can be obtained by using a hemodynamic analysis. This is an invasive procedure, and animals such as the mouse and rat must be anesthetized during measurement. The instrumentation for this measurement consists of a computerized oscilloscope, a transducer, and digital analyzers for the different digitalized signals obtained from recordings. Surgical procedures include the following: A midline incision (1 to 2 cm) in the throat area external to the trachea is made and the right and left sternohyoid muscles are pulled apart using forceps with serrated tips. A small opening in the trachea is made for the insertion of a PE-100 catheter to ensure a patent airway. The catheter is cut to a length of the animal's trachea to maintain normal resistance to airflow. The common right carotid artery, which is buried under the sternocleidomastoid muscle immediately adjacent to the sternohyoid muscles of the tracheae, is then isolated. The common artery is tied on one end at the branch point of the internal and external parts (toward the head) to prevent the backflow of blood from the peripheral vasculature of the head. The other end of the artery (toward the body) is clamped using a vascular clamp to occlude blood flow from the heart. A small incision is then made in the artery for the insertion of a hand-stretched PE-50 catheter. The catheter preparation and transducer connection are then made ready. The catheter is slowly advanced through the common carotid artery, the ascending aorta, and into the left ventricle. When a waveform characteristic of ventricular pressure is achieved, the wound is covered to minimize liquid evaporation. The animal is then allowed to stabilize for 20 to 30 minutes before recording the waveform for up to 2 hours. At the end of each experiment, the chest is opened to confirm the presence of the catheter inside the left ventricle.

### Transducer Connection

The PE-50 catheter to be inserted into the carotid artery must first be prepared by soaking it in heparanized saline for 24 hours. This prevents clot formation on the outside of the catheter and facilitates its insertion into the artery. One end of the catheter is cut to produce a beveled end for insertion into the artery. The transducer is prepared by placing a three-way stopcock on both ends. The stopcock and the transducer are filled with normal saline. A needle for insertion into the catheter is prepared by cutting the sharp end off. The resulting needle is only 1.0 cm long. The needle hub is inserted into the blunt end of the catheter and is connected on the other end to the three-way stopcock.

### Stress Test for Myocardial Function

Isoproterenol is a reagent used to generate β-adrenergic stimulation of the left ventricle function. The right femoral vein is cannulated with PE-10 tubing and connected to a Harvard Apparatus/22 microinjection pump for the infusion of isoproterenol. After a 30- to 45-minute stabilization period, isoproterenol is infused for 1 minute at a rate of 0.02 mL/min (1.6 ng isoproterenol per minute per g body weight). Myocardial functional changes in response to isoproterenol are recorded immediately after the infusion for 20 to 30 minutes, at which time myocardial functions are recovered to baseline.

The parameters to be obtained by the hemodynamic analysis include the end diastolic pressure (EDP), a measurement of ventricular wall compliance; ventricle minimum diastolic pressure (VMDP); maximum ventricular dP/dt (MAX dP/dt), an index for the mechanical ability of the heart to generate force for ejection of blood from the ventricle; maximum ventricular negative dP/dt (MAX –dP/dt); heart rate (HR); duration of contraction (DCON); time constant of pressure decay (tau); ventricular peak systolic pressure (VPSP); and duration of 1/2 relaxation (1/2R).

The maximum rate of rise in intraventricular pressure during ventricular contraction (MAX dP/dt) is an index frequently used to assess the mechanical ability of the heart to generate force for the ejection of blood from the ventricle. This parameter is increased with inotropic intervention such as treatment with digitalis glycosides, as well as an increase in heart rate that reflects the augmented level of contractility associated with tachycardia [31]. Furthermore, an increase in preload (elevated end-diastolic volume and pressure) and afterload (increased aortic diastolic pressure) increases MAX dP/dt [31]. Under normal physiological conditions, the heart responds to β-adrenergic stimulation by increasing heart rate in addition to an increase in MAX dP/dt. If the MAX dP/dt does not increase in response to β-adrenergic stimulation, it suggests some degree of heart failure. The MAX dP/dt is thus an indicator for systolic function of the heart. The changes in this parameter together with the VPSP provide a complementary indication of cardiac systolic dysfunction.

Diastolic dysfunction is characterized by slowed or incomplete relaxation, abnormal left ventricle filling, and altered passive elastic properties of the heart [49,50]. The increase in the time constant of pressure decay (tau) and changes in MAX –dP/dt are indicative of diastolic dysfunction. The end diastolic pressure is a measurement of ventricular wall compliance; if the EDP increases, it suggests either increased peripheral blood pressure or stiffness of the myocardial muscle due to cardiomyopathy.

## QT PROLONGATION AND SUDDEN CARDIAC DEATH

The recognition of QT prolongation and its associated adverse effects on the heart has been a major focus in drug discovery and development and environmental cardiac toxicity in the last decade. Many cardiac and noncardiac drugs have been found to cause QT prolongation and torsade de pointes (TdP) and have been removed from the market or relabeled for restricted use. It has been known for a long time that quinidine causes sudden cardiac death; however, the severe and lethal side effect of QT prolongation did not attract sufficient attention until the last decade due to a lack of knowledge and experimental approaches available to develop a comprehensive understanding of QT prolongation. QT prolongation is now better understood, and new regulatory guidelines for a battery of preclinical tests to assess new drugs for potential QT liability in humans have been recommended.

### Definition of QT Prolongation

A simple definition of QT prolongation is that the length of the QT interval observed on a typical electrocardiogram is prolonged. Clinically, long QT syndrome is defined when the QT interval is longer than 460 msec; however, TdP occurs with an average increase in the QT interval of approximately 200 msec (a normal QT interval is about 300 msec). A human study has found that TdP did not occur with a QT interval shorter than 500 msec in the cases studied [51]. In general, the long QT syndrome can be divided into two classes: congenital and acquired. Congenital long QT syndrome is rare, and the acquired syndrome is the major concern of drug cardiac toxicity in pharmaceutical discovery and development.

## Molecular Basis of QT Prolongation

Prolongation of QT intervals on electrocardiograms is caused by prolongation of the action potential of ventricular myocytes. For the cardiac action potential, phase 0 represents depolarization of myocytes (Figure 32.2), and the depolarization of all ventricular myocytes is measurable as the QRS complex on the electrocardiogram (Figure 32.3). Phase 1 of the cardiac action potential is recognized as a partial repolarization of the membrane due to inactivation of cardiac sodium channels and activation of transit outward potassium channels. Phase 2 of the action potential is generated primarily by slowly decreasing inward calcium currents through L-type calcium channels and gradually increasing outward currents through several types of potassium channels. This phase is sensitive to small changes in ion currents and is a critical determinant of the duration of the action potential. At this point, the cardiac cycle of the electrocardiogram has returned to baseline. Phase 3 of the cardiac action potential represents myocardial cell repolarization due to outward potassium currents. Two critical potassium channels terminate the plateau phase (phase 2) and initiate the final repolarization (phase 3): $I_{kr}$, which is the rapidly activating delayed rectifier potassium current, and $I_{ks}$, which is the slowly activating delayed rectifier potassium current. The repolarization phase correlates with the T wave on the electrocardiogram; therefore, the duration of the QT interval is related to the length of ventricular action potentials.

A reduction in net outward current and an increase in inward current are potential contributors to the prolongation of cardiac action potential, as reflected by QT prolongation on the electrocardiogram. Although many channels are potentially involved in the prolongation of the cardiac action potential, current studies have identified three important channels that play a critical role in the plateau phase (phase 2) of the cardiac action potential: sodium inward channels and potassium outward channels ($I_{kr}$ and $I_{ks}$).

*Sodium channel dysfunction* in congenital long QT syndrome is related to mutations in the *SCN5A* gene that encodes the α subunits of sodium channels. Mutational analyses have found 14 distinct mutations of *SCN5A* associated with long QT syndrome [52]. It has been hypothesized that gain-of-function mutations in *SCN5A* would cause long QT syndrome because reopening of the sodium channels during the plateau phase of action potential, even in a small inward current, would lengthen the duration of the cardiac action potential. Sodium channel inactivation immediately following depolarization (phase 1) is important for the transition to phase 2 of the action potential. A mutation of *SCN5A* has been found to destabilize the inactivation gate [53]. Activation of these mutant sodium channels is normal, and the rate of inactivation appears slightly faster than normal, but these mutant channels can reopen during the plateau phase of the action potential, leading to a prolonged plateau phase.

The $I_{kr}$ potassium channels critically affect the length of the plateau phase of the cardiac action potential. The human *ether-a-go-go*-related gene (*hERG*) is expressed primarily in the heart and encodes the α-subunit of the cardiac $I_{kr}$ potassium channel. The 94 mutations of *hERG* that have been identified represent 45% of the total number of mutations related to long QT syndrome to date [52]. The *hERG* α-subunits assemble with MiRP1 β-subunits to form cardiac $I_{kr}$ channels. The $I_{kr}$ potassium channel is one of the two channels primarily responsible for termination of the plateau phase of the action potential. During the repolarization of the action potential, the $I_{kr}$ channels open, resulting in an increase in the magnitude of $I_{kr}$ current during the first half of phase 3 repolarization. Many *hERG* mutations occur around the membrane-spanning domains and the pore region of the channel. Most of these mutations have a loss-of-function effect, and many long-QT-syndrome-associated mutations in *hERG* are missense mutations, which lead to a dominant negative effect on $I_{kr}$ channels because the functional $I_{kr}$ potassium channels are composed of heteromultimers, including several *hERG* subunits. Therefore, the loss-of-function mutations in *hERG* make a critical contribution to the long QT syndrome due to the prolonged plateau phase of cardiac potential.

The $I_{ks}$ potassium channel is the other one of the two channels primarily responsible for the termination of the plateau phase of the action potential. The $I_{ks}$ potassium channel is assembled from *KVLQT1* α-subunits and the *minK* β-subunits. Two molecular mechanisms possibly account for reduced *KVLQT1* function in the long QT syndrome [54]. First, intragenic deletions of one *KVLQT1* allele result in synthesis of abnormal α-subunits that do not assemble with normal subunits, leading to a 50% reduction in the number of the functional channels. Second, missense mutations result in synthesis of *KVLQT1* subunits with structural abnormalities that can assemble with normal subunits. Channels formed from the mutant *KVLQT1* subunits have reduced or no function. Both of these mutations result in a dominant negative effect. Interestingly, both *KVLQT1* and *minK* are expressed in the inner ear, where the channels function to produce a potassium-rich fluid known as *endolymph* that bathes the organ of Corti, the cochlear organ responsible for hearing. Individuals with Jervell and Lange–Nielsen syndrome have homozygous mutations of *KVLQT1* or *minK* and no functional $I_{ks}$ channels. These individuals have severe arrhythmia susceptibility and congenital neural deafness.

In summary, the molecular basis of QT prolongation observed from electrocardiogram is the prolongation of cardiac action potential. In this regard, the inward sodium channels and outward potassium channels play important roles in increasing the length of the plateau phase of action potential. Congenital long QT syndrome is related to gain-of-function mutations in sodium channels or loss-of-function mutations in potassium channels. Acquired long QT syndrome is also related to altered function of these channels; however, many other factors affect the phenotype of long QT syndrome and the clinical manifestations.

## Torsade de Pointes and Sudden Cardiac Death

The abnormalities of different channels in different regions of the heart at varying levels result in channel dysfunction with regional variability. The regional abnormalities of cardiac repolarization or conductance provide a substrate for arrhythmia. Under these conditions, arrhythmia is induced if a trigger mechanism is implanted. The trigger for arrhythmia in the long QT syndrome is believed to be spontaneous secondary depolarization that arises during or just following the action potential plateau. This small action potential is referred to as *early afterdepolarization*, which occurs preferentially in M cells and Purkinje cells due to reactivation of the L-type calcium channels or activation of the sodium–calcium exchange current. When the spontaneous depolarization is accompanied by a marked increase in dispersion of repolarization, the likelihood of an arrhythmia being triggered is increased. Once triggered, the arrhythmia is maintained by a regenerative circuit of electrical activity around relatively unexcitable tissue, a phenomenon known as *reentry*. The development of multiple reentrant circuits within the heart causes ventricular arrhythmia, or TdP, leading to sudden cardiac death.

## Parameters Affecting QT Prolongation and Torsadogenesis

Alterations in the function of cardiac channels, or cardiac channelopathies, occur at the cellular level; however, electrotonic cell-to-cell coupling influences the dispersion of repolarization. If myocardial cells with intrinsically different durations of action potential are well coupled, electrotonic current flow attenuates differences in the action potential duration. Many factors affect the clinical manifestations of QT prolongation and torsadogenesis. Genetic polymorphisms and female gender are two distinct risk factors. The mechanism of the polymorphisms and the rationale for the high susceptibility of females to QT prolongation and torsadogenesis have yet to be determined. The following factors are better characterized.

### Drugs and Environmental Toxicants

Drug-induced QT prolongation is a major acquired long QT syndrome. Selective blockers of potassium channels, including class III antiarrhythmics, have been developed for the treatment of various atrial arrhythmias; however, these drugs predictably produce the long QT syndrome, which is sufficient to cause TdP in 5 to 7% of recipients. Environmental exposure to particulate matter pollution in air is a risk factor for QT prolongation in the elderly, children, and individuals with compromised hearts.

### Disturbances in Ion Homeostasis

Hypokalemia in combination with torsadogenic drugs is a well-recognized risk factor for QT prolongation and TdP. It is also shown that sodium supplementation can diminish the long QT syndrome due to the gain-of-function mutations in sodium channels. Stress-induced $Ca^{2+}$ overload in myocardial cells increases the likelihood of arrhythmia. The electrode imbalance exerts a greater effect on compromised hearts.

### Abnormal Gap Junction

Gap-junction-mediated intercellular communication is essential in the propagation of electrical impulse in the heart. The gap junction is composed of connexons, as described earlier. Under normal conditions, the gap junction electrotonic current flow attenuates the differences in action potential duration of myocardial cells. Toxicological exposures cause damage to connexons, leading to disruption of electrotonic cell-to-cell coupling; thus, differences in the action potential duration would be significant, particularly under the influence of torsadogenic drugs or conditions.

### Myocardial Ischemic Injury

Acute myocardial ischemia can cause immediate arrhythmia due to disturbance in ionic homeostasis, which is often transient; however, acute ischemia induces myocardial infarction that can lead to blockage of cardiac conductance. Under conditions of myocardial infarction, the areas separated by the scar tissue would be uncoupled, making differences in the duration of the action potentials of myocardial cells in various regions apparent. The infarct heart thus is more susceptible to drug-induced QT prolongation and TdP.

### Cardiac Hypertrophy

Purkinje fibers are derived from myogenic precursors during embryonic development. The normal distribution of Purkinje fibers in the myocardium is proportional to the mass of the heart. Cardiac hypertrophy resulting from the hypertrophic growth of cardiac myocytes would lead to unbalanced distribution of Purkinje fibers in the remodeling heart. The conduction of pacemaker potentials would thus be interrupted.

### Myocardial Fibrosis

Dilated cardiomyopathy in alcoholics often involves myocardial fibrosis, which simulates the effect of myocardial infarction on electrical conduction in the heart and blockage of cardiac conductance.

### Heart Failure

Most individuals with failing hearts die suddenly of cardiac arrhythmias. Heart failure presents a common, acquired form of the long QT syndrome. In human heart failure, selective downregulation of two potassium channels ( $I_{to1}$ and $I_{k1}$) has been shown to be involved in action potential prolongation. The $I_{to1}$ current is involved in phase 1 of action potential and opposes the depolarization. The increase in depolarization may be adaptive in the short term because it provides more time for excitation–contraction coupling, thus mitigating the decrease in cardiac output. However, downregulation of potassium channels becomes maladaptive in the long term because it predisposes the individual to early afterdepolarization, inhomogeneous repolarization, and polymorphic ventricular tachycardia.

## BIOMARKERS: APPLICATION AND LIMITATION

Myocardial injury can be divided into two major classes: structural and nonstructural injuries. The structural damage of the heart includes cell death and the associated histopathological changes such as myocardial infarction. Functional deficits often accompany the structural injury. Nonstructural damage represents functional deficits without apparent structural alterations. Myocardial adaptation to intrinsic and extrinsic stress leading to myocardial structural changes such as hypertrophy should be in the category of structural damage because the progression of hypertrophy leads to heart failure in which cell death is a major determining factor. Myocardial structural changes and functional alterations can be indirectly measured by echocardiography and electrocardiograms in combination with stress testing. The data generated from these measurements can be considered in a broad sense as biomarkers; however, in clinical practice and experimental approach, biomarkers are considered to be indexes of myocardial injury measured from blood samples. The fundamental principle of the biomarkers is that molecules that are released from the myocardium under various injury conditions are readily detectable from blood samples.

### Validation of Biomarkers

For a biomarker to be indicative of myocardial damage, an important question to address is what characterizes a valid biomarker. In 2000, an Expert Working Group (EWG) on biomarkers of drug-induced cardiac toxicity was established under the Advisory Committee for Pharmaceutical Sciences of the Center for Drug Evaluation and Research of the U.S. Food and Drug Administration. The report from this EWG has summarized the characteristics of ideal cardiac toxic injury biomarkers [55]. These characteristics include cardiac specificity, sensitivity, predictive value, robustness, ability to bridge preclinical to clinical, being a noninvasive procedure, and accessibility. These characteristics have been adopted as standards for the development and validation of biomarkers of myocardial injury.

### Availability of Biomarkers

Currently, validated biomarkers that are included in clinical diagnostic testing guidelines are all related to myocardial structural injury. Developing biomarkers for nonstructural injury is most challenging and demands the utilization of more advanced technologies such as functional genomics and proteomics. In addition, currently available biomarkers have limitations, although they are useful.

#### Creatine Kinase

Three major creatine kinase (CK) isoenzymes have been identified. CK-MM is the principal form in skeletal muscle, CK-MB presents in myocardium in which CK-MM is also found, and CK-BB is the predominant form in brain and kidney. Elevation of serum CK-MB is considered a reasonably specific marker of acute myocardial infarction.

#### Myoglobin

Myoglobin is found in all muscle types, and its value as a biomarker of myocardial injury is based on the fact that serum concentrations of myoglobin increase rapidly following myocardial tissue injury, with peak values observed 1 to 4 hours after acute myocardial infarction. Elevation of serum myoglobin is likely reflective of the extent of myocardial damage.

#### B-Type Natriuretic Peptide

B-type natriuretic peptide (BNP) is a cardiac neurohormone secreted by the ventricular myocardium in response to volume and pressure overload, and the release of BNP appears to be directly correlated with the degree of ventricular wall tension. BNP is now accepted as a biomarker for congestive heart failure and is included in the European guideline for the diagnosis of chronic heart failure.

#### C-Reactive Protein

The acute-phase reactant C-reactive protein (CRP) is a marker of systemic and vascular inflammation that appears to predict future cardiac events in asymptomatic individuals. In particular, inflammation has been shown to play a pivotal role in the inception, progression, and destabilization of atheromas; thus, the predictive value of CRP in the prognosis of coronary heart disease has been proposed. The measurement of CRP appears to provide additional prognostic information when cTnT is measured at the same time.

### Cardiac Troponins

Cardiac troponin T (cTnT) and I (cTnI) are constituents of the myofilaments and expressed exclusively in cardiomyocytes, so they have absolute myocardial tissue specificity. In healthy persons, serum cTnT or cTnI are rarely detectable; therefore, any measurable concentrations of serum cTnT or cTnI reflect irreversible myocardial injury such as myocardial infarction. Clinical experience has led to the recommendation that cTn measurement be considered the gold standard for the diagnosis of acute myocardial infarction.

### Biomarker Applications and Limitations

All of the biomarkers described above have been used as indices of myocardial injury in clinical practice and experimental studies, and the same use continues. The major concern of most of the biomarkers is their specificity. CK-MB is present in small quantities in skeletal muscle and other tissues, thus elevations of CK-MB occur in some diseases involving skeletal muscle injury. Myoglobin is found in all muscle types, and the concentrations of myoglobin vary significantly between species and even within species. BNP has been proposed for use as a prognostic indicator of disease progression and outcome of congestive heart failure; however, the actual utility of this biomarker remains untested. BNP is also indicated in the counter-regulation of heart hypertrophy; thus, it is important to understand the changes in serum BNP concentrations as a function of time in the transition from cardiac hypertrophy to heart failure. Higher levels of BNP do not necessarily indicate greater severity of the heart disease, so more study is needed. CRP is a biomarker of inflammation, and its use in evaluating myocardial injury is supplementary to other tests that offer greater predictive value.

Considering all of the limitations above, more reliable biomarkers are needed. A significant advance in the development and validation of biomarkers for myocardial injury is the promising clinical experience with cTn, which has absolute myocardial tissue specificity and high sensitivity. It is now accepted by the clinical community that cardiac troponins can be used as the biomarkers of choice for assessing myocardial damage in humans. Their preclinical value for monitoring drug cardiac toxicity and in drug development still must be evaluated.

## REFERENCES


1. Robbins, J. (2004): Genetic modification of the heart: exploring necessity and sufficiency in the past 10 years. *J. Mol. Cell. Cardiol.*, 36:643–652.
2. Olson, E. N. (2004): A decade of discoveries in cardiac biology. *Nat. Med.*, 10:467-74.
3. Berenji, K., Drazner, M. H., Rothermel, B. A., and Hill, J. A. (2005): Does load-induced ventricular hypertrophy progress to systolic heart failure? *Am. J. Physiol. Heart Circ. Physiol.*, 289:H8–H16.
4. Dorn, G. W. and Force, T. (2005): Protein kinase cascades in the regulation of cardiac hypertrophy. *J. Clin. Invest.*, 115:527–537.
5. van Empel, V. P. and De Windt, L. J. (2004): Myocyte hypertrophy and apoptosis: a balancing act. *Cardiovasc. Res.*, 63:487–499.
6. de Simone, G. (2003): Left ventricular geometry and hypotension in end-stage renal disease: a mechanical perspective. *J. Am. Soc. Nephrol.*, 14:2421–2427.
7. Kass, D. A., Saavedra, W. F., and Sabbah, H. N. (2004): Reverse remodeling and enhanced inotropic reserve from the cardiac support device in experimental cardiac failure. *J. Card. Fail.*, 10:S215–S219.
8. Richey, P. A. and Brown, S. P. (1998): Pathological versus physiological left ventricular hypertrophy: a review. *J. Sports Sci.*, 16:129–141.
9. Piano, M. R., Bondmass, M., and Schwertz, D. W. (1998): The molecular and cellular pathophysiology of heart failure. *Heart Lung*, 27:3–19.
10. Goldsmith, S. and From, A. H. (1980): Arsenic-induced atypical ventricular tachycardia. *N. Engl. J. Med.*, 303: 1096–1098.
11. Van Vleet, J. F., Amstutz, H. E., Weirich, W. E., Rebar, A. H., and Ferrans, V. J. (1983): Clinical, clinicopathologic, and pathologic alterations in acute monensin toxicosis in cattle. *Am. J. Vet. Res.*, 44:2133–2144.
12. Litwak, K. N., McMahan, A., Lott, K. A., Lott, L. E., and Koenig, S. C. (2005): Monensin toxicosis in the domestic bovine calf: a large animal model of cardiac dysfunction. *Contemp. Top. Lab. Anim. Sci.*, 44:45–49.
13. Dockery, D. W. (2001): Epidemiologic evidence of cardiovascular effects of particulate air pollution. *Environ. Health Perspect.*, 109(Suppl. 4):483–436.
14. Gordon, T. and Reibman, J. (2000): Cardiovascular toxicity of inhaled ambient particulate matter. *Toxicol. Sci.*, 56:2–4.
15. Ventura-Clapier, R., Garnier, A., and Veksler, V. (2004): Energy metabolism in heart failure. *J. Physiol.*, 555:1–13.
16. Ingwall, J. S. and Weiss, R. G. (2004): Is the failing heart energy starved? On using chemical energy to support cardiac function. *Circ. Res.*, 95:135–145.
17. Neubauer, S., Horn, M., Cramer, M., Harre, K., Newell, J. B., Peters, W., Pabst, T., Ertl, G., Hahn, D., Ingwall, J. S., and Kochsiek, K. (1997): Myocardial phosphocreatine-to-ATP ratio is a predictor of mortality in patients with dilated cardiomyopathy. *Circulation*, 96:2190–2196.
18. Bittl, J. A. and Ingwall, J. S. (1985): Reaction rates of creatine kinase and ATP synthesis in the isolated rat heart. A $^{31}$P NMR magnetization transfer study. *J. Biol. Chem.*, 260:3512–3517.

19. Saupe, K. W., Spindler, M., Hopkins, J. C., Shen, W., and Ingwall, J. S. (2000): Kinetic, thermodynamic, and developmental consequences of deleting creatine kinase isoenzymes from the heart: reaction kinetics of the creatine kinase isoenzymes in the intact heart. *J. Biol. Chem.*, 275:19742–19746.
20. Wallimann, T., Dolder, M., Schlattner, U., Eder, M., Hornemann, T., O'Gorman, E., Ruck, A., and Brdiczka, D. (1998): Some new aspects of creatine kinase (CK): compartmentation, structure, function and regulation for cellular and mitochondrial bioenergetics and physiology. *Biofactors*, 8:229–234.
21. Neubauer, S. et al. (1999): Downregulation of the Na(+)-creatine cotransporter in failing human myocardium and in experimental heart failure. *Circulation*, 100:1847–1850.
22. Huss, J. M. and Kelly, D. P. (2005): Mitochondrial energy metabolism in heart failure: a question of balance. *J. Clin. Invest.*, 115:547–555.
23. Shipp, J. C., Opie, L. H., and Challoner, D. (1961): Fatty acid and glucose metabolism in the perfused heart. *Nature*, 189:1018–1019.
24. Wisneski, J. A., Gertz, E. W., Neese, R. A., and Mayr, M. (1987): Myocardial metabolism of free fatty acids: studies with $^{14}$C-labeled substrates in humans. *J. Clin. Invest.*, 79:359–366.
25. van Bilsen, M., Smeets, P. J., Gilde, A. J., and van der Vusse, G. J. (2004): Metabolic remodelling of the failing heart: the cardiac burn-out syndrome? *Cardiovasc. Res.*, 61:218–226.
26. Sun, X., Zhou, Z., and Kang, Y. J. (2001): Attenuation of doxorubicin chronic toxicity in metallothionein-overexpressing transgenic mouse heart. *Cancer Res.*, 61:3382–3387.
27. Bertazzoli, C., Bellini, O., Magrini, U., and Tosana, M. G. (1979): Quantitative experimental evaluation of adriamycin cardiotoxicity in the mouse. *Cancer Treat. Rep.*, 63:1877–1883.
28. Kang, Y. J., Chen, Y., Yu, A., Voss-McCowan, M., and Epstein, P. N. (1997): Overexpression of metallothionein in the heart of transgenic mice suppresses doxorubicin cardiotoxicity. *J. Clin. Invest.*, 100:1501–1506.
29. Gottlieb, R. A., Burleson, K. O., Kloner, R. A., Babior, B. M., and Engler, R. L. (1994): Reperfusion injury induces apoptosis in rabbit cardiomyocytes. *J. Clin. Invest.*, 94:1621–1628.
30. Olivetti, G. et al. (1997): Apoptosis in the failing human heart. *N. Engl. J. Med.*, 336:1131–1141.
31. Kang, Y. J., Zhou, Z. X., Wang, G. W., Buridi, A., and Klein, J. B. (2000): Suppression by metallothionein of doxorubicin-induced cardiomyocyte apoptosis through inhibition of p38 mitogen-activated protein kinases. *J. Biol. Chem.*, 275:13690–13698.
32. Wang, G. W., Klein, J. B., and Kang, Y. J. (2001): Metallothionein inhibits doxorubicin-induced mitochondrial cytochrome c release and caspase-3 activation in cardiomyocytes. *J. Pharmacol. Exp. Ther.*, 298:461–468.
33. EL-Sherif, L., Wang, G. W., and Kang, Y. J. (2000): Suppression of cadmium-induced apoptosis in metallothionein-overexpressing transgenic mouse cardiac myocytes. *FASEB J.*, 14:1193.
34. Kang, Y. J., Zhou, Z. X., Wu, H., Wang, G. W., Saari, J. T., and Klein, J. B. (2000): Metallothionein inhibits myocardial apoptosis in copper-deficient mice: role of atrial natriuretic peptide. *Lab. Invest.*, 80:745–757.
35. Kajstura, J., Cheng, W., Reiss, K., Clark, W. A., Sonnenblick, E. H., Krajewski, S., Reed, J. C., Olivetti, G., and Anversa, P. (1996): Apoptotic and necrotic myocyte cell deaths are independent contributing variables of infarct size in rats. *Lab. Invest.*, 74:86–107.
36. Eliot, R. S., Clayton, F. C., Pieper, G. M., and Todd, G. L. (1977): Influence of environmental stress on pathogenesis of sudden cardiac death. *Fed. Proc.*, 36:1719–1724.
37. Yaoita, H., Ogawa, K., Maehara, K., and Maruyama, Y. (2000): Apoptosis in relevant clinical situations: contribution of apoptosis in myocardial infarction. *Cardiovasc. Res.*, 45:630–641.
38. Wyllie, A. H. (1994): Death from inside out: an overview. *Philos. Trans. R. Soc. Lond. B Biol. Sci.*, 345:237–241.
39. Leist, M., Single, B., Castoldi, A. F., Kuhnle, S., and Nicotera, P. (1997): Intracellular adenosine triphosphate (ATP) concentration: a switch in the decision between apoptosis and necrosis. *J. Exp. Med.*, 185:1481–1486.
40. Didenko, V. V., Tunstead, J. R., and Hornsby, P. J. (1998): Biotin-labeled hairpin oligonucleotides: probes to detect double-strand breaks in DNA in apoptotic cells. *Am. J. Pathol.*, 152:897–902.
41. Frustaci, A., Kajstura, J., Chimenti, C., Jakoniuk, I., Leri, A., Maseri, A., Nadal-Ginard, B., and Anversa, P. (2000): Myocardial cell death in human diabetes. *Circ. Res.*, 87: 1123–1132.
42. Anversa, P. (2000): Myocyte death in the pathological heart. *Circ. Res.*, 86:121–124.
43. Guerra, S., Leri, A., Wang, X., Finato, N., Di Loreto, C., Beltrami, C. A., Kajstura, J., and Anversa, P. (1999): Myocyte death in the failing human heart is gender dependent. *Circ. Res.*, 85:856–866.
44. Yamashita, K., Kajstura, J., Discher, D. J., Wasserlauf, B. J., Bishopric, N. H., Anversa, P., and Webster, K. A. (2001): Reperfusion-activated Akt kinase prevents apoptosis in transgenic mouse hearts overexpressing insulin-like growth factor-1. *Circ. Res.*, 88:609–614.
45. Frey, N. and Olson, E. N. (2003): Cardiac hypertrophy: the good, the bad, and the ugly. *Annu. Rev. Physiol.*, 65:45–79.
46. Zou, Y. et al. (2001): Calcineurin plays a critical role in the development of pressure overload-induced cardiac hypertrophy. *Circulation*, 104:97–101.
47. Hill, J. A., Karimi, M., Kutschke, W., Davisson, R. L., Zimmerman, K., Wang, Z., Kerber, R. E., and Weiss, R. M. (2000): Cardiac hypertrophy is not a required compensatory response to short-term pressure overload. *Circulation*, 101:2863–2869.
48. Esposito, G. et al. (2002): Genetic alterations that inhibit in vivo pressure-overload hypertrophy prevent cardiac dysfunction despite increased wall stress. *Circulation*, 105:85–92.
49. Matter, C., Nagel, E., Stuber, M., Boesiger, P., and Hess, O. M. (1996): Assessment of systolic and diastolic LV function by MR myocardial tagging. *Basic Res. Cardiol.* 91(Suppl. 2):23–28.

50. Grossman, W. (1991): Diastolic dysfunction in congestive heart failure. *N. Engl. J. Med.*, 325:1557–1564.
51. Joshi, A., Dimino, T., Vohra, Y., Cui, C., and Yan, G. X. (2004): Preclinical strategies to assess QT liability and torsadogenic potential of new drugs: the role of experimental models. *J. Electrocardiol.*, 37(Suppl.):7–14.
52. Splawski, I. et al. (2000): Spectrum of mutations in long-QT syndrome genes: *KVLQT1*, *hERG*, *SCN5A*, *KCNE1*, and *KCNE2*. *Circulation*, 102:1178–1185.
53. Bennett, P. B., Yazawa, K., Makita, N., George, Jr., A. L. (1995): Molecular mechanism for an inherited cardiac arrhythmia. *Nature*, 376:683–685.
54. Wollnik, B., Schroeder, B. C., Kubisch, C., Esperer, H. D., Wieacker, P., and Jentsch, T. J. (1997): Pathophysiological mechanisms of dominant and recessive KVLQT1 $K^+$ channel mutations found in inherited cardiac arrhythmias. *Hum. Mol. Genet.*, 6:1943–1949.
55. Wallace, K. B., Hausner, E., Herman, E., Holt, G. D., MacGregor, J. T., Metz, A. L., Murphy, E., Rosenblum, I. Y., Sistare, F. D., and York, M. J. (2004): Serum troponins as biomarkers of drug-induced cardiac toxicity. *Toxicol. Pathol.*, 32:106–121.

# Notes

# 33 Assessment of Male Reproductive Toxicity*

*Sally D. Perreault, Gary R. Klinefelter, and Eric Clegg*

## CONTENTS



## INTRODUCTION

Reproduction is a complex process requiring exquisite integration of signals along the hypothalamic–pituitary–gonadal axis in both males and females. To be successful, functional gametes must be produced, matured, released, and transported effectively. Specific behavioral repertoires also must be executed precisely to ensure mating and fertilization with optimal timing. Our current state of knowledge about the physiological, biochemical, and

* This document has been reviewed in accordance with U.S. Environmental Protection Agency policy and approved for publication. Mention of trade names or commercial products does not constitute endorsement or recommendation for use.

molecular orchestration of reproduction is captured in many excellent reference volumes [1–5]. Based on this understanding, male reproductive physiology would be expected to be impacted by chemicals that act through a variety of mechanisms, including disruption of the cell cycle or chromosome movement in dividing cells (spermatogonia) and those undergoing meiotic processes (spermatocytes); damage to DNA, especially during DNA repair-deficient stages (condensed spermatids); and interference with endocrine/paracrine signaling that is vital to the regulation and integration of reproductive function.

Much of the current research in reproductive physiology is motivated by the widely recognized need to understand both fertility and infertility in humans and wildlife so as to develop effective technologies for both contraception and treatment of infertility. Part of this need is understanding the extent to which chemicals in the environment or pharmaceuticals used to treat diseases may negatively impact reproduction. Such understanding is essential if protective actions by regulatory agencies are to be based on sound science. The goal of this chapter is to summarize how information on the potential for chemicals to interfere with male reproductive capability is obtained and used by regulatory agencies, industry, and others. Importantly, this topic includes not only the ability to produce fertile spermatozoa but also to determine if they are able to transmit a healthy, intact, genome to the ensuing embryo.

The incidence of infertility is high in humans relative to many other species. It is estimated that 8 to 15% of couples seek medical treatment for infertility [6–9]. Although infertility is a function of the couple, in a sizable number of cases the infertility can be attributable to deficiencies in the male [7,8]. In addition, male-mediated effects on offspring have been demonstrated [10–19]. The potential for the male to contribute to reproductive failure and adverse pregnancy outcomes is significant, but the causes of male infertility are only partially understood. We now know that genetic alterations (e.g., aneuploidy), chromosomal aberrations (e.g., translocations), and gene polymorphisms account for a significant proportion of male infertility cases [15–17]. Furthermore, recent evidence confirms that male reproductive potential, as reflected by semen quality, also declines with age and can be negatively impacted by self-inflicted exposures to cigarette smoke, alcohol, and drugs [18,19]. Against this backdrop of factors, it is difficult to decipher the contribution of chemicals in the environment to the etiology of male infertility, which remains largely unexplained in the majority of individual cases. Despite such difficulty, it is imperative to elucidate the contribution of environmental exposures because, unlike genetically based infertility, chemically induced infertility is presumably preventable.

The assessment of environmental risk factors is further complicated by the growing awareness that exposures during fetal and neonatal development have the potential to impact reproductive function during adulthood [20–22]. Accordingly, it is critical to consider the impact of exposures at all periods of the life cycle, not just on adults; therefore, our reproductive effect test protocols and epidemiological study designs must be optimized to identify critical windows of exposure and developmental bases of adult disease.

Whereas numerous environmental, occupational, and therapeutic agents have been identified as male reproductive toxicants in test species such as mice and rats, relatively few have been shown to cause similar effects in human males. This discrepancy is likely due to the reduced resolving power that is inherent in many human studies rather than lower susceptibility for humans. Agents that have been shown to cause male reproductive effects in humans include heavy metals, chemotherapeutic agents, radiation, pesticides such as dibromochloropropane and ethylene dibromide, and other chemicals, such as carbon disulfide, chloroprene, and 2-ethoxyethanol [23–25].

More is known about chemicals that are capable of causing adverse male reproductive system effects in test species such as rats and mice where experimental protocols can be designed to control the exposure and dose response precisely [26]. The toxicology literature is rich in reviews of many studies relating male reproductive effects to exposures that target male germ cells at specific stages of development, the supporting Sertoli cells, and, importantly, the endocrine support of spermatogenesis and sperm maturation [3,4,27–35]. The outcomes of such exposures have included not only reduced fertility but also embryo/fetal loss, birth defects, cancer, and other postnatal structural or functional deficits.

Of particular relevance regarding animal toxicology studies is the growing awareness that exposures to chemicals that mimic or interfere with androgen and other endocrine signaling during fetal development can significantly alter male reproductive organ and tract development, sometimes with far-reaching impacts such as decreased sperm production or increased risk of testicular cancer later in life [20–22]. This literature indicates that, for certain agents, the developing male reproductive system could be the first affected or the most sensitive target organ.

Furthermore, concerns about human male reproductive health have been elevated by reports of downward secular trends in sperm counts, as well as increases in incidences of male reproductive tract malformations (e.g., hypospadias and testicular maldescent) and testicular cancer observed during the latter half of the twentieth century [36–39]. It has been suggested that exposures during reproductive system development to environmental agents that mimic or antagonize endogenous hormones may be causally related to such effects and that low sperm counts are only part of a syndrome of testicular dysfunction in humans that also may include reproductive tract malformations such as hypospadias, cryptorchidism, and predisposition to testicular cancer [40,41]. These concerns prompted the U.S. Environmental Protection Agency (EPA) and international partners to design and launch a comprehensive, tiered testing program to screen for endocrine-active and endocrine-disrupting chemicals [42].

In extrapolating effects observed in animal test species to the human situation, it is important to consider that fertility of the human male may be particularly susceptible to agents that reduce the number or quality of sperm produced. Whereas in some strains of mice and rats sperm production must be reduced by up to 90% before effects on fertility are seen, at least with routine mating procedures [43,44], the distribution of number of normal sperm for human males appears to be much closer to the threshold for reduced fertility. With that being the case, smaller decreases in sperm production in men could have serious consequences on their reproductive potential. If the number of normal sperm per ejaculate is sufficiently low, fertilization is unlikely and an infertile condition exists. The incidence of infertility in men has been considered to increase at sperm concentrations below $20 \times 10^6$ sperm per milliliter of ejaculate [45]; nevertheless, some men with low sperm concentrations but otherwise healthy sperm are able to achieve conception, and many subfertile men have concentrations greater than $20 \times 10^6$, illustrating the importance of sperm quality. Results from a recent prospective study indicate that human conception rate may begin to decline below sperm concentrations of $60 \times 10^6$ sperm per milliliter [46]; the average sperm concentration of ejaculates in that study of 200 couples trying to achieve pregnancy was $65 \times 10^6$ with 36- to 48-hour abstinence intervals. It is reasonable to assume, therefore, that reductions in sperm production by a toxic agent may further decrease the human male reproductive potential.

To ascertain the relationships between environmental exposures and male reproductive risks and thereby inform the risk assessment process, a sufficient amount and type of experimental data and a systematic approach to evaluating those data are needed. The approach utilized by the U.S. EPA for male reproductive toxicity risk assessment [47] is presented in this chapter, with particular emphasis on a critical analysis of the strategies and endpoints that are available for male reproductive risk assessment.

Although the focus of this chapter is on the assessment of reproductive risk in the male, it is important to point out that the concept of exposure being limited only to the male parent is most appropriate when applied to occupational or clinical settings. On the other hand, as one shifts to considering effects of exposures in the general environment, the likelihood increases that both parents experience common exposures, although dose, duration, and periodicity may vary. Examples where this is the case include ambient outdoor and indoor air pollutants, drinking water contaminants, residential and recreational pesticide contact, and potential exposure from neighborhood waste disposal sites. In these instances, the ability to discriminate between a male or female contribution to a reproductive problem or adverse developmental outcome may be difficult and inappropriate. Because a pregnancy that produces a healthy, normal child is the result of the reproductive competence of both parents, reproductive toxicity risk assessments must consider the couple as the unit for evaluation. Studies that focus solely on the contribution of one parent (paternal or maternal) to reproductive failure resulting from environmental exposures, although valuable for identifying the more sensitive gender and for elucidating modes of toxicant action, may either miss or substantially underestimate the true reproductive risk.

## RISK ASSESSMENT IN MALE REPRODUCTIVE TOXICOLOGY

### Overview of the Risk Assessment Process

A paradigm for the risk assessment process has been described in detail in two publications prepared by the National Academy of Sciences [48,49]. Although devised primarily for cancer risk assessment, many of the components also apply to the assessment of noncancer health effects such as reproductive toxicity. The major components of that paradigm are (1) hazard identification, (2) dose-response assessment, (3) exposure assessment, and (4) risk characterization. The EPA's Guidelines for Reproductive Toxicity Risk Assessment [47] have modified the first two components to hazard characterization followed by quantitative dose–response analysis for risk assessments involving nonlinear, low-dose extrapolation with reproductive effects.

### Hazard Characterization and Quantitative Dose–Response Assessment

#### Laboratory Protocols

Testing protocols describe the procedures to be used to provide data for risk assessments. The quality and usefulness of those data are dependent on the design and conduct of the tests, including endpoint selection and resolving power. The most widely accepted comprehensive protocol is the multigeneration reproductive test [50]. As a screen, it detects reproductive toxicity and provides dose–response information but may not provide all of the information required for a comprehensive risk assessment. This protocol was revised in 1996 to add specific sperm measures (motility, morphology, and numbers) and is continually being evaluated for its effectiveness. It is noteworthy that recent recommendations would modify it further with the goal of enhancing its sensitivity to detect reproductive effects while expanding its scope to detect developmental immuno- and developmental neurotoxicity [51]. The latter changes would also reduce the cost and number of animals necessary for noncancer health effects testing. The EPA has harmonized the multigenerational test protocol with those from other federal agencies and international organizations to achieve uniformity of approaches at the international level [52]. Results of this screen may indicate that modifications or follow-up testing should be conducted to determine the reversibility of adverse effects or elucidate mechanism of toxic action [26,53]. Acute exposure protocols may be used to pin down the time of onset of an effect or identify the pathogenesis of a lesion.

## Selection of Species

It is intuitively obvious that information from human studies is desirable when estimating specific exposure levels below which there is no appreciable risk; however, human data based on defined exposures to single chemicals are rarely available, so it is necessary to extrapolate human risk from data derived from animal test data. Generally, the rat is the species of choice and the default species for reproductive toxicity testing [47,50]; however, pharmacokinetic and mechanistic information may indicate that the rat is not an appropriate model for a specific chemical, in which case a more appropriate species should be selected. In any case, confidence in the results of testing for male reproductive toxicity is increased when multiple species have been examined. Advantages of the rat as a test species include the availability of an extensive reproductive toxicity database for this species, uniformity of reproductive endpoints within strain, and consistently efficient reproductive performance. In addition, the basic mechanisms underlying male reproductive function in the rat are well researched and reasonably representative of those in human males.

For a second mammalian test species, the rabbit has specific advantages that make it a good choice. First, ejaculated semen can be collected from bucks using an artificial vagina, allowing longitudinal assessment of semen quality. Second, by collecting semen (as opposed to epididymal sperm), alterations in the accessory sex gland secretions can be assessed, and levels of xenobiotics present in seminal fluid can be measured. Finally, ejaculated semen samples can also be used for artificial insemination. Use of artificial insemination with a limited number of sperm can be a highly effective strategy for detecting adverse effects on the sperm fertilizing ability in rabbits or rats.

Under some circumstances, data from other mammalian or nonmammalian species may be appropriate for incorporation into human health risk assessments; for example, mice in which specific genes have been knocked out may make excellent models for elucidating the mechanisms of toxicant action. Thus, use of other species is likely to become increasingly important as the ability to use mechanistic and molecular genetic information increases.

## Dose Selection and Duration of Dosing

To ensure adequate detection of toxicity, the use of relatively high dose levels and a sufficient array of endpoints is recommended. This decreases the likelihood of missing an effect. For toxicity testing, the test guidelines specify that the highest dose should produce systemic toxicity but not mortality. Dose–response assessment requires the generation of dose–response curves that adequately describe the increments in degree of effect as well as any changes in pattern of endpoints affected with changing dose level. Dose–response data should also include sufficiently low dose levels such that a low level of response or no effect is produced. Spacing between doses is especially critical in dose–response assessment. If the gaps between dose levels are too large, the estimate of the lowest observed-adverse-effect level (LOAEL) could be too high, and the no-observed-adverse-effect level (NOAEL) could be too conservative. Adequate coverage of a range of low doses is a critical issue in screening for endocrine disrupting chemicals [42].

The multigeneration test protocol was originally designed mainly to detect effects on fertility. Adverse effects of an agent on the spermatogenic process may not be observed in sperm or semen evaluations or in fertility until a substantial time after initiation of treatment. Damage that is limited to spermatogonial stem cells, for example, would not appear in cauda epididymal sperm or in ejaculates for 8 to 14 weeks, depending on the species examined. To allow effects on spermatogonial stem cells to be expressed in all evaluations of cauda epididymal or ejaculated sperm in subchronic studies, treatment of adult males should be continued for a minimum of six cycles of the seminiferous epithelium [54] prior to mating or sample collection. For the more commonly used species, one cycle of the seminiferous epithelium requires the following number of days: rat, 12.9; mouse, 8.6; rabbit, 10.7; rhesus monkey, 9.5; human, 16.0 [54]. Therefore, treatment for six cycles of the seminiferous epithelium for the test species requires from 52 to 82 days to ensure that all possible adverse effects are expressed in each endpoint observed. This recommendation assumes that levels and cumulative effects of the agent at the sites of attack reach steady state within one cycle of the seminiferous epithelium after initiation of treatment. If that assumption is not valid for an agent, the treatment period may need to be extended accordingly. In studies using shorter dosing periods, a prolonged follow-up may be necessary to detect effects on the earlier stages of spermatogenesis or to determine the persistence of an effect. In these situations, knowledge of the relevant pharmacokinetic and pharmacodynamic data can facilitate selection of dose levels and treatment duration. Equally important is proper timing of examination of treated animals relative to initiation and termination of exposure to the agent.

**TABLE 33.1**
**Reproductive Toxicity Data from Restricted Mating Study[a]**

| Dose (mg/kg/day) | Testis Weight (g) | Spermatid Count ($10^6$) | Cauda Sperm Count ($10^6$) | Fertility[b] | |
|---|---|---|---|---|---|
| | | | | 1× (%) | >3× (%) |
| 0 | 1.78 | 149.7 | 143.3 | 60 | 80 |
| 450 | 1.23 | 32.4 | 28.4 | 22 | 72 |

[a] Rats were treated with ethylene glycol monoethyl ether for 7 weeks, given mating experience, and then allowed either one copulation followed by separation or at least three copulations by overnight cohabitation with a female in estrus.

[b] Fertility index (number pregnant/number mated).

## Length of Mating Period

In fertility testing, pairs of animals are cohabited for periods of time sufficient to ensure conception in healthy animals. This is generally 2 weeks in rats, which allows for at least two opportunities for normally cycling females to conceive. Under those conditions, a large impact on sperm production is often necessary before an adverse effect on fertility can be detected, and reduced fertility in a male may be masked due to the multiple mating opportunities during each estrus and the multiple estrous periods. This is considered by some to be a weakness of the test protocol. During cohabitation, females should be examined daily for presence of seminal plugs or by vaginal lavage for evidence of mating. Females are usually separated from the male on the day following mating. This practice limits mating to one estrus but still allows numerous copulations during that estrus. As a result, an adequate number of sperm may be ejaculated to ensure fertility even when the males have reduced sperm production as a consequence of treatment. A hypothesis that restricting the number of copulations would increase the sensitivity of fertility testing was tested by Clegg and Zenick [55]. Male rats were exposed to a well-documented spermatotoxicant, ethoxyethanol (EE; 0 or 450 mg/kg/day), by gavage for 7 weeks to depress sperm counts and testis weight (Table 33.1). Each control and EE-treated male was mated initially, in a counterbalanced design, to a female in proestrus, with either a single mating or a minimum of three matings allowed. Three days later, each male was mated under the alternate condition. Despite the extreme reduction in sperm counts in EE males, no differences in fertility were observed relative to controls with multiple matings; however, a marked decrease in fertility was seen in the EE group when only a single copulation was allowed (Table 33.1). These results suggest that the sensitivity of breeding protocols may be enhanced by experimentally limiting the number of copulations.

The observation and control of number of copulations is facilitated by maintaining the animals on a reverse light–dark schedule so estrus and matings occur during normal working hours. If sexually experienced rats are used, copulatory behavior can be rapidly and accurately monitored, with several males observed simultaneously. Details of this mating procedure have been published previously [56,57].

## Number of Animals

The number of animals per dose group in a toxicology study is determined by the number of animals expected to survive and yield data, the expected variation between animals in the endpoints to be examined, the magnitude of effect to be detected, and the level of probability selected for statistical significance. The number of animals required per treatment should be calculated by standard statistical methods as part of the study design process [58]. Estimates of the coefficients of variation for some parameters used for tests of the male rat reproductive system have been reported by several authors [59–64]. In general, when multiple endpoints are used in a study of male reproductive toxicity, 20 males per treatment should be sufficient to detect effects; however, in tests designed to evaluate fertility, it is often necessary to start with more males per treatment group to obtain 20 pregnancies per treatment. Some protocols specify mating of two females per male. In such cases, it is important to note that this practice does not double the sample size for statistical analysis; the male must be the unit of measure in the analysis, and the use of the number of pregnant females as the unit of analysis would inflate sample size artificially.

# Testing Protocols

## Single- and Multigeneration Reproduction Tests

Comprehensive reproductive toxicity studies in laboratory animals generally involve continuous exposure to a test substance and evaluation of reproductive capability for one or more generations. The objective is to detect effects on the integrated reproductive process as well as to study effects on the individual reproductive organs.

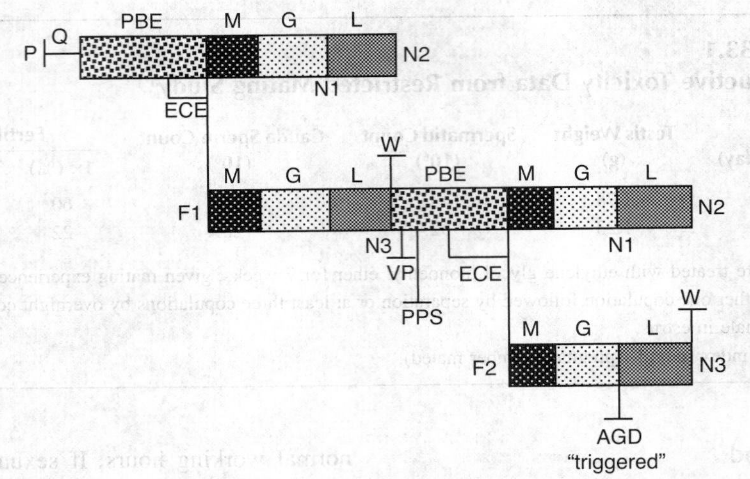


**FIGURE 33.1** A schematic depicting the EPA's harmonized multigeneration reproduction test. Q = quarantine; PBE = prebreeding exposure; M = a 2-week mating; G = gestation; L = lactation; VP = vaginal patency monitored in F1 females from postnatal day 22; PPS = preputial separation monitored in F1 males from postnatal day 35; W = weaning on postnatal day 21; N1 = necropsy of the parental or F1 males (assessments include organ weights, histology, and sperm measures); N2 = necropsy of the parental or F1 females (assessments include organ weights and histology); N3 = necropsy of three weanlings per sex per litter (assessment is a macroscopic evaluation, and if indicated organs are weighed and preserved for histology); ECE = a 3-week evaluation estrous cyclicity; AGD = anogenital distance measured in the F2 pups at birth if pubertal landmarks are affected in the F1.

The single-generation reproduction test evaluates effects of subchronic exposure of peripubertal and adult animals on reproductive organs and performance. In the multigeneration reproduction protocol, both sexes of the parental generation (P) are exposed for about 10 weeks (rats) prior to breeding, and dosing is continued during pregnancy such that the F1 and F2 offspring are exposed continuously *in utero* from conception until birth and during the preweaning period. F1 offspring exposures are continued beyond puberty. This allows detection of effects that occur from exposures throughout development, including the peripubertal and young adult phases. Because the parental and subsequent filial generations have different exposure histories, reproductive effects seen in any particular generation are not necessarily comparable with those of another generation. Also, successive litters from the same parents cannot be considered as replicates because of factors such as continuing exposure of the parents, increased parental age, sexual experience, and parity of the females.

According to the EPA's multigeneration protocol illustrated in Figure 33.1, rats are dosed for 8 to 10 weeks after initiation at 5 to 8 weeks of age (i.e., soon after puberty). This allows effects on gametogenesis and sperm maturation in the epididymis to be expressed and increases the likelihood of detecting histologic lesions in the testis and epididymis. Three dose levels plus one or more control groups are usually included. Enough males and females are mated to ensure 20 pregnancies per dose group for each generation. Animals producing the first generation of offspring are considered the parental (P) generation, and the subsequent offspring generations are designated filial generations (e.g., F1, F2). Only the P generation is mated in a single-generation test, while both the P and F1 generations are mated in the standard two-generation reproduction test illustrated.

Cohabitation is terminated when evidence of mating is detected. Females continue to be exposed during gestation and lactation. In the two-generation reproduction test, randomly selected F1 male and female offspring continue to be exposed after weaning (day 21) and through the mating period. Treatment of mated F1 females is continued throughout gestation and lactation. More than one litter may be produced from either P or F1 animals. Depending on the route of exposure of lactating females, it is important to consider that offspring may be exposed to a chemical by ingestion of maternal feed or water (diet or drinking water studies), by licking of exposed fur (inhalation study), by contact with treated skin (dermal study), or by coprophagia, as well as via the milk.

In single- and multigeneration reproduction tests, reproductive endpoints evaluated in P and F generations include visual examination of the reproductive organs, weights and histopathology of the pituitary (both sexes), testes, epididymides, male accessory sex glands, uterus, ovaries, and vagina. The parental and F1 animals are evaluated to obtain the number and quality of sperm produced, estrous cycle normality, number of ovarian primordial follicles, and pubertal landmarks associated with normal development of the reproductive system, including the age at vaginal opening in females and the age at preputial separation in males. In addition, anogenital distance (AGD), an endocrine-sensitive endpoint in pups, may be triggered in the F2 animals if developmental effects are observed in the F1 animals. Based on value-added evaluations, recent recommendations are to include AGD in an augmented test design and to increase the number of F1 and F2 pups maintained through weaning, puberty and adulthood [65]. Evaluation of reproductive performance is standard for P and F1. Both male and female mating and fertility indices are calculated. Litters (and often individual pups) are weighed at birth and examined for the number of live and dead offspring, gender, gross abnormalities, and growth and survival to weaning.

Inclusion of testicular histopathology and sperm evaluations helps risk assessors judge whether effects on fertility can be attributed specifically to the male or the female, in which case the most sensitive gender is used for defining the NOAEL and setting safe exposure limits; however, identification of effects in one sex does not exclude the possibility that both sexes may have been affected adversely. Data from matings of treated males with untreated females and *vice versa* (cross-over matings) may be necessary to separate sex-specific effects.

An EPA workshop has considered the relative merits of one- vs. two-generation reproductive effects studies [66]. The participants concluded that a one-generation study is insufficient to identify all potential reproductive toxicants, because it would exclude detection of effects caused by prenatal and postnatal exposures (including the prepubertal period), as well as effects on germ cells that could be transmitted to and expressed in the next generation. A one-generation test might also miss adverse effects with delayed or latent onset because of the shorter duration of exposure for the P generation. These limitations are shared with the shorter term screening protocols described below. Because of these limitations, a comprehensive reproductive risk assessment should include results from a two-generation test or its equivalent.

In studies where parental and offspring generations are evaluated, additional risk assessment issues arise with regard to the relationships of reproductive outcomes across generations. Increasing vulnerability of subsequent generations is sometimes, but not always, observed. Qualitative predictions of increased risk of the filial generations could be strengthened by knowledge of the reproductive effects in the adult, the likelihood of bioaccumulation of the agent, and the potential for increased sensitivity resulting from exposure during critical periods of development [67]. In addition to the sensitivity of the developing reproductive system to endocrine-active agents [36], many of the detoxifying enzyme systems and renal excretory mechanisms develop neonatally in the rodent, contributing to the increased vulnerability of generations exposed during this critical period [68]. Qualitative predictions of the increased risk of the filial generations could be strengthened by knowledge of the nature of the reproductive effects in the adult, the metabolic pathways and the likelihood of bioaccumulation of the agent, and the potential for increased sensitivity resulting from exposure during critical periods of development [69]. An instructive example of the use and interpretation of multigenerational test data to evaluate the potential for effects to worsen across generations (or not) is provided by the Center for the Evaluation of Risks to Human Reproduction in its evaluation of bromopropane [70]. This and related reports are publicly available at https://cerhr.niehs.nih.gov.

Several recent reports suggest that exposures during early phases of testis or prostate differentiation and development may alter the epigenetic programming of the organ, resulting in altered function during adulthood, including increased predisposition to cancer [71,72]. Such alterations may potentially be transmissible across subsequent generations, even in unexposed offspring. Such observations may impact both testing strategies and risk assessment approaches in the future.

On the other hand, subsequent generations may be less affected than the F1 offspring. This can occur when the F1 and F2 animals represent survivors who are (or become) more resistant to the agent than the P generation. Therefore, results between generations or between sequential litters within a generation should not necessarily be compared directly. Significant adverse effects in any generation should be considered cause for concern unless inconsistencies in the data indicate otherwise.

A review of 20 positive multigeneration reproduction studies has provided some insights on the relationship of male toxicity to effects on offspring within a given generation and the relationships of reproductive outcomes across generations [73]:

- The presence of toxicity in the adult male (P), reproductive or otherwise, was not a prerequisite for the occurrence of effects on offspring.
- Approximately one half of the studies were classified as "positive-increasing." In these cases, the second-generation (F1) animals exhibited effects that were more severe than those in the first generation or occurred at equivalent or lower doses.
- The increasing toxicity across generations is consistent for chemicals that bioaccumulate; however, exposure of sequential generations that involve different developmental stages (P vs. F1 adults) might also contribute to differential effects across generations.
- The multigeneration reproduction test does not address the issue of reversibility; however, inclusion of additional mated pairs in a study can provide additional animals for a reversibility test or developmental toxicity evaluation.

## Subchronic Test Protocols

Subchronic toxicity tests may have been conducted before a detailed reproduction study is initiated. In the subchronic toxicity test with rats, exposure usually begins at 6 to 8 weeks of age and is continued for 90 days. The initiation of exposure at 8 weeks of age (compared with 6) and exposure for approximately 90 days allows the animals to reach a more mature stage of sexual development and ensures an adequate length of dosing for observation of effects on the reproductive organs with most agents. Dosing is often done orally (i.e., gavage, in diet, or in drinking water) but may be by inhalation or dermal application. Animals are monitored for clinical signs throughout the test and are necropsied at the end of dosing, without evaluating reproductive function. The endpoints evaluated for the male reproductive system include visual examination of the reproductive organs, plus weights and histopathology for the testes, epididymides, and accessory sex glands. For the females, endpoints may include visual examination of the reproductive organs, uterine and ovarian weights, and histopathology of the vagina, uterus, cervix, ovaries, and mammary glands.

Scientists in the National Toxicology Program have examined the feasibility and value of incorporating some basic measures of reproductive toxicity into the protocols of their standard 13-week, prechronic toxicity studies to serve as a reproductive screening battery. The sperm morphology and vaginal cytology examination (SMVCE) includes evaluations of epididymal sperm motility, sperm concentration, sperm head morphology, and reproductive organ weights in the males and average estrous cycle length and relative frequency of different estrous cycle stages in the females. Data were collected at the end of 50 13-week studies, of which 25 were conducted in mice and the remainder in rats [62]. In this evaluation, reproductive organ weights and sperm motility appeared to be the most statistically powerful endpoints. Seven days appeared to be an inadequate time frame for evaluating stages of the estrous cycle. Based on the results of these studies, the authors recommended that multiple endpoints of spermatotoxicity be evaluated in screening tests. This test may be useful to identify an agent as a potential reproductive hazard but usually does not provide information about the integrated function of the reproductive systems (sexual behavior, fertility, and pregnancy outcomes) nor does it include effects of the agent on immature animals.

## Specific Tests for Endocrine-Disruptive Chemicals

A battery of short-term *in vitro* and *in vivo* tests has been proposed to fulfill congressional mandates to screen for endocrine disrupting chemicals [74] (full details available at http://www.epa.gov/scipoly/oscpendo/index.htm). The *in vitro* tests screen for chemicals that bind estrogen and androgen receptors or affect steroidogenesis by altering the steroidogenic enzymes, including aromatase. Short-term *in vivo* tests relevant to male reproductive effects detect chemicals that alter androgen action; for example, the Hershberger assay is designed to detect environmental antiandrogens (or androgens) and is based on exposing young castrated male rats to the compounds in question and weighing the prostate gland (an androgen-dependent organ) 5 to 7 days later. A male pubertal assay is designed to detect a wider array of endocrine disruptors by exposing male rats from weaning through puberty and then evaluating reproductive and thyroid organ weights and hormones, as well as assessing pubertal indices. A slightly longer protocol would initiate dosing during pregnancy and continue through puberty. This assay would also detect reproductive tract malformations such as cleft phallus, ectopic testes, retained nipples, and altered AGD. Finally efforts are underway to ascertain whether the addition of a few endpoints to the harmonized reproductive toxicity test, described above, would ensure that it, too, would detect endocrine disruptors [51]. These assays are undergoing international validation in conjunction with the Organization for Economic Cooperation and Development (OECD). Although a full discussion of these protocols is beyond the scope of this chapter, information on the history and progress of this effort (including detailed protocols) can be found at www.epa.gov/scipoly/oscpendo/index.htm.

## Short-Term Reproductive Protocols

Although short-term tests (i.e., less than one full spermatogenic cycle for the species being evaluated) have been proposed to screen chemicals for testicular toxicity, the risk of false negatives due to insufficient duration of exposure is high. A serious limitation, especially when using them for chemicals with unknown toxicity, is that effects of exposures during development would not be evaluated; however, these tests may be appropriate when prior information exists about the target organ or if a chemical is suspected to be a testicular toxicant by structure–activity relationships. Short-term tests are also of value, if not essential, in the identification of target sites, affected cell types, and mechanisms of toxicity, delineation of which is often impossible after subchronic treatment. Because a wide variety of short-term tests have been used to assess reproductive toxicity, only a few examples are presented here.

### *Protocol for Toxicity in the Epididymis*

This protocol was developed to identify toxicants that alter the structure or function of the epididymis within 5 days of toxicant exposure [75]. Based on the known transit rate of rat sperm through the epididymis, sperm that are within the proximal region of the epididymis at the onset of exposure (day 1) can be recovered from the proximal cauda epididymis, the first site in which fertile sperm can be found, 4 days later (day 5). By limiting the exposure period in this manner, sperm that would be within the testis at the onset of exposure are precluded from assessment. With toxicants that compromise testosterone production, it is possible to implant testosterone-filled Silastic® capsules to clamp serum testosterone at control levels. Also, the efferent ducts can be ligated to prevent toxicant-induced perturbations in testicular fluid from altering the epididymis. This protocol has been used to identify epididymal toxicity resulting from exposure to chloroethylmethanesulfonate [76], as well as epichlorohydrin and hydroxyflutamide [77]. To detect antiandrogenic effects from a chemical such as hydroxyflutamide, however, it is necessary to castrate the animals to lower endogenous androgen levels to the extent that competition of test chemical for the androgen receptor is effective. To prevent the loss of androgen-dependent function within the epididymis following castration, a testosterone-filled Silastic® capsule can be implanted to maintain androgen status at a level that maintains epididymal sperm maturation over 4 days [78]. A similar short exposure design has been used to identify sperm defects associated with cyclophosphamide-induced post-implantation loss [79], as well as infertility of epididymal sperm that results from the metabolic inhibitors ornidazole [80] and α-chlorohydrin [81].

Importantly, one manifestation that seems to be common to the epididymal toxicants ethane dimethane sulfonate (EDS), chloroethylmethanesulfonate, hydroxyflutamide, and epichlorohydrin, as well as developmental exposure to 2,3,7,8-tetrachlorodibenzo-*p*-dioxin (TCDD) and methoxychlor, is reduced epididymal sperm number without any concomitant reduction in testicular sperm number. This suggests that epididymal sperm transit is accelerated by certain chemical exposures. Indeed, when the short-term epididymal toxicity protocol was modified to test this hypothesis, accelerated transit was demonstrated [77]. The acceleration appeared to be independent of androgen status but was significantly correlated to several constitutive epididymal proteins. Accelerated sperm transit in humans, which have a much shorter period of epididymal transit, could have adverse effects on the process of sperm maturation and numbers of sperm ejaculated.

### Spermatoxicity Test Protocol

Linder et al. [82] originally proposed a short-duration test to screen chemicals for spermatotoxicity in structure–activity studies or to set priorities for chemicals requiring further evaluation. Depending on the duration of dosing and the day on which necropsies are performed, the protocol may cover a period of up to 2.5 weeks (e.g., dose for 5 days, necropsy 14 days later). The 14-day period allows for spermatids that are compromised at the onset of exposure to appear in the epididymis. In a validation study, groups of male rats were dosed for 1 to 5 days with 14 chemicals shown to produce minimal testicular effects in subchronic studies, and necropsies were performed 2 to 3 days and 13 to 14 days after dosing was terminated. Reproductive organ weights (testis, epididymis, seminal vesicle, and prostate), sperm counts (testicular and epididymal), and sperm motion parameters (computer-assisted sperm analysis, or CASA) were measured. Both the testes and epididymides were subjected to rigorous histopathologic evaluation. Spermatotoxicity was detected for the ten most potent testicular toxicants. Results for the other four chemicals, which were judged to be minimally toxic in a subchronic test, were essentially negative; thus, chemicals that produce moderate to severe damage to germ cells in latter stages of spermatogenesis are detectable with this short-duration test. More recently, this protocol has been abbreviated to encompass a 14-day exposure with a necropsy following administration of the last dose. With this, it has been established that the disubstituted haloacetic acid by-products of drinking water disinfection (i.e., dibromoacetic acid [83], dichloroacetic acid [84], and bromochloroacetic acid [85]) produce lesions in the latter stages of spermatogenesis. The histopathologic profile of these insults includes the formation of atypical residual bodies, fusion of sperm, and delayed spermiation.

### Thirty-Five Day Reproductive Assessment

Harris et al. [86] initiated a 21-day reproductive and developmental toxicity screen that has subsequently been revised to a 35-day screen. This screen has been utilized increasingly to identify potential reproductive toxicants among the growing list of putative endocrine disruptors and byproducts of drinking water disinfection. In the male portion of this protocol, animals (mice or rats) are dosed from study day (SD) 5 to 35 with one of three dosages of test chemical; the highest dose is selected as a dose that begins to inhibit drinking water or food consumption. These males are mated from SD 12 to 17 to females that are dosed with chemical from SD 1 to 35. At necropsy, testis and epididymal weights are obtained, and cauda epididymal sperm concentration and motility are determined. In addition, the testis is subjected to a thorough histopathologic evaluation. This reproductive screen is likely to detect potent testicular toxicants, but subtle effects on early stages of spermatogenesis and effects on other organ systems (epididymis, central nervous system, sex accessory glands, and pituitary) may go undetected unless manifested rapidly.

### OECD ReproTox Test Protocol

The Organization for Economic Cooperation and Development has developed a guideline protocol, ReproTox, to rapidly screen high priority chemicals for safety assessment [87]. The protocol is designed to generate information on systemic toxic effects with repeated dosing along with data on reproductive and developmental toxicity in both sexes. Tanaka et al. [88] have examined the ability of this protocol to confirm the effects of cyclophosphamide reported in the published literature. In their study, male rats were dosed for 42 to 43 days, including 14 days prior to mating. After 14 days of treatment, each male rat was paired with a treated female for 14 days or until sperm were detected in the vaginal smear. Males were necropsied on days 43 or 44, and thymus, liver, kidneys, spleen, testes, and epididymal weights were measured. In addition, the brain, heart, liver, kidneys, adrenal glands, spleen, bone marrow, testes, and epididymides were subjected to histopathologic examination. The authors found that the ReproTox protocol identified most of the known toxicological properties of this chemical, except for the adverse effects on spermatogenesis and fertility.

*Dominant Lethal Test Protocol*

The dominant lethal test is intended to detect mutagenic effects in the spermatogenic process that are lethal to the embryo or fetus. A review of this test has been published as part of the EPA's Gene-Tox program [89]. Dominant lethal protocols may utilize acute dosing (1 to 5 days) followed by serial matings with one or two females per male per week for the duration of the spermatogenic process. An alternative protocol may utilize subchronic dosing for the duration of the spermatogenic process followed by matings. Females are monitored for evidence of mating, sacrificed at approximately mid-gestation, and examined for incidence of pre- and post-implantation loss. The acute exposure protocol of the standard dominant lethal test, combined with serial mating, may allow identification of the spermatogenic cell types that are affected; however, acute dosing may not produce adverse effects at levels as low as with subchronic dosing because of factors such as bioaccumulation. Information from such studies can be useful for identifying site and potential mechanism of action and, thus, facilitate design of subsequent studies.

## Endpoints for Evaluating Male Reproductive Toxicity

The following sections describe various endpoints that can reflect male reproductive toxicity and their use in risk assessment [26,47]. A comprehensive assessment of male reproductive toxicity requires information on multiple endpoints that are capable of detecting the range of potential adverse effects. These should include measures of the primary functions of fertility and reproductive behavior. Because the usual measures of fertility in rodents have limited sensitivity, endpoints should also be included that are capable of detecting effects on components of the male reproductive system that support those functions (e.g., production of normal spermatozoa and normal differentiation of the reproductive tract and external genitalia).

Alterations in these reproductive endpoints may be the result of direct or indirect toxicity to the male reproductive system. In either case, the exposure to the agent has caused a reproductive effect and there may be cause for concern. Careful evaluation of the dose–response curves for the various target organs and effects may provide insight into whether the reproductive effects are independent of these other toxicities. Seldom, however, is such a judgment possible. Estimating the dose levels at which these various target organ effects occur has significance in predicting the effects of anticipated human exposure. Also, the likelihood that reproductive toxicity will be present in the absence of other systemic effects can be better characterized.

Statistical analyses are important in determining the effects of a particular agent, but the biological significance of data is most important. When many endpoints are investigated, statistically significant differences may occur by chance. On the other hand, apparent trends with dose may be relevant biologically even though pair-wise comparisons do not indicate a statistically significant effect. In the following discussion, endpoints are identified in which significant changes may be considered adverse, but concordance of results and known biology should be considered in interpreting all results. All effects that may be considered as adverse are appropriate for use in establishing a NOAEL, LOAEL, or benchmark dose.

Although the measures discussed in this section may detect impairment to the various components of the reproductive process, they do not discriminate effectively between nonmutagenic and mutagenic mechanisms. If the effects seen in evaluation of male reproductive endpoints are the result of mutagenic events (e.g., interaction with DNA), then there is the potential for transmissible genetic damage.

To facilitate discussion, the endpoints of reproductive toxicity can be separated into three categories: couple-mediated, female-specific, and male-specific. Couple-mediated endpoints are those in which both sexes can have a contributing role if both partners are exposed. In this chapter, couple-mediated endpoints are included along with male-specific endpoints because male exposures may result in effects on those endpoints.

The discussions of endpoints and the factors influencing results that are presented in this section are directed to evaluation and interpretation of results with test species. Many of those endpoints require invasive techniques that preclude routine use with humans; however, in some instances, related endpoints that can be used with humans are identified.

### Couple-Mediated Endpoints

Breeding studies with test species are a major source of data on reproductive toxicants. Evaluations of fertility and pregnancy outcomes provide measures of the functional

**TABLE 33.2**
**Couple-Mediated Endpoints of Reproductive Toxicity**

| Multigeneration Studies | Other Reproductive Endpoints |
|---|---|
| Mating rate, time to mating (time to pregnancy[a]) | Ovulation rate |
| Pregnancy rate[a] | Fertilization rate |
| Delivery rate | Preimplantation loss |
| Gestation length[a] | Implantation number |
| Litter size (total and live) | Postimplantation loss[a] |
| Number of live and dead offspring (fetal death rate[a]) | Internal malformations and variations[a] |
| Offspring gender[a] (sex ratio) | Postnatal structural and functional development[a] |
| Birth weight[a] | |
| Postnatal weights[a] | |
| Offspring survival[a] | |
| External malformations and variations[a] | |
| Offspring reproduction[a] | |

[a] Endpoints that can be obtained with humans.

consequences of reproductive injury. Measures of fertility and pregnancy outcome that are often obtained are presented in Table 33.2. Many of the endpoints reflect developmental toxicity that may be male mediated. Some of the endpoints identified above are used to calculate ratios or indices [47,50]. Definitions of some of these indices in published literature vary substantially, which can make interstudy comparisons difficult. Also, the calculation of an index may be influenced by the test design; therefore, it is important that the methods used to calculate indices be specified. Some commonly reported indices are listed in Table 33.3.

Because the reproductive testing often entails the cohabitation of treated males with treated females, the influences of the gender-specific impact on changes in fertility or other reproductive outcome (e.g., reduced survival) may not be readily discriminated. Assignment to the male of at least some of the responsibility for a toxic effect on fertility may be possible from evaluation of data on other reproductive measures (organ weights, histopathology, sperm measures, cleft phallus). Data on mating behavior could also clarify effects on fertility by indicating whether or not the males were behaviorally competent. Finally, if male measures do indeed suggest that a male component does exist, subsequent mating with untreated females is recommended for confirmation.

If evaluation of mating success is included, useful data would include confirmation of the day of insemination (i.e., sperm plugs or sperm-positive vaginal lavages), plus analysis of the length of time required for each animal to achieve successful mating (time to mating). Although evidence of mating is not synonymous with successful impregnation and does not preclude undetected matings, such data could provide a more complete evaluation of reproductive competence.

Evaluations of time to mating might also help detect the presence of subfertility. Exposure to a reproductive toxicant may not produce a total absence of fertility (i.e., sterility), but rather a condition of subfertility seen as an increased time to conception. (Subfertility may also be reflected as a reduction in litter size in polytocous species.) The assessment of time to mating in reproductive studies might indicate the potential for increased time to conception in humans. Unfortunately, time to pregnancy is rarely evaluated in epidemiology studies.

As noted earlier, dominant lethal assays, in which the female is sacrificed in mid- to late pregnancy, may also produce reproductive data. Endpoints examined from dominant lethal tests often include mating and fertility ratios and estimates of preimplantation loss: [(number of corpora lutea – number of implantation sites)/number of corpora lutea] and postimplantation loss [(number of implantation sites – number of fetuses)/number of implantation sites]. The occurrence of pre- or postimplantation loss is often considered to provide sufficient evidence that the agent has gained access to the reproductive organs and has compromised fertilizing ability or induced mutagenic damage to the sperm, respectively. Such data were informative, for example, in the CERHR's report on acrylamide [90], which is available at http://cerhr.niehs.nih.gov/chemicals/acrylamide-eval.html.

A genotoxic basis for postimplantation loss (i.e., mutations in sperm) is accepted widely; however, current methods of assessing preimplantation loss provide little distinction between contributions of DNA or chromosome damage in sperm that cause embryo or fetal death and nonmutagenic factors that result in failure of fertilization (e.g., inadequate numbers of qualitatively normal and motile sperm, failure in sperm transport or ovum penetration). To distinguish between fertilization failure and early pregnancy loss, additional experimental approaches are

**TABLE 33.3**
**Selected Indices That May Be Calculated from Endpoints of Reproductive Toxicity in Test Species**

**Mating Index**

$$\frac{\text{Number of males or females mating}}{\text{Number of males or females cohabited}} \times 100$$

*Note:* Mating is used to indicate that evidence of copulation (observation or other evidence of ejaculation such as vaginal plug or sperm in vaginal smear) was obtained.

**Fertility Index**

$$\frac{\text{Number of cohabited females becoming pregnant}}{\text{Number of unpregnant couples cohabited}} \times 100$$

*Note:* Because both sexes are often exposed to an agent, distinction between sexes often is not possible. If responsibility for an effect can be clearly assigned to one sex (as when treated animals are mated with controls), then a female or male fertility index could be useful.

**Gestation (Pregnancy) Index**

$$\frac{\text{Number of females delivering live young}}{\text{Number of females with evidence of pregnancy}} \times 100$$

**Live Birth Index**

$$\frac{\text{Number of live offspring}}{\text{Number of offspring delivered}} \times 100$$

**Sex Ratio**

$$\frac{\text{Number of male offspring}}{\text{Number of female offspring}} \times 100$$

**4-Day Survival Index (Viability Index)**

$$\frac{\text{Number of live offspring at lactation day 4}}{\text{Number of live offspring delivered}} \times 100$$

*Note:* This definition assumes that no standardization of litter size is done until after the day 4 determination is completed.

**Lactation Index (Weaning Index)**

$$\frac{\text{Number of live offspring at day 21}}{\text{Number of live offspring born}} \times 100$$

*Note:* If litters were standardized to equalize the number of offspring per litter, the number of offspring after standardization should be used instead of the number born alive. When no standardization is done, the measure is called the *weaning index*. When standardization is done, the measure is called the *lactation index*.

**Preweaning Index**

$$\text{Number of offspring born} - \frac{\text{Number of offspring weaned}}{\text{Number of live offspring born}} \times 100$$

*Note:* If litters were standardized to equalize the number of offspring per litter, then the number of offspring remaining after standardization should be used instead of the number born.

**TABLE 33.4**
**Male-Specific Endpoints of Reproductive Toxicity**

| | |
|---|---|
| Organ weights | Testes, epididymides, seminal vesicles, prostate, pituitary |
| Visual examination and histopathology | Testes, epididymides, seminal vesicles, prostate, pituitary |
| Sperm evaluation[a] | Sperm number (count) and quality (morphology, motility) |
| Sexual behavior[a] | Mounts, intromissions, ejaculations |
| Hormone levels[a] | Luteinizing hormone, follicle stimulating hormone, testosterone, estrogen, prolactin |
| Developmental effects | Testis descent,[a] preputial separation, sperm production,[a] anogenital distance, structure of external genitalia[a] |

[a] Reproductive endpoints that can be obtained or estimated relatively noninvasively with humans.

needed [91,92]; for example, oocytes can be recovered from the oviduct the day after mating and examined to evaluate fertilization directly [81,93], and the zygotes can be cultured *in vitro* to the blastocyst stage to evaluate preimplantation developmental potential [94].

Animal data on reproductive success (or the lack thereof) are difficult to verify in human populations. First, humans are characterized by low rates of conception; thus, an insufficient number of pregnancies may occur in an exposed population to provide sufficient power to detect an effect. Moreover, both partners may be exposed to the toxicant, making it more difficult to ascribe reproductive failure solely to the male. Studies of gender-specific occupational work groups may provide some clarification as to the male's contribution. Such studies also provide the opportunity to study individuals with higher exposures than those encountered in the general population.

Data on fertility potential plus other measures of reproductive outcomes provide the most comprehensive and direct insight into reproductive capability. Fertility assessments are limited by their insensitivity as measures of reproductive injury. As noted earlier, normal males of most test species produce numbers of qualitatively normal sperm that greatly exceed the minimum requirements for fertility as evaluated in current protocols. However, human males appear to function nearer to the threshold for the number of normal sperm needed to ensure reproductive competence. This difference between test species and humans means that data that fail to demonstrate an effect on fertility in a test species should not be the basis for concluding that the test agent poses no reproductive hazard to fertility in humans. In such instances, data from additional reproductive endpoints may provide clarification and should be examined.

A number of investigators have used the strategy proposed by Amann [95] that employs artificial insemination (AI) with a limited number of sperm from both control and treated males. Artificial insemination by *in utero* insemination does indeed increase the sensitivity of detection for toxicant-induced decreases in sperm quality [78,83,96–98]. Moreover, Robl and Dzuik [99] have successfully used *in utero* insemination in three strains of mice. Dose–response curves for fertility as a function of number of sperm inseminated intracervically have been developed for both mice [99] and rats [78]. Robl and Dziuk [99] found that the $ED_{50}$ sperm concentrations for fertility varied among the strains: $1.5 \times 10^6$ (DBA/2N), $3 \times 10^6$ (CF1), and $6.3 \times 10^6$ (C57BL/6N) sperm. Klinefelter et al. [78] reported that the $ED_{50}$ for rat sperm was $2.5 \times 10^6$ inseminated per uterine horn. However, an insemination dose of $5.0 \times 10^6$ sperm per uterine horn was selected to evaluate toxicant-induced alterations in fertility, as this concentration lies within the linear (i.e., sensitive) portion of the sperm dose–response curve and provides optimal control fertility (i.e., 75%); $2.5 \times 10^6$ sperm results in suboptimal control fertility and $10 \times 10^6$ sperm results in 100% fertility with no enhancement of sensitivity. Rather than recovering zygotes to calculate fertilization rates, these studies used vasectomized (sterile) males to cervically stimulate synchronized females and thereby prime them for implantation. Then, the number of postimplantation implants was determined on gestation day 9 or 20 without any decrease in control fertility levels [97,98]. These studies have been used to demonstrate the potential for by-products of drinking water disinfection to produce low experimental dose alterations in fertility, as well as identify a novel sperm protein biomarker of fertility (SP22).

## Male-Specific Endpoints

The following sections describe various male-specific endpoints of reproductive toxicity that can be obtained (Table 33.4). Guidance is presented for interpretation of results involving these endpoints and their use in risk assessment. Effects are identified that should be considered as adverse reproductive effects if significantly different from controls.

### *Body Weight and Reproductive Organ Weights*

Monitoring body weight during treatment provides an index of the general health status of the animals, and such information may be important for the interpretation of reproductive effects. Depression in body weight or reduction in weight gain may reflect a variety of responses, including rejection of chemical-containing food or water because of reduced palatability, treatment-induced anorexia, or systemic toxicity. Less than severe reductions in adult body weight induced by restricted nutrition have shown little effect on the male reproductive organs or on male reproductive function [100,101]. When a meaningful biologic relationship between a body weight decline and a significant effect on the male reproductive system is not apparent, it is not appropriate to dismiss significant alteration of the male reproductive system as secondary to the occurrence of nonreproductive toxicity. Unless additional data provide the needed clarification, alteration in a reproductive measure that would otherwise be considered adverse should still be considered as an adverse male reproductive effect in the presence of mild to moderate body weight changes. In the presence of severe body weight depression or other severe systemic debilitation, it should be noted that an adverse effect on a reproductive endpoint occurred, but the effect may have resulted from a more generalized toxic effect.

The male reproductive organs for which weights may be useful for reproductive risk assessment in adults include the testes, epididymides, pituitary gland, seminal vesicles (with coagulating glands), and prostate. Organ weight data may be presented as both absolute weights and as relative weights (i.e., organ weight to body weight ratios). Organ weight data may also be reported relative to brain weight because, subsequent to development, the weight of the brain usually remains quite stable [102]. Evaluation of data on absolute organ weights is important, because a decrease in a reproductive organ weight may occur that was not necessarily related to a reduction in body weight gain. The ratio of organ weight to body weight may show no significant difference if both body weight and organ weight change in the same direction, masking a potential organ weight effect.

Normal testis weight varies only modestly within a given test species [60]. This relatively low interanimal variability suggests that absolute testis weight should be a precise indicator of gonadal injury; however, damage to the testes may be detected as a weight change only at doses higher than those required to produce significant effects in other measures of gonadal status [103–105]. This contradiction may arise from several factors, including a delay before cell deaths are reflected in a weight decrease (due to preceding edema and inflammation, cellular infiltration) or Leydig cell hyperplasia. Blockage of the efferent ducts by cells sloughed from the germinal epithelium or the efferent ducts themselves can lead to an increase in testis weight due to fluid accumulation [106,107], an effect that could offset the effect of depletion of the germinal epithelium on testis weight. Thus, although testis weight measurements may not reflect certain adverse testicular effects and do not indicate the nature of an effect, a significant increase or decrease is indicative of an adverse effect.

Pituitary gland weight can provide valuable insight into the reproductive status of the animal; however, the pituitary contains cell types that are responsible for the regulation of a variety of physiologic functions, including some that are separate from reproduction, so changes in pituitary weight may not necessarily reflect reproductive impairment. If weight changes are observed, gonadotroph-specific histopathologic evaluations may be useful in identifying the affected cell types. This information may be used then to judge whether the observed effect on the pituitary is related to reproductive system function and therefore an adverse reproductive effect.

Prostate and seminal vesicle weights are androgen-dependent and may reflect changes in the animal's endocrine status or testicular function. Separation of the seminal vesicles and coagulating gland (dorsal prostate) is difficult in rodents; however, the seminal vesicle and prostate can be separated and results may be reported for these glands separately or together, with or without their secretory fluids. Differential loss of secretory fluids prior to weighing could produce artifactual weights. Because the seminal vesicles and prostate may respond differently to an agent (endocrine dependency and developmental susceptibility differ), more information may be gained if the weights are examined separately.

### *Adverse Effects*

Significant changes in absolute or relative male reproductive organ weights may constitute an adverse reproductive effect. Such changes also may provide a basis for obtaining additional information on the reproductive toxicity of that agent, but significant changes in other important endpoints that are related to reproductive function may not be reflected in organ weight data. For this reason, lack of an organ weight effect should not be used to negate significant changes in other endpoints that may be more sensitive.

### *Histopathologic Evaluations*

Histopathologic evaluations of test animal tissues have a prominent role in male reproductive risk assessment. Organs that are often evaluated include the testes, epididymides, prostate, seminal vesicles (often including coagulating glands), and pituitary. Tissues from lower dose exposures are often not examined histologically if the high dose produced no difference from controls. Histologic evaluations can be especially useful by: (1) providing a relatively sensitive indicator of damage; (2)

providing information on toxicity from a variety of protocols; (3) with short-term dosing, providing information on site (including target cells) and extent of toxicity; and (4) indicating the potential for recovery.

The quality of the information presented from histologic analyses of spermatogenesis is improved by proper fixation and embedding of testicular tissue. With adequately prepared tissue, a description of the nature and background level of lesions in control tissue, whether preparation-induced or otherwise, can facilitate interpreting the nature and extent of the lesions observed in tissues obtained from exposed animals [108–111].

Many histopathologic evaluations of the testis only detect lesions if the germinal epithelium is severely depleted or degenerating, if multinucleated giant cells are obvious, or if sloughed cells are present in the tubule lumen. More subtle lesions, such as retained spermatids or missing germ cell types that can significantly affect the number of sperm being released normally into the tubule lumen may not be detected when less adequate methods of tissue preparation are used. Also, familiarity with the detailed morphology of the testis and the kinetics of spermatogenesis of each test species can assist in the identification of less obvious lesions that may accompany lower dose exposures or lesions that result from short-term exposure [110]. Several approaches for qualitative or quantitative assessment of testicular tissue are available that can assist in the identification of less obvious lesions that may accompany lower dose exposures, including use of the technique of staging. A text is available [110] that provides extensive information on tissue preparation, examination, and interpretation of observations for normal and high-resolution histology of the germinal epithelium of rats, mice, and dogs. Included is guidance for identification and quantification of the various cell types and associations for each stage of the spermatogenic cycle. Also, a decision-tree scheme for staging with the rat has been published [112].

Cell staging is based on the examination of cross-sections of the seminiferous tubules. The proliferation and differentiation of germ cells is a highly ordered, time-locked process; thus, the temporal and spatial relationships of the different spermatogenic cell types can be defined for the different stages of the spermatogenic cycle (Figure 33.2) [110,112]. Based on differences in light absorption of tubules containing different stages, transillumination has also proven useful for isolating specific cell stages for subsequent biochemical analyses [113]. Knowledge of the cytoarchitecture of the testis can allow identification of specific lesions that have resulted from toxicity to the germ cell at a given stage of development.

Quantification of cell staging may include analysis of frequency distributions of the cell stages present or the proportion of tubules that have identifiable stages with all expected cell types [110,112,114]. The cell-staging approach is being applied more frequently in the evaluation of environmental agents [114–121]. It is not, however, required for routine toxicology testing.

*Morphometry* is a term applied to a variety of specialized techniques to obtain quantitative data on cellular or organelle characteristics [103,110,122]. The methods may be applied to measure diameters, areas or volumes of testicular compartments (e.g., tubular versus interstitial), specific cell types, or subcellular structures. Cell counts may also be obtained as well as ratios of one cell type to another (e.g., pachytene spermatocytes:Sertoli cell). Although not required in test protocols, these specialized methods can be used in mechanistic studies, which, in turn, can inform the risk assessment. Aside from an evaluation of the seminiferous epithelium, the interstitium of the testis also warrants a careful evaluation. Some reproductive toxicants can result in Leydig cell hyperplasia, which may or may not progress to adenoma formation. While most, if not all, of the discussion to date regarding Leydig cell hyperplasia and its association with Leydig cell adenoma formation has centered around mechanisms of toxicant-induced hyperplasia in the adult testis [123,124], it is possible that the fetal or prepubertal Leydig cell is uniquely capable of becoming hyperplastic and that the mechanisms of initiation may differ. Transgenic mice that are deficient in anti-Mullerian hormone exhibit Leydig cell hyperplasia [125]. In this regard, gestational exposures to dibutyl hexyl phthalate and dibutyl phthalate have been associated with an increased postnatal incidence of Leydig cell hyperplasia [126] and adenoma formation [127,128], respectively.

Although adenoma formation does bear clinical significance, the relevance of Leydig cell hyperplasia alone, as a reproductive effect in test species, remains controversial [124]. Numerous modes of action have been postulated for induction of Leydig cell hyperplasia, including androgen receptor antagonists, testosterone biosynthesis inhibitors, aromatase inhibitors, and estrogen agonists [124]. Each of these modes of action would result in a change in the testosterone to estradiol ratio within the Leydig cell. Even though the importance of this ratio has been demonstrated in recent years [129], its direct association with Leydig cell hyperplasia has not been demonstrated; moreover, few studies have demonstrated Leydig cell number in a definitive fashion [128,130]. There is really no way to confirm an increase number of Leydig cells in the testis without performing a thorough morphometric analysis in which the number of Leydig cells are enumerated using Leydig cell-specific probes (i.e., antibodies that recognize steroidogenic enzymes specific to Leydig cells).

*In utero* exposure to the commercially important environmental contaminant di(*n*-butyl) phthalate (DBP) has been linked to fetal Leydig cell hyperplasia [131,132]. Using the Leydig cell marker 3beta hydroxysteroid dehydrogenase and the Sertoli cell marker anti-Mullerian hor-

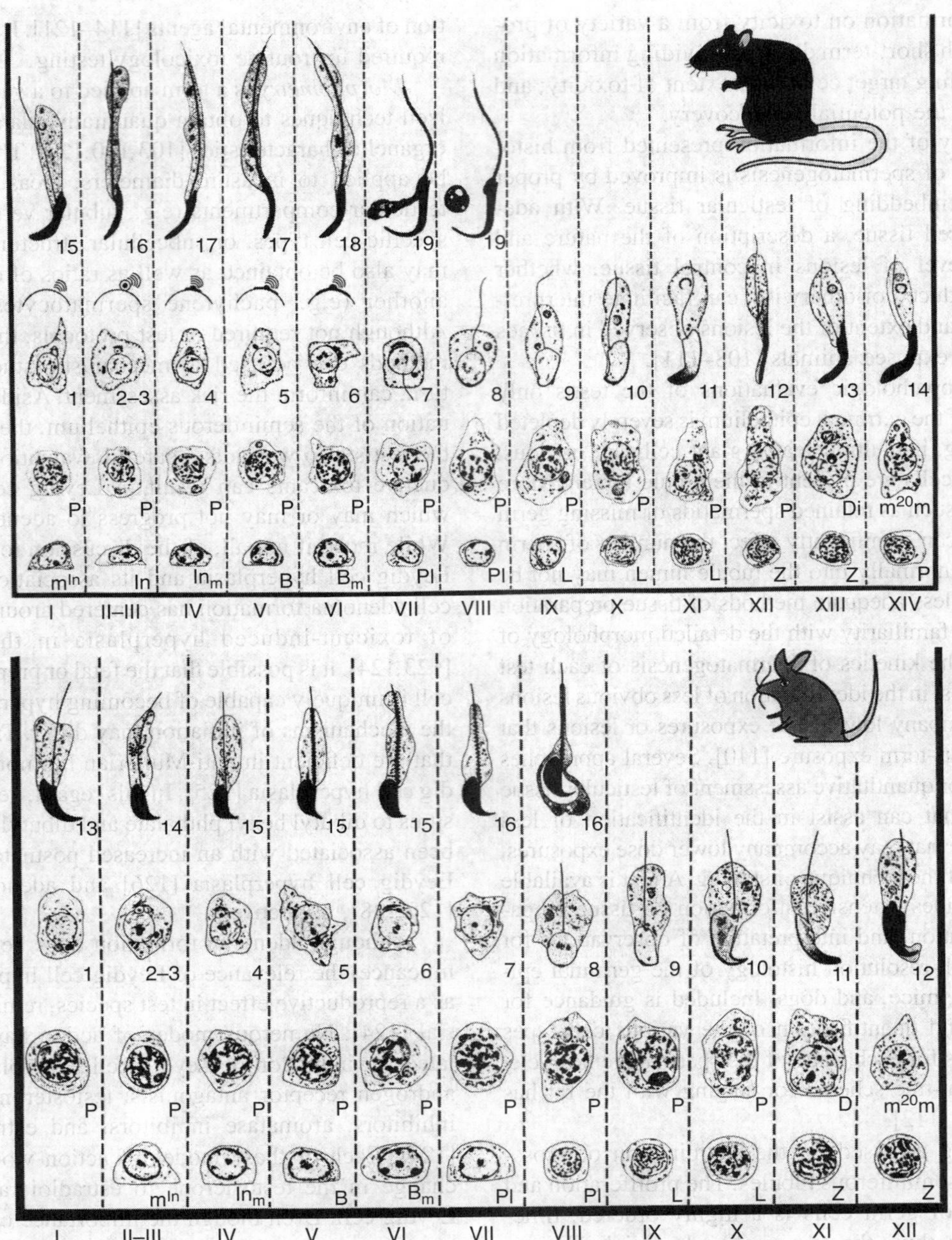


**FIGURE 33.2** Cycle maps of spermatogenesis for (top) the rat and (bottom) the mouse. The vertical columns, designated by Roman numerals depict cell associations (stages). In the scheme provided, stages II and III are combined into a single stage called II–III. The developmental progression of a cell is followed horizontally until the right-hand border of the cycle map is reached. The cell progression continues at the left of the cycle map one row up. The cycle map ends with the completion of spermiation. The symbols used designate specific phases of cell development. (From Russell, L.D. et al., *Histological and Histopathological Evaluation of the Testes*, Cache River Press, Clearwater, FL, 1990. Reprinted with permission of Cache River Press.)

mone, recent studies by Mahood et al. [133,134] now suggest that the Leydig cell aggregates that appear in the DBP exposed fetal testis represent clusters of smaller Leydig cells and trapped Sertoli cells rather than an increase in Leydig cell number. Many of these Leydig cells appear to remain within dysgenic seminiferous tubules found in the DBP-exposed testes through adulthood. The decrease in Leydig cell size itself may account for the observed reductions in intratesticular testosterone following exposure to DBP *in utero* [135,136], but genomic and steroidogenic profiling has indicated that multiple Leydig cell steroidogenic enzymes are compromised by DBP exposure [135]. Although a morphometric study of Leydig cell populations at adulthood following *in utero* exposure has not yet been performed, Leydig cell hyperplasia may exist as demonstrated following exposure to di(2-ethylhexyl) phthalate (DEHP) [128] and may possibly result from decreased testosterone biosynthesis in the fetal Leydig cell. If studied adequately, Leydig cell hyperplasia may become one of the outcomes currently associated with the testicular dysgenesis syndrome (TDS) in humans (i.e., decreased sperm quantity/quality, hypospadias, cryptorchidism, and testicular germ cell cancer). The outcomes of phthalate exposure in rodents frequently parallel those of TDS [137,138]. Testicular dysgenesis has been linked to abnormal male reproductive development or masculinization which is now known to be dependent on testosterone as well as anti-Mullerian hormone and insulin-like factor 3 (Insl3) [137]. Indeed, Insl3 mRNA and protein have recently been shown to be decreased in cryptorchid testes of rats exposed to selected phthalate esters during gestation [139,140]. Cryptorchid testes frequently manifest testicular carcinoma *in situ* (CIS) cells prior to frank formation of germ cell tumors. Recently, light and electron microscopic evaluations have revealed that the driving force in the progression to CIS cells is the chemical nature of the toxicant insult and not the abdominal location of the testis itself [141]. It is interesting that those chemicals that were shown to produce a progression from cryptorchid to CIS-cell-containing testes—1,1-dichloro-2,2-*bis*(*p*-chlorophenyl) ethylene (DDE) and estradiol—would each alter the normal testosterone:estrogen ratio during reproductive development.

The basic morphology of other male reproductive organs (e.g., epididymides, accessory sex glands, and pituitary) has been described as well as the histopathologic alterations that may accompany certain disease states [142–144]. Compared with the testes, less is known about structural changes in these tissues that are associated with exposure to toxic agents. With the epididymides and accessory sex glands, histologic evaluation is usually limited to the height and possibly the integrity of the secretory epithelium. Evaluation should include information on the caput, corpus, and cauda segments of the epididymis. The presence of debris and sloughed cells in the epididymal lumen is a valuable indicator of damage to the germinal epithelium or the excurrent ducts. The presence of lesions such as sperm granulomas, leukocyte infiltration (inflammation), or the absence of clear cells in the cauda epididymal epithelium should be noted. Information from examinations of the pituitary should include evaluation of the morphology of the cell types that produce the gonadotropins and prolactin.

Historically, the degree to which histopathologic effects are quantified has been limited to classifying animals, within dose groups, as either affected or not affected using qualitative criteria. Little effort was made to quantify the extent of injury, and procedures for such classifications were not applied uniformly [93]. This situation has prompted increased attention to the development of improved methods and more uniform approaches for evaluating the seminiferous epithelium [109–111]. These efforts have reinforced the importance of quantifying the extent of histopathologic damage per individual and have established high-quality histopathology as a sensitive and value-added endpoint in reproductive toxicity testing.

With proper tissue preparation and analysis, data from histopathologic evaluations provide a relatively sensitive tool that is useful for the detection of low-dose effects. Furthermore, changes in testis histology provide insights regarding the sites and mechanisms of action for the agent on that reproductive organ. When similar targets or mechanisms exist in humans, the basis for interspecies extrapolation is strengthened. Depending on the experimental design, information can also be obtained that may allow prediction of the eventual extent of injury and degree of recovery in that species and humans [145].

### *Adverse Effects*

Significant and biologically meaningful histopathologic damage in excess of the level seen in control tissue of any of the male reproductive organs should be considered an adverse reproductive effect. Significant histopathologic damage in the pituitary should be considered as an adverse effect, but they should be shown to involve cells that control gonadotropin or prolactin production to be considered a reproductive effect. Although thorough histopathologic evaluations that fail to reveal any treatment-related effects may be quite convincing, consideration should be given to the possible presence of other testicular or epididymal effects that are not detected histologically (e.g., genetic damage to the germ cell, decreased sperm motility) but may affect reproductive function.

## Sperm Evaluations

The EPA harmonized reproductive test guidelines [50] call for evaluation of epididymal (or vas) sperm number, sperm morphology, and sperm motility. Data on these parameters allow more adequate estimation of the number of normal sperm, a parameter that is likely to be more informative than sperm numbers alone. Although effects on sperm production can be reflected in other measures such as testicular spermatid count or cauda epididymal weight, no surrogate measures are adequate to reflect direct effects, including those that may occur after the sperm have left the testes, on sperm morphology or motility. Furthermore, similar data can be obtained noninvasively from human ejaculates, thus enhancing the ability to confirm effects seen in test species or to detect effects in humans. Note, however, that fundamental differences exist between epididymal sperm measures in rodents, which are relatively consistent among animals of the same strain, and comparable measures in human semen, which are highly variable both between and within men [46]. Brief descriptions of these measures in rats are provided below, followed by a discussion of the use of various sperm measures in male reproductive risk assessment.

*Sperm Number*

Measures of sperm concentration (count) have been the most frequently reported semen variable in the literature on humans [146]. Sperm number or sperm concentration from test species may be derived from ejaculated, epididymal, or testicular samples [147]. Of the common test species, ejaculates (sperm suspended in seminal fluid) can only be obtained readily from rabbits or dogs; however, ejaculates can be recovered from the reproductive tracts of mated females of other species including the rat [56]. Indeed, by using ovariectomized, hormonally primed, receptive females and limiting the breeding time, serial ejaculated samples can be evaluated in rats [56,148]. Using this approach, sperm are recovered from the genital tract of the receptive female soon after mating; for example, 97% of the sperm are found in the uterus within 15 minutes of coitus [149]. By using frequent mating, Hurtt and Zenick [150] were able to deplete ejaculated sperm counts and increase the sensitivity of breeding to detect toxicant-induced effects. More recently, enumeration of ejaculated sperm numbers has provided data implicating toxicant action at the level of mating behavior rather than decreased epididymal sperm numbers [151]. In this study, chemical denervation by guanethidine produced infertility that was linked to insufficient deposition of sperm upon mating rather than to alterations in sperm quality coincident with aging in the epididymis.

Ejaculated sperm number from any species is influenced by several variables, including the length of abstinence and the ability to obtain the entire ejaculate. Intra- and interindividual variation is often high but is reduced somewhat if ejaculates are collected at regular intervals from the same male, as can be done with the rabbit [152]. Likewise, repeated measures study designs in epidemiology studies can improve detection sensitivity so fewer subjects are required [153]. When a pre-exposure baseline is obtained for each man, then changes during exposure or recovery can be better defined because each individual serves as his own control or baseline [154].

In toxicity testing using rats, epididymal sperm are sampled, typically from the cauda region, but the samples for sperm motility and morphology may be derived also from the vas deferens. Automation of counting has improved the efficiency with which this endpoint can be obtained [155]. It has been customary to express the sperm count in relation to the weight of the cauda epididymis; however, because sperm contribute to epididymal weight, expression of the data as a ratio may actually mask declines in sperm number. The inclusion of data on absolute sperm counts can improve resolution. As is true for ejaculated sperm counts, epididymal sperm counts are influenced directly by level of sexual activity [150,156].

Sperm production data may be derived from counts of the distinctive elongated spermatid nuclei that remain after homogenization of testes in a detergent-containing medium [43,156–158]. The elongated spermatid counts are a measure of sperm production from the stem cells and their ensuing survival through spermatocytogenesis and spermiogenesis [43,150]. If evaluation was conducted when the effect of a lesion would be reflected adequately in the spermatid count, then spermatid count may serve as a substitute for quantitative histologic analysis of sperm production [110]; however, spermatid counts may be misleading when the duration of exposure is shorter than the time required for a lesion to be fully expressed in the spermatid count. Also, spermatid counts reported from some laboratories have large coefficients of variation that may reduce the statistical power and thus the usefulness of that measure.

The ability to detect a decrease in testicular sperm production may be enhanced if homogenization-resistant spermatid counts are obtained, but spermatid enumerations reflect only the integrity of spermatogenic processes within the testes. Post-testicular effects or toxicity expressed as alterations in motility, morphology, viability, fragility, and other properties of sperm can be determined only from epididymal, vas deferens, or ejaculated samples.

*Sperm Morphology*

Sperm morphology refers to structural aspects of sperm and can be evaluated in cauda epididymal, vas deferens, or ejaculated samples. A thorough morphologic evaluation identifies abnormalities in the sperm head and flagellum. Because of the suggested relationship between an agent's mutagenicity and its ability to induce abnormal sperm, sperm head morphology has been a frequently reported sperm variable in toxicology studies on test species [159]. The tendency has been to conclude that increased incidence of sperm head malformations reflects germ cell mutagenicity; however, not every mutagen induces sperm head abnormalities, and other nonmutagenic chemicals may alter sperm head morphology. For example, microtubule poisons may cause increases in abnormal sperm head incidence, presumably by interfering with spermiogenesis, a microtubule-dependent process [160]. Sperm morphology also may be altered due to degeneration subsequent to cell death; thus, the link between sperm morphology and mutagenicity is not consistent.

An increase in abnormal sperm morphology has been considered evidence that the agent has gained access to the germ cells [161]. Exposure of males to toxic agents may lead to sperm abnormalities in their progeny [62,162,163]; however, transmissible germ cell mutations might exist in the absence of any morphologic indicator such as abnormal sperm. The relationships between these morphologic alterations and other karyotypic changes remain uncertain [164].

The traditional approach to characterizing morphology in toxicologic testing has relied on subjective categorization of sperm head, midpiece, and tail defects in either stained preparations by bright field microscopy [165] or fixed, unstained preparations by phase contrast microscopy [69,147]. Such approaches may be adequate for mice and rats with their distinctly angular head shapes. Because human sperm exhibit considerable heterogeneity of structure and categorizing normal sperm involves subjectivity, the World Health Organization (WHO) has provided consensus guidance on classification of sperm morphology [45] and emphasizes that reference values relating this outcome to fertility depend largely on the strictness with which a normal sperm is defined. Data that categorize the types of abnormalities observed and quantify the frequencies of their occurrences may provide more information in a toxicology or epidemiology study than simply quantifying the percentage of normal sperm.

In comparison with other sperm measures, sperm morphology is a relatively stable endpoint in humans and test species. This feature may enhance its use in the detection of spermatotoxic events. The majority of studies in test species and humans have suggested that abnormally shaped sperm may not reach the oviduct or participate in fertilization [166,167]. The implication is that the greater the number of abnormal sperm or the smaller the number of normal sperm in the ejaculate, the greater the probability of reduced fertility. A prospective human male fertility study [46] has reported that the number of normal sperm (by strict criteria [168] in ejaculates was highly related to the probability of couples achieving pregnancy within one year.

*Sperm Motility*

The biochemical environments in the testes and epididymides are highly regulated to ensure proper development and maturation of the sperm and the acquisition of critical functional characteristics such as progressive motility and the potential to fertilize [169]. With chemical exposures, perturbation of this balance may occur, producing alterations in sperm properties such as motility. Chemicals (e.g., epichlorohydrin) have been identified that selectively affect epididymal sperm motility and also reduce fertility [170]. Rat sperm motility is an established male reproductive toxicology endpoint [171–174], and sperm motility assessments are an integral part of reproductive toxicity testing [50]. Motility measurements are typically obtained on epididymal rat spermatozoa, and sperm recovered from the vas deferens can be similarly analyzed [175].

Motility estimates may be obtained on ejaculated, vas deferens, or cauda epididymal samples. Standardized methods are necessary because motility is influenced by a number of experimental variables, including abstinence interval, method of sample collection and handling, elapsed time between sampling and observation, the temperature at which the sample is stored and analyzed, the extent of sperm dilution, the nature of the dilution medium, and the microscopic chamber employed for the observations [147,176–180].

Sperm motility can be evaluated in real time under phase contrast microscopy, or sperm images can be recorded and stored in video or digital format and analyzed later, either manually or by CASA [173,174, 181–185]. For manual assessments, the percentage of motile and progressively motile sperm can be estimated and a simple scale used to describe the vigor of the sperm motion.

The application of video and digital technology to sperm analysis allows a more detailed and presumably more objective evaluation of sperm motion, including information about the individual sperm tracks. It also provides a permanent record of the sperm tracks that can be reanalyzed as necessary (manually or computer-assisted). With computer-assisted technology, information about sperm velocity (straight-line and curvilinear) as well as the amplitude and frequency of the track is obtained rapidly and efficiently on large numbers of sperm [181]. Using this technology, chemically induced alterations in sperm motion have been detected [76,186,187], and such changes have been related to the fertility of the exposed animals [171,188,189]. These studies indicate that significant reductions in sperm velocity are associated with infertility, even when the percentage of motile sperm may not be affected. The ability to distinguish between the proportion of sperm showing any type of motion and those with progressive motility is important [147]. CASA can be used to provide an objective classification of progressively motile sperm by setting user-defined thresholds for one or more CASA outcomes based on the respective motion characteristics of sperm from control animals [173,174]. The computer then calculates the percentage of motile sperm that exceed these thresholds.

## Relationships between Semen Quality Measures and Fertility

Changes in endpoints that measure effects on spermatogenesis (i.e., histopathology, homogenization-resistant spermatid numbers) and sperm maturation (i.e., epididymal sperm number, motility, and morphology) have been related to fertility in several test species, but the ability to predict infertility from these data (in the absence of fertility data) for test species is not reliable. This is in part due to the observation, in both test species and humans, that fertility is dependent not only on having adequate numbers of sperm but also on the degree to which those sperm are qualitatively normal. If sperm quality is high, sperm number must be reduced substantially before fertility by natural

**TABLE 33.5**
**Relative Sensitivity of Testicular and Epididymal Sperm Parameters for the Rat**

| Parameter | Coefficient of Variation | Percent Difference Detected ($N$ = 10) | ($N$ = 15) | ($N$ = 20) |
|---|---|---|---|---|
| Testis weight | 4.65 | 6.08 | 4.96 | 4.31 |
| Epididymal weight | 9.40 | 12.30 | 10.03 | 8.70 |
| Sperm production rate[a] | 16.56 | 21.67 | 17.68 | 15.33 |
| Sperm count/g epididymis | 29.32 | 38.36 | 31.30 | 27.14 |
| Percent motile | 16.00 | 20.93 | 17.09 | 14.82 |
| Swimming speed | 7.24 | 9.47 | 7.73 | 6.70 |
| Percent normal morphology[b] | 2.70 | 3.53 | 2.88 | 2.49 |

[a] Sperm rate production = spermatid enumeration/rat.
[b] Derived from literature on Long–Evans and Sprague–Dawley rats.

*Source:* Data from Blazak, W.F. et al., *Fundam. Appl. Toxicol.*, 5, 1097–1103, 1985.

mating is affected. Similarly, if sperm numbers are normal in rodents, a relatively large effect on sperm motility is required before fertility is affected. Again, this is because rodents and other test species produce an excess of qualitatively normal sperm. In the presence of adequate numbers of sperm, average sperm velocity must be reduced substantially before fertility is affected [78]. Nevertheless, fertility in other species may be impaired by smaller changes in both number and motility (or other qualitative characteristics). Thus, relatively modest reductions in sperm number or quality may not cause infertility in species with relatively robust reproductive characteristics, but these changes can be predictive of infertility at higher exposure levels or in species that do not have the same level of excess sperm production.

In humans, the distribution of sperm counts for fertile and infertile men overlap, with the mean for fertile men being higher [190]. Recent prospective studies [46,191] have examined the influence of varying semen quality on human male fertility. Those studies have shown that the number of sperm, percent motile sperm, and percent normal sperm were all significantly associated with fertility, with the number of morphologically normal sperm using strict criteria being particularly strongly associated with fertility. These results with humans suggest that more specific biomarkers of sperm function are needed.

For semen quality measures to be of value in predicting effects on male fertility, it is necessary that the variances associated with their application be sufficiently low to have adequate resolving power. The relative variability associated with a number of indices of spermatotoxicity in the rat is indicated in Table 33.5 [60]. These data, for the most part, are in close agreement with values reported elsewhere in the literature [63,64] and reflect the percent difference that can be detected statistically for a change in a given measure as a function of sample size. Similar data on endpoint variability, including a variety of motility measures, are presented in Table 33.6.

In summary, sperm measures add considerable information and sensitivity to standard fertility measures in reproductive toxicity testing. Significant changes in any of these measures, even if modest in magnitude, may be considered adverse and used to set the no-effect or benchmark dose, particularly when they predict more severe effects, including infertility, at higher doses. Such changes are proving useful as critical endpoints for risk assessment; for example, sperm motility and morphology helped inform detailed assessments of several male reproductive toxicants (e.g., acrylamide and 1- and 2-bromopropane) by the National Institute of Environmental Health Sciences (NIEHS) Center for the Evaluation of Risks to Human Reproduction [70,90].

## Additional Markers of Sperm Function and Integrity

The functional capacity of sperm can be evaluated *in vivo* by recovering eggs at the appropriate time after copulation (species-dependent) and determining whether fertilization and normal initial development (timing of cleavage divisions) occurred [82,92,192]. Alternatively, sperm can be collected and cultured *in vitro* under capacitating conditions and then cocultured with eggs. *In vitro* fertilization (IVF) assays have been used for years to study basic mechanisms of sperm maturation and function, but only recently have such methods been proposed for use in toxicology studies [172,193], where they may be applied after either *in vitro* or *in vivo* exposures. For example, the latter approach has been used to evaluate sperm function in rats exposed acutely to the testicular toxicant 1,3 dini-

**TABLE 33.6**
**Variability in Testicular and Sperm Motility Parameters in Fischer 344 Rats**

| Indicator | $N^a$ | Range | Mean ± SD | $CV^b$ (%) |
|---|---|---|---|---|
| Testes weight (g) | 30 | 2.3–2.9 | 2.6 ± .01 | 5.61 |
| Sperm production rate (×10⁶) | | | | |
| Per testis per day | 30 | 14.7–25.6 | 22.3 ± 2.7 | 12.09 |
| Per g testis per day | 30 | 14.6–21.1 | 17.8 ± 1.5 | 8.54 |
| Cauda epididymal sperm number (×10⁶) | | | | |
| Per cauda epididymis[c] | 50 | 55.2–153.2 | 86.1 ± 20.8 | 24.16 |
| Per g cauda epididymis | 30 | 307.1–972.9 | 696.5 ± 151.9 | 21.81 |
| Cauda epididymal sperm motility | | | | |
| Motile cells (%) | 50 | 39–82 | 61 ± 9 | 14.72 |
| Curvilinear velocity (μm/sec) | 50 | 98.3–156.5 | 128.3 ± 11.4 | 8.88 |
| Straight-line velocity (μm/sec) | 50 | 41.8–93.5 | 69.2 ± 10.2 | 14.68 |
| Linearity | 50 | 4.2–6.3 | 5.4 ± 0.5 | 9.45 |

[a] Number of animals examined for each indicator.
[b] Coefficient of variation = [(SD/mean) × 100].
[c] Regions 5, 6A, and 6B of the cauda epididymis.

*Source:* Working, P.K. and Hurtt, M.E., *J. Androl.*, 8, 330–337, 1987. With permission.

trobenzene [194]. Although labor intensive, such methods may prove valuable to test hypotheses regarding toxicant-induced effects on sperm function specifically.

The extrapolation of *in vitro* fertilization data to predict the *in vivo* condition requires certain considerations. The ease with which *in vitro* fertilization can be achieved varies across species, being readily accomplished in the mouse and hamster but rather difficult in the rat. Such test systems do not reflect the dynamic role played by the female reproductive tract in sperm transport, survival, and capacitation prior to fertilization. The conditions required to achieve successful fertilization *in vitro* may, in some instances, bear little resemblance to the state that exists *in vivo* [195,196]; for example, IVF typically employs far higher sperm concentrations than those found at the site of fertilization *in vivo*.

Another direct indicator of sperm function is the ability to undergo the acrosome reaction after *in vitro* incubation under conditions that support sperm capacitation. Fluorescent probes can be used to label the acrosome and permit determination of acrosomal status [174]; for example, the plant lectin, *Pisum sativum* agglutinin, can be used to identify acrosome-intact sperm [197]. In combination with viability stains, live acrosome-reacted sperm can be distinguished from live acrosome-intact sperm and from dead sperm that have shed their acrosomes and the percentages of each type quantified using flow cytometry.

Recent research has identified specific sperm proteins associated with fertility; for example, the sperm protein SP22 has been proposed as a biomarker of effect on fertility [75,98,198]. Diminutions in this protein have been linked to toxicant-induced effects, particularly chemicals thought to alter epididymal function such as EDS, chloroethanemethanesulfonate, epichlorohydrin, α-chlorohydrin, 6-chloro-6-deoxyglucose, ornidazole, and hydroxyflutamide [198]. More recently, it has been demonstrated that testicular toxicants such as brominated haloacids [98] also produce alterations in the protein profiles of rat epididymal or ejaculated rabbit sperm. From this work, SP22 levels have been associated with fertility and changes in fertility after exposure to toxicants [97,99–202]. In some cases, the toxicant-induced diminutions in the levels of this 28-kD membrane protein proved to be more sensitive than other more typical measures of sperm quality (i.e., sperm motion and sperm morphology). Because SP22 actually originates in the testis, it is a feasible biomarker of effect following exposure to both testicular and epididymal toxicants. Antibodies to recombinant SP22 are now being used in an ELISA [202] to screen extracts of sperm from human and laboratory animals alike for chemical-induced alterations in fertility.

As mentioned earlier, however, the ability of sperm to fertilize an egg does not guarantee fertility. The sperm genome must also be intact for embryonic/fetal development to proceed normally. A variety of assays for DNA and chromosome damage in sperm are being used in human epidemiology and rodent toxicology studies. Of these, the sperm chromatin structure assay has been most widely used [203,204]. This assay is based on the properties of the fluorescent, metachromatic dye acridine orange (AO), which fluoresces green when intercalated into normal double-stranded DNA but red when bound to denatured DNA (or RNA). In the conduct of this assay, sperm are first acid denatured, then stained with AO and examined using flow cytometry. The relative abundance of red fluorescence is an indication of abnormal chromatin structure and DNA damage; increased percentages of sperm with excess red fluorescence have been correlated with infertility in humans [205]. Other tests for detecting DNA damage in human and rodent sperm, including the comet assay, the TUNEL assay, and measurement of oxidative DNA adduct sites, are being increasingly used to evaluate the effects of environmental contaminants on the genetic integrity of sperm [91,204]. New assays for detecting aneuploidy and chromosome breakage in sperm are also beginning to be applied in epidemiology and toxicology studies. These are based on the use of chromosome-specific fluorescent probes detected by fluorescence *in situ* hybridization (FISH) methods [206]. For example, exposure to air pollution was recently reported to be associated with increased incidence of an extra Y chromosome [207], and a similar effect was found to be associated with smoking [208]. Because these methods are also being developed for use with rat sperm [209], it should be possible in the near future to design studies to compare responses in humans and rats for this endpoint.

## *In Vitro* Test Systems

The following discussion highlights *in vitro* methods in use in male reproductive toxicology. *In vitro* methods have been proposed with two goals in mind: either as screens that would assist in prioritizing chemicals for *in vivo* testing or as specific tests for use in elucidating modes and mechanisms of toxicant action. For the latter purpose, it is important to consider linkage to a specific tissue or cell type that has been demonstrated to be altered by *in vivo* toxicant exposure. Even if appropriate preliminary *in vivo* data are obtained, one may fail to demonstrate an effect *in vitro*. This may reflect the need for a toxicant to interact *in vivo* with another tissue or cell type or to form an active metabolite; thus, it is advisable to ascertain both the metabolic fate and dosimetry of a toxicant and metabolites within the putative target tissue.

Today, given the rising practical and ethical concerns associated with the use of large numbers of animals for research and the growing number of environmental chemicals requiring risk assessment, there is an increased emphasis on the use of cell lines for screening potential reproductive toxicants. Progress in this direction has been limited because cell lines frequently fail to retain specific aspects of normal differentiated function. Nevertheless, they may prove to have utility in screening assays as long as provisions are made to test positive chemicals further using either a primary cell culture or an *in vivo* test system. It is also important to recognize that a chemical that tests negative in a cell line screen may actually test positive in a primary cell culture or *in vivo* system.

### Leydig Cells *In Vitro*

Today, a variety of well-accepted methods are available to identify and characterize Leydig cell toxicity *in vitro*. The choice of method depends largely on the available *in vivo* data, the tissue availability, and the sensitivity desired. Klinefelter et al. [210,211] used a highly purified Leydig cell preparation to establish the dose–response, intracellular site of action, and morphological integrity following *in vitro* EDS exposure. Results from this study demonstrated clearly that the steroidogenic lesion induced by either *in vivo* or *in vitro* EDS exposure exists between the second messenger cyclic AMP and cytochrome P450 side-chain cleavage enzyme. In another study, highly purified Leydig cells were incubated with increasing concentrations of EDS and with either a maximally stimulating concentration of luteinizing hormone (LH) or with [$^{35}$S]methionine to discriminate between functional and cytotoxic effects [212]. Similarly purified Leydig cells were used recently to study the interaction between Sertoli cells and Leydig cells during *in vitro* exposure to tri-*o*-cresyl phosphate (TOCP). This approach demonstrated that Leydig cells must first metabolize the toxicant for subsequent Sertoli cell toxicity to occur [213].

To evaluate the ability of the active metabolite of methoxychlor to produce differentiation-dependent alterations in the ability of the Leydig cell to produce testosterone during postnatal development, purified Leydig cells were isolated from rats on postnatal days 21, 35, and 90, when progenitor, immature, and adult Leydig cell populations, respectively, are present within the testis [214]. Results indicated that the prepubertal Leydig cell was more sensitive than the adult Leydig cell to disrupted cholesterol side-chain cleavage activity and cholesterol mobilization. Another relatively simple approach involves the use of decapsulated testicular parenchyma incubated with or without LH stimulation to identify alterations in testosterone biosynthetic ability under stimulated or basal conditions, respectively. These incubations can be performed using parenchyma derived from testes exposed *in vivo* or *in vitro*. An advantage is that data can be easily obtained on a per animal basis, an important consideration when using species such as the rabbit and hamster which contain relatively few steroidogenically active Leydig cells following enzymatic dispersion. This technique has been used to identify chemical effects on both LH-stimulated and basal testosterone production. The effects of EDS on LH-stimulated testosterone production were compared in the rat and hamster [215]. Testosterone production was linear over time in the testis incubations of both species, and the results demonstrated that the hamster Leydig cell is far less sensitive than the rat Leydig cell to the cytotoxic effects of EDS. Chloroethylmethanesulfonate is an example of a toxicant that resulted in significantly reduced basal testosterone production [76].

## Seminiferous Tubule Culture

Toxicology studies that have demonstrated a disruption in the process of spermatogenesis have implicated virtually every cell type in the seminiferous tubule. A toxicant may perturb either one of the early germ cell types in the basal compartment of the seminiferous epithelium or one of the more mature germ cell types in the adluminal compartment. To affect the more advanced germ cells, a toxicant might either exert its effects directly by passing through the blood–testis barrier or indirectly by perturbing Sertoli cell function.

The cell–cell interactions that occur normally within the seminiferous tubule, particularly the interactions between Sertoli cells and germ cells [216], justify the use of *in vitro* models in which the whole seminiferous tubule is studied. Seminiferous tubule cultures from rats have been used to study the *in vitro* effects of 1,3-dinitrobenzene and methoxyacetic acid (MAA). These agents are toxic to Sertoli cells and pachytene spermatocytes, respectively [217]. The amount of inhibin secreted by the seminiferous tubules *in vitro* was significantly increased by both compounds. More importantly, the stimulation in inhibin secretion was seen in rats exposed *in vivo* to doses resulting in intratesticular toxicant concentrations that approximated the effective *in vitro* concentrations. The seminiferous tubule culture reported by Allenby et al. [217] formed the basis of a series of elegant studies designed to determine the role germ cells play in the expression of androgen-dependent protein secretion [218,219]. Seminiferous tubules were isolated 4, 18, and 30 days following exposure to MAA when pachytene spermatocytes, round spermatids, and elongating spermatids, respectively, would be selectively depleted. Specific proteins were secreted selectively by these different germ cells. The effects of androgen depletion were assessed in the latter study 4 days following exposure to EDS, with or without exogenous androgen supplementation. Together these studies present evidence that specific androgen-dependent proteins are synthesized by seminiferous tubules when different germ cell types are present. This may explain why it has been impossible to demonstrate androgen-dependent proteins in isolated Sertoli cell cultures. It is clear from this work that isolated seminiferous tubules can be used to identify *in vitro* effects on the seminiferous epithelium if a toxicant disrupts spermatogenesis *in vivo*.

Short-term seminiferous tubule culture has also been employed to elucidate modes of action of certain drinking water disinfection byproducts—namely, the disubstituted haloacids—that have been identified as male reproductive toxicants in the rat. Low-dose effects after *in vivo* exposure include abnormal spermatid head morphology, delayed spermiation, and spermatid fusion, implicating a defect during spermiogenesis that could be mediated through the Sertoli cell or directly through the late germ cells. Stage-isolated seminiferous tubules from adult rats following a combination of *ex vivo* and *in vitro* exposures were used to explore the possibility that requisite Sertoli cell–germ cell interactions might be altered via changes in protein synthesis and secretion [220]. At least three specific proteins were diminished by either *in vivo* or *in vitro* exposure to dibromoacetic acid.

In the study by Allenby et al. [217] mentioned previously, a comparison was made between the response of optimized cultures of isolated seminiferous tubules and cultures of isolated Sertoli cells. The results clearly showed that inhibin production by isolated Sertoli cells was smaller and more variable in response to both toxicants. This may be attributed to the fact that the isolated Sertoli cells were derived from immature Sertoli cells. It would be useful to know whether the response of Sertoli cells from adult rats would better approximate the response of the isolated seminiferous tubules. Immature cells have been used in most of the Sertoli cell culture/toxicological studies to date, although the effects of various phthalate esters have been compared in cultures of Sertoli cells obtained from both immature and young adult rats [221]. Because differentiated function (i.e., transferrin, androgen binding protein, and inhibin secretion) of the Sertoli cell changes with sexual maturity [222,223], an *in vitro* study should use Sertoli cells that are at the same ontogenic stage as those affected in the *in vivo* studies. This is important if *in vitro* data are to be useful in risk assessment. Recent advances in culturing adult Sertoli cells may provide promising *in vitro* models for male reproductive toxicology research [224,225]. For this and any system using testicular cells [226] or combinations of cells *in vitro*, it will be important to demonstrate the extent to which differentiated endpoints are maintained and can be evaluated following both *in vivo* and *in vitro* exposures.

The use of cell lines representing cells in a less differentiated state holds more promise for screening, particularly as it applies to perturbation in development of the male reproductive system. Testicular cells from neonatal rats and novel approaches for culturing cells on matrices have been used recently to generate Sertoli cell–germ cell cocultures [227]. In theory, these undifferentiated cells should be maintained well in culture and be capable of responding to developmental male reproductive toxicants in a manner similar to that observed *in vivo*. A challenge will be to demonstrate whether functional (i.e., cell–cell interactions) and molecular (i.e., transcriptional and proteomic) alterations can be evaluated in neonatal Sertoli cells and spermatogonia following *in vivo* insult.

Advances in stem cell biology may lead to other approaches, including the use of pluripotent spermatogonial stem cells [228] at different stages of differentiation. Likewise, advances in bioengineering may produce novel systems wherein testis development can be re-enacted and studied with respect to toxicant response. For example, dissociated immature testis cells from rats were recently reported to develop into structures resembling immature seminiferous tubules when they were grown in an extracellular matrix gel and then xenographed into a rat. Encouragingly, the xenographs became vascularized and the tubular structures exhibited lumena and a few putative spermatogonia, as well as an interstitium containing putative Leydig cells [229].

### Epididymis

Because the epididymis is the organ that confers fertilizing ability on maturing sperm, it is important to develop methods that identify epididymal toxicity following *in vivo* exposure and confirm any direct toxicant action on the epididymis following *in vitro* exposure. To establish direct toxicant action, a culture system must be used that is capable of maintaining facets of normal epididymal sperm maturation. This means that the culture system must: (1) preserve the morphological integrity of both epididymal epithelial cells and cocultured sperm, (2) support protein synthesis by the epididymal epithelial cells and the association of the secreted protein with cocultured sperm, and (3) promote the acquisition of progressive motility by the cocultured sperm. An *in vitro* model has been developed that meets these criteria. Using that model, the exposure of epididymal epithelial cells and sperm to EDS *in vitro* resulted in a significant decline in protein secretion by the epididymal epithelial cells during coculture and a decreased association of specific secretory proteins with the cocultured sperm. Moreover, EDS exposure resulted in a dose-dependent decline in the progressive motility of cocultured sperm [230]. It was concluded that EDS acts directly on epididymal epithelial cells to disrupt protein secretion and this, in turn, disrupts facets of sperm maturation. Because the biomarker of fertility (SP22) was discovered using a protocol to identify toxicity within the epididymis and is now known to be expressed in the testis around the time of round spermatid formation [199], it is likely that a protein secreted by the epididymal epithelium serves to stabilize SP22 on the sperm surface. To explore this possibility, epididymal epithelial cell–sperm cocultures will be useful. Recently, a human epididymal epithelial cell–sperm coculture approach has been successfully adopted [231]. Finally, because it is known that gestational exposures to a variety of environmental chemicals perturb reproductive development, including differentiation of the Wolffian ducts, it may be worthwhile to investigate the critical events (e.g., gene/protein expression) that imprint the development of the epididymis. For this, a system for the isolation and culture of the Wolffian ducts will be required.

## Paternally Mediated Effects on Offspring

The concept is well accepted that exposure of a female to toxic chemicals during gestation or lactation may produce death, structural abnormalities, growth alteration, or postnatal functional deficits in her offspring. Sufficient data now exist with a variety of agents to conclude that male-only exposure can also produce deleterious effects in offspring [10,13,232,233]. Agents for which such adverse effects in test species have been reported include lead [234], diethylstilbestrol [235], urethane [236], cyclophosphamide [237–246], marijuana [247], and opiates [248]. Although a number of human studies have reported associations between a variety of paternal occupations and the occurrence of birth defects or childhood cancer [249–255], others have failed to observe such relationships [256–258]. Paternally mediated effects include pre- and postimplantation loss, growth and behavioral deficits, and malformations. A large proportion of the chemicals reported to cause paternally mediated effects have genotoxic activity and are considered to exert this effect via transmissible genetic or epigenetic alterations [259–260]. Low doses of cyclophosphamide have resulted in induction of single-strand DNA breaks during rat spermatogenesis that, due in part to the absence of subsequent DNA repair capability, remain at fertilization [79]. The results of such damage have been observed in F2 generation offspring [261] and may result from epigenetic changes evident as early as the pronuclear stage of development [262].

Other mechanisms of induction of paternally mediated effects also are possible. Xenobiotics present in seminal plasma or bound to the fertilizing sperm could be introduced into the female genital tract or into the oocyte directly and might also interfere with fertilization or early development. With humans, the possibility exists that a parent could transport the toxic agent from the work environment to the home (e.g., on work clothes), exposing other adults or children. Further work is needed to clarify the extent to which paternal exposures may be associated with adverse effects on offspring. Regardless, if an agent is identified in test species or in humans as causing a paternally mediated adverse effect on offspring, the effect should be considered an adverse reproductive effect.

## Sexual Behavior

Sexual behavior is a complex process involving neural and endocrine components of the central and peripheral nervous systems and the reproductive system. For humans, interactions of personality, social, and experiential factors also influence the initiation and performance of these behaviors. Similar factors may exist in other species, but they are more controlled by standardized laboratory conditions; however, the perturbation of sexual behavior in animals suggests the potential for similar effects on humans. Consistent with this position are data on CNS-active drugs that have been shown to disrupt sexual behavior in both animals and humans [263].

Although the functional components of sexual performance can be quantified in rats [264], direct evaluation of this behavior is not typically done in most breeding studies; rather, the presence of copulatory plugs or sperm-positive vaginal lavages has been interpreted as indirect evidence of successful mating. These markers do not demonstrate that male performance necessarily resulted in adequate sexual stimulation of the female. In rats, the degree of sexual preparedness of the female partner can strongly influence the site of semen deposition and subsequent sperm transport in her genital tract [265]. Failure of the female to achieve sufficient stimulation may adversely influence these processes, thereby reducing the probability of successful impregnation. Such a mating failure would be reflected in the fertility index as reduced fertility and could erroneously be attributed to a spermatotoxic effect. Other aspects of current breeding protocols exist that may serve to mask a decline in the fertility potential of a given male.

The need to evaluate directly sexual behavior routinely for all suspected reproductive toxicants is questionable. Likely candidates may be agents reported to exert neurotoxic effects, as several neurotoxicants have also been shown to produce disruptions in copulatory behavior (e.g., trichloroethylene [266], carbon disulfide [267], acrylamide [268]). Chemicals possessing or suspected to possess androgenic or estrogenic properties (or antagonistic properties) also are potential candidates for the evaluation of copulatory behavior, separate from effects on reproductive organs (e.g., chlorinated hydrocarbon pesticides).

## Pharmacokinetic Considerations

Pharmacokinetic data are most useful for risk assessment when administration of the test agent has been done by the routes expected for human exposure. Differences in metabolic fates at the site of entry, absorption rate into the blood, initial absorption into the portal vs. systemic blood, and lipophilic properties can markedly affect the amount, form, and time course in which a toxic agent is delivered to a target site. Several major factors influence the pharmacokinetics of a given agent as related to gonadal toxicity, including: (1) the existence of a blood–testis barrier that may restrict access of a compound to the adluminal compartment of the seminiferous tubules, and (2) the metabolic capability (including DNA repair) of the different compartments of the testis and other reproductive system organs that determine the eventual disposition of the agent.

The reproductive organs appear to have a wide range of metabolic capabilities directed at both steroid and xenobiotic metabolism. These properties have been best characterized for the testis. The distributions of these enzymes and cytochrome P450 levels (multiple forms) in the testis differ between interstitial and germ cell compartments. Aryl hydrocarbon hydroxylase activity and cytochrome P-450 levels in interstitial tissue are approximately twice as high as those in the seminiferous tubules. On the other hand, activities of some of the detoxifying enzymes in the germ cells such as epoxide hydralase and glutathione transferase are nearly double those in the interstitial compartment [269]. High levels of glutathione transferase are seen in the neonatal rodent testis and rapidly approach adult levels [270], suggesting an early capacity to detoxify electrophilic agents.

The protective role of glutathione should not be underestimated, as it may serve to prevent interactions between reactive electrophiles and critical cellular proteins and nucleic acids. For germ cell mutagens such as ethylene oxide, ethyl methanesulfonate, and acrylamide, the level of mutagenic response may be directly related to the rate of glutathione depletion [271–273]. Moreover, the concurrent exposure of an individual to a germ cell mutagen and a nonmutagenic, glutathione depletor could significantly lower the dose–response threshold of the former. This has been demonstrated for the induction of dominant lethal mutations by ethyl methanesulfonate administered in combination with buthionine sulfoximine, a glutathione depletor [274]. The sensitivity of different species to germ cell mutagens may also, in part, be a function of the concentration of glutathione [271,275].

The majority of pharmacokinetic studies have incompletely characterized the distribution of toxic agents and their subsequent metabolic fate within the testis. Generalizations based on hepatic metabolism are inadequate because the metabolic profile for a given agent may differ between the liver and the testes [276]. As an example, the isoenzyme patterns for glutathione transferase are markedly different for these two systems [277]. Detailed interspecies comparisons of the metabolic capabilities of the testis have not been conducted.

Attention should be directed toward delineating the relationship between the pharmacokinetic fate of an agent in the testes and the occurrence of spermatotoxicity. Of primary importance is determining the relationships between different exposure conditions (acute, intermittent, subchronic, chronic), pharmacokinetic status (e.g., bioaccumulation, steady state), and the nature of the response (i.e., transient, static, or progressive) as a function of site or mechanism of action (e.g., stem cell, mature sperm). Understanding these interactions is critical to more accurately assess the risks for different human exposure situations as well as to predict the degree of injury associated with prolonged exposures in humans. Such predictions currently must be based upon test animal data from different exposure protocols.

### Factoring Reversibility into Male Reproductive Risk Assessment

When an agent has been identified as a male reproductive toxicant, it may be of interest to determine whether or not the effects are reversible [53]. In the spermatogenic process, extensive damage to the spermatogonial stem cells is known to produce an irreversible effect. Even though an agent may affect only spermatocytes or spermatids at low dose levels, stem cells are often affected at higher doses. Damage to certain other cell types in the male reproductive system may also result in irreversible effects that are of concern (e.g., Sertoli cells).

When recovery from stem cell damage is possible, the duration of the recovery period is determined by the time for regeneration (for stem cells) and repopulation of the affected spermatogenic cell type. To that must be added the time required for appearance of those cells as sperm in an ejaculate. The time required for these events varies with species, pharmacokinetic properties of the agent, the extent to which the stem cell population has been destroyed, and the degree of sublethal toxicity inflicted on the stem cells or Sertoli cells. When the stem cell population has been partially destroyed, humans require longer to attain the same degree of recovery than mice [278].

The design of a protocol to study reversibility of effects on spermatogenesis requires assessment of degree of damage at intervals after cessation of treatment. In the absence of the ability to monitor ejaculates in a longitudinal design, necropsy of satellite groups at intervals during the recovery phase is necessary to specify the time required for recovery. Even with the ability to obtain ejaculates, data on testis parameters (e.g., histopathology, spermatid count) at the end of dosing and at intervals during recovery are useful in determining the potential for and progress of recovery.

Under some conditions, the level of concern associated with a reversible male reproductive effect might be less than with an irreversible effect; however, that stance is not necessarily justified for several reasons. First, reversibility assumes a discontinuation in exposure or a decrease in exposure below a critical (threshold) level. Thus, the assessment of reversibility must consider the exposure conditions. Second, an agent that produces a reversible effect with a low exposure level may produce an irreversible effect at a higher exposure level. Third, the potential for reversibility may vary greatly between individuals. Individuals who border on subfertility may have a reduced capability to compensate for spermatotoxicity; therefore, the extent of an effect on fertility may be greater in those individuals and the probability of full recovery less likely. Fourth, exposures that occur prior to puberty may produce effects that are not reversible, even if they would be reversible in adults. Finally, even if the effect is fully reversible, a period of infertility may be disruptive to family and career planning, as well as psychological health. Unless those factors described above have been considered carefully and judged to be of lesser significance compared to other considerations for that agent, the same level of concern should be given to an apparent reversible effect as to an irreversible male reproductive effect.

## THE FUTURE OF MALE REPRODUCTIVE TOXICOLOGY

The field of male reproductive toxicology and risk assessment has been evolving rapidly since reports of adverse effects of pesticides such as dibromochloropropane became evident in the 1970s [25]. Our understanding of modes of action of chemicals on male (and female) reproductive health has expanded with the application of cell and molecular approaches, including genomics and proteomics. The need to develop better interfacing between toxicology and epidemiology has become widely accepted [279], as has the importance of measuring contaminants in the environment and in human tissues and fluids to build more precise models of exposure–effect relationships. To this end, the Centers for Disease Control and Prevention has added many reproductive toxicants and endocrine disruptors to its list of substances measured in Americans and listed in the National Report on Human Exposure to Environmental Chemicals (see www.cdc.gov/exposurereport). The importance of understanding pathways of reproductive toxicity has grown in scope to include not only the relationships between exposure and infertility in adults but also with the impact of exposure during embryonic and fetal development on increased risks of abnormal reproductive function, cancer, and other diseases later in life. U.S. and international agencies and their industrial partners have recognized and are acting upon the need for more practical testing approaches for reproductive effects, both with respect to the efficiency and sensitivity of multigenerational test protocols [51] and the critical need for improved tools to screen the large number of existing chemicals and growing list of newly introduced substances. Although chemicals are usually listed and regulated singly, their occurrence as components of highly complex mixtures is recognized—for example, disinfection by-products in our drinking water, various products that are widely used by consumers (e.g., pharmaceuticals and personal care products), antibiotics and growth promoters (and their metabolites) used in agriculturally important species, and the rapidly growing list of nanomaterials introduced into our environment. The genomics/proteomics revolution is also identifying polymorphisms in genes that regulate reproduction and responses to toxicant exposure. This information should eventually help us understand interindividual differences in susceptibility to toxicants. These are the challenges to be faced by reproductive toxicologists in the new millennium.

## QUESTIONS

1. Why is it necessary to evaluate so many different endpoints to assess adult male reproductive toxicity? What are the limitations of the harmonized multigeneration test protocol and what changes or new endpoints might strengthen it?
2. Why is data from a one-generation or subchronic toxicity test insufficient to support a comprehensive male reproductive risk assessment?
3. What cellular and molecular mechanisms are thought to be responsible for male-mediated transmissible (transgenerational) effects?
4. Under what testing conditions is it particularly important to use the technique of staging for testis histopathologic examinations? Why?
5. What *in vitro* approaches hold the most promise for advancing our understanding of toxic mechanisms? Why?

## REFERENCES


1. Knobil, E., Plant, T. M., Pfaff, D. W., Challis, J. R. G., deKretser, D. M., Richards, J. S., and Wassarman, P. M., *Knobil and Neill's Physiology of Reproduction*, 3rd ed., Academic Press, New York, 2006.
2. Knobil, E. and Neill, J. D., *Encyclopedia of Reproduction*, Academic Press, New York, 1998.
3. Chapin, R. E., Stevens, J. T., Hughes, C. L., Kelce, W. R., Hess, R. A., and Daston, G. P., Endocrine modulation of reproduction, *Fundam. Appl. Toxicol.*, 29, 1–17, 1996.
4. Johnson, L., Welsh, T. H., and Wilker, C. E., Anatomy and physiology of the male reproductive system and potential targets for toxicants, in *Reproductive and Endocrine Toxicology*, Boekelheide, K., Chapin, R. E., Hoyer, P. B., and Harris, C., Eds., Elsevier, New York, 1998, pp. 5–62.
5. McPhaul, M. J., The biology of the male reproductive tract, in *Reproductive and Developmental Toxicology*, Korach, K. S., Ed., Marcel Dekker, New York, 1998, pp. 475–508.
6. Mosher, W. D. and Pratt, W. F., Fecundity and infertility in the United States: incidence and trends, *Nature*, 296, 575–577, 1991.
7. WHO, Towards more objectivity in diagnosis and management of male infertility, *Int. J. Androl.*, 7(Suppl.), 1–53, 1997.
8. de Kretser, D. M. and Baker, H. W. G., Infertility in men: recent advances and continuing controversies, *J. Clin. Endocrinol. Metab.*, 84, 3443–3450, 1999.
9. Thonneau, P., Marchand, S., Tallec, A., Ferial, M. L., Ducot, B., Lansac, J., Lopes, P., Tabaste, J. M., and Spira, A., Incidence and main causes of infertility in a resident population (1,850,000) of three French regions (1988–1989), *Hum. Reprod.*, 6, 811–816, 1991.
10. Davis, D. L., Friedler, G., Mattison, D., and Morris, R., Male-mediated teratogenesis and other reproductive effects: biologic and epidemiologic findings and a plea for clinical research, *Reprod. Toxicol.*, 6, 289–292, 1992.
11. Friedler, G., Paternal exposures: impact on reproductive and developmental outcome, an overview, *Pharmacol. Biochem. Behav.*, 55, 691–700, 1996.
12. Olshan, A. F. and Faustman, E. M., Male-mediated developmental toxicity, *Reprod. Toxicol.*, 7, 191–202, 1993.
13. Savitz, D. A., Sonnenfeld, N. L., and Olshan, A. F., Review of epidemiologic studies of paternal occupational exposure and spontaneous abortion, *Am. J. Indust. Med.*, 25, 361–383, 1994.
14. Robaire B. and Hales B. F., Eds., *Advances in Male Mediated Developmental Toxicity*, Kluwer Academic/Plenum Press, New York, 2003.
15. Cram, D. S., O'Bryan, M. K., and deKretser, D. M., Male infertility genetics: the future, *J. Androl.*, 22, 738–746, 2001.
16. Layman, L. C., Human gene mutations causing infertility, *J. Med. Genet.*, 39, 153–161, 2002.
17. Nishimune, Y. and Tanaka, H., Infertility caused by polymorphisms or mutations in spermatogenesis-specific genes, *J. Androl.*, 27, 326–334, 2006.
18. Eskenazi, B., Wyrobek, A. J., Sloter, E., Kidd, S. A., Moore, L., Young, S., and Moore, D., The association of age and semen quality in healthy men, *Hum. Reprod.*, 18, 447–454, 2003.
19. Wyrobek, A. J., Eskenazi, B., Young, S., Arnheim, N., Tiemann-Boege, I., Jabs, E. W., Glaser, R. L., Pearson, F. S., and Evenson, D., Advancing age has differential effects on DNA damage, chromatin integrity, gene mutations, and aneuploidies in sperm, *Proc. Natl. Acad. Sci U.S.A.*, 103(25), 9601–9606, 2006.
20. Gray, L. E., Ostby, J., Furr, J., Wolf, C. J., Lambright, C., Parks, L. et al., Effects of environmental antiandrogens on reproductive development in experimental animals, *Hum. Reprod. Update*, 7, 248–264, 2001.
21. Gray, Jr., L. E., Wilson, V. S., Stoker, T., Lambright, C., Furr, J., Noriega, N., Howdeshell, K., Ankley, G. T., and Guillette, L., Adverse effects of environmental antiandrogens and androgens on reproductive development in mammals, *Int. J. Androl.*, 29, 96–104, 2006.
22. Sharpe, R. M., Pathways of endocrine disruption during male sexual differentiation and masculinization., *Best Pract. Res. Clin. Endocrinol. Metab.*, 20, 91–110, 2006.
23. Schrader, S., Male reproductive toxicants, in *CRC Handbook of Human Toxicology*, Massaro, E. J., Ed., CRC Press, Boca Raton, FL, 1997, pp. 961–980.
24. Sever, L. E., Arbuckle, T. E., and Sweeney, A., Reproductive and developmental effects of occupational pesticide exposure: the epidemiological evidence, *Occup. Med.*, 12, 305–325, 1997.
25. Lawson C. C., Schnorr, T. M., Daston, G. P., Grajewski, B., Marcus, M., McDiarmid, M., Murono, E., Perreault, S. D., Shelby, M., and Schrader, S. M., An occupational research agenda for the third millennium, *Environ. Health Persp.*, 111, 584–592, 2003.
26. Klinefelter, G. R. and Gray, L. E., The clinical relevance of animal models: animal studies that assess the potential for drugs and environmental agents to cause reproductive disorders in humans, in *Reproductive Toxicology and Infertility*, Scialli, A. R. and Zinaman, M. J., Eds., McGraw-Hill, New York, 1993, pp. 219–282.
27. Chapin, R. E., Germ cells as targets for toxicants, in *Comprehensive Toxicology*, Boekelheide, K., Chapin, R. E., Hoyer, P. B., and Harris, C., Eds., Elsevier, New York, 1997, pp. 139–150.

28. Lamb, J. C. and Foster, P. M. D., *Physiology and Toxicology of Male Reproduction*, Academic Press, New York, 1988.
29. Lewis, J. R., *Reproductively Active Chemicals: A Reference Guide*, Van Nostrand Reinhold, New York, 1991.
30. Li, L.-H. and Heindel, J. J., Sertoli cell toxicants, in *Reproductive and Developmental Toxicology*, Korach, K. S., Ed., Marcel Dekker, New York, 1998, 655–691.
31. Perreault, S. D., The mature spermaozoon as a target for reproductive toxicants, in *Comprehensive Toxicology*, Boekelheide, K., Chapin, R. E., Hoyer, P. B., and Harris, C., Eds., Elsevier Science, New York, 1997, pp. 165–179.
32. Peterson, R. E., Cooke, P. S., Kelce, W. R., and Gray, L. E., Environmental endocrine disruptors, in *Comprehensive Toxicology*, Boekelheide, K., Chapin, R. E., Hoyer, P. B., and Harris, C., Eds., Elsevier Science, New York, 1997, 181–192.
33. Richburg, J. H., Boekelheide, K., and Blanchard, K. T., The Sertoli cell as a target for toxicants, in *Comprehensive Toxicology*, Boekelheide, K., Chapin, R. E., Hoyer, P. B., and Harris, C., Eds., Elsevier Science, New York, 1997, pp. 127–150.
34. Witorsch, R. J., *Reproductive Toxicology*, Raven Press, New York, 1995.
35. Working, P. K., *Toxicology of the Male and Female Reproductive Systems*, Hemisphere, New York, 1989.
36. Crisp, T. M., Clegg, E. D., Cooper, R. L., Wood, W. P., Anderson, D. G., Baetcke, K. P., Hoffmann, J. L., Morrow, M. S., Rodier, D. J., Schaeffer, J. E., Touart, L. W., Zeeman, M. G., and Patel, Y. M., Environmental endocrine disruption: an effects assessment and analysis, *Environ. Health Perspect.*, 106, 11–56, 1998.
37. National Academy of Science, *Hormonally Active Agents in the Environment*, National Academy Press, Washington, D.C., 1999, pp. 1–414.
38. Toppari, J., Larsen, J. C., Christiansen, P., Giwercman, A., Grandjean, P., Guillett, L. J., Jegou, B., Jensen, T. K., Jouannet, P., Keiding, N., Leffers, H., McLachlan, J. A., Meyer, O., Muller, J., Rajpert-De Meyts, E., Scheike, T., Sharpe, R., Sumpter, J., and Skakkebaek, N. E., Male reproductive health and environmental xenoestrogens, *Environ. Health Perspect.*, 104, 741–803, 1996.
39. Giwercman, A. and Bonde, J. P., Declining male fertility and environmental factors, *Endocrinol. Metab. Clin. North Am.*, 27, 807–830, 1998.
40. Bay, K., Asklund, C., Skakkebaek, N. E. and Andersson, A. M., Testicular dysgenesis syndrome: a possible role of endocrine disruptors, *Best Pract. Res. Clin. Endocrinol. Metab.*, 20, 77–90, 2006.
41. Sharpe, R. M., The 'oestrogen hypothesis': where do we stand now? *Int. J. Androl.*, 26, 2–15, 2003.
42. U.S. EPA, *Endocrine Disruptor Screening and Testing Advisory Committee (EDSTAC) Final Report*, U.S. Environmental Protection Agency, Washington, D.C., 1999 (http://www.epa.gov/scipoly/oscpendo/edspoverview/finalrpt.htm).
43. Meistrich, M. L., Quantitative correlation between testicular stem cell survival, sperm production, and fertility in the mouse after treatment with different cytotoxic agents, *J. Androl.*, 3, 56–68, 1982.
44. Robaire, B., Smith, S., and Hales, B. F., Suppression of spermatogenesis by testosterone in adult male rats: effect on fertility, pregnancy outcome and progeny, *Biol. Reprod.*, 31, 221–230, 1984.
45. World Health Organization, *WHO Laboratory Manual for the Examination of Human Semen and Sperm-Cervical Mucus Interaction*, 4th ed., Cambridge University Press, Cambridge, U.K., 1999.
46. Zinaman, M. J., Brown, C. C., Selevan, S. G., and Clegg, E. D., Semen quality and human fertility: a prospective study with healthy couples, *J. Androl.*, 21, 145–153, 2000.
47. U.S. EPA, Guidelines for reproductive toxicity risk assessment, *Fed. Reg.*, 61, 56274–56322, 1996.
48. National Research Council, *Risk Assessment in the Federal Government*, National Academy Press, Washington, D.C., 1983.
49. National Research Council, *Science and Judgement in Risk Assessment*, National Academy Press, Washington, D.C., 1994.
50. U.S. EPA, *Health Effects Test Guidelines: Reproduction and Fertility Effects*, EPA 712-C-98-208, OPPTS 870.3800, Office of Prevention, Pesticides, and Toxic Substances, U.S. Environmental Protection Agency, Washington, D.C., 1998.
51. Cooper, R. L., Lamb IV, J. C., Barlow, S. M., Bentley, K., Brady, A. M., Doerrer, N. A., Eisenbrandt, D. L., Fenner-Crisp, P. A., Hines, R. N., Irvine, L. F. H., Kimmel, C. A., Koeter, H., Li, A. A., and Makris, S. L., A tiered approach to life stages testing for agricultural chemical safety assessment, *Crit. Rev. Toxicol.*, 36, 69–98, 2006.
52. Parker R. M., Testing for reproductive toxicity, In: *Developmental and Reproductive Toxicology: A Practical Approach*, 2nd ed., Hood, R. D., Ed., Taylor & Francis, Boca Raton, FL, 2006, pp. 425–488.
53. Scialli, A. R. and Clegg, E. D., *Reversibility in Testicular Toxicity Assessment*, CRC Press, Boca Raton, FL, 1992.
54. Galbraith, W. M., Voytek, P., and Ryon, M. S., Assessment of risks to human reproduction and development of the human conceptus from exposure to environmental substances, in *Advances in Modern Environmental Toxicology*, Christian, M. S., Galbraith, W. M., Voytek, P., and Mehlman, M. A., Eds., Princeton Scientific, Princeton, NJ, 1983, pp. 41–153.
55. Clegg, E. D. and Zenick, H., Restricting mating trials enhanced the detection of ethoxyethanol-induced fertility impairment in rats, *Toxicologist*, 8, 19, 1988.
56. Zenick, H., Blackburn, K., Hope, E., and Baldwin, D. J., Evaluating male reproductive toxicity in rodents: a new animal model, *Teratogen. Carcinogen. Mutagen.*, 4, 109–128, 1984.
57. Zenick, H. and Goeden, H., Evaluation of copulatory behavior and sperm in rats: role in reproductive risk assessment, in *Physiology and Toxicology of Male Reproduction*, Lamb, J. C. and Foster, P. M. D., Eds., Academic Press, New York, 1987, pp. 174–197.
58. Berndtson, W. E. and Thompson, T. L., Age as a factor influencing the power and sensitivity of experiments for assessing body weight, testis size, and spermatogenesis in rats, *J. Androl.*, 11, 325–335, 1990.

59. Berndtson, W. E. and Clegg, E. D., Developing improved strategies to determine male reproductive risk from environmental toxins, *Theriogenology*, 38, 223–237, 1992.
60. Blazak, W. F., Ernst, T. L., and Stewart, B. E., Potential indicators of reproductive toxicity: testicular sperm production and epididymal sperm number, transit time, and motility in Fischer 344 rats, *Fundam. Appl. Toxicol.*, 5, 1097–1103, 1985.
61. Gray, L. E., Ostby, J., Ferrell, J., Sigmon, R., Cooper, R., Linder, R., Rehnberg, G., Goldman, J. M., and Laskey, J., Correlation of sperm and endocrine measures with reproductive success in rodents, in *Sperm Measures and Reproductive Success: Institute for Health Policy Analysis Forum on Science, Health, and Environmental Risk*, Alan R. Liss, New York, 1989, pp. 193–209.
62. Morrissey, R. E., Schwetz, B. A., Lamb, J. C., Ross, M. D., Teague, J. L., and Morris, R. W., Evaluation of rodent sperm, vaginal cytology, and reproductive organ weight data from National Toxicology Program 13-week studies, *Fundam. Appl. Toxicol.*, 11, 343–358, 1988.
63. Working, P. K. and Hurtt, M. E., Computerized videomicrographic analysis of rat sperm motility, *J. Androl.*, 8, 330–337, 1987.
64. Zenick, H. and Clegg, E. D., Issues in risk assessment in male reproductive toxicology, *J. Am. Coll. Toxicol.*, 5, 249–259, 1986.
65. Gray, Jr., L. E., Wilson, V., Noriega, N., Lambright, C., Furr, J., Stoker, T. E., Laws, S. C., Goldman, J., Cooper, R. L., and Foster, P. M. D., Use of laboratory rat as a model in endocrine disruptor screening and testing, *ILAR*, 45, 425–437, 2004.
66. Francis, E. Z. and Kimmel, G. L., Proceedings of the workshop on one- vs. two-generation reproductive effects studies, *J. Am. Coll. Toxicol.*, 7, 911–925, 1988.
67. Gray, Jr., L. E., Delayed effects on reproduction following exposure to toxic chemicals during critical periods of development, in *Aging and Environmental Toxicology: Biological and Behavioral Perspectives*, Cooper, R. L., Goldman, J. M., and Harbin, T. J., Eds., The Johns Hopkins University Press, Baltimore, MD, 1991, pp. 183–210.
68. Park, D. V., Development of detoxication mechanisms in the neonate, in *Toxicology and the Newborn*, Kacew, S. and Reason, M. J., Eds., Elsevier, New York, 1984, pp. 1–32.
69. Gray, L. E., Chemical-induced alterations of sexual differentiation: a review of effects in humans and rodents, in *Advances in Modern Environmental Toxicology*, Colburn, T. and Clement, C., Eds., Princeton Scientific, Princeton, NJ, 1992, pp. 203–230.
70. Boekelheide, K., Darney, S. P., Daston, G. P., David, R. M., Luderer, U., Olshan, A. F., Sanderson, W. T., Willhite, C. C., and Woskie, S., NTP Center for the Evaluation of Risks to Human Reproduction Bromopropane Expert Panel, NTP-CERHR Expert Panel report on the reproductive and developmental toxicity of 2-bromopropane, *Reprod. Toxicol.*, 18, 189–217, 2004.
71. Anway, M. D., Cupp, A. S., Uzumcu, M., and Skinner, M. K., Epigenetic transgenerational actions of endocrine disruptors and male fertility, *Science*, 308, 1466–1469, 2005.
72. Ho, S. M., Tang, W. Y., Belmonte de Frausto, J., and Prins, G. S., Developmental exposure to estradiol and bisphenol A increases susceptibility to prostate carcinogenesis and epigenetically regulates phosphodiesterase type 4 variant 4, *Cancer Res.*, 66:5624–5632, 2006.
73. Christian, M. S., A critical review of multigenerational studies, *J. Am. Coll. Toxicol.*, 5, 161–180, 1986.
74. Laws, S. C., Stoker, T. E., Goldman, J. M., Wilson, V., Gray, Jr., L. E., and Cooper, R. L., The U.S. EPA Endocrine Disruptor Screening Program: *in vitro* and *in vivo* mammalian tier 1 screening assays, in *Developmental and Reproductive Toxicology: A Practical Approach*, 2nd ed., Hood, R. D., Ed., Taylor & Francis, Boca Raton, FL, 2006, pp. 489–523.
75. Klinefelter, G. R., Laskey, J., Roberts, N. L., Slott, V. L., and Suarez, J., Multiple effects of ethane dimethanesulfonate on the epididymis of adult rats, *Toxicol. Appl. Pharmacol.*, 105, 271–287, 1990.
76. Klinefelter, G. R., Laskey, J., Kelce, W. R., Ferrell, J., Roberts, N. L., Suarez, J., and Slott, V. L., Chloroethylmethanesulfonate-induced effects on the epididymis seem unrelated to altered Leydig cell function, *Biol. Reprod.*, 51, 82–91, 1994.
77. Klinefelter, G. R. and Suarez, J., Toxicant-induced acceleration of epididymal sperm transit: androgen-dependent proteins may be involved, *Reprod. Toxicol.*, 11, 511–519, 1997.
78. Klinefelter, G. R., Laskey, J., Perreault, S. D., Ferrell, J., Jeffay, S. C., Suarez, J., and Roberts, N. L., The ethane dimethanesulfonate-induced decrease in the fertilizing ability of cauda epididymal sperm is independent of the testis, *J. Androl.*, 15, 318–327, 1994.
79. Qui, J., Hales, B. F., and Robaire, B., Adverse effects of cyclophosphamide on progeny outcome can be mediated through post-testicular mechanisms in the rat, *Biol. Reprod.*, 46, 926–931, 1992.
80. Wagenfeld, A., Ching-Hei, Y., Strupat, K., and Cooper, T. G., Shedding of a rat epididymal sperm protein associated with infertility induced by ornidazole and a-chlorohydrin, *Biol. Reprod.*, 58, 1257–1265, 1998.
81. Slott, V. L., Jeffay, S. C., Dyer, C. J., Barbee, R. R., and Perreault, S. D., Sperm motion predicts fertility in male hamsters treated with alpha-chlorohydrin, *J. Androl.*, 18, 708–716, 1997.
82. Linder, R., Strader, L. F., Slott, V. L., and Suarez, J., Endpoints of spermatotoxicity in the rat after short duration exposures to fourteen reproductive toxicants, *Reprod. Toxicol.*, 6, 491–505, 1992.
83. Linder, R., Klinefelter, G. R., Strader, L. F., Narotsky, M. G., Suarez, J., Roberts, N. L., and Perreault, S. D., Dibromoacetic acid affects reproductive competence and sperm quality in the male rat, *Fundam. Appl. Toxicol.*, 28, 9–17, 1995.
84. Linder, R., Klinefelter, G. R., Strader, L. F., Suarez, J., and Roberts, N. L., Spermatotoxicity of dichloroacetic acid, *Reprod. Toxicol.*, 11, 681–688, 1997.
85. Strader, L. F., Suarez, J., Roberts, N. L., and Klinefelter, G. R., Spermatotoxicity of bromochloroacetic acid (BCA) in the rat after 14 daily doses, *Toxicologist*, 47, 511A, 1998.

86. Harris, M. W., Chapin, R. E., Lockhart, A. C., Jokinen, M. P., Allen, J. D., and Haskins, E. A., Assessment of a short-term reproductive and developmental toxicity screen, *Fundam. Appl. Toxicol.*, 19, 186–196, 1992.
87. OECD, *Guideline for Testing of Chemicals: Extended Steering Group Document 3*, Organization for Economic Cooperation and Development, Brussels, Belgium, 1990.
88. Tanaka, S., Kawashima, K., Naito, K., Usami, M., Nakadate, M., Imaida, K., Takahashi, M., Hayashi, Y., Kurokawa, Y., and Tobe, M., Combined repeat dose and reproductive/developmental toxicity screening test (OECD): familiarization using cyclophosphamide, *Fundam. Appl. Toxicol.*, 18, 89–95, 1992.
89. Green, S., Auletta, A., Fabricant, R., Kapp, M., Sheu, C., Springer, J., and Whitfield, B., Current status of bioassays in genetic toxicology: the dominant lethal test, *Mutat. Res.*, 154, 49–67, 1985.
90. Manson, J., Brabec, M. J., Buelke-Sam, J., Carlson, G. P., Chapin, R. E., Favor, J. B., Fischer, L. J., Hattis, D., Lees, P. S., Perreault-Darney, S., Rutledge, J., Smith, T. J., Tice, R. R., and Working, P., NTP-CERHR expert panel report on the reproductive and developmental toxicity of acrylamide, *Birth Defects Res. B Dev. Reprod. Toxicol.*, 74, 17–113, 2005.
91. Perreault, S. D., Aitken, R. J., Baker, H. W. G., Evenson, D. P., Huszar, G., Irvine, D. S., Morris, I. D., Morris, R. A., Robbins, W. A., Sakkas, D., Spano, M., and Wyrobek, A. J., Integrating new tests of sperm genetic integrity into semen analysis: breakout group discussion, *Adv. Exp. Med. Biol.*, 518, 253–268, 2003.
92. Sublet, V., Zenick, H., and Smith, M. K., Factors associated with reduced fertility and implantation rates in females mated to acrylamide-treated rats, *Toxicology*, 55, 53–67, 1989.
93. Linder, R., Strader, L. F., Barbee, R. R., Rehnberg, G., and Perreault, S. D., Reproductive toxicity of a single dose of 1,3-dinitrobenzene in two ages of young adult male rats, *Fundam. Appl. Toxicol.*, 14, 284–298, 1990.
94. Goldstein, L. S., Use of an *in vitro* technique to detect mutations induced by antineoplastic drugs in mouse germ cells, *Cancer Treat. Rep.*, 68, 855–858, 1984.
95. Amann, R. P., Detection of alterations in testicular and epididymal function in laboratory animals, *Environ. Health Perspect.*, 70, 149–158, 1986.
96. Cukierski, M. A., Sina, J. L., Prahalada, S., and Robertson, R. T., Effects of seminal vesicle and coagulating gland ablation on fertility in rats, *Reprod. Toxicol.*, 5, 347–352, 1991.
97. Klinefelter, G. R., Laskey, J., Ferrell, J., Suarez, J., and Roberts, N. L., Discriminant analysis indicates a single sperm protein (SP22) is predictive of fertility following toxicant exposure, *J. Androl.*, 18, 139–150, 1997.
98. Klinefelter, G. R., Suarez, J., Roberts, N. L., and Strader, L. F., The sperm biomarker SP22 is highly correlated with infertility resulting from the testicular toxicant bromochloroacetic acid, *Biol. Reprod.*, 60, 187A, 1999.
99. Robl, J. M. and Dziuk, P. J., Influence of the concentration of sperm on the percentage of eggs fertilized for three strains of mice, *Gamete Res.*, 10, 415–422, 1984.
100. Chapin, R. E., Gulati, D. K., Barnes, L. H., and Teague, J. L., The effects of feed restriction on reproductive function in Sprague–Dawley rats, *Fundam. Appl. Toxicol.*, 20, 23–29, 1993.
101. Chapin, R. E., Gulati, D. K., Fail, P. A., Hope, E., Russell, S. R., Heindel, J. J., George, J. D., Grizzle, T. B., and Teague, J. L., The effects of feed restriction on reproductive function in Swiss CD-1 mice, *Fundam. Appl. Toxicol.*, 20, 15–22, 1993.
102. Stevens, K. R. and Gallo, M. A., Practical considerations in the conduct of chronic toxicity studies, in *Principles and Methods of Toxicology*, Hayes, A. W., Ed., Raven Press, New York, 1989, pp. 237–250.
103. Berndtson, W. E., Methods for quantifying mammalian spermatogenesis: a review, *J. Anim. Sci.*, 44, 818–833, 1977.
104. Foote, R. H., Schermerhorn, E. C., and Simkin, M. E., Measurement of semen quality, fertility, and reproductive hormones to assess dibromochloropropane (DBCP) effects in live rabbits, *Fundam. Appl. Toxicol.*, 6, 637, 1986.
105. Ku, W. W., Chapin, R. E., Wine, R. N., and Gladen, B. C., Testicular toxicity of boric acid (BA): relationship of dose to lesion development and recovery in the F344 rat, *Reprod. Toxicol.*, 7, 305–319, 1993.
106. Hess, R. A., Moore, B. J., Forrer, J., Linder, R., and Abuel-Atta, A. A., The fungicide Benomyl (methyl 1-(butylcarbamoyl)-2-benzimidazolecarbamate) causes testicular dysfunction by inducing the sloughing of germ cells and occlusion of efferent ductules, *Fundam. Appl. Toxicol.*, 17, 733–745, 1991.
107. Nakai, M., Moore, B. J., and Hess, R. A., Epithelial reorganization and irregular growth following carbendazim-induced injury of the efferent ductules of the rat testis. *Anat. Rec.*, 235, 51–60, 1993.
108. Chapin, R. E., Morphologic evaluation of seminiferous epithelium of the testis, in *Physiology and Toxicology of Male Reproduction*, Lamb, J. C. and Foster, P. M. D., Eds., Academic Press, New York, 1988, pp. 155–177.
109. Hess, R. A. and Moore, B. J., Histological methods for evaluation of the testis, in *Methods in Toxicology: Male Reproductive Toxicology*, Chapin, R. E. and Heindel, J. J., Eds., Academic Press, San Diego, CA, 1993, pp. 52–85.
110. Russell, L. D., Ettlin, R., Sinha Hikim, A. P., and Clegg, E. D., *Histological and Histopathological Evaluation of the Testes*, Cache River Press, Clearwater, FL, 1990.
111. Creasy, D. M., Evaluation of testicular toxicology: a synopsis and discussion of the recommendations proposed by the Society of Toxicologic Pathology, *Birth Defects Res. B Dev. Reprod. Toxicol.*, 68, 408–415, 2003.
112. Hess, R. A., Quantitative and qualitative characteristics of the stages and transitions in the cycle of the rat seminiferous epithelium: light microscopic observations of perfusion-fixed and plastic-embedded testes, *Biol. Reprod.*, 43, 525–542, 1990.
113. Parvinen, M. and Vanha-Pettula, T., Identification and enzymatic quantification of the stages of the seminiferous epithelial wave in the rat, *Anat. Rec.*, 174, 435–449, 1972.
114. Chapin, R. E., Dutton, S. L., Ross, M. D., Sumrell, B. M. and Lamb, J. C., The effects of ethylene glycol monomethyl ether on testicular histology in F344 rats, *J. Androl.* 5, 369–380, 1984.

115. Chapin, R. E., White, R. D., Morgan, K. T., and Bus, J. S., Studies of lesions induced in the testis and epididymis of F-344 rats by inhaled methyl chloride, *Toxicol. Appl. Pharmacol.*, 76, 328–343, 1984.
116. Creasy, D. M., Ford, G. R., and Gray, T. J., The morphogenesis of cyclohexylamine-induced testicular atrophy in the rat: *in vivo* and *in vitro* studies, *Exp. Mol. Pathol.*, 52, 155–169, 1990.
117. Creasy, D. M., Foster, J. R., and Foster, P. M., The morphological development of di(*n*-pentyl) phthalate induced testicular atrophy in the rat, *J. Pathol.*, 139, 309–321, 1983.
118. Creasy, D. M., Jones, H. B., Beech, L. M., and Gray, T. J., The effects of two testicular toxins on the ultrastructural morphology of mixed cultures of Sertoli and germ cells: a comparison with *in vivo* effects, *Fund. Chem. Toxicol.*, 24, 655–656, 1986.
119. Hess, R. A., Linder, R., Strader, L. F., and Perreault, S. D., Acute and long term sequelae of 1,3-dinitrobenzene on male reproduction in the rat. II. Quantitative and qualitative histopathology of the testis, *J. Androl.*, 9, 327–342, 1988.
120. Somkuti, S. G., Lapadula, D. M., Chapin, R. E., and Abou-Donia, M. B., Light and electron microscopic evidence of tri-*O*-cresyl phosphate (TOCP)-mediated testicular toxicity in Fischer 344 rats, *Toxicol. Appl. Pharmacol.*, 107, 35–46, 1991.
121. Treinen, K. A. and Chapin, R. E., Development of testicular lesions in F344 rats after treatment with boric acid, *Toxicol. Appl. Pharmacol.*, 107, 325–335, 1991.
122. Mori, H. and Christensen, A. K., Morphometric analysis of Leydig cells in the normal rat testis, *J. Cell Biol.*, 84, 340–354, 1980.
123. Clegg, E. D., Cook, J. C., Chapin, R. E., Foster, P. M., and Daston, G. P., Leydig cell hyperplasia and adenoma formation: mechanisms and relevance to humans, *Reprod. Toxicol.*, 11, 107–121, 1997.
124. Cook, J. C., Klinefelter, G. R., Hardisty, J. F., Sharpe, R. M., and Foster, P. M., Rodent Leydig cell tumorigenesis: a review of the physiology, pathology, mechanisms, and relevance to humans, *Crit Rev. Toxicol.*, 29, 169–261, 1999.
125. Racine, C., Rey, R., Forest, M. G., Louis, F., Ferre, A., Huhtaniemi, I., Josso, N., and diClemente, N., Receptors for anti-Mullerian hormone on Leydig cells are responsible for its effects on steroidogenesis and cell differentiation, *Proc. Natl. Acad. Sci. U.S.A.*, 95, 594–599, 1998.
126. Parks, L. G., Ostby, J., Lambright, C. R., Abbott, B. D., Klinefelter, G. R., and Gray, L. E., Perinatal butyl benzyl phthalate (BBP) and *bis*(2-ethylhexyl) phthalate (DEHP) exposures induce antiandrogenic effects in Sprague–Dawley rats, *Biol. Reprod.*, 60, 191A, 1999.
127. Mylchreest, E., Sar, M., Cattley, R. C., and Foster, P. M. D., Disruption of androgen-regulated male reproductive development by di(*n*-butyl) phthalate during late gestation in rats is different from flutamide, *Toxicol. Appl. Pharmacol.*, 156, 81–95, 1999.
128. Akingbemi, B. T., Renshan, G., Klinefelter, G. R., Zirkin, B. R., and Hardy, M. P., Phthalate-induced Leydig cell hyperplasia is associated with multiple endocrine disturbances, *PNAS*, 101, 775–780, 2004.
129. Rivas A., Fisher, J. S., McKinnell C., Atanassova, N., and Sharpe, R. M., Induction of reproductive tract developmental abnormalities in the male rat by lowering androgen production or action in combination with a low dose of diethylstilbestrol: evidence for importance of the androgen-estrogen balance. *Endocrinology*, 143, 4797–4808, 2002.
130. Tarka-Leeds, D. K., Suarez, J. D., Roberts, N. L., Rogers, J. M., and Hardy, M. P., Gestational exposure to ethane dimethanesulfonate permanently alters reproductive competence in the CD-1 mouse, *Biol. Reprod.*, 69, 959–967, 2003.
131. Parks, L. G., Ostby, J. S., Lambright, C. R., Abbott, B. D., Klinefelter, G. R., Barlow, N. J., and Gray, Jr., L. E., The plasticizer diethylhexyl phthalate induces malformations by decreasing fetal testosterone synthesis during sexual differentiation in the male rat, *Toxicol. Sci.*, 58, 339–349, 2000.
132. Mylchreest, E., Wallace, D. G., Cattley, R. C., and Foster, P. M. D., Dose-dependent alterations in androgen-regulated male reproductive development in rats exposed to di(*n*-butyl) phthalate during late gestation, *Toxicol. Sci.*, 55, 143–151, 2000.
133. Mahood, I. K., Hallmark, N., McKinnell, C., Walker, M., Fisher, J., and Sharpe, R. M., Abnormal Leydig cell aggregation in the fetal testis of rats exposed to di(*n*-butyl) phthalate and its possible role in testicular dysgenesis, *Endocrinology*, 146, 613–623, 2005.
134. Mahood, I. K., McKinnell, C., Walker, M., Hallmark, N., Scott, H., Fisher, J., Rivas, A., Hartung, S., Ivell, R., Mason, J. I., and Sharpe, R. M., Cellular origins of testicular dysgenesis in rats exposed *in utero* to di(*n*-butyl) phthalate, *Int. J. Androl.*, 29, 148–154, 2006.
135. Lehmann, K. P., Phillips, S., Sar, M., Foster, P. M. D., and Gaido, K. W., Dose-dependent alterations in gene expression and testosterone synthesis in the fetal testes of male rats exposed to di(*n*-butyl) phthalate, *Toxicol. Sci.*, 81, 60–68, 2004.
136. Barlow, N. J., Phillips, S. L., Wallace, D. G., Sar, M., Gaido, K. W., and Foster, P. M. D., Quantitative changes in gene expression in fetal rat testes following exposure to di(*n*-butyl) phthalate, *Toxicol. Sci.*, 73, 431–451, 2003.
137. Sharpe, R. M., Pathways of endocrine disruption during male sexual differentiation and masculinisation., *Best Pract. Res. Clin. Endrocrinol. Metab.*, 20, 91–110, 2006.
138. Fisher, J. S., Environmental anti-androgens and male reproductive health: focus on phthalates and testicular dysgenesis syndrome, *Soc. Reprod. Fertil.*, 1470–1626, 2004.
139. Wilson, V. S., Lambright, C., Furr, J., Ostby, J., Wood, C., Held, G., and Gray, Jr., L. E., Phthalate ester-induced gubernacular lesions are associated with reduced insl3 gene expression in the fetal rat testis, *Toxicol. Lett.*, 146, 207–215, 2004.
140. McKinnell C., Sharpe, R. M., Mahood, K., Hallmark, N., Scott, H., Ivell, R., Staub, C., Jegou, B., Haag, F., Koch-Nolte, F., and Hartung, S., Expression of insulin-like factor 3 protein in the rat testis during fetal and postnatal development and in relation to cryptorchidism induced by in utero exposure to di(*n*-butyl) phthalate. *Endocrinology*, 146, 4536–4544, 2005.

141. Veeramachaneni, D. N. R., Germ cell atypia in undescended testes hinges on the aetiology of cryptorchidism but not the abdominal location per se, *Int. J. Androl.*, 29, 235–240, 2006.
142. Fawcett, D. W., *Bloom and Fawcett: A Textbook of Histology*, W.B. Saunders, Philadelphia, PA, 1986.
143. Haschek, W. M. and Rousseaux, C. G., *Handbook of Toxicologic Pathology*, Academic Press, New York, 1991.
144. Jones, T. C. and Mohr, U., *Genital System*, Springer-Verlag, New York, 1987.
145. Russell, L. D., Normal testicular structure and methods of evaluation under experimental and disruptive conditions, in *Reproductive and Developmental Toxicity of Metals*, Clarkson, T. W., Nordberg, G. F., and Sager, P. R., Eds., Plenum Press, New York, 1983, pp. 227–252.
146. Wyrobek, A. J., Gordon, L. A., Burkhart, J. G., Francis, M. W., Kapp, R. W., Letz, G., Malling, H. V., Topham, J. C., and Whorton, D. M., An evaluation of the mouse sperm morphology test and other sperm tests in nonhuman mammals, *Mutat. Res.*, 115, 1–72, 1983.
147. Seed, J., Chapin, R. E., Clegg, E. D., Dostal, L. A., Foote, R. H., Hurtt, M. E., Klinefelter, G. R., Makris, S. L., Perreault, S. D., Schrader, S., Seyler, D., Sprando, R., Trienen, K. A., Veeramachaneni, D. N. R., and Wise, L. D., Methods for assessing sperm motility, morphology, and counts in the rat, rabbit, and dog: a consensus report, *Reprod. Toxicol.*, 10, 237–244, 1996.
148. Ratnasooriya, W. D., A simplified method for measuring ejaculated sperm content of male rats, *J. Pharmacol. Meth.*, 2, 379–381, 1979.
149. Carballada, R. and Esponda, P., Role of fluid from seminal vesicles and coagulating glands in sperm transport into the uterus and fertility in rats, *J. Reprod. Fertil.*, 95, 639–648, 1992.
150. Hurtt, M. E. and Zenick, H., Decreasing epididymal sperm reserves enhances the detection of ethoxyethanol-induced spermatotoxicity, *Fundam. Appl. Toxicol.*, 7, 348–353, 1986.
151. Kempinas, W. D., Suarez, J. D., Roberts, N. L., Strader, L. F., Ferrell, J. M., Goldman, J. M., Narotsky, M. G., Perreault, S. D., and Klinefelter, G. R., Fertility of rat epididymal sperm after chemically and surgically induced sympathectomy, *Biol. Reprod.*, 59, 897–904, 1998.
152. Williams, J., Gladen, B. C., Schrader, S. M., Turner, T. W., Phelps, J. L., and Chapin, R. E., Semen analysis and fertility assessment in rabbits: statistical power and design considerations for toxicology studies, *Fundam. Appl. Toxicol.*, 15, 651–665, 1990.
153. Wyrobek, A. J., Watchmaker, G., and Gordon, L., An evaluation of sperm tests as indicators of germ-cell damage in men exposed to chemical or physical agents, in *Reproduction: The New Frontier in Occupational and Environmental Health Research*, Lockey, J. E., Lemasters, G. K., and Keye, W. R., Eds., Alan R. Liss, New York, 1984, pp. 385–407.
154. Rubes, J., Selevan, S. G., Zudova, D., Zudova, Z., Evenson, D. P., and Perreault, S. D., Exposure to episodic air pollution is associated with increased DNA fragmentation in human sperm without other changes in semen quality, *Hum. Reprod.*, 20, 2776–2783, 2005.
155. Strader, L. F., Linder, R. E., and Perreault, S. D., Comparison of rat epididymal sperm counts by IVOS HTM-IDENT and hemacytometer, *Reprod. Toxicol.*, 10, 529–533, 1996.
156. Amann, R. P., A critical review of methods for evaluation of spermatogenesis from seminal characteristics, *J. Androl.*, 2, 37–58, 1981.
157. Blazak, W. F., Treinen, K. A., and Juniewicz, P. E., Application of testicular sperm head counts in the assessment of male reproductive toxicity, in *Methods in Toxicology: Male Reproductive Toxicology*, R. E. Chapin, R. E. and Heindel, J. J., Eds., Academic Press, San Diego, 1993, pp. 86–94.
158. Cassidy, S. L., Dix, K. M., and Jenkins, T., Evaluation of a testicular sperm head counting technique using rats exposed to dimethoxyethyl phthalate (DMEP), glycerol alpha-monochlorohydrin (GMCH), epichlorohydrin (ECH), formaldehyde (FA), or methyl methanesulphonate (MMS), *Arch. Toxicol.*, 53, 71–78, 1983.
159. Wyrobek, A. J., Gordon, L. A., Burkhart, J. G., Francis, M. W., Kapp, R. W., Letz, G., Malling, H. V., Topham, J. C., and Whorton, D. M., An evaluation of human sperm as indicators of chemically induced alterations of spermatogenic function, *Mutat. Res.*, 115, 73–148, 1983.
160. Russell, L. D., Malone, J. P., and McCurdy, D. S., Effect of microtubule disrupting agents, colchicine and vinblastine, on seminiferous tubule structure in the rat, *Tiss. Cell*, 13, 349–367, 1981.
161. U.S. EPA, Guidelines for mutagenicity risk assessment, *Fed. Reg.*, 51, 34006–34012, 1986.
162. Hugenholtz, A. P. and Bruce, W. R., Radiation induction of mutations affecting sperm morphology in mice, *Mutat. Res.*, 107, 177–185, 1983.
163. Wyrobek, A. J. and Bruce, W. R., The induction of sperm-shape abnormalities in mice and humans, in *Chemical Mutagens: Principles and Methods for Their Detection*, Hollander, A. and de Serres, F. J., Eds., Plenum Press, New York, 1978.
164. de Boer, P., van der Hoeven, F. A., and Chardon, J. A. P., The production, morphology, karyotypes and transport of spermatozoa from tertiary trisomic mice and the consequences for egg fertilization, *J. Reprod. Fertil.*, 48, 249–256, 1976.
165. Filler, R., Methods for evaluation of rat epididymal sperm morphology, in *Methods in Toxicology: Male Reproductive Toxicology*, Chapin, R. E. and Heindel, J. J., Eds., Academic Press, San Diego, CA, 1993, pp. 334–343.
166. Nestor, A. and Handel, M. A., The transport of morphologically abnormal sperm in the female reproductive tract of mice, *Gamete Res.*, 10, 119–125, 1984.
167. Redi, C. A., Garagna, S., Pellicciari, C., Manfredi-Romanini, M. G., Capanna, E., Winking, H., and Gropp, A., Spermatozoa of chromosomally heterozygous mice and their fate in male and female genital tracts, *Gamete Res.*, 9, 273–286, 1984.
168. Menkveld, R., Stander, F. S. H., Kotze, T. J., Kruger, T. F., and VanZyl, J. A., The evaluation of morphological characteristics of human spermatozoa according to stricter criteria, *Hum. Reprod.*, 5, 586–592, 1990.

169. Robaire, B., Hinton, B. T., and Orgebin-Crist, M.-C., The epididymis, In: *Knobil and Neill's Physiology of Reproduction*, 3rd ed., Knobil, E., Plant, T. M., Pfaff, D. W., Challis, J. R. G., deKretser, D. M., Richards, J. S. and Wassarman, P. M., Eds., Academic Press, New York, 2006, pp. 1071–1148.
170. Toth, G. P., Stober, J. A., Zenick, H., Read, E. J., Christ, S. A., and Smith, M. K., Correlation of sperm motion parameters with fertility in rats treated subchronically with epichlorohydrin, *J. Androl.*, 12, 54–61, 1991.
171. Toth, G. P. et al., The automated analysis of rat sperm motility following subchronic epichlorohydrin administration: methodologic and statistical considerations, *J. Androl.*, 10, 401–415, 1989.
172. Perreault, S. D., Gamete toxicology: the impact of new technologies, in *Reproductive and Developmental Toxicology*, Korach, K., Ed., Marcel Dekker, New York, 1998, pp. 635–654.
173. Perreault, S. D., Smart use of computer-aided sperm analysis (CASA) to characterize sperm motion, in *The Epididymis, From Molecules to Clinical Practice*, Robaire, B. and Hinton, B. H., Eds., Kluwer, New York, 2002, pp. 459–471.
174. Perreault, S. D. and Cancel, A., Significance of incorporating measures of sperm production and function into rat toxicology studies, *Reproduction*, 121, 207–216, 2001.
175. Dostal, L. A., Faber, C. K., and Zandee, J., Sperm motion parameters in vas deferens and cauda epididymal rat sperm, *Reprod. Toxicol.*, 10, 231–235, 1996.
176. Chapin, R. E., Filler, R. S., Gulati, D., Heindel, J. J., Katz, D. F., Mebus, C. A., Obasaju, F., Perreault, S. D., Russell, S. R., and Schrader, S., Methods for assessing rat sperm motility, *Reprod. Toxicol.*, 6, 267–273, 1992.
177. Schrader, S., Chapin, R. E., Clegg, E. D., Davis, R. O., Fourcroy, J. L., Katz, D. F., Rothman, S. A., Toth, G., Turner, T. W., and Zinaman, M. J., Laboratory methods for assessing human semen in epidemiologic studies: a consensus report, *Reprod. Toxicol.*, 6, 275–279, 1992.
178. Slott, V. L., Suarez, J., and Perreault, S. D., Rat sperm motility analysis: methodologic considerations, *Reprod. Toxicol.*, 5, 449–458, 1991.
179. Toth, G. P., Stober, J. A., George, E. L., Read, E. J., and Smith, M. K., Sources of variation in the computer-assisted motion analysis of rat epididymal sperm, *Reprod. Toxicol.*, 5, 487–495, 1991.
180. Weir, P. J. and Rumberger, D., Isolation of rat sperm from the vas deferens for sperm motion analysis, *Reprod. Toxicol.*, 9, 327–330, 1995.
181. Boyers, S. P., Davis, R. O., and Katz, D. F., Automated semen analysis, *Curr. Prob. Obstet. Gynecol. Fertil.*, 12, 173–200, 1989.
182. Linder, R., Hess, R. A., and Strader, L. F., Testicular toxicity and infertility in male rats treated with 1,3-dinitrobenzene, *J. Toxicol. Environ. Health*, 19, 477–489, 1986.
183. Slott, V. L. and Perreault, S. D., Computer-assisted sperm analysis of rodent epididymal sperm motility using the Hamilton–Thorne motility analyzer, in *Methods in Toxicology: Male Reproductive Toxicology*, Chapin, R. E. and Heindel, J. J., Eds., Academic Press, San Diego, CA, 1993, pp. 319–333.
184. Toth, G. P., Zenick, H., and Smith, M. K., Effects of epichlorohydrin on male and female reproduction in Long–Evans rats, *Fundam. Appl. Toxicol.*, 13, 16–25, 1989.
185. Yeung, C. H., Oberlander, G., and Cooper, T. G., Characterization of the motility of maturing rat spermatozoa by computer-aided objective measurement, *J. Reprod. Fertil.*, 96, 427–441, 1992.
186. Slott, V. L., Suarez, J., Simmons, J. E., and Perreault, S. D., Acute inhalation exposure to epichlorohydrin transiently decreases rat sperm velocity, *Fundam. Appl. Toxicol.*, 15, 597–606, 1990.
187. Toth, G. P., Wang, S. R., McCarthy, H., Tocco, D. R., and Smith, M. K., Effects of three male reproductive toxicants on rat cauda epididymal sperm motion, *Reprod. Toxicol.*, 6, 507–515, 1992.
188. Oberlander, G., Yeung, C. H., and Cooper, T. G., Induction of reversible infertility in male rats by oral ornidazole and its effects on sperm motility and epididymal secretions, *J. Reprod. Fertil.*, 100, 551–559, 1994.
189. Slott, V. L., Jeffay, S. C., Suarez, J., Barbee, R. R., and Perreault, S. D., Synchronous assessment of sperm motility and fertilizing ability in the hamster following treatment with alpha-chlorohydrin, *J. Androl.*, 16, 523–535, 1995.
190. Meistrich, M. L. and Brown, C. C., Estimation of the increased risk of human infertility from alterations in semen characteristics. *Fertil. Steril.*, 40, 220–230, 1983.
191. Bonde, J. P., Ernst, E., Jensen, T. K., Hjollund, N. H., Kolstad, H., Henriksen, T. B., Scheike, T., Giwercman, A., Olsen, J., and Skakkebaek, N. E., Relation between semen quality and fertility: a population-based study of 430 first-pregnancy planners, *Lancet*, 352, 1172–1177, 1998.
192. Perreault. S. D., Distinguishing between fertilization failure and early pregnancy loss when identifying male-mediated adverse pregnancy outcomes, *Adv. Exp. Med. Biol.*, 518, 189–198, 2003.
193. Darney, S. P., *In vitro* assessment of gamete integrity, in *In Vitro Toxicology: Mechanisms and New Technology*, Goldberg, A. M., Ed., Mary Ann Liebert, New York, 1991, pp. 63–75.
194. Holloway, A. J., Moore, H. D. M., and Foster, P. M. D., The use of *in vitro* fertilization to detect reductions in the fertility of male rats exposed to 1,3-dinitrobenzene, *Fundam. Appl. Toxicol.*, 14, 113–122, 1990.
195. Goeden, H. and Zenick, H., Influence of the uterine environment on rat sperm motility and swimming speed, *J. Exp. Zool.*, 233, 247–251, 1985.
196. Shalgi, R., Developmental capacity of rat embryos produced by *in vivo* or *in vitro* fertilization, *Gamete Res.*, 10, 77–82, 1984.
197. Graham, J. K., Kunze, E., and Hammerstedt, R. H., Analysis of sperm cell viability, acrosomal integrity, and mitochondrial function using flow cytometry, *Biol. Reprod.*, 43, 55–64, 1990.
198. Klinefelter, G. R. and Hess, R. A., Toxicology of the male excurrent ducts and accessory glands, in *Reproductive and Developmental Toxicology*, Korach, K. S., Ed., Marcel Dekker, New York, 1998, pp. 553–591.

199. Klinefelter, G. R. and Welch, J. E., The saga of a male fertility protein (SP22), *Ann. Rev. Biomed. Sci.*, 1, 145–184, 1999.
200. Klinefelter, G. R., Welch, J. E., Perreault, S. D., Moore, H. D., Zucker, R. M., Suarez, J. D., Roberts, N. L., Bobseine, K., and Jeffay, S. C., Localization of the sperm protein SP22 and inhibition of fertility *in vivo* and *in vitro*, *J. Androl.*, 23, 48–63, 2002.
201. Klinefelter, G. R., Strader, L. F., Suarez, J. D., and Roberts, N. L., Bromochloroacetic acid exerts qualitative effects on rat sperm: implications for a novel biomarker, *Toxicol. Sci.*, 68, 164–173, 2002.
202. Kaydos, E. H., Suarez, J. D., Roberts, N. L., Bobseine, K., Laskey, J. L., and Klinefelter, G. R., Haloacid induced alterations in fertility and the sperm biomarker SP22 in the rat are additive: validation of an ELISA, *Toxicol. Sci.*, 81, 419–429, 2004.
203. Evenson, D. and Jost, L., Sperm chromatin structure assay: DNA denaturability, in *Methods in Cell Biology*, Darzynkiewicz, L. and Robinson, J. P., Eds., Academic Press, New York, 1994, pp. 159–176.
204. Evenson, D. P., Larson, K. L., and Jost, L. K., Sperm chromatin structure assay: its clinical use for detecting sperm DNA fragmentation in male infertility and comparisons with other techniques, *J. Androl.*, 23, 25–43, 2002.
205. Evenson, D., Jost, L., Marshall, D., Zinaman, M. J., Clegg, E. D., Purvis, K., deAngelis, P., and Claussen, O. P., Utility of the sperm chromatin structure assay as a diagnostic and prognostic tool in the human fertility clinic, *Hum. Reprod.*, 14, 1039–1049, 1999.
206. Baumgartner, A., Van Hummelen, P., Lowe, X. R., Adler, I. D., and Wyrobek, A. J., Numerical and structural chromosomal abnormalities detected in human sperm with a combination of multicolor FISH assays, *Environ. Mol. Mutagen.*, 33, 49–58, 1999.
207. Robbins, W. A., Rubes, J., Selevan, S. G., and Perreault, S. D., Air pollution and sperm aneuploidy in healthy young men, *Environ. Epidemiol. Toxicol.*, 1, 125–131, 1999.
208. Rubes, J., Lowe, X. R., Moore, D., Perreault, S. D., Slott, V. L., Evenson, D., Selevan, S. G., and Wyrobek, A. J., Smoking cigarettes is associated with increased sperm disomy in teenage men, *Fertil. Steril.*, 70, 715–723, 1998.
209. Lowe, X. R., de Stoppelaar, J. M., Bishop, J. B., Cassel, M., Hoebee, B., Moore, D., and Wyrobek, A. J., Epididymal sperm aneuploidies in three strains of rats detected by multicolor fluorescence *in situ* hybridization, *Environ. Molec. Mutagen.*, 31, 125–132, 1998.
210. Klinefelter, G. R., Laskey, J., and Roberts, N. L., *In vitro/in vivo* effects of ethane dimethanesulfonate on Leydig cells of adult rats, *Toxicol. Appl. Pharmacol.*, 107, 460–471, 1991.
211. Klinefelter, G. R., Hall, P. F., and Ewing, L. L., Effect of luteinizing hormone deprivation *in situ* on steroidogenesis of rat Leydig cells purified by a multistep procedure, *Biol. Reprod.*, 36, 769–783, 1987.
212. Kelce, W. R., Zirkin, B. R., and Ewing, L. L., Immature rat Leydig cells are intrinsically less sensitive than adult Leydig cells to ethane dimethanesulfonate, *Toxicol. Appl. Pharmacol.*, 111, 189–200, 1991.
213. Chapin, R. E., Phelps, J. L., Somkuti, S. G., Heindel, J. J., and Burka, L. T., The interaction of Sertoli and Leydig cells in the testicular toxicity of tri-*o*-cresyl phosphate, *Toxicol. Appl. Pharmacol.*, 104, 483–495, 1990.
214. Akingbemi, B. T., Ge, R. S., Klinefelter, G. R., Gunsalus, G. L., and Hardy, M. P., A metabolite of methoxychlor, 2,2-*bis*(*p*-hydroxyphenyl)-1,1,1-trichloroethane, reduces testosterone biosynthesis in rat Leydig cells through suppression of steady-state messenger ribonucleic acid levels of the cholesterol side-chain cleavage enzyme, *Biol. Reprod.*, 62, 571–578, 2000.
215. Gray, Jr., L. E., Klinefelter, G. R., Kelce, W. R., Laskey, J., Ostby, J., Marshall, R., and Ewing, L. L., Hamster Leydig cells are less sensitive to ethane dimethane sulphonate when compared to rat Leydig cells both *in vivo* and *in vitro*, *Toxicol. Appl. Pharmacol.*, 130, 248–256, 1995.
216. Janecki, A., Jakubowiak, A., and Steinberger, A., Effect of germ cells on vectorial secretion of androgen binding protein and transferrin by immature rat Sertoli cells, *J. Androl.*, 9, 126–132, 1988.
217. Allenby, G., Foster, P. M., and Sharpe, R. M., Evaluation of changes in the secretion of immunoactive inhibin by adult rat seminiferous tubules *in vitro* as an indicator of early toxicant action on spermatogenesis, *Fundam. Appl. Toxicol.*, 16, 710–724, 1991.
218. McKinnell, C. and Sharpe, R. M., The role of specific germ cell types in modulation of the secretion of androgen-regulated proteins (ARPs) by stage VI–VIII seminiferous tubules from the adult rat, *Mol. Cell Endocrinol.*, 83, 219–231, 1992.
219. Sharpe, R. M., Maddocks, S., Millar, M., Kerr, J. B., Saunders, P. T. K., and McKinnell, C., Testosterone and spermatogenesis: identification of stage-specific, androgen-regulated proteins secreted by adult rat seminiferous tubules, *J. Androl.*, 13, 172–184, 1992.
220. Holmes, M., Suarez, J., and Klinefelter, G. R., Dibromoacetic acid perturbs protein synthesis in adult rat seminiferous tubules, *Biol. Reprod.*, 60, 146A, 1999.
221. Heindel, J. J. and Powell, C. J., Phthalate ester effects on rat Sertoli cell function *in vitro*: effects of phthalate side chain and age of animal, *Toxicol. Appl. Pharmacol.*, 115, 116–123, 1992.
222. Castellon, E., Janecki, A., and Steinberger, A., Influence of germ cells on Sertoli cell secretory activity in direct and indirect co-culture with Sertoli cells from rats of different ages, *Mol. Cell Endocrinol.*, 64, 169–178, 1989.
223. Wright, W. W., Zabludoff, S. D., Erickson-Lawrence, M., and Karzai, A. W., Germ-cell–Sertoli cell interactions, *Ann. N.Y. Acad. Sci.*, 564, 173–185, 1989.
224. Anway, M. D., Folmer, J., Wright, W. W., and Zirkin, B. R., Isolation of Sertoli cells from adult rat testes: an approach to *ex vivo* studies of Sertoli cell function, *Biol Reprod.*, 68, 996–1002, 2003.
225. Anway, M. D., Show, M. D., and Zirkin, B. R., Protein C inhibitor expression by adult rat Sertoli cells: effect of testosterone withdrawal and replacement, *J. Androl.*, 26, 578–585, 2005.
226. Rahman, N. A. and Huhtaniemi, I. T., Testicular cell lines *Mol. Cell Endocrinol.*, 228, 53–65, 2004.

227. Yu, X., Sidhu, J. S., Hong, S., and Faustman, E. M., Essential role of extracellular matrix (ECM) overlay in establishing the functional integrity of primary neonatal rat sertoli cell/gonocyte co-cultures: an improved *in vitro* model for assessment of male reproductive toxicity, *Toxicol. Sci.*, 84, 378–393, 2005.
228. Guan, K., Nayernia, K., Maier, L. S., Wagner, S., Dressel, R., Lee, J. H., Nolte, J., Wolf, F., Li, M., Engel, W., and Hasenfuss, G., Pluripotency of spermatogonial stem cells from adult mouse testis, *Nature*, 440, 1199, 2006.
229. Gassei, K., Schlatt, S., and Ehmche, J., *De novo* morphogenesis of seminiferous tubules from dissociated immature rat testicular cells in xenografts, *J. Androl.*, 27, 611–618, 2006.
230. Klinefelter, G. R., Roberts, N. L., and Suarez, J., Direct effects of ethane dimethanesulphonate on epididymal function in adult rats: an *in vitro* demonstration, *J. Androl.*, 13, 409–421, 1992.
231. Moore, H. D., Curry, M. R., Penfold, L. M., and Pryor, J. P., The culture of human epididymal epithelium and *in vitro* maturation of epididymal spermatozoa, *Fertil. Steril.*, 58, 776–783, 1992.
232. Colie, C. F., Male mediated teratogenesis, *Reprod. Toxicol.*, 7, 3–9, 1993.
233. Qui, J., Hales, B. F., and Robaire, B., Damage to rat spermatozoal DNA after chronic cyclophosphamide exposure, *Biol. Reprod.*, 53, 1465–1473, 1995.
234. Brady, M., Herrera, Y., and Zenick, H., Influence of parental lead exposure on subsequent learning ability in offspring, *Pharmacol. Biochem. Behav.*, 3, 561–565, 1975.
235. Turusov, V. S., Trukhanova, L. S., Parfenov, Y. D., and Tomatis, L., Occurrence of tumours in the descendents of CBA male mice prenatally treated with diethylstilbestrol, *Int. J. Cancer*, 50, 131–135, 1992.
236. Nomura, T., Parental exposure to x-rays and chemicals induces heritable tumors and anomalies in mice, *Nature*, 296, 575–577, 1982.
237. Adams, P. M., Fabricant, J. D., and Legator, M. S., Cyclophosphamide-induced spermatogenic effects detected in the F1 generation by behavioral testing, *Science*, 211, 80–82, 1981.
238. Adams, P. M., Fanini, D., and Legator, M. S., Neurobehavioral effects of paternal drug exposure on the development of offspring, in *Functional Teratogenesis*, Fujii, T. and Adams, P. M., Eds., Teikyo University Press, 1987, pp. 147–156.
239. Auroux, M. R., Dulioust, E. M., Nawar, N. Y., and Yacoub, S. G., Antimitotic drugs (cyclophosphamide and vinblastine) in the male rat: deaths and behavioral abnormalities in the offspring, *J. Androl.*, 7, 378–386, 1986.
240. Hales, B. F. and Robaire, B., Reversibility of effects of chronic paternal exposure to cyclophosphamide on pregnancy outcome in rats, *Mutat. Res.*, 229, 129–134, 1990.
241. Hales, B. F., Smith, S., and Robaire, B., Cyclophosphamide in the seminal fluid of treated males: transmission to females by mating and effect on pregnancy outcome, *Toxicol. Appl. Pharmacol.*, 84, 423–430, 1986.
242. Jenkinson, P. C. and Anderson, D., Malformed foetuses and karyotype abnormalities in the offspring of cyclophosphamide and allyl alcohol-treated male rats, *Mutat. Res.*, 229, 173–184, 1990.
243. Kelly, S. M., Robaire, B., and Hales, B. F., Paternal cyclophosphamide treatment causes postimplantation loss via inner cell mass-specific cell death, *Teratology*, 45, 313–318, 1992.
244. Trasler, J. M., Hales, B. F., and Robaire, B., Paternal cyclophosphamide treatment of rats causes fetal loss and malformations without affecting male fertility, *Nature*, 316, 144–146, 1985.
245. Trasler, J. M., Hales, B. F., and Robaire, B., Chronic low dose cyclophosphamide treatment of adult male rats: effect on fertility, pregnancy outcome and progeny, *Biol. Reprod.*, 34, 275–283, 1986.
246. Trasler, J. M., Hales, B. F., and Robaire, B., A time course study of paternal cyclophosphamide treatment in rats: effects on pregnancy outcome and the male reproductive and hematologic systems, *Biol. Reprod.*, 37, 317–326, 1987.
247. Dalterio, S. L., Steger, R. W., and Bartke, A., Maternal or paternal exposure to cannabinoids affects central neurotransmitter levels and reproductive function in male offspring, in *The Cannabinoids: Chemical, Pharmacologic, and Therapeutic Aspects*, Agurell, S., Dewey, W. L., and Willette, R. E., Eds., Academic Press, New York, 2005, pp. 411–425.
248. Friedler, G. and Wheeling, H. S., Behavioral effects in offspring of males injected with opioids prior to mating, *Pharmacol. Biochem. Behav.*, 11, 23–28, 1979.
249. Aschengrau, A. and Monson, R. R., Paternal military service in Vietnam and the risk of late adverse pregnancy outcomes, *Am. J. Public Health*, 80, 1218–1224, 1990.
250. Federick, J., Anencephalus in the Oxford record linkage study area, *Child Neurol.*, 18, 643–656, 1976.
251. Hemminki, K., Saloniemi, I., and Salonen, T., Childhood cancer and paternal occupation in Finland, *J. Epidemiol. Commun. Health*, 35, 11–15, 1981.
252. Johnson, C. C., Annegers, J. F., Frankowski, R. F., Spitz, M. R., and Buffler, P. A., Childhood nervous system tumors: an evaluation of the association with paternal occupational exposure to hydrocarbons, *Am. J. Epidemiol.*, 126, 605–613, 1987.
253. Kluwe, W. M., Weber, H., Greenwell, A., and Harrington, F., Initial and residual toxicity following acute exposure of developing male rats to dibromochloropropane, *Toxicol. Appl. Pharmacol.*, 79, 54–68, 1985.
254. Peters, J. M., Preston-Martin, S., and Yu, M. C., Brain tumors in children and occupational exposure of the parents, *Science*, 213, 235–237, 1981.
255. Polednak, A. P. and Janerich, D. T., Use of available record systems in epidemiologic studies of reproductive toxicology, *Am. J. Indust. Med.*, 4, 329–348, 1983.
256. Hemminki, K., Mutanen, P., Luoma, K., and Saloniemi, I., Congenital malformations by the parental occupation in Finland, *Int. Arch. Occup. Environ. Health*, 46, 93–98, 1980.
257. Papier, C. M., Parental occupation and congenital malformations in a series of 35,000 births in Israel, *Prog. Clin. Biol. Res.*, 163, 291–294, 1985.
258. Zack, M., Cannon, S., Lloyd, D., Heath, C. W., Falleta, J. M., Jones, B., Housworth, J., and Crowley, S., Cancer in children of parents exposed to hydrocarbon-related industries and occupations, *Am. J. Epidemiol.*, 3, 329–336, 1980.

259. Hales, B. F. and Robaire, B., Paternally mediated effects on development, in *Handbook of Developmental Toxicology*, Hood, R. D., Ed., CRC Press, Boca Raton, FL, 1996, pp. 91–107.
260. Hales, B. F. and Robaire, B., Paternally mediated effects on development, in *Developmental and Reproductive Toxicology: A Practical Approach*, 2nd ed., Hood, R. D., Ed., CRC Press, Boca Raton, FL, 2006, pp. 125–145.
261. Hales, B. F., Crosman, K., and Robaire, B., Increased post-implantation loss and malformations among the F2 progeny of male rats chronically treated with cyclophosphamide, *Teratology*, 45, 671–678, 1992.
262. Barton, T. S., Robaire, B., and Hales, B. F., Epigenetic programming in the preimplantation rat embryo is disrupted by chronic paternal cyclophosphamide exposure, *Proc. Natl. Acad. Sci. U.S.A.*, 102, 7865–7870, 2005.
263. Rubin, H. B. and Henson, D. E., Effects of drugs on male sexual function, in *Advances in Behavioral Pharmacology*, Academic Press, New York, 1979, pp. 65–86.
264. Dewsbury, D. A., A quantitative description of the behavior of rats during copulation, *Behavior*, 29, 154–178, 1967.
265. Adler, N. T. and Toner, J. P., The effect of copulatory behavior on sperm transport and fertility in rats, in *Behavioral and Neuroendocrine Perspective*, Komisaruk, B. R., Siegel, H. I., Chang, M. F., and Feder, H. H., Eds., New York Academy of Science, New York, 1986, pp. 21–32.
266. Nelson, J. L. and Zenick, H., The effect of trichloroethylene on male sexual behavior: possible opiate role, *Neurobehav. Toxicol. Teratol.*, 8, 441–445, 1986.
267. Zenick, H., Blackburn, K., Hope, E., and Baldwin, D. J., An assessment of the copulatory, endocrinologic, and spermatotoxic effects of carbon disulfide exposure in rats, *Toxicol. Appl. Pharmacol.*, 73, 275–283, 1984.
268. Zenick, H., Hope, E., and Smith, K., Reproductive toxicity associated with acrylamide treatment in male and female rats, *J. Toxicol. Environ. Health*, 17, 457–472, 1986.
269. Mukhtar, H., Philpot, R. M., Lee, I. P., and Bend, J. R., Developmental aspects of epoxide-metabolizing enzyme activities in adrenals, ovaries, and testes of the rat, in *Developmental Toxicology of Energy Related Pollutants*, Mahlum, D. D., Sikov, M. R., Hackett, P. L., and Andrew, F. D., Eds., Technical Information Center, U.S. Department of Energy, Springfield, VA, 1978, pp. 89–104.
270. Grosshans, K. and Calvin, H. I., Estimation of glutathione in purified populations of mouse testis germ cells, *Biol. Reprod.*, 33, 1197–1205, 1985.
271. McKelvey, J. A. and Zemaitis, M. A., The effects of ethylene oxide (EO) exposure on tissue glutathione levels in rats and mice, *Drug Chem. Toxicol.*, 9, 51–66, 1986.
272. Smith, M. K., Zenick, H., Preston, R. J., George, E. L., and Long, R. E., Dominant lethal effects of subchronic acrylamide administration in the male Long–Evans rat, *Mutat. Res.*, 173, 273–277, 1986.
273. Teaf, C. M., Harbison, R. D., and Bishop, J. B., Germ cell mutagenesis and GSH depression in reproductive tissue of the F-344 rat induced by ethyl methanesulfonate, *Mutat. Res.*, 144, 93–98, 1985.
274. Bishop, J. B. and Teaf, C. M., *A Dominant Lethal Mutation Study in Male F-344 Rat: Effects of Ethyl Methane Sulfonate Alone and in Combination with Agents Perturbing the Glutathione System in Reproductive Tissue, Experiment 6298 and 6314*, NCTR Final Report, National Center for Toxicological Research, U.S. Food and Drug Administration, Washington, D.C., 1985.
275. Gandy, J., Bates, H. K., Conder, L. A., and Harbison, R. D., Effects of reproductive tract glutathione enhancement and depletion on ethyl methanesulfonate-induced dominant lethal mutations in Sprague–Dawley rats, *Teratogen. Carcinogen. Mutagen.*, 12, 61–70, 1992.
276. Lee, I. P. and Nagayama, J., Metabolism of benzo(*a*)pyrene by the isolated perfused rat testis, *Cancer Res.*, 40, 3297–3303, 1980.
277. Guthenberg, B., Alin, P., and Mannervik, B., Glutathione transferases in rat testis, *Acta Chem. Scand.*, 261, 1983.
278. Meistrich, M. L. and Samuels, R. C., Reduction in sperm levels after testicular irradiation of the mouse: a comparison with man, *Rad. Res.*, 102, 138–147, 1985.
279. Lawson, C. L., Grajewski, B., Daston, G. P., Frazier, L., Lynch, D., McDiarmid, M., Murono, E., Perreault, S. D., Robbins, W., Shelby, M., and Whelan, E. A. Implementing a national occupational reproductive research agenda: decade one and beyond, *Environ. Health Persp.*, 114, 435–341, 2006.

# 34 Test Methods for Assessing Female Reproductive and Developmental Toxicology

*Mildred S. Christian*

## CONTENTS

## PURPOSE

The principal subject of this chapter is the presentation of practical methods used in performing reproductive (female) and developmental toxicity studies. In general, the methods reported are those used in studies conducted for regulatory use—that is, in the process of identifying the safe use of pharmaceuticals (including small molecules and biologics), chemicals, pesticides, fungicides, direct and indirect food additives, and devices. Although many additional *in vitro* and elegant *in vivo* methods exist for screening agents and to identify mechanisms, the methods described in this chapter are generally limited to *in vivo* tests conducted for regulatory use in standard laboratory species. Tests used to evaluate reproductive toxicity in male animals are addressed elsewhere in this book, as are *in vitro* methods used to evaluate development, genomic methodology, and *in vivo* behavioral and functional tests. Mechanistic studies employing biotechnology techniques and methods for identifying the pharmacodynamics and kinetics of specific exposures are alluded to but are also described in detail in other chapters.

## INTRODUCTION

Although diseases and some agents, such as alcohol, historically had been considered to possibly affect reproductive performance or a given pregnancy, before 1961 human birth defects were usually considered spontaneous or hereditary events [236]. Jackson's definitive 1925 review [95] of the effects of malnutrition identified potential reduction in fertility, reduced birth weight, abortion, or death of the conceptus, but not congenital malformation, as possible outcomes. The relationship between maternal measles and congenital cataracts in humans was the first case of a disease being clearly identified as affecting the outcome of a human pregnancy [73], but this causal relationship was considered unique. Only after World War II, when the effects of malnutrition in Leningrad and Holland were reported [3,195], was it determined that maternal malnutrition could result in congenital malformations in humans as well as other unfavorable outcomes, including reproductive failure, abnormal or absent menstrual cycles, increased abortion, small birth weight, and reduced postnatal survival.

Animal research in the first half of the twentieth century generally focused on enhancing the reproductive performance of animals used for food and the adverse effects of malnutrition on these animals. Although a 1921 anecdotal report identified a congenital malformation associated with maternal malnutrition (rudimentary limbs in four piglets born to a sow fed a deficient diet in a study of fat-soluble factor [241]), it was not until Hale's studies of vitamin A deficiency in pigs [76,77] that it was definitively demonstrated that the observed anophthalmia was associated with a maternal dietary deficiency (vitamin A). Congenital malformations in a conventional laboratory species, the rat, were first reported in 1940 and 1941 by Warkany and Nelson [234,235], who were also studying agents that might affect nutrition and subsequently affect reproduction. Once it was clearly proven that mammalian maternal organisms did not prevent teratogenic effects, considerable research into teratogenic agents and susceptibility to these agents followed [236].

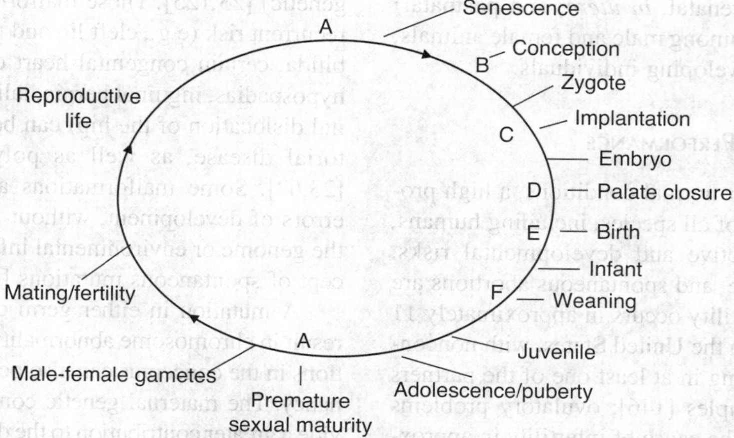


**FIGURE 34.1** Reproductive and developmental effects.

Regulatory research and revision and development of testing guidelines were directly related to three human tragedies that resulted from *in utero* exposure to a drug: (1) 1961, congenital malformations [122,124]; (2) early 1970s, cancer [85]; and (3) 1976, behavioral and functional alterations [115]. The first event, the thalidomide tragedy, completely changed perceptions regarding the consumption of a medicine and potential adverse outcomes of a pregnancy. The second event, cancer resulting from *in utero* exposure to diethystilbestrol (DES), raised additional concerns regarding adverse effects that were not evident until after puberty. The third, Minamata disease, resulted in the addition of tests for postnatal behavioral changes to the regulations.

As the result of increased concern, including regulatory requirements, a large body of research developed associated with morphological changes resulting from teratogenic insult during the first trimester of gestation [240]. Many investigators continue to assume that susceptibility to teratogenesis (i.e., gross anatomical malformation) is limited to the period of organogenesis [104,239]. It is now clearly evident, however, that the "all or nothing" law of recovery or death does not always apply to preimplantation embryos (i.e., surviving embryos may demonstrate growth retardation, malformation, and/or functional impairment), that both the preimplantation and fetal stages of development are also susceptible to toxic insult (i.e., toxic exposure may result in death, retarded growth, malformation, and/or functional impairment), and that many systems continue to develop after birth (e.g., lungs, immune system, reproductive organs).

Despite theoretical and practical efforts to separate exposures and effects in the maternal animal from those of the developing conceptus, it remains axiomatic that three dynamic populations are at risk: the male, the female, and the conceptus [30]. The very existence of any given pregnancy, as well as the outcome of that pregnancy, is dependent on dynamic complex interactive systems. As shown in Figure 34.1, each pregnancy is dependent on the entire reproductive process, including the development, genetic makeup, and nutritional status of the parental generation, as well as the genetic makeup, nutritional status during growth and development, and exposure of the offspring to perturbations in the macro and micro environments. Perturbations include, but are not limited to, primary and secondary toxic effects resulting from exposures to xenobiotics. Both sensitivity and response are sex, age, dose, and tissue dependent.

## NORMAL REPRODUCTION AND DEVELOPMENT

Reproduction and production of offspring can occur multiple times in an individual's life within the interval beginning at puberty and ending at reproductive senility. During this process, two haploid chromosomes, one from each sex, are joined to produce a diploid state in a new individual. In most mammalian species, phases of the reproductive cycle during which development of a new individual occurs principally take place within the uterus of the female. This development of the conceptus from the fertilized ovum to birth is a complex process during which extremely rapid cell proliferation occurs, apoptosis is normal and required, and regulatory genes produce necessary products that are associated with cancers if produced in adult life. As noted by Klinefelter and Gray [110], the multiplicity of xenobiotic exposures, lesion sites, and potential reproductive disorders precludes simple screening of suspected reproductive toxicants in humans, because the effects and the agents causing reproductive dysfunctions or altered development are often impossible to predict. The entire reproductive process is interactive, with the affected sex, organ, tissue, or cell having continually changing sensitivities in terms of dose, duration, and timing (age) of insult (prenatal, *in utero*, and postnatal) and involves interactions among male and female animals, maternal animals, and developing individuals.

## Normal Reproductive Performance

As described below, under normal conditions, a high proportion of the population of all species, including humans, has remarkable reproductive and developmental risks. Many couples are infertile, and spontaneous abortions are frequent in humans. Infertility occurs in approximately 11 million married couples in the United States, with noncontraceptive sterility occurring in at least one of the partners in 3 million of these couples [146]; ovulatory problems have been reported to be the cause of infertility in approximately 40% of all infertile couples [199]. Hertig [86] estimated that as many as 50% of all fertilized ova are lost within the first 3 weeks of human development. The World Health Organization [241] estimated that 15% of all clinically recognizable pregnancies end in a spontaneous abortion, and that 50 to 60% of the spontaneously aborted fetuses have chromosomal abnormalities [12,192]. Prematurity occurs in approximately 7% of births (i.e., birth before the 37th week of gestation), and the most common developmental abnormality in humans is low birth weight (2.5 kg or less) [90,156], a finding often associated with functional and neurological defects and malformations. Of the approximately 3 million infants born annually, 13.1 per 1000 die before one year of age [152], and 2 to 3% of these liveborn infants are identified as having a congenital malformation within the first year postnatal [51,83]. Although technical improvements for identifying defects will increase this number, based on the current relatively crude criteria approximately 16% of live births have major or minor malformations that become apparent after the first year [35].

The cause of approximately 65 to 75% of human congenital malformations is unknown [15,17,23,64,83]. In the early 1900s, genetic causes were believed to be the predominant cause, with the remaining birth defects being unsolved clinical problems. In 1976, Fraser [62] published the multifactorial/threshold hypothesis, which involved modulation of a continuum of genetic characteristics by intrinsic and extrinsic (environmental) factors. Modulating factors include, but are not limited to, placental blood flow, placental transport, site of implantation, maternal disease states, infections, drugs, chemicals, and spontaneous errors of development. Genetic causes may continue to predominate if Fraser's multifactorial/threshold hypothesis is modified to include altered genetic expression and interaction of environmental factors, although the three traditional categories for the etiology of congenital malformations (i.e., unknown, genetic, and environmental factors) continue to be used. A large proportion of these congenital malformations is likely to be due to two or more genetic loci (polygenetic) [23,125]. Those malformations with an increased recurrent risk (e.g., cleft lip and palate, anencephaly, spina bifida, certain congenital heart diseases, pyloric stenosis, hypospadias, inguinal hernia, talipes equinovarus, congenital dislocation of the hip) can be categorized as multifactorial disease, as well as polygenic inherited disease [23,62]. Some malformations are probably spontaneous errors of development, without apparent abnormalities of the genome or environmental influence, similar to the concept of spontaneous mutations [14,62].

A mutation in either germ cell before conception can result in chromosome abnormalities in the conceptus. Mutations in the conceptus can also occur during any given pregnancy. The maternal genetic contribution is known to provide a greater contribution to the development of the embryo, although maternal health, nutrition, and toxic exposures can also affect the outcome of a given pregnancy. With few exceptions, toxic exposures before implantation generally result in death of the embryo or no observable effect. Toxic exposures during the first trimester of pregnancy (embryogenesis) are those most likely to produce gross morphological changes, and those during later stages of pregnancy (fetogenesis) and postpartum are those more likely to be associated with retarded growth, functional alterations and reduced capacity of the various systems, including the cardiovascular, endocrine, reproductive, pulmonary, central nervous system (CNS), urinary tract, gastrointestinal, and immune systems [28,31,64,99,100,105,155,185]. Thus, the development of an organism is a life-long cycle characterized by changes in size, biochemistry, physiology, form, and functionality [102,108,179].

## Human Epidemiology Studies

Epidemiology has the potential to play central and imperative roles regarding identifying how developmental and reproductive toxicology affects human populations. This is particularly apparent when one considers that, regardless of the results of studies in animals, for the purpose of human reproductive and developmental risks the human is the ultimate species of interest [16]. As noted by Erickson [54], several inherent problems exist in performing human epidemiology studies. First, they are time consuming and expensive, especially when dealing with rare disease states, as are most conditions that could conceivably result from reproductive or developmental toxicity. Second, reported findings are often based on anecdotal information, and, logically, it is essentially impossible to prove the absence of an association. Third, studies may be either experimental (interventional) or observational, but the more scientifically useful type (interventional) is essentially precluded by ethical concerns that disallow direct testing of women of child-bearing potential; thus, epidemiology studies in the area of reproduction and development usually are observational in type (i.e., evaluation of events in two groups without intervention or random assignment of subjects to treatments). As a result, epidemiology studies usually involve statistical analyses of medical data to identify associations between toxic agents and adverse outcomes, resulting in identification of relative risks that may or may not identify causal relationships.

The major types of study designs considered useful are studies of group characteristics, cohort studies, case-control studies, and intervention studies. Ongoing population surveillance and maintenance of registries are also commonly practiced, although there are remarkable differences in the quality of the data collected [98,121]. The greatest difficulty is in obtaining adequate sample sizes; for example, to detect a 3.2-fold increase in major malformations occurring at a frequency of 3%, a sample size of at least 300 live births is required [51,127]. Manson and Kang [127] cite that frequent outcomes, such as spontaneous abortion, which occurs at a 15% incidence, would require a population of only 50 pregnancies to detect a 3-fold increase. The importance of perceived risk vs. actual risk is evident for the most well-studied risk of human exposure: environmental radiation. Despite well-characterized risks of exposure to environmental radiation [18], the perceived risk of any exposure remains high, perpetuating the nonscientific approach often taken regarding potential adverse effect of exposure to any agent during pregnancy. Because of the difficulties and complexities associated with human clinical and epidemiological studies and the desire to prevent potential human exposures to agents at levels producing adverse effects, animal studies are those most frequently used to detect potential reproductive or developmental toxins. Such studies are, for the most part, the subject of this chapter.

## FEMALE REPRODUCTIVE TOXICOLOGY

### Maturation of the Female Reproductive System

#### Gametogenesis and Ovulation

Each sex (male and female) of most multicellular animals produces specialized cells (gametes), which are joined (fertilization) to form a new individual (zygote/conceptus). The mammalian ovulatory cycle includes multiple interrelated events involving folliculogenesis, ovulation, and preparation of the reproductive tract for fertilization and implantation leading to pregnancy. Ovulation is the central event in the ovulatory cycle [13]. Ovulation results from interaction among multiple feedback systems, including the diencephalon, especially the hypothalamic regions, the anterior pituitary and the ovary. The hypothalamus releases gonadotropin-releasing hormone (GnRH), and through this process regulates anterior pituitary production and secretion of gonadotropin hormones, including luteinizing hormone (LH) and follicle-stimulating hormone (FSH). LH and FSH released from the anterior pituitary and transported to the ovary initiate and maintain ovarian follicle growth. Figure 34.2 demonstrates the progression of the ovulatory cycle in humans.

The initial phase of positive feedback is that in which the hypothalamic–anterior pituitary axis component signals the ovary to initiate growth of the follicles. Phase two of the cascade is initiated by the mature ovarian follicle, which signals readiness to ovulate through production and secretion of estradiol and progesterone. Estradiol and progesterone secreted by the ovary initiate a surge of GnRH, which is followed by a surge of the ovulating-inducing hormones, LH and FSH, which provide the stimulus initiating the cascade of events in the ovary ultimately resulting in ovulation [13].

Ovulation of a fertilizable ovum (oocyte, egg, or female gamete) requires formation of a corpus luteum and growth, maturation, and differentiation of three cell types: the germ cell (oocyte), granulosa cells, and thecal endocrine cells. Each of these cell types is susceptible to toxic effects, as are also the three major processes that occur during development of the mature oocyte: (1) mitosis of oogonia and granulosa cells during follicular growth; (2) meiosis of oogonia to form oocytes; and (3) differentiation of granulosa cells and theca cells, which allows a response to a surge of luteinizing hormone and subsequent ovulation.

The inseparability of the reproductive process and development of the conceptus is most apparent when one considers that the female germ cells and follicles are formed during prenatal life. Primordial germ cells are first detectable in the yolk sac at 3 weeks of human development. These cells undergo mitosis, migrate to the urogenital ridge, populate the indifferent gonad, and then differentiate into oogonia or prespermatogonia. Approximately 1700 germ cells migrate to the gonads in a human embryo. These increase to approximately 600,0000 germ cells by 2 months of gestation and peak at approximately 7,000,000 germ cells by 5 months of gestation. Oogonia begin to enter meiosis at month 3, and all oogonia are in early prophase I by the end of month 5, at which time the oogonia are termed *primary oocytes* [71]. From this point onward, atresia results in a decline of oocyte numbers to less than 1,000,000 by birth, with continued reductions in numbers throughout the remainder of the reproductive life of the woman through atresia and ovulation [133].

The four stages of meiotic divisions are prophase, metaphase, anaphase, and telophase. The first meiotic division (prophase) occurs during the fetal or neonatal period. Within 8 weeks after birth, human oocytes enter a resting phase (diakinesis), in which they stay until puberty begins [10]. Sex differentiation and ovarian germ cell development occur at different developmental ages in various mammalian species, as shown in Table 34.1.

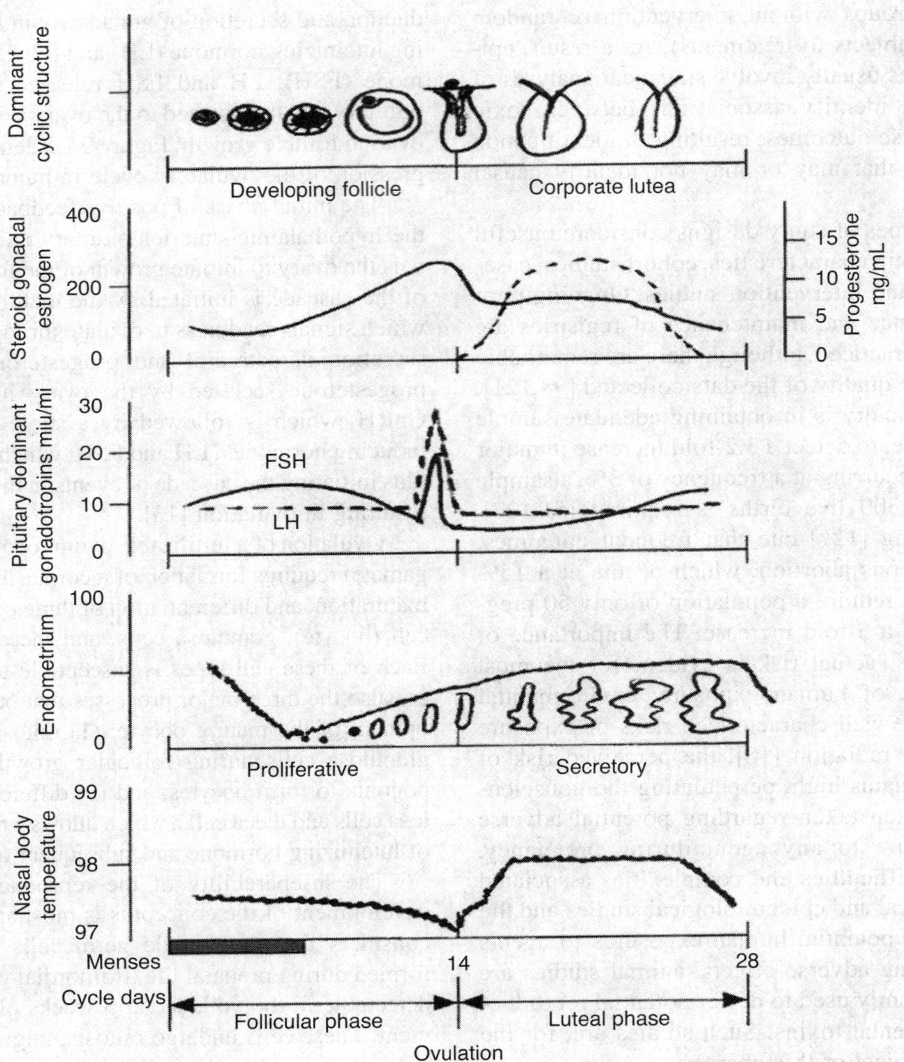


**FIGURE 34.2** Progression of ovulatory cycle in humans; development of the dominant follicle of pituitary gonadotropins, endometrial proliferation, and basal body temperature are depicted. (Adapted from Haney, A.F., in *Endocrine Toxicology*, Thomas, J. A. et al., Eds., Raven Press, New York, 1985, pp. 181–210; Mattison, D.R. and Ross, G.T., in *Methods for Assessing the Effects of Chemicals on Reproductive Functions*, Vouk, V.B. and Sheehan, P.J., Eds., Wiley, New York, 1983, pp. 217–246.)

## Atresia

As noted previously, the majority of germ cells are lost to normal physiological degeneration (atresia), which occurs during the oogonial and primary oocyte stages. Approximately 60% of the germ cells in a human fetus are lost between 5 months of gestation and birth, with three distinct waves of oogonial degeneration occurring. One wave affects oogonia in mitosis (final interphase), and the other two affect oocytes in the pachytene and diplotene stages of prophase I. This normal apoptosis is synchronous in oogonia connected by cytoplasmic bridges. After the meiotic prophase, simultaneous atresia no longer occurs, although individual oocytes spontaneously degenerate at all stages of development.

## Folliculogenesis

Surviving oocytes in prophase are surrounded by granulosa cells and begin to form follicles (folliculogenesis). Initially, there is a period of gonadotropin-independent growth when very early stages of follicle development occur without LH or FSH support. After the early stages of gonadotropin independence, follicular growth becomes dependent on the continuous presence of gonadotropins. Nongrowing follicles remain as primary oocytes within unilamellar follicles during prepubertal and reproductive periods. This provides a pool from which groups (cohorts) of small follicles are recruited for further maturation to preovulatory or Graafian follicles. Follicular growth occurs daily and is characterized by three events: oocyte

**TABLE 34.1**
**Ovarian Germ Cell Development in Mammalian Species**

| Species | Days of Gestation | Gonadal Sex Differentiation | Initiation of Meiosis | Completion of Oogenesis | Arrest of Meiosis |
|---|---|---|---|---|---|
| Mouse | 19 | 12 | 13 | 16 | (5) |
| Rat | 21 | 13–14 | 17 | 19 | (5) |
| Hamster | 16 | 11–12 | (1) | (5) | (9) |
| Rabbit | 31 | 15–16 | (1) | (10) | (21) |
| Monkey (rhesus) | 165 | 38 | 56 | 165 | Newborn |
| Human | 270 | 40–42 | 84 | 150 | Newborn |

*Note:* Numbers in parentheses indicate days of gestation or postnatal age. Completion of oogenesis refers to transformation of all oogonia to primary oocytes.

*Source:* Adapted from Gondos, B., in *The Vertebrate Ovary*, Jones, R.E., Ed., Plenum Press, New York, 1978, pp. 83–120; Manson, J.M. and Kang, Y.J., in *Principles and Methods of Toxicology*, 3rd ed., Hayes, A.W., Ed., Raven Press, New York, 1994, pp. 989–1037.

enlargement, transition of granulosa cells from flattened to rounded configuration, and formation of the zona pellucida, an extracellular matrix present between the granulosa cells and oocyte and comprised of a complex protein–carbohydrate extracellular matrix. The first stage is characteristic of small type 2 follicles [168] that enter the pool of committed growing follicles. What triggers follicular growth remains unidentified, although it is known that follicle-stimulating hormone (FSH) and luteinizing hormone (LH) are not involved [175]. Follicle growth requires five events: continued oocyte enlargement, rapid proliferation and increase in layers of granulosa cells, formation of basal lamina (extracellular matrix external to outer layer of granulosa cells), organization of endocrine thecal cells around a basal lamina, and formation of the antrum (fluid-filled cavity within the follicle). The majority of type 2 follicles grow into the preantral stage (types 5 and 6). A large surge of gonadotropins in the cycle preceding ovulation results in selection of a few type 7 and 8 follicles (antral stage) from a pool of large preantral follicles.

The preovulatory surge in gonadotropin includes an increase in LH that stimulates conversion of progesterone to androstenedione in theca cells [176]; the androstenedione is then converted to estradiol in granulosa cells. The estradiol secreted by the growing follicles, in conjunction with FSH, effects differentiation of granulosa cells. Differentiation of the granulosa cells includes increased cellular content of FSH and LH receptors, increased aromatase activity, cholesterol side-chain cleavage, and prostaglandin synthetase activity [175,176], which regulates the synthesis of the prostaglandins required for ovulation. Only follicles that can produce estradiol progress to preovulatory follicles. It should be noted that this scheme is well established in the rat ovary, but there is some question as to the importance of estrogen in the proliferation of granulosa cells in the primate ovary [13]. Atresia can be induced by any agent that inhibits either theca cell function (ability to synthesize androstenedione) or granulosa cell function (synthesis and action of estradiol). Atresia may also be caused by agents that alter gonadotropin receptor content or the functional coupling of the receptor to adenylate cyclase, because FSH and LH act via cycle adenosine monophosphate (AMP).

After the rise in FSH and LH, primary oocytes in preovulatory follicles continue to form secondary oocytes through the first meiotic division and remain in metaphase of the second division. The first polar body, containing half of the chromosomes that were present in the primary oocyte, is extruded. In addition, as ovulation nears, the follicle vascularizes and swells on the ovary's surface, becoming a macroscopically visible blister-like protuberance. The secondary oocyte is ovulated at metaphase II and stays in this stage until fertilization. At fertilization, the second meiotic division is completed, the second polar body extruded, and the female pronucleus formed. Male and female pronuclei combine at fertilization to regain the diploid state [55].

## Fertilization

Fertilization (union of a spermatozoon, or sperm, and an oocyte, or egg) occurs in the female reproductive tract in birds and mammals. Restoring the diploid number of chromosomes determines the genetic sex of the zygote and initiates cleavage (rapid mitotic division). It should be noted that maternal and paternal genomes are not functionally equivalent in their contributions to the zygotic genome due to imprinting. The process of imprinting occurs during gametogenesis, conferring differential expressivity to certain allelic genes, depending on whether they originated from the male or the female [78]. Although imprinting is not well understood and there are no documented examples of toxicant effects on this process, such could conceivably play a role in paternally mediated developmental toxicity.

### Implantation/Luteinization

Following ovulation, vascularization of the granulosa cell layer occurs, and granulosa cells are transformed into luteal cells, which produce the progesterone required for preparation of the endometrial lining of the uterus for implantation of the conceptus [79]. In the absence of fertilization, luteinization occurs. This process includes degeneration of the ovulated oocyte, continued LH stimulation, and luteinizing of the empty follicle into a corpus luteum, which secretes progesterone. The process continues to occur throughout reproductive life until all primordial follicles are depleted or menopause occurs.

### Corpus Luteum

The function of the corpus luteum is dependent on LH, and withdrawal of LH leads to luteal failure, decreased estrogen and progesterone secretion, and failure to maintain the pregnancy. In a normal cycle without fertilization, corpus luteal failure occurs after approximately 10 days of functioning. This failure is believed to be associated with an increase in activity of prostaglandin $F_{2\alpha}$ ($PGF_{2\alpha}$) [63]. $PGF_{2\alpha}$ is produced by the uterus in some animals, but the importance of endogenous prostaglandin from the uterus in the primate uterus has not been established, although exogenously administered $PGF_{2\alpha}$ is luteolytic in primates. When the oocyte is fertilized, the corpus luteum is maintained by secretion of human chorionic gonadotropin (hCG), which is an LH-like molecule synthesized by the trophoblastic tissue of the embryo. Under hCG stimulation, progesterone synthesis continues in the corpus luteum until this steroid is principally produced by the placenta. The corpus luteum also provides an important source of relaxin (oxytocin), GnRH-related molecules, and growth factors that function in pregnancy and parturition [13].

### Ovulatory Cycles

Ovulatory cycles vary widely in laboratory animals, farm animals, nonhuman primates, and humans [13] and among humans [84,209]. Two major categories exist: animals with spontaneous ovulation and animals with ovulation induced by mating. Spontaneous ovulators include laboratory rats, hamsters, mice, guinea pigs, sheep, pigs, rhesus monkeys, baboons, and humans, with cycle lengths ranging from 3 to 33 days. Mice, rats, and hamsters develop a functional corpus luteum only if mated, with vaginal stimulation resulting in pseudopregnancy or pregnancy. The remaining species always form an active corpus luteum that secretes progestoerone and is functional for 10 to 15 days. Reflex or induced ovulators include rabbits, cats, ferrets, short-tailed shrews, and voles. Mechanical or coital stimulation in these species results in a gonadotropin surge (primarily LH) within 1 to 2 hours that results in ovulation of follicles that have reached maturity in the ovary. The FSH secreted in the surge may result in development of new ovarian follicles for estradiol production, which is essential for corpus luteum function.

At least two other neuroendocrine signals are important in regulating spontaneous ovulatory cycles: circadian rhythms in the gonadotropin surge and seasonal variation. Humans and guinea pigs appear to be the species least affected by circadian rhythms and seasonal variations, although some circadian variations in LH levels have been reported in women during the follicular phase of the cycle [13]. In rats, the suprachiasmatic nucleus (SCN) in the hypothalamus has been identified as the site of signal transduction by light–dark cycles, and lesioning of this site terminates the estrous cycle, as well as exposure to constant light. Seasonal reproduction is also known to be associated with the duration of the photoperiod, and for most laboratory species a minimum of 10 hours of light is required for reproduction to occur. In addition, diet (in particular, phytoestrogen content) is known to affect ovulatory cycles and fertility (phytoestrogens are naturally occurring diphenols found in many plants and have structural and functional similarities to 17β-estradiol) [84,111,112].

## Ovarian Toxicology

During each menstrual cycle in humans, oocyte maturation, folliculogenesis, ovulation, and corpus luteum formation occur. These processes are under the influence of gonadotropins that act on the ovarian cells to initiate morphologic changes, steroidogenesis, and induction of various receptors. Many of these events also occur in rodents, although factors other than gonadotropins appear to be involved in the initial stimulation of growth of the resting follicle; however, once the follicle is committed to growth, selection of antral, preovulatory follicles depends on the LH surge preceding the cycle [127].

Evaluation of follicular alterations is difficult, because a sexually mature ovary contains a diverse population of resting, maturing, and mature follicles, and the female has only one cluster of follicles that is selected for maturation in each cycle. In addition, the oogonia enter meiosis during fetal life, after a specified number of mitotic divisions, and there is no mechanism to replace oocytes. Thus, the presence of reduced numbers of oocytes, which may occur in the fetus or result from subsequent destruction, has the potential to reduce reproductive life in the female. If the oocytes are subject to irreversible DNA damage, mutagenic manifestations may occur if the damaged oocyte is fertilized. During adulthood, each of the three components of the follicle may be uniquely susceptible to specific toxicants [130,132]; for example, a xenobiotic agent might increase the progesterone/estradiol or testosterone/estradiol ratios in the granulosa cell stimulated by FSH or alter steroidogenesis and androgen production by the thecal cells, thus delaying follicle maturation and ovulation. DNA in the oocyte can be damaged during maturation and fertilization of the oocyte, depending on the agent [68], and such damage is the probable mechanism by which antineoplastic agents deplete the oocyte pool and result in premature menopause in women [130].

## Reproductive Endocrinology and Toxicologic Interactions

For an in-depth review of endocrine disruptive chemicals, a subject that has become a field in and of itself, suggested reading includes Stoker et al. [204].

### Interaction of Reproductive Hormones and Target Tissues

Estrogens increase oviduct secretions and muscular contractions, actions antagonized by progesterone. Thus, cyclic patterns of estrogen and progesterone levels that occur during the menstrual cycle in primates and humans play important roles in regulating sperm transport through the cervix, into the uterine lumen, and to the oviduct [120]. Of 150 to 300 million sperm in an ejaculate, less than 500 reach the site of fertilization, with the greatest obstacle to sperm transport being the cervix and uterotubal junction, where most sperm loss occurs. Mammalian sperm undergo changes in the female reproductive tract to allow penetration of the ovum. This change, or capacitation, is normally induced by secretions in the female genital tract and requires one to several hours for completion. Capacitated sperm penetrate the layers of the granulosa cells and bind to a major glycoprotein (ZP3) in the zona pellucida, which is the main barrier to fertilization in mammals. Sperm binding to ZP3 induce the acrosomal reaction, releasing proteases and hyaluronidase, which are essential enzymes for sperm penetration of the zona. After penetration, contact of the sperm and oocyte membrane triggers the cortical reaction in the egg, releasing enzymes that act on the zona and oocyte membrane to prevent further binding and entry of sperm [120].

After fertilization, the ovum divides and slowly moves down the oviduct. Decreased motility in the oviduct is associated with rising progesterone levels secreted by the corpus luteum and prevents the ovum from prematurely reaching the uterus. Estrogen treatment during ovum migration accelerates movement of the embryo so it prematurely reaches the uterus, as occurs with use of the "morning-after" pill, where a synthetic estrogen, diethylstilbestrol (DES), is administered for 5 days. Similar effects have been observed with agents such as methoxychlor that mimic estrogens [45]. Before implantation, which begins on day 7 and is completed on day 12 in humans [66], the embryo floats free in the uterus, nourished by endometrial gland secretions, which are under the control of progesterone. Combined estrogen and progesterone actions signal the uterine endometrium to prepare for implantation of the blastocyst and the embryo invading the uterine endometrium. Implantation triggers a decidualization reaction in several rodent species (the placental sign) when the uterine stromal cells are actively converted to decidual cells. In humans, decidualization of the uterine lining normally occurs during the luteal phase of the menstrual cycle, whether or not implantation occurs. Decidualization of the uterine lining contributes the maternal portion of the placenta [120].

Estrogen and progesterone synthesis and secretion during normal human pregnancy are a cooperative effort of the mother, fetus, and placenta. Extremely large amounts of estrogens (estradiol, estrone, estriol) are produced during pregnancy, with the placenta being the primary source. Estriol excretion is a biomarker used to monitor fetal well-being, because one of the precursors for estriol synthesis by the placenta is dehydroepiandrosterone, which is produced by the fetal adrenal gland [49]. In addition, the placenta can synthesize estrogens *de novo,* although the exact role of the high levels of estrogen present during pregnancy is unclear [191].

### Hormonal Interaction with the Placenta

The placenta also secretes human placental lactogen (HPL), which is similar to human growth hormone and human prolactin; it does not appear to promote body growth, may be involved in fetal nutrition, and is probably involved in preparation of the mammary gland for lactation. This protein usually appears in the blood approximately 2 months after fertilization, increases until parturition, and provides a biomarker to monitor placental size and growth during pregnancy [120,206].

## Mammary Development During Pregnancy

Mammary development in women is controlled by prolactin, which is secreted by the anterior pituitary gland. Prolactin levels increase during pregnancy from about 2 months after fertilization until parturition [119]. Although the role of prolactin during pregnancy is unclear, it probably contributes to the profound mammary development that occurs during pregnancy.

## Parturition

The exact mechanisms that initiate parturition in humans are unknown; however, most current evidence suggests that estrogen, progesterone, oxytocin, and prostaglandins may all be involved [25,196]. Estrogen and oxytocin stimulate uterine contractility, which increases and becomes more coordinated as labor approaches. Progesterone antagonizes the effects of progesterone and oxytocin, and declining levels of progesterone may trigger initiation of uterine contractions, although peripheral progesterone concentrations do not fall before the onset of labor. It is hypothesized that a fall in tissue levels of progesterone could occur before detection of peripheral changes and that local prostaglandin production may also be involved in initiating labor by direct or indirect means [25].

The role of oxytocin in labor is also unclear, although it is important once labor begins. Stretching of the uterine cervix and vagina stimulates the reflex release of oxytocin from the posterior pituitary, resulting in increased cyclical uterine contractions and stretching, facilitating the delivery process. The increase in oxytocin receptors near term may be the cause of the increased sensitivity of the myometrial response to oxytocin [198]. Physical factors also regulate myometrial activity in the human, of which the most important are the increase in uterine volume associated with fetal and placental growth resulting in stimulation of myometrial contractions and the inhibition of myometrial contractions by progesterone. Estrogens, oxytocin, and $PGF_{2\alpha}$ are stimulatory hormones additionally contributing to uterine contractility and inhibited by progesterone, which blocks estrogen action, oxytocin receptor production, and prostaglandin release. Autonomic innervation also contributes to increased or inhibited myometrial activity. Drugs, such as alcohol, can directly act on uterine smooth muscle or indirectly inhibit oxytocin release [120].

In summary, parturition in humans is usually effected by three major factors: myometrial stretch, maturation of the fetal adrenal, and sensitivity to oxytocin. Maturation of the fetal adrenal late in pregnancy causes increased secretion of cortisol, promoting myometrial contractility. Myometrial contractile activity is also stimulated by stretching of the uterine smooth muscle fibers as the result of growth of the fetus and placenta and by increased sensitivity to oxytocin as the result of increased numbers of oxytocin receptors.

## Lactation

Various hormones also modulate the development and secretory capacity of the mammary gland. Prolactin is the primary hormone responsible for mammary growth. Estrogen works indirectly on the mammary gland via promoting prolactin synthesis and secretion by the anterior pituitary. Gonadotropin hormones are also involved, in that they stimulate ovarian estrogen secretion. Estrogens also directly stimulate mammary gland growth, but only in the presence of prolactin.

Development of the breast is a major event at the onset of puberty; milk secretion is not usually initiated until after parturition and is inhibited by estrogen and progesterone secreted by the placenta, antagonizing prolactin's ability to stimulate milk secretion [97]. Estrogens are the primary ovarian hormones responsible for mammary gland development and growth, although progesterone is important for alveolar development in some animals; however, no evidence suggests that progesterone is required for alveolar development in humans [210]. The milk secretion stimulatory effect of prolactin is different from the growth promoting effects of prolactin. Initiation of lactation normally involves withdrawal of an estrogen and progesterone block to the stimulatory effect of prolactin on milk secretion [120].

Milk secretion is initiated by suckling of the nipple; continued secretion requires continued suckling stimulus and the associated rapid release of prolactin from the pituitary gland. The prolactin release as the result of suckling does not cause the milk release at the time of suckling, but it is responsible for the release of milk (milk letdown) during the subsequent nonsuckling interval. The young obtain this milk at the next suckling. Normal milk secretion also requires other hormones, including adrenocorticotropic hormone (ACTH), provided the adrenals are intact. Insulin, growth hormone, parathyroid hormone, and thyroid hormones are also interactive stimulators of milk secretion [210]. Milk letdown at suckling is a phenomenon caused by the suckling-induced release of oxytocin and contraction of the myoepithelial cells [97]. In humans, this is a conditioned reflex in response to the presence or crying of a baby. Release of oxytocin apparently can be inhibited by higher levels of the CNS, because emotional disturbances or pain during nursing can reduce milk release.

Both LH and FSH secretion are reduced during lactation, apparently as the result of suckling [126], and lactating women may have a diminished response to GnRH, which suggests that the inhibition occurs at the level of the pituitary. The ovaries also seem refractory to gonadotropic stimulation during lactation, although the mechanisms resulting in lactational amenorrhea are not currently understood. Several conditions are known to result in abnormal lactation associated with increased prolactin levels; for example, pituitary tumors can cause abnormal lactation as the result of increased prolactin levels. Hypothalamic lesions, abnormal afferent neural input into the hypothalamus, and tranquilizers can also raise prolactin secretion sufficiently to result in galactorrhea [97].

## Test Systems and Endpoints for Detection of Toxic Effects on Female Reproductive Capacity

### Ovarian Cyclicity

Despite the commonality of many features of ovarian cyclicity in humans and rodents, distinct differences exist that should be considered in designing toxicologic studies [110]; for example, in most mammals, follicular growth continues through the luteal phase. Luteolysis is believed to be mediated by uterine prostaglandins in rats and nonhuman primates. In human females, follicular growth does not continue through the luteal phase, and estrogen, rather than prostaglandins, appears to be the luteolytic agent [110]. Elger et al. [53] noted other important differences among species; for example, progesterone production by the corpora lutea is maintained throughout pregnancy in the rat, but in humans and guinea pigs placental progesterone contributes significantly to maintenance of pregnancy during mid- and late gestation. In rats, progesterone levels must be remarkably reduced to induce parturition, but humans and guinea pigs have high progesterone levels until parturition. In rats, deciduoma formation is highly dependent on a correct progesterone/estrogen ratio, and abortion occurs when the serum estrogen level is increased, but humans and guinea pigs are estrogen resistant. In contrast, prostaglandins have abortifacient properties in humans and guinea pigs. The hormonal similarities between guinea pigs and humans, in addition to the greater similarities in timing of fetal *in utero* development, suggest that the guinea pig may be a better model for human risk assessment than the rat, although this model has not been as extensively studied as the rat for evaluations of toxic effects in mid- and late pregnancy. Normal ovarian function depends on the appropriate interactions of various compartments of the ovary and changes that occur in these individual compartments during the cycle. *In vivo* adverse effects may result from active intermediates or metabolites or through modulation of hypothalamic-pituitary axis; such effects would not be identified by *in vitro* assays. In addition, *in vitro* screening techniques are appealing for use in assessment of effects on the ovary; effects identified in *in vitro* systems must be confirmed using *in vivo* methods.

### Reproductive Behavior

Female reproductive behavior in vertebrates is dependent on the ovarian hormones estrogen and progestins [8,240]. Many normal, neural, and molecular mechanisms for female reproductive behaviors in the rat have been widely investigated and, on this basis, the rat is frequently considered to provide a model system to study CNS function and behavioral development. Unfortunately, little is known about the role of these processes in human and nonhuman primate reproductive behaviors.

Although not the model most used for basic research, female monkeys show clear changes in behavior during the menstrual cycle and subsequent to injection of steroid hormones. Copulation in nonhuman primates is restricted to the periovulation period. Similar findings have not been made in women, although the absence of clear findings may reflect methodological problems inherent in the conduct of such studies [170]. It can be assumed that steroid hormone behavioral mechanisms are present in humans; however, other social and environmental factors appear to be more important. Humans clearly have sexual behaviors that begin and are dependent on steroid hormone secretion during puberty, and perceptual, and sensory behaviors vary within the menstrual cycle [65], as well as the susceptibility to irritability, anxiety and depression during the premenstrual period [181]. Although it has been reported that women exposed to DES prenatally had reduced female sexual behavior [137] and that some masculization of girls can be induced by androgens prenatally [52], these studies remain equivocal.

Maternal behaviors are also hormonally dependent in the rat but have not been well investigated in nonhuman primates, although steroid hormones may facilitate maternal responsiveness to alien newborns in monkeys [157]. The early postpartum period in humans is known to be an interval of emotional alterations in many women and severe disturbances in some women [182]. Human and rodent studies suggest that serum prolactin may affect postpartum hostility [157].

### Alteration of Hypothalamic–Pituitary Axis/Ovarian Feedback

As is evident in Table 34.2, effects on female reproduction are generally studied by evaluation of many endpoints also examined in developmental toxicity. Toxic effects may be produced by targeting the hypothalamic pituitary axis, the ovary, or any point in the complex feedback system. Examples of some agents known to adversely affect female reproductive performance are cited in Table 34.3. Altered hypothalamic and pituitary secretions have been demonstrated to adversely affect fertility and estrous cyclicity in female rats [110]. The mechanism by which such effects can occur can sometimes be identified through the use of *in vitro* perfusion systems. These can be used to evaluate specific functions of an organ system independently from other endocrine organs and sometimes provide less variable data. For example, Middleton et al. [139] demonstrated inhibition of ovarian estradiol synthesis by R151775, a triazole fungicide, with resultant delay in the LH surge and ovulation. Further study of this fungicide found that pituitaries from rats treated at midday on diestrus 2 had LH production after GnRH stimulation [141]. This chemical was later identified as an aromatase inhibitor that prevented ovarian synthesis of estradiol [110].

**TABLE 34.2**
**Scheme for Identifying Effects on Female Reproductive Process**

| Step | Treatment | Activity | Potential Adverse Effects |
|---|---|---|---|
| 1 | Pretreatment (14 days) | Identify estrous cycling | Abnormal cycling; if incidence is abnormally high, consider environmental factors |
| 2 | Treatment (14 days precohabitation) | Identify estrous cycling | Altered from pretreatment; assume hormonal changes |
| 3 | Treatment through cohabitation and gestation | Cohabit females with untreated males (1:1). Observe mating behavior (receptivity) and fertility (persistent diestrus, sperm in smear or plug; sperm/plug = day 0 of gestation [GD 0]). | Reduced or absent copulatory behavior; irregular estrous cycling |
| 4 | Treatment through day before killed | Kill half of pregnant animals/group at GD 21 (preselect, to preclude biasing results based on mating performance). Observe for gross lesions, corpora lutea, pregnancy, implantation numbers, early and late resorptions and live and dead fetuses, fetal body weights, and sexes. | Reduced corpora lutea, implantation sites, live fetuses (timing indicates when effect occurred); reduced fetal body weight; altered sex ratios |
| 5 | Treatment through delivery and lactation | Allow remaining half of animals in group to deliver. Observe durations of gestation and parturition, maternal behavior peripartum, pup number, viability, body weight, sex and morphology at birth; pup viability, growth, clinical signs, and interaction with dam to weaning; maternal implantation sites and gross lesions. | Reduced implantation sites; altered duration of gestation; altered duration of parturition; reduced maternal care/lactation; reduced total litter size, live litter size, and/or pup viability; reduced or increased pup weight; altered pup sex ratio; altered pup morphology; reduced pup growth and viability; altered function |

*Source:* Adapted from Chapin, R.E. and Heindel, J.J., in *Methods in Toxicology*. Vol. 3B. *Female Reproductive Toxicology*, Heindel, J.J. and Chapin, R.E., Eds., Academic Press, San Diego, CA, 1993, pp. 1–15.

## Ovarian Morphometry

Ovarian compartments likely to be affected by xenobiotic exposure and of interest include oocyte number and follicular development, ovulation, estrous cycling, fertility, and maintenance of pregnancy. Models used to evaluate dose- and stage-dependent, as well as age-dependent, effects of xenobiotic exposures of the ovary generally utilize morphometry, which can be performed either as part of general toxicity evaluations or within the framework of the reproductive toxicity studies. Although the original methods were extremely time consuming and required that serial sections of the ovary be made and numbers of follicles [170] or oocyte size [192] evaluated, current screening methods generally are restricted to every 10th or 20th section. Unfortunately, although these methods have been incorporated into some guidelines, they produce highly variable results that should not be used as the sole endpoint in a risk assessment.

## Corpora Lutea Count and Preimplantation Loss

Evidence of toxic activity at the ovarian level in *in vivo* studies is most frequently based on identification of reduced numbers of corpora lutea, often in combination with preimplantation loss, although these methods are not appropriate or equivalent for all species. An example of the complex nature of the ovarian endpoints is provided here with regard to calculation of preimplantation loss in developmental toxicity studies in rabbits (see Table 34.4). Preimplantation loss reflects the number of eggs ovulated, fertilized, and implanted, as well as the receptivity of the uterus. Rabbits are reflex (induced) ovulators, requiring stimulation of the cervix, which is essentially absent when artificial insemination procedures are used for breeding. An intravenous injection of human chorionic gonadotropin (hCG) is generally administered to compensate for reduced natural cervical stimulation during mating (hCG injection is less frequently used in natural mating practices). Artificial insemination procedures result in greater variability in ovulation, numbers of corpora lutea, and preimplantation loss values than natural mating because of several factors.

First, release of ova depends on the number of mature follicles present when ovulation is induced by the stimulation of mating (natural mating) or injected hCG. One major consequence of hCG priming of the female is the

**TABLE 34.3**
**Examples of Agents Producing Female Reproductive Toxicity**

| General Mechanism | Potential Activity | Examples |
|---|---|---|
| Altered puberty and estrous or menstrual cycling | Altered ovarian activity and hypothalamic/pituitary feedback | Alcohol, *o,p′*-DDT, isoflavones |
| Impaired ovulation | Altered endocrine signal | *o,p′*-DDT, kepone |
| Altered mating behavior | Altered modulation | β-Endorphin, naloxone |
| Altered gamete or embryo transport | Increased urine contractions | Progesterone, diethylstilbestrol, estrogens, methoxychlor |
| Suboptimal endometrial environment | — | Estrogens, DDT, methoxychlor, kepone, EGME |
| Ovarian toxicity, oocyte destruction, atresia | Mimic structure of naturally occurring hormones; general chemical reactivity | Alkylating agents, chemotherapeutic agents (e.g., prednisone, vincristine, vinblastine, 6-mercaptopurine, radiation, methotrexate, adriamycin), alcohol, polycyclic aromatic hydrocarbons, 4-vincyclohexene, cyclophosphamide. |
| Altered steroid synthesis | Inhibition of steroidogenic enzymes | Aminoglutethimide, 3-methoxybenzidine, cyanoketone, estrogens, azastene, danazol, spironolactone epostane traizole fungicide (L151775) |
| Antagonized steroid action | Inhibited steroid activity | Clominiphene citrate, cimetidine, spironolactone, opioid peptides |
| Inhibition of gonadotropins | Alterations at hypothalamic level | Marijuana (Δ-9-tetrahydrocannabinol) |
| Altered maternal behavior/lactation | Alterations at hypothalamic level | Tranquilizers |
| Impaired lactation | Altered prolactin levels | Tranquilizers |

*Source:* Data from Cassidy et al. [24], Cummings and Perreault [45], Elger et al. [53], Haney [79], Heindel et al. [81], and Pfaff and Schwartz-Giblin [170].

possibility of superovulation, where a large number of eggs are ovulated, including ones not completely mature; these eggs cannot be fertilized or implanted, resulting in a large percentage of preimplantation loss [211].

Second, fertilization depends on multiple factors, including the quality and quantity of both the eggs and the sperm, as well as the timing of priming, ovulation, and insemination. Artificial insemination generally is performed using one introduction of diluted sperm into the primed female's vagina, with 4 to 20 does inseminated with sperm from the same buck. The angle of insertion of the insemination tube, handling of the female, and degree of priming of the female all affect the number of eggs fertilized and implanted. In contrast, in natural mating, generally only one buck inseminates any female on one day (bucks may inseminate multiple females used in the same study, but generally rabbit breeders allow matings of only the same pair on any single breeding day). Natural breeding may involve one or more intromissions and inseminations, generally increasing the number of sperm inseminated, cervical stimulation of the female, associated hormonal changes and ovulation of appropriately aged eggs, and the number of egg fertilized and implanting.

Third, natural patterns are associated with the number of corpora lutea ovulated and the success of a specific pregnancy. Feussner et al. [60] published data for 1463 control group rabbits artificially inseminated at Argus (now Charles River Laboratories, Preclinical Services) in 98 studies from 1980 through 1989. Preimplantation loss was increased in does with a high number (15 or more corpora lutea, 90 does) or low number (7 or fewer corpora lutea, 117 does) of corpora lutea (numbers presumably associated with the development of the eggs in the ovary and degree of ovulation induced by hormonal stimulation). Preimplantation loss averaged 36.8% for the does with high numbers of corpora lutea, presumably reflecting superovulation of these animals resulting in the ovulation of immature follicles. Preimplantation loss in the 177 does with low numbers of corpora lutea averaged 33.9%, numbers that probably resulted from either insufficient maturity of ovarian follicles or inadequate stimulation of the ovary by intravenously injected hCG. Preimplantation loss for the entire population averaged 27.8%, a value very similar to that reported for a smaller population by Tyl and Marr [211].

**TABLE 34.4**
**Comparative Reproductive and Placental Parameters**

| Parameter | Mice | Rats | Rabbits | Hamsters | Guinea Pigs | Rhesus Monkeys | Humans |
|---|---|---|---|---|---|---|---|
| Estrous cycle (days) | 3–9 | 4–6 | None | 4 | 16 | 28 (menstrual cycle) | 28 (menstrual cycle) |
| Ovulation stimulus | Spontaneous | Spontaneous | Coitus | Spontaneous | Spontaneous | Spontaneous | Spontaneous |
| Uterus | Bicornuate | Bicornuate | Duplex | Bicornuate | Bicornuate | Simplex | Simplex |
| Implantation (days) | 4.5–5 | 5.5–6 | 7–7.5 | 4.5–6 | 6–6.5 | 9 | 5–7.5 |
| Implantation type | Early: eccentric; late: interstitial | Early: eccentric; late: interstitial | Superficial | Interstitial | Interstitial | Superficial | Interstitial |
| Fetal membranes in placental type | Early: inverted yolk sac; definitive chorioallantoic | Early: inverted yolk sac; definitive chorioallantoic | Early: inverted yolk sac; definitive chorioallantoic | Early: inverted yolk sac; definitive chorioallantoic | Early: inverted yolk sac; definitive chorioallantoic | Chorioallantoic | Chorioallantoic |
| Placental shape | Discoid | Discoid | Discoid | Discoid | Discoid | Bidiscoid | Discoid |
| Internal placental shape | Labyrinthine | Labyrinthine | Labyrinthine | Labyrinthine | Labyrinthine | Villous | Villous |
| Placental relation to maternal tissues | Hemotrichorial | Hemotrichorial | Hemodichorial | Hemotrichorial | Hemomonochorial | Hemomonochorial | Hemomonochorial |
| Duration of gestation (days) | 19 | 22 | 30–32 | 15.5 | 67–68 | 166 | 266 |
| Expected number of offspring | 4–12 | 6–14 | 6–12 | 5–12 | 3–4 | 1 | 1 |

*Source:* DeSesso, J.M., in *Handbook of Developmental Toxicology,* Hood, R.D., Ed., CRC Press, Boca Raton, FL, 1997, pp. 111–174. With permission.

## DEVELOPMENTAL TOXICOLOGY

As early as 1973, Wilson [239] expressed the concept of developmental toxicology in his statement: "The unborn is not only the embryo and the fetus *in utero* but also the as yet unconceived individuals whose potential for future development is represented in the germ cells residing in the parental gonads. Also of interest … are those who have already been born but who are incompletely developed." This giant in the field of teratology also expanded the definition of *teratology* (study of congenital malformations) to the "study of the adverse effects of environment on developing systems, that is, on germ cells, embryos, fetuses, and immature postnatal individuals" and stated that "a more comprehensive definition is that teratology is the science dealing with the causes, mechanisms, and manifestations of developmental deviations of either structural or functional nature" [239]. Based on these definitions, developmental toxicology can be considered to be the study of the entire reproductive process, with special emphasis on the developing conceptus but including the development and function of that conceptus throughout its entire lifespan. Wilson's basic principles remain valid. All investigators in the field should be familiar with their details (to comply with current usage, *teratogenesis* has been made synonymous with *developmental toxicity*):

- Susceptibility to developmental toxicity depends on the genotype of the conceptus and the manner in which this interacts with adverse environmental factors.
- Susceptibility to developmental toxicity varies with the developmental stage at the time of exposure to an adverse influence.
- Developmental toxins act in specific ways (mechanisms) on developing cells and tissues to initiate sequences of abnormal developmental events (pathogenesis).
- The access of adverse influences to developing tissues depends on the nature of the influence (agent).
- The four manifestations of deviant development are death, malformation, growth retardation, and functional deficit.
- Manifestations of deviant development increase in frequency and degree as dosage increases, from the no-effect to the totally lethal level.

Schmidt and Johnson [186] provided an elegant refinement of Wilson's principles, including their statement that, while "teratogenesis *per se* has its primary focus on embryogenesis, developmental biology, and molecular genetics, the field of developmental toxicology also involves basic principles of toxicokinetics, dose–response relationships, target organ toxicity and exposure assessment." Thus, for many investigators, the fields of embryology and teratology have been essentially replaced by molecular cell biology. We now use transgenic animal models [113], consider gene therapy a practical possibility [11,44], and are approaching deciphering the human genome. Despite these advances and the common use of molecular cell biology techniques by academic and pharmaceutical laboratories for mechanistic studies, such techniques are not commonly used in the first line screening to identify doses causing general and specific toxicity in normal animals and their offspring, the principle subject of this chapter.

## NORMAL DEVELOPMENT

The following information is provided as an overview of the complex process of normal development and also as an introduction to differences in developmental events in laboratory animal species and humans. For additional information on specific facets discussed, the reader is referred to reviews by DeSesso [47], Klinefelter and Gray [110], Rogers and Kavlock [179], and Shield and Mirkes [189], as well as to basic texts on embryology. Much of the information below is excerpted from these sources. For the interrelationship of the reproductive process in the maternal animal, *vis-à-vis* the developing conceptus, the reader is referred to the prior sections in this chapter. As noted by Beltrame and Giavini [9], development is characterized by changes in size, biochemistry and physiology, and form and functionality. The overall process is orchestrated by a cascade of gene transcription regulating factors, the first of which are present in the egg before fertilization. These factors activate regulatory genes in the embryonic genome, with sequential gene activation continuing throughout development.

### Development of the Conceptus

#### Preimplantation

From sperm penetration to first cleavage requires approximately 12 hours in most laboratory species. As the fertilized oocyte (zygote) travels down the fallopian tube to the uterus, cleavage continues with growth to the morula stage. During this time, the zygote is surrounded by an acellular mucopolysaccharide layer, the zona pellucida, which previously prevented sperm penetration and now prevents premature implanting. In most mammals, the zona pellucida disappears at the blastocyst stage, and the morula cavitates between 5 and 8 days of gestation. The preimplantation embryo has remarkable regulative (restorative) growth potential [197], and it has been shown that one cell from an eight-celled rabbit embryo can produce a normal offspring [144]. The preimplantation period is generally identified as a period during which toxic insult generally results in embryo death or absence of effect

**TABLE 34.5**
**Historical Values for Mean Percent Preimplantation Loss in Hra:(NZW)SPF Rabbits**

| Facility and Mating Procedure | Number of Studies | Time Period | Number of Does | Mean ± Standard Deviation (%) | Range per Study (%) |
|---|---|---|---|---|---|
| *Artificially inseminated hCG-primed rabbits* | | | | | |
| RTI [211] | Not reported | Not reported | 88 | 30.90 ± 2.9 | Not reported |
| Argus Research Laboratories [60] | 98 | 1980–1989 | 1463 | 17.8 | 8.2–38.4 |
| *Naturally mated rabbits* | | | | | |
| RTI [211] | Not reported | Not reported | 117 | 11.34 ± 1.49 | Not reported |
| Argus Research Laboratories (historical control data) | 20 | 1998–1999 | 378 | 13.0 ± 7.8 | 3.7–30.0 |

because of the regenerative powers of the embryo [240], although it has been shown that preimplantation exposures to some agents affect the ultimate growth and development of the embryo (e.g., actinomycin D and methotrexate, among others) [147,200].

## Implantation and Placentation

Cavitation is followed by attachment of the blastocyst to the uterine wall (nidation) and subsequent invasion of the uterine wall (implantation) by the syncytiotrophoblast, which erodes the endometrium. Placental circulation is subsequently established [208]. Each blastocyst includes two different cell populations: the outer layer (trophoblast), which becomes the placenta and fetal membranes, and the inner cell mass (cluster of cells within the blastocyst), which becomes the embryo. Although sometimes not considered in the overall development of the embryo, the trophoblast serves an important function: The extraembryonic membranes protect the conceptus, while the placenta provides a means to supply nutrients and remove metabolic waste. Multiple diverse mechanisms exist for the placental transport of molecules, including both simple diffusion and carrier-mediated mechanisms (active transport, facilitated diffusion, receptor-mediated endocytosis). The placenta is clearly not a barrier, but rather a means by which a substance in the maternal system can be transported at some rate and by some mechanism into the embryo. Remarkable differences exist across species with respect to the layers of embryonic and maternal tissues interposed between the respective circulations, as well as differences in the duration and functions of the yolk sac among species which can affect the rate and access of a test material to a conceptus. Comparative reproductive and placental parameters are provided in Table 34.5.

## Embryogenesis

The rapid growth of the conceptus continues through embryonic, fetal, and neonatal stages. The duration of these stages differ in various species, as shown in Table 34.6 [19]. The period of embryogenesis (organogenesis) is generally identified as the interval between implantation (and formation of the neural plate in the ectoderm) and closure of the hard palate. Most organ systems are formed during this period, requiring cell proliferation, cell migration,

**TABLE 34.6**
**Timing of Early Development in Some Mammalian Species**

| | Times of Early Development (Days from Ovulation) | | | |
|---|---|---|---|---|
| Mammal | Blastocyst Formation | Implantation | Organogenesis Period | Length of Gestation |
| Mice | 3–4 | 4–5 | 6–15 | 19 |
| Rats | 3–4 | 5–6 | 6–15 | 22 |
| Rabbits | 3–4 | 7–8 | 6–18 | 33 |
| Sheep | 6–7 | 17–18 | 14–36 | 150 |
| Monkeys (rhesus) | 5–7 | 9–11 | 20–45 | 184 |
| Humans | 5–8 | 8–13 | 21–56 | 267 |

*Source:* Adapted from Brinster, R.L., in *Methods for Detection of Environmental Agents That Produce Congenital Defects*, Shepard, T.H. et al., Eds., Elsevier, New York, 1975, pp. 113–124.

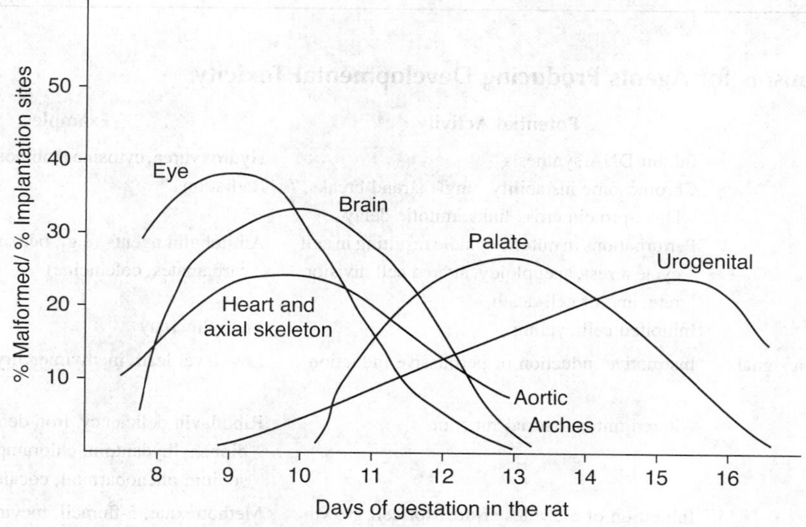


**FIGURE 34.3** Hypothetical pattern of susceptibility of embryonic organs to teratogenic insult. (Adapted from Wilson, J.G. and Warkany, J., Eds., *Teratology: Principles and Techniques*, University of Chicago Press, Chicago, IL, 1965.)

cell–cell interactions, and morphogenetic tissue remodeling. Each forming structure has a period of maximum susceptibility, with peak susceptibility to insult coinciding with the time when key developmental events occur in these structures. Wilson's classic diagram (Figure 34.3) [238] demonstrates the pattern for the rat embryo. Shenefelt [187] demonstrated the varying susceptibilities in the hamster; Rogers et al. [178] provided similar observations in the mouse, and Hoar and Salem [87] in the guinea pig. *In toto*, these investigations demonstrate that peak sensitivities may not only differ for a given tissue or organ but also with the administered dose, reverting to Wilson's concepts [238–240]. In addition, the same insult may affect the growth of concurrently developing systems. Thus, insult during the period of organogenesis is most likely to result in gross structural malformations [238,239]; however, all endpoints of developmental toxicology have been shown to be affected, although the dose–response patterns for agents often differ (see Figure 34.4), and the interrelationships of the responses often confound apparent dose dependency. In general, the type of agent that is of most concern is that causing malformation at doses that are lower than those associated with growth retardation or lethality. As discussed later, regardless of the dose–response, current practice in risk assessment is to base the developmental toxicity effect level on the lowest of the four potential endpoints adversely affected, whether the effect occurs alone or in combination.

## Fetogenesis

The fetal period is characterized by tissue differentiation, growth, and physiological maturation. Almost all organs are present and grossly recognizable, and further development of these organs proceeds with the fetus attaining required functions before birth. These include fine structure morphogenesis (e.g., synaptogenesis, neural outgrowth, branching morphogenesis of the bronchial tree and renal cortical tubules) and biochemical maturation (e.g., induction of tissue-specific enzymes and structural proteins). Insult during the period of fetogenesis is most likely to affect growth and functional maturation of the central nervous system, reproductive organs (including behavioral and motor deficits and reductions in fertility),

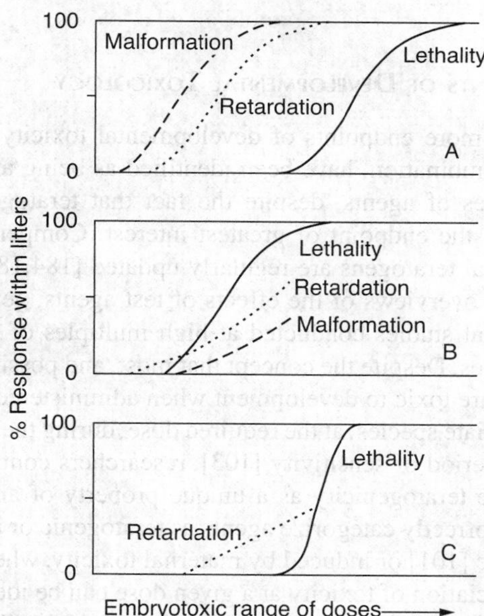


**FIGURE 34.4** Dose–response patterns (A–C) for different types of developmental toxicant. (Adapted from Neubert, D. et al., *Curr. Top. Pathol.*, 69, 241–331, 1980.)

**TABLE 34.7**
**Examples of Mechanisms for Agents Producing Developmental Toxicity**

| General Mechanism | Potential Activity | Examples | Ref. |
|---|---|---|---|
| Mitotic interference | Inhibit DNA synthesis | Hydroxyurea, cytosine arabinoside, 5-fluroiacil | [58] |
| | Chromosome instability, single strand breaks, DNA-protein cross-links, mitotic delay | Radiation | [18] |
| | Perturbations in mitotic spindle resulting in cell cycle arrest, aneuploidy, altered cell division rate, and/or cell death | Antitubulin agents (e.g., benzimidazoke carbamates, colchicine) | [58] |
| | Inhibited cell cycling | Methylmercuy | [58] |
| Altered membrane function/signal transduction | Instructive induction or permissive induction | Low level lead, methylmercury | [58] |
| Altered energy sources | Altered mitochondrial function | Riboflavin deficiency, iron deficiency, diphenylhydantoin, chloramphenicol, sodium phenobarbital, cocaine | [58] |
| Enzyme inhibition | Inhibition of enzymes critical for cell growth and proliferation | Methotrexate, 5-fluracil, mevinolin | [58] |
| Nucleic acid interaction | Interference with normal synthesis and functioning of DNA and RNA | Hydroxyurea, cytosine arabinodide | [58] |
| Mutations | Alternation of DNA nucleotide sequence | Ionizing radiation, alkylating agents, aromatic amines | [58] |
| Altered gene expression | Induction or repression of transcription | Retinoicacid, dioxin, cocaine | [114] |
| | | Phenytoin | [67] |
| Programmed cell death (apoptosis) | Alterations of normal apoptosis | Retinoic acid, dioxin, DNA-damaging agents | [58] |

the pulmonary system, and the immune system. Although gross structural changes can occur during the fetal period, such observations are generally secondary to deformations (changes in previously normal structures, such as clubbed or bent limbs), rather than malformations (abnormal growth).

## Endpoints of Developmental Toxicology

One or more endpoints of developmental toxicity, alone or in combination, have been identified as being affected by scores of agents, despite the fact that teratogenicity remains the endpoint of greatest interest. Compendiums of animal teratogens are regularly updated [184,188] and provide overviews of the effects of test agents, generally in animal studies conducted at high multiples of human exposures. Despite the concept that most, and possibly all, agents are toxic to development when administered to the appropriate species, at the required dose, during the appropriate period of sensitivity [103], researchers continue to perceive teratogenicity as a unique property of an agent and incorrectly categorize agents as teratogenic or nonteratogenic [101] or induced by maternal toxicity, when only an association of toxicity at a given dose can be identified on the basis of the study design [46,89,106,107]. It is hoped that current readers will understand that agents resulting in toxic effects in the maternal and paternal animals may be expected to result in adverse effects on the developing conceptuses from which the parents but not necessarily the conceptuses may recover, and that cross-species extrapolation of the affected developmental endpoint does not necessarily occur because of multiple timing, exposure, and species-specific differences.

## Mechanisms of Developmental Toxicity

As early as 1977, Wilson [240] identified six general categories for mechanisms resulting in developmental toxicity: mitotic interference, altered membrane function or signal transduction, altered energy sources, enzyme inhibition, altered nucleic acid synthesis, and mutations. The current increased knowledge regarding molecular mechanisms of normal development has led to the addition of perturbations in gene expression and programmed cell death to Wilson's general categories [58]. Examples of developmental toxins known to act by one or more of these mechanisms are identified in Table 34.7.

### Apoptosis

Because of the relatively recent interest in apoptosis as a mechanism in carcinogenicity, the following information is provided to illustrate the occurrence of the phenomenon from conception to death. Although not previously emphasized, for almost 50 years, controlled cell death (apoptosis) has been recognized to be as important to development of the conceptus as are cell proliferation and differentiation [70,75,183,189]. Apoptosis occurs in the developing embryo and in normal healthy adult tissues, as well as in many pathological settings. It is genetically directed and usually requires ongoing protein synthesis [2]; it continues through life and provides the central mechanism for the removal of surplus, unwanted, damaged, or aged cells. It is characterized by ultrastructural and biochemical features, including cytoplasmic and nuclear condensation, formation of membrane-bound apoptotic bodies, oligonucleosomal DNA degradation into micronuclei, and cytolysis into condensed apoptotic bodies. Apoptosis is not generally associated with an inflammatory cell infiltrate, a hallmark of necrosis [59]. In contrast, changes associated with necrosis include cellular swelling, organelle dysfunction, mitochondrial collapse, and ultimately cellular disintegration, release of the cellular contents in the extracellular milieu, and a marked host inflammatory reaction [11].

Apoptosis occurs in normal tissue turnover, organ involution after withdrawal of trophic hormones, distinct cells after deprivation of growth factors or specific stimuli, and cells that have undergone sublethal damage [2,75,233]. Interference with the interrelated cell functions that result in normal growth during embryogenesis can result in abnormalities during embryogenesis as well as cancers in adult life. For example, exaggerated or defective apoptosis can result in developmental abnormalities, such as cleft palate, neural tube defects, phocomelia, and hypospadias, or a likelihood of tumorous growth, as occurred in daughters of mothers administered DES during pregnancy [233]. Among the multiple genetic events associated with tumor development are activation of antiapoptotic oncogenes, such as the *BCL*-2 family, and inactivation of apoptosis-inducing tumor suppressor genes, including the retinoblastoma (*Rb*), *p53*, *BAX*, and *BCL*-2 genes [59,167,187,235].

Apoptosis occurs in almost every tissue during development [36,140], including during palate formation [190,230], body sculpting (e.g., digit formation [91]), gastrointestinal development [233], sexual organ development and gamete formation and number [2,5,22,75,96, 135,140,233], and homeostasis of normal tissues, especially the gastrointestinal tract, immune system, and skin [2,233]. Establishment of normal craniofacial pattern requires apoptosis of the neural crest [72], with apoptosis perhaps being most well studied in development of the nervous system. It is critical in the development of both neuronal and non-neuronal cells in the peripheral and central nervous systems [21,148] and occurs both pre- and postnatally. As noted by Mazarakis et al. [135], apoptosis is seen in the developing nervous system as early as neural tube formation and persists throughout terminal differentiation of the neural network, with more than 50% of the neurons being lost during development. These cell deaths occur in structures as diverse as sensory and autonomic ganglia, cranial motor nuclei and spinal motor neuron pools, the retina, various brainstem nuclei, and the cerebral cortex during development of most regions of the central and peripheral nervous systems. Apoptosis is evident during transformation of the neural plate into the neural tube, in the suboptic death center at the junction between the diencephalon and telencephalon, and in the ventral midline of the embryonic retina and optic stalk and is also a common phenomenon in the CNS during normal postnatal development [151].

Several mechanisms appear responsible for survival of developing neurons, including retrograde transport of trophic factors, of which nerve growth factor (NGF) is the best characterized [148]. Unchecked apoptosis and failure of the apoptotic pathway contribute to nervous system pathology in both the developing embryo [151] and the adult [148] and can be caused by viruses, metabolic stress, damage to cell structures, and drug, chemical, and physical insults [75]. Dysregulation of apoptosis has been shown to result in neurodegeneration and tumorigenesis [167]. Failure to undergo apoptosis is also associated with low-grade follicular lymphoma [2].

Studies have identified an association between apoptosis and tumor suppressor genes [233,237]; for example, the retinoblastoma gene (*Rb*) and *p53* genes are mutated in a wide range of sporadic human tumor types, and germline mutations in *Rb* and *p53* result in a greatly increased cancer risk. The *Rb* gene encodes a nuclear phosphoprotein believed to act through protein–protein interactions as a regulator of the $G_1$ to S phase of cell-cycle transition. The *p53* gene encodes a transcriptional regulator, has been linked to cell cycle arrest and the induction of apoptosis [237], and has been termed the "guardian of the genome" because it plays an essential role in surveillance of DNA damage and regulation of the cell cycle and a regulatory role in apoptosis.

The interrelationship of oncogenes, tumor suppressor genes, apoptosis, and normal growth and development is being actively studied but has not yet been defined [167]; for example, apoptosis in the neural crest rhombomeres 3 and 5 and in the interdigital mesenchyme appears to involve homeobox gene *MSX*-2 and the signaling molecule BMP-4 [72]. Although the 5′-located *HOXd* genes are reported to be involved in patterning of the digital rays, Hurle et al. [91] found that experimental inhibition of interdigital cell death and formation of extra digits was not accompanied by a modified expression of these genes nor precocious modification of *MSX*-1 and *MSX*-2 gene expression, even though these two genes exhibit a domain of expression relatively coincident with the interdigital cell death zones, indicating involvement of other unidentified genes. Studies of the *BCL*-2 gene identified a contradictory phenomenon. Although the *BCL*-2 gene is not expressed in areas of cell death in the developing limb but is expressed in the digital rays, transgenic mice with disruption of the *BCL*-2 gene have normal limbs. Transgenic embryo models have also been used to demonstrate that proper regulation of the *p53* tumor suppressor gene is necessary for normal morphogenesis, that too little *p53* makes the embryo susceptible to neural defects associated with instability of the genome or chromosomal damage, and that excessive *p53* may result in cell death abnormalities, particularly in the eye [113].

### Structure–Activity and Agent Interaction Considerations

Although the reason for many investigations into developmental toxicity has been to establish a relationship between structure and human teratogens, such as thalidomide, as occurred for the phthalic acids, or valproic acid, as occurred for di-(2-ethylhexyl) phthalate (DEHP), structure–activity relationships remain weak for elucidating mechanisms of developmental toxicity [61]. Considerable interest also lies in the potential for interactions of agents by the same or different mechanisms. It is known that the potential for interaction of multiple chemicals exists for developmental and other toxicities and that such interactions may enhance or reduce general toxicity of the agents and their potential developmental toxicity, although not necessarily equally or by the same mechanism. For example, early on it was found that caffeine potentiates the activity of multiple agents [176]; although the initial emphasis was on the pharmacologic activity of the putative agent (in this case, the vasoconstrictive effect of caffeine [48]), studies now tend to investigate the interaction of the agent with gene expression. The U.S. Environmental Protection Agency (EPA) has an active program for evaluating multiple chemical combinations that are potential water contaminants, often using modified reproductive or developmental toxicity protocols [69,74,82,149].

## METHODS USED IN REPRODUCTIVE (FEMALE) AND DEVELOPMENTAL TOXICOLOGY

### Testing Procedures and Guidelines for Regulatory Use

The purpose of this section is to provide an overview of the methods used in the collection and interpretation of data obtained from reproductive and developmental toxicity studies conducted for regulatory use. The methods described are those used in our laboratory and were developed from the literature and practical experience. Because it was not always practical to provide in-depth details, useful publications have been referenced when possible. Multiple laboratory species are used in these types of studies; however, for practical reasons, most studies are performed using rats, rabbits, mice, or hamsters. In general, the described techniques can be applied to ferrets, guinea pigs, mini-pigs, dogs, and nonhuman primates by using species-specific considerations. As previously noted, there are often special reasons to use alternative species; for example, the prolonged *in utero* development and comparable *in utero* CNS development in the guinea pig theoretically make this species ideal for evaluation of neural development. Important caveats are that studies conducted for regulatory use should provide evidence that the agent is pharmacologically or toxicologically active in the species; that it is absorbed, if administered orally or topically; and, if possible, that it is similarly handled metabolically by the test species and humans.

The International Conference for Harmonization (ICH) guideline on the detection of toxicity to reproduction for medicinal products [93] has been accepted by the U.S. Food and Drug Administration (FDA) [227,228], the European Union, and Japan. The guideline addresses all elements of reproductive and developmental toxicity testing and provides an overview of the various study designs and endpoints evaluated in the regulatory testing of pharmaceuticals, biotechnology products [226], indirect food additives and devices [228], chemicals [29,222,223,224], pesticides, fungicides, and rodenticides [29,216–218]. As harmonization progresses it is likely that the guidelines will become more, rather than less similar. For example, the ICH guideline [229,230] replaces the former guidelines issued by the U.S. FDA Bureau of Drugs [226], EU countries [40,56], and Japan [142,143,207], and the Office of Prevention, Pesticides, and Toxic Substances (OPPTS) guidelines replace the former EPA Toxic Substances Control Act (TSCA) guidelines [29,222–224] and Federal Insecticide, Fungicide, and Rodenticide Act (FIFRA) guidelines [29,216–218], and harmonization efforts are currently ongoing between the EPA and the Organization for Economic Cooperation and Development (OECD).

Comprehensive comparisons of the ICH guideline with other guidelines have been published by Christian and Hoberman [33] and Collins [39]. For complete details, it is recommended that the documents be obtained; they are available on the Internet and from the various regulatory publishing houses. Practical updates on regulatory issues, shared experiences, and access to historical databases can be obtained through access to the websites cited in Table 34.8. A listing of the majority of the regulatory guidelines currently in use is provided in Table 34.9.

For the purposes of this chapter, the ICH guideline [93] was used as the reference for identifying the various stages and interrelationships of female reproductive functions and development of the offspring. The guideline segments the reproductive cycle into six ICH stages (ICH stages A through F) that may be tested separately or in combination, generally using a one-day overlap of treatment. Because many publications describe tests using now-outdated terms (e.g., segment I, II, and III tests), Table 34.10 is provided for use in the identification of comparable intervals in the various guidelines and in academic research.

**TABLE 34.8**
**Useful Websites**

| Organization | Website |
|---|---|
| American College of Toxicology | http://actox.org |
| Congenital Anomalies (Journal of the Japanese Teratology Society) | http://www.blackwellpublishing.com |
| Environmental Protection Agency | http://www.epa.gov (search OPPTS) |
| European Teratology Society | http://www.etsoc.com |
| FDA-CDER | http://www.fda.gov/cder/guidance/index.htm (a front page for guidance documents; click on ICH) |
| FDA-CFSAN | http://vm.cfsan.fda.gov/~redbook/red-toct.html |
| International Federation of Teratology | http:// www.ifts-atlas.org |
| Societies (IFTS) Atlas of Developmental Abnormalities | http:// www.ifts-atlas.org |
| Mid-Atlantic Reproduction and Teratology Society (MARTA) | http://www.e-marta.org (provides link to Teratology Society) |
| Midwest Teratology Association | http://www.midwest-teratology.org (provides link to Teratology Society) |
| National Institute of Environmental Health Sciences (NIEHS) | http://www.niehs.nih.gov |
| Neurobehavioral Teratology Society | http://nbts.org |
| *Reproductive Toxicology* (fee charged for online subscription | http://www.elsevier.com |
| Society of Toxicology (SOT) | http://www.toxicology.org |
| Teratology Society | http://www.teratology.org |
| TERIS (Teratogen Information System: $1000/yr subscription price) | http://depts.washington.edu/~terisweb/ |

**TABLE 34.9**
**Current Regulatory Guidelines: ICH, FDA, EPA, and OECD**

| | |
|---|---|
| **Medical Agents** | |
| ICH [93] | Detection of Toxicity to Reproduction for Medicinal Products (proposed rule endorsed by the ICH Steering Committee at Step 4 of ICH Process, 1993) |
| ICH [94] | Detection of Toxicity to Reproduction for Medicinal Products (proposed rule endorsed by the ICH Steering Committee at Step 4 of ICH Process, 1995) |
| FDA [27] | ICH: Guideline on detection of toxicity to reproduction for medicinal products, *Fed. Reg.*, 59(183), 1994 |
| FDA [230] | ICH: Guideline on detection of toxicity to reproduction for medicinal products: addendum on toxicity to male fertility, *Fed. Reg.*, 61(67), 1996 |
| **Indirect Food Additives** | |
| FDA, Bureau of Foods [227] | *Toxicological Principles and Procedures for Priority Based Assessment of Food Additives (Redbook)*; guidelines for reproduction testing with a teratology phase |
| FDA, Center for Food Safety and Applied Nutrition [227] | *Toxicological Principles for the Safety Assessment of Direct Food Additives and Color Additives Used in Food (Red Book II)*; guidelines for reproduction and developmental toxicity studies |
| FDA, Center for Food Safety and Applied Nutrition [227] | *Toxicological Principles and Procedures for Priority Based Assessment of Food Additives (Redbook)* |
| **Chemicals** | |
| EPA, OPPTS [219] | *Health Effects Test Guidelines: Prenatal Developmental Toxicity Study*, OPPTS 870.3700, August, 1998 |
| EPA, OPPTS [220] | *Health Effects Test Guidelines: Reproduction and Fertility Effects*, OPPTS 870.3800, August, 1998 |
| EPA, OPPTS [221] | *Health Effects Test Guidelines: Prenatal Developmental Toxicity Study*, OPPTS 870.6300, August, 1998 |
| EPA, Risk Assessment Forum [214] | Guidelines for developmental toxicity risk assessment, *Fed. Reg.*, 56(234), 63798-63826, 1991 |
| EPA [215] | *Guidelines for Developmental Toxicity Risk Assessment*, NTIS PB No. PB97-100098, Sept. 5, 1996 |
| OECD [160] | *OECD Guidelines for Testing of Chemicals: Teratogenicity*, Section 4, No. 414, adopted May 12, 1981 |
| OECD [161] | *OECD Guidelines for Testing of Chemicals: One-Generation Reproduction Toxicity*, Section 4, No. 415, adopted May 26, 1983 |
| OECD [162] | *OECD Guidelines for Testing of Chemicals: Two-Generation Reproduction Toxicity Study*, Section 4, No. 416, adopted May 26, 1983 |
| OECD [164] | *OECD Guidelines for Testing of Chemicals: Combined Repeated Dose Toxicity Study with the Reproduction/Developmental Toxicity Screening Test*, Section 4, No. 422, adopted March 22, 1996 |

**TABLE 34.10**
**Comparison of ICH Stages and Study Types with Similar Observations Made[a]**

| ICH Stage | FDA Guidelines[b] | Great Britain and EEC Guidelines[c] | Japanese Guidelines[d] | EPA OPPTS, OECD, and FDA Redbook Guidelines[e] | Alternative or Additional Evaluations[f] |
|---|---|---|---|---|---|
| A. *Premating to conception*: Reproductive functions in adult animals, including development and maturation of gametes, mating behavior, and fertilization | *Segment I* | *Segment I* | *Segment I* | *Multigeneration*<br>*One-generation* | *Continuous breeding*<br>*Modified Chernoff–Kavlock* |
| B. *Conception to implantation*: Reproductive functions in the adult female; preimplantation and implantation stages of the conceptus | *Segment I* | *Segment I* | *Segment I* | *Multigeneration*<br>*One-generation*<br>*Developmental toxicity* | *Continuous breeding*<br>*Modified Chernoff–Kavlock* |
| C. *Implantation to closure of the hard palate*: Adult female reproductive functions and development of the embryo through major organ formation | *Segment I*<br>*Segment II* | *Segment I*<br>*Segment II* | *Segment II* | *Multigeneration*<br>*One-generation*<br>*Developmental toxicity*<br>*Developmental neurotoxicity* | *Continuous breeding*<br>*Modified Chernoff–Kavlock* |
| D. *Closure of the hard palate to the end of pregnancy*: Adult female reproductive function; fetal development and growth; organ development and growth | *Segment I*<br>Segment II<br>*Segment III* | *Segment I*<br>Segment II | *Segment II* | *Multigeneration*<br>*One-generation*<br>*Developmental toxicity*<br>*Developmental neurotoxicity* | *Continuous breeding*<br>*Modified Chernoff–Kavlock* |
| E. *Birth to weaning*: Adult female reproduction function; adaptation of the neonate to extrauterine life, including preweaning development and growth (postnatal age optimally based on postcoital age) | *Segment I*<br>*Segment III*<br>*Pediatric* | *Segment I*<br>Segment II<br>*Segment III* | Segment II<br>*Segment III* | *Multigeneration*<br>*One-generation*<br>*Developmental toxicity*<br>*Developmental neurotoxicity* | *Modified Chernoff–Kavlock* |
| F. *Weaning to sexual maturity*: (Pediatric evaluation when treated); postweaning development and growth; adaptation to independent life; and attainment of full sexual development | *Pediatric* | Segment I | Segment II<br>Segment III | *Multigeneration*<br>*Developmental neurotoxicity*<br>*Developmental immunotoxicity* | |

[a] Italicized information indicates treatment interval.
[b] Data from [226].
[c] Data from Committee on the Safety of Medicines [40] and European Economic Community [56].
[d] Data from Ministry of Health and Welfare [142,143] and Tanimura et al. [207].
[e] Data from OECD [160,164], U.S. EPA [217–219], U.S FDA [227–229].
[f] Data from Chernoff and Kavlock [27], Hardin [80], Lamb [116], Morrissey et al. [145], and Narotsky et al. [149].

The ICH guideline cites recommendations that were often unidentified in other guidelines, including use of: (1) scientific justification of flexible study designs, (2) kinetics, (3) expanded male reproductive toxicity evaluations, (4) a requirement for mechanistic studies, and (5) essentially equal emphasis on all endpoints of developmental toxicity (death, malformation, reduced weight, functional or behavioral alterations), rather than emphasizing malformation as the most important outcome. Flexible testing strategies are to be based on: (1) anticipated drug use, especially in relation to reproduction; (2) the form of the substance and routes of administration intended for humans; and (3) consideration of existing data on toxicity, pharmacodynamics, kinetics, and similarity to other compounds in structure and activity. *In toto*, the purpose of these studies is to identify any effect of an active substance on mammalian reproduction, to compare this effect with all other pharmacologic and toxicologic data for the agent and ultimately to determine whether the human risk for reproductive and developmental effects is the same as, increased, or reduced in comparison with the risks of other toxic effects of the agent. As should occur for all toxicologic observations, additional pertinent information should be considered before the results of the animals studies are extrapolated to humans. This information includes human exposure considerations, comparative kinetics, and the mechanism of the toxic effect.

Because this chapter emphasizes toxic effects on the female animal and her offspring (through completion of the lactation period), sections of the ICH guideline relevant to male reproductive performance and postnatal development of the offspring are discussed only when relevant to the primary subject area. Assessments of male reproductive effects, functional and behavioral effects in the offspring, and pediatric effects are more fully described elsewhere.

## General Procedures for Pharmaceuticals

The ICH stages are shown in Figure 34.1. Conduct of a study including two full reproductive cycles is equivalent to a two-generation study. As noted earlier, rodents and rabbits are the most frequently used laboratory animals. Common protocol designs used for evaluation of these species are provided in Figure 34.5 through Figure 34.8. The most common practice used in developing pharmaceuticals is to evaluate ICH stage A (premating to conception) and stage B (conception to implantation); treatment of the female animals is initiated 2 weeks before cohabitation and continues through day 7 of gestation (day 0 of gestation = sperm in smear), a protocol comparable to the formerly used Japanese segment I design [142]. Many investigators continue treatment through ICH stage C (implantation to closure of the hard palate; in rats, days 6/7 through 17 of gestation). ICH stage C is also evaluated in a nonrodent species (generally the rabbit, treatment days 6/7 through 19 of gestation, where day 0 of gestation = mating or insemination) early in product development, although depending on the proposed use of the agent some investigators have chosen to delay this test until after the early clinical trials are completed (metabolism, kinetics, safety, and efficacy). Based on the proposed use of the agent, studies are also conducted to evaluate ICH stage C (implantation to closure of the hard palate), stage D (closure of the hard palate to parturition; rats, day 17 through 22 of gestation), stage E (birth to weaning; rats, day 0/1 through 21 postnatal), and stage F (without drug exposure from weaning to sexual maturity; rats, day 22 through 60 postnatal), a design comparable to a combination of the former Japanese segment II and segment III designs. When these studies provide a second evaluation of stage C (embryogenesis), the female animals are usually permitted to naturally deliver all offspring, and the fetal evaluations not repeated, although the results of the fetal evaluations should be referenced in the discussion of the results of these studies.

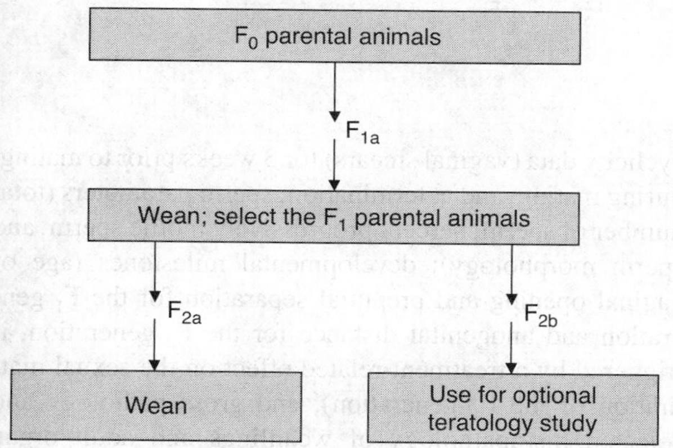


**FIGURE 34.5** Three-generation study in rats.

## General Procedures for Indirect Food Additives

The FDA *Redbook 2000* [227] allows conduct of a teratology study in rodents within a multigeneration study (see Figure 34.5). Current common practice is to evaluate two generations, one generation per litter, although continuous breeding and developmental toxicology studies may be conducted concurrently. When such occurs, exposure is continual in the parental generation from before mating and in the dams throughout gestation. The usual treatment period in the multigeneration study before mating of the first parental generation ($P_0/F_0$, 8 to 11 weeks before cohabitation in male rats and 2 weeks before mating in female rats). It continues through mating, gestation, and lactation and then in the selected weanlings identified as the second ($P_1/F_1$) gestation. These guidelines now include optional procedures for additional litters per generation, additional generations, tests for developmental toxicity effects, and neurotoxicity and immunotoxicity screening. It is acceptable to conduct the developmental toxicity studies in rats and rabbits separately. The current requirements identify the treatment period cited in the 1966 FDA pharmaceutical testing requirements [226]; however, studies conducted in conformance with the FDA ICH guideline [229], with treatment on days 6/7 through 17 in rats and on days 7 through 19 in rabbits, are acceptable.

## General Procedures Specific to EPA OPPTS Protocols

Developmental toxicity studies are required in two species [217–219]. Treatment is administered from implantation to delivery (rats, gestation days 6 through 20; rabbits, days 6 through 29), with 20 pregnant animals per group assigned to the study. It is acceptable to provide treatment throughout the entire gestation period (rats, gestation days 0 through 20; rabbits, days 0 through 29). The EPA multi-

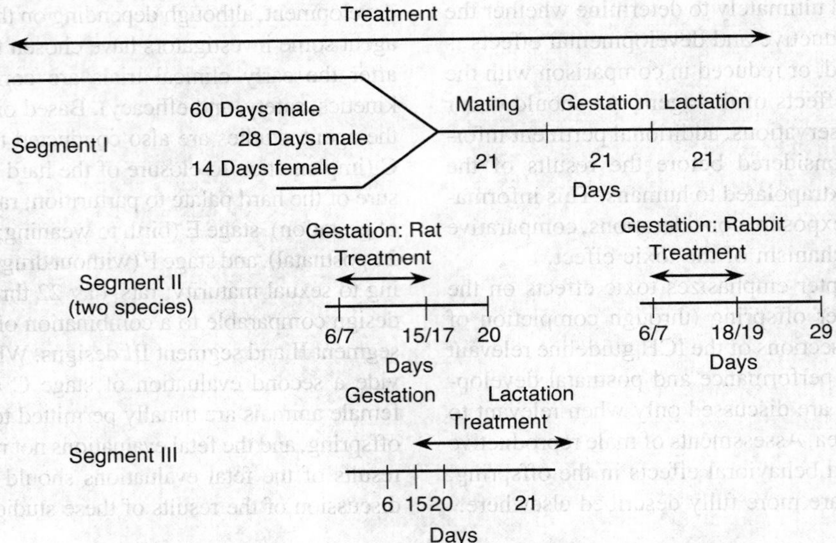


**FIGURE 34.6** Three segments, FDA.

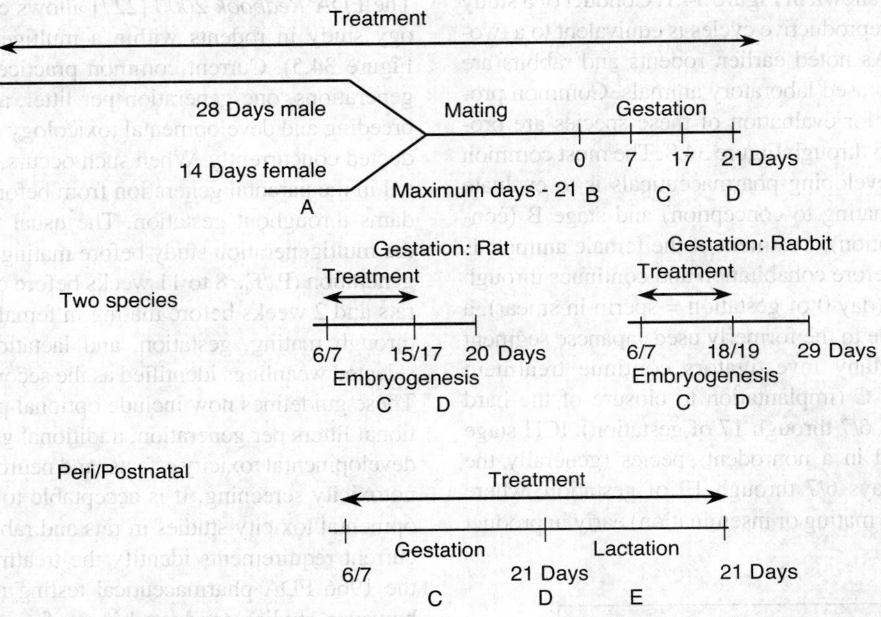


**FIGURE 34.7** Study designs illustrating ICH stages.

generation (two-generation) protocol has multiple endpoints and incorporates many endpoints traditionally performed in pharmaceutical development, although differing from the ICH guideline in that the protocols are set rather than open to scientific judgment regarding design. Treatment of the parental generation ($P_0/F_0$ male and female rats) is initiated 10 weeks before mating and continues through a 3-week mating period (the male rats can be terminated after the cohabitation period). Treatment of the $PP_0/F_0$ and $F_1$ male and female rats continues until termination. Additional endpoints assessed include estrous cyclicity data (vaginal smears) for 3 weeks prior to mating, during mating, and at termination; sperm parameters (total number of sperm, percent progressively motile sperm, and sperm morphology); developmental milestones (age of vaginal opening and preputial separation for the $F_1$ generation and anogenital distance for the $F_2$ generation, if triggered by a treatment-related effect on the sexual maturation of the $F_1$ generation); and gross pathology and selected histopathology of weanlings and adult organ weight data (uterus; ovaries; testes; one epididymis, including the total weight and cauda epididymal weight;

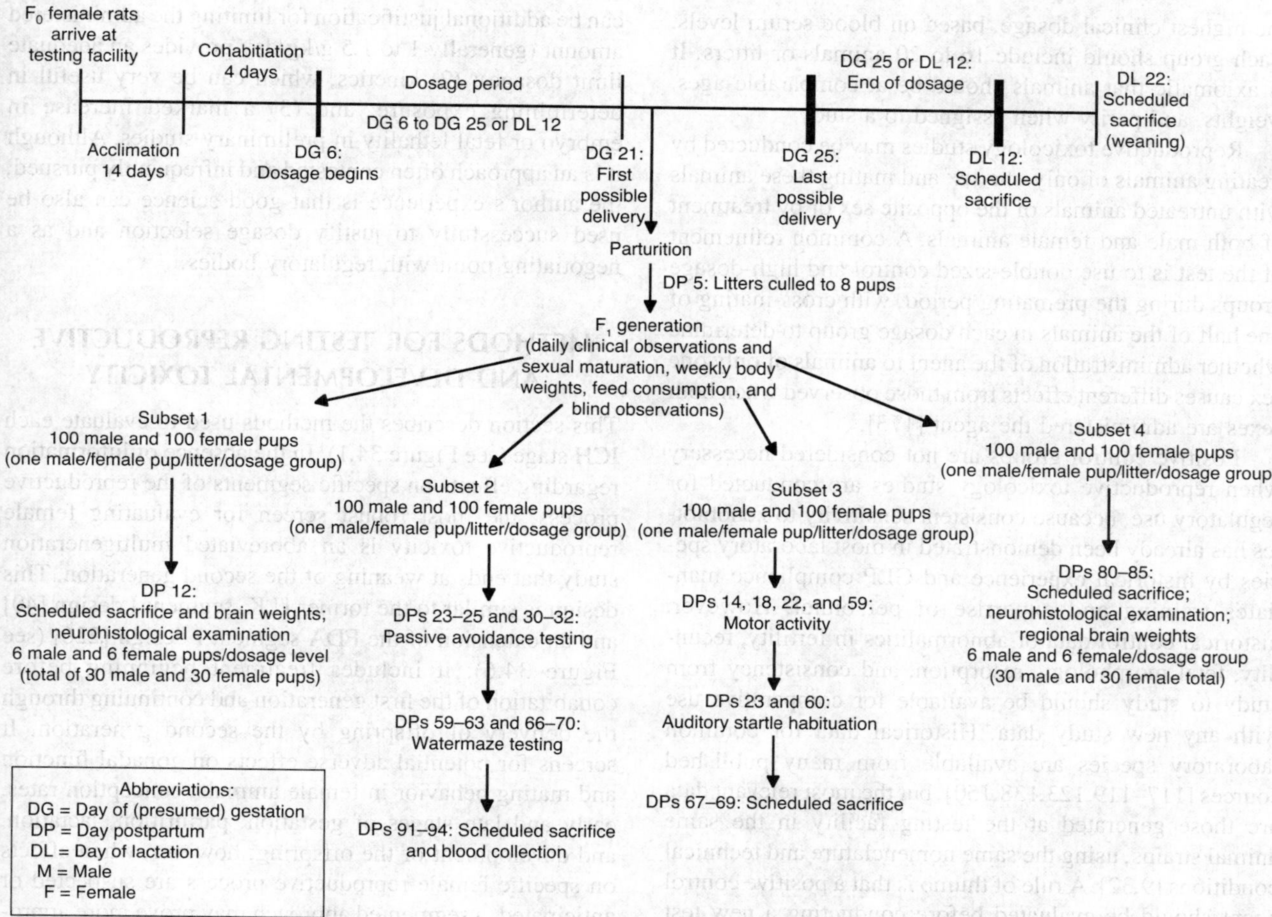


**FIGURE 34.8** Neurobehavioral developmental toxicity study in rats.

seminal vesicles with coagulating glands and fluids; prostate; brain; liver; kidneys; adrenal glands; spleen; thymus; and all known target organs). Histologic evaluation of the ovary is to include a minimum of ten sections, randomly selected from one completely sectioned ovary per female in the control and high-dosage groups. Examination of the intact epididymis is to include the caput, corpus, and cauda regions. For the $F_1$ generation, organ weight data are to be collected for one pup per sex per litter from all dose groups for the ovaries, testes, brain, liver, kidney, adrenal glands, spleen, thymus, and known target organs, with histopathology triggered by treatment-related effects in the $F_1$ high-dosage group histopathology.

The EPA developmental neurotoxicology protocol [221] may be incorporated within the multigeneration study design or conducted separately. A potential schematic for this very complex study is presented as Figure 34.8. In this study, treatment occurs from implantation through postnatal development of the majority of the CNS (rats, gestation day 6 through lactation day 12), although treatment can be continued through lactation to day 21, when weaning occurs.

## General Considerations Regarding Reproductive/Developmental Toxicology Studies Conducted for Submission to Regulatory Agencies

These studies should be conducted in conformance with appropriate good laboratory practice (GLP) regulations [57,165,213,212,225]. Whenever possible, the same species and strain of animal should be used as tested in other toxicology studies and in the kinetics studies. The kinetics, pharmacological, and toxicological data should support the position that the selected species provides a model relevant for use in human safety assessment. Reproductive toxicology studies are generally conducted in one species (usually the rat), with some subsegments (e.g., ovarian toxicity, testicular toxicity) obtained from companion subchronic studies in the same species and strain. Developmental toxicology studies are generally conducted in two species, one rodent (usually the rat) and one nonrodent (usually the rabbit). Each study usually includes a vehicle control group and three or more groups administered dosages of the test agent at arithmetic multiples of the highest clinical dosage. Ideally, the low dosage is a multiple of the highest clinical dosage, based on blood serum levels. Each group should include 16 to 20 animals or litters. It is axiomatic that animals should be at comparable ages, weights, and parity when assigned to a study.

Reproductive toxicology studies may be conducted by treating animals of only one sex and mating these animals with untreated animals of the opposite sex or by treatment of both male and female animals. A common refinement of the test is to use double-sized control and high-dosage groups during the premating period, with cross-mating of one half of the animals in each dosage group to determine whether administration of the agent to animals of only one sex causes different effects from those observed when both sexes are administered the agent [173].

Positive control groups are not considered necessary when reproductive toxicology studies are conducted for regulatory use, because consistent sensitivity to xenobiotics has already been demonstrated in most laboratory species by historical experience and GLP compliance mandates training and expertise of personnel. However, historical control data of abnormalities in fertility, fecundity, fetal morphology, resorption, and consistency from study to study should be available for comparative use with any new study data. Historical data for common laboratory species are available from many published sources [117–119,123,138,150], but the most relevant data are those generated at the testing facility in the same animal strains, using the same nomenclature and technical conditions [9,32]. A rule of thumb is that a positive control agent should be evaluated before conducting a new test paradigm or examining a new endpoint to document appropriate training and expertise.

In general, regulatory studies require that the high dosage be maternally or developmentally toxic and that the low dosage should be a no-observed-adverse-effect level (NOAEL). In the past, depending on the regulatory body, no more than 10% mortality and/or 10% reduction in maternal body weight gain were usually considered sufficient to identify a maternally toxic dosage. Conformance with these criteria and the associated testing of excessively high dosages resulted in both maternal toxicity and developmental toxicity [136]. In contrast, the ICH guideline indicates that the high dosage should produce minimal maternal (adult) toxicity and be selected on the basis of data from all available studies, including pharmacologic, acute and subchronic toxicity, and kinetic studies. These guidelines allow evidence of maternal toxicity to include reduced or increased weight gain, specific target organ toxicity, changes in hematology or clinical chemistry parameters, and exaggerated pharmacological response, which may or may not be reflected as marked clinical reactions (e.g., sedation, convulsions). Additional justification for high dosage selection include: (1) physicochemical properties of the test substance or dosage formulation which, in combination with the route of administration, can be additional justification for limiting the administered amount (generally 1 to 1.5 g/kg/day provides an adequate limit dosage); (2) kinetics, which can be very useful in determining exposure; and (3) a marked increase in embryo or fetal lethality in preliminary studies. Although it is an approach often contested and infrequently pursued, the author's experience is that good science can also be used successfully to justify dosage selection and as a negotiating point with regulatory bodies.

## METHODS FOR TESTING REPRODUCTIVE AND DEVELOPMENTAL TOXICITY

This section describes the methods used to evaluate each ICH stage (see Figure 34.1). In the absence of information regarding effects on specific segments of the reproductive process, the most robust screen for evaluating female reproductive toxicity is an abbreviated multigeneration study that ends at weaning of the second generation. This design is similar to the former U.K. segment I design [40] and an extension of the FDA segment I design [226] (see Figure 34.6); it includes treatment beginning before cohabitation of the first generation and continuing through the delivery of offspring by the second generation. It screens for potential adverse effects on gonadal function and mating behavior in female animals, conception rates, early and late stages of gestation, parturition, lactation, and development of the offspring; however, when effects on specific female reproductive process are suspected or anticipated, a segmented approach may prove more appropriate. For example, evidence of LH inhibition in a pharmacology study, ovarian atresia in a chronic study, or abnormal estrous cycling in a subchronic toxicity study or preliminary reproductive toxicology screen may be predictive of potential reductions in mating and fertility in the parental generation and preclude production of adequate numbers of litters for further evaluation. Common protocols for use in evaluation of the various ICH segments are identified in Figure 34.7.

### ICH Stage A

ICH stage A includes premating to conception (reproductive functions in adult animals, including development and maturation of gametes, mating behavior, and fertilization).

#### *Evaluation of the Ovary*

##### *Weight*

Ovarian weights are recorded using the same techniques as those generally used for recording organ weights. The tissue should be trimmed and weighed close to the time of its removal from the animal to avoid dehydration. An alternative method is to weigh the ovaries after trimming and fixation.

*Follicle Number and Size*

Follicle number and size can be identified by histological evaluation of cross-sections of the ovary [171,194]. This procedure is extremely time consuming and expensive (30 to 60 sections per ovary for mice and rats, resulting in 300 and 600 sections per group of 10 animals, respectively). A shorter screening method is described by Plowchalk [172] for potential use when indicated by observations in companion studies (e.g., pharmacologic action or reduced ovarian weight or atresia). After fixation of the ovaries in Bouin's solution, to prevent undue shrinkage, standard histologic techniques are used to prepare 6.0-μm serial sections. Five random sections are evaluated for primordial, growing, and antral follicles. Although this method reduces the statistical power of the assay, it allows assessment of the loss of specific follicle types and changes in component volumes.

*Corpora Lutea Number*

Details regarding gross evaluation of corpora lutea at Caesarean sectioning of the female animal are provided later in this section. As previously discussed, the number of corpora lutea can be affected by many factors, including enhanced ovarian atresia, injection of hormones (a procedure common in mating rabbits), handling (especially in reflex ovulators), and litter size and hormonal feedback mechanisms, some of which are species specific [88,211]. As a result, identification of preimplantation loss on the basis of comparison of the number of implants at Caesarean sectioning of the dams is often quite imprecise, especially in mice and rabbits. Generally, in these types of studies, comparison of litter sizes is a more representative indicator of preimplantation loss. As regression of the corpora lutea occurs during lactation, inclusion of this parameter in postlactational evaluations is unnecessary.

*Hormone Integrity and Function*

Direct measurements of hormone levels are usually second-level evaluations. These can most effectively be performed using commercially available antibody kits or by sending samples to contract laboratories for evaluation. Unfortunately, little historical experience exists with laboratory animal species. The expected background levels relevant to the specific strain and source should be developed before using these evaluations in a regulatory setting.

### *Evaluation of the Uterus*

*Weight*

Uterine weight measurements are an historic method of measuring potential estrogenicity of a test material. Both juvenile and ovariectimized models are used, with the intact model demonstrating both primary and secondary effects and the ovariectimized model identifying uterine-specific effects. Despite the use and value of this assay for pharmacologic testing since 1939 [4], no standardized method is currently available [158,159,174], although in 2005 the Endrocrine Disruptor Methods Validation Advisory Committee (EDMVAC) recommended its acceptance, after an extensive validation process, for use in screening environmental contaminants. Regardless of the duration of the treatment period, responses can vary with the strain, source, and specific population and age tested, requiring characterization of historical data at the testing laboratory. The common use of only ten animals per group often results in statistical significance between groups when outlier values (increased and reduced uterine weights) are included, with inappropriate identification of false positives and negatives [24]. Because of the sensitivity of the assay, the screen is a good predictor of absence of effect. When a positive effect is observed, it should be further characterized. When uterine weights are evaluated in animals in subchronic or chronic tests, it is helpful to identify the estrous cycle stage shortly before sacrifice, as cyclicity greatly affects uterine weight. The estrous stage identified from a vaginal lavage and the uterine estrous stage identified by histopathology will differ by approximately half a day.

### *Evaluation of Estrous Cycling*

Initial reproductive toxicity screening studies are usually performed in rodent species, with the female animals being treated for the first 14 days before cohabitation and then until mating occurs. Ideally, sexually mature female rodents are evaluated for estrous cycling [41] for 14 days before treatment to establish a baseline for regularity of cycling. Animals that irregularly cycle should be excluded from evaluation before the treatment phase. They can then be evaluated daily for one or more comparable intervals during the study to identify potential effects and possible reversal of effects. Animals housed in the same room tend to have synchronous estrous cycles, with resultant cyclic matings. Repeated use of this evaluation can be used to identify the onset of reproductive senility in chronic toxicity studies.

Estrous cycling in rodent species is evaluated by obtaining a sample of cells from the vagina by lavage and examining the cytology. The regularity of the light cycle should be monitored, and vaginal smears should be collected from the animals at approximately the same time each day. Care should be taken in handling the animals to prevent pseudopregnancy. It is also important to avoid contamination of the saline and pipette to prevent potential infection of the animal and inappropriate sample collection. Vaginal lavage samples are prepared from one or two drops of saline, delivered into and removed from the vagina with a dropping pipette. A new (clean) pipette should be used for each animal, and the tip of the pipette should always point downward when it contains a sample to avoid contaminating the bulb. After the pipette is removed from the vagina, the contents are delivered onto either a clean glass slide (identified with the animal number) or a ring slide containing designated areas for the placement of each sample. The smear of the vaginal contents is then examined (wet and unstained) by using a microscope at a power of 100 to 200×. Ring slides should not be used for vaginal smears taken during cohabitation because it is difficult to wash sperm off of this type of slide. After all daily smears are collected, the saline used in preparing the smears should be evaluated for potential contamination, because carryover of sperm from an inseminated female could result in incorrect identification of mating.

The cellular characteristics of vaginal smears reflect structural changes of vaginal epithelium and follow a predictable course during the estrous cycle. The stage of the estrous cycle is determined by recognition of the predominant cell type present in the smear at each daily evaluation. An estrous cycle in rats and mice is typically 4 to 5 days and may be influenced by multiple factors, including light, temperature, humidity, noise, nutrition, and social relationship. In general, cycles are more easily influenced in mice than in rats.

Cycling changes are usually divided into four stages: estrus, metestrus, diestrus, and proestrus (Figure 34.9). Estrus lasts approximately 10 to 15 hours in rats and approximately 21 hours in mice; it is the only stage in the cycle when the female will copulate with the male. As shown in Figure 34.9a, cornified squamous epithelial cells are the dominant cell type in the vaginal smear and are few in number or absent. The squamous epithelial cells vary from flat (occur singly during early estrus) to curled (occur in large sheets in late estrus). Metestrus lasts from 6 to 14 hours in rats and approximately 22 hours in mice. It is characterized by the presence of numerous cornified epithelial cells together with irregularly shaped nucleated epithelial cells. Leukocytes are present in considerable numbers (Figure 34.9b). Diestrus lasts 60 to 70 hours (approximately half of the entire cycle) in rats and 22 to 33 hours in mice. It is dominated by leukocytes, although some nucleated epithelial cells may be present. In late diestrus, the nucleated epithelial cells become more spherical, and occasional cornified cells and erythrocytes are present (Figure 34.9c). Proestrus has a duration of 12 to 18 hours in rats and approximately 21 hours in mice and denotes the beginning of the next cycle. The vaginal smear is dominated by numerous nucleated epithelial cells, usually arranged in grape-like clusters (Figure 34.9d).

The duration of an estrus cycle is calculated for each animal based on the number of days between estrus stages per interval observed; these values are then averaged. When high variability is present in the averages or there are animals with persistent diestrus or estrus (more than six days), the abnormally cycling animals should be excluded from analyses and separately cited. Such findings may reflect environmental or handling problems but may also be associated with effects of the test agent and should be evaluated for dose dependency.

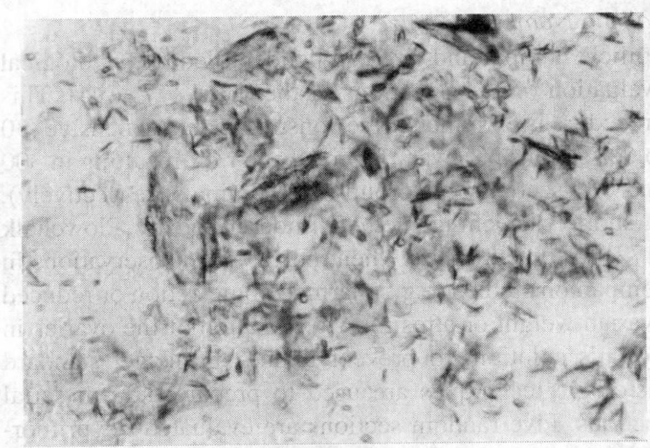

**FIGURE 34.9a** Estrus cycle stage 1: estrus.

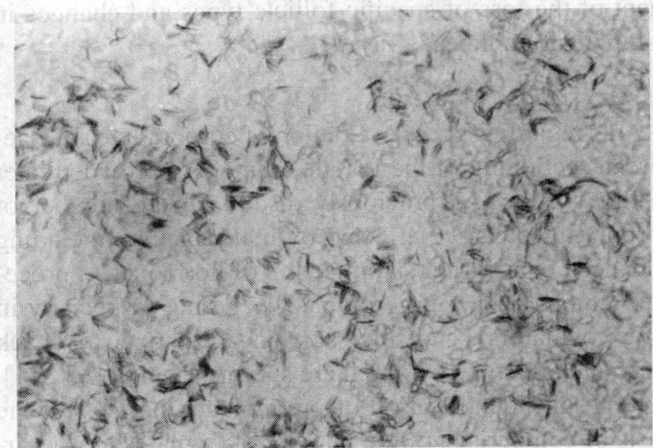

**FIGURE 34.9b** Estrus cycle stage 2: metestrus.

### *Evaluation of Mating Behavior and Fertilization*

Techniques for evaluating mating performance and fertility are used both for impregnating animals for use in developmental toxicity studies and when mating performance is evaluated in a reproductive toxicology evaluation. In the latter case, treatment usually is begun in the female animal at least two weeks before cohabitation and continued until necropsy. The timing of necropsy is dependent on the endpoints to be evaluated, occurring at some interval within the gestation period. If the purpose of the screen is to observe mating behavior and any associated impairment of fertility or implantation, necropsy is usually performed at 13 to 15 days of presumed gestation. For alternative uses, treatment continues to the day before termination, either approximately one day before expected parturition with Caesarean sectioning and evaluation of uterine contents or at completion of a 21- to 28-day lactation period with evaluation of uterine contents and litter sizes and viability.

In general, a mating ratio of one male per female is ideal for use in rodents, because it allows easy identification of the sire and dam, provides a history of reproductive performance, and excludes potential reductions in performance associated with excess use of the same male. The

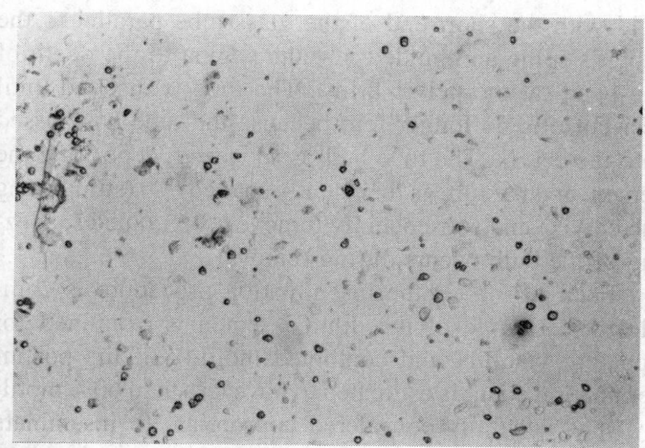

**FIGURE 34.9c** Estrus cycle stage 3: diestrus.

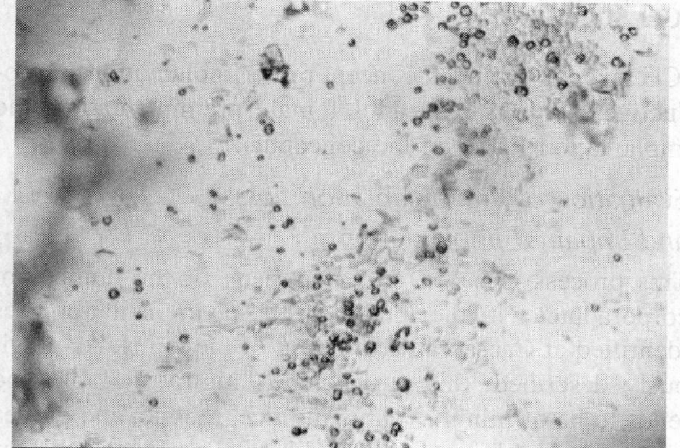

**FIGURE 34.9d** Estrus cycle stage 4: proestrus.

most effective way to produce pregnancies is to pair the animals when the female is in proestrus (usually late in the afternoon, depending on the light cycle used), as it will soon enter estrus and be receptive. However, assuming a large population of animals is available (approximately 125 breeder males and 125 virgin females), sufficient numbers (approximately 100 mated females) can be obtained within 4 to 5 days simply by cohabiting the entire population simultaneously.

Although estrous cycling is not usually evaluated when animals are bred for use in developmental toxicity studies, this parameter is very helpful in identifying potential effects in female animals in reproductive toxicity evaluations, when perturbations in estrous cycling may predict reduced female receptivity and fertility. Male mating performance should be considered in evaluation of any observed effects, as altered estrous cycling during cohabitation and reduced female fertility may reflect male-mediated effects (e.g., naïveté, general health, altered/reduced mating performance, testicular lesions, abnormal sperm).

Rats and mice generally have multiple intromissions in one copulatory interval. If there is an indication that aberrant copulatory behavior may be present, the animals can be observed (videotaped) and graded for number of copulations, number of intromissions, duration of intromission, and expected female receptivity (lordosis). Mating (copulation) is confirmed by the presence of spermatozoa in the contents of the vaginal smear or the observation of a copulatory plug *in situ*; these findings designate day 0 of presumed gestation. It is prudent to note when less than ten sperm are present in a smear, because this information can assist in interpreting mating and fertility data. For mice, the presence of an expelled copulatory plug in the pan is often considered adequate proof of mating, although only approximately 85% of the mice identified in this fashion actually become pregnant.

Female mating performance can be measured by several endpoints. These include identifying the number of days in cohabitation before mating, the number of mated (inseminated) female animals per group, and pregnancy incidences based on the total population per group (% pregnant/number cohabited) and on the inseminated population per group (% pregnant/number inseminated). The number of females assigned to an alternative male and the duration of cohabitation with the second male should always be identified. Care must be taken to ensure that all parameters based on group values are calculated on the basis of observations for the initial cohort pair, and data regarding any subsequent matings should be identified as such.

### *Artificial Insemination Procedures*

Artificial insemination is an alternative method of mating rabbits. It has the advantages of better control over fewer male breeders and reductions in possible cross-infection and technical time requirements. The downside is a slightly reduced pregnancy rate and limited genetic background of the sires, although this can prove a benefit should a malformation be traceable to the sire. The methodology, although not the subject of this chapter, can be easily adapted for use in male reproductive toxicity evaluations. Semen is collected in an artificial vagina that is lined with a condom from which the tip has been cut off. A collection cone is attached to the artificial vagina, which is then filled with warm water (approximately 50°C). The condom is then lubricated with a small amount of petroleum jelly (do not use a condom lubricated with a spermicide). The teaser female is introduced into the male's cage, and the semen is collected when the buck attempts copulation with the doe.

Average spermatozoa concentration during a collection is 1.5 to 3.0 × $10^8$/mL from a mature male breeder rabbit. Sperm concentration is determined by evaluation in a Neubauer RBC counting chamber to ensure appropriate quality. The mucus is removed, and the volume recorded. A specimen sample is drawn to the 0.5 mark in an RBC counting pipette and then diluted with warm saline to the highest mark on the pipette, resulting in a dilution of 1:200. The pipette is shaken for thorough mixing, several drops are discarded from the tip of the pipette, and an appropriate amount of the sample is placed onto the Neubauer counting chamber. The number of sperm and the estimated percentage of motile sperm are then counted in the four corner squares and one of the middle squares of the lined area.

After determining that the sample has live, apparently normal spermatozoa, a count is calculated as follows:

Number of live spermatozoa × 1.0 × $10^7$ × semen volume = spermatozoa count

The standard sperm concentration used for insemination is 6.0 × $10^6$ (0.6 × $10^7$) spermatozoa per 0.25 mL. This concentration is obtained by using the following formula:

Spermatozoa count/2.4 (dilution constant) – mL semen = mL saline to add

The measured volume of saline is then used to dilute the semen sample by adding a portion of the saline to the collection cone. A glass bottle that has been warmed with normal saline is emptied, and the diluted semen in the collection cone is poured into the warmed bottle; this procedure is repeated several times, using the remaining saline from the graduated cylinder.

Before insemination, each doe is given an intravenous injection (ear vein) of 20 U.S.P. units/kg human chorionic gonadotropin (hCG) to induce ovulation (larger dosages tend to result in superovulation and do not increase fertility or litter sizes). hCG is supplied as a powder with 10,000 U.S.P. units per vial. These 10,000 U.S.P. units are diluted with normal saline to a volume of 50 mL (200 U.S.P. units/mL). The same volume of diluted hCG is separated into labeled sterilized vials, one for each day of insemination (usually four or five consecutive days) and maintained refrigerated after preparation. At this concentration, each injection volume is 0.10 mL/kg of body weight (e.g., a rabbit weighing 5.0 kg is administered 0.50 mL for a dosage of 20 U.S.P. units/kg, which is a total dose of 100 U.S.P. units/animal).

Approximately 3 to 6 hours after hCG-induced ovulation of the female rabbits, each is inseminated with approximately 0.25 mL of diluted semen. To do so, the technician gently restrains the doe by holding it between his or her knees, with the rabbit's head downward and facing away from the technician. A glass insemination tube with an inside diameter of 4 mm and length of 7.5 inches is used. The tube is bent at a 45° angle 1.5 inches from one end; a rubber bulb covers the opposite end. The tube is filled to the bend (which is the point at which it will contain 0.25 mL) with the diluted semen (this volume is a close approximation). The tail of the rabbit is pulled toward the technician, exposing the rabbit's vulva, which should be pink and glistening (evidence of receptivity). The glass tube containing 0.25 mL of diluted semen is held with the short end of the glass tube parallel to the rabbit's spine and guided into the vagina to the depth of the bend (at the pelvic brim). The tube is inserted until resistance is no longer felt, because the tube has passed into the cervix. The tube bulb is squeezed, depositing the semen, and the tube is then gently rotated 90° (stimulating the cervix) and subsequently removed (continued squeezing of the bulb occurs during removal).

Each day, after the insemination procedure is completed, a sample of the diluted semen is examined for apparent viability and continued motility. If the postinsemination count is estimated to have less than 60% motility, it is generally considered appropriate to inseminate the does again.

## ICH Stage B

ICH stage B includes conception to implantation (reproductive functions in the adult female, preimplantation, and implantation stages of the conceptus).

### *Evaluation of Preimplantation Loss and Impaired Implantation*

This process is based on comparison of the number of corpora lutea with the number of uterine implantation sites identified at Caesarean sectioning of the dams. As previously described, this parameter is highly variable and tends to be of minimal value in mice, rabbits, and guinea pigs; it is highly associated with fetal survival and hormonal feedback in rats.

## ICH Stages B, C, and D

ICH stage B includes conception to implantation (reproductive functions in the adult female, preimplantation, and implantation stages of the conceptus); ICH stage C, implantation to closure of the hard palate (adult female reproductive functions and development of the embryo through major organ formation); ICH stage D, closure of the hard palate to the end of pregnancy (adult female reproductive function, fetal development, and growth and organ development and growth).

### *Caesarean Sectioning Procedures*

When the purpose of the reproductive or developmental toxicity study is to screen for preimplantation loss and early implantation, the animals are killed near the beginning of the fetal period, as this allows easy identification of viability based on a beating heart. In a full developmental toxicity screen, the animals are killed on the day before or the day parturition is expected, to preclude effects associated with delivery complications (e.g., prolonged gestation and associated pup mortality and maternal cannibalization of stillborn or malformed pups). Use of time-mated animals tends to increase the variability in

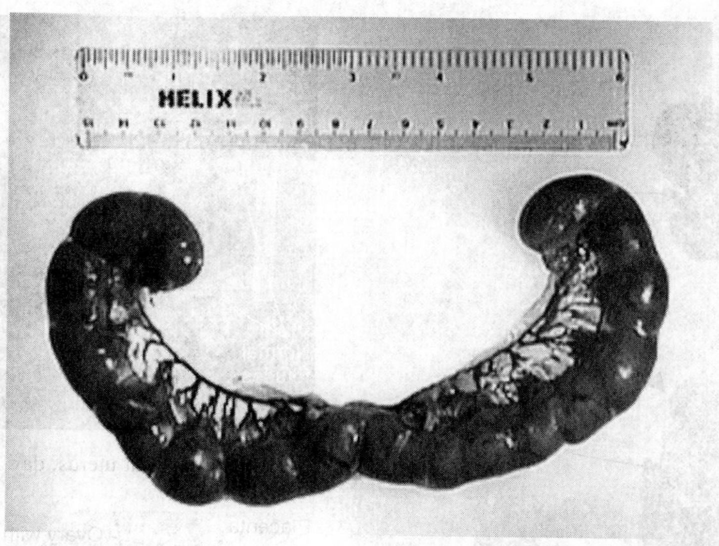


**FIGURE 34.10** Rat, day 20 of gestation uterus.

gestational ages and viability of the conceptuses and should be considered as a potential confounding factor in data evaluation. Regardless of the day of gestation, balanced termination of animals across groups should occur to preclude differences associated with gestational ages. Termination of large populations of animals on the day parturition is expected tends to result in some dams delivering litters. This is particularly common in mice and hamsters and occurs occasionally in rabbits. Such events produce complications in data capture because the animals can be included in some analyses (e.g., pregnancy incidence, implantation numbers) but should be excluded from others (e.g., fetal body weights, fetal ossification site observations). One benefit of termination of the dam on the day parturition is expected is that there is less variability in the degree of skeletal ossification. General practice is to terminate mice, rats, rabbits, and hamsters on days 18, 20 or 21, 29, and 15 of gestation, respectively.

### Gross Necropsy of Maternal Animals

Euthanasia of the dams is easily accomplished by asphyxiation with carbon dioxide gas, and animals are terminated at intervals to ensure that the time between death and necropsy is no more than 20 minutes. After external examination, the animal is placed on its dorsal surface and an incision made through the skin and musculature, extending from the inguinal area to the anterior portion of the neck. The abdominal viscera are exposed by making an incision through the abdominal muscles and peritoneum up to the sternum. The thoracic viscera are exposed by making an incision along one or both sides of the sternum up to the level of the clavicles.

Gross examination of the thoracic, abdominal, and pelvic viscera of the dam and retention of gross lesions in neutral buffered 10% formalin is a standard procedure that often provides evidence of maternal effects. When required, the brain may also be examined. In some cases, target organs, such as the liver, kidneys, and brain, are weighed and retained for further evaluation. It is also common to collect maternal blood, placentas, and embryos or fetuses for use in pharmacokinetics evaluations at Caesarean sectioning. Obviously, data from nonpregnant animals should be excluded in the pharmacokinetics studies relevant to pregnancy.

According to current ICH guidelines, pharmacokinetic studies should be conducted in pregnant animals to determine if pharmacokinetic differences occur in the pregnant state vs. the nonpregnant state and whether the fetus is exposed to appreciable amounts of the test material. These studies are usually conducted twice (first and last dosage) during pregnancy in rats and rabbits and are most frequently performed as separate studies to preclude the occurrence of developmental abnormalities that may be associated with the stress of multiple blood collections or unusually high dosages of test material in the dams assigned to full reproduction or developmental toxicity studies. Although it is occasionally appropriate to perform detailed kinetics studies, this is not usually done. Common practice in the kinetics animals is to identify only pregnancy (presence of live conceptuses) to avoid potential legal or liability actions based on the identification of possible fetal malformations associated with non-drug-related manipulations of the dam.

### Evaluation of Ovaries and Uterus

Evaluation of the uterus and its contents can be performed at any time during the gestation period, depending on the purpose of the evaluation (see Figure 34.10). The uterus with attached ovaries is removed from the abdominal cavity and placed on a dissection blotting

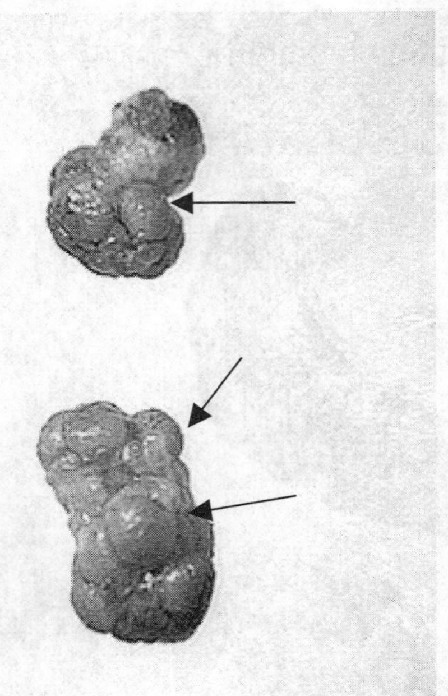

FIGURE 34.11a Rat ovaries, corpora lutea.

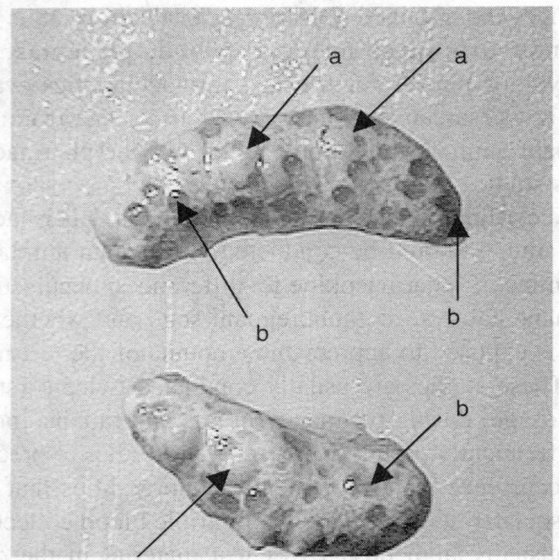


FIGURE 34.11b Rabbit ovaries: (a) corpora lutea, (b) empty follicles.

paper. The number of corpora lutea in each ovary is counted (with ovaries attached to oviduct), and this number is compared with the number of implantation sites, which can be identified through the uterine wall. Figure 34.11 shows rat and rabbit ovaries with corpora lutea. The uterus of female rodents that do not appear pregnant can be examined effectively by either pressing the uterus between glass plates and examining it for the presence of implantation sites or staining it with ammonium sulfide [180].

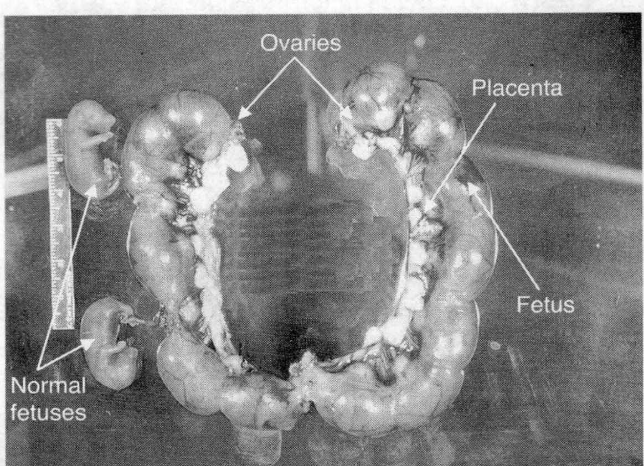


FIGURE 34.12a Rat uterus, day 20 of gestation.

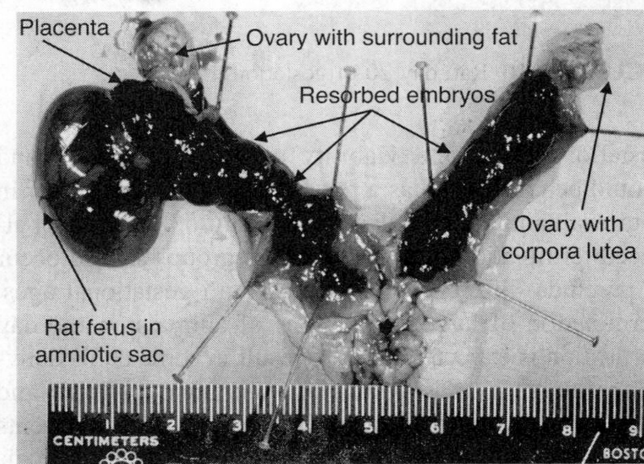


FIGURE 34.12b Rat uterine contents, day 20 of gestation.

*Evaluation of Uterine Weight*

When a uterine weight is desired, the right ovary is removed, and the left ovary remains on the oviduct of the uterine horn (for identification of right and left uterine horns). An alternative method is to remove both ovaries and attach a hemostat (weight tared before weighing) to the right uterine horn for identification. Care is taken throughout the process not to cut the uterus or express any fluid. The same procedure can be used to evaluate nonpregnant uterine weights of rats or mice used in assays for estrogenicity, as described previously.

*Evaluation of Uterine Contents*

After weighing, the uterine horns are opened along the greater curvature, and the number and distribution of implantation sites are recorded (Figure 34.12). Each implantation site is consecutively numbered in both uterine horns, beginning with the ovarian end of the right horn, counting toward the cervix, continuing with the ovarian end of the left uterine horn, and again counting toward the cervix. Each implantation site is described as:

- An early or late resorption—Early or late resorption is defined as a conceptus in which it is not grossly evident that organogenesis has occurred; a late resorption is defined as a fetus (day 16 of developmental age or later in the rat) in which it is grossly evident that organogenesis has occurred.
- A live or dead fetus—A live fetus is defined as one that responds to stimuli; a dead fetus is defined as a term fetus not demonstrating marked to extreme autolysis. Fetuses with marked to extreme autolysis are considered to be late resorptions.
- Viable or nonviable embryos (day 10, 13, or 15 of gestation for rats)—A viable embryo is oval or crescent shaped, pink, firm, and enclosed in an amniotic sac filled with clear fluid. A nonviable embryo is amorphous, small, pale pink to tan or deep red to black, soft, and enclosed in an amniotic sac filled with clear or cloudy fluid. It is not necessary to open the uterine horns to determine viability of day 10 or day 13 rat embryos.

Excess blood and amniotic fluid are removed from the fetus by rinsing or blotting on absorbent paper. Individual fetuses are deposited in multicompartmented boxes, each compartment of which contains a tag that ultimately will be tied to the fetus and will remain with the fetus throughout processing, evaluation, and archiving. The tag cites the study number, dam number, uterine placement of the fetus, and appropriate fixative to be used for further processing, if required. When evaluations are to be made without the investigator's knowledge of the dosage group of the fetus (i.e., blind), the box containing the fetus is given a white label bearing only the study number and animal number and bears a Caesarean sectioning processing tag. Although some investigators believe that blind evaluations enhance the quality of the data, this procedure is not always valid. In many cases, knowledge regarding the specimen evaluated enhances the ability of an experienced investigator to identify alterations associated with treatment. In addition, conducting evaluations under blinded conditions increases the probability of error in data processing and storage of the specimens. It also increases the difficulties of data collection, should an automated data collection system not be available.

### *Gross Examination and Weight of Placenta*

The placenta and attached amniotic sac containing the fetus are removed from the uterine horn. The placenta is separated from the amniotic sac, and the placenta and fetus are placed in the compartmentalized box for the litter. Each placenta is observed for gross changes (e.g., white spots, areas of necrosis, abnormal size). Individual weights are sometimes taken when an estrogenic effect is suspected, although the relatively high variability in placental weights associated with fetal size and blood status at maternal death makes this endpoint relatively variable.

### *Fetal Evaluations*

Experience and expertise are pivotal requirements for accurate evaluations of fetus specimens. Technical training should include evaluation of a positive control group (e.g., aspirin or vitamin A), as well as evaluation of several untreated control group litters (co-evaluation of two or three studies with an experienced investigator), particularly for soft tissue and skeletal evaluations, to ensure comparable levels of expertise when multiple technicians perform these evaluations in the same laboratory. It is also recommended that identified alterations be confirmed by a senior-level investigator to ensure comparable evaluation within a study as well as across studies. To the extent possible, observations made during gross external examination of the fetus should be confirmed in soft tissue and skeletal examinations and the related changes described. Because the degree of ossification increases very rapidly shortly before birth, any errors regarding day 0 of gestation for the dam should result in exclusion of the litter from summarizations of delays in ossification.

### *Gross External Fetal Examination*

Examples of typical externally identifiable abnormalities in rat fetuses are provided in Figure 34.13. All fetuses are examined for gross external alterations, externally sexed, weighed, and tagged. It should be noted that no truly adequate method for fetal euthanasia exists. Because of ethical animal care and use concerns, multiple new methods have been developed (e.g., chilling, carbon dioxide, oral administration of barbiturates). Each has associated problems, the most frequent of which is prolonged time between administration and fetal death, inadvertent production of artifacts (intrathoracic and intraperitoneal injections of barbiturates with associated deformations and occasional internal bleeding or apparent hemorrhage tracts) and potential employee health concerns associated with increased access to barbiturates in the laboratory. The investigator should be aware of the potential for artifact production, as many artifacts have been incorrectly diagnosed as findings by inexperienced technicians. The methods described in the following paragraphs are those generally used in our laboratory, although alternative methods are sometimes used.

Each fetus is euthanized by intraperitoneal injection of pentobarbital or euthanasia solution (preferred method of euthanasia for fetuses assigned to skeletal evaluation). Decapitation may be used and is preferred for fetuses assigned to evaluation using Wilson's free-hand sectioning technique [238] for the head and visceral dissection of the body [202]. Hypothermia may be used alone or may be combined with one of the above methods to facilitate death.

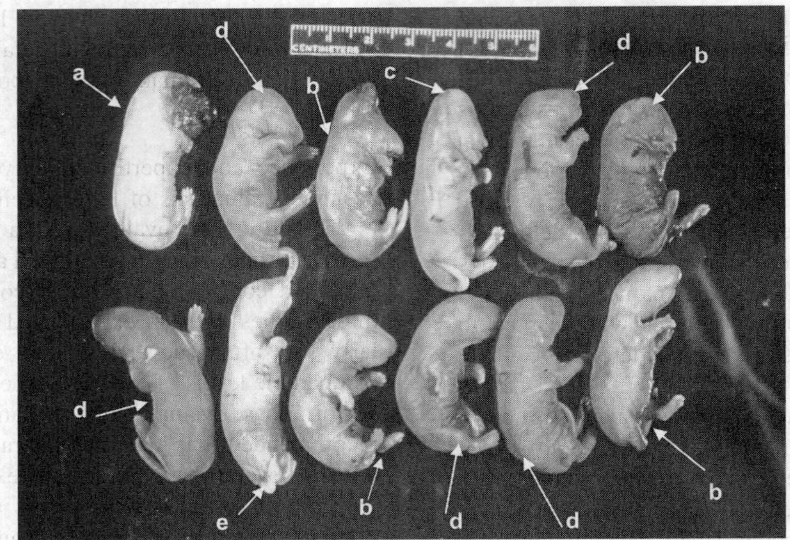


**FIGURE 34.13a** Rat fetuses, day 20 of gestation, example of differential effects on litter: a, dead, macerated; b, malformed; c, dead, compressed; d, live, externally normal, e, dead.

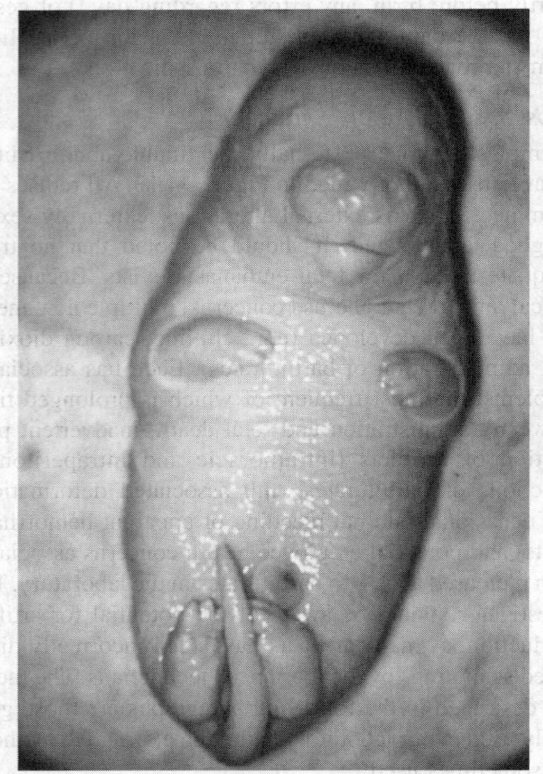

**FIGURE 34.13b** Rat fetus, day 20 of gestation, anasarca (edema of entire body); note apparent clubbing of limbs.

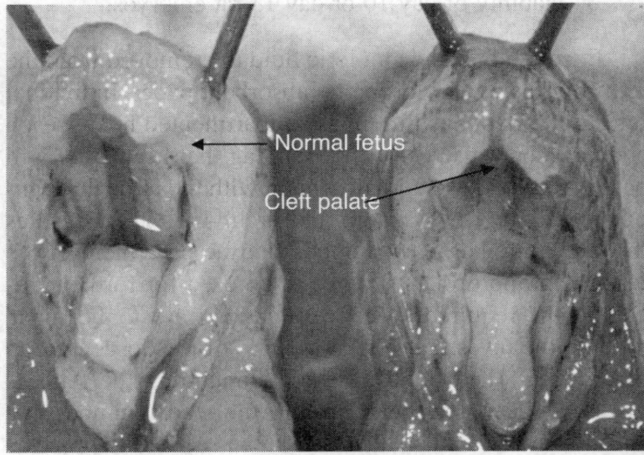


**FIGURE 34.13c** Rat fetuses, day 20 of gestation, cleft palate.

After completion of the gross external fetal examinations, fetuses are placed in an appropriate fixative for subsequent processing, when soft tissue evaluation is to be performed using Wilson's free-hand sectioning technique. Should all fetuses be evaluated for both soft tissue and skeletal examinations, dissection procedures such as those described by Staples [202] and Stuckhardt and Poppe [205] are performed before fixation of tissues and processing for skeletal examination.

The general practice for rabbits and other large mammals with relatively small litter sizes is to evaluate all fetuses for gross external, soft tissue, and skeletal alterations; for small rodents (rats, mice, and hamsters) with relatively large litter sizes, the usual practice is to evaluate half of the fetuses in each litter for soft tissue alterations and the remaining fetuses for skeletal alterations. Although some investigators recommend random assignment of fetuses to either evaluation, this procedure complicates an already technically difficult process. The most common practice is to assign fetuses to skeletal or soft tissue evaluation on an every-other-fetus basis, beginning with the first fetus at the ovarian position of the right

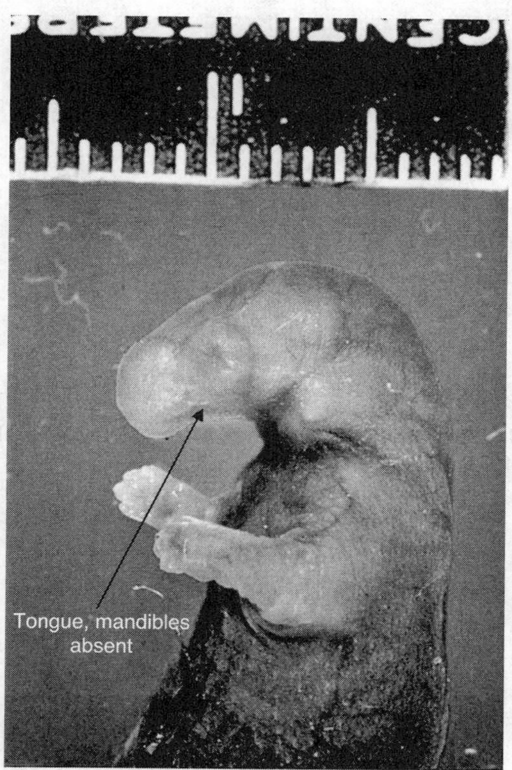


**FIGURE 34.13d** Mouse fetus, day 18 of gestation, agmathia (absence of tongue and mandibles).

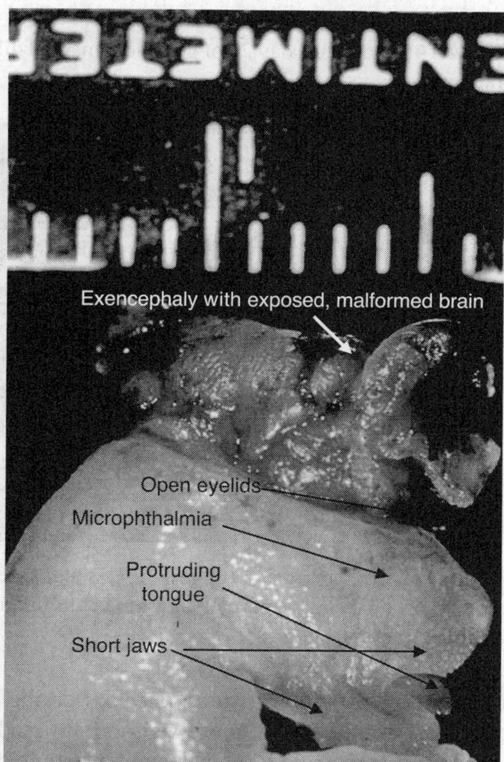


**FIGURE 34.13e** Rat fetus, day 20 of gestation, exencephaly, short jaws, protruding tongue, microphthalmia.

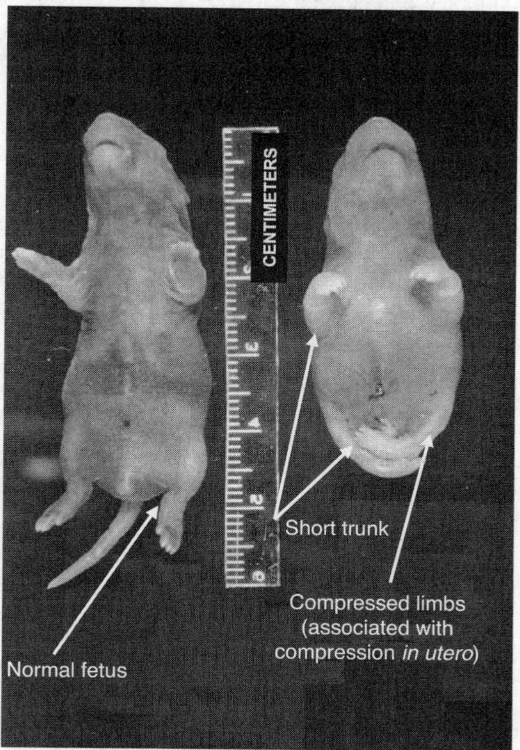


**FIGURE 34.13f** Rat fetuses, day 20 of gestation, normal fetus and fetus with short trunk and compressed limbs.

uterine horn and ending with the last fetus at the vaginal position of the left uterine horn. The first fetus is assigned to skeletal evaluation, although use of the reverse procedure would not affect the overall outcome of the evaluation. Occasionally, it may be desirable to evaluate a fetus with a specific gross external alteration to the alternate fixative to achieve a more meaningful evaluation.

### Soft Tissue Evaluation

All nonrodent fetuses are evaluated for soft tissue alterations by using gross dissection techniques. Soft tissue evaluation in rodents can be performed by using either Wilson's free-hand cross-sectioning [238] or gross dissection [202]. Examples of normal sections and a few abnormalities in rat fetuses are provided as Figure 34.14. Although debate has occurred as to which method is optimal, both are currently considered to be equivalent. The method selected should be based on the laboratory's personnel resources, training, historical experience, and scheduling requirements. Several of the senior and technical personnel in our laboratory had the honor of being trained by Dr. Wilson or Dr. Staples in their respective techniques, and this author is particularly indebted to them both for access to their figures and descriptions of the procedures, some of which are presented in this volume.

As noted previously, approximately half of the rodent fetuses in each litter are generally assigned to soft tissue

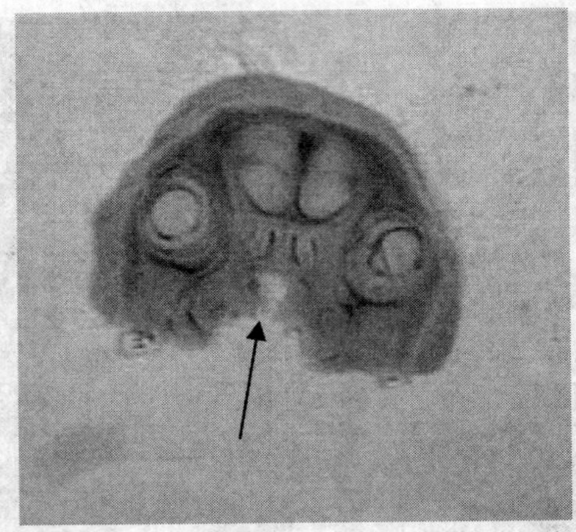

**FIGURE 34.14a** Rat fetus, day 20 of gestation, Bouin's fixation, cleft palate.

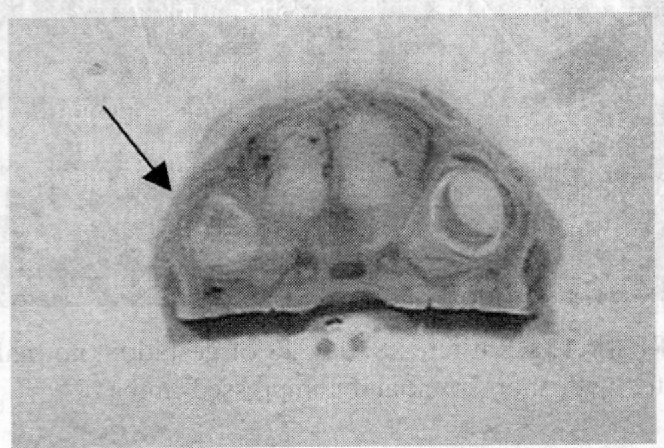

**FIGURE 34.14b** Rat fetus, day 20 of gestation, unilateral microphthalmia.

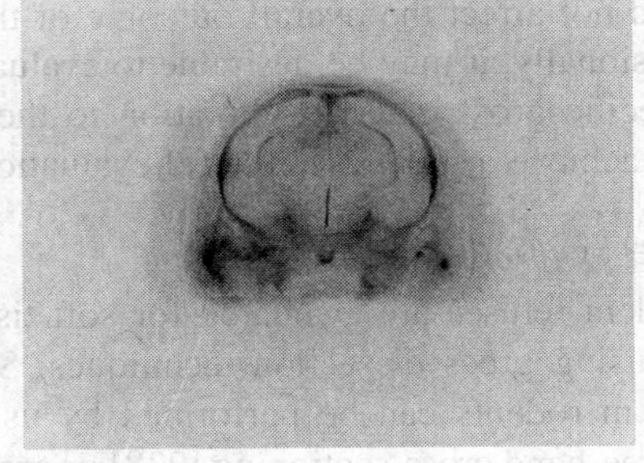

**FIGURE 34.14c** Rat fetus, day 20 of gestation, Bouin's fixation, normal brain.

evaluation after gross external evaluation. Fixation of these specimens in Bouin's solution and subsequent evaluation by Wilson's method [238], rather than dissection of the fetuses directly after Caesarean delivery, allows the use of fewer personnel and reduces the time required to perform all observations associated with Caesarean sectioning. Evaluation of the Bouin's-fixed fetuses can occur at any time after hardening and decalcification of the specimens (a process usually requiring approximately a week). In the absence of high incidences of multiple abnormalities in the fetuses, an experienced individual can generally evaluate all specimens

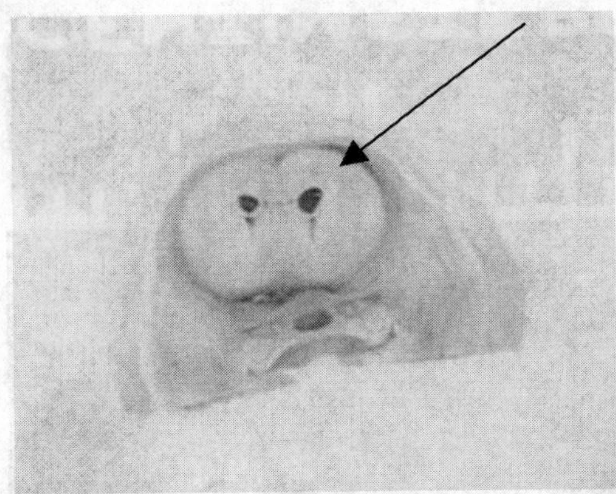

**FIGURE 34.14d** Rat fetus, day 20 of gestation, slight dilation lateral ventricles in brain.

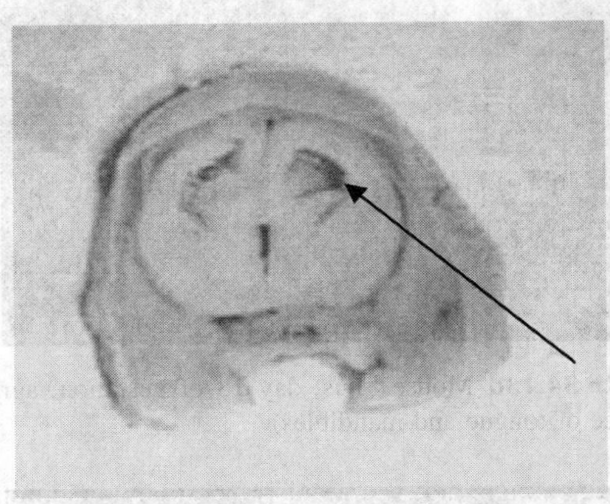

**FIGURE 34.14e** Rat fetus, day 20 of gestation, marked dilation of lateral ventricles in brain.

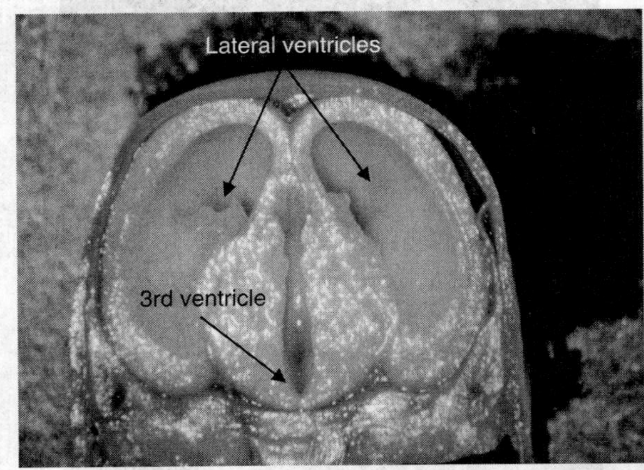


**FIGURE 34.14f** Rat fetus, day 20 of gestation, Bouin's fixation, extreme dilation of lateral ventricles and moderate dilation of third ventricle in brain.

(approximately 1000) from a study over 5 working days. The cross-sections are easily preserved and are ready for processing for histopathological evaluation, if required. The disadvantages of this method are that it tends to require more intensive training and experience than fresh dissection. This

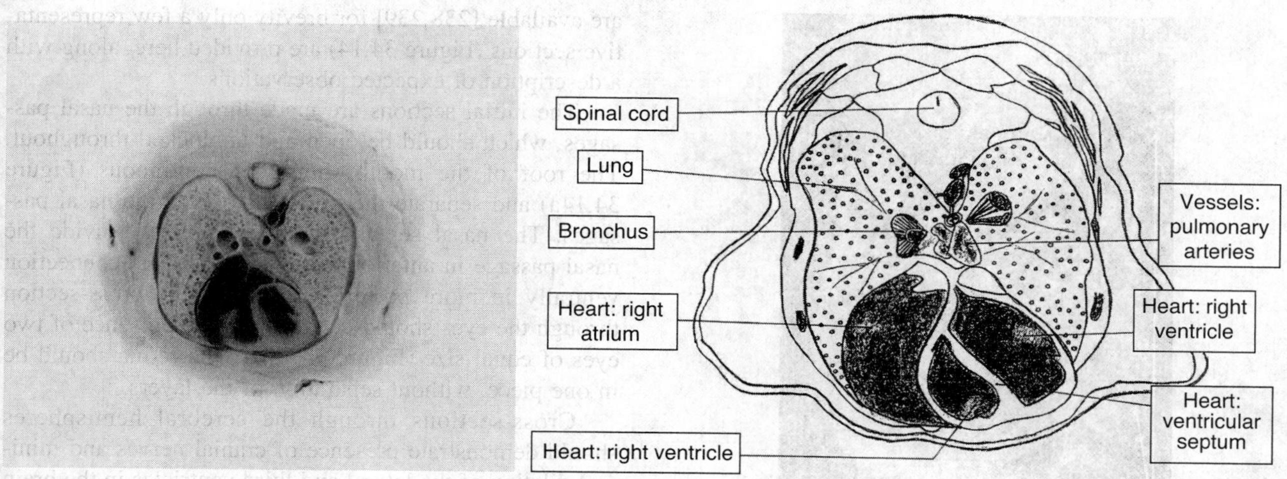


**FIGURE 34.14g** Rat fetus, day 20 of gestation, Bouin's fixation, major organs around the heart.

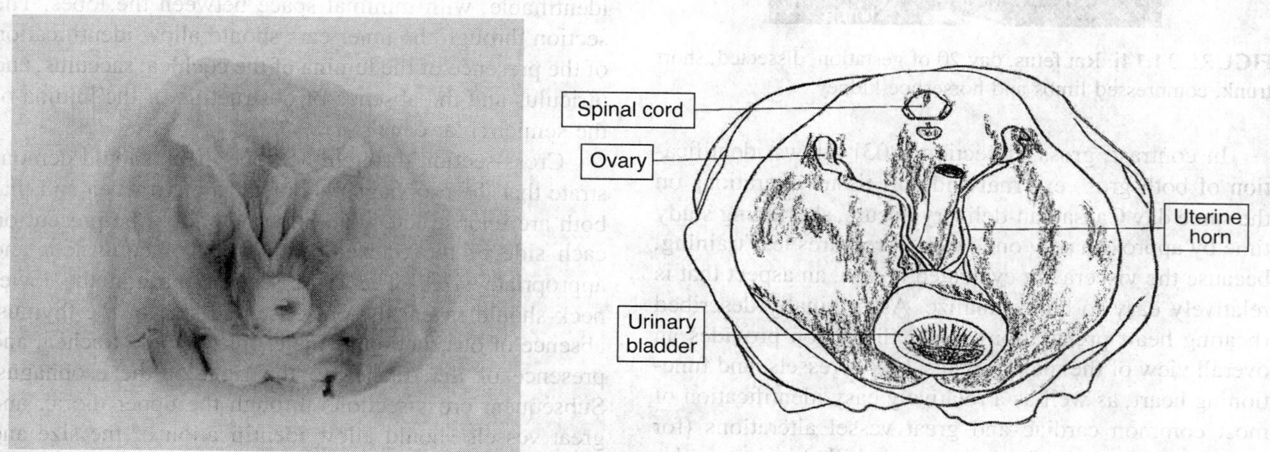


**FIGURE 34.14h** Rat fetus, day 20 of gestation, Bouin's fixation, normal male reproductive organs.

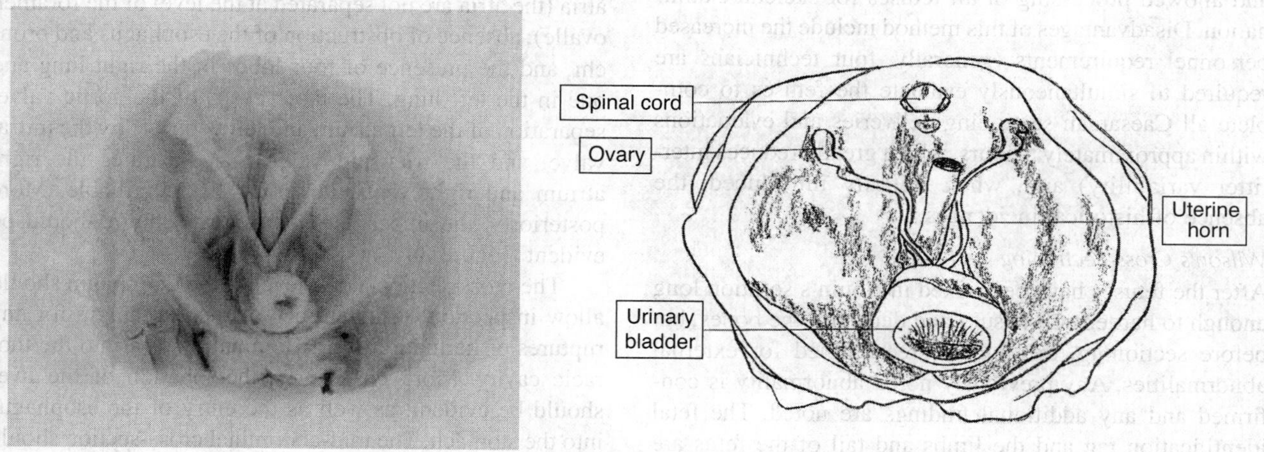


**FIGURE 34.14i** Rat fetus, day 20 of gestation, Bouin's fixation, normal female reproductive organs.

is because the evaluator must have an extensive knowledge of normal appearance of each cross-section and must be aware of potential artifacts associated with fixation (e.g., potential shrinkage, tracks from slicing), biased slicing, and small differences in the thickness of the slices and the tissue present associated with individual fetal size.

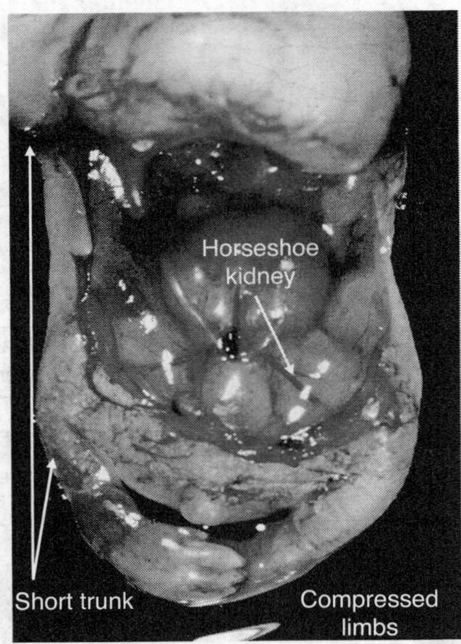


**FIGURE 34.14j** Rat fetus, day 20 of gestation, dissected, short trunk, compressed limbs and horseshoe kidney.

In contrast, gross dissection [203] allows identification of both gross external and soft tissue alterations on the same day Caesarean-delivery occurs, shortening study time by approximately one week. It requires less training, because the viscera are evaluated *in situ*, an aspect that is relatively easy to conceptualize. As originally described (beating heart preparation), gross dissection provides an overall view of the major organs, blood vessels, and functioning heart, as well as a relatively easy identification of most common cardiac and great vessel alterations (for humane reasons, the fetuses are now killed before evaluation). It also eliminates artifacts associated with fixation and allowed processing of all fetuses for skeletal examination. Disadvantages of this method include the increased personnel requirements (generally, four technicians are required to simultaneously evaluate the fetuses to complete all Caesarean sectioning deliveries and evaluations within approximately 5 hours, which greatly reduces interlitter variability) and, when initially introduced, the absence of historical incidences.

*Wilson's Cross-Sectioning Technique*

After the fetuses have been fixed in Bouin's solution long enough to harden soft tissues and decalcify the bones, but before sectioning, the fetus is reexamined for external abnormalities. Any previously noted abnormality is confirmed and any additional findings are noted. The fetal identification tag and the limbs and tail of the fetus are then removed, and the fetus is cross-sectioned. The actual number of cross-sections made is dependent on the size of the fetus. The sections are kept moist with Bouin's solution, and each section is examined with the aid of a stereo microscope at 7 to 40× magnification. Because so many diagrams and photographs of Wilson's procedure are available [238,239] for brevity only a few representative sections (Figure 34.14) are provided here, along with a description of expected observations.

The initial sections are made through the nasal passages, which should be open and unblocked throughout. The roof of the mouth should be continuous (Figure 34.14a) and separate the oral cavity from the nasal passages. The nasal septum should completely divide the nasal passage in anterior sections but lose its connection ventrally in more posterior sections. The cross-section through the eyes should demonstrate the presence of two eyes of equal size (Figure 34.14b). The retina should be in one piece, without separation of the layers.

Cross-sections through the cerebral hemispheres should demonstrate presence of cranial nerves and minimal dilation of the lateral and third ventricles in the brain (Figure 34.14c–f). Both lobes of the pituitary should be identifiable, with minimal space between the lobes. The section through the inner ears should allow identification of the presence of the lumina of the cochlea, sacculus, and utriculus and the absence of obstruction of the lumina of the semicircular canals.

Cross-section through the upper neck should demonstrate that the esophagus is dorsal to the trachea and that both are unobstructed. The thyroid should be present on each side of the trachea, with lobes of equivalent and appropriate sizes. The cross-section through the lower neck should reveal the presence and size of the thymus, absence of obstruction of the esophagus and trachea, and presence of the trachea to the right of the esophagus. Subsequent cross-sections through the upper thorax and great vessels should allow identification of the size and orientation of the great vessels, the size of the atria (they should not be enlarged), separation of the left and right atria (the atria are not separated at the level of the foramen ovalle), absence of obstruction of the esophagus and bronchi, and the presence of four lobes in the right lung and one in the left lung. The three cusps of the aortic valve, separation of the left atrium and left ventricle by the mitral valve, and the tricuspid valve that separates the right atrium and right ventricle should be identifiable. More posteriorly, the intact interventricular septum should be evident (Figure 34.14g).

The cross-section at the level of the diaphragm should allow inspection of the surface of the diaphragm for any ruptures or herniations of abdominal viscera into the thoracic cavity. More posteriorly, the lobation of the liver should be evident, as well as the entry of the esophagus into the stomach. The mid-abdominal cross-section should be evaluated for any unusual patches or raised areas in the mucosa of the stomach. More posteriorly, it should be noted that the right kidney is slightly anterior to the left kidney; the size and shape of the kidneys and each pelvis should be checked for enlargement and papillary development and the spleen evaluated for size, shape, and texture.

If the adrenal glands are not transected, they should be examined separately and sectioned; normal adrenal glands appear as a single solid mass of tissue.

The pelvic area should be examined for the ureters, located in the connective tissue of the lower back, and any deviations in the course or diameter noted. The sex of each fetus should be confirmed (internally (Figure 34.14h,i), and any discrepancy in the externally identified sex noted. Bilateral testes and ovaries should be present, with the testes descended.

*Staples' Dissection Technique*

Unless also assigned to skeletal examination, the head of the rodent fetus is removed and stored in Bouin's solution for subsequent examination after cross-sectioning, as described by Wilson [238] or Barrow and Taylor [7]. The fetus is placed on a board, and the limbs are restrained with elastic bands or dissection pins. Visceral examinations may be made using a binocular microscope or a magnifying lens and light, when necessary (for the larger nonrodent fetuses, a microscope is not generally required). A longitudinal cut is made extending from below the umbilicus through the midline of the trunk. The diaphragm is examined for intactness, after which the cut is continued along one side of the sternum, exposing the thoracic viscera. The trachea is detached from the surrounding tissues, and the supporting ligaments are separated from the viscera, after which the trachea, esophagus, and thymus are examined. The thymus is removed, and the heart and great vessels are examined for shape and position. The main arterial branches above the heart are examined, but subbranches of these major vessels are not explored because of the extensive variability present in normal fetuses (Figure 34.15a). At the anterior portion of the heart, the semilunar values are present at the base of the pulmonary truncus. Also present are the ascending arch of the aorta, the innominate artery, the right carotid artery, the right subclavian artery, the left carotid artery, and the left subclavian artery. The heart is then pulled over to the right, revealing the pulmonary arteries leading to the lungs and the ductus arteriosis.

The heart is cut and examined for internal alterations (Figure 34.15b,c), as follows. The first cut is made anteriorly from the apex of the heart, enters the right ventricle, and exits from the pulmonary truncus, without cutting the dorsal musculature of the heart. This cut reveals the papillary muscles, the tricuspid valve, and the three semilunar valves. In a beating heart preparation, a functional septal defect could be easily detected because blood would spurt through the septum. A second cut is made through the left ventricle, which will cross over the first incision and enter the ascending aorta. This cut will reveal the bicuspid valve, the papillary muscles, and the three semilunar values.

The lungs, liver, stomach, pancreas, spleen, and gallbladder (when appropriate) are examined for color, size, shape, position, and appropriate lobation. The color, size, and position of adrenals, kidneys, ureters, intestines, bladder, and genitalia are then evaluated (the rectum should contain meconium) (Figure 34.14j). Particular attention is given to evaluation of the reproductive organs for gross integrity, size, shape, and position. During this procedure, gender is confirmed for rodent species and identified for rabbits (ferrets can be sexed either as fresh or fixed specimens). The ureters are observed for hydroureter, tortuousness, and obstruction, and it is noted whether urine is present in the urinary bladder. The kidneys are sectioned for a detailed examination of infrastructures (size of each renal pelvis, size and appearance of the renal papilla). The patency of the anus can be checked by use of a hair.

When the head is examined for skeletal alterations, the skull and brain may be sectioned at the level of the frontal–parietal suture and the brain examined *in situ*. Alternatively, the head or brain may be retained in Bouin's solution for later sectioning, as previously described. The eyes are removed and examined for color, size, and shape (nonrodent species).

*Evisceration of Fetuses*

Fetuses are eviscerated to aid in clearing and staining of the skeleton. It is helpful to also remove the skin from nonrodent fetuses. Small scissors are used to make a longitudinal cut in rodent fetuses. This cut extends from below the umbilicus through the midline of the trunk and along one side of the sternum, severing costal cartilage, but not ribs, and avoiding the clavicle. Before further processing, the externally identified fetal sex is confirmed internally (evidence of testicles or uterine horns and ovaries). Similar procedures are then used to evaluate both rodent and nonrodent specimens, although small forceps are used for the rodent fetuses, and blunt forceps may be used for the larger nonrodent fetuses. Evisceration is performed by inserting the forceps into the thoracic cavity and pulling downward on the trachea and esophagus, removing these organs and the lungs and heart (thoracic viscera) as a unit. The diaphragm is gripped and pulled downward, and abdominal viscera (liver, kidneys, intestines) are removed. Any remaining viscera in the abdomen or pelvis are then removed. Throughout this process, care must be taken to prevent damaging the skeleton. For nonrodent fetuses, blunt forceps are used to remove as much of the skin and subcutaneous fat as possible, although skin can remain on the paws, tail, and snout to avoid damaging the underlying bones. Any damage occurring during processing should be noted to prevent the artifact from being potentially incorrectly identified as a skeletal anomaly. After verifying that the fetal identification tag is secure, that the eyes have been removed, and that the head was sectioned, the eviscerated, skinned fetus is returned to the holding tray, containing 95 to 99% isopropyl alcohol or 70 to 95% ethanol.

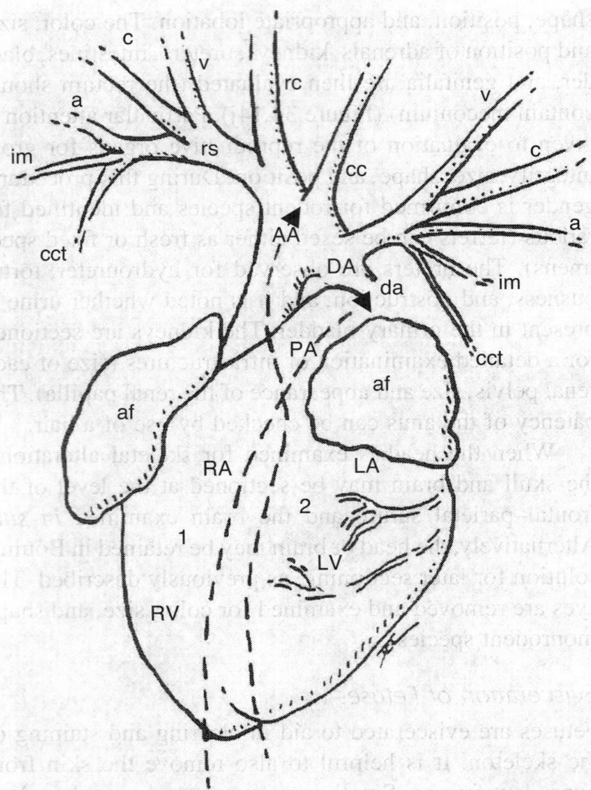


| | | | |
|---|---|---|---|
| AA | ascending aorta | LV | left ventricle |
| af | auricular flap | PA | pulminary artery |
| cc | common carotid artery | RA | right atrium |
| DA | descending aorta | rc | right carotid artery |
| da | ductus arteriousus | rs | right subclavian artery |
| i | innominate artery | RV | right ventricle |
| LA | left atrium | | |

**FIGURE 34.15a** Major arteries above the heart.

*Staining of Fetal Skeletons*

Recently, there has been an emphasis on the use of double-staining to identify changes in cartilage in term fetuses. In practice, this procedure has not been found to greatly assist in evaluation of late fetal development, because most of the observed changes are of a transient nature. Retaining the cartilage by not overprocessing during clearing allows detection of whether apparent absence of an ossification site normally present is a developmental delay or a valid absence of the site. Double-staining may be of use if one is studying development of the skeleton earlier than the day before expected parturition or a specific site (e.g., development of the cervical vertebrae) is being investigated. Although not described here, several references are available for double-staining [92,128,169]. A description of age-related fetal observations in the rat skeleton can be found in Marr et al. [129].

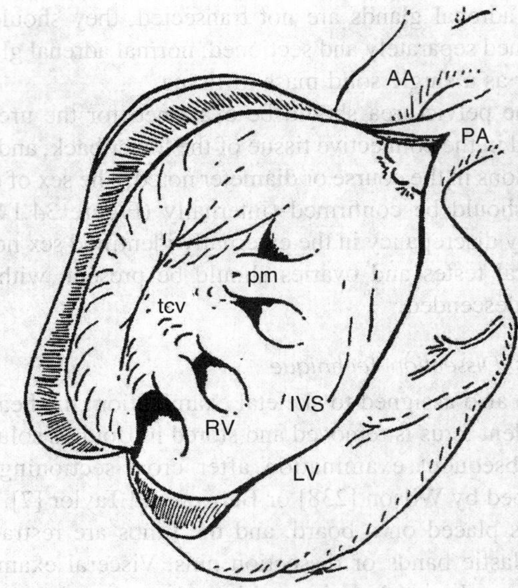


| | |
|---|---|
| AA | ascending aorta |
| IVS | interventricular septum |
| LV | left ventricle |
| PA | pulmonary artery |
| pm | papillary muscle |
| RV | right ventricle |
| tcv | tricuspid valve |

**FIGURE 34.15b** First heart cut.

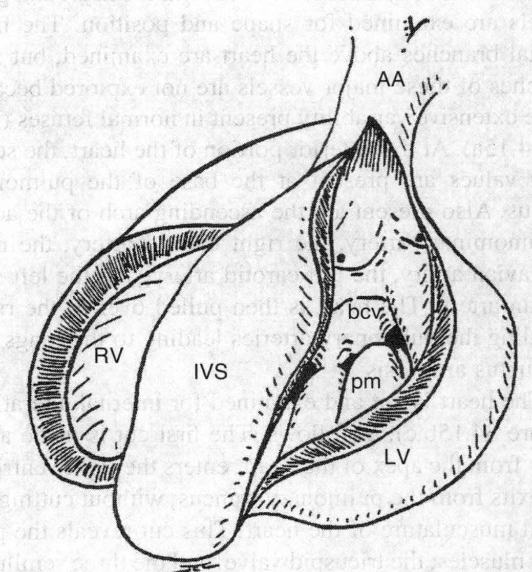


| | |
|---|---|
| AA | ascending aorta |
| bcv | bicuspid valve |
| IVS | interventricular septum |
| LV | left ventricle |
| pm | papillary muscle |
| RV | right ventricle |

**FIGURE 34.15c** Second heart cut.

A 1% KOH solution is recommended for clearing rodent fetuses and a 2% KOH solution for nonrodent fetuses. Although higher concentrations may be used, the speed of maceration is associated with fetal size, temperature, and lighting conditions and agitation, and small specimens can be easily and inadvertently dissolved. For optimal skeletal processing, minor changes should be made in any maceration or staining process, depending on individual specimens. Inspection should be ongoing and the speed of the process modified depending on the observed degree of completion of the maceration and staining of the individual specimen. It goes without saying that it is necessary to have full dissolution of the KOH pellets into a true solution. KOH is generally prepared or obtained in 1-gallon containers, and alizarin red S is added (50.0 mg/1 gallon of 1% KOH solution) to stain the skeletons.

Several concentrations of glycerine are required to clear and preserve the stained skeletal preparations during storage. As supplied by the manufacture, glycerine is 99.5% pure. This glycerine should be diluted to 20, 40, 60, and 80% concentrations so the macerated, stained fetuses can be gradually brought up to the 80% or 99.5% glycerine concentration used for storage. Too rapid an increase of the glycerine concentration will result in compression of the fetal skull. Processing can be performed using compartmentalized polystyrene utility boxes or alternative containers that do not react with the KOH or alizarin. Several commercial units are now available, although they are not recommended for use in staining very small fetuses, which should be closely observed during the processing to prevent inadvertent damage.

The procedure described is a modification of the method of Staples and Schnell [201] in which eviscerated, skinned (if appropriate) fetuses are fixed in plastic bottles filled with 99% isopropanol or 95% ethanol. Fixation of rodent fetuses generally requires at least 7 days, and nonrodent fetuses may require at least 14 days for fixation. After the alcohol fixation period, the fetuses are placed into individual compartmentalized plastic boxes and any remaining alcohol drained away. Each compartment is then filled with 1% (rodent) or 2% (nonrodent) KOH solution, and the fetuses are allowed to macerate for approximately 24 hours. The KOH solution is then drained and replaced with 1% KOH solution containing the alizarin red S stain, and the fetuses are allowed to remain in the stain for approximately 24 hours, after which this solution is drained and replaced with a fresh 1% KOH solution in which the specimens remain for another 24 hours. The KOH solution is again drained, and the rodent fetuses are cleared with progressively higher concentrations of glycerine (20, 40, 60, and 80% for rodents; 20, 40, and 80% for nonrodents). After completion of this process, the fetuses are stored, ultimately for archiving, in plastic jars to which 80% glycerine and a few crystals of thymol, a preservative, have been added.

*Examination of Fetal Skeletons*

We have found that calculation of ossification site averages for each litter is the most appropriate unit of evaluation for the hyoid (body), vertebrae (cervical, thoracic, lumbar, sacral, and caudal), ribs (pairs), sternum (manubrium, xiphoid, and sternal centra between these sites), forepaws (carpals, metacarpals, phalanges), and hindpaws (tarsals, metatarsals, and phalanges); it provides an easy method for identifying delays in ossification. Separate averages for the manubrium, four or five sternebrae, and xiphoid better identify reduced ossification than citing retarded sternal ossification by site, to the exclusion of the interrelationship of the biological pattern of sternal development. Delays in ossification identified by using this method should be correlated with other evidence of retarded ossification that may be observed (e.g., a reduction in the average for the litter average number of ossified sternebrae per fetus should be compared with nonsequential sternal ossification). This method also allows association of increases in rib numbers (supernumerary ribs) with increases in thoracic vertebrae and reductions in lumbar vertebrae, thus eliminating identification of supernumerary thoracic ribs as rudimentary, full, unilateral, or bilateral, which often obscures an increase in supernumerary ribs because the counts are recorded as separate findings rather than an overall response.

As stated previously, fetal processing (evisceration, skinning, clearing, and staining) and examination require considerable manipulation of the fetuses, and any artifacts (e.g., breakage or removal of a bone) should be recorded during the processing procedures, because they can be misidentified as malformations by inexperienced investigators. During many years of consulting, it has been the author's experience that many "major malformations" are often processing artifacts; for example, one agent was incorrectly labeled as a "teratogen" because multiple fetuses were missing portions of the digits. Reexamination of the digits identified that the small fetuses had the ends of the digits removed as the result of being caught in the screening of the cover of the tray used in the staining process and were associated with poor technique. Unfortunately, to convince regulatory reviewers that this reported malformation was an artifact, it was necessary to repeat the study, as well as to provide photographs that clearly demonstrated tissue loss.

Terminology remains a serious problem in the field of developmental toxicology, especially in the area of skeletal assessment. Because definitions differ among laboratories, comparative historical data or a glossary should be provided so consistent terminology is, at the least, used in the same laboratory [32]. It is highly recommended that technical personnel describe the observations using clear statements and that any summarization of associated alterations and classification as malformation or variation, should this archaic practice be used, occur only during higher level evaluations to prevent inappropriate summarization and statistical analyses. For example, the technical observations of fusion of the eighth and ninth right ribs, right arches, and right aspects of the centra in the eighth and ninth thoracic vertebrae, as well as asymmetric and bifid centra in the seventh thoracic vertebra, should be classified at the supervisory and data summarization levels as an interrelated vertebral-rib malformation, not as separate findings. The bifid thoracic vertebral centra in this malformation are part of the overall malformation, not a variation; however, bifid centra in the absence of other vertebral changes would generally be classified as a variation. Our practical definition of "absent" is that neither bone nor cartilage is present in an expected site—for example, when inadvertently removed in processing or associated with inappropriate development (e.g., anencephaly is associated with absence of the bones of the calvaria).

Terms more relevant to human syndromology should be avoided when examining laboratory specimens, as they are often alarming to reviewers inexperienced in animal observations (e.g., frequently physicians) and often inappropriate. Interrelated alterations (external, soft tissue, and skeletal) in the same fetus should be described as one entity; for example, external doming of the skull, dilation of the lateral and third ventricles in the brain, and small holes in the skull bones and enlargement of the fontanelles represent one malformation, not multiple malformations. Unfortunately, current use of computer data entry, proscribed terms, and the inability of naïve investigators to identify related changes in multiple organ systems often result in inappropriate conclusions, statistical findings that are not biologically relevant, and inappropriate assignment of site-specific findings to bimodal categorization such as malformation and variation. Biologically based data interpretation and summarization are highly recommended.

*Procedures*

To perform a skeletal examination, each fetus is removed from the container holding the processed litter, checked for identification, and then placed in a Petri dish and examined using a light source and magnification (5 to 10×). When fetuses are to be evaluated without the investigator's knowledge of the dosage group (i.e., blind), a white label bearing only the study number, maternal animal number, and fixative is affixed to the container holding the fetuses (see also the section on evaluation of uterine contents). Rat, mouse, hamster, and ferret fetuses are examined using a binocular microscope. All nonrodent fetuses (except ferrets) are examined using a magnifying light.

Each skeletal examination proceeds systematically from head to tail (refer to the diagrams and comparable photographs by Ms. Donna W. Lewis of our laboratory provided in Figure 34.16). Representative abnormalities observed in day 20 of gestation rat fetuses are provided as Figure 34.17. Representative abnormalities observed in day 29 of gestation rabbit fetuses are provided as Figure 34.18. The skull (Figure 34.16a–c) is examined for size, shape, and extent of ossification; each skull bone is examined for ossification appropriate to the specimen's gestational age. A listing of the skull bones is provided below:

| **Singular** | **Plural** |
|---|---|
| Nasal | Nasals |
| Frontal | Frontals |
| Parietal | Parietals |
| Interparietal | Interparietals |
| Supraoccipital | Supraoccipitals |
| Exoccipital | Exoccipitals |
| Premaxilla | Premaxillae |
| Maxilla | Maxillae |
| Zygomatic | Zygomatics |
| Squamosal | Squamosals |
| Eye socket | Eye sockets |
| Mandibula (mandible) | Mandibulae (mandibles) |
| Hyoid (ala) | Hyoids (alae) |
| Basisphenoid | Basisphenoids |
| Basioccipital | Basioccipitals |
| Tympanic ring | Tympanic rings |

If the fetus has a domed skull associated with marked or extreme dilation of the lateral ventricles in the brain or if it is small and has retarded ossification, the anterior and posterior fontanelles are often enlarged and the parietals frequently contain holes. In such cases, the affected fontanelle and degree of dilation should be noted (1, slight; 2, moderate; 3, marked; 4, extreme enlargement of the fontanelle). When holes are present (Figure 34.18a), the number, location, and approximate size of each hole should be noted, as should be intra- and interossification sites (Figure 34.18b).

The hyoid bone (Figure 34.16c) is observed to be present (1) or not ossified (0). Rabbit fetuses frequently have angulation of the hyoid alae (Figure 34.18d); the hyoid is not ossified (cartilage is present) in rat fetuses at 20 or 21 days of gestation (Figure 34.16c). Appropriate alignment and closure of the upper and lower jaws (mandibles and maxillae) should be present (teeth should be present in rabbit fetuses). The size and shape of each eye socket should be checked and correlated with reported findings of microphthalmia or apparent anophthalmia.

The vertebral column is next examined (see Figure 34.16d–k); each vertebra consists of three parts (the centrum and two arches) and is evaluated to determine whether areas other than those expected to have minor developmental alterations demonstrate changes. The total number of vertebrae present and the relationship of the subsections to the ribs and pelvis are considered to identify when there is an anterior–posterior shift in the number of thoracic or lumbar vertebrae. Although some laboratories identify only the number of presacral

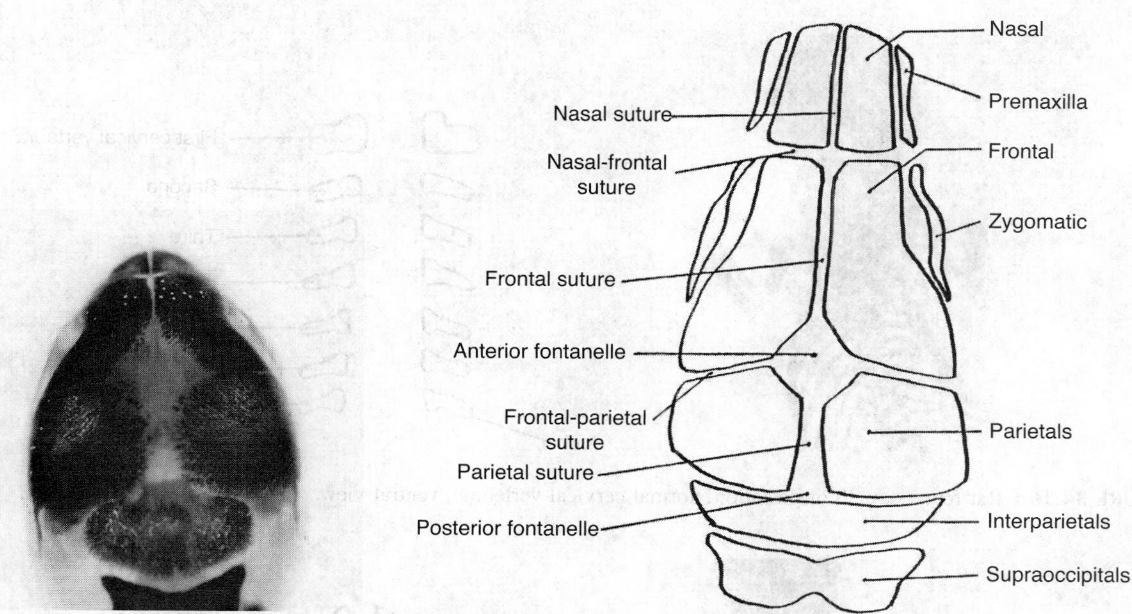


**FIGURE 34.16a** Rat fetus, day 20 of gestation, normal skull, dorsal view.

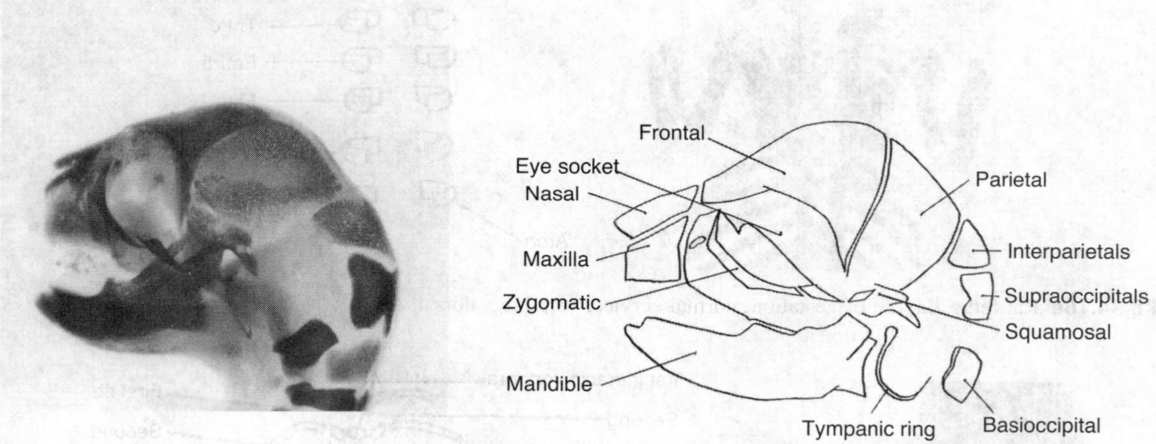


**FIGURE 34.16b** Rat fetus, day 20 of gestation, normal skull, lateral view.

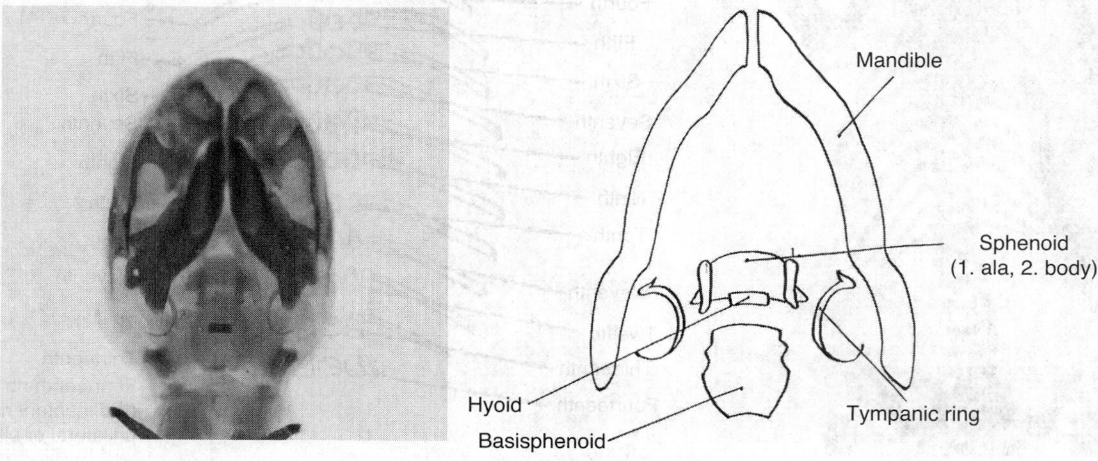


**FIGURE 34.16c** Rat fetus, day 20 of gestation, normal skull, ventral view.

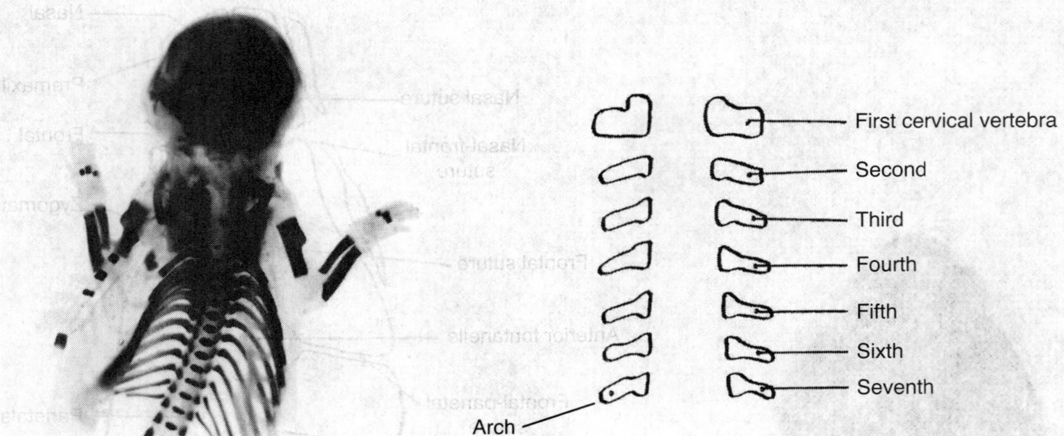


**FIGURE 34.16d** Rat fetus, day 20 of gestation, normal cervical vertebrae, ventral view.

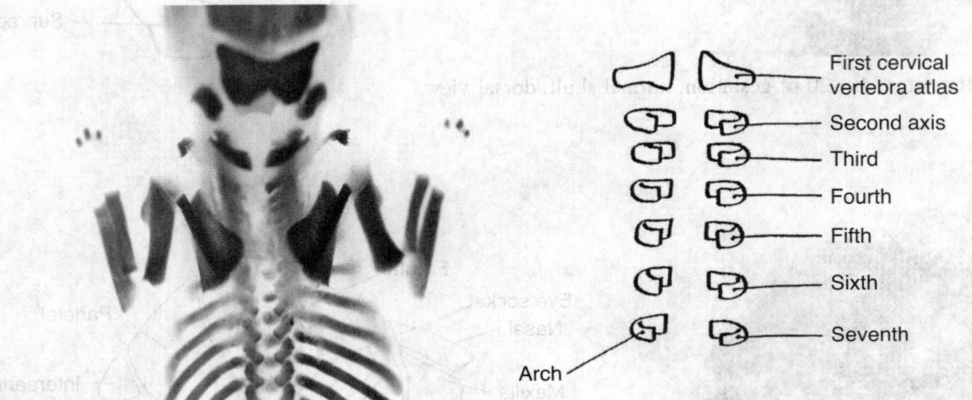


**FIGURE 34.16e** Rat fetus, day 20 of gestation, normal cervical vertebrae, dorsal view.

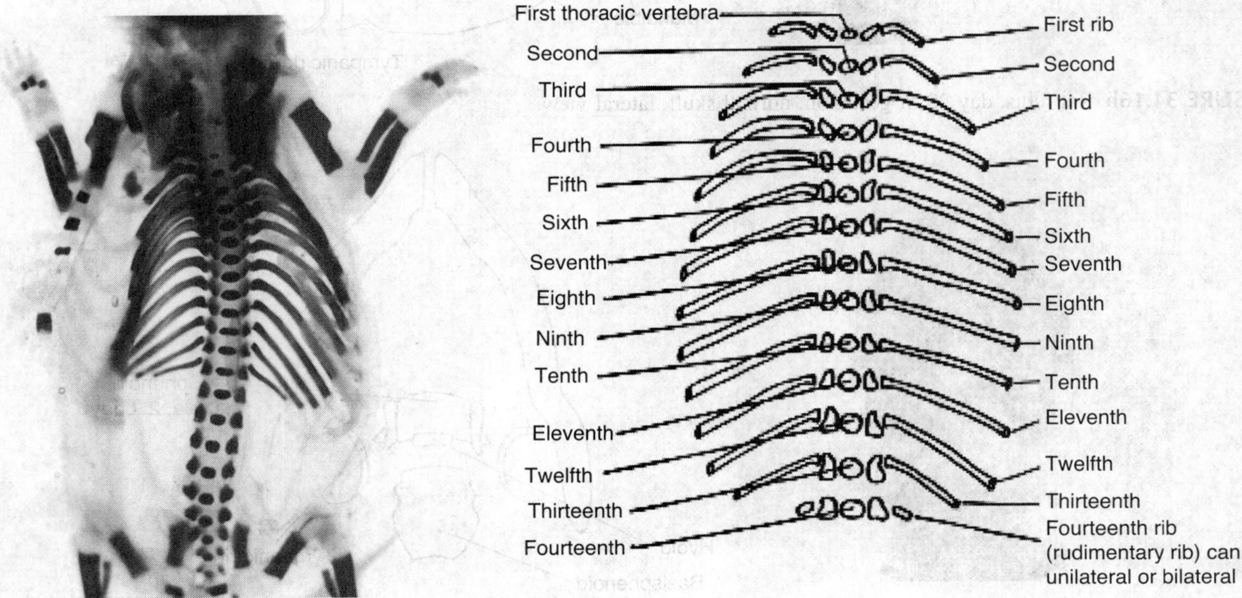


**FIGURE 34.16f** Rat fetus, day 20 of gestation, normal thoracic vertebrae and ribs, ventral view.

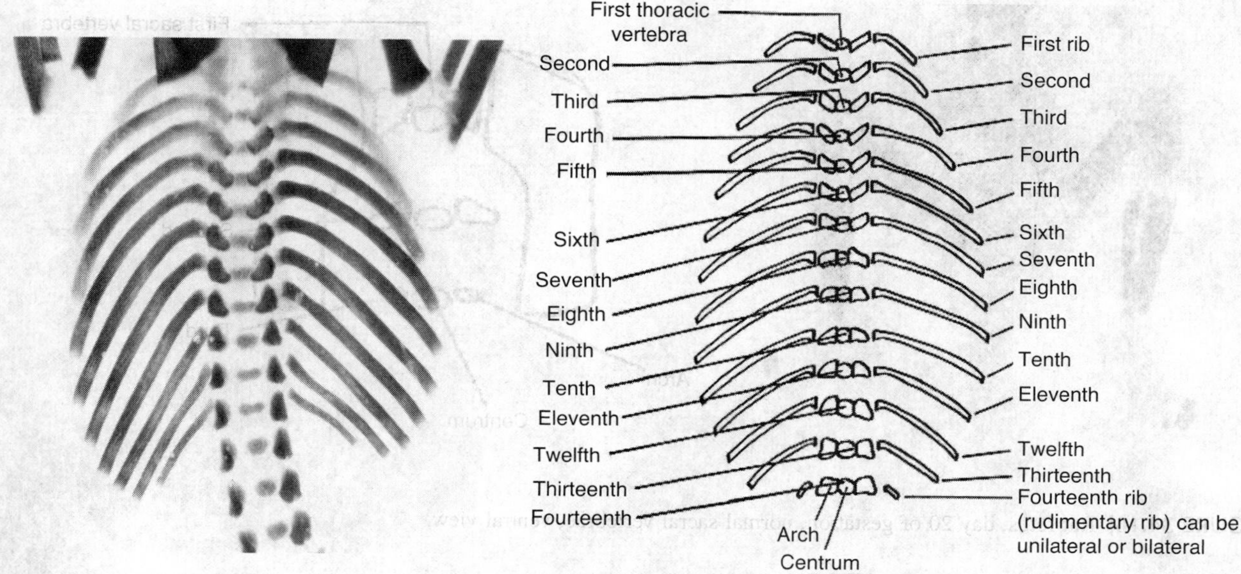


**FIGURE 34.16g** Rat fetus, day 20 of gestation, normal thoracic vertebrae and ribs, dorsal view.

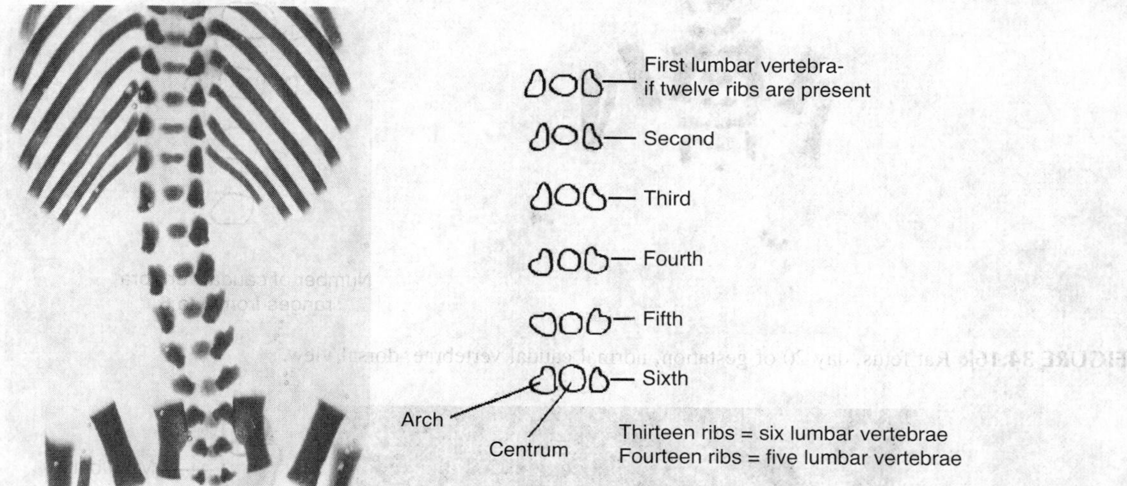


**FIGURE 34.16h** Rat fetus, day 20 of gestation, normal lumbar vertebrae, ventral view.

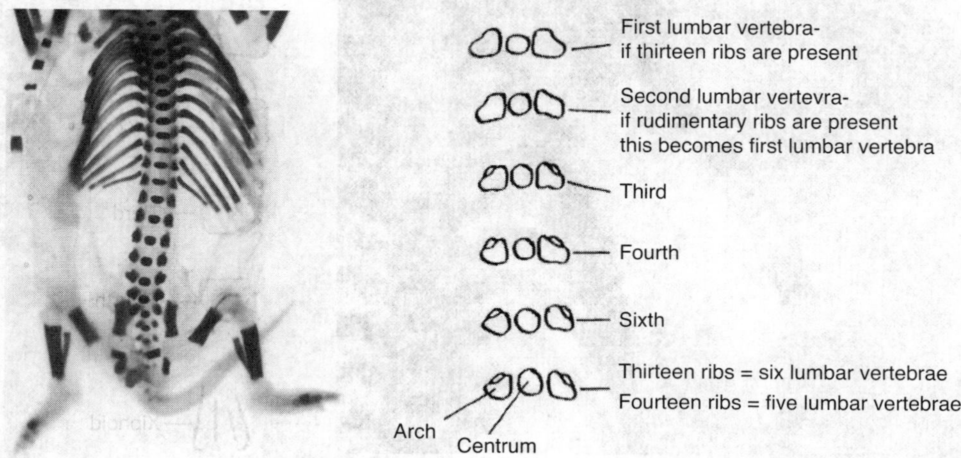


**FIGURE 34.16i** Rat fetus, day 20 of gestation, normal lumbar vertebrae, dorsal view.

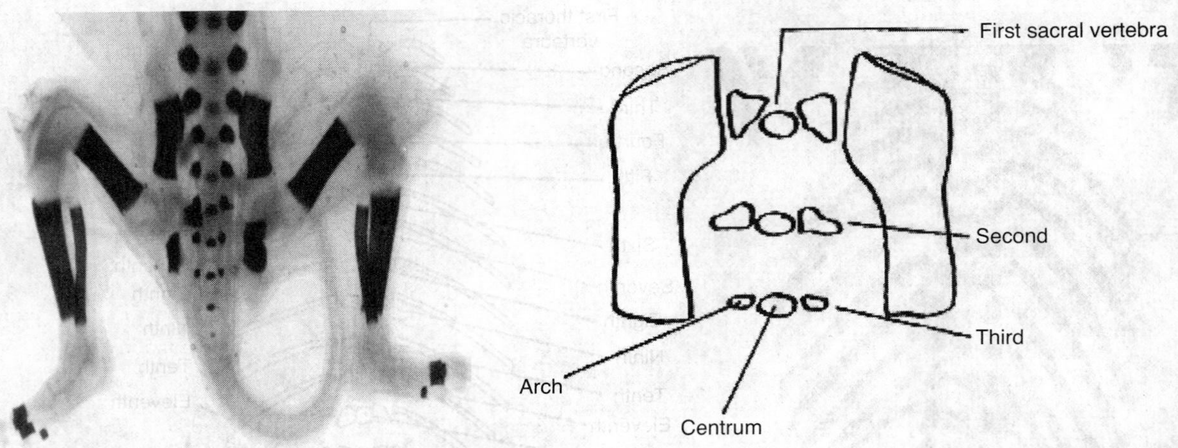


**FIGURE 34.16j** Rat fetus, day 20 of gestation, normal sacral vertebrae, ventral view.

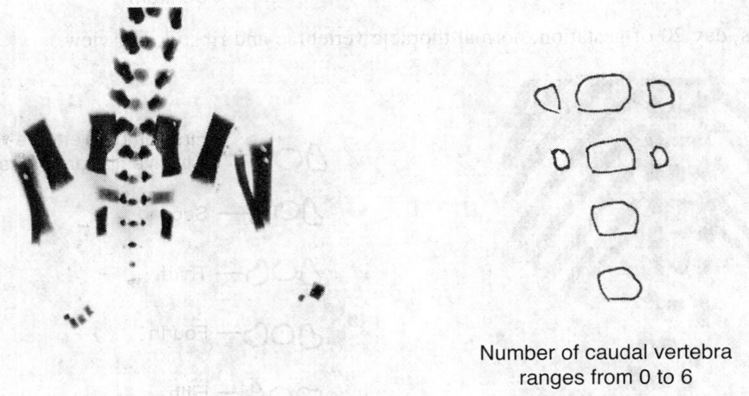

Number of caudal vertebra
ranges from 0 to 6


**FIGURE 34.16k** Rat fetus, day 20 of gestation, normal caudal vertebrae, dorsal view.

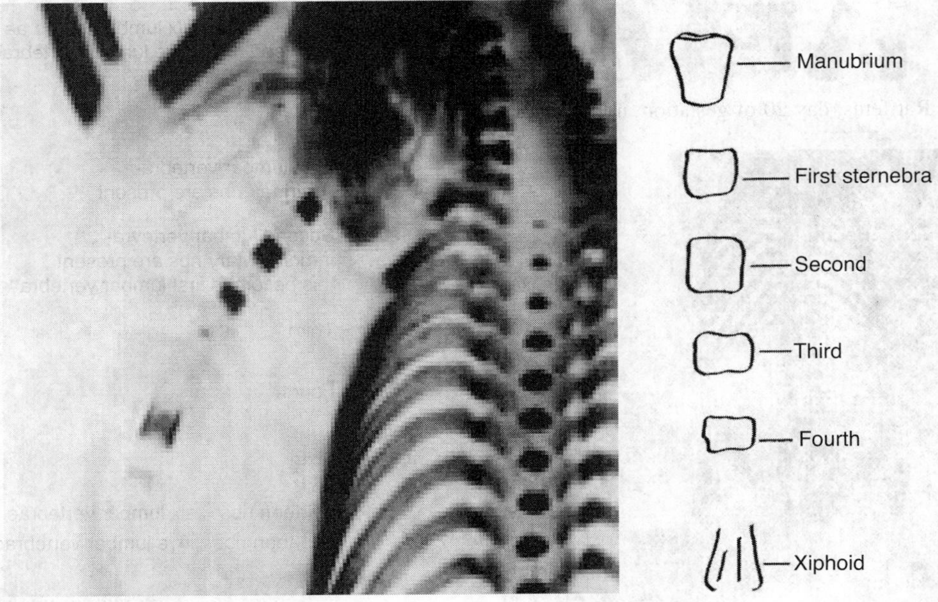


**FIGURE 34.16l** Rat fetus, day 20 of gestation, normal sternebrae, ventral view.

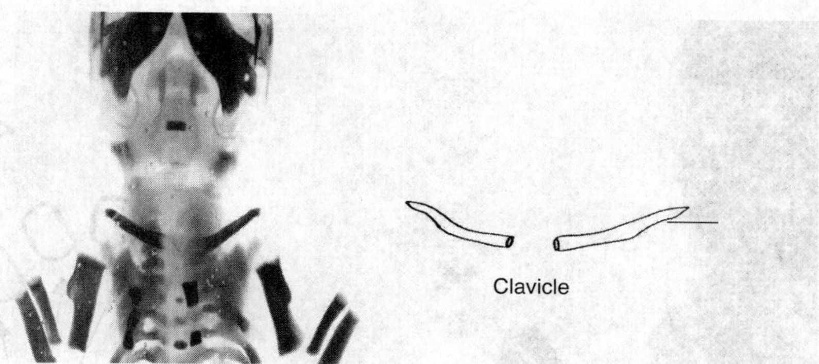


FIGURE 34.16m Rat fetus, day 20 of gestation, normal claviculae, ventral view.

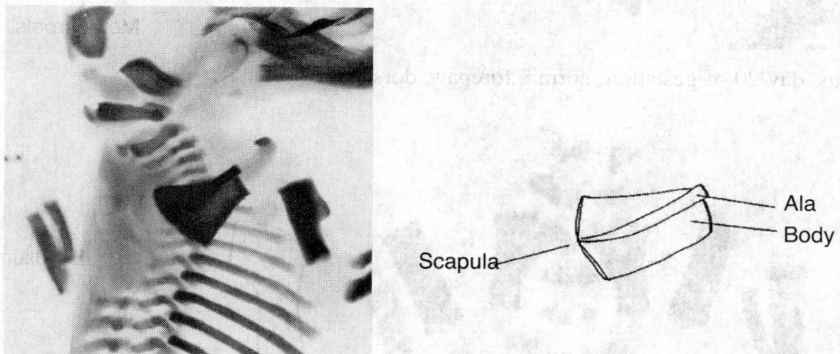


FIGURE 34.16n Rat fetus, day 20 of gestation, normal scapula, dorsal view.

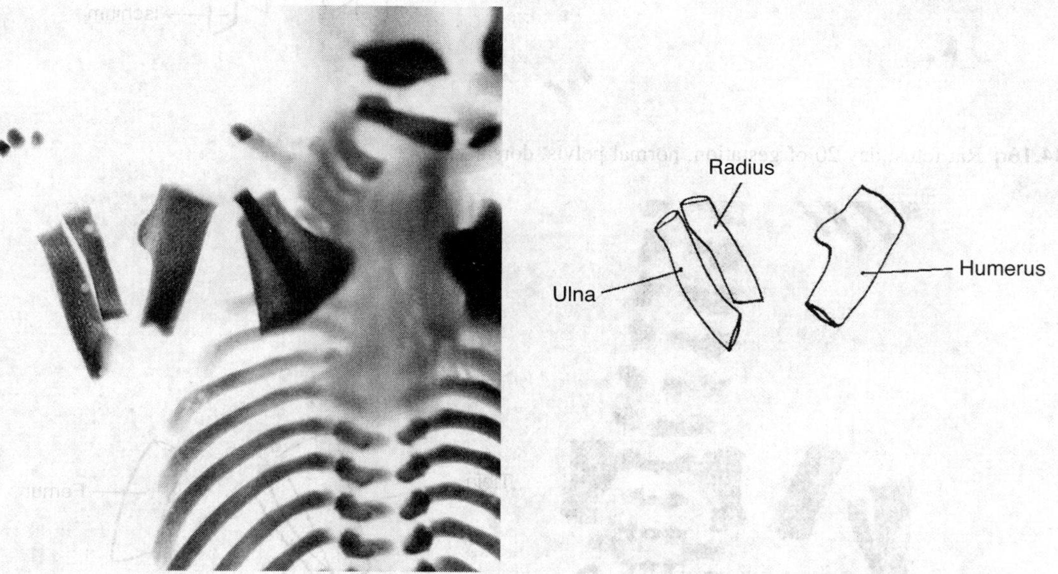


FIGURE 34.16o Rat fetus, day 20 of gestation, forelimb, dorsal view.

vertebrae, we have found that this method does not allow for correlation of increases in supernumerary thoracic ribs and the related increase in thoracic vertebrae and decrease in lumbar vertebrae. Occasionally, a dose-dependent increase in both thoracic and lumbar vertebrae occurs which is usually a more severe expression of an increase in only thoracic ribs. Use of the following working definition for numbers of vertebrae improves comparisons across studies in the same species:

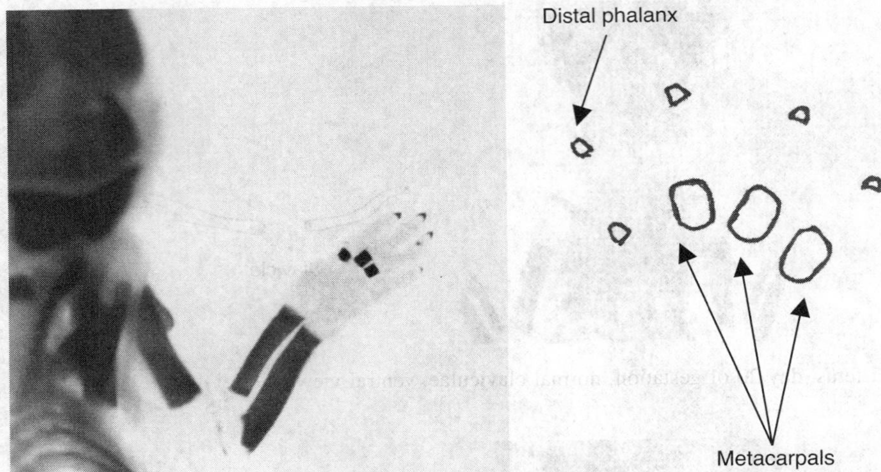


**FIGURE 34.16p** Rat fetus, day 20 of gestation, normal forepaw, dorsal view.

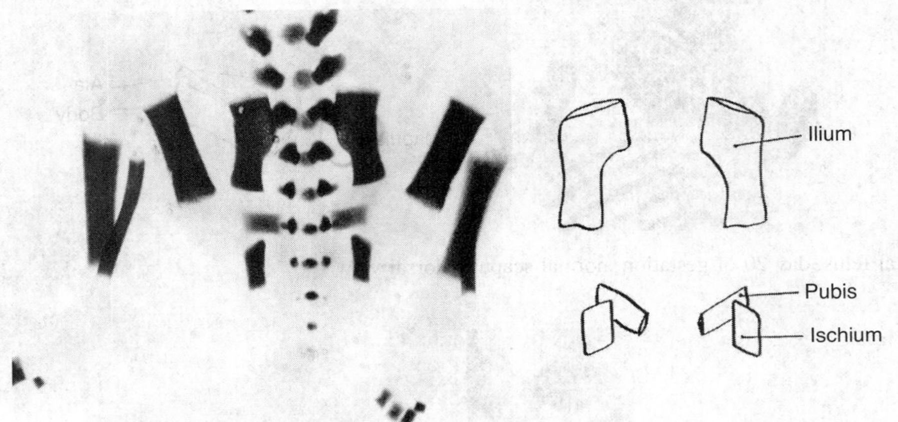


**FIGURE 34.16q** Rat fetus, day 20 of gestation, normal pelvis, dorsal view.

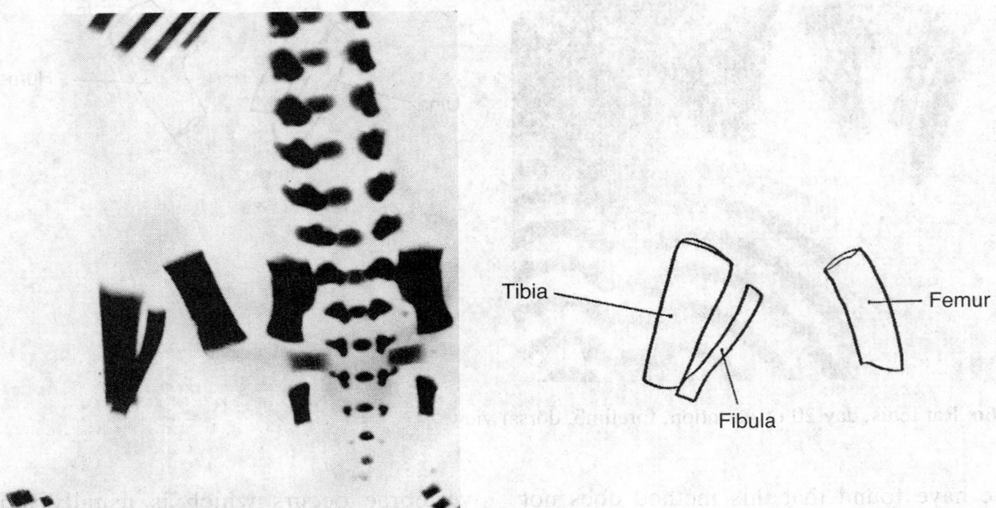


**FIGURE 34.16r** Rat fetus, day 20 of gestation, normal hindlimb, dorsal view.

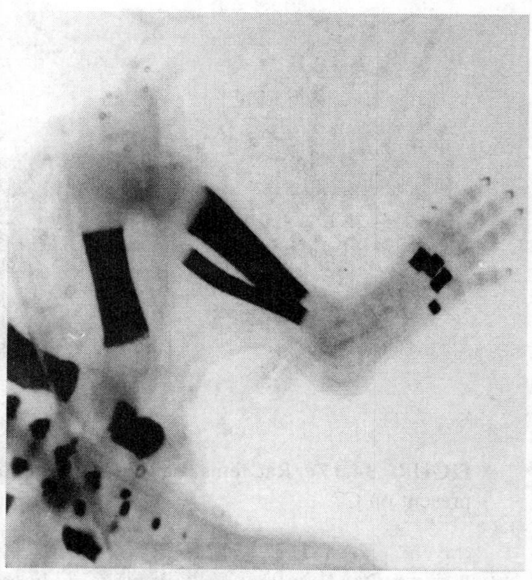

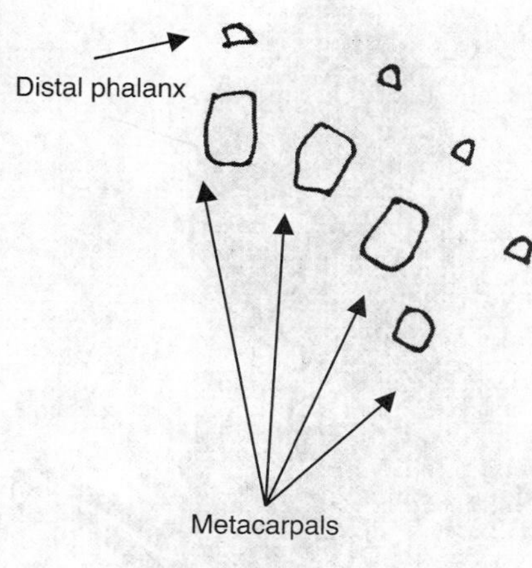


**FIGURE 34.16s** Rat fetus, day 20 of gestation, normal hindpaw, dorsal view.

- *Cervical vertebrae*—7 (C1 to C7) (see Figure 34.16d,e)
- *Thoracic vertebrae*—Rat, 13 or 14 (T1 to T13 or T14); rabbit, 12 or 13 (T1 to T12 or T13) (see Figure 34.16f,g)
- *Lumbar vertebrae*—Rat, 5 or 6 (L1 to L5 or L6); rabbit, 6 or 7 (L1 to L6 or L7) (see Figure 34.16h,i)
- *Sacral vertebrae*—3 (S1 to S3); the vertebrae present between the limits of the ilia are counted (see Figure 34.16j)
- *Caudal vertebrae*—Number varies (CA1 to CAX); any ossification point is counted (see Figure 34.16k)

Based on our working definition, the number of thoracic vertebrae is based on the number of thoracic ribs. The number of thoracic vertebrae and number of thoracic ribs are equal; when supernumerary thoracic ribs are present, the last thoracic ribs may be unilateral or bilateral. As previously noted, this counting procedure simplifies and more clearly identifies the presence of supernumerary thoracic ribs and associated changes. Using this system, the normal sum of the lumbar and thoracic vertebrae is normally 19 (normal rat fetus, 13 thoracic vertebrae + 6 lumbar vertebrae = 19; normal rabbit fetus, 12 thoracic vertebrae + 7 lumbar vertebrae = 19).

Although adult rats and rabbits have four sacral vertebrae, the fetal specimens do not have complete development of vertebral arches or the ilia for use as landmarks. To simplify identification, our working definition for fetal specimens is that sacral vertebrae are those present between the cephalic and caudal limits of the ilia, generally resulting in identification of three ossified sacral vertebrae. It is presumed that one of the vertebrae identified as caudal in the fetal specimen will be identified as sacral in the mature animal. This procedure eliminates the need to identify uneven ossification of the iliac crest and also more clearly identifies when an increase in thoracic–lumbar vertebrae has occurred.

An alternative method used in many laboratories is counting the total number of presacral vertebrae, normally 26 (rat fetus, 7 + 13 + 6 = 26; rabbit fetus, 7 + 12 + 7 = 26), and then identifying the number of fetuses with additional presacral vertebrae. This method is considered inferior because it fails to correlate the relationship of increased thoracic ribs with the number of vertebrae, and frequently, when remarkable increases occur, such as the presence of 28 presacral vertebrae, it incorrectly identifies the number of fetuses with 27 presacral vertebra, because those with 28 are incorrectly excluded from this category in subsequently statistical analyses. This method can serve well when it is biologically based and overall changes are integrated; however, experience has shown that this seldom occurs.

Any alterations in vertebral ossification are identified. This includes a bifid (split) centrum (Figure 34.17d and 34.18i); absent, asymmetric, or small centrum (Figure 34.17e,j); unilateral (left, right, or bilateral) ossification of a centrum or arch (usually identified as a hemivertebrae with associated changes in the vertebrae and ribs) (Figure 34.17f and Figure 34.18f,g); and small arches, open arches, and fused centra or arches of vertebrae (Figure 34.17e,j and Figure 34.18g).

The ribs are examined and counted after assessment of the vertebral column. Ribs on a cervical vertebra are difficult to see and are often quite small (Figure 34.17c). Cervical ribs are most frequently found on the seventh cervical

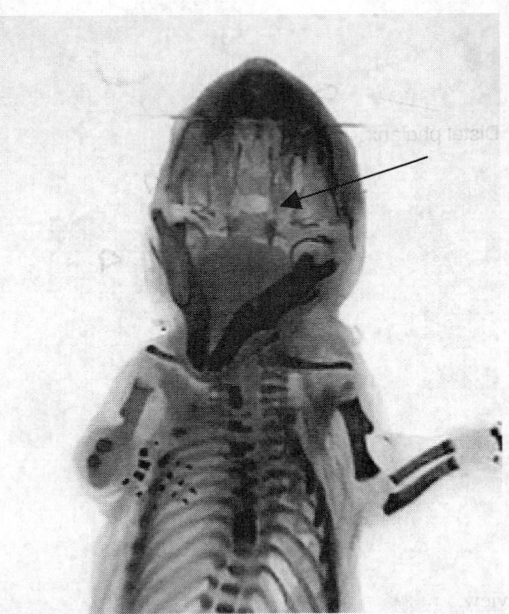

**FIGURE 34.17a** Rat fetus, day 20 of gestation, incomplete ossification of palantine shelves.

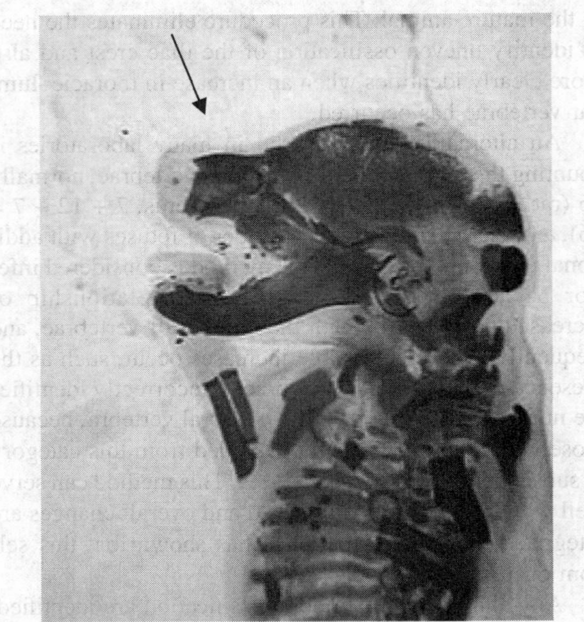

**FIGURE 34.17b** Rat fetus, day 20 of gestation, short nasal bones.

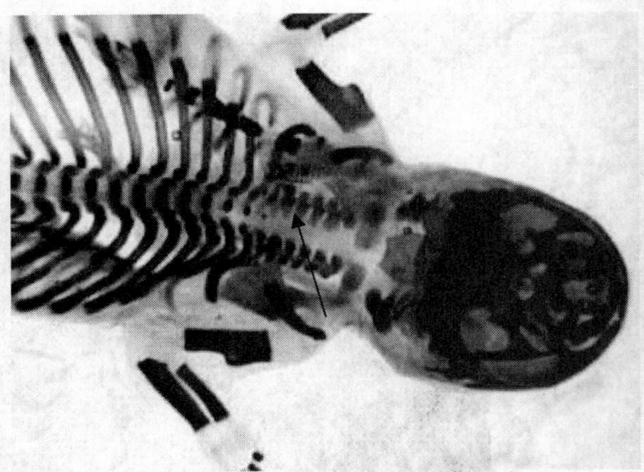

**FIGURE 34.17c** Rat fetus, day 20 of gestation, cervical ribs present on C7.

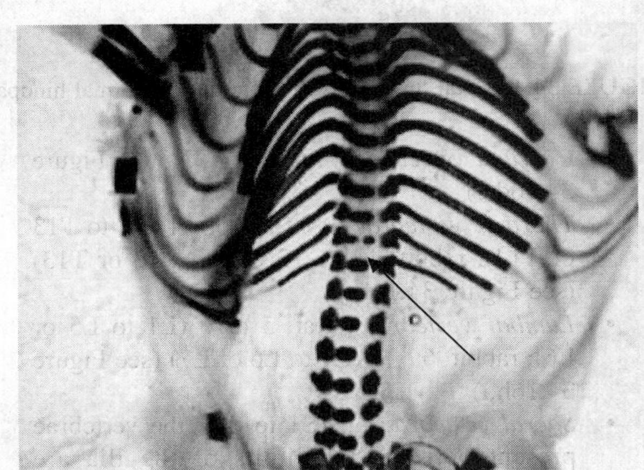

**FIGURE 34.17d** Rat fetus, day 20 of gestation, bifid centrum, T12.

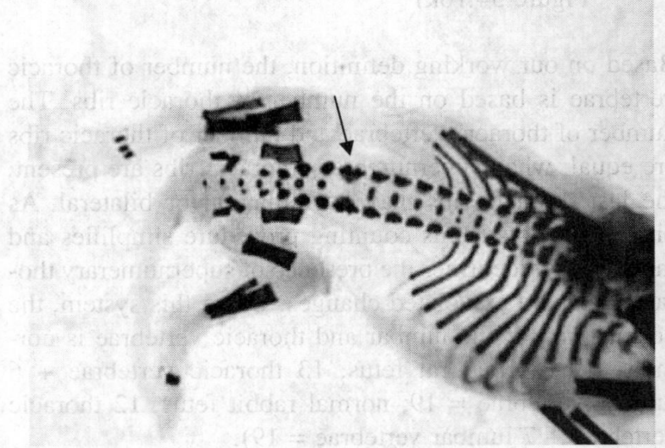

**FIGURE 34.17e** Rat fetus, day 20 of gestation, day 20 of gestation, fused arches and absent centra of L4 and L5.

vertebra (C7) and may be unilateral or bilateral. Although tabulated as ribs in summary presentations, they are excluded from thoracic rib counts and averages. The total number of thoracic ribs (regardless of size) on each side is identified, with the highest number of thoracic ribs on either side defining the number of thoracic vertebrae. For example, 13 ribs bilaterally (13 rib pairs) = 13 thoracic vertebrae; 14 ribs, bilaterally (14 rib pairs) = 14 thoracic vertebrae; 13 ribs, right, 14 ribs, left (13.5 rib pairs) = 14 thoracic vertebrae. Any alterations in rib ossification are noted, such as a thickened area of ossification (Figure 34.18e); waviness (Figure 34.17i); and splitting (branching), fusion, or misalignment (Figure 34.17k,m, and Figure 34.18h). To the

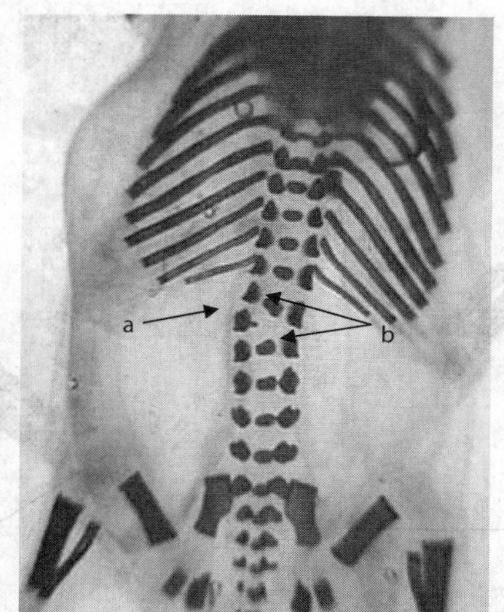


**FIGURE 34.17f** Rat fetus, day 20 of gestation, interrelated malformations of lumbar vertebrae: L2 present as a left hemivertebra with asymmetry of centra of L1 and L3.

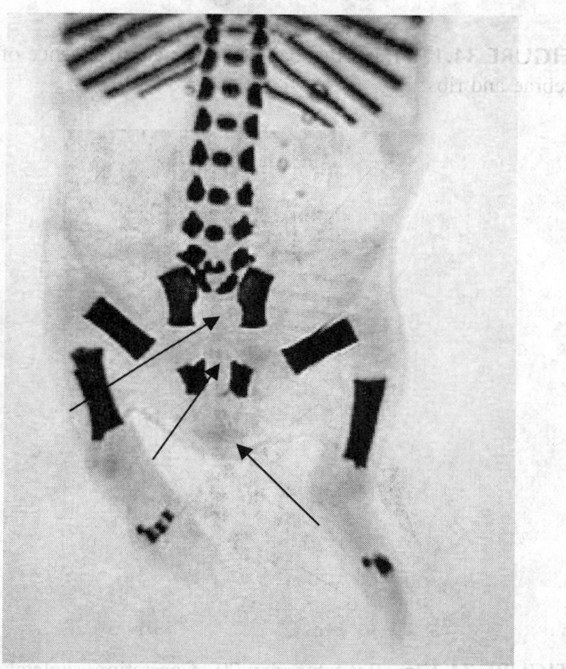

**FIGURE 34.17g** Rat fetus, day 20 of gestation, skeletal malformations associated with absence of tail (absence of lumbar, sacral and caudal vertebrae).

extent possible, interrelated vertebral-rib malformations are identified and this malformation counted as one finding (Figure 34.18f,g). Abnormal rib numbers (unilaterally increased or decreased) associated with increased or absent ribs and the presence of a hemivertebra are excluded from the litter average. Altered ribs are noted by number (T1 to T13), affected area of ossification (proximal, beside the vertebral column; medial, middle of the ossified area; distal,

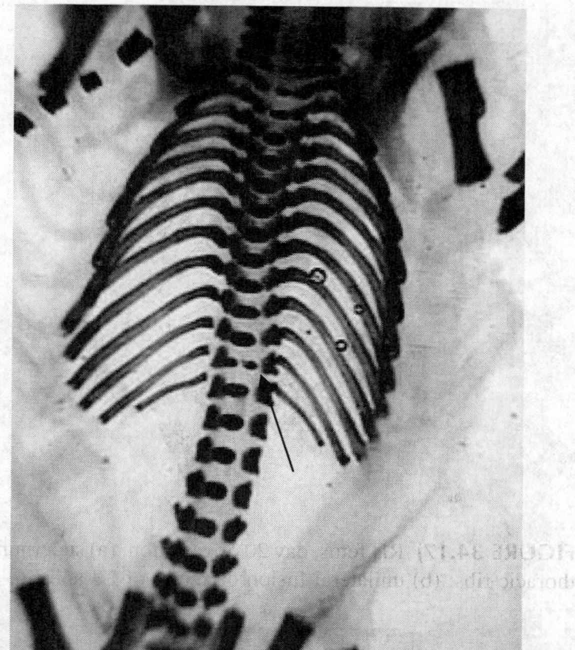

**FIGURE 34.17h** Rat fetus, day 20 of gestation, bifid vertebral centrum.

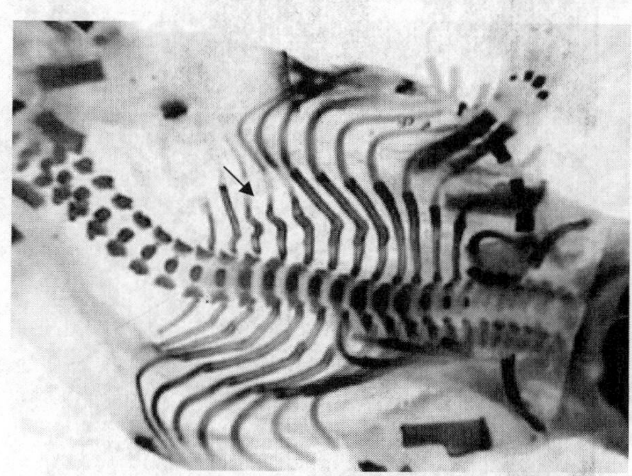

**FIGURE 34.17i** Rat fetus, day 20 of gestation, wavy, hypoplastic thoracic ribs.

near costochondral junction), and whether the alteration is unilateral (left or right) or bilateral.

Evaluation of the sternum then occurs (Figure 34.16l,n and Figure 34.18m–o). The adult sternum usually consists of six or seven ossification sites (manubrium, four or five sternal centers and xiphoid). For the fetal examination, the manubrium and xiphoid are counted (0, no ossification is present; 1, any degree of ossification) followed by the intermediate sternal ossification sites, which may have degrees of delayed ossification (incompletely ossified or not ossified). For these sites, incompletely ossified centra are included in the count and unossified sites are excluded. Any alterations in sternal ossification are noted; for example,

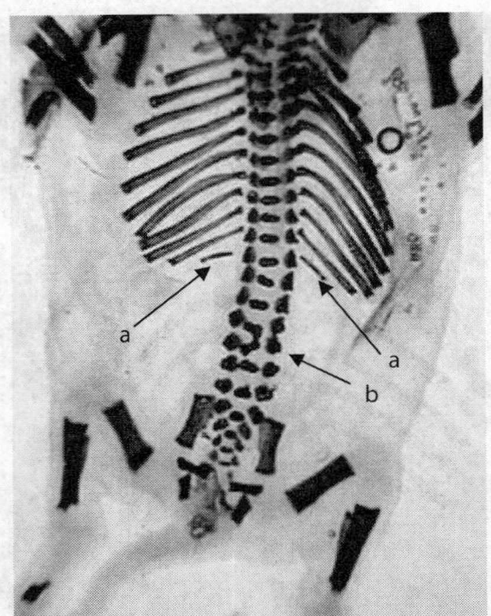


**FIGURE 34.17j** Rat fetus, day 20 of gestation: (a) supernumary thoracic ribs; (b) unilateral fusion of centra of L3 and L4.

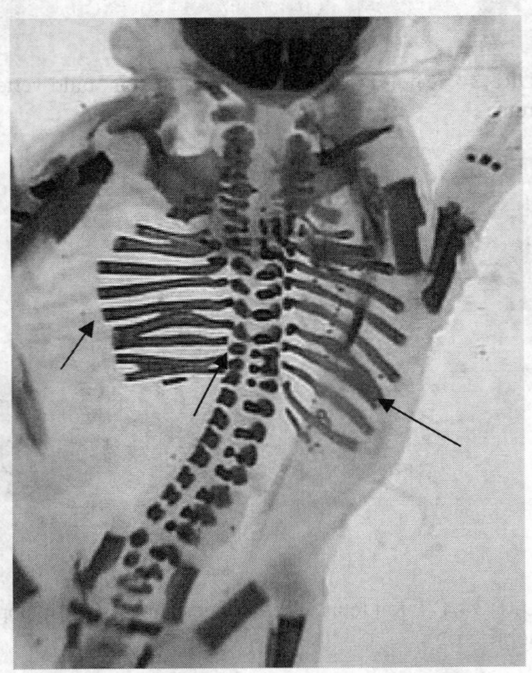

**FIGURE 34.17k** Rat fetus, day 20 of gestation, multiple interrelated vertebral/rib fusions.

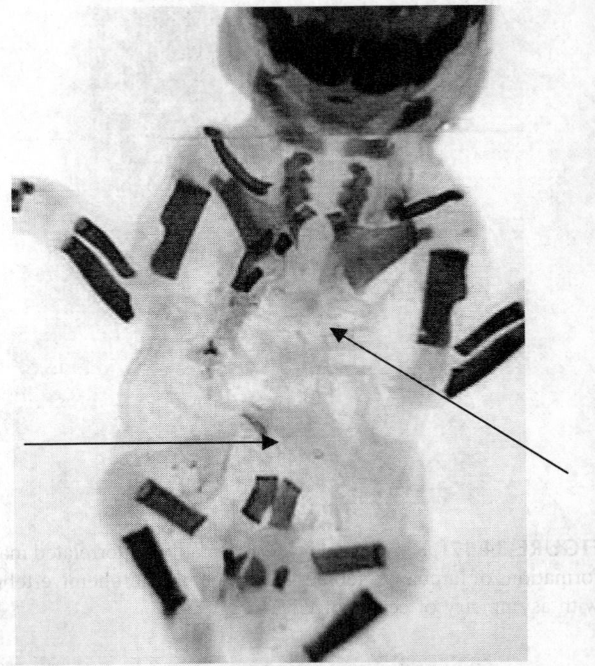

**FIGURE 34.17l** Rat fetus, day 20 of gestation, absence of vertebrae and ribs.

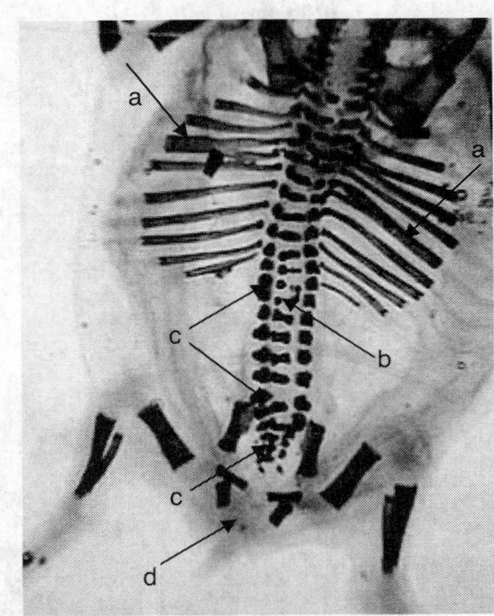


**FIGURE 34.17m** Rat fetus, day 20 of gestation, skeletal alterations associated with short trunk and short tail: (a) fused ribs; (b) bifid unilateral/incomplete ossification of vertebral centra; (c) fused vertebral arches and centra; (d) absent caudal vertebrae.

sternal ossification between the manubrium and manubrium is nonsequential, and centra 2 and 3 normally ossify before center 1. Thus, delayed ossification (i.e., centrum 1 unossified and centra 2, 3, and 4 ossified) should be identified and incompletely ossified and not ossified sites noted individually. Incomplete ossification of the last (third, fourth, or fifth) centrum between the manubrium and xiphoid is not recorded, because it is considered normal and identified as a delay based on the number of ossified sites present (i.e., two, three, or four).

The pectoral girdle consists of two claviclae and two scapulae (Figure 34.16m,n). Any alterations in ossification, such as an irregular shape, waviness, bending, or small size, are noted.

Each forelimb (Figure 34.16o,p) consists of long bones (humerus, radius, and ulna) and the bones of the forepaws (metacarpals and phalanges). The number of carpals, metacarpals, and phalanges in each digit are

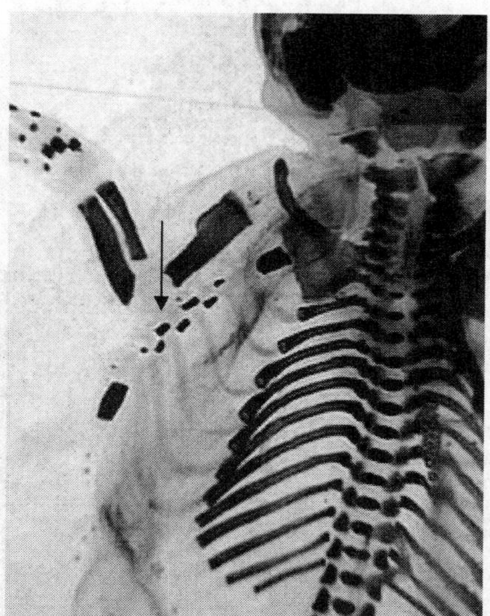

**FIGURE 34.17n** Rat fetus, day 20 of gestation, asymmetric, incompletely ossified sternebrae.

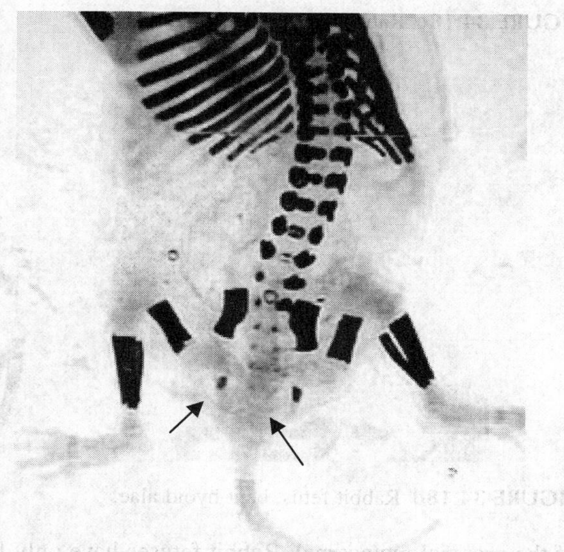

**FIGURE 34.17o** Rat fetus, day 20 of gestation, incomplete ossification of pubes and absent ossification of ischia.

counted, without separate identification of incomplete ossification. Averaging these values identifies retarded ossification on a litter basis. Each paw is evaluated separately and between-paw differences noted. The carpals are usually not ossified, but there are normally four or five metacarpals present in each forepaw. Because the preaxial metacarpal is often not ossified in fetuses, the presence of only four metacarpals indicates delayed ossification of the preaxial metacarpal, unless the finding is associated with absence of the pollex (first preaxial digit), a relatively common finding in some rabbit strains.

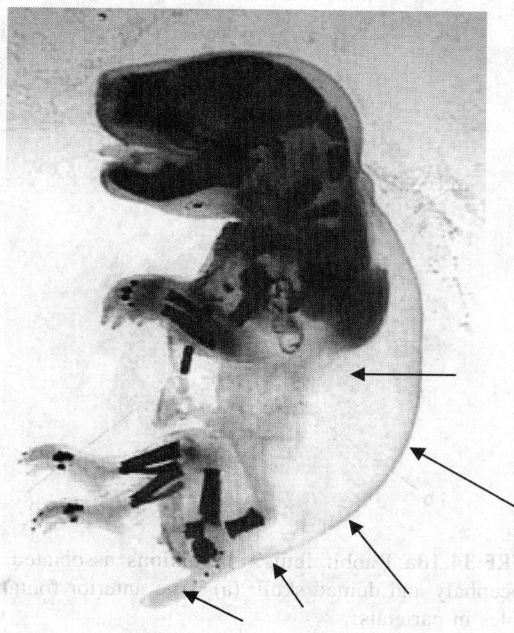

**FIGURE 34.17p** Rat fetus, day 20 of gestation, absence of thoracic vertebrae, ribs, lumbar, sacral and most caudal vertebrae.

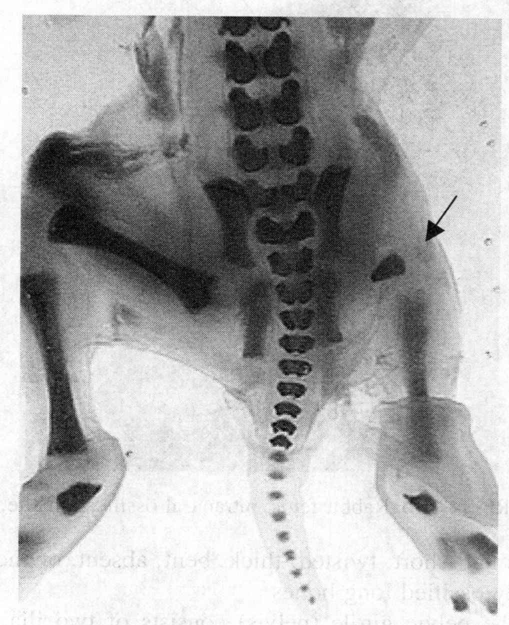

**FIGURE 34.17q** Rat fetus, day 20 of gestation, short femur.

Each forepaw in most rodent and nonrodent species has five digits. The digit count begins at the preaxial digit, after which the phalanges in each digit are counted, including the distal phalanx (claw). Day 20 rat fetuses normally have one, one, one, one, and one ossified phalanges, and day 29 rabbit fetuses normally have two, three, three, three, and three ossified phalanges in the five respective digits. Counts excluded from the ossification site averages are noted, such as fused or absent

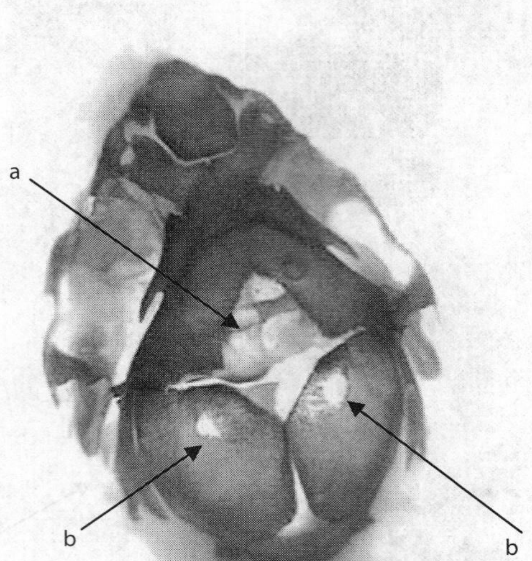


**FIGURE 34.18a** Rabbit fetus, alternations associated with hydrocephaly and domed skull: (a) large anterior fontanelle; (b) holes in parietals.

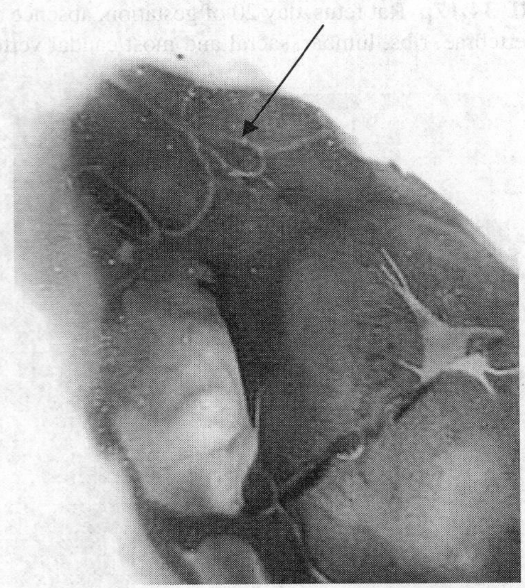

**FIGURE 34.18b** Rabbit fetus, intranasal ossification site.

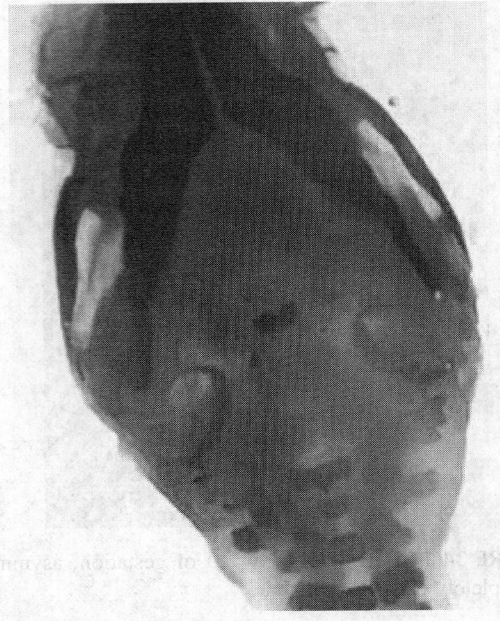

**FIGURE 34.18c** Rabbit fetus, absent hyoid alae.

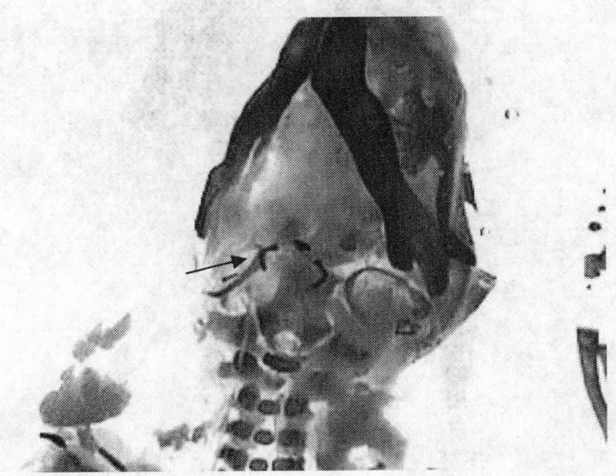

**FIGURE 34.18d** Rabbit fetus, bent hyoid alae.

digits and short, twisted, thick, bent, absent, or incompletely ossified long bones.

The pelvic girdle (pelvis) consists of two ilia, two ischia, and two pubes (Figure 34.16q). Any alterations in ossification are recorded, including incomplete, absent, or abnormal ossification (Figure 34.17o).

Each hindlimb (Figure 34.16r,s) consists of long bones (tibia and fibula) and the bones of the hindpaws (tarsals, metatarsals, and phalanges). Using the methods described to count the bones in the forepaws, the numbers of tarsals, metatarsals, and phalanges in each digit are counted. Normally, four metatarsals are observable. Rat fetuses have five hindpaw digits; when four metatarsals are present, the finding indicates delayed ossification of the preaxial metacarpal. Rabbit fetuses have only four hindpaw digits and four metacarpal bones. Beginning with the preaxial digit, the hindpaw phalanges are counted across. Normal day 20 rat fetuses have one, one, one, one, and one phalanges and normal day 29 rabbit fetuses have three, three, three, and three phalanges, including the distal phalanx, or claw (Figure 34.18p). Examples of reasons for the exclusion of count values from averages include fused or absent digits or short, thick, and bent or absent long bones (Figure 34.17q).

Our practice is to have a senior-level scientist confirm alterations identified at the technical level, with random evaluations of additional sites or specimens made to test for laboratory consistency across studies. These confirmations generally exclude sites included in ossification site averages and common variations in skull ossification (rabbits).

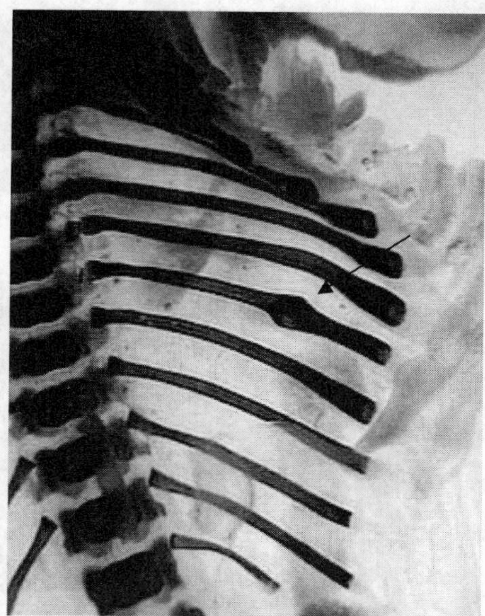

**FIGURE 34.18e** Rabbit fetus, thickened area of ossification in rib.

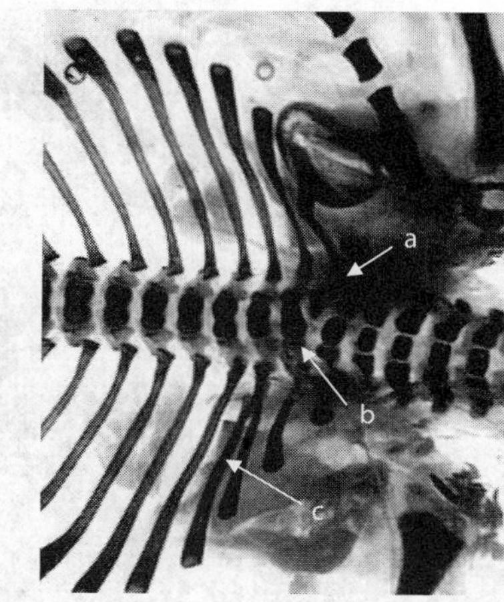


**FIGURE 34.18f** Rabbit fetus, interrelated vertebral/rib malformation: (a) T2 present as right hemivertebra with unilateral ossification of centrum and rib; (b) T3 asymmetric centrum; (c) unilateral extra rib present between right ribs 3 and 4.

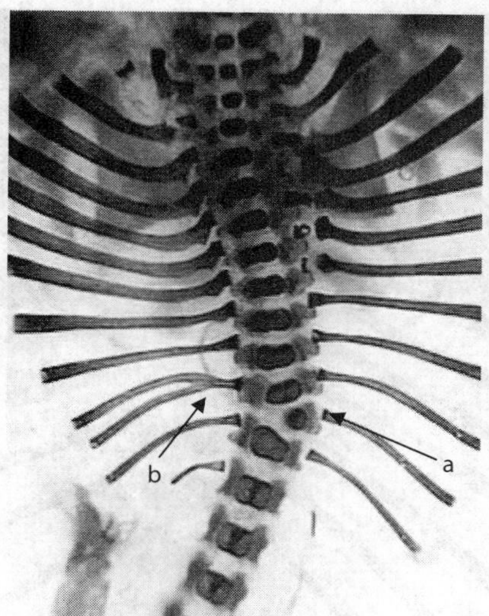


**FIGURE 34.18g** Rabbit fetus, Interrelated vertebral/rib malformations: (a) T12 present as left hemivertebra with unilateral centrum and rib; (b) forked right 10th rib, compensating for absence of 11th right rib.

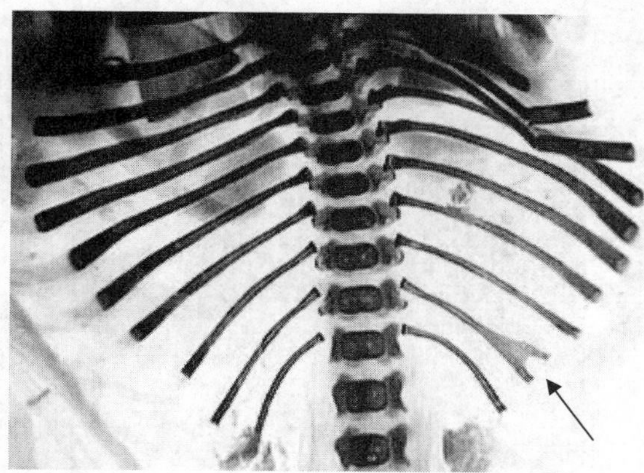

**FIGURE 34.18h** Rabbit fetus, forked rib.

## ICH Stage E

ICH stage E includes birth to weaning (adult female reproduction function, adaptation of the neonate to extrauterine life, including preweaning development and growth); postnatal age is optimally based on postcoital age.

### Observation and Timing of Parturition

Parturition, lactation, maternal–pup interaction, and pup growth and development until weaning are monitored during ICH stage E. For some classes of compounds (e.g., NSAIDS, sedatives), gestation will be prolonged and insufficient numbers of deliveries will occur during the normal 10- or 12-hour light interval. The identified pharmacologic activity of the agent, in combination with a screening study, generally will identify whether the duration of gestation or parturition is affected and overnight observations will be needed. When continual observations are made (i.e., 24 hours/day), modification of the light cycle will lengthen the duration of gestation after approximately 3 days. Although this can be prevented by conducting the evaluations under a red light, this is sometimes considered to be detrimental to employee health and safety.

Presumed pregnant animals should be housed in a nesting box with litter, beginning approximately 2 days before the expected day of delivery (hanging cages can be used before this time). They should be observed periodically for signs indicating the onset of parturition, such as stretching, visible uterine contractions, vaginal bleeding, and the presence of pups or placentas in the nesting box.

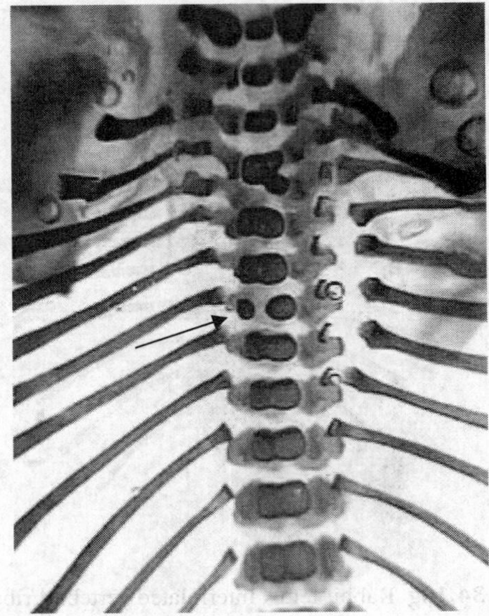

**FIGURE 34.18i** Rabbit fetus, bifid centrum in T6.

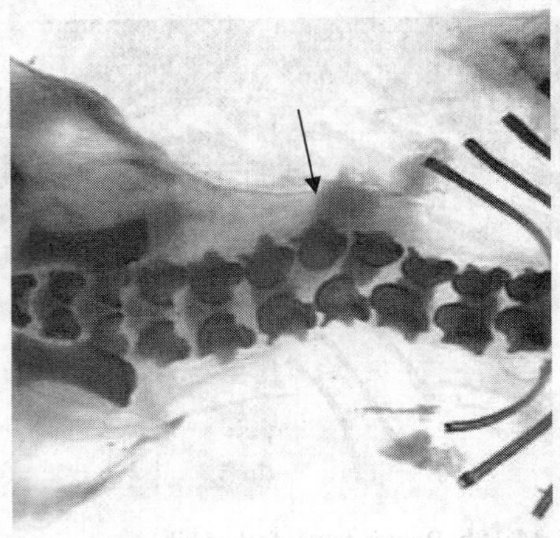

**FIGURE 34.18j** Rabbit fetus, lumbar hemivertebra with associated scoliosis.

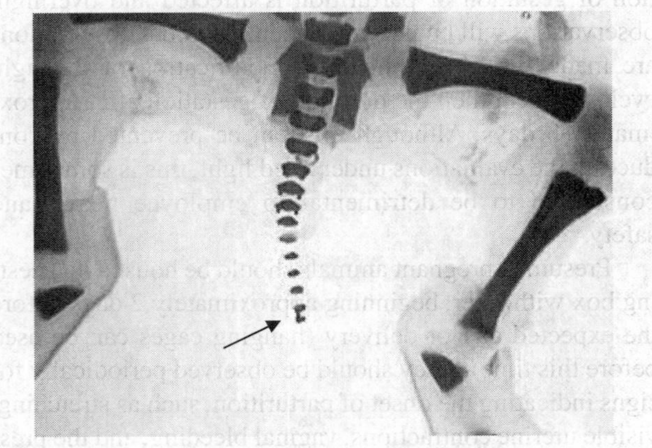

**FIGURE 34.18k** Rabbit fetus, fused caudal vertebrae.

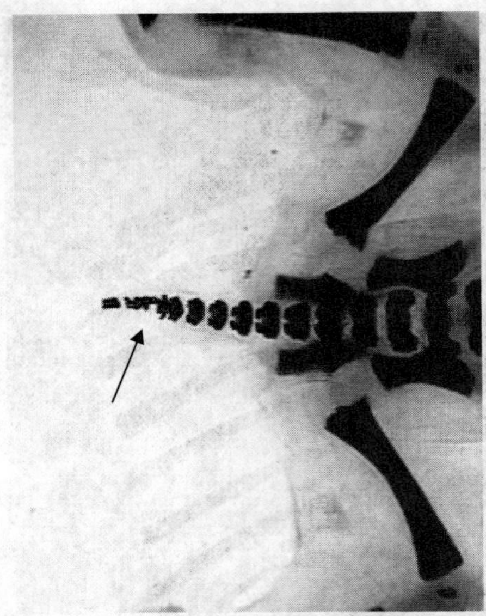

**FIGURE 34.18l** Rabbit fetus, short tail with associated fusion of caudal vertebrae.

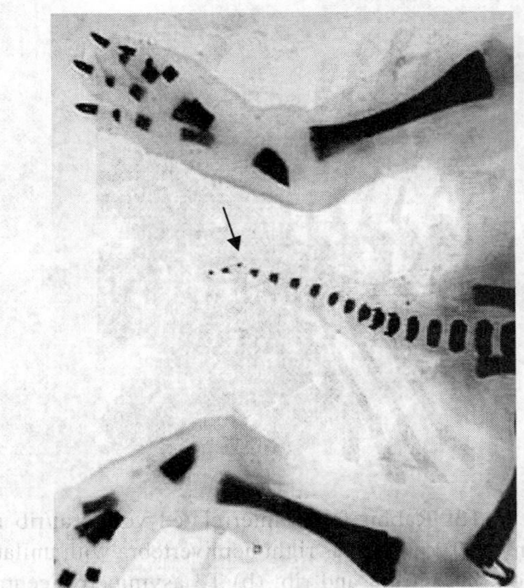

**FIGURE 34.18m** Rabbit fetus, caudal vertebra misaligned.

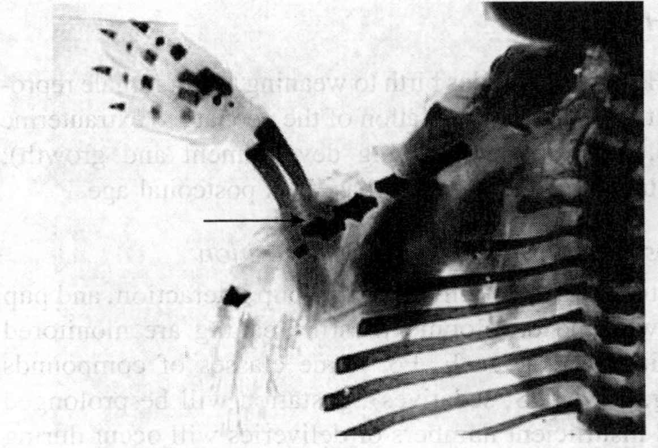

**FIGURE 34.18n** Rabbit fetus, asymmetric fused sternebrae.

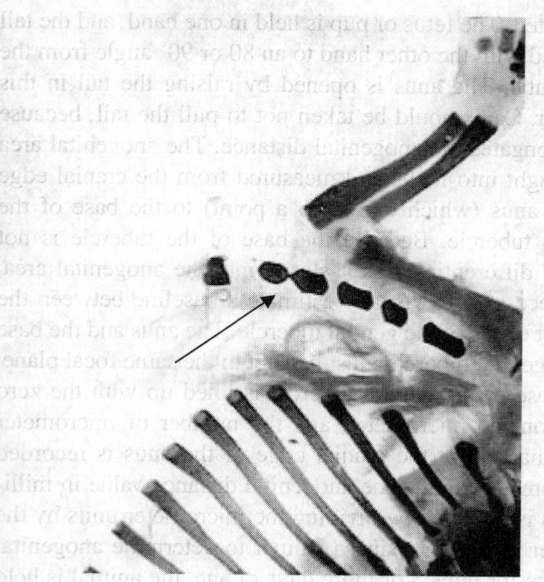

**FIGURE 34.18o** Rabbit fetus, fused sternebrae 3 and 4.

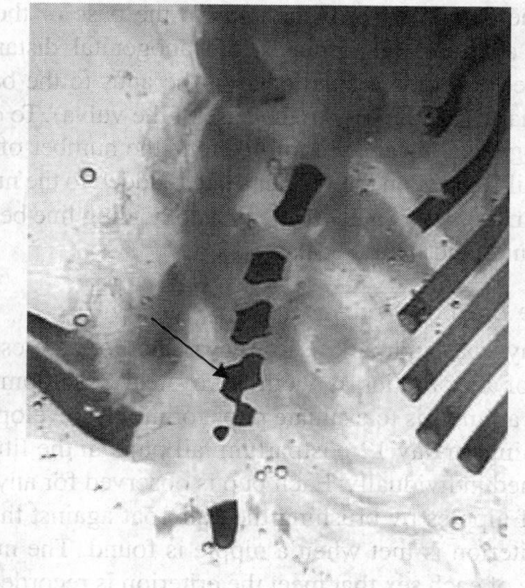

**FIGURE 34.18p** Rabbit fetus, fused asymmetric sternal centra.

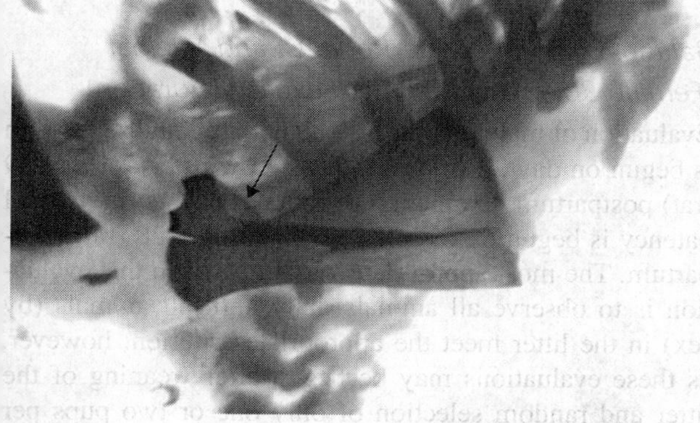

**FIGURE 34.18q** Rabbit fetus, bent scapular ala.

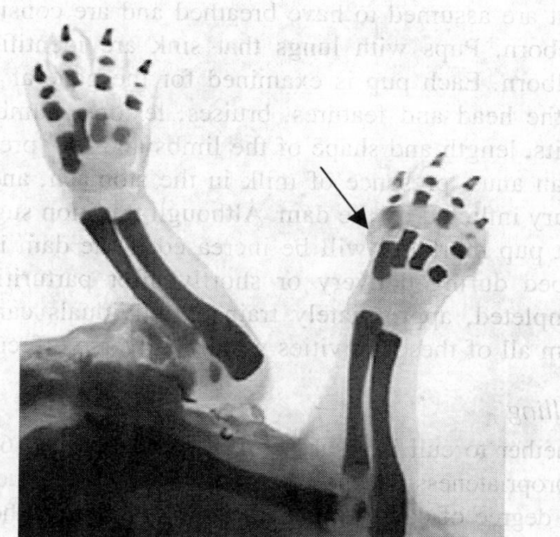

**FIGURE 34.18r** Rabbit fetus, absent pollex.

After parturition begins, observations are made every few minutes for delivery of a pup (the dam may be gently lifted to verify the presence of pups), and the delivery time for each pup is recorded. Delivery complications (dystocia) and abnormal maternal behavior (e.g., failure to remove the placenta, groom and nest the litter, mutilization or cannibalization of pups) are noted. Pups that appear stillborn or die shortly after birth are removed from the nesting box to preclude cannibalization by the dam. Dams in the process of delivery are generally not treated until parturition is completed to prevent overdosing of the animal (rats generally lose 100 g or more from the beginning to end of parturition, which would remarkably affect a dosage adjusted for body-weight changes).

The day on which delivery is completed is functionally defined as day 0 or 1 postpartum, depending on the laboratory's historical use (our laboratory uses day 1). Appropriate maternal behavior is determined on the basis of examinations for maternal and pup nesting behavior. Criteria evaluated include pup appearance (clean and warm), presence of a nest in which the pups are grouped together, and evidence of nursing activity or milk in the pup stomach. Should a dam die during delivery or during the lactation period, no further observations are made on the litter.

### *Evaluation of Pups at Birth*

The pups are further evaluated after removal from the nesting box. Gender is identified on the basis of observed anogenital distance (longer in male than in female pups). For studies in which endocrine perturbations are expected, these observations are made using calipers. Any gross physical alterations are identified, and the viability and weight of each pup are determined. Dead pups are necropsied, and sections of their lungs are placed in containers of water. Pups with lungs that float are assumed to have breathed and are considered liveborn. Pups with lungs that sink are identified as stillborn. Each pup is examined for the general shape of the head and features, bruises, lesions, number of digits, length and shape of the limbs and tail, presence of an anus, presence of milk in the stomach, and any injury inflicted by the dam. Although tradition suggests that pup mortality will be increased if the dam is disturbed during delivery or shortly after parturition is completed, appropriately trained individuals can perform all of these activities without adverse effects.

*Culling*

Whether to cull is a subject of some debate [1,166]. Its appropriateness depends on the purpose of the study and the degree of control required for evaluation of the endpoint in question. Although culling is a common practice and is required in some guidelines, this practice does increase the variability in litter values for viability and weight gain and has the potential to obscure late-occurring effects because pups are removed from evaluation. It goes without saying that these comments also apply to the selection of animals for continued evaluation in multigeneration studies.

## ICH Stages E and F

ICH stage E includes postnatal development to weaning (adult female reproduction function, adaptation of the neonate to extrauterine life, including preweaning development and growth); postnatal age is optimally based on postcoital age. ICH stage F includes postweaning development of reproductive organs to puberty (pediatric evaluation when treated; postweaning development and growth, adaptation to independent life, and attainment of full sexual development).

*Anogenital Distance*

With the advent of enhanced concern regarding estrogenic agents, many investigators have incorporated determination of anogenital distance at Caesarean delivery of fetuses or birth of pups [37]. Anogenital distance measurements of fetuses and pups through 4 days of age can easily and accurately be made using a micrometer and a stereomicroscope. Measurements taken on pups, day 5 of age or later, are made by using a caliper. For fetuses and pups up to 4 days of age, the evaluation is made as follows: A micrometer is placed into one of the eyepieces of a stereomicroscope and calibrated so the magnification is approximately 10×. The entire length of the micrometer is aligned with 10 mm on a ruler and the ruler is brought into focus. The zero micrometer mark is aligned with the left side of the 0-mm mark and the zoom magnification is adjusted until the 100-μm mark is at the beginning of the 10-mm mark. The zoom magnification may not be changed once the microscope is calibrated. The fetus or pup is held in one hand, and the tail is raised with the other hand to an 80 or 90° angle from the horizontal. The anus is opened by raising the tail in this manner. Care should be taken not to pull the tail, because this elongates the anogenital distance. The anogenital area is brought into focus and measured from the cranial edge of the anus (which comes to a point) to the base of the genital tubercle. Because the base of the tubercle is not clearly differentiated as it slopes into the anogenital area, it is necessary to visually estimate a baseline between the distinct edges of the genital tubercle. The anus and the base of the genital tubercle must be kept in the same focal plane. The base of the genital tubercle is lined up with the zero mark on the micrometer, and the number of micrometer units that fall at the cranial edge of the anus is recorded (micrometer units). The anogenital distance value in millimeters is obtained by dividing the micrometer units by the number 10. When using a caliper to determine anogenital distance for pups 5 or more days of age, the animal is held by the tail, keeping the tail at an 80 to 90° angle from the horizontal. For males, the arms of the caliper are aligned from the cranial edge of the anus to the base of the anogenital aperture. For females, the anogenital distance is measured from the cranial edge of the anus to the base of the urinary aperture (*not* to the base of the vulva). To obtain the anogenital distance in millimeters, the number of complete millimeters on the vernier scale is added to the number of millimeters indicated by the first coinciding line between the vernier and metric scales.

*Nipple Evaluation*

One physical landmark used to evaluate both the estrogenicity of xenobiotics as well as physical development of female animals is to evaluate pups for nipple development. Beginning on day 12 postpartum, all pups in the litter are examined individually. Each pup is observed for any presence of nipples by brushing the hair coat against the nap. The criterion is met when a nipple is found. The number of pups of each sex that meet the criterion is recorded over the total number of pups of each sex tested. Testing continues until all animals identified as female have at least one nipple identified.

*Balano–Preputial (Male Rodents) and Vaginal (Female Rodents) Opening (Sexual Maturation)*

Evaluation of male rodents for balano–preputial separation is begun on day 22 (mouse), day 27 (hamster), or day 39 (rat) postpartum. Evaluation of female rodents for vaginal patency is begun on day 21 (mouse) or day 28 (rat) postpartum. The most appropriate way to perform this evaluation is to observe all animals daily until all animals (by sex) in the litter meet the appropriate criterion; however, as these evaluations may be begun after weaning of the litter and random selection of only one or two pups per sex has been made for continued evaluation, this process

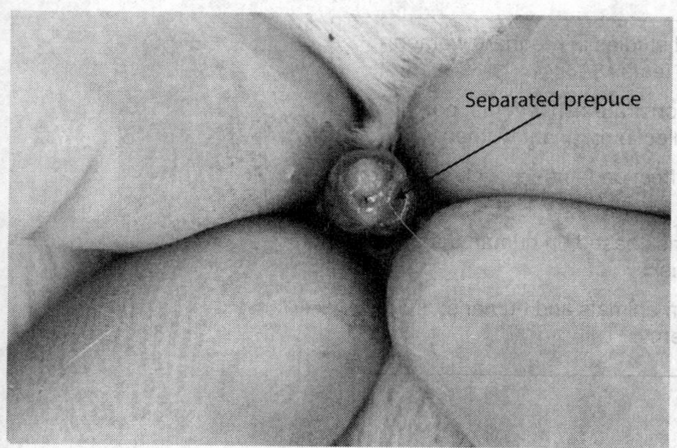


**FIGURE 34.19** Rat pup, preputial separation.

has the potential to skew data distribution. When differences occur on the basis of individual animal evaluations, litter analyses should be made. Evaluations are performed while the male or female rodent is held in a supine position. For male rodents, gentle digital pressure is applied to the sides of the rodent's prepuce. The criterion for balano–preputial opening is met when the prepuce completely retracts from the head of the penis (see Figure 34.19) and when the prepuce remains attached along the shaft of the penis but is not attached to the opening of the urethra.

## RISK ASSESSMENT

Assessing the human risk of reproductive and developmental toxicities associated with exposure to pharmaceutical and chemical agents is a complex process involving the conduct of animal and, if possible, human studies, evaluating all relevant data and then extrapolating the risk to humans. Although the endpoints that are used to identify reproductive and developmental toxicity are the same for both environmental and pharmaceutical agents, the methods used to identify the risk of exposure to these agents differ, based on the regulatory agency involved. A summary of current methodologies used by the FDA (food, drugs), the EPA (chemicals), and California (Proposition 65) will serve to illustrate the differences in assessing reproductive or developmental risk for humans.

### U.S. Environmental Protection Agency

Guidelines for reproductive toxicity risk assessment were issued by the U.S. EPA in 1996 [215]. Those for developmental toxicity were issued in 1991 [214]. These procedures provide guidance for interpreting, analyzing, and using the data from reprotoxicity and developmental toxicity studies conducted in conformance with the then issued guidelines, as well as information to be used in interpreting evaluations of epidemiologic data, sperm production, reproductive endocrine system function, sexual behavior, and female reproductive cycle normality. Risk assessment is defined by the National Research Council [153] as consisting of several components: hazard identification, dose–response assessment, exposure assessment, and risk characterization. The EPA uses these components as key determinants in their procedure for risk assessment.

*Hazard identification* includes evaluation of all available experimental animal and human data to determine whether and under what conditions reproductive toxicity is caused in a specific species. This information can then be characterized as to whether it is sufficient or insufficient to use in risk assessment. The recommended approach is to include an evaluation of dose–response relationships and the route, timing, and duration of exposure in studies used to identify a hazard. As noted by Kimmel et al. [109], determining a hazard is often dependent upon whether a dose–response relationship is present; thus, information important in comparing the toxicity of a chemical to potential human exposures is included as part of the exposure assessment, minimizing inappropriate labeling of chemicals as "reproductive toxicants" or "developmental toxicants" on a purely qualitative basis.

Quantitative *dose–response analysis* includes determining a no-observed-adverse-effect level (NOAEL) and the lowest-observed-adverse-effect level (LOAEL) for each endpoint and study. When the data are sufficient, a benchmark dose, an alternative method considered to provide a more quantitative dose–response evaluation, may be used [42,43,109]. This method accounts for the variability in the data and the slope of the dose–response curve when calculating the reference dose for developmental toxicity (RfD) or chronic exposure (RfC). When reproductive toxicity occurs at the lowest toxic dose, the RfD or RfC can be derived using either the NOAEL or the benchmark dose divided by uncertainty factors, which are used to account for interspecies differences in response, intraspecies variability, and database deficiencies. Uncertainty factors traditionally have been in units of 10, although mechanistic and pharmacokinetics evaluations can be used to reduce these values. Recently, the EPA interpreted the 1996 Food Quality Protection Act (Public Law 104-170 [231]) and proposed an additional safety factor of 10 for use when there is evidence of developmental toxicity.

*Exposure assessment* includes the populations potentially exposed or actually exposed to an agent, as well as the type, magnitude, frequency, and duration of these exposures, some of which are unique for reproductive toxicity exposure assessments; thus, exposures during certain critical points in the reproductive process may affect the outcomes observed in humans.

*Risk characterization* results in a statement of the potential for human risk and the consequences of exposure as the result of integration of the hazard characterization, quantitative dose–response analysis, and human exposure estimates. Included in this characterization are the strengths

A: Animal studies and well-controlled studies in pregnant women failed to demonstrate a risk to the fetus.

B: Animal studies have failed to demonstrate risk to fetus; no adequate and well-controlled studies in pregnant women.

C: Animal studies showed adverse effect on fetus; no well-controlled human studies.

D: Positive evidence of human fetal risk based on human data, but potential drug benefit outweighs risk.

X: Studies show fetal abnormalities in animals and humans; drug is contraindicated in pregnant women.

**FIGURE 34.20** FDA labeling requirements: pregnancy categories.

and weaknesses of each component of the risk assessment, as well as major assumptions, scientific judgments, and, when possible, qualitative descriptions and quantitative estimates of uncertainties. In 1995 [20], the U.S. EPA issued a new risk characterization policy and guidance refining the principles of the earlier 1992 policy [215].

Finally, risk assessment, in conjunction with risk management, is used in the overall regulatory process, which includes risk characterization, directives of the enabling regulatory legislation, and other factors regarding whether and the degree to which exposure to an agent should be controlled. Risk management decisions involve socioeconomic, technical, and political factors that extend beyond scientific considerations; for example, acceptability of the margin of exposure (MOE) is a risk management decision, although the scientific bases used to generate this value fall under risk assessment.

## U.S. Food and Drug Administration

The U.S. Food and Drug Administration considers risk on a risk–benefit basis; that is, the beneficial uses of the agent are compared with the risk of its use. The conclusion regarding this risk–benefit consideration is published in the labeling (the label is the official FDA approved package insert of a drug or biologic). This system was established by law in 1979 and categorizes risk by assigning a pregnancy label category to drugs and biologics (see Figure 34.20). However, a new proposal, *Considerations in the Integration of Study Results for the Assessment of Concern for Human Reproductive and Developmental Toxicities*, is currently being developed into a guidance document in accordance with the FDA's Good Guidance Practices (62 FR 8961; February 27, 1997).

At present, the FDA is reviewing these procedures and considering changing the labeling of drugs and biologics from categories (A to D, X) to text that would include each of the seven potentially affected reproductive and developmental endpoints. Conclusions regarding the animal data are to be based on an integrated analysis of all pharmacology, toxicology, and pharmacokinetics data [232]. It is to be performed using the "wedge," a tool that is designed to assess each endpoint and the strength of its signal (effect), ultimately resulting in a quantification of the risk for human use of the agent. This decision tree process is to be used to integrate clinical and nonclinical data for therapeutic agents (drugs or biologics) and to attempt to assess the probability or likelihood of an adverse reproductive outcome in humans. It reduces the previously existing emphasis on malformations as the most important adverse outcome and is independent of the type of response and its reversibility; however, the nature of the risk, along with the severity and reversibility of the response, remain as considerations in the clinical management of therapeutic use (i.e., risk vs. benefit).

In the new proposal, *positive signals* are broadly classified as reproductive or developmental toxicities. The three subclasses of reproductive toxicity—fertility and fecundity, parturition, and lactation—include both structural and functional changes affecting the reproductive competence of male and female animals in the parental ($F_0$) generation. Developmental toxicity has four subclasses—developmental mortality, dysmorphogenesis, alterations to growth, and functional toxicity—which are actually more complex descriptions of the four endpoints originally described by Wilson [238]. Each signal, regardless of type or whether it is present in a reproductive or developmental or general toxicity study, is independently evaluated by using three flowcharts (see Figure 34.21 to Figure 34.23) to estimate the concern for human risks. Flowchart A is used to determine, if animal or human reproductive or developmental studies were conducted, whether the test system and routes were relevant and whether there are any positive signals. If there are no positive signals in any of the studies addressing a particular reproductive or developmental endpoint, the decision tree evaluation of risk ends with flowchart B; however, when positive signals exist for reproductive or developmental endpoints, six factors—signal strength part A, signal strength part B, pharmacodynamics, metabolic/toxicologic concordance, relative exposure, and class alerts—must be considered (see Figure 34.23).

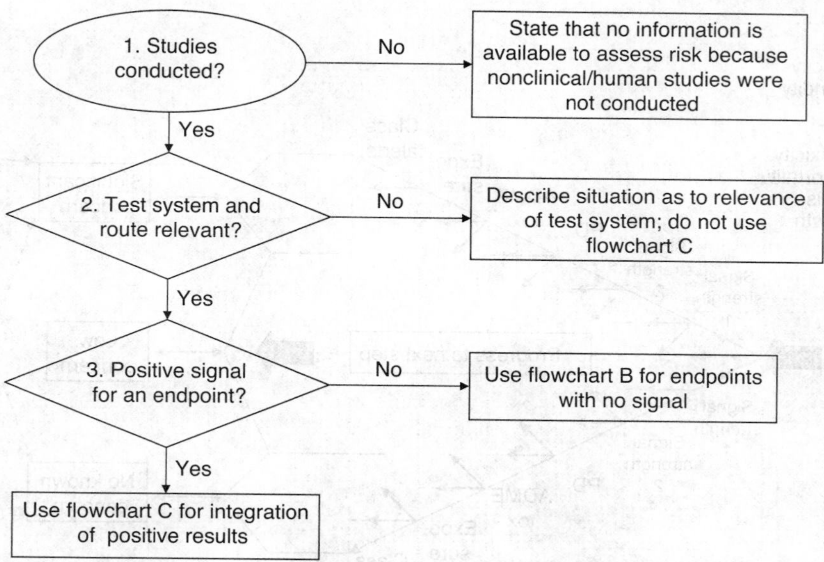


**FIGURE 34.21** Flowchart A: overall decision tree for evaluation of reproduction/developmental toxicity risk (Wedge Document, www.FDA.gov; June, 1999).

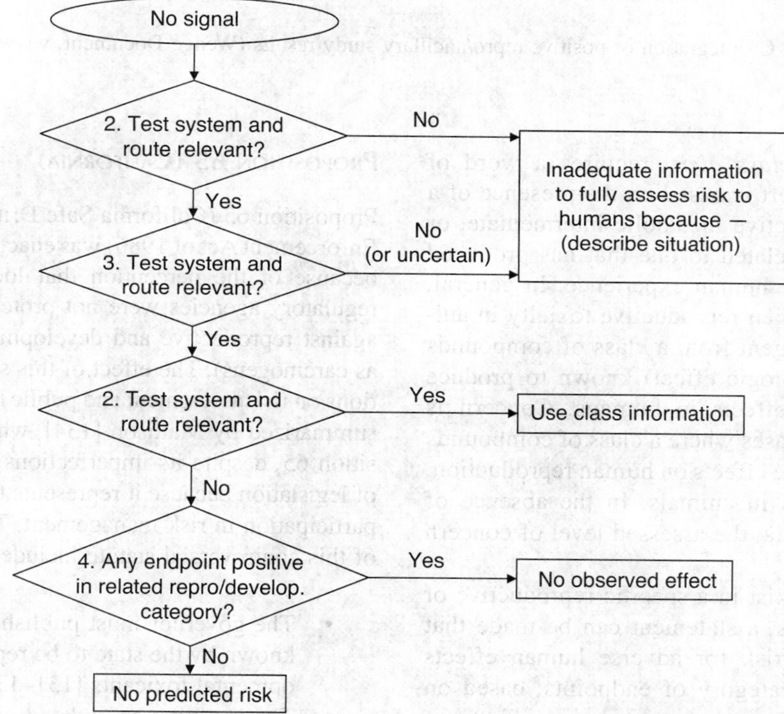


**FIGURE 34.22** Flowchart B: decision tree for endpoints with no signal (Wedge Document, www.FDA.gov; June, 1999).

*Signal strength* has six contributory elements that should be evaluated independently and integrated into the overall risk evaluation. Cross-species concordance, multiplicity of effects, and adverse effects as a function of time are assessed in part A; maternal toxicity, dose–response relationship, and rare effects are evaluated in part B. Unitary values of +1, −1, or 0 are assigned to each of the six factors, if the factor is perceived as increasing, decreasing, or having no effect on the level of risk. Conclusions regarding the six factors are summed to produce an overall level of concern for human reproductive risk.

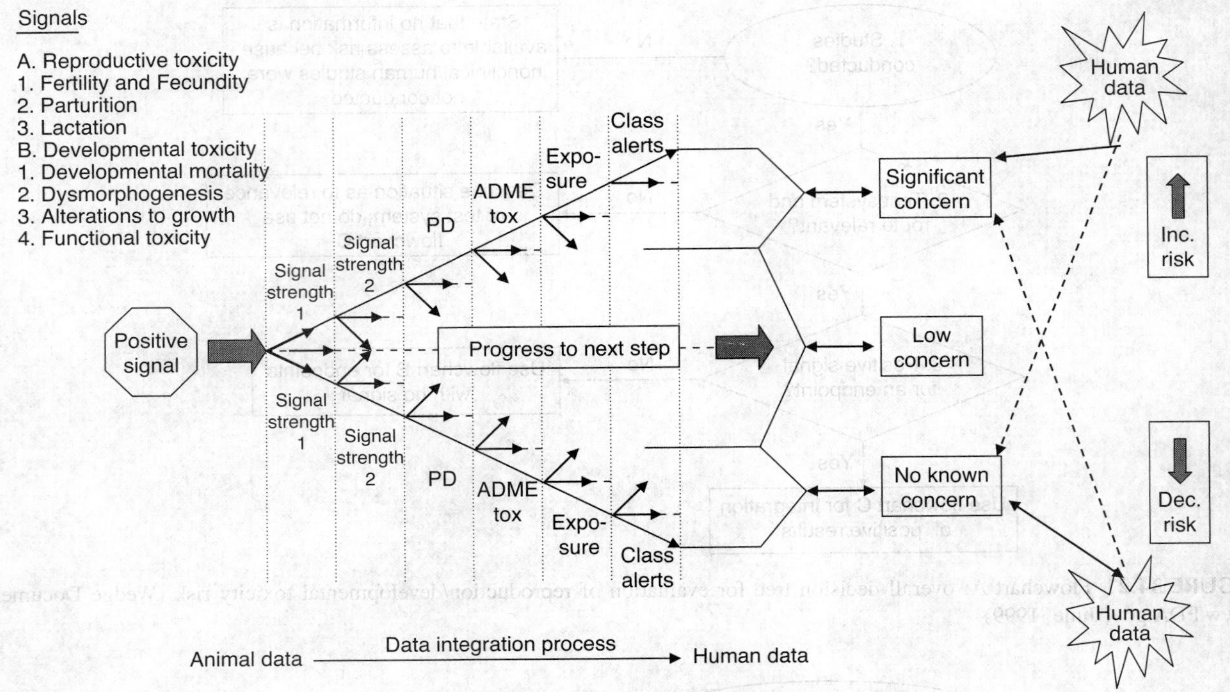


**FIGURE 34.23** Flowchart C: integration of positive repro/ancillary study results (Wedge Document, www.FDA.gov; June, 1999).

The sixth factor, *class alert*, requires a word of explanation. A class alert is based on the presence of a chemical structure, reactive metabolic intermediate, or pharmacologic effect related to one that has produced adverse effects in prior human experience. In general, concern is increased when reproductive toxicity in animals is caused by an agent from a class of compounds (structure or pharmacologic effect) known to produce adverse reproductive effects in humans. Concern is reduced only in those cases where a class of compounds has produced no adverse effects on human reproduction, despite adverse effects in animals. In the absence of human reproduction data, the assessed level of concern remains unchanged.

When no signals exist in a specific reproductive or developmental subclass, a statement can be made that there is no predicted risk for adverse human effects regarding the named category of endpoints, based on the animals studies; however, when adverse effects on other endpoints within the same category are identified, the evaluation should state that there was no effect on the incidence of the specific endpoint in animal studies. To summarize the reproductive and developmental toxicity of a therapeutic accurately and concisely in one or two paragraphs of a package insert is almost impossible; therefore, it will be interesting to see if the "wedge" concept will withstand the test of time and effectively replace the current category system.

## Proposition 65 (California)

Proposition 65 (California Safe Drinking Water and Toxic Enforcement Act of 1986) was enacted by California voters because of the perception that local, state, and federal regulatory agencies were not protecting them adequately against reproductive and developmental hazards (as well as carcinogens). The effect of this statute, and its implications on risk assessment and public health, are thoughtfully summarized by Mattison [134], who believes that Proposition 65, despite its imperfections, is an important piece of legislation because it represents the beginning of public participation in risk management. The basic requirements of this controversial statute include the following:

- The governor must publish a list of chemicals known by the state to be reproductive or developmental toxicants [131–133].
- The list must be updated on a regular basis, at least annually.
- Identification of a chemical as a reproductive or developmental toxicant under Proposition 65 requires that human exposure be no more than 1/1000 of the NOAEL.
- Chemicals listed under Proposition 65 are identified as producing male reproductive toxicity, female reproductive toxicity, or developmental toxicity. Reproductive toxicants are defined as

impairing male or female fecundity or decreasing couple specific factors and are reflected by impaired fertility; developmental toxicants impair the structural or functional integrity of the developing embryo or fetus.

Chemicals were selected for the original master list by (1) literature reports, (2) expert opinion, (3) evaluations by regulatory agencies, and (4) a modified air pollutant prioritization screen. The chemicals were then prioritized according to potential for human exposure and perception of hazard for human reproduction and development. Substances that were placed on the priority listing (42 of 162 chemicals on the master list) were then subjected to evaluation by the California Department of Health Services and considered by the Science Advisory Panel before ultimately appearing on the Proposition 65 listing [50].

Proposition 65 has been included in this section on risk assessment not only because of its novel role in public health but also because it serves to illustrate the problems and difficulties associated with human reproductive and developmental risk assessment. For example, the ability to assess and characterize the reproductive and developmental health of a community is often limited by a lack of data and a lack of consensus as to the adverse effects of a specific chemical. Evidence cited by Mattison [134] indicates that of the estimated 90,000 chemicals in commerce in the United States, only 4000 have been tested in animals for reproductive or developmental toxicity, and up to one third of the tested substances may be reproductive or developmental toxicants. A study by the National Research Council [153] also concluded that only a small fraction of the chemicals tested contained enough toxicity information for reproductive and developmental hazard identification. This data gap is reason for concern, despite the findings by Barlow and Sullivan [5] that show that less than half of the animal reproductive or developmental toxicants produce reproductive or developmental toxicity in humans.

According to most guidelines, hazard assessment and dose–response analysis are the primary principles of risk assessment, and they depend on the results of laboratory animal studies; yet, the predictiveness of animal studies is itself a problem and depends to a great extent on the expertise with which these studies are conducted and interpreted. Thus, experiments with inappropriate dosage regimens or lack of pharmacokinetics considerations may provide entirely misleading risk assessments, while studies with insufficient numbers of animals or effects that are incorrectly classified preclude meaningful interpretation due to insufficient statistical power upon which to make accurate predictions.

## CONCLUSION

Human risk assessment for reproductive and developmental toxicity is a complex procedure that has yet to be perfected, regardless of whether the assessment is for environmental chemicals, pharmaceutical products, or physical phenomena. Many procedures have been tried over the years and refinements in methodologies (e.g., the "benchmark dose" and the "wedge") will continue to be made. It is hoped that computerized characterizations of potential structure–activity relationships for reproductive and developmental toxicity will someday emerge, although to date this methodology has not been successful in detecting reproductive or developmental toxicants. Regardless of the method used for risk assessment, the endpoints used to characterize reproductive and developmental toxicity remain unchanged, and every effort should be made to ensure the accuracy of these basic components of risk assessment and thereby prevent the compounding of errors that are all too frequent in the prediction of risk.

## ACKNOWLEDGMENTS

Special thanks are due to Dr. Robert E. Staples, for the use of his diagrams and procedures; to Dr. James G. Wilson, for access to his procedures; to Mrs. Donna W. Lewis, for her diagrams and photographs; to Dr. Robert M. Diener, for his photographs and editorial assistance; and to Mrs. Marjorie B. Vargo, for her assistance in developing and compiling this chapter. I would also like to thank Charles River Laboratories (formerly Argus Research Laboratories, Inc.) for the use of their SOPs.

## QUESTIONS

1. Design the testing needs for a new drug entity that is to be used as a contraceptive agent.
2. A female employee was exposed to insecticide fumes used during office cleaning procedures. Nine months later she delivers a child with a cleft palate. Determine whether there a potential causal relationship between the insecticide fumes and the birth defect in the child.
3. A chemical company wants to develop a new herbicide. Design the testing paradigm that should be run.
4. A woman falls down the steps and breaks her leg. As part of the hospital examination, she is x-rayed and it is discovered that she is about 2 months pregnant. She has read that exposure to radiation sometimes results in microcephaly and calls her physician to schedule an abortion. What counsel should the physician provide?
5. A new drug entity results in increased hematopoiesis and is administered weekly to humans. What are the expected adverse effects of pharmacotoxic doses of this agent in women of childbearing potential and their offspring? Design a testing paradigm that addresses these endpoints and results in registration of the agent.

## REFERENCES


1. Agnish, N. D. and Keller, K. A. (1997): The rationale for culling of rodent litters. *Fund. Appl. Toxicol.*, 38:2–6.
2. Alison, M. R. and Sarraf, C. E. (1992): Apoptosis: a gene-directed programme of cell death. *J. Roy. Coll. Phys. Lond.*, 26:25–35.
3. Antonov, A. N. (1947): Children born during the siege of Leningrad in 1942, *J. Pediatr.*, 30:250.
4. Astwood, E. B. (1939): Changes in the weight and water content of the uterus of the normal adult rat. *The Amer. Journal of Physiol.*, 126: 162–170.
5. Baker, T. G. (1963): A quantitative and cytological study of germ cells in the human ovaries. *Proc. Roy. Soc. Lond. Biol.*, 158:417–433.
6. Barlow, S. M., Sullivan, F. M. (1982): *Reproductive Hazards of Industrial Chemicals: and Evaluation of Animal and Human Data*. Academic Press, New York.
7. Barrow, M. V. and Taylor, W. J. (1969): A rapid method for detecting malformations in rat fetuses. *J. Morph.*, 127: 291–306.
8. Beach, F. (1948): *Hormones and Behavior.* Hoeber, New York.
9. Beltrame, D. and Giavini, E. (1990): Morphological abnormalities in experimental teratology: need for a standardization of the current terminology. *Cong. Anom.*, 3: 187–195.
10. Biggers, J. D. (1975): Oogenesis. In: *Gynecologic Endocrinology*, edited by J. J. Gold, pp. 612–620. Harper & Row, New York.
11. Bold. R. J., Termuhlen, P. M., and McConkey, D. J. (1997): Apoptosis, cancer and cancer therapy. *Surg. Oncol.*, 6:133–142.
12. Boué, J., Boué, A., and Lazar, P. (1975): Retrospective and prospective epidemiological studies of 1500 karyotyped spontaneous abortions. *Teratology*, 12:11–26.
13. Brann, D. W., Mills, T. M., and Mahesh, V. B. (1995). Female reproduction: the ovulatory cycle. In: *Reproductive Toxicology*, 2nd ed., edited by R. J. Witorsch, pp. 23–44. Raven Press, New York.
14. Brent, R. L. (1964): Drug testing in animals for teratogenic effects: thalidomide in the pregnant rat. *J. Pediatr.*, 64:762–770.
15. Brent, R. L. (1976): Environmental factors: miscellaneous. In: *Prevention of Embryonic, Fetal and Perinatal Disease*, DHEW Publ. No. NIH 76-853, edited by R. L. Brent and M. I. Harris, pp. 211–218. National Institutes of Health, Bethesda, MD.
16. Brent, R. L. (1980). The prediction of human diseases from laboratory and animal tests for teratogenicity, carcinogenicity and mutagenicity. In*: Controversies in Therapeutics*, edited by L. Lasagna. W.B. Saunders, New York.
17. Brent, R. L. (1985): The magnitude of the problem of congenital malformations. In: *Prevention of Physical and Mental Congenital Defects*. Part A. *The Scope of the Problem*, edited by M. Marois, pp. 55–68. Alan R. Liss, New York.
18. Brent, R. L. (1999): Utilization of developmental basic science principles in the evaluation of reproductive risks from pre- and postconception environmental radiation exposures. *Teratology*, 59:182–204.
19. Brinster, R. L. (1975): Teratogen testing using preimplantatoin mammalian embryos. In: *Methods for Detection of Environmental Agents That Produce Congenital Defects*, edited by T. H. Shepard, J. R. Miller, and M. Marois, pp. 113–124. Elsevier, New York.
20 Browner, C. M. (1995): *EPA Risk Characterization Program* [memorandum]. U.S. Environmental Protection Agency, Washington, D.C.
21. Burek, M. J. and Oppenheim, R. W. (1996): Programmed cell death in the developing nervous system. *Brain Pathol.*, 6:427–446.
22. Byskov, A. G. (1978). Follicular atresia. In: *The Vertebrate Ovary*, edited by R. E. Joes, pp. 533–562. Plenum Press, New York.
23. Carter, C. O. (1976): Genetics of common single malformations. *Br. Med. Bull.*, 32:21–26.
24. Cassidy, A., Bingham, S., and Setchell, K. (1996): Biological effects of a diet of soy protein rich in isoflavones on the menstrual cycle of premenopausal women. *Am. J. Clin. Nutr.*, 60:333–340.
25. Challis, J. R. G. and Olson, D. M. (1988): Parturition. In: *The Physiology of Reproduction*, Vol. 2, edited by E. Knobil and J. D. Neill, pp. 2177–2216. Raven Press, New York.
26. Chapin, R. E. and Heindel, J. J. (1993): Introduction. In: *Methods In Toxicology*. Vol. 3B. *Female Reproductive Toxicology*, edited by J. J. Heindel and R. E. Chapin, pp. 1–15. Academic Press, San Diego, CA.
27. Chernoff, N. and Kavlock, R. J. (1982): An *in vivo* teratology screen utilizing pregnant mice. *Toxicol. Environ. Health*, 10:541–550.
28. Christian, M. S. (1978): Postnatal Functional Teratology Resulting From Fetal Insult, Ph.D. dissertation. Thomas Jefferson University, Philadelphia, PA.
29. Christian, M. S. and Voytek, P. E. (1982): *In vivo* reproductive and mutagenicity tests. In: *A Guide to General Toxicology*, edited by F. Homburger, J. A. Hayes, and E. W. Pelikan, pp. 294–325. Karger, Basel.
30. Christian, M. S. (1983): Statement of problem. In: *Advances in Modern Environmental Toxicology*. Vol. 3. *Assessment of Reproductive and Teratogenic Hazards*, edited by M. S. Christian, W. M. Galbraith, P. Voytek, and M. A. Mehlman, pp. 1–4. Princeton Scientific, Princeton, NJ.
31. Christian, M. S. (1983): Postnatal alterations of gastrointestinal physiology, hematology, clinical chemistry, and other non-CNS parameters. In: *Handbook of Experimental Pharmacology: Teratogenesis and Reproductive Toxicology*, Vol. 65, edited by E. M. Johnson and D. M. Kochhar, pp. 263–286. Springer-Verlag, Berlin.
32. Christian, M. S. (1993): Problems in developmental toxicology caused by incorrectly used terminology. *J. Am. Coll. Toxicol.*, 12:323–328.
33. Christian, M. S. and Hoberman, A. M. (1996): Perspectives on the U.S., EEC, and Japanese developmental toxicity guidelines. In: *Handbook of Developmental Toxicology*, edited by R. D. Hood, pp. 551–596. CRC Press, Boca Raton, FL.
34. Christian, M. S., Hoberman, A. M. Bachmann, S., and Hellwig, J. (1998): Variability in the uterotrophic response assay (an *in vivo* estrogenic response assay) in untreated control and positive control (DES-DP, 2.5 μg/kg, bid) Wistar and Sprague–Dawley rats. *Drug Chem. Toxicol.*, 21(Suppl. 1):51–100.

35. Chung, C. S. and Myrianthopoulos, N. C. (1975): Factors affecting risks of congenital malformations. *Birth Defects Orig. Artic. Ser.*, 11(10):23–38.
36. Clarke, P. G. H. (1990): Developmental cell death: morphological diversity and multiple mechanisms. *Anat. Embryol.*, 181:195–213.
37. Clark, R. L., Anderson, C. A., Prahalada, S., Robertson, R. T., Lochry, E. A., Leonard, Y. M., Stevens, J. L., and Hoberman, A. M. (1993): Critical developmental periods for effects on male rat genitalia induced by finasteride, a 5α-reductase inhibitor. *Toxicol. Appl. Pharmacol.*, 119: 34–40.
38. Collins, T. F. X. (1978): Reproduction and teratology guidelines: review of deliberations by the National Toxicology Advisory Committee's Reproduction Panel. *J. Environ. Pathol. Toxicol.* 2:141–147.
39. Collins, T. F. X., Sprando, R. L., Hansen, D. L., Schackelford, M. E., and Welsh, J. J. (1998): Testing guidelines for evaluation of reproductive and developmental toxicity of food additives in females. *Int. J. Toxicol.*, 17:299–325.
40. Committee on the Safety of Medicines. (1974): *Notes for Guidance on Reproduction Studies*. Department of Health and Social Security, Great Britain.
41. Cooper, R. L., Goldman, J. M., and Vandenbergh, J. G. (1993): Monitoring of the estrous cycle in the laboratory rodent by vaginal lavage. In: *Methods in Toxicology.* Vol. 3B. *Female Reproductive Toxicology*, edited by J. J. Heindel and R. E. Chapin, pp. 45–56. Academic Press, San Diego, CA.
42. Crump, K. S. (1984): A new method for determining allowable daily intakes. *Fund. Appl. Toxicol.*, 4:854–871.
43. Crump, K. S. (1995): Calculation of benchmark doses from continuous data. *Risk Analysis*, 15:79–89.
44. Culver, K. W. (1994): *Gene Therapy: A Handbook for Physicians*. Mary Ann Liebert, New York.
45. Cummings, A. M. and Perreault, S. D. (1990): Methoxychlor accelerates embryo transport through the rat reproductive tract. *Toxicol. Appl. Pharmacol.*, 102:110–116.
46. Daston, G. P. (1994): Relationship between maternal and developmental toxicity. In: *Developmental Toxicology*, 2nd ed., edited by C. A. Kimmel and J. Buekle-Sam, pp. 189–212. Raven Press, New York.
47. DeSesso, J. M. (1997): Comparative embryology. In: *Handbook of Developmental Toxicology*, edited by R. D. Hood, pp. 111–174. CRC Press, Boca Raton, FL.
48. Dews, P. B., Ed. (1984): *Caffeine*. Springer-Verlag, Berlin.
49. Diczfalusy, E. (1964): Endocrine functions of the human feto-placental unit. *Fed. Proc.*, 23:791–798.
50. Donald, J. M., Monserrat, L. E., Hooper, K., Book, S. A., and Chernoff, G. F. (1991): Prioritizing candidate reproductive/developmental toxicants for evaluations. *Reprod. Toxicol.*, 5:99–108.
51. Edmonds, L., Hatch, M., Holmes, L., Kline, J., Letz, G., Levin, B., Miller, R., Shrout, P., Stein, Z., Warburton, D., Weinstock, M., Whorton, R. D., and Wyrobek, A. (1981): Report of panel II: guidelines for reproductive studies in exposed human populations. In: *Guidelines for Studies of Human Populations Exposed to Mutagenic and Reproductive Hazards*, edited by A. D. Bloom, pp. 37–110. March of Dimes Birth Defects Foundations, White Plains, NY.
52. Ehrhardt, A. A. and Meyer-Bahlburg, H. F. L. (1981): Effects of prenatal sex hormones on gender-related behavior. *Science*, 211:1312–1318.
53. Elger, W., Beier, S., and Faehnrich, M. (1990): Interference with hormonal control of rodent reproduction and its implications for human risk assessment. In: *Proc. of the Fifth Int. Congress of Toxicology*, edited by G. N. Valans, J. Sims, F. M. Sullivan, and P. Turner, pp. 445–456. Taylor & Francis, New York.
54. Erickson J. D. (1981): Epidemiology and developmental toxicology. In: *Developmental Toxicology*, edited by C. A. Kimmel and J. Buelke-Sam, pp. 289–301. Raven Press, New York.
55. Espey, L. L. (1978): Ovulation: In: *The Vertebrate Ovary*, edited by R. E. Jones, pp. 503–532. Plenum Press, New York.
56. European Economic Community. (1988): *The Rules Governing Medicinal Products in the European Community.* Vol. III. *Guidelines on the Quality, Safety and Efficacy of Medicinal Products for Human Use*. Office of Official Publications of the European Communities, Brussels, Belgium.
57. European Economic Community. (1989): Council decision on 28 July 1989 on the acceptance by the European Economic Community of an OECD decision/recommendation on compliance with principles of good laboratory practice. *Off. J. Eur. Commun. Legis.*, 32(L315):1–17.
58. Faustman, E. M., Ponce, R. A., Seeley, M. R. and Whittaker, S. G. (1997): Experimental approaches to evaluate mechanisms of developmental toxicity. In: *Handbook of Developmental Toxicology*, edited by R. D. Hood, pp. 13–41. CRC Press, Boca Raton, FL.
59. Favrot, M., Coll, J.-L., Louis, N., and Negoescu, A. (1998): Cell death and cancer: replacement of apoptotic genes and inactivation of death suppressor genes in therapy. *Gene Ther.*, 5:728–739.
60. Feussner, E. L., Lightkep, G. E., Hennesy, R. A., Hoberman, A. M., and Christian, M. S. (1992): A decade of rabbit fertility data: study of historical control animals. *Teratology*, 46:349–365.
61. Francis, B. M., Metcalf, R. L., Lewis, P. A., and Chernoff, N. (1999): Maternal and developmental toxicity of halogenated 4′-nitrodiphenyl ethers in mice. *Teratology*, 59: 69–80.
62. Fraser, F. C. (1976): The multifactorial/threshold concept-uses and misuses. *Teratology*, 14:267–280.
63. Fritz, M. A. and Fitz, T. A. (1991). The functional microscopic anatomy of the corpus luteum: the 'small cell–large cell' controversy. *Clin. Obstet. Gynecol.*, 34:144–156.
64. Fujii, T. and Adams, P. M. (1987): *Functional Teratogenesis: Functional Effects on the Offspring After Parental Drug Exposure*. Teikyo University Press, Tokyo, Japan.
65. Gandelman, R. (1983): Gonadal hormones and sensory function. *Neurosci. Behav. Rev.*, 7:1–17.
66. Garside, D. A., Charlton, A., and Heath, K. J. (1996): Establishing the timing of implantation in the Harlan Porcellus Dutch and New Zealand white rabbit and the Han Wistar rat. *Regul. Toxicol. Pharmacol.*, 23:69–73.
67. Gelineau-Van Waes, J., Bennett, G. D., and Finnell, R. H. (1999): Phenytoin-induced alterations in craniofacial gene expression. *Teratology*, 59:23–34.

68. Generoso, W. M. (1980): Repair in fertilized eggs of mice and its role in the production of chromosomal aberrations. In: *DNA Repair and Mutagenesis in Eukaryotes*, edited by W. M. Generoso, M. S. Shelby, and F. J. De Serres, pp. 389–410. Plenum Press, New York.
69. George, J. D., Fail, P. A., Grizzle, T. B., Heindel, J. J., and Chapin, R. E. (1990): *Mixed Chemicals (MIX): Reproduction and Fertility Assessment in Swiss (CD-1) Mice When Administered in the Drinking Water, Final Study Report*. National Technical Information Service, Springfield, VA.
70. Glücksman, A. (1951): Cell deaths in normal vertebrate ontogeny. *Biol. Rev. Cambridge Philos. Soc.*, 26:59–86.
71. Gondos, B. (1978): Oogonia and oocytes in mammals. In: *The Vertebrate Ovary*, edited by R. E. Jones, pp. 83–120. Plenum Press, New York.
72. Graham, A., Koentges, G., and Lumsden, A. (1996): A review: neural crest apoptosis and the establishment of craniofacial pattern—an honorable death. *Mol. Cell. Neurosci.*, 8:76–83.
73. Gregg, N. M. (1941): Congenital cataract following German measles in the mother. *Trans. Ophthalmol. Soc. Aust.*, 3:35–46.
74. Gulati, D. K., Barnes, L. H., Chapin, R. E., and Heindel, J. (1991): *Final Report on the Reproductive Toxicity of a Complex Mixture of Groundwater Contaminants in Sprague–Dawley Rats*. National Technical Information Service, Springfield, VA.
75. Haanen, C. and Vermes, I. (1996): Apoptosis: programmed cell death in fetal development. *Eur. J. Obstetr. Gynecol. Reprod. Biol.*, 64:129–133.
76. Hale, F. (1933): Pigs born without eyeballs. *J. Hered.*, 24:105.
77. Hale, F. (1935): The relation of vitamin A to anophthalmos in pigs. *Am. J. Ophthalmol.*, 18:1087.
78. Hall, J. G. (1990): Genetic imprinting: review and relevance to human diseases. *Ann. Hum. Genet.*, 46:857–873.
79. Haney, A. F. (1985): Effects of toxic agents on ovarian function. In: *Endocrine Toxicology*, edited by J. A. Thomas et al., pp. 181–210. Raven Press, New York.
80. Hardin, B. D. (1987): Evaluation of the Chernoff/Kavlock test for developmental toxicity. *Teratogen. Carcinogen. Mutagen.* 7:1–127.
81. Heindel, J. J., Thomford, P. J., and Mattison, D. R. (1989): Histological assessment of ovarian follicle number as a screen for ovarian toxicity. In: *Growth Factors and the Ovary*, edited by A. N. Hirshfield, pp. 421–425. Plenum Press, New York.
82. Heindel, J. J., George, J. D., and Fail, P. A. (1990): *Final Report on the Reproductive Toxicity of a Chemical Mixture in CD-1 Swiss Mice*. National Technical Information Service, Springfield, VA.
83. Heinonen, O. P. et al. (1977): *Birth Defects and Drugs in Pregnancy*, pp. 127, 450. PSG, Littleton, MA.
84. Henderson, B. E., Ross, R. K., Judd, H. L., Krailok, M. D., and Pike, M. C. (1985): Do regular ovulatory cycles increase breast cancer risk? *Cancer*, 56:1206–1208.
85. Herbst, A. L., Ulfelder, H., and Poskanzer, D. C. (1971): Adenocarcinoma of the vagina: association of maternal stilbestrol therapy with tumor appearance in young women. *New Engl. J. Med.*, 284:878–881.
86. Hertig, A. T. (1967): The overall problem in man. In: *Comparative Aspects of Reproductive Failure*, edited by K. Benirschke, pp. 11–41. Springer-Verlag, Berlin.
87. Hoar, R. M. and Salem, A. J. (1961): Time of teratogenic action of trypan blue in guinea pigs. *Anat. Rec.*, 141: 173–182.
88. Hoar, R. M. (1969): Resorption in guinea pigs as estimated by counting corpora lutea: the problem of twinning. *Teratology*, 2:187–190.
89. Hood, R. D. and Miller, D. B. (1997): Maternally mediated effects on development. In: *Handbook of Developmental Toxicology*, edited by R. Hood, pp. 61–90. CRC Press, Boca Raton, FL.
90. Hull, D., Dobbing, J., Miller, R. W., Naftolin, F., Ounsted, M., D., Rehder, H., Robinson, J. S., Tuge, C., and Usher, R. H. (1978): Definition, epidemiology, identification of abnormal fetal growth: group report. In: *Abnormal Fetal Growth: Biological Bases and Consequences*, edited by F. Naftolin, pp. 69–83. Dahlem Konferenzen, Berlin.
91. Hurle, J. M., Ros, M. A., Garcia-Martinez, V., Macias, D., and Gañan, Y. (1995): Cell death in the embryonic developing limb. *Scan. Microsc.*, 9:519–534.
92. Inouye, M. (1976): Differential staining of cartilage and bone in fetal mouse skeleton by alcian blue and alizarin red S. *Cong. Anom.*, 16:171–173.
93. ICH. (1994): Harmonised tripartite guideline: detection of toxicity to reproduction for medicinal products, proposed rule endorsed by the ICH Steering Committee at Step 4 of the ICH process, June 24, 1993. In: *Proc. of the Second Int. Conf. on Harmonization*, edited by P. F. D'Arcy and D. W. G. Harron, pp. 557–586. Greystone Books, Antrim, N. Ireland.
94. ICH. (1996): Harmonised tripartite guideline: male fertility studies in reproductive toxicology. In: *Proc. of the Third Int. Conf. on Harmonization*, edited by P. F. D'Arcy and D. W. G. Harron, pp. 245–252. Greystone Books, Antrim, N. Ireland.
95. Jackson, C. M. (1925): *Effects of Inanition and Malnutrition Upon Growth and Structure*, P. Blakiston's Son & Co., Philadelphia, PA.
96. Jacobson, M. D., Weil, M., and Roff, M. C. (1997): Programmed cell death in animal development. *Cell Press*, 88:347–354.
97. Jaffe, R. B., Ed. (1981): *Prolactin*. Elsevier, New York.
98. Joffe, M. (1985): Biases in research on reproduction and women's work. *Int. J. Epidemiol.*, 4:118–123.
99. Johnson, E. M. (1964). A histologic study of postnatal vitamin $B_{12}$ deficiency in the rat. *Am. J. Pathol.*, 44:73–83.
100. Johnson, E. M. and Armenti, V. T. (1978): Postnatal effects of prenatal insult on lung development in the rat. *Anat. Rec.*, 190:432–433.
101. Johnson, E. M. and Christian, M. S. (1984): When is a teratology study not an evaluation of teratogenicity?, *J. Am. Coll. Toxicol.*, 3:431–434.
102. Johnson, E. M. (1986): The scientific basis for multigeneration safety evaluation, *J. Am. Coll. Toxicol.*, 5:197–201.
103. Karnofsky, D. A. (1965): Mechanism of action of certain growth-inhibiting drugs. In: *Teratology Principles and Techniques*, edited by J. G. Wilson and J. Warkany, pp. 185–194. University of Chicago Press, Chicago, IL.

104. Kalter, H. (1968): *Teratology of the Central Nervous System*. University of Chicago Press, Chicago, Chicago, IL.
105. Kavlock, R. J. and Grabowski, C. T., Eds. (1983): Abnormal functional development of the heart, lungs, and kidneys: approaches to functional teratology. *Prog. Clin. Biol. Res.*, 140:1–387.
106. Khera, K. S. (1984): Maternal toxicity: a possible factor in fetal malformations in mice. *Teratology*, 29:411–416.
107. Khera, K. S. (1985): Maternal toxicity: a possible etiologic factor in embryo-fetal deaths and fetal malformations in rodent-rabbit species. *Teratology*, 31:129–153.
108. Kimmel, C. A. (1981): A profile of developmental toxicity. In: *Developmental Toxicology*, edited by C. A. Kimmel and J. Buelke-Sam, pp. 321–331. Raven Press, New York.
109. Kimmel, C. A. (1990): Quantitative approaches to human risk assessment for noncancer health effects. *Neurotoxicology*, 11:189–98.
110. Klinefelter, G. and Gray, Jr., L. E. (1993): The clinical relevance of animal models: animal studies that assess the potential for drugs and environmental agents to cause reproductive disorders in humans. In: *Reproductive Toxicology and Infertility*, edited by A. R. Scialli and M. J. Zinaman, pp. 219–282. McGraw-Hill, New York.
111. Knight, D. C. and Eden, J. A. (1995): Phytoestrogens: a short review. *Maturitas*, 22:167–175.
112. Knight, D. C. and Eden, J. A. (1996): A review of clinical effects of phytoestrogens. *Obstet. Gynecol.*, 87:897–904.
113. Knudsen, T. B. and Wubah, J. A. (1998): Transgenic animal models: functional analysis of developmental toxicity as illustrated with the p53 suppressor model. In: *Handbook of Developmental Neurotoxicology*, edited by W. Slikker, Jr., and L. W. Chang, pp. 209–221. Academic Press, New York.
114. Koebbe, M. J., Golden, J. A., Bennett, G., and Finnell, R. H. (1999): Effects of prenatal cocaine exposure on embryonic expression of *Sonic Hedgehog*. *Teratology*, 59:12–19.
115. Koos, B. J. and Longo, L. (1976): Mercury toxicity in pregnant women, fetus and newborn infant. *Am. J. Obstet. Gynecol.*, 126:390–409.
116. Lamb, J. C. (1985): Reproductive toxicity testing: evaluating and developing new systems, *J. Am. Coll. Toxicol.*, 4:163–171.
117. Lang, P. L. (1988): *Embryo and Fetal Developmental Toxicity (Teratology) Control Data in the Charles River Crl:CD7BR Rat*. Charles River Laboratories, Wilmington, MA (database provided by Argus Research Laboratories, Inc., now Charles River Laboratories, Preclinical Services).
118. Lang, P. L. (1993): *Historical Control Data for Development and Reproductive Toxicity Studies Using the New Zealand White Rabbit*, compiled by MARTA (Middle Atlantic Reproduction and Teratology Association). Charles River Laboratories, Wilmington, MA (database provided by Argus Research Laboratories, Inc., now Charles River Laboratories, Preclinical Services).
119. Lang, P. L. (1993): *Historical Control Data for Development and Reproductive Toxicity Studies Using the Charles River Crl:CD7BR Rat*, compiled by MARTA (Middle Atlantic Reproduction and Teratology Association). Charles River Laboratories, Wilmington, MA (database provided by Argus Research Laboratories, Inc., now Charles River Laboratories, Preclinical Services).
120. Leavitt, W. W. (1995): The female reproductive system during pregnancy, parturition, and lactation. In: *Reproductive Toxicology*, 2nd ed., edited by R. J. Witorsch, pp. 45–72. Raven Press, New York.
121. Lemasters, G. K. and Pinney, S. M. (1989): Employment status as a confounder when assessing occupational exposures and spontaneous abortion. *J. Clin. Epidemiol.*, 42:975–981.
122. Lenz, W. (1961): Kindliche micebildungen nach medikament wahrend der draviditat? *Deutsch. Med. Wochenschr.*, 86:2555.
123. Lewis, E. M., Barnett, Jr., J. F., Freshwater, L., Hoberman, A. M., and Christian, M. S. (2002): Sexual maturation data for CRL Sprague–Dawley rats: criteria and confounding factors. *Drug Chem. Toxicol.*, 25:437–458.
124. McBride, W. G. (1961): Thalidomide and congenital abnormalities. *Lancet*, 2:1358.
125. McLaughlin, J. A. (1977): Prenatal exposure to diethylstilbestrol in mice: toxicological studies. *J. Toxicol. Environ. Health*, 2:527–537.
126. McNeilly, A. S. (1988): Suckling and the control of gonadotropin secretion. In: *The Physiology of Reproduction*, Vol. 2, edited by E. Knobil and J. D. Neill, pp. 2323–2349. Raven Press, New York.
127. Manson, J. M. and Kang, Y. J. (1994): Test methods for assessing female reproductive and developmental toxicology. In: *Principles and Methods of Toxicology*, 3rd ed., edited by A. W. Hayes, pp. 989–1037. Raven Press, New York.
128. Marr, M. C., Myers, C. B., George, J. D., and Price, C. J. (1988): Comparison of single and double staining for evaluation of skeletal development: the effects of ethylene glycol (EG) in CD rats. *Teratology*, 37:476.
129. Marr, M. C., Price, C. J., Myers, C. B., and Morrissey, E. (1992): Developmental states of the CD® (Sprague-Dawley) rat skeleton after maternal exposure to ethylene glycol. *Teratology*, 46: 169–181.
130. Mattison, D. R. and Ross, G. T. (1983): Laboratory methods for evaluating and predicting specific reproductive dysfunctions: oogenesis and ovulation. In: *Methods for Assessing the Effects of Chemicals on Reproductive Functions*, edited by V. B. Vouk and P. J. Sheehan, pp. 217–246. Wiley, New York.
131. Mattison, D. R., Kochar, D. M., and Rao, K. S. (1989): Criteria for identifying and listing substances shown to cause developmental toxicity under California's Proposition 65. *Reprod Toxicol.*, 3:3–12.
132. Mattison, D. R., Plowchalk, D. R. Meadows, MJ., Al-Juburi, A. Z., Gandy, J., and Malek, A. (1990): Reproductive toxicity: male and female reproductive systems as targets for chemical injury. *Med. Clin. North Am.*, 74:391–411.
133. Mattison, D. R., Working, P. K., Blazak, W. F., Hughes, C. L. jr., Killinger, J. M., Olive, D. L., Rao, K. S., Hanson, J. W., and Kochar, D. M. (1991): Reply to: when scientists become policy makers: shaping hazard identification under Proposition 65. *Reprod. Toxicol.*, 5:175–178.
134. Mattison, D. R. (1992): Protecting reproductive and developmental health under Proposition 65: public health approaches to knowledge, imperfect knowledge, and the absence of knowledge. *Reprod. Toxicol.*, 6:1–7.

135. Mazarakis, N. D., Edwards, A. D., and Mehmet, H. (1997): Apoptosis in neural development and disease. *Arch. Dis. Child.*, 77:F165–F170.
136. Mermelstein, R., Morrow, P. E., and Christian, M. S. (1994): Letter to the editor: organ or system overload and its regulatory implications. *J. Am. Coll. Toxicol.*, 13: 143–147.
137. Meyer-Bahlburg, H. F. L., Ehrhardt, A. A., Feldman, J. F., Rosen, L. R., Veridiano, N. P., and Zimmerman, I. (1985): Sexual activity level and sexual functioning in women prenatally exposed to diethylstilbestrol. *Psychosom. Med.*, 47:497–511.
138. Midwest Teratology Association. (1994): *Historical Control Project: External and Visceral Malformations, 1988–1992. Sprague–Dawley CD® Rats, New Zealand White Rabbits.*
139. Middleton, M. C., Milne, C. M., Moreland, D., and Hasmall, R. L. (1986): Ovulation in rats is delayed by a substituted triazole. *Toxicol. Appl. Pharmacol.*, 83:230–239.
140. Milligan, C. E. and Schwartz, L. M. (1997): Programmed cell death during animal development. *Br. Med. Bull.*, 52: 570–590.
141. Milne, C. M., Hasmall, R. L., Russell, A., Watson, S. C., Vaughan, Z., and Middleton, M. C. (1987): Reduced estradiol production by a substituted triazole results in delayed ovulation in rats. *Toxicol. Appl. Pharmacol.*, 90:426–435.
142. Ministry of Health and Welfare. (1975) *On Animal Experimental Methods for Testing the Effects of Drugs on Reproduction*, Notification No. 529 of the Pharmaceutical Affairs Bureau, Ministry of Health and Welfare, Tokyo, Japan.
143. Ministry of Health and Welfare. (1984): *Information on the Guidelines of Toxicity Studies Required for Applications for Approval to Manufacture (Import) Drugs*, Notification No. 118 of the Pharmaceutical Affairs Bureau, Ministry of Health and Welfare, Tokyo, Japan.
144. Moore, N. W., Adams, C. E., and Rowson, L. E. A. (1968): Developmental potential of single blastomeres of the rabbit egg. *J. Reprod. Fertil.*, 17:527–531.
145. Morrissey, R. E., Lamb, 4th, J. C., Morris, R. W., Chapin, R. E., Gulati, D. K., and Heindel, J. J. (1989): Results and evaluation of 48 continuous breeding reproduction studies conducted in mice. *Fund. Appl. Toxicol.*, 13:747–777.
146. Mosher, W. E. (1985): Reproductive impairments in the United States, 1965–1982. *Demography*, 22:415–430.
147. Mukherjee, A. B., Chan, M., Waite, R., Metzger, M. I., and Yaffee, S. J. (1975): Inhibition of RNA synthesis by acetyl salicylate and actinomycin-D during early development in the mouse. *Pediat. Res.*, 9:652–657.
148. Narayanan, V. (1997): Apoptosis in development and disease of the nervous system. 1. Naturally occurring cell death in the developing nervous system. *Pediatr. Neurol.*, 16:9–13.
149. Narotsky, M. G., Weller, E. A., Chinchilli, V. M., and Kavlock, R. J. (1995): Nonadditive developmental toxicity in mixtures of trichloroethylene, di(2-ethylhexyl) phthalate, and heptachlor in a 5 × 5 × 5 design. *Fund. Appl. Toxicol.*, 27:203–216.
150. Nakatsuka, T., Horimoto, M., Ito, M., Matsubara, Y., Akaike, M., and Ariyuki, F. (1997): Japan Pharmaceutical Manufacturers Association (JMPA) survey on background control data of developmental and reproductive toxicity studies in rats, rabbits and mice. *Cong. Anom.*, 37:47–138.
151. Naruse, I. and Keino, H. (1995): Apoptosis in the developing CNS. *Prog. Neurobiol.*, 47:135–155.
152. NCHS. (1980): *Births, Marriages, Divorces and Deaths for 1979: Monthly Vital Statistics Report*. National Center for Health Statistics, U.S. Department of Health, Education, and Welfare, Washington, D.C.
153. National Research Council. (1985): *Toxicity Testing: Strategies To Determine Needs and Priorities*. National Academy Press, Washington, D.C.
154. Neubert, D., Barrach, H. J., and Merker, H. J. (1980): Drug-induced damage to the embryo or fetus: molecular and multilateral approach to prenatal toxicology. *Curr. Top. Pathol.*, 69:241–331.
155. Newman, L. M. and Johnson, E. M. (1986): Teratogen-induced decrements of postnatal functional capacity. *J. Am. Coll. Toxicol.*, 5:517–524.
156. Niswander, K. R. and Gordon, M., Eds. (1972): *The Women and Their Pregnancies: The Collaborative Perinatal Study of the National Institute of Neurological Diseases and Stroke*. W.B. Saunders, Philadelphia, PA.
157. Numan, M. (1988): Maternal behavior. In: *The Physiology of Reproduction*, edited by E. Knobil and J. Neill, pp. 1569–1645. Raven Press, New York.
158. O'Connor, J. C., Cook. J. C., Craven, S. c., VanPelt, C. S., and Obourn, J. D. (1996): An *in vivo* battery for identifying endocrine modulators that are estrogenic or dopamine regulators. *Fund. Appl. Toxicol.*, 33:182–195.
159. Odum, J., Lefevre, P. A., Tittensor, S., Paton, D., Routledge, E. J., Beresfor, N. A., Sumpter, J. P., and Ashby, J. (1997): The rodent uterotrophic assay: critical protocol features, studies with nonyl phenols, and comparison with a yeast estrogenicity assay. *Regul. Toxicol. Pharmacol.*, 25: 176–188.
160. OECD. (1981): *Guidelines for Testing of Chemicals: Teratogenicity*, Section 4, No. 414. adopted May 12, 1981. Organization for Economic Cooperation and Development, Brussels, Belgium.
161. OECD. (1983): *OECD Guidelines for Testing of Chemicals: One-Generation Reproduction Toxicity*, Section 4, No. 415. adopted May 26, 1983. Organization for Economic Cooperation and Development, Brussels, Belgium.
162. OECD. (1983): *OECD Guidelines for Testing of Chemicals: Two-Generation Reproduction Toxicity Study*, Section 4, No. 416, adopted May 26, 1983. Organization for Economic Cooperation and Development, Brussels, Belgium.
163. OECD. (1995): *OECD Guidelines for Testing of Chemicals: Reproduction/Developmental Toxicity Screening Test*, Section 4, No. 421, adopted July 27, 1995. Organization for Economic Cooperation and Development, Brussels, Belgium.
164. OECD. (1996): *OECD Guidelines for Testing of Chemicals: Combined Repeated Dose Toxicity Study with the Reproduction/Developmental Toxicity Screening Test*, Section 4, No. 422, adopted March 22, 1996. Organization for Economic Cooperation and Development, Brussels, Belgium.
165. OECD. (1998): *The Revised OECD Principles of Good Laboratory Practices*, C(97)186/Final. Organization for Economic Cooperation and Development, Brussels, Belgium.

166. Palmer, A. K. and Ulbrich, B. C. (1997): The cult of culling. *Fund. Appl. Toxicol.*, 38:7–22.
167. Pan, H., Yin, C. and Van Dyke, T. (1997): Apoptosis and cancer mechanisms. *Cancer Surv.*, 29:305–327.
168. Pederson, T. and Peters, H. (1968): Proposal for a classification of oocytes and follicles in the mouse ovary. *J. Reprod. Fertil.*, 17:555–557.
169. Peters, P. W. J. (1977): Double staining of fetal skeletons for cartilage and bone. In: *Methods in Prenatal Toxicology*, edited by D. Neubert, H. J. Merker, and T. E. Kwasigroch, pp. 153–154. Georg Thieme, Stuttgart.
170. Pfaff, D. W. and Schwartz-Giblin, S. (1988): Cellular mechanisms of female reproductive behaviors. In: *The Physiology of Reproduction*, edited by E. Knobil and J. Neill, pp. 1987–1568. Raven Press, New York.
171. Plowchalk, D. R. and Mattison, D. R. (1991): Phosphoramide mustard is responsible for the ovarian toxicity of cyclophosphamide. *Toxicol. Appl. Pharmacol.*, 107: 472–481.
172. Plowchalk, D. R., Smith, B. J., and Mattison, D. R. (1993): Assessment of toxicity to the ovary using follicle quantitation and morphometrics. In: *Methods in Toxicology*. Vol. 3B. *Female Reproductive Toxicology*, edited by J. J. Heindel and R. E. Chapin, pp. 57–68. Academic Press, San Diego, CA.
173. PMA. (1981): *PMA Guidelines: Reproduction, Teratology and Pediatrics*. Pharmaceutical Manufacturers Association, Washington, D.C.
174. Reel, J. R., Lamb, 4th, J. C., and Neal, B. H. (1996): Survey and assessment of mammalian estrogen biological assays for hazard characterization. *Fund. Appl. Toxicol.*, 34: 288–305.
175. Richards, J. S. (1980): Maturation of ovarian follicles. *Physiol. Ref.*, 60:51–89.
176. Richards, J. S. and Bogvich, K. (1980): Development of gonadotropin receptors during follicular growth. In: *Functional Correlates of Hormone Receptors in Reproduction*, edited by Maresh, V. B. et al., pp. 223–244. Elsevier/North Holland, Amsterdam.
177. Ritter, E. J., Scott, Jr., W. J., Randall, J. L., and Ritter, J. M. (1987): Teratogenicity of di(2-ethylhexyl) phthalate, 2-ethylhexanol, 2-ethylhexanoic acid, and valproic acid, and potentiation by caffeine. *Teratology*, 35:41–46.
178. Rogers, J. M., Barbee, B. D., and Rehnberg, B. F. (1993): Critical periods of sensitivity for the developmental toxicity of inhaled methanol. *Teratology*, 47:395A.
179. Rogers, J. M. and Kavlock, R. J. (1996): Developmental toxicology. In: *Casarett & Doull's Toxicology: The Basic Science of Poisons*, 5th ed., edited by C. D. Klassen, pp. 301–331. McGraw-Hill, New York.
180. Salewski, E. (1964): Färbemethode zum makroskopischen Nachweis von Implantationsstellen am Uterus der Ratte. *Arch. Pathol. Exp. Pharmakol.*, 247:367.
181. Sanders, S. A. and Reinisch, J. M. (1985): Behavioral effects on humans of progesterone-related compounds during development and in the adult. In: *Current Topics in Neuroendocrinology: Actions of Progesterone on the Brain*, Vol. 5, edited by D. Ganten and D. Pfaff, pp. 175–205. Springer-Verlag, Berlin.
182. Sandler, M., Ed. (1985): *Mental Illness in Pregnancy and the Puerperium*. Oxford University Press, New York.
183. Sauders, Jr., J. W. (1966): Death in embryonic systems. *Science*, 154:604–612.
184. Schardein, J. L. (1992): *Chemically Induced Birth Defects*. 2nd ed., Marcel Dekker, New York.
185. Schmidt, R. R. (1984): Altered development of immunocompetence following prenatal or combined prenatal–postnatal insult: a timely review. *J. Am. Coll Toxicol.*, 3:57–72.
186. Schmidt, R. R. and Johnson, E. M. (1997): Principles of teratology. In: *Handbook of Developmental Toxicology*, edited by R. D. Hood, pp. 3–12. CRC Press, Boca Raton, FL.
187. Shenefelt, R. E. (1972): Morphogenesis of malformations in hamsters caused by retinoic acid: relation to dose and stage at treatment. *Teratology*, 5:103–118.
188. Shepard, T. H. (1995): *Catalog of Teratogenic Agents*, 8th ed. The John Hopkins University Press, Baltimore, MD.
189. Shield, M. A. and Mirkes, P. E. (1998): Apoptosis. In: *Handbook of Developmental Neurotoxicology*, edited by W. Slikker, Jr., and L. W. Chang, pp. 159–188. Academic Press, New York
190. Shuler, C. F. (1995): Programmed cell death and cell transformation in craniofacial development. *Crit. Rev. Oral Biol. Med.*, 6:202–217.
191. Siiteri, P. K. (1966). Placental endocrine biosynthesis during human pregnancy. *J. Clin. Endocrinol.*, 26:751–761.
192. Simpson, J. L. (1980): Genes, chromosomes and reproductive failure. *Fertil. Steril.*, 33:107–116.
193. Smith, B. J., Mattison, D. R., and Sipes, I. G. (1990): The role of epoxidation in 4-vinylcyclohexene-induced ovarian toxicity. *Toxicol. Appl. Pharmacol.*, 105:372–381.
194. Smith, B. J. Plowchalk, D. R., Sipes, I. G., and Mattison, D. R. (1991): Comparison of random and serial sections in assessment of ovarian toxicity. *Reprod. Toxicol.*, 5: 379–383.
195. Smith, C. A. (1947): Effects of maternal undernutrition upon the newborn infant in Holland (1944–1945). *J. Pediatr.*, 30:229.
196. Smith, R. (1999): The timing of birth. *Sci. Am.*, 280:68–75.
197. Snow, M. H. L. and Tam, P. P. L. (1979): Is compensatory growth a complicating factor in mouse teratology? *Nature*, 279:555–557.
198. Soloff, M. S. (1989): Endocrine control of parturition. In: *Biology of the Uterus*, 2nd ed., edited by R. M. Wynn and W. P. Jollie, pp. 559–607. Plenum Press, New York.
199. Speroff, L., Glass R. H., and Kase, N. G. (1983): Investigation of the infertile couple. In: *Clinical Gynecologic Endocrinology and Infertility*, 3rd ed., pp. 467–492. Williams & Wilkins, Baltimore, MD.
200. Spielmann, H. (1987). Analysis of embryotoxic effects in preimplantation embryos. In: *The Mammalian Preimplantation Embryo*, edited by B. D. Bavister, pp. 309–331. Plenum Press, New York.
201. Staples, R. E and Schnell, V. L. (1964): Refinement in rapid clearing technique in the KOH-alizarin red s method for fetal bone. *Stain Technol.*, 29:61–63.
202. Staples, R. E. (1974): Detection of visceral alterations in mammalian fetuses. *Teratology*, 9(3):A37–A38.

203. Staples, R. E. (1993): *Staples Technique for Evaluation of Fetal Soft Tissue*, course presented for Center for Professional Advancement, New Brunsick, NJ.
204. Stoker, T. E., Parks, L. G., Gray, E. G., and Cooper, R. L. (2000): Endocrine-disrupting chemicals: prepubertal exposures and effects on sexual maturation and thyroid function in the male rat. A focus on the EDSTAC recommendations. *Crit. Rev. Toxicol*., 30:297–252.
205. Stuckhardt, J. L. and Poppe, S. M. (1984): Fresh visceral examination of rat and rabbit fetuses used in teratogenicity testing. *Teratogen. Carcinogen. Mutagen*., 4:181–188.
206. Talamantes, F. and Ogren, L. (1988): The placenta as an endocrine organ: polypeptides. In: *The Physiology of Reproduction*, Vol. 2, edited by E. Knobil, E. and J. D. Neill, pp. 2093–2144. Raven Press, New York.
207. Tanimura, T., Kameyama, Y., Shiota, K., Tanaka, S., Matsumoto, N., and Mizutani, M. (1989): *Report on the Review of the Guidelines for Studies of the Effect of Drugs on Reproduction*, Notification No. 118. Pharmaceutical Affairs Bureau, Ministry of Health and Welfare, Tokyo, Japan.
208. Thomas, J. (1996): Toxic responses of the reproductive system. In: *Casarett & Doull's Toxicology: The Basic Science of Poisons*, 5th ed., edited by C. D. Klaassen, pp. 547–581. McGraw-Hill, New York.
209. Treolar, A. E., Boynton, R. E., Behn, B. G., and Brown, B. W. (1967): Variation of the human menstrual cycle through reproductive life. *Int. J. Fertil*. 12:77–126.
210. Tucker, H. A. (1988): Lactation and its hormonal control. In: *The Physiology of Reproduction*, Vol. 2, edited by E. Knobil and J. D. Neill, pp. 2235–2263, Raven Press, New York.
211. Tyl, R. W. and Marr, M. C. (1996): Developmental toxicity testing: methodology. In: *Handbook of Developmental Toxicology*, edited by R. Hood, pp. 175–225. CRC Press, Boca Raton, FL.
212. USEPA. (1996; updated 2007): *Federal Insecticide, Fungicide, and Rodenticide Act (FIFRA): Good Laboratory Practice Standards—Final Rule*, 40 CFR Part 160. U.S. Environmental Protection Agency, Washington, D.C.
213. USEPA. (1976): *Toxic Substances Control Act (TSCA): Good Laboratory Practice Standards, Final Rule*, 40 CFR Part 792. U.S. Environmental Protection Agency, Washington, D.C.
214. USEPA. (1991): Guidelines for developmental toxicity risk assessment. *Fed. Reg*., 56(234):63798–63826.
215. USEPA. (1996). *Guidelines for Reproductive Toxicity Risk Assessment*, NTIS PB No. PB97-100098. U.S. Environmental Protection Agency, Washington, D.C.
216. USEPA. (1985): *Hazard Evaluation Division Standard Evaluation Procedure: Teratology Studies*, EPA-540/9-85-018. Office of Pesticide Programs, U.S. Environmental Protection Agency, Washington, D.C.
217. USEPA. (1991): *Pesticide Assessment Guideline, Subdivision F, Hazard Evaluation: Human and Domestic Animals*. Addendum 10. *Neurotoxicity*. Health Effects Division, Office of Pesticide Programs. U.S. Environmental Protection Agency, Washington, D.C.
218. USEPA. (1993): *Health Effects Division Draft Standard Evaluation Procedure: Developmental Toxicity Studies*. Office of Prevention, Pesticides, and Toxic Substances (OPPTS), U.S. Environmental Protection Agency, Washington, D.C.
219. USEPA. 1998): *Health Effects Test Guidelines: Prenatal Developmental Toxicity Study*, OPPTS 870.3700. Office of Prevention, Pesticides, and Toxic Substances, U.S. Environmental Protection Agency, Washington, D.C.
220. USEPA. (1998): *Health Effects Test Guidelines: Reproduction and Fertility Effects*, OPPTS 870.3800. Office of Prevention, Pesticides, and Toxic Substances, U.S. Environmental Protection Agency, Washington, D.C.
221. USEPA. (1998): *Health Effects Test Guidelines: Developmental Neurotoxicity Study*, OPPTS 870.6300. Office of Prevention, Pesticides, and Toxic Substances, U.S. Environmental Protection Agency, Washington, D.C.
222. USEPA. (1985): Subpart E - Specific Organ/Tissue Toxicity, No. 798. 4900: Developmental Toxicity Study. 40 CFR Part 798 - Toxic Substances Control Act Test Guidelines: Final Rules. Fed. Reg. 50:39433–39434.
223. USEPA. (1985): Subpart E, Specific Organ/Tissue Toxicity, No. 798.4700, Reproduction and Fertility Effects; 40 CFR Part 798, Toxic Substances Control Act Test Guidelines: Final Rules. *Fed. Reg*., 50:39432–39433.
224. USEPA. (1997): *Toxic Substances Control Act Test Guidelines. (TSCA): Final Rule*, 40 CFR Part 799, OPPTS-42193, FRL-5719–5. U.S. Environmental Protection Agency, Washington, D.C.
225. USFDA. (1984): *Good Laboratory Practice Regulations: Final Rule*, 21 CFR Part 58. U.S. Food and Drug Administration, Washington, D.C.
226. USFDA. (1966): *Guidelines for Reproduction Studies for Safety Evaluation of Drugs for Human Use*. U.S. Food and Drug Administration, Washington, D.C.
227. USFDA. (2000): *Toxicological Principles for the Safety Assessment of Food Ingredients* (*Redbook 2000*). Office of Food Additive Safety, U.S. Food and Drug Administration, Washington, D.C. (http://www.cfsan.fda.gov/~redbook/red-toca.html).
228. USFDA (1993): Draft: toxicological principles for the safety assessment of direct food additives and color additives used in food (Redbook II). In: *Guidelines for Reproduction and Developmental Toxicity Studies*, pp. 123–134. Center for Food Safety and Applied Nutrition, U.S. Food and Drug Administration, Washington, D.C.
229. USFDA (1994): International Conference on Harmonization: guideline on detection of toxicity to reproduction for medicinal products. *Fed. Reg*., 59(183):48746–48752.
230. USFDA (1996): International Conference on Harmonization: guideline on detection of toxicity to reproduction for medicinal products, addendum on toxicity to male fertility. *Fed. Reg*., 61(67).
231. USFDA (1996): Food Quality Protection Act (FQPA), Public Law 104-170.
232. USFDA (1999): FDA/industry meeting on preclinical assessment of reproductive toxicity, June 24, 1999 (http://www. fdagov/cder/meeting/advcomm/repro-tox 62499.htm).
233. Vermes, I. and Haanen, C. (1994): Apoptosis in programmed cell death in health and disease. *Adv. Clin. Chem*., 31:177–246.
234. Warkany, J. and Nelson, R. C. (1940): Appearance of skeletal abnormalities in the offspring of rats reared on a deficient diet. *Science*, 92:383–384.

235. Warkany, J. and Nelson, R. C. (1941): Skeletal abnormalities in offspring of rats reared on deficient diets, *Anat. Rec.*, 79:83–100.
236 Warkany, J. (1965): Development of experimental mammalian teratology. In: *Teratology: Principles and Techniques*, edited by J. G. Wilson and J. Warkany, pp. 1–20. University of Chicago Press, Chicago, IL.
237. Williams, B. O., Morgenbesser, S. D., DePinho, R. A., and Jacks, T. (1994): Tumorigenic and developmental effects of combined germ-line mutations in *Rb* and *p*53. *Cold Spring Harbor Symp. Quant. Biol.*, 49:449–447.
238. Wilson, J. G. and J. Warkany, J., Eds. (1965): *Teratology: Principles and Techniques*, lectures and demonstrations given at the First Workshop in Teratology, University of Florida, February 2–8, 1964. University of Chicago Press, Chicago, IL.
239. Wilson, J. G. (1973): *Environment and Birth Defects*. Academic Press, New York.
240. Wilson, J. G. and Fraser, F. C., Eds. (1977): *Handbook of Teratology*, Vols. 1–4. Plenum Press, New York.
241. WHO. (1970): *Spontaneous and Induced Abortion*, WHO Tech. Rep. Ser. No. 461, World Health Organization, Geneva.
242. Young, W. C. (1961): The hormones and mating behavior. In: *Sex and Internal Secretions*, 3rd ed., Vol. 2, edited by W. C. Young, pp. 1173–1239. Williams & Wilkins, Baltimore, MD.
243. Zilva, S. S., Goulding, J., Brummond, J. C., and Coward, K. H. (1921): The relation of the fat-soluble factor to rickets and growth in pigs. *Biochem. J.*, 15:427–437.

# Notes

# 35 Hormone Assays and Endocrine Function

*Robert W. Kapp, Jr., and John A. Thomas*

**CONTENTS**

## INTRODUCTION

All physiological responses are essentially controlled by two primary monitoring/control systems: (1) the nervous system, which is electrical in nature and controls processes requiring immediate attention, such as breathing, skeletal muscular movements, and heartbeat, and (2) the endocrine system, which is chemical in nature and controls slower processes, such as cell growth and development, metabolism, and sexual function. Both systems are critical to the body's function, and they frequently work together to allow the body to function at its most optimal level.

The purpose of this chapter is examine the endocrine system of physiological control. The endocrine system basically consists of ductless glands or cells that secrete chemical messengers that are released into the bloodstream or local cellular fluid. Hormone receptors strategically placed throughout the body recognize and ultimately react to the presence of the hormone and affect a biological change. Hormones are usually very specific in nature, binding to a specific receptor in much the same way a key fits into a lock. There may be many keys present at the location of the locks, but only one fits and creates a hormonally activated receptor; therefore, even though hormones reach all parts of the body, only target cells with compatible receptors are equipped to fully respond. When a receptor and a hormone bind, the receptor carries out the instructions of the hormone by either altering the existing proteins of the cell or building new proteins. These actions create hormonally mediated chemical reactions throughout the body.

In conjunction with the nervous system, the endocrine system integrates many different processes that allow multicellular organisms to maintain homeostasis. It is estimated that that may be as many as 50 hormones in mammals [117]. Several drugs and chemicals are known to affect the endocrine system [108]. Perturbation of endocrine homeostasis often produces consequences that can lead to metabolic derangements, developmental abnormalities, and sexual or reproductive dysfunction. *Endocrine disruptors*, including environmental estrogens, is a contemporary term used to describe such effects. Endocrine toxicology involves the study of chemicals and drugs that disrupt endocrine processes, leading to either augmentation or inhibition of a physiologic response [111]. In a broad sense, endocrine toxicology encompasses reproductive toxicology (see Chapters 33 and 34) [128]. In addition, endocrine toxicology can also include metabolic derangements that result from toxic injury to nonendocrine systems; for example, renal or hepatic toxicity may alter the rate of hormone catabolism. This chapter reviews endocrine physiology, morphology, and endocrine function as it relates to toxicology. Another area of interest is endocrine pharmacology, which involves the therapeutic and diagnostic use of hormones and other nonhormonally related agents. It can also encompass endocrine toxicology, as many nonhormonal medicinal products possess side effects that can alter the biochemical activity of organs of internal secretion [105]. Importantly, molecular biology continues to provide a number of techniques that are very useful in endocrinology [26].

Although some endocrine organs appear to be more sensitive or vulnerable to toxicologic agents than others, hormone–target organ interrelations often lead to a substance causing multiple disruptions in the hormonal balance of the organism. It is not uncommon to witness chemically induced changes in gonadal function along with alterations in thyroid gland activity. In some species, chemically induced changes in sex steroids can affect pancreatic secretion of insulin. Chemically induced stress leading to increased secretion of glucocorticoids can also affect insulin secretion but more importantly can affect adrenocorticotropin (ACTH) levels and hence alter the pituitary–adrenal axis.

The thalidomide tragedy of the 1960s led to the formulation of toxicologic testing guidelines for the field of teratology beginning as early as 1962. No such guidelines have been invoked for the endocrine system. Pesticide-induced sterility (e.g., DBCP) first reported in 1977 [125] eventually led to concern about the deleterious actions of such substances in both male and female reproductive systems [28,56]. The inherent estrogenicity of *o,p′*-dichlorodiphenyldichloroethane (*o,p*-DDD) can affect the reproductive system, and concern has been expressed about the possible relation between diethylstilbestrol (DES) and the incidence of vaginal and cervical cancer [95]; thus, numerous examples of chemically induced changes in the endocrine system have been reported. Because many of these agents represent reproductive hazards in the workplace, the endocrine system falls prey to many such agents. Polyhalogenated biphenyls, dibenzodioxins, and dibenzofurans can interfere with thyroid hormone metabolism. Certain insecticides and pesticides can be toxic to the pancreatic β cell.

Agents toxic to the endocrine system reach a relatively small number of individuals at the site of manufacture, which can be controlled, but they gain access to entire communities when, for example, toxic wastes are released into the environment. Large geographic regions are exposed to natural goitrogens, such as resorcinol, that are derived from regional coal and shale deposits. Likewise, radioisotopes of iodine released during atmospheric nuclear weapons testing or reactor accidents can be disseminated over very large areas in detectable, if not clearly toxic, concentrations throughout the world [8].

Conceptually, the endocrine system is vulnerable to chemical toxicity at multiple points. Most, if not all tissues, are target organs of one or more hormones and are influenced in some fashion by endocrine substances. A vast number of critical biological processes, for example, are regulated by the endocrine system, including brain and nervous system development, development of the reproductive system and secondary sex characteristics at puberty, metabolism, and blood sugar maintenance. In fact, major physiological events from conception and infancy through adolescence to old age are generally mediated by large numbers of various hormones working together with the nervous system to achieve this immensely complex task. Because of structural similarity to certain hormones, some toxic substances interfere with hormone metabolism or subsequent actions at the receptor site. Still other toxic agents interfere directly with the glands that synthesize and secrete hormones. Compounding the chemical vulnerability of the endocrine system are the tier and feedback systems of many hormones.

The effects of thyroid hormone on a target organ are dependent on the quantity of thyroid hormone secreted by the thyroid. Thyroid secretion is regulated by the serum concentration of the tropic hormone, thyroid-stimulating hormone (TSH), which is secreted by the pituitary thyrotroph cell. The function of the thyrotroph cell is likewise influenced by another tropic hormone, thyrotropin-releasing hormone (TRH). Multilevel regulatory systems for thyroid and other hormones are counter-regulated in a negative system by the serum concentration of the hormone. A transient excess of circulating thyroid hormone would, for example, feed back upon the hypothalamus and pituitary thyrotroph to bring about a corrective reduction in levels of TRH and TSH secretion.

In effect, the glands producing the regulatory hormones are also target organs of the primary hormone; hence, every cell in the endocrine system is chemically linked to every other cell within the sphere of influence. Given this complexity of the endocrine system, it is not difficult to understand how the similar endocrine toxicities of relatively diverse compounds may, in fact, represent common manifestations of interventions of the compound at very different points in a hormone's pathways of production, regulation, and action. The effects can be widespread and can be life threatening if they do not function in synchrony.

Interestingly, chemically induced changes in the endocrine system are not always considered undesirable. Indeed, with what is now a very common type of birth control, synthetic steroids are therapeutically directed to inhibit pituitary gonadotropins, thereby providing a chemical method of birth control for millions of women. Chemical (or drug) suppression of target organ hormone secretion can also be therapeutically useful in such organs as the thyroid gland and the adrenal gland.

## HISTORICAL PERSPECTIVE

Claude Bernard first broached the subject of internal secretions as early as 1855 in his research on the pancreas. His concept of a stable *milieu intérieur*, or the internal environment, in spite of the constantly changing exterior environment was critical in understanding of maintenance of body functions [11]. Ivan Pavlov's early experiments on classical conditioning in dogs provided additional groundwork for the concept of internal control of bodily functions; however, the digestive processes described by Pavlov were thought to be controlled solely by the nervous system. In 1902, William Bayliss and Ernest Starling were able to identify secretin in experiments in which they were able to chemically increase pancreatic secretions in dogs. They were first to use the term *hormone* (from the Greek word *horman*, meaning "to set in motion") to describe secretin as a chemical messenger [126]. Their experiments revealed that secretin could cause changes in physiology at a distant site from the site of entry after being transported through the circulatory system.

In 1915, Walter Cannon revealed the connection between the endocrine glands and emotions such as fear, pain, hunger, and rage [15]. Later that year, Edward Kendall isolated thyroxin, which is the active component of the thyroid and for which he was awarded the Nobel Prize in 1950 [59]. Frederick Banting, J.J. McLeod, and Charles Best were awarded the Nobel Prize in 1923 for their work in 1921 on isolating insulin in the control of carbohydrate metabolism [7]. In 1953, Vincent du Vigneaud synthesized oxytocin and other posterior pituitary hormones for which he was awarded the Nobel Prize in 1955 [31,49,84]. In 1959, Rosalyn Yalow and Solomon Berson first published a method to detect minute amounts of hormones in humans with the radioimmunoassay (RIA) [129]. They were awarded the Nobel Prize in 1977 for their work on the development and application of radioimmunoassays of peptide hormones [84]. Georges J.F. Köhler and César Milstein published work in 1975 [63] on theories concerning the specificity in development and control of the immune system and the discovery of the principle for production of monoclonal antibodies and were awarded the Nobel Prize for this groundbreaking work in 1984 [84]. The nervous and chemical control of bodily functions works intimately in concert with the immune system to maintain a delicate internal balance that is collectively termed *homeostasis*. Momentous events in scientific history have led to our current understanding of this field. The importance of these findings to mankind cannot be minimized in light of this worldwide recognition of their importance to the health and welfare of humans. The focus of this chapter is to review testing assays in an effort to understand the chemical control exerted by the endocrine system.

## GENERAL PRINCIPLES

The basic foundations of the endocrine system include glands, the hormones they produce, and the receptors at the site of action. The fundamental concept of the endocrine system is that endocrine cells release a hormone that is transported to a receptor site in the target tissue, where the hormone binds and subsequently exerts its biologic effect. In a traditional sense, endocrine systems have been portrayed as encompassing those tissues that release a hormone that is transported through the bloodstream to a target tissue, resulting in an effect at some point distant from the original tissue. This has been found to be a simplistic definition, as many variations have been discovered. New techniques in tissue cell culture and molecular biology have permitted the identification of several intercellular signaling pathways that do not export hormones in the traditional way via the general circulation to target tissue; for example, paracrine effects are produced when an effector cell releases a hormone that acts on adjacent target cells to produce a local effect. Many examples of paracrine systems can be found among various growth factors and inflammatory mediators (such as arachadonic acid metabolites, complement factors, cytokines, clotting factors, and somatostatin). Autocrine systems occur when a particular cell type releases a hormone that can act on the same cell to augment a particular response. Examples of autocrine systems include the cytokine interleukin-1 in monocytes and peripheral membrane protein in T-cells.

Several classification schemes have been developed for hormones. One classifies hormones according to their source, another classifies them according to their solubility characteristics in water or fat, and yet another classifies them according to their chemical composition [9,89]. Examining hormones based on their chemical composition also leads to several sets of data. One grouping that is routinely utilized in the endocrine literature divides hormones into one of six general chemical categories (see Table 35.1). These include biogenic hormones, polypeptide hormones, protein hormones, steroid/sterol hormones, thyroid hormones, and fatty-acid-derived hormones.

**TABLE 35.1**
**Categories of Hormones**

Biogenic amines (e.g., epinephrine)
Polypeptides (e.g., thyroid-releasing hormone)
Proteins (e.g., insulin, growth hormone)
Steroids (e.g., estrogens, androgens)
Thyroid hormones (e.g., thyroxine)
Fatty acids (e.g., prostaglandins)

*Source:* Adapted from Gardner, D.G. and Nissenson, R. A., in *Basic and Clinical Endocrinology*, 7th ed., Greenspan, F.S. and Gardner, D.G., Eds., McGraw-Hill, New York, 2004, pp. 61–84; Porterfield, S.P., Ed., *Endocrine Physiology*, 2nd ed., Mosby, St. Louis, MO, 2001.

Amine hormones include catecholamines (notably, epinephrine, norepinephrine, and dopamine) as well as serotonin and derivatives of tryptophan, tyrosine, or glutamic acid. They are hydrophilic in nature and are related to compounds found in the nervous system that can exert local effects in tissues such as the gastrointestinal tract. Biogenic amines are small, biologically active, modified, single amino acids that are stored in vesicles prior to their release into the circulation. They act primarily as neurotransmitters and are capable of affecting mental functioning. In contrast to most polypeptide hormones, stores of biogenic amines can be quickly restored by rapid synthesis.

Polypeptide hormones are composed of short amino acid chains; protein hormones are composed of many amino acids and polypeptides together in a single molecule and are generally hydrophilic in nature. These types of hormones can range in length from three amino acids (as in the case of thyrotropin-releasing hormone) to several hundred amino acids (e.g., growth hormone). They can be composed of two or more subunits (e.g., gonadotropins) or may be linked by disulfide bonds (e.g., insulin). These hormones are synthesized in the endoplasmic reticulum and subsequently transferred to the Golgi apparatus for inclusion in secretory vesicles. Other posttranslational modifications, such as glycosylation, may occur that can affect biologic activity. Following synthesis in an endocrine cell, hormones may be stored in vesicles prior to stimulation of the endocrine system. Patterns of peptide and protein hormone secretion can be either pulsatile (regulated) or basal (constitutive).

Steroid and sterol hormones are derived from cholesterol, contain a cyclopentanoperhydrophenanthrene ring (four ring) structure, and are hydrophobic in nature. Steroid hormones are generally synthesized in the adrenal cortex, the placenta, and the gonads and are further characterized by being immediately secreted by these glands. Because they are fat soluble, these hormones can readily cross cell membranes and have intracellular receptors. When they have entered the cytoplasm, these steroid hormones bind to a specific receptor (e.g., $\alpha$ and $\beta$ estrogen receptors), a large metalloprotein. Further binding occurs that permits the steroid to enter the nucleus. In the nucleus, the steroid–receptor ligand complex binds to specific DNA sequences and induces transcription of its target genes. Examples of steroid hormones include the sex steroid hormones (e.g., testosterone, estrogen, and progesterone), corticosteroids, mineralocorticoids, vitamin D, and retinoic acid. Recent evidence suggests that megalin, a member of the low-density lipoprotein receptor superfamily of endocytic proteins, is an important facilitator of steroid entry into cells [19].

Thyroid hormones are an intermediate-molecular-sized group of compounds that fall in between biogenic amines and short polypeptides. Thyroxine is produced by attaching iodine atoms to the ring structures of tyrosine molecules. Thyroxine contains four iodine atoms. They are composed of two iodinated tyrosine residues, which are converted to triiodothyronine by deiodinases to enhance biologic activity.

Fatty acid hormones or eicosanoids are generally derived from polyunsaturated fatty acids such as linoleic acid and phospholipids, with the most prominent of these being arachadonic acid. They are synthesized throughout the body, and their primary effects are paracrine and autocrine in nature through both surface and nuclear receptors [89]. These hormones are effective at minute concentrations and are typically active for 5 to 10 seconds. These hormones include prostaglandins, prostacyclin, leukotrienes and thromboxanes, lipoxins, isoprostanoids, and hydroxylated fatty acids.

Transport of hormones from endocrine tissues to target tissues frequently involves carrier plasma proteins (e.g., sex-hormone-binding protein [SHBP]) that bind the hormones with high affinity and specificity while the hormone is transported within the circulatory system. Although many of the carrier proteins possess high specificity and affinity, they often possess low capacity, so the availability of binding sites is limited. Other nonspecific carrier proteins, such as albumin, can bind hormones with a high capacity but possess low specificity and characteristics unlike biogenic amines, steroid, thyroid, and some polypeptide hormones, which have specific binding proteins. Some hormones such as protein hormones are transported free in the blood, while steroid and thyroid hormones are transported bound to plasma proteins.

After a hormone is transported to the target tissue, it must then interact with a receptor within the target tissue to exert a biologic effect. An important concept of hormone action is that only the free, unbound hormone can interact with its receptor to exert a biologic effect in the particular target tissue; therefore, the hormone must become unbound from any carrier protein before it can interact with a receptor. Hormonal signal transduction occurs when the hormone interacts with the target tissue receptor, which in turn produces an intracellular change in the target tissue. Mechanisms of signal transduction depend on whether the target tissue receptor is bound to the cell membrane surface or is present in the cytoplasm.

To permit exposure of the required amount of hormone, a delicate equilibrium exists among the bound hormone ($B$), the plasma protein ($P$), and unbound or free hormone ($F$), which is expressed in the following equation [89]:

$$F \times P = [B] = ([F] \times [P]) \div [B]$$

Not only is the free hormone the critical moiety for activation, but it is also critical for deactivation in feedback control. Signal transduction by cell-membrane-bound receptors is usually either via intracellular second messengers (e.g., cyclic AMP, calcium, or phosphatidylinositol metabolites) or through mechanisms such as phosphorylation of serine, threonine, or tyrosine residues of intracellular kinases and other enzymes [72]. Most polypeptide hormones interact with cell membrane surface receptors, although there is some evidence to suggest that hormonal receptor internalization may play a role in some processes. In contrast, steroid and thyroid hormones interact with intracellular receptors. Transport into the cell may be aided by cell membrane transporters (e.g., megalin), and once inside the cell the steroid or thyroid hormone receptor complex is transported to the nucleus, where it may modulate gene expression by binding to certain DNA regulatory sequences.

## CONTROL OF ENDOCRINE ACTIVITY

Regulation of hormonal activity is critical to all biological systems in their quest for biological homeostasis. Overall, the endocrine system utilizes two major control systems to respond to physiological changes: (1) automatic internal controls from the various concentrations of the hormones themselves, and (2) nervous system override of the self-regulating chemical system. The primary function of the nervous system override is to respond to stimuli (i.e., fight or flight situations) and produce a hormonal readjustment. The automatic and continuous self-regulation is a complex physiologic hierarchy of chemically mediated controls that modulates the homeostasis of the endocrine system through a series of feedback loops.

The maintenance of biological systems involves negative feedback loops where the output of a hormone controls the input in a simple inverse relationship. As the concentration of the hormone increases in the system, the amount of input into the system is decreased, in much the same way as a thermostat regulates room temperature. As the temperature increases, the thermostat turns off the

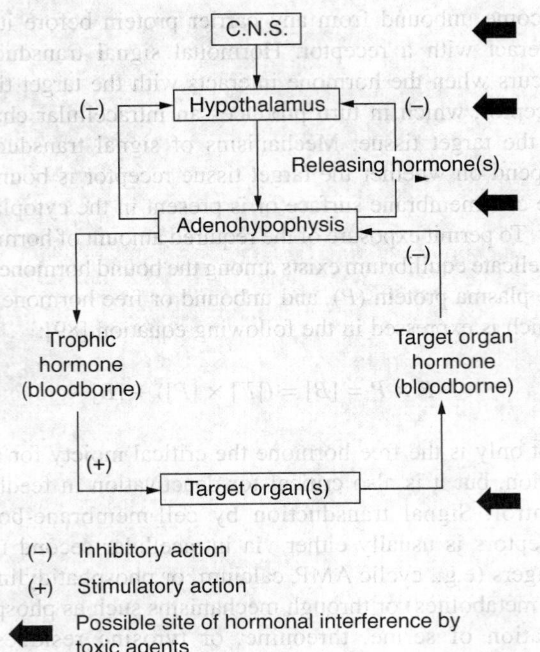


**FIGURE 35.1** Relationship between adenohypophyseal–hypothalmic axis and hormone target organs.

heating unit until the temperature drops to a predetermined point, whereupon the signal is sent to the heating unit to increase heat once again. Such would be an example of simple hormonal regulation where the involvement includes a single endocrine gland. Complex hormonal regulation is another level of control involving the anterior pituitary (adenohypophysis) as well as the target receptors. The anterior pituitary is frequently referred to as the *master gland* because of its control over a number of other glands and target cells. In this case, one endocrine gland controls the hormone release of another which ultimately regulates another distant target. An example of this type of control is the mechanism involving the pituitary gland, the adrenal cortex, and the gonads. The anterior pituitary releases hormones that stimulate the adrenal cortex, which, in turn, releases hormones that effect a change at the gonads. The secretory releases of both the adrenal cortex and the gonads feed back to decrease the hormonal secretion of the anterior pituitary. The overall effect of negative feedback loops is that hormone levels are maintained within the narrow range the body needs to maintain normal homeostasis. A negative feedback loop inhibits an endocrine pathway, whereas a positive feedback loop enhances or augments an endocrine response (see Figure 35.1). Positive feedback relationships, where hormone levels are actually increased by elevating levels of a hormone (i.e., ovulation), are also operants.

The concentration of a hormone at the target cells is generally determined by three factors: (1) the rate of hormone production, (2) the rate of hormone delivery, and (3) the rate of hormone elimination. The rate of hormone production is affected by many factors, and is primarily controlled by feedback loops. Feedback loops can be further modulated by other endocrine systems that are pulsatile, cyclical, or stimulated through other mechanisms. The integration of various feedback loops gives rise to a complex cascade of endocrine responses to particular stimuli. The rate of hormone delivery is affected by the rate of blood flow to the target receptor. The rate of elimination is affected by two factors: (1) the body's ability to metabolize and excrete the hormone from the bloodstream and through renal elimination, and (2) the physiological rate of degradation of the hormone. Some hormones have short biological half-lives, and others have long biological half-lives. Protein binding can affect a hormone residence time in the blood.

## ENDOCRINE DISRUPTION

Several drugs and chemicals are known to affect the endocrine system [41,108]. Perturbation of endocrine homeostasis often produces consequences that can lead to metabolic derangements, developmental abnormalities, and sexual or reproductive dysfunction. The U.S. Environmental Protection Agency (EPA) has defined an endocrine disruptor as "an exogenous agent that interferes with the production, release, transport, metabolism, binding, action, or elimination of natural hormones in the body responsible for the maintenance of homeostasis and the regulation of developmental processes" [57,118]. Endocrine disruption has become a subject of intense scientific research, in large part due to the publication of *Our Stolen Future* in 1996 [23]. The book and its findings became sensationalized and politicized, which may or may not have been warranted; however, with the recognition of the effects of estrogen-like environmental contaminants, logically, the concern extends to many types of contaminants that might in some way effect unwanted changes to the endocrine system. Such exposures might result in ill-timed maturation, development, growth, or regulation, resulting in the reduced ability of an organism to cope with external changes or birth defects if the changes occurred during gestation. Indeed, the EPA has indicated that at least four categories of adverse effects might be linked to endocrine disruptors, including cancer, immunological effects, neurological effects, and reproductive and developmental effects involving the adrenals, thyroid, pituitary, and gonads [116]. An Endocrine Disruptor Screening and Testing Advisory Committee (EDSTAC) was charged with developing "a screening program, using appropriate validated test systems and other scientifically relevant information, to determine whether certain substances may have an effect in humans that is similar to an effect produced by a naturally occurring estrogen, or other such endocrine effect as the Administrator may designate" [115,118,119].

Evidence indicates that a number of chemicals can perturb the endocrine systems of animals in a laboratory setting, where one can precisely measure the changes in a controlled environment. There is also substantial evidence that birds [12], dolphins [2], alligators [44], and other wildlife [4,37,77] exposed to high levels of various environmental contaminants manifest adverse developmental and reproductive effects. Although the logical extension of this concept is that it could produce similar effects in humans, the data are not clear and are scientifically controversial. [57,115]. Because the effects on humans are so poorly understood, the EPA is performing due diligence in assessing these potential relationships on the environment and as it relates to humans with the Endocrine Disruptor Screening Program. The EPA states: "The seriousness of the endocrine disruptor hypothesis and the many scientific uncertainties associated with the issue are sufficient to warrant a coordinated federal research effort" [114]. Future developments in screening tests for endocrine disruption will lead to additional hormone assays being added to the standard hormonal assays.

## BIOCHEMICAL ASSESSMENT

Many reliable endocrine methods are available to the toxicologist [110]. Molecular genetics has become increasingly important in reproductive endocrinology [69]. The diagnosis of endocrine disorders as well as an understanding of the mechanisms of hormonal action were significantly advanced when methods became available to measure hormones in the blood [97]. Until about the 1960s, the measurement of hormones usually involved *in vivo* bioassays. The discovery of monoclonal antibody techniques revolutionized the measurement of hormones. Subsequently, antibody-based competitive protein-binding assays (i.e., radioimmunoassays) were perfected with the availability of radioactive, or tracer, hormones.

Evaluation of endocrine function is an important aspect in determining the mechanism of endocrine toxicity [9]. Measurement of hormone and hormonal metabolite levels in the blood or urine can reveal important information regarding the site of endocrine toxicity. These sites can include the endocrine tissue itself (e.g., the site of hormone synthesis and release), the endocrine target tissue (e.g., the hormone receptor and intracellular signaling pathways), and the mechanism of transport from the endocrine tissue to the target cell receptor to exert an endocrine effect (e.g., the hormone carrier proteins). Furthermore, toxic actions on feedback control mechanisms can augment or inhibit endocrine responses.

There are limitations to the biochemical assessment of the endocrine system. For example, factors that may have modulatory effects on endocrine target cells cannot always be identified by simple evaluation of blood hormone levels. For this reason, biochemical assessment of the endocrine system frequently offers a macroscopic survey of toxicology mechanisms. Determination of pathophysiology and molecular endocrine toxicity is limited due to the complex interplay of the numerous and frequently redundant regulatory and counterregulatory feedback loops [36]. Biochemical assessment of physiologic conditions can be confounded by several external factors, such as stress, extreme temperatures, or changes in the diurnal variation of sleep patterns. Similarly, the biochemical assessment can be confounded by internal factors such as the female menstrual cycle in both the reproductive and nonreproductive organs. Finally, the sensitivity and specificity of hormonal testing vary greatly among different hormonal systems.

### *IN VIVO* BIOASSAYS

An *in vivo* bioassay determines the biological activity of a hormone by noting its effect on a live animal or isolated organ preparation compared to a known standard preparation. Bioassays provide a means of assessing endocrine status, and, although more accurate measurement of hormonal levels can now be achieved with radioimmunoassay, bioassays are sometimes useful when radioimmunoassay is unable to distinguish active from inactive hormonal metabolites or precursors [103]. Historically, bioassays have used hypoglycemia to measure insulin, bone growth to measure growth hormone, and ovarian weight change to measure gonadotropins. The main drawbacks to *in vivo* bioassays are that they are frequently insensitive and nonspecific, and comparisons to the dose–response curve for a standard preparation are imprecise.

### *IN VITRO* BIOASSAYS

*In vitro* bioassays employ endocrine-responsive tissue cell culture lines that can assess the amount of biologically active hormone in sera. The hormonal activity of the sera can be determined by measuring a cellular response to a hormone. Classically, hormonally induced changes in adenylate cyclase activity (with changes in cAMP levels) have been used, but more recently changes in intracellular calcium levels, phosphoinositol metabolites, and protein phosphorylation have been used to assess hormonal activity. Other *in vitro* bioassays can examine changes more distal to the receptor and signal transduction mechanisms and note changes in enzymatic activity or steroidogenesis. Finally, some *in vitro* bioassays can assess the mitogenic responses to a given hormone. A shortcoming to *in vitro* bioassays is that there may be coexisting stimulatory or inhibitory substances in the sera that can confound the observed response. For this reason, it is difficult to design a valid bioassay without using fractionated serum, and full characterization of an endocrine-responsive cell line is required.

## Radioimmunoassay

Measurement of endocrine values was revolutionized in 1959 when Yalow and Berson published their work on the radioimmunoassay (RIA) to detect nanomolar concentrations of hormones in human subjects [129]. Immunoassays are based on the fact that specific proteins can distinguish three-dimensional structures at the molecular level. When specific proteins are produced by a biological organism, their specificity can be discriminated and tightly bound to those molecules in a highly complex mixture providing a means to measure these proteins in a controlled situation. Radioimmunoassays use radioactive isotopes to identify these specific proteins. Classic bioassays were vastly surpassed in sensitivity, specificity, and facilitation by RIA. In addition, RIA allows measurement of biological materials not previously detectable by chromatographic or spectrophotometric techniques [66]. RIAs can be used to measure hormones that cannot be radiolabeled to detectable levels *in vivo*. They are also used for hormones that cannot fix complement when bound to antibodies, or they can be used to identify cross-reacting antigens that compete and bind with the antibody.

Generally, immunoassays can be divided into two categories: competitive inhibition and noncompetitive inhibition. Competitive assays use a single specific antibody type that is fixed to a surface and a corresponding analog of the protein to carry the label. The protein in the sample competes with the labeled analog for binding positions on the antibody. Once the unbound analog is separated, the amount of label remaining is inversely related to the amount of bound protein. Competitive inhibition of radiolabeled hormone antibody binding by unlabeled hormone (either as a standard or an unknown mixture) is the principle of most RIAs. A standard curve for measuring antigen (hormone) binding to antibody is constructed by placing known amounts of radiolabeled antigen and the antibody into a set of test tubes. Varying amounts of unlabeled antigen are added to the test tubes. Antigen–antibody complexes are separated from the antigen, and the amount of radioactivity from each sample is measured to detect how much unlabeled antigen is bound to the antibody. Smaller amounts of radiolabeled antigen–antibody complexes are present in the fractions containing higher amounts of unlabeled antigen. Usually, a standard curve is constructed that measures the percent of radiolabeled antigen bound with the concentration of unlabeled antigen present.

Noncompetitive assays utilize two specific antibodies to sandwich the protein being examined. One antibody (termed the *capture protein*) is immobilized to a surface and a second (termed the *label protein*) carries the label. In the assay, the protein being examined is bound simultaneously by both the capture and label proteins. At completion of the assay, the unbound label protein is separated and measured; the remaining label is measured and is directly proportional to the protein concentration in the sample. Sandwich assays are generally limited to those proteins of sufficient size to be able to bind two materials simultaneously, typically proteins and microorganisms. Competitive assays are compatible with a wide variety of proteins and are used for the majority of low-molecular-weight proteins. Although several methods exist for the separation of antigen–antibody complexes, two methods are most commonly employed in RIAs. The first, the double-antibody technique, precipitates antigen–antibody complexes out of solution by utilizing a second antibody, which binds to the first antibody. Although other means of antigen–antibody precipitation exist, they can sometimes chemically alter antigen–antibody binding properties. The drawback to the double-antibody technique is its expense, which makes this technique uneconomical for RIA screening procedures.

Another commonly used method is the dextran-coated activated charcoal technique. Addition of dextran-coated activated charcoal to the sample followed by immediate centrifugation absorbs free antigen and leaves antigen–antibody complexes in the supernatant fraction. Although it is economical, a drawback to this technique is that it works best only when the molecular weight of the antigen is 30 kDa or less. Also, sufficient carrier protein must be present to prevent adsorption of unbound antibody.

Once a standard curve has been constructed, the RIA can determine the concentration of hormone in a sample (usually plasma or urine). The values of hormone levels are usually accurate using the RIA, but certain factors (e.g., pH or ionic strength) can affect antigen binding to the antibody; thus, controlled and similar conditions must be used when assessing the standard and the sample.

Difficulties with RIAs include a lack of specificity. This problem is usually due to nonspecific cross-reactivity of the antibody. Another complication is that, because the assays require the use of radioactive materials by definition, the laboratory must be regulated by the Nuclear Regulatory Commission (NRC). Despite the more complex and involved RIA and monoclonal antibody methodologies, they are of immense value for measuring various tropic hormones. RIA represents an analytical approach of great sensitivity, and such techniques have been applied to more than 200 biological substances, many of which cannot be assessed by other techniques. Unlike bioassays that often require large amounts of tissue (or blood), the greater sensitivity of the RIAs or monoclonal antibody techniques allows the use of smaller samples of biological fluids. Some of these RIA methodologies are more useful than others, and their usefulness to some extent depends on the degree of hormonal cross-reactions or, in the case of monoclonal antibody methods, their degree of sensitivity.

## Enzyme-Linked Immunosorbent Assay

A second generation of radioimmunoassay technology was developed that were procedurally less cumbersome. The enzyme-linked immunosorbent assay (ELISA) is comparable to the immunoradiometric assay except that an enzyme tag is attached to the antibody instead of a radioactive label. This change eliminated the complex handling and disposal issues involved with radioisotopes as mandated by the NRC. ELISAs produce an end product that can be assessed with a spectrophotometer. The hormone is bound to the enzyme-labeled antibody, and the excess antibody is removed for immunoradiometric assays. After excess antibody has been removed or the second antibody containing the enzyme has been added (two-site assay), the substrate and cofactors necessary are added to visualize and record enzyme activity. The level of hormone present is directly related to the level of enzymatic activity. The sensitivity of the ELISAs can be enhanced by increasing the incubation time for producing substrate. Sometimes the substrate formed may yield a color change so that detection of the hormone being measured can be determined visually.

## Immunoradiometric Assays

Immunoradiometric assays (IRMAs) are like RIAs in that a radiolabeled substance is used in an antibody–antigen reaction. The radioactive label, however, is attached to the antibody instead of the hormone. This assay is based upon the reversible and noncovalent binding of an antigen by a specific antibody labeled with a radioisotope. Further, excess of antibody, rather than limited quantity, is present in the assay. All of the unknown antigen becomes bound in IRMA, rather than just a portion as in RIA. IRMA assays are more sensitive. In the one-site assay, the excess antibody that is not bound to the sample is removed by addition of a precipitating binder. In the two-site assay (i.e., sandwich technique), a hormone with at least two antibody-binding sites is adsorbed onto a solid phase, to which one of the antibodies is firmly attached (either the walls of the assay tube itself or beads that are added to the patient sample in assay buffer). After binding to the antibody is completed, a second antibody labeled with $^{125}I$ is added to the assay. This antibody reacts with the second antibody-binding site to form the sandwich, which is composed of antibody-hormone-labeled antibody. In contrast to RIA and similar competitive protein-binding assays, the amount of hormone present is directly proportional to the amount of radioactivity measured in the assay [10].

## Enzyme-Multiplied Immunoassay Technique

Using enzyme-multiplied immunoassay technique (EMIT) assays, enzyme tags replace the radiolabels; however, the antibody binding alters the enzyme characteristics, allowing for measurement of hormone without separating the bound and free components (i.e., homogeneous assay). EMIT assays are used to monitor urine for drugs, but because of a lack of sensitivity they have not been used to assess hormones. No extraction is required, and the assay can be completed within a few minutes [33]. The enzyme is attached to the hormone or drug being tested. This enzyme-labeled antigen is incubated with the sample and with antibody to the hormone or drug. Binding of the antibody to the enzyme-linked hormone either physically blocks the active site of the enzyme or changes the protein conformation so the enzyme is no longer active. When antibody binding has occurred, the enzyme substrate and cofactor are added and the enzyme activity can be measured. If the sample contains hormone or drug, it will compete with enzyme-linked hormones for antibody binding. The enzyme will not be blocked by the antibody, and more enzyme activity will be measurable.

## Monoclonal Antibodies

Many hormones can now be assessed using monoclonal antibody (mAb) techniques. Because hormones possess a number of antigenic determinants, it is possible to produce antisera containing a variety of polyclonal antibodies that recognize and bind many parts of the hormone. Antibodies against hormones have been used in many types of RIAs and radioreceptor assays, but polyclonal antisera can create some nonspecificity problems such as cross-reactivity and variation in binding affinity. It is oftentimes desirable, therefore, to produce a group of antibodies that selectively bind to a specific region of the hormone (i.e., antigenic determinant). In the past, investigators produced antisera to antigenic determinants of the hormone by cleaving the hormone and immunizing an animal with the fragment of the hormone containing the antigenic determinant of interest (e.g., ACTH immunization with the 24-amino-acid *N*-terminal end of the hormone). This approach solved some problems with cross-reactivity of antisera with other similar antigenic determinants, but problems were still associated with the heterogeneous collection of antibodies found in polyclonal antisera.

In 1975, Köhler and Milstein [63] first discovered monoclonal antibodies and devised an immunological method for producing large quantities of monoclonal antibodies that could be targeted to specific proteins. The production of monoclonal antibodies offered investigators a homogeneous collection of antibodies that could selectively bind to a specific antigenic determinant with the same affinity. In addition to protein isolation and diagnostic techniques, monoclonal antibodies have contributed greatly to the development of RIAs (see Figure 35.2). Because propagation of a single antibody-producing cell *in vitro* does not occur, Köhler and Milstein used spleen cells from an immunized mouse and fused them with myeloma cells (malignant lymphocytes) in the presence

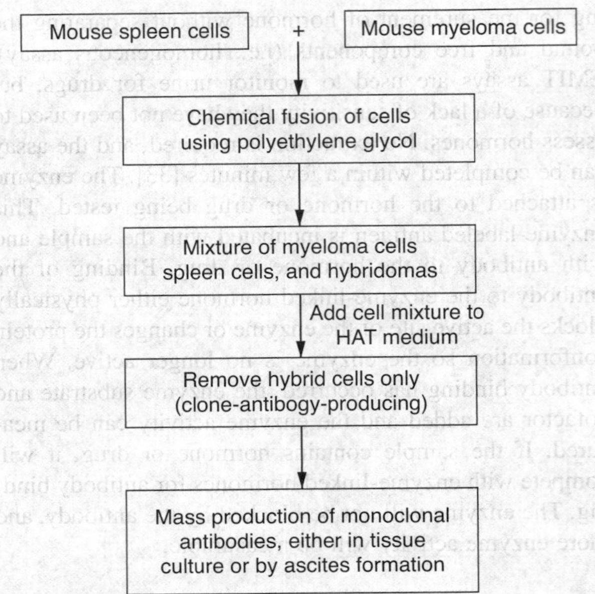


**FIGURE 35.2** Diagram for producing monoclonal antibodies.

of a reagent that causes the cells to fuse (e.g., Sendai virus of polythylene glycol). The fused cells or hybridomas share characteristics of their antibody-producing cells from the spleen by continuously synthesizing identical antibodies and could be infinitely propagated *in vitro* using tissue culture techniques.

Unfused spleen cells do not survive in tissue culture media. To separate unfused myeloma cells from hybridomas, the cells are placed in hypoxanthine, aminopterin, and thymidine (HAT) medium. Myeloma cells used in the production of monoclonal antibodies lack the enzyme hypoxanthine-guanine-phosphoribosyltransferase (HGPRT), whereas hybridomas contain the enzyme (contributed by fused spleen cells). Because the main pathway of DNA synthesis is blocked by aminopterin, only cells containing HGPRT can utilize hypoxanthine and synthesize DNA to propagate. Thus, unfused myeloma cells are eliminated in HAT medium because they lack HGPRT. The hybridomas propagate to further isolate and separate variant myeloma cells that can overcome the aminopterin block (using thymidine kinase and exogenous thymidine), the cells are subjected to 5-bromodeoxyuridine, a pyrimidine analog that kills cells using the pathway (see Figure 35.2). When the hybridomas have been isolated from unfused cells, they are separated into individual colonies and the type of antibody they produce is characterized. Identification of a hybridoma clone that is producing a specific antibody to the antigen of interest allows for harvesting large quantities of the monoclonal antibody.

Although monoclonal antibodies offer a highly sensitive, specific method for detecting antigen, sometimes increasing monoclonal antibody specificity compromises affinity of the antibody for the antigen. In addition, complement fixation is usually reduced and costs are high for preparing and maintaining hybridomas that produce monoclonal antibodies (Table 35.2). Monoclonal antibody techniques do provide a means of producing a specific antibody for binding antigen; this technique is useful for studying protein structure relations (or alterations) and has been used for devising specific RIAs.

**TABLE 35.2**
**Advantages and Disadvantages of Monoclonal Antibodies Compared to Polyclonal Antisera**

| Advantages | Disadvantages |
|---|---|
| Sensitivity | Overly specific |
| Quantities available | Decreased affinity |
| Immunologically defined | Diminished complement fixation |
| Detection of neoantigens on cell membrane | Labor-intense; high cost |

*Source:* Adapted from Srikanta, S. et al., *Diabetes*, 34, 300, 1985.

## Gene Expression

The effect of an endocrine modulator or disruptor can influence gene expression via transcriptional control. By increasing or decreasing the rate of gene transcription, one can effectively modify gene expression. By quantitating the amount of messenger ribonucleic acid (mRNA) for a given gene, it is possible to measure the effect on transcriptional gene expression [99]. Because mRNA represents a relatively small fraction of RNAs in the cell, methods have been developed that are sensitive and specific for detecting changes in mRNA levels and can distinguish between transfer ribonucleic acid (tRNA) and ribosomal ribonucleic acid (rRNA). Northern blots employ total or poly-A-enriched RNA that is electrophoretically separated on an agarose gel that is transferred to a matrix (e.g., nitrocellulose), where it is then allowed to hybridize to label an RNA segment that can be utilized for a complementary nucleotide sequence (antisense riboprobe) of interest. Although the northern blot is a standard molecular biology technique to examine gene expression via RNA, it is sometimes not sensitive for low-abundance mRNAs or specific enough for RNAs derived from closely related genes or alternative splicing. RNase protection (solution hybridization) assays offer increased sensitivity and specificity over the northern blot but are technically more difficult. In this case, a specific radiolabeled antisense riboprobe is permitted to hybridize with the RNA for several hours before digesting with RNase, which digests only single-stranded but not hybridized double-stranded RNA. The sample is then electrophoresed on a denaturing gel and subjected to autoradiography to detect specific bands (indicating the presence of protected RNA:RNA hybrids during RNase digestion). Quantitative reverse transcriptase–polymerase chain reaction (RT-PCR) is sometimes required when a very small amount of mRNA must be amplified so analyses can be made. In this process, the data-bearing RNA is reverse transcribed into its complementary DNA, which subsequently replicates the mRNA in question using the polymerase chain reaction.

Subtractive hybridization can be used to identify mRNAs that are differentially expressed between tissue or cell types under different hormonal or environmental conditions [93]. Subtractive hybridization is a method that allows for the rapid isolation of differentially distributed nucleic acids (differentially expressed, differentially present, or differentially arranged in two or more different sources). The substrates in the process are nucleic acids (tester component) containing specifically expressed sequences that are to be extracted (target component) and nucleic acids that are utilized for comparison of the target molecules (driver component). The tester is combined with the driver to form a tester–driver hybrid, which is the nontarget component. Tester–tester hybrids and single-stranded tester molecules are also formed which represent the fractions amplifying the target molecules. After hybridization is complete, target and nontarget fractions are separated and quantified. More specifically, poly A + RNA is isolated from the two sources, and the cell or tissue expressing the mRNA of interest is converted into radiolabeled cDNAs (complementary DNA is DNA synthesized from an mRNA template) using reverse transcriptase. After removing the RNA for the cDNA, the poly A + RNA from the second cell or tissue source is exhaustively hybridized to the radiolabeled cDNAs, and unhybridized cDNA is isolated and utilized to screen a cDNA library constructed from the cells or tissues expressing the sequences of interest. Another technique for RNA fingerprinting includes differential display. Following reverse transcription of cDNA, selected segments of the transcripts are amplified, and the sequence of the transcript represented by the differentially displayed PCR end product is identified by utilizing multiple polymerase chain reaction primer pairs.

## Polymerase Chain Reaction

The polymerase chain reaction is another molecular biology technique for enzymatically replicating specific regions of DNA or cDNA template without using a living organism. The technique allows a small amount of the DNA molecule to be amplified many times by using a pair of oligonucleotide primers (usually ranging from 10 to 40 kbp in length, flanking a DNA sequence of interest, approximately 100 to 500 bp in length), a thermostable DNA polymerase (an enzyme that catalyzes the elongation of DNA from a template), deoxynucleotide triphosphates (dNTPs), and a reaction buffer containing magnesium [54].

**TABLE 35.3**
**Selected Basic and Clinical Applications of the Polymerase Chain Reaction (PCR)**

| |
|---|
| *Basic:* |
| DNA sequencing |
| Identification of DNA polymorphisms |
| cDNA cloning |
| DNA mutagenesis |
| Expression and quantitation of mRNA |
| *In vivo* footprinting (DNA–protein interactions) |
| *Clinical:* |
| Detection of infectious agents |
| Diagnosis of gene and chromosome defects |
| HLA haplotyping |
| Legal and forensic applications |

The process consists of 25 to 35 cycles. Each cycle of PCR is comprised of three steps:

1. A denaturing step, where a double-stranded DNA is broken apart at the hydrogen bonds that connect the two DNA strands; primers and DNA strands become singled stranded and are activated by a thermostable polymerase used to check for the presence or absence of a gene by amplifying a DNA fragment termed a *Taq polymerase*.
2. An annealing step, where the primers can bind to the template DNA.
3. An extension step (extension of the annealed primer Taq polymerase) at a rate usually about 1 kilobase pair (kbp)/minute, where 1 kilobase pair equals 1000 nucleotides.

The PCR cycles are performed in a machine called a *DNA thermal cycler*, which can be adjusted to optimize the amplification of DNA, depending on the length and nature of the DNA sequence, the source and quality of the DNA template, and the application of the procedure by heating and cooling the internal reaction tubes within a precise temperature range for each reaction step. Other critical parameters include the ionic conditions (most notably, $Mg^{2+}$ concentration), pH, and purity of reagents (i.e., absence of DNA contamination). DNA templates can be derived from genomic DNA (human, animal, bacterial, or viral) or RNA that has been converted to a DNA template using reverse transcriptase (RT-PCR). This latter technique has been quantitatively adapted for measuring small amounts of mRNA (quantitative RT-PCR) or detecting differences in the amount of expressed mRNA (differential display). PCR can be employed to detect mRNA for hormones, growth factors, polypeptides, receptors, and other proteins involved with the endocrine system (Table 35.3).

## *In Situ* Hybridization and Immunohistochemistry

*In situ* hybridization is a type of hybridization that utilizes labeled nucleic acid probes (either DNA or RNA) which permits the detection and localization of mRNA in tissue samples with labeled nucleic acid probes (DNA or RNA). Tissue samples (or cultured cells) are treated to increase permeability, thus allowing the desired probe to enter the cell. The samples are fixed, embedded, and thinly sectioned prior to hybridization. Hybridization is permitted at elevated temperatures, and the excess probe is then removed. A complementary probe is labeled with an antigenic, fluorescent, or radioactive tag to precisely locate and quantify the probe remaining in the tissue. Radiolabeled probes work well for abundant mRNAs and can be detected using autoradiography. Other methods of signal detection include fluorescence, such as fluorescent *in situ* hybridization (FISH), which is a cytogenetic technique used to detect and localize specific DNA sequences on chromosomes. Fluorescence microscopy is used to localize specific DNA sequences on chromosomes. For example, a probe is constructed and tagged with fluorophores for specific targets. A chromosome preparation is prepared with the chromosomes fixed to a glass substrate whereupon the probe is applied to the DNA of the chromosome and allowed to hybridize. Unbound probe is removed. If a very limited amount of probe remains, fluorescent-tagged antibodies or streptavidin are bound to the tagged molecules, resulting in amplified fluorescence. This preparation is embedded and susequently examined under the fluoroscence microscope. Low-abundance mRNAs can be detected by *in situ* PCR, which amplifies the amount of nucleic acid target in the tissue. This technique is particularly useful for detecting hormone or receptor subtypes that are biologically unique by combining immunogenic features with related biological molecules. Generally, probe specificity parameters are first tested by northern blot analysis.

Immunochemistry refers to a process of detecting and localizing gene products (e.g., proteins) in cells of a tissue section utilizing antibodies [121]. The antibody is tagged with a color-producing tag, which can include such materials as alkaline phosphatase or horseradish peroxidase. Primary monoclonal or polyclonal antibodies can be used to bind specific epitopes of a hormone or receptor, which can then be visualized with a secondary marker, such as fluorochrome-conjugated secondary antibody or streptavidin–biotin labels. Selection of a fluorochrome depends on microscopy wavelength and filters, the stability of the signal, the type of the tissue being examined, and whether or not double-labeling is necessary. Nonfluorescent markers, such as immunoperoxidase or immunogold conjugates, are suitable for bright-field microscopy and offer increased stability over fluorochrome conjugates. The technique is widely used in basic research to help characterize the distribution and localization of biomarkers in different parts of a tissue. Some specific markers are known for specific cancers, such as [126]:

- Carcinoembryonic antigen (CEA), a marker for colon cancer
- CD15 and CD30, markers for Hodgkin's disease (cluster of differentiation [CD] molecules are recognized by specific sets of antibodies, used to identify the cell type, stage of differentiation, and activity of a cell)
- Prostate-specific antigen (PSA), a marker for prostate cancer
- CD117, a marker for gastrointestinal stromal tumors (GISTs)

Immunohistochemistry is a powerful complement to *in situ* hybridization, but limitations include the lack of antibody specificity, denaturation of the antigen during fixation, and cell membrane permeabilization, which refers to the changes that must occur in the plasma membrane to allow for the entry of impermeable fluorescent probes and the binding of antibodies to their respective antigens.

## Analysis of Signal Transduction Pathways

Biological signal transduction is a process by which an organism converts one type of signal (stimuli) to another type of signal. The process by which biochemical events transition from one signal to another is generally referred to as *signal transduction*. The process has several common characteristics, including the fact that these reactions are enzymatic in nature, they occur within a few fractions of a second, and they are mediated by second messengers (low-weight diffusible molecules that are used to relay signals within a cell).

The effects of a specific compound on hormone receptor signal transduction can be analyzed, depending on the nature of the signaling pathway and the generation of second messengers. Although a comprehensive review of signal transduction is beyond the scope of this chapter, several common endocrine signal transduction mechanisms are summarized in Table 35.4. A variety of techniques, including high-performance liquid chromatography (HPLC),

**TABLE 35.4**
**Common Endocrine Signal Transduction Mechanisms**

| |
|---|
| Activation of adenylate cyclase (cyclic AMP) |
| Activation of guanylate cyclase (cyclic GMP) |
| Activation of intracellular kinases |
|   Serine/threonine kinases |
|   Tyrosine kinases |
| Activation of phospholipases (phosphoinositides) |
| Release of intracellular calcium |
| Receptor-operated ion channels |

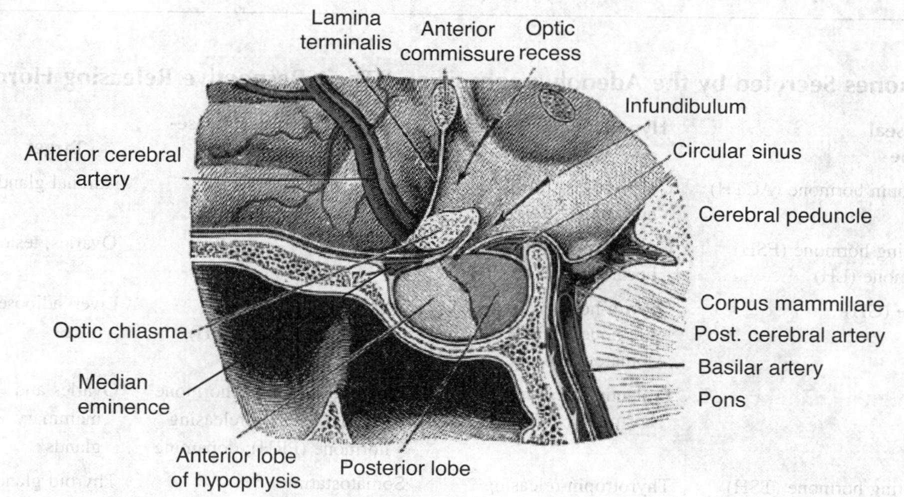


**FIGURE 35.3** Relationship of pituitary gland to surrounding anatomical structures.

western immunoblotting, and protein kinase assays, can be utilized to examine the effects of a toxicant on a hormone signaling pathway. In addition to affecting hormone receptor activation, some compounds can inhibit inactivation of signal transduction pathways, such as phosphodiesterase inhibitors (e.g., caffeine and theophylline). Drugs such as sildenafil and tadalafil are selective inhibitors of type V phosphodiesterase, which degrades the cyclic guanosine monophosphate (cGMP) in the corpus cavernosum, leading to vascular engorgement in erectile dysfunction [34].

Not all substances result in an effect on signal transduction; phytoestrogens are defined as "any plant substance or metabolite that induces biological responses in vertebrates and can mimic or modulate the actions of endogenous oestrogens usually by binding to oestrogen receptors" [53]. They may be flavonoids that have structural similarities to estradiol. Despite the paucity of research on these materials, some data indicate that phytoestrogens help the body balance hormone levels. Also, phytoestrogens have been identified as having a role in some types of cancers, although it is not clear if this is due to the phytoestrogens or eating patterns [3].

## PITUITARY GLAND

### Overview

The pituitary gland, which includes the neurohypophysis and the adenohypophysis, is located at the base of the brain in a small cavity called the *sella turcica* ("Turkish saddle"), which is a saddle-shaped depression in the sphenoid bone at the base of the human skull (see Figure 35.3) [76]. The pituitary secretes several tropic hormones that regulate the activities of cells within distant endocrine glands. The hypothalamic–pituitary axis is the union formed by the interaction between the hypothalamus and pituitary gland. It exerts control over many parts of the endocrine system. This axis functions by interacting between the nervous and endocrine systems; the nervous system regulates the endocrine system, and endocrine activity modulates the activity of the central nervous system. The pituitary is physically attached to the brain by the hypophyseal stalk connected through the median eminence. The median eminence is integral to the hypophyseal portal system, which connects the hypothalamus with the anterior lobe of the pituitary gland. The anterior lobe of the pituitary (i.e., adenohypophysis) receives releasing and inhibitory hormones on their way to the medium eminence, where they collect before entering the portal system. The hypothalamus and the anterior pituitary secrete a number of important hormones. (Table 35.5). The posterior pituitary (i.e., neurohypophysis) is primarily a conduit for the hypothalamus. It does not produce its own hormones; instead, it stores and releases the hormones oxytocin and vasopressin into the blood stream.

### Pituitary and Target Organ Relationships

To understand the toxicology of the endocrine system, it is necessary to grasp the concept of hormonal feedback systems (see Figure 35.1). Although measurement of specific hormone levels might not be feasible or economical for the general toxicologic screening of a substance, some bioassays and microscopic techniques do yield useful information. The reduction in animal growth rates, although in most instances caused by diminished nutritional intake, might be due to the suppression of pituitary growth hormone secretion. Similarly, a decrease in testicular weight following the administration of certain chemicals can be due to interference with pituitary gonadotropins and not a simple direct toxic effect upon the primary target [106,107].

**TABLE 35.5**
**Major Hormones Secreted by the Adenohypophysis and Their Respective Releasing Hormones**

| Adenohypophyseal Tropic Hormone | Hypothalamic Releasing Factors | Hypothalamic Release-Inhibiting Factors | Target | Effect |
|---|---|---|---|---|
| Adrenocorticotropin hormone (ACTH) | Corticotropin-releasing hormone (CRH) | — | Adrenal gland | Secretion of glucocorticoids |
| Follicle-stimulating hormone (FSH) Luteinizing hormone (LH) | Gonadotropin-releasing hormones (FRH/LRH) | — | Ovaries, testes | Reproductive growth |
| Growth hormone (GH) Somatotropin | Growth-hormone-releasing hormone (GHRH); somatotropin | Somatotropin release-inhibiting factor (SRH) | Liver, adipose tissue | Promotes growth; lipid metabolism |
| Prolactin (Prl) | Prolactin-releasing hormone (PRH) | Prolactin-inhibitory hormone (PIH); prolactin-releasing hormone (PRH); dopamine | Ovaries and mammary glands | Secretion of estrogens and progesterone; milk production |
| Thyroid-stimulating hormone (TSH) | Thyrotropin-releasing hormone | Somatostatin | Thyroid gland | Secretion of thyroid hormones |
| Melanocyte-stimulating hormone (MSH) | Melanocyte-stimulating release hormone (MRH) | Melanocyte-inhibiting hormone (MIH) | Melanocytes | Synthesis of melanin |

*Source:* Adapted from Gardner, D.G. and Nissenson, R.A., in *Basic and Clinical Endocrinology*, 7th ed., Greenspan, F.S. and Gardner, D.G., Eds., McGraw-Hill, New York, 2004, pp. 61–84; Porterfield, S.P., Ed., *Endocrine Physiology*, 2nd ed., Mosby, St. Louis, MO, 2001.

Chemically induced changes that affect pituitary–target organ relationships seldom are manifested after a single administration of a toxic substance; rather, compounds that have the potential to exert deleterious effects on the endocrine system ordinarily require multiple administrations and longer durations of time before hormonally complex events can occur. Whereas chemically induced stress could provoke a rapid response in catecholamine secretion and production of glucocorticoids, other hormonal responses would not be as immediate. For those chemicals that initiate the induction of hepatic microsomal enzyme systems that affect hormone metabolism (i.e., catabolism), it may be a week or longer before any changes are detected in the endocrine system. Even those chemicals or drugs purposely designed to suppress a particular hormone target organ secretion (e.g., antithyroidal agents) may exhibit an onset of action of several days.

Many steps are involved in the regulation of target organs by the endocrine system (see Figure 35.1). Note that at no less than four possible sites can hormonal messages be disrupted by various toxic agents. Chemicals, including certain classes of therapeutic drugs, can interfere with the release of tropic hormones or can affect their synthesis. Still other toxic agents can exert disruptive actions on the central nervous system or at the hypothalamus or target organ itself in the biosynthesis of target organ hormone regulatory secretions. Thus, various chemicals have several sites of action on adenohypophyseal–target organ feedback systems.

The sites of a chemical's action may differ in their sensitivity. Target organs such as the gonads are frequently sensitive to toxic substances, particularly because rapidly dividing cells (e.g., spermatogonia) are often vulnerable to chemicals and drugs. Furthermore, stress can affect the secretory activity of certain of the hypothalamic-releasing hormones and hence alter pituitary–target organ relationships (e.g., ACTH, prolactin). Often, toxic agents bind to circulating blood proteins and alter the ratio of free to bound target organ hormones. Such changes in binding also can modify the pituitary–target organ relationship. Numerous target organs, then, can be affected by chemical perturbation.

## ADENOHYPOPHYSIS

The adenohypophysis, also called the *anterior pituitary*, comprises the anterior lobe of the pituitary gland (see Figure 35.3). It is modulated by the hypothalamus. This part of the pituitary secretes hormones that regulate processes such as growth, stress, and reproduction. The stimulus that regulates the secretion of hormones from the anterior pituitary originates in the hypothalmus and is transported to the anterior pituiary via the hypophyseal portal system. The major hormones of the adenohypophysis and their hypothalamic-releasing hormones and targets, as well as effects, are shown in Table 35.5. Because tropic hormones are either protein or glycoproteins with half-lives of 20 minutes to 6 hours, they cannot be measured by standard spectrophotometric procedures. These hormones must be either bioassayed or measured using radioimmunoassays or monoclonal antibody techniques. Although bioassays may be useful for certain of the adenohypophyseal hormones, such tests are often inaccurate or have been replaced by more sensitive methods. Bioassays, however, might be employed when there is only a secondary interest in determining if a particular toxicologic agent is affecting tropic hormone levels. Sometimes a target organ that is known to be directly influenced by a particular tropic hormone can be measured and hence provide some general insight into the nature of the chemically induced alterations in the endocrine system. Target organs whose secretions (e.g., estrogen, testosterone, $T_4$) act back upon the adenohypophysis are usually low-molecular-weight hormones. The glycoproteins include thyroid-stimulating hormone (TSH), luteinizing hormone (LH), follicle-stimulating hormone (FSH), and human chorionic gonadotropin (hCG). Each hormone is a heterodimer of two noncovalently associated subunits α and β, which are encoded by separate genes situated on different chromosomes. It is the β subunit that generally confers the unique biological specificity of each hormone. This biological activity is dependent on the intact dimers; free subunits are biologically inactive.

## MEASUREMENT OF ANTERIOR PITUITARY HORMONES

### Adrenocorticotropin

Adrenocorticotropin (ACTH) stimulates the growth and the maintenance of the adrenal gland, in addition to stimulating the adrenal cortex to secrete cortical steroids. Many pathologic states can alter ACTH secretion. Stress, caused by a variety of environmental or chemical stimuli, including surgery, trauma, infection, hypoxia, and anxiety, can cause a rapid elevation in ACTH blood levels. ACTH and cortisol exhibit diurnal variations, with the highest levels occurring in the morning (about 8:00 a.m.) and the lowest levels in late afternoon. The hypothalamic–pituitary–adrenal axis has many modulators, many of which are mediated through the central nervous system and stimulate or inhibit the secretion of ACTH. Several methods are available for measuring ACTH [80], but most measurements have indirectly assessed adrenal gland secretions. A highly sensitive IRMA ACTH assay has been very useful in diagnosing adrenal deficiencies [112]. Gravimetric assay of adrenal glands represents one of the simplest methods for indirectly evaluating ACTH activity. This assay uses hypophysectomized animals; injections of ACTH-like material can enhance the weight of the adrenal glands. ACTH stimulates increases in plasma cortisol and corticosterone and elevates urinary 17-hydroxycorticosteroids and 17-ketosteroids; these steroid levels can also be used to assess ACTH. ACTH causes involution of the thymus gland and deposition of hepatic glycogen and leads to a decrease in circulating eosinophils in hypophysectomized rodents. ACTH can also cause depletion of adrenal ascorbic acid. In addition to those assays for ACTH that rely on adrenal gland responses (see section on adrenal glands), radioligand–receptor assays have been developed for ACTH. Often, cortisol levels are used to assess ACTH. Cortisol can be determined using commercially available antibody-coated tube RIA kits.

### Thyroid-Stimulating Hormone

Thyroid-stimulating hormone (TSH) is a glycoprotein that is synthesized and secreted by the thyrotrope cells of the anterior pituitary gland. This hormone regulates the growth and proliferation of cells of the thyroid gland. In this case, the hypothalamus produces thyroid-releasing hormone, which subsequently stimulates the anterior pituitary to release TSH. TSH, in turn, stimulates the thyroid gland to secrete triiodothyronine ($T_3$) and thyroxine ($T_4$). These tyrosine-based hormones function to increase the basal metabolic rate and make changes to the rate of protein, carbohydrate, and fat synthesis. $T_3$ and $T_4$ (thyroid) as well as somatostatin (hypothalamus) negatively feedback to inhibit TSH production. The TSH receptor site is complex and involves an extracellular domain as well as a transmembrane component [65]. TSH assays have employed the uptake of $^{32}P$ in the thyroid glands of experimental animals. Like ACTH and for routine toxicologic assessment of TSH, tests often involve the measurement of target organ secretory responses; hence, the evaluation of TSH in routine toxicologic experiments often employs measurement of the thyroid hormones $T_3$ and $T_4$ (see the later section on the thyroid gland). Serum thyroid hormone concentrations alone do not explain the variability and severity of the range of symptoms observed in thyrotoxicosis [83].

Many chemicals, drugs, environmental factors, and pathologic states can affect thyroid hormone secretion [21]. Certain foodstuffs and plants contain chemicals (goitrogens) that can act as antithyroidal agents. Most of these conditions or factors, however, seem to affect thyroid gland activity itself rather than impinge upon TSH secretion.

Immunochemistry assays are also available for measuring TSH [47,101]. A liquid-phase, two-site immunoradiometric assay (IRMA) has been described for human TSH (hTSH). IRMA is based on the simultaneous addition of affinity purified sheep anti-hTSH IgG-$^{125}I$ and rabbit anti-hTSH antiserum. This assay is specific for hTSH and exhibits no cross-reactivity with other pituitary glycoprotein [87]; thus, the IRMA assays for TSH may be more specific than current RIAs for TSH. The one-step IRMA method involves the use of monospecific antibody against two immunogenic sites on the TSH molecule [92]. Other assays for thyrotropin involve a combination of bioluminescence and immunoassay techniques [98].

### Growth Hormone (Somatotropin)

Growth hormone (GH) is also referred to as *somatropin*, *somatotropin*, or *somatotrophin*. Human growth hormone is a peptide composed of 191 amino acids with a molecular weight of 22,000 daltons. The structure contains four helices that precisely interact with the GH receptor. GH is secreted by the somatotrope cells found

**TABLE 35.6**
**Effects of Various Factors on Growth Hormone (GH) Levels**

| Increased GH Levels | Decreased GH Levels |
|---|---|
| Apomorphine | Dietary carbohydrate and glucocorticoids |
| Arginine[a] | Sleep |
| Clonidine[a] | Somatostatin |
| a-Deoxyglucose | |
| Dietary protein | |
| L-DOPA[a] | |
| Endorphins | |
| Enkephalins | |
| Epinephrine | |
| Estradiol | |
| Exercise | |
| Insulin (hypoglycemia)[a] | |
| Norepinephrine | |
| Prostaglandins | |
| Serotonin | |
| Stress | |
| Substance P | |
| TRH | |
| Vasopressin | |

[a] Used as a provocative test for the diagnosis of GH disorders.

in the anterior pituitary. It has been found that transcription factor Pit-1 stimulates the development of somatotrope cell and subsequently the production of GH. Growth-hormone-releasing hormone (GHRH) released from the hypothalamus promotes the secretion of GH, while somatostatin from the periventricular nucleus inhibits GH production, as does the concentration of GH and insulin-like growth factor 1 (IGF-1), a protein similar in structure to insulin that is produced by target cells and the liver and is involved in the regulation of metabolism. GH exerts its actions on a variety of cells to stimulate lipolysis, protein anabolism, and hyperglycemia. Its actions on bone and cartilage are mediated primarily through IGF-1. A deficiency in GH leads to a reduction in the incorporation of amino acids into protein. GH causes a marked stimulation of cartilaginous growth at the epiphyses of long bones.

A constant flow of stimulating and inhibiting peptides enhances or suppresses the rate of secretion of GH. In addition, many physiological events effect changes in GH secretion. Many of the agents and physiological factors that can affect the rate of GH secretion are listed in Table 35.6. Hypoglycemia or insulin can cause a sudden and dramatic increase in serum GH. Starvation can affect GH levels, while cold, stress, or surgical trauma can lead to an increase in serum GH. Drugs and chemicals that affect catecholamine neurotransmission and the autonomic nervous system can influence GH secretion.

It has been shown that most of the physiologically critical GH secretions generally occur as several large surges over the course of the day that may extend from 10 to 30 minutes, with the largest GH peak occurring during the first hour of sleep. The frequencies are most pronounced (8 peaks per 24 hours) and the levels of GH are highest during early childhood and at puberty during adolescence, whereas the mature adult averages approximately 5 peaks per 24 hours.

Growth hormone has been assessed using bioassays. Some GH assays simply use a 10-day body weight gain test in hypophysectomized female rats. The hormone has also been assayed by measuring the width of the tibial epiphysial growth plate. Sensitive RIAs have been developed for experimental animals (e.g., rat) and for humans. Human GH concentrations can be measured with double-antibody RIAs. Immunocytologic assays for GH can also be performed on Epon-embedded, semithin sections of tissue using the avidin–biotin–peroxidase complex technique [52]. Microdissection techniques [70] and other *in vitro* assays for the release of GH have been employed [5]; in addition, an ultrasensitive immunofluometric assay [91] and a chemiluminescence-based GH assay [120] are now available. Whether GH is assessed using bioassays or by the more accurate and sensitive RIA procedures, the experimental design of either acute or chronic toxicity tests must closely monitor the nutritional and behavioral status of the animals. Several toxic agents can affect dietary intake and hence reduce body weight. Experimental designs using paired-feeding protocols are often necessary for interpreting GH activity in any well-designed toxicology protocol.

## Gonadotropins (FSH and LH)

Gonadotropins are hormones secreted by the gonadotrope cells located in the anterior pituitary gland. The two primary gonadotropins are the follicle-stimulating hormone (FSH) and luteinizing hormone (LH), both of which are glycoproteins that contain $\alpha$ and $\beta$ peptide subunits. The $\alpha$ units in both hormones are essentially identical, but there is some functional and immunologic specificity with the $\beta$ subunits of the gonadotropins, including human chorionic gonadotropin (hCG), which is produced by the placenta during pregnancy.

Gonadotropin receptors are found on the surface of the gonads, and these are connected to the G-protein system. G-proteins are guanine nucleotide binding proteins, a group of specific proteins involved in second-messenger feedback systems. Second messengers are used in signal transduction to relay signals within a particular cell in response to an external signal received by a transmembrane receptor. These signals are relayed by cyclic AMP when the receptor is bound by the G-protein.

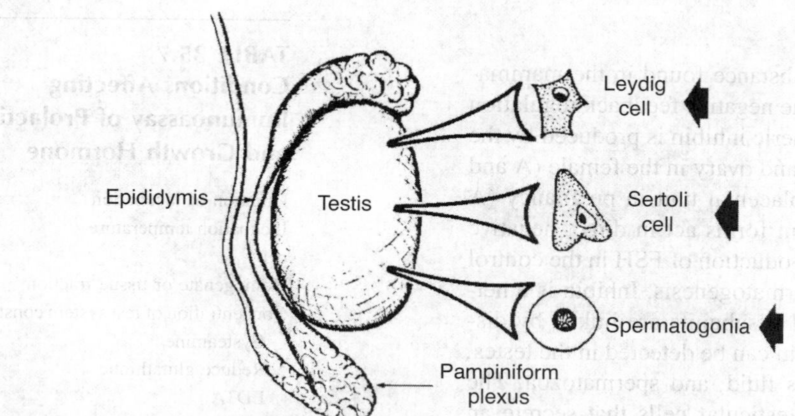


**FIGURE 35.4** Some possible sites of action of selected gonadotoxins in the testes.

The critical gonadotropins, FSH and LH, are secreted from the adenohypophysis as directed by the gonadotropin-releasing hormone from the hypothalamus. Their principal stimulatory actions and primary targets are the testes or ovaries. FSH stimulates follicular development in the ovary and spermatogenesis in the testes. LH, also referred to as *interstitial-cell-stimulating hormone* (ICSH) in the male, causes luteinization of the ovary and stimulates androgen production in testicular Leydig cells. The gonads contain receptors for LH and FSH which are involved in intracellular signaling [72].

In the human female, blood levels of FSH and LH vary according to the phase of the menstrual cycle. In human males, although some diurnal fluctuation in FSH and LH may occur, blood levels are noncyclic. Unfortunately, there is a paucity of information about the direct effects of chemicals and drugs on the ovary [45,51]; indeed, evaluating gonadotropin secretion in women is more complex than in males from a toxicologic standpoint, as the various secretory or cellular processes of the male gonad are more readily evaluated. Figure 35.4 shows the major cellular sites of testicular toxicants. In general, the Leydig cells (or interstitial cells) are comparatively resistant to chemical insult; however, the process of steroidogenesis is quite sensitive and can be used as an indicator of male sex steroid (e.g., testosterone, dihydrotestosterone) activity (see later section on gonads). Methods have been devised to examine isolated Leydig cell cultures as well as testicular perfusion systems in an effort to study androgen secretion by the male gonad [5]. Spermatogenesis, however, is sensitive to chemical insult. Testicular sections (or biopsy) can be used to assess the degree of spermatogenic arrest as evidenced by the presence (or absence) of sterile or partially sterile seminiferous tubules. In domestic and laboratory animals, electroejaculation methods and the use of an artificial vagina can be used for collecting semen for examination of sperm counts, morphology, and motility.

In response to FSH, Sertoli cells in the seminiferous tubules secrete a number of proteinaceous substances, including androgen-binding protein (ABP), which binds testosterone within the testes presumably to maintain high levels of testosterone at the site of the developing germ cells. ABP has been used as an indicator of toxicologic insult following the administration of potentially damaging chemicals or drugs. FSH or LH levels can be assessed by direct measurement of the hormones or by indirect assays that reflect an influence of these hormones on their target tissues. RIAs are available to measure gonadotropins.

Earlier methods in classical endocrinology studies often employed bioassays. FSH has been bioassayed by assessing its ability to enhance ovarian weight in immature rats treated with a placental gonadotropin. A sensitive assay for LH once employed ovarian ascorbic acid depletion. The bioassay of gonadotropins has largely been abandoned because of low sensitivity and expense. The sensitivity of RIAs offers substantial advantages over bioassays, even though the biological and immunological activities do not always correlate.

Indirect measurement of gonadotropins in women often includes the measurement of ovarian steroids or the study of vaginal cytology (see later section on gonads). In men, indirect assays include the measurement of androgens or the histological assessment of spermatogenesis. Chromosome studies to rule out Klinefelter's syndrome or Turner's syndrome can also be of some value when assessing hypogonadism for pituitary gonadal activity. Localization of FSH and LH and their receptors can be detected using immunoperoxidase techniques.

It is important that the appropriate separation system be employed when measuring FSH [29]. Solid-phase RIAs have been developed that measure free $\alpha$ subunits of pituitary glycoprotein hormones [90]. Semiquantitative assays for LH are also available [18].

### INHIBIN (A AND B)

Inhibin, a proteinaceous substance found in the mammalian gonad, is involved in the negative feedback regulation of FSH secretion [42]. Dimeric inhibin is produced by the testes in the male (B only) and ovary in the female (A and B), together with the fetoplacental unit in pregnancy (A only). These various inhibin forms act in direct negative feedback on the pituitary production of FSH in the control of folliculogenesis and spermatogenesis. Inhibin is a heterodimer, where the α and β subunits are linked by disulfide bonds. In men, inhibin can be detected in the testes, seminal plasma, rete testes fluid, and spermatozoa. The Sertoli cells are the only testicular cells that secrete an inhibin-like substance referred to as *Sertoli cell factor* (SCF). Inhibin, as well as SCF, can suppress the pituitary secretion of FSH. Inhibin (A or B) is synthesized in granulosa cells, and production is stimulated by FSH and by estradiol. The biological activities of various inhibin preparations, including SCF, can be assessed *in vitro* using pituitary cell cultures [96,46]. This *in vitro* assay consists of evaluating the degree of suppression of basal FSH release following an incubation with the test material relative to that of a control culture.

### PROLACTIN

Prolactin is a protein hormone whose amino acid composition is quite similar to growth hormone. It is synthesized and secreted by lactotrope cells in the adenohypophysis. Prolactin is a single-chain polypeptide structure of 199 amino acids resulting in a molecular weight of about 24,000 daltons. Interestingly, it is also produced in decidua and the breast tissue. Prolactin secretion is initiated in the hypothalamus by Pit-1 transcription factor, which binds to the prolactin genes at several sites. Estrogen also enhances the growth of lactotrope cells. Dopamine suppresses the production of prolactin and itself is controlled by estrogen, which inhibits the release of dopamine. Prolactin causes initiation and maintenance of lactation in women. It has no known physiological function in men. In rodents, prolactin maintains the corpus luteum. In many species, milk ejection cannot be produced by suckling unless prolactin first stimulates the myoepithelial cells of the mammary glands.

Prior to the development of sensitive RIAs for prolactin, it was bioassayed. Prolactin has the ability to stimulate the crop sac of the pigeon. Prolactin can also be identified in tissues using *in vitro* immunoassays, but a number of test system conditions *in vitro* can affect the detectable levels of prolactin (Table 35.7) [73]. It is important to determine the optimal conditions for assessing tissue prolactin. Many drugs, chemicals or even some physiological conditions can affect blood levels of prolactin (Table 35.8). Several of the actions of these agents are mediated by dopaminergic mechanisms. Prolactin secretion can be increased by physical exercise, coitus, suckling, and surgical stress. Such factors must be taken into consideration when assessing prolactin levels in toxicologic protocols.

**TABLE 35.7**
**Conditions Affecting Immunoassay of Prolactin and Growth Hormone**

Duration of incubation
Incubation temperature
pH
Homogenate or tissue fraction
Concentration of test system constituents:
  Cysteamine
  Reduce glutathione
  EDTA
  Urea
  Sodium dodecyl sulfate
  Iodoacetate

## POSTERIOR PITUITARY

The posterior pituitary gland, or neurohypophysis, comprises the posterior lobe of the pituitary gland (see Figure 35.3). It consists primarily of neuronal axon nerve fiber projections from the hypothalamus. The hormones of the posterior pituitary are produced in the hypothalamus and are transported through the hypophyseal vein to the posterior pituitary, where they are stored until needed for use. The posterior pituitary is not a gland in the conventional sense, but rather an extension of the hypothalamus from the supraoptic and paraventricular nuclei of the hypothalamus.

### NEUROHYPOPHYSEAL PEPTIDES

The posterior pituitary contains a group of peptides known as neurophysins, as well as oxytocin and vasopressin (antidiuretic hormone [ADH]). The neurophysins appear to be synthesized in the same hypothalamic neurons as the nonapeptides oxytocin and vasopressin. Whereas the physiological function of oxytocin and vasopressin are well established, less is known about the biological function of the neurophysins. It appears that oxytocin and vasopressin are packaged with the protein neurophysin into discrete granules, which then move down the supraoptic and paraventricular nuclei of the hypothalamus axon and are stored in the posterior pituitary. When the hypothalamus is stimulated, these hormones are then released into the bloodstream.

The principal physiological function of vasopressin is conservation of fluids, and it exerts its action on the renal tubule leading to the reabsorption of water. The release of

**TABLE 35.8**
**Effect of Various Agents on Blood Prolactin Levels**[a]

| Increase Prolactin Levels | Decrease Prolactin Levels |
|---|---|
| Atropine | Acetylcholine |
| Chlorpromazine | Apomorphine |
| Diethyl ether | Bromocriptine |
| β-Endorphin | Bulimia |
| Estrogens | Dopamine |
| Haloperidol | L-DOPA |
| Histamine | β-Hydroxy-GABA |
| Hypothyroidism | Iproniazid |
| Met-enkephalin | Somatostatin |
| Methyl-DOPA | Thyroxine (and $T_3$) |
| α-Methyl-*p*-tyrosine | |
| Nicotine | |
| Opiates | |
| Perphenazine (and other phenothiazines) | |
| Prolactinoma | |
| Prostaglandin E | |
| Reserpine | |
| Supiride | |
| Tricyclic antidepressants | |
| TRH | |
| Vasopressin | |

[a] Response may vary quantitatively depending on the dose and the particular species.

vasopressin is mediated by neural impulses from osmoreceptors located in the hypothalamus. The principal physiological function of oxytocin is to stimulate uterine smooth musculature and to aid in the process of lactation. Oxytocin is released in response to suckling, and it may also play a role in parturition.

Drugs, other hormones, and physiological state can affect the secretion of vasopressin or oxytocin (Table 35.9). Nonspecific stress also can stimulate the release of ADH and hence is an important variable to consider in any toxicologic protocol involved with monitoring water balance. Some chemicals and drugs affect the central release of posterior pituitary hormones at the source, whereas others may block their peripheral actions. Physiological factors also affect the secretory rate of oxytocin, and considerable differences exist among species. Suckling and mammary duct dilation and cervical distension lead to enhanced secretion of oxytocin. Estrogens and pregnancy enhance the sensitivity of the uterine smooth muscle to oxytocin.

## Oxytocin

Oxytocin is a oligopeptide that contains 9 amino acids (termed a nonapeptide) and has a molecular weight of 1007 daltons. Oxytocin is produced by the oxytocin-producing magnocellular neurosecretory cells located within the supraoptic and paraventricular nuclei. The oxytocin neurons affect corticotropin-releasing hormone (CRH). The oxytocin receptor is a G-protein-coupled receptor, a protein family of transmembrane receptors that transduce extracellular signals into intracellular signals and requires magnesium and cholesterol. Oxytocin is released after stimulation of the nipples and distention of the vagina and cervix to prepare the female for birthing and breastfeeding. It is also released in males and females during orgasm and has been shown to be involved in bonding between individuals.

In females, the three critical actions involving oxytocin are [60]:

- Uterine contraction, which is critical for cervical dilation in late labor and for blood-clotting mechanisms with parturition
- Letdown reflex, which occurs during lactation and permits the milk to accumulate in the collecting chambers for the infant to suckle (suckling activity at the nipple stimulates increased secretion of oxytocin)
- Maternal behavior, which appears to have an oxytocin chemical component in some female animals such as voles and sheep

In males the actions of oxytocin include [39]:

- Facilitation of sperm transport with orgasm (which releases oxytocin into the bloodstream)
- Enhanced sexual arousal

**TABLE 35.9**
**Effects of Various Factors on the Release or Action of Neurohypophyseal Hormones**

| Enhanced Release | Inhibited Release | Blocked Peripheral Action |
|---|---|---|
| | **Vasopressin** | |
| Acetylcholine | Ethanol | Lithium |
| Nicotine | Caffeine | β-Adrenergic agonists |
| α-Adrenergic agonists | | Tetracyclines |
| Vincristine | | |
| Clofibrate | | |
| Stress | | |
| Increased plasma osmolality | | |
| | **Oxytocin** | |
| Prostaglandin $E_2$ | Ethanol | Propranolol |
| Prostaglandin $F_{1\alpha}$ | Methalibure | Vasopressin analogs |
| Suckling | | Oxytocin analogs |
| Mammary duct dilation | | |
| Distention of the cervix | | |

General actions of oxytocin that affect both sexes include:

- Secretion into the bloodstream at orgasm in both males and females [17]
- Reduced anxiety, blood pressure, and cortisol levels, as well as mediation of fight-or-flight-behavior
- Increased tolerance to pain
- Stimulation of sodium excretion from kidneys (natriuresis)
- Possible cardiac development involvement (specifically, by enhancing cardiomyocyte development) [55]

Oxytocin may be measured by RIA or bioassay. A solid-phase RIA for the direct measure of plasma oxytocin has been developed that is rapid, relatively sensitive, and reproducible and does not require the prior extraction of plasma samples [14]. Because oxytocin can stimulate the contraction of smooth muscles, several bioassays have been developed using isolated muscle strips. Oxytocin can be assayed employing the mammotonic activity of the hormone on strips of lactating rat mammary gland. The increment of tension developed by the muscle strip is used as an index of oxytocic activity. A four-point assay can be carried out using USP posterior pituitary standard as a reference.

### Vasopressin (Antidiuretic Hormone)

Vasopressin is an oligopeptide that contains 9 amino acids (termed a nonapeptide) and has a molecular weight of about 1084 daltons. Vasopressin is produced by the vasopressin-producing magnocellular neurosecretory cells located within the supraoptic and paraventricular nuclei. Interestingly, vasopressin is also secreted from the parvocellular neuroendocrine neurons found in the paraventricular nucleus, which transports the hormone to the anterior pituitary, where it becomes a releasing factor for ACTH. Vasopressin is also released into the brain by neurons of the suprachiasmatic nucleus of the hypothalamus, which generates a circadian rhythm of neuronal activity, ultimately regulating many different body functions over a 24-hour period. In addition, vasopressin released into the brain has been shown to be involved with delayed reflexes, memory, aggression, blood pressure modulation, and temperature regulation.

Peripheral actions of vasopressin primarily include activities at three receptor sites:

- V1a, which is located in various tissues and is involved with vasoconstriction, gluconeogenesis in the liver, platelet aggregation, and release of factor VIII
- V1b, which is located in the anterior pituitary and is involved with corticotropin secretion
- V2, which is located primarily in the distal convoluted tubules and collecting ducts of the kidneys, where it is involved with the control of free water reabsorption in the cortical collecting ducts of the renal medulla

Antidiuretic hormone can be measured by RIA or bioassay. With bioassay, ADH or extracts of posterior pituitary tissue can be measured for their antidiuretic activity in the rat using ether anesthesia and a constant water load; the diminution in urine flow is an index of antidiuretic activity. Unanesthetized trained dogs also have been employed in the bioassay, but stress must be minimized because it can reduce urinary output and hence the ADH assay. Simple and rapid radioimmunoassays for vasopressin are available [68]. Because vasopressin is present in biological fluids in low concentrations (picograms), tests must be sensitive. It ordinarily has been difficult to measure basal levels or small fluctuations resulting from particular experimental designs.

## THYROID GLAND

The thyroid gland is located on the anterior of the spinal chord between vertebrae C-5 and T-1 (see Figure 35.5). It is one of the largest endocrine glands, weighing about 20 g in the adult. It is butterfly shaped, with the wings extending over each side of the trachea. Iodine is selectively absorbed by follicles in the thyroid gland. These follicles consist of thyroid epithelial cells whose purpose is to secrete triiodothyronine ($T_3$) and thyroxine ($T_4$). In addition, the follicles contain a protein known as *thyroglobulin*. Parafollicular cells (C cells) are found throughout the gland; these cells secrete calcitonin, which participates in the regulation of vitamin D, bone mineralization, and the reduction of calcium absorption.

The secretory process of the thyroid gland is modulated by the adenohypophyseal–hypothalamic system and is mediated by thyroid-stimulating hormone (TSH). TSH

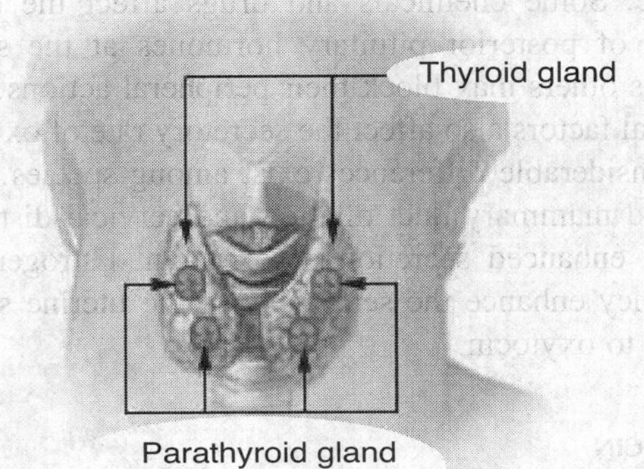


**FIGURE 35.5** Parathyroid and thyroid glands. (Courtesy of U.S. National Cancer Institute Surveillance, Epidemiology, and End Results [SEER] Program.)

stimulates the thyroid gland, leading to increased secretion and release of $T_3$ and $T_4$. The initial step in the synthesis of these thyroid hormones is the uptake of iodide (derived from dietary iodine) into the follicular cells of the thyroid in response to the adenohypophyseal hormone thyrotropin (TSH) (Figure 35.6). Internalized iodide is oxidized, possibly through a free-radical mechanism, and then chemically combined with the tyrosine components of a protein, thyroglobulin, to form either monoiodotyrosyl or diiodotyrosyl residues. Two molecules of the latter can combine to form triiodothyronine. As much as 40% of the $T_4$ is converted to $T_3$ by the liver, spleen, and kidney; hence, $T_3$ is many more times active than $T_4$, although the ratio can be altered by certain physiological states [85]. $T_3$ and $T_4$ remain incorporated in the thyroglobulin, which is stored in the follicular colloid material of the gland, until their release in response to TSH. The release, which involves lysosomes, results from the proteolytic cleavage of thyroglobulin into the thyroid hormones and component amino acids. Monoiodotyrosine and diiodotyrosine also are released at this stage; however, before reaching the circulation, they are enzymatically degraded, liberating iodide, which is eventually reincorporated into protein by the gland. Normally, thyroglobulin does not reach the circulation but remains inside the thyroid cell.

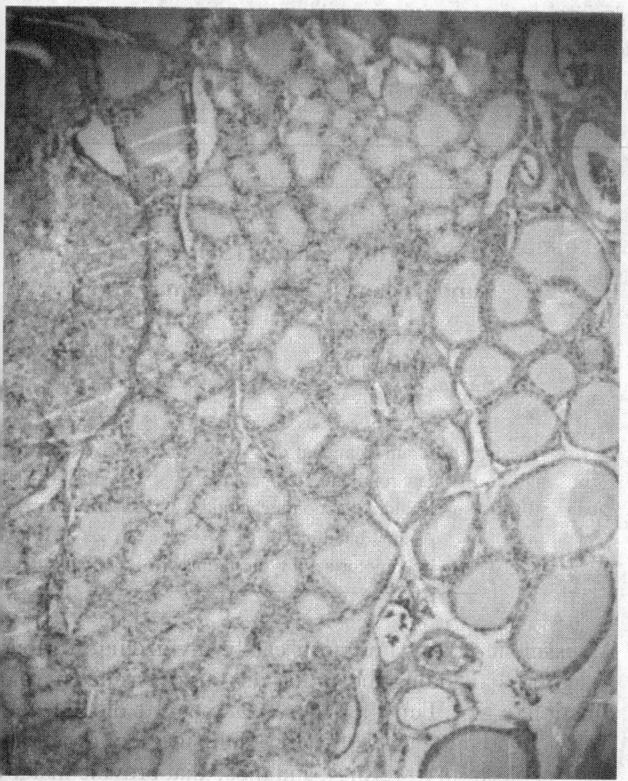

FIGURE 35.6 Histology of rat thyroid follicular cells (original magnification, 125×).

Once released into the circulation, $T_3$ and $T_4$ are transported in the plasma bound to specific proteins. Although there are species differences in the protein-binding patterns of the thyroid hormones, the primary binding protein in humans is thyroxine-binding globulin (TBG). TBG is an acidic glycoprotein (molecular weight 40,000) that binds $T_4$ with a relatively high binding affinity and $T_3$ with a lower binding affinity. A second transport protein, thyroxine-binding prealbumin, although present in higher amounts than thyroxine-binding globulin, has a lower binding affinity for the thyroid hormones and is considered of secondary physiologic importance. In humans and most other mammals, the thyroid hormones can also bind to albumin following occupation of the higher-affinity binding sites. As a consequence of this plasma protein binding, less than 0.1% of the total plasma thyroid hormones exist in a free or unbound form. Care must be exercised when monitoring thyroid function in a species such as the rat, which does not possess TBG (a high-affinity binding protein) and therefore has lower plasma levels of protein-bound thyroid hormone. Furthermore, many factors can affect the measurement of rodent thyroid hormones (see Table 35.12). Also, it is the free hormone that is available for degradation, and the plasma half-life for $T_4$ would be longer in a species with a TBG than in a species without the protein. The $T_4$ plasma half-life in the human, which has a TBG, is 5 to 9 days. In the rat, which does not have a TBG, the $T_4$ plasma half-life is 12 to 24 hours.

The thyroid gland can be affected by various disease states (Table 35.10). Antithyroidal drugs can interfere with the biosynthesis of thyroid hormones. Some agents interfere with the uptake of iodine and some act by inhibiting thyroid peroxidase (TPO), resulting in a reduction in $T_3$ and $T_4$ levels. Furthermore, several other chemicals can affect thyroid hormone secretion (Table 35.11). Industrial or environmental agents that affect thyroid function in humans typically do so by interfering with the intrathyroidal synthesis or secretion of thyroid hormone (i.e., primary hypothyroidism) [7]. Polychlorinated biphenyls (PCBs) affect thyroid hormone metabolism. Amiodarone, an antiarrhythmic drug, is an iodinated benzofuran derivative with a chemical structure similar to thyroxine that may cause hyperthyroidism. PCBs are goitrogenic in some animals. 2,3,7,8-Tetrachlorodibenzo-*p*-dioxin (TCDD) can reduce $T_4$ levels in experimental animals, and DDT can produce avian hypothyroidism. Hydroxyphenols (e.g., resorcinol) and hydroxypyridines have been shown to inhibit TPO; likewise, phthalates, also known as plasticizers, can be degraded by certain bacteria producing dihydroxybenzoic acid (DHBA). DHBA can also inhibit TPO [7]. In addition to the various drugs, chemicals, and environmental pollutants that can affect the thyroid gland, other factors can influence the measurement of TSH, $T_3$, and $T_4$ levels; see Table 35.12.

### TABLE 35.10
### Diseases of the Thyroid Gland

| Type of Effect | Disease | Specific Condition |
|---|---|---|
| Hypothyroidism (myxedema) | Hashimoto's thyroiditis or thyroiditis | Autoimmune disorder where antibodies destroy thyroid cells |
| | Ord's thyroiditis | Atrophic autoimmune disorder common in Europe |
| | Postpartum thyroiditis | Inflammation and dysfunction of thyroid after delivery, due to antibodies |
| | Silent thyroiditis | A variant of chronic autoimmune thyroiditis similar to postpartum thyroiditis |
| | Postoperative thyroidism | Development of hypothyroidism upon surgical removal of the thyroid |
| | Acute and subacute thyroiditis | Inflammation and dysfunction of the thyroid |
| | Iatrogenic hypothyroidism | Dysfunction of the thyroid after exposure to irradiation |
| Hyperthyroidism (thyrotoxicosis) | Graves' disease | Autoimmune disorder where antibodies stimulate the thyroid cells to overproduce |
| | Thyroid storm | An acute, life-threatening, thyroid-hormone-induced hypermetabolic state |
| | Toxic thyroid nodule | Autonomously functioning thyroid nodules that secrete excess thyroid hormone |
| | Toxic nodular struma (Plummer's disease) | Enlarged thyroid gland containing small, rounded masses called *nodules* that secrete excess thyroid hormone |
| | Hashitoxicosis | Autoimmune disorder where thyroid cells are sporadically stimulated to overproduce |
| | De Quervain thyroiditis | Viral infection that initially causes over secretion and then under secretion of thyroid hormone |
| | Iatrogenic hyperthyroidism | Overtreatment of hypothyroidism with thyroid hormone replacement |
| Additional thyroid pathology | Goiter | Swelling of the thyroid gland because of lack of iodine (endemic); toxic goiter (from inflammation, neoplasm); nontoxic goiter (caused by autoimmune or chemical reaction); diffuse/multinodular goiter (spread throughout the thyroid) |
| | Thyroid adenoma | Usually benign tumors of the follicular epithelium of the thyroid gland |
| | Thyroid cancer | Cancer in one of the following forms: papillary, follicular, medullary, or anaplastic |
| | Cretinism (congenital hypothyroidism) | A congenital defect that results in an underdevelopment of the thyroid; thought to be a result of iodine deficiency |

*Source:* Data from Greenspan, F.S., in *Basic and Clinical Endocrinology*, 7th ed., Greenspan, F.S. and Gardner, D.G., Eds., McGraw-Hill, New York, 2004, pp. 215–294; Wikidepia, http://en.wikipedia.org, 2006.

### TABLE 35.11
### Chemicals Producing Abnormal Thyroid Function

| Blocks Iodide Trapping | Blocks Iodide Oxidation | Mechanism Not Established |
|---|---|---|
| Chlorate | Amphenone | Acetazolamide |
| Hypochlorite | Carbimazole | Chlorpromazine |
| Iodate | Cobalt | Chlortrimeton |
| Nitrate | Methimazole | Thiopental |
| Perchlorate | *p*-Aminosalicylate | Tolbutamide |
| Thiocyanate | Phenylbutazone | |
| | Phenylindanedione | |
| | Propylthiouracil | |
| | Resorcinol | |

### TABLE 35.12
### Factors Influencing TSH, $T_3$, and $T_4$ Levels in Rat Plasma

| |
|---|
| Sex of animal |
| Age of animal |
| Time of day |
| Stage of estrous cycle |
| Strain of animal |
| Environmental temperature |
| Blood collection technique |
| Animal handling |
| Locomotor activity of animal |

## Measurement of Thyroid Hormones

Assays for thyroid hormones are used to monitor and confirm the extent of hyper- or hypothyroidism. A large number of laboratory tests are available that can be used to assess the function of the thyroid gland (Table 35.13). Assessment of laboratory tests can be divided into various categories, including those that directly measure thyroid function (e.g., $^{131}$I uptake), those that measure blood levels (free vs. bound), those that measure metabolic actions (e.g., cholesterol lowering), and those that provide insight into the modulations of thyroid function by the adenohypophysis (e.g., thyroid suppression test).

**TABLE 35.13**
**Laboratory Assessment of the Thyroid**

Tests assessing metabolic effect of thyroid:
- Serum cholesterol
- Basal metabolic rate
- Cardiac rate (i.e., systolic time intervals)

Tests assessing thyroid function:
- TRH stimulation
- Thyroid suppression test
- Serum TSH
- $^{131}I$ uptake; $T_3$ uptake

Tests assessing blood levels (free and bound):
- Serum total $T_3$ and total $T_4$
- Serum free (unbound) $T_4$
- Reverse $T_3$ ($rT_3$)
- Thyroxine-binding globulin (TBG)

Radioimmunoassays can be used to determine serum $T_3$ and $T_4$ levels. The RIA is based on the binding of endogenous hormone to a specific antibody, thereby displacing a proportional amount of radiolabeled hormone from that antibody. Obviously, the hormone-binding proteins in the sample (e.g., thyroxine-binding globulin, thyroxine-binding prealbumin, and albumin) tend to compete with the antibody for the hormone. This interference can be prevented by extracting the serum prior to the assay or adding chemical agents to block the binding of hormone to binding proteins. Another difficulty in measuring $T_3$ or $T_4$ by RIA is that the hormones are small and similar in structure; hence, it has been difficult to produce a specific antibody for $T_3$ and $T_4$ without the need for prior chromatographic separation.

Both $T_4$ and $T_3$ can be measured using a competitive protein-binding assay that is based on the displacement of radiolabeled hormone from thyroxine-binding globulin. The amount of label displaced is proportional to the amount of hormone added to the assay in the sample serum. $T_4$ can be effectively measured using the competitive protein-binding assay, but the hormone must first be extracted from the sample serum to eliminate any interference from endogenous binding proteins, which would tend to compete with the binding protein used in the assay. Non-isotopic immunotechniques can also be used to measure thyroid hormones (e.g., EMIT). The principles of these tests are comparable to RIAs except that the labeled analyte may be an enzyme, as in ELISA [43]. EMIT is an enzymatic method that does not require separation of the free and bound portions of the hormone.

## PARATHYROID GLANDS

Parathyroid glands are small glands located in the neck and situated on the posterior surface of the thyroid gland. They produce parathyroid hormone (PTH) (see Figure 35.5). The densely packed parathyroid gland contains two primary types of cells: oxyphil cells and chief cells. Oxyphil cells usually appear in clusters and contain large numbers of mitochondria. The specific function of oxyphil cells is unknown. The chief cells synthesize PTH in the rough endoplasmic reticulum; it is packaged by the Golgi apparatus and then stored in secretory granules. PTH is synthesized as a prohormone consisting of 115 amino acids, whereas PTH itself contains 84 amino acids. The balance of endogenous calcium is maintained by several intrinsic factors that modulate the remodeling of bone and the absorption and excretion of calcium and phosphorus homeostasis. PTH, vitamin D, and calcitonin are the principal factors involved in calcium homeokinesis [16,61]. About 98% of endogenous calcium resides in skeletal tissue, with the remainder sequestered in soft tissues and extracellular fluids. About 50% of serum calcium exists in an ionized form; ionized calcium is the biologically active form of the cation. Hypocalcemia activates calcium-sensing receptors which induces the release of PTH which, in turn, stimulates osteoclasts to break down bone and release calcium into the blood. Although the direct involvement of PTH, calcitonin, and vitamin D are important in modulating calcium, other hormones such as thyroxine, growth hormone, estrogens, and corticosteroids contribute to the maintenance of calcium homeostasis [61].

Parathyroid hormone was originally bioassayed using a hormone-induced elevation in serum calcium in dogs. Several commonly used *in vivo* assays employ the measurement of serum calcium in parathyroidectomized rats or in calcium-injected chicks or quail. A commonly used *in vitro* assay is based on determining the activation of renal adenylate cyclase in response to PTH. PTH can be measured by immunoassay. The majority of assays are directed against the more stable *C*-terminal fraction [9]. Assays for this *N*-terminal fragment are available and are better suited to detect rapid fluctuations in PTH levels; however, the *C*-terminal assay is the method of choice for assessing abnormal PTH function particularly with concomitant hypercalcemia [27, 43]. A monoclonal antibody assay for human vitamin-D-binding protein has also been reported [88]. The anti-human vitamin-D-binding protein antibodies cross-react with monkey and pig vitamin-D-binding protein, but not with vitamin-D-binding protein from the rat, mouse, or chicken.

## ADRENAL GLANDS

Also known as *suprarenal glands*, the adrenal glands are triangular-shaped glands located on the anteriosuperior aspect of both kidneys. They are found at the level of the T-12 vertebra; they have their own blood supply through the adrenal arteries and measure about 1/2 by 3 inches in humans (see Figure 35.7). The adrenal gland consists of an outer layer (i.e., adrenal cortex) and an inner layer (i.e.,

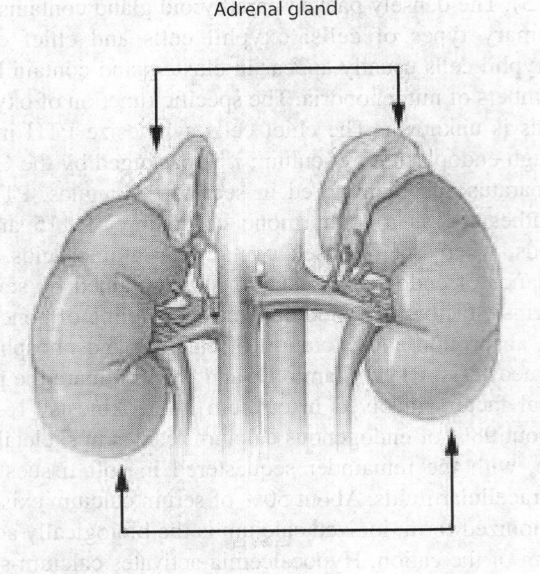


**FIGURE 35.7** Relationship of adrenal glands to the kidneys. (Courtesy of U.S. National Cancer Institute Surveillance, Epidemiology, and End Results [SEER] Program.)

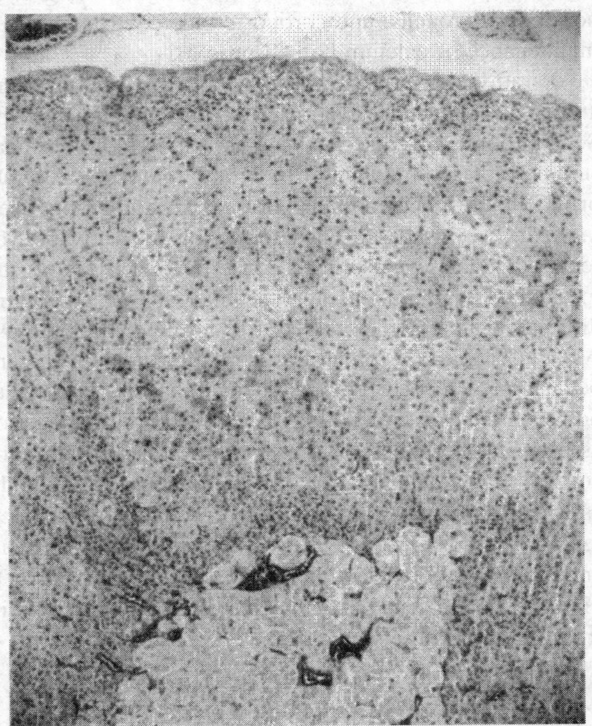

**FIGURE 35.8** Histology of rat adrenal cortex (outer area) and medulla (inner area) (original magnification, 126×).

adrenal medulla). The inner and outer layers can be readily distinguished by histologic preparations (Figure 35.8). These glands are regulated by both the hormonal and neural systems. The adrenal medulla is the primary source of the catecholamine hormones (epinephrine and norepinephrine) and responds to sympathetic nerve stimulation. The release of these hormones produces systemic effects resembling generalized sympathetic nerve stimulation. The adrenal cortex, on the other hand, produces two major groups of hormones—namely, mineralocorticoids and glucocorticoids. The hypothalamic–pituitary–adrenal axis controls reactions to stress and functions in regulating various body processes such as digestion and the immune system.

## Adrenal Medulla

The adrenal medulla is composed primarily of chromaffin cells, which secrete epinephrine, norepinephrine, and enkephalin, which are critial to the fight-or-flight response. The adrenal medulla can synthesize tyrosine into epinephrine and norepinephrine. The adrenal medulla is considered to be a ganglion of the sympthetic nervous system, as it is derived from neuronic neural crest and from modified neurons. In response to imminent danger, medullary cells release epinephrine and norepinephrine into the bloodstream at a ratio of about 70:30. This release causes an immediate increase in the heart rate, vasoconstriction, and metabolism.

## Adrenal Cortex

The adrenal cortex is composed of three tissue layers. The external layer, the *zona glomerulosa*, secretes mineralocorticoids such as aldosterone as part of the renin–angiotensin system regulating blood pressure. The intermediate layer, the *zona fasciculata*, produces cortisol, which is involved in the response to stress, increased blood pressure, hyperglycemia, and suppression of the immune system. ACTH causes corticol-producing cells to release glucocorticoids as part of the hypothalamic–pituitary–adrenal axis. The inner layer of the adrenal cortex is the *zona reticularis*. The cells of the zona reticularis are arranged in a network that has the same functions as cells of the zona fasciculata. Evidence suggests that the zona reticularis is the primary source of glucocorticoids and adrenal androgens. The zona fasciculata becomes involved only in very severe stress or after prolonged stimulation.

In humans, the main mineralocorticoid is aldosterone, although deoxycorticosterone (DOCA) also exhibits mineralocorticoid activity; however, DOCA potency is reported to be only 1/30th that of aldosterone. Cortisol, the major glucocorticoid produced in humans, also exhibits a small amount of mineralocorticoid activity. The basic physiological effect of aldosterone is on extracellular fluid volume. It promotes the renal reabsorption of sodium in the ascending portion of the loop of Henle, the distal tubule, and the collecting tubule by acting on the mineralocorticoid receptor. Sodium reabsorption is accompanied by reabsorption of the chloride anion. In addition to stimulating sodium reabsorption, aldosterone enhances the urinary excretion of potassium and hydrogen ions. The increased elimination of hydrogen ions can lead to alkalosis and an increased extracellular content of bicarbonate ions, which, when combined with an increased extracellular sodium and chloride content, promotes tubular reabsorption of water [127].

The adrenal gland is essential for life, especially the salt-retaining properties of the mineralocorticoids. In the absence of mineralocorticoids, the extracellular fluid potassium concentration rises, and sodium and chloride contents fall. Total lack of aldosterone secretion can cause the urinary elimination of 20% of the total body sodium in one day. The salt elimination can cause a life-threatening reduction in the extracellular and blood volume which, if untreated, leads to diminished cardiac output and death.

The major glucocorticoid secreted in humans is cortisol, or hydrocortisone, although both corticosterone and cortisone possess some glucocorticoid activity. The three major systems regulated by the glucocorticoids are carbohydrate, protein, and fat metabolism. With regard to carbohydrate metabolism, the glucocorticoids stimulate gluconeogenesis and decrease glucose utilization by the cells leading to hyperglycemia. Cortisol also breaks down protein and lipids which increases blood glucose, with a resulting increase in glycogen in the liver [35]. The glucocorticoids also produce a marked reduction in cellular protein content. An exception to this protein catabolic action is the liver, where protein content increases as does the production of plasma protein by the liver.

Cortisol is an essential chemical in the restoration of homeostasis after exposure to stress. Among other things, it promotes gluconeogenesis by inhibiting insulin. Glucocorticoids probably interfere with the transport of amino acids into extrahepatic cells, and this action combined with continuing protein catabolism in these cells produces an increase in plasma amino acid content. Increased plasma amino acid levels and their subsequent transport into the liver promote gluconeogenesis (i.e., the conversion of amino acids to glucose). The glucocorticoids also promote mobilization of fatty acids from adipose tissue which raises plasma fatty acid levels. This effect, plus increased oxidation of fatty acids in the cells, is probably involved in the metabolic conversion from glucose utilization to fatty acid utilization as a source of energy during periods of stress.

Corticosteroids can bind to plasma protein fractions. A corticosteroid-binding globulin (CBG) has a high affinity but a low binding capacity. Plasma albumin has a low affinity and a relatively high binding capacity. Under physiological conditions, most of the hormone is bound to CBG.

It is now recognized that many drugs and chemicals can produce lesions of either the adrenal medulla or the adrenal cortex (Table 35.14). Still other agents can specifically inhibit particular enzymatic steps involved in adrenal steroidogenesis (Figure 35.9). Not only have metyrapone, aminoglutethimide, and *o,p*′-DDD been used as agents to aid in the diagnosis of adrenal cortical dysfunction, but some have also been used as adrenolytic drugs in treating inoperable adrenal adenocarcinomas.

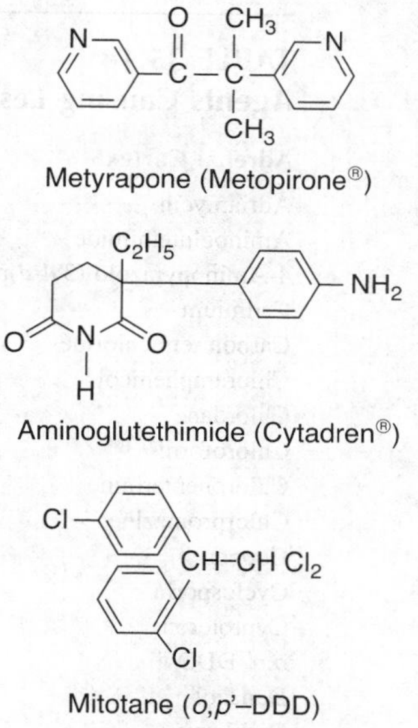

**FIGURE 35.9** Agents that inhibit adrenocortical steroid biosynthesis.

## Measurement of Corticosteroids

Several methods have been used to measure plasma and urinary aldosterone levels. Among them are double-isotope dilution techniques, gas–liquid chromatographic techniques, and RIAs [10]. A competitive immunoassay for cortisol reportedly is based on a capillary electrophoresis and laser-induced fluorescence technique [94]. Several analytical procedures have been used for the quantitation of glucocorticoids. One of the earliest procedures involved a colorimetric reaction between the glucocorticoid and a phenylhydrazine reagent. The Porter–Silber method and Zimmerman reactions are colorimetric assays used for measuring corticoids. Both cortisol, which is the primary glucocorticoid in humans, and corticosterone, which is the primary glucocorticoid in the rat, have also been measured using a competitive protein-binding assay that takes advantage of the binding affinity of the glucocorticoid-binding globulin found in the plasma. Many drugs can interfere with the measurement of corticosteroids and ketosteroids; see Table 35.15. With the development of immunochemical techniques, it was found that plasma glucocorticoids could be effectively measured using the RIA. A complicating factor was the existence of competition between the endogenous glucocorticoid-binding globulin and the added antibody for the radioligand.

**TABLE 35.14**
**Agents Causing Lesions to the Adrenal Gland**

| Adrenal Cortex | Adrenal Medulla |
|---|---|
| Adriamycin | Acrylonitrile |
| Aminoglutethimide | ACTH |
| 4-Aminopyrazolo (3,4-*d*)pyrimidine | Alloxan |
| Cadmium | Blocadren |
| Carbon tetrachloride | Chlordecone |
| Chloramphenicol | *o*-Chlorobenzylidine |
| Chlordane | Malononitrile |
| Chloroform | Cysteamine |
| Chlorphentermine | Dichloromethane |
| Chlorpromazine | 7,12-Dimethylbenzanthracene |
| Copper | Estrogens |
| Cyclosporin | Growth hormone |
| Cyproterone | Interleukin-2 |
| *o,p′*-DDD | Lactitol |
| Danazol | Lactose |
| Dichlorvos | Malathion |
| 7,12-Dimethylbenzanthracene | Mannitol |
| Etomidate | 1-Methyl-4-phenyl-1,2,3,6-tetrahydropyrine (MPTP) |
| 5-Fluorouracil | Neuroleptics |
| Kepone | Nicotine |
| Ketoconazole | Pyrazole |
| Nicotine | Reserpine |
| Phenobarbital | Retinol acetate |
| Polychlorinated biphenyls | Sorbitol |
| Spironolactone | Thiouracil |
| Tamoxifen | Thyroid hormones |
| TCDD | Thyroid-stimulating hormone |
| Tetrahydrocannabinol | 1,1,2-Trichloroethane |
| Toxaphene | Xylitol |

*Source:* Colby, H.D. and Longhurst, P.A., in *Endocrine Toxicology*, Atterwill, C.K. and Flack, J.D., Eds., Cambridge University Press, Cambridge, U.K., 1992. With permission.

Stress can produce a rapid increase in ACTH production, which in turn promotes the secretion of adrenocortical hormones; thus, the conditions under which the animal is prepared for blood sampling can modify the corticoid levels. It has been demonstrated that under ether anesthesia, blood corticoid levels are higher than those levels found after pentobarbital administration, which are higher than those measured following decapitation. Likewise, the time elapsed following the administration of pentobarbital influences blood hormone levels. Thus, stress can be a major problem affecting the interpretation of results. An additional related problem is that fairly large volumes of blood must be drawn for adrenocortical hormone measurement.

Adrenocortical function can be effectively monitored by measuring corticosteroid levels in 24-hour urine samples. Major advantages to this approach are that: (1) the animals can be housed stress-free in metabolic cages during toxicity testing, and (2) serial measurements can be conducted in the same animal by noninvasive techniques without sacrificing the animal. This approach cannot be used to measure urinary aldosterone in the rat because the primary site of degradation for aldosterone is the liver, and only about 1% of the secreted aldosterone is excreted unchanged in the urine. Nevertheless, relative changes in the urinary corticosteroid patterns could be of value as a screen for adrenotoxicity during chronic toxicity studies. RIA is the method of choice for the determination of urinary corticosteroids.

## GONADS

### Male Sex Hormones (Androgens)

Androgens control and maintain masculine characteristics. These actions are generally mediated by binding to intracellular steroid receptors that specifically bind testosterone and dihydrotestosterone. Testosterone is a produced from cholestorol in the Leydig cells (Figure 35.4) of the male testes and, to a much lesser degree, in the thecal cells of the ovaries. Small amounts of testosterone are produced in both males and females by the zona reticulosa of the adrenal glands.

**TABLE 35.15**
**Drugs That Interfere with the Measurement of Corticosteroids and Ketosteroids**

| |
|---|
| Antibiotics/antibacterial agents: |
| Nalidixic acid |
| Sulfamerazine |
| Triacetyloleandomycin |
| Sedatives/tranquilizers: |
| Chloral hydrate |
| Chlordiazepoxide |
| Chlorpromazine |
| Ethinamate |
| Meprobamate |
| Phenaglycodol |
| Reserpine |
| Monoamine oxidase inhibitors: |
| Etryptamine |
| Oral hypoglycemic agents: |
| Acetohexamide |
| Miscellaneous drugs: |
| Colchicine |
| Phenytoin (DPH) |
| Quinidine |
| Quinine |
| Spironolactone |

**TABLE 35.16**
**Endpoints Used in Assessment of the Male Reproductive System**

| |
|---|
| Sperm count |
| Sperm motility (reflectospermiograph) |
| Sperm head morphology |
| Testicular morphology |
| Sperm production rates |
| Epididymal sperm numbers and transit time |
| Spermatogenesis (dual-parameter flow cytometry) |
| Sperm membrane integrity: |
| Viability (eosin Y exclusion |
| Hypo-osmotic swelling |
| Nuclear maturity: |
| Acid aniline blue stain |
| Nuclear chromatin decondensation (SDS) |
| Acrosome assessment: |
| Normal intact acrosomes |
| Acrosin activity |
| Objective motility assessment: |
| Linearity (VSL/VCL) |
| ALH |
| Sperm–oocyte interaction: |
| Sperm–zona pellucida binding ratio |
| Sperm–oolemma binding ratio |
| Serum follicle-stimulating hormone and luteinizing hormone |
| Serum testosterone and dihydrotestosterone |
| Gravimetric response (e.g., prostate, seminal vesicles) |
| Sex accessory organ biochemical constitutions (e.g., fructose) |
| Hemizona assay (HZA) |

*Source:* Thomas, J.A., in *Casarett & Doull's Toxicology: The Basic Science of Poisons*, 5th ed., Klaassen, C.D., Ed., Pergamon Press, New York, 1996, pp. 547–581. With permission.

Testosterone is the primary male sex hormone. It acts as a virilizing agent and promotes secondary male sex characteristics and anabolic effects, such as increased bone and muscle strength and size and inhibition of estrogenic effects. In addition, testosterone enhances libido, immune function, normal bone growth, and general well-being. Testosterone binds to sex-hormone-binding globulin (SHBG), which is a glycoprotein that specifically binds testosterone and estradiol. The exact role of SHBG is not clear; however, it is thought that SHBG in concert with other carrier proteins provides a dynamic equilibrium between free and bound androgens in response to fluctuations in the secretion of androgens [58].

Testosterone not bound to SHBG can bind to androgen receptors or can be reduced to 5α-dihydrotestosterone (DHT) by 5α-reductase. Interestingly, DHT is a more potent agonist than testosterone. Once activated, the androgen receptors are transported to the nucleus, where they bind to specific areas of chromosomes to form hormone response elements (HREs), which ultimately enhance the appropriate gene transcription activity and result in anabolic or virilizing protein production. Anabolic steroids (e.g., 19-norsteroids) bind the androgen receptor, resulting in increased protein synthesis, and simultaneously block the effects of cortisol, thus reducing the rate of catabolism, especially of the muscle mass.

Male reproductive toxicology endpoints can be assessed in several ways (Table 35.16). The components of normal human male semen are listed in Table 35.17. Androgens can be chemically measured,or androgen-dependent end organs can be used to determine the endocrine status of male sex hormones. Androgens and other steroids can be biotransformed in several anatomical sites (Table 35.18).

Some cell types of the testes are more sensitive to chemical insult than are others. The various subpopulations of cells within the testes include the germ cells, the Sertoli cells, and the Leydig cells (interstitial cells) (Figure 35.4). Cross-sections of the mammalian testes reveal the seminiferous tubules containing the germinal epithelium and the Sertoli cells. Leydig cells lie outside these tubules and are located in the interstitium (Figure 35.10).

Gonadotoxicants may act directly on the testes, indirectly via the central nervous system (CNS), or by a combination of the two. Relative to the germinal epithelium, the Leydig cells are not as sensitive to the toxic effects of chemicals. A pig Leydig cell culture system has been devised to evaluate testicular toxicity [13]. This *in vitro* system can test the ability of a compound to inhibit

**TABLE 35.17**
**Characteristics of Normal Human Male Semen**

| Factor | Measurement |
|---|---|
| Appearance | Homogeneous, gray-opalescent ejaculate |
| Volume | >2 mL |
| Consistency | Not viscous |
| Liquefaction (conversion to liquid) | Complete within 60 minutes |
| Concentration | >20 million sperm per mL |
| Total count | >40 million sperm per ejaculate |
| Motility | >50% at 1 hour |
| pH | >7.2 |
| Morphology | >30% normal shape (WHO criteria) |
| | >14% normal shape (Krueger criteria) |
| White blood cells | <1 million per mL |

*Source:* ASRM, *Patient Fact Sheet*, American Society for Reproductive Medicine, Birmingham, AL, 2001. With permission.

**TABLE 35.18**
**Possible Sites of Action of Agents Affecting the Reproductive System**

| Anatomical Site | Endocrine Effects |
|---|---|
| *Central nervous system* | |
| Cerebral cortex | Altered secretion of follicle-stimulating hormone (FSH) and luteinizing hormone (LH) |
| Median eminence | Altered releasing hormone secretion |
| Adenohypopysis | Changes in gonadotropin secretion |
| *Peripheral target organs* | |
| Ovary | Altered secretion of estrogens |
| Testes | Altered secretion of androgens |
| Liver | Increased catabolism of steroids |
| Kidney | Increased excretion of steroids |

*Source:* Thomas, J.A., in *Casarett & Doull's Toxicology: The Basic Science of Poisons*, 5th ed., Klaassen, C.D., Ed., Pergamon Press, New York, 1996, pp. 547–581. With permission.

steroidogenesis. Some physiochemical characteristics of gonadotoxicants are shown in Table 35.19. The toxicity of a particular gonadotoxicant depends to some extent on its ability to penetrate the blood–testes barrier, which is the barrier formed by tight connections between the Sertoli cells. In younger animals, the testicular gap junctions are still relatively permeable, and the germinal epithelium may be more vulnerable than in the adult testes.

Androgens are steroids that can be measured fluorometrically, by chromatography, and by RIA (Figure 35.11). They also can be bioassayed using various biological responses such as the stimulation of accessory sex organ weights in castrated animals (e.g., prostate gland or seminal vesicle weights). Still other androgen assessments can be made by measuring male sex accessory gland constituents (e.g., seminal vesicle fructose levels and zinc concentrations).

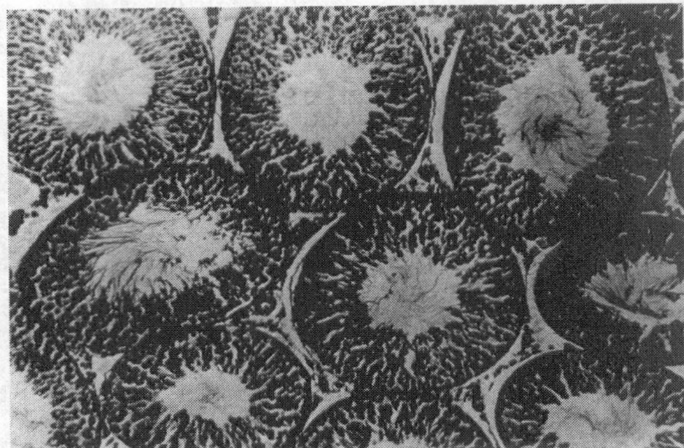

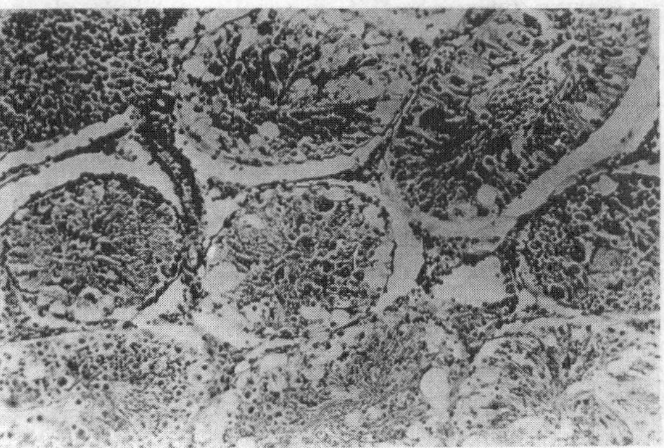

**FIGURE 35.10** Normal rat testes (top) and chemically induced sterility in rat testes (bottom). Note the partially sterile seminiferous tubules in lower panel.

**TABLE 35.19**
**Some Physicochemical Characteristics of Gonadotoxicants**

- Usually lipophilic
- Avidity/affinity for androgen receptor
- Can permeate the testes–blood barrier
- Diverse chemical structures
- Molecular weight frequently less than 400
- Propensity for rapidly dividing cells

Androgens possess three primary biological actions: (1) virilizing or masculinizing actions, (2) protein anabolic or myotropic actions, and (3) antiestrogenic properties. All of these biological effects can be bioassayed in experimental animals. Rodents are the most commonly used animal for the bioassay of androgens. The actions of male sex hormones (i.e., their masculinizing actions) are usually bioassayed on the basis of gravimetric responses in sex accessory organs. Generally, either the rat seminal vesicle or ventral lobes of the prostate gland are used to bioassay androgens (Figure 35.12). For the bioassay, rats are castrated, and the sex accessory glands are allowed to regress for about 7 days. Using

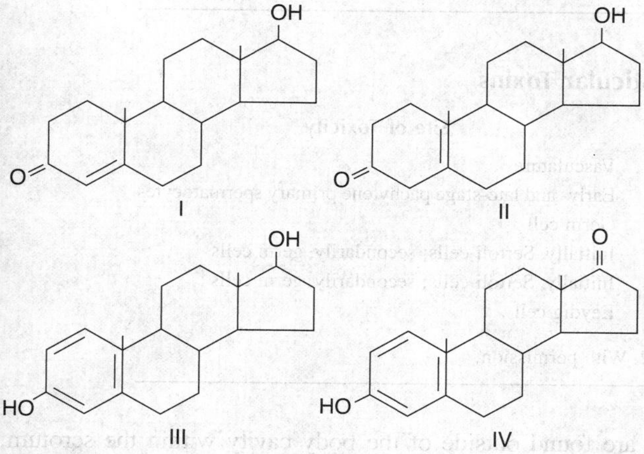

**FIGURE 35.11** I. Testosterone. II. Dihydrotestosterone. III. Estradiol-17β. IV. Estrone.

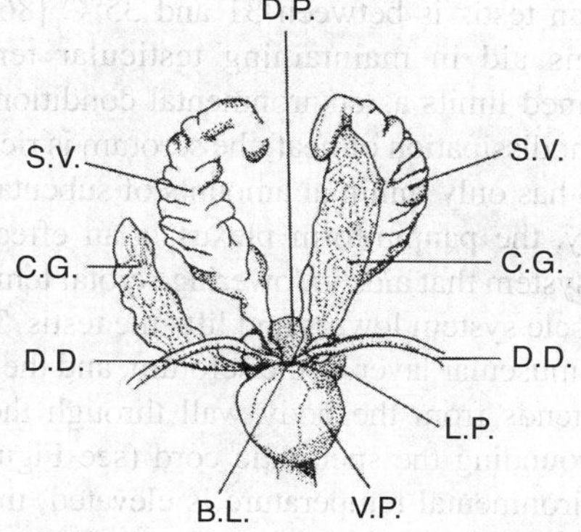


**FIGURE 35.12** Anatomical components of rodent sex accessory glands. D.D., ductus deferens; B.I., bladder; V.P., ventrtal prostate; L.P., lateral prostate; C.G., coagulating gland (also called anterior prostate); S.V., seminal vesicle; D.P., dorsal prostate.

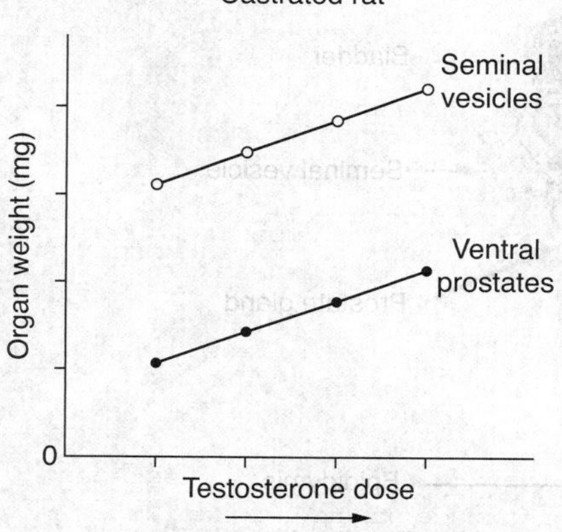


**FIGURE 35.13** Bioassay of adrogens using rat seminal vesicles or ventral prostate glands from castrated rats.

testosterone as a standard, castrated rats are injected for several days, the animals are killed, and either the seminal vesicles (empty) or the ventral prostate glands are removed and weighed (Figure 35.13). The data reveal that organ weights of both the seminal vesicles and the ventral prostates increase directly proportional to the increase in concentration of injected testosterone. Sometimes immature rats are used for the bioassay instead of castrate animals.

A classical, but now obsolete, bioassay used to assess androgenic activity was the weight or the size of the comb of the castrated rooster or immature cockerel. The direct application of testosterone (or other androgenic substances) to the comb causes it to increase in size and weight. The rat levator ani muscle weight has been used to bioassay the protein anabolic actions of androgens. The levator ani bioassay was developed following observations of the myotropic and nitrogen-retaining activities of certain androgenic steroids. This test lacks specificity as a measure of protein anabolic activity. The levator ani test ordinarily uses immature castrated rats that have been treated for 1 week with the steroid. In addition to androgen bioassay methods using rodent sex accessory organs and muscles, as well as bird combs, male sex hormones, which exert a renotropic action in mice, can be utilized. Androgens can stimulate the growth of the kidneys in castrated or immature mice.

Chemical indicator tests also have been used to assess androgen activity. Sex accessory organs contain several biochemical constituents that are androgen dependent and quantifiable [75]; for example, sex accessory fructose decreases following castration and can be restored by androgen administration. Sex accessory fructose has been used as a sensitive chemical indicator for testosterone and other androgens. Fructose can be measured spectrophotometrically by several colorimetric reactions. Similarly, sex accessory organ citric acid can be used to assess androgenic activity. These chemical indicator tests for androgens are more sensitive than the gravimetric responses used in bioassay procedures. A number of different endpoints to assess male reproductive function have been proposed for food and color additives [100].

Urinary creatine profiles have been used as a chemical indicator for testicular toxicity [82]. Several chemicals have been assessed for their testicular toxicity, as evidenced by increases in urinary creatine (Table 35.20). Creatine is associated primarily with cells of the seminiferous epithelium. Several chemicals can destroy the germ cells, as reflected by creatinuria. Chemicals that destroy primarily Leydig cells do not lead to creatinuria; for example, ethane dimethane sulfonate (EDS) exerts a direct cytotoxic action on the Leydig cells. Creatinuria is related to testicular degeneration, and a substantial portion of the testicular creatine is associated with the cells of the seminiferous epithelium.

Testosterone levels have been determined using double-isotope derivative methods, gas–liquid chromatography, and fluorometric procedures. Although these testosterone methods are specific and sensitive, elaborate purification is essential for accuracy, and routine application is often time

**TABLE 35.20**
**Urinary Creatine: A Test for Cell-Specific Testicular Toxins**

| Agent | Urinary Creatine | Site of Toxicity |
|---|---|---|
| Cadmium (Cd) | ↑ | Vasculature |
| 2-Methoxyethanol (2-ME) | ↑ | Early- and late-stage pachytene primary spermatocytes |
| Methoxyacetic acid (MAA) | ↑ | Germ cell |
| di-(*n*-Pentyl) phthalate (DPP) | ↑ | Initially, Sertoli cells; secondarily, germ cells |
| Dinitrobenzene (DNB) | ↑ | Initially, Sertoli cells; secondarily, germ cells |
| Ethane dimethane sulphonate (EDS) | ↔ | Leydig cell |

*Source:* Moore, N.P. et al., *Arch. Toxicol.*, 66, 435–442, 1992. With permission.

consuming. Competitive protein binding assays and RIAs are available. In the RIA, antiserum against testosterone can be produced in rabbits immunized with testosterone-3-oxime-beef serum albumin. Biological samples must undergo solvent extraction and be eluted on microcolumns. If the extraction and purification of the samples are appropriately carried out, the RIA of testosterone is highly sensitive and accurate; it is capable of detecting testosterone in nanogram to picogram amounts. Recently, a method has been described for assessing androgens under field or remote site conditions [50]. Monoclonal antibody techniques can be used for assessing androgen receptors.

## Male Infertility/Testicular Toxicity

It is estimated that 30 to 40% of infertility in humans is due to the male partner. The etiology of male infertility may be genetic or nongenetic. There may also be psychological aspects to male infertility. Infertility may or may not be due to direct toxic insults to the testicles. In mammals, the testes are found outside of the body cavity within the scrotum. Because spermatogenesis cannot occur at body temperature, scrotal temperature is a few degrees lower than internal body temperature. The physiological temperature of the adult human testis is between 31 and 35°C [86]. Several mechanisms aid in maintaining testicular temperature within defined limits as environmental conditions change. For efficient dissipation of heat, the scrotum is rich in sweat glands and has only minimal amounts of subcutaneous fat. Importantly, the panpiniform plexus is an effective heat-exchange system that aids in lowering scrotal temperatures. A two-muscle system lowers and lifts the testis. The tunica dartos is a muscular layer of the scrotum, and the cremaster muscle extends from the body wall through the inguinal canal, surrounding the spermatic cord (see Figure 35.14). When environmental temperature is elevated, the muscles relax in unison, and the testes are lowered from the body, permitting the escape of excess heat. Under cold conditions, the muscles contract, pulling the testes toward the warmth generated by the body.

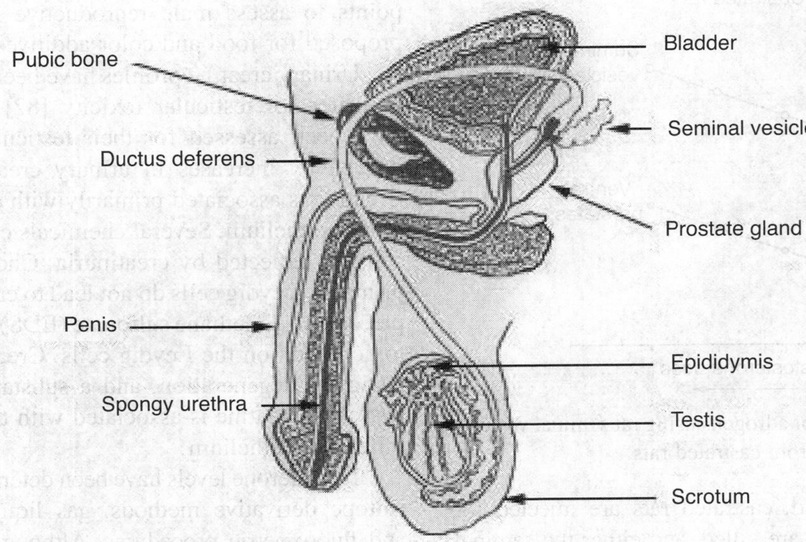


**FIGURE 35.14** Male reproductive tract. (Courtesy of U.S. National Cancer Institute Surveillance, Epidemiology, and End Results [SEER] Program.)

**TABLE 35.21**
**Potentially Useful Tests of Male Reproductive Toxicity for Laboratory Animals and Humans**

| | |
|---|---|
| *Testis* | *Fertility* |
| Size *in situ* | Ratio of exposed to pregnant females |
| Weight | Number embryos or young per pregnant female |
| Spermatid reserves | Ratio of viable embryos to corpora lutea |
| Gross and histologic evaluation | Number of 2- to 8-cell eggs |
| Nonfunctional tubules (%) | Number of unfertilized eggs; abnormal eggs |
| Tubules with lumen sperm (%) | Sperm per ovum |
| Tubule diameter | |
| Counts of leptotene spermatocytes | *In vitro* |
| | Incubation of sperm in agent |
| *Sperm motility* | Hamster egg penetration test |
| Time-exposure photography | |
| Multiple-exposure photography | *Semen* |
| Cinemicrography | Total volume |
| Videomicrography | Gel-free volume |
| Sperm membrane characteristics | Sperm concentration |
| Evaluation of sperm metabolism | Total sperm/ejaculate |
| Fluorescent Y bodies in spermatozoa | Total sperm/day of abstinence |
| Flow cytometry of spermatozoa | Sperm motility, visual (%) |
| Karyotyping human sperm proneuclei | Sperm motility, videotape (% and velocity) |
| Cervical mucus penetration test | Gross sperm morphology |
| | Detailed sperm morphology spermatozoa |
| *Endocrine* | |
| Luteinizing hormone | *Accessory sex glands* |
| Follicle stimulating hormone | Histology |
| Gonadotropin-releasing hormone | Gravimetric |
| | |
| *Epididymis* | *Other tests considered* |
| Weight and histology | Tonometric measurement of testicular consistency |
| Number of sperm in distal half | Qualitative testicular histology |
| Motility of sperm, distal end (%) | Stage of cycle at which spermiation occurs |
| Gross sperm morphology, distal end (%) | Quantitative testicular histology |
| Detailed sperm morphology, distal end (%) | |
| Biochemical assays | |

*Source:* Thomas, J.A., in *Casarett & Doull's Toxicology: The Basic Science of Poisons*, 5th ed., Klaassen, C.D., Ed., Pergamon Press, New York, 1996, pp. 547–581. With permission.

Numerous factors may contribute to male infertility and encompass both neural and hormonal elements. The hormonal factor can be affected by gonadotoxicants, which cause spermatogenic arrest and degeneration of seminiferous tubules, leading to infertility (i.e., low sperm count). Industrial chemicals can cause sterile tubules and result in infertility [104]; examples include dibromochloropropane (DBCP) [27] and alkylating agents (e.g., nitrogen mustards) [38]. Also, x-rays [22] and cryptorchidism (undescended testes) [48] can lead to infertility.

Many useful tests can aid in determining reproductive success, but such tests must be undertaken in both the male (Table 35.21) and the female (Table 35.22). Although the measurement of fertility is imprecise, it does have the advantage of integrating all reproductive functions from both the male and the female [124]. Testicular histology is not quantitative but does provide some information about gonadal physiology (or pathology).

Due to transgene technology, several animal models are now available for studying male sterility [62]. Genetic aberrations are evident not only in the processes involved in spermatogenesis but also in anatomical defects observed in sex accessory organs. Gonadotropin-releasing hormone (GnRH) cell lines have been developed and can produce a transgenic mouse with hypothalamic hypogonadism [122]; hence, transgenic mouse models can be used to study the mechanisms of spermatogenic arrest as well as altered neuroendocrine systems that interfere with gonadotropin secretion.

**TABLE 35.22**
**Potentially Useful Tests of Female Reproductive Toxicity**

*Ovary*
Organ weight
Histology
Number of oocytes
Rate of follicular atresia
Follicular steroidogenesis
Follicular maturation
Oocyte maturation
Ovulation
Luteal function

*Cervix/vulva/vagina*
Cytology
Histology
Mucus production
Mucus quality (sperm penetration test)

*Oviduct*
Histology
Gamete transport
Fertilization
Transport of early embryo

*In vitro*
*In vitro* fertilization of superovulated eggs, either exposed to chemical in culture or from treated females

*Uterus*
Cytology and histology
Luminal fluid analysis (xenobiotics, proteins)
Decidual response
Dysfunctional bleeding

*Fertility*
Ratio of exposed to pregnant females
Number of embryos or young per pregnant female
Ratio of viable embryos to corpora lutea
Number of 2- to 8-cell eggs
Number of unfertilized eggs; abnormal eggs
Number of corpora lutea

*Hypothalamus*
Histology
Altered synthesis and release of neurotransmitters, neuromodulators, and neurohormones

*Pituitary*
Histology
Altered synthesis and release of tropic hormones

*Endocrine*
Gonadotropin
Chorionic gonadotropin levels
Estrogen and progesterone

*Source:* Thomas, J.A., in *Casarett & Doull's Toxicology: The Basic Science of Poisons*, 5th ed., Klaassen, C.D., Ed., Pergamon Press, New York, 1996, pp. 547–581. With permission.

## Female Sex Hormones (Estrogens)

Estrogens are biosynthesized primarily by theca interna cells of the developing follicles in the ovaries, the corpus luteum, the placenta, and, in lesser amounts, the adrenal cortex. Secondary sources such as the breasts, liver, and adrenals are an important source of estrogens in postmenopausal women. Cholesterol is converted to pregnenolone by the enzyme P450 side-chain cleavage; another enzyme, CYP17A, converts pregnenolone to 17α-hydroxypregnenolone and then to dehydroepiandrosterone (DHEA), which is also a precursor steroid that is converted to estrogens and estradiol. Cholesterol is also synthesized to androstenedione in the theca interna cells which is subsequently absorbed into the nearby granulosa cells of the ovary. In granulosa cells, androstenedione is converted primarily to estradiol-17β. Evidence suggests that estrone is also secreted in the process and can be further metabolized to estradiol by hepatic enzymes (see Figure 35.15). Of these three main naturally occurring estrogens, estradiol-17β is undoubtedly the most potent. All of these naturally occurring estrogens

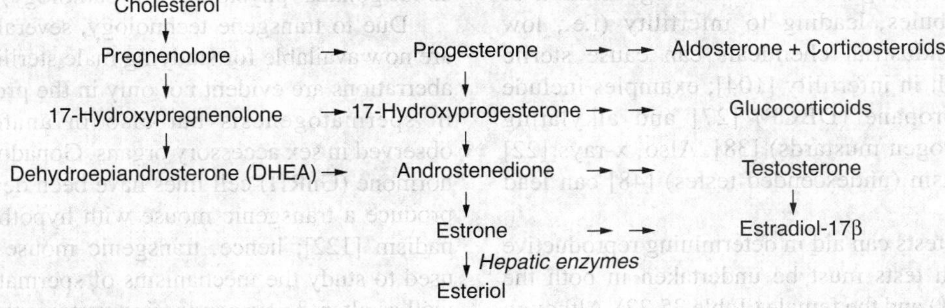


**FIGURE 35.15** Biosynthesis of sex steroids.

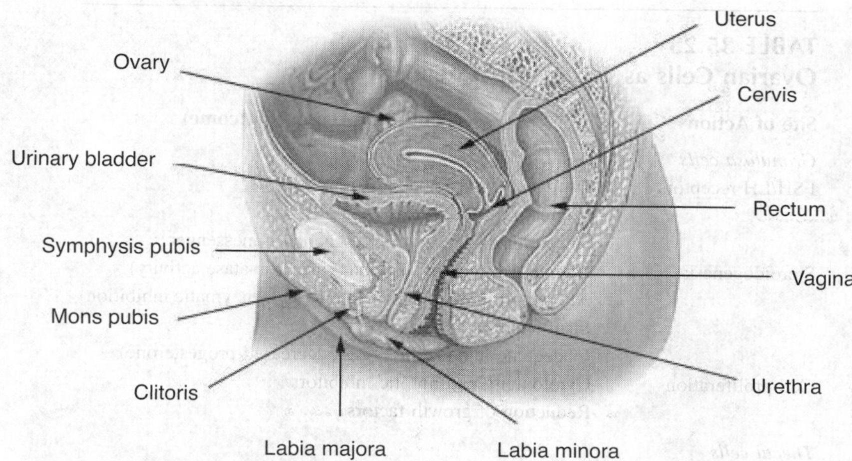


**FIGURE 35.16** Female reproductive tract. (Courtesy of U.S. National Cancer Institute Surveillance, Epidemiology, and End Results [SEER] Program.)

are steroids. Synthetic estrogens such as diethylstilbestrol are not steroids. Although steroidal and nonsteroidal estrogens can be bioassayed using similar tests, their different molecular structures do not allow them to be measured by similar chemical methodologies. Testing guidelines for evaluating reproductive and developmental toxicity in the female have recently undergone some revisions [25]. The two methods commonly once used to bioassay estrogenic hormones involve either histological changes in the vaginal epithelium or an increase in the weight of the rodent uterus. Both assays require the use of ovariectomized animals.

The mouse vaginal smear bioassay uses ovariectomized animals with different standard doses of estradiol as controls. To establish that the animals are responding, vaginal smears are characterized by nucleated epithelial cells or cornified cells. Vaginal responses characterized by epithelial cornification are considered positive responders to estradiol. Estrogen bioassays also include using an increase in uterine weight in the ovariectomized rat or mouse. Uterine weight falls precipitously after ovariectomy. The reduced uterine weight can be restored by daily injections of estradiol. Ovariectomized mice are injected subcutaneously and killed, and the uteri are removed and weighed. Like the vaginal smear bioassay, at least two dilutions of the unknown substance are run concurrently with the estradiol dose–response curve. The immature chick can also be used to assess estrogenic potency. The immature chick bioassay is based on estrogens stimulating oviduct weights.

These bioassays (i.e., vaginal cornification, uterine weights, and oviduct weights) are quite sensitive to estrogen but have largely been abandoned for newer and more expedient and accurate assays. Several useful tests can be utilized in determining reproductive toxicity in the female reproductive system (Table 35.22). Estrogens can be measured colorimetrically, by RIA, by monoclonal assays, and by other immunotechniques. Estrogen receptors can also be measured by monoclonal antibody techniques that employ enzyme immunoassays (EIAs). Likewise, progesterone receptors can be measured with EIAs [30]. The evaluation of techniques for the detection of functional estrogenicity has received renewed attention [64].

### Female Infertility/Ovarian Toxicity

In females, the ovaries are homologous to the testicles in the male. They are located internally in the female on the posterior wall of the pelvis lateral to the position of the uterus and are structurally supported by three ligaments: the suspensory ligament, the broad ligament, and the ovarian ligament. The ovaries are close to but not attached to the fallopian tubes, which transport the ovum as it is expelled from the ovary to the uterus (Figure 35.16). Toxicants can affect reproductive outcomes at a number of points in the female reproductive process through their direct effects on the ovaries, ovulation, fertilization, implantation, and the developing fetus.

A paucity of research data exists regarding ovarian toxicants. Direct effects on the ovary can produce a reduction in ovarian hormonal secretion, resulting in a change in the LH and FSH levels as a result of reduced negative feedback signals to the pituitary and hypothalamus. Direct ovarian toxicants could also hinder the production of follicles which could result in early menopause or result in a disruption of menses [51]. Reproductive toxicants produce changes in ovarian functions at different levels. Some drugs are known to be toxic to the ovaries, such as nitrogen mustards, chlorambucil, cyclophosphamide, busulfan, vinblastine, and polycyclic aromatic hydrocarbons [79];

**TABLE 35.23**
**Ovarian Cells as Targets for Chemical Injury**

| Site of Action | Mechanism of Action (Outcome) |
|---|---|
| *Granulosa cells* | |
| FSH/LH receptors | Decreased receptor population |
| | Competition for receptors |
| | Uncoupling of receptors to secondary messenger |
| Steroidogenesis | Altered estrogen production (e.g., aromatase activity) |
| | Altered progesterone production (e.g., enzymatic inhibition) |
| | Insufficient androgens |
| | Inadequate luteinization (e.g., decreased progesterone) |
| Cell proliferation | Cytotoxicity and mitotic inhibitors |
| | Reduction of growth factors |
| *Thecal cells* | |
| LH receptors | Decreased receptor population |
| | Competition for receptors |
| | Uncoupling of receptors to secondary messengers |
| Steroidogenesis | Inhibition of enzymes (e.g., decreased androgens) |
| | Insufficient substrate for granulosa cells |
| Cell proliferation | Cytotoxicity and mitotic inhibitors |
| | Disrupted migration of stroma |
| | Reduction of growth factors |

*Source:* Mattison, D.R. et al., *Med. Clin. North Am.*, 74, 391–411, 1990. With permission.

hence, the ovary is clearly vulnerable to chemical injury [67,78]. Both the granulosa cells and the thecal cells are targets for chemical injury (Table 35.23). Chemicals that inhibit gonadotropin secretion, that damage gonadotropin receptors, or that uncouple the receptor from other molecules necessary for hormone action would be expected to adversely affect granulosa cells. Thecal cells provide precursors for granulosa cell steroidogenesis. Xenobiotics can alter either granulosa cells or thecal cells. The oocytes themselves are also targets for chemical insult or injury. Alkylating agents, lead, and mercury can be destructive to the mammalian oocyte [51,78].

## PANCREAS

In humans, the pancreas is a retroperitoneal lobulated organ located posterior to the stomach and in close association with the duodenum, approximately 1 inch in height and 5 inches in length and weighing up to about 100 g (see Figure 35.17) [130]. The pancreas has two primary

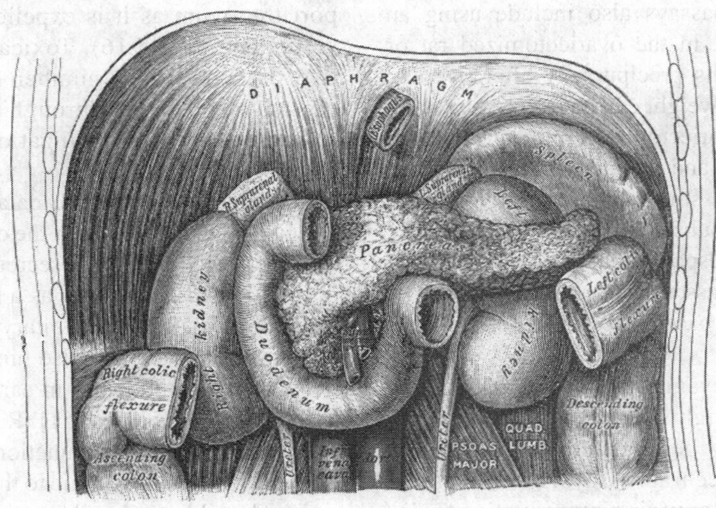


**FIGURE 35.17** Anatomical relationship of pancreas to adjacent organs.

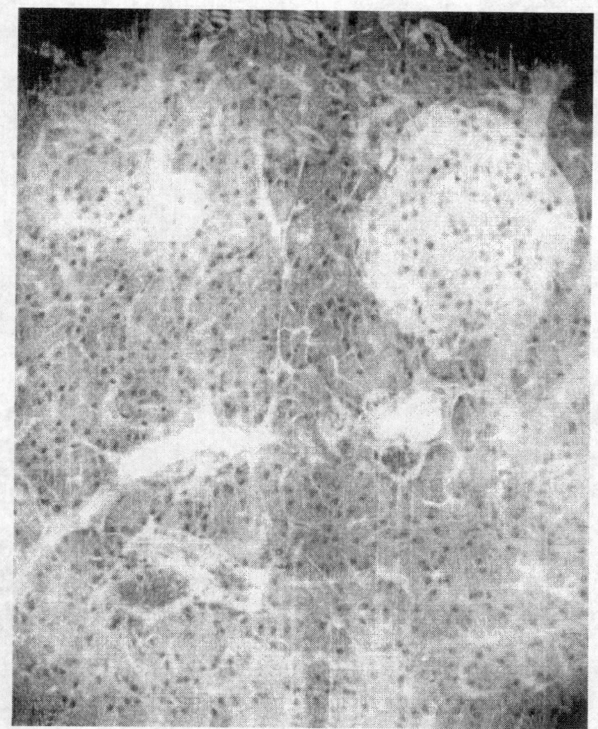

**FIGURE 35.18** Histology of rat pancreas. Note the acini ducts (original magnification, 256×).

functions: (1) The exocrine pancreas produces enzymes for digesting lipids and proteins that break down various categories of digestible foods from asini cells, and (2) the endocrine pancreas secretes hormones from clusters of glandular epithelial cells called *pancreatic islets* (islets of Langerhans) that produce the hormones glucagon, insulin, and somatostatin (see Chapter 31).

The exocrine pancreas constitutes the vast majority of the pancreas and contains the acini cells. The acini cell secretions collect in the acini ducts, which communicate directly with the intralobular ducts, eventually draining into the major pancreatic duct or the duct of Wirsung. This duct subsequently unites with the common bile duct, which enters the upper end of the duodenum in a duct common to both organs termed the *hepatopancreatic ampulla* or the *ampulla of Vater*. At this juncture, the exocrine pancreatic secretions enter the duodenum and exert their effects upon the intestinal chyme. The morphologic arrangement of the rat pancreas is depicted in Figure 35.18.

The endocrine pancreas secretes two hormones important in the regulation of carbohydrate metabolism: insulin and glucagon. Insulin is the primary regulator in carbohydrate homeostasis, as it controls blood glucose levels. Additionally, insulin regulates muscle tone, the uptake of amino acids and electrolytes, and the release of triglycerides. By binding to specific hepatocyte receptors, glucagon also assists in the regulation of the level of glucose in the blood, resulting in: (1) the release of stored glycogen as glucose into the bloodstream (termed *glycogenolysis*), and (2) the synthesis of additional glucose for release into the bloodstream (termed *gluconeogenesis*). Glucogon stimulates the release as well as the synthesis of glucose in response to hypoglycemia. These hormones are synthesized in the islets of Langerhans by α cells (glucagon) and β cells (insulin). Somatostatin is considered an inhibitory hormone that can retard the release of growth-hormone-releasing factor (GHRF), growth hormone, thyroid-stimulating hormone, insulin, and glycogen. It also suppresses the release of gastrointestinal hormones such as gastrin, cholecystokinin, secretin, motilin, gastric inhibitory polypeptide, and enteroglucagon and generally slows the rate of smooth muscle contractions and the flow of blood into the small intestine.

A concern when testing for toxicity is the potential for the compound to interfere with the normal functioning of the pancreatic β cells. The indicator that often alerts the toxicologist to a possible pancreatic side-effect is hyperglycemia. Although this increased blood glucose level would usually be detected during routine clinical chemistry analyses, additional tests may be required to pinpoint specific pancreatic toxicity.

The interactions between insulin and carbohydrate, fat, and protein metabolism represent some of the complexities seen in diabetes mellitus. Hyperglycemia may result from impaired utilization of glucose by cells due principally to the insufficient production of insulin. The failure of glucose to penetrate adipose tissue mobilizes fat, producing a rise in the free fatty acid and triglyceride content of plasma and the triglyceride content of the liver. A diabetic fatty liver can occur from the absence of lipoprotein synthesis due to accelerated gluconeogenesis. If glucose oxidation is impaired, fatty acids form the major source of energy. This condition generates an excess of intermediary metabolites, collectively described as ketone bodies (acetone, acetoacetic acid, and β-hydroxybutyric acid), which can lead to metabolic acidosis.

Hyperglycemia also can lead to the presence of glucose in the urine (glycosuria) when the blood glucose levels exceed the renal threshold of approximately 180 mg/dL. At lower levels, all the filtered glucose is normally reabsorbed by the renal tubules. Blood glucose levels increased to the point of glycosuria also can be caused by emotional stress and the concomitant release of glucose from liver glycogen in response to epinephrine. Several drugs can affect blood glucose levels and can either increase or antagonize glucose levels (Table 35.24). Glycosuria also can be the consequence of impaired renal tubular function caused by compounds such as the glycoside phlorizin [20]. Renal glycosuria can produce an osmotic diuretic effect that leads to dehydration and polydipsia. Glycogenolysis and gluconeogenesis are increased in diabetes mellitus, generating glucose, which in turn increases blood glucose levels.

**TABLE 35.24**
**Interaction of Drugs with Oral Hypoglycemic Agents and Insulin**

| Enhanced Effect[a] | Antagonist Effect |
|---|---|
| Anabolic steroids | Acetazolamide |
| Chloramphenicol | Corticosteroids |
| Dihydroxycoumarin | Diuretics (e.g., chlorthalidone, ethracrynic acid, furosemide, thiazides, triamterene) |
| Ethanol | |
| Guanethidine | |
| K+ salts | D-Thyroxine |
| MAO inhibitors | Epinephrine |
| Oxytetracycline | Marijuana |
| Phenylbutazone | Oral contraceptives |
| Phenyramidol | Phenothiazines |
| Probenecid | Phenytoin (DPH) |
| Propranolol | |
| Salicylates | |
| Sulfinpyrazone | |
| Sulfonamides | |

[a] Greater hypoglycemic action.

**TABLE 35.25**
**Genetically Altered Rodents Used To Study Diabetes Mellitus**

| | |
|---|---|
| *Diabetes-susceptible strains* | |
| Mice | C57BLKsJ |
| | DBA/2J |
| | SWR/J |
| | C3H.SW/SnJ |
| | C3HeB/FeCHp (males only) |
| | CBA/Lt (males only) |
| | NOD (IDDM) |
| Rats | BB (IDDM) |
| | BHE |
| | BHE/cdb |
| *Diabetes-resistant strains* | |
| Mice | C57BL/6J |
| | 129/J |
| | Ma/MYJ |

*Source:* Thomas, J.A., in *Biotechnology and Safety Assessment*, Thomas, J.A. and Myers, L.A., Eds., Raven Press, New York, 1993. With permission.

Experimental diabetes mellitus can be produced by destroying β cell function with alloxan or streptozocin (Figure 35.19). In addition to chemically induced destruction of β cells by alloxan and streptozocin, these chemicals can be diabetogenic [32,131]. Alloxan, a cyclic urea analog, can produce permanent hyperglycemia in the rabbit. Streptozotocin, a methylnitroso–urea analog, has generally replaced alloxan to produce experimental insulin-dependent diabetes in laboratory animals. Both agents destroy pancreatic β cells, but streptozotocin may involve the alkylation of critical cell components. Chlorozotocin, an analog of streptozocin, is lethal to β cells in culture. Vacor (pyriminil), cyproheptadine, and pentamidine are all capable of causing pancreatic dysfunction. In addition to chemically induced experimental models for diabetes mellitus, several genetic strains are either diabetes susceptible or diabetes resistant Table 35.25); thus, experimental animal models for diabetes mellitus may exploit genetic susceptibility or may involve the chemically

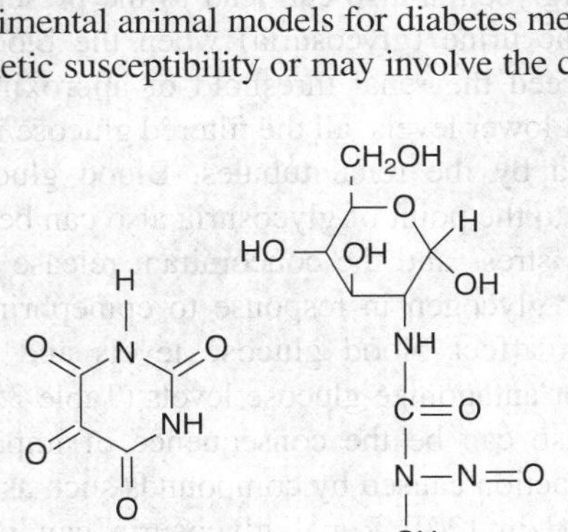

**FIGURE 35.19** Agents that inhibit pancreatic insulin and glucagon secretion.

induced destruction of β cells [71]. Several inbred strains of mice are sensitive to *db* gene-induced diabetes, with sexual dimorphism in some inbred strains emphasizing the relationship between the obesity gene and sex. A major genetic regulator of inbred-strain diabetogenic sensitivity is gender related. In the rat, the BHE strain is an excellent animal model for the study of non-insulin-dependent diabetes mellitus (NIDDM). The Cdb:BHE stock is a subline of the parent BHE stock. The nonobese diabetic (NOD) mouse is an ideal animal model of insulin-dependent diabetes mellitus [113]. Spontaneous diabetes in the BioBreeding (BB) rat, like human type 1 diabetes mellitus, results from the destruction of the pancreatic islets by autoreactive T lymphocytes recognizing β-cell-specific antigens.

Chemical insult is not unique to the endocrine pancreas; indeed, many agents can cause acute pancreatitis (Table 35.26). Side-effects resulting from a number of cytotoxic agents (e.g., colchicine), antibiotics (e.g., tetracyclines), antibacterial drugs (e.g., sulfonamides), and diuretics (e.g., thiazides) can produce irritation and inflammation of pancreatic cells.

## Measurement of Insulin

Before RIAs, insulin was bioassayed using the rabbit hypoglycemia test. The biologic potencies of porcine, bovine, and human insulin were assessed using the rabbit bioassay. Insulin was the first protein hormone to be measured by RIA. Although the RIA techniques that are available are similar with regard to the interaction between antigen and antibody, numerous variations exist in the methods

**TABLE 35.26**
**Drugs Implicated in Acute Pancreatitis**

| *Definite* | *Possible* |
|---|---|
| Azathiorine | Anticholinesterase |
| Cisplatin | Bumetanide |
| Colaspase (L-asparaginase) | Carbamazepine |
| Corticosteroids | Chlorthalidone |
| Frusemide | Clonidine |
| Lisinopril | Colchicine |
| Sulfonamides | Co-trimoxaz (furosemide) |
| Tetracycline | Cyclosporin |
| Thiazides | Cytarabine (cytosine arabinoside) |
| | Diazoxide |
| *Probable* | Enalapril |
| Cimetidine | ERCP contrast media |
| Diazinon | Ergotamine |
| Estrogens | Ethacrynic acid |
| Indomethacin | Isoniazid |
| Fonofos | Isotretinoin (13-*cis*-retinoic acid) |
| Mefenamic acid | Mercaptopurine |
| Opiates | Methyldopa |
| Paracetamol (acetaminophen) | Metronidazole |
| Phenformin | Nitrofurantoin |
| Valproic acid | Oxphenbutazone |
| | Piroxicam |
| | Procainamide |
| | Rifampicin |
| | Salicylates |
| | Sulindac |

*Source:* Banerjee, A.K. et al., *Med. Toxicol. Adverse Drug Exp.*, 4, 186–198, 1989. With permission.

for separating free insulin from antibody-bound insulin: gel filtration, salt precipitation, alcohol precipitation, precipitation with anti-gamma-globulin serum, and absorption on anion-exchange resin, cellulose, dextran-coated charcoal, antibody-coated tubes, or Sephadex-coupled antibodies. A radioreceptor assay has been developed for quantitation of plasma insulin levels, which may also be determined by ELISA. Generally, ELISA has a number of drawbacks that preclude its routine use for the measurement of insulin, including a lower sensitivity than is seen with RIA. A hyperglycemic clamp can be used to assess insulin secretion and insulin sensitivity [81]. Insulin receptors can be isolated from a variety of sources, including placenta, rat liver plasma membranes, human lymphocytes, and guinea pig kidney.

### Glucose Tolerance Tests

The glucose tolerance test (GTT), useful for evaluating endocrine pancreatic function, is based on the compensatory regulation of blood glucose levels by insulin following ingestion of a glucose load. The glucose load also can be administered orally or intravenously. If 50 g of glucose is taken orally, the blood glucose level rises rapidly for approximately 30 to 60 minutes and then falls rapidly to obtain fasting levels by 2 to 3 hours. In situations where the synthesis or release of insulin is insufficient, ingestion of the glucose load leads to an excessive rise in blood glucose levels, followed by a slow, gradual decline to preingestion levels. Abnormal glucose tolerance curves are evident in diabetes mellitus but may also be abnormal in other pathologic states. For example, elevated GTT values can indicate any one of the following conditions or diseases: acromegaly, chronic renal failure, Cushing's syndrome, stress, hyperthyroidism, pancreatic cancer, pancreatitis, or the use of various drugs (e.g., tricyclic antidepressants, epinephrine, diuretics, lithium). On the other hand, decreased GTT response may indicate any one of the following conditions or diseases: adrenal insufficiency, liver disease, hypothyroidism, hypopituitarism, insulinomas, or the use of acetaminophen and anabolic steroids [126]. Glucose oxidase methods involving colorimetric reactions are routinely used to measure blood sugar. Capillary blood glucose can be used to monitor diabetes mellitus [27].

### Cytological Evaluation of Pancreatic Islet Cells

Alpha and beta cells can be differentiated using an aldehyde–fuchsin stain (Scott stain) following fixation of the tissue in Bouin's solution. The $\alpha$ cells stain light, and the $\beta$ cells stain dark, permitting calculation of the ratio of $\alpha$ cell to $\beta$ cells. Pseudoisocyanic staining permits direct demonstration of insulin in the $\beta$ cells. The reaction involves the development of $SO_2$ groups formed by the oxidative splitting of the disulfide bridges of insulin with potassium permanganate. Organ culture techniques for studying pancreatic islets have been investigated [74]. The availability of suitably characterized dispersed islet cell preparations offers another *in vitro* test system to examine the effects of various drugs and chemicals on the pancreas [123]. Monoclonal antibody methods are also available to assay for islet cell antibodies [102].

## QUESTIONS

*Essay*

1. Briefly diagram and describe the hypothalamic–pituitary–gonadal axis.
2. Outline the biosynthesis of thyroid hormones and the influence of TRF and TSH.
3. Describe factors which modulate secretions of the endocrine pancreas.

*Multiple choice*

4. Which serum protein(s) can specifically bind to hormones?
   a. SHBG
   b. CBG
   c. TBG
   d. All of the above

5. Which gonadal cell(s) produce testosterone?
   a. Germ cells
   b. Sertoli cells
   c. Leydig cells
   d. Endothelial cells
6. Suppression of ACTH occurs with:
   a. Dexamethasone (answer)
   b. TSH
   c. Triiodothyronone
   d. None of the above
7. Destruction of pancreatic beta cells are most likely to occur following:
   a. FSH administration
   b. Streptomycin
   c. Prl administration
   d. None of the above
8. Biosynthesis of adrenocortical steroids can be inhibited by:
   a. Mitotane (*o,p'*-DDD)
   b. Aminogluthethimide
   c. Metyrapone
   d. All of the above
9. Castration can lead to the following endocrine modifications:
   a. Elevation of blood levels of FSH and LH
   b. Reduced levels of testosterone
   c. Atrophy of sex accessory organs
   d. All of the above
10. Infertility may be due to:
    a. Failure to ovulate
    b. Reduced sperm count
    c. Absence of germinal epithelium
    d. All of the above

*True/false*

11. Monoclonal antibody techniques represent a very sensitive method for determining certain hormones.
12. Bioassays have largely been replaced by newer and more reliable hormone tests.

## REFERENCES


1. ASRM. (2001): *Patient Fact Sheet*. American Society for Reproductive Medicine, Birmingham, AL (www.asrm.org).
2. Aguilar, A. and Borrell, A. (1994): Abnormally high polychlorinated biphenyl levels in striped dolphins (*Stenella coeruleoalba*) affected by the 1990–1992 Mediterranean epizootic. *Sci. Tot. Environ.*, 154:237–247.
3. Adlercreutz H. (2002): Phyto-oestrogens and cancer. *Lancet Oncol.*, 3(6):364–373.
4. Aulerich, R., Ringer, R., and Iwamoto, S. (1973): Reproductive failure and mortality in mink fed on great lakes fish. *J. Reprod. Fertil.*, 19(Suppl.):365–376.
5. Badger, T. M., Millard, W. J., McCormick, G. F., Bowers, C. Y., Martin, J. B. (1984): The effects of growth hormone (GH)-releasing peptides on GH secretion in perfused pituitary cells of adult male rats. *Endocrinology*, 115:1432–1438.
6. Banerjee, A. K., Patel, K. J., and Grainger, S. L. (1989): Drug-induced acute pancreatitis: a critical review. *Med. Toxicol. Adverse Drug Exp.*, 4:186–198.
7. Banting, F.G., Best, C. H., and MacLeod, J. J. R. (1922): Internal secretions of the pancreas, *Am. J. Physiol.*, 59:479
8. Barsono, C. P. and Thomas, J. A. (1992): Endocrine disorders of occupational and environmental origin. *Occup. Med.*, 7:479–502.
9. Baxter, J. D., Ribiero, R. C. J., and Webb, P. (2004): Introduction to endocrinology. In: *Basic and Clinical Endocrinology*, 7th ed., edited by F. S. Greenspan and D. G. Gardner, pp. 1–37. McGraw-Hill, New York.
10. Bennett, B. D. and Wells, D. J. (1992): Endocrinology. In: *Clinical Chemistry*, 2nd ed., edited by M. L. Bishop J. L. Duben-Engelkirk, and E. P. Fody, pp. 317–352. Lippincott, Philadelphia, PA.
11. Bernard, C. (1865/1927/1949): *An Introduction to the Study of Experimental Medicine*. Macmillan, New York.
12. Broley, C. (1958): The plight of the American bald eagle. *Audubon Mag.*, 60:162–163.
13. Brun, H. P., Leonard, J. F., Moronvalle, V., Caillaud, J. M., Melcion, C., and Cordier, A. (1991): Pig Leydig cell culture: a useful *in vitro* test for evaluating the testicular toxicity of compounds. *Toxicol Appl. Pharmacol.*, 108:307–320.
14. Burd, J. M., Weightman, D. R., and Baylis, P. H. (1985): Solid phase radioimmunoassay for direct measurement of human plasma oxytocin. *J. Immunoassay*, 6:227–243.
15. Cannon, W. C. (1915): *Bodily Changes in Pain, Hunger, Fear and Rage*. Appleton, New York.
16. Capen, C. C. (1992): Pathophysiology and xenobiotic toxicity of parathyroid glands. In: *Endocrine Toxicology*, edited by C. K. Atterwill and J. D. Flack. Cambridge University Press, Cambridge, U.K.
17. Carmichael, M. S., Humbert, R., Dixen, J., Palmisano, G., Greenleaf, W., and Davidson, J. M. (1987): Plasma oxytocin increases in the human sexual response. *J. Clin. Endocrinol. Metab.*, 64:27–31.
18. Chiu, T. T., Tam, P. P., and Mao, K. R. (1990): Evaluation of a semiquantitative urinary LH assay for ovulation detection. *Int. J. Fertil.*, 35:120–124.
19. Christensen, E. I. and Birn, H. (2002): Megalin and cubulin: multifunctional endocytic receptors. *Nat. Rev. Mol. Cell Biol.*, 3:282–268
20. Christopher, M. J., Rantzau, C., McConell, G., Kemp, B. E., and Alford, F. P. (2005): Prevailing hyperglycemia is critical in the regulation of glucose metabolism during exercise in poorly controlled alloxan-diabetic dogs. *J. Appl. Physiol.*, 98:930–939.
21. Clark F. and Hutton, C. W. (1985): The effect of drugs upon the assessment of thyroid function. *Adverse Drug React. Acute Poison. Rev.*, 4:59–81.
22. Clifton, D. K. and Bremner, W. J. (1983): The effect of testicular x-irradiation on spermatogenesis in man: a comparison with the mouse. *Int. J. Andro.*, 4:387–392.
23. Colborn, T., Dumanoski, D., and Myers, J.P. (1996): *Our Stolen Future*. Penguin Books, New York.
24. Colby, H. D. and Longhurst, P. A. (1992): Toxicology of the adrenal gland. In: *Endocrine Toxicology*, edited by C. K. Atterwill and J.D. Flack. Cambridge University Press, Cambridge, U.K.

25. Collins, T. F. X., Sprando, R. L., Hansen, D. L., Shackelford, M. E., and Welsh, J. J. (1998): Testing guidelines for evaluation of reproductive and developmental toxicity of food additives in females. *Int. J. Toxicol.*, 17:299–325.
26. Davis, J. R. E. (1996): Molecular biology techniques in endocrinology. *Clin. Endocrinol.*, 45:125–133.
27. Davis, M. and Walker, E. A. (1992): Capillary blood glucose monitoring for clinical decision making. *Lab. Med.*, 23:591–598.
28. IRIS. (1991): *1,2-Dibromo-3-Chloropropane (DBCP)*, CASRN 96-12-8. Integrated Risk Information System, U.S. Environmental Protection Agency, Washington, D.C. (http://www.epa.gov/IRIS/subst/0414.htm).
29. Desai, M. P., Khatkhatay, M. I., Sankolli, G. M., and Joshi, U. M. (1991): Importance of selection of separation system in the development of enzyme immunoassay: an experience with follicle stimulating hormone (FSH) assay. *J. Immunoassay*, 12:83–98.
30. DiFronzo, G., Miodini, P., Brivio, M., Cappelletti, V., Coradino, D., Granata, G., and Ronchi, E. (1986): Comparison of immunochemical and radioligand binding assays for estrogen receptors in human breast tumors. *Cancer Res.*, 46:4278s–4281s.
31. Du Vigneaud, V., Ressler, C., and Tripett, S. (1953): The sequence of amino acids in oxytocin, with a proposal for the structure of oxytocin. *J. Biol. Chem.*, 205(2):949–957.
32. Fischer, L. J. (1985): Drugs and chemicals that produce diabetes. *TIPS*, 2:72–75.
33. Foltz, R. L., Fentiman, A. F., and Flotz, R. B. (1980): *GC/MS Assays for Abused Drugs in Body Fluids*, NIDA Research Monograph 32. National Institute on Drug Abuse, U.S. Department of Health and Human Services, Rockville, MD.
34. Francis, S. H. and Corbin, J. D. (1999): Cyclic nucleotide-dependent protein kinases: intracellular receptors for cAMP and cGMP action. *CRC Crit. Rev. Clin. Lab Sci.*, 36(4):275–328.
35. Freeman, S. (2004): *Biological Science*, 2nd ed. Prentice-Hall, Upper Saddle River, NJ.
36. Gardner, D. G. and Nissenson, R. A. (2004): Mechanisms of hormone action. In: *Basic and Clinical Endocrinology*, 7th ed., edited by F. S. Greenspan and D. G. Gardner, pp. 61–84. McGraw-Hill, New York.
37. Gilbertson, M., Kubiak, T., Ludwig, J., and Fox, G. (1991): Great Lakes embryo mortality, edema, and deformities syndrome (GLEMEDS) in colonial fish-eating birds: similarity to chick edema disease, *J. Toxicol. Environ. Health*, 33(4):455–520.
38. Gilman, A. (1963): The initial clinical trial of nitrogen mustard. *Am. J. Surg.*, 105:574–578.
39. Gimpl, G. and Fahrenholz, F. (2001): The oxytocin receptor system: structure, function, and regulation. *Physiol. Rev.*, 81(2):629–683.
40. Greenspan, F. S. (2004): The thyroid gland. *Basic and Clinical Endocrinology*, 7th ed., edited by F. S. Greenspan and D. G. Gardner, pp. 215–294. McGraw-Hill, New York.
41. Gornall, A. G., Luxton, A. W., and Bhavnini, B. R. (1986): Endocrine disorders. In: *Applied Biochemistry of Clinical Disorders*, edited by A. G. Gornall, pp. 285–358. Lippincott, Philadelphia, PA.
42. Groome, N. P., Illingworth, P. J., O'Brien, M., Cooke, I., Ganesan, T. S., Baird, D. T., and McNeilly, A. (1994): Detection of dimeric inhibin throughout the human menstrual cycle by two-site enzyme immunoassay. *Clin. Endocrinol.*, 40:717–723.
43. Guiles, H. J. (1992): Thyroid function. In: *Clinical Chemistry*, 2nd ed., edited by M. L. Bishop et al., pp. 509–525. Lippincott, Philadelphia, PA.
44. Guillette, L., Gross, T., Gross, D., Rooney, A., and Percival, H. (1995): Gonadal steroidogenesis *in vitro* from juvenile alligators obtained from contaminated or control lakes, *Environ. Health Perspect.*, 103(4):31–36.
45. Haney, A. F. (1985): Effects of toxic agents on ovarian function. In: *Target Organ Toxicology Series*, edited by J. A. Thomas, K. S. Korach, and J. A. McLachlan, pp. 181–193. Raven Press, New York.
46. Hasegawa, Y., Miyamoto, K., Iwamura, S., and Igarashi, M. (1988): Changes in serum concentrations of inhibin in cyclic pigs. *J. Endocrinol.*, 118:211–219.
47. Hermann, G. A., Sugiura, H. T., and Krumm, R. P. (1986): Comparison of thyrotropin assays by relative operating characteristic analysis. *Arch. Pathol. Lab. Med.*, 110: 21–25.
48. Herzog, B., Hadziselimovic, F., and Strebel, C. (1987): Primary and secondary testicular atrophy, *Eur. J. Pediatr.*, 146(Suppl. 2):S53–S55.
49. RCP. (2005): *Hormone Timeline*. Royal College of Physicians, London, (http://rcplondon.ac.uk/heritage/hormones/index.htm).
50. Howe, C. J. and Handelsman, D. J. (1997): Use of filter paper for sample collection and transport in steroid pharmacology. *Clin. Chem.*, 43:1408–1415.
51. Hoyer, P. (2006): Impact of metals on ovarian function. In: *Metals, Fertility and Reproductive Toxicity*, edited by M. Golab, pp. 93–116. Taylor & Francis, Boca Raton, FL.
52. Hsu, S. M., Raine, L., and Fanger, H. (1981): A comparative study of the peroxidase-antiperoxidase method for studying polypeptide antibodies. *Am. J. Clin. Pathol.*, 75:734.
53. Hughes, I. and Woods, H. F. (2003): *Phytoestrogens and Health*. Committee on Toxicity of Chemicals in Food, Consumer Products, and the Environment (COT) Working Group on Phytoestrogens, U.K. Food Standards Agency, Crown Copyright FSA/0826/0503 (http://www.food.gov.uk/multimedia/pdfs/phytoreport0503).
54. Innis, M. A., Gelfand, D. H., Sninsky, J. J., and White, T. J. (1990): *PCR Protocols: A Guide to Methods and Applications*. Academic Press, San Diego, CA.
55. Jankowski, M., Danalache, B., Wang, D., Bhat, P., Hajjar, F., Marcinkiewicz, M., Paquin, J., McCann, S. M., and Gutkowska, J. (2004): Oxytocin in cardiac ontogeny. *Proc. Natl. Acad. Sci. U.S.A.*, 101(35):13074–13079.
56. Kapp, Jr., R.W., Picciano, D. J., and Jacobson, C. B. (1979): Y-chromosomal nondisjunction in dibromochloropropane-exposed workmen. *Mutat Res.*, 64(1):47–51.
57. Kavlock, R. J. (1999): Overview of endocrine disruptor research activity in the United States, *Chemosphere*, 39(8): 1227–1236.
58. Kelly, J. A. and Vankrieken, L. (1997): *Sex Hormone Binding Globulin and the Assessment of Androgen Status*. Diagnostic Products Corporation, Los Angeles, CA.

59. Kendall, E. C. (1971): *Cortisone: Memoirs of a Hormone Hunter*, Charles Scribner's Sons, New York.
60. Kendrick, K. M. (2005): *The Neurobiology of Social Bonds*. British Society for Neuroendocrinology, London (http://neuroendo.org.uk/index.php/content/view/34/11/).
61. Klee, D. G., Kao, P. C., and Heath III, H. (1988): Hypercalcemia. *Endocrinol. Metab. Clin. North Am.*, 17:573–600.
62. Kobayashi, E., Kunieda, T., Ikadai, H., Imamichi, T., and Matsumoto, K. (1992): Genetic profiles of 11 inbred rat strains at 25 biochemical marker loci and five RFLP loci. *Lab. Anim. Sci.*, 42:86–88.
63. Köhler, G. and Milstein, C. (1975): Continuous cultures of fused cells secreting antibody of predefined specificity. *Nature*, 256:495–497.
64. Korach, K. S. and McLachlan, J. A. (1995): Techniques for detection of estrogenicity. *Environ. Health Perspect.*, 103:5–8.
65. Kosugi, S., Sugawa, H., and Mori, T. (1996): TSH receptor and LH receptor. *Endocr. J.*, 43:595–604.
66. Krumm, R. (1994): Radioimmunoassay: a proven performer in the bio lab. *Scientist*, 8(10):17–23.
67. Kulkarni, A. P. (2006): The role of xenobiotic metabolism in developmental and reproductive toxicity. In: *Developmental and Reproductive Toxicology: A Practical Approach*, 2nd ed., edited by R. D. Hood, pp. 525–570. Taylor & Francis, Boca Raton, FL.
68. LaRose, P., Ong, H., and Du Souich, P. (1985): Simple and rapid radioimmunoassay for the routine determination of vasopressin in plasma. *Clin. Biochem.*, 18:357–362.
69. Layman, L. C. and McDonough, P. G. (1995): Molecular genetics in reproductive endocrinology. In: *Reproductive Medicine and Surgery*, edited by E. E. Wallach and H. A. Zacur. Mosby, St. Louis, MO.
70. Leidy, J. W. and Robbins, R. J. (1986): Regional distribution of human growth hormone-releasing hormone in the human hypothalamus by radioimmunoassay. *J. Clin. Endocrinol. Metab.*, 62:372.
71. Leiter, E. H., (1989): The genetics of diabetes susceptibility in mice. *FASEB J.*, 3:2231–2241.
72. Leung, P. C. K. and Steele, G. L. (1992): Intracellular signaling in the gonads. *Endocr. Rev.*, 13:476–498.
73. Lorsenson, M. Y. (1985): *In vitro* conditions modify immunoassayability of bovine pituitary prolactin and growth hormone: insights into their secretory granule storage forms. *Endocrinology*, 116:1399–1407.
74. Mandel, T. E., Hoffman, L., Colier, S., Carter, W., and Koulmanda, M. (1982): Organ culture of fetal mouse and fetal human pancreatic islets for allografting. *Diabetes*, 3:39–47.
75. Mann T. (1964): *The Biochemistry of Semen and of the Male Reproductive Tract*. Wiley, New York.
76. Marieb, E.N. (2004): Human Anatomy and Physiology, 6th ed., edited by E. M. Marieb, p. 209. Pearson Education, Upper Saddle River, NJ.
77. Mason, C., Ford, T., and Last, N. (1986): Organochlorine residues in British otters. *Bull. Environ. Contam. Toxicol.*, 36:656–661.
78. Mattison, D. R., Plowchalk, D. R., Meadows, M. J., Al-Juburi, A. Z., Gandy, J., and Malek, A. (1990): Reproductive toxicity: male and female reproductive systems as targets for chemical injury. *Med. Clin. North Am.*, 74:391–411.
79. Mattison, D. R., Shiromizu, K., and Nightinggale, M. S. (1985): The role of metabolic activation in gonadal and gamete toxicity. In: *Occupational Hazards and Reproduction*, edited by K. Hemminki, M. Sorsa, and H. Vainio. Hemisphere, Washington, D.C.
80. May, M. E. and Carey, R. M. (1985): Rapid ACTH test in practice. *Am. J. Med.*, 79:679–684.
81. Mitrakou, A., Vuorinen-Markkola, H., Raptis, G., Toft, I., Mokan, M., Strumph, P., Pimenta, W., Veneman, T., Jansen, T., and Bolli, G. (1992): Simultaneous assessment of insulin secretion and insulin sensitivity using a hyperglycemic clamp. *Clin. Endocrinol. Metab.*, 75:379–382.
82. Moore, N. P., Creasy, D. M., Gray, T. J. B., and Timbrell, J. A. (1992): Urinary creatine profiles after administration of cell-specific testicular toxicants to the rat. *Arch. Toxicol.*, 66:435–442.
83. Motomura, K. and Brent, G. A. (1998): Mechanisms of thyroid hormone action. *Endo. Metab. Clin. North Am.*, 27:1–23.
84. Nobel Prize website, http://nobelprize.org/.
85. Nussey, S. and Whitehead, S. (2001): The thyroid gland. In: *Endocrinology: An Integrated Approach*, edited by S. Nussey and S. Whitehead. BIOS Scientific, Oxford.
86. Partsch, C. J., Aukamp, M., and Sippell, W. G. (2000): Scrotal temperature is increased in disposable plastic lined nappies. *Arch. Dis. Child.*, 83:364–368.
87. Piaditis, G. P., Hodgkinson, S., McLean, C., and Lowry, P. J. (1985): Thyroid stimulating hormone. *J. Immunoassay*, 6:299–319.
88. Pierce, E. A., Dame, M. C., Bouillon, R., Van Baelen, H., DeLuca, H. F. (1985): Monoclonal antibodies to human vitamin D-binding protein. *Proc. Natl. Acad. Sci. U.S.A.*, 82:8429–8433.
89. Porterfield, S. P., Ed. (2001): *Endocrine Physiology*, 2nd ed. Mosby, St. Louis, MO.
90. Preissner, C. M., Klee, G. G., Scheithauer, B. W., and Abboud, C. F. (1990): Free alpha subunit of the pituitary glycoprotein hormones: measurement in serum and tissue of patients with pituitary tumors. *Am. J. Clin. Pathol.*, 94: 417–421.
91. Root, A. W., Duckett, G. E., Geiszler, J. E., Hu, C. S., and Bercu, B. B. (1997): Evaluation of the clinical utility of the ultrasensitive immunofluometric assay for growth hormone (GH) and of the cortisol secretory pattern in prediction of the linear growth response to treatment with GH. *J. Pediatr. Endocrinol. Metab.*, 10:3–10.
92. Rosenfeld, L. and Blum, M. (1986): Immunoradiometric (IRMA) assay for thyrotropin (TSH) should replace the RIA method in the clinical laboratory. *Clin. Chem.*, 32:1.
93. Sambrook, J., Fritsch, E. F., and Maniatis, T. (1989): *Molecular Cloning: A Laboratory Manual*, 2nd ed. Cold Spring Harbor Laboratory Press, Cold Spring Harbor, NY.
94. Schmalzing, D., Nashabeh, W., Yao, X. W., Mhatre, R., Regnier, F. E., Afeyan, N. B., and Fuchs, M. (1995): Capillary electrophoresis-based immunoassay for cortisol in serum. *Anal. Chem.*, 67:606–612.
95. Schrager, S. and Potter, B. E. (2004): Diethylstilbestrol exposure, *Am. Fam. Phys.*, 69:2395–2400.
96. Seethalakshmi, L., Steinberger, A., and Steinberger, E. (1984): Pituitary binding of $^3$H-labeled Sertoli cell factor *in vitro*: a potential radioreceptor assay for inhibin. *Endocrinology*, 115:1289–1294.

97. Segre, G. V. and Brown, E. N. (1998): Measurement of hormones. In: *Williams Textbook of Endocrinology*, 9th ed., edited by J. D. Wilson D. W. Foster, H. M. Kronenberg, and P. R. Larsen, pp. 43–54. W.B. Saunders, Philadelphia, PA.
98. Sgoutas, D. S., Tuten, T. E., Verras, A. A., Love, A., and Barton, E. G. (1995): AquaLite bioluminescence assay of thyrotropin in serum elevated. *Clin. Chem.*, 41:1637–1643.
99. Shupnik, M. A. (1995): Measurement of gene transcription and messenger RNA. In: *Molecular Endocrinology: Basic Concepts and Clinical Correlations*, edited by B. D. Weintraub, pp. 41–58. Raven Press, New York.
100. Sprando, R. L. and Collins, T. F. X. (1998): Testing guidelines for evaluation of food additives' effects on male reproduction. *Int. J. Toxicol.*, 17:327–336.
101. Squire, C. R. and Fraser, W. D. (1995): Thyroid stimulating hormone measurement using a third generation immunometric assay. *Ann. Clin. Biochem.*, 32:307–313.
102. Srikanta, S., Rabizadeh, A., Omar, M. A. K., and Eisenbarth, G. S. (1985): Assay for islet cell antibodies. *Diabetes*, 34:300.
103. Stites, D. P., Stobo, J. D., Fudenberg, H. H., and Wells, J. V. (1987): *Basic and Clinical Immunology*, 6th ed. Lange Medical, Los Angeles, CA.
104. Thomas, J. A. (1981): Reproductive hazards and environmental chemicals. *J. Toxic Subst.*, 2:318.
105. Thomas, J. A. and Keenan, E. J. (1986): Drugs affecting the endocrine system. In: *Principles of Endocrine Pharmacology*, edited by J. A. Thomas and E. J. Keenan. Plenum Press, New York.
106. Thomas, J. A. and Ballantyne, B. (1990): Occupational reproductive risks: sources, surveillance, and testing. *J. Occup. Med.*, 32:547–553.
107. Thomas, J. A. (1995): Gonadal-specific metal toxicology. In: *Metal Toxicology*, edited by R. Goyer, pp. 413–446. Academic Press, San Diego, CA.
108. Thomas, J. A. (1996): Toxic responses of the reproductive system. In: *Casarett & Doull's Toxicology: The Basic Science of Poisons*, 5th ed., edited by C. D. Klaassen, pp. 547–581. Pergamon Press, New York.
109. Thomas, J. A. (1993): Transgenic animal models and genetically altered species. In: *Biotechnology and Safety Assessment*, edited by J. A. Thomas and L. A. Myers. Raven Press, New York.
110. Thomas, J A., Ed. (1996): *Endocrine Methods*, pp. 1–447. Academic Press, San Diego, CA.
111. Thomas, J. A. and Colby, H. D., Eds. (1996): *Endocrine Toxicology*, 2nd ed. Taylor & Francis, New York.
112. Thronton, P. S. et al. (1994): The new highly sensitive adrenocorticotropin assay improves detection of patients with partial adrenocorticotropin deficiency in a short-term metyrapone test. *J. Pediatr. Endocrinol. Metab.*, 7:317–324.
113. Tochino, Y., Kanaya, T., and Makino, S. (1982): Genetics of NOD mice. *Excepta Medica*, 44:285–291.
114. USEPA. (2006): *Endocrine Disruptor Research Initiative*, U.S. Environmental Protection Agency, Washington, D.C. (http://www.epa.gov/endocrine/index.html).
115. USEPA. (1996): *Endocrine Disruption Screening Program: Endocrine Primer*, U.S. Environmental Protection Agency, Washington, D.C. (http://www.epa.gov/scipoly/oscpendo/edspoverview/primer.htm#1).
116. USEPA. (1996): *Report on the Health and Ecological Effects of Endocrine Disrupting Chemicals: A Framework for Planning*, U.S. Environmental Protection Agency, Washington, D.C. (http://www.epa.gov/endocrine/Pubs/framewrk.pdf).
117. USEPA. (2006): *What Are Endocrine Disruptors?* Endocrine Disruptor Screening Program, U.S. Environmental Protection Agency, Washington, D.C. (http://www.epa.gov/scipoly/oscpendo/edspoverview/whatare.htm).
118. USEPA. (1998): *Endocrine Disruptor Screening and Testing Advisory Committee (EDSTAC) Final Report*, U.S. Environmental Protection Agency, Washington, D.C. (http://www.epa.gov/scipoly/oscpendo/edspoverview/finalrpt.htm).
119. USEPA. (2006): *History*. Endocrine Disruptor Screening Program, U.S. Environmental Protection Agency, Washington, D.C. (http://www.epa.gov/scipoly/oscpendo/edspoverview/primer.htm#4).
120. Veldhuis, J. D., Liem, A. Y., South, S., Weltman, A., Weltman, J., Clemmons, D. A., Abbott, R., Mulligan, T., Johnson, M. L., Pincus, S., Straume, M., and Iramanesh, A. (1995): Differential impact of age, sex steroid hormones, and obesity on basal versus pulsatile growth hormone secretion in men as assessed in an ultrasensitive chemiluminescence assay. *J Clin. Endocrinol. Metab.*, 80: 3209–3222.
121. Watkins, S. (1998): Immunohistochemistry. In: *Current Protocols of Molecular Biology*, edited by F. M. Ausubel et al., pp. 1–13. John Wiley & Sons, New York.
122. Weiner, R. I., Wetsel, W., and Goldsmith, P. (1992): Gonadotropin-releasing hormone neuronal cell lines. *Front. Neuroendocrinol.*, 13:95–119.
123. Weir, G. C., Halban, P. A., Wollheim, C. B., Orci, L., and Renold, A. E. (1984): Dispersed adult rat pancreatic islet cells in culture): A, B, and D cell function. *Metabolism*, 33:447–453.
124. Wenk, R. E. (1992): Reproductive medicine and the clinical laboratory. *Clin. Lab. Med.*, 12(3):393–653.
125. Whorton, D., Krauss, R. M., Marshall, S., and Milby, T. H. (1977): Infertility in male pesticide workers. *Lancet*, 2:1259–1260.
126. Wikidepia. (2006): http://en.wikipedia.org/.
127. Williams, J. S. and Williams, G. H. (2003): 50th anniversary of aldosterone. *J. Clin. Endocrinol. Metab.*, 88(6):2364–2372.
128. Witorsch, R. J., Ed. (1995): *Reproductive Toxicology*, 2nd ed. Raven Press, New York.
129. Yalow, R. S. and Berson, S, A. (1959): Assay of plasma insulin in human subjects by immunological methods. *Nature (Lond.)*, 184:1648–1649.
130. Yamaguchi, K., Masahiko, M., and Tanaka, M. (2006): Gross anatomy of the pancreas. In: *Toxicology of the Pancreas*, edited by P. M. Pour. Taylor & Francis, Boca Raton, FL.
131. Yoon, J. W. (1990): The role of viruses and environmental factors in the induction of diabetes. *Curr. Top. Microbiol. Immunol.*, 164:95–123.

# Notes

# 36 Immunotoxicology: Effects of and Response to Drugs and Chemicals

*Jack H. Dean, Robert V. House, and Michael I. Luster*

## CONTENTS

## INTRODUCTION

The immune system is a complex multicellular organ system consisting of granulocytes, macrophages, lymphocytes, and dendritic cells with various functions and phenotypic characteristics, as well as various soluble mediators. These cells are of hemopoietic origin and, in adults, are found in the peripheral blood, lymphatic fluid, and organized lymphoid tissues, including bone marrow, spleen, thymus, lymph nodes, tonsils, and mucosa-associated lymphoid tissue. The immune system is in a constant state of self-renewal involving cell proliferation, differentiation, activation, and maturation. It exists to defend the body against invasion by infectious and opportunistic microorganisms and spontaneously arising neoplasia. This network of cells and soluble factors is highly regulated and interdependent, must discriminate self from non-self, and can react to non-self with many different (pleiotropic) defensive responses [1]. In addition, this immune system occasionally develops a response to a chemical or its metabolite that binds to or alters a host protein, resulting in allergy or autoimmune disease. The immune systems of experimental animals, although exhibiting some obvious differences from that of humans, are still sufficiently similar that data obtained from lower species are instructive of a potential human response [2].

## IMMUNE MECHANISMS RESPONSIBLE FOR HOST DEFENSE

The host defense functions of the immune system are provided by two major mechanisms: a *nonspecific* (innate) mechanism that does not require prior sensitization with the inducing agent to elicit a response, and a *specific* (adaptive) mechanism directed against the eliciting agent to which the individual has been previously sensitized. Penetration of the skin or mucosal defense barriers by an invading microorganism results in nonspecific reactions by phagocytic cells (granulocytes and macrophages [MØ]). If the microorganism is not controlled and persists, specific responses involving antibody production and the induction of effector lymphocytes follow. The effector lymphocytes respond through cytokine mediators to seek out and destroy the invading microorganism. Both antibody-producing lymphocyte responses (B-lymphocytes or B-cells) and thymus-dependent lymphocyte responses (T-lymphocytes or T-cells) are triggered by the presentation of foreign antigen to appropriate lymphocytes by dendritic cells, macrophages, or other antigen-presenting cells (APCs). Following antigen-induced activation, B-cells proliferate and differentiate into plasma cells (PCs), with the support of T-helper 2 (Th2) cells, which produce large quantities of antigen-specific immunoglobulins (antibodies). Antibodies enter the plasma, where they bind the foreign material and neutralize, lyse, or facilitate phagocytosis of the agent. Antibody–antigen interactions are expanded by actions of the complement (C′) system and other inflammatory mediators (e.g., prostaglandins and leukotrienes). Another population of T-cells, referred to as cytotoxic T-cells, proliferate and recognize viral infected cells that they can destroy before viral replication is complete.

### INNATE AND ADAPTIVE MECHANISMS OF IMMUNITY

Two categories of phagocytic leukocyte—the polymorphonuclear phagocyte (PMN), or granulocyte, and the mononuclear phagocyte, or MØ—are involved with nonspecific mechanisms of host resistance. Both cell types originate from myeloid progenitor cells in the bone marrow and normally pass through several maturation stages before entering the bloodstream. PMNs readily traverse blood vessels and provide the primary defense against infectious agents. The inflammation associated with a splinter is typical of a nonspecific PMN and MØ response. Both PMNs and MØ exhibit phagocytic activity toward foreign material, especially MØ in the presence of specific opsonic antibodies and C′, and can destroy most microorganisms. MØ are recruited to the site in the event that PMNs either cannot contain or are destroyed by the infectious agent, as is the case with certain bacteria (e.g., *Listeria*). MØ can be activated to a state of enhanced bactericidal or tumoricidal activity by soluble lymphocyte products (e.g., cytokines) produced by T-lymphocytes sensitized to the invading microbe.

The immune responses that characterize adaptive host defenses represent a series of complex events that occur following the introduction of foreign antigenic material into the body. The two major types of specific immune response are (1) *cell-mediated immunity* (CMI), which is initiated by specifically sensitized T-cells and is generally associated with delayed-type hypersensitivity (DTH), rejection of tumors or foreign grafts, and resistance to viral agents; and (2) *humoral immunity* (HI), which involves the production of antibodies by PCs following sensitization to a specific antigen and is important in resistance to extracellular pathogens.

## Organization, Differentiation, and Function of Primary Lymphoid Tissue

The cellular elements of the immune system arise from *pluripotent stem cells*, a unique group of unspecialized cells that have self-renewal capacity. These cells are found in the blood islands of the embryonic yolk sac and in the liver of the fetus during fetal development, and later in the bone marrow. The pluripotent stem cell differentiates along several pathways, giving rise to erythrocytes, myeloid series cells (i.e., MØ and PMNs), megakaryocytes (platelets), or lymphocytes. Maturation generally occurs within the bone marrow, although lymphoid progenitor cells are disseminated through the blood and lymphatic vessels to the primary lymphoid organs where they undergo further differentiation under the influence of the humoral microenvironment of these organs (Figure 36.1).

The *primary lymphoid organs* include the *thymus* in all vertebrates and the bursa of Fabricius (in birds) or *bursa-equivalent tissue* in other vertebrates; the latter is believed to be bone marrow and gut-associated lymphoid tissue in mammals (Table 36.1). Primary lymphoid organs are lymphoepithelial in origin and are derived from ectoendodermal junctional tissue in association with gut epithelium. During the beginning of the second half of embryogenesis (days 12 to 13 in the mouse), stem cells migrate into the epithelia of the thymus and bursa-equivalent areas, where they differentiate independently of antigenic stimulation into immunocompetent T- and B-cells, respectively (Figure 36.1). The thymus, which is derived embryologically from the third and fourth pharyngeal pouches, is an organization of lymphoid tissue located in the chest, above the heart. Thymus development occurs during the sixth week of embryological development in humans and day 9 of gestation in the mouse. The thymus reaches its maximum size at birth or shortly thereafter in most mammals and then begins a slow involution that is complete between the ages of 5 and 15 years in humans.

Histologically, the thymus consists of multiple lobules, each lobule containing a cortex (outer) and a medulla (inner) (Figure 36.2). Lymphocyte precursors from bone marrow proliferate in the cortex of the lobules and then migrate to the medulla. In the medulla, they further differentiate, under the influence of thymic epithelium and hormonal factors, into mature T-lymphocytes before emigrating to secondary lymphoid tissues. The neonatal/postnatal thymus has a significant endocrine function supported by nonlymphoid thymic epithelium cells. These cells produce a family of thymic hormones essential for T-lymphocyte maturation and differentiation. In contrast, B-cell differentiation occurs in the bursa of Fabricius in birds, a lymphoepithelial organ that develops from a diverticulum of the posterior wall of the cloaca. The thymus is divided into a medullary region, containing lymphoid follicles and a cortical region (Figure 36.2). The mammalian bursa equivalent is believed to be the fetal liver, neonatal spleen, gut-associated lymphoid tissue, and adult bone marrow, depending on age. Mature B-lymphocytes migrate from the bursa-equivalent tissue to populate the B-dependent areas of the secondary lymphoid tissues.

Neonatal removal or chemical destruction of primary lymphoid organs prior to the maturation of lymphocytes into T- or B-cells or prior to their population of secondary peripheral lymphoid tissue dramatically depresses the immunological capacity of the host. However, removal of these same organs in adults has little influence on immunological capacity. In addition, neonatal thymectomy in mammals dramatically impairs the development of CMI but does not generally influence the generation of immunoglobulin-producing cells involved in antibody-mediated immunity (unless they strictly require T-lymphocyte help for the induction of antibody production). In contrast to the removal of primary lymphoid organs, removal of secondary lymphoid organs does not inhibit the development of immune competence, although it may suppress the magnitude or alter the tissue location of the responsive cells.

## Organization and Function of Secondary Lymphoid Tissue

The organization and function of secondary lymphoid organs are extremely important for immune competence and host defense (Table 36.1). The organized areas of secondary lymphoid tissue are the spleen, lymph nodes, gut-associated lymphoid tissue (GALT), and bronchial-associated lymphoid tissue (BALT). The anatomical organization of these tissues provides a microenvironment for functional development of lymphoid cells and vital immune responses.

Lymph nodes are discrete, organized secondary lymphoid organs that serve as filtering devices for lymphatic fluid [3]. Lymph nodes are divided structurally into three areas: cortex, paracortex, and medulla (Figure 36.3) and are served by several afferent lymphatic vessels that collect lymphatic fluid (lymph) from distal tissue sites. Because lymph may contain foreign antigens, the efferent lymphatic vessel, which drains lymph from the node, con-

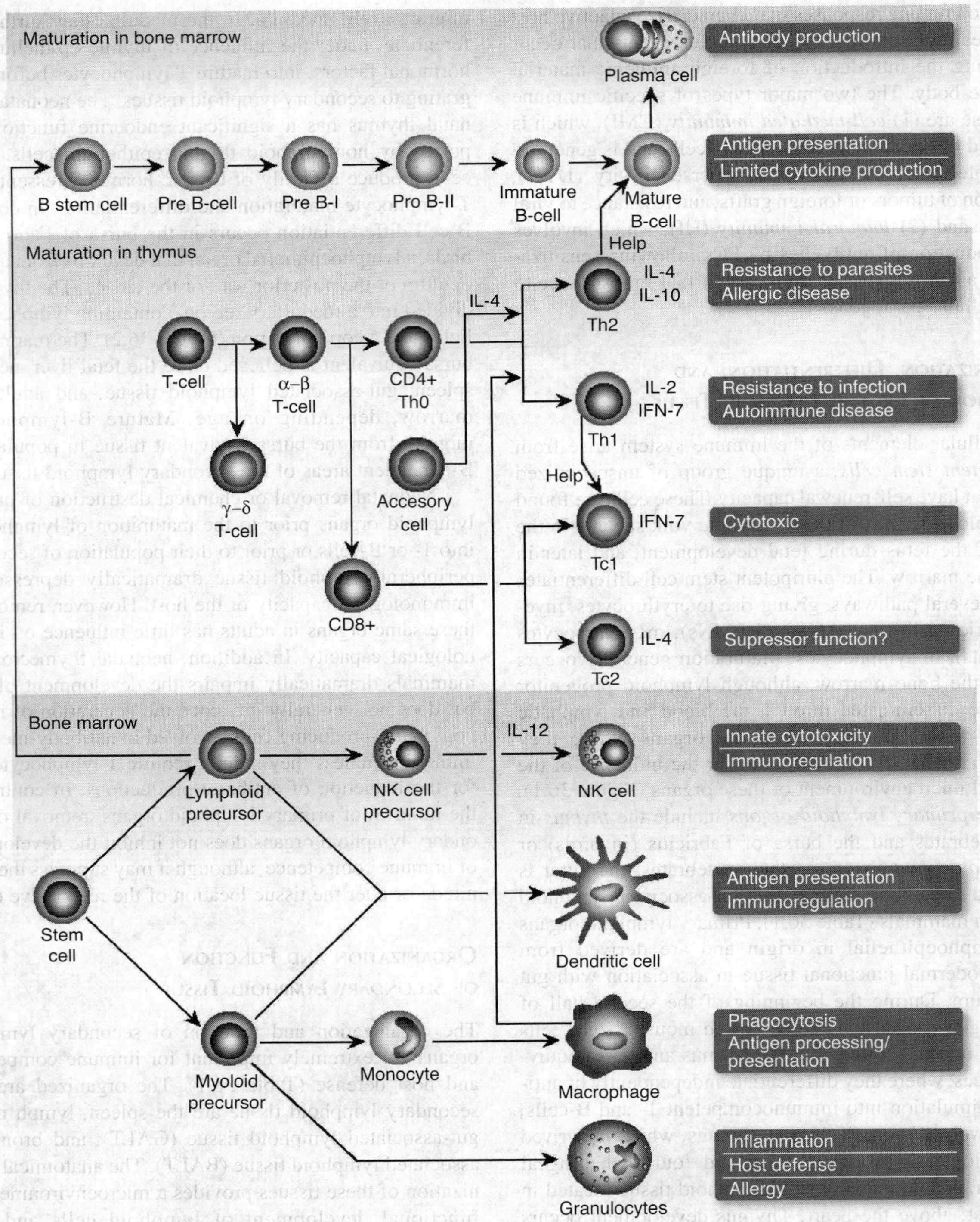


**FIGURE 36.1** Origins and interactions of cells of the immune system.

tains antibodies, cytokines, and lymphocytes produced in response to foreign antigenic stimulation occurring within the node. The cortex, located underneath the subcapsular sinus, receives the afferent lymph and serves as the major site of B-lymphocyte localization. The cortex consists of a narrow rim of small lymphocytes in the absence of antigenic stimulation. Also located in the cortex are aggregations of small lymphocytes, termed *lymphoid follicles*, which contain dendritic reticulum cells capable of retaining antigens on their plasma membranes. When lymphocytes within the lymphoid follicles are stimulated by antigens they proliferate, giving rise to dense aggregations of

**TABLE 36.1**
**Organization and Characterization of Primary and Secondary Lymphoid Organs**

| | Primary Lymphoid Organs | Secondary Lymphoid Organs |
|---|---|---|
| Generation and maturation of cells | Thymus<br>Bursa of Fabricius (avians)<br>Fetal liver (mammals)<br>Adult bone marrow | Spleen<br>Lymph nodes<br>Gut-associated lymphoid tissue<br>Bronchial-associated lymphoid tissue |
| Embryonic origin | Ectoendodermal junction | Mesoderm |
| Development | Thymus: mouse, days 9–10; humans, week 6 | — |
| | Bursa equivalent: mouse, days 10–13; humans, week 10 | — |
| Lymphoid cell proliferation | Independent of antigenic stimulation | Dependent on antigenic stimulation |
| Function | Generation and maturation of cells | Expansion of cells after antigenic stimulation |
| Cells repopulating after depletion | Stem cells only | Differentiated lymphocytes |
| Effect of depletion | Persistent effect if stem cells targeted | Likely to regenerate |

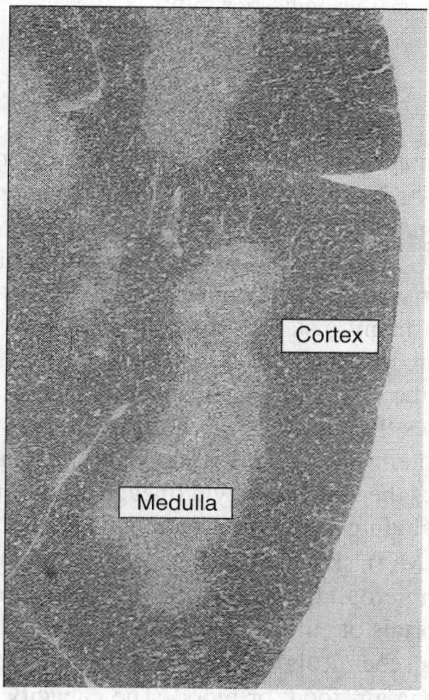


**FIGURE 36.2** Photomicrograph of a rat thymus with a densely populated cortex and a less densely populated medulla (H&E-stained section).

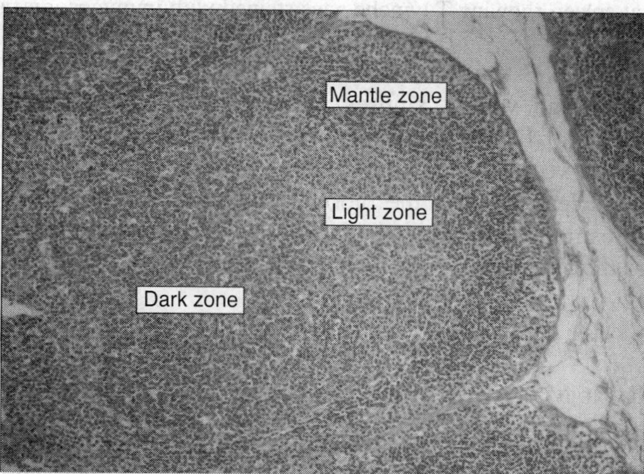


**FIGURE 36.3** Photomicrograph of a rat lymph node showing a secondary lymphoid follicle including a mantle zone and a germinal center with a dark and light zone (H&E-stained section).

lymphocytes, termed *germinal centers*. These germinal centers serve as sites for differentiation of B-lymphocytes into PCs capable of antibody production. Following antigenic stimulation, germinal centers are easily detectable as spherical or ovoid structures containing many large and medium-sized lymphocytes, predominantly B-lymphocytes. The paracortex, lying between the cortex and the medulla, is composed predominantly of T-cells and is a major site of MØ and T-cell interaction. Neonatal thymectomy or lymphocyte depletion by cytolytic drugs reduces the production of paracortical lymphocytes, leading to depressed immune capacity. In addition, the paracortex contains a specialized blood vasculature, termed *postcapillary venules*, that serves as a point of entry for recirculating lymphocytes from the bloodstream. The medulla of the lymph node is composed primarily of networks of cords and sinuses; it serves as an effective filter for removing particulate material from lymphatic fluid. Following antigenic stimulation, the majority of antibody that is produced comes from the PCs found within these medullary cords of the nodes.

The spleen is the major filter of bloodborne antigens and the site of immunological response to these antigens [4]. In addition, the spleen is a site of extramedullary hematopoiesis (non-bone-marrow-derived red blood cell production) and responsible for the removal of damaged blood cells. The two major histological regions within the spleen are the red pulp and the white pulp containing IgM-staining B-cells and T-cells (Figure 36.4). These areas have been named for their colors in a freshly cut spleen. The white pulp consists of numerous white blood cell aggregates and lymphoid follicles. The red pulp contains

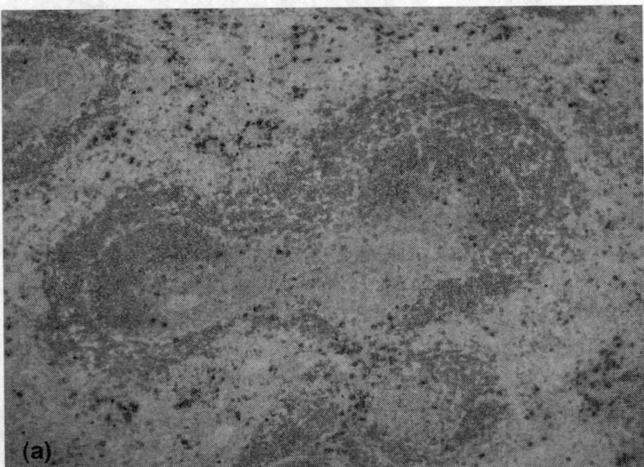


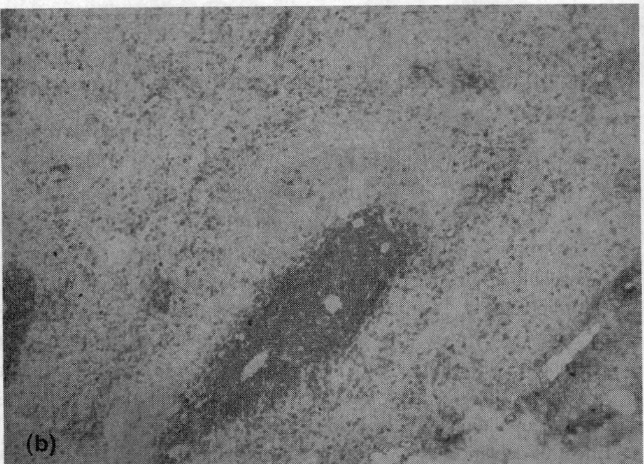


**FIGURE 36.4** (a) Photomicrograph of rat spleen showing IgM staining cells (original magnification, 50×). (b) Photomicrograph of rat spleen showing T-lymphocytes stained with immunoconjugated antibody (original magnification, 50×).

cords and venous sinuses analogous to the medullary region of lymph nodes. The spleen has no afferent lymphatic vessels; thus, all antigenic material or cells enter the spleen through the blood vasculature. The marginal sinus in the spleen is structurally and functionally similar to the subcapsular sinus of the lymph node.

## IMMUNE FUNCTION AND RESPONSES

### Bone Marrow

The bone marrow functions as a primary lymphoid organ and serves as the principal source of uncommitted stem cells, including both myeloid and erythroid precursor cells. The bone marrow architecture is highly organized and complex, consisting of a matrix or cellular stroma derived from local mesenchymal cells, as well as cells of hemopoietic parenchyma that are descendants of circulating stem cells. The bone marrow matrix consists of reticular–dendritic cells, fibroblast-like cells, and immune cells within the bone marrow microenvironment. Bone marrow stem and stromal cells have been shown to possess a significant capacity for metabolic activation because they contain cytochromes of the P450 and P448 families, as well as peroxidases, and can generate reactive oxygen species, which could also activate xenobiotics via oxidant-dependent mechanisms [5]. This metabolic activity is thought to contribute to the sensitivity of bone marrow elements to toxicants such as benzene, which is extensively metabolized within the bone marrow. In light of the cell proliferation and differentiation occurring within the marrow, this tissue is also one of the most sensitive tissues to drugs or chemicals affecting cell division. Dose-limiting bone marrow toxicities are a significant problem with antiproliferative drugs, including cytotoxic agents, antifolates, AIDS therapeutics, and certain cytokines [6,7].

### Mononuclear Phagocytic System

Whether an antigen induces CMI, antibody production, or both depends on the physical and chemical characteristics of the antigen, the mode of presentation of the antigen to lymphocytes, the pattern of antigen distribution within lymphoid tissue, and the molecular configuration of the antigen. In many instances, antigen is initially phagocytized and processed by APCs. Antigenic peptides are transported to the cell surface, where they are presented to lymphocytes through cell surface interactions via specific surface proteins (e.g., class II molecule antigens).

Cells of the MØ/monocyte lineage are found in many tissues, including liver (Kupffer cells), lung (alveolar and interstitial MØ), skin (Langerhans cells), and brain (astrocytes and microglia). These cells, because of their proximity to portals of entry, are often the first cells to interact with drugs, chemicals, and physical agents entering the organism via air, food, or blood. The capacity of cells of the mononuclear phagocytic system (formerly known as the reticuloendothelial system) to carry out these functions is associated with their state of activation, which in turn is a function of both endogenous (e.g., interferon-gamma [IFN-γ]) and exogenous (e.g., bacterial lipopolysaccharide [LPS]) stimuli. Responsive MØ obtained from the peritoneal cavity are relatively quiescent and require extracellular signals or "priming," followed by a second signal induced by triggers such as LPS, to be fully activated.

Although the mononuclear phagocyte system is designed to protect the host, once a xenobiotic has gained entry, extensive or persistent tissue damage can result, mediated, in part, by MØ products. Silicosis and asbestosis are two examples of diseases where MØ mediators are involved in the pathology [8]. In this condition, tissue damage from an environmental or infectious agent results in the influx of phagocytic cells, mainly PMNs. In most

**TABLE 36.2**
**Biological Properties of Mammalian Immunoglobulin Classes**

| Class | Serum Concentration (mg/dL) | Molecular Weight | Placental Transfer | Half-Life (days) | Biological Function | Abnormalities |
|---|---|---|---|---|---|---|
| IgG | 670 ± 33 | 150,000 | + | 23 | Primarily synthesized during secondary immune response; readily diffuses into extravascular tissue; fixes complement | Increased in liver disease chronic infection; reduced in B-cell depression |
| IgM | 61 ± 5 | 890,000 | – | 5 | Produced early in immune response; isoagglutinins; fixes complement | Increased in infection; reduced in B-cell depression |
| IgA | 40 ± 4 | 170,000 | – | 6 | Major immunoglobulin in seromucous secretions | Increased in liver disease; increased or decreased in sinopulmonary infection |
| IgD | — | 150,000 | – | 2.8 | Lymphocyte receptor | Decreased following thymectomy |
| IgE | 0.02 | 196,000 | – | 1.5 | Mediator of allergic reactions and atopic diseases | Increased in parasitic and allergic diseases, homocytotropic |

instances, these cells effectively eliminate the agents by digesting them in internal vacuoles; however, if the foreign material persists (as, for example, silica crystals or asbestos fibers), a chronic inflammatory process ensues in which PMNs are replaced by monocytes/MØ as the predominant effector cells. These cells release a variety of reactive molecules (e.g., cytokines, nitric oxide, amines, lipid mediators) that damage tissue, in addition to recruiting other inflammatory cells into the local environment. This sometimes leads to the development of a granuloma, a collection of inflammatory cells surrounded by fibrotic tissue [9].

## Humoral Immunity

The principal function of B-lymphocytes is production of specific antibody in response to antigenic stimulation. B-cells recognize antigen via a specific receptor, comprised of membrane immunoglobulins associated with accessory proteins either directly or in the presence of an APC. Binding of the receptor with its cognate antigen triggers transmembrane signaling, leading to activation of the B-cell. The antigen is subsequently internalized, where it is processed and associated with class II major histocompatibility complex (MHC) molecules. Antigen-derived peptides, along with MHC proteins, are then transferred to the cell surface, where they are free to interact with helper T-cells.

Within 3 to 5 days following antigen exposure, this T-/B-cell interaction results in the B-lymphocytes differentiating into blast cells, then into immature PCs, and finally into antibody-secreting PCs. The establishment of humoral immunity is characterized by an early rise in IgM antibody titer in the serum, followed several days later by the appearance of IgG antibodies. During this differentiation process, some of the lymphocytes develop into long-lived or memory cells (sensitized but non-blast cells), so subsequent antigen encounters result in an enhanced (secondary) response. This secondary response is characterized by a shorter latency for antibody appearance, as well as an increased affinity and synthesis of IgG antibodies. Antibody molecules react with specific antigenic determinants (epitopes) on their target, facilitating its removal (e.g., lysis, clearance, or enhanced phagocytosis).

Based on chemical structure and biological function, the five classes of antibody molecules in mammals are IgM, IgG, IgA, IgD, and IgE; some of the physical and biological characteristics of each of these classes are listed in Table 36.2. Antibodies operate via several mechanisms to protect the host from infectious agents. Some of these mechanisms include virus neutralization, in which antibodies bind and prevent virus particles from infecting target cells; opsonization, the process by which antibody molecules react with infectious agents and thus enhance their phagocytosis; and antibody-dependent cellular cytotoxicity, the process whereby antibody-coated target cells are killed by Fc receptor-bearing lymphocytes.

Of increasing interest is the concept that naturally occurring IgM antibodies (that is, antibodies that are secreted in the absence of antigen stimulation) may play an important role in immune surveillance against neoplasia [10,11]. This concept has not as yet been explored in the context of immunotoxicology but may be a contributory factor in decreased resistance to tumors following immunotoxic insult.

## Cell-Mediated Immunity

Cell-mediated immunity, often referred to as *T-cell-mediated immunity*, refers broadly to any host resistance mechanism in which cellular elements play a direct role and which is part of the acquired arm of immunity. This is in comparison to humoral immunity, in which there are certainly cellular interactions but the final host resistance products are soluble factors such as antibody. A number of host defenses are mediated directly by cells, including MØ-mediated cytolysis, antibody-dependent cellular cytotoxicity, and natural killer (NK) cell cytotoxicity although cytotoxic T-lymphocytes (CTLs) usually predominate, particularly in the destruction of virus-infected cells.

Functions associated with CMI are commonly considered the province of T-lymphocytes, although immune cells (e.g., B-cells and MØ) as well as non-immune cells (e.g., fibroblasts and dendritic cells) contribute to the development of CMI. As the primary effector cell in CMI, the T-cell represents one of the most complex and multifunctional immune cells. Antigens that generally elicit CMI include tissue-associated antigens, chemicals and drugs that covalently bind to autologous proteins, and antigenic determinants on persistent intracellular microorganisms. The route of exposure also plays a major role in the type of response generated; for example, sheep erythrocytes elicit antibody production (but not CMI) when injected intravenously in humans, but elicit both when injected intracutaneously. The induction of CMI proceeds when small lymphocytes differentiate into large pyroninophilic cells and ultimately divide, giving rise to cells responsible for effector function as well as immunological memory. In contrast to HI, which is more effective against extracellular pathogens, CMI helps protect against intracellular bacteria, viruses and neoplasia and is responsible for graft rejection.

T-cells can differentiate into populations responsible for either regulatory or effector function. For example, regulatory and inducer T-cell functions are provided by CD3/CD4+ T-helper cells. The T-helper function facilitates antibody responses by B-cells and assists in other T-cell responses. For most antigens, B-cells require assistance from T-cells for differentiation into PCs. T-helper cells are integral in the B-cell response by participating in two distinct mechanisms: (1) major-histocompatibility-locus-restricted B- and T-cell collaborations, and (2) cytokine-mediated differentiation. Helper function is a result of interactions between surface molecules on T-helper cells and B-cells, as well as the production and secretion of immunoregulatory cytokines.

Effector functions take the form of cytotoxic activity (CD3/CD8 phenotype), manifested by CTLs. These cells are able to specifically lyse target cells via the release of various bioactive molecules. Another effector function is the ability of T-cells to mediate suppressor activity for both T- and B-cell responses. Suppressor activity is also mediated by cells bearing the CD3/CD8 phenotype, although recent studies suggest that this activity may be the result, at least in part, of differential cytokine production by this population (Figure 36.1). This responsibility for both helper and suppressor activities indicates the crucial role of T-cells in normal immune function.

## The T-Helper 1/T-Helper 2 Cell Paradigm

An important conceptual breakthrough in immunology was the finding that two major populations of T-helper cells exist that have different, sometimes opposing, functions. Mosmann et al. [12] first established the concept by demonstrating that cloned murine T-cells exhibited differential patterns of cytokine production. One population, designated T-helper 1 cells (Th1), was found to produce interleukin-2 (IL-2), IFN-γ, and lymphotoxin. The second major population (designated Th2 cells) produces IL-4, IL-5, IL-10, and IL-13. Both populations of T-cells produce IL-3, granulocyte–macrophage-colony-stimulating factor (GM-CSF) and tumor necrosis factor (TNF). Later, a third population, Th0, was described and was found to exhibit an intermediate pattern of cytokine production. These cells are less well defined but may be an early precursor of Th1 and Th2, or, alternatively, they may represent an intermediate stage in development of the other two populations.

Although there were initial doubts that human T-cells followed this paradigm, it is now known that a similar paradigm exists for human T-cells [13]. The major differences appear to be in the profile of cytokine production, cytokine response (e.g., human Th1 and Th2 proliferate in response to IL-4 while only Th2 cells proliferate in the presence of IL-4 in rodents), and cytolytic potential. Despite these differences, the human and rodent systems are similar enough to make experimental rodent models meaningful for understanding the human immune response.

Recent studies suggest that Th1 and Th2 cells may not necessarily represent distinct lineages descending from a common precursor but rather may be seen as points in a continuum; for example, development of each population is influenced by type, location, and concentration of eliciting antigen. More important may be the cytokine milieu (Figure 36.1). For example, the cytokines IL-12 and IFN-γ-inducing factor (from MØ) and IFN-γ (from NK cells) drive the development of Th1 cells, whereas IL-4 (from the ill-defined "T-accessory" cell, mast cells, or other sources) drives the development of Th2 cells [14]. Interestingly, IL-4-driven development of Th2 appears to take precedence over IL-12-induced Th1 production; this may have ramifications in the etiology of some disease states.

The Th1/Th2 paradigm is important for immunotoxicology in that certain immunopathology has been associated with the predominance of one helper cell type over another, particularly in human disease states; for example, Th1 polarization has been associated with organ-specific autoimmune diseases such as multiple sclerosis and Hashimoto's thyroiditis, whereas systemic autoimmune conditions, such as rheumatoid arthritis and Sjögren's syndrome, lack a clear T-cell polarization [15].

On the other hand, strong Th2-type responses appear to result in many hypersensitivity disorders, including asthma. The extent to which cytokine polarization contributes to these pathologies, as opposed to being a sequel of other mechanisms, remains to be elucidated. It is possible, however, that assignment of Th1/Th2 patterns may eventually become much more important when designing and performing mechanistic immunotoxicology studies [16].

## Natural Killer Cells

Natural killer (NK) cells are a population of non-B-, non-T-lymphocytes that exhibit cytotoxicity toward a variety of target cells, including tumor cells and virally infected cells. NK cells express a unique panel of cell surface markers (e.g., asialo GM1) and are morphologically distinct, being larger than other lymphocytes. In addition, they contain numerous granules, leading to their designation as large granular lymphocytes (LGLs) [17,18]. Unlike CMI, NK-cell-mediated cellular cytotoxicity is MHC unrestricted and does not require prior exposure to the target. NK cells have been seen principally as mediators of so-called "immune surveillance" [19], resulting in the concept of a constant removal of spontaneously arising neoplastic cells [20,21]. In fact, the standard methodology for assessing NK cell function relies on the *in vitro* lyses of tumor target cells; however, NK cells are more likely to play a role in resisting the progression and metastatic spread of tumors once they develop, rather than preventing initiation [22].

In contrast to previous models in which NK cells were considered independent of the acquired immune response, recent studies have revealed an important role for these cells in the induction and regulation of acquired immunity [23–25]. NK cells respond to, and produce, key immunoregulatory cytokines and thus play an important role in the normal immune response. In fact, studies of individuals with NK cell deficiency states, most of which are associated with single gene mutations, have helped identify a role for NK cells in defense against human infectious disease. A resounding theme of NK cell deficiencies is susceptibility to herpes viruses [26–28].

Related to NK cells are the NKT cells, a population of CD1d+ lymphocytes that exhibit properties of both NK cells and T-cells [29,30]. These cells appear to have primarily immunoregulatory functions and serve in the coordination of functions between the innate and adaptive immune responses [31]. To date, this cell population has not been a target of immunotoxicology assessment, although it is conceivable that these cells would be as likely a target as NK cells. Moreover, NKT cells have been associated with immunotoxicologically important endpoints, including resistance to infection and autoimmunity [32,33].

## Other Immunoregulatory Circuits

### Cytokines

Cytokines are glycoproteins that are generally produced in response to cellular activation. Most cytokines studied to date have multiple and overlapping actions, and they frequently function via cascading mechanisms referred to as the *cytokine network*, interacting with each other both synergistically and antagonistically. Two important features of cytokines are that they usually act at a local level and they are rapidly cleared from the circulation. This combination of features helps ensure that cytokines remain compartmentalized, undoubtedly an important consideration given the potent bioactivity of these molecules [34,35]. Cytokines serve as immune system mediators and regulators. They are produced proportionally by T-helper lymphocytes but are not exclusive to the immune system; in fact, some cytokines are phylogenetically ancient and highly conserved. Furthermore, both IL-1 and TNF are intrinsically involved in apoptosis and cellular proliferation, both fundamental biological processes. Thus, cytokines should be recognized for their role as conveyers of bioinformation, rather than as simple effector molecules involved in a single physiological process such as immunity and host resistance. For convenience, cytokines may be grouped into several classes (Table 36.3). These classifications are necessarily arbitrary due to the overlapping activity of these molecules. An updated listing of all known cytokines may be found at www.copewithcytokines.de.cgi.

### Chemokines

Another related group of molecules is the chemokines. Chemokines are small peptide molecules that, like cytokines, were originally associated with the immune system but which now are recognized as being produced by almost all cells of the body and involved in a multitude of biological functions (Table 36.3). Chemokines play many roles, including modulation of the Th1/Th2 balance associated with autoimmunity and hypersensitivity [36], as mediators of allergic inflammation [37], and modulation of the function of leukocytes in disease states such as rheumatoid arthritis and asthma [38].

### The Immune/Nervous/Endocrine System Axis

It has been recognized for some time that the nervous, immune, and endocrine systems, rather than being separate in function and structure, share many features and appear to cross-regulate each other's function [39–43]. The ramification for immunotoxicology of the existence of this neuro–immune–endocrine axis is that xenobiotics can affect the immune response indirectly by affecting

**TABLE 36.3**
**The Major Classes of Cytokines and Chemokines**

| Class | Members | Functions |
|---|---|---|
| Interleukins (IL) | IL-1 (α and β), IL-1 receptor antagonist, IL-2, IL-4, IL-5, IL-6, IL-7, IL-9–IL-30 | Primarily immunoregulatory; act on immune system cells (generally lymphocytes) in either stimulatory or inhibitory fashion |
| Colony-stimulating factors (CSF) | Granulocyte (G-CSF), macrophage (M-CSF), granulocyte/macrophage (GM-CSF), IL-3, MEG-CSF | Involved in the proliferation of leukocyte progenitors; GM-CSF and IL-3 share certain immunoregulatory functions |
| Interferon (IFN) | IFN-α, IFN-β, IFN-δ, IFN-ω, IFN-γ | |
| Type I | | Primarily antiviral activity; some immunoregulatory functions |
| Type II | | Primarily immunoregulatory |
| Tumor necrosis factors (TNF) ligand superfamily | TNF-α, TNF-β, TNF-related apoptosis-inducing ligand (TRAIL); CD27 ligand, CD30 ligand, CD40 ligand, CD95 ligand, FAS ligand | Immunoregulatory activities; antitumor effector functions; apoptosis growth regulation |
| Hematopoietins | Stem cell factor, stem cell growth factor, erythropoietin, thrombopoietin | Involved in the regulation of bone marrow function and the production of hematopoietic cells |
| Miscellaneous | Oncostatin M, leukemia inhibitory factor, transforming growth factors | Various pleiotropic functions |
| Chemokines | | |
| C | Lymphotactin | Chemotactic for lymphocytes |
| CC | C10, eotaxin, I-309, leukotactin-1, MARC, MCP, MIP-1, MIP-3, MPIF-1, PARC, RANTES, TARC, TECK | Primarily active on monocytes/macrophages |
| CXC | IL-8, 6Ckine (Exodus), BLC, CINC-1, CINC-2, CRG, ENA-78, gro, KC, MIG, MIP-2, NAP-2 | Primarily active on neutrophils and T-cells |
| CX3C | Fractalkine | Modulates calcium flux; involved in cell adhesion |

other organ systems; conversely, modulation of the immune system may have secondary effects on other organ systems. To date, these interactions have not been extensively studied as they relate to immunotoxicology.

## CONCEPTS AND APPROACHES FOR DETECTING IMMUNOMODULATION IN EXPERIMENTAL ANIMALS

### Basic Considerations in Study Design

The value of incorporating immunological assessment in the toxicological evaluation of drugs, chemicals, and biologicals in human risk assessment has been increasingly accepted by regulatory agencies. These data are now more frequently used in the overall safety assessment process of drugs and to help determine acceptable exposure levels to chemicals. For example, the U.S. Environmental Protection Agency (EPA) has used immunotoxicity data for establishing the reference doses (RfDs) or reference concentrations (RfCs) for a number of compounds, including 1,1,2-trichloromethane, 2,4-dichlorophenol, benzene, and dibutyltinoxide. Also, the Agency for Toxic Substances and Disease Registry has derived "minimum risk levels" for arsenic, dieldrin, nickel, 1,2-dichloroethane, and 2,4-dichlorophenol using immune endpoints. At the time of this writing, the EPA is preparing test guidelines for immunotoxicity testing. Both the U.S. Food and Drug Administration (FDA) and the European Agency for the Evaluation of Medicinal Products (EMEA) have issued guidelines for the evaluation of drug candidates. Currently, the International Committee on Harmonization (ICH) has taken on standardizing immunotoxicology testing guidelines for the safety assessment of drug candidates across the major geographic areas affected.

In general, effects observed in the immune system may be due to the general properties of the chemical (e.g., interaction with macromolecules) and amplified by the complex nature of the immune system, such as antigen recognition and processing; cell cooperation, regulation, and amplification; cell activation, proliferation, and differentiation; and the production of regulatory mediators (lymphokines or cytokines) by the various cell types involved. Because of this complexity, the initial strategy among immunologists working in toxicology and safety assessment was to select and apply a tiered panel of assays to identify immunosuppressive or immunostimulatory agents in laboratory rodents [44–46]. Although the configurations of these testing panels vary by laboratory and species, they generally include measures for: (1) altered lymphoid organ weights and histomorphology; (2) quantitative changes in cellularity of lymphoid tissue, peripheral blood leukocytes, and bone marrow; (3) impairment of cell function at the effector or regulatory level; and (4) increased susceptibility to experimental challenge with infectious agents or transplantable tumors cells [47].

There are a number of advantages and limitations to using such test panels. For example, although the sensitivity of these methods to detect immune system changes is well recognized in animal models, it is often difficult to establish the clinical significance of subtle immune changes on the development of neoplasia or infectious diseases, particularly in humans. Furthermore, some of the tests require invasive procedures such as immunization or are not feasible or ethical for inclusion in human studies; thus, they limit the potential for animal–human comparisons, although several recent immunotoxicology studies in humans have included primary immunization (e.g., hepatitis B) as a test measure. In this respect, assays that require *in vivo* primary antigenic challenge are generally accepted as the most sensitive and predictive of all immune function tests.

A variety of factors must be considered when evaluating the potential of an environmental agent or drug to adversely influence the immune system. Assessment requires validation of the endpoints to be measured (quality control and biological relevance), as well as the knowledgeable selection of animal models and exposure parameters and consideration of general toxicological parameters, including metabolism, distribution, and toxicokinetics. The treatment protocol should take into account the potential route and level of exposure expected in humans, the biophysical properties of the agent (e.g., protein binding, bioavailability, and toxicokinetics of the agent), as well as information on the mechanism of action of the agent. The dose selected should attempt to establish a clear dose–response curve as well as a no-observable-effect level (NOEL) up to or near the maximum tolerated dose (MTD). Although in some instances it might be beneficial to include a dose level that induces overt toxicity, any immune change observed at such a dose level should be interpreted cautiously, as severe stress and malnutrition are reported to impair immune responsiveness. It is often recommended that the highest dose used be considerably lower than a dose producing severe weight loss. Although laboratories routinely employ three dose levels, dose range-finding studies are recommended prior to a full-scale immunotoxicology evaluation.

The selected exposure route should parallel the most probable route of exposure in humans, which is most frequently oral but it could be respiratory, dermal, parenteral, subcutaneous, or intraperitoneal. Because the major routes of exposure (i.e., skin, lung, or gut) are also associated with local immunity, attention has been directed in some laboratories to the development of methodology for the assessment of local immune responses, particularly in the lung.

Selection of the most appropriate animal model for immunotoxicology studies has also been a matter of much discussion. Ideally, toxicity testing should be performed in a species that: (1) will elicit chemical-related pharmacology and toxicities similar to those anticipated in humans. For most immunosuppressive therapeutics, rodent data for target organ toxicities and comparability of the immunosuppressive doses have generally been predictive of the clinical observations. Exceptions to the predictive value of rodent immunotoxicology data are seen infrequently but have occurred (e.g., glucocorticoids are lympholytic in rodents but not in primates [48,49]). Although certain compounds may exhibit different pharmacokinetic properties in rodents than in humans, rodents still appear to be the most appropriate animal model for examining the immunotoxicity of small molecules [50]. This statement is based on the established similarities of toxicological profiles across these models and when comparing host susceptibility challenge and immune function data. Although exceptions do exist [51,52], for the most part, when studied using a similar experimental design, mice and rats have provided comparable immunotoxicology results.

The quantitative and qualitative response of an experimental animal to an immunotoxic agent can be influenced by its genetic composition (genotype), indicating a need to consider not only species but in some cases also strain. Rao et al. [53] described two approaches to the selection of genotypes for rodent toxicity studies. The first approach is to select a strain where the genotype is representative of the animal species, in the hope that the choice will also exhibit sensitivities similar to humans. This can best be accomplished by using randomly bred rodents; however, due to the variability in immune responses associated with outbred animals, it may be necessary to use a greater number of animals to identify a sensitive population. A second approach attempts to identify genotypes that are uniquely suitable for evaluation of a specific class of chemicals. This requires considerable knowledge of the mechanisms of toxicity for the particular compound. One compromise would be to use F1 hybrids that contain the stability, phenotypic uniformity, and background information of an inbred animal and yet have heterozygosis. At present, it is impossible to determine how applicable these conclusions will be for immunotoxic compounds with different immune profiles; however, as more comparative analyses become available, the ability to accurately estimate the potential clinical effects from immunological tests in animals should improve.

**TABLE 36.4**
**Examples of Drugs and Chemicals Associated with Immunosuppression**

| | | |
|---|---|---|
| Pharmaceuticals | Cytoreductive agents | Transplantation drugs |
| | Opiates | Therapeutic immunosuppressant |
| | Antibiotics | AIDS therapeutics |
| Industrial chemicals | Organic solvents | Halogenated aromatic hydrocarbons |
| | Polychlorinated biphenyls | Polycyclic aromatic hydrocarbons |
| | Glycol ethers | |
| Environmental agents | Heavy metals | Air pollutants |
| | Ultraviolet light | Dusts (silica, asbestos) |
| | Pesticides | |
| Recreational adjuncts | Ethanol | Tobacco (smoke) |
| | Cannabinoids | Opiates |
| | Cocaine | |

## Immunosuppression

### Introduction and Fundamental Concepts

Based on the preceding discussion of the important role that the immune system plays in protection of the host from infectious organisms and incipient neoplasia, it is logical to expect that disruption of this system following exposure to xenobiotics would have serious consequences. A large body of information has developed demonstrating that xenobiotic exposure can produce immune suppression and altered host resistance in experimental animals (Table 36.4) following acute or chronic exposure. Although only a limited number of reports document immune dysfunction following human exposure to xenobiotics, clinical data with a number of agents appear to demonstrate that immunotoxicity in rodents may form the basis for human risk assessment [54]. Given the complexity of the immune system (both natural and acquired) and the many potential target cells and molecules, it is impractical to enumerate all of the potential targets of immunosuppressive agents. For this reason, several immune function assays have been developed and validated for evaluating immunotoxicity. These techniques and approaches are discussed in the following sections.

### Techniques for Assessing Immunosuppression

The basic approach to immunotoxicity testing, as it is currently practiced, is based on the work of Luster et al. [55–57]. This early work established the concept of the tier approach, where test materials are evaluated for effects on the immune system using a biphasic system of descriptive and functional assays. This group of tests provides a fairly comprehensive evaluation of immune structure and function (Table 36.5). The tests included in tier 1 (screening) have varied depending on the individual laboratory conducting the studies or the recommending agency; however, in most cases, tier 1 includes tests that can at least be incorporated into a standard 28-day safety study, including routine hematology, selected organ weights (spleen, thymus), and histology of lymphoid organs. Some agencies request assessment of the T-dependent antibody response (TDAR) using sheep red blood cell, keyhole limpet hemocyanin, or other T-dependent antigen to monitor an active immune response to the antigen as part of a tier 1 evaluation. This requires the use of additional animals and specialized technical training. Tier 1 assays may also include assessment of NK cell activity and immunophenotyping, which does not require additional animals.

In situations where a significant effect was observed in one of the tier 1 tests, the nature of the immune defect can be confirmed by using tier 2 (comprehensive) tests. These include immunopathology (quantitation of T- and B-cell numbers), enumeration of IgG antibody response for HI, functional assessment of CMI using the CTL or DTH assays, and assessment of natural immunity using MØ function assays. In addition, tier 2 testing often includes host resistance challenge assays (bacterial, viral, parasite, or transplantable tumor) as a measure of immune function of the whole animal.

Given the high predictive value of certain of these assays for immunotoxicity assessment, as well as the time and expense involved in performing the entire battery of tier 1 and tier 2 tests, many practicing immunotoxicologists now use the TDAR assay, in conjunction with more routine toxicological tests such as lymphoid organ weights and histomorphology, as an initial assessment for potential immunotoxicity of drugs or chemicals. In many cases, measurement of NK cell activity is also included in this initial assessment, as alterations in this effector of natural immunity would not normally be detected using the other assays. In the following section we will, therefore, concentrate on these particular assays.

**TABLE 36.5**
**Assays Commonly Employed To Assess Immunosuppression in Laboratory Animals**

| Tier | Rodent | Nonhuman Primate |
|---|---|---|
| Initial assessment (tier 1) | Hematology | Hematology |
| | Bone marrow histomorphology | — |
| | Lymphoid organ weight and histomorphology | — |
| | Primary antibody response | Serum Ig level |
| | NK cell activity | NK cell activity |
| | Surface marker analysis | Surface marker analysis |
| Advanced assessment (tier 2) | CTL or DTH | — |
| | MØ function | MØ function |
| | Apoptosis | Apoptosis |
| | Cytokine analysis | Cytokine analysis |
| | Host resistance assays | — |

*Abbreviations:* CTL, cytotoxic T-lymphocyte; DTH, delayed-type hypersensitivity; Ig, immunoglobulin; NK, natural killer; —, not routinely performed.

Recently, a work group was formed under the auspices of the National Toxicology Program to examine the utility of extended histopathology of critical immune organs, originally described by Kuper et al. [58] for identifying immunotoxic chemicals [59]. The results from these studies suggest that, although not as sensitive as functional assays, extended histopathology of the immune system provides a reasonable level of accuracy for identifying immunotoxic chemicals, provided the level of stringency used to score histological lesions allows for detection of true positives while minimizing false positives.

The following tests are commonly performed using the B6C3F$_1$ mouse or the Fischer 344 rat, although they are readily applicable to other rodent strains. Fewer well-developed immunotoxicology assays are available for nonhuman primates at present, and even fewer tests are available for use with canines.

### *Initial Assessment*

#### *Routine Tests and Extended Immunopathology*

A hemogram (complete blood count and differential) and examination of lymphoid organ weights, including the spleen, thymus, and lymph node, are useful for assessing the immunotoxic activity of a material [60–62]. Because of the structural division of the spleen and lymph nodes into thymus-dependent and thymus-independent compartments, immunocytochemical staining may indicate preferential effects for T- or B-cells, although this is seldom conducted. Likewise, microscopic examination of the thymus may reveal a compound that affects thymocyte viability, although careful measurement of thymus weight is easier and provides equal, if not more, sensitivity. As indicated earlier, extended histopathology provides considerably more information than routine histopathology and does not involve any specialized stains. This involves a semiquantitative assessment to estimate histological changes within different anatomical compartments of the spleen, thymus, and lymph node and considers changes in cell density and anatomical compartment size [58,59]. Five endpoints are examined in the spleen: cellularity of periarteriolar lymphoid sheaths (PALSs), lymphoid follicles, marginal zone, red pulp, and the number of germinal centers. Three endpoints are examined in the thymus: cortex cellularity, medullary cellularity, and the cortico/medullary ratio. Four endpoints are evaluated in the lymph node: grade of cellularity in the follicles, paracortical areas, medullary cords, and sinuses.

#### *T-Cell Dependent Antibody Response*

Within a few days following *in vivo* injection of a foreign antigen, antibody molecules of the IgM class are produced and released from PCs into the systemic circulation. The antibody-forming cell (AFC) assay, alternatively referred to as the plaque-forming cell (PFC) assay, quantitates the production of specific antibody, either indirectly in cell culture or in the circulation via standard ELISA techniques or through enumeration of antibody-producing cells in the spleen following a primary antigenic stimulus such as sheep red blood cells (SRBCs). Although TDAR is, strictly speaking, a measure of B-cell function rather than T-cell function, it is an excellent functional parameter to examine, as this response requires cognate cell interaction and regulation by MØ, T-cells, and soluble regulatory molecules such as cytokines. The primary antibody response is currently measured using either a PFC assay [63] or an ELISA [64]. The steps involved in this assay are illustrated in Figure 36.5.

### IgM Plaque Assay

*Materials and reagents required:*

- Earle's balanced salt solution (EBSS) supplemented with 25-m*M* HEPES buffer
- SRBCs in Alsever's solution
- Guinea pig complement (GPC′)
- Dulbecco's phosphate-buffered saline (DPBS)
- DEAE dextran (30 mg/mL in saline, pH 6.9)
- Bacto-agar
- Petri dishes and cover slips

*Procedure:*

1. Four days prior to assay, immunize animals with an intravenous injection of washed SRBCs in sterile saline. Recommended inocula are approximately $1 \times 10^8$ SRBCs for mice and approximately $2 \times 10^8$ SRBCs for rats.
2. Euthanize the animals, remove the spleens, and prepare a single-cell splenocyte suspension in EBSS. Prepare two dilutions of the cell suspension in EBSS.
3. Wash SRBC three times by centrifugation. After the final wash, retain approximately 100 μL of SRBC and then adjust the remaining cells to a final density of 10% in EBSS. Add GPC′ to the reserved SRBCs, mix well, and hold on ice until needed.
4. Prepare a solution containing 0.5% agar in DPBS, add DEAE dextran (1.6 mL stock solution per 100 mL agar), and mix. Dispense the agar in 0.35-mL aliquots into polypropylene culture tubes, and maintain these tubes at 45°C.
5. For the assay, each tube contains 0.35 mL agar solution, 100 μL cell dilutions, and 25 μL GPC′. Add SRBCs first and then the cell suspension; immediately remove the tube from the water bath. Add the GPC′ and mix the contents of the tube. Dispense the contents into a Petri dish and then drop the cover slip so an even layer of fluid forms underneath.
6. Incubate the plates at 37°C for approximately 3 hours, and enumerate the plaques. While the plates are incubating, determine cell number and viability of the original splenocyte suspensions.
7. Calculate the results as follows:
   a. Plaques counted under each cover slip × 10 × dilution factor = PFCs/mL of the original cell suspension (as 0.1 mL of the cell dilution is counted)
   b. PFCs/mL × volume of original cell suspension = PFC/spleen
   c. PFCs/mL/number of viable cells/mL = PFC/$10^6$ viable splenocytes

### Anti-SRBC IgM ELISA

*Materials and reagents required:*

- SRBC in Alsever's solution
- Horseradish peroxidase (HRP)-conjugated, affinity-purified goat anti-mouse/anti-rat IgM antibody
- Peroxidase substrate (2-azino-*bis*-3-ethylbenzthiazoline-6-sulfonic acid [ABTS])
- ABTS buffer (phosphate–urea–hydrogen peroxide)
- Phosphate-buffered saline (PBS)
- 96-well microplates
- General reagents and supplies for ELISA

*Procedure:*

1. Immunize mice or rats with SRBCs as for the plaque assay. Five days (mice) or 6 days (rats) later, obtain serum from both immunized and naïve animals. Pool each as appropriate to use as standards or controls and freeze at –20°C.
2. Treat mice or rats with test material and vehicle (and a positive control, if necessary). On day 5 or 6, respectively, post-treatment, obtain serum from animals. *Note:* If serum is collected via the retro orbital sinus, additional samples may be collected later for time-course studies.
3. Prepare SRBC membrane antigens by lyses and solubilization [64]. This antigen serves as the capture reagent in the ELISA.
4. Obtain anti-SRBC IgM monoclonal antibodies to use as standards. *Note:* Anti-SRBC must be of the appropriate species depending on the test animal (e.g., mouse or rat).
5. Dilute membrane antigen to 1.0 μg/mL in PBS and coat the wells of the microplates at approximately 4°C using 125 μL/well of the antigen preparation.
6. On the day of assay, wash the plates three times with 0.01% Tween®-20 in water. Block any unbound sites on the plates by incubating the plates with 200 μL/well of PBS per 0.05% Tween®-20, 3% bovine serum albumin, or 3% powdered milk.
7. Prepare serial twofold dilutions of test sera and antibody standards. Add to the plates and incubate for at least 1 hour at room temperature.
8. Wash the plates three times, then add HRP-conjugated secondary antibody. Incubate for at least 1 hour at room temperature and then wash the plates three times.

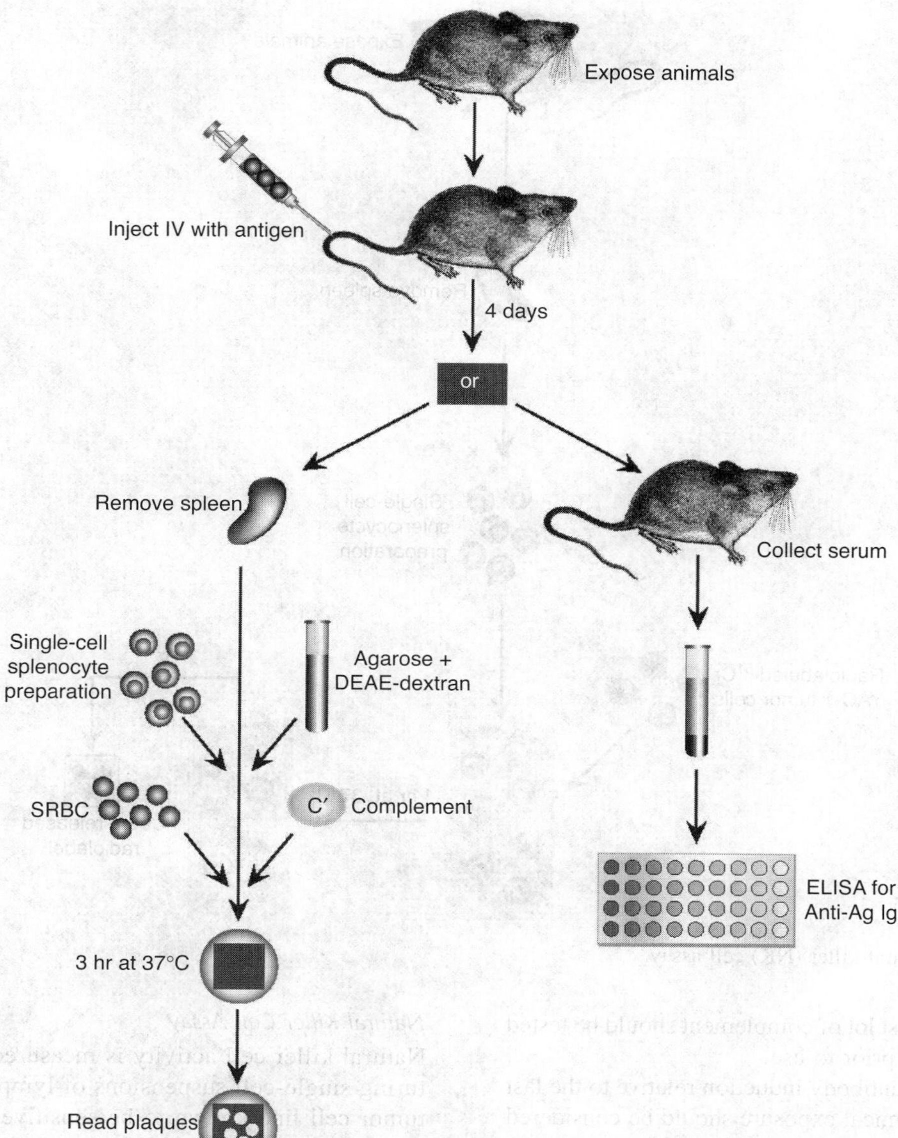


**FIGURE 36.5** T-dependent antibody response (TDAR) assay.

9. Add peroxidase substrate (ABTS) and incubate the plates at room temperature for 45 minutes. Stop the reaction by adding 3% oxalic acid to all wells.
10. Read the plates at 405 nm, and calculate the results based on curves prepared using the antibody standards.

*Positive control:* Cyclophosphamide is routinely used as a positive immunosuppression control for the AFC assay (either plaque or ELISA format). For mice, cyclophosphamide is administered intraperitoneally at 80 mg/kg once approximately 24 hours prior to euthanasia. For rats, it is given intraperitoneally at 20 to 25 mg/kg daily for 4 to 5 days prior to euthanasia.

*Notes:*

1. The antibody-forming cell response varies depending on the day of analysis following immunization. Each species and strain should also be evaluated for the optimum response, although for intravenous injection the optimum assay period is usually 4 days following immunization.
2. The dose and route of antigen exposure alter the peak AFC response. Intravenous injections shift the optimum response to an earlier time, whereas an intraperitoneal injection delays the peak response.

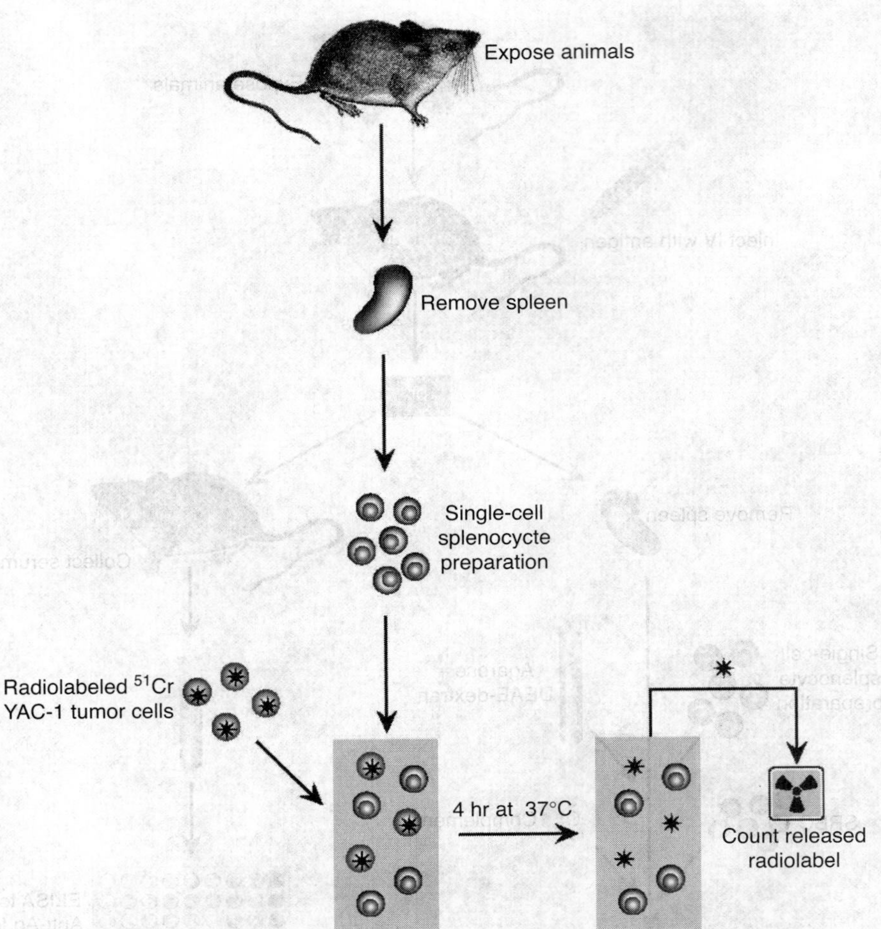


**FIGURE 36.6** Natural killer (NK) cell assay.

3. Each new test lot of complement should be tested and titrated prior to use.
4. The day of antibody induction relative to the last dose of chemical exposure should be considered when designing a study.
5. SRBC membrane antigens are generally prepared individually by a laboratory. Monoclonal and polyclonal antibodies specific for SRBCs are commercially available from a variety of sources.
6. The direct comparability of results between the plaque assay and the ELISA is the source of some discussion. The AFC assay measures antibody production in one organ (spleen) only, whereas the ELISA is a measure of systemic antibody production, which may have a different time course.
7. The current trend in TDAR assessment favors use of an ELISA against keyhole limpet hemocyanin (KLH), a widely used T-dependent antigen. Methodology for standardizing this technique and applying it to immunotoxicology assessment has been published [65–67].

*Natural Killer Cell Assay*

Natural killer cell activity is measured *in vitro* by culturing single-cell suspensions of lymphoid cells with a tumor cell line known to be sensitive to NK-mediated cytotoxicity. The target cells are radiolabeled prior to the assay; thus, any cells that have been lysed will release their radioactivity into the culture medium, where it can subsequently be quantitated. The procedure described below is modified from the micro-culture method described by Reynolds and Herberman [68] and is the standard approach for immunotoxicity assessment. The procedure for this assay is illustrated in Figure 36.6.

*Materials and reagents required:*

- RPMI-1640 culture medium supplemented with 25-m*M* HEPES buffer, 10% fetal bovine serum (FBS), 2-m*M* L-glutamine, and 50 μg/mL gentamicin
- Fetal bovine serum (FBS)
- DPBS
- Wash solution (DPBS/1% FBS)

- YAC-1 cell line (for rodent NK evaluation; ATCC #TIB 160) or K562 cell line (for primate NK evaluation; ATCC #243) maintained in log-phase growth in the culture medium described above
- 96-well, round-bottom, microculture plates
- 0.1% solution of Triton® X-100 in distilled $H_2O$
- $^{51}Cr$ as sodium chromate in sterile saline; specific activity of 200 to 500 mCi/mg
- Supernatant collection system

*Procedure:*

1. Prepare a single-cell suspension of the effector spleen cells, and adjust to a density of $5 \times 10^6$ viable cells per mL in culture medium.
2. Prepare two serial 1:3 dilutions of the cell suspension in culture medium. Dispense 100 μL of each dilution in quadruplicate wells of 96-well, round-bottom microculture plates.
3. Centrifuge a log-phase culture of target cells and suspend the cell pellet in 0.5 mL FBS. Add 200 μL $^{51}Cr$ to the cells, mix well, and incubate at 37°C for 1 hour. Wash the cells three times.
4. Suspend the target cells in culture medium, determine cell number and viability, and adjust the cells to a final density of $5 \times 10^4$ viable cells per mL in culture medium. Add the target cells to all wells in a volume of 100 μL/well. Include a row containing 100 μL target cell suspension and 100 μL culture medium per well (spontaneous release) and one row consisting of 100 μL target cell suspension and 100 μL 0.1% Triton® X-100 per well (total release).
5. Incubate the plates at 37°C, 5% $CO_2$, for 4 hours.
6. Harvest supernatant fractions either manually or by using a semiautomatic harvesting system. Quantitate radiolabel released into the supernatant fractions in a gamma counter, and determine percent cytolysis using the following formula: percent cytolysis = (experimental release − spontaneous release)/(total release − spontaneous release) × 100.

*Positive controls:*

- Immunosuppression control—Unless the laboratory has extensive experience, a positive suppression control of the NK response should be included. Approximately 24 to 78 hours prior to euthanasia, a separate group of animals is injected intravenously with an optimum concentration of anti-asialo GM1 antibody. The exact amount to be given will vary from lot to lot and between suppliers. Treatment with an optimum dose of anti-asialo GM1 will result in an essentially complete abrogation of the NK response in rodents [69].
- Immunostimulation control—In some cases, it may be useful to include a positive control for NK cell augmentation. Although cytokines (IL-2 and IFN-γ) can enhance this response both *in vivo* and *in vitro*, an equally efficient, and more economical and reproducible, option is the use of interferon inducers such as polyinosinic:polycytidilic acid (poly I:C) [23]. Poly I:C is administered intraperitoneally at a concentration of 100 μg/mouse or 500 to 1000 μg/rat approximately 24 hours prior to assay.

*Notes:*

1. Natural killer activity is highest in young mice, declining after 12 weeks of age. Basal NK activity may be highly variable or undetectable in mice over 20 weeks old.
2. The target cells must be in log-phase growth to achieve adequate labeling with $^{51}Cr$. In addition, the target cell lines should be assessed for *Mycoplasma* contamination at periodic intervals.
3. The assessment of NK cell activity has been utilized most extensively in rodents and primates. For instances where evaluation of canine NK cell function would be useful, modified techniques have been published [70].
4. For laboratories unable or unwilling to use radioisotopes, flow cytometry serves as a useful alternative [71]. A full comparison has not been made between these alternative methodologies and the standard chromium release assay, although a study by Motzer et al. [72] found little advantage in flow cytometry in comparison to the standard radiolabel assays.

### *Phenotypic Analysis of Cell Surface Markers by Flow Cytometry*

Both pros and cons exist for including immunophenotypic analysis as part of a screen. It offers a rapid, sensitive, and quantitative measure of a heterogeneous cell population in blood or cell suspensions prepared from lymphoid organs. The major concern is that the test is not well validated, particularly with respect to sensitivity or how less profound changes relate to disease. For mechanistic studies, however, they are of unquestionable relevance, especially when considering the vast array of well-defined surface markers now available for rodent and human immune cells. The technique involves treating cells with monoclonal antibodies covalently bound to different fluorochromes. These antibodies recognize surface antigens, referred to as *clusters of differentiation* (CDs), unique to different cell types. The availability of fluorochromes, which emit light at different wavelengths following excitation, combined with flow cytometers that are capable of performing multiple color analysis, provides a rapid and effective method of analyzing cell types. The most commonly examined CDs in the mouse are those that recognize pan T-cells (CD90 and TCR complex), T-helper cells (CD4), T-suppressor cells (CD8), and pan B-cells (CD45R/B220 or CD19).

*Materials and reagents required:*

1. Prepare single-cell suspensions from the spleen (Ficoll-separated and whole blood have both been used occasionally). For washing and staining, use DPBS (0.01 *M*).
2. Centrifuge conjugated reagents at $15 \times 10^3$ *g* to remove aggregates.
3. Pipette desired concentration of antibody or control sera in 50-μL volumes to a small test tube or 96-well microtiter plates. For two-color analysis, both conjugated antibodies can be added together.
4. Add $10^6$ viable cells in a volume of 50 μL to the test tube or well containing the antibody.
5. Incubate 30 minutes on ice in the presence of 0.1% sodium azide.
6. Wash with 2.0-mL (if in tubes) or two 100-μL (if in wells) washes.
7. Suspend to a volume of 1 to $2 \times 10^6$ cells per mL in cold PBS containing 0.1% sodium azide, and perform analysis.
8. Cell fluorescence and integrity can be preserved for up to 5 days by rapidly suspending the cell pellet in 50 μL cold PBS containing 1% paraformaldehyde.

### Mechanistic Immunotoxicology Assays

From its beginnings, the discipline of immunotoxicology has constantly evolved and incorporated new techniques and paradigms to understand the nature of immunomodulation. The assays described previously in this section allow one to make a first-pass evaluation of drugs and chemical agents for generalized immunotoxicity; that is, the assays will indicate that the immune system had been perturbed, although the cellular and molecular mechanism involved will not necessarily be obvious. These assays are valuable for quick and relatively accurate identification of toxic agents. On the other hand, the tools and concepts of immunotoxicology are increasingly being utilized as research tools to understand the function of the immune system; for example, it may be useful to know not only whether or not an agent modulates the immune response, but why. This is especially important in the discovery and development of pharmaceutical agents, where therapeutic manipulation of the immune system may be a desirable goal. In response to these novel applications of immunotoxicology, assays are needed that will allow us to determine the precise mechanism of immunomodulation.

The methodology for tier 2 assays such as T-cell mediated immune function (CTL/DTH), MØ function, IgG antibody cell forming response, and host resistance models has been reviewed in detail elsewhere and will not be reiterated here [73–76]. These assays are still valuable tools for understanding the mechanistic basis of immunotoxicity. In addition to these assays, several other methodologies are now being included in the immunotoxicology armamentarium.

#### Assessment of Apoptosis

Apoptosis (programmed cell death) is increasingly recognized as a fundamental process in both health and disease states, including the response to toxic insult [77]. Apoptosis plays a vital role in the immune response, regulating the number and action of immune cells such as lymphocytes [78,79]. Given the important role apoptosis plays in normal regulation of the immune system, as well as its implication in immune-related disease, it is a logical and potentially valuable endpoint for mechanistic immunotoxicology evaluation [80]. Apoptosis has been found to play an important role in the immunotoxicity of a number of compounds, including organotins [81], polychlorinated biphenyls [82], methylmercury [83], and 2,3,7,8-tetrachlorodibenzo-*p*-dioxin [84]. Numerous techniques are available for assessing apoptosis, including analysis of DNA degradation, flow cytometry, morphological analysis, 3′-OH end labeling, and endonuclease analysis [85]. More recently, a number of ELISAs have become available for assessing apoptosis; these ELISAs are based on the detection of Bcl-2 or histone-associated DNA fragments. The ELISA format offers a number of benefits over the other techniques, including rapidity, simplicity, and cost effectiveness.

#### Cytokine Analysis

As described previously, cytokines represent an important mechanism not only for regulating the function of the immune system but also for linking the immune system with other organ systems. Early studies employing cytokine analysis in immunotoxicology studies were more descriptive [86,87]; however, as the intricacies of the cytokine–chemokine network become better understood, these assays are allowing us to assess the mechanisms responsible for a variety of immunomodulatory effects. As an example, a variety of nonbiological agents have been described that either specifically or nonspecifically alter cytokine production. These agents act via myriad mechanisms, including direct toxicity to cytokine-producing cells (cyclophosphamide); inhibition of cytokine production (cyclosporin, FK506, pentoxifylline); inhibition of cytokine release (pentamidine); induction of immunosuppressive factors (leflunomide); alterations in cellular homeostasis (tenidap); alterations in cellular activational or transcriptional mechanisms (thalidomide); alteration of cell cycle progression (Rapamune®); and miscellaneous or undefined mechanisms (glucocorticoids, phosphodiesterase isozyme inhibitors, metalloproteinase inhibitors, and p38 kinase inhibitors) [88]. Currently, four major types of cytokine assays are in use: bioassays, immunoassays, mRNA gene expression, and flow cytometry. What may be termed the *hybrid assay* is sometimes used; it employs elements of two or more of the main assay categories and molecular biology assays to examine cytokines and cytokine receptors [89]. Each of these assay types exhibits advantages and disadvantages, and no one type of assay is best suited for all applications. The type of assay chosen is subjective and will depend on the capabilities of the laboratory, as well as the type of information to be gained.

## Allergy and Autoimmune Reactions to Xenobiotics

### Introduction and Fundamental Concepts

Immune reactions to xenobiotics (e.g., industrial and naturally occurring chemicals, metals, and drugs) can give rise to allergy and autoimmunity; these reactions are frequent and can encompass a broad spectrum of diseases and organs. To decrease the health risks associated with exposure to these agents, it is important to understand the pathogenic mechanisms involved and to identify human subjects at risk of developing these reactions prior to serious clinical manifestations.

Allergic or hypersensitivity reactions, which include allergic asthma, food and environmental allergies, and allergic contact dermatitis, are common and costly health problems in the United States afflicting a large part of the population. Individuals with potential occupational exposure (e.g., agricultural or manufacturing workers) are at a higher risk than the general public for development of respiratory and cutaneous contact hypersensitivity to chemicals. The indirect costs, such as wages lost because of illness, are estimated to be in excess of $1 billion annually for asthma alone, and more than 35 million work days are lost to sickness each year [90]. Food allergy affects 5 to 7.5% of all children and 1 to 2% of adults. Industrial processes utilize many materials capable of inducing occupational immunological lung disease or contact hypersensitivity in workers and thus must be rigorously controlled to ensure worker safety. Studies in the metal-refining industry, for example, suggested that many workers regularly exposed to the complex salts of platinum develop disorders of the respiratory tract. A study of workers exposed to toluene diisocyanate (TDI), a substance used in the manufacture of polyurethane, revealed that 5% of those surveyed developed occupational asthma to TDI.

Drug allergy is also a significant problem and among the most common causes for new pharmaceuticals being withdrawn from the market after they are released. This adverse immune reactivity is not well predicted from the current battery of preclinical safety assessment methods [51]. A distinction should be made between immunologically mediated adverse drug reactions that represent true drug allergy reactions and those that are autoimmune in nature. In drug allergy, the adverse reaction occurs as a consequence of an immune response directed against the drug or a metabolite. Examples of drug allergy include immediate anaphylactic reactions mediated by IgE antibody, delayed systemic reactions (IgG), or cell-mediated reactions. Drugs are unique in that they can provoke allergic and autoimmune reactions against blood cells, including erythrocytes and platelets, as well as a variety of other antigens including the haptenated drug. Usually the reaction is directed against the drug or drug metabolites, in which case it is necessary for both the drug and the antibody to be present to produce the allergic or autoimmune reaction. In addition to producing allergic manifestations or pathology under certain conditions, drug-specific antibodies can also alter the pharmacokinetics and clearance of the drug in plasma. Thus, drug-induced allergic reactions can come in many forms, producing either allergic or autoimmune phenomena.

Allergy (acute hypersensitivity) is a pathological state resulting from prior sensitization to a specific molecule or structurally related compound. Because of the clinical and pathological similarities between xenobiotics-induced allergy, autoimmunity, and graft-vs.-host reactions, there is considerable evidence that these reactions are initiated and maintained by sensitized T-cells [91]. The major difficulty in trying to study these immune reactions is the fact that the ultimate neoantigen formed by the chemical or drug is often unknown. Based on our understanding, it is assumed that allergic reactions to most xenobiotics involve the formation of protein adducts (i.e., hapten–carrier conjugates). Metal ions react by oxidizing proteins or forming stable protein–metal chelates through multiple binding points with several amino acid side chains in the protein. A second mechanism for xenobiotics not reactive enough to bind covalently to proteins is through the direct antigen recognition by $\gamma$-$\delta$ and $\alpha$-$\beta$ T-cells. This hypothesis is supported by evidence of human $\gamma$-$\delta$ T-cells specific for the drug lidocain [92] and human CD8+ $\alpha$-$\beta$ T-cells that recognize pollen-derived carbohydrate antigen [93].

There appears to be multiple mechanisms by which xenobiotics can become sensitizers. Highly reactive organic compounds most often act as electrophiles and react with protein nucleophilic groups such as thiol, amino, and hydroxyl groups to form hapten–protein conjugates (reviewed by Griem et al. [94]). Examples of reactive hap-

**TABLE 36.6**
**Materials Associated with Contact, Food or Respiratory Allergy**

| | | |
|---|---|---|
| Pharmaceuticals | Phenylglycine acid chloride | Ampicilline |
| | Piperazine | Spiramycin |
| | Amprolium hydrochloride | Antibiotic dust |
| | Antihistamines | Quinidine |
| | Anesthetics | Plasma substitutes |
| Foodstuffs | Castor bean | Pancreatic extracts |
| | Green coffee bean | Grain and flour |
| | Papain | Molds |
| | Tree nuts | Shellfish |
| | Peanuts | |
| Industrial chemicals | Ethylenediamine | Phthalic anhydride |
| | Diisocyanates (TMI, HDI, MDI) | Trimellitic anhydride |
| | Metallic salts | |
| Miscellaneous | Wood dusts | Animal products |
| | Latex proteins | Fragrance components |
| | Flour | Detergent enzymes (subtisilin) |

tenic chemicals that frequently lead to sensitization after dermal or inhalation exposure are TDI, trimellitic anhydride, phthalic anhydride, benzoquinone, formaldehyde, hexyl cinnamic aldehyde, ethylene oxide, dinitrochlorobenzene, picryl chloride, penicillins, and D-penicillamine. In contrast, most xenobiotics are unable to bind to proteins until they are converted to reactive intermediates and are thus considered prohaptens.

The liver is the main organ for metabolic conversion of chemicals and drugs; in spite of this, immune reactions to liver or its constituents are relatively rare. However, such reactions, although rare, have been observed, as seen with autoimmune hepatitis that results from chronic treatment with the diuretic tienilic acid, and are believed to result from the production of autoantibodies against the cytochrome P450 isozyme 2C9, the enzyme that converts this prohapten [95]. Two other examples of drug-induced autoimmune hepatitis involving a similar mechanism where autoantibodies are raised against the enzyme converting the prohapten to a reactive metabolite are hepatitis produced by the anesthetic halothane (anti-CYPE1 [96]) and by dihydralazine (anti-CYP1A2 [97]).

Extrahepatic metabolism appears to play a more critical role in hypersensitivity reactions in prohaptens conversion, and the skin is a good example of an organ with metabolic conversion potential via dendritic Langerhans cells (isoenzyme CYP1A) or inflammatory leukocytes (myeloperoxidase [MPO], prostaglandin H synthase, and CYPs). Examples of prohaptens converted in the skin include the polyaromatic hydrocarbon dimethylbenz(*a*)anthracene that is converted to a potent haptenic sensitizer [98] and urushiol, a mixture of allergenic 2-alkyl and 3-alkenyl catechols from poison ivy and poison oak that are oxidized to reactive *o*-quinones that elicit specific T-cell responses [99].

Exposure to any of a number of industrial chemicals and drugs has been associated with the development of allergic or hypersensitivity reactions (Table 36.6). In addition, in some cases the reactive metabolite of the xenobiotics acting as a hapten and producing the adverse immune response has been identified in rodents, humans, or both (Table 36.7) [94].

What is not well understood is why only a few of the hapten–protein conjugates formed in the body induce clinical allergy or autoimmunity. Administrative route, dose, and genetic background clearly all play an important role in this process. Cutaneous sensitization of mice to the hapten dinitroflurobenzene (DNFB) does not occur if the animals are orally pretreated with DNFB [100]. This is believed to be due to the induction of both hapten-specific CD8+ T-cells and CD4+ suppressor T-cells. In the case of dose, subsensitizing doses of oxazolone induce tolerance while higher doses sensitize. Genetic background further complicates our ability to predict, as gold thiomalate induces a Th2-like response in Brown Norway rats, but not in Lewis rats. This induction of autoimmunity in Brown Norway rats is explained by a polymorphism in a putative enhancer element in the second intron of the IL-4 gene [101].

Autoimmunity is a multifactorial pathological process comprised of at least two processes. Initially, an immune response is initiated to normal components of the host, and, second, a pathological condition may ensue in which the response causes structural or functional damage (e.g., pathology). An autoimmune response does not necessarily reflect disease, although it is a prerequisite for disease to occur. The autoimmune response induced can be cellular in nature, mediated by CD4 or CD8 T-cells, or, more often, it arises from antibody, mediated by specific B-cells. The most common autoimmune diseases are rheumatoid arthritis and those associated with the thyroid, such as

**TABLE 36.7**
**Examples of Allergic Reactions to Xenobiotics That Involve Reactive Metabolites**

| Parent Compound | Allergic Reaction Observed | Candidate Metabolite |
|---|---|---|
| Dihydralazine | Drug-induced lupus, autoimmune hepatitis | Hydralazine radical |
| Gold thiomalate | Dermatitis, glomerulonephritis | Gold (III) |
| Halothane | Autoimmune hepatitis | Trifluoracetylchloride |
| Practolol | Oculomucocutaneous syndrome | Practolol epoxide |
| Procainamide | Drug-induced lupus | *N*-hydroxyprocainamide |
| Propylthiouracil | Vasculitis, drug-induced lupus | Propyluracilsulfonic acid |
| *p*-Phenylenediamine | Contract dermatitis | Bandrowski's base |
| Tienilic acid | Autoimmune hepatitis | Thiophene sulfoxide |
| Urushiol | Contact dermatitis | 3-pentadecyl-*o*-quinone |

Graves' disease [102]. In total, they represent a significant and chronic morbidity problem; recent estimates suggest that 1 in 31 individuals in the United States is affected, with women at 2.7 times greater risk than men [102].

What is the underlying problem in autoimmunity and autoimmune disease? It is well understood that the immune system is able to recognize and produce a response against foreign material but is tolerant towards the body's own constituents. Unfortunately, immunological tolerance to self is not absolute; autoantibodies can be produced experimentally and are seen in a variety of human autoimmunity diseases (e.g., rheumatoid arthritis, systemic lupus erythematosus, and myasthenia gravis). The mechanisms responsible for conversion from an autoimmune response to autoimmune disease are not clear. It is believed that the failure of any one of several immune processes can lead to the development of autoimmune disease; however, the key process involves the loss of self-tolerance, such as the missed deletion or activation of autoreactive lymphocyte precursors. This process may be exacerbated by altered immunoregulation, such as overexpression of the immunoregulatory cytokine IL-4 or underexpression of IFN-$\gamma$. Autoimmunity may also occur in the absence of an aberration in the immune system, such as when microbial agents express cryptic determinants [103]. Although autoimmunity is a disease of the immune system, nonimmunological genetic and epigenetic factors play a major role in disease development; for example, autoimmunity is influenced strongly by infectious agents, stress, and diet (epigenetic), as well as polymorphisms in the T-cell receptor and drug-metabolizing phenotypes (genetic). The association of autoimmune diseases with certain MHC haplotypes, such as HLA-DR3 in systemic lupus, is striking. A detailed description of the potential mechanisms and the influential factors leading to autoimmune disease is beyond the scope of this section, and the reader is referred to several excellent reviews [103–106].

In drug-induced autoimmunity, the adverse reaction results from an immune response directed against the body's own tissue, subcellular, or cellular components. A large variety of drugs with a molecular weight of less than 1000 Da and from a variety of different chemical classes have been shown to induce autoimmunity. These include aromatic amines (procainamide, practolol), hydrazines (hydralazine), hydantoins (phenytoin, mephenytoin, ethotoin, nitrofurantoin), thioureylenes (methimazole, propylthiouracil), oxazolidinediones (trimethadione, paramethadione), succinimides (ethosuximide, methsuximide, phensuximide), dibenzazepines (carbamazepine), phenothiazines (chlorpromazine), sulfonamides (sulfasalazine, sulfadiazine), pyrazolones (phenylbutazone), amino acids (D-penicillamine, captopril, methyldopa), allyl amines (zimeldine, halothane), and certain metal salts (mercuric chloride, gold salts).

The most common examples of drugs that have produced autoimmune disease are those that cause hematological disorders such as neutropenia, thrombocytopenia, and immune hemolysis; they include a variety of antibiotics as well as anticonvulsants, such as phenytoin (Table 36.8). Approximately 10 to 20% of patients receiving procainamide and 5 to 20% receiving hydralazine develop drug-induced systemic lupus erythrematosus (SLE) [107].

Although not as well demonstrated as for drugs, considerable evidence exists that autoimmunity can also be induced by substances found in food or the environment. Regarding food consumption, strong associations have been found to exist between the consumption of iodine and autoimmune thyroiditis, L-5-hydroxytryptophan and scleroderma, and alfalfa seeds and SLE. Exposure to occupational agents has also been linked to autoimmune diseases. Scleroderma-like skin diseases can result from exposure to vinyl chloride, silica, or aniline derivatives, the latter presumably the active agent resulting in the "toxic oil syndrome" [108]. Agents such as heavy metals, nitrofurantoin, and organic solvents such as trichloroethylene are associated with SLE or glomerulonephritis. Like their idiopathic counterparts, xenobiotic-induced autoimmune diseases are also associated with certain genetic backgrounds. Individuals with the low acetylator phenotypes are associated with drug-induced SLE. The relative

**TABLE 36.8**
**Examples of Drugs and Chemicals Implicated in Autoimmune Disease**

| Pathology | Agents | |
|---|---|---|
| Systemic lupus erythematosus/ immune complex glomerulonephritis | Hydralazine | Heavy metals |
| | Penicillamine | Isoniazid |
| | Chlorpromazine | Organic solvents |
| | Anticonvulsants | Procainamide |
| | Alfalfa sprouts (L-canavanine) | |
| Hemolytic anemia | Methyldopa | Diphenylhydantein |
| | Penicillin | Interferon-alpha |
| | Mefenamic acid | Sulfa |
| Thrombocytopenia | Acetazolamide | *p*-Amionsalicylic acid |
| | Chlorothiazide | Rifampin |
| | Gold salts | Quinidine |
| Scleroderma-like disease | Vinyl chloride | L-Tryptophan |
| | Silica | |
| Pemphigus | Penicillamine | — |
| Thyroiditis | Polychlorinated biphenyls | Lithium |
| | Iodine | IL-2 |

risk for developing autoimmunity from gold salts increases 32-fold in individuals who possess the HLA-DR3 allele. Likewise, experimental studies of mercury-induced autoimmunity in the Brown Norway rat and B-10 mice suggests the same genetic influences apply in animals as in humans [109].

## Techniques for Assessing Contact and Respiratory Allergy

Given the potential economic and medical importance of hypersensitivity, the importance of sensitive and reliable assays for the detection of sensitizing potential for drugs and chemicals is obvious. For over 100 years, the guinea pig has served as the principal model for allergic reactions in humans, as they demonstrate many similarities in their response to pulmonary hypersensitivity (e.g., response to histamine, demonstration of immediate and delayed allergic reactions), as well as dermal hypersensitivity. In addition, the lightly pigmented skin of albino guinea pigs and their relatively small size and docile nature make them manageable model animals; thus, they have traditionally been used for assessing the human safety of drugs, as well as other chemicals, and for contact and respiratory sensitization. Based on the specific experimental needs at hand, a variety of modifications have been described. In this section, we will discuss only two, the Buehler assay and the guinea pig maximization test, which are probably the two most widely used guinea pig tests for risk evaluation of contact sensitization [110].

The mouse has been developed as an alternative model to the guinea pig. The impetus for this development has been the mouse's small size and reduced cost, more thoroughly understood immune system, and the perceived need of a more quantitative endpoint than the subjective degree of erythema that is the hallmark of most guinea pig assays. Of the various mouse models developed, including the mouse ear swelling test (MEST) [111], the murine local lymph node assay (LLNA) first described by Kimber and Weisenberger [112] has undergone extensive validation studies [113] and is rapidly replacing guinea pig tests as the method of choice for identifying contact sensitizers and, thus, will be described in detail in this section.

Considerable efforts have been devoted to establishing relatively simple tests to identify respiratory and food allergens. Guinea pig models have been used historically for the assessment of potential respiratory sensitizers [114,115]. The interested reader is directed to a number of excellent reviews on the use of these assays in risk assessment and drug development [115–118]. In addition, murine models for assessment of respiratory sensitizing potential, such as cytokine fingerprinting and the mouse IgE test, are being examined for its utility [119–121]. Likewise, models are being developed to identify food allergens, including the development of IgE antibody in Brown Norway rats or BALB/c mice and a transgenic mouse strain engineered to produce class II HLA molecules (reviewed by Tryphonas et al. [122]).

### *Buehler Assay*

The Buehler assay [123] was originally developed to evaluate strong and moderate contact sensitizers, leaving only negative or weakly positive compounds for testing in humans. The hallmark of the Buehler assay was the use of an occlusive patch to enhance or exaggerate exposure to test materials. This method also has the advantage of using an exposure method similar to that encountered in human exposure. The specific technique for performing this assay is involved and is only summarized below. A detailed description of the assay has been provided by Buehler [123,124].

*Materials and reagents required:*

- Young adult albino guinea pigs (Dunkin–Hartley strain)
- Guinea pig restrainers
- Patch delivery system (e.g., Hilltop chambers, Webril patch, PMP patch)
- Dental dam

*Procedure:*

1. On the day before induction exposure, clip the guinea pig's fur. Expose the skin to the selected test dose using a patch delivery system and then restrain the animal using a combination of guinea pig restrainer and dental dam. Duration of exposure, number of inductions, and induction regimens may vary but generally entail three 6-hour induction exposures with an interval of 5 to 9 days.
2. Approximately 2 weeks after the last induction exposure, the animals are exposed to the test material again, but at a different skin site that has not been previously exposed. Again, the timing and duration of exposure may vary.
3. If necessary, an animal may be rechallenged with test material between 6 and 14 days following the primary challenge.
4. The day after the challenge or rechallenge, depilate the animals using a commercial depilatory; 2 hours later the animals are ready to be scored.
5. Results are scored as 0 (no reaction), ± (slight patchy erythema), 1 (slight but confluent erythema), 2 (moderate erythema), and 3 (severe erythema, with or without edema). Scoring should be performed at 24 and 48 hours after the challenge or rechallenge.
6. Scores of 1 or greater in the test group usually indicate that the test material is a sensitizer, if the scores in control animals are less than 1. Results of the challenge and rechallenge should be expressed as both incidence and severity.

*Positive control:* Dinitrochlorobenzene (DNCB) has been the traditional positive control for this assay. Suggested test concentrations are 0.3% DNCB in ethanol for induction and two concentrations of DNCB in acetone (0.05% and 0.01%) for challenge. Inclusion of two different challenge doses provides for a range of response [124].

*Notes:*

1. An irritation screen is performed prior to the actual test. This assay requires an induction concentration that will not produce severe irritation or toxicity. The concentration used for challenge should produce only a slight degree of irritation.
2. Proper technique is essential to the success of this assay; in particular, animal restraint, test material occlusion, and consistency in scoring are all critical aspects. Whenever possible, it would be advisable to learn these techniques from a laboratory that has already demonstrated success with the assay.
3. Due to the relatively small number of animals and the nature of the readout, statistical analysis has not generally been practical for this assay; rather, hazard assessment has been defined in terms of threshold levels.
4. It is important to maintain occlusion for the entire duration of the exposure in this assay. Proper use of the restraining devices is considered to be critical to obtaining consistent, meaningful results. Proper clipping and depilation of the animals are also important variables.

### Guinea Pig Maximization Assay

The maximization assay was described by Magnusson and Kligman in 1964 [125] and was developed to maximize the sensitivity of guinea pig tests. This was accomplished by intradermal injections of test material, inclusion of Freund's complete adjuvant (FCA), and the use of a pretreatment to irritate the skin at the site of exposure. These treatments enhanced the chance of a test material to penetrate the skin and subsequently produce allergic contact dermatitis [126]. Perhaps even more so than the Buehler assay, the maximization assay is technically detailed, and only the basics of its performance are summarized below.

*Materials and reagents required:*

- Young adult albino guinea pigs (Dunkin–Hartley strain)
- Freund's complete adjuvant
- Hilltop chambers or PMP patches
- Hypoallergenic tape and elastic wrap
- Dental dam

*Procedure:*

1. Similar to the Buehler assay, a preliminary irritation/toxicity screen must be performed to determine the highest concentration of test material that can be tested. Both intradermal injection and occlusive patch tests are performed.

2. For the induction step, the animal's fur is clipped over the back on either side of the spine, and test material is injected intradermally in a volume of 0.1 mL. Each animal receives a total of six injections: two each of diluted adjuvant, two each of adjuvant containing test material, and two each of test material in vehicle. Control animals are treated similarly, but do not receive test material. Let the animals rest for 6 days.
3. If the test material being used is not an irritant, the injection sites should be treated with 10% sodium lauryl sulfate in petrolatum under an occluded patch for 24 hours. This step may be skipped if the test material is a known irritant.
4. Place 0.8 mL of test material on a PMP patch (booster patch) and place this patch over the injection sites. Cover with dental dam and wrap the animal with elastic tape. Control animals should be treated with vehicle only. Remove the booster patch after 48 hours.
5. Challenge the animals 10 days later by exposing to test material (or vehicle) under an occlusive patch (PMP patch or Hilltop chamber). Wrap the animals with dental dam and elastic tape. *Note:* Animals must be clipped prior to challenge, as the fur will have grown back.
6. Remove the patches 24 hours later. Approximately 21 hours later, remove any remaining fur with depilatory. Grade the reactions approximately 24 and 48 hours after the challenge patches are removed.
7. Animals may be rechallenged with the same test material at the same or a different site with 2 weeks of the primary challenge. Naïve test sites should be used on the animals.
8. The grading of this assay is similar to that of the Buehler assay.

*Positive control:* The positive control material generally used in the maximization test is 1-chloro-2,4-dinitrobenzene, at a concentration of 0.1% in a vehicle of propylene glycol for the intradermal injections. The booster patch incorporates 0.1% (w/v) of test material in an 80:20 ethanol/water (v/v) vehicle.

### Murine Local Lymph Node Assay

In contrast to the guinea pig tests described previously, sensitizing activity is assessed as a function of events occurring in the induction, rather than the elicitation, phase of sensitization and thus does not utilize a secondary (challenge) exposure to the test material. The test involves quantifying cell proliferation in the draining lymph node following topical application of the test material. Guinea pig contact sensitization models have been used for assessing the potential of compounds to induce hypersensitivity reactions for some time, but they have certain drawbacks. The endpoints are subjective and require skilled technicians to evaluate the intensity of the reaction; in addition, this subjectivity precludes the use of statistical analysis. Moreover, the assays are relatively expensive and time consuming, and animal welfare issues are associated with the use of an adjuvant. Although none of these issues alone is a major detriment, together they led to the search for an alternative method.

In 1989, Kimber and Weisenberger [112] reported the development of an alternative approach to assess potential contact hypersensitivity using the mouse as a model system. This model is known as the local lymph node assay (LLNA). The LLNA exhibits a number of advantages over guinea pig assays, including more quantitative and objective endpoints, insensitivity to colored compounds, reduced turnaround time and cost, no need for specialized reagents or materials (adjuvant, wrapping material), and reduced animal welfare concerns.

The biological basis of the LLNA is simple. Test materials are applied epicutaneously to the dorsal surface of the pinnae (Figure 36.7). The test agent is transported from the skin by Langerhans cells to the draining (i.e., local) lymph node, where it is presented to T-lymphocytes. Contact sensitizers induce proliferation of these T-cells. By radiolabeling these proliferating cells *in situ* using a radioactive tracer or a direct cell count, it is possible to determine the degree of proliferation induced. It is important to note that the LLNA does not utilize a secondary exposure (challenge) to the test material, as do guinea pig assays; thus, the LLNA differs fundamentally in that it evaluates only the induction phase of the hypersensitivity response.

The LLNA has been the subject of numerous international validation studies, including a round of international validation studies employing a standardized protocol [127–129]. The LLNA validation data were evaluated by the Interagency Coordinating Committee on the Validation of Alternative Methods (ICCVAM), and the assay was found to provide an equivalent prediction of the risk for human contact dermatitis when compared to guinea pig assays [130,131]; in addition, the method was robust, sensitive, and reproducible. More recently, efforts have been made to define how the LLNA can be used to measure the relative potency of contact allergens [132]. The following protocol is the standard procedure recommended by the ICCVAM Working Group (IWG).

*Materials and reagents required:*

- Female CBA/J mice, 6 to 9 weeks old at initiation of assay
- Tritiated thymidine ([$^3$H]TdR), specific activity 5 to 10 Ci/m*M*
- Phosphate-buffered saline (PBS)
- Nylon mesh (100-μm opening size)

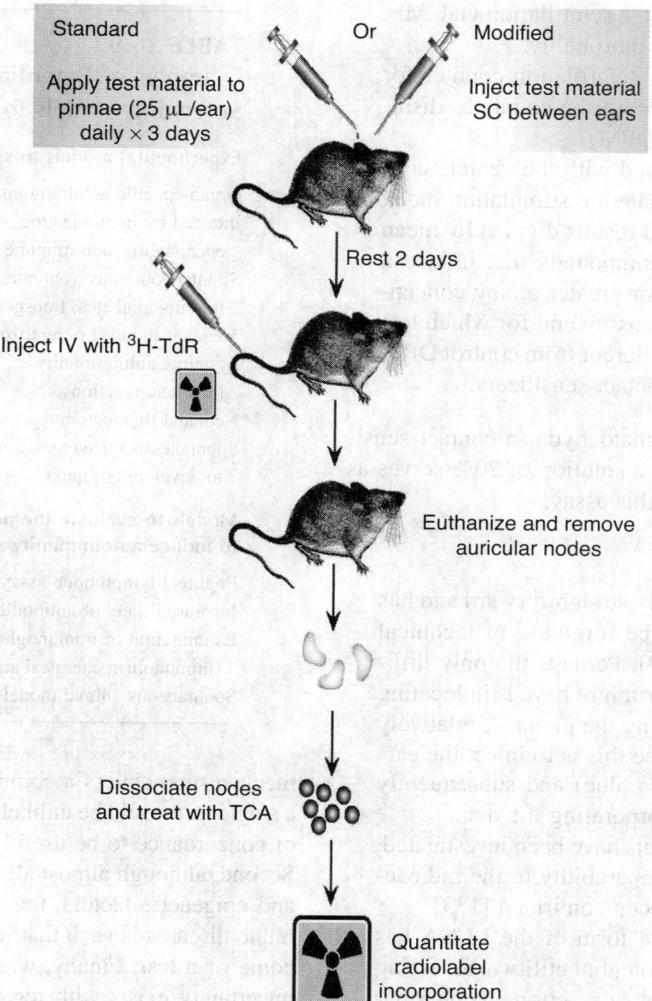


**FIGURE 36.7** Murine local lymph node assay.

- 15-mL conical capped polypropylene centrifuge tubes
- 5% (w/v) trichloroacetic acid (TCA)
- Scintillation vials and scintillation cocktail

*Procedure:*

1. Apply vehicle, test compound, or positive control compound to the dorsum of each pinna (25 μL per ear), ensuring that the vehicle is evenly distributed on the pinna. Dose the animals daily for 3 consecutive days.
2. Rest the mice for 2 days, then inject each mouse intravenously with 20 μCi of [$^3$H]TdR in saline.
3. Five hours after the [$^3$H]TdR injection, euthanize the mice by $CO_2$ and remove the lymph nodes draining the ear. Place the nodes from individual mice in culture tubes containing 4 mL of PBS.
4. Transfer the nodes from the culture tubes to Petri dishes containing a 1-inch square of nylon mesh. Gently rub the lymph node cells through the mesh, transfer the cell suspension back to the tube, and allow it to settle for approximately 5 minutes.
5. Transfer the cell suspension to a 15-mL centrifuge tube containing 6 mL of PBS, taking care not to transfer the sedimented debris. Centrifuge the tubes for 10 minutes at approximately 200 *g*. Wash the cells a second time in PBS.
6. After the second wash, suspend the cell pellet in 3 mL of 5% TCA and incubate at approximately 4°C for approximately 18 hours.
7. Centrifuge the cell suspensions at approximately 200 *g* for 10 minutes, discard the supernatant, and suspend the pellet in 1 mL of 5% TCA. Transfer this suspension to a scintillation vial. Rinse the culture tubes with an additional 1 mL

of TCA, and add this to the scintillation vial. Mix the contents of the vial thoroughly.

8. Count the samples in a scintillation counter for 5 to 10 minutes, and record the counts as disintegrations per minute (DPM).
9. Using the results obtained with the vehicle controls as baseline, calculate the stimulation index (i.e., mean experimental results divided by mean control results). Test compounds that induce a stimulation index of 3 or greater at any concentration evaluated in this assay and for which test DPM are statistically different from control DPM are considered to be contact sensitizers.

*Positive control:* Hexylcinnamaldehyde (a contact sensitizer of moderate activity) in a solution of 20% serves as a useful positive control for this assay.

*Notes:*

1. The LLNA is technically straightforward and has been demonstrated to be forgiving of technical modifications [127,129]. Perhaps the only difficulty a new investigator might have is in locating the lymph nodes draining the pinna. A relatively simple way to determine this is to inject the ears with a dye (e.g., Evans blue) and subsequently identify the nodes incorporating the dye.
2. Nonradioactive endpoints have been investigated, but at present their comparability to the radioactive endpoint has not been confirmed [133].
3. Recently, an alternative form of the LLNA has been evaluated for its potential utility in assessing immune-mediated adverse reactions to pharmaceuticals. This method differs from the standard LLNA technique in that the test material is injected subcutaneously between the ears, rather than being applied topically. Initial collaborative studies have shown some promise of this technique for its intended use, although much work remains to be done before the assay can be considered predictive [134].

## Techniques for Assessing Autoimmunity

Although the general consensus within the toxicology community is that there is a major need to screen drugs or chemical agents for their potential to induce autoimmunity, suitable validated models do not, as yet, exist. This is despite the fact that a large number of experimental animals and *in vitro* models are available to study the mechanisms of autoimmune disease (Table 36.9). The primary reason for the lack of validated assays probably stems from the complexity of the disease. First, and as mentioned earlier, autoimmune disease is not one disease but a group of over 25 diseases affecting distinct organs, often through different mechanisms. Unless a common early process is identified, a single test would be unlikely to provide an adequate degree of concordance to be useful for predictive risk assessment. Second, although almost all diseases are affected by genetic and epigenetic factors, the degree of influence in autoimmune diseases is such that it could drastically alter the outcome of a test. Finally, when using animal models, some uncertainty exists with regard to what actually constitutes autoimmunity. This is also reflected by a lack of well-defined diagnostic tests for identifying autoimmune disease in humans. Despite these challenges, attempts to develop predictive screening assays for detecting xenobiotic-induced autoimmunity have been undertaken in several laboratories. Currently, four screening approaches that are clearly different from those used for mechanistic studies have been suggested, each having received varying levels of attention. These include: (1) monitoring changes in the frequency or rate of autoimmune disease using autoimmune prone rodents [135], (2) identifying immunoglobulin complexes or immunoglobulin deposits using immunohistological procedures, (3) monitoring for increased levels of serum autoantibodies, and (4) the use of the popliteal lymph node assay (PLNA) with reporter antigens [136].

The successful use of exacerbating disease by chemical exposure in autoimmune-prone rodent species has been illustrated by administering streptozotocin in diabetic mice [137] or $HgCl_2$ in glomerulonephritis-prone rodents [138]. Less studied has been the monitoring of immunoglobulin deposits or autoantibody production [139]. The approach that has received the most attention is the PLNA with

**TABLE 36.9**
**Examples of Potential Experimental and Screening Models for Autoimmunity**

| |
|---|
| **Experimental models to study autoimmunity** |
| Organ-specific autoimmunity |
| Induced by immunization (experimental autoimmune encephalitis, autoimmune arthritis) |
| Spontaneous mice (nonobese diabetic, which develop immune diabetes; transgenics) |
| Toxicant-induced (streptozotocin, cadmium) |
| Systemic autoimmunity |
| Allogeneic reactions |
| Neonatal thymectomy |
| Spontaneous mice (New Zealand mixed, which are prone to developing lupus) |
| **Models to evaluate the potential of xenobiotics to induce autoimmunity** |
| Popliteal lymph node assay with reporter antigens |
| Increased titers of antibodies to self constituents |
| Examination of immunoglobulin complexes/deposits (immunohistochemical staining for immune complexes) |
| Spontaneous animal models |

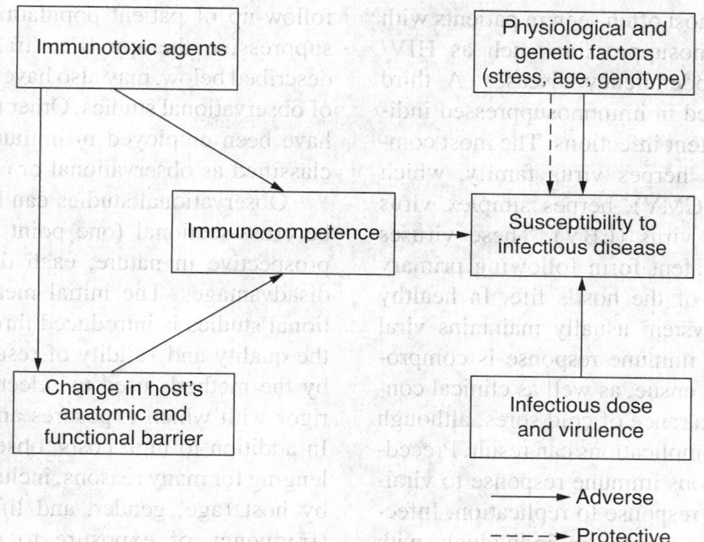


**FIGURE 36.8** Relationship between altered immune responses and frequency of severity of human disease.

reporter antigens [136]. In this model, the autoimmunogenicity of a chemical, believed to mimic the graft-vs.-host reaction, is determined by its ability to stimulate specific IgG responses to TNP–Ficoll and TNP–ovalbumin in the popliteal lymph node. Although more an indicator for adjuvancy than for disease, it has proven valuable as a first-tier screen, as it is independent of the nature of the neoantigens and eliminates many of the potential genetic confounders. In this assay, the test compound is coinjected with 10 μg TNP–Ficoll or 10 μg TNP–ovalbumin subcutaneously into the right hind paw of BALB/c mice. The amount of test substance injected can be equimolar to a related compound or, if known, a concentration demonstrated to be stimulatory in the PLNA. Seven days following treatments, the thickness of the paw is measured using a micrometer and the draining popliteal lymph node is isolated. Specific antibody-forming cells from the PLN are quantitated by any one of several methods such as ELISPOT. For memory responses, mice are similarly treated and then challenged in the right paw with 10 μg antigen 4 to 5 weeks after the primary immunization, and antibody-producing cells from the PLN are determined 6 days later. Serum samples can also be collected weekly following the primary immunization and serum antibodies determined using commercial procedures.

## EVALUATION OF IMMUNOLOGICAL CHANGES IN HUMANS

### FUNDAMENTAL CONCEPTS

Although infectious disease is the most obvious adverse consequence of immunotoxicity (immunosuppression), the etiology, progression, and severity of a much broader range of disorders, including certain cancers, allergy, and autoimmune disease, can result. Establishing the quantitative relationship between altered immune responses and frequency or severity of diseases in human populations is challenging, as many factors may contribute [140]. Figure 36.8 illustrates how the appearance, progression, and outcome of infectious disease depend on the virulence of the organism, infectious dose (number of organisms required to produce illness), integrity of the host's anatomical and functional barriers, and overall immunocompetence of an individual, which, in turn, is affected by genetics as well as age, gender, use of certain medications, drug or alcohol use, smoking history, stress, and nutritional status. These factors probably account for most of the variability reported in mean immune values, which in some instances demonstrate two standard deviations.

Infectious agents associated with immunodeficiency disorders include community-acquired (common), opportunistic, and latent pathogens. Community-acquired pathogens, such as influenza virus, occur in the general population at frequencies associated with their infectious nature. The respiratory system is the most vulnerable target for common pathogens. Upper respiratory infections occur in all age groups but, due to their lack of immunocompetence, can be more severe in the very young and very old. Although influenza is responsible for more morbidity and mortality than any other infectious agent in recorded history [141], the low individual rates of infections in the general population (only one or two episodes in an individual per year), combined with underreporting, make changes in infection rates difficult to detect in prospective epidemiological studies. Although infections with common pathogens occur in the healthy population, opportunistic infections, such as *Candida albicans*, *Mycobacterium avium* complex (MAC), and

*Pneumocystis carinii*, are most often seen in patients with profound forms of immunosuppression such as HIV/AIDS or primary immunodeficiency diseases. A third group of infections observed in immunosuppressed individuals is reactivation of latent infections. The most common agents are from the herpes virus family, which include cytomegalovirus (CMV), herpes simplex virus (HSV), and Epstein–Barr virus (EBV). These viruses remain in the tissue in a latent form following primary infection for the duration of the host's life. In healthy individuals, the immune system usually maintains viral latency. When the cellular immune response is compromised, viral replication can ensue, as well as clinical consequences such as the appearance of cold sores, although in some instances severe complications can result. Preceding viral activation a vigorous immune response to viral-specific antigens occurs in response to replication. Infections with these viruses can occur in individuals with secondary immunodeficiency disorders where mild to moderate immunosuppression exists [142].

Immunodeficiency is also associated with an increased incidence of certain virally induced tumors, such as non-Hodgkin's lymphomas (NHLs) and tumors of the skin [143]. In contrast to cancers of internal organs, in particular those in the lung and liver that are often induced by chemical carcinogens, virus-induced cancers are more immunogenic and, thus, more likely influenced by immunological factors. Suppression of cell-mediated immunity has been associated with higher incidences of skin cancers, leukemias, and lymphoproliferative disorders in transplant patients, whereas Kaposi's sarcoma and EBV-associated B-cell lymphomas are associated with severe immunosuppression as seen in patients with AIDS.

## General Testing Considerations and Approach

The design of human studies can range from controlled clinical trials to large, population-based observational studies. Clinical studies offer advantages in that exposure parameters of interest can often be controlled (e.g., clinical drug trials, chamber studies of inhaled toxicants, challenge studies of adenovirus infection) and outcomes can be prospectively monitored; however, a disadvantage is that ethical considerations provide little opportunity for exposure with suspected toxic chemicals. Furthermore, studies with extensive biological monitoring and functional immune tests can be expensive, and exposures as well as outcomes of interest may be difficult to study in the available time frame as study participants are not typically available for long-term exposures or extended follow-up. For the purpose of obtaining data for immunotoxicological risk assessment, clinical studies are particularly useful as they provide an opportunity to collect data on the frequency of infections or the level of immune response to vaccines. Variations on this type of study design might include follow-up of patient populations administered immunosuppressive therapy (i.e., transplant patients) that, as described below, may also have many of the characteristics of observational studies. Other types of human studies that have been employed in immunotoxicology are typically classified as observational or epidemiological.

Observational studies can be of varying size and can be cross-sectional (one point in time), retrospective, or prospective in nature; each design has advantages and disadvantages. The initial means of control in observational studies is introduced through the study design, and the quality and validity of results can be greatly affected by the methods used to select the study sample and the rigor with which exposures and outcomes are measured. In addition to high costs, observational studies are challenging for many reasons, including potential confounding by host (age, gender, and lifestyle) and environmental (frequency of exposure to chemicals and infectious agents) factors. A secondary measure of control in observational studies is through the use of multivariable analysis techniques (e.g., regression modeling), providing the sample size and information on potential confounders are sufficient. Overall, well-designed epidemiological studies (e.g., absence of selection bias, exposure or outcome misclassification, control of confounding) can contribute valuable information to the assessment of risk due to immunotoxic exposures. Because of concern about individual variation, a confirmatory evaluation and a cross-sectional or longitudinal study design can be employed using randomized normal, nonexposed individuals. It is also imperative to obtain a careful medical history that covers the clinical features of immune dysfunction and accurate exposure assessments.

Testing for primary immunodeficiency diseases is normally undertaken by a stepwise (tiered) approach [144] and is usually initiated because the patient has a history of excessive infectious disease episodes. Initial screening tests include measurements of general parameters, such as complete blood counts, serum immunoglobulin levels, chest radiograph of the thymus (for adolescents), and delayed hypersensitivity tests. In immunotoxicology, this screening paradigm may be appropriate for clinical studies or in individuals where significant immunosuppression is suspected; however, this approach is unlikely to detect subtle immune changes, and there are no currently validated tests for determining immunotoxic effects in human epidemiological studies. A systematic approach based on simple screening procedures followed by more specialized tests of immune function should provide the best overall assessment. This approach should include the functional evaluation of cellular immunity (T-cell), an antibody response (B-cell), and nonspecific resistance (e.g., PMN function). Recently, it has been suggested that immunization through vaccination to elicit a primary immune response may be the best criterion to establish immuno

**TABLE 36.10**
**Classification of Immune Assessment Tests for Humans**

| Basic Tests | General Indicators | Procedures |
|---|---|---|
| Should be included with general health panels along with immune status questionnaire | Assay methods are standardized among laboratories.<br>Results are clinically interpretable.<br>Reference ranges are established. | Complete blood count and differential<br>Acute-phase proteins (C-reactive protein)<br>Humoral immunity—serum IgG, IgA, IgM levels<br>Cell-mediated immunity—delayed-type skin test |
| **Confirmatory Tests** | **More Specific Immune Tests** | **Procedures** |
| Should be included when indicated by clinical findings or prior test results | Assay methods are less standardized.<br>Results are difficult to interpret.<br>Reference ranges are less well established. | Surface-marker analysis—assessment of phenotypes for major lymphocyte subsets (CD3, CD4, CD8, CD2)<br>Humoral immunity—primary antibody response to immunogen; total serum IgE; secondary antibody response to proteins and polysaccharide antigens<br>Nonspecific—auto-antibodies (ANA, DNA, RA, mitochondria); granulocyte/leukocyte function (oxidative burst)<br>Serum sample banked for additional analysis |

toxicity [145]. Most investigations have included one or more of the following: complete blood cell counts (CBCs) and differential immunophenotypic analyses, quantification of serum immunoglobulin levels, and subject's recall of infectious disease frequency.

## Tests Commonly Conducted in Humans

### Complete Blood Count and Differential

Complete white blood cell counts (WBCs) and differentials on all individuals whose immune status is being evaluated are routinely incorporated into a tier 1 panel (Table 36.10). The data should be expressed as absolute leukocyte count for each cell type which is the total WBC multiplied by the differential percentage for that cell type. Higher absolute lymphocyte counts should be expected in children than adults and in certain ethnic groups. Absolute lymphocyte counts consistently below 1500/mm$^3$ are indicative of lymphocytopenia and may signal a defect in the T-cell system. Lymphocytopenia can be associated with primary immune deficiency disease but can also occur secondary to viral infections, malnutrition, severe stress, autoimmune diseases, and hematopoietic malignancy. When lymphocytopenia is repeatedly observed, a bone marrow biopsy is recommended as an important adjunct for exclusion of other diseases and for identification of normal plasma cells, pre-B-cells, or diagnosis of bone marrow depression or dysplasia. Individuals with lymphocytopenia should be reevaluated and further assessed for changes in CMI. Lymphocytosis can be caused by chronic infections or allergic reactions. Monocytosis is often associated with stress, infections, and hematologic disorders. Eosinophilia can be caused by allergic reactions, parasitism, skin diseases, neoplasia, and adrenocortical dysfunction.

### Immunoglobulin Concentrations

Serum concentration of the major immunoglobulin classes IgG, IgM, and IgA can be readily measured in epidemiological studies. Several standardized laboratory methods and reagents are available for measuring these major classes of immunoglobulin. These methods include single-radial diffusion, double diffusion in agar gel, immunoelectrodiffusion, radioimmunoassay, enzyme-linked immunosorbent assay, and automated laser nephelometry. Serum Ig concentrations vary with age, ethnicity, geographical location, gender, and environment; thus, appropriate norms must be used with any type of population assessment. Patients with primary immunodeficiency disease can manifest a profound decrease in all Ig classes or only in a single class or subclass and are associated with increases in infectious disease frequency. The adverse health effects associated small changes in immunoglobulin levels is unclear.

#### *Delayed-Type Skin Testing*

Skin testing is a commonly used procedure (see Table 36.10) to assess cellular immune competence, as delayed cutaneous hypersensitivity, a localized immunological skin response, depends on functional T-cells and the production of inflammatory cytokines. Antigens commonly employed to elicit a positive skin response include purified protein derivative (PPD) of mumps, trichophyton, *Candida*, tetanus, or diphtheria. These antigens usually are employed in a panel and are administered by intradermal injection at the appropriate dilution. Skin responses are read at 48 and 72 hours for maximal diameter of erythema and induration. The test is not considered very sensitive unless severe immunosuppression is suspected, which is unlikely to occur.

## Specific Antibody Assessment

Recently, efforts have been made to quantify the response to vaccines, either by measurement of antibody titers (HI) or lymphocyte proliferation tests (CMI) to specific vaccine epitopes following vaccination as an indicator of immune function. To use this test in children, the study must be conducted in specified age groups and serum samples collected at specific times following vaccination to help limit variability. Several investigations have examined associations between chronic psychological stress and immune function in adults using hepatitis B, influenza virus, or pneumococcal vaccine responses [146,147]. More recently, Sleijffers et al. [148] demonstrated a decreased response to hepatitis vaccination in a subgroup of students exposed to ultraviolet radiation in association with polymorphisms in genes that control inflammatory cytokines. An excellent review on the use of vaccination in immunotoxicity testing has recently been published [149].

Although quantifying vaccination responses for clinical diagnoses has not been validated, the ability to detect changes in populations with moderate degrees of immunosuppression, likely associated with the above examples, suggest a level of sensitivity that would unlikely be achieved with more commonly employed clinical tests such as immunophenotyping or determination of immunoglobulin levels. Historical values for normal vaccine responses in children can be found; for example, Swartz et al. [150] monitored changes in antibody titers in a large group of children following diphtheria–tetanus–pertussis (DTP) vaccination from birth to 8 years of age. Antibody titers to diphtheria, while significantly elevated shortly after the primary and secondary (booster) immunizations, were almost undetectable at 3 years of age. These data are applicable to epidemiological studies, as they provide an indication of the antibody decay rates and the variability in titers that may be expected to occur within a normal population.

## Phenotypic Analysis by Flow Cytometry

Enumerating cell-surface markers (CDs) on lymphoid cells by flow cytometry has provided considerable information on the ontogeny and activation state of the human immune system in children and adults, as well as assisting in the clinical diagnosis for immunological and hematopoietic disorders [151]. Specific CD markers have been identified for almost all lymphoid cell populations and subpopulations, as well as for specific stages of cell differentiation and activation. In contrast to adults, children in the first few years of life have a much larger number of total lymphocytes, and both the percentage and numbers of leukocyte populations can vary significantly during critical periods of development. Age-related differences in immunophenotypic profiles in children were recently addressed by the Pediatric AIDS Clinical Trials Group, sponsored by the National Institute of Allergy and Infectious Diseases and National Institute of Child Health and Human Development [152]. In this study, lymphocyte subsets were phenotyped in 807 normal children ranging from birth to 18 years of age. Despite efforts to control for inter- and intralaboratory methodological differences, the variance within each age group was significant, often exceeding twofold, even when discarding the highest and lowest tenth percentile. Certainly, not all of the variability in human immunophenotyping studies is related to technical variability, as both genetic and environmental influences play even more significant roles [151]. Nonetheless, this database may prove useful for epidemiological studies in developmental immunotoxicology, as it provides not only extensive reference values but can assist in developing appropriate study designs.

Immunophenotyping is conducted using flow cytometers that have multiple photomultiplier tubes (four or more) and are capable of measuring three color fluorescence, 90° light scatter, and forward light scatter. When highly specific fluor-labeled monoclonal antibodies are used in these instruments, very quantitative measurements can be made of T-, B-, and T-cell subsets. The most commonly used procedure for processing peripheral blood samples for immunofluorescence is first to stain an aliquot of whole blood with fluorescent-conjugated monoclonal antibodies and then to lyse the erythrocytes. The proportion of circulating T-cells is then determined by immunofluorescence with fluor-labeled CD2 or CD3 monoclonal antibodies in a flow cytometer. Normally, T-cells constitute 55 to 80% of peripheral blood lymphocytes. Normal values reported for absolute numbers of circulating T-cells are 590 to 3090/mm$^3$ for individuals greater than 18 months of age [153]. If an immune defect is suspected, the ratio of CD4 to CD8 T-cells can also be beneficial. Although this method is quite quantitative, the ability of this test method to detect subtle immune changes in populations of individuals has recently been challenged [154].

## Nonspecific Measurements

### *Neutrophil Function*

The measurement of nitroblue tetrazolium dye reduction by actively phagocytosing PMNs is a method that has been historically utilized when a PMN defect is suspected. Flow cytometric methods are also available [155].

### *Autoantibodies*

It is often stated that the immune system is established on a principle of self/non-self recognition. In some cases, tolerance to self-antigens breaks down and autoantibodies are produced that can be manifested as autoimmune disease. Antibodies to cellular components and nuclear antigens (ANA, DNA, mitochondria) and to rheumatoid factor (RA) and their frequency in a population may reflect an immune alteration. Standardized diagnostic kits are available to detect the presence of these autoantibodies in sera. Finally, it is a good practice to establish a freezer bank of an aliquot of each test subject's sera for later evaluation when new research questions or test methods are developed.

## RISK ASSESSMENT CONSIDERATIONS

### Immunosuppression

Data used in risk assessment for immunotoxicology are derived primarily from animal toxicology studies. When adequate data are available, epidemiological or controlled clinical exposure studies take precedence. The results obtained from *in vitro* studies or from structure–activity relationship (SAR) or mechanistic investigations are used normally as supportive information. Mechanistic studies, however, are important, as without them the tenfold classical defaults in the risk assessment process, such as inter-individual variability and species differences, are assumed valid [156]. In toxicology, human clinical studies, of course, represent the gold standard. Questionnaires offer some value, particularly as they provide information on reportable diseases such as autoimmunity. They provide less utility in studies of immunosuppression.

Because of the complexity of the immune system, the initial strategy devised by immunologists working in toxicology was to select and apply tiers of assays to identify hazardous agents. Among these are consideration of induced changes in the weight, composition, and histology of lymphoid organs; immunophenotyping, generally performed by cytometric analysis; and various functional assays designed specifically to evaluate B-cell, T-cell, MØ, and NK cell function to *in vitro* or *in vivo* antigen challenge. These tests were usually accompanied by an additional tier that included host resistance assays to help establish whether the immune changes observed translated into increased susceptibility to infectious or malignant diseases. This testing battery has been conducted in a number of laboratories and the results analyzed with the objective to improve the testing configuration to accurately identify immunosuppressive chemicals with the least number of tests and to help establish the quantitative relationship between immune function and host resistance [56,57].

Although a number of limitations exist, the following conclusions were drawn from these analyses: First, examination of only two or three immune parameters is required to accurately predict an immunosuppressive agent; in particular, results from the T-cell dependent antibody response appears to provide excellent concordance. Second, altered host resistance is closely associated with immune function, although changes in immune function often occur at lower dose levels. Third, no single immune test is predictive for altered host resistance, although some tests show relatively good concordance (>70%). Fourth, logistic and standard modeling, using a single data set indicated most immune function-host resistance relationships follow a linear-quadratic model rather than a true threshold. This would suggest that even very small changes in the immune system can alter host resistance, although due to the frank nature of most susceptibility tests a large group size might be required to achieve statistical significance. At an individual level, then, small changes in immune function would likely have little impact in combating infectious disease; however, such changes may have a significant impact at the population level, particularly in groups that are already at increased risk, such as the aged or the very young. Identifying changes in the frequency or severity of infectious diseases, particularly from community acquired pathogens, can be difficult for several reasons. For example, as noted earlier, although influenza is responsible for greater morbidity and mortality than any other infectious agent in recorded history, the low individual rates of common infections in the general population (only one or two episodes in an individual per year), combined with under-reporting, make it difficult to detect changes in infection rates.

In contrast to profound immunosuppression, such as that which occurs in patients with HIV/AIDS or primary immunodeficiency diseases, exposure to immunotoxic chemicals or drugs may only result in mild to moderate levels of immunosuppression (e.g., 20 to 30% decrease in white blood cell counts). Data, while limited, are available suggesting that mild to moderate immunosuppression can lead to an increase in infectious disease. The types of infections that occur tend to result from communicable infections (e.g., upper respiratory) or latent viruses (e.g., herpes cold sores), rather than opportunistic infections such as *Pnuemocyctis carinii* pneumonia. These infections can be, but usually are not, life threatening, except in certain susceptible populations, such as the elderly. Opportunistic infections, in contrast, are more prevalent in individuals where severe forms of immunosuppression are present, such as primary immunodeficiency diseases or HIV/AIDS.

### Hypersensitivity/Allergy

As ethical considerations usually prevent the use of human patch testing to establish the potential of agents to induce allergic contact dermatitis, animal models, particularly the Buehler occluded and Magnuson–Kligman maximization tests in guinea pigs, have been used as predictive tests. Several graded doses of antigen may be examined simultaneously, and comparing skin reactions in individual animals can generate an entire dose–response curve; however, it is expensive to purchase as well as maintain guinea pigs, few inbred strains are available, and immunological reagents are not widely available. Furthermore, there is some suggestion, although never fully substantiated, that these models are overly sensitive when compared to studies in humans and may thus present false positives.

A more quantitative and objective assay than the guinea pig tests, the LLNA [114] has successfully undergone a series of exercises that supports its use as a standalone test to assess allergic contact hypersensitivity. Although the strengths and weaknesses of this test have been discussed earlier, it is important to note that, much like tests for immunosuppression [55–57], the assay has undergone a series of examinations for technical refinement and to assess inter- and intralaboratory reproducibility as well as the relative sensitivity and specificity, referred to as *concordance* [128,132,157]. Concordance for a new assay should be established with previously used test models as well as available human data. Such data for the LLNA [130,131] indicate that the LLNA is highly comparable to guinea pig tests (concordance almost 90%), but only about 70% accurate when compared directly to human studies. As this is similar to results obtained when guinea pig tests are compared to human studies, in terms of risk assessment, the LLNA can be used in lieu of guinea pig tests, but an alternative assay that could provide higher concordance with humans would be desirable.

In contrast to predictive tests for allergic contact hypersensitivity, the identification of proteins and chemicals to induce respiratory hypersensitivity is in its infancy. As these tests are difficult to undertake, often involving respiratory exposure and lung function tests, efforts to develop and validate new methods are limited. Although the guinea pig has significant immunological differences compared to humans (e.g., IgG1 vs. IgE reagenic antibodies), it appears to be a predictive model for humans given the limited comparative data available and has been used to test for high- and low-molecular-weight sensitizers. This test requires a systemic or inhalation sensitization phase and an aerosol challenge, and both immediate and delayed-onset responses are measured, although this does not distinguish between nonspecific pulmonary hyperreactivity and specific immune responses [158]. The latter can be established by examining sera for the presence of reagenic antibodies. An IgE test has been proposed for the prospective identification of chemical respiratory allergens in the mouse [159].

### Autoimmunity

General agreement exists among the regulatory and pharmaceutical communities that predictive tests for autoimmunity or systemic allergy are in most need of development to improve risk assessment in immunotoxicology [51]. Although many models exist to study autoimmune processes, they do not readily lend themselves to use in risk assessment, as they do not consider the multifactorial nature of the disease. To improve the risk assessment process, screening models must be developed and validated that not only incorporate mechanistic information into the assessment process but also allow for consideration of the genetic, physiological, and environmental influences that lead to the loss of self-tolerance, autoimmune disease, or systemic allergy. Despite the challenges in developing such screening tests, the considerable amount of data generated by immunologists and pharmacologists pertaining to basic mechanisms of chemically induced autoimmune diseases have provided a conceptual framework that allows the establishment of potential structure–activity relationships. These structure–activity relationships are by no means definitive, and, as the database increases, no doubt some will not be supported while others will be added. In all cases, a basic understanding of immunological and pharmacological processes supports these relationships; for example, estrogens are known to be a major factor in classical autoimmune diseases, presumably due to their ability to stimulate certain components of the immune system [160], so agents with estrogenic activity may be of concern.

Laboratory studies have also shown that thymolytic chemicals, such as cyclophosphamide and cyclosporin A, can induce autoimmunity when given neonatally by altering normal patterns of autoreactive T-cell deletion [161]. In this respect, the thymus has been shown to be a target for many toxic chemicals. As in the case of halothane, chemicals that form protein adducts or damage tissue in such a way as to allow expression of cryptic determinants would provide novel host antigens that could now be recognized by T-cells. Agents that have adjuvant activity or biologicals that stimulate certain cytokines may shift the balance of Th1 and Th2 cells and allow exacerbation of preexisting autoimmune disease [162]. Common features associated with many drugs that induce autoimmune diseases are that they serve as myeloperoxidase substrates or cause changes in methylation [163]. The explanation for the latter association is less clear but may require identification of the specific antigenic epitopes responsible for the autoimmune response. In the case of the association with myeloperoxidase substrates, it has been suggested that many of the chemicals require metabolism in proximity to immune cells to be antigenic, and immune cells such as monocytes contain high levels of myeloperoxidase.

## REGULATORY GUIDANCE FOR ASSESSMENT OF IMMUNOTOXICITY

### Overview

Some of the earliest codified immunotoxicology test guidelines were developed to augment toxicological assessment of pesticides. In 1996, the Office of Prevention, Pesticides, and Toxic Substances (OPPTS) of the U.S. Environmental Protection Agency (EPA) published guidelines entitled *Biochemicals Test Guidelines: OPPTS*

*880.3550 Immunotoxicity* [164] that described the preferred study design for evaluating potential immunotoxicity in biochemical pesticides. The panel of tests included in this guideline includes standard toxicology tests as well as immune function tests. A second document published concurrently (*Biochemicals Test Guidelines: OPPTS 880.3800 Immune Response* [165]) provided rationale for testing, detailed explanations for testing strategies, and additional mechanistic tests, including host resistance and bone marrow function.

Whereas immunotoxicity evaluation encompassed by the 880 series of guidelines would arguably detect any type of immunotoxicity, its breadth would probably render it tremendously expensive and time consuming. In 1998, the EPA followed up with *Health Effects Test Guidelines: OPPTS 870.7800 Immunotoxicity* [166], which described immunotoxicology testing for non-biochemical agents that would be regulated by the EPA. This document provides descriptions of both why and how, with a much more abbreviated panel of testing to be performed. Whereas the 880 series of immunotoxicology guidelines is probably excessive, the testing approach mandated by 870.7800 has stood up well over the years and reflects the more limited, case-by-case approach currently favored. Most notably, the functional assessment is pared down to T-dependent antibody formation (plaque assay), NK cell function, and quantitation of T- and B-cells; this combination is derived from the early work of Luster et al. [55–57] demonstrating the greatest predictivity of known immunotoxicants using these three assays. This study design described in this document is amenable for testing a wide range of industrial and environmental chemicals.

In Europe, the Organization for Economic Cooperation and Development (OECD) regulates testing of chemicals for toxicity. OECD Guideline 407 (*Repeated Dose 28-Day Oral Toxicity Study I Rodents*), while not specific for immunotoxicology, includes a variety of toxicological endpoints that can provide early evidence of immune system alterations. Missing, however, are any functional assays to directly measure any immune deficit.

## PHARMACEUTICALS

In Europe, safety testing of pharmaceuticals is regulated by the Committee for Proprietary Medicinal Products (CPMP). In 2000, the CPMP published a *Note for Guidance on Repeated Dose Toxicity* (CPMP/SWP/1042/99) [169]. Although the primary purpose of this document was to describe an overall approach to safety testing of pharmaceuticals, it was important as the first guidance document mandating specific immunotoxicology screening for pharmaceuticals. An appendix describes a staged evaluation, emphasizing that information gained in standard toxicology studies is useful as a primary indicator for immunotoxicity. Functional tests were incorporated to gain additional information. The choice of assays to be used and combinations of functional tests came from National Toxicology Program publications. As the first published document requiring immunotoxicology evaluation, CPMP/SWP/1042/99 predictably was met with some confusion and resistance. The Drug Information Association sponsored a workshop in The Netherlands in 2001 that allayed some of the confusion. A summary of findings from this workshop was published [170].

In the United States, the safety testing of small-molecule pharmaceuticals is the purview of the U.S. Food and Drug Administration (FDA) Center for Drug Evaluation and Research (CDER). In 2002, CDER released a long-awaited document entitled *Guidance for Industry: Immunotoxicology Evaluation of Investigational New Drugs* [167]. This document is arguably the most comprehensive of any published guidance and described a diversity of adverse events including immunosuppression, immunogenicity, hypersensitivity, autoimmunity, and adverse immunostimulation. The document also suggested methodology for evaluating each of these types. Like the CPMP document, the CDER guidance advocated the use of information derived from standard repeated-dose toxicity studies to provide early evidence of immunotoxicity, with subsequent evaluations to be rationally designed to use a minimum of animals, but it did not mandate animal testing [168]. Differences in how the U.S. and E.U. guidance to immunotoxicology assessment was interpreted have led to intense discussion regarding the best approach for performing this assessment on a routine basis [171–174].

In an effort to provide greater consistency, the International Committee on Harmonization (ICH) initiated an effort in 2003 to harmonize immunotoxicology testing across the major geographic areas. This working group made recommendations on nonclinical testing approaches to identify compounds that have the potential to be immunotoxic, provided guidance on a weight-of-evidence decision-making approach for immunotoxicity, and defined immunotoxicity as suppression or enhancement of the immune response. It was recommended that information gained from the standard toxicity studies (e.g., hematological changes, changes in immune system organ weights and/or histology, changes in serum immunoglobulins, an increased incidence of infections or tumors) should be viewed as potential signs of immunosuppression in the absence of other plausible causes. The assays proposed included the TDAR, immunophenotyping, NK, host-resistance studies, macrophage and neutrophil function, and CMI. This guidance was finalized at step 4 in 2005, and a meeting report was issued [175]. Although specific immunotoxicity testing was not mandated, if weight-of-evidence evaluation indicates a need, it gives regulatory agencies authority to request such studies. It also indicated that drug allergy and autoimmunity issues cannot be adequately addressed without validated methods.

### Biologicals

Biologicals (therapeutics derived by biotechnology) present a unique challenge for immunotoxicity assessment. Many of these agents, such as cytokines and other immunomodulatory molecules, are intended to therapeutically modulate the immune response; therefore, it can be difficult to differentiate between the agent's efficacy and a truly adverse reaction. Second, because many of these agents are proteins or peptides, their introduction into a host often triggers an immune response directed against the molecule itself; this can lead to alterations in pharmacodynamics or to other adverse reactions. Thus, development of appropriate guidance on testing these agents is problematic. One approach is promulgated by the ICH in *Preclinical Safety Evaluation of Biotechnology-Derived Pharmaceuticals S6* [176]. This document includes sections on immunogenicity (as described above) as well as a brief mention of immunotoxicity studies. In short, the S6 document recognizes the inappropriateness of a structured tier approach, opting instead for the careful design of screening studies followed by mechanistic studies to clarify any potential evidence of immunotoxicity. Specific techniques and approaches are not described in ICH S6. Safety evaluation of biological drugs is regulated in the United States by the FDA's Center for Biologics Evaluation and Research (CBER). To date, CBER has not promulgated any written guidance on immunotoxicology; the reason for this lack of written guidance is the extreme diversity of biological therapeutics, which makes it difficult to design a standardized testing approach. Instead, CBER's approach to addressing potential immunotoxicology has been case by case and has followed suggestions provided in ICH S6.

### Vaccines

Along with certain biologicals, vaccines present a challenge for immunotoxicological evaluation, as they are specifically designed to induce an immune response, a situation deemed undesirable (or potentially so) for most of the other agents described in this review. Because the methodology to evaluate the desirable immunomodulation produced by vaccine is well established, the concern of regulatory agencies is the propensity of these therapeutics to produced undesired or deleterious effects on the immune system.

European regulation of vaccines is described in *Note for Guidance on Preclinical Pharmacological and Toxicological Testing of Vaccines*, published by the CPMP [177]. In this document, immunotoxicology is to be considered during toxicology testing. In particular, vaccines should be considered for their immunological effect on toxicity, such as antibody complex formation, release of cytokines, induction of hypersensitivity reactions (either directly or indirectly), and association with autoimmunity. No specifics are described for methods or approaches; rather, each vaccine is to be evaluated on a case-by-case basis.

The FDA's CBER is tasked with regulating vaccines in the United States. One of the primary documents describing vaccine studies is *Guidance for Industry for the Evaluation of Combination Vaccines for Preventable Diseases: Production, Testing, and Clinical Studies* [178]. Animal immunogenicity is covered in detail in the document, although immunotoxicity is not specified as an area of concern. On the other hand, CBER's *Considerations for Reproductive Toxicity Studies for Preventive Vaccines for Infectious Disease Indications* [179], although intended primarily to assess effects of vaccination on reproductive function (e.g., generalized toxicity such as fetal malformations), acknowledges the potential immunological reactions resulting from the vaccination process to exert unintended consequences. No specific guidance is provided on methods or approaches to be used in this evaluation.

### Devices and Radiological Agents

It has been recognized by the FDA that immunotoxicity may result not only from chemical or biological agents that dynamically interact with humans' physiology such as small-molecule drugs or biological agents but also from medical devices that contact the body externally (via skin or mucosa) or internally (implantable devices) or by external communication to the blood or tissue. Thus, the FDA's Center for Devices and Radiological Health (CDRH) published the guidance entitled *Guidance for Industry and FDA Reviewers: Immunotoxicology Testing Guidance* in 1999 [180], which addresses testing for medical devices. This guidance is based on General Program Memorandum G95-1, an FDA-modified version of International Standard ISO-10993 (*Biological Evaluation of Medical Devices-Part 1: Evaluation and Testing*). The Immunotoxicology Testing Guidance provides detailed guidance for determining when immunotoxicity testing should be performed (including a flowchart and numerous tables) but does not provide details on which methods should be employed or for overall study design. Anderson and Langone [181] provided explanatory details on the use of this guidance.

### Hypersensitivity

Although much attention is paid to immunosuppression (low immune response) in the majority of guidance documents, hypersensitivity (hyperactive immune response) is the most common type of immunomodulation resulting from exposure to xenobiotics. Due to the acknowledged frequency of this occurrence, as well as the multiplicity of testing methods that have been developed complete coverage of this condition is not included here however, one method for assessing hypersensitivity ha taken priority in assessing contact hypersensitivity—namely, the murine local lymph node assay. Detailed explanations of this assay and its use are covered in the OECD 429 guideline *Skin Sensitization: Local Lymph Node Assay* and the U.S. EPA document *OPPTS 870.2600 Skin Sensitization* [182].

## FRONTIER TECHNOLOGIES IN EXPERIMENTAL IMMUNOTOXICOLOGY

### Knockout/Transgenic Animals

Significant information regarding immunomodulation has been gained by the study of experiments of nature—that is, naturally occurring immune deficiencies [183]; however, the technology for specifically engineering mutations in the immune system of laboratory animals has resulted in the ability to evaluate various perturbations of the immune response. The promise of this technology for immunotoxicology was first described by Lovik [184], and several recent uses of this technology for investigational immunotoxicology have been described [185].

### Nanotechnology

An emerging safety concern is the emergence of successful implementation of nanotechnology, the manufacture of materials and devices at the molecular scale. It is unknown at present whether these new materials will exhibit unanticipated toxicology, although several groups have already begun raising potential concerns [186,187]. One of the most immediate concerns will probably be effects on respiratory function following inhalation of nanoscale particles [188,189], although other immune toxicities, such as immune reactions to medical devices, are possible [186].

### The Interface Between Innate and Adaptive Immunity

The innate immune system is often described as being primitive, but it is increasingly being recognized that the innate and adaptive systems represent a continuum of highly interactive and often complementary protective mechanisms. A number of effectors form the bridge between these systems, including the Toll-like receptors [190], γ-δ T-cells [191], NK cells, dendritic cells [192,193], T-regulatory cells (TREGs) [194] and Fas signals [195] to name but a few. As a better understanding of these complex regulatory circuits develops, it is becoming apparent that this system is responsible for immunomodulation of both normal [196,197] and pathologic processes, such as autoimmune disease [198]. Although very little mechanistic immunotoxicology has thus far evaluated the role of this system, its importance to human health makes it an ideal target for future research.

## CONCLUSIONS AND FUTURE DIRECTIONS

The discipline of immunotoxicology has grown in importance in toxicology since its inception in the mid-1970s. It has progressed from the early identification of chemicals that may cause immunosuppression/modulation and allergic contact dermatitis to the validation of sensitive and quantitative assays that serve as biomarkers of immune system alterations in animals and humans. More recently, academic, industrial, and government scientists have taken a more mechanistic approach to define how therapeutic and environmental agents alter immune function at a cellular and molecular level. Immunotoxicity data derived from experimental and human immunosuppression and hypersensitivity studies play an increasing role in establishing health standards and defining permissible levels of toxic chemical exposure in humans. What is still needed is better correlation between animal data with known immunotoxicants and epidemiological or clinical studies to ascertain the predictive value of the immune evaluation methods for human populations that may be occupationally or environmentally exposed. Well-controlled studies are still needed in human subjects exposed to environmental chemicals to establish concretely the relationship between documented exposure and immune-mediated effects. For pharmaceuticals where exposure is well documented, correlation of the prediction value of animal studies for immunological alterations (e.g., immunosuppression or allergy) in humans is more clearly defined.

## QUESTIONS

1. The immune system exists to protect the body against:
   a. Specific invading pathogens and microorganisms
   b. Neoplastic cells
   c. Non-self antigens
   d. Transplanted foreign antigen
   e. All of the above (answer)
   f. A, B, and D
2. Which of the following statements is false?
   a. Macrophages and leukocytes are types of phagocytic cells derived from the bone marrow.
   b. Cell-mediated immunity represents a type of nonspecific immune response. (answer)
   c. The two major mechanisms of immunity are nonspecific and specific.
   d. Humoral immunity is associated with the production of antibody.
   e. Pluripotent stem cells are found in the bone marrow and give rise to megakaryocytes and lymphocytes.

3. Which of the following statements is true?
   a. The primary lymphoid organs are represented by the thymus and bursa-equivalent tissues. (answer)
   b. Lymphoid tissue is derived from ectoendodermal junctional tissue.
   c. Secondary lymphoid tissue is found in the spleen, lymph nodes, gut-associated lymphoid tissue (GALT), and bronchial-associated lymphoid tissue (BALT).
   d. a and c.
   e. All of the above.
4. Autoimmunity is best defined as:
   a. An immune response to normal components of the host (answer)
   b. Being mediated by IgE
   c. Being best measured in guinea pigs
   d. Reflecting a single organ
5. Immunotoxicity assessment is most often conducted using:
   a. Epidemiology studies
   b. *In vitro* studies
   c. Animal studies (answer)
   d. Combinations of SAR and clinical trials
6. Chemical- and drug-induced autoimmunity differ from their idiopathic counterparts in that they:
   a. Usually remit when the drug is withdrawn (answer)
   b. Only target the kidney
   c. Only target blood elements
   d. Are more common in females
7. Validation of animal models for immunotoxicology studies requires:
   a. Laboratory validation
   b. Establishment of specificity
   c. Establishment of sensitivity
   d. Reproducibility
   e. All of the above (answer)
8. The most appropriate animal model for evaluating immunotoxicity appears to be:
   a. Rodents (answer)
   b. Mini-pigs
   c. Guinea pigs
   d. Nonhuman primates
9. Allergic reactions to drugs may result from:
   a. Direct antigenicity of the drug moiety (answer)
   b. Activation of complement proteins
   c. Haptenation of self proteins by the drug or a metabolite (answer)
   d. Bone marrow ablation
10. Macrophages are an important potential target of immunotoxicants because:
    a. They are capable of metabolizing xenobiotics.
    b. They are potent immunoregulatory cells.
    c. They secrete large quantities of inflammatory antibodies.
    d. a and b
    e. b and c (answer)

## REFERENCES


1. Paul, W. E. (1999): *Fundamental Immunology*, 4th ed. Lippincott, Philadelphia, PA.
2. Haley, P. J. (2003): Species differences in the structure and function of the immune system. *Toxicology*, 188(1):49–71.
3. von Andrian, U. H. and Mempel, T. R. (2003): Homing and cellular traffic in lymph nodes. *Nat. Rev. Immunol.*, 3:867–878.
4. Mebius, R. E. and Kraal, G. (2005): Structure and function of the spleen. *Nat. Rev. Immunol.*, 5: 606–616.
5. Twerdok, L. E. and Trush, M. A. (1988): Neutrophil derived oxidants as mediators of chemical activation in bone marrow. *Chem. Biol. Int.*, 65:261–273.
6. Greenberger, J. S. (1991): Toxic effects on the hematopoietic microenvironment. *Exp. Hematol.*, 19:1101–1109.
7. Rosenthal, G. J. and Kowolenko, M. (1994): Immunotoxicological manifestations of AIDS therapeutics. In: *Immunotoxicology and Immunopharmacology*, 2nd ed., edited by J. H. Dean, M. I. Luster, A. E. Munson, and I. Kimber, pp. 249–265. Raven Press, New York.
8. Warheit, D. B. and Hesteberg, T. W. (1994): Asbestos and other fibers in the lung. In: *Immunotoxicology and Immunopharmacology*, 2nd ed., edited by J. H. Dean, M. I. Luster, A. E. Munson, and I. Kimber, pp. 363–376. Raven Press, New York.
9. Rosenberg, H. F. and Gallin, J. I. (1999): Inflammation. In: *Fundamental Immunology*, 4th ed., edited by W. E. Paul, pp. 1051–1066. Lippincott, Philadelphia, PA.
10. Vollmers, H. P. and Brandlein, S. (2005): The 'early birds': natural IgM antibodies and immune surveillance. *Histol. Histopathol.*, 20(3):927–937.
11. Vollmers, H. P. and Brandlein, S. (2005): Death by stress: natural IgM-induced apoptosis. *Methods Find. Exp. Clin. Pharmacol.*, 27(3):185–191.
12. Mosmann, T. R., Cherwinski, H., Bond, M. W., Giedlin, M. A., and Coffman, R. L. (1986): Two types of murine helper T cell clone. I. Definition according to profiles of lymphokine activities and secreted proteins. *J. Immunol.*, 136:2348–2357.
13. Romagnani, S. (1995): Biology of human Th1 and Th2 cells. *J. Clin. Immunol.*, 15:121–129.
14. O'Garra, A. (1998): Cytokines induce the development of functionally heterogeneous T helper cell subsets. *Immunity*, 8:275–283.
15. Del Prete, G. (1998): The concept of type-1 and type-2 helper T cells and their cytokines in humans. *Intern. Rev. Immunol.*, 16:427–455.
16. Selgrade, M. K., Lawrence, D. A., Ullrich, S. E., Gilmour, M. I., Schuyler, M. R., and Kimber, I. (1997): Modulation of T-helper cell populations: potential mechanisms of respiratory hypersensitivity and immune suppression. *Toxicol. Appl. Pharmacol.*, 145:218–29.

17. Smyth, M. J., Cretney, E., Kelly, J. M., Westwood, J. A., Street, S. E., Yagita, H., Takeda, K., van Dommelen, S. L., Degli-Esposti, M. A., and Hayakawa, Y. (2005): Activation of NK cell cytotoxicity. *Mol. Immunol.*, 42(4): 501–510.
18. Lotzova, E. (1993): Definition and functions of natural killer cells. *Nat. Immun.*, 12:169–176.
19. Whiteside, T. L. and Herberman, R. B. (1995): The role of natural killer cells in immune surveillance of cancer. *Curr. Opin. Immunol.*, 7(5):704–710.
20. Penn, I. (1985): Neoplastic consequences of immunosuppression. In: *Immunotoxicology and Immunopharmacology*, edited by J. H. Dean, A. Munson, M. I. Luster, and H. Amos. Raven Press, New York.
21. Pross, H. F. and Lotzová, E. (1993): Role of natural killer cells in cancer. *Nat. Immun.*, 12:279–292.
22. Herberman, R. B. (2001): Immunotherapy. In: *Clinical Oncology*, edited by R. E. Lenhard, Jr., et al., pp. 215–223. American Cancer Society, Atlanta, GA.
23. Kos, F. J. (1998): Regulation of adaptive immunity by natural killer cells. *Immunol. Res.*, 17:303–312.
24. Lanier, L. L., Corliss, B., and Phillips, J. H. (1997): Arousal and inhibition of human NK cells. *Immunol. Rev.*, 155:145–154.
25. Naume, B. and Espevik, T. (1994): Immunoregulatory effects of cytokines on natural killer cells. *Scand. J. Immunol.*, 40:128–134.
26. Orange, J. S., Human natural killer cell deficiencies and susceptibility to infection, *Microbes Infect.*, 4, 1545, 2002.
27. See, D. M., Khemka, P., Sahl, L., Bui, T., and Tilles, J. G. (1997): The role of natural killer cells in viral infections. *Scand. J. Immunol.*, 46:217–224.
28. Tay, C. H., Szomolanyi-Tsuda, E., and Welsh, R. M. (1998): Control of infections by NK cells. *Curr. Top. Microbiol. Immunol.*, 230:193–220.
29. Papamichail, M., Perez, S. A., Gritzapis, A. D., and Baxevanis, C. N. (2004): Natural killer lymphocytes: biology, development, and function. *Cancer Immunol. Immunother.*, 53(3):176–186.
30. Kronenberg, M. (2005): Toward an understanding of NKT cell biology: progress and paradoxes. *Annu. Rev. Immunol.*, 23:877–900.
31. Godfrey, D. I. and Kronenberg, M. (2004): Going both ways: immune regulation via CD1d-dependent NKT cells. *J. Clin. Invest.*, 114(10):1379–1388.
32. Hammond, K. J. and Kronenberg, M. (2003): Natural killer T cells: natural or unnatural regulators of autoimmunity? *Curr. Opin. Immunol.*, 15(6):683–689.
33. David, T., Thomas, C., Zaccone, P., Dunne, D. W., and Cooke, A. (2004): The impact of infection on the incidence of autoimmune disease. *Curr. Top. Med. Chem.*, 4(5):521–529.
34. Dinarello, C. A. (1997): Role of pro- and anti-inflammatory cytokines during inflammation: experimental and clinical findings. *J. Biol. Reg. Homeostat. Agents*, 11:91–103.
35. Lunney, J. K. (1998): Cytokines orchestrating the immune response. *Rev. Sci. Tech. Off. Int. Epiz.*, 17:84–94.
36. Montovani, A., Allavena, P., Vecchi, A., and Sozzani, S. (1998): Chemokines and chemokine receptors during activation and deactivation of monocytic and dendritic cells and in amplification of Th1 versus Th2 responses. *Int. J. Clin. Lab. Res.*, 28:77–82.
37. Bacon, K. B. and Schall, T. J. (1996): Chemokines as mediators of allergic inflammation. *Int. Arch. Allergy Immunol.*, 109:97–109.
38. Taub, D. D. (1996): Chemokine-leukocyte interactions: the voodoo that they do so well. *Cytokine Growth Factor Rev.*, 7: 355–376.
39. Fuchs, B. A. and Sanders, V. M. (1994): The role of brain–immune interactions in immunotoxicology. *CRC Crit. Rev. Toxicol.*, 24:151–176.
40. Savino, W., Arzt, E., and Dardenne, M. (1999): Immunoneuroendocrine connectivity: the paradigm of the thymus–hypothalamus–pituitary axis. *Neuroimmunomodulation*, 6(1–2):126–136.
41. Friedman, E. M. and Lawrence, D. A. (2002): Environmental stress mediates changes in neuroimmunological interactions. *Toxicol Sci.*, 67(1):4–10.
42. Weigent, D. A. and Blalock, J. E. (1995): Associations between the neuroendocrine and immune systems. *J. Leukoc. Biol.*, 57:137–150.
43. Haskó, G. and Szabó, C. (1998): Regulation of cytokine and chemokine production by transmitters and co-transmitters of the autonomic nervous system. *Biochem. Pharmacol.*, 56:1079–1087.
44. Dean, J. H., Padarathsingh, M. L., and Jeffells, T. R. (1979): Assessment of immunobiological effects induced by chemicals, drugs and food additives. I. Tier testing and screening approach. *Drug Chem. Toxicol.*, 2:5–17.
45. National Research Council. (1992): *Biologic Markers in Immunotoxicology*. National Academy Press, Washington, D.C.
46. Vos, J. G. (1980): Immunotoxicity assessment: screening and function studies. *Arch. Toxicol.*, 4(Suppl.):95–108.
47. Germolec, D. R. (2004): Selectivity and predictivity in immunotoxicity testing: immune endpoints and disease resistance, *Toxicol. Lett.*, 149:109.
48. Haynes, R. C. and Murad, F. (1985): Adrenocortical steroids and their synthetic analogs: inhibitors of adrenocortical steroid biosynthesis. In: *Goodman and Gilman's Pharmacological Basis of Therapeutics*, edited by A. G. Gilman, L. S. Goodman, T. W. Rall, and F. Murad, pp. 1459–1489. Macmillan, New York.
49. Luster, M. I., Germolec, D. R., Clark, G., Wiegand, G., and Rosenthal, G. J. (1988): Selective effects of 2,3,7,8-tetrachlorodibenzo-*p*-dioxin and corticosteroid on *in vitro* activation, proliferation and differentiation of murine B-lymphocytes. *J. Immunol.*, 140:928–935.
50. Olson, H., Betton, G., Robinson, D., Thomas, K., Monro, A., Kolaja, G., Lilly, P., Sanders, J., Sipes, G., Bracken, W., Dorato, M., Van Deun, K., Smith, P., Berger, B., and Heller, A. (2000): Concordance of the toxicity of pharmaceuticals in humans and in animals. *Regul. Toxicol. Pharmacol.*, 32(1):56–67.
51. Dean, J. H., Hincks, J. R., and Remandet, B. (1998): Immunotoxicology assessment in the pharmaceutical industry. *Toxicol. Lett.*, 102–103:247–255.
52. Smialowicz, R. J., DeVito, M. J., Riddle, M. M., Williams, W. C., and Birnbaum, L. S. (1997): Opposite effects of 2,2′,4,4′,5,5′- hexachlorobiphenyl and 2,3,7,8-tetrachlorodibenzo-*p*-dioxin on the antibody response to sheep erythrocytes in mice. *Fundam. Appl. Toxicol.*, 37: 141–149.

53. Rao, G. N., Birnbaum, L. S., Collins, J. J., Tennant, R. W., and Skow, L. C. (1988): Mouse strains for chemical carcinogenicity studies: overview of workshop. *Fund. Appl. Toxicol.*, 10:385–394.
54. Vos, J. G. and Van Loveren, H. (1998): Experimental studies on immunosuppression: how do they predict for man? *Toxicology*, 129:13–26.
55. Luster, M. I., Munson, A. E., Thomas, P. T., Holsapple, M. P., Fenters, J. D., White, Jr., K. L., Lauer, L. D., Germolec, D. R, Rosenthal, G. J., and Dean, J. H. (1988): Development of a testing battery to assess chemical-induced immunotoxicity: National Toxicology Program's guidelines for immunotoxicity evaluation in mice. *Fund. Appl. Toxicol.*, 10:2–19.
56. Luster, M. I, Portier, C., Pait, D. G., White, Jr., K. L., Gennings, C., Munson, A. E., and Rosenthal, G. J. (1992): Risk assessment in immunotoxicology. I. Sensitivity and predictability of immune tests. *Fund. Appl. Toxicol.*, 18:200–210.
57. Luster, M. I., Portier, C., and Pait, D. G. (1993): Risk assessment in immunotoxicology. II. Relationships between immune and host resistance tests. *Fund. Appl. Toxicol.*, 21:71–82.
58. Kuper, C. F., Harleman, J. H., Richter-Reichelm, H. B., and Vos, J. G. (2000): Histopathologic approaches to detect changes indicative of immunotoxicity, *Toxicol. Pathol.*, 28, 454.
59. Germolec, D. R., Kashon, M., Nyska, A., Kuper, C. F., Portier, C., Kommineni, C., Johnson, K. A., and Luster, M. I. (2004): The accuracy of extended histopathology to detect immunotoxic chemicals, *Toxicol. Sci.*, 82:504–514.
60. Basketter, D. A., Bremmer, J. N., Buckley, P., Kammuller, M. E., Kawabata, T., Kimber, I., Loveless, S. E., Magda, S., Stringer, D. A., and Vohr, H.-W. (1995): Pathology considerations for, and subsequent risk assessment of, chemicals identified as immunosuppressive in routine toxicology. *Food Chem. Toxicol.*, 33:239–243.
61. Gopinath, C. (1996): Pathology of toxic effects on the immune system. *Inflamm. Res.*, 45:S74–S78.
62. Schuurman, H.-J., Kuper, C. F., and Vos, J. G. (1994): Histopathology of the immune system as a tool to assess immunotoxicity. *Toxicology*, 86:187–212.
63. Cunningham, A. J. and Szenberg, A. (1968): Further improvement in the plaque technique for detecting single antibody-forming cells. *Immunology*, 14:599–600.
64. Temple, L., Butterworth, L., Kawabata, T. T., Munson, A. E., and White, K. L. (1995): ELISA to measure SRBC specific serum IgM: method and data evaluation. In: *Methods in Immunotoxicology*, Vol. 1, edited by G. R. Burleson, J. H. Dean, and A. E. Munson, pp. 137–157. Wiley-Liss, New York.
65. Gore, E. R., Gower, J., Kurali, E., Sui, J. L., Bynum, J., Ennulat, D., and Herzyk, D. J. (2005): Primary antibody response to keyhole limpet hemocyanin in rat as a model for immunotoxicity evaluation. *Toxicology*, 197(1):23–35.
66. Ulrich, P., Paul, G., Perentes, E., Mahl, A., and Roman, D. (2004): Validation of immune function testing during a 4-week oral toxicity study with FK506. *Toxicol Lett.*, 149(1–3):123–131.
67. Shkedy, Z., Straetemans, R., Molenberghs, G., Desmidt, M., Vinken, P., Goeminne, N., Coussement, W., Van Den Poel, B., and Bijnens, L. (2005): Modeling anti-KLH ELISA data using two-stage and mixed effects models in support of immunotoxicological studies. *J. Biopharm. Stat.*, 15(2):205–223.
68. Reynolds, C. W. and Herberman, R. B. (1981): *In vitro* augmentation of rat natural killer (NK) cell activity. *J. Immunol.*, 126:1581–1585.
69. Habu, S., Fukui, H., Shimamura, K., Kasai, M., Nagai, Y., Okumura, K., and Tamaoki, N. (1981): *In vivo* effects of anti-asialo GM1. I. Reduction of NK activity and enhancement of transplanted tumor growth in nude mice. *J. Immunol.*, 127:34–38.
70. Knapp, D. W., Leibnitz, R. R., DeNicola, D. B., Turek, J. J., Teclaw, R., Shaffer, L., and Chan, T. C. K. (1993): Measurement of NK activity in effector cells purified from canine peripheral lymphocytes. *Vet. Immunol. Immunopathol.*, 35:239–251.
71. Marcusson-Stahl, M. and Cederbrant, K. (2003): A flow-cytometric NK-cytotoxicity assay adapted for use in rat repeated dose toxicity studies. *Toxicology*, 193(3):269–279.
72. Motzer, S. A., Tsuji, J., Hertig, V., Johnston, S. K., and Scanlan, J. (2003): Natural killer cell cytotoxicity: a methods analysis of $^{51}$chromium release versus flow cytometry. *Biol. Res. Nurs.*, 5(2):142–152.
73. Burleson, G. R., Dean, J. H., and Munson, A. E., Eds. (1995): *Methods in Immunotoxicology*, Vols. 1 and 2. Wiley-Liss, New York.
74. House, R. V. (1997): Immunotoxicology methods. In: *Handbook of Human Toxicology*, edited by E. J. Massaro, pp. 677–708. CRC Press, Boca Raton, FL.
75. Smialowicz, R. J. and Holsapple, M. P., Eds. (1996): *Experimental Immunotoxicology.* CRC Press, Boca Raton, FL.
76. Thomas, P. T. and House, R. V. (1995): Preclinical immunotoxicity assessment. In: *CRC Handbook of Toxicology*, edited by M. J. Derelanko and M. A. Hollinger, pp. 293–316. CRC Press, Boca Raton, FL.
77. Corcoran, G. B., Fix, L., Jones, D. P., Moslen, M. T., Nicotera, P., Oberhammer, F. A., and Buttyan, R. (1994): Apoptosis: molecular point control in toxicity. *Toxicol. Appl. Pharmacol.*, 128:169–181.
78. Howie, S. E., Harrison, D. J., and Wyllie, A. H. (1994): Lymphocyte apoptosis: mechanisms and implications in disease. *Immunol. Rev.*, 142:141–156.
79. Mountz, J. D., Zhou, T., Wu, J., Wang, W., Su, X., and Cheng, J. (1995): Regulation of apoptosis in immune cells. *J. Clin. Immunol.*, 15:1–16.
80. Pallardy, M., Kerdine, S., and Lebrec, H. (1998): Testing strategies in immunotoxicology. *Toxicol. Lett.*, 102–103: 257–260.
81. Pieters, R. H., Bol, M., and Penninks, A. H. (1994): Immunotoxic organotins as possible model compounds in studying apoptosis and thymocyte differentiation. *Toxicology*, 91:189–202.
82. Yoo, B. S., Jung, K. H., Hana, S. B., and Kim, H. M. (1997): Apoptosis-mediated immunotoxicity of polychlorinated biphenyls (PCBs) in murine splenocytes. *Toxicol. Lett.*, 91:83–89.

83. Shenker, B. J., Guo, T. L., and Shapiro, I. M. (1998): Low-level methylmercury exposure causes human T-cells to undergo apoptosis: evidence of mitochondrial dysfunction. *Environ. Res.*, 77:149–159.
84. Kamath, A. B., Nagarkatti, P. S., and Nagarkatti, M. (1998): Characterization of phenotypic alterations induced by 2,3,7,8-tetrachlorodibenzo-*p*-dioxin on thymocytes *in vivo* and its effect on apoptosis. *Toxicol. Appl. Pharmacol.*, 150:117–124.
85. Sgonc, R. and Wick, G. (1994): Methods for the detection of apoptosis. *Int. Arch. Allergy Immunol.*, 105:327–332.
86. House, R. V., Lauer, L. D., Murray, M. J., and Dean. J. H. (1987): Suppression of T-helper cell function in mice following exposure to the carcinogen 7,12-dimethylbenz(*a*)anthracene and its restoration by interleukin 2. *Int. J. Immunopharmacol.*, 9:95–87.
87. Lyte, M. and Bick, P. H. (1986): Modulation of interleukin I production by macrophages following benzo(*a*)pyrene exposure. *Int. J. Immunopharmacol.*, 8:377–381.
88. House, R. V. (1999): The theory and practice of cytokine assessment in immunotoxicology. *Methods*, 19(1):17–27.
89. Vandebriel, R. J., Van Loveren, H., and Meredith, C. (1998): Altered cytokine (receptor) mRNA expression as a tool in immunotoxicology. *Toxicology*, 130:43–67.
90. Young, P. (1980): *Asthma and Allergies: An Optimistic Future*, based on report on the Task Force on Asthma and the Other Allergic Diseases, NIH Publ. No. M388. U. S. Government Printing Office, Washington, D.C.
91. Goldman, M, Druet, P., and Gleichmann, E. (1991): TH2 cells in systemic autoimmunity: insights from allogeneic diseases and chemically induced autoimmunity. *Immunol. Today*, 12:223–227.
92. Zanni, M. P., Mauri-Hellweg, D., Brander, C. et al (1995): Characterization of lidocaine-specific T cells. *J. Immunol.*, 158:1139–1148
93. Corinti, S., DePalma, R., Fontana, A., Gagliardi, C., Pini, C., and Sallusto, F. (1997): Major histocompatibility complex-independent recognition of a distinctive pollen antigen, most likely a carbohydrate, by human CD8+ alpha/beta T cells. *J. Exp. Med.*, 186:899–908
94. Griem, P., Wulferink, M., Sachs, B., Gonzalez, J., Gleichmann, E. (1998): Allergic and autoimmune reactions to xenobiotics: how do they arise? *Immunol. Today*, 19:133–141.
95. Lecoeur, S., Gautier, J. C., Belloc, C., Gauggre, A., and Beaune, P. H. (1996): Use of heterologous expression systems to study autoimmune drug-induced hepatitis. *Methods Enzymol.*, 272:76–85.
96. Eliasson, E. and Kenna, J. G. (1996): Cytochrome P450 2E1 is a cell surface autoantigen in halothane hepatitis. *Mol. Pharmacol.*, 50:573–582.
97. Bourdi, M., Tinel, M., Beaune, P. H., and Pessayre, D. (1994): Interactions of dihydralazine with cytochromes P4501A: a possible explanation for the appearance of anti-cytochrome P4501A2 autoantibodies. *Mol. Pharmacol.*, 45:1287–1295.
98. Anderson, C., Hehr, A., Robbins, R. et al. (1995): Metabolic requirements for induction of contact hypersensitivity to immunotoxic polyaromatic hydrocarbons. *J. Immunol.*, 155:3530–3537.
99. Schmidt, R. J., Khan, L., and Chung, L. Y. (1990): Are free radicals and not quinones the haptenic species derived from urushiols and other contact allergenic mono- and dihydric alkylbenzenes? The significance of NADH, glutathione, and redox cycling in the skin. *Arch. Dermatol. Res.*, 282(1):56–64.
100. Bour, H., Peyron, E., Gaucherand, M. et al. (1995): Major histocompatibility complex class I-restricted CD8+ T cells and class II-restricted CD4+ T cells, respectively, mediate and regulate contact sensitivity to dinitrofluorobenzene. *Eur. J. Immunol.*, 25:3006–3010.
101. Kermarrec, N., Dubay, C., DeGouyon, B. et al. (1996): Serum IgE concentration and other immune manifestations of treatment with gold salts are linked to the MHC and IL4 regions in the rat. *Genomics*, 31:111–114.
102. Jacobson, D. L., Gange, S. J., Rose, N. R., and Graham, N. M. H. (1997): Epidemiology and estimated population burden of selected autoimmune diseases in the United States. *Clin. Immunol. Immunopathol.*, 84:223–243.
103. Sercarz, E. E., Lehmann, P. V., and Ametani, A. (1993): Dominance and crypticity of T cell antigenic determinants. *Annu. Rev. Immunol.*, 11:729–766.
104. Liblan, R. S., Singer, S. M., and McDevitt, H. O. (1995): Th1 and Th2 CD4+ T cells in the pathogenesis of organ-specific autoimmune diseases. *Immunol. Today*, 16:3–8.
105. Rose, N. R. and Caturegli, P. P. (1997): Autoimmune diseases of humans. In: *Comprehensive Toxicology*, edited by D. Lawrence, pp. 381–390. Elsevier, New York.
106. Theofilopoulus, A. N. (1995): The basis for autoimmunity. Part II. Genetic predisposition. *Immunol. Today*, 16:150–159.
107. Bigazzi, P. E. (1995): *Autoimmunity Caused by Xenobiotics*, presented at the 4th Summer School in Immunotoxicology, October 18–20, 1995, Aix-les-Bains, France.
108. Kammuller, M. E., Bloksma, N., and Seinen, W. (1988): Chemical-induced autoimmune reactions and Spanish toxic oil syndrome: focus on hydantoins and related compounds. *Clin. Toxicol.*, 26:157–174.
109. Pelletier, L., Ramanathan, S., and Druet, P. (1997): Autoimmune models. In: *Comprehensive Toxicology*, edited by D. Lawrence, pp. 365–380. Elsevier, New York.
110. Maurer, T., Arthur, A., and Bentley, P. (1994): Guinea-pig contact sensitization assays. *Toxicology*, 93:47–54.
111. Gad, S. C. (1994): The mouse ear swelling test (MEST) in the 1990s. *Toxicology*, 93:33–46.
112. Kimber, I. and Weisenberger, C. (1989): A murine local lymph node assay for identification of contact allergens. *Arch. Toxicol.*, 63:274–282.
113. Gerberick, G. F., Ryan, C. A., Kimber, I., Dearman, R. J., Lea, L. J., and Basketter, D. A. (2000): Local lymph node assay: validation assessment for regulatory purposes. *Am. J. Contact Dermat.*, 11(1):3–18.
114. Sarlo, K. and Karol, M. H. (1994): Guinea pig predictive tests for respiratory allergy. In: *Immunotoxicology and Immunopharmacology*, 2nd ed., edited by J. H. Dean, M. I. Luster, A. E. Munson, and I. Kimber, pp. 703–720. Raven Press, New York.
115. Verdier, F., Chazal, I., and Descotes, J. (1994): Anaphylaxis models in the guinea pig. *Toxicology*, 93:55–61.

116. Choquet-Kastylevsky, G. and Descotes, J. (1998): Value of animal models for predicting hypersensitivity to medicinal products. *Toxicology*, 129:27–35.
117. Maurer, T. (1996): Guinea pig predictive tests. In: *Toxicology of Contact Hypersensitivity*, edited by I. Kimber and T. Maurer, pp. 107–126. Taylor & Francis, London.
118. Vial, T. and Descotes, J. (1994): Contact sensitization assays in guinea pigs: are they predictive of the potential for systemic allergic reactions? *Toxicology*, 93:63–75.
119. Dearman, R. J., Basketter, D. A., Blaikie, L., Clark, E. D., Hilton, J., House, R. V., Ladics, G. S., Loveless, S. E., Mattis, C., Sailstad, D. M., Sarlo, K., Selgrade, M. K., and Kimber, I. (1998): The mouse IgE test: inter-laboratory evaluation and comparison of BALB/c and C57BL/6 strain mice. *Toxicol. Meth.*, 8:69–85.
120. Sarlo, K., Dearman, R. J., and Kimber, I. (2005): Guinea pig, mouse and rat models for safety assessment of protein allergenicity. In: *Investigative Immunotoxicology*, edited by H. Tryphonas, M. Fournier, B. R. Blakley, J. E. G. Smits, and P. Brousseay, pp 278–289. Taylor & Francis, Boca Raton, FL.
121. Johnson, V. J., Matheson, J. M., and Luster, M. I. (2004): Animal models for diisocyanate asthma: answers for lingering questions. *Curr. Opin. Allergy Clin. Immunol.*, 4:105–110.
122. Tryphonas, H., Arvanitakis, G., Vavasour, E., and Bondy, G. (2003): Animal models to detect allergenicity to foods and genetically modified products: workshop summary. *Environ. Health Persp.*, 111:221–222.
123. Buehler, E. V. (1965): Delayed contact hypersensitivity in the guinea pig. *Arch. Dermatol.*, 91:171.
124. Buehler, E. V. (1995): Prospective testing for delayed contact hypersensitivity in guinea pigs: the Buehler method. In: *Methods in Immunotoxicology*, Vol. 2, edited by G. R. Burleson, J. H. Dean, and A. E. Munson, pp. 343–356. Wiley-Liss, New York.
125. Magnusson, B. and Kligman, A. M. (1964): The identification of contact allergens by animal assay: the maximization test. *J. Invest. Dermatol.*, 52:268.
126. Hiles, R. A. (1988): Predicting hypersensitivity responses. In: *Product Safety Evaluation Handbook*, edited by S. C. Gad, pp. 107–142. Marcel Dekker, New York.
127. Kimber, I., Hilton, J., Dearman, R. J., Gerberick, G. F., Ryan, C. A., Basketter, D. A., Scholes, E. W., Ladics, G. S., Loveless, S. E., House, R. V., and Guy, A. (1995): An international evaluation of the murine local lymph node assay and comparison of modified procedures. *Toxicology*, 103:63–73.
128. Kimber, I., Hilton, J., Dearman, R. J., Gerberick, G. F., Ryan, C. A., Basketter, D. A., Lea, L., House, R. V., Ladics, G. S., Loveless, S. E., and Hastings, K. L. (1998): Assessment of the skin sensitization potential of topical medicaments using the local lymph node assay: an interlaboratory evaluation. *J. Toxicol. Environ. Health*, 53:563–579.
129. Loveless, S. E., Ladics, G. S., Gerberick, G. F., Ryan, C. A., Basketter, D. A., Scholes, E. W., House, R. V., Hilton, J., Dearman, R. J., and Kimber, I. (1996): Further evaluation of the local lymph node assay in the final phase of an international collaborative trial. *Toxicology*, 108:141–152.
130. Dean, J. H., Twerdok, L. E., Tice, R. R., Sailstad, D. M., Hattan, D. G., and Stokes, W. S. (2001): ICCVAM evaluation of the murine local lymph node assay: conclusions and recommendations of an independent scientific peer review panel. *Regul. Toxicol. Pharmacol.*, 34(3):258–273.
131. National Toxicology Program. (1999): *The Murine Local Lymph Node Assay: A Test Method for Assessing the Allergic Contact Dermatitis Potential of Chemicals/Compounds*. NIH Publ. No. 99-4494. National Institutes of Health, Bethesda, MD.
132. Kimber, I., Basketter, D. A., Butler, M., Gamer, A., Garrigue, J. L., Gerberick, G. F., Newsome, C., Steiling, W., and Vohr, H. W. (2003): Classification of contact allergens according to potency: proposals. *Food Chem. Toxicol.*, 41(12):1799–1809.
133. Ehling, G., Hecht, M., Heusener, A., Huesler, J., Gamer, A. O., van Loveren, H., Maurer, T., Riecke, K., Ullmann, L., Ulrich, P., Vandebriel, R., and Vohr, H. W. (2005): An European inter-laboratory validation of alternative endpoints of the murine local lymph node assay: second round. *Toxicology*, 212(1):69–79.
134. Weaver, J. L., Chapdelaine, J. M., Descotes, J., Germolec, D., Holsapple, M., House, R. V., Lebrec, H., Meade, J., Pieters, R., Hastings, K. L., and Dean, J. H. (2005): Evaluation of a lymph node proliferation assay for its ability to detect pharmaceuticals with potential to cause immune-mediated drug reactions. *J. Immunotoxicol.*, 2(1):11–20.
135. Lai, H. and Forster, M. J. (1991): Autoimmune mice as models for discovery of drugs against age-related dementia. *Drug Devel. Res.*, 24:1–27.
136. Albers, R., Broeders, A., van der Pijl, A., Seinen, W., and Pieters, R. (1997): The use of reporter antigens in the popliteal lymph node assay to assess immunomodulation by chemicals. *Toxicol. Appl. Pharmacol.*, 143:102–109.
137. Leiter, E. H. (1982): Multiple low-dose streptozotocin-induced hyperglycemia and insulitis in C57BL mice: influence of inbred background, sex and thymus. *Proc. Natl. Acad. Sci. U.S.A.*, 79:630–634.
138. Hultman, P., Turley, S. J., Enestrom, S., Lindh, A., and Pollard, K. M. (1996): Murine genotype influences the specificity, magnitude and persistence of murine mercury-induced autoimmunity. *J. Autoimmun.*, 9:139–149.
139. Kilburn, K. H. and Warshaw, R. H. (1992): Prevalence of symptoms of systemic lupus erythematosus (SLE) and of fluorescent antinuclear antibodies associated with chronic exposure to trichloroethylene and other chemicals in well water. *Environ. Res.*, 57:1–9.
140. Morris, Jr., J. G. and Potter, M. (1997): Emergence of new pathogens as a function of changes in host susceptibility. *Emerg. Infect. Dis.*, 3:435.
141. Patriarca, P. A. (1994): A randomized controlled trial of influenza vaccine in the elderly: scientific scrutiny and ethical responsibility. *JAMA*, 272:1700–1701.
142. Luster, M. I., Germolec, D. R., Parks, C. G., Blancifort L., Kashon, M., and Luebke, R. W. (2005): Are changes in the immune system predictive of clinical disease. In *Investigative Immunotoxicology*, edited by H. Tryphonas M. Fournier, B. R. Blakely, J. E. G. Smits, and P. Brousseau pp. 165–182. Taylor & Francis, Boca Raton, FL.

143. Penn, I. (2000): Post-transplant malignancy: the role of immunosuppression. *Drug Saf.*, 23:101.
144. Noroski, L. M. and Shearer, W. T. (1998): Screening for primary immunodeficiencies in the clinical immunology laboratory. *Clin. Immunol. Immunopathol.*, 86(3):237–245.
145. van Loveren, H., Germolec, D., Koren, H. S., Luster, M. I., Nolan, C., Repetto, R., Smith, E., Vos, J. G., and Vogt, R. F. (1999): Report of the Bilthoven Symposium: advancement of epidemiological studies in assessing the human health effects of immunotoxic agents in the environment and the workplace. *Biomarkers* 4:135–157.
146. Kiecolt-Glaser, J. K., Glaser, R., Gravenstein, S., Malarkey, W. B., and Sheridan, J. (1996): Chronic stress alters the immune response to influenza virus vaccine in older adults. *Proc. Natl. Acad. Sci. U. S.A.*, 93:3043–3047.
147. Kiecolt-Glaser, J. K., McGuire, L., Robles, T. F., and Glaser, R. (2002): Psychoneuroimmunology: psychological influences on immune function and health. *J. Consult. Clin. Psychol.*, 70:537–547.
148. Sleijffers, A., Yucesoy, B., Kashon, M., Garssen, J., De Gruijl, F. R., Boland, G. J., Van Hattum, J., Luster, M. I., and Van Loveren, H. (2003): Cytokine polymorphisms play a role in susceptibility to ultraviolet B-induced modulation of immune responses after hepatitis B vaccination. *J. Immunol.*, 170:3423–3428.
149. van Loveren, H., van Amsterdam, J. G., Vandebriel, R. J., Kimman, T. G., Rumke, H. C., Steerenberg, P. S., and Vos, J. G. (2001): Vaccine-induced antibody responses as parameters of the influence of endogenous and environmental factors. *Environ. Health Perspect.*, 109: 757–764.
150. Swartz, T. A., Saliou, P., Catznelson, E., Blondeau, C., Gil, I., Peled, T., Havkin, O., and Fletcher, M. (2003): Immune response to a diphtheria and tetanus toxoid administration in a three-dose diphtheria tetanus whole-cell pertussis/enhanced inactivated poliovirus vaccination schedule: a 7-year follow up. *Eur. J. Epidemiol.*, 18:827–833.
151. Marti, G. E., Zenger, V. E., Vogt, R., and Gaigalas, A. (2002): Quantitative flow cytometry: history, practice, theory, consensus, inter-laboratory variation and present status. *Cytotherapy*, 4, 97–98
152. Shearer, W. T., Rosenblatt, H. M., Gelman, R. S., Oyomopito, R., Plaeger, S., Stiehm, E. R., Wara, D. W., Douglas, S. D., Luzuriaga, K., McFarland, E. J., Yogev, R., Rathore, M. H., Levy, W., Graham, B. L., and Spector, S. A. (2003): Lymphocyte subsets in healthy children from birth through 18 years of age: the Pediatric AIDS Clinical Trials Group P1009 study. *J. Allergy Clin. Immunol.*, 112:973–980.
153. Fleisher, T. A., Luckasen, J. R., Sabad, A., Gehrtz, R. C., and Kersey, J. H. (1975): T and B lymphocyte subpopulations in children. *Pediatrics*, 55:162–165.
154. Immunotoxicity Technical Committee. (1999): *Application of Flow Cytometry to Immunotoxicity Testing: Summary of a Workshop. Report from an October 1997 Workshop.* ILSI HESI, Washington, D.C.
155. van Eeden, S. F., Klut, M. E., Walker, B. A., and Hogg, J. C. (1999): The use of flow cytometry to measure neutrophil function. *J. Immunol. Meth.*, 232(1–2):23–43.
156. Scala, R. (1991): Risk assessment. In: *Casarett & Doull's Toxicology: The Basic Science of Poisons*, 4th ed., edited by M. Amdur, J. Doull, and C. Klaassen, pp. 985–996. Pergamon Press, Elmsford, NY.
157. Kimber, I. and Basksetter, D. A. (1992): The murine local lymph node assay: a commentary on collaborative studies and new directions. *Food Chem. Toxicol.*, 30:165–169.
158. Karol, M. H. (1988): The development of an animal model for TDI asthma. *Bull. Eur. Physiopath. Respir.*, 23:571–576.
159. Dearman, R. J., Basketter, D. A., and Kimber, I. (1992): Variable effects of chemical allergens on serum IgE concentration in mice: preliminary evaluation of a novel approach to the identification of respiratory sensitizers. *J. Appl. Toxicol.*, 12:317–323.
160. Homo-Delarche, F., Fitzpatrick, F., Christeff, N., Nunez E. A., Bach, J. F., and Dardenne, M. (1991): Sex steroids, glucocorticoids, stress and autoimmunity. *J. Steroid Biochem. Mol. Biol.*, 40:619–637.
161. Sakaguchi, S. and Sakaguchi, N. (1989): Organ-specific autoimmune disease induced in mice by elimination of T cell subsets: neonatal administration of cyclosporin A causes autoimmune disease. *J. Immunol.*, 142: 471–480.
162. Chazerain, P., Meyer, O., and Kahn, M. F. (1992): Rheumatoid arthritis-like disease after alpha-interferon therapy. *Ann. Intern. Med.*, 116:427–439.
163. Greim, P., Gleichmann, E., and Shaw, C. F. (1997): Chemically induced allergy and autoimmunity: what do T cells react against? In: *Comprehensive Toxicology*, edited by D. Lawrence, pp. 324–338. Elsevier, New York.
164. U.S. EPA. (1996): *Biochemicals Test Guidelines: OPPTS 880.3550 Immunotoxicity*. U.S. Environmental Protection Agency, Washington, D.C.
165. U.S. EPA. (1996): *Biochemicals Test Guidelines: OPPTS 880.3800 Immune Response*. U.S. Environmental Protection Agency, Washington, D.C. (http://www.epa.gov/docs/OPPTS_Harmonized/880_Biochemicals_Test_Guidelines/Series/880–3800.pdf).
166. U.S. EPA. (1996): *Biochemicals Test Guidelines: OPPTS 880., 7800 Immunotoxicity*. U.S. Environmental Protection Agency, Washington, D.C. (http://www.epa.gov/docs/OPPTS_Harmonized/870_Health_Effects_Test_Guidelines/Drafts/870–7800.pdf).
167. CDER. (2002): *Guidance for Industry: Immunotoxicology Evaluation of Investigational New Drugs*. Center for Drug Evaluation and Research, U.S. Department of Health and Human Services, U.S. Food and Drug Administration, Washington, D.C.
168. Hastings, K. L. (2002): Implications of the new FDA/CDER immunotoxicology guidance for drugs. *Int. Immunopharmacol.*, 2(11):1613–1618.
169. EMEA. (2000): Note for Guidance on Repeated Dose Toxicity, CPMP/SWP/1042/99. The European Agency for the Evaluation of Medicinal Products, London (http://www.emea.eu.int/pdfs/human/swp/104299en.pdf).
170. Putman, E. et al. (2002): Assessment of the immunotoxic potential of human pharmaceuticals: a workshop report, *Drug Inform. J.*, 36:417–427.

171. Snodlin, D. J. (2004): Regulatory immunotoxicology: does the published evidence support mandatory nonclinical immune function screening in drug development? *Regul. Toxicol. Pharmacol.*, 40(3):336–355.
172. Ryle, P. R. (2005): Justification for routine screening of pharmaceutical products in immune function tests: a review of the recommendations of Putman et al. (2003): *Fundam. Clin. Pharmacol.*, 19(3):317–322; discussion, 329–330.
173. Descotes, J. (2005): Immunotoxicology: role in the safety assessment of drugs. *Drug Saf.*, 28(2):127–136.
174. van der Laan, J. W. and van Loveren, H. (2005): Immune function testing of human pharmaceuticals: regulatory overshoot? *Expert Opin. Drug Saf.*, 4(1):1–5.
175. Weaver, J. L. et al. (2007): Meeting report: development of the ICH guidelines for immunotoxicology evaluation of pharmaceuticals using a survey of industry practices, *J. Immunotoxicol.*, in press.
176. U.S. FDA (1997): *Guidance for Industry: S6 Preclinical Safety Evaluation of Biotechnology-Derived Pharmaceuticals*. Center for Drug Evaluation and Research, U.S. Food and Drug Administration, Washington, D.C. (http://www.fda.gov/cder/guidance/1859fnl.pdf).
177. EMEA. (1997): *Note for Guidance on Preclinical Pharmacological Testing of Vaccines*, CPMP/SWP/465/95. The European Agency for the Evaluation of Medicinal Products, London (http://www.emea.eu.int/pdfs/human/swp/046595en.pdf).
178. U.S. FDA. (1997): *Guidance for Industry for the Evaluation of Combination Vaccines for Preventable Diseases: Production, Testing and Clinical Studies*. Center for Biologics Evaluation and Research, U.S. Food and Drug Administration, Washington, D.C. (http://www.fda.gov/cber/gdlns/combvacc.pdf).
179. U.S. FDA. (2000): *Guidance for Industry: Considerations for Reproductive Toxicity Studies for Preventive Vaccines for Infectious Disease Indications*. Center for Biologics Evaluation and Research, U.S. Food and Drug Administration, Washington, D.C. (http://www.fda.gov/cber/gdlns/reprotox.htm).
180. U.S. FDA. (1999): *Guidance for Industry and FDA Reviewers: Immunotoxicology Testing Guidance*. Center for Devices and Radiological Health, U.S. Food and Drug Administration, Washington, D.C. (http://www.fda.gov/cdrh/ost/ostggp/immunotox.html).
181. Anderson, J. M. and Langone, J. J. (1999): Issues and perspectives on the biocompatibility and immunotoxicity evaluation of implanted controlled release systems. *J. Controlled Rel.*, 57:107–113.
182. U.S. EPA. (1998): *Health Effects Test Guidelines: Skin Sensitization*, OPPTS 870.2600. U.S. Environmental Protection Agency, Washington, D.C. (http://www.epa.gov/docs/OPPTS_Harmonized/870_Health_Effects_Test_Guidelines/Series/870-2600.pdf).
183. Fischer, A., Cavazzana-Calvo, M., De Saint Basile, G., DeVillartay, J. P., Di Santo, J. P.,Hivroz, C., Rieux-Laucat, F., and Le Deist, F. (1997): Naturally occurring primary deficiencies of the immune system. *Annu. Rev. Immunol.*, 15:93–124.
184. Lovik, M. (1997): Mutant and transgenic mice in immunotoxicology: an introduction. *Toxicology*, 119(1):65–76.
185. House, R. V. (2005): Transgenic rodent models in immunotoxicology. In: *Investigative Immunotoxicology*, edited by H. Tryphonas, M. Fournier, B. Blakley, J. Smits, and P. Brousseau, pp. 345–362. CRC Press, Boca Raton, FL.
186. Shetty, R. C. (2005): Potential pitfalls of nanotechnology in its applications to medicine: immune incompatibility of nanodevices. *Med. Hypoth.*, 65(5):998–999.
187. Thomas, K. and Sayre, P. (2005): Research strategies for safety evaluation of nanomaterials. Part I. Evaluating the human health implications of exposure to nanoscale materials. *Toxicol. Sci.*, 87(2):316–321.
188. Donaldson, K., Stone, V., Tran, C. L., Kreyling, W., and Borm, P. J. (2004): Nanotoxicology. *Occup. Environ. Med.*, 61(9):727–728.
189. Oberdorster, G., Oberdorster, E., and Oberdorster, J. (2005): Nanotoxicology: an emerging discipline evolving from studies of ultrafine particles. *Environ. Health Perspect.*, 113(7):823–839.
190. Pasare, C. and Medzhitov, R. (2005): Toll-like receptors: linking innate and adaptive immunity. *Adv. Exp. Med. Biol.*, 560:11–18.
191. Holtmeier, W. and Kabelitz, D. (2005): Gamma delta T cells link innate and adaptive immune responses. *Chem. Immunol. Allergy*, 86:151–183.
192. Jakob, T., Traidl-Hoffmann, C., and Behrendt, H. (2002): Dendritic cells: the link between innate and adaptive immunity in allergy. *Curr. Allergy Asthma Rep.*, 2(2):93–95.
193. Hemmi, H. and Akira, S. (2005): TLR signalling and the function of dendritic cells. *Chem. Immunol. Allergy*, 86:120–135.
194. Kubo, T., Hatton, R. D., Oliver, J., Liu, X., Elson, C. O., and Weaver, C. T. (2004): Regulatory T cell suppression and anergy are differentially regulated by proinflammatory cytokines produced by TLR-activated dendritic cells. *J. Immunol.*, 173(12):7249–7258.
195. Guo, Z., Zhang, M., Tang, H., and Cao, X. (2005): Fas signal links innate and adaptive immunity by promoting dendritic-cell secretion of CC and CXC chemokines. *Blood*, 106(6):2033–2041.
196. Peng, G., Guo, Z., Kiniwa, Y., Voo, K. S., Peng, W., Fu, T., Wang, D. Y., Li, Y., Wang, H. Y., and Wang, R. F. (2005): Toll-like receptor 8-mediated reversal of CD4+ regulatory T cell function. *Science*, 309(5739):1380–1384.
197. Liew, F. Y., Xu, D., Brint, E. K., and O'Neill, L. A. (2005): Negative regulation of toll-like receptor-mediated immune responses. *Nat. Rev. Immunol.*, 5(6):446–458.
198. Rifkin, I. R., Leadbetter, E. A., Busconi, L., Viglianti, G., and Marshak-Rothstein, A. (2005): Toll-like receptors, endogenous ligands, and systemic autoimmune disease. *Immunol. Rev.*, 204:27–42.

# 37 Assessment of Behavioral Toxicity

*Deborah A. Cory-Slechta and Bernard Weiss*

## CONTENTS

To survive, organisms must be sensitive to events occurring in their environments and respond appropriately. At the most elementary level, organisms must avoid hazards such as predators and other threats, must secure food, and, for species survival, must pursue reproduction. The nervous system is the site at which such transactions with the environment are processed. The nervous system also governs endogenous transactions such as controlling neuroendocrine secretions and carries on commerce with the immune system, but such functions are processed in the background, so to speak. The integrity of the nervous system, paramount to both individual and species survival, is reflected predominantly by the integrity of behavior.

This chapter describes methods and issues related to the assessment of behavioral toxicity. Its complexities are significant because of the multiple dimensions and behavioral domains that comprise the human behavioral repertoire. Primary among these are motoric, sensory, and cognitive (e.g., learning, memory, attention) domains, each of which can range from simple to highly complex levels of function. Further, human behavior requires integration across these various domains. It is for this reason that *in vitro* methods are unlikely to ever provide sufficient alternative approaches for behavioral toxicology [96], particularly when it is clear that even subtle functional differences can significantly influence neurochemical and neurophysiological outcomes.

Often, the first clues of toxicity to humans may be subjective disturbances such as nervousness (chlordecone) or personality changes (manganese), succeeded by more overt signs such as tremor (chlordecone) or akinesia (manganese). Regulatory standards aim for exposure levels providing enough of a margin to preclude even these preliminary, nonspecific symptoms [262]. In contrast, early animal laboratory investigations tended to adopt exposure levels likely to evoke a clearly visible toxic response, a strategy requisite to develop and validate methods responsive to neurotoxic agents and to acquire corresponding information about mechanisms of toxicity applicable to lower exposure levels. Since then, animal studies have expanded to focus on lower levels and cumulative exposures and associated behavioral mechanisms, and both human and experimental studies address issues such as long-term effects of toxicants as well as reversibility of adverse effects.

In addition to providing a measure of functional competence, behavioral outcomes often give guidance to the underlying neurochemical, neurobiological, and histopathological substrates, an understanding continually increased by parallel advances in the fields of psychology and neuroscience. For example, elevated motor activity may indicate actions on specific neurotransmitter systems; an inability to distinguish geometric form would point to the visual cortex as a possible site of damage.

Given that behavioral functions can range from simple to highly complex, the assessment of behavioral toxicity often proceeds in stages. The first stage often examines dose–response relationships based on systematic observations of responses such as those included in functional observation batteries that generally evaluate simple and innate behaviors. While alerting to potential behavioral toxicity, such evaluations typically provide little information about the specificity of a functional deficit [172], which requires more complex assays that provide greater precision, thus yielding more specific and quantitative information to permit better prediction of possible adverse effects in humans.

The chapter proceeds from descriptions of the simpler behavioral techniques generally applied in the earliest steps of hazard identification to more complex approaches designed to clarify specific deficits and to understand both behavioral and neurobiological mechanisms. The techniques and paradigms described include those utilized in an experimental context, many of which are appropriate across species, including humans, with appropriate parametric modifications. In fact, some approaches were originally implemented in human subjects and subsequently modified for experimental animal use. This discussion is followed by a presentation of behavioral techniques that have evolved from clinical neuropsychological assessments, designed to identify disease and dysfunction, that may not have an experimental analog, being utilized to

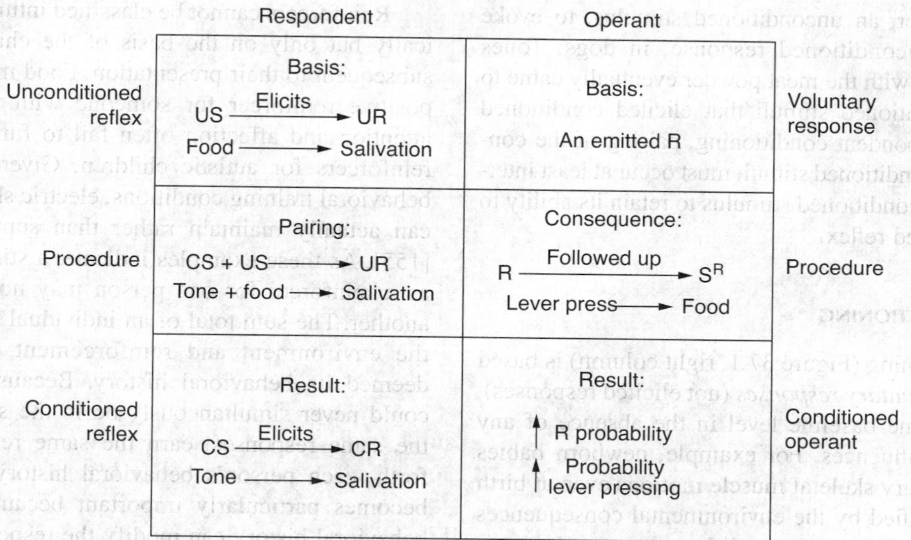


**FIGURE 37.1** The two types of behavioral conditioning are illustrated. Respondent conditioning (left column) is based on unconditioned (innate) reflexes. In respondent conditioning, an unconditioned stimulus elicits an unconditioned response. When an initially neutral stimulus is paired with this unconditioned stimulus (procedure), it acquires eliciting properties; that is, it becomes a conditioned stimulus capable of eliciting a conditioned reflex. In the example, elicitation of salivation by food is an unconditioned reflex. When a tone is appropriately paired with the food (procedure), its subsequent presentation alone will come to elicit salivation. Operant conditioning (right column) is based on voluntary responses emitted by the organism. If these responses are followed by a reinforcing stimulus (procedure), the frequency of the response will subsequently increase (result). In the example, pressing of a lever by an organism results in food reward (procedure), thereby increasing the frequency of lever pressing. If reinforcement is withheld (extinction) or a punishing stimulus is presented, response frequency will decline.

evaluate effects of chemical exposures in human populations. The chapter aims to provide enough familiarity with these techniques to allow a more than elementary understanding of behavioral toxicology. Inevitably, new techniques and modifications of existing ones will continue to proliferate, not only in behavioral toxicology itself but also in fields that share common borders, such as the study of neurodegenerative diseases. Two prior reports [175,187] aimed at the impact of neurobehavioral toxicants on public health demonstrate how behavior serves as the sentinel for less accessible endpoints. This widening scope of behavioral toxicology is certain to alter how it is practiced.

## ORIGINS OF BEHAVIOR

The behavioral repertoire varies along two dimensions: origin and modifiability. At one end of such a continuum are behaviors designated as innate or hardwired which include unconditioned reflexes, fixed-action patterns, and instinctive behaviors that do not require learning. The components of such innate behaviors may be unique to a particular species and may be insensitive to modification by the environment. For some such behaviors, once evoked, the responses continue to completion even in the absence of the appropriate environmental substrates. On the other extreme are learned or acquired behaviors that are subject to modification by the environment. Such behaviors are based on voluntary emitted responses, which are increased or decreased in strength, altered, or refined in a dynamic manner across time by environmental consequences through operant or respondent conditioning. Falling between these two extremes are instinctive behaviors such as fixed-action patterns that are modifiable by environmental circumstances. The extent to which the behavioral repertoire of an organism is comprised of learned vs. unlearned behaviors is dependent on the species, with a tendency for learned behaviors predominating at the higher end of the phylogenetic continuum.

### RESPONDENT CONDITIONING

In respondent conditioning, as shown in Figure 37.1 (left column), an *unconditioned reflex* elicited (evoked) by an unconditioned stimulus is the basis for a new conditioned respondent. The term *unconditioned* signifies the innate characteristics of the stimulus and response comprising the reflex. When the unconditioned stimulus is repeatedly paired with a neutral stimulus (*procedure*), the neutral stimulus comes to acquire eliciting properties similar to those of the unconditioned stimulus; that is, the neutral stimulus becomes a conditioned stimulus that can evoke a *conditioned reflex* that typically, though not uniformly, resembles the unconditioned response. Figure 37.1 uses the classic example from the Russian physiologist I.P. Pavlov, who used meat powder, an unconditioned stimulus, to evoke salivation, an unconditioned response, in dogs. Tones repeatedly paired with the meat powder eventually came to function as conditioned stimuli that elicited conditioned salivation. In respondent conditioning, pairing of the conditioned and unconditioned stimuli must occur at least intermittently for the conditioned stimulus to retain its ability to elicit a conditioned reflex.

### Operant Conditioning

Operant conditioning (Figure 37.1, right column) is based on emitted or *voluntary responses* (not elicited responses), that occur at some baseline level in the absence of any environmental influences. For example, newborn babies engage in voluntary skeletal muscle motions even at birth that can be modified by the environmental consequences that follow them (*procedure*) and thus serve as the basis of later, more coordinated movements and precise motor functions (operant behaviors). Operant conditioning is the process by which the frequency or strength of an operant (voluntary) response is modified by the consequences of that behavior. Figure 37.1 depicts an example of a lever press response by a slightly hungry rat resulting in food as a reward that increases the future probability of the lever press.

### Reinforcement

Reinforcement is the presentation of a stimulus contingent upon a response that results in an *increase* in the future frequency of that response. Reinforcers may be either positive or negative, a distinction that is purely procedural (Table 37.1). Specifically, a *positive* reinforcer strengthens a response; that is, it increases its probability by its *presentation*, such as the delivery of food to a hungry organism or the presentation of money for work performed. With *negative* reinforcement, a response is strengthened by the *removal* or diminution of a stimulus event or the prevention of its onset. Here, common examples used experimentally include cessation of electric shock or loud noise contingent upon a response; a human example might be the prevention of a parent's scolding by cleaning one's room. Shock avoidance and escape procedures fit into this category.

**TABLE 37.1**
**Consequences of Responding**

| Stimulus Conditions | Change in Response Strength: Increased | Change in Response Strength: Decreased |
|---|---|---|
| Presentation | Positive reinforcement | Punishment |
| Withdrawal | Negative reinforcement | Extinction |

Reinforcers cannot be classified intuitively or categorically but only on the basis of the change in behavior subsequent to their presentation. Food may not serve as a positive reinforcer for someone with the stomach flu; attention and affection often fail to function as positive reinforcers for autistic children. Given the appropriate behavioral training conditions, electric shock presentation can actually maintain rather than suppress responding [155]. As these examples indicate, a stimulus that serves as a reinforcer for one person may not so function for another. The sum total of an individual's interaction with the environment and reinforcement contingencies is deemed its behavioral history. Because all individuals could never simultaneously be in the same place, make the same responses, earn the same reinforcers, and so forth, each person's behavioral history is unique. This becomes particularly important because differences in behavioral history can modify the response to drugs and, thus, perhaps to toxicants as well, an important possibility that as yet has received little experimental attention.

Reinforcers also are defined as primary (unlearned, $S^R$) or secondary (conditioned, learned, $S^r$). Primary reinforcers function as effective reinforcers without any prior experience or conditioning. Examples include food, water, and the opportunity to engage in sexual behavior. Conditioned reinforcers are stimuli that *acquire* their reinforcing efficacy through pairing with other established reinforcers. Money serves as an extremely effective generalized conditioned reinforcer because of its pairing with many other reinforcers, both primary and conditioned. Conditioned reinforcers frequently are generated in an experimental setting by repeatedly pairing an initially neutral stimulus, such as a light flash or tone, with a primary reinforcer, such as food delivery. After repeated pairings, the light stimulus acquires conditioned reinforcing properties of its own and can thereby maintain substantial operant responding itself. To retain its reinforcing efficacy, however, the conditioned reinforcer must be paired at least occasionally with the primary reinforcer.

### Punishment and Extinction

Two operant procedures that *decrease* the frequency of learned responses are punishment and extinction. Table 37.1 summarizes these terms, with respect both to the procedures involved and to the subsequent behavioral outcome. In *punishment*, a stimulus presented after a response decreases the frequency of that response. Scolding a child who has drawn on the wallpaper may constitute an example. Again, the stimulus must be classified only on the basis of the subsequent change in response frequency; if a promptly delivered scolding fails to decrease the frequency of writing on the wallpaper, then by definition it was not an effective punishing stimulus. In a clinical context, the presentation of ammonia to the nose contingent

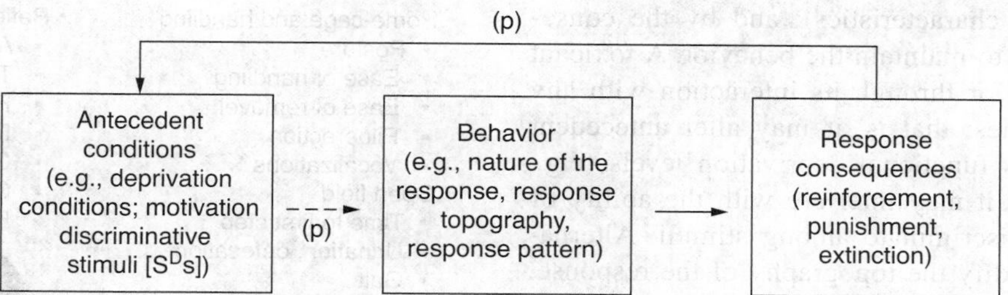


**FIGURE 37.2** The three-term, contingency-describing operant behavior. In the presence of antecedent conditions, operant responses are followed by consequences. The nature of the consequence determines the future probability (frequency) of the response and will also alter the strength of the antecedent stimuli (discriminative stimuli) controlling the response. A toxicant may act by altering the antecedent conditions (e.g., functional deprivation level or motivation of the subject) or the efficacy of relevant discriminative stimuli. A toxicant may act by altering the characteristics of the response itself (e.g., its topography or duration). A toxicant may act to alter the response consequences (e.g., change the perceived magnitude of reward or punishment). Understanding how a toxicant interacts with these factors defines the behavioral mechanism of effect.

on self-abusive behavior often decreases the frequency of this self-injurious response sometimes observed in autistic children. In *extinction*, the reinforcer for a response is withheld and the frequency of the response eventually (usually after an initial increase) declines. Withdrawing social attention from a child who has thrown a temper tantrum will in many cases decrease the future incidence of tantrums, albeit after an initial increase in frequency.

### Discriminative Stimuli

A stimulus in the presence of which an operant response is repeatedly reinforced comes to acquire *stimulus control* over the response; in the presence of this and related stimuli, the probability of the response is increased. This stimulus that defines the occasion on which an operant response is followed by reinforcement is called a *discriminative stimulus* ($S^D$). For example, the sight or smell of freshly baked cookies may be an $S^D$ that sets the occasion for the response of reaching into the cookie jar to be reinforced with a cookie, whereas an empty cookie jar would be unlikely to occasion such a response. Similarly, a red light and a stop sign are both discriminative stimuli that control the responses involved in stopping a vehicle. Braking at other times (e.g., at a green light) may have disastrous consequences. It is important to remember that discriminative stimuli do not elicit or evoke responses as do unconditioned and conditioned stimuli in respondent conditioning procedures; they merely, but importantly, indicate the likelihood or probability of reinforcement for a given operant response and thereby influence response probability.

The response of an organism may be explicitly reinforced in the presence of one stimulus ($S^+$), such as a red light, while reinforcement is explicitly withheld in the presence of another stimulus, the $S^-$ (or S), such as a green light. Using such a *discrimination procedure*, the organism soon comes to respond in a certain way only when the red light is on (i.e., in the presence of the $S^+$), such that the behavior is said to be under stimulus control, as noted above. When responding generalizes to other stimuli of the same or related stimulus classes, such as when a young child refers to all grown males as "Daddy," responding has generalized across the stimulus class "adult male"; that is, *stimulus generalization* has occurred. These processes of stimulus control and stimulus generalization are important aspects of such behavioral functions as concept formation and learning to learn.

In summary, a three-term contingency describes operant conditioning, as illustrated in Figure 37.2 [173]. In the presence of antecedent conditions that include discriminative stimuli, operant behaviors occur and are followed by consequent events (response consequence) that alter the future probability or frequency of that response and the control by the antecedent discriminative stimulus ($S^D$) over the response. It should be noted that discriminative and reinforcing stimuli need not be external or visible/audible environmental events; internal stimuli, whether internal physiological changes such as headaches or our own nonvocal verbal behavior (thinking), can acquire important behavioral functions, including serving as both discriminative stimuli and conditioned reinforcers.

Precise and operational definitions are critical in behavior because a toxic agent, like a brain lesion, may affect any stage in this three-term contingency. A goal of behavioral toxicology is to understand how toxic agents alter function or, more precisely, the behavioral mechanisms of action of the compound. As discussed by Thompson and Schuster [248] and Thompson and Boren [244] in the context of behavioral pharmacology, attaining such a goal depends on a thorough knowledge of the variables that control behavior, just as the biologist must understand a biological system. Behavior is influenced by multiple factors: by antecedent factors, such as deprivation or motivational state; by the nature of the response and response parameters (i.e., its topography or physical characteristics); and by the consequences that serve to maintain the behavior. A toxicant may change behavior through its interaction with any or all such variables; that is, it may alter antecedent conditions such as functional deprivation levels (e.g., induce nausea), or it may interfere with the ability of the organism to discriminate among stimuli. Alternatively, it could modify the topography of the response, such as by evoking incoordination, or it could interfere with the ability to associate the discriminative stimulus with the contingency, and so forth. Understanding the exact components of behavioral processes affected by toxic agents assists in understanding the underlying behavioral mechanisms of action and offers guidance as to the associated neurobiological substrates of effect.

## SCREENING BATTERIES

Screening batteries are used for evaluating the potential neurotoxicity of new or existing chemicals. These have typically included two behavioral components: a functional observational battery (FOB) and motor activity. The behavioral tests utilized for such purposes are often referred to as *apical tests* because they require the integrated function of several organ systems, including the nervous system. Often a distinction is made between what are deemed naturalistic behaviors, such as those scored on functional observation batteries, and complex learned behaviors. But, all behaviors are naturalistic; none exceeds the bounds of biological possibility. For many naturalistic behaviors, there is simply a lack of understanding of the controlling variables rather than an absence of their sensitivity to environmental control. Even as apparently natural and spontaneous a behavior as self-grooming by monkeys can be brought under experimental control by reinforcing (rewarding) it with food [117].

From the standpoint of screening and hazard identification, two points related to the interpretation of FOB and motor activity studies deserve mention. First, effects observed in response to toxicant exposure in such batteries can either be the result of a direct effect of the toxicant on the nervous system or be secondary to changes in other systems, as such apical testes rely on the functional integrity of multiple systems. Under some circumstances, the fact that the toxic effect is ultimately expressed in behavior may minimize the importance of the source of the effect. A second point is that the concurrent presence of body weight loss or decline in food or water intake does not necessarily indicate that behavioral changes observed in an FOB or in locomotor activity are the result of malaise or sickness, as these measures may change independently of each other. Certain agents, such as volatile organic solvents, tend to enhance motor activity at low concentrations, so the problem of confounding with malaise is negligible with such results [285].

- Home-cage and handling
  - Posture
  - Ease of handling
  - Ease of removel
  - Piloerection
  - Vocalizations
- Open field
  - Time to first step
  - Urination, defecation
  - Gait
  - Bizarre behavior
  - Rearing behavior
- Reflex and physiological
  - Approach response
  - Touch response
  - Finger snap response
  - Righting reflex
  - Grip strength
  - Catalepsy
  - Forelimb grip strength

**FIGURE 37.3** Typical measures used in many functional observational batteries (FOBs) include aspects of home-cage and handling, behavior in an open field and various reflex and physiological responses. Measurement of locomotor activity is also often included in an FOB. (From Moser, V.C. et al., *Fundam. Appl. Toxicol.*, 11, 189–206, 1988. With permission.)

### FUNCTIONAL OBSERVATIONAL BATTERIES

Figure 37.3 depicts the component tests of the functional observational battery developed by Moser et al. [172], which includes an array of measures of both unconditioned operant and respondent behaviors. Such batteries have been shown to exhibit utility for screening potential neurotoxicity (i.e., hazard identification and elaboration), as shown in Figure 37.4. As the figure shows, components of the FOB directed toward cholinergic functions exhibited sensitivity to the effects of the anticholinesterase carbaryl, whereas few such signs of cholinergic disturbances were evident in the presence of the non-anticholinesterase pesticide chlordimeform. Further discussion of FOB measures by domain is provided by Moser et al. [172] and by Baird et al. [11].

### MOTOR ACTIVITY

Motor activity is one component of motor function [249]. Generally described as an unconditioned behavior, motor activity exists at some baseline (operant) level and is a complex behavioral class that includes numerous components such as ambulation, rearing, grooming, and sniffing, all of which are environmentally influenced. Toxicants may alter motor activity by affecting any or all of its component behaviors. Numerous devices have been designed to measure motor activity. These vary in complexity and in the specific component of motor activity that they measure. One frequently employed device is the figure-8 maze, which consists of a series of interconnected alleys converging on a central open area. Motor activity is detected by photobeams, and an activity count is registered each time a photobeam is interrupted by the animal. In a device such as the open field (Figure 37.5), motor activity is quantified by counting the number of squares entered by the animal within some prescribed period of

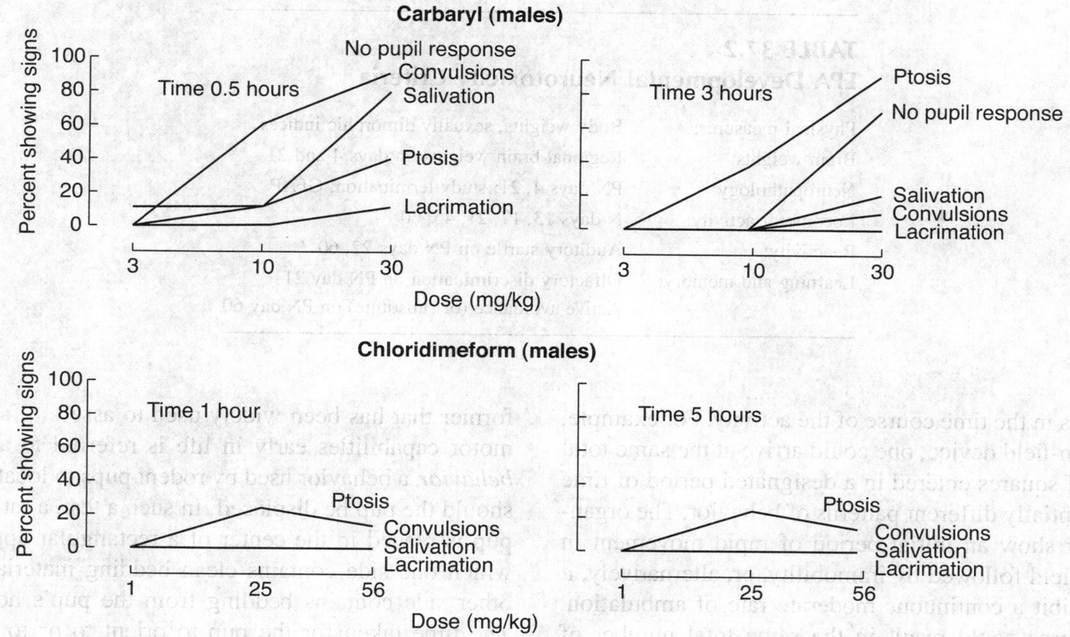


**FIGURE 37.4** Autonomic cholinergically mediated measures included in a functional observational battery are increased in response to increasing doses of the anticholinesterase carbaryl (top row), as might be predicted, but not to the non-anticholinesterase pesticide chlordimeform. (From Moser, V.C. et al., *Fundam. Appl. Toxicol.*, 11, 189–206, 1988. With permission.)

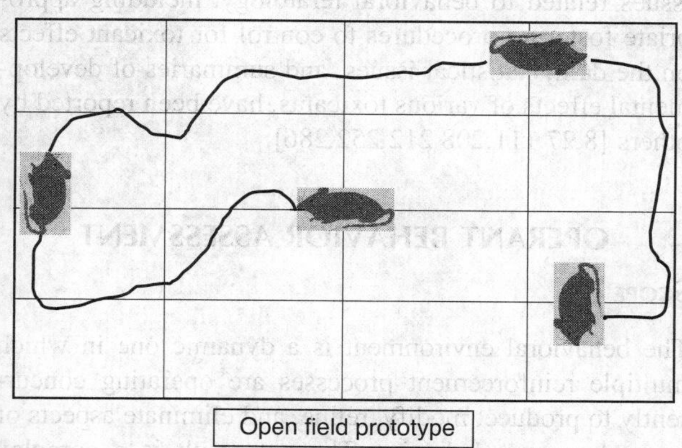


**FIGURE 37.5** Open field for the measurement of locomotor activity. In nonautomated procedures, the number of movements per unit time are recorded by an observer. Automated devices typically rely on computerized measurement of photobeam breaks by the movement of the subject to access locomotor activity. Photobeams are typically positioned so as to measure aspects of both horizontal ambulation and rearing in rodents.

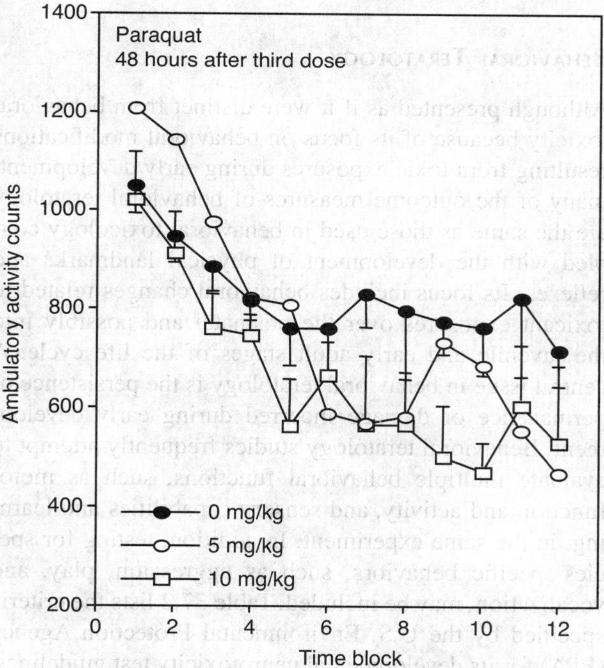


**FIGURE 37.6** Decreases in ambulatory motor activity counts over the course of a 60-minute behavioral test session in adult C57BL/6 mice injected i.p. with saline or paraquat (5 mg/kg or 10 mg/kg) one time per week for a total of 3 weeks. Motor activity was measured in an automated device using photobeam breaks. (From Brooks, A.I. et al., *Brain Res.* 823, 1–10, 1999. With permission.)

time. Figure 37.6 plots the motor activity of mice, as measured in an open field, following exposure to the herbicide paraquat [25]. Over the years such devices have become increasingly automated and generally locate photobeam devices in a manner that permits the detection, in a time-course fashion, of different types of motor activity, such as ambulation vs. rearing. This is important, because reliance on total counts may obscure toxicant-induced

**TABLE 37.2**
**EPA Developmental Neurotoxicity Criteria**

| | |
|---|---|
| Physical measures: | Body weights, sexually dimorphic indices |
| Brain weights: | Regional brain weights on days 4 and 21 |
| Neuropathology: | PN days 4, 21; study termination; GFAP |
| Locomotor activity: | N days 13, 17, 21, 45, 60 |
| Reactivity: | Auditory startle on PN days 22, 60 |
| Learning and memory: | Olfactory discrimination on PN day 21 |
| | Active avoidance (or substitute) on PN day 60 |

differences in the time course of the activity. For example, in the open-field device, one could arrive at the same total number of squares entered in a designated period of time via substantially different patterns of behavior. The organism might show an initial period of rapid movement in the open field followed by immobility, or, alternatively, it could exhibit a continuous moderate rate of ambulation. Both patterns could result in the same total number of squares entered, even though the underlying patterns of behavior are quite distinct. Thus, in the absence of a more detailed analysis of patterns of activity, a toxicant that alters motor activity primarily by altering its distribution in time could go undetected.

### Behavioral Teratology

Although presented as if it were distinct from behavioral toxicity because of its focus on behavioral modifications resulting from toxic exposures during early development, many of the outcome measures of behavioral teratology are the same as those used in behavioral toxicology coupled with the development of physical landmarks and reflexes. Its focus includes behavioral changes related to toxicant exposures over the postnatal and possibly into the juvenile and early adult stages of the life cycle. A central issue in behavioral teratology is the persistence or permanence of damage incurred during early development. Behavioral teratology studies frequently attempt to evaluate multiple behavioral functions, such as motor function and activity, and sensory capabilities and learning, in the same experiment. In addition, testing for species-specific behaviors, such as aggression, play, and vocalization, may be included. Table 37.2 lists the criteria specified by the U.S. Environmental Protection Agency (EPA) for its developmental neurotoxicity test guidelines.

What distinguishes behavioral teratology is that testing during the infant, postnatal, and juvenile periods of development sometimes requires modifications of procedures utilized with adults or even the development of new paradigms. In other cases, behavioral paradigms identical to those used in more mature subjects may be employed, albeit with parametric modifications. An example of the former that has been widely used to assess olfactory and motor capabilities early in life is referred to as *homing behavior,* a behavior used by rodent pups to locate the nest should the pup be displaced. In such a test, a rat or mouse pup is placed in the center of a rectangular apparatus in which one side contains clean bedding material and the other side contains bedding from the pup's home cage. The time taken for the pup to orient to or to reach the home-cage bedding constitutes the dependent variable of interest. Because this performance depends on both olfactory capabilities and the development of appropriate motor skills, it represents a type of apical evaluation [112]. Other issues related to behavioral teratology, including appropriate fostering procedures to control for toxicant effects on the dam, statistical issues, and summaries of developmental effects of various toxicants, have been reported by others [8,27,111,208,212,252,286].

## OPERANT BEHAVIOR ASSESSMENT

### Scope

The behavioral environment is a dynamic one in which multiple reinforcement processes are operating concurrently to produce, modify, refine, and eliminate aspects of learned operant behavior. The net result is a complex behavioral repertoire that includes a multitude of different behavioral domains and processes, any or all of which may be the target of a toxic agent. Enumerating all such possibilities is beyond the scope of such a chapter and, indeed, constitutes a subject matter having spawned numerous volumes of its own. Consequently, this chapter focuses on six particular operant behavioral domains that have evolved as primary interest areas in neurotoxicology: motor function, sensory behavior, learning, memory, schedule-controlled behavior, and stimulus and reinforcing properties of toxicants. The techniques and corresponding apparati designed to evaluate the entirety of these behavioral domains are numerous and cannot be fully enumerated here. This chapter therefore emphasizes those procedures that are the most widely used, as well as those that exemplify the range and scope of behavioral technologies.

**FIGURE 37.7** The prototypical operant chamber (Skinner box) for use with the rat is equipped with three response levers (left, center, right) and a bank of lights above each lever. Also included are a sonalert for delivering tones, a speaker for the delivery of auditory stimuli (top right), and a houselight for use as a visual stimulus. Food deliveries used as rewards are delivered from a feeder outside of the operant chamber through a plastic tube into a pellet trough located below the middle lever. The rat can be seen pressing the left lever.

## Apparatus

The basic requirements for an assessment of complex operant behavior must include a defined or specified unit of behavior to serve as the designated response. An environmental consequence or reinforcing event appropriate to the species and experimental parameters is arranged to follow the designated response. For some types of experiments, environmental stimuli of different modalities (e.g., visual, auditory) that can be varied along certain dimensions may be used as discriminative stimuli and conditioned reinforcers. The choice of the operant response, reinforcing consequence, and external or environmental stimuli should be congruent with the physical and behavioral capabilities of the experimental species. Teaching the chimpanzee to vocalize speech sounds proved impossible because of the physical limitations of its vocal apparatus. When the operant response was changed instead to hand signing, experimental studies of the acquisition of verbal behavior proceeded successfully (reviewed by Ristau and Robbins [209] and Terrace [240]). Similarly, under certain conditions, visual stimuli may prove inappropriate for rodents, because of the poor visual capabilities of these species, which, instead, depend heavily upon olfactory information in their normal environment [106].

A broad spectrum of reinforcers has been used in behavioral experiments. The choice depends on the species and the aim of the particular experiment. Reinforcers can include the delivery of food or liquid reinforcers such as water, fruit juice, or saccharin solutions; the opportunity to engage in wheel running, sexual behavior, or aggressive behavior; the production of heat in a cold environment; the opportunity to self-administer a drug such as cocaine or electrical brain stimulation; and escaping from or avoiding the onset of electric shock. Such a breadth of choice in a standard experimental arrangement allows the experimenter enormous flexibility, such as to compare heat and food reinforcement in metabolically impaired animals [131].

Many experiments on complex behavior are carried out in operant chambers typified by that illustrated in Figure 37.7. In conventional chambers, response devices typically include levers (rat, mouse, monkey, human), disks (pigeons, monkeys, humans), running wheels (mouse, rat), and cones for snout insertion (mouse, rat, guinea pig). Typically, the execution of the response is defined by the closure of a switch or electrical circuit that can be recorded and acted on by a computer or by electromechanical devices. The operant chamber may contain multiple response devices, usually but not necessarily of the same type. In addition, stimuli of different modalities can be delivered to the chamber by lights, loudspeakers, and such, and the operant chamber may be adapted to utilize different reinforcers. Equally important are the ease and flexibility available to program experimental contingencies in the operant chamber and the precision and resolution with which data can be collected. Computer technology allows responses and other events to be measured and stored sequentially with millisecond resolution. For these reasons, an operant chamber equipped to deliver stimuli of various modalities, containing multiple response devices, and offering a choice of reinforcement delivery systems provides the maximal flexibility and versatility for behavioral studies. It similarly permits a continuum in the pursuit of scientific questions directed by research findings as opposed to research questions that are dictated by the nature of the equipment, particularly in cases where the equipment has been specifically designed to measure a specific behavioral function.

Over the years, however, different types of devices, some automated and others not, have been used to measure specific aspects of behavior, some commercially available, others designed to meet the requirements of a particular experiment. This has been the case particularly for the assessment of learning and memory functions, where mazes have been used, including the more traditional mazes such as the T maze and L maze; the radial-arm maze, which consists of a central area from which eight or more arms radiate like spokes; and, more recently, water mazes. The Wisconsin General Test Apparatus (WGTA) was used to study learning in primates, and modifications have been made to accommodate other experimental species. The absence of automation of some such devices can require a far greater expenditure of personnel time and, ultimately greater costs to carry out experiments and also introduces the possibility of experimental bias. Although such devices can certainly provide in some cases a reasonable alternative to an operant chamber, as noted earlier, their limited utility for addressing a wide variety of other behavioral functions must also be considered when apparatus decisions are made and where resources are limited.

## Shaping an Operant Response

Before implementing a behavioral experiment, an effective reinforcer must be identified, one that can be presented immediately after the designated response occurs, because delayed reinforcement is less effective. A procedure known as *magazine training* is used to ensure the adequacy of the reinforcement contingency in an operant chamber. Magazine training consists of the presentation, at intermittent intervals, of the reinforcing stimulus, such as a food pellet, independently of behavior. After repeated presentations of food pellets in this manner, a slightly food deprived organism will reliably approach the feeder and ingest the food whenever the noise generated by the operation of the feeder occurs. A stimulus such as a light is frequently paired with the food delivery and, along with the noise generated by the feeder, comes to serve as a conditioned reinforcer. Such conditioned reinforcers are especially important because they provide the immediate reinforcement for responding given the delay necessarily imposed by the time required for the organism to approach and ingest the food. For some reinforcers, no such training is required. For example, electric shock is used in escape and avoidance procedures in the absence of any prior training of subjects.

Once a reliable reinforcer has been established, shaping of the designated response can proceed. At this point, the reinforcing stimulus is presented only when the organism emits responses that more and more closely resemble the final designated response, a process of reinforcing successive approximations to the desired response. For example, in the case of a lever press response, reinforcer delivery might first be contingent on touching the lever, then putting two paws on the lever, and eventually applying sufficient force to displace the lever downward and close the associated electric circuit. Reynolds [199] discusses these techniques in greater detail. In some cases, magazine training and shaping procedures can be automated [52], a major advantage when a large number of experimental subjects must be used.

Shaping procedures also may be used in other types of apparatus to train the initial response; for example, in the WGTA, the monkey is first shaped to reach out and retrieve food cups on the shelf in front of the enclosures. For other devices, such as mazes and running wheels, little overt shaping may be necessary, as exploring alleys and running usually occur spontaneously at a high enough frequency (high operant level) in rodents to ensure contact with the reinforcement contingency.

Once the organism is emitting the designated response at an adequate rate, it may still be necessary to progress through a series of training conditions before imposing a complex behavioral paradigm of experimental interest, just as a dancer must learn to connect the different steps of a choreographed performance. Likewise, a beginning reader could hardly be expected to tackle the plays of Shakespeare. In a similar way, requiring an organism to emit 100 responses for each food delivery will likely require intervening sessions with smaller response requirements to build response strength and to prevent the behavior from undergoing extinction. It is the shaping process that creates new behaviors and pieces together responses to produce increasingly complex chains of behavior.

An accumulating literature attests to the fact that many of the specific behavioral procedures described here are successfully used with both mice and rats, although differences between the two and between strains of each in operant level behavior may sometimes necessitate alterations in approach. In operant chambers, for example, higher overall activity levels may mean that high levels of lever pressing occur in the absence of any shaping; in this case, additional procedures must be implemented to bring the high rates of behavior under appropriate control of the operant contingencies, such as time outs contingent on excess responding, specific implementation of differential reinforcement for low rates of responding, or the use of response devices such as nose cones rather than levers (see Figure 37.27). Further, many of the behavioral paradigms described in this chapter have been more broadly used across human and nonhuman species with appropriate parametric and response device modifications.

## MOTOR FUNCTION

Visible indications of toxic processes in the nervous system often occur as abnormal movements, impaired coordination, slowing of responses, and complaints of weakness. Because any reduction in the capacity for coordinated movement reduces an organism's ability to cope with the demands of its environment, even subtle defects will influence how effectively it functions. Learned motor skills play an especially salient role in human activities. Even apart from the advanced skills of the surgeon or violinist, which are the culmination of years of practice, consider how much we rely on proficiencies in writing and driving as part of our daily activities.

Despite many clear examples, the full scope of contributions to movement disorders by neurotoxic chemicals remains vague. Part of the vagueness arises because the etiology may lie buried many years in the past. For example, understanding whether pesticide exposures serve as risk factors for Parkinson's disease [97], as suggested by experimental models and epidemiological evidence, is hampered by the inability to identify specific past exposures. The emergence of clinical signs may also reflect the diminished ability of compensatory mechanisms during advanced age to overcome the effects of earlier damage—that is, silent damage [269].

Clear connections between exposure and motoric dysfunction have been established for several metals. Wrist drop afflicted many painters who were occupationally exposed to lead pigments. The cardinal sign of mercury vapor neurotoxicity is excessive tremor. One of the primary signs of methylmercury poisoning is ataxia. Manganese miners display a condition best described as dystonia but with some features of Parkinson's disease. Insecticides are designed as neurotoxicants, and the organophosphorous compounds produce axonopathies that impair both motor and sensory function. The industrial chemical acryamide also induces both motor and sensory neuropathies. Certain organic solvents produce central nervous system damage expressed as motor dysfunction. Many of the chemical classes described in this volume, in fact, even those not classified primarily as neurotoxicants, can induce motor disorders.

The control of posture and movement is anatomically organized in the central nervous system as a collection of diverse motor centers arranged hierarchically from the least integrative, at the level of the spinal cord, through the basal ganglia and cerebellum, to the ultimate level of the cerebral cortex. Weaving through this basic hierarchical structure is a web of enormously complex pathways connecting the various motor centers and involving both afferent and efferent transmission. The total system depends on the functional integrity of many different components, all of which seem to present unique opportunities for the actions of toxic agents.

As this implies, the sources of movement disorders may arise from many different sites and processes. Some may be due to damage in the peripheral nervous system, although most originate in central structures. Some are an expression of cell damage or loss, whereas others reflect neurochemical abnormalities. Some are manifested clinically as relatively blunt abnormalities, such as ataxia; others are as subtle as the inability to grasp small objects with thumb and forefinger. Because they represent a leading clinical problem, the neural basis of movement disorders is the subject of a vast literature.

The basic properties of movement, however, subsume only a few fundamental dimensions described by mass, time, and displacement, expressed in measures such as force, duration, velocity, acceleration, momentum, amplitude, accuracy, and patterning in time. Screening batteries often rely on single global indices, such as spontaneous locomotor activity, that are influenced by many variables independent of motor capacity and do not encompass the full spectrum of motor functions. Some incorporate more specific assays of motor function such as the ability to maintain balance on a rotating rod. Because organisms effect change by producing patterns of muscular contraction, those measures of function likely to prove most sensitive to toxic impairment will typically be based on careful analyses of such patterns and will reflect the integration of multiple systems that yield complex movements. Some measures used to evaluate gait and balance have been described, such as the quantification of Parkinson's disease outcomes in clinical studies and their analogs in the animal laboratory—i.e., studies of changes in the walking patterns of rats treated with organophosphate compounds [278], or the kind of kinematic analysis undertaken by Cohen and Gans [39] to describe the patterns of rat locomotion in running wheels. Kulig and Lammers [128] reviewed a wide range of techniques currently in use to investigate motor dysfunction.

The focus of this chapter is on operant behavioral measures of motor function, as much human motor behavior is skilled, learned, and voluntary. Learned skills probably offer the most useful baselines for such an assessment because of the flexibility they offer to the experimenter for precise specification of the form of the response. Because of the inherent capacity to compensate for incipient difficulties, however, they are especially challenging to evaluate. For either learned or unlearned motor functions, it seems likely that detection of early subclinical toxic effects will require a detailed quantitative analysis of movement topography.

### Response Duration

The time occupied by a particular component of a movement can be studied either as an intrinsic variable, arising indirectly from contingencies applied to another response component such as force, or as an explicit variable specified directly by the experimenter. Especially in the latter instance, the stimuli governing the response are almost wholly proprioceptive, arising from sources such as muscle spindle receptors. Assume a situation in which the required operant response is a lever press, and the reinforcement contingency specifies that response durations must exceed *t* seconds to produce reinforcement. If no external stimuli that change systematically with time are presented, the primary stimuli available to the organism with respect to response duration are those arising from muscle and joint receptors. Because such proprioceptors are key elements in the control of movement, measures of response duration could prove useful in evaluating whether these proprioceptors have been damaged by agents such as acrylamide [142].

Shaping long-duration responses is a fairly straightforward process, as described earlier. Once the designated response, such as lever pressing, has been learned, the additional requirement of maintaining it for a specified duration can be imposed. Early in training, the duration should be short enough that a substantial proportion of responses meet the criterion for reward. The duration, as in all shaping procedures, can then be raised gradually until the final value is attained. Both minimal and maximal durations can be imposed, so a band of durations comes to serve as the criterion. Stevenson and Clayton [230] trained rats to hold down a lever for at least 40 consecutive seconds, after which a white noise signal was sounded. Releasing the bar in the presence of the noise turned it off and triggered a feeder in the operant chamber to deliver a pellet of food.

Relatively few experiments with neurotoxic agents have sought to exploit such possibilities. Cory-Slechta et al. [48] trained rats to respond on a schedule that reinforced only durations above a specified minimum value. After preliminary training, during which the rats were first trained to press a lever for food reinforcement, the experimenters imposed a schedule based on differential reinforcement of response duration. Each lever press that exceeded a specified duration was followed by food pellet delivery. These durations ranged from 0.5 seconds at the beginning to 6.0 seconds during later training. Lead treatment reduced durations and also expanded within-group variability.

Several experiments with dogs have relied on response duration as the primary measure of performance. In one, dogs were trained to press a button with their snouts for a food reward [265]. The schedule did not specify response duration directly; instead, it specified that reinforcement would be delivered when 60 seconds of responding had been accumulated. The dogs adjusted to this contingency by pressing the button between 10 and 20 times for each reinforcement; short responses predominated. Response durations were longer under control conditions than after amphetamine, pentobarbital, or ethanol. Concurrent administration of pentobarbital or ethanol with amphetamine shortened durations even more sharply. In later experiments [259,260], response duration was made the criterion for a titration schedule; that is, the duration required for reinforcement changed in accordance with the dog's performance. Each session began with a specified minimum duration of 0.25 seconds. Each time the dog (this time pressing its snout against a panel) ended a response that exceeded the current minimum, the required duration was raised to a higher level. In one variant of the program, the criterion also fell after a series of unsuccessful responses. The dogs learned the progression and, over the course of each session, emitted longer and longer responses. Amphetamine reduced this rate of rise, and alpha-methylparatyrosine ($\alpha$-MPT), a tyrosine hydroxylase inhibitor, lengthened it, even at doses as small as 3.12 mg/kg (Figure 37.8). Other studies with $\alpha$-MPT, such as those based on avoidance performance, have identified it as a source of performance degradation. Under the conditions of the study just described, it might be viewed as a source of performance enhancement and provides an example of how much influence the chosen parameters of a behavioral assay exert on the results and interpretation of an experiment.

## Precision

In the absence of clear visual cues, movement precision is also guided largely by proprioceptive information. As an example, Falk and associates [76] trained rats to exert forces within a specified range (15 to 20 g) for a duration of 1.5 seconds. The rats responded on a lever that transmitted an electrical signal proportional to the applied force and were reinforced by food pellets for responses meeting the joint force–duration criterion. Training was carried out by approaching the joint criteria of force and duration in small increments and required only 5- to 11-hour-long training sessions. With this system, Falk demonstrated that several common central nervous system drugs exerted unique effects on this form of discriminative motor control; for example, the relative amount of time spent within the specified force band varied with the dose of amphetamine and pentobarbital, declining as dose was raised.

A similar device, adapted for monkeys, enabled Preston et al. [195] to study deficits in fine motor control produced by various drugs. One side of a standard metal cage was modified with the addition of a Plexiglas® tube through which the monkey could extend its arm and make contact with a conically shaped response device connected to a force transducer. The monkeys were trained to emit forces between 25 and 40 g for a continuous 3 seconds to obtain small quantities of water. Various indices of performance demonstrated that the acute effects of methamphetamine, similar to the effects described earlier for rats, impaired performance after repeated high-dose treatment, the pattern of intake adopted by amphetamine abusers.

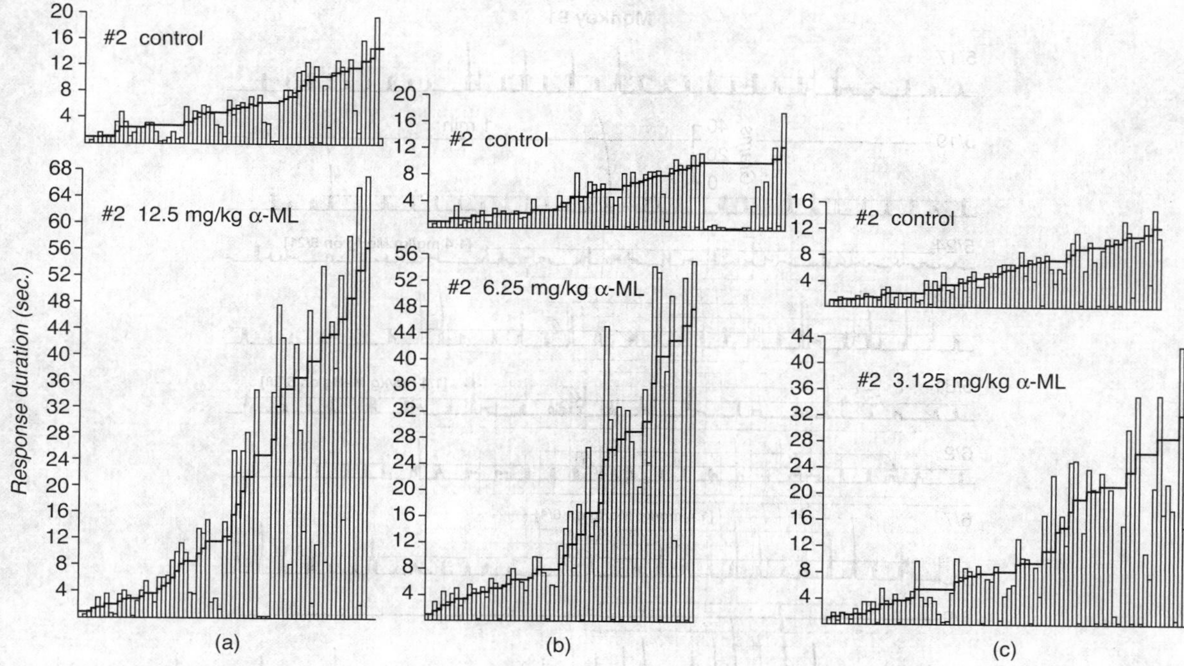


**FIGURE 37.8** Complex duration discrimination in Basenji dogs. The dogs were trained to press a panel with their snouts. An experimental session began with a predetermined duration criterion of 0.25 seconds. Any response duration above the criterion produced delivery of reinforcement (dry dog food) and elevated the criterion by 25% of the difference between the old criterion and the new duration that had exceeded it. This proportion was chosen empirically. If the criterion were raised too quickly, the behavior would be lost, as it often is in shaping procedures that advance so rapidly that they lose contact with the behavior. Each panel shows response durations (seconds) on the *y*-axis in relation to session time on the *x*-axis. Under control conditions, the dogs learned to produce longer and longer durations through the session. The compound α-methyltyrosine, which inhibits the rate-limiting step in the synthesis of catecholamines, enhanced the rate at which successive response durations rose, even at doses many times lower than those used in neuropharmacology experiments. (Adapted from Weiss, B., in *International Symposium on Amphetamine and Related Compounds*, Costa, E. and Garratini, S., Eds., Raven Press, New York, 1970, pp. 797–812; Weiss, B., in *Recent Developments in the Quantification of Steady-State Operant Behavior*, Bradshaw, C.M., Ed., Elsevier, Amsterdam, 1981, pp. 249–265.)

In another adaptation of this system [261], monkeys were trained to insert a paw through a slot and to touch a Lucite® plate connected to a strain gauge. The strain gauge output was transmitted to an amplifier, the output of which was coupled to the analog-to-digital converter inputs of a digital computer. The analog-to-digital converter transforms the continuously varying electrical signal from the amplifier into a form that can be processed by the computer. The computer was used to specify upper and lower bounds of the force that defined the response. With this system, the gradual degradation of fine motor control produced by methylmercury could be traced. The typical pattern before treatment consisted of a precise emission of force within the 25- to 40-g range specified as the response criterion, maintenance in that range for the required 2 seconds, and then release. After treatment, the first sign of coordination difficulty was overshoot at the beginning of the response—that is, a transient force above the specified ceiling (Figure 37.9). This early indication of impaired motor control was followed, after further dosing, by an increasing inability to maintain the force within the prescribed limits. Elsner [68] devised a somewhat similar system to study motor performance in rats, aiming to construct a model reflecting some of the characteristics of attention deficit disorder.

Because rodents remain the most widely used species for routine toxicological testing, additional techniques for the measurement of movement precision in rats and mice, in addition to those mentioned above, are especially appealing. All require some investment of time or instrumentation but can yield critical information about the dimensions of motor control. Lesion experiments designed to clarify the functional domain of certain brain areas are a useful source of approaches, and Newland [183] has compiled a table showing which basic tests have shown motor effects of lesions in cerebral cortex, basal ganglia, cerebellum, peripheral nerves, and the neuromuscular junction. In a combination of operant and observational methods, Whishaw et al. [272] examined the impact of motor cortex lesions in rats on their method of grasping food pellets. Before surgery, the rats were reduced to 90% of their initial body weight and trained to eat in a special

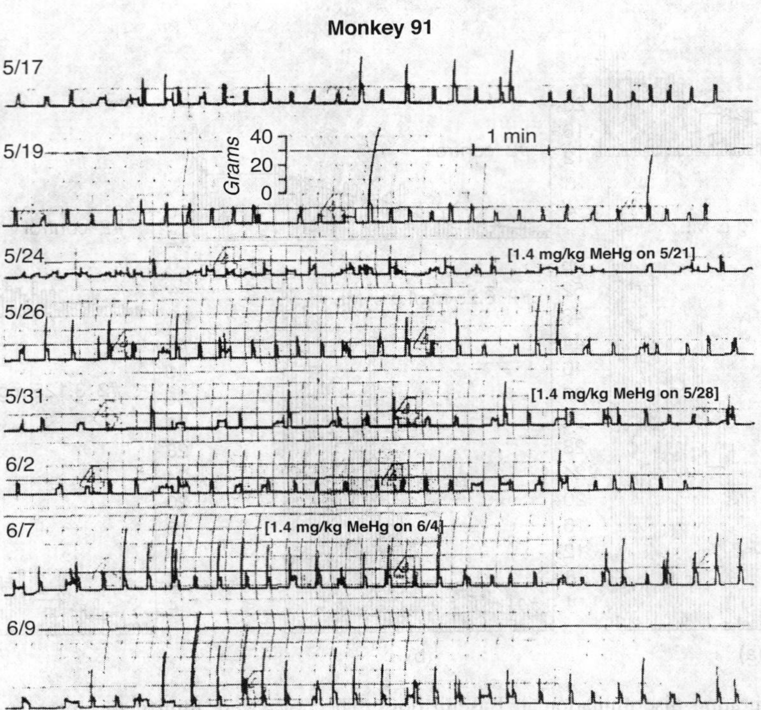


**FIGURE 37.9** Performance on a motor discrimination task of a monkey (*Saimiri sciurea*) treated with methylmercury as labeled in this figure. The monkey was trained to insert a paw into a slot and to apply pressure on a force-sensitive plate connected to a computer. Applied forces of 25 to 40 g maintained continuously for 2 seconds triggered delivery of a sucrose pellet. The top two tracings show baseline performance, which was characterized by a few overshoots or undershoots. After three weekly doses of methylmercury, most responses were initiated by sharp overshoots before settling into the reinforced zone. (From Weiss, B., in *Neurobiology of the Trace Elements*, Vol. 2, *Neurotoxicology and Neuropharmacology*, Dreosti, I.E. and Smith, R.M., Eds., Humana Press, Clifton, NJ, 1983, pp. 1–50. With permission.)

filming box designed to capture their movements from various perspectives. Video records then were analyzed by a dance notation system adapted to describe animal motor behavior [94]. Such an approach, applied frame by frame, required considerable patience but revealed a spectrum of impairments that were not grossly obvious. A later study [273] used the technique to evaluate motor skills required to reach for food located on a shelf and to manipulate and eat pieces of pasta. Both the pyramidal tract and the red nucleus are also involved in skilled movements, but lesions in these sites suggested a greater role for the pyramidal tract in guiding limb movements.

Another technique designed to evaluate forelimb motor impairment was described by Schrimsher and Reier [221] in an evaluation of cervical spinal cord injury. Rats were trained in a special apparatus to reach into a recessed tray to retrieve a food pellet. Training occurred 2 weeks after body weight had stabilized on a food deprivation schedule. Videotape recordings provided the raw data for analysis. Hypometria turned out to be the primary disability, and it remained permanent in most of the rats. Note that automation of procedures such as the two just discussed, with the potential to yield more accessible and quantitative data, requires little additional training of the animals [80].

Another situation devised for testing skilled movement in rats is the staircase test [170], in which rats reach down from a central platform to retrieve food pellets located on each step of two adjacent staircases. The pellets are placed on the staircase and presented bilaterally at seven graded stages of reaching difficulty to provide measures of side bias, maximum forelimb extension, and grasping skill for each paw. Performance is measured by the number of food pellets obtained.

Wolthuis and Vanwersch [278] designed an ingenious procedure for recording and analyzing coordination deficits in rats. A traditional running wheel was modified with the addition of flanges spaced at 30° intervals around the interior surface. The wheel was driven at specified rates by a motor. The ability of rats to step from flange to flange as the wheel rotated was recorded on videotape and the analysis was performed by automated measures of the placement of the hind feet, which had been coded by color so appropriate software could be used to measure positions in the video image. With this system, the experimenters could demonstrate that organophosphate treatment produced coordination deficits not readily detectable by common early screening techniques such as the rotarod and hindlimb foot splay.

Running wheels also served Cohen and Gans [39] in studies aimed at providing a detailed analysis of rat locomotion, given that the motor activity of rats is frequently monitored to determine the effects of physiological and behavioral variables. Adult male rats were trained to run in an activity wheel. Photoflood lights provided the discriminative stimulus for running, and mild electrical stimuli delivered through an electrode provided a source of aversive reinforcement to encourage running. Electrodes implanted in forelimb muscles allowed the investigators to monitor muscle activity at the same time movement was being filmed. In an analysis of this functional morphology, running could be shown to represent a "complex and highly adaptable grouping of motor sequences." The rat's forelimb was shown to act as a steering, propulsive, and supportive device with complicated temporal relationships defining their joint actions. Studies such as these underscore the fact that simple locomotor activity, which has been adopted widely as a component of screening batteries for neurotoxicology, or assays such as hindlimb foot splay offer a misleading, perhaps even deceptive, simplicity and sometimes a lack of sensitivity with respect to motor function deficits.

## Tremor

Excessive limb tremor accompanies many neurological disorders, is a product of many drug treatments, and can result from exposure to many different classes of toxicants [81]. Tremor is associated with exposure to metals such as mercury and manganese; to insecticides such as chlordecone, dieldrin, and the organophosphates; and to solvents such as carbon disulfide. In a survey, Anger [2] noted tremor as a response to 177 chemicals or chemical classes. Newland [178] describes some of the characteristics of abnormal tremor associated with lesions or chemical exposures. Because of its pervasiveness as a marker of nervous system dysfunction, many techniques have been developed to measure tremor. The availability of inexpensive digital computers and appropriate programs makes the task of recording and analyzing tremor much simpler now than in the past. Because abnormal tremor is a marker for many different neurological syndromes and can be induced by damage at many sites in the nervous system, experimenters have to be wary of ascribing too much specificity to it.

Tremor is one of the cardinal signs of excessive mercury vapor exposure. Wood et al. [284] studied several women exposed to vapor in a factory workroom devoted to pipette calibration. When first seen clinically, the women exhibited visible tremor. After a prolonged absence from the factory, the pathological tremor faded. To follow the course of recovery, and especially to relate it to diminished blood mercury levels, a system was devised to quantify tremor using a strain gauge to which was attached a Lucite® slot in which the patient placed her finger. She was instructed to maintain a force within a range designated by lights to mark the upper (40 g) and lower (10 g) bounds. The output of the strain gauge was transmitted to the analog-to-digital converter of a digital computer for processing to express the continuously varying electrical signal corresponding to the output of the strain gauge in the form of tremor frequency. Because physiological tremor is composed of many different frequencies, the analysis is designed to yield their relative contributions to the total signal. The algorithm by which this analysis is performed is called a *fast Fourier transform*.

Fast Fourier transforms demonstrated two features of the tremor induced by mercury vapor exposure (Figure 37.10). First, as expected, the total amount of tremor was greater than any seen with normal subjects. Second, and unexpected, the distribution of the relative amounts attributable to different frequencies showed multiple modes, indicating a very complex signal whose component could not be seen clinically. Eventually, the amplitude of the tremor returned to normal levels, and at the same time the multiple modes collapsed into one dominant peak. Later, Langolf et al. [129] used a similar technique to monitor workers in a chloralkali processing plant who were exposed to mercury from the massive electrodes used in such plants. They found that as urine mercury rose, the distribution of tremor frequencies began to show multiple modes. With removal from exposure and a parallel decline in urinary concentrations of mercury, the secondary modes declined.

Several later studies have confirmed the usefulness of this method, called *power spectral analysis*, for monitoring workers exposed to mercury. Chapman et al. [37] compared battery workers exposed to mercury with controls by instructing them to maintain a steady force on a displacement transducer. All of the battery workers were judged asymptomatic after interviews and clinical examinations; however, differences in power spectra were able to separate exposed from control workers. One distinguishing feature consisted of peak frequencies in the spectrum. The exposed workers generally showed a displacement toward the higher frequencies. Chapman et al. [37] noted similar findings in workers exposed to carbon disulfide in the grain industry and concluded that frequency differences more effectively indicate neurotoxicity than amplitude differences. The investigators noted that such subtle changes in tremor characteristics are not apparent in clinical examinations and are seen at relatively low exposure levels.

One of the most easily recognized adverse effects of ethanol is incoordination. Newland and Weiss [180] arranged a situation to try to quantify some of the less overt consequences of ethanol consumption (Figure 37.11). They trained squirrel monkeys to grip a rod attached to the hub of a rotary transformer whose output voltage was proportional to angular displacement. The monkeys received juice reinforcement for maintaining the rod within 15° of hori-

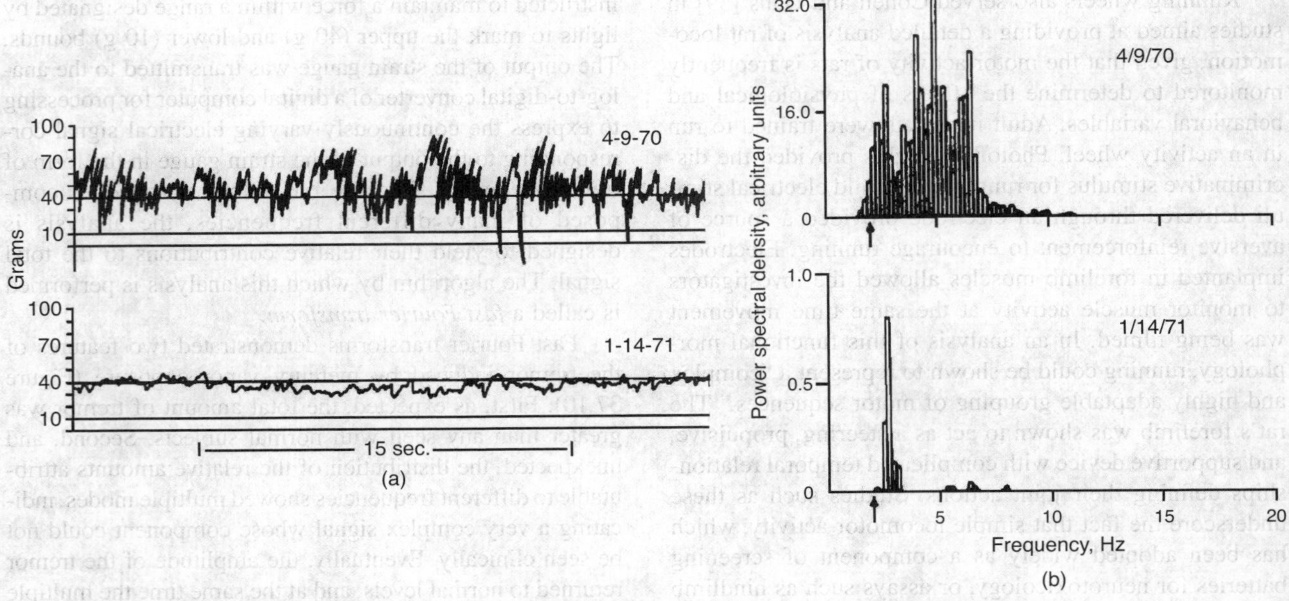


**FIGURE 37.10** (a) Tremor tracings from a female worker chronically exposed to mercury vapor in a factory area devoted to pipette calibration. The worker rested a forefinger in a Lucite® slot attached to a strain gauge and was requested to maintain a force between 10 and 40 g, as signaled by lights. Strain-gauge output was amplified and transmitted to a digital computer for analysis. The upper tracing was recorded shortly after the worker entered the hospital for treatment. The lower tracing was recorded 9 months later, with no intervening workplace exposure and a marked decrease of mercury blood levels. (b) Power spectral density plots of tremor corresponding to racings shown in A. These plots show the amount of total power (variance) contributed by each component frequency in the tremor spectrum and were calculated by fast Fourier analysis. They emphasize the feature of tremor, such as multiple modes, that cannot be evaluated by ordinary clinical examination. (From Wood, R.W. et al., *Arch. Environ. Health*, 26, 249–252, 1973. With permission.)

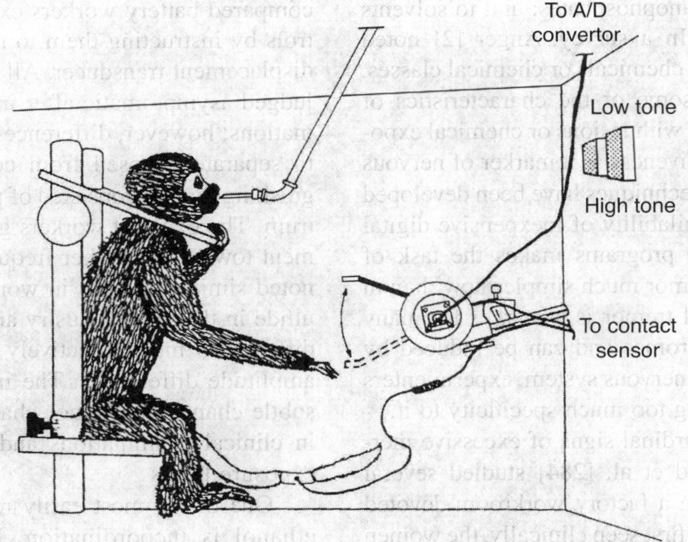


**FIGURE 37.11** Depiction of a squirrel monkey (*Saimiri sciurea*) responding on a tremor assessment apparatus. The bar was insulated everywhere except the handle to limit response topography. A contact sensor circuit between the bar and the monkey's tail closed when the bar was gripped. The position of the bar provided a continuous electrical signal that was transformed into a digital code by the analog–digital converter input of the computer that controlled the experimental contingencies. When the bar was held in range, a tone sounded. If held for a sufficient duration, a high-frequency tone burst sounded; on a random ration 2 schedule of reinforcement, a pulse of fruit juice was delivered to the monkey. (From Newland, M.C. and Weiss, B., *J. Stud. Alcohol*, 52, 492–499, 1991. With permission.)

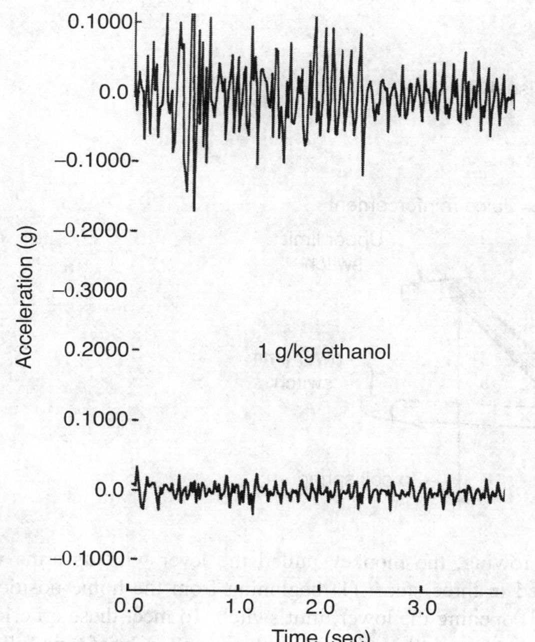


**FIGURE 37.12** Tracings of bar position (see Figure 37.11) after vehicle and after the oral administration of 1 g/kg ethanol to monkey SM834. Each tracing represents a 5.12-second sample of the second derivative of the bar position. The ordinate is expressed as *g* values, or acceleration due to gravity. Differentiation of the position signal emphasizes frequencies important for tremor measures and facilitates comparison with the literature, which typically records tremor with accelerometers. The contribution of slow or low-frequency undulations and position drift is accordingly reduced. Ethanol reduced peak-to-peak excursions in acceleration. (From Newland, M.C. and Weiss, B., *J. Stud. Alcohol*, 52, 492–499, 1991. With permission.)

zontal for 8 seconds. The output voltage was sampled for 5.12 seconds, and tremor was measured using power spectral analysis [178]. As Figure 37.12 shows, ethanol greatly reduced the amplitude of tremor, even at doses of 0.25 g/kg. It also tended to flatten the spectral distributions. Procedures such as those developed by Falk [76] and Fowler et al. [81] and described earlier can be used to obtain tremor indices in rats. The neuroleptic haloperidol, which can induce a pseudoparkinsonism syndrome in patients, also elevates forelimb tremor in rats. As the above examples illustrate, tremor measures may be sensitive to disturbances of motor function that are not clinically apparent; however, as Beuter and de Geoffroy [19] emphasize, the data yielded by tremor measurement techniques depend on the apparatus, the procedure, the data analysis techniques, and other variables, and all should be weighed when comparing different reports.

Simple tests of motor function almost always are included in screening batteries for humans. One of the most common is finger-tapping rate. Subjects or patients are asked to tap a key or button as rapidly as possible during a test period of several seconds. Both the preferred and alternate hand are usually tested. Chaffin and Miller [36] included both finger and toe tapping in a test battery applied to chloralkali workers exposed to metallic mercury and were able to show that performance correlated with exposure. A more complex version of the tapping task, developed originally by Fitts [79], was included as part of a test battery by Maizlish et al. [146]. In this version, the subjects tapped between two copper plates with a stylus connected to a timer so the intervals between taps could be recorded. In addition, variations in difficulty were introduced by changing the size of or distance between the plates. With a similar technique, Sanes et al. [215] found slower rates in patients with Parkinson's disease than in controls.

Tests developed originally for assessing manual dexterity in workers also have been adopted for human behavioral toxicology studies. The Santa Ana test requires the subject to remove pegs from holes in a board, turn them 180°, and reinsert them. The number rotated correctly within a specified time is taken as the score. It is similar to a test known as the *grooved pegboard*. Hanninen [103] first used the test in a study of workers exposed to carbon disulfide and then included it in the screening battery developed at the Finnish National Institute of Occupational Health. The Purdue Pegboard, also developed to assess manual dexterity in factory workers, requires subjects to place pins in a series of holes, but they can then be asked to place collars or washers on the pins. Baker et al. [13] included it in a test battery and note that it is one of the tests deemed acceptable for occupational studies by a World Health Organization (WHO)–National Institute for Occupational Safety and Health (NIOSH) expert committee.

Tests borrowed from the experimental psychology laboratory are frequently adopted by investigators. Studies of manganese-exposed workers by Roels et al. [210] tried to assess tremor by using a device that required the subject to hold a stylus in a hole without touching the sides. Small-diameter holes offer a greater challenge than large-diameter holes. Each contact is recorded as the completion of an electrical circuit. Workers with higher urinary values of manganese produced a greater incidence of contacts. A much more extensive test battery was applied to another population of workers exposed to manganese [113]; these data also indicate subclinical deficits in such populations. Both of these studies suggest that the methods used by Newland and Weiss [179] to study manganese neurotoxicity in monkeys might be directed to human assessment.

## Strength

Complaints of weakness appear after exposure to acetylcholinesterase inhibitors, manganese, and other agents. In fact, weakness is one of the most frequent subjective indices of neurotoxicity. Although simple procedures for assessing strength in rodents, such as forcing them to pull against a spring (grip strength), have been devised, com-

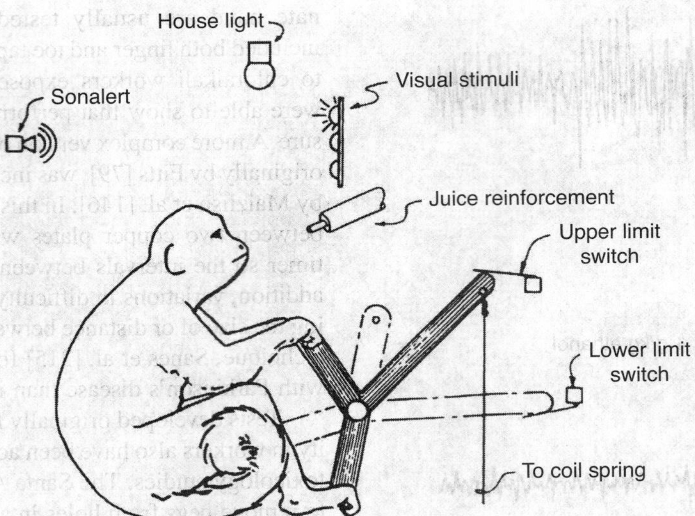


**FIGURE 37.13** Apparatus for testing strength and endurance. As in rowing, the monkey pulled the lever with its arms while simultaneously thrusting with its legs. A complete response was defined in three steps: (1) beginning from the home position by closing the lower limit switch, (2) closing the upper limit switch, and (3) opening the lower limit switch. To meet these criteria, the monkey had to move the response device through an arc length of 10 cm against a 40-N (4-kg) circular spring. A brief tone followed each complete response. The visual stimulus panel indicated the schedule component in effect. (From Newland, M.C. and Weiss, B., *Pharmacol. Biochem. Behav.*, 36, 381–387, 1990. With permission.)

plex learned performance offers more direct answers to the kinds of questions likely to reflect human complaints or observations, such as the relationship between blood pressure and effort and greater sensitivity to subclinical effects [61]. With an apparatus similar to a rowing machine, requiring simultaneous applications of force by the legs as well as the arms (Figure 37.13), Newland and Weiss [182] found that low doses of *d*-amphetamine and L-DOPA reduced rates when the behavior was maintained by multiple fixed-ratio (FR)/fixed-interval (FI) schedules. Domperiodone, a peripheral dopamine-blocking agent, counteracted the effects of L-DOPA, suggesting that at least some of these effects arose peripherally. These techniques are readily adaptable to questions about complaints about weakness and fatigue arising from other agents and, because they are based on apparatus developed for human physical conditioning, could play a role in human toxic assessments as well.

Lead at high doses, such as observed in painters exposed to lead pigments, produces neuropathies such as the syndrome of wrist drop due to radial nerve damage. Newland et al. [184] showed that prenatal lead exposure, at levels (21 to 760 μg/dL) insufficient to induce overt motor dysfunction, can cause deficiencies in strength. Squirrel monkey subjects, at 3 to 7 years of age, were trained to pull a T-shaped bar against a 1-kg spring through a distance of 1 cm. On FR schedules, which tend to evoke high rates of responding, the lead-exposed monkeys showed a higher incidence of incomplete responses than control monkeys. The authors concluded that *in utero* lead exposure at these levels produces subtle impairments in motor function detectable years after birth.

Studies such as those just described indicate how corresponding questions in toxicology may be approached. In its advanced stages, manganese intoxication results in signs sometimes interpreted as Parkinsonism but, as argued by Barbeau [14], more closely correspond to dystonia. In its earlier stages, feelings of weakness and excessive fatigue predominate. As with many questions of neurotoxicity, tracking its progression provides the key to identification of underlying neural substrates and appropriate measures on which to base quantitative risk estimates. The exercise device described earlier [182] offers such a means to track progression [181]. After training, monkeys were followed during a sequence of manganese or vehicle treatments. Performance measures included number of missed or incomplete responses, response durations, and interresponse times (IRTs). Figure 37.14, depicting a history of more than 400 days, shows the effect on one index: number of misses, or failures to pull through the complete required arc. After the first manganese treatment (10 mg/kg i.v.), the frequency of misses on the FR component jumped sharply. Although this index occasionally drifted downward, it never returned to a pretreatment baseline. Additional manganese treatments produced further variability. Overt signs, such as tremor and dystonias, appeared late in the course of the treatment. Even with the cumulative dosing regimen adopted for this experiment, total exposure amounted to a small fraction of the doses reported by Suzuki et al. [234] to induce overt signs of manganese toxicity.

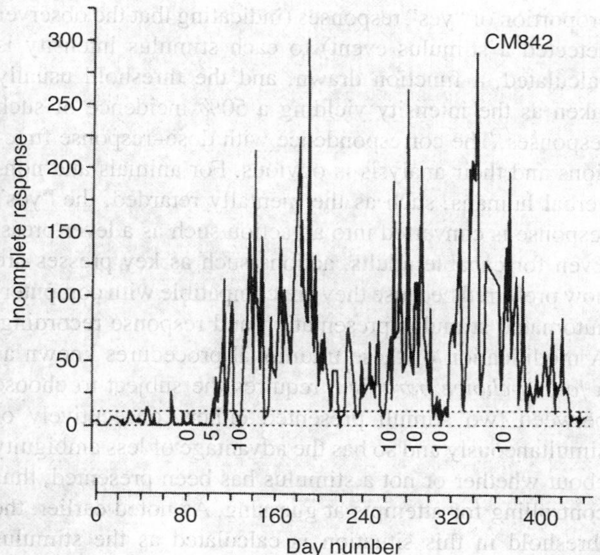


**FIGURE 37.14** Performance of a Cebus monkey on a device shown in Figure 37.35 during a course of manganese treatments. Incomplete responses designate those defaulting on the criteria listed in Figure 37.35, typically because of failure to operate the upper limit switch. The performance charted was maintained by the fixed-ratio (FR) 20 component of the FR fixed-interval multiple schedule of reinforcement. Manganese chloride was administered intravenously in doses of 5 or 10 mg/kg at times indicated on the baseline; saline infusions are represented by zeros. Dotted lines show the 5th and 95th percentiles calculated from baseline sessions. (From Newland, M.C. and Weiss, B., *Toxicol. Appl. Pharmacol*, 113, 87–97, 1992. With permission.)

## SENSORY FUNCTION

Almost all behavioral processes depend ultimately on an organism receiving information from its environment. Many behavioral deficits can be traced to disturbances in the way this information is received or processed. Moreover, sensory systems seem to be special targets of certain agents. Acrylamide and methylmercury toxicity are characterized during their earliest stages by loss of sensitivity to touch. Both agents also damage the visual system, although by different mechanisms. A large number of agents, in fact, impair visual function, sometimes so subtly that human victims may be unaware of the deficit. Individuals with defective color vision or areas of scotoma often are detected only on clinical examination. Hearing is degraded by toluene, lead, methylmercury, salicylates, certain antibiotics and diuretics, and noise. Cadmium exposure and chronic solvent exposure have been reported to impair the sense of smell. Behavioral assessments of sensory function can be especially useful because they provide an integrated evaluation of the entire system, starting at the receptors and progressing through the intermediate to the final processing stages of the central nervous system. Furthermore, it may prove misleading to rely mainly on histopathology as the primary index of sensory system toxicity. The intactness of the visual system, for example, cannot be judged simply by examining the retina with a fundascope or by the gross histological appearance of the visual pathways. The ability to discriminate colors requires a verbal response in humans and some form of motor response in trained animals.

### PSYCHOPHYSICS

Psychophysics is a branch of psychology that studies the relationships between sensory stimuli and behavior [90]. One of its central themes is the concept of a sensory threshold, or the limits of sensitivity for a particular sensory modality, defined as the minimum energy necessary to produce a sensation. Even with the trained observers who typically served as subjects in such experiments in early years, variation from trial to trial in sensitivity was apparent, leading to the development of procedures to cope with these fluctuations and to obtain precise estimates of thresholds. These methods have been refined and expanded since their introduction.

A crucial ingredient of sensory testing is precise specification of the stimulus and its properties. Such descriptions are often lacking in much of behavioral toxicology, as in the reaction-time literature discussed later in this chapter. Light and sound stimuli are used without any apparent reports of their parameters, such as brightness and loudness, despite evidence, extending to the beginning of the century, that reaction time latencies are related directly to stimulus amplitude. Many of the devices marketed for sensory testing are also deficient; rather than providing direct measures of stimulus qualities such as vibratory stimulus amplitude, they simply use a dial setting on a potentiometer, say, for specification.

The behavioral measurement of sensory function in laboratory animals also embraces a long history and was one of the earliest topics addressed by experimental psychologists and physiologists, including Pavlov. The development of operant behavior technology contributed major advances to the precision with which sensory function could be determined and to a marked elevation in the status of animal psychophysics. As Stebbins [228] noted, "Training and testing procedures based on the principles of operant conditioning have shown that animals can report on their sensory capabilities in as precise and reliable a fashion as humans." The basic technique in animal psychophysics is to reinforce responses to specified physical characteristics of the stimulus or stimuli. These stimuli serve no eliciting role. They exert control over behavior because they are associated with a particular history of reinforcement. This feature is of special relevance for extrapolation to humans, as many measurement techniques are applied to animals and humans with equal facility.

**TABLE 37.3**
**Basic Psychophysical Methods**

| | |
|---|---|
| Method of limits: | Series of stimulus intensities ascend and descend from above and below thresholds, respectively. |
| Method of constant stimuli: | Equally spaced stimulus values are presented in random sequence. |
| Method of adjustment: | Stimulus intensity is varied by the observer to exceed and dip below detection limits. |
| Adaptive methods: | Stimulus intensity rises with failure of detection and falls with correct detection. |

Two principles are pertinent to the current chapter. First, all psychophysical experiments whether conducted with animals or humans, are based on a discriminative response by the subject; lever presses and verbal responses happen to be functionally equivalent. Second, a threshold is a statistical concept; it does not imply an absolute physical limit, and its value depends on the methods used to estimate it.

## Psychophysical Methods

In most instances, the aim of psychophysical testing is to derive a threshold. During the extensive history of psychophysics, many procedures were developed for the collection of data (Table 37.3 and Table 37.4). Most of these have been effectively translated into procedures that can be used with animals. As noted above, thresholds are not absolute values but statistical functions of the techniques used to derive them. For example, if the subject is called on to designate the presence or absence of a stimulus, the threshold is typically defined as the stimulus value at which it is detected on 50% of the trials. If the subject is required to discriminate a variable stimulus from a standard, the threshold is typically defined as the magnitude of the variable stimulus that can be differentiated from the standard on 75% of the presentations.

The traditional method of constant stimuli presents the observer with between five and nine different stimulus values, each tested repeatedly in a random sequence. The proportion of "yes" responses (indicating that the observer detected a stimulus event) to each stimulus intensity is calculated, a function drawn, and the threshold usually taken as the intensity yielding a 50% incidence of such responses. The correspondence with dose–response functions and their analysis is obvious. For animals and nonverbal humans, such as the mentally retarded, the "yes" response is converted into an action such as a lever press. Even for capable adults, actions such as key presses are now preferred because they are compatible with computer-automated stimulus presentation and response recording. A modification of these traditional procedures known as a *forced-choice paradigm* requires the subject to choose between two stimuli presented either consecutively or simultaneously and so has the advantage of less ambiguity about whether or not a stimulus has been presented, thus controlling for attempts at guessing. As noted earlier, the threshold in this situation is calculated as the stimulus magnitude corresponding to 75% correct detections, because 50% incidence corresponds to chance. In the method of limits, the observer is presented with a series of stimulus intensities that begin either well above or well below the presumed threshold. The limits of detection are approached in steps of either diminishing or increasing stimulus values. With alternating descending and ascending series of stimulus values, the transition intensities between "yes" and "no" responses in both directions become the value from which a threshold value is computed. An ascending series, for example, begins with a stimulus intensity well below the limits of detection. After each trial on which the observer fails to report detection, the intensity of the next stimulus is raised by a prescribed amount. Once the observer reports detection, the series is ended and the intensity recorded. A descending series is conducted in the same way, except that it begins with a stimulus intensity well above the detection limit. Thresholds are typically calculated as the average value or midpoint at which a transition between detection success and failure occurs.

**TABLE 37.4**
**Response Requirements for Psychophysical Assays**

| | |
|---|---|
| Forced-choice procedures: | Subject must choose among alternatives. |
| Yes–no procedures: | Subject responds if stimulus is detected, refrains if it is not. |
| Rating procedures: | Subject reports likelihood of presence or absence using rating scale. |

In one variation of the method of limits, called variously the *up-and-down*, *staircase*, or *titration* method, stimulus intensity is modulated in accordance with the responses of the subject, so the threshold can be tracked continuously. A correct detection lowers the next stimulus amplitude, and an incorrect response raises it (Figure 37.15). This is the most efficient system for calculating thresholds and was a clever variation introduced into audiometry by von Békesy [250]. In the latter procedure, intensity rose or fell continuously. As the subject held down a switch, intensity rose; when it was released at the point where the subject could detect the stimulus, it reversed in direction. In this way, as intensity rose and fell, auditory thresholds for different frequencies could be traced continuously by a graphic recorder. For both human and animal testing, such continuous responses typically are

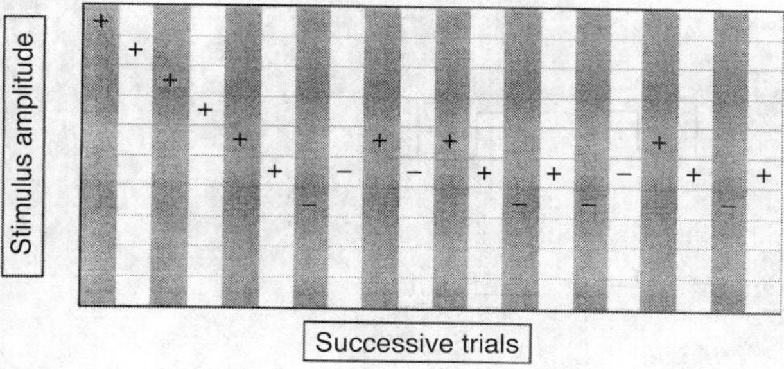


**FIGURE 37.15** Depiction of up–down (titration) psychophysical procedure. Stimulus amplitude typically begins at an intensity above the detection limit. Each failure (–) increases the intensity setting (stimulus amplitude) for the next trial or stimulus presentation.

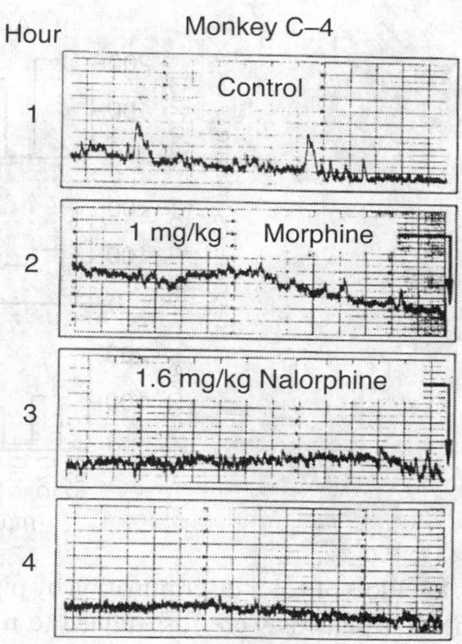


**FIGURE 37.16** Panel 1 shows a tracing of current amplitudes (aversive thresholds) maintained by monkeys. The current was delivered to the soles of the feet by electrodes. Every 2 seconds the current rose by a small increment. Each lever press by the monkey produced an equivalent decrement in the current. A steady tracing then reflects the level of current tolerated by the monkey. Panel 2 shows the effects of 1 mg/kg morphine (time runs from left to right); the aversive threshold rose markedly. After the administration of 1.6 mg/kg nalorphine, an opiate antagonist, the amplitude of the aversive threshold was reduced to control levels, indicating that nalorphine antagonized the analgesic properties of morphine. (From Weiss, B. and Laties, V.G., in *Animal Psychophysics: The Design and Conduct of Sensory Experiments*, Stebbins, W.C., Ed., Appleton-Century-Crofts, New York, 1970, pp. 185–210. With permission.)

replaced by discrete actions such as key presses, a feature that makes it easily adapted for animal and computer-based testing. Computer control also allows the use of parameter estimation by sequential testing (PEST) procedures; these vary the sizes of the steps in up–down methods so the threshold can be tracked to increasingly narrow boundaries.

Because titration procedures are designed so the amplitude of the stimulus is governed by the behavior of the subject, they proved especially suitable for studies of analgesia [266], providing a means for tracing the impact over time of drugs such as morphine and aspirin on thresholds of sensitivity to shock delivery (Figure 37.16). Alternative techniques would have required suprathreshold aversive stimuli to be applied to the animals, with inevitable behavioral disruption and unnecessary pain.

The theory of signal detection [236], first proposed in the 1950s, conceives of the situation as one in which the task of the observer is to distinguish or detect the presence of a signal or stimulus against a background of random activity or noise. Techniques derived from the theory permit the experimenter to take into account variables such as guessing habits or biases on the part of the observer that lead to false detections. Unlike classical psychophysics, these techniques can make explicit the consequences of correct detections and false reports, such as in the form of monetary gains and losses.

In all psychophysical experiments, precise specification of stimulus attributes is essential [148,152], a facet that may require complex equipment and procedures and thus partly explains why psychophysics remains a relatively neglected component of toxicology. Animal psychophysics introduces further complications, deriving mostly from the extended training sometimes required to extract precise and reliable data. Such investments in time and apparatus are worthwhile when the results are coupled closely to problems of human toxic exposure. Circumstances in which animal and human data require reconciliation are likely to generate questions that mandate the application of complex procedures for sensory assessment. For this reason, as in the other sections, the more advanced techniques are featured. Numerous descriptions of simpler approaches, such as reflex measures [53], have already appeared in the toxicology literature, whereas the more complex procedures and the contexts in which they confer special advantages are less abundant.

## Specific Sensory Systems

### Vision

Vision is the sensory system that has received the most attention from toxicology, where it is represented by at least two books [99,163] and chapters in textbooks [194]. Toxic reactions can take place at many different sites in the visual system. Grant [99] has compiled a list of over 2800 agents that can induce visual system toxicity. Corrosive chemicals can damage the cornea, certain drugs can induce cataract formation in the iris, many different chemicals can damage the retina, and even that part of the cerebral cortex subserving vision can be a toxic target. Vision is unique in the way that information is transmitted

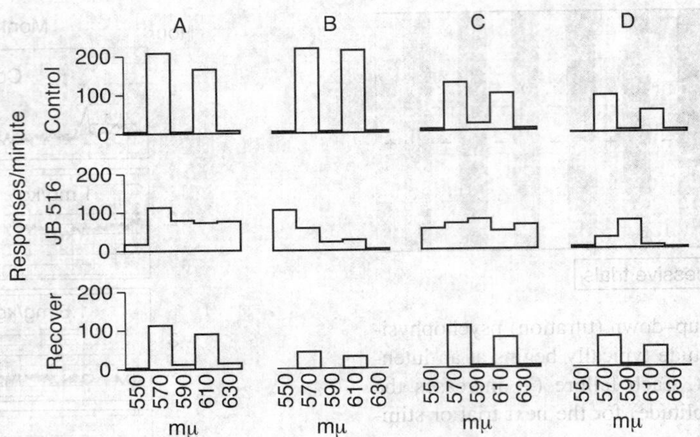


**FIGURE 37.17** Plots of color discrimination by pigeons after chronic administration of the monoamine oxidase inhibitor pheniprazine (JR 516). The four pigeons were first trained to respond differentially to wavelengths of 570 and 610 nm, as shown in the top row by rates of pecking (*y*-axis) at keys illuminated by different colors under control conditions (*x*-axis). Pheniprazine treatment (middle row) gradually obliterated the discrimination, and the pigeons pecked at the keys regardless of their color. Following drug withdrawal (bottom row), discrimination performance tended to recover. (From Hanson, H.M. et al., *Toxicol. Appl. Pharmacol.*, 6, 690–695, 1964. With permission.)

along the system. The representation of the visual field on the retina, which contains the light receptors and constitutes what has been labeled a retinotopic map, preserves an analogous spatial distribution in the neuronal pathways ascending from the periphery to the final cortical map. Although ancillary influences may act on these pathways and the distribution of information in the cortex involves secondary projections, parallel and topographic segregation in the primary projection areas is preserved with remarkable fidelity. This feature of the visual system makes testing simple and complicated at the same time.

Color sensitivity, for example, is localized to the center of the visual field (the fovea), in receptor elements called *cones*, which are coded chemically to respond to light of different wavelengths. Deficits in color discrimination arise when the function of these receptors is impaired. Humans, however, are often unaware of such deficits because they learn to compensate by relying on brightness or other contextual cues, which is why specialized tests are used to detect color blindness. An illustration of how easily color discrimination deficits may be masked was provided by the experience with the monoamine oxidase inhibitor pheniprazine, which had been prescribed to treat depression. The treated patients developed red–green color blindness, a toxic effect of which neither the patients nor their physicians were aware. Only later, when normal visual function could be recovered in a relatively small fraction of the patients, did clinical investigators learn of this problem. Once they did, it was possible to demonstrate a corresponding effect in pigeons, whose color discriminative capacity is close to that of humans.

Hanson et al. [104] trained pigeons to peck a translucent disk on which different colors could be projected. The birds were reinforced with food for pecking the disk when it was illuminated by green and orange stimuli but not by blue, yellow, and red stimuli. Discriminative responding quickly appeared; few responses were made in the presence of the negative stimuli. After about a month of daily administration of pheniprazine, however, discriminative capacity was lost, with pigeons responding equally to all stimuli. (The experimenters took care to match brightness so it could not be used as a cue.) When treatment with pheniprazine ceased, discriminative ability gradually recovered (Figure 37.17). This fairly simple discrimination procedure could have been used to preclude a serious, irreversible effect in humans had it been implemented.

Color vision deficits are recognized as one outcome of occupational exposure to organic solvents. Earlier, Raitta et al. [198] had noted such deficits among workers in the viscose rayon industry, where they are exposed to carbon disulfide, by using the Farnsworth–Munsell 100 Hue Panel, which requires subjects to arrange a series of 85 reference caps representing incremental changes in hue. Color vision function is based on the ability of the subject to place the color caps in order of hue. Modifications of the 100 Hue Test include the Farnsworth–Munsell Dichotomous D-15 Color Test, requiring the subject to arrange 15 numbered disks with different hues, and the Lanthony D-15d desaturated hue panel, claimed to be more sensitive. Geller and Hudnell [87] have noted the high incidence of errors sometimes made by control subjects and suggested revised test protocols to improve its diagnostic value.

Numerous studies, based on such techniques, have now been published in support of the claim that workplace organic solvent exposure impairs color vision. They implicate styrene, toluene, tetrachlorethylene, xylene, methyl

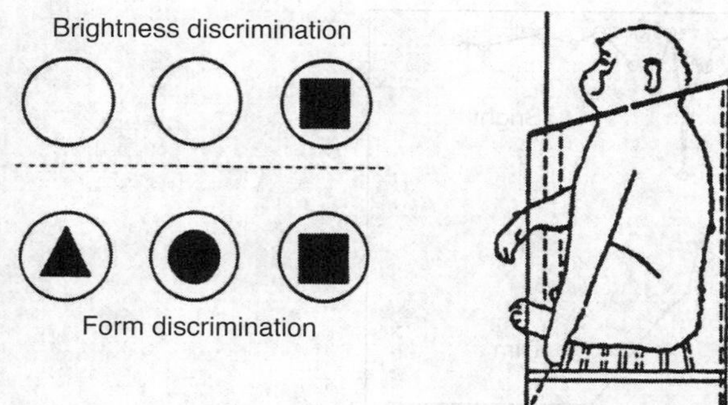


**FIGURE 37.18** Testing geometric form discrimination in monkeys exposed to methylmercury. The monkey faced a panel of three disks. When examined for the ability to distinguish shape, the monkey was required to press the disk on which the square was projected. The position of the square varied randomly from trial to trial. When examined for its ability to perform simple brightness discrimination, the monkey was required simply to indicate which disk contained the square. (On the figure, the light and dark areas are reversed for ease of presentation.) Correct responses were reinforced by pulses of fruit juice. (Adapted from Evans, H.L., in *Proceedings of the International Conference on Heavy Metals in the Environment*, Vol. 3, Hutchinson, T.C., Ed., University of Toronto Institute of Environmental Studies, Toronto, 1978, pp. 241–256; Evans, H.L. et al., *Fed. Proc.*, 34, 1858–1867, 1975.)

ethyl ketone, and others [32,93,160]. A common finding in these studies is blue–yellow confusion, but red–green confusion seems to be a more specific marker for solvents.

Color vision deficits due to solvents are likely due to retinal damage or dysfunction. Toxicants that damage visual pathways at upstream sites in the central nervous system create other kinds of deficits that depend on the site of damage. Methylmercury is a potent central nervous system poison that in primates, including humans, tends to be most lethal to nerve cells buried deep in the folds of the cortex. The peripheral projections of the visual fields lie within the medial portions of the occipital cortex along the calcarine fissure, which is a principal site of damage induced by methylmercury. Humans who have undergone serious exposure show constriction of the visual fields, sometimes progressing to severe tunnel vision and, occasionally, blindness. Korogi et al. [126] found, by magnetic resonance imaging, considerable atrophy in several brain regions, including the calcarine fissure, of Japanese victims of methylmercury poisoning. Evans et al. [72,74] traced this progression in monkeys and related it to exposure and tissue levels of methylmercury (Figure 37.18). Monkeys were first accustomed to perching in a primate test chair. During testing, they faced a panel containing three Lucite® disks illuminated from behind by three geometric forms—a square, a circle, and a triangle. Pressing any one of the disks closed a circuit and allowed the experimenter to record the source of the response and to arrange certain consequences as a result. In this experiment, the monkeys were reinforced with a small amount of fruit juice for pressing the key with the square.

The reasons for the choice of a geometric form discrimination to trace the progression of methylmercury toxicity illustrate how animal psychophysics is applied to toxicological questions. If methylmercury preferentially damages cortical cells receiving projections from the peripheral areas of the visual fields, then visual discriminations at low luminances should be differentially impaired. The periphery is represented in the retina by visual elements called *rods*, which are sensitive to low light levels. The cones in the center of the field, which are responsible for color vision and for fine acuity, function at high light levels. To detect a differential effect in the central and peripheral visual fields, the forms were illuminated on different occasions with a range of luminance values, the lowest of which made the forms visible only after the monkey had remained in the dark for at least 10 minutes so the rods had become adapted. Monkeys treated chronically with methylmercury began to show deficits in visual function revealed earliest by diminished accuracy on the form discrimination at the lowest luminances. Only much later did damage progress far enough in the cortex to produce deficits in discrimination at the higher luminances (Figure 37.19).

The primary health concerns aroused by methylmercury arise from its effects on brain development. Because visual deficits are such a conspicuous feature of adult poisoning, Rice and Gilbert [204] characterized visual function in two groups of monkeys (*Macaca fascicularis*) exposed developmentally to methylmercury. One group was dosed from birth onward with 50 μg/kg/day. A second group was exposed *in utero* by dosing the mother with 10, 25, or 50 μg/kg/day and was then exposed postnatally until 4.0 to 4.5 years of age with the same dose the mother had received. The authors based their assessment on what is now conceived by vision scientists to be the principles on which the visual system functions. They view the visual system as basically a frequency analyzer, responding to variations in both time and space. Variations in time are exemplified by flickering light sources. Spatial variations are illustrated by almost all natural scenes containing different textures. Low spatial frequencies are represented by objects with relatively broad features, such as faces. High spatial frequencies correspond to fine details, such as the print on the page. Figure 37.20 shows the kinds of variations in a visual display defined by spatial frequency. Spatial and temporal visual function was tested in both groups.

For spatial testing, subjects viewed a display composed of gratings, or alternating light and dark bars. These bars were not sharply defined stripes but varied sinusoidally so the highest (lightest) luminance in the modulated signal represented the peak and the lowest (darkest) luminance the trough. In correspondence with

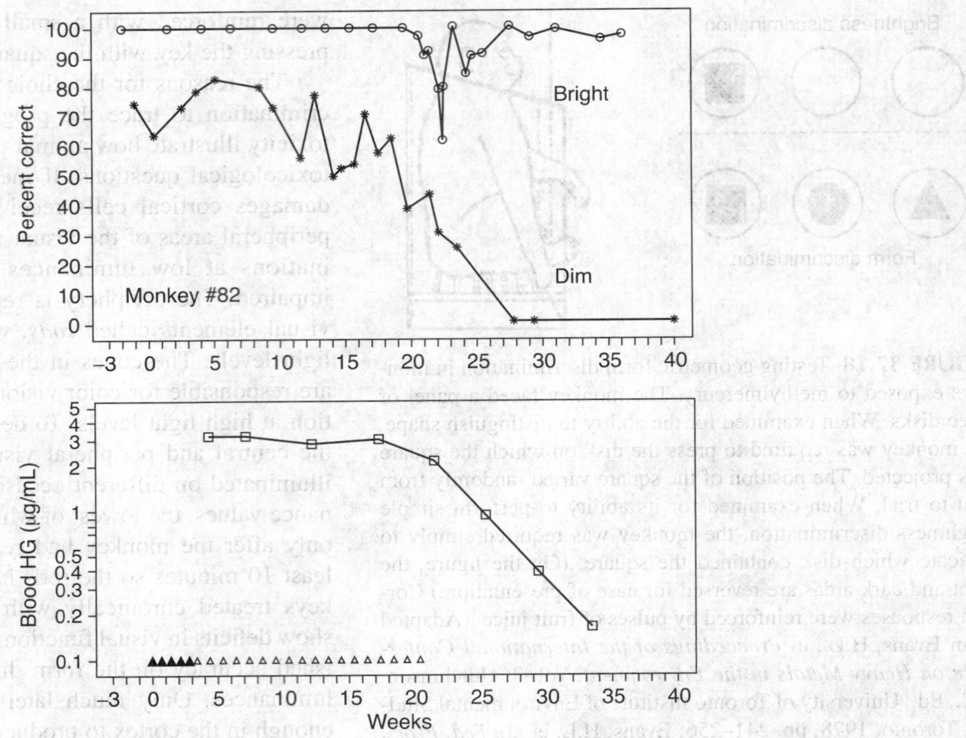


**FIGURE 37.19** Form accuracy (top plot) of a monkey (*Macaca speciosa*) treated with methylmercury (see Figure 37.18). The methylmercury was administered in doses of 0.5 mg/kg on occasions marked by small triangles on the tower graph Resulting blood levels are traced by the connected squares (bottom plot). Performance under dim (scotopic) target luminances deteriorated before performance under bright (photopic) tuminances and eventually reached zero, where it remained even after treatment ceased. Note the sharp fall in photopic accuracy beginning at about week 20. At that time, methylmercury administration was discontinued and was followed by recovery, but performances at scotopic luminances continued to deteriorate, suggesting that sensitive assays might help identify toxicity at a stage at which damage is still reversible. (Adapted from Evans, H.L., in *Proceedings of the International Conference on Heavy Metals in the Environment*, Vol. 3, Hutchinson, T.C., Ed., University of Toronto Institute of Environmental Studies, Toronto, 1978, pp. 241–256; Evans, H.L. et al., *Fed. Proc.*, 34, 1858–1867, 1975.)

**FIGURE 37.20** Spatial contrast model. Spatial frequency decreases from top to bottom, and contrast decreases from right to left. A conventional contrast sensitivity plot would be based on the display of a grating with a specified visual angle and would indicate the contrast level at which the grating appears uniform.

temporal stimuli, the variables comprise spatial frequency (bar width), mean luminance, and contrast between peak and trough. High spatial frequencies are represented by narrow bars, which require intact visual acuity, and low frequencies by wide bars.

To access temporal acuity, the subject is asked to indicate whether a target, such as an oscilloscope screen, is flickering or steady. His or her ability to do so is measured by raising the frequency of flicker to a rate at which it no longer appears to vary. The current approach varies both depth of modulation, or the difference in luminance between the brightest and dimmest extremes of the light source, and the mean luminance around which these values fluctuate sinusoidally [162]. Some neuropsychological test batteries [110] continue to rely on the old method, designed to provide a measure called *critical flicker frequency* (CFF), which typically flashes a bright light source on and off at different rates. The CFF is defined as that frequency at which the subject reports a shift from a flickering to a steady stimulus.

Earlier, Rice and Gilbert [202] had reported spatial contrast sensitivity deficits in the group dosed only postnatally. Monkeys exposed both *in utero* and postnatally exhibited deficits in both high- and low-luminance spatial sensitivity. They also exhibited deficits in low-frequency, high-luminance temporal discriminations (flickering light), while low-luminance temporal vision was superior to that of control monkeys. Those monkeys exposed from birth displayed superior low-luminance temporal vision and no impairment of high-luminance temporal vision. Constriction of the visual fields was not detected. The authors suggest that the pattern of visual deficits produced by methylmercury exposure during development differs from that seen in the adult and speculate that the developing visual system may be able to remodel in response to early damage by a neurotoxic agent.

Similar approaches indicate that the visual system is subject to damage from the axonopathic agent acrylamide. Until the report by Merigan et al. [164], it had been assumed that acrylamide neurotoxicity, expressed predominantly as a central–peripheral distal axonopathy, was largely reversible. Moreover, none of the previous publications, including those reporting cases of human poisonings, had mentioned visual deficits. Merigan et al. [164–166], in a series of experiments with acrylamide, trained monkeys to choose between two targets presented on two oscilloscopes. The monkeys sat in a special test stand positioned before a response panel that supported two pushbuttons and a spout for juice delivery. Before a test trial began, both screens were illuminated evenly. At the onset of a trial, a tone sounded and the test stimulus appeared on one of the oscilloscopes. For acuity testing, the test stimulus was a vertical grating. For tests of temporal resolution, it was a flickering screen. If the monkey pressed the pushbutton corresponding to the position of the test screen (left or right), it received a juice reward. Each response terminated the trial. An adjusting or titration procedure governed the stimulus parameters. Correct responses made the gratings finer on the next trial during acuity testing and raised flicker rate during flicker fusion testing. Incorrect responses drove the stimulus values in the other direction. The positions of the variable and steady targets shifted randomly from trial to trial and stimulus position and characteristics were controlled by a digital computer.

Treatment with acrylamide continued until the appearance of overt toxic signs, then it was stopped and recovery monitored. All measures of function recovered except for visual acuity, defined as the ability to resolve gratings at the highest contrast (Figure 37.21) which evidenced only partial recovery. A subsequent series of histological studies [69,70,143] demonstrated that the source of this deficit was the destruction of a class of cells in the retina that project to a particular site in the lateral geniculate nucleus of the midbrain, which in turn projects to the visual areas

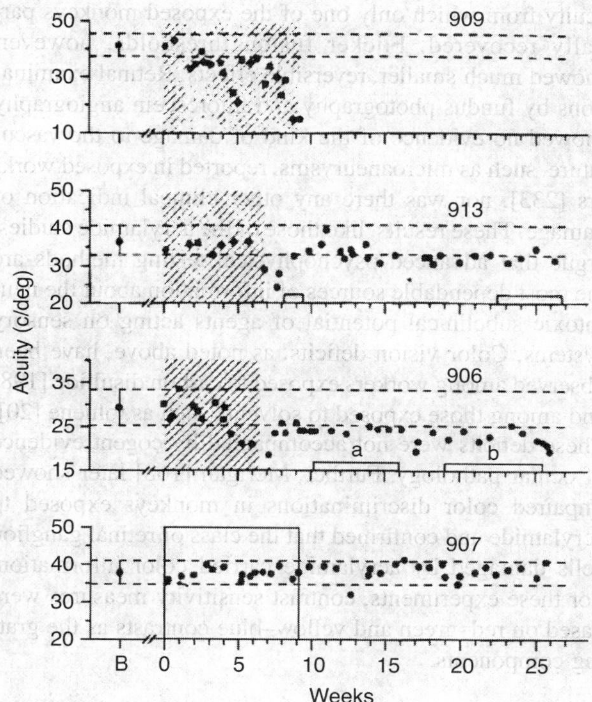


**FIGURE 37.21** Visual acuity plots of four monkeys (*Macaca nemestrina*) dosed with acrylamide. The monkeys faced two high-resolution oscilloscopes during testing. One displayed a grating and the other a uniform field, according to a randomly chosen sequence. For expressing acuity, grating contrast, which describes the differences between the brightest and dimmest areas of the vertical grating pattern, remained constant at 0 55. Spatial frequency, based on the width of light and dark bars of the grating and expressed as cycles per degree of visual angle, was varied in steps according to an up–down procedure to obtain a threshold for the monkey's ability to distinguish a grating from a uniform display. Hatched areas show the period of acrylamide treatment. Acrylamide (10 mg/kg) was administered five times weekly until ataxia appeared. Monkey 907 received only sham treatment. Part B shows the mean of eight baseline sessions (±2 SD), which was extended across the duration of the experiment (dashed tines). The intervals marked *a* and *b* for monkeys 913 and 906 designate collection of contrast sensitivity measures. Monkey 909 was sacrificed at the end of dosing for neuropathology. Monkeys 913 and 906, despite some recovery, failed to reach predosing acuity. (From Merigan, W.H. et al., *Invest. Ophthalmol. Vis. Sci.*, 26, 309–316, 1985. With permission.)

of the cerebral cortex. The degree of persisting functional impairment noted in some of these studies, however, might go unnoticed by many, if not most, people, just like color blindness.

Merigan et al. [167] and Eskin et al. [71] adopted a similar approach to study the visual toxicity of carbon disulfide. Monkeys were exposed in inhalation chambers to 256 ppm for 6 hours daily, 5 times each week, for 7 weeks. The visual acuity thresholds of the two exposed monkeys indicated severe functional losses after about 5 weeks. Further testing revealed a 7- to 10-fold loss of acuity from which only one of the exposed monkeys partially recovered. Flicker fusion thresholds, however, showed much smaller, reversible effects. Retinal examinations by fundus photography and fluorescein angiography showed no evidence of the kind of damage to the vasculature, such as microaneurysms, reported in exposed workers [233], nor was there any other clinical indication of damage. These results, like those of the acrylamide studies, argue that advanced psychophysical testing methods are the most dependable sources of information about the neurotoxic subclinical potential of agents acting on sensory systems. Color vision deficits, as noted above, have been observed among workers exposed to carbon disulfide [198] and among those exposed to solvents such as toluene [20]. These deficits were not accompanied by cogent evidence of ocular pathology. Further, Merigan [168] later showed impaired color discriminations in monkeys exposed to acrylamide and confirmed that the class of retinal ganglion cells damaged by acrylamide carried color information. For these experiments, contrast sensitivity measures were based on red–green and yellow–blue contrasts as the grating components.

## Hearing

The most frequent cause of hearing loss after aging is exposure to excessive noise. Several classes of drugs also impair hearing. Some (e.g., salicylates) produce transient effects such as tinnitus. Others, such as the aminoglycoside antibiotics (streptomycin and kanamycin), damage the hair cells in the cochlea, where the mechanical movements of sound are transformed into nerve impulses. Loop diuretics and quinine, especially with prolonged administration, can cause similar damage. It is now standard practice for such classes of drugs to be tested for auditory-system pathology, but behavioral testing may detect impairment at a stage when the cessation of treatment leads to recovery.

Stebbins and his coworkers [229] have produced an extensive body of data on ototoxicity from which two important conclusions have emerged. First, they demonstrated that hearing can be assessed by many different behavioral methods and in many different species. Second, they established correlations between histopathology and the results of behavioral testing that yield valuable information about how the auditory system works. In this work, a trial begins by pressing a key or contact-sensitive plate, a response that activates a light. At a variable time after light onset, the acoustic stimulus, usually a pure tone, is presented. If the subject responds during the tone, the reinforcer is delivered immediately. Premature releases are followed by termination of the trial and a delay of 6 to 10 seconds before the next trial. Trials on which no tone is presented ("catch" trials) are interspersed among the tone trials to estimate the subject's tendency to guess. Training is continued until guessing is reduced to a low, stable rate. Both the method of constant stimuli and the up–down or titration method have been used in these studies.

Measurements of relative loudness, which involve stimulus intensities well above threshold, are made with a modification of these methods. The animals are trained by differential reinforcement, to respond quickly, say within 500 msec. This is basically a reaction-time situation, and, as noted earlier, because greater stimulus intensities produce shorter response latencies, loudness, which is a subjective variable, can be measured in animals. For example, response latency in monkeys falls from about 900 msec at a sound pressure level of 10 dB to about 200 ms at a level of 90 dB. With these techniques, in species as diverse as macaque monkeys, cats, guinea pigs, and chinchillas, these investigators have been able to demonstrate the effects of aminoglycoside antibiotics on the progression of hearing loss and both the temporary and permanent consequences of exposure to noise. The earliest effects are seen at the high frequencies, a typical finding. Mattson et al. [147] note that such specificity makes it unlikely that ototoxicity would be detected with the usual elementary screening techniques because their stimulus dimensions tend not to be precisely described.

With continued treatment, losses extend to lower and lower frequencies. Such a progression has its morphological counterpart in the hair cells attached to the basilar membrane, which stretches along the winding spiral structure of the cochlea. If the cochlea is examined at an intermediate stage of hearing loss, when high- but not low-frequency discrimination is impaired, it shows damage to hair cells in the lower half of the cochlea, where high frequencies are represented, but no damage to the upper half of the cochlea, where the basilar membrane responds maximally to low frequencies. Given such correlations as a guide, it has become possible to largely confirm the assertion that frequency is coded according to location along the basilar membrane from the base to the apex of the cochlea.

Lead and methylmercury have both been documented as ototoxicants. Schwartz and Otto [219,220] reported that thresholds to a 2-kHz tone rose almost linearly with blood lead levels in subjects 14 to 19 years of age. This function is based on conventional audiograms from more that 4000 individuals surveyed in the second National Health and Nutrition Examination Survey (NHANES II). In another analysis, based on a subset of Hispanic subjects between 6 and 19 years of age, an increase in blood lead level from 6 to 18 μg/dL was associated with a 2-dB loss of hearing at all frequencies tested. Although a difference of such magnitude is seemingly of little consequence to an individual, it is significant to a population because of the great magnification it undergoes at the extremes of the distribution [261].

Complaints of hearing difficulties were voiced by individuals from Minamata, Japan, who were heavily exposed to methylmercury, but intensive testing beyond conventional audiometry has not been carried out. Rice and Gilbert [205] surveyed auditory function in *Macaca fascicularis* monkeys exposed to methylmercury from birth to 7 years of age. The testing was conducted at 14 years of age, after the monkeys were trained on an up–down detection procedure, as described earlier. Exposed monkeys generally exhibited deficits at frequencies above 10 kHz. The age at which these deficits began to appear is not known because auditory testing did not begin until long after exposure had ended. Another group of monkeys was exposed from gestation to 4 years of age [206]. As in the postnatal group, both mothers and offspring received daily doses of 0, 10, 25, or 50 μg/kg. Pure-tone detection thresholds were determined at 11 and 19 years of age by the up–down psychophysical procedure. The degree of deterioration of hearing thresholds between 11 to 19 years of age in the treated monkeys greatly exceeded that observed in the controls and supports speculations that accelerated impairment of function during aging is one possible consequence of developmental exposure to neurotoxicants.

Other instrumental techniques have also been applied to questions of sensory function. A series of studies by Pryor and associates, for example, using conditioned avoidance responding, established that hearing loss could be induced in rats by chronic exposure to common organic solvents such as toluene [196]. Rats were trained to jump and grasp a pole upon presentation of a tone to avoid an aversive electric shock delivered through a grid floor in the chamber.

Most of the testing methods described above command extensive resources and time for their execution and limit the number of subjects that can be evaluated, and as such they are not suitable for inclusion in a preliminary screening battery designed to identify potential hazards. Their advantage lies in their ability to respond to questions arising at later stages of evaluation. Even so, there is constant speculation about how these methods might be replaced by less expensive and lengthy approaches. One attractive alternative is the reflex modulation paradigm. It is based on the ability of a low-intensity stimulus, delivered before a stimulus intense enough to elicit a startle response, to modify the amplitude of that response [116]. The startle stimulus is typically a loud sound, and the response is measured in rats by confining them to a platform connected to a force or acceleration transducer. If a prepulse, such as a soft sound, appears at an appropriate interval, such as 80 msec before the startle stimulus, the amplitude of the startle stimulus will be reduced. If the prepulse stimuli consist of pure tones of various frequencies and energies, it is possible to chart what in essence is a conventional audiogram because the amount of reduction is directly related to prepulse amplitude.

Applications of the auditory startle response technique have established the potency of solvents as ototoxicants; for example, rats exposed to trichloroethylene at 4000 ppm for 6 hours a day for 5 days showed elevated thresholds at 8 and 16 kHz, perhaps due to a loss of spinal ganglion cells [77]. Startle also can be inhibited by cutaneous and visual stimuli and by brief gaps in a continuous auditory stimulus. The reflex modification technique, nevertheless, is limited to relatively simple questions. It would be unsuitable for plotting contrast sensitivity functions, such as those reported by Merigan and his coworkers in studies of acrylamide and carbon disulfide, or for evaluating disorders of color sensitivity. The latter demand an extremely subtle form of stimulus discrimination that could be achieved only by highly trained and motivated subjects. What seems most reasonable is to include reflex modulation in preliminary screening batteries and to reserve the complex psychophysical methods for advanced questions.

## Somesthesis

The skin contains a heterogeneous population of receptors whose central representation is equally diverse. Mechanoreceptors are specialized to respond to deformations of the skin. Some respond on the basis of depth of displacement, some on the basis of velocity, and others on the basis of acceleration. Still others are activated by the movement of hairs. Additional receptors are responsive to temperature. Free nerve endings are thought to subserve pain. Pathways transmitting information from the skin travel to the central nervous system through the peripheral nerves and then, except for the head region, penetrate the spinal cord through the dorsal roots. The head region is supplied by the 12 cranial nerves. At the level of the cerebral cortex, the skin surface is represented in accordance with the density of receptors in various areas, so the cortical areas devoted to the face and hands greatly exceed those devoted to the back, buttocks, and other low-density areas. A corresponding nonspecific system, which is less defined and which includes the reticular formation of the midbrain, makes a wider range of connections as it ascends and is thought to serve an arousal function.

Most disturbances of somesthesis are attributed to impaired function of the peripheral nerves (i.e., peripheral neuropathy), although one prototypical agent, acrylamide, apparently exerts its earliest peripheral effects on the

**TABLE 37.5**
**Chemicals Impairing Vibration Sensitivity**

| | |
|---|---|
| Acrylamide | Triorthocresyl phosphate |
| *N*-Hexane | Lead |
| Carbon disulfide | Cyanide |
| Arsenic | Chlorobiphenyl |
| Methylmercury | Methyl bromide |
| Thallium | Methamidophos |
| Methyl-*n*-butyl ketone | |

acceleration receptors known as Pacinian corpuscles. Damage anywhere along the pathways from receptor to cortex may produce such disturbances, but methylmercury is the only poison clearly documented to be involved at the cortical level. A wide array of chemical classes, as listed in Table 37.5, seem to produce their somatosensory effects primarily through peripheral nerve damage, although some act as well at subcortical and cortical levels. This list includes metals, solvents, organophosphates, and other chemicals. A more complete list would include drugs from many different therapeutic classes.

In assessing the scope of such a dysfunction, histology will be of limited utility when the basic question is the risk to the health of a specific population such as workers. Moreover, the typical clinical neurological examination is generally inadequate because its aim is to uncover frank disease. The clinician may prick the skin with a pin, draw a wisp of cotton across the skin, or ask the patient to judge whether a tuning fork is vibrating. Even the most controlled of these stimuli, the tuning fork, is a crude device held in a hand that itself has an oscillation of much greater amplitude than is discriminable at the skin. The threshold for vibration detection at a frequency of, for example, 150 Hz lies close to 1 μm and requires precise instrumentation, calibrated to 0.1 μm, to determine. The simple vibration devices marketed by several producers and often used in clinical studies are too gross to yield useful data; moreover, they tend to be used with improper psychophysical procedures [148,152]. A major flaw is their lack of direct specification of stimulus characteristics, particularly amplitude. Some of the commercial units express stimulus output as volts, which refers to the setting on a dial; others use unspecified units that are equally arbitrary.

Despite the requirements for precision, however, vibration sensitivity testing has particular utility as a method for assessing sensory neuropathy. One is its underlying structural and functional diversity. It is thought to be mediated by at least two sets of large-diameter myelinated fibers originating from two kinds of receptors: the Pacinian corpuscles, presumed to be maximally responsive to stimulus frequencies from about 100 to 400 Hz, and Meissner corpuscles, maximally responsive to frequencies of about 50 Hz and below. As in the visual and auditory systems, then, vibration sensitivity can be evaluated by determining detection thresholds across a broad range of frequencies.

Monkeys and humans possess about the same degree of sensitivity to vibration and can be tested in essentially the same way. Maurissen and Weiss [149] tested monkeys perched in a primate chair, with one hand restrained by a plasticene mold fitted individually for each monkey. An electromagnetic vibrator drove a rod in contact with the middle finger and was positioned to indent the skin at a constant depth in the absence of vibration. It is essential to use a vibrator powerful enough to maintain its calibrated amplitude when opposing the mechanical resistance of the skin. (One flaw of some commercial instruments advertised for vibration testing is their sensitivity to pressure, which renders them unsuitable for precise measurement of displacement.) The monkey's other hand was free to respond on a key. The frequency and amplitude of stimuli were determined by function generators whose outputs, in turn, were governed by parameters entered into a computer program that controlled the sequence of experimental events and that stored the results.

One experiment with this system sought to trace the onset and development of sensory neuropathy induced by the axonopathic agent acrylamide [151] in *Macaca nemestrina* monkeys. During training, they learned to press and release the key to obtain juice, succeeded by a phase in which they learned to wait for the onset of vibration delivered to the other hand before releasing the key. In the final protocol, a tone signaled the beginning of a trial. The monkey then held down the key. After a foreperiod of variable length, a vibratory stimulus was presented. If the key were released during this period, juice delivery followed. On catch trials, inserted to discourage and compensate for random guessing and on which no vibration occurred, the monkey earned juice delivery by holding down the key during the tone and releasing it only after the tone ceased. Correct responses lowered the amplitude of the stimulus on the next trial and incorrect responses raised it, as in the vision experiments described previously.

Figure 37.22 shows the results of this experiment. After stable baselines were established, monkeys were given 10 mg/kg acrylamide monomer in apple juice on weekdays during the treatment period. In addition to testing two vibration frequencies, 40 and 150 Hz, the monkeys were also tested, in a separate but identical apparatus, for electrical sensitivity. In this apparatus, the rod in contact with the monkey's finger was an electrode through which a 60-Hz electrical stimulus was delivered. As during vibration testing, correct detections lowered the amplitude of the stimulus on the next trial, and incorrect responses raised it. Gross motor deficits were assessed with an apparatus that allowed an observer to record the length of time required to retrieve marshmallow from a mesh grid. Body weights and the occurrence of standard clinical signs also were recorded.

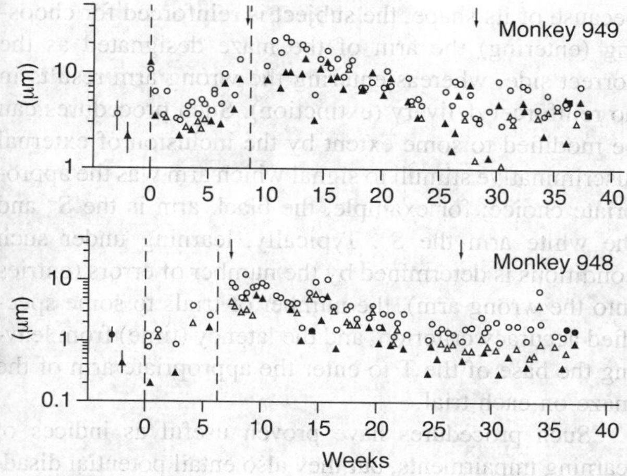


**FIGURE 37.22** Vibration sensitivity of two monkeys (*Macaca nemestrina*) assessed by determining thresholds to stimulation by a vibrating rod applied to a finger. The rod was driven by an electromagnetic vibrator controlled by a voltage determined by the computer program that managed the experiment. The monkey's paw was maintained in a stable position by a plasticene cast. The subjects were trained to press a telegraph key with the unrestrained hand when a vibratory stimulus was detected and to respond only at the end of the trial period when a stimulus was not detected. Correct responses were reinforced by the delivery of fruit juice. Baseline thresholds (±2 SD) are plotted in the area preceding acrylamide dosing at week 0 for vibration frequencies of 40 Hz (circles) and 150 Hz (triangles). Dashed vertical lines indicate the beginning and end of acrylamide treatment (10 mg/kg, five times weekly). Thresholds, given as vibration amplitude in grams on the ordinate, rose slowly for several weeks during treatment, then slowly declined. (From Maurissen, J.P.J. et al., *Toxicol. Appl. Pharmacol.*, 71, 266–279, 1983. With permission.)

As the figure shows, vibration thresholds rose soon after dosing began. Once dosing ended, they fell slowly after several weeks. Electrical sensitivity remained stable, presumably because it is subserved by small, unmyelinated fibers that are not as sensitive to acrylamide as the large, myelinated nerve fibers that carry information from the Pacinian corpuscles and other receptors that respond to vibration. A second course of acrylamide treatment, following the return to baseline sensitivity, duplicated the results obtained during the first course [150]. What is especially notable about both sets of data is the absolute sensitivity to vibration. Note the baseline thresholds, which lie in the range of a few micrometers. As noted earlier, measuring such thresholds, which is required if early intervention is sought, requires instrumentation that can be calibrated to within tenths of micrometers.

Rice and Gilbert [207] used this method to determine the effects of methylmercury and lead on vibration sensitivity. Methylmercury-exposed monkeys consisted of the same group described previously in discussions of visual and auditory function. Lead-exposed monkeys had been treated throughout life. Paresthesias comprise the earliest indication of methylmercury neurotoxicity in adults, and peripheral neuropathies are a recognized consequence of lead exposure. Of five monkeys dosed with methylmercury from birth to 7 years of age, four exhibited reduced sensitivity to vibration when tested at 18 years of age. Two monkeys dosed prenatally through 4 years of age exhibited reduced sensitivity at 15 years of age, while two monkeys treated with higher doses showed little impairment. Lifetime lead exposure produced ambiguous but suggestive indications of impairment.

Quantitative sensory assessments are becoming more routine in clinical neurology because of the recognition that the traditional clinical methods are insensitive [101], and vibration sensitivity is one of the more common indices. Quantitative measures are especially useful for monitoring patients treated with drugs, such as cancer chemotherapeutic agents, that induce peripheral neuropathies, so treatment can be interrupted before permanent damage is inflicted. Despite the availability of instrumentation capable of the precise control of amplitude and frequency, however, considerable misunderstanding remains about the appropriate procedures, including confusion as to the difference between psychophysical methods for the presentation of stimuli and the type of responses with which they are used. Maurissen [152] describes some of the typical errors made by investigators.

### Smell

Olfaction is a primitive sense in the context of evolution. Its representation in the brain takes a different course than the senses already described because the olfactory pathways bypass the thalamus and travel directly to the piriform cortex. Recognition of odorants by the olfactory receptors comprises the first stage in odor discrimination, a process that is beginning to yield to a molecular understanding of odorant recognition. Only then do they link to subcortical structures in the limbic system, which includes the amygdala and hypothalamus. Because the limbic system is associated with behaviors such as those involved in reproduction, olfaction carries a crucial responsibility in species as well as in individual adaptation to the environment. Even in humans, the sense of smell, although not critical in meeting most environmental challenges, is nevertheless the source of both pleasant and unpleasant stimuli that contribute to the quality of life. Moreover, some chemicals announce the approach of dangerous ambient levels by stimulating olfactory receptors, so diminished smell sensitivity might pose a danger.

Disorders of olfaction recently entered the argument about whether or not chronic, low-level exposure to volatile organic solvents inflicts neurotoxic consequences. Schwartz et al. [218] examined workers in the paint industry who had been exposed to a variety of solvents, all in settings well enough regulated that exposure concentrations did not exceed the Threshold Limit Values (TLVs®). The workers were tested by asking them to identify standardized odor patches produced by the Monell Center at the University of Pennsylvania. Workers exhibited a lowered discriminative capacity. Such a test does not afford precise control over parameters such as concentration; it is a fairly blunt instrument designed for screening large groups. Precise control is achievable in both animal and human studies, although it requires specialized instrumentation, as shown by recent attempts to provide adequate precision for clinical use [127]. In animal studies, operant behavior, based on the use of odor as a discriminative stimulus, has proven successful [226]. In the presence of one odor, responses are reinforced with food delivery; in the presence of another odor or a neutral stimulus, responses are not reinforced. Similar experimental arrangements are discussed in a later section (Stimulus Properties of Chemicals). Differences in responding serve to index the ability of the subject to distinguish the specific odors at varying concentrations. A special reason for not neglecting olfaction is its exquisite elaboration in rodents, the most common laboratory species. Odor discrimination learning occurs rapidly in rats [67] and also seems sensitive to toxic intervention.

## LEARNING

Learning refers to the process of behavioral adaptation to changes in environmental contingencies—that is, behavior in transition. As Laties [130] pointed out, "There can be as many ways of studying learning as there are ways of confronting organisms with changed reinforcement contingencies and then watching them adapt to the new contingencies." The procedures and apparatus devoted to the study of learning are so diverse and numerous that an adequate description of their domain is beyond the scope of this chapter. As previously noted, this chapter focuses on approaches that have been most frequently utilized, as well as those most promising for illuminating behavioral and neurobiological mechanisms of toxicity. It also highlights some of the important methodological issues involved in interpretation of changes in learning. Recent reviews describe in greater detail the effects of specific toxicants on learning, as well as other procedural issue [31,65,108,251].

### Simple Mazes

The evaluation of learning processes has frequently been based on procedures involving a response choice from which measures including accuracy and latency can be derived. Before the advent of more sophisticated technologies, the assessment of learning often relied on the use of simple mazes. For example, in a T maze, so named because of its shape, the subject is reinforced for choosing (entering) the arm of the maze designated as the correct side, whereas entering the wrong arm results in no reinforcer delivery (extinction). Such procedures can be modified to some extent by the inclusion of external discriminative stimuli to signal which arm was the appropriate choice; for example, the black arm is the $S^+$ and the white arm the $S^-$. Typically, learning under such conditions is determined by the number of errors (entries into the wrong arm), the number of trials to some specified accuracy criterion, and the latency (time) from leaving the base of the T to enter the appropriate arm of the maze on each trial.

Such procedures have proven useful as indices of learning impairments, but they also entail potential disadvantages and caveats with respect to interpretation of outcome. Because such devices are not easily automated, they are personnel intensive; for example, after each trial the animal must be replaced in the start box by the experimenter for initiation of the next trial. The necessity of handling animals and experimenter intrusion into the recording of data introduce the possibility of experimenter bias, unless procedures are carried out by personnel blind to all experimental conditions. Additionally, demands on personnel time by nonautomated versions of mazes markedly increase the operational costs of such procedures and may limit the number of experiments that can be undertaken. Objections to purchasing fully automated equipment such as operant chambers, based on the expense of instrumentation, often neglect to account for personnel costs incurred by procedures relying on human intervention. This disadvantage applies to other nonautomated instrumentation in general.

Moreover, simple mazes represent relatively simple learning tasks, decreasing sensitivity to drug, or toxicant effects. The relative rapidity with which learning occurs also renders such baselines ineffectual for assessing leaning deficits following exposures to toxicants that have a delayed onset of action or for detecting chronic effects or for tracking reversibility of any observed learning impairments. Finally, and perhaps most importantly, these dependent variables may be influenced by changes in other behavioral processes, limiting the ability to define a learning impairment; for example, the time to traverse the maze can be affected by changes in motor function. Increases in time to reach the correct arm via, for example, motor dysfunction, can also delay time to reward, itself a variable known to retard learning. Olfactory cues left by experimental animals can be shown under some conditions to influence the performance of animals tested later. Such difficulties in controlling variables known to influence behavior may result in replication failures both within and across laboratories. Interpretation of a change in behavior as a true learning impairment, therefore, imposes the need to eliminate these potential confounds.

## Radial-Arm Maze

The radial-arm maze is a more complex learning task (Figure 37.23) that can be partially automated. It consists of a central area from which, typically, eight arms radiate like spokes, and a single food pellet is available in each arm. The subject then has access to eight reinforcer deliveries, one in each of the eight arms radiating from the central compartment. The accuracy, efficiency, and speed with which the organism learns to retrieve all eight food deliveries constitute the data of interest, with maximal efficiency requiring only eight arm entries to collect all eight food pellets. The radial-arm maze has been shown to be sensitive to the effects of a variety of toxic agents and prenatal insults. For example, Walsh et al. [253] reported an impairment of reacquisition of radial-arm maze performance in rats that had been treated with trimethyltin, a toxicant that damages the hippocampus, a brain structure thought to play a prominent role in learning and memory functions. Because it presents a more complex problem to the animal, the radial-arm maze has advantages relative to the simpler mazes described earlier, with respect to sensitivity to detecting toxicant effects. Modification of the standard procedures used with the radial-arm maze can make it amenable to evaluation of repeated learning and thus of long-term effects, delayed onset effects, and reversibility. Peele and Baron [193], for example, accomplished this by baiting only four of the eight arms; the particular four arms that were baited changed during each successive experimental test session. Nevertheless, in parallel with the simpler maze procedures is the possibility that changes in the baseline may reflect not changes in learning processes *per se* but rather the effects of the toxicant on motor function, sensory capabilities, etc. that must be accounted for in the interpretation of behavioral changes.

## Water Mazes

The water maze is a learning paradigm based on negative reinforcement procedures. A rodent is placed in a large pool of water that has been made opaque; escape from the water is possible only by finding a platform submerged just under the surface so as to be invisible. Again, dependent variables include the number of trials to learn where the platform is hidden and the latency to find the hidden platform on each trial. The water maze has been employed widely in behavioral toxicology, in neuroscience, and in aging studies because of its ostensible simplicity and the lack of a training requirement or any food deprivation procedures. Performance on the maze has been reported to be influenced by a variety of manipulations, including lesions, toxicants, drugs, and aging. The procedure typically is not automated, rendering it a personnel-intensive approach to learning. Moreover, in the configuration described, it represents a relatively simple learning assay

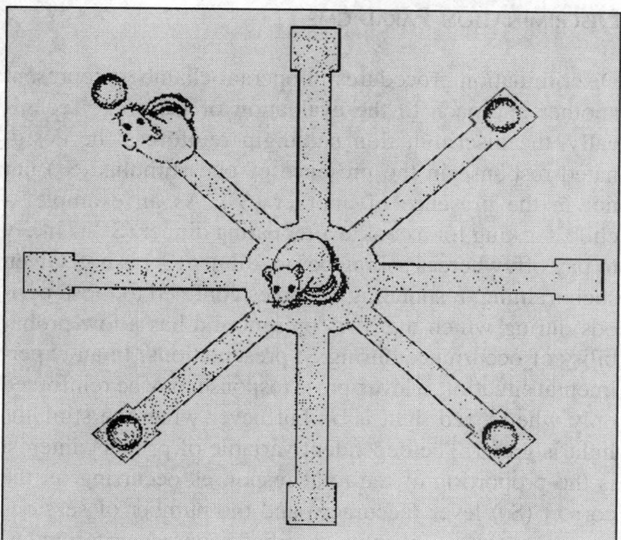

**FIGURE 37.23** A prototypical radial-arm maze for use with a rodent. The subject is typically placed in the center arena from which some number of arms radiate. Food reward is available under different configurations in various arms. Indices of performance include latencies to obtain rewards, number of arms visited to obtain rewards, and errors (e.g., returning to unbaited arms or arms in which the reward had already been obtained).

and thus may be of limited sensitivity and utility for studies of delayed onset of effect, reversibility, chronic exposure, etc. This simplicity can be addressed to some extent by movement of the platform to new locations requiring learning of a new spatial route for escape.

Like the maze procedures described earlier, it is important to note that the ability to define a selective effect on learning may be confounded by concurrent changes in other behavioral functions. These include impaired motor function, where, for example, a weaker animal will have to exert more effort to swim, a highly effortful response. Body temperature is also a determinant of performance, as has been demonstrated in aging animals. Changes of as little as 1°C in the water temperature may significantly alter performance, making the response more stressful and effortful [141]. Sensory alterations could impair the ability to utilize environmental stimuli to guide the response, and studies have shown that olfactory cues [158] in the water from prior subjects can influence the performance of animals tested subsequently. Such possibilities require the use of additional probes or experiments (e.g., cued platform, swim speed, endurance) for evaluation of a true learning alteration. Furthermore, in some studies, floaters are identified who must be eliminated from the experiment, thus enhancing the possibility of experimental bias. Although frequently employed, the use of motor activity to rule out motoric deficits as contributing to latency differences in a water maze is unlikely to be adequate, given that swimming represents a far more effortful response than simple ambulation.

## Discrimination Paradigms

Discrimination procedures in operant chambers represent another approach to the evaluation of learning. Technically, the discrimination paradigm reinforces the designated response in the presence of one stimulus ($S^+$) but not in the presence of another ($S^-$). As an example, a child's asking for a cookie after eating dinner ($S^+$) is likely to pay off, whereas asking before dinner ($S^-$) is not. Given such training, responding becomes confined to those periods during which the $S^+$ is present and has a low probability of occurrence during $S^-$ presentations. In an experimental situation, a lever-press response may be reinforced only when a red light is on but never when the stimulus light is green. The dependent variable of primary interest is the proportion of the total responses occurring on the correct ($S^+$) lever (accuracy) and the number of sessions or session time until some specified accuracy criterion has been achieved. Discrimination paradigms are of two types: simultaneous or successive. In a simultaneous discrimination, both stimuli are presented at the same time; if only a single response device is available, responding on this device is reinforced during $S^+$ presentations but not during $S^-$ presentations in a successive discrimination.

Discrimination paradigms carried out in operant chambers as shown in Figure 37.7 offer the distinct advantage of being conducted as free operant rather than discrete trial procedures such as must be used in the maze techniques described earlier. In trials procedures, the time between each trial or opportunity to respond (i.e., the inter-trial interval) is determined by the experimenter, and a trial ends with a designated response. The necessity for an inter-trial interval is imposed by the requirement of removing the animal from the reinforcement delivery site back to the start box. In free operant procedures, a response by the organism initiates a trial and no inter-trial interval is necessarily imposed between responses. The advantage of this approach is that the subject's rate of responding can be used as a potential index of motivational and motor effects of a treatment, which may contribute to any presumed effect on discrimination learning (see also later discussion). Another advantage of the free operant procedure is that initiation of the trial by the subject rather than by the experimenter ensures the subject's attention to the relevant environmental stimuli and improves accuracy [66].

A relatively simple operant discrimination procedure is exemplified by the study of Hastings et al. [107] examining the impact of neonatal lead exposure of rats in which a simultaneous visual discrimination task using lights as stimuli proved sensitive to lead exposure. In this discrete trial procedure, a trial was initiated by the insertion of two levers into the operant chamber. The light above one of these levers was illuminated, signaling it as the correct lever ($S^+$), whereas the other light was not illuminated ($S^-$). The lever associated with $S^+$ varied randomly from trial to trial. Criterion performance, defined as 90% correct responses during a daily test session, was reached significantly more slowly by lead-exposed rats.

An example of the kinds of complex problems that may be designed to confront the organism is a conditional discrimination procedure shown by Rice [200] to be impaired in monkeys following lead exposure. One of three disks (the sample) was illuminated with one of three colors on this delayed matching-to-sample paradigm. A press on this sample disk darkened it. After a delay period, which was constant under some conditions and variable under others, all three disks were illuminated, each with one of three colors. If the monkey pressed the disk with the color that had been presented on the sample disk (i.e., if it matched the previously presented sample), a fruit-juice reward was delivered. The discrimination is described as conditional because the correct response for any trial is conditional on the sample stimulus for that trial. In this experiment, differences between control and lead-treated monkeys were not observed in the initial acquisition of the behavior but appeared when delays were imposed between the conditional stimulus and matching stimuli, with shorter delay values in treated monkeys impairing performance to a greater extent than in controls. In this matching-to-sample paradigm, the response to the sample (conditional) stimulus itself is called an *observing response*; requiring a response to this sample stimulus ensures that the organism is attending to the relevant stimuli when a trial or experimental sequence begins and improves the accuracy of performance [66]. A variant of the matching-to-sample procedure, the oddity paradigm, requires the organism to choose the stimulus that does not match the previously presented sample.

Typically, a specified criterion defines learning in discrimination procedures, such as 8 correct responses in a block of 10 trials. In a discrete trials experiment with both a control and a treated group, the group mean total number of trials to reach the specific criterion can be compared. Using the free operant procedure, the proportion of the total responses occurring during the $S^+$ presentation (i.e., the correct responses) may be contrasted between the two groups. Alternatively, an organism can be used as its own control, in which case the accuracy of performance or the rate of learning can be compared before and after treatment. This approach has been made possible by the development of procedures that allow the repeated measurement of learning (discussed later).

## Repeated Learning

As previously pointed out, one major limitation of many of the simpler procedures described is that, once the correct response has been learned, performance rather than learning is being measured. To pursue issues such as time course of a toxicant's effects on learning, chronic toxicity, or reversibility of toxicant effects, such procedures offer limited utility. Some, however, can be modified to allow repeated measurements of learning. In a discrimination reversal task, for example, acquisition of the original discrimination can be followed by a reversal of the $S^+$ and the $S^-$; that is, the stimulus–reward contingency is reversed until the learning criterion is met for the new discrimination problem. Multiple reversals of this sort can be carried out and behavioral adaptation evaluated by measuring the number of trials to criterion on each reversal. After a number of such reversals, however, subjects may acquire the concept of "reversal" and effectively learn the discrimination in a single trial [105]. Bushnell and Bowman [29] found that monkeys exposed to lead showed an increase in the number of trials to criterion and in the number of errors on the first reversal of discrimination problem, but no effect on six subsequent reversals. Likewise, water maze procedures can be modified to permit repeated measurement of learning by moving the location of the submerged platform after the subject has successfully learned its initial location.

An automated paradigm designed specifically to provide a measure of repeated learning was developed for human subjects by Boren [21,22]. This procedure requires the subject to learn a new response sequence or response chain during each experimental session and has since been widely used to study the effects of drugs on learning. In an experiment with pigeons [242], three stimulus keys in a chamber were all illuminated simultaneously by one of four colors. The pigeon's task was to peck the correct key in the presence of each color—for example, if the key was yellow, peck the left key; if green, peck right; if red, peck center; if white, peck right. In such a case, if the sequence of light presentations was yellow, green, red, and white, then the correct sequence of responses would be left, right, center, and right. The association between color and key position, however, changed with each successive experimental session, and the subject was required to learn a new four-response sequence. Each correctly completed sequence was followed by food delivery. With training, the number of errors (incorrect responses at any point in the sequence) per session stabilizes, yielding a steady baseline from which drug or toxicant effects on learning can be assessed repeatedly. Thompson and Moerschbacher [243] have studied the effects of several drugs on the baseline in nonhuman primates.

Paule and McMillan [191] used a variant of this procedure to track the time course of trimethyltin (TMT) effects on learning in rats. In their incremental repeated acquisition procedure, the sequence length was incremented during the course of a session from one- to five-member sequences. By separating certain types of error classes in their analyses, they were better able to understand the particular behavioral processes affected by TMT. Their observations indicated a differential time course for TMT effects on various classes of errors and showed that acquisition of early responses in the sequence was disrupted to a greater degree than the later stages of the sequences, suggesting an effect on learning, while the recall necessary for the longer sequences remained intact.

Another variant of the repeated acquisition procedure devised for rodents by Cory-Slechta and colleagues [40,42,49] extended the explanatory power of this procedure. Through microanalyses of response patterns and patterns of errors, it could be shown that different drugs could achieve what appeared to be a similar effect (namely, decreases in overall accuracy) through very different patterns of errors. For example, the noncompetitive glutamatergic antagonist MK-801 decreased overall accuracy on the repeated acquisition paradigm by increasing the frequency of preservative (repetitive) errors, while scopolamine, a cholinergic muscarinic antagonist, increased the frequency of errors composed of incorrectly skipping forward or backward through the sequence.

Before classifying the effects of a toxicant as one on learning, however, the critical issue of specificity raised earlier must be addressed. Changes in accuracy on learning paradigms may well arise from nonspecific behavioral deficits such as changes in motor activity or function, sensory capacity, motivation, or other unrelated factors, as noted previously. Studies have shown that aging-related impairments in performance by rats in a water maze can be attributed to hypothermia induced by the temperature of the water and inefficient body-temperature regulation [141]. Odor trials can likewise affect performance in the water maze [158]. The use of odor trials as a confounder of learning deficits in conventional mazes is a well-documented phenomenon. Diminished motivation could increase latency in a maze, increasing the delay to reward, which, by itself would slow the rate of learning, as noted above. For the maze procedures described earlier, the potential influence of changes in motor, sensory, and motivational functions and other influences such as temperature-related effects must be ruled out using an independent procedure—that is, by carrying out additional experiments or probes.

One approach that has been widely adopted for assessing the contribution of such nonspecific effects to learning with the repeated acquisition paradigm is to alternate the repeated learning component with a performance component during each experimental session (technically, a multiple schedule of reinforcement; discussed below). The learning or acquisition component, as described previously, requires the organism to learn a new response sequence during each successive experimental session. In contrast, the response sequence reinforced during the performance component remains *constant* across sessions, so the organism is simply performing an already learned or rote response. By alternating the learning and performance

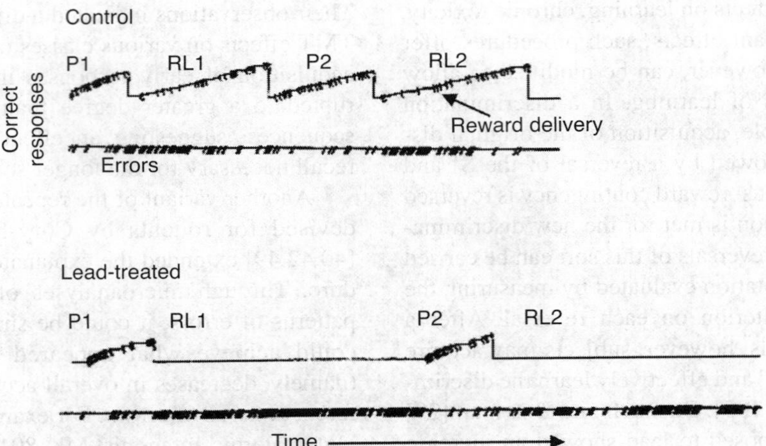


**FIGURE 37.24** Cumulative records of performance on a multiple schedule of repeated learning and performance in a rat exposed chronically to distilled water (top record) vs. 250 ppm lead acetate drinking solutions over the course of a behavioral session from left to right. The multiple schedule involved a repeated learning component (RL), which required learning a new three-response sequence during each successive experimental session. These alternated with a performance (P) component in which the three-response sequence remained constant across sessions. The top tracing of each record shows the correct responses, which cumulate vertically. Each pip on the correct response curve depicts the delivery of a reinforcer for correctly completing the sequence of three responses required for reward. The pen resets to the baseline with each transition between the performance and repeated learning components. The bottom tracing for each subject shows errors that occurred during the components. The lead-treated rat earned virtually no food deliveries (i.e., completed no correct sequences) during the repeated learning components of the session despite normal performance under conditions where no learning was required (P component). This was not due to a lack of responding, as the rat emitted hundreds of errors during this period. (Adapted from Cory-Slechta, D.A., unpublished data; Cory-Slechta, D.A. et al., *Behav. Brain Res.*, 102, 181–194, 1999.)

components during each experimental session, such as after every tenth reinforcer delivery, the experimenter can separate, during the same session for each individual animal, drug- or toxicant-induced changes in learning from nonspecific changes in performance or motivation. This is based on the premise that if a compound selectively affects learning, then decreases in accuracy will only be manifest during the learning (acquisition) component of the schedule, as no learning is required in the performance component. Because intact motor, sensory, and motivational processes are required in both components, impairments of these functions would result in decreases in accuracy in both the learning and performance components.

This approach demonstrated selective effects of lead on learning as illustrated in an extreme case in Figure 37.24 [41], which compares the behavior of a typical control rat to that of a rat exposed to 250 ppm lead acetate from weaning. The particular session depicted began with a performance (P) component, which was followed sequentially by the learning component (A), a return to the performance component, and a final presentation of the learning component, with these components alternating after 25 reinforcer deliveries. Reinforcement followed each completion of a correct sequence. Both rats had received extensive training on the schedule. The control rat (top record) evidenced high levels of accuracy in the first P component, generating a high rate of reinforcement delivery and a relatively low rate of errors, as expected. After the onset of the learning component (A), indicated to the subject by a change in illumination in the operant chamber, the rat was required to learn a new three-response sequence. As anticipated, the error rate was initially high, producing a lower rate of reinforcement delivery. By the end of the learning component, however, the rate of errors had begun to decline and the rat was earning more food deliveries as it gradually began to learn the correct sequence. The acquisition process was even more pronounced during the second presentation of the learning component, as the rat continued to learn the correct sequence for this particular session. The bottom record of Figure 37.24 shows the rather dramatic impact of lead exposure that was selective for the learning component of the schedule. As can be seen during this particular session, the lead-exposed rat earned virtually no food deliveries during the learning component, emitting hundreds of incorrect responses over the course of both presentations of the learning component, even while exhibiting high accuracy levels during both presentations of the P component, rapidly earning all 25 available reinforcers. By undertaking a microanalysis of the various patterns of responses and classes of errors, it was determined that the detrimental effect of lead on learning occurred via an increase in response perseveration—that is, repetitive responding on the same lever or repetitive iteration of the same sequence of responses during the learning components.

Any learning assay also must consider the values of the experimental parameters and nature of the problem selected, because such factors influence the sensitivity of the task to disruption by chemical agents. For example, Winneke et al. [276] investigated the effects of lead exposure on the acquisition by rats of both a form and a size discrimination. The form discrimination required rats to distinguish between vertical and horizontal stripes. On the average, only 8 training sessions were required by rats to reach criterion accuracy, and the procedure did not differentiate the lead-exposed from the control groups. In contrast, size discrimination, in which a small circle had to be distinguished from a larger circle, proved to be a much more difficult problem, requiring more than 20 training sessions; it revealed a substantial impairment due to lead. A similar effect was reported by Carlson et al. [34] in lead-exposed sheep. Such data demonstrate the important role of task complexity and difficulty in modulating sensitivity to disruption by chemical agents.

Often the degree of difficulty of the task can be equated with the degree of stimulus control over the performance—that is, the strength of the stimuli controlling the response; for example, Laties et al. [133] required rats to press a fixed number of times on one of two levers, then respond once on the alternate lever for reinforcement. Completion of an insufficient number of responses on the first lever before switching to the alternate lever reset the response requirement. During some parts of the session, a light and tone signaled that the response criterion on the first lever had been met; in other parts of the session, no external stimulus signaled the completion of the response requirement. As would be expected, accuracy of performance was superior during the signaled components of the session, and much higher doses of *d*-amphetamine [133] or toluene [283] were required to disrupt the signaled ($S^D$) than the unsignaled performance (unlabeled, no $S^D$), as can be seen in Figure 37.25. A similar example is exemplified by the data presented in Figure 37.24. Responding on the performance component of the multiple repeated learning and performance schedule represents a far less difficult discrimination and maintains higher overall levels of accuracy. These findings indicate that behavior strongly controlled by environmental stimuli, such as is the case during the signaled component of Laties et al [133], is less easily disrupted than behavior under weaker stimulus control. Thus, the degree of stimulus control over behavior can markedly influence its susceptibility to the effects of drugs of toxicants.

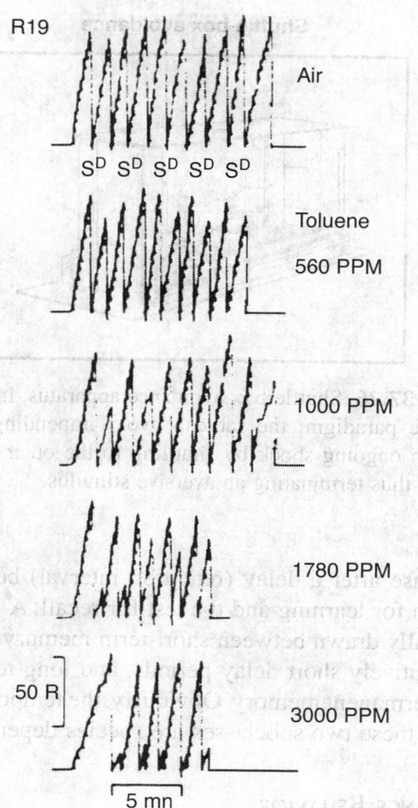


**FIGURE 37.25** Behavior of rats on a fixed consecutive number (FCN) schedule which required subjects to complete a specified number of responses on one lever before a single response on the second lever would produce reinforcement. This was carried out in a multiple schedule format; in some FCN presentations, a discriminative stimulus was provided to signal to the subject that the response requirement on the first lever had been met (labeled as $S^D$ components), whereas in the alternate components, no such stimulus was provided. Accuracy levels were higher in the $S^D$ component, and behavior in this component was also significantly less disrupted by toluene exposure. (Adapted from Laties, V.G. et al., *Psychopharmacology*, 75, 277–282, 1981; Wood, R.W. et al., *Toxicol. Appl. Pharmacol.*, 68, 462–472, 1983.)

## MEMORY

*Memory* is a term that has been used both to describe and to attempt to account for the behavior of remembering, or the influence on behavior of previous events. Experiences are said to be stored in memory, in encoded forms, such as neural engrams or, more recently, as electrical fields, and later recalled from storage by some type of retrieval system. Many theorists have adopted the vocabulary of computer technology to describe memory processes, despite the lack of evidence of operational correspondence. In a behavioral analysis, memory may be best understood and experimentally approached in terms of stimulus control. Remembering is really another way to assert that the probability of certain learned responses is increased. Heise [108] referred to the fact that, while learning is "manifested behaviorally by acquisition, an enduring change in behavior, ... memory may be defined as the preservation of the learned behavior over time." Experimentally, memory is indicated by the accuracy of

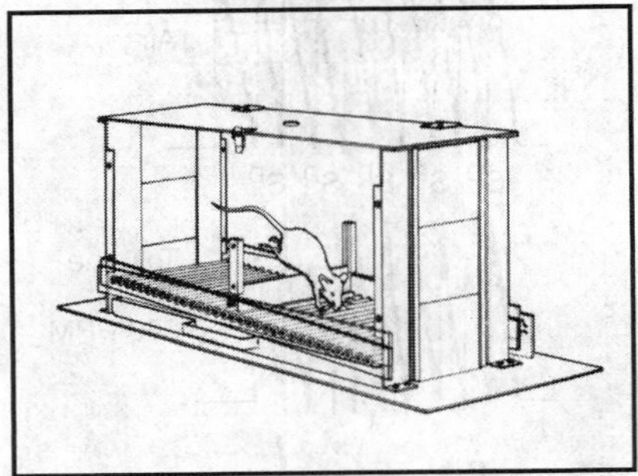


**FIGURE 37.26** Shuttle-box avoidance apparatus. In this active avoidance paradigm, the rat can avoid impending shock or escape an ongoing shock by shuttling to the other side of the chamber, thus terminating an aversive stimulus.

a response after a delay (retention interval) between the occasion for learning and the test for recall. A distinction is typically drawn between short-term memory, occurring over relatively short delay periods, and long-term or virtually permanent memory. Obviously, the temporal parameters of these two subclasses are species dependent.

## Avoidance Behavior

As noted earlier, the various techniques devised to assess memory typically involve the measurement of response accuracy after various delays. An effect of a drug or chemical on memory is suggested by an increased impairment of accuracy with increasing delay values (retention intervals). Avoidance of an aversive stimulus, such as electric shock, is one frequently used measure of both learning and memory. Several different variants of these procedures have been employed; some are discrete trial procedures, and others are free operant procedures. The frequently used passive avoidance paradigm assesses the tendency of the subject to avoid the site of previous shock delivery. A mouse that avoids returning to a normally preferred dark chamber where it previously has been shocked is said to be exhibiting a memory of that event. Active avoidance paradigms require a specified response to be emitted to postpone an impending shock onset. In some cases, the apparatus used is a two-compartment chamber in which the subject switches compartments at the appearance of a stimulus that signals impending shock (Figure 37.26).

Although useful as a screen for potential memory dysfunction, it is important to note that there are also interpretation issues associated with such approaches, especially the passive avoidance paradigm. As with learning, a drug or toxicant can alter performance in memory paradigms through indirect changes in non-mnemonic functions. For example, motivation to avoid shock can be produced by differences in shock sensitivity prior to the training phase in passive avoidance paradigms. If the toxicant decreases shock sensitivity *per se*, then treated animals will be less motivated to subsequently avoid the shocked compartment during the memory testing phase, as shock level is an important determinant of subsequent memory. Similarly, in active avoidance paradigms, toxicant-induced decreases in shock sensitivity would diminish response rates to avoid impending shocks. The drug or toxicant may also modify activity levels (e.g., induce hypo- or hyperactivity), thus increasing the probability of the subject returning to the shocked compartment independently of remembering in the passive avoidance paradigm. Furthermore, state-dependent learning may alter subsequent retention for passive avoidance training if the animal is trained during drug or toxicant exposure but retested under nonexposed conditions. In other words, the response may have an altered probability of recurrence during the retest simply because the environmental stimulus conditions (nondrugged) differ from stimulus conditions during training (drugged) and thus may be independent of memory. Such possibilities must then be sorted out in additional experiments that probe these confounding interpretations. Other problems that can be encountered with these simple avoidance paradigms include the fact that variability of response among animals tends to be high, often necessitating large groups of subjects. Additionally, some rats fail to learn such procedures and may be discarded as slow learners, biasing the sample population for unknown reasons.

## Delayed Alternation

Delayed alternation requires a subject to alternate responses on each of two response devices such as levers or nose cones for reinforcement (Figure 37.27) [50]; that is, a response on device A followed by a response on device B produces a reinforcer, a return to device A, another reinforcer, and so on. After initial training, a delay requirement is interposed between the alternating responses, so only alternations separated by at least the required delay interval are reinforced, rendering this a memory paradigm. Generally, the delay value varies randomly from trial to trial during an experimental session, allowing the assessment of accuracy by delay (retention interval) information from each experimental session. A response occurring before the end of the delay period typically resets the time requirement, with the particular delay values specified being dependent on the experimental species. Importantly, on some trials, there is no delay. This no-delay condition is critical for evaluating the contribution of non-memory-related changes in behavior to any presumed memory impairment. Since no memory is required in the absence of a delay, accuracy values should not decline in response to treatment in the no-delay condition. Typically, one also expects to see a greater decline in accuracy with increasing delay values in drug- or toxicant-treated groups. Using this

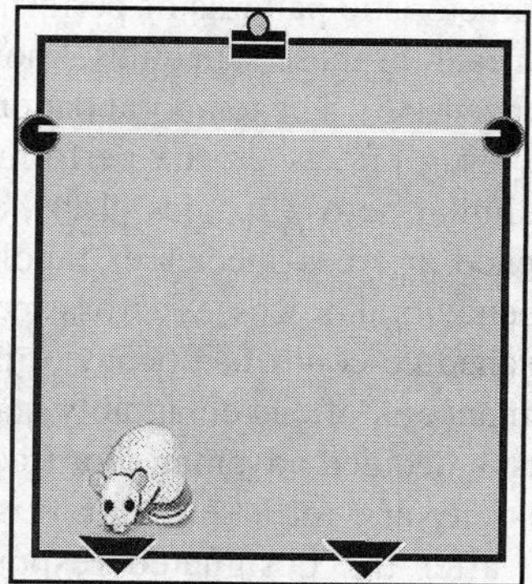


FIGURE 37.27 Schematic of the delayed alternation procedure. This paradigm requires the subject to alternate responses between two response manipulanda for reward (in this case, to nose poke first on the right manipulanda and then on the left manipulanda for reinforcement). The next reinforcement delivery then requires a switch back to the right manipulanda. For the assessment of memory function, a delay between these response opportunities is imposed, requiring the subject to recall on which manipulanda the previous response occurred and thus alternate accordingly. This delay value typically varies from trial to trial during an experimental session so a complete function of accuracy by delay length (retention interval) can be generated for each experimental session.

procedure, Bushnell [29] reported that TMT could produce deficits in memory processes in rats, with TMT accelerating the decline in the delay interval function relative to control but not affecting accuracy on the zero-second delay trials. These data are consistent with the notion of a selective impairment of remembering.

### Delayed Matching-to-Sample

The matching-to-sample procedure described earlier becomes a memory paradigm when delay intervals are imposed between the presentation of the sample stimulus and the later presentation of the matching stimuli options. As with delayed response procedures, a range of delay values is presented in a semi-random manner throughout the course of an experimental session, allowing the collection of accuracy by delay interval information in every session. Taylor and Evans [237] reported impairments in delayed matching-to-sample performance in pigeons at an exposure concentration of toluene that maintains self-administration behavior (discussed later) in the primate [280]. This impairment was not interpreted as an effect on memory, however, because accuracy was significantly impaired even in the no-delay condition.

As in the case for learning, then, changes in performance on a memory task may have little to do with memory but may instead be indirectly influenced by other nonspecific behavioral effects such as changes in motivation, arousal, sensation, and perception. For example, a toxicant that increased the rate of responding might produce premature responding to the comparison stimuli in a matching-to-sample task and thereby increase error rates. Alternatively, decreased rates of responding might increase the latency to make the response in a choice situation and thus increase the functional delay interval. One efficient way to assess the contribution of nonspecific effects in free operant memory procedures is to examine changes in the rate of responding as well as changes in accuracy in the presence of a no-delay condition. Nonspecific effects on levels of arousal or motivation may be reflected in altered response rates. The no-delay condition permits drug or toxicant effects not related to memory to be determined, as no remembering is required at the zero-second delay. A further condition supporting the interpretation of a true memory impairment is that the magnitude of the decrease in accuracy increases with increasing delay value relative to controls [108] and that similar effects of the toxic agent can be demonstrated in other memory paradigms using other stimulus and response conditions.

## SCHEDULE-CONTROLLED BEHAVIOR

Every experimental procedure using operant behavior is based on the principle that behavior is generated, altered, refined, and eliminated by its consequences. Seldom, however, is each and every instance of a specific behavior in the natural environment followed by reinforcement. Continuous or invariant reinforcement is infrequent; instead, intermittent reinforcement is the rule. Paychecks are typically distributed on a weekly, semiweekly, or even monthly basis. Not every visit to the mailbox will be rewarded by the arrival of the letter we are awaiting. Often a string or chain of responses may be emitted before there is a reward, much as the pianist finishes playing an entire piece of music before the audience applauds. Similarly, the child may correctly put together several puzzle pieces or even the entire puzzle before the parent praises the child. In the wild, predation and foraging behavior are certainly maintained under conditions of intermittent reinforcement. Besides the economy achieved by intermittent presentation of reinforcement, behavior maintained under conditions of intermittent reinforcement is actually considerably more robust than that maintained by continuous reinforcement [120]. For example, a response that has been reinforced on every occurrence declines much more rapidly during an extinction procedure (withholding of reinforcement) than does behavior that has been intermittently reinforced. Put another way, continuously reinforced behavior is much less resistant to extinction than that maintained by intermittent payoff. Many parents have learned, to their distress, how occasional reinforcement of a temper tantrum (failure to ignore it) may subsequently increase the magnitude and persistence of the behavior [274].

In the human environment, reinforcement schedules—that is, the nature of the rules by which reinforcement is allocated—may be complex. The laboratory offers the experimenter direct control over these contingencies and thereby allows a more precise analysis of the ways in which the scheduling of reinforcements controls various aspects of responding, including its temporal distribution, force, rate, resistance to extinction, and so on. The study of reinforcement schedules is a discipline in itself [78,91,217]. Of most relevance to toxicology, an extensive body of literature has accrued describing the effects of a wide variety of central nervous system drugs on schedule-controlled behavior [118,121,123,124,154,156,222,244,248]. This permits comparison of toxicant effects on schedule-controlled responding to those of central nervous system compounds whose mechanisms of action are generally well understood. The effects of numerous toxicants on schedule-controlled behavior have since begun to be enumerated [45–47,92,145,201,270].

Schedules of reinforcement are especially important to behavioral toxicology, as reinforcement schedules govern the rate and pattern of behavioral responding involved in different behavioral processes [46]. The rate of learning, for example, may well be influenced by the reinforcement schedule according to which the reinforcer is presented. The response may be only slowly acquired if initially reinforced only infrequently. Whether the response is learned at all may depend on the strength of competing responses, the magnitude of reinforcement, and the concurrent availability of competing schedules of reinforcement. In addition, changes in schedule-controlled behavior may actually underlie behavioral changes that are attributed to impairments of other behavioral processes. Decreases in rates of responding produced by a toxicant may be incorrectly interpreted as a decline in the rate of learning in a discrimination procedure. Schedules of reinforcement can also be used as an index of learning processes *per se*—that is, whether treated subjects learn the behavioral patterns characteristic of the particular schedule under study and whether changes in parametric values of the schedule produce corresponding changes in schedule-controlled performance at that same rate in treated and control subjects. To truly understand the behavioral effects of a toxicant, then, requires an understanding of its impact on schedule-controlled behavior, as behavior occurs at given rates and in particular pattern over time.

### Simple Schedules

Although a virtually limitless number of ways to schedule reinforcement delivery is possible, most schedules are based on either number of responses or time. Four simple schedules have been defined: the fixed-interval (FI) and the variable-interval (VI), both of which are time-based reinforcement schedules, and the fixed-ratio (FR) and the variable-ratio (VR), both of which are response-based schedules. Characteristic patterns of performance eventually emerge on each of these schedules, known as steady-state performance [245]. The temporal pattern of responding as well as drug effects on the performance are frequently very similar across species [124], a feature providing confidence in cross-species extrapolation and the continuity of behavioral processes across species. Another advantage of schedule-controlled behavior is that the pattern of performance is often remarkably stable over prolonged periods, a decided advantage for tracking the progression of toxicity and for assessing reversibility.

Typically, after the designated response has been shaped, the schedule of interest is then imposed directly, although it may reach its final parametric value only after a series of gradual changes from the initial parametric values. For example, in studying FR performance in a rat, the experimenter will most likely carry out several sessions at lower FR values before imposing a final value as high as 100. This prevents "ratio strain" (irregular pauses occurring between responses on an FR schedule [78] associated with high FR values) and prevents extinction of the response.

Interval schedules stipulate that a certain amount of time must elapse following a previous response before the next occurrence of the response will be reinforced. On an FI schedule, the time period is constant and typically is measured from the previously reinforced response. For example, an FI 5-minute schedule reinforces the first response occurring at least 5 minutes after the preceding reinforced response; responses during the 5-minute interval have no programmed consequence. The FI schedule typically generates a characteristic scalloped pattern of performance, as shown in Figure 37.28, which consists of a pause after reinforcement delivery (designated as the post-reinforcement pause time, or PRP) followed by a progressive increase in the rate of responding (running rate) as the time for the next food delivery approaches. In the human environment, studying for a scheduled exam has features resembling performance maintained by an FI schedule: Normally, little or no studying occurs early in the semester, but as the time for the midterm approaches the rate of studying begins to escalate and is highest right before the exam. The species generality of the scalloped pattern of FI performance is amply demonstrated in Figure 37.29 across species, type of response, and type of reinforcer. Human FI performance shows a similar scallop under many, although not all, conditions [132,241]. Sensitivity of the FI schedule to a variety of toxicants has been demonstrated [47].

The typical, though grossest, measure of schedule-controlled behavior is absolute or overall response rate, which is simply the total number of responses divided by total session time. Often, a full understanding of chemical modification requires a microanalysis to dissect the behavior into more elementary components. Like the separation of

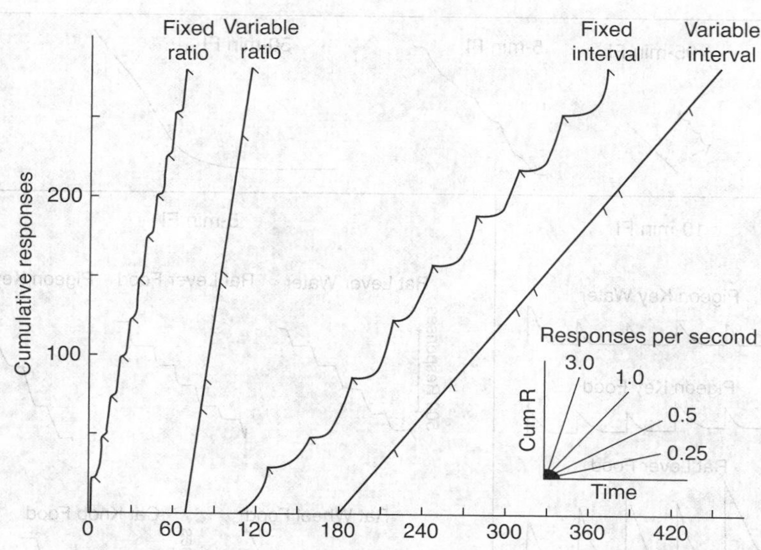


**FIGURE 37.28** Characteristic behavior on each of the four simple schedules of reinforcement: the fixed ratio (FR), variable ratio (VR), fixed interval (FI), and variable interval (VI). In these records, responses cumulate vertically, and the slope of the line indicates the rate of responding; examples of slope-rate comparisons are shown in the inset. Each pip on the response curve shows where a reinforcer was delivered according to schedule contingencies. The FR schedule generates a very high rate of responding, with short pauses (indicated by horizontal periods on the curve) following each reinforcer delivery. The VR schedule generates an even higher rate of responding with little or no pausing after reinforcement, as the very next response may produce reward. The FI schedule produces a scalloped pattern of responding characterized by pauses after reinforcement delivery followed by a gradually increasing rate of responding as the time of the next available reinforcer delivery approaches. The VI schedule generates a very stable moderate rate of responding with almost no pausing, as the time to the next available reinforcement opportunity is unpredictable. (From Seiden, L.S. and Dykstra, L.A., *Psychopharmacology: A Biochemical and Behavioral Approach,* Van Nostrand Reinhold, New York, 1977. With permission.)

various classes of errors in a learning or memory experiment, schedule-controlled performance can be differentiated into component parts, which permits a more precise understanding of the manner in which the schedule controls performance. Such an analysis can suggest the possible behavioral processes that are altered by a chemical and point to directions for further analyses and manipulation.

Fixed-interval performance provides an example. The schedule has been the focus of much experimental study in part because it exemplifies temporal control over behavior. As previously described, FI performance is characterized by a pause after reinforcement delivery (Figure 37.28 and Figure 37.29). The length of this PRP time (time from food delivery to the first response in the next interval) depends on the length of the fixed interval [224]: the longer the interval, the longer the pause. The PRP is followed by a long period of shorter pauses, interspersed with short bursts of responding. As the interval progresses, the long pauses cease and are replaced by alternating periods of moderate and high response rates until reinforcement delivery [89]. The rate of responding during an interval, after the PRP is subtracted out, is referred to as the *running rate*. With the aid of a computer, the actual times between each successive response, the interresponse times (IRTs), can be collected, a frequency distribution of various length IRTs generated, and sequential patterns of IRTs reconstructed. The characteristic scalloped pattern of performance is usually described by one of two measures. The index of curvature [83] specifies the extent to which the scallop deviates from a straight line (a constant rate of responding throughout the interval), and the quarter-life [109] is defined as that proportion of the interval required for the first 25% of the total responses in the interval. It is these microanalyses that may permit detection of a toxicant effect and understanding of its behavioral mechanism of action. In the case of low-level lead exposure in rats, for example, the proportion of short IRTs (0.5 seconds or less) is consistently increased by lead exposure (Figure 37.30), even while increases in overall response rate may be less impressive [51]. A similar effect has been reported in chronically exposed monkeys [201].

The type of effect produced by lead exposure on FI schedule-controlled behavior might be viewed as a loss of discriminative control by the schedule [44,47]. The behavioral mechanism might be explained as a failure of the FI schedule to exert temporal control over behavior or as a failure to learn to discriminate interval length. This hypothesis earned further support from two additional experiments. In one study [48], rats were required to hold down the lever for 3 seconds for each food delivery. Control rat

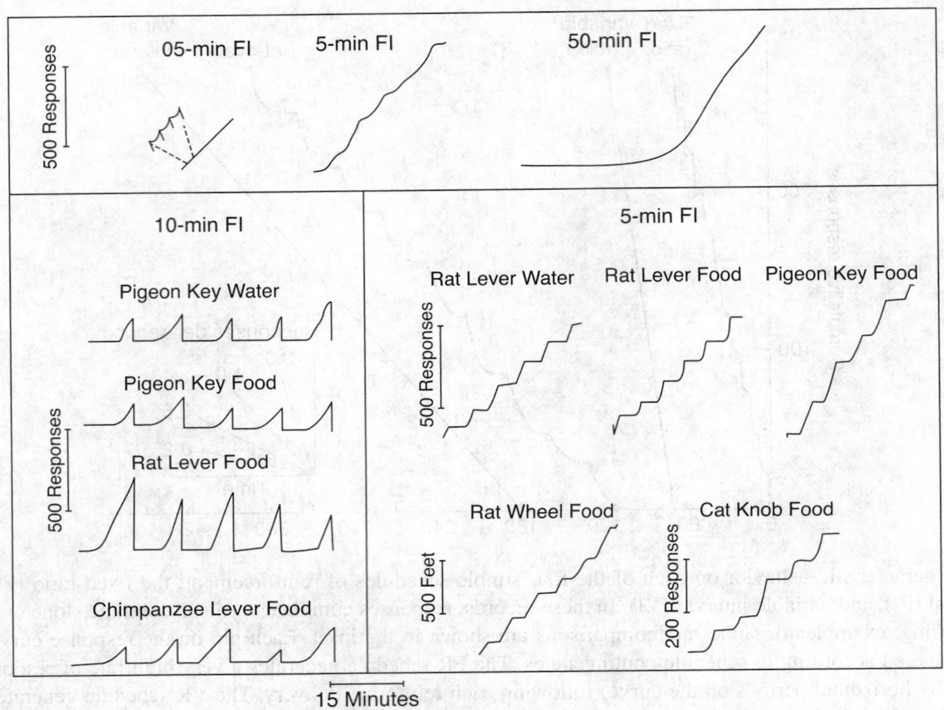


**FIGURE 37.29** Species generality of characteristic fixed interval (FI) schedule-controlled behavior. Responses cumulate vertically in these records, and time is represented horizontally. The top panel shows the behavior of an individual pigeon pecking a key for food at three different FI values; the general scalloped pattern persists despite the 100-fold change in time value. Performance on an FI 5-minute schedule is shown in the bottom right panel. As can be seen, FI performance is remarkably similar for different species (rat, pigeon, cat) with different responses (lever press, key peck, wheel running, and knob pushing) and different reinforcers (water, food). The bottom left panel shows performance on an FI 10-minute schedule of reinforcement; in this case, the pen reset to the baseline after each reinforcer delivery. Again, comparable FI performance occurs across species, response, and reinforcers. (From Kelleher, R. and Morse, W.H. et al., *Ergebnisse der Physiol.*, 60, 1–56, 1968. With permission.)

median response durations were between 2.5 and 3 seconds, while some lead-exposed rats exhibited a much higher proportion of response durations too short to produce reinforcement, even after a tone stimulus was subsequently added to signal that the required duration had been met. In another experiment [203], reinforcement depended on separation of responses by at least 30 seconds (a schedule known as DRL, or differential reinforcement of low rate). Lead-treated monkeys acquired the performance more slowly than controls, as indicated by a higher frequency of non-reinforced responses during the initial sessions. These findings with lead are consistent with other hypotheses regarding its behavioral mechanisms of action.

In contrast to lead, certain pesticide classes alter both response rate and the temporal patterning of FI behavior, suggesting actions through different behavioral and neurobiological mechanisms [47]. Both oranochlorine pesticides [28] and formamidine compounds [134] have been shown to decrease rates of responding on the FI schedule and to disrupt the normal temporal pattern of responding. An example is provided in the cumulative records shown in Figure 37.31.

When relevant information about a toxic agent is scarce, the FI schedule offers several distinct advantages as an early test for behavioral toxicity. Because the schedule reinforces only the first response after the end of the interval, and because responses during the interval have no programmed consequence, response rates during the interval itself can vary quite broadly before they are sufficient to alter the frequency of reinforcement. This feature may explain in part why FI performance is often more sensitive to drugs than ratio-based schedules of reinforcement [58,122,225] on which a decrease in the rate of responding necessarily decreases reinforcement frequency (discussed later).

The patterns of behavior by the three other simple reinforcement schedules are shown in Figure 37.28. On a variable-interval (VI) schedule, the intervals between reinforcement availability are determined on the basis of time, with the specific value varying from interval to interval and the mean of those values indicated by the schedule parameter value. For example, on a VI 30-second schedule, the specified interval between reinforcement opportunities will vary from one reinforcement delivery to the next, but the average of all the values will be 30 seconds.

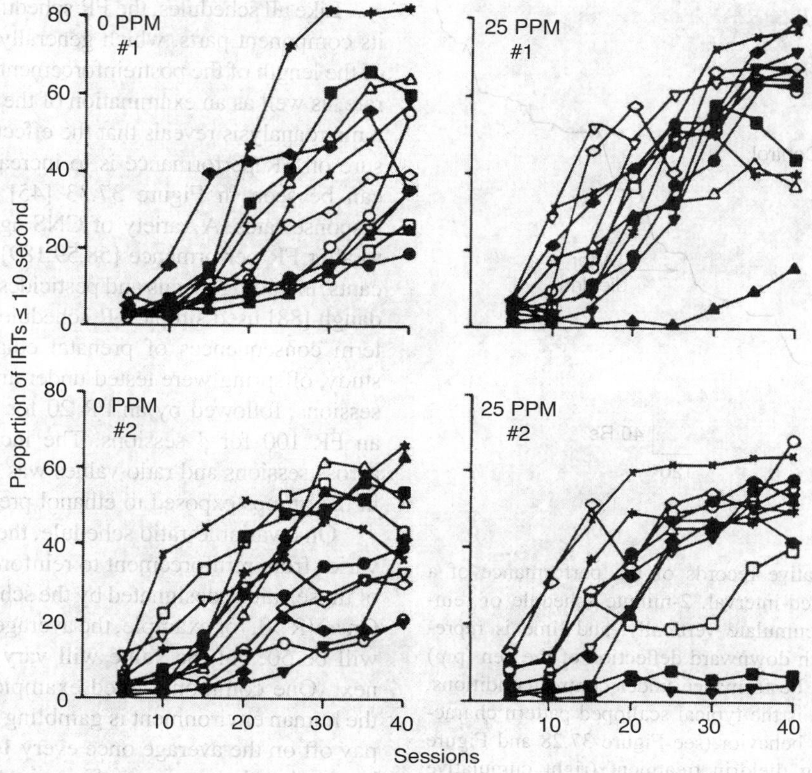


**FIGURE 37.30** Proportion of interresponse times (IRTs) less than or equal to 1.0 second over the course of 40 experimental sessions on a fixed-interval, 1-minute schedule of food reinforcement. Each curve represents the performance of an individual control (left panels) of lead-exposed rat (25 ppm lead acetate in drinking water from weaning; right panels). The top row shows results of the first experiment; row 2 plots show results from a replication experiment. Lead increases overall response rates by increasing the proportion of short IRTs. (From Cory-Slechta, D.A. et al., *Toxicol. Appl. Pharmacol.*, 78, 291–299, 1985. With permission.)

As indicated by Figure 37.28, the VI schedule generates a moderate but steady rate of responding, with little pausing evident after the reinforcement, consistent with the lack of predictability of reinforcement availability. Reinforcement may be available immediately after the previous reinforcer delivery or may be delayed. Reynolds [199] cites as an example a busy signal on the telephone, with the caller continuing to make the response that is reinforced (by a ringing sound) on a VI schedule because of the variable length of telephone conversations. The steady persistent pattern of responding on the VI schedule would suggest its utility as a baseline for detecting toxic effects; however, little information in this regard is currently available, even though alterations in VI performance certainly are produced by central nervous system agents [59].

Ratio schedules require a specified number of responses for reinforcement. On the fixed-ratio schedule, the requirement remains constant; for example, each completion of 100 responses produces food delivery on an FR 100 schedule. The piecework system used early in U.S. history serves as a classic example; salespeople working exclusively on a commission basis is another. The FR schedule typically generates a pattern of performance in which there is a characteristic pause after food delivery, the length of which is related to ratio size, followed by a very rapid rate of responding until the ratio requirement is completed. The high, constant response rate on the FR schedule is the result of the relation between rate of reinforcement and rate of responding; the faster the ratio is completed, the sooner reinforcement delivery occurs. As a result, short IRTs tend to be differentially reinforced, amplifying the relationship. Figure 37.32 [98] plots the frequency distribution of IRTs for a pigeon maintained on an FR 30 schedule of reinforcement on the left and shows that short IRTs predominate. As shown in the right panel, a dose of 1.0 mg/kg of *d*-amphetamine produces marked changes in the IRT distribution, including a shift of the main peak to the right and a decline in the number of very short IRTs. As mentioned earlier, the rate of responding on the FI schedule can vary widely before affecting the frequency of reinforcement delivery; however, decreases in response rate on the FR schedule necessarily decrease the rate of reinforcement. It is precisely this difference in the contingencies controlling the two performances that generates the stark contrasts in FI and FR schedule-controlled performance.

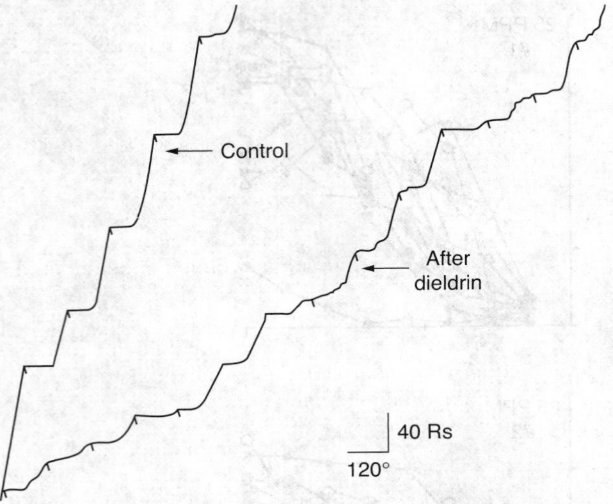


**FIGURE 37.31** Cumulative records of the performance of a Japanese quail on a fixed-interval, 2-minute schedule of reinforcement. Responses cumulate vertically, and time is represented horizontally. Each downward deflection of the pen (pip) indicates food delivery. Performance under control conditions, shown in the left panel, is the typical scalloped pattern characteristic of fixed-interval behavior (see Figure 37.28 and Figure 37.29). After 5.0-mg/kg dieldrin treatment (right cumulative record), the response rate is decreased, as indicated by the shallower slope of the curve, and the typical pause after reinforcement delivery is disrupted. (From Burt, G.A., in *Behavioral Toxicology*, Weiss, B. and Laties, V.G., Eds., Plenum Press, New York, 1975, pp. 241–263. With permission.)

Like all schedules, the FR schedule can be analyzed into its component parts, which generally include measurement of the length of the postreinforcement pause and the running rate, as well as an examination of the IRT distribution. Such a microanalysis reveals that the effect of chronic lead exposure on FR performance is to increase the median IRT, as can be seen in Figure 37.33 [45] and thus to decrease response rates. A variety of CNS agents have been shown to alter FR performance [58,59,189], as have various toxicants, including metals and pesticides [47]. Gentry and Middaugh [88] used simple FR schedules to examine the long-term consequences of prenatal ethanol exposure. In that study, offspring were tested under an FR 1 schedule for 10 sessions, followed by an FR 20 for 9 sessions, and finally an FR 100 for 4 sessions. The increase in response rate across sessions and ratio values was significantly depressed in the groups exposed to ethanol prenatally.

On a variable-ratio schedule, the response requirement varies from reinforcement to reinforcement, with the mean of those values designated by the schedule parameter value. On a VR 50, for example, the average response requirement will be 50, but the value will vary from one ratio to the next. One commonly cited example of a VR schedule in the human environment is gambling. The slot machine may pay off on the average once every 100 plays, but the number of plays between payoffs varies unpredictably. Another example includes the sale of real estate. As can be seen with gamblers, the VR schedule generates very high and consistent rates of responding with little or no pausing (Figure 37.28). Both the VR and the FR schedule generate

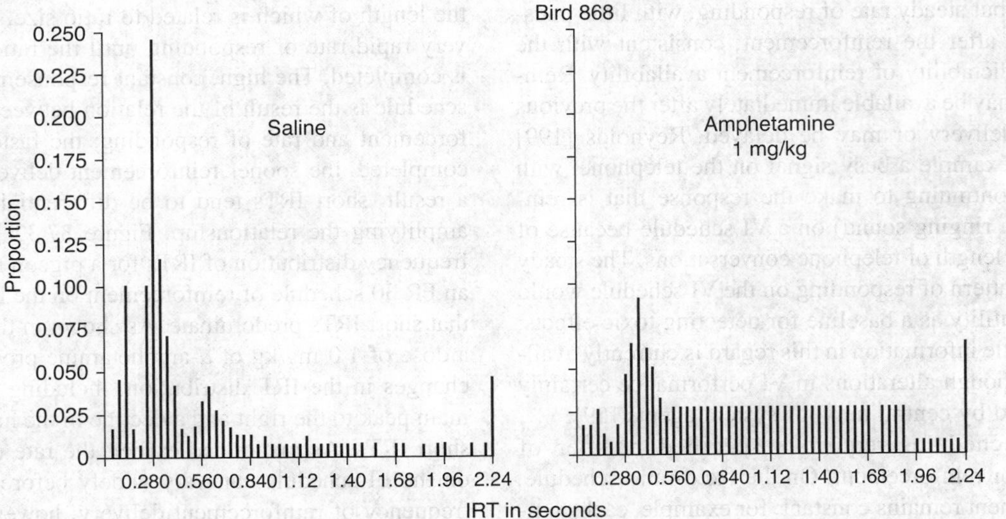


**FIGURE 37.32** Histogram of interresponse times (IRTs) from the performance of a pigeon on a fixed-ratio (FR) 30 schedule of food reinforcement. The temporal resolution was 40 msec/bin on the abscissa, and the distribution was based on a total of 1200 key-peck responses. Performance under saline control conditions is shown in the left panel. After treatment with 1.0 mg/kg *d*-amphetamine the distribution of IRTs changes with a decrease in the proportion of the shortest IRT bins and an increase in the proportion of slightly longer IRTs. (From Gott, C.T. and Weiss, B., *J. Exp. Anal. Behav.*, 18, 481–497, 1972. With permission.)

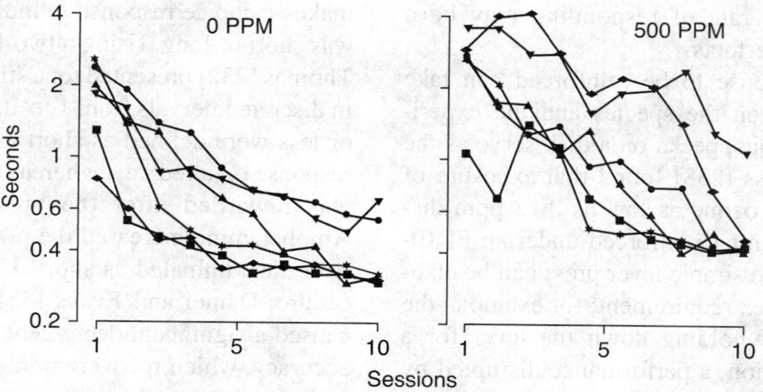


**FIGURE 37.33** Median interresponse time (IRT) values (seconds) for individual control rats (left panel) and rats treated chronically with 500 ppm lead in drinking water from weaning (right panel) over the course of the first 10 experimental sessions on a fixed-ratio (FR) 5 schedule of food reinforcement. Each data point on each curve represents the median IRT value of a subject for the indicated session. Median IRTs of lead-treated rats were considerably longer than controls, accounting for the decreased overall response rates observed in lead-exposed animals. (From Cory-Slechta, D.A., *Neurotoxicol. Teratol.*, 8, 237–244, 1986. With permission.)

very high rates of responding, but they differ in the characteristic pause after reinforcement, seen only on the FR. This difference in response pattern between the two schedules derives from the fairly constant amount of time required to complete the fixed number of responses on the FR schedule. One hypothesis is that reinforcement delivery then becomes a stimulus associated with a subsequent period of non-reinforcement or extinction, decreasing response probability, as it represents the earliest part of such a temporal interval. In contrast, reinforcement opportunity on the VR schedule is unpredictable. A reinforcement may follow after any number of responses since the preceding food delivery; thus, reinforcement delivery itself does not become a stimulus that indicates absence of reinforcement availability. The VR schedule is comparable to the VI schedule in that both maintain fairly constant rates of responding with little or no pausing after reinforcement. The VR, however, generally maintains the higher rates of responding of the two because the faster the ratio is completed, the sooner reinforcement delivery occurs. On the VI schedule, higher rates of responding cannot accelerate reinforcement availability. Sensitivity of the VR schedule to toxic agents remains relatively unexplored.

The differences in performance on these four simple schedules of reinforcement thus reflect the very different contingencies of reinforcement. Comparing the effects of a chemical agent on the various schedules can provide a better understanding of the mechanisms by which drugs or toxicants modify behavior or the behavioral mechanisms by which changes occur. For example, suppression of response rate on all schedules might suggest a nonspecific effect of a treatment on antecedent factors, such as the motivational level of the subject (i.e., an alteration in functional deprivation conditions). An effect specific to a schedule would implicate the unique contingency of that schedule for further study.

Several lines of evidence [60,124] indicate that the type of consequence is less important in determining the behavioral effect of a chemical than is the schedule according to which such consequences are presented. Kelleher and Morse [122] found that both amphetamine and chlorpromazine had similar effects on behavior maintained on a given schedule (a multiple FI–FR, described later), regardless of whether it was maintained by food reinforcement or escape from electric shock. Weiss and Laties [264] studied the effects of amphetamine, chlorpromazine, and pentobarbital on the behavioral regulation of temperature. Shaved rats, placed in a cold compartment, could warm themselves by responding on a lever that turned on a heat lamp for a short period of time. Amphetamine increased the frequency of responding even while elevating skin temperature above normal. Despite accelerating heat loss, chlorpromazine decreased the rate of turning on the heat lamp.

Another variable of importance in determining drug or toxicant effect is the baseline rate of responding [59]. Many compounds, including stimulants, barbiturates, minor tranquilizers, and opiates, have been found to increase the length of short IRTs and to decrease the length of long IRTs on certain reinforcement schedules. In other words, many agents increase low rates of responding and decrease higher rates of responding. For amphetamine, such effects have been noted in several species and across a wide variety of reinforcement schedules [62]. Thus, on an FI schedule, low rates of responding early in the interval may be increased, while the higher rates of responding occurring just prior to reinforcement may be decreased, leading to a loss of the scalloped pattern of responding characteristic of FI performance. One notable exception to the rate-dependency phenomenon is responding suppressed by punishment, which may be even further decreased by amphetamine

[86]. Thus, the baseline rate of responding may be a determinant of toxicant effects.

The designated response to be reinforced can take many forms, depending on the species and the experimenter's goals. For pigeons, pecks on a disk serve as the operant. Tepper and Weiss [238] found that exposure of rats to concentrations of ozone as low as 0.12 ppm disrupted operant wheel running reinforced under an FI 10-minute schedule. Also, the simple lever press can be elaborated into a more complex requirement; for example, the response may consist of holding down the lever for a specified minimum duration, a performance disrupted by chronic lead exposure [48]. A force requirement can also be specified, or the animal may be required to emit a series of responses with different topographies, such as a wheel run followed by a lever press. Another kind of complex response with an extensive literature in behavioral pharmacology specifies that only lever presses separated by a specified interval of time will be reinforced (i.e., the differential reinforcement of the low rate schedule described earlier, which makes the pause part of the operant response).

Temporally based schedules of reinforcement also have been used to study processes analogous to retention (memory). Consider the FI schedule of reinforcement. Food delivery is programmed to follow the first response occurring after a specified interval has elapsed since the preceding food delivery. Responses emitted before the end of the interval carry no penalty. The DRL schedule previously described is another type of interval schedule but does prescribe a penalty for early responding in stipulating a minimum interval (and sometimes a maximum) between successive responses. Neither schedule conventionally provides external stimuli, such as lights or tones, to indicate the passage of time. Both schedules generate distinct temporal patterns of responding. FI schedules foster patterns in which little or no responding occurs at the beginning of the interval, which has become a discriminative stimulus for the absence of reinforcement availability, but as the time of the next food delivery approaches the rate of responding increases. Subjects experienced on the DRL schedule learn to separate successive responses by enough time to earn a high percentage of reinforcements. Increases in response rate on a DRL schedule, such as found by Colotla et al. [43] to be produced by solvent exposure, then decrease the rate of reinforcement.

One criticism leveled against the use of temporally defined schedules of reinforcement in a memory context is the inability to provide a within-session manipulation of the temporal intervals to be discriminated. Although multiple FI–FI schedules (discussed later) represent an alternative approach [78], more direct psychophysical techniques for estimation of time intervals are available [54]. In some duration estimation procedures, for example, a stimulus of specified length is presented and the subject makes a choice response to indicate whether the stimulus was short or long. Using a two-key procedure, Stubbs and Thomas [232] presented tone stimuli to pigeons that varied in discrete intervals from 1 to 10 seconds. Tones 5 seconds or less were defined as short and were reinforced after a response on a red key, whereas responding on a green key was rewarded after the presentation of long tones. Amphetamine increased the proportion of long tones that were discriminated as short. Using a variant of this procedure, Daniel and Evans [55] reported that acrylamide caused a significant decrement in duration discrimination accuracy, which recovered only gradually.

## Complex Schedules

A major advantage of schedule-controlled performance is the flexibility of schedule combinations, so more than one baseline can be studied concurrently in the same subject. This approach compounds the amount of information obtained in an experiment. In a multiple-schedule format, the most common combination, component schedules alternate during the course of an experimental session. Frequently, FI and FR schedules are used as the component schedules because of the marked differences in the contingencies of reinforcement on these two schedules (described earlier). On a multiple FI–FR, the two schedules may alternate either on the basis of time (e.g., every 15 minutes) or after a certain number of reinforcers have been delivered (e.g., switch to the other schedule after every 10 reinforcer deliveries). Different external stimuli are used to indicate to the organism which schedule component is in effect at a given time; for example, for pigeons, a red light might be illuminated throughout each FI component and a green light used to signal the FR component. After some training, each light serves as a discriminative stimulus controlling the performance typical of the reinforcement schedule with which it is associated.

The multiple FI–FR schedule of reinforcement has been used by several investigators to study a range of toxicants, such as mercury vapor [9] and methyl *n*-butyl ketone [5]. In those studies, differential sensitivity of the two schedules was not demonstrated, indicating that these toxicants, at the doses used, produced a generalized impairment of performance. In contrast, Levine [139] demonstrated a greater sensitivity of FI than of FR performance in pigeons exposed to carbon disulfide, a difference similar to that reported for many drugs (Figure 37.34). A decline in FI rate, as indicated by the lower slope of the cumulative record, occurred after a single 8-hour exposure (day 1), whereas the concurrent FR performance remained intact. A 2-day, 8-hour exposure to carbon disulfide was required to disrupt FR responding. Similarly, Leander and MacPhail [134] studied the effects of the pesticide chlordimeform on a multiple FI 1-minute–FR

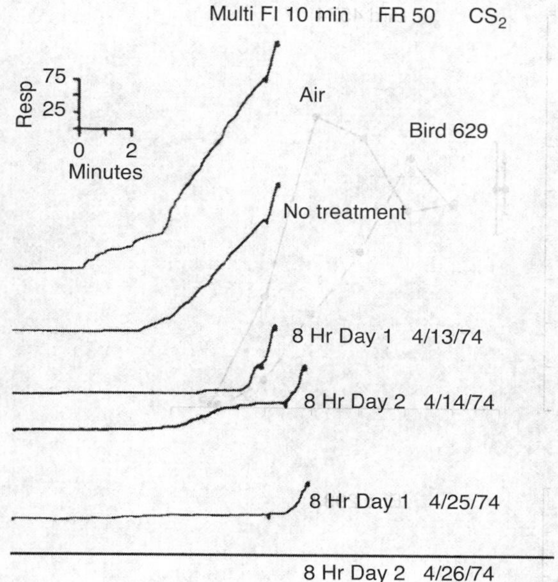


**FIGURE 37.34** Cumulative records of the performance of a pigeon on a multiple fixed-interval (FI) 10-minute fixed-ratio (FR) 50 schedule of food reinforcement. Different stimuli signaled whether the FI or FR component was active. Records show a single FI followed by a single FR from the middle of each behavioral session with the exposures to vehicle or carbon disulfide as indicated. The diagonal slashes (pips) on each curve indicate reinforcement delivery. The first pip shows reinforcement delivery for the FI component; the second, for the FR component. FI performance was disrupted even by a single 8-hour carbon difsulfide exposure, whereas FR performance disruption only occurred after two 8-hour exposures. (From Levine, T.E., *J. Pharmacol. Exp. Ther.*, 199, 669–678, 1976. With permission.)

30 schedule of reinforcement (Figure 37.35) and noted reductions in FI performance at doses lower than those required to disrupt FR performance. Wenger et al. [271] tracked the consequences of trimethyltin exposure in C57B/6N mice responding on a multiple FR 30–FI 10-minute schedule of reinforcement. Rates of responding on the FI schedule had increased substantially within 3 hours of the administration of TMT, whereas FR performance was as yet unchanged. Markedly divergent effects of TMT on these two schedules were observed 5 to 9 days after injection, with substantial rate increases on the FI and decreases on the FR. Such divergent drug or toxicant effects reflect the different reinforcement contingencies of the two-component schedules and the very different behavioral performances they produce. These examples illustrate how the schedule of reinforcement itself may be a powerful determinant of the effect of a chemical agent.

A mixed schedule of reinforcement operates identically to a multiple schedule, but without external stimuli to indicate which schedule component is in effect [78,199]. The only stimuli available to the subject, then, come from its own behavior in relation to the reinforcement schedule contingencies; thus, comparisons of drug or toxicant effects on multiple vs. mixed schedules permit an assessment of the role of discriminative stimulus control over behavior. In that context, Leander and McMillan [135] compared the effects of chlorpromazine on the performance of pigeons under a multiple FR 30–FI 5-minute schedule and a mixed FR 30–FI 5-minute schedule of reinforcement. The contingencies of reinforcement were identical on the multiple and the mixed schedules; they differed only in the extent of stimulus control. On the multiple schedule, blue and red key lights were illuminated during the FR and FI components, respectively, while a white key light was illuminated during both components of the mixed schedule. A dose of 3.0 mg/kg chlorpromazine decreased FR response rates on the mixed schedule, whereas a dose of 100 mg/kg was required to produce an equivalent decrease in FR response rates on the multiple schedule. These results again demonstrate the modulation of the effects of a chemical agent by environmental conditions—in this case, the degree and nature of the stimulus control over the performance. A toxicological example is provided by work using a fixed consecutive number (FCN) schedule of reinforcement [73]. On this baseline, pigeons were required to peck eight or nine times on one key and then one time on a second key for reinforcement. Methylmercury exposure shortened the run of responses on the first key below the required level. When an external stimulus was added to signal the completion of the eight to nine required responses, however, the effects of methylmercury were eliminated. A return of the effects of methylmercury on the FCN baseline re-emerged when the external stimulus was again removed, a further example illustrating how strong external stimulus control can overcome a toxicant-induced discriminative deficit.

A chained schedule of reinforcement also has different stimuli associated with each component of the schedule, but it requires the completion of the entire sequence of schedule components for reinforcement delivery. For example, on a chained FI 5-minute–FR 30 schedule, an external stimulus first signals the FI component and the subject is then required to complete the FI with a response after 5 minutes elapses. This event produces a change in the external stimulus, which acts both as a conditioned reinforcer ($S^r$) for the FI performance that preceded it and as a discriminative stimulus ($S^D$) for the FR schedule component that will follow. After completing the FR requirement, the reinforcer is delivered, and the chain starts again. A tandem schedule of reinforcement is equivalent to the chained schedule, but no external stimuli indicate which component is in effect.

Wood et al. [283] compared performances on FR 8–FR 1 chained and tandem schedules to determine whether the behavioral effects of toluene were modulated by stimulus control. Both schedules required the completion of the FR 8 on one lever, followed by a single response (FR 1) on

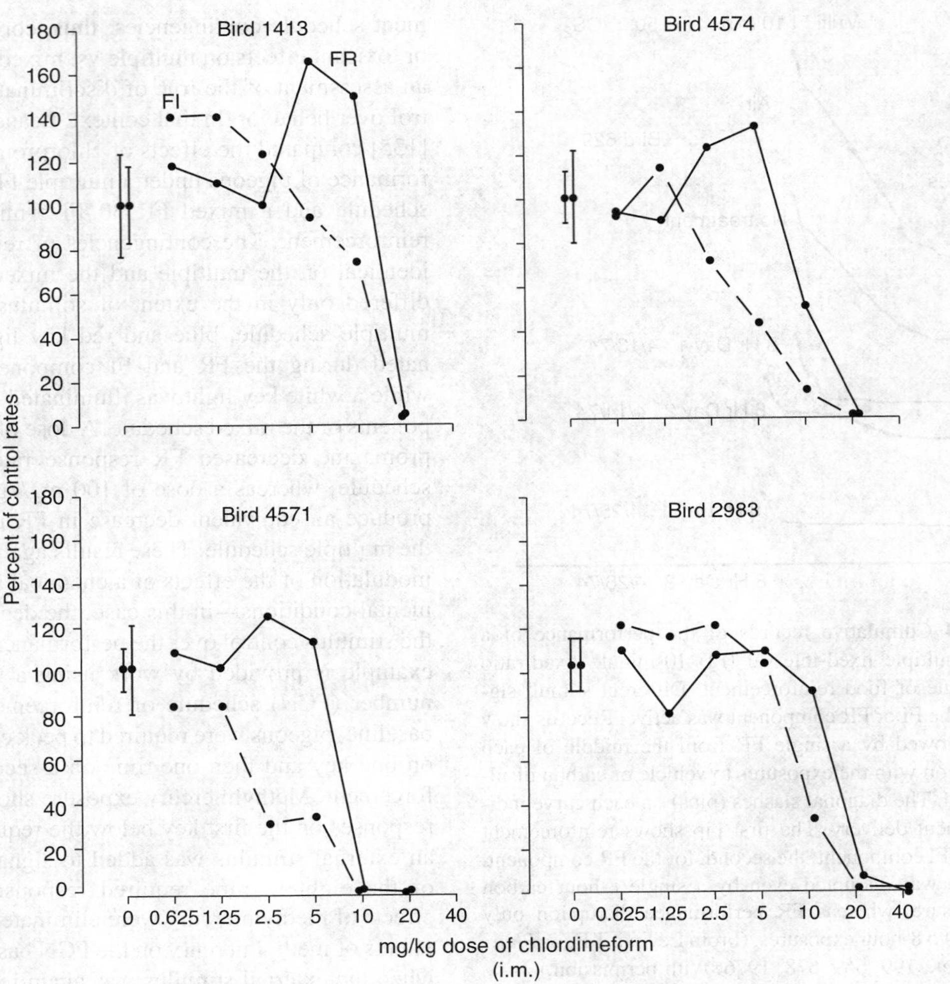


**FIGURE 37.35** Effects of acute administration of chlordimeform on a multiple fixed-interval (FI), fixed-ratio (FR) schedule of food reinforcement in pigeons. Changes in response rates (total responses divided by session time) are plotted in relation to increasing dose of chlordimeform separately for each component and for each of four subjects. Points and brackets above 0 indicate the mean ± S.E. under control conditions. As is evident, overall response rates (ordinate) decreased with increasing dose of chlordimeform (abscissa). FI performance was generally more sensitive, however, as it declined at lower doses than those required to disrupt FR performance. (From Leander, J.D. and Macphail, R.C., *Neurotoxicol. Teratol.*, 2, 315–321, 1980. With permission.)

the other lever for food delivery. During the chained schedule, the light above the first lever served as the discriminative stimulus for the FR 8 component. Completion of the FR 8 activated a tone and illuminated the light above the second lever, signaling the subject that a response on the alternate lever would now result in reinforcement delivery. The first of these two stimuli, then, served as the discriminative stimulus ($S^D$) for FR 8 performance, while the second acted as both a conditioned reinforcer for FR 8 performance and the discriminative stimulus for the FR 1 component, with completion of the latter followed by food delivery. On the tandem schedule, only the house light was illuminated, and it remained illuminated during both the FR 8 and the FR 1 components. Clear differential effects of toluene were observed under these conditions, with performance on the chained schedule, which was under stronger external stimulus control, disrupted far less than the unsignaled tandem performance (see Figure 37.25).

In the complexity of the natural environment, we are routinely faced with many contingent relationships simultaneously, sometimes requiring choices among them. Concurrent schedules, an experimental simulation of such circumstances, facilitate a study of choice by making it possible to select among concurrently operative reinforcement schedules. In an experimental chamber, each such choice would be associated with a separate response device [289]. The choice of component schedules used will, of course, depend on the experimental question of interest. In addition to securing data about preferences for the available reinforcement options, concurrent schedules permit analyses of the behavior of switching between schedule options, a category of responding under the control of the contingencies governing choice. Under some conditions [35], schedule-controlled performance maintained by concurrent schedules differs from the performance observed under conventional schedule conditions.

Despite the wealth of information about complex behavior such as choice yielded by concurrent schedules, few studies of chemical effects on such baselines have as yet been conducted. Newland et al. [184–186] used concurrent schedules of reinforcement to examine the effects of *in utero* exposure to lead, methylmercury, and elemental mercury vapor in squirrel monkeys. Random-interval (RI) schedules were programmed on two response levers, with the reinforcement density always being richer on one of the levers. RI schedules are similar to VI schedules, as the intervals between available reinforcement deliveries are truly random and the RI schedule parameter represent the average of these values. Under steady-state conditions, the behavior of the control monkeys was indeed sensitive to the differences between the two levers in reinforcement density, as indicated by the relative distribution of responses across the levers; that is, monkeys allocated proportionately more time to the response lever associated with the greater reinforcement density. In contrast, the behavior of monkeys exposed to methylmercury and elemental mercury and with blood lead values in excess of 40 μg/dL was more biased and less sensitive to the differences in reinforcement rates. Further, as exemplified by the results with mercury vapor, when the relative reinforcement densities of the two levers were switched, as shown in Figure 37.36, control monkeys gradually shifted to the new, richer lever, whereas the offspring of females exposed to mercury vapor changed more slowly, or not at all, or even in the wrong direction. These effects led the authors to suggest that one behavioral mechanism by which these exposures altered behavior was by causing insensitivity to alterations in reinforcement contingencies.

## STIMULUS PROPERTIES OF CHEMICALS

Chemical agents or drugs themselves may function as unconditioned stimuli, as discriminative stimuli, and as reinforcing stimuli [246]. These roles correspond to those of conventional exteroceptive stimuli in controlling behavior. As reinforcing stimuli, they may pose the problem of dependence and the allied problem of abuse. Chemicals can function as discriminative stimuli in drug discrimination paradigms to provide information about the central nervous system properties of the agent and its neurochemical mediation. Chemicals have also been shown to have the capacity to act as positive reinforcing stimuli in a drug abuse context, as well as negative reinforcing stimuli that provoke avoidance behavior.

### CHEMICAL AS DISCRIMINATIVE STIMULI

The capacity of a chemical to act as a discriminative stimulus and the neurochemical mediation of its central nervous system effects can be evaluated by a simple discrimination procedure referred to as the *drug discrimination*

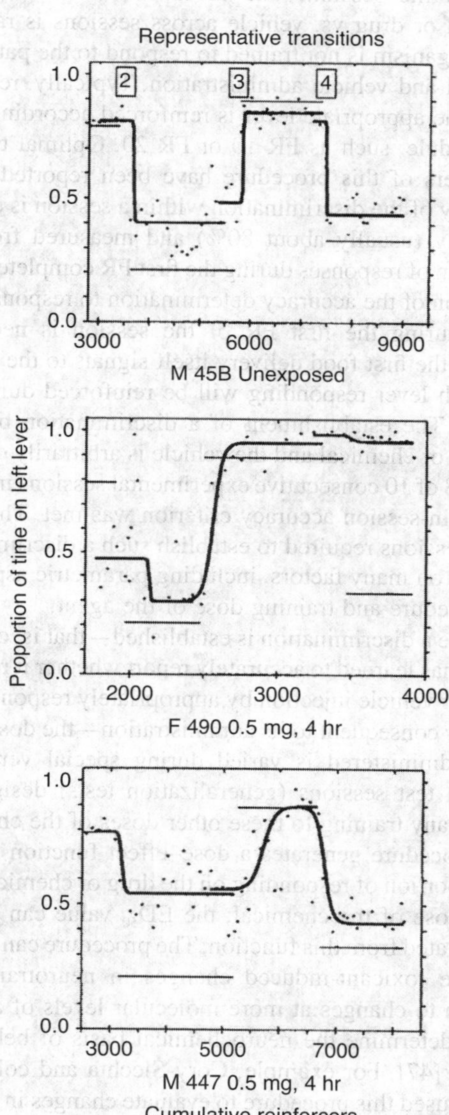


**FIGURE 37.36** Plot of transition performance in one control monkey (M 458) and two monkeys whose mothers were exposed to mercury vapor during gestation. Proportion of time on the left lever should track the proportion of reinforcements (shown by the unconnected horizontal lines) on that lever. The response allocations (proportion of time on the left lever) of the control monkey shifted rapidly when the proportion of reinforcers programmed for the left lever was changed. The two exposed monkeys proved less sensitive to the change in contingencies. The heavy black line shows the fit of a logistic equation to the plotted points. (From Newland, M.C. et al., *Toxicol. Appl. Pharmacol.*, 139, 374–386, 1996. With permission.)

*paradigm*. Before an experimental training session, an organism is injected with, or exposed to, a specified dose of an agent (drug, toxicant) or an appropriate vehicle. Subsequently, responding on one of two levers is reinforced if the vehicle was administered prior to the session but on the alternate lever if drug or toxicant administration preceded the session. The order of administering the chemical or drug vs. vehicle across sessions is random, so the organism is not trained to respond to the pattern of chemical and vehicle administration. Typically, responding on the appropriate lever is reinforced according to an FR schedule, such as FR 10 or FR 20. Optimal training parameters of this procedure have been reported [188]. Accuracy of the discrimination within a session is defined arbitrarily (usually about 80%) and measured from the allocation of responses during the first FR completed. This restriction of the accuracy determination to response allocation during the first FR of the session is necessary because the first food delivery itself signals to the subject on which lever responding will be reinforced during the session. The establishment of a discrimination between the drug or chemical and the vehicle is arbitrarily defined, such as 8 of 10 consecutive experimental sessions in which the within-session accuracy criterion was met. The number of sessions required to establish such a discrimination depends on many factors, including parametric aspects of the procedure and training dose of the agent.

Once a discrimination is established—that is, once the subject has learned to accurately report whether it received a drug or vehicle injection by appropriately responding on the lever consequent to its administration—the dose of the agent administered is varied during special very short duration test sessions (generalization tests) designed to prevent any training to these other doses of the chemical. This procedure generates a dose–effect function relating the proportion of responding on the drug or chemical lever to the dose of the chemical; the $ED_{50}$ value can then be extrapolated from this function. The procedure can be used to relate toxicant-induced changes in neurotransmitter function to changes at more molecular levels of analysis and to determine the neurochemical basis of behavioral toxicity [47]. For example, Cory-Slechta and colleagues [47,52] used this procedure to evaluate changes in dopaminergic and glutamatergic function in relation to lead exposures. Rats were trained to discriminate either a $D_2$ dopamine agonist from saline or a $D_1$ agonist from saline using standard drug discrimination procedures. When the dose–effect curves depicting the proportion of drug lever responding to various doses of these agonists were determined, the $ED_{50}$ values for postweaning (see Figure 37.37) and postnatally lead-exposed rats were significantly shifted to the left of those of control, consistent with dopaminergic supersensitivity. A previous study also has shown a lead-induced subsensitivity to a *d*-amphetamine drug stimulus [288].

### Chemicals as Positive Reinforcers

Operant techniques for experimentally evaluating the efficacy of drugs as reinforcers were developed in the 1960s for both rat [258] and monkey [247]. Animals are often equipped with intravenous catheters attached to infusion pumps through which specified amounts of the drug are administered when response requirements are met by the organism. The development of such procedures established the capability for an experimental analysis of drug dependence, and the resulting literature demonstrated unequivocally the correspondence between most of the substances self-administered experimentally and those abused by humans. Consequently, these techniques are commonly relied on today to evaluate the abuse liability of newly synthesized compounds. In some cases, animals are first trained to self-administer cocaine. The test compound is then substituted for cocaine, and its efficacy in maintaining responding is determined. Abuse potential is not, however, limited to drugs. Volatile materials and aerosols frequently encountered in occupational settings and also commercially available are likewise subject to abuse by inhalation. Using such procedures in squirrel monkeys, Wood demonstrated self-administration of toluene [280] and nitrous oxide [282], two substances that are also abused by humans. Other toxic compounds reported to engender self-administration in humans include *n*-hexane, gasoline, vinyl chloride, and other organic solvents and volatile agents.

### Chemicals as Negative Reinforcers

Drugs or chemicals may also function as negative reinforcers and maintain escape or avoidance behavior. As shown in Figure 37.38, for example, mice will respond to terminate the flow of ammonia through a chamber [279]. Response latency and incidence are directly related to the concentration of ammonia. In other studies, similar aversive properties of ozone [239], acetic acid [212], and formaldehyde [281] have been demonstrated. Also, Tepper and Weiss [238] found that rats working under an FR 20 schedule of reinforcement made fewer responses to release a brake on their running wheels, an avoidance of exercise that increased ozone flow into the lungs. Thus, such techniques can be used to assess both the pleasurable and aversive/irritating properties of a chemical, properties previously considered subjective and consequently barred to experimental evaluation.

## HUMAN BEHAVIORAL TESTING

With the recognition that toxic exposures in the workplace might lead to functional disturbances too subtle to be detected by ordinary clinical or neurological evaluation, a growing body of research based on clinical psychological

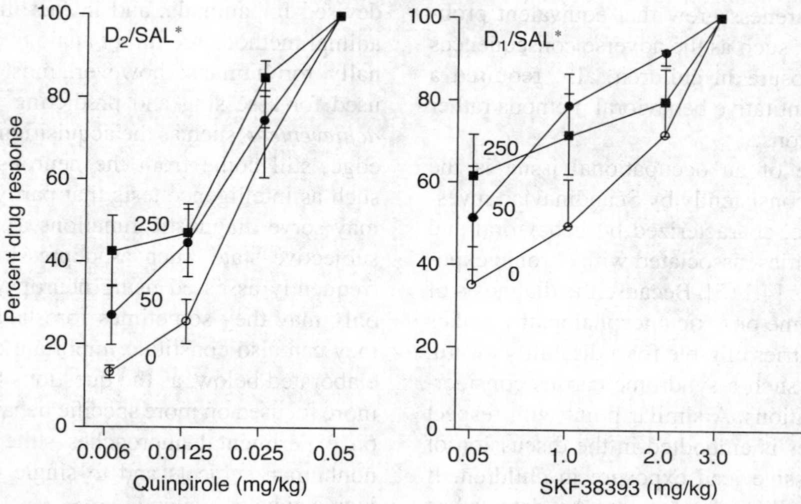


**FIGURE 37.37** Generalization dose–effect curves for control (0 ppm; open circles), 50 ppm (filled circles) and 250 ppm (filled squares) lead-exposed groups trained to discriminate a dopamine $D_2$ receptor agonist from saline (left panel) or a dopamine $D_1$ receptor agonist from saline (right panel) using a drug discrimination paradigm that reinforced responding on one lever if the session was preceded by saline administration and on an alternate lever if dopamine agonist administration preceded the session. Test sessions involved administration of lower doses of the agonist. Each data point shows a group mean ± S.E. based on $n = 10$. The ability to discriminate drug from saline declines with decreasing dose of the drug as expected. These dose–effect curves were shifted to the left following lead treatment, indicating supersensitivity to the dopamine agonists. (From Cory-Slechta, D.A. and Widzowski, D.V., *Brain Res.*, 553, 65–74, 1991. With permission.)

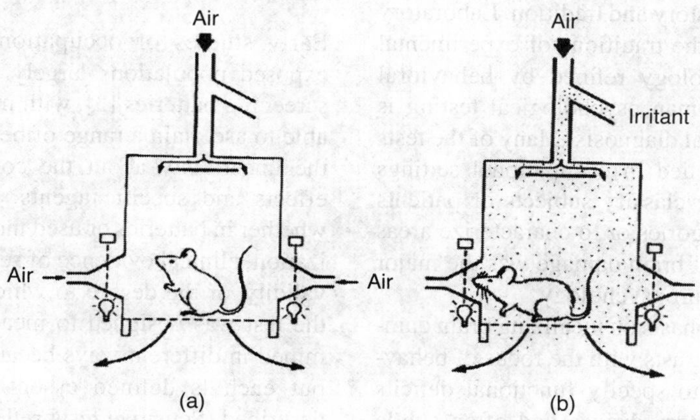


**FIGURE 37.38** Schematic drawing of an exposure chamber used to study the irritant properties of compounds. Panel A shows the situation before irritant delivery; panel B, during irritant delivery. The chamber atmosphere was introduced at the top, and a baffle ensured even mixing. The mouse stood on a perforated stainless-steel platform though which the atmosphere exhausted. The irritant was added to the dilution air immediately above the chamber. Delivery of the irritant could be terminated by a nose poke, as shown in panel B, which interrupted a light beam and was then recorded. Only one of the sensors shown was the active sensor for terminating irritant deliveries; the other served to measure the specificity of any behavioral changes. When the irritant exposure was terminated, either by a response or the end of the trial, a stream of clean humidified air was delivered through each cone to minimize the delay of irritant termination after a response had occurred. (From Wood, R.W., *Toxicol Appl. Pharmacol.*, 50, 157–162, 1979. With permission.)

testing methods began to develop in the early 1970s. Gamberale et al. [85, p. 359] cogently described the rationale for behavioral testing:

> The growing interest in the measurement of performance is most probably due to the sensitivity shown by these methods in unveiling changes in the human organism that otherwise would not be detected. By now, the evidence that these changes are some of the earliest indicators of the occurrence of health effects has become unequivocal. As a consequence, the measurement of performance has come to be regarded by many as a device of major importance for monitoring hazards to health and safety in the work environment.

At the same time, awareness grew that equivalent problems in the community, such as the adverse consequences of excessive lead exposure in children, also required a greater reliance on quantitative behavioral methods rather than clinical examinations.

A salient example of an occupational issue is the claim, advanced most consistently by Scandinavian investigators, of a syndrome, characterized by behavioral and neurological abnormalities, associated with chronic exposure to organic solvents [4,115]. Because the diagnosis of organic solvent syndrome or toxic encephalopathy makes a worker in those countries eligible for a disability award, official recognition of such a syndrome carries considerable economic implications. A similar point with respect to environmental issues is embodied in the discussion of what constitutes excessive lead exposure in children. If lowered scores on intelligence tests can be detected at blood lead concentrations of 10 or even <10 μg/dL, then sources of environmental lead must be much more stringently controlled.

Neuropsychological test methods, confined previously to clinic or laboratory settings, have been increasingly adapted to serve in both field and population studies. Although in principle both clinical neuropsychology and experimental psychology flow from common tenets, there are marked differences in history and tradition. Laboratory animal testing is based on the traditions of experimental psychology and the technology refined by behavioral pharmacology. Much of human psychological testing is rooted in questions of clinical diagnosis. Many of the tests that were most widely applied in occupational settings were designed to originally classify subjects or patients into various diagnostic categories or to characterize areas of deficit. The assessment of brain damage was the major focus of the specialty of neuropsychology.

Neuropsychology's emphasis on localizing brain damage and on focal effects contrasts with the focus of behavioral toxicology, which is to specify functional deficits that generally arise from more diffuse and often subtle deficits. The recent surge of interest by neuropsychology in degenerative diseases such as Huntington's disease and Alzheimer's disease, which also involve diffuse damage, heightens its congruence with toxicology, however, and should help provide additional assessment tools. Brandt and Buttes [23] note, for example, that performance on the arithmetic, digit span, and digit–symbol subtests of the Wechsler scale, discussed later, may indicate modifications of concentration and freedom from distraction rather than a direct impairment of memory function during the early stages of Huntington's disease.

Both standard clinical neuropsychological methods and those developed in experimental psychology have now come to be used in occupational and environmental studies. Advanced tests of sensory and motor function in humans, as described earlier, are closer in design to those devised for animals, and in fact many of the laboratory animal methods are direct analogs of those created originally for humans; however, most techniques currently used for assessing and predicting what is designated as *achievement*, such as the acquisition of specialized knowledge, still come from the neuropsychological tradition, such as intelligence tests that partly attain such aims and may serve diagnostic functions as well. Disturbances of subjective state, such as depression and irritability, are frequently assessed using neuropsychological tools. Not only may they sometimes foreshadow overt toxicity, but they can also constitute important data in themselves. As elaborated below, as the questions being pursued become more focused on more specific behavioral deficits, reliance on experimental approaches suited to both human and nonhuman subjects and to single behavioral domains is increasing.

Certain functional variables have assumed a special role in human testing, partly because of the frequency with which they have been reported, partly because tools for their assessment are available, and partly because they are deemed critical qualities. Only a few of the methods are reviewed here, and they are meant to serve as illustrations.

## Test Choice

Early studies of occupationally and environmentally exposed populations largely involved the utilization of screening batteries [3], with multiple different tests available to ascertain a range of behavioral functions based on the uncertainty about the correlations between specific effects and specific agents. Criteria for test selection, whether in batteries or used individually, include standardization, clinical evidence of sensitivity, selectivity and test validity, or the degree to which a test score reflects what the test was designed to measure. Validity can be determined in different ways because of its many dimensions, but each is defined essentially by the criterion just described. Sensitive tests reflect earlier manifestations of toxicity or efficacy at low doses; tests with a high degree of specificity are less responsive to conditions not relevant to those the test was designed to detect. Another criterion is reliability, or reproducibility. A test that fails to yield consistent results from one administration to the next will bury true differences in excessive variability. A related problem is standardization, which refers to the population from which the norms for certain kinds of tests are derived. The original Stanford–Binet intelligence test, for example, was developed and standardized on a white, middle-class sample of children, so the intelligence quotient (IQ) scores calculated from it probably are misleading when the test is given to black children from other social strata. The mode of framing the question will also help guide choice. If the aim is screening or determining whether toxicity is detectable at a particular exposure level, functional breadth is accorded more importance than selectivity. If the aim is to clarify the nature of the functional deficits or their mechanisms, selectivity will receive greater emphasis.

It is important to note that sensitivity cannot be judged on an absolute scale in human studies, however. Only relative sensitivity, derived from a comparison of different measures, is within our grasp. Unlike the situation in animal studies, we rarely are in a position to exert control over exposure levels, studies with human volunteers being the exception. If one test instrument discriminates an exposed from an unexposed population more definitively than another, we may term it more sensitive. Such results do not indicate whether it could identify a population with even lower exposures. Further, the exposure level at which the effects are observed is not necessarily the lowest level of effect. One approach, advocated by Kennedy et al. [125], calibrates the test battery by relating deficits in computerized performance tests to graded doses of alcohol.

Accumulated experience, coupled with theory and tradition, has led most investigators to select tests for batteries that encompass a fairly common set of behavioral domains. These may include tests of memory, of simple motor function, of complex cognitive performance, of attention and vigilance, and of mood or subjective state. Not all categories may appear in any single study, because some investigations are directed at one or a small cluster of functions. A common choice for inclusion in a test battery has been the Wechsler Adult Intelligence Scale (WAIS), an attractive feature of which is its relatively careful standardization [255] and large body of accumulated experience. The WAIS is actually a battery of tests in itself and was designed to assess a broad sample of functions. It consists of 11 subtests assigned to verbal and performance categories to yield two major subscores (Table 37.6). This division is somewhat arbitrary because the two categories show considerable functional overlap and high correlations. Many investigators have chosen certain WAIS subtests rather than the complete assemblage for incorporation into their own repertoire of tests.

**TABLE 37.6**
**Subtests of the Wechsler Adult Intelligence Scale (WAIS)–Revised**

| | |
|---|---|
| **Verbal** | |
| Information: | Questions of a general nature |
| Comprehension: | Interpretation, judgment |
| Arithmetic: | Numerical calculations |
| Similarities: | Comparisons of nouns |
| Digit span: | Repetition of digit sequences |
| Vocabulary: | Word definitions |
| **Performance** | |
| Digit–symbol: | Symbol–number coding |
| Picture completion: | Identify missing portions |
| Block design: | Duplication of patterns |
| Picture arrangement: | Construct narrative sequence |
| Object assembly: | Assemble jigsaw puzzle |

**TABLE 37.7**
**WHO Neurobehavioral Core Test Battery**

| Functional Domain | Core Test |
|---|---|
| Motor speed, steadiness | Aiming; placing dots in circle |
| Attention | Simple reaction time |
| Perceptual–motor | WAIS digit–symbol |
| Manual dexterity | Santa Ana test |
| Visual memory | Benton Visual Retention Test |
| Auditory memory | WAIS digit span |

**TABLE 37.8**
**Neurobehavioral Evaluation System (NES3)**

| | |
|---|---|
| Vocabulary | Tracing |
| Continuous Performance Test, Letters | Symbol–digit |
| Continuous Performance Test, Animal | Pattern recognition |
| Auditory digit span | Line orientation matching |
| Visual span | Incomplete figures |
| Paced Auditory Serial Addition Test | List learning |
| Sequencing B | List delayed recall |
| Diamond naming | Pattern memory |
| Finger tapping | Profile of Mood States |

Vocabulary tests are examples of the verbal category. An example of the performance category is the digit–symbol substitution test of the WAIS, which requires entering symbols that are paired with numbers (digits) according to a specified code into appropriate blanks on a form. All of the subscales can be combined to yield a full-scale IQ score that correlates highly with academic achievement. The full-scale score, because it is a global score composed of the 11 subscale scores, is unlikely to provide much information about the precise areas of impaired function arising from exposure to a particular toxic agent. For this reason, some investigators have simply adopted certain subscales of the WAIS to assay specific functions; however, the pattern itself is often useful for diagnosis, and techniques have been developed to explicitly relate the profile of performance to diagnosis [213].

Two of the most widely used test batteries in worksite research are shown in Table 37.7 and Table 37.8. The World Health Organization (WHO) Neurobehavioral Core Test Battery consists of a small number of components selected because of broad use and because they were not reliant on instruments or techniques that would be unavailable in developing countries. To verify the applicability of the WHO battery in different settings, Anger [3] compared performance in ten countries representing diverse populations. In general, the results from country to country were remarkably consistent. A major objection to the WHO battery [85] is that its emphasis on traditional manual and paper-and-pencil tests, largely to avoid complicated technical equipment, especially computer-based administration and scoring, diminishes its sensitivity to neurotoxic effects.

The Neurobehavioral Evaluation System (NES), devised originally by Letz and Baker [136] to exploit the potential of computer administration and scoring [13], has been the most widely used of the current batteries (Table 37.8). It embodies three main categories of tests: psychomotor, memory and learning, and cognitive (executive function). It is a much more ambitious battery than the WHO collection and has been translated into several languages other than English. In its current form (NES3), it includes 18 tests of functions such as reaction time, motor coordination, and simple cognition [137]. Other groups have devised batteries of their own, based on personal preferences, theories, or unique aims, but the NES and WHO formats offer the most extensive databases.

Even the more advanced batteries, however, exclude many of the higher level functions such as the types of complex monitoring and decision processes required of aircraft pilots. One development in computerized testing, pioneered by Anger and colleagues [6], uses both computer technology and operant behavior principles to deliver instructions to subjects in novel ways. Reliance upon language to deliver instructions can be an impediment to the use of neurobehavioral tests internationally because of the problem of translation. To avoid such problems, a set of preliminary instructions is provided so the subject's responses are operantly shaped by techniques such as successive approximation to the criterion response to conform to the test requirements. After the subject masters the test procedure, he or she is then ready for the actual test. Even under these circumstances, which are designed to minimize educational and cultural contributions, such variables still exert a pronounced effect. Another study [7] compared subjects of European descent with American Indian, African-American, and Latin-American populations on two consensus neurotoxicity test batteries. One was the WHO collection (Table 37.7), and the other one was assembled by the Agency for Toxic Substances and Disease Registry (ATSDR). Education accounted for the most variance in the tests studied, followed by cultural group. Years of education and cultural group had 13 to 25% shared variance on the cognitive tests, suggesting that these factors should be controlled during the design phase of a study rather than in the statistical analysis. The authors stress that failure to adequately control and analyze these variables could lead to inaccurate conclusions about the association between poor performance and neurotoxic insult.

Another battery certain to gain increasing use in behavioral evaluation is the Cambridge Neuropsychological Testing Automated Battery (CANTAB), a computer-automated battery of neuropsychological tests designed for accurate, sensitive cognitive assessment [82]. It includes tests of memory, attention, and executive function administered via a touch screen, many identical to those used in the experimental psychology domain. Studies carried out to date have shown its sensitivity to a variety of neurodegenerative diseases and behavioral dysfunctions, including Alzheimer's, Parkinson's, and Huntington's diseases; attention deficit disorder; autism; depression; and schizophrenia, as well as age-related deficits in cognition. Moreover, its utility extends from children as young as 4 year old through seniors, and its potential for use in neurotoxicology has been suggested [82]. Recent studies demonstrate the sensitivity of CANTAB in demonstrating deficits in spatial memory and reversal learning in lead-exposed children as young as 4 to 5 years of age [33]. Indeed, this battery shows promise for elaborating specific behavioral deficits related to chemical exposures in human studies. The fact that many of these tests are the same as those used in an extensive experimental animal literature allows extrapolation from a rich base of experimental studies that includes neurobiological and neurochemical substrates of specific functional deficits.

## Memory

Complaints of impaired memory surface frequently in workers exposed to neurotoxic agents and, according to a survey by Anger [3], are the functions most often assessed in work-site research. Memory disorders also appear in adults with focal brain damage and in patients suffering from degenerative neurological diseases. Given the prominence of such complaints in patients and the central role accorded memory in psychological theory and in neuroscience, it is not surprising that numerous tests and techniques are available for the study of this function. A frequent distinction is made between immediate or working or short-term memory and remote or reference or long-term memory. Such distinctions, and the more elaborate ones offered by current workers, often are difficult to apply to specific experimental conditions, and it has even been argued that the term *memory* itself has grown too vague to be useful scientifically Often, terms borrowed from computer technology, such as *storage* and *retrieval*, are alleged to account for memory. But, if memory, as noted earlier, is defined by responses based on earlier experiences, this more neutral and empirical definition is a better platform from which to launch useful experiments, and the examples that follow are presented from an empirical standpoint for that reason.

One widely used test is the WAIS assay of short-term memory known as *digit span*. Traditionally, the examiner calls out a series of numbers that the patient or subject is asked to repeat. Smith et al. [227], like many other current investigators, adapted the digit-span technique to the computer and presented the numbers, one at a time, on a display terminal. Their subject population consisted of workers from two mercury-cell chloralkali plants whose urinary mercury concentrations had been monitored repeatedly. The lists of digits were presented with ascending length, beginning with three digits. If the worker then recited the list correctly, a list with four digits was presented, and so on. Errors on two successive presentations of the same list length halted the test, and the worker's score was noted as the length of the previously correct list. No relationship with mercury excretion was observed, primarily because the standard WAIS procedure proved to be unreliable, yielding a correlation between successive administrations of 0.36. To overcome this inherent flaw, the investigators modified the procedure so they could use probit analysis to estimate the 50% threshold span. With this procedure, the reliability coefficient rose to 0.85, and regression analysis showed a significant correlation between this measure of digit span and urinary mercury concentration.

Hanninen [103] also noted that the digit-span subtest of the WAIS can produce ambivalent results; however, Baker et al. [12], in a study of foundry workers exposed to lead, obtained a significant relationship between blood lead values and performance on a version of the digit-span test that required the subjects to repeat the list backward. It is possible that this result arose not so much from the inability of workers to remember the list but from intruding factors such as the fatigue, depression, and confusion detected by an inventory of subjective state. Complaints of cognitive difficulties and lowered scores on memory tests may also stem from sources such as depression and the inability to concentrate [140]. These potential confounds need to be addressed in interpreting behavioral outcomes. Variations of the digit-span procedure that are said to test verbal memory include letter span and word span. One contribution made possible by computer-based batteries is a restructuring of tests such as digit span to offer adaptive procedures in which the presentation of test items is contingent on the performance of the subject, as in programmed instruction.

In the Benton Visual Retention Test, a test of nonverbal memory, a common component of many neuropsychological test batteries [13], the subject is asked to reproduce a three-figure design. Scoring systems have been developed to quantify fidelity and to correlate types of distortions with different kinds of brain damage. The test is sensitive to brain lesions that lead to neglect of one side of the body [169], but such deficits are uncommon in neurotoxicology. Another variant, the Graham–Kendall Memory-for-Designs Test [140] consists of a series of 15 geometric designs presented to the patient one at a time for 5 seconds. The patient is then asked to copy the design, and the reproduction is scored for various kinds of errors.

The ability to memorize lists of words or nonsense syllables is commonly assessed in neuropsychological test batteries. The Wechsler Memory Scale® [254] contains a paired associate learning task consisting of 10 word pairs. Some, such as baby–cries, are relatively simple; others, such as cabbage–pen, are less transparent. The list is read three times. After each repetition, a test trial is conducted during which the examiner calls out the first word and the subject is required to say the second word of the pair. This kind of learning paradigm has a long history in experimental psychology, and many different kinds of items, including pictures and nonsense syllables, have been used as materials. Adverse effects of food dyes on young children were demonstrated by the paired-associates learning of number and zoo (animal) combinations [235].

As part of their performance battery used to test lead workers, Williamson and Teo [275] also examined paired-associates learning of five pairs of three-letter words. The subject was required to write down the second member of the pair after the first member of the pair was presented. This sequence was repeated until the subject was able to identify all five pairs. Lead-exposed workers recalled fewer items than controls after the first presentation (mean of 0.72 vs. 2.19), required more trials to reach criterion (mean of 3.74 vs. 2.81), and had a higher incidence of individuals failing to reach criterion (64.2% vs. 13.8%).

Although it may have been sensible in the beginning to choose methods for assessing memory function that were based on widespread diagnostic acceptance, new concepts, data, and techniques are emerging at a rapid rate. Sahakian et al. [214] relied on complex matching-to-sample stimuli, presented by computer displays in the CANTAB battery, to evaluate memory deficits in Alzheimer's disease. As previously noted, these emerging trends are bound to influence behavioral toxicology and eventually to displace or augment the traditional approaches that so far have dominated the literature.

## Vigilance and Attention

These two categories are often jointly assessed, and performance on such tasks assumes that the required motor responses are reasonably intact. Highway driving and piloting aircraft are situations that require competence in both. In other situations, vigilance predominates. Some examples are provided by power stations and chemical process plants where remaining alert enough to respond to infrequent or slowly changing signals is critical and is challenged by shift-work schedules that induce chronic sleep disruption. In the laboratory, these functions can be examined in detail because of the available instrumentation and time. In field testing, where simpler, often portable equipment is mandated and where workers must be allowed time from their jobs, performance assessment is usually directed toward global screening for adverse effects rather than determining definitive answers about the parameters of dysfunction.

Reaction-time measures are relatively simple to implement and generate straightforward values that do not require elaborate clinical interpretation. Basic reaction-time paradigms can be modified to provide an immense range of complexity in the assessment of many kinds of stimulus variables. A frequent variation is to compare reaction times under conditions where alterations in responding are required in accordance with the location or other properties of the stimulus. Reaction-time measures seem to reflect both acute and subchronic toxicity resulting from volatile organic solvent exposure. Experiments with workers exposed to styrene, a solvent widely used in fabricating fiberglass boat hulls, called for the experimenters to visit the plant on Friday, familiarize the workers with the reaction-time situation, and deliver a questionnaire about work history and personal habits [38]. On the succeeding Monday, simple reaction times were measured with a portable device at the beginning and end of the shift, and urine was collected by each subject until bedtime. Blood samples were drawn at the end of the shift. Following the suggestions of Gamberale and Kjellberg [84], the rate of stimulus presentation was fairly high (16 trials per minutes), a decision based on findings that such high rates lead to a quicker decline in performance.

Men exposed to styrene began the shift with slower reaction times than did men in a reference group, but by the end of the day the groups did not differ. Within the exposed group, however, those men with the highest blood levels of styrene showed no improvement and displayed the greatest deterioration in ratings of alertness and exhaustion. In a subsequent study, the investigators observed substantial correlations between urinary mandelic acid (a styrene metabolite) and reaction-time performance on Monday morning, indicating that, even with a 60-hour period away from the plant, the effects of exposure were still evident in performance. Exposure levels, it should be noted, did not exceed the British Threshold Limit Value (TLV®) of 100 ppm.

A study with volunteers exposed to 1,1,1-trichloroethane (methyl chloroform, or MC) examined effects on both simple and complex reaction time [144]. Simple reaction time was measured as the time required to respond to a stimulus light presented about eight times per minute during a 5-minute test period. For complex reaction time measures, the subject faced a panel with four lights at the corners of a square and was instructed to respond on one of four buttons beneath the panel, arranged in the same way. The stimuli were flashed at a high rate; only 120 msec separated the next stimulus presentation from the previous response. The same arrangement had been used successfully to study the effects of sleep deprivation, antihistamines, alcohol, and styrene. All subjects were exposed to all treatments. These consisted of 0 ppm, 175 ppm (950 mg/m$^3$), and 350 ppm (1900 mg/m$^3$) of MC. The experimental sessions lasted for 2 hours. MC exposure lengthened both simple reaction time and complex reaction time for those responses made on the correct button.

Needleman et al. [177] demonstrated the sensitivity of reaction time measures to an index of cumulative lead exposure in children by comparing levels of dentine lead in the highest and lowest deciles of a distribution obtained from two Boston suburbs based on more than 2000 children. They used simple reaction times but varied the interval between the ready signal, which alerts the subject to the next stimulus, and the presentation of the stimulus. On two blocks of trials, the interval was specified as 3 seconds; on two other blocks, it was specified as 12 seconds. These values were chosen to probe for the possible influence of distractibility. The longer interval produced longer reaction times in both groups, but for both intervals the children with the higher lead levels responded more slowly than those with the lower levels. A subsequent study by Yule et al. [287] confirmed these results in British children with even lower lead levels, while a study from Germany adds another dimension of support to these findings [277]. In addition to confirming the relationship between tooth lead and reaction time performance, these investigators also observed that, when the children were categorized by social class, the data of those from the less advantaged group showed an even greater correlation with tooth lead than the group as a whole.

Despite a large number of publications featuring or including such measures, however, the power of reaction time measures to detect neurotoxicity remains only superficially exploited. Gamberale et al. [85] and Iregren and Gamberale [114] demonstrated that, with computerized testing and analysis, the sensitivity of such measures can be enhanced and can even contribute to the differentiation of various diagnostic groups, such as those suffering from solvent-induced deficits. One of the limitations of reaction-time measures is the common practice of ignoring stimulus properties such as intensity. Almost never is the brightness of a stimulus light or the properties of a stimulus tone reported, even in studies of agents known to affect vision or hearing. Because reaction time is sensitive enough to stimulus intensity to be used as a psychophysical measure [171], ignoring stimulus characteristics can lead to the suspicion of confounding.

Vigilance includes elements of both motor performance and reaction time but is the subject of a considerable literature of its own. The prototype is the clock test devised by Mackworth, which has served as the criterion task in many studies with drugs [263]. It is arranged so a pointer on a clock face moves 1° every second, except for occasional deflections of two steps. The subject's task is to respond to these infrequent events by pressing a key. Typically, the frequency of detections falls sharply after

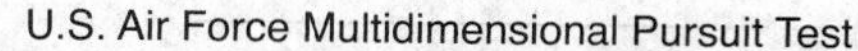

**FIGURE 37.39** U.S. Air Force Multidimensional Pursuit Test. The test was devised during World War II for aviation cadet selection. The four dial pointers drifted randomly from their null positions. The subject's task was to use the joystick, throttle level, and rudder pedals to restore and maintain the pointers at the null positions simultaneously. (From Melton, A.W., Ed., *Apparatus Tests*, Report No. 4, AAF Aviation Psychology Research Reports, U.S. Government Printing Office, Washington, D.C., 1947.)

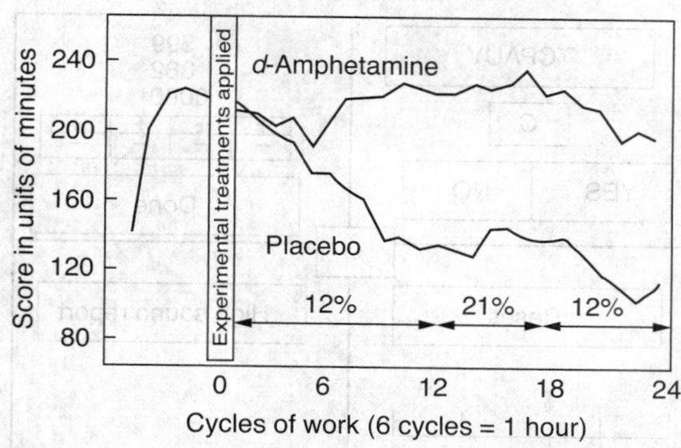


**FIGURE 37.40** U.S. Air Force Multidimensional Pursuit Test performance (see Figure 37.39) as a function of oxygen level (12 or 21%) and *d*-amphetamine administration (5 mg). Treatments were administered after 1 hour of practice. Score refers to the times during which all four dial pointers were held simultaneously at the null positions. (From Payne, R.P. and Hauty, G.T., *J. Aviat. Med.*, 26, 382–389, 1955. With permission.)

the first 30 minutes but can be counteracted by administering drugs, such as the amphetamines, that promote alertness. Dick et al. [63] included the clock test in a battery to evaluate experimental exposures to toluene, methyl ethyl ketone, and ethanol. Exposure to 100 ppm toluene (the TLV) for 4 hours lowered the proportion of correct detections. Ethanol also impaired performance at a dose of 0.8 ml/kg.

More complex vigilance performance, which may demand a certain amount of coordination skills, also offers unexplored potential. In a series of classic studies, Payne and Hauty [192] reported the effects of various drugs and parametric variations on the School of Aviation Medicine Multidimensional Pursuit Test [159], which was designed originally as a tool for pilot selection. (Figure 37.39). The subject was seated in a simulated aircraft cockpit and asked to monitor four dials. The pointers on the dials drifted irregularly from their null positions, and the subject's task was to restore them to null by manipulating a control stick, rudder pedals, and a throttle control. With a typical group of subjects, total time at null reached asymptotic values after about 1 hour of practice, but with further testing began a gradual decline. The rate of decline was enhanced by drugs with sedative actions and retarded by drugs such as the amphetamines, even at remarkably low dose levels (Figure 37.40). The unusual sensitivity of this task to drugs suggests that it, or a contemporary analog based on computer technology, deserves to be evaluated in behavioral toxicology.

The Multidimensional Pursuit Test is not simply watchkeeping, because it also embodies the type of visually directed motor performance called tracking. Mackay et al. [144] examined the acute effects of carbon monoxide on a tracking task. Subjects controlled the position of a cross displayed on an oscilloscope screen by moving a joystick and were required to maintain the cross within a square target that moved continuously around the screen in random patterns. During a single test session, they performed on five 1-minute trials. Carbon monoxide significantly increased error amplitude and decreased time on target during the 2-hour exposure period. Tracking tasks are now included in a number of test batteries; for example, one of the tests used in studies of methylmercury-exposed children required the subjects to follow a sine-wave display by appropriate movements of a joystick [100].

The quantitative methods developed by engineers for evaluating systems in which operators track complex displays, coupled with promising results from the experiments just described and several conducted with patients diagnosed with Parkinson's disease [80], indicate that complex tracking performance could play a more useful role in behavioral toxicology than it has in the past. Particularly with tracking tasks that require predictive adjustments, as would be the case if the subject were instructed to minimize the lag in following a target moving according to a function such as simple harmonic movement, the impact of neurotoxicants on what are called *executive processes* might be assessed more directly.

More typically than not, performance demands occur in settings requiring more than one task to be monitored and responded to concurrently. Driving, piloting, air traffic control, and scanning and adjusting medical monitoring equipment provide examples. Most neurobehavioral test systems strive to examine one function at a time, making extrapolation to multitasking situations somewhat uncertain. A study conducted by Rahill et al. [197] employed a multitasking scheme to evaluate the performance effects of exposure to 100 ppm of toluene (see Figure 37.41).

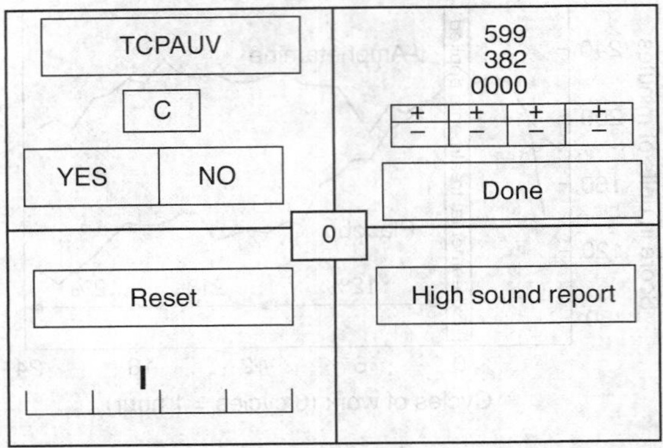


**FIGURE 37.41** Multitasking performance test (SYNWORK). (Top left) A series of letters appears on the screen, then it is removed. In the box below, letters are flashed in sequence, and the subject is required to indicate whether or not the letter belongs to the original set. (Top right) The subject calculates the sum of the two upper numbers and, using the mouse, scans through the numerals 0 to 9 to place them in the appropriate positions in the third number. (Lower left) The cursor drifts toward the end of the scale, and the subject is required, by mouse clicks, to prevent it from reaching the end. (Lower right) A series of tones is presented through a headset. Intermittently, a louder tone is inserted. The subject is asked to respond when the louder tone is detected. The box in the center of the screen displays the subject's numerical score for that session. The SYNWORK task proved sensitive to toluene exposure. (From Rahill, A.A. et al., *Aviat. Space Environ. Med.*, 67, 640–647, 1996. With permission.)

Significantly lower composite performance scores were obtained by the subjects during toluene exposure.

### Subjective State

Certain clusters of symptoms seem to be characteristic of different toxic exposures. Mercury vapor induces a cluster so well known that it earned the label *erethism* (Table 37.9). Chronic solvent exposure is associated with complaints of tiredness, depression, and confusion. Roels et al. [210] and Mergler and Baldwin [161] both found excessive fatigue to dominate the list of subjective complaints among manganese workers. Measurement of such vague symptoms is not a task to be undertaken without an appreciation of the principles of psychometrics. The earlier discussion of validity and reliability probably applies more directly to symptom measurement than to almost any other criterion because accessory criteria are so difficult to define. Memory, for example, can be assessed by enough different techniques to provide some index of consistency. A complaint of depression is not as readily confirmed. To simply construct a questionnaire based on what the investigator believes are the most conspicuous symptoms associated with a particular exposure is a virtual guarantee of uninterpretable data. For this reason, neuropsychologists rely on standardized instruments such as the Beck Depression Inventory [16].

**TABLE 37.9**
**Symptoms of Erethism**

| |
|---|
| Hyperirritability |
| Blushes easily |
| Labile temperament |
| Avoids friends, public places |
| Timid, shy |
| Depressed, despondent |
| Insomnia |
| Fatigue |

An example of the arduous process required to design and develop a quantitative measure of specific symptoms is described by Goldberg [95], who aimed to construct a scale sensitive to minor psychiatric illness. The evolution of the questionnaire proceeded through several steps: selection of items, including careful editing of wording; decisions about the form of the response—that is, whether to ask for a "yes"–"no" response or to ask for some form of rating such as a scale from "infrequent" to "always"; selecting criterion groups and then determining which items of the inventory discriminate among them (a measure of validity); and assessing the consistency of responses within individuals (reliability). Some of these stages may be repeated and items added, deleted, or modified. Some items may carry an excessive burden of social desirability and may not be answered with total candor. Several other well-designed inventories of subjective state are available that also offer considerable data about validity and reliability. Some were constructed, such as the Goldberg questionnaire, to detect mild psychiatric dysfunction.

Some were developed to measure the acute effects of drugs or to follow the course of drug therapy. The Profile of Mood States (POMS) is a self-rating scale designed to measure subjective responses to various therapeutic or experimental maneuvers [157]. It consists of 65 adjectives that the subject judges on a 5-point rating scale that has been tested on both psychiatric patients and normals. A factor analysis yielded the following major factors: tension–anxiety, depression–dejection, anxiety–hostility, vigor–activity, fatigue–inertia, and confusion–bewilderment. Baker et al. [12] included the POMS in a battery given to lead foundry workers and observed marked elevations in several of these factors in workers with the higher blood lead levels. The SCL-90 [57] was devised as a self-report rating scale for psychiatric outpatients, and, like those already referred to, provides a set of dimensions (factors) extracted from the entire 90-item scale. Its designers cite it as particularly useful in clinical drug trials undertaken to evaluate psychoactive agents. Anyone contemplating the framing of an *ad hoc* questionnaire should note the care taken in the construction of the SCL-90. For example, to make the wording of the items accessible to

**TABLE 37.10**
**Items on the Taylor Manifest Anxiety Scale Sensitive to Differences in Organophosphate Insecticide Exposure**

I work under a great deal of tension.
My sleep is fitful and disturbed.
I have periods of such restlessness that I cannot sit long in a chair.
I believe that I am no more nervous than most other people.
I feel anxiety about something or someone almost all the time.
I am usually calm and not easily upset.
It makes me nervous to have to wait.
I practically never blush.

the widest variety of subjects, the designers of the inventory turned to the Thorndike–Lorge *Word Book*, which provides frequency counts obtained from materials such as newspapers for 30,000 words. It allowed them to equate the vocabulary levels of the nine factors and to select the most basic word levels to express the item accurately.

One example of the data that may be extracted with a properly designed inventory was provided by Levin et al. [138]. They compared 24 control subjects matched by age and education with 24 subjects who had been exposed to organophosphate insecticides in recent weeks. The subjects were evaluated by a structured interview, a depression inventory, and the Taylor Manifest Anxiety Scale consisting of 50 true–false items derived from the Minnesota Multiphasic Personality Inventory, which enjoys wide use in psychiatric diagnosis. These items include both psychological and somatic manifestations of anxiety. The structured interview and the depression scale yielded no differences between the two groups, but several items on the Taylor scale provided a clear separation. These are listed in Table 37.10 and support the view that organophosphate exposure may induce subclinical psychological effects that would not be detected by conventional medical surveillance or even by inventories and questionnaires that did not contain specific kinds of symptom items. The 13 commercial applicators in the exposed group (the others were farmers) accounted for the bulk of the difference; their plasma cholinesterase levels, however, fell within the normal range, suggesting that this exposure measure is not an accurate reflection of central cholinergic status.

A more ominous legacy of organophosphate exposure is now undergoing examination. Earlier probes of the aftermath of acute organophosphate poisoning had suggested persisting neurotoxic consequences, but Savage et al. [216] were the first to combine well-documented exposures and neuropsychological measures. They reported on 100 individuals who had experienced a poisoning episode one year earlier and on 100 matched controls. Although clinical chemistries and medical examinations indicated normal function in the exposed group, psychological testing indicated residual deficits as measured by several WAIS subtests and other indices (Table 37.11). Subsequent reports by others have confirmed these findings [211].

**TABLE 37.11**
**Persistent Neurological Sequelae of Acute Organophosphate Poisoning**

| Test Name | 100 Cases | 100 Controls | *p*-Value |
|---|---|---|---|
| WAIS verbal IQ | 105.40 | 111.86 | .001 |
| WAIS performance | 108.41 | 110.13 | .242 |
| WAIS full scale | 107.50 | 111.77 | .001 |
| Impairment rating | 1.07 | 0.91 | .001 |
| Halstead Index | 0.30 | 0.23 | .020 |
| Pegboard | 148.34 | 137.96 | .002 |
| Card sorting | 17.07 | 12.91 | .001 |

## Developmental Assessment

Most of the tests described earlier were devised to assess adults. Many questions about behavioral toxicity, however, arise from the possible impact on early development and involve testing during infancy or childhood, which has been the domain of behavioral teratology. Methylmercury, lead, and polychlorinated biphenyls (PCBs) are examples of substances closely connected with developmental disturbances. Although the essential functions that should be evaluated remain the same throughout the life cycle, early developmental stages can pose several unique problems. One is that young children have not yet acquired or developed the full behavioral and functional repertoire of adults. Another is that early childhood is a period of rapid change with marked individual differences in rate of development, one of the factors that dilutes the ability of tests administered at that time to predict later capacities. For these and associated reasons, investigations of developmental toxicity have to adopt approaches that sometimes differ markedly from those suitable for adults. Infancy is the most difficult stage to evaluate because of the plasticity of infant behavior.

The most common scale of infant development is the Brazelton Neonatal Behavioral Assessment Scale [24], which usually is administered within the first 3 days after birth. It consists of two subscales. One includes items reflecting habituation, motor activity and control, stimulus responsiveness, and similar responses among 27 behavioral items. The other is comprised of 21 neurological items based on elicited reflexes. Although Brazelton scores are not highly correlated with tests of later function, they have proven capable of reflecting the effects of obstetrical medications [223].

One of the best examples of how the Brazelton scale might be used in a study of developmental toxicology is provided by Streissguth et al. [231], who reported their findings on a cohort of about 500 children, selected at birth, to determine the consequences of various levels of maternal alcohol consumption. The mothers of these children were interviewed during the fifth month of pregnancy to obtain data on this and other relevant drug practices, as well as on social, educational, and demographic variables. Habituation in the infant was a special focus of this investigation because of observations indicating that neonates with a clinically overt fetal alcohol syndrome failed to display a response decrement to repeated auditory stimulation. A factor analysis of scores (a procedure that reduced the 27 original Brazelton scores to a smaller number of dependent variables) yielded 6 independent factors. After a multiple-regression analysis, which adjusted for possibly confounding contributions from smoking, caffeine, maternal nutrition, and so on, maternal alcohol consumption during pregnancy remained a significant determinant of poorer habituation and increased low arousal in the infants. The care taken by these investigators to ensure that the examiners were kept uninformed of the other data and the efforts made to train and preserve consistency in examiner ratings can serve as a model for such studies.

The ultimate question for investigators is how well measures obtained during infancy predict later outcomes such as intelligence test scores or school status. One of the more promising techniques, the Fagan Test of Infant Intelligence, measures visual recognition memory [75]. The infant is held so it faces a display with two screens, behind which an observer can record the amount of time the infant's gaze is directed to each screen. On an individual trial, a particular visual stimulus, such as a face, is projected on one of the screens. After the infant has had time to examine the stimulus and after a brief interval, the previous stimulus and a new one are presented simultaneously. The observer measures the amount of time spent by the infant in examining the novel stimulus compared to the one displayed earlier. Infants who score higher on later developmental tests such as the Stanford–Binet tend to spend less time viewing the old stimulus. In studies exploring the developmental neurotoxicity of PCBs, the Fagan test revealed a significant relationship between umbilical cord PCB levels (attributed to maternal consumption of tainted Great Lakes fish) and psychological development [119]. Gunderson et al. [102] found that infant monkeys exposed prenatally to methylmercury spent more time than controls gazing at the previously displayed stimulus.

The Fagan test has also been used in epidemiological studies of methylmercury neurotoxicity. Myers et al. [174], in a prospective study of more than 700 children, examined the relationship between measures of child development and prenatal exposure to methylmercury in the Seychelle Islands of the Indian Ocean. They found no association between the maternal hair mercury level during pregnancy, a marker of maternal exposure, and scores on the Fagan test at 6 months of age.

For later stages of infancy, encompassing a range from 2 to 30 months of age, the Bayley Scales of Infant Development [15] are among the most popular because of the care taken in their standardization and their ability to be used for repeated testing. The Bayley items are arranged chronologically and divided into three subscales: mental, motor, and behavioral. The mental scale is composed of 163 items, ranging from responses to visual and auditory stimuli to abilities such as naming objects. A notable application to toxicology was suggested in studies relating cord blood lead concentrations to performance, summarized by Bellinger et al. [18]. At 6, 12, 18, and 24 months of age, even after correcting statistically for many potential confounding variables such as maternal age and intelligence, a significant relationship between cord blood lead and scores on the mental development index emerged. Children in the group with the highest cord blood lead values (a mean of 14.6 μg/dL, compared to 6.5 for the middle group and 1.8 for the low group) attained the lowest scores at all ages tested. At 24 months of age, the difference between the lowest and highest groups reached 8%, although by that time all three groups had converged toward the same mean blood level of about 7 μg/dL. Similar findings have since emerged from other prospective studies in Cincinnati [64] and Australia [10].

At later ages, the number and variety of instruments available for psychological assessment are overwhelming. Most investigators choosing a battery of tests will typically include one of the intelligence tests designed for children. The Wechsler Intelligence Scale for Children–Revised (WISC-R) [257] is the dominant choice for children older than 6 years of age. In addition to its demonstrated reliability, it is attractive because, like the WAIS, it provides separate verbal and performance subscales that include several individual tasks. These offer a profile of the child's abilities across a variety of functional components. For younger children, the Wechsler Preschool and Primary Scale of Intelligence (WPPSI) [256] yields the same subscale advantages as the WISC and is basically an extension of the WISC for ages 4 to 6.5 years. Another instrument suitable for younger children, the McCarthy Scales of Children's Abilities [153], also extends into early school years, spanning 2.5 to 8.5 years of age. It provides five separate scales that can be combined into what is called a *general cognitive index*, similar to an IQ. All three of these, plus the older Stanford–Binet, have been used to assess the effects of lead. Needleman et al. [177] included the WISC-R in their landmark study, which indicated lowered performance on IQ tests in those children whose tooth lead concentrations fell at the upper portions of the distribution. Their results have since been confirmed by other investigators, including some from outside the United States who also used the WISC-R or an adaptation of it to measure intellectual function. This is one illustration of how reliance on a standardized test allowed different investigations to be combined with some degree of confidence.

**TABLE 37.12**
**Agency for Toxic Substances and Disease Registry Test Battery**

| | |
|---|---|
| Vineland Adaptive Behavior Scales | Kaufman Brief Intelligence Test |
| Parenting Stress Index | Story memory |
| Personality Inventory for Children | Finger tapping |
| Henderson Environmental Learning Process Scale | Divided attention test |
| Family Resources Scale | Visual–motor integration |
| Visual acuity, contrast sensitivity | Purdue Pegboard |
| Contrast sensitivity pretest | Verbal cancellation |
| Vibration II | Story memory delay |

Needleman [176] and Bellinger and Needleman [17], in two companion articles, discuss the strategy for using such scores in a context where many potential confounding variables should be considered. Some examples, besides obvious factors such as race, birth weight, and length of gestation, are maternal age, birth order, parental education, family social class, and quality of the rearing environment. Even when these and other variables were compensated for in the statistical analysis, cord blood lead values remained significantly associated with reduced scores on the mental development index of the Bayley scales. Parallel conclusions have been drawn by other groups.

As in adult epidemiological studies, most investigations try to assemble comprehensive test batteries to try to determine relationships between neurotoxic exposure across a broad spectrum of functional outcomes. For the evaluation of the cohort of another large prospective methylmercury study [100], in the Faroe Islands of the North Atlantic, investigators chose a neuropsychological test battery for 7-year-olds consisting of the following components: finger tapping, hand–eye coordination, Tactual Performance Test (shape discrimination when blindfolded), Continuous Performance Test (vigilance and reaction time), Wechsler Intelligence Scale for Children–Revised (digit spans, similarities, and block designs), Bender Visual Motor Gestalt Test, Boston Naming Test, California Verbal Learning Test (children), and a pictorial analog of the Profile of Mood States survey. They also included a chart that is used to measure visual contrast sensitivity and several neurophysiological measures. Their analyses pointed to mercury-related deficits primarily in the domains of language, attention, and memory, but they also observed deficits in visuospatial and motor functions. Subsequent studies of this cohort have indicated that coexposures to PCBs from consumption of whale blubber [26] could be contributing to these deficits.

Evaluations of the Seychelles cohort, where PCBs are not present, have so far failed to demonstrate any adverse developmental outcomes due to methylmercury [56]. For this assessment, the test battery included the McCarthy Scales of Children's Abilities (the General Cognitive Index), the Preschool Language Scale for measuring expressive and receptive language, two subtests of the Woodcock–Johnson series designed to measure reading and arithmetic achievement, the Bender–Gestalt test, and the Child Behavior Checklist designed to assess social and adaptive behavior.

These two groups of investigators selected somewhat different arrays of instruments for these two parallel investigations because they were pursuing somewhat different questions and because they were evaluating different age ranges and stages of maturity. All investigators agree that no one test battery will be suitable for all purposes or populations. Selection will be guided by the questions asked, the population studied, and the investigator's inclinations. Often, designers of test batteries attempt to provide researchers with a menu of tests, so to speak. One example is the battery assembled [1] to assist the Agency for Toxic Substances and Disease Registry (ATSDR), which is responsible for evaluating health risks at hazardous waste sites in the United States. The prototype test battery consists of the components shown in Table 37.12. The entire battery, because it is designed for screening populations rather than for the intensive appraisal of small groups or individuals, requires only about 1 hour for children. The first five items refer to information secured from the parent.

## EMERGING ISSUES

Like other components of contemporary toxicology, behavioral toxicology is undergoing almost continuous change. Its literature has grown almost exponentially, and the range of techniques applied to its problems has expanded over a broader spectrum of behavior and neuroscience. Compare, for example, the first U.S. book on the topic [267] with a comprehensive survey published almost 20 years later [268]. The range of topics is not only broader but now more firmly based on a comprehensive literature. Especially for human assessment, however, greater efforts are necessary to extend the range, relevance, and specificity of test procedures. Although standardization and an extensive history of use are advantages, they should not limit investigators. Advances in test development are at least as crucial to behavioral toxicology as the analysis of toxic effects themselves. Contrast, for example, the relative crudeness of sensory and motor testing in most test batteries with the precise, elegant methods described in earlier sections of this chapter that have been applied to animals.

Even cognitive function, which has received the bulk of attention from neuropsychology, is only superficially addressed in most test batteries. Among the more promising advances in this area are those stemming from the translation of rigorous laboratory test procedures into practical test methods suitable for clinical and even field evaluations, such as CANTAB, described previously [82,214], and an operant battery for measuring complex performance in children [190] that, like the items on intelligence tests, reflects developmental maturity. Both have the advantage of direct comparability with the techniques used for assessment in animals with the associated literature on neurobiological substrates. In addition, because of the flexibility afforded by computer technology, both can introduce increasingly difficult versions of the same basic test to avoid ceilings on performance. Finally, because they do not require verbal responses, problems of language translation and illiteracy can be bypassed. As more data have become available, the ability to more precisely specify hypotheses with respect to predicted functional deficits should become possible, allowing the use of increasingly sophisticated and specific methods in both human and experimental studies.

## ACKNOWLEDGMENTS

The preparation of this chapter was supported, in part, by grants ES01247, ES08109, ES08958, ES05903, and ES05017 from the National Institute of Environmental Health Sciences.

## QUESTIONS

1. New behavior can be generated in two basic ways: operant conditioning and respondent conditioning. They differ in (a) the type of response; (b) the conditioning procedure; and (c) the criteria for judging the strength of the conditioned response. How would you use each type of conditioning to evaluate impaired learning capacity? What potential confounds must be considered?
2. As a test of memory difficulties, design a procedure to test whether a rat can remember which of two levers (a left and a right, say) it had pressed 10 seconds earlier.
3. Workers in a machine shop, where tools are constantly being degreased, have been complaining that, away from work, their friends and families are telling them that they are confused about identifying colors. Could there be some validity to their complaints? How would you test them?
4. The Food Quality Protection Act of 1996 is a product of growing concern over vulnerability of children to environmental chemicals. What components of a neuropsychological test battery might be used to evaluate whether children living in a community located near a waste dump might be suffering adverse neurobehavioral effects?

## REFERENCES

1. Amler, R. W., Giberini, M., Lybarger, J. A., Hall, A., Kakolweski, K., Phifer, B. L., and Olsen, K. L., Selective approaches to basic neurobehavioral testing of children in environmental health studies. *Neurotoxicol. Teratol.*, 18, 429–434, 1996.
2. Anger, W. K., Worksite behavioral research: results, sensitivity methods, test batteries and the transition from laboratory data to human health. *Neurotoxicology*, 11, 629–720, 1990.
3. Anger, W. K., Assessment of neurotoxicity in humans, in *Neurotoxicology*, Tilson, H. A. and Mitchell, C. L., Eds., Raven Press, New York, 1992, pp. 368–386.
4. Anger, W. K. and Johnson, B. L., Chemicals affecting behavior, in *Neurotoxicity of Industrial and Commercial Chemicals*, O'Donoghue, J., Ed., CRC Press, Boca Raton, FL, 1985, pp. 51–148.
5. Anger, W. K. and Lynch, D. W., The effect of methyl *n*-butyl ketone on response rates of rats performing on a multiple schedule of reinforcement. *Environ. Res.*, 14, 204–211, 1977.
6. Anger, W. K., Rohlman, D. S., Sizemore, O. J., Kovera, C. A., Gibertini, M., and Ger, J., Human behavioral assessment in neurotoxicology: producing appropriate test performance with written and shaping instructions. *Neurotoxicol. Teratol.*, 18, 371–379, 1996.
7. Anger, W. K., Sizemore, O. J., Grossmann, J., Glasser, J. A., Letz, R., and Bowler, R., Human neurobehavioral research methods: impact of subject variables. *Environ. Res.*, 73, 18–41, 1997.
8. Annau, Z. and Eccles, C. U., Prenatal exposure, in *Neurobehavioral Toxicology*, Annau, Z., Ed., The Johns Hopkins University Press, Baltimore, MD, 1986, pp. 153–169.
9. Armstrong, R. D., Leach, L. J., Belluscio, P. R., Maynard, E. A., and Hodge, H. C., Behavioral changes in the pigeon following inhalation of mercury vapor. *Am. Indust. Hyg. Assoc. J.*, 24, 366–375, 1963.
10. Baghurst, P. A., McMichael, A. J., Wigg, N. R., Harleu, J. P., Niles, C. A., Dimse, G. E., and Berkey, C. S., Environmental exposure to lead and children's intelligence at the age of seven years. *N. Engl. J. Med.*, 327, 1279–1284, 1992.
11. Baird, S. J., Catalano, P. J., Ryan, L. M., and Evans, J. S., Evaluation of effect profiles: functional observational battery outcomes. *Fundam. Appl. Toxicol.*, 40, 37–51, 1997.
12. Baker, E. L., Feldman, R. C., White, R. A., Vimpani, G. V., Robertson, E. F., Roberts, R. J., and Tong, S. L., Occupational lead neurotoxicity: a behavioral and electrophysiological evaluation. Study design and year one results. *Br. J. Indust. Med.*, 41, 352–361, 1984.

13. Baker, B. L., Letz, R. B., Fidler, A. T., Shalat, S., Plantamura, D., and Lyndon, M., A computer-based neurobehavioral evaluation system for occupational and environmental epidemiology: methodology and validation studies. *Neurobehav. Toxicol. Teratol.*, 7, 369–377, 1985.
14. Barbeau, A., Manganese and extrapyramidal disorders. *Neurotoxicology*, 5, 13–26, 1984.
15. Bayley, N., *Manual for the Bayley Scales of Infant Development*, Psychological Corp., Berkeley, CA, 1969.
16. Beck, A. T., *Beck Depression Inventory*, Psychological Corp., San Antonio, TX, 1978.
17. Bellinger, D. and Needleman, H. L., Prenatal and early postnatal exposure to lead: developmental effects, correlates, and implications. *Int. J. Mental Health*, 14, 78–111, 1985.
18. Bellinger, D., Leviton, A., Waternaux, C., Needleman, H., and Rabinowitz, M., Longitudinal analyses of prenatal and postnatal lead exposure and early cognitive development. *N. Engl. J. Med.*, 316, 1037–1043, 1987.
19. Beuter, A. and de Geoffroy, A., Can tremor be used to measure the effect of chronic mercury exposure in human subjects? *Neurotoxicology*, 17, 213–227, 1996.
20. Blain, L., Lagace, J. P., and Mergler, D., Sensitivity and specificity of the Lanthony desaturated D-15 panel to assess chromatic discrimination loss among solvent-exposed workers, in *Neurobehavioral Methods in Occupational and Environmental Health*, Regional Office for Europe, World Health Organization, Copenhagen, 1985, pp. 105–109.
21. Boren, J. J., The Study of Performance-Enhancing Drugs with Respect to Acquisition Technique, paper presented at the Eastern Psychological Association, Boston, April, 1967.
22. Boren, J. J. and Devine, D. D., The repeated acquisition of behavioral chains. *J. Exp. Anal. Behav.*, 11, 651–660, 1968.
23. Brandt, J. and Butters, N., The neuropsychology of Huntington's disease. *Trends Neurosci.*, 9, 118–120, 1986.
24. Brazelton, T. B., *Neonatal Behavioral Assessment Scale*, monograph of the National Spastics Society, Lippincott, Philadelphia, PA, 1973.
25. Brooks, A. I., Chadwick, C. A., Gelbard, H. A., Cory-Slechta, D. A., and Federoff, H. J., Paraquat elicited neurobehavioral syndrome caused by dopaminergic neuron loss. *Brain Res.*, 823, 1–10, 1999.
26. Budtz-Jorgensen, E., Keiding, N., Grandjean, P. and Weihe, P., Estimation of health effects of prenatal methylmercury exposure using structural equation models. *Environ. Health*, 1, 2, 2002.
27. Buelke-Sam, J., Kimmel, C. A., and Adams, J., Design considerations in screening for behavioral teratogens: results of the collaborative behavior teratology study. *Neurotoxicol. Teratol.*, 7, 537–789, 1987.
28. Burt, G. A., Use of behavioral techniques in the assessment of environmental contaminants. In: *Behavioral Toxicology*, Weiss, B. and Laties, V. G., Eds., Plenum Press, New York, 1975, pp. 241–263.
29. Bushnell, P. J., Effects of delay, intertrial interval, delay behavior and trimethyltin on spatial delayed response in rats. *Neurotoxicol. Teratol.*, 10, 237–244, 1988.
30. Bushnell, P. J. and Bowman, R. E., Reversal learning deficits in young monkeys exposed to lead. *Pharmacol. Biochem. Behav.*, 10, 733–742, 1979.
31. Cabe, P. A. and Eckerman, D. A., Assessment of learning and memory dysfunction in agent-exposed animals, in *Nervous System Toxicology*, Mitchell, C. L., Ed., Raven Press, New York, 1982, pp. 133–198.
32. Campagna, D., Gobba, F., Mergler, D., Moreau, T., Galassi, C., Cavalleri, A., and Huel, G., Color vision loss among styrene-exposed workers neurotoxicological threshold assessment. *Neurotoxicology*, 17, 367–373, 1996.
33. Canfield, R. L., Gendle, M. H., and Cory-Slechta, D. A., Impaired neuropsychological functioning in lead-exposed children. *Dev. Neuropsychol.*, 26, 513–540, 2004.
34. Carlson, R. L., Van Gelder, C. A., Karas, C. C., and Buck, W. B., Slowed learning in lambs prenatally exposed to lead. *Arch. Environ. Health*, 29, 154–156, 1974.
35. Catania, A. C., Concurrent operants, in *Operant Behavior: Areas of Research and Application*, Honig, W. K., Ed., Appleton-Century-Crofts, New York, 1966, pp. 213–270.
36. Chaffin, D. B. and Miller, J. M., Behavioral and neurological evaluation of workers exposed to inorganic mercury, in *Behavioral Toxicology*, Xintaras, C., Johnson, B. L., and deGroat, I., Eds., Publ. No., 74-126, U.S. Department of Health, Education, and Welfare, Washington, D.C., 1974, pp. 214–239.
37. Chapman, L. J., Sauter, S. L., Henning, R. A., Dodson, V. N., Reddan, W. C., and Matthews, C. C., Differences in frequency of finger tremor in otherwise asymptomatic mercury workers. *Br. J. Indust. Med.*, 47, 838–843, 1990.
38. Cherry, N., Venables, H., and Waldron, H. A., The use of reaction times in solvent exposure, in *Neurobehavioral Methods in Occupational Health*, Cilioli, R., Cassitto, M. C., and Foa, V., Eds., Pergamon Press, New York, 1983, pp. 191–195.
39. Cohen, A. H. and Gans, C., Muscle activity in rat locomotion: movement analysis and electromyography of the flexors and extensors of the elbow. *J. Morphol.*, 146, 177–196, 1975.
40. Cohn, J. and Cory-Slechta, D. A., Differential effects in rats of MK-801, NMDA and scopolamine on learning in a four-member repeated acquisition paradigm. *Behav. Pharmacol.*, 3, 403–413, 1992.
41. Cohn, J. C., Cox, C., and Cory-Slechta, D. A., The effects of lead exposure on learning in a multiple repeated acquisition and performance schedule. *Neurotoxicology*, 14, 329–346, 1993.
42. Cohn, J. C., Ziriax, J., Cox, C., and Cory-Slechta, D. A., Comparison of the error patterns produced by scopolamine and MK-801 on a repeated acquisition and repeated transition baseline. *Psychopharmacology*, 107, 243–254, 1992.
43. Colotla, V., Bautista, S., Lorenzana-Jiminez, M., and Rodriguez, R., Effects of solvents on schedule-controlled behavior. *Neurotoxicol. Teratol.*, 1(Suppl.), 113–118, 1979.
44. Cory-Slechta, D. A., The behavioral toxicity of lead: problems and perspectives, in *Advances in Behavioral Pharmacology*, Vol., 4, Thompson, T., Dews, P. B., and Barrett, J. E., Eds., Academic Press, New York, 1984, pp. 211–256.

45. Cory-Slechta, D. A., Prolonged lead exposure and fixed-ratio performance. *Neurotoxicol. Teratol.*, 8, 237–244, 1986.
46. Cory-Slechta, D. A., Schedule-controlled behavior in neurotoxicology, in *Neurotoxicology*, Tilson, H. A. and Mitchell, C. L., Eds., Raven Press, New York, 1992, pp. 271–294.
47. Cory-Slechta, D. A., Neurotoxicant-induced changes in schedule-controlled behavior, in *Basic Principles of Neurotoxicology*, Chang, L., Ed., Marcel Dekker, New York, 1994, pp. 313–344.
48. Cory-Slechta, D. A., Bissen, S. T., Young, A. M., and Thompson, T., Chronic postweaning lead exposure and response duration performance. *Toxicol. Appl. Pharmacol.*, 60, 78–84, 1981.
49. Cory-Slechta, D. A., O'Mara, D. J., and Brockel, B. J., Learning versus performance impairments following regional administration of MK-801 into nucleus accumbens and dorsomedial striatum. *Behav. Brain Res.*, 102, 181–94, 1999.
50. Cory-Slechta, D. A., Pokora, M. J., and Widzowski, D. V., Behavioral manifestations of prolonged lead exposure initiated at different stages of the life cycle. II. Delayed spatial alternation. *Neurotoxicology*, 12, 761–76, 1991.
51. Cory-Slechta, D. A., Weiss, B., and Cox, C., Performance and exposure indices of rats exposed to low concentrations of lead. *Toxicol. Appl. Pharmacol.*, 78, 291–299, 1985.
52. Cory-Slechta, D. A. and Widzowski, D. V., Low-level lead exposure increases sensitivity to the stimulus properties of dopamine $D_1$ and $D_2$ agonists. *Brain Res.*, 553, 65–74, 1991.
53. Crofton, K. M., Reflex modification and the assessment of sensory function, in *Neurotoxicology*, Tilson, H. A. and Mitchell, C., Eds., Raven Press, New York, 1992, pp. 181–211.
54. Daniel, S. A. and Evans, H. L., Discriminative behavior as an index of toxicity, in *Advances in Behavioral Pharmacology*, Vol. 4, Thompson, T., Dews, P. B., and Barrett, J. E., Eds., Academic Press, New York, 1984, pp. 257–283.
55. Daniel, S. A. and Evans, H. L., Effects of acrylamide on multiple behavioral endpoints in the pigeon. *Neurotoxicol. Teratol.*, 7, 267–273, 1985.
56. Davidson, P. W., Myers, G. J., Cox, C., Axtell, C., Shamlaye, C., Sloane-Reeves, J., Cernichiari, E., Needham, L., Choi, A., Wang, Y., Berlin, M., and Clarkson, T. W., Effects of prenatal and postnatal methylmercury exposure from fish consumption on neurodevelopment: outcomes at 66 months of age in Seychelles Child Development Study. *JAMA*, 280, 701–707, 1998.
57. Derogatis, L.R., Lipman, R. S., Rickels, K., Uhlenhuth, E. H., and Covi, L., The Hopkins Symptom Checklist (HSCL): a self-report system inventory. *Behav. Sci.*, 19, 1–15, 1974.
58. Dews, P. B., Studies on behavior. I. Differential sensitivity to pentobarbital of pecking performance in pigeons depending on schedule of reward. *J. Pharmacol.*, 113, 393–401, 1955.
59. Dews, P. B., Studies on behavior. IV. Stimulant actions of methamphetamine. *J. Pharm. Exp. Ther.*, 122, 137–147, 1958.
60. Dews, P. B. and Morse, W. H., Behavioral pharmacology. *Annu. Rev. Pharmacol.*, 1, 145–174, 1961.
61. Dews, P. B. and Herd, J. A., Behavioral activities and cardiovascular functions: effects of hexamethonium on cardiovascular changes during strong sustained static work in rhesus monkeys. *J. Pharmacol. Exp. Ther.*, 189, 12–23, 1974.
62. Dews, P. B. and Wenger, C. R., Rate-dependency of the behavioral effects of amphetamine, in *Advances in Behavioral Pharmacology*, Vol. 1, Thompson, T. and Dews, P. B., Eds., Academic Press, New York, 1977, pp. 167–229.
63. Dick, R. B., Setzer, J. V., Wait, R., Hayden, M. B., Taylor, B. J., Tolos, B., and Puts-Anderson, V., Effects of acute exposure of toluene and methyl ethyl ketone on psychomotor performance. *Int. Arch. Occup. Environ. Health*, 54, 91–109, 1984.
64. Dietrich, K. N., Kraftt, K. M., Bornschein, R. L., Hammond, P. B., Berger, O., Succop, P. A., and Bier, M., Low-level fetal lead exposure effect on neurobehavioral development in early infancy. *Pediatrics*, 80, 721–730, 1987.
65. Eckerman, D. A. and Bushnell, P. J., The neurotoxicology of cognition: attention, learning, and memory, in *Neurotoxicology*, Tilson, H. A. and Mitchell, C. L., Eds., Raven Press, New York, 1992, pp. 213–270.
66. Eckerman, D. A., Lanson, R. N., and Cummung, W., Acquisition and maintenance of matching without a required observing response. *J. Exp. Anal. Behav.*, 11, 435–442, 1968.
67. Eichenbaum, H. and Otto, T., Odor-guided learning and memory in rats: is it "special?" *Trends Neurosci.*, 16, 22–24, 1993.
68. Elsner, J., Tactile–kinesthatic system of rats as an animal model for minimal brain dysfunction. *Arch. Toxicol.*, 65, 465–473, 1991.
69. Eskin, T. A., Lapham, L. W., Maurissen, J. P. J., and Mergian, W. H., Acrylamide effects on the macaque visual system. *Invest. Ophthalmol. Vis. Sci.*, 26, 317–329, 1985.
70. Eskin, T. A. and Mergian, W. H., Selective acrylamide induced degeneration of color opponent ganglion cells in macaques. *Brain Res.*, 378, 379–384, 1986.
71. Eskin, T. A., Mergian, W. H., and Wood, R. W., Carbon disulfide effects on the visual system. II. Retinogeniculate degeneration. *Invest. Ophthalmol. Vis. Sci.*, 29, 519–527, 1988.
72. Evans, H. L., Early methylmercury signs revealed in visual tests, in *Proceedings of the International Conference on Heavy Metals in the Environment*, Vol. 3, Hutchinson, T. C., Ed., University of Toronto Institute of Environmental Studies, Toronto, 1978, pp. 241–256.
73. Evans, H. L., Scopolamine effects on visual discrimination: modifications related to stimulus control. *J. Pharmacol. Exp. Ther.*, 195, 105–113, 1975.
74. Evans, H. L., Laties, V. G., and Weiss, B., Behavioral effects of mercury and methylmercury. *Fed. Proc.*, 34, 1858–1867, 1975.
75. Fagan, J. F. and McGrath, S. K., Infant recognition memory and later intelligence. *Intelligence*, 5, 121–130, 1981.
76. Falk, J. L., Drug effects on discriminative motor control. *Physiol. Behav.*, 4, 421–427, 1969.
77. Fechter, L. D., Liu, Y., Herr, D. W., and Crofton, K. M., Trichloroethylene ototoxicity: evidence for a cochlear origin. *Toxicol. Sci.*, 42, 28–35, 1998.

78. Ferster, C. B. and Skinner, B. F., *Schedules of Reinforcement*, Prentice Hall, Englewood Cliffs, NJ, 1957.
79. Fitts, P. M., The information capacity of the human motor system in controlling the amplitude of movement. *J. Exp. Psychol.*, 47, 381–391, 1954.
80. Flowers, K., Ballistic and corrective movement on an aiming task: intention tremor and parkinsonian movements compared. *Neurology*, 25, 413–421, 1976.
81. Fowler, S. C., Gramling, S. E., and Liao, R.-M., Effects of pinnozide on emitted force, duration and rate of operant response maintained at low and high levels of required force. *Pharmacol. Biochem. Behav.*, 25, 615–622, 1986.
82. Fray, P. J. and Robbins, T. W., CANTAB battery: proposed utility in neurotoxicology. *Neurotoxicol. Teratol.*, 18, 499–504, 1996.
83. Fry, W., Kelleher, R. T., and Cook, L., A mathematical index of performances on fixed-interval schedules of reinforcement. *J. Exp. Anal. Behav.*, 3, 193–199, 1960.
84. Gamberale, F. and Kjellberg, A., Behavioral performance assessment as a biological control of occupational exposure to neurotoxic substances, in *Neurobehavioral Methods in Occupational Health*, Gilioli, R., Cassitto, M. C., and Foa, V., Eds., Pergamon Press, New York, 1983, pp. 111–121.
85. Gamberale, F., Iregren, A., and Kjellberg, A., Computerized performance testing in neurotoxicology: why, what, how and whereto? In: *Behavioral Measures of Neurotoxicity*, Russell, R. W., Flattau, P. E., and Pope, A. M., Eds., National Academy Press, Washington, D.C., 1990, pp. 359–394.
86. Geller, I. and Seifter, J., The effects of meprobamate, barbiturates, *d*-amphetamine and promazine on experimentally induced conflict in the rat. *Psychopharmacology*, 1, 482–492, 1960.
87. Geller, A. M. and Hudnell, H. K., Critical issues in the use and analysis of the Lanthony desaturate color vision test. *Neurotoxicol. Teratol.*, 19, 455–465, 1997.
88. Gentry, G. D. and Middaugh, L. D., Prenatal ethanol weakens the efficacy of reinforcers for adult mice. *Teratology*, 37, 135–144, 1988.
89. Gentry, D. G., Weiss, B., and Laties, V. G., The microanalysis of fixed-interval responding. *J. Exp. Anal. Behav.*, 39, 327–343, 1983.
90. Gescheider, G. A., *Psychophysics: Method and Theory*, Lawrence Erlbaum Associates, Hillsdale, NJ, 1976.
91. Gilbert, R. M. and Millenson, J. R., *Reinforcement: Behavioral Analyses*, Academic Press, New York, 1972.
92. Glowa, J. R., Behavioral effects of volatile organic solvents, in *Behavioral Pharmacology: The Current Status*, Seiden, L. D. and Balster, L. S., Eds., Alan R. Liss, New York, 1985, pp. 537–552.
93. Gobba, F., Righi, E., Fantuzzi, G., Predieri, G., Cavazzuti, L., and Aggazzotti, G., Two-year evolution of perchloroethylene-induced color-vision loss. *Arch. Environ. Health*, 53, 196–198, 1998.
94. Golani, I., Wolgin, D. L., and Teitelbaum, P., A proposed natural geometry of recovery from akinesia in the lateral hypothalamic rat. *Brain Res.*, 641, 237–267, 1979.
95. Goldberg, D. P., *The Detection of Psychiatric Illness by Questionnaire*, Oxford University Press, London, 1972.
96. Goldberg, L., Charting a course for cell culture alternatives to animal testing. *Fundam. Appl. Toxicol.*, 6, 607–617, 1986.
97. Gorell, J. M., Johnson, C. C., Rybicki, B. A., Peterson, E. L., and Richardson, R. J., The risk of Parkinson's disease with exposure to pesticides, farming, well water, and rural living. *Neurology*, 50, 1346–1350, 1998.
98. Gott, C. T., and Weiss, B., The development of fixed-ratio performance under the influence of ribonucleic acid. *J. Exp. Anal. Behav.*, 18, 481–497, 1972.
99. Grant, W. M., *Toxicology of the Eye*, 3rd ed., Charles C Thomas, Springfield, IL, 1986.
100. Grandjean, P., Weihe, P., White, R. F., Debes, F., Araki, S., Yokoyama, K., Murata, K., Sorensen, N., Dahl, R., and Jorgensen, P. J., Cognitive deficit in 7-year-old children with prenatal exposure to methylmercury. *Neurotoxicol. Teratol.*, 19, 417–428, 1997.
101. Grant-Webster, K. S., Gunderson, V. M., and Burbacher, T. M., Behavioral assessment of young nonhuman primates: perceptual–cognitive development. *Neurotoxicol. Teratol.*, 12, 543–546, 1990.
102. Gunderson, V. M., Grant, K. S., Barbacher, T. M., Fagan, J. F., and Mottet, U. K., The effect of low-level prenatal methylmercury exposure on visual recognition memory in infant crab-eating macaques. *Child. Dev.*, 57, 1076–1083, 1986.
103. Hanninen, H., Psychological test batteries: new trends and developments, in *Neurobehavioral Methods in Occupational Health*, Gillioli, R., Cassitto, M. G., and Foa, V., Eds., Pergamon Press, New York, 1983, pp. 123–129.
104. Hanson, H. M., Witoslawski, J. J., and Campbell, E. H., Reversible disruption of a wavelength discrimination in pigeons following administration of pheniprazine. *Toxicol. Appl. Pharmacol.*, 6, 690–695, 1964.
105. Harlow, H. F., The formation of learning pets. *Psychol. Rev.*, 56, 51–65, 1949.
106. Hastings, L., Sensory neurotoxicology: use of the olfactory system in the assessment of toxicity. *Neurotoxicol. Teratol.*, 12, 455–459, 1990.
107. Hastings, L., Cooper, G. P., Bornschein, R. L., and Michaelson, I. A., Behavioral deficits in adult rats following neonatal lead exposure. *Neurotoxicol. Teratol.*, 1, 227–231, 1979.
108. Heise, G. A., Behavioral methods for measuring effects of drugs on learning and memory in animals. *Med. Res. Rev.*, 4, 535–558, 1984.
109. Hernstein, R. J. and Morse, W. H., Effects of pentobarbital on intermittently reinforced behavior. *Science*, 125, 929–931, 1957.
110. Hindmarch, I., Alford, C., Barwell, F., and Kerr, J. S., Measuring the side effects of psychotropics: the behavioral toxicity of antidepressants. *J. Psychopharmacol.*, 6, 198–203, 1992.
111. Holson, R. R. and Buelke-Sam, J., Design and analysis issues in developmental neurotoxicology: papers from a symposium on experimental design and statistical analysis. *Neurotoxicol. Teratol.*, 14, 197–228, 1992.
112. Infurna, R. and Weiss, B., Neonatal behavioral toxicity in rats following prenatal exposure to methanol. *Teratology*, 33, 259–265, 1986.

113. Iregren, A., Psychological test performance in foundry workers exposed to low levels of manganese. *Neurotoxicol. Teratol.*, 12, 673–675, 1990.
114. Iregren, A. and Gamberale, F., Human behavioral toxicology: central nervous effects of low-dose exposure to neurotoxic substances in the work environment. *Scand. J. Work. Environ. Health*, 16(Suppl. 1), 17–25, 1990.
115. Iregren, A., Gamberale, F., and Kjellberg, A., SPES: a psychological test system to diagnose environmental hazards. Swedish Performance Evaluation System. *Neurotoxicol. Teratol.*, 18, 485–91, 1996.
116. Ison, J. R. and Hoffman, H. S., Reflex modification in the domain of startle. II. The anomalous history of a robust and ubiquitous phenomenon. *Psychol. Bull.*, 94, 3–17, 1983.
117. Iversen, I. H., Ragnarsdottir, A., and Randrup, K. I., Operant conditioning of autogrooming in vervet monkeys (*Ciropithecus aethiops*). *J. Exp. Anal. Behav.*, 42, 189–191, 1984.
118. Iversen, S. D. and Iversen, L. L., *Behavioral Pharmacology*, 2nd ed., Oxford University Press, New York, 1981.
119. Jacobson, J. L., Jacobson, S. W., and Humphrey, H. E., Effects of exposure to PCBs and related compounds on growth and activity in children. *Neurotoxicol. Teratol.*, 12, 319–326, 1990.
120. Jenkins, W. O. and Stanley, J. C., Partial reinforcement: a review and critique. *Psychol. Bull.*, 47, 193–234, 1950.
121. Special issue on behavioral pharmacology. *J. Exp. Anal. Behav.*, 56, 167–423, 1991.
122. Kelleher, R. T. and Morse, W. H., Escape behavior and punished behavior. *Fed. Proc.*, 23, 808–817, 1964.
123. Kelleher, R. and Morse, W. H., Determinants of the specificity of behavioral effects of drugs. *Ergebnisse der Physiol.*, 60, 1–56, 1968.
124. Kelleher, R. T. and Morse, W. H., Determinants of the behavioral effects of drugs. In: *Importance of Fundamental Principles of Drug Evaluation*, Tedeschi, D. H. and Tedeschi, R. E., Eds., Raven Press, New York, 1968, pp. 383–405.
125. Kennedy, R. S., Turnage, J. J., and Lane N. E., Development of surrogate methodologies for operational performance measurement: empirical studies. *Human Perf.*, 10, 251–282, 1997.
126. Korogi, Y., Takahashi, M., Okajima, T., and Eto, K., MR findings of Minamata disease-organic mercury poisoning. *J. Magn. Reson. Imaging*, 8, 308–16, 1998.
127. Kremer, B., Klimek, L., and Mosges R., Clinical validation of a new olfactory test. *Eur. Arch. Otorinolaryngol.*, 255, 355–58, 1998.
128. Kulig, B. M. and Lammers, J. H. M., Assessment of neurotoxicant-induced effects on motor function, in *Neurotoxicology*, Tilson, H. A. and Mitchell, C. L., Eds., Raven Press, New York, 1992, pp. 147–179.
129. Langolf, G. D., Chaffin, D. B., Henderson, R., and Whittle, H. P., Evaluation of workers exposed to elemental mercury using quantitative test of tremor and neuromuscular functions. *Am. Indust. Hyg. Assoc. J.*, 39, 976–984, 1978.
130. Laties, V. G., How operant conditioning can contribute to behavioral toxicology. *Environ. Health Perspect.*, 26, 29–35, 1978.
131. Laties, V. G. and Weiss, B., Thyroid state and working for heat in the cold. *Am. J. Physiol.*, 197, 1028–1034, 1959.
132. Laties, V. G. and Weiss, B., Effects of a concurrent task on fixed-interval responding in humans. *J. Exp. Anal. Behav.*, 6, 431–436, 1963.
133. Laties, V. G., Wood, R. W., and Rees, D. C., Stimulus control and the effects of *d*-amphetamine in the rat. *Psychopharmacology*, 75, 277–282, 1981.
134. Leander, J. D. and MacPhail, R. C., Effect of chlordimeform (a formamidine pesticide) on schedule-controlled responding pigeons. *Neurotoxicol. Teratol.*, 2, 315–321, 1980.
135. Leander, J. D. and McMillian, D. E., Rate-dependent effects of drugs. I. Comparisons of *d*-amphetamine, pentobarbital and chlorpromazine on multiple and mixed schedules. *J. Pharmacol. Exp. Ther.*, 188, 726–739, 1974.
136. Letz, R. and Baker, E. L., Computer-administered neurobehavioral testing in occupational health. *Semin. Occup. Med.*, 1, 197–203, 1986.
137. Letz, R., DiIorio, C. K., Shafer, P. O., Yeager, K. A., Schomer, D. L., and Henry, T. R., Further standardization of some NES3 tests. *Neurotoxicology*, 24, 491–501, 2003.
138. Levin, H. S., Rodnitzky, R. L., and Mick, D. L., Anxiety associated with exposure to organophosphate compounds. *Arch. Gen. Psychiatry*, 33, 225–228, 1976.
139. Levine, T. E., Effects of carbon disulfide and FLA-63 on operant behavior of pigeons. *J. Pharmacol. Exp. Ther.*, 199, 669–678, 1976.
140. Lezak, M. D., *Neuropsychological Assessment*. Oxford University Press, New York, 1976.
141. Lindner, M. D. and Gribkoff, V. K., Relationship between performance in the Morris water task, visual acuity, and thermoregulatory function in aged F-344 rats. *Behav. Brain Res.*, 45, 445–455, 1991.
142. Lowndes, H. E., Baker, T., Cho, E. S., and Jortner, B., Position sensitivity of de-efferented muscle spindles in experimental acrylamide neuropathy. *J. Pharmacol. Exp. Ther.*, 205, 40–48, 1978.
143. Lynch, J. J., Silveira, L. C., Perry, V. H., and Merigan, W. H., Visual effects of damage to P ganglion cells in macaques. *Vis. Neurosci.*, 8, 575–582, 1992.
144. Mackay, C. J., Campbell, L., Samuel, A. M., Alderman, K. J., Idzikowslei, C., Wilson, H. K., and Gompertz, D., Behavioral changes during exposure to 1,1,1-trichloroethane: time-course and relationship to blood solvent levels. *Am. J. Indust. Med.*, 11, 223–239, 1987.
145. MacPhail, R. C., Effects of pesticides on schedule-controlled behavior, in *Behavioral Pharmacology: The Current Status*, Seiden, L. S. and Balster, R. L., Alan R. Liss, New York, 1985, pp. 519–536.
146. Maizlish, N. A., Langolf, C. D., Whitehead, L. W., Fine, L. J., Albers, J. W., Goldberg, J., and Smith, P., Behavioral evaluation of workers exposed to mixtures of organic solvents. *Br. J. Indust. Med.*, 42, 579–590, 1985.
147. Mattson, J. L., Boyes, W. K., and Ross, J. F., Incorporating evoked potentials into neurotoxicity test schemes, in *Neurotoxicology*, Tilson, H. A. and Mitchell, C. M., Eds., Raven Press, New York, 1992, pp.125–145.
148. Maurissen, J. P. J., Quantitative sensory assessment in toxicology and occupational medicine: applications, theory, and critical appraisal. *Toxicol. Lett.*, 43, 321–343, 1988.

149. Maurissen, J. P. J. and Weiss, B., Vibration sensitivity as an index of somatosensory function, in *Experimental and Clinical Neurotoxicity*, Spencer, P. S. and Schaumbers, H. H., Eds., Williams & Wilkins, Baltimore, MD, 1980, pp. 767–774.
150. Maurissen, J. P. J., Weiss, B., and Cox, C., Vibration sensitivity recovery after a second course of acrylamide intoxication. *Fundam. Appl. Toxicol.*, 15, 93–98, 1990.
151. Maurissen, J. P. J., Weiss, B., and Davis, H. T., Somatosensory thresholds in monkeys exposed to acrylamide. *Toxicol. Appl. Pharmacol.*, 71, 266–279, 1983.
152. Maurissen, J. P. J., Neurobehavioral methods for the evaluation of sensory functions, in *Neurotoxicology. Approaches and Methods*, Chang, L. W. and Slikker, W., Eds., Academic Press, San Diego, CA, 1995, pp. 239–264.
153. McCarthy, D., *The Manual for the McCarthy Scales of Children's Abilities*, Psychological Corp., New York, 1972.
154. McKearney, J. W. and Barrett, J. E., Schedule-controlled behavior and the effects of drugs, in *Contemporary Research in Behavioral Pharmacology*, Blackman, D. B. and Sanger, D. J., Eds., Plenum Press, New York, 1978, pp. 1–68.
155. McKearney, J. W., Interrelations among prior experience and current conditions in the determination of behavior and the effects of drugs, in *Advances in Behavioral Pharmacology*, Vol., 2, Thompson, T. and Dews, P. B., Eds., Academic Press, New York, 1979, pp. 39–64.
156. McMillan, D. B. and Leander, J. D., Effects of drugs on schedule-controlled behavior, in *Behavioral Pharmacology*, Click, S. D. and Goldfard, J., Eds., Mosby, St. Louis, MO, 1976, pp. 85–139.
157. McNair, D. M. and Kahn, R. J., Self-assessment of cognitive deficits, in *Assessment in Geriatric Psychopharmacology*, Crook, T., Ferris, S., and Bartus, E., Eds., Mark Powley Associates, New Canaan, CT, 1983, pp. 13–143.
158. Means, L. W., Alexander, S. R., and O'Neal, M. F., Those cheating rats: male and female rats use odor trails in a water-escape 'working memory' task. *Behav. Neural Biol.*, 58, 144–151, 1982.
159. Melton, A. W., Ed., *Apparatus Tests*, Report No. 4, AAF Aviation Psychology Research Reports, U.S. Government Printing Office, Washington, D.C., 1947.
160. Mergler, D. and Blain, L., Assessing color vision loss among solvent-exposed workers. *Am. J. Indust. Med.*, 12, 195–203, 1987.
161. Mergler, D. and Baldwin, M., Early manifestations of manganese neurotoxicity in humans: an update. *Environ. Res.*, 73, 92–100, 1997.
162. Merigan, W. H., Effects of toxicants on visual systems. *Neurobehav. Toxicol.*, 1(Suppl. 1), 15–22, 1979.
163. Merigan, W. H. and Weiss, B., Eds., *Neurotoxicity of the Visual System*, Raven Press, New York, 1980.
164. Merigan, W. H., Barkdoll, B., and Maurissen, J. P. J., Acrylamide-induced visual impairment in primates. *Toxicol. Appl. Pharmacol.*, 62, 342–345, 1982.
165. Merigan, W. H., Barkdoll, B., Maurissen, J. P. J., Eskin, T. A., and Lapham, L. W., Acrylamide effects on the macaque visual system. I. Psychophysics and electrophysiology. *Invest. Ophthalmol. Vis. Sci.*, 26, 309–316, 1985.
166. Merigan, W. H. and Eskin, T. A., Spatio-temporal vision of macaques with severe loss of P-3 retinal ganglion cells. *Vision Res.*, 26, 1751–1761, 1986.
167. Merigan, W. H., Wood, R. W., Zehl, D., and Eskin, T. A., Carbon disulfide effects on the visual system. I. Visual thresholds and ophthalmoscopy. *Invest. Ophthalmol. Vis. Sci.*, 29, 512–518, 1988.
168. Merigan, W. H., Chromatic and achromatic vision of macaques: role of the P pathway. *J. Neurosci.*, 9, 776–783, 1989.
169. Mesulam, M.-M., *Principles of Behavioral Neurology*. F.A. Davis, Philadelphia, PA, 1986.
170. Montoya, C. P., Campbell, H. L., Pemberton, K. D., and Dunnett, S. B., The 'staircase test': a measure of independent forelimb reaching and grasping abilities in rats. *J. Neurosci. Meth.*, 36, 219–228, 1991.
171. Moody, D. B., Reaction time as an index of sensory function, in *Animal Psychophysics: The Design and Conduct of Sensory Experiments*, Stebbins, W. C., Ed., Appleton-Century-Crofts, New York, 1970, pp. 277–302.
172. Moser, V. C., McCormick, J. P., Creason, J. P., and MacPhail, R. C., Comparison of chlordimeform and carbaryl using a functional observation battery. *Fundam. Appl. Toxicol.*, 11, 189–206, 1988.
173. Moxley, R., Graphics for three-term contingencies. *Behav. Anal.*, 5, 45–51, 1982.
174. Myers, G. J., Marsh, D. O., Davidson, P. W., Cox, C., Shamlaye, C. F., Tanner, M., Choi, A., Cernichiari, E., Choisy, O., and Clarkson, T. W., Main neurodevelopmental study of Seychellois children following *in utero* exposure to methylmercury from a maternal fish diet: outcome at six months. *Neurotoxicology*, 16, 653–664, 1995.
175. National Research Council Committee on Neurotoxicology and Models for Assessing Risk, *Environmental Neurotoxicology*, National Academy Press, Washington, D.C., 1992.
176. Needleman, H. L., The neurobehavioral effects of low-level exposure to lead in childhood. *Int. J. Mental Health*, 14, 64–77, 1985.
177. Needleman, H. L., Gunnoe, C., Leviton, A., Reed, R., Peresie, H., Maher, C., and Barrett, P., Deficits in psychological and classroom performance of children with elevated dentine lead levels. *N. Engl. J. Med.*, 300, 689–695, 1979.
178. Newland, M. C., Quantification of motor function in toxicology. *Toxicol. Lett.*, 43, 295– 319, 1988.
179. Newland, M. C., Ceckler, T. L., Kordower, J. H., and Weiss, B., Visualizing manganese in the primate basal ganglia with magnetic resonance imaging. *Exp. Neurol.*, 106, 251–258, 1989.
180. Newland, M. C. and Weiss, B., Ethanol's effects on tremor and positioning in squirrel monkeys. *J. Stud. Alcohol*, 52, 492–499, 1991.
181. Newland, M. C. and Weiss, B., Persistent effects of manganese on effortful responding and their relationship to manganese accumulation in the primate globus pallidus. *Toxicol. Appl. Pharmacol*, 113, 87–97, 1992.
182. Newland, M. C. and Weiss, B., Drug effects on an effortful operant: pentobarbital and amphetamine. *Pharmacol. Biochem. Behav.*, 36, 381–387, 1990.

183. Newland, M. C., Motor function and the physical properties of the operant: applications to screening and advanced techniques, in *Neurotoxicology: Approaches and Methods*, Chang, L., Ed., Academic Press, San Diego, CA, 1995, pp. 265–299.
184. Newland, M. C., Yezhou, S., Logdberg, B., and Berlin, M., *In utero* lead exposure in squirrel monkeys: motor effects seen with schedule-controlled behavior. *Neurotoxicol. Teratol.*, 18, 33–40, 1996.
185. Newland, M. C., Yezhou, S., Logdberg, B., and Berlin, M., Prolonged behavioral effects of in utero exposure to lead or methylmercury: reduced sensitivity to changes in reinforcement contingencies during behavioral transitions and in steady state. *Toxicol. Appl. Pharmacol.*, 126, 6–15, 1994.
186. Newland, M. C., Warfvinge, K., and Berlin, M., Behavioral consequences of *in utero* exposure to mercury vapor: alterations in lever-press durations and learning in squirrel monkeys. *Toxicol. Appl. Pharmacol.*, 139, 374–386, 1996.
187. OTA, *Neurotoxicity: Identifying and Controlling Poisons of the Nervous System*, OTA-BA-436, Office of Technology Assessment, U.S. Congress, U.S. Government Printing Office, Washington, D.C., 1990.
188. Overton, D. A. and Hayes, M. W., Optimal training parameters in the two-bar fixed ratio drug discrimination task. *Pharmacol. Biochem. Behav.*, 21, 19–28, 1984.
189. Owen, J. B., The influence of *dl*-, *d*-, and *l*-amphetamine and *d*-methamphetamine on a fixed ratio schedule. *J. Exp. Anal. Behav.*, 3, 293–310, 1960.
190. Paule, M. C., Cranmer, J. M., Wilkins, J. D., Stern, H. P., and Hoffman, B. L., Quantitation of complex brain function in children: preliminary evaluation using a nonhuman primate behavioral test battery. *Neurotoxicology*, 9, 367–378, 1988.
191. Paule, M. C. and McMillan, D. B., Effects of trimethyltin on incremental repeated acquisition. *Neurotoxicol. Teratol.*, 8, 245–253, 1986.
192. Payne, R. P. and Hauty, G. T., Factors affecting the endurance of psychomotor skill. *J. Aviat. Med.*, 26, 382–389, 1955.
193. Peele, D. B. and Baron, S. P., Effects of scopolamine on repeated acquisition of radial-arm maze performance by rats. *J. Exp. Anal. Behav.*, 49, 275–290, 1988.
194. Potts, A. M. and Conasun, L. M., Toxic response of the eye, in *Casarett & Doull's Toxicology: The Basic Science of Poisons*, 2nd ed., Doull, J., Klaassen, C. D., and Amdur, M. O., Eds., McGraw-Hill, New York, 1980, pp. 275–310.
195. Preston, K. L., Schuster, C. R., and Seiden, L. S., Methamphetamine, physostigmine, atropine and mecamylamine: effects on force lever performance. *Pharmacol. Biochem. Behav.*, 23, 781–788, 1985.
196. Pryor, C. T., Rebert, C. S., Dickinson, J., and Feeney, B., Factors affecting toluene-induced ototoxicity in rats. *Neurobehav. Toxicol. Teratol.*, 6, 223–238, 1984.
197. Rahill, A. A., Weiss, B., Morrow, P. E., Frampton, M. W., Cox, C., Gibb, R., Gelein, R., Speers, D., and Utell, M. J., Human performance during exposure to toluene. *Aviat. Space Environ. Med.*, 67, 640–647, 1996.
198. Raitta, C., Impaired colour discrimination among viscose rayon workers exposed to carbon disulphide. *J. Occup. Med.*, 23, 589–592, 1981.
199. Reynolds, G. S., *A Primer of Operant Conditioning*. Scott Foresman, Glenview, IL, 1968.
200. Rice, D. C., Behavioral deficit (delayed matching to sample) in monkeys exposed from birth to low levels of lead. *Toxicol. Appl. Pharmacol.*, 75, 337–345, 1984.
201. Rice, D. C., Testing effects of toxicants on sensory system function by operant methodology, in *Neurobehavioral Toxicity: Analysis and Interpretation*, Weiss, B. and O'Donoghue, J., Raven Press, New York, 1994.
202. Rice, D. C. and Gilbert, S. G., Early chronic low-level methylmercury poisoning in monkeys impairs spatial vision. *Science*, 216, 759–761, 1982.
203. Rice, D. C. and Gilbert, S. G., Low lead exposure from birth produces behavioral toxicity (DRL) in monkeys. *Toxicol. Appl. Pharmacol.*, 80, 421–426, 1985.
204. Rice, D. C. and Gilbert, S. G., Effects of developmental exposure to methylmercury on spatial and temporal visual function in monkeys. *Toxicol. Appl. Pharmacol.*, 102, 151–163, 1990.
205. Rice, D. C. and Gilbert, S. G., Exposure to methylmercury from birth to adulthood impairs high-frequency hearing in monkeys. *Toxicol. Appl. Pharmacol.*, 115, 6–10, 1992.
206. Rice, D. C., Age-related increase in auditory impairment in monkeys exposed *in utero* plus postnatally to methylmercury. *Toxicol. Sci.*, 44, 191–196, 1998.
207. Rice, D. C. and Gilbert, S. G., Effects of developmental methylmercury exposure or lifetime lead exposure on vibration sensitivity function in monkeys. *Toxicol. Appl. Pharmacol.*, 134, 161–169, 1995.
208. Riley, E. P. and Voorhees, C. V., *Handbook of Behavioral Teratology*, Plenum Press, New York, 1986.
209. Ristau, C. A. and Robbins, D., Language in the great apes: a critical review. *Adv. Study Behav.*, 12, 141–255, 1982.
210. Roels, H., Lauwerys, R., Buchet, J. P., Genet, P., Sarhan, M. J., Hanotiau, I., deFays, M., Bernard, A., and Stanescu, D., Epidemiological survey among workers exposed to manganese: effects on lung, central nervous system, and some biological indices. *Am. J. Indust. Med.*, 11, 307–327, 1987.
211. Rosenstock, L., Keifer, M., Daniell, W. E., McConnell, R., and Claypoole, K., Chronic central nervous system effects of acute organophosphate pesticide intoxication: the Pesticide Health Effects Study Group, *Lancet*, 338, 223–227, 1991.
212. Ruppert, P. H., Postnatal exposure, in *Neurobehavioral Toxicology*, Annau, Z., Ed., The Johns Hopkins University Press, Baltimore, MD, 1986, pp. 170–192.
213. Russell, E. W., Neuringer, C., and Goldstein, G., *Assessment of Brain Damage: A Neuropsychological Key Approach*, Wiley Interscience, New York, 1970.
214. Sahakian, B. J., Morris, R. G., Evenden, J. L., Heald, A., Levy, R., Philpot, M., and Robbins, T. W., A comparative study of visuospatial memory and learning in Alzheimer type dementia and Parkinson's disease. *Brain*, 111, 695–718, 1988.

215. Sanes, J. N., Colburn, T. R., and Morgan, N. T., Behavioral motor evaluation for neurotoxicity screening. *Neurobehav. Toxicol. Teratol.*, 7, 329–337, 1985.
216. Savage, E. P., Keefe, T. J., Mounce, L. M., Heaton, R. K., Lewis, J. A., and Burcar, P. J., Chronic neurological sequelae of acute organophosphate poisoning. *Arch. Environ. Health*, 43, 38–45, 1988.
217. Schonenfeld, W. N., *The Theory of Reinforcement Schedules*, Appleton-Century-Crofts, New York, 1970.
218. Schwartz, B. S., Ford, D. P., Bolla, K. I., Agnew, J., Rothman, N., and Bleecker, M. L., Solvent-associated decrements in olfactory function in paint manufacturing workers. *Am. J. Indust. Med.*, 18, 697–706, 1990.
219. Schwartz, J. and Otto, D., Blood lead, hearing thresholds, and neurobehavioral development in children and youth. *Arch. Environ. Health*, 42, 153–160, 1987.
220. Schwartz, J. and Otto, D., Lead and minor hearing impairment. *Arch. Environ. Health*, 46, 300–305, 1991.
221. Schrimsher, G. W. and Reier, P. J., Forelimb motor performance following cervical spinal cord contusion injury in the rat. *Exp. Neurol.*, 117, 287–298, 1992.
222. Seiden, L. S. and Dykstra, L. A., *Psychopharmacology: A Biochemical and Behavioral Approach*, Van Nostrand Reinhold, New York, 1977.
223. Sepkoski, C. M., Lester, B. M., Ostheimer, G. W., and Brazelton, T. B., The effects of maternal epidural anesthesia on neonatal behavior during the first month. *Dev. Med. Child. Neurol.*, 34, 1072–1080, 1992.
224. Shull, R. L., The postreinforcement pause: some implications for the correlational law of effect, in *Reinforcement and the Organization of Behavior*, Vol. 1, Zeiler, M. D. and Harzem, P., Eds., John Wiley & Sons, New York, 1979, pp. 193–222.
225. Smith, C. B., Effects of *d*-amphetamine upon operant behavior of pigeons: enhancement by reserpine. *J. Pharmacol. Exp. Ther.*, 146, 167–174, 1964.
226. Smith, J., Conditioned suppression as an animal psychophysical technique, in *Animal Psychophysics: The Design and Conduct of Sensory Experiments*, Stebbins, W. C., Ed., Appleton-Century-Crofts, New York, 1970, pp. 25–159.
227. Smith, P. J., Langolf, G. D., and Goldberg, J., Effects of occupational exposure to elemental mercury on short term memory. *Br. J. Med.*, 40, 413–419, 1983.
228. Stebbins, W. C., Principles of animal psychophysics, in *Animal Psychophysics: The Design and Conduct of Sensory Experiments*, Stebbins, W. C., Ed., Appleton-Century-Crofts, New York, 1970, pp. 1–19.
229. Stebbins, W. C. and Moody, D. B., Comparative behavioral toxicology. *Neurobehav. Toxicol.*, 1, 33–44, 1979.
230. Stevenson, J. G. and Clayton, F. L., A response duration schedule: effects of training, extinction, and deprivation. *J. Exp. Anal. Behav.*, 13, 359–367, 1970.
231. Streissguth, A. P., Barr, H. M., and Martin, D. C., Maternal alcohol use and neonatal habituation assessed with the Brazelton scale. *Child Dev.*, 54, 1109–1118, 1983.
232. Stubbs, D. A. and Thomas, J. R., Discrimination of stimulus duration and *d*-amphetamine in pigeons: a psychophysical analysis. *Psychopharmacology*, 36, 313–322, 1974.
233. Sugimoto, K. and Goto, S., Retinopathy in chronic carbon disulfide exposure, in *Neurotoxicity of the Visual System*, Merigan, W. H. and Weiss, B., Eds., Raven Press, New York, 1980, pp. 55–71.
234. Suzuki, Y., Mouri, T., Suzuki, Y., Nishiyama, K., Fujii, N., and Yano, H., Study of subacute toxicity of manganese dioxide in monkeys. *Tokushima J. Exp. Med.*, 22, 5–10, 1975.
235. Swanson, J. M. and Kinsbourne, M., Food dyes impair performance of hyperactive children on a laboratory learning test. *Science*, 207, 1485–1487, 1980.
236. Swets, J. A., The science of choosing the right decision threshold in high-stakes diagnostics. *Am. Psychol.*, 47, 522–532, 1992.
237. Taylor, J. D. and Evans, H. L., Effects of toluene inhalation on behavior and expired carbon dioxide in macaque monkeys. *Toxicol. Appl. Pharmacol.*, 80, 487–495, 1985.
238. Tepper, J. S. and Weiss, B., Determinants of behavioral response with ozone exposure. *J. Appl. Physiol.*, 60, 868–875, 1986.
239. Tepper, J. S. and Wood, R. W., Behavioral evaluation of the irritating properties of ozone. *Toxicol. Appl. Pharmacol.*, 78, 404–411, 1985.
240. Terrace, H. S., In the beginning was the name. *Am. Psychol.*, 40, 1011–1028, 1985.
241. Tewes, P. A., and Fischman, M. W., Effects of *d*-amphetamine and diazepam on fixed- interval, fixed ratio responding in humans. *J. Pharmacol Exp. Ther.*, 221, 373–383, 1982.
242. Thompson, D. M., Repeated acquisition of response sequences: stimulus control and drugs. *J. Exp. Anal. Behav.*, 23, 429–436, 1975.
243. Thompson, D. M. and Moerschbaecher, J. M., Drug effects on repeated acquisition. In: *Advances in Behavioral Pharmacology*, Vol. 2, Thompson, T. and Dews, P. B., Eds., Academic Press, New York, 1979, pp. 229–260.
244. Thompson, T. and Boren, J. J., Operant behavioral pharmacology, in *Handbook of Operant Behavior*, Honig, W. K. and Staddon, J. E. R., Eds., Prentice Hall, Englewood Cliffs, NJ, 1977, pp. 540–569.
245. Thompson, T. and Grabowski, J. C., *Reinforcement Schedules and Multioperant Analysis*, Appleton-Century-Crofts, New York, 1972.
246. Thompson, T. and Pickens, R., *Stimulus Properties of Drugs*, Appleton-Century-Crofts, New York, 1971.
247. Thompson, T. and Schuster, C. R., Morphine self-administration, food-reinforced and avoidance behavior in rhesus monkeys. *Psychopharmacology*, 5, 87–94, 1964.
248. Thompson, T. and Schuster, C. R., *Behavioral Pharmacology*, Prentice Hall, Englewood Cliffs, NJ, 1968.
249. Tilson, H. A., Behavioral indices of neurotoxicity: what can be measured? *Neurotoxicol. Teratol.*, 9, 427–444, 1987.
250. von Békesy, G., A new audiometer. *Acta Otolaryngol. (Stockh.)*, 35, 411–422, 1947.
251. Vorhees, C. V., Methods for detecting long-term CNS dysfunction after prenatal exposure to neurotoxins. *Drug Chem. Toxicol.*, 20, 387–399, 1997.

252. Vorhees, C. V., Developmental neurotoxicology, in *Neurotoxicology*, Tilson, H. A. and Mitchell, C. L., Eds., Raven Press, New York, 1992, pp. 295–330.
253. Walsh, T. J., Miller, D. B., and Dyer, R. S., Trimethyltin, a selective limbic system neurotoxicant, impairs radial-arm maze performance. *Neurotoxicol. Teratol.*, 4, 177–183, 1982.
254. Wechsler, D., A standardized memory scale for clinical use. *J. Psychol.*, 19, 87–95, 1945.
255. Wechsler, D., *Wechsler Adult Intelligence Scale Manual*, Psychological Corp., New York, 1955.
256. Wechsler, D., *Wechsler Preschool and Primary Scale of Intelligence*, Psychological Corp., New York, 1967.
257. Wechsler, D., *Manual for the Wechsler Intelligence Scale for Children–Revised*, Psychological Corp., New York, 1974.
258. Weeks, J. R., Experimental morphine addiction: method for automatic intravenous injections in unrestrained rats. *Science*, 138, 143–144, 1962.
259. Weiss, B., Amphetamine and the temporal structure of behavior, in *International Symposium on Amphetamine and Related Compounds*, Costa, E. and Garratini, S., Eds., Raven Press, New York, 1970, pp. 797–812.
260. Weiss, B., Microproperties of operant behavior as aspects of toxicity, in *Recent Developments in the Quantification of Steady-State Operant Behavior*, Bradshaw, C. M., Ed., Elsevier, Amsterdam, 1981, pp. 249–265.
261. Weiss, B., Behavioral toxicology of heavy metals, in *Neurobiology of the Trace Elements*, Vol. 2, *Neurotoxicology and Neuropharmacology*, Dreosti, I. E. and Smith, R. M., Humana Press, Clifton, NJ, 1983, pp. 1–50.
262. Weiss, B., Neurobehavioral toxicity as a basis for risk assessment. *Trends Pharmacol. Sci.*, 9, 59–62, 1988.
263. Weiss, B. and Laties, V. C., Enhancement of human performance by caffeine and the amphetamines. *Pharmacol. Rev.*, 14, 1–36, 1962.
264. Weiss, B. and Laties, V. G., Effects of amphetamine, chlorpromazine and pentobarbital on behavioral thermoregulation. *J. Pharmacol. Exp. Ther.*, 140, 1–7, 1963.
265. Weiss, B. and Laties, V. G., Effects of amphetamine, chlorpromazine, pentobarbital, and ethanol on operant response duration. *J. Pharmacol. Exp. Ther.*, 144, 17–23, 1964.
266. Weiss, B. and Laties, V. G., The psychophysics of pain and analgesia in animals, in: *Animal Psychophysics: The Design and Conduct of Sensory Experiments*, Stebbins, W. C., Ed., Appleton-Century-Crofts, New York, 1970, pp. 185–210.
267. Weiss, B. and Laties, V. G., *Behavioral Toxicology*. Plenum Press, New York, 1975.
268. Weiss, B. and Elsner, J., Eds., Risk assessment for neurobehavioral toxicity. *Environ. Health Perspect.*, 104(Suppl. 2), 171–412, 1996.
269. Weiss, B. and Reuhl, K., Delayed neurotoxicity: a silent toxicity, in *Handbook of Neurotoxicology. Approaches and Methods for Neurotoxicology*, Chang, L., Ed., Marcel Dekker, New York, 1994, pp. 765–784.
270. Wenger, C. R., The effects of trialkyl tin compounds on schedule-controlled behavior, in *Behavioral Pharmacology: The Current Status*, Seiden, L. S. and Balster, R. L., Eds., Alan R. Liss, New York, 1985, pp. 503– 518.
271. Wenger, C. R., McMillan, D. B., and Chang, L. W., Behavioral effects of trimethyltin on two strains of mice. II. Multiple fixed-ratio, fixed interval. *Toxicol. Appl. Pharmacol.*, 73, 89–96, 1984.
272. Whishaw, I. Q., Dringenberg, H. C., and Pellis, S. M., Spontaneous forelimb grasping in free feeding by rats: motor cortex aids limb and digit positioning. *Behav. Brain Res.*, 48, 113–125, 1992.
273. Whishaw, I. Q., Gorny, B., and Sarna, J., Paw and limb use in skilled and spontaneous reaching after pyramidal tract, red nucleus and combined lesions in the rat: behavioral and anatomical dissociations. *Behav. Brain Res.*, 93, 167–183, 1998.
274. Williams, C. D., The elimination of tantrum behavior by extinction. *J. Abnorm. Soc. Psychol.*, 59, 269, 1959.
275. Williamson, A. M. and Teo, R. K. C., Neurobehavioral effects of occupational exposure to lead. *Br. J. Indust. Med.*, 43, 374–380, 1986.
276. Winneke, C., Brockhaus, A., and Baltissen, R., Neurobehavioral and systemic effects of long-term blood lead elevation in rats. I. Discrimination learning and open-field behavior. *Arch. Toxicol.*, 37, 247–263, 1977.
277. Winneke, C. and Kraemer, U., Neuropsychological effects of lead in children: interactions with social background variables. *Neuropsychobiology*, 11, 195–202, 1984.
278. Wolthuis, O. L. and Vanwersch, R. A. P., Behavioral changes in the rat after low doses of cholinesterase inhibitors. *Fundam. Appl. Toxicol.*, 5, 5195–5208, 1984.
279. Wood, R. W., Behavioral evaluation of sensory irritation evoked by ammonia. *Toxicol. Appl. Pharmacol.*, 50, 157–162, 1979.
280. Wood, R. W., Stimulus properties of inhaled substances: an update, in *Nervous System Toxicology*, Mitchell, C. L., Ed., Raven Press, New York, 1982, pp. 199–212.
281. Wood, R. W. and Coleman, J. B., Behavioral evaluation of the irritant properties of formaldehyde. *Toxicologist*, 4, 119, 1984.
282. Wood, R. W., Grubman, J., and Weiss, B., Nitrous oxide self-administration by the squirrel monkey. *J. Pharmacol. Exp. Ther.*, 202, 491–499, 1977.
283. Wood, R. W., Rees, D. C., and Laties, V. G., Behavioral effects of toluene are modulated by stimulus control. *Toxicol. Appl. Pharmacol.*, 68, 462–472, 1983.
284. Wood, R. W., Weiss, A. B., and Weiss, B., Hand tremor induced by industrial exposure to inorganic mercury. *Arch. Environ. Health*, 26, 249–252, 1973.
285. Wood, R. W. and Colotla, V. A., Biphasic changes in mouse motor activity during exposure to toluene. *Fundam. Appl. Toxicol.*, 14, 6–14, 1990.
286. Yanai, J., *Neurobehavioral Teratology*, Elsevier, Amsterdam, 1984.
287. Yule, W., Lansdown, R., Miller, I. B., and Urbanowicz, M., The relationship between blood lead concentrations, intelligence, and attainment in a school population: a pilot study. *Dev. Med. Child Neurol.*, 23, 567–576, 1981.
288. Zenick, H. and Goldsmith, M., Drug discrimination learning in lead-exposed rats. *Science*, 212, 569–571, 1983.
289. Ziriax, J. M., Snyder, J. R., Newland, M. C., and Weiss, B., *d*-Amphetamine modifies the pattern of concurrent behavior. *Exper. Clin. Psychopharmacol.*, 1, 1–12, 1993.

# 38 Application of Isolated Organ Perfusion Techniques in Toxicology

*Harihara M. Mehendale*

## CONTENTS

## INTRODUCTION

The concept of employing organ perfusion for physiological and biochemical studies is not new. Early accounts of organ perfusion techniques in biochemical and physiological studies may be found in the descriptions of Baglioni [24] and Muller [226]. More detailed accounts of the historical development of organ perfusion techniques may be found in the works of Brodie [43] and Embden and Glaüssner [84]. Systematic early developments in organ perfusion techniques have been reviewed by Skutul [310] and Kapfhammer [148]. Increasing interest in the application of the techniques of perfusing isolated organs and tissues in biochemical and physiological investigations is apparent in the later works of Ross [287], Diczfalusy [77], and Ritchie and Hardcastle [283]. The interest in the application of isolated perfused organ techniques in studies on toxicological mechanisms is apparent in the increasing number of reviews appearing in the literature [24,63,64, 211–213,221,290,332]. Some investigators have advanced the term "*ex vivo* perfusion" to refer to the technique of perfusing and maintaining functional isolated organs outside the animal body.

The rationale for experimental studies using perfused organs and for trying to improve the technology of organ perfusion lies in the following physiological and biochemical considerations. We recognize homeostasis to be the outcome of many simultaneously occurring, interacting complex processes. It is recognized that when many simultaneously occurring processes interact, they may collectively take on functional properties that cannot be perceived in any of the individual component processes. In endocrine and metabolic systems, a basic experimental question arises at the organ level: How does the organ's uptake or output of some toxic substance depend on the composition of the arterial blood reaching the organ? The technique of organ perfusion can, if certain conditions are met, permit one to make controlled concentration changes in the perfusing blood while observing the time course of the organ's response in terms of its uptake or output of one or more substances. This kind of experimentation has several advantages. Mathematical models based on some convenient equations can be constructed relating the nature of a toxic compound and its concentration in blood. The nature of the substances produced as a result of the biotransformation can be investigated and related to the specific organ, and the role of that organ in converting a substance into a more harmful, less harmful, or biochemically inert species can be effectively evaluated. Quantitative input on these pathways of biotransformation into the appropriate mathematical models can lead to the understanding and development of predictable values.

In simplest terms, the essentials of isolated organ perfusion techniques include: (1) the desirability of separating individual organs from the whole animal to permit the study of one in the absence of complex interaction by others; (2) the need for the tissue to be physiologically compatible to the *in vivo* situation; and (3) the desire to simulate the natural circulation through an organ. In the latter regard, the composition of the medium changes constantly, as in the whole body, at least with respect to the experimental toxic substance. Not all substances reenter the perfusion medium, nor are all substances entirely removed by the organ. Finally, an analytical study of the organ itself can be undertaken; artificial means of stimulating organ function may aid in magnifying the physiological role of, or effect on, the organ, thus enabling determination of such an interaction. Although use of autologous whole blood would be considered ideal for homeostatic mechanisms, partial or complete substitution with artificial media is often necessary for technical reasons; nonetheless, an attempt is made to maintain the cell structure and function by following as many viability criteria as possible.

### Choice of Donor Animal

A variety of factors may influence the choice of organ donor animals in perfused organ studies. Often, the nature of the particular problem being investigated is the determining factor when choosing the experimental animal. Susceptibility or refractoriness to the toxic agent to be tested and the presence or absence of biotransformation pathways govern the selection of a particular species and often a particular strain. Other factors may influence the final selection of the experimental animal as well. Availability of pertinent background information in a particular animal model may compel the investigator in favor of that species in the interest of savings in time and resources. Availability, cost of animals, and maintenance or unique genetic characteristics may become important considerations. The rat has been most popular in this regard as a donor for perfusion experiments; thus, perfused heart [15,202,225,232], liver [42,203,215,218], lung [52,183,247–248,252,260], kidney [25,28,187,188], brain [10,90, 166,330], and pancreas [115,152,253,254,358] obtained from the rat have been employed by many investigators for a variety of biochemical studies. Larger animals employed in perfusion experiments include the cat [11,71,102,155,255], dog [11,83,139, 175,348,357], rabbit [11,59,175,184,211,236,252,274, 276,285], and monkey [11]. Among the larger animals employed for perfusion experiments are calves [185], chickens [200], dogs [126], goats [201], pigs [82], and sheep [82,325].

All small animals have the disadvantage of small blood vessels, which pose difficulties in surgical procedures. In many instances, the experimenter is interested in using autologous blood for perfusion, and small animals may not yield sufficient blood supply to prime the perfusion apparatus. The trend has been to utilize either diluted or reconstituted blood or a completely artificial perfusion medium composed of natural or synthetic ingredients. Nevertheless, in a few instances, the necessity of using autologous whole blood as a perfusate essentially eliminates the use of small experimental animals in perfusion experiments. Additional factors to be considered are the volume and number of the perfusate samples required to carry out necessary analytical tests during the course of the experiment. These difficulties are clearly overcome by using large experimental animals. Large animals, however, have the distinct disadvantage of increased cost, on the one hand, and requisite chemicals, equipment, space, and other supplies on the other. Use of expensive isotopically labeled chemicals or limited availability of valuable samples of newly synthesized or isolated test drugs may make it necessary to restrict the perfusion experiments to organs from smaller animals.

Large or small, other considerations may also be important. Much fat, particularly in the abdominal areas of large animals or old small animals, contributes to surgical difficulties. The presence or absence of some tissues, such as the gallbladder, which is absent in rat and present in the rabbit, might be an additional consideration. For the purpose of acquainting oneself with the techniques of organ perfusion, the size of the animal *per se* matters little; however, considerations of economy of space, equipment, apparatus size, and animal costs might make the choice of a small animal a prudent one. In view of these and other considerations discussed above, whenever possible we consider the rat the animal of choice; however, it is important to bear in mind that most procedural and other technical considerations remain the same with minor modifications when a particular perfusion technique is intended to be applied to a large animal or to an animal of similar, larger, or smaller size.

### *In Situ* and Isolated Organ Perfusion System

Isolated organ perfusion may be defined as the maintenance of an organ in vascular isolation from the rest of the tissues and organs of the body by mechanically assisted circulation of a suitable fluid through its vascular bed. In most cases, special apparatus is acquired, and each investigator has invariably adopted an individual approach to solving the technical problems associated with maintaining a particular organ in viable condition for a particular toxicological investigation. The resulting scattered literature and many technical variations introduced to the techniques of organ perfusion have to a large degree contributed to the difficulty of a newcomer to the field of isolated organ perfusion to readily adopt the application of these techniques toward investigating the special problems of toxicology. Often, it is difficult to assess the merits of the available methods. Given the complexity of the problem, there is often reluctance on the part of some toxicologists to embark on isolated perfused organ studies, even in those areas of biochemical toxicology, where these techniques offer unique and definitive advantages over other *in vitro* or *in vivo* techniques. Establishment of a set of standards for each organ perfusion system by an internationally composed committee might alleviate many of the problems arising out of infinitely varied perfusion techniques and methodology introduced by individual investigators in a scattered body of diverse literature.

One rather obvious prerequisite for isolated organ perfusion studies is that the organ to be perfused is capable of vascular isolation from the neighboring tissues, although physical isolation is not obligatory. A separate vascular bed is sufficient to ensure that only one tissue or organ is perfused in isolation. For example, an organ may be perfused in isolation but may remain *in situ*, as in the case of lung [180], liver [33,125], intestine [150,257], or kidney [25]. A principal advantage of perfusing the organ *in situ* is the time saved in surgical removal of the organ, thereby reducing the time of interrupted perfusion. An additional advantage is that it reduces physical damage to the organ that may be inflicted during surgical removal of the organ.

Alternatively, the organ to be perfused may be physically isolated from the animal, as in the case of lung [143,238], liver [45,203,218], kidney [186,235], heart [225,232,309], intestine [81], and pancreas [115]. The principal advantage of physically isolating an organ is the elimination of interactions between the organ being perfused and other tissues and organs present in the body. Although the vascular bed of the organ being perfused may be totally isolated, the endogenous substances being secreted may seep out of other tissues and organs and might come in physical contact with the organ being perfused, resulting in uncontrollable or even unanticipated interactions. Similarly, the compound being studied in the perfused organ may be secreted and absorbed by other surrounding tissues and organs by physical contact and hence may introduce experimental errors in the quantitative and qualitative aspects of the disposition of the toxic chemical being studied. Accuracy and reliability of mass-balance studies of toxic chemicals and hence the accountability of the test chemical and possible metabolites are vastly superior with isolated organ perfusion systems.

## Advantages of Isolated Perfused Organ Techniques

Isolated perfused organ preparations offer several advantages over experimentation with intact animals. Perfusion experiments lend themselves to a definitive evaluation of the role of a particular organ or tissue in the disposition of endogenous or exogenous chemicals. Although experimentation with whole-animal preparations may provide clues implicating a possible role of a particular organ in regulating the levels of a test toxin or an endogenous substance in response to a toxin, decisive conclusions may not be feasible. A case in point is provided by the studies [92] that reported the presence of γ-aminobutyrate in the lungs of rats and mice after these animals were injected with radioactive putrescine. This finding cannot be taken as conclusive evidence for the formation of γ-aminobutyrate in the lungs, as it could have been transported from other sites of synthesis. Isolated, ventilated, and perfused lungs [276] and other tissue preparations were used to examine if rat and rabbit lungs were capable of metabolizing putrescine to γ-aminobutyrate [249], establishing that the lungs of these species are devoid of the diamine oxidase necessary for this metabolism. Isolated perfused organ studies provide opportunities to decisively ascertain or reject such possibilities.

Unlike the *in vitro* homogenate preparations, intact organ perfusion studies allow the experimenter to retain the structural and functional integrity of the organ in question during such experiments. Unlike in the intact animal, perfusion experiments allow the experimenter to retain control over several experimental parameters (e.g., perfusion pressure and blood flow); in the intact animals, these measurements are likely to change during the course of an experiment, especially in response to administration of the experimental toxic chemical. The concentrations of endogenous or exogenous stimulatory substances and other factors can be under experimental control in isolated perfused organ studies. The isolated perfused organ would lend itself to a broader range of concentrations of the experimental drug to be used in the study; that is, concentrations of drug at which the intact animal would not be expected to survive can be tested in isolated perfused organs. Determination of an accurate and complete mass-balance of the toxic chemical in question is possible throughout the perfusion experiment, as the compound must be in the perfusate or the tissue or must be excreted via excretory fluids such as bile and urine. Binding of the test drug to the glassware, tubing, and other components of the perfusion apparatus may occur, but this possibility can be explored in blank experiments, in which the perfusion experiment is conducted without the organ, from which appropriate correction factors are derived; moreover, removal of such interfering factors is often technically feasible. Another advantage of perfusion studies is the availability of large blood or perfusate samples; thus, complete qualitative and quantitative analyses of minor and major biotransformation products of the test compound are feasible, as the volume of perfusate used in these experiments can be controlled. A further advantage of perfusion experiments in comparison with whole-animal experiments is the feasibility of tests with smaller quantities of toxic chemicals. This point is particularly noteworthy, as limitations of either the availability of small quantities of the toxins or the cost of isotopically labeled newly synthesized compounds can be formidable.

Another advantage of perfused organ studies is the maintenance of appropriate membrane barriers, not only between vascular and parenchymal sides but also between individual cells; hence, the natural constraints of intact organs are retained throughout the experimental duration. Evidence has made it clear that one may not be able to predict the qualitative and quantitative aspects of biotransformation of a test drug by intact organ based only on the results of *in vitro* experiments with homogenate preparation [208]. Factors governing the generation and availability of cofactors and transport of substrate to the site of biotransformation influence the final results in the intact perfused organ [136,208,331]. These factors can remain operative in perfusion studies unlike with other *in vitro* techniques, thereby enabling realistic extrapolation of the results to *in vivo* situations. Finally, no matter how determined, experimental results have to be interpreted and extrapolated to the *in vivo* situation, where intact organs interact continuously; such interpretation and extrapolation are made easier by use of intact perfused organs in toxicological investigations.

Furthermore, the cell-to-cell interactions are preserved in an intact perfused organ, which might be either missing or at least compromised in isolated cells or other *in vitro* incubations. It is known that gap junctioning plays an important role in the regulation of cellular and tissue homeostatic mechanisms. These would be preserved in intact perfused organs. The collagenase trypsin or other proteases used in procedures to isolate cells might alter the plasma membrane, thereby altering the permeability and even receptor characteristics of isolated cells; for example, in freshly prepared hepatocytes, glutathione levels are only half of the normal values. Some essential and critical differences between the tissue slice experiments and perfusion studies are also of interest in this regard. Whereas the perfusion of intact organs allows entry of the chemical through the endothelium, which would be representative of what happens in the intact animal, tissue slice incubations permit entry of chemicals directly into the parenchymal cells through a direct contact. Studies using tissue slice incubations might not represent the *in vivo* situation, as some chemicals may not be taken up through the endothelial barrier altogether or be taken up to a small extent. Hepatocytes or liver slices incubated with the calcium channel blockers do not have any influence on cellular calcium, whereas the perfused liver is responsive to these same calcium blockers. The latter findings would be clearly more representative of the *in vivo* situation than the former.

## Limitations

The current state of the art allows maintenance of isolated perfused organ preparations with adequate physiological and biochemical integrity for only short periods of time. Clinically, advances have allowed maintenance of the kidney for several days for later physical transplantation in patients. These procedures require subambient temperatures to preserve organ function. Such techniques are not generally useful in toxicological studies, as maintenance of the organ at optimal functional level is a prerequisite for most toxicological studies; hence, the principal limitation imposed by the isolated perfused organ preparations is the short duration of study. Critical and vital organ functions deteriorate in isolated perfused organs with time; for example, isolated perfused lung preparations can be maintained for a maximum of only 4 hours [204]. Often, it is not possible to determine the effect of therapeutic agents on the lung tissue in such a short period of time. Similarly, isolated perfused liver preparations cannot be maintained for longer periods without compromising liver function [131,172,252]. A practical consideration of interest in this connection may also be the level of expertise required for setting up the perfusion experiments. Setting up and conducting successful perfusion experiments requires specially trained personnel in all aspects of the surgical procedures as well as the technical aspects of associated instrumentation. The unavailability of such personnel requires that the investigator allow valuable time for training.

Often a principal argument in favor of isolated perfused organ studies is the maintenance of natural membrane barriers, the integrity of the intact cells, and the complex and dynamic interrelations between individual and groups of cells. For certain studies, this argument may represent a principal limitation. The complexity of a whole organ deprives the toxicologist of access to individual reactions that occur within the organ. Compartments and permeability barriers may prevent substrates and test drugs from exerting effects that are known to manifest when the particular toxic agent is allowed direct access to the enzyme or organelle of interest. *In vitro* experiments with homogenate preparations and tissue slice preparations would be the obvious choice of techniques when dissection of individual transport processes and biotransformation reactions is the principal objective. The size of a single experiment and time required to perform it may make organ perfusion far less efficient from the point of time and resources than *in vitro* preparations that demonstrate the same effects with lesser investments of time and resources. Another consideration is the availability of the experimental tissue or organs. Although access to valuable human tissues might be available, such access might be infrequent, and, in any event, the available tissue would be limiting. Clearly, isolated cell techniques or other *in vitro* techniques have the advantages of maximizing the use of such invaluable experimental material when designing and carrying out studies of utmost relevance. Schimmel and Knobil [300] pointed out the greater efficiency of establishing an experimental fact with tissue slices than with isolated perfused organ. Finally, despite all the refined techniques of maintaining the organ *in vitro* in as near a normal state as possible, the resulting preparation may differ in some highly significant manner from the organ *in vivo*, limiting the interpretation and application of results obtained in the organ perfusion system.

## Scope of This Chapter

The principal purpose of this chapter is to acquaint the reader with generalized principles and methods for the isolation and perfusion of selected organs from experimental animals. Accordingly, the following discussion represents simplified and often idealized procedures, so a relative novice could undertake application of perfusion techniques to toxicological investigations. Sufficient references are included to serve as focal points for the benefit of those who are already engaged in perfusion of isolated organs and who seek advanced information. The principles governing the choice of isolated perfused organ studies, the experimental protocols (including the composition of perfusion media, duration of perfusion, considerations of recirculation, or single-pass circulation), and other related aspects are available in the examples of perfused organ studies employed in toxicology studies.

Thorough considerations of pumps and mechanical devices [33] used in the perfusion of organs, the application of pharmacokinetic concepts to isolated perfused organs [252], and the mathematical considerations of single-pass studies in isolated organs [78,225] are available elsewhere. Similarly, use of radiolabeled microspheres to determine intraorgan and regional blood flows [253] may also be considered in isolated perfused organ studies. Excellent reviews on the application of perfusion techniques to biochemical studies [247], to the studies of reproductive endocrinology [76], and to other general consideration of perfusion techniques, especially in large experimental animals [245], are available elsewhere for those seeking additional and detailed considerations of organ perfusion techniques.

## METHODOLOGY

### Isolated Perfused Heart

Since Langendorff [175] described a procedure for isolation and perfusion of dog or rabbit heart, heart has been the model for studying the effects of toxic agents on metabolism in muscle. Numerous investigators have demonstrated the stability and versatility of the preparation [9,16,36,43,59,225,251]. The original method of Langendorff [175] has survived with only minor modifications and remains the standard preparation on which toxicological studies are performed today [9,20,22,75,86,98,153, 158,160,161,174,179,196,200,294,309]. The use of isolated heart preparations for toxicity studies has been recently reviewed [9]. A wide variety of species have been employed for isolated perfused heart preparations. Chicken [200], ferret [158], guinea pig [196], rabbit [22,99] and rat [86,153,174,179,294,309,356] are representative examples.

Two procedures have been used for isolated perfused heart preparations. One uses the aortic perfusion described by Langendorff [175], and the other uses atrial perfusion, in which the left atrium is cannulated. The Langendorff preparation perfuses the muscle of both ventricles, although only the left ventricle produces any tension by contracting against the closed aortic valve. The atrial perfusion method gives a "working heart" preparation: It allows the left ventricle to fill, which results in normal systolic and diastolic cycles. The perfusion described by Morgan et al. [225] and later improvised by Neely et al. [232] provides for the working heart circulation [202,304]. Several studies have compared the aortic perfusion of Langendorff and atrial perfusion of working heart preparations. One considerable advantage in the working heart preparation is the ease and accuracy with which myocardial performance can be quantified. In the Langendorff preparation, coronary flow and heart rate are the only available physical parameters of function. The "working heart," on the other hand, has aortic output as a quantifiable parameter of function, although it is dependent on the heart achieving an aortic pressure of at least 100 cm$H_2O$. Anoxia, lack of substrate, poison, and drugs may be qualitatively and quantitatively tested on this basis. The linearity of oxygen and substrate uptake provides an additional means of assessing function in this preparation and, together with the easily observed abnormalities of cardiac rhythm, which may also indicate failure of the preparation [85], allows the working heart preparation to be easily and accurately assessed [174,202,304].

Despite the completeness of description and advantages of the working heart, however, investigators in this field have experienced many difficulties establishing a viable working heart preparation for significant lengths of experimental time. Hence, although quantitative differences do occur in many biochemical and functional measurements, the working and nonworking heart preparations are similar in many respects [164]. Although with certain experimental protocols it is necessary to employ both preparations, it is doubtful if it will be necessary to carry out all toxicological tests on the heart in both Langendorff and working heart preparations. Because the former is so much simpler, its use will continue to be popular for some time to come.

#### Aortic Perfusion

*Apparatus*

The heart is suspended in a water-jacketed cylindrical chamber, 3 cm in diameter and 20 cm in length, with a coarse sintered-glass filter disk sealed into the lower portion (Figure 38.1). The aortic cannula is mounted in a Teflon® stopper through which pass the gas inlet tube and an outlet vent for the excess gases. The inflowing gas is delivered by means of a fine plastic tube extending into a small pool of perfusion medium that collects on the surface of the sintered-glass filter. After passing through the filter, medium is recirculated by a peristaltic pump. This pumping arrangement results in a waveform applied to the heart, but it can be considerably dampened by passing the perfusate first through a bubble trap containing 1 to 2 mL of air so the characteristics of the pump waveform may be eliminated.

*Surgical Procedure*

Rats weighing approximately 250 to 300 g can be used for obtaining isolated heart preparations. The donor animal is killed by decapitation. It may be desirable to heparinize the animal by administering heparin intraperitoneally (5 mg)

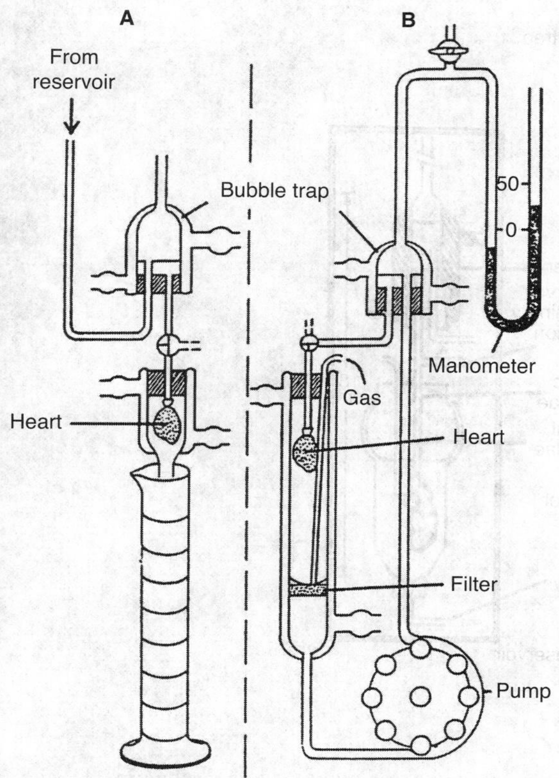


**FIGURE 38.1** Langendorff heart perfusion: (A) single-pass perfusion; (B) recirculating perfusion. (From Manabe, S. et al., *Toxican*, 29, 787–790, 1991. With permission.)

one hour before decapitation to prevent formation of large clots. Within 20 seconds after decapitation, the thorax is widely opened by incisions to remove the anterior wall. The heart is removed immediately by means of a scissor cut across the great vessels about 5 mm from the heart. Earlier investigators [225] allowed a 1- to 2-minute cooling period by immersing the heart in ice-cold Krebs–Ringer buffer solution. This step is not necessarily desirable; in fact, greater speed may help reduce the likelihood of anoxic damage to the heart. Ischemia and reperfusion are known to give rise to the formation of toxic free-radical species that adversely affect the quality of the heart preparations [149,163,356]. After blotting the heart with filter paper and weighing it, a small glass cannula (or polyethylene tubing [PE-200]) can be inserted into the aorta, with the tip of the cannula positioned just above the semilunar valves. Instantaneously beating heart is then perfused through the aortic cannula. The heart preparation thus established may be allowed to equilibrate for a period of 15 to 20 minutes by means of a setup in which the perfusate is allowed to recirculate (Figure 37.1). Oxygenation of the perfusate is accomplished using a mixture of oxygen and carbon dioxide ($O_2/CO_2$, 95:5) that is humidified by passing through water at 37°C before entering the apparatus to prevent loss of perfusate volume by evaporation.

## Atrial Perfusion of the Working Heart

### *Apparatus*

The apparatus for the atrial perfusion of heart is shown in Figure 38.2 [232]. The glass components of the working heart perfusion apparatus are a double-jacketed oxygenator, a 100-cm bubble trap, a mixing column, connecting pieces, condensers, a 100-cm reservoir, a heart chamber, and a second bottle trap. The filter used is a closed Millipore® system adapted from that used in ultrafiltration; for example, a Millipore® inline filter holder may be used. Whenever possible, tubing components are composed of glass; connections between glass tubings are made with flexible tubing, with the exception of the compression tubing used in the roller pumps, for which siliconized medical-grade tubing may be used. The apparatus is assembled as illustrated in Figure 38.2.

### *Surgical Procedure*

A rat weighing approximately 250 g is lightly anesthetized with ether. Without preparation of the skin, the first incision is made with large, pointed scissors around the lower margin of the ribs. The first cut is made across the upper abdomen, taking care not to injure the liver. This cut is extended laterally, and for this purpose the rat may be held in the left hand. The point of the scissors is then directed toward the head, and a single cut is made through the layers of the thorax, including the ribs up to the clavicle. Cuts are made on the left side and the right side, producing a free flap of the anterior chest, which is removed with a cut transversely at the level of the second rib. At this time, the heart is fully exposed. The heart is grasped firmly between the thumb and forefinger of the left hand and lifted, drawing it to the animal's right. This step exposes the pulmonary veins and the site of entry into the left atrium. The point of the scissors is passed horizontally and behind the left atrium, pulmonary veins, and aortic arch. Before closing the blades to cut, the instrument is drawn well over the animal's left to leave the maximum length of pulmonary vein in continuity with the atrium. After cutting the pulmonary veins, the scissors are pointed downward and away from the operator, and the aorta is cut about 0.5 cm from the ventricle. This length is adequate for cannulation. If the cut is made too near the heart, a hole is made in the left atrium rather than through the pulmonary veins, which makes cannulation difficult.

The heart is thus removed and transferred rapidly to ice-cold Krebs–Ringer buffer, and aortic cannulation is carried out. The cannula is filled to the tip with oxygenated medium, making sure to avoid entrapping air bubbles. The aorta is grasped from opposite sides in two pairs of fine curved forceps and gently lifted over the straight cannula. The cannula may be retained by a single ligature around the aorta. It is important to watch that the tip of the cannula

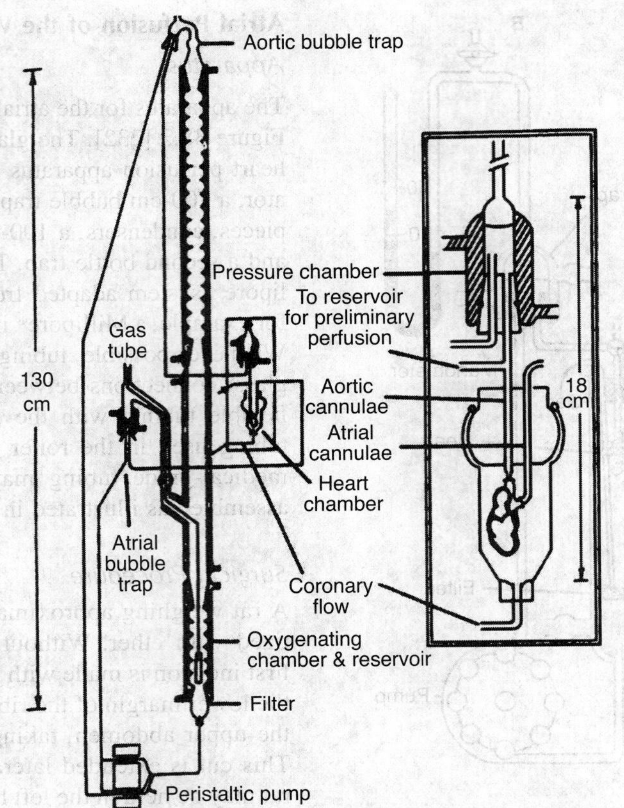


**FIGURE 38.2** Working rat heart preparation. Completely assembled apparatus is shown on the left. Heart chamber, cannula assembly, and pressure chamber are enlarged and illustrated on the right. The apparatus consists of the following components: *Heart chamber and cannula assembly*—The male portion of a size 35 ball joint is made of Teflon® and holds two stainless steel cannulas (0.134-inch o.d.) grooved to hold the ligatures. A tip of 0.109-inch o.d. tubing is soldered into the aortic cannula. The female portion of a 35/25 size ball joint is adapted for use as a heart chamber and held in place with a pinch clamp. *Aortic and atrial bubble traps and pressure chamber*—These parts are made from female portions of 14/35 standard taper joints. Male plugs are made of Teflon® and fitted with neoprene O-rings to facilitate sealing and removal of the stoppers. A side arm is sealed onto the arterial bubble trap, extended with Silastic® tubing, and connected to the central portion of the apparatus by an 18/9 size ball joint. A side arm for the aortic bubble trap can be made from the male portion of a size 15 ball joint. This is connected to an adapter (size 28 ball, 29/42 standard taper), which in turn can be fitted into the top condenser of the central portion of the apparatus. *Oxygenating chamber*—Three condensers with size 29/42 standard taper connections make up the central portion of the apparatus. Both the top condenser (60 cm long) and the bottom one (20 cm long) are Ful-Jak Allihn condensers. An additional condenser can be used as the middle portion. It receives the overflow from the atrial bubble trap and coronary effluent from the heart chamber. A coarse-porosity, sintered glass filter is fitted into the bottom of the oxygenating chamber. A peristaltic pump is used to transport the blood to the top of the oxygenator via glass tubings. The shaded portion of the glass apparatus represents a double-jacket arrangement for circulating warm water to facilitate warming the perfusate. Fluid in bubble traps and the pressure chamber is indicated by stippling. (From Masuda, Y. and Yamamori, Y., *Jap. J. Pharmacol.*, 56, 143–150, 1991. With permission.)

does not rest on the aortic valve because it is too far down the aorta. A flow of the medium at maximal rate is begun as soon as the ligature is tied, ending the unavoidable period of operative myocardial anoxia. The heart should begin to beat within 10 to 15 seconds of starting the perfusion.

Atrial cannulation may be conducted at leisure, as the heart is fully perfused via the aorta and is no longer at risk of anoxia. The atrial wall is grasped by forceps as it lies on either side of the cannula, and by turning the wrist outward the wall is slightly averted. The atrium is then drawn over the cannula itself. The cannula is positioned and retained in this position by means of a ligature. The final step is to incise the right ventricle and its pulmonary trunk, which allows drainage of the medium that has accumulated in the right side of the heart from the minor coronary veins; otherwise, such accumulation results in poor cardiac function. The incision is made with pointed scissors through the base of the ventricle and at the origin of the pulmonary trunk.

## Perfusion Media

A variety of perfusion media with minor modifications have been used to perfuse isolated heart preparations. Use of whole blood [99,261], Krebs–Henseleit bicarbonate buffer medium [225,232], and Krebs–Ringer buffered solution [16,59] has been reported for maintaining successful perfused heart preparations. Many advantages have been noted for the use of whole blood to perfuse isolated heart preparations [99]; however, whenever it has been used, circulation of whole blood through the isolated heart was maintained using support animals [88,222]. In these preparations, the circulation from the isolated heart enters the circulation of the support animal via the right jugular venous cannula and exits the support animal via the left carotid artery [99]. Although satisfactory preparations are obtained, such isolated preparations may not adapt to all types of toxicological investigation. The principal limitation arises from introducing the support animal to the perfusion circuit. The support animal would clear the test chemical, making it difficult to correlate the effects of test toxins on cardiac function and to evaluate possible myocardial biotransformation; hence, artificial perfusates have been used most often in biochemical and toxicological investigations.

Various investigators have introduced minor variations in individual ion composition of the particular medium used in their own experimental setups. The medium of choice, however, seems to be the Krebs–Ringer bicarbonate (KRB)-buffered solution of the following composition: NaCl, 119 m*M*; KCl, 4.75 m*M*; $CaCl_2{\cdot}2H_2O$, 2.54 m*M*; $KH_2PO_4$, 1.19 m*M*; $MgSO_4{\cdot}7H_2O$, 1.19 m*M*; $NaHCO_3$, 25.0 m*M*; glucose, 5.5 m*M*. The solution may be sterilized by ultrafiltration to avoid microbial contamination. If autoclaving is used, the addition of glucose is withheld until later. Equilibration with $O_2/CO_2$ (95:5) for at least 30 minutes before using the perfusate increases the viability of the perfused preparation as well as the buffering capacity of the perfusate during perfusion. The pH of the medium is adjusted to 7.4 by means of a dilute solution of NaOH.

## Viability Criteria

A number of functional, biochemical, and histological determinations have been used to ensure the viability and validity of the perfused heart preparation. Physical measurements of cardiac function include electrocardiogram (ECG) recordings of the left ventricular pressure and left ventricular end-diastolic pressure and the perfusion pressure of the isolated heart preparations [99]. Other functional parameters that can be continuously recorded include the heart rate, isometric systolic tension, and coronary flow [16]. Aronson and Serlick [16] examined the effect of perfusion time on a number of these parameters (Table 38.1). Changes in the heart rate were evident 15 minutes after the end of the initial equilibration period; however, after being perfused for as long as 4 hours, the heart rate was approximately 86% of the initial level, indicating the usefulness of the perfused heart preparation in toxicological studies. In contrast to the heart beat, coronary flow and isometric systolic tension (Table 38.1) appear to be more sensitive, as indicated by their decrease with perfusion time. A number of biochemical parameters can be examined in heart preparations as indices of viability (Table 38.2). Glycogen concentration was not significantly decreased until 3 hours but decreased to 33% of the zero time control values after 4 hours of perfusion. Likewise, adenosine-5-triphosphate (ATP) concentrations remained stable for 3 hours but significantly decreased in hearts perfused for 4 hours. Creatine phosphate concentrations showed the greatest change during the first hour of perfusion; they diminished to 51% of the amount present at zero time in the control group. After perfusion for 4 hours, the creatine phosphate content of the tissue was 45% of the zero time control value. Histochemical evaluation of the heart preparations for viability can be useful initially in establishing the perfusion methodology [16]. Tissues fixed by classical histological methods can be examined for any evidence of inflammatory infiltrate, edema, or hemorrhage.

To develop a better model of isolated perfused heart, a new apparatus of coronary artery cannula fixed in aortic tube was developed for continuous, normothermic perfusion and compared to the Casalis apparatus with cold ischemia [8]. All the continuous perfusion experimental hearts resumed a spontaneous heart beat and stabilized earlier than the control hearts without the need of defibrillator or pacemaker, indicating no reperfusion injury on the heart. None of the experimental hearts showed fibrillation nor stopped beating during the entire experiment, whereas the control hearts fibrillated. Two control hearts stopped beating, and only one of the two survived with the help of pacemaker. The coronary systolic, diastolic, and mean pressures were more stable with low variation in the experimental hearts than the cold ischemic control hearts. The hearts maintained with the new technique consumed more oxygen than the control hearts, indicating greater cardiac output [8]. According to these results, the continuous normothermic perfusion method by the new cannula, even though with a short period of hypothermic perfusion, provided better myocardial protection than the cold ischemia technique [8].

## Applications

Although isolated perfused heart preparations have been used in a variety of ways to study the mechanism and interaction of various drugs and hormones [15–20,75, 109,118], as well as in biochemical investigations [149,286], these

**TABLE 38.1**
**Effect of Perfusion on Heart Rate, Coronary Flow, and Isometric Systolic Tension in the Isolated Perfused Rat Heart[a]**

| Perfusion Time[b] (min) | Spontaneous No.[c] | Heart Rate ± SE (beats/min) | Coronary Flow ± SE (mL/min) | Isometric Systolic Tension (g) |
|---|---|---|---|---|
| 0 | 32 | 285 ± 5.9 | 7.9 ± 0.4 | 13.8 ± 0.7 |
| 15 | 26 | 280 ± 5.2 | 7.8 ± 0.4 | 15.0 ± 0.9[d] |
| 30 | 26 | 270 ± 5.3[d] | 7.7 ± 0.4 | 14.3 ± 0.9 |
| 45 | 26 | 267 ± 5.0[d] | 7.8 ± 0.4 | 13.6 ± 0.9 |
| 60 | 26 | 270 ± 5.8[d] | 7.9 ± 0.5 | 12.6 ± 1.0[d] |
| 75 | 19 | 266 ± 6.0[d] | 7.9 ± 0.6 | 12.3 ± 1.2[d] |
| 90 | 19 | 263 ± 6.7[d] | 7.5 ± 0.6 | 11.4 ± 1.2[d] |
| 105 | 19 | 263 ± 6.7[d] | 7.3 ± 0.6 | 10.6 ± 1.2[d] |
| 120 | 19 | 262 ± 6.4[d] | 7.0 ± 0.5[d] | 9.6 ± 1.2[d] |
| 135 | 12 | 272 ± 8.6 | 7.2 ± 0.6 | 8.2 ± 1.4[d] |
| 150 | 12 | 272 ± 10.1 | 7.1 ± 0.5 | 7.7 ± 1.4[d] |
| 165 | 12 | 275 ± 11.0 | 7.3 ± 0.7[d] | 7.0 ± 1.4[d] |
| 180 | 12 | 265 ± 9.7[d] | 6.9 ± 0.5[d] | 6.4 ± 1.3[d] |
| 195 | 6 | 255 ± 10.2[d] | 6.2 ± 0.2[d] | 6.8 ± 2.0[d] |
| 210 | 6 | 245 ± 9.2[d] | 6.1 ± 0.6 | 6.0 ± 1.8[d] |
| 225 | 6 | 245 ± 9.2[d] | 5.7 ± 0.5[d] | 5.3 ± 1.7[d] |
| 240 | 6 | 245 ± 9.2[d] | 5.2 ± 0.5[d] | 4.8 ± 1.3[d] |

[a] Hearts obtained from untreated normal male animals.
[b] Duration of perfusion after initial 15-minute equilibration period.
[c] Total number of measurements.
[d] Significant ($p < 0.05$) compared to zero perfusion time by paired variate Student's t-test.

*Source:* Aronson, C.E., *Arch. Int. Pharmacodyn.*, 222, 351–360, 1976. With permission.

preparations are not generally used to screen drugs or toxic chemicals for potential cardiotoxicity. However, efforts have been directed toward using such preparations for toxicological investigations [20,74,98,99, 155,161,163,202, 272,273,304,356]; for example, Autian [20] stressed the need for the development of reliable *in vitro* systems with which to screen large numbers of drugs and emphasized the advantages of using isolated perfused heart preparations. Gad et al. [87] used isolated perfused rat heart preparations to examine the inhibitory actions of the prooxidant butylated hydroxytoluene (BHT) on cardiac function. Cardiac contractility was depressed and leakage of creatine phosphate into the perfusate was found within 30 minutes of perfusion when BHT was included in the perfusion medium in concentrations ranging from 1 to 500 mg/L; thus, these investigations provided direct evidence that BHT depresses contractility and causes cellular damage of isolated heart as measured by the leakage of creatine phosphate from the myocardium.

Drug effects on myocardial contractile function are obviously of considerable practical importance for the toxicologist. The basic mechanism of such actions must reside at some point in the metabolism of cardiac muscle. Interference in the liberation of energy of the metabolic fuels utilized by the myocardial tissue may be implicated in many of the toxicological effects induced by drugs and other toxic chemicals. For example, chlorpromazine (50 ng/mL) decreased isometric systolic tension and prolonged the QT interval in the ECG of the perfused rat heart [20]. Fructose-1,6-diphosphate and pyruvate concentrations were elevated, suggesting aldolase inhibition and decreased pyruvate utilization. Anesthetic drugs that produce reversible depression of myocardial contractile function in a dose-dependent fashion have been shown to interfere with many of the mechanisms involving the generation and utilization of cellular energy in the heart tissue [216]. Using ischemia and the reperfusion injury model, 3-m*M* fructose-1,6-bisphosphate was recently shown to also preserve higher concentrations of high energy bonds in the heart tissue, indicating the protective effects of this substrate [179]. Use of isolated perfused heart preparations would be a valid and useful technique in the toxicological investigations related to drug-induced heart disease [20,21,72].

De Wildt and Speijers [74] studied the effects of dietary rapeseed oil and pure erucic acid on the mechanical behavior of the isolated rat hearts after 24 to 26 weeks. Neither compound caused any effects on the ECG changes in comparison to the control sunflower seed oil diet. After inotropic intervention, only the rapeseed oil group showed less contractile reserve capacity. The authors concluded that a

**TABLE 38.2**
**Effect of Perfusion Time on Metabolite Concentrations in the Isolated Perfused Rat Heart[a]**

| Metabolite | Concentration (μm/g ± SE[b]) at Various Perfusion Times[c] | | | | |
|---|---|---|---|---|---|
| | 0 hr | 1 hr | 2 hr | 3 hr | 4 hr |
| Glycogen | 14.23 ± 1.25 | 11.83 ± 1.60 | 9.59 ± 1.60 | 5.44 ± 0.93[d] | 4.75 ± 0.39[d] |
| D-Glucose-1-phosphate | 0.0214 ± 0.0154 | 0.0142 ± 0.0141 | 0.0155 ± 0.0141 | 0.0018 ± 0 | 0.0062 ± 0 |
| D-Glucose-6-phosphate | 0.0336 ± 0.0141 | 0.0480 ± 0.0109 | 0.0343 ± 0 | 0.0410 ± 0.0063 | 0.0532 ± 0.0134 |
| D-Fructose-6-phosphate | 0.0252 ± 0.0134 | 0.0096 ± 0 | 0[d] | 0.0018 ± 0[d] | 0.0052 ± 0 |
| Fructose-1,6-diphosphate | 0.0380 ± 0.0161 | 0.0656 ± 0.0236 | 0.0248 ± 0.0091 | 0.0204 ± 0.0109 | 0.0258 ± 0.0118 |
| D-Glyceraldehyde-3-phosphate | 0.0750 ± 0.0218 | 0.0908 ± 0.0148 | 0.0626 ± 0.0276 | 0.0680 ± 0.0271 | 0.0732 ± 0.0373 |
| Dihydroxyacetone phosphate | 0.0438 ± 0.0271 | 0.0238 ± 0.0089 | 0.0320 ± 0.0091 | 0.0572 ± 0.0209 | 0.0134 ± 0.0077 |
| Total triose phosphate | 0.1188 ± 0.0223 | 0.1146 ± 0.0161 | 0.0946 ± 0.0294 | 0.1246 ± 0.0209 | 0.0866 ± 0.0313 |
| L-(–)-Glycerol phosphate | 0.1892 ± 0.0588 | 0.1764 ± 0.0331 | 0.2175 ± 0.0764 | 0.2326 ± 0.0399 | 0.4018 ± 0.1011 |
| Pyruvate | 0.0206 ± 0.0173 | 0.0146 ± 0.0099 | 0.0166 ± 0.0107 | 0 | 0.0300 ± 0.0299 |
| L-(+)-Lactate | 1.1458 ± 0.3296 | 0.6068 ± 0.1193 | 0.7733 ± 0.1505 | 0.6020 ± 0.0470 | 0.8980 ± 0.1961 |
| Adenosine-5′-triphosphate | 2.8882 ± 0.1258 | 2.4058 ± 0.2271 | 2.6448 ± 0.1670 | 2.3692 ± 0.1846 | 1.7400 ± 0.1290[d] |
| Adenosine-5′-diphosphate | 0.1370 ± 0.0199 | 0.1688 ± 0.0503 | 0.1613 ± 0.0244 | 0.2118 ± 0.0354 | 0.2138 ± 0.0519 |
| Adenosine-5′-monophosphate | 0.0544 ± 0.0118 | 0.0776 ± 0.0138 | 0.0441 ± 0.0091 | 0.0364 ± 0.0126 | 0.0808 ± 0.0099 |
| Total nucleotides | 3.0796 ± 0.1174 | 2.6522 ± 0.1902 | 2.8503 ± 0.1650 | 2.6174 ± 0.1980 | 2.0346 ± 0.1140[d] |
| Creatine phosphate | 3.7180 ± 0.1710 | 1.8866 ± 0.3085[d] | 2.3355 ± 0.2002[d] | 1.4494 ± 0.0885[d] | 1.6766 ± 0.1698[d] |

[a] Hearts were obtained from untreated control animals. Results are from five or six hearts in each group.
[b] Expressed per gram of tissue (wet weight).
[c] Duration of perfusion after initial 15-minute equilibration period.
[d] Significant at 5% level when compared to zero perfusion time by an independent Student's t-test.

*Source:* Angevine, L.S. and Mehendale, H.M., *Am. Rev. Respir. Dis.*, 122, 891–898, 1980. With permission.

fat-rich diet might result in reduced myocardial function during a state of energy demand and that the erucic acid effects must be on the peripheral vascular system.

A number of studies have examined the role of oxyradicals in myocardial damage due to ischemia and reperfusion using perfused heart preparations [86,88, 160,163,179,229,356]. Koster et al. [163] demonstrated the release of malonaldehyde (MDA) in the coronary effluent of hearts perfused with cumene peroxide (0.5 m*M*), indicating a susceptibility of the coronary vascular tissue preradical-induced lipid peroxidation. The possible role of oxygen free radicals in the development of reperfusion arrhythmias was investigated using a 10-minute period of coronary ligation followed by reperfusion in the isolated rat heart [356]. Glutathione (GSH) and a combination of superoxide dismutase, catalase, and mannitol reduced the incidence of reperfusion-induced ventricular fibrillation when given just prior to reperfusion. Because these oxyradical scavengers protected against the reperfusion-induced myocardial injury, such experiments have been employed to implicate the role of oxyradicals in reperfusion-induced arrhythmias [356]. Ferrari et al. [88] induced ischemia in isolated, perfused rabbit hearts by reducing the coronary flow from 25 to 1 mL/min for 90 minutes. The effects of postischemic reperfusion were also followed for 30 minutes. These studies provided evidence that severe ischemia induces a reduction of the protective mechanisms, such as GSH and protein SH groups and mitochondrial superoxide dismutase (SOD) activity. Reperfusion induces a massive release of reduced GSH, oxidized GSH (GSSG), and creatine phosphokinase (CPK), leading to a further decrease in tissue content of these important protective mechanisms which in turn leads to loss of mechanical function. Interestingly, the CPK leakage associated with ischemia and reperfusion of the heart is augmented by arachidonic acid [149]. In these studies, the recovery of contractility was also suppressed by arachidonic acid (10 mg/mL). Arachidonic acid augmentation was protected by antioxidant vitamin E, indomethacin (a prostaglandin synthesis inhibitor), and nordihydroguarietic acid (a lipoxygenase inhibitor), observations that are consistent with the preradical mediation of ischemia reperfusion injury of the myocardium. The studies demonstrated that isolated, perfused heart preparations have significantly advanced our understanding of the oxyradical involvement in myocardial injury.

Akahira et al. [5] employed the Langendorff preparation of rat heart to study the effect of prazosin, an alpha-1 selective adrenergic antagonist, on $H_2O_2$-induced cardiotoxicity. $H_2O_2$ (600 μ*M*) produced an increase in the left ventricular end-diastolic pressure, a decrease in ATP, and an increase in melondialdehyde levels. The mechanical and metabolic derangements were attenuated by 2.5-, 5-, and 10-μ*M* prazosin, while the melondialdehyde formation was attenuated by 5- and 10-μ*M* prazosin. Prazosin (up to 10 μ*M*) had no adverse effects on the normal $H_2O_2$-nontreated perfused hearts. Nazeyrollas et al. [231] used the Langendorff constant-pressure isolated rat hearts to test the protective effects of amifostine, known to possess free-radical scavenging properties, on doxorubicin cardiotoxicity. Amifostine at $10^{-5}$ and $10^{-4}$ *M* concentrations induced coronary dilation and protected against the cardiotoxic effects of doxorubicin ($2.5 \times 10^{-5}$ *M*).

Isolated, perfused heart preparations have also been employed to study mechanisms of specific toxic chemicals [22,86,153,160,174,200,196,202,228,229,294,309]. McDonough et al. [202] employed the perfused working rat hearts to assess the intrinsic function after challenge with lethal or nonlethal doses of endotoxin to the rats. They concluded that the myocardial reserve was compromised by *in vivo* administration of endotoxin in a dose-dependent fashion. Bunc et al. [49] demonstrated the direct cardiotoxic effect of equinatoxin II, a protein extracted from the sea anemone *Actinia equina*.

Nahas and Trouve [228] employed Langendorff perfused rat heart preparations to assess the dose–response effects of natural cannabinoids on heart rate, coronary flow, and supraaortic differential pressure (ΔP). $\Delta^9$-Tetrahydrocannabinol produced a biphasic increase in heart rate without any change in coronary flow, but ΔP was increased. The toxicity of acetone [236], acrolein [271], allylamine [309], chronic alcohol consumption [274], dantrolene [108], cocaine [145], digitalis [155], doxorubicin [75,294], thioridazine-5-sulfoxide [118], *n*-hexane [153,273], and nonionic as well as monomeric contrast media [22] has also been studied using perfused heart preparations.

For many toxicological studies, it is necessary to determine the perfusate concentration of the experimental drugs with time. It is also necessary to determine the appearance and disappearance kinetics of possible metabolites of the test toxin. As is true in the case of most perfused organ studies, these experiments can be carried out employing either recirculating perfusate or a single-pass mode. In most experimental protocols, it is prudent to determine in advance which mode of circulation would be utilized. It will also be necessary to determine the size of the perfusate sample required to carry out the analyses intended and the time course and total duration of the experiment. As indicated in Table 38.1, many enzymatic determinations can be carried out in the perfusate as well as in the heart tissue itself.

The influence of propafenone associated with propofol on myocardial contractility (dP/dt and heart rate), coronary flow, and the incidence of arrhythmia in isolated rat hearts was investigated [133]. Arrhythmogenic effects of propafenone (pro-arrhythmia) were verified in 50% of hearts receiving 100 μg propafenone. In the association with propofol, no significant difference occurred, and arrhythmias (pro-arrhythmic effect) were observed in 40% of the hearts.

## Isolated Perfused Liver

The isolated perfused liver preparation has enjoyed long-standing, sustained attention in terms of functionally preserved isolated organs for biochemical, pharmacological, and toxicological studies. The view of Miller et al. [218] concerning the perfused liver is that functional performance of the liver cells is best studied in the isolated liver perfused with continuously oxygenated whole, homologous blood under closely approximating physiological conditions. Earlier attempts to obtain viable, perfused liver preparations were marked by many problems, and in most cases the failure can be attributed to several factors [218]: use of aqueous perfusion media such as Ringer's solution in place of whole blood; lack of adequate filtering devices to remove tiny fibrin clots that plug the hepatic circulatory system; and the relative unavailability of effective, nontoxic anticoagulants such as heparin, which results in ominous failure to carry out perfusion for more than 1 to 2 hours when whole blood is used as the perfusate. We have now witnessed the increased use of isolated liver preparations in a variety of pharmacological and toxicological investigations. Much credit for developing and standardizing isolated perfused liver techniques belongs to Miller et al. [218] and Schimassek [297].

The liver has two sources of blood supply. The portal vein supplies 80% and the hepatic artery 20% of the blood flow. Most workers have ignored the hepatic arterial supply when perfusing the rat liver. Although experiments confirm that the liver functions normally even in the absence of perfusion through the hepatic artery, techniques are now available that allow this small blood vessel to be perfused at normal arterial pressure, either under hydrostatic conditions [288] or by direct pumping [266]. By and large, the rat has been the animal of choice for isolated perfused liver preparations [3,5,45,46,55,56,193,194,197,213,218,300,323,326,327,331,333], although livers from other experimental animals have been perfused for various investigations [11,56,82,212,346].

### Apparatus

In the liver perfusion technique of Miller et al. [218], the liver is perfused at a constant hydrostatic pressure using diluted rat blood as the perfusion medium, which passes through a glass multibulb oxygenator. The perfusate enters the portal vein via a filter and drains from the liver through the inferior vena cava either via an indwelling cannula or through a free cut of the vena cava; it collects in a bottom reservoir. From the reservoir it is pumped to the top of the oxygenator, which it enters through a filter. The liver is removed from the animal after the cannulation procedure, and perfusion is conducted by connecting the portal cannula of the liver to the preprimed circulation of the apparatus housed in an enclosed liver chamber.

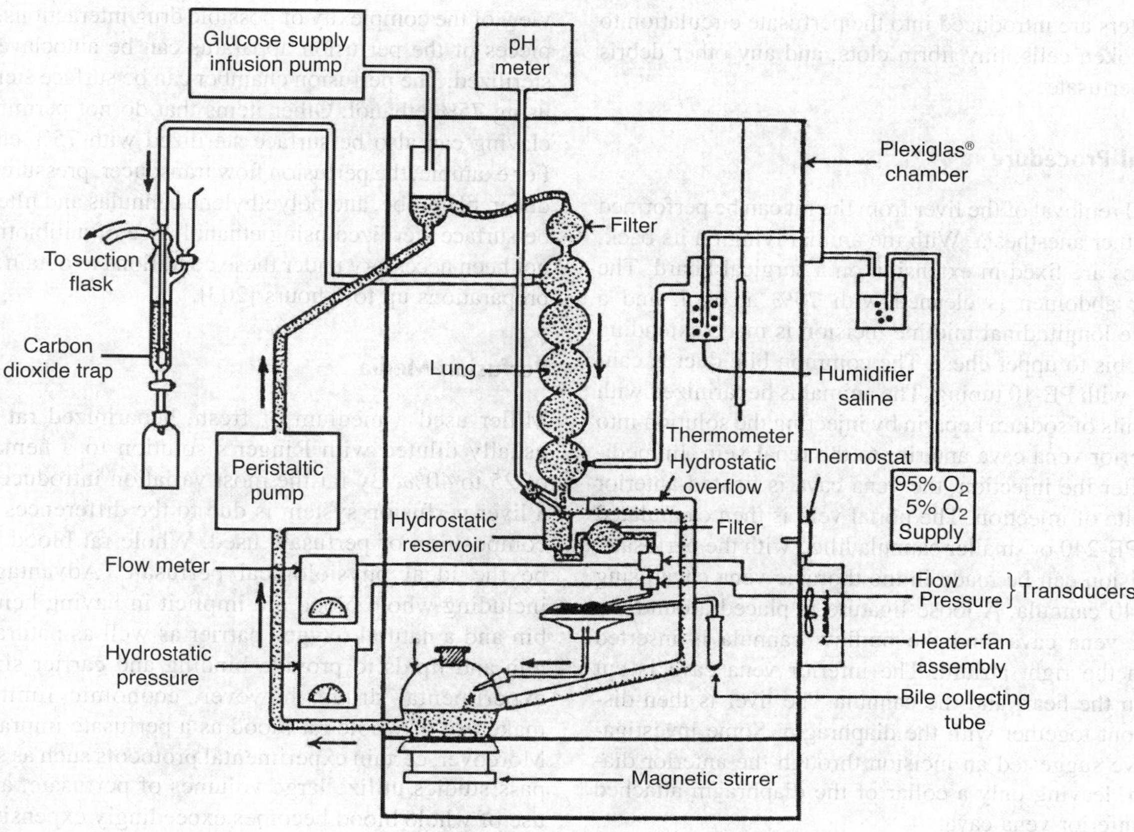


**FIGURE 38.3** Liver perfusion apparatus modified from that described by Miller et al. [181]. Most of the circulation carrying the perfusate is housed inside a Plexiglas® chamber fitted with a thermostatically controlled heater–fan assembly, which is connected to the top of the chamber to recirculate warm air from the chamber. In the center of the chamber, a pair of vertical cylindrical rods (not shown) is fitted to facilitate securing various items inside the chamber by means of appropriate clamps. The bottom reservoir, liver platform, flow and pressure transducers, filter, multibulb glass lung hydrostatic reservoir, humidifier, and upper reservoir to dampen the peristaltic wave of the perfusate are held by means of suitable clamps on these vertical rods. The liver platform has a ground-glass joint that can fit the top of the bottom reservoir. A different arrangement is shown here that allows sampling the effusate directly from the liver by disconnecting the side-arm connection of the bottom reservoir. The inside of the chamber is accessible via two overlapping sliding doors (not shown) in the front of the chamber. (From Lambert, C. et al., *Cardiovasc. Res.*, 24, 653–658, 1990. With permission.)

The apparatus designed by Miller has been almost universally adopted by workers using this technique, and it is also available commercially. The liver perfusion apparatus used in our laboratory [203] is based on Miller's [218] description and is illustrated in Figure 38.3. The perfusion chamber is kept warm by means of a heater–fan assembly that is thermostatically controlled. The original description of Miller et al. [218] noted the use of a heating coil that traversed the inner surface of the chamber, but cleaning the chamber becomes difficult with this arrangement; hence, heating may be accomplished by installing a highly efficient heating coil inside a box that would also house a fan [203].

The glass multibulb oxygenator can be easily put together by connecting a series of 100-mL round-bottomed flasks [203]. At the bottom of the multibulb glass oxygenator, an inlet is provided for oxygen supply, and at the top of the oxygenator a side outlet is provided for the carbon dioxide to escape. The bottom reservoir is provided with a number of side arms and a principal opening at the top to connect the platform that supports the liver. A magnetic stirrer can be introduced into the bottom reservoir, and a magnetic stirring device can be placed below the reservoir under the chamber. The perfusate can be pumped by means of a peristaltic pump to avoid hemolysis of red blood cells when either whole blood or diluted blood is used as the perfusate. $O_2/CO_2$ (95:5) is passed through a water trap to humidify the gas mixture. The expired gas escaping from the top of the multibulb glass lung is passed through a carbon dioxide trapping device. This phase is especially useful in studies in which either labeled carbon dioxide or other volatile metabolic products are expected to be formed. Efficient filtration of the blood perfusion medium is considered to be crucial for successful perfusion. Two disks of Lucite® are compressed together to hold a disk of white silk (100 × 150 mesh per inch), and two such filters are introduced into the perfusate circulation to clear broken cells, tiny fibrin clots, and any other debris in the perfusate.

## Surgical Procedure

Surgical removal of the liver from the rat can be performed under ether anesthesia. With the animal lying on its back, the limbs are fixed in extension on a surgical board. The anterior abdomen is cleaned with 75% alcohol, and a ventricle longitudinal midline incision is made extending from pubis to upper chest. The common bile duct is cannulated with PE-10 tubing. The animal is heparinized with 1000 units of sodium heparin by injecting the solution into the inferior vena cava anterior to the renal vein. Immediately after the injection, the vena cava is ligated anterior to the site of injection. The portal vein is then cannulated with a PE-240 or smaller cannula filled with the perfusate. An incision can be made in the thoracic vena cava using a PE-240 cannula. A loose ligature is placed around the inferior vena cava, and the outflow cannula is inserted through the right atrium. The inferior vena cava is cut between the heart and the cannula The liver is then dissected out together with the diaphragm. Some investigators have suggested an incision through the anterior diaphragm, leaving only a collar of the diaphragm attached to the inferior vena cava.

The liver with or without the diaphragm is then lifted free of the abdomen together with its cannulas and transferred to a warm saline bath. Immersing the liver in the warm saline facilitates proper orientation of the lobes as well as cleaning the blood clots and debris that may be on the surface of the liver. Removing the liver and subsequent handling require skill and should be accomplished with minimum handling of the organ itself. After ensuring that the lobes of the liver are properly oriented, the liver may be attached to the perfusion apparatus by connecting the portal cannula to the circulating perfusate. The liver is placed on the platform, making sure the outflow cannula is let down through the central porthole of the platform extending into the neck of the bottom reservoir. Proper orientation of the common bile duct cannula ensures unhindered bile flow.

An equilibration period of 30 minutes is generally sufficient to establish proper perfusion flow and for the liver to recover from the brief period of anoxia it underwent during the surgical procedure. Glucose can be infused throughout the perfusion to replenish the glucose utilized by the liver. This procedure also allows replacing any fluid losses due to evaporation of the perfusate in the heated chamber [193].

Earlier investigators have used antibiotics to prevent bacterial growth in the perfusate during the course of the perfusion [218,298]. In many toxicological studies, it is important to keep the perfusion system free of drugs in view of the complexity of possible drug interactions. Most pieces of the perfusion apparatus can be autoclaved and sterilized. The perfusion chamber can be surface sterilized using 75% ethanol. Other items that do not permit autoclaving can also be surface sterilized with 75% ethanol. For example, the perfusion flow transducer, pressure transducer, pH probe, and polyethylene cannulas and filters can be surface sterilized using ethanol. Use of antibiotics has not been necessary under these conditions to obtain viable preparations up to 6 hours [203].

## Perfusion Media

Miller used a medium of fresh, heparinized rat blood usually diluted with Ringer's solution to a hematocrit of 25 to 40%. By far the most variation introduced into a liver perfusion system is due to the differences in the composition of perfusate used. Whole rat blood would be the ideal physiological perfusate. Advantages of including whole blood are implicit in having hemoglobin and a natural oxygen carrier as well as natural protein and lipids to provide binding and carrier sites for experimental drugs; however, economic limitations make use of whole rat blood as a perfusate impractical. Moreover, certain experimental protocols such as single-pass studies utilize large volumes of perfusate, and the use of whole blood becomes exceedingly expensive and impractical. The following blood perfusate has the advantages of containing rat blood as well as being economical, as it is mixed with two parts of KRB-buffered solution (pH 7.4). The Krebs–Ringer bicarbonate solution includes the following (in g/L): NaCl, 6.896; KCl, 0.354; $CaCl_2 \cdot 2H_2O$, 0.373; $KH_2PO_4$, 0.162; $NaHCO_3$, 2.1; $MgSO_4 \cdot 7H_2O$, 0.293; bovine serum albumin (BSA) Cohn fraction V, 45; and glucose, 0.901. The solution is adjusted to pH 7.4 with 1-*N* NaOH solution. Freshly collected heparinized whole rat blood is mixed with this solution to obtain a 30% blood perfusate. The required volume of glucose solution (20% solution) is added to the perfusate to obtain a final glucose concentration of 3.2 g/L. The advantages of this type of medium are the physiological nature and high oxygen-supplying capacity. Disadvantages include the poorly defined nature of the blood with respect to the presence of hormones and substrates and the difficulty of collecting blood from small animals. To circumvent the problem of uncertainty of hormones and substrates, some investigators have used bovine [298–300] or human [125] erythrocytes, which can be included in the perfusate after washing.

Triner et al. [337] demonstrated that isolated perfused liver preparations can be supported using artificial oxygen carriers such as fluorocarbon emulsions emulsified in electrolyte buffer solution. The use of fluorocarbon emulsion in the perfusion of isolated organs would have a number of advantages over erythrocyte suspension. Avoidance of possible antigenicity from nonautologous erythrocytes and simpler, more standardized, less expensive preparations of perfusion media are among the advantages. In experiments comparing the adequacy of fluorocarbon emulsions to replace erythrocytes in the perfusate, three kinds of fluorocarbon were evaluated [112,243,337]. Urea nitrogen, glucose, sodium, potassium, and alanine aminotransferase in the medium and incorporation of $^{14}$C-lysine into the liver proteins were found to be either normal or above normal compared to perfusate containing erythrocytes [243] when fluosol-43 was used as the fluorocarbon oxygen carrier. Using the fluorocarbon FC-47 emulsified in KRB-buffered solution, Goodman et al. [112] found that oxygen consumption, alanine aminotransferase gluconeogenesis, production of lactate and ketone bodies, and hepatic ATP concentrations were no different than when buffer or a medium containing suspended erythrocytes was used as perfusate. The cytosolic and mitochondrial redox states, as indicated by the hepatic lactate/pyruvate and β-hydroxybutyrate/acetic acid ratios, respectively, were also the same whether the medium contained erythrocytes or FC-47 [112].

Because of the varieties of perfusion media used for liver perfusion by various investigators, it is prudent to select the most appropriate perfusion medium for a particular experimental protocol [30,42,203,213,218, 230,331]. Many variables include the presence or absence of bovine or other serum albumin, erythrocytes, buffering agents, amino acids, glucose, vitamins, antibiotics, and heparin. These conditions in combination may or may not be appropriate for the particular experimental design. A study designed to examine the utilization of externally provided GSH may not be valid if BSA is included in the perfusion medium [144]. Reduced GSH (GSSG) is oxidized rapidly ($t_{1/2}$ 10 minutes) in KRB medium containing BSA [144]. Replacing the albumin with high-molecular-weight dextran was found to preserve the GSSG. A perfusion system designed to study $Ca^{2+}$ fluxes from the liver may not contain albumin or hemoglobin, as the ion-selective electrode used for measuring the change in perfusate $Ca^{2+}$ might not be compatible with these constituents [213].

## Viability Criteria

Certain viability criteria can be readily used to evaluate the perfused liver. For example, bile flow can be used as a viability criterion. Here, 1 to 1.5 μL/min/g of liver can be expected from a normal rat liver, which can change depending on the experimental conditions. Bile flow can be expected to drop with the time of perfusion, as the endogenous bile acid pool would be depleted during bile collection. Some investigators have used an infusion of sodium taurocholate in the perfusate to maintain the bile flow throughout the perfusion time.

Another easily recognizable viability criterion is the perfusion flow rate. At a hydrostatic pressure of 15 to 20 cm$H_2O$, a flow rate of up to 60 mL/min can be obtained using diluted (30%) blood as perfusate. Even after allowing for a lower viscosity of the perfusate, this flow rate would be judged to be beyond the normal physiological range; hence, the perfusion flow rate should be controlled by means of a suitable clamp placed between the portal cannula and the hydrostatic reservoir. Once a stable flow is attained, the flow rate through the liver can be used as a readily available criterion for evaluating the viability of the organ.

Another easily detectable criterion is visual examination of the liver. Inadequately perfused liver gives a reddish appearance, indicating anoxia, as well as a blotchy appearance on the surface of the liver. Often, if the liver is not secured on the platform, the liver moves in such a way as to impede proper, continuous, uniform perfusion through the organ. The liver may move such that the flow of the liver through the outflow cannula may be impeded, resulting in swelling of the liver. Visually, this problem is easily recognized by the tensile and anoxic appearance of one or more lobes of the liver.

Oxygen consumption by the liver can be used as another criterion for viability. For example, Schimassek [297] reported that oxygen consumption by the perfused liver was 2.2 nmol/min/g of tissue after the 30-minute equilibration period, and it was maintained at this level thereafter. Oxygen consumption can be measured by following the oxygen tension of the perfusate before it enters the liver and sampling the oxygen content of the perfusate after it effuses from the liver [203].

A number of biochemical parameters can be used for ascertaining the viability of the perfused liver. Schimassek [298] has shown conclusively that the isolated perfused livers under standard conditions have glycolytic intermediate concentrations, a respiratory quotient, and adenine nucleotide levels close to those found in the liver fresh out of the animal (Table 38.3). Such determinations in the isolated perfused liver preparation have allowed several investigators to determine normal conditions of perfusion. A variety of other biochemical parameters have also been determined in isolated perfused liver preparations to establish the physiological validity of using such a preparation [170]. Miller et al. [218] determined the incorporation of $^{14}$C-lysine into hepatic proteins as a biochemical index of optimum macromolecular synthetic activity during perfusion. Bock et al. [38] measured a number of biochemical parameters associated with the microsomal mixed function oxidase (MFO) and cytochrome P450 system and found satisfactory preservation of the hepatic MFO system after 4 hours of perfusion when erythrocyte-containing perfusate was used. Biliary excretion of sulfobromophthalein (BSP) and indocyanine green has also been useful as a measure of the functional status of the isolated perfused

**TABLE 38.3**
**Substrate Content of Isolated, Perfused Rat Liver after Various Times of Perfusion**

| Metabolite | *In Vivo* | Content (μmol wet weight of liver) Before Perfusion | After 30-Minute Perfusion | After 120-Minute Perfusion |
|---|---|---|---|---|
| Lactate (L) | 1450 | 12,000 | 3570 | 3400 |
| Pyruvate (P) | 145 | 50 | 277 | 345 |
| α-Glycero-P (G) | 253 | 1,560 | 304 | 450 |
| Diammonium phosphate (D) | 38 | 21 | 45 | 67 |
| Malate (M) | 443 | 750 | 281 | 280 |
| Oxaloacetate (O) | 7 | <1 | 4.2 | 4 |
| Fructose-1,6-diphosphate (FDP) | 22 | 17 | 20 | 41 |
| Fructose-6-phosphate | 75 | — | — | 41 |
| Glucose-6-phosphate | 370 | 986 | 141 | 206 |
| Glucose | 8600 | 25,900 | 11,600 | 11,000 |
| Glycogen | 340,000 | 320,000 | 214,000 | 185,000 |
| Adenosine monophosphate (AMP) | 300 | — | 209 | 280 |
| Adenosine diphosphate (ADP) | 900 | 1853 | 684 | 620 |
| Adenosine triphosphate (ATP) | 2900 | 710 | 2270 | 2060 |
| L/P | 10 | 239 | 13 | 10 |
| G/D | 6.7 | 72 | 6.7 | 6.5 |
| M/O | 64 | — | 70 | 70 |
| ATP/ADP | 3.3 | 0.4 | 3.3 | 3.3 |

*Source:* Adapted from Thurman, R.G. et al., *Rev. Biochem. Toxicol.*, 1, 249–286, 1979.

liver. The disadvantage of using these markers for functional status is that the same perfused livers cannot be used for toxicological investigations after establishing that these livers are indeed viable. These tests are helpful for evaluating isolated perfused liver preparations, establishing the procedure, and subsequently evaluating the functional status of the liver preparations after treating with an experimental toxic agent. Phenolphthalein glucuronide (PG) has been used as a marker of biliary excretory function [213]. The advantages of using PG are that it requires no further metabolism, it can be used at low concentrations, and it is sensitive for detecting hepatobiliary dysfunction and so can be used in livers being perfused for toxicological investigations [207,213].

Although histological examination of perfused livers is not done routinely, the technique is useful for setting up a perfused preparation. Several authors [42,293,298] have reported results of histological examinations of perfused livers, indicating the general usefulness of morphological examination in ascertaining the viability of perfused liver preparations. However, one should be aware that morphological examinations may be of limited usefulness, as cells that appear abnormal morphologically may exhibit normal cellular function; conversely, normal-appearing cells may exhibit abnormal cellular functions [42,218]. Oomen and Chamalarun [250] reported an excellent correlation between biochemical and histological parameters in their isolated perfused rat liver preparations.

## Applications

Depending on the experimental protocol, the isolated perfused liver preparation can be used in either a single-pass or a recirculating mode [203,213,335]. Many examples can be cited for the use of isolated perfused liver preparations in the evaluation of toxic responses to a variety of chemicals. It is instructive to consider a few examples of the use of isolated perfused liver preparations in toxicological investigations [29,43,46,100,138,181,197,198,205,207,213,217,221,230,233,271,278,282,295,312,326,327,329,331,343,346]. Only a few of these will be discussed here. Rice et al. [282] employed isolated perfused rat liver preparations to determine the effect of carbon tetrachloride ($CCl_4$) administered *in vivo* on the hemodynamics of the liver. Portal blood pressure and flow were recorded in perfused livers from either control or treated animals. The study concluded that, although the primary lesion caused by $CCl_4$ was hepatocellular damage, subsequent effects included increased vascular resistance and enhanced response to norepinephrine. Masuda et al. [197] employed the perfused liver to examine if bromotrichloromethane-induced cell necrosis can be dissociated from lipid peroxidation. Their study provided histological evidence for dissociation of lipid peroxidation from hepatocellular necrosis induced by this halomethane. Ambs et al. [7] investigated acute and chronic toxicity of aromatic amines in the isolated perfused rat liver. 2-Acetyl amino-fluorine (2-AAF) and its principal metabolites were not toxic in the range of concentrations from 200 to 400 μ*M* in a 2-hour exposure to perfused male Wistar rat livers; however, *N*-acetyl 2-AAF was severely toxic. Chronic effects of feeding 2-AAF (0.02% in the diet) for up to 12 weeks were also studied. Excretion of glutathione in bile was drastically reduced after 5 or more weeks, increasingly less glucose was released in the perfusate, and $O_2$ consumption was constantly increased by 20% after 3 weeks of 2-AAF feeding. The authors suggest that these effects are adaptive responses to the toxicity of 2-AAF and may be related to the promoting properties of this carcinogen. Iwamoto et al. [138] used perfused rat liver to establish the decreased intrinsic hepatic clearance of propranolol in $CCl_4$-injured liver. Another study used perfused liver to investigate the hepatic elimination of galactose in $CCl_4$-injured liver [343]. Bullock et al. [48] utilized isolated perfused rat liver preparations to examine the effects of two fungal toxins (i.e., sporidesmin and icterogenin) on the mechanisms of bile secretion. Electron microscopic examination of livers perfused with these two toxins indicated that the cholestatic reaction was due to changes in canalicular membranes, which included extrusion of material into the canalicular lumen and aggregation of lysosomes in the cytoplasm. Abraham et al. [1] examined the effect of hyperoxia on lysosomal enzymes using perfused liver preparations. Colantoni et al. [62] investigated reoxygenation injury in isolated perfused rat livers.

Employing 60 minutes of hypoxia followed by 30 minutes of reoxygenation, liver injury assessed as lactate dehydrogenase (LDH) release, protein, carbonyl content, and melondialdehyde production was significantly decreased in livers perfused with 2-m*M* salicylate. In another study, cold ischemia–reperfusion injury was investigated using isolated perfused rat liver preparations [168]. The objectives were to investigate whether the inactivation of Kupffer cells by gadolinium chloride modulates cold ischemia–reperfusion liver injury and whether cold storage of rat liver involves injury to biliary epithelial cells. The authors concluded that cold ischemia–reperfusion liver injury of rat liver is mediated by both Kupffer cell-dependent and -independent mechanisms, and cold storage of rat liver induces functional impairment of biliary epithelial cells.

Radwan and Henschler [271] employed isolated liver preparations to study the uptake and metabolism of the hepatocarcinogen vinyl chloride. Using erythrocyte-suspension perfusion medium, they found that the solubility of vinyl chloride stayed constant at concentrations of 50 to 25,000 ppm. The amount metabolized, as determined by the difference between vinyl chloride concentration before and after passage through the liver, was found to stay constant at 14.6% of the 50 to 25,000 ppm concentrations. Ethanol (12 m*M*) and pyrazole (200 m*M*) decreased the metabolism of vinyl chloride which was also modified by prior exposure of the animals to other inducing and inhibiting agents. It was concluded that vinyl chloride underwent a metabolic transformation via mixed function oxidation to reactive metabolites. The above study can be taken to represent how perfused liver preparations can be used to determine the metabolism of even volatile substances.

Lemaster et al. [181] used perfused liver to study hypoxic hepatocellular injury and concluded that shedding of cytoplasmic fragments resulting from the blebbing of centrilobular hepatocytes may represent a basis for the appearance of hepatic enzymes in the sera of patients with liver disease. Nastainczyk and Ullrich [230] used isolated perfused rat liver preparations to examine the effect of hypoxia on the metabolism of halothane. The study concluded that halothane is biotransformed via reductive *in vivo* metabolism to reactive intermediates when the oxygen concentration of the perfusate drops below a critical level (about 50 m*M*). These authors employed whole-organ spectrophotometry of isolated perfused livers to establish that a complex of macromolecules and halothane is formed under slightly hypoxic conditions and that metyrapone, an inhibitor of MFO reactions, abolished the formation of this complex.

Use of isolated perfused liver preparations in the metabolism of toxic substances as well as the effect of the toxic substances on hepatic function are illustrated by a series of studies in which the hepatobiliary function was examined after the animals had been exposed to toxic chemicals [205–207]. In these studies, the effect of exposure to chlorocarbon pesticides, mirex, and chlordecone (Kepone®) was examined in isolated perfused liver preparations. Biliary excretion of anionic model compounds, BSP, and imipramine was examined. These studies also illustrated the utility of isolated liver preparations in studying the biotransformation of chemicals. By assaying a series of perfusate samples as well as liver tissue at the end of a perfusion study, the metabolism of imipramine by control as well as pretreated livers was followed. Although both chlordecone and mirex were known to be inducers of hepatic MFOs, these experiments revealed that the biliary excretion of endogenously formed metabolites of imipramine was suppressed by prior exposure to the above chlorocarbons. The pattern of imipramine metabolism indicated that the suppressed biliary excretory function was not related to alterations in metabolism of imipramine.

Other experiments in which readily excretable polar metabolites of imipramine were introduced into the perfusate of control and treated liver preparations indicated that biliary excretion of these metabolites was also hindered by prior exposure to mirex and chlordecone [205,206]; thus, such experimental manipulations using isolated liver preparations were useful for evaluating the role of drug metabolism in hepatobiliary function. Isolated liver preparations can also be utilized to examine the role of hepatic uptake and metabolism in the disposition of toxic chemicals [204]. In these studies, the uptake, metabolism, and biliary excretion of polychlorinated biphenyls were examined using isolated perfused rat liver preparations. Similar preparations were useful in discovering the inhibitory effect of mirex on the biliary excretion of polar metabolites of monochlorobiphenyl [206].

Choo et al. [60] employed isolated perfused rat liver in single-pass perfusion mode to study if certain drug-metabolizing pathways are affected more than the others in hepatic cirrhosis. Using *p*-nitrophenol as a substrate for glucuronidation and sulfation and *d*-propranolol as a substrate for oxidative metabolism, the authors concluded that, in cirrhosis, oxidative metabolism and sulfation are significantly impaired but glucuronidation is spared. The decreased sulfation is attributed to a decrease in sulfotransferase as well as to decreased cofactor (PAPS) synthesis. Hoffmann et al. [132] examined $^{14}$C-phenol in isolated perfused mouse livers. These studies were conducted to examine if the metabolic fate of phenol produced during benzene metabolism was different in the absence of benzene. Administration of benzene produces bone marrow depression, whereas administration of phenol, a major metabolite of benzene, does not. Mouse livers were perfused orthograde (portal vein to central vein) or retrograde (central vein to portal vein) to investigate the metabolic zonation of enzymes involved in phenol hydroxylation and conjugation. It was found that a larger percentage of radioactivity released from the liver was unconjugated hydroquinone after benzene perfusion than after phenol perfusion. Enzymes involved in the *p*-hydroxylation of phenol are located nearer the central vein than those involved in conjugation. The amount of radioactivity covalently bound to liver macromolecules was measured after each perfusion and was determined to be the amount of hydroquinone glucuronide detected in the perfusate samples.

In other studies, the pharmacokinetics [128] and acute toxic action of diclofenac [344] and troglitazone [269] were studied in isolated perfused rat livers. Livers were perfused for 2 hours with Krebs–Henseleit bicarbonate buffer (250 mL) containing either 10.75 mg or 1.075 mg of diclofenac, representing 100 and 10 times the therapeutic levels, respectively. At the higher concentration, liver injury (LDH release) was observed at 90 minutes; neither alanine aminotransferase (ALT) nor aspartate aminotransferase (AST) was released. The lower concentration did not elicit liver injury. The authors concluded that diclofenac at therapeutic levels is unlikely to have any direct toxic effects, not even at 100 times the therapeutic levels.

Thiazolidinedione, a compound used as an antidiabetic agent in type 2 (adult-onset) diabetes, is known to enhance insulin action and reduce plasma glucose concentrations when administered clinically to type 2 diabetic patients. Preininger et al. [269] investigated the acute actions of troglitazone (0.61 and 3.15 μ*M*) on hepatic glucose and lactate flux, bile secretion, and portal pressure under basal or insulin- or glucagon-stimulated conditions in isolated perfused rat livers. During BSA-free perfusion, high doses of troglitazone increased basal, but inhibited glucagon-stimulated, incremental glucose production by 75% vs. control. Low-dose troglitazone did not enhance the inhibitory effects of insulin on glucagon-stimulated glucose production but rapidly increased LDH release and portal venous pressure. The authors concluded that troglitazone exerts both insulin-like and non-insulin-like hepatic effects which are blunted by the presence of albumin due to binding to albumin.

In another study, Villanueva et al. [344] investigated the effects of bile acids on the transport of cisplatin in perfused rat liver. Enhancing the excretion of the cytostatic drug, cisplatin from the body is envisaged as a way of decreasing cisplatin toxicity. The objective of the study was to investigate if bile acids could be used to enhance biliary excretion of cisplatin by urodeoxycholic acid (a highly choleretic acid), glycocholic acid, and chenodeoxycholic acid, the latter two being the micelle-forming bile acids. The authors concluded that, even though bile acids (1 m*M*) induce an enhancement in the transport of cisplatin from the hepatocyte to bile, the net excretion of cisplatin in the bile is reduced.

Hadasova et al. [117] investigated the influence of immunosuppression on *O*-demethylation of dextromethorphan in isolated perfused rat liver. They examined the effect of cyclophosphamide and dexamethasone on the CYP2D1-dependent metabolism of dextromethorphan in isolated perfused rat liver from male Wistar rats. Although cyclophosphamide and dextromethorphan both increased the *O*-demethylation of dextromethorphan to dextrorphan, only cyclophosphamide caused significant changes in the pharmacokinetic properties characteristic of increased rate conversion of dextromethorphan to dextrorphan. These findings suggest that CYP2D-dependent metabolism might be promoted in immunosuppressant therapy of autoimmune and other diseases.

Thurman et al. [331] studied the kinetics of *p*-nitroanisole *O*-demethylation in hemoglobin-free perfused rat liver preparations. Using the isolated liver, these investigators were able to demonstrate that the rates of *p*-nitroanisole metabolism were linear for 30 minutes in normal livers and for 1 to 2 minutes in phenobarbital-induced livers. This reduced rate of metabolism could be reversed by infusing additional glucose, suggesting an intimate relation between drug and carbohydrate metabolism in the intact liver. Alteration in the rate of *p*-nitroanisole metabolism with various inducing agents of the MFO system produced parallel changes in rates of hepatic lactase production, reflecting the action of *p*-nitrophenol to uncouple oxidative phosphorylation; thus, these investigators were able to demonstrate that the reduction in the rate

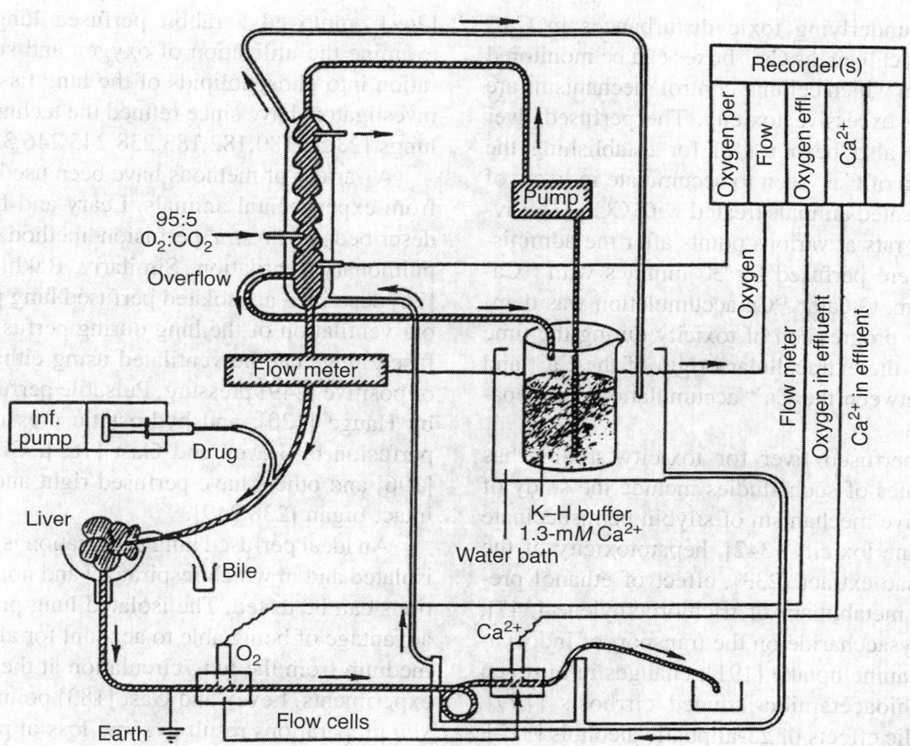


**FIGURE 38.4** Liver perfusion apparatus designed for on-line measurement of perfusate $Ca^{2+}$ levels. (From Levin, N.W. et al., *S. Afr. J. Med. Sci.*, 30, 78–79, 1965. With permission.)

of *p*-nitroanisole metabolism in induced liver preparations was due to reduced availability of NADPH for MFO-catalyzed substrate oxidation. Using a similar noncirculating liver perfusion system, Belinsky et al. [30] employed trypan blue to investigate the regiospecific hepatotoxic response of the liver. Periportal hepatocellular injury after allyl alcohol infusion to isolated perfused liver was readily evident from the stained nuclei only in the periportal zone. Takano et al. [323] described a technique similar to the one described by Thurman et al. [331], in which the dynamic effects of environmental agents on the hepatic drug-metabolizing system and on the energy metabolism could be monitored in the perfused liver. Such experiments with intact liver preparations have provided valuable insights into what might be occurring in terms of drug metabolism under *in vivo* conditions. These results demonstrated that, despite the induced status of the liver, enhanced drug metabolism may not necessarily be the end result, as other factors (e.g., availability of cofactors) might become limiting and hence limit the quantitative aspects of drug metabolism.

There has been significant interest in understanding the role of $Ca^{2+}$ in chemical toxicity [295]. Although it is generally agreed that a rising cytosolic $Ca^{2+}$ level is detrimental to the cell, the source of the increased $Ca^{2+}$ has been strongly debated [295,312]. Because of the inherent disadvantages of working with isolated cells and organelles, intact perfused liver preparations offer the most suitable model. Such a perfusion setup, used by Mehendale et al. [213], is illustrated in Figure 38.4. Infusion of menadione elicited an increased oxygen utilization by the liver, followed by a decrease in the perfusate $Ca^{2+}$. Hepatic accumulation of $Ca^{2+}$ was accompanied by stimulation of cytosolic phosphorylase a activity, indicating a rise in the cytosolic $Ca^{2+}$ levels. A gradual recovery of perfusate $Ca^{2+}$ to base levels was observed after cessation of menadione infusion. Leakage of LDH into the perfusate followed $Ca^{2+}$ uptake, which was not accompanied by a decrease in reduced pyridine nucleotide or ATP level in the liver, as evidenced by measurements either during maximal $Ca^{2+}$ uptake or after recovery. However, $Ca^{2+}$ uptake was correlated with decreased GSH and increased GSSG levels in the liver, both of which reversed during recovery from $Ca^{2+}$ uptake. The amount of protein-bound mixed disulfides showed a striking relation to $Ca^{2+}$ uptake, reaching a maximal value during $Ca^{2+}$ uptake and reversing toward normal during recovery from $Ca^{2+}$ accumulation. Depletion of hepatic GSH with prior diethylmaleate treatment resulted in increased $Ca^{2+}$ accumulation during menadione infusion. These findings suggested that menadione-induced $Ca^{2+}$ uptake is due to plasma membrane dysfunction as a result of loss of protein thiol groups, which are critical for maintaining the plasma membrane $Ca^{2+}$ extrusion mechanism. This perfused liver model is particularly useful for studying the mechanisms underlying toxic disturbances in $Ca^{2+}$ homeostasis in intact liver, as $Ca^{2+}$ fluxes can be monitored under conditions in which cellular control mechanisms are not obliterated by excessive toxicity. The perfused liver preparations have also been useful for establishing the extracellular origin of $Ca^{2+}$ seen to accumulate in livers of chlordecone-pretreated animals treated with $CCl_4$ [3]. Livers obtained from rats at various points after the administration of $CCl_4$ were perfused for 30 minutes with $^{45}Ca$-containing medium. Greater $^{45}Ca$ accumulation was demonstrated with the progression of toxicity during the time course, indicating the extracellular origin of the $Ca^{2+}$ and the association between the $Ca^{2+}$ accumulation and hepatocellular toxicity.

Use of the perfused liver for toxicity studies has increased. Examples of such studies include the study of the hepatoprotective mechanism of silybin hemisuccinate on phenylhydrazine toxicity [342], hepatotoxicity of the hornet's venom sac extract [233], effect of ethanol pretreatment on the metabolism of trichloroethylene [347], effects of lipopolysaccharide on the transport of indocyanine green and alanine uptake [191], changes in nitrogen metabolism in thioacetamide-induced cirrhosis [199], toxic and metabolic effects of 23-aliphatic alcohols [317], dose-dependent effects of acute lindane treatment on Kupffer cell function [343], hepatotoxicity of gossypol [193], hepatic uptake of the anticancer drug mitomycin [194], the mechanism of chlorpromazine-induced cholestasis [6], effects of vanadate on glucose output by the liver [197], galactosamine hepatotoxicity [278,346], metabolism of acetaminophen by liver after overdosage [265], lipid peroxidation associated with acetaminophen toxicity [293], isolation and characterization of the metabolites of T-2 toxin [100], and the hepatotoxicity of several cytotoxic agents [217]. One report [221] described a technique of *in vivo* isolated perfusion of the rat liver to study the effect of hyperthermochemotherapy with 5-fluorouracil for possible clinical application in treating unresectable liver cancer in patients. An example of the use of isolated perfused liver in evaluating the pharmacokinetics of drugs such as propranolol is also available [87]. In recent studies, the uptake and excretion of taurocholate by isolated perfused neonatal sheep liver [114] and kinetic modeling of the slow dissociation of bromosulfophthalein from albumin in perfused rat liver and toxicological implications [95] have been examined.

## Isolated Perfused Lung

The heart–lung preparation of Knowlton and Starling [159] has been used extensively to study the respiratory functions of the lung in small and large animals; however, refinement of the isolated perfused lung preparation technique was expedited only after the nonrespiratory functions of the lungs were recognized. Popjak and Beeckmans [262] employed a rabbit perfused lung preparation to examine the utilization of oxygen and substrate incorporation into phospholipids of the lung tissue. A number of investigators have since refined the technique of perfusing lungs [23,52,180,182,183,238,245,246,316].

A variety of methods have been used to perfuse lungs from experimental animals. Leary and Ledingham [180] described an *in situ* perfusion method with or without pulmonary ventilation. Similarly, Bakhle and coworkers [26] described an isolated perfused lung preparation without ventilation of the lung during perfusion. Isolated perfused lungs can be ventilated using either negative [238] or positive [249] pressure. Pulsatile perfusion was utilized by Hauge [120], and hydrostatic pressure was used for perfusion by Levey and Gast [183]. Gillis and Iwasawa [106] and others have perfused right and left lungs as an intact organ [238,249].

An ideal perfused lung preparation is one that is totally isolated and in which respiratory and nonrespiratory functions can be tested. The isolated lung preparation has the advantage of being able to account for all of the perfusion medium from the lung circulation at the end of perfusion experiments. Levey and Gast [183] pointed out that the *in situ* preparations result in some loss of perfusion medium through collateral vessels supplying the chest wall, and some fluid may be lost through exudation from the lung surface. Ventilation of the lung is essential for toxicological investigations in which maintaining a physiological route of gas exchange is important. Ventilation is an integral part of lung function; hence, nonventilating lung preparations represent less than desirable conditions, regardless of whether respiratory or nonrespiratory functions are being investigated. Retaining the ability to test certain experimental drugs through the gaseous phase to simulate inhalational exposure would also be an additional advantage of maintaining a ventilating perfused lung preparation [209]. Two types of isolated lung preparation are described here, one that utilizes negative-pressure ventilation and another that utilizes positive-pressure ventilation.

### Apparatus

The perfusion system developed by Niemeier and Bingham [238], with small modifications currently in use [12,52,70,111,211,246], is described here (Figure 38.5). The apparatus consists of a combination of pumps for ventilation, a peristaltic pump to drive the perfusate, an assembly of tubing for carrying the perfusate to and from the lung, and an artificial thorax kept warm by heated, circulating water. The thorax is made up of double-jacketed thick glass provided with an airtight lid. Perfusate flow to the lung is maintained from the upper reservoir connected to the central porthole of the lid. The bottom of the reservoir has an opening through which the perfusate can be directed to the peristaltic pump. A small-

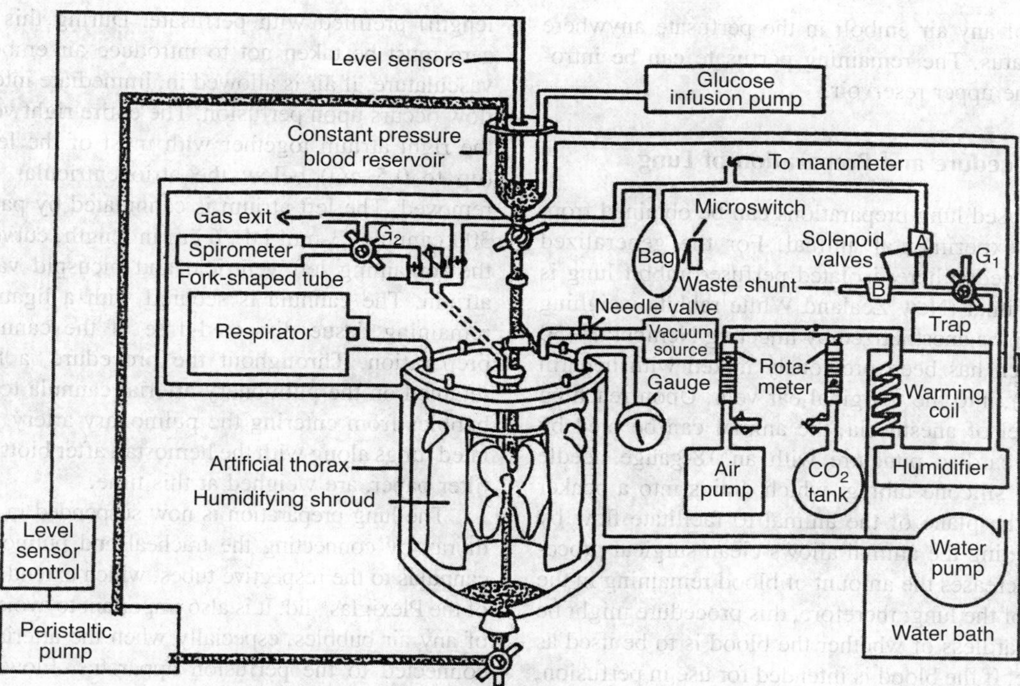


**FIGURE 38.5** Isolated perfused lung apparatus. This apparatus, originally developed by Niemeier and Bingham [198] for perfusion of rabbit lung, uses alternating negative pressure for ventilation. The apparatus can be scaled down to perfuse lungs from smaller animals (rats, guinea pigs), using essentially the same procedure. (From Mehendale, H.M., *Toxicol. Appl. Pharmacol.*, 36, 369–381, 1976. With permission.)

animal respirator and a vacuum pump are connected to two of the portholes on the lid; these parts provide alternating negative pressure as a means of ventilating the perfused lung preparation. A Magnehelic® gauge or a simple manometer can be connected to another porthole on the lid to monitor the operating negative pressure in the thorax. At the center of the lid are two portholes, one for a tracheal cannula and another for a pulmonary arterial cannula. The upper reservoir, also double-jacketed, connects to the central pulmonary arterial cannula by means of a stopcock arrangement and via pressure and flow transducers. The tracheal cannula is connected to a source of $O_2/CO_2$ (95:5), which is filtered and humidified by bubbling through warm saline. By means of appropriate one-way valves, provisions are made for inspiration as well as expiration of the ventilating lung. A spirometer is connected to measure the inspiration volume.

The perfusate from the upper reservoir passes through the assembly of transducers into the lung via the pulmonary arterial cannula. The perfusate empties into the bottom of the reservoir and is led to a peristaltic pump, which delivers it to the upper reservoir. A level-sensing controller device can be introduced at the upper reservoir to maintain a constant level of the perfusate in the reservoir so as to provide constant hydrostatic pressure. This automatic sensing device regulates the peristaltic pump, maintaining a designated level of perfusate in the upper reservoir. An infusion pump can be used to infuse glucose, which replaces the glucose utilized by the perfusing lung. In the upper reservoir, a pH probe can be installed to monitor the pH of the circulating perfusate. The waterbath, equipped with a heater and a circulating water pump, provides a means for maintaining the thorax and the upper reservoir at physiological temperature (37°C).

Almost all components that come in direct contact with the perfusate are composed of glass, with the exception of small pieces of medical-grade Silastic® tubing used in the peristaltic pump, which requires flexible tubing. The glass components and flexible tubing of the apparatus are coated with Siliclad® or a similar liquid silicone to avoid binding of experimental drugs to the tubing used for transporting the perfusate. The lid (thoracic roof) has a number of portholes and is composed of Plexiglas® fitted with a rubber O-ring and sealed with silicone high-vacuum grease; the lid is held in place on the ground-glass rim of the thorax by appropriate clamps. The small-animal respirator is connected in reverse to the porthole on the lid for the purpose of generating alternating negative pressure in combination with the vacuum pump. Prior to the surgical procedure to remove the lung, the apparatus is thoroughly cleaned, assembled, and rinsed with physiological saline. A measured volume of perfusate can be introduced into the upper reservoir, and the entire perfusate line can be primed with the perfusate. Precaution is taken to avoid entrapment of any air emboli in the perfusate anywhere in the apparatus. The remaining perfusate can be introduced into the upper reservoir.

### Surgical Procedure and Preparation of Lung

Isolated perfused lung preparations can be obtained from almost any experimental animal. For the generalized description needed here, isolated perfused rabbit lung is used. The animal (New Zealand White rabbits weighing 2 to 3 kg) can be anesthetized by injecting Nembutal® (50 mg/kg), which has been previously mixed with heparin (1000 IU/kg), into the marginal ear vein. Upon reaching a proper level of anesthesia, the animal can be bled by means of a cardiac puncture with an 18-gauge needle connected to silicone tubing, which drains into a beaker held below the plane of the animal to facilitate flow by gravity. Bleeding the animal allows clean surgical procedures and decreases the amount of blood remaining in the vasculature of the lung; therefore, this procedure might be followed regardless of whether the blood is to be used as the perfusate. If the blood is intended for use in perfusion, it is collected in a heparinized container and immediately stirred. (Blood can also be used as a perfusate upon proper filtration.) When carrying out cardiac puncture, care is taken to enter between the sixth and seventh ribs next to the sternum so as to not damage the lungs.

A midline incision is made from the neck to the abdomen to expose the trachea and rib cage. The liver is retracted and the sternum grasped by means of a curved hemostat. The sternum is lifted upward to facilitate inflation of the lungs, at which time the trachea can be clamped by means of a hemostat to entrap the proper amount of oxygen in the lungs. An incision in the diaphragm at the midline area facilitates cutting the diaphragm on both sides along the rib cage. The rib cage can be cut laterally on both sides, making sure that the lung tissue is detached from the roof of the rib cage. The lungs and heart are thus exposed through a midline sternotomy, and the rib cage is retracted. The lungs and heart can be removed from the animal *en bloc* and transferred to a Petri dish containing warm saline. All the subsequent operations can be carried out while the lungs rest in this Petri dish.

The trachea is first cannulated using PE-300 tubing (PE-200 for the rat). The lungs can then be ventilated artificially by means of a 100-cc syringe attached to the tracheal cannula by silicone tubing. Alternatively, the tracheal cannula can be attached to a small-animal ventilator so the lungs can be ventilated during the subsequent cannulation procedure. The trachea, lungs, and heart are dissected free from their attachments, connective tissue, and other extraneous material, taking care not to puncture the lungs, and then are rinsed with warm physiological saline. The pericardium is removed and the pulmonary artery cannulated with a PE-300 cannula (3-mm i.d., 5 cm in length) prefilled with perfusate. During this procedure, care must be taken not to introduce air emboli into the vasculature; if air is allowed in, immediate interruption of flow occurs upon perfusion. The entire right ventricle and the right atrium together with most of the left ventricle (up to 0.5 cm) below the atrioventricular septum are removed. The left atrium is cannulated by passing a PE-300 cannula (3-mm i.d., 6 cm in length, curved) through the remaining left ventricle and bicuspid valves to the atrium. The cannula is secured with a ligature and the remaining tissue dissected free of the cannulated lung preparation. Throughout the procedure, a hemostat is retained on the pulmonary arterial cannula to prevent air bubbles from entering the pulmonary artery. The cannulated lungs along with the hemostat, after blotting dry with filter paper, are weighed at this time.

The lung preparation is now suspended in the artificial thorax by connecting the tracheal and pulmonary arterial cannulas to the respective tubes, which extend to the inside of the Plexiglas® lid. It is also important to avoid entrapment of any air bubbles, especially when the arterial cannula is connected to the perfusion apparatus. Flow is resumed through the arterial cannula after the lid is closed, and the perfusate line from the pump is connected to the top of the upper reservoir. Perfusion can be slowly established at this time, ensuring that no bubbles pass through into the pulmonary artery. The pumps can be activated to inflate the lung. The lungs are inflated by applying a negative pressure of 25 to 30 cm$H_2O$ by activating the vacuum pump and increasing the vacuum. When the collapsed lungs are inflated to a desired level, because the respirator is already turned on, the alternating negative–positive pressure automatically ventilates the lung. In the rabbit, the frequency is kept at 50 respirations per minute. In the rat, respiration can be maintained at approximately 60 respirations per minute. When the ventilation cycle is initiated, the apparatus is automatic, and perfusion and ventilation continue throughout the experiment. The perfusion flow rate increases quickly upon inflation of the lung and may steadily increase until steady-state ventilation is established. If the automatic level sensor is in operation, monitoring the perfusion flow rate is not necessary; however, if such an arrangement is not available, care must be exercised to maintain perfusate in the upper reservoir by manual control of the peristaltic pump. The lungs are usually allowed to equilibrate in the perfusion apparatus over a period of 15 to 20 minutes.

### Positive-Pressure Ventilation Procedure

The perfusion apparatus developed by O'Neil and Tierney [249] can be used to illustrate the perfusion of lungs from smaller animals and by using a positive-pressure ventilation procedure. The apparatus and assembly are illustrated in Figure 38.6. The lungs are ventilated directly by means of an animal respirator that is attached to the tracheal

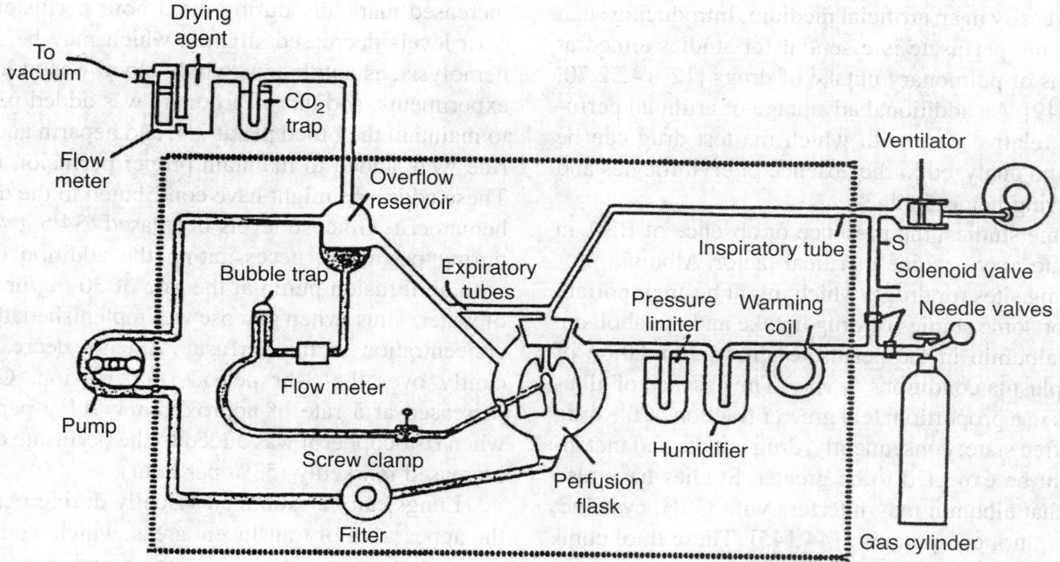


**FIGURE 38.6** Isolated perfused lung preparation of O'Neil and Tierney [208]. This apparatus is used to perfuse lungs with positive-pressure ventilation. It can be scaled up to perfuse lungs from large animals. S, solenoid value; O, needle valve. (From Menaouar, A et al., *Eur. Respir. J.*, 10, 1100–1107, 1997. With permission.)

cannula. This procedure eliminates the additional vacuum pump that is required in the negative alternating-pressure ventilation method. Two selenoid valves (Figure 38.6) are used to direct gas flow during the respiratory cycle. A tidal volume of approximately 3.5 mL is obtained for the rat and is adjusted to provide a maximum transpulmonary pressure of 12 $cmH_2O$. The end-expiratory pressure is set at 3.5 $cmH_2O$ to keep the lung from collapsing during expiration. The rat lungs are perfused in this apparatus at a frequency of 13 respirations per minute. The procedure for isolating and cannulating the lung is essentially the same as described for the rabbit lung preparation, with the exception of smaller diameter PE cannulas. The procedure for maintaining isolated perfused rat lung using positive-pressure ventilation has been described adequately by O'Neil and Tierney [249] and Young [359], whose papers may be consulted for additional details. Lungs can also be perfused using a setup similar to the one described for the negative-pressure ventilation procedure with simple modifications. The vacuum pump is deleted, and the ventilator is connected directly to the trachea in a forward direction instead of the reverse direction used in the negative-pressure ventilation procedure. Camus and Mehendale [52] may be consulted for other details regarding this procedure.

## Perfusion Media

Nicolaysen [236] studied the effect of perfusate composition on edema development and found whole blood to be the most suitable perfusate. The perfusion procedure described by Niemeier and Bingham [238] included the use of autologous whole blood as a perfusate. This perfusate was practical in the case of a rabbit, as 100 mL or more of whole blood can be obtained by cardiac puncture from a rabbit weighing 3 kg or more. Blood is collected in a heparinized container and filtered to remove any debris, dead cells, or small blood dots. The blood is heparinized once again prior to circulation in the apparatus to avoid clotting. Using autologous whole blood is impractical, however, for perfusing rat lungs or lungs of other small animals. Blood may have to be collected from several animals to supply an adequate volume of perfusate for a single perfusion experiment; furthermore, for certain experimental protocols such as single-pass experiments, several liters of perfusate may be required. For these reasons, the use of artificially constituted medium as perfusate has been popular.

A widely used lung perfusate is the one described by Junod [146], which has the following composition in millimolar concentrations: NaCl, 118; KCl, 4.75; $CaCl_2$, 2.54; $KH_2PO_4$, 1.19; $MgSO_4$, 1.19; and $NaHCO_3$, 25. BSA is added at a concentration of 4.5 g/L, and the final pH of the solution is adjusted to 7.4 with 1-*N* NaOH, so the final $Na^+$ concentration of the standard medium is 161 m*M*. The medium is equilibrated with $O_2/CO_2$ (95:5) prior to priming the apparatus. This perfusate has been used by a number of investigators [52,60,95,182,183,211–215,218]. The standard perfusion medium is KRB buffer solution, described by Umbreit et al. [340]; it contains 5 m*M* of glucose and 4.5% BSA (Cohn fraction V). Several advantages of using this artificial medium as a perfusate can be cited. First, in the one-pass perfusion experiments, large volumes of the perfusate are often used [262] and using whole blood as a perfusate is not economical. Second, manipulations of ionic changes in the perfusate can be introduced easily in an artificial medium. Introducing such changes in the perfusate is essential for studies aimed at mechanisms of pulmonary uptake of drugs [12–14,52,70, 146,245–249]. An additional advantage of artificial perfusate is the relative ease with which the test drug can be extracted and analyzed in the absence of erythrocytes and any interfering hemoglobin.

For some studies, the presence or absence of BSA in the perfusate may become a critical factor. Albumin provides binding sites for drugs, which might be an important criterion for some studies of drug uptake and metabolism. Including albumin in the perfusate allows simulation of the blood plasma conditions *in vivo*. The absence of albumin results in a proportionately greater fraction of the drug being in a free state; consequently, drug uptake and metabolism might be expected to be greater. Studies have also indicated that albumin may interfere with GSH, cysteine, and similar thiol compounds [144,145]. These thiol compounds are oxidized rapidly in the presence of albumin in KRB buffer [144,145]. Replacing albumin with high-molecular-weight dextran ameliorated this particular difficulty. Other variations of the perfusion media include the use of 4.5% Ficoll® 70 instead of BSA [81] and HEPES buffer with only 2% BSA [31].

A combination of either autologous or mixed whole blood and the KRB buffer artificial medium as a perfusate can also be used for perfusing lung preparations. Such a preparation has the advantage of including the natural constituents of blood so natural binding sites can be provided for the test drug. Differences in the uptake of drugs by the lung have been found between whole blood and artificial medium used as perfusate [12,252]. Perfluorocarbons have not been used in the perfusate in isolated perfused lung preparations, although their successful usage in perfusing livers [112,243,297,298] suggests that fluorocarbon emulsions would also adequately support the isolated lung preparations. In the isolated lung preparations ventilated by positive pressure, Young [359] utilized KRB-buffered solution containing glucose and BSA, similar to the preparation described above. By and large, this erythrocyte-free medium has been widely used as a perfusate for maintaining viable preparations of isolated perfused lungs.

## Viability Criteria

Niemeier and Bingham [238] measured a number of biochemical parameters in the circulating perfusate as well as in the lung to ascertain the viability of isolated perfused rabbit lung preparations. The concentrations of blood urea nitrogen (BUN), $Ca^{2+}$, albumin, total protein, and pyruvate in the perfusate changed little throughout the perfusion (Table 38.4). Inorganic phosphate, uric acid, lactic acid, and total bilirubin increased moderately during perfusion. LDH and serum glutamic oxaloacetic transaminase (SGOT) activities, as well as plasma hemoglobin, increased markedly during the 3-hour perfusion. Hematocrit levels decreased slightly, which may be a result of hemolysis, as autologous whole blood was used. In these experiments, sodium bicarbonate was added periodically to maintain the blood pH at 7.4, and heparin and epinephrine were added to maintain proper perfusion flow rates. These additions might have contributed to the decrease in hematocrit. Glucose levels decreased (34.5 ± 4.1 mg/hr) during perfusion, necessitating the addition of glucose with an infusion pump at the rate of 30 mg/hr in 0.3 mL of water. Thus, when glucose was replenished, the glucose concentration in the perfusate did not decrease significantly over the 3-hr period of perfusion. Cholesterol increased at a rate of approximately 11% per hour, but when α-tocopherol was added to the perfusate cholesterol increased markedly (53% per hour).

Lungs can be examined visually during perfusion for the appearance of translucent areas, which would indicate development of edema. Niemeier and Bingham [238] found that the lungs gained weight an average of 2.8% per hour over the 3-hour perfusion with autologous whole blood. Histopathological examination revealed no edema after 3 hours of perfusion, and the integrity of the pulmonary ultrastructure was well preserved [238,252].

Often, after prolonged perfusion of lungs, the lung preparations deteriorate, with the concomitant development of edematous areas characterized by a translucent appearance of the lung surface. After continued perfusion, the lung does not inflate and deflate with each respiratory cycle, and surfactant material might appear in the tracheal cannula. If the perfusion is continued after the lung appears to be edematous, large and copious flows of the surfactant material continue to appear through the tracheal cannula. Maintenance of perfusate pH can be a problem. Niemeier and Bingham [238] maintained the pH by adding 1-m*M* sodium bicarbonate to the perfusate; however, they used room air mixed with 5% $CO_2$ to ventilate the rabbit lung preparations. If $O_2/CO_2$ (95:5) is used for ventilating the isolated lungs and the lungs are properly inflated, no problem is encountered in maintaining physiological pH of the perfusate [212].

An additional criterion of viability of the lung preparation is the evaluation of drug-metabolizing enzymes in the perfused lung [178]. Various investigators have found that, after 2 to 3 hours of perfusion, microsomal drug-metabolizing activity of the lung remains unaltered, indicating satisfactory preservation of microsomal MFO activity of perfused lung preparation. ATP levels were measured in rat lungs perfused for 90 minutes by a positive-pressure ventilation procedure [359] and were found to be at or above the ATP levels in nonperfused lungs. By contrast, ATP content was decreased in lung slice incubations. Perfusion of rat [31] and rabbit [81,144] lungs results in only a slight decrease in the lung content of GSH, indicating that lungs can maintain GSH levels during perfusion.

**TABLE 38.4**
**Biochemical Changes and Physiological Values in the Isolated Perfused Lung**

| Parameter | Average Concentration in Plasma Prior to Perfusion | Average Change in Concentration per Hour |
|---|---|---|
| Calcium (mg/dL) | 13.8 ± 0.8 | ↓0.15 ± 0.29 |
| Inorganic phosphate (mg/dL) | 4.1 ± 0.4 | ↑0.78 ± 0.27 |
| Glucose (mg/dL) adding 30 mg/hr | 236 ± 35 | ↓34.5 ± 4.1 ± 2.3 |
| BUN (mg/dL) | 17.5 ± 3.2 | ↑0.14 ± 0.22 |
| Uric acid (mg/dL) | 0.62 ± 0.18 | ↑0.20 ± 0.12 |
| Cholesterol (mg/dL) with vitamin E | 37 ± 13 | ↑4.2 ± 0.7 |
| | | ↑19.5 ± 7.1 |
| Total protein (g/dL) | 5.7 ± 0.4 | ↑0.10 ± 0.15 |
| Albumin (g/dL) | 0.53 ± 0.07 | ↓0.04 ± 0.03 |
| Total bilirubin (mg/dL) | 0.14 ± 0.06 | ↑0.12 ± 0.16 |
| Alkaline phosphatase (mU/mL) with vitamin E | 55 ± 36 | ↑4.8 ± 2.1 |
| | | ↑0.6 ± 1.1 |
| Lactate dehydrogenase (mU/mL) | 135 ± 42 | ↑485 ± 201 |
| Serum glutamic oxaloacetic transaminase (SGOT) (mU/mL) | 58 ± 23 | ↑121 ± 58 |
| Plasma hemoglobin (mg/dL) | 0.19 ± 0.12 | ↑2.3 ± 0.6 |
| Lactic acid (mg/dL) | 173 ± 17 | ↑19.8 ± 5.2 |
| Pyruvic acid (mg/dL) | 1.19 ± 0.06 | n.c.[a] ± 0.01 |
| Hematocrit (%) | 35.0 ± 5.0 | ↓1.63 ± 0.34 |
| Weight gain (% hr) | ↑2.81 ± 1.36 | — |
| Blood flow (mL/min) | 160 | — |
| $pO_2$ (mmHg), typical values | 118 ± 6; 121 ± 10 | — |
| $pCO_2$ (mmHg), typical values | 39 ± 4; 34 ± 4 | — |
| pH range | 7.35–7.45 | — |
| Tidal volume (mL) | | |
| Typical values | 11.7 ± 0.3; 11.0 ± 0.4 | — |
| Normal values | 23.9 ± 5.5 (16) | — |

[a] No change.

*Source:* Post, C. and Hede, A.R., *Biochem. Pharmacol.*, 31, 353–358, 1982. With permission.

## Applications

Isolated perfused lung preparations have been used for a variety of studies, including drug uptake and metabolism and the disposition of various pharmacological and toxicological agents [212]. Although rats and rabbits appear to be the choice of animals, studies with other species such as mouse [147], guinea pig [143], and other species can be found in the literature. Rhoades [281] described a technique of utilizing isolated perfused lungs ventilated by positive pressure for evaluating the effect of various gaseous environments on pulmonary biochemistry. Block and Cannon [37] investigated the effect of anoxia or hyperoxia on the ability of lungs to clear endogenous and exogenous chemicals. Use of isolated lung preparations has resulted in significant contributions to our understanding of the pulmonary role in uptake and metabolism of a variety of xenobiotics [39,52,57,70,137,177,210,209–213,237,238,341,354].

Examples of toxicological investigations using isolated perfused lung preparations include studies on the pulmonary uptake and disposition of aldrin and dieldrin [210,211]; uptake of the herbicide paraquat [57]; uptake and metabolism of trichloroethylene [69]; pharmacokinetics and toxicity of doxorubicin [23]; metabolism and toxicity of naphthalene oxide [147]; metabolism of nitroaromatics [334]; uptake, reactivity, and impairment of lung function by diisocyanates [151,176]; nitrofurantoin toxicity [32]; $H_2O_2$ toxicity [116]; chlorine gas toxicity [215]; modulation of 4-ipomeanol activation in the lung [336]; effect of xanthine oxidase-induced lung injury on removal of 5-hydroxy-tryptamine by the lung [65]; and uptake and metabolism of benzo(*a*)pyrene [34,341]. Although this list is by no means a survey of toxicological investigations utilizing isolated perfused lung preparations, these preparations have been useful for determining the pulmonary contribution to the disposition of toxic chemicals.

Perfused lung preparations have been useful for demonstrating epoxidase activity in the lung tissue [208,211,212]. Aldrin is a cyclodiene pesticide that is readily epoxidized to dieldrin and in the liver can be further metabolized by epoxide hydrase to dihydrodiol metabolites. In the lung, aldrin could be readily epoxidized to dieldrin, which is a stable epoxide and can be quantitated as a measure of aldrin epoxidase activity. This finding represents the first direct demonstration of epoxidase capability of lung tissue. Aldrin is thus metabolized to dieldrin irrespective of whether it enters via airways or through the vascular system, and the metabolite appears in the perfusate rapidly [208]. In these studies, it was demonstrated that intact perfused lung preparations were able to turn over aldrin to dieldrin at a lesser rate than the *in vitro* incubations of lung homogenate preparations. These studies serve to illustrate the utility of perfusing intact organs to realistically evaluate the biochemical metabolic contribution by the organ to the metabolism and disposition of a test chemical. The turnover of aldrin to dieldrin was four to seven times greater *in vitro* preparations than in *ex vivo* preparations using perfused intact lungs [208]. Although the reason for such a discrepancy between *in vitro* and intact organ perfusion systems is incompletely understood, Itakura et al. [136] demonstrated that the availability of necessary cofactors might be limiting in intact perfused lung preparations. They observed that demethylation of *p*-nitroanisole by the perfused rabbit lung preparations was limited by the availability of the cofactor NADPH, the generation of which could be stimulated by introducing glucose to the perfusion medium. Even in the presence of excessive glucose in the perfusate, aldrin epoxidation, a reaction requiring NADPH, was saturable [211], suggesting that the cofactor availability can be limiting in intact lungs, an observation that would not be apparent from the idealized *in vitro* incubations.

Perfused lung preparations of rat and rabbit [245–248] have also been helpful in identifying a flavin monooxygenase capable of *N*-oxidizing chlorpromazine and imipramine. Most interestingly, these studies established that, although the rat lung is capable of *N*-oxidation of both of these substrates, rabbit lung is devoid of this activity. Curiously, rabbit lung does contain a flavin monooxygenase capable of *N*-oxidizing *N,N*-dimethylaniline [210]. These studies also established the absence of any significant cytochrome P450-mediated metabolism of these substrates. Furthermore, the remarkable species differences between rats and rabbits in pulmonary flavin monooxygenase activities became apparent from these perfusion studies [210,245–248].

Lesire et al. [182] investigated the toxicokinetics of parathion and paraoxon, their metabolic activity, and cholinesterase inhibition in guinea pig and rabbit lungs using single-pass perfusion. Although lungs of both species extracted both compounds from the perfusate, the extraction ratio was higher for guinea pig lungs. Cytochrome P450-related lung metabolic activity, inhibited by the inclusion of piperonyl butoxide, mediates the activation of parathion to paraoxon, as was demonstrated in the lungs of both species. Cytochrome P450 activity was required for maximum inhibition of lung acetylcholinesterase activity.

The study of Dalbey and Bingham [69] illustrates the utility of isolated perfused lung preparations for studying the uptake and disposition of gaseous toxic substances. They studied the uptake and metabolism of trichloroethylene in isolated perfused rat lung preparations by introducing trichloroethylene vapors through the trachea. Perfused rat lung preparations metabolized trichloroethylene to trichloroethanol, and guinea pig lungs were even more active in the metabolism of trichloroethylene to the alcohol. In the future, it should be anticipated that more experimentation will be carried out utilizing isolated perfused lung preparations for such toxicological investigations [209].

Kennedy et al. [151] and Lastbom et al. [176] studied the reactivity and lung impairment by diisocyanates in isolated perfused guinea pig lungs. Perfused lungs were exposed to $^{14}$C-labeled toluene diisocyanate (TDI) at 0.2 and 0.7 ppm. KRB buffer only, with or without guinea pig albumin, human albumin, or diluted guinea pig plasma, was used to perfuse the lungs. The rate of TDI uptake was dependent on the TDI concentration and composition of the perfusate. The percentage of conjugated products was higher when diluted guinea pig plasma was used as the perfusate (15% in buffer only vs. 45% with diluted plasma). The authors concluded that perfused lungs could serve as a useful model to study the molecular mechanism of isocyanate-induced lung disease and metabolic activity. In another study [176], isolated perfused ventilated guinea pig lungs were exposed to 3.5 and 11 mg of hexamethylene diisocyanate (HDI). A dose-dependent decrease was observed in both airway conductance and compliance, but no effects were noted on the pulmonary circulation. The reduction in lung function (at the 11-mg/m$^3$ exposure) was abolished when 100-$\mu M$ diclofenac (a cyclooxygenase inhibitor) was added to the perfusate. The thromboxane $A_2$ antagonist L-670,596 (20 $\mu M$) exerted only a partial protective effect. The authors concluded that HDI-induced brochoconstriction is mediated via arachidonic acid release and thromboxane formation in isolated perfused guinea pig lungs.

The use of isolated perfused lung preparations to study the metabolism of carcinogenic chemicals such as benzo(*a*)pyrene is another example of how perfused lung preparations have furthered advancements in this area. Bingham and associates [34] studied the metabolism of benzo(*a*)pyrene and reported the formation of several metabolites of this chemical in the lung, including carcinogenic reactive epoxide and dihydrodiol metabolites. In addition to containing the necessary enzyme systems for carrying out the oxidative metabolism of these toxic chemicals, lungs contain enzyme systems that catalyze phase II reactions such as epoxide hydrase, glucuronyl transferase, and glutathione *S*-transferase using 1-chloro-2,4-dinitrobenzene as the substrate, which can be demonstrated using intact perfused lung preparations [71,341].

The use of isolated lung preparations for determining other physiological effects of toxic chemicals mediated via the endogenous hormone system is illustrated by the studies of Seiler et al. [305]. They examined the effect of certain anorexic agents, including chlorphentermine, on the clearance of 5-hydroxytryptamine (5-HT; serotonin) and observed that the anorexic agents enhanced the vasoconstrictor effects of 5-HT in the pulmonary circulation. Subsequent studies by Angevine and Mehendale in perfused rabbit [13] and rat [14] lungs revealed the mechanism. Chlorphentermine was found to inhibit the pulmonary deactivation of 5-HT [13,14,210], consequently allowing the action of the vasoactive 5-HT to prevail. Another example of the use of isolated perfused lung preparations to evaluate the effect of foreign toxic chemicals on pulmonary clearance of endogenous chemicals is the study of Gillis and Roth [107] in which they examined the turnover of endogenous hormones (5-HT and norepinephrine).

Isolated perfused rabbit lungs were employed by Dunbar et al. [81] to examine the GSH status of the lung upon perfusion with 420-μ*M* paraquat or nitrofurantoin. Significant increases in lung GSSG were observed, which provided evidence for the preoxidant nature of the lung injury caused by these agents. Possible utilization of externally provided GSH by rat [31] and rabbit [129] lungs was studied by determining the GSH status of the isolated, ventilated, and perfused lungs with or without prior GSH depletion. The rat lungs appeared to be able to utilize external GSH [31], whereas the rabbit lungs failed to do so [144]. The question of whether externally provided GSH can be utilized by the lung tissue is of significance for many toxicological considerations. The role of the GSH redox cycle as a defense system against $H_2O_2$-induced prostanoid formation and vasoconstriction was investigated in perfused rabbit lungs [303].

Bernard et al. [32] studied the toxicology of nitrofurantoin in the isolated perfused rat lung. Nitrofurantoin induced a decrease in tissue levels of glutathione but not protein thiols by the end of 3-hour perfusion. Tissue levels of angiotensin-converting enzyme were not decreased. Electron microscopic analysis of the tissue revealed detachment of endothelial cells from the basement membrane, which may account for the edematogenic weight gain. The edema was matched by an increase in protein content of the alveolar lavage fluid. Coinfusion of penicillamine, *N*-acetylcysteine, or *N*-(2-mercaptopropionyl)-glycine failed to mitigate nitrofurantoin-induced edema. Allopurinol, an inhibitor of xanthine oxidase and a metal chelator, significantly decreased lung weight gain but did not prevent the loss of glutathione. The authors concluded that organ function was compromised more than the individual cells, and the allopurinol might be useful in modulating nitrofurantoin pulmonary toxicity. Oxyradicals generated by xanthine–xanthine oxidase also interfere with the endothelial removal of 5-HT in perfused rabbit lungs [65]. The question of hydroxyl radical (·OH) involvement in granulocyte-mediated oxidant lung injury was examined in perfused rat lungs using dimethylthiourea (DMTU) as the radical scavenger [95]. When isolated rat lungs were perfused with polymorphonuclear neutrophils (PMNs) activated by phorbal myristate acetate (PMA) to produce ·OH, lung weights were increased significantly. Because the increase in lung weights was preventable by 10-m*M* DMTU, the findings are supportive of the role for the ·OH radical in acute granulocyte-mediated lung injury. Recently, Weissmann et al. [350] established a protocol for the measurement of intravascular reactive oxygen species (ROS) release from isolated buffer-perfused and ventilated rabbit and mouse lungs, combining lung perfusion with the spin probe 1-hydroxy-3-carboxy-2,2,5,5-tetramethylpyrrolidine (CPH) and electron spin resonance (ESR) spectroscopy [350]. This technique was employed to characterize hypoxia-dependent ROS release, with specific attention paid to NADPH-oxidase-dependent superoxide formation as a possible vasoconstrictor pathway. The perfusion of isolated lungs with CPH is suitable for the detection of intravascular ROS release by ESR spectroscopy. The authors employed this technique to demonstrate that (1) PMA-induced vasoconstriction is caused directly by superoxide generated from NADPH oxidases and (2) this pathway is pronounced in hypoxia. NADPH oxidases thus may contribute to the hypoxia-dependent regulation of pulmonary vascular tone.

Evidence for a protective role of intact human erythrocytes against $H_2O_2$-mediated damage [295] and the finding that DMTU treatment might be helpful for treating acute edematous lung injury such as seen in adult respiratory distress syndrome have provided some insights into the origin and mechanism of such lung injury. Habib and Clements [116] employed isolated perfused rat lungs to investigate if treatment with $H_2O_2$ would result in measurable changes in exhaled ethane during early stages of capillary leak. Exhaled ethane was not increased when perfused with 0.25-m*M* $H_2O_2$ in albumin containing KRB perfusate. $H_2O_2$ caused a small but significant increase in capillary permeability coefficient, and the wet weight/dry weight ratio was increased. The authors concluded that small amounts of $H_2O_2$ may increase pulmonary capillary permeability without affecting exhaled ethane.

Additional examples of the use of perfused lungs are available [39,53,121,172,264,302]. These studies include the effects of asbestos [53], chlorine gas [215], staphylococcal α-toxin [302], and haloalkanes [264] on the pulmonary metabolism of vasoactive substances. Additional examples of applications of the perfused lung preparations include pulmonary uptake and release of morphine [70,121], metabolism and macromolecular binding of the carcinogenic nitropyrene [39], and cystamine uptake and metabolism [307]. Lafranconi et al. [173] employed perfused rat lungs to investigate the toxic effects of equinatoxin, a peptide of 147 amino acid residues isolated from the venom of the sea anemone *Actinia equina*. In this study, equinatoxin was found to adversely affect fluid regulation in the lung tissue, suggesting that it might become an important tool for the investigation of fluid regulation in the lung.

Activation of cytokines after reexpansion of collapsed lung during one-lung ventilation (OLV) has not been thoroughly investigated. Funakoshi et al. [97] investigated the effects of reexpansion of the collapsed lung on pulmonary edema formation and proinflammatory cytokine expression. Proinflammatory cytokines were upregulated upon reexpansion and ventilation after short-period lung collapse, although no changes were noted in pulmonary capillary permeability [97]. In recent investigations, isolated perfused rat lung was used to study the permeability and translocation of superfine and nanoparticles [214]. Permeability of the lung barrier to ultrafine particles or nanoparticles is controlled both at the epithelial and endothelial level. Conditions that affect this barrier function, such as inflammation, may affect translocation of nanoparticles [214]. Using isolated perfused murine lungs, Song et al. [314] determined the role of aquaporin-4 (AQP4) in airspace-to-capillary water transport by comparing water permeability in wild-type mice and transgenic null mice lacking AQP1, AQP4, or AQP1 and AQP4 together. A simple gravimetric method was established to quantify osmosis and filtration in intact mouse lung and provide direct evidence for a contribution of the distal airways to airspace-to-capillary water transport. Conhaim et al. [68] employed isolated perfused rat lungs to study the influence of the thromboxane receptor analog W-46619 on microvascular circulatory effects. These results suggest that vasoconstriction occurred in the microvessels themselves, which are much smaller vessels than those previously thought to be capable of vasoconstriction. Although the recirculation apparatus was illustrated above, simple modifications allow one to perform experiments using a single-pass perfusion system. A number of investigators have utilized single-pass perfusion to investigate the mechanisms of drug uptake and release from the lung [106]; for example, Junod [146] examined the mechanisms of pulmonary uptake of imipramine using a single-pass mode of perfusion. Likewise, single-pass kinetics were used to determine the uptake, metabolism, and release mechanism of aldrin and dieldrin in the rabbit perfused lung preparations [211]. Single-pass experiments can be expensive in terms of both materials (albumin used for preparing the perfusate) and the technical assistance required to conduct these experiments. An ideal single-pass experiment requires the assistance of three individuals to work in swift coordination and usually involves analyzing a large number of samples; hence, the information to be gained from such experimental protocols must be weighed against the expense involved. Often, equally valuable information results from recirculating perfusion experiments at a fraction of the effort and resources expended. For additional information, the reader may wish to refer to Roth [290] for a review of the methodology and applications of perfused lung preparations.

## Isolated Perfused Kidney

Although a number of methods have been devised for obtaining perfused kidney preparations from various mammalian species, the use of isolated kidney preparations in toxicological studies has been infrequent. Historically, most workers in this field have been concerned with the study of autoregulation, excretion, and reabsorption functions of the organ [20,26]. Isolated perfused kidney preparations have been used in such studies from the rat [25,28,41,73,162,287], dog [19,296,348,357], rabbit [92,285], pig [47], monkey and sheep [325], and human [169]. Considerable variation has existed among the various investigators concerning the specific techniques used for perfusing kidneys. The kidney may be perfused via the dorsal aorta or the renal artery. Pulsatile or nonpulsatile perfusate flow can be used, and perfusion pressure may be exerted either with the pump [124] or by means of a hydrostatic pressure head [28].

As with other perfused organ systems, the most important variable has been the perfusion medium, and a considerable range of varying compositions has now been tested. Argument persists in the literature relating to whether pulsatile flow is preferred as a simulation of the *in vivo* situation [287]. A second major problem has been vasoconstriction and deterioration of the kidney preparations associated with the use of whole blood as a perfusion medium. By and large, to circumvent this problem, investigators have used diluted blood, a reconstituted blood perfusate, or an erythrocyte-free medium containing various electrolytes, glucose, and albumin. Most investigators are content with the use of a KRB-buffered solution containing albumin and glucose. As in the case of other organs, kidneys can also be perfused either *in situ* or in total isolation in chambers that can be kept warm for normothermic conditions. A further variation can be in the mode of perfusion; kidneys can be perfused either using recirculating perfusate or in a single-pass mode.

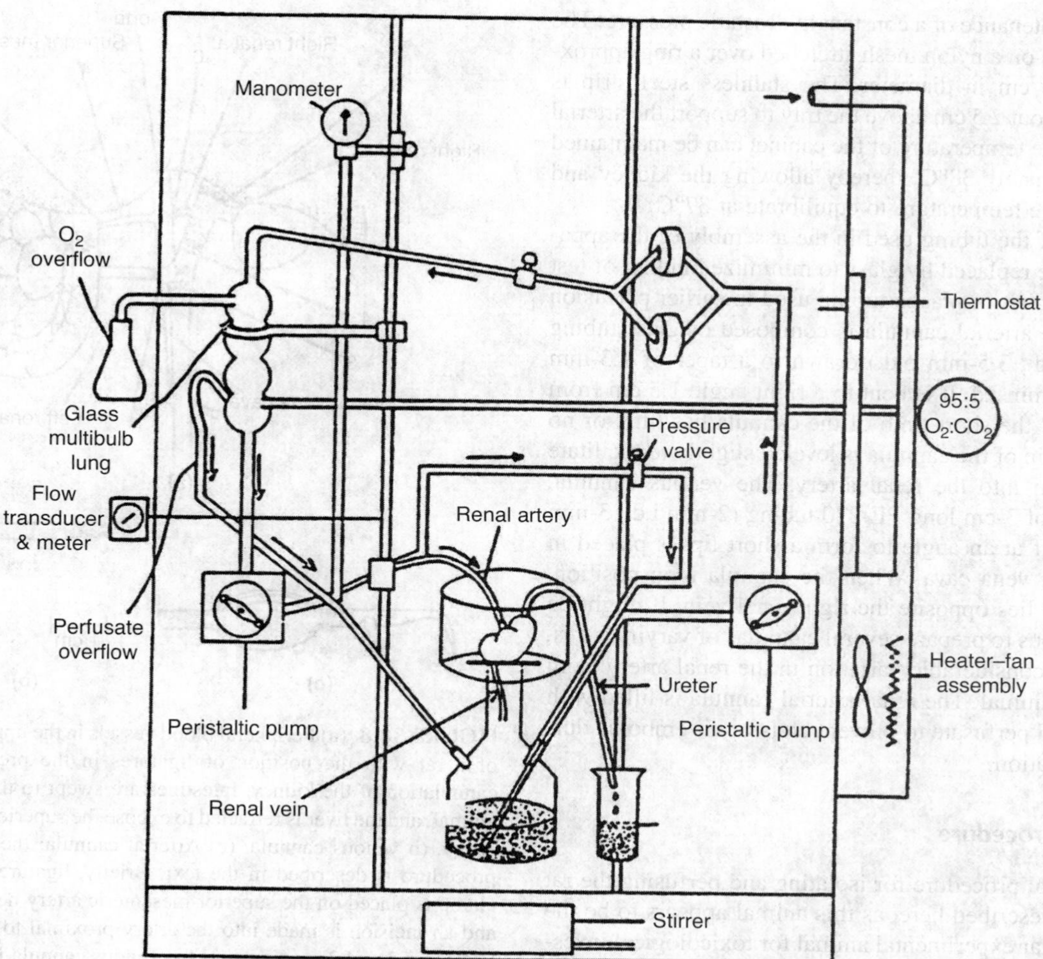


**FIGURE 38.7** Isolated perfused kidney apparatus [20]. The entire perfusion assembly is housed in a Plexiglas® chamber fitted with a heater–fan assembly [168]. The circulation of the perfusate is accomplished by two peristaltic pumps; one directs the perfusate via a Millipore® filter device into a multibulb glass lung for oxygenation, and the other directs the perfusate to the renal artery at a desired perfusion pressure. The perfusion pressure can be adjusted by imposing a clamp on the arterial diverticulum, which can be opened to reduce perfusion pressure or closed to increase it. The perfusion pressure is monitored by placing a manometer in the system. An outflow arrangement is provided at the bottom of the lung, and this tube drains the excess oxygenated fluid into the central reservoir, which also receives the venous flow from the kidney. Additions or sampling can be done in the central reservoir. The perfused kidney rests on a platform made up of a Petri dish with a central porthole for the venous flow. A muslin cloth stretched across the mouth of the Petri dish and held by a rubber band provides the platform for the perfused kidney.

## Apparatus

The apparatus used for perfusing kidney preparations is similar to the one described for perfusing livers (Figure 38.7). It includes the outer chamber used for perfusing the liver and is fitted with a heater–fan assembly for maintaining the desired temperature inside the chamber. A multibulb glass lung can be used to oxygenate the perfusate. Perfusate is led to a roller type or peristaltic pump and is transported to the top of the oxygenator via glass tubing. A filter placed between the lung and the glass tubing ensures trapping fat droplets and any other particulate material, including cell debris, from the perfusate before it enters the kidney. A major difference between the liver and kidney is the higher (120 cm) hydrostatic pressure used for perfusing the kidney. Oxygenated blood by means of glass tubing then enters the renal arterial cannula. A set of perfusion pressure and flow transducers can be placed between the arterial cannula and the glass tubing to measure the perfusion pressure and flow rates. Similarly, electrodes can be placed before and after the kidney to monitor oxygen levels of the perfusate. A pH probe can be placed anywhere in the circulation to monitor pH continuously. The effusate from the kidney is guided back to the reservoir to complete the recirculation. An overflow arrangement from the hydrostatic reservoir to the central reservoir allows maintenance of a constant hydrostatic pressure. The kidney rests on a nylon mesh stretched over a ring approximately 7.5 cm in diameter. The stainless steel strip is mounted about 2.5 cm above the tray to support the arterial cannula. The temperature of the cabinet can be maintained at approximately 38°C, thereby allowing the kidney and the perfusate temperature to equilibrate at 37°C.

Most of the tubing used in the assembly of the apparatus can be replaced by glass to minimize binding of test drugs to rubber or plastic tubing used in earlier perfusion setups. The arterial cannula is composed of glass tubing (2.8-mm i.d., 3.5-mm o.d.) drawn to a taper of 1.3-mm o.d. and 1-mm i.d. It is bent to a right angle 1.5 cm from the tip, and the short limb of the cannula has little or no taper. The tip of the cannula is leveled slightly to facilitate its insertion into the renal artery. The venous cannula, consisting of 3-cm long PE-270 tubing (2-mm i.d., 3-mm o.d.) cut off at an angle to form a short tip, is placed in the inferior vena cava. When the cannula is in position, its opening lies opposite the right renal vein. It might be advantageous to prepare several cannulas of varying sizes, as there is considerable variation in the renal artery from animal to animal. The renal arterial cannula is filled with heparinized perfusate to prevent clots or air embolus during cannulation.

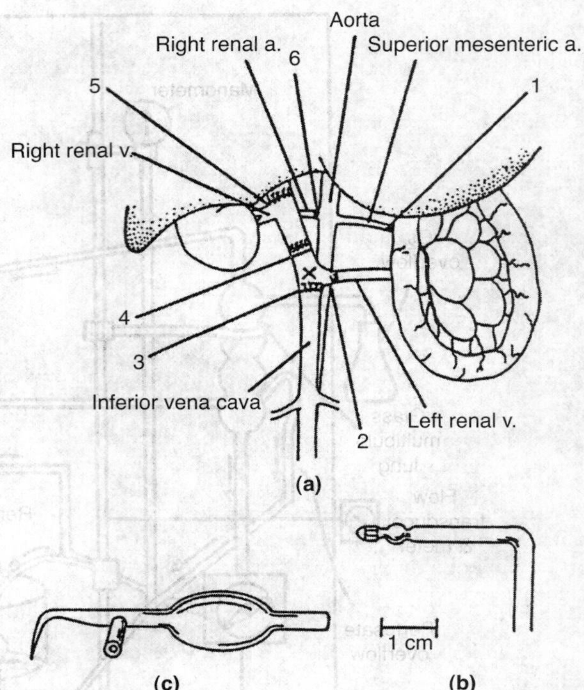


**FIGURE 38.8** (a) Peripheral blood vessels in the upper abdomen of a rat with the position of ligatures in the preparation for cannulation of the kidney. Intestines are swept to the left of the animal, and the liver is retracted to expose the superior mesenteric artery. (b venous cannula. (c) Arterial cannula; the cannulation procedure is described in the text. Briefly, ligature 1 is tied, a clamp is placed on the superior mesenteric artery near its origin, and an incision is made into the artery proximal to the ligature. Ligatures 2 and 3 are tied, and the venous cannula is introduced into inferior vena cava at X; ligature 4 is tied around the venous cannula. The right renal artery is cannulated via the superior mesenteric artery, and ligations 5 and 6 are completed. (From Kopp, S.J. et al., *Toxicol. Appl. Pharmacol.*, 82, 200–210, 1986. With permission.)

## Surgical Procedure

The surgical procedure for isolating and perfusing the rat kidney is described here, as this animal appears to be the most popular experimental animal for toxicological investigation [64,234,287,288]. Kidneys can be surgically removed from rats weighing 300 to 400 g, preferably starved overnight to decrease rates of gluconeogenesis and possibly reduce perinephric fat. After anesthetizing the rat with an injection of pentobarbital (50 mg/kg), an abdominal incision is made in the midline and extended laterally. The intestines can be swept to the animal's left to facilitate the next steps. Because of the anatomical advantage of the mesenteric artery arising from the aorta at the same level as the renal artery, the right kidney is used for perfusion. This technique facilitates passing the cannula through the aorta into the renal artery with loss of little blood and no interruption of blood flow to the kidney. Figure 38.8 illustrates the principal blood vessels encountered in the cannulation and surgical preparation of the right rat kidney.

To expose the major abdominal vessels and the right kidney, fat and perivascular tissue are cleared away by teasing the tissues around the blood vessels. The adrenal branch of the right renal artery is ligated, and loose ligatures are placed around the vessels as follows: one on the inferior vena cava just distal to the liver, one on the aorta above the mesenteric artery, two on the mesenteric artery near the aorta separated by 0.5 cm, and three on the inferior vena cava, one between the right and left renal vein, one distal to the left renal vein, and one further down on the inferior vena cava. Finally, a ligature is placed on the left renal vein. The ureter is cannulated by means of PE-10 tubing, and a ligature is placed around the ureter to hold the cannula in place.

The animal is heparinized by injecting approximately 200 units of heparin into the inferior vena cava, after which the opening in the wall of the vein is closed by means of a ligature passed over the point of the injecting needle. After tying the ligature on the left renal vein, the venous cannula is inserted into the inferior vena cava and tied in place by means of the two upper ligatures on the inferior vena cava. The cannula is turned such that the opening lies opposite the right renal vein. The other end of the cannula can be temporarily closed by a loose plug of tissue paper. The distal ligature on the mesenteric artery is tied, the artery is grasped at its origin with fine curved forceps, and an incision is made on the wall. The cannula filled with perfusion medium is inserted and passed to meet the forceps, which are then removed. The tip of the cannula is advanced into the aorta and then into the renal artery, which takes off on the opposite side of the aorta, allowing the perfusion medium to flow to the kidney. The cannula is tied in place by means of the anterior ligature on the renal artery and the ligature on the mesenteric artery as well.

With all the cannulas intact and in place, the kidney is surgically removed. The isolated kidney is transferred to the kidney platform in the perfusion chamber, and perfusion is resumed by connecting the arterial cannula to the perfusion flow of the preprimed apparatus. With practice, the total time required for the surgical procedure can be reduced to approximately 15 minutes, starting from the initial midline incision.

## Perfusion Media

As indicated earlier, the perfusion medium of choice appears to be a cell-free perfusion fluid with adequate buffering capacity and containing salts, glucose, and albumin [25,41,186,234,288]. The perfusion medium described earlier for liver and lung, containing KRB-buffered solution, appears to be satisfactory for perfusing kidney preparations. Earlier attempts to utilize blood as a natural perfusate have met with problems relating to vasoconstriction, whether the blood was defibrinated [348] or heparinized [239]. Another cause of impaired renal flow was found to be embolization of fat droplets, especially at later stages of perfusion using whole blood [239] as a perfusate. An additional disadvantage of using whole blood as a perfusate is the unavailability of a requisite volume of blood, especially in small experimental animals such as the rat. A volume of 100 mL of perfusate is often necessary to conduct an isolated perfused kidney experiment, and several animals would be required to obtain the volume of blood needed for perfusing the rat kidney. In addition to the economic considerations, other problems may be encountered when mixing blood from several animals. One is the possibility of immunological interactions in blood pooled from several animals. Finally, in certain experiments, it is essential to have one-pass circulation through the kidney. This method requires greater volumes of perfusate, and using blood would be impractical in such studies.

A major advance in the development of isolated perfused kidney preparations was made with the use of artificial, cell-free perfusion fluids such as buffered saline solutions supplemented with serum albumin [25,41,63, 186,234,289] or macromolecular plasma substitutes [96,301,319,349]. A principal disadvantage of using a perfusion fluid devoid of red blood cells is the relatively high perfusate flow rates observed under these conditions [25] and the relative anoxia due to the limitation of oxygen-carrying capacity of the cell-free medium [96]. Second, an absence of natural components of blood in perfusion medium devoid of whole blood may alter the disposition of the test chemical. Although KRB-buffered solution containing glucose and albumin can be used for perfusing kidneys, investigators have found it necessary to dialyze the BSA (Cohn fraction V) to obtain satisfactory kidney preparations that are being perfused for longer durations [187]. Third, Millipore® filters must be used to filter out any cellular debris that may enter the circulation. Fourth, higher flow rates are necessary to ensure an adequate oxygen supply to the kidney when an artificial medium is used for perfusion.

In addition to the composition for artificial perfusate given above, Fonteles et al. [92] found that the addition of GSH (500 mg/L) to the perfusate prevented depletion of endogenous cortical and medullary GSH. Furthermore, GSH supplementation of the perfusate decreased renal vascular resistance and increased perfusate flow. GSH extraction studies revealed a progressive decrease in renal extraction with time, ranging from complete extraction at 10 minutes to a value of 38% at 60 minutes [92]. Including GSH in the perfusate might be especially relevant for toxicological investigations in view of the reports that many toxic agents deplete endogenous GSH levels in various tissues. Thus far, only kidney has been shown to be able to use external GSH to any significant extent. As was pointed out earlier, rat lung is reportedly able to utilize external GSH [28] but rabbit lung cannot [144].

## Viability Criteria

To date most of the isolated perfused kidney studies have dealt with the mechanism of autoregulation and physiological functions of the kidney. Because of such a background on which perfused kidney preparations have been refined, techniques for evaluating the adequacy and viability of the perfused kidney are abundant. Renal blood flow can be measured by placing a flow transducer prior to the kidney. Alternatively, a flow transducer can also be placed after the perfusate exits the kidney. Blood flow through the kidney can reach 6 to 7 mL/min/g of tissue, depending on perfusion pressures varying from 90 to 130 mmHg [239]. With the artificial perfusate, flow through the kidney can reach as much as 30 to 60 mL/min/g of tissue. One problem associated with using perfusion flow rate as an index of viability is not knowing the intrarenal distribution of the flow. The regional distribution of the blood flow within the kidney can change markedly with artificial perfusion, with a striking increase in flow to the medulla and inner cortex [292]. This fact is especially important for any toxicological investigations, as blood flow through an organ can be either shunted or altered in other ways as a result of toxic action of a test drug. Radiolabeled microspheres (10-15 μm) introduced into the circulation of a perfused organ may be used to assess the regional distribution of flow through the organ. Detailed accounts of such techniques are available [292].

Another criterion used for viability of perfused kidney preparations is the glomerular filtration rate (GFR). During the initial phase of perfusion, the GFR is at a lower limit (46 to 50 mL/min/100 g) of the normal range [239,348]. Better values have been obtained with difibrinated blood at 69 mL/min/100 g tissue. GFR decreases progressively after 2 to 3 hours of perfusion. Often, when blood is used as a perfusate, this impairment has been attributed to the embolization of fat droplets in the glomeruli [348]. In unsuccessful experiments, increases in the weight of the kidney are often due to retention of fluid in the tubules and interstitial edema [239,348].

Urine concentration and excretion of water are other criteria used for establishing the viability of perfused kidney preparations. During the first period of up to 1 hour, urine may reach an osmolality of 800 mOsm, which is later reduced (60 to 150 mOsm) due to increased urine flow. The water diuresis in the kidney preparation can be suppressed by administration of vasopressin in the perfusate. After 3 to 4 hours, the urine concentration approaches isotonicity [169,348,357]. When observed, loss of urine concentrating power is probably due to medullary edema and disturbances of deep cortical and medulla blood flow [239]. Sodium excretion can be used as an additional parameter of viability. In contrast to frequent statements in the literature, the fractional reabsorption of sodium can be normal in perfused kidney preparations. In Berndt's experiments [unpublished data], sodium excretion exceeded 0.4% of the filtered load after 2 hours of perfusion using the KRB-buffer type of artificial medium. After prolonged perfusion, however, sodium rejection may develop unless the perfusate is replaced which may be due to accumulation of metabolic end products such as ammonia in the perfusate [169,348].

Acidification of urine is yet another functional viability criterion used for evaluating perfused kidney preparations. The loss of ability to excrete acidic urine represents a major functional abnormality of isolated kidney [169,348]. An additional functional parameter that can be applied to perfused kidney preparations is the determination of insulin as well as *para*-aminohippurate (PAH) clearance values. A disadvantage of using these clearances as determinants of viability is that, depending on the experimental conditions, these tests may or may not be compatible with the original intended use of the perfused preparation. Hence, many functional tests may have to be carried out infrequently rather than routinely as an internal check of the perfused preparation.

Finally, the kidney preparations can be sampled for electron microscopy as well as light microscopy, and morphological examination can be used as a determinant of functional abnormality. However, routine morphological examination at the light microscopic or ultrastructural level might not be practical for several reasons. Whether the preparation was viable cannot be determined until after the perfusion experiment. Facilities or expertise for routine morphological examination may not be available or, when available, might be prohibitively expensive for routine use in experimental work. Often, the validity of considering morphological alterations to be indicative of functional abnormality is questionable because other functional parameters might be optimal despite the morphological alterations at the cellular or subcellular level. Conversely, despite the normal morphological appearance, distinct functional aberrations may be observed. At least in part, such discrepancies can be explained on the basis of the relatively short time required for a functional abnormality to be detected, whereas longer periods may be required for the observed morphological alterations to develop and *vice versa*. At any rate, as pointed out earlier for the liver and lungs, morphological observation should be helpful initially for establishing a viable perfused preparation in any laboratory [92,348,357]. Also, it is useful for examining the effects of toxic chemicals on perfused kidney preparations [64,88].

## Applications

Historically, the isolated perfused kidney preparations have not been utilized in pharmacological and toxicological investigations despite the availability of refined techniques for some time [186,239]. The bulk of the isolated perfused kidney work can be seen in the physiological literature. One reason might be increased attention devoted to establishment of the physiological parameters such as GFR, autoregulation of blood flow through the kidney, absorption–reabsorption mechanisms in the kidney tubules, and hormonal regulation of renal tubular functions; however, the utility of isolated perfused kidney preparations can be demonstrated by the nature of the information that can be obtained using this technique [209]. For example, the control of GFR by an endogenously released humoral factor was definitively demonstrated using isolated kidney preparations [239].

Using isolated erythrocyte-free, perfused rat kidney preparations, Schureck et al. [301] demonstrated that sodium reabsorption can be increased by including glucose as the sole energy source of the kidney. The nature of glucose handling by the kidney has also been investigated using perfused kidney [41]. Because the tubular transport maximum ($T_m$) for glucose is proportional to GFR in the perfused kidney preparations, it was concluded that the $T_m$ for glucose is not controlled by extrarenal factors.

Many physiological parameters of kidney function have been studied and are understood, and in many cases the renal control mechanism has been confirmed using isolated perfused kidney preparations [239]. Use of the perfused kidney to investigate the toxicological mechanisms has increased [64,141,186,219,220,289,308,315, 319,322,354], and these and other applications of the perfused kidney are briefly discussed here.

The study of GSH extraction from the circulating perfusate by isolated perfused kidney preparations is an example of a biochemical study that can be useful in toxicological investigations [92]. Including GSH in the circulating perfusate at a concentration of 500 mg/L resulted in the preservation of cortical and medullary GSH. Perfusion without the addition of GSH consistently resulted in depletion of tissue levels of this important tripeptide. In addition, including GSH in the perfusate resulted in decreased renal vascular resistance and increased perfusion flow. Whether the increased flow is due to intrarenal alterations in flow patterns is unclear. It appears that GSH storage in the kidney can level off, as indicated by the above study, in which complete extraction of added GSH was seen at 10 minutes; this uptake was reduced to 38% of administered GSH at 60 minutes. These studies of Fonteles et al. [92] indicated a high affinity of rabbit kidney for GSH and a relatively large net reabsorption of this important tripeptide. In view of the many toxicological molecular events being related to alterations in the endogenous pools of GSH, this observation might be important in toxicological studies using isolated perfused kidney preparations.

Another example of using the isolated perfused kidney in toxicological investigations lies in the studies of Dovrak et al. [79], in which they examined the effects of high doses of methylprednisolone on the isolated perfused dog kidney. They noted several histological changes in the kidneys perfused with methylprednisolone for 20 hours or longer. The primary changes consisted of necrosis of capillary loops, inclusion of eosinophilic material in Bowman's space, thickening of the basement membrane, and endothelial cell damage. Arterial changes consisted primarily of inclusion of afferent arterioles with dense eosinophilic material. Tubular changes consisted of inclusion of tubular lumens and damage to tubular epithelial cells. These studies demonstrated that administration of high doses of methylprednisolone can produce irreversible hemodynamic and histological changes in the kidney. Trumper et al. [338] investigated effects of different concentrations of acetaminophen (*N*-acetyl-*p*-aminophenol; APAP) on renal function in isolated perfused rat kidney. Changes in the fractional excretion of sodium, water, and glucose and in the glomerular filtration rate were measured. APAP (10 m*M*) increased the fractional excretion of water (72%), sodium (79%), and glucose (55%) and this increase was associated with a decrease in GFR. Prostaglandin $E_2$ ($PGE_2$) prevented the decrease in GFR and glucose reabsorption but did not change the fraction of water or glucose excretion induced by APAP. Verapamil prevented the glomerular but not tubular effects of APAP. The authors concluded that APAP exerts a direct effect on isolated perfused kidney, affecting hemodynamic and tubular functions, and that the latter are not a consequence of hemodynamic alterations. Aiba et al. [4] investigated the renal handling of tobramycin (TOB), an aminoglycoside antibiotic using a single-pass isolated perfused rat kidney. At tracer concentrations (7.4 m*M*), 32% of TOB remained in the lumen, but no TOB was found in the vein. This ratio of luminal uptake was reduced in a dose-dependent manner. Other aminoglycosides such as gentamycin inhibit this uptake, but tetramethyl ammonium and glucosamine had no effect. Alkalinization of urine led to a 67% decrease of the TOB uptake. This indicated that TOB was mainly taken up by the epithelial cells from the luminal site and that this uptake process was saturable and specific for aminoglycosides, which have more than one cationic group.

Three studies [108,188,284] have investigated the toxicology of cyclosporin A (CsA) in isolated perfused kidney. Roby and Shaw [284] studied the acute effects of CsA and its metabolites. Intralipid was used as a vehicle for CsA because ethanol, methanol, or Cremophor® caused significant effects on GFR. Intralipid enhanced the effects of CsA25-fold, giving CsA dose responses comparable to human kidneys. This enhanced effect was due to vasoconstriction, not vasoobstruction, and was specific to CsA as enhancement of norepinephrine with intralipid did not occur. Primary metabolites (M1, M17, and M21) caused decreases in GFR comparable to or slightly less than the modest declines in GFR. Because CsA metabolites in human blood often exceed CsA concentration, the study suggests that CsA metabolites may contribute substantially to CsA nephrotoxicity. Longoni et al. [188] investigated the protective effects of L-propionyl carnitine (LPC) against CsA nephrotoxicity; their work with isolated perfused kidney revealed that LPC protects against the toxic lipid peroxidation phenomenon induced by CsA. In additional studies, Giovannini et al. [108] found that LPC can restore CsA-induced decreases in intracellular ATP levels to normal. At the same time, a decrease in the increased vascular resistance was not noted; therefore, it was suggested that the protective effects of LPC included correcting biochemical alterations induced by CsA, explaining the pathogenesis of renal damage induced by CsA.

Herrero et al. [127] employed isolated perfused kidney preparations to evaluate a preservation solution containing fructose-1,6-diphosphate and mannitol (F-M) with the University of Wisconsin solutions used to preserve kidneys intended for human transplantation. The F-M flushing solution contained fructose-1,6-diphosphate (1 g/dL) and mannitol (2 g/ dL) during cold ischemia. The kidneys were stored in hypothermia for 4 and 18 hours after initial flushing with the solution being tested and then reperfused at 37°C in an isolated perfused circulation for 90 minutes with KRB containing 4.5% albumin. Plasma flow rate (PFR), renal vascular resistance (RVR), urine flow rate (UFR), GFR, fractional sodium reabsorption (FRNa), and net sodium reabsorption (TNa) were studied during perfusion. Conventional histology and tissue malondialdehyde levels were also evaluated. After 4 and 18 hours of cold ischemia, GFR, FRNa, and TNa were better and conventional histology worse in F-M than Euro-Collins (EC) flushed kidneys. After 18 hours of cold ischemia, F-M flushed kidneys were better than the University of Wisconsin (UW) flushed kidneys although after 4 hours they did not differ. After 18 hours, malondialdehyde was lower in F-M compared to EC or UW, although after 4 hours there were no differences. The authors concluded that the newly developed flushing solution (F-M) showed promising results in renal preservation and that its ability to preserve is at least as good as UW solution as assessed by isolated perfused kidney preparations.

Summerfield et al. [318] utilized isolated perfused kidney preparations to examine conjugating reactions involved in the elimination of certain bile acids in urine. These investigators demonstrated that lithocholic and chenodeoxycholic acids can be metabolized by the perfused kidney to their monosulfate conjugates; the disulfate metabolites of these bile acids were not detected in the urine. These findings supported the hypothesis that renal synthesis of monosulfate conjugates may account for at least some of the bile acid sulfates present in urine in the cholestatic syndrome of humans. The results further suggested that in chemically induced hepatic injury the kidney may be able to conjugate some of the bile acids, and they demonstrated the presence of sufficient biochemical machinery within the renal tissue for conjugating endogenous substrates. The experiments also demonstrated the possibility of conjugation of foreign chemicals in the renal tissue, facilitating their elimination in urine.

Jaffe et al. [141] employed perfused rabbit kidney to evaluate the genotoxic potential of *S*-(*trans*-1,2-dichlorovinyl)-L-cysteine (DCVC). The proposed mechanism of renal toxicity of this and other vinyl cysteine conjugates is activation by β-lyase, an enzyme in the brush-border membrane of renal tubular epithelial cells. This enzyme converts the halogenated vinyl cysteine conjugate to reactive thiovinyl intermediates, which alkylate subcellular macromolecules such as DNA. In these rabbit kidney perfusion studies, a dose-dependent (0.01 to 1.00 m*M* DCVC) effect of DCVC on DNA single-strand breaks was demonstrated.

The study of Tark et al. [324] may be cited as an example of yet another use of isolated perfused kidney in toxicological investigations. These investigators examined substrate metabolism in the isolated perfused dog kidney and established that free fatty acids (FFAs) and glucose serve as significant substrates for providing energy for sodium transport in the kidney. Second, their studies suggested that glucose may substitute for FFAs as an energy source at times when FFAs in the circulation are decreased. The effect of toxic chemicals in the renal circulation on substrate metabolism in the kidney can be examined using the methods described by Tark et al. [324]. Johannesen et al. [143] employed perfused kidney preparations to study the renal energy metabolism inhibitor 2,4-dinitrophenol.

The use of isolated perfused kidney preparations in toxicological investigations has increased [64,162,186, 219,220,289,308,319,322,354]. Perfused rat kidney has been used to investigate the mechanism of gentamicin toxicity [64,219,220]. Mitchell et al. [219] established that a specific effect of gentamicin on potassium secretion is responsible for clinically observed hypokalemia during gentamicin therapy. This study and those of Hook and associates [64] have established that reduced water and electrolyte reabsorption were the earliest effects of gentamicin on the kidney. Perfusion of the rat kidneys with gentamicin induced a dose-dependent decrease in reabsorption and metabolism of lysozyme [64]. The basis of the sex differences in renal toxicity was the subject of additional inquiry [220]. The sex differences could not be demonstrated in the perfused rat kidneys exposed to gentamicin, suggesting that the sex differences in the susceptibility to this antibiotic *in vivo* might be due to some extrarenal factors. Koschier et al. [140] perfused rat kidneys to study the effects of *bis*-(*p*-chlorophenyl) acetic acid (DDA), the principal water-soluble metabolite of dichlorodiphenyltrichloroethane (DDT). At a concentration of 1-m*M* DDA in the perfusate, the GFR, urine volume, and fractional excretion of sodium were decreased, suggesting a direct action of DDA on nephron function. Sumpio et al. [319] employed isolated perfused rat kidneys to characterize the renal toxicity of *cis*-diaminedichloroplatinum (CDDP) and to determine if treatment with ATP–$MgCl_2$ could prevent or reduce the nephrotoxic effect of CDDP. After 2 hours of perfusion, CDDP (100 μg/mL) treatment led to marked inhibition of protein reabsorption with only a minimal decrease in sodium and water reabsorption. Despite a marked diuresis, GFR was not significantly altered. Posttreatment with ATP–$MgCl_2$ (2 m*M*) led to partial alleviation of the nephrotoxic effect of CDDP; however, after 1 hour of perfusion, simultaneous treatment with ATP–$MgCl_2$ (0.3-m*M*) fully protected the protein reabsorptive capacity of CDDP-treated kidneys. Because the CDDP-induced toxicity simulates the acute renal failure seen clinically, these findings are suggestive of a new therapeutic modality for clinical management.

Using isolated perfused kidney, Adin et al. [2] showed that bilirubin treatment resulted in significant improvements in renal vascular resistance, urine output, glomerular filtration rate, tubular function, and mitochondrial integrity after ischemia–reperfusion injury (IRI). Beneficial effects on organ viability were achieved most consistently with a dose of 10-μ*M* bilirubin. The authors concluded that the protective effects of hemeoxygenase-1 activity during IRI in the kidney are mediated, at least in part, by bilirubin and that pretreatment with micromolar doses of bilirubin may offer a simple and inexpensive method to improve renal function after IRI [2].

## Isolated Perfused Brain

Brain does not lend itself to simple and totally isolated perfusion, and all the preparations to date include more or less extraneural tissue. Perfusion has found little place in many of the biochemical and toxicological studies with brain tissue, as the technique presents great difficulty even when effort is not made to exclude extraneural tissue. In addition, the brain is heterogeneous, and the contribution by both neuronal and nonneuronal tissue of the brain to drug uptake and turnover in perfusion causes difficulties in duplicating the results. The blood–brain barrier is one aspect of brain metabolism in particular that remains noticeably obscure, and it has been the subject of studies with perfused brain preparations [10,90,102,105,330]. The heterogeneity of the preparation together with the many neural tissues represented in the brain make perfusion a somewhat less valuable technique for toxicological investigations than other individual organs. Nevertheless, several investigators have been able to maintain a viable perfused brain preparation that can be utilized for drug metabolism and investigations on the effects of toxic chemicals that may adversely affect the central nervous system. The difficulty of maintaining the isolated perfused brain preparation coupled with the readily available *in vitro* and *in vivo* techniques have resulted in underutilization of perfused brain preparations for pharmacological and toxicological investigations.

Comparatively simple techniques of perfusing the rat brain are described here. The perfused rat head preparation of Thompson et al. [330], in which the entire head is perfused with no attempt to limit the circulation to the brain, is not described in detail; however, such a technique may be useful in some toxicological investigations and may be considered before setting up more difficult preparations. Far more elaborate and therefore technically more difficult is the preparation described by Andjus et al. [10], which attempts to exclude the muscle of the head and neck from the perfusion circuit; this method is described because of the obvious superiority of the technique. The preparation is based on the more elaborate perfusion technique developed in the cat by Geiger and Magnes [102] and that described by Chute and Smith [61], which may be consulted if larger animal models are suitable to meet the particular need. Readers may refer to White [351] for a total isolated, vascularly perfused monkey brain preparation or to Gilboe et al. [105] for a dog brain perfusion preparation.

### Apparatus

Figure 38.9 shows the schematics of the equipment used for oxygenating the venous drainage from the brain and perfusion of the isolated rat brain described by Andjus et al. [10]. The bubble oxygenator and reservoir are made from two disposable plastic drug administration sets combined and fitted with connectors and plastic tubing. A small volume of recirculating fluid pumped by a peristaltic pump is used to perfuse the brain. The perfused brain preparation is held in a funnel from which the effluent perfusate drips to the oxygenator and passes through a filter to complete one circulation. If a single-pass circulation is desired, the peristaltic pump is disconnected, and a reservoir of perfusate is used at a height sufficient to provide satisfactory perfusion pressure and flow.

### Surgical Procedures

After the animal is anesthetized using a proper anesthetic agent, both common carotid arteries are exposed, and the trachea is intubated via a tracheostomy. The animal is heparinized by injecting 500 IU of sodium heparin through the jugular vein, which is then ligated by means of a suture. The external carotid and pterygopalatine arteries are ligated. In the rat the pterygopalatine artery is a branch of the internal carotid artery, which supplies blood to various extracranial structures. A plastic cannula filled with perfusion fluid is then inserted into each common carotid, advanced into the internal carotid artery, and tied in position so its tip is near the origin of the previously ligated pterygopalatine artery. The arrangement of the vessels and the cannulas is illustrated in Figure 38.9.

A slow perfusion is initiated, and the skin and the muscles of the head, face, and neck are then removed together with the mandible. A sturdy ligature is placed around the vertebral column, which is transected just below the ligature. The transected vertebral canal is packed with a cotton or tissue paper plug, and the canal is sealed with melted wax. The completed preparation consists of the skull and its contents with the upper cervical vertebrae and small remnants of muscle tissue attached. Toward the end of the surgical procedure, the perfusion flow is gradually increased. After severing the vertebral column, the perfusion flow rate is adjusted to a desired value (1.4 mL/min). The cannulated preparation is then mounted above the collecting funnel. The entire preparation and the apparatus can be housed in a heating chamber to facilitate maintenance of the proper temperature. The chamber used for liver preparations [203,218] may be used for this purpose.

Thompson et al. [330] described a technique of *in situ* perfusion of the rat head. They used an apparatus reminiscent of that of Miller et al. [218]. The aortic arch of the rat is perfused, and the inferior vena cava is cannulated

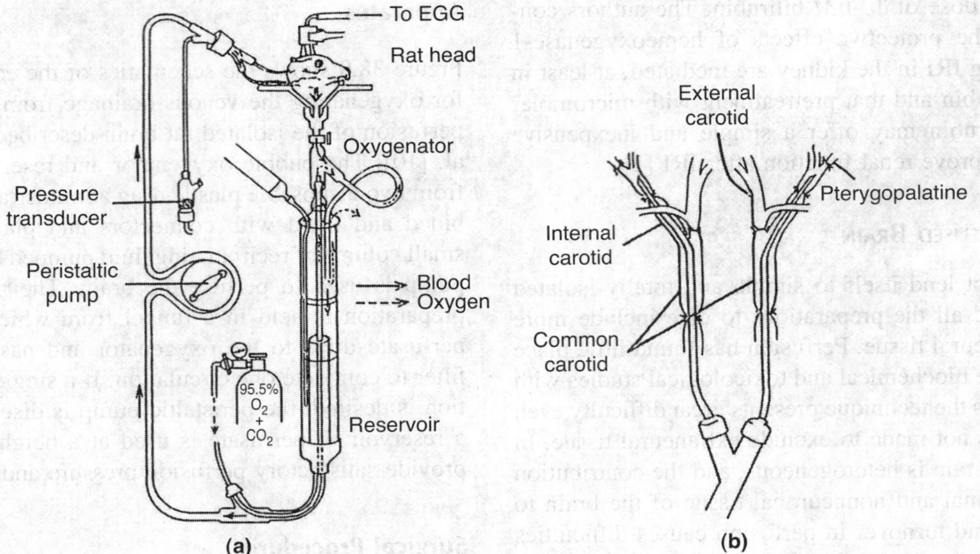


**FIGURE 38.9** (a) Perfusion system for the isolated perfused rat brain preparation. Venous blood from the preparation flows by gravity into the collecting funnel and then is dispersed by the stream of gas ($O_2/CO_2$, 95:5) into short segments and carried through the side branch of the oxygenator into the reservoir. The stream of gas is prevented from escaping through the collecting funnel by adjusting the rate of flow of gas and the screw clamp just below the funnel. Fixed in the lower part of the reservoir is a metal- mesh filter. For perfusion without recirculation, the blood from the collecting funnel drains into a bottle, and fresh blood is added at the top of the oxygenator as needed. The peristaltic pump has a continuously variable speed with maximum delivery of 2 mL/min. The perfusion pressure is monitored by a pressure transducer. Substances can be added to or samples taken from the reservoir by means of the syringe and attached tubing shown to the left of the oxygenator. The solid arrows show the direction of flow of blood, the dashed arrows the direction of flow of gas. (b) Relation of the perfusion cannulas to the cannulated and adjacent arteries. (From Akerboom, T. et al., *Hepatology*, 13, 216–221, 1991. With permission.)

for the outflow. Recirculation of the perfusion medium, made up of diluted rat blood, maintains the preparation for up to 3 hours. The isolated perfused rat brain preparation described by Andjus et al. [10] can be maintained for 2 hours with satisfactory CNS function. Geiger and Magnes [102] described an isolated perfused brain from the cat that is similar to the preparation described above for the rat. The Geiger–Magnes preparation has been used for various studies by other investigators; for example, Barrett et al. [27] used it to evaluate the effect of a number of centrally acting drugs on cat brain. In addition, Otsuki et al. [255] used the cat brain preparation to evaluate the suitability of using various perfusion media.

## Perfusion Media

Andjus et al. [10] used pooled rat blood obtained from several rats as the perfusion medium. However, they found that when blood was used it did not support the spontaneous electrical activity of the isolated brain. After several trials they concluded that an artificial perfusion fluid similar to that described by Geiger [101] should be used.

The fluid portion of the artificial perfusate is KRB buffer solution, described earlier for perfusion of the liver, lung, and kidney. BSA (Cohn fraction V) is dissolved in distilled water and deionized by passing it through a column of Amberlite® MB-3 before mixing it with the KRB-buffered solution. Erythrocytes obtained from dog blood were washed and used in this perfusion fluid. The cells are washed four times with cold, buffered (0.01-*M*, pH 7.4) isotonic sodium chloride solution and finally with isotonic KRB-buffered solution prior to use. The final perfusate contains 7 to 8% BSA, a hematocrit of 20 to 25%, and pH adjusted to 7.3. The perfusate is always best freshly prepared just prior to use. The glucose concentration of the blood for perfusion is adjusted to 200 mg/dL by adding a requisite volume of 5% glucose in normal saline solution.

The perfusate and the isolated brain preparation described by Andjus et al. [10] remained at room temperature (23 to 27°C) throughout the experiment. Most of the perfusion experiments were conducted at 25°C; however, maintaining the entire perfusion at body temperature should be considered. It can be easily done by employing a heated chamber such as the one used for perfusing the liver [197,211]. The total volume of perfusate is 100 mL, and perfusion is carried out at a constant arterial pressure of 100 to 120 mmHg. The perfusion rate is 3 to 5 mL/min and should be maintained at this rate throughout the experiment. The temperature of the brain can be maintained at 30°C using a heated chamber as indicated above.

Otsuki et al. [255] compared a number of perfusion media using KRB buffer as the solution and either low-molecular-weight (40,000) or high-molecular-weight (70,000) dextran as the substituent for BSA. They also compared the effect of including a number of amino acids in the perfusion medium. Their study concluded that dextran could replace BSA in the perfusion medium and that the low-molecular-weight dextran was superior to the high-molecular-weight dextran. Furthermore, including glutamic acid in the perfusate along with low-molecular-weight dextran improved functional performance of the isolated perfused cat brain.

## Viability Criteria

Spontaneous electrical activity was recorded by Andjus et al. [10] as a functional parameter of brain activity. The isolated rat brain preparation had spontaneous electroencephalographic (EEG) activity that persisted as long as 5 hours with single-pass perfusion and about 2 hours when recirculating perfusion was conducted. Although the reason for the shorter period of viability in a recirculation is not well understood, it is possible that endogenous biochemical products of intermediary metabolism may accumulate in the recirculating perfusion, to the detriment of the perfused brain. Addition of pentylenetetrazol to the perfusing blood evoked characteristic EEG signs of convulsive activity either before or after spontaneous EEG activity had ceased. Otsuki et al. [255] carried out a similar study using EEG measurements to compare the suitability of altering the perfusion medium to perfuse cat brain. Response to loud sounds can also be observed by EEG recordings and can be used as a viability criterion. After perfusion for 5 hours, Andjus et al. [10] reported that the rat brain was nonresponsive to sound, indicating deterioration of the preparation. Thus, although EEG recordings indicated viability of the rat brain preparations, response to loud noise was lost after 5 hours of perfusion [10], suggesting that not all of the criteria are satisfied, especially during a perfusion that lasts several hours. Gilboe and associates [105] have used oxygen utilization and EEG pattern as satisfactory criteria for dog brain perfusion.

An additional criterion for viability is the rate of glucose utilization by the isolated brain preparation. It can be measured as the decrease in glucose concentration in the perfusing blood and should be linear during the first hour of the experiment; later, it tends to decrease with the deterioration of the preparation. Lactate accumulates in the circulating perfusate, but the rate at which it accumulates does not appear to be directly related to glucose utilization [10]. Otsuki et al. [255] also used rates of glucose utilization as a measure of viability for evaluating the various perfusion media to support isolated perfused cat brain preparations.

## Applications

Although the use of isolated perfused brain preparations for toxicological investigations has been infrequent, examples can be cited where isolated perfused brain preparations have been used for such studies. The principal reason for the relatively infrequent use of isolated perfused brain preparations seems to be the relatively slow development of techniques [10,27,102,255]. Hein et al. [123] investigated the effect of thiopental anesthesia on the energy metabolism of the isolated perfused rat brain. They perfused the rat brain in the presence and absence of 5- to 15-m$M$ thiopental and investigated glucose turnover as well as a number of indicators of the glycolytic pathway. They noted that glucose uptake by the brain preparation was increased when thiopental was included in the perfusate; however, the glycolytic pathway remained inhibited, indicating the sensitivity of the glycolytic pathway to thiopental. They also noted that this effect was not mediated via hindered uptake of glucose. In another study, Dirks et al. [77] investigated the effect of piracetam and methohexital on rat brain energy metabolism and concluded that piracetam had no acute effect on energy metabolism and methohexital protected the rat brain from ischemic effects. These reports demonstrated the usefulness of the perfused brain preparation for studying intermediary metabolism [77,165] of the brain as affected by the presence of toxic drugs or chemicals in the circulating perfusion medium. Similar preparations should be useful for studying the effect of centrally acting toxic chemicals such as industrial solvents and gaseous substances as well as other neurotoxins; for example, Krieglstein and Stock [166] described the effect of chloral hydrate and trichloroethanol on cerebral intermediary metabolism. Fink et al. [89] used perfused rat brain to study the central activity of valtrate, following the EEG activity of brains perfused with or without valtrate. When Fitzpatrick and Gilboe [90] used perfused dog brain to study the effects of nitrous oxide on cerebrovascular tone, oxygen consumption, and EEG, they observed that nitrous oxide reduces cerebral vascular tone but exhibits no effects on central oxygen metabolism.

The applicability of isolated rat brain preparations to study central neurological mechanisms can be illustrated by the study of Kilbinger and Krieglstein [154]. They found that physostigmine caused a rise in the acetylcholine concentration of both the isolated perfused rat brain and the rat brain *in vivo*. Oxotremorine, on the other hand, produced an increase in acetylcholine content in the brain *in vivo* but was ineffective in the isolated rat brain at the same dosage. These investigations were carried out by recording the EEG as well as determining perfusate and brain levels of acetylcholine. These studies suggested the feasibility of using brain preparations to evaluate the effect of toxic chemicals on alterations of endogenous neurohormones and neurotransmitters.

Isolated guinea pig brain preparation was employed to investigate the influence of chronic amphetamine intoxication on the permeability of the blood–brain barrier (BBB) to inert and polar molecules [274]. This study illustrates the use of isolated brain preparations to evaluate the effect of chronic drug treatments on vascular permeability. Multiple-time brain analysis was used to study the effects of chronic (14 and 20 days) amphetamine intoxication (5 mg/kg daily, i.p.) on the kinetics of $^{14}$C-mannitol and $^{3}$H-polyethylene glycol (PEG) in the forebrain guinea pigs. Regardless of their molecular weight or lipophilicity, these molecules were transferred across the BBB progressively with the amphetamine treatment. The opening of the BBB was associated with changes in behavior (increased locomotor activity, stereotypy, hypervigilance, social withdrawal, and loss of weight) with 14- and 20-day treated animals. At 7 and 28 days after the withdrawal of the amphetamine treatment, behavioral manifestations were absent and the opening of the BBB to these inert molecules was not different from that in normal animals [274].

Although not isolated, *in situ* perfused rat brain was used to study the drug metabolizing enzymes. Chikaoka and Tamura [58] employed the *in situ* perfused rat brain to study deethylation of 7-ethoxycoumarin (7-EC). Infusion of 7-EC through an internal carotid artery resulted in the formation of 7-hydroxycoumarin (7-HC) and its conjugates in the effluent perfusate collected from the superior vena cava. The rate of formation of products was 200 nmol/hr/g when 130 μM 7-EC was infused. This value was 100 times higher than the rate predicted from brain microsomal activity. Induction of drug metabolizing enzymes by phenobarbital and β-naphthoflavone increased the deethylation activity in the perfused brain, just as in the perfused liver.

The usefulness of brain perfusion for bridging *in vitro* and *in vivo* studies was demonstrated by the studies of Inagaki and Tamura [136]. They employed a cannulation procedure for *in situ* preparation of a functionally intact, hemoglobin-free isolated perfused rat brain without interrupting the brain circulation. The spontaneous distribution (8 to 25 Hz) and amplitude of a spontaneous electroencephalogram (EEG, 50 μV) were kept within the normal ranges for up to 4 hours after perfusion was started. The preparation gave characteristic flash-evoked EEG responses through the eyes. Administration of bicuculline elicited an epileptic seizure similar to that in normal rats. Redox behaviors of cytochrome oxidase a and a3 in the brain were identical to those observed in the brain tissue.

Yet another example can be found in the studies of Gibbs et al. [104]. The bilateral *in situ* guinea pig brain perfusion method, linked to high-performance liquid chromatography analyses, was used to examine (–)-2′-deoxy-3′-thiacytidine (3TC) uptake into brain and cerebrospinal fluid (CSF) simultaneously. The influence of transport inhibitors and additional nucleoside analogs on this uptake was investigated. 3TC movement across the blood–CSF barrier was examined in more detail by the isolated choroid plexus model. 3TC movement across the brain barriers and subsequent accumulation in the brain and CSF was low; however, 3TC uptake from blood into the choroid plexus (a potential CNS target for HIV treatment) was significant and was facilitated by a digoxin-sensitive transporter. Another transporter was identified that removed 3TC from the choroid plexus. Abacavir, 2′,3′-didehydro-3′-deoxythymidine, and 3′-azido-3′-deoxythymidine did not interact with 3TC at either of the brain barriers to affect CNS concentrations of 3TC; however, a significant interaction between 3TC and 2′,3′-dideoxyinosine was observed at the choroid plexus, and it may prove beneficial to select drug combinations where no such interaction is indicated [104].

In addition to the studies mentioned above, the technique of perfusion has also been used to aid histological fixation [256,333] of brains from experimental animals. For this purpose, the arch of the aorta is perfused with a balanced salt solution followed by a fixative. A distinct advantage of the perfusion technique to fix the brain preparation for histological examination is the delivery of oxygen through the oxygenated perfusion medium to the brain to achieve better preservation of the tissue.

Although the isolated perfused brain preparations described to date include extraneural tissue and many types of nerve centers within the brain itself, the perfusion technique should nevertheless be valuable for examining the specific effects of centrally active toxic chemicals. The state of the art in this area has developed to a degree to which the technique should prove useful in toxicological investigations.

## Isolated Perfused Intestine

Over the years, several attempts have been made to study the mammalian intestine as an isolated tissue sustained by a vascular perfusion. Mainly because of the nature of the tissue itself, use of the term "perfusion" has been ambiguous in the field of intestinal research. Intraluminal circulation of fluid for the purpose of studying the transport of small molecules across the intestinal mucosa has often been referred to as *intestinal perfusion* [311,313]. Intraluminal perfusion experiments can be performed *in vivo* for subacute toxicity studies [189] or for shorter periods of time [280], or *in vitro* with excised segments of intestine [191]. A full range of experimental techniques is available for work with intestinal tissue, and the topic has been reviewed [258], including the technique of vascular perfusion [140]. No single experimental technique provides information about all phases of the absorptive processes involved in the removal of a substance from the lumen of the small intestine, its transport across the intestinal wall, and its entry into either blood or the lymphatic circulation. Techniques such as intraluminal perfusion and everted sac

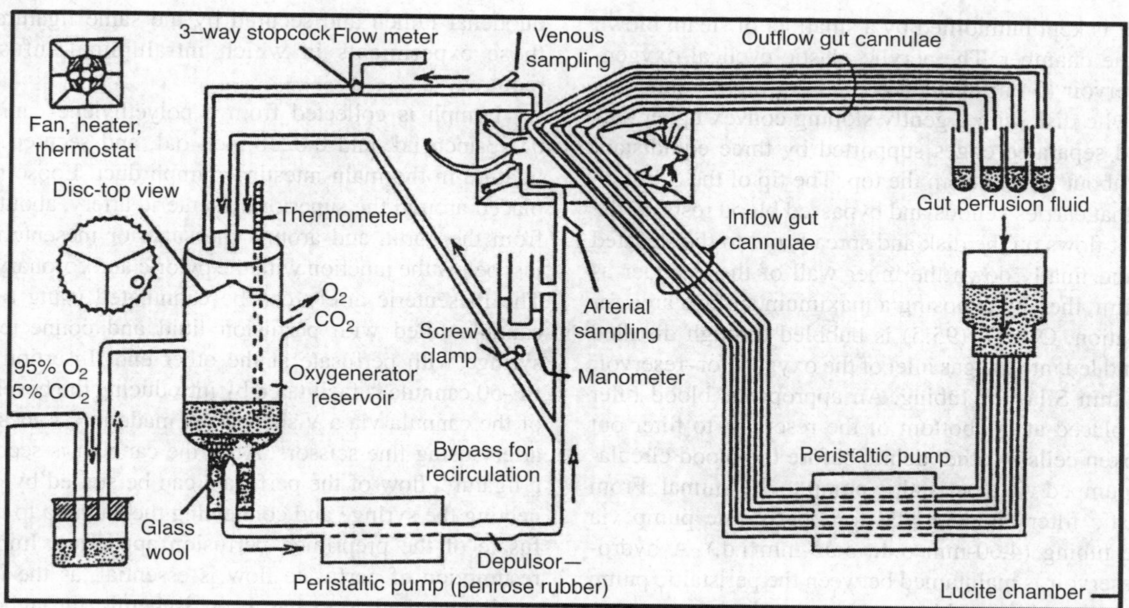


**FIGURE 38.10** Apparatus for vascular perfusion of small intestine. (From Hidaka, T. et al., *Arch. Toxicol.*, 64, 103–108, 1990. With permission.)

described by Wilson and Wiseman [353] are discussed elsewhere in this book. This particular discussion deals with the vascular perfusion of small intestine.

Isolated vascular perfusion of intestine has not been prominent in gastrointestinal research because of the many problems encountered when trying to obtain sustained viability [355]. It is clear that vascular resistance, spasmodic bowel contractions, tissue edema, and progressive destruction of the mucosal epithelium were the principal problems encountered during earlier attempts to establish a vascularly perfused intestinal preparation. Second, because most investigators were concerned with the mechanisms of intestinal absorption and were dealing with the mucosal layer, these investigators found it convenient to use intraluminal perfusion with fluid containing the experimental drug to study the intestinal absorption. In addition to supplying the experimental drugs through the intraluminal fluid, such fluid could also carry the necessary oxygen for supporting the mucosal layer [40]; however, arguments can be made for developing a viable, vascularly perfused intestinal preparation [150,355]. The role played by other layers of intestinal wall, such as the muscle, can be studied in a vascularly perfused intestinal preparation. Also, the disadvantages of everted gut preparation, in which only transport across the wall is studied (rather than transport into the blood circulation), are overcome by utilizing an isolated vascularly perfused preparation. Thus, anatomically and functionally distinct compartments (e.g., lumen, lymph, bloodstream) are kept separated and can be studied individually. These arguments and the need to separate intestinal tissue to describe those absorptive, distributive, and metabolic functions of the tissue fully justify the development and use of the isolated, vascularly perfused intestinal preparation.

As has been observed for other isolated perfused organs, intestine can be perfused *in situ* [150,257,355] or in complete isolation [50,93,131]. In addition, the perfused preparation can be used with an intraluminal flow maintained in the natural direction of peristalsis [80,150] or without the intraluminal flow [76,84]. If the intestine is perfused without the intraluminal flow, two dynamic compartments (perfusion fluid and lymph) can be sampled for an experimental test chemical in addition to the intestinal tissue itself. If intraluminal flow is maintained, however, three dynamic compartments (perfusion fluid, lymph, and intraluminal fluid) and the static compartment of intestine tissue can be sampled for an experimental test chemical. For most practical purposes, maintaining intraluminal flow would introduce another dynamic variability in the experimental condition so analysis and interpretation of the experimental results from analyzing all four compartments would be difficult if not impossible. The following description of an isolated perfused intestinal preparation is based on the procedure of Kavin et al. [150], later improvised by Windmueller et al. [355].

### Apparatus

An acrylic plastic box (75 cm high, 60 cm wide, 50 cm deep) houses the apparatus (Figure 38.10). Temperature in the chamber is maintained at 37°C by means of a heater and fan assembly that is thermostatically controlled. The chamber is kept humidified by a small jet of steam blown inside the chamber. The acrylic plastic cyclical oxygenator-reservoir (7 cm in diameter, 15 cm high) contains a thin acrylic disk with a gently sloping convex upper surface and separated edges supported by three equidistant flanges about 3.7 cm from the top. The tip of the Silastic® tubing that carries venous and bypassed blood rests on the disk so it flows on the disk and spreads out to the serrated edges and finally down the inner wall of the cylinder as a thin film, thereby exposing a maximum surface area for oxygenation. $O_2/CO_2$ (95:5) is bubbled through distilled water and led into the gas inlet of the oxygenator–reservoir via the thin Silastic® tubing. An appropriate blood filter can be placed at the bottom of the reservoir to filter out any broken cells or other debris before the blood circulation is pumped via a peristaltic pump to the animal. From the plastic filter, blood is led to a peristaltic pump via silicone tubing (4.60-mm o.d., 3.35-mm i.d.). A hydrostatic reservoir is maintained between the peristaltic pump and the animal to provide a constant perfusion pressure. A bypass flow from the hydrostatic reservoir to the central reservoir facilitates maintaining a constant hydrostatic head. Blood flow is determined by means of a transducer connected to the venous circuit by a three-way stopcock. Blood is sampled from the arterial and venous channel via polyethylene tubing, using Y connections near the arterial and venous cannulas; thus, blood is in contact with Silastic® tubing, a hydrostatic reservoir, and polyethylene connections (Figure 38.10). Most of the silicone tubing could be replaced by glass in the apparatus, and use of the silicone tubing should be restricted to glass tube connections and the peristaltic pump. If intraluminal flow is desired in the experimental protocol, an infusion pump can be used to introduce a fluid, with or without the test chemical, into the intestinal lumen. A peristaltic pump can be used for this purpose, giving a peristaltic wave motion for the intraluminal flow. Fluid effusing out of the intestine can be collected as desired or recirculated after sampling.

## Perfusion Procedure

The rat is anesthetized with a mixture of ether and oxygen or an intraperitoneal injection of pentobarbital (50 mg/kg). Through an L-shaped abdominal incision, the small intestine, cecum, and proximal large intestine are gently exteriorized and supported on a plastic platform that is covered with gauze soaked in warm 0.9% NaCl. The intestine is covered with saline-soaked gauze in a plastic sheet and kept at 37°C by means of a heat lamp. Both ends of the intestine are ligated using an appropriate suture: proximally at the duodenum about 1.5 cm from the pylorus and distally near the midpoint of the descending colon. Included in the proximal ligature are the common bile duct and the mesentery between the duodenum and superior mesenteric vein. A PE-90 cannula is placed in the duodenal lumen and secured by the same ligature as in those experiments in which intraluminal infusion is required.

Lymph is collected from a polyethylene cannula of 0.023-inch i.d. and 0.0238-inch o.d. and secured with a ligature in the main intestinal lymph duct. Loose ties are placed around the superior mesenteric artery, about 5 mm from the aorta, and around the superior mesenteric vein just below the junction with the pyloric and coronary veins. The mesenteric artery can be cannulated using a PE-50 cannula filled with perfusion fluid and connected to a syringe with perfusate at the other end. Insertion of the PE-50 cannula is facilitated by introducing the beveled end of the cannula via a V-shaped cut made in the mesenteric artery using fine scissors. After the cannula is secured by a ligature, flow of the perfusate can be started by disconnecting the syringe and connecting the cannula to the perfusate of the preprimed perfusion apparatus. Immediate resumption of perfusate flow is essential, as the isolated small intestinal loop has been ischemic throughout the surgical procedure. At this point, a flow of 8 to 9 mL/min would ensure adequate oxygen supply to the tissue.

The rat is then exsanguinated by severing the left jugular vein and carotid artery. The superior mesenteric vein is cannulated by means of another PE-50 cannula in a manner analogous to the above procedure and secured by means of a suitable suture. The venous effluent is recycled through the perfusion apparatus to the oxygenator-reservoir. Once recirculation of the perfusate is established, arterial flow is increased until an arterial pressure of approximately 95 mmHg is reached. The venous pressure is adjusted to 150 $mmH_2O$ by an adjustable clamp on the outflow cannula. With experience, total surgical time can be minimized to 30 minutes or less. All of these procedures are easier with large animals; the isolated, vascularly perfused canine intestine is described in the literature [50,131,242].

## Perfusion Media

The semiartificial perfusion medium consists of 80 to 120 mL fresh heparinized rat blood drawn by abdominal aortic puncture from an ether-anesthetized animal [355]. The blood is heparinized with 10,000 to 25,000 IU of sodium heparin. Windmueller et al. [355] used antibiotics in the blood (penicillin G, streptomycin), but often use of these compounds is undesirable as they may interfere with the test drug. If the perfusion chamber and all of the components of the apparatus are sterilized, antibiotics have not been necessary for the isolation and perfusion of organs such as the lung and liver [203,212,213]. Use of antibiotics should be avoided unless it becomes critical in a particular experimental protocol. Norepinephrine is used continuously at a rate of 1.0 to 2.2 mL/min as a 0.153-mg/mL solution to alleviate the increased vascular resistance noted by earlier investigators. A glucocorticoid (dexamethasone, $6 \times 10^{-7}$ *M*) is added in a single dose as a solution (25 mg/dL) in the perfusate to the reservoir. A combination of the glucocorticoid and norepinephrine aids in maintaining the low vascular resistance and improved tissue preservation.

The perfusate described for other perfusate organs containing KRB-buffered solution with glucose and BSA can be used for perfusing intestinal preparations with satisfactory results. Kavin et al. [150] utilized a similar perfusate with low-molecular-weight dextran instead of albumin in their perfusion fluid. The addition of norepinephrine and a glucocorticoid to this perfusate would make it comparable to the preparation of Windmueller et al. [309]. The fluid for intraluminal flow is 0.9% saline containing glucose (220 m*M*) and sodium taurocholate (10 m*M*) infused at 2.6 mL/hr. Regardless of whether intraluminal flow is intended, use of the KRB-buffered solution with glucose and albumin or low-molecular-weight dextran should be considered for toxicological investigations. The perfusion flow rate of 16 to 20 mL/min is obtained under these conditions at an arterial pressure of 95 mmHg. *In vivo* flow in rats was found to be 8 to 10 mL/min [355], approximately half of that observed with the heparinized blood as perfusate. This rate is an indication of low vascular resistance; when the intestinal vasculature is isolated, this resistance may persist during the remainder of the perfusion period.

## Viability Criteria

Histological examination of the intestine after 5 hours of perfusion [355] indicated that the integrity of the vascularly perfused tissue was well preserved. Cross-sections through the duodenum showed that the brush border of the intestinal wall and the base epithelium of the duodenum were well preserved after 5 hours of perfusion [150,355]. Additional parameters used for determining the viability of vascularly perfused intestinal preparations include the perfusion flow rate, uptake of glucose by tissue, and production of lymph and continued satisfactory oxygen consumption. In general, the preparation can be maintained viable for at least 5 hours, as indicated by its gross and microscopic appearance and by the continued oxygen consumption, pelistaltic motility, water transport, and vascular responsiveness to norepinephrine. The perfused intestine is capable of glucose transport, and in most experiments lymph flow continues without reduction for 5 hours [355]. The rate of lymph flow is increased by infusing intraluminal fluid and by increasing venous pressure. An additional parameter used for evaluating the perfused intestinal preparation is fat transport and lipoprotein biosynthesis [355]. Morphological examination of the intestinal tissue by light and electron microscopy [150,355] is useful for ascertaining viability.

## Applications

The effects of bolus intraarterial doses of heroin and other stimulant drugs were studied [242] in vascularly perfused isolated segments of canine dog small intestine. Heroin caused dose-related increases in intraluminal pressure similar to those caused by morphine. Perfusion with Krebs bicarbonate solution containing naloxone selectively abolished intestinal responses to heroin. Perfusion with cinanserin, a 5-HT antagonist, decreased intestinal responses to 5-HT and heroin without affecting responses to dimethylphenyl–piperazinium or bethanechol. Atropine antagonized the contractile responses. The authors concluded that heroin interacts with a conventional opiate receptor in the intestine and that the intestinal stimulatory effect of heroin is mediated by the release of endogenous 5-HT, which activates intramural cholinergic neurons. In other studies, perfused canine intestines were used to evaluate the proposed *k* receptor agonists such as ethylketoxyclazocine, nalorphine, and bremazocine [131] and concluded that *k* receptors did not appear to be involved in the contractile response of the canine small intestine to opioids.

The effects of hydrolytic products of food digestion on jejunal mucosal injury and restitution were assessed in anesthetized rats [167]. Mucosal epithelial integrity was continuously monitored by measuring the blood-to-lumen clearance of $^{51}$Cr-labeled ethylenediaminetetraacetic acid (EDTA). Perfusion of the lumen with hydrolyzed casein (3%) or glucose (150 m*M*) did not affect $^{51}$Cr-EDTA clearance compared with saline controls. By contrast, perfusion with emulsified lipids (20-m*M* sodium taurocholate and 10- to 40-m*M* oleic acid) increased $^{51}$Cr-EDTA clearance in a dose-dependent manner. This increase in clearance of $^{51}$Cr-EDTA could be restored to normal if infusion of the lipid emulsion was terminated and saline infusion was resumed. Histological evaluation of jejunal mucosa indicated that the epithelial lining of the villous tips was damaged during lipid infusion and that restitution of the lining occurred within 50 minutes after the resumption of saline perfusion. Because the concentrations of the nutrients used in this study were similar to those measured in postprandial chyme, these findings suggest that the intestinal epithelium is injured and restitutes during the normal course of digestion and absorption of a meal.

Isolated small bowel from the duodenojejunal junction with intact superior mesenteric vein (SMV) and artery (SMA) was perfused with KRB solution intraluminally in rat [262]. After all the branches from the aorta and portal vein were cannulated and perfused with Krebs–dextran solution, the isolated perfused bowel was transferred immediately to a chamber with constant temperature and humidity. After B4 endotoxin (Difco™) was added to intravascular perfusate, changes in endotoxin perfused bowel were compared to the control bowel perfused without the endotoxin. Active transport of D-glucose was decreased. Entry of water into the gut was not affected. Lactic acid level in the intravascular fluid was significantly high, which correlated with the decreased pH of this fluid. Although oxygen consumption was not changed, carbon dioxide accumulation in the intravascular fluid was significant. The intestinal perfusion technique was employed to demonstrate a carrier-mediated process in the elimination of specific optical isomer of ofloxacin, a fluoroquinolone [270]. When intestinal elimination of ofloxacin was studied using *in situ* isolated intestines [270], it was shown that elimination was favored for the R-(+) form of the molecule. Similar results were found with another fluoroquinolone, ciprofloxacin, in previous work. P-glycoprotein appears to be involved in the intestinal elimination of fluoroquinolones in rats.

Isolated vascularly perfused intestinal preparations have been used infrequently in toxicological investigations. It is not readily apparent as to why this is so, but difficulty developing the techniques necessary to maintain a viable perfused preparation and the requirement for a thoroughly elaborate set of equipment must be deterrents to the use of this technique. The preparation should prove especially useful in studies relating to drug absorption and the interaction of drugs relative to intestinal absorption and transport mechanisms. It might also be useful for evaluating the role of intestinal metabolism and overall disposition of various drugs and toxic chemicals. The preparation should also prove useful for testing the reported intestinal elimination of a number of chlorinated hydrocarbon compounds via the luminal surface. All of these aspects relate to toxicology of a particular test drug. Finally, the effect of toxic chemicals on the absorption of nutrients, the generation and preservation of the mucosal cell lining, and various drug-metabolizing enzymes can also be evaluated using the perfused intestinal preparations. The effect of toxic chemicals on endogenous biochemical parameters that are related to the intermediary metabolism of the intestine itself can also be investigated.

## Isolated Perfused Pancreas

The pancreas is a highly vascular endocrine organ in which the anatomy of the blood supply lends itself to isolated vascular perfusion. Compared to the isolated islet incubation [171], pancreatic slices [185], and tissue fragments [66], the superfusion method of Burr et al. [57] for the isolated perfused pancreas is the most satisfactory and ideal approach to study the interrelations between pancreatic and other hormones as well as the effect of toxic chemicals on pancreatic function. The perfused pancreas is preferred over all the other tissue preparations for these studies in view of the following advantages offered by the isolated perfused pancreas preparation [190]. Superficial and deep islets are equally provided with oxygen, which is constantly replenished, and the substrates and effectors arrive at the cell in a physiologically normal way. The islets remain in an anatomical relation with other cells and tissues of the organs, including blood vessels and nerves. Only extrinsic nerve control is lost, and many intrinsic factors that actively regulate the function of the gland may be preserved in an isolated perfused pancreatic preparation. Maintenance of the cellular integrity of the tissue can be confirmed at the end of the experiment by light and electron microscopic examination. It is also possible to have a control and experimental period of study in the same pancreas preparation; such controls in other *in vitro* systems require multiple incubations.

Of greater significance might be the ability to prevent the exocrine secretions of the pancreas from coming back into contact with the islet cells, as cannulation of the segment of duodenum into which the exocrine secretions drain allows collection of these secretions separately. Thus, the enzymes and other secretions produced are kept separate from the cells that produce them, and the well-known digestive effect of pancreatic enzymes on the pancreatic tissue is avoided. Despite the overwhelming superiority of vascularly perfused pancreatic preparations, there has been a lag in the development of perfusion techniques with this endocrine tissue. This lag can be attributed to the technical difficulties encountered with surgical preparations and the heterogeneity of the tissue in the isolated pancreatic preparation.

The location of the pancreatic duct entering the duodenum might be of some significance, especially in studies related to exocrine function such as protein synthesis. The pancreatic ducts are multiple, and their entry into the intestine is variable [54]. Doerr and Becker [78] reported the main pancreatic duct as emptying into the bile duct rather than into the duodenum in the rat. In the guinea pig and rabbit [78], the main pancreatic duct enters the duodenal tract. The general rule is that in carnivores the ducts empty near or into the common bile duct, whereas in herbivores they empty more distally into the duodenum.

A limited literature is available on the technique of isolated perfused pancreatic preparation, although excellent descriptions of feline [54,83] and canine [126,140, 157] pancreas preparations are available. The principal descriptions are those of Grodsky et al. [115], Sussman et al. [320], and Loubatieres et al. [190]. The following is a description of a combination of the better features of all three perfusion preparations.

### Apparatus

The apparatus used by Sussman et al. [320] is adequate for perfusion of the isolated pancreas and is described here. The chamber of the perfusion apparatus is based on the original description of that used by Miller et al. [218]. The details of the apparatus are given in Figure 38.11. Perfusate is pumped from the reservoir through Silastic®

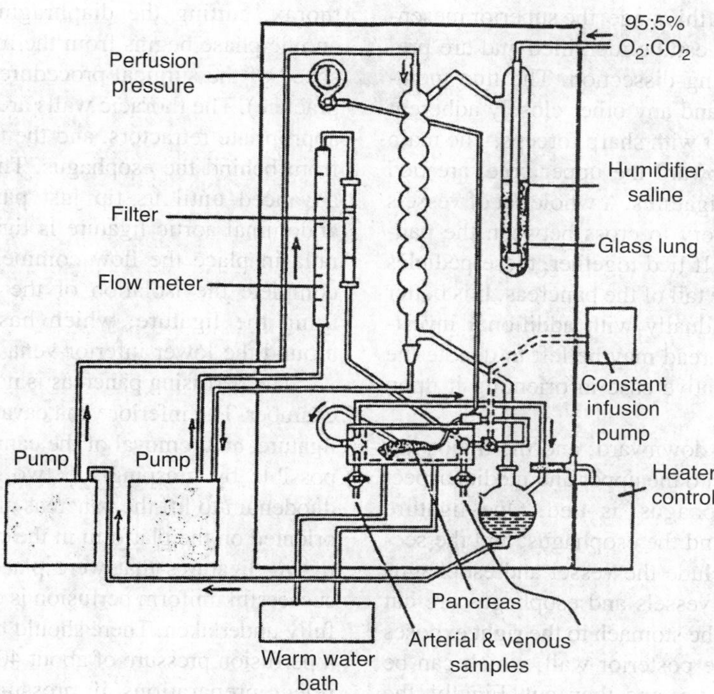


**FIGURE 38.11** Apparatus for perfusion of an isolated pancreas preparation, described by Sussman et al. [273]. It is a modification of the apparatus described by Miller et al. [181] and adapted to the perfusion of the rat pancreas. (From Rhoades, R.A., *Environ. Health Perspect.*, 16, 73–75, 1976. with permission.)

tubing to a multibulb glass lung for oxygenation. The perfusate is then pumped by means of a peristaltic pump through a flow transducer, after which it courses through a filter to the enclosed perfusion chamber. The perfusate enters the pancreatic circulation through the arterial cannula, and the pressure is measured in millimeters of mercury using an aneroid manometer. The arterial blood may be sampled through another, similar arrangement (Figure 38.11), where $O_2/CO_2$ (95:5) is directed to the lung after bubbling it through a humidifier. The pancreas is kept moist with either the same solution used as perfusate or normal saline. The pancreas is perfused using KRB-buffered solution (pH 7.4) containing glucose and BSA, as described for the liver, which is circulated through the glass lung for oxygenation. A pediatric plastic cannula (1.5-mm o.d.) is used to cannulate the arterial side with an attachment for measuring intraluminal pressure. The venous cannula is PE tubing (2-mm o.d.). A duodenal cannula similar to the one described earlier for perfusion of the intestine would be adequate.

## Surgical Procedure

The animal may be starved overnight to facilitate the operation by depleting the omental fat. Although this method facilitates surgical procedures, the investigator may wish to avoid starving the animal as starvation is known to cause increased toxicity of some chemicals, which might also mean that pancreatic effects are altered. The animal is anesthetized by appropriate means (e.g., pentobarbital, 50 mg/kg i.p.). A lateral incision and a midline incision through the skin and linea alba and through the skin over the inferior thorax are made as described for the surgical procedure to isolate the liver. The intestine is moved to the animal's left and covered with a wet saline gauze, and the aorta with the superior mesenteric branch is clearly made visible. The descending colon is easily identified as it remains attached to the lower posterior abdominal wall by a short mesentery. The fine layer of connective tissue is cut along the length of the colon in a plane in which no blood vessels are found, thereby mobilizing the lower gut, which is later removed. Using blunt-end dissection scissors, the pancreas is separated from the overlying colon. The jejunum is ligated and severed just behind its pancreatic attachment. This ligature will be useful for orientation later. The distal jejunum has a copious blood supply, and its vessels are tied for the last 0.5 inches of the distal segment to facilitate its later removal.

At this time, the superior mesenteric vein, which runs into the portal vein, can be seen over the surface of the pancreas. The superior mesenteric vein is tied just below the pancreas with double ligatures and cut between them. Any other attachments of the pancreas to the descending colon are ligated and cut. At this stage, the superior mesenteric artery and celiac axis can be identified and are preserved through the following dissection. The fine membrane covering the spleen and any other closely adherent tissue is carefully picked off with sharp forceps. The main splenic vessels, entering toward the upper pole, are tied twice and cut between the ligatures. A whole set of vessels curves through the mesentery to cross between the pancreatic tail and the spleen. If tied together, these pedicles can bunch the vessels in the tail of the pancreas. It is better to tie each of them individually with additional investments of time. A marker thread may be left to denote the tail of the pancreas for relative ease in orienting it upon its isolation.

The stomach is pulled downward, and the major left gastric artery, which runs into the upper and medial aspect of stomach near the esophagus, is tied. One ligature encloses both the vessels and the esophagus, and the second and third ligatures include the vessel and esophagus separately higher up. The vessels and esophagus are cut between ligatures. Lifting the stomach to the right exposes vascular connections to the posterior wall, which can be tied, using a single distal tie, and then cut. Finally, the pylorus is tied twice and cut between the ligatures, releasing the stomach, which can now be checked. At this stage, the animal can be heparinized by means of an injection via the inferior vena cava. The right renal pedicle (containing artery, vein, and ureter) is cleared with blunt dissection and a ligature is passed behind the artery and vein with curved forceps. Double ligatures allow the vessels to be cut and the kidney removed.

By careful dissection, the aorta is exposed at this level to clear it from the inferior vena cava between the left renal artery and superior mesenteric artery. A ligature passed behind the aorta at this level can now be tied. Three loose ligatures are passed around the portal vein as it leaves the pancreas for the liver. The uppermost ligatures should include all structures of the portal tract (vein, hepatic artery, bile duct), whereas the other two include only the vein. Cannulation is delayed until the aorta has been cleared along its length, preserving the celiac and mesenteric branches. This step entails passing ligatures around and tying the lumbar arteries. The aorta is free from the inferior vena cava from the level of the diaphragm to the ligature below the left renal artery. Loose ligatures are positioned around the aorta just below the diaphragm, avoiding the origin of the celiac artery.

The portal vein can be cannulated in a retrograde manner by tying the uppermost ligatures first and making an incision in the anterior wall of the vein. The cannula (made from PE-140 to PE-100 tubing) is filled with heparinized perfusion medium before insertion. When flow through a portal cannula is ensured, aortic cannulation is performed. The rat is turned around with its head toward the technician, and a midline incision is made through the thorax, cutting the diaphragm through the edge. The anoxic phase begins from the moment of entering, so the speed of the surgical procedure is critical (it comes with practice). The thoracic walls are spread apart by means of appropriate retractors, and the thoracic aorta is separated from behind the esophagus. The cannula is inserted and advanced until its tip just passes the diaphragm. The abdominal aortic ligature is tightened, and with the cannula in place the flow commences. The final step is to complete the isolation of the pancreatic circulation by tying the ligature, which has already been prepared, around the lower inferior vena cava.

The perfusing pancreas is now transported to the organ chamber. The inferior vena cava is cut distal to the last tied ligature, and removal of the cannulated pancreas should be possible by grasping the two cannulas and the loop of duodenum to lift the pancreas clear of the rat. The organ is oriented on the platform in the perfusion chamber by identifying ligatures that were placed during the preparation; successful uniform perfusion is obtained if this step is carefully undertaken. There should be a flow of 2 mL/min with a perfusion pressure of about 40 mmHg. It may be wise to reject preparations if pressures above 80 mmHg are obtained, as such pressure usually results in further deterioration of the perfused pancreatic preparation. The entire surgical procedure takes roughly 60 to 75 minutes, assuming familiarity with the surgical techniques [67].

For the method of Grodsky et al. [101], perfusion of the pancreas is performed together with perfusion of the stomach, spleen, and duodenum as a unit; they are then removed from the rat through a vertical abdominal incision and transferred to the prewarmed chamber. The celiac axis is cannulated for the inflow and the portal vein for the outflow. A peristaltic pump supplies the perfusion medium to the celiac axis at a pressure of 40 to 100 mmHg, and the flow is adjusted to 10 mL/min. The principal disadvantage of Grodsky's procedure is that it involves perfusion of additional tissues so ascribing a particular biochemical function to the pancreas would be more difficult when perfusion is used. The principal advantage is that the surgical procedure is much simpler. Procedures for the cannulation and perfusion of feline and canine pancreas are available. Figure 38.12 shows the constant perfusion system of the blood-perfused canine pancreas [139].

## Perfusion Media

Grodsky et al. [101] used whole rat blood mixed with an artificial medium as a perfusate, but they, as well as others who have utilized whole blood, encountered hemolysis of the red blood cells, which resulted in complications of the perfusion. Hence, the use of artificial media has become more popular for perfusion of the pancreas. The medium containing 4% dextran in Krebs–Henseleit

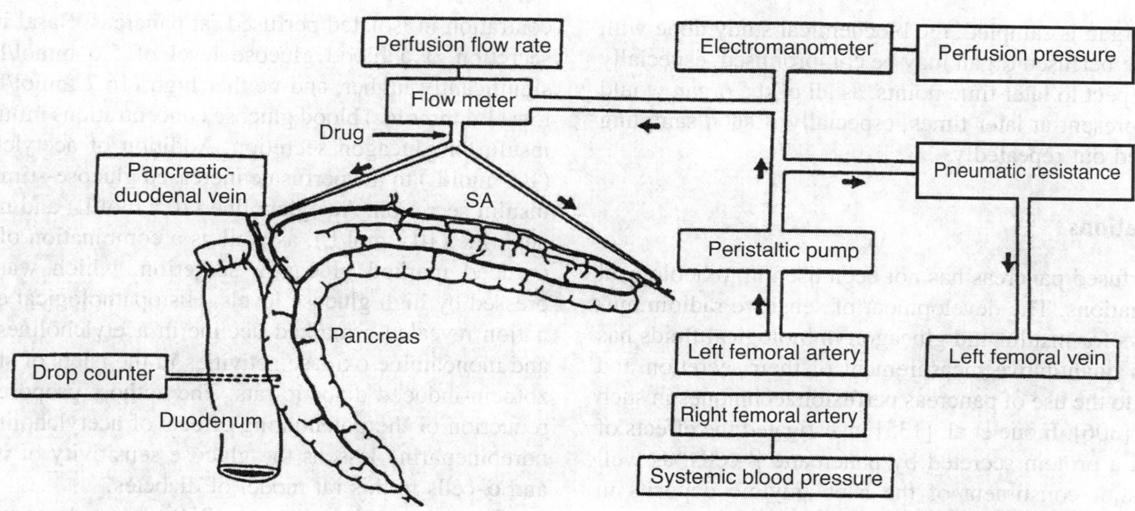


**FIGURE 38.12** Constant perfusion system of the blood–perfused canine pancreas preparation. Arterial cannulas are inserted into the pancreaticoduodenal (PDA) and splenic (SA) arteries. (From Hadasova, E. et al., *Exp. Toxicol. Pathol.*, 51, 330–334, 1999. With permission.)

medium prepared and gassed with $O_2/CO_2$ (95:5) is satisfactory. This medium is the standard one mentioned in the subsequent work of Grodsky as well as by many other investigators. Alternatively, 4% human serum albumin or BSA can be used instead of dextran with no apparent difference in insulin production, medium flow, and rate of circulation of perfusion [67]. The pH is adjusted to 7.4, and the addition of sodium bicarbonate and adequate bubbling with the $O_2/CO_2$ ensure maintenance of the appropriate pH. If the pH seems to fluctuate during the experiment, the addition of sodium bicarbonate solution to the perfusate helps stabilize its pH.

Khayambashi and Lyman [152] used a medium containing fresh rat plasma diluted 1:1 with Ringer's saline or O.9% saline with satisfactory results. Costiner et al. [67] used diluted heparinized rat blood as perfusate to support satisfactory pancreatic function. Advantages of perfusing pancreas with whole blood have been pointed out [119], and the advantage of the larger animal size has been noted by several investigators when using autologous blood [44,119,140,157].

## Viability Criteria

Several parameters of physiological and biochemical function were reported by Grodsky et al. [101]. Histological examinations at the end of the perfusion period indicated that well-preserved granules diminished in number roughly in proportion to the measured release of insulin into the medium [101]. Oxygen consumption by the perfused pancreas can be used as a comparatively rough indicator of viability. However, in the preparation of Grodsky et al. [101], because other tissues are involved in the perfusion circulation, oxygen consumption may not be an exclusive indicator of viability of the perfused pancreas. In the preparation of Sussman and Vaughn [320], because less peripheral tissue is involved, oxygen consumption might represent a better viability criterion. In any case, production of insulin by the preparation, consumption of glucose, and analysis of the exocrine secretion collected through the duodenal cannula can be used as adequate criteria for the viability of the perfused pancreas. The pH of the circulating perfusate can be monitored and adjusted if necessary by regulating the $O_2/CO_2$ when oxygenating the perfusate. Second, appropriate amounts of sodium bicarbonate solution can be added to adjust the pH of the circulating perfusate.

Amylase formation has been shown to be linear for 40 minutes in the preparation used by Khayambashi and Lyman [152], and an amylase assay might be a reasonable parameter for viability; however, they observed that the rate of amylase production fell beyond 40 minutes despite the linearity of the perfusion flow rate. This decrease may be a reflection of the discharge of existing enzymes rather than the inability for the isolated perfused organ to synthesize enzymes. In the work of Khayambashi and Lyman [152], the composition of the perfusion medium was demonstrated to have a definite effect on the secretion of the amylase type of enzymes by the pancreas; hence, such a criterion may have to be established in conjunction with other parameters of perfused preparation. Sampling of the tissue to demonstrate effects that may be reflected in the medium or related to a particular treatment of perfusion is clearly important in the assessment of the perfused organ. In this connection, it may be pointed out that the dendritic form of the rat pancreas lends itself to sampling when a single arm of the pancreas is tied off and perfusion is continued in the remaining tissue. If, however, a portion of the organ is sampled, the biochemical study done with the same perfused organ may be compromised, especially with respect to later time points, as all of the organ would not be present at later times, especially if such sampling is carried out repeatedly.

## Applications

The perfused pancreas has not been used in toxicological investigations. The development of sensitive radioimmunoassays for insulin and glucagon in biological fluids has enabled quantitative measurement of their secretion and has led to the use of pancreas perfusion techniques in such studies [306]. Inoue et al. [135] investigated the effects of amylin, a protein secreted by pancreatic β-cells, as well as a major constituent of the islet amyloid deposits in patients with non-insulin-dependent diabetes mellitus, on the release of insulin and glucagon. Amylin (100 n*M*) did not alter glucose-stimulated secretion of insulin but significantly inhibited arginine-stimulated secretion of insulin. Amylin did not alter the release of glucagon from perfused rat pancreas in response to 16.7-m*M* glucose and 10-m*M* arginine. These findings suggest that amylin may modulate the secretion of insulin from pancreatic β-cells. Clearly, the perfused pancreas can be used to evaluate the effect of drugs and drug interactions that affect pancreatic function. Drugs not having, exclusive action on the pancreas may mutually interact within the body to produce an action on pancreatic function [259]. Yelich [358] studied the effect of epinephrine on insulin and glucagon secretion from the endotoxic rat pancreas. Pancreases from control and endotoxin treated rats were perfused with 240 mg/dL glucose in the presence or absence of 13.6-n*M* epinephrine. The absolute ability of epinephrine to inhibit glucose-induced immunoreactive glucagon secretion was similar for both control and endotoxic pancreases; however, because endotoxic pancreases hypersecrete insulin, the relative ability of epinephrine to inhibit insulin secretion was reduced in endotoxic pancreas. Although epinephrine did not appreciably alter immunoreactive glucagon secretion, it prevented a progressive decrease in its secretion. The results partially explain endotoxic-induced hyperinsulinemia and also demonstrate a possible role for epinephrine with regard to the production of elevated glucagon levels during endotoxicosis. Blech et al. [35] tested the *in vitro* and *in vivo* effects of gold thioglucose on insulin and glucagon secretion of the isolated perfused Wistar rat pancreas. The authors concluded that gold thioglucose reacts primarily on the hypothalamus and modulates the reactivity of the pancreas in a permanent manner via the nervous system. To elucidate the mechanism of insensitivity of hormone secretion to glucose in streptozotocin-induced diabetic rat islets, Ito et al. [137] investigated the effects of acetylcholine and borepinephrine on insulin and glucagon secretion to changes in glucose concentration in isolated perfused rat pancreas. Basal insulin secretion at a blood glucose level of 5.6 mmol/L was significantly higher, and neither high (16.7 mmol/L) nor low (1.4 mmol/L) blood glucose concentrations influenced insulin or glucagon secretion. Addition of acetylcholine ($10^{-6}$ mol/L) to the perfusate increased glucose-stimulated insulin secretion. Acetylcholine ($10^{-6}$ mol/L) and norepinephrine ($10^{-7}$ mol/L), as well as a combination of both, induced marked glucagon secretion, which was suppressed by high glucose levels. Histopathological examination revealed a marked decline in acetylcholinesterase and monoamine oxidase activities in the islets of streptozotocin-induced diabetic rats. The authors proposed that reduction of the potentiating effects of acetylcholine and norepinephrine lessens the glucose sensitivity of islet β- and α-cells in this rat model of diabetes.

Peterson and Fujimoto [259] have demonstrated increased pancreatic secretory activity after exposure to such agents as ethanol and $CCl_4$. The mechanism of such enhanced pancreatic secretory activity is not understood. Use of a perfused pancreatic preparation obtained from control and treated animals may be useful in the elucidation of the underlying mechanisms. Nordback et al. [240] investigated the role of acetaldehyde in the pathogenesis of acute alcoholic pancreatitis. They tested their hypothesis that acute alcoholic pancreatitis may be initiated by acetaldehyde in the presence of active xanthine oxidase in their experiments in which xanthine dehydrogenase was converted to xanthine oxidase by a period of ischemia and the infusion of acetaldehyde (250 mg/hour after 2 hours of ischemia). The authors concluded that in the presence of active xanthine oxidase acetaldehyde can initiate acute alcoholic pancreatitis and that oxic oxygen metabolites play an important intermediary role.

Examples of the use of isolated perfused pancreas for a number of pharmacological [18,113,115,254], physiological [83,254], and toxicological [44,157] studies are available. The studies of Kimura et al. [157] provide one example of the utility of the perfused pancreas in mechanistic toxicological investigations. Although steroid administration has long been suspected of causing acute pancreatitis, clinical and experimental data have failed to firmly establish the association or to uncover a pathogenic mechanism. The acute effects of large doses of methylprednisolone on the pancreas were evaluated utilizing an isolated, perfused, canine pancreas [157]. Using a dose of 200 mg of methylprednisolone, no significant differences were observed between the control and steroid-treated preparations over a 4-hour perfusion period. When the dose of methylprednisolone was increased to 400 mg, again there were no significant differences in gross appearance, weight gain, or serum amylase during a 3-hour perfusion period; however, pancreatic secretion was initially depressed in the steroid-treated preparations. Following a maximal secretory stimulus (secretin), secretion markedly increased during the fourth hour of perfusion but again was significantly less in the steroid-treated glands. Viscosity of pancreatic secretions was significantly increased in the steroid-treated glands. These studies suggest a mild inhibitory effect of steroids on pancreatic secretion, which might be mediated through an increase in viscosity. Another example is provided by the studies of Broe and Cameron [44].

Controlled clinical trials have documented the development of acute pancreatitis in 5% of patients receiving azathioprine for Crohn's disease, by far the highest incidence of drug-induced pancreatitis recorded to date. The isolated, perfused canine pancreas was used to evaluate the effects of azathioprine on the pancreas. No significant changes in gross appearance, weight, or serum amylase were observed in azathioprine-treated glands compared to controls. Azathioprine administration, however, resulted in a twofold increase in secretory volume and bicarbonate output as well as a profound depression of trypsin output compared to controls. These preliminary studies demonstrate that azathioprine has a marked effect on pancreatic function in this model [46].

## ACKNOWLEDGMENTS

The author is the recipient of the 1988 Burroughs Wellcome Toxicology Scholar Award and support from the Kitty DeGree Eminent Scholar's Endowment and Enhancement fund. This effort was supported in part by a U.S. Public Health Service grant (HL-20622) from the National Heart, Lung, and Blood Institute; by a U.S. Environmental Protection Agency grant (R-811072); and by a Starter grant from The Burroughs Wellcome Fund. The excellent assistance of Lillian Brown in the preparation of this manuscript is greatly appreciated. The fine contributions of my former and present colleagues included in the work referenced in this manuscript are also appreciated.

## QUESTIONS

1. Describe a technique for preparing an isolated perfused ventilated lung using positive pressure and negative pressure for ventilation. Which blood vessels are cannulated and why?
2. Describe the isolated perfused liver preparation. What are the advantages and disadvantages of *in situ* and isolated liver perfusion techniques?
3. Substantiate the statement that Langendorff preparations are useful for mechanistic studies in toxicology.
4. Why are isolated perfused organ techniques useful in toxicological investigations? Compare and contrast advantages and disadvantages of isolated vascularly perfused organ techniques and isolated cell incubations.
5. Describe three toxicological studies using isolated perfused brain preparations.
6. Using suitable examples, illustrate how isolated perfused kidneys are useful in investigating mechanisms of toxicity.
7. Describe an isolated perfused pancreatic preparation. Design a study to investigate the mechanism of pancreatotoxic action of a suspected toxicant.
8. Design an investigation of toxicity of an intestinal toxicant using an isolated perfused rat intestine preparation.
9. What are the advantages and disadvantages of using artificial medium vs. whole or diluted blood perfusate for toxicokinetic perfusion studies with isolated perfused organs?
10. Compare and contrast the use of one-pass vs. recirculating perfusion systems for toxicokinetic and mechanistic studies.

## REFERENCES


1. Abraham, R., Dawson, W., Grasso, P., and Goldberg, L. (1968): Lysosomal changes associated with hyperoxia in the isolated perfused rat liver. *Exp. Mol. Pathol.*, 8: 370–387.
2. Adin, C.A., Croker, B.P., and Agarwal, A. (2005): Protective effects of exogenous bilirubin on ischemia-reperfusion injury in the isolated, perfused rat kidney. *Am. J. Physiol. Renal Physiol.*, 289:F778–F784.
3. Agarwal, A.K. and Mehendale, H.M. (1986): Effect of chlordecone on carbon tetrachloride induced increase in calcium uptake in isolated perfused rat liver. *Toxicol. Appl. Pharmacol.*, 83:342–348.
4. Aiba, T., Itoga, Y., Shimizu, H., Tanigawara, Y., and Hori, R. (1994): Renal handling of tobramycin in the isolated perfused rat kidney. *J. Pharm. Sci.*, 83:723–726.
5. Akahira, M., Hara, A., Hashizume, H., Nakamura, M., and Abiko, Y. (1998): Protective effect of prazocin on the hydrogen peroxide-induced derangements in the isolated perfused rat heart. *Life Sci.*, 62:1755–1766.
6. Akerboom, T., Schneider, I., Von Dahl, S., and Sies, H. (1991): Cholestasis and changes of portal pressure caused by chlorpromazine in the perfused rat liver. *Hepatology*, 13:216–221.
7. Ambs, S. and Neumann, H. G. (1996): Acute and chronic toxicity of aromatic amines studied in the isolated perfused rat liver. *Toxicol. Appl. Pharmacol.*, 139:186–194.
8. An, Mi-Y. et al. (2002): Development and evaluation of a new apparatus for continuous perfusion of isolated perfused pig heart. *J. Vet. Sci.*, 3:219–232.
9. Anderson, P.G., Digerness, S.B., Sklar, J.L., and Boor, P.J. (1990): Use of the isolated perfused heart for evaluation of cardiac toxicity. *Toxicol. Pathol.*, 18:497–510.
10. Andjus, R.K., Suhara, K., and Stoviter, H.A. (1967): An isolated, perfused rat brain preparation, its spontaneous and stimulated activity. *J. Appl. Physiol.*, 22:1033–1039.

11. Andrews, W.H.H., Hecker, R., and Maegraith, B.G. (1956): The action of adrenaline, noradrenaline, acetylcholine and histamine on the perfused liver of the monkey, cat and rabbit. *J. Physiol. (Lond.)*, 132:509–521.
12. Angevine, L.S. and Mehendale, H.M. (1980): Chlorphentermine uptake by isolated perfused rabbit lung. *Toxicol. Appl. Pharmacol.*, 52:336–346.
13. Angevine, L.S. and Mehendale, H.M. (1980): Effect of chlorphentermine on the pulmonary disposition of 5-hydroxytryptamine in the isolated perfused rat lung. *Am. Rev. Respir. Dis.*, 122:891–898.
14. Angevine, L.S. and Mehendale, H.M. (1982): Effect of chlorphentermine treatment on 5-hydroxytryptamine disposition in the isolated perfused rat lung. *Fundam. Appl. Toxicol.*, 21:306–312.
15. Aronson, C.E. (1976): Effects of thyroxine pretreatment and calcium on the isolated, perfused rat heart. *Arch. Int. Pharmacodyn.*, 222:351–360.
16. Aronson, C.E. and Serlick, E.R. (1976): Effects of prolonged perfusion time on the isolated perfused rat heart. *Toxicol. Appl. Pharmacol.*, 38:479–488.
17. Aronson, C.E. and Serlick, E.R. (1977): Effects of chlorpromazine on the isolated rat heart. *Toxicol. Appl. Pharmacol.*, 39:157–176.
18. Arredondo, A.A., Chaudhuri, B., Kar, R., Crist, K.A., Thomford, M.R., and Chaudhuri, P.K. (1990): Isolated perfusion of pancreas with mitomycin C. *Am. J. Surg.*, 159:569–574.
19. Auda, S.P., Kesner, L., Butt, K., and Dountz, S.L. (1975): Continuous single pass perfusion of the isolated kidney. *Trans. Am. Soc. Artif. Intern. Organs*, 21:84–88.
20. Autian, J. (1975): *In vitro* toxicity testing gains strength. *Forum Adv. Toxicol.*, 8:1–2.
21. Aviado, D.M. (1975): Drug action, reaction, and interaction. II. Teratogenic cardiopathies. *J. Clin. Pharmacol.*, 15: 641–655.
22. Baath, L. and Almen, T. (1989): Reduction of the risk of ventricular fibrillation in the isolated rabbit heart by small additions of electrolytes to non-ionic monomeric contrast media. *Acta. Radiol.*, 30:327–333.
23. Baciewicz, F.A., Arredondo, M., Chaudhuri, B., Crist, K.A., Basilius, D., Bandyopadhyaah, S., Thomford, N.R., and Chaudhuri, P.K. (1991): Pharmacokinetics and toxicity of isolated perfusion of lung with doxorubicin. *J. Surg. Res.*, 50:124–128.
24. Baglioni, D.E. (1910): Stoffwecheseluntershuganagen and Uberlebenden organen. *Han. Biol. Arb. Meth.*, 3:364.
25. Bahlmann, J., Giebisch, G., Ochwady, B., and Schoeppe, W. (1967): Micropuncture study of isolated perfused rat kidney. *Am. J. Physiol.*, 221:77–82.
26. Bakhle, T.S., Reynard, A.M., and Vane, J.R. (1969): Metabolism of the angiotensins in isolated perfused tissues. *Nature*, 222:956–958.
27. Barrett, J.P., Ingenito, A.J., and Procita, L. (1969): A brain perfusion technique adapted for the study of drugs which may affect the peripheral circulation through a central action. *J. Pharmacol. Exp. Ther.*, 170:199–209.
28. Bauman, A.W., Clarkson, T.W., and Miles, E.M. (1963): Functional evaluation of isolated perfused rat kidney. *J. Appl. Physiol.*, 18:1239–1246.
29. Bautista, A.P. and Spitzer, J.J. (1990): Superoxide anion generation by *in situ* perfused rat liver: effect of *in vivo* endotoxin. *Am. J. Physiol.*, 259:G907–G912.
30. Belinsky, S.A., Popp, J.A., Kauffman, F.C., and Thurman, R.G. (1984): Trypan blue uptake as a new method to investigate hepatotoxicity in periportal and pericentral regions of the liver lobule: studies with allyl alcohol in the perfused liver. *J. Pharmacol. Exp. Ther.*, 230:755–760.
31. Berggren, M., Dawson, J., and Moldéus, P. (1984): Glutathione biosynthesis in the isolated perfused rat lung: utilization of extracellular glutathione. *FEBS Lett.*, 176: 189–192.
32. Bernard, C. E., Magid, A. A., Yen, T. S., and Hoener, B. A. (1997): Mitigation of nitrofurantoin-induced toxicity in the perfused rat lung. *Hum. Exp. Toxicol.*, 16:727–732.
33. Bernstein, E.F. (1971): Evaluation of mechanical systems used in perfusion. In: *Perfusion Techniques*, edited by E. Diszfalusy, pp. 44–73. Karolinska Institute, Stockholm.
34. Bingham, E., Warshawsky, D., and Niemeier, R. W. (1978): Metabolism of benzo(*a*)pyrene in the isolated perfused rabbit lung following *N*-dodecane inhalation. In: *Carcinogenesis: Mechanisms of Tumor Production and Carcinogenesis*, Vol. 2, edited by T.J. Slaga, A. Sivak, and R.K. Boutwell, pp. 509–516. Raven Press, New York.
35. Blech, W., Bierwolf, B., Weiss, I., and Ziegler, M. (1986): *In vitro* and *in vivo* effect of gold thioglucose on the insulin and glucagon secretion of the isolated perfused rat pancreas. *Biomed. Biochim. Acta*, 45:507–522.
36. Bleehan, N.M. and Fisher, R.B. (1954): The action of insulin on the isolated rat heart. *J. Physiol. (Lond.)*, 123: 260–276.
37. Block, E.R. and Cannon, J.K. (1978): Effect of oxygen exposure on lung clearance of amines. *Lung*, 155:287–295.
38. Bock, K.W., Forhling, W., and Scholte, W. (1972): Activity and stability of microsomal mixed function oxidase and NAD glycohydrolase in isolated perfused rat liver. *Naunyn Schmiedebergs Arch. Pharmacol.*, 273:193–203.
39. Bond, J.A. and Mauderly, J.L. (1984): Metabolism and macromolecular covalent binding of [$^{14}$C]-1-nitropyrene in isolated perfused and ventilated rat lungs. *Cancer Res.*, 44: 3924–3929.
40. Boyd, C.A.R., Parsons, D.S., and Thomas, A.V. (1968): The presence of $K^+$-dependent phosphatase in intestinal epithelial cell brush borders isolated by a new method. *Biochim. Biophys. Acta*, 150:723–726.
41. Bowman, R.H. and Maack, T. (1972): Glucose transport by the isolated rat kidney. *Am. J. Physiol.*, 222:1499–1504.
42. Brauer, R.W., Pessotti, R.L., and Pizzolato, P. (1951): Isolated rat liver preparation: bile production and other basic properties. *Proc. Soc. Exp. Biol. Med.*, 78:174–185.
43. Brodie, T.G. (1903): The perfusion of surviving organs. *J. Physiol. (Lond.)*, 29:266–275.
44. Broe, P.J. and Cameron, J.L. (1983): Azathioprine and acute pancreatitis: studies with an isolated perfused canine pancreas. *J. Surg. Res.*, 34:159–163.
45. Brown, P.C., Thurman, R.G., Belinsky, S.A., and Kauffman, F.C. (1991): Effect of allyl alcohol on xanthine dehydrogenase activity in the perfused rat liver. *Toxicol. Lett.*, 58:1–6.

46. Bruck, R., Prigozin, H., Krepel, Z., Rotenberg, P., Schechter, Y., and Bar-Meir, S. (1991): Vanadate inhibits glucose output from isolated perfused rat liver. *Hepatology*, 14:540–544.
47. Brull, L. and Louis-Bar, D. (1957): Toxicity of artificially circulated heparinized blood on the kidney. *Arch. Int. Physiol. Biochem.*, 65:470–476.
48. Bullock, G., Eakins, M.N., Sawyer, B.C., and Slater, T.F. (1974): Studies on bile secretion with the aid of the isolated perfused rat liver. I. Inhibitory action of sporidesmin and icterogenin. *Proc. R. Soc. Lond.*, 186:333–356.
49. Bunc, M., Drevensek, G., Budinha, M., and Suput, D. (1999): Effects of equinatoxin II from *Actinia equina* (L.) on isolated rat heart: the role of direct cardiotoxic effects in equitoxin II lethality. *Toxicon*, 37:109–123.
50. Burks, T. F. (1974): Vascularly perfused isolated perfused intestine. In: *Proceedings of the Fourth International Symposium on Gastrointestinal Motility*, edited by Daniel, E.E. Mitchell Press, Vancouver.
51. Burr, I.M., Stauffacher, W., Balant, L., Renold, A.E., and Grodsky, G. (1969): Dynamic aspects of proinsulin release from perfused rat pancreas. *Lancet*, 2:882–883.
52. Camus, Ph. and Mehendale, H.M. (1986): Pulmonary sequestration of amiodarone and desethylamiodarone. *J. Pharmacol. Exp. Ther.*, 237:867–873.
53. Cardieux, A., Masse, S., and Sirois, P. (1983): Effect of asbestos in the metabolism of vasoactive substances in isolated perfused guinea pig lungs. *Environ. Health Perspect.*, 51:287–291.
54. Case, R.M., Harper, A.A., and Scratcherd, T. (1968): Water and electrolyte secretion by the perfused cat pancreas. *J. Physiol. (Lond.)*, 196:133–149.
55. Cesarone, C.F., Fugassa, E., Gallo, G., Voci, A., and Orunesu, M. (1984): Collagenase perfusion rat liver induces DNA damage and DNA repair in hepatocytes. *Mutat. Res.*, 141:113–116.
56. Chapman, N.D., Saint George, S., and Ishida, T. (1960): Small volume perfusion system of the isolated rat liver. *J. Appl. Physiol.*, 15:128–136.
57. Charles, J.M., Abou-Donia, M.B., and Menzel, D.B. (1978): Absorption of paraquat and diquat from the airways of the perfused rat lung. *Toxicology*, 9:59–67.
58. Chikaoka, Y. and Tamura, M. (1991): 7-Ethoxycoumarin deethylation activity in perfused isolated rat brain. *J. Biochem.(Tokyo)*, 125:634–640.
59. Chiong, M.A., Berenzy, G.M., and Winton, T.L. (1978): Metabolism of the isolated perfused rabbit heart. I. Responses to anoxia and reoxygenation. II. Energy stores. *Can. J. Physiol. Pharmacol.*, 56:844–856.
60. Choo, E. F., Angus, P. W., and Morgan, D. J. (1999): Effect of cirrhosis on sulphation by the isolated perfused liver. *J. Hepatol.*, 30:498–502.
61. Chuté, A.L. and Smyth, D.H. (1939): Metabolism of the isolated perfused cat's brain. *Q. J. Exp. Physiol.*, 29: 379–394.
62. Colantoni, A., de Maria, N., Caraceni, P., Bernardi, M., Floyd, R. A., and Van Thiel, D. H. (1998): Prevention of reoxygenation injury by sodium salicylate in isolated-perfused rat liver. *Free Radic. Biol. Med.*, 25:87–94.
63. Cohrs, P., Jaffe, R., and Meesen, H. (1958): *Pathologie der Laboratoriumstiere*, Vol. 1. Springer-Verlag, Berlin.
64. Cojocel, C., Docius, N., Maita, K., Smith, J.H., and Hook, J.B. (1984): Renal ultrastructural and biochemical injuries induced by aminoglycosides. *Environ. Health Perspect.*, 57:293–299.
65. Cook, D.R., Howell, R.E., and Gillis, C.N. (1982): Xanthine oxidase-induced lung injury inhibits removal of 5-hydroxytryptamine from the pulmonary circulation. *Anesth. Analg.*, 61:666–670.
66. Coore, H.G. and Randle, P.J. (1964): Regulation of insulin secretion studied with pieces of rabbit pancreas incubated *in vitro*. *Biochem. J.*, 93:66–78.
67. Costiner, E., Ghiea, D., Simionescu, L., and Oprescu, M. (1975): Modified technique of perfusion of isolated rat pancreas tested by insulin release after glucose administration. *Endocrinol. Exp.*, 9:197–204.
68. Conhaim, R.L., Watson, K.E., Heisey, D.M., Leverson, G.E., and Harms, B.A. (2004): Thromboxane receptor analog, W-46619, redistributes pulmonary microvascular perfusion in isolated rat lungs. *J. Appl. Physiol.*, 96: 245–252.
69. Dalbey, W. and Bingham, E. (1978): Metabolism of trichloroethylene by the isolated perfused lung. *Toxicol. Appl. Pharmacol.*, 43:267–277.
70. Davis, M.E. and Mehendale, H.M. (1979): Absence of metabolism of morphine during accumulation by isolated perfused rabbit lung. *Drug Metab. Dispos.*, 7:425–428.
71. Dawson, J.R., Vapakangas, K., Jernstrom, B., and Moldeus, P. (1984): Glutathione conjugation by isolated lung cells and isolated perfused lung: effect of external glutathione. *Eur. J. Biochem.*, 138:439–443.
72. Deglin, S.M., Deglin, J.M., and Chung, E.K. (1977): Drug-induced cardiovascular diseases. *Drugs*, 14:29–40.
73. DeMello, G. and Maack, T. (1976): Nephron function of the isolated perfused rat kidney. *Am. J. Physiol.*, 231: 1699–1707.
74. De Wildt, D.J. and Speijers, G.J.A. (1984): Influence of dietary rapeseed oil and erucic acid upon myocardial performance and hemodynamics in the rat. *Toxicol. Appl. Pharmacol.*, 74:99–108.
75. De Wildt, D.J., De Jong, Y., Hillen, F.C., Steerenberg, P.A., and Van Hoesel, Q.G.C.M. (1985): Cardiovascular effects of doxorubicin-induced toxicity in the intact Lou/M Wsl rat and isolated heart preparations. *J. Pharmacol. Exp. Ther.*, 235:234–240.
76. Diczfalusy, E. (1971): *Perfusion Techniques*. Karolinska Institute, Stockholm.
77. Dirks, V.B., Seibert, A., Sperling, G., and Krieglstein, J. (1984): Comparison of the effects of piracetam and methohexital on brain energy metabolism. *Azneimittelforschung*, 34:258–266.
78. Doerr, W. and Becker, V. (1958): Bauchspeicheldruse (pancreas): In: *Pathologie der Laboratariumstierre*, Vol. 1, edited by P. Cohrs, R. Jaffe, and Hm. Messen, p. 130. Springer-Verlag, Berlin.
79. Dovrak, K.J., Braun, W.E., Magnusson, M.O., Stowe, N.T., and Banowsky, L.H.W. (1976): Effect of high methylprednisolone on the isolated perfused canine kidney. *Transplantation*, 21:149–157.

80. Dubois, R.S., Vaughn, G.D., and Roy, C.C. (1968): Isolated rat small intestine with intact circulation. In: *Organ Perfusion and Perservation*, edited by J.C. Norman, pp. 863–868. Appleton-Century-Crofts, New York.
81. Dunbar, J.R., Delucia, A.J., and Bryant, L.R. (1984): Glutathione status of isolated rabbit lungs: effects on nitrofurantoin and paraquat perfusion with normoxic and hyperoxic ventilation. *Biochem. Pharmacol.*, 33:1343–1348.
82. Eisman, B., Knipe, P., McColl, H., and Orloff, M.J. (1961): Isolated liver perfusion for reducing blood ammonia. *Arch. Surg.*, 83:356–363.
83. Elisha, E.E., Hutson, D., and Seratcherd, T. (1984): The direct inhibition of pancreatic electrolyte secretion by noradrenaline in the isolated perfused cat pancreas. *J. Physiol. (Lond.)*, 351:77–85.
84. Embden, G. and Glaüssner, K. (1902): Uber den ortder atherschwefelsaurebildung im tierkorpen. *Hoffmlisters Beituug 2 Chem. Phys. Band.*, 1:310–327.
85. Enser, M.B., Kunz, F., Borenstanjn, J., Opie, L.H., and Robinson, D.S. (1967): Metabolism of triglyceride fatty acids by perfused rat heart. *Biochem. J.*, 104:306–317.
86. Esser, E., Loschen, G., and Flohe, L. (1991): Phorbol ester cardiotoxicity: are $O_2$ radicals involved? *Arch. Toxicol.*, 65: 335–339.
87. Evans, G.H., Wilkinson, G.R., and Shand, D.G. (1973): The disposition of propranolol. IV. A dominant role of tissue uptake in the dose dependent extraction of propranolol by the perfused rat liver. *J. Pharmacol. Exp. Ther.*, 186:447–456.
88. Ferrari, R., Ceconi, C., Currello, S., Guarnieri, C., Caderara, C.M., Albertini, A., and Visioli, O. (1985): Oxygen-mediated myocardial damage during ischemia and reperfusion: role of the cellular defenses against oxygen toxicity. *J. Mol. Cell Cardiol.*, 17:937–945.
89. Fink, V.C., Holzl, J., Riegger, H., and Kriegelstein, J. (1984): Effects of valtrate on the EEG of the isolated perfused rat brain. *Azneimittelforschung*, 34:170–174.
90. Fitzpatrick, Jr., J.H. and Gilboe, D.D. (1982): Effects of nitrous oxide on the cerebrovascular tone, oxygen metabolism, and electroencephalogram of the isolated perfused canine brain. *Anesthesiology*, 57:480–484.
91. Fogel, W.A., Bieganski, T., Schayer, R.W., and Maslinski, C. (1981): Involvement of diamine oxidase in catabolism of $^{14}$C-putrescine in mice *in vivo* with special reference to the formation of $\gamma$-aminobutyric acid. *Agents Actions*, 11: 679–684.
92. Fonteles, M.C., Pillion, D.J., Jeske, A.H., and Leibach, F.H. (1976): Extraction of glutathione by the isolated perfused rabbit kidney. *J. Surg. Res.*, 21:169–174.
93. Forth, W. (1968): Eisen und Kobalt-Resorption am perfundierten Dendarmasegment. In: *3 Konfder Gesellschaft fur Biologische Chemie*, edited by W. Staib and R. Scholz, pp. 242–254. Springer-Verlag, Berlin.
94. Foy, B. D. Toxopeus, C., and Frazier, J. M. (1999): Kinetic modeling of slow dissociation of bromosulphothalein from albumin in perfused rat liver: toxicological implications. *Toxicol. Sci.*, 50:20–29.
95. Fox, R.B. (1984): Prevention of granulocyte-mediated oxidant injury in rats by a hydroxyl radical scavenger, dimethylthiourea. *J. Clin. Invest.*, 74:1456–1464.
96. Franke, H. and Weiss, C. (1976): The $O_2$ supply of the isolated cell-free perfused rat kidney. *Adv. Exp. Med. Biol.*, 75:425–432.
97. Funakoshi, T., Ishibe, Y., Okazaki, N., Miura, K., Liu, R., Nagai, S., and Minami, Y. (2004): Effect of re-expansion after short-period lung collapse on pulmonary capillary permeability and pro-inflammatory cytokine gene expression in isolated rabbit lungs. *Br. J. Anaesth.* 92: 558–563.
98. Gad, S.C., Leslie, S.W., and Acosta, D. (1979): Inhibitory actions of butylated hydroxytoluene on isolated ileal, atrial and perfused heart preparations. *Toxicol. Appl. Pharmacol.*, 49:45–52.
99. Gamble, W.J., Conn, P.A., Edalji-Kumer, A., Pleuge, R., and Monroe, R.G. (1970): Myocardial oxygen consumption of blood-perfused, isolated supported rat heart. *Am. J. Physiol.*, 219:604–612.
100. Gareis, M., Hashem, A., Bauer, J., and Gedek, B. (1986): Identification of glucuronide metabolites of T-2 toxin and diacetoxyscerpinol in the bile of isolated perfused rat liver. *Toxicol. Appl. Pharmacol.*, 84:168–172.
101. Geiger, A. (1958): Correlation of brain metabolism and function by the use of a brain perfusion *in situ*. *Physiol. Rev.*, 38:1–20.
102. Geiger, A. and Magnes, J. (1947): The isolation of the cerebral circulation and the perfusion of the brain in the living cat. *Am. J. Physiol.*, 149:517–537.
103. Gerber, G.B. and Remy-Defraigne, J. (1966): DNA metabolism in perfused organs. II. Incorporation in DNA and catabolism of thymidine at different levels of substrate by normal and x-irradiated liver and intestine. *Arch. Int. Physiol. Biochem.*, 74:785–794.
104. Gibbs, J.E., Rashid, T., and Thomas, S.A. (2003): Effects of transport inhibitors and additional anti-HIV drugs on the movement of lamivudine (3TC) across the guinea pig brain barriers. *J. Pharmacol. Exp. Ther.*, 306: 1035–1041.
105. Gilboe, D.D., Betz, A.L., and Langebartel, D.A. (1973): A guide for the isolation of the canine brain. *J. Appl. Physiol.*, 34:534–537.
106. Gillis, C.N. and Iwasawa, Y. (1972): Technique for measurement of norepinephrine and 5-hydroxytryptamine uptake by rabbit lung. *J. Appl. Physiol.*, 33:404–408.
107. Gillis, C.N. and Roth, J.A. (1976): Pulmonary disposition of circulating vasoactive hormones. *Biochem. Pharmacol.*, 25:2547–2553.
108. Giovannini, L., Migliori, M., De Pietro, S., Taccola, D., Panichi, V., Bertelli, A. A., and Bertelli, A. (1999): L-Propionyl carnitine reduces toxicity correlated to cyclosporine-induced intracellular ATP concentrations. *Drugs Exp. Clin. Res.*, 25:173–177.
109. Gollan, F. and McDermott, J. (1979): Effect of skeletal muscle relaxant dantrolene sodium on the isolated, perfused heart. *Proc. Soc. Exp. Biol. Med.*, 160:42–45.
110. Gonzalez-Martin, G., Dominguez, A. R., and Guevara, A. (1997): Pharmacokinetics and hepatotoxicity of diclofenac using an isolated perfused rat liver. *Biomed. Pharmacother.*, 5:170–175.
111. Gonmori, K., Prasada Rao, K.S., and Mehendale, H.M. (1986): Pulmonary synthesis of 5-hydroxytryptamine in isolated perfused rabbit and rat lung preparations. *Exp. Lung Res.*, 11:295–306.

112. Goodman, M.N., Parilla, R., and Toews, C.J. (1973): Influence of fluorocarbon emulsions on hepatic metabolism in perfused rat liver. *Am. J. Physiol.*, 225:1384–1388.
113. Goto, Y., Seino, Y., Note, S., and Imura, H. (1980): The dual effect of alloxan modulated by 3-*O*-methylglucose or somatostatin on insulin secretion in the isolated perfused rat pancreas. *Horm. Metab. Res.*, 12:140–143.
114. Gow, P. J. et al. (1999): Uptake and excretion of sodium taurocholate by the isolated perfused neonatal sheep liver. *J. Pharm. Sci.*, 88:445–449.
115. Grodsky, G.M., Batts, A.A., Bennett, L.L., Veella, C., McWilliams, N.B., and Smith, D.F. (1963): Effects of carbohydrates on secretion of insulin from isolated rat pancreas. *Am. J. Physiol.*, 205:638–644.
116. Habib, M. P. and Clements, N. C. (1995): Effects of low-dose hydrogen peroxide in the isolated perfused rat lung. *Exp. Lung Res.*, 21:95–112.
117. Hadasova, E., Charvatova, Z., Nerusilova, K., Hykosova, M., and Zelenkova, O. (1999): Influence of pretreatment with immunosuppressants on *O*-demethylation of dextro-methorphan in isolated perfused rat liver. *Exp. Toxicol. Pathol.*, 51:330–334.
118. Hale, Jr., P.W. and Poklis, A. (1984): Thioridazine-5-sulfoxide cardiotoxicity in the isolated, perfused rat heart. *Toxicol. Lett.*, 21:1–8.
119. Hashimoto, K., Satoh, S., and Takeuchi, O. (1971): Effect of dopamine on pancreatic secretion in the dog. *Br. J. Pharmacol.*, 43:739–746.
120. Hauge, A. (1968): Conditions governing the pressor response to ventilation hypoxia in isolated perfused rat lungs. *Acta Physiol. Scand.*, 72:33–44.
121. Heaton, J.D., McAnalley, B.H., Gardiner, T.H., and Johnson, A.R. (1982): Uptake and release of $^{14}$C-morphine by pulmonary endothelium and cultured pulmonary endothelial cells. *Gen. Pharmacol.*, 13:105–110.
122. Heavner, J. E., Shi, B., Inners-McBride, K., Asimakis, G., Wang, M. J., and McIntyre, D. C. (1998): Cocaine cardiotoxicity differs markedly in isolated hearts of two strains of rats exhibiting phenolic differences in sensitivity to seizures. *Life Sci.*, 63:625–633.
123. Hein, H., Krieglstein, J., and Stock, R. (1975): The effects of increased glucose supply and thiopental anesthesia on energy metabolism of the isolated perfused rat brain. *Naunyn Schmiedebergs Arch. Pharmacol.*, 289:399–407.
124. Hemingway, A. (1931): Some observations on the perfusion of the isolated kidney by a pump. *J. Physiol. (Lond.)*, 71:201–213.
125. Hems, R., Ross, B.D., Berry, M.N., and Krebs, H.A. (1966): Gluconeogenesis in the perfused rat liver. *Biochem. J.*, 101:284–292.
126. Herman-Taylor, J. (1973): The isolated perfused canine pancreas. In: *Isolated Organ Perfusion*, edited by H.D. Ritchie and J.D. Hardcastle, pp. 171–190. University Park Press, Baltimore, MD.
127. Herrero, I., Torras, J., Carrera, M., Castells, A., Pasto, L., Gil-Vernet, S., Alsina, J., and Grinyo, J. M. (1995): Evaluation of a preservation solution containing fructose-1,6-diphosphate and mannitol using isolated perfused rat kidney: comparison with Euro-Collins and University of Wisconsin solutions. *Nephrol. Dial. Transplant.*, 10:519–526.
128. Hidaka, T., Furuno, H., Inokuchi, T., and Ogura, R. (1990): Effects of diethyl maleate (DEM), a glutathione depletor, on prostaglandin synthesis in the isolated perfused spleen of rabbits. *Arch. Toxicol.*, 64:103–108.
129. Hilliker, K.S., Imlay, M., and Roth, R.A. (1984): Effects of monocrotaline treatment on norepinephrine removal by isolated, perfused rat lungs. *Biochem. Pharmacol.*, 33: 2692–2695.
130. Hilliker, K.S. and Roth, R.A. (1985): Injury to the isolated, perfused lung by exposure *in vitro* to monocrotaline pyrrole. *Exp. Lung Res.*, 8:201–212.
131. Hirning, L.D., Porreca, F., and Burks, T.F. (1985): Mu, but not kappa, opioid agonists induce contractions of the canine small intestine *ex vivo*. *Eur. J. Pharmacol.*, 109:49–54.
132. Hoffmann, M.J., Ji, S., Hedli, C.C., and Snyder, R. (1999): Metabolism of ($^{14}$C) phenol in the isolated perfused liver. *Toxicol. Sci.*, 49:40–47.
133. Horizonte, B. and Brazil, M.G. (2003): Effects of propafenone associated with propofol on myocardial contractivity, heart rate, coronary flow, and the incidence of arrhythmia in isolated hearts of rats. *Arqi. Bras. Cardiol.*, 82:88–93.
134. Inagaki, M. and Yamura, M. (1993): Preparation and optical characteristics of hemoglobin-free isolated perfused rat head *in situ*. *J. Biochem. (Tokyo)*, 113:650–657.
135. Inoue, K., Hiramatsu, S., Hisatomi, A., Umeda, F., and Nawata, H. (1993): Effects of amylin on the release of insulin and glucagon from the perfused rat pancreas. *Horm. Metab. Res.*, 25:135–137.
136. Itakura, N., Fisher, A.B., and Thurman, R.G. (1977): Cytochrome P450-linked *p*-nitroanisole *O*-demethylation in the perfused lung. *J. Appl. Physiol.*, 43:238–245.
137. Ito, K., Hirose, H., Maruyama, H., Fukamachi, S., Tashiro, Y., and Saruta, T. (1995): Neurotransmitters partially restore glucose sensitivity of insulin and glucagon secretion from perfused streptozotocin-induced diabetic rat pancreas. *Diabetologia*, 38:1276–1284.
138. Iwamoto, K., Watanabe, J., Araki, K., Satoh, M., and Deguchi, N. (1985): Reduced hepatic clearance of propranolol induced by chronic carbon tetrachloride treatment in rats. *J. Pharmacol. Exp. Ther.*, 234:470–475.
139. Iwatsuki, K., Ikeda, K., and Chiba, S. (1982): Effects of nitroprusside on pancreatic juice secretion in the blood perfused canine pancreas. *Eur. J. Pharmacol.*, 79:53–60.
140. Jacobs, F.A. (1968): Continuous radioactivity monitoring of perfusion in the small intestine of the intact animal. *Adv. Tracer Methodol.*, 4:255–272.
141. Jaffe, D.R., Hassal, C.D., Gandolfi, A.J., and Brendel, K. (1985): Production of DNA single strand breaks in rabbit renal tissue after exposure to 1,2-dichlorovinylcysteine. *Toxicology*, 35:25–33.
142. Johannesen, J., Lie, M., and Kiil, F. (1977): Renal energy metabolism and sodium reabsorption after 2,4-nitrophenol administration. *Am. J. Physiol.*, 233:207–217.
143. Johnson, A., Phillips, P., Hocking, D., Tsan, M.F., and Ferro, T. (1989): Protein kinase inhibitor prevents pulmonary edema in response to $H_2O_2$. *Am. J. Physiol.*, 256:H1012–1022.
144. Joshi, U.M., Dumas, M., and Mehendale, H.M. (1986): Glutathione turnover in perfused rabbit lung. *Biochem. Pharmacol.*, 35(19):3409–3412.

145. Joshi, U.M., Prasada Rao, K.S., and Mehendale, H.M. (1987): Glutathione status in constituted physiological fluid containing albumin. *Int. J. Biochem*., 19:1129–1135.
146. Junod, A.F. (1972): Accumulation of $^{14}$C-imipramine in isolated perfused rat lungs. *J. Pharmacol. Exp. Ther.*, 183: 182–187.
147. Kanekal, S., Plopper, C., Morin, D., and Buckpitt, A. (1991): Metabolism and cytotoxicity of naphthalene oxide in the isolated perfused mouse lung. *J. Pharmacol. Exp. Ther.*, 256:391–401.
148. Kapfhammer, J. (1927): Die leber im Stoffwechzel. In: *Handbuch der Biochemie*, Vol. 9, 2nd ed., edited by K. Oppenheimer, pp. 98–150. G. Fischer, Jena.
149. Karmazyn, M. and Moffat, M.P. (1985): Toxic properties of arachidonic acid in normal, ischemic and reperfused hearts: indirect evidence for free radical involvement. *Prostaglandins Leukotrienes Med*., 17:251–264.
150. Kavin, H., Levin, N.W., and Stanley, M.M. (1967): Isolated perfused rat small bowel-technique: studies of viability, glucose absorption. *J. Appl. Physiol*., 22:604–611.
151. Kennedy, A. L., Lastbom, L., Skarping, G., Dalene, M., Ryrfeldt, A., Moldeus, P., and Brown W. E. (1995): Analysis of the reactivity of ($^{14}$C) toluene diisocyanate (TDI) in an isolated, perfused lung model. *Chem. Biol. Interact.*, 98:167–183.
152. Khayambashi, H. and Lyman, R.L. (1969): Secretion of rat pancreas perfused with plasma from rats fed soybean trypsin inhibitor. *Am. J. Physiol*., 217:646–651.
153. Khedun, S.M., Maharaj, B., Leary, W.P., and Lockett, C.J. (1992): The effect of hexane on the ventricular fibrillation threshold of the perfused rat heart. *Toxicology*, 71: 145–150.
154. Kilbinger, H. and Krieglstein, J. (1974): Applicability of the isolated perfused rat brain for studying central cholinergic mechanisms. *Naunyn Schmiedebergs. Arch. Pharmacol*., 285:407–411.
155. Kim, D.-H. and Akera, T. (1984): Effects of myocardial hypoxia on digitalis-induced toxicity in the isolated heart of guinea pigs and cats. *Eur. J. Pharmacol*., 104:303–312.
156. Kim, J.H. and Miller, K.L. (1969): The functional significance of changes in activity of the enzymes, tryptophan pyrrolase and tyrosine transaminase after induction in intact rats and isolated perfused rat liver. *J. Biol. Chem*., 244:1410–1416.
157. Kimura, T., Zuidema, G.D., and Cameron, J.L. (1979): Steroid administration and acute pancreatitis studies with an isolated, perfused canine pancreas. *Surgery*, 85:520–524.
158. Kitakaze, M., Weisman, H.F., and Marban, E. (1988): Contractile dysfunction and ATP depletion after transient calcium overload in perfused ferret hearts. *Circulation*, 77: 689–695.
159. Knowlton, F.P. and Starling, E.H. (1912): The influence of variations in temperature and blood pressure on the performance of the isolated mammalian heart. *J. Physiol. (Lond.)*, 44:206–219.
160. Komai, T., Yamamoto, F., Tanaka, K., Ichikawa, H., Shibata, T., Koide, A., Nakashima, N., Ohashi, T., and Kawashima, Y (1991): Harmful effects of inotropic agents on myocardial protection. *Ann. Thorac. Surg*., 52:927–933.
161. Kopp, S.J., Daar, A.A., Prentice, R.C., Tow, J.P., and Feliksik, J.M. (1986): $^{31}$P-NMR studies of the intact perfused rat heart: a novel analytical approach for determining functional-metabolic correlates, temporal relationships, and intracellular actions of cardiotoxic chemicals nondestructively in an intact organ model. *Toxicol. Appl. Pharmacol*., 82:200–210.
162. Koschier, F.J., Gigliotti, P.J., and Hong, S.K. (1980): The effect of *bis*(*p*-chlorophenyl) acetic acid on the renal function on the rat. *J. Environ. Pathol. Toxicol*., 4:209–217.
163. Koster, J. F., Slee, R.G., Essed, C.E., and Stam, H. (1985): Studies on canine hydroperoxide-induced lipid peroxidation in the isolated perfused rat heart. *J. Mol. Cell. Cardiol*., 17:701–708.
164. Kraupp, O., Adler-Kastner, L., Niessner, H., and Plank, B. (1967): The effects of starvation and acute and chronic alloxan diabetes on myocardial substrate levels and on liver glycogen in the rat *in vivo*. *Eur. J. Biochem*., 2:197–214.
165. Krieglstein, G., Krieglstein, J., and Stock, R. (1972): Suitability of the perfused rat brain for studying effects on cerebral metabolism. *Naunyn Schmiedebergs. Arch. Pharmacol*., 275:124–134.
166. Krieglstein, J. and Stock, R. (1973): Comparative study of the effects of chloral hydrate and trichloroethanol on cerebral metabolism. *Naunyn Schmiedebergs. Arch. Pharmacol*., 277:323–332.
167. Krietys, P.R., Specian, R.D., Grisham, M.B., and Tso, P. (1991): Tejunal mucosal injury and restitution: role of hydrolytic products of food digestion. *Am. J. Physiol*., 261:G384–G391.
168. Kukan, M., Vajdova, K., Horecky, J., Nagyova, A., Mehendale, H. M., and Trnovec, T. (1997): Effects of blockade of Kupffer cells by gadolinium chloride on hepatobiliary function in cold ischemia-reperfusion injury of rat liver. *Hepatology*, 26:1250–1257.
169. Kulatilake, A.E. (1967): Isolated perfusion of canine and human kidneys. *Br. J. Surg*., 54:877–882.
170. Kvetina, J. and Guaitani, A. (1969): A versatile method for the *in vitro* perfusion of isolated organs of rats and mice with particular reference to liver. *Pharmacology*, 2:65–81.
171. Lacy, P.E. and Kostianowsky, M. (1967): Method for isolation of intact islets of Langerhans from the rat pancreas. *Diabetes*, 16:35–39.
172. Lafranconi, W.M., Ferlan, I., Russell, F.E., and Huxtable, R.J. (1984): The action of equinatoxin, a peptide from the venom of the sea anemone, *Actinia equina*, on the isolated lung. *Toxicology*, 22:347–352.
173. Lafranconi, W.M. and Huxtable, R.J. (1984): Hepatic metabolism and pulmonary toxicity of monocrotaline using isolated perfused liver and lung. *Biochem. Pharmacol*., 33:2479–2484.
174. Lambert, C., Mossiat, C., Tanniere, Z.M., Maupoil, V., and Rochette, L. (1990): Antiarrhythmic effect of amiodarone on doxorubicin acute toxicity working rat hearts *Cardiovasc. Res*., 24:653–658.
175. Langendorff, O. (1895): Untersuchungen am Uberleberden Saügertierzen. *Pflügers Arch. ges. Physiol*., 61:291–332.
176. Lastbom, L., Skarping, G., Moldeus, P., and Ryrfeldt, A. (1997): Hexamethylene diisocyanate (HDI)-induced lung impairment studies in isolated perfused and ventilated guinea pig lungs. *Pharmacol. Toxicol*., 81:85–89.

177. Law, F.C.P. (1978): Metabolism and disposition of 4′-tetrahydrocannabinol by the isolated perfused rabbit lung. *Drug Metab. Dispos.*, 6:154–163.
178. Law, F.C.P., Eling, T.E., Bend, J.R., and Fouts, J.R. (1974): Metabolism of xenobiotics by the isolated perfused lung. *Drug Metab. Dispos.*, 2:433–442.
179. Lazzarino, G., Tavazzi, B., DiPierro, D., and Giardina, B. (1992): Ischemia and reperfusion: effect of fructose 1,6-bisphosphate. *Free Radic. Res. Commun.*, 16:325–329.
180. Leary, W.P.P. and Ledingham, J.G. (1969): Removal of angiotensin by isolated perfused organs of the rat. *Nature*, 222:959–960.
181. Lemaster, J.J., Ji, S., Stemkowski, C.J., and Thurman, R.G. (1983): Hypoxic hepatocellular injury. *Pharmacol. Biochem. Behav.*, 18:455–459.
182. Lessire, F., Gustin, P., Delaunois, A., Bloden, S., Nemmar, A., Vargas, M., and Ansay, M. (1996): Relationship between parathion and paraoxontoxicokinetics, lung metabolic activity, and cholinesterase inhibition guinea pig and rabbit lungs. *Toxicol. Appl. Pharmacol.*, 138: 201–210.
183. Levey, S. and Gast, R. (1966): Isolated perfused rat lung preparation. *J. Appl. Physiol.*, 21:313–316.
184. Levin, N.W., Ryan, W.G., Hayashi, J., and Kark, R.M. (1965): Studies on the isolated perfused rabbit kidney. *S. Afr. J. Med. Sci.*, 30:78–79.
185. Light, A. and Simpson, M.S. (1966): Studies on the biosynthesis of insulin. I. The paper chromatographic isolation of $^{14}$C-labeled insulin from calf-pancreas slices. *Biochem. Biophys. Acta*, 20:251–261.
186. Linas, S.L., Whittenburg, D., and Repine, J.E. (1991): Role of neutrophil derived oxidants and elastase in lipopolysaccharide-mediated renal injury. *Kidney Int.*, 39:618–623.
187. Little, J.R. and Cohen, J.J. (1974): Effect of albumin concentration on function of isolated perfused rat kidney. *Am. J. Physiol.*, 226:512–517.
188. Longonoi, B., Giovannini, L., Migliori, M., Bertelli, A. A., and Bertelli, A. (1999): Cyclosporine-induced lipid peroxidation and propionyl carnitine protective effect. *Int. J. Tissue React.*, 21:7–11.
189. Lorenz-Meyer, H., Roth, H., Elsässer, P., and Hahn, R. (1985): Cytotoxicity of lectins on rat intestinal mucosa enhanced by neuraminidase. *Eur. J. Clin. Invest.*, 15: 227–234.
190. Loubatieres, A., Mariani, M.M., Chapal, J., and Portal, A. (1967): Action penatrice de faibles doses d'adrenaline et de noradrenaline sur l'insulino secretion etadiee sur le pancreas isole et perfuse du rat. *C. R. Soc. Biol. (Paris)*, 161:2578–2586.
191. Lund, M., Kang, L., Tygstrup, N., Wolkoff, A. W., and Ott, P. (1999): Effects of LPS on transport of indocyanine green and alanine uptake in perfused rat liver. *Am. J. Physiol.*, 277:G91–100.
192. Lyons, D.E., Beery, J.T., Lyons, S.A., and Taylor, S.L. (1983): Cadaverine and aminoguanidine potentiate the uptake of histamine *in vitro* in perfused intestinal segments of rats. *Toxicol. Appl. Pharmacol.*, 70:445–458.
193. Manabe, S., Nuber, D.C., and Lin, Y.C. (1991): Zone-specific hepatotoxicity of gossypol in perfused rat liver. *Toxicon*, 29:787–790.
194. Marinelli, A., Pons, D.H., Vreeken, J.A., Nagessen, S.K., Kuppen, P.J., Tjadew, U.R., and Van deVelde, C.J. (1991): High mitomycin C concentration in tumor tissue can be achieved by isolated liver perfusion in rats. *Cancer Chemother. Pharmacol.*, 28:109–114.
195. Martinus, A. M.,Monreiro, H. S., Junior, E. O., Menezes, D. B., and Fonteles, M. C. (1998): Effects of *Crotalus durissus casacavella* venum in the isolated rat kidney. *Toxicon*, 36:1441–1450.
196. Masini, E., Giannella, E., Palmerani, B., Pistelli, A., Gambassi, F., and Mannaioni, P. F. (1989): Free radicals induce ischemia-reperfusion injury and histamine release in the isolated guinea pig heart. *Int. Arch. Appl. Immunol.*, 88: 132–133.
197. Masuda, Y. and Yamamori, Y. (1991): Histological evidence for dissociation of lipid peroxidation and cell necrosis in bromotrichloromethane hepatotoxicity in the perfused rat liver. *Jpn. J. Pharmacol.*, 56:143–150.
198. Masuda, Y. and Nakamura, Y. (1990): Effects of oxygen deficiency and calcium omission on carbon tetrachloride hepatotoxicity in isolated perfused livers from phenobarbital-pretreated rats. *Biochem. Pharmacol.*, 40: 1865–1876.
199. Masumi, S., Moriyama, M., Kannan, Y., Ohta, M., Koshitani, O., Sawamoto, O., and Sugano, T. (1999): Changes in hepatic metabolism in isolated perfused liver during the development of thioacetamide-induced cirrhosis in rats. *Toxicology*, 135:21–31.
200. McCallum, T., Badylak, S.F., Van-Vleet, J.F., and Reed, R.M. (1989): Furazolidone-induced injury in the isolated perfused chicken heart. *Am. J. Vet. Res.*, 50:1183–1185.
201. McCarthy, R.D., Shaw, J.C., and Lakshmanan, S. (1958): Metabolism of volatile fatty acids by the perfused goat liver. *Proc. Soc. Exp. Biol. Med.*, 99:560–564.
202. McDonough, K.H., Brumfield, B.A., and Lang, C.H. (1986): *In vitro* myocardial performance after lethal and nonlethal doses of endotoxin. *Am. J. Physiol.*, 250: H240–H246.
203. Mehendale, H.M. (1976): Uptake and disposition of chlorinated biphenyls by isolated perfused rat liver. *Drug Metab. Dispos.*, 4:124–132.
204. Mehendale, H.M. (1976): Effect of preexposure to kepone on the biliary excretion of polychlorinated biphenyl compounds. *Toxicol. Appl. Pharmacol.*, 36:369–381.
205. Mehendale, H.M. (1977): Mirex-induced impairment of hepatobiliary function: suppressed biliary excretion of imipramine and sulfobromophthalein. *Drug Metab. Dispos.*, 5:56–62.
206. Mehendale, H.M. (1977): Effect of preexposure to kepone on the biliary excretion of imipramine and sulfobrophthalein. *Toxicol. Appl. Pharmacol.*, 40:247–259.
207. Mehendale, H.M. (1978): Pesticide-induced modification of hepatobiliary function; hexachlorobenzene, DDT and toxaphene. *Food Cosmet. Toxicol.*, 16:19–25.
208. Mehendale, H.M. (1980): Aldrin epoxidase activity in the developing rabbit lung. *Pediatr. Res.*, 14:282–285.
209. Mehendale, H.M. (1982): Use of isolated perfused lung in determining pulmonary disposition and potential toxicological significance of inhaled environmental pollutants. *J. Environ. Toxicol. Chem.*, 1:231–244.

210. Mehendale, H.M. (1984): Pulmonary disposition of pneumophilic agents and possible relationship to pulmonary hypertension. *Fed. Proc.*, 43:2586–2591.
211. Mehendale, H.M. and El-Bassiouni, E.A. (1975): Uptake and disposition of aldrin and dieldrin by isolated perfused rabbit lung. *Drug Metab. Dispos.*, 3:543–556.
212. Mehendale, H.M. et al. (1981): The isolated perfused lung: a critical evaluation. *Toxicology*, 21:1–36.
213. Mehendale, H.M., Svensson, S.A., Baldi, C., and Orrenius, S. (1985): Accumulation of $Ca^{2+}$ induced by cytotoxic levels of menadione in the isolated perfused rat liver. *Eur. J. Biochem.*, 149:201–206.
214. Meiring, J., Borm, P.J.A., Bagate, K., Semmler, M., Seitz, J., Takenaka, S., and Kreyling, W.G. (2005): The influence of hydrogen peroxide and histamine on lung permeability and translocation of iridium nanoparticles in the isolated perfused rat lung. *Part. Fibre Toxicol.*, 2: 3–15.
215. Menaouar, A., Anglade, D., Baussand, P., Pelloux, A., Corboz, M., Lantuejeul, S., Benchetrit, G., and Grimbert, F. A. (1997): Chlorine gas induced acute lung injury in isolated rabbit lung. *Eur. Respir. J.*, 10:1100–1107.
216. Merin, R.G. (1978): Myocardial metabolism for the toxicologist. *Environ. Health Perspect.*, 26:169–174.
217. Merker, G., Helling, H.J., Krahl, M., and Aigner, K. (1983): Ultrastructural changes in the dog liver cell after isolated liver perfusion with various cytotoxins. *Recent Results Cancer Res.*, 86:103–109.
218. Miller, L.L., Bly, C.G., Berry, M.N., and Krebs, H.A. (1951): The dominant role of the liver in plasma protein synthesis: a direct study of the isolated perfused rat liver with the aid of lysine-$^{14}C$. *J. Exp. Med.*, 94:431–453.
219. Mitchell, C.J., Bullock, S., and Ross, B.D. (1977): Renal handling of gentamicin and other antibiotics by the isolated perfused rat kidney: mechanism of nephrotoxicity. *J. Antimicrob. Chemother.*, 3:593–600.
220. Miura, K. et al. (1985): Effects of gentamicin on renal function in isolated perfused kidneys from male and female rats. *Toxicol. Lett.*, 26:15–18.
221. Miyazaki, M., Makowka, L., Falk, R.E., Falk, W., Venturi, D., Ambus, U., and Falk, J.A. (1983): Hyperthermochemotherapeutic *in vivo* isolated perfusion of the rat liver. *Cancer*, 51:1254–1260.
222. Monroe, R.G., Larfarge, C.G., Gamble, W.J., Honda, S., and Kevy, S.W. (1968): Ventricular performance and coronary flow of isolated hearts when perfused through isolated lungs and membrane oxygenators. In: *Organ Perfusion and Preservations*, edited by J.C. Norman, pp. 779–979. Appleton-Century-Crofts, New York.
223. Monteiro, H. S., Lima, A. A., and Fonteles, M. C. (1999): Glomerular effects of cholera toxin in isolated perfused rat kidney: a potential role for platelet activating factor. *Pharmacol. Toxicol.*, 85:105–110.
224. Moore, G.K. and Hook, J.B. (1978): Hemodynamic effects of furosemide in isolated perfused rat kidneys. *Proc. Soc. Exp. Biol. Med.*, 158:354–358.
225. Morgan, H.E., Henderson, M.J., Regen, D.M., and Park, C.R. (1961): Regulation of glucose uptake in muscle. I. The effect of insulin and anoxia on glucose transport and phosphorylation in isolated perfused heart of normal rats. *J. Biol. Chem.*, 235:253.
226. Muller, F. (1910): Dir kunslitche Durchbluntung resp. durchspulung von orgen. *Handb. biol. Arb.-Meth.*, 3:327.
227. Nakai, T. (1984): Toxic effects of endotoxin perfusion on isolated rat bowel. *Nippon Geba Gakkai. Zasshi.*, 85: 370–377.
228. Nahas, G. and Trouve, R. (1985): Effects of interactions of natural cannabinoids on the isolated heart. *Proc. Soc. Exp. Biol. Med.*, 180:312–316.
229. Nakazawa, H., Arroyo, C.M., Ichimori, K., Saigusa, Y., Minezaki, K.K., and Pronai, L. (1991): The demonstration of DMPO superoxide adduct upon reperfusion using a law nontoxic concentration. *Free Radic. Res. Commun.*, 14:297–302.
230. Nastainczyk, W. and Ullrich, V. (1978): Effect of oxygen concentration of the reaction of halothane with cytochrome P-450 in liver microsomes and isolated perfused rat liver. *Biochem. Pharmacol.*, 27:387–392.
231. Nazeyrollas, P., Prevost, A., Baccard, N., Manot, L., Devillier, P., and Millart, H. (1999): Effects of amifostine on perfused isolated rat heart and on acute doxorubicin-induced cardiotoxicity. *Cancer Chemother. Pharmcol.*, 43: 227–232.
232. Neely, J.R., Liebermiester, H., Battersby, E.J., and Morgan, H.E. (1967): Effect of pressure development on oxygen consumption by isolated rat heart. *Am. J. Physiol.*, 212: 804–814.
233. Neuman, M.G., Eshchar, J., Cotariu, D., Ben-Sason, R., Ziv, E., Baron, H., and Ishay, J.S. (1985): Hepatotoxicity of hornet's venom sac extract in isolated perfused rat liver. *Acta Pharmacol. Toxicol. (Copenh.)*, 56:133–138.
234. Newton, J.E. and Hook, J.B. (1981): Isolated perfused kidney. *Meth. Enzymol.*, 77:94–105.
235. Nichiitsutsuiji-Uwo, J.M., Ross, B.D., and Dribs, H.A. (1967): Metabolic activation of the isolated perfused rat kidney. *Biochem. J.*, 103:852–862.
236. Nicolaysen, G. (1971): Perfusate qualities and spontaneous edema formation in an isolated perfused lung preparation. *Acta Physiol. Scand.*, 83:563–570.
237. Niemeier, R.W. (1976): Isolated perfused rabbit lung: a critical appraisal. *Environ. Health Perspect.*, 16:67–71.
238. Niemeier, R.W. and Bingham, E. (1972): An isolated perfused lung preparation for metabolic studies. *Life Sci.*, 11: 807–820.
239. Nizet, A. (1975): The isolated perfused kidney: possibilities, limitations and results. *Kidney Int.*, 7:1–11.
240. Nordback, I.H., MacGowan, S., Potter, J.J., and Cameron, J.L. (1991): The role of acetaldehyde in the pathogenesis of acute alcoholic pancreatitis. *Ann. Surg.*, 214:671–678.
241. Norman, J.C., Ed. (1978): *Organ Perfusion and Perservation*. Appleton-Century-Crofts, New York.
242. Northway, M.O. and Burks, T.F. (1979): Indirect intestinal stimulatory effects of heroin: direct action on opiate receptors. *Eur. J. Pharmacol.*, 59:237–243.
243. Novakova, V., Birke, G., Plantin, L.O., and Wretland, A. (1976): A perfluorochemical oxygen carrier (Fluosol-43) in a synthetic medium used for perfusion of isolated rat liver. *Acta Physiol. Scand.*, 98:356–365.
244. O'Brien, R.F., Makarski, J.S., and Rounds, S. (1985): Studies on the mechanism of decreased angiotensin. I. Conversion in rat lungs injured with alpha-naphthylthiourea. *Exp. Lung Res.*, 8:243–259.

245. Ohmiya, Y., Angevine, L.S., and Mehendale, H.M. (1983): Effect of drug-induced phospholipidosis on pulmonary disposition of pneumophilic drugs. *Drug Metab. Dispos.*, 11:25–30.
246. Ohmiya, Y. and Mehendale, H.M. (1979): Uptake and accumulation of chlorpromazine in the isolated perfused rabbit lung. *Drug Metab. Dispos.*, 7:442–443.
247. Ohmiya, Y. and Mehendale, H.M. (1980): *N*-Oxidation of imipramine by isolated perfused rat and rabbit lung. *Life Sci.*, 26: 1411–1421.
248. Ohmiya, Y. and Mehendale, H.M. (1980): Uptake and metabolism of chlorpromazine by rat and rabbit lungs. *Drug Metab. Dispos.*, 8:313–318.
249. O'Neil, J.J. and Tierney, F. (1974): Rat lung metabolism, glucose utilization by isolated perfused lungs and tissue slices. *Am. J. Physiol.*, 226:867–873.
250. Oomen, H.A.P.C. and Chamalarun, R.A.F.M. (1971): Correlation between histological and biochemical parameters of isolated perfused rat liver. *Virchows Arch.*, 8:243–251.
251. Opie, L.H. (1965): Coronary flow rate and perfusion pressure as determinants of mechanical function and oxidative metabolism of isolated perfused rat heart. *J. Physiol. (Lond.)*, 180:529–541.
252. Orton, T.C., Anderson, M.W., Pickett, R.D., Eling, T.E., and Fouts, J.R. (1973): Xenobiotic accumulation and metabolism by isolated perfused rabbit lungs. *J. Pharmacol. Exp. Ther.*, 186:482–497.
253. Otsuki, M., Nakamura, T., Okabayashi, Y., Oka, T., Fuji, M., and Baba, S. (1985): Comparative inhibitory effects of pirenzapine and atropine on cholinergic stimulation of exocrine and endocrine rat pancreas. *Gastroenterology*, 89:408–414.
254. Otsuki, M., Sakamoto, C., Ohki, A., Akabayashi, Y., Suehiro, I., and Baba, S. (1983): Effect of acarbose on exocrine and endocrine pancreatic function in the rat. *Diabetologia*, 24:445–448.
255. Otsuki, S., Watanabe, S., Morimistu, J., and Edamatsu, N. (1967): Regulatory effects of blood constituents on the function and metabolism of the cat brain perfusion experiments. *Acta Med. Okayama*, 21:279–296.
256. Palay, S.L., McGee-Russell, S.M., Gordon, S., and Grillo, M.A. (1962): Fixation of neural tissues for electron microscopy by perfusion with solutions of osmium tetroxide. *J. Cell Biol.*, 12:385–410.
257. Pang, K.S., Yuen, V., Fayz, S., Tekopple, J.M., and Mulder, G.J. (1986): Absorption and metabolism of acetaminophen by the *in situ* perfused rat small intestine preparation. *Drug Metab. Dispos.*, 14:102–111.
258. Parsons, D.D. and Prichards, J.S. (1968): A preparation of perfused small intestine for the study of absorption in amphibia. *J. Physiol. (Lond.)*, 198:405–434.
259. Peterson, R.E. and Fujimoto, J.M. (1976): Increased 'bile duct-pancreatic fluid' flow in rats pretreated with carbon tetrachloride. *Toxicol. Appl. Pharmacol.*, 35:29–39.
260. Pickett, R.D., Anderson, M.W., Orton, T.C., and Eling, T.E. (1975): The pharmaco-dynamics of 5-hydroxytryptamine uptake and metabolism by the isolated perfused rabbit lung. *J. Pharmacol. Exp. Ther.*, 194:545–553.
261. Pitzele, S., Sze, S., and Dosell, A.R.C. (1971): Hypothermic plasma perfusion of the isolated heart. *Surgery*, 70:407–412.
262. Popjak, G. and Beeckmans, M. (1950): Extra-hepatic lipid synthesis. *Biochem. J.*, 47:233–238.
263. Post, C., Anderson, R.G.G., Ryfeldt, A., and Nilsson, E. (1978): Transport and binding of lidocaine by lung slices and perfused lung of rats. *Acta Pharmacol. Toxicol. (Copenh.)*, 43:156–163.
264. Post, C. and Hede, A.R. (1982): Trichloroethylene and halothane inhibit uptake of 5-hydroxytryptamine in the isolated perfused rat lung. *Biochem. Pharmacol.*, 31: 353–358.
265. Poulsen, H.E., Lerche, A., and Skovgaard, L.T. (1985): Acetaminophen metabolism by the perfused rat liver twelve hours after acetaminophen overdose. *Biochem. Pharmacol.*, 34:3729–3733.
266. Powis, G. (1970): Perfusion of rat liver with blood: transmitter overflows and gluco-neogenesis. *Proc. R. Soc. Lond.*, B174:503–515.
267. Prasada Rao, K. S., and Mehendale, H. M. (1987): Precursor utilization of 5-hydroxytryptophan for 5-hydroxytryptamine biosynthesis in isolated perfused rabbit and rat lungs. *Can. J. Physiol. Pharmacol.*, 65:2117–2123.
268. Prasada Rao, K.S. and Mehendale, H.M. (1988): Precursor utilization of [$^{14}$C]-L-tryptophan and [$^{14}$C]-5-hydroxytryptophan for pulmonary biosynthesis of [$^{14}$C]-5-hydroxythryptamine: a review. *Indian J. Pharmacol.*, 18(4):186–196.
269. Preininger, K., Stingl, H., Englisch, R., Furnsinn, C., Graf, J., Waldhausl, W., and Roden, M. (1999): Acute troglitazone action in isolated perfused rat liver. *Br. J. Pharmacol.*, 126:372–378.
270. Rabbaa, H., Dautrey, S., Colas-Linhart, N., Carbon, C., and Farinotti, R. (1996): Intestinal elimination of ofloxacin enantiomers in the rat: evidence of a carrier-mediated process. *Antimicrob. Agents Chemother.*, 40:2126–2130.
271. Radwan, Z. and Henschler, D. (1977): Uptake and rate of metabolism of vinyl chloride by the isolated perfused rat liver preparation. *Int. Arch. Occup. Environ. Health*, 40: 101–110.
272. Raje, R.R. (1980): *In vitro* toxicity of acetone using coronary perfusion in isolated rabbit heart. *Drug Chem. Toxicol.*, 3:333–342.
273. Raje, R.R. (1983): *In vitro* toxicity of *n*-hexane and 2,5-hexanedione using isolated perfused rabbit heart. *J. Toxicol. Environ. Health*, 11:879–884.
274. Rakic, L.M., Zlokovic, B.V., Davson, H., Segal, M.B., Begley, D.J., Lipovac, M.N., and Mitrovic, D.M. (1989): Chronic amphetamine intoxication and the blood–brain barrier permeability to inert polar molecules studied in the vascularity perfused guinea pig brain. *J. Neurol. Sci.*, 94: 41–50.
275. Rao, M.M. and Elmslie, R.G. (1970): A modified technique of isolated pancreatic perfusion. *J. Surg. Res.*, 10:357–362.
276. Rao, S.B. and Mehendale, H.M. (1987): Uptake and disposition of putrescine, spermidine, and spermine by isolated perfused rabbit lungs. *Drug Metab. Dispos.*, 15:189–194.
277. Rao, S.B., Rao, K.S.P., and Mehendale, H.M. (1986): Absence of diamine oxidase activity from rabbit and rat lungs. *Biochem. J.*, 234:733–736.
278. Rasenack, J., Koch, H.K., Lesch, R., and Decker, K. (1980): Hepatotoxicity of D-galactosamine in isolated perfused rat liver. *Exp. Mol. Pathol.*, 32:264–275.

279. Redgeld, F. A., Hofman, G. A., van de Loo, P. G., Koster, A. S., and Noordhoek, J., (1991): Nephrotoxicity of the glutathione conjugate of menadione (2-methyl-1,4-naphthoquinone) in the isolated perfused rat kidney: role of metabolism by gamma-glutamyltranspeptidase and probenecid-sensitive transport. *J. Pharmacol. Exp. Ther.*, 256:665–669.
280. Reichelderfer, M., Pero, B., Lorenzsonn, V., and Olsen, W.A. (1984): Magnesium sulfate-induced water secretion in hamster small intestine. *Proc. Soc. Exp. Biol. Med.*, 176:8–13.
281. Rhoades, R.A. (1976): Perfused lung preparation for studying altered gaseous environments. *Environ. Health Perspect.*, 16:73–75.
282. Rice, A.J., Roberts, R.J., and Plaa, G.L. (1967): The effect of carbon tetrachloride administered *in vivo* on the hemodynamics of the isolated perfused rat liver. *Toxicol. Appl. Pharmacol.*, 11:422–431.
283. Ritchie, H.D. and Hardcastle, J.D. (1973): *Isolated Organ Perfusion*. University Park Press, Baltimore, MD.
284. Roby, K. A. and Shaw, L. M. (1993): Effects of cyclosporine and its metabolites in the isolated perfused rat kidney. *J. Am. Soc. Nephrol.*, 4:168–177.
285. Rosenfeld, S., Sellers, A.L., and Katz, J. (1959): Development of an isolated perfused rabbit kidney. *Am. J. Physiol.*, 196:115–159.
286. Ross, B.D. (1972): *Perfusion Techniques in Biochemistry*. Clarendon Press, Oxford.
287. Ross, B.D., Epstein, F.H., and Leaf, A. (1973): Sodium reabsorption in the perfused rat kidney. *Am. J. Physiol.*, 225:1165–1171.
288. Ross, B.D., Hems, R., and Krebs, H.A. (1967): The rate of gluconeogenesis from various precursors in the perfused rat liver. *Biochem. J.*, 102:942–951.
289. Rossi, N.F., Churchill, P.C., McDonald, F.D., and Ellis, V.R. (1989): Mechanism of cyclosporine A-induced renal vasoconstriction in the rat. *J. Pharmacol. Exp. Ther.*, 250:896–901.
290. Roth, J.A. (1979): Use of the isolated perfused lung in biochemical toxicology. *Rev. Biochem. Toxicol.*, 1:287–310.
291. Rowland, M. (1972): Application of clearance concepts to some literature data on drug metabolism in the isolated perfused liver preparation and *in vivo*. *Eur. J. Pharmacol.*, 17:352–356.
292. Rudolph, A.M. and Hyemann, M.A. (1971): Measurement of flow in perfused organs using microsphere techniques. In: *Perfusion Techniques*, edited by E. Diczfalusy, pp. 112–117. Karolinska Institute, Stockholm.
293. Ryoo, H. and Tarver, H. (1968): Studies on plasma protein synthesis with a new liver perfusion apparatus. *Proc. Soc. Exp. Biol. Med.*, 128:760–772.
294. Sato, Y., Eddy, L., and Hochstein, P. (1991): Comparative cardiotoxicity of doxorubicin and a morpholino-anthracycline derivative. *Biochem. Pharmacol.*, 42:2283–2287.
295. Schanne, F.A.X., Kane, A.B., Young, E.E., and Farber, J.L. (1979): Calcium dependence of toxic cell death: a final common pathway. *Science*, 206:700–702.
296. Schermann, J., Stowe, N., Yarimizu, S., Magnusson, M., and Tingwald, G. (1977): Feedback control of glomerular filtration rate in isolated, blood-perfused dog kidneys. *Am. J. Physiol.*, 223:217–224.
297. Schimassek, H. (1962): Perfusion of rat liver with a semi-synthetic medium and control of liver function. *Life Sci.*, 1:629–637.
298. Schimassek, H. (1963): Metabolite des kohlendydrastoffwechels der isoliert perfundierten rattenleber. *Biochem. Z.*, 336:460–467.
299. Schimassek, H. and Gerok, W. (1965): Control of the levels of free amino acids in plasma by the liver. *Biochem. Z.*, 343:407–415.
300. Schimmel, R.J. and Knobil, E. (1969): Role of free fatty acid in stimulation of gluconeogenesis during fasting. *Am. J. Physiol.*, 217:1803–1808.
301. Schureck, J., Brecht, J.P., Lofert, H., and Hierholzer, K. (1975): The basic requirements for the function of the isolated cell free perfused rat kidney. *Pflügers Arch.*, 354: 349–365.
302. Seeger, W., Bauer, M., and Bhakdi, S. (1984): Staphylococcal alpha-toxin elicits hypertension in isolated rabbit lungs: evidence for thromboxane formation and the role of extracellular calcium. *J. Clin. Invest.*, 74:849–858.
303. Seeger, W., Suttrop, N., Schmidt, F., and Neuhof, H. (1986): The glutathione redox cycle a defense system against hydrogen-peroxide induced prostanoid formation and vasoconstriction in rabbit lungs. *Am. Rev. Respir. Dis.*, 133:1029–1036.
304. Segel, L.D., Rendig, S.V., and Mason, D.T. (1979): Left ventricular dysfunction of isolated working rat hearts after chronic alcohol consumption. *Cardiovasc. Res.*, 13: 136–146.
305. Seiler, K.U., Tamm, G., and Wasserman, O. (1974): On the role of serotonin in the pathogenesis of pulmonary hypertension induced by anorectic drugs: an experimental study in the isolated perfused rat lung. *Clin. Exp. Pharmacol. Physiol.*, 1:463–471.
306. Seiver, B.R. and Whitney, J.E. (1967): Biosynthesis of insulin by the isolated perfused dog pancreas. *Diabetes*, 16:647–651.
307. Sharma, R., Kodavanti, U.P., Smith L.L., and Mehendale, H.M. (1995): The uptake and metabolism of cystamine and taurine by isolated ventilated and perfused rat and rabbit lungs. *Int. J. Biochem. Cell Biol.*, 27:655–664.
308. Silva, P., Rosen, S., Spokes, K., and Epstein, F.H. (1991): Effect of glycine on medullary thick ascending limb injury in perfused kidneys. *Kidney Int.*, 39:653–658.
309. Sklar, J.L., Anderson, P.G., and Boor, P.J. (1991): Allylamine and acrolein toxicity in perfused rat hearts. *Toxicol. Appl. Pharmacol.*, 107:535–544.
310. Skutul, K. (1908): Uber durchstromunsapporate. *Pflügers. Arch.*, 123:249–273.
311. Sladen, G.E. (1968): Perfusion studies in relation to intestinal absorption. *Gut*, 9:624–628.
312. Smith, M.T., Thor, H., and Orrenius, S. (1981): Toxic injury to isolated hepatocytes is not dependent on extracellular calcium. *Science*, 213:1257–1259.
313. Soergel, K.H. (1971): Intestinal perfusion studies: values, pitfalls, and limitations. *Gastroenterology*, 61:261–263.
314. Song, Y., Ma, T., Matthay, M.A., and Verkman, A.S. (2001): Role of aquaporin-4 in airspace to capillary water permeability in intact mouse lung measured by a novel gravimetric method. *J. Gen. Physiol.*, 115:17–27.

315. Southard, J.H., Senzig, K.A., Hoffman, R.M., and Belzer, F.O. (1980): Toxicity of oxygen to mitochondrial respiratory activity in hypothermically perfused canine kidneys. *Transplantation*, 29:459–461.
316. Sperling, F. and Marcus, W.L. (1968): Turpentine-induced histological changes in isolated rat and guinea-pig lungs. *Arch. Int. Pharmacodyn. Ther.*, 175:330–338.
317. Strubelt, O., Deters, M., Pentz, R., Siegers, C. P., and Younes, M. (1999): The toxic and metabolic effects of 23 aliphatic alcohols in the isolated perfused rat liver. *Toxicol. Sci.*, 49:133–142.
318. Summerfield, J.A., Gollan, J.L., and Billing, B.H. (1976): Synthesis of bile acid monophosphates by the isolated perfused rat kidney. *Biochem. J.*, 156:339–345.
319. Sumpio, B.E., Chandry, I.H., and Baue, A.E. (1985): Reduction of the drug-induced nephrotoxicity by ATP-$MgCl_2$. I. Effects of the *cis*-diaminedichloro-platinum-treated isolated perfused kidneys. *J. Surg. Res.*, 38: 429–437.
320. Sussman, K.E. and Vaughn, G.D. (1967): Insulin release after ACTH, glucagon and adenosine 3′,5′-phosphate (cyclic AMP) in the perfused isolated rat pancreas. *Diabetes*, 16:449–454.
321. Sussman, K.E., Vaughn, G.D., and Timmer, R.F. (1966): An *in vitro* method for studying insulin secretion in perfused rat pancreas. *Metabolism*, 15:466–476.
322. Takano, T., Nakata, K., Kawakami, T., Miyazaki, Y., Murakami, M., Seo, Y., and Suzuki, E. (1991): Validation of a toxicity testing model by evaluating oxygen supply and energy state in the isolated perfused rat kidney. *J. Pharmacol. Meth.*, 25:195–204.
323. Takano, T., Miyazaki, Y., and Motohashi, Y. (1983): A method to evaluate the dynamic effects of environmental chemical agents on intracellular functions: the real time observations of changes in the spectra of mitochondrial cytochromes, cytochrome P-450, and catalase and in the fluorescence or reduced pyridine nucleotides in perfused rat liver. *Jpn. J. Hyg.*, 38:649–656.
324. Tark, M., Randall, Jr., H.M., and Hoffer, T.L. (1976): Substrate metabolism in the isolated perfused kidney. *Invest. Urol.*, 14:132–136.
325. Telander, R.L. (1964): Prolonged monothermic perfusion of the isolated primate and sheep kidney. *Surg. Gynecol. Obstet.*, 118:347–353.
326. te Kopelle, J.M., Keller, B.J., Caldwell-Kenkel, J.C., Lemasters, J.J., and Thurman, R.G. (1991): Effect of hepatotoxic chemicals and hypoxia on hepatic nonparenchymal cells: impairment of phagocytosis by Kupffer cells and disruption of the endothelium in rat livers perfused with colloidal carbon. *Toxicol. Appl. Pharmacol.*, 110:20–30.
327. Teo, S. and Vore, M. (1991): Mirex inhibits bile acid secretory function *in vivo* and in the isolated perfused rat liver. *Toxicol. Appl. Pharmacol.*, 109:161–170.
328. Terao, N. and Shen, D.D. (1985): Reduced extraction of *l*-propranolol by perfused rat liver in the presence of uremic blood. *J. Pharmacol. Exp. Ther.*, 233:277–284.
329. Thelen, M. and Wendel, A. (1983): Drug-induced lipid peroxidation in mice. V. Ethane production and glutathione release in the isolated liver upon perfusion with acetaminophen. *Biochem. Pharmacol.*, 32:1701–1706.
330. Thompson, A.M., Cavert, H.M., and Lifson, N. (1968): A rat head-perfusion technique developed for the study of brain uptake of materials. *J. Appl. Physiol.*, 24:407–411.
331. Thurman, R.G., Marazzo, D.P., Jones, L.S., and Kauffman, F.C. (1977): The continuous kinetic determination of *p*-nitroanisole *O*-demethylation in hemoglobin-free perfused rat liver. *J. Pharmacol. Exp. Ther.*, 201:498–506.
332. Thurman, R.G., Reinke, L.A., and Kauffman, F.C. (1979): The isolated perfused liver: a model to define biochemical mechanisms of chemical toxicity. *Rev. Biochem. Toxicol.*, 1:249–286.
333. Torack, R.M. (1969): Sodium demonstration in rat cerebellum following perfusion hydroxyadipaldehyde antimonate. *Acta Neuropathol. (Berl.)*, 12:173–182.
334. Tornquist, S., Moller, L., Gabrielsson, J., Gustafsson, J.A., and Toftgard, R. (1990): 2-Nitrofluorene metabolism in the rat lung: pharmacokinetic and metabolic effects of beta-naphthoflavone treatment. *Carcinogenesis*, 11:1249–1254.
335. Toth, K.M., Clifford, D.P., Berger, E.M., White, C.W., and Repine, J.E. (1984): Intact human erythrocytes prevent hydrogen peroxide-mediated damage to isolated perfused rat lungs and cultured bovine pulmonary artery endothelial cells. *J. Clin. Invest.*, 74:292–295.
336. Trela, B.A., Carlson, G.P., Turek, J., Rebar, A., and Mathews, J.M. (1989): Effect of carbon monoxide on the cytochrome P-450-mediated activation of 4-ipomeanol by the isolated perfused rabbit lung. *J. Toxicol. Environ. Health*, 27:341–350.
337. Triner, L., Verosky, M., Habif, D.V., and Nahas, G.G. (1970): Perfusion of isolated liver with fluorocarbon emulsions. *Fed. Proc.*, 29:1778–1781.
338. Trumper, L., Monasterolo, L. A.,Ochoa, E., and Elias, M. M. (1995): Tubular effects of acetaminophen in the isolated perfused kidney. *Arch. Toxicol.,* 69:248–252.
339. Tsuji, M. and Nakajima, T. (1978): Studies on the formation of γ-aminobutyric acid from putrescine in rat organic purification of its synthesis enzyme from rat intestine. *J. Biochem. (Tokyo)*, 83:1407–1420.
340. Umbreit, W.W., Burris, R.H., and Stauffer, J.F. (1972): *Manometric Biochemical Techniques*, 5th ed., p. 146. Burguess, Minneapolis, MN.
341. Vähäkangas, K., Nevasaari, K., Pelkonen, O., and Karki, N.T. (1977): The metabolism of benzo(*a*)pyrene in isolated perfused lungs from variously treated rats. *Acta Pharmacol. Toxicol. (Copenh.)*, 41:129–140.
342. Valenzuela, A. and Guerra, R. (1985): Protective effect of the flavonoid silybin dihemisuccinate on the toxicity of phenyl hydrazine on rat liver. *FEBS Lett.*, 181:292–294.
343. Videla, L. A., Troncoso, P., Arisi, A. C., and Junquiera, V. B. (1997): Dose-dependent effects of acute lindane treatment on Kupffer cell function assessed in the isolated perfused rat liver. *Xenobiotica*, 27:747–757.
344. Villanueva, G. R., Mendoza, M. E., el-Mir, M. Y., Herrera, M. C., and Marin, J. J. (1997): Effect of bile acids on hepatobiliary transport of cisplatin by perfused rat liver. *Pharmacol. Toxicol.*, 80:111–117.
345. Vilstrap, H. (1983): Effects of acute carbon tetrachloride intoxication on kinetics or galactose elimination by perfused rat livers. *Scand. J. Clin. Lab. Invest.*, 43:127–131.

346. Wang, J.F. and Wendel, A. (1990): Studies on the hepatotoxicity of galactosamine/endotoxin on galactosamine/TNF in the perfused mouse liver. *Biochem. Pharmacol.*, 39:267–270.
347. Watanabe, M., Takano, T., and Nakamura, K. (1998): Effect of ethanol on the metabolism of trichloroethylene in the perfused rat liver. *J. Toxicol. Environ. Health.*, 55: 287–305.
348. Waugh, W.A., and Kubo, T. (1959): Development of an isolated perfused dog kidney with improved function. *Am. J. Physiol.*, 217:227–290.
349. Welbourne, T.C. (1974): Ammonia production and pathways of glutamine metabolism in the isolated perfused rat kidney. *Am. J. Physiol.*, 226:544–548.
350. Weissmann, N., Kuzkaya, N., Fuchs, B., Tiyerili, V., Schafer, R.W., Schutte, H., Ghofrani, H.A., Schermuly, R.T., Schudt, C., Sydykov, A., Egemnazarow, B., Seeger, W., and Grimminger, F. (2005): Detection of reactive oxygen species in isolated perfused lungs by electron spin resonance spectroscopy. *Resp. Res.*, 6:86–100.
351. White, R.J. (1971): Preparation and mechanical perfusion of the isolated monkey brain. In: *Perfusion Techniques*, edited by E. Diczfalusy, pp. 200–216. Karolinska Institute, Stockholm.
352. Willinger, C. C., Moschen, I., Kulmer, S., and Pfaller, W. (1995): The effect of sodium fluoride at prophylactic and toxic doses on renal structure and function in the isolated perfused rat kidney. *Toxicology*, 95:55–71.
353. Wilson, T.H. and Wiseman, G. (1954): Use of sacs of everted small intestine for study of transference of substances from the mucosal to serosal surface. *J. Physiol. (Lond.)*, 123:116–125.
354. Winchell, R.J. and Halasz, N.A. (1989): Lack of effect of oxygen-radical scavenging systems in the preserved-reperfused rabbit kidney. *Transplantation*, 48:393–396.
355. Windmueller, H.G., Spaeth, A.E., and Ganote, C.E. (1970): Vascular perfusion of isolated rat gut: norepinephrine and glucocorticoid requirement. *Am. J. Physiol.*, 218:197–204.
356. Woodward, B. and Zakaria, M.N.M. (1985): Effect of some free radical scavengers on reperfusion induced arrhythmias in the isolated rat heart. *J. Mol. Cell. Cardiol.*, 17:485–493.
357. Yamamoto, K., Hasegawa, T., and Ueda, J. (1968): Renin secretion in the perfused dog kidney. *Jpn. J. Pharmacol.*, 18:1–8.
358. Yelich, M.R. (1993): The effect of epinephrine on insulin and glucagon secretion from the endotoxic rat pancreas. *Pancreas*, 4:450–458.
359. Young, S.L. (1976): An isolated perfused rat lung preparation. *Environ. Health Perspect.*, 16:61–66.
360. Youngman, R.C., Klugo, R.C., Cruickshank, R.D., and Cerny, J.C. (1976): A technique for isolated *in vivo* renal perfusion. *Invest. Urol.*, 14:187–190.
361. Zamlauski,-Tucker, M.J., Morris, M.E., and Springate, J.E. (1994): Ifosfamide metabolite chloroacetaldehyde causes Fanconi syndrome in the perfused rat kidney. *Toxicol. Appl. Pharmacol.*, 129:170–175.

# 39 Organelles as Tools in Toxicology

*Bruce A. Fowler, Mary L. Haasch, Katherine S. Squibb, and A. Wallace Hayes*

## CONTENTS

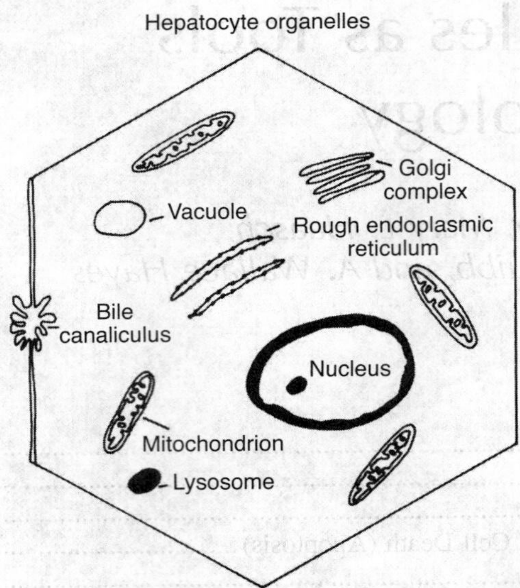


**FIGURE 39.1** Illustration of a hepatocyte showing nucleus, mitochondria, lysosomes, peroxisomes, rough and smooth endoplasmic reticulum, and Golgi apparatus.

Cells are composed of a number of organelle compartments that play crucial roles in facilitating metabolic processes essential to cellular viability (Figure 39.1). The effects of many toxic agents on cells are mediated via damage to one or more of these specialized subcellular compartments. Specific organelle systems may become damaged by toxic agents when they perform a primary role in the metabolism of a particular toxicant, when a toxicant is stored intracellularly, or as a result of an inherent sensitivity of some essential biochemical pathways in the organelle to perturbation. In terms of understanding the mechanisms of cellular toxicity, it is clear that evaluation of organelles as basic units of subcellular function may provide useful insights into the basis of toxicant action. It also should be obvious that the ability to detect damage within particular organelle systems depends on the sensitivity and nature of the parameters measured. The following discussion examines some of the current ultrastructural and biochemical methods available for evaluation of specific organelles and reviews some of the ways in which these techniques have aided understanding the mechanisms of toxicity. A critical examination of these techniques is presented to aid the reader in assessing the potential value of a given procedure for delineating information about a specific toxic process.

## MITOCHONDRIA

Mitochondria are essential organelles that play an important role in cell metabolism by mediating a number of metabolic functions (Figure 39.2). Enzymes involved in energy production, carbohydrate metabolism, heme biosynthesis, and the urea cycle are found in this organelle.

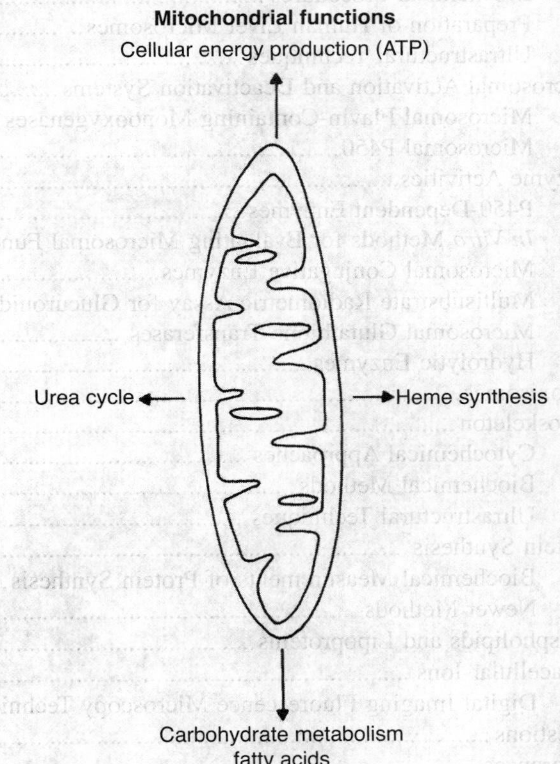


**FIGURE 39.2** Illustration of a mitochondrion showing oute[r] membrane, inner membrane with infoldings (cristae), an[d] matrix. This multifunctional organelle has enzyme systems tha[t] are concerned with the production of ATP, carbohydrate metab[-]olism, heme biosynthesis, and the urea cycle and that are spe[-]cifically localized in the outer membrane, inner membrane, o[r] matrix. (From Fowler, B.A. et al., *Environ. Health Perspect.*, 19[…] 197–204, 1977.)

These enzymes are not distributed randomly within the mitochondria but are localized within specific subcompartments, such as the outer and inner membranes, intermembrane space, and matrix (Table 39.1). In terms of the effects of toxicants on this organelle, it is important to understand the relationship between particular metabolic functions and the physical integrity of the mitochondrion as a structure, because *in vivo* biochemical perturbations frequently result directly from structural damage. The following examination of ultrastructural and biochemical methods for mitochondrial evaluations uses examples of some well-known toxicants to illustrate how each technique aided in understanding the mechanisms of toxicity. More recently, demonstration of the role of the mitochondria in releasing molecular factors involved in the initiation of apoptosis has added new insights into how this organelle may modulate basic processes of cell death. Over the last 5 years, great progress has been made in understanding the major roles played by this organelle in regulating the processes of cell injury and cell death through the regulation and control of a number of molecular factors and proteases which appear to be central factors in modulating programmed cell death or apoptosis.

## Ultrastructural Techniques

### Fixation and Embedding

Preservation of mitochondria within intact cells is carried out routinely by rapid chemical fixation using glutaraldehyde or glutaraldehyde-formaldehyde-based fixatives. Tissues may be either placed in these fixatives or perfused via the blood vasculature for optimal preservation of cellular structure. Electron density is imparted to the mitochondrial membranes by postfixation in a 1% solution of osmium tetroxide ($OsO_4$) followed by dehydration in a graded series of alcohol from 70 to 100%. Dehydrated tissues are then placed in solutions of propylene oxide and embedded in plastic resins, such as Epon®. A stepwise routine procedure for fixation and embedding of tissues for electron microscopy is as follows:

1. Place 1-mm$^3$ tissue blocks in fixative—2% glutaraldehyde, 2.6% formaldehyde in 0.07-*M* cacodylate buffer (pH 7.4) and 3% sucrose—for 2 hours in a refrigerator.
2. Decant fixative and place blocks in cacodylate buffer overnight in a refrigerator.
3. Postfix blocks in 1% $OsO_4$ (*caution:* volatile toxicant) in 0.1-*M* phosphate buffer (pH 7.4) for 2 hours and then decant in a fume hood.
4. Dehydrate tissue blocks in 70%, 90%, 95% (two changes), and 100% alcohol at room temperature for 15 minutes at each step.
5. Decant final 100% alcohol solution and place blocks in two changes of propylene oxide.
6. Place blocks in 50:50 propylene oxide–plastic resin mixture overnight to infiltrate tissue blocks.
7. Place tissue blocks in final plastic resin mixture and embed in Teflon® capsules.
8. Place in curing oven (60°C) to harden plastic before sectioning.

### Ultrastructural Morphometry

This technique, which is essentially an approach to quantifying the dimensions of organelle compartments within intact cells based on evaluation of their surface area in a large number of electron micrographs, has been reviewed extensively [2–5]. The method may be employed to determine the overall volume of organelles, such as mitochondria, within cells (volume density), but determinations of mitochondrial membrane surface area (surface density) and numbers of mitochondria (numerical density) require the application of correction factors [2–4]. The specific steps in this technique, as well as the equations necessary for evaluation of generated data, are given in articles by Weibel and Paumgartner [5] and Williams [6] and will not be repeated here. The application of morphometry to the evaluation of mitochondria following *in vivo* exposure to arsenate [7,8], cortisone [9], methylmercury [10], and vitamin E deficiency [11] has been employed successfully to document increases or decreases in this organelle system and the relationship of these effects to observed biochemical changes.

The primary value of ultrastructural morphometry in delineating toxic mechanisms for organelles, such as the mitochondrion, rests with the ability to quantitatively assess changes in mitochondrial structure within the intact cell. Such data have proved to be invaluable not only in interpreting the results of biochemical studies on this organelle [12] but also in suggesting new and more integrative hypotheses that consider change in the biochemical functionality of the mitochondrion in relation to concomitant chemical-induced alterations in other organelle systems (e.g., the endoplasmic reticulum) within the same cells [13]. In other words, this technique has provided a rigorous approach for assessing whether changes are occurring in more than one organelle system, thereby minimizing the possibility of erroneously concluding that a chemical is acting at only one site within a target cell population, which is one of the great pitfalls in contemporary mechanistic toxicology. It should be noted that the application of this technique to toxicology is extremely labor intensive and requires a serious commitment of resources for successful use [14,15]. This aspect requires serious consideration by those contemplating use of this powerful morphological technique.

**TABLE 39.1**
**Activity and Functions of Enzymes and Proteins in Mitochondria**

| Enzyme Activities and Proteins | Function |
|---|---|
| *Outer membrane* | |
| Fatty acid CoA synthetase | Fatty acid metabolism |
| Monoamine oxidase | Catecholamine metabolism |
| *Inter-membrane space* | |
| I-AAA proteases | Regulation of mitochondrial protein metabolism |
| *Inner membrane* | |
| NADH oxidase | Transport |
| Succinic dehydrogenase | TCA substrate |
| β-Hydroxybutyrate dehydrogenase | Oxidation |
| $Mg^{2+}$-APTase | Ion transport |
| Coproporphyrinogen oxidase | — |
| Ferrochelatase | Heme biosynthesis |
| ALA synthetase | — |
| Anion and cation transport systems | Mitochondrial conformation |
| m-AAA proteases | Mitochondrial protein metabolism |
| Cytochrome *c* | Mitochondrial induction of apoptosis |
| *Matrix* | |
| Pyruvate dehydrogenase complex | — |
| Malate dehydrogenase | — |
| Isocitric dehydrogenase | Intermediary metabolism |
| Citrate synthetase | — |
| Fumarase | — |
| Glutamic dehydrogenase | — |
| Glutamic transminases | Ammonia metabolism |
| Ornithine carbamoyltransferase | Urea synthesis |
| Carbamoyl $PO_4$ synthetase | — |
| HSP 60 | Protein refolding |
| Lon/PIM1 proteases | Regulation of mitochondrial protein metabolism |
| Clp proteases | — |

## Ultrastructural Evaluation of Mitochondrial Fractions

Evaluation of mitochondria from tissues following homogenization and isolation in sucrose (discussed later) by electron microscopy provides one method for evaluating the purity of the samples and the degree of structural integrity. This technique also has been employed to examine changes in mitochondrial conformational behavior during respiration following *in vitro* exposure to uncoupling agents such as dinitrophenol [16] or *in vivo* following exposure to lead [17] or arsenate [8]. The technique essentially involves using the chemical fixation and embedding process described previously to processed pellets of mitochondria and other organelles. A more quantitative approach to evaluation of isolated organelles has been described by Deter [18].

## Negative Staining of Isolated Mitochondria

The technique of negative staining [19] involves uranyl acetate, phosphotungstic acid, or ammonium molybdate to stain the Formvar™ grid backing so isolated organelles, such as mitochondria, stand out against the dark background. This method has proved to be extremely useful for high-resolution microscopy studies of mitochondrial membrane preparations and has been used to evaluate changes in mitochondrial membranes following *in vitro* exposure of these organelles to uncoupling agents [20]. A general outline of this technique follows, and a more complete discussion is given elsewhere [19]. As with the other morphological techniques discussed in this section, the main value of this procedure rests with providing correlative structural information about changes in the internal structure of this organelle in relation to biochemical alterations within intramitochondrial compartments [15].

### *Technique*

1. Isolate mitochondria (see Figure 39.3).
2. Final dilution of mitochondria to 60 mg protein per mL.
3. Pipet sample on the Formvar™-coated grids and allow to dry in covered dish.

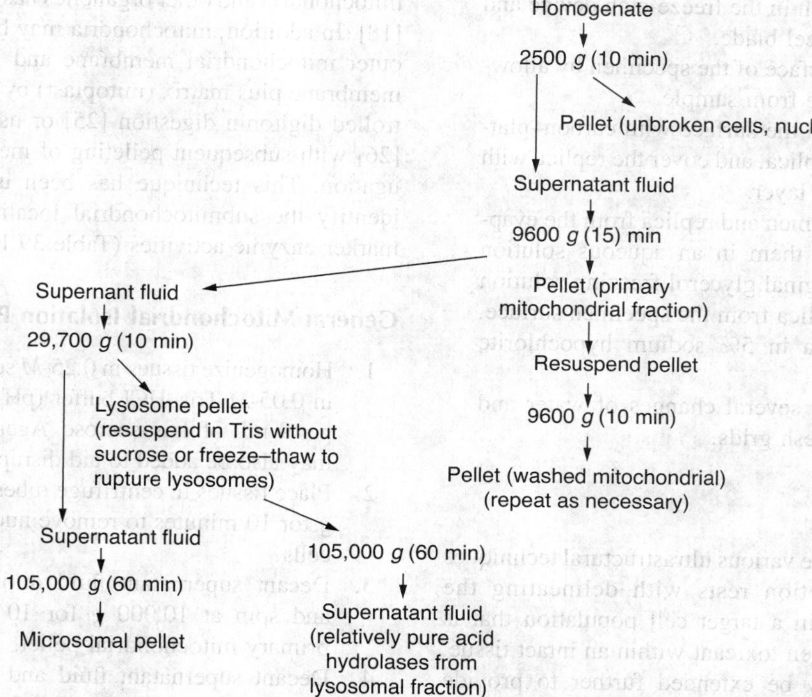


**FIGURE 39.3** Standard isolation procedure for mitochondria and other organelles, such as lysosomes and microsomes by differential centrifugation in 0.25-*M* sucrose and 0.05-*M* Tris buffer (pH 7.4).

4. Cover grids with a drop of negative stain (pH 7.4) at 1 to 2% concentration.
5. Blot excess stain from edge of grid with filter paper.
6. Examine sample with transmission electron microscope.

## Scanning Electron Microscopy

Application of scanning electron microscopy to evaluation of mitochondrial conformational behavior or the conformational behavior of the intact inner membrane (mitoplast) has been employed by Andrews and Hackenbrock [21] to confirm findings obtained by transmission electron microscopy.

### *Technique*

1. Place mitochondrial or mitoplast samples (1 to 2 mg protein per milliliter) on Formvar™-coated grids, and cover each grid with one drop of 2% glutaraldehyde in 0.1-*M* phosphate buffer (pH 7.4).
2. Place the grids in perforated vials and dehydrate in acetone; follow with critical-point drying.
3. Coat samples with a 150-Å layer of palladium–gold in a vacuum evaporator.
4. Examine samples in a scanning electron microscope.

## Freeze-Etch Analyses

The technique of freeze-etching has been employed to study the three-dimensional structure of mitochondrial membranes during different energy states [22,23] and their relationship to localization of protein complexes within the membrane. This method essentially involves chemical fixation and rapid freezing of biological samples in Freon® prior to fracturing in a freeze-etch device [24]. The fracture plane is believed to cleave primarily across the hydrophobic regions of the membranes, thereby exposing both inner and outer surfaces. The surfaces are then sputter-coated with metals, such as platinum and carbon, to form replicas that are floated off the tissue and collected on standard electron microscopy grids for evaluation in a transmission electron microscope. A detailed examination of the techniques and known artifacts has been given elsewhere [24]. To date, this technique has not been applied to evaluation of toxicant action on mitochondria.

### *Technique*

1. Fix tissue in the glutaraldehyde fixative described previously.
2. Incubate tissue in 10 to 20% glycerol until tissue is impregnated.
3. Place the specimen on a specimen carrier and immerse in Freon® 22 cooled to −165°C with liquid nitrogen.

4. Place the specimen in the freeze-etch device and fracture with a steel blade.
5. Etch-clean the surface of the specimen by allowing ice to sublime from sample.
6. Shadow specimen the surface with carbon–platinum to form a replica, and cover the replica with a carbon backing layer.
7. Remove the specimen and replica from the evaporator and place them in an aqueous solution similar to the original glycerol freezing solution to release the replica from the specimen surface.
8. Clean the replica in 5% sodium hypochlorite solution.
9. Rinse replicas in several changes of water and mount on 150-mesh grids.

### Overall Assessment

The primary value of the various ultrastructural techniques described in this section rests with delineating the organelle system within a target cell population that is being affected by a given toxicant within an intact tissue. These techniques may be extended further to provide information about changes in organelle infrastructure that may point to a molecular site of action, which may be approached by biochemical techniques. In this regard, these techniques also have proved highly useful in interpreting results of biochemical studies on various organelle systems following either *in vivo* or *in vitro* chemical treatment. Thus, although these procedures do not by themselves delineate mechanisms, they are extremely valuable techniques for detecting molecular sites of action and correctly interpreting biochemical studies that examine chemical mechanisms at those loci.

## Biochemical Procedures

A variety of biochemical parameters can be used to assess the effects of toxicants on mitochondrial function. In part, the effectiveness of these techniques depends on the procedures used to isolate mitochondria prior to evaluation. A relatively standard procedure is given in Figure 39.3 that essentially involves homogenizing in Tris (0.05-*M*)–sucrose (0.25-*M*) with subsequent pelleting of mitochondria by centrifugation. In addition to this basic procedure, resuspension and recentrifugation may be used to wash the mitochondria and to remove contamination by microsomes. In the process of reducing mitochondrial contamination, it should be noted that mitochondria from different tissues vary in their sensitivity to physical damage or chelating agents such as ethylenediaminetetraacetic acid (EDTA). This means that care must be taken to separate toxicant effects on these organelles from other effects derived from the isolation procedures. A more complete examination of problems encountered in the isolation of mitochondria and other organelles has been given by Deter [18]. In addition, mitochondria may be separated into the outer mitochondrial membrane and inner mitochondrial membrane plus matrix (mitoplast) by treatment with controlled digitonin digestion [25] or use of a pressure cell [26] with subsequent pelleting of membranes by centrifugation. This technique has been used successfully to identify the submitochondrial localization of a host of marker enzyme activities (Table 39.1).

### General Mitochondrial Isolation Procedures

1. Homogenize tissues in 0.25-*M* sucrose or mannitol in 0.05-*M* Tris–HCl buffer (pH 7.4) at 1 g tissue per 9 mL of Tris–sucrose. Agents such as EDTA may also be added to aid disruption of cells.
2. Place tissues in centrifuge tubes and spin at 2500 *g* for 10 minutes to remove nuclei and unbroken cells.
3. Decant supernatant fluid into centrifuge tubes and spin at 10,000 *g* for 10 minutes to form primary mitochondrial pellet.
4. Decant supernatant fluid and gently resuspend pellet in 10 mL Tris–sucrose for washing. Recentrifuge pellet and decant supernatant fluid. This washing cycle may be repeated a number of times, depending on the tissue involved and the degree of mitochondrial purity desired.
5. Resuspend the final mitochondrial pellet (1 mL Tris–sucrose per 1 g of original sample).

### Separation of Outer and Inner Mitochondrial Membranes

1. Place washed mitochondria (30 to 60 mg protein per mL) in a precooled French pressure cell and subject to 1500 psi. Extruded material is taken up in an equal volume of double-strength medium and centrifuged at 12,100 *g* for 10 minutes.
2. Resuspend the resultant pellet in the previous volume and recentrifuge at 12,100 *g* for 10 minutes.
3. Combine supernatant material from the above pellets and centrifuge at 27,100 *g* for 10 minutes.
4. Centrifuge supernatant fluid from this pellet at 144,000 *g* for 90 minutes to obtain the outer membrane (pellet) and intermembrane fraction (supernatant fluid).

### Respiratory Function

One of the primary functions of mitochondria within intact cells is the oxidation of substrates with subsequent generation of adenosine triphosphate (ATP). Two major classes of oxidizable substrates are capable of causing electron flow through the mitochondrial electron transport chain. The first of these involves those substrates (pyruvate, malate, and *p*-hydroxybutyrate) that use nicotinamide adenine dinucleotide (NAD) as an acceptor of protons and is capable of generating 3 mol of ATP per molecule oxidized. Succinate, the other substrate type, generates 2 mol of ATP per molecule oxidized. Methods employed for the evaluation of mitochondrial respiratory function include Warburg respirometry and the oxygen electrode; each measures oxygen consumption by mitochondria in the presence of oxidizable substrates. The advantage of the first type of measurement rests with its ability to measure oxygen consumption within intact tissue slices, whereas the latter is capable of detecting changes in respiration during different states of respiration.

*Technique (Oxygen Electrode)*

1. Isolated mitochondria in Tris–sucrose medium (10 to 20 mg protein per mL) are placed into a 1- to 3-mL oxygen electrode cell with stirrer containing a reaction mixture composed of 40-m*M* Tris–HCl (pH 7.5), 5-m*M* $K_2HPO_4$, 5-m*M* $MgSO_4$, and 100-m*M* KCl with 1 to 2 mg mitochondrial protein per milliliter.
2. A stable recorder baseline is obtained and initial state 4 respiration is initiated by adding succinate or NAD-linked substrates to yield a final concentration in the cell of 5 m*M*.
3. After 1 to 2 minutes of state 4 respiration, state 3 respiration is initiated by adding 2 to 5 μmol adenosine diphosphate (ADP).
4. Following complete use of the added ADP, a return of state 4 respiration will be observed.
5. Respiratory control ratios (RCRs) are calculated by dividing the state 3 rate by the state 4 rate. ADP:O ratios are calculated by dividing the amount of ADP added by the calculated amount of oxygen consumed, as described by Estabrook [27].

As an approach to the toxicity assessment of mitochondria, respiratory function is an essential index of mitochondrial function that is easily damaged by many toxic agents. Toxic trace metals, such as arsenic [1,7,8], lead [17], mercury [10,28], and cadmium [29], inhibit mitochondrial respiration. For lead and arsenic, this inhibition is relatively specific for NAD-linked substrates, such as pyruvate or malate [1,8,17]. This process is believed to be due to inhibition of mitochondrial dehydrogenases for these substrates, which are located in the mitochondrial matrix. In addition, alteration of mitochondrial conformation behavior has been reported in relation to these phenomena [8,17], indicating that the well-known energy-linked transformation [16,22,30] of these organelles also is altered. Organic toxicants,

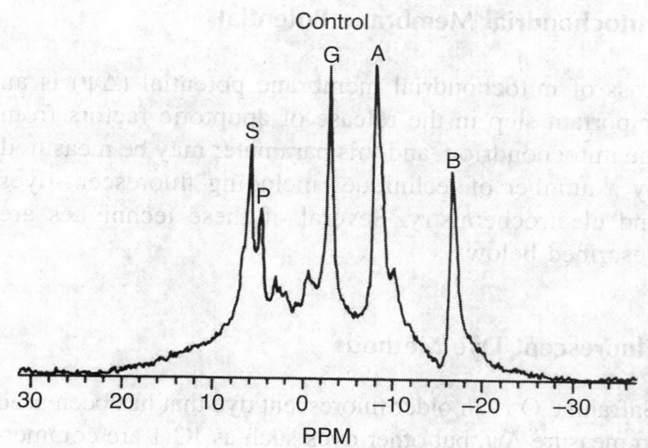


**FIGURE 39.4** *In vivo* $^{31}$P-NMR spectra from a control rat liver showing three ATP peaks (G, A, B) and inorganic phosphorus (P) and sugar (S) compounds. (Courtesy of Dr. Benjamin Chen.)

such as pesticides [31] and others [32,33], also damage mitochondrial respiratory function, leading to diminished production of ATP.

Obviously, as the primary energy source for most cells, mitochondrial respiration and ATP generation are essential to cell survival. Although impairment of mitochondrial respiration implies the reduced availability of ATP for maintaining essential cellular processes, it should be noted that quantification of cellular ATP levels is essential to confirming such a mechanism, as ATP appears to be present in excess within cells. Chemical methods for quantifying ATP require an extraction process, which in our experience usually added to the variability in these measurements by such procedures. Such methodological problems increase standard deviations, thus reducing the ability to discriminate effects between treatment groups on a statistical basis. More recently, the advent of *in vivo* $^{31}$P-nuclear magnetic resonance (NMR) spectroscopy has permitted more specific measurement of the three ATP resonances as well as inorganic phosphorus, NAD, and sugar phosphates in other phosphorylated chemical species (Figure 39.4). A major advantage of this technique is the ability to monitor changes in ATP concentrations in major target organs, such as the liver, in real time without killing an animal by placing an NMR surface coil over the organ of interest [34] while the anesthetized animal rests inside the large-bore NMR magnet. We have employed this technique [35] to study the effects of acute arsenite ($As^{+3}$) treatment on hepatic ATP content following a single intravenous dose. The data demonstrate not only the expected decrease in hepatic ATP and rise in inorganic phosphate (Pi) but also the attendant increased phosphorylation of several other chemical species. These latter events would never have been detected via simple extraction and measurement of ATP.

## Mitochondrial Membrane Potential

Loss of mitochondrial membrane potential ($\Delta\psi$) is an important step in the release of apoptotic factors from the mitochondrion, and this parameter may be measured by a number of techniques including fluorescent dyes and electrochemistry. Several of these techniques are described below.

## Fluorescent Dye Methods

Safranine O is an older fluorescent dye that has been used to measure $\Delta\psi$, but other dyes such as JC-1 are commercially available from Molecular Probes. Electrodes sensitive to the lipophilic cation TPP+ have also been used effectively for this purpose.

### *Safranine O Technique*

1. Isolate mitochondria as described above and wash twice.
2. Incubate with the electrogenic dye safranine O ± succinate or ATP stimulation.
3. Measure difference spectra at 554 and 524 nm with a double-beam spectrophotometer at 25°C [36].

### *JC-1 Technique*

The JC-1 technique [37] utilizes fluorescence-activated cell sorting (FACS) to measure changes in the fluorescence of the JC-1 dye (Molecular Probes) in intact isolated cells at various time points after a challenge with $H_2O_2$. Fluorescence is measured at 530 nm and 585 nm, depending on the $\Delta\psi$ of the dye. The generated data are analyzed by commercially available software.

### *Electrochemistry (TPP+)*

This method uses an electrode which is sensitive to the lipophilic cation TPP+ [38]. TPP+ is rapidly accumulated by energized mitochondria and TPP+ is released into the incubation media as the $\Delta\psi$ decreases thereby presenting a measure of the state of the membrane energy.

## Carbohydrate Metabolism

Many of the enzymes involved in intermediary metabolism are localized in the mitochondrial matrix. Dehydrogenases for pyruvate, malate, and glutamate are localized in this portion of the organelle. A typical assay procedure for malate dehydrogenase has been described extensively elsewhere [39]. Toxicant damage to this aspect of mitochondrial function has been demonstrated for agents such as arsenic [1,7,8,40] and methylmercury [41].

## Heme Biosynthesis

Three of the key enzymes in the heme biosynthesis pathway are localized in the mitochondrion and are associated with the inner mitochondrial membrane. Ferrochelatase, coproporphyrinogen oxidase, and δ-aminolevulinic acid synthetase are highly sensitive to the action of toxic trace metals [42–46], with resultant increases in the urinary excretion of porphyrin precursors that have proved to be useful biological indicators of toxicity. Assay procedures for these mitochondrial enzymes also have been described extensively [42–46] and will not be described here.

The value of measuring mitochondrial heme biosynthetic pathway enzymes rests with determining enzymatic mechanisms for specific chemical-induced porphyrinuria patterns, which have widespread use as biological indicators for both organic [47] and inorganic [12] chemicals. Measurement of these enzymes in target tissues such as the liver [43,45,47] and kidney [42,44] has provided valuable insight into the tissue source of the excreted porphyrins, which are among the most useful biological indicators available for chemical exposure and toxicity. If other parameters of mitochondrial structure and function are measured [1,7,8,10,36] along with these enzyme activities, then a rather complete picture of the nature and mechanism of the mitochondrial toxicity emerges. In other words, both the biochemical mechanism and tissue/organelle localization of the chemically induced injury are identified. Because these events usually precede the onset of overt clinical disease, they offer the prospect of detecting target tissue toxicity at an early stage.

## Mitochondrial Protein Synthesis

Studies on the synthesis of mitochondrial proteins have been reviewed extensively [48] and generally may be regarded as divisible into two categories: structural and enzymatic. Beattie [49] showed that these two categories of protein could be separated biochemically on the basis of solubility in dilute acetic acid into (1) proteins synthesized within the mitochondria for structural purposes, and (2) those enzymes synthesized outside the mitochondria in the endoplasmic reticulum with subsequent incorporation into the mitochondria.

Mitochondrial protein synthesis studies are essential for determining whether changes in the specific activities of mitochondrial marker enzymes following *in vivo* chemical exposure are the result of a direct chemical-enzyme interaction or a change in the synthesis of that enzyme, or both. We have found these studies of value for interpreting structural changes in mitochondria delineated by ultrastructural morphometry. Thus, protein synthesis studies in this organelle are extremely valuable for interpreting the results of morphological and other biochemical studies.

Application of this technique to studies of toxicology has shown that prolonged *in vivo* exposure of fetal rat liver mitochondria to methylmercury produced preferential suppression of membrane but not enzymatic protein synthesis [10]. In contrast, exposure of adult rats to arsenate [8] produced an increased synthesis of both protein compartments and morphometric increases in the surface density of the inner mitochondrial membrane. The changes in protein synthesis were associated with increases in the specific activities of the mitochondrial marker enzymes monoamine oxidase, cytochrome oxidase, and $Mg^{2+}$-ATPase.

*Technique*

1. Give the rat an intraperitoneal injection of [$^{14}C$]-leucine (20 μCi) and kill it 10 minutes later.
2. Excise liver tissue and isolate mitochondria as described previously.
3. Place isolated mitochondria in 1.4% acetic acid in capped ultracentrifuge tubes and shake for 30 minutes in the cold (4°C).
4. Centrifuge tubes at 90,000 *g* for 1 hour to pellet acid-insoluble proteins. Rinse pellets with ice-cold water and suspend in 0.4-*N* NaOH followed by shaking at 37°C in an incubator until the material is dissolved.
5. Pipet the supernatant fluid into new centrifuge tubes and neutralize the solution while shaking with 2-*N* NaOH.
6. Centrifuge solutions at 105,000 *g* for 1 hour to pellet acid-soluble proteins.
7. Wash pellet in ice-cold water and suspend in 0.4-*N* NaOH followed by shaking at 37°C in an incubator until the material is dissolved.
8. Pipet 0.2 mL of each fraction into counting vials, add 20 mL of scintillation fluid, shake, and count in a liquid scintillation counter.

## Conformation Behavior

The technique of following mitochondrial swelling and contraction by measurement of light scattering in a spectrophotometer was developed by Tedeschi and Harris [50]. This method is based on the increased optical density of mitochondria in a contracted state and decreased density in a swollen or orthodox configuration due to cation influx. Agents such as arsenic [8] and phosphate [51] produce detectable alterations of swelling and contraction behavior that can be detected by measurement of light scattering at 520 nm. Data from these studies are useful functional tests of mitochondrial membrane integrity following *in vivo* or *in vitro* exposure to a chemical agent. The technique is also useful for discriminating between high- and low-amplitude mitochondrial swelling.

*Technique*

1. Place a solution of 0.12-*M* KCl in 0.02-*M* Tris–Cl (pH 7.4) in spectrophotometer cuvettes, and add isolated mitochondria to a final concentration of 2 mg/mL.
2. Measure mitochondrial swelling as a decrease in optical density at 520 nm with time; maximal swelling usually is achieved within 15 minutes with liver mitochondria.
3. Initiate contraction of the mitochondria after about 15 minutes by adding $Mg^{2+}$-ATP (5 m*M*), which produces a corresponding increase in the optical density of the sample to near its original reading.

## Ion Translocation by Specific Ion Electrodes

During mitochondrial respiration or changes in conformation, the transport of $H^+$, $Na^+$, $K^+$, or $Ca^{2+}$ occurs [52–54]. Movement of these cations between isolated mitochondria and the surrounding medium may be monitored by specific ion electrodes, as described in a review by Pressman [54], which contains specific details for application of this technique. Application of this approach to measuring mitochondrial membrane functionality following exposure to mercurials [55,56] and lead [57] has provided useful information about the nature of mercury–mitochondrial membrane interactions. Other studies have demonstrated the energy-dependent mitochondrial uptake of arsenic. Data from such investigations are highly useful in more specifically delineating which ion transport systems are altered by agents that affect mitochondrial membrane integrity.

## Mitochondria and Programmed Cell Death (Apoptosis)

In recent years, it has become clear that the mitochondria contain a number of factors that regulate the process of programmed cell death or apoptosis [58,59]. This process appears to be controlled by a series of proteases known as *caspases*, whose activities are regulated by a number of mitochondrial proteins, such as cytochrome *c*, which binds to a cytoplasmic protein identified as Apaf-1; the combination of proteins forms a regulatory complex, which in turn activates caspase 9, which in turn activates other caspases and results in the degradation of genomic DNA via DNAase activation. More recently [60,61], studies have shown that mitochondria can release an apoptosis-inducing factor (AIF) from the intermembrane space. This protein has sequence homology with bacterial oxidoreductases. This factor is also capable of activating nuclear DNAase activity and producing the characteristic DNA laddering on sucrose gels, which is one hallmark of apoptosis.

As noted above, mitochondrial release of cytochrome *c* from the intermembrane space is an important step for the induction of apoptosis. It is generally recognized that cytochrome *c* release from the intermembrane space may occur by two general mechanisms: (1) $Ca^{2+}$-induced swelling of the mitochondrial inner membrane with subsequent rupture of the mitochondrial outer membrane and cytochrome *c* release, or (2) Bcl-2-family-mediated opening of the outer mitochondrial membrane permeability transition pore and release of cytochrome *c* into the extramitochondrial space. Studies by Orrenius and coworkers [62,63] have demonstrated that this second mechanism is a two-step process that also involves cleavage of the cardiolipin–cytochrome *c* linkage in the mitochondrial inner membrane. The complexity of these findings indicates the need for a variety of multidisciplinary techniques for determining the roles of loss of mitochrondrial respiratory function, decreases in mitochondrial membrane potential, reactive oxygen species (ROS) formation in relation to $Ca^{2+}$ influxes, Fe mobilization from intracellular storage sites, and the release of cytochrome *c* from the intermembrane space via cleavage of the linkage to inner membrane cardiolipin. AIF released from the mitochondria under a variety of conditions results in activation of the caspase system and initiation of apoptosis or necrosis. The techniques described below are intended to serve as a summary of some of the methods for measuring various aspects of the roles mitochondria may play in cell death processes.

### The Lon Family of Proteases

The Lon proteases are ATP-dependent proteases found in bacteria, and related proteases have been found in the matrix of mitochondria (Table 39.1). These proteases are essential for mitochondrial respiration and maintenance of mitochondrial genomic integrity. Alteration of these enzymes by damage to the mitochondria results in increased cell injury.

#### *Technique*

Mitochondria are isolated as indicated previously and assayed according to the method of Wang et al. [64] using $^{3}$H-alpha casein as the substrate. Protease activity is assayed at 37°C using a 50-m*M* Tris–HCl buffer at pH 8.0 with 10-m*M* $MgCl_2$ and 1-m*M* dithiothreitol in the presence or absence of 4-m*M* ATP. The unit of activity is defined as the degradation of 1 µg of casein/hr released into a tricarboxylic acid (TCA)-soluble form.

### The CIp Family of Proteases

The CIp family of proteases represents a second class of ATP-dependent mitochondrial proteases that also appear to operate to regulate protein turnover in the mitochondrial matrix compartment [65]. These proteins have both protease and chaperone-like activities that may be related to the yeast HSP 78.

### The AAA Family of Proteases

This AAA family of ATP-dependent proteases is localized in the mitochondrial inner membrane; they exist in two main forms: m-AAA protease and i-AAA protease. The m-AAA protease acts on newly synthesized protein products at the mitochondrial inner membrane, whereas the i-AAA protease operates in the mitochondrial inter-membrane space.

### Mitochondrial Release of Cytochrome *c*

Release of cytochrome *c*, a major trigger for initiation of apoptosis, from the mitochondrial inner membrane is a two-step process that involves the release of cytochrome *c* from electrostatic binding to cardiolipin in the inner membrane and then a permeabilization of the mitochondrial outer membrane known as the membrane permeability transition (MPT) which results in the release of cytochrome *c* into the extramitochondrial space, resulting in interactions with a number of other pro-apoptotic effector proteins. Cytochrome *c* may be measured by several methods, which are discussed below [66,67].

#### *Method 1 Technique*

1. Cell homogenate or cell supernatant following centrifugation at 15,000 *g* for 15 minutes in the presence of 3-µ*M* rotenone, 0.8-µ*M*, and 6-µ*M* myxothiazole are incubated with potassium ferricyanide (0.1-m*M*) to oxidize reduced cytochrome *c* and then with KCN (1-m*M*) to inhibit cytochrome oxidase.
2. The reduction of cytochrome *c* (E550 = 21 m*M*/cm) is obtained by adding a few grains of sodium dithionite.
3. Increase in absorption was measured using a dual wavelength spectrophotometer (548 to 540 nm).

#### *Method 2 Technique*

1. Mitochondria supernatants are evaluated for cytochrome *c* content by high-performance liquid chromatography (HPLC) using a 5-mm C4 reversed-phase column (150 × 4.5 mm) with an ultraviolet detection system set at 393 nm.
2. A 20% acetonitrile–60% acetonitrile gradient is established in water containing trifluroacetic acid (0.1% v/v). A flow rate of 1 mL/min and a 12-minute runtime are utilized to elute cytochrome *c*, which is monitored at 393 nm.

### Immunological Method: Western Blots for Cytochrome *c*

This assay uses western blots and anticytochrome *c* antibodies diluted 1:2500 [38].

*Technique*

1. Samples of mitochondrial pellets and mitochondrial pellet supernatants are isolated following excision of liver tissue by homogenization in manitol (210-m*M*)–sucrose (70-m*M*) in a 5-m*M* HEPES buffer (pH 7.5) containing 1-m*M* EDTA.
2. The homogenates are initially centrifuged at 600 *g* for 8 minutes at 4°C. The supernatant is decanted and the pellet resuspended in the above buffer without EDTA. This supernatant is then centrifuged at 5500 *g* for 15 minutes to produce the primary mitochondrial pellet, which is then resuspended in the above buffer to a protein concentration of 80 to 100mg/mL.
3. For western blot electrophoresis, the samples are mixed with the loading buffer described by Laemmli [131], boiled for 5 minutes, and loaded onto 15% polyacrylamide gels (30 μg of protein) for one-dimensional electrophoresis at 130 V.
4. The gels are removed from the glass plates and the protein bands electroblotted onto nitrocellulose membranes for 2 hours at 100 V.
5. The membranes are blocked with nonfat milk in physiologically buffered saline at room temperature and incubated overnight with the diluted anticytochrome *c* antibody.
6. The membranes are rinsed and then incubated with a secondary horseradish peroxidase conjugated antibody (1:10,000 dilution).
7. The membranes are washed and the bands visualized by enhanced chemiluminescence followed by computerized quantitative scanning using commercially available software.

## Apotosis-Inducing Factor

Apotosis-inducing factor is another pro-apoptotic protein, which, when released from the mitochondrion, stimulates activation of the executioner caspase 3. AIF is released from the mitochondria by increased permeability of the mitochondrial outer membrane (MPT). Recent data suggest that caspase 2 may be involved in the mitochondrial release of AIF. AIF is measured using commercially available monoclonal antibodies and may be detected by western blot technique as described above for cytochrome *c* or by immunofluorescence imaging, as described below [68].

*Immunofluorescence Microscopy Technique*

1. Cells of interest are grown on sterile glass coverslips and treated with the apoptosis-inducing agent under study. After treatment, the cells are washed in physiological saline and fixed in 2% paraformaldehyde in physiological saline for 15 minutes.
2. The cells are then washed twice with physiological saline. They are subsequently permeabilized in a blocking buffer containing 1% bovine serum albumin, 1% goat serum, 0.1% saponin, 1-m*M* $CaCl_2$, 1-m*M* $MgCl_2$, and 2-m*M* $NaV_2O_5$ in physiological saline for 1 hour.
3. The cells are then incubated with mouse monoclonal anti-AIF antibody (1:200 dilution) at 37°C for 1 hour in a humidified incubator. At the end of the incubation period, the cells are washed 3 times in a washing buffer containing 1% serum albumin and 0.1% saponin in physiological saline.
4. The cells are next incubated in a humidified incubator for 1 hour with a 1:500 dilution of a fluor-conjugated goat anti-mouse secondary antibody. The coverslips are then mounted on slides and visualized using a fluorescent light microscope.

## Bcl-2 Family Proteins: Bid, Bax, Bad, Bak

The Bcl-2 family of apoptotic regulatory proteins includes a number of major proteins. The major anti-apoptotic proteins that have been identified to date are Bcl-2 and Bcl-xL [69,70], which regulate the development of the mitochondrial mediated apoptotic response as negative effectors at the level of the mitochondrial MPT. Positive pro-apoptotic proteins of interest include Bax [71], Bad [72], and Bid [73]. Others, such as Bak, may also play roles in the cell death process, and all of them may be studied using commercially available antibodies and the western blotting techniques described above.

## Voltage-Dependent Anion Channel

The voltage-dependent anion channel (VDAC) is another key element in mitochondrial-dependent apoptosis that is modulated by pro-apotopic elements such as Bax and Bak. Interactions between the Bcl-2 family of proteins and this mitochondrial outer membrane channel may be studied using anti-VDAC antibodies and the western blot techniques discussed above [74,75].

## Adenine Nucleotide Translocase

Mitochondrial adenine nucleotide translocase (ANT) exists as three isoforms (ANT1, ANT2, ANT3) in different tissues and interacts with other cellular proteins to form the mitochondrial MPT, which is directly involved in the release of cytochrome *c* into the cellular cytoplasm and resulting activation of the apoptotic machinery. The release of cytochrome *c* is prevented in part by anti-apoptotic Bcl2 and Bcl-XL, while pro-apoptotic Bax and Bak operate in the opposite direction [75]. As with the other proteins that regulate the apoptotic response, commercially available antibodies against ANT may be used to study the behavior of this protein under a variety of conditions via western blot analysis using the techniques described above.

### ROS Formation and Roles of the FeS Clusters in Mitochondrial Aconitase

Numerous studies have demonstrated roles of ROS formed subsequent to inhibition of mitochondrial respiration by toxic agents as key triggering agents for initiation of the apoptotic response. Studies from a number of laboratories [76–78] have shown that ROS ($O_2$) produced by mitochondrial respiration can inhibit mitochondrial aconitase and liberate $Fe^{2+}$ from the FeS clusters of this enzyme. This results in the intracellular availability of $Fe^{2+}$, which can catalyze the Fenton reaction, resulting in the further formation of highly reactive OH· as well as other ROS species. The measurement of mitochondrial/cytosolic aconitase activity [79] is thus a potentially useful marker for intracellular ROS bioavailability.

*Mitochondrial Aconitase Activity Technique*

1. Mitochondria are isolated as described above (0.35 mg protein per mL) and incubated for 20 minutes at 37°C in a 3.5-mL assay buffer containing 120-m*M* KCl, 5-m*M* $KH_2PO_4$, 3-m*M* HEPES buffer, 1-m*M* ethylene glycol tetraacetic acid (EGTA), 1-m*M* $MgCl_2$, 0.2-m*M* NADP, 2 units of isocitrate dehydrogenase, and 5-m*M* citrate (pH.7.2).
2. The incubation mixture is run in duplicate ± xanthine plus xanthine oxidase (0.01 units/3.5 mL). At the end of the 20-minute incubation, superoxide dismutase (45 units) is added and the reaction started by addition of Triton X100 (0.12% v/v).
3. The formation of NADPH is followed at 340 nm by ultraviolet spectrophotometry.

## LYSOSOMES

Lysosomes are spherical structures ranging from 25 nm to 1 μm in diameter that play a central role in the storage and catabolism of many endogenous and exogenous substances. Biochemically, these organelles are characterized by the presence of acid hydrolases. Lysosomes synthesized in the Golgi apparatus are filled with enzymes synthesized on the rough endoplasmic reticulum. Initially formed primary lysosomes fuse with endocytic or autophagic vesicles or cellular organelles to become larger secondary lysosomes, which either breakdown the engulfed material or fuse with the cellular membrane to eject their contents from the cell. The enzyme content of lysosomes can vary in different cell types. To date, more than 30 lysosomal acid hydrolases have been characterized. To understand the impact of toxicants on this organelle system, ultrastructural and biochemical techniques have been developed that discern the various categories of lysosomes and the impact of toxicants on lysosome function and structure.

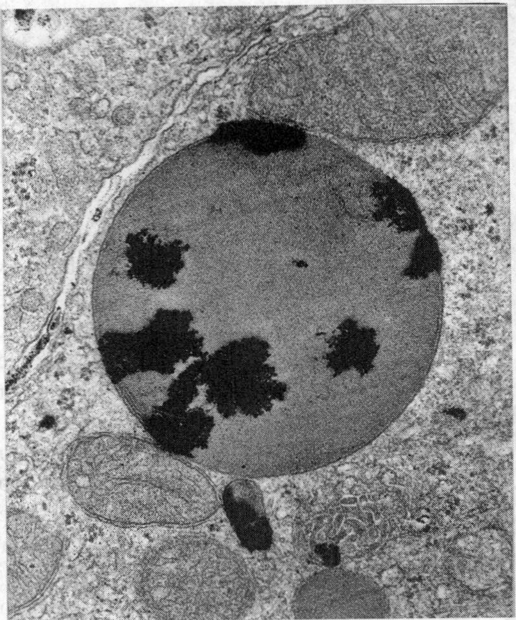

**FIGURE 39.5** Cytochemical demonstration of acid phosphatase activity at the electron microscope level showing positive (dense lead phosphate precipitate) over a secondary lysosome in a renal proximal tubule cell of a rat exposed to methylmercury in its drinking water. (Original magnification, 46,000×.)

## Ultrastructural Techniques

### Cytochemistry

Active lysosomes (secondary lysosomes) may be cytochemically distinguished from inactive (teleolysosomes) or autophagic vacuoles by the presence of acid phosphatase activity (Figure 39.5). This technique gives a clear demonstration of this enzyme's activity, provided the development time of the reaction is monitored carefully to minimize spurious or nonspecific deposition of lead–phosphate reaction product. Although the technique is largely qualitative, it does provide essential information for delineating lysosomes from other subcellular structures.

*Technique for Histochemical Determination of Acid Phosphatase*

1. Remove tissue under light ether anesthesia.
2. Cut into thick (2- to 3-mm) slices on plate of dental wax.
3. Fix at approximately 4°C for 2 to 3 hours in 2.5% glutaraldehyde in 0.1-*M* sodium cacodylate buffer containing 7.5% sucrose (final pH 7.1) or standard glutaraldehyde–formaldehyde fixative described previously.

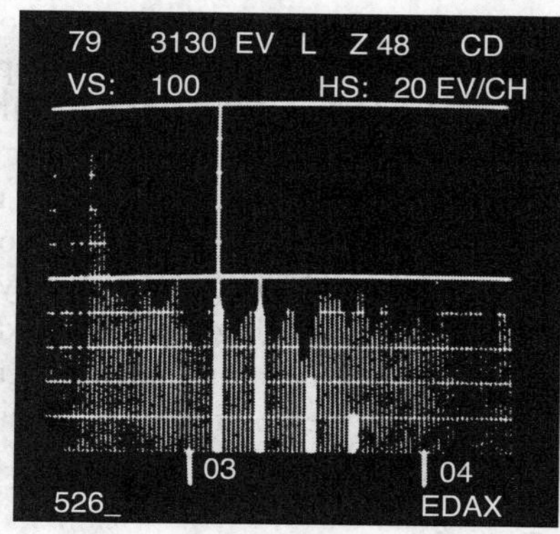


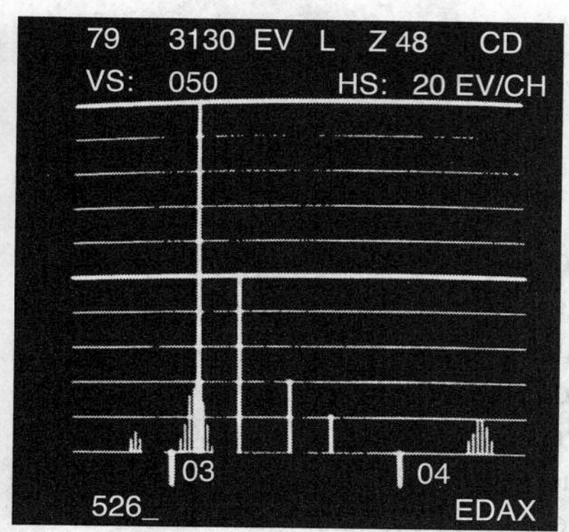


**FIGURE 39.6** Energy-dispersive x-ray spectrum from a renal proximal tubule lysosome of a rat injected with 0.6 mg/kg cadmium (Cd) as Cd–metallothionein before (right) and after (left) background subtraction. The presence of a Cd Lα x-ray peak (3.13 keV) is indicated by the first vertical marker bar.

4. Rinse slices in cold sodium cacodylate buffer (pH 7.4) containing 0.33-*M* sucrose.
5. Transfer slices to stage of tissue chopper and cut ten 50-μm sections.
6. Collect in cold sodium cacodylate (pH 7.4) containing 0.33-*M* sucrose.
7. Rinse 20 minutes to 2 hours in two changes of sucrose buffer.
8. Warm Gomori medium to 60°C for 1 hour; cool to room temperature for 4 minutes; filter through one piece of Whatman #l.
9. Incubate sections 15 minutes to 2 hours at 37°C in medium (depending on reactivity of tissue).
10. Rinse twice for 1 minute in cold 0.05-*M* acetate buffer (pH 5.0) containing 7.5% sucrose and 4% formaldehyde.
11. For light-microscopic monitoring of reaction development, expose sections to $(NH_4)_2S$ (two drops of 45% $(NH_4)_2S$ in 10 mL $H_2O$). Transfer to glass slides and mount in water-soluble embedding medium.
12. For electron microscopy, postfix for 30 to 60 minutes in 1% $OsO_4$ in acetate–veronal buffer (pH 7.4) containing 49-mg/mL sucrose. Rapidly dehydrate beginning with 70% ethanol. Embed in plastic resin as described previously.

*Gomori Medium*

1. 0.12 g $Pb(NO_3)_2$
2. 100 mL 0.05-*M* sodium acetate buffer (pH 5.0) containing 7.5% sucrose
3. 10 mL of 3% sodium-β-glycerophosphate or cytidine monophosphate (CMP) added slowly, with gentle mixing

*Glutaraldehyde Fixative*

1. 0.2-*M* cacodylate buffer (pH 7.4), 97.5 mL, containing 7.5% sucrose
2. Ultrapure glutaraldehyde (70%), 2.5 mL

*Buffer Rinse*

3. 0.1-*M* cacodylate buffer (pH 7.4) containing 0.33-*M* sucrose

*Formaldehyde Rinse*

1. 0.05-*M* sodium acetate buffer (pH 5.0), 90 mL
2. 37% Formaldehyde (formalin solution), 10 mL
3. Sucrose, 7.5 g

*Acetate Buffer*

1. 15 mL 1-*N* HCl
2. 50 mL 1-*N* sodium acetate
3. pH adjusted to 5.0; diluted to 1300 mL

## Localization of Substances within Lysosomes

Several ultrastructural techniques are available for demonstrating the presence of particular substances within lysosomes of intact cells. X-ray microanalysis (Figure 39.6) has been used by several investigators [80–83] to demonstrate the presence of toxic trace metals within lysosomes following *in vivo* exposure. This method essentially uses the focused electron beam of an electron microscope to displace orbital electrons from the atoms present in the sample with the resultant generation of characteristic x-rays from within the sample that are separated by wavelength- or energy-dispersive techniques. Major problems with the technique for the analysis of biological samples are related to extraction or translocation of elements during tissue processing [84], volatilization of elements by specimen heating [80], and detection of elements within biological thin sections due to insufficient excitation or low concentrations of the elements within the tissue [84]. The obvious chief advantage to this technique is that it provides a clear means of placing the toxic element of concern in structures such as lysosomes within target cell populations, thus providing evidence that could not be generated readily by subcellular fractionation studies. A more complete description of this technique and the available instrumentation has been provided by Hall [84].

*Technique*

1. Section blocks of tissue embedded for electron microscopy (as previously described) at 2500 Å or less using an ultramicrotome. Place on carbon-coated grids made of carbon, beryllium, or some other element with x-ray emission lines different from those in the sample to be analyzed.
2. Place the sample grid in the specimen holder of a transmission or scanning electron microscope fitted with energy-dispersive or wavelength-dispersive spectrometers.
3. Perform x-ray microanalysis of lysosomes (or other organelles of interest) by condensing the electron beam onto the site to be analyzed and monitoring the elemental x-rays generated.
4. Problems associated with extraction of elements from the tissue during fixation, dehydration, and embedding may be circumvented to some degree by cryosectioning frozen samples and utilizing liquid-nitrogen-cooled cold stages.
5. Vaporization of elements by specimen-heating from the condensed electron beam may be dealt with to some degree by altering the accelerating the voltage of the electron microscope and reducing the counting times.

Autoradiography of compounds labeled with $^{125}I$ or $^{3}H$ is another sensitive tool that requires great care in application due to translocation of the label and insufficient grain development. This technique has been applied successfully to the detection of proteins within lysosomes of intact cells to show uptake into this cellular compartment. At the light-microscope level, lysosomal uptake of fluorescent dyes has been demonstrated by fluorescence microscopy [85–88]. Histochemical staining methods [89] also have been used to demonstrate lysosomal uptake of metals in cells of metal-exposed animals. These techniques, like x-ray microanalysis, provide useful approaches for localizing chemicals of interest in lysosomes of target cell populations.

### Ultrastructural Identification of Lysosomal Storage Disorders

Phospholipidosis, a lipid storage disorder that can occur following exposure to a variety of cationic amphiphillic drugs, is characterized by the formation of lamellar myelin-like bodies in lysosomes [90,91]. Inhibition of lysosomal phospholipase activity is thought to be a primary cause of this disorder; however, recent DNA microarray studies suggest that genes involved in lysosomal enzyme transport and phospholipid and cholesterol biosynthesis are also altered [91]. Although a rapid screening method utilizing a Nile red fluorescent stain with a high affinity for lipids has been reported for detecting toxicant-induced phospholipidosis [86], the most reliable method for identifying this lysosomal storage disorder is by transmission electron microscopy, which provides a means of quantifying the increased number and size of lysosomes containing myelin figures (electron-dense deposits and membranous structures arranged in whorled arrays) [91].

*Technique*

1. Fix pelleted cells in 1% glutaraldehyde for 2 hours, wash with sodium phosphate buffer, and postfix with 2% osmium tetroxide for 2 hours.
2. Following dehydration in increasing concentrations of alcohol, fix the cell pellet in epoxy resin.
3. Cut ultrathin sections (80 nm) using an ultramicrotome, and double-stain sections with uranyl acetate and lead acetate.
4. Observe cells using an electron microscope to score pathological changes indicative of phospholipidosis based on the following scale: 0, no evidence of lamellar myelin-like bodies in lysosomes; 1, slight evidence of lamellar myelin-like bodies in lysosomes; 2, moderate evidence of lamellar myelin-like bodies in lysosomes; or 3, severe evidence of lamellar myelin-like bodies in lysosomes.

### Ultrastructural Morphometry

For quantifying *in vivo* changes in the lysosomal compartment, ultrastructural morphometry has been employed to evaluate changes in the lysosome system with prolonged methylmercury exposure [92], age [93], and cadmium metallothionein exposure [94]. Application of this method to lysosomes is subject to some of the same constraints and limitations noted previously for mitochondria.

## Biochemical Procedures

Isolation procedures for lysosomes by centrifugation are given in Figure 39.3 and have been described extensively elsewhere [90,95]. Changes in lysosomal sedimentation characteristics have been reported [18,90] following loading with organic and inorganic compounds. This effect, as well as alterations of lysosomal membrane stability [87,88,96], should be considered carefully when isolating lysosomes in toxicity studies. In addition, consideration also should be given to distribution of lysosomes within different cell types within a given organ because not all cells will be affected equally.

## Lysosomal Protein Degradation

Protein degradation by lysosomes has been monitored by following release of $^{125}I$ from labeled protein following either *in vitro* [97–100] or *in vivo* [101] incubations. *In vitro* exposure of lysosomes to agents such as toxic metals [97–99,101] or mycotoxins [98] has been found to alter the ability of lysosomes to perform this basic function.

## Marker Enzyme Assays

Measurement of the various acid hydrolase activities found in lysosomes is another means for assessing lysosome functionality. As noted in Figure 39.3, these assays frequently are performed on lysed lysosomes so activities of the lysosomal enzymes may be more clearly separated from those present in the microsomal fraction. Marker enzymes frequently measured are the cathepsins A, B, C, and D [95]; acid phosphatase; aryl sulfatase; glycosidases; and acid RNAase. Exposure of animals by intravenous injection of protein [102,103] activates a number of the above enzymes in kidney lysosomes. As with the protein degradation procedure, these assays provide essential information about changes in lysosome functionality following chemical exposure. The metabolic consequences of lysosomal enzyme inhibition are varied but may include proteinuria or a number of lysosomal storage diseases that involve inhibition of lipid metabolizing enzymes such as phospholipase $A_1$ and acid sphingomyelinase [90,91].

### *Acid Phosphatase Assay Technique*

1. Technique requires 0.1 mL 0.004-*M* citrate buffer (pH 4.8), 0.1 mL *p*-nitrophenylphosphate (100 mg/25 mL; kept frozen), 0.1 mL enzyme extract.
2. Incubate at 37°C for various times.
3. Stop reaction by adding 5 mL of 0.2-*M* glycine (pH 10.4).
4. Centrifuge in tabletop centrifuge for 5 minutes.
5. Read optical density at 405 nm.
6. Prepare standard curve of $OD_{405}$ vs. nanomoles *p*-nitrophenol per 5.4 mL of reaction mixture.
7. Report specific activity in terms of nmol/min/mg.

### *Cathepsin D Assay Technique*

1. Technique requires 1.0 mL 0.2-*M* acetate (pH 4.5) and 0.5 mL 2% hemoglobin.
2. Add 0.5 mL enzyme solution.
3. Incubate at 37°C for 1 hour.
4. Stop reaction by adding 8 mL 5% TCA.
5. Centrifuge at 1400 rpm for 5 minutes.
6. Read absorbance at 280 nm.
7. Report specific activity as $OD_{280}$/mg protein.

### *RNAase Assay Technique*

1. Technique requires 0.2 mL 0.03-*M* acetate, 0.15-*M* NaCl (pH 5.8), 0.1 mL enzyme solution, 0.1 mL $H_2O$, and 0.2 mL 1% RNA.
2. Shake the reaction mixture at 37°C for 20 minutes.
3. After incubation, place the tubes in ice.
4. To precipitate the protein and RNA, add 0.9 mL of a mixture of 10 volumes of 76% ethanol in 1 *N* HCl and 1 volume of 0.75% uranyl acetate in 2.5 *N* $HClO_4$.
5. After allowing the mixture to stand for 10 minutes, centrifuge at 1000 *g* for 10 minutes.
6. Read the absorbance of a 1:10 dilution of each supernatant fraction at 260 nm.
7. Report specific activity as $OD_{260}$/min/mg.
8. *Caution:* The RNA is unstable and must be prepared just before use to prevent high readings in the blank.

### *Glucuronidase Assay Using p-Nitrophenyl-β-D-Glucuronide as Substrate*

1. Technique requires 1.0 mL 0.2-*M* acetate (pH 5.0), 0.6 mL *p*-nitrophenyl-β-D-glucuronide (15 m*M*), 0.2 mL lysosomal protein.
2. Incubate at 37°C for 15 to 30 minutes.
3. Stop the reaction by adding 3 mL 0.2-*M* glycine (pH 10.4).
4. Centrifuge in a tabletop centrifuge for 5 minutes.
5. Read at 405 nm.
6. Obtain nanomoles of *p*-nitrophenol from the standard curve.
7. Report the activity in terms of nmol/min/mg of protein.

## Lysosomal Membrane Integrity

Recent studies indicate that alterations in lysosomal membrane stability play a role in triggering apoptosis-related mitochondrial pathways [87,88]. Membrane integrity can be measured by evaluation of the translocation of substances such as fluoroscein isothiocyanate (FITC)-conjugated dextran or acridine orange out of lysosomes due either to lysosomal membrane rupture or to an inability to maintain the lysosomal–cytosolic pH gradient. The technique utilizing acridine orange is based on a change in fluorescence of acridine orange from red when excited by blue light, when it is highly concentrated in the lysosomes, to green, when it is released from lysosomes to the cytoplasm due to a loss of membrane integrity. The following technique involves direct observation of cells by confocal microscopy and a quantitative measurement of a change in emission ratio in comparison to controls [88].

*Technique*

1. Preload cells with acridine orange (5 μg/mL) in complete medium for 15 minutes at 37°C, followed by rinsing with media.
2. Incubate cells with toxicant in Hank's Balanced Salt Solution (HBSS). (*Note:* Preliminary tests should be conducted to ensure that the toxicant being tested does not interfere with acridine orange fluorescence). The lysosomotropic detergent *O*-methylserine dodecylamide hydrochloride (MSDH) can be used as a positive control.
3. At appropriate times, rinse off the toxicant or MSDH using ice-cold phosphate-buffered saline (PBS) supplemented with 3.6-m*M* $CaCl_2$ and 33-m*M* $MgSO_4$; immediately observe the cells using an confocal microscope coupled to a fluorescence scanner set at an excitation wavelength of 488 nm and emission wavelengths of 520 nm and 615 nm.
4. For quantitative analysis, 530 nm/620 nm emission ratios can be compared between control and treated cells using a microplate fluorometer set at an excitation wavelength of 485 nm.

### Lysosomal Accumulation of Drug Compounds

Accumulation of drug compounds by lysosomes can be an important determinant of drug action and potential interactions between pharmacological agents. Use of specific inhibitors can demonstrate lysosomal accumulation and provide some information on mechanisms of accumulation. Inhibition of acidotrophic sequestration can be demonstrated using ammonium chloride or the proton ionophore monensin, which increases the internal pH of lysosomes, or with the vaculolar ATPase inhibitor bafilomycin $A_1$ [90,104].

## ENDOPLASMIC RETICULUM

The endoplasmic reticulum (ER) is comprised of a complex, net-like pattern of membranes or cisternae (sac-like structures) found throughout the cytoplasmic matrix as a continuation of the outer nuclear membrane. The ER forms a large surface area for organizing synthesis of proteins and lipids and controlling chemical reactions. Although the ER contains many folds, it is comprised of only one sheet enclosing one area, termed the *lumen*, and part of the same space as that inside the nuclear envelope. The space of the lumen is separate from the cytoplasm and may take up more than 10% of the total volume of a cell. The ER allows molecules to be selectively transferred between the lumen, the cytoplasm, and the nucleus, thus facilitating efficient information processing to maintain cellular health. Functionally, the ER produces the proteins and lipids for all of the other cellular organelles and moves proteins and carbohydrates to the Golgi apparatus, plasma membranes, peroxisomes, lysosomes, or wherever required.

Two distinct forms of endoplasmic reticulum have been characterized by histological as well as biochemical and centrifugal techniques. The rough endoplasmic reticulum (RER) is a series of flattened sacs consisting of a complex of granular basophilic membranes distinguished by extensive ribosomal units on the outer surface of the membrane. In mammals, the RER forms layered stacks of cisternae. In leukocytes, the RER produces antibodies; in pancreatic cells, insulin. The RER also functions in protein folding and modification and controls the quality of proteins. Proteins synthesized in the RER are quickly transported to the smooth endoplasmic reticulum (SER) and packaged into vesicles for transport to the Golgi. The SER is essentially agranular and is formed of a myriad of branching interconnecting tubules extending to all areas of the cytoplasmic matrix. The RER and SER are interconnected. In general, protein synthesis occurs in the RER, whereas the SER functions in the synthesis of membrane proteins, lipids, steroid hormones, and fatty acids. In addition, the SER has a role in carbohydrate metabolism, calcium storage and transport of cellular nutrients. In muscle cells, the SER helps in muscle contraction. In the brain, the SER synthesizes hormones. Both the RER and SER function in biotransformation as evidenced by the detection of bioactivation/detoxification enzyme systems in both fractions.

The heterogeneous microsomal fraction (centrifugal fraction containing fragmented SER or microsomes) is the primary location for the cytochromes P450, flavin-containing monooxygenases, and NADPH cytochrome P450 reductase and cytochrome $b_5$ reductase as resident proteins in the ER. Microsomes are used to assess the capacity of the cell to bioactivate or detoxify a variety of foreign chemicals, as well as some endogenous compounds, including the steroid hormones. The role that microsomal enzymes play in chemical toxicity is extremely difficult to evaluate, as a single enzyme system might activate or deactivate a chemical depending on the molecular structure of the chemical, animal age (state of development or differentiation), site of metabolism (as related to organ-specific toxicity), and interactions with other chemicals (potentiation or antagonism). All of these variables contribute to the ever-changing complement of enzymes acting on the chemical of interest. Much of the information presented in this section deals with hepatic microsomal function because of the wealth of methodological information available for this tissue; however, the relative lack of time spent on the extrahepatic tissues should not detract from the contribution of extrahepatic pathways to pharmacokinetics and toxic reactions.

## Biochemical Procedures

### Standard Method for Preparation of Microsomes

The most common method used to prepare microsomes from a variety of tissues involves tissue and cell disruption followed by differential centrifugation. A general procedure for rat liver follows.

*Technique*

1. Remove, mince, and wash the liver in 1.15% KCl and keep on ice. Homogenize the cleaned tissue in 3 volumes of 0.25-*M* sucrose, 50-m*M* potassium phosphate (pH 7.4), and 1-m*M* EDTA. Homogenization is accomplished by using six strokes in a motor-driven Potter-Elvehjem glass homogenizer with smooth Teflon® pestle.
2. Remove nuclei and cell debris by centrifugation at 670 *g* for 10 minutes.
3. Remove mitochondria by centrifugation of the 670-*g* supernatant fluid at 10,000 *g* for 15 minutes. Carefully remove the postmitochondrial supernatant fluid to avoid the mitochondrial layer that may appear slightly lighter in color and close to the pellet. Surface lipids should also be avoided. The 10,000-*g* pellet can be homogenized by hand and centrifuged again to avoid loss of microsomes in the unbroken cells and sedimented microsomal vesicles. This fraction may be added to the postmitochondrial supernatant fluid previously obtained.
4. Pellet the microsomes by centrifugation of the postmitochondrial supernatant fluid at 105,000 *g* for 60 minutes.
5. Wash the resulting pellet once with homogenization buffer, and resuspend by homogenization so 1.0 mL of microsomal suspension contains material from 0.5 g liver (wet weight); pellet the suspension again at 105,000 *g* for 60 minutes.
6. Resuspend the final pellet by homogenization in a buffer of 100-m*M* potassium phosphate, 20% glycerol, and 1-m*M* EDTA (pH 7.4) so 1.0 mL of microsomal suspension contains material from 0.5 g liver (wet weight). Alternatively, the protein can be measured and the suspension volume manipulated to result in a known amount of protein per volume (e.g., 30 mg microsomal protein per mL).

Each part of the procedure should be carefully evaluated during method development if changes from the standard procedure are to be performed because of the need to examine a new tissue type or for any other reason; for example, depending on the tissue, microsomal fragments may pellet with the nuclear or mitochondrial fraction. Researchers have mistakenly reported the subcellular distribution of microsomal enzymes because of tissue differences in fragmentation of the endoplasmic reticulum; therefore, preliminary experimentation must include a thorough examination of the effect of various disruption techniques (e.g., homogenization apparatus, sonication) on disruption of the endoplasmic reticulum. In addition, certain tissues may require the use of protease inhibitors or other buffer components to maintain protein activity or to produce the best differential separation. Because of the variety of microsomal preparation methods available, the optimum method for each specific application should be determined empirically with the guidance of peer-reviewed publications and other available reference materials.

### Calcium Aggregation Method for Preparation of Microsomes

An alternative method to prepare microsomes from rat liver involves the aggregation of microsomes following the addition of $Ca^{2+}$ ions to the postmitochondrial supernatant fluid [105–107]. In addition to eliminating the need for an ultracentrifuge, this method greatly reduces the time required to prepare microsomes. The procedure is outlined as follows.

*Technique*

1. Remove, mince, and homogenize the liver in 10-m*M* Tris–HCl containing 250-m*M* sucrose (pH 7.4) to make a 20% (w/v) mixture. Homogenization is accomplished by using six strokes in a motor-driven Potter-Elvehjem glass homogenizer with smooth Teflon® pestle.
2. Remove nuclei and cell debris by centrifugation at 670 *g* for 10 minutes.
3. Remove mitochondria by centrifugation of the 670-*g* supernatant fluid at 10,000 *g* for 15 minutes.
4. Add solid $CaCl_2$ to the postmitochondrial supernatant fluid to achieve a final concentration of 8 m*M*. Stir the suspension and pellet the microsomes by centrifugation at 25,000 *g* for 15 minutes.
5. Resuspend the microsomal pellet in 150-m*M* KCl and 10-m*M* Tris–HCl (pH 7.4), and centrifuge at 25,000 *g* for 15 minutes to produce a microsomal pellet.

Many studies have compared the activities of microsomal enzymes prepared by these two methods. In general, specific activities of rat liver microsomes were similar in preparations derived by either method [107]; however, the calcium aggregation method cannot be applied to all tissues or species. Researchers have found markedly different enzyme activities in preparations derived by the two methods as a function of the species and tissue. These findings emphasize the need to determine whether the calcium aggregation method is a viable one before applying it to preparation of microsomes from a source other than rat liver.

Although many xenobiotics induce the specific activity of microsomal enzymes, these chemicals generally do not cause great changes in total microsomal protein content; however, some changes do occur in the relative distribution of SER and RER, and in the SER/RER specific activity ratios of enzymes. For example, the potent inducing agent 2,3,7,8-tetrachlorodibenzo-*p*-dioxin (TCDD) reduces the SER/RER activity ratio for aminopyrine demethylation, benzo(*a*)pyrene hydroxylation, *p*-nitrophenol glucuronidation, and microsomal protein [108]. These biochemical and pharmacological changes are associated with concomitant alterations in the cellular distribution of SER and RER in hepatocytes following exposure to TCDD as well as to a wide range of organohalogens, some of which are hepatotoxic. The following discontinuous sucrose gradient method is commonly used to isolate SER from RER in liver [109].

*Technique*

1. Homogenize the liver in 0.25-*M* sucrose to make a 20% (w/v) mixture.
2. Prepare the postmitochondrial supernatant fluid as described earlier in this section.
3. Add 2.0 mL of 1.3-*M* sucrose (not containing CsCl) to a centrifuge tube.
4. Layer 0.5 mL of 0.6-*M* sucrose (also not containing CsCl) on the heavy sucrose.
5. Bring the postmitochondrial supernatant fluid to 15 m*M* with respect to CsCl, and layer 4.0 mL of the suspension above the 0.6-*M* sucrose. Centrifuge the three-layered system at 105,000 *g* for 90 minutes. The RER forms a pellet at the bottom of the centrifuge tube and the SER forms a band at the top of the 1.3-*M* sucrose.
6. Aspirate off and pellet the SER fraction by dilution with buffer and centrifugation at 105,000 *g* for 60 minutes.

Several methods are available to further differentially fractionate SER and RER, including, discontinuous sucrose gradients, rate-differential centrifugation, and isopycnic density gradient centrifugation [109].

## Preparation of Human Liver Microsomes

Human liver samples may be obtained from organ donors through the National Disease Research Interchange (NDRI) and should be considered biohazardous and handled accordingly. Samples should be removed within 30 minutes of death and microsomes prepared immediately, or, alternatively, samples can be quick-frozen in liquid nitrogen and stored at –80°C until microsomes can be prepared [110].

*Technique*

1. Slowly thaw samples at 4°C in 100-m*M* Tris–HCl buffer (pH 7.4) containing 100-m*M* KCl, 1-m*M* EDTA, and 1-m*M* phenylmethylsulfonyl fluoride (PMSF).
2. Cut into small pieces and homogenize in 4 volumes of the same buffer using two 40-second bursts in a prechilled Waring® blender.
3. Pass the homogenate through cheesecloth or glass wool to remove capsule fibers.
4. Homogenize the filtrate using four strokes in a motor-driven Teflon®/glass Potter-Elvehjem homogenizer.
5. Centrifuge at 10,000 *g* for 30 minutes at 4°C.
6. Filter the supernatant through cheesecloth or glass wool to remove lipid (which may be quite substantial).
7. Centrifuge at 100,000 *g* for 90 minutes at 4°C.
8. Resuspend the pellet in 100-m*M* sodium pyrophosphate buffer (pH 7.4) in a volume equal to the volume of the original homogenate (second step).
9. Centrifuge at 100,000 *g* for 60 minutes at 4°C.
10. Resuspend the pellet in 50-m*M* potassium phosphate ($KPO_4$) buffer (pH 7.4) containing 0.1-m*M* EDTA, 0.1-m*M* dithiothreitol (DTT), and 20% glycerol. Freeze at –80°C or use immediately.

## Ultrastructural Techniques

### Morphometry

The surface area or more precisely the surface density ($S_v$) of smooth and rough endoplasmic reticulum may be estimated by application of morphometric techniques to intact cells [50,94,111]. This approach has been used to quantify changes in the endoplasmic reticulum of hepatocytes following exposure of rats to phenobarbital [112]. Studies of this type provide useful *in situ* correlations with the biochemical evaluation of microsomal enzyme preparations (described later), in addition to providing a means for estimating membrane recoveries from intact cells [2,3]. The labor-intensive nature of the technique discussed earlier in reference to mitochondrial preparation also applies to surface-density measurements of the endoplasmic reticulum. Despite this, the data generated have proved extremely useful in interpreting changes in microsomal enzyme activities produced by metals such as indium [13] and thallium [13,42,113].

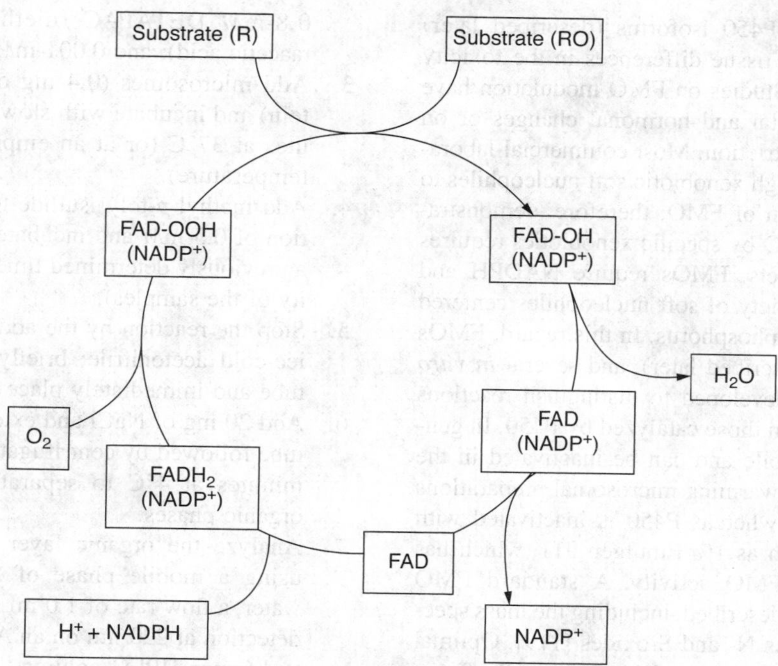


**FIGURE 39.7** Catalytic cycle and mechanism of action of flavin-containing monooxygenase (FMO): A single point of contact between the xenobiotic or endogenous substrate (R) and the terminal oxygen of the hydroperoxyflavin is all that is required for the formation of the oxygenated product (RO) by oxygen transfer to the nucleophile and immediate release of the product. The remaining steps, which do not require the presence of (R), simply regenerate the enzyme-bound oxygenating agent from NADPH and oxygen. (Adapted from Ziegler, D. M., *Annu. Rev. Pharmacol. Toxicol.*, 33, 179–199, 1993.)

## Evaluation of Microsomal Fractions

Ultrastructural examination of microsomal fractions may be conducted to assess the purity of the preparation in a manner similar to that previously described for mitochondria. This examination is highly recommended when examining a new tissue or organism for which characterization has not yet been done. Empirical optimization of the most efficient centrifugal method and buffer systems unique to the specimen should be a part of the characterization. Fixation, dehydration, and embedding procedures for microsomal pellets are essentially similar to those used for other organelle fractions. The value of these procedures for assessing relative microsomal purity and determining the efficiency of RER and SER centrifugal separations cannot be understated.

# MICROSOMAL ACTIVATION AND DEACTIVATION SYSTEMS

## Microsomal Flavin-Containing Monooxygenases

In humans, the liver, kidney, and lung contain one or more flavin adenine dinucleotide (FAD)-containing monooxygenases (FMOs) that belong to a family of proteins that are important in the NADPH-dependent metabolism of hundreds of exogenous compounds (Figure 39.7) [114,115]. Endogenous substrates of FMO include cysteamine, cystamine, and trimethylamine. FMOs catalyze the oxidation of nucleophilic tertiary amines to *N*-oxides, secondary amines to hydroxylamines, oximes, and sulfur- or phosphorus-containing xenobiotics *S*- and *P*-oxides, respectively. Hydrazines, iodides, selenides, and boron-containing compounds are also substrates for FMOs. FMOs are generally associated with detoxication reactions, although several sulfur-containing xenobiotics (thiols, thioamides, 2-mercaptoimidazoles, thiocarbamates, thiocarbamides) are oxygenated to electrophilic reactive intermediates. Humans and other mammals express five different FMO isoforms (FMO1 to FMO5) in a species- and tissue-specific manner. Each FMO isoform appears to have arisen from one ancestral gene family, with the five different members sharing 52% or more amino acid sequence identity. The major form of FMOs in human liver has been designated FMO3 and is only 52 to 57% identical to animal liver FMO1, whereas rabbit lung FMO2 is 55% identical to other FMOs. The various forms of FMOs are distinct gene products with different physical properties and substrate specificities [116–118]. The substrate stereoselectivity of *N*- or *S*-oxide product formation has been used in the development of isoform-selective catalytic methods. Species and tissue differences in the relative expression of

FMO and cytochrome P450 isoforms (described later) account for species and tissue differences in the toxicity of several xenobiotics. Studies on FMO modulation have focused on developmental and hormonal changes or on decreases by dietary restriction. Most commercial laboratory chows contain enough xenobiotic soft nucleophiles to cause maximal induction of FMO; therefore, demonstration of induction of FMO by specific xenobiotics requires specially formulated diets. FMOs require NADPH and oxygen to oxidize a variety of soft nucleophiles centered on nitrogen, sulfur, and phosphorus. In this regard, FMOs are similar to P450 (described later), and several *in vitro* techniques have been developed to distinguish reactions catalyzed by FMOs from those catalyzed by P450. In general, FMOs are heat labile and can be inactivated in the absence of NADPH by warming microsomal preparations to 50°C for 1 minute, whereas P450 is inactivated with nonionic detergent, such as 1% Emulgen 911, which has a minimal effect on FMO activity. A standard FMO enzyme assay has been described, including the mass spectral properties of various N- and S-oxides [119]. Optimal conditions for these assays should be empirically determined for each FMO source and particular substrate.

### FMO Enzyme Assays

*Technique*

1. Place 0.8 to 1.6 mg of microsomal protein in buffer (50-m*M* potassium phosphate, pH.8.4; 0.5-m*M* $NADP^+$; 2.0-m*M* glucose-6-phosphate; 1 IU glucose-6-phosphate dehydrogenase; 0.8-m*M* diethylenetriaminepentaacetic acid [DETAPAC]), and equilibrate for 1 minute at 33°C.
2. Initiate reaction by the addition of substrate (various tertiary amines and sulfides) and incubate at 33°C for various time intervals.
3. Stop the reaction by the addition of 3 volumes of ice-cold methylene chloride, mix thoroughly, centrifuge to separate the aqueous and organic fractions, and analyze for product by HPLC or other appropriate method.

Alternatively, FMO activity can be assessed using the following modified protocol [120]:

1. Obtain a microsomal source of FMO and use the described method to empirically determine optimal protein concentrations and time and temperature dependence of the reaction.
2. In a total reaction volume of 0.25 mL, add buffer: 50-m*M* potassium phosphate (pH 8.4) containing an NADPH-generating system (0.5-m*M* NADP, 2.0-m*M* glucose-6-phosphate, 2 IU glucose-6-phosphate dehydrogenase per mL of reaction); 0.8-m*M* DETAPAC (diethylenetriaminepentaacetic acid); and 0.001-m*M* FAD.
3. Add microsomes (0.4 mg of microsomal protein) and incubate with slow shaking for 2 minutes at 37°C (or at an empirically determined temperature).
4. Add methyl-*p*-tolyl sulfide to a final concentration of 0.5 m*M* and incubate for 40 minutes (or a previously determined time based on the activity of the samples).
5. Stop the reaction by the addition of 0.70 mL of ice-cold acetonitrile; briefly vortex the sample tube and immediately place on ice.
6. Add 20 mg of NaCl and extensively vortex each tube followed by centrifugation at 2000 *g* for 10 minutes at 4°C to separate the aqueous and organic phases.
7. Analyze the organic layer by isocratic HPLC using a mobile phase of 50% acetonitrile in water, a flow rate of 1.0 mL/min, and ultraviolet detection at 220 nm on an Alltech C8 RSIL 250 × 4.6-mm HPLC column. The products of the reaction—methyl-*p*-tolyl sulfoxide and the parent compound, methyl-*p*-tolyl sulfide—are quantified by comparison of the peak heights relative to those of reference standards of these two compounds.

### Colorimetric FMO Assay

A convenient colorimetric method for measuring the activity of FMO in crude tissue fractions has been developed [121]. The assay measures the thiourea-dependent oxidation of thiocholine. An alternative to preparing microsomes is to dilute the tissue homogenate about fivefold, followed by centrifugation at 40,000 *g* for 45 minutes and resuspension of the pellet in 0.25-*M* sucrose containing 0.05-*M* potassium phosphate (pH 7.5). The assay can be performed without further tissue processing. *Note:* To perform the measurements at pH 8.4, change the phosphate buffer to 0.03-*M* pyrophosphate and 0.1-*M* glycine

*Technique*

1. Remove tissue and immediately place in 0.25-*M* sucrose on ice; mince and rinse several times to remove excess blood.
2. Homogenize with a glass–Teflon® homogenizer in 6 volumes of 0.25-*M* sucrose containing 0.05-*M* potassium phosphate (pH 7.5) and 0.1-m*M* butylated hydroxy toluene.
3. Prepare microsomes by differential centrifugation and wash once and resuspend in 0.25-*M* sucrose containing 0.05-*M* potassium phosphate to the original volume of the homogenate aliquot.

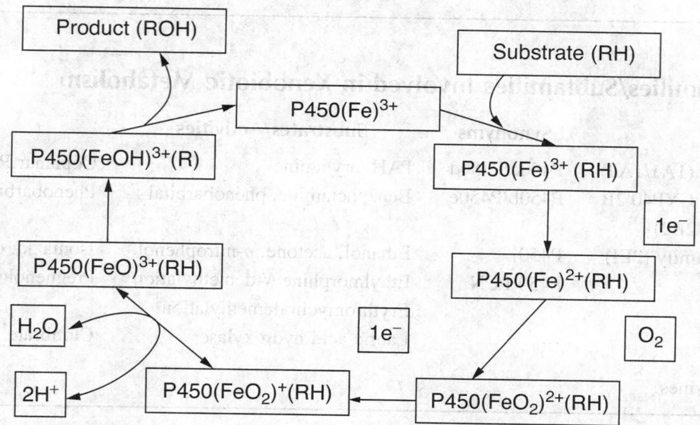


**FIGURE 39.8** Catalytic cycle and mechanism of action of cytochrome P450. The integral iron atom of the heme group at the active site is represented by $Fe^{2+}$ or $Fe^{3+}$, whereas RH represents the substrate and ROH the monooxygenation product of the reaction. This scheme is simplified for clarity. (Adapted from Parkinson, A., in *Casarett & Doull's Toxicology: The Basic Science of Poisons*, Klaassen, C.D., Ed., McGraw-Hill, New York, 1996, pp. 113–186; Porter, T.D. and Coon, M.J., *Biochem. J.*, 158, 289–294, 1991.)

4. Prepare an open 10-mL Erlenmeyer flask to contain 0.1-*M* potassium phosphate (pH 7.5), 0.25-m*M* $NADP^+$, 2.5-m*M* glucose-6-phosphate, sufficient glucose-6-phosphate dehydrogenase to reduce 1 µmol $NADP^+$ per min·mL, 80- to 160-µ*M* thiocholine, 100 units catalase, and 2-m*M* benzylimidazole, 0.4-m*M* EDTA (the last three items are added to minimize formation, accumulation, and metal-catalyzed oxidation of thiols, respectively) for a final total volume of 2.5 mL in a 37°C metabolic shaker for each sample.
5. Equilibrate for 4 to 6 minutes and add the microsomal source (0.5 to 10 mg protein in no more than 0.2 mL).
6. Equilibrate for an additional minute and add thiourea to a final concentration of 1.2 m*M* to start the reaction.
7. Withdraw 0.4 mL aliquots at 0, 3, 6, 9, and 12 minutes and transfer to tubes on ice containing 0.04 mL of 3.0-*M* trichloroacetic acid. After all the aliquots have been collected, separate the precipitated protein by centrifugation and transfer 0.35 mL of clear supernatant fluid to tubes containing 1 mL of 1.0-*M* phosphate (pH 7.5), 0.6 mL of water, and 0.05 mL of 10-m*M* DTNB.
8. Measure the concentration of thiocholine in each aliquot at 412 nm using a millimolar absorptivity of 13.6 $cm^{-1}$ for 5-thio-2-nitrobenzoate.

The reduction of thiocholine disulfide can be determined by measuring the rate of thiocholine formation in aliquots of the completed reaction mixture containing 200 to 300 µ*M* thiocholine disulfide in place of thiourea and thiocholine.

The rate constant for thiocholine disulfide reduction by glutathione (GSH) can be determined by the following procedure: In a 1-mL cuvette maintained at 37°C, add the reaction mixture consisting of 0.1-*M* potassium phosphate (pH 7.5), 150-µM NADPH, 1.0-m*M* EDTA, 18 units crystalline glutathione reductase, and 100- to 500-µ*M* GSH. Record the reaction velocity after adding 100-, 200-, 300-, 400-, and 500-µ*M* thiocholine disulfide and record the absorbance for 2 to 3 minutes after each addition.

## Microsomal P450

The cytochromes P450 are not in fact cytochromes in the true sense but belong to a large family of heme thiolate proteins consisting of multiple isozymes associated with the endoplasmic reticulum. Originally named for their absorption maximum at 450 nm (when bound with CO and reduced), these proteins serve as terminal oxidases in the membrane-bound electron transport system involved in biotransformation. Collectively, these heme protein catalysts participate in a variety of oxidative reactions with numerous lipophilic xenobiotics and many endogenous substrates, such as bile acids, fatty acids, eicosanoids, and steroids. These biotransformation reactions include epoxide formation, *N*-hydroxylation, *C*-hydroxylation, *N*-dealkylation, *O*-dealkylation, S–O exchange, deamination, and S/N oxidations. For an in-depth overview of the versatility and regulation of P450, see the review by Ioannides [122].

The function of the cytochrome P450 system involves a series of sequential reactions (Figure 39.8): (1) The substrate combines with the oxidized P450 to form a substrate–ferric heme complex; (2) flavoprotein NADPH–P450 reductase mediates a reduction of this complex; (3) oxygen reacts with the reduced hemeprotein to form an

**TABLE 39.2**
**Selected Gene Families/Subfamilies Involved in Xenobiotic Metabolism**

| Gene Family | Synonyms | Substrates/Activities | Inducers |
|---|---|---|---|
| CYP450 1 gene family (1A1/1A2) | P450c/P450d | PAH, arylamine | Coplanar PCBs/PAHs |
| CYP450 2 gene family; CYP40 2B gene subfamily (2B1/2B2) | P450b/P450e | Benzphetamine, phenobarbital | Phenobarbital, noncoplanar PCBs |
| CYP450 2E gene subfamily (2EI) | P450j | Ethanol, acetone, *p*-nitrophenol | Isoniazid, ethanol |
| CYP450 3 gene family | P450 PCN | Ethylmorphine *N*-demethylation[a] Erythromycin demethylation[a] | Pregnenolone 16α-carbonitrile |
| CPY450 4 gene family | — | Lauric acid hydroxylase[a] | Clofibrate |

[a] Activities of the enzymes.

oxycytochrome–P450 substrate complex; (4) this complex accepts another electron from NADPH, activating oxygen for interaction with the organic substrate; (5) these reactions result in the introduction of one oxygen atom into the substrate, and the other is reduced into water; and (6) the oxygenated substrate dissociates, regenerating the oxidized form of P450. As biological catalysts, P450 hemeproteins are unique in their multiplicity of isozymes, substrates, reactions, and regulatory mechanisms. Although individual isozymes may exhibit substrate overlap, each isozyme exhibits a unique overall profile of substrate selectivity. The substrate–heme binding domain appears to be structurally variable so even a minor amino acid substitution at the critical positions will define the altered specificity. Another unique feature of P450 hemeproteins is the loosely isozyme-specific inducibility by a variety of chemicals, including phenobarbital, 3-methylcholanthrene, isosafrole, ethanol, clofibrate, and pregnenolone 16α-carbonitrile. The relative abundance of individual P450 isozymes is dependent on prior chemical exposure history and other factors, such as sex, age, species, and strain.

The nomenclature for individual isozymes is critical to understanding the scientific literature concerning P450-mediated biotransformation, regulation, and induction. Early on, almost every major laboratory developed its own nomenclature; for example, for what is now referred to as the CYP1A1 isozyme, the terms P450c, P450BNF, P450mc, PCB P448L, $P_1450$, P450MC, and P450 isozyme 6 were all used in the literature. The current nomenclature system was developed to alleviate confusion and is based on alignment of amino acid sequence data from proteins or deduced from cDNAs. According to a 2001 entry to the Cytochrome P450 Home Page website (http://drnelson.utmem.edu/famcount.html), the P450 gene superfamily consists of 265 gene families, 18 of which exist in all mammals examined, with 43 subfamilies represented. Within a given family, the P450 protein sequences are more than 40% identical. Mammalian subfamilies are always more than 55% identical, but inclusion of other species drops this value to more than 46% identical. The gene is designated by the capitalized and italicized root symbol "*CYP*" (CYtochrome P450) for human and all other species except mouse and *Drosophila*, which are designated by "*Cyp*," followed by an Arabic number denoting the family, a letter designating the subfamily (when two or more exist), and an Arabic numeral representing the individual gene within the subfamily (e.g., *CYP1A1*). The cDNAs, mRNAs, and enzymes in all species (including mouse) are denoted by all capital letters without italics or hyphens (e.g., CYP1A1) [125]. The primary hepatic drug metabolism enzymes currently comprise gene families 1 through 4. A brief outline of P450 nomenclature is given in Table 39.2.

The net action of the P450-dependent monooxygenase system is several-fold: (1) Lipophilic compounds can be metabolized to more polar and therefore more easily excretable molecules; (2) lipophilic compounds may be metabolized such that the compounds are substrates for phase II reactions, such as glucuronidation, sulfation, glutathione conjugation, amino acid conjugation, and acetylation (these in turn are generally more polar); and (3) some P450 biotransformation products are electrophilic, reactive, and highly toxic metabolites.

Several methods are available to quantitatively or qualitatively study P450. Analytically, this system may be approached spectrophotometrically, enzymatically, immunologically, or by examination of specific gene or transcription products. Each approach has inherent strengths and weaknesses, as well as optimal applications. The approach selected is largely a function of the experimental goals, the biochemical level one is interested in (transcription, translation, activity), and the degree of specificity required.

## Determination of P450 by Difference Spectra

A classical method of measuring the P450 content of a microsomal preparation utilizes the appearance of an absorbance band at 450 or 448 nm for the CO adduct of the reduced cytochrome [126]. Animals induced with phenobarbital demonstrate P450 absorbance maxima at 450 nm, whereas those exposed to 3-methylcholanthrene exhibit a hypochromic shift to 448 nm in the CO-reduced spectrum. Early on, other compounds that elicited an absorbance maxima at 450 nm were referred to as phenobarbital-type inducers, whereas those with maxima at 448 nm were of the methylcholanthrene type [127]. These types of findings, along with enzymatic data, formed the early basis for the concept of P450 enzyme heterogeneity. Current applications of difference spectra are less oriented toward P450 heterogeneity as more specific immunological and molecular methodologies are available. Difference spectra, however, are useful for determining total P450 content (inducible and constitutive), examining substrate interactions with P450, and investigating mechanisms of inhibition of mixed-function oxidation reactions. Depending on the substrate or inhibitor, several types of spectral interactions can be detected *in vitro*: type I spectral change, which is characterized by a peak at 385 nm and a trough at 420 nm; reverse type I spectral change, which is the mirror image of type I spectral change; and type II spectral change, which is characterized by a broad trough between 390 and 410 nm and a peak between 425 and 435 nm [107]. Heterogeneous microsomal preparations in which more than one isozyme interacts with the substrate or inhibitor may produce intermediate spectra; therefore, it is recommended that purified isozymes be used to determine specific interactions. The procedure for determining P450 by difference spectra requires a spectrophotometer with reference and sample locations and is performed as follows.

*Technique*

1. Dilute the microsomal suspension with the homogenizing buffer to achieve a protein concentration of approximately 1.5 mg/mL.
2. Determine the baseline of equal light absorbance by placing equal volumes of diluted microsomal suspension into two cuvettes.
3. Gently gas only the sample cuvette with CO (approximately 30 seconds) and record the spectrum that quantifies oxyhemoglobin contamination.
4. Add sodium dithionite (about 1 mg solid) to the sample cuvette and record the difference spectrum of the CO adduct of the reduced P450.
5. Add sodium dithionite to the reference cuvette and record the difference spectrum of the CO complex of reduced P450 minus the spectral contribution of reduced P450.
6. Convert the change in absorbance at 450 nm relative to 490 nm to P450 concentration using a millimolar extinction coefficient of 91 m*M*/cm.

### Techniques for Identification and Quantification of P450 Isozymes

Techniques for the purification of microsomal P450 isozymes have been reviewed elsewhere [128,129]. The large number of purified and characterized isozymes has allowed for a variety of inter-isozyme comparisons and reconstitution studies [130]. Purified P450 proteins have also led to the production of specific antibodies providing an important set of reagents for the study of P450.

Several techniques have been employed to assess the purity of isolated P450 isozymes. Sodium dodecyl sulfate–polyacrylamide gel electrophoresis (SDS-PAGE) [131], at a resolution of about 5 kDa, has demonstrated the multiplicity of the P450 proteins. Certain isozymes, however, have nearly identical molecular weights, which limits the use of stained SDS-PAGE gels in examining isozyme purity. Techniques such as two-dimensional isoelectric focusing [132] have more clearly delineated highly homologous isozymes and also have provided insights into microheterogeneity resulting from polymorphisms. Other techniques, such as peptide mapping [133] and the use of monospecific antibodies, have provided further help in the examination and analysis of purified P450 isozymes [134].

Immunodetection of microsomal protein with specific antibodies is a technique widely used for the identification and quantification of individual P450 isozymes. This approach is contingent on the specificity of the antibodies [135]. This specificity may be lost on isolation of the antigen if it is not purified to homogeneity or if the recognition sites of the antibodies are found in more than one protein. Both polyclonal and monoclonal antibodies may lack specificity [135,136]. Although monoclonal antibodies (mAbs) are highly specific for a single protein epitope, if that epitope exists on another P450, the mAb will cross-react with all isozymes containing the epitope. Conversely, polyclonal antibodies (pAbs) are a group of antibodies directed at a variety of epitopes (antigenic sites). Some pAbs may be specific for the isozyme of interest, whereas others may be directed toward epitopes on other isozymes. Conversely, pAbs are often useful for investigating species other than the one from which the isozyme was purified and against which the antibody was prepared, as a pAb preparation potentially provides recognition of related isozymes. A panel of techniques is required to screen the specificity of antibodies. Included in this list are enzyme-linked immunosorbent assay (ELISA), inhibition of enzymatic activity, immunodiffusion, immunoaffinity chromatography, and immunoblotting [135]. These techniques can provide information on cross-reactions and specificity with purified isozymes and microsomal systems.

When the specificity of an antibody has been verified, the antibody may be used for a variety of applications, including: (1) as a reagent in recombinant DNA technology;

(2) for immunoinhibition studies, as a means to examine xenobiotic metabolism; (3) for immunocytochemistry, as a means to localize P450 proteins; and (4) as a method for immunoquantification of P450 isozymes. The following immunodetection technique encompasses gel electrophoresis, electrophoretic transfer to a membrane, and one of the many available staining techniques. It should be noted that some manufacturers have integrated systems for the entire process, including high-quality control precast gels, and all of the necessary buffers, thus eliminating much of the trial and error usually involved in achieving immunodetection success.

## SDS-PAGE of Microsomal Proteins

The SDS-PAGE technique as described by Laemmli [131] is the most common electrophoretic technique used for the separation of P450 proteins. In this technique, proteins are separated by molecular weight using polyacrylamide gels of varying acrylamide/*N,N′*-methylene-bisacrylamide (Bis) percentages in an overlying stacking and lower resolving gel configuration. The stacking gel serves to concentrate the sample upon entry into the resolving gel. Cross-linking of the acrylamide and Bis upon initiation of polymerization with tetramethylethylene diamine (TEMED) and ammonium persulfate results in a characteristic porosity. Changing the acrylamide concentration, generally between 4 and 20%, results in differing mobility of low- and high-molecular-weight proteins. For most applications, P450 separation can be accomplished with polyacrylamide concentrations of 3 to 4% and 7.5 to 10% for stacking and resolving gels, respectively. These percentages may vary depending on conditions and the application. A modification of the Laemmli [131] procedure follows.

*Technique*

1. Be sure electrophoresis apparatus, spacers, and glass plates are clean and dry; set up the casting apparatus, avoiding contamination of glass plates with bare hands.
2. The following reagents should be mixed together for formulations of resolving gel: Tris–HCl (pH 8.8; 0.375 *M*); 7.5% acrylamide (w/v); 0.02% bisacrylamide (w/v); and 0.10% SDS (w/v)
3. Degas this solution under vacuum for 20 minutes.
4. Add ammonium persulfate (0.05%) and TEMED (0.04%; v/v) to the components of the resolving gel (see step 2).
5. Following gentle mixing without introducing bubbles, pipet resolving gel components along the internal edge of the apparatus and glass plates. Avoid the introduction of bubbles in delivery of the acrylamide solution. Cover gel with water and allow 1 hour to polymerize (at 20°C).
6. Remove water from the top of the polymerized resolving gel.
7. After placement of the sample comb for the stacking gel, follow the same procedures as in the steps 2 through 5, using the following reagents: Tris–HCl (pH 6.8; 0.125 *M*); 4.0% acrylamide (w/v); 0.01% bisacrylamide (w/v); 0.10% SDS (w/v); 0.05% ammonium persulfate (w/v); and 0:05% TEMED (v/v). Cover with water for polymerization.
8. Dilute microsomal samples 1:5 or at a minimum 1:4 with sample buffer as formulated below: Tris–HCl (pH 6.8; 62-m*M*); 2.0% SDS (w/v); 10% glycerol (v/v); 5% 2-mercaptoethanol; and 0.001% bromophenol blue (w/v). *Note:* Bromophenol blue marks the migration front of the proteins but may be omitted for immunodetection (western blotting), as it will pass through the membrane during electrophoretic transfer. For immunodetection, pyronin Y takes the place of the bromophenol blue as a dye front marker and is retained on the membrane as a check of transfer efficiency. Pyronin Y at 0.05% is diluted 1:20 with sample buffer just prior to dilution of the microsomal samples.
9. Heat buffered samples at 95°C for 2 to 4 minutes to denature the proteins.
10. Remove the comb from the stacking gel. Wash wells with electrode buffer (see step 12), taking care to maintain well integrity.
11. Under a layer of electrode buffer, place 10 to 50 µg of microsomal protein in each well.
12. Place gels in the electrophoresis apparatus and fill with electrode buffer: 25-m*M* Tris, 192-m*M* glycine, and 0.10% SDS (w/v).
13. With running conditions set according to apparatus instructions, run times range from 1 to 4 hours for large gels or approximately 45 minutes for minigels.

*Note:* Acrylamide is neurotoxic. Care should be taken to avoid dermal contact or inhalation.

## Immunodetection of P450 Proteins

This procedure electrophoretically transfers the separated proteins from the SDS-PAGE gel to a membrane matrix, such as nitrocellulose or polyvinylidene difluoride (PVDF). The choice of membrane depends on the characteristics of the detection system and the type of analysis to be performed. Nitrocellulose is the most generally applicable and, with appropriate blocking (discussed later), is compatible with most detection systems, whereas PVDF is often used when multiple probes with different antibodies will be done with the same membrane as it is much more durable than nitrocellulose. PVDF is also used when specific proteins will be submitted for automated solid phase protein sequencing. Membranes are also available that do not require blocking steps or that are optimized for the desired detection system. The membrane specifications of each manufacturer should be consulted to meet the needs of the investigation. The following method is modified from the procedure of Burnette [137,138].

*Electrophoretic Transfer Buffer*

- 25-m*M* Tris (pH 8.3)
- 192-m*M* glycine
- 20% methanol (v/v)

*Electrophoretic Transfer Technique*

Specific manufacturers' directions should be followed, but a general method follows:

1. Equilibrate gel approximately 30 minutes in transfer buffer to remove buffer salts and detergents from gel and allow for methanol-dependent shrinking of the gel.
2. With gloved hands, cut nitrocellulose (0.45 μm or less) to the size of the gel or use precut membranes. Wet membrane for 30 minutes in transfer buffer, being careful not to trap air bubbles. PVDF membranes must be wet with methanol prior to buffer equilibration.
3. Soak filter paper and fiber pads, removing any trapped air bubbles from pads.
4. Place presoaked fiber pad on the cathode panel of cassette. Place filter paper saturated with transfer buffer on top of the fiber pad.
5. Flood the filter paper with transfer buffer, and place the gel on the filter paper, aligning the gel with the other components.
6. Saturate the gel surface with transfer buffer and place equilibrated nitrocellulose on the gel in a manner to exclude air bubbles from between nitrocellulose and gel (bubbles will prevent transfer).
7. Cover nitrocellulose with buffer. Complete the sandwich sequentially with saturated filter paper and fiber sponge. Close cassette and place the cathode panel toward the cathode in the buffer chamber. Protein transfer occurs in a cathode-to-anode direction.
8. Fill the transblot chamber with transfer buffer. Maintain the temperature at 4°C for the duration of the transfer. For rapid miniblot transfer, allow 1 hour at a constant 100 V for transfer of the P450 isozymes. Larger formats will require different settings of 200 to 400 mÅ (for 1 to 2 hours) as common constant amperage settings.

*Immunological Detection Technique*

A wide variety of immunodetection techniques are available. In general, for most procedures, the membrane is incubated first with gelatin, casein, or albumin to block nonspecific binding sites. The membrane is then placed in an antibody solution composed of the polyclonal or monoclonal antibody specific to the P450 isozyme or other enzyme of interest (primary antibody). After a suitable incubation period, the membrane is washed. At this step, the nitrocellulose may be exposed to protein A or protein G conjugates (bacterial proteins bound to some detection modality, such as a radiolabel or enzyme) that bind directly to the Fc region (non-antigen-binding fragments) of the primary antibody. Alternatively, a common technique uses a secondary antibody specific for the species IgG of the primary antibody. This secondary antibody is conjugated to an enzyme or metal colloid complex (such as gold) that mediates the detection [139,140]. Other approaches require further specific colorimetric or chemiluminescent (such as with the antibody enzyme conjugates horseradish peroxidase or alkaline phosphatase), radiographic ([$^{125}$I] protein A), enhanced (silver enhancement of gold colloids), or fluorescent development. The choice of the detection technique will depend on the required sensitivity and the available imaging equipment as well as the availability of the secondary antibody conjugate. Chemiluminescent/fluorescent detection is often the method of choice, and detection reagents are available from a variety of vendors. Each detection technique offers inherent advantages and disadvantages that should be evaluated in terms of the particular purpose for using the method.

The choice of method, instrumentation, and quantification approach and software for determining the immunologically detected protein levels are all equally important in the validity of the data. Extreme care must be taken at each step in the process for the data to be reliable. As noted above, numerous choices are available for the method of detection, and just as many vendors for imaging equipment and image acquisition. Many of these instruments are well supported with specific directions and methods for achieving the best results with each particular detection method. Following the manufacturer's guidelines can eliminate numerous problems. Software for densitometry is available for MacOS (http://rsb.info.nih.gov/ nih-image/default.html) or for Linux, MacOS, or Windows (http://rsbweb.nih.gov/ij/). Alternatively, the software provided with the imaging equipment can be used according to the manufacturer's directions. Although a discussion of the intricacies and ethics of image acquisition and manipulation is beyond the scope of this section, it is emphasized that the integrity of the process and of the data is the responsibility of the investigator.

# ENZYME ACTIVITIES

## P450-Dependent Enzymes

### Phenoxazone-*O*-Dealkylation

The *O*-dealkylation of phenoxazone ethers is a commonly used method to examine P450 enzyme activity and induction; methoxy [141], ethoxy [142], pentoxy [143], and benzyloxyphenoxazones [144] have received the greatest attention in regard to characterization and use. P450-mediated *O*-dealkylation of these –R groups for each of these substrates results in the formation of resorufin. This metabolic product can be assayed directly by fluorimetric techniques. Studies with ethoxyphenoxazone (EROD assay) and pentoxyphenoxazone (PROD assay) have indicated that a fair degree of selectivity exists between these substrates and the phenobarbital and polyaromatic hydrocarbon (PAH)-inducible isozymes; for example, 6- and 283-fold inductions were demonstrated in a group of phenobarbitone-exposed rats, while β-naphthoflavone (BNF) administration resulted in 74- and 8-fold inductions when assayed with ethoxyphenoxazone and pentoxyphenoxazone, respectively [144]. Studies with the purified CYP1A1 isozyme further indicate a substrate selectivity for ethoxyphenoxazone, whereas pentoxyphenoxazone is metabolized preferentially by the CYP2B1 isozyme [144]. Other phenoxazone substrates, such as benzyloxyphenoxazone, appear to be much less specific. Although the *O*-dealkylation of benzyloxyphenoxazone is induced preferentially by isosafrole (43-fold), it is also a substrate for the phenobarbital and 3-methylcholanthrene-inducible isozymes. The assay method is nearly the same for each of the aforementioned phenoxazone ethers [144].

*Technique*

*Reagents*

- 0.1-*M* phosphate buffer (Na–K salts; pH 7.6)
- 50-m*M* NADPH (in phosphate buffer)
- 1-m*M* substrate in dimethylsulfoxide (DMSO): methoxyphenoxazone, ethoxyphenoxazone, pentoxyphenoxazone, and benzyloxyphenoxazone
- 10- or 25-μ*M* resorufin in DMSO

*Fluorimeter settings*

- Excitation and emission slit width: 5 to 10 nm
- Excitation wavelength: 530 nm
- Emission wavelength: 585 nm

*Method*

*To cuvette:*

1. Add 5-μ*M* of substrate (10 μL of stock).
2. Add 20 to 200 μg of microsomal protein (depending on substrate and reaction rate).
3. Use phosphate buffer to bring reaction volume to 1990 μL.
4. Incubate reaction mixture 1 to 2 minutes at 37°C.
5. Add 10 μL of NADPH stock solution to start reaction (mix).
6. Allow reaction to run at least several minutes; record linear reaction rate on chart recorder.
7. Add 10 μL of resorufin standard to cuvette and record increase (used as calibration standard).

*Calculations:*

1. Rate of increase in fluorescence (*X* units/min).
2. *Y* n*M* of resorufin = *t* fluorescence units.

$$\text{n}M\text{/unit} = \frac{Y\ \text{n}M\ \text{resorufin}}{t\ \text{fluorescence units}}$$

3. *X* units/min × n*M*/unit = n*M*/min
4. $\text{n}M\text{/min} \times \frac{1}{[\text{protein}]} = \text{n}M\text{/min/mg protein}$

*Notes*

- Narrow bandpass slits (4 to 5 nm) on the fluorimeter will reduce the effect of microsomal turbidity on excitation light scatter.
- Preparation of phenoxazones and resorufin in DMSO will give these reagents greater stability; store both stocks in the dark.

### Aminopyrine-*N*-Demethylation

One of the most common assays is oxidative demethylation using aminopyrine, ethylmorphine, or benzphetamine as the substrate. This method measures the production of formaldehyde, which is an intermediate in oxidative demethylation reactions [112]. This assay is particularly useful with microsomes from animals induced with phenobarbital or other cytochrome CYP2B inducers. The assay for aminopyrine as the substrate is conducted as follows:

1. Add buffer (50-m*M* Tris–HCl, pH 7.5, 1.5 mL) to incubation tube.
2. Add saturating levels of NADPH (3.1-m*M*) in 0.5 mL Tris buffer; the incubation medium is made 20-m*M* with respect to $MgCl_2$.
3. Add aminopyrine to achieve a substrate concentration of 2.5-m*M*.
4. After prewarming the incubation contents to 37°C, initiate the reaction by adding approximately 0.5 to 1.5 mg microsomal protein.
5. After a 10-minute incubation, stop the reaction by adding 1.0 mL 10% trichloroacetic acid.
6. Sediment the protein by low-speed centrifugation and add 2 mL of supernatant fluid to 1 mL of NASH reagent (2-*M* ammonium acetate, 0.05-*M* acetic acid, 0.02-*M* acetylacetone).

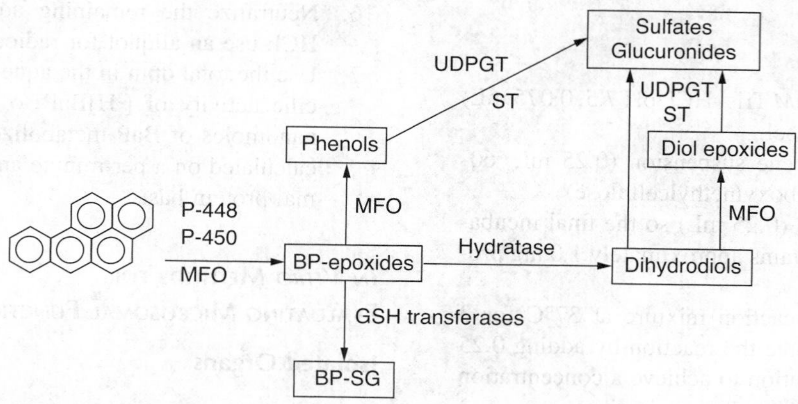


**FIGURE 39.9** Metabolic pathways for benzo(*a*)pyrene. BP, benzo(*a*)pyrene; MFO, mixed function oxidase; GSH, glutathionine; UDPGT, UDP-glucuronyl transferase; ST, sulfotransferase; BP-SG, glutathione conjugate of benzo(*a*)pyrene. (From Kleinow, K.M. et al., *Toxicol. Appl. Pharmacol.*, 104, 367–374, 1990. With permission.)

7. Heat the solution for 8 minutes at 60°C. Measure the formaldehyde concentration at 405 nm and read against a standard curve. Blank values are obtained by omitting microsomes from the incubation medium.

The reaction rate is linear with respect to time and microsomal protein under these incubation conditions, although each investigator should assess these parameters as part of preliminary investigations. Ethylmorphine and benzphetamine demethylation rates can be determined using the same method. A radiolabel assay has been devised for aminopyrine demethylation that can be used when increased sensitivity is required to detect low enzyme activity as a function of tissue, developmental stage, toxicity, or disease state [145].

### Benzo(*a*)pyrene Hydroxylase (Aryl Hydrocarbon Hydroxylase)

Benzo(*a*)pyrene (BaP), a polycyclic hydrocarbon, is metabolized by P450 as well as by a variety of conjugative enzymes (Figure 39.9). In this biotransformation process, several polar water-soluble and carcinogenic metabolites (epoxides) are formed. One of the first steps in this process is catalyzed by benzo(*a*)pyrene hydroxylase, a microsomal-bound monooxygenase. More than 12 oxygen-containing polar metabolites are formed from this action, including 3-hydroxy and 9-hydroxy BaP. This enzyme system is inducible by a variety of compounds, including both phenobarbital and 3-methylcholanthrene. Assays for measuring benzo(*a*)pyrene hydroxylase activity have been developed based on fluorimetric detection [147–149] or by isotopic detection through use of radiolabeled substrates [150–152]. Often, these techniques, while differing in detection modality, also differ in the number of metabolites available for detection. This is a function of the extraction protocol [152]. Each of the techniques, however, takes advantage of the aqueous solubility of the metabolites vs. the hydrophobicity of the parent benzo(*a*)pyrene. The nature of these assays requires pure BaP substrate to reduce background noise. BaP is highly photosensitive and can undergo chemical or radiochemical breakdown upon prolonged storage under nitrogen at 70°C. A general extraction procedure for purification of benzo(*a*)pyrene follows [152].

*Technique*

1. Dissolve up to 50 mg of BaP or [$^3$H]BaP in 100 mL of hexane.
2. Extract the hexane/BaP solution 5 times with 50 mL of aqueous 1-*M* KCl/DMSO (65/85; v/v); parent BaP should be retained in the hexane fraction that can subsequently be blown down under nitrogen.

Another purification method uses a neutral alumina resin and extraction with hexanes and toluene [153]. An example with slight modification follows [154].

*Technique*

1. Load a slurry of alumina resin (30 g) and hexane/methylene chloride (85/15) onto a column.
2. Apply BaP dissolved in hexane/methylene chloride (85/15) to the surface of the column bed.
3. Collect fractions through a solvent series of 85/15, 80/20, and 75/25 hexane/methylene chloride.
4. Clear the column with a 50/50 hexane/methylene chloride solution.

### Fluorimetric Method

*Technique*

1. Add buffer (50-m*M* Tris–HCl, pH 7.5, 0.075 mL) to incubation vessel.
2. Add benzo(*a*)pyrene suspension (0.25 mL, 60-m*M*, in 2.5% carboxymethylcellulose).
3. Add microsomes (0.25 mL) so the final incubation medium contains approximately 1.0 mg protein per mL.
4. Equilibrate the reaction mixture at 37°C for 3 minutes and initiate the reaction by adding 0.25 mL NADPH solution to achieve a concentration of 3.1 m*M* in the incubation medium.
5. After 10 minutes, stop the reaction by adding 2 mL ice-cold acetone.
6. Add hexane (20 mL) to 45-mL shaking tubes (stoppered or screw-capped). Wash the incubation mixtures into the shaking tubes with water (three times with 0.5 mL).
7. Shake the tubes 10 minutes and store overnight at 4°C or freeze.
8. Centrifuge the tubes for 15 minutes at 600 *g*.
9. Transfer the hexane layer (upper 15 mL) by pipette to a clean 45-mL shaking tube.
10. Add NaOH (5 mL, 0.1-*M*) and shake the tubes for 10 minutes followed by centrifugation for 10 minutes.
11. Read fluorescence of the aqueous layer (excitation, 400 nm; emission, 525 nm). Determine the concentration of phenolic metabolite by using 3-hydroxy-benzo(*a*)pyrene as the standard.

### Radiometric Method

The following is a radiometric benzo(*a*)pyrene hydroxylase assay [152].

*Technique*

1. Add Tris–HCl (pH 7.6; 50-m*M*), $MgCl_2$ (5-m*M*), NADP (0.37-m*M*), glucose-6-phosphate dehydrogenase (1 IU/mL), bovine serum albumin (0.8 mg/mL), [$^3$H]BaP (0.08-m*M*; $2 \times 10^5$ dpm) in acetone, and glucose-6-phosphate (2.5-m*M*). All concentrations listed are final concentrations in a volume of 0.5 mL.
2. Incubate reaction mixture for approximately 3 minutes at 37°C.
3. Initiate the reaction with the addition of appropriate amounts of microsomal protein.
4. After a 10-minute incubation, stop the reaction with 1 mL KOH (0.15-*M* in 85% DMSO).
5. Extract the parent substrate in the aqueous phase with hexane (5 mL). Centrifuge the sample to form a clear interphase and remove the hexane. Repeat this step.
6. Neutralize the remaining aqueous phase with HCl; use an aliquot for radioactivity counting.
7. Use the total dpm in the aqueous phase and specific activity of [$^3$H]BaP to calculate the total nanomoles of BaP metabolized. This amount is calculated on a per-minute and per-mg microsomal protein basis.

## *In Vitro* Methods for Evaluating Microsomal Function

### Isolated Organs

Although the use of isolated microsomes offers many advantages in the characterization of individual enzyme systems and quantification of the response of these systems to inducers, inhibitors, or repressors, it is difficult to develop a good pharmacokinetic model using such preparations. Accordingly, many investigators have used isolated perfused organs to evaluate the complex interrelationships among heterogeneous cell types, different metabolic pathways, variations in substrate concentrations, and time–course relationships. This system represents an open metabolic system capable of generating important information on steady-state pharmacokinetics. Several organ systems have been used, including the liver, lung, intestine, kidney, and testis. Detailed methods for conducting isolated organ studies have been reviewed [155–159].

### Isolated Cells

Isolated cells often are selected as an experimental model to study microsomal function because they provide a reasonable intermediate between perfused organ systems and preparation of organelle reconstituted systems. Isolated hepatocytes may be used to investigate the activity and products of complex bioactivation/detoxification enzyme systems. In addition, isolated cells allow one to investigate, in a more precise way, organelle interactions that might qualitatively or quantitatively alter the metabolic capacity of microsomal enzyme systems. References 157 to 161 detail and summarize procedures for evaluating metabolism and toxicity in isolated cells. Of particular importance for the use of isolated cells are the requirements that the cells remain viable for a sufficient time to evaluate a biochemical or pharmacological parameter, and that the isolated cells retain the characteristics and functions present *in vivo*. Cell viability commonly is determined by the trypan blue exclusion test or by leakage of cytosolic enzymes (indication of membrane dysfunction or damage), such as lactate dehydrogenase [163]. Following the maintenance of isolated hepatocytes for relatively long periods of time (more than 2 days), liver cells have been reported to revert to a more fetal form [165,166]. The relative contribution of fetal-type cells can be monitored by biochemical indicators, such as α-fetoprotein, alkaline phosphatase, and γ-glutamyl transpeptidase. The techniques that have been used to stabilize the function of P450 in culture include altering the culture matrix [165], supplying additives [167], or coculture with other cell types [168]. Although most studies are conducted on liver cells, other organ systems can be investigated by analogous techniques.

## Microsomal Conjugative Enzymes

In biotransformation, conjugation reactions refer to the enzymatically mediated addition of endogenous molecules to xenobiotics. Endogenous compounds undergoing conjugative biotransformation include sulfate, amino acids, acetyl groups, glutathione, and glucuronide. Conjugation occurs at suitable functional groups preexisting on the acceptor molecule or introduced by phase I metabolism. A number of these conjugation reactions are primarily cytosolic, such as acetylation and sulfate or amino acid conjugation. Other reactions, such as glucuronidation and glutathione conjugation, are entirely or partially microsomal. This section deals with those conjugative enzymes of microsomal origin.

### Uridine Diphosphate Glucuronosyltransferase

Uridine diphosphate glucuronosyltransferase (UDP-GT, or UGT) is a family of inducible microsomal isozymes associated with the liver, kidney, intestine, lung, and olfactory epithelium. These isozymes catalyze glucuronidation, the transfer of glucuronic acid from the high-energy nucleotide UDP-glucuronic acid (UDP-GA) to an electronegative group on a wide variety of endogenous and xenobiotic substrates. Endogenous substrates include steroid hormones, bilirubin, biogenic amines, and fat-soluble vitamins. Xenobiotic substrates include drugs, carcinogens, plant metabolites, and other environmental pollutants. Conjugation with alcohols, phenols, and carboxylic acids yield *O*-glucuronides, whereas *N*-glucuronides are formed with carbamates, amides, and amines. Substrates such as thiocarbamates and mercaptans form *S*-glucuronides.

Endogenous and xenobiotic compounds may serve as substrates for UGT with or without prior metabolism by P450. Conjugation without other biotransformation steps is probably the rule rather than the exception, as many xenobiotics (e.g., morphine, naphthols, phenols) exist in forms suitable as substrates for UGT. Structurally these ~50- to 60-kDa UGT isozymes exhibit a transmembrane region with a hydrophobic segment bordered on both sides by highly charged amino acid residues [169,170]. Sequence, enzymatic, and antibody studies suggest that the active site of UGT is in the lumen of the endoplasmic reticulum. Cytosolic substrates and donors, such as UDP-glucuronic acid, appear to require transmembrane transport to the active site. Translocases embedded in the endoplasmic reticulum membrane are believed to be involved in this process. Substrates formed with the action of membrane-bound P450 and associated reductases appear to be vectored to the closely located UGT.

Multiple forms of UGT appear to be present in most of the species thus far examined [171]. The isolation and purification of UGTs have been reviewed [172]. The family of enzymes has grown large enough to necessitate regular updates in the nomenclature [173]. The regulation of expression of this gene family is complex, giving rise to different isozyme patterns during development and following induction by xenobiotics. A tenfold difference in the rates of drug glucuronidation is common in a healthy human population. Whether the variation is due to age, disease state, exposure to xenobiotics, or genetic background is yet to be definitively determined [171]; however, significant advancement has been made in the determination of genetic polymorphisms and possible effects on disease and drug metabolism [174–176]. UGT expressed in rat and humans belongs to two gene families (UGT1 and UGT2), differing from each other by more than 50%. In humans, UGT1 consists of 5 exons and has a unique gene structure. There are 13 exon 1s from UGT1A1 to UGT1A13P, and exons 2 to 5 are used in common for all mRNAs expressed from the gene. Each isoform of UGT1 results from differential splicing of exon 1s to common exons 2 to 5 and has a unique spectrum of substrate specificity. In contrast, the genes of the UGT2 family consist of 6 exons, and all the enzymes have an individual set of exons 1 to 6. In UGT1, there are no reports of polymorphism in the common exons, although a number of polymorphisms have been reported for exon 1s. The mutations of UGT1A1 cause hereditary unconjugated hyperbilirubinemias, Crigler–Najjar syndrome type I and type II, and Gilbert syndrome. UGT1A1 has two major polymorphisms: a missense mutation of G71R and an insertion mutation of the TATA box. The prevalence of Gilbert syndrome is attributed to these polymorphisms. Because UGT1A1 metabolizes not only bilirubin but also hormones and drugs, the mutations could be involved in carcinogenesis and adverse drug reactions. Recent studies also revealed a widespread presence of diverse polymorphisms in other isoforms of UGT1 as well as the UGT2 family, including UGT1A6, UGTG1A7, UGT1A8, UGT1A10, UGT2B4, UGT2B7, and UGT2B15. The incidences and types of the polymorphisms for these enzymes are quite different in region and ethnic groups. Understanding of these polymorphisms is essential for the prevention of adverse effects of a considerable number of drugs and to predict cancer risks [176]. In addition, the regulation of gene expression of the UGTs is complex. Several transcription factors involved in the regulation of the UGT genes have been identified. These include factors such as hepatocyte nuclear factor 1, CAAT-enhancer binding protein, octamer transcription factor 1, and Pbx2, which appear to control the constitutive levels of UGT in tissues and organs. In addition, UGT gene expression is also modulated by hormones, drugs, and other foreign chemicals through the action of proteins that bind or sense the presence of these chemicals. These proteins include the Ah receptor, members of the nuclear receptor superfamily (such as CAR and PXR), and transcription factors that respond to stress [177].

Glucuronidation generally detoxifies xenobiotics and potentially toxic endogenous compounds, such as bilirubin, and is therefore considered beneficial; however, glucuronidation of steroid hormone D-rings causes cholestasis, and induced glucuronidation may lead to abnormal decreases in serum thyroid hormone levels [178–180]. In addition, glucuronidation can represent an important step in xenobiotic toxicity, such as with aromatic amine-mediated bladder cancer or colon tumor formation [176].

Considerable interlaboratory variations of *in vitro* activities exist in the assay of microsomal UGTs. In part, this may be due to the dependence of these enzymes on the phospholipid environment for activity and the extremely labile nature of UGT, which exhibits a latency in activity that is expressed upon membrane disruption. Lubrol-PX® and other detergents have been demonstrated to effectively release latent UGT activity. Accurate quantification of these activities requires measurement in the presence of optimal concentrations of detergent [169]. This often necessitates a detergent titration curve to determine optimal transferase activity and incubation of detergent with the membrane preparation prior to use in enzyme assays. It has been reported with microsomal pellets (20 mg/mL) resuspended in 0.25-m*M* sucrose/5-m*M* HEPES (pH 7.4) that this latency, which may be as high as 95%, may be retained for as long as 2 months at –80°C [169]. It has been postulated that disruption of the membrane barrier by detergents allows free access of the rate-limited donor substrate (UDP-GA) and reveals the full catalytic potential of the transferases.

Glucuronidation may be assayed by a variety of colorimetric, fluorometric, and radioisotopic methodologies. Many of these methods are substrate specific and require specific methods or specifically labeled substrates. Other methods, best described as multisubstrate assays, are based on quantification of the donated glucuronic acid moiety or by indirect reaction products. Often, the former assays are based on the use of UDP-[$^{14}$C]GA to produce labeled glucuronides that are separated by chromatography and quantified by liquid scintillation counting [181]. Assay adaptations including high-throughput format [182], measurement of recombinant enzyme activity [183], and high-sensitivity assays [184] have been developed. Indirect reactions have been described that link the UGT reaction to NADH/NAD conversion through pyruvate kinase and lactate dehydrogenase [185].

### Colorimetric Assay for *p*-Nitrophenol Glucuronidation

A reproducible simple colorimetric assay has been presented as a standard method for measuring *p*-nitrophenol glucuronidation [186].

*Technique*

1. Incubation mixture (total volume of 0.45 mL):
   - Liver microsomes (1 mg protein/mL) (0.10 mL)
   - 1-*M* Tris–HCl (pH 7.4) (0.05ml)
   - Sodium cholate (0.25% w/v) or Triton® X-100 (0.25% w/v) (0.02 mL)
   - 50-m*M* $MgCl_2$ (0.05 mL)
   - $H_2O$ (0.18 mL)
   - 5-m*M* p-Nitrophenol (0.05 mL)
2. Preincubate mixture for 2 minutes at 37°C.
3. Add 30-m*M* UDP-GA (0.05 mL) to start the reaction; incubate for 10 to 30 minutes at 37°C.
4. Stop the reaction and precipitate protein with 1 mL 5% trichloroacetic acid.
5. Centrifuge to clear protein.
6. Add 0.25 mL 2-*M* NaOH to 1 mL of supernatant fluid.
7. Read supernatant OD at 405 nm.
8. *p*-Nitrophenol extinction coefficient is $18.1 \times 10^3$ $cm^2$/mol at pH > 10.

### Single-Substrate Glucuronosyltransferase Assays (Radiometric)

Radiolabeled single-substrate assays that have been developed for determination of UDP-GT activity often use radiolabeled substrates and some means to separate unreacted substrate from the respective glucuronide conjugates. The first of the methods described is a rapid radiometric method developed by Lucier and McDaniel [187]. This method is particularly useful for the study of steroid conjugation reactions.

*Rapid Radiometric Glucuronosyltransferase Assay*

The incubation system for glucuronidation measurements is added to liquid scintillation vials and consists of the following: 1.2 mL 75-m*M* Tris–HCl buffer (pH 7.4), 1.0-μ*M* UDP-GA, 10.0-μ*M* unlabeled substrate in 50 μL methanol, and $1 \times 10^5$ dpm-labeled substrate in 50 μL methanol. This volume of methanol is used to ensure substrate solubilization and has no apparent effect on glucuronosyltransferase activity. The incubation contents are warmed at 37°C for 3 minutes, and then 0.4 to 0.6 mg microsomal protein is added. The incubation period is approximately 10 minutes. The reaction is stopped by the addition of 10 mL nonaqueous scintillation fluid. Samples are capped and shaken for 10 seconds on a vortex mixer, and radioactivity is counted in the same vials in which the incubation reactions were performed. Addition of the toluene-based scintillation fluid results in a two-phase mixture (toluene on top and the aqueous fraction on the bottom). Unreacted substrate partitions into the toluene, and this radioactivity is detected in a liquid scintillation counter equipped with an automatic quench analyzer. Glucuronides remain in the aqueous fraction, and because $^{14}C$ and $^{3}H$ in a water medium do not scintillate, radioactivity associated with glucuronides is not detectable by liquid scintillation spectrometry. This phenomenon enables steroid glucuronidation rates to be measured by substrate disappearance. Blank values are obtained by omitting UDP-GA from the reaction media. The incubation blanks represent 0% activity and correct for the amount of substrate remaining in the aqueous fraction (incubation medium). Glucuronides detected after addition of scintillation fluid to incubation media reflect the amount of radioactivity detected after 100% glucuronidation of substrate. Enzyme activity using 300 nmol substrate in the incubation medium is expressed by the equation:

$$1 - \frac{\text{Radioactivity after incubation} - \left[\text{Radioactivity after incubation} \times \begin{array}{c}\text{Fraction of respective}\\ \beta\text{-D-glucuronide detected}\end{array}\right]}{\text{Radioactivity in black}}$$

$$\times\ 300\ \text{nmol} = \text{nmol substrate conjugated}$$

This assay procedure is applicable to a wide range of substrates and enzyme reactions in which the polarity of the product is significantly different from the substrates.

### *1-[1-$^{14}$C]-Naphthol Glucuronidation*

A commonly used assay for glucuronidation is similar to the technique described by Hazelton et al. [146]. This technique uses radiolabeled 1-[1-$^{14}$C]naphthol as the substrate and removes unreacted naphthol from the glucuronide conjugate by a chloroform extraction. The method is simple and sensitive.

*Technique*

1. Incubation mixture for liver microsomes:
   - *Native*—Microsome pellet is suspended in 0.25-*M* sucrose (0.33 g equivalent wet weight of liver/mL 0.25-*M* sucrose). This is then diluted with an equal volume of 0.25-*M* sucrose.
   - *Activated*—Similar to native microsomes; however, an equal volume of sucrose contains 16-m*M* CHAPS and is agitated for 20 minutes at 4°C.
2. Incubation mixture: 0.2-*M* Tris–HCl (pH 7.5), 10-m*M* $MgCl_2$, 2.2-m*M* saccharic acid-1,4-lactone, 0.5-m*M* 1-naphthol, and 0.04 μCi 1-[1$^{14}$C]naphthol at 3°C. Total volume is 0.5 mL.
3. Add UDP-GA to start the reaction such that the final concentration is 4 m*M*.
4. Terminate the reactions at the desired time (10 minutes) by the addition of an equal volume of ice-cold ethanol.
5. Remove unreacted 1-naphthol with an 8-mL chloroform extraction.
6. Determine the remaining radioactivity in the aqueous phase in a liquid scintillation counter.

### *Multisubstrate Glucuronosyltransferase Assay Using UDP-[$^{14}$C]GA*

Assay of glucuronosyltransferase activity can be accomplished using a multiple-substrate approach. This type of approach allows examination of multiple substrate interactions with the various isozymes for definition of UGT isozyme catalytic boundaries. One such method, through the use of HPLC and incubation with UDP-[$^{14}$C]GA, directly measures $^{14}$C-labeled glucuronides of a range of model substrates [180]. This method, as described later, has been used successfully with a variety of aglycone substrates, including 1-naphthol, 4-nitrophenol, androsterone, phenol, phenolphthalein, 2-hydroxybiphenyl, 4-hydroxybiphenyl, menthol, diethylstilbestrol, 1-hydroxypenzo(*a*)pyrene, 3-hydroxybenzo(*a*)pyrene, and 9-hydroxybenzo(*a*)pyrene. The glucuronides of these substrates were all resolved from precursors upon HPLC with this system.

*Technique*

*Incubation:*

1. The incubation mixture is comprised of 50-m*M* Tris–maleate buffer, 10-m*M* $MgCl_2$ (pH 7.4), 2.7-m*M* UDP-GA, UDP-[$^{14}$C]GA (0.25 μCi), and substrate such that the substrate is delivered in 5 μL of carrier vehicle and the final concentration is 1 m*M*. The final volume of the reaction mixture is 100 μL.
2. Add the detergent Lubrol-PX® such that the detergent-to-protein ratio is 0.25.
3. Preincubate the mixture for 2 minutes at 37°C. Start the reaction by adding 50 μL of microsomal protein (0.1 to 0.5 mg).
4. Incubate the reaction 10 to 30 minutes with gentle shaking at 37°C. Time should be regulated such that the reaction rate is linear and no more than 10% of the UDP-[$^{14}$C]GA is used during the reaction.
5. Stop the reaction with 475 μL of 100% ethanol at −15°C.
6. Centrifuge in a microfuge at 16,000 *g* for 3 minutes.
7. Filter supernatant fraction through a 0.45-μm filter.
8. Store samples up to 1 week in liquid nitrogen.

*HPLC analysis:* Analyses are performed on a Partisil® 5 polar amino-cyano (PAC) bonded phase column (4.5 mm × 26 cm). The mobile phase consists of a linear gradient (1.5 mL/min) with acetonitrile and the ion pairing agent tetrabutylammonium hydrogen sulfate (TBAHS). Initially, the gradient runs from 100% acetonitrile to 100% 0.01-*M* TBAHS over 20 minutes. The gradient is then maintained at 100% 0.01-*M* TBAHS for 25 minutes before returning to 100% acetonitrile over 10 minutes. Radioactivity measurements of HPLC eluent are taken at 20-second intervals over the 45-minute gradient.

### Multisubstrate Radiometric Assay for Glucuronidation in Intact Cells

In addition to microsomal preparations, glucuronidation may be studied in intact cells, such as those in tissue culture. As described earlier, glucuronidation is often studied with radiolabeled UDP-[$^{14}$C]GA as the measurable moiety in microsomal preparations. These types of studies are not possible in culture, as the radiolabeled UDP-[$^{14}$C]GA is not cell membrane permeable in intact cells. A novel method bypasses this limitation by labeling GA in the cell by using permeable [$^{14}$C]fructose as a GA precursor [188]. In this technique, donor animals are fasted so endogenous glucuronic acid is depleted. Because GA is derived from both glycogenolysis and gluconeogenesis, use of radiolabeled gluconeogenic substrates such as [$^{14}$C]fructose allows the radiolabel to be incorporated into UDP-GA and any subsequent glucuronides. The specific activity of the glucuronide is similar to that of the substrate. This method incorporates three major steps following cell procurement: (1) incubation, (2) HPLC of radiolabeled glucuronides, and (3) specific activity determinations. The method has been applied to the glucuronidation of a variety of substrates, including 4-nitrophenol and 1-naphthol.

#### Technique

*Incubation:*

1. To cell suspensions prepared from fasted animals, add 1-naphthol or 4-nitrophenol substrate dissolved in dimethylsulfoxide (final DMSO concentration is 1% v/v).
2. Add 200 μ*M* [$^{14}$C]fructose (0.5% μCi).
3. Gas with 95% $O_2$ and 5% $CO_2$, and cap vials.
4. Incubate mixture for 20 to 90 minutes in a shaking water bath at 37°C.
5. Centrifuge (12,000 *g*, 30 seconds) through 250 μL of silicone oil (Dow Corning® 550 dinonylphthalate 2:1 v/v) to terminate incubation and separate cells from medium.
6. Heat supernatant fraction in boiling water bath for 5 minutes. Remove denatured protein by centrifugation at 12,000 *g* for 1 minute.
7. Store for glucuronide analysis.

*HPLC analysis:*

1. Carry out analysis on a 4.5 mm × 25-cm Partisil® 5 PAC column.
2. The mobile phase consists of a linear gradient from 100% acetonitrile to 67% 0.01-*M* TBAHS over 20 minutes. The gradient is maintained at 67% 0.01-*M* TBAHS for 5 minutes before conditions are returned to 100% acetonitrile over the subsequent 10 minutes (1-mL/min flow rate).
3. Collect HPLC fractions at 0.5-minute intervals.
4. Identify glucuronide peaks by hydrolysis with glucuronidase and comparison to nontreated chromatogram.
5. Remove 200 μL of HPLC fraction.
6. Add 40 μL 0.5-*M* sodium phosphate buffer (pH 7) containing 1000 U P-glucuronidase (*Escherichia coli*).
7. Incubate for 2 hours at 37°C.
8. Assay 0.1 mL of incubation mixture by HPLC, identifying glucuronide peaks by their disappearance.

*Specific activity determination of [$^{14}$C]glucose:*

1. Take 0.1 mL of boiled and centrifuged supernatant fraction from the incubation procedure. Add 0.9 mL of 100-m*M* Tris buffer, with 1-m*M* magnesium acetate, 1.7-m*M* NAD$^+$, and 1.1-m*M* ATP.
2. Incubate mixture at 37°C in disposable cuvettes. Measure change in absorbance at 340 nm following addition of 1 U hexokinase per 1 U glucose-6-phosphate dehydrogenase, as well as after the addition of 0.1 U of 6-phosphogluconate dehydrogenase. Glucose concentrations are calculated using $6.22 \times 10^3$ *M*/cm as the extinction coefficient for NADH (total glucose). Hexokinase converts glucose to glucose-6-phosphate. Glucose-6-phosphate dehydrogenase then converts glucose-6-phosphate to 6-phosphoglucono-δ-lactone, which is hydrolyzed to 6-phosphogluconate. This substrate is converted by 6-phosphogluconate dehydrogenase to ribulose 5-phosphate, liberating $CO_2$.
3. Under identical conditions, incubate another aliquot in a 20-mL stoppered vial with a center well. Add enzymes by syringe through stopper.
4. Add 0.1 mL of 20% $HClO_4$ (v/v) to end the incubation and drive off $^{14}CO_2$.
5. The $^{14}CO_2$ released is absorbed by 2-phenethylamine/methanol (0.2 ml, 1:1 v/v) on a filter paper strip in center well. Allow $^{14}CO_2$ absorption to proceed 1 hour with shaking at room temperature.
6. Remove wells to vials containing scintillation fluid for liquid scintillation counting.
7. Radioactivity in the original glucose is calculated as being six times that in the single carbon released as $^{14}CO_2$.

## Microsomal Glutathione Transferases

Glutathione transferase (GST), a multigene family of isozymes, catalyzes the nucleophilic attack of glutathione on a variety of electrophilic compounds. Besides detoxifying electrophilic xenobiotics, such as chemical carcinogens, environmental pollutants, and antitumor agents, GSTs also inactivate endogenous α,β-unsaturated aldehydes, quinines, epoxides, and hydroperoxides formed as secondary metabolites during oxidative stress [189]. GST biosynthesis reactions include leukotrienes, prostaglandins, testosterone, and progesterone [189]. The metabolism of endogenous lipid mediators influences diverse signaling pathways leading to many biological consequences [189]. GSTs detoxify xenobiotics through the mercapturic acid pathway catalyzing the first of four steps required for the synthesis of mercapturic acids [189]. As part of an integrated defense strategy, GST effectiveness is dependent on the synthesis of GSH and on the activity of transporters to remove the glutathione conjugates from the cell [189]. The conjugation of glutathione to xenobiotics most often leads to the formation of less reactive products that are readily excreted; however, in a few instances, the glutathione conjugate is more reactive than the parent compound, such as short-chain alkyl halides that contain two functional groups or the solvent dichloromethane and the 1,2-dihalo-ethanes, resulting in bioactivation [189].

These enzymes may be found both in the soluble or membrane-bound cell fractions. The majority of cytosolic GST isozymes are found in the cytoplasm of the cell, but mouse and human alpha-class GST4-4 can associate with mitochondria and membranes as can mouse GSTM1-1 [187]. Mouse, rat, and human possess only a single mitochondrial kappa GST, a distinct type of GST more bacterial in nature that is also present in peroxisomes [189]. The cytosolic and mitochondrial GST enzymes are only distantly related. The third family of GSTs, microsomal GSTs that bear no structural resemblance to the cytosolic and mitochondrial GSTs, is now referred to as membrane-associated proteins in eicosanoid and glutathione (MAPEG) metabolism [189]. All three families, however, contain members that catalyze the conjugation of glutathione with 1-chloro-2,4-dinitrobenzene (CDNB) and exhibit glutathione peroxidase activity toward cumene hydroperoxide (CuOOH) [189]. Assays of glutathione transferase activity have been accomplished spectrophotometrically. This procedure is facilitated by substrates that change optical absorbance upon glutathione conjugation. A common procedure is to assay glutathione transferase activity with CDNB [190,191].

### Technique

*Reagents:*

- 10-m*M* glutathione
- 20-m*M* 1-chloro-2,4-dinitrobenzene in ethanol
- 0.1-*M* potassium phosphate buffer
- Triton® X-100

*Assay conditions:*

- Final reaction volume 2 mL
- 30°C reaction temperature

*Incubation mixture:*

- 1-m*M* 1-chloro-2,4-dinitrobenzene/ethanol (50 μL)
- Microsomes, washed twice with 0.15-*M* Tris–HCl (pH 8.0) to remove cytosolic contamination (20 to 100 μL, dependent on activity and protein content)
- 0.1-*M* potassium phosphate buffer (pH 6.5; minimizes nonenzymatic reactions); balance of reaction volume to 1 mL
- 0.1 to 0.2% Triton® X-100 (2 to 4 μL)
- 1-m*M* glutathione (100 μL), added to start reaction

*Analysis:*

1. Read absorbance at 340 nm.
2. Molar extinction coefficient is 9.6-m*M*/cm.

## Hydrolytic Enzymes

The endoplasmic reticulum also contains a variety of hydrolytic enzymes that exhibit a dual localization in that they are also active in lysosomes. These enzymes include β-D-glucuronidase (performs the reverse reaction of UDP-glucuronidase by liberating the free aglycone from β-D-glucuronic acid conjugates), acid and alkaline phosphatases, and aryl sulfatase. The activity of β-glucuronidase may play a role in the accumulation, targeting, and disposition of drugs [192]. Because β-glucuronidase is expressed in cell lysosomes, in some larger tumors β-glucuronidase is found at high levels in necrotic tissue areas. Several glucuronide prodrugs have been synthesized that can be activated by β-glucuronidase, mediating drug release specifically at the tumor sites [193]. In liver, a significant fraction of β-glucuronidase (30 to 50%) is not found in the lysosome but is localized to the endoplasmic reticulum [194]. Hepatic microsomal β-glucuronidase appears likely to influence the biliary excretion and hence the hepatic elimination of endogenous and xenobiotic substrates (e.g., carcinogens) which undergo hepatic glucuronidation [195]. Organophosphate compounds are known to cause the selective release of rat liver microsomal β-glucuronidase into plasma [196]. Hydrolytic enzymes, such as β-glucuronidase, can be detected by histochemical or biochemical methods. One simple method for measuring mammalian β-glucuronidase activity is as follows.

### Technique

1. The incubation medium contains 1.0-m*M* substrate (β-D-glucuronide of *p*-nitrophenol, 4-methyl-umbelliferone, or phenolphthalein) in 50-m*M* acetate buffer (pH 4.5).

2. Warm the incubation mixture for 3 minutes at 37°C.
3. Initiate the reaction by adding 1.0 mg microsomal protein.
4. After a 10-minute incubation period, stop the reaction by adding 5 mL glycine buffer (see *p*-nitrophenol glucuronidation assay).
5. Remove protein by low-speed centrifugation and measure the formation of the product spectrophotometrically (*p*-nitrophenol, 405 nm; phenolphthalein, 550 nm; and methyl-umbelliferone, 365 nm).

The value of measuring conjugation–deconjugation enzyme activities rests with assessing the potential capacity of a cell or organ to deactivate or activate potentially toxic reactive intermediates. Although these measurements are an indirect approach to assessing tissue and cell susceptibility to these highly toxic chemical species, the data generated from past studies have proved highly useful for predicting cellular potential for injury.

## PEROXISOMES

The peroxisomes are single, membrane-bound cytoplasmic organelles present in the cells of animals, plants, fungi, and protozoa. With the exception of mature erythrocytes, peroxisomes appear to be present in all eukaryotic cells. Studies by de Duve and coworkers [197] have shown that rat liver peroxisomes contain both hydrogen-peroxide-generating oxidase enzymes and hydrogen-peroxide-degrading catalase, causing the breakdown of hydrogen peroxide to oxygen and water. Peroxisomes are involved in lipid, sterol, and purine metabolism, as well as peroxidative detoxification. Peroxisomes contain a complete fatty acid β-oxidation cycle. In addition, peroxisomes are known to proliferate in response to both natural ligands, such as fatty acids, and a group of chemicals known as peroxisome proliferators or peroxisome proliferating agents (PPAs). PPAs are a very structurally diverse group of chemicals that include hypolipidemic agents, phthalate esters, perfluorocarboxy acids, solvents, and phenoxyacetic acid herbicides. All are considered to be nongenotoxic carcinogens because when tested they fail to cause DNA damage directly. Peroxisome proliferation consists of gross hepatomegaly associated with both hypertrophy and hyperplasia, with hepatocytes exhibiting a marked proliferation of peroxisomes and the endoplasmic reticulum. These morphological changes are associated with changes in fatty acid β-oxidation, lipid metabolism, and carcinogenesis in mice and rats [198,199]. Extrahepatic tissues, including the kidney, intestine, testis, and adipose tissue, have also been shown to be susceptible to peroxisomal proliferation, but peroxisomes are most abundant in the liver and kidney. Peroxisomal proliferation has been linked with reproductive and developmental toxicity, as well as hepatic and testicular cancers [200,201]. A marked species-specific sensitivity to peroxisomal proliferation has been noted, with humans and apes considered refractory species, whereas rats and mice are sensitive species; however, even in humans, peroxisomal dysfunction can result in severe disease states, including Zellweger syndrome, adrenoleukodystrophy, and Refsum syndrome [202].

By far the majority of peroxisomal enzymes are lipid-metabolizing enzymes. Over 40 different enzymes have been identified, including enzymes involved in fatty acid activation, elongation, and β-oxidation; the oxidation of polyunsaturated fatty acids, carnitine acyltransferases, and acyl-CoA hydrolases; bile acid synthesis; cholesterol and dolichol synthesis; glycerolipid synthesis; purine catabolism; polyamine catabolism; amino acid and glyoxylate metabolism; oxygen metabolism; reactive oxygen species; and nucleotide binding proteins [203,204]. The levels of the peroxisomal β-oxidation enzymes, such as acyl-CoA oxidase and the microsomal enzyme P450 4A1, are elevated between 10- and 30-fold in response to PPA administration. This is paralleled by an increase in the respective mRNAs which is observed as early as 2 hours after administration. This coordinate and rapid increase in gene expression suggests a common mechanism of induction and peroxisome proliferation.

The peroxisome proliferator-activated receptors (PPARs) are transcriptional factors belonging to the ligand-activated nuclear receptor superfamily. PPARs are ubiquitously expressed throughout the body. When activated by endogenously secreted prostaglandins and fatty acids, they initiate transcription of an array of genes that are involved in energy homeostasis. So far, three major types have been identified, namely PPAR-α (NR1C1), PPAR-β/δ (NR1C2), and PPAR-γ (NR1C3). PPAR-α and PPAR-γ are crucial for lipid and glucose metabolism, respectively. Although limited information is available on PPAR-β biological functions, recent studies have shown that PPAR-β also regulates glucose metabolism and fatty acid oxidation. The discovery of PPAR-α agonists such as fibrates and PPAR-γ agonists such as thiazolidinediones has led to recognition of the mechanisms involved in ameliorating the adverse effects of chronic disorders such as atherosclerosis and diabetes. In addition, PPARs are also involved in the regulation of various types of tumors, inflammation, cardiovascular diseases, and infertility. The importance of these transcription factors in physiology and pathophysiology has instigated considerable research in this field [205].

Studies have shown that all of the PPARs form heterodimers with RXRα (retinoic acid X receptor; X = 9-*cis*-retinoic acid) via the ligand-binding domain (LBD). Heterodimerization facilitates binding to the DNA PPAR response element (PPRE) consisting of a direct repeat (DR)-1 of the hexanucleotide AGGTCA separated by a single nucleotide spacer. Binding of the heterodimer to the PPRE results in transcriptional activation of genes containing the appropriate regulatory elements for activation by the specific PPAR [205].

PPARγ is expressed from three different promoters yielding three different isoforms with different tissue specificity. Whereas PPAR$\gamma_1$ is found in a broad range of tissues, PPAR$\gamma_2$ is restricted to adipose tissue, and PPAR$\gamma_3$ is highly expressed in macrophages, the large intestine, and white adipose tissue [203]. Endogenous ligands of PPARγ include polyunsaturated fatty acids, while exogenous ligands are most notably the thiazolidinediones; a triterpenoid, bisphenol diglycidyl ether; some synthetic chemicals; and several natural products, including the flavonoids kampferol and apigenin. PPARγ is involved in adipocyte differentiation, insulin sensitization, inflammation, and atherosclerosis [203].

PPARβ/δ is expressed in a wide range of cells and tissues, with higher expression in brain, adipose tissue, and skin. Endogenous ligands include fatty acids, while exogenous ligands include some synthetic chemicals and nonsteroidal antiinflammatory drugs (NSAIDS). Functionally, PPARβ/δ is involved in hyperlipidemia, inflammation, atherosclerosis, obesity, and fertility and has several functions in the nervous system and muscle metabolism [205,206].

PPARα serves as a receptor for PPAs and is required for both peroxisomal proliferation and the induction of certain peroxisomal and microsomal enzymes [207], as evidenced by the absence of response when PPARα expression is disabled by targeted disruption [208]. PPARα has received the most attention related to activation by xenobiotics and exogenous ligands include several phthalate esters, solvents, herbicides, pesticides and drugs like the fibrates, aspirin and NSAIDS. At least 15 different genes are activated in response to PPAs or elevated levels of fatty acids and other endogenous compounds. PPARα is involved in lipid metabolism, inflammation and atherosclerosis [205].

The investigation of the myriad functional and structural features of peroxisomes requires the preparation of highly purified organelle fractions separated from microsomes and other contaminants, such as lysosomes and mitochondria. This is hampered by the fragility of peroxisomes and their relative paucity (2.5% of the total protein) within the cell; thus, separation conditions must maintain the mechanical, hydrostatic and osmotic integrity of the sample. Generally, three steps are employed in the purification of peroxisomes: (1) homogenization or tissue disruption, (2) subfractionation by differential centrifugation, and (3) isolation of the purified fraction by density gradient centrifugation (Figure 39.10). Purified peroxisomes can be used for enzyme assays, which are too numerous and varied to be covered in detail here, or for immunodetection of peroxisome-associated proteins using the methods covered under SDS-PAGE of microsomal fractions for P450. Several commercially available antibodies recognize catalase, PMP70, and any proteins containing certain peroxisomal membrane-targeting sequences. Fractions may also be observed for purity or structural features using electron microscopic evaluations to check for mitochondria or by staining with 3,3′-diaminobenzidine, which produces an electron-dense reaction product with catalase found throughout the peroxisome. A method for the isolation of highly purified peroxisomes from rat liver follows [207].

### *Technique*

*Reagents:*

- Homogenization buffer—150-m*M* sucrose, 6.717-m*M* morpholinopropase sulfonic acid (MOP; free acid), 1-m*M* EDTA, and 0.9% saline (NaCl)
- Gradient buffer—6.717-m*M* MOPS, 1-m*M* EDTA, and 1 mL of ethanol per liter, adjusted to pH 7.2 with NaOH. May be stored at 4°C. Just prior to use, per 100 mL add 0.2 mL of 0.1-*M* PMSF, 0.1 mL of 1-*M* ε-aminocaproic acid, and 20 μL of 1-*M* DTT.
- Metrizamide solutions—stock solution, 60% w/v (at 15°C density = 1.328 g/mL; 40%, 1.218 g/mL; 20%, 1.108 g/mL) in gradient buffer. To make one gradient, take 3.76, 3.38, 3.53, 2.06, and 3.2 mL of the stock solution and bring to total volumes of 10, 7, 6, 3, and 4 mL, respectively, using gradient buffer. Resultant densities are 1.12, 1.155, 1.19, 1.225, and 1.26.

*Isolation of purified peroxisomes:*

1. Anesthetize the animal.
2. Open the abdominal cavity and perfuse the liver with 0.9% saline via the portal vein until all the blood is removed.
3. Remove the liver, cut into small pieces, and add 3 mL/g tissue (wet liver weight) of the homogenization buffer. Place into ice-cold Potter–Elvehjem homogenizer and homogenize with a single up-and-down stroke using a motorized, loose fitting pestle.
4. Decant into a 50-mL centrifuge tube and centrifuge at 70 *g* for 10 minutes in a refrigerated centrifuge.
5. Collect the supernatant fraction and resuspend the pellet in 2 mL/g of ice-cold homogenization buffer, rehomogenize as before, and spin again under the same conditions. Collect the second supernatant fraction and combine it with the first to make up the postnuclear supernate. The remaining pellet is the nuclear fraction.
6. Centrifuge the postnuclear supernate at 1950 *g* for 10 minutes in a refrigerated centrifuge.

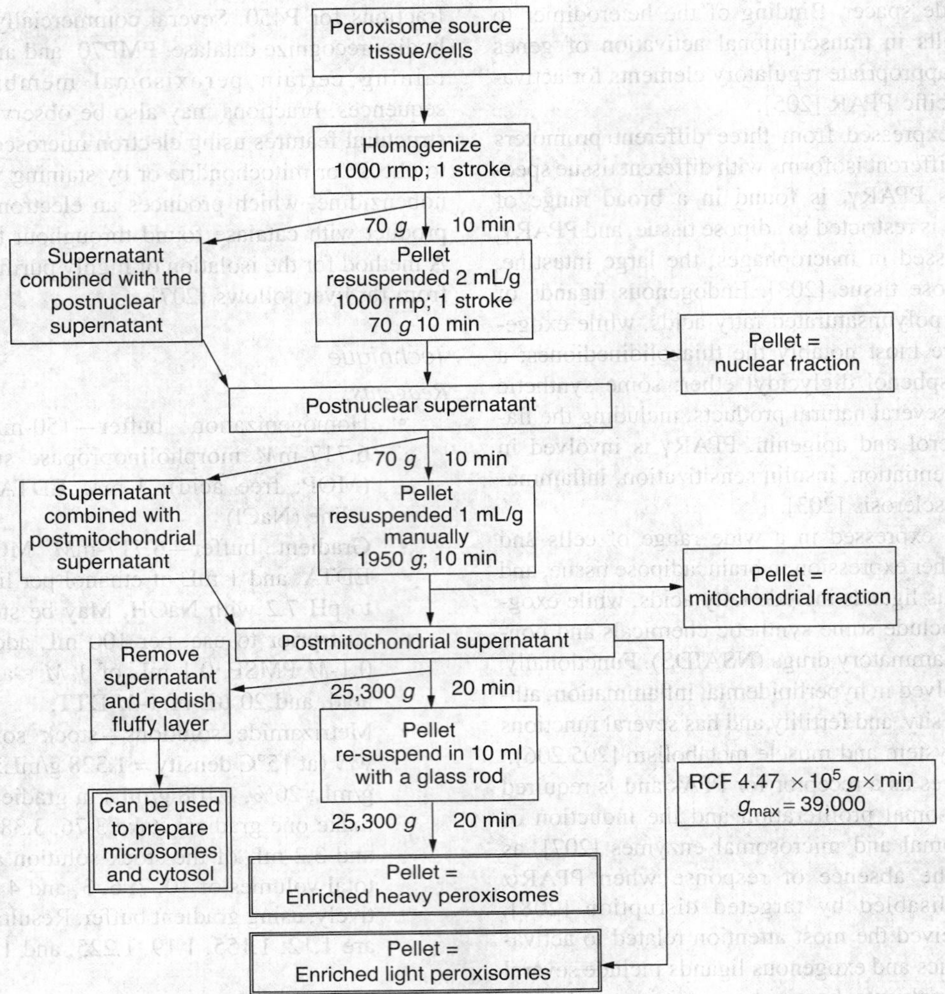


**FIGURE 39.10** Flow chart of the peroxisome isolation method. Rat liver peroxisomes prepared by differential centrifugation are purified to more than 95% of the total protein content with contributions of about 2% each from mitochondria and microsomes and 1% from lysosomes. Peroxisomal fractions prepared by density-gradient centrifugation are highly purified (98–99%), as confirmed by electron microscopy. (Adapted from Grimaldi, P.A., *Biochimie*, 87, 5–8, 2005.)

7. Collect the supernate and resuspend the pellet manually in 1 mL/g of ice-cold homogenization buffer and spin again under the same conditions. Collect the second supernate and combine with the first to make up the postmitochondrial supernate. The final pellet contains the majority of mitochondria, large microsomal sheets, and some remaining nuclei.
8. Centrifuge the postmitochondrial supernate at 25,300 *g* for 20 minutes and, using suction, remove the supernate and the reddish fluffy layer. Resuspend the pellet in about 10 mL of ice-cold homogenization buffer using a glass rod and centrifuge again at 25,300 *g* for 15 minutes. Resuspend the final pellet (enriched heavy peroxisomes) in 5 mL of ice-cold homogenization buffer using a glass rod. The supernate may be used to prepare a microsomal fraction with the supernatant fraction from that process constituting the cytosolic fraction. The postmitochondrial supernate may also be centrifuged at an integrated relative centrifugal force (RCF) of $4.47 \times 10^5$ $g \times$ minutes ($g_{max}$ = 39,000), and the resulting pellet represents the enriched light peroxisomes. For highly purified peroxisomes, complete the density gradient fractionation described.

*Isolation of highly purified peroxisomes*

1. Layer sequentially 4, 3, 6, 7, and 10 mL of the metrizamide solutions described above (1.26 to 1.12 g/mL) in a 40-mL Quick-Seal™ polyallomer centrifuge tube (Beckman™) to form a discontinuous gradient.
2. Freeze the gradient in liquid nitrogen and store at –20°C. Prior to use, thaw the gradient quickly

at room temperature using a metallic stand to transform the discontinuous gradient into one with an exponential profile.
3. Layer 5 mL of the appropriate peroxisomal fraction on top of the thawed gradient and seal the tube.
4. Centrifuge the gradient in a vertical-type rotor at an integrated RCF of $1.256 \times 10^6$ $g \times$ minutes ($g_{max}$ = 39,000) using slow acceleration and deceleration. Under these conditions, the heavy peroxisomes band at 1.23 to 1.24 g/ mL and the light peroxisomes band at 1.20 to 1.21 g/mL.
5. Recover the peroxisomes using a fraction collector or by puncturing the tube and aspirating from the bottom using a syringe. Store the fractions at −80°C until used for electron microscopy, enzyme assays, or immunodetection.
6. To remove metrizamide, which interferes with some enzyme activities (e.g., urate oxidase) or with protein determinations using the Lowry method, dilute the peroxisomal fraction containing heavy or light peroxisomes about 10-fold with homogenization buffer followed by centrifugation at 25,000 or 39,000 *g*, respectively, to pellet the peroxisomes.

## CYTOSKELETON

In recent years, there has been a growing appreciation of the roles played by the cytoskeleton and related proteins in numerous functions of the cell. In particular, the importance of the cytoskeleton in regulation of pinocytosis via clathrin-coated vesicles for both neuronal synaptosomic vesicle recycling [210] and other endocytic functions of other cell types, such as those of the kidney tubular epithelium [211]. This endocytic pathway has been shown [94] to be disrupted by nephrotoxic agents such as cadmium following administration of cadmium metallothionein (CdMT) *in vivo* and it has been hypothesized that $Cd^{2+}$-mediated disruption of the cytoskeleton [209] is a primary underlying mechanism following release of the $Cd^{2+}$ ions by lysosomal proteolysis of the CdMT complex. This organelle system may play an important role in mechanisms of chemical toxicity via its central place in endocytotic mechanisms and the intracellular trafficking of vesicles. An excellent overview of the various important roles played by actin and microtubules in the cytoskeleton in endocytosis in polarized cells, such as kidney cells, has been published by Apodaca [213]. Numerous cytochemical and biochemical methods are available for studying the various aspects of the cytoskeleton, but the following review focuses only on a few of the more commonly utilized approaches for actin, microtubules, and the central acting GTPase dynamin 2.

### Cytochemical Approaches

#### Phalloidin Binding

The protein phalloidin binds to F-actin in the cytoskeleton with great affinity, and the conjugation of fluorescent dyes such as Rhodamine Green™ or Oregon Green® [210] has allowed visualization of the cytoskeleton during recycling of synaptosomic vesicles in neuronal cell types or kidney proximal tubule cells during absorption of albumin [214].

#### Dynamin

Dynamin (Dyn2) is a GTPase that is essential for endocytosis via an interaction with actin [215,216]. This protein and others appear to regulate endocytosis via direct interactions with F-actin. Dynamin also forms lockwasher-type collars around the necks of clathrin-coated invaginations of the cell membrane, and it is believed that a conformational change occurs in dynamin following GTP cleavage that results in endocytic vesicle formation [217]. This protein thus plays a major role in the formation of endocytic vesicles. In addition, dynamin has been shown to regulate and to be a component of actin comets, which are actin-based motile vesicles that migrate within living cells along actin filaments [216]; thus, dynamin also plays an important role in the intracellular vesicular trafficking of vesicles. An excellent overview of the various forms of endocytosis and the roles of dynamin in this important cellular process has been published by Connor and Schimd [218]. This protein may be visualized within intact cells via commercially available fluorescent-tagged antibodies or by the western blot technique described earlier.

### Biochemical Methods

#### Taxol® Binding to Microtubules

Taxol® is an anticancer drug that interferes with the division of cancer cells via binding to the tubulin subunits [219]. Conjugation fluorscein dyes that bind to Taxol® to form the fluorescent taxoids (Flutax-1 and Flutax-2) have been demonstrated to allow assessment of the kinetics of Taxol® binding to microtubules [219]. These dyes bind to the microtubules with Ka bands on the order of $10^7$ $M^{-1}$.

*Technique*

1. Incubate Flutax-1 and Flutax-2 with samples of cross-linked microtubules for 1 hour over a concentration range of the dyes and at various temperatures.
2. Centrifuge the samples for 20 minutes at 50,000 rpm in an ultracentrifuge.
3. Save the supernatants and re-suspend the pellets in 10-m*M* phosphate buffer at pH 7.0 with 1% SDS. Dilute the supernatants and solubilized pellets 1:5 in the buffer.

4. Measure fluorescence in a spectrofluorometer with excitation and emission wavelengths of 492 and 522, respectively, using 5-mm excitation and emission slits.
5. Determine concentrations of the bound dye by using appropriate Flutax standards.

### Cytochemical Method

The cytoskeletal system in cells is composed of three main types of fibrils: microtubules (25 nm in diameter), which consist primarily of the protein tubulin; microfilaments (6 nm in diameter), which are composed primarily of actin; and intermediate filaments (10 nm in diameter), which consist of different tissue-specific proteins [215,216]. This system of fibrillar structures criss-crosses the cytoplasm of eukaryotic cells to form an internal network that changes dynamically in response to external and internal stimuli. This network functions to maintain the internal structure and shape of individual cells and also is involved in cell processes that involve movement such as exocytosis and endocytosis, organelle transport, protoplasmic streaming, and locomotion, as well as cellular polarity, cell adhesion, and cell division [222–227]

#### Ultrastructural Techniques

Visualization of proteins within microtubules and microfilaments can be accomplished by double antibody procedures in which antibodies to specific cytoskeletal proteins are reacted with cytoskeletal preparations. A second antibody that has been tagged with either a fluorescent dye molecule or a label visible by electron microscopy (e.g., colloid gold) is then allowed to react with the first immunoglobulin, and the structures are visualized by fluorescence or electron microscopy [228,229]. The following fluorescence microscopic technique was developed by Zhao et al. [230].

*Technique*

1. Wash cells grown on coverslips (12 mm × 12 mm) with a microtubule stabilizing buffer: PM2G (0.1-*M* PIPES, 1 MM $MgSO_4$, 2-*M* glycerol, and 2-m*M* EGTA; pH 6.9).
2. Fix cells with 3.7% formaldehyde in PM2G for 30 minutes and wash with 0.1-*M* glycine in PBS (pH 7.4) for 5 minutes.
3. Extract cells by incubating in 0.3% Nonidet™ P40 (NP40) in PBS for 10 minutes, followed by two washes with PBS.
4. Incubate the cytoskeletal preparations with a mixture containing NBD-phallacidin (which labels microfilaments directly) and diluted rabbit antitubulin antiserum (the primary antibody) for 30 minutes in a moist chamber at 37°C.
5. Wash thoroughly by dipping in PBS, drain, and incubate the coverslips with Rhodamine-conjugated goat and anti-rabbit IgG (the secondary antibody) for 30 minutes in a moist chamber at 37°C.
6. Wash coverslips thoroughly in PBS followed by distilled water; drain, and mount (with cells side down) on glass slides with Glevatol mounting solution.
7. Store slides flat for 24 hours and then examine using a fluorescence microscope.

Both metal ions [229,231–235] and organic compounds [230,236] have been shown to affect microtubule and microfilament organization in cells. Due to the fact that most nonviable cells are removed by the many extractions and washings involved in this procedure, results observed on the final slides will reflect early effects on cytoskeletal elements and not those that reflect alterations due to cell death.

## PROTEIN SYNTHESIS

Although much of the pharmacological and toxicological research on microsomes focuses on their role in metabolic activation and deactivation reactions, the main function of this organelle involves protein synthesis. Increasingly, researchers are attempting to identify sensitive biochemical indicators of toxicity (usually proteins) that have predictive and diagnostic value and also provide insight into the mechanisms of toxic actions of these chemicals. Although a survey of the genetics and molecular biology of protein synthesis is not in the scope of this chapter, some general useful methods can be applied to investigations on the effects of chemicals on overall or specific protein synthesis. These methods involve injection of radiolabeled amino acids into intact animals (usually tail vein), and the incorporation of radiolabel into specific organelles or specific proteins is then determined. Obviously, protein purification or immunochemical procedures must be undertaken to assess synthesis and degradation of individual proteins. When conducting such pulse-label experiments, it is essential that samples are taken over a wide range of sample periods so meaningful and valid conclusions can be made concerning alterations either in synthetic or degradative phases for a particular protein. For example, if a toxic chemical selectively alters protein degradation and a pulse label is conducted during the synthetic phase, then the investigator could miss critical information. If possible, it is desirable to conduct pulse-label experiments with carbon rather than tritium to avoid nonspecific redistribution of the radiolabel.

Another method, two-dimensional gel electrophoresis, is being used to determine protein synthesis profiles [238]. This procedure involves administration of high-

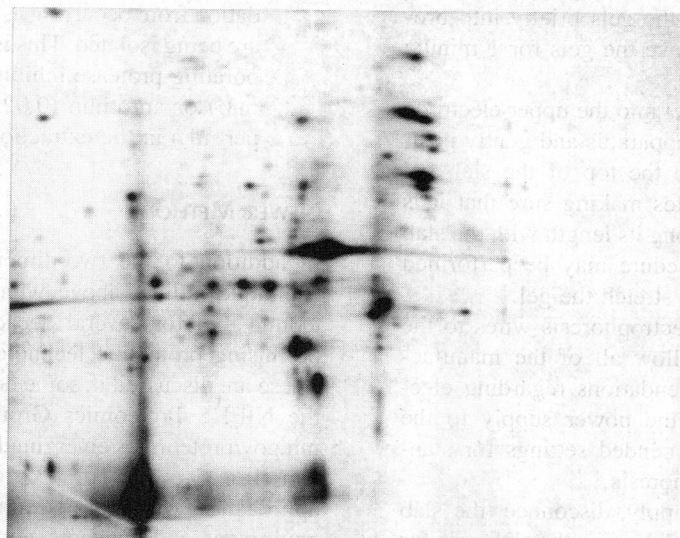

**FIGURE 39.11** $^{35}$S methionine autoradiogram of a two-dimensional gel showing numerous gene product spots.

specific-activity [$^{35}$S]methionine to intact animals or isolated cells, and the subsequent electrophoretic mapping of labeled proteins by SDS in one direction and isoelectric focusing in the second direction. This procedure is now commonly used in studies investigating the role of specific protein synthesis tissue differences in response to chemical exposure and stress proteins, as well as in development and differentiation, and it can resolve as many as 1000 proteins (Figure 39.11).

*Technique*

The method described below is adapted from Spector et al. [238]; the technique is based on classical methods. It should be noted that commercially available, premade immobilized pH gradient (IPG) strips with a wide variety of isoelectric ranges are available, as are precast polyacrylamide gels with a wide range of polyacrylamide percentages, to expedite implementation of this technique. Cells, tissues or isolated organelle fractions are solubilized in sodium dodecyl sulfate and urea using the method of Laemmli [131]. For purposes of comparing control to chemical- or drug-treated samples, it is important to load equivalent amounts of radioactivity (e.g., $^{35}$S-methionine) or protein onto the isoelectric focusing gels/IPG strips so direct treated-to-control comparison may be made.

1. Prepare the isoelectric focusing gels in vertical casting capillary tubes (3-mm diameter) using a commercially available apparatus and prepared isoelectric focusing gel stock solution with the desired isoelectric focusing range generated by ampholines in the gel stock. Initiate polymerization by addition of 10% ammonium persulfate.
2. Leave approximately 1 cm at the top of the capillary tubes for addition of protein sample and allow the gels to polymerize at 30°C for 60 minutes.
3. Place the gels in their capillary tubes in the isolectric focusing apparatus and fill the upper and lower chambers with 0.1-*N* NaOH and the lower chamber with 0.01-*M* $H_3PO_4$ electrode buffers.
4. Apply about 3 μL of overlay buffer to each tube; connect the wires to the power supply and prefocus the gels to provide the desired pH gradient (using a starting amperage of 100 amps/gel and continue until the voltage has reached 1000 V).
5. Apply the protein sample (5 to 15 μL) with a fine pipet tip or microsyringe directly onto the top of the isolectric focusing gel and allow the overlay buffer to flow over.
6. Run the isolectric focusing gels overnight according to the instructions of the manufacturer for the isolectric focusing apparatus.
7. Turn off the power; remove the gels from the apparatus and place them on ice prior to loading on the slab gels to minimize diffusion of the protein bands.
8. Use precast slab gels or those prepared as described above for microsomal proteins using the Laemmli procedure [131].
9. Gently push the isolectric focusing gels from the capillary tubes into equilibration buffer using a fine glass rod. Gently flushing around the gel while it is in the capillary tube with equilibration buffer using a syringe prior to pushing will also help to minimize chances of breaking the gel during removal.

10. Dip the acidic end of the gels briefly into bromophenol blue and shake the gels for 1 minute at 37°C.
11. Rapidly place the IF gel into the upper electrode buffer of the slab gel apparatus and gently position the tube gel onto the top of the slab gel between the glass plates making sure that it is making full contact along its length with the slab gel. This seating procedure may be performed with a spatula. Do not stretch the gel.
12. Attach the slab gel electrophoresis wires to the power supply, and follow all of the manufacturer's safety recommendations regarding electrophoresis. Turn on the power supply to the manufacturer's recommended settings for standard slab gel electrophoresis.
13. Turn off the power supply, disconnect the slab gel electrophoresis, and remove the slab gels for further processing (e.g., autoradiography) or staining once the bromophenol blue tracking dye band is near the bottom of the gel.

## Biochemical Measurements of Protein Synthesis

The synthesis of specific groups of proteins, such as cytoskeletal protein synthesis, can be measured by determining the amount of radiolabel incorporated into isolated fractions of cytoskeletal extracts from cells exposed to radioactively labeled amino acids [232,236].

### *Technique*

1. Plate cells at $1 \times 10^5$ cells/60 minutes and grow to subconfluency.
2. Replace medium with a serum-free Eagle's basal medium containing 1/20th of the normal concentration of amino acids plus $^3$H-labeled amino acid mixture at 1.5 μCi/mL.
3. Incubate the cells for 1 hour. Remove the labeled medium, rinse twice, gently scrape the cells from the plate, and centrifuge.
4. Wash the cell pellets once with PM2G buffer (0.1-*M* PIPES, 1-m*M* $MgSO_4$, 2-*M* glycerol, and 2-m*M* EDTA; pH 6.9). Extract cell pellets with 0.3% NP-40 in PM2G buffer containing 0.5-m*M* phenylmethyl sulfonyl fluoride (PMSF), microcentrifuge to pellet the cytoskeletal fractions, and wash twice with PM buffer (0.1-*M* PIPES and 1-m*M* $MgSO_4$; pH 6.9) containing 0.5-m*M* PMSF.
5. Extract the cytoskeletal fraction pellet with PM buffer containing 5-m*M* $CaCl_2$ and 0.5-m*M* PMSF to depolymerize the microtubules and release the cytoskeletal proteins.
6. Count an aliquot of the protein extract in a liquid scintillation counter and normalize to total cell protein. It is important to prevent protein degradation from occurring as the cytoskeletal proteins are being isolated. This is accomplished by incorporating protease inhibitors such as PMSF (0.5-m*M*) or aprotinin (0.02 trypsin inhibitory units per mL) in the extraction and wash buffers.

## Newer Methods

In addition to the two-dimensional gel electrophoresis technique outlined above, which has been the primary proteomic tool for several decades, a number of new and promising proteomic technologies have been developed. These are discussed in some detail updated periodically on the NIEHS Proteomics Group website (http://dir.niehs.nih.gov/proteomics/emerging.htm). At present, the major limiting factors for exploitation of these newer approaches, which are summarized below, are costs for equipment and reagents and the need for specialized training such that these techniques are limited to a relatively limited number of facilities with the resources required to maintain the equipment and personnel to conduct research. These techniques include fluorescence difference two-dimensional gel electrophoresis, isotope-coded affinity tagging (ICAT™), matrix-assisted laser desorption ionization–time-of-flight (MALDI-TOF) and surface-enhanced laser desorption ionization–time-of-flight (SELDI-TOF) mass spectrometry, commercially available protein arrays, and laser-capture microdissection. The following section provides an overview of each of these newer proteomic approaches discussed on the NIEHS Proteomics Group website and briefly lists the advantages and disadvantages of each. Detailed technical protocols are given for each method on the website and hence are not be presented here.

### Fluorescence Difference Two-Dimensional Gel Electrophoresis

This technique attempts to speed up the standard two-dimensional gel method as outlined above by incubating protein samples with fluorescent dyes prior to two-dimensional gel electrophoresis which allows several samples (>3) to be visualized on a single two-dimensional gel. The fluorescent dyes used are designated Cy2, Cy3, and Cy5, and they have different excitation wavelengths. The combined images are merged and image analysis software is used to sort out relative differences in expression patterns. This method is sensitive and linear over a large range of protein concentrations. The main limitations of this technique reside in costs for the needed equipment and dyes.

### Isotope-Coded Affinity Tagging (ICAT™)

This is a very specialized technique that employs the differential labeling of the cysteine residues of proteins with either hydrogen- or deuterium-based tags followed by liquid chromatography and mass spectrometry to separate the proteins. The technique requires some expertise to ensure the uniformity of protein labeling and to determine the degree of protein recovery from the liquid chromatography and proteins that contain cysteine residues. This technique works well for making comparisons of known proteins with specific characteristics.

### Matrix-Assisted Laser Desorption Ionization–Time-of-Flight and Surface-Enhanced Laser Desorption Ionization–Time-of-Flight Mass Spectrometry

The MALDI-TOF method separates peptides as either intact proteins or peptide fragments based on the time of flight in a mass spectrometer and mass and charge properties. A limitation of this approach is that it does not identify specific proteins but only size and charge patterns. Linkages of the MALDI-TOF approach with one-dimensional SDS-PAGE followed by LC–MS/MS analysis or amino acid sequencers capable of analyzing small samples would be a great advancement in that it would permit specific identification of detected peptides [239,240]. The SELDI-TOF method utilizes metallic chips with differently charged surfaces (positive or negative, lipophillic or hydrophyllic) or surfaces with antibodies or antibody fragments. These are generically termed *bait surfaces*, and they differentially bind various proteins based on the chemical nature of the proteins of interest, so protein groups of particular interest are preselected. Biological protein solutions of interest are applied and then washed to remove nonspecifically or weakly bound proteins, leaving only the more specifically bound proteins. The chips are then laser desorbed and the released proteins subject to mass spectrometry. The working protein molecular mass range is given as 1 to 300 kDa. Limitations of the technique are that it would obviously only select for those proteins with characteristics that allow them to bind to the metallic chip surface, and the mass spectrometry, while providing a unique fingerprint, would only yield information on the masses of detected proteins rather than specific protein identities. As with the MALDI-TOF approach, the linkage of this technology with systems capable of providing protein identification would greatly increase the value of the data generated.

### Commercially Available Protein Arrays

This technology is analogous to the DNA array technology except that specific antibodies bound to the surface of chips or capillary tubes are utilized to bind proteins and captured proteins are detected by fluorescence imaging technology. An obvious limitation of this technology is the availability of specific antibodies with enough affinity to retain the proteins of interest on the chip for detection. In addition to antibodies, oligonucleotides may bind proteins of interest via ultraviolet photo-cross-linking or DNA probes that bind to proteins of interest thorough the mRNA encoding for these proteins. Again, these approaches are targeted to specific proteins by prior knowledge of those proteins that are of particular interest. On the other hand, the printing of proteins in a microarray format on glass slides [241] is one approach for increasing the number of proteins which may be analyzed at one time.

### Laser Capture Microdissection

Another potentially useful technique is that of laser capture microdissection, which enables the selective laser dissection of a particular sample of tissue from a tumor or apoptotic area and normal tissue for comparison by two-dimensional gel electrophoresis [242]. The technique uses a thermosensitive polymer film to collect cells dissected from an area of interest by an infrared laser. The collected cells are transferred to tubes and released from the polymer film by an extraction buffer prior to two-dimensional gel electrophoresis as discussed above. Individual protein spots may be visualized by silver staining or fluorescent dyes followed by computerized image analysis. This technique obviously requires some equipment investment and a high degree of technical training to be used on a routine basis. This technique has also been used in combination with reverse-transcriptase polymerase chain reaction to identify specific changes in gene expression patterns [243].

## PHOSPHOLIPIDS AND LIPOPROTEINS

Phospholipids and lipoproteins provide important permeability properties to membrane structures such as the endoplasmic reticulum. These membrane phospholipids and lipoproteins are synthesized on the endoplasmic reticulum and often play an integral role in enzyme activity by regulating the membrane environment of enzymes that are imbedded in the endoplasmic reticulum. The production of certain forms of lipoproteins seems to be associated with specific toxic and disease states; for example, the relative amount of very-low-density lipoproteins produced by the liver appears to play a critical role in the development and susceptibility of cardiovascular disease. Phospholipid and lipoprotein synthesis can be studied by measuring $^{32}P$ and $^{14}C$ incorporation into these compounds as outlined earlier.

## INTRACELLULAR IONS

Intracellular ions such as $Na^+$, $K^+$, $Ca^{2+}$, $Mg^{2+}$, and $H^+$ play important regulatory roles in normal cell processes and in the response of cells to injury. Sequential alterations in cellular ion concentrations have been shown to be an integral part of cellular repair and regeneration processes and in the sequelae of events leadings to cell death [244–246]. Direct effects of chemicals on ion regulation also can occur through direct effects on energy metabolism, plasma membrane integrity, and ion translocation systems. Recent advances in instrumentation have made it possible to measure changes in ion concentrations in living cells following exposure to toxic compounds. Fluorescent probes sensitive to various ions are currently available and have been used to study effects of metals [247], oxidative stress [248], and chemical ischemia [246]. These probes include Quin-2 AM and Fura-2 AM for [$Ca^{2+}$ ], Mag-Fura-2 AM for [$Mg^{2+}$]$_i$, SBIF for $Na^+$, PBFI for $K^+$, BCECF for pH, and rhodamine-123 for mitochondrial membrane potential. Initial studies of cellular ion concentrations were conducted in cell suspensions in which average changes occurring in cell populations were measured. These studies used standard fluorimeters to measure total fluorescence of the sample [249]. With the development of digital imaging fluorescence microscopy (DIFM) coupled with image analysis, however, it has become possible to quantitate ions such as [$Ca^{2+}$]$_i$ within individual cells and in subcellular compartments [250,251].

## Digital Imaging Fluorescence Microscopy Technique

1. Monolayer cell cultures grown in specially prepared dishes, with a window on the bottom of the plate [250], are loaded with 3 $\mu M$ Fura-2 AM for 30 minutes at 37°C by adding 6 µL of 1-m$M$ Fura-2 AM in dry DMSO to culture dishes containing 2 mL of medium (final concentration of DMSO is 0.3%). After loading, the Fura-2-containing medium is removed and the cells are washed three times with fresh medium.
2. Fura-2-loaded cell cultures are placed on a heated (37°C) stage and examined with an inverted microscope equipped with ultraviolet optics. Changes in [$Ca^{2+}$]$_i$ are observed by alternating 340- and 380-nm excitation produced by a chopper-based xenon light source coupled through a quartz bifurcated fiberoptic to the microscope. The microscope is equipped with an Opelco KS1380 intensifier and a Dage 65 videocamera.
3. Fluorescent images are acquired at an alternating rate of 5 frame-pairs/second and are processed by a Tracor Northern 8502 image analyzer. The fluorescent images are corrected for camera and intensifier variations by subtracting background images collected when the port is closed. The images collected with the 340-nm excitation are divided on a pixel-by-pixel basis by the images collected with 380-nm excitation, and the mean value of pixel ratios for each ratioed image is obtained by histogram analysis.
4. Calibration of the ratioed images is accomplished using a calcium ionophore such as ionomycin or 4-bromo-A23187 to give fully $Ca^{2+}$-saturated Fura-2 [250]. The $Ca^{2+}$ ion concentrations are calculated using the formula:

$$\text{nM Ca}^{2+} = K_d(R - R_{min})/(R_{max} - R)$$

where $K_d$ = 224 nM; $R_{max}$ is the ratio obtained in a cell treated with a Ca ionophore; and $R_{min}$ is calculated from the ratio ($R$) obtained with normal cells in which [$Ca^{2+}$] was assumed to be 100 nm [253].

## QUESTIONS

1. What are some of the advantages and disadvantages of studying the effects of toxic agents on subcellular organelles vs. whole cells and intact organisms?
2. What measurements could one make to determine the purity of a sample of lysosomes isolated by differential centrifugation?
3. Why would it be advantageous to use both immunodetection and enzymatic assay measurements to study the effect of toxic chemicals on microsomal P450 enzymes?
4. What might be a possible benefit of not having a particular FMO isozyme?
5. Peroxisomes produce both catalase and hydrogen peroxide. What might be the possible adverse effects to a cell that could result from a toxic agent that perturbs production of either hydrogen peroxide or catalase activity?
6. What functional changes would lead you to suspect that a toxic agent was targeting the cytoskeleton and how could you assess this?
7. The combined ultrastructural and biochemical techniques described in this chapter provide powerful approaches to understanding mechanisms of cell injury from chemicals by helping to localize the primary organelle systems involved. Cell death via apoptosis or necrosis are major areas of current interest. How would you use the techniques described in this chapter to further basic knowledge about the underlying mechanisms which determine whether a cell dies by apoptosis or necrosis?

## REFERENCES


1. Fowler, B.A., Woods, J.S., and Schiller, C.M. (1977): Ultrastructural and biochemical effects of prolonged oral arsenic exposure on liver mitochondria of rats. *Environ. Health Perspect.*, 19:197–204.
2. Blouin, A., Bolender, R. P., and Weibel, E. R. (1977): Distribution of organelles and membranes between hepatocytes and nonhepatocytes in the rat liver parenchyma: a stereological study. *J. Cell Biol.*, 72:441–455.
3. Bolender, R.P. et al. (1978): Intergrated stereological and biochemical studies of hepatocytic membranes. I. Membrane recoveries in subcellular fractions. *J. Cell Biol.*, 77:565–583.
4. Weibel, E.R. et al. (1969): Correlated morphometric and biochemical studies on the liver cell. I. Morphometric model, stereologic methods, and normal morphometric data for rat liver. *J. Cell Biol.*, 42:68–91.
5. Weibel, E.R. and Paumgartner, D. (1979): Integrated stereo1ogica1 and biochemical studies on hepatocyte membranes. II. Correction of section thickness effect and volume and surface density estimates. *J. Cell Biol.*, 77: 584.
6. Williams, M.A. (1977): Stereological techniques. In: *Practical Methods in Electron Microscopy*, Vol. 6, edited by A.M. Glauert, pp. 1–84. North-Holland, Amsterdam.
7. Fowler, B. A. and Woods, J. S. (1979): The effects of prolonged oral arsenate exposure on liver mitochondria of mice: morphometric and biochemical studies. *Toxicol. Appl. Pharmacol.*, 50:177–187.
8. Fowler, B.A., Woods, J.S., and Schiller, C.M. (1979): Studies of hepatic mitochondrial structure and function: morphometric and biochemical evaluation of *in vivo* perturbation by arsenate. *Lab. Invest.*, 41:313–320.
9. Wiener, J. et al. (1968): A quantitative description of cortisone-induced alterations in the ultrastructure of rat liver parenchymal cells. *J. Cell Biol.*, 37:47–62.
10. Fowler, B.A. and Woods, J.S. (1977): The transplacental toxicity of methyl mercury to fetal rat liver mitochondria: morphometric and biochemical studies. *Lab. Invest.*, 36:122–130.
11. Frigg, M. and Rohr, H.P. (1976): Ultrastructural and stereological study on the effects of vitamin E on liver mitochondrial membranes. *Exp. Mol. Pathol.*, 24:236–243.
12. Fowler, B.A. and Woods, J.S. (1987): Metal and metalloid-induced porphyrinurias: relationship to cell injury. *Ann. N.Y. Acad. Sci.*, 514:172–182.
13. Fowler, B. A., Kardish, R. M., and Woods, J. S. (1983): Alteration of hepatic microsomal structure and function by indium chloride: ultrastructural, morphometric, and biochemical studies. *Lab. Invest.*, 48:471–478.
14. Fowler, B.A. (1983): The role of ultrastructural techniques in understanding mechanisms of metal-induced nephrotoxicity. *Fed. Proc.*, 42:2957–2964.
15. Fowler, B.A. (1980): Ultrasturctural morphometric/biochemical assessment of cellular toxicity. In: Proceedings of the Symposium on the "Scientific Basis of Toxicity Assessment," edited by H. P. Witschi, pp. 211–218. Elsevier, Amsterdam.
16. Hackenbrock, C.R., and Caplan, A. I. (1969): Ion-induced ultrastructural transformations in isolated mitochondria: the energized uptake of calcium. *J. Cell Biol.*, 42:221–234.
17. Goyer, R.A. and Krall, R. (1969): Ultrastructural transformation in mitochondria isolated from kidneys of normal and lead intoxicated rats. *J. Cell Biol.*, 41:393–400.
18. Deter, R.L. (1973): Electron microscopic evaluation of subcellular fractions obtained by ultracentrifugation. In: *Principles and Techniques of Electron Microscopy: Biological Applications*, Vol. 3, edited by M.A. Hayat, pp. 199–235. Van Nostrand Reinhold, New York.
19. Haschemeyer, R.H. and Myers, R.T. (1973): Negative staining. In: *Principles and Techniques of Electron Microscopy: Biological Applications*, Vol. 2, edited by M.A. Hayat, pp. 99–147. Van Nostrand Reinhold, New York.
20. Muscatello, U. et al. (1975): Configurational changes in isolated rat liver mitochondria as revealed by negative staining. III. Modifications caused by uncoupling agents. *J. Ultrastruct. Res.*, 52:2–12.
21. Andrews, P.M. and Hackenbrock, C.R. (1975): A scanning and stereographic ultrastructural analysis of the isolated inner mitochondrial membrane during change in metabolic activity. *Exp. Cell Res.*, 90(1):127–136.
22. Hackenbrock, C.R. (1972): Energy-linked ultrastructural transformations in isolated liver mitochondria and mitop1asts: preservation of configurations by freeze-cleaning compared to chemical fixations. *J. Cell Biol.*, 53: 450–465.
23. Lang, R.D.A. and Bronk, J.R. (1978): A study of rapid mitochondrial structural changes *in vitro* by spray-freeze-etching. *J. Cell Biol.*, 77:134–147.
24. Koehler, J.K. (1973): The freeze-etching technique. In: *Principles and Techniques of Electron Microscopy: Biological Applicaitons*, Vol. 2, edited by M.A. Hayat, pp. 51–98. Van Nostrand Reinhold, New York.
25. Schnaitman, C., Erwin, V.G., and Greenawalt, J.W. (1967): The submitochondrial localization of monoamine oxidase, an enzymatic marker for the outer membrane of rat liver mitochondria. *J. Cell Biol.*, 32:719–735.
26. Greenwalt, J.W. (1979): Survey and update of outer and inner mitochondrial membrane separation. In: *Methods in Enzymology*, edited by S. Fleischer and L. Packer, pp. 88–98. Academic Press, New York.
27. Estabrook, R. W. (1967): Mitochondrial respiratory control and the polarographic measurement of ADP:Q ratios. In: *Methods in Enzymology*, Vol. 10, edited by S. Colowick and N.O. Kaplan, pp. 41–47. Academic Press, New York.
28. Fowler, B.A. and Woods, J.S. (1977): Ultrastructural and biochemical changes in renal mitochondria during chronic oral methyl mercury exposure: the relationship to renal function. *Exp. Mol. Pathol.*, 27:403–412.
29. Jacobs, E.E. et al. (1956): Uncoupling of oxidative phosphorylation by cadmium ion. *J. Biol. Chem.*, 223:147–156.
30. Hackenbrock, C.R. (1972): States of activity and structure in mitochondrial membranes. *Ann. N.Y. Acad. Sci.*, 195: 492–505.
31. Nelson, B. D. (1975): The action of cyclodiene pesticides on oxidative phosphorylation in rat liver mitochondria. *Biochem. Pharmacol.*, 24:1485–1490.

32. Cederbaum, A.I., Lieber, C.S., and Rubin, E. (1974): The effect of acetaldehyde on mitochondrial function. *Arch. Biochem. Biophys.*, 161:26–39.
33. Inouye, B. et al. (1978): Effects of phthalate esters on mitochondrial oxidative phosphorylation in the rat. *Toxicol. Appl. Pharmacol.*, 43:189–198.
34. London, R.E. et al. (1985): An approach to NMR studies of the metabolism of internal organs using surface coils. *J. Biochem. Biophys. Meth.*, 11:21–29.
35. Chen, B. et al. (1986): *In vivo* $^{31}$P nuclear magnetic resonance studies of arsenite-induced changes in hepatic phosphate levels. *Biochem. Biophys. Res. Commun.*, 139(1):228–234.
36. Halestrap, A.P. et al. (1997): Oxidative stress, thiol reagents, and membrane potential modulate the mitochondrial permeability transition by affecting nucleotide binding to the adenine nucleotide translocase. *J. Biol. Chem.*, 272(6):3346–3354.
37. Santos, J.H. et al. (2003): Cell sorting experiments link persistent mitochondrial DNA damage with loss of mitochondrial membrane potential and apoptotic cell death. *J. Biol. Chem.*, 278(3):1728–1734.
38. Ott, M. (2002): Cytochrome *c* release from mitochondria proceeds by a two-step process. *Proc. Natl. Acad. Sci. U.S.A.*, 99(3):1259–1263.
39. Ochoa, S. (1955): Malic dehydrogenase from pig heart. In: *Methods in Enzymology*, Vol. 1, edited by S. Colowick and N.O. Kaplan, pp. 735–739. Academic Press, New York.
40. Frenkel, R. and Cobo-Frenkel, A. (1973): Differential characteristics of the cytosol and mitochondrial isozymes of malic enzyme from bovine brain: effects of dicarboxylic acids and sulfhydryl reagents. *Arch. Biochem. Biophys.*, 158:323–330.
41. Magnaval, R., Batti, R., and Thiessard, J. (1975): Methyl mercury effect on rat liver mitochondrial dehydrogenase. *Experientia*, 31:406–407.
42. Woods, J.S., Eaton, D.L., and Lukens, C.B. (1984): Studies of porphyrin metabolism in the kidney: effects of trace metals and glutathione on renal uroporphyrinogen decarboxylase. *Mol. Pharmacol.*, 26:336–341.
43. Woods, J.S. and Fowler, B.A. (1977): Effects of chronic arsenic exposure on hematopoietic function in adult mammalian liver. *Environ. Health Perspect.*, 19:209–213.
44. Woods, J.S. and Fowler, B.A. (1977): Renal porphyrinuria during chronic methyl mercury exposure. *J. Lab Clin. Med.*, 90:266–272.
45. Woods, J.S. and Fowler, B.A. (1978): Altered regulation of mammalian hepatic heme biosynthesis and urinary porphyrin excretion during prolonged exposure to sodium arsenate. *Toxicol. Appl. Pharmacol.*, 43:361–371.
46. Woods, J.S. and Fowler, B.A. (1987): Metal alterations of uroporphyrinogen decarboxylase and coproporphyrinogen oxidase. *Ann. N.Y. Acad. Sci.*, 514:55–64.
47. Strik, J.J. (1987): Porphyrins in urine as an indication of exposure to chlorinated hydrocarbons. *Ann. N.Y. Acad. Sci.*, 514:219–221.
48. Chua, N.H. and Schmidt, G.W. (1979): Transport of proteins into mitochondria and chloroplasts. *J. Cell Biol.*, 8:461–483.
49. Beattie, D.S. (1968): Studies on the biogenesis of mitochondrial protein components in rat liver slices. *J. Biol. Chem.*, 243:4027–4033.
50. Tedeschi, H. and Harris, D. L. (1958). Some observations on the photometric estimation of mitochondrial volume. *Biochim. Biophys. Acta*, 28:392–402.
51. Matlib, M.A. and Srere, P.A. (1976): Oxidative properties of swollen rat liver mitochondria. *Arch. Biochem. Biophys.*, 174:705–712.
52. Mintz, H.A. et al. (1967): Morphological and biochemical studies of isolated mitochondria from fetal neonatal and adult liver and from neoplastic tissues. *J. Cell Biol.*, 34:513–525.
53. Papa, S. et al. (1973): Mechanisms of respiration-driven proton translocation by the inner membrane. *Biochem. Biophys. Acta*, 292:20–28.
54. Pressman, B.C. (1967): Biological applications of ion-specific glass electrodes. In: *Methods in Enzymology*, Vol. 10, edited by S. Colowick and N.O. Kaplan, pp. 714–726. Academic Press, New York.
55. Bogucka, K. and Wojtczak, L. (1979): On the mechanisms of mercurial-induced permeability of the mitochondrial membrane to K+, *FEBS Lett.*, 100(2):301–304.
56. Lee, M.J., Harris, R.A., and Green, D.E. (1969): Action of fluorescein mercuric acetate upon mitochondrial-energized processes. *Biochem. Biophys. Res. Commun.*, 36:937–946.
57. Parr, D.R. and Harris, E.J. (1976): The effect of lead on the calcium-handling capacity of rat heart mitochondria. *Biochem. J.*, 158:289–294.
58. Lenaz, G. (2001): The mitochondrial production of reactive oxygen species: mechanisms and implications in human pathology. *IUBMB Life*, 52:159–164.
59. Fariss, M.W. et al. (2005): Role of mitochondria in toxic oxidative stress. *Mol. Interv.*, 5(2):94–111.
60. Liu, X. et al. (1996): Induction of apoptotic program in cell-free extracts: requirement for dATP and cytochrome *c*. *Cell*, 86:147–156.
61. Wallace, D.C. (1999): Mitochondrial diseases in man and mouse. *Science*, 283:1482–1488.
62. Iverson, S.L. and Orrenius, S. (2004): The cardiolipin–cytochrome *c* interaction and the mitochondrial regulation of apoptosis. *Arch. Biochem. Biophys.*, 423(1):37–46.
63. Samali, A. and Orrenius, S. (1998): Heat shock proteins: regulators of stress response and apoptosis. *Cell Stress Chaperones*, 3(4):228–236.
64. Wang, N. et al. (1993): A human mitochondrial ATP-dependent protease that is highly homologous to bacterial Lon protease. *Proc. Natl. Acad. U.S.A.*, 90:11247–11251.
65. Langer, T. and Neupert, W. (1996): Regulated protein degradation in mitochondria. *Experientia*, 52:1069–1076.
66. Atlante A. et al. (2000): Cytochrome *c* is released from mitochondria in a reactive oxygen species (ROS)-dependent fashion and can operate as a ROS scavenger and as a respiratory substrate in cerebellar neurons undergoing excitotoxic death. *J. Biol. Chem.*, 275(47): 37159–37166.
67. Petrosillo, G. et al. (2004): $Ca^{2+}$-induced reactive oxygen species production promotes cytochrome *c* release from rat liver mitochondria via mitochondrial permeability transition (MPT)-dependent and MPT-independent mechanisms: role of cardiolipin. *J. Biol. Chem.*, 279(51): 53103–53108.

68. Seth, R. et al. (2005): p53-Dependent caspase-2 activation in mitochondrial release of apoptosis-inducing factor and its role in renal tubular epithelial cell injury. *J. Biol. Chem.*, 280(35):31230–31239.
69. Finucane, D.M. et al. (1999): Bax-induced caspase activation and apoptosis via cytochrome *c* release from mitochondria is inhibitable by Bcl-xL. *J. Biol. Chem.*, 274(4):2225–2233.
70. Zhang, W., Shi, H.Y., and Zhang, M. (2005): Maspin overexpression modulates tumor cell apoptosis through the regulation of Bcl-2 family proteins, *BMC Cancer*, 5(1):50.
71. Adachi, M. et al. (2004): BAX interacts with the voltage-dependent anion channel and mediates ethanol-induced apoptosis in rat hepatocytes. *Am. J. Physiol. Gastrointest. Liver Physiol.*, 287(3):G695–G705.
72. Hashimoto, A. et al. (2005): BAD detects coincidence of $G_2$/M phase and growth factor deprivation to regulate apoptosis. *J. Biol. Chem.*, 280(28):26225–26232.
73. Kulik, G. et al. (2001): Tumor necrosis factor alpha induces BID cleavage and bypasses angtiapoptotic signals in prostate cancer LNCaP cells. *Cancer Res.*, 61(6):2713–2719.
74. Shimizu, S. et al. (2001): Essential role of voltage-dependent anion channel is various forms of apoptosis in mammalian cells. *J. Cell Biol.*, 152(2):237–250.
75. Zamora, M. et al. (2004): Recruitment of NF-κB into mitochondria is involved in adenine nucleotide translocase 1(ANT1)-induced apoptosis. *J. Biol. Chem.*, 279(37): 38415–38423.
76. Flint, D.H. et al. (1993): The inactivation of Fe–S cluster containing hydro-lyases by superoxide. *J. Biol. Chem.*, 268(30):22369–22376.
77. Vasquez-Vivar, J., Kalyanaraman, B., and Kennedy, M.C. (2000): Mitochondrial aconitase is a source of hydroxyl radical. *J. Biol. Chem.*, 275(19):14064–14069.
78. Gardner, P.R. (1997): Superoxide-driven aconitase Fe–S center cycling. *Biosci. Rep.*, 17(1):33–42.
79. Echtay, K.S. et al. (2002): Superoxide activates mitochondrial uncoupling protein 2 from the matrix side. *J. Biol. Chem.*, 277(49):47129–47135.
80. Carmichael, N.G. and Fowler, B.A. (1979): Effects of separate and combined chronic mercuric chloride and sodium selenate administration in rats: histological, ultrastructural and x-ray microanalytical studies of liver and kidney. *J. Environ. Pathol. Toxicol.*, 3: 399–4l2.
81. Fowler, B.A. et al. (1974): Mercury uptake by renal lysosomes of rats ingesting methyl-mercury hydroxide: ultrastructural observations and energy dispersive x-ray analysis. *Arch. Pathol.*, 98:297–301.
82. Fowler, B.A. and Nordberg, G.F. (1978): The renal toxicity of cadmium metallothionein: morphometric and x-ray microanalytical studies. *Toxicol. Appl. Pharmacol.*, 46:609–623.
83. Goldfisher, S. (1965): The localization of copper in the pericanalicular granules (lysosomes) of liver in Wilson's disease (hepatolenticular degeneration (1)). *Ariz. J. Pathol.*, 46:977–983.
84. Hall, T.A. (1971): The microprobe assay of chemical elements. In: *Physical Techniques in Biological Research*, Vol. IA, edited by G. Oster, pp. 157–275. Academic Press, New York.
85. Allison, A.C. and Young, M.R. (1964): Uptake of dyes and drugs by living cells in culture. *Life Sci.*, 3:1407–1414.
86. Casartelli, A. et al. (2003): A cell-based approach for the early assessment of the phospholipidogenic potential in pharmaceutical research and drug development. *Cell Biol. Toxicol.*, 19:161–176.
87. Thibodeau, M.S. et al. (2004): Silica-induced apoptosis in mouse alveolar macrophages is initiated by lysosomal enzyme activity. *Toxicol. Sci.*, 80:34–48.
88. Servais, H. et al. (2005): Gentamicin-induced apoptosis in LLC-PK1 cells: involvement of lysosomes and mitochondria. *Toxicol. Appl. Pharmacol.*, 206:321–333.
89. Brun, A. and Brunk, U. (1970): Histochemical indications for lysosomal localization of heavy metals in normal rat brain and liver. *J. Histochem. Cytochem.*, 18:820–827.
90. Nassogne, M.-C. et al. (2004): Cocaine induces a mixed lysosomal lipidosis in cultured fibroblasts, by inactivation of acid sphingomyelinase and inhibition of phospholipase *Appl. Toxicol. Appl. Pharmacol.*, 194:101–110.
91. Sawada, H., Takami, K., and Asahi, S. (2005). A toxicogenomic approach to drug induced phospholipidosis: analysis of its induction mechanism and establishment of a novel *in vitro* screening system. *Toxicol. Sci.*, 83: 282–292.
92. Fowler, B.A. et al. (1975): The. effects of chronic oral methyl mercury exposure on the lysosome system of rat kidney: morphometric and biochemical studies. *Lab. Invest.*, 32:313–322.
93. Conney, A.M. (1967): Pharmacological implications of microsomal enzyme induction. *Pharmacol. Rev.*, 19:317–366.
94. Squibb, K.S., Pritchard, J.B., and Fowler, B.A. (1984): Cadmium metallothionein nephropathy: ultrastructural/biochemical alterations and intracellular cadmium binding. *J. Pharmacol. Exp. Ther.*, 228:311–321.
95. Shibko, S. and Tappel, A.L. (1965): Rat kidney lysosomes: isolation and properties. *Biochem. J.*, 95:731–741.
96. Lauwerys, R. and Buchet, J. P. (1972): Study on the mechanism of lysosome labilization by inorganic mercury *in vitro*. *Eur. J. Biochem.*, 26:535–542.
97. Mego, J.L. and Barnes, J. (1973): Inhibition of heterolysosome formation and function in mouse kidneys by injection of mercuric chloride. *Biochem. Pharmacol.*, 22:373–381.
98. Mego, J.L. and Hayes, A.W. (1973): Effects of fungal toxins on uptake and degradation of formaldehyde-treated $^{125}$I-albumin in mouse liver phagolysosomes. *Biochem. Pharmacol.*, 22:3275–3286.
99. Mego, J.L. and Cain, J.A. (1973): The effect of carbon tetrachloride on lysosome function in kidneys and livers of mice. Biochem. Biophys. Acta, 297:343–345.
100. Mego, J.L. and Cain, J.A. (1975): An effect of cadmium on heterolysosome formation and function in mice. *Biochem. Pharmacol.*, 24:1227–1232.
101. Madsen, K.M. and Christensen, E.L. (1978): Effects of mercury on lysosomal protein digestion in the kidney proximal tubule. *Lab. Invest.*, 38:165–174.
102. Maack, T. (1967): Changes in the activity of acid hydrolases during reabsorption of lysozyme. *J. Cell Biol.*, 35:268–273.

103. Maack, T., Mackensie, D.D.S., and Kinter, W.D. (1971): Intracellular pathways of renal reabsorption of lysozyme. *Am. J. Physiol.*, 221:1609–1616.
104. Daniel, W.A. and Wojcikowski, J. (1999): The role of lysosomes in the cellular distribution of thioridazine and potential drug interactions. *Toxicol. Appl. Pharmacol.*, 158: 115–124.
105. Kupfer, D., and Levin, E. (1972): Monooxygenase drug metabolizing activity in $CaCl_2$ aggregated hepatic microsomes from rat liver. *Biochem. Biophys. Res. Commun.*, 47:611–8.
106. Schenkman, J.B. and Cinti, D.L. (1972): Hepatic mixed function oxidase activity in rapidly prepared microsomes. *Life Sci.*, 11:247–257.
107. Schenkman, J.B. and Cinti, D.L. (1978): Preparation of microsomes with calcium. *Meth. Enzymol.*, 52:83–89.
108. Lucier, G.W. et al. (1973): TCDD-induced changes in rat liver microsomal enzymes. *Environ. Health Perspect.*, 5: 199–209.
109. Dallner, G. (1978): Isolation of microsomal subfractions by use of density gradients. *Meth. Enzymol.*, 52:71–82.
110. Raucy, J. L., and Lasker, J. M. (1991): Isolation of P450 enzymes from human liver. *Meth. Enzymol.*, 206:577–587.
111. Wakabayashi, M. (1970): p-Glucuronidase in metabolic hydrolysis. In: *Metabolic Conjugation and Metabolic Hydrolysis*, edited by W. Fishman, pp. 520–592. Academic Press, New York.
112. Stäubli, W., Hess, R., and Weibel, E.R. (1969): Correlated morphometric and biochemical studies on the liver cell. II. Effects of phenobarbital on rat hepatocytes. *J. Cell Biol.*, 42:92–112.
113. Woods, J.S. and Fowler, B.A. (1986): Alteration of hepatocellular structure and function by thallium chloride: ultrastructural, morphometric, and biochemical studies. *Toxicol. Appl. Pharmacol.*, 83:218–229.
114. Cashman, J.R. (1995): Structural and catalytic properties of the mammalian flavin-containing monooxygenase. *Chem. Res. Toxicol.*, 8:166–181.
115. Ziegler, D.M. (1993): Recent studies on the structure and function of multisubstrate flavin-containing monooxygenases. *Annu. Rev. Pharmacol. Toxicol.*, 33:179–199.
116. Cashman, J.R. and Zhang, J. (2002): Interindividual differences of human flavin-containing monooxygenase. 3. Genetic polymorphisms and functional variation. *Drug Metab. Dispos.*, 30:1043–1052.
117. Hines, R.N. et al. (1994): The mammalian flavin-containing monooxygenases: molecular characterization and regulation of expression. *Toxicol. Appl. Pharmacol.*, 125:1–6.
118. Ziegler, D.M. (2002): An overview of the mechanism, substrate specificities, and structure of FMOs. *Drug Metab. Rev.*, 34:503–11.
119. Lomri, N., Yang, Z., and Cashman, J.R. (1993): Regio- and stereoselective oxygenations by adult human liver flavin-containing monooxygenase 3. Comparison with forms 1 and 2. *Chem. Res. Toxicol.*, 6:800–807.
120. Brunelle, A. et al. (1997): Characterization of two human flavin-containing monooxygenase (form 3): enzymes expressed in escherichia coli as maltose binding protein fusions. *Drug Metab. Dispos.*, 25:1001–1007.
121. Guo, W.X. and Ziegler, D.M. (1991): Estimation of flavin-containing monooxygenase activities in crude tissue preparations by thiourea-dependent oxidation of thiocholine. *Anal. Biochem.*, 198:143–148.
122. Ioannides, C. (1996): *Cytochromes P450: Metabolic and Toxicological Aspects*. CRC Press, Boca Raton, FL.
123. Parkinson, A. (1996): Biotransformation of xenobiotics. In: *Casarett & Doull's Toxicology: The Basic Science of Poisons*, edited by C.D. Klaassen, pp. 113–186. McGraw-Hill, New York.
124. Porter, T.D. and Coon, M.J. (1991): The effect of lead on the calcium-handling capacity of rat heart mitochondria. *Biochem. J.*, 158:289–194.
125. Nelson, D.R. et al. (1996): P450 superfamily: update on new sequences, gene mapping, accession numbers and nomenclature. *Pharmacogenetics*, 6:1–42.
126. Omura, T. and Sato, R. (1964): The carbon monoxide-binding pigment of liver microsomes. I. Evidence for its hemoprotein nature. *J. Biol. Chem.*, 239:2370–2378.
127. Goldstein, J.A. et al. (1977): Separation of pure polychlorinated biphenyl isomers into two types of inducers on the basis of induction of cytochrome P-450 or P-448. *Chem. Biol. Interact.*, 17:69–87.
128. Ryan, D., Lu, A. Y., and Levin, W. (1978): Purification of cytochrome P-450 and P-448 from rat liver microsomes. *Meth. Enzymol.*, 52:117–123.
129. Ryan, D.E. and Levin, W. (1990): Purification and characterization of hepatic microsomal cytochrome P-450. *Pharmacol. Ther.*, 45:153–239.
130. Guengerich, F.P. et al. (1982): Purification and characterization of liver microsomal cytochromes P-450: electrophoretic, spectral, catalytic, and immunochemical properties and inducibility of eight isozymes isolated from rats treated with phenobarbital or beta-naphthoflavone. *Biochemistry*, 21:6019–6030.
131. Laemmli, U.K. (1970): Cleavage of structural proteins during the assembly of the head of bacteriophage T4. *Nature*, 227:680–685.
132. Vlasuk, G.P. and Walz, F.G. (1980): Liver endoplasmic reticulum polypeptides resolved by two-dimensional gel electrophoresis. *Anal. Biochem.*, 105:112–120.
133. Cleveland, D.W. et al. (1977): Peptide mapping by limited proteolysis in sodium dodecyl sulfate and analysis by gel electrophoresis. *J. Biol. Chem.*, 252:1102–1106.
134. Ryan, D.E. et al. (1985): Characterization of a major form of rat hepatic microsomal cytochrome P-450 induced by isoniazid. *J. Biol. Chem.*, 260:6385–6393.
135. Thomas, P.E. et al. (1986): Antibodies as probes of cytochrome P450 isozymes. In: *Biological Reactive Intermediates*. Vol. III. *Mechanisms of Action in Models and Human Disease*, edited by J. Kocsis et al., pp. 95–106. Plenum Press, New York.
136. Friedman, F.K. et al. (1986): Monoclonal antibody-directed analysis of cytochrome P-450. *Adv. Exp. Med. Biol.*, 197:145–154.
137. Burnette, W.N. (1981): 'Western blotting': electrophoretic transfer of proteins from sodium dodecyl sulfate–polyacrylamide gels to unmodified nitrocellulose and radiographic detection with antibody and radioiodinated protein A. *Anal. Biochem.*, 112:195–203.

138. Kleinow, K.M. et al. (1990): Interaction of carbon tetrachloride with beta-naphthoflavone-mediated cytochrome P450 induction in winter flounder (*Pseudopleuronectes americanus*). *Toxicol. Appl. Pharmacol.*, 104:367–374.
139. Hsu, Y.H. (1984): Immunogold for detection of antigen on nitrocellulose paper. *Anal. Biochem.*, 142:221–225.
140. Surek, B. and Latzko, E. (1984): Visualization of antigenic proteins blotted onto nitrocellulose using the immuno-gold-staining (IGS) method. *Biochem. Biophys. Res. Commun.*, 121:284–289.
141. Mayer, R.T. et al. (1977): Methoxyresorufin as a substrate for the fluorometric assay of insect microsomal *O*-dealkylases. *Pest. Biochem. Physiol.*, 7:349–354.
142. Burke, M.D. and Mayer, R.T. (1974): Ethoxyresorufin: direct fluorimetric assay of a microsomal *O*-dealkylation which is preferentially inducible by 3-methylcholanthrene. *Drug Metab. Dispos.*, 2:583–588.
143. Lubet, R.A. et al. (1985): Dealkylation of pentoxyresorufin: a rapid and sensitive assay for measuring induction of cytochrome(s) P-450 by phenobarbital and other xenobiotics in the rat. *Arch. Biochem. Biophys.*, 238:43–48.
144. Burke, M.D. et al. (1985): Ethoxy-, pentoxy- and benzyloxyphenoxazones and homologues: a series of substrates to distinguish between different induced cytochromes P-450. *Biochem. Pharmacol.*, 34:3337–3345.
145. Poland, A.P. and Nebert, D.W. (1973): A sensitive radiometric assay of aminopyrine *N*-demethylation. *J. Pharmacol. Exp. Ther.*, 184:269–277.
146. Hazelton, G.A., Hjelle, J.J., and Klaasen, C.D. (1985): Effects of butylated hydroxyanisole on hepatic glucuronidation capacity in mice. *Toxicol. Appl. Pharmacol.*, 78:280–290.
147. Dehnen, W., Tomingas, R., and Roos, J. (1973): A modified method for the assay of benzo(*a*)pyrene hydroxylase. *Anal. Biochem.*, 53:373–383.
148. Nebert, D.W. and Gelboin, H.V. (1968): Substrate-inducible microsomal aryl hydroxylase in mammalian cell culture. I. Assay and properties of induced enzyme. *J. Biol. Chem.*, 243:6242–6249.
149. Nebert, D.W. and Gielen, J.E. (1972): Genetic regulation of aryl hydrocarbon hydroxylase induction in the mouse. *Fed. Proc.*, 31:1315–1325.
150. DePierre, J.W. et al. (1975): A reliable, sensitive, and convenient radioactive assay for benzpyrene monooxygenase. *Anal. Biochem.*, 63:470–484.
151. Selkirk, J.K. et al. (1974): High-pressure liquid chromatographic analysis of benzo(alpha)pyrene metabolism and covalent binding and the mechanism of action of 7,8-benzoflavone and 1,2-epoxy-3,3,3-trichloropropane. *Cancer Res.*, 34:3474–3480.
152. Van Cantfort, J., De Graeve, J., and Gielen, J. E. (1977): Radioactive assay for aryl hydrocarbon hydroxylase: improved method and biological importance. *Biochem. Biophys. Res. Commun.*, 79:505–512.
153. James, M.O. et al. (1995): Biotransformation, hepatopancreas DNA binding and pharmacokinetics of benzo(*a*)pyrene after oral and parenteral administration to the American lobster, *Homarus americanus*. *Chem. Biol. Interact.*, 95:141–160.
154. Selkirk, J.K., Croy, R.G., and Gelboin, H.V. (1975): Isolation by high pressure liquid chromatography and characterization of benzo(*a*)pyrene-4,5-epoxide as a metabolite of benzo(*a*)pyrene. *Arch. Biochem. Biophys.*, 168:322–326.
155. Cross, D.M. and Bayliss, M.K. (2000): A commentary on the use of hepatocytes in drug metabolism studies during drug discovery and development. *Drug Metab. Rev.*, 32:219–240.
156. de Graaf, I.A.M. et al. (2002): Comparison of *in vitro* preparations for semi-quantitative prediction of *in vivo* drug metabolism. *Drug Metab. Dispos.*, 30:1129–1136.
157. Renwick, A.B. et al. (2000): Differential maintenance of cytochrome P450 enzymes in cultured precision-cut human liver slices. *Drug Metab. Dispos.*, 28:1202–1209.
158. Sies, H. (1978): The use of perfusion of liver and other organs for the study of microsomal electron-transport and cytochrome P-450 systems. *Meth. Enzymol.*, 52:48–59.
159. Thohan, S. et al. (2001): Tissue slices revisited: evaluation and development of a short-term incubation for integrated drug metabolism. *Drug Metab. Dispos.*, 29:1337–1342.
160. Gebhardt, R. et al. (2003): New hepatocyte *in vitro* systems for drug metabolism: metabolic capacity and recommendations for application in basic research and drug development, standard operation procedures. *Drug Metab. Rev.*, 35:145–213.
161. Gomez-Lechon, M.J. et al. (2004): Human hepatocytes in primary culture: the choice to investigate drug metabolism in man. *Curr. Drug Metab.*, 5:443–462.
162. Grisham, J.W., Charlton, R.K., and Kaufman, D.G. (1978): *In vitro* assay of cytotoxicity with cultured liver: accomplishments and possibilities. *Environ. Health Perspect.*, 25:161–171.
163. Moldeus, P., Hogberg, J., and Orrenius, S. (1978): Isolation and use of liver cells. *Meth. Enzymol.*, 52:60–71.
164. Parkinson, A. et al. (2004): The effects of gender, age, ethnicity, and liver cirrhosis on cytochrome P450 enzyme activity in human liver microsomes and inducibility in cultured human hepatocytes. *Toxicol. Appl. Pharmacol.*, 199:193–209.
165. Guzelian, P.S. et al. (1988): Sex change in cytochrome P-450 phenotype by growth hormone treatment of adult rat hepatocytes maintained in a culture system on matrigel. *PNAS*, 85:9783–9787.
166. Sirica, A. et al. (1979): Fetal phenotypic expression by adult rat hepatocytes on collagen gel/nylon meshes. *Proc. Natl. Acad. Sci. U.S.A.*, 76:283–287.
167. Engelmann, G.L., Staecker, J.L., and Richardson, A.G. (1987): Effect of sodium butyrate on primary cultures of adult rat hepatocytes. *In Vitro Cell Dev. Biol.*, 23:86–92.
168. Utesch, D., Molitor, E., Platt, K.L., and Oesch, F. (1991): Differential stabilization of cytochrome P-450 isoenzymes in primary cultures of adult rat liver parenchymal cells. *In Vitro Cell Dev. Biol.*, 27A:858–863.
169. Burchell, B. and Coughtrie, M.W. (1989): UDP-glucuronosyltransferases. *Pharmacol. Ther.*, 43:261–289.
170. Iyanagi, T., Watanabe, T., and Uchiyama, Y. (1989): The 3-methylcholanthrene-inducible UDP-glucuronosyltransferase deficiency in the hyperbilirubinemic rat (Gunn rat) is caused by a –1 frameshift mutation. *J. Biol. Chem.*, 264:21302–21307.

171. Tephly, T.R. and Burchell, B. (1990): UDP-glucuronosyltransferases: a family of detoxifying enzymes. *Trends Pharmacol. Sci.*, 11:276–279.
172. Tephly, T.R. (1990): Isolation and purification of UDP-glucuronosyltransferases. *Chem. Res. Toxicol.*, 3:509–516.
173. Burchell, B. et al. (1991): The UDP glucuronosyltransferase gene superfamily: suggested nomenclature based on evolutionary divergence. *DNA Cell Biol.*, 10:487–494.
174. Burchell, B. (2003): Genetic variation of human UDP-glucuronosyltransferase: implications in disease and drug glucuronidation. *Am. J. Pharmacogen.*, 3:37–52.
175. Guillemette, C. (2003): Pharmacogenomics of human UDP-glucuronosyltransferase enzymes. *Pharmacogen. J.*, 3:136–158.
176. Maruo, Y. et al. (2005): Polymorphism of UDP-glucuronosyltransferase and drug metabolism. *Curr. Drug Metab.*, 6:91–99.
177. Mackenzie, P.I. et al. (2003): Regulation of UDP glucuronosyltransferase genes. *Curr. Drug Metab.*, 4:249–257.
178. Curran, P.G. and DeGroot, L.J. (1991): The effect of hepatic enzyme-inducing drugs on thyroid hormones and the thyroid gland. *Endocr. Rev.*, 12:135–150.
179. Qatanani, M., Zhang, J., and Moore, D.D. (2005): Role of the constitutive androstane receptor in xenobiotic-induced thyroid hormone metabolism. *Endocrinology*, 46(3):995–1002.
180. McClain, R.M. (1989): The significance of hepatic microsomal enzyme induction and altered thyroid function in rats: implications for thyroid gland neoplasia. *Toxicol. Pathol.*, 17:294–306.
181. Coughtrie, M.W., Burchell, B., and Bend, J.R. (1986): A general assay for UDP-glucuronosyltransferase activity using polar amino-cyano stationary phase HPLC and UDP[U-$^{14}$C]glucuronic acid. *Anal. Biochem.*, 159: 198–205.
182. Di Marco, A. et al. (2005): Determination of drug glucuronidation and UDP-glucuronosyltransferase selectivity using a 96-well radiometric assay. *Drug Metab. Dispos.*, 33:812–819.
183. Trubetskoy, O.V., and Shaw, P.M. (1999): A fluorescent assay amenable to measuring production of beta-D-glucuronides produced from recombinant UDP-glycosyl transferase enzymes. *Drug Metab. Dispos.*, 27:555–557.
184. Luukkanen, L. et al. (2001): Glucuronidation of 1-hydroxypyrene by human liver microsomes and human UDP-glucuronosyltransferases UGT1A6, UGT1A7, and UGT1A9: development of a high-sensitivity glucuronidation assay for human tissue. *Drug Metab. Dispos.*, 29:1096–1101.
185. Mulder, G.J. and van Doorn, A.B. (1975): A rapid NAD$^+$-linked assay for microsomal uridine diphosphate glucuronyltransferase of rat liver and some observations on substrate specificity of the enzyme. *Biochem. J.*, 151:131–140.
186. Bock, K.W. et al. (1983): UDP-glucuronosyltransferase activities: guidelines for consistent interim terminology and assay conditions. *Biochem. Pharmacol.*, 32:953–955.
187. Lucier, G.W. and McDaniel, O.S. (1977): Steroid and nonsteroid UDP glucuronyltransferase: glucuronidation of synthetic estrogens as steroids. *J. Steroid Biochem.*, 8:867–872.
188. Dawson, J., Knowles, R.G., and Pogson, C.I. (1992): Measurement of glucuronidation by isolated rat liver cells using [$^{14}$C]fructose. *Biochem. Pharmacol.*, 43:971–978.
189. Hayes, J.D., Flanagan, J.U., and Jowsey, I.R. (2005): Glutathione transferases. *Annu. Rev. Pharmacol. Toxicol.*, 45:51–88.
190. Habig, W.H., Pabst, M.J., and Jakoby, W.B. (1974): Glutathione *S*-transferases: the first enzymatic step in mercapturic acid formation. *J. Biol. Chem.*, 249:7130–7139.
191. Morgenstern, R. and DePierre, J.W. (1983): Microsomal glutathione transferase: purification in unactivated form and further characterization of the activation process, substrate specificity and amino acid composition. *Eur. J. Biochem.*, 134:591–597.
192. Sperker, B., Backman, J.T., and Kroemer, H.K. (1997): The role of beta-glucuronidase in drug disposition and drug targeting in humans. *Clin. Pharmacokinet.*, 33:18–31.
193. de Graaf, M. et al. (2002): Beta-glucuronidase-mediated drug release. *Curr. Pharm. Des.*, 8:1391–1403.
194. Li, H. et al. (1990): The propeptide of beta-glucuronidase: further evidence of its involvement in compartmentalization of beta-glucuronidase and sequence similarity with portions of the reactive site region of the serpin superfamily. *J. Biol. Chem.*, 265:14732–14735.
195. Whiting, J.F. et al. (1993): Deconjugation of bilirubin-IX alpha glucuronides: a physiologic role of hepatic microsomal beta-glucuronidase. *J. Biol. Chem.*, 268:23197–23201.
196. Nishimura, Y., Kato, K., and Himeno, M. (1995): Biochemical characterization of liver microsomal, Golgi, lysosomal, and serum beta-glucuronidases in dibutyl phosphate-treated rats. *J. Biochem. (Tokyo)*, 118:56–66.
197. de Duve, C. (1996): The peroxisome in retrospect. *Ann. N.Y. Acad. Sci.*, 804:1–10
198. Desvergne, B. and Wahli, W. (1999): Peroxisome proliferator-activated receptors: nuclear control of metabolism. *Endocr. Rev.*, 20:649–688.
199. Lake, B.G. (1995): Peroxisome proliferation: current mechanisms relating to nongenotoxic carcinogenesis. *Toxicol. Lett.*, 82/83:673–681.
200. Cook, J.C. et al. (1992): Induction of Leydig cell adenomas by ammonium perfluorooctanoate: a possible endocrine-related mechanism. *Toxicol. Appl. Pharmacol.*, 113:209–217.
201. Vanden Heuvel, J.P. (1996): Perfluorodecanoic acid as a useful pharmacologic tool for the study of peroxisome proliferation. *Gen. Pharmacol.*, 27:1123–1129.
202. Moser, H.W. (2000): Molecular genetics of peroxisomal disorders. *Front. Biosci.*, 5:D298–D306.
203. Singh, I. (1997): Biochemistry of peroxisomes in health and disease. *Mol. Cell Biochem.*, 167:1–29.
204. Wanders, R.J., Van Grunsven, E.G., and Jansen, G.A. (2000): Lipid metabolism in peroxisomes: enzymology, functions and dysfunctions of the fatty acid alpha- and beta-oxidation systems in humans. *Biochem. Soc. Trans.*, 28:141–149.
205. Kota, B.P., Huang, T.H., and Roufogalis, B.D. (2005): An overview on biological mechanisms of PPARs. *Pharmacol. Res.*, 51:85–94.
206. Grimaldi, P.A. (2005): Regulatory role of peroxisome proliferator-activated receptor delta (PPAR delta): in muscle metabolism. A new target for metabolic syndrome treatment? *Biochimie*, 87:5–8.

207. Corton, J.C., Lapinskas, P.J., and Gonzalez, F.J. (2000): Central role of PPARalpha in the mechanism of action of hepatocarcinogenic peroxisome proliferators. *Mutat. Res.*, 448:139–151.
208. Lee, S.S. et al. (1995): Targeted disruption of the alpha isoform of the peroxisome proliferator-activated receptor gene in mice results in abolishment of the pleiotropic effects of peroxisome proliferators. *Mol. Cell. Biol.*, 15:3012–3022.
209. Völkl, A. and Fahimi, H.D. (1998): Isolation of peroxisomes. In: *Cell Biology: A Laboratory Manual*, Vol. 2, edited by J.E. Celis, pp. 87–92. Academic Press, San Diego, CA.
210. Shupliakov, O. et al. (2002): Impaired recycling of synaptic vesicles after acute perturbation of the presynaptic actin cytoskeleton. *Proc. Natl. Acad. Sci. U.S.A.*, 99(22):14476–14481.
211. Gottlieb, T.A. et al. (1993): Actin microfilaments play a critical role in endocytosis at the apical, but not the basolateral surface of polarized epithelial cells. *J. Cell Biol.*, 120(3):695–710.
212. Fowler, B.A. (1989): Biological roles of high affinity metal-binding proteins in mediating cell injury. *Comm. Toxicol.*, 3:27–46.
213. Apodaca, G. (2001): Endoctytic traffic in polarized epithelial cells: role of the actin and microtubule cytoskeleton. *Traffic*, 2:149–159.
214. Hryciw, D.H., Pollock, C.A., and Poronnik, P. (2005): PKC-alpha-mediated remodeling of the actin cytoskeleton is involved in constitutive albumin uptake by proximal tubule cells. *Am. J. Physiol. Renal Physiol.*, 288(6): F1227–F1235.
215. Lee. E. and De Camilli, P. (2002): Dynamin at actin tails. *Proc. Natl. Acad. Sci. U.S.A.*, 99(1):167–172.
216. Orth, J.D. et al. (2002): The large GTPase dynamin regulates actin comet formation and movement in living cells. *Proc. Natl Acad. Sci. U.S.A.*, 99(1):167–172.
217. Takei, K. et al. (1996): The synaptic vesicle cycle: a single vesicle budding step involving clathrin and dynamin. *J. Cell Biol.*, 133(6):1237–1250.
218. Conner, S.D. and Schmid, S.L. (2003): Regulated portals of entry into the cell. *Nature*, 422(6927):37–44.
219. Diaz, J.F. et al. (2000): Molecular recognition of Taxol by microtubules. *J. Biol. Chem.*, 275(34):26265–26276.
220. Kreis, T. and Vale, R. (1994): *Guidebook to the Extracellular Matrix and Adhesion Proteins*. Oxford University Press, New York.
221. Shay, J.W. (1986): *Cell and Molecular Biology of the Cytoskeleton*. Plenum Press, New York.
222. Alberts, B., Bray, D., Lewis, J., Raff, M., Roberts, K., and Watson, J.D., Eds. (1983): *Molecular Biology of the Cell*, pp. 549–668. Garland Press, New York.
223. Fuchs, E. and Cleveland, D.W. (1998): A structural scaffolding of intermediate filaments in health and disease. *Science*, 279:514–519.
224. Hall, A. (1998): Rho GTPases and the actin cytoskeleton. *Science*, 279:509–514.
225. Hirokawa, N. (1998): Kinesin and dynein superfamily proteins and the mechanism of organelle transport. *Science*, 279:519–526.
226. Kreis, T. and Vale, R. (1993): *Guidebook to the Cytoskeletal and Motor Proteins*. Oxford University Press, New York.
227. Mermall, V., Post, P.L., and Mooseker, M.S. (1998): Unconventional myosins in cell movement, membrane traffic, and signal transduction. *Science*, 279:527–533.
228. Bershadsky, A.D. and Vasiliev, J.M. (1988): *Cytoskeleton*. Plenum Press, New York.
229. Graff, R.D. et al. (1997): Altered sensitivity of post translationally modified microtubules to methylmercury in differentiating embryonal carcinoma-derived neurons. *Toxicol. Appl. Pharmacol.*, 144:215–224.
230. Zhao, Y., Li, W., and Chou, I.N. (1987): Cytoskeletal perturbation induced by herbicides, 2,4-dichlorophenoxyacetic acid (2,4-D) and 2,4,5-trichlorophenoxyacetic acid (2,4,5-T). *J. Toxicol. Environ. Health*, 20:11–26.
231. Elliget, K.A., Phelps, P.C., and Trump, B.F. (1991): $HgCl_2$-induced alteration of actin filaments in cultured primary rat proximal tubule epithelial cells labeled with fluorescein phalloidin. *Cell Biol. Toxicol.*, 7:263–280.
232. Li, W. and Chou, I.N. (1992): Effects of sodium arsenite on the cytoskeleton and cellular glutathione levels in cultured cells. *Toxicol. Appl. Pharmacol.*, 114:132–139.
233. Mills, J.W. and Ferm, V.H. (1989): Effect of cadmium on F-actin and microtubu1es of Madin–Darby canine kidney cells. *Toxicol. Appl. Pharmacol.*, 101:245–254.
234. Mills, J.W. et al. (1992): Zinc alters actin filaments in Madin–Darby canine kidney cells. *Toxicol. Appl. Pharmacol.*, 116:92–100.
235. Prozialeck, W.C. and Niewenhuis, R.J. (1991): Cadmium ($Cd^{2+}$) disrupts intercellular junctions and actin filaments in LLC-PK1 cells. *Toxicol. Appl. Pharmacol.*, 107:81–97.
236. Li, W., Zhao, Y., and Chou, I.N. (1987): Paraquat-induced cytoskeleta1 injury in cultured cells. *Toxicol. Appl. Pharmacol.*, 91:96–106.
237. O'Farrell, P.H. (1975): High-resolution two-dimensional electrophoresis of proteins. *J. Biol. Chem.*, 250: 4007–4021.
238. Spector, D.L., Goldman, R.D., and Leinwand, L.A. (1998): Two-dimensional gel electrophoresis of proteins. In: *Cells: A Laboratory Manual*, edited by D.L. Spector et al., pp. 58.1–58.9. Cold Spring Harbor Press, New York.
239. Zimmerman, L.J. et al. (2005): Identification of protein fragments as pattern features in MALDI: MS analyses of serum. *J. Proteome Res.*, 4(5):1672–1680.
240. Petricoin, E.F. et al. (2002): Serum proteomic patterns for detection of prostate cancer. *J. Natl. Cancer Inst.*, 94(20):1576–1578.
241. MacBeath, G. and Schreiber, S.L. (2000): Printing proteins as microarrays for high-throughput function determination. *Science*, 289(5485):1760–1763.
242. Craven, R.A. et al. (2002): Laser capture microdissection and two-dimensional polyacrylamide gel electrophoresis. *Am. J. Pathol.*, 160(3):815–822.
243. Vincent, V.A.M. et al. (2002): *J. Neurosci. Res.*, 69(5): 578–586.
244. Trump, B.F. and Berezesky, I.K. (1989): Cell injury and cell death: the role of ion deregulation. *Comm. Toxicol.*, 3:47–67.

245. Trump, B.F. et al. (1990): Cell toxicity and ion regulation in the kidney and bronchus: an hypothesis. In: *Basic Science in Toxicology: Proceedings of the Vth International Congress of Toxicology*, edited by G.N. Volans, J. Sims, F. Sullivan, and P. Turner, pp. 636–650. Taylor & Francis, London.
246. Trump, B.F. et al. (1992): The role of cytosolic calcium ($[Ca^{2+}]_I$) in injury and recovery from anoxia and ischemia. *Md. Med. J.*, 41:301–304.
247. Smith, M.W., Phelps, P.C., and Trump, B.F. (1991): Cytosolic $Ca^{2+}$ deregulation and blebbing after $HgCl_2$ injury to cultured rabbit proximal tubule cells as determined by digital imaging microscopy. *Proc. Natl. Acad. Sci. U.S.A.*, 88:4926–4930.
248. Nitta, N. et al. (1989): The effects of oxidative stress on rat proximal tubular epithelium (PTE): a role for cytosolic cadmium ($Ca^{2+}$). *Circ. Shock*, 27:333–334.
249. Ambudkar, I.S. et al. (1988): Extracellular $Ca^{2+}$-dependent elevation in cytosolic $Ca^{2+}$ potentiates $HgCl_2$-induced renal proximal tubular cell damage. *Toxicol. Indust. Health*, 4(1):107–123.
250. Swann, J.D. et al. (1991): Oxidative injury induces influx-dependent changes in intracellular calcium homeostasis. *Toxicol. Pathol.*, 19:128–137.
251. Trump, B.F., Berezesky, I.K., and Morris, A.C. (1989): New technique for the assessment of cellular injury: digital imaging fluorescence microscopy (DIFM). In: *The Applications of Histochemistry to Toxicology*, edited by J. Baker and P. Bach. Chapman & Hall, London.
252. Grynkiewicz, G., Poenie, M., and Tsien, R.Y. (1985): A new generation of $Ca^{2+}$ indicators with greatly improved fluorescence properties. *J. Biol. Chem.*, 260:3440–3450.
253. Smith, M.W. et al. (1987): $HgCl_2$-induced changes in cytosolic $Ca^{2+}$ of cultured rabbit renal tubular cells. *Biochem. Biophys. Acta*, 931:130–142.

# 40 Analysis and Characterization of Enzymes and Nucleic Acids

F. Peter Guengerich and Cheryl J. Bartleson

## CONTENTS

## INTRODUCTION

Twenty-five years have elapsed since one of us prepared this chapter for the first edition [1]. The book has been very successful, and the original chapter has been revised three times since then [2–4]. Many things have changed since the first version. Toxicology has become a more mechanistic field. In this fifth edition, the general discussion has been extensively modified. An additional author has been included to help with coverage of some of the areas. Some of the basic methods have been retained, because they are still utilized widely, without extensive changes, or they serve as prototypes for other assays. Several of the procedures have been changed in this fifth version to reflect perceived needs in readers' laboratories. In some cases, detailed protocols have been deleted because commercial manufacturers routinely supply detailed instructions; in those cases some general guidance about the basis of the system will be provided. We will also refer readers to some details previously covered in earlier editions of the chapter and book.

## BIOACTIVATION AND DETOXICATION

What can be learned by studying enzymes? The general paradigm for the relationship of enzymes to toxicology issues is shown in Figure 40.1. A chemical can be directly transformed to an inactive product. An alternative enzymatic reaction is conversion to a "biological enzymatic intermediate," such as the epoxide shown in the example. This reactive compound can be hydrolyzed or conjugated (possibly enzymatically in either case), yielding detoxicated product. Some of the reactive product may also react with tissue nucleophiles. Most of the reactions are with nontargets, which is good, although these products obscure the searches toward understanding mechanisms. The reactions with targets for toxicity are the detrimental ones. Nearly 50 years have elapsed since the concept of ultimate toxic forms was first developed [5]. Such activation is now widely accepted as the first step leading to the toxic and carcinogenic effects of many chemicals.

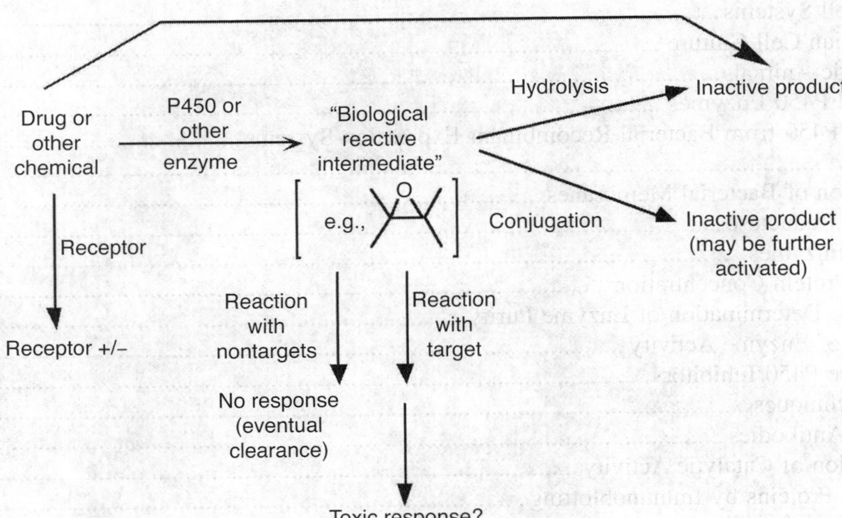


**FIGURE 40.1** General paradigm for the role of enzymatic bioactivation and detoxication.

The classification of the enzymes of interest is somewhat subjective, but the list provided in Figure 40.2 would probably be generally agreed on, with some possible changes. The list is based on the work of Jakoby [7] and was utilized in a later monograph [6]. The reader is referred to this list. Many of these enzymes exist in medium to large gene families and are inducible. This review will be focused on the P450 enzymes as general models for many of these enzymes, although many of the assays presented should, at least in principle, be useful for many of these enzyme systems.

The basic principles of the chemistry involved in toxicity are relatively straightforward. Most can be explained by either (1) the irreversible reaction of an electrophile with a tissue nucleophile or (2) propagation of free radicals (Figure 40.3). The enzymatic steps leading to the generation of electrophile are myriad but have been extensively studied [8,9]. The following principles govern these reactions [10,11]: (1) The basic reactions involve simple chemistry (e.g., reactions of electrophiles with nucleophiles, free radical propagation). (2) The first metabolic product or the most obvious chemical prospect may not be the reactant. (3) The stability of reactive products influences site and patterns of damage; a short time ($t_{1/2}$ = 1 second) may be considerable in a cell, and some reactive molecules are long lived ($t_{1/2}$ of minutes to hours). (4) *In vitro* systems are models, good for elucidating details; however, only some results, but not all, apply *in vivo*. (5) The dose is an issue (or more properly *the* issue) that can be traced to Paracelsus [12]. (6) Covalent binding can be an index of toxicity, but exceptions exist even after considerations of dose. Other issues (in addition to covalent binding) are receptor-mediated events (especially signaling), ability to repair (DNA and protein) damage, cell proliferation, and immune responses.

Events leading to radical initiation generally involve some type of one-electron chemistry, and radical propagation is often the result of free metal ions in biological systems. An important component of the radical chemistry is oxidative damage, resulting from the production of partially reduced oxygen species. In many cases, the effects of oxidative damage are not readily distinguished from alkylation and reactions of electrophiles. These and other events, including direct interactions of either parent chemicals or products with cellular receptor systems, are elaborated in the complex, more global view of the current knowledge of events related to toxicity presented in Figure 40.4 [11,13]. Covalent binding of metabolites is only one of the ways of disrupting the networks in a cell. Some direct effects on receptors can produce similar outcomes, in some cases. The numbers indicate the events that would be useful to monitor: (1) Genotoxicity tests are relatively well established and have been used in preliminary screening for more than 30 years. (2) Many receptor assays are now routinely done (e.g., AhR, PPARα); most of the guilt is by association in that many of the details related to toxicity are still vague. (3) Recently, assays have been developed to monitor the reaction of electrophiles with small thiols, with the hope of predicting toxicity in medium to high-throughput assays [8]. (4) The real goal is to develop biomarkers and assays that will predict events related to toxicity; some assays are in use, but this area still has the greatest potential.

Some examples of bioactivation processes follow. The first is the classic case of acetaminophen (paracetamol). Although this analgesic and antipyretic is used safely every day by many people, overdoses (accidents, attempted suicides) result in a large fraction of the cases of liver failure. The chemistry related to acetaminophen is rather simple (Figure 40.5). Small amounts are efficiently conjugated by uridine diphosphate glucuronosyltransferases (UGTs) or sulfotransferases to yield products that are readily eliminated; however, cytochrome P450 enzymes (P450s) are involved in minor oxidation pathways. The catecholic product does not appear to be a problem or the iminoquinone reacts with glutathione (GSH) and blocks its reaction with proteins. Thus, at low concentrations of acetaminophen, very little tissue damage results; however, when the concentration of acetaminophen increases, the sulfation and glucuronidation pathways are overwhelmed, and, in some cases, GSH is depleted and more of the iminoquinone reacts with tissues [16]. The extent of covalent binding is well correlated with parameters of toxicity [17]. Interestingly, the *meta* isomer of acetaminophen is much less toxic but demonstrates the same level of total covalent binding [18,19]. The protein targets for the two isomers are not identical [20], but exactly which targets for acetaminophen are relevant to toxicity, if any are, is still not clear.

The second example is the polycyclic aromatic hydrocarbon benzo(*a*)pyrene (Figure 40.6). This carcinogen, known since the 1930s [21], is oxidized to an epoxide. This first epoxide is not unusually reactive and is readily hydrolyzed by microsomal epoxide hydrolase to yield a dihydrodiol. The dihydrodiol is an excellent substrate for P450s. The resulting diol epoxide is quite reactive and can react with $H_2O$ directly, be conjugated by GSH transferases (GSTs), or react with DNA.

One example, provided in Figure 40.7, is the metabolism of aflatoxin $B_1$ ($AFB_1$). Over the course of a decade of research in this and other laboratories, the enzymes involved in nearly every step of $AFB_1$ metabolism have been identified, and the numerical value of the catalytic efficiency, $k_{cat}/K_m$, is given in Figure 40.7. For nonenzymatic processes, the estimated second-order rate constant is given. These are the parameters that should be operative at the low $AFB_1$ concentrations encountered in relevant human exposures, although some caveats must be considered when substrate concentrations begin to approximate

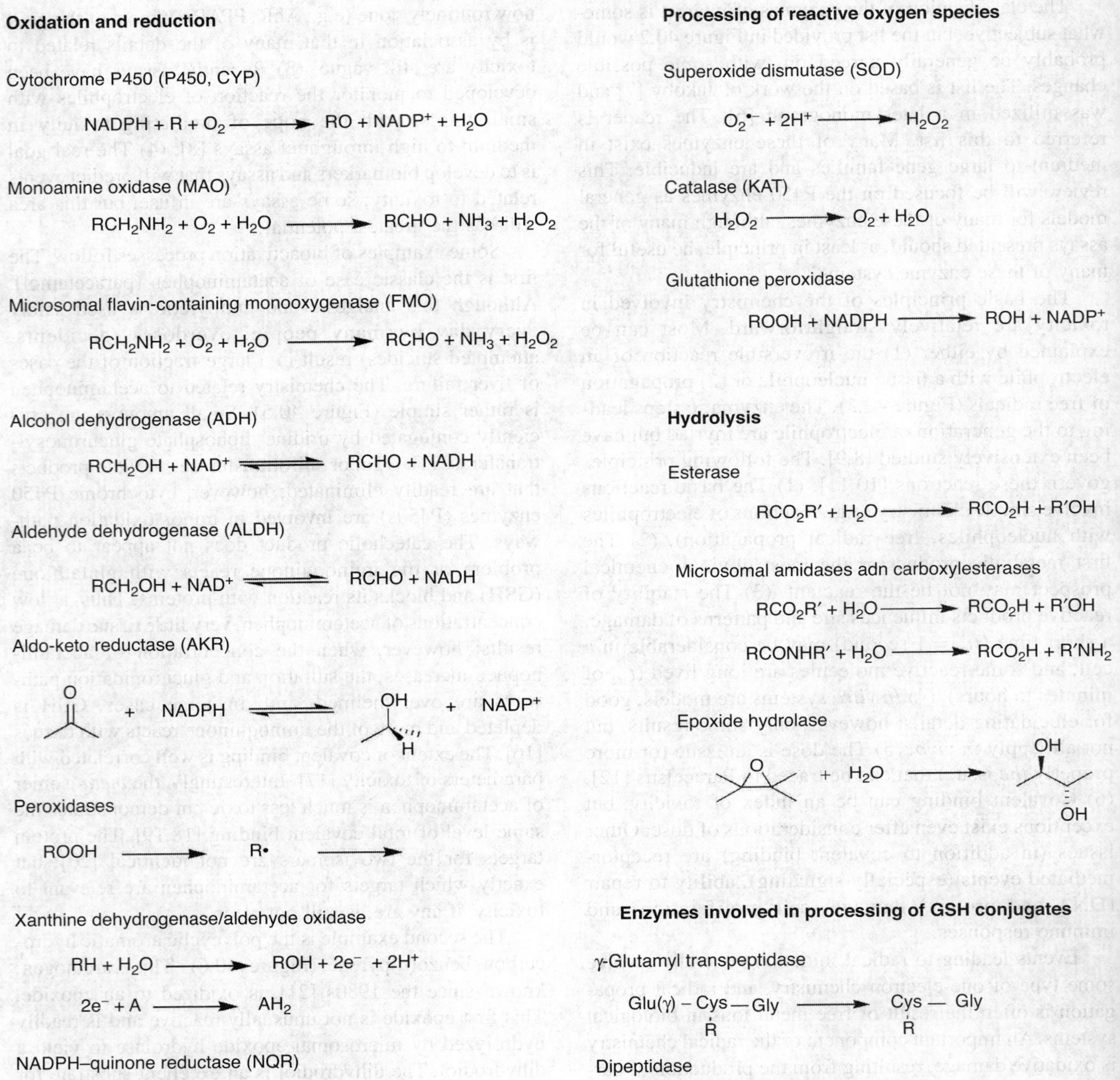

**Oxidation and reduction**

Cytochrome P450 (P450, CYP)

$NADPH + R + O_2 \longrightarrow RO + NADP^+ + H_2O$

Monoamine oxidase (MAO)

$RCH_2NH_2 + O_2 + H_2O \longrightarrow RCHO + NH_3 + H_2O_2$

Microsomal flavin-containing monooxygenase (FMO)

$RCH_2NH_2 + O_2 + H_2O \longrightarrow RCHO + NH_3 + H_2O_2$

Alcohol dehydrogenase (ADH)

$RCH_2OH + NAD^+ \rightleftharpoons RCHO + NADH$

Aldehyde dehydrogenase (ALDH)

$RCH_2OH + NAD^+ \rightleftharpoons RCHO + NADH$

Aldo-keto reductase (AKR)

O + NADPH ⇌ OH H + $NADP^+$

Peroxidases

ROOH ⟶ R• ⟶

Xanthine dehydrogenase/aldehyde oxidase

$RH + H_2O \longrightarrow ROH + 2e^- + 2H^+$

$2e^- + A \longrightarrow AH_2$

NADPH–quinone reductase (NQR)

O O + NAD(P)H ⟶ OH OH + $NAD(P)^+$

NADPH–P450 reductase

$R + NADPH \longrightarrow R^{\bullet +} + NADP^+$

**Processing of reactive oxygen species**

Superoxide dismutase (SOD)

$O_2^{\bullet -} + 2H^+ \longrightarrow H_2O_2$

Catalase (KAT)

$H_2O_2 \longrightarrow O_2 + H_2O$

Glutathione peroxidase

$ROOH + NADPH \longrightarrow ROH + NADP^+$

**Hydrolysis**

Esterase

$RCO_2R' + H_2O \longrightarrow RCO_2H + R'OH$

Microsomal amidases adn carboxylesterases

$RCO_2R' + H_2O \longrightarrow RCO_2H + R'OH$

$RCONHR' + H_2O \longrightarrow RCO_2H + R'NH_2$

Epoxide hydrolase

+ $H_2O$ ⟶ OH OH

**Enzymes involved in processing of GSH conjugates**

γ-Glutamyl transpeptidase

Glu(γ) – Cys — Gly (R) ⟶ Cys — Gly (R)

Dipeptidase

Cys — Gly (R) ⟶ Cys– R

Cys conjugate acetyl transferase

Cys—R + AcSCoA ⟶ NAcCys — R

Cys conjugate β-lyase

R′ S $NH_2$ $CO_2H$ ⟶ O $CO_2H$ + $NH_3$ + RSH

**FIGURE 40.2** Enzymes involved in metabolism and relevant to toxicology. (Adapted from Guengerich, F.P., in *Comprehensive Toxicology*, Vol. 3, *Biotransformation*, Guengerich, F.P., Ed., Elsevier, Oxford, 1997, pp. 1-6.)

**Conjugation**

GSH transferases (GST)

$GSH + RX \longrightarrow GSR + (H)X$

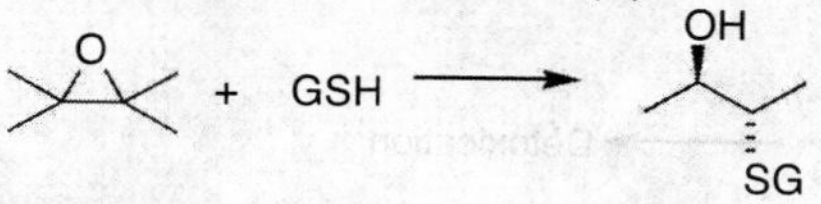

+ GSH ⟶ (OH, SG)

Sulfotransferases (SULT)

$ROH$ + 3′-phosphoadenosine 5′-phosphosulfate ⟶ $R\text{-}OSO_3^-$

*N*-Acetytransferase (NAT)

$RNH_2 + AcSCoA \longrightarrow RNHAc$

$RNHOH + AcSCoA \longrightarrow RNHOAc \longrightarrow RNH^+$

Methyltransferases

$RNH_2$ + S-AdenosylMet ⟶ RNHMe

ROH + S-AdenosylMet ⟶ ROMe

RSH + S-AdenosylMet ⟶ RSMe

Enzymes involved in formation of amide bonds

$RCO_2H + ATP + CoASH \longrightarrow RCOSCoA + AMP + PPi$

$RCOSCoA + NH_2CHR'CO_2H$ (glysine, taurine) ⟶ $RCONHCHR'CO_2H + CoASH$ (hippuric acid)

UDP glucuronosyl transferase (UGT)

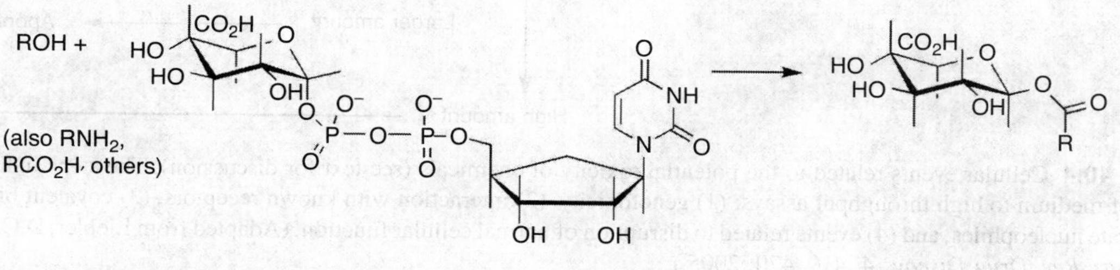

ROH + (also $RNH_2$, $RCO_2H$, others) (UDP-glucuronic acid: $CO_2H$, HO, HO, OH, O, $O^-$, P, O, NH, N, OH, OH) ⟶ ($CO_2H$, HO, HO, OH, O, R)

Rhodaneses and enzymes involved in the sulfane pool

$S + CN^- \longrightarrow SCN^-$

**FIGURE 40.2 (cont.)**

$X^{\delta+}$ + $Y^{\delta-}$ ⟶ X – Y

Electrophile (from drug) Nucleophile (from tissue)

Free radical propagation: $R\cdot + R'H \longrightarrow RH + R'\cdot \xrightarrow{(O_2)} R'O_2\cdot$

$\cdots \longleftarrow RO \longleftarrow R'O_2H \longleftarrow R'H$

**FIGURE 40.3** Basic chemical reactions involved in chemical toxicity: reaction of an electrophile with a nucleophile; free radical propagation.

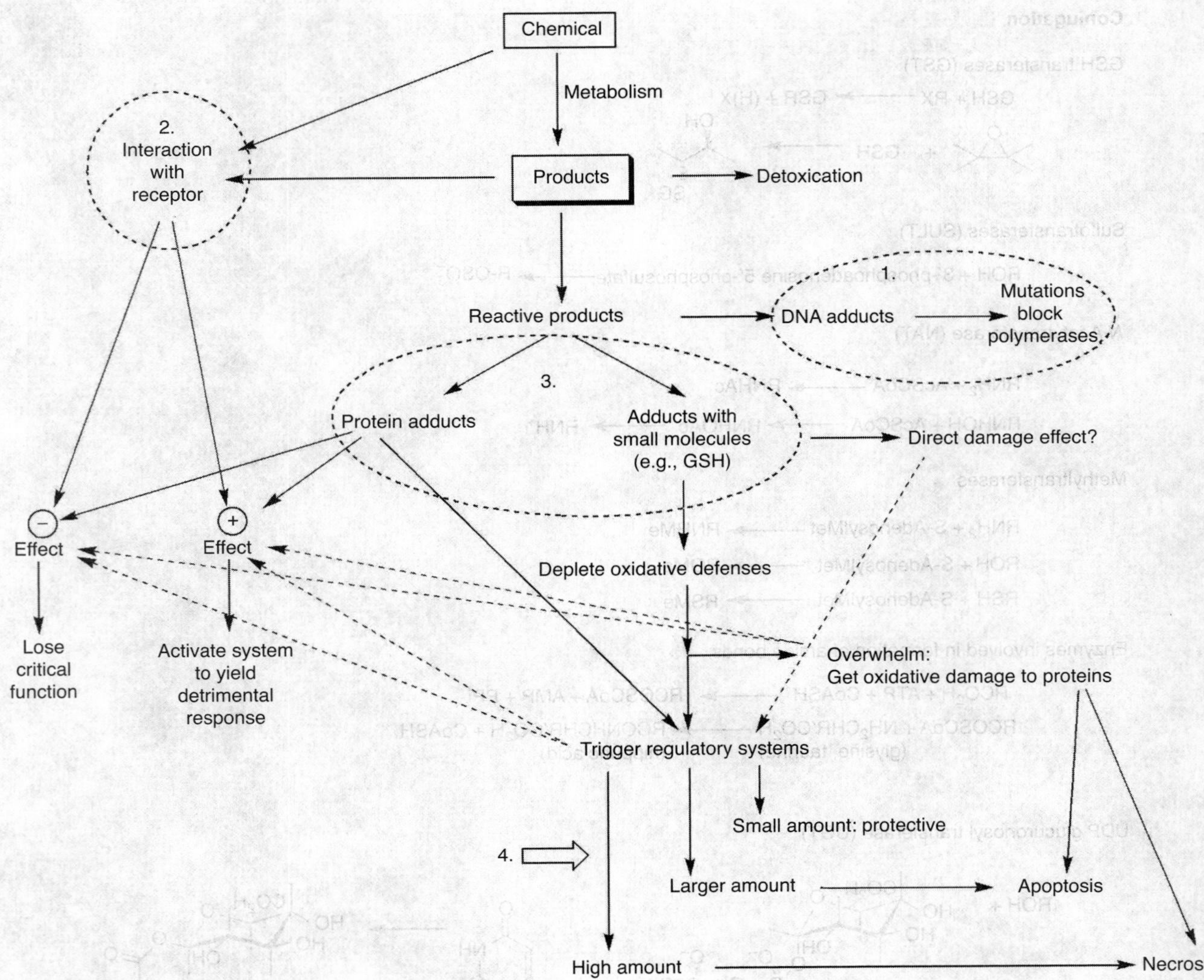


**FIGURE 40.4** Cellular events related to the potential toxicity of chemicals (see text for discussion). The steps indicate potential for the use of medium to high throughput assays: (1) genotoxicity, (2) interaction with known receptors, (3) covalent binding to proteins or surrogate nucleophiles, and (4) events related to disruption of normal cellular function. (Adapted from Liebler, D.C. and Guengerich, F.P., *Nature Rev. Drug Discov.*, 4, 410–420, 2005.)

the enzyme concentrations. As we consider this work in a global text, P450s can activate or detoxicate $AFB_1$ (Figure 40.7). GST M1 has a protective role, although it is not as effective as in some species (e.g., mice). Some aldo–keto reductases protect against protein (but not DNA) damage. Epoxide hydrolase is too inherently slow to provide much catalytic protection of DNA from the 8,9-*exo* epoxide, although enough of the enzyme might scavenge the epoxide, acting as a reagent.

Knowledge about the overall pathway leads to logical intervention strategies; for example, chlorophyllin can be used to absorb $AFB_1$, the initial substrate [24]. Oltipraz inhibits P450 oxidation [25]. Oltipraz and other inducers (Keap1 modifiers) induce GSTs and aldehyde–keto reductase (AKR) enzymes involved in detoxication [26].

## P450 AS A PARADIGM FOR ENZYMES INVOLVED IN TOXICOLOGY

By far the most common mechanism for bioactivation of chemicals is mixed-function oxidation by P450. This enzyme system was discovered in the late 1950s and has attracted a great deal of interest ever since. A number of enzyme forms exist within each species; some are tissue specific. The different forms preferentially oxidize different substrates, and this specificity contributes to the preferential bioactivation and detoxication of chemicals by different enzyme forms.

With a general catalytic mechanism involving abstraction of electrons or hydrogen atoms followed by oxygen rebound [27], one can explain the apparently diverse oxidative reactions catalyzed by P450 which can be classified

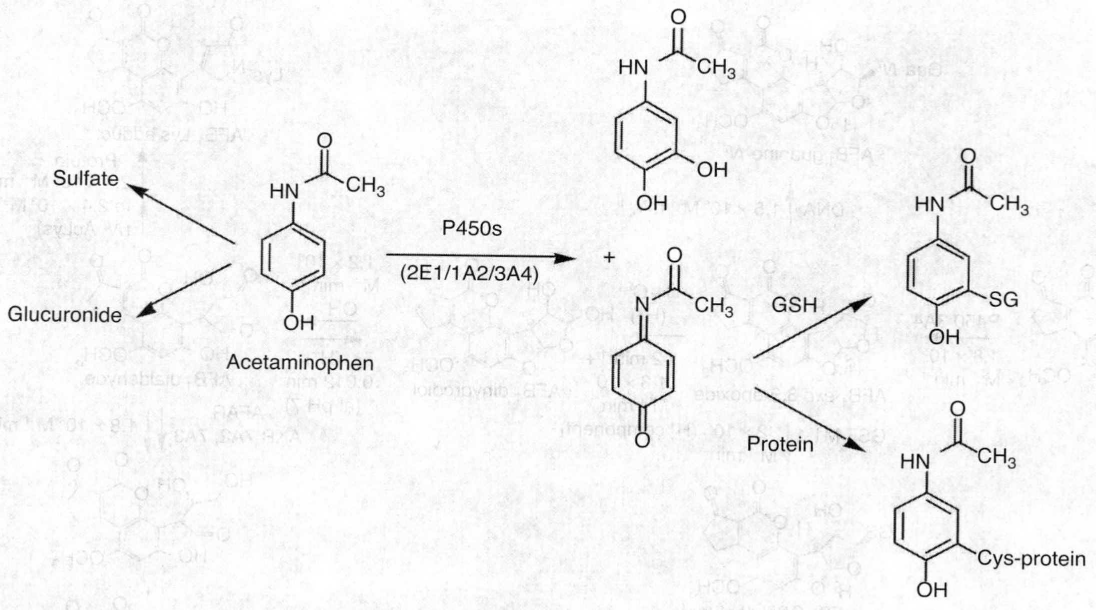

**FIGURE 40.5** Pathways involved in metabolism of acetaminophen. The conjugates are formed by the action of sulfotransferases and UGTs. For reference to the oxidation, see Harvison et al. [14] and Patten et al. [15].

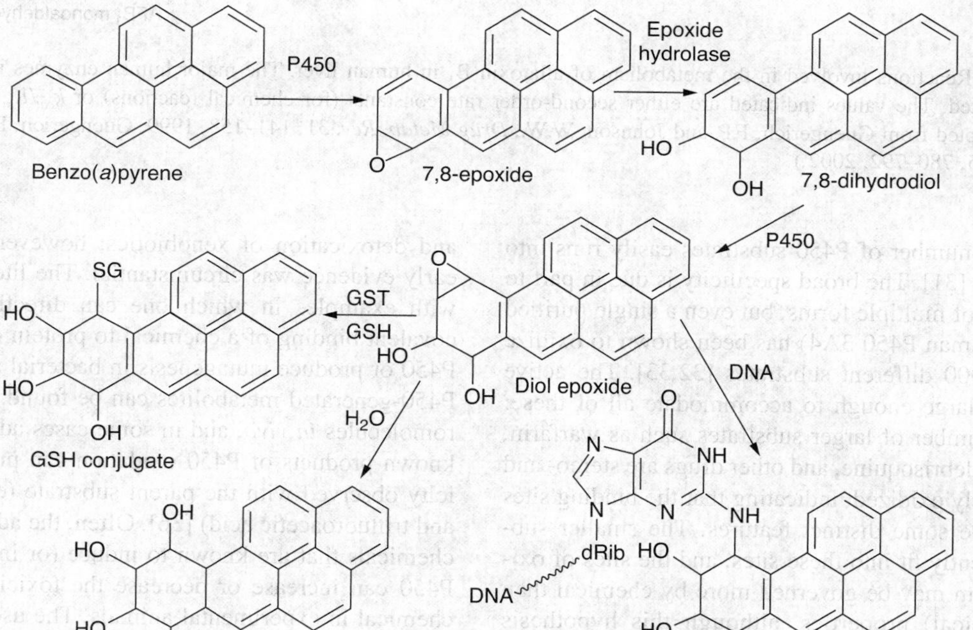

**FIGURE 40.6** Major reactions involved in the conversion of benzo(*a*)pyrene to a reactive diol epoxide, plus reactions of the diol epoxide, plus reactions of the diol epoxide with $H_2O$, DNA, and GSH.

as carbon hydroxylation, heteroatom oxygenation, heteroatom release (dealkylation), epoxidation, oxidative group migration, and other related events (Figure 40.8) [9,28]. These basic mechanisms can also explain the suicidal inactivation observed with substrates such as olefins, cyclopropylamines, and aminobenztriazole; the variation of structural features has allowed the selective inactivation of individual enzymes by mechanism-based (suicide) inhibitors [29,30]. P450 also reduces some compounds such as azo dyes, $CCl_4$, and *N*-oxides [9,31].

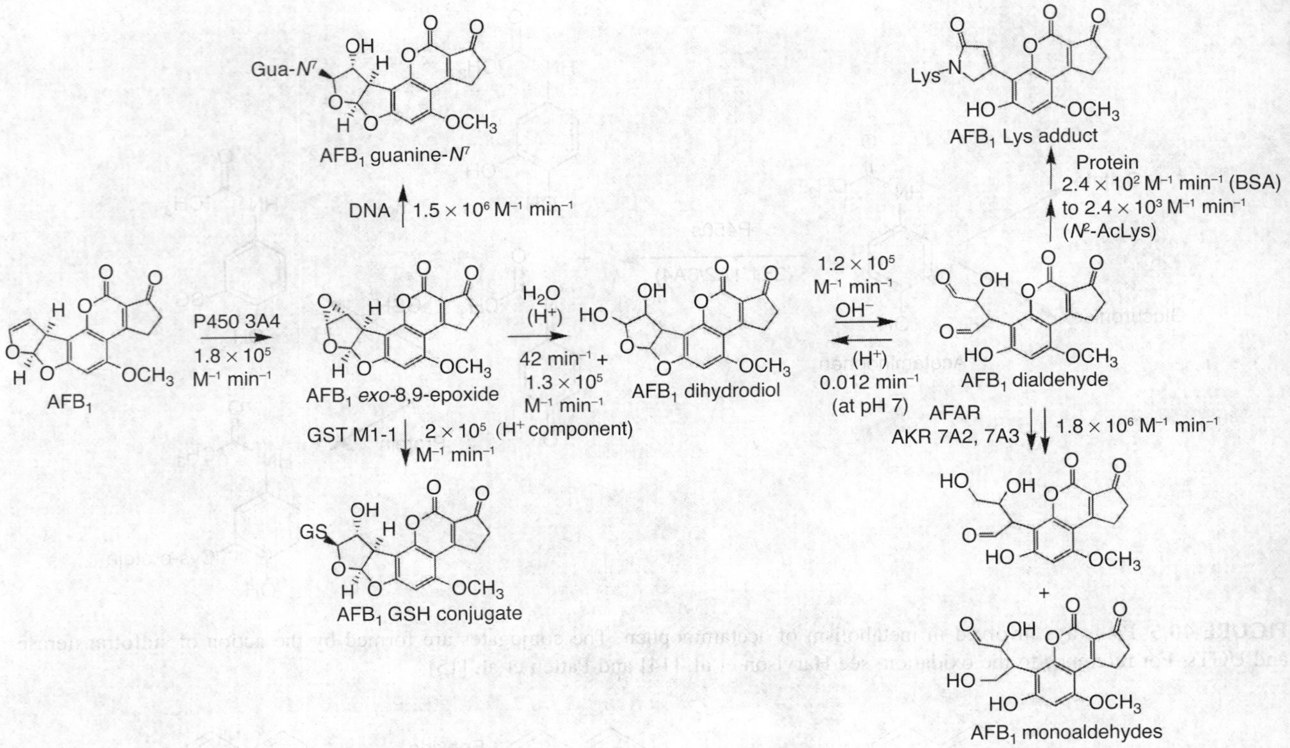

**FIGURE 40.7** Reactions involved in the metabolism of aflatoxin $B_1$ in human liver. The major human enzymes involved in each step are indicated. The values indicated are either second-order rate constants (for chemical reactions) or $k_{cat}/K_m$ for the purified enzymes. (Adapted from Guengerich, F.P. and Johnson, W.W., *Drug Metab. Rev.* 31, 141–158, 1999; Guengerich, F.P. et al., *Chem. Res. Toxicol.* 15, 780–792, 2002.)

The total number of P450 substrates easily runs into the thousands [31]. The broad specificity is due in part to the existence of multiple forms, but even a single purified form (e.g., human P450 3A4) has been shown to oxidize more than 1000 different substrates [32,33]. The active site must be large enough to accommodate all of these; however, a number of larger substrates such as warfarin, testosterone, debrisoquine, and other drugs are stereo- and regioselectively oxidized, indicating that the binding sites do really have some distinct features. The smaller substrates apparently fit into these sites, and the sites of oxidation on them may be governed more by chemical than spatial (physical) properties, although this hypothesis must be tested further.

One view is that most P450 enzymes exist for the metabolism of specific endogenous substrates such as fatty acids, steroids, and eicosanoids; however, others feel that the purpose of these enzymes is the clearance of ingested foreign chemicals (e.g., terpenes, alkaloids, pyrolysis products) [34]. There is validity in both viewpoints, and good cases for individual enzymes with each function may be found (Table 40.1).

One of the major reasons for the widespread study of P450 is that these enzymes are involved in the activation and detoxication of xenobiotics; however, much of the early evidence was circumstantial. The literature is filled with examples in which one can directly demonstrate covalent binding of a chemical to protein or DNA with a P450 or produce mutagenesis in bacterial strains. Known P450-generated metabolites can be found bound to macromolecules *in vivo*, and in some cases administration of known products of P450 oxidation can produce the toxicity observed with the parent substrate (e.g., fluoroxene and trifluoroacetic acid) [36]. Often, the administration of chemicals that are known to induce (or inhibit) forms of P450 can increase or decrease the toxicity of a certain chemical in experimental animals. The usefulness of this approach is limited because of the ability of these chemicals to alter the levels of several cytochromes P450 concomitantly, to alter the content of other enzymes, and to affect physiological parameters (e.g., blood flow) that contribute to endpoints under consideration. The problem of not knowing exactly *how* metabolites exert toxic effects also obfuscates the problem. Although P450s do appear to play a role in the generation of ultimate toxicants and carcinogens, efforts to show the importance of changes in the composition of individual forms on the effects are still not very clear.

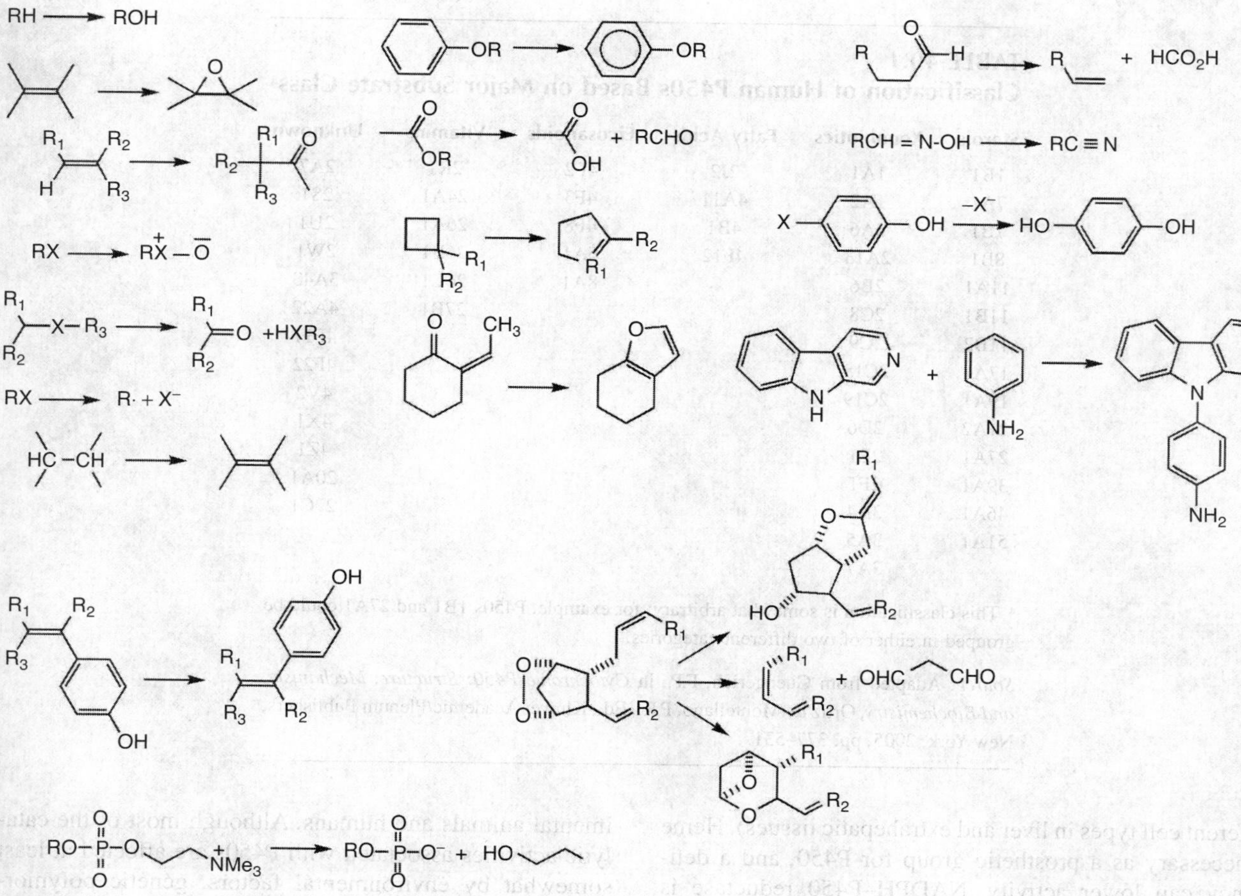

**FIGURE 40.8** Some P450 reactions. (Adapted from Guengerich, F.P., *Chem. Res. Toxicol.* 14, 611–650, 2001.)

Today no doubt exists concerning the existence of distinct forms of P450 in experimental animals or humans. At this time, 83 gene products have been characterized in rats and 57 in humans (Table 40.1) [37]. See the websites http://drnelson.utmem.edu/CytochromeP450.html and http://www.imm.ki.se/CYPalleles/ for current information regarding P450 genes and sequences. All of the human P450 genes are known. The purified proteins differ in electrophoretic properties, immunochemical aspects, primary sequence, and other criteria, including catalytic specificity (Table 40.2). The assignment of the proteins as separate gene products has been done using recombinant DNA. cDNAs have been cloned and used to establish the amino acid sequences of several thousand P450 enzymes in different species. In addition, many genomic DNA sequences have been established. There is no evidence to support an earlier view that gene translocations have made P450 a huge supergene family like the immunoglobulins.

In some cases considerable conservation of P450 structure is found between species. Small differences among the P450 proteins, however, can generate large differences in catalytic activity and substrate specificity, as shown in several notable examples. Lindberg and Negishi [38] showed that the change of a single residue of mouse P450 2A5 could change its catalytic selectivity quite dramatically.

Many human P450 enzymes have been purified [32,39] and even more have been characterized using recombinant methods [32]. These have been shown to have specificity in catalyzing the oxidation of drugs and other chemicals. In every case, some structural similarity with certain P450 forms isolated from experimental animals has been shown; however, this similarity can be misleading in ascertaining catalytic specificity in some cases. As in the case of experimental animal models, immunochemical methods have been of use in establishing the roles and catalytic specificities of the human P450s. Probes for P450s from experimental animals have been used to identify cDNAs for human P450s and derive their sequences.

Overall catalytic activity can be influenced at several levels [40,41]. Cofactor supply can be important: NADPH levels can be altered by starvation, and oxygen gradients exist in the liver. The different P450 forms also are localized preferentially in different regions of the liver (and in

**TABLE 40.1**
**Classification of Human P450s Based on Major Substrate Class[a]**

| Sterols | Xenobiotics | Fatty Acids | Eicosanoids | Vitamins | Unknown |
|---|---|---|---|---|---|
| 1B1 | 1A1 | 2J2 | 4F2 | 2R1 | 2A7 |
| 7A1 | 1A2 | 4A11 | 4F3 | 24A1 | 2S1 |
| 7B1 | 2A6 | 4B1 | 4F8 | 26A1 | 2U1 |
| 8B1 | 2A13 | 4F12 | 5A1 | 26B1 | 2W1 |
| 11A1 | 2B6 | | 8A1 | 26C1 | 3A43 |
| 11B1 | 2C8 | | | 27B1 | 4A22 |
| 11B2 | 2C9 | | | | 4F11 |
| 17A1 | 2C18 | | | | 4F22 |
| 19A1 | 2C19 | | | | 4V2 |
| 21A2 | 2D6 | | | | 4X1 |
| 27A1 | 2E1 | | | | 4Z1 |
| 39A1 | 2F1 | | | | 20A1 |
| 46A1 | 3A4 | | | | 27C1 |
| 51A1 | 3A5 | | | | |
| | 3A7 | | | | |

[a] This classification is somewhat arbitrary; for example, P450s 1B1 and 27A1 could be grouped in either of two different categories.

*Source:* Adapted from Guengerich, F.P., in *Cytochrome P450: Structure, Mechanism, and Biochemistry*, Ortiz de Montellano, P.R., Ed., Kluwer Academic/Plenum Publishers, New York, 2005, pp. 377–531.

different cell types in liver and extrahepatic tissues). Heme is necessary as a prosthetic group for P450, and a deficiency can lower activity. NADPH–P450 reductase is present at a level an order of magnitude lower than P450 and is needed for activity. While reconstituted P450 systems are stimulated by phospholipid, *in vivo* changes in lipid composition probably do not have any significant effect in most cases [42] with the very notable exception of the P450 3A family enzymes [43,44].

In experimental animals, the major way in which activities, or at least rather specific catalytic activities, are modulated is via changes in the amounts of individual P450 forms present [45,46]. Such changes have been clearly demonstrated to involve *de novo* synthesis and often involve specific increases in rates of nuclear DNA transcription. The list of P450 inducers is nearly as long as the list of substrates. Only in a few cases does one see induction of only one P450; more commonly several are induced. In some cases, the level of one of these proteins is depressed while others are induced [47,48]. Several of the rodent liver cytochromes P450 are also regulated by steroid hormones [49,50]. Most of the P450 proteins have similar half-lives (about 24 hours); however, some evidence exists for a stabilizing effect of certain classic inducers on mRNA stability. In the case of most cytochromes P450, evidence exists for intracellular receptors that bind the inducing ligands [51,52].

Another aspect of regulation is polymorphism, which is observed with certain catalytic activities in both experimental animals and humans. Although most of the catalytic activities associated with P450s are affected at least somewhat by environmental factors, genetic polymorphisms are distinct and not so readily influenced by other factors. These can be produced by variations in structural genes (coding or regulatory regions) or by variations in other proteins which regulate expression. Several polymorphisms have been mapped to chromosomes in rats and mice, and structural details underlying the polymorphisms have been elucidated in many cases. In humans, genetic polymorphisms in the metabolism of certain drugs have been identified. Correlations have been made between individuals expressing individual phenotypes (or genotypes) and susceptibility to chemical carcinogenesis. Low activity of the steroidogenic P450s can lead to lethality or debilitating diseases [35,53,54]. Still another issue with human P450 is inhibition, which can be a major issue in drug–drug interactions (Table 40.3).

Although the cytochromes P450 have received considerable attention, one should realize that the existence of isozymes (or more appropriately, *enzymes*) in a family is not unusual nor is a multigene family. These probably occur with many enzymes, including some others that are involved in other aspects of metabolism of xenobiotic chemicals; for example, at least 19 different forms of human glutathione-*S*-transferase exist and have been shown to be distinct gene products [59,60]. Again, the catalytic specificities of these enzyme forms differ, and many are under differential regulatory control. Consider-

**TABLE 40.2**
**Marker Activities for Some Human P450s Involved in Toxicology Studies**

| P450 | Tissue Sites | Typical Reaction |
|---|---|---|
| 1A1 | Lung, several extrahepatic sites, peripheral blood cells | Benzo(*a*)pyrene 3-hydroxylation |
| 1A2 | Liver | Caffeine $N^3$-demethylation |
| 1B1 | Many extrahepatic sites, including lung and kidney | 17β-Estradiol 4-hydroxylation |
| 2A6 | Liver, lung, and several extrahepatic sites | Coumarin 7-hydroxylation |
| 2A13 | Nasal tissue | Activation of 4-(methylnitrosamino)-1-(3-pyridyl)-1-butanone (NNK) |
| 2B6 | Liver, lung | (*S*)-Mephenytoin *N*-demethylation |
| 2C8 | Liver | Taxol 6α-hydroxylation |
| 2C9 | Liver | Tobutamide methyl hydroxylation |
| 2C19 | Liver | (*S*)-Mephenytoin 4′-hydroxylation |
| 2D6 | Liver | Debrisoquine 4-hydroxylation |
| 2E1 | Liver, lung, other tissues | Chlorzoxazone 6-hydroxylation |
| 3A4 | Liver, small intestine | Testosterone 6β-hydroxylation |
| 3A5 | Liver, lung | Testosterone 6β -hydroxylation |
| 3A7 | Fetal liver | Testosterone 6β -hydroxylation |
| 4A11 | Liver | Fatty acid ω-hydroxylation |

*Source:* Adapted from Guengerich, F.P., in *Cytochrome P450: Structure, Mechanism, and Biochemistry*, Ortiz de Montellano, P.R., Ed., Kluwer Academic/Plenum Publishers, New York, 2005, pp. 377–531.

**TABLE 40.3**
**Useful Selective Inhibitors of Human P450 Enzymes**

| | |
|---|---|
| P450 1A1 | 7,8-Benzoflavone (but see references 55 and 57 regarding P450 1A2) |
| | Ellipticine |
| | 1-(1-Propynyl)pyrene |
| | 2-(1-Propynyl)phenanthrene |
| P450 1A2 | 7,8-Benzoflavone |
| | Furafylline |
| | Fluvoxamine |
| P450 1B1 | 7,8-Benzoflavone |
| | 2-Ethynylpyrene |
| P450 2A6 | Diethyldithiocarbamate (see reference 58) |
| P450 2C9 | Sulfaphenazole |
| | Tienilic acid |
| P450 2D6 | Quinidine |
| P450 2E1 | Aminoacetonitrile |
| | 4-Methylpyrazole |
| | Diethyldithiocarbamate (see reference 58) |
| P450 3A4 | Troleandomycin |
| | Ketoconazole |
| | Gestodene |

*Source:* Data from Correia and Ortiz de Montellano [30], Shimada et al. [55], and Newton et al. [56].

able evidence supports the view that many forms of UGT exist [61], and the epoxide hydrolases are distinct gene products [62]. Distinct forms of microsomal flavin-containing monooxygenase are found in different tissues, specifically the liver and lung [63,64]; however, in some of the other instances only one gene appears to be involved (e.g., NADPH–cytochrome P450 reductase) [65].

## ROLES OF P450S AND OTHER ENZYMES IN BIOACTIVATION AND DETOXICATION

As mentioned above, P450-catalyzed oxidation can result in either the bioactivation or detoxication of a potential toxicant. In general, reduction reactions catalyzed by either P450 or NADPH–P450 usually lead to more reactive products.

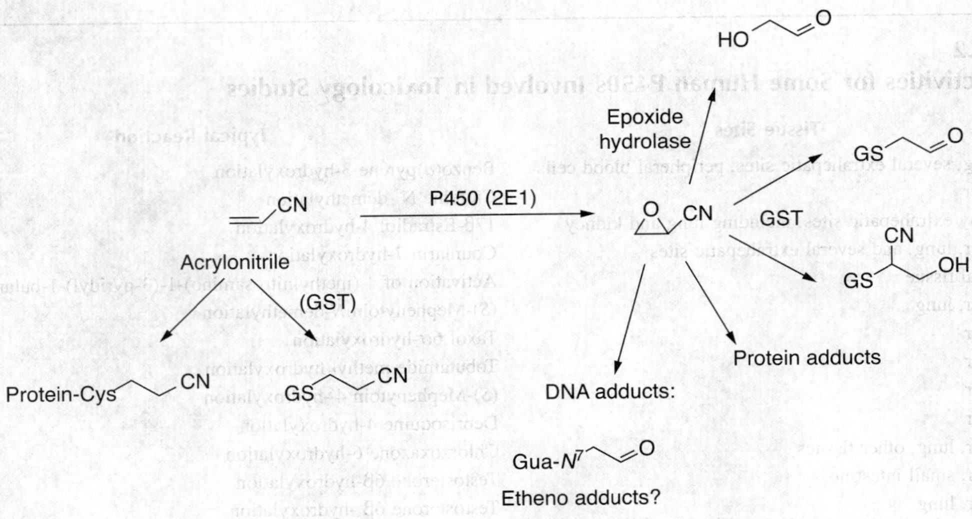

FIGURE 40.9 Proposed scheme for metabolism of acrylonitrile. (Adapted from Geiger, L.E. et al., *Cancer Res.*, 43, 3080–3087, 1983.)

Oxidations by other enzymes (i.e., microsomal flavin-containing monooxygenase, alcohol, and aldehyde dehydrogenases) can also result in either bioactivation or detoxication; for example, oxidation of allylic alcohols by alcohol dehydrogenase yields acrolein derivatives, which react rapidly with soft nucleophiles. In classical drug metabolism, oxidation–reduction reactions have been termed *phase I*, and conjugation reactions, which usually follow after oxidation or reduction, are termed *phase II* [66]. This classification, however, is outdated and frankly misleading [67], and it will not be used here. The majority of these conjugation processes detoxicate chemicals; therefore, increases in the concentrations of the proteins that catalyze these reactions or increases in the concentrations of cofactors (cosubstrate) tend to render an organism at decreased risk to protoxicants. Many exceptions to this generalism can be found, however; for example, epoxide hydrolase action on benzo(*a*)pyrene 7,8-oxide leads to formation of a substrate that is efficiently converted to 7,8-dihydroxy-7,8-dihydro-9,10-oxo-benzo(*a*)pyrene (the dihydrodiol), which reacts rapidly with DNA and is a potent mutagen and carcinogen (Figure 40.6). GST, as indicated later, can activate *vic*-dihaloalkanes to yield DNA damage. Glucuronides are formed by the action of UGT on hydroxylamines, which can break down in the acidic environment of the bladder to release nitrenium ions to alkylate DNA. Thus, we see that metabolic transformations must be viewed in a global manner to put the importance of individual steps into context.

What can studies on metabolic transformations tell us about the toxicity of chemicals? Comparing the actions of a series of small industrial compounds provides some examples:

$CH_2{=}CH{-}CN$
$CH_2{=}CH{-}Cl$
$Br{-}CH_2{-}CH_2{-}Br$

The first, acrylonitrile, is acutely toxic and also causes several types of general toxicity problems when administered at high doses in chronic studies (e.g., nausea, weight loss, gastric disturbances). The compound is not particularly carcinogenic, causing only tumors of the forestomach, brain, and possibly Zymbal's gland at high doses. These actions can be understood when the various pathways for acrylonitrile are measured using *in vitro* assays. Acrylonitrile reacts rapidly and nonenzymatically with sulfhydryls, both in proteins and in GSH (Figure 40.9). Conjugation with GSH is the major fate of acrylonitrile and renders it innocuous. Reaction with proteins is considerable and probably accounts for the toxic effects of acrylonitrile. About 10% of acrylonitrile is oxidized by P450 to its epoxide, which can (1) release cyanide (which does not appear to play a role in toxicity) or be conjugated with (2) GSH, (3) alkylate proteins, or (4) alkylate nucleic acids. The extent of the latter reaction does not appear to be very great, consistent with the relatively low tumorigenic potential of acrylonitrile [68–70].

Vinyl chloride appears similar in structure to acrylonitrile at first glance but behaves quite differently. Only very high doses are acutely toxic, and this toxicity is probably unrelated to metabolism; however, vinyl chloride is carcinogenic, causing a peculiar hemangiosarcoma that is almost unique to vinyl chloride production workers and can be reproduced in laboratory animals. Unlike acrylonitrile, vinyl chloride does not react directly with thiols, and its metabolism proceeds strictly through oxidation. The epoxide 1-chlorooxirane (2-chloroethylene oxide) can react with nucleic acids to form several lesions, including 1,$N^6$-ethenoadenine, $N^3$,4-ethenocytosine, 1,$N^2$-ethenoguanine, 7-hydroxy-1,$N^2$-ethanoguanine (5,6,7,9-tetrahydro-7-hydroxy-9-oxoimidazo[1,2-*a*]purine), $N^2$,3-ethenoguanine, and $N^7$-(2-oxoethyl)guanine [71]. Which of these is most

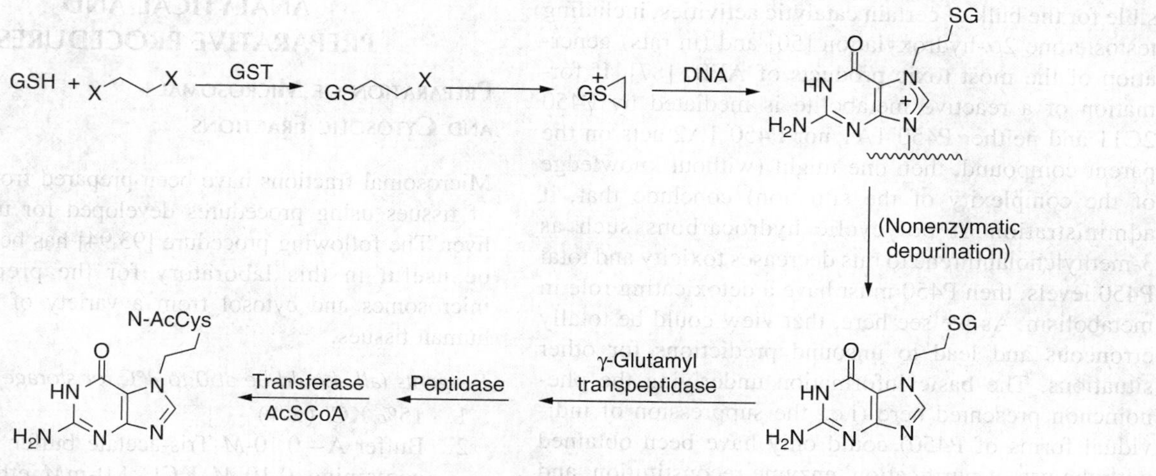

**FIGURE 40.10** Scheme depicting formation of the major DNA adduct from ethylene dihalides and the degradation of the adduct. (Adapted from Inskeep, P.B. et al., *Cancer Res.*, 46, 2839–2844, 1986; Kim, D.H. and Guengerich, F.P., *Cancer Res.*, 49, 5843–5851, 1989.)

intimately related to tumorigenesis is yet unclear, although evidence favors the etheno adducts [72,73]. The epoxide also spontaneously rearranges to form 2-chloroacetaldehyde [74–76], which is more like acrylonitrile, reacting rapidly and nonenzymatically with GSH and protein thiols. It reacts only slowly with nucleic acids and is probably not relevant to tumor initiation. The major site of vinyl chloride oxidation is the parenchymal cells of the liver; however, hepatic tumors originate in the reticuloendothelial cells, which have little if any oxidation capacity. A possible explanation is that the epoxide is formed in the parenchymal cells and is stable enough to migrate to other cells (some experimental evidence supports this view [77]); the differential susceptivity to the alkylating agent may be explained by variations in rates of DNA adduct repair among the cell types.

The next compound to consider in this series is ethylene dibromide (1,2-dibromoethane). This compound causes kidney toxicity and is carcinogenic at a number of sites. Oxidation by P450 (especially P450 2E1 [78]) yields 2-bromoacetaldehyde, which behaves in the same way as 2-chloroacetaldehyde (*vide supra*) and depletes sulfhydryls. GSH-transferase-catalyzed conjugation of ethylene dibromide with GSH also occurs; the ratio of ethylene dibromide metabolized through the oxidative and conjugative pathways is ~4:1 in rats [79]. In this case, the GSH conjugate is unstable, however, because the leaving group is still present (Br) (Figure 40.10). Nonenzymatic dehydrohalogenation produces an episulfonium (thiiranium) ion [82,83], which also has several fates. If it is hydrolyzed, *S*-(2-hydroxyethyl)GSH is formed, and this innocuous product is degraded and excreted. The putative episulfonium ion can also react with another GSH to form the ethylene–*bis* GSH adduct, which is also innocuous. Another possibility is elimination to yield GSSG (oxidized GSH) plus ethylene, another mode of detoxication [84]; however, another reaction (of the episulfonium ion) occurs with DNA to yield *S*-(2-[$N^7$-guanyl)ethyl]GSH as the major product [85]. This GSH pathway appears to be related to carcinogenesis because *in vitro* DNA binding and mutagenesis are much more dependent upon cytosolic than microsomal enzymes, and *in vivo* studies also support this view [80,86]. Thus, we see here that GSH conjugation can become a major bioactivation pathway.

The above three compounds share some apparent features of similarity, yet further analysis indicates that they differ widely in terms of their chemical properties, the manner in which they are handled by the body, and the biological effects that are exerted. Much of our understanding of these chemicals has come from *in vitro* studies using assays of the type that are described here, but what can we learn from studies that focus on identification and quantitation of individual enzymes?

One example involves the suppression of a particular P450 in rat liver. Polycyclic hydrocarbons such as 3-methylcholanthrene, β-naphthoflavone, and isosafrole induce increased synthesis of at least two forms of P450 in rat liver: P450 1A1 and P450 1A2. Increases in microsomal catalytic activities following administration of such compounds has generally been held to support the involvement of these inducible forms in a particular transformation, and in many *in vivo* studies alterations in the acute toxicity of compounds by these inducers have been interpreted in the same terms. The levels of a particular form of P450, P450 2C11 (measured with a specific antibody), are decreased when compounds are given to rats that induce P450 1A1 and P450 1A2 (also measured immunochemically) [48]. The decrease is as much as tenfold when certain polybrominated biphenyl congeners are administered to rats [47]. P450 2C11 is male specific [49,50] and is responsible for the bulk of certain catalytic activities, including testosterone 2α-hydroxylation [50] and (in rats) generation of the most toxic products of $AFB_1$ [87]. If formation of a reactive metabolite is mediated by P450 2C11 and neither P450 1A1 nor P450 1A2 acts on the parent compound, then one might (without knowledge of the complexity of the situation) conclude that, if administration of polycyclic hydrocarbons such as 3-methylcholanthrene to rats decreases toxicity and total P450 levels, then P450 must have a detoxicating role in metabolism. As we see here, that view could be totally erroneous and lead to unsound predictions for other situations. The basic information underlying the phenomenon presented here (i.e., the suppression of individual forms of P450) could only have been obtained with the use of purification, enzyme reconstitution, and immunochemical techniques.

Do the identification and assay of individual enzyme forms have any relevance in clinical settings? The answer is "yes," and several examples will be given from the realm of drug toxicity and therapeutic effectiveness. The antituberculosis drug rifampicin is a potent enzyme inducer and appears to increase the P450 forms that catalyze the A-ring hydroxylation of 17α-ethynylestradiol, the major estrogenic component of oral contraceptives. Such oxidation renders the drug ineffective, and cases have been reported where rifampicin administration to women has led to unanticipated pregnancies [88]. In other clinical cases, genetic deficiency in P450 2D6 has led to the accumulation of certain drugs and the production of undesirable side effects, such as the neuropathy associated with perhexiline- and captopril-induced agranulocytosis [89]. The suggestion has been made that dangers associated with chemicals in the environment may be affected by some of the same factors that influence drug clearance; for example, it has been suggested that individuals lacking P450 2D6 are less prone to tumors related to $AFB_1$ and cigarette smoking [90,91], although the findings are controversial. The molecular basis of the P450 2D6 polymorphism is now known [92]. We can now understand some interindividual variations in response to potentially toxic chemicals at the level of specific sequence changes. Toward this end, methods in enzymology are necessary and need to be applied in the field of toxicology.

For many of the enzymes considered here, a number of purification and heterologous expression techniques have been developed independently, and the reader is referred to the original literature for details. In describing the general assay procedures for use with microsomal and purified fractions, an effort has been made to deal with some of those most commonly used in the authors' and other laboratories. Selected examples of different types of assay procedures are given, and many of these can be adapted to other uses.

## ANALYTICAL AND PREPARATIVE PROCEDURES

### Preparation of Microsomal and Cytosolic Fractions

Microsomal fractions have been prepared from a variety of tissues using procedures developed for use with rat liver. The following procedure [93,94] has been found to be useful in this laboratory for the preparation of microsomes and cytosol from a variety of animal and human tissues.

*Reagents (all should be at 0 to 4°C for storage and use):*

1. 15% KCl (w/v)
2. Buffer A—0.10-*M* Tris-acetate buffer (pH 7.4) containing 0.10-*M* KCl, 1.0-m*M* ethylenediamine tetraacetic acid (EDTA), and 20-m*M* butylated hydroxytoluene (BHT)
3. Buffer B—0.10-*M* potassium pyrophosphate buffer (pH 7.4) containing 1.0-m*M* EDTA and 20-m*M* BHT
4. Buffer C—10-m*M* Tris-acetate buffer (pH 7.4) containing 1.0-m*M* EDTA and 20% glycerol (w/v)

Rats are killed by $CO_2$ asphyxiation in a closed chamber, in line with current animal care regulations. Livers are excised and placed in cold 1.15% KCl. All subsequent steps are carried out at 0 to 4°C. The livers are trimmed of debris and washed with 1.15% KCl; if one desires, hemoglobin contamination can be lowered by perfusing livers with KCl via the portal vein. The livers are blotted and weighed, placed in 4 times that weight of buffer A, and minced with a scissors. The method of homogenization depends on the scale of the preparation. If only a few livers are used, a mechanically driven Teflon®–glass homogenizer (4 to 5 vertical passes) is preferred. For larger preparations, two 40-second bursts in a Waring® blender are more efficient.

The homogenate is centrifuged at $10^4$ *g* for 20 minutes, and the supernatant is saved. If the yield of microsomes is a factor, the precipitate can be homogenized in buffer A again and recentrifuged to obtain additional supernatant. The supernatant is centrifuged for 60 minutes at $10^5$ *g* ($3.5 \times 10^4$ rpm in a Beckman 45 Ti rotor) to yield a microsomal pellet. After discarding the supernatant, a volume of buffer B equal to that of the discarded supernatant is added, and microsomes are removed from the clear glycogen pellet by gentle swirling or, if necessary, with the use of a rubber policeman. The suspended microsomes are homogenized with 4 passes of a mechanically driven Teflon®–glass homogenizer and recentrifuged at $10^5$ *g* for 60 minutes; the resulting pellets are homogenized and recentrifuged (60 minutes at $10^5$ *g*). The pellet is homogenized (with four strokes of the Teflon®–glass system) in a minimum volume of buffer C (to give 20 to 50 mg protein per mL) and stored at –20°C or –70°C.

Several comments are in order. BHT and EDTA are added to retard lipid peroxidation, and the pyrophosphate buffer is useful in removing hemoglobin and nucleic acids [94]. If proteases are a potential problem, as is often the case in extrahepatic tissues, phenylmethylsulfonyl fluoride (PMSF) or other protease inhibitors can be used. PMSF is unstable in water; a stock 0.10-*M* solution should be prepared in absolute ethanol or *n*-propanol, stored at –20°C, and added to buffers to give a final concentration of 0.10 m*M* immediately prior to their use. The use of dithiothreitol has also been reported to be necessary for the preparation of functional rat colon microsomes [95].

Buffers containing 0.25-*M* sucrose can be substituted for buffers A and B in the procedure. Some older procedures can be applied if an ultracentrifuge is not available. Precipitation of microsomes can be done at much lower speeds when 8-m*M* $CaCl_2$ is added to buffers [96]. Alternatively, microsomes have been isolated using gel exclusion chromatography [97,98]. More sophisticated techniques are available for the separation of rough and smooth endoplasmic reticulum and Golgi apparatus fractions. Some workers prefer to store microsomes as frozen pellets. For many enzyme activities, microsomes are functional for at least several months when stored either as pellets or frozen in buffer C.

Essentially identical procedures are routinely used to prepare microsomes and cytosol from livers of other animals, including humans [99]. In work with human samples, it is advisable to have individual samples tested for HIV and hepatitis B (and other forms of hepatitis) if at all possible before proceeding. Personnel should be immunized against hepatitis B. When handling tissues, personnel should handle samples as if viruses such as HIV or hepatitis might be present (even a viral test could be in error). Good hygiene is essential, and some key practices used in this laboratory include: (1) delivery of all residual tissue materials to the infectious diseases division of the institution for incineration or other disposal; (2) carrying out early steps that produce aerosols (homogenization, balancing of tubes containing crude fractions) in a fume hood; (3) absolutely *no* mouth pipetting; (4) prompt disposal of blotters over which all work has been done; (5) use of disposable plastic gloves and other protective clothing; (6) disinfection of glassware, knives, etc. in an appropriate detergent (e.g., O-Syl disinfectant detergent; National Laboratories, Lehn and Fink Industrial Products, Montvale, NJ; contains *o*-benzyl-*p*-chlorophenol-*o*-phenylphenol and isopropanol); and (7) above all, use of common sense in handling potentially dangerous material.

The same procedures used for livers of various animals may be adapted to extrahepatic tissues, although these are usually more resistant to homogenization. Cutting devices (e.g., Tekmar Tissuemizer®) may be used, although caution is advised if catalytic activity is to be measured in samples. The effects of such procedures should be checked.

## Assay of P450

The most generally used method is that of Omura and Sato [100], which utilizes the reduced-CO vs. reduced difference spectrum. The procedure used in this laboratory is outlined below.

*Reagents*

1. 0.10-*M* potassium phosphate buffer (pH 7.4) containing 1.0-m*M* EDTA, 20% glycerol (v/v), 0.50% sodium cholate (w/v), and 0.40% Triton® N-101 (Sigma Chemical Co.; St. Louis, MO), Emulgen™ 913 (Kao-Atlas, Tokyo), or equivalent detergent (w/v)
2. $Na_2S_2O_4$ (sodium dithionite, sodium hydrosulfite), reagent-grade (keep bottle tightly closed when not in use)
3. CO gas, reagent-purity; store and use in fume hood

Microsomes (or other preparations) are added to the buffer to give a final concentration of 0.05 to 5 μ*M* P450, mixed, and divided into two 1.0-mL glass or (disposable) polystyrene cuvettes (10-mm pathlength). The sample cuvette is saturated with 40 to 60 bubbles of CO at a rate of about 1 bubble per second. A baseline is recorded between 400 to 500 nm using a split-beam spectrophotometer. A few crystals of $Na_2S_2O_4$ (1 to 2 mg) are added to each cuvette; the cuvettes are covered with parafilm, inverted several times to mix the $Na_2S_2O_4$, and placed in the spectrophotometer again after checking for liquid on the sides of the cuvettes; alternatively, a plumping device, such as Add-a-Mixer (NSG Precisions Cells; Farmingdale, NY) can be used to mix the contents, without removing the cuvette from the chamber. Spectra are recorded (400 to 500 nm) until the 450-nm peak reaches a maximum.

The $A_{490}$ (isosbestic point) serves as a reference point. P450 content is determined as follows (Figure 40.11):

$$[(A_{450-490})_{observed} - (A_{450-490})_{baseline}]/0.091 = \text{nmol P450/mL}$$

Cytochrome P420 represents denatured forms of P450 and is determined using the following formulae:

$$(\text{nmol P450/mL}) \times (-0.041) = (A_{420-490})_{theoretical}$$

$$(A_{420-490})_{observed} - (A_{420-490})_{theoretical} - (A_{420-490})_{baseline}/0.110 = \text{nmol cytochrome P420/mL}$$

The extinction coefficient ($\Delta\varepsilon_{450-490}$) of 91-m$M^{-1}$ cm$^{-1}$ has been verified using highly purified rat and rabbit liver P450 preparations [101,102]. The second set of formulae is based on the observation that P450 has an extinction coefficient of –41 m$M^{-1}$ cm$^{-1}$ ($\Delta\varepsilon_{420-490}$) in the difference spectrum (i.e., the $A_{420}$ of the reduced-CO complex is less than the $A_{420}$ of reduced P450) [103].

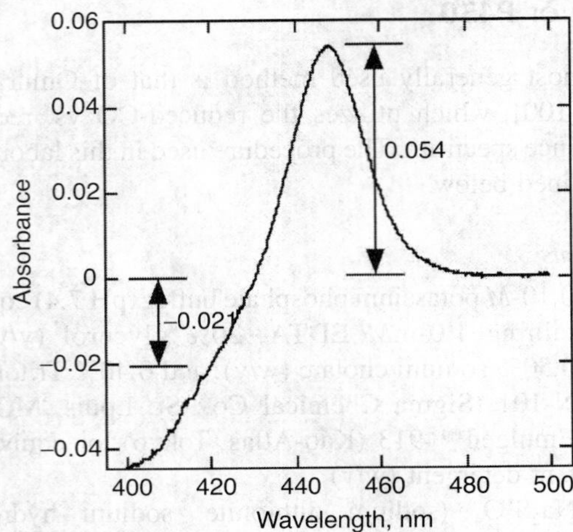


**FIGURE 40.11** Calculation of P450 and P420 concentrations. A sample of rat liver P450 1A1 was diluted 10-fold with 0.1-*M* potassium phosphate buffer (pH 7.7) containing 1-m*M* EDTA, 40% glycerol (v/v), 0.2% Emulgen 913 (w/v), and 0.5% sodium cholate (w/v). The sample was divided into two cuvettes, and one was saturated with CO gas. The two cuvettes were balanced in an Amino DW2a/OLIS spectrophotometer using the automatic baseline correction mode. The corrected baseline was recorded. After the addition of $Na_2S_2O_4$ as indicated in the text, the final difference spectrum was obtained. The calculations are as follows: $0.054 \div 0.091 = 0.59$ nmol P450/mL; $0.59 \times (-0.041) = -0.024$ ($A_{420}$); $-0.024 - (-0.021) = 0.003$; $0.003 \div 0.110 = 0.03$ nmol cytochrome P420/mL; $0.59 \times 10 = 5.9$ nmol P450/mL; $0.03 \times 10 = 0.3$ nmol cytochrome P420/mL. See text for further discussion.

Whereas rat liver microsomes can be routinely prepared with minimal hemoglobin contamination, this is not the case for many other preparations. The basic procedure of Matsubara et al. [104] for assaying P450 in liver homogenates is then useful. In this method, two cuvettes are prepared as before, but both are equilibrated with CO and the baseline is recorded. $Na_2S_2O_4$ is added *only* to the sample cuvette to obtain a reduced-CO vs. oxidized-CO difference spectrum; the extinction coefficient ($\Delta\varepsilon_{450-490}$) is 106 m$M^{-1}$ cm$^{-1}$. Distinguishing between methemoglobin and cytochrome P420 is difficult, although Johannesen and DePierre [105] have reported that methemoglobin can be specifically reduced by ascorbate and phenazinemethosulfate.

Detergents are routinely used in the assay of P450 in this laboratory, as these solubilize the microsomal membranes to reduce light scattering and do not denature P450 in the presence of glycerol [94]. The buffer also helps prevent settling of any insoluble particles; however, some particular P450 proteins may not necessarily be stable in the presence of these detergents [43]. If one desires to carry out determinations in the absence of detergents, a spectrophotometer should be used that is capable of handling turbid solutions. The limit of detection of P450 in extrahepatic tissues is influenced more by the presence of hemoglobin than instrumental considerations. In our own laboratory, we have used Varian® 635M and Cary 14, 210, and 219 spectrophotometers for such measurements in the past. Other instruments are also suitable. The Aminco™ DW-2a instrument has historically been popular among investigators, and we currently use a computer-updated version (On-Line Instrument Systems; Bogart, GA) in our own laboratory; this is particularly useful with bacterial cells because of the ability to handle turbid samples.

## Assay of NADPH-Cytochrome *c* Reduction

The enzyme NADPH–P450 reductase is conveniently measured by its NADPH-cytochrome *c* reduction activity [106].

*Reagents*

1. Horse heart cytochrome *c* (0.50-m*M*) in 10-m*M* potassium phosphate buffer (pH 7.7)
2. 0.30-M potassium phosphate buffer (pH 7.7)
3. 10-m*M* NADPH (fairly stable for <7 days at 4°C in the dark; however, for the most accurate work, solutions should be prepared fresh daily)

Pipette 80 µL cytochrome *c* solution, the enzyme sample, and sufficient 0.30-*M* phosphate buffer in a 1.0-mL cuvette (10-mm pathlength) to bring the total volume to 0.99 mL. The components are mixed and preincubated at 30°C (room temperature is also acceptable for most measurements) in a recording spectrophotometer; the recorder is adjusted to zero absorbance at 550 nm (full scale, 1.0; slit width, 1.0 nm if possible, because of the narrow band being observed). After recording the baseline for 3 minutes, 10 µL of NADPH is added and $A_{550}$ is followed for about 3 minutes. Activity is calculated as follows:

$$\frac{\Delta A_{550}\ \text{min}^{-1}}{0.021} = \text{nmol cytochrome } c \text{ reduced min}^{-1}$$

An amount of enzyme should be used such that the initial $\Delta A_{550}$ does not exceed 0.2 min$^{-1}$. The assay is an indirect measure of NADPH–P450 reductase activity; measurement of the actual NADPH–P450 reductase activity requires anaerobic conditions and rapid reaction techniques [106]. A small peptide of the reductase is required for efficient reduction of P450 but not cytochrome *c*; if no proteolysis has occurred, the two activities are closely correlated [107]. NADPH–cytochrome *c* reduction activity is stimulated by the high salt concentration used in the assay [106]. If activity is assayed in the presence of mitochondria (or in bacteria), KCN (1.0-m*M*) may be added as a precautionary measure to block nonmicrosomal activity. The reduction of several other compounds, such as dichlorophenolindophenol, ferricyanide [107], or a tetrazolium dye [108] may also be used to assay activity. Spectrophotometers with automatic sample positioners may be used to carry out multiple assays simultaneously.

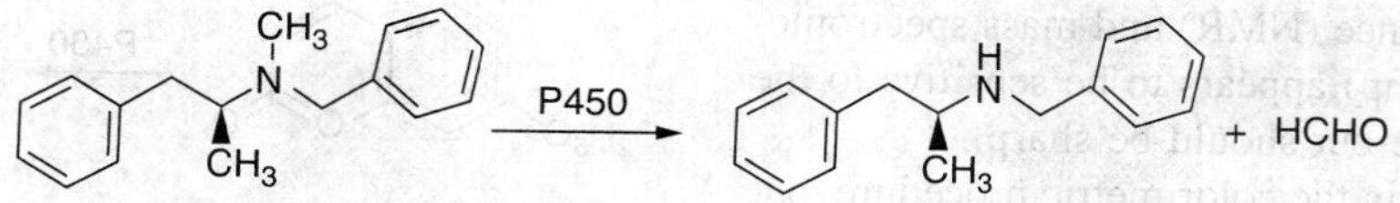

FIGURE 40.12 *d*-Benzphetamine *N*-demethylation.

## P450-Linked Activities

Some classic and typical assays are presented, along with prototypes of different types of methods and approaches.

### Benzphetamine *N*-Demethylation

The assay of this activity (Figure 40.12) is popular because of its ease and sensitivity, especially with microsomes prepared from animals treated with barbiturates or similar inducers. At least three assays can be used. HCHO is released during the reaction and forms the basis for the first two of the assays.

#### *Colorimetric Measurement of HCHO*

For studies on colorimetric measurement of HCHO and other carbonyls, see references 109 to 111.

*Reagents*

1. 1.0-*M* potassium phosphate buffer (pH 7.7)
2. 10-m*M* $NADP^+$
3. 1 mg = $10^3$ IU yeast glucose-6-phosphate dehydrogenase/mL (dissolved in 10-m*M* Tris-acetate buffer, pH 7.7, containing 1.0-m*M* EDTA and 20% glycerol [v/v])
4. 0.10-*M* glucose-6-phosphate
5. NADPH-generating system: Mix 25 parts reagent 2, 1 part reagent 3, and 50 parts reagent 4 in an amount sufficient to supply 0.15 mL per mL of total reaction; store (on ice) during the day of use and then discard. This mixture may have to be preincubated to the reaction temperature prior to addition; otherwise, keep it on ice during the day of use, then discard. *Note:* The system may be easily checked before use, which is a good idea when using new reagents or doing a large number of experiments: Add 0.10 mL of the reaction buffer, 0.05 mL of reagent 2, 0.10 mL of reagent 4, and 0.75 mL $H_2O$ to a cuvette and mix. Place this in a spectrophotometer and adjust the baseline to zero. Add 2 μL of reagent 3, preferably using a plumping device (see P450 assay above) and record the change in $A_{340}$. The system should reach an $A_{340}$ of 2.0 within ~60 seconds.
6. 10-m*M* *d*-benzphetamine·HCl
7. 17% aqueous $HClO_4$ (1/4 dilution of concentrated $HClO_4$)
8. Nash reagent (300 g $NH_4CH_4CO_2$, 4.0 mL acetylacetone [2,4-pentanedione], and 6.0 mL glacial $CH_3CO_2H$ per liter)

Incubations are carried out in 1.27-mL total volume (prior to addition of reagent 5) and include 0.5 to 2 mg microsomal protein, 50-m*M* phosphate buffer, and 1.0-m*M* benzphetamine·HCl (this should be added after the other components from a 10-m*M* aqueous stock). Tubes are preincubated for 3 minutes at 37°C and then 0.23 mL of reagent 5, the NADPH-generating system, to start incubations. After shaking the tubes (150 rpm) for 10 minutes at 37°C, the incubations are stopped by the addition of 0.50 mL of 17% $HClO_4$ and chilled on ice for 5 to 10 minutes. (Because of the short incubation time, it is convenient to start and stop individual tubes every 10 seconds.) Tubes are centrifuged at $3 \times 10^3$ *g* for 5 minutes, and 1.0 mL of each supernatant is transferred to a new tube. To each of these tubes is added 0.40 mL of Nash reagent. The tubes are heated at 60 to 70°C for 20 minutes in a water bath (covered to prevent evaporation). The tubes are cooled in tap water and $A_{412}$ values are read vs. a water blank. Both minus benzphetamine and minus NADPH-generating system blanks should be included; the $A_{412}$ values for these should be similar and are subtracted from the experimental values. A standard curve can be prepared using HCHO; we find that such curves routinely yield factors of 460 to 480 which, when multiplied by net $A_{412}$, give the total nmol of HCHO produced. If microcuvettes are used for reading $A_{412}$, the entire procedure can be scaled down tenfold if an appropriate spectrophotometer is available. The same procedure is generally applicable to many substrates that release HCHO as a consequence of oxidation by P450.

#### *Radiometric Extraction of $H^{14}CHO$*

This procedure requires the use of [*N*-methyl-$^{14}$C]-benzphetamine but offers increased sensitivity [102,112]. This material can be synthesized from *d*-benzylamphetamine. The synthesis has been carried out in this laboratory as follows [113].

Benzylamphetamine (free base, 1.0 mmol, 225 mg) is stirred with 1.1 mmol of $^{14}CH_3I$ (2 mCi, 156 mg) and 1.1 mmol of $K_2CO_3$ (152 mg) in 30 mL of $(CH_3)_2CO$ overnight. The solvent is removed *in vacuo* and the residue is suspended in water. The pH is adjusted to >10 with $K_2CO_3$ if necessary; the solution is extracted three times with $CH_2Cl_2$. The $CH_2Cl_2$ layers are combined, dried with anhydrous $Na_2SO_4$, and saturated with dry HCl gas. Solvent is removed *in vacuo*, and the residue is crystallized from ethyl acetate to give [*N*-methyl-$^{14}$C] benzphetamine·HCl in ~50% yield. Purity can be checked by nuclear magnetic resonance (NMR) and mass spectrometry; the melting point (mp) appears to be sensitive to the crystallization procedure but should be sharp.

Assays are set up as in the colorimetric procedure, but the volume is reduced to 0.75 mL and the protein concentration may be reduced to fit the situation. Incubations are stopped by the addition of 0.25 mL 1-*M* NaOH and 5.0 mL $CHCl_3$ (or $CH_2Cl_2$). Tubes are mixed using a vortex device and centrifuged ($3 \times 10^3$ *g* for 5 minutes). The aqueous upper layer is transferred to a clean tube, 5.0 mL of $CHCl_3$ is added, and the mixing, centrifugation, and transfer steps are repeated. The above step is repeated once more, and a 0.50-mL aliquot of the aqueous phase is transferred to a mini-scintillation vial. The contents are neutralized by the addition of 0.50 mL of 0.10-*M* sodium citrate buffer (pH 6.5) to which has been added 0.060-*M* HCl; 5 mL of a water-miscible liquid scintillation cocktail is added. Vials are capped, mixed, and counted (10 minutes will usually produce satisfactory counting deviation). Blanks contain all components except NADPH or protein.

The efficiency of extraction is >95%. HCHO remains in the aqueous phase and residual substrate is extracted in the $CHCl_3$ layers at the basic pH. A similar procedure may be used in the assay of any substrate that can be labeled with a labeled methyl group that is released following oxidation. An alternative method involves trapping the labeled HCHO product as the dimedone derivative [114] or as the 2,4-dinitrophenylhydrazone derivative and analysis by high-performance liquid chromatography (HPLC) [115].

### *Enhancement of NADPH Oxidation or $O_2$ Uptake*

Because of high rates of endogenous oxidase activity, these procedures are more commonly used with reconstituted enzyme systems than with microsomes [116,117]. In the oxygen electrode procedure, experiments are set up as before and a background rate of $O_2$ uptake is observed. The differences in the rates obtained with substrates are measured; each nmol of $O_2$ consumed corresponds to 1 nmol of substrate metabolized [117]. The NADPH oxidation assay is carried out in a similar way. The NADPH-generating system is deleted. Incubations, containing all components except NADPH, are preincubated for 3 minutes at 37°C in 1.0-mL cuvettes in a recording spectrophotometer set at 340 nm (1.0 or 2.0 full-scale absorbance). NADPH (15 μL of a 10-m*M* solution, prepared the same day) is added and the rate of decrease in $A_{340}$ is observed. Blank incubations contain all components except benzphetamine. Rates are determined by dividing net $\Delta A_{340}$ $min^{-1}$ by 0.00622 to obtain nmol NADPH oxidized/min. The benzphetamine demethylation rate determined by this procedure has been reported to be identical to that obtained by HCHO assay under some conditions [118] but not others [117] and should be checked before routine use.

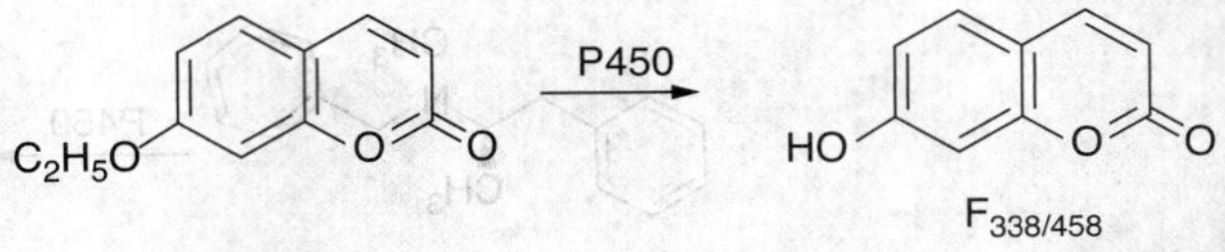

$C_2H_5O$ — P450 → HO — $F_{338/458}$

**FIGURE 40.13** 7-Ethoxycoumarin *O*-deethylation.

## Fluoresence

### *7-Ethoxycoumarin O-Deethylation*

This reaction (Figure 40.13) is an example of a fluorescence assay involving extraction; it is very sensitive, convenient, and applicable to a wide variety of samples [119,120].

*Reagents*

1. 30-m*M* 7-ethoxycoumarin (Aldrich Chemical Co.; Milwaukee, WI), dissolved in $CH_3OH$ (avoid exposure to light)
2. 1.0-*M* potassium phosphate buffer (pH 7.4)
3. 10-m*M* $NADP^+$
4. $10^3$ IU yeast glucose-6-phosphate dehydrogenase/mL (dissolved in 10-m*M* Tris-acetate buffer, pH 7.7, containing 1-m*M* EDTA and 20% glycerol [v/v])
5. 0.10-*M* glucose-6-phosphate
6. NADPH-generating system: Mix 25 parts reagent 2, 1 part reagent 3, and 50 parts reagent 4 in an amount sufficient to supply 0.15 mL per mL of total reaction; refer to the earlier section on benzphetamine *N*-demethylation regarding checking the system.
7. 0.20-*M* sodium borate (pH 9.6)
8. 1.0-m*M* 7-hydroxycoumarin (Aldrich), dissolved in an aqueous solution of 0.10-*N* NaOH and 0.10-*M* NaCl (prepare fresh solution each day; avoid exposure to light)

An appropriate amount of enzyme is placed in a test tube along with 50-m*M* phosphate buffer, 0.30-m*M* 7-ethoxycoumarin, and water to bring the volume to 0.90 mL. After 3 minutes of preincubation at 37°C, 0.15 mL of reagent 6, the NADPH-generating system, is added per mL to start incubations. (As in the case of benzphetamine, reactions are conveniently started and stopped each 10 to 15 seconds.) After 5 to 10 minutes, incubations are stopped by the addition of 0.10 mL of 2.0-*M* HCl and 2.0 mL $CHCl_3$. Tubes are mixed and centrifuged for 5 minutes at $3 \times 10^3$ *g*. One mL of the lower $CHCl_3$ phase (containing both substrate and product) is transferred to a clean tube, and 2.0 mL of 0.20-*M* sodium borate buffer is added. The tubes are mixed and centrifuged 5 minutes at $3 \times 10^3$ *g*. The upper phase, containing the phenolic product, is transferred to a new tube, and fluorescence is read vs. a standard curve in a fluorimeter with the excitation wavelength set at 338 nm and the emission wavelength set at 458 nm.

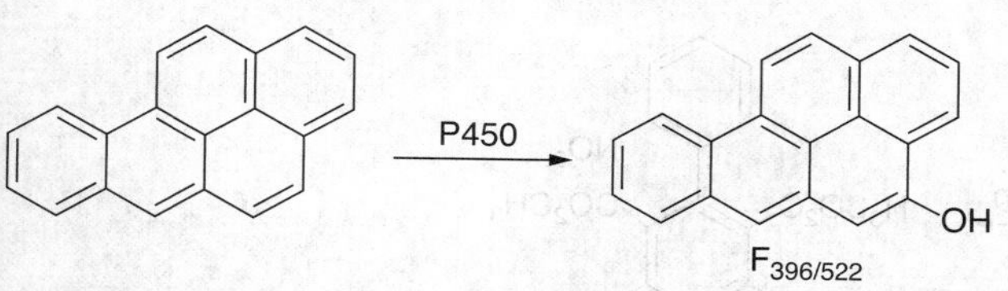

P450

OH

$F_{396/522}$

**FIGURE 40.14** Benzo(*a*)pyrene 3-hydroxylation.

*Benzo(a)pyrene Hydroxylation*

This assay has been widely used because of its sensitivity, the widespread occurrence of this activity, and the interest in carcinogenic aspects of benzo(*a*)pyrene. This substrate is carcinogenic and light sensitive and gives rise to many metabolites. The following procedure measures primarily the 3-hydroxy derivative (Figure 40.14) and, to a lesser extent, 9-hydroxybenzo(*a*)pyrene [121].

*Reagents*

1. 1.0-*M* potassium phosphate buffer (pH 7.4)
2. 10-m*M* $NADP^+$
3. $10^3$ IU yeast glucose-6-phosphate dehydrogenase/mL (dissolved in 10-m*M* Tris-acetate buffer, pH 7.7, containing 1.0-m*M* EDTA and 20% glycerol [v/v])
4. 0.10-*M* glucose-6-phosphate
5. NADPH-generating system: Mix 25 parts reagent 2, 1 part reagent 3, and 50 parts reagent 4, in an amount sufficient to supply 0.15 mL per mL of total reaction; refer to the earlier section on benzphetamine *N*-demethylation regarding checking the system.
6. 8.0-m*M* benzo(*a*)pyrene, dissolved in $(CH_3)_2CO$
7. 6.0-m*M* quinine, dissolved in 0.10 N $H_2SO_4$
8. 1.0-m*M* 3-hydroxybenzo(*a*)pyrene (this material can be obtained from the National Cancer Institute Chemical Carcinogen Reference Repository, c/o Midwest Research Institute, Kansas City, MO)

An appropriate amount of enzyme is placed in a test tube along with 50-m*M* phosphate buffer (pH 7.4), 80-μ*M* benzo(*a*)pyrene, and sufficient water to bring the total volume to 1.0 mL. (All procedures should be carried out in dim light or under yellow light or reactions may be done in amber glass vials; appropriate precautions should be taken to prevent exposure of skin to benzo(*a*)pyrene or its metabolites, and when solid material is being handled precautions should be taken to avoid breathing the dust.) After 3 minutes of preincubation at 37°C, 0.15 mL of reagent 4, the NADPH-generating system, is added per mL to initiate reactions (this is conveniently done every 10 to 15 seconds). After 5 to 10 minutes, reactions are stopped by the addition of 1.0 mL of cold acetone and mixed. Hexane (3.25 mL) is added and mixing is repeated. Two mL of the upper layer is transferred to a clean tube with a pipette and 4.0 mL of 1.0-*N* NaOH is added to this. After vortex mixing and centrifugation for 5 minutes at $3 \times 10^3$ *g*, the aqueous phase is carefully transferred to a clean tube.

Fluorescence is read with an excitation wavelength of 396 nm and emission wavelength of 522 nm. A standard curve is prepared in 0.10-*N* NaOH using 3-hydroxybenzo(*a*)pyrene. Because solutions of the standard are unstable, a convenient method involves setting up the standard curve, changing the wavelength settings to 350 nm (excitation) and 450 nm (emission) without adjusting other settings, and preparing a standard curve using serial dilutions of quinine sulfate. The quinine sulfate can then be used as a secondary standard in subsequent experiments, and levels of 3-hydroxybenzo(*a*)pyrene can be calculated by reference to the original curves.

An alternative procedure devised by Dehnen et al. [122] is comparable in terms of convenience and sensitivity. Incubations are set up as before (1.0-mL volume) and stopped by the addition of 2.3 mL of a fresh aqueous mixture of 1.0-m*M* EDTA, 10% Triton® X-100 (w/v), and 1.2% $(C_2H_5)_3N$ (w/v). After mixing, tubes are capped until fluorescence measurements are made. The excitation wavelength is set at 435 nm, and fluorescence emission is scanned (and a chart recorded) between 450 and 650 nm. Residual substrate gives a large peak near the 450-nm region of the chart, and the product appears as a peak at 522 nm. A baseline is drawn between the trough (between benzo(*a*)pyrene and 3-hydroxybenzo(*a*)pyrene) and the point at which the 3-hydroxybenzo(*a*)pyrene peak tails into a baseline at about 575 nm. The distance between this slanted baseline and the top of the 522-nm peak can be calibrated against the quinine sulfate curve prepared as described above.

Other benzo(*a*)pyrene metabolism assays can be carried out with radioactive substrate to measure total polar metabolites [123], individual metabolites (after separation by HPLC) [124–126], or metabolites covalently bound to protein or added nucleic acids [127].

## High-Performance Liquid Chromatography

*Nifedipine Oxidation*

A description of this assay is added to serve as an example of how HPLC methods may be utilized very effectively. Nifedipine is a widely used calcium channel blocker and oxidation by P450 renders it inactive [128]. Nifedipine oxidation (from the dihydropyridine to the pyridine product; see Figure 40.15) is measured in the following manner. Note that *all* incubation, extraction, and other handling of samples is done in amber vials because of the light sensitivity of nifedipine solutions; amber glass vials are available from the Pierce Chemical Co. (Rockford, IL).

Typical incubations include liver microsomes containing 10 to 100 pmol of P450, 0.10-*M* potassium phosphate buffer (pH 7.85), and 0.20-m*M* nifedipine (added

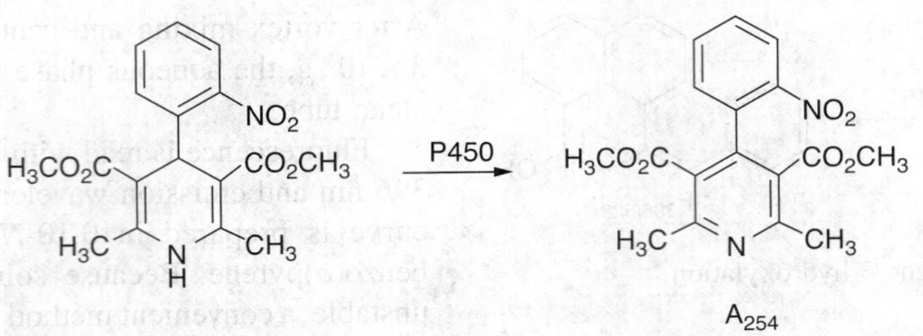

FIGURE 40.15 Oxidation of nifedipine to the pyridine derivative. (Adapted from Guengerich, F.P. et al., *J. Biol. Chem.*, 261, 5051–5060, 1986.)

from a stock solution of 20-m*M* in $CH_3OH$) in a final volume of 0.50 mL. The components are equilibrated for 3 minutes at 37°C, and the reaction is initiated by the addition of an NADPH-generating system consisting of (final concentrations) 10-m*M* glucose-6-phosphate, 0.50-m*M* $NADP^+$, and 1.0 IU yeast glucose-6-phosphate dehydrogenase per mL (refer to earlier section on benzphetamine *N*-demethylation regarding using and checking the system). The reaction proceeds for 10 minutes at 37°C and is then quenched by the addition of 2.0 mL of $CH_2Cl_2$; 100 μL of 1-*M* $Na_2CO_3$ buffer (pH 10.5, 1-*M*) containing 2.0-*M* NaCl is then added to each vial. The contents of each vial are mixed using a vortex device, and the two layers are separated by centrifugation at 3 × $10^3$ *g* for 10 minutes. From each lower organic layer, 1.4 mL is transferred to an amber Reacti-Vial™ (Pierce Chemical Co.; Rockford, IL). The total extract is reduced to dryness at 23°C under an $N_2$ stream. The residue is dissolved in 50 μL of $CH_3OH$, and 20 μL is injected onto an octyldecylsilane ($C_{18}$) reversed-phase HPLC column (e.g., 6.2 mm × 80 mm; Mac-Modd, Chadds Ford, PA) placed in series following a 1-cm-long octyldecylsilane guard column and 0.2-mm filter. The column is eluted with an isocratic mixture of 64% $CH_3OH$–36% $H_2O$ (v/v) at a flow rate of 3.0 mL/min. Detection is at 254 nm (found to be optimal by previous scanning). Quantitation is usually done with external standards and by the use of peak heights. Alternatively, nitrendipine or another dihydropyridine [129,130] can be used as an internal standard. Typically, a 20-ng sample of the metabolite (59 pmol) yields a maximal $A_{254}$ of about 0.015 under these conditions (in the HPLC effluent). A typical chromatogram resulting from injection of a human liver microsomal incubation extract is shown in Figure 40.16.

Assay conditions were optimized with a human liver microsomal sample [128]. Product formation was linear up to a time of 20 minutes, the pH optimum was 7.85 (Tris-HCl yielded lower rates than did potassium phosphate buffer), the rate of product formation per unit enzyme was constant over a range of 5 to 1000 pmol P450/mL, and a substrate concentration of 200 μ*M* was optimal ($K_m$ ~10 μ*M*; substrate inhibition observed at concentrations >500 μ*M*; no evidence for multiphasic behavior observed over the concentrations range of 2 to 1000 μ*M*).

When purified P450 fractions are assayed for activity, the microsomes are replaced with 20 to 100 pmol of P450, 250 pmol of rabbit NADPH–P450 reductase, 250 pmol cytochrome $b_5$, and 15 nmol of L-α-1,2-dilauroyl-*sn*-glyceryl-3-phosphocholine. These components are mixed and then incubated for 10 minutes at 23°C prior to addition of other materials. Examination of experimental conditions indicated that the NADPH–P450 reductase, substrate, and phospholipid concentrations used are optimal; however, product formation is not linear for more than 5 minutes. The use of alternate phospholipids has been found to improve the activity [44].

The separation of the product from the substrate is very efficient with the reversed-phase system used. The only solvent components needed are $CH_3OH$ and $H_2O$, thus eliminating any problems with salts. The product elutes before the substrate, enhancing sensitivity of the assay. It is not necessary to have more than baseline separation, and interfering peaks are absent; thus, a short column can be used with a high flow rate and low back pressure. The total HPLC time for each assay can be less than 3 minutes. The efficiency of analysis could also be improved with the use of an automated injector system. Analysis time can also be reduced if incubations are only deproteinized and not extracted prior to HPLC analysis, but the sensitivity would be reduced.

### *Chlorzoxazone 6-Hydroxylation*

Another HPLC-based assay of P450 activity involves measurement of the 6-hydroxylation of chlorzoxazone (Figure 40.17). This reaction has been reported to be highly selective for P450 2E1 in human liver [78,131]. The assay offers considerable advantages with regard to sensitivity, reliability, and specificity over many other traditional assays for the function of P450 2E1 assays (e.g., *N,N*-dimethylnitrosamine *N*-demethylation and 4-nitrophenol 2-hydroxylation). Chlorzoxazone is used as a drug [131] and this general procedure can be used to measure pharmacokinetic parameters in humans to assess their relative levels of P450 2E1 [132,133].

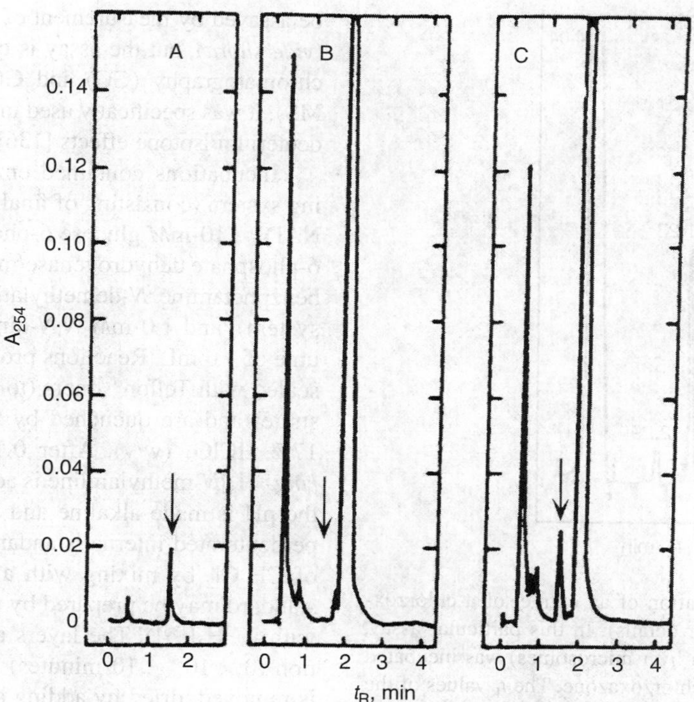


**FIGURE 40.16** HPLC separation of the oxidized (pyridine) product of nifipidine oxidation from human liver microsomes. A typical 10-minute incubation with 100 pmol of human liver microsomal P450 was done and 20 μL of the total extract was chromatographed as described. The $t_R$ of the nifedipine oxidation product is indicated with an arrow. (A) Authentic standard of the metabolite (200 ng); (B) extract of an incubation devoid of NADPH; (C) complete incubation. (Adapted from Guengerich, F.P. et al., *J. Biol. Chem.*, 261, 5051–5060, 1986.)

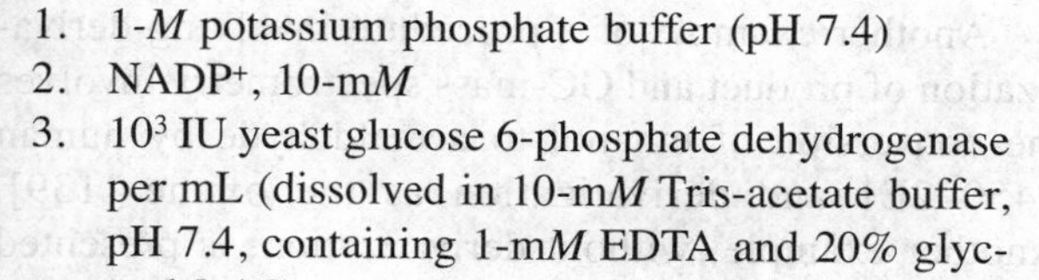

**FIGURE 40.17** Chlorzoxazone 6-hydroxylation.

*Reagents*

1. 1-*M* potassium phosphate buffer (pH 7.4)
2. NADP+, 10-m*M*
3. $10^3$ IU yeast glucose 6-phosphate dehydrogenase per mL (dissolved in 10-m*M* Tris-acetate buffer, pH 7.4, containing 1-m*M* EDTA and 20% glycerol [v/v])
4. 0.1-*M* glucose 6-phosphate
5. NADPH-generating system—mix 25 parts reagent 2, 1 part reagent 3, and 50 parts reagent 4 in an amount sufficient to supply 0.15 mL per mL of total reaction; refer to earlier section on benzphetamine *N*-demethylation regarding checking the system)
6. 50-m*M* chlorzoxazone in 60-m*M* KOH (prepare fresh daily and store on ice)—dissolve 51 mg chlorzoxazone (Sigma Chemical Co.; St. Louis, MO) in 0.36 mL 1.0-*M* KOH by vigorous mixing with a vortex device or sonic bath, add 0.64 mL $H_2O$, continue mixing until dissolved, and then add 5.0 mL $H_2O$; avoid the introduction of organic solvents
7. 6-Hydroxychlorzoxazone (standard product; see Peter et al. [131] for synthesis)
8. Either 2-benzoxazolinone (Aldrich Chemical Co.; Milwaukee, WI) or 5-flurobenzoxazolinone dissolved in 2.0% aqueous propylene glycol 400 (internal standard)

An appropriate amount of enzyme is placed in a test tube along with 50-m*M* potassium phosphate buffer (pH 7.4) and 0.50-m*M* chlorzoxazone (final incubation volume, 0.50 mL). After 3 minutes of preincubation at 37°C, 0.15 mL of reagent 5 (the NADPH-generating system) is added per mL to start incubations. After 10 to 20 minutes at 37°C, reactions are stopped by the addition of a 25 μL of aqueous 43% $H_3PO_4$ (w/v). An appropriate amount of the internal standard (5.0 nmol) is also added to each tube, followed by 2.0 mL of $CH_2Cl_2$. The contents of each tube are mixed using a vortex device and the layers are separated by brief centrifugation ($3 \times 10^3$ *g*, 10 minutes). An aliquot (1.6 mL) of each lower $CH_2Cl_2$ layer is transferred with a pipette to a clean test tube or conical vial (Reacti-

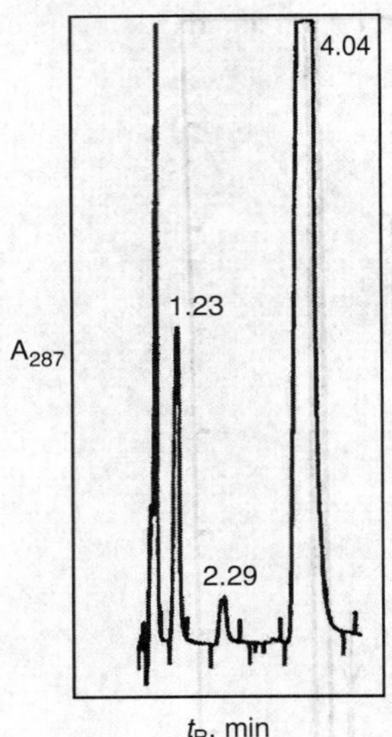


**FIGURE 40.18** HPLC separation of an extract of a chlorzoxazone incubation (see text for details). In this particular assay, 0.27 nmol of human P450 (in liver microsomes) was incubated for 10 minutes with 0.5 m*M* chlorzoxazone. The $t_R$ values of the individual peaks are designated 1.23 minutes, 6-hydroxychlorzoxazone; 2.29 minutes, 5-fluorobenzoxazolinone; 4.04 minutes, chlorzoxazone. (Adapted from Peter, R. et al., *Chem. Res. Toxicol.*, 3, 566–573, 1990.)

Vial™; Pierce Chemical Co., Rockford, IL), and the solvent is removed under an $N_2$ stream at 23°C. The residue is dissolved in 50 μL of $CH_3CN$ (mixing on a vortex device), and 20 μL is injected onto a 6.2 mm × 80-mm Zorbax® octylsilane ($C_8$) HPLC column (3-mm; MacModd, Chadds Ford, PA) utilizing a solvent mixture of 32% $CH_3CN$ (v/v) and 0.5% $H_3PO_4$ (w/v) in $H_2O$, with ultraviolet (UV) detection at 287 nm (flow rate, 3.5 mL/min) (Figure 40.18). Under these conditions, each assay requires ≤5 minutes HPLC time. For more complex samples the $CH_3CN$ concentration may have to be decreased to move the produce peak away from the solvent front. It is reasonable to use as little as 5 pmol of microsomal P450 in this assay and obtain reliable results.

## Gas Chromatography–Mass Spectrometry

### *N,N-Dimethylaniline N-Demethylation*

Amine *N*-dealkylations are catalyzed by a number of different oxidases and are important in the disposition of drugs and other chemicals. The P450-dependent reactions have historically been of the most interest, although such reactions can also be catalyzed by peroxidases, flavoproteins, and other enzymes as well [134,135]. The mechanism of the reaction has been a matter of considerable interest, and the reader is referred to recent work on this topic [136,137]. The *N*-demethylation of *N,N*-dimethylaniline (Figure 40.19) can be assayed by measurement of HCHO by other procedures (*vide supra*), but the assay is typical of many used for gas chromatography (GC) and GC–mass spectrometry (GC-MS). It was specifically used in our work to measure kinetic deuterium isotope effects [136].

Incubations contained enzyme, an NADPH-generating system (consisting of final concentrations of 0.50-m*M* $NADP^+$, 10-m*M* glucose 6-phosphate, and 1.0 IU glucose 6-phosphate dehydrogenase/mL; refer to earlier section on benzphetamine *N*-demethylation regarding checking the system), and 1.0-m*M* *N,N*-dimethylaniline in a final volume of 1.0 mL. Reactions proceed for 10 minutes in vials sealed with Teflon® liners (to prevent evaporation of substrate) and are quenched by the addition of 0.40 mL of 17% $HClO_4$ (w/v). After 0.10 mL of 270-μ*M* [$N^2$-$H_3$, *ring*-$^2H_5$]*N*-methylaniline is added as an internal standard, the pH is made alkaline and the substrate, products, and perdeuterated internal standard are extracted into 2.0 mL of $CH_2Cl_2$ by mixing with a vortex device (the internal standard may be prepared by reaction of [*ring*-$^2H_5$]aniline with [$C^2H_3I^{136}$]). The layers are separated by centrifugation (3 × $10^3$ *g*, 10 minutes) and the lower $CH_2Cl_2$ layer is removed, dried by adding anhydrous $Na_2SO_4$ and mixing, and then transferred to a clean vial. $(CF_3CO)_2O$ (50 μL) is added to each vial. The vials are sealed with caps and Teflon® liners and allowed to stand overnight at 4°C to form trifluoroacetanilide.

Most of the $CH_2Cl_2$ is removed under a stream of $N_2$ (≤30°C) and aliquots (~2 μL) are injected onto a 0.5 mm × 3-m SPB-1 capillary GC column connected to a mass spectrometer operating in the chemical ionization mode with He as the carrier gas and $CH_4$ as the ionization gas. The column temperature is programmed from 100 to 280°C at 20°/min. The ions at *m/z* 205 and 214 ($[M+H]^+$) are monitored for the product and internal standard. Mixtures of varying ratios of trifluoroacetanilide and the perdeutero derivative are used to prepare a standard curve.

Another example of a procedure involving derivatization of product and GC-mass spectrometry involves the conversion of ethanol to acetaldehyde by human P450 2E1 and derivatization as the oxime [139]. Another example, without derivatization, is presented for ethyl carbamate. In some cases, it is impractical to use a heavy isotope as an internal standard, and product analogs must be considered. There are many procedures for GC-based methods involving flame ionization, electron capture, and other means of detection. Electron capture can be a very sensitive assay procedure when halides or nitro groups are present. The use of capillary GC columns has made packed columns nearly obsolete because of the superior resolution; however, the need to use very small injection volumes requires the use of internal standards except in cases where head space analysis is done.

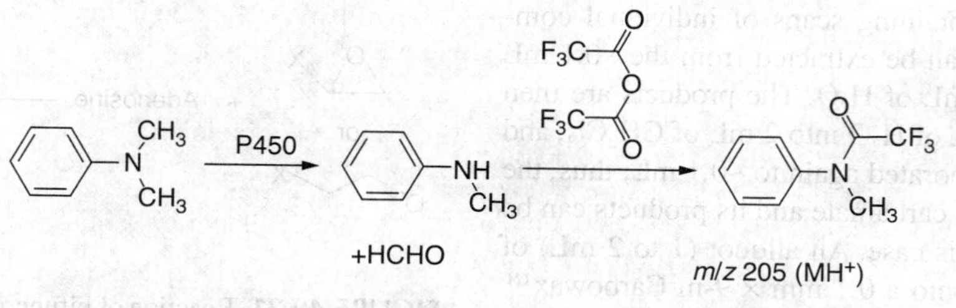

FIGURE 40.19 *N,N*-Dimethylaniline *N*-demethylation assay. (Adapted from Okazaki, O. and Guengerich, F.P., *J. Biol. Chem.*, 268, 1546–1552, 1993; Miwa, G.T. et al., *J. Biol. Chem.*, 258, 14445–14449, 1983.)

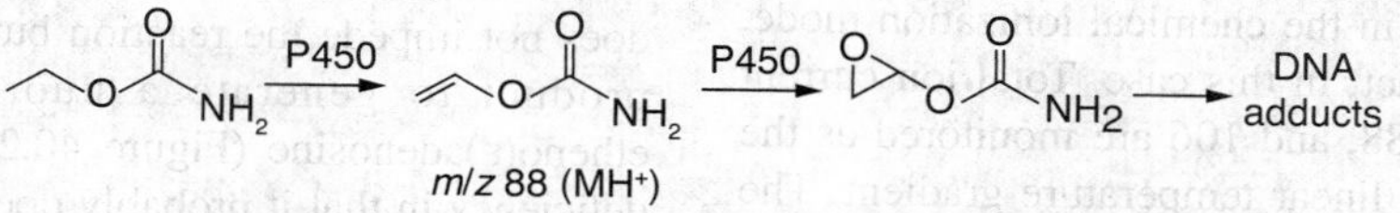

FIGURE 40.20 Desaturation of ethyl carbamate and epoxidation of vinyl carbamate. (Adapted from Guengerich, F.P. and Kim, D.-H., *Chem. Res. Toxicol.*, 4, 413–421, 1991.)

### *Ethyl Carbamate Desaturation*

Ethyl carbamate (urethane) desaturation is another example of a reaction with an assay involving GC-mass spectrometry. The desaturation of ethyl carbamate to vinyl carbamate (Figure 40.20) had been suggested by Dahl et al. [141,142], but direct evidence for this view had been difficult to obtain. Now it is realized that the oxidation of ethyl carbamate to vinyl carbamate is a very slow process and that the succeeding step, epoxide formation, is ~400 times faster (and is catalyzed by the same P450 enzyme) [140]; thus, the steady-state level of vinyl carbamate is very low. The procedure described below can be used to estimate levels of several ethyl carbamate metabolites—vinyl carbamate, 2-hydroxyethyl carbamate, and *N*-hydroxyethyl carbamate [140]. In the reference cited, a major use was the qualitative demonstration of these products. With this procedure, it was possible to detect one part product in $10^4$ parts substrate.

*Reagents*

1. 1.0-*M* potassium phosphate buffer (pH 7.4)
2. $NADP^+$, 10-m*M*
3. $10^3$ IU yeast glucose 6-phosphate dehydrogenase/mL (dissolved in 10-m*M* Tris-acetate buffer, pH 7.4, containing 1.0-m*M* EDTA and 20% glycerol [v/v])
4. 0.10-*M* glucose 6-phosphate
5. NADPH-generating system: Mix 25 parts reagent 2, 1 part reagent 3, and 50 parts reagent 4 in an amount sufficient to supply 0.15 mL per mL of total reaction; refer to earlier section on benzphetamine *N*-demethylation regarding checking the system.
6. 0.10-*M* ethyl carbamate (urethane, Aldrich Chemical Co., Milwaukee, WI; purify by sublimation—use house vacuum or a water aspirator to achieve ~15 mmHg)
7. 2-Hydroethyl carbamate (Pfaltz and Bauer, Stamford, CT)
8. *N*-Hydroxyethyl carbamate (Aldrich)
9. Vinyl carbamate [141,143]
10. Methyl carbamate (Aldrich; purify by sublimation before use)

*Note:* These carbamates, particularly ethyl and vinyl, should be handled with care. They are carcinogenic and volatile. Use fume hoods, closed vials, and appropriate protective clothing.

The oxidation of ethyl carbamate is catalyzed primarily by P450 2E1, so care should be taken to avoid adding organic solvents, which are inhibitory [144]. The reaction is slow and a high concentration of microsomal protein is needed, so it is useful to remove glycerol (or sucrose) from microsomes by dialysis for 1 to 2 hours vs. 0.10-*M* potassium phosphate buffer (pH 7.4) at 4°C just prior to use. Microsomal protein (10 to 15 mg protein/mL) is mixed in a vial with 100-m*M* potassium or an equivalent amount of a reconstituted P450 system), 10-m*M* ethyl carbamate, and an NADPH-generating system consisting of (final concentrations of) 0.50-m*M* $NADP^+$, 1 IU glucose 6-phosphate dehydrogenase per mL, and 10-m*M* glucose 6-phosphate (refer to earlier section on benzphetamine *N*-demethylation regarding checking the system). The total volume is 1.0 mL. Each vial is sealed with a cap and Teflon® liner and incubated for 2 hours at 37°C.

Methyl carbamate is then added as an internal standard and the products are extracted twice into 1 mL of $CH_2Cl_2$ by mixing on a vortex device, and the layers are separated by centrifugation (3 × $10^3$ *g*, 10 minutes). The lower $CH_2Cl_2$ layer is removed and concentrated to ≤0.1 mL under an $N_2$ stream at ≤23°C, taking care to avoid concentration to dryness. Either this extract can be analyzed directed by GC-mass spectrometry or, if a cleaner sample is required (e.g., for obtaining scans of individual compounds), the products can be extracted from the ~0.1 mL $CH_2Cl_2$ sample into 1 mL of $H_2O$. The products are then extracted from the 1 mL of $H_2O$ into 2 mL of $CH_2Cl_2$, and the $CH_2Cl_2$ layer is evaporated again to ~0.1 mL; thus, the water solubility of ethyl carbamate and its products can be used to advantage in this case. An aliquot (1 to 2 mL) of the extract is injected onto a 0.1 mm × 9-m Carbowax™ 20-*M* capillary GC column with the effluent directed into a mass spectrometer operating in the chemical ionization mode with $CH_4$ as the carrier gas. The selectivity and sensitivity are enhanced in the chemical ionization mode, relative to electron impact, in this case. Total ion current and the ions at *m/z* 74, 88, and 106 are monitored as the column is heated with a linear temperature gradient. The actual $t_R$ values depend on the rate of heating, gas flow rate, etc. Under typical operating conditions methyl carbamate (*m/z* 74) elutes at ~4 minutes, vinyl carbamate (*m/z* 88) at 6.8 minutes, ethyl carbamate (*m/z* 90, not monitored) at 8.0 minutes, ethyl *N*-hydroxycarbamate (*m/z* 106) at 10.4 minutes, and 2-hydroxyethyl carbamate (*m/z* 106) at 12.9 minutes [140]. Standard curves may be prepared by plotting the ratio of detector response relative to methyl carbamate vs. the concentration of each product; such curves are linear over several orders of magnitude for each of the oxidation products under consideration [140].

The steady-state concentration of vinyl carbamate ([VC]) can be expressed as a function of the ethyl carbamate concentration ([EC]):

$$[\mathrm{VC}] = \frac{(k_1 K_2)}{V_2}[\mathrm{EC}]$$

where $k_1$ is the rate constant for oxidation of ethyl carbamate to vinyl carbamate, and $V_2$ and $K_2$ are the parameters $V_{max}$ and $K_m$ for the oxidation of vinyl carbamate to form the epoxide (estimated from the rate of formation of 1,$N^6$-ethenoadenoxine) [140]. Thus, the steady-state concentration should be relatively independent of the concentration of enzyme, if both desaturation and epoxidation are catalyzed by the same enzyme (in this case, P450 2E1) [140].

### HPLC-Fluorescence Assay: 1,$N^6$-Ethenoadenosine Formation

Fluorescence assays have long been regarded for their high sensitivity [145]. The discrimination of compounds provided by the use of selective excitation and emission wavelengths may be further enhanced by coupling to HPLC. In many cases, the enzyme product has intrinsic fluorescence and may be monitored. In other cases, postcolumn derivatization can be used to render products fluorescent. The situation described here is somewhat different. A compound is present in the enzyme mixture (adenosine) that does not impede the reaction but reacts with the reaction product to generate a fluorescent product: 1,$N^6$-etheno(ε)adenosine (Figure 40.21). The procedure has a deficiency in that it probably does not serve as a quantitative trap for any of the compounds under consideration. On the other hand, it is of considerable relevance to studies in toxicology and chemical carcinogenesis in that the same reaction serves to modify DNA to introduce a potentially mutagenic lesion. The procedure described here is based on our own use [78,140] of methods described by Leithauser et al. [143] and Rinkus and Legator [146].

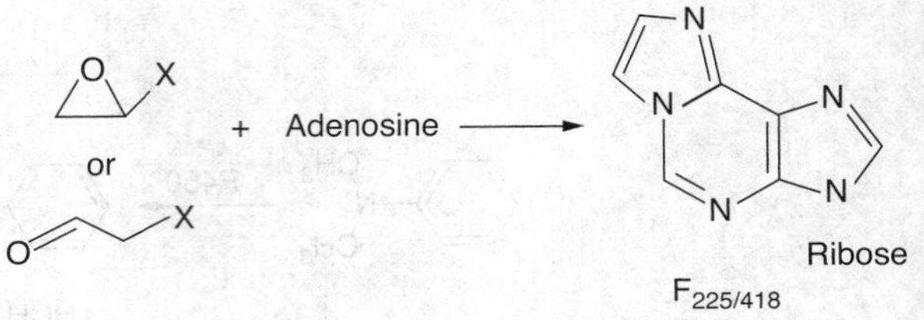

**FIGURE 40.21** Reaction of bifunctional electrophiles with adenosine.

*Reagents*

1. 1.0-*M* potassium phosphate buffer (pH 7.4)
2. 50-m*M* adenosine, dissolved in $H_2O$ (heat in a bath of hot tap water or sonicate to dissolve)
3. Appropriate substrate, dissolved in $H_2O$ (not organic solvent at ~10× the concentration used in the assay)
4. 10-m*M* $NADP^+$
5. $10^3$ IU yeast glucose 6-phosphate dehydrogenase/mL (dissolved in 10-m*M* Tris-acetate buffer, pH 7.7, containing 1.0-m*M* EDTA and 20% glycerol [v/v])
6. 0.1-*M* glucose-6-phosphate
7. NADPH-generating system: Mix 25 parts reagent 4, 1 part reagent 5, and 50 parts reagent 6 in an amount sufficient to supply 0.15 mL per mL of total reaction; refer to earlier section on benzphetamine *N*-demethylation regarding checking the system.
8. Standard solutions of 1,$N^6$-ethenoadenosine dissolved in $H_2O$
9. 0.60-*M* $ZnSO_4$ (in $H_2O$)

An appropriate amount of microsomes or purified enzyme system is placed in a glass vial (generally to provide ~2 mg microsomal protein per mL). Reagents are added to give 50-m*M* potassium phosphate buffer (pH 7.4), 0.50-m*M* $NADP^+$, 1.0 IU glucose-6-phosphate dehydrogenase per mL, and the substrate. Most of the substrates that generate products reacting with adenosine to produce 1,$N^6$-ethenoadenosine are small organic compounds (e.g., ethylene dibromide, ethylene dichloride, vinyl chloride, vinyl bromide, acrylonitrile, vinyl carbamate, ethyl carbamate, and some nitrosamines) [78]. These are usually substrates for P450 2E1 enzymes; because organic solvents are competitive inhibitors, they should be avoided [144]. Although these compounds are considered to be organic solvents, most are actually reasonably soluble in $H_2O$ [78] (see the *Merck Index* and various handbooks of physical constants of chemicals). Many of these are rather volatile, so the vials may have to be sealed with screw caps and Teflon® liners (rubber septa absorb these chemicals and should be avoided). Gases such as vinyl chloride can be added directly to the head space of such vials.

Incubations are initiated by the addition of 0.15 mL of reagent 7, the NADPH-generating system; shake at 37°C for a specified period of time. With some of the slower reactions (e.g., ethyl carbamate) we [140] and others [143] have found that reactions can proceed in a linear manner for up to 3 hours. Reactions are stopped by the addition of $ZnSO_4$ to 30-m*M*. The protein is precipitated by centrifugation at $3 \times 10^3$ *g* for 10 minutes, and aliquots of the supernatant are injected onto an HPLC system (octadecylsilane [$C_{18}$], equilibrated with 11% $CH_3CN$ [v/v] in 50-m*M* $NH_4CH_3CO_2$ [pH 5.0]). With a 4.6 mm × 250-mm or a 6.2 mm × 80-mm column, as much as 250 µL of material may be injected; at a flow rate of 1.0 mL/min, the $t_R$ of 1,$N^6$-ethenoadenosine is typically ~8 minutes. The effluent passes through the flow cell of a spectrofluorimeter designed for such monitoring, with the excitation wavelength set at 225 nm and using a 418-nm emission filter. The 1,$N^6$-ethenoadenosine peak follows a large negative peak of residual adenosine (Figure 40.22). If enough separation is not seen, then the concentration of $CH_3CN$ should be reduced. As little as 0.5 pmol of 1,$N^6$-ethenoadenosine can be detected using this method.

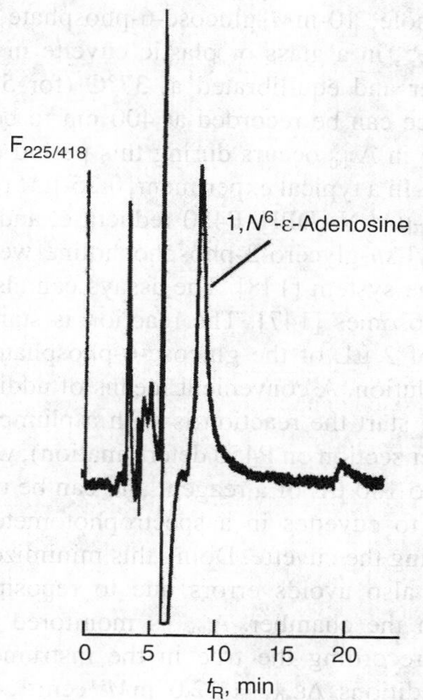


**FIGURE 40.22** Formation of 1,$N^6$-ethenoadenosine in a microsomal incubation. The incubation included 5 mg liver microsomal protein (prepared from isoniazid-treated rats)/mL, 2.0-m*M* *N*-methyl, *N*-vinylnitrosamine, 10-m*M* adenosine, and an NADPH-generating system. After an incubation time of 120 minutes, a 100-µL aliquot was analyzed as described in the text. (Adapted from Guengerich, F.P. et al., *Chem. Res. Toxicol.*, 4, 168–179, 1991.)

$OCH_3$ / $NO_2$ —P450→ [$OCH_2OH$ / $NO_2$] → $O^-$ / $NO_2$ ($A_{400}$)

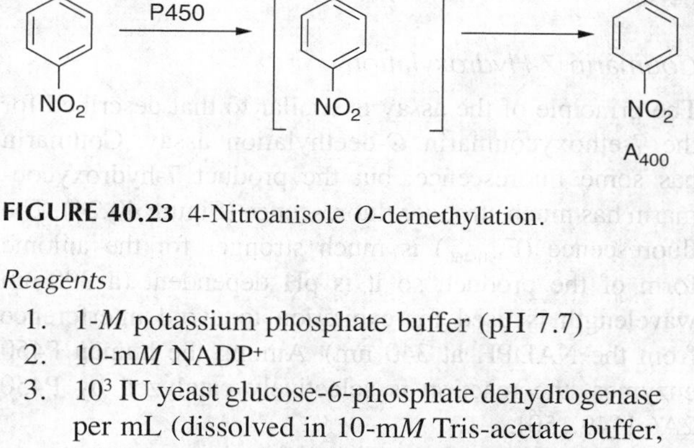

**FIGURE 40.23** 4-Nitroanisole *O*-demethylation.

## Continuous Assays

Two examples of P450-based assays are provided as examples: coumarin 7-hydroxylation and 4-nitroanisole *O*-demethylation. The continuous monitoring of a reaction, when possible, allows the investigator to determine exactly what part of the enzyme assay data to establish linearity (i.e., when the enzyme is functioning under steady-state conditions and before substrate depletion, product inhibition, etc. are issues). The change in absorbance is used to calculate a rate, using the known extinction coefficient for the product. In the case of fluorescence, a known amount of the product must be used to calibrate the instrument.

### *4-Nitroanisole O-Demethylation*

This assay has been used in P450 assays for many years; it uses the (yellow) absorbance of the product as the basis (Figure 40.23) [147,148].

*Reagents*

1. 1-*M* potassium phosphate buffer (pH 7.7)
2. 10-m*M* $NADP^+$
3. $10^3$ IU yeast glucose-6-phosphate dehydrogenase per mL (dissolved in 10-m*M* Tris-acetate buffer, pH 7.7, containing 1.0-m*M* EDTA and 20% glycerol [v/v])
4. 0.10-*M* glucose-6-phosphate
5. 10-m*M* 4-nitroanisole (in $CH_3CN$)

In this assay, the product 4-nitrophenol has a broad absorbance band at 400 nm. The absorbance is only seen for the anionic form of the product, so the apparent extinction coefficient is pH sensitive.

An enzyme system containing P450 and NADPH–P450 reductase is incubated in 1.0 mL of 0.1-*M* potassium phosphate buffer (pH 7.7), along with 100-µ*M*

4-nitroanisole, 10-m$M$ glucose-6-phosphate, and 150-µ$M$ $NADP^+$, in a glass or plastic cuvette in a spectrophotometer and equilibrated at 37°C (for 5 minutes). Absorbance can be recorded at 400 nm to be sure that no change in $A_{400}$ occurs during this period due to turbidity, etc. In a typical experiment, 0.25-µ$M$ rabbit P450 1A2, 0.50-µ$M$ NADPH–P450 reductase, and 50-µ$M$ L-α-dilauroyl-*sn*-glycero-3-phosphocholine were used as the enzyme system [148]. The assays can also be done with microsomes [147]. The reaction is started by the addition of 2 µL of the glucose-6-phosphate dehydrogenase solution. A convenient means of adding the last reagent to start the reaction is with a plumping device (see earlier section on P450 determination), which holds up to 50 to 100 µL of a reagent and can be used to add materials to cuvettes in a spectrophotometer, without withdrawing the cuvette. Doing this minimizes the dead time and also avoids errors due to repositioning the cuvette in the chamber. $A_{400}$ is monitored for 1 to 2 minutes, recording the true in the instrument. Under these conditions $\Delta\varepsilon_{400}$ = 12.0 m$M^{-1}$ cm$^{-1}$, so an $A_{400}$ change of 0.012 corresponds to 1 nmol of product (in a 1.0-mL cuvette, with an optical pathlength of 10 mm).

A convenient means of analyzing the results is to do the experiments in a spectrophotometer connected to a computer. In this laboratory, the assays are usually done with a Cary 14-OLIS instrument. The collected data can be handled to use only the linear portions of the traces and to fit the lines with regression analysis. The disappearance of substrate can also be monitored at 314 nm ($\Delta\varepsilon_{314}$ = 10.4 m$M^{-1}$ cm$^{-1}$) [148].

### *Coumarin 7-Hydroxylation*

The principle of the assay is similar to that described for the 7-ethoxycoumarin *O*-deethylation assay. Coumarin has some fluorescence, but the product 7-hydroxycoumarin has much stronger fluorescence (Figure 40.24). The fluorescence ($F_{390/460}$) is much stronger for the anionic form of the product, so it is pH dependent (a 390-nm wavelength is used for excitation to avoid interference from the NADPH at 340 nm). Among the human P450 enzymes, the reaction is selectively catalyzed by P450 2A6 [149,150].

*Reagents*

1. 1-$M$ potassium phosphate buffer (pH 7.7)
2. 10-m$M$ $NADP^+$
3. $10^3$ IU yeast glucose-6-phosphate dehydrogenase/mL (dissolved in 10-m$M$ Tris-acetate buffer, pH 7.7, containing 1.0-m$M$ EDTA and 20% glycerol [v/v])
4. 0.10-$M$ glucose-6-phosphate
5. 5-m$M$ coumarin, dissolved in $H_2O$
6. 100-µ$M$ 7-hydroxycoumarin, dissolved in $H_2O$

P450 2A6 $F_{380/460}$

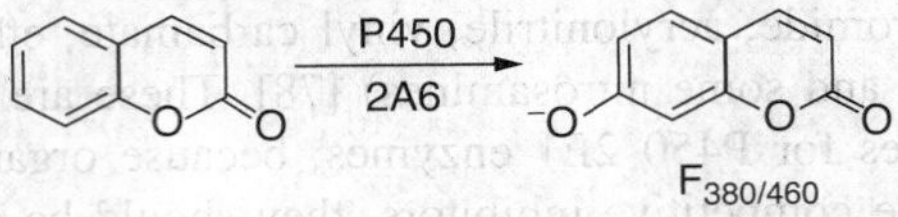

**FIGURE 40.24** Coumarin 7-hydroxylation.

P450 2A6 $F_{380/460}$

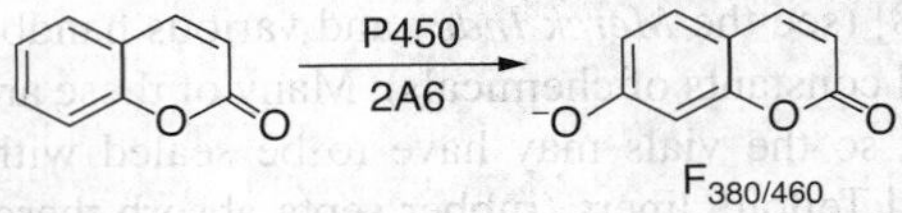

**FIGURE 40.25** Continuous fluorescence traces for 7-hydroxylation of coumarin by P450 2A6. An NADPH-generating system was used in both cases. (Adapted from Yun, C.-H. et al., *J. Biol. Chem.*, 280, 12279–12291, 2005.)

An enzyme system containing P450 2A6 and NADPH–P450 reductase is incubated in 3.0 mL of 0.10-$M$ potassium phosphate buffer (pH 7.7) and equilibrated at 37°C for 3 to 5 minutes in the compartment of the spectrofluorimeter, along with 50-µ$M$ coumarin, 10-m$M$ glucose-6-phosphate, and 150-µ$M$ $NADP^+$. The fluorescence excitation wavelength is set to 390 nm and the emission wavelength is 460 nm. The reading can be adjusted to zero with some instruments at this point, and the fluorescence output should be stable. The reaction is initiated by the addition of 30 µL of the glucose-6-phosphate dehydrogenase solution, and fluorescence ($F_{390/460}$) is monitored continuously for 2 to 3 minutes (Figure 40.25). The data are calibrated by adding a known amount of 7-hydroxycoumarin (also using a plumping device) and recording the change in $F_{390/460}$. A computer-linked spectrofluorimeter is useful for recording and analyzing the data. In this laboratory, an OLIS DM-45 instrument is routinely used.

## HPLC-Mass Spectrometry: Midazolam Hydroxylation

Midazolam has been increasing used as *in vivo* probe for P450 3A4 activity because its metabolism is essentially complete and is totally mediated by P450 3A4 [152].

*Reagents*

1. β-glucuronidase type H1 from *Helix pomatia* (if sample is plasma)
2. Diazepam
3. 2-$M$ bicine buffer saturated with NaCl (pH 9.3)
4. Methyl-*t*-butyl ether containing 10% ethyl acetate
5. 20-m$M$ $NH_4CH_3CO_2$
6. 20-m$M$ $NH_4CH_3CO_2$ in $CH_3CN$

Midazolam and the 1′-hydroxy product (Figure 40.26) are determined via HPLC–tandem mass spectrometry. One mL (200 units) of β-glucuronidase type H1 from *Helix pomatia* (Sigma) is added to 1 mL of plasma and

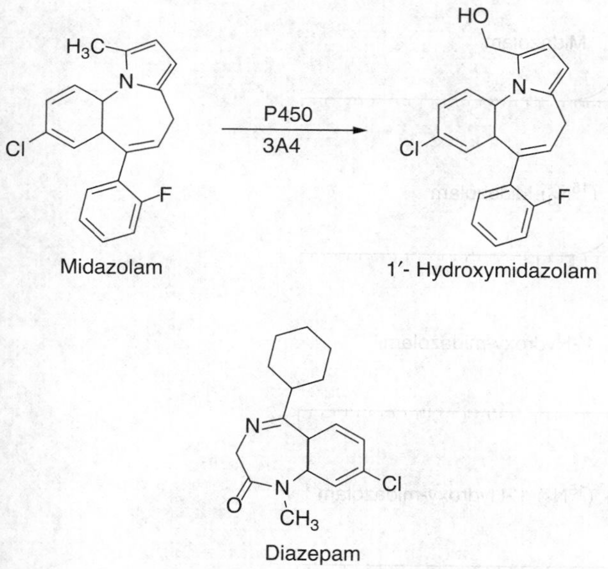

Midazolam

P450
3A4

1′- Hydroxymidazolam

Diazepam

**FIGURE 40.26** Midazolam hydroxylation by P450 3A4 to yield 1′-hydroxymidazolam; internal standard diazepam. (Adapted from Wandel, C. et al., *Clin. Pharmacol. Ther.*, 68, 82–91, 2000.)

diluted with 0.55 mL of $H_2O$ containing 50 ng of diazepam. Diazepam is used as the internal standard for the study. The mixture is incubated at room temperature for 20 to 24 hours, and 1 mL of 2-*M* bicine buffer saturated with NaCl (pH 9.3) is then added. The mixture is extracted with 6 mL of methyl-*t*-butyl ether containing 10% ethyl acetate (v/v). After centrifugation, the organic layer is evaporated to dryness under an $N_2$ stream at 60°C. Samples are dissolved in 250 μL of an equal mixture of mobile phase A and mobile phase B. Samples are briefly vortexed and incubated for 2 minutes and then filtered through 2-mm Costar® Spin-X® centrifuge filters. A 60-μL aliquot is chromatographed. Gradient chromatography is performed with a 4.6 mm × 50-mm Zorbax™ XDB $C_8$ octylsilane column using a flow rate of 0.35 mL/min. Mobile phase A consists of 20-m*M* $NH_4CH_3CO_2$ in $H_2O$, and mobile phase B is 20-m*M* $NH_4CH_3CO_2$ in 90% $CH_3CN$ (v/v). The gradient is initiated with equal parts of both phases and then linearly changed over 6 minutes to 10% phase A/90% phase B (v/v). The gradient is then returned to the original 50:50 mixture and equilibrated for 2 minutes prior to the next injection.

A Finnegan 7000 TSQ instrument with an APCI source interfaced with a Waters HPLC system is used to perform LC-MS-MS analysis. The APCI settings are as follows: capillary temperature, 200°C; vaporizer temperature, 500°C; current, 5 mA. Collision energy is 27 eV with a collision gas pressure of 2.5 mTorr for MS-MS. Selected reaction monitoring (SRM) and retention characteristics of the various compounds are as follows: [$^{15}N_0$]-midazolam, 326 to 291 *m/z* and 4.4 minutes; [$^{15}N_0$]-1′-hydroxymidazolam, 342 to 324 *m/z* and 3.2 minutes; diazepam, 285 to 193 *m/z* and 5.4 minutes; [$^{15}N_3$]-midazolam, 329 to 294 *m/z*; and [$^{15}N_3$]-1′-hydroxymidazolam, 345 to 327 *m/z* and 3.2 minutes [152]. A sample chromatogram is shown in Figure 40.27.

## Covalent Binding

### *Binding of Metabolites to Protein*

The irreversible binding of reaction products to protein was observed 60 years ago [5]. Although no consensus currently exists regarding the importance of individual targets or how such modification of proteins actually leads to death of cells, *in vitro* binding of chemicals to protein provides an index of bioactivation processes and can be useful in the characterization of reactive intermediates.

In general, the enzyme system under investigation is incubated with the radioactive substrate for a fixed amount of time, during which the rate of production of species binding covalently should remain constant. In practice, this is usually less than 1 hour. Incubations are terminated and binding to protein is measured. Several approaches are available for measuring binding.

One method involves precipitation of the protein with organic solvent ($C_2H_5OH$ or $CH_3OH$, >2 volumes) or aqueous $Cl_3CCO_2H$ (5% final concentration, w/v) and collection of the pelleted material after centrifugation in any case ($10^4$ *g*, 15 minutes). The sensitivity of the residual substrate to the solvent must be considered, as well as the solubility. The supernatant is decanted and more acid or solvent is added; the protein pellet is washed by vigorous mixing or homogenization. We have found that carrying out the entire procedure in stainless-steel centrifuge tubes is convenient because the vessels can be centrifuged in a Sorvall® SS-34 or SA-600 rotor and homogenized with a Sorvall® Omni-Mixer (DuPont Instruments, Wilmington, DE) without the need to transfer contents. The process of homogenization, centrifugation, and decantation of supernatant is repeated several times until significant radioactivity no longer appears in the wash fractions. At that point, the protein samples are dried by heating at 60°C for 2 hours. The protein is dissolved in 1.0-*N* NaOH (about 1 to 2 mL) for 1 hour at 60°C. Insoluble material is removed by centrifugation and the protein in an aliquot of *each* sample is measured by the Lowry et al. [153] or Pierce bicinchoninic acid method, because the recovery is variable. A larger aliquot of each sample is added to 5 to 10 mL of a scintillation cocktail capable of holding water, and chemiluminescence is allowed to decay overnight at room temperature in the dark prior to counting. The results are expressed in terms of nmol adduct per mg protein, with subtraction of values obtained with an inactive enzyme (e.g., without NADPH in the case of mixed-func-

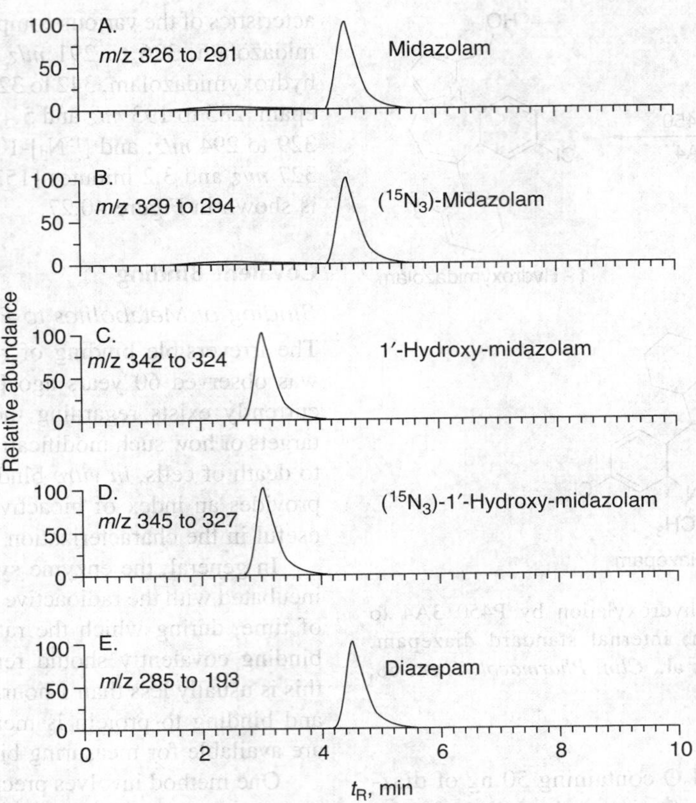


**FIGURE 40.27** A sample chromatogram of the midazolam hydroxylation assay with P450 3A4 [152]. Selected reaction monitoring combined with HPLC: (A) midazolam, *m/z* 326 to 291, 4.4 min; (B) [$^{15}N_3$]-midazolam, *m/z* 329 to 294; (C) 1′-hydroxymidazolam, *m/z* 342 to 324, 3.2 min; (D) [$^{15}N_3$]-1′-hydroxymidazolam, *m/z* 345 to 327; (E) diazepam, *m/z* 285 to 193.

tion oxidases). Another approach is extensive dialysis of protein samples against buffer containing sodium dodecyl sulfate (SDS) [154]. We have utilized this method with hydrophilic materials such as acrylonitrile [68–70] but have not obtained as reliable results with more hydrophobic materials.

A third method involves the adsorption of protein to glass fiber filters, such as those used for *in vitro* protein translation experiments [155]. These disks can be washed with organic solvents in a shaking device to remove unbound material. We have usually used 5 to 8 changes of the wash solvent, with wash times of 30 minutes per cycle (usually with $C_2H_5OH$ as the solvent) [76,78]. The capacity of these filters is limited, so such an approach may not be satisfactory if considerable amounts of protein must be used; however, if satisfactory sensitivity can be achieved with the use of submilligram amounts of protein, then the method is considerably easier than the other approaches. In our experience with trichloroethylene, we have found that recovery of protein on the filters is nearly quantitative when ≤1 mg protein is used (checks on binding can be done with radiolabeled proteins) [76,78].

### *Binding of Metabolites to DNA*

Like the binding of reaction products to protein, binding to DNA provides an index of bioactivation. One must remember that binding to naked DNA in such a system may be much higher than *in vitro* or in cells. Binding to protein is not to be equated with nucleic acid binding; notable examples demonstrate that different enzymes and pathways can be involved in the generation of the various types of adducts [156]. Calf thymus DNA has been used in many *in vitro* binding experiments. Herring sperm DNA is considerably less expensive and much more soluble; its fragmented nature should not really cause a problem in this type of work. In general, DNA is added to incubations containing a radioactive substrate at a concentration of ~2 mg/mL, with other components as used in standard procedures (described below). Several methods can be used to purify the DNA for measurement of bound adducts, and the choice depends upon the situation.

One approach involves initial extraction of the aqueous solution with $H_2O$-saturated butanol to remove small compounds. Centrifugation ($3 \times 10^3$ *g*, 10 minutes) is used to separate the layers after each step. The DNA solution is mixed with an equal volume of phenol solution; liquid phenol is first washed sequentially with equal volumes of 1.0-*M*, 0.50-*M*, and 0.02-*M* Tris-HCl (pH 7.4), and then 1 g of 8-hydroxyquinoline is added per liter of phenol. After mixing (with a vortex device), the layers are separated by centrifugation; DNA remains in the lower phase. The step is repeated one or two more times. NaCl is added to the DNA layer to a concentration of 0.10 *M*. The DNA is precipitated by the addition of 5 volumes of cold (–20°C) $C_2H_5OH$ and recovered by centrifugation; the material, dried under a $N_2$ stream, can be dissolved in any of a number of low ionic strength buffers, although shaking or other agitation may be necessary.

Another procedure that has been used as an alternative to the butanol and phenol extractions involves the sedimentation of DNA by ultracentrifugation [157]. This process is usually done following the enzymatic incubation: SDS is added to a final concentration of 1% w/v (from a stock 10% solution). The incubation buffer should not contain potassium ions because potassium dodecyl sulfate is rather insoluble. The preparations are centrifuged at $10^5$ *g* for 16 hours at 20°C to pellet the DNA. The recovery is generally good (>80%); most proteins remain soluble and are decanted in the supernatant. Although the procedure is limited by the availability of rotor places in the ultracentrifuge, the effort involved in manipulations is minimal.

Sometimes these procedures do not completely remove unbound materials. A useful procedure for further purification is hydroxylapatite chromatography. A method used in this laboratory is outlined here. A sample containing 1 mg of DNA and 1 g of dry hydroxylapatite (DNA-grade; Calbiochem, San Diego, CA) is suspended in 5-m*M* sodium phosphate buffer (pH 6.8) and swirled to hydrate the particles. The suspension is evacuated (at 40°C) to remove gas bubbles using an aspirator and poured into a column (1.6-cm diameter), where 5-m*M* sodium phosphate buffer (pH 6.8) is pumped through the column at a flow rate of ~2 mL/min using a peristaltic pump. The DNA is applied to the column (using the pump), and the column is sequentially eluted with 100-, 200-, and 300-m*M* sodium phosphate buffers (pH 6.8). The eluate is monitored at 254 nm, and the elution buffer is not changed until $A_{254}$ has decreased nearly to the baseline (eluted fractions are collected). Proteins are eluted with 100-m*M* phosphate, RNA with 200-m*M* phosphate, and DNA with 300-m*M* phosphate. Aliquots of fractions can be assayed for radioactivity after mixing with a cocktail containing a detergent and designed for aqueous samples. When large volumes of water and high phosphate concentrations are used, the conditions required for formation of a stable gel should be checked carefully beforehand.

If concentration of peak fractions is necessary, this is conveniently achieved by removal of the salt from the pooled samples by extensive dialysis and subsequent lyophilization. In some cases, further treatment of DNA with RNase (heating at 80°C for 10 minutes prior to use to destroy DNase) or pronase may be desired, with subsequent recovery of DNA by phenol extraction and ethanol precipitation. Analyses for RNA can be done using the orcinol procedure [158]. Protein can be measured with any of several assays or, if necessary, by amino acid hydrolysis [70]. (*Note:* Purine decomposition leads to abnormally high levels of glycine.) DNA itself can be estimated using the diphenylamine procedure [158], Hoechst 33258 dye with a fluorimeter [159], or the approximate relationship of $E_{260}^{1\%} = 200\ cm^{-1}$. If major DNA adducts have been identified for a particular substrate, then the best approach is to hydrolyze the DNA and measure the adduct by a specific method, e.g., chromatography or immunoassay.

### Other Enzyme Assays

Many of the types of enzyme assays have been covered in the examples provided here; the reader is referred to Table 35.1 of the last version of this chapter [4] for more. Other types of assays that are not covered here include the use of a bacterial mutagenic or other genotoxicity endpoint [160,161]and radioisotope characterization (i.e., adding a radiolabeled compound to derivatize the product) [162,163].

## STRUCTURAL ELUCIDATION OF ENZYME REACTION PRODUCTS AND DEVELOPMENT OF ASSAYS

One common problem that occurs when an enzyme acts on a drug or other chemical is that a new, undefined compound results. Of course, this can happen not only with a purified enzyme but also with crude systems. In the past, the elucidation of the structure of a new metabolite was often a heroic task, and a considerable amount of material was required. Today, the process is much more efficient, and in many situations there are demands to identify new products quickly and develop assays for further quantitative analysis. Such situations are now routine in the pharmaceutical industry, and in our own experience we are of the opinion that the rapid identification of reaction products will gain even more importance in academic laboratories.

A very typical situation is when a researcher incubates a chemical with an enzyme and analyzes the extracted products by HPLC, finding one or more new products as judged by the appearance of UV response. The challenge is to identify these. First, consider the HPLC system. The majority of work is done with reversed-phase systems (e.g., $C_8$, $C_{18}$, phenyl), but not all. The effluent from any system can probably be used with online UV–visible detection, but the system may be incompatible for other spectroscopy; therefore, there is a preference for avoiding normal-phase, ion-exchange, and ion-pairing systems. Some of the salts and organic solvent may be incompatible with a mass spectrometer. The researcher will also need to decide if spectra will be acquired online or offline. In the latter case, the peaks of interest are collected manually and processed. Manual collection is preferable to use of a fraction collector, unless major separations are achieved. If manual collection is done, it works best with a detector having a rapid response (e.g., a simple detector with a strip-chart recorder), because many computer-based detectors have time delays for averaging and processing signals. Early in the work, the researcher will also have to decide what scale will be needed for the separation and accumulation. If a semipreparative HPLC column is needed (typically 10 × 250 mm), then the change should be made immediately to define the operating conditions. Typically, the flow rate for a 10 × 250-mm reversed-phase HPLC column is about 4 mL/min, which will give a separation similar to a classical analytical column (4.6 mm × 150–250 mm) operating at 1.5 to 2 mL/min. In some cases, we have used 25 mm × 250-mm columns, operating at 10 mL/min [164]. The analytical columns are usually very appropriate for doing online UV–visible spectroscopy and mass spectrometry. If fractions are to be collected offline for NMR, then the larger columns may be in order.

Online UV–visible spectroscopy is reasonably straightforward. In this laboratory, we utilize systems based on both diode arrays (Hewlett-Packard) and rapid-scanning monochromators (Thermo). The two systems each have their own advantages in handling and displaying data. With both, software can be used to subtract the solvent absorbance and any peaks contributed by other eluting chemicals (e.g., not completely resolved). If the sample is too concentrated, the spectrum will not be valid (e.g., if the absorbance is greater than 1 or 2), and a more dilute sample should be analyzed. With regard to the interpretation of spectra, some well-established general rules can be of use in the interpretation of spectra [165–167]. The data should be stored in files on the hard drive of the computer, backed up (preferably to a CD rather than a magnetic diskette), and printed to a version that can be appropriately saved in a notebook. As always, all relevant information necessary to repeat the work should be recorded.

Online liquid chromatography (LC)–mass spectrometry is similar, in terms of the HPLC. In contrast to HPLC/UV–visible spectroscopy, HPLC columns larger than analytical cannot be used. In some cases, microbore columns are preferred. A review of mass spectrometry and complete description of all relevant aspects are beyond the scope of this review; however, almost all work of this sort will begin with either electrospray or atmospheric pressure chemical ionization, in either the positive or negative ion mode. The preferred solvents are those composed of $H_2O$, $CH_3OH$, or $CH_3CN$ and some $NH_4CH_3CO_2$ as a similar compatible buffer salt. With a new compound, one will have to collect full-scan data because the *m/z* values are unknown. To monitor the presence of all eluted chemicals, one should follow a UV absorbance signal (from an inline detector) as the total ion current as a function of timed. The detected peaks can then be scrutinized to recover full spectra.

The interpretation of mass spectra is beyond the scope of this chapter. One should have some intuition as to what *m/z* values are realistic from knowledge of the substrate. The molecular ion will be manifested as $[M+H]^+$ or $[M-H]^-$ in electrospray (M is the molecular mass), depending on whether the positive or negative ion mode is used. One simple guide for chemicals containing the common atoms dealt with in this type of research is that an odd number for M indicates an odd number of nitrogen atoms (1, 3, 5, …), and an even number indicates either none or an even number (2, 4, 6, …). It is possible to utilize a higher resolution instrument to acquire greater mass accuracy, but linking such instruments (e.g., magnetic sector, Fourier transform–ion cyclotron) to HPLC is not as common. With a resolution of $>1/10^4$, 4 decimal places are recorded and the atomic composition of a molecule can be established. It is possible to collect samples and submit them for elemental analysis or exact mass determination, but the sensitivity is usually not as good.

In some cases, the structure of a compound can be established with reasonable certainty just from the fragmentation pattern. Modern mass spectrometers allow the fragmentation of individual ions, which can be quite useful; however, with complex molecules and in the absence of work done with stable isotopes, establishing structures of unknown compounds only by UV–visible and mass spectrometry methods is usually not a definitive process.

Nuclear magnetic resonance spectroscopy is a very powerful approach to determination of chemical structures. Two basic types of information can be obtained: through-bond coupling (which atoms are attached to each other) and through-space coupling (for our purposes, stereochemistry). The simplest through-based method is a simple one-dimensional $^1H$ NMR spectrum. More complex two-dimensional methods require more sample or longer acquisition times. Some of these establish the linkages of protons and $^{13}C$ atoms, in a manner based on the protons and thus are much more sensitive than direct analysis of the $^{13}C$ signals. Examples we have utilized in this laboratory include the COESY, HSQC, HMBC, and CIGAR methods. Two-dimensional through-space methods we have utilized with small molecules are NOESY and ROESY [168,169].

Nuclear magnetic resonance spectra of small molecules are usually run in deuterated solvents to avoid artifacts introduced by signal suppression methods. If HPLC peaks are collected offline, then the chemical of interest can often be extracted from the (aqueous) elute into an organic solvent (e.g., $CH_2Cl_2$) and concentrated for NMR analysis. If compounds cannot be extracted (e.g., nucleosides), one approach is to use a volatile combination of buffer salts (e.g., $NH_4HCO_2$, $NH_4CH_3CO_2$, *N*-ethylmorpholine acetate) and remove the buffer by repeated lyophilization. Remove solvents and $H_2O$ as rigorously as possible to avoid interfering peaks. With small amounts of a chemical, the limitation to sensitivity is usually not the amount of the chemical itself but the extent of contamination by solvents, $H_2O$, grease, and other contaminants. Spectra are usually recorded on instruments with field strengths of 300 to 500 MHz. The inherent sensitivity increases as a function of the square of the field strength, so an 800-MHz instrument should provide fourfold greater sensitivity than a 400-MHz system. Other factors for sensitivity are the sample volume (smaller is better) and the probe temperature. Cryoprobe technology provides greater sensitivity because of a square-root relationship with the absolute temperature.

A relatively recent development is the implementation of combined LC-NMR systems (Figure 40.28). In true online systems, one of the two HPLC buffers is usually deuterated and the elution is paused when peaks are eluted (into the NMR probe) to collect spectra for longer time periods. A new development involves automated robotic systems for collecting designated HPLC peaks (detected by UV–visible or mass spectral measurements). The designated effluent is directed on to a solid-phase extraction (SPE) cartridge. (The run can be repeated to apply a multiple load.) Each SPE cartridge, loaded with a peak from the HPLC run, is then washed with solvent to remove any salts and the HPLC solvent, then a small amount of deuterated organic solvent is applied to each SPE cartridge to elute the bound compound into a 30-μL capillary tube. These capillaries, all filled offline, are then moved to the NMR spectrometry, where spectra can be recorded for the desired length of time.

The amount of material required to obtain useful spectra in each of these approaches is dependent on the particular compound and its physical characteristics. UV–visible spectra can be collected with absorbance maxima of <0.01, and the sensitivity will depend upon the dimension of the HPLC column and the detector cell. Subnanomolar amounts can be characterized in many cases, depending on the chromophore. Mass spectrometry has a similar detection limit, depending on the ionization characteristics of the compound, the HPLC column bore, and the details of the specific mass spectrometer. Our own experience has been mainly with ThermoFinnigan electrospray instruments, and we have found that the newer Quantum series has an ~50-fold greater sensitivity than the older TSQ7000 series. NMR is usually the limiting factor in analyses of this type, depending on the situation. If only a one-dimensional $^1H$ NMR is needed then, in principle, the UV–visible, mass, and NMR spectra can be collected with sample in the low-microgram range.

One powerful approach to spectral analysis is the use of online systems (e.g., liquid chromatography coupled to spectroscopy). A recent example from this laboratory, presented in Figure 40.28, involves the characterization of an unknown oxidation product of testosterone by recombinant P450 3A4. Testosterone was incubated with 0.1 nmol of the P450 (plus NADPH–P450 reductase) for 5 minutes, and the substrate and products were extracted and applied to an HPLC column. The effluent was split, with a minor fraction directed to a UV detector and a mass spectrometer. On the basis of the UV profile, fractions were collected (robotically) on small SPE cartridges. In this work, three HPLC runs were pooled before further processing. The SPE columns were switched and washed to elute solvent and any salts and then eluted with a small amount of deuterated solvent into small capillary tubes. All of this processing can be done offline from the NMR spectrometer. The capillary tube containing the peak of interest was then used for NMR analysis—in this case, one-dimensional ($^1H$), COSY ($^1H$–$^1H$), HSQC, and NOESY ($^1H$). Most of the $^1H$ and $^{13}C$ resonances were already known from the literature on testosterone. The ultraviolet spectrum (not shown) was very similar to that of testosterone, suggesting little change in the double bonds that constitute the chromophore. The mass spectrum confirmed the addition of one oxygen atom (+16). The chemical shift (δ) of the proton associated with the new product (relative to the substrate testosterone) suggested that it was adjacent to either an allylic carbon or a carbon α to a carbonyl (i.e., C-1 or C-7). Inspection of the COSY spectra indicates a lack of connection to C8, arguing against a hydroxyl at C-7 and in favor of C-1. The HSQC spectrum supports this view. The stereochemistry was easily established as 1β with the NOESY spectra, after considering the three-dimensional possibilities for the molecule (Figure 40.28). If the H-7 proton was in the β-position, it would be expected to show a very strong cross peak with the (three) protons of the C-19 methyl group. It does not; therefore, the H-7 proton is α (and the hydroxyl group is β). The NOESY spectrum shows the expected coupling to H-2α,β, H-9, and H-11 (Figure 40.28). Thus, the structure was established as 1β-hydroxytestosterone with ~6 μg of the product [169].

In some cases, the above information may not be sufficient to assign the structure. Other spectroscopy may be in order. Infrared (IR) spectra can be useful, particularly if the presence of certain diagnostic groups is suspected [166]. Circular dichroism (CD) spectra can be useful in assigning stereochemistry [170,171]. In the example shown here with 1β-hydroxytestosterone, an exact mass was obtained. As mentioned earlier, this is not always done as a first course to characterizing a molecule but can be

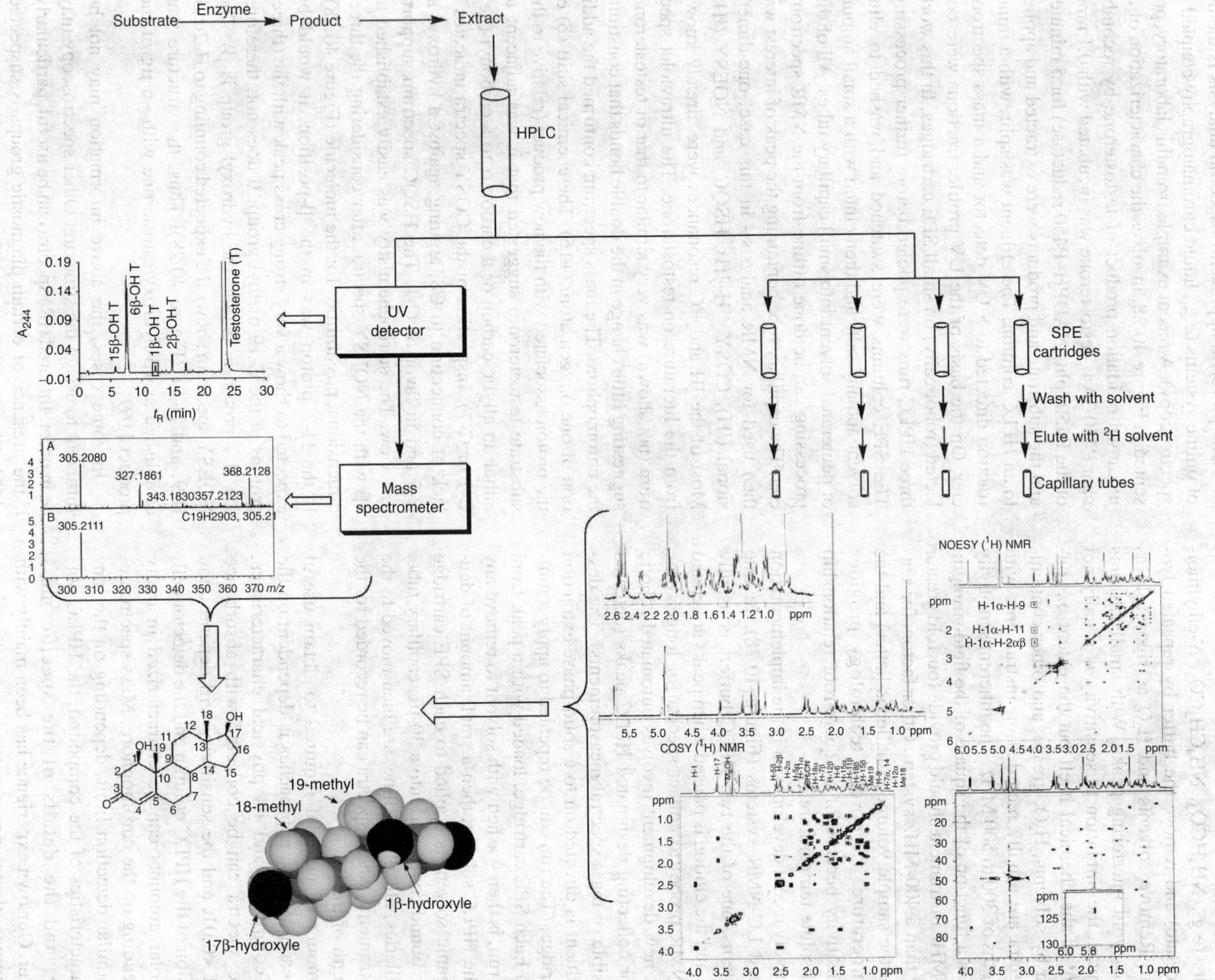


**FIGURE 40.28** Determination of a structure of a reaction product by online LC-SPE-MS-NMR. The UV detector can also be used to obtain the UV spectrum of the peak of interest (indicated with box, spectrum not shown). The data are from the characterization of 1β-hydroxytestosterone as an oxidation product of testosterone by P450 3A4. (Adapted from Krauser, J.A. et al., *Eur. J. Biochem.*, 271, 3962–3969, 2004.)

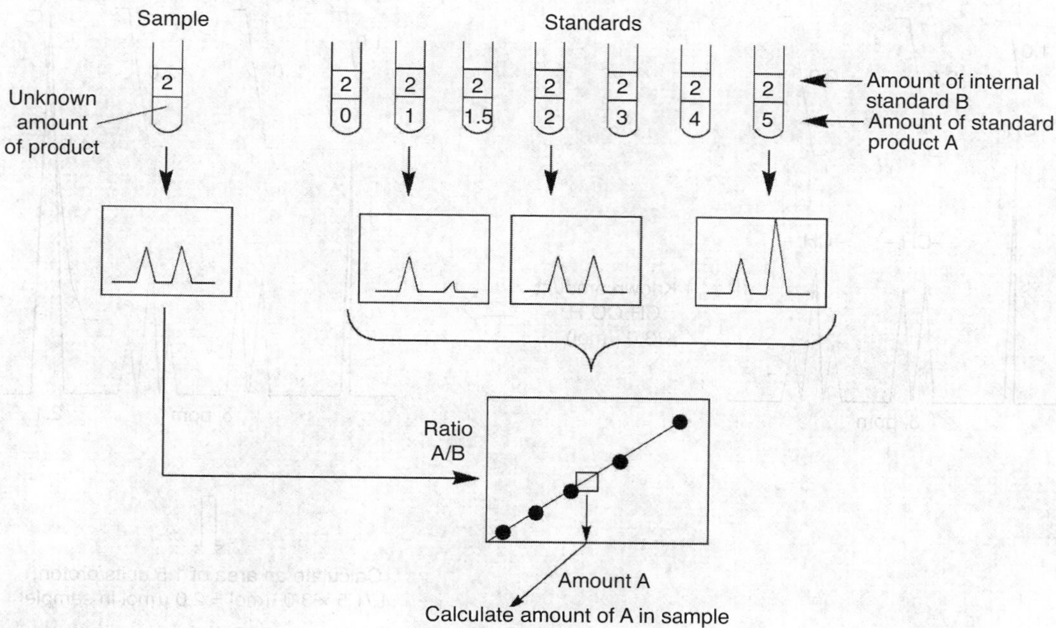


**FIGURE 40.29** General approach to use of an internal standard (closely related compound) for quantitative HPLC analysis.

very important in distinguishing among possibilities after obtaining a lower resolution spectrum. Another approach required sometimes involves the use of more NMR spectra, especially two-dimensional spectra [168,172].

A common next step in work of this type is development of an assay. One preferred approach is to synthesize the reaction product and use it as a standard, preparing stocks by gravimetric methods. Typically, an assay would be set up using HPLC; in the case cited here (Figure 40.28), the product was first separated by this method. Detection could be by UV or fluorescence spectroscopy or mass spectrometry. A common approach is to use an internal standard (i.e., a closely related compound that is still distinguished). This is added after the enzyme reaction, prior to the extraction step, to control for product recovery. If the analysis utilizes mass spectrometry, then ideally the internal standard will be the same compound with heavy atoms (≥2 preferred) covalently attached ($^{13}C$ and $^{15}N$ are preferred) to avoid isotopic ($^{2}H$) effects on the HPLC $t_R$. In other cases, a compound differing slightly but having similar physical properties (extraction, ionization) is derived. The principle of internal standard curves is presented in Figure 40.29.

One problem that arises in many cases is that of quantifying a very small amount of a product that cannot be synthesized; also, the amount available may be low for gravimetric determination. One option is the use of a good microbalance. Another, depending on the structure, is use of the UV–visible extinction coefficient of a similar molecule with the same basic chromophore, in the absence of interference problems. Another method of quantifying small amounts of source materials is to use $^{1}H$ NMR spectra, which are absolute in their signal intensity. The choice depends on the interfering signals; an example of a compound devoid of signals in the aliphatic region is $CH_3CO_2H$, which gives a sharp singlet (δ 1.91) in DMSO (Figure 40.30) [173,174].

An extensive discussion of all possible assays is beyond the scope of this article, and some representative assays are mentioned earlier. If interference is a problem (due to contaminating ions at the $m/z$ of interest), a suitable assay may be set up by monitoring the loss of a specific fragment characteristic of the analyte of interest.

## HETEROLOGOUS EXPRESSION OF ENZYMES

The overall approach to handling enzymes has changed dramatically since the chapter was published in the first edition of this book [1]. In today's world, the genomes of humans and several of the principal experimental systems are known. In such systems, the isolation of an enzyme from a tissue or a microorganism is rare. Most enzymes are now prepared by heterologous expression methods. The advantages are simple: (1) The amount of enzyme that can be obtained is much less of an issue; (2) the expression system is usually less expensive to work with; (3) in principle, all human enzymes are accessible; (4) individual amino acids can be replaced to study their roles; and (5) purification is easier, because the starting concentration of the enzyme of interest is much higher, the presence of interfering, closely related enzymes (e.g., other P450s) can be

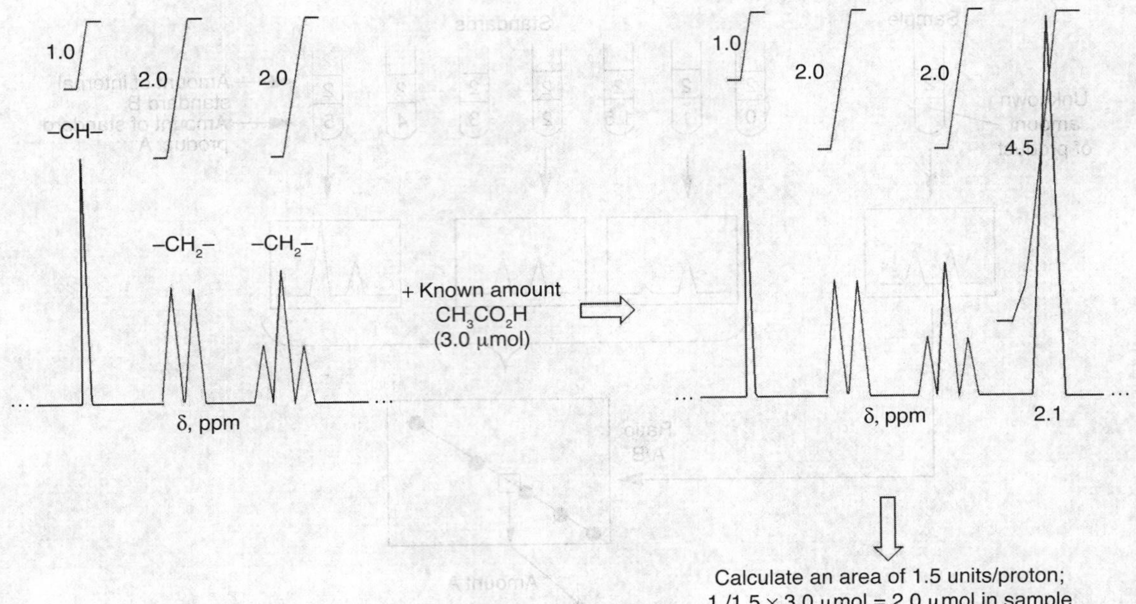


**FIGURE 40.30** Use of NMR and an internal standard ($CH_3CO_2H$) to establish the concentration of a small amount of a new product. (Adapted from Gottlieb, H.E. et al., *J. Org. Chem.*, 62, 7512–7515, 1997; Wu, Z. et al., *Chem. Biodiversity*, 2, 51–65, 2005.)

avoided, and many proteins can be expressed with innocuous tags that will facilitate isolation. A thorough treatment of heterologous expressions is beyond the scope of this test, and the reader is referred to other work, including some compilations [175–177], original work, and commercial literature; however, the point should be made that most practicing biomedical scientists (and certainly all enzymologists) must be able to utilize recombinant DNA methods and heterologous expression in today's scientific environment.

### Considerations

Many methods for expression are available, and many scientists have personal favorites. What are some of the issues in deciding which system to use? Here are some issues to consider:

- *Amount needed*—Some scientific problems require large amounts of enzyme; for example, x-ray crystallography and some physical techniques require considerable amounts, at least milligrams of purified protein.
- *Posttranslation modification*—If the protein of interest is known to require posttranslational modification (other than incorporation of heme, flavin, and metal ions), then eukaryotic expression systems are in order. The disadvantage is that these systems generally have lower yields. Another issue is that some posttranslation systems differ in lower and higher eukaryotes.
- *Difficulty of reconstitution*—Multiprotein systems can be difficult to reconstitute; for example, if a protein has multiple subunits, coexpression of the components in a single system may facilitate the interaction, and the proteins may be isolated as a complex.
- *Need for a specific cellular system*—In some applications, the interest is not in using a purified enzyme but in doing studies within the context of a cellular environment because of the endpoints to be examined; for example, in this laboratory we have utilized bacteria to report genotoxicity, in both the analysis of chemicals [161,178] and doing random mutagenesis/molecular breeding [179]. Others have used mammalian cell systems to evaluate toxicity [180]; thus, the complexity of the system depends upon the project. An extension of this question is whether a transgenic animal will be used instead of a cellular system. Obviously, many mammalian responses can be detected only in such complex settings.
- *Use of protein tags*—The purification of proteins is facilitated by the attachment of tags to either end of the protein. In work in this laboratory, we have often employed a $(His)_5$ tag at the C-terminus [181,182]; N-terminal tags can be employed in some cases [183], although that is the first end of the protein synthesized, and it may be less tolerant to disruption, particularly in bacteria [184]. In bacteria, the N-terminus can often be

modified to improve protein expression, if there is no effect on the properties of the protein. An alternative approach is to add larger peptide units to expressed proteins and utilize chromatography systems that will bind the tag. Some of the larger tags can be removed with selective proteases, if the interference is a problem.

## Heterologous Expression Systems Available for Use

As indicated earlier, these systems will be described only briefly, mainly in terms of the advantages and limitations of each, moving from the simplest to the more complex.

### *In Vitro* Translation

A simple, low-cost instrument can be purchased (e.g., Roche), and this approach can be very useful if only small amounts of protein are required. The need to optimize organisms for expression efficiency is avoided. The reagents (aside from the cDNA) are commercially available and are expensive for large-scale work but reasonable for small projects.

### Bacteria

Bacteria have many advantages and remain the preferred vehicle in this laboratory [177]. Most work is with *Escherichia coli*, although sometimes *Salmonella typhimurium* [185–187] and *Bacillus subtilis* [175] have applications. Bacterial expression is generally fast (1- to 2-day culture) and relatively inexpensive and gives high yields. Many vectors are available. The systems can be employed for various screens done in the cells [177,185,186]. Scale-up to high-volume production in fermentors is most easily accomplished with bacterial systems. The limitations for bacterial expression are the need for posttranslational modification and problems in the proper folding of source proteins. In addition, many proteins express better if the N-terminus is modified, for various reasons [177,188]. Thus, some trials are usually necessary with several constructs before proceeding with productive expression. If expression of proteins is too rapid in bacteria, then misfolding can occur. Sometimes, the misfolded proteins are accumulated in inclusion bodies, which are particles of damaged protein (devoid of prosthetic groups). Several approaches can be used to minimize this problem, including the use of weaker promoters, lower incubation temperatures, and coexpression of chaperone proteins. For the expression of human P450s in *E. coli*, the addition of chloramphenicol and ethanol has been helpful in some cases. For the P450s that are more difficult to express, the addition of heme precursors has been helpful, probably by providing a prosthetic group for the protein to wrap itself around.

### Yeast

Most work is done with *Saccharomyces cerevisiae*, but other yeasts have some advantages [189]. Yeast expression of P450s has been done in this laboratory in the past [190] and is still done in other groups [191,192]. In addition to yeasts, some expression has been done in the false yeast *Pichia pastoris* [193,194]. Yeasts are eukaryotes and have advantages in terms of posttranslational processing capability and capacity for proper protein folding. Numerous vectors are available for expression, and generally less manipulation of sequences is necessary for expression. *S. cerevisiae* does have an NADPH–P450 reductase and cytochrome $b_5$, which couple with some mammalian P450s but not all [195,196]. The expression levels achieved in yeast are generally lower than in bacteria. The recombinant DNA manipulations needed to prepare vectors are done in bacteria and then transferred to yeast. The expression times are longer than in bacteria. A major disadvantage of yeast expression is the difficulty in breaking yeast cells for enzyme purification. Breaking cells with only (strong) mechanical methods is not very inefficient, and enzymes such as yeast lytic enzymes are required.

### Insect Cell Systems

Systems based on baculovirus infections are useful [197,198]. The advantages and limitations of these systems are similar to those for yeast. Enzymes are not required to break the cells, but the costs of media are considerable. High levels of expression can be achieved, although the expression volume is usually a limitation because of the cost of media. As with yeast, posttranslational modification occurs; it may or may not be identical to what occurs in mammalian system. Baculovirus-based expression of P450s and other heme proteins does require the addition of heme. As a general rule, baculovirus is a reliable means of expressing proteins that are recalcitrant to expression in other systems. These expression systems are usually used in spinner cultures, which do not require rigorous aseptic conditions. Insect cell culture is possible in shaking cultures (e.g., Fernbach flasks) although such methods have been used less. Large-scale cultures are possible although expensive.

### Mammalian Cell Culture

COS-1 cells have a monkey kidney origin and have been used extensively for many proteins [199]. Other systems used for enzymes of relevance to toxicology include TKK lymphoblasts and V-79 cells, which have been used in toxicity and mutation assays. HEK cells have been used for higher level expression, usually in roller bottles. In some cases, these latter systems have been used to generate enough material for structural biology studies, although this is not a trivial or inexpensive process. Another variation is vaccinia virus systems, which are used to produce proteins in mammalian cells [200,201]. These systems have been used for the synthesis of drug-metabolizing enzymes and transporters, with higher levels of expression than some of the other mammalian systems. The mammalian cell systems have the advantages of not requiring much optimization and rather faithfully doing posttranslational modification. They also provide good models for cell-based assays. The disadvantages are the need for cell culture facilities, high cost, and difficulties in large-scale work.

### Transgenic Animals

A detailed discussion is far beyond the scope of this chapter [202]. Transgenic animals can be considered one type of heterologous expression, although they are usually not used to harvest the protein except in certain commercial settings with large animals and biopharmaceuticals (proteins used as drugs); however, transgenic mice are commonly used to address hypotheses in *in vivo* settings, in terms of both gain and loss of function (gene knock-in and knock-out animals, respectively). Obviously, these are expensive experiments. The ease of developing transgenic mice has improved in recent years; the cost of maintaining the animals is still a problem. They do provide a reasonably appropriate setting for analyzing the functions of proteins *in vivo*.

## Use of Purified P450 Enzymes

Many individual forms of P450 have now been purified and characterized from a number of species, including humans. A complete discussion of these preparations is beyond the scope of this article, and the reader is referred to other references [39,45,46,48,203,204]. For a discussion of the current recommended nomenclature for the P450 enzymes see Nebert et al. [205] and the website http://drnelson.utmem.edu/CytochromeP450.html. One of the reasons for purifying individual forms of P450 is to determine which individual forms are involved in particular reactions. This task has been made somewhat easier in the light of current knowledge concerning P450s.

With rat liver systems, the first step is the development of an *in vitro* assay for the particular activity under consideration. This has to be developed with liver microsomes, and the sensitivity should be optimized, along with conditions such as pH, time, and protein and substrate concentrations. The next step involves comparison of the rates of oxidation with microsomes isolated from (untreated) male and female rats and male rats treated with various inducing agents. A considerable body of knowledge now exists regarding the effects of gender and inducing agents on individual P450 forms, and this information can be used to advantage [46,48-50,206]; for example, adult male rats contain P450 2C11 and P450 3A2 but not P450 2C12. Phenobarbital administration induces P450 2B1, P450 2C6, P450 2B2, and P450 3A1. Pregnenolone 16α-carbonitrile and dexamethasone induce only P450 3A1 (and any closely related forms). Levels of P450 2C11 are suppressed by the administration of any of several of the typical inducers, particularly polycyclic aromatic hydrocarbons. From information obtained in such experiments one can begin to hypothesize which forms are involved. The next step involves examination of the relative abilities of individual forms of P450 to catalyze the reaction, if these are available. Reconstitution conditions are described elsewhere. When specific inhibitors are known for individual forms of P450, such as 7,8-benzoflavone (α-naphthoflavone) for rat P450 1A1 and P450 1A2, these can also be used to advantage [30,207]. Another step involves using specific antibodies with microsomal preparations to determine which will inhibit the reaction. This approach is discussed later in this chapter. These results should agree with and confirm those obtained in the above experiments.

The strategy is slightly different for studies with human enzymes, as induction and gender patterns are not involved. If some results are available on the catalytic selectivity of rat or other animal P450 enzymes, this information may be of some use in predicting human enzymes, although there are many cases of catalytic selectivity jumping subfamilies when making such comparisons, particularly in the P450 2C family [208,209].

One way to proceed is to first use diagnostic chemical inhibitors (Table 40.2) [30,207]. Because human samples are known to vary considerably in their P450 composition, it is risky to rely only on a single human sample, particularly if there is little prior knowledge about its behavior. Usually, several samples (≥3) should be used in initial screens. An alternative is to use a single experiment with microsomes prodded on the basis of protein content (≥10 samples). The same strategy applies to human hepatocytes (here the number of samples available may be low).

Another useful strategy with microsomes is correlation analysis [210,211]. A set of microsomal samples is compared for the new activity under consideration and also marker activities diagnostic of individual P450 enzymes (Table 40.2 and Figure 40.31). In general, ≥10 samples should be used. Correlation analysis can be done with either linear or Spearman rank methods [210,213]. In principle, the parameter $r^2$ estimates the fraction of the variation accounted for by the relationship between two variables. Values of $r^2 \geq 0.5$ are generally judged to be significant (but $p$ is dictated by sample size).

An additional approach is to utilize purified or, more likely, recombinant enzymes to assay the activity. If this is done with a chemical, there should be some appreciation of the plasma or tissue concentrations relevant to the prob-

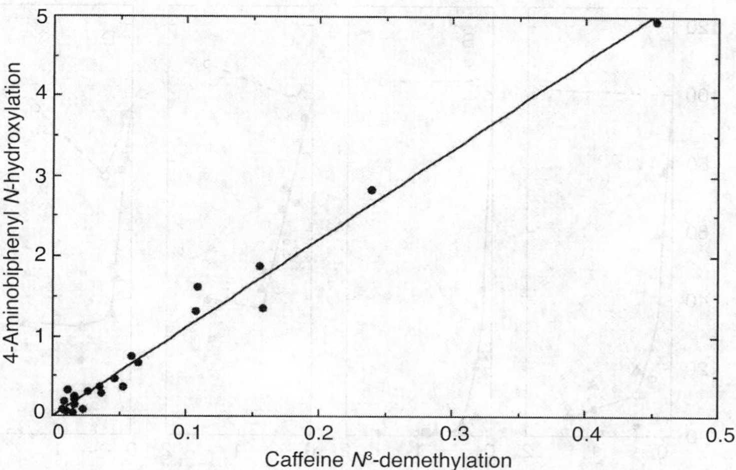


**FIGURE 40.31** Correlation of caffeine $N^3$-demethylation and 4-aminobiphenyl *N*-hydroxylation in different human liver microsomal samples, indicative of P450 1A2 variation and involvement in both reactions. All rates are expressed in nmol product formed/min/mg microsomal protein. (Adapted from Butler, M.A. et al., *Proc. Natl. Acad. Sci. U.S.A.*, 86, 7696–7700, 1989.)

lem under consideration when a substrate concentration is selected. An improved approach involves measuring activity at several substrate concentrations and estimating $V_{max}$ (or, more appropriately, $k_{cat}$, which is $V_{max}$ divided by the enzyme concentration) and $K_m$.

Inhibitory antibodies can also be useful. In principle, the fraction of inhibition one sees when one adds an antibody (to a particular P450) to microsomes is the fraction of that reaction catalyzed by the P450 to which the antibody was raised (Figure 40.32). There are several confounding problems. One is that antibodies generally do not distinguish between P450 subfamily members, unless extensive cross-adsorption is done with the antibody. Despite early success in the area [214,215], not all anti-P450 antibodies are inhibitory. Further, even antibodies raised against recombinant P450 proteins have sometimes showed reactions with similar epitopes in different P450 families [216]. This cross-reactivity can vary with the individual rabbits. Although polyclonal antibodies raised against purified P450 enzymes have been highly useful in development of the field, there are two prospects for improvements. One is the use of monoclonal antibodies, which have already been exploited for some uses [217–220]. The other, more recent, is the use of antipeptide antibodies. Some of these have shown impressive specificity [221,222], although sustained immunoinhibitory potency may still be a problem. An area for possible development in this area is phage display methods [223].

Two general points need to be made about the use of such information (some points are also made about the use of such information, and some points are made later regarding heterologous expression systems). First, many of the parameters often cited for the *in vitro* systems used today are inappropriate, particularly with the recombinant systems. The expression of an activity of a recombinant system in terms of mg protein is not particularly helpful because no information is available about the level of expression. The concentration of the enzyme under consideration should be determined and used in the normalization. Such an analysis may be difficult if an immunochemical method is needed, as apoprotein or inactive enzyme will also be measured. Another common problem in P450 work (and with most of the other enzymes of interest here) is the attachment of too much significance to the $K_m$ value itself. Contrary to what is often conveyed in many pharmacology and toxicology texts, $K_m$ does not generally note substrate affinity unless specifically proven so. The parameter $K_m$ is an operational term, simply denoting the substrate concentration at which half-maximal velocity occurs [134,224]. It is a complex collection of microscopic rate constants, which usually are not known. In the case of P450 2E1, $K_m$ can be considered a function of $k_{cat}$, with $k_{cat}$ appearing as a dependent variable [139].

Second, the ratio of $k_{cat}/K_m$ (or $V_{max}/K_m$) is considered the most appropriate estimate of enzyme efficiency by enzymologists [224]. In a plot of *v* vs. *S* for an enzyme, this is the tangent to the plot at low substrate concentrations and has units of $mM^{-1}$ $sec^{-1}$ (or the equivalent); however, $k_{cat}/K_m$ (or $V_{max}/K_m$) is *not* intrinsic clearance, which is an *in vivo* parameter with a distinct meaning. If blood flow is not rate limiting and there are no complicating factors due to transport, then intrinsic clearance within a given organ might be a direct function of $k_{cat}/K_m$ [225]; however, the more proper term for the ratio $k_{cat}/K_m$ is simply *enzyme efficiency*. This is not to demean efforts at physiologically based pharmacokinetic modeling, which can be very useful [224,226], but these models must not make inappropriate assumptions about what they incorporate in the way of meanings of parameters, any more than they would incorporate the wrong pathways.

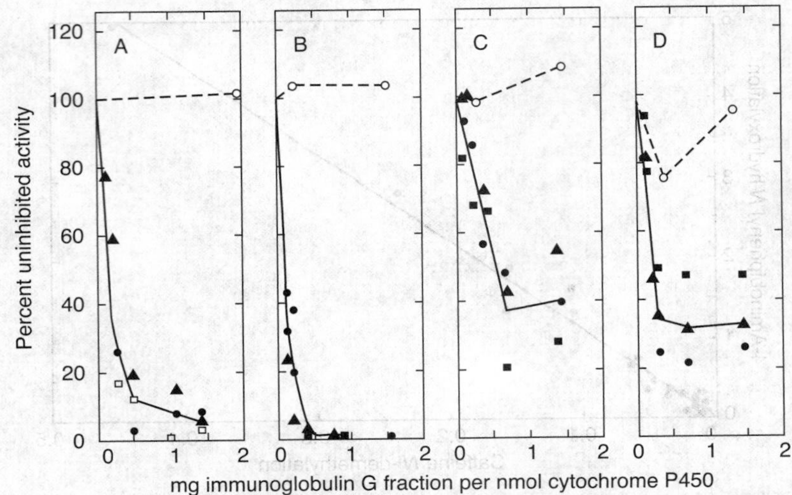


**FIGURE 40.32** Inhibition of mixed-function oxidation of debrisoquine, sparteine, encainide, and propranolol in human liver microsomes by anti-rat P450 2D1. Incubations were carried out with microsomes prepared from liver samples 17 (△), 25 (□), 31 (●), 32 (▲), 34 (■), 72 (◇), and 86 (◆), which contained 0.32, 0.55, 0.77, 0.50, 0.55, 0.61, and 0.30 nmol P450/mg protein, respectively. The microsomes were incubated with indicated amounts of anti-P450 2D1 (IgG fraction) for 30 minutes at 23°C in the appropriate buffer and other incubation components were then added. Results are expressed as percent of the control activity (obtained in the absence of antibody) for (A) debrisoquine 4-hydroxylation, (B) formation of $\Delta^5$-dehydrosparteine from sparteine, (C) encainide *O*-demethylation, and (D) propranolol 4-hydroxylation. The solid lines are drawn connecting the means of the values obtained with the various samples, and the broken lines connect the means of the values (○) obtained with IgG prepared from preimmune antisera (individual values not shown). Control activities of debrisoquine 4-hydroxylation were 0.16, 0.060, and 0.19 pmol/min/nmol P450 for samples 25, 31, and 32, respectively. Control activities of $\Delta^5$-dehydrosparteine formation were 26.6, 20.4, and 2.0 pmol/min/nmol P450 for samples 31, 32, and 34, respectively. Control activities of encainide *O*-demethylation were 0.044, 0.072, and 0.039 nmol/min/nmol P450 for samples 31, 32, and 34, respectively. Control activities of propranolol 4-hydroxylase were 0.17, 0.12, and 0.081 nmol/min/nmol P450 for samples 31, 32, and 34, respectively. (From Distlerath, L.M. and Guengerich, F.P., *Proc. Natl. Acad. Sci. U.S.A.*, 81, 7348–7352, 1984. With permission.)

## Purification of P450 from Bacterial Recombinant Expression Systems

### Overview

A procedure is outlined here that is used for purification of P450 3A4 from *Escherichia coli* membranes in this laboratory. Details of construction of the expression plasmid (NF14) are presented elsewhere [227,228], including the addition of the C-terminal *penta*-histidine, $(His)_5$, tag [181,229]. This chapter is an updated version of an earlier procedure published in this series [230]. The major change has been the incorporation of the $(His)_5$ tag and the use of metal-affinity chromatography (Figure 40.33).

Attachment of a *oligo*-His region, usually at the N- or C-terminus, has been used to facilitate protein purification [231]. The free His residues (usually $(His)_4$, $(His)_5$, or $(His)_6$) can chelate $Ni^{2+}$ and similar metal ions; thus, a $Ni^{2+}$-chelate affinity column can be used for rapid purification [232]. Such approaches have been used with P450s, with His tags at either the C-terminus [233,234] or N-terminus [183,235]. Detergent is needed to solubilize the membranes and keep the proteins disaggregated during chromatography, and the detergent must be removed in a subsequent step.

In our early work on the purification of *Escherichia coli*-expressed human P450s, the procedures consisted mainly of ion-exchange chromatography methods [227,228,230, 236–239]. The change to metal-affinity methods was made for two reasons: (1) the metal-affinity approaches can be used to reduce the need for nonionic detergents or to facilitate removal of these (nonionic detergents can yield artifacts and are even substrates [240]), and (2) some P450 mutants are relatively unstable and their purification requires more rapid methods [241,242]. An example is provided in which an ion-exchange step is used prior to metal-affinity chromatography [181]. In some cases, the step can be omitted and solubilized P450 preparations can be used directly for metal-affinity chromatography (e.g., P450s 1A2 and 2D6 [183,241–244]). In these cases, highly purified P450s can be prepared without the need for earlier steps.

*Reagents*

1. 1.0-*M* Tris-acetate buffer (pH 7.6)
2. Sucrose
3. Sodium EDTA, 0.10-m*M* in $H_2O$, adjusted to pH 7.5
4. Lysozyme, 50 mg/mL
5. 1.0-*M* potassium phosphate buffer (pH 7.4)

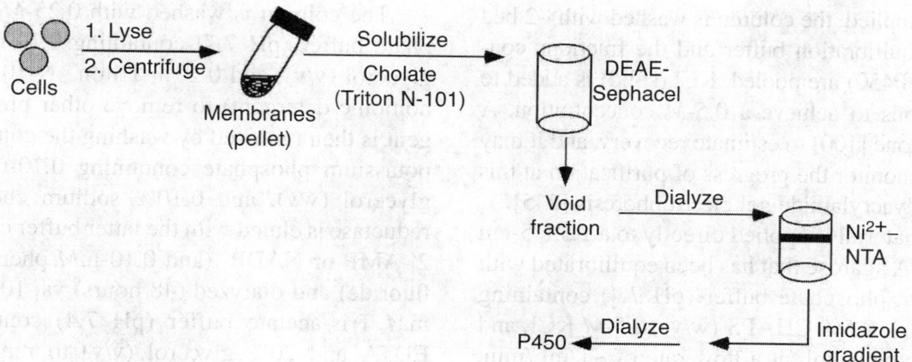


**FIGURE 40.33** Purification of recombinant human P450 from *Escherichia coli* membranes utilizing nickel affinity chromatography. (From Guengerich, F.P. and Martin, M.V., in *Methods in Molecular Genetics, Cytochrome P450 Protocols*, Phillips, I.R. and Shephard, E., Eds., Academic Press, Orlando, FL, 2005. With permission.)

6. Magnesium acetate
7. Glycerol
8. 2-Mercaptoethanol
9. Phenylmethylsulfonyl fluoride (PMSF), 0.1-*M* in *n*-propanol (stored at –20°C)
10. Leupeptin, 0.20-m*M* in $H_2O$
11. Bestatin, 0.1-m*M* in $H_2O$
12. Aprotinin, 4 U/mL
13. CHAPS (3-[(3-cholamidopropyl)dimethylammonio]-1-propanesulfonic acid), Sol-Grade® (Anatrace; Maumee, OH)
14. Sodium EDTA, 0.10-m*M* in $H_2O$, adjusted to pH 7.5
15. Diethylaminoethyl (DEAE)-Sepharose® (Amersham Biosciences; Piscataway, NJ), adjusted to pH 7.4
16. Ni-nitrilotriacetate (NTA) agarose (QIAGEN, Valencia, CA); an alternative is a $Co^{2+}$-based matrix, such as the Talon® system (Clontech/BD Biosciences; Palo Alto, CA), although we have not used this in our own work
17. Dithiothreitol (DTT)

## Preparation of Bacterial Membranes

Bacterial cells are recovered by centrifugation ($4 \times 10^3$ *g*, 15 minutes) and resuspended in 0.10-*M* Tris-acetate buffer (pH 7.6) containing 0.50-*M* sucrose and 0.5-m*M* EDTA at a concentration of ~70 mg wet cells per mL. The suspension is diluted twofold with $H_2O$, and lysozyme is added to 0.1 mg/mL. The suspension is gently shaken and incubated on ice for 30 minutes to hydrolyze the outer membranes. The resulting spheroplasts are recovered by centrifugation at $4 \times 10^3$ *g* for 15 minutes; the pellet is resuspended in buffer at a concentration of ~0.5 g/mL. This buffer for contains 0.10-*M* potassium phosphate (pH 7.4), 6-m*M* magnesium acetate, 20% glycerol (v/v), and 10-m*M* 2-mercaptoethanol. At this point, the preparation can be stored frozen at –70°C until further use [228].

The frozen spheroplasts are thawed in a water bath at room temperature. During the thawing process, protease inhibitors are added to the spheroplasts to the following final concentrations: PMSF, 1.0-m*M*; leupeptin, 2.0 μ*M*; bestatin, 1.0 μ*M*; aprotinin, 0.04 U/mL. Cells, in a Rosette cell packed in ice, are lysed for approximately 15 minutes (at ~70% full power) using a Branson sonicator or until the cell lysate is void of clumps. The lysate is subjected to centrifugation ($10^4$ *g*, 20 minutes), and the pellet is discarded. The supernatant is then centrifuged at $10^5$ *g* for 90 minutes (e.g., $3.5 \times 10^4$ rpm in a Beckman 45 Ti rotor). The pelleted membranes are resuspended in a minimum volume of 100-m*M* potassium phosphate buffer (pH 7.4) containing 20% glycerol (v/v), 6-m*M* magnesium acetate, and 10-m*M* 2-mercaptoethanol. The suspension is quickly frozen in liquid $N_2$, and stored at –70°C (unless further purification is begun immediately).

The *Escherichia coli* membranes prepared as above are diluted to a 20-m*M* potassium phosphate concentration by addition of 4 volumes of a solution of 20% glycerol (v/v), 1.25% CHAPS (w/v), and 10-m*M* 2-mercaptoethanol. Buffer, containing 20-m*M* potassium phosphate (pH 7.4), 10-m*M* 2-mercaptoethanol, 20% glycerol (v/v), and 1% CHAPS (w/v), is added, as necessary, to dilute to a protein concentration of ~2.0 mg/mL. Some P450s are not solubilized well with only an ionic detergent (CHAPS, cholate, etc.) and require the addition of a nonionic detergent (e.g., Triton® N-101, Emulgen™ 911 or 913, Tergitol® NP-10) [230]. P450s in this category that we have encountered in this laboratory include P450 2A6 and 2D6. The mixture is stirred gently for 2 to 4 hours and centrifuged at $10^5$ *g* for 30 minutes; the pellet is discarded.

The resulting supernatant is applied to a 2.5 × 10-cm column of DEAE-Sepharose® (suitable for ~1000 mL of solubilized membranes) that has been equilibrated with 20-m*M* potassium phosphate buffer (pH 7.4) containing 10-m*M* 2-mercaptoethanol, 20% glycerol (v/v), and 1% CHAPS (w/v), and 10-mL fractions are collected. After all of the sample has been applied, the column is washed with ~2 bed volumes of the equilibration buffer and the fractions containing red color (P450) are pooled. KCl (solid) is added to the pooled fractions to achieve a 0.5-*M* concentration. A spectral assay is done [100] to estimate recovery, and it may be useful to also monitor the progress of purification at this point by SDS–polyacrylamide gel electrophoresis [245].

The pooled material is applied directly to a 1.5 × 5-cm column of Ni-NTA agarose that has been equilibrated with 20-m*M* potassium phosphate buffer (pH 7.4) containing 20% glycerol (v/v), 0.5% CHAPS (w/v), 0.5-*M* KCl, and 10-m*M* 2-mercaptoethanol, at a flow rate of ~1 mL/min. Most of the brown color should be adsorbed to the column. The column is washed with 10 column volumes of the equilibration buffer. The column is then washed with 10 bed volumes of 20-m*M* potassium phosphate buffer (pH 7.4) containing 20% glycerol (v/v), 0.5-*M* KCl, and 10-m*M* 2-mercaptoethanol. The protein (P450) is then eluted from the column with 20-m*M* potassium phosphate buffer (pH 7.4) containing 20% glycerol (v/v), 0.5-*M* KCl, 10-m*M* 2-mercaptoethanol, and 400-m*M* imidazole, collecting 5-mL fractions. The pH of the elution buffer is rechecked after addition of the imidazole and is adjusted to 7.4 with 43% (or more dilute) $H_3PO_4$. The fractions are analyzed for $A_{417}$ and by SDS–polyacrylamide gel electrophoresis, taking care to dilute samples to avoid potassium dodecyl sulfate precipitates. The P450 fractions (as judged by $A_{417}$) that are homogeneous (>95%) as judged by SDS–polyacrylamide gel electrophoresis are pooled and dialyzed extensively (4×, >6 hours each time) with 100-m*M* potassium phosphate buffer (pH 7.4) containing 20% glycerol (v/v), 0.1-m*M* EDTA, and 0.1-m*M* DTT, either before or after concentration with an Amicon® ultrafiltration system and a PM-30 membrane. The concentration of P450 is estimated spectrally [100].

## NADPH–P450 Reductase

Affinity procedures have been described by Yasukochi and Masters [246] and Strobel and Dignam [247] for purification of this enzyme from rat and hog liver using detergent extraction from microsomes, DEAE-cellulose chromatography, and 2′,5′-ADP- or NADP+-agarose affinity chromatography. 2′,5′-ADP-agarose is commercially available from Pharmacia. Ion-exchange chromatography, or some other initial purification procedure, is often used to facilitate binding of the reductase to the affinity column [246]. Alternatively, the *n*-octylamino–Sepharose® 4B reductase fraction from rabbit, rat, or human liver P450 purification procedures [120,248–253] can be directly applied to the affinity column. In our own laboratory, we now produce the rat or human reductase in *Escherichia coli* and isolate it from the membranes, using the methods of Hanna et al. [254]. These expression vector systems are freely available upon request.

The column is washed with 0.25-*M* potassium phosphate buffer (pH 7.7) containing 0.10-m*M* EDTA, 20% glycerol (v/v), and 0.20% Triton® N-101 (w/v) (or other nonionic detergent) to remove other proteins. The detergent is then removed by washing the column with 30-m*M* potassium phosphate containing 0.10-m*M* EDTA, 20% glycerol (v/v), and 0.10% sodium cholate (w/v). The reductase is eluted with the latter buffer containing 10-m*M* 2′-AMP or NADP+ (and 0.10-m*M* phenylmethylsulfonyl fluoride) and dialyzed (48 hours) vs. 100 volumes of 10-m*M* Tris-acetate buffer (pH 7.4) containing 0.10-m*M* EDTA and 20% glycerol (v/v) to remove cholate and 2′-AMP (or NADP+) [48,250,255].

This procedure has been used to obtain NADPH–P450 reductase in yields as high as 50%; specific activities for cytochrome *c* reduction range from 40 to 70 mmol/min/mg protein [246,247,255]. Spectra show the absence of nonflavin components; the ratios of $A_{455}$ to $A_{380}$ range from 1.10 to 1.15 [250,255]. The apparent monomeric $M_r$ is 74 kDa (the protein sequence has been deduced from cDNA and the actual mass is somewhat different [246]). Sometimes, proteolysis is a problem and results in cleavage of a peptide necessary for activity toward P450 (but not cytochrome *c*). This problem can be avoided by adding phenylmethylsulfonyl fluoride (from a stock ethanolic solution) to 0.10 m*M* to buffers immediately prior to use to inhibit serine-active proteases. Some preparations appear to lose some flavin mononucleotide (FMN), as evidenced by stimulation by 10-μ*M* FMN. This problem can be minimized by including 1 mM concentrations of FMN in buffers and minimizing exposure to light during dialysis [251].

# METHODS FOR USE OF ENZYMES

## Estimation of Protein Concentration

Estimates of protein concentrations are useful in a number of settings, ranging from standardization of crude assays for the purpose of repeatability to precisely establishing the amount of an enzyme being used in a reaction. In the course of monitoring proteins during purifications, following the absorbance at 280 nm is a reasonable approach; however, even this can be a nontrivial issue if absorbing materials are present in buffers; for example, alkyl phenyl ether detergents (e.g., Triton®, Emulgen™) are problems. At least five methods are used to estimate protein concentrations in individual samples. The first three are colorimetric assays and are used widely.

The Lowry method is based on the use of Folin–Ciocolteau reagent [153] and yields a blue color, based on reactions with tryptophan and tyrosine (which still remains vague after all these years). A detailed protocol was published in this chapter in the previous edition [4]. The assay is reasonably rapid and sensitive. A major limitation is the sensitivity to interference by contaminating materials (e.g., detergent, buffer salts). The biurette assay is an older method, using complexation of $Cu^{2+}$ ions with the amide bonds. Compared with the Lowry assay, this method has less interference by buffer anomalies but is also much less sensitive. The Bradford method involves a color change of the dye Coomassie® Brilliant Blue G upon binding to protein [256]. This assay is convenient and has sensitivity comparable to the Lowry method; however, as might be suspected, any extraneous hydrophobic material will interfere (especially detergents).

Another method, the bicinchoninic acid (BCA) assay [257], is even more sensitive than the Lowry method and generally has fewer interference problems. The reagents can be purchased (Pierce; Rockford, IL), and the manufacturer provides a detailed description of the method. It is technically simpler than the Lowry assay in that only a single reagent is used and the timing of additions is less critical; this makes the procedure more adaptable in microassay systems. The sensitivity is increased at higher temperatures (e.g., 50°C), and detection of submicrogram amounts of protein is not unrealistic.

All of the above four systems are based on phenomena that are not well understood and, even if the basis were established, a real issue is the protein-to-protein variability of the colorimetric response. Thus, an assay used to establish the concentration of a protein by one of these methods can often show twofold error. This inaccuracy may not be an issue in some settings but could be critical in others, such as establishing the stoichiometry of binding of a substrate or prosthetic group or the extent of a kinetic burst measurement [258].

Two methods are accurate when the concentration of an individual protein must be known with great accuracy. One approach is to calculate the expected extinction coefficient ($E^{1\%}_{260}$) from the content of tryptophan and aromatic amino acids in the sequence. The websites http://www.biomol.net/en/tools/proteinextinction.htm and http://www.basic.nwu.edu/biotools/ProteinCalc.html can be utilized. This approach can provide a reasonable value. Extinction coefficients at the lower ultraviolet wavelengths (i.e., 210 and 215 nm) are based on the amide bonds and might be expected to be accurate. They are more sensitive [259]; however, the far ultraviolet region is also very sensitive to interference by contaminants, including some buffer components.

The other accurate method is probably the best, although it is more laborious and expensive enough to restrict use to critical applications. The method is a throwback to a method referred to as *quantitative amino analysis*, which is used in protein chemistry. The protein is carefully hydrolyzed (usually 6-*M* HCl, 24 hours at 110°C) to its component amino acids. These are derivatized, usually with phenylisothiocyanate, and analyzed by HPLC. The amounts of each amino acid are determined by comparisons with external standards processed in the same way (using the integrals of the $A_{254}$ peaks). Alternatively, the derivatization can be done with a fluorescent reagent to increase the sensitivity. The hydrolysis yields of tryptophan and cysteine are not ideal, but the deviation due to these rarer amino acids is usually not significant enough to limit the application. The amount of the protein can be calculated from knowledge of the amounts of each amino acid and the known sequence of each protein [258]. In one application, three component DNA polymerase δ protein subunits were resolved by SDS–polyacrylamide gel electrophoresis and the amounts of each were determined by quantitative amino acid analysis [260].

## Methods for the Determination of Enzyme Purity

As in the case of other chemicals, no single technique can be used to establish purity; moreover, purity is always defined as a limit of given impurities in a given analytical system, and one can argue that nothing is really pure; however, some techniques are more useful than others in ruling out heterogeneity, and they are discussed here. Some of the classical considerations about purity have been relaxed by the ability to express in recombinant systems, in that related enzyme family members are not expressed. For some purposes, though, proteins must still hold to strict criteria (e.g., antibody production, certain enzymology experiments):

1. An obvious criterion of homogeneity is the absence of suspected contaminants; for example, P450 preparations should be devoid of NADPH-cytochrome *c* reduction activity, and epoxide hydrolase preparations should be devoid of heme. Of course, such impurities must always be defined in terms of detectable limits of contamination. A point to be considered here is that an experimental situation may call for the absence of lipid or detergent contamination as well as protein contamination, and the limit of such impurity must be determined. Phospholipid can be determined by thoroughly dialyzing the enzyme vs. Tris buffer and then $H_2O$ to remove soluble phosphates; lipids are extracted as described by Bligh and Dyer [261], and phosphate is determined according to Chen et al. [262]. Ethylene-oxide-based detergents (including Emulgen™ 911 and 913, Renex™ 690, Triton® N-101, and Lubrol® PX) can be extracted and assayed as described by Garewal [263] and subsequently modified by Goldstein and Blecher [264]. Alternatively, detergents can also be assayed by HPLC [240].
2. Specific activity or specific content of the isolated enzyme is a guide to follow in purification; for example, P450 preparations should contain *x*

nmol of P450 per mg protein, where $x = 10^6 \div$ subunit $M_r$ (i.e., $x$ = 16–22), and NADPH–P450 reductase preparations should catalyze the reduction of 40 to 70 mmol cytochrome *c* per min per mg protein under optimal conditions. However, such measurements are dependent on the accuracy of the protein estimation, which may be a problem. Specific activities are sometimes more variable than expected. In the case of P450, some forms may exist *in vivo* without a full complement of heme [265] (these are issues regarding prosthetic group loss that involve expressed proteins as well as those purified from tissues). These general guidelines about activity, etc. are useful when evaluating purity.

3. SDS–polyacrylamide gel electrophoresis [245] is routinely used as the main criterion for homogeneity. This technique has been quite useful in determination of homogeneity; moreover, the subunit $M_r$ estimates appear to be reasonably valid for P450 [113,120,266] and NADPH–P450 reductase [65,267]. The reader should remember, however, that even this powerful technique has its limitations. Evidence has been presented that different microsomal enzymes cannot always be distinguished by this technique [48,268]. Furthermore, the results obtained with this technique are rather dependent upon the exact procedure used, and the methods vary in resolving abilities [48,120]. Finally, apparent $M_r$ values also vary depending on the procedure and the standards used, and the reader is cautioned to compare results from different laboratories carefully and allow for as much as 3- to 4-kDa differences. Further, it is well established that in some cases such a change may result from a single amino acid replacement regardless of the effect on the true $M_r$. Isoelectric focusing offers a great potential for resolution of enzymes and has been used in studying cytochromes P450; however, a number of artifacts can be encountered in the use of this methodology, at least when used without specialized procedures [203,269–272]. Staining of electrophoretograms is usually done with protein stains. In the absence of SDS, NADPH–P450 reductase can be stained using tetrazolium dyes [273], and P450 can be stained using benzidine derivatives and $H_2O_2$ [274,275]. The latter method has also been used to tentatively identify P450 in SDS–polyacrylamide gel electrophoresis, although much of the heme leaves the enzyme, even in the absence of reducing agents. While others have claimed that heme does not bind to other proteins, Thomas et al. [275] found that heme was bound to albumin; thus, one must be cautious in interpreting data involving such a technique. Immunochemical techniques have been developed for the identification of P450 separated from microsomal membranes by SDS–polyacrylamide gel electrophoresis; these methods proved useful in answering a number of questions about P450 induction.
4. N-Terminal amino acid analysis should produce single residues at each step of automated Edman degradation for homogeneous proteins [266,276]. There has been considerable development in this area, and it is now common to be able to do this work with samples of <10 pmol [277,278].
5. A variety of immunochemical criteria have been used to assess homogeneity. An antibody should produce a single precipitin line when diffused against the antigen, if the antigen is homogeneous. Immunoelectrophoresis of a crude preparation should show only the antigen. Cross-reactivity of related proteins does exist, and many of the isolated microsomal and even recombinant proteins do not necessarily induce the production of monospecific antibodies [216].
6. Hydrodynamic criteria have been used in the past to assess the homogeneity of the various isolated microsomal proteins; because all of these proteins tend to aggregate, velocity and equilibrium studies carried out to examine homogeneity must be done in the presence of detergents or other strong denaturants. These techniques are not very commonly used, but they are of great value in ascertaining aggregation properties.

## Reconstitution of Enzyme Activity

Early work on reconstitution of mixed-function oxidase activity was reviewed by Lu and West [111]. The following general statements can be made. The optimum rate of enzyme activity, based on P450, is obtained when NADPH–P450 reductase is present at an equimolar concentration or slight excess. Phospholipid enhances the rates of most activities; this phospholipid can be in the form of a microsomal extract or synthetic L-α-1,2-dilauroyl-*sn*-glyceryo-3-phosphocholine. Some nonionic detergents will partially replace phosphatidylcholine at low concentrations [279–281]. The activity toward some substrates can be further enhanced by small amounts of cholate or deoxycholate [111]. The role of phospholipid is not completely understood, but a dual role has been postulated [282]: L-α-1,2-dilauroyl-*sn*-glyceryo-3-phosphocholine increased the affinity of a rabbit liver P450 for *both* organic substrate and NADPH–P450 reductase; all four components are complexed during catalysis. Several investigators have studied synthetic liposomal systems;

however, no such system has been prepared to date that is more active in hydroxylation than a system reconstituted in the presence of subcritical micelle concentration levels of phospholipid. Neither cytochrome $b_5$ nor NADH is required for activity in many reactions, although some are definitely enhanced considerably by cytochrome $b_5$. The effect of the phospholipid appears to be kinetic and can, at least in some cases, be overcome with high protein concentrations and extended preincubation conditions [42]; however, see also Halvorson et al. [43] and Imaoka et al. [44].

A basic procedure for reconstituting mixed-function oxidase activity is outlined below. Equimolar concentrations of P450 and NADPH–P450 reductase (both of which have been stripped of excess detergent by treatment with beads or calcium phosphate gel or hydroxylapatite [94]) are first mixed in the presence of 40-m*M* sonicated L-α-1,2-dilauroyl-*sn*-glyceryo-3-phosphocholine, and, after 5 minutes, an appropriate buffer (i.e., 50- to 100-m*M* potassium phosphate or other buffer, pH 7.0 to 7.7) is added plus a sufficient volume of water. The substrate is then added, preferably in water if possible. If not, the substrate should be dissolved in acetone, DMSO, or $CH_3OH$ such that the final concentration of organic solvent is $\leq 1\%$ (v/v) for the enzymes. Some forms of P450 oxidize these compounds or are inhibited by them (e.g., P450 2E1), and caution must be exercised [144,283]. The system is pre-equilibrated at 37°C for 3 to 5 minutes, and the reaction is initiated by the addition of NADPH (0.15 to 0.5-m*M*) or, preferably, an NADPH-generating system. (The organic substrate should not be the last addition, as prior addition of NADPH will result in generation of $H_2O_2$, which can destroy P450 [255].) The length of the time for which the reaction is linear depends on the substrate; in general, rapidly metabolized substrates do not give long periods of linearity, and some substrates are converted to metabolites that destroy P450 rapidly.

Flavin-containing monooxygenase activity requires only the single protein, of course, but the enzyme is dramatically less heat stable than others [284–286]. The enzyme is stabilized by pyridine nucleotides, so NADPH should be added to the enzyme before incubation at 37°C, and then the organic substrate should be added. NADPH–P450 reductase has other enzyme activities in the absence of P450; many of these require no special conditions but may be stimulated, as is the case for lipid peroxidation, by high salt concentrations.

### Use of Selective P450 Inhibitors

Chemical inhibitors offer considerable potential in the discrimination of enzymes involved in reactions. Unlike antibodies, they can readily be obtained by many laboratories by chemical synthesis or, in many cases, purchased directly. Further, they have the advantage that they can be used *in vivo* in many cases, even in humans. The approach of using chemical inhibitors with crude enzyme preparations to discern catalytic specificity has been most highly developed for the P450 enzymes (Figure 40.28). P450 inhibitors have been known for some time but many of the early compounds, such as SKF-525A (*N,N*-diethylaminoethyl-2,2-diphenyl valerate) and metyrapone, are only partially selective. Newer and more specific compounds are now available. A list of those used for human P450 enzymes is presented in Table 40.3 [30,56].

A few comments are in order. First, the specificity tends to carry over within each enzyme family (1A, 2A, 2B, 2C, ...) between species, although some differences can be noted; for example, quinine is a better inhibitor of rat P450 2D enzymes than its diastereomer quinidine, while the opposite is true in humans. Some inhibitors are highly effective with more than one enzyme (e.g., diethyldithiocarbamate). Some inhibitors are irreversible and act by modifying the protein or prosthetic group. Ketoconazole is able to inhibit several P450 enzymes, but many of its *in vivo* effects can probably be attributed to P450 3A4. The same is probably true of cimetidine [287].

## IMMUNOCHEMICAL TECHNIQUES

Antibodies have been raised to P450, NADPH–P450 reductase, epoxide hydrolase, and many of the other enzymes considered here; for example, these antibodies have been used to show the involvement of P450 [213–215,288–293] and its reductase [273,294] in a number of reactions. Antibodies have also been used to examine the homogeneity of isolated enzyme fractions, the multiplicity of enzymes in microsomes, and the amounts of individual enzyme forms in microsomal preparations [47,48,268,288,293,295–299]. The topical location of the enzymes in microsomal membranes has also been studied with immunological techniques [214,288,300], as have several aspects of enzyme biosynthesis. Antibodies have also been used to study the localization of the enzymes in various sections of individual organs [301,302].

### Preparation of Antibodies

All three of the above enzymes are rather antigenic, and antibodies have been raised in rabbits using less than 50 mg of protein. Sheep, goats, and guinea pigs have also been used for antibody production. A number of different immunization schedules can be used, depending on the animal and the dose. Antisera can be used in some procedures, but immunoglobulin G (IgG) fractions are necessary in some applications [215,288]. To prepare these fractions, antisera are heated 20 minutes at 56°C and centrifuged at $10^4$ *g* for 10 minutes. The supernatants are mixed with equal volumes of 50% $(NH_4)_2SO_4$ (w/v) and recentrifuged. The pellets are then washed with 25%

$(NH_4)_2SO_4$ (w/v) to remove most of the color and then dissolved in 10-m*M* potassium phosphate buffer (pH 8.0) and dialyzed against the same buffer (at 4°C). The dialysates are passed through columns of DEAE–cellulose equilibrated with the same buffer. The void volume fractions (measured by absorbance at 280 nm) are pooled, retreated with $(NH_4)_2SO_4$ as above to remove color if necessary, concentrated by ultrafiltration (up to 50 mg/mL), and stored at –20°C. A number of alternative procedures are available that use, for example, Protein A affinity columns.

An alternative approach to using the purified protein of interest is to utilize peptides to generate antibodies. This approach has become more popular with P450s, as well as many other enzymes, for several reasons. First, the intact protein does not have to be expressed or isolated, although it may be necessary for use as a standard in either defining specificity or quantifying the protein. Second, the ability to use only part of the protein provides an opportunity to choose sequences that are specific for a particular protein compared to closely related forms. Some caution (and experimental work) is necessary because antibodies can show reactivity with less closely related proteins; for example, some antibodies recognize P450 proteins from multiple gene families [216]. Finally, the purification of peptides is generally more straightforward than proteins, in that they are smaller and can be readily subjected to methods such as reversed-phase HPLC with organic solvents and acidic conditions; thus, the prospects of including contaminating antigens are fewer. Many commercial vendors supply peptides already purified. The quality of these peptides is important, and the purity and identity can be readily confirmed by methods such as capillary electrophoresis and matrix-assisted laser desorption ionization/time-of-flight mass spectrometry. Peptides are generally conjugated to a carrier protein (e.g., keyhole limpet hemocyanin) before injection into animals.

Another approach involves the production and use of monoclonal antibodies. This methodology, originally developed by Milstein [303], involves initial injection of animals and utilization of spleen cells, either in *in vitro* or in intact animals for the hybrodoma production. An advantage of using a monoclonal antibody is that it is only one of a complex mixture of antibodies that normally is generated in the polyclonal response; therefore, no problems should result from the variability one often sees with polyclonal antibodies (i.e., with variation in the individual antibodies produced among different animals and at different times) [216].

Another approach avoids the need to use animals and has other advantages. Large phage display libraries exist (e.g., $10^9$ antibody molecules). These systems are bacteria based and can be readily screened; for example, blots containing the protein of interest (e.g., a particular P450 and other proteins that one wishes to avoid recognition of) can be used with such a library. The phage that selectively binds can be recovered and cultured in bacteria. Recovery can be expedited using epitope tags. Further, such approaches can be used to select antibodies using unpurified systems; for example, one can screen an existing phage display library for antibodies (or, more correctly, $F_c$ chains) with a set of cells that is hypothesized to contain a specific protein and another set that is not. If a difference is observed, in terms of which $F_c$ chains are selected, then that $F_c$ chain can be recovered and used as a reagent to either monitor purification of the protein or to immunopurify the protein directly.

The list of immunochemical techniques that can be used with such antibodies is quite lengthy and the reader is referred to texts on the general subject [304]; the procedures include double-diffusion analysis, radial diffusion quantitation, inhibition of enzyme activity, complement fixation, radioimmune assay, crossed gel electrophoresis, immunoprecipitation, immunoaffinity column chromatography, and immunohistochemical localization. These techniques should continue to be useful in future studies of the roles of individual forms of these microsomal enzymes in various processes.

### Immunoinhibition of Catalytic Activity

In this approach one adds an antibody to an enzyme preparation and determines if that antibody preparation can block catalytic activity (Figure 40.32). This approach is most useful with a crude enzyme system, such as a subcellular organelle preparation. Thus, one can ask what fraction of total activity is the result of the enzyme specifically recognized by the antibody (if the antibody completely inhibits the activity of the antigen itself). Such an approach has been useful in a number of cases in our own laboratory [203,213,305,306], as well as in many others.

Antibodies to some proteins tend to be more inhibitory than others. In general, polyclonal antibodies raised against cytochromes P450 are usually inhibitory; however, inhibitory antibodies have not been reported for microsomal flavin-containing monooxygenase and only occasionally has inhibitory anti-epoxide hydrolase been prepared [307]. Apparently, antibodies raised against GSTs are not inhibitory. Some of the difference may be due to the size of the substrate or accessory protein that interacts with the antigen in catalysis. Thus, one would expect binding of an antibody to block binding of very large substrates (namely, other proteins) more readily than smaller compounds. In support of this view, anti-NADPH–P450 reductase blocks reduction of cytochrome *c* but not ferricyanide or neotetrazolium blue [273]. Another general trend is that only a limited fraction of monoclonal antibodies are inhibitory [308,309].

Analyzing for antibody inhibition is a relatively straightforward process. In general, the enzyme preparation of interest is mixed with the antibody and incubated for 20 minutes at room temperature. Other components are then added, and catalytic activity is measured in the usual manner. A good way to properly assess enzyme inhibition is to run several incubations, varying the amount of antibody and holding the amount of enzyme constant. Parallel assays should be done in which a nonimmune antibody preparation, prepared in the same way, is added at the same levels to the enzyme preparation of interest (Figure 40.32). (Alternatively, one can mix varying ratios of immune and nonimmune antibodies with each aliquot of enzyme, maintaining a constant *total* amount of antibody added.) Most assays of this type are done with immunoglobulin G (IgG) antibody fractions. Serum and ascites fluid contain other materials that can cause nonspecific inhibition, but, if the antibody titer is very high (with regard to inhibition) or if the catalytic assay is so sensitive that little antibody is needed for inhibition, then such crude materials may be used.

In general little can be said about inhibition <15% of the total unless enough careful replicates are done and the difference between immune and nonimmune serum incubates is reproducible, concentration dependent, and statistically significant. To the first approximation the percentage of inhibition is a reflection of the fraction of the total catalytic activity in the preparation that is due to the protein that reacts with the antibody. The antibody should completely inhibit the purified enzyme itself; however, for this analysis to be valid, the possibility exists that noninhibitory antibodies may hinder the binding of inhibitory antibodies and total inhibition may never be achieved.

## Quantitation of Proteins by Immunoblotting

In many cases, the absolute concentration of a particular protein in a sample is derived apart from its catalytic activity. The most direct way to obtain such measurements is with the use of specific antibodies. A variety of immunochemical techniques is available for use, including various types of radioimmunoassays (RIAs) and enzyme-linked immunosorbent assays (ELISAs) [310]; however, knowledge concerning the specificity of the antigen–antibody reaction must be available. Probably the single most reliable technique for evaluating specificity is coupled SDS–polyacrylamide gel electrophoresis/immunoperoxidase staining, or immunoblotting (which often goes by the slang term *western blotting*), where a crude mixture of protein is separated by electrophoresis, and the resolved proteins are transferred to a thin sheet of nitrocellulose paper, where they can be detected after binding antibodies and antibodies coupled to enzymes with chromogenic substrates. In our early studies with this system, we found that the intensity of the staining of protein bands was proportional to the amount of antigen applied and that such a procedure could be utilized in making quantitative measurements (Figure 40.34). We still continue to use such a system to quantify many proteins, for several reasons. Under appropriate conditions, the method is accurate and quite sensitive. It provides a check on the specificity of antigen–antibody interaction in each individual antigen sample and provides data even when cross-reactive materials are present (if they can be resolved in a single electrophoretic dimension). The method is relatively rapid and straightforward, and even when new systems are explored little optimization is required.

Samples of roughly 5 μg of microsomes or cellular homogenate protein are solubilized by heating with SDS and 2-mercaptoethanol. The samples are electrophoresed in a typical system based on the procedure of Laemmli [245]—a slab gel is used with up to 25 samples. Five or six lanes are used to prepare a standard curve for each gel. The lanes contain, for example, 0.5, 1, 2, 3, 5, and 10 pmol of the purified antigen. Crude protein samples to be analyzed are loaded into the wells for the other lanes. Typically, 1 to 50 μg of microsomal protein might be loaded per well for analysis of P450 proteins. Protein samples are dissolved in a mixture of 63-m*M* Tris-HCl buffer (pH 6.8) containing 10% glycerol (v/v), 1.0% SDS (w/v), 0.001% pyronin Y (w/v), and 5.0% 2-mercaptoethanol (v/v) and are heated for 60 seconds at 95°C. Aliquots are loaded into the wells of a 1.5 mm × 16 cm × 20 cm gel (e.g., Bio-Rad; Richmond, CA). The separating gel is poured from a mixture of 0.375-*M* Tris-HCl buffer (pH 8.8) containing 7.5% acrylamide (w/v), 0.03% tetramethylethylenediamine (TEMED) (w/v), 0.10% SDS (w/v), and 0.0425% $(NH_4)_2S_2O_8$ (w/v). The stacking gel, in which the wells are formed, is poured from a mixture of 0.14-*M* Tris-HCl buffer (pH 6.8) containing 3.5% acrylamide (w/v), 0.057% TEMED (v/v), 0.65% sucrose (w/v), 0.11% SDS (w/v), and 0.045% $(NH_4)_2S_2O_8$ (w/v). The electrode buffer (pH 8.3) contains 190-m*M* glycine, 25-m*M* Tris, and 0.10% SDS (w/v). Power is applied to the system at a constant current setting of 25 mA per gel to move the samples through the separating gel. The electrophoresis takes ~4 hours. When the pink dye front has moved to within about 1 cm of the edge of the gel, the power is turned off and the system is separated. One of the glass plates on the gel is removed, and water is sprinkled on the surface. A wetted piece of nitrocellulose paper (0.45 mm; Scheicher and Schull, Keene, NH) is laid over the wet gel. Care should be taken (and enough water used) to avoid trapping air bubbles. Two sheets of Whatman #3 paper, prewetted with water, are laid over the nitrocellulose. The glass plate is removed from the other side of the gel and replaced by a wet sheet of Whatman #3 paper. The entire sandwich is placed between two wet sponges and then enclosed between the two electrode baffles of an electrotransfer apparatus, with the nitrocellulose closer to the anode than the cathode. The apparatus, with the gel

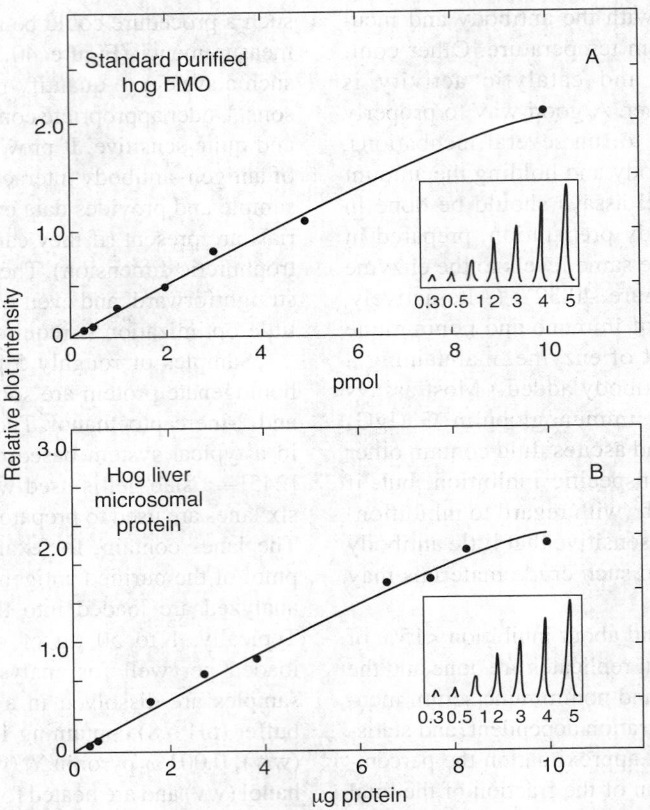


**FIGURE 40.34** Immunoelectrophoresis and densitometry of flavin-containing monooxygenase (FMO) in purified samples and porcine liver microsomes [268]. (A) Area under the densitometric peak as a function of the amount of purified porcine liver FMO used for electrophoresis; the inset shows the actual densitometric traces. (B) Area under the densitometric peak as a function of the amount of porcine liver microsomal protein used for electrophoresis; the inset shows the actual densitometric traces.

and nitrocellulose between the baffles, is filled with 25-m*M* Tris-HCl buffer (pH 8.2) containing 190-m*M* glycine and 20% $CH_3OH$ (v/v). Constant current (400 mA for 1 hour or 200 mA for 2 hours) is applied to the system. If a commercial electroblotting apparatus is not available, household sponges (e.g., Brillo®; Purex Corp., Lakewood, CA) and a pair of stainless-steel plates (attached to electrodes) can be substituted, with the device held by rubber bands and immersed in a beaker. Satisfactory results can be obtained [311,312], although the cathode plate tends to pit and corrode, especially if not washed thoroughly.

After the blotting operation, the polyacrylamide gel and the filter papers are discarded, and the nitrocellulose sheet is placed in 25 mL of a solution of phosphate-buffered saline (PBS; 10-m*M* potassium phosphate buffer, pH 7.4, containing 0.9% NaCl [w/v]) containing 0.50% Tween® 80 (w/v). The sheet is conveniently placed in a plastic box of only slightly larger dimensions (13 × 18 × 3 cm) with a lid. The gel is shaken in a 37°C water bath for 30 minutes to block reactive sites on the nitrocellulose sheet with serum proteins so antibodies will not be bound in subsequent steps. After the blocking step, the nitrocellulose sheet is washed twice with 60 mL of PBS at room temperature. In practice, the box is rocked on a platform rocker (Bellco Glass, Vineland, NJ) for 5 minutes, the buffer is decanted, and 60 mL of fresh PBS is added each time.

In the next step, 25 mL of PBS containing 0.50% Tween® 80 (v/v) and an appropriate dilution of the antiserum of choice is poured into the box over the nitrocellulose sheet. The system is shaken (or rocked) at 37°C for 30 minutes and then overnight at 4°C; alternatively, 37°C for 2 hours is usually satisfactory. Typical antisera dilutions range from 1/100 to 1/2000 (some monoclonal antibodies have been successfully used at $1/10^6$ dilutions of ascites fluid). The nitrocellulose sheet is then washed six times (5 minutes each, room temperature) with 60 mL of PBS. The next addition is 25 mL of PBS containing 0.50% Tween® 80 (w/v) and 0.20% (v/v) goat anti-rabbit IgG antiserum (if the primary antiserum was made in rabbits). This solution is rocked or shaken with the nitrocellulose sheet at room temperature for 30 minutes, and the sheet is then washed again six times with PBS as before. The next addition to the sheet is 25 mL of a solution of PBS containing 0.50% Tween® 80 (w/v) and 0.20% (v/v) horseradish peroxidase–rabbit anti-horseradish peroxidase complex (Miles Laboratories,

Elkhart, IN). The sheet is rocked in this solution at room temperature for 30 minutes and washed six times with PBS as before.

Development of the stain is done in the following manner. 4-Chloro-1-naphthol (32 mg) is dissolved in 12 mL of $CH_3OH$ and diluted with 60 mL of PBS. $H_2O_2$ (120 mL of a 30% solution) is added, and the solution is poured over the nitrocellulose sheet; bands usually appear within a few minutes. The solution is removed; the nitrocellulose sheet is washed three times with PBS and twice with $H_2O$. Sheets can be dried between two layers of Whatman #3 filter paper, with a uniform weight applied. When the gel is dry (within 1 to 2 hours), the bands can be scanned using a densitometer. The integrals are used to construct standard curves and estimate the amount of antigen in each sample (Figure 40.34).

In practice, a standard curve is constructed on each nitrocellulose sheet. An additional way to reduce error is to include an internal standard in each protein sample. Equine alcohol liver dehydrogenase can be used for this purpose, adding 0.2 mg to each sample prior to electrophoresis. The buffer containing the primary antisera is fortified with a 1/500 dilution of rabbit antisera raised against equine alcohol dehydrogenase. When the nitrocellulose sheets are visualized, the P450 band in the 50- to 60-kDa region is accompanied by a second band migrating with apparent $M_r$ of 43 kDa. The integrals of both bands are obtained from the densitometer. The ratio of the areas of the two bands can be compared to the ratios found with the standard antigen samples.

Even if the antigen–antibody system is not specific enough to visualize only a single electrophoretic band, useful information can be obtained if the different antigens are electrophoretically separable. For example, rat P450s 1A1 and 1A2 usually cross-react but can be separated and quantified [47,313]. The same situation exists with human P450 2C enzymes [306].

Since the original immunoblotting work was done [312], a number of variations of the procedure have been reported. Many of these are cited in subsequent reviews [314,315]; for example, different additives can be used in the buffers for blocking the sheets. Nylon membranes, such as Zeta-Probe™ (Bio-Rad; Richmond, CA), have increased capacity and can be used to increase the sensitivity of the methods. *Staphylococcus aureus* Protein A conjugates can be bound to the primary antibody. Other enzymes such as alkaline phosphate can replace peroxidase; alternatively, $^{125}$I-labeled antibodies can be used with autoradiography, as described in the original Towbin et al. paper [312]. Another alternative is the use of luminescent substrates, which are much more sensitive. These can be visualized with autoradiography sheets. If monoclonal antibodies are used, the methods must be adapted by including a step with rabbit anti-mouse immunoglobulin G [308,309]; many monoclonal antibodies give poor responses in this system because the individual epitopes do not have appropriate affinity. In our own laboratory, we have applied this approach to rat and human microsomal epoxide hydrolase, rat NADPH–P450 reductase, several different forms of rat and human P450, and flavin-containing monooxygenase. The method can be utilized with cells [316] or tissue homogenates [268] as well as with subcellular organelles.

## RECOMBINANT DNA TECHNIQUES

### mRNA Isolation

Analysis of mRNA levels provides insight into regulatory mechanisms. In many cases, regulation of enzymatic activity occurs primarily at the level of transcription, and mRNA levels are well correlated with protein levels; however, this is not always the case, and notable exceptions have been documented with some P450 enzymes [317,318]. Techniques of mRNA isolation and handling have become routine and may be mastered without considerable difficulty. Moreover, the generation of highly specific probes can be easily achieved by synthesis of oligonucleotides or long probes developed by polymerase chain reaction (PCR) technology, in contrast to the production of antigens and antibodies for analysis of protein levels. The most widely used procedure for RNA isolation is that of Chomczynski and Sacchi [319]. It is rapid, reliable, and useful with large numbers of samples. First and foremost, one of the critical aspects of working with RNA is to avoid RNase. Any traces of this protein on glassware, dust, or fingertips will be devastating. Disposable gloves must be worn during all steps, and all glassware and plasticware should be treated as described below and stored specifically for RNA procedures. Several commercial kits for total RNA isolation are currently available based on the guanidine method. These kits do offer the ease of ready-made reagents or a single mono-solution combining all the initial reagents but at much higher costs.

*Reagents for RNA Isolation*

1. Stock guanidine thiocyanate—250 g of guanidine thiocyanate (Fluka™) is dissolved (at 65°C) in a mixture of 293 mL $H_2O$, 18 mL of 0.75 sodium citrate buffer (pH 7.0), and 26 mL of 10% sarcosyl (w/v). This solution can be stored ≥3 months at room temperature.
2. Denaturing solution—0.36 mL of β-mercaptoethanol is added per 50 mL of the solution just described. Add before use; can be stored 1 month at room temperature if necessary.
3. $H_2O$-saturated phenol—Nucleic-acid-grade phenol is saturated with $H_2O$. The solution may be stored for less than 1 month at 4°C. This may be purchased ready-made from a variety of sources.
4. 2-*M* sodium acetate (pH 4.0)

5. $CHCl_3$/isoamyl alcohol (49:1, v/v) or bromochloropropane [320]
6. 100% isopropanol
7. 75% ethanol

All glassware and plastic ware to be utilized should be previously treated with a solution of diethylpyrocarbonate (DEPC) and autoclaved, preferably in individually self-sealed pouches. Autoclaving does not inactivate all RNases. Glassware can be baked at 300°C for 4 hours. Some types of plastic ware may be washed with chloroform as an alternative to DEPC [321].

If possible, solutions should be shaken with 0.1% DEPC (w/v) and then autoclaved to destroy excess DEPC, which can react with nucleic acids. For chemicals that react with DEPC (e.g., Tris buffers) or that cannot be autoclaved, sterile filtration through 0.2-μm Nalge filters may reduce potential RNase contamination. Presterilized disposable pipettes, pipet tips, filter units, etc. may be used directly if individually wrapped as supplied. Whenever possible, all chemicals to be used in RNA work should be reserved for this purpose only and weighed out only with the use of DEPC-treated spatulas.

This procedure may be used for 100 mg of tissue or an equivalent amount of cells, $10^7$ cells [321]. The tissue is removed from the animal and minced on ice. If frozen material is utilized, the tissue should be ground with a mortal and pestle in liquid nitrogen. It is important to note that substantial RNA degradation may occur during this step. Homogenize the material in 10 mL of denaturing solution per g of tissue with either a Teflon®–glass homogenizer or a Tissuemizer® (Tek-Mar; Cincinnati, OH). Transfer homogenate to either a 25-mL glass Corex tube or a smaller polypropylene centrifuge tube. To this are added, sequentially, per 1.0 g of tissue, 1.0 mL of 2-*M* sodium acetate (pH 4.0; mix thoroughly by inversion), 10 mL of water-saturated phenol repeat mixing, and 2 mL of the $CHCl_3$–isoamylalcohol or bromochloropropane. The final mixture is mixed vigorously for 10 seconds and incubated on ice for 15 minutes. Centrifuge the suspension at $10^4$ *g* for 20 minutes at 4°C and transfer the aqueous (upper) RNA containing phase to a new tube. Precipitate the RNA by the addition of 1 volume of 100% isopropanol. Incubate for 30 minutes at –20°C and then centrifuge at $10^4$ *g* for 20 minutes at 4°C. Discard the supernatant. Redissolve the RNA pellet in 0.3 mL of the denaturing solution, and transfer to a microcentrifuge tube. Reprecipitate the RNA with 1 volume of 100% isopropanol for 30 minutes at –20°C. Centrifuge for 10 minutes at $10^3$ g and discard the supernatant. Resuspend this RNA pellet in 75% ethanol, and collect the pellet as before. Dry the RNA pellet under vacuum for 5 minutes. Redissolve the RNA pellet in 100 to 200 mL of DEPC-treated $H_2O$ and incubate 10 to 15 minutes at 55 to 60°C. Store the final RNA at –70°C. The total RNA concentration may be estimated by measuring the absorbance at 260 nm ($A_{260}$ = 1.0 for a solution of 40 μg/mL).

*Reagents for mRNA Preparation*

1. Oligo (dT)–cellulose, available from a number of suppliers
2. Binding buffer—10-m*M* Tris-HCl (pH 7.5), 0.5-*M* NaCl, 1-m*M* EDTA, and 0.5 % SDS (v/v); prepare a 2× binding buffer as well
3. Wash buffer—10-m*M* Tris-HCl (pH 7.5) containing 0.1-*M* NaCl and 1-m*M* EDTA
4. Elution buffer—10-m*M* Tris-HCl (pH 7.5) containing 1-m*M* EDTA
5. 3-*M* sodium acetate
6. Microfuge spin columns Ultra-free MC 0.45 mm filter units (Waters, Bedford, MA)
7. 100% ethanol

As described in the previous section, all reagents and glassware must be meticulously treated and subsequently handled to avoid contamination with RNases.

The basic procedure can be varied depending on the scale [322]. Oligo (dT)-cellulose (0.1 g/mg total RNA) is suspended in 1 to 5 mL of binding buffer and equilibrated for 60 minutes at room temperature with gentle agitation. The slurry is then pelleted and resuspended in fresh binding buffer, rewashed, and finally resuspended in binding buffer (50 mg/mL). The RNA is suspended at 1 mg/mL in elution buffer and heated at 65°C for 5 minutes. The sample is chilled on ice and diluted with an equal volume of 2× binding buffer. The RNA sample can be added to equilibrated oligo (dT)-cellulose from which excess binding buffer has been removed by a brief spin immediately prior to loading. RNA is incubated with the oligo (dT)-cellulose for 15 minutes at room temperature with gentle rocking. The sample can then be loaded into a DEPC-treated microfuge spin column. The flow-through (void) fraction should be collected by a brief spin and reapplied to the column, repeat. The column is washed with 5 to 10 volumes of binding buffer followed by 5 volumes of wash buffer. The bound RNA is eluted with 2 to 3 column volumes of elution buffer adjusted to 0.5-*M* NaCl with 2× binding buffer; the entire procedure is repeated (i.e., binding, washing, elution). RNA in the final sample is recovered via ethanol precipitation by adding 0.1 volumes of ethanol and centrifuging. We do not recommend reuse of the oligo (dT) resin.

## Northern and Southern Blotting

Specific DNA and RNA sequences can be detected by blotting and hybridization, referred to as Southern and northern blotting [323] (*northern* and *western blotting* are slang derived from use of a method developed by Southern [323] for DNA). Northern blotting differs from Southern blotting primarily in the initial gel fractionation step [324].

RNA molecules are single stranded and thus could form secondary structures. They must be electrophoresed under denaturing conditions for good separations to occur. Denaturation is achieved by the addition of formaldehyde to the gel and buffers or by treating the RNA with glyoxal and DMSO. Previous precautions mentioned for dealing with RNA apply here. All materials must be handled meticulously, and all glassware should be DEPC treated and baked. A gel tank should be set aside for RNA work. You should not utilize a DNA tank for RNA work.

*Reagents for Electrophoresis and Blotting [326]*

1. 5× MOPS (3-(*N*-morpholino) propanesulfonic acid) running buffer and 1× 0.2-*M* MOPS buffer (pH 7.0) with 50-m*M* sodium acetate and 5.0-m*M* EDTA
2. 37% formaldehyde (12.3-*M*; pH > 4)
3. Deionized formamide—If formamide has a yellowish color, it can be deionized by adding 5 g of mixed-bed ion exchange resin (e.g., Bio-Rad AG 501-X8) per 100 mL; stir 1 hour at room temperature and filter through Whatman #1 filter paper. Formamide is a teratogen; handle with care [324].
4. Loading buffer—1-m*M* EDTA (pH 8.0), 0.25% (w/v) bromophenol blue, 0.25% (w/v) xylene cyanol, 50% (v/v) glycerol; store up to 3 months at room temperature.
5. 1× SSC buffer (prepare 20×, 10×, and 0.25×, as well), 15-m*M* sodium citrate (pH 7.0), 150-m*M* NaCl
6. Ethidium bromide, 5 mg/mL
7. Prehybridization buffer—Mix 60 mL formamide, 3 mL sonicated solution of 10 mg/mL salmon sperm DNA, 12 mL of 100× Denhardt's solution (2.0% Ficoll® [w/v], 2.0% polyvinylpyrrolidine [w/v], and 2.0% bovine serum albumin [w/v]), 6 mL of a solution of 1 mg/mL polyA, 1.2 mL of 10% SDS (w/v), 30 mL of 20× SSC buffer, and 7 to 8 mL DEPC-treated $H_2O$.
8. SDS, 10% (w/v)

To pour a 20 × 20-cm gel, boil 3 g of agarose in 186 mL of DEPC-treated $H_2O$ until dissolved and then cool to 60°C. Add 60 mL of 5× MOPS buffer and 54 mL of 37% formaldehyde. Pour the gel into the gel box with the comb such that the slots are 1 mm above the horizontal surface. Once the gel has set, remove the comb and add enough 1× MOPS buffer to completely cover the gel.

Each sample should contain 2 µL of 5× MOPS buffer, 3.5 µL of 37% formaldehyde, and ≤20 µg RNA in 5 µL. Mix using a vortex, centrifuge in a microfuge for 5 to 10 seconds, and incubate 15 minutes at 55°C or 5 minutes at 65°C. Add 2 µL loading buffer to every sample and spin again for 5 to 10 seconds in a microfuge. Load samples onto the gel and run at constant voltage of 5 V/cm until the bromophenol blue band has migrated halfway to two thirds down the gel. It can be advantageous to run duplicate sets of gels. In this way, one gel may be stained with ethidium bromide and the other used for transfer. For staining, place the gel in a pan, cover with 20× MOPS, and add a few drops of the stock ethidium bromide solution. Incubate for 40 minutes and then examine gel on a UV transilluminator to visualize bands and photograph. Two sharp bands, the 18S and 28S rRNA, should appear if total RNA was used in the electrophoresis.

The gel to be transferred should be rinsed several times with $H_2O$. It is necessary to remove the formaldehyde, as it can hinder retention of RNA to nitrocellulose and nylon membranes [324]. Replace the $H_2O$ with 500 mL of 20× SSC. A piece of nitrocellulose is cut 3 mm smaller on each side than the gel. Wet this 1 minute in $H_2O$ followed by 5 to 10 minutes in 20× SSC buffer. Cut 5 sheets of Whatman 3MM paper 3 mm smaller on each side than the nitrocellulose. To prepare a wick, cut a piece of chromatography paper 2 cm larger than the gel with 4 tabs extending out 10 cm on each side. Place a glass plate, slightly larger than the gel, on top of a large buffer reservoir. Place one wick on every side of the glass plate, hanging down into the 20× SSC buffer reservoir. Remove any air bubbles trapped between the wick and the glass plate with a glass rod. At this point, carefully place the gel on top of the wicks, taking care to remove any air bubbles. Cover the four edges of the gel with either plastic wrap or parafilm to prevent wicking at the edges as the gel contracts. Place the nitrocellulose sheet on top of the gel, then cover the sheet with the Whatman papers, one at a time, taking care to remove air bubbles after the addition of each new sheet. Cover the stack with 5 cm of paper towels cut to the same size followed by another glass plate. Add a weight to hold everything in place. Transfer will occur overnight.

Remove all the materials to expose the nitrocellulose sheet. Mark the wells front to back. It is best to use pencil, as ink will wash off in subsequent washes. If nitrocellulose was used, place it between two sheets of Whatman paper and bake for 2 hours at 80°C. Nylon membranes may be baked as described, or wrap the dry membrane in UV-transparent plastic wrap, such as Saran™ wrap (polyvinylidene chloride), and place the RNA side down on a UV transilluminator (254 nm), and irradiate for the appropriate length of time. For hybridization, place the blot in a sealable bag with 1 mL formamide prehybridization solution per 10 $cm^2$ of membrane. Rotate the bag and incubate it at 37 to 42°C for 2 hours to overnight. An appropriate probe might be (1) nick translation of a cDNA fragment, (2) random labeling, (3) PCR, or (4) oligonucleotide synthesis followed by $^{32}P$-end labeling. For additional information, please consult the current literature. Longer probes with higher levels of incorporation are more sensitive, while shorter probes can be more specific when optimized for differences. The probe is mixed with hybridization solution and heated at 95°C for 10 minutes. Hybridization will occur overnight at the same temperature used for prehybridization. On the next day, open the bag or box and remove the filter. Wash the filter twice, 1 to 2 minutes each, with 2× SSC buffer at room temperature. Two similar washes follow, 45 minutes each, with 1× SSC containing 0.10% SDS (w/v). The hybridization can be changed by varying the temperature of the hybridization, or wash, and the salt concentration of the wash buffer to increase the stringency of the wash and the specificity of hybridization [327]. Expose the filter to film. The time required for an appropriate exposure varies with the specific activity of the probe, stringency of the hybridization, and concentration of the RNA and DNA. Previously, most work has been done with $^{32}P$-labeled probes. Alternative procedures utilize $^{33}P$, avidin/biotin, immunochemical, and luminescence procedures (consult commercial vendors). These procedures might have increased use due to decreased costs and increased restrictions on the disposal of radioactive waste.

*Reagents for Southern Gels*

1. 0.25-*M* HCl
2. Denaturation solution—1.5-*M* NaCl, 0.5-*M* NaOH
3. Neutralization solution—1.5-*M* NaCl, 0.5-*M* NaOH (pH 7.0)
4. 20× and 2× SSC

Once the DNA samples have been digested with the restriction enzymes of choice, run an agarose gel with the appropriate markers and stain with ethidium bromide. Photograph the gel with a ruler alongside so the band positions may be compared to the membrane after hybridization. Rinse the gel in distilled water and place it in a dish containing 10 gel volumes of 0.25-*M* HCl. Shake slowly for 30 minutes at room temperature. Remove the HCl and rinse with distilled $H_2O$. Add 10 gel volumes of denaturation solution and shake for 20 minutes at room temperature. Repeat this step one time. After the second incubation, rinse the gel and incubate 2 times; shake 20 minutes with 10 gel volumes of neutralization solution [328]. The setup for a Southern transfer is exactly the same as for the northern transfer. Please refer to the discussion on northern sections regarding how to assemble the gel and membrane for transfer. After the DNA has been transferred to the membrane, it must be immobilized. The protocol from here on is the same as that for a northern procedure. Place the nitrocellulose membrane between two sheets of Whatman 3MM paper and bake for 2 hours at 80°C. Nylon membrane may be UV irradiated. Hybridization of probes can proceed as described for the northern procedure.

## Plasmid DNA Isolation (Minipreps)

Alkaline lysis is the most common procedure used for minipreps [329,330]. This procedure is simple, and multiple samples can be run at the same time. Many commercial vendors offer kits based on this procedure. Plasmid DNA is isolated from small amounts of plasmid-containing bacteria. The bacteria are lysed using SDS and NaOH. The SDS denatures the bacterial proteins, and the NaOH denatures the chromosomal and plasmid DNA. This solution is neutralized with potassium acetate. The covalently closed plasmid DNA reanneals rapidly. The chromosomal DNA and proteins precipitate and are removed by centrifuging the sample. The plasmid DNA is recovered from the supernatant by ethanol precipitation [331].

*Reagents for DNA Isolation*

1. Luria–Bertani (LB) media with an appropriate antibiotic (i.e., selective for your plasmid)
2. Cell resuspension buffer—50-m*M* glucose, 25-m*M* Tris-Cl (pH 8.0), 10-m*M* EDTA; sterilize and store at 4°C
3. Cell lysis buffer—0.2-*M* NaOH, 1% (w/v) SDS; prepare immediately prior to use.
4. Cell neutralization buffer—5-*M* potassium acetate (pH 4.8); start with 29.5 mL glacial acetic acid and add KOH pellets until pH 4.8 is reached. Add $H_2O$ to 100 mL. Do not autoclave. Store at room temperature.
5. 95% and 70% ethanol

Inoculate a single colony in 5 mL of LB$^+$ antibiotic and grow overnight. Spin down 1.5 mL of the cells in a microcentrifuge 20 seconds at maximum speed. Remove the supernatant, and add 100 μL of the cell resuspension buffer. Incubate for 5 minutes. It is important that the cells are completely resuspended. Add 200 μL of the cell lysis buffer and mix by tapping the tubes (by finger). Place tubes on ice for 5 minutes. Next, add 150 μL of cell neutralization buffer and mix. Incubate on ice for 5 minutes. Precipitate the chromosomal DNA and cellular debris by spinning in a microfuge for 3 minutes. Transfer supernatant to a new tube and add 0.8 mL of 95% ethanol. Incubate for 2 minutes at room temperature. The DNA can be precipitated by microcentrifuging for 1 minute. Carefully remove the supernatant and add 1 mL of 70% ethanol to wash the pellet. Dry the pellet under vacuum. The DNA pellet may be resuspended in 30 μL of sterile $H_2O$ and stored at –20 or –70°C. Some procedures call for storing the DNA in TE buffer (10-m*M* Tris-Cl, pH 8.0; 1-m*M* EDTA); however, TE can interfere with sequencing reactions. DNA concentration can be determined by measuring $A_{260}$. A solution with an $A_{260}$ of 1.0 contains 50 μg/mL of DNA [332]. Plasmids can be maintained for a very brief amount of time, 2 to 4 weeks, on selective media plates by storing them at 4°C. Permanent storage should involved storage of the isolated plasmid in addition to storage of the bacterial strain. Bacterial strains can be stored by growing the cells to saturation in the presence of the selective antibiotic. An equal volume of bacteria is added to sterile 100% glycerol and quick frozen in liquid nitrogen. Cells may then be stored at –70°C. For recovery, simply streak out a sample on selective media and grow.

## Polymerase Chain Reaction

The basic principles of the method were developed in 1986 and are relatively straightforward [333]. The process provides the opportunity for considerable innovation in the application to a wide variety of research problems. The reader is advised to consult the current literature as only the basic points are mentioned here. The double-stranded DNA is heat denatured, and the two primers complementary to the ends of the target segment are annealed at a temperature close to the temperature of the oligomers and then extended at a temperature optimal for the polymerase used. One set of these three consecutive steps is referred to as a *cycle*; thus, the amount of the original sequence of interest is expanded in a geometric progression. The process is greatly facilitated by the use of heat-stable polymerase isolated from thermophilic bacterium; such polymerases function most effectively at high temperatures (72 to 78°C). Many cycles can be carried out with the same enzyme. The cycle times are relatively short, and commercial instruments can be programmed to cycle through steps automatically. DNA segments can be amplified $10^5$- to $10^9$-fold. Heat-stable polymerases of very high fidelity are available and reduce errors.

A typical PCR reaction would be performed on the 100 µL scale in 0.5 mL sterile thin-walled microcentrifuge tubes. There is considerable potential to generate false positives, and it is imperative to minimize laboratory errors. Reagent purity is important and avoiding contamination at every step is crucial. Carryover of amplified sequences contributes to the majority of false positives; appropriate precautions include physical separation of pre- and post-PCR reaction and aliquoting reagents to minimize the number of repeated samplings. Positive-placement pipettes are the best choice for PCR work. Some other general precautions include changing gloves frequently, careful uncapping of samples to prevent aerosol formation, addition of nonsample components (buffer, nucleotides, primers) to the reaction mixture prior to DNA addition, and absolutely the use of new pipette tips every time.

The PCR sample may be single- or double-stranded DNA or RNA. If the starting sample is RNA, reverse transcriptase is used to first prepare cDNA prior to conventional PCR (i.e., RT-PCR). PCR primers are oligonucleotides 20 to 30 bases long that are complementary to sequences defining the 3′ ends of the complementary template strands. Several computer programs can assist in primer design. Keep in mind that computer design is not foolproof. Utilizing these programs will aid in detecting primer pairs with intra- or intermolecular complementarities (primer dimers). Ideally, 40 to 60% G+C content is recommended without long stretches of either base. Many of these computer programs will analyze the primer secondary structures and calculate the $T_m$ values. The $T_m$ values for both primers should be well matched. Internal secondary structures (e.g., hairpin loops) should be avoided in primers. Nontemplate-complementary 5′ extensions may be added to primers to allow a variety of useful operations on the PCR product (e.g., the addition of restriction enzyme sites), without significant perturbation to the amplification.

Optimal annealing temperature must be determined empirically. A good annealing temperature from which to begin optimizing is 5 to 10°C below the $T_m$ of the primers. Another critical parameter for PCR is the $MgCl_2$ concentration needed in the reaction to generate maximum product. Titration of $Mg^{2+}$ over a range of 1.5 to 4 m*M* is recommended to find the concentration producing the highest yield of product. The fidelity of *Taq* polymerase decreases in the presence of high $Mg^{2+}$. *Pfu* polymerase has a higher fidelity rate compared to *Taq* (Stratagene). This multifunctional, thermostable enzyme possess both 5′-3′ DNA polymerase activity and 3′-5′ exonuclease activity. This latter activity results in a 12-fold increase in fidelity over *Taq* polymerase. A typical PCR reaction performed in 100 µL would contain dATP, dCTP, dGTP, dTTP (200 µ*M* each), primer (0.25 µ*M*), template (0.1 to 500 ng), and 10 µL of a stock buffer containing 0.20-*M* Tris-HCl (pH 8.75), 0.10-*M* KCl, 0.10-*M* $(NH_4)_2SO_4$, 20-m*M* $MgCl_2$, 1.0% Triton® X-100 (w/v), and 1.0 mg nuclease-free bovine serum albumin per mL (w/v). The tubes are mixed gently and centrifuged briefly. It is important to note that some protocols call for the addition of mineral oil. If the thermocycler being utilized has a heated lid, then mineral oil is not needed in the reaction; if the thermocycler does not have a heated lid, add 50 µL of mineral oil. Heat samples to 95°C for 5 minutes and then cool to the desired annealing temperature to allow the primers to anneal to the template DNA. Add *Pfu* polymerase (2.5 units). Allow primer extension to proceed at 75°C. Repeat the cycles to achieve adequate amplification. Check PCR products by electrophoresis, with detection by ethidium bromide staining or hybridization to labeled probes as appropriate. An alternative to ethidium bromide staining is the SYBR Gold™ nucleic acid gel stain from Molecular Probes. This stain is 25 to 100 times more sensitive than ethidium bromide, is easier to use, and permits optimization of 10- to 100-fold lower starting template copy number [334].

## Site-Directed Mutagenesis

Site-directed mutagenesis has been defined as "any of various techniques by which defined mutations can be made *in vitro* in a cloned DNA" [335]. The Kunkel method of site-directed mutagenesis is a classic method for introducing mutations into a DNA sequence. This method of site-directed mutagenesis by deoxyuridine incorporation depends on the host strain to degrade template DNA that contains uracil in place of thymidine [336,337]. A number of dUTPs are incorporated into the template strand in place of dTTP in a host that lacks dUTPase (i.e., *dut*$^-$) and uracil *N*-deglycosidase (i.e., *ung*$^-$) activities. Uracil itself is not mutagenic and base pairs with adenine. Under normal cellular conditions, dUTPase degrades deoxyuridine, and uracil *N*-deglycosidase removes any incorporated uracil. Postmutation replication in a *dut*$^+$*ung*$^+$ *Escherichia coli* strain is then used to degrade the nontarget strand DNA. This approach requires that single-stranded DNA be utilized so only one strand contains the uracils that are susceptible to degradation.

A more convenient method for site-directed mutagenesis is based on PCR using the selection method of DAM methylation. This is commonly sold in a commercial kit available from Stratagene known as QuickChange®; however, the procedure can be accomplished without the purchase of a kit. Everything needed is available individually. It is not necessary to isolate single-stranded DNA. The entire process is PCR based, utilizing *Pfu* polymerase. After the PCR has finished, the products are digested with the restriction enzyme *Dpn*I. This restriction enzyme is specific for DAM-methylated $G^{Me6}ATC$ sequences [338]. The wild-type template will be digested by *Dpn*I, and the PCR-generated mutant DNA will be not be degraded by this enzyme; therefore, the reaction is enriched for the mutant DNA population. Multiple mutations can be introduced using the *Dpn*I method. A selection feature one can incorporate through the primer design in the addition of other unique restriction enzyme sites (e.g., unique to your DNA sequence). This method works well with moderately sized plasmids (<8 kb) and eliminates the need for subcloning.

Another type of mutagenesis is cassette mutagenesis. In the basic form of cassette mutagenesis, a small section of DNA is removed from the wild-type gene and is replaced with a synthetic segment the carries one or more mutations [339]. Cassette mutagenesis can be utilized to produce multiple mutations within a target zone. One efficient way to generate the pool of oligonucleotides is to synthesize one strand of the cassette with equal mixtures of all four nucleotides in the first two positions and an equal mixture of G and C in the third [340]. This composition will result in all 20 amino acids [341]. The second strand for the cassette is synthesized with inosine at each target position. Inosine will base pair with all four bases [342]. The two strands are annealed and the population of cassettes is ligated into a vector.

## Random Mutagenesis, Directed Evolution, and Staggered Extension Process

Random mutagenesis and subsequent screening, also termed *directed evolution* or *molecular breeding*, has become a widely used tool in protein engineering and has been increasingly applied to investigate structure–function relationships of proteins. The idea of directed evolution of biomolecules *in vitro* on a molecular level was first introduced by pioneering work of Spiegelman for nucleic acids, and later Eigen and Kauffman proposed a theory of molecular evolution [343–346]. Arnold's group was among the first to apply the principles of molecular evolution to improvement of enzyme biocatalysis [347]. A series of random mutagenesis and screening of the generated mutant libraries could produce more stable and active variant enzymes in unusual environments (i.e., organic solvents). Later, Stemmer developed a method, called *DNA shuffling*, which allows *in vitro* recombination of mutants to eliminate deleterious mutations and combine beneficial mutations [348].

The simplest way to create sequence diversity is random point mutagenesis using error-prone PCR. The mutation rate is typically adjusted to a low rate of one or two amino acid changes per protein, allowing for exhaustive screening of the created library [343]. If the number of amino acid substitutions were increased, the number of possible variants numbers would exceed current high-throughput screening capabilities (libraries with $10^4$ to $10^6$ members for enzyme screens), and a large fraction of the library would contain nonfunctional variants because beneficial mutations are rare and combinations of beneficial mutations are extremely rare [349]. Error-prone PCR reactions typically contain a higher concentration of $MgCl_2$ (7-m*M*) compared to the basic PCR reaction (1.5-m*M*). $MnCl_2$ can also be added to increase the variation of ratios of nucleotides incorporated in the reaction [350]. More recent approaches include the use of a newly developed error-prone polymerase, Mutazyme® (Stratagene), which can be adjusted to achieve desired mutation frequencies simply through changes in the DNA template concentration [242].

In contrast to error-prone PCR, whole-scale plasmid PCR amplification using complementary degenerate primers is useful, particularly to achieve construction of mutant libraries that are randomized within a limited target zone [351,352]; however, a major challenge in random mutagenesis is to establish a screen that is sensitive to the properties of interest. The best screens are high throughput, to increase the likelihood that useful clones will be found; sufficiently sensitive, to allow the isolation of altered activity clones early in the process; reproducible, to allow one to find small improvements; robust, meaning that the signal afforded by active clones is not depended on difficult to control environmental variables; and sensitive to the desired functions [353].

*In vitro* PCR-based recombination can be utilized to shuffle segments from homologous DNA sequences to produce highly mosaic chimeric sequences [353]. The staggered extension process, referred to as *StEP recombination*, is based on cross-hybridization of growing gene fragments during polymerase catalyzed primer extension [354]. After denaturation, primers anneal and extend in a step whose brief duration and suboptimal extension temperature limit primer extension. The partially extended primers randomly anneal to different parent sequences throughout multiple cycles, creating novel recombinants [353]. The process is illustrated in Figure 40.35. Full-length product can be amplified by PCR, depending on the yield of the StEP reaction.

## GENOTOXICITY ASSAYS LINKED TO METABOLISM

One use of bacterial P450 systems is the development of convenient genotoxicity assays. This area has been reviewed elsewhere, including the expression of *Salmonella typhimurium* [177,186]. More recently, *Escherichia coli* systems have been developed where the endpoint is *lac* protrophy instead of the usual *his* in the Ames test. A bacterial *N*-acetyltransferase can be concurrently expressed to increased sensitivity to aryl and heterocyclic amines (Figure 40.36). Such systems have been shown to respond to concentrations of some aryl amines as low as the classical *S. typhimurium* systems [161]. These systems offer several advantages. In other work, it has been possible to express rat and human GSH transferases along with *N*-acetyltransferase in *S. typhimurium* for use in genotoxicity assays [185,355,356]. More recently, several human P450s have been coexpressed with NADPH–P450 reductase in *S. typhimurium* [357]. These systems have considerable potential, as bacterial systems will continue to be the mainstay of high-throughput, first-check genotoxicity screens. In these systems, reactive products are generated with the cells and more closely approximate the normal cellular situation.

Some examples are the *Salmonella typhimurium umu* test (or the similar *Escherichia coli* chromotest) in which the reporter gene is linked to a promoter that responds to a cascade emanating from damage to bacterial DNA [160,358]. These systems often have faster readout than colony-counting systems (e.g., Ames test) with their colorimetric endpoints, which are often seen within a few hours. The Ames test relies on the reversion of a variety of different *S. typhimurium* mutants with histidine-dependent growth requirements to prototrophy by a chemical carcinogen. All of the strains contain two additional mutations that disrupt excision repair and cause loss of the lipopolysaccharide barrier [359]. In the *S. typhimurium umu* test, DNA damage (as a result of modification of DNA by chemical carcinogens or their metabolic products) invokes the SOS response. The SOS response initiates proteolytic cleavage of the LexA protein by the activated RecA protease and ultimately results in β-galactosidase activity that can be colorimetrically quantified [160].

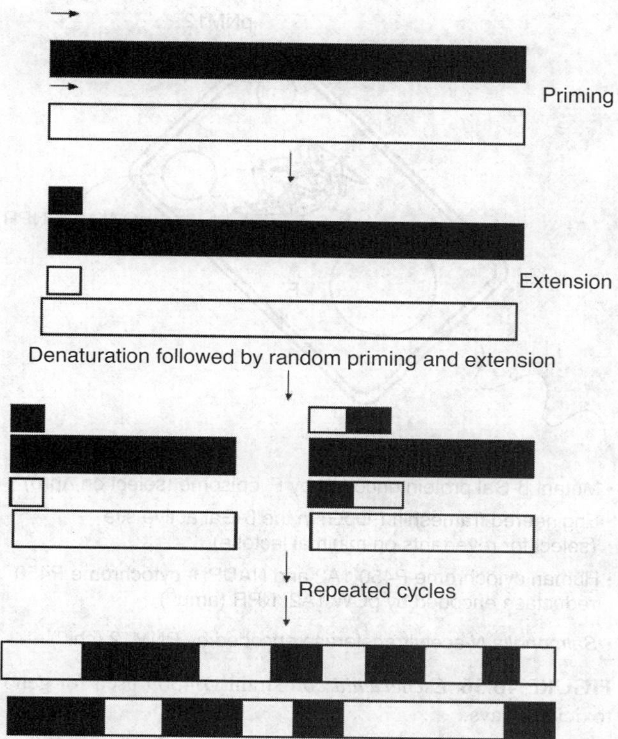


**FIGURE 40.35** StEP recombination illustrated for two gene templates. Only one primer and single strands of the two genes (open and solid blocks) are shown for simplicity. During priming, oligonucleotide primers anneal to denatured template. Short fragments are produced by brief polymerase catalyzed extension that is interrupted by denaturation. Subsequent random annealing-abbreviated extension cycles produce randomly primed templates. This results in the eventual extension of full-length chimeric genes. (Adapted from Arnold, F.H. and Georgiou, G., *Directed Enzyme Evolution, Screening and Selection Methods*, Humana Press, Totowa, NJ, 2003.)

Briefly described, a chemical is incubated with a metabolic system capable of converting it to a reactive product, in the presence of *Salmonella typhimurium* cells containing the plasmid pSK1002. This can be liver microsomes or a recombinant P450 system expressed within the cells. Formation of DNA adducts blocks replication of the modified strand and leaves regions of single-stranded DNA, to which the protein LexA binds. The protease LexA is then activated and cleaves RecA, leading to more events in a cascade and activating more than 30 genes. One of these is *umu*C, which codes for a translesion polymerase that the bacteria use to bypass DNA adducts and replicate. A reporter plasmid in the cell (pSK1002) contains the regulatory region of the *umu* gene attached to *lac*Z; thus, generation of DNA adducts has the overall effect of activating this *umu*-based plasmid and producing the products β-galactosidase. Production of β-galactosi-

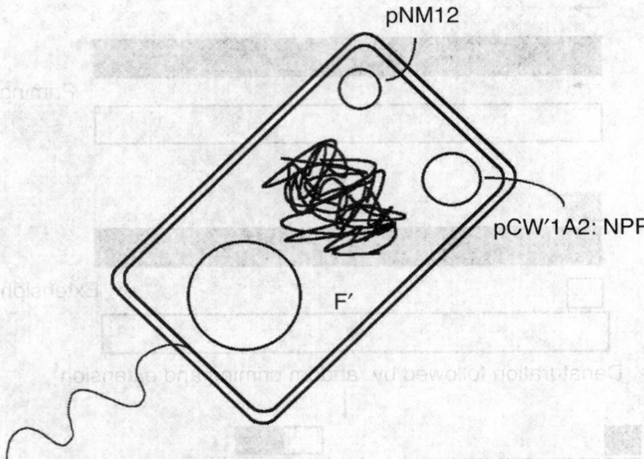


• Mutant β-Gal protein encoded by F′ episome (select on Δpro)
• Engineered frameshift (-CpG) in the β-Gal active site (select for revertants on minimal lactose)
• Human cytochrome P450 1A2 and NADPH–cytochrome P450 reductase encoded by pCW′1A2: NPR ($amp^R$)
• *Salmonella N*-acetyltransferase encoded by PNM12 ($chl^R$)

**FIGURE 40.36** *Escherichia coli* strain DJ4309 used for genotoxicity assays.

dase can be quantified by lysing cells and using a colorimetric assay (Figure 40.37). The assay can be done in microtiter plates.

The *Escherichia coli* chromotest is very similar, although unlike the previous examples this system utilizes only one strain. This test also relies on the cellular SOS response system. In this case, cell division is tied into the assay, so survival of the host is not important [360]. The *E. coli* strain has the same mutation that renders the strain lipopolysaccharide deficient. This allows for better diffusion of chemicals through the outer membrane [361]. Several steps occur between the reaction of a carcinogen with DNA and a resulting mutation. In this laboratory we have recently utilized two different systems to study DNA repair. One system is the human p53-driven *Ade* reporter system in *Saccharomyces cerevisiae*. Cells with the wild-type p53 express Can1, and these cells are able to take up canavanine, which is cytotoxic and limits growth on selective media. Cells with mutant p53 are unable to express Ade2 and accumulate an adenine metabolite intermediate, and the colonies are red in color [362]. The second system utilizes resistance to the antibiotic rifampicin in *E. coli* through mutations in the *rpo*B gene as a result of carcinogen exposure [363,364]. Utilization of this system enabled the identification of the direct inactivation of $O^6$-alklyguanine–DNA alkyltransferase (AGT) and formation of AGT–$Cys^{145}$–$CH_2$–DNA adducts by which $CH_2Br_2$ (1,2-dibromoethane) might cause damage to the genome [364]. Additional work resulted in identification of another $CH_2Br_2$ DNA adduct, guanine–$N^7$ alkylation, and a potential second site that was not identified [363].

## CONCLUSIONS

Enzymes have important roles in the metabolism and toxicology of many chemicals. Important examples are known with drugs and with environmental chemicals, and the principles are the same. Some background has been presented here, along with some detailed procedures for working with these systems. Obviously, the examples are not comprehensive, and they can be modified to make them more useful for particular needs. The need for better understanding of the biochemical and molecular events involved in toxicology (Figure 40.4) will continue for some time, and we hope that this information will facilitate some of the studies in the area.

## ACKNOWLEDGMENTS

The authors are supported by U.S. Public Health Service grants R37 CA90426, R01 ES10375, R01 ES10546, and P30 ES00267. Thanks are extended to G. R. Wilkinson for providing assay methods for midazolam, to M. V. Martin for assistance in describing P450 purification, and to K. Trisler for assistance in preparation of the manuscript.

## QUESTIONS

1. Most assays for competitive inhibition of cytochrome P450 enzymes involve direct studies with assays of substrate oxidation. How could you develop a high-throughput screening strategy for identifying high-affinity inhibitors of a particular cytochrome P450 (e.g., 2D6) without the need to do assays of catalytic activity?
2. A new chemical of interest induces total hepatic cytochrome P450 (as measured spectrally) and is hepatotoxic (in rats). Are the two phenomena necessarily related? How would you establish the relationship?
3. An investigator tells you that he has found that a particular recombinant human cytochrome P450 enzyme catalyzes a reaction of interest to you. What other pieces of evidence are needed to put this information into the perspective of how much this enzyme contributes to the process in human liver?
4. The pathways shown on the next page have been extended to demonstrate an example of situations often encountered in practical situations. Provide reasonable stepwise pathways to the products indicated. The number of steps is not specified. For each step, provide the necessary cosubstrate(s) and name of the enzyme (if enzymatic, only group name; no need for individual form).

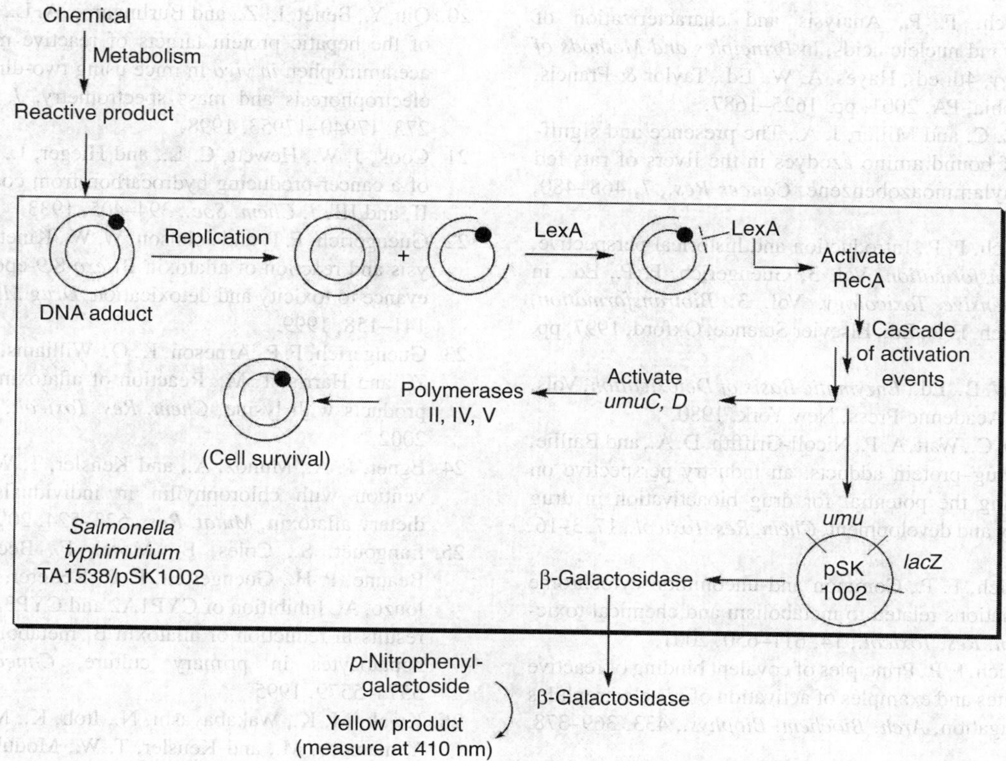


**FIGURE 40.37** Use of bacterial SOS response to identify genotoxins with the *umu* response. A chemical is incubated with a metabolic system capable of converting it to a reactive product in the presence of *Salmonella typhimurium* cells containing the plasmid pSK1002. Formation of DNA adducts blocks replication of the modified strand and leaves regions of single-stranded DNA, to which the protein LexA binds. The protease LexA is then activated and cleaves RecA, leading to more events in a cascade and activating more than 30 genes. One of these is *umu*C, which codes for a translesion polymerase that the bacteria use to bypass DNA adducts and replicate. A reporter plasmid in the cell (pSK1002) contains the regulatory region of the *umu* gene attached to *lac*Z. Thus, generation of DNA adducts has the overall effect of activating this *umu*-based plasmid and producing the products β-galactosidase. The production of β-galactosidase can be quantified by lysing cells and using a colorimetric assay.

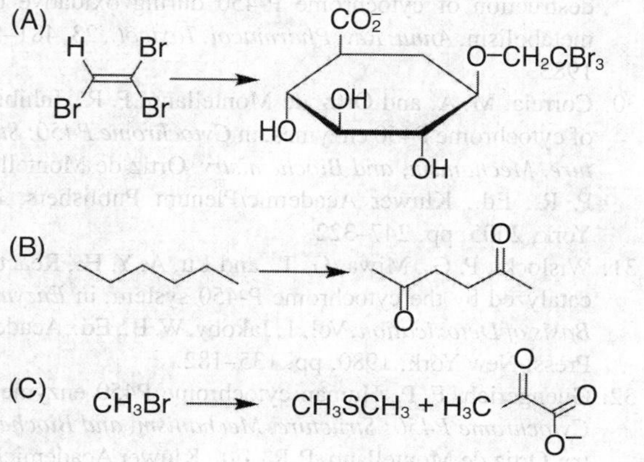

(A) H, Br, Br, Br → $CO_2^-$, O–$CH_2CBr_3$, OH, HO, OH

(B) → O, O

(C) $CH_3Br$ → $CH_3SCH_3 + H_3C$ (O, O, $O^-$)

5. Your research group is interested in generating multiple mutants in a specific area of your enzyme of interest (a region of ten amino acids suspected to be the active site). Outline your plan to generate these mutations and analyze their effects.

6. You have encountered an interesting enzyme in your research and wish to further study its function. This enzyme requires glycosylation for activity. How does this influence your choice of expression systems?

## REFERENCES

1. Guengerich, F. P., Microsomal enzymes involved in toxicology: analysis and separation, in *Principles and Methods of Toxicology*, 1st ed., Hayes, A. W., Ed., Raven Press, New York, 1982, pp. 609–634.
2. Guengerich, F. P., Analysis and characterization of enzymes, in *Principles and Methods of Toxicology*, 2nd ed., Hayes, A. W., Ed., Raven Press, New York, 1989, pp. 777–814.
3. Guengerich, F. P., Analysis and characterization of enzymes, in *Principles and Methods of Toxicology*, 3rd ed., Hayes, A. W., Ed., Raven Press, New York, 1994, pp. 1259–1313.

4. Guengerich, F. P., Analysis and characterization of enzymes and nucleic acids, in *Principles and Methods of Toxicology*, 4th ed., Hayes, A. W., Ed., Taylor & Francis, Philadelphia, PA, 2001, pp. 1625–1687.
5. Miller, E. C. and Miller, J. A., The presence and significance of bound amino azodyes in the livers of rats fed *p*-dimethylaminoazobenzene, *Cancer Res.*, 7, 468–480, 1947.
6. Guengerich, F. P., Introduction and historical perspective, in *Biotransformation*, Vol. 3, Guengerich, F. P., Ed., in *Comprehensive Toxicology*, Vol. 3, *Biotransformation*, Guengerich, F. P., Ed., Elsevier Science, Oxford, 1997, pp. 1–6.
7. Jakoby, W. B., Ed., *Enzymatic Basis of Detoxication*, Vols. 1 and 2, Academic Press, New York, 1980.
8. Evans, D. C., Watt, A. P., Nicoll-Griffith, D. A., and Baillie, T. A., Drug–protein adducts: an industry perspective on minimizing the potential for drug bioactivation in drug discovery and development, *Chem. Res. Toxicol.*, 17, 3–16, 2004.
9. Guengerich, F. P., Common and uncommon cytochrome P450 reactions related to metabolism and chemical toxicity, *Chem. Res. Toxicol.*, 14, 611–650, 2001.
10. Guengerich, F. P., Principles of covalent binding of reactive metabolites and examples of activation of *bis*-electrophiles by conjugation, *Arch. Biochem. Biophys.*, 433, 369–378, 2005.
11. Liebler, D. C. and Guengerich, F. P., Mechanisms of drug-induced cell damage, *Nature Rev. Drug Discov.*, 4, 410–420, 2005.
12. Borzelleca, J. F., Profiles in toxicology: Paracelsus, herald of modern toxicology, *Toxicol. Sci.*, 53, 2–4, 2000.
13. Guengerich, F. P., Life and times in biochemical toxicology, *Int. J. Toxicol.*, 24, 1–17, 2005.
14. Harvison, P. J., Guengerich, F. P., Rashed, M. S., and Nelson, S. D., Cytochrome P-450 isozyme selectivity in the oxidation of acetaminophen, *Chem. Res. Toxicol.*, 1, 47–52, 1988.
15. Patten, C. J., Thomas, P. E., Guy, R. L., Lee, M., Gonzalez, F. J., Guengerich, F. P., and Yang, C. S., Cytochrome P450 enzymes involved in acetaminophen activation by rat and human liver microsomes and their kinetics, *Chem. Res. Toxicol.*, 6, 511–518, 1993.
16. Mitchell, J. R., Jollow, D. J., Potter, W. Z., Davis, D. C., Gillette, J. R., and Brodie, B. B., Acetaminophen-induced hepatic necrosis. I. Role of drug metabolism, *J. Pharmacol. Exp. Ther.*, 187, 185–194, 1973.
17. Jollow, D. J., Mitchell, J. R., Potter, W. Z., Davis, D. C., Gillette, J. R., and Brodie, B. B., Acetaminophen-induced hepatic necrosis. II. Role of covalent binding *in vivo*, *J. Pharmacol. Exp. Ther.*, 187, 195–202, 1973.
18. Streeter, A. J., Bjorge, S. M., Axworthy, D. B., Nelson, S. D., and Baillie, T. A., The microsomal metabolism and site of covalent binding to protein of 3′-hydroxyacetanilide, a nonhepatotoxic positional isomer of acetaminophen, *Drug Metab. Dispos.*, 12, 565–576, 1984.
19. Roberts, S. A., Price, V. F., and Jollow, D. J., Acetaminophen structure-toxicity studies: *in vivo* covalent binding of a nonhepatotoxic analog, 3-hydroxyacetanilide, *Toxicol. Appl. Pharmacol.*, 105, 195–208, 1990.
20. Qiu, Y., Benet, L. Z., and Burlingame, A. L., Identification of the hepatic protein targets of reactive metabolites of acetaminophen *in vivo* in mice using two-dimensional gel electrophoresis and mass spectrometry, *J. Biol. Chem.*, 273, 17940–17953, 1998.
21. Cook, J. W., Hewett, C. L., and Hieger, I., The isolation of a cancer-producing hydrocarbon from coal tar, Parts I, II, and III, *J. Chem. Soc.*, 394–405, 1933.
22. Guengerich, F. P. and Johnson, W. W., Kinetics of hydrolysis and reaction of aflatoxin $B_1$ *exo*-8,9-epoxide and relevance to toxicity and detoxication, *Drug Metab. Rev.*, 31, 141–158, 1999.
23. Guengerich, F. P., Arneson, K. O., Williams, K. M., Deng, Z., and Harris, T. M., Reaction of aflatoxin $B_1$ oxidation products with lysine, *Chem. Res. Toxicol.*, 15, 780–792, 2002.
24. Egner, P. A., Munoz, A., and Kensler, T. W., Chemoprevention with chlorophyllin in individuals exposed to dietary aflatoxin, *Mutat. Res.*, 523–524, 209–216, 2003.
25. Langouët, S., Coles, B., Morel, F., Becquemont, L., Beaune, P. H., Guengerich, F. P., Ketterer, B., and Guillouzo, A., Inhibition of CYP1A2 and CYP3A4 by oltipraz results in reduction of aflatoxin $B_1$ metabolism in human hepatocytes in primary culture, *Cancer Res.*, 55, 5574–5579, 1995.
26. Kwak, M. K., Wakabayashi, N., Itoh, K., Motohashi, H., Yamamoto, M., and Kensler, T. W., Modulation of gene expression by cancer chemopreventive dithiolethiones through the Keap1–Nrf2 pathway: identification of novel gene clusters for cell survival, *J. Biol. Chem.*, 278, 8135–8145, 2003.
27. Guengerich, F. P. and Macdonald, T. L., Chemical mechanisms of catalysis by cytochromes P-450: a unified view, *Acct. Chem. Res.*, 17, 9–16, 1984.
28. Ortiz De Montellano, P. R., Ed., *Cytochrome P450: Structure, Mechanism, and Biochemistry*, 3rd ed., Kluwer Academic/Plenum Publishers, New York, 2005.
29. Ortiz de Montellano, P. R. and Correia, M. A., Suicidal destruction of cytochrome P-450 during oxidative drug metabolism, *Annu. Rev. Pharmacol. Toxicol.*, 23, 481–503, 1983.
30. Correia, M. A. and Ortiz de Montellano, P. R., Inhibition of cytochrome P450 enzymes, in *Cytochrome P450: Structure, Mechanism, and Biochemistry*, Ortiz de Montellano, P. R., Ed., Kluwer Academic/Plenum Publishers, New York, 2005, pp. 247–322.
31. Wislocki, P. G., Miwa, G. T., and Lu, A. Y. H., Reactions catalyzed by the cytochrome P-450 system, in *Enzymatic Basis of Detoxication*, Vol. 1, Jakoby, W. B., Ed., Academic Press, New York, 1980, pp. 135–182.
32. Guengerich, F. P., Human cytochrome P450 enzymes, in *Cytochrome P450: Structure, Mechanism, and Biochemistry*, Ortiz de Montellano, P. R., Ed., Kluwer Academic/Plenum Publishers, New York, 2005, pp. 377–531.
33. Rendic, S., Summary of information on human CYP enzymes: human P450 metabolism data, *Drug Metab. Rev.*, 34, 83–448, 2002.
34. Jakoby, W. B., Detoxication enzymes, in *Enzymatic Basis of Detoxication, Vol. 1*, Jakoby, W. B., Ed., Academic Press, New York, 1980, pp. 1–6.

35. Guengerich, F. P., Human cytochrome P450 enzymes, in *Cytochrome P450: Structure, Mechanism, and Biochemistry*, Ortiz de Montellano, P. R., Ed., Kluwer Academic/Plenum Publishers, New York, 2005, pp. 377–531.
36. Guengerich, F. P. and Liebler, D. C., Enzymatic activation of chemicals to toxic metabolites, *CRC Crit. Rev. Toxicol.*, 14, 259–307, 1985.
37. Nelson, D. R., Koymans, L., Kamataki, T., Stegeman, J. J., Feyereisen, R., Waxman, D. J., Waterman, M. R., Gotoh, O., Coon, M. J., Estabrook, R. W., Gunsalus, I. C., and Nebert, D. W., P450 superfamily: update on new sequences, gene mapping, accession numbers, and nomenclature, *Pharmacogenetics*, 6, 1–42, 1996.
38. Lindberg, R. L. P. and Negishi, M., Alteration of mouse cytochrome $P450_{coh}$ substrate specificity by mutation of a single amino-acid residue, *Nature*, 339, 632–634, 1989.
39. Distlerath, L. M. and Guengerich, F. P., Enzymology of human liver cytochromes P-450, in *Mammalian Cytochromes P-450*, Vol. 1, Guengerich, F. P., Ed., CRC Press, Boca Raton, FL, 1987, pp. 133–198.
40. Guengerich, F. P., Effects of nutritive factors on metabolic processes involving bioactivation and detoxication of chemicals, *Annu. Rev. Nutr.*, 4, 207–231, 1984.
41. Guengerich, F. P., Human cytochrome P-450 3A4: regulation and role in drug metabolism, *Annu. Rev. Pharmacol. Toxicol.*, 39, 1–17, 1999.
42. Müller-Enoch, D., Churchill, P., Fleischer, S., and Guengerich, F. P., Interaction of liver microsomal cytochrome P-450 and NADPH-cytochrome P-450 reductase in the presence and absence of lipid, *J. Biol. Chem.*, 259, 8174–8182, 1984.
43. Halvorson, M., Greenway, D., Eberhart, D., Fitzgerald, K., and Parkinson, A., Reconstitution of testosterone oxidation by purified rat cytochrome P450p (IIIA1), *Arch. Biochem. Biophys.*, 277, 166–180, 1990.
44. Imaoka, S., Imai, Y., Shimada, T., and Funae, Y., Role of phospholipids in reconstituted cytochrome P450 3A forms and mechanism of their activation of catalytic activity, *Biochemistry*, 31, 6063–6069, 1992.
45. Nebert, D. W., Nelson, D. R., Coon, M. J., Estabrook, R. W., Feyereisen, R., Fujii-Kuriyama, Y., Gonzalez, F. J., Guengerich, F. P., Gunsalus, I. C., Johnson, E. F. et al., The P450 superfamily: update on new sequences, gene mapping, and recommended nomenclature, *DNA Cell Biol.*, 10, 397–398, 1991.
46. Guengerich, F. P., Cytochrome P-450 enzymes and drug metabolism, in *Progress in Drug Metabolism*, Vol. 10, Bridges, J. W., Chasseaud, L. F., and Gibson, G. G., Eds., Taylor & Francis, London, 1987, pp. 1–54.
47. Dannan, G. A., Guengerich, F. P., Kaminsky, L. S., and Aust, S. D., Regulation of cytochrome P-450: immunochemical quantitation of eight isozymes in liver microsomes of rats treated with polybrominated biphenyl congeners, *J. Biol. Chem.*, 258, 1282–1288, 1983.
48. Guengerich, F. P. et al., Purification and characterization of liver microsomal cytochromes P-450: electrophoretic, spectral, catalytic, and immunochemical properties and inducibility of eight isozymes isolated from rats treated with phenobarbital or β-naphthoflavone, *Biochemistry* 21, 6019–6030, 1982.
49. Dannan, G. A., Waxman, D. J., and Guengerich, F. P., Hormonal regulation of rat liver microsomal enzymes: role of gonadal steroids in programming, maintenance, and suppression of $\Delta^4$-steroid 5α-reductase, flavin-containing monooxygenase, and sex-specific cytochromes P-450, *J. Biol. Chem.*, 261, 10728–10735, 1986.
50. Waxman, D. J., Dannan, G. A., and Guengerich, F. P., Regulation of rat hepatic cytochrome P-450: age-dependent expression, hormonal imprinting, and xenobiotic inducibility of sex-specific isoenzymes, *Biochemistry*, 24, 4409–4417, 1985.
51. Waterman, M. R. and Guengerich, F. P., Enzyme regulation, in *Comprehensive Toxicology*, Vol. 3, *Biotransformation*, Guengerich, F. P., Ed., Elsevier Science, Oxford, 1997, pp. 7–14.
52. Williams, S. N., Dunham, E., and Bradfield, C. A., Induction of P450 enzymes: receptors, in *Cytochrome P450: Structure, Mechanism, and Biochemistry*, Ortiz De Montellano, P. R., Ed., Kluwer Academic/Plenum Publishers, New York, 2005, pp. 323–346.
53. White, P. C., New, M. I., and Dupont, B., HLA-linked congenital adrenal hyperplasia results from a defective gene encoding a cytochrome P-450 specific for steroid 21-hydroxylation, *Proc. Natl. Acad. Sci. U.S.A.*, 81, 7505–7509, 1984.
54. Keeney, D. S. and Waterman, M. R., Regulation of steroid hydroxylase gene expression: importance to physiology and disease, *Pharmacol. Ther.*, 58, 301–317, 1993.
55. Shimada, T., Yamazaki, H., Foroozesch, M., Hopkins, N. E., Alworth, W. L., and Guengerich, F. P., Selectivity of polycyclic inhibitors for human cytochromes P450 1A1, 1A2, and 1B1, *Chem. Res. Toxicol.*, 11, 1048–1056, 1998.
56. Newton, D. J., Wang, R. W., and Lu, A. Y. H., Cytochrome P450 inhibitors: evaluation of specificities in the *in vitro* metabolism of therapeutic agents by human liver microsomes, *Drug Metab. Dispos.*, 23, 154–158, 1994.
57. McManus, M. E., Burgess, W. M., Veronese, M. E., Huggett, A., Quattrochi, L. C., and Tukey, R. H., Metabolism of 2-acetylaminofluorene and benzo(*a*)pyrene and activation of food-derived heterocyclic amine mutagens by human cytochromes P-450, *Cancer Res.*, 50, 3367–3376, 1990.
58. Yamazaki, H., Inui, Y., Yun, C.-H., Mimura, M., Guengerich, F. P., and Shimada, T., Cytochrome P450 2E1 and 2A6 enzymes as major catalysts for metabolic activation of *N*-nitrosodialkylamines and tobacco-related nitrosamines in human liver microsomes, *Carcinogenesis*, 13, 1789–1794, 1992.
59. Armstrong, R. N., Structure, catalytic mechanism, and evolution of the glutathione transferases, *Chem. Res. Toxicol.*, 10, 2–18, 1997.
60. Hayes, J. D., Flanagan, J. U., and Jowsey, I. R., Glutathione transferases, *Annu. Rev. Pharmacol. Toxicol.*, 45, 51–88, 2005.
61. Burchell, B., McGurk, K., Brierly, C. H., and Clarke, D. J., UDP-glucuronosyltransferases, in *Comprehensive Toxicology*, Vol. 3, *Biotransformation*, Guengerich, F. P., Ed., Oxford, 1997, pp. 401–435.
62. Hammock, B. D., Grant, D. F., and Storms, D. H., Epoxide hydrolases, in *Comprehensive Toxicology*, Vol. 3, *Biotransformation*, Guengerich, F. P., Ed., Oxford, 1997, pp. 283–305.

63. Williams, D. E., Hale, S. E., Muerhoff, A. S., and Masters, B. S. S., Rabbit lung flavin-containing monooxygenase: purification, characterization, and induction during pregnancy, *Mol. Pharmacol.*, 28, 381–390, 1985.
64. Cashman, J. R., Monoamine oxidase and flavin-containing monooxygenases, in *Comprehensive Toxicology*, Vol. 3, *Biotransformation*, Guengerich, F. P., Ed., Oxford, 1997, pp. 69–96.
65. Porter, T. D. and Kasper, C. B., Coding nucleotide sequence of rat NADPH-cytochrome P-450 oxidoreductase cDNA and identification of flavin-binding domains, *Proc. Natl. Acad. Sci. U.S.A.*, 82, 973–977, 1985.
66. Williams, R. T., *Detoxication Mechanisms*, 2nd ed., Wiley, New York, 1959.
67. Josephy, P. D., Guengerich, F. P., and Miners, J. O., Phase 1 and phase 2 drug metabolism: terminology that we should phase out, *Drug Metab. Rev.*, 37, 579–584, 2005.
68. Geiger, L. E., Hogy, L. L., and Guengerich, F. P., Metabolism of acrylonitrile by isolated rat hepatocytes, *Cancer Res.*, 43, 3080–3087, 1983.
69. Guengerich, F. P., Geiger, L. E., Hogy, L. L., and Wright, P. L., *In vitro* metabolism of acrylonitrile to 2-cyanoethylene oxide, reaction with glutathione, and irreversible binding to proteins and nucleic acids, *Cancer Res.*, 41, 4925–4933, 1981.
70. Hogy, L. L. and Guengerich, F. P., *In vivo* interaction of acrylonitrile and 2-cyanoethylene oxide with DNA in rats, *Cancer Res.*, 46, 3932–3938, 1986.
71. Müller, M., Belas, F. J., Blair, I. A., and Guengerich, F. P., Analysis of 1,$N^2$-ethenoguanine and 5,6,7,9-tetrahydro-7-hydroxy-9-oxoimidazo[1,2-*a*]purine in DNA treated with 2-chlorooxirane by high-performance liquid chromatography/mass spectrometry and comparison of amounts with other adducts, *Chem. Res. Toxicol.*, 10, 242–247, 1997.
72. Basu, A. K., Wood, M. L., Niedernhofer, L. J., Ramos, L. A., and Essigmann, J. M., Mutagenic and genotoxic effects of three vinyl chloride-induced DNA lesions: 1,$N^6$-ethenoadenine, 3,$N^4$-ethenocytosine, and 4-amino-5-(imidazol-2-yl)imidazole, *Biochemistry*, 32, 12793–12801, 1993.
73. Langouët, S., Mican, A. N., Müller, M., Fink, S. P., Marnett, L. J., Muhle, S. A., and Guengerich, F. P., Misincorporation of nucleotides opposite 5-membered exocyclic ring guanine derivatives by *Escherichia coli* polymerases *in vitro* and *in vivo*: 1,$N^2$-ethenoguanine, 5,6,7,9-tetrahydro-9-oxoimidazo[1,2-*a*]purine, and 5,6,7,9-tetrahydro-7-hydroxy-9-oxoimidazo[1,2-*a*]purine, *Biochemistry*, 37, 5184–5193, 1998.
74. Guengerich, F. P., Crawford, Jr., W. M., and Watanabe, P. G., Activation of vinyl chloride to covalently bound metabolites: roles of 2-chloroethylene oxide and 2-chloroacetaldehyde, *Biochemistry*, 18, 5177–5182, 1979.
75. Liebler, D. C. and Guengerich, F. P., Olefin oxidation by cytochrome P-450: evidence for group migration in catalytic intermediates formed with vinylidene chloride and *trans*-1-phenyl-1-butene, *Biochemistry*, 22, 5482–5489, 1983.
76. Miller, R. E. and Guengerich, F. P., Metabolism of trichloroethylene in isolated hepatocytes, microsomes, and reconstituted enzyme systems containing cytochrome P-450, *Cancer Res.*, 43, 1145–1152, 1983.
77. Guengerich, F. P., Mason, P. S., Stott, W. T., Fox, T. R., and Watanabe, P. G., Roles of 2-haloethylene oxides and 2-haloacetaldehydes derived from vinyl bromide and vinyl chloride in irreversible binding to protein and DNA, *Cancer Res.*, 41, 4391–4398, 1981.
78. Guengerich, F. P., Kim, D.-H., and Iwasaki, M., Role of human cytochrome P-450 IIE1 in the oxidation of many low molecular weight cancer suspects, *Chem. Res. Toxicol.*, 4, 168–179, 1991.
79. van Bladeren, P. J., Breimer, D. D., van Huijgevoort, J. A. T. C. M., Vermeulen, N. P. E., and van der Gen, A., The metabolic formation of *N*-acetyl-*S*-2-hydroxyethyl-L-cysteine from tetradeutero-1,2-dibromoethane: relative importance of oxidation and glutathione conjugation *in vivo*, *Biochem. Pharmacol.*, 30, 2499–2502, 1981.
80. Inskeep, P. B., Koga, N., Cmarik, J. L., and Guengerich, F. P., Covalent binding of 1,2-dihaloalkanes to DNA and stability of the major DNA adduct, *S*-[2-($N^7$-guanyl)ethyl]glutathione, *Cancer Res.*, 46, 2839–2844, 1986.
81. Kim, D. H. and Guengerich, F. P., Excretion of the mercapturic acid *S*-[2-($N^7$-guanyl)ethyl]-*N*-acetylcysteine in urine following administration of ethylene dibromide to rats, *Cancer Res.*, 49, 5843–5851, 1989.
82. Peterson, L. A., Harris, T. M., and Guengerich, F. P., Evidence for an episulfonium ion intermediate in the formation of *S*-[2-($N^7$-guanyl)ethyl]glutathione in DNA, *J. Am. Chem. Soc.*, 110, 3284–3291, 1988.
83. Guengerich, F. P., Activation of dihaloalkanes by thiol-dependent mechanisms, *J. Biochem. Mol. Biol.*, 36, 20–27, 2003.
84. Cmarik, J. L., Inskeep, P. B., Meyer, D. J., Meredith, M. J., Ketterer, B., and Guengerich, F. P., Selectivity of rat and human glutathione S-transferases in activation of ethylene dibromide by glutathione conjugation and DNA binding and induction of unscheduled DNA synthesis in human hepatocytes, *Cancer Res.*, 50, 2747–2752, 1990.
85. Ozawa, N. and Guengerich, F. P., Evidence for formation of an *S*-[2-($N^7$-guanyl)ethyl]glutathione adduct in glutathione-mediated binding of 1,2-dibromoethane to DNA, *Proc. Natl. Acad. Sci. U.S.A.*, 80, 5266–5270, 1983.
86. Koga, N., Inskeep, P. B., Harris, T. M., and Guengerich, F. P., *S*-[2-($N^7$-guanyl)ethyl]glutathione, the major DNA adduct formed from 1,2-dibromoethane, *Biochemistry*, 25, 2192–2198, 1986.
87. Shimada, T., Nakamura, S., Imaoka, S., and Funae, Y., Genotoxic and mutagenic activation of aflatoxin $B_1$ by constitutive forms of cytochrome P-450 in rat liver microsomes, *Toxicol. Appl. Pharmacol.*, 91, 13–21, 1987.
88. Bolt, H. M., Metabolism of estrogens: natural and synthetic, *Pharmacol. Ther.*, 4, 155–181, 1979.
89. Shah, R. R., Oates, N. S., Idle, J. R., Smith, R. L., Dayer, P., Courvoisier, F., Balant, L., and Fabre, J., Beta-blockers and drug oxidation status, *Lancet*, 508–509, 1982.
90. Ayesh, R., Idle, J. R., Ritchie, J. C., Crothers, M. J., and Hetzel, M. R., Metabolic oxidation phenotypes as markers for susceptibility to lung cancer, *Nature*, 312, 169–170, 1984.
91. Idle, J. R., Mahgoub, A., Sloan, T. P., Smith, R. L., Mbanefo, C. O., and Bababunmi, E. A., Some observations on the oxidation phenotype status of Nigerian patients presenting with cancer, *Cancer Lett.*, 11, 331–338, 1981.

92. Gonzalez, F. J. and Meyer, U. A., Molecular genetics of the debrisoquin-sparteine polymorphism, *Clin. Pharmacol. Ther.*, 50, 233–238, 1991.
93. Guengerich, F. P., Studies on the activation of a model furan compound: toxicity and covalent binding of 2-(*N*-ethylcarbamoylhydroxymethyl)furan, *Biochem. Pharmacol.*, 26, 1909–1915, 1977.
94. van der Hoeven, T. A. and Coon, M. J., Preparation and properties of partially purified cytochrome P-450 and reduced nicotinamide adenine dinucleotide phosphate-cytochrome P-450 reductase from rabbit liver microsomes, *J. Biol. Chem.*, 249, 6302–6310, 1974.
95. Fang, W. F. and Strobel, H. W., The drug and carcinogen metabolism system of rat colon microsomes, *Arch. Biochem. Biophys.*, 186, 128–138, 1978.
96. Cinti, D. L., Moldeus, P., and Schenkman, J. B., Kinetic parameters of drug-metabolizing enzymes in $Ca^{+2}$-sedimented microsomes from rat liver, *Biochem. Pharmacol.*, 21, 3249–3256, 1972.
97. Jernström, B., Capdevila, J., Jakobsson, S., and Orrenius, S., Solubilization and partial purification of cytochrome P-450 from rat lung microsomes, *Biochem. Biophys. Res. Commun.*, 64, 814–822, 1975.
98. Taugen, O., Jonasson, J., and Orrenius, S., Isolation of rat liver microsomes by gel filtration, *Anal. Biochem.*, 54, 597–603, 1973.
99. Wang, P. P., Beaune, P., Kaminsky, L. S., Dannan, G. A., Kadlubar, F. F., Larrey, D., and Guengerich, F. P., Purification and characterization of six cytochrome P-450 isozymes from human liver microsomes, *Biochemistry*, 22, 5375–5383, 1983.
100. Omura, T. and Sato, R., The carbon monoxide-binding pigment of liver microsomes. I. Evidence for its hemoprotein nature, *J. Biol. Chem.*, 239, 2370–2378, 1964.
101. Haugen, D. A. and Coon, M. J., Properties of electrophoretically homogenous phenobarbital-inducible and β-naphthoflavone-inducible forms of liver microsomal cytochrome P-450, *J. Biol. Chem.*, 251, 7929–7939, 1976.
102. Ryan, D., Lu, A. Y. H., West, S., and Levin, W., Multiple forms of cytochrome P-450 in phenobarbital- and 3-methylcholanthrene-treated rats, *J. Biol. Chem.*, 250, 2157–2163, 1975.
103. Omura, T. and Sato, R., Isolation of cytochromes P-450 and P-420, *Meth. Enzymol.*, 10, 556–561, 1967.
104. Matsubara, T., Koike, M., Touchi, A., Tochino, Y., and Sugeno, K., Quantitative determination of cytochrome P-450 in rat liver homogenate, *Anal. Biochem.*, 75, 596–603, 1976.
105. Johannesen, K. A. M. and DePierre, J. W., Measurements of cytochrome P-450 in the presence of large amounts of contaminating hemoglobin and methemoglobin, *Anal. Biochem.*, 86, 725–732, 1978.
106. Phillips, A. H. and Langdon, R. G., Hepatic triphosphopyridine nucleotide-cytochrome *c* reductase: isolation, characterization, and kinetic studies, *J. Biol. Chem.*, 237, 2652–2660, 1962.
107. Vermilion, J. L. and Coon, M. J., Purified liver microsomal NADPH-cytochrome P-450 reductase: spectral characterization of oxidation-reduction states, *J. Biol. Chem.*, 253, 2694–2704, 1978.
108. Roerig, D. L., Mascaro, Jr., L., and Aust, S. D., Microsomal electron transport: tetrazolium reduction by rat liver microsomal NADPH–cytochrome *c* reductase, *Arch. Biochem. Biophys.*, 153, 475–479, 1972.
109. Nash, T., The colorimetric estimation of formaldehyde by means of the Hantzsch reaction, *Biochem. J.*, 55, 416–421, 1953.
110. Cochin, J. and Axelrod, J., Biochemical and pharmacological changes in the rat following chronic administration of morphine, nalorphine, and normorphine, *J. Pharmacol. Exp. Ther.*, 125, 105–110, 1959.
111. Lu, A. Y. H. and West, S. B., Reconstituted mammalian mixed-function oxidases: requirements, specificities and other properties, *Pharmacol. Ther.*, 2, 337–358, 1978.
112. Guengerich, F. P., Ballou, D. P., and Coon, M. J., Purified liver microsomal cytochrome P-450: electron-accepting properties and oxidation–reduction potential, *J. Biol. Chem.*, 250, 7405–7414, 1975.
113. Guengerich, F. P. and Holladay, L. A., Hydrodynamic characterization of highly purified and functionally active liver microsomal cytochrome P-450, *Biochemistry*, 18, 5442–5449, 1979.
114. Thomas, P. E., Bandiera, S., Maines, S. L., Ryan, D. E., and Levin, W., Regulation of cytochrome P-450j, a high-affinity *N*-nitrosodimethylamine demethylase, in rat hepatic microsomes, *Biochemistry*, 26, 2280–2289, 1987.
115. Yoo, J. S. H., Guengerich, F. P., and Yang, C. S., Metabolism of *N*-nitrosodialkylamines by human liver microsomes, *Cancer Res.*, 48, 1499–1504, 1988.
116. Lu, A. Y. H., Strobel, H. W., and Coon, M. J., Properties of a solubilized form of the cytochrome P-450-containing mixed-function oxidase of liver microsomes, *Mol. Pharmacol.*, 6, 213–220, 1970.
117. Nordblom, G. D. and Coon, M. J., Hydrogen peroxide formation and stoichiometry of hydroxylation reactions catalyzed by highly purified liver microsomal cytochrome P-450, *Arch. Biochem. Biophys.*, 180, 343–347, 1977.
118. Kamataki, T., Lin, M. L., Belcher, D. H., and Neal, R. A., Studies of the metabolism of parathion with an apparently homogeneous preparation of rabbit liver cytochrome P-450, *Drug Metab. Dispos.*, 4, 180–189, 1976.
119. Greenlee, W. F. and Poland, A., An improved assay of 7-ethoxycoumarin *O*-deethylase activity: induction of hepatic enzyme activity in C57BL/6J and DBA/2J mice by phenobarbital, 3-methylcholanthrene and 2,3,7,8-tetrachlorodibenzo-*p*-dioxin, *J. Pharmacol. Exp. Ther.*, 205, 596–605, 1978.
120. Guengerich, F. P., Separation and purification of multiple forms of microsomal cytochrome P-450: partial characterization of three apparently homogeneous cytochromes P-450 prepared from livers of phenobarbital- and 3-methylcholanthrene-treated rats, *J. Biol. Chem.*, 253, 7931–7939, 1978.
121. Nebert, D. W. and Gelboin, H. V., Substrate-inducible microsomal arylhydroxylase in mammalian cell culture: assay and properties of induced enzyme, *J. Biol. Chem.*, 243, 6242–6249, 1968.
122. Dehnen, W., Tomingas, R., and Roos, J., A modified method for the assay of benzo(*a*)pyrene hydroxylase, *Anal. Biochem.*, 53, 373–383, 1973.

123. DePierre, J. W., Johannesen, K. A. M., Moron, M. S., and Seidegård, J., Radioactive assay of aryl hydrocarbon monooxygenase and epoxide hydrase, *Meth. Enzymol.*, 52, 412–418, 1978.
124. Selkirk, J. K., Croy, R. G., Roller, P. P., and Gelboin, H. V., High-pressure liquid chromatographic analysis of benzo(*a*)pyrene metabolism and covalent binding and the mechanism of action of 7,8-benzoflavone and 1,2-epoxy-3,3,3-trichloropropane, *Cancer Res.*, 34, 3474–3480, 1974.
125. Thakker, D. R., Yagi, H., and Jerina, D. M., Analysis of polycyclic aromatic hydrocarbons and their metabolites by high-pressure liquid chromatography, *Meth. Enzymol.*, 52, 279–296, 1978.
126. Bauer, E., Guo, Z., Ueng, Y.-F., Bell, L. C., and Guengerich, F. P., Oxidation of benzo(*a*)pyrene by recombinant human cytochrome P450 enzymes, *Chem. Res. Toxicol.*, 8, 136–142, 1995.
127. Deutsch, J., Leutz, J. C., Yang, S. K., Gelboin, H. V., Chiang, Y. L., Vatsis, K. P., and Coon, M. J., Regio- and stereoselectivity of various forms of purified cytochrome P-450 in the metabolism of benzo(*a*)pyrene and (–)*trans*-7,8-dihydroxy-7,8-dihydrobenzo(*a*)pyrene as shown by product formation and binding to DNA, *Proc. Natl. Acad. Sci. U.S.A.*, 75, 3123–3127, 1978.
128. Guengerich, F. P., Martin, M. V., Beaune, P. H., Kremers, P., Wolff, T., and Waxman, D. J., Characterization of rat and human liver microsomal cytochrome P-450 forms involved in nifedipine oxidation, a prototype for genetic polymorphism in oxidative drug metabolism, *J. Biol. Chem.*, 261, 5051–5060, 1986.
129. Böcker, R. H. and Guengerich, F. P., Oxidation of 4-aryl- and 4-alkyl-substituted 2,6-dimethyl-3,5-*bis*(alkoxycarbonyl)-1,4-dihydropyridines by human liver microsomes and immunochemical evidence for the involvement of a form of cytochrome P-450, *J. Med. Chem.*, 29, 1596–1603, 1986.
130. Guengerich, F. P., Brian, W. R., Iwasaki, M., Sari, M.-A., Bäärnhielm, C., and Berntsson, P., Oxidation of dihydropyridine calcium channel blockers and analogues by human liver cytochrome P-450 IIIA4, *J. Med. Chem.*, 34, 1838–1844, 1991.
131. Peter, R., Böcker, R. G., Beaune, P. H., Iwasaki, M., Guengerich, F. P., and Yang, C.-S., Hydroxylation of chlorzoxazone as a specific probe for human liver cytochrome P-450 IIE1, *Chem. Res. Toxicol.*, 3, 566–573, 1990.
132. O'Shea, D., Davis, S. N., Kim, R. B., and Wilkinson, G. R., Effect of fasting and obesity in humans on the 6-hydroxylation of chloroxazone: a putative probe of CYP2E1 activity, *Clin. Pharmacol. Ther.*, 56, 359–367, 1994.
133. Kim, R. B., Yamazaki, H., Mimura, M., Shimada, T., Guengerich, F. P., Chiba, K., Ishizaki, T., and Wilkinson, G. R., Chlorzoxazone 6-hydroxylation in Japanese and Caucasians: *in vitro* and *in vivo* differences, *J. Pharmacol. Exp. Ther.*, 279, 4–11, 1996.
134. Walsh, C., *Enzymatic Reaction Mechanisms*, W.H. Freeman, San Francisco, CA, 1979.
135. Guengerich, F. P., Enzymatic oxidation of xenobiotic chemicals, *CRC Crit. Rev. Biochem. Mol. Biol.*, 25, 97–153, 1990.
136. Okazaki, O. and Guengerich, F. P., Evidence for specific base catalysis in *N*-dealkylation reactions catalyzed by cytochrome P450 and chloroperoxidase: differences in rates of deprotonation of aminium radicals as an explanation for high kinetic hydrogen isotope effects observed with peroxidases, *J. Biol. Chem.*, 268, 1546–1552, 1993.
137. Guengerich, F. P., Yun, C.-H., and Macdonald, T. L., Evidence for a one-electron oxidation mechanism in *N*-dealkylation of *N,N*-dialkylanilines by cytochrome P450 2B1: kinetic hydrogen isotope effects, linear free energy relationships, comparisons with horseradish peroxidase, and studies with oxygen surrogates, *J. Biol. Chem.*, 271, 27321–27329, 1996.
138. Miwa, G. T., Walsh, J. S., Kedderis, G. L., and Hollenberg, P. F., The use of intramolecular isotope effects to distinguish between deprotonation and hydrogen atom abstraction mechanisms in cytochrome P-450- and peroxidase-catalyzed *N*-demethylation reactions, *J. Biol. Chem.*, 258, 14445–14449, 1983.
139. Bell, L. C. and Guengerich, F. P., Oxidation kinetics of ethanol by human cytochrome P450 2E1: rate-limiting product release accounts for effects of isotopic hydrogen substitution and cytochrome $b_5$ on steady-state kinetics, *J. Biol. Chem.*, 272, 29643–29651, 1997.
140. Guengerich, F. P. and Kim, D.-H., Enzymatic oxidation of ethyl carbamate to vinyl carbamate and its role as an intermediate in the formation of 1,$N^6$-ethenoadenosine, *Chem. Res. Toxicol.*, 4, 413–421, 1991.
141. Dahl, G. A., Miller, J. A., and Miller, E. C., Vinyl carbamate as a promutagen and a more carcinogenic analog of ethyl carbamate, *Cancer Res.*, 38, 3793–3804, 1978.
142. Dahl, G. A., Miller, E. C., and Miller, J. A., Comparative carcinogenicities and mutagenicities of vinyl carbamate, ethyl carbamate, and ethyl *N*-hydroxycarbamate, *Cancer Res.*, 40, 1194–1203, 1980.
143. Leithauser, M. T., Liem, A., Stewart, B. C., Miller, E. C., and Miller, J. A., 1,$N^6$-Ethenoadenosine formation, mutagenicity and murine tumor induction as indicators of the generation of an electrophilic epoxide metabolite of the closely related carcinogens ethyl carbamate (urethane) and vinyl carbamate, *Carcinogenesis*, 11, 463–473, 1990.
144. Yoo, J. S. H., Cheung, R. J., Patten, C. J., Wade, D., and Yang, C. S., Nature of *N*-nitrosodimethylamine demethylase and its inhibitors, *Cancer Res.*, 47, 3378–3383, 1987.
145. Udenfriend, S., *Fluoresence Assay in Biology and Medicine,* Academic Press, New York, 1969.
146. Rinkus, S. J. and Legator, M. S., Fluorometric assay using high-pressure liquid chromatography for the microsomal metabolism of certain substituted aliphatic to 1,$N^6$-ethenoadenine-forming metabolites, *Anal. Biochem.*, 150, 379–393, 1985.
147. Mitoma, C., Yasuda, D. M., Tagg, J., and Tanabe, M., Effect of deuteration of the O–$CH_3$ group on the enzymic demethylation of *o*-nitroanisole, *Biochim. Biophys. Acta*, 136, 566–567, 1967.
148. Miller, G. P. and Guengerich, F. P., Binding and oxidation of alkyl 4-nitrophenyl ethers by rabbit cytochrome P450 1A2: evidence for two binding sites, *Biochemistry*, 40, 7262–7272, 2001.

149. Yamano, S., Tatsuno, J., and Gonzalez, F. J., The *CYP2A3* gene product catalyzes coumarin 7-hydroxylation in human liver microsomes, *Biochemistry*, 29, 1322–1329, 1990.
150. Yun, C.-H., Shimada, T., and Guengerich, F. P., Purification and characterization of human liver microsomal cytochrome P-450 2A6, *Mol. Pharmacol.*, 40, 679–685, 1991.
151. Yun, C.-H., Kim, K.-H., Calcutt, M. W., and Guengerich, F. P., Kinetic analysis of oxidation of coumarins by human cytochrome P450 2A6, *J. Biol. Chem.*, 280, 12279–12291, 2005.
152. Wandel, C., Witte, J. S., Hall, J. M., Stein, C. M., Wood, A. J., and Wilkinson, G. R., CYP3A activity in African American and European American men: population differences and functional effect of the CYP3A4*1B5′-promoter region polymorphism, *Clin. Pharmacol. Ther.*, 68, 82–91, 2000.
153. Lowry, O. H., Rosebrough, N. J., Farr, A. L., and Randall, R. J., Protein measurement with the Folin phenol reagent, *J. Biol. Chem.*, 243, 1331–1332, 1951.
154. Sun, J. D. and Dent, J. G., A new method for measuring covalent binding of chemicals to cellular macromolecules, *Chem. Biol. Interact.*, 32, 41–61, 1980.
155. Wallin, H., Schelin, C., Tunek, A., and Jergil, B., A rapid and sensitive method for determination of covalent binding of benzo(*a*)pyrene to proteins, *Chem. Biol. Interact.*, 38, 109–118, 1981.
156. Guengerich, F. P., Crawford, Jr., W. M., Domoradzki, J. Y., Macdonald, T. L., and Watanabe, P. G., *In vitro* activation of 1,2-dichloroethane by microsomal and cytosolic enzymes, *Toxicol. Appl. Pharmacol.*, 55, 303–317, 1980.
157. Inskeep, P. B. and Guengerich, F. P., Glutathione-mediated binding of dibromoalkanes to DNA: specificity of rat glutathione *S*-transferases and dibromoalkane structure, *Carcinogenesis*, 5, 805–808, 1984.
158. Dische, Z., Color reactions of nucleic acid components, in *The Nucleic Acids*, Vol. 1, Chargoff, E. and Davidson, J. N. Academic Press, New York, 1955, pp. 285–305.
159. Cerasone, C. F., Bolognesi, C., and Santi, L., Improved microfluorometric DNA determinations in biological material using 33258 Hoechst, *Anal. Biochem.*, 100, 188–197, 1979.
160. Shimada, T., Oda, Y., Yamazaki, H., Mimura, M., and Guengerich, F. P., SOS function tests for studies of chemical carcinogenesis in *Salmonella typhimurium* TA 1535/pSK1002, NM2009, and NM3009, in *Methods in Molecular Genetics*, Vol. 5, *Gene and Chromosome Analysis*, Adolph, K. W., Ed., Academic Press, Orlando, FL, 1994, pp. 342–355.
161. Josephy, P. D., Evans, D. H., Parikh, A., and Guengerich, F. P., Metabolic activation of aromatic amine mutagens by simultaneous expression of human cytochrome P450 1A2, NADPH-cytochrome P450 reductase, and *N*-acetyltransferase in *Escherichia coli*, *Chem. Res. Toxicol.*, 11, 70–74, 1998.
162. Randerath, K., Reddy, M. V., and Gupta, R. C., $^{32}$P-labeling test for DNA damage, *Proc. Natl. Acad. Sci. U.S.A.*, 78, 6126–6129, 1981.
163. Sheabar, F. Z., Morningstar, M. L., and Wogan, G. N., Adduct detection by acylation with [$^{35}$S]methionine: analysis of DNA adducts of 4–aminobiphenyl, *Proc. Natl. Acad. Sci. U.S.A.*, 91, 1696–1700, 1994.
164. Guengerich, F. P., Sorrells, J. L., Schmitt, S., Krauser, J. A., Aryal, P., and Meijer, L., Generation of new protein kinase inhibitors utilizing cytochrome P450 mutant enzymes for indigoid coupling, *J. Med. Chem.*, 43, 3236–3241, 2004.
165. Jaffé, H. H. and Orchin, M., *Theory and Applications of Ultraviolet Spectroscopy*, John Wiley & Sons, New York, 1962.
166. Silverstein, R. M., Bassler, G. C., and Morrill, T. C., *Spectrometric Identification of Organic Compounds*, 5th ed., John Wiley & Sons, New York, 1991.
167. Dyer, J. R., *Applications of Absorption Spectroscopy of Organic Compounds*, Prentice-Hall, Englewood Cliffs, NJ, 1965.
168. Guengerich, F. P. et al., Structure of aflatoxin $B_1$ dialdehyde adduct formed from reaction with methylamine, *Chem. Res. Toxicol.*, 15, 793–798, 2002.
169. Krauser, J. A., Voehler, M., Tseng, L.-H., Schefer, A. B., Godejohann, M., and Guengerich, F. P., 1β-Hydroxylation of testosterone by human cytochrome P450 3A4, *Eur. J. Biochem.*, 271, 3962–3969, 2004.
170. Woody, R. W., Circular dichroism, *Methods Enzymol.*, 246, 34–71, 1995.
171. Meyring, M., Muhlbacher, J., Messer, K., Kastner-Pustet, N., Bringmann, G., Mannschreck, A., and Blaschke, G., *In vitro* biotransformation of (*R*)- and (*S*)-thalidomide: application of circular dichroism spectroscopy to the stereochemical characterization of the hydroxylated metabolites, *Anal. Chem.*, 74, 3726–3735, 2002.
172. Rabenstein, D. L., NMR spectroscopy: past and present, *Anal. Chem.*, 73, 214A–223A, 2001.
173. Gottlieb, H. E., Kotlyar, V., and Nudelman, A., NMR chemical shifts of common laboratory solvents as trace impurities, *J. Org. Chem.*, 62, 7512–7515, 1997.
174. Wu, Z., Aryal, P., Lozach, O., Meijer, L., and Guengerich, F. P., Biosynthesis of new indigoid inhibitors of protein kinases using recombinant cytochrome P450 2A6, *Chem. Biodiversity*, 2, 51–65, 2005.
175. Goeddel, D. W., Ed., *Methods in Enzymology*, Vol. 185, *Gene Expression Technology*, Academic Press, San Diego, CA, 1990.
176. Waterman, M. R. and Johnson, E. F., *Methods in Enzymology*, Vol. 206, *Cytochrome P450*, Academic Press, San Diego, CA, 1991.
177. Guengerich, F. P., Gillam, E. M. J., and Shimada, T., New applications of bacterial systems to problems in toxicology, *CRC Crit. Rev. Toxicol.*, 26, 551–583, 1996.
178. Langouët, S., Furge, L. L., Kerriguy, N., Nakamura, K., Guillouzo, A., and Guengerich, F. P., Inhibition of human cytochrome P450 enzymes by 1,2-dithiole-3-thione, oltipraz and its derivatives, and sulforaphane, *Chem. Res. Toxicol.*, 13, 245–252, 2000.
179. Parikh, A., Josephy, P. D., and Guengerich, F. P., Selection and characterization of human cytochrome P450 1A2 mutants with altered catalytic properties, *Biochemistry*, 38, 5283–5289, 1999.

180. Jensen, K. G., Poulsen, H. E., Doehmer, J., and Loft, S., Paracetamol-induced spindle disturbances in V79 cells with and without expression of human CYP1A2, *Pharmacol. Toxicol.*, 78, 224–228, 1996.
181. Hosea, N. A., Miller, G. P., and Guengerich, F. P., Elucidation of distinct binding sites for cytochrome P450 3A4, *Biochemistry*, 39, 5929–5939, 2000.
182. Guengerich, F. P. and Martin, M. V., Purification of cytochrome P450: products of bacterial recombinant expression systems, in *Methods in Molecular Genetics, Cytochrome P450 Protocols*, Phillips, I. R. and Shephard, E., Ed., Academic Press, Orlando, FL, 2005.
183. Hanna, I. H., Dawling, S., Roodi, N., Guengerich, F. P., and Parl, F., Cytochrome P450 *1B1* (*CYP1B1*) pharmacogenetics: association of polymorphisms with functional differences in estrogen hydroxylation activity, *Cancer Res.*, 60, 3440–3444, 2000.
184. Barnes, H. J., Arlotto, M. P., and Waterman, M. R., Expression and enzymatic activity of recombinant cytochrome P450 17α-hydroxylase in *Escherichia coli*, *Proc. Natl. Acad. Sci. U.S.A.*, 88, 5597–5601, 1991.
185. Thier, R., Pemble, S. E., Taylor, J. B., Humphreys, W. G., Persmark, M., Ketterer, B., and Guengerich, F. P., Expression of mammalian glutathione *S*-transferase 5-5 in *Salmonella typhimurium* TA1535 leads to base-pair mutations upon exposure to dihalomethanes, *Proc. Natl. Acad. Sci. U.S.A.*, 90, 8576–8580, 1993.
186. Josephy, P. D., DeBruin, L. S., Lord, H. L., Oak, J., Evans, D. H., Guo, Z., Dong, M.-S., and Guengerich, F. P., Bioactivation of aromatic amines by recombinant human cytochrome P450 1A2 expressed in bacteria: a substitute for mammalian tissue preparations in mutagenicity testing, *Cancer Res.*, 55, 799–802, 1995.
187. Suzuki, A., Kushida, H., Iwata, H., Watanabe, M., Nohmi, T., Fujita, K., Gonzalez, F. J., and Kamataki, T., Establishment of a *Salmonella* tester strain highly sensitive to mutagenic heterocyclic amines, *Cancer Res.*, 58, 1833–1838, 1998.
188. Barnes, H. J., Maximizing expression of eukaryotic cytochrome P450s in *Escherichia coli*, *Methods Enzymol.*, 272, 3–14, 1996.
189. Gellissen, G., Janowicz, Z. A., Weydemann, U., Melber, K., Strasser, A. W., and Hollenberg, C. P., High-level expression of foreign genes in *Hansenula polymorpha*, *Biotechnol. Adv.*, 10, 179–189, 1992.
190. Guengerich, F. P., Brian, W. R., Sari, M.-A., and Ross, J. T., Expression of mammalian cytochrome P450 enzymes using yeast-based vectors, *Meth. Enzymol.*, 206, 130–145, 1991.
191. Loeper, J., Louérat-Oriou, B., Duport, C., and Pompon, D., Yeast expressed cytochrome P450 2D6 (CYP2D6) exposed on the external face of plasma membrane is functionally competent, *Mol. Pharmacol.*, 54, 8–13, 1998.
192. Marques-Soares, C., Dijols, S., Macherey, A.-C., Wester, M. R., Johnson, E. F., Dansette, P. M., and Mansuy, D., Sulfaphenazole derivatives as tools for comparing cytochrome P450 2C5 and human cytochrome P450 2Cs: identification of a new high affinity substrate common to those CYP 2C enzymes, *Biochemistry* 42, 6363–6369, 2003.
193. Newton-Vinson, P., Hubalek, F., and Edmondson, D. E., High-level expression of human liver monoamine oxidase B in *Pichia pastoris*, *Prot. Express. Purif.*, 20, 334–345, 2000.
194. Reddy, R. G., Yoshimoto, T., Yamamoto, S., and Marnett, L. J., Expression, purification, and characterization of porcine leukocyte 12-lipoxygenase produced in the methyltrophic yeast, *Pichia pastoris*, *Biochem. Biophys. Res. Commun.*, 205, 381–388, 1994.
195. Brian, W. R., Srivastava, P. K., Umbenhauer, D. R., Lloyd, R. S., and Guengerich, F. P., Expression of a human liver cytochrome P-450 protein with tolbutamide hydroxylase activity in *Saccharomyces cerevisiae*, *Biochemistry* 28, 4993–4999, 1989.
196. Brian, W. R., Sari, M.-A., Iwasaki, M., Shimada, T., Kaminsky, L. S., and Guengerich, F. P., Catalytic activities of human liver cytochrome P-450 IIIA4 expressed in *Saccharomyces cerevisiae*, *Biochemistry* 29, 11280–11292, 1990.
197. Miller, L. K., Baculoviruses as gene expression vectors, *Annu. Rev. Microbiol.*, 42, 177–199, 1988.
198. Tamura, S., Korzekwa, K. R., Kimura, S., Gelboin, H. V., and Gonzalez, F. J., Baculovirus-mediated expression and functional characterization of human NADPH-P450 oxidoreductase, *Arch. Biochem. Biophys.*, 293, 219–223, 1992.
199. Zuber, M. X., Simpson, E. R., and Waterman, M. R., Expression of bovine 17α-hydroxylase cytochrome P-450 cDNA in nonsteroidogenic (COS 1) cells, *Science*, 234, 1258–1261, 1986.
200. Gonzalez, F. J., Aoyama, T., and Gelboin, H. V., Expression of mammalian cytochrome P450 using vaccinia virus, *Meth. Enzymol.*, 206, 85–92, 1991.
201. Kimchi-Sarfaty, C., Gribar, J. J., and Gottesman, M. M., Functional characterization of coding polymorphisms in the human *MDR*1 gene using a vaccinia virus expression system, *Mol. Pharmacol.*, 62, 1–6, 2002.
202. Gonzalez, F. J. and Kimura, S., Study of P450 function using gene knockout and transgenic mice, *Arch. Biochem. Biophys.*, 409, 153–158, 2003.
203. Guengerich, F. P., Enzymology of rat liver cytochromes P-450, in *Mammalian Cytochromes P-450*, Vol. 1, Guengerich, F. P., Ed., CRC Press, Boca Raton, FL, 1987, pp. 1–54.
204. Guengerich, F. P., Distlerath, L. M., Reilly, P. E. B., Wolff, T., Shimada, T., Umbenhauer, D. R., and Martin, M. V., Human liver cytochromes P-450 involved in polymorphisms of drug oxidation, *Xenobiotica*, 16, 367–378, 1986.
205. Nebert, D. W., Nelson, D. R., Coon, M. J., Estabrook, R. W., Feyereisen, R., Fujii-Kuriyama, Y., Gonzalez, F. J., Guengerich, F. P., Gunsalus, I. C., Johnson, E. F., Loper, J. C., Sato, R., Waterman, M. R., and Waxman, D. J., The P450 superfamily: update on new sequences, gene mapping, and recommended nomenclature, *DNA Cell Biol.*, 10, 1–14, 1991.
206. Guengerich, F. P., Hosea, N. A., Parikh, A., Bell-Parikh, L. C., Johnson, W. W., Gillam, E. M. J., and Shimada, T., Twenty years of biochemistry of human P450s: purification, expression, mechanism, and relevance to drugs, *Drug Metab. Dispos.*, 26, 1175–1178, 1998.

207. Halpert, J. R., Guengerich, F. P., Bend, J. R., and Correia, M. A., Selective inhibitors of cytochromes P450, *Toxicol. Appl. Pharmacol.*, 125, 163–175, 1994.
208. Shimada, T. and Guengerich, F. P., Participation of a rat liver cytochrome P-450 induced by pregnenolone 16α-carbonitrile and other compounds in the 4-hydroxylation of mephenytoin, *Mol. Pharmacol.*, 28, 215–219, 1985.
209. Guengerich, F. P., Comparisons of catalytic selectivity of cytochrome P450 subfamily members from different species, *Chem. Biol. Interact.*, 106, 161–182, 1997.
210. Beaune, P., Kremers, P. G., Kaminsky, L. S., de Graeve, J., and Guengerich, F. P., Comparison of monooxygenase activities and cytochrome P-450 isozyme concentrations in human liver microsomes, *Drug Metab. Dispos.*, 14, 437–442, 1986.
211. Guengerich, F. P. and Shimada, T., Oxidation of toxic and carcinogenic chemicals by human cytochrome P-450 enzymes, *Chem. Res. Toxicol.*, 4, 391–407, 1991.
212. Butler, M. A., Iwasaki, M., Guengerich, F. P., and Kadlubar, F. F., Human cytochrome P-$450_{PA}$ (P-450IA2), the phenacetin *O*-deethylase, is primarily responsible for the hepatic 3–demethylation of caffeine and *N*-oxidation of carcinogenic arylamines, *Proc. Natl. Acad. Sci. U.S.A.*, 86, 7696–7700, 1989.
213. Distlerath, L. M. and Guengerich, F. P., Characterization of a human liver cytochrome P-450 involved in the oxidation of debrisoquine and other drugs using antibodies raised to the analogous rat enzyme, *Proc. Natl. Acad. Sci. U.S.A.*, 81, 7348–7352, 1984.
214. Thomas, P. E., Lu, A. Y. H., West, S. B., Ryan, D., Miwa, G. T., and Levin, W., Accessibility of cytochrome P450 in microsomal membranes: inhibition of metabolism by antibodies to cytochrome P450, *Mol. Pharmacol.*, 13, 819–831, 1977.
215. Kaminsky, L. S., Fasco, M. J., and Guengerich, F. P., Production and application of antibodies to rat liver cytochrome P-450, *Meth. Enzymol.*, 74, 262–272, 1981.
216. Soucek, P., Martin, M. V., Ueng, Y.-F., and Guengerich, F. P., Identification of a common epitope near the conserved heme-binding region with polyclonal antibodies raised against cytochrome P450 family 2 proteins, *Biochemistry*, 34, 16013–16021, 1995.
217. Fujino, T., Park, S. S., West, D., and Gelboin, H. V., Phenotyping of cytochromes P-450 in human tissues with monoclonal antibodies, *Proc. Natl. Acad. Sci. U.S.A.*, 79, 3682–3686, 1982.
218. Gelboin, H. V., Cytochrome P450 and monoclonal antibodies, *Pharmacol. Rev.*, 45, 413–453, 1993.
219. Thomas, P. E., Reidy, J., Reik, L. M., Ryan, D. E., Koop, D. R., and Levin, W., Use of monoclonal antibody probes against rat hepatic cytochromes P-450c and P-450d to detect immunochemically related isozymes in liver microsomes from different species, *Arch. Biochem. Biophys.*, 235, 239–253, 1984.
220. Reubi, I., Griffin, K. J., Raucy, J. L., and Johnson, E. F., Three monoclonal antibodies to rabbit microsomal cytochrome P-450 1 recognize distinct epitopes that are shared to different degrees among other electrophoretic types of cytochrome P-450, *J. Biol. Chem.*, 259, 5887–5892, 1984.
221. Cribb, A., Nuss, C., and Wang, R., Antipeptide antibodies against overlapping sequences differentially inhibit human CYP2D6, *Drug Metab. Dispos.*, 23, 671–675, 1995.
222. Wang, R. W. and Lu, A. Y. H., Inhibitory anti-peptide antibody against human CYP3A4, *Drug Metab. Dispos.*, 25, 762–767, 1997.
223. Baca, M., Scanlan, T. S., Stephenson, R. C., and Wells, J. A., Phage display of a catalytic antibody to optimize affinity for transition-state analog binding, *Proc. Natl. Acad. Sci. U.S.A.*, 94, 10063–10068, 1997.
224. Kyte, J., *Mechanism in Protein Chemistry*, Garland, New York, 1995.
225. Renwick, A. G., Toxicokinetics: pharmacokinetics in toxicology, in *Principles and Methods of Toxicology*, 3rd ed., Hayes, A. W., Ed., Raven Press, New York, 1994, pp. 101–147.
226. Andersen, M. E., Clewell, H. J., and Frederick, C. B., Applying simulation modeling to problems in toxicology and risk assessment: a short perspective, *Toxicol. Appl. Pharmacol.*, 133, 181–187, 1995.
227. Gillam, E. M. J., Baba, T., Kim, B.-R., Ohmori, S., and Guengerich, F. P., Expression of modified human cytochrome P450 3A4 in *Escherichia coli* and purification and reconstitution of the enzyme, *Arch. Biochem. Biophys.*, 305, 123–131, 1993.
228. Guengerich, F. P., Martin, M. V., Guo, Z., and Chun, Y.-J., Purification of recombinant human cytochrome P450 enzymes expressed in bacteria, *Meth. Enzymol.*, 272, 35–44, 1996.
229. Domanski, T. L., Liu, J., Harlow, G. R., and Halpert, J. R., Analysis of four residues within substrate recognition site 4 of human cytochrome P450 3A4: role in steroid hydroxylase activity and α-naphthoflavone stimulation, *Arch. Biochem. Biophys.*, 350, 223–232, 1998.
230. Guengerich, F. P., Hosea, N. A., and Martin, M. V., Purification of P450s. Products of bacterial recombinant systems, in *Methods in Molecular Genetics*, Vol. 107, *Cytochrome P450 Protocols*, Phillips, I. R. and Shephard, E., Eds., Academic Press, Orlando, FL, 1998, pp. 77–83.
231. Porath, J., Immobilized metal ion affinity chromatography, *Prot. Express. Purif.*, 3, 263–281, 1992.
232. Porath, J., Carlsson, J., Olsson, I., and Belfrage, G., Metal chelate affinity chromatography, a new approach to protein fractionation, *Nature*, 258, 598–599, 1975.
233. Jenkins, C. M. and Waterman, M. R., Flavodoxin and NADPH-flavodoxin reductase from *Escherichia coli* support bovine cytochrome P450c17 hydroxylase activities, *J. Biol. Chem.*, 269, 27401–27408, 1994.
234. Imai, T., Globerman, H., Gertner, J. M., Kagawa, N., and Waterman, M. R., Expression and purification of functional human 17alpha-hydroxylase/17,20-lyase (P450c17) in *Escherichia coli*. Use of this system for study of a novel form of combined 17alpha-hydroxylase/17,20-lyase deficiency, *J. Biol. Chem.*, 268, 19681–19689, 1993.
235. Kempf, A., Zanger, U. M., and Meyer, U. A., Truncated human P450 2D6: expression in *Escherichia coli*: $Ni^{2+}$-chelate affinity purification, and characterization of solubility and aggregation, *Arch. Biochem. Biophys.*, 321, 277–288, 1995.

236. Sandhu, P., Baba, T., and Guengerich, F. P., Expression of modified cytochrome P450 2C10 (2C9) in *Escherichia coli*, purification, and reconstitution of catalytic activity, *Arch. Biochem. Biophys.*, 306, 443–450, 1993.
237. Gillam, E. M. J., Guo, Z., and Guengerich, F. P., Expression of modified human cytochrome P450 2E1 in *Escherichia coli*, purification, and spectral and catalytic properties, *Arch. Biochem. Biophys.*, 312, 59–66, 1994.
238. Gillam, E. M. J., Guo, Z., Ueng, Y.-F., Yamazaki, H., Cock, I., Reilly, P. E. B., Hooper, W. D., and Guengerich, F. P., Expression of cytochrome P450 3A5 in *Escherichia coli*: effects of 5′ modifications, purification, spectral characterization, reconstitution conditions, and catalytic activities, *Arch. Biochem. Biophys.*, 317, 374–384, 1995.
239. Guo, Z., Gillam, E. M. J., Ohmori, S., Tukey, R. H., and Guengerich, F. P., Expression of modified human cytochrome P450 1A1 in *Escherichia coli*: effects of 5′ substitution, stabilization, purification, spectral characterization, and catalytic properties, *Arch. Biochem. Biophys.*, 312, 436–446, 1994.
240. Hosea, N. A. and Guengerich, F. P., Oxidation of non-ionic detergents by cytochrome P450 enzymes, *Arch. Biochem. Biophys.*, 353, 365–373, 1998.
241. Yun, C.-H., Miller, G. P., and Guengerich, F. P., Rate-determining steps in phenacetin oxidations by human cytochrome P450 1A2 and selected mutants, *Biochemistry*, 39, 11319–11329, 2000.
242. Kim, D. and Guengerich, F. P., Selection of human cytochrome P450 1A2 mutants with selectivity enhanced catalytic activity for heterocyclic amine *N*-hydroxylation, *Biochemistry*, 43, 981–988, 2004.
243. Hanna, I. H., Kim, M.-S., and Guengerich, F. P., Heterologous expression of cytochrome P450 2D6 mutants, electron transfer, and catalysis of bufuralol hydroxylation: the role of aspartate 301 in structural integrity, *Arch. Biochem. Biophys.*, 393, 255–261, 2001.
244. Chun, Y.-J., Kim, S., Kim, D., Lee, S.-K., and Guengerich, F. P., A new selective and potent inhibitor of human cytochrome P450 1B1 and its application to antimutagenesis, *Cancer Res.*, 61, 8164–8170, 2001.
245. Laemmli, U. K., Cleavage of structural proteins during the assembly of the head of bacteriophage $T_4$, *Nature*, 227, 680–685, 1970.
246. Yasukochi, Y. and Masters, B. S. S., Some properties of a detergent-solubilized NADPH-cytochrome *c* (cytochrome P-450) reductase purified by biospecific affinity chromatography, *J. Biol. Chem.*, 251, 5337–5344, 1976.
247. Strobel, H. W. and Dignam, J. D., Purification and properties of NADPH-cytochrome P-450 reductase, *Meth. Enzymol.*, 52, 89–96, 1978.
248. Guengerich, F. P., Separation and purification of multiple forms of microsomal cytochrome P-450: activities of different forms of cytochrome P-450 towards several compounds of environmental interest, *J. Biol. Chem.*, 252, 3970–3979, 1977.
249. Guengerich, F. P., Preparation and properties of highly purified cytochrome P-450 and NADPH-cytochrome P-450 reductase from pulmonary microsomes of untreated rabbits, *Mol. Pharmacol.*, 13, 911–923, 1977.
250. Guengerich, F. P. and Martin, M. V., Purification of cytochrome P-450, NADPH-cytochrome P-450 reductase, and epoxide hydratase from a single preparation of rat liver microsomes, *Arch. Biochem. Biophys.*, 205, 365–379, 1980.
251. Imai, Y., The use of 8-aminooctyl sepharose for the separation of some components of the hepatic microsomal electron transfer system, *J. Biochem.*, 80, 267–276, 1976.
252. Imai, Y. and Sato, R., A gel-electrophoretically homogeneous preparation of cytochrome P-450 from liver microsomes of phenobarbital-pretreated rabbits, *Biochem. Biophys. Res. Commun.*, 60, 8–14, 1974.
253. Wang, P., Mason, P. S., and Guengerich, F. P., Purification of human liver cytochrome P-450 and comparison to the enzyme isolated from rat liver, *Arch. Biochem. Biophys.*, 199, 206–219, 1980.
254. Hanna, I. H., Teiber, J. F., Kokones, K. L., and Hollenberg, P. F., Role of the alanine at position 363 of cytochrome P450 2B2 in influencing the NADPH- and hydroperoxide-supported activities, *Arch. Biochem. Biophys.*, 350, 324–332, 1998.
255. Guengerich, F. P., Destruction of heme and hemoproteins mediated by liver microsomal reduced nicotinamide adenine dinucleotide phosphate-cytochrome P-450 reductase, *Biochemistry*, 17, 3633–3639, 1978.
256. Bradford, M. M., A rapid and sensitive method for the quantitation of microgram quantities of protein utilizing the principle of protein-dye binding, *Anal. Biochem.*, 72, 248–254, 1976.
257. Brown, R. E., Jarvis, K. L., and Hyland, K. J., Protein measurement using bicinchoninic acid: elimination of interfering substances, *Anal. Biochem.*, 180, 136–139, 1989.
258. Furge, L. L. and Guengerich, F. P., Explanation of pre-steady-state kinetics and decreased burst amplitude of HIV-1 reverse transcriptase at sites of modified DNA bases with an additional non-productive enzyme-DNA-nucleotide complex, *Biochemistry*, 38, 4818–4825, 1999.
259. Segel, I. H., *Biochemical Calculations: How To Solve Mathematical Problems in General Biochemistry*, 2nd ed., John Wiley & Sons, New York, 1976.
260. Einolf, H. J. and Guengerich, F. P., Kinetic analysis of nucleotide incorporation by mammalian DNA polymerase δ, *J. Biol. Chem.*, 275, 16316–16322, 2000.
261. Bligh, E. G. and Dyer, W. J., A rapid method of total lipid extraction and purification, *Can. J. Biochem. Physiol.*, 37, 911–917, 1959.
262. Chen, Jr., P. S., Toribara, T. Y., and Warner, H., Microdetermination of phosphorus, *Anal. Chem.*, 28, 1756–1758, 1956.
263. Garewal, H. S., A procedure for the estimation of microgram quantities of Triton X-100, *Anal. Biochem.*, 54, 319–324, 1973.
264. Goldstein, S. and Blecher, M., The spectrophotometric assay for the polyethoxy nonionic detergents in membrane extracts: a critique, *Anal. Biochem.*, 64, 130–135, 1975.
265. Sadano, H. and Omura, T., Reversible transfer of heme between different molecular species of microsome-bound cytochrome P-450 in rat liver, *Biochem. Biophys. Res. Commun.*, 116, 1013–1019, 1983.

266. Black, S. D. and Coon, M. J., Comparative structures of P-450 cytochromes, in *Cytochrome P-450*, Ortiz de Montellano, P. R., Ed., Plenum, New York, 1986, pp. 161–216.
267. Knapp, J. A., Dignam, J. D., and Strobel, H. W., NADPH–cytochrome P-450 reductase: circular dichroism and physical studies, *J. Biol. Chem.*, 252, 437–443, 1977.
268. Dannan, G. A. and Guengerich, F. P., Immunochemical comparison and quantitation of microsomal flavin-containing monooxygenases in various hog, mouse, rat, rabbit, dog, and human tissues, *Mol. Pharmacol.*, 22, 787–794, 1982.
269. Guengerich, F. P., Artifacts in isoelectric focusing of the microsomal enzymes cytochrome P-450 and NADPH-cytochrome P-450 reductase, *Biochim. Biophys. Acta*, 577, 132–141, 1979.
270. Guengerich, F. P., Isolation and purification of cytochrome P-450, and the existence of multiple forms, *Pharmacol. Ther.*, 6, 99–121, 1979.
271. O'Farrell, P. Z., Goodman, H. M., and O'Farrell, P. H., High resolution two-dimensional electrophoresis of basic as well as acidic proteins, *Cell*, 12, 1133–1142, 1977.
272. Vlasuk, G. P. and Walz, Jr., F. G., Liver endoplasmic reticulum polypeptides resolved by two-dimensional gel electrophoresis, *Anal. Biochem.*, 105, 112–120, 1980.
273. Guengerich, F. P., Wang, P., and Mason, P. S., Immunological comparison of rat, rabbit, and human liver NADPH–cytochrome P-450 reductases, *Biochemistry*, 20, 2379–2385, 1981.
274. Guengerich, F. P., Wang, P., Mason, P. S., and Mitchell, M. B., Immunological comparison of rat, rabbit, and human microsomal cytochromes P-450, *Biochemistry*, 20, 2370–2378, 1981.
275. Thomas, P. E., Ryan, D., and Levin, W., An improved staining procedure for the detection of the peroxidase activity of cytochrome P-450 on sodium dodecyl sulfate polyacrylamide gels, *Anal. Biochem.*, 75, 168–176, 1976.
276. Shively, J. E., Reverse-phase HPLC isolation and microsequence analysis, in *Methods of Protein Microcharacterization*, Shively, J. E., Ed., Humana Press, Clifton, NJ, 1986, pp. 41–87.
277. Shimada, T., Wunsch, R. M., Hanna, I. H., Sutter, T. R., Guengerich, F. P., and Gillam, E. M. J., Recombinant human cytochrome P450 1B1 expression in *Escherichia coli*, *Arch. Biochem. Biophys.*, 357, 111–120, 1998.
278. Dong, M.-S., Bell, L. C., Guo, Z., Phillips, D. R., Blair, I. A., and Guengerich, F. P., Identification of retained *N*-formylmethionine in bacterial recombinant cytochrome P450 proteins with the N-terminal sequence MALLLAVFL...: roles of residues 3–5 in retention and membrane topology, *Biochemistry*, 35, 10031–10040, 1996.
279. Lu, A. Y. H. and Levin, W., The resolution and reconstitution of the liver microsomal hydroxylation system, *Biochim. Biophys. Acta*, 344, 205–240, 1974.
280. Lu, A. Y. H., Levin, W., and Kuntzman, R., Reconstituted liver microsomal enzyme system that hydroxylates drugs, other foreign compounds and endogenous substrates. VII. Stimulation of benzphetamine *N*-demethylation by lipid and detergent, *Biochem. Biophys. Res. Commun.*, 60, 266–272, 1974.
281. Lu, A. Y. H., Strobel, H. W., and Coon, M. J., Hydroxylation of benzphetamine and other drugs by a solubilized form of cytochrome P-450 from liver microsomes: lipid requirement for drug demethylation, *Biochem. Biophys. Res. Commun.*, 36, 545–551, 1969.
282. French, J. S., Guengerich, F. P., and Coon, M. J., Interactions of cytochrome P-450, NADPH–cytochrome P-450 reductase, phospholipid, and substrate in the reconstituted liver microsomal enzyme system, *J. Biol. Chem.*, 255, 4112–4119, 1980.
283. Chauret, N., Gauthier, A., and Nicoll-Griffith, D. A., Effect of common organic solvents on *in vitro* cytochrome P450-mediated metabolic activities in human liver microsomes, *Drug Metab. Dispos.*, 26, 1–4, 1998.
284. Ziegler, D. M., Flavin-containing monooxygenases: catalytic mechanism and substrate specificities, *Drug Metab. Rev.*, 19, 1–32, 1988.
285. Ziegler, D. M., Recent studies on the structure and function of multisubstrate flavin-containing monooxygenases, *Annu. Rev. Pharmacol. Toxicol.*, 33, 179–199, 1993.
286. Ziegler, D. M., Microsomal flavin-containing monooxygenase: oxygenation of nucleophilic nitrogen and sulfur compounds, in *Enzymatic Basis of Detoxication*, Vol. 1, Jakoby, W. B., Ed., Academic Press, New York, 1980, pp. 201–227.
287. Knodell, R. G., Browne, D., Gwodz, G. P., Brian, W. R., and Guengerich, F. P., Differential inhibition of human liver cytochromes P-450 by cimetidine, *Gastroenterology* 101, 1680–1691, 1991.
288. Dean, W. L. and Coon, M. J., Immunochemical studies on two electrophoretically homogeneous forms of rabbit liver microsomal cytochrome P-450: P-$450_{LM2}$ and P-$450_{LM4}$, *J. Biol. Chem.*, 252, 3255–3261, 1977.
289. Johnson, E. F. and Muller-Eberhard, U., Multiple forms of cytochrome P-450: resolution and purification of rabbit liver aryl hydrocarbon hydroxylase, *Biochem. Biophys. Res. Commun.*, 76, 644–651, 1977.
290. Kamataki, T., Belcher, D. H., and Neal, R. A., Studies of the metabolism of diethyl *p*-nitrophenyl phosphorothionate (parathion) and benzphetamine using an apparently homogeneous preparation of rat liver cytochrome P-450: effect of a cytochrome P-450 antibody preparation, *Mol. Pharmacol.*, 12, 921–932, 1976.
291. Kaminsky, L. S., Fasco, M. J., and Guengerich, F. P., Comparison of different forms of liver, kidney, and lung microsomal cytochrome P-450 by immunological inhibition of regio- and stereoselective metabolism of warfarin, *J. Biol. Chem.*, 254, 9657–9662, 1979.
292. Kaminsky, L. S., Fasco, M. J., and Guengerich, F. P., Comparison of different forms of purified cytochrome P-450 from rat liver by immunological inhibition of regio- and stereoselective metabolism of warfarin, *J. Biol. Chem.*, 255, 85–91, 1980.
293. Thomas, P. E., Koreniowski, D., Ryan, D., and Levin, W., Preparation of monospecific antibodies against two forms of rat liver cytochrome P-450 and quantitation of these antigens in microsomes, *Arch. Biochem. Biophys.*, 192, 524–532, 1979.

294. Masters, B. S. S., Baron, J., Taylor, W. E., Isaacson, E. L., and LoSpalluto, J., Immunochemical studies on electron transport chains involving cytochrome P-450. I. Effects of antibodies to pig liver microsomal reduced triphosphopyridine nucleotide–cytochrome *c* reductase and the non-heme iron protein from bovine adrenocortical mitochondria, *J. Biol. Chem.*, 246, 4143–4150, 1971.
295. Bentley, P. and Oesch, F., Purification of rat liver epoxide hydratase to apparent homogeneity, *FEBS Lett.*, 59, 291–295, 1975.
296. Johnson, E. F. and Muller-Eberhard, U., Resolution of two forms of cytochrome P-450 from liver microsomes of rabbits treated with 2,3,7,8-tetrachlorodibenzo-*p*-dioxin, *J. Biol. Chem.*, 252, 2839–2845, 1977.
297. Johnson, E. F. and Muller-Eberhard, U., Purification of the major cytochrome P-450 of liver microsomes from rabbits treated with 2,3,7,8-tetrachlorodibenzo-*p*-dioxin (TCDD), *Biochem. Biophys. Res. Commun.*, 76, 652–659, 1977.
298. Knowles, R. G. and Burchell, B., A simple method for purification of epoxide hydratase from rat liver, *Biochem. J.*, 163, 381–383, 1977.
299. Ryan, D. E., Thomas, P. E., Korzeniowski, D., and Levin, W., Separation and characterization of highly purified forms of liver microsomal cytochrome P-450 from rats treated with polychlorinated biphenyls, phenobarbital, and 3-methylcholanthrene, *J. Biol. Chem.*, 254, 1365–1374, 1979.
300. De Lemos-Chiarandini, C., Frey, A. B., Sabatini, D. D., and Kreibich, G., Determination of the membrane topology of the phenobarbital-inducible rat liver cytochrome P-450 isoenzyme PB-4 using site-specific antibodies, *J. Cell Biol.*, 104, 209–219, 1987.
301. Kawabata, T. T., Guengerich, F. P., and Baron, J., An immunohistochemical study on the localization and distribution of epoxide hydrolase within livers of untreated rats, *Mol. Pharmacol.*, 20, 709–714, 1981.
302. Shen, J., Moy, J. A., Green, M. D., Guengerich, F. P., and Baron, J., Immunohistochemical demonstration of β-naphthoflavone-inducible cytochrome P450 1A1/1A2 in rat intrahepatic biliary epithelial cells, *Hepatology*, 27, 1483–1491, 1998.
303. Milstein, C., From antibody diversity to monoclonal antibodies, *Eur. J. Biochem.*, 118, 429–436, 1981.
304. Gill III, T. J., The chemistry of antigens and its influence on immunogenicity, in *Imunogenicity*, Borek, F., Ed., Elsevier, New York, 1972, pp. 5–44.
305. Distlerath, L. M., Reilly, P. E. B., Martin, M. V., Davis, G. G., Wilkinson, G. R., and Guengerich, F. P., Purification and characterization of the human liver cytochromes P-450 involved in debrisoquine 4-hydroxylation and phenacetin *O*-deethylation, two prototypes for genetic polymorphism in oxidative drug metabolism, *J. Biol. Chem.*, 260, 9057–9067, 1985.
306. Shimada, T., Misono, K. S., and Guengerich, F. P., Human liver microsomal cytochrome P-450 mephenytoin 4-hydroxylase, a prototype of genetic polymorphism in oxidative drug metabolism: purification and characterization of two similar forms involved in the reaction, *J. Biol. Chem.*, 261, 909–921, 1986.
307. Oesch, F. and Bentley, P., Antibodies against homogeneous epoxide hydratase provide evidence for a single enzyme hydrating styrene oxide and benzo(*a*)pyrene 4,5-oxide, *Nature*, 259, 53–55, 1976.
308. Park, S. S., Fujino, T., Miller, H., Guengerich, F. P., and Gelboin, H. V., Monoclonal antibodies to phenobarbital-induced rat liver cytochrome P-450, *Biochem. Pharmacol.*, 33, 2071–2081, 1984.
309. Park, S. S., Fujino, T., West, D., Guengerich, F. P., and Gelboin, H. V., Monoclonal antibodies that inhibit enzyme activity of 3-methylcholanthrene-induced cytochrome P-450, *Cancer Res.*, 42, 1798–1808, 1982.
310. Paye, M., Beaune, P., Kremers, P., Frankinet-Collignon, C., Guengerich, F. P., Goujon, F., and Gielen, J., Quantification of two cytochrome P-450 isoenzymes by an enzyme-linked immunosorbent assay (ELISA), *Biochem. Biophys. Res. Commun.*, 122, 137–142, 1984.
311. Guengerich, F. P., Wang, P., and Davidson, N. K., Estimation of isozymes of microsomal cytochrome P-450 in rats, rabbits, and humans using immunochemical staining coupled with sodium dodecyl sulfate-polyacrylamide gel electrophoresis, *Biochemistry*, 21, 1698–1706, 1982.
312. Towbin, H., Staehelin, T., and Gordon, J., Electrophoretic transfer of proteins from polyacrylamide gels to nitrocellulose sheets: procedure and some applications, *Proc. Natl. Acad. Sci. U.S.A.*, 76, 4350–4354, 1979.
313. Shaw, P. M., Reiss, A., Adesnik, M., Nebert, D. W., Schembri, J., and Jaiswal, A. K., The human dioxin-inducible NAD(P)H:quinone oxidoreductase cDNA-encoded protein expressed in COS-1 cells is identical to diaphorase 4, *Eur. J. Biochem.*, 195, 171–176, 1991.
314. Gershoni, J. M. and Palade, G. E., Protein blotting: principles and applications, *Anal. Biochem.*, 131, 1–15, 1983.
315. Towbin, H. and Gordon, J., Immunoblotting and dot immunobinding-current status and outlook, *J. Immunol. Meth.*, 72, 313–340, 1984.
316. Steward, A. R., Dannan, G. A., Guzelian, P. S., and Guengerich, F. P., Changes in the concentration of seven forms of cytochrome P-450 in primary cultures of adult rat hepatocytes, *Mol. Pharmacol.*, 27, 125–132, 1985.
317. Simmons, D. L., McQuiddy, P., and Kasper, C. B., Induction of the hepatic mixed-function oxidase system by synthetic glucocorticoids: transcriptional and post-transcriptional regulation, *J. Biol. Chem.*, 262, 326–332, 1987.
318. Song, B. J., Gelboin, H. V., Park, S. S., Yang, C. S., and Gonzalez, F. J., Complementary DNA and protein sequences of ethanol-inducible rat and human cytochrome P-450s: transcriptional and post-transcriptional regulation of the rat enzyme, *J. Biol. Chem.*, 261, 16689–16697, 1986.
319. Chomczynski, P. and Sacchi, N., Single-step method of RNA isolation by acid guanidinium thiocyanate-phenol-chloroform extraction, *Anal. Biochem.*, 162, 156–159, 1987.
320. Chomczynski, P. and Mackey, K., Substitution of chloroform by bromo-chloropropane in the single-step method of RNA isolation, *Anal. Biochem.*, 225, 163–164, 1995.
321. Kingston, R. E., Chomczynski, P., and Sacchi, N., Guanidine methods for total RNA preparation, *Curr. Protocols Mol. Biol.*, 4.2.1–4.2.9, 1996.

322. Jacobson, A., Purification and fractionation of poly(A)+ RNA, *Meth. Enzymol.*, 152, -32768, 1987.
323. Southern, E. M., Detection of specific sequences among DNA fragments separated by gel electrophoresis, *J. Mol. Biol.*, 98, 503–517, 1975.
324. Brown, T., Mackey, K., and Du, T., Analysis of RNA by northern and slot blot hybridization, *Curr. Protocols Mol. Biol.*, 4.9.1–4.9.19, 2004.
325. Ogden, R. C. and Adams, D. A., Electrophoresis in agarose and acrylamide gels, *Meth. Enzymol.*, 152, 61–87, 1987.
326. Wahl, G. M., Meinkoth, J. L., and Kimmel, A. R., Northern and Southern blots, *Meth. Enzymol.*, 572, 581, 1987.
327. Bork, R. W., Muto, T., Beaune, P. H., Srivastava, P. K., Lloyd, R. S., and Guengerich, F. P., Characterization of mRNA species related to human liver cytochrome P-450 nifedipine oxidase and the regulation of catalytic activity, *J. Biol. Chem.*, 264, 910–919, 1989.
328. Brown, T., Analysis of DNA sequences by blotting and hybridization, *Curr. Protocols Mol. Biol.*, 2.9.1–2.9.15, 1999.
329. Birnboim, H. C. and Doly, J., A rapid alkaline extraction procedure for screening recombinant plasmid DNA, *Nucl. Acids Res.*, 7, 1513–1523, 1979.
330. Birnboim, H. C., A rapid alkaline extraction method for the isolation of plasmid DNA, *Meth. Enzymol.*, 100, 243–255, 1983.
331. Engelbrecht, J., Brent, R., and Kaderbhai, M. A., Mini-preps of plasmid DNA, *Curr. Protocols Mol. Biol.*, 1.6.1–1.6.10, 1991.
332. Sambrook, J. and Russell, D. W., *Molecular Cloning: a Laboratory Manual*, 3rd ed., Cold Spring Harbor Laboratory Press, Cold Spring Harbor, NY, 2001.
333. Mullis, K., Faloona, F., Scharf, S., Saiki, R., Horn, G., and Erlich, H., Specific enzymatic amplification of DNA *in vitro*: the polymerase chain reaction, *Cold Spring Harbor Symp. Quant. Biol.*, 51(Pt. 1), 263–273, 1986.
334. Kramer, M. and Coen, D. M., Enzymatic amplification of DNA by PCR: standard procedures and optimization, *Curr. Protocols Mol. Biol.*, 15.11.11–15.11.14, 2001.
335. Kendew, J., *The Encyclopedia of Molecular Biology*, Blackwell Science, London, 1994.
336. Kunkel, T. A., Rapid and efficient site-specific mutagenesis without phenotypic selection, *Proc. Natl. Acad. Sci. U.S.A.*, 82, 488–492, 1985.
337. Kunkel, T. A., Roberts, J. D., and Zakour, R. A., Rapid and efficient site-specific mutagenesis without phenotypic selection, *Meth. Enzymol.*, 154, 367–382, 1987.
338. Vovis, G. F. and Lacks, S., Complementary action of restriction enzymes endo R-*Dpn*I and endo R-*Dpn*II on bacteriophage f1 DNA, *J. Mol. Biol.*, 115, 525–538, 1977.
339. Wells, J. A., Vasser, M., and Powers, D. B., Cassette mutagenesis: an efficient method for generation of multiple mutations on defined sites., *Gene*, 34, 315–323., 1985.
340. Reidhaar-Olson, J. R. and Sauer, R. T., Combinatorial cassette mutagenesis as a probe of the informational content of protein sequences., *Science*, 241, 53–57, 1988.
341. Reidhaar-Olson, J. F., Bowie, J. U., Breyer, R. M., Hu, J. C., Knight, K. L., Lim, W. A., Mossing, M. C., Parsell, D. A., Shoemaker, K. R., and Sauer, R. T., Random mutagenesis of protein sequences using oligonucleotide cassettes, *Meth. Enzymol.*, 208, 564–586., 1991.
342. Martin, F. H., Castro, M. M., Aboul-ela, F., and Tinoco, Jr., I., Base pairing involving deoxyinosine: implications for probe design, *Nucl. Acids Res.*, 13, 8927–8938., 1985.
343. Schmidt-Dannert, C., Directed evolution of single proteins, metabolic pathways, and viruses, *Biochemistry*, 40, 13125–13136, 2001.
344. Mills, D. R., Peterson, R. L., and Spiegelman, S., An extracellular Darwinian experiment with a self-duplicating nucleic acid molecule, *Proc. Natl. Acad. Sci. U.S.A.*, 58, 217–224, 1967.
345. Eigen, M., New concepts for dealing with the evolution of nucleic acids, *Cold Spring Harbor Symp. Quant. Biol.*, 52, 307–320, 1987.
346. Kauffman, S., *The Origins of Order: Self-Organization and Selection in Evolution,* Oxford University Press, New York, 1993.
347. Chen, K. and Arnold, F. H., Tuning the activity of an enzyme for unusual environments: sequential random mutagenesis of subtilisin E for catalysis in dimethylformamide, *Proc. Natl. Acad. Sci. U.S.A.*, 90, 5618–5622, 1993.
348. Stemmer, W. P., Rapid evolution of a protein *in vitro* by DNA shuffling, *Nature*, 370, 389–391, 1994.
349. Arnold, F. H., Design by directed evolution, *Acct. Chem. Res.*, 31, 125–131, 1998.
350. Cirino, P. C., Mayer, K. M., and Umeno, D., Generating mutant libraries using error-prone PCR, *Meth. Mol. Biol.*, 231, 3–9, 2003.
351. Kim, D. and Guengerich, F. P., Random mutagenesis by whole-plasmid PCR amplification, *Meth. Mol. Biol.*, 192, 241–245, 2002.
352. Parikh, A. and Guengerich, F. P., Random mutagenesis via whole plasmid PCR amplification, *BioTechniques*, 24, 428–431, 1998.
353. Arnold, F. H. and Georgiou, G., *Directed Enzyme Evolution, Screening and Selection Methods*, Humana Press, Totowa, NJ, 2003.
354. Zhao, H., Giver, L., Shao, Z., Affholter, J. A., and Arnold, F. H., Molecular evolution by staggered extension process (StEP) *in vitro* recombination, *Nat. Biotechnol.*, 16, 258–261, 1998.
355. Evans-Storms, R. B. and Cidlowski, J. A., Regulation of apoptosis by steroid hormones, *J. Steroid Biochem. Mol. Biol.*, 53, 1–8, 1995.
356. Simula, T. P., Glancey, M. J., and Wolf, C. R., Human glutathione *S*-transferase-expressing *Salmonella typhimurium* tester strains to study the activation/detoxification of mutagenic compounds: studies with halogenated compounds, aromatic amines and aflatoxin $B_1$, *Carcinogenesis*, 14, 1371–1376, 1993.
357. Yamazaki, Y., Fujita, K.-I., Nakayama, K., Suzuki, A., Nakamura, K., Yamazaki, H., and Kamataki, T., Establishment of ten strains of genetically engineered *Salmonella typhimurium* TA1538 each co-expressing a form of human cytochrome P450 with NADPH-cytochrome P450 reductase sensitive to various promutagens, *Mutat. Res.*, 562, 151–162, 2004.
358. White, P. A. and Rasmussen, J. B., SOS chromotest results in a broader context: empirical relationships between genotoxic potency, mutagenic potency, and carcinogenic potency, *Environ. Mol. Mutagen.*, 27, 270–305, 1996.

359. Ames, B. N., McCann, J., and Yamasaki, E., Methods for detecting carcinogens and mutagens with the *Salmonella*/mammalian-microsome mutagenicity test, *Mutat. Res.*, 31, 347–364, 1975.
360. Quillardet, P., Huisman, O., D'ari, R., and Hofnung, M., SOS chromotest, a direct assay of induction of an SOS function in *Escherichia coli* K-12 to measure genotoxicity, *Proc. Natl. Acad. Sci. U.S.A.*, 79, 5971–5975, 1982.
361. Quillardet, P. and Hofnung, M., The SOS chromotest, a colorimetric bacterial assay for genotoxins: procedures, *Mutat. Res.*, 147, 65–78, 1985.
362. Valadez, J. G. and Guengerich, F. P., *S*-(2-chloroethyl)glutathione-generated p53 mutation spectra are influenced by differential repair rates more than sites of initial DNA damage, *J. Biol. Chem.*, 279, 13435–13446, 2004.
363. Liu, L., Hachey, D. L., Valadez, J. G., Williams, K. M., Guengerich, F. P., Loktionova, N. A., Kanugula, S., and Pegg, A. E., Characterization of a mutagenic DNA adduct formed from 1,2-dibromoethane by $O^6$-alkylguanine-DNA alkyltransferase, *J. Biol. Chem.*, 279, 4250–4259, 2004.
364. Liu, L., Williams, K. M., Guengerich, F. P., and Pegg, A. E., $O^6$-Alkylguanine-DNA alkyltransferase has opposing effects in modulating the genotoxicity of dibromomethane and bromomethyl acetate, *Chem. Res. Toxicol.*, 17, 742–752, 2004.

# 41 Modern Instrumental Methods for Studying Mechanisms of Toxicology

*Peter A. Crooks, David R. Worthen, Gary D. Byrd, J. Donald deBethizy, and William S. Caldwell*

## CONTENTS

## INTRODUCTION

The mechanisms underlying an organism's response to a toxic insult are usually complex, involving the toxicant itself, metabolites derived from it, and numerous tissue-derived endogenous compounds. The ability to monitor the chemical changes that result from intoxication is critical to understanding these mechanisms. Our ability to monitor chemical changes associated with intoxication has often limited our mechanistic understanding of toxicity. The development of gas chromatography in the 1950s and 1960s extended the lower range for detecting chemical changes in organisms and in the environment. The realization that exposure to chemicals such as DDT was widespread and that these chemicals could concentrate in the food chain fueled the development of modern toxicology. This early breakthrough in instrumental analysis led to the development of a host of powerful instruments and techniques that have profoundly increased our ability to define mechanisms of toxicity. Because each instrumental method has both strengths and limitations, it is important for toxicologists to be aware of available analytical methods and to understand the types of studies for which each is suited.

This chapter is not intended to be a comprehensive survey of modern instrumental methods; rather, it focuses on those techniques that are likely to be most useful to toxicologists. Some of these techniques, such as mass spectrometry and nuclear magnetic resonance spectroscopy, are widely used in toxicology. Others, such as near-infrared spectroscopy, are not as commonly used but have proven extremely useful for toxicological studies. Recent advances in the important areas of *in vivo* toxicology and metabolomics further underscore the importance of these analytical methods in toxicological research. Emphasis has been placed on practical applications, not in-depth theoretical discussion. This chapter serves as a starting point for further study. The interested reader is encouraged to consult the cited references to gain a more in-depth understanding of modern instrumental methods.

## MASS SPECTROMETRY

Mass spectrometry is an analytical technique that determines the mass of ionized molecules and their fragments and adducts. It is perhaps the most generally applicable tool in chemical analysis and the method of choice for many specific analytical procedures. In addition to its application as a universal detector in qualitative analysis, mass spectrometry can be fine-tuned to quantify trace amounts of specific components in complex mixtures. Mass spectrometers are often part of a hyphenated analytical system where chromatography (usually gas or liquid) precedes mass spectrometric detection. Mass spectrometry has impacted biomedical, environmental, and toxicological research, including the study of toxic substances and their fate in the body. Considerable progress has been made in toxicology using mass spectrometry as an analytical tool. This section looks at the basic principles of mass spectrometry, new trends in the field, and some examples where mass spectrometry has been applied to toxicology studies such as elucidation of detoxification mechanisms and exposure assessment. This is not a comprehensive review of recent breakthroughs in the area as much as it is a starting point for the curious student of the art. A good text covering routine techniques in mass spectrometry with consideration of biochemical applications has been written by Johnstone and Rose [96].

## Mass Spectrometers and Mass Spectra

There are many different types of mass spectrometers [47], but certain features are common to them all. Ions, unlike neutral compounds, can be manipulated by electromagnetic forces such that a separation by mass-to-charge ($m/z$) ratio is possible. A mass spectrometer is a device that creates ions from sample molecules, separates them by $m/z$, and determines the number of ions at each particular ratio. The mass spectrum is a plot of $m/z$ ratio on the $x$-axis vs. relative ion abundances (RA) on the $y$-axis. The mass spectrum can give molecular weight and structural information about the compound that was ionized. Ions are normally produced with single charges ($z = 1$) so the $m/z$ ratio gives the mass in Daltons (Da) of intact molecules and fragments directly. A simple model of a molecule AB undergoing ionization is shown below:

$$AB \rightarrow \text{Ionization process} \rightarrow AB^{+\bullet} + A^{+} + B^{+}$$

The ionization process in this case produces a molecular ion $AB^{+\bullet}$ (usually designated as $M^{+\bullet}$) that is the sample molecule with an electron removed. Often, enough energy is imparted in the process so that a portion of the molecular ions decompose to form various stable fragment ions as shown by $A^{+}$ and $B^{+}$ above. The mass spectrum shows the mass and abundance of each of these fragments (see Figure 41.1). The spectrum is highly characteristic of a particular compound and has been likened to a fingerprint that can be used for identification purposes. Mass spectra of compounds with known structures are continuously reported and compiled so patterns of fragmentation are

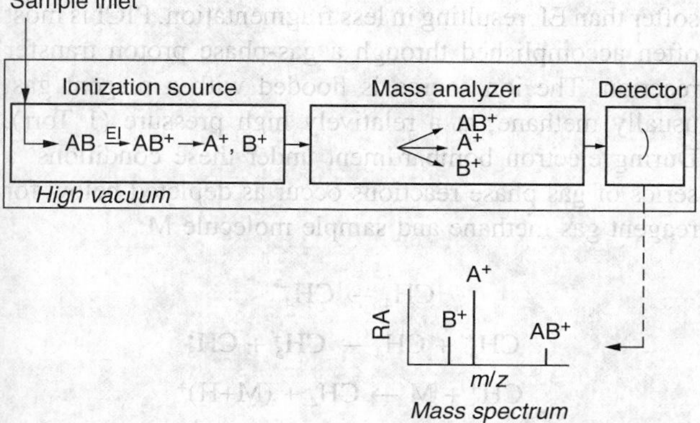


**FIGURE 41.1** Schematic of a basic mass spectrometer system interfaced to an inlet.

established. This knowledge is useful in interpreting spectra of unknown compounds. In addition, computerized searches of mass spectral databases can facilitate compound identification. The sensitivity of conventional mass spectrometers permits good spectra to be obtained on less than 10 ng of material, so, although the ionization process destroys the sample, the mass spectrometer requires very small amounts of material for analysis.

## Instrument Design

### Basic Configuration

The many different types of mass spectrometers vary in size and complexity from small bench-top units to large multiple-sector machines [47]. Nevertheless, mass spectrometers can all be broken down into a few basic components and their functions: sample introduction, ion production, mass analysis, and ion detection. These components are shown schematically in Figure 41.1. The hardware components from ion source to detector are housed in a high-vacuum chamber to provide the necessary mean free path for ions to travel from the source to the detector. Typical vacuum for a commercial mass spectrometer is less than $10^{-6}$ Torr.

### Inlet Systems

Samples are introduced into the mass spectrometer vacuum housing via some type of inlet. The inlet selected depends on the sample, the nature of the sample matrix, and the type of ionization desired. Many mass spectrometers are designed with multiple inlets to accommodate a wide variety of samples with minimal time required for instrument reconfiguration. For stable solid materials, the simplest means of introduction is to place the sample onto a long probe that is inserted directly into the ion source through a valve. This is commonly referred to as a *direct-inlet probe* (DIP). Heating the probe volatilizes the sample into the gas phase where it can be ionized easily, usually by energetic electrons. The DIP method is very sensitive and can produce spectra from submicrogram quantities of material; however, some materials such as glutathione conjugates are not easily volatilized and decompose on the probe. Analytes that occur in mixtures may require some separation before DIP analysis. Although some separation of components in time is achieved by heating the probe slowly, it is a crude means of separation and works best on reasonably pure samples. Probes are also used in conjunction with other ionization techniques where they serve merely to place the sample in a position to be ionized. Examples of this would be *fast atom bombardment* (FAB) or *laser desorption* (LD), which are described later in this chapter.

Chromatographic separation techniques coupled to the source add another dimension of analysis, and most mass spectrometer systems used in biological sciences are configured in this manner. *Gas chromatography–mass spectrometry* (GC–MS) has worked very well in this regard, making GC–MS perhaps the most common hyphenated method for performing organic analysis [137,134]. Although applications with packed gas chromatography columns are still reported with various types of interfaces, most GC–MS is performed with fused silica capillary columns. These columns use gas flows of 1 to 2 mL/min, and the ends can be placed directly into the ionization source. High-resolution chromatography of reasonably volatile and thermally stable samples is possible with these columns.

Many compounds will not pass through a gas chromatograph, and *liquid chromatography-mass spectrometry* (LC–MS) has been developed as an alternative method for introducing analytes dissolved in solution into the mass spectrometer. In fact, because biological samples are usually aqueous based and include thermally labile and polar substances, LC–MS now rivals GC–MS in popularity. This is due to the development and refinement of many reliable interfaces in the last two decades. *High-performance liquid chromatography* (HPLC) methods using a variety of columns and flow rates have used mass spectrometry as a detector. General procedures are well documented for LC–MS techniques [231,222].

Another chromatographic technique used with mass spectrometry is *capillary electrophoresis* (CE) [105]. A high electrical field in a small-diameter capillary tube filled with aqueous solution is used to separate charged compounds. CE offers the advantages of high resolution, low sample consumption, and short analysis times. It is well suited to the analysis of ions in solution. Interfacing to the mass spectrometer is often accomplished through electrospray ionization or continuous flow-fast atom bombardment, which are described below.

## Ionization Sources for Volatile Compounds

This discussion of ionization sources is divided into two parts based on the volatility of the compound of interest. It might well be divided into the same parts based on the two popular chromatography methods interfaced to mass spectrometers: GC and LC. Because mass spectrometry was originally limited to volatile samples, these types of ionization sources are considered conventional. Volatile samples are introduced directly into the ion source using a leak valve for gases, a heated DIP described above for solids, or a GC. This section describes methods used in these types of applications.

The most established means of ionization for volatile samples is *electron impact* (EI), in which a sample is volatilized into the gas phase and passes through a beam of energetic (70 eV) electrons boiled from a filament. The process may be written as:

$$M + e^- \rightarrow M^{+\bullet} + 2e^-$$

The high-energy electron displaces an electron from molecule M, which may remain intact as the molecular ion $M^{+\bullet}$ and thus provide direct molecular weight information. The energetics of the process can also cause fragmentation of the molecule. Fragmentation patterns are related to the structure of the molecule and thus can be interpreted as representative of that compound. As an example, Figure 41.2 shows the EI mass spectrum of nicotine identified in an extract of urine from a smoker. The highest mass ion in the spectrum is *m/z* 162, which is the molecular ion. The most abundant peak in the spectrum (referred to as the base peak) is *m/z* 84 and results from cleavage of the bond between the two rings with the charge remaining with the pyrrolidine ring.

Electron impact has several advantages. Both fragment and molecular ions are produced in most EI spectra. The mass spectra are fairly reproducible from one instrument to the next, which makes it possible to match sample spectra to reference EI mass spectra with some certainty. Several large mass spectral databases of compounds are available that can be readily searched against an EI spectrum produced on most types of mass spectrometers. Some disadvantages of EI include the occasional lack of a molecular ion for some compounds, the difficulty of distinguishing between mass spectra of isomers, and limited application to samples with sufficient gas phase volatility and thermal stability.

An alternative ionization method called *chemical ionization* (CI) can be used in cases where the molecular ion is weak or not present in the EI mass spectrum. The process derives its name from the use of gas-phase chemical reactions to produce ions from the sample [146]. The two types of chemical ionization are *positive ion chemical ionization* (PICI or PCI) and *negative ion chemical ionization* (NICI or NCI). In both cases, the ionization is softer than EI, resulting in less fragmentation. PICI is most often accomplished through a gas-phase proton transfer reaction. The ion source is flooded with a reagent gas, usually methane, at a relatively high pressure (1 Torr). During electron bombardment under these conditions, a series of gas phase reactions occur as depicted below for reagent gas methane and sample molecule M:

$$CH_4 \rightarrow CH_4^{+\bullet}$$

$$CH_4^{+\bullet} + CH_4 \rightarrow CH_5^+ + CH_3^\bullet$$

$$CH_5^+ + M \rightarrow CH_4 + (M{+}H)^+$$

Methane molecular ions formed in the high-pressure source collide with neutral reagent molecules ($CH_4$) and produce protonated methane. $CH_5^+$ acts as a strong Brönsted acid and

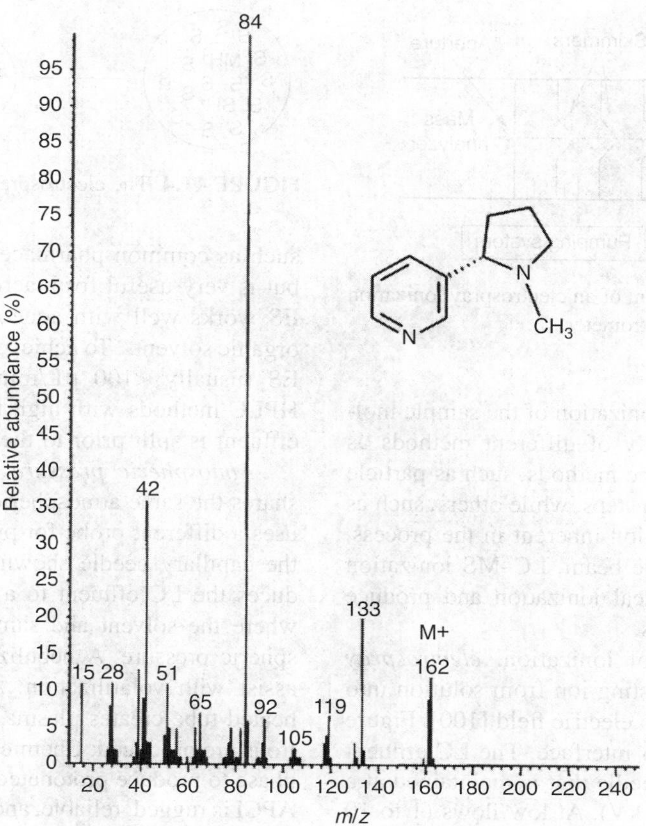


**FIGURE 41.2** Electron impact mass spectrum of nicotine; data acquired on a quadrupole GC–MS system.

transfers a proton to the sample molecule to produce the protonated molecular adduct $(M+H)^+$. PICI works best for samples with a relatively high proton affinity such as those containing a heteroatom. Fewer fragment ions and more abundant molecular ions characterize PICI mass spectra. In addition, sensitivity is enhanced relative to EI. The higher source pressure used, however, contaminates the instrument more rapidly and results in more frequent maintenance. Also, some modifications to the source are usually required when switching from EI to CI operation.

Negative ion chemical ionization sources operate under high-pressure conditions similar to those for PICI with voltages switched to detect negative ions. The underlying ionization processes, however, are somewhat different. Several processes can occur to form negative ions. A common method is *electron capture*, which uses the reagent gas as a buffer to reduce the energy of electrons such that a molecule with a suitable electron affinity can capture an electron as shown below:

$$M + CH_4 + e^- \rightarrow M^{-\bullet} + CH_4$$

This process is rather selective, as not every molecule will readily form a stable $M^{-\bullet}$. Another process is adduct formation where a background anion such as $Cl^-$ will attach to a molecule to form $(M+Cl)^-$. Numerous ion–molecule reactions are possible, and some can be rather complex. Proton abstraction is common and works well for samples with an acidic proton. Negative ion formation can be enhanced using derivatization to form an analog with a high electron affinity. The sensitivity and selectivity of such assays can be very high; for example, one group reports conversion of nicotine using heptafluorobutyric anhydride to a stable electron scavenger derivative that can be detected at the femtogram ($10^{-15}$g) level on column [48].

## Ionization Sources for Nonvolatile Compounds

This section describes ionization sources for compounds that cannot be volatilized sufficiently for EI ionization, particularly those that decompose upon heating. These include very polar or very large molecules such as those encountered in biological samples. For these types of samples, LC is preferred over GC for chromatographic separation. A number of interfaces deal with samples in solution such as those that emerge from an HPLC column. Before introduction into the high vacuum of the mass spectrometer, most of the solvent molecules must be removed and the sample molecules ionized. Heat, nebulization with gas, and differential pumping are techniques

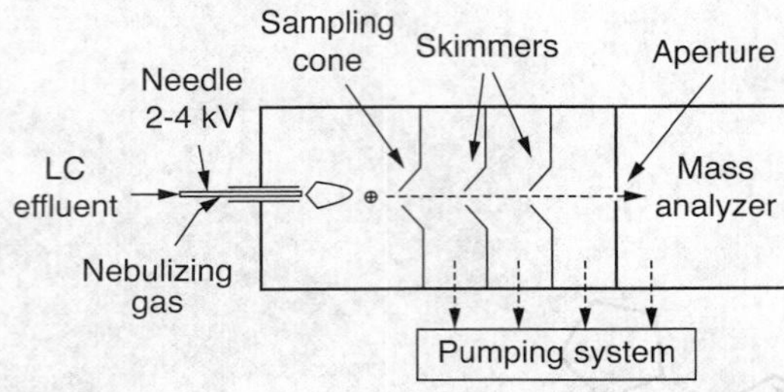


**FIGURE 41.3** Schematic diagram of an electrospray ionization source interfaced to a mass spectrometer.

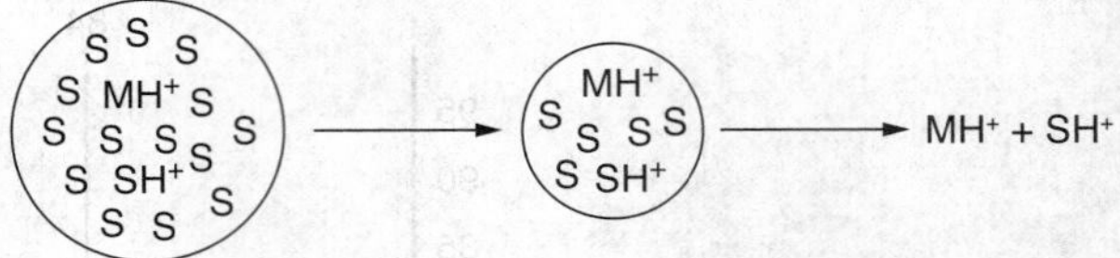


**FIGURE 41.4** The electrospray ionization process.

used to remove the solvent. Ionization of the sample molecules can occur by a variety of different methods as described in this section. Some methods, such as particle beam, have distinct ionization steps, while others, such as electrospray, have ion formation inherent in the process. With the exception of particle beam, LC–MS ionization methods are soft like chemical ionization and produce mostly molecular adduct ions.

An intriguing method of ionization, *electrospray* (ES) simply transfers an existing ion from solution into the gas phase by using a high electric field [100]. Figure 41.3 is a schematic of an ES interface. The LC effluent passes through a capillary needle that is maintained at a high applied voltage (2 to 5 kV). At low flows (1 to 10 μL/minute), the high electrical field at the tip produces a mist of charged droplets at atmospheric pressure. For higher flow rates, a nebulizing gas delivered coaxially to the needle assists production of the mist. As evaporation decreases droplet size, ions are ejected as depicted for sample molecules M and solvent molecules S in Figure 41.4. Of course, the analyte must already exist as an ion in solution, but this is easily accomplished by adjusting the pH. Positive ions are shown in the figure, but the process works equally well for negative ions. The ions are swept through a sampling cone and passed through a differential pumping system to remove solvent and nebulizing gas molecules before the ions enter the mass analyzer through a tiny aperture. ES is a very gentle ionization method that can be used with either small or large molecules. Figure 41.5 shows an ES spectrum of the glucuronide conjugate of cotinine, a phase II metabolite of nicotine. This thermally labile compound produces $(M+H)^+$ as the base peak at *m*/*z* 353. In comparison, a thermospray mass spectrum (discussed below) of this same compound yields mostly the protonated aglycone fragment [29,30]. An interesting feature of ES is the ability to produce ions with multiple charges. Because mass spectra are plotted with *m*/*z*, this permits the available mass range of the analyzer to extend to thousands of Daltons; for example, a compound with $(M+H)^+$ at *m*/*z* 5000 would have $(M+5H)^{+5}$ at nominal *m*/*z* 1001. This feature is less useful for small molecules such as common pharmaceuticals and their metabolites but is very useful for macromolecules such as proteins. ES works well with aqueous mobile phases and some organic solvents. To achieve the low flows necessary with ES (usually <100 μL/min) when using conventional HPLC methods with high flows (0.5 to 2 mL/min), the effluent is split prior to the interface.

*Atmospheric pressure chemical ionization* (APCI) shares the same atmospheric pressure interface as ES but uses a different probe for producing ions [37]. Instead of the capillary needle shown in Figure 41.3, APCI introduces the LC effluent to a heated tube (400 to 550°C), where the solvent and samples are volatilized at atmospheric pressure. A nebulizing gas (nitrogen) is used to assist with volatilization. A discharge needle after the heated tube creates plasma where reagent ions and electrons are produced. Chemical ionization occurs in the gas phase to produce protonated and other molecular adducts. APCI is rugged, reliable, and very sensitive due to efficient ionization, especially for molecules with heteroatoms. The rapid heating does produce some thermal degradation of labile species. APCI can accommodate HPLC flows of 2 mL/min and works well with most types of solvents.

*Desorption electrospray ionization* (DESI) is an exciting ionization technique that eliminates the need for a bulk solvent, as it is used in ES, while allowing researchers to prepare MS samples at ambient conditions, as in APCI [174]. The DESI experiment is performed by spraying a stream of charged solvent particles across an analyte surface while collecting the deflected droplets in a portable, charged wand, which serves as an atmospheric pressure ion transfer inlet into the MS. In this way, analytes may be collected from the surface of an object, such as a briefcase, a floor tile, or an article of clothing, which is suspected of being contaminated with toxins, drugs, or other materials of interest. The portability and ease of sample handling make DESI a particularly promising technique for remote environmental testing, field forensics, and counterterrorism. Growing interest in these toxicological disciplines continues to foster the development of portable field MS units for remote sensing [10].

One of the first widely applied interfaces developed for LC–MS was *thermospray* (TS) [18]. The LC effluent passes through a narrow metal tube whose tip is maintained at a high temperature (100 to 200°C) by resistive heating. The heated liquid volatizes and forms a spray in a reduced pressure chamber. Ions in solution are desorbed

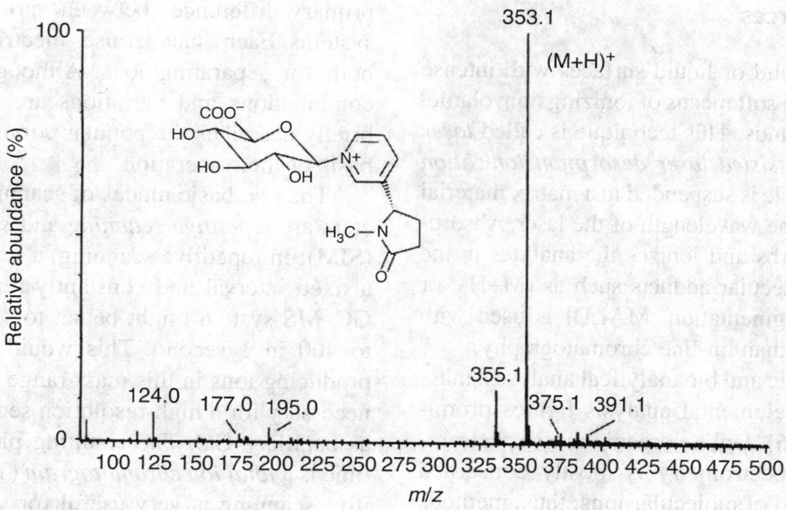


**FIGURE 41.5** Electrospray mass spectrum of cotinine-*N*-glucuronide taken with a quadrupole mass analyzer.

from the shrinking droplet in the spray; some gas-phase chemical ionization reactions also occur. The LC effluent must contain an ionizing buffer such as ammonium acetate to produce ions, and precise temperature control of the probe is crucial for TS operation. An optional filament placed near the probe tip can be used to increase ionization in some cases. The spray passes a conical aperture where ions are directed into the mass analyzer. TS works best with aqueous mobile phases such as those used in reversed-phase HPLC. Numerous applications of TS LC–MS have been published [231]. The strengths of TS are its simplicity and broad applicability to many types of samples. Like APCI, however, the high temperatures required for volatilization can degrade thermally labile samples. TS applications have decreased in the past few years due primarily to the success of APCI and ES techniques discussed previously.

The *particle beam* (PB) interface is not an ionization source so much as it is a solvent removal system utilized prior to introducing the sample into a conventional EI source [33,221]. The LC effluent is first mixed with a gas (usually helium) to create an aerosol in a heated chamber. Selective evaporation and removal of solvent leave a beam of sample molecules that passes through a series of apertures into the mass spectrometer source. There, ionization may occur by conventional EI or chemical ionization techniques. Unlike other LC–MS interfaces, PB produces EI spectra, which are reproducible and searchable by computer; this makes PB invaluable for identification. PB is limited to small molecular weight samples (<1000 Da), and sensitivity can be low for certain applications. Sensitivity may be enhanced by using volatile organic buffers [12]. PB maintains a small market niche for certain industrial and environmental applications.

*Continuous-flow fast atom bombardment* (CFFAB), sometimes referred to as *dynamic FAB* or *liquid secondary ion mass spectrometry* (SIMS), allows the LC effluent to flow directly across the target area of a FAB probe (34). The target is bombarded with fast atoms or ions and the sample is ionized by a desorption process. The LC effluent enters the probe at a very low flow rate (<10 μL/min). It is mixed with a matrix material such as glycerol to facilitate the ionization process. Like ES described above, it is a very gentle method of ionization. CFFAB is usually used on sector instruments with capillary LC columns for separation. CFFAB is not preferred as a general method because ionization of the sample depends on the matrix. It is good for certain classes of compounds such as peptides.

For all of these interfaces, an important consideration in coupling an HPLC method to a mass spectrometer is mobile-phase compatibility. Popular LC–MS mobile-phase solvents are water, methanol, and acetonitrile. They are good solvents for a wide range of samples, and their low molecular weights reduce background ions in the mass spectrometer. Organic modifiers such as triethylamine, which are often used to improve chromatography, should be avoided as they can suppress ionization of sample molecules. Although many of the interface/ionization sources listed here work best with a buffer at low concentrations (<20 m*M*), it must be a volatile organic buffer such as ammonium formate or ammonium acetate that will not leave deposits that block the apertures leading to the mass analyzer. Some recent work has been done on two-stage orthogonal sampling that allows the use of harsh mobile phases such as potassium phosphate buffers (e.g., the commercial Z-Spray interface offered by Micromass, Manchester, U.K.). This is extremely useful when adapting existing HPLC methods to a mass spectrometer.

## Other Ionization Sources

The bombardment of solid or liquid surfaces with intense light from a laser offers a soft means of ionizing nonvolatile, thermally labile compounds. This technique is called *laser desorption*. In *matrix-assisted laser desorption ionization* (MALDI) [99], the sample is suspended in a matrix material that absorbs light near the wavelength of the laser. Absorption of the energy desorbs and ionizes the analytes in the matrix, resulting in molecular adducts such as $(M+H)^+$ or $(M+Na)^+$ with little fragmentation. MALDI is used with isolated samples rather than on-line chromatography.

In addition to organic and bioanalytical analyses, mass spectrometry used for elemental analysis figures prominently in toxicology [86]. Unlike organic mass spectrometry, inorganic mass spectrometry is mostly concerned with atomic ions instead of molecular ions; thus, methods have been developed for producing atomic ions of various elements from samples. A common ionization method used in inorganic mass spectrometry is *thermal ionization* that takes place when an atom or molecule interacts with a heated surface. Samples are deposited directly on a filament and heated. Another method is the use of *inductively coupled plasma* (ICP), where the sample is volatilized, atomized, and ionized in a few milliseconds at plasma temperatures of 7000 to 8000 K.

An interesting field of expanding applications is *accelerator mass spectrometry* (AMS). Unlike the ion sources described above, AMS is a high-energy (MeV) nuclear physics technique that uses a Van de Graaf accelerator to measure very small amounts of rare and long-lived isotopes [7,211]. For $^{14}C$ analysis, the sample is converted to graphite powder by oxidation followed by reduction. A cesium ion gun bombards the sample, and the negative carbon ions produced are selected from the source and accelerated at 3 to 10 MeV. They pass through a thin foil or gas where electrons are stripped from $C^-$ to create positive ions. These ions are then accelerated to several MeV, selected by a mass analyzer (quadrupole or magnet), and detected. AMS was first applied to geochemical and archeological samples but has seen growing biomedical applications, particularly in the monitoring of $^{14}C$-labeled drugs. The high sensitivity of this method ($10^{-18}$ mole) permits low doses of labeled materials to be monitored in humans. Given its high sensitivity and specificity, AMS has been particularly useful for *in vivo* studies involving minute doses of xenobiotics. The kinetics and dynamics of xenobiotic transport, metabolism, and receptor binding studies in living organisms, including humans, have been extensively explored using AMS [210].

## Mass Analyzers

Mass analysis is the process by which a mixture of ions is separated according to their *m/z* ratios. Several types of mass analyzers are commonly used, and these form the primary differences between various mass spectrometer systems. Each analyzer uses electric or magnetic fields or both for separating ions. Although many experimental combinations and variations are available, this section briefly describes the popular commercial models and the basis of their operation.

The two basic modes of scanning for mass spectrometers are *repetitive scanning* and *selected ion monitoring* (SIM). In repetitive scanning, a mass range is covered in a fixed interval and constantly repeated; for example, a GC–MS system might be set to scan repetitively *m/z* 40 to 400 in 1 second. This would detect any compound producing ions in this mass range every second, which is necessary for a high-resolution separation technique such as capillary GC. The resulting plot of all responses vs. time is a *total ion chromatogram* (TIC). Wide-range repetitive scanning is very useful for screening samples such as body fluid extracts for all types of compounds present. The mass range may be adjusted to detect only signals of interest or to exclude interferences such as low-molecular-weight solvents. This procedure offers less sensitivity than SIM but provides much more information. In addition, ion chromatograms can be produced from the stored data file. Ion chromatograms are reconstructed responses for a particular ion and are useful in quantitation to exclude interferences in complex chromatograms. For quantifying specific analytes it is useful to restrict the ions detected to those of interest. Operating in the SIM mode, where the analyzer scans discontinuously, rapidly switching between only a few ions does this. Because more dwell time is spent on these ions, sensitivity is increased on the order of 10 to 100 times. This technique ignores ions from background materials and is useful in quantifying analytes in fairly complex matrices. Multiple-stage analyzers are capable of several more modes of scanning, and these are described later in a separate section.

The earliest mass spectrometers used a *magnetic sector* to deflect ion currents into a detector system. Ions accelerated into a large electromagnet sector of radius *R* by a high voltage (*V*) are dispersed by the magnetic field (**B**) according to:

$$m/z = (\text{constant})\frac{\mathbf{B}^2 R^2}{V} \tag{41.1}$$

A spectrum is normally produced by varying **B** and monitoring the ions with a detector at the end to convert the impinging ions to a signal. Early magnetic sector instruments were large, heavy, expensive, and relatively slow in scanning a large *m/z* range. When used in combination with an electrostatic sector, these instruments can provide high mass resolution.

The most popular analyzer system used today is the quadrupole mass analyzer. These compact systems are relatively inexpensive and durable and can scan very rapidly. The quadrupole assembly consists of four parallel rods arranged equidistant around a central axis. DC and radiofrequency (RF) voltages are applied to opposite pairs of rods to form a fluctuating electric field. Ions formed in the source are directed and focused into this field by electrostatic lenses. Only ions with a particular *m/z* ratio as determined by the DC and RF field strength can pass through to the detector; thus, the analyzer scans for particular ions by ramping these parameters. Typically, a full mass spectrum of *m/z* 40 to 400 can be taken in less than 1 second with good sensitivity. A detector is located at the end of the quadrupole assembly.

By using quadrupole analyzers in series (*multiple-stage quadrupoles*), tandem mass spectrometry [24] is possible. Although sector instruments with magnetic and electrostatic fields can be used as stages, quadrupoles are more often used for tandem mass spectrometry. These systems operate by selecting a particular ion from the first stage analyzer and sending it into a second stage where the ion collides with an inert gas (such as argon) to produce further fragmentation. A third stage analyzes the resulting fragments, thus producing a mass spectrum from an ion in a mass spectrum. This technique is referred to as *MS/MS* or *collision-induced dissociation* (CID) and is often used with soft ionization such as CI, ES, or APCI to produce fragments from molecular adducts. An example is given below for a molecular adduct produced by PICI:

| **Source** | **Analyzer 1** | **Collision Cell** | **Analyzer 2** |
|---|---|---|---|
| M + $CH^{5+}$ → | $(M+H)^+$ → | Argon (eV) → | $A^+$, $B^+$ |
| | (Precursor) | | (Products) |

The selected ion is usually referred to as the *precursor* or *parent*, and its fragments as *products* or *daughters*. Precursors and products can be linked to establish fragmentation pathways, thereby producing additional structural information. MS/MS systems can also be set to monitor specific fragmentation pathways, a process known as *multiple reaction monitoring* (MRM). Setting the mass analyzer for a specific transition creates a very specific detector that can monitor an analyte in very complex samples. In this regard, it is analogous to selected ion monitoring. Although the cost of an MS/MS system is much more than for a single-stage mass spectrometer system, the increased structural information and selectivity of these instruments are very attractive. They have become workhorse systems for identifying and quantifying metabolites, particularly when interfaced to LC.

A more recent development in mass analyzers that is increasingly found in analytical laboratories is the *ion trap* [41]. These mass analyzers are a type of three-dimensional quadrupole composed of a doughnut-shaped ring and two end caps. Ions are contained in the circular electromagnet by a RF field until swept into a detector by an applied RF voltage-amplitude ramp. Ions are ejected according to their resonance energy, which is related to their mass, and a spectrum is thus produced. A time sequence of events produces ions in the trap and then ramps the RF field to detect specific masses. Both GC–MS and LC–MS ion trap systems are commercially available. When coupled directly to a GC, ions are formed by EI just as for a quadrupole system, and switching between EI and CI modes is simple. For LC–MS, ions are usually injected into the trap from an external source such as ES. Ion traps are similar to quadrupoles in sensitivity, size, and cost, but they are also capable of MS/MS analysis. MS/MS in the ion trap is performed in a timed series of events by colliding the ions while trapped either with themselves or a collision gas. The product ions are then detected. In addition, MS/MS on the product ions may be performed such that MS/MS/MS/... ($MS^n$) is possible. For LC–MS interfaces that produce molecular adducts, the ion trap is a less expensive option over the triple stage systems described above to perform CID analyses.

Somewhat related to the ion trap but longer in use is Fourier transform mass spectrometry (FT–MS) [45]. Ions are produced by EI in a cubic cell consisting of opposing pairs of trapping plates, transmitting plates, and receiving plates. Ions may also be injected into the source from an external ionization source such as ES. A high constant magnetic field and electrostatic trapping plates trap the ions formed. Each ion undergoes cyclotron motion at a frequency determined by its *m/z* ratio. All the ions are excited (resulting in increased radius of motion) simultaneously by applying a burst of RF energy over a range of frequencies corresponding to the cyclotron frequencies of the ions. The ions are detected simultaneously by measuring the image current induced on the detection plates of the cell. The frequencies form a beat pattern and, by Fourier transformation, the individual cyclotron frequencies of the ions are determined and the *m/z* values produced. Many operational aspects of FT–MS are similar to FT–NMR. High mass resolution (discussed below) is possible with these systems, as is $MS^n$ described for the ion trap above.

Time-of-flight (TOF) analyzers work on the simple principle that ion velocity is mass dependent [217]. A high voltage sent down a tube accelerates ionized sample ions, and the different *m/z* ratios are separated in time. The arrival time at the detector is based on *m/z*, and a mass spectrum is produced. Naturally, scan speed is very fast and mass range is unlimited. Coupled with MALDI ionization sources described above, they have been applied to a variety of studies in biomedical research that require the mass analysis of very large molecules. TOF mass spectrometer systems have shown a growing popularity recently due to techniques that have improved their performance, including mass resolution. TOF has been recently interfaced with LC and GC systems [84].

High-resolution mass analyzers can measure masses to within a thousandth of a mass unit (0.001 Da) or better. The utility of this feature is based on the fact that atomic masses, although close to integral values, are not truly integral. Although $^{12}C$ is assigned 12.000000 Da, $^{16}O$ is actually 15.994915 Da, $^{1}H$ is 1.007825 Da, $^{14}N$ is 14.003074 Da, and so on. Thus, high-resolution mass measurements can lead to determination of elemental composition; for example, the exact mass of nicotine ($C_{10}H_{14}N_2$) is 162.115699. Using a high-resolution mass spectrometer, its molecular ion would be distinguishable from that of another compound with a nominal $m/z$ 162 but different molecular formula such as $p$-nitrobenzyl cyanide ($C_8H_6N_2O_2$ = 162.042928). As detectors, high-resolution mass spectrometers offer more selectivity than low-resolution mass analyzers by monitoring very narrow mass ranges and thus eliminating potential interferences. Quadrupole analyzers are not capable of resolving power much in excess of 0.1 mass unit and cannot produce high-resolution spectra. The most commonly used high-resolution system is a double-focusing instrument with an electrostatic sector to first select ions of a specific kinetic energy before analysis by magnetic sector. As mentioned earlier, FT–MS and some TOF systems are capable of high-resolution mass spectra.

### Ion Monitoring

Conventional mass spectrometer systems (quadrupoles and magnetic sector instruments) focus the resolved beams of ions from the mass analyzer onto a detector. The most common detector is an electron multiplier, which is a series of electrodes. When an ion impinges on the first electrode it releases a shower of electrons that impact a second electrode and so on. The cascading effect produces gains on the order of $10^6$. Such high gain produces sensitivity, so a complete mass spectrum can be produced from a few nanograms or less of material.

## Interpretation of Mass Spectra

Interpretation of mass spectra involves correlating the plots of ion abundance vs. $m/z$ ratio with structure. For unknowns, it is best performed with all auxiliary information possible regarding a compound such as its origin and preparation, chromatographic behavior, and spectra from ultraviolet, infrared, nuclear magnetic resonance, etc. to assign structure with confidence. In this section, basic interpretation of EI mass spectra is reviewed followed by a discussion of mass spectra produced by softer ionization techniques.

### EI Mass Spectra

Fragmentation in EI mass spectrometry can be complex but rich in information if correctly interpreted [135]. It remains difficult to interpret all features in a mass spectrum, but a few simple rules can be used to glean useful information. The most significant datum in identifying an unknown compound is its molecular weight; thus, assignment of the molecular ion in the spectrum is the most important step in its interpretation. The molecular ion must be the highest mass ion in the spectrum and fragment ions must be logical neutral losses from this ion. Keep in mind that not all compounds will give a molecular ion. Fragmentation peaks in the spectrum result from one or more cleavages that may or may not involve rearrangements. The particular fragments produced are related to the strength and chemical nature of the bonds that held the fragment to the rest of the molecule; thus, an understanding of organic chemistry is useful in assigning fragment ion structures and the associated neutral species lost. MacLafferty's text [135] remains a popular text on the interpretation of mass spectra. EI mass spectra are very reproducible on different instruments, and assistance is available in the form of libraries of mass spectra that are searchable by computer. These are included with most data systems and are very straightforward to use. The data system suggests the best matches for your compound and provides some number indicating its confidence in the assignment. These libraries are useful for quickly identifying known compounds and suggesting identities for others. The ability to create your own library is an option in most data packages.

### Soft Ionization Mass Spectra

These mass spectra are those produced by less energetic ionization methods such as CI, MALDI, and LC–MS interfaces such as ES and APCI. They are all marked by the appearance of molecular adducts such as $(M+H)^+$. Other common adducts depend on the sample matrix and may include $(M+Na)^+$ and $(M+K)^+$. When using ammonium buffers in LC–MS, $(M+NH_4)^+$ is common along with adducts formed with solvent molecules such as $(M+H_2O+H)^+$ or $(M+CH_3CN+H)^+$. Also, dimers such as $(2M+H)^+$ may occur, especially when higher concentrations of analyte are present. These spectra serve primarily to verify the molecular weight of the compound.

### CID Mass Spectra

Mass spectra produced by MS/MS are greatly influenced by different parameters such as collision energy and collision cell pressure [24]. These conditions can vary from one compound to the next and also from one instrument to the next for the same compound; thus, CID spectra are less quantitatively reproducible than EI mass spectra, yet CID spectra remain extremely useful for qualitative structure elucidation. Common neutral losses such as $H_2O$ for hydroxyl groups or loss of glucuronic acid from a glucuronide conjugate are very informative and complement the soft ionization techniques.

## Experimental Design

With such a variety of instruments and methods available, some thought must be given to designing a mass spectrometric analysis. This section provides some basic guidance for acquiring qualitative and quantitative information for a sample.

### Method Selection

The nature of the sample determines the best approach; for example, if the sample is a large protein it is unlikely to volatilize intact from a DIP or pass through a GC. Another consideration is the complexity of the sample. Purified samples may not require chromatography but mixtures do. GC–MS and LC–MS are established techniques that perform separation and mass analysis and can rapidly identify unknowns and assess the relative composition of a sample. Samples that are thermally labile or nonvolatile must be ionized with a soft ionization method such as ES or APCI. MS/MS techniques are useful for providing structural information from soft ionization techniques. If high mass resolution is required, the mass analyzer must be a double-focusing sector instrument, an FT–MS, or one of the new TOF systems.

### Qualitative Analysis

Qualitative analysis refers to correctly identifying an unknown substance. Although interpretation of mass spectra has already been discussed, other considerations must be mentioned. Many compounds have similar mass spectra, and a single mass spectrum cannot give certain identification in every case. For example, at least three polycyclic aromatic hydrocarbons have a molecular weight of 252 (benzo(*a*)pyrene, benzo(*e*)pyrene, and perylene), and all have similar EI mass spectra. In these situations, using a separation technique such as GC or LC prior to mass analysis can assist identification. When an unknown gives a matching spectrum and coelutes with an authentic standard, then confidence in its identification is high. When no authentic standard is available, the use of two or more chromatographic techniques and ionization methods can be used to enhance confidence in the identity of an unknown. It should be emphasized that, even though mass spectrometry is a powerful tool in determining the identity of an unknown, auxiliary techniques outside of mass spectrometry should be used whenever possible to provide increased confidence in the identification of a compound.

### Quantitative Analysis

The number of ions detected in the mass spectrometer is proportional to the amount of material introduced to the source, making quantitative analysis possible. Using standard analytical calibration principles, mass spectrometers are capable of very precise, accurate, and sensitive determinations of analytes. The response of the mass spectrometer is calibrated using a series of standards that contain known amounts of the analyte or analytes of interest. External standard calibration plots are possible but are prone to errors due to instrument response variation through the course of analyzing several standards and samples.

A preferred calibration system is the use of internal standards where a known amount of a reference compound is added to the samples and also to the standards. A response ratio of the analyte to an internal standard is produced which can partially offset variations in sample preparation and instrument sensitivity. The internal standard should be added to the sample prior to any mixing, extraction, derivatization, enzyme treatment, or chromatography to account for any sample losses during these processes. The instrument is usually calibrated with a series of standards that contain a fixed amount of the internal standard and varying amounts of the analyte of interest. The range of analyte concentrations in the standards must cover the values expected for the samples. A calibration plot can be produced from the response ratio vs. analyte concentration. Analyte concentration can be obtained from this plot by interpolation. Most analysts prefer linear curves because the data are easier to process; however, many analytical software packages are available that will fit curves to nonlinear plots. Good texts are available that provide rigorous coverage of calibration plots, calculation of concentration, and limits of detection and quantitation [138,201].

A special case in quantitative mass spectrometry is the use of a labeled analog of the analyte as an internal standard in a procedure known as *isotope dilution*. Isotope dilution provides a more accurate and precise means of calibration than either external standards or conventional internal standard methods. The isotopically labeled analog has physical, chromatographic, and mass spectral properties that are nearly identical to those of the analyte and thus compensates in like manner for any losses due to sample preparation or instrument variability [44]. Stable isotopes (such as $^{2}H$ and $^{13}C$) are preferred to radioactive isotopes because of reduced hazards of handling the samples.

### Sample Preparation

Investigations in toxicology involve the analysis of biological samples from *in vivo* and *in vitro* sources. Such samples include urine, blood, saliva, tissue extracts, and cell suspensions. The use of specific detection methods such as SIM or MRM coupled with chromatography permits these very complex samples to be analyzed directly with LC–MS; sample preparation is minimal (filter and inject, for example). As trace analysis becomes important, however, many samples require a preconcentration step prior to analysis so adequate limits of detection can be obtained. Isolation and purification techniques commonly employed are solid- or liquid-phase extraction, preparatory HPLC, thin-layer chromatography (TLC), or a combination of these. GC–MS analyses require the sample to be volatile and thermally stable. If the analyte of interest does not meet these requirements, derivatization of an analyte can assist in its analysis. Derivatization can enhance sensitivity, selectivity, and chromatographic performance in some cases. Thus, it should be considered as an alternative when difficulties with an analyte are discovered. Ideally, the mass spectrum of the derivative will provide an intense molecular ion or unique fragment ion. A common derivatization choice is trimethylsilation of hydroxyl or amine groups to increase volatility and thermal stability. Several commercial trimethylsilating reagents are available such as *bis*-trimethylsilylacetamide (BSA).

#### Applications

Mass spectrometry analysis is extensively employed in general toxicological disciplines, such as clinical toxicology, forensics, and metabolomics. This versatile technique is also highly specific, as it has been used for the study of specific toxins, including heavy metals, aromatic amines, natural products, and DNA adduct formers. In the emergency clinical setting, the presentation of a highly intoxicated patient requires prompt blood and body fluid analysis, as one or a number of common intoxicating agents may be involved. Van Hee and coworkers [206] developed a method for rapidly screening body fluids for a series of intoxicating agents and their characteristic metabolites. A 20-μL sample of blood, plasma, or urine is derivatized with a trimethylsilyl agent, mixed with a 1,3-propylene glycol internal standard, and analyzed by GC/EI–MS. γ-Hydroxybutyric acid, a common "date-rape" drug, and several other potential intoxicants and metabolites were readily separated and quantified using this method. These compounds were detectable at body fluid concentrations of less than 1 mg/L, and the entire process, from derivatization to analysis, took less than 30 minutes.

Metabolomics is the exciting discipline of studying changes in the production and relative concentration of small-molecular-weight endogenous biochemicals and metabolites in living organisms before and after exposure to xenobiotics or pathological processes [59]. Metabolomic fingerprinting enables the toxicologist to identify likely xenobiotic exposure, monitor environmental stressors, and predict chemical toxicity. Dynamic metabolomic profiles consist of complex mixtures of very similar compounds that are constantly changing in terms of composition and relative amount. Accordingly, the precision and extreme sensitivity of MS are very useful in metabolomic studies, and examples of this application are numerous. The complex, dynamically changing metabolomic profile of the urine of toxin-exposed rats was elucidated and quantified in detail using an integrated HPLC–MS–electrochemical system [69]. Both toxic metabolites and discrete biochemical changes associated with histopathological changes were amenable to study. The chronic effects of heavy metal toxicity were detected and characterized by analyzing the metabolomic profile of several chemical families of endogenous biochemicals in rat urine [112]. Amines, carboxylates, and sulfated, glucuronidated, and glycosidic secondary metabolites were readily discernible using this integrated method. The time-dependent expression of certain analytes may allow researchers to monitor responses to specific toxins.

#### Determination of Toxic Metals Using Stable Isotope Dilution

Cadmium is another toxic heavy metal derived from a variety of industrial and environmental sources. Aggarwal et al. [1] have developed a stable isotope dilution GC–MS method for determining cadmium exposure. Their method determined cadmium in urine by using the chelating agent lithium *bis*(trifluoroethyl)dithiocarbamate (Li(FDEDTC)), to form the chelate $Cd(FDEDTC)_2$, which was analyzed by GC–MS with EI ionization. $M^{+\bullet}$ was monitored for quantitation. The internal standard used was $^{106}CdO$. Cd determination in urine was possible with this method at the 10-μg/L level with good precision and accuracy. The researchers note that the isotope dilution technique provides freedom from matrix effects so that precision and accuracy are not affected by incomplete recovery.

#### Determination of Aromatic Amines in Human Milk Using GC–MS

Aromatic amines come from several environmental pollution sources and have been associated with breast cancer in humans. The presence of these compounds in human breast milk was demonstrated in a straightforward manner by DeBruin et al. [55] using GC–MS. The method used *solid-phase microextraction* (SPME) [56] to sample the headspace over a heated milk sample. The SPME fiber was then inserted into the injector port of a GC–MS and desorbed onto the column. Aromatic amines identified were aniline, *o*-toluidine, and *N*-methylaniline. Two internal standards, aniline-$d_5$ and *o*-toluidine-$d_9$, were added to the milk samples prior to preparation. For quantitation, a quadrupole mass spectrometer was operated in the SIM mode with a base ion for each analyte selected. A qualifying ion was also used to confirm the identity. Using a calibration curve constructed from standards made in bovine milk, the authors quantified the three analytes in milk samples at the sub-ppb level.

### Serum Cotinine by LC–MS/MS Using APCI

Serum cotinine is often used as a biomarker of exposure to nicotine. If methods have sufficient limits of detection, exposure to extremely low amounts of nicotine as in environmental tobacco smoke can be assessed. The analysis of low concentrations of analytes in complex matrices may be approached in different ways. One approach seeks to perform extensive sample clean-up to present a relatively clean sample to the instrument. Although this is labor intensive and requires meticulous attention to handling of the sample, it permits the analytical system to operate longer and more reproducibly between servicing. A good example of this approach is the determination of cotinine in serum by the method of Bernert et al. [15]. They desired to measure exposure to nicotine for large numbers of samples from both smokers and nonsmokers. A rapid, sensitive LC–MS/MS method with APCI was developed for serum cotinine that utilized sample extraction and concentration. Although sample preparation has many steps, the results of this method are impressive. The HPLC method had a retention time of less than 1 minute for the analyte and internal standard, which permitted samples to be injected every 2 minutes. Under routine operation, 100 samples a day were analyzed. A detection limit of 0.05 ng/mL was achieved which was sufficient to monitor serum cotinine in nonsmokers exposed to environmental tobacco smoke.

Sample preparation began with addition of the internal standard, methyl-$D_3$-cotinine, to 1 mL serum. The sample was acidified and centrifuged to remove protein. The supernatant was basified and extracted with methylene chloride, dried, redissolved for transferal to a microvial, evaporated to dryness again, and finally reconstituted in 20 μL of toluene. For LC–MS/MS analysis, 10 μL were injected onto a 4.6-mm × 3-cm $C_{18}$ column for an isocratic separation with a flow rate of 1 mL/minute of 80% methanol and 20% 2-m*M* ammonium acetate. The effluent was introduced to an APCI source on a triple-stage quadrupole mass spectrometer set to operate in the MRM mode. Ions monitored were $m/z$ 177 → $m/z$ 80 for cotinine and $m/z$ 180 → $m/z$ 80 for the internal standard. For a confirming transition for cotinine, $m/z$ 177 → $m/z$ 98 was also monitored.

The key features of this method were sample extraction and concentration followed by very specific detection. Recoveries were 60 to 70%, and most background materials were removed. Reducing the final volume to 0.02 mL resulted in a >10-fold concentration of the sample. Highly specific MRM analysis and focusing on only one analyte reduced chromatography requirements so retention time could be minimal, and a simple isocratic method that did not require equilibration could be used. Thousands of serum samples were analyzed by this method but only for cotinine. The mass spectral parameters of this method could be modified to include nicotine as it could be extracted by this same procedure. Extension to other more polar metabolites of nicotine such as *trans*-3′-hydroxycotinine would be problematic as the method now exists, because they are more difficult than cotinine to efficiently extract into an organic phase.

### Determination of Nicotine and Several Metabolites in Smokers' Urine Using LC–MS with Thermospray

A different approach is to perform minimal sample preparation and let the specificity and sensitivity of the instrument work to measure the analyte. Keeping the sample close to its original state minimizes chances of analyte loss prior to detection, especially when numerous analytes are involved. An example of such a method is that of McManus et al. [136]. Though neither as rapid nor sensitive as the method above, it was ambitious in the scope of analytes it sought to determine. The objective of this method was to determine as many known urinary metabolites of nicotine as possible to assess nicotine exposure. The method used TS interfaced to a single quadrupole instrument. It permitted direct injection of urine from smokers and used a 10-minute reversed-phase HPLC gradient method for separation. The $(M+H)^+$ values for numerous known nicotine metabolites were detected by SIM, and standards were used to establish retention times. Sample preparation was minimal with addition of methyl-$D_3$-cotinine as internal standard and filtration through a 0.2-μ*M* mesh. Because no extractions were involved in this approach, extraction efficiency was not an issue, and all identified metabolites could be reasonably determined. Only 4 of 17 known metabolites were detected in smokers' urine initially (*trans*-3′-hydroxycotinine, cotinine, nicotine-*N*′-oxide, and norcotinine).

The strength of this method is the determination of nicotine and several metabolites in a single run. By determining multiple metabolites at the same time, a better indication of nicotine absorption is provided, as some individuals may produce more of one metabolite over another in their urine. The limit of detection of approximately 20 ng/mL makes this method useful for smokers but not for nonsmokers. Also, undiscovered minor nicotine metabolites would be difficult to find by this approach. The time and resources saved by this approach are offset partially by the introduction of many contaminants to the analytical system that damage columns and foul ionization sources so performance is eventually compromised. Later applications of the method to urine samples from smokers used enzymatic hydrolysis to indirectly account for glucuronide conjugates of nicotine, cotinine, and *trans*-3′-hydroxycotinine [26–28]. Further modification of this method has been reported using slightly faster chromatography and APCI tandem mass spectrometry [25]. MRM detection of six metabolites was applied to saliva and serum samples from smokers in addition to urine.

### Determination of DNA Adducts

DNA adducts are believed to be biological dosimeters for exposure to chemical carcinogens. This is a developing area of mass spectrometry research, and some impressive work has been done using radiolabeled compounds administered to animals and humans [70]. Using AMS to detect $^{14}C$ has lowered detection limits to the point that $^{14}C$-labeled carcinogens may be safely administered to humans at realistic exposure levels. Turtelbaum et al. [205] have studied the administration of $^{14}C$-labeled heterocyclic amines to rodents in their diet at 100 ng/kg/day to mimic human exposure. Following isolation and purification of the DNA, adducts were detectable after 24 hours in the liver and kidney. These data suggested that DNA adducts are formed at human exposure levels and are indicative of dose for this particular compound.

Another elegant approach for monitoring DNA adducts using two different LC–MS methods has been described by Siethoff et al. [187]. One of the problems with quantifying DNA adducts accurately is the lack of standards. This group has used ICP with a high-resolution sector mass spectrometer coupled to an HPLC to resolve the different adducts following digestion of the DNA sample to its monophosphate units. Each unit contains a phosphorous that is converted to $P^+$ by ICP. This ion occurs at *m/z* 30.97 but must be resolved from $^{15}N^{16}OH^+$ and $^{14}N^{16}OH^+$ interferences in the plasma background using high-resolution mass discrimination. Because ICP completely atomizes molecules, any phosphorous-containing molecule such as phosphoric acid may be used as an internal standard for quantitative purposes. Thus, ICP–MS is used to locate and quantify the DNA adducts. The authors then used the same LC method connected to a mass spectrometer with an ES source to elucidate the structures of the adducts from the molecular adduct $(M–H)^-$.

### The Rationale for Choosing Mass Spectrometry for an Analysis

Contemporary mass spectrometry combines specificity and sensitivity and gives highly reliable analytical results. Femtomolar determinations are commonplace. The unique fingerprint identification provided by mass spectra gives a great deal of confidence that an analysis has been performed correctly. Improvements in inlet systems, ionization sources such as ES and APCI, and mass analyzers such as multiple-stage quadrupoles for tandem mass spectrometry have greatly expanded the applicability and portability of mass spectrometry. LC–MS is the workhorse in modern bioanalytical laboratories. It is the method of choice for toxicokinetic and pharmacokinetic studies and is commonly used for metabolite characterization, quantification and metabolomic profiling. Its high sensitivity makes it ideal for quantifying exposure to environmental toxins. Nevertheless, the analyst must keep in mind that mass spectrometry is a technique that destroys the sample. This is particularly important in cases where sample sizes are very small and result from time consuming and labor intensive isolation and purification methods. Mass spectrometers are expensive, although the price of benchtop systems is becoming attractive. Newer techniques in mass spectrometry are very powerful but do not work well for all samples; for example, if a sample could be one of several possible positional isomers, mass spectrometry may be inadequate for the unequivocal assignment of structure. Despite these limitations, mass spectrometry is critical to studying mechanisms of toxicity.

## NUCLEAR MAGNETIC RESONANCE SPECTROSCOPY

In 1946, Purcell, Torrey, and Pound of Harvard University and Bloch, Hansen, and Packard of Stanford University independently detected nuclear magnetic resonance (NMR) effects of the hydrogen nucleus in paraffin wax (Purcell) and water (Bloch). These first observations of nuclear magnetic resonance won the Nobel Prize in Physics for Bloch and Purcell in 1952. In 1951, Packard reported that the NMR spectrum of ethanol consisted of three distinct resonances and that these resonances arose from the three different chemical environments for the hydrogen nuclei in the molecule ($CH_3$, $CH_2$, and OH). This discovery quickly caught the attention of organic chemists who realized that NMR spectroscopy could be used as a tool for determining chemical structure [101,164]. NMR spectroscopy has since emerged as the most powerful technique for structural characterization of organic compounds. NMR is the method of choice for determining the identity of xenobiotic metabolites if they can be isolated in microgram to milligram quantities. In this regard, it is often used in conjunction with chromatographic and mass spectral methods. It can also be used to probe the structures of xenobiotic adducts to biological macromolecules such as DNA and proteins. NMR is used to quantify xenobiotic exposure and changes in endogenous compounds in response to such exposure, and it is widely employed in metabolomic studies.

Due to its selectivity and the wealth of information that is available from NMR spectroscopy, it has found increasing application in pharmacology, toxicology, and biomedical research. Recent advances in NMR technology have enabled researchers and clinicians to study chemical and physiological changes within cell suspensions, isolated organs, and living organisms in a noninvasive manner. *In vivo* spectroscopic techniques have been used to measure cellular metabolism, intracellular pH, cytosolic sodium, magnesium and calcium levels, organ damage in response to toxicants, and a host of other toxicologically relevant phenomena.

Nuclear magnetic resonance spectroscopy is a very powerful tool for studying mechanisms of toxicology. Toxicologists should be familiar with the strengths and limitations of NMR spectroscopy and the types of studies for which it is suited. The following sections include an overview of the theory of NMR and numerous examples of its application in toxicology and related disciplines. Although a detailed mathematical treatment of magnetic resonance is beyond the scope of this chapter, mathematical formulae will be used to describe and clarify the phenomena. It is assumed that the reader has encountered traditional proton and carbon NMR spectroscopy in an introductory organic chemistry course, and very little discussion of the interpretation of NMR spectra appears. A number of extremely good texts on the theory, application, and interpretation of NMR are available. The interested reader might refer to them for more detail [57,60,164, 169,175].

## Basic Theory of NMR Spectroscopy

### Nuclear Spin

An understanding of the NMR phenomenon begins with a look at the atomic nucleus. Nuclei are composed of protons and (except for $^1$H) neutrons; therefore, all atomic nuclei carry a positive charge. Many nuclei also rotate about the nuclear axis, and the angular momentum of this spinning charge is described in terms of the spin quantum number, *I*. Spin numbers have values of 0, 1/2, 1, 3/2, and so forth. For our purposes, the most important spin-active nuclei are $^1$H, $^{13}$C, $^{31}$P, and $^{19}$F, all of which have $I = 1/2$.

### Behavior of Nuclei in an External Magnetic Field

Because a spinning nucleus ($I > 0$) is a charge in motion, it generates a magnetic field and behaves like a small bar magnet with a magnetic moment (μ). Just as a bar magnet will align itself with an external magnetic field, so too will any nucleus with $I > 0$. For $I = 1/2$, the nuclei can adopt one of two orientations relative to the external magnetic field. They may be aligned with the field (low-energy state) or against the field (high-energy state).

A nucleus with $I > 0$ exhibits another important property when placed in an external magnetic field. That property is precessional motion. We have all observed the behavior of a spinning top. As it spins, the axis of the top slowly revolves around the vertical. The top is said to be precessing around the vertical axis, and this type of behavior is called *precessional motion*. A nucleus spinning in an external magnetic field also exhibits precessional motion, with the magnetic moment (μ) precessing around the axis of the applied magnetic field $\mathbf{B}_o$ (Figure 41.6). The frequency at which the magnetic moment precesses about $\mathbf{B}_o$ is called the precessional frequency, ν. While the spinning frequency of any given nucleus is constant, ν varies directly with the strength of the external magnetic field $\mathbf{B}_o$:

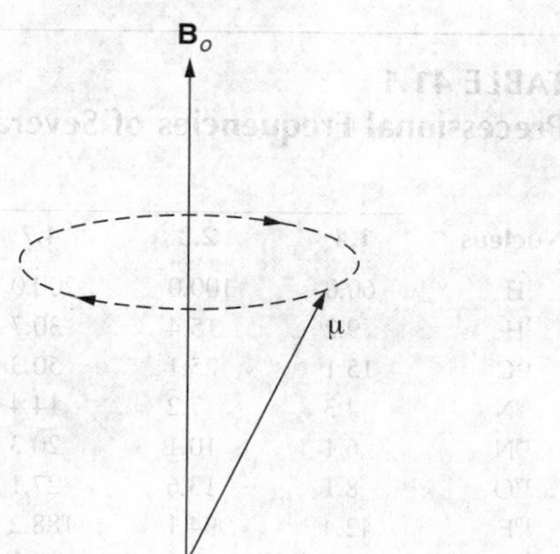


**FIGURE 41.6** Magnetic moment (μ) of a spinning nucleus precessing about the applied magnetic field ($\mathbf{B}_o$) with precessional frequency ν.

$$\nu \propto \mathbf{B}_o$$

This proportionality is the most fundamental relationship of NMR spectroscopy and leads to the fundamental NMR equation:

$$\nu = \frac{\gamma \mathbf{B}_o}{2\pi} \tag{41.2}$$

The proportionality constant (γ) is the magnetogyric ratio and is a fundamental nuclear constant. It is related to the nuclear magnetic moment (μ) and the spin quantum number (*I*):

$$\gamma = \frac{2\pi\mu}{hI} \tag{41.3}$$

where *h* is Planck's constant.

A proton (hydrogen nucleus) has a magnetogyric ratio of $2.6752 \times 10^8$ radians s$^{-1}$ T$^{-1}$; therefore, in a magnetic field of 7.1 tesla (7.1 T), its precessional frequency is approximately 300 MHz. A proton in a magnetic field of 14.1 T will have $\nu \approx 600$ MHz. The magnetogyric ratio of $^{13}$C is $6.727 \times 10^7$ radians s$^{-1}$ T$^{-1}$, so in a field of 7.1 T, $\nu \approx 75$ MHz. Table 41.1 lists the precessional frequencies at selected field strengths for several common magnetic nuclei. NMR spectrometers are most frequently classified by the precessional frequency of the $^1$H nucleus at the field strength of the magnet; for example, a spectrometer with a field strength of 7.1 T is commonly referred to as a 300-MHz NMR.

A spin 1/2 nucleus in a magnetic field will have a precessional frequency (ν) and will be able to adopt one of two orientations, or spin states, relative to $\mathbf{B}_o$. It will

**TABLE 41.1**
**Precessional Frequencies of Several Common Nuclei at Selected Field Strengths**

| | Precessional Frequency (MHz) at Field Strength (T) | | | | | | | | | |
|---|---|---|---|---|---|---|---|---|---|---|
| **Nucleus** | **1.4** | **2.3** | **4.7** | **7.1** | **9.4** | **11.7** | **14.1** | **17.6** | **18.8** | **21.1** |
| $^{1}H$ | 60.0 | 100.0 | 200.0 | 300.0 | 400.0 | 500.0 | 600.00 | 750.0 | 800.0 | 900.0 |
| $^{2}H$ | 9.2 | 15.4 | 30.7 | 46.1 | 61.4 | 76.8 | 92.1 | 115.1 | 122.8 | 138.2 |
| $^{13}C$ | 15.1 | 25.1 | 50.3 | 75.4 | 100.6 | 125.7 | 150.9 | 188.6 | 201.2 | 226.3 |
| $^{14}N$ | 4.3 | 7.2 | 14.4 | 21.7 | 28.9 | 36.1 | 43.3 | 54.1 | 57.8 | 65.0 |
| $^{15}N$ | 6.1 | 10.1 | 20.3 | 30.4 | 40.5 | 50.7 | 60.8 | 76.0 | 81.0 | 91.2 |
| $^{17}O$ | 8.1 | 13.6 | 27.1 | 40.7 | 54.2 | 67.8 | 81.3 | 101.6 | 108.4 | 122.0 |
| $^{19}F$ | 42.4 | 94.1 | 188.2 | 282.2 | 376.3 | 470.4 | 564.5 | 705.6 | 752.6 | 846.7 |
| $^{23}Na$ | 15.9 | 26.5 | 52.9 | 79.4 | 105.8 | 132.3 | 158.7 | 198.4 | 211.6 | 238.1 |
| $^{31}P$ | 24.3 | 40.5 | 81.0 | 121.4 | 161.9 | 202.4 | 242.9 | 303.6 | 323.8 | 364.4 |

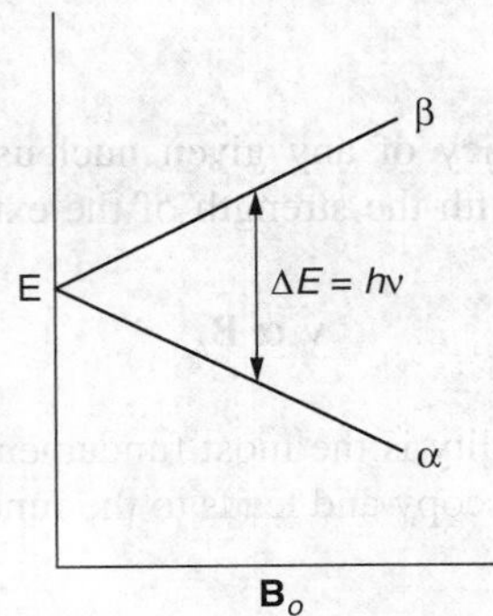


**FIGURE 41.7** Energy level diagram for a spin 1/2 nucleus in a magnetic field $\mathbf{B}_o$.

be aligned with (parallel) or against (antiparallel) the external field. The parallel orientation (α) is a low-energy state for the system, and the antiparallel orientation (β) is a high-energy state (Figure 41.7). A nucleus precessing in the parallel orientation can absorb energy (electromagnetic radiation) and be excited to the high-energy, antiparallel orientation. The precessing nucleus will only absorb electromagnetic radiation with a frequency equal to ν. Because precessional frequencies are on the order of MHz, the electromagnetic radiation that will excite a precessing nucleus is in the radiofrequency range. When the applied RF equals ν, the precessing nucleus and the RF are said to be in resonance, hence the name nuclear magnetic resonance. The frequency at which resonance occurs for a given nucleus is called its *resonance frequency.*

## Relaxation Times

When a nucleus has absorbed energy and is excited to the high-energy spin state, it will tend to lose energy and return to the low-energy state. In addition to direct reemission of RF, there are two radiationless processes by which nuclei can exchange energy with their environment. These relaxation processes are a direct result of interaction of the nucleus with some electromagnetic vector in the local environment. The nucleus is surrounded by solvent molecules, and energy can be transferred to the solvent or other nearby atoms in a process called *spin-lattice* relaxation. The spin-lattice relaxation time ($T_1$) depends on such factors as temperature and solvent viscosity, with higher temperature and lower solvent viscosity slowing $T_1$ relaxation.

The nucleus can also transfer energy to nearby nuclei in a process called *spin–spin* relaxation. In spin–spin relaxation, one nucleus loses energy and the other one gains energy so there is no net change in the populations of the two spin states. The spin–spin relaxation time ($T_2$) depends on molecular mobility, and nuclei in large molecules that have highly constrained molecular motions have very efficient spin–spin relaxation (short $T_2$). The magnitude of $T_1$ and $T_2$ determines the line widths of NMR spectral lines; short relaxation times lead to broad lines, and long relaxation times lead to sharp lines. This means that NMR spectra obtained in viscous solvents (short $T_1$) will have broader lines than those obtained in nonviscous solvents. Also, NMR spectra of biological matrices such as plasma or cell suspensions will show many very broad resonances from proteins, nucleic acids, and other macromolecules that have very short $T_2$ relaxation times. These broad resonances can obscure the signals of small molecules of interest such as xenobiotic metabolites. A variety of NMR techniques have been developed that enable the observation of low-molecular-weight compounds in the presence of macromolecules [170].

## Chemical Shifts

In the simplest NMR experiment, *continuous wave NMR* (CW–NMR), a sample is placed in a strong, uniform magnetic field and the nuclei in the sample process with their characteristic frequency ν. The sample is irradiated with radiowaves of increasing frequency, and when the radiofrequency equals ν, the nuclei are excited to the high-energy state and RF is absorbed. This absorption is recorded as the NMR spectrum. If all nuclei of a given type ($^1$H, for example) precessed with exactly the same frequency, a NMR spectrum would convey very little structural information. It would consist of a single line. Fortunately, the exact value of ν for a given nucleus depends on the chemical environment of that nucleus.

So far we have discussed the magnetic properties of spinning nuclei, but atoms and molecules also contain electrons. Under the influence of an external magnetic field, electrons will circulate and generate their own magnetic field that will tend to oppose $\mathbf{B}_o$. This results in a magnetic shielding of nearby nuclei, which slightly shifts their precessional frequencies. The density of the circulating electron cloud depends greatly on its chemical environment so different nuclei in a molecule will experience different degrees of shielding and hence will have different values of ν. The shift in ν depends on the chemical environment of the nucleus, and for this reason it was given the name *chemical shift*.

Relative to the resonance frequency, chemical shifts are quite small. For protons in a field of 7.1 T (ν ≈ 300 MHz), the chemical shifts cover a range of about 4000 Hz. The absolute value of chemical shifts in frequency units depends on the strength of the applied magnetic field. The greater the value of $\mathbf{B}_o$, the greater the chemical shift range. This explains, in part, the desire to move to higher and higher field strengths for NMR spectroscopy. The higher the field, the greater the resolution of chemical shifts.

Chemical shifts are rarely (if ever) expressed in absolute frequency units; instead, they are measured relative to a standard reference compound. For $^1$H and $^{13}$C NMR, the universally accepted reference is tetramethylsilane (TMS). It is commonly added to samples as an internal standard at concentrations <1%. TMS has several important properties that make it a near ideal reference standard. It contains 12 magnetically equivalent protons and 4 magnetically equivalent carbons, so it gives a single, intense peak in the $^1$H and $^{13}$C NMR spectra. It is chemically inert, soluble in most organic solvents, and volatile (b.p. = 27°C) so it can be removed easily from the sample after analysis. The protons and carbons of TMS absorb at a lower frequency (are more shielded) than those of almost all other organic compounds, so their chemical shifts can be arbitrarily set to 0 Hz and most other chemical shifts measured relative to them will be positive. TMS is not soluble in water, so for aqueous solutions a suitable water soluble internal standard must be chosen. The most widely used internal standard for aqueous samples is the sodium salt of 3-(trimethylsilyl)-propanoic acid (TSP).

Because chemical shifts vary with the strength of the applied magnetic field, their values (in Hz) vary from one NMR spectrometer to the next. Protons that absorb at 300 Hz relative to TMS in a field of 7.1 T (300-MHz instrument) will absorb at 100 Hz on a 2.3-T (100-MHz) instrument. To avoid confusion and enable comparisons of chemical shifts from instrument to instrument, chemical shifts are expressed in dimensionless units designated δ. The chemical shift in δ units is defined according to the following relationship:

$$\delta = \frac{\nu_s - \nu_{std}}{\text{Operating frequency}} \times 10^6 \text{ ppm} \qquad (41.4)$$

where $\nu_s$ and $\nu_{std}$ are the resonance frequencies of the sample and standard, respectively.

## Spin–Spin Coupling

Whereas a great deal of structural information may be obtained from the chemical shifts of nuclei in a molecule, even more information is available from a NMR spectrum. Nuclei that are closely connected through bonding electrons are influenced by the spin state of their neighbors. Consider two closely connected protons, $H_a$ and $H_b$:

$H_a$ $H_b$
| |
C — C

The bonding electrons in the $H_a$–C bond tend to align their spins with the spin of $H_a$, the C–C bonding electrons tend to align with the spin of the $H_a$–C bonding electrons, and the $H_b$–C bonding electrons tend to align with the spin of the C–C bonding electrons. Finally, $H_b$ tends to align its spin with the spin of the $H_b$–C bonding electrons. In this way, the spin state of $H_a$ directly influences the spin state of $H_b$. Because $H_a$ can have two spin states, parallel or antiparallel, the resonance line of $H_b$ is split into two closely spaced lines, a so-called *doublet*. $H_b$ influences the spin state of $H_a$ in exactly the same way, so $H_a$ also appears as a doublet. The spacing between the lines in the $H_a$ doublet is the same as that in the $H_b$ doublet. Figure 41.8 illustrates the appearance of the $H_a$ and $H_b$ NMR signals. This phenomenon, called *spin–spin coupling* (or simply *coupling*), results in characteristic splitting of NMR sig-

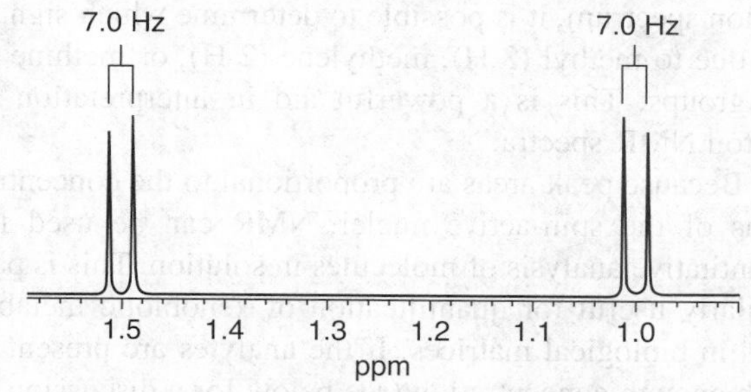


**FIGURE 41.8** Proton nuclear magnetic resonance spectrum of the $H_a$/$H_b$ spin system. Note that coupling constant $J$ = 7.0 Hz.

nals which is dependent on the number of neighboring spin-active nuclei, their geometrical arrangement, and the number of bonds between them. Coupling is usually not important beyond four bonds, and magnetically equivalent nuclei do not show coupling. Unlike chemical shifts, coupling is independent of $\mathbf{B}_o$.

The number of lines into which a NMR signal is split depends on the number of adjacent nuclei. The general rule is that *n* adjacent spin-active nuclei will split a signal into $2nI + 1$ lines. For spin 1/2 nuclei such as $^1H$, two adjacent nuclei will split a signal into three lines (a triplet), three adjacent nuclei will split the signal into four lines (a quartet), and so forth. The relative line intensities within a multiplet are determined by the coefficients in the binomial expansion. A triplet will have intensities of 1:2:1, a quartet will have intensities 1:3:3:1, and so on. The separation between spectral lines due to coupling is the coupling constant (*J*) (Figure 41.8). The magnitude of a coupling constant depends on the number of intervening bonds and the bond geometry. The coupling constants for protons that are *trans* to one another in a rigid molecule are greater than those for protons that are *cis*. Spin–spin coupling is most important in proton NMR spectroscopy, as most organic molecules have adjacent protons and show extensive coupling. In $^{13}C$ NMR spectroscopy, proton coupling to the carbons is often undesirable because it leads to very complex spectra and reduced signal-to-noise ratio. For these reasons, it is usually suppressed by irradiating over the entire proton resonance frequency range to saturate the proton transitions. This so-called decoupling produces carbon spectra made up of only singlets.

## Integral Areas

The area under a peak in a NMR spectrum is proportional to the total number of nuclei giving rise to that signal; in other words, the area of a peak can be used to gain information about the concentrations of the spin-active nuclei. All modern NMR spectrometers are capable of integrating peak areas, and the resulting integrals can be plotted on the spectrum. In the case of a multiplet, the total area of the signal is integrated. By comparing integrals within a proton spectrum, it is possible to determine which signals are due to methyl (3 H), methylene (2 H), or methine (1 H) groups. This is a powerful aid in interpretation of proton NMR spectra.

Because peak areas are proportional to the concentrations of the spin-active nuclei, NMR can be used for quantitative analysis of molecules in solution. This is particularly useful for quantification of xenobiotic metabolites in biological matrices. If the analytes are present in high enough concentration (see below for a discussion of NMR sensitivity) and signals due to the presence of the analytes of interest can be distinguished in the spectrum, then very little sample preparation is required. For quantitative analysis, analyte integrals are compared to integrals of an internal standard that is added to the sample at a known concentration. Quantitative analysis by NMR requires very careful calibration and attention to spectral acquisition parameters so it should be carried out cautiously. Excellent discussions of quantitative analysis by NMR are available [164,203].

For heteronuclear NMR (such as carbon) where broadband proton decoupling is used, the peak areas do not always give an accurate estimate of the concentration of spin-active nuclei. Decoupling gives rise to nuclear Overhauser effects that complicate the interpretation of integral areas. For this reason, carbon spectra are rarely integrated; however, spectral acquisition parameters can be adjusted to correct for these effects if accurate integrals are required from a carbon (or other heteronuclear) NMR spectrum.

## Sensitivity

For nuclei in an external magnetic field, the energy difference between spin states is quite small. This means that the two spin states are almost equally populated at room temperature, with the population of the low-energy state exceeding that of the high-energy state by only about 0.001%. The intensity of absorption (and hence the sensitivity of NMR) is proportional to the number of nuclei absorbing RF energy. As the population difference between spin states increases, there will be more nuclei to absorb RF energy, and the sensitivity will increase. From Equation 41.3, we see that the energy difference between spin states ($\Delta E = h\nu$) is proportional to the magnetogyric ratio ($\gamma$). This means that the sensitivity of NMR depends on the magnitude of $\gamma$. Detection of nuclei with relatively large magnetogyric ratios (such as $^1H$ and $^{19}F$) will be fairly sensitive. The sensitivity of NMR also depends on the natural abundance of the spin-active nucleus under observation. Nuclei with a high natural abundance will be detected with greater sensitivity. Table 41.2 lists natural abundance, spin quantum number, magnetogyric ratio, and sensitivity for some selected nuclei.

The energy difference between spin states ($\Delta E = h\nu$) is also proportional to the magnetic field strength $\mathbf{B}_o$. One way to increase the sensitivity of NMR is to increase the field strength. This approach has been somewhat successful; however, even at a field strength of 14.1 T, the energy difference between spin states is only on the order of $10^{-4}$ kJ mol$^{-1}$. NMR is (and is likely to remain) less sensitive than optical techniques such as electronic absorption spectroscopy, where $\Delta E$ is considerably larger. Even with the most sensitive high field instruments, tens to thousands of micrograms of sample are required.

**TABLE 41.2**
**Selected NMR Properties of Several Common Nuclei**

| Nucleus | Natural Abundance (%) | Spin Quantum Number I | Magnetogyric Ratio ($\times 10^{-7}$ radians $s^{-1}$ $T^{-1}$) | Relative Sensitivity at Constant Field |
|---|---|---|---|---|
| $^{1}H$ | 99.985 | 1/2 | 26.752 | 1.000 |
| $^{2}H$ | 0.015 | 1 | 4.107 | $9.65 \times 10^{-3}$ |
| $^{13}C$ | 1.108 | 1/2 | 6.727 | 0.016 |
| $^{14}N$ | 99.635 | 1 | 1.933 | $1.01 \times 10^{-3}$ |
| $^{15}N$ | 0.365 | 1/2 | 2.711 | $1.04 \times 10^{-3}$ |
| $^{17}O$ | 0.037 | 5/2 | 3.627 | 0.029 |
| $^{19}F$ | 100 | 1/2 | 25.167 | 0.834 |
| $^{23}Na$ | 100 | 3/2 | 7.076 | 0.093 |
| $^{31}P$ | 100 | 1/2 | 10.829 | 0.067 |

## FT–NMR

Because the energy difference between spin states is small, NMR signals are invariably weak. In fact, they are often only slightly more intense than the background noise caused by the electronics of the NMR spectrometer. To improve the signal-to-noise ratio (S/N), several spectra can be obtained and the resulting collection of data averaged. The NMR signals in these spectra would always occur at the same frequencies; however, the random noise would not. The desired signals would build up over time relative to the noise and the S/N would increase. It can be shown that, for *n* repetitions, the signal increases by a factor of *n* and the noise increases by a factor of $\sqrt{2}$ . As a result, signal averaging increases the S/N by a factor of $\sqrt{2}$ .

If signal averaging were applied to the CW–NMR experiment described previously, the spectrum would be scanned repetitively over the frequency range of interest and the resulting spectra averaged. This would result in an increase in S/N, but the time required for the experiment would be quite long. Consider a proton spectrum covering 10 ppm at a field strength of 2.3 T. The sweep width (frequency range) of this spectrum is 1000 Hz. A sweep rate of 1 Hz/s is required to achieve a spectral resolution of 1 Hz so it would take 1000 seconds to acquire one spectrum. Increasing the S/N by a factor of 4 requires 16 repetitions ( $\sqrt{2}$ ), so this experiment would take almost 4-1/2 hours. You can see that signal averaging in CW–NMR is not very practical. How, then, do we achieve a significant increase in the S/N in a reasonable amount of time?

The solution to this problem is found in a technique called *pulse NMR*. In pulse NMR, the sample is irradiated with a pulse of RF energy containing all the frequencies required to excite the nuclei of interest. This pulse is applied for a very short period of time (on the order of µsec) and the nuclei are allowed to relax to their equilibrium state with the emission of all the frequencies previously absorbed by the nuclei. This signal decays over time and is called *free induction decay* (FID). The FID contains all the spectral information including chemical shifts, coupling constants, and intensity. To convert this information from the time domain to the frequency domain, which is the normal mode for a NMR spectrum, the FID is subjected to a mathematical operation called *Fourier transformation*. To achieve signal averaging, a series of FIDs are acquired and averaged prior to Fourier transformation. The resulting FT–NMR spectrum is equivalent to a CW spectrum and is obtained in considerably less time. Figure 41.9 shows the FID and resulting proton FT–NMR spectrum of a 0.1% solution of ethylbenzene. For this spectrum, 64 repetitions, or transients, were acquired, requiring a total time of 4.7 minutes. Note the large TMS peak at 0 ppm and the presence of an impurity peak at about 1.55 ppm. Figure 41.10 shows the effect on the S/N of increasing the number of transients. These proton spectra were obtained on the same ethylbenzene sample used to generate Figure 41.9 and only the quartet is plotted.

All modern NMR spectrometers are of the FT type. FT–NMR has allowed insensitive nuclei such as $^{13}C$ to be studied routinely and sensitive nuclei such as $^{1}H$ and $^{19}F$ to be studied at much lower concentrations than previously possible. It has also opened up a multitude of powerful new multipulse and two-dimensional NMR experiments, some of which will be discussed in a later section.

### Instrument Design

Although NMR spectrometer design and construction vary from one manufacturer to the next, all spectrometers share the same basic components. Figure 41.11 shows a schematic representation of the basic features of a high-field NMR spectrometer. The heart of the spectrometer is the NMR probe that sits inside the superconducting magnet. The probe contains coils that transmit the RF pulses to the sample and receive the NMR signals that are emitted. Specialized probes for a variety of NMR experiments are available, and selection of the proper probe for the partic-

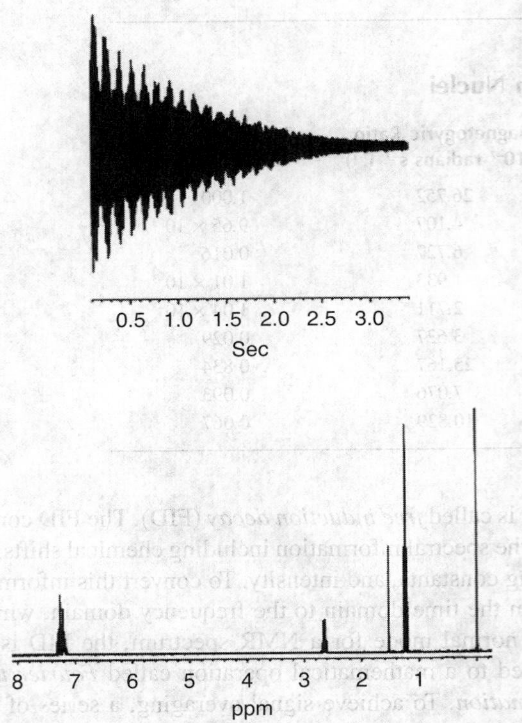


**FIGURE 41.9** Free induction decay and proton nuclear magnetic resonance spectrum of a 0.1% solution of ethylbenzene in $CDCl_3$; data were collected at a frequency of 300 MHz and a field strength of 7.1 T.

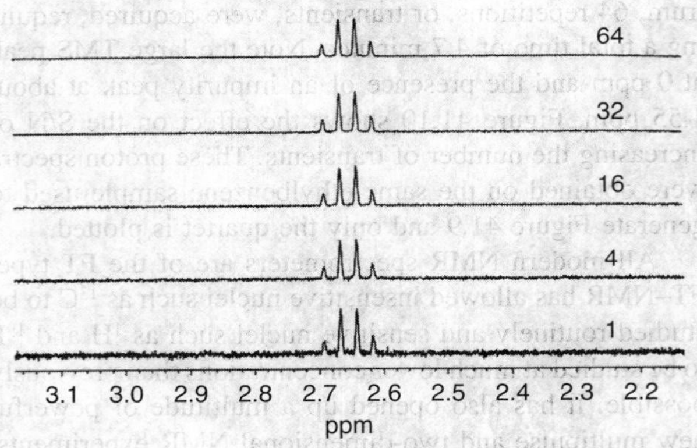


**FIGURE 41.10** The effect of increasing the number of transients on the S/N ratio of the proton nuclear magnetic resonance spectrum of ethylbenzene.

ular experiment is crucial to its success. They are exchangeable, and most NMR laboratories have more than one type. For *in vivo* spectroscopy, the probe may be surface coils placed in close proximity to the organ under study. The spectrometer contains a RF source for generating RF pulses and a RF detector for receiving and amplifying the NMR signal. A computer that is also used for storing, processing, and displaying data controls the RF source and detector.

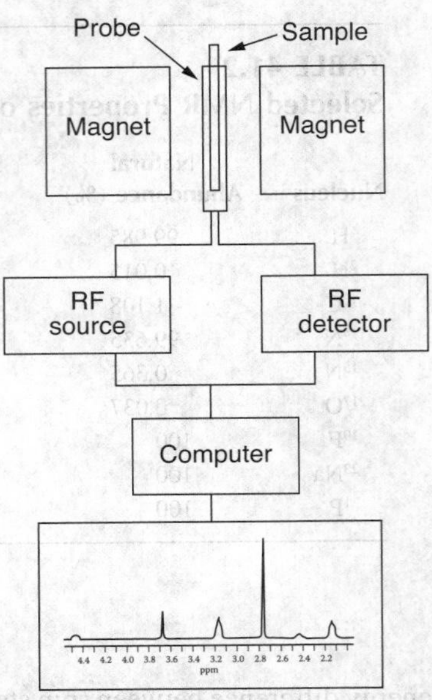


**FIGURE 41.11** Schematic representation of a nuclear magnetic resonance spectrometer.

## Experimental Design

The proper design of toxicological NMR studies depends on a number of factors. The first consideration when designing a NMR study is the nature of the sample. If the sample is a solution or cell suspension or can be dissolved in a suitable solvent, then a host of liquid-state NMR experiments is possible. If the sample is alive (perfused organs, laboratory animals, or humans) then *in vivo* spectroscopy using surface coils or imaging techniques is required. The choice of nuclei, type of NMR experiment, probe, solvent (if liquid-state analysis is required), and method of sample preparation will affect the outcome of the study. This section provides some guidance to the researcher in designing NMR studies. Particular emphasis is placed on liquid-state methods, as spectrometers for liquid-state analyses are more readily available than those for *in vivo* spectroscopy, and liquid-state methods are by far the most commonly used for toxicology research.

### Choice of Nucleus

The most useful NMR active nuclei for toxicology studies are $^{1}H$, $^{13}C$, $^{31}P$, and $^{19}F$. When the effects of heavy metal exposure are of interest, nuclei such as $^{199}Hg$ or $^{113}Cd$ can sometimes be used. Biologically derived samples are often complex mixtures, and NMR signals from the matrix can cause significant interference. This is particularly troublesome for $^{1}H$ and $^{13}C$, as these nuclei are ubiquitous. With proper sample cleanup and utilization of appropriate NMR techniques, these matrix effects can be overcome. $^{1}H$ is the most sensitive NMR active nucleus (in absolute terms) and is present in almost all organic compounds. If care is taken to minimize signal overlap from the matrix, it can provide qualitative and quantitative information on the metabolism and biochemical effects of xenobiotics. $^{1}H$ spectra can be obtained for compounds present in concentrations greater than about 50 to 100 $\mu M$ in a reasonable amount of time. For these reasons, $^{1}H$ is the most commonly used NMR active nucleus for studying mechanisms of toxicology. A number of studies of crude urine or urine extracts have been reported; however, biological fluids are usually concentrated and fractionated by HPLC prior to $^{1}H$ NMR analysis. In some cases, two-dimensional NMR experiments such as COSY (discussed below) can adequately resolve signals of interest with little or no sample cleanup.

$^{13}C$ is of very low natural abundance (about 1%), and its magnetogyric ratio is four times lower than $^{1}H$. As a result, its absolute sensitivity is over 5000 times less than $^{1}H$. Despite this low sensitivity, $^{13}C$ NMR spectra can be obtained for compounds present at concentrations >10 m$M$ in a reasonable amount of time. As in $^{1}H$ NMR studies, samples are most often concentrated and purified by HPLC prior to analysis. In some cases, isotopic enrichment with $^{13}C$ has been used to increase sensitivity. Enrichment to 90 atom-% $^{13}C$ results in a 90-fold increase in sensitivity, rendering direct analysis of crude urine for low-molecular-weight metabolites feasible. Combined with two-dimensional NMR experiments such as INADEQUATE, isotopic enrichment with $^{13}C$ is a powerful tool for metabolite characterization. Noninvasive *in vivo* $^{13}C$ NMR spectroscopy has been used to study the pharmacokinetics of $^{13}C$-enriched xenobiotics in the rat [145]. The high cost and limited availability of $^{13}C$ enriched materials are the main disadvantages to isotopic enrichment.

$^{31}P$ is a sensitive nucleus with a 100% natural abundance. As such, it is an attractive candidate for NMR studies. $^{31}P$ is one of the most commonly used nuclei for *in vivo* NMR spectroscopy and has been used extensively to determine the effects of xenobiotics on *in vivo* energy metabolism and intracellular pH [127]. It is much less useful for studies of xenobiotic metabolism, as very few xenobiotic metabolites contain phosphorous.

$^{19}F$ is almost as sensitive as $^{1}H$ and, because it is not normally found in biological samples, $^{19}F$ NMR signals from the matrix are not a problem. If the xenobiotic of interest contains fluorine, useful information such as quantitative determination of fluorinated metabolites and excretion rates can be obtained from $^{19}F$ NMR. Most fluorinated compounds contain only a few fluorine atoms (one to three), so structural characterization of metabolites by $^{19}F$ NMR is rare. $^{19}F$ is a very good tracer for *in vitro* and *in vivo* spectral determination of organ perfusion and has been used to determine cerebral, hepatic, and muscular blood flow [147]. Using fluorophenols as model toxins, the microbial degradation of environmental pollutants has been examined in a detailed metabolomic study using $^{19}F$ NMR [19]. In practice, NMR studies often utilize more than one type of NMR-active nucleus. Characterization of metabolites is best accomplished with more than one type of NMR experiment. Typically, both one-dimensional $^{1}H$ and $^{13}C$ spectra and a variety of two-dimensional experiments are used.

### Choice of NMR Experiments

Although a detailed discussion of modern NMR experiments is far beyond the scope of this chapter, several excellent books on the subject are available [57,60,133,175]. A brief description of those experiments that are likely to be most useful for studying mechanisms of toxicology is appropriate. NMR experiments can be divided into two main categories: one-dimensional and multidimensional. One-dimensional NMR experiments are of the type discussed in the earlier section on basic theory. The output of a one-dimensional experiment is a spectrum of intensity vs. frequency (or chemical shift in $\delta$ units). The $^{1}H$ spectrum of Figure 41.9 is a very good example. In principle, one-dimensional spectra can be obtained for any spin-active nucleus. One-dimensional spectra are the type used most often for quantitative analysis.

For structural characterization of organic compounds, one-dimensional proton and carbon spectra are almost always used. These spectra contain information about the chemical environments of the hydrogen and carbon atoms in the molecule. The proton spectrum, with its coupling information, also helps in assigning connectivities, as coupling constants and peak multiplicity can assist the spectroscopist in determining which hydrogens are coupled. Coupling constants also help in assigning geometry in rigid systems (*cis* or *trans* isomers, for example). Although a great deal of structural information can be obtained from these one-dimensional spectra, it is often useful to obtain two-dimensional spectra as well.

For two-dimensional experiments, two or more RF pulses with variable delay times between them are used to excite the nuclei. The variable delay times introduce a second time domain and, after two-dimensional Fourier transformation, a spectrum with two frequency domains is obtained. No attempt to explain the details of two-dimensional NMR is made here; the interested reader should refer to other texts for such detail [133]. The two-dimensional experiments, which are likely to be the most useful in toxicology research, are used to investigate spin–spin coupling by correlation spectroscopy. COSY spectra reveal $^{1}H$–$^{1}H$ spin–spin correlations and generate all connectivities between coupled protons. HETCOR (or HMQC and HMBC) spectra reveal proton–heteroatom correlations (usually $^{1}H$–$^{13}C$ correlations) and generate connectivities between protons and heteroatoms, such as carbon. INADEQUATE spectra (which are used much less frequently and most often for $^{13}C$-enriched compounds) are used to generate connectivities between carbons. Two-dimensional experiments are very useful for resolving signals of interest from overlapping matrix signals in biological samples, and examples of their application in metabolism studies are presented in a later section.

### Instrument Considerations

Ideally, the instrument with the highest available field strength (frequency) should be used to obtain maximum resolution and sensitivity. Commercial instruments with a field strength of 900 MHz are available. NMR but advances in magnet design should make even higher field instruments possible. Price must be considered when purchasing an instrument. Fortunately, most academic NMR laboratories make instrument time available free or for a reasonable hourly rate.

In toxicology research, sample size is often limited. It may take many hours (or weeks) to isolate a few hundred micrograms of a xenobiotic metabolite, so sensitivity is a major concern. For a given field strength, the component that has the greatest influence on sensitivity is the probe. The most common probes are designed for 5-mm sample tubes and in many cases will give adequate sensitivity. Where sample size is limited, the smallest possible probe should be used. Microprobes designed for 3-mm sample tubes are available, and a 1.7-mm probe has been developed [132].

Probes are tuned to the resonance frequency of the nucleus under observation. Selectively tuned probes are designed to work over a very narrow frequency range. They are selected based on the frequency of the nucleus under observation. Selectively tuned probes for proton may be tuned slightly lower to observe fluorine but can never be tuned low enough to observe carbon. Broadband probes can be tuned over a very broad frequency range (a factor of ten from lowest to highest) so one broadband probe can be used to observe most NMR active nuclei. Although this sounds very attractive, there is a tradeoff when using such a probe. Broadband probes are inherently less sensitive than selectively tuned probes. The S/N with a broadband probe will be at least a factor of two less than with a selectively tuned probe. For maximum sensitivity, it is best to use selectively tuned probes.

A powerful method for heteronuclear correlation experiments, called *indirect* or *reverse detection* [50,131], considerably increases the sensitivity of heteronuclear correlation experiments. It takes advantage of the sensitivity of proton NMR and the selectivity of heteronuclear NMR. It requires irradiation at the heteroatom frequency and observation at the proton frequency. Probes specifically designed for indirect detection are available in 5-, 3-, and 1.7-mm sizes. Using a 1.7-mm indirect detection probe reduces the sample size required for heteronuclear correlation experiments from ~50 μmol for a 5-mm probe to <0.05 μmol [132].

### Sample Preparation

Interference from the sample matrix and the dilute nature of many biological samples often require a purification or preconcentration step prior to NMR analysis. HPLC, solid- or liquid-phase extraction, and TLC have all been used to purify xenobiotic metabolites. The method of choice will depend on the nature of the matrix, the chemical properties of the metabolites, and the sample size. If organic solvents are used for the purification, it is very important to remove as much of the solvent as possible by evaporation prior to NMR analysis. Residual solvents can add very large signals to proton and carbon spectra. These large signals often obscure significant portions of the spectra and, if large enough, can make detection of small signals difficult or impossible due to the limited dynamic range of most spectrometers.

Liquid-state NMR requires the sample to be dissolved in a suitable solvent. Solvents used for most NMR analyses are deuterated; that is, they contain deuterium ($^2$H) instead of protium ($^1$H). Deuterium is required for the spectrometer frequency lock, which corrects for field drift during spectral acquisition. For proton NMR, deuterated solvents are also desirable as they cut down on the intense solvent resonance that would appear in the presence of protiated solvents. Common NMR solvents, commercially available in >99 atom-% deuterium, are $CDCl_3$, $CD_2Cl_2$, $d_4$-methanol, $d_6$-DMSO, $d_6$-acetone, and $D_2O$. If samples in $H_2O$ are to be analyzed, a small amount of $D_2O$ (10 to 20%) should be added for frequency locking, and a suitable solvent suppression method (discussed later) should be used.

When the sample has dissolved in a suitable solvent, it should be filtered to remove particulate matter. Even small particles can result in a loss of resolution, so filtration is particularly important when closely spaced signals or very small couplings are to be observed.

### Solvent Suppression

Many biological samples are aqueous solutions. Acquisition of $^1$H NMR spectra of such samples is complicated by the fact that the solvent water protons are present at a concentration of 110 *M* and the analytes of interest are often present in submillimolar concentrations. As a result, weak solute signals are often obscured or undetectable in the presence of the very large solvent signal. It is essential to attenuate the water signal to observe weak signals in the $^1$H NMR spectra of biological samples. A variety of so-called solvent suppression methods have been developed to reduce or eliminate the water peak in $^1$H NMR spectra [91].

The simplest solvent suppression method is the removal of $H_2O$ by lyophilization. The sample is lyophilized to dryness and redissolved in $D_2O$. Typically this is repeated several times to remove the last traces of $H_2O$

prior to spectral acquisition. In practice, it is extremely difficult to eliminate the water peak completely because lyophilized biological samples such as urine are often quite hygroscopic and the sample invariably picks up moisture from the air. This method does, however, provide adequate reduction in the size of the water peak for most purposes. An additional advantage to this approach is that dilute samples can be concentrated to increase overall sensitivity.

Several instrumental methods are available for solvent suppression. Perhaps the simplest of these involves irradiation at the solvent resonance frequency to saturate the solvent signal. This irradiation may be continuous or it may be gated off during the pulse and acquisition. Suppression ratios of 1000 are possible for water. Unfortunately, peaks close to it are often attenuated or distorted. To suppress the water peak with less distortion of other resonances, selective relaxation methods based on $T_1$ relaxation differences between solvent and solute, such as the *water elimination Fourier transform* (WEFT) method, have been used. Although adequate water suppression is often achieved with WEFT, if signals from the molecule of interest have slow relaxation times their intensities may be attenuated. If accurate integration is required for signals with a range of relaxation times, other solvent suppression methods should be used.

A powerful method for water suppression takes advantage of rapid proton exchange between water and an added chemical agent such as ammonium chloride. Rapid proton exchange greatly reduces $T_2$ for the water signal, which is suppressed using a special pulse sequence called *spin–echo*. This pulse sequence allows the water signal to relax but still permits detection of the desired solute resonances that have longer $T_2$ relaxation times. This method, *water attenuation by $T_2$ relaxation* (WATR), can reduce the intensity of the water signal by factors of $>10^4$. WATR was employed in the determination of benzene and *N*-nitrosodimethylamine in aqueous solution by 500-MHz proton NMR with limits of detection of 35 and 510 ng/mL, respectively [68].

Recent advances in the theory and technology of NMR have led to the development of a host of pulse sequences designed specifically for solvent suppression. Fortunately, modern NMR computer control systems contain software for all common solvent suppression pulse sequences, and they are now routinely used with a minimum of operator involvement. A very powerful method, based on the two-dimensional experiment NOESY, is called NOESYPRESAT. This method results in attenuation of the water signal by a factor of $10^5$ or more. Using NOESYPRESAT (for one-dimensional spectra), a series of two-dimensional experiments, and a 750-MHz spectrometer, Nicholson et al. [111] were able to assign over 150 resonances in the $^1H$ spectrum of human blood plasma diluted with 10% $D_2O$.

## Applications

### Identification of Metabolites

Nuclear magnetic resonance is often critical for elucidating the structure of unknown organic compounds. Caldwell et al. [30] considered both one-dimensional and two-dimensional NMR spectra to fully characterize (S)-(–)-cotinine *N*-glucuronide, a previously unidentified metabolite of nicotine. The structure of this glucuronide was determined using one-dimensional proton and carbon and two-dimensional COSY and HETCOR spectra. The one-dimensional spectra (Figure 41.12) were consistent with the proposed structure. The COSY spectrum shown in Figure 41.13A was used to establish connectivities of the coupled protons enabling the complete assignment of the proton resonances in the one-dimensional spectrum. In COSY spectra, the contours that fall on the diagonal correspond to the one-dimensional spectrum except they are seen as if the observer were looking down from above the spectrum. Contours that fall off the diagonal result from spin–spin coupling, so they reveal correlations between coupled protons. COSY spectra are usually presented with the one-dimensional spectrum plotted on the top to facilitate interpretation. The HETCOR spectrum shown in Figure 41.13B was used to establish proton–carbon connectivities and enabled the complete assignment of the carbon resonances in the one-dimensional carbon spectrum. HETCOR spectra may be thought of as a carbon spectrum plotted against a proton spectrum. Contours that appear in the HETCOR result from $^1H$–$^{13}C$ coupling and reveal proton–carbon connectivities—that is, which protons are bound to which carbons. This HETCOR spectrum is presented with the carbon spectrum (projection) plotted on the vertical axis and the proton spectrum plotted on the horizontal axis.

Perhaps the most common applications of NMR in toxicology research involve characterization of xenobiotic metabolites. Usually, NMR spectra of the purified metabolites are used in conjunction with mass spectral characterization and are included as final confirmation of the proposed structure. Such applications of NMR have contributed to the understanding of xenobiotic detoxification mechanisms. NMR analysis of crude or partially purified biological samples, although not frequently used, can be useful for metabolic studies. Typically, one-dimensional proton NMR spectra of crude biological samples are very complex, with many overlapping signals. Nicholson and Wilson [151] have demonstrated the utility of two-dimensional COSY NMR spectroscopy of crude urine for the simplification of such complex spectra. They collected urine from a human volunteer before and 3 hours after ingestion of 1 g paracetamol. The urine was lyophilized, reconstituted in $D_2O$, and examined directly by one-dimensional proton and COSY NMR. The one-dimensional spectrum revealed considerable signal overlap in the aromatic region; however, in the COSY spectrum, well-

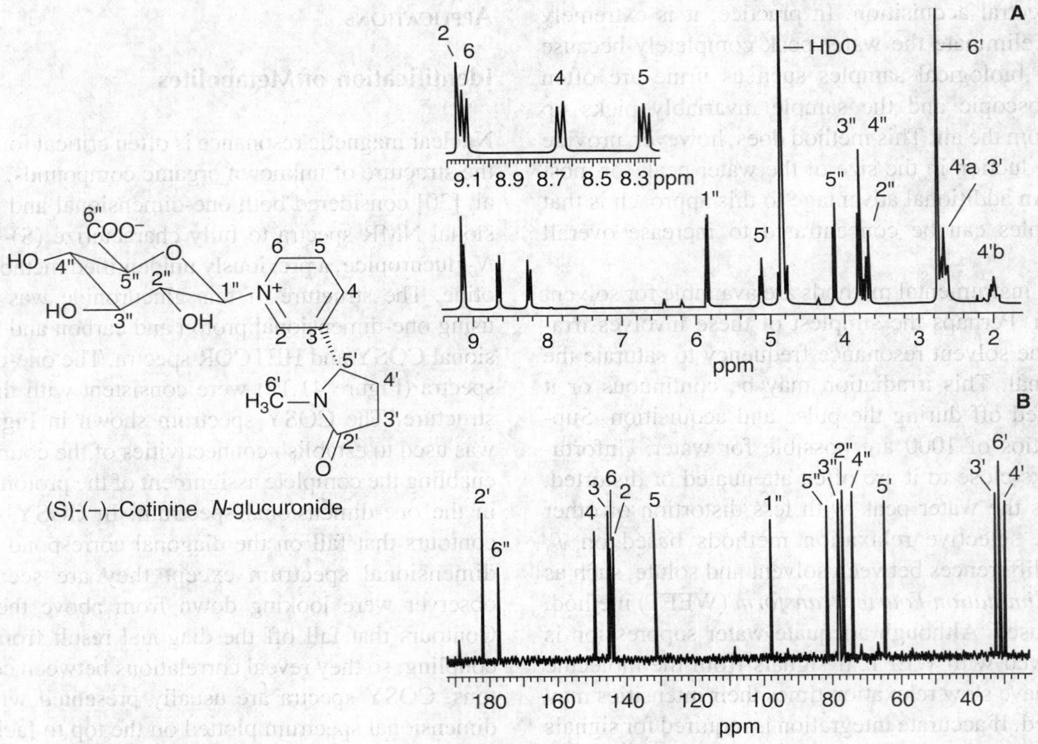


**FIGURE 41.12** (A) One-dimensional proton and (B) carbon nuclear magnetic resonance spectra of (S)-(–)-cotinine *N*-glucuronide.

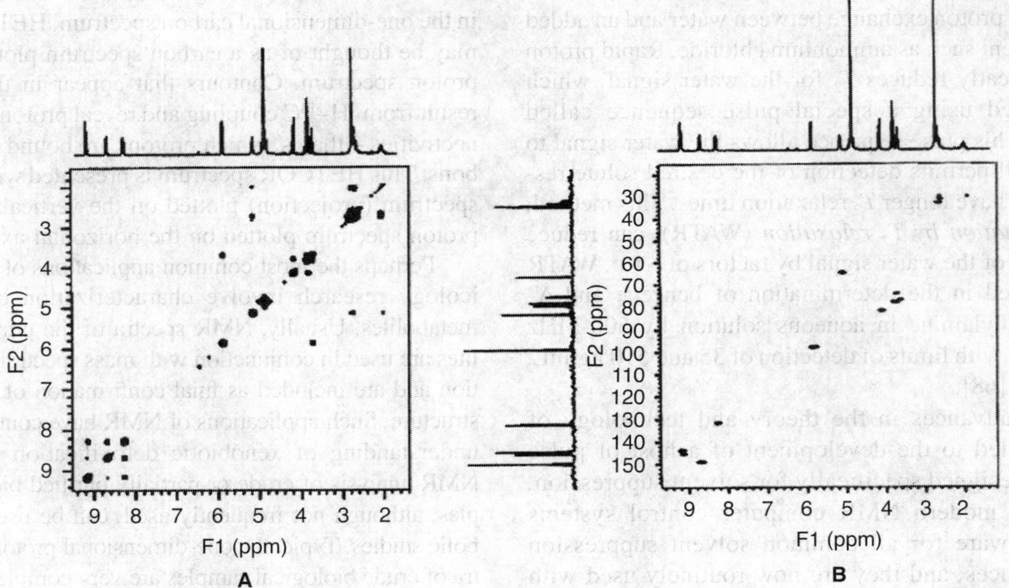


**FIGURE 41.13** (A) COSY spectrum and (B) HETCOR spectrum of (S)-(–)-cotinine *N*-glucuronide.

resolved cross peaks for five paracetamol metabolites were immediately evident. They pointed out that two-dimensional techniques are costly in terms of instrument time but can be used quite successfully when a great deal of metabolic information is required from a few samples or spectral assignments are difficult from the one-dimensional spectra alone.

When NMR spectroscopy is coupled with partial purification, structural characterization is often possible from one-dimensional spectra alone. Solid-phase extraction is a good method for the partial purification of biological samples on a moderately large scale and has been applied successfully to the characterization of ibuprofen metabo-

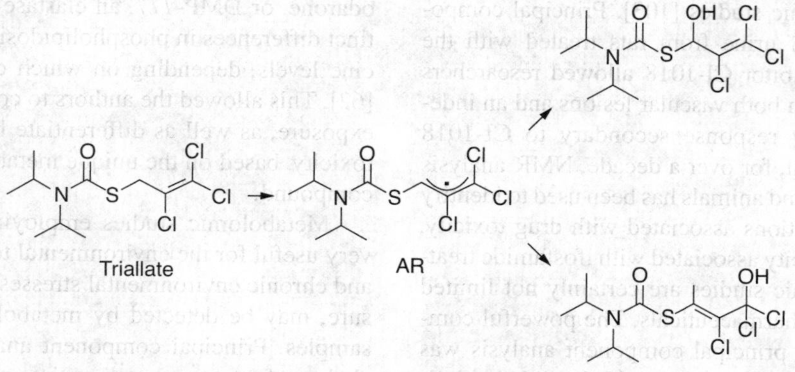

**FIGURE 41.14** Cytochrome P450 catalyzed oxidation of triallate showing the allylic radical (AR) intermediate.

lites by one-dimensional proton and carbon NMR [151]. For this study, a normal, healthy male volunteer was administered 400 mg ibuprofen and urine was collected predose, at 0 to 2 hours, and at 2 to 4 hours. The urine was lyophilized and reconstituted in $D_2O$, and one-dimensional proton spectra were acquired. The samples were next applied to a C-18 solid-phase extraction column and eluted with a stepwise gradient of increasing methanol concentration. Fractions were collected, the solvent was evaporated, and the residue was redissolved in $D_2O$. Spectra of each fraction were acquired to follow the progress of metabolite elution. Using this method, three ibuprofen metabolites, including the glucuronide, were recovered essentially free of impurities.

## Mechanistic Studies of Xenobiotic Metabolism

To elucidate the mechanism of xenobiotic metabolism, detailed structural characterization of metabolites is required. Often, such detail is only available from NMR spectroscopic analysis. Because many metabolites are present in very small quantities, extremely sensitive NMR methods are required. A type of indirect detection called *proton-detected heteronuclear multiple quantum coherence* (HMQC) NMR provides detailed structural information (proton–carbon connectivities) with very good sensitivity. Using a novel one-dimensional application of HMQC, Hackett et al. [79,80] have studied the microsomal hydroxylation of the herbicide triallate. They determined that cytochrome-P450-catalyzed oxidation of triallate leads to the formation of an intermediate *allylic radical* (AR) that undergoes rearrangement leading to hydroxylation at two different positions (Figure 41.14). The structural characterization that led to such detailed mechanistic understanding was carried out on 20 to 45 μg of material isolated and purified by HPLC after microsomal incubation of $^{13}C$-labeled triallate. Hackett et al. [79,80] used a 500-MHz NMR equipped with a 5-mm indirect detection probe. Even greater sensitivity could be realized with a higher field NMR or a 1.7-mm probe. Their work has demonstrated the utility of one-dimensional HMQC for mechanistic studies when sample size is quite limited and no other suitable analytical method is available.

## Biochemical Changes Associated with Xenobiotic Toxicity

Exposure to toxins invariably leads to certain biochemical changes. These include changes in the urinary excretion of endogenous compounds such as carbohydrates, amino acids, and carboxylic acids. Mercury is known to accumulate in the kidneys of experimental animals after injection, causing damage to the proximal tubular epithelium and severe kidney failure. Mercury-induced nephrotoxicity is characterized by increased urinary excretion of amino acids, glucose, calcium, phosphate, bicarbonate, and low-molecular-weight proteins. Nicholson et al. [150] have studied changes in urinary and plasma levels of a large number of low-molecular-weight compounds in rats exposed to mercuric chloride by proton NMR and correlated these changes with histopathology and enzyme excretion. They quantified metabolites in untreated urine and plasma and observed dose-dependent decreases in urinary excretion of creatinine and citrate and increases in glucose, glycine, alanine, α-ketoglutarate, succinate, and acetate. They observed increases in plasma levels of lactate and creatinine. The observed changes were consistent with $Hg^{2+}$ inhibition of certain citric acid cycle enzymes and intracellular, tubular acidosis. Their NMR data provided not only a sensitive measure of $Hg^{2+}$-induced nephrotoxicity but mechanistic insight as well. Similar quantitative NMR techniques may also prove useful for studying the mechanism of action of other toxins.

## Metabolomics

The systematic evaluation of biochemical responses to intoxication is the foundation of metabolomics. NMR spectroscopy of biological fluids and tissues, coupled with pattern recognition and chemometric analysis, is often essential to metabolomic studies [103]. Principal component NMR analysis of urine from rats treated with the phosphodiesterase inhibitor CI-1018 allowed researchers to differentiate between both vascular lesions and an independent inflammatory response secondary to CI-1018 treatment [189]. Indeed, for over a decade, NMR analysis of urine from humans and animals has been used to identify the metabolic perturbations associated with drug toxicity, such as the nephrotoxicity associated with ifosfamide treatment [66]. Metabolomic studies are certainly not limited to the study of toxic pharmaceuticals. The powerful combination of NMR and principal component analysis was employed in the development of a rapid throughput analysis of urine from rats treated with common hepato- and nephrotoxins [173]. Metabolomic analysis of urine components readily and reliably detected both the onset and reversal of toxicity associated with carbon tetrachloride, 4-aminophenol, and other compounds.

Hydrazine, a model hepatotoxin, is an excellent example of a compound that may be systematically studied using NMR-based metabolomics. The onset and recovery from hydrazine-induced liver toxicity in rats was easily monitored in a time- and dose-dependent manner using NMR analysis of urine, coupled with pattern recognition and pattern classification algorithms [195]. Similar analysis of urinary biomarkers, including lactate, acetate, taurine, and β-alanine, allowed researchers to distinguish between different rat strains and their biochemical response to treatment with hydrazine [89]. Predictive models for classifying the type of intoxication and the strain of intoxicated rat were highly accurate and reproducible. Interspecies variations in the biochemical responses to hydrazine intoxication are also readily discernible using NMR-based metabolomics. Urinary metabolomic analysis of hydrazine metabolites and biochemical markers, including citrulline and trimethylamine-*N*-oxide, was very effective for detecting and differentiating between hydrazine toxicity in rats and mice [20].

Lipids are a particularly useful group of metabolomic markers for toxicological studies. Often, changes in lipid profiles in biological matrices are both pronounced and readily detectable by NMR after intoxication. Metabolomic studies of urinary dicarboxylic aciduria, which combined NMR with multivariate statistical analysis, were very useful for detecting and characterizing the mechanism of liver toxicity attributed to MrkA, an experimental therapeutic [143]. The effect of rosiglitazone, an antidiabetic compound, on several metabolic pathways was characterized by NMR metabolomics. Changes in liver–plasma lipid exchange, *de novo* fatty acid synthesis, lipid levels in the peroxisome, cardiac lipid metabolism and lipid metabolism in adipose tissue were all detected using this method, which served as a phenotypic model for assessing clinical response to the drug [216]. Metabolomic analysis of urine from rats treated with chloroquine, amiodarone, or DMP-777, an elastase inhibitor, showed distinct differences in phospholipidosis and phenylacetylglycine levels, depending on which drug was administered [62]. This allowed the authors to correctly determine drug exposure, as well as differentiate between tissue-specific toxicity, based on the unique metabolomic profile of each compound.

Metabolomic studies employing NMR may also be very useful for the environmental toxicologist. Both acute and chronic environmental stresses, including toxin exposure, may be detected by metabolomic analysis of field samples. Principal component analysis of extracts from abalone foot muscle, digestive gland, and hemolymph allowed toxicologists to distinguish between the biochemical profiles of healthy, stunted, and diseased abalone, an important commercial shellfish [209]. In addition to providing a means of assessing the presence of toxins and other environmental stressors, this metabolomic approach also allowed the researchers to diagnose withering syndrome, a common disease in abalone. High-resolution NMR was employed in a metabolomic study of earthworms treated with 3-fluoro-4-nitro phenol, a model toxin. Changes in the relative concentrations of organic acids (in particular, acetate and malonate) served as a reliable marker of toxin exposure [22].

As discussed, the intensity and resolution of NMR spectra may be compromised by a number of factors, including pH, solvent effects, overlapping peaks, and the like. These may limit the reliability and robustness of NMR-based metabolomics. Accordingly, several specialized systems and mathematical approaches have been developed to overcome these limitations. Variations in peak position might be overcome by employing orthogonal projection, back-scale plots, and weighted variable peak position data in NMR plot analysis [43]. Two-dimensional J-resolved spectra, combined with precision spectral preprocessing and logarithmic transformation, afford proton-decoupled projected one-dimensional spectra with reduced peak congestion and more accurate integration of metabolomic peaks [208]. Magic-angle spinning NMR is a more sophisticated technique that has been successfully employed for identifying and correlating metabolomic analytes with histopathological changes in thioacetamide toxicity [215]. Advances in NMR technology for metabolomic applications have been reviewed by Griffin [76]. Advances in the mathematical approaches used to interpret metabolomic NMR spectra include advanced pattern recognition algorithms, hierarchical cluster analysis, and nearest neighbor classification systems [11,75].

## *In Vivo* Spectroscopy and Imaging

Advances in NMR theory and hardware have enabled the study of morphological and metabolic changes in isolated organs and whole animals and humans based on NMR principles. Magnetic resonance imaging (MRI), widely used in clinical medicine, relies mainly on the detection of hydrogen nuclei in water and fat to construct high-resolution images. The contrast in these images results from different $T_1$ and $T_2$ relaxation times for hydrogen nuclei in different tissue environments. *In vivo* magnetic resonance spectroscopy (MRS) is used to obtain spectral information (such as chemical shift and intensity) on chemical compounds within living tissues and can be used to monitor metabolic changes resulting from disease or xenobiotic exposure. MRI and MRS are based on the same principles as liquid-state NMR—that is, the behavior of nuclei in a magnetic field under the influence of RF pulses—but the hardware, pulse sequences, and data processing are somewhat different. Several good reviews describe these differences in detail [8,142,200]

*In vivo* $^{31}P$ MRS has been used to monitor energy metabolism in tumors in laboratory animals during growth [197] and following hyperthermia [207], treatment with interleukin-1$\alpha$ [46], and endocrine therapy [196]. In many cases, the changes seen in phosphorous metabolites (ATP, ADP, PCr, $P_i$, $\beta$-nucleoside triphosphate, phospholipids) correlate well with tumor growth. These results demonstrate the potential for noninvasive monitoring of tumor development by *in vivo* NMR spectroscopy.

Both MRI and MRS have been used to study liver damage induced by hepatotoxic halocarbons. Locke and Brauer [122] monitored the response of rat liver *in situ* to bromobenzene by proton MRI and $^{31}P$ MRS. They found that a sublethal dose of bromobenzene induced acute hepatic edema and decreased energy metabolism. Both effects had an onset of 15 to 20 hours and were maximal at 25 to 60 hours. These effects were blocked by Trolox C, a potent inhibitor of lipid peroxidation.

*In vivo* NMR spectroscopy has tremendous potential as a tool for studying mechanisms of toxicology. MRI and MRS techniques provide detailed information on the response of specific organs to toxicants and can also be used to monitor xenobiotic metabolism *in vivo*. In addition, they could greatly reduce the number of animals required for toxicology studies, as a single animal could be followed over an extended period of time to monitor internal changes. As spectrometers for *in vivo* NMR become more readily available to the practicing toxicologist, *in vivo* NMR spectroscopy will almost certainly become the method of choice for many toxicology studies.

## LC–NMR

Hyphenated techniques coupling mass spectrometry and various separation methods such as GC, LC, and CE are standard analytical tools widely used in toxicology research. Advances in the design of NMR probes and in methods for adequate solvent suppression together with the availability of high-field NMR spectrometers have enabled the coupling of NMR with LC. This hyphenated method takes advantage of high-performance liquid chromatographic separations and the wealth of structural information provided by one-dimensional and two-dimensional NMR spectra. This section provides a brief overview of LC–NMR instrumentation, methods, and applications in toxicology research. For more detailed information, the interested reader is referred to several very good reviews of the topic [119–121,149].

The complex nature of most biological samples makes direct analysis by NMR spectroscopy difficult. Spectra of crude biological samples typically contain many overlapping resonances, which greatly complicates their interpretation. One approach to the simplification of such samples is the removal of endogenous components and the separation of the compounds of interest. The use of solid-phase extraction followed by one-dimensional NMR analysis (SPE–NMR) for the isolation and characterization of ibuprofen metabolites from human urine has already been mentioned [151]. Although SPE–NMR can be quite effective, it suffers from several limitations. Solid-phase extraction is a relatively low-resolution separation method. It is inadequate for very complex mixtures or mixtures of very similar compounds. SPE–NMR is a tedious and time-consuming technique involving the collection of samples, solvent removal, and reconstitution of samples in an appropriate NMR solvent. It also requires a relatively large amount of sample. Although this is usually not a problem for the analysis of human urine, it can be limiting in some circumstances. These limitations are largely removed by hyphenation of NMR with HPLC.

Although reports of successful LC–NMR experiments date back to the late 1970s, the limited sensitivity of NMR and the technical hurdles associated with suppression of signals from the protonated solvents commonly used for LC greatly limited the utility of LC–NMR. With the advent of micro NMR probes, the greater availability of high-field NMR spectrometers (>300 MHz) with increased dynamic range, and the development of truly effective solvent suppression methods, LC–NMR has come into its own as a widely used analytical technique. Hardware and software making LC–NMR a relatively routine method are now commercially available from major vendors.

Although the configurations of LC–NMR systems vary from vendor to vendor and laboratory to laboratory, all systems share the same basic components. The design of a typical LC–NMR system is shown schematically in Figure 41.15. The LC pumps, injector, column, and detector are all standard equipment. From the detector the flow enters the LC–NMR interface, which contains components for flow control and peak sampling. The LC–NMR interface can send the flow directly to the NMR probe, divert the flow to waste, or store peaks detected by the inline detector for subsequent NMR analysis. The NMR

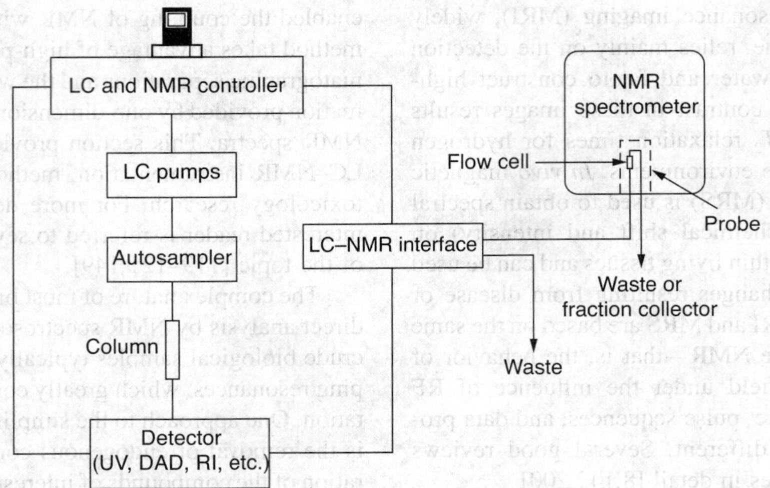


**FIGURE 41.15** Schematic representation of an LC–NMR system.

probe contains a flow cell that replaces the standard glass NMR tube and typically has a volume between about 50 and 250 µL. The probe can be of the selectively tuned type for direct detection of $^1H$, $^{19}F$, $^{31}P$, etc., or it can be of the indirect detection type. Broadband probes are inherently less sensitive than selectively tuned or indirect probes and therefore are rarely, if ever, used for LC–NMR. The most commonly used LC–NMR probe is the dual $^1H/^{13}C$ indirect probe. Instruments configured as shown in Figure 41.15 can be used in the continuous-flow mode if peaks eluted from the LC column pass through the LC–NMR interface to the flow cell for detection in real time. This so-called online LC–NMR works well for fairly concentrated samples (mass detection limits >10 µg in the flow cell), but because the residence time in the flow cell is limited, it is difficult to obtain more than 32 to 64 transients, and two-dimensional experiments are not possible. When higher resolution or sensitivity is required, the LC–NMR interface can be programmed to stop the flow once a peak of interest enters the probe. Such stop-flow or static LC–NMR offers the opportunity to acquire as many transients as necessary for adequate S/N in one-dimensional experiments and is compatible with two-dimensional experiments as well. Using stop-flow, samples of ~1 µg or less can be analyzed. A variation of the stop-flow technique involves programming the LC–NMR interface to store peaks of interest in sample loops and then pass them one at a time to the probe. Once a peak enters the probe, the flow is stopped and the peak is scanned.

A major development in the evolution of routine LC–NMR was in the area of solvent suppression. Because most LC separations of biological samples depend on reversed-phase columns, they utilize solvent mixtures containing water and protonated organic solvents such as methanol or acetonitrile. The most common LC–NMR solvent mixtures are $D_2O$/acetonitrile or $D_2O$/methanol. The use of $D_2O$ reduces the effective concentration of solvent protons and provides a source of deuterium for the instrument lock. Because $D_2O$ is relatively inexpensive, it is a very good alternative to $H_2O$. As discussed previously, the use of protonated solvents (such as acetonitrile) requires attenuation of the solvent signals to observe the weak signals from solute molecules. Fortunately, LC–NMR control systems contain computer software for solvent suppression pulse sequences such as NOESYPRESAT. The suppression of signals from mixed solvents is now a routine operation requiring minimal operator involvement.

Major NMR vendors offer systems for fully automated LC–NMR. Naturally, such systems are capable of automated online LC–NMR, but they are also capable of automated stop-flow LC–NMR. A typical system can:

- Auto-detect a LC peak.
- Transfer the peak to the flow cell.
- Stop the flow.
- Shim the NMR magnet.
- Optimize solvent suppression.
- Acquire the spectrum.
- Restart the LC pump.
- Send the peak to waste or a fraction collector.
- Repeat the procedure for any desired peaks in the chromatogram.

Such a system can also automatically store peaks of interest in the LC–NMR interface for subsequent analysis.

The most common applications of LC–NMR involve the characterization of xenobiotic metabolites. An early example, reported by Spraul et al. [194], was the characterization of ibuprofen metabolites in human urine. For this study, a normal, healthy male volunteer was administered 400 mg ibuprofen and urine was collected for the period from 0 to 4 hours after dosing. The urine was lyophilized and reconstituted in a $D_2O$ $d_3$-acetonitrile mixture prior to LC–NMR analysis. Gradient-elution LC was accomplished with (1) potassium dihydrogen phosphate in $D_2O$, and (2) acetonitrile. A linear gradient from 2 to 45% acetonitrile at a flow rate of 1 mL/min was used for stop-flow analysis. For online analysis, a flow rate of 0.5 mL/min was used. The 500-MHz NMR was equipped with a commercial selectively tuned $^1H$ flow probe with a 60-μL flow cell, and solvent suppression was accomplished using the NOESYPRESAT pulse sequence. For online analysis, 16 transients were collected for each peak, giving a time resolution of 12 seconds. For stop-flow analysis, both one-dimensional (256 or 512 transients) and two-dimensional TOCSY experiments were performed. Using online analysis, Spraul et al. [194] were able to identify five metabolites, including three glucuronide conjugates. The stop-flow analyses provided high-resolution one-dimensional $^1H$ spectra and two-dimensional TOCSY spectra that permitted unambiguous characterization of the five metabolites. On-column detection limits were 10 μg for the online analyses and 1 μg for the stop-flow. Using LC–NMR, the authors were able to identify one metabolite, the dicarboxylic acid metabolite of ibuprofen, which was not observed using SPE–NMR.

Glucuronic acid conjugates of carboxylic acids can undergo an isomerization reaction known as *acyl migration*, where the carboxylic acid moiety migrates from the 1 position of the glucuronic acid to the 2, 3, or 4 positions. Such isomerizations can occur *in vivo* and *in vitro* at physiological pH and under mildly alkaline conditions. The resulting acyl migrated positional isomers can exist as either β or α anomers. Lenz et al. [118] used LC–NMR to characterize the glucuronic acid conjugates of the nonsteroidal antiinflammatory drug 6,11-dihydro-11-oxodibenz(*b*,*e*)oxepin-2-acetic acid in human urine and study the pH dependence of the acyl migration reaction in that matrix. They used a 400-MHz NMR with a 120-μL flow cell in the stop-flow mode to obtain $^1H$ spectra of the α and β anomers of all possible acyl migration products. For one metabolite, they used a 600-MHz NMR to obtain higher resolution spectra that aided in full characterization. This work is noteworthy in that the α and β anomers were not separated by the LC and yet using NMR as a detector allowed the unambiguous characterization of the metabolite anomers. The high-resolution and rich information content of NMR spectra can often be used to characterize closely related metabolites that coelute in the liquid chromatogram.

The use of very-high-field NMR spectrometers (800 MHz) greatly increases spectral resolution and can aid tremendously in metabolite characterization by LC–NMR. Sidelmann et al. [186] used SPE–NMR at 400 MHz and stop-flow LC–NMR at 800 MHz to identify the major phase II metabolites of tolfenamic acid in human urine. They identified glucuronide conjugates of the parent compound and five metabolites, the first report of the direct identification of these phase II metabolites in biofluids. The 800-MHz NMR was particularly useful in determining the exact position of hydroxylation of the aromatic ring of tolfenamic acid. The increase in spectral dispersion obtained at ultra-high field provided very high-resolution one-dimensional $^1H$ spectra, thus permitting unambiguous assignments.

Liquid chromatography NMR has also been applied to studies of *in vitro* microsomal metabolism. Corcoran et al. [49] identified the metabolites of two phenoxypyridines obtained from incubation with rat microsomes using stop-flow LC–NMR at 750 MHz. They were able to characterize one metabolite that was 6% of the total and another that was only 0.6%. The unequivocal identification of these metabolites without the use of radiolabeled substrates or synthetic metabolite standards demonstrated that LC–NMR can be used for metabolite characterization in *in vitro* systems and could be used in a high throughput mode for lead optimization in drug discovery.

In an effort to further reduce the sample size required for hyphenated separation NMR techniques, some laboratories have explored microbore or capillary LC as well as other separation methods such as capillary electrophoresis (CE) and capillary electrochromatography (CEC) coupled to NMR. Wu et al. [229] developed a theoretical model for predicting the signal-to-noise ratio as the NMR flow-cell volume is scaled down. Their model predicted only a twofold reduction in S/N for a 400-fold reduction in flow-cell volume. They constructed a 50-nL flow cell by wrapping narrow-gauge copper wire around a fused silica capillary column. Using this microflow cell and a 300-MHz spectrometer, they realized a mass limit of detection of ~1 μg on-column in an online microbore LC–NMR analysis. The most serious limitation to their approach was the relatively broad line widths in the online NMR spectra. When the analysis was performed in the stop-flow mode, they were able to optimize the NMR shims and other instrument parameters and reduce the line widths to <1 Hz. As a result, two-dimensional experiments such as COSY and NOESY were possible.

Pusecker et al. [168] constructed a 240-nL capillary flow cell coupled to a packed fused silica microbore column. This column could be used for CE, CEC or capillary HPLC separations. Using a 600-MHz NMR they realized a mass limit of detection of ~300 ng in the stop-flow mode. They also reported that in the online mode the limit of detection was adequate for all three micro separation techniques. A major advantage to this system is the ease of changing from one separation method to another. When this micro flow cell was used in an online CE analysis of paracetamol metabolites in an extract of human urine (at 600 MHz) the mass limit of detection was ~10 ng [180].

Online CEC analysis afforded a similar mass limit of detection. Identification of metabolites by NMR could be accomplished with nanoliter sample volumes. Improvements in flow cell and probe design are likely to reduce limits of detection even further, and micro separation methods coupled to online NMR detection will almost certainly become routine tools for metabolite identification. Metabolomic studies are routinely performed using NMR-LC systems; for example, a high-throughput metabolomic profiling system that used a short monolithic column, a rapid gradient, and a high flow rate to analyze large numbers of urine samples was recently reported [159].

### LC–NMR–MS

The hyphenated techniques of LC–MS and LC–NMR are both very powerful tools for characterization of xenobiotic metabolites but they both suffer from unavoidable limitations. Neither technique can, by itself, provide the unequivocal assignment of chemical structure in all cases. Mass spectrometry often cannot distinguish between positional isomers but NMR is very good for this application. Certain functional groups do not contain NMR active nuclei and are invisible in NMR spectra. An example of a NMR-invisible functional group is the sulfate group of sulfate conjugates. In such cases, mass spectrometry can be used to characterize the sample. Because MS and NMR are complementary techniques and both have been successfully hyphenated with LC, it was inevitable that these techniques would be merged to form LC–NMR–MS. The first report of successful LC–NMR–MS appeared in 1995 [166] and was soon followed by reports of its application for characterization and quantification of xenobiotic metabolites [23,53,90,176–178,185,224]. The advantages of LC–NMR–MS are obvious in these reports. By capitalizing on the strengths of NMR and MS in one method, it affords rapid identification of unknown compounds in complex mixtures such as urine in a single chromatographic run. With the increasing availability of LC–NMR instruments that can be coupled to mass spectrometers, more reports of LC–NMR–MS in toxicology research are sure to appear.

### Limitations

Although NMR spectroscopy is a very powerful tool for studying mechanisms of toxicology, it is subject to a number of significant limitations. Perhaps the greatest limitation of NMR is its sensitivity. As discussed above, even the most sensitive spectrometers require nanograms of sample for NMR analysis. Advances in magnet, flow-cell, and probe design have greatly reduced the amount of sample required for NMR, but it would be unrealistic to assume that NMR will be as sensitive an analytical technique as mass spectrometry in the foreseeable future. Another limitation of NMR is cost and, hence, availability. NMR spectrometers are quite expensive, and their price goes up dramatically with field strength. High-field spectrometers are not common laboratory instruments, so instrument time must be purchased from a local (or regional) NMR laboratory or acquired through collaboration. NMR spectroscopy is a specialized discipline requiring a significant amount of expertise. Operation of NMR spectrometers and interpretation of spectra are not straightforward and require a great deal of training and experience. A toxicologist lacking this training and experience would do well to establish an active collaboration with a NMR spectroscopist to carry out sophisticated NMR studies.

## ELECTRON PARAMAGNETIC RESONANCE SPECTROSCOPY

A free radical is a chemical species that contains an unpaired electron. Extremely reactive, free radicals play a role in the mechanisms of tissue injury and toxicity of many chemicals [5,81,144]. Free radicals are often formed during xenobiotic metabolism by enzymes such as the cytochromes P450 and peroxidases. In the Haber–Weiss reaction, superoxide forms the extremely reactive hydroxide radical (HO·) in the presence of ferrous ion. The toxic effects of iron overload have been attributed to the hydroxyl radical. Free radicals are also involved in lipid peroxidation, leading to LDL oxidation and atherosclerosis. Because free radicals are so reactive, they have short lifetimes and are present at very low concentrations in biological systems; consequently, they are usually difficult to detect. Although indirect methods such as product analysis, inhibition by antioxidants, and other radical scavengers and photolysis have been used to detect free radicals in biological systems, these methods provide little information on the nature of the radical. *Electron paramagnetic resonance* (EPR) spectroscopy, also known as *electron spin resonance* (ESR) spectroscopy, is the most versatile and information-rich method for free radical analysis. The following sections provide a concise discussion of the theory of EPR spectroscopy, as well as several examples of its application in toxicology research. The basic phenomenon of magnetic resonance is common to both EPR and NMR; thus, the reader should be familiar with the preceding theoretical discussion of NMR spectroscopy. More detailed descriptions of the theory and instrumentation of EPR are available [3,4,163].

### Basic Theory of EPR Spectroscopy

Like a proton, an electron rotates about its axis and has a magnetic moment ($\mu$). In an applied magnetic field, $\mathbf{B}_o$, this magnetic moment will precess around the axis of $\mathbf{B}_o$ with a characteristic precessional frequency ($\upsilon$). Because the spin quantum number ($S$) of an electron is 1/2, it exists in two energy states in a magnetic field. The energy difference ($E$) between these two spin states is given by:

$$E = h\upsilon = gB_o \frac{eh}{4\pi m_e c} \quad (41.5)$$

where $m_e$ is the electron mass, $e$ is the electronic charge, $c$ is the speed of light, and $h$ is Planck's constant. The proportionality constant ($g$), the spectroscopic splitting factor, is the ratio of the magnetic moment to the angular momentum and has a value of 2.0022319 for an unbound electron.

The exact precessional frequency of an electron in a radical and the position of its resonance signal depend on its chemical environment. In NMR spectroscopy, the position of the resonance signal is denoted by the chemical shift ($\delta$). In EPR, resonance positions are expressed as $g$ values. Tables of the $g$ values of common radicals are available [72]. Because the magnetic moment of an electron is approximately 700 times greater than that of a proton, the energy difference between spin states is correspondingly higher in EPR than in NMR. Accordingly, EPR is more sensitive than NMR, and EPR spectra can be obtained from radicals present in low micromolar concentrations. The larger energy difference between radical spin states also means that EPR requires more energy than NMR spectroscopy. The frequencies used for EPR are in the microwave region of the electromagnetic spectrum. In a typical EPR spectrometer operating at a field strength of 0.34 T, the precessional frequency of an electron is approximately 9.5 GHz.

Electron paramagnetic resonance signals are typically much broader than NMR signals. Assignment of EPR resonance positions ($g$ values) is facilitated by plotting first-derivative traces. Although NMR spectra are plotted as absorption vs. frequency, EPR spectra are plotted as the rate of change of absorption vs. frequency. These two types of spectral traces are shown in Figure 41.16. Because radicals have only one unpaired electron, the EPR spectrum of a single radical species will generate only one signal. If a sample contains more than one radical species, multiple signals will appear. The total area under a peak in an EPR spectrum is proportional to the number of unpaired electrons in a sample associated with that peak. For quantification of unknown radical concentrations in a sample, direct comparison to a radical standard of known concentration is made. Common radical standards include diphenylpicrylhydrazyl (DPPH), which contains 1.53 × $10^{21}$ unpaired electrons per gram.

In a radical, the unpaired electron is not associated with only one atom. The electron spin is distributed over several atoms in the radical and may interact with any spin-active nuclei with which it is associated. The interaction of electron and nuclear spins leads to spin–spin coupling called *hyperfine splitting*, similar to the coupling seen in NMR. An EPR signal will be split into $2nI + 1$ peaks, where $n$ is the number of equivalent nuclei of spin $I$. The hyperfine splitting constant is denoted $a_i$, where $i$ is the atomic symbol of the nucleus to which the electron is coupled. Hyperfine EPR splitting constants are typically measured in gauss (G) or in hertz (Hz). Figure 41.17 shows a computer-simulated spectrum of the methyl radical. The unpaired electron is coupled to three equivalent protons, so the signal appears as a quartet. The magnitude of the hyperfine splitting constant is directly proportional to the electron spin density at the coupled nucleus. It is thus related to the probability of finding the unpaired electron associated with the coupled nucleus. Hyperfine splitting constants are often applied in the interpretation EPR spectra. Tables of $a_i$ values are available for reference [72].

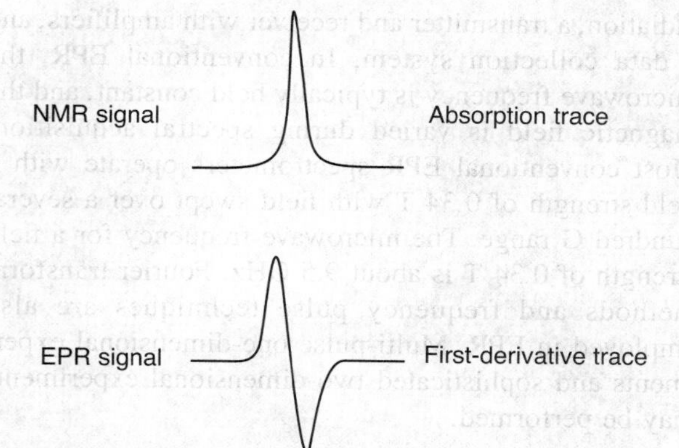


**FIGURE 41.16** Nuclear magnetic resonance spectrum plotted as an absorption trace and an EPR spectrum plotted as a first derivative trace.

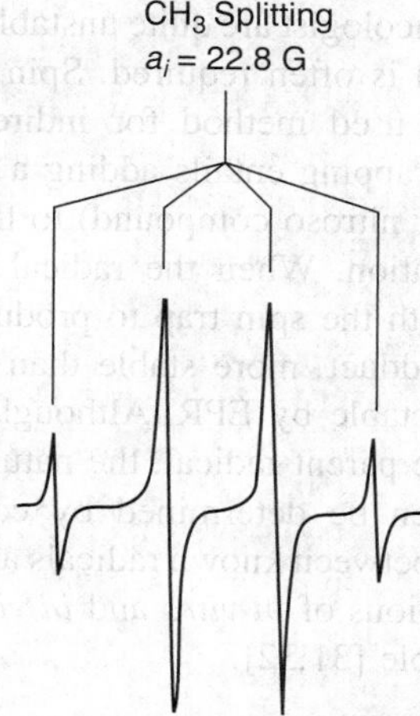


**FIGURE 41.17** Computer-simulated EPR spectrum of the methyl radical showing the 22.8 G hyperfine splitting caused by coupling of the unpaired electron to the three spin 1/2 protons.

## Instrument Design

Electron paramagnetic resonance spectrometers contain the same basic components as NMR spectrometers: a probe, a magnet, a source of electromagnetic microwave radiation, a transmitter and receiver with amplifiers, and a data collection system. In conventional EPR, the microwave frequency is typically held constant, and the magnetic field is varied during spectral acquisition. Most conventional EPR spectrometers operate with a field strength of 0.34 T with field swept over a several hundred G range. The microwave frequency for a field strength of 0.34 T is about 9.5 GHz. Fourier transform methods and frequency pulse techniques are also employed in EPR. Multi-pulse one-dimensional experiments and sophisticated two-dimensional experiments may be performed.

### Spin Trapping

Some free radicals are stable enough in solution to be directly detected by EPR; however, most free radicals of interest to the toxicologist are quite unstable, so an indirect detection method is often required. Spin trapping is the most commonly used method for indirect free radical detection. Spin trapping entails adding a spin trap (typically a nitrone or nitroso compound) to the sample prior to radical generation. When the radical is generated it rapidly reacts with the spin trap to produce a secondary radical or spin adduct, more stable than the parent free radical and detectable by EPR. Although the detectable species is not the parent radical, the nature of the parent radical may often be determined by comparison with model reactions between known radicals and the spin trap. Detailed descriptions of *in vitro* and *in vivo* EPR experiments are available [31,32].

### Applications

#### Model Studies

The EPR study of appropriate model systems in aqueous solutions may yield important insights into related *in vivo* radical reactions and pathways. Such model systems are often used to study free radical processes thought to play a role in radical toxicity. Compounds containing cobalt or other transition metals have been linked to free radical generation by reaction of the metal ions with lipid hydroperoxides. An examination of the *in vitro* reaction of Co(II) with $H_2O_2$ and lipid hydroperoxides in the presence of various biological ligands and the spin trap DMPO revealed that, in the presence of sulfhydryl-containing molecules, Co(II) can react with $H_2O_2$ and lipid hydroperoxides to generate free radicals [183]. This observation may be important for understanding Co(II)-mediated toxicity and carcinogenicity. High-field EPR and site-directed spin labels have been used to study the structure and conformational dynamics of toxic proteins, thereby providing detailed information on transient intermediates and their relationship to biological action [141]. D-Mannitol has neuroprotectant properties and reduces sensory neurological disturbances associated with ciguatera intoxication. *In vitro* EPR analysis of rat neurons exposed to the toxin and several potential neuroprotectants and antioxidants revealed that toxin-induced changes in neuronal excitability could not be attributed to toxin-mediated hydroxyl radical generation alone. The authors suggest that the neuroprotectant effects of D-mannitol are more complex than simple osmotic reduction in neuronal swelling and may be due to reduced toxin association with ion channels [16].

#### *In Vivo* Spin Trapping

Direct evidence for *in vivo* free radical generation in response to xenobiotic exposure may be obtained with EPR spin-trapping techniques. Ozone exposure is believed to cause pulmonary damage as a result of lipid peroxidation in lung tissue. Kennedy and colleagues [102] employed a spin trap with EPR in ozone-exposed rats and detected a spin adduct in lipid extracts of lung tissue obtained from exposed animals. The concentration of the radical spin adduct in lung extracts correlated well with ozone exposure, suggesting both free radical production and toxicity in ozone-exposed lungs.

Nitric oxide is an important endogenous metabolite, vasoactive substance, and neurotransmitter in many organisms. At higher concentrations, nitric oxide may have toxic effects, and its production may be induced by exposure to xenobiotics. Nitric oxide contains an unpaired electron and is a relatively stable free radical. Nitric oxide binds to Fe(II)-containing molecules such as hemoglobin and endogenous iron–sulfur proteins, thereby forming stable complexes readily detectable by EPR. EPR is useful for studying nitric oxide–Fe(II) complexes in iron containing biomolecules generated in response to environmental toxins and disease. The administration of an endogenous spin trap has also been extensively employed for EPR studies. Kubrina and coworkers [104] used an Fe(II)–diethyldithiocarbamate (Fe–DETC) complex, which forms a characteristic radical spectrum with nitric oxide, to clearly demonstrate that nitric oxide originates from the guanidine nitrogens of L-arginine *in vivo*. $^{15}$N-labeled L-arginine (L-guanidineimino-$^{15}$N-arginine) was administered to rats along with the Fe–DETC complex and lipopolysaccharide. EPR analysis of the livers from sacrificed animals revealed a very characteristic doublet hyperfine splitting associated with $^{15}$NO complexed with the iron. This study clearly demonstrated the application of both spin trapping and stable isotope labeling in EPR while providing the first direct evidence that L-arginine is the ultimate precursor of endogenous nitric oxide.

### EPR Imaging

Sensitive and specific, EPR is readily adaptable to *in vitro* and *in vivo* imaging of free radicals in biological samples [193]. Measurements of *in vivo* blood oxygenation in rats [108] and *in vivo* imaging of kidneys [107] and tumor heterogeneity and oxygenation in mice [109] are some examples. A murine model of septic shock has been developed using EPR for *in vivo* monitoring of real-time nitric oxide generation secondary to intoxication with bacterial lipopolysaccharide [92]. The authors employed both surgically implanted *in situ* radical probes and a soluble systemic probe with EPR to monitor and localize both oxygen and nitric oxide. *In vivo* EPR techniques are also extensively employed in the pharmaceutical industry. Pharmacokinetic and pharmacodynamic interactions, as well as the metabolism of nitroxides and metals, have been measured noninvasively *in vivo* with EPR [126]. An *in vivo* evaluation of adriamycin nephropathy in rats [156] employed EPR with a nitroxide radical to monitor the reducing ability of renal tissues in adriamycin-exposed rats. The decay rate of the EPR signal intensity correlated with renal reducing activity and the extent of drug-induced nephropathy.

### Limitations

Electron paramagnetic resonance analysis in toxicology is limited by the very nature of the species under study, the free radical. Unstable free radicals are generally detected indirectly by spin-trapping techniques. The spin trap itself may have toxic effects independent of the toxic mechanisms under study. Because spin adducts are secondary radicals, it is not always possible to determine the structure of the parent radical by EPR analysis of the spin adduct. EPR is nonetheless a very important tool for studying free radical mechanisms of toxicity and may be combined with other analytical techniques, such as stable isotope labeling. EPR remains the method of choice for detecting and monitoring free radicals both *in vitro* and *in vivo*.

## ULTRAVIOLET AND VISIBLE SPECTROPHOTOMETRY

### Principles

Ultraviolet and visible (UV–VIS ) absorption spectrophotometry have been the principal methods of chemical analysis for many years. They involve the measurement of light absorption by substances in the wavelength region from 190 to 380 nm for UV absorption and 380 to 900 nm for visible light absorption [52,61,182,192]. Light absorption in these regions arises from electronic transitions within the molecule. The frequency of absorption depends on the energy difference between the normal or

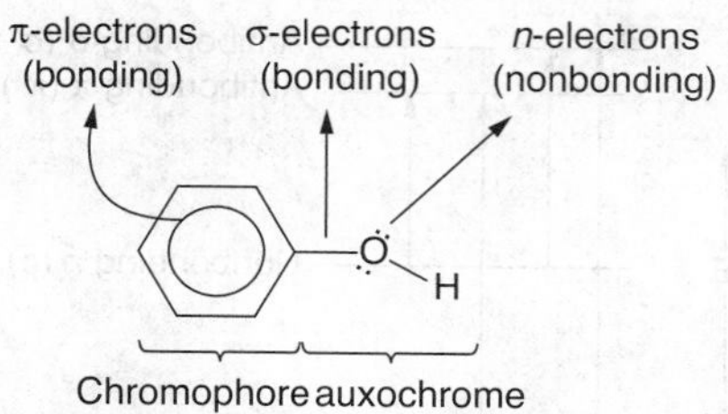


**FIGURE 41.18** Some basic terminologies in ultraviolet spectrophotometry.

ground state of an electron vs. that of the excited state (higher energy level). Absorption of UV or visible light is also accompanied by vibrational and rotational transitions that result in relatively broad bands characteristic of UV–VIS spectra.

A molecule containing electrons in σ, π, and *n*-orbitals (see Figure 41.18) may absorb light energy and be promoted from the ground state to higher energy states. Antibonding orbitals (σ* and π*) exist in the excited state for the bonding electrons, and *n* electrons may be associated in the ground state with heteroatoms that do not participate in bonding yet can absorb energy and be promoted to either σ* or π* orbitals. From a practical consideration, the π → π* and *n* → π* transitions are of most utility because these transitions occur in the useful range (200 to 750 nm) of the UV–VIS spectrum; σ → σ* transitions require more energy and usually occur at wavelengths of less than 200 nm. Compounds that contain nonbonding electrons on oxygen, nitrogen, sulfur, or halogen atoms can undergo *n* → σ* transitions; however, these are of lower energy than σ → σ* transitions. The absorption due to *n* → σ* transitions is of limited utility because it is very weak and in most cases occurs at wavelengths too short to be easily measured on conventional instruments—for example, trimethylamine (λ, 277 nm; ε, 227) and methanol (λ, 183 nm; ε, 150). Molecules that contain oxygen, nitrogen, sulfur or halogen atoms usually show an intense absorption, known as *end absorption*, around 200 nm due to *n* → σ* transitions.

The energy required for the σ → σ* transition is very high; consequently, compounds in which all valence shell electrons are involved in single-bond formation, such as saturated hydrocarbons, do not show absorption in the ordinary ultraviolet region. An exception is cyclopropane, which shows a wavelength of maximum absorption ($\lambda_{max}$) of about 190 nm (propane shows $\lambda_{max}$ about 135 nm). Transitions to antibonding π* orbitals are associated with unsaturated centers in the molecule such as alkenyl, carbonyl, imino, and azo groups. These transitions are of relatively lower energy requirements and occur in the useful part of the UV spectrum—for example, C=O: ~285 nm (low intensity, *n* → π*) and 185 nm (high intensity, π → π*). The π → π* transitions lie between *n* → σ*

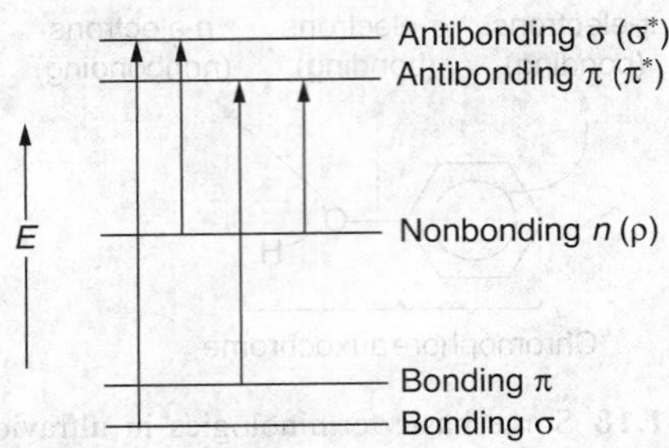


**FIGURE 41.19** Summary of electronic energy transitions.

and $n \rightarrow \pi^*$ transitions in terms of energy content. Figure 41.19 illustrates in a nonempirical manner the relative electronic excitation energies for the above electronic transitions.

## Quantitative Aspects of UV–VIS Spectrophotometry and the Beer–Lambert Law

When UV light traverses a cell containing an absorbing solute dissolved in a suitable solvent, the light intensity is diminished by reflection at the inner and outer surfaces of the cell, by light scattering by any particles in the solution, or by absorption of light by the molecules of the solute. The intensity of the light absorbed can be expressed as:

$$I_{absorbed} = I_O - I_T \tag{41.6}$$

where $I_O$ is the original intensity incident on the cell and $I_T$ is the reduced intensity transmitted from the cell. The transmittance ($T$) is the ratio $I_T/I_O$ and the percent transmittance ($\%T$) is given by:

$$\%T = \frac{100 I_T}{I_O} \tag{41.7}$$

The *absorbance* ($A$) is the common logarithm of the reciprocal of $T$:

$$A = \log \frac{I_O}{I_T} \tag{41.8}$$

It can be shown that the intensity of a beam of parallel monochromatic radiation decreases exponentially as it passes through a medium of homogeneous thickness. Or, alternatively, the absorbance is proportional to the pathlength ($b$) of the solution. This is the basis of Lambert's law.

Beer's law states that the intensity of a beam of parallel monochromatic radiation decreases exponentially with the number of absorbing molecules, or more simply the absorbance is proportional to the concentration ($c$). A combination of the two laws yields the Beer–Lambert law:

$$A = \log \frac{I_O}{I_T} = abc \tag{41.9}$$

The proportionality constant ($a$) is the absorptivity. The name and value of $a$ depend on the units of concentration. When $c$ is in moles per liter, the constant is called *molar absorptivity* or the *molar extinction coefficient* (ε). Thus:

$$A = \varepsilon bc \tag{41.10}$$

The molar absorptivity at a specified wavelength of a compound in solution is the absorbance at that wavelength of a 1 mol per liter solution in a 1-cm cell. The units of ε are therefore liter mol$^{-1}$ cm$^{-1}$. Expressing the absorptivity in terms of a 1-mol-per-liter solution facilitates the comparison of the light-absorbing abilities of compounds with widely differing molecular weights. Substances that have ε values less than 100 are weakly absorbing; those with ε values above 10,000 are intensely absorbing. Many absorbing xenobiotics and drugs have ε values at their wavelengths of maximum absorption of $10^{3.5}$ to $10^{4.5}$.

Another form of the Beer–Lambert proportionality constant is the *specific absorbance*, which is the absorbance of a specified concentration in a cell of specified pathlength. The most common form is the $A$(1%, 1 cm) value, which is the absorbance of a 1 g/100 mL (1% w/v) solution in a 1-cm cell. The Beer–Lambert equation therefore takes the form:

$$A = A_{1\text{cm}}^{1\%} bc \tag{41.11}$$

where $c$ is in g/100 mL and $b$ is in cm.

Whenever an analyte is involved in an equilibrium, such as protonation or deprotonation, tautomerism, dimerization, or complex formation, the material added to the solution will be distributed among several forms and the apparent concentration (amount of material dissolved/volume) will not be proportional to the actual concentration of the parent substance. A deviation for Beer's law will be observed under these circumstances unless the absorptivity is identical for all the species present or the equilibrium is controlled in some manner. If only two species are present and their spectra are not too different, useful measurements following Beer's law can be made by measuring at the isosbestic point rather than at $\lambda_{max}$. The isosbestic point is the wavelength at which the UV spectra of the two species cross when measured at equal molarities or, equivalently, the wavelength at which their molar absorptivities are equal. Deviations arising from acid–base equilibria can be avoided by carefully buffering the solutions because the ratio of protonated to deprotonated analyte will be constant at constant pH. A variety of other equilibria can be controlled in a similar manner.

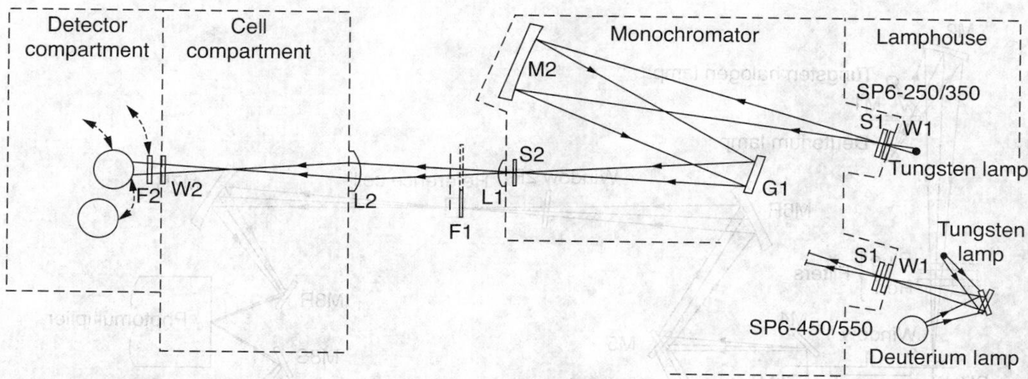


**FIGURE 41.20** Optical diagram of a single-beam ultraviolet spectrophotometer. Abbreviations: F, filter; G, grating; L, lens; M, mirror; S, slit; W, window. (Courtesy of Thermo Fisher Scientific, Inc., Waltham, MA.)

Absorption spectra of compounds with conjugated chromophores or aromatic moieties in their structure show maxima shifts to longer wavelengths (bathochromic shifts) when compared to the wavelength of individual chromophores. This is due to increased stability of the π-electron system which requires less energy for the $\pi \rightarrow \pi^*$ transition. This bathochromic shift is usually accompanied by an increase in intensity of the absorption (a hyperchromic shift).

## Instrument Design

### Single-Beam Spectrophotometers

The arrangement of the components in a commercially available, single-beam, UV–VIS spectrophotometer is shown in Figure 41.20. The essential characteristic is that the light travels in a single continuous optical path between the light source and the detector. Single-beam instruments are relatively inexpensive and are satisfactory when many samples are being assayed by a simple measurement of absorbance at the same wavelength. A major disadvantage is the need to reset the 100% transmission value at each wavelength to compensate for the large variation of intensity of light from the lamp at each wavelength.

### Double-Beam Spectrophotometers

In this type of instrument (Figure 41.21), the monochromatic light is split by a rapidly rotating beam chopper into two beams that are alternately directed in rapid succession through a cell containing the sample and one containing only solvent. If there is greater absorption of light in the sample cell than in the reference cell, the recombination of the beams at the detector produces a pulsating current that is converted into two direct-current voltages proportional to the light intensities $I_0$ and $I_T$ transmitted by the reference solution and the sample solution, respectively. The ratio of voltages is recorded as a transmission. Double-beam optics, therefore, automatically compensate for variation of $I_0$ with wavelength. Recording spectrophotometers are double-beam instruments equipped with a wavelength scanning device which enables rapid automatic scanning of spectra.

## Solvents and Sample Conditions

### Solvents

The solvent of choice in ultraviolet and visible spectrophotometry is determined by several factors, including the solubility of the analyte and the absorption of the solvent at the wavelength employed for analysis. The solvent should be available in a purity-grade suitable for carrying out spectrophotometric work, devoid of contaminants that are either fluorescent or absorb at the analytical wavelength. Moisture-sensitive compounds require nonhygroscopic solvents that are easily dried. Recovery of the analyte after analysis requires the use of more volatile solvents.

It is important to note that the analyte may be sensitive to pH changes; for example, acid–base equilibria, tautomerism, complex formation, and other equilibria are often pH dependent. In such systems, then, a strongly acidic or basic solvent may be indicated to ensure that the analyte is present in solution as a single species. Nonabsorbing buffers may also be used in UV analysis for this purpose.

The exact wavelength of a particular electronic transition depends not only on the chromophore but also on the solvent, on substituents present on the chromophore, and on chromophore geometry. The solvent effect arises because solvation alters the electronic energy levels of a chromophore, and the degree of solvation is frequently different for the ground and excited states. If the ground state is solvated more strongly than the excited state, the energy difference between the levels is increased. The increase in energy difference is reflected in a shift of the absorbance

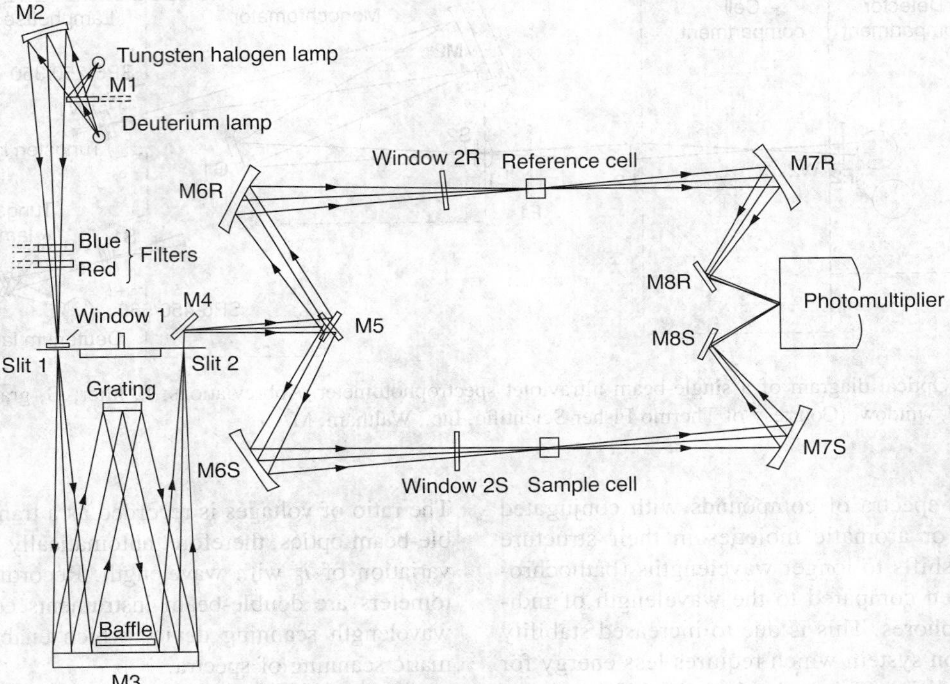


**FIGURE 41.21** Optical diagram of a double-beam UV–visible spectrophotometer. (Courtesy of Thermo Fisher Scientific, Inc., Waltham, MA.)

to shorter wavelengths (hypsochromic or blue shift) than those observed in the gas phase where there is no solvation. If the excited state is solvated more strongly, the energy difference decreases and absorbance is shifted to a longer wavelength (bathochromic or red shift). Absorption due to $n \rightarrow \sigma^*$ and $n \rightarrow \pi^*$ transitions are usually shifted to shorter wavelengths in more polar solvents.

If a group is more polar in the ground state than in the excited state, the nonbonding electrons in the ground state are stabilized (relative to the excited state) by hydrogen bonding and other electrostatic interactions with a polar solvent. The absorption is shifted to shorter wavelengths (higher energy) with increasing solvent polarity. Conversely, if the group is more polar in the excited state, the nonbonding electrons of the excited state are stabilized (relative to the ground state) by interaction with a polar solvent, and the absorption is shifted to longer wavelengths (lower energy) with increasing solvent polarity. Thus, polar solvents generally shift the $n \rightarrow \pi^*$ and $n \rightarrow \sigma^*$ bands to shorter wavelength and the $\pi \rightarrow \pi^*$ band to longer wavelength.

## Cells

Ultraviolet–visible cells (also called *cuvettes*) may be made of glass (for use down to about 360 nm) or silica. Disposable plastic cells are also available. They usually have a transmission cut-off at about 320 nm and may not be suitable for high-precision work. Modern cells are fused and may be square, rectangular, or (rarely) cylindrical in section. Silica is substantially transparent between about 190 nm and 1000 nm, and special grades extend this range downward to below 180 nm and upward to about 2000 nm. For precision work and operation near the wavelength limits, the use of these purer silicas is recommended.

## Path Length and Concentration

Optimum accuracy and precision in UV spectrophotometric analyses are obtained when the absorbance is about 0.9; however, in practice, absorbencies in the range of 0.3 to 1.5 are sufficiently reliable, and the combination of cell pathlength and concentration of analyte should be adjusted to give an absorbance within this range.

## Correlation of Molecular Structure to UV–VIS Absorption

An isolated functional group not in conjugation with any other group is said to be a *chromophore* if it exhibits absorption of a characteristic nature in the ultraviolet or visible region. If a series of compounds has the same functional group and no complicating factors are present, all of the compounds will generally absorb at nearly the same wavelength and will have nearly the same molar extinction coefficient. Thus, the spectrum of a compound, when compared to published spectra for known compounds, can be a valuable aid in determining the functional groups present in the molecule.

Auxochrome*s* are groups that do not in themselves show selective absorption above 200 nm but which, when attached to a given chromophoric system, usually cause a shift in the absorption to longer wavelength and an increased intensity of the absorption peak. Common auxochromic groups are hydroxyl, amino, sulfhydryl (and their derivatives), and some of the halogens. These groups all contain nonbonding electrons; transitions involving these *n* electrons are responsible for these effects; for example, the absorption band at the longest wavelength of *trans-p-*ethoxyazobenzene is shifted 65 nm to longer wavelengths and is about twice as intense as that of the corresponding band of *trans*-azobenzene. Benzene shows $\lambda_{max}$ 255 nm and $\varepsilon$ 230, and aniline shows $\lambda_{max}$ 280 nm and $\varepsilon$ 1430. (Interestingly, the anilinium ion, which has no nonbonding electrons, shows $\lambda_{max}$ 254 nm and $\varepsilon$ 160.) Some functional groups that do not contain nonbonding electrons, such as alkyl groups, can also be considered as auxochromes due to weak inductive or hyperconjugative effects.

If two or more chromophoric groups are present in a molecule and they are separated by two or more single bonds, the effect on the spectrum is usually additive, as there is little electronic interaction between isolated chromophoric groups. However, if two chromophoric groups are separated by only one single bond (a conjugated system), a large effect on the spectrum results because the $\pi$ electron system is spread over at least four atomic centers. When two chromophoric groups are conjugated, the high-intensity ($\pi \rightarrow \pi^*$ transitions) absorption band is generally shifted 15 to 45 nm to a longer wavelength as compared to the unconjugated chromophore.

Many colorimetric analyses developed for UV-absorbing drugs, xenobiotics, and metabolites have been based on the formation of an analyte-specific, multiple conjugated chromophoric system that readily absorbs in the visible range of the spectrum. This helps to avoid interfering UV absorption from impurities in the sample observed when the underivitized analyte is analyzed.

Compounds containing an extensively conjugated chromophore will appear colored to the eye if they absorb above 400 nm. As UV absorption peaks are frequently broad, the absorption of a peak with a $\lambda_{max}$ of approximately 350 nm will generally extend into the visible region. Usually, if a compound appears to be colored, it will contain not less than four and usually five or more conjugated chromophoric and auxochromic groups.

Substitution of aromatic chromophores with auxochromic groups is worthy of mention. When benzene is substituted by halogen or alkyl groups, only a slight shift with a small increase in extinction coefficient is seen; however, substitution by groups carrying nonbonding electrons or $\pi$ electrons (e.g., –OH, $-NH_2$, –CHO) causes a pronounced wavelength shift and a greatly intensified absorption relative to benzene.

In aromatic and conjugated structures where an auxochromic group may function as an acid or base, the effect of pH on the absorption spectrum of the conjugated system can be qualitatively useful; for example, the conversion of phenol (PhOH) to the phenolate ion ($PhO^-$) by addition of base results in an additional electron pair and a formal negative charge being located on the auxochromic group. The interaction of these electrons with the conjugated system results in a bathochromic–hyperchromic shift when compared to the neutral phenol spectrum. Adjusting the pH of the solution reverses this process, such that the phenol is regenerated. The conversion of aniline ($PhNH_2$) to the anilinium ion ($PhNH_3^+$) by lowering the pH of the medium results in a hypsochromic–hypochromic shift that is reversible by increasing the pH of the medium.

In polyaromatic compounds, as the number of fused rings increases, the absorption band is shifted to longer wavelengths. Extensively conjugated polyaromatics, such as naphthacene ($\lambda_{max}$, 480 nm; $\varepsilon$, 11,000–yellow) and pentacene ($\lambda_{max}$, 580 nm; $\varepsilon$, 12,000–blue), absorb in the visible region of the spectrum. The spectra of simple heterocyclic aromatic compounds such as pyridine, pyrrole, indole, furan, thiophene, and their derivatives generally resemble the spectra of analogous benzenoid or naphthalenoid structures.

## UV Spectrophotometry: Direct and Indirect Methods

When carrying out a quantitative spectrophotometric assay, the analyte is dissolved in a solvent that is transparent in the wavelength region to be examined. The wavelength normally selected is that at which the analyte exhibits maximum absorption ($\lambda_{max}$). The usual procedure is to obtain the absorbance value of the solution under non-scanning conditions (i.e., with the monochromator set at the analytical wavelength). Alternatively, if a recording double-beam spectrophotometer is used, the absorbance may be read from a recording of the spectrum. This latter procedure is generally utilized for qualitative purposes and in assays in which absorbances at more than one wavelength are required.

The measurement of absorbance is generally carried out using one of three methods:

1. Comparison with a standard absorptivity value
2. Use of a calibration curve
3. Single- or double-point standardization

Use of a standard absorptivity value is generally restricted to compounds that exhibit broad absorption bands and are not significantly affected by variation in instrumental parameters. An available value obviates the need to prepare a standard solution for absorptivity determination if the reference analyte is difficult to obtain.

The use of a calibration curve is a common procedure for carrying out quantitative spectrophotometric assays. The absorption of four or more standard solutions of the reference compound at concentrations above and below the expected concentration of the analyte is determined. A concentration vs. absorbance graph is then constructed. The concentration of the analyte in the sample solution is then read from the graph as the concentration corresponding to the absorbance of the solution. Calibration data are essential if the absorbance has a nonlinear relationship with concentration, if it is necessary to confirm the proportionality of absorbance as a function of concentration, or if the absorbance or linearity is dependent on the assay conditions. In certain visible spectrophotometric assays of colorless substances, performed by converting the analyte to a colored derivative by heating it with one or more reagents, slight variation of assay conditions, such as pH, temperature, and time of heating, may cause a significant variation in absorbance. Here, experimentally derived calibration data are required for each set of samples.

Single- or double-point standardization is often used in place of a calibration curve. In the single-point procedure, the absorbance of a solution of the sample and that of a standard solution of the reference substance (the concentration should be close to that of the sample solution) are determined. The concentration of the test compound is calculated as follows:

$$C_{test} = \frac{A_{test} \times C_{std}}{A_{std}} \quad (41.12)$$

where $C_{test}$ and $C_{std}$ are concentrations in the sample and standard solutions respectively, and $A_{test}$ and $A_{std}$ are the absorbances of the sample and standard solutions, respectively. This method is best suited for those compounds that obey Beer's law and for which a reference standard of acceptable purity is readily available.

Occasionally, a linear but nonproportional relationship between concentration and absorbance occurs, which is indicated by a significant positive or negative intercept in a Beer's law plot. A two-point bracketing standardization is therefore required to determine the concentration of the sample solutions. The concentration of one of the standard solutions is greater than that of the sample, while the other standard solution has a lower concentration than the sample. The concentration of the substance in the sample solution is given by the equation:

$$C_{test} = \frac{\left(A_{test} - A_{std_1}\right)\left(C_{std_1} - C_{std_2}\right) + C_{std_1}\left(A_{std_1} - A_{std_2}\right)}{A_{std_1} - A_{std_2}} \quad (41.13)$$

where $std_1$ and $std_2$ refer to the more concentrated standard and the less concentrated standard, respectively.

Direct spectrophotometric analysis of a xenobiotic or metabolite may not be possible for several reasons. The natural absorption of the analyte may occur at too low a wavelength to be useful, the molar absorptivity may be too small to give the required sensitivity, or other materials contaminating the analyte may absorb at the same wavelength. These problems can be overcome in many cases by chemical modification of the analyte to change its absorption characteristics. Some examples of useful derivitization procedures for spectrophotometric utility include: (1) diazotization and coupling of primary aromatic amines, (2) condensation reactions (e.g., between amines or hydrazines and carbonyl compounds), (3) reduction of tetrazolium salts in the present of an α-ketol group (–CHOH–C=O), (4) ion pairing of amines with ionized acidic dyes, (5) oxidation of the side chains of weakly absorbing compounds containing an aromatic ring, and (6) metal–ligand complexation. It is also possible to measure a substance by the change in absorbance when a chromophore is destroyed.

## Difference Spectrophotometry

This method of spectrophotometric analysis is useful for obtaining selective and accurate analytical data on solutions of analytes containing absorbing interferants. Basically, the technique measures a difference absorbance ($\Delta A$) between two equimolar solutions of the analyte in different chemical forms that exhibit different spectral characteristics. The method is valid provided that *reproducible* changes are induced in the spectrum of the analyte by the addition of one or more reagents, and that the absorbance of the interfering substance is not altered by the reagents.

The simplest and most commonly employed technique for altering the spectral properties of the analyte is the adjustment of the pH by means of aqueous solutions of acid, alkali, or buffers. The UV–visible absorption spectra of many substances containing ionizable functional groups, including phenols, aromatic carboxylic acids, and amines, are dependent on the state of ionization of the functional groups and consequently on the pH of the solution.

The pHs chosen must quantitatively form single species with at least 99% spectral purity. This can be achieved with monofunctional analytes (e.g., aromatic amines or aromatic carboxylic acids and phenols) by simply working at a pH at least 2 pKa units above the pKa of the analyte. The difference spectrum is obtained by placing one form of the analyte in the sample cell of the spectrophotometer and the other form in the reference cell and plotting the observed absorbance against wavelength. The value of difference spectrophotometry is that it provides a reference solution that contains both analyte and interfering substances in the same concentrations but at a pH different

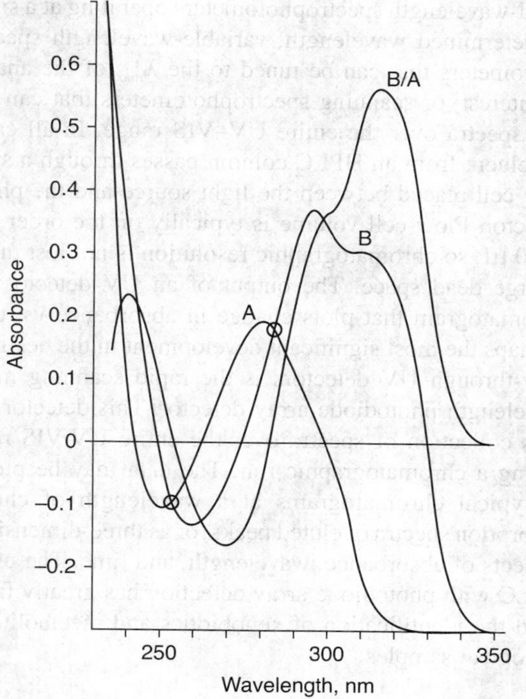


**FIGURE 41.22** Conventional and difference spectra of benzthiazide (15 g/mL) in (A) acidic solution and (B) basic solution, and the difference spectrum of basic vs. acidic solutions (B/A). (From Doyle, T.D. and Fazzari, F.R., *J. Pharm. Sci.*, 63, 1921–1926, 1974. With permission.)

from that of the analyte solution in the sample cell. Interferants present in the sample should not be affected by the pH changes, and their contribution to the total absorbance is therefore canceled.

The absorption spectra of the drug benthiazide is shown in Figure 41.22 in both acidic and basic media [58]. The difference absorption spectrum is plotted as the difference in absorbance between the basic solution and the acidic solution against pH. The spectrum may be generated automatically using a double-beam recording spectrophotometer with the basic solution in the sample cell and the acidic solution in the reference cell. At 255 nm and 287 nm, both solutions have identical absorbance and consequently exhibit zero difference absorbance. Such wavelengths of equal absorptivity are the isosbestic or isoabsorptive points. Above 287 nm, the basic solution absorbs more intensely than the acidic solution and the $\Delta A$ is positive. Between 255 and 287 nm, $\Delta A$ has a negative value. A maximum in the difference spectrum occurs at 313 nm and a minimum at 271 nm. The absorbance difference at these two wavelengths is termed the *amplitude*. A plot of absorbance differences vs. drug concentration will obey Beer's law and can be used for quantitative determinations. The isosbestic points are useful indicators of whether the absorbance of interferants in the sample is

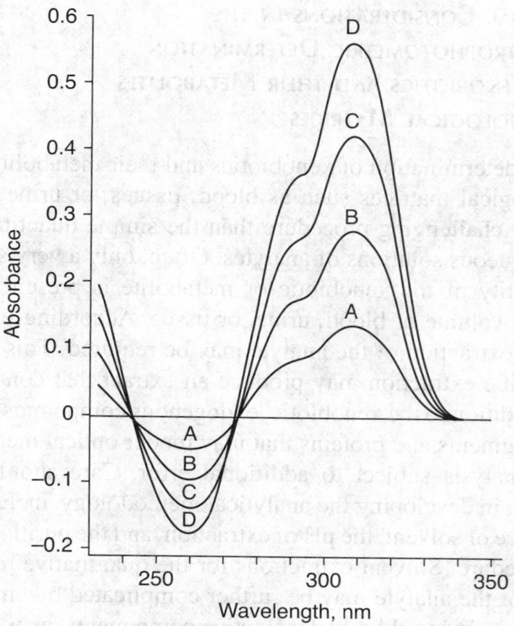


**FIGURE 41.23** Difference spectra used to verify Beer's law for benthiazide. Plots are basic vs. acid difference spectra for concentrations of (A) 5, (B) 10, (C) 15, and (D) 20 g/mL. (From Doyle, T.D. and Fazzari, F.R., *J. Pharm. Sci.*, 63, 1921–1926, 1974. With permission.)

affecting the measurement of the absorbance of the drug. Figure 41.23 shows the isosbestic points for benzthiazide generated from several concentrations of the analyte. Any sample containing the drug or any standard curve concentration should show zero absorbance at the isosbestic wavelengths unless an interfering compound is present. If this is not observed, an alternative assay procedure or removal of the interferant is indicated.

Difference spectrophotometry is useful in the analysis of macromolecules such as proteins, peptides, and nucleic acids, as conformational effects are likely to be accompanied by changes in the environment of aromatic residues which will in turn lead to small wavelength shifts as well as other perturbations. A difference spectrum in a protein may be generated by unfolding or by proteolytic degradation, as well as by more localized changes engendered by conformational adjustments or subunit association or dissociation, or, in favorable cases, the binding of a ligand, as well as a change in bulk solvent. The latter effect can be achieved simply by adding a benign perturbant, such as glycerol, sucrose, or $D_2O$, or by a change of temperature below the region of denaturation. Such a change in solvent character will in general affect only those aromatic residues in contact with solvent and therefore provides a means of determining, at least in a semiquantitative way, what proportion of aromatic residues are so exposed and following any change in this degree of exposure.

## Special Considerations in the Spectrophotometric Determination of Xenobiotics and their Metabolites in Biological Matrices

The determination of xenobiotics and their metabolites in biological matrices such as blood, tissues, or urine, is a more challenging procedure than the simple quantitation of aqueous solutions of analytes. Often, only a very small quantity of the xenobiotic or metabolite is present in a large volume of blood, urine, or tissue. Accordingly, solvent extraction of the analyte may be required. This nonspecific extraction may produce an extract that contains, in addition to the xenobiotic, endogenous compounds such as pigments and proteins that may render optical methods of analysis subject to additional error. Care should be taken in developing the analytical methodology, including choice of solvent, the pH of extraction, and the purification procedure. Solvent extractions for the quantitative recovery of the analyte may be further complicated by emulsification induced by surface-active components, as well as managing large volumes of solvent for evaporation. These problems may be ameliorated by utilizing a large solvent-to-sample ratio and by carrying out control analyses with biological matrices spiked with the analyte. Extraction procedures must account for xenobiotics that may be present both free and as polar, water-soluble conjugates such as glucuronide or sulfate. Another complicating factor for many drug molecules is protein binding. This can lead to poor recoveries of the drug, thereby necessitating the inclusion of a protein denaturation step in the overall extraction procedure.

Direct spectrophotometric procedures, even after purification of the sample by solvent extraction or chromatographic cleanup, often lack the sensitivity and selectivity required for the assay of low concentrations of drugs that are found in body fluids after the administration of therapeutic doses. Modified spectrophotometric techniques, however, such as those involving chemical derivatization or difference spectrophotometry, are sufficiently discriminating and sensitive for the assay of many drugs and other xenobiotics.

## Flow-Through UV Detection

Many of these difficulties associated with the spectrophotometric determination of xenobiotics and metabolites in biological matrices can be overcome by using a tandem separation method with flow-through UV detection. Capillary electrophoresis, affinity columns, size-exclusion chromatography, and HPLC are routinely employed in flow-through UV detection systems. Indeed, because many xenobiotics exhibit characteristic UV spectra, the most common HPLC detector for toxicological applications is a UV spectrophotometer. UV detectors may be fixed-wavelength spectrophotometers operating at a single predetermined wavelength, variable-wavelength spectrophotometers that can be tuned to the $\lambda_{max}$ of the analyte of interest, or scanning spectrophotometers that can collect spectra over the entire UV–VIS range. In all cases, the eluent from an HPLC column passes through a small flow cell placed between the light source and the photodetector. Flow-cell volume is typically on the order of 1 to 10 μL so chromatographic resolution is not lost due to a large dead space. The output of an UV detector is a chromatogram that plots change in absorbance vs. time. Perhaps the most significant development in the design of flow-through UV detectors is the rapid scanning multiwavelength photodiode array detector. This detector permits collection of spectra over the entire UV-VIS range during a chromatographic run. The data may be plotted as typical chromatograms at a wavelength of choice, absorption spectra of eluted peaks, or as three-dimensional projects of absorbance, wavelength, and time. The use of HPLC with photodiode array detection has greatly facilitated the identification of xenobiotics and metabolites in biological samples.

## Applications

Changes in the production of glycolytic metabolites may be useful biomarkers for monitoring pathophysiology, toxic injury, and clinical therapeutics. A metabolomic analysis of glycolysis cycle metabolites in human erythrocytes was developed using a tandem capillary electrophoresis–UV analytical method [129]. The nine components of the glycolytic metabolome of 22 volunteers were readily and reproducibly assayed in micromolar quantities with minimal error. Temporal changes in the microbial metabolome may also be monitored using tandem UV methods. Over 400 fungal metabolites, including important mycotoxins, have been analyzed and identified in culture using chromatographic UV–MS systems [152].

Martin and colleagues [130] developed an *in vivo* visible spectrophotometric assay to simultaneously monitor the nitric oxide, hemoglobin, and deoxyhemoglobin content in discrete areas of the rat brain. The overlapping absorption of spectra of these oxidized hemoglobin chromophores has complicated their analytical resolution in biological matrices. Here, the combination of visible spectroscopy with least-squares mathematical analysis allowed the researchers to study the interaction of the important biomolecule and potential toxin, nitric oxide, with hemoglobin status in brain tissue.

Tandem HPLC–phased diode array analysis has simplified the rapid detection and identification of drugs, toxins, and their metabolites in a variety of systems. This may be particularly useful for the clinical toxicologist [165]. When integrated with a database of known retention parameters and UV spectra, tandem HPLC–phased diode array analysis

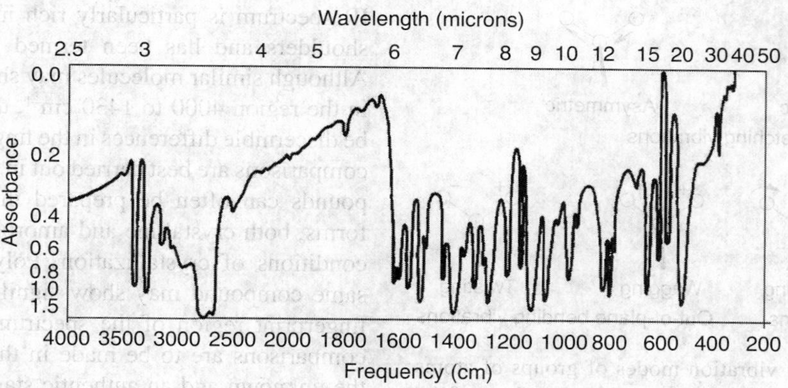


**FIGURE 41.24** Infrared spectrum of sulfacetamide.

may be one of the most effective techniques for the rapid, systematic toxicological analysis of biological samples.

Ultraviolet spectroscopy is often combined with other detection methods for specialized toxicological analyses. Domoic acid is a potent neurotoxin that causes amnesic shellfish poisoning in humans who have ingested contaminated shellfish. The presence of domoic acid in bivalve molluscs was determined for the first time in commercial shellfish harvested in Ireland using a liquid chromatography–photodiode array UV–MS system [94]. Another application of tandem UV analysis in the area of food toxicology was conducted using size-exclusion chromatography with UV and MS detection for the determination of toxic metals in edible mushrooms [230]. The association of silver, arsenic, cadmium, mercury, lead, and tin with the high- and low-molecular-weight fractions of edible mushroom components was strongly influenced by mushroom species and growth medium composition.

## INFRARED SPECTROSCOPY

### Basic Technique

Infrared (IR) spectroscopy [157] deals with the absorption of electromagnetic radiation in the wavelength range 0.8 μm (800 nm) to 1000 μm (1 mm). This range can be subdivided into the near-IR region (0.8 to 2 μm), the middle or fundamental IR region (2 to 15 μm), and the far-IR region (15 to 1000 μm). The fundamental IR region is the one that provides the greatest amount of information for the elucidation of molecular structure. Most IR spectrophotometers are designed to carry out measurements in this wavelength range.

When a molecule absorbs IR radiation, changes in the vibrational energy in the ground state of the molecule occur. All but the simplest of molecules have a large number of accessible vibrational and rotational energy levels with a correspondingly large number of allowed transitions. The energies of these transitions are strongly influenced by molecular structure; thus, IR spectra are highly structured, with unique features that are ideally suited for the identification of xenobiotics, drugs, and other organic substances. A typical IR spectrum of the antibacterial drug sulfacetamide is illustrated in Figure 41.24.

For energy to be transferred from the IR source to the molecule, the frequency of vibration of both must coincide, and energy transfer must be accompanied by a change in the dipole moment of the molecule. Molecules that contain certain symmetry elements will display more simplified spectra; for example, the C=C stretching vibration of ethylene and the symmetrical C–H stretching of the four C–H bonds of methane do not produce an absorption band in the IR region. For nonlinear polyatomic molecules, the number of fundamental modes of vibration is $3n - 6$ ($3n - 5$ for linear molecules), where $n$ is the number of atoms. Certain groups within a molecule, such as –OH, –C=O, $-NH_2$, –CN, and –CC–, have characteristic absorption frequencies known as *group frequencies*. These frequencies are generally independent of the structure of the rest of the molecule and can therefore be used diagnostically to confirm the presence of the functionality in a molecule of unknown structure.

In practice, it is usually not possible to observe the calculated number of peaks in the spectrum of a known compound. This may be due to the superimposition or coalescing of absorptions that are too close to be resolved. A fundamental band may be too weak to be observed. Alternatively, additional (nonfundamental) bands may be observed that are either overtones and harmonics that occur with greatly reduced intensity or combination and difference bands.

Molecules have two types of fundamental vibrations. Stretching occurs when the distance between two atoms increases or decreases but the atoms remain in the same bond axis. Polyatomic molecules may produce in-phase (symmetric) or out-of-phase (asymmetric) stretching vibrations (see Figure 41.25). Bending (or deformation) occurs when the position of the atom changes relative to the original bond axis.

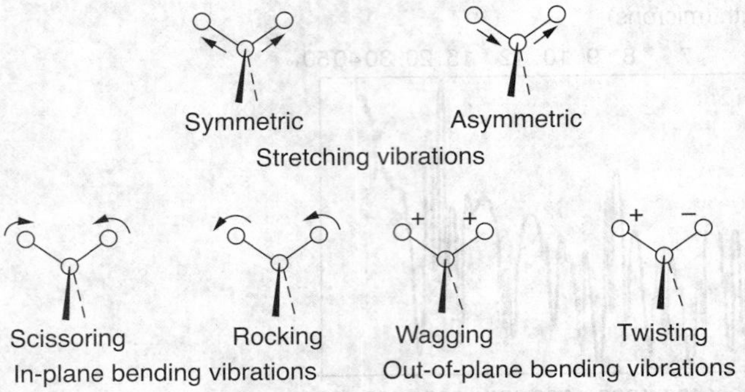

Symmetric Asymmetric
Stretching vibrations

Scissoring Rocking Wagging Twisting
In-plane bending vibrations Out-of-plane bending vibrations

**FIGURE 41.25** Infrared vibration modes of groups of atoms (+ and – refer to vibrations above and below the plane of the paper, respectively).

The stretching frequency (ν) of a bond is related to the masses of the two atoms involved ($M_a$ and $M_b$, in g), the velocity of light ($c$), and the force constant of the bond ($k$, in dynes cm$^{-1}$). An approximate value for the stretching frequency can be calculated from the following equation:

$$\nu \text{ (in cm}^{-1}) = \frac{1}{2\pi c}\sqrt{\frac{k}{M_a M_b/(M_a + M_b)}} \quad (41.14)$$

Note that the force constants for sp$^3$, sp$^2$, and sp bonds have values of 5, 10, and $15 \times 10^5$ dynes cm$^{-1}$, respectively.

Infrared spectra are usually recorded as plots of sample absorbance (or percent transmittance) vs. wavelength or frequency in reciprocal cm (wavenumbers). The relationship between wavelength and wavenumber is:

$$\text{wavenumber (cm}^{-1}) = \frac{1 \times 10^4}{\lambda\ (\mu\text{m})} \quad (41.15)$$

Unlike UV and visible spectra, IR spectra are by convention plotted with zero absorbance (or 100% transmittance) at the top of the spectrum—that is, in an inverted mode relative to UV–visible spectra.

## Qualitative Uses and Interpretation of Spectra

The complexity in the IR spectra of organic molecules is a valuable tool that can be used for the unambiguous identification of unknown compounds if an authentic standard is available. If all the bands in the IR spectrum of the unknown structure are identical in all respects (i.e., in their wavenumber value and their relative intensity) when compared with an IR spectrum of an authentic standard, then the two compounds are identical. The region 1430 to 910 cm$^{-1}$ contains many absorptions caused by bending vibrations as well as absorptions caused by C–C, C–O, and C–N stretching vibrations. As a molecule has many more bending vibrations than stretching vibrations, this region of the IR spectrum is particularly rich in absorption bands and shoulders and has been termed the *fingerprint region*. Although similar molecules may show very similar spectra in the region 4000 to 1430 cm$^{-1}$, there will nearly always be discernible differences in the fingerprint region. Spectral comparisons are best carried out in the solution state. Compounds can often be prepared in different polymorphic forms, both crystalline and amorphous, depending on the conditions of crystallization. Polymorphic forms of the same compound may show significant differences in the fingerprint region of the spectrum; therefore, if spectral comparisons are to be made in the solid state, then both the unknown and an authentic standard should be recrystallized from a specific solvent in the same manner.

A second important use of qualitative IR spectroscopy is that it gives structural information about an unknown molecule. The previously mentioned group frequencies, together with frequencies of other characteristic bands, can be utilized in the form of comprehensive frequency correlation charts which have been compiled for easy reference [72]. This type of compilation is invaluable as a means of confirming the presence or absence of a particular functionality in an unknown structure. In this respect, the region between 4000 and 1500 cm$^{-1}$ is probably easier to interpret than that between 1500 and 650 cm$^{-1}$, as the latter includes many skeletal vibrations, typical of molecules as a whole, which are not diagnostic.

## Instrument Considerations

### Dispersive IR Spectrophotometers

The arrangement of a typical double-beam recording IR spectrophotometer is shown in Figure 41.26. The beam chopper, which is a rotating mirror, permits radiation passing alternately through sample cell and solvent cell to reach the IR detector. The difference in absorption by solute and solvent is measured as an alternating electric current from the thermocouple. The system operates on the optical null principle with the recorder pen linked mechanically to a "comb" (not shown), which is placed across the solvent cell beam and moved by a servomechanism to reduce or increase the solvent cell-beam intensity. The servomechanism is actuated by the amplified thermocouple output to make the solvent beam intensity equal to the solution beam intensity, which reduces the detector output to zero or the null point. The spectrum can be scanned through the various wavelengths by rotation of the prism in synchronization with the motion of the recorder drum or chart. Most modern instruments now utilize a diffraction grating in place of the prism. Sodium chloride prisms can be used for the entire region from 4000 to 650 cm$^{-1}$ but suffer from the disadvantage of low resolution at 4000 to 2500 cm$^{-1}$. The use of a grating monochromator provides a better overall resolution throughout the range 4000 to 625 cm$^{-1}$.

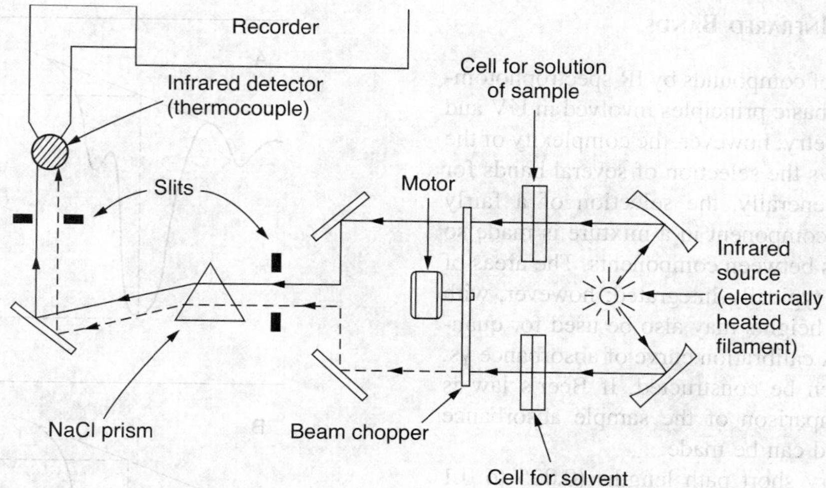


**FIGURE 41.26** Schematic representation of a double-beam recording infrared spectrophotometer.

## Fourier Transform IR Spectrometers

Fourier transform (FT) IR spectrometers are widely used in chemical analysis [2,77]. The FT–IR spectrometer is built around an interferometer rather than a monochromator (see Figure 41.27). The beam from the light source passes through the chopper and is collimated and directed to the beam splitter via mirror C. The beam splitter is a half-silvered mirror that reflects 50% of the incident light onto movable mirror F and allows 50% to pass through fixed mirror E. The beams reflected from mirrors F and E are then combined at the beam splitter so an interference pattern results, and this is focused by mirror G onto detector J. The sample to be analyzed is located in the combined beams between spectral filter H and the detector.

The interferogram of the source is obtained by driving the moveable mirror over a fixed distance and determining the interference pattern as a function of path length difference traveled by the light beams in the two arms of the interferometer. Each frequency of light that passes through the interferometer produces its own interference pattern. Because all of the frequencies generated are observed all at once by the detector, the spectrum is said to be multiplexed, and no grating is required to disperse the radiation. The increased energy reaching the detector allows the relatively rapid accumulation of data, so repetitive scans can be carried out.

Extensive data analysis must be undertaken to extract the frequency information from the interferogram. The principle computation required is Fourier transformation of the interferogram. Because data that have been transformed to the frequency domain are available in digital form, manipulations such as substitution of solvent background and smoothing are simplified.

Fourier transform–IR analysis offers considerable advantages over conventional prism or grating instruments. The S/N under the conditions of the FT experiment is increased from the conventional apparatus by $N^{1/2}$, where $N$ is the number of resolution elements in the spectrum; thus, for a 1000-cm$^{-1}$ scan with 2-cm$^{-1}$ resolution, the S/N advantage is $(1000/2)^{1/2}$, or about 22. These high S/N values permit the accurate subtraction of backgrounds due to liquid $H_2O$. Accordingly, one principle application of FT–IR spectrometry is in the acquisition of protein spectra in aqueous media. In terms of sample preparation and analysis time, FT–IR is considered to be a truly high-throughput analytical technique [59].

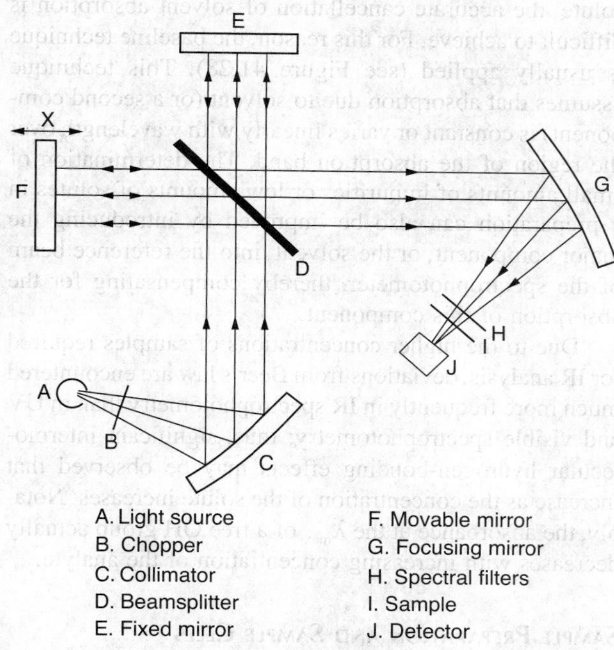


**FIGURE 41.27** Layout of the optics of a Fourier transform infrared spectrometer.

## Quantification of Infrared Bands

Quantitative analysis of compounds by IR spectrophotometry utilizes the same basic principles involved in UV and visible spectrophotometry; however, the complexity of the spectra obtained allows the selection of several bands for quantitative work. Generally, the selection of a fairly strong band for each component in a mixture is made so no interference occurs between components. The areas of absorption bands are typically integrated; however, with sharp IR bands, peak heights may also be used for quantitative calculations. A calibration curve of absorbance vs. concentration can then be constructed. If Beer's law is obeyed, a direct comparison of the sample absorbance with that of a standard can be made.

In IR studies, very short path lengths (0.025 to 0.1 mm) are used, and solution concentrations of analytes in are often in the 10% region. All solvents absorb in some part of the IR spectrum. Given this high concentration of solute, the accurate cancellation of solvent absorption is difficult to achieve. For this reason, the baseline technique is usually applied (see Figure 41.28). This technique assumes that absorption due to solvent (or a second component) is constant or varies linearly with wavelength over the region of the absorption band. The determination of small amounts of impurities or low amounts of solutes in a preparation can also be improved by introducing the major component, or the solvent, into the reference beam of the spectrophotometer, thereby compensating for the absorption of this component.

Due to the higher concentrations of samples required for IR analysis, deviations from Beer's law are encountered much more frequently in IR spectrophotometry than in UV and visible spectrophotometry; thus, significant intermolecular hydrogen-bonding effects may be observed that increase as the concentration of the solute increases. Notably, the absorbance at the $\lambda_{max}$ of a free OH group actually decreases with increasing concentration of the analyte.

## Sample Preparation and Sample Cells

Although IR spectra have been obtained on materials in every physical state from solids to gases, most analyses are carried out on the neat liquid analyte, solutions of the analyte in organic solvents, suspensions, Nujol suspensions, and KBr discs. Solution spectra are preferred for quantitative analysis, as errors arising from nonhomogeneous samples and pathlengths are minimized. KBr discs are often used for qualitative analysis because subtle structural differences such as polymorphism can often be observed in this medium.

Cells for quantitative IR analysis of solutions consist of a pair of sodium chloride plates with a thin metal spacer between them. This unit is housed in a metal frame along with protective washers. The plate spacing is ~$10^{-2}$ cm. Considerable care in handling and storing the IR cells should be exercised, because the windows are easily fogged by traces of moisture and are easily scratched. When not in use, cells should be stored in a desiccator.

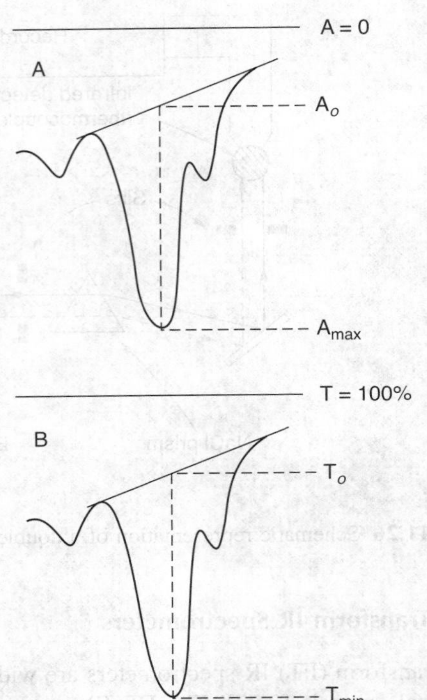


**FIGURE 41.28** Measurement of infrared band intensities using the baseline technique. (A) When the scale is linear in absorbance, peak absorbance = $A_{max} - A_o$. (B) When the scale is linear in transmittance, peak absorbance = $-\log(T_{min}/T_o)$.

Neat liquids and Nujol suspensions or mulls for qualitative analysis are usually placed as a drop on an unmounted circular sodium chloride plate, and a second plate is pressed on top until the liquid is spread into a thin film with a thickness on the order of $1 \times 10^{-4}$ to $50 \times 10^{-4}$ cm. The plates are held in a frame while the spectrum is scanned. Nujol mulls are prepared by grinding the solid analyte in a small mortar with mineral oil until a milky emulsion is obtained. Other suspending agents such a perfluorokerosene, hexachlorobutadiene, or other heavy liquids can also be used. The spectrum obtained with mulls will consist of bands from the analyte superimposed upon bands from the mulling agent. The C–H regions of the spectrum will be obscured when mineral oil mulls are used, and the C–F or C–Cl regions will be obscured when the halogenated mulls are used.

Qualitative analysis of solid analytes is typically performed using KBr discs. Here, the analyte (0.3 to 9 mg) is ground with 300 mg of spectral-quality KBr (~400 mesh). Between 100 and 300 mg of the grind is then pressed into pellets or discs at between 20,000 and 100,000 psig using a stainless steel die and a vacuum pump. The finished pellet is transparent and produces

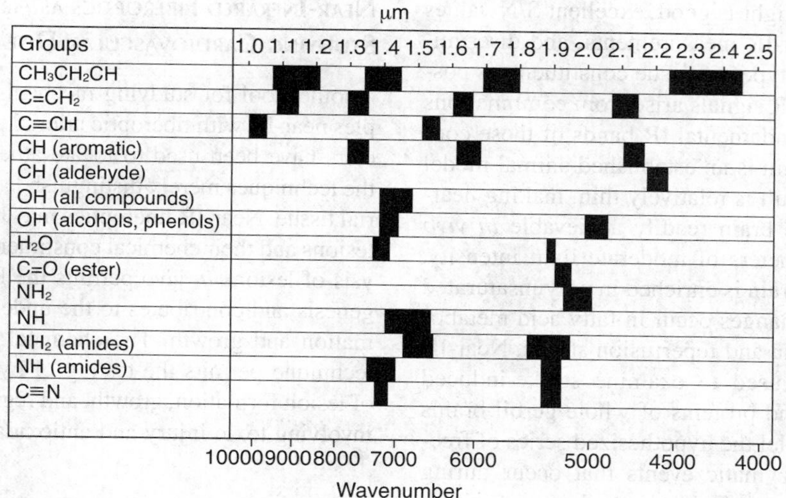


FIGURE 41.29 Characteristic group frequencies in the near-infrared spectrum.

excellent spectra. Unlike mulling agents, KBr does not contribute extraneous bands in the IR spectrum; however, KBr is slightly hygroscopic and may pick up moisture during the disc preparation. This will lead to characteristic water bands at 3300 and 1640 $cm^{-1}$.

## INFRARED SPECTROSCOPY: APPLICATIONS

High signal-to-noise spectra and sophisticated data reduction techniques have made FT–IR a versatile, high-throughput analytical technique. Advances in the acquisition of FT–IR spectra from preparations such as monolayers, tissues, and samples *in situ* include fiberoptical waveguides for beam handling and remote field sensing. Metabolic fingerprinting, a process where crude metabolite mixtures are rapidly screened and analyzed by multivariate mathematical models, is central to metabolomic studies. FT–IR is critical to these studies, wherein mixed sample profiles are correlated to biological origin, metabolic status, and exposure to chemical or environmental stimuli [59].

Near-infrared spectroscopy is particularly useful as a noninvasive *in vivo* analytical tool. Virtually every organic compound has a near-IR spectrum that can be measured. Near-IR spectra consist of overtones and combinations of fundamental mid-IR bands. These IR spectra are characterized by good penetration of light into tissues and a high capacity to identify organic compounds. Figure 41.29 illustrates some of the structural moieties that can be determined by near-IR and their absorption characteristics.

The application of *in vivo* FT–IR microspectroscopy in the diagnosis of disease has been reviewed [98,115]. FT–IR spectroscopy is used for studying human arteries, cancers and tumors, and brain tissues from stroke and Alzheimer's patients [116,117,128]. Wetzel et al. [219] utilized FT–IR microspectroscopy for deuteration studies investigating the metabolic activity in various layers of the cerebellum. Some specific examples of the use of FT–IR spectroscopy in disease diagnosis and toxicological research follow.

### Analysis of Solid Human Tumor Cells by FT–IR Microspectroscopy

Early detection of human tumors often allows for a higher survival rate. Cells from normal and neoplastic human lung tissue have been analyzed by means of FT–IR microspectroscopy [13]. Reliable spectra that can differentiate between normal and neoplastic cells may be obtained. Neoplastic cells show an increase in the intensity of the bands corresponding mainly to the $PO_2^-$ symmetrical and asymmetrical vibrations (1080 to 1540 $cm^{-1}$) of DNA compared to normal cells. This analytical method may be useful for recognizing early neoplastic transformation that is usually not possible with traditional procedures. Similar results have been observed in the FT–IR spectra of sections of normal and malignant human colon tissues and in other malignant tissues such as stomach, esophagus, skin, liver, cervix, and vagina [227]. The results suggest that, in cancerous tissues, most $PO_2^-$ groups become hydrogen bonded and the intermolecular packing among $PO_2^-$ groups becomes closer. Nucleic acids may be the molecules primarily responsible for the observed changes in the $\nu_s(PO_2^-)$ and $\nu_{as}(PO_2^-)$ bands.

### Analysis of Stroke-Induced Changes by Near-Infrared Spectroscopy

Historically, near-IR spectroscopy has been used to monitor fat and protein in agricultural products. Near-IR spectroscopy is also well suited for animal work because tissue penetration by near-IR light is good, excellent S/N values can be obtained in near-IR measurements, and discrimination between various types of tissue constituents is possible because the near-IR signals arise from combinations and overtones of the fundamental IR bands of these constituents. The gerbil brain is an established animal model of stroke. The gerbil skull is relatively thin, making near-IR spectroscopy of the brain readily achievable *in vivo* with common spectrometers of moderate light intensity. In addition, the gerbil brain is enriched in polyunsaturated fatty acids. Dramatic changes occur in fatty acid metabolism during the ischemia and reperfusion stages. Near-IR techniques have been used to examine stroke-induced changes in the lipids and proteins of whole gerbil brains [36]. The changes parallel the hypothesized series of free-radical and altered enzymatic events that occur during transient ischemia and reperfusion in the human brain.

To understand the early changes in lipid and protein metabolism following stroke and trauma, animal models have been developed to recreate these changes. The examination of whole brains has been made possible by a combination of hardware modifications and mathematical techniques designed to make the sample presentation to the spectrometer quite reproducible. A refrigerated sample compartment with dry nitrogen purge can be constructed for analysis of whole frozen brains. The cooled compartment enables repeated scans of frozen brains to be collected over time without thawing. The spectrophotometer itself can be purged with dry nitrogen gas, eliminating spectral artifacts associated with lipid/protein oxidation and atmospheric water vapor or other gases. A geometric noise filter removes spectral variations arising from positional variation of the brain. The BEST and extended BEST algorithms, which scale spectral vectors in multidimensional hyperspace with a directional probability, can be used with a supercomputer to analyze the spectra collected.

In addition to changes observed in stroke, age-related changes occur in the polyunsaturated fatty acid pool and in the state of protein oxidation within the central nervous system. Near-IR spectral analysis has many applications to aging and stroke research, including:

- Determination of age from brain spectra
- Prediction of short-term memory deficit from the spectra of injured brains
- Simultaneous multicomponent analysis of lipids and proteins
- Quantification of edema
- Transcranial scanning of the brain *in vivo*

Near-IR scanning of brains *in vivo* simplifies the testing of antiepileptic drug candidates by reducing the number of subjects required, by allowing each subject to be used as its own control, and by eliminating variance due to outlier subjects, such as those that have had a stroke before the experiment.

## Near-Infrared Fiberoptics as Arterial Probes for Studying Cardiovascular Disease

Another tool for studying mechanisms of toxicology couples near-IR with fiberoptic arterial probes. Fiberoptic catheters have been used to locate atherosclerotic lesions, but the techniques merely distinguish lesions from healthy arterial tissue. Near-IR fiberoptics may be used to spatially map lesions and their chemical constituents [39]. Chemical analysis of lesions *in vivo* permits the kinetic study of atherogenesis and contributes to the understanding of lesion formation and growth. The chemical imaging power of this technique permits the testing of hypothetical mechanisms of lesion formation, growth, and regression, including those involving toxic injury and antioxidant protection.

## Toxicological Studies

Infrared spectroscopy is extensively employed in toxicological research. The toxic effects of selenium exposure on the external cell membranes of living bacteria may be efficiently analyzed by FT–IR directly in the biomass [65]. FT–IR spectroscopy was employed to assess the hepatotoxic effects of a single dose of carbon tetrachloride in rat liver [51]. Dynamic chemical changes in liver samples were monitored over time by analyzing the IR spectra in the lipid (1800 to 1000 $cm^{-1}$) and C–H stretching (3000 to 2400 $cm^{-1}$) regions and comparing them to untreated controls. Changes in band shape and intensity were correlated with toxicity. To develop a high-throughput metabolomic toxicology assay, researchers hypothesized that FT–IR could be used to detect differences between urine collected from rats treated with lipopolysaccharide, a potent inflammatory agent, and urine obtained from untreated controls [82]. In addition, cotreatment with ranitidine, a drug often associated with idiosyncratic hepatotoxicity, was performed to determine whether or not the FT–IR method would be useful for predicting this idiosyncratic toxicity. The results of this pilot study suggest that similar methods might be applied for the rapid metabolomic screening of drug toxicity. In the occupational and environmental toxicology settings, FT–IR analysis may be useful for the continuous monitoring of the presence of airborne toxins. Over a 2-year period, 39 selected air toxins were measured at 11 different petrochemical sites using open-path FT–IR [40]. The data enabled the researchers to calculate the hazard indices for both acute and chronic health effects attributable to these toxins over time and by location, thereby estimating potential health risks to workers.

## Specialized Instrumentation

Several modified approaches for performing IR spectroscopic imaging microscopy have been developed. One instrument integrates an acoustic-optic tunable filter

(AOTF) and charge-coupled device (CCD) detector with an infinity-corrected microscope for operation in the near-IR spectral regions [204]. Images at moderate spectral resolution (2 nm) and high spatial resolution (1 μm) can be rapidly collected. Data can be presented with 128 × 128 pixels. The CDD is a true imaging detector, with wavelength selection provided by the AOTF and quartz tungsten halogen lamp to create a tunable source. The instrument can be utilized for both absorption and reflectance spectroscopies.

Synchrotron FT–IR (SFT–IR) microscopy, which does not require sample mounting or specialized preparation, has been used to probe the molecular responses to toxic injuries in intact, living cells [88]. SFT–IR allows the observation of several types of molecular responses and lesion development induced by stressors such as radiation and toxic compounds in the same cell over time. Real-time changes in nucleic acids and proteins in live human cells treated with dioxin were studied using this technique.

Portable FT–IR units are available and broadly deployed for field use. Potential applications include the rapid detection of toxins and suspected bioterror agents at remote locations. To evaluate the utility of these portable systems for the field-based characterization of biological threats, a series of peptide and nonpeptide toxins were analyzed and compared to libraries of known spectra [181]. In the case of pure compounds, this spectral searching technique allowed the researchers to discriminate between aflatoxin, mycotoxin, and strychnine at a 99% confidence level. Nonpeptide toxins were readily identified in chemical mixtures, and peptides such as ricin were correctly identified in mixtures at a 95% confidence level. FT–IR analysis is being rapidly developed as a versatile tool for onsite identification of chemicals, toxins, and biological threats.

## RAMAN SPECTROSCOPY

Raman spectroscopy is closely related to infrared spectroscopy, in that the information about molecular vibrational frequencies provided by the latter technique is of the same kind as that provided by the Raman vibrational spectrum [35,157]. In molecules with a center of symmetry, however, vibrational transitions that are allowed in the IR spectrum are forbidden in the Raman effect, and *vice versa*, providing useful information about molecular symmetry. Structurally symmetrical diatomic molecules such as $H_2$ and $O_2$ are also electrically symmetrical and do not give IR absorption spectra. These molecules do afford Raman spectra due to excitation of symmetrical vibrations. In a molecule such as tetrachloroethylene ($CCl_2{=}CCl_2$), the double-bond stretching frequency is symmetrical and the molecule does not show a double-bond stretching frequency in the IR spectrum; however,

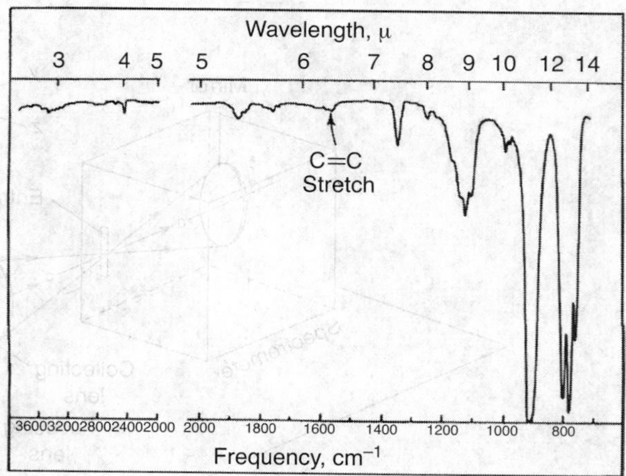


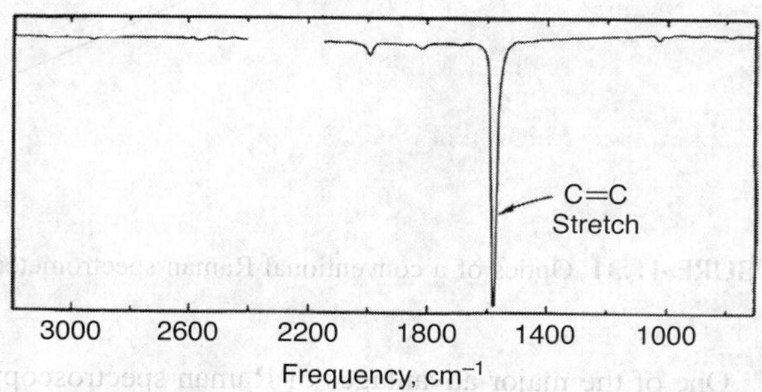


**FIGURE 41.30** Infrared (top) and Raman (bottom) spectra of tetrachloroethylene.

this vibration appears strongly in the Raman spectrum of tetrachloroethylene and provides evidence of a symmetrical structure (see Figure 41.30). Thus, the two techniques are complementary to each other.

The Raman effect is a scattering process in which the interaction between photon and the molecule occurs in a very short period of time, and the Raman peaks obtained correspond to photons that have bounced inelastically off the molecule. The Raman spectrum arises as a result of the light photons being captured momentarily by molecules in the sample and giving up (or gaining) small increments of energy through changes in the molecular vibrational and rotational energies before being emitted as scattered light. These changes in vibrational and rotational energies result in changes in wavelength of the incident light. The convention in Raman spectra is to quote the positions of vibrational peaks as the difference between the absolute wavenumbers of the exciting line and the absolute wavenumbers of the resulting scattered photons. The Raman effect is extremely weak, and only a minute portion of the incident photons are useful emergent photons; thus, relatively high-power lasers must be used to create a high photon flux. Sophisticated optical and electronic equipment is also required to detect the scattered photons.

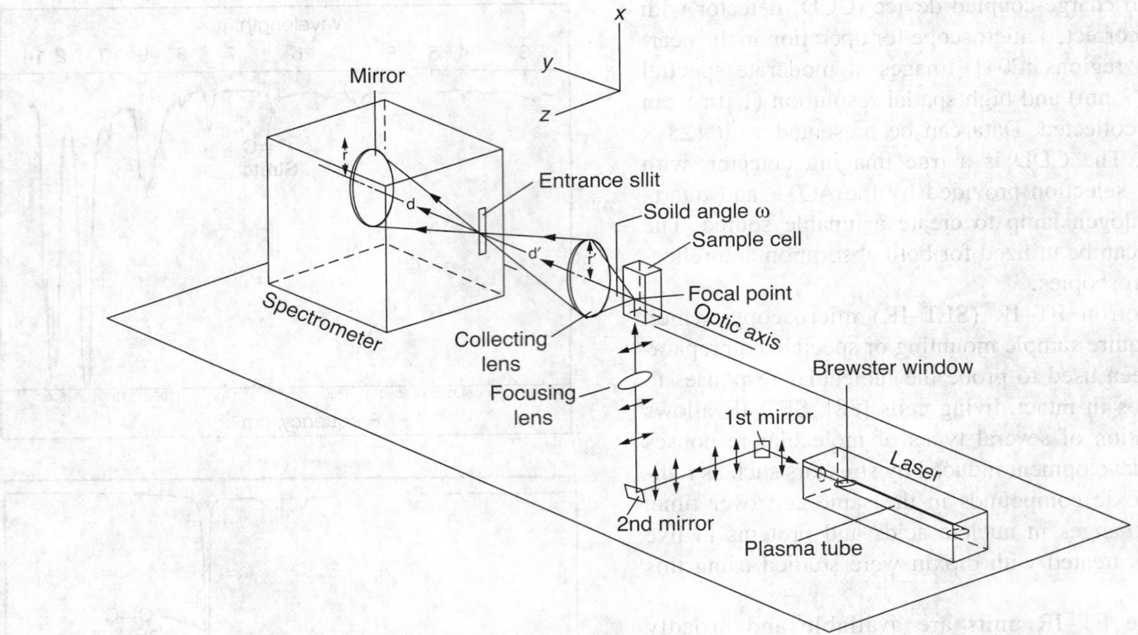


**FIGURE 41.31** Optics of a conventional Raman spectrometer.

One of the major advantages of Raman spectroscopy is that spectra may be obtained for molecules in aqueous solutions, as water has a weak Raman spectrum that interferes only minimally with the spectrum of the solute. Analyte concentrations in the range of 0.1 to 0.01 *M* in water are normally used. In resonance Raman spectroscopy, concentrations of chromophoric molecules in the range $10^{-4}$ to $10^{-6}$ *M* can be used, making this technique particularly useful for biochemical studies.

## Basic Optics of the Raman Experiment

In a standard Raman experiment, intense monochromatic radiation provided by a continuous-wave laser is focused onto or into the sample. Some of the resultant scattered light is collected by optics and directed to a dispersing system that is usually a monochromator. The monochromator separates the scattered light on the basis of frequency, and these frequencies are then detected and recorded either by single-channel (scanning) or multichannel detection. Figure 41.31 illustrates the optics involved in a conventional Raman experiment.

## Sampling Techniques and Problems

Samples for Raman spectroscopy may be examined in any physical state. Liquid samples are usually measured in a quartz (1-cm) cuvette similar to the type used in fluorescence spectroscopy; however, because the incident laser beam travels longitudinally down the length of the liquid column, the cell bottom must be transparent. Capillary cells are often used for biological samples, especially when material is limited in availability. Single crystals and fibers can be analyzed by mounting on a goniometer head. Solid crystalline or polycrystalline materials can be pressed into pellets, prepared as KBr discs, or packed into capillary microprobes. Samples in the form of thin films can also be examined. Potential problems include the breakdown of photolabile analytes during laser irradiation. This can be reduced or eliminated for liquid samples by utilizing a spinning cell, a cell in which the liquid sample is continually moved through the laser beam, or a cell in which the liquid is continually stirred with a magnetic stirrer. These procedures reduce the build-up of degradation products usually observed with static sample analysis.

In Raman spectroscopy, it is important for the analytes to be optically homogeneous, especially in the case of biological samples. Particulate matter in solutions should be removed either by centrifugation or filtration; otherwise, hot spots may occur in the sample on irradiation, leading to possible degradation. Luminescence, which may often obliterate the Raman effect, may occur if the sample or an impurity in the sample has a chromophore. Luminescence can often be reduced or eliminated by changing the wavelength of excitation or adding a quenching agent such as KI.

## Protein Conformation Determination

Protein molecules are classical examples of the application of Raman spectroscopy to biomolecules. This technique can probe structural details such as average peptide backbone conformation, as well as the side chains of some amino acids, such as tyrosine and tryptophan. Protein spectra are usually obtained in the 450- to 650-nm region. UV-excited resonance Raman spectra of proteins containing aromatic amino acids of interest is often conducted with excitation below 300 nm. The normal Raman spectrum of proteins contains diagnostic amide I (C=O stretch) and amide III (N–H in-plane bending) bands that can be utilized to characterize the secondary structure of the protein or peptide backbone. Table 41.3 gives approximate positions for the amide I and amide III bands in both the IR and Raman spectra of various polypeptide conformations. Characterization of the secondary structure of a protein depends on the determination of characteristic amide I and III frequencies in the Raman spectrum for α-helical, β-sheet, and random protein conformations. This is often achieved by using polypeptide models and proteins of known conformation. Figure 41.32 illustrates the Raman spectra, run in water, of native and denatured ribonuclease A and shows the change in the amide III band in the disordered protein [42].

## Resonance Raman Labels

Resonance Raman labels are chromophores that have been carefully designed to mimic natural biochemical components and that are themselves biologically active molecules. They provide detailed vibrational and electronic spectral data when in the vicinity of a biologically important site. Extrinsic protein-bound chromophores, such as methyl orange bound to bovine serum albumin, have been studied in detail. Other systems that have been studied include protein–ligand interactions, such as drug–enzyme and hapten–antibody complexes where the drug or hapten is the chromophoric resonance Raman label. This method is useful for studying enzyme–substrate complexes and can provide vibrational spectra of the substrate during enzyme catalysis.

**TABLE 41.3**
**Approximate Positions ($cm^{-1}$) of the Most Intense Amide I and Amide III Bands in Raman Spectra and the Amide I Band in IR Spectra for Various Polypeptide Conformations**

| | Amide I | | Amide III |
|---|---|---|---|
| Conformation | Raman | Infrared | Raman |
| α-Helix | 1645–1660 | 1650.00 | 1265–1300 |
| β-Sheet | 1665–1680 | 1632.00 | 1230–1240 |
| Unordered | 1660–1670 | 1658.00 | 1240–1260 |

## RNA and DNA Structural Analysis

Analysis of polynucleotides by Raman spectroscopy affords bands (~30) primarily attributable to purine or pyrimidine ring modes. In addition, the phosphate group shows interesting features in the spectrum. The sugar moieties in DNA and RNA molecules and the related polynucleotides are poor Raman scatterers. From the Raman spectrum it is possible to obtain a semiquantitative estimate of the relative population of bases in the polynucleotide molecule. Base protonation in DNA has been studied using Raman spectroscopy. Metal–nucleotide binding has also been investigated. The mode of binding of ions such as $Ca^{2+}$, $Mg^{2+}$, $Co^{2+}$, $Cu^{2+}$, and $Mg^{2+}$ to adenosine triphosphate over a wide pH range has also been studied. Raman spectroscopy can detect the disruption of base-pairing and base-stacking interactions [113]. In Figure 41.33, raising the temperature of poly(rA)·poly(rU) from 32°C to 82°C results in thermal disruption of the helix. Important features in the difference spectra are the radical changes in the carbonyl region (1650 to 1700 $cm^{-1}$) due to the disruption of H bonding in the Watson–Crick mode. Phosphate backbone conformation in nucleic acids has also been studied by monitoring the –O–P–O– symmetrical

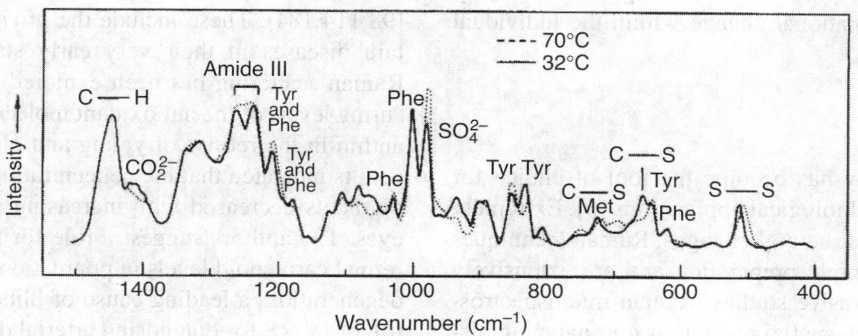


**FIGURE 41.32** Raman spectra of native and denatured ribonuclease A, at 32° and 70°, after correction for the water background and being normalized to the intensity of the methylene deformation mode at 1447 $cm^{-1}$. Protein concentrations of about 10% were used with typical spectral conditions of 488-nm excitation, 200-mW power, and 7 $cm^{-1}$ spectral slit. (From Chen, M.C. and Lord, R.C., *Biochemistry*, 15, 1889–1897, 1976. With permission.)

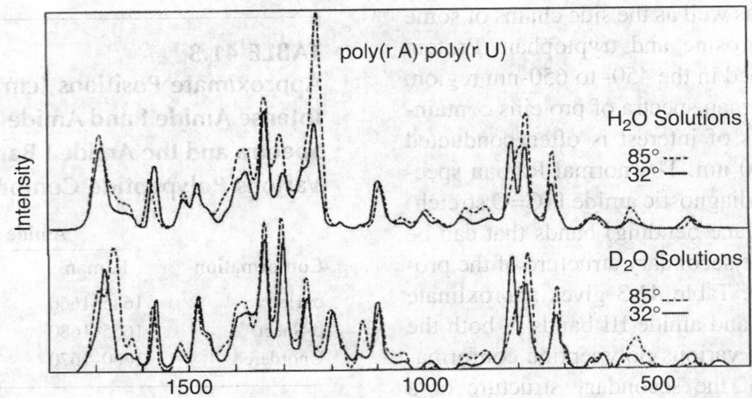


**FIGURE 41.33** Raman spectra of $H_2O$ and $D_2O$ solutions of poly(rA)·poly(rU) at 32° and 85°C. The background of Raman scattering by the solvent has been subtracted from each spectrum. (From Lafleur, L. et al., *Biopolymers*, 11, 2423–2437, 1972. With permission.)

stretching vibration (800 $cm^{-1}$) and the $–PO^{2-}$ symmetrical stretch motion (1100 $cm^{-1}$). The interaction of nucleic acids with proteins is an area of active study; for example, the stabilizing effect of the viral capsids on the secondary structure of the viral RNA has been investigated. Also, Raman data on DNA–histone interactions have indicated that the sites of DNA–protein interactions are probably located in the grooves running along DNA.

## Lipids and Membranes

Raman spectroscopy is ideally suited for the study of lipids and membranes and has some advantages over other techniques. The analysis time frame of fractions of a picosecond provides instant snapshots, thereby eliminating the line broadening commonly seen in magnetic resonance spectra. No probe molecule is required, and both gel and liquid-crystal hydrocarbon regions can be monitored. The most useful regions are 1000 to 1150 $cm^{-1}$ (C–C accordion stretch) and the 2800 to 3000 $cm^{-1}$ (C–H stretching mode). Transitions from gel to liquid-crystal in membranes are indicated by marked changes in the bands assigned to the C–C stretch mode, and the C–H stretch region is extremely sensitive to conformational change within the individual fatty acid chains.

## Applications

Raman spectroscopy has become the tool of choice for many chemical and biological applications [6]. Extremely sensitive to minute structural changes, Raman techniques require minimal sample preparation and are extensively employed in noninvasive studies. Raman microspectroscopy also offers high spatial resolution for analyzing discrete regions of biological matrices. These characteristics make Raman spectroscopy particularly useful for exploring biological processes in living tissues and in individual living cells and organelles [167]. Continuous Raman spectroscopy monitoring of living A549 human lung cells was employed to evaluate the effects of nonionic surfactant exposure on biochemical processes and the sequence of events in cell death. The molecular mechanisms of cell death were strongly associated with specific decreases in Raman peaks associated with nucleic acids and proteins [154]. Western blotting analysis supported the conclusions derived from the Raman spectroscopy data. Raman spectroscopy analysis of specific, time-dependent biochemical changes might be applied to tissue engineering and high-throughput toxicity screening.

Evans and colleagues [63] combined coherent anti-Stokes Raman scattering (CARS) with video-rate microscopy for real-time monitoring of dynamic processes in lipid-rich tissues of living mice. CARS imaging of the methylene stretching vibrational band afforded exceptional subcellular resolution in tissues such as adipocytes, corneocytes, and the sebaceous glands. Important processes such as *in vivo* cellular transport, xenobiotic diffusion, and transcorneal absorption might be monitored using the CARS method.

Several reports on the biomedical applications of *in vivo* Raman spectroscopy (IV-RS) have been published [98,114,184]. These include the *in vivo* diagnosis of certain diseases in their very early stages [184]. *In vivo* Raman scattering has been explored as a means of measuring levels of the antioxidant molecules lutein and zeaxanthin in the retinas of young and older adults [71]. The results indicated that the concentration of these protective pigments decreased with increasing age, even in normal eyes. The authors suggest a role for Raman screening of retinal carotenoid levels in populations at risk for macular degeneration, a leading cause of blindness. The potential use of IV-RS for diagnosing arterial disease and cancer in gynecological tissues, soft tissues, breast, colon, bladder, and brain has been discussed by Manoharan et al. [128]. Lawson et al. [114] reviewed the application of Raman spectroscopy to the study of human arteries, tumors, gallstones, hair, and nails. Brain tissues have also been investigated using FT–Raman spectroscopy, and spectra from the cerebral cortex, cerebral white matter, caudate–putamen, thalamus, synaptosomal fraction, and myelin fractions have been recorded [139,140]. Brain tumors have also been studied using FT–Raman spectroscopy (139).

Ong et al. [155] studied the substantia nigra of the rostral mid-brain in monkeys using Raman microspectroscopy to determine differences between white and gray matter. As compared to the laser spot size of 20 µm in IR microspectroscopy, the 1-µm laser spot size in Raman microspectroscopy afforded much greater spatial resolution in sample analysis. The white and gray matter could be clearly distinguished and their relative proportions evaluated from Raman frequencies in the 3000-$cm^{-1}$ region.

Singer and coworkers [188] evaluated Raman confocal microscopy as a means of assessing environmental pollutant bioavailability and toxicity in a cell culture system. Phenanthrene, dodecane, 3-chlorobiphenyl, and pentachlorophenol (four common environmental pollutants) were detected and quantified using this potential pollution bioassay.

Along with IR spectrometers, portable Raman spectrometers have also been developed for remote field use. Raman analysis with portable instrumentation was used in a blind field test to analyze a matrix of 58 unknowns for comparison with a standard hazardous materials response library of known spectra [83]. Despite the inclusion of multiple solvents and compounds with uncatalogued spectra, over 97% of the samples were correctly identified with no false positives. Free-base crack cocaine and its hydrochloride salt were readily identified and distinguished from common street adulterants using fiberoptic Raman spectroscopy [38].

Raman spectroscopy systems have been developed for the detection of toxins associated with bioterror, including chemical and microbial weapons. The capacity of a human-lung-cell-based Raman spectroscopy system to detect and differentiate between sulfur mustard and ricin exposure was assessed noninvasively over time [153]. Differences in cellular Raman spectra correlated with characteristic changes in cellular biochemistry and structure. Damaged cells were detected with high sensitivity and specificity, and toxic agent identification was reasonably accurate for ricin (71.4%) and sulfur mustard (88.6%).

## ISOTOPIC LABELING

The use of isotopes to study the fate of xenobiotics *in vitro* and *in vivo* developed initially with the use of radioisotopes. The development of methods for detecting and quantifying energy emitted from radioisotopes led to synthetic xenobiotics containing radioactive isotopes that could be tracked or traced in organisms. The development of sophisticated NMR and mass spectral methods has made the use of stable isotopes for studying the fate of xenobiotics fairly common. Stable isotopes are popular because they often yield a great deal of structural information about xenobiotic metabolites without the hazards to the organism (often humans) inherent in radioisotopes. The following sections provide an introduction and overview of isotope methods. More comprehensive information is available [97,106,220,226].

### Radioisotopes

Most elements exist as several isotopes that vary in atomic weight. Some of these isotopes are radioactive, spontaneously decaying to form an atom of another element. This decay is accompanied by the emission of radiation. The radiation emitted is of three distinct types: alpha (α), beta (β), and gamma (γ). Alpha particles are actually helium nuclei ($^4He$), β particles are electrons, and γ rays are high-energy electromagnetic radiation. Isotopes that emit β particles, so-called β emitters, are less hazardous to laboratory workers than γ emitters because β particles do not possess sufficient energy to penetrate the skin. β emitters are only hazardous if ingested or where they otherwise come in contact with cells. In contrast, γ emitters are more hazardous to laboratory workers because γ rays are highly energetic and can easily penetrate the skin. Special precautions such as lead shielding are often required when working with γ emitters.

Radionuclides are quantified in terms of their *specific activity*, which is the radioactivity per unit mass of material. Specific activity can be given in Curies per mole (Ci $mol^{-1}$) or Bequerels per mole (Bq $mol^{-1}$). A Bequerel is defined as 1 disintegration per second (dps) and 1 Curie = $3.7 \times 10^{10}$ dps; thus, 1 Ci = $3.7 \times 10^{10}$ Bq. The most commonly used radionuclides are $^3H$ and $^{14}C$. Both of these isotopes are weak β emitters that emit fast-moving electrons that can penetrate up to 50 cm in air and up to 0.5 cm in aluminum. Both $^3H$- and $^{14}C$-labeled compounds can be handled easily and safely with the use of glass containers, protective gloves, and safety glasses. $^{14}C$ β particles carry significantly more energy ($\beta^-$ = 155 KeV) than $^3H$ β particles ($\beta^-$ = 18.6 KeV). This difference enables mixtures of radionuclides (e.g., double-labeled $^3H/^{14}C$-containing compounds) to be measured simultaneously (see later discussion). Although $^{14}C$ is a more energetic nuclide and therefore a more easily detected tracer atom than $^3H$, its maximum specific activity when incorporated into a drug molecule is only 62.4 mCi $milliatom^{-1}$ of carbon, compared to a maximum specific activity for $^3H$ of 29.1 Ci $milliatom^{-1}$ of hydrogen.

The energy emitted when radioactive isotopes decay is easily traced, as the radionuclide itself is not metabolically altered by the biological system under study. The radioisotopes most commonly used in biological systems are shown in Table 41.4.

**TABLE 41.4**
**Radioactive Isotopes Commonly Used To Study the Fate of Xenobiotics**

| Atomic No. | Element | Atomic Weight | Half-Life | Radiation (MeV) |
|---|---|---|---|---|
| 1 | Hydrogen | 3 | 12.33 yr | $\beta^-$ (0.019) |
| 6 | Carbon | 11 | 20.4 min | $\beta^+$ (0.96) |
| | | 14 | 5730 yr | $\beta^-$ (0.156) |
| 15 | Phosphorus | 32 | 14.28 d | $\beta^-$ (1.71) |
| | | 33 | 25.3 d | $\beta^-$ (0.25) |
| 16 | Sulfur | 35 | 87.5 d | $\beta^-$ (0.167) |
| 17 | Chlorine | 36 | $3 \times 10^5$ yr | $\beta^-$ (0.71) |
| 37 | Rubidium | 87 | $4.8 \times 10^{10}$ yr | $\beta^-$ (0.272) |
| 53 | Iodine | 125 | 60.14 d | $\gamma$ (0.035) |
| | | 131 | 8.040 d | $\beta^-$ (0.607, 0.336) |
| | | | | $\gamma$ (0.080, 0.284, 0.364, 0.637, 0.723) |

## Xenobiotic Disposition Studies Using Radiolabeled Tracers

The most common use of radiolabeled xenobiotics in toxicology is the study of the fate of a chemical in animal models. These studies are often referred to as absorption, distribution, metabolism, and excretion (ADME). The successful ADME study requires:

- A radiolabeled xenobiotic with the most appropriate isotope and position of the label
- An analytical method for separating the parent compound and its hypothesized metabolites
- Methods for quantifying the amount of radioactivity

### Choice and Location of Label

The radioisotope of choice for xenobiotic ADME studies is carbon-14 ($^{14}C$) because it is a radioactive form of the element that forms the backbone of most xenobiotics. In addition, $^{14}C$ has a very long half-life (over 5000 years), and, as a weak β emitter, it poses fewer health risks to laboratory workers. The position of the label must be carefully selected to ensure that the label is not lost upon metabolism. Researchers using $^{14}C$ will often uniformly label one of the aromatic rings of a xenobiotic if it is thought not to undergo ring opening during metabolism. These positions are preferable to $^{14}C$ labeling in a methyl group attached to an oxygen or a nitrogen, because both of these carbons will undergo demethylation reactions catalyzed either by cytochrome P450 or flavin-containing monooxygenase. Sometimes there are limitations for placing the label, based on synthetic concerns. Every effort should be made to locate the label in a chemically and metabolically stable position.

Tritium ($^3H$), the radioactive form of hydrogen, is often used as a tracer because of the high specific activity that can be obtained with this isotope. High specific activity increases the researcher's ability to detect smaller amounts of the xenobiotic. Receptor binding assays typically use tritiated ligands for this reason; however, the use of tritium has some shortcomings, including a relatively short half-life (12.33 years) that requires adjustment of the specific activity over time. In addition, because tritium is an isotope of hydrogen it undergoes exchange with nonradioactive protium ($^1H$) in solvents. This so-called *solvent exchange* must be taken into account when positioning a label on the xenobiotic. Positions that readily undergo solvent exchange are not good candidates for labeling. These positions are referred to as *labile positions* (e.g., protons attached to oxygen, nitrogen, or sulfur atoms are labile). Tritium may also be lost due to metabolism of the xenobiotic, and an experienced investigator will usually avoid inserting labels at positions in the molecule where they may be lost during biotransformation; for example, $^3H$ labeling at a carbon adjacent to a heteroatom or at a hydroxylation site on a phenyl ring or a $^3H$-labeled *N*-methyl group, which may often be lost by oxidation, should generally be avoided.

### Radiochemical Purity

The purity of labeled compounds is usually critical to the success of an experiment. When following (tracing) a radiolabeled compound, there is no specificity to the radioactivity emitted by the radioactive isotope. Chemical impurities containing the radioisotope will be indistinguishable from the xenobiotic of interest; therefore, the original material administered to the organism should be of the highest purity possible to ensure that the radioactivity detected is derived from the parent compound. Radiochemical purity should be confirmed by a suitable method, such as radiochromatography (discussed later), prior to using a radiolabeled compound for an ADME study.

In addition to radiochemical purity, enantiomeric purity may also be important. Many biological processes are stereoselective, with either the S or R enantiomer of a biologically active compound having significant activity. That is, only one of the stereoisomers of a compound is active in the system under study; for example, the (S)-(–) enantiomer of nicotine is responsible for most of its pharmacological activity. The affinity of the (R)-(+) enantiomer for high-affinity nicotinic acetylcholine receptors is 60-fold less than the (S)-(–) enantiomer. With the recognition that many toxic responses are receptor mediated, the enantiomeric purity of radiochemicals has become more important.

The solvent selected for storage of a radiolabeled compound is an important consideration; for example, radiolabeled peptides and proteins that have been stored in water may undergo extensive degradation. Solvent exchange can also be a problem for tritiated compounds, as described above. The selection of a solvent is usually a compromise between adequate solubility and a minimum of inherent reactivity with the radiolabeled compound.

## Route of Administration

A primary consideration when conducting an ADME study with radiolabeled xenobiotics is the route of administration. The xenobiotic is usually administered to the animal model by a route appropriate to either the anticipated major route of exposure to the species of interest (usually humans) or by a route of administration that is consistent with the goals of the research; for example, if the xenobiotic of interest is a component of a consumer product applied to the skin, then the appropriate route of delivery may be dermal administration. If, however, the objective of the ADME experiment is to determine the pharmacokinetics of the xenobiotic in blood and tissues, then the intravenous route of administration would be chosen over the dermal route. The solubility of the compound in an appropriate delivery vehicle must also be considered when deciding on a route because many vehicles are not compatible with intravenous administration.

### Liquid Scintillation Spectrometry

Total xenobiotic-derived radioactivity can be determined in all excreta including urine, feces, and expired air to quantify the routes of excretion for a xenobiotic and its metabolites. Total radioactivity in aqueous samples is relatively easy to quantify using liquid scintillation spectrometry. This method for quantifying radioactivity involves mixing an aliquot of the sample with a *liquid scintillation mixture* (LSM) that uses an organic compound or mixture of compounds that are scintillators, compounds that give off light when they absorb radioactive energy. Traditionally these scintillators were dissolved in toluene-based mixtures that enabled counting both organic and aqueous samples. For aqueous samples, a typical mixture of 2 parts sample to 8 parts LSM formed a gel that was counted. These LSMs could disperse no more than about 20% water. Newer LSMs that are less hazardous, can disperse more water, and can be discarded down the sanitary sewer have replaced toluene-based LSMs for most applications. The mixture of the radioactive sample and the liquid scintillation mixture is placed in special vials that are highly efficient at passing light without absorbing radioactive energy. The vials are counted by placing them in a liquid scintillation counter, which uses two opposed photomultiplier tubes to detect light emitting events triggered by the radioactive decay of the isotope. Coincidence circuitry is used to separate random events from radioactive-isotope-driven events. The counting region is lined with lead to shield the vial from extraneous environmental radiation. The sample is compared to a blank or background vial that contains everything but the radioisotope-containing sample. The radioactive counts from the background vial must be subtracted from the sample to obtain net radioactivity.

The liquid scintillation counter expresses data in counts per minute (CPMs) which must be corrected to disintegrations per minute (DPMs) to account for inefficiencies in capturing all of the radioactive energy emitted. The method employed in most liquid scintillation counters today involves the use of external standard calibration. Here, a radioactive source housed in the instrument is placed automatically near the vial containing the sample, and the photomultiplier tubes detect the resulting light emitted from the scintillator within the sample vial. The CPMs detected by the instrument are then automatically compared to the known DPMs of the external standard source and a counting efficiency is determined. The counting efficiencies for $^{14}C$ are usually much higher than for the less energetic β emitter tritium.

In addition to the inefficiencies in transferring the energy of radioactive decay into light energy, there is also *quenching* of the light emitted from the scintillator. Many solvents and biological molecules can quench the light emitted by the scintillator, so the researcher must correct for the amount of quenching within the sample by comparing the expected DPMs to a quench curve and determining the actual DPMs present in the quenched sample. This is usually accomplished automatically by the liquid scintillation counter by first running a set of vials containing a known amount of radioactivity with increasing amounts of a quenching agent added to each vial. The resulting quench curve is stored in the counter memory and used to determine the DPMs of the sample.

Radioactive samples containing scintillant are often sensitive to external light, especially sunlight, which results in excitation and abnormally high CPMs on analysis. This phenomenon, called *chemiluminescence*, is usually more of a problem with $^3$H-containing rather than $^{14}$C-containing radiolabeled samples and can usually be minimized or eliminated by storing scintillation fluids and samples containing scintillant in the dark for 30 to 60 minutes before analysis.

## Combustion Techniques

Many samples are not amenable to liquid scintillation counting because they are either solids or otherwise incompatible with LSMs. Total radioactivity in tissue samples and fecal samples can be determined by combustion of the sample to carbon dioxide and water. $^{14}$C in the original sample is converted to $^{14}CO_2$ and tritium is converted to tritiated water. Combustion instruments automatically combust samples on a platinum electrode covered with a glass chimney. The products of combustion are swept away by an airstream and trapped in appropriate solvents. Liquid scintillation mixture is added and sample radioactivity is measured in a liquid scintillation counter.

## Autoradiography

Autoradiography is the production of an image by the emission of radioactive decay energy from a radionuclide. It provides a qualitative visual image of the tissue distribution of a xenobiotic and has been used to provide quantitative distribution data, as well. $^{14}$C is probably the most often used radionuclide for autoradiography because of its long half-life and relatively high energy. Tritium is used when greater resolution is required because it is a weaker emitter. Radioisotopes of nitrogen and oxygen do not have sufficiently long half-lives to permit development of an image. Iodine, sulfur, chlorine, and phosphorous are also used for autoradiography.

Whole-body autoradiography is an excellent technique for visualizing the distribution of xenobiotic-derived radioactivity in an animal model. This technique involves administration of the xenobiotic to the animal; following anesthesia, the animal is quick frozen by immersion in hexane or acetone with dry ice. The time interval between administration and freezing must be selected based on some knowledge of the rate of elimination of the compound. After complete freezing, the animal is placed in a block of carboxymethylcellulose ice on the stage of a microtome, an instrument that uses a sharp blade to shave thin sections off one side of the block. Sections varying in thickness from 5 to approximately 80 μm are captured from the microtome blade using an adhesive matrix. The section is then placed on x-ray film and stored in a freezer while the radiation emitted by the isotope exposes the film. The resulting pictures (autoradiograms) illustrate where in the body the xenobiotic-derived radioactivity has distributed. Highly perfused organs such as the kidney, liver, and heart are the first to light up with the compound [214]. Over time it becomes obvious which tissues sequester the compound.

Digital imaging techniques have enabled quantitative analysis of whole-body autoradiograms providing absolute concentrations of xenobiotic-derived radioactivity in tissues. Zane et al. [232] have shown that autoradiography with quantitative digital image analysis compared extremely well with the more traditional combustion method. For this comparison, they treated rats with $^{14}$C-labeled CGS 18102A and sectioned the animals as described above. They also took 16 tissue samples from each animal for combustion analysis. The sections were placed on x-ray film along with a series of calibration standards of known radioactivity. After 3 weeks of film exposure, the autoradiograms were analyzed with a digital imaging system. The concentrations obtained from digital analysis of the autoradiograms compared very well to those obtained by combustion analysis. Jacob et al. [93] used quantitative whole-body autoradiography (QWBA) with digital image analysis to study the transplacental uptake and covalent binding of [$^{14}$C]-chloroacetonitrile (CAN) to various tissues in normal and glutathione-depleted pregnant mice. Their finding that covalent binding of CAN to maternal and fetal tissue was elevated in glutathione-depleted mice suggested a modulatory role of glutathione in CAN distribution and transplacental toxicity and demonstrated the utility of QWBA in mechanistic studies.

Although x-ray film provides excellent resolution and has been used successfully for QWBA, it is not necessarily the medium of choice. Exposure times for x-ray film are typically weeks to months. Also, x-ray film often does not provide a linear calibration over the wide range of radioactivity (optical densities) common in QWBA. To avoid these limitations, storage phosphor screens are commonly employed for QWBA instead of x-ray film. Storage phosphor screens trap the energy released by radioactive samples. The energy is stored by the phosphor screen and is only released when the screen is scanned with a laser beam. The released energy appears in the form of blue light, which is detected by a photomultiplier tube and converted to a digital signal that is stored in a computer file. To quantify the levels of radioactivity in the whole body sections, a series of standards are analyzed and the response of each is determined. A regression equation, derived for the standard responses, is used to determine the absolute concentration of radioactivity in the whole body sections.

The autoradiograms can also be visualized by converting the digital data to a graphics file and printing the results. Commercially available hardware and software are available for QWBA using storage phosphor

**TABLE 41.5**
**Tissue Concentrations of Total Radioactivity After Single Oral Administrations of [$^{14}$C]-RJR-2403 to Male Albino Rats**

| Tissue | 30 minutes | 4 hours |
|---|---|---|
| Eye | 1.01 | 0.67 |
| Heart | 5.39 | 2.46 |
| Kidney | 39.17 | 39.15 |
| Large intestine wall | 3.28 | Not measured |
| Liver | 26.12 | 17.53 |
| Lung | 6.49 | 3.52 |
| Spleen | 5.18 | 4.96 |
| Stomach wall | 15.38 | 7.53 |

*Note:* Target dose, 15 mg/kg; results are expressed as μg equivalents per g.

screens. Storage phosphor screens overcome the major limitations of x-ray film. They have a linear response over a 5 log-unit range and can be developed in hours to days. It has been estimated [223] that a 2-week exposure to x-ray film is equivalent to a 24-hour exposure to a storage phosphor screen. For a more comprehensive introduction to storage phosphors for QWBA, the reader is referred to the work of Wilson and Kraus [223] and Herman and Chay [85].

An interesting application of storage phosphor QWBA is illustrated in Figure 41.34. For this study, Sprague–Dawley albino rats received [$^{14}$C]-RJR-2403 at a dose of 15 mg/kg by gastric gavage. Animals were euthanized at specific time points and sectioned (30 μm) as described above. Sections were exposed to storage phosphor screens for 7 days, and data were collected using a PhosporImager SF (Molecular Dynamics). Figure 41.34A shows an autoradiogram obtained at 30 minutes, and Figure 41.34B shows an autoradiogram obtained at 4 hours after dosing. Table 41.5 lists the tissue concentrations by tissue for each time point. Note that radioactivity was extensively absorbed and distributed throughout the tissues of the rat at 30 minutes after the dose. Tissue concentrations were highest in the kidney and liver. High levels were also detected in the stomach, heart, lung, and spleen. By 4 hours post-dose, tissue concentrations decreased considerably, with high levels seen in tissues associated with metabolism and excretion such as the liver and kidney.

## Analytical Methods for Determining Chemical Form

Total radioactivity data are of limited value because of their lack of specificity. When major routes of excretion have been determined, it is important to characterize the chemical form of the excreted radioactivity. As an exam-

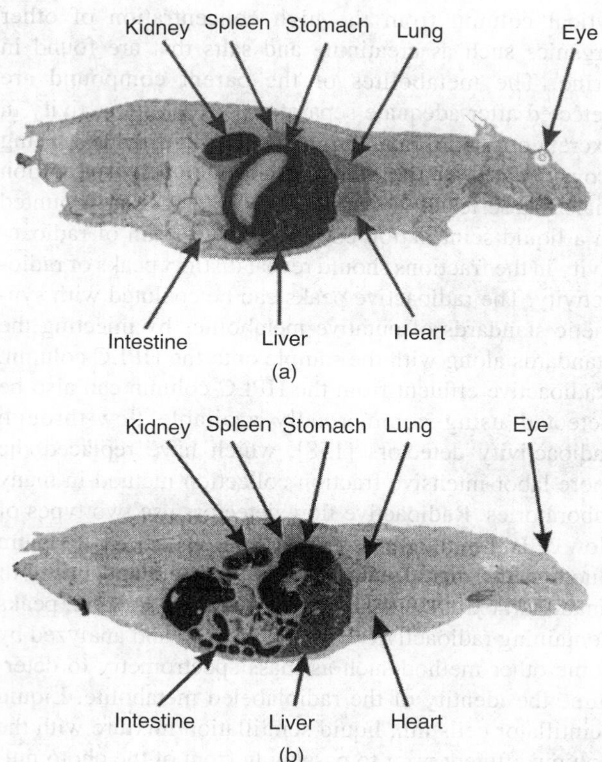


**FIGURE 41.34** Quantitative whole-body autoradiograms of male albino rats after administration of [$^{14}$C]-RJR-2403 at a dose of 15 mg/kg by gastric gavage. (A) 30 minutes after dosing, (B) 4 hours after dosing.

ple, expired $CO_2$ can be selectively trapped in base such as sodium hydroxide or ethanolamine, thereby providing insight into how extensively a xenobiotic is metabolized. Acrylic acid and ethyl acrylate, two monomers used extensively in the plastics industry, are rapidly metabolized, and 70% of the compounds are eliminated from the body as $CO_2$ via normal catabolic biochemical pathways that degrade the carbon skeleton of the molecule [54]. On the other hand, many xenobiotics are metabolized to polar compounds by selective metabolism of functional groups on the molecule without catabolic degradation of the carbon skeleton. High-performance liquid chromatography is useful for quantifying metabolites in biological matrices because metabolites that are relatively hydrophilic and nonvolatile can be chromatographed easily. Other methods for separating metabolites from the parent compound include LC–MS, LC–NMR, thin-layer chromatography, capillary zone electrophoresis, chemical reaction interface mass spectrometry, and gas–liquid chromatography.

The simplest approach to characterizing the radioactivity in excreta is to inject urine directly onto an HPLC. Usually the urine is filtered first by passing it through a centrifugal ultrafilter with a small pore size (0.2 to 0.45 μm). HPLC guard columns are used to protect the analytical column from the high concentration of other organics such as creatinine and salts that are found in urine. The metabolites of the parent compound are detected after adequate separation of the radioactivity in excreta on the HPLC column. Radioactive peaks eluting from the HPLC column can be collected in scintillation vials as a series of fractions and the radioactivity counted in a liquid scintillation counter. A histogram of radioactivity in the fractions should reveal distinct peaks of radioactivity. The radioactive peaks can be coeluted with synthetic standards of putative metabolites by injecting the standards along with the sample onto the HPLC column. Radioactive effluent from the HPLC column can also be detected using commercially available flow-through radioactivity detectors [158], which have replaced the more labor-intensive fraction collection method in many laboratories. Radioactive flow detectors use two types of flow cells. Solid scintillator cells use glass beads, calcium fluoride, or yttrium silicate and permit the recovery of unadulterated effluent. Using this type of detector, peaks containing radioactivity can be collected and analyzed by some other method such as mass spectrometry to determine the identity of the radiolabeled metabolite. Liquid scintillator cells mix liquid scintillation mixture with the column effluent prior to passing in front of the photomultiplier tubes where the radioactivity is detected. The mixture of effluent and LSM must be discarded as radioactive waste. The choice of the type and volume of the flow cell depends on the isotope and specific activity of the label. Optimization of these parameters permits rapid and accurate determinations.

## Double-Label Techniques

The use of doubly labeled compounds in metabolic experiments is often necessary to obtain in-depth information on metabolic pathways and biotransformation mechanisms; thus, one part of a xenobiotic molecule may be labeled with $^{14}C$, and another group in the molecule may contain a $^{3}H$ label. Because $^{3}H$ and $^{14}C$ have different maximal β energies (see earlier discussion), this allows the quantitative measurement of both $^{3}H$ and $^{14}C$ within the same sample. Similarly, other β-emitting nuclides, such as $^{35}S$, can also be measured in the presence of $^{3}H$ or $^{14}C$. The determination depends on the fact that there will be a region in the energy spectrum of the mixture where the β particles from only one of the nuclides will be present. By measuring this region and comparing it with a nuclide standard, the appropriate isotope content in the mixture can be obtained. Some very elegant experiments have been conducted using doubly labeled xenobiotics. Pool and Crooks [162] used $^{3}H$ and $^{14}C$ to determine the *in vivo* stability of (R)-(+)-[$^{3}H$–N′-$CH_3$; $^{14}C$–N–$CH_3$]-*N*-methylnicotinium ion, a primary nicotine metabolite in the guinea pig.

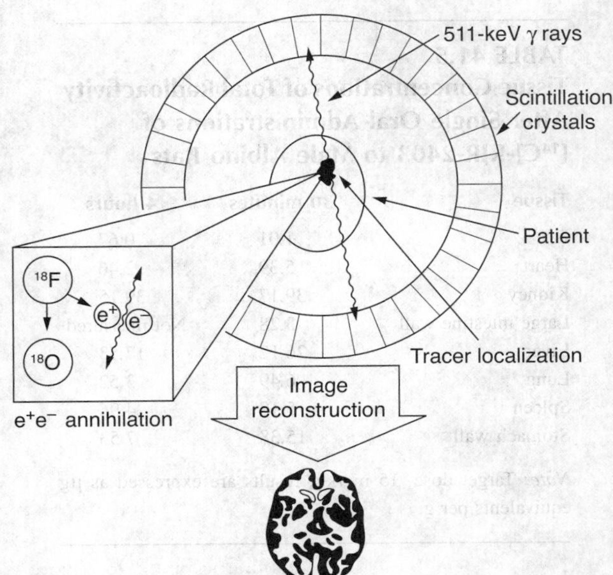


**FIGURE 41.35** Key features of a PET scan: radiopharmaceutical localization, radioisotope decay by positron emission, and immediate positron-electron annihilation; detection of body-penetrating, 511-keV annihilation radiation by an external circular assay of scintillation crystals; reconstructed image of radioactivity distribution. (From Feliu, A.L., *J. Chem. Educ.*, 65, 655–660, 1988. With permission.)

## Positron Emission Tomography

Positron emission tomography (PET) was introduced in the early 1970s as a noninvasive diagnostic technique to study *in vivo* physiological processes in both animals and humans [171,202]. The process is an imaging technique that provides quantitative, regional measurements, and kinetics of specific biochemical and physiological processes in living animals or human subjects. PET is similar to x-ray computerized axial tomography and magnetic resonance imaging in that images of cross-sectional slices of the body are produced. The technique involves the use of a substance with the desired biological activity containing a positron-emitting radioactive isotope. Positron-emitting nuclides are neutron deficient compared to their stable isotopes and decay by spontaneous conversion of a proton to a neutron. This conversion is accompanied by release of a positron, which travels a small distance before encountering an electron, resulting in antimatter–matter annihilation. This annihilation event releases energy in the form of two 511-KeV tissue penetrating γ-ray photons radiating at approximately 180° from one another. A circular array of scanners consisting of scintillation crystals arranged so opposing crystals are grouped in coincidence circuits are placed around the subject to detect the paired γ rays as they simultaneously arrive on opposite sides (see Figure 41.35) [64]. The PET scanner's coincidence circuits enable the localization of the source of each annihilation. A computer

**TABLE 41.6**
**Positron-Emitting Isotopes Commonly Used in PET**

| Nuclide | Half-Life (min) | Carrier-Free Specific Activity (Ci/mol) |
|---|---|---|
| $^{11}$C | 20.40 | $9.22 \times 10^3$ |
| $^{13}$N | 9.96 | $1.89 \times 10^4$ |
| $^{15}$O | 2.07 | $9.10 \times 10^4$ |
| $^{18}$F | 109.70 | $1.7 \times 10^3$ |

then uses this information to reconstruct an image of radionuclide distribution within the body. Several million coincidences may be assimilated during a 1- to 15-minute scan interval. In this way, a tomographic image can be obtained, illustrating the spatial distribution of the radionuclide. When images are recorded at appropriate intervals after the administration of the radionuclide, quantitative measurements reflecting the dynamic process under study can be obtained. Images can also be color coded to show differences in the levels of activity from one time point to the next.

## Positron-Emitting Isotopes

The radionuclides most commonly used in PET are carbon-11, nitrogen-13, oxygen-15, and fluorine-18. All of these isotopes decay exclusively by positron emission, producing readily detectable tissue-penetrating γ-ray photons, but have relatively short half-lives (see Table 41.6). Their brief existence means that they must be manufactured close to the detection site, and, more importantly, fast chemical reaction techniques and isolation procedures must be developed to ensure the production of a useful radionuclide. In addition, the use of such labeled compounds is limited to biochemical processes with rapid rates of turnover.

The production of positron-emitting isotopes is normally carried out using onsite medical cyclotrons immediately prior to use, via the transmutation of stable isotopes; for example, $^{11}$C is usually prepared by irradiating nitrogen or boron with accelerated protons, $^{15}$O by irradiating nitrogen with deuterons, and $^{13}$N by irradiating nitrogen with deuterons or irradiating carbon with deuterons. In the case of the longer lived $^{18}$F, stocks may be obtained from a regional distribution center. At sites remote from an available cyclotron, a radioactive generator system may be used to derive short-lived isotopes from longer lived radioactive nuclides as they decay.

Although in some cases the positron-emitting isotope can be used directly in its elemental form (e.g., $^{15}$O can be directly used in metabolic studies, and $^{13}$N is routinely used for studying lung ventilation), its incorporation into an unlabeled precursor molecule represents a formidable challenge for the imaginative synthetic chemist. In addition to the usual risk of working with body-penetrating radiation, the chemist must design suitable synthetic procedures that will proceed rapidly and in high radiochemical yield, avoid unintentional dilution of the radionuclide by stable carrier, and avoid difficult and time-consuming separation procedures. The number of steps in the synthetic process should be minimal to avoid working with large quantities of radioactivity and to ensure compatibility with the half-life of the isotope.

A description of the successful synthesis of L-[$^{13}$N]-tyrosine serves to illustrate the above points [64]. A mixture of [$^{13}$N]-nitrate and [$^{13}$N]-nitrite obtained from an onsite cyclotron is converted to [$^{13}$N]-$NH_3$ by reduction with Devarda's alloy/NaOH reagent. The product is collected by distillation and immediately reacted with α-ketoglutarate and NADH. The reaction mixture is then passed through a column containing glutamate dehydrogenase immobilized on a solid support to give [$^{13}$N]-glutamate. Passage of this product down a second column containing immobilized glutamate oxaloacetate transaminase transfers the [$^{13}$N]-$NH_2$ group from glutamate to *p*-hydroxyphenylpyruvate to give L-[$^{13}$N]-tyrosine in radiochemically pure form and 28% overall radiochemical yield. The total synthesis time is 25 minutes.

## Instrument Design

A functional PET unit consists of a data-acquisition system and a data-processing computer. The acquisition system incorporates the radiation detectors, their associated circuitry, and in some designs a mechanical system that imparts a small motion to the detectors to obtain better sampling. Data from the acquisition system are assimilated in a computer, and a display system for immediate viewing and recording of the image is available, with interactive capabilities for image analysis.

## Specific Applications

Positron emission tomography is an extremely versatile method for probing fundamental biochemical processes in living systems. Important applications include metabolism, drug disposition and pharmacokinetics, and physiological and neurochemical mechanisms. In the brain, PET has been used to measure regional blood flow [179], glucose utilization, blood volume, oxygen utilization, the oxygen extraction ratio, the permeability-surface area product for water, acid–base chemistry, protein synthesis, and the characterization of dopamine $D_2$, benzodiazepine, and opiate receptors. PET measurement of regional rates of cerebral glucose metabolism utilizes [$^{18}$F]-2-fluoro-2-deoxyglucose (FdG) as the tracer component [74,160]. This procedure can quantitatively determine glucose utilization, because FdG, unlike dG, cannot be biotransformed further after its initial conversion to FdG-6-phosphate (FdG-6P) by hexokinase, due to the absence of an hydroxyl group at C-2; thus, FdG-6P (and dG-6P) are metabolically trapped within the cell. Regional accumulation of FdG radioactivity is therefore proportional to blood glucose and is a sensitive index of brain function.

Regional blood flow and glucose utilization are established markers of local neuronal activity, and PET measurement of these parameters has been used to identify regions of the brain that are activated during visual, auditory, and somatic sensations; limb and eye movements; speech; semantic associations; and pathologic and normal forms of anxiety. Such parameters are also being used to identify regional brain abnormalities in patients with schizophrenia, panic disorders, epilepsy, Parkinson's and Huntington's diseases, obsessive–compulsive disorder, and clinical depression. An FdG–PET study in patients with acquired immune deficiency syndrome (AIDS) has focused on improving the diagnosis of the AIDS dementia complex (ADC), the major neurological outcome of AIDS that results in behavioral, cognitive, and motor disturbances [64]. In this study, progressive development of ADC correlated well with regional rates of cerebral glucose metabolism, particularly in the basal ganglia and thalamus. PET studies have also indicated that patients with Alzheimer's disease exhibit characteristic patterns of cerebral blood flow and glucose utilization; glucose metabolism in such patients is significantly reduced in the posterial parietal temporal region of the brain, compared to control patients [95]. PET images of cerebral glucose utilization have also been used to localize the epileptogenic foci in patients with intractable partial epilepsy.

The use of PET with radiolabeled synthetic drugs or radiolabeled analogs of naturally occurring neurotransmitters may lead to more reliable diagnoses of neurological disorders, perhaps even presymptomatic diagnosis. *N*-methyl-[$^{18}$F]-spiroperidol (NMSP), a neuroleptic drug with high affinity for dopamine-$D_2$ receptors, has been used to visualize brain receptors *in vivo* in both normal and schizophrenic patients. The effectiveness of the anti-Parkinson drug, L-deprenyl, as a monoamine oxidase inhibitor in the brain has been verified from PET studies with $^{11}$C-labeled L-deprenyl [161]. Ricaurte and colleagues [172] employed PET imaging of a $^{11}$C-labeled serotonin transporter ligand to assess the long-term neurotoxicity of the recreational drug MDMA (Ecstasy) in humans. Here, reduced serotonergic receptor binding positively correlated with the extent of previous MDMA use.

Other studies [123,124] have shown that [$^{18}$F]-4′-fluoroclebopride ([$^{18}$F]-FCP) is a most promising candidate ligand for studying $D_2$ receptors with PET, and [$^{18}$F]-(+)-fluorobenzyltrozamicol ([$^{18}$F]-(+)-FBT) is a suitable ligand for studying cholinergic terminal density with PET via the vesicular acetylcholine transporter (125). [$^{18}$F]-FCP has been used in PET studies designed to measure synaptic dopamine levels and $D_2$ receptor numbers in Rhesus monkeys after treatment with psychostimulants such as (–)-cocaine, *d*-amphetamine, methylphenidate, and *d*-methamphetamine [123]. In these studies, psychostimulants caused an increase in the rate of washout of [$^{18}$F]-FCP from the basal ganglia. *d*-Methamphetamine and *d*-amphetamine had the greatest effect on washout kinetics of [$^{18}$F]-FCP relative to (–)-cocaine and methylphenidate, which was consistent with their ability to elevate synaptic dopamine levels. Thus, challenge studies with [$^{18}$F]-FCP may be a useful technique for studying the dynamics of interaction between psychostimulant-induced increases in synaptic dopamine and postsynaptic $D_2$ receptors.

Positron emission tomography and [$^{18}$F]-FCP analysis have been utilized in chronic stress studies. Chronic stress results in heightened synaptic dopaminergic levels and a concomitant downregulation of $D_2$ receptors, analogous to the downregulation of dopaminergic receptors found following chronic cocaine exposure. Experiments with [$^{18}$F]-FCP in dominant and subordinate cynomologus monkeys showed a clearly greater uptake of the ligand in the basal ganglia by the dominant monkeys [73]. The data provide strong evidence that stimuli controlling behavior/physiological consequences (stress) have a neurochemical correlate that can be imaged with PET techniques and suggest that chronic stress results in heightened synaptic dopaminergic levels and a concomitant downregulation of $D_2$ receptors. Previous PET studies have found downregulation in dopaminergic receptors in human cocaine addicts [212,213] and alcoholics [87]. PET may also be a powerful tool in addressing the etiology of complex behavioral disorders associated with stress.

[$^{18}$F]-(+)-fluorobenzyltrozamicol has high affinity for the vascular acetylcholine transporter and low affinity for $\sigma_1$ and $\sigma_2$ receptors [125]. Its high uptake in the basal ganglia and reversible binding kinetics *in vivo* indicate that this tracer can provide quantitative measurements of vesicular acetylcholine transporter function *in vivo* with PET imaging. The vesicular acetylcholine transporter is a marker for cholinergic nerve terminals and may be directly related to cholinergic neuronal density. An important goal in PET studies is relating cholinergic neuronal density to cognitive performance. Such studies may be able to correlate an individual's change in behavioral performance (i.e., cognition) with a change in (reduction in) cholinergic function.

Positron emission tomography scanning of the torso is routinely employed in the diagnostic evaluation of patients with certain disorders of the heart. Modeling metabolic processes in the heart is more complex than in the brain, because the heart draws energy from several substrate pools, such as fatty acids, carbohydrates, and lactate. FdG–PET studies and blood flow [$^{13}$N]-ammonia–PET measurements have been used to diagnose coronary artery disease with better than 95% accuracy. Other studies have utilized [$^{11}$C]-palmitate, which is accumulated homogeneously by the heart muscle, to diagnose cardiovascular abnormalities [64]. Whole-body copper flux following transgastric injection was assessed in Long–Evans Cinnamon (LEC) rats with a defective copper transporter using $^{64}$Cu and PET imaging [17]. The authors concluded that increased biliary secretion may not provide sufficient protection against high doses of orally administered copper, whereas the trapping of copper in the upper gastrointestinal tract may be critical to reducing toxicity.

Positron emission tomography images of glucose utilization and amino acid uptake are particularly useful for examining patients with a variety of tumors throughout the body. PET images of labeled estradiol uptake are used to examine patients with primary and metastatic breast tumors.

An established technique in diagnostic medicine, PET is now widely used for neuroscientific, pharmacological and toxicological studies *in vivo*. Two major obstacles to more general use of this technique are its cost effectiveness and the availability of appropriate radiotracers. The synthesis of novel positron-emitting drug molecules remains a challenge for the radiochemist. Nevertheless, PET has advantages over other functional imaging techniques such as single-photon emission tomography (SPECT). Unlike magnetic resonance imaging, it has the potential to measure characteristics of biological compounds, such as neurotransmitters and neuroreceptors, that exist in minute concentrations. Improvements in the methodology for determining anatomical localization, data analysis, and the performance of PET systems are ongoing. This important analytical tool continues to offer unique opportunities for researchers in the study of physiology, biochemistry, pharmacology, and toxicology.

## Stable Isotopes

Most elements exist as several isotopes that vary in atomic weight. Some of the heavy isotopes are stable and do not decay with the emission of radiation. These so-called *stable isotopes* typically have a low natural abundance and chemical and physical properties that are nearly identical to those of their more abundant counterparts. Carbon exists naturally as a mixture of isotopes, $^{12}$C, $^{13}$C, and $^{14}$C. $^{12}$C is the most abundant isotope of carbon (98.89%); $^{13}$C, the stable isotope of carbon, is much less abundant (1.11%); and $^{14}$C is a radioisotope of carbon with a natural abundance so low as to be insignificant (except for its special application for carbon dating). Hydrogen also exists as a mixture of isotopes, $^{1}$H, $^{2}$H, and $^{3}$H. $^{1}$H (protium) is the most abundant isotope of hydrogen (99.985%); $^{2}$H (deuterium), the stable isotope of hydrogen, is much less abundant (0.015%); and $^{3}$H (tritium) is a radioactive isotope of hydrogen with insignificant natural abundance. Other stable isotopes that are useful for toxicology research include $^{15}$N (0.37%), $^{17}$O (0.037%), and $^{18}$O (0.204%). A wide variety of chemicals specifically labeled with $^{2}$H, $^{13}$C, $^{15}$N, $^{17}$O, and $^{18}$O at high atom-% enrichments are commercially available.

Stable isotopes with low natural abundance make ideal tracers for studying mechanism of toxicology. Because they are stable, they pose no significant health risk to the researcher or the experimental subject. No special license is required for the acquisition and use of stable isotopes, and no special precautions are necessary for their safe handling (other than the normal precautions used for handling the unlabeled compound). Because their chemical and physical properties are very similar to those of the more abundant isotope, their behavior *in vivo* is also quite similar. Compounds enriched in a stable isotope can be studied in the presence of natural abundance compounds with minimal background interference. Finally, a wealth of mechanistic information is available from stable isotope studies through the application of modern magnetic resonance and mass spectral techniques.

### Choice and Location of Label

The stable isotope of choice for a toxicology study depends on the nature of the study (ADME, exposure assessment, mechanism of toxicity) and the instrumental method that will be used for analysis. The most common stable isotopes for toxicology research are $^{13}$C, $^{2}$H, $^{15}$N, and $^{18}$O. Stable isotopes of heavy metals such as $^{199}$Hg and $^{206}$Pb have been used to study mechanisms of toxicity and clearance. If NMR is used for analysis, it is best to use a spin 1/2 isotope for maximum sensitivity. For $^{13}$C-NMR analysis, uniform labeling offers distinct advantages in terms of metabolite characterization as will be discussed further below. For mass spectral analyses, the stable isotope that will give the greatest mass difference between natural abundance and labeled analytes is appropriate. Multiple sites of labeling are quite common and provide greater mass discrimination. If GC–MS is used, the optimum mass difference is 2 to 4 u. A greater mass difference can lead to altered retention times and introduce uncertainty in quantitative analysis. Unlike radioisotopes where enrichments are quite low, stable isotopic enrichments are typically high, at least 90 atom-%. In general, the factors that govern the position of a radioisotopic label (stability and exchangeability) also govern the position of a stable isotopic label.

## Applications

### Pharmacokinetics and Metabolism of Isotopically Labeled Nicotine

Due to the widespread use of tobacco products, the pharmacokinetics of nicotine and its metabolites have been the subject of much study. Early studies focused on nicotine and its major metabolites cotinine and nicotine-1′-oxide. Kyerematen et al. [111] reported a pharmacokinetic study of nicotine and 12 of its metabolites in the rat. They administered $^{14}C$-labeled nicotine in serial intraarterial doses and collected blood for up to 30 hours and urine up to 120 hours. Nicotine and 12 of its metabolites were quantified by HPLC with radiometric detection. One previously unidentified metabolite, collected from the HPLC and analyzed by EI mass spectrometry, was identified as allohydroxy-demethylcotinine. Plasma nicotine half-life, total body clearance, and apparent volume of distribution and half-lives of urinary excretion of cotinine, cotinine-*N*-oxide, and allohydroxy-demethylcotinine were all determined. Radiolabeled nicotine afforded convenient determination of nicotine and 12 of its metabolites with high sensitivity, demonstrating the power of this method for the study of nicotine pharmacokinetics.

Kyerematen et al. [110] also determined the disposition of radiolabeled nicotine and 8 of its metabolites in humans. The use of radiolabeled nicotine enabled detection of plasma and urinary nicotine and metabolites from a very low dose (190 μg per subject) so a comparison of smokers and nonsmokers was possible. Two new urinary metabolites, 3′-hydroxycotinine glucuronide and demethylcotinine $\Delta^{2',3'}$-enamine, were identified by mass spectral analyses.

The ability to distinguish labeled from unlabeled nicotine by GC–MS enabled the study of nicotine pharmacokinetics and bioavailability in humans [14]. After smoking a cigarette, smokers received nicotine labeled with deuterium in both 3′ positions (previously shown to be stable to metabolism). Plasma levels of both labeled and unlabeled nicotine were monitored by GC–MS with selected ion monitoring. The sensitivity of the analytical method and the specificity of the labeled nicotine that eliminated background interference from unlabeled nicotine gave high-quality pharmacokinetic data.

Metabolomic analysis is also facilitated using these techniques. At the microbial scale, a mixture of uniformly labeled $^{13}C$-labeled metabolites was extracted from yeast grown on $^{13}C$-labeled substrates [228]. The labeled metabolites were then used as internal standards for the LC–MS analysis of the glycolytic and TCA cycle metabolites produced in yeast grown under varied conditions. The precision and efficiency of cellular extraction, sample processing, and quantification of metabolite recovery for each metabolite were significantly improved when using labeled metabolites as internal standards.

### Characterization of Urinary Metabolites

Detection of $^{13}C$ resonances of natural abundance xenobiotic metabolites in a complex matrix such as urine is often difficult due to interference from endogenous compounds. Appropriate use of $^{13}C$-enriched xenobiotics can facilitate the characterization of metabolites in whole urine by $^{13}C$ NMR spectroscopy with little or no sample cleanup. In higher organisms, the characterization and quantification of dynamic renal xenobiotic metabolism have been extensively explored using $^{13}C$-labeled compounds coupled with NMR and other analytical methods [9].

Isotopically labeled $^{13}C$-benzene was employed to distinguish benzene metabolites produced from known benzene exposure from background levels typically present in the urine [218]. Human volunteers were exposed to an environmentally relevant dose of benzene, 40 ± 10 ppb, for 2 hours. Benzene metabolites, readily discernible from unlabeled background metabolites, were then quantified using GC–MS. The percentage of muconic acid metabolites excreted was higher than that reported after exposure to occupational levels. The authors suggest that such stable isotope studies may be useful for validating toxicokinetic models and occupational exposure risk extrapolation.

Sumner et al. [198] studied the metabolism of [1,2,3-$^{13}C$]-acrylamide in rats and mice using one-dimensional and two-dimensional $^{13}C$ NMR spectroscopic analysis of urine. The animals received the labeled xenobiotic orally. Urine was collected for 24 hours and then centrifuged, and $D_2O$ was added to a final concentration of 15%. INADEQUATE spectra were used to correlate all carbon signals. HET2DJ spectra revealed the number of protons attached to each carbon for all metabolites. These spectral data, along with calculated chemical shifts of proposed metabolites and spectra of synthetic standards, were used to identify five urinary metabolites of acrylamide. Quantification of the urinary output of acrylamide and the five metabolites was accomplished by integration of the carbons signals in one-dimensional spectra relative to dioxane, an internal standard. Approximately 50% of the administered dose of acrylamide was recovered as the five metabolites, confirming a previous study wherein 62% of the administered dose of $^{14}C$-labeled acrylamide was recovered in urine within 24 hours. Stable isotope labeling and $^{13}C$ NMR analysis enabled the characterization and quantification of xenobiotic metabolites in a complex matrix without the need for tedious sample cleanup or chromatographic separation.

### Determination of Toxic Metals Using Stable Isotope Dilution Mass Spectrometry

Stable isotopes are widely employed in elemental mass spectrometry studies of toxic elements. Smith et al. [190,191] used stable lead isotopes to study the effectiveness of a chelating agent (DMSA) in removing lead from skeletal vs. soft tissue and the redistribution of lead in other tissues. Lead has four naturally occurring isotopes: $^{204}Pb$ (1.4%), $^{206}Pb$ (24.1%), $^{207}Pb$ (22.1%), and $^{208}Pb$ (52.4%). By enriching drinking water given to rats with $^{206}Pb$, ingested lead was distinguishable from the endogenous lead already present in the animals. Here, the mass spectrometer used thermal emission to produce ions and measure the isotopic ratios, thereby quantifying the $^{206}$Pb dose.

Cadmium is another highly toxic metal. Environmental and industrial cadmium exposure is of great concern. In contrast to the previous thermal emission mass spectrometric method for lead determination, a stable isotope dilution GC–MS method was employed for assessing cadmium exposure. Urinary cadmium was determined by using the chelating agent lithium *bis*(trifluoroethyl)dithiocarbamate (Li(FDEDTC)) to form the chelate $Cd(FDEDTC)_2$. The chelate was then analyzed by GC–MS with EI ionization, where $M^{+\cdot}$ was monitored for quantitation. Precise, accurate cadmium detection in urine at the 10-mg/L level was achieved. This isotope dilution technique eliminated matrix effects so precision and accuracy were not affected by incomplete recovery.

## INTEGRATION OF TECHNIQUES IN TOXICOLOGY

Mechanisms of xenobiotic toxicity are usually complex, involving not only the parent compound but also a broad array of metabolites and tissue derived endogenous compounds. All of these materials are produced in a dynamic process as the intoxicated organism responds to the toxic insult and attempts to maintain homeostasis. Modern instrumental methods are powerful tools that have their greatest utility when they are used in conjunction with one another in a synergistic fashion to characterize materials in complex mixtures. Numerous examples are available in the literature of the use of multiple instrumental techniques to describe a variety of different chemical entities in complex mixtures.

The forensic toxicologist often conducts hair analysis to assess both acute and chronic exposure to drugs and toxins. Newborn hair analysis may be particularly suited for studying both fetal and maternal exposure to xenobiotics. Tagliaro [199] and colleagues assayed thyroxine concentration in newborn hair by first extracting the hair, fractioning it with HPLC, and then using a radioimmunoassay to quantify the analyte. The amphetamine and alkaloid drug content in hair samples was similarly assessed by derivatizing the amines in alkaline hair extracts to produce trichloroethylcarbamates, followed by combined GC–MS analysis of the derivatives [67]. Gustafson et al. [78] simultaneously determined plasma levels of THC and two of its metabolites, first by enzymatically cleaving their glucuronide conjugates, followed by solid phase extraction, separation, and, finally, quantification by GC–MS. Given the large number of biomarkers and dynamic concentration ranges normally present in a metabolomic fingerprint, integrated separation and analytical techniques are commonly employed for mapping these complex mixtures. A method for multicomponent analysis has been developed wherein rat urinary metabolite profiles correlated with toxin exposure may be accurately determined by combining HPLC separation with simultaneous electrochemical redox array and MS analysis to identify intoxicated animals [69].

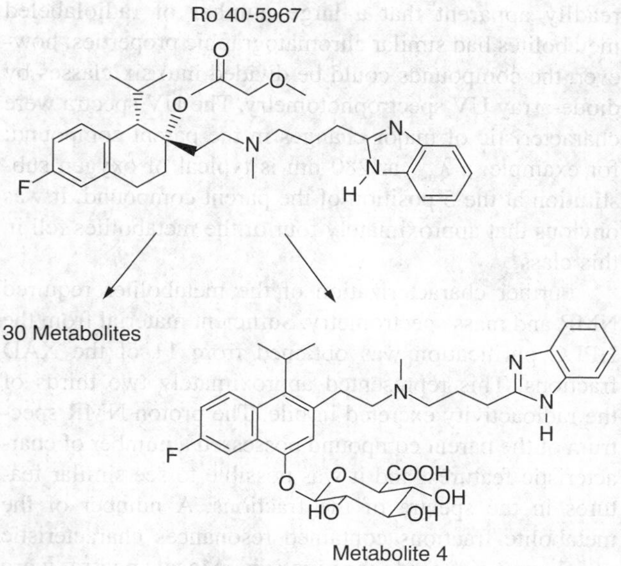

**FIGURE 41.36** Metabolism of the calcium antagonist Ro 40-5967 in the rat; characterization of metabolite 4 exemplifies the integration of modern instrumental techniques to study mechanisms of toxicology.

In one exemplary study, many of the techniques that have been described in this chapter were used to confirm the identity of many compounds. Wiltshire et al. [225] used diode-array ultraviolet spectrophotometry, mass spectrometry, NMR, radiolabeled tracers, open column chromatography, HPLC, and TLC to characterize the complex metabolic pathway of a calcium antagonist (Ro 40-5967) in the rat (Figure 41.36). A classic xenobiotic disposition study was conducted using Ro 40-5967 labeled with $^{14}$C. Approximately 80% of the radiolabeled dose was recovered in all excreta including bile, urine, and feces; 37% of the dose was excreted in the bile so chemical characterization of the biliary radioactivity was carried out. Bile was partially purified by passing it through an open column of Amberlite XAD-2 resin. The fractions collected from the open column were evaporated to dryness and partitioned between ethyl acetate and water. Two of the fractions contained an insoluble gum that was taken up in ethanol. All fractions (aqueous and organic) were taken to dryness and then dissolved in aqueous methanol. These fractions were subjected to three sequential reversed-phase HPLC purification steps to obtain material that was suitable for further characterization. It was readily apparent that a large number of radiolabeled metabolites had similar chromatographic properties; however, the compounds could be divided into six classes by diode-array UV spectrophotometry. The UV spectra were characteristic of major changes in the parent compound; for example, a $\lambda_{max}$ at 280 nm is typical of oxygen substitution at the 5 position of the parent compound. It was obvious that approximately four of the metabolites fell in this class.

Further characterization of the metabolites required NMR and mass spectrometry. Sufficient material from the HPLC purification was obtained from 11 of the XAD fractions. This represented approximately two thirds of the radioactivity excreted in bile. The proton NMR spectrum of the parent compound possessed a number of characteristic features, and it was possible to see similar features in the spectra of the fractions. A number of the metabolite fractions contained resonances characteristic of glucuronic acid conjugation. Mass spectra were obtained from 45 subfractions of the original XAD fractions by thermospray LC–MS. As with the proton NMR spectra, features in the mass spectra were characteristic of the parent compound; for example, *m/z* 406 was characteristic of the loss of the side chain.

Rather than discussing the entire characterization that involved the description of 31 metabolites, it is instructive to describe the characterization of just one of the metabolites. Metabolite 4 was the major biliary metabolite and equivalent to about 25% of the total biliary radioactivity. It had a class 3 UV spectrum, indicating the presence of a substituted naphthyl group. Remember that the parent compound contains a tetrahydronaphthyl ring system where one of the rings of naphthylene is saturated. The very strong double peak at 240 to 250 nm is characteristic of a fully aromatic substituted naphthylene.

The proton NMR spectrum of this metabolite showed the presence of glucuronic acid as well as loss of the ester side chain. Evidence that the asymmetric center of the isopropyl side chain had been lost was obtained from a shift in the methyl resonances of the isopropyl group. Aromatization of the saturated ring was obvious by the addition of an aromatic singlet at 7.1 ppm. The metabolite structure was deduced to be an oxyglucuronide derivative of the parent compound in which the alicyclic ring had been aromatized. This was consistent with a mass spectrum containing a protonated molecular ion at *m/z* = 596, loss of the benzimidazole group (*m/z* = 438), and loss of glucuronic acid (*m/z* = 262).

This integrated application of instrumental techniques permitted a thorough characterization of metabolites that varied greatly in chemical structure and polarity. A total of 31 biliary metabolites, representing 67% of the excreted dose of Ro 40-5967, were characterized. Only through the integration of modern instrumental methods could this complex metabolic pathway be thoroughly elucidated.

## QUESTIONS

1. Name four ways that mass spectrometry has been used to assess exposure to xenobiotics.

   *Answer:* Mass spectrometry has been used to determine metabolites of xenobiotics in biological fluids, to analyze the metabolomic profile of heavy metal exposure, to determine DNA adducts formed as a result of toxin exposure, and for the rapid identification of intoxicants.

2. An unknown compound in an organic extract of urine is sufficiently volatile to pass through a GC–MS system but fragments extensively under EI such that no $M^{+\bullet}$ is observed. Name at least two alternative approaches to determine the molecular weight of this compound using mass spectrometry.

   *Answer:* (a) A GC–MS system with PICI could be used to form $(M+H)^+$. (b) An HPLC method could be developed and the material analyzed by LC–MS using a soft ionization method such as ES or APCI. (c) If the compound contained a hydroxyl or amine group, derivatization with a common trimethylsilation reagent could form a compound with a more stable molecular ion.

3. An LC–MS method with an ES ionization source runs a binary gradient with mobile phase 1 = 2-m*M* ammonium acetate and mobile phase 2 = 90% methanol:10% acetonitrile. There is a large background ion in the system at *m/z* 59. CID on this ion produces a fragment at *m/z* 42. What are likely assignments for these ions?

   *Answer:* The *m/z* 42 is protonated acetonitrile. The mass difference of 17 Da (59 – 42 ) suggests ammonia ($NH_3$) as a neutral loss; thus, *m/z* 59 is likely an ammonium adduct of acetonitrile, $(CH_3CN)(NH_4)^+$, produced from background mobile-phase components.

4. EI mass spectrometers have several ions constantly present in the background mass spectrum at *m/z* 18, 28, 32, 40, and 44. Identify these components.

   *Answer:* These are all components of air with moisture present: *m/z* 18 = $H_2O$; *m/z* 28 = $N_2$, *m/z* 32 = $O_2$, *m/z* 40 = Ar, and *m/z* 44 = $CO_2$.

5. Using positive ES, a large protein is ionized under acidic conditions that create an $(M+4H)^{4+}$ ion in the mass spectrum at *m/z* 712.2. What is the molecular weight of this compound?

   *Answer:* Remember that four protons have been attached to this molecule, so $m/z = 712.2 = (M+4H)/4, \rightarrow M = (712.2 \times 4) - 4; M = 2844.8$ Da.

6. $CDCl_3$ is a commonly used NMR solvent. Carbon NMR spectra of samples dissolved in $CDCl_3$ contain a residual solvent peak that is split due to $^{13}C$–$^{2}H$ coupling.
   a. Into how many lines is this signal split?

      *Answer:* $^{2}H$ has a spin quantum number ($I$) of 1. Applying the $2nI + 1$ rule, the number of lines is $(2 \times 1 \times 1) + 1 = 3$.
   b. In most cases, carbon spectra are acquired with hydrogen decoupling. Why isn't the $^{13}C$–$^{2}H$ multiplet reduced to a singlet under the conditions of this decoupling?

      *Answer:* Decoupling is applied at the $^{1}H$ (protium) frequency. Deuterium ($^{2}H$) has a very different precessional frequency; therefore, it is not decoupled in the carbon NMR spectrum.
7. Why is $^{13}C$–$^{13}C$ coupling rarely seen in carbon NMR spectra?

   *Answer:* Coupling is only seen for closely spaced nuclei (fewer than four to five bonds). To see $^{13}C$–$^{13}C$ coupling, two $^{13}C$ atoms would have to be in close proximity. The natural abundance of $^{13}C$ is only 1%, so the likelihood of two $^{13}C$ nuclei being in close proximity is very low. Carbon spectra of molecules composed of natural abundance carbon do not show $^{13}C$–$^{13}C$ coupling. $^{13}C$–$^{13}C$ coupling is only seen with isotopically enriched molecules.
8. Sensitivity in NMR depends on what factors?

   *Answer:* (a) Magnetic field strength; (b) magnetogyric ratio; and (c) natural abundance of the spin-active nucleus.
9. Describe three approaches for increasing sensitivity for a given nucleus in NMR?

   *Answer:* (a) Using a higher field-strength instrument (magnet); (b) using a smaller NMR probe or a selectively tuned probe; and (c) signal averaging (FT-NMR).
10. Why are chemical shifts expressed in dimensionless units designated δ?

    *Answer:* Chemical shifts vary with the strength of the applied magnetic field. To enable comparisons of chemical shifts from one instrument to the next, chemical shifts are expressed in δ units.
11. What factors limit time resolution in online LC–NMR?

    *Answer:* (a) HPLC mobile-phase flow rate; (b) total time required to obtain one transient (scan); and (c) total number of transients acquired per LC peak.
12. Why is EPR inherently more sensitive than NMR?

    *Answer:* The magnetic moment of an electron is 700 times greater than that of a proton. This means that the energy difference between spin states is greater for electrons than for protons. This greater energy difference leads to higher sensitivity. It also means that the energy required for an EPR transition is higher than that required for a NMR transition. NMR uses energy in the radiofrequency range; EPR uses energy in the microwave range.
13. The EPR spectrum of the Fe–DETC complex with $^{15}NO$ shows doublet hyperfine splitting.
    a. Why?

       *Answer:* $^{15}N$ has a spin quantum number ($I$) of 1/2. Applying the $2nI + 1$ rule, ($n = 1$, $I = 1/2$), the number of lines is 2 (doublet).
    b. How was this used to distinguish between NO produced from $^{15}N$-labeled arginine and NO produced from endogenous arginine?

       *Answer:* $^{14}N$ has a spin quantum number of 1; therefore, it splits electron signals into three lines (triplet hyperfine splitting). Endogenously formed NO is derived from natural abundance arginine. Natural abundance nitrogen is 99.6% $^{14}N$, so it leads to triplet hyperfine splitting. Any doublet splitting results from NO produced from added $^{15}N$-labeled arginine.
14. Quantitative whole-body autoradiography (QWBA) is becoming a very popular method for determining xenobiotic tissue distribution.
    a. What is the major limitation of this technique?

       *Answer:* QWBA is used to determine total radioactivity. It cannot distinguish between different chemical forms of radioactivity; in other words, it cannot distinguish between xenobiotic metabolites. It determines the sum of parent compound plus metabolites and, therefore, yields no information on xenobiotic metabolism.
    b. In practice, how can this limitation be overcome?

       *Answer:* To overcome this limitation, follow-up studies must be performed. QWBA is used to determine the distribution of a xenobiotic. QWBA data reveal the tissues containing xenobiotic-derived radioactivity. The follow-up studies involve administration of the xenobiotic (radiolabeled or natural abundance) and isolation of tissues containing the parent xenobiotic and its metabolites.

Extraction of these tissues provides solutions that can be analyzed for metabolites by suitable separation techniques such as LC–MS or LC–NMR.

15. What effect, if any, would you expect to observe in the UV absorbance spectrum of an ethanol solution of each of the following compounds: phenol, aniline, and *trans*-cinnamic acid, after the addition of a drop of 1-*N* aqueous sodium hydroxide solution to the UV cuvette containing the solution of each compound?

    *Answer:* (a) Phenol—Addition of 1-*N* NaOH will convert phenol (PhOH) to the phenolate ion ($PhO^-$), leading to a formal negative charge on the auxochromic group attached to the phenyl ring chromophore. The interaction of these electrons with the conjugated phenyl ring produces a bathochromic–hyperchromic shift; that is, the absorption band of phenol shifts to a longer wavelength, and the extinction coefficient increases in value. (b) Aniline—Addition of 1-*N* NaOH will have no significant effect on the absorption spectrum of aniline, as the alkaline medium will not alter the structure of the molecule. (c) *trans*-Cinnamic acid—Addition of 1-*N* NaOH will generate the carboxylate ion of this organic acid ($R–COO^-$). As with phenol, this will result in a bathochromic–hyperchromic shift in the absorption band of *trans*-cinnamic acid.

16. Under what circumstances may the UV absorbance of organic molecules deviate from Beer's law?

    *Answer:* The UV absorbance of organic molecules may not obey Beer's law if (a) protonation or deprotonation, (b) tautomerization, (c) dimerization, or (4) complex formation of the analyte occurs in solution.

17. Even relatively simple, low-molecular-weight compounds exhibit an abundance of absorption bands in their IR spectrum. What factors influence the number and value of the IR absorption bands in organic molecules?

    *Answer:* The number of fundamental modes of vibrations (i.e., stretching and bending vibrations) is $(3n - 6)$ for nonlinear polyatomic molecules and $(3n - 5)$ for linear molecules, where $n$ is the number of atoms in the molecule. The factors that influence the value of IR absorption frequencies of chemical bonds in organic molecules are (a) the masses of the two atoms in the bond, (b) the velocity of light, and (c) the force constant of the bond.

18. Discuss the usefulness of group frequencies in the IR spectrum of unknown organic molecules.

    *Answer:* Group frequencies are characteristic absorption frequencies of certain functionalities within an organic molecule which can be used diagnostically to confirm the presence (or absence) of the functionality in an unknown molecular structure. Examples of such functionalities are C=O, C≡N, C≡C, –OH, and $–NH_2$. The frequencies of these structural moieties are usually independent of the structure of the rest of the molecule.

19. What is one of the major advantages of using Raman spectroscopy compared to IR spectroscopy for studying biochemical processes?

    *Answer:* One of the major advantages of using Raman spectroscopy in biochemical studies is that spectra can be obtained for molecules in aqueous solution, as water exhibits a weak Raman spectrum that interferes only minimally with the spectrum of the analyte.

20. Which general class of nuclides is useful in PET scanning, and which isotopes are commonly utilized?

    *Answer:* Position-emitting radioactive nuclides, which are neutron deficient, are used in PET scanning. These isotopes are short lived and decay by spontaneous conversion of a proton to a neutron. This process is accompanied by release of a positron, which collides with an electron, resulting in annihilation and release of energy in the form of two tissue penetrating γ-ray photons. Common positron-emitting isotopes used in PET scanning are $^{11}C$, $^{13}N$, $^{15}O$, and $^{18}F$. These nuclides have half-lives ranging from 2 to 110 minutes.

## REFERENCES


1. Aggarwal, S. K., Orth, R. G., Wendling, J., Kinter, M., and Herold, D. A. (1993): Isotope dilution gas chromatography/mass spectrometry for cadmium determination in urine. *J. Anal. Toxicol.*, 17:5–10.
2. Alben, J. O. and Fiamingo, F. G. (1984): Fourier transform infrared spectroscopy. In: *Optical Techniques in Biological Research*, edited by D. L. Rousseau, pp. 133–179. Academic Press, New York.
3. Alger, R. S. (1968): *Electron Paramagnetic Resonance, Techniques, and Applications*. Wiley-Interscience, New York.
4. Assenheim, H. M. (1967): *Introduction to Electron Spin Resonance*. Plenum Press, New York.
5. Aust, S. D., Chignell, C. F., Bray, T. M., Kalyanaraman, B., and Mason, R. P. (1993): Contemporary issues in toxicology: free radicals in toxicology. *Toxicol. Appl. Pharmacol.*, 120:168–178.

6. Baena, J. R. and Lendl, B. (2004): Raman spectroscopy in chemical bioanalysis. *Curr. Opin. Chem. Biol.* 8:534–539.
7. Barker, J. and Garner, R. C. (1999): Biomedical applications of accelerator mass spectrometry-isotope dilution measurements at the level of the atom. *Rapid Commun. Mass Spectrom.*, 13:285–293.
8. Baudouin, C. J., Bryant, D. J., Collins, A. G. et al. (1990): Aspects of chemical shift imaging which illustrate the cross-fertilization of methods and techniques in *in vivo* NMR imaging and spectroscopy. *Phil. Trans. R. Soc. Lond.*, 333:545–559.
9. Baverel, G., Conjard, A., Chauvin, M.-F., Vercoutere, B., Vittorelli, A., Dubourg, L., Gauthier, C., Michoudet, C., Durozard, D., and Martin, G. (2003): Carbon-13 NMR spectroscopy: a powerful tool for studying renal metabolism. *Biochimie*, 85:863–871.
10. Beaugrand, C. and Kostelitz, S. (2003): Field detection of chemical and biological compounds with mass spectrometry. *Spectra Analyse*, 32:26–30.
11. Beckonert, O. et al. (2003): NMR-based metabonomic toxicity classification: hierarchical cluster analysis and *k*-nearest-neighbour approaches. *Anal. Chim. Acta*, 490:3–15.
12. Bellar, T. A., Behymer, T. D., and Budde, W. L. (1990): Investigation of enhanced ion abundances from a carrier process in high-performance liquid chromatography particle beam mass spectrometry. *J. Am. Soc. Mass Spectrom.*, 1:92–98.
13. Benedetti, E., Teodori, L., Trinca, M. L. et al. (1990): A new approach to the study of human solid tumor cells by means of FT–IR microspectroscopy. *Appl. Spectrosc.*, 44:1276–1281.
14. Benowitz, N. L., Jacob III, P., Denaro, C., and Jenkins, R. (1991): Stable isotope studies of nicotine kinetics and bioavailability. *Clin. Pharmacol. Ther.*, 49:270–277.
15. Bernert, Jr., J. T., Turner, W. E., Pirkle, J. L. et al. (1997): Development and validation of sensitive method for determination of serum cotinine in smokers and nonsmokers by liquid chromatography/atmospheric pressure ionization tandem mass spectrometry. *Clin. Chem.*, 43: 2281–2291.
16. Birinyi-Strachan, L. C., Davies M. J., Lewis R. J., and Nicholson, G. M. (2005): Neuroprotectant effects of isoosmolar D-mannitol to prevent Pacific ciguatoxin-1 induced alterations in neuronal excitability: a comparison with other osmotic agents and free radical scavengers. *Neuropharmacology*, 49:669–686.
17. Bissig, K. D., Honer, M., Zimmermann, K., Summer, K. H., and Solioz, M. (2005): Whole animal copper flux assessed by positron emission tomography in the Long–Evans Cinnamon rat: a feasibility study. *Biometals*, 18:83–88.
18. Blakely, C. R., Carmody, J. J., and Vestal, M. L. (1980): A new liquid chromatograph/mass spectrometer interface using crossed-beam techniques. *Adv. Mass Spectrom.*, 8B:1616–1623.
19. Boersma, M. G., Solyanikova, I. P., Van Berkel, W. J. H., Vervoort, J., Golovleva, L. A., and Rietjens, I. M. (2001): F-19 NMR metabolomics for the elucidation of microbial degradation pathways of fluorophenols. *J. Indust. Microbiol. Biotechnol.*, 26:22–34.
20. Bollard M. E., Keun, H. C., Beckonert, O., Ebbels, T. M. D., Antti, H., Nicholis, A. W., Shockcor, J. P., Cantor, G. H., Stevens, G., Lindon, J. C., Holmes, E., and Nicholson, J. K. (2005): Comparative metabonomics of differential hydrazine toxicity in the rat and mouse. *Toxicol. Appl. Pharmacol.*, 204:135–151.
21. Brackett, D. J., Wallis, G., Wilson, M. F., and McCay, P. B. (1998): Spin trapping and electron paramagnetic resonance spectroscopy. *Meth. Mol. Biol.*, 108:15–25.
22. Bundy, J. G., Osborn, D., Weeks, J. M., Lindon, J. C., and Nicholson, J. K. (2001): An NMR-based metabonomic approach to the investigation of coelomic fluid biochemistry in earthworms under toxic stress. *FEBS Lett.*, 500:31–35
23. Burton, K. I., Everett, J. R., Newman, M. J., Pullen, F. S., Richards, D. S., and Swanson, A. G. (1997): Online liquid chromatography coupled with high field NMR and mass spectrometry (LC–NMR–MS): a new technique for drug metabolite structure elucidation. *J. Pharm. Biomed. Anal.*, 15:1903–1912.
24. Busch, K. L., Glish, G. L., and McLuckey, S. A. (1988): *Mass Spectrometry/Mass Spectrometry: Techniques and Applications of Tandem Mass Spectrometry.* VCH Publishers, New York.
25. Byrd, G. D. (1996): LC–MS/MS method for profiling nicotine and its metabolites in biological fluids. In: *Proc. of the 44th Annual Conf. on Mass Spectrometry and Allied Topics*, Portland, OR, May 12–16.
26. Byrd, G. D., Chang, K. M., Greene, J. M., and deBethizy, J. D. (1992): Evidence for urinary excretion of glucuronide conjugates of nicotine, cotinine, and *trans*-3′-hydroxycotinine in smokers. *Drug Metab. Dispos.*, 20:192–197.
27. Byrd, G. D., Davis, R. A., Caldwell, W. S., Robinson, J. H., and deBethizy, J. D. (1998): A further study of FTC yield and nicotine absorption in smokers. *Psychopharmacology*, 139:291–299.
28. Byrd, G. D., Robinson, J. H., Caldwell, W. S., and deBethizy, J. D. (1995): Comparison of measured and FTC-predicted nicotine uptake in smokers. *Psychopharmacology*, 122:95–103.
29. Byrd, G. D., Uhrig, M. S., deBethizy, J. D., Caldwell, W. S., Crooks, P. A., Ravard, A., and Riggs, R. M. (1994): Direct determination of cotinine-*N*-glucuronide in urine using thermospray liquid chromatography/mass spectrometry. *Biol. Mass Spectrom.*, 23:103–107.
30. Caldwell, W. S., Greene, J. M., Byrd, G. D. et al. (1992): Characterization of the glucuronide conjugate of cotinine: a previously unidentified major metabolite of nicotine in smokers' urine. *Chem. Res. Toxicol.*, 5:280–285.
31. Cammack, R. and Shergill, J. K. (1998): Biomedical applications of EPR spectroscopy. In: *Modern Applications of EPR/ESR: From Biophysics to Material Science. Proceedings of the First Asia-Pacific EPR/ESR Symposium*, edited by C. Z. Rudowicz, P. K. N. Yu, and H. Hiraoka, pp. 66–73. Springer-Verlag, Singapore.
32. Cammack, R., Shergill, J. K., Inalsingh, V. A., and Hughes, M. N. (1998): Applications of electron paramagnetic resonance spectroscopy to study interactions of iron proteins in cells with nitric oxide. *Spectrochim. Acta, Part A*, 54:2393–2402.

33. Cappiello, A. (1996): Is particle beam an up-to-the-date LC–MS interface? *Mass Spectrom. Rev.*, 15:283–296.
34. Caprioli, R. M. (1990): Continuous-flow fast atom bombardment mass spectrometry. *Anal. Chem.*, 62:477A–485A.
35. Carey, P. R. (1982): *Biochemical Applications of Raman and Resonance Raman Spectroscopies*. Academic Press, New York.
36. Carney, J. M., Landrum, W., Mayes, L., Zou, Y., and Lodder, R. A. (1993): Near-IR spectrophotometric monitoring of stroke-related changes in the protein and lipid composition of whole gerbil brains. *Anal. Chem.*, 65: 1305–1313.
37. Carroll, D. I., Dzidic, I., Horning, E. C., and Stillwell, R. N. (1981): Atmospheric pressure ionization mass spectrometry. *Appl. Spectrosc. Rev.*, 17:337–406.
38. Carter J. C., Brewer W. E., and Angel, S. M. (2000): Raman spectroscopy for the *in situ* identification of cocaine and selected adulterants. *Appl. Spectrosc.*, 54:1876–1881.
39. Cassis, L. A. and Lodder, R. A. (1993): Near-IR imaging of atheromas in living arterial tissue. *Anal. Chem.*, 65:1247–1256.
40. Chan, C. C., Shie, R. H., Chang, T. Y., and Tsai, D. H. (2006): Workers' exposures and potential health risks to air toxics in a petrochemical complex assessed by improved methodology. *Int. Arch. Occup. Environ. Health*, 79:135–142.
41. Charles, M. J. and Glish, G. L. (1995): Review of modern ion trap research. In: *Practical Aspects of Ion Trap Mass Spectrometry*, Vol. III, edited by R. E. March, and J. F. J. Todd, pp. 89–118. CRC Press, Boca Raton, FL.
42. Chen, M. C. and Lord, R. C. (1976): Laser Raman spectroscopic studies of the thermal unfolding of ribonuclease A. *Biochemistry*, 15:1889–1897.
43. Cloarec, O., Dumas, M. E., Trygg, J., Craig, A., Barton, R. H., Lindon, J. C., Nicholson, J. K., and Holmes, E. (2005): Evaluation of the orthogonal projection on latent structure model limitations caused by chemical shift variability and improved visualization of biomarker changes in H-1 NMR spectroscopic metabonomic studies. *Anal. Chem.*, 77:517–526.
44. Colby, B. N., Ryan, P. W., and Wilkinson, J. E. (1983): Strategies for compound identification and quantification using fused silica capillary column GC/MS. *J. High Resolut. Chromatogr. Chromatogr. Commun.*, 6:72–76.
45. Comisarow, M. B. and Marshall, A. G. (1996): The early development of Fourier transformation cyclotron resonance (FT-ICR) spectroscopy. *J. Mass Spectrom.*, 6:581–585.
46. Constantinidis, I., Braunschweiger, P. G., Wehrle, J. P. et al. (1989): $^{31}$P-Nuclear magnetic resonance studies of the effect of recombinant human interleukin 1$\alpha$ on the bioenergetics of RIF-1 tumors. *Cancer Res.*, 49:6379–6382.
47. Cooks, R. G., Hole II, S. H., Morand, K. L., and Lammert, S. A. (1992): Mass spectrometers: instrumentation. *Int. J. Mass Spectrom. Ion Proc.*, 118/119:1–36.
48. Cooper, D. A. and Moore, J. M. (1993): Femtogram on-column detection of nicotine by isotope dilution gas chromatography/negative ion detection mass spectrometry. *Biol. Mass Spectrom.*, 22:590–559.
49. Corcoran, O., Spraul, M., Hofmann, M., Ismail, I. M., Lindon, J. C., and Nicholson, J. K. (1997): 750 MHz HPLC–NMR spectroscopic identification of rat microsomal metabolites of phenoxypyridines. *J. Pharm. Biomed. Anal.*, 16:481–489.
50. Crouch, R. C. and Martin, G. E. (1992): Micro inverse-detection: a powerful technique for natural product structure elucidation. *J. Nat. Prod.*, 55:1343–1347.
51. Crupi, V., Majolino, D., Migliardo, P., Mondello, M. R., Pergolizzi, S., and Venuti, V. (2004): FT–IR spectroscopy for the detection of liver damage. *Spectroscopy*, 18:67–73
52. Davidson, A. G. (1988): Ultraviolet and visible absorption spectroscopy. In: *Practical Pharmaceutical Chemistry*, Part 2, 4th ed., edited by A. H. Beckett, and J. B. Stanlake, pp. 275–337. Athlone Press, London.
53. Dear, G. J., Ayrton, J., Plumb, R., Sweatman, B. C., Ismail, I. M., Fraser, I. J., and Mutch, P. J. (1998): A rapid and efficient approach to metabolite identification using nuclear magnetic resonance spectroscopy, liquid chromatography/mass spectrometry and liquid chromatography/nuclear magnetic resonance spectroscopy/sequential mass spectrometry. *Rapid Commun. Mass Spectrom.*, 12:2023–2030.
54. deBethizy, J. D., Udinsky, J. R., Scribner, H. E., and Frederick, C. B. (1987): The disposition and metabolism of acrylic acid and ethyl acrylate in male Sprague–Dawley rats. *Fund. Appl. Toxicol.*, 8:549–561.
55. DeBruin, L. S., Pawliszyn, J. B., and Josephy, P. D. (1999): Detection of monocyclic aromatic amines, possible mammary carcinogens, in human milk. *Chem. Res. Toxicol.*, 12:78–82.
56. DeBruin, L. S., Josephy, P. D., and Pawliszyn, J. B. (1998): Solid-phase microextraction of monocyclic aromatic amines from biological fluids. *Anal. Chem.* 70:1986–1992.
57. Derome, A. E. (1987): *Modern NMR Techniques for Chemistry Research*. Pergamon Press, Oxford.
58. Doyle, T. D. and Fazzari, F. R. (1974): Determination of drugs in dosage forms by difference spectrophotometry. *J. Pharm. Sci.*, 63:1921–1926.
59. Dunn, W. B., Bailey, N. J. C., and Johnson, H. E. (2005): Measuring the metabolome: current analytical technologies. *Analyst*, 130:606–625.
60. Dybowski, C. and Lichter, R. L., Eds. (1987): *NMR Spectroscopy Techniques*. Marcel Dekker, New York.
61. Dyer, J. R. (1965): *Applications of Absorption Spectroscopy of Organic Compounds*. Prentice-Hall, Englewood Cliffs, NJ.
62. Espina, J. R., Shockcor, J. P., Herron, W. J., Car, B. D., Contel, N. R., Ciaccio, P. J., Lindon, J. C., Holmes, E., and Nicholson, J. K. (2001): Detection of *in vivo* biomarkers of phospholipidosis using NMR-based metabonomic approaches. *Magnet. Res. Chem.*, 39:559–565.
63. Evans, C. L. et al. (2005): Chemical imaging of tissue in vivo with video-rate coherent anti-Stokes Raman scattering microscopy. *PNAS*, 102:16807–16812.
64. Feliu, A. L. (1988): The role of chemistry in positron emission tomography. *J. Chem. Educ.*, 65:655–660.
65. Feo, J. C., Castro, M. A., Robles, L. C., and Aller, A. J. (2004): Fourier transform infrared spectroscopic study of the interactions of selenium species with living bacterial cells. *Anal. Bioanal. Chem.*, 378:1601–1607.

66. Foxall, P. J. D., Lenz, E. M., Lindon, J. C., Neild, G. H., Wilson, I. D., and Nicholson, J. K. (1996): Nuclear magnetic resonance and high-performance liquid chromatography nuclear magnetic resonance studies on the toxicity and metabolism of ifosfamide. *Therap. Drug Monit.*, 18:498–505.
67. Frison, G., Tedeschi, L., Favretto, D., Reheman, A., and Ferrara, S. D. (2005): Gas chromatography/mass spectrometry determination of amphetamine-related drugs and ephedrines in plasma, urine and hair samples after derivatization with 2,2,2-trichloroethyl chloroformate. *Rapid. Commun. Mass Spectrom.*, 19:919–927.
68. Fulton, D. B., Sayer, B. G., Bain, A. D., and Malle, H. V. (1992): Detection and determination of dilute, low molecular weight organic compounds in water by 500-MHz proton nuclear magnetic resonance spectroscopy. *Anal. Chem.*, 64:349–353.
69. Gamache, P. H., Meyer, D. F., Granger, M. C., and Acworth, I. N. (2004): Metabolomic applications of electrochemistry/mass spectrometry. *J. Am. Soc. Mass Spectrom.*, 15:1717–1726.
70. Garner, R. C. (1998): The role of DNA adducts in chemical carcinogenesis. *Mut. Res.*, 402:67–75.
71. Gellermann, W., Ermakov, I. V., Ermakova, M. R., McClane, R. W., Zhao, D. Y. and Bernstein, P. S. (2002): *In vivo* resonant Raman measurement of macular carotenoid pigments in the young and the aging human retina. *J. Opt. Soc. Am. A Opt. Image Sci. Vis.*, 19:1172–1186.
72. Gordon, A. J. and Ford, R. A. (1972): *The Chemist's Companion: A Handbook of Practical Data, Techniques, and References*. John Wiley & Sons, New York.
73. Grant, A. G., Shively, C. A., Nader, M. A., Ehrenkaufer, R. L., Line, S. W., Morton, T. E., Gage, H. D. and Mach, R. H. (1998): Effect of social status on striatal dopamine $D_2$ receptor binding characteristics in cynomolgus monkeys assessed with positron emission tomography. *Synapse*, 29:80–83.
74. Greitz, T., Ingvar, D. H., and Widen, L., Eds. (1985): *Metabolism of the Human Brain Studied with Positron Emission Tomography*. Raven Press, New York.
75. Griffin, J. L. (2003): Metabonomics: NMR spectroscopy and pattern recognition analysis of body fluids and tissues for characterisation of xenobiotic toxicity and disease diagnosis. *Curr. Opin. Chem. Biol.*, 7:648–654.
76. Griffin, J. L. and Bollard, M. E. (2004): Metabonomics: its potential as a tool in toxicology for safety assessment and data integration. *Curr. Drug Metab.*, 5:389–398.
77. Griffiths, P. R. and DeHaseth, J. A. (1986): *Fourier Transform Infrared Spectroscopy*. Wily Interscience, New York.
78. Gustafson, R. A., Moolchan, E. T., Barnes, A., Levine, B., and Huestis, M. A. (2003): Validated method for the simultaneous determination of Delta(9)-tetrahydrocannabinol (THC), 11-hydroxy-THC and 11-nor-9-carboxy-THC in human plasma using solid phase extraction and gas chromatography-mass spectrometry with positive chemical ionization. *J. Chromatogr. B Anal. Technol. Biomed. Life Sci.*, 798:145–154.
79. Hackett, A. G., Kotyk, J. J., Fujiwara, H., and Logusch, E. W. (1990): Identification of a unique glutathione conjugate of trichloroacrolein using heteronuclear multiple quantum coherence $^{13}C$ nuclear magnetic resonance spectroscopy. *J. Am. Chem. Soc.*, 112:3669–3671.
80. Hackett, A. G., Kotyk, J. J., Fujiwara, H., and Logusch, E. W. (1991): Microsomal hydroxylation of triallate: identification of a 2-chloroacrylate glutathione conjugate using heteronuclear multiple quantum coherence NMR spectroscopy. *Drug Metab. Dispos.*, 19:1163–1165.
81. Halliwell, B. and Gutteridge, J. M. C. (1985): *Free Radicals in Biology and Medicine*. Clarendon Press, Oxford.
82. Harrigan, G. G., LaPlante, R. H., Cosma, G. N., Cockerell, G., Goodacre, R., Maddox, J. F., Luyendyk, J. P., Ganey, P. E., and Roth, R. A. (2004): Application of high-throughput Fourier-transform infrared spectroscopy in toxicology studies: contribution to a study on the development of an animal model for idiosyncratic toxicity. *Toxicol. Lett.*, 146:197–205.
83. Harvey, S. D., Vucelick, M. E., Lee, R. N., and Wright, B. W. (2002): Blind field test evaluation of Raman spectroscopy as a forensic tool. *Forensic Sci. Int.*, 125:12–21.
84. Haufler, R. E., and Kerley, E. L. (1997): Miniaturized time of flight mass spectrometer for high-speed applications. In: *Proc. of the 45th Annual Conf. on Mass Spectrometry and Allied Topics*, Palm Springs, CA, June 1–5.
85. Herman, J. L. and Chay, S. H. (1998): Quantitative whole-body autoradiography in pregnant rabbits to determine fetal exposure of potential teratogenic compounds. *J. Pharmacol. Toxicol. Methods*, 39:29–33.
86. Heumann, K. G. (1987): Trace determination and isotopic analysis of the elements in life sciences by mass spectrometry. *Biomed. Mass Spectrom.*, 12:477–488.
87. Hietala, J., West, C., Syvalahti, E., Nagren, K., Lehikoinen, P., Sonninen, P., and Ruotsalainen, U. (1994): Striatal $D_2$ dopamine receptor binding characteristics *in vivo* in patients with alcohol dependence. *Psychopharmacology*, 116:285–290.
88. Holman, H. Y. N., Bjornstad, K. A., and McNamara, M. P. (2002): Synchrotron infrared spectromicroscopy as a novel bioanalytical microprobe for individual living cells: cytotoxicity considerations. *J. Biomed. Opt.*, 7:417–424.
89. Holmes, E., Nicholls, A. W., Lindon, J. C., Connor, S. C., Connelly, J. C., Haselden, J. N., Damment, S. J., Spraul, M., Neidig, P., and Nicholson, J. K. (2000): Chemometric models for toxicity classification based on NMR spectra of biofluids. *Chem. Res. Toxicol.*, 13:471–478.
90. Holt, R. M., Newman, M. J., Pullen, F. S., Richards, D. S., and Swanson, A. G. (1997): High-performance liquid chromatography/NMR spectrometry/mass spectrometry: further advances in hyphenated technology. *J. Mass Spectrom.*, 32:64–70.
91. Hore, P. J. (1989): Solvent suppression. In: *Methods in Enzymology*, Vol. 176, edited by N. J. Oppenheimer and T. L. James, pp. 64–77. Academic Press, New York.
92. Jackson, S. K., Thomas, M. P., Smith, S., Madhani, M., Rogers, S. C., and James, P. E. (2004.): *In vivo* EPR spectroscopy: biomedical and potential diagnostic applications. *Faraday Discuss.*, 126:103–117.
93. Jacob, S., Abdel-Aziz, A. H., Shouman, S. A., and Ahmed, A. E. (1998): Effect of glutathione modulation on the distribution and transplacental uptake of 2-[$^{14}C$]-chloroacetonitrile (CAN) quantitative whole-body autoradiographic study in pregnant mice. *Toxicol. Indust. Health*, 14:533–546.

94. James, K. J., Gillman, M., Amandi, M. F., Lopez-Rivera, A., Puente, P. F., Lehane, M., Mitrovic, S., and Furey, A. (2005): Amnesic shellfish poisoning toxins in bivalve molluscs in Ireland. *Toxicon*, 46:852–858.
95. Johnson, K. A., Holman, L., Rosen, T. J., Nagel, J. S., English, R. J., and Growdon, J. H. (1990): Iofetame I-123 single photon emission computed tomography is accurate in the diagnosis of Alzheimer's disease. *Arch. Intern. Med.*, 150:752–756.
96. Johnstone, R. A. W. and Rose, M. E. (1996): *Mass Spectrometry for Chemists and Biochemists*, 2nd ed. Cambridge University Press, Cambridge, U.K.
97. Jones, J. R., Ed. (1988): *Isotopes: Essential Chemistry and Applications II*. Royal Society of Chemistry, London.
98. Kalasinsky, V. F. (1996): Biomedical applications of infrared and Raman microscopy. *Appl. Spectrosc. Rev.*, 31:193–249.
99. Kaufmann, R. (1995): Matrix-assisted laser desorption ionization (MALDI) mass spectrometry: a novel analytical tool in molecular biology and biotechnology. *J. Biotechnol.*, 41:155–175.
100. Kebarle, P. and Liang, T. (1993): From ions in solution to ions in the gas phase. *Anal. Chem.*, 65:972A–986A.
101. Kemp, W. (1991): Nuclear magnetic resonance spectroscopy. In: *Organic Spectroscopy*, pp. 101–241. W.H. Freeman, New York.
102. Kennedy, C. H., Hatch, G. E., Slade, R., and Mason, R. P. (1992): Application of the EPR spin-trapping technique to the detection of radicals produced *in vivo* during inhalation exposure of rats to ozone. *Toxicol. Appl. Pharmacol.*, 114:41–46.
103. Keun, H. C., Ebbels, T. M. D., Antti, H., Bollard, M. E., Beckonert, O., Schlotterbeck, G., Senn, H., Niederhauser, U., Holmes, E., Lindon, J. C., and Nicholson, J. K. (2002): Analytical reproducibility in H-1 NMR-based metabonomic urinalysis. *Chem. Res. Toxicol.*, 15:1380–1386.
104. Kubrina, L. N., Caldwell, W. S., Mordvintcev, P. I., Malenkova, I. V., and Vanin, A. F. (1992): EPR evidence for nitric oxide production from guanidino nitrogens of L-arginine in animal tissues *in vivo*. *Biochim. Biophys. Acta*, 1099:233–237.
105. Kuhr, W. G. (1990) Capillary electrophoresis. *Anal. Chem.*, 62:403R–414R.
106. Kuntzman, R. (1981): Applications of tracer techniques in drug metabolism studies. In: *Fundamentals of Drug Metabolism and Drug Disposition*, edited by B. N. La Du, H. G. Mandel, and E. L. Way, pp. 489–504. Robert E. Krieger Publishing, Malabar, FL.
107. Kuppusamy, P., Wang, P., Chzhan, M., and Zweier, J. L. (1997): High resolution electron paramagnetic resonance imaging of biological samples with a single line paramagnetic label. *Magn. Reson. Med.*, 37:479–483.
108. Kuppusamy, P., Shankar, R. A., and Zweier, J. L. (1998): *In vivo* measurement of arterial and venous oxygenation in the rat using 3D spectral-spatial electron paramagnetic resonance imaging. *Phys. Med. Biol.*, 43:1837–1844.
109. Kuppusamy, P., Afeworki, M., Shankar, R. A., Coffin, D., Krishna, M. C., Hahn, S. M., Mitchell, J. B., and Zweier, J. L. (1998): *In vivo* electron paramagnetic resonance imaging of tumor heterogeneity and oxygenation in a murine model. *Cancer Res.*, 58:1562–1568.
110. Kyerematen, G. A., Morgan, M. L., Chattopadhyay, B., deBethizy, J. D., and Vesell, E. S. (1990) Disposition of nicotine and eight metabolites in smokers and nonsmokers: identification in smokers of two metabolites that are longer lived than cotinine. *Clin. Pharmacol. Ther.*, 48: 641–651.
111. Kyerematen, G. A., Taylor, L. H., deBethizy, J. D., and Vesell, E. S. (1988): Pharmacokinetics of nicotine and 12 metabolites in the rat: application of a new radiometric high performance liquid chromatography assay. *Drug Metab. Dispos.*, 16:125–129.
112. Lafaye, A., Junot, C., Ramounet-Le Gall, B., Fritsch, P., Tabet, J.-C., and Ezan, E. (2003): Metabolite profiling in rat urine by liquid chromatography/electrospray ion trap mass spectrometry: application to the study of heavy metal toxicity. *Rapid Commun. Mass Spectrom.*, 17:2541–2549.
113. Lafleur, L., Rice, J., and Thomas, G. J. (1972): Raman studies of nucleic acids. VII. Poly(A)·poly(U) and poly(G)·poly(C). *Biopolymers.*, 11:2423–2437.
114. Lawson, E. E., Barry, B. W., Williams, A. C., and Edwards, H. G. M. (1997): Biomedical applications of Raman spectroscopy. *J. Raman Spectrosc.*, 28:111–117.
115. LeVine, S. M. and Wetzel, D. L. (1993): Analysis of brain tissue by FT–IR microspectroscopy. *Appl. Spectrosc. Rev.*, 28:385–412.
116. LeVine, S. M. and Wetzel, D. L. (1994): *In situ* chemical analysis from frozen tissue sections by Fourier transform infrared microspectroscopy: examination of white matter exposed to extravasated blood in rat brain. *Amer. J. Pathol.*, 145:1041–1047.
117. LeVine, S. M. and Wetzel, D. L. (1994): *In situ* chemical analysis of brain tissue by Fourier transform infrared microspectroscopy. *Neuroprotocols*, 5:72–79.
118. Lenz, E. M., Greatbanks, D., Wilson, I. D., Spraul, M., Hofmann, M., Troke, J., Lindon, J. C., and Nicholson, J. K. (1996): Direct characterization of drug glucuronide isomers in human urine by HPLC–NMR spectroscopy: application to the positional isomers of 6,11-dihydro-11-oxodibenz[*b,e*]oxepin-2-acetic acid glucuronide. *Anal. Chem.*, 68:2832–2837.
119. Lindon, J. C., Nicholson, J. K., and Wilson, I. D. (1996): The development and application of coupled HPLC–NMR spectroscopy. *Adv. Chromatogr.*, 36:315–382.
120. Lindon, J. C., Nicholson, J. K., and Wilson, I. D. (1996): Direct coupling of chromatographic separations to NMR spectroscopy. *Prog. Nucl. Magn. Reson. Spectrosc.*, 29:1–49.
121. Lindon, J. C., Nicholson, J. K., Sidelmann, U. G., and Wilson, I. D. (1997): Directly coupled HPLC–NMR and its application to drug metabolism. *Drug Metab. Rev.*, 29:705–746.
122. Locke, S. J. and Brauer, M. (1991) The response of the rat liver *in situ* to bromobenzene: *in vivo* proton magnetic resonance imaging and $^{31}$P magnetic resonance spectroscopy studies. *Toxicol. Appl. Pharmacol.*, 110:416–428.
123. Mach, R. H., Nader, M. A., Ehrenkaufer, R. L. E., Line, S. W., Smith, C. R., Gage, H. D., and Morton, T. E. (1997): Use of positron emission tomography to study the dynamics of psychostimulant-induced dopamine release. *Pharmacol. Biochem. Behav.*, 57:477–486.

124. Mach, R. H., Nader, M. A., Ehrenkaufer, R. L. E., Line, S. W., Smith, C. R., Luedtke, R. R., Kung, M.-P., Kung, H. F., Lyons, D., and Morton, T. E. (1996): Comparison of two fluorine-18 labeled benzamide derivatives that bind reversibly to dopamine $D_2$ receptors. *Synapse*, 24: 322–333.
125. Mach, R. H., Voytko, M. L., Ehrenkaufer, R. L. E., Nader, M. A., Tobin, J. R., Efange, S. M. N., Parsons, S. M., Gage, H. D., Smith, C. R., and Morton, T. E. (1997): Imaging of cholinergic terminals using the radiotracer [$^{18}$F](+)-4-fluorobenzyltrozamicol: *in vitro* binding studies and positron emission tomography studies in nonhuman primates. *Synapse*, 25:368–380.
126. Maeder, K. and Gallez, B. (2003): Pharmaceutical applications of *in vivo* EPR. *Biol. Magnet. Reson.*, 18:515–545.
127. Malhotra, D. and Shapiro, J. I. (1993): Nuclear magnetic resonance measurements of intracellular pH: biomedical implications. *Concepts Magn. Reson.*, 5:123–150.
128. Manoharan, R., Wang, Y., and Feld, M. S. (1996): Histochemical analysis of biological tissues using Raman spectroscopy. *Spectrochim. Acta Part A*, 52:215–249.
129. Markuszewski, M. J., Szczykowska, M., Siluk, D., and Kaliszan, R. (2005): Human red blood cells targeted metabolome analysis of glycolysis cycle metabolites by capillary electrophoresis using an indirect photometric detection method. *J. Pharm. Biomed. Anal.*, 39: 636–642.
130. Martin, F. A., Rojas-Diaz, D., Luis-Garcia, L., Gonzalez-Mora, J. L., and Castellano, M. A. (2004): Simultaneous monitoring of nitric oxide, oxyhemoglobin and deoxyhemoglobin from small areas of the rat brain by *in vivo* visible spectroscopy and a least-square approach. *J. Neurosci. Meth.*, 140(1–2):75–80.
131. Martin, G. E. and Crouch, R. C. (1991): Inverse-detected two-dimensional NMR methods: applications in natural products chemistry. *J. Nat. Prod.*, 54:1–70.
132. Martin, G. E., Guido, J. E., Robins, R. H., Sharaf, M. H. M., Schiff, Jr., P. L., and Tackie, A. N. (1998): Submicro inverse-detection gradient NMR: a powerful new way of conducting structure elucidation studies with $<0.05$ μmol samples. *J. Nat. Prod.*, 61:555–559.
133. Martin, G. E. and Zektzer, A. S. (1988): *Two-Dimensional NMR Methods for Establishing Molecular Connectivity*. VCH Publishers, New York.
134. McFadden, W. A. (1973): *Techniques of Combined Gas Chromatography/Mass Spectrometry*. John Wiley & Sons, New York.
135. McLafferty, F. W. (1980): *Interpretation of Mass Spectra*, 3rd ed., University Science Books, Mill Valley.
136. McManus, K. T., deBethizy, J. D., Garteiz, D. A., Kyerematen, G. A., and Vessel, E. S. (1990) A new quantitative thermospray LC–MS method for nicotine and its metabolites in biological fluids. *J. Chromatogr. Sci.*, 28:510–516.
137. McMaster, M. and McMaster, C. (1998): *GC/MS: A Practical User's Guide*. Wiley-VCH, New York.
138. Miller, J. C. and Miller, J. N. (1988): *Statistics for Analytical Chemistry*, 2nd ed. Ellis Horwood, Chichester.
139. Mizuno, A., Hayashi, T., Tashibu, K., Maraishi, S., Kawauchi, K., and Ozaki, Y. (1992): Near-infrared FT-Raman spectra of the rat brain tissue. *Neurosci. Lett.*, 141:47–52.
140. Mizuno, A., Kitajima, H., Kawauchi, K., Muraishi, S., and Ozaki, Y. (1994): Near-infrared Fourier transform Raman spectroscopic study of human brain tissues and tumors. *J. Raman Spectrosc.*, 25:25–29.
141. Mobius, K., Savitsky, A., Wegener, C., Plato, M., Fuchs, M. Schnegg, A. Dubinskii, A. A., Grishin, Y. A., Grigor'ev, I. A., Kuehn, M., Duche, D., Zimmermann, H., and Steinhoff, H.-J. (2005): Combining high-field EPR with site-directed spin labeling reveals unique information on proteins in action. *Mag. Reson. Chem.*, 43: S4–S19.
142. Moonen, C. T. W., van Zijl, P. C. M., Frank, J. A., Le Bihan, D., and Becker, E. D. (1990): Functional magnetic resonance imaging in medicine and physiology. *Science.*, 250:53–61.
143. Mortishire-Smith, R. J., Skiles, G. L., Lawrence, J. W., Spence, S., Nicholls, A. W., Johnson, B. A., and Nicholson, J. K. (2004): Use of metabonomics to identify impaired fatty acid metabolism as the mechanism of a drug-induced toxicity. *Chem. Res. Toxicol.*, 17:165–173.
144. Moslen, M. T. and Smith, C. V., Eds. (1992): *Free Radical Mechanisms of Tissue Injury*. CRC Press, Boca Raton, FL.
145. Muller, H. J., Lanens, D., de Cock Buning, T. J. et al. (1992): Noninvasive *in vivo* $^{13}$C-NMR spectroscopy in the rat to study the pharmacokinetics of $^{13}$C-labeled xenobiotics. *Drug Metab. Dispos.*, 20:507–509.
146. Munson, M. S. B. and Field, F. H. (1966): Chemical ionization mass spectrometry. I. General introduction. *J. Am. Chem. Soc.*, 88:2621–2630.
147. Neil, J. J. (1991): The use of freely diffusible, NMR-detectable tracers for measuring organ perfusion. *Concepts Magn. Reson.*, 3:1–12.
148. Nicholson, J. K., Foxall, P. J. D., Spraul, M., Farrant, R. D., and Lindon, J. C. (1995): 750 MHz $^{1}$H and $^{1}$H-$^{13}$C NMR spectroscopy of human blood plasma. *Anal. Chem.*, 67:793–811.
149. Nicholson, J. K., Holmes, E., Sidelmann, U., Lindon, J. C., and Wilson, I. D. (1996): HPLC–NMR spectroscopy: a powerful tool for the investigation of drug metabolism and metabolite reactivity. *Pharm. Sci.*, 2:127–130.
150. Nicholson, J. K., Timbrell, J. A., and Sadler, P. J. (1985): Proton NMR spectra of urine as indicators of renal damage: mercury-induced nephrotoxicity in rats. *Mol. Pharmacol.*, 27:644–651.
151. Nicholson, J. K. and Wilson, I. D. (1987): High resolution nuclear magnetic resonance spectroscopy of biological samples as an aid to drug development. *Prog. Drug Res.*, 31:427–479.
152. Nielsen, K. F. and Smedsgaard, J. (2003): Fungal metabolite screening: database of 474 mycotoxins and fungal metabolites for dereplication by standardised liquid chromatography-UV-mass spectrometry methodology. *J. Chromatogr. A*, 1002:111–136.
153. Notingher, I., Green, C., Dyer, C., Perkins, E., Hopkins, N., Lindsay, C., and Hench, L. L. (2004): Discrimination between ricin and sulphur mustard toxicity in vitro using Raman spectroscopy. *J. R. Soc. Interface*, 1:79–90.
154. Notingher, I., Selvakumaran, J., and Hench, L. L. (2004): New detection system for toxic agents based on continuous spectroscopic monitoring of living cells. *Biosens. Bioelectr.*, 20:780–789

155. Ong, C. W., Shen, Z. X., He, Y., Lee, T., and Tang, S. H. (1999): Raman microspectroscopy of the brain tissues in the substantra nigra and MPTP-induced Parkinson's disease. *J. Raman Spectrosc.*, 30:91–96.
156. Oteki, T., Nagase, S., Yokoyama, H., Ohya, H., Akatsuka, T., Takao, T., Mika, U., Hirayama, A., and Koyama, A. (2005): Evaluation of adriamycin nephropathy by an *in vivo* electron paramagnetic resonance. *Biochem. Biophys. Res. Commun.*, 332:326–331.
157. Parker F. (1983): *Applications of Infrared, Raman and Resonance Raman Spectroscopy in Biochemistry.* Plenum Press, New York.
158. Parvez, H., Reich, A., Lucas-Reich, S., and Parvez, S. (1988): *Flow-Through Radioactivity Detection in HPLC.* VSP, Utrecht, The Netherlands.
159. Pham-Tuan, H., Kaskavelis, L., Daykin, C. A., and Janssen, H.-G. (2003): Method development in high-performance liquid chromatography for high-throughput profiling and metabonomic studies of biofluid samples. *J. Chromatogr. B Anal. Tech. Biomed. Life Sci.*, 789:283–301.
160. Phelps, M. E. and Mazziotta, J. C. (1985): Positron emission tomography: human brain function and biochemistry. *Science*, 228:779–809.
161. Philips, M., Mazziotta, J. C., and Schelbert, H. R., Eds. (1986): *Positron Emission Autoradiography: Principles and Applications for the Brain and Heart.* Raven Press, New York.
162. Pool, W. F. and Crooks, P. A. (1988): Biotransformation of primary nicotine metabolites: metabolism of R-(+)-[$^3$H-$N'$-$CH_3$,$^{14}$C-$N$-$CH_3$] *N*-methylnicotinium acetate: the use of double isotope studies to determine the *in-vivo* stability of the *N*-methyl groups of *N*-methylnicotinium ion. *J. Pharm. Pharmacol.*, 40:758–762.
163. Poole, C. P. (1967): *Electron Spin Resonance, a Comprehensive Treatise on Experimental Techniques.* Wiley-Interscience, New York.
164. Popov, A. I. and Hallenga, K., Eds. (1991): *Modern NMR Techniques and Their Application in Chemistry.* Marcel Dekker, New York.
165. Pragst, F., Herzler, M., and Erxleben, B. T. (2004): Systematic toxicological analysis by high-performance liquid chromatography with diode array detection (HPLC-DAD). *Clin. Chem. Lab. Med.*, 42:1325–1340.
166. Pullen, F. S., Swanson, A. G., Newman, M. J., and Richards, D. S. (1995): Online liquid chromatography/nuclear magnetic resonance mass spectrometry: a powerful spectroscopic tool for the analysis of mixtures of pharmaceutical interest. *Rapid Commun. Mass Spectrom.*, 9:1003–1006.
167. Puppels, G. J., de Mui, F. F. M., Otto, C., Greve, J., Robert-Nicoud, D., Arndt-Jovin, D. J., and Jovin, T. M. (1990): Studying single living cells and chromosomes by confocal Raman microspectroscopy. *Nature (Lond.)*, 347:301–303.
168. Pusecker, K., Schewitz, J., Gfrörer, P., Tseng, L.-H., Albert, K., and Bayer, E. (1998): On-line coupling of capillary electrochromatography, capillary electrophoresis, and capillary HPLC with nuclear magnetic resonance spectroscopy. *Anal. Chem.*, 70:3280–3285.
169. Rabenstein, D. L. and Guo, W. (1988) Nuclear magnetic resonance spectroscopy. *Anal. Chem.*, 60:1R–28R.
170. Rabenstein, D. L., Millis, K. K., and Strauss, E. J. (1988): Proton NMR spectroscopy of human blood plasma and red blood cells. *Anal. Chem.*, 60:1380–1390.
171. Reiman, E. M. and Mintan, M. A. (1990): Positron emission tomography. *Arch. Intern. Med.*, 150:729–731.
172. Ricaurte, G. A., McCann, U. D., Szabo, Z., and Scheffel, U. (2000): Toxicodynamics and long-term toxicity of the recreational drug, 3,4-methylenedioxymethamphetamine (MDMA, 'Ecstasy'). *Toxicol. Lett.*, 112:143–146.
173. Robertson, D. G., Reily, M. D., Sigler, R. E., Wells, D. F., Paterson, D. A., and Braden, T. K. (2000): Metabonomics: evaluation of nuclear magnetic resonance (NMR) and pattern recognition technology for rapid *in vivo* screening of liver and kidney toxicants. *Toxicol. Sci.*, 57:326–337.
174. Russo, E. (2005): Mass spectrometry goes offsite. *Scientist*, 19:32.
175. Sanders, J. K. M. and Hunter, B. K. (1987): *Modern NMR Spectroscopy: A Guide for Chemists.* Oxford University Press, Oxford.
176. Scarfe, G. B., Wilson, I. D., Spraul, M., Hofmann, M., Braumann, U., Lindon, J. C., and Nicholson, J. K. (1997): Application of directly coupled high-performance liquid chromatography-nuclear magnetic resonance-mass spectrometry to the detection and characterization of the metabolites of 2-bromo-4-trifluoromethylaniline in rat urine. *Anal. Commun.*, 34:37–39.
177. Scarfe, G. B., Wright, B., Clayton, E., Taylor, S., Wilson, I. D., Lindon, J. C., and Nicholson, J. K. (1998): $^{19}$F-NMR and directly coupled HPLC–NMR–MS investigations into the metabolism of 2-bromo-4-trifluoromethylaniline in rat: a urinary excretion balance study without the use of radiolabeling. *Xenobiotica*, 28:373–388.
178. Scarfe, G. B., Wright, B., Clayton, E., Taylor, S., Wilson, I. D., Lindon, J. C., and Nicholson, J. K. (1999): Quantitative studies on the urinary metabolic fate of 2–chloro-4–trifluoromethylaniline in the rat using $^{19}$F-NMR spectroscopy and directly coupled HPLC–NMR–MS. *Xenobiotica*, 29:77–91.
179. Schelbert, H. R. (1985): Positron emission tomography: assessment of myocardial blood flow and metabolism. *Circulation*. 72(Suppl. IV):122–133.
180. Schewitz, J., Gfrörer, P., Pusecker, K., Tseng, L-H., Albert, K., Bayer, E., Wilson, I. D., Bailey, N. J., Scarfe, G. B., Nicholson, J. K., and Lindon, J. C. (1998): Directly coupled CZE-NMR and CEC-NMR spectroscopy for metabolite analysis: paracetamol metabolites in human urine. *Analyst*, 123:2835–2837.
181. Schiering, D. W., Walton, R. B., Brown, C. W., Norman, M. L., Brewer, J., and Scott, J. (2004): Toward the characterization of biological toxins using field-based FT–IR spectroscopic instrumentation. *Proc. SPIE Int. Soc. Opt. Eng.*, 585:21–32.
182. Schirmer, R. E. (1982): *Modern Methods of Pharmaceutical Analysis.* Vol. 1. CRC Press, Boca Raton, FL.
183. Shi, X., Dalal, N. S., and Kasprzak, K. S. (1993): Generation of free radicals from model lipid hydroperoxides and $H_2O_2$ by Co(II) in the presence of cysteinyl and histidyl chelators. *Chem. Res. Toxicol.*, 6:277–283.
184. Shim, M. G. and Wilson, B. C. (1997): Development of an *in vivo* Raman spectroscopic system for diagnostic applications. *J. Raman Spectrosc.*, 28:131–142.

185. Shockcor, J. P., Unger, S. H., Wilson, I. D., Foxall, P. J. D., Nicholson, J. K., and Lindon, J. C. (1996): Combined HPLC, NMR spectroscopy, and ion-trap mass spectrometry with application to the detection and characterization of xenobiotic and endogenous metabolites in human urine. *Anal. Chem.*, 68:4431–4435.
186. Sidelmann, U. G., Braumann, U., Hofmann, M., Spraul, M., Lindon, J. C., Nicholson, J. K., and Hansen, S. H. (1997): Directly coupled 800 MHz HPLC–NMR spectroscopy of urine and its application to the identification of the major phase II metabolites of tolfenamic acid. *Anal. Chem.*, 69:607–612.
187. Siethoff, C., Feldmann, I., Jakubowski, N., and Linscheid, M. (1999): Quantitative determination of DNA adducts using liquid chromatography/electrospray ionization mass spectrometry and liquid chromatography/high resolution inductively coupled plasma mass spectrometry. *J. Mass Spectrom.*, 34:421–426.
188. Singer, A. C., Huang, W. E., Helm, J., and Thompson, I. P. (2005): Insight into pollutant bioavailability and toxicity using Raman confocal microscopy. *J. Microbiol. Meth.*, 60:417–422.
189. Slim, R. M., Robertson, D. G., Albassam, M., Reily, M. D., Robosky, L., and Dethloff, L. A. (2002). Effect of dexamethasone on the metabonomics profile associated with phosphodiesterase inhibitor-induced vascular lesions in rats. *Toxicol. Appl. Pharmacol.*, 183:108–116.
190. Smith, D. R. and Flegal, A. R. (1992): Stable isotopic tracers of lead mobilized by DMSA chelation in low lead-exposed rats. *Toxicol. Appl. Pharmacol.*, 116:85–91.
191. Smith, D. R. et al. (1992): Stable isotope labeling of lead compartments in rats with ultra-low lead concentrations. *Environ. Res.*, 57:190–207.
192. Smith, R. V. and Stewart, J. T. (1981): *Textbook of Biopharmaceutical Analysis*. Leo & Febiger, Philadelphia, PA.
193. Sotgiu, A., Colacicchi, S., Placidi, G., and Alecci, M. (1997): Water soluble free radicals as biologically responsive agents in electron paramagnetic resonance imaging. *Cell. Mol. Biol.*, 43:813–823.
194. Spraul, M., Hofmann, M., Dvortsak, P., Nicholson, J. K., and Wilson, I. D. (1993): High-performance liquid chromatography coupled to high-field proton nuclear magnetic resonance spectroscopy: application to the urinary metabolites of ibuprofen. *Anal. Chem.*, 65:327–330.
195. Stoyanova, R., Nicholson, J. K., Lindon, J. C., and Brown, T. R. (2004): Sample classification based on Bayesian spectral decomposition of metabonomic NMR data sets. *Anal. Chem.*, 76:3666–3674.
196. Stubbs, M., Coombes, R. C., Griffiths, J. R., Maxwell, R. J., Rodrigues, L. M., and Gusterson, B. A. (1990): $^{31}$P-NMR spectroscopy and histological studies of the response of rat mammary tumours to endocrine therapy. *Br. J. Cancer*, 61:258–262.
197. Stubbs, M. et al. (1989): Growth studies of subcutaneous rat tumours: comparison of $^{31}$P-NMR spectroscopy, acid extracts and histology. *Br. J. Cancer*, 60:701–707.
198. Sumner, S. C. J. et al. (1992): Characterization and quantitation of urinary metabolites of [1,2,3-$^{13}$C] acrylamide in rats and mice using $^{13}$C nuclear magnetic resonance spectroscopy. *Chem. Res. Toxicol.*, 5:81–89.
199. Tagliaro, F., Camilot, M., Valentini, R., Mengarda, F., Antoniazzi, F., and Tato, L. (1998). Determination of thyroxine in the hair of newborns by radioimmunoassay with high-performance liquid chromatographic confirmation. *J. Chromatogr. B Biomed. Sci. Appl.*, 716:77–82.
200. Talangala, S. L. and Lowe, I. J. (1991): Introduction to magnetic resonance imaging. *Concepts Magn. Reson.*, 3:145–159.
201. Taylor, J. K. (1987): *Quality Assurance of Chemical Measurements*. Lewis Publishers, Chelsea, MI.
202. Ter-Pogossian, M. M., Raichle, M. E., and Sobel, B. E. (1980): Positron-emission tomography. *Sci. Am.*, 243:171–181.
203. Traficante, D. D. (1992): Optimum tip angle and relaxation delay for quantitative analysis. *Concepts Magn. Reson.*, 4:153–160.
204. Treado, P. J., Levin, I. W., and Lewis, E. N. (1992): Near-infrared acousto-optic filtered spectroscopic microscopy: a solid-state approach to chemical imaging. *Appl. Spectrosc.*, 46:553–559.
205. Turtelbaum, K. W., Frantz, C. E., Creek, M. R., Vogel, J. S., Shen, N., and Fultz, E. (1993): DNA adducts in model systems and humans. *J. Cell Biochem.*, 17F:138–148.
206. Van Hee, P., Neels, H., De Doncker, M., Vrydags, N., Schatteman, K., Uyttenbroeck, W., Hamers, N., Himpe, D., and Lambert, W. (2004): Analysis of gamma-hydroxybutyric acid, DL-lactic acid, glycolic acid, ethylene glycol and other glycols in body fluids by a direct injection gas chromatography-mass spectrometry assay for wide. *Clin. Chem. Lab. Med.*, 42:1341–1345.
207. Vaupel, P., Okunieff, P., and Neuringer, L. J. (1990): *In vivo* $^{31}$P-NMR spectroscopy of murine tumours before and after localized hyperthermia. *Int. J. Hyperthermia*, 6:15–31.
208. Viant, M. R. (2003). Improved methods for the acquisition and interpretation of NMR metabolomic data. *Biochem. Biophys. Res. Commun.*, 310:943–48.
209. Viant, M. R., Rosenblum, E. S., and Tieerdema, R. S. (2003): NMR-based metabolomics: a powerful approach for characterizing the effects of environmental stressors on organism health. *Environ. Sci. Technol.*, 37:4982–4989.
210. Vogel, J. S. (2005): Accelerator mass spectrometry for quantitative *in vivo* tracing. *BioTechniques*, 38:S25–S29.
211. Vogel, J. S., Turteltaub, K. W., Finkel, R., and Nelson, D. E. (1995): Accelerator mass spectrometry. *Anal. Chem.*, 67:353A–359A.
212. Volkow, N. D., Fowler, J. S., Wang, G. J., Hitzemann, R., Logan, J., Schlyer, D. J., Dewey, S. L., and Wolf, A. P. (1993): Decreased dopamine $D_2$ receptor availability is associated with reduced frontal metabolism in cocaine abusers. *Synapse*, 14:169–177.
213. Volkow, N. D., Fowler, J. S., Wolf, A. P. et al. (1990): Effects of chronic cocaine abuse on postsynaptic dopamine receptors. *Am. J. Psychol.*, 147:719–724.
214. Waddell, W. J. (1981): Autoradiography in drug disposition studies. In: *Fundamentals of Drug Metabolism and Drug Disposition*, edited by B. N. La Du, H. G. Mandel, and E. L. Way, pp. 505–514. Robert E. Krieger Publishing, Malabar, FL.

215. Waters, N. J., Waterfield, C. J., Farrant, R. D., Holmes, E., and Nicholson, J. K. (2005): Metabonomic deconvolution of embedded toxicity: application to thioacetamide hepato- and nephrotoxicity. *Chem. Res. Toxicol.*, 18:639–654.
216. Watkins, S. M., Reifsnyder, P. R., Pan, H. J., German, J. B., and Leiter, E. H. (2002): Lipid metabolome-wide effects of the PPARgamma agonist rosiglitazone. *J. Lipid Res.*, 43:1809–1817.
217. Weickhardt, C., Moritz, F., and Grotemeyer, J. (1996): Time-of-flight mass spectrometry: state-of-the-art in chemical analysis and molecular science. *Mass Spectrom. Rev.*, 15:139–162.
218. Weisel, C. P., Park, S., Pyo, H., Mohan, K., and Witz, G. (2003): Use of stable isotopically labeled benzene to evaluate environmental exposures. *J. Expo. Anal. Environ. Epidemiol.*, 13:393–402.
219. Wetzel, D. L., Slatkin, D. N., and LeVine, S. M. (1998): FT–IR microspectroscopic detection of metabolically deuterated compounds in the rat cerebellum: a novel approach for the study of brain metabolism. *Cell. Mol. Biol.*, 44:15–27.
220. Whateley, T. L. (1988): Radiochemistry and radiopharmaceuticals. In: *Practical Pharmaceutical Chemistry*, Part 2, 4th ed., edited by A. H. Beckett, and J. B. Stanlake, pp. 501–534. Athlone Press, London.
221. Willoughby, R. C. and Browner, R. F. (1984): Monodispersed aerosol generation interface for coupling liquid chromatography with mass spectrometry. *Anal. Chem.*, 56:2626–2631.
222. Willoughby, R., Sheehan, E., and Mitrovich, S. (1998): *A Global View of LC–MS*. Global View, Pittsburgh, PA.
223. Wilson, A. G. E. and Kraus, L. J. (1995): Application of direct analytic and storage phosphor techniques in quantitating whole-body autoradiography data. *Toxicol. Meth.*, 5:15–20.
224. Wilson, I. D., Lindon, J. C., and Nicholson, J. K. (1998): Liquid chromatography directly and jointly combined with nuclear magnetic resonance spectroscopy and mass spectrometry. *LC-GC*, 16:842–852.
225. Wiltshire, H. R., Harris, S. R., Prior, K. J., Kozlowski, U. M., and Worth, E. (1992): Metabolism of calcium antagonist Ro 40-5967: a case history of the use of diode-array UV spectroscopy and thermospray-mass spectrometry in the elucidation of a complex metabolic pathway. *Xenobiotica*, 22:837–857.
226. Wolfe, R. R. (1992): *Radioactive and Stable Isotope Tracers in Biomedicine: Principles and Practice of Kinetic Analysis*. Wiley-Liss, New York.
227. Wong, P. T. T. and Rigas, B. (1990) Infrared spectra of microtome sections of human colon tissues. *Appl. Spectrosc.*, 44:1715–1720.
228. Wu, L., Mashego, M. R., van Dam, J. C., Proell, A. M., Vinke, J. L., Ras, C., van Winden, W. A., van Gulik, W. M., and Heijnen, J. J. (2005): Quantitative analysis of the microbial metabolome by isotope dilution mass spectrometry using uniformly $^{13}$C-labeled cell extracts as internal standards. *Anal. Biochem.*, 336:164–171.
229. Wu, N., Webb, A., Peck, T. L., and Sweedler, J. V. (1995): On-line NMR detection of amino acids and peptides in microbore LC. *Anal. Chem.*, 67:3101–3107.
230. Wuilloud, R. G., Kannamkumarath, S. S., and Caruso J. A. (2004): Speciation of essential and toxic elements in edible mushrooms: size-exclusion chromatography separation with on-line UV-inductively coupled plasma mass spectrometry detection. *J. Organomet. Chem.*, 18:156–165.
231. Yergey, A. L., Edmonds, C. G., Lewis, I. A. S., and Vestal, M. L. (1990): *Liquid Chromatography/Mass Spectrometry, Techniques and Applications*. Plenum Press, New York.
232. Zane, P. A. et al. (1997): Validation of procedures for quantitative whole-body autoradiography using digital imaging. *J. Pharm. Sci.*, 86:733–738.

# 42 Methods in Environmental Toxicology

*Michael A. Lewis, Anne Fairbrother, and Robert E. Menzer*

## CONTENTS

Most testing of chemicals for their toxic effects has traditionally focused on concerns regarding the safety of humans. Such testing, however, invariably is conducted using surrogate species that are supposed to approximate the responses of the human. Over the years, a body of literature has accumulated that provides a measure of confidence that such an approach, although with appropriate safety factors applied, has indeed provided data that can be extrapolated for the protection of human health. Most of that literature is based on testing chemicals with a variety of laboratory animal species that can be bred, reared, and maintained with confidence that results obtained will be comparable from one testing situation to another. Mice, rats, rabbits, dogs, and occasionally primates are the principal species used.

Two important considerations have emerged in the discipline of toxicology that demand an extension of the limited testing protocols of the past. First, we have discovered that some animal species more closely mimic the human response than traditionally used laboratory animals [109]. Second, we recognize today the important implications for human health in the response to xenobiotics of wild animals in their own environments. Thus, the subdiscipline of ecotoxicology has emerged with its research emphasis on bioindicators of ecosystem health.

By expanding the number of species tested in assessing the toxicology of a chemical, one is able to gain considerable insight into its mechanism of action, biodegradability, organ-specific toxicity, and acute and potential chronic effects. The expansion of comparative toxicology from reliance on laboratory mammals to the inclusion of wild mammals, fish, birds, and some invertebrates is highly desirable to better understand the range of responses to a chemical in its interactions with the various target systems possible in different animals. With advances in understanding the physiology and biochemistry of different species, the possibility of a better understanding of toxic responses is enhanced. With the inclusion of additional species in toxicity testing has come the need for the development of protocols to standardize approaches for the use of such new species. In this chapter, we provide some principles and examples of the development of such protocols.

## ENVIRONMENTAL BEHAVIOR OF XENOBIOTICS

When testing the toxicity of chemicals to laboratory animals, the chemical is usually fed to or applied directly to the animal or directly into the medium in which the animal is living; however, when using wild mammals, fish, birds, or invertebrates, testing protocols must be designed to take into account the effects of the transport and fate of the chemical in the natural environment. Thus, the physical properties and chemical behavior of the test substance are important factors that must be understood. The most comprehensive and useful treatise on this subject is that of Lyman et al. [100], updated by Boethling and Mackay [28].

### Water Solubility and Lipophilicity

The most significant determinant of the transport and fate of a chemical in the environment is its water solubility. Highly soluble chemicals are transported through the hydrologic cycle and thus may be found widely distributed at great distances from their points of introduction into the environment. Conversely, hydrophobic compounds will tend to be more static and move little through the hydrologic cycle. Generally, the more water soluble the chemical, the less lipophilic, the less sorbed to soils and sediments, and the less bioconcentrated it will be. Water solubility is defined as the maximum amount of a chemical that may be dissolved in a given quantity of pure water at a particular temperature. It is important to note that even chemicals described in tables of physical constants as very insoluble may have sufficient water solubility to have a significant impact on their behavior in the environment. It is particularly important to note that metals and other inorganics with low water solubility may be converted to more water-soluble forms in the environment. A derivative property of the water solubility and lipophilicity of a chemical is its octanol–water partition coefficient ($K_{OW}$ or $P$), frequently reported as log $K_{OW}$. $K_{OW}$ is defined as the ratio of the concentrations of a chemical in the water phase and the $n$-octanol phase after a chemical is equilibrated between equal volumes of the two solvents. It is a key property of a chemical for environmental considerations, as it is related to soil and sediment adsorption and bioconcentration of a chemical in aquatic organisms. In designing studies to assess toxicity of a chemical in model ecosystems, such as microcosms or mesocosms, the octanol/water partition coefficient must be known or experimentally determined so the behavior of the chemical in the system can be predicted and the system be designed appropriately.

### Soil Adsorption

The extent of partitioning of a chemical between the solid and solution phases of a water-saturated soil or sediment is described by the soil sorption coefficient ($K$

or $K_d$) [100,156]. It is determined experimentally using the Freundlich equation:

$$x/m = KC^{1/n}$$

where $x/m$ is the μg of chemical adsorbed per g of soil, $C$ is the μg of chemical per mL of solution, and $K$ and $n$ are constants for a particular soil type. The value of $n$ must be experimentally determined, but is frequently assumed to be 1. $K_{OC}$, the soil sorption constant, is determined from $K$ by dividing by the percent organic carbon in the soil and multiplying the result by 100. This constant is observed to be relatively independent of the type of soil or sediment and is the value most frequently used to describe the adsorption of a chemical to soil or sediment. Consideration of the soil adsorption of a chemical is extremely important for the proper design of ecological test systems.

## Atmospheric Particulate Matter

When pollutant-contaminated soils become airborne, chemicals sorbed to the resulting particulate matter must be considered in some testing protocols. Such sorbed chemicals may be bioavailable in biological systems. Soil-derived particulate matter is composed of a mineral fraction and an organic matter fraction. Affinity of organic chemicals for the mineral fraction is believed to be low, but hydrophobic organic chemicals may have a high affinity for the organic portion [28]. Once particulate matter is suspended in the atmosphere, gas/particle partitioning of adsorbed chemicals is a function of the vapor pressure of the chemical. Chemicals with a vapor pressure $<10^{-6}$ atm will be primarily in the particle phase, while those with a vapor pressure >1 atm will partition primarily to the gas phase [28]. Thus, under some testing conditions, the air quality of the test system must be controlled.

## Vaporization

The vaporization of a chemical from a solid surface or a solution is an important mass transfer process. Factors that control volatilization are diffusivity of the chemical, its water solubility, vapor pressure, the Henry's law constant, and temperature. Diffusivity is the rate of diffusion of a chemical through a medium and depends on the nature of the chemical itself and the nature of the medium through which it moves. Vapor pressure is the tendency of a liquid to change from the liquid to the gaseous state and is highly dependent on temperature.

The air–water interface is most important in environmental analyses, so the question of the ability of a chemical to diffuse across that interface is significant in evaluating its environmental behavior. The tendency of a chemical to escape from solution is described by the Henry's law constant. Henry's law states that the solubility of a gas in a liquid is directly proportional to the pressure of the gas above the liquid at equilibrium: $H = P/C$, where $C$ is measured in mol/m$^3$ and $P$ is in atm; that is, $H$ is the ratio of the saturation vapor pressure and the water solubility of the chemical. Units of $H$ are most often reported as atm-m$^3$/mole; however, if $C$ is expressed in moles/L and $P$ is expressed in moles/m$^3$, $H$ is dimensionless. Under these circumstances, the Henry's law constant is sometimes referred to as the air/water partition coefficient.

Movement of chemicals in the environment occurs in the vapor state as well as in solution or adsorbed to particulate matter. The most important consideration in evaluating the extent of such movement is the Henry's law constant, because in the environment most chemicals will eventually be found in water solution, and their tendency to move into the air will have an impact on their potential for toxic effects in organisms which may be exposed. In the design of toxicology test methods, such considerations must be taken into account. Their importance is illustrated in Figure 42.1, in which the relation of the Henry's law constant of a chemical to its volatility characteristics is depicted.

## Bioaccumulation

Certain chemicals accumulate in organisms exposed to them in their environments. There are now many well-known cases of chemicals being concentrated in food webs to the point that toxic effects are exhibited in organisms that may have had no direct exposure to the chemical itself at its point of application, the insecticide DDT and other chlorinated hydrocarbons being examples. The tendency of a chemical to be more concentrated in an organism than the concentration in its environment is described by its bioconcentration factor (BCF). It is calculated by dividing the concentration of the chemical in the organism (wet weight) by the concentration of the chemical in the air or water in which it lives. The units in both the numerator and denominator must be the same (e.g., μg/g). The BCF can be estimated from the water solubility of a chemical, $K_{OW}$, or $K_{OC}$. A number of empirical equations specifically useful for calculating BCF for particular chemical groups will be found in Lyman et al. [100]. The estimates of BCF should be used in designing toxicity studies to understand the degree to which the test chemical will be taken up and stored in the organisms being used in tests; however, bioaccumulation does not take into account the ability of an organism to metabolize the chemical, so the use of BCF without considering biodegradation may yield misleading conclusions.

## Biodegradation

As chemicals move in the environment, they are subjected to break down, primarily by microorganisms, in a process known as *biodegradation*. This process represents a sig-

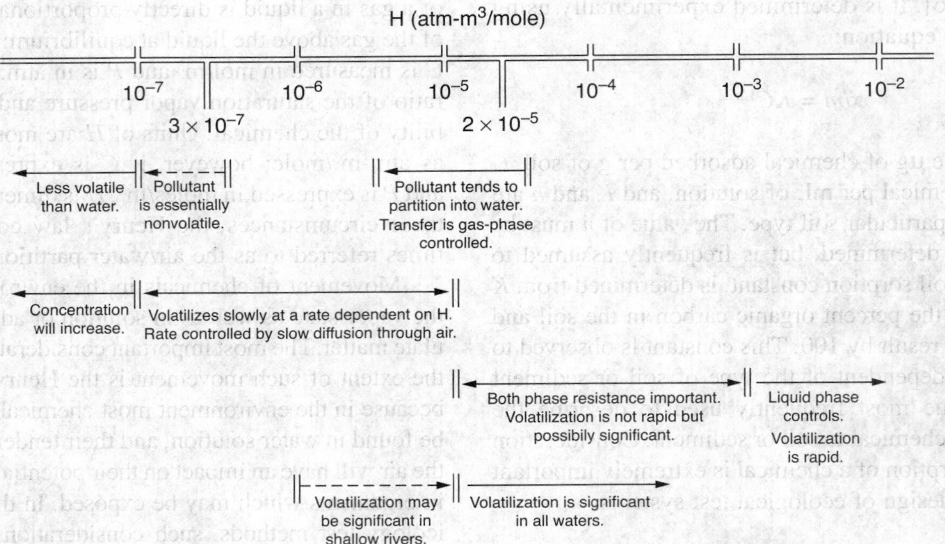


**FIGURE 42.1** Volatility characteristics associated with various ranges of Henry's law constant. (From Smith, J.H. et al., *Environ. Sci. Technol.*, 14, 1332, 1980; reproduced in Lymen et al. [100]. With permission.)

nificant loss mechanism in soil, sediments, and aquatic systems which leads ultimately to mineralization of the compound—that is, degradation to carbon dioxide, water, and the inorganic forms of other elements they may contain. Microorganisms are the primary converters of complex organic chemicals to inorganic substances. In many instances, higher organisms are able to metabolize compounds, but they generally play a less significant role in environmental systems. Microorganisms are generally the first agents in biodegradation, converting compounds into the simpler forms that are more susceptible to degradation by higher organisms.

Almost all degradative reactions in the environment are oxidative, reductive, hydrolytic, or conjugative. Biodegradation can take place in virtually any environmental situation, aerobic or anaerobic; thus, in a test system to assess the toxicity of a chemical, the potential for biodegradation must always be considered regardless of the presence or absence of oxygen. Photochemical degradation is often also a factor in the environment.

When testing chemicals in anything more complex than a single organism (i.e., the microcosms, mesocosms, etc. discussed later in this chapter), the medium, whether water, sediment, soil, or a combination of these, plays an important role in the behavior of the system. Organic matter in the medium strongly influences its microbial density, and microorganisms may comprise as much as 80% of the biomass of soil. The microbial community in turn determines how stable a xenobiotic chemical will be in the system.

Organic compounds can be divided into four groups according to their biodegradability: (1) usable immediately by an exposed organism as a nutrient or energy source, (2) usable by microorganisms following an acclimation period, and (3) degraded slowly or not at all. In the fourth group, compounds are subject to cometabolic degradation wherein a compound that does not provide a nutrient or energy source for the degrading organism is broken down in conjunction with the degradation of other substances. Ideally, when evaluating the effect of xenobiotics on complex systems, one needs to consider all aspects of the biology and chemistry of all of the organisms and chemicals present. The best design of such test systems demands the most complete knowledge possible of all the potential interactions of chemicals and organisms in the system.

Before presenting and discussing the various specific test methods used in environmental toxicology, it is important to note that there are several significant differences between the practice of toxicology using laboratory animals and methods that use wild animals, often located in their usual or at least simulated environmental conditions. Most importantly, the goal of environmental toxicology is considerably broader and more varied, in that, while the assessment of the impact of a test chemical on human health is important, an understanding of the impact of a chemical on whole ecosystems as well as their component parts is the paramount consideration. Furthermore, in ecotoxicology the test organisms frequently are the actual targets for a chemical pollutant and the test methodology thus utilizes the endpoint subjects of concern, rather than a surrogate animal model. Finally, these species live in a variable environment subject to seasonal physical and chemical changes. These changes can affect toxicity; consequently, toxic effects are more difficult to predict than in mammalian toxicology.

**TABLE 42.1**
**Estimated Number of Species of Fauna and Flora in the Gulf of Mexico**

| Taxa | No. of Species |
|---|---|
| Microalgae | 30,000 |
| Sea grasses | 7 |
| Molluscs | 500 |
| Polychaetes | 600 |
| Oligochaetes | 200 |
| Echinoderms | 400 |
| Cnidarians | 600 |
| Sponges | 100 |
| Fishes | 800 |
| Marine animals | 32 |

*Source:* Gore, R.H., *The Gulf of Mexico*, Pineapple Press, Sarasota, FL, 1992. With permission.

The challenge of environmental toxicology is to protect ecosystems and their component parts. In general, mechanisms of toxicity, test methods and their limitations, and the delivered dose to the target organism are better understood and more easily controlled in mammalian toxicology. Methodologies for testing chemicals in ecosystems have developed rapidly in recent years, and the methods presented in the following pages provide approaches on which to build the definitive and precise techniques for refining ecotoxicology in the future.

## TESTING OF AQUATIC ORGANISMS

Complex ecosystems such as rivers, lakes, wetlands, and estuaries contain unique biota that may be represented by several thousand species, such as for the Gulf of Mexico (Table 42.1). The indigenous flora and fauna may be exposed to a variety of potentially toxic chemicals, which, in most cases, result from anthropogenic activities. As a result, environmental damage may occur. The study of the adverse effects of chemicals on freshwater and saltwater biota and on the ecosystems that contain them defines the aquatic toxicology discipline.

Aquatic toxicology differs from mammalian toxicology in several aspects. The primary goal of aquatic toxicology is to assess the effect of toxicants on the many diverse species, populations, and communities of plants and animals inhabiting saltwater and freshwater environments. The biota are usually cold-blooded, and the naturally variable physical and chemical characteristics of the aquatic habitat have a considerable effect on their sensitivity to toxicants. The aquatic test species of interest, unlike in mammalian studies, can be used directly. The objective of mammalian toxicology is to assess effects on humans whose sensitivity to toxicants is less affected by their environment than aquatic organisms. The dose of the toxicant used in mammalian toxicology can be measured more accurately, the mechanisms of toxic action are better understood, and the test methods are more established.

Various species of aquatic life have been used in toxicity experiments for over 130 years. One of the earliest reported studies was conducted with fish in 1863 [139], and the first proposed standard test species was the goldfish in 1917 [144]. Toxicity tests have been conducted with increasing frequency since the 1960s due to the numerous environmental regulations that have been enacted requiring their use. An additional reason is the increasing availability of standardized test methods, the first of which were published in 1960 for animal test species and 1970 for algae. Hunn [79] provides additional detail on the development of aquatic toxicology as a field of study.

Many toxicity test methods are available for use [4,13]. They differ in their cost, precision, complexity, and the skill needed to conduct them; nevertheless, their objectives are similar. They are conducted to determine the relative potency among chemicals and the relative susceptibility among different species and life stages, as well as to identify environmental variables that influence the overall outcome of exposure. Toxicity tests are conducted frequently to meet regulatory guidelines for the use and discharge of commercial chemicals and effluents [162,163,179]. In addition, toxicity results are used to assess water and sediment quality and to support the development of numerical, effects-based water quality standards and sediment quality benchmarks to protect aquatic life.

Aquatic toxicologists do not use all of the available toxicity tests to determine the effects for any single toxicant; instead, a tiered approach is used to provide a systematic and comprehensive process of deriving the toxicity data needed to assess the environmental hazard of a chemical. This approach consists of conducting short-term screening tests prior to using longer term studies that are more complex. This sequential evaluation provides an efficient use of resources and tends to eliminate unnecessary testing. The criteria used to determine the appropriate level of testing have been discussed in detail [9]. The decision points and testing phase depend on the quality and quantity of data required for the test substance of interest. The types of toxicity tests used in this tiered approach are discussed briefly below.

## SINGLE-SPECIES TOXICITY TESTS

### METHODOLOGIES

The two basic types of aquatic single-species toxicity tests are acute and chronic. Acute toxicity tests have been the workhorse of aquatic toxicologists for many years. These

**TABLE 42.2**
**The Availability of Toxicity Data for High-Volume Commercial Chemicals**

| | |
|---|---|
| Acute toxicity | 90% |
| Repeat dose toxicity | 30% |
| Carcinogenicity | 20% |
| Mutagenicity | 50% |
| Reproductive toxicity | 10% |
| Teratogenicity | 30% |
| Acute toxicity (fish or daphnids) | 50% |
| Short-term toxicity (green algae) | 5% |
| Effects on soil organisms | <5% |

*Source:* Van der Zandt, P.T.J. and Van Leeuwen, C.J., *A Proposal for Priority Setting of Existing Chemical Substances*, proposal prepared for the Directorate General for Environment, Nuclear Safety and Civil Protection of the Commission of the European Communities, Directorate General for Environmental Protection, The Hague, The Netherlands, 1992.

tests are relatively simple, have short durations, and are cost effective. As a result, a large historical data base exists for many chemicals, including detergent surfactants, trace metals, pesticides, and various other organic compounds [89,90,104,189].

Acute toxicity tests are most often used to screen toxicity quickly or to determine the relative sensitivities of different test species. Mortality is the effect monitored during the usual test duration of 48 hours (invertebrates) or 96 hours (fish). In a typical acute toxicity test, 5 to 10 organisms are exposed under static conditions in glass test beakers to five test concentrations. A control is included. The test concentrations and control are conducted in triplicate. Daily observations are made on survival, and dead organisms are removed. At test termination, the concentration that kills 50% of the test organisms ($LC_{50}$ value) is determined using probit analysis or graphical interpolation. Unlike in chronic toxicity tests, there is no test solution renewal, the organisms are unfed, and there is no analytical verification of the test concentrations. Furthermore, cumulative, chronic, and sublethal effects of a chemical are not evaluated in acute toxicity tests, although behavioral changes and lesions caused by a chemical can be observed.

Chronic toxicity tests are more complex and time consuming than acute studies and for these reasons are conducted less frequently (Table 42.2). The methodologies for these tests differ considerably from acute tests because they are designed for the specific life histories of the various test species. Chronic toxicity tests may be for a full life-cycle (egg–egg), partial life cycle (embryo–larval), and partial life history (egg–death). Full life-cycle tests are uncommon with fish due to the long durations that are necessary (0.5 to 2 years). Partial life-cycle tests with fish can be as long as 60 days. The early life-stage of fish (embryo/larvae) is usually the most sensitive period; consequently, partial life-cycle tests are often used as surrogates for the full life-cycle studies. Chronic tests may be conducted for more than one complete life-cycle if algal and invertebrate species are used, because their life cycles are shorter than those of fishes. Lethal and sublethal effects that are monitored in chronic toxicity studies include changes in growth, reproduction, behavior, physiology, and histology.

Toxicity tests may be static, continuous flow, recirculating, or static renewal based on the toxicant dosing technique. Static and flow-through procedures are more widely used in toxicity tests conducted with pure chemicals and animal test species. Chronic toxicity tests conducted with effluents are usually static renewal; those with algae, static. There is no change or renewal of the test substance and dilution water in a static test. This design is the simplest and least expensive; however, the toxicant concentrations may decrease due to adsorption, uptake, volatilization, and biodegradation. The test solutions and dilution water are renewed periodically, usually daily in a static-renewal test. In a continuous-flow test, the dilution water and test substance are continuously renewed. The exposure concentrations remain fairly constant, and the dose–response relationship can be well defined.

Various aquatic toxicity test methods have been published for single species, and many have been standardized (Table 42.3). Some are required to control the environmental entry of toxic industrial chemicals, pesticides, new drugs, and wastewaters (Table 42.4). Test method development is an ongoing process, however, which continues to improve these methods by identifying more sensitive test species and effect parameters. These efforts often result in alternative study designs; for example, Blaise [26] and Wells [196] summarize several of the microbiotests developed to provide toxicity data more quickly. In addition, the use of genotoxicity as an effect parameter is becoming more common. The use of genotoxicity as a diagnostic indicator of environmental contamination provides insight on the potential occurrence of enhanced mutation rates and alterations in the gene pool of organisms. Mutagenicity tests in bacterial and cell cultures, cytogenetic tests, and DNA tests for strand breaks and unscheduled synthesis have been used to screen the genotoxic effects of treated wastewaters, pesticides, sediment extracts, and dredged sediments [40,76].

### Experimental Conditions

In general terms, toxicity tests are conducted in a laboratory controlled for light and temperature. The test solutions containing the test species are monitored for pH, temperature, dissolved oxygen, and hardness. The test organisms are exposed for a predetermined duration that varies depending upon the type of test and test species. Daily

**TABLE 42.3**
**Standard Toxicity Test Methods Available from Standards-Writing Organizations**

| ASTM[a] | APHA et al.[b] |
|---|---|
| Phytoplankton | Bacterial bioluminescence |
| Algal growth potential | Phytoplankton |
| Static tests with microalgae | Plants |
| Bioluminescence dinoflagellates | Microalgae |
| Plants | Duckweed |
| Freshwater emergent macrophytes | Emergent vascular plants |
| Seaweeds | Ciliated protozoans |
| Duckweeds | Invertebrates |
| Early seedling growth | Rotifers |
| Invertebrates | Annelids |
| Rotifers | *Daphnia* spp. |
| Polychaete annelids | *Ceriodaphnia* spp. |
| Bivalve molluscs | Mysids |
| *Daphnia magna* | Decapods |
| Saltwater mysids | Aquatic insects |
| *Ceriodaphnia dubia* | Echinoderms |
| Echinoid embryos | Mollusks |
| Vertebrates | Vertebrates |
| Fish | Fish |
| Macroinvertebrates | |
| Amphibians | |
| Freshwater microcosm | |
| Fish behavior | |

[a] American Society for Testing and Materials [14].
[b] American Public Health Association, American Water Works Association, Water Pollution Control Federation [4].

observations on lethal and sublethal effects are made, and several calculations such as the $LC_{50}$ value and the highest no-observed-effect concentration (NOEC) and lowest first-observed-effect concentration (LOEC) are determined based on the most sensitive effect parameter of interest. Although toxicity tests have similarities, as discussed below, interspecific sensitivities, instrumentation, and methods influence the outcomes and utility of the results.

## Test Chambers

The types of test chambers used in toxicity tests depend on the test species. Various sizes of beakers, aquaria, jars, bowls, and Petri dishes have been used. The test chambers are usually constructed of materials such as glass, Teflon®, and certain plastics that minimize leaching of potential toxicants and adsorption of the test substance.

## Test Concentrations

The chemical concentrations used in an acute toxicity test are routinely based on results obtained from a pretest or range-finding test. There are no standard guidelines for conducting these preliminary tests. Generally, 5 to 10 organisms are exposed to several test concentrations that are usually an order of magnitude apart. The dilution water and exposure conditions (i.e., water temperature, salinity, hardness, and pH) in range-finding tests are usually similar to those in the definitive test. The test concentration range for a chronic test is based on the results of an acute test conducted prior to the chronic test.

The test organisms are exposed in the definitive test to five or more concentrations of the test compounds chosen in a geometric progression. The test concentrations and controls are usually replicated at least threefold. The test compound is added to the dilution water, which may be well water, reconstituted water, dechlorinated tap water, uncontaminated river water, or natural or artificial seawater. The dilution water is well aerated and undesirable organisms are removed, usually by filtration, before use.

An organic solvent is used to dissolve test compounds with minimal water solubility. Several have been used and include triethylene glycol, dimethyl sulfoxide, acetone, and dimethyl formamide. The $LC_{50}$ values for these solvents are between 9000 and 92,500 mg/L [141]. The concentration of the solvent in the test water should not exceed 0.5 ml/L or should not be greater than 1/1000 of the $LC_{50}$ value of the solvent. When a solvent is used, a solvent control is included in the study.

Toxicant delivery systems are used to deliver, on a once-through basis, the various test concentrations to the test chambers in continuous-flow toxicity tests. The serial proportional diluter (Figure 42.2) is the most common design used to mix the dilution water with the test substance to produce the desired test concentrations. The construction materials in toxicant delivery systems, such as for the test chambers, should not be rubber or certain plastics to prevent leaching of potentially toxic substances. The test concentrations are analytically confirmed during chronic toxicity tests. Analyses are performed at least weekly for each test concentration and control for tests of 7-day duration or longer. In chronic tests of shorter duration, analyses are usually conducted on alternate days. Analytical verification of the test concentrations in range-finding and static acute toxicity tests is seldom performed, and the results from these tests are generally based on nominal concentrations of the test substance.

## Test Species

Historically, animal test species have been used more frequently than plant species and freshwater species more frequently than marine species. This trend can be seen in Table 42.5. Most toxicity tests are conducted with single cultured test species such as those listed in Table 42.6. The more commonly used freshwater species, particularly

**TABLE 42.4**
**Toxicity Tests Conducted with Aquatic Species Required for Regulatory Control of Industrial Chemicals, Pesticides, New Drugs, and Treated Municipal and Industrial Effluents**

| TSCA (Industrial Chemicals) and FIFRA (Pesticides)[a] | FDA (New Drugs)[b] | OECD (New Chemicals)[c] | NPDES (Effluents)[d] |
|---|---|---|---|
| Aquatic invertebrate acute toxicity | Algal growth inhibition | Algal growth inhibition test | Acute toxicity |
| Gammarid acute toxicity test | *Daphnia magna* acute toxicity | *Daphnia magna* acute immobilization test | *Ceriodaphnia dubia* |
| Oyster acute toxicity test | *Daphnia magna* chronic toxicity | Fish, acute toxicity test | *Daphnia magna* and *D. dubia* |
| Mysid acute toxicity test | *Hyallela azteca* acute toxicity | Fish, prolonged toxicity test | *Pimephales promelas* |
| Penaeid acute toxicity test | Fish early life stage | Fish, early life stage test | *Oncorhynchus mykiss* |
| Bivalve acute toxicity test | Fish chronic toxicity | *Daphnia magna* reproduction test | *Mysidopsis bahia* |
| Fish acute toxicity test | Sediment invertebrate species toxicity | Fish, short-term toxicity test | *Cyprinodon variegatus* |
| Daphnid chronic toxicity test | | *Lemna* growth inhibition test | *Menidia beryllina* |
| Mysid chronic toxicity test | | Bioconcentration flow-through fish test | Chronic toxicity |
| Fish life stage | | | *Pimephales promelas* |
| Fish early life stage | | | *Ceriodaphnia dubia* |
| Whole sediment acute toxicity | | | *Pseudokirchneriella subcapitata* |
| Chironomid sediment toxicity test | | | *Cyprinodon variegatus* |
| Aquatic food chain transfer | | | *Menidia beryllina* |
| Generic freshwater microcosm test | | | *Arbacia punctulata* |
| Field testing for aquatic organisms | | | *Champia parvula* |
| Aquatic plant toxicity test using *Lemna* spp. | | | |
| Algal toxicity | | | |

[a] U.S. Environmental Protection Agency Toxic Substances Control Act and U.S. Environmental Protection Agency Federal Insecticide, Fungicide, Rodenticide Act [162,163].
[b] U.S. Food and Drug Administration [187].
[c] Organization for Economic Cooperation and Development [121].
[d] National Pollutant Discharge Elimination System [179,180].

in tests used for regulatory compliance, are fathead minnows (*Pimephales promelas*), daphnid species (*Daphnia magna*, *Ceriodaphnia dubia*), and a green alga (*Pseudokirchneriella subcapitata*). Common marine species include sheepshead minnows (*Cyprinodon variegatus*), mysids (*Americamysis bahia*, formerly *Mysidopsis bahia*), and diatoms (*Skeletonema costatum*). The species in Table 42.6 were selected based on several criteria, primarily ease of culture, commercial availability, and size. The test species are acclimated for a specific time prior to testing to eliminate diseased organisms and to acclimate the organisms to the test conditions. Generally, a minimum of 10 animals are exposed in static and static-renewal tests and 20 in a flow-through test to each test concentration and control. The recommended loading density for the test species is between 0.5 and 0.8 g/L in static tests and 1 and 10 g/L in continuous-flow tests.

Historically, environmental toxicology has centered on determining the effects of anthropogenic chemicals on organisms inhabiting temperate zones. Tropical marine ecosystems have been considered relatively pristine; however, as a result of resource exploitation that began in the 1960s, the coastal waters of the tropics and subtropics have experienced increased sedimentation, nutrient loading, and chemical contaminants. The impact of these events on highly productive ecosystems such as mangrove forests, seagrasses, and coral reefs are relatively unknown; consequently, there is an increased focus on developing toxicity tests and toxicity data for tropical plant and animal species. A review of these efforts and examples of the data have been published [140].

Reference toxicants are often used to determine the health of the test species. There is no widely used reference toxicant; several that have been used include sodium dodecyl sulfate (anionic surfactant), sodium chloride, sodium pentachlorophenol, and cadmium chloride.

Sensitivity is a criterion that is also used in the choice of a test species. The sensitivities of the species in Table 42.6 relative to one another as well as to indigenous flora and fauna in the ecosystem are a matter of scientific debate. No single animal or plant test species is consistently most sensitive to toxicants or most reliable for extrapolation to all other organisms. Because toxicity is species specific, acute toxicity tests are conducted first with a variety of freshwater and marine test species to determine the most sensitive plant and animal. These sensitive species are then used in all subsequent chronic testing.

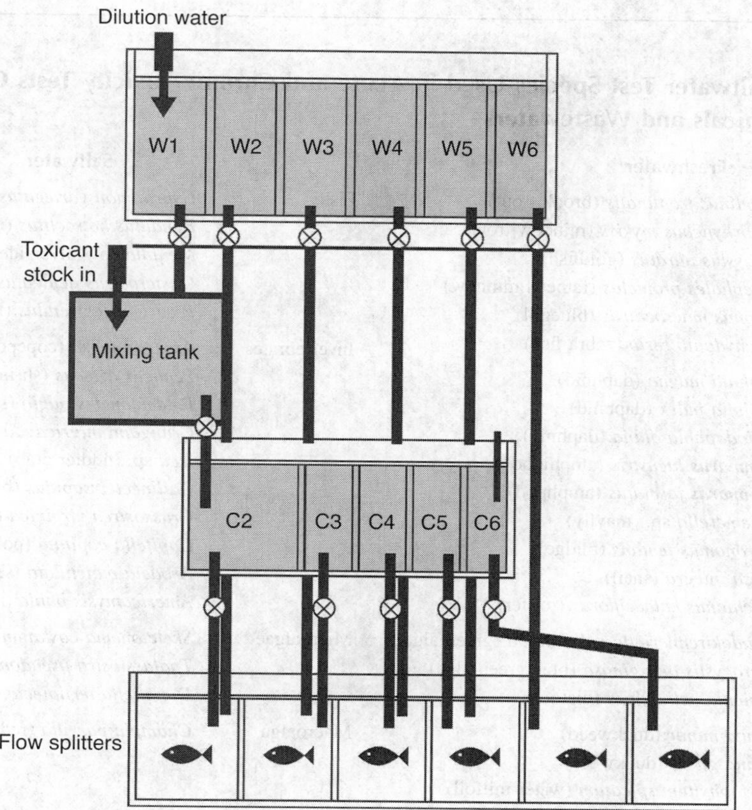


**FIGURE 42.2** Diagram of a proportional diluter designed to ensure that five concentrations of a toxicant and a control treatment are delivered to the test species in a toxicity test. (Adapted from Landis, W.G. and Yu, M.H., *Introduction to Environmental Toxicology*, Lewis Publishers, Boca Raton, FL, 1995.)

## Calculations

The results of acute toxicity tests are reported usually as the $LC_{50}$ value and the corresponding 95% confidence interval. These calculations are determined using one of several statistical methods that are discussed by Stephan [154], such as probit analysis and moving average interpolation. Probit analysis [51] is the most commonly used statistical method to determine $LC_{50}$ values. Graphical interpolation can be used also to estimate the $LC_{50}$ value where the proportion of deaths versus the test concentration is plotted for each observation time. The no-observed-effect concentration and lowest-observed-effect concentration are the usual calculations reported from chronic toxicity tests. The NOEC is the highest concentration in which the measured effect is not statistically different from that of the control. The LOEC is the lowest concentration at which a statistically significant effect occurred. These concentrations are based on the most sensitive effect parameter, such as hatchability, growth, and reproduction. The statistical procedure for these calculations combines the use of analysis of variance techniques and multiple comparison tests. In some cases, the maximum acceptable toxic concentration (MATC) is reported from chronic toxicity results. The MATC is a concentration ($x$) that is within the range of the NOEC – LOEC, NOEC $< x <$ LOEC. The first-effect concentration can be expressed as the geometric mean of the two terms.

**TABLE 42.5**
**More Commonly Reported Test Species in the AQUIRE Database**

| Common Name | Species | Percentage of Reported Data (%) |
|---|---|---|
| Rainbow trout | *Oncorhynchus mykiss* | 8.6 |
| Bluegill | *Lepomis macrochirus* | 5.6 |
| Fathead minnow | *Pimephales promelas* | 4.9 |
| Water flea | *Daphnia magna* | 3.9 |
| Carp | *Cyprinius carpio* | 2.4 |
| Coho salmon | *Oncorhynchus kisutch* | 2.1 |
| Goldfish | *Carassius auratus* | 1.9 |
| Channel catfish | *Ictalurus punctatus* | 1.7 |
| Mosquito fish | *Gambusia affinis* | 1.5 |
| Brook trout | *Salvelinus fontinalis* | 1.0 |
| Green alga | *Scenedesmus quadricauda* | 0.9 |

**TABLE 42.6**
**Freshwater and Saltwater Test Species Used in Acute and Chronic Toxicity Tests Conducted with Commercial Chemicals and Wastewaters**

| | Freshwater | | Saltwater |
|---|---|---|---|
| Fish | *Salvelinus frontinalis* (brook trout) | Fish | *Cyprinodon variegatus* (sheepshead minnow) |
| | *Oncorhynchus mykiss* (rainbow trout) | | *Fundulus hetroclitus* (mummichog) |
| | *Carassius auratus* (goldfish) | | *Menidia* sp. (silverside) |
| | *Pimephales promelas* (fathead minnow) | | *Gasterosteus aculeatus* (threespine stickleback) |
| | *Lepomis macrochirus* (bluegill) | | *Pleuronectes vetulus* (English sole) |
| | *Brachydanio rerio* (zebra fish) | Invertebrates | *Arcartia tonsa* (copepod) |
| Invertebrates | *Daphnia magna* (daphnid) | | *Penaeus aztecus* (shrimp) |
| | *Daphnia pulex* (daphnid) | | *Palaemonetes pugio* (shrimp) |
| | *Ceriodaphnia dubia* (daphnid) | | *Crangdon migricauda* (shrimp) |
| | *Gammarus lacustris* (amphipod) | | *Uca* sp. (fiddler crab) |
| | *Gammarus fasciatus* (amphpod) | | *Callinectes sapidus* (blue crab) |
| | *Ephemerella* sp. (mayfly) | | *Crassostrea virginica* (oyster) |
| | *Chironomus tentans* (midge) | | *Capitella capitata* (polychaete) |
| | *Physa integra* (snail) | | *Arbacia punctulata* (sea urchin) |
| | *Brachionus calccciflorus* (rotifer) | | *Americamysis bahia (mysid)* |
| Microalgae | *Pseudokirchneriella subcapitata* (green algae) | Microalgae | *Skeletonema costatum* (diatom) |
| | *Microcystis aeruginosa* (blue-green alga) | | *Thalassiosira pseudonana* (diatom) |
| | *Navicula pelliculosa* (diatom) | | *Dunaliella tertiolecta* (flagellate) |
| Vascular plants | *Lemna minor* (duckweed) | Macroalga | *Champia parvula* (red macroalga) |
| | *Lemna gibba* (duckweed) | | |
| | *Myriophyllum spicatum* (water milfoil) | | |
| | *Ceratophyllum demersum* (coontail) | | |

### Variability/Precision

Laboratory toxicity tests conducted with freshwater and saltwater species are considered relatively precise and reliable based on current information from inter and intralaboratory comparisons of toxicity results. Generally, the $LC_{50}$ values from acute toxicity tests conducted under similar experimental conditions vary less than threefold. This has been observed for metals, effluents, reference toxicants, and different organic compounds [37,58,63,109a,134,138, 170,171]. Coefficients of variation (CVs) for acute and chronic toxicity tests conducted with daphnid species and chemicals and effluents are between 27 and 39% [42,63,149]. The CV values for several reference toxicants and acute daphnid studies ranged between 10 and 72% [92] and from 47 to 83% for chronic toxicity tests with algae [180]. Anderson and Norberg-King [17] reported the CV values for a variety of freshwater and marine species and found minimal variation. They concluded that biological tests can be conducted with a precision similar to that for chemical-specific measurements.

### Data Utility

One measure of the usefulness of the different types of toxicity tests is their frequency of use. Based on a review of the premanufacture notification (PMN) submissions under the Toxic Substances Control Act (TSCA), the more frequently conducted tests are, in decreasing order, acute fish, acute invertebrates, algal toxicity, chronic toxicity, and terrestrial toxicity. In addition, several judgments on the value of toxicity tests to the scientific and regulatory communities have been reported [3,59,101]. The acute toxicity test was judged the most useful for commercial chemicals, with tests monitoring behavioral, histological, and physiological effects being less useful. Toxicity tests with daphnids (*Daphnia magna* and *Ceriodaphnia dubia*) were judged to be of more value in monitoring sediment toxicity than those with other species, and early life-stage tests were judged most useful to determine the toxicity of refinery wastewaters.

## MULTISPECIES TOXICITY TESTS

The results of the standard acute and chronic single-species toxicity tests conducted in the laboratory cannot be used alone to predict effects on natural populations, communities, and ecosystems. First, the cultured species in laboratory tests are often different from those found inhabiting most ecosystems and conditions, and the size of the test species, its life stage, and nutritional state, among others, can have an effect on toxicity. Second, the laboratory tests conducted under controlled conditions cannot duplicate the complex interacting physical and chemical conditions of

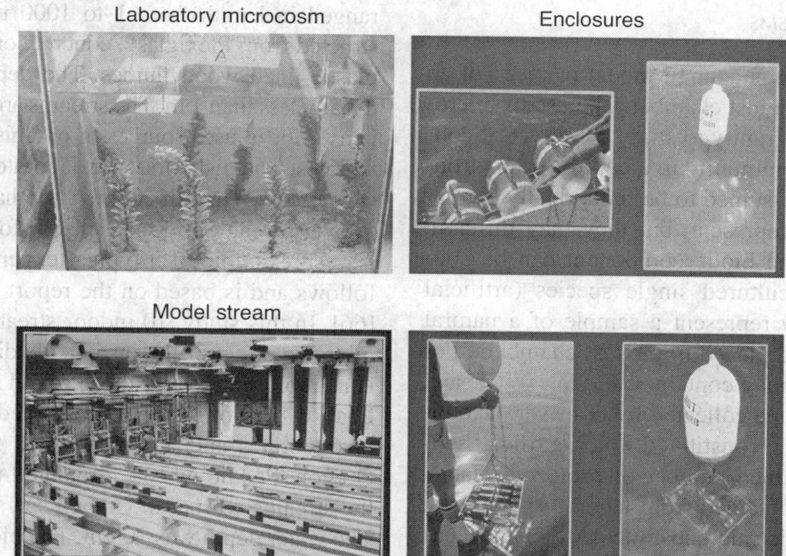


**FIGURE 42.3** Examples of experimental designs used to expose either multiple test species simultaneously or natural biotic assemblages to toxicants. Test durations ranged from 3 hours to 21 days.

ecosystems, such as seasonal changes in water temperature, dissolved oxygen, salinity, and suspended solids. Third, aquatic species are usually exposed simultaneously to numerous potential toxicants (mixtures). Although the toxicities of binary and ternary mixtures have been evaluated for some chemicals in laboratory toxicity tests, the resultant information has predictive limitations.

Because of the deficiencies of single-species toxicity tests, multispecies toxicity tests have been developed to address ecosystem structural and functional processes [34,35]. These tests include the use of laboratory microcosms and mesocosms such as outdoor ponds, experimental streams, and enclosures (Figure 42.3). Standardized procedures for these tests, with the exception of a freshwater microcosm, do not exist. They can be conducted indoors and outdoors with plant and animal species obtained from laboratory cultures or collected from natural sources. The toxic effects, in addition to those determined in single-species tests, are determined for structural parameters, such as community similarity, diversity, and density, and functional parameters, such as community respiration and photosynthesis. Effects on these parameters are reported usually as the NOEC and LOEC. A brief description of the major types of multispecies toxicity tests follows, and their advantages and disadvantages relative to single-species studies are summarized in Table 42.7. More detail concerning specific test conditions, replicability, and regulatory usefulness is available from a number of reviews [25,34,67,87,166].

**TABLE 42.7**
**Advantages and Disadvantages of Single-Species and Multispecies Toxicity Tests**

| | Single-Species | Multispecies |
|---|---|---|
| Advantages | Simple | Conducted under more realistic conditions |
| | Cost effective | Can simultaneously study different trophic levels |
| | Standard protocols available | Can be indigenous species |
| | Existing databases | Can study effect of environmental modification |
| | Rapid | |
| | Reproducible | |
| Disadvantages | Responses of only individual determined | Not cost effective |
| | Use nonindigenous species | Standard protocols largely absent |
| | Controlled experimental conditions | Poor replication |
| | Fate processes ignored | Adaptation not considered |
| | Recovery rate not considered | Stability of exposure concentration uncertain |
| | Cumulative effects not studied | Data interpretation difficult |
| | Ecological interactions not considered | |

### Laboratory Microcosms

A laboratory microcosm is a small model or piece of an ecosystem contained in a test chamber. The use of microcosms provides an opportunity to study the effect of contaminants on a biotic community in a controlled environment. Microcosms are assumed to be functionally similar to the ecosystem they represent, but they may differ in origin and structure. The biotic component can be constructed from several cultured single species (artificial microcosms), or it may represent a sample of a natural ecosystem collected and placed in the test chambers. For example, a microcosm may contain sediment, water, and indigenous flora and fauna collected from a river, lake, or pond, or it may contain reconstituted water, artificial sediment, and a predetermined number of protozoa, plants, invertebrates, and fish obtained from laboratory cultures. The standardized aquatic microcosm (SAM) method has a set density of 15 species and is the only standardized microcosm test method that is currently available [15]. In this method, 24 3-L microcosms are used, and they can be aquaria. They contain a well-defined medium of trace metals and vitamins and 200 g of sand (sediment) to which chitin and cellulose are added. The tests are conducted in the laboratory at 22°C with a 12-hour period of light at 80 $\mu E/m^2/s$. Ten species of freshwater algae and five species of freshwater invertebrates are added to the microcosms, which are intended to represent a new pond ecosystem. The test species are exposed to three test concentrations of the contaminant for 63 days during which observations are made on a variety of parameters, including algal biovolume, species diversity and density, and dissolved oxygen concentrations. The toxicant concentrations are renewed periodically during the study, and biological and chemical measurements are taken weekly or biweekly depending on the parameter.

### Outdoor Ponds

Outdoor ponds have been used to investigate the fate and effects of several chemicals, but primarily pesticides, on aquatic life [31,166]. There is no universally accepted test design, and various types of plant and animal life have been used in these systems. Ponds of various sizes and shapes have been utilized. The depth is usually 1 m or less. Volumes have ranged from 10 to 650 $m^3$.

### Experimental Streams

Experimental streams have been used to assess the effects of thermal effluents, nutrients, metals, insecticides, and municipal and industrial effluents on natural biota in studies ranging in duration from several days to several years. Experimental streams may be flow through or circulating and may be located indoors or outdoors. The streams have ranged in length from 1 to 1000 m and have been constructed from Plexiglas®, concrete or aluminum troughs, or plastic-lined wood flumes. The depth of most streams is usually less than 1 m. The streams are colonized with organisms prior to use. Sources of organisms (periphyton, invertebrates, and fish) may be laboratory cultures or natural ecosystems. Time of colonization has varied from a month to over a year. See Kosinski [84] for additional detail.

An example of a typical experimental stream design follows and is based on the report of Hansen and Garton [66]. In this study, 10 indoor streams were used to determine the effect of the insecticide diflubenzuron on organisms representative of those found in a freshwater stream. Each 6.1-m-long stream contained natural substrate and invertebrates and algae collected from a small creek in Oregon. Paddle wheels circulated the well water through the streams with a midstream surface current of 30 cm/s. The water depth was 10 cm. The biota in the streams were allowed to equilibrate for 3 months prior to the 5-month exposure to the toxicant. Eight streams were paired to deliver the four test concentrations, and the remaining pair served as the control. Light intensity was 200 fc, with a natural diurnal light cycle being used. The insecticide was continuously delivered to the streams, and effects on biomass and diversity of plant and animal life were determined monthly. Water chemistry such as pH, hardness, and temperature were determined at least weekly.

### Enclosures

Enclosures isolate a portion of an ecosystem that can be dosed with the toxicant in such a way that significant contamination beyond the limits of the enclosure will not occur. The size, shape, and volume of enclosures used have varied considerably. In freshwater experiments, the volumes have ranged from 8 to 3,000,000 liters. Plastic tubes, carboys, plastic bags, and limnocorrals have been used to enclose the biota. The effects of oil, mercury, phenolic compounds, acidification, and pesticides, among other toxicants, have been investigated in studies ranging up to 3 months.

## SINGLE-SPECIES–MULTISPECIES TOXICITY COMPARISONS

The results of several multispecies tests have been compared to those of single-species tests conducted with the same toxicant; for example, the results from the experimental stream study of Hansen and Garton [66], discussed previously, were compared to those for single-species chronic toxicity tests conducted with fish, daphnids, algae, and snails. These single, cultured species were more than an order of magnitude more sensitive than the stream communities to the insecticide diflubenzuron. Other comparisons of this type are discussed by Boyle [31] and Cairns and coworkers [34,35]. In most cases, the outcomes of these comparisons have been species and compound specific; thus, the use of toxicity test results for single species to predict effects on natural biotic assemblages without supporting data requires careful consideration. The validation of laboratory-derived toxicity data continues to receive high priority by aquatic toxicologists.

## PREDICTIVE TOXICOLOGY

The large number of chemicals combined with the many different regulations, the cost of conducting test batteries, and the number of species to be protected limit the full use of standard toxicity tests to only a relatively few chemicals. In the absence of such data, methodologies have been developed to predict chemical toxicities. These methodologies include the use of quantitative structure–activity relationships [191] and techniques to predict acute toxicity of chemicals from one species to others [18] and to predict chronic toxicity from acute toxicity [46]. The acute-to-chronic ratio (ACR) is used to predict chronic toxicity when only acute toxicity data are available for a chemical. ACR values have been reported for a variety of chemicals, particularly metals and pesticides, and most are 25 or less. Kenaga and Moolenaar [82], Maki [102], and LeBlanc [88] provide equations that can be used to predict the acute toxicity of pesticides, metals, and nonpesticide organics to freshwater and marine algae, invertebrates, and fish based on interspecies relationships in sensitivity.

Computational toxicology is a relatively new concept that offers an alternative to traditional animal-intensive approaches used in toxicological testing [181]. It utilizes mathematical and computer models and combines techniques from computational chemistry, molecular biology, and systems biology to improve prioritization of data requirements and risk assessments for toxic chemicals. The value of computational toxicology as a predictive tool is still to be determined.

The results of single-species toxicity tests are often used to predict effects in natural ecosystems with little being known about the accuracy of the prediction. In some cases, multispecies toxicity test results are available and can be used with more confidence. In other cases, conservative correction factors are used to compensate for the limitations of the toxicity data. A variety of these correction techniques have been reported [117,190] and their usefulness reviewed [67,115]. One of the more simple and common methods is the use of numerical safety or uncertainty factors to estimate *safe*, *concern*, or *risk concentrations* [43,165]. Laboratory-derived chronic toxicity results (usually the NOEC) are divided by these factors (1000, 100, and 10, respectively) to determine the concentration that, if exceeded in the ecosystem, may represent an environmental risk. The magnitude of the factor used depends on the quantity and quality of the toxicity data available for the toxicant of interest. The greater the quality and quantity of data, the smaller the safety factor used. The technical validity and magnitude of these factors are largely unproven, although they are in general use; consequently, this issue continues to be a subject of considerable debate within the scientific and regulatory communities.

An additional tool related to differences in interspecific sensitivities that has received considerable attention is the use of species sensitivity distributions (SSD). A SSD is a cumulative frequency distribution (usually a log-normal or Weibull function) of the sensitivities of a set of species to a chemical of interest. The available toxicity data (e.g., NOAEC or $EC_{50}$ values) are assumed to be a representative sample of the entire variation of sensitivity of all species to the chemical. A concentration of the chemical of concern that will be protective of a desired percentage of all species can be calculated from the frequency distribution in a manner similar to determination of a benchmark dose. Generally, the value selected is the lower confidence interval of the 5th or 10th percentile of the distribution. A modification of this approach is used for development of water quality criteria, and SSDs are frequently used in environmental risk assessments and remediation. A detailed description of SSDs and their applications and limitations is available in Posthuma et al. [143].

## SEDIMENT TOXICITY TESTS

In the past, toxicity tests have been conducted primarily with water-column-dwelling or planktonic organisms, with the objective of controlling water pollution; however, it has been realized that sediments act as reservoirs for chemicals that can adversely affect benthic aquatic life and, at times, also affect planktonic life [65]. This concern has led to the development of a variety of assessment techniques [1], including the use of sediment quality guidelines to protect benthic or bottom-dwelling life [39,97]. Test methods have been developed to support the derivation of these guidelines and to support other related regulatory activities, such as the disposal of dredged materials [167].

Sediment toxicity tests conducted in the laboratory with a variety of single species of freshwater and marine benthic organisms were reviewed by Traunsperger and Drews [158]. Several toxicity tests and methods for the collection and preparation of sediments used in toxicity tests have been standardized (Table 42.8). Additional test guidelines for marine sediment [72] and freshwater sediment [33,59] have also been reported. Most tests conducted to date have been acute and of 10-day duration or less. Toxicity tests are conducted usually with the whole sediment (solid phase) or pore water [38]. An example of a whole-sediment acute toxicity test is shown in Figure 42.4. In this static test, a burrowing amphipod or an epibenthic mysid is exposed to undiluted field-collected contaminated sediment, and mortality is recorded after 10 days of expo-

## TABLE 42.8
**Examples of Standardized Toxicity Tests Available for Sediments**

| | |
|---|---|
| USEPA [173,177,178] | Whole Sediment Acute Toxicity Invertebrates |
| | Chironomid Sediment Toxicity Test |
| | Methods for Measuring the Toxicity and Bioaccumulation of Sediment-Associated Contaminants with Freshwater Invertebrates |
| | Methods for Assessing the Chronic Toxicity of Marine and Estuarine Sediment-Associated Contaminants with the Amphipod, *Leptocheirus plumulosus* |
| ASTM [15] | Standard Guide for Conduction of 10-Day Static Sediment Toxicity Tests with Marine and Estuarine Amphipods |
| | Standard Guide for Collection, Storage, Characterization, and Manipulation of Sediments for Toxicological Testing |
| | Standard Guide for Designing Biological Test with Sediments |
| | Standard Test Methods for Measuring the Toxicity of Sediment-Associated Contaminants with Freshwater Invertebrates |
| | Standard Guide for Conduction of Sediment Toxicity Tests with Marine and Estuarine Polychaetous Annelids |
| | Standard Guide for Determination of Bioaccumulation of Sediment-Associated Contaminants by Benthic Invertebrates |
| APHA et al. [4] | Test Procedures Using the Marine Polychaete *Neanthes arenaceodentata* |
| | Test Procedures Using the Freshwater and Marine Oligochaetes *Pristina leidyi*, *Tubifex tubifex*, and *Lumbriculus variegatus* |
| | Sediment Test Procedures for Marine Bivalves |
| | Sediment Porewater Toxicity |

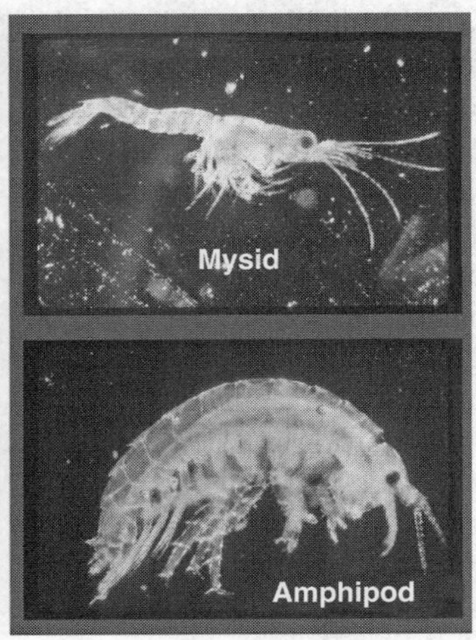


**FIGURE 42.4** Example of a whole-sediment acute toxicity test conducted for 10 days with epibenthic mysid and burrowing amphipod tests species.

sure. The mortality of amphipods is obvious by the reduction in their burrowing entrances (Figure 42.5).

A chronic toxicity test has been developed for marine sediments [178]. The results from such a test conducted with a sediment collected below a pulp mill outfall are shown in Table 42.9. Survival, growth, and reproductive effects of the burrowing amphipod *Leptocheirus plumulosus* were determined at the end of the 28-day toxicity test. Significant effects on the test species were noted at contaminated sediment concentrations of 10% (young production, fertility), 50% (survival), and 100% (length).

Reliable test methods for contaminated sediments have only recently become available, and the test method development process continues. Several issues remain to be resolved. Among the more important of these are validation of the single-species test results and the identification of sensitive species and response parameters for whole sediments and their associated pore waters.

## EFFLUENT TOXICITY TESTS

Toxicity tests are used in the National Pollutant Discharge Elimination System (NPDES) permitting process to determine the toxicities of municipal and industrial effluents prior to discharge on aquatic life [62,179,180]. A summary of the experimental conditions for test methodologies using freshwater species appears in Table 42.10. The methodologies differ slightly from those used for pure chemicals; for example, the choice of the dilution water

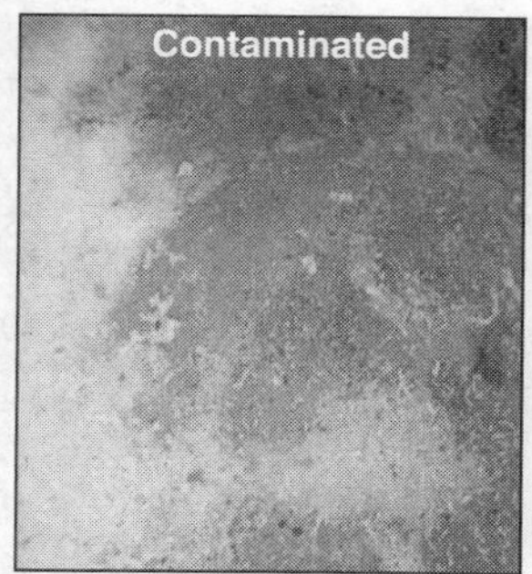


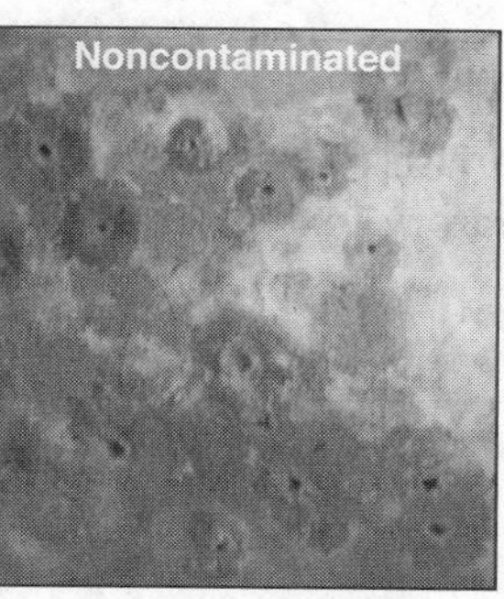


**FIGURE 42.5** Burrowing entrances of infaunal amphipods exposed for 10 days in an acute toxicity test to contaminated and noncontaminated estuarine sediments. Lack of entrances indicates mortality.

**TABLE 42.9**
**Results of a 28-Day Chronic Sediment Toxicity Test Conducted with the Marine Amphipod *Leptocheirus plumulosus* and Whole Sediment Collected Below a Pulp Mill and Plant**

| Test Concentration[a] | Mean Survival (%) | Young Production (Number of Juveniles) | Fertility | Mean Length (mm) |
|---|---|---|---|---|
| Control | 97 | 91 | 59 | 5.83 |
| 10 | 93 | 32 | 31 | 5.31 |
| 25 | 92 | 30 | 19 | 5.45 |
| 50 | 83 | 8 | 6 | 4.95 |
| 100 | 9 | 0 | 0 | 3.06 |

[a] Contaminated sediment was diluted with a noncontaminated reference sediment to obtain test concentrations.

and the effluent collection technique used are important considerations. In most cases, water collected from the receiving water above the outfall is used for dilution, and composite samples of treated effluent are used. The test species (algae, invertebrates, or fish) are usually exposed to five effluent dilutions for 4 to 7 days. The tests are static renewal except those for algae, which are static. The calculations reported are the $LC_{50}$ value, the NOEC, and the LOEC, which are expressed as percent effluent. The specific causes of toxicity in the effluent can be identified by using a toxicity identification evaluation (TIE) that consists of comparative toxicity testing and chemical fractionation techniques [169].

**TABLE 42.10**
**Several Experimental Conditions for Acute Toxicity Tests Conducted with Effluents and Freshwater Fish, Invertebrates, and Algae**

| Experimental Condition | Fish (*Pimephales promelas*) | Invertebrate (*Ceriodaphnia dubia*) | Alga (*Pseudokirchneriella subcapitata*) |
|---|---|---|---|
| Test type | Static nonrenewal, static renewal, or flow-through | Static nonrenewal, static renewal, or flow-through | Static nonrenewal |
| Test duration | 24, 48, or 96 hr | 24, 48, or 96 hr | 96 hr |
| Temperature | 20 ± 1°C or 25 ± 1°C | 20 ± 1°C or 25 ± 1°C | 25 ± 1°C |
| Light Intensity | 10–20 μE/m²/s (50–100 ft-c) | 10–20 μE/m²/s (50–100 ft-c) | 86 ± 8.6 μE/m²/s |
| Photoperiod | 16 hr light, 8 hr darkness | 16 hr light, 8 hr darkness | Continuous illumination |
| Test chamber size | 250 mL | 30 mL | 125 mL or 250 mL |
| Test solution volume | 200 mL | 15 mL | 50 mL or 100 mL |
| Renewal of test solutions | After 48 hr | After 48 hr | None |
| Age of test organisms | 1–14 days; ≤24-hr range in age | Less than 24 hr old | 4–7 days |
| No. organisms per test chamber | 10 for effluent and receiving water tests | 5 for effluent and receiving water tests | 1 × 10⁴ cells/mL |
| No. of replicate chambers per concentration | 2 for effluent tests, 4 for receiving water tests | 4 for effluent and receiving water tests | 3 to 4 |
| No. of organisms per concentration | 20 for effluent tests, 40 for receiving water tests | 20 for effluent and receiving water tests | 3 × 10⁴ |
| Test concentrations | 5 and a control | 5 and a control | 5 and a control |
| Endpoint | Mortality | Mortality | Growth (cell counts, biomass chlorophyll fluorescence) |

*Source:* USEPA, *Methods for Measuring the Acute Toxicity of Effluents and Receiving Waters to Freshwater and Marine Organisms*, 5th ed., EPA/821/R-02/012, Office of Water, U.S. Environmental Protection Agency, Washington, D.C., 2002.

**TABLE 42.11**
**Response of a Diatom (*Dunaliella tertiolecta*), a Fish (*Cyprinodon variegatus*), and an Invertebrate (*Mysidopsis bahia*) Exposed in Laboratory Toxicity Tests to Treated Municipal Wastewater (Test Durations 4 to 7 Days)**

| Species | Wastewater Concentration (%) | Effect Parameters | | | | | | |
|---|---|---|---|---|---|---|---|---|
| | | Growth (% of Control) | | | | | | |
| Diatom: | Control | 100 | | | | | | |
| | 6 | 160 | | | | | | |
| | 12 | 144 | | | | | | |
| | 25 | 222 | | | | | | |
| | 50 | 178 | | | | | | |
| | 100 | 756 | | | | | | |
| | **Wastewater Concentration (%)** | **Mortality (%)** | **Mean Weight (mg)** | | | | | |
| Fish | Control | 0 | 1.00 | | | | | |
| | 6 | 0 | 1.19 | | | | | |
| | 12 | 0 | 0.84 | | | | | |
| | 25 | 2 | 1.01 | | | | | |
| | 50 | 2 | 0.90 | | | | | |
| | 100 | 2 | 0.90 | | | | | |
| | **Wastewater Concentration (%)** | **Mortality (%)** | **Mean Weight (mg)** | **Reproductive Maturity** | | | | |
| | | | | **Eggs** | **No Eggs** | **Immature** | **Mature** | |
| Invertebrate | Control | 2 | 0.21 | 8 | 7 | 17 | 7 | |
| | 6 | 4 | 0.30 | 8 | 7 | 18 | 6 | |
| | 12 | 12 | 0.27 | 3 | 9 | 14 | 9 | |
| | 25 | 8 | 0.32 | 6 | 7 | 17 | 7 | |
| | 50 | 20 | 0.19 | 2 | 3 | 13 | 10 | |
| | 100 | 63 | 0.25 | 1 | 5 | 9 | 19 | |

The results of three wastewater toxicity tests are shown in Table 42.11. Three saltwater test species were exposed to a municipal wastewater for 4 to 7 days. The interspecific differences in response are obvious. It can be seen that the wastewater stimulated algal growth at a diluted concentration as low as 6%. The same wastewater concentration, however, was not toxic to fish but did have an adverse effect on an estuarine invertebrate (mysid). For this species, 63% mortality was observed after exposure to the 100% wastewater concentration.

## PHYTOTOXICITY TO AQUATIC PLANTS

The majority of aquatic toxicity tests have been conducted with animal test species because they were once thought to be more sensitive than plants. This generalization is not true [89], and recently phytotoxicity tests have been more commonly conducted although with a limited number of algal species and even fewer species of vascular plants.

Several standardized methods are available to determine the phytotoxic effects of chemicals and effluents (Table 42.12). Species of microalgae have been used more frequently than vascular species. The freshwater algal species most frequently used has been the green microalga *Pseudokirchneriella subcapitata*, for which a relatively large database exists. Marine species used include the diatom *Skeletonema costatum*, the flagellate *Dunaliella tertiolecta*, and the red macroalga *Champia parvula*.

Most toxicity tests conducted with algae are chronic, 3 to 4 days in duration, although exposures can be less than 1 day if effects on photosynthesis are measured. These static exposures occur in a liquid nutrient-enriched medium under conditions of controlled pH, temperature, and light (Figure 42.6). Inhibitory and stimulatory effects on population growth are monitored during the exponential growth phase. Five test concentrations and a control are included in each study. The most common calculation reported is the 96-hr $IC_{50}$ value (concentration that reduces the parameter of interest 50%), but algistatic (completely stops growth) and algicidal (lethal) concentrations have also been reported. In addition, the $SC_{20}$ concentration (stimulatory concentration) is reported if growth stimulation is

**TABLE 42.12**
**Examples of Experimental Conditions in Several Phytotoxicity Tests Conducted with Micro and Macro Algae, Duckweed, and Emergent Vascular Plants**

| Test Type | Duration (days) | No. of Test Concentrations | Test Species | No. of Replicates | Temperature (°C) | Light Intensity |
|---|---|---|---|---|---|---|
| *Algae* | | | | | | |
| Static | 3 | 5 | *Pseudokirchneriella subcapitata* (F) | 3 | 21–25 | 120 μE/m²/s |
| | | | *Scenedesmus quadricauda* (F) | | | |
| | | | *Chlorella vulgaris* (F) | | | |
| Static | 4 | 5 | *Scenedesmus subspicatus* (F) | 3 | 20–24 | 30–90 μmol/m²/s |
| | | | *Microcystis aeruginosa* (F) | | | |
| | | | *Anabaena flos-aquae* (F) | | | |
| | | | *Navicula pelliculosa* (F) | | | |
| | | | *Skeletonema costatum* (M) | | | |
| | | | *Dunaliella tertiolecta* (M) | | | |
| *Duckweed* | | | | | | |
| Static | 4 | 5 | *Lemna gibba* (F) | 4 | 25 ± 2 | 2150–4300 lux |
| Static | 7 | 3–5 | *Lemna minor* (F) | 3 | 25 ± 2 | 6200–6700 lux |
| *Emergent/submersed vascular plants* | | | | | | |
| Static-renewal | 14 | 5 | *Spartina pectinata* (E) | 5 | 20–30 | 150–200 μmol/m²/s |
| | | | *Scirpus acutus* (E) | | | |
| Static | 14 | 5 | *Myriophyllum sibiricum* (F) | 5 | 20–25 | 100–150 μmol/m²/s |

*Note:* F, freshwater; M, marine; E, estuarine.

*Source:* Methods are from APHA et al. [4], ASTM [15], and OECD [121].

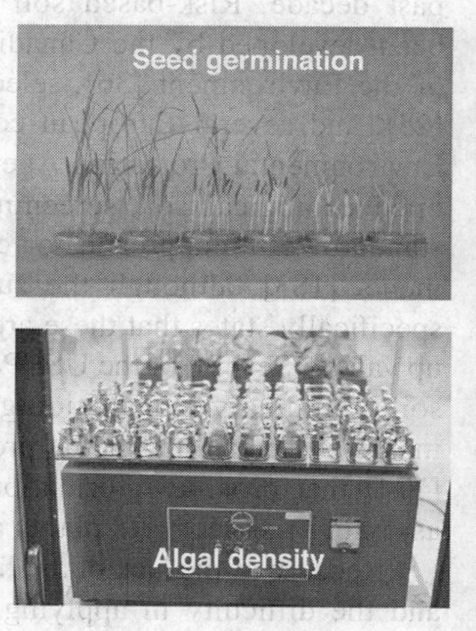


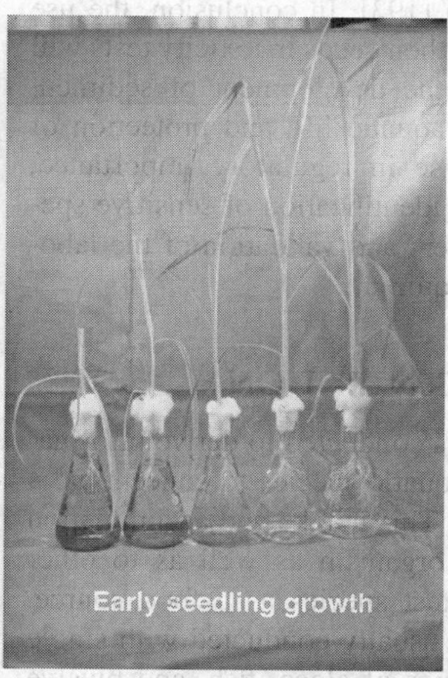


**FIGURE 42.6** Examples of toxicity tests conducted in the laboratory for 4 to 28 days with microalgae and rooted vascular plants. Toxicants include commercial chemicals, effluents, and contaminated sediments.

observed. The $SC_{20}$ value represents that concentration that increases algal growth 20% above that of the algal population in the control. Additional information on the use of algae in toxicity tests is available from Walsh [192], Thursby et al. [157], and Nyholm and Kallqvist [116].

Freshwater floating and rooted vascular plants have been used more frequently in toxicity tests than marine species. The duckweeds have been the more commonly used freshwater species due to their small size and rapid growth. Several published methods are available for *Lemna minor* and *L. gibba* [4,13]. Tests with these species are usually 4 to 14 days in duration, during which effects on frond number and chlorophyll content are monitored. The results are expressed as an $EC_{50}$ value and the NOEC.

**TABLE 42.13**
**Examples of Bioaccumulation Tests**

| Test | Reference |
|---|---|
| Bioaccumulation Tests with Whole Sediments (Dredged Materials) | USEPA/U.S. Army Corps of Engineers [167] |
| Oyster Bioconcentration Factor | USEPA [173] |
| Fish Bioconcentration Factor | |
| Methods for Measuring the Toxicity and Bioaccumulation of Sediment-Associated Contaminants with Freshwater Invertebrates | USEPA [177] |
| Bioconcentration: Flow-Through Fish Test | OECD [121] |
| Practice for Conducting Bioconcentration Tests with Fishes and Saltwater Bivalve Molluscs | ASTM [15] |

As for algae, the tests are conducted in a nutrient-enriched medium. The test chambers can be jars, plastic cups, test tubes, or Erlenmeyer flasks.

The use of rooted vascular plants as test species in toxicity tests has been less frequent than for algae and duckweed due to their large size, slow growth, and, until recently, lack of standardized methods. Standardized tests are now available for a few species (Table 42.12). In addition, other experimental techniques are available [53,153]. Several of these describe the use of seeds and seedlings of macrophytic vegetation to assess the toxicities of chemicals, and effluents (Figure 42.6); for example, seedlings of several wetland plants have been used to determine the effects of contaminated sediments in studies of 21 to 28 days' duration [193]. In conclusion, the use of rooted macrophytes and their seeds in toxicity tests will increase in the future as the development of sediment quality criteria to protect aquatic life and protection of wetland vegetation increase in regulatory importance; however, for this to occur, identification of sensitive species and response parameters and validation of the laboratory-derived results is required.

## BIOCONCENTRATION

A bioconcentration study is conducted to derive information on the ability of an aquatic species to concentrate a toxicant in its tissues [64]. This uptake and accumulation can be hazardous to the organism as well as to other aquatic life utilizing the test species as a food source. Bioconcentration tests are usually conducted with single chemicals and single species of algae, fish, and bivalve molluscs. A variety of fishes have been used, including the fathead minnow, bluegill, rainbow trout, and sheepshead minnow, and several species of oysters, scallops, and mussels. Several test methodologies are available to estimate the bioconcentration potential of a compound (Table 42.13). Typically, the test species is exposed to the toxicant for an uptake and depuration phase. A control is included in which the test species is not exposed to the toxicant. The uptake phase is usually for 28 days or until a steady state is attained. The depuration period lasts until the concentration in the test species is 10% of the steady-state concentration in the tissue. During both phases, the test water and test species are analyzed daily for the test chemical. All results from a bioconcentration study are based on measured concentrations. The uptake rate, depuration rate, and the bioconcentration factor are typically reported. The relevance of tissue-concentrated chemicals to the survival of the organism and to ecosystem dynamics is an issue that has received and continues to receive significant scientific attention [41,80,172,176].

## TERRESTRIAL SYSTEMS

Standardization of toxicity testing for environmental problems in terrestrial systems has matured during the past decade. Risk-based soil quality guidelines have been developed by the Canadian Council of Ministers of the Environment [36], selected Canadian provinces [25], and several European countries [184]. The U.S. Environmental Protection Agency (USEPA) has developed ecological soil screening levels (Eco-SSLs) to standardize the approach for contaminated site assessments [183], although the supporting documentation specifically states that these are not to be used as clean-up values. Several of the USEPA regions have developed soil preliminary remediation goals (PRGs) for contaminated site clean-up for the protection of human health [174] that provide information and some guidance to assessing risks to wild plants and animals, as well.

Because of the lack of applicable soil quality criteria, and the difficulty in applying such criteria to specific sites within the heterogeneous soil ecosystem, terrestrial environmental toxicology still relies on standardized toxicity tests for assessing toxicity of new and existing chemicals to plants, wildlife, and soil ecosystem functions. These tests fall into one of two major categories: (1) *a priori* toxicity tests of single chemicals proposed for use or disposal in terrestrial systems, and (2) site-specific assessments of extant contamination. Both approaches utilize the same suite of tests for determining plant, wildlife, and soil toxicity and rely, to some extent, on data generated for human toxicology studies although

**TABLE 42.14**
**Extrapolation Factors for the Median and One-Sided 95% Left Confidence Limit for the Log-Logistic Distribution of Bird and Mammal Toxicity Values**

| Sample Size | Median Estimate | | One-Sided Left Confidence Limits | |
|---|---|---|---|---|
| | Birds | Mammals | Birds | Mammals |
| 1 | 5.7 | 3.8 | 32.9 | 14.9 |
| 2 | 5.7 | 3.8 | 19.6 | 10.0 |
| 3 | 5.7 | 3.8 | 15.6 | 8.4 |
| 4 | 5.7 | 3.8 | 13.7 | 7.6 |
| 5 | 5.7 | 3.8 | 12.4 | 7.0 |

*Source:* Luttik, R. and Aldenberg, T., *Environ. Toxicol. Chem.*, 16, 1785, 1997. With permission.

they differ significantly in their risk assessment approach [49]. All of the tests discussed below in detail have standard operating procedures published by the USEPA, the American Society for Testing and Materials (ASTM), the European Union (EU), or the Organization for Economic Cooperation and Development (OECD). Many of these tests are under discussion for revision to incorporate up-to-date statistical designs and measurement of potential disruption of the endocrine system [122]; however, at the time of this writing, no revisions have been formally accepted to the protocols as they are presented here. Table 42.14 provides extrapolation factors for all of the terrestrial toxicity tests for which detailed protocols currently are available. Additional tests are being developed, and, where appropriate, references are provided to direct the reader desiring further information.

One major challenge remaining in terrestrial toxicology is extrapolation of results from test species to species of concern that have not been tested. Given the large number of terrestrial biota (mammals, birds, reptiles, amphibians, invertebrates, and soil microbes) it will never be possible to test all species, so protection levels must be established based on some method of extrapolation from the data at hand. Two methods currently are in use: (1) Calculate a fifth percentile of the distribution of species mean toxicity values [2,4], or (2) use a fixed extrapolation factor on the lowest available value. If toxicity endpoints (e.g., $LD_{50}$ values) are available for fewer than five species, the method of Luttik and Aldenberg [99] may be used to estimate the fifth percentile from the available data by applying the extrapolation factors derived for small samples (Table 42.15) to the geometric mean of the species tested. This method is based on pooled log-normal standard deviations. If more than one toxicity value is available for a species, the geometric mean of all available values is used as the input for the species. Further information on statistical analysis of ecotoxicity data is available in OECD [129].

## AVIAN TESTS

### Laboratory Tests

Single chemical tests for acute, subacute, and chronic reproductive toxicity have accepted standard methods. The USEPA requires tests with bobwhite quail (*Colinus virginianus*) and mallards (*Anas platyrhynchos*) at a minimum for pesticide registration, so the largest database of toxicity information and testing experience for such compounds is available for these species, although relatively large databases also exist for Japanese quail (*Coturnix coturnix japonica*) (required by the EU for pesticide registration) and ring-necked pheasant (*Pahianus colchicus*) [70,71]. The test methods, however, are equally applicable to all other avian species, provided appropriate adjustments are made in caging and feeding regimens [77,150].

The acute toxicity test ($LD_{50}$) is required when determining effects of ingestion of pesticide granules, seed treatments, or baits [161] or for comparative toxicology [77,150]. A modified version of this test has been developed for testing potential toxicity of alternative nontoxic shot pellets developed for waterfowl hunting [185]. The $LD_{50}$ test [162] requires the use of adult birds (e.g., ≥16 weeks old for bobwhite and mallards; > 6 weeks old for Japanese quail). Within a given test, all birds should be from the same hatch unless the test is being conducted with wild birds. All birds should be of uniform weight and size and absent of obvious signs of disease. Birds should be housed under acceptable animal husbandry practices with a 10:14 light:dark (L:D) cycle to avoid inducing a reproductive state. Animals should be acclimated to the test environment for a minimum of 2 weeks prior to starting the test. For pesticide testing, a minimum of six birds per dose level is required, with the doses arranged by geometric progression such that at least one dose will be below the estimated $LD_{50}$, one dose above, and one control group (administered carrier only). For nontoxic shot studies, 10 birds of each sex are dosed with

**TABLE 42.15**
**Standardized Ecotoxicological Effect Tests for Terrestrial Organisms**

| | Guideline Number | | | | |
|---|---|---|---|---|---|
| | OECD[a] | EU[b] | OPPTS[c] | Year[d] | Title |
| *Terrestrial vertebrates* | | | | | |
| Birds | 223 | — | 850.2100 | 1984 | Avian acute oral toxicity |
| | 205 | — | 850.2200 | 1984 | Avian short-term dietary toxicity |
| | 206 | — | 850.2300 | 1984 | Avian reproduction |
| | — | — | 850.2400 | 1982 | Wild mammal acute toxicity |
| | — | — | 850.2500 | 1996 | Field testing for terrestrial wildlife |
| *Terrestrial invertebrates* | | | | | |
| Bees | 213 | C.16 | 850.3020 | 1998 | Honeybees, acute oral toxicity test |
| | 214 | C.17 | 850.3030 | 1998 | Honeybees, acute contact toxicity test |
| | — | — | 850.3040 | 1982 | Field testing for pollinators |
| Others | ISO 11267[e] | — | — | 1999 | Collembola (springtail) reproduction test |
| | 207 | — | 850.6200 | 1984 | Earthworm (sub)acute lab test |
| | 222 | — | — | 2004 | Earthworm reproduction test |
| | 220 | — | — | 2004 | Enchytraeid (potworm) reproduction test |
| *Terrestrial plants* | | | | | |
| | — | — | 850.4100 | 1982 | Seedling emergence, tier I |
| | — | — | 850.4225 | 1982 | Seedling emergence, tier II |
| | 208 | — | 850.4230 | 1984 | Early seedling growth toxicity test |
| | 227 | — | 850.4150 | 1996 | Vegetative vigor test |
| | — | — | 850.4200 | 1982 | Seed germination/root elongation toxicity test |
| | — | — | 850.2450 | 1996 | Terrestrial (soil-core) microcosm test |
| | — | — | 850.4300 | 1982 | Terrestrial plants field study |
| *Soil microorganisms* | | | | | |
| | 216 | C.21 | — | 2000 | Nitrogen transformation test |
| | 217 | C.22 | — | 2000 | Carbon transformation test |

[a] OECD, Organization for Economic Cooperation and Development; guidelines can be purchased from http://www.oecd.org/document/62/0,2340,en_2649_34377_2348862_1_1_1_1,00.html.
[b] EU, European Union; guidelines can be downloaded from http://ecb.jrc.it/testing-methods/.
[c] OPPTS, USEPA's Office of Pollution Prevention and Toxics; guidelines can be downloaded from http://www.epa.gov/opptsfrs/OPPTS_Harmonized/850_Ecological_Effects_Test_Guidelines/Drafts/.
[d] Earliest year published by OECD, EU, or OPPTS.
[e] ISO, International Organization for Standardization. Guideline can be purchased from http://www.iso.org/iso/en/CatalogueDetailPage.CatalogueDetail?CSNUMBER=19245.

8 no. 4 shot, and a similar negative control (steel shot) and positive control (lead shot) also are dosed. Birds should be fasted for 15 hours prior to exposure; at all other times, they should be allowed *ad libitum* access to food and water. Chemical or shot should be administered through oral intubation into the crop or proventriculus, preferably in an encapsulated form, as birds frequently regurgitate liquid formulations or small pellets. The test animals should be observed for a minimum of 14 (pesticide) or 30 (shot) days after exposure, and all signs of intoxication and number and time of mortality recorded. For pesticide studies, if more than one bird from the control group dies during the 14-day period, the test may be considered invalid. Necropsies and histopathological studies of all birds that die during the test are encouraged. For studies of alternative shot, measurement of hematocrit, hemoglobin concentration, and other specified blood chemistries is required on days 15 and 30 after exposure, and histopathological analysis is required of liver and kidney at test termination. Additionally, analytical chemistry analysis of these or other tissues may be required to ascertain residue concentrations. The $LD_{50}$ value is calculated, along with the 95% confidence interval, using probit analysis [51] or another acceptable statistical method. At the time of this writing, the OECD is reviewing a proposed revised draft guideline that will significantly reduce the number of birds required for $LD_{50}$ determination [128]. This approach uses several repeated stages of testing to approximate the $LD_{50}$ and then to refine its estimate to within acceptable limits of statistical error. For those

instances where the slope of the dose–response curve is needed as well as the $LD_{50}$, the design provides for additional stages and requires the use of more animals.

The subacute avian toxicity test (also known as the dietary test or $LC_{50}$ assay) is conducted using 10- to 14-day-old bobwhite or 5- to 10-day-old mallards, although the test is not limited to these species [6,18,24,162]. Birds used in the test should be either wild birds or pen-reared birds that are phenotypically indistinguishable from wild birds, preferably from colonies with known breeding history. If possible, all test birds should be from a single hatch, and only birds free of obvious injury and disease should be used. Birds must be kept in brooders of appropriate dimensions and temperature (about 35°C for bobwhite and mallards). The standard protocol does not require a certain lighting regimen; however, because the length of the photoperiod influences daily food consumption, it is recommended that a 10:14 or 12:12 L:D cycle be maintained. Water should be available *ad libitum*. A standard commercial game bird diet in mash or crumble should be used for mallard and bobwhite tests; appropriate formulations should be developed as needed for other species. If possible, the test material should be added to the diet without the use of a vehicle. If a vehicle is needed, water is preferred but reagent-grade evaporative material (acetone, methylene chloride) may be used if necessary and completely evaporated at room temperature prior to feeding. Other acceptable vehicles include table-grade corn oil, propylene glycol, carboxymethylcellulose, and gum arabic.

The toxicant is added to feed in a ratio of two parts of solution to 98 parts of feed by weight, and subsampling should be done to confirm uniformity of mixing. Large batches may be mixed in mechanical food mixers or similar devices. It is encouraged that feed batches be analyzed and actual concentrations be reported along with the nominal value. The chemical should be administered in at least four concentrations spaced geometrically to produce mortality ranging from 10 to 90%. A concurrent control group and a vehicle control group (if appropriate) are required. A minimum of 10 birds is required for each dose and control group. Birds should be acclimated to the testing conditions and feed prior to presentation of treated feed. Birds are exposed to treated feed for 5 days, followed by an observation period on clean feed for at least 3 days or until mortality ceases. There must be at least 72 consecutive hours without treatment-related mortality, and control mortality may not exceed 10% during this period for the test to be considered valid. Throughout the test, all signs of intoxication should be recorded, including time of onset and duration. Times of all mortalities must be recorded. Estimates of average food consumption must be made for each pen of birds, with weighing of feed occurring at the beginning and end of the pretreatment, treatment, and observation periods. Provisions for minimizing spillage should be reported. Necropsies and histopathological examination of all dead birds are suggested. The $LC_{50}$ and 95% confidence interval must be reported, along with the method used to determine these statistics.

The avian dietary test may also be used as an indication of repellency of the chemical [98]. Frequently, sufficiently severe reduction in food consumption by the chicks occurs due to inherent properties of the chemical (e.g., bad taste) to induce starvation by the end of the 5-day treatment period, significantly confounding the results of the dietary test. Thus, the $LC_{50}$ should be calculated based on measured concentration of chemical in the feed and actual amount of food consumed (the daily dietary dose, or DDD) or alternative test designs should be employed [129]. Repellency (the inherent property of a test substance that causes a reduction in feed ingestion) is calculated as the percent reduction in food consumption or the highest dose where no food reduction is observed. A standard guidance document on testing for avoidance behavior has been proposed [130] that uses adult birds presented with two dietary choices (treated vs. nontreated) to avoid confounding the data with starvation-induced mortality and to present a more ecologically realistic scenario. This approach, however, remains controversial, as several experts argue that more than two feed treatments are needed to generate realistic worst-case scenarios for exposure potential.

Avian reproductive effects and chronic toxicity endpoints are determined according to methods described by the USEPA [108,162] and OECD [119] and in ASTM E1062 [7]. The U.S. Fish and Wildlife Service [185] has modified these protocols for the testing of alternative nontoxic shot to include a cold stress (ambient temperatures of 6.6 to 4.4°C) and dietary stress (a diet for mallards consisting of whole kernel corn). Both these tests are one-generation studies. A two-generation avian reproduction test guideline is currently in draft form, in support of second tier testing of potential endocrine disrupting chemicals [131]. Changes to the one-generation protocol also have been under discussion, to reduce the length of the study and increase the statistical power of the test.

As with the subacute toxicity test, birds in the avian reproduction study should be pen reared and phenotypically indistinguishable from wild birds. Additionally, it is recommended that the birds come from a colony that has maintained breeding records and that all birds be free of obvious disease or injury. Bobwhite quail should be at least 16 weeks old at the beginning of the test period, and mallards should be at least 7 months of age. Ages of other bird species should correspond to known times of reproductive maturity. Birds should be acquired in a quiescent reproductive state, so onset of egg laying occurs in the test facility. Birds should be acclimated to the test environment for 2 to 6 weeks prior to the presentation of treated feed. Birds must be maintained in pens or rooms that conform to good husbandry practices [111], with minimum space being defined as the ability to stand upright and stretch the wings to the full extent possible. Birds may be housed as one pair per pen or in groups (one male and two females per pen for bobwhite or two males and five females for mallards). Generally, tests are conducted indoors; however, for some birds (e.g., kestrels), it is more appropriate to house them outside, and the alternative shot study allows for this practice if the study is conducted in a cold weather environment. Photoperiod must be rigorously controlled, as the onset and maintenance of reproductive activity are determined by the length of the light period. During the acclimation period and for the first 8 weeks of a test conducted indoors, the photoperiod should be set at 7 or 8 hours of light per day. The photoperiod should then be increased to 16 hours of light per day, preferably by gradually increasing the day length over a 2-week period, and should be maintained at this level for the remainder of the test. Lights should have an intensity of approximately 65 lux (6 fc) at each cage. For outdoor test environments, the tests should be initiated according to the phenology of the species tested, such that presentation of treated feed begins approximately 10 weeks prior to the anticipated onset of egg laying.

Feed preparation should follow the same guidelines discussed above for the subacute test, with a minimum of one control and two test doses. Both test doses should be at sublethal concentrations and frequently represent known environmental concentrations. If a no-observed-adverse-effect level (NOAEL) determination is required, then three concentrations arranged in a geometric progression should be used in addition to the control group. The test chemical should be administered for at least 10 weeks prior to the onset of egg laying, and administration should continue until all control pens have produced 25 eggs, or 6 weeks after 50% of the control hens have laid one egg. For studies of nontoxic shot, birds are orally gavaged with 1 no. 4 lead pellet (positive control) or 8 no. 4 pellets of either the test shot or steel shot (negative control), with a minimum of 4 pairs in the lead treatment group and 20 pairs in the other treatment groups. Shot is administered on 0, 30, 60, and 90 days after the start of the study.

Food and water should be presented *ad libitum* for the entire test period, and food consumption should be recorded at least biweekly throughout the study. Eggs should be collected daily, marked according to the pen from which they were collected, and stored at 16°C and 65% relative humidity. All eggs should be set in incubators once a week at a temperature and humidity suitable for the species being tested. Parental incubation and rearing of chicks may be used if suitable artificial husbandry parameters cannot be determined. Eggs should be candled on day 0 to look for cracks and on days 11 and 18 (bobwhite) or 14 and 21 (mallard) to determine fertility and embryo survival. Eggs collected on one day of weeks 1, 3, 5, 7, and 9 of the egg-laying period should be opened at the equator, the contents washed out, and shells dried and measured for the thickness of the shell plus the membranes to the nearest 0.01 mm. Eggs should be transferred to a hatcher one day before expected pipping (day 21 for bobwhite and 23 for mallard). Hatchability for each egg batch is recorded and chicks are placed in brooders for 14 days and fed a control diet *ad libitum*. Survival at the end of 14 days is recorded. All birds should be weighed weekly prior to the onset of egg laying and at the termination of the experiment; handling should be minimized during the egg-laying period to reduce disturbance. In addition to weight changes, reported endpoints include daily egg production per pen; type and frequency of abnormal eggs (including cracks and other gross defects); number of incubated eggs that are fertile (fertility); the number of fertile eggs that produce hatchlings that completely free themselves from the shell (hatchability); the number of normal young surviving to 14 days of age; weight of young at 1 day and 14 days of age; eggshell thickness; and chemical residue in tissues of adults, chicks, and eggs. Alternative shot studies also require measures of hematocrit, hemoglobin concentration, and appropriate blood chemistries. Standard appropriate statistical analyses should be conducted to determine if treated birds differ from controls or to ascertain the dose-response relationship.

## Field Tests

Methods for examining effects of toxic substances on birds under field conditions have been developed by the USEPA's Office of Pesticide Programs for use in obtaining pesticide registration data. Generally, field studies to determine effects of specific toxic substances are conducted only if preliminary laboratory studies indicate a potential hazard (e.g., high acute toxicity) and have been developed primarily for pesticide registration testing; however, these techniques are being adopted for use in determining individual level and population level effects of existing environmental pollution, such as at hazardous waste sites [195]. Methods for field surveys of bird density, community composition, nesting success, and other ecologically relevant endpoints have a long history of use in biological assessments [30] but only recently have become widely used for ecotoxicological assessments. These methods are beyond the scope of this chapter and will not be discussed further here; of more direct relevance are the standardized controlled field studies described below.

The starling (*Sturnus vulgaris*) nest box study [83] is one means of introducing standardized testing procedures into a field study and can easily be adapted for most cavity-nesting passerines. Nest boxes encourage birds to nest within a study site and establish a large local population with readily accessible nests from which chemical-induced reproductive, behavioral, and biochemical perturbations can be documented. The full-scale test is a 3-year procedure. The first field season serves as a pilot study (and allows birds to locate and colonize the nest boxes) in which 30 nest boxes are erected on each of 12 fields and occupancy and reproduction are documented. Pesticide application occurs in the definitive study during years two and three in a paired block design (i.e., each treated field is paired with a similar control field). The type of fields selected depends on the pesticide (or other chemical) to be applied and the bird species of interest; for example, starlings prefer grasslands (e.g., hay fields), which should be selected in preference over other field types when using these birds as a representative species. Study fields are 16 hectares in size and should be located next to similarly cropped fields to keep starlings foraging on the test sites. Study sites must be located at least 3 km from each other to reduce the possibility that starlings will leave their initial study site and forage in another study area.

Nest boxes are constructed from utility grade lumber and measure 29 × 23 × 38 cm ($l \times w \times h$) (dimensions will vary, depending on the study species). The back portion of the box that attaches to the post is 58 cm in length, centered so 10 cm extends above and below the box. The box is attached to the top of a 10-cm × 10-cm × 3.67-m ($l \times w \times h$) post with aluminum sheet metal tubing placed around the post starting immediately below the box and extending down about 75 cm to reduce predation. Nest boxes must be placed in the fields so crop culture can occur with minimal disturbance; therefore, boxes are placed in a single row down the middle of the field in the same orientation as the crop rows. Boxes are separated by 10 m and erected at least 2 months before the beginning of the breeding season.

Following the placement of the nest boxes, 7 days are allowed to pass before first observations are made. This allows time for the starlings to locate and utilize the nest boxes without disturbance. During the initial nest-building stage (by males), observations are made every 4 days. When the females arrive and during the mating period, observations are made every 3 days. When eggs have been observed in the nest, observations are made every other day. Daily observations are made from the time hatching begins until fledging occurs. During all monitoring periods, observations are made between 11:30 a.m. and 3:00 p.m., the time of least activity by the birds. Following is the reproductive information observed or calculated from nest box observations:

- Date of nest box selection (based on the presence of some nesting material)
- Notes on the development of the nest (e.g., amount of nest building material, quality of the nest, presence of a nest cup)
- Date of the laying of the first egg (the previous day is designated the date of nest completion)
- Interval of egg laying
- Clutch size
- Number of eggs missing or broken
- Number of eggs that hatch
- Date of egg hatch
- Number of missing or dead nestlings
- Weight of 16-day-old nestlings (g)
- Number of fledglings on each day
- Date of fledging

Additional information collected includes weather data (maximum and minimum daily temperature, precipitation events, wind speed), observations of behaviors of adult starlings, and presence of predators in the study area. The study continues until all eggs and nestlings from the second clutch (starlings generally produce two clutches of eggs per breeding season) have hatched and fledged or died and at least 2 weeks have passed without any more eggs being laid. If none of the eggs hatches but they remain in the nest box, the study is terminated when the last box occupied prior to the chemical application has been monitored for 12 days after the laying of the final egg in the clutch. All nestlings that are found dead in the box are analyzed for brain cholinesterase activity, liver enzyme induction, and tissue residues to verify the cause of death (choice of assay is dependent on the chemical being studied).

If the study being conducted is a pesticide safety study, the application rate of the pesticide should be the highest concentration proposed for registration. The timing of application is dependent on the age class of starling nestlings. The initial (or single) application to the fields should be performed when there is the greatest number of 1- to 14-day-old nestlings in the boxes. Pesticide application rates are verified through analysis of vegetation, spray cards, and subsampling of the pesticide holding tanks. For granular formulations, the deposition rate of the granules from the dispensing apparatus is measured. About 50 samples per field are required for an accurate representation of application rate.

The reproductive success of the starlings (as summarized by the number of fledglings per field) should be analyzed initially using a two-way analysis of variance (ANOVA) randomized block design, with the number of fledglings per field as the dependent variable, to determine differences among treatment groups. It is advisable to transform data using a square root transformation prior to analysis. If this initial test shows significant differences, then inferential statistics should be used to identify which parameters were affected.

Guidance for how to design and perform avian field studies for toxicity determinations is provided by Fite et al. [52]. General guidelines are given for two kinds of studies: a general screening study to detect acute toxic effects resulting in mortality or obvious behavioral changes and a detailed, definitive study to quantify the magnitude of acute mortality, determine and measure reproductive impairment, and integrate indirect effects (e.g., food reduction) that may influence the long-term survival of the subpopulation under study. The species and location for the study are determined primarily by the known or proposed location of the pollutant and the population densities and species diversity of the bird communities. The number of replicate study plots (both control and treatment) required to make a determination of effect is based on the expected probability of occurrence of effect and the known variability of the system. True replicates are required, and the tendency toward pseudoreplication must be avoided [78]. Many statistical techniques are available for determining the minimum number of replicates required [57].

Information on population density, age, and sex structure, as well as survival data, can be acquired through a range of methods, including the use of mark–recapture techniques, radiotelemetry [32], line transects (for carcass searches), captures per unit effort, or counts of animal signs [30]. Various models are available that utilize these types of data to determine the probability of survival per unit time and intrinsic population growth rates [85,136]. Reproductive success (from time of initiation of egg laying to fledging of young) is determined through visual observations of nests (egg and fledging counts), radiotelemetry of young, and behavioral observations [32,105].

In addition to documenting direct toxic effects to the species of concern, information also is gathered during a field study about the distribution and persistence of the chemicals in the environment. Relating harmful effects on animals to the degree of environmental contamination is dependent on establishing that a toxic dose has reached either the animal or another species upon which its survival depends. Such information is collected through determining the tissue residues of the chemicals or their metabolites or through the use of markers of exposure, such as cholinesterase activity changes following exposure to organophosphorus or carbamate insecticides or δ-aminolevulinic acid dehydratase (ALA-D) inhibition due to lead exposure. Both ALA-D and cholinesterase inhibition are legally acceptable indicators of harmful effects of exposure to environmental pollutants [112]. Other biomarker responses have been developed and used in wildlife exposure studies for risk assessment purposes (e.g., cytochrome P450 induction) but are not yet standardized or legally acceptable endpoints [137]. For persistent, bioaccumulative compounds, residues in food sources and the lower portion of the food web of the species of concern also should be measured to determine bioaccumulation and bioconcentration factors. It is important to recognize that, for most chemicals, bioaccumulation is not a linear function of exposure concentration but rather attenuates at higher doses. Additional data about the effects of the chemical on the food source, competitors, or predators of the species of concern also should be gathered to ascertain whether indirect effects of food reduction or changes in community composition may be affecting the endpoint species in addition to (or instead of) direct toxicology effects. These data must be interpreted in the context of what is known about the natural history and normal phenology of the species being studied [47].

## Mammalian Tests

In environmental toxicology, mammalian effects data generally are gathered as part of human health impact studies. Both the USEPA [162] and ASTM [5,8] have published general guidelines for conducting acute and subacute toxicity studies with wild mammals and reference the avian and laboratory mammal protocols for more detailed approaches. For large or relatively scarce species, a single-dose regimen may be used where a group of three or more test animals is exposed to the chemical at a dose thought to be representative of expected environmental exposures and observed for at least 10 days for signs of intoxication. Additional doses may be tested sequentially if needed to develop a dose–response curve. This follows the method described for estimating acute oral toxicity in rats by the up–down method [8].

Efficacy testing of anticoagulants in rodents (particularly rats and mice) is well developed, as these animals are the target species [5]. The tests present the animals with both contaminated feed and control feed simultaneously for a free-choice scenario, using a single concentration of chemical in the feed. Exposure times vary from 8 to 20 days, depending on the species being tested, with a 7-day postexposure observation period. The higher the mortality, the greater the efficacy of the product.

Detailed protocols have also been developed for mink (*Mustela vison*) and the European ferret (*Mustela purotius furo*) as representative carnivores due to concern about compounds that biomagnify in the aquatic or terrestrial food chain, respectively and therefore might be expected to cause secondary poisoning due to high concentrations in the tissues of the target species. Both of these animals can be successfully propagated in the laboratory, and stocks of known genetic origin are available. Additionally, the mink has been shown to be extremely sensitive to polychlorinated biphenyls [20], polybrominated biphenyls [21], hexachlorobenzene [27], aflatoxins [29], and 2,3,7,8-tetrachlorodibenzo-*p*-dioxin [73]. Three protocols are available: dietary $LC_{50}$, reproduction, and secondary toxicity [147]. For all three tests, animals should be of mature body weight (about 18 to 20 weeks of age). Either sex can be used in the subacute dietary and secondary toxicity tests; however, the two sexes must be treated as separate subgroups due to significant size differences. Test animals may be obtained from commercial sources or reared in the laboratory and should be free from obvious disease or injury. It is recommended that mink be vaccinated against canine distemper, virus enteritis, infectious pneumonia, and botulism and that ferrets be vaccinated against canine distemper and botulism. Although space requirements for most carnivores have not been standardized, adherence to the guidelines of the Fur Farm Animal Welfare Coalition [54] should provide adequate husbandry guidance. Cages measuring 61 × 76 × 46 cm ($l \times w \times h$) have proven adequate for housing individual mink or ferrets. Solid dividers should be used between adjoining cages to reduce aggression, and a nest box containing straw, shredded wood, or marsh hay must be provided for females in reproduction tests prior to parturition. No specific photoperiod is required, but the day length should not be altered from that in which the animals have previously been reared. A minimum of 7 days of acclimation is required during which time food consumption should be measured and an initial body weight determined. Diets must be formulated to meet the requirements of the test species. Suggested composition of mink diets is provided by the National Research Council [113] and by Ringer et al. [147]. Fresh food and water must be provided daily *ad libitum*. Test diets are prepared by mixing the chemical directly into the feed or by dissolving or suspending in a solvent or carrier prior to mixing in the feed. If a solvent or carrier is used, it must also be added to the control group diet. If a volatile solvent is used (acetone or hexane), the diet should be air dried to evaporate the solvent prior to feeding. Sufficient diet should be mixed to provide feed for the entire exposure period and frozen in aliquots sufficient for one or two days' feeding. All diets should be analyzed after mixing to determine actual concentrations and homogeneity of the test substance in the diet.

## Subacute Dietary Toxicity Protocol for Ferrets or Mink

The doses for a definitive $LC_{50}$ study [147] should be arranged in a geometric progression with the highest dose set so an animal will consume the equivalent of an $LD_{50}$ dose in one day's feed. The test concentrations should then be spaced to achieve at least two concentrations, yielding between 10 and 90% mortality. This generally can be achieved with four to six different concentrations. A minimum of eight animals per dose is required. If an $LD_{50}$ value is not available, one may be determined from a range-finding procedure. A geometrically spaced series of doses is administered by gavage to two animals per dose, and the $LD_{50}$ is determined to be the dose at which one or two animals die during a 1-week observation period. Animals should be treated by gavage with a 3-inch, 14-gauge, curved, stainless-steel animal feeding needle. If a lethal dose is not found, the highest dietary concentration should be set at 5000 mg/kg. It is recommended that the highest concentration to be tested is fed to two to four animals for several days prior to the start of the test to be sure that it is palatable. If it is not, the highest concentration should be reduced to a level that will be eaten. The test diet is then presented for 28 days. A withdrawal period when clean feed is given should be included if animals are still exhibiting signs of intoxication at the end of the exposure period but should not exceed 14 days. Body weights should be recorded at the beginning of the treatment period and weekly thereafter. Feed consumption should also be measured weekly. Mortality, behavioral abnormalities, and other signs of toxicity should be recorded daily. A test is considered invalid if more than 12.5% of the control animals die. It is recommended that a dietary concentration group be removed from the test when food consumption values indicate that feed consumption has dropped to less than 10% of control values after the first 2 weeks of measurements or if animals lose 30% of their initial body weight. It is recommended that necropsies be performed on all animals at the time of death and on all test animals euthanized at the end of the observation period. Gross and histopathological examinations of all major organs should be conducted, including measurement of organ weight. Comparison of body and organ weight changes and feed consumption between control and treatment groups should be made by analysis of variance procedures with a posterior comparison of groups conducted by Dunnett's method [44]. The $LC_{50}$ value is calculated by probit analysis or another standard method.

## Reproduction Test Protocol for Ferrets or Mink

This test [147] primarily measures female reproductive effects because the male is used only for the period of time needed for insemination of the female. Proven breeders should be used if possible; otherwise, nulliparous animals may be used. A minimum of 12 females per treatment group is required. One male is needed for every three females in each treatment group; males should not service females in more than one treatment group. If proven breeders are used, the number of females per group can be reduced to eight and the number of males to two. Males are left with the females only for the duration of breeding. A minimum of two test concentrations and one control group is required. The highest dose must (1) produce an effect, (2) contain at least 1000 mg/kg, or (3) be at least 100 times higher than the known or expected environmental concentration. Animals are fed test diets for 8 weeks prior to breeding; during breeding, gestation, and parturition; and for 3 weeks of lactation (approximately 23 weeks in total). The test will be longer for mink than for ferrets, as mink exhibit a variable delay in implantation of fertilized ova and ferrets do not. The gestation period for mink can range from 42 to 60 days, whereas the ferret has a more constant gestation period of 42 days. Under natural conditions, mating attempts begin at the first of March for mink and at the end of April for ferrets.

To breed mink, a female is presented to a male (in his cage) and, if receptive, is allowed to mate. If not receptive, the female is removed and a mating attempt is tried again in 4 days. Successful mating is verified by the presence of viable spermatozoa in a vaginal aspiration taken after copulation. Following a successful mating, the female is given a second opportunity to mate to the same male either the next day or (preferably) 8 days later. In breeding ferrets, females are presented to males when in estrus (determined by extent of vulvar swelling) and left with the male overnight. They are not given the opportunity for additional matings.

When the breeding period is over, animals should be left undisturbed except for daily feedings. Body weights should be measured weekly only during the 8 weeks prior to breeding. Feed consumption should be measured weekly throughout the duration of the test. Observations on behavioral changes, mortality, or other signs of toxicity should be recorded daily. During parturition (up to 3 weeks in length), females should be checked daily for newborn. Number, weight, and sex of all newborns are recorded within 24 hours of birth. It may be necessary to remove the female from the nest box to check for newborns. Offspring should be weighed again at 3 weeks of age. After this time, they will begin eating the adult diet and should be fully weaned at 6 weeks of age, at which time the test is terminated. At the end of the test, all males and at least an equal number of females are euthanized [16] and necropsies performed for gross and histopathological examinations. A test is considered invalid if more than 20% of the control animals die during the test. The following reproductive parameters must be reported:

- *Length of gestation*—Time, in days, from last confirmed mating until parturition
- *Number whelped, not whelped*—Number of females giving birth or not giving birth in a treatment group per number of females with confirmed matings; number whelped includes those females that die during parturition
- *Live newborns per females whelped*—Average number of live newborns produced by all females that give birth in a treatment group; this does not include females that die during whelping
- *Average birth weight*—Average weight within 24 hours of birth of all live newborns within a treatment group
- *Average litter weight*—Average weight of all litters (live newborns only) within each treatment group
- *Percent newborn survival to 3 weeks*—Number of live newborn in a treatment group surviving to 21 days of age, expressed as a percentage of all live births in the treatment group
- *Average body weight at 3 weeks of age*—Average weight of all live newborns in a treatment group on the 21st day after birth
- *Total newborns per female whelped*—Average number of all newborn (alive and dead) produced by all females that give birth in a treatment group (including those females that die during whelping)
- *Percent newborn survival to 6 weeks*—Identical to 21-day survival but extended to 42 days
- *Average 6-week body weight*—Identical to 21-day weights but measured at 42 days of age

Number whelped and percent survival should be analyzed for differences among treatment groups using contingency tables and Bonferroni's chi-square test or similar statistical procedures. The remaining variables can be compared among treatment groups by analysis of variance techniques with Dunnett's method for comparison to controls.

## Secondary Toxicity Protocol for Ferrets or Mink

This protocol [147] is used to determine the comparative toxicity of both a parent compound and its metabolites administered in the diet, such as might be ingested when consuming contaminated prey. The toxicity of the parent compound is determined following the procedures detailed above for subacute dietary toxicity testing ($LC_{50}$). Secondary toxicity testing is conducted using the same protocol with the exception that the contaminated feed consists of mink and ferret prey items that have themselves been fed the test substance. Prey animals can be any species readily consumed by the test animals, including fish (salmon, perch, alewife, sucker, carp, and bloater chubs), birds (chickens), and mammals (beef, nutria, rabbits, voles, pocket gophers, rats, and mice). Prey animals may be contaminated by dietary, inhalation, or dermal routes and should be exposed to the same test substance (same source and lot number) as fed to the mink or ferrets in the primary toxicity test. The concentration of the test substance to which the prey are exposed should be sufficient to generate tissue residues of parent compound or metabolite at levels known to cause 10 to 90% mortality of the mink or ferret. This may need to be determined through a range-finder test with the prey species. The final body burden in the prey should allow for dilution of the tissues by the remainder of the dietary ingredients provided and for dilution by noncontaminated portions of the prey; for example, if 10 mg/kg causes a 50% lethality in the mink or ferret and a diet consisting of 40% prey tissue is being fed, then a prey body burden of 25 mg/kg is required to yield the final dietary concentration of 10 mg/kg. Prey animals that are not killed by the test substance should be euthanized in a manner that will not interfere with the test results. The gastrointestinal (GI) tract may be removed from the carcass if the test substance is known to accumulate in tissues. This will reduce the probability of direct poisoning from consumption of undigested material in the upper GI tract; however, some compounds do not readily accumulate in tissues (e.g., organophosphorus insecticides), and poisoning of carnivore species occurs primarily from ingestion of material in the GI tract. In these instances, the GI tract may be left intact. All prey animal carcasses should be frozen until being fed to the mink or ferret.

## Honey Bees

Honey bees (*Apis mellifera*) are important economically, due to their use for pollination of many crops (e.g., fruit trees). Methods have been developed to evaluate the effects of pesticides on honey bees [124,125,164], and these tests have been adapted to determinations of environmental pollution *in situ* [195]. Toxicity to bees of residues on foliage is conducted using individual worker bees of uniform age [164]. The compound of concern is applied to alfalfa foliage in the field and allowed to weather under natural conditions. Worker bees—honey bees, the alfalfa leafcutting bee *Megachile rotundata*, or the alkali bee *Nomia melanderi*—are collected from the frames of established colonies, and 50 to 100 individuals are introduced into each test cage. Cages are constructed by cutting wire screen into strips 46 × 5 cm long and stapling the ends to form a cylinder. Tops and bottoms of 150- × 15-mm plastic Petri plates serve as the tops and bottoms of the cages. Fifty to100 honey bees, 20 to 40 leafcutting bees, or 15 to 30 alkali bees should be placed in each cage. Ambient temperature during the 24-hour test period should be 24 to 26°C (honey bee) or 29 to 31°C (leafcutting bee and alkali bee). Bees are fed during the test by providing cotton squares (5 × 5 cm) soaked with 50% sugar syrup and placed under the treated foliage.

The test compound is applied to 0.01-acre alfalfa plots in a randomized block design at applications simulating known environmental concentrations or proposed use patterns. The test may be designed to test a single dose; preferably, a geometrically spaced range of doses is used to determine an $LC_{50}$ value. Foliage is harvested at 3, 8, and 24 hours after application. If greater than 25% mortality of bees occurs when exposed to 24-hour foliage, sampling should continue at 24-hour intervals until mortality resulting from exposure to treated foliage is not significantly different from control groups. Foliage is chopped and mixed and 500 mL introduced to each cage. Mortality is determined after 24 hours of exposure to the treated foliage. At least three cages of bees must be used per replicate, and each treatment, including controls, must be replicated at least three times.

The honey bee acute contact $LD_{50}$ test is based on the protocol developed by Atkins et al. [19] for screening pesticide dust and follows the same husbandry procedures detailed above. Worker bees of uniform age are exposed to the test substance through a dusting procedure. Twenty bees are transferred to the dusting cage and dusted for a period of 15 seconds. The dusted bees are then put in holding cages and observed for mortality over the next 24 hours.

The honey bee subacute test [164] is designed to test the effects of a chemical on the colony as a unit. The study is intended to identify those chemicals that may cause adverse reproductive, behavioral, or other subacute effects that can be brought back to the hive by exposed foragers. This test is less developed than the tests described above for individual bee toxicity and, consequently, will not be detailed here. The general approach involves exposure of intact bee colonies to the test substance through feeding in pollen or sugar syrup (there is no consensus as to which component of the diet should contain the test compound). Through caging or location, the colonies are restricted to feeding only on the treated food provided. Periodic observations are made for 42 days to 4 months (42 days is the approximate time needed to complete two brood cycles). Observations include the number of eggs, open brood or sealed brood, gross colony weight, estimated adult population, and amount of honey in storage. Other suggested observations are the presence or absence of disease, discoloration, desiccation, egg and larvae abnormalities, and morphological or behavioral abnormalities in adults.

## Soil Invertebrates

Soil toxicity is an important parameter in terrestrial environmental assessments. Direct measures of chemical contamination of the soil can be, and frequently are, made following traditional analytical chemistry methods; however, these methods are costly and time consuming and do not provide information about the toxicity of the soil to terrestrial organisms, particularly because synergistic, antagonistic, and additive effects of the complex mix of compounds found in contaminated soils can occur. Furthermore, it has become recognized that chemicals in soil become bound to soil particles during the aging process and become progressively less bioavailable [95]. Chemical assays that use harsh extraction procedures to extract all the chemical from the soil do not provide a realistic estimate of exposure to soil organisms. Models are being developed to calculate pore water concentrations for varying soil types (especially with reference to pH, organic matter, and clay content) to more closely estimate actual exposure regimes. Even these methods cannot account for biological adaptations, behaviors, and microenvironment alterations; therefore, bioassays have been developed to provide direct measures of toxicity.

Plant germination and growth tests are the most commonly used methods for determination of soil toxicity and are described later in this chapter. The amphibian bioassay

(FETAX) described below can be conducted utilizing eluates from contaminated soils. A solid-phase version of the aquatic MICROTOX test has been developed and is used occasionally. Soil microbial function assays are available for nitrogen and carbon transformation process [93,126,127] and are described briefly below. Analysis of the soil transcriptome (i.e., transfer RNA) is being studied as a sensitive and more realistic *in situ* method for measuring functional changes in the microbial community.

Soil invertebrates play a pivotal role in the terrestrial ecosystem, providing functions such as decomposition of organic matter as well as providing a prey base for many higher order organisms. A well-developed, but infrequently applied, test determines an $LC_{50}$ in soil for harvester ants (*Pogonomyrmex owyheei*) [55]. Crickets (*Acheta domesticus*), isopods (e.g., *Porcellio scaber*), millipedes (*Brachydesmus superus*), centipeds (*Lithobius mutabilis*), Collembola (e.g., *Folsomia candida*), staphylinids (e.g., *Philonthus cognatus*), oribatid mites (e.g., *Platynothrus peltifer*), and soil nematodes also have been studied as possible toxicity test organisms for soil toxicity determinations [96,194], but none of these protocols has yet been accepted by the OECD, EU, or United States for regulatory purposes [107]. The soil invertebrates most frequently used for toxicity testing are the oligochaetes: the potworm (Enchytraeids, such as *Cognettia sphagnetorum*) and earthworms (e.g., Lumbricidae, such as *Eisenia foetida*). Standard assays are available for earthworm survival, reproduction, and avoidance behaviors [123,132] and a standard protocol was recently adopted by the OECD for Enchytraeid reproduction assays [133]. Both of these types of worms are found in nearly every soil type (except for xeric soils) and play an important role in soil decomposition [69]. Furthermore, they are composed of a high percentage of lipids and readily bioaccumulate contaminants from soils. As they are a significant food item for many vertebrates, they often are an important first step in movement of soil contaminants into the aboveground food web.

## Earthworm Tests

The earthworm survival test [11,45,61] uses adult (>60 days old, 300 to 500 mg, with clitellum) *Eisenia foetida* grown in a single culture chamber. Worms may be purchased from a commercial source or acquired from the environment. The identification must be verified [50] to ensure that the species is *E. foetida*. Test soils are first homogenized using a blender and then mixed with artificial soil (10% 2.36-mm-screened sphagnum peat, 20% colloidal kaolinite clay, and 70%-grade 70 silica sand) to prepare 700 g each of a geometric series of test soil concentrations (e.g., 100, 50, 25, 12.5, 6.25, 3.13% w/w) plus a 100% artificial soil control. The total amount for each concentration is blended to ensure even mixing prior to dividing into aliquots for the test. After mixing, the soils are hydrated to 75% water-holding capacity. Standard 1-pint canning jars with screw-top lids and rings are used as test chambers. Three replicate chambers are filled with 200 g (dry weight) of soil for each dilution. Ten earthworms are placed on the soil surface, the jars are capped, and they are incubated at 20±2°C under continuous light (540 to 1080 lux) for 14 days. Worms are not fed during the test. Soil pH is measured at the beginning and end of the test, and the temperature of the environmental control chamber is continuously monitored. The total organic carbon of the test soils should be measured in one test jar for each test concentration and the control. Jars are examined daily for dead worms (worms are considered dead when they do not respond to a gentle touch on their front end). At the end of the 14-day period, the soil is emptied into a tray and the jar thoroughly searched for worms. Dead worms decay very rapidly so all ten worms in a container may not be accounted for during the test and it is assumed that all missing worms have died. The percent mortality for each concentration is determined, and the $EC_{50}$ for mortality is calculated by probit analysis. Mortality is defined as a lack of response to a gentle mechanical stimulus (e.g., a touch with a small glass rod). Loss of earthworm biomass and behavioral and morphological endpoints such as coiling, segmental swellings or constrictions, lesions, rigidness, and flaccidness also can be used as toxicity endpoints.

Growth and reproduction are biological endpoints in earthworm tests of longer duration, generally 140 days [11]. Test conditions are the same as described for the 14-day study, although worms must be fed for any study with duration greater than 28 days. Control of pH, temperature, and soil moisture content are very important, as variations in these environmental parameters have been shown to significantly affect the outcome of earthworm reproduction studies. Mortality and other sublethal observations are made at least weekly. At the conclusion of the test, the containers are emptied, and the number of adult worms remaining is tabulated. The number of cocoons formed, cocoon mass, number and growth of young worms, and rate of clitellum development are measured as reproductive endpoints.

## Enchytraeid Tests

The potworm, *Cognettia sphagnetorum*, is readily obtained from the field and is easy to culture in the laboratory on a 75-vol% Sphagnum peat substrate mixed into 25-vol% LUFA 2.2 soil (LUFA is a natural soil type found in Europe that contains ~3.9% organic matter and 3.5% clay and has a pH of 5.8) [96]. Animals of similar size (i.e., with the same number of segments) are used in each study. A single study is conducted to determine percent mortality and effects on growth rate of adults and fragments. Potworms reproduce asexually, resulting in the production of three fragments: head, middle, and tail. The middle and tail fragments develop a functional mouth within 2 to 3 weeks, and all three fragments continue to add new segments. After another 2 weeks of adding segments, the new worms are indistinguishable from older specimens. Worms that have >30 segments are considered to be adult. Growth is defined as the increase in number of segments over the test period, reproductive success is quantified as the mean number of fragments that survive and add segments over the test period, and the fragmentation rate is defined as the average number of fragments produced per day. The chemicals of interest are dissolved in water or an organic solvent and mixed with the soil substrate, and the soil is wetted to 80% water-holding capacity (WHC) 24 hours before introducing the worms. For chemicals dissolved in organic solvent, the chemical is first mixed with a small amount of sand to allow the solvent to evaporate and then mixed into the final batch of soil, As a food source is required due to the long duration of the test, 1% algae (*Pleurococcus* spp.) or 0.2% baker's yeast is added to the soil substrate prior to putting ~2 g of soil into each test tube. Animals are maintained one per test tube at 15±1°C, with a relative humidity in the test chamber of 75% under constant light, after first counting the number of segments present. At weekly intervals for 10 weeks, the tubes are examined for the number and size of fragments and number and size of unfragmented worms. Animals are transferred to vials with fresh soil and food (one worm or fragment per vial) prior to returning them to the incubation chamber. Copper chloride is used as a positive control (reference); a test is considered valid if the growth rate in control animals is >1 segment/week and there is <10% mortality of controls.

## Soil Microorganism Tests

Soil microbes (i.e., fungi, bacteria, actinomycetes, and protozoa) make up 80% of the living matter in soil and contribute significantly to essential soil function such as decomposition of organic matter, nutrient cycle, and soil aggregation [107]. Because of the diversity of species and the importance of microbial functions, tests for the effects of chemicals on soil microbes use functional endpoints rather than effects to individuals or population structure. Two standardized tests are available: nitrogen transformation [127] and carbon transformation [126], but significant controversy remains over the repeatability and relevance of these and other soil function assays [131]. The nitrogen transformation test adds a nitrogenous substrate to reference or contaminated soils brought in from the field (or clean soils spiked with chemical). Soils are incubated for 28 days and then analyzed for the production of mineralized N (i.e., amount of ammonium, nitrite, and nitrate production). The carbon transformation test is similar and measures the soil respiration rate (carbon dioxide production). A dilution series of five soil concentrations of the chemicals of concerned is preferred, and at least a 20% difference between reference and treated soils is required to specify that a significant effect has occurred.

### Amphibians

The frog embryo teratogenesis assay–*Xenopus* (FETAX) is a standardized bioassay developed for obtaining data on the developmental toxicity of test materials. Amphibians usually are associated with wetlands, many of which are impacted from chemical pollution due to their distribution in low-lying areas of the landscape and their connection to the surface and subsurface hydrology [91]. FETAX primarily is used as a measure of toxicity to amphibians of environmental samples (water, sediments, and soil eluates); however, results from this bioassay have an 85% correspondence with results from mammalian developmental toxicity tests so the information can be extrapolated to other classes of animals with a high degree of certainty. The standard protocol, as described below, is a laboratory assay. A field application (by putting the egg masses in porous containers submerged in the water at the study site and following all other standard procedures) is being developed [94].

In FETAX [10], a range-finding and three replicate tests are performed. Each test includes both a negative control (no test material added) and a positive control (6-aminonicotinamide). The 96-hour $LC_{50}$ and 96-hour $EC_{50}$ (malformation) are determined, and the teratogenic index (TI) is calculated by dividing the $LC_{50}$ by the $EC_{50}$. The FETAX protocol is designed to use embryos of the South African clawed frog *Xenopus laevis* (Daudin). Other North American species can be used and are listed in the appendix of the protocol, although breeding times and methods would have to be adjusted to produce an appropriate egg mass for the test. For *Xenopus*, adult frogs that are proven breeders should be purchased and maintained in single pairs. The frogs should be bred in the same water in which the test is to be conducted (natural, nonchlorinated water known to be free of contaminants) and fed ground adult beef liver three times per week. Temperature should be kept at 23±3°C on a 12:12 L:D photoperiod. Breeding is induced in the males and females by injection into the dorsal lymph sac of 250 to 500 and 500 to 1000 IU, respectively, of human chorionic gonadotropin. Egg deposition usually occurs 9 to 12 hours later. The eggs should be inspected for fertility, which must be >75% for the egg mass to be used in a test. Eggs are then separated from the jelly coat by gently swirling them for 1 to 3 minutes in a 2% w/v L-cystein solution prepared in FETAX solution (625 mg NaCl, 96 mg $NaHCO_3$, 30 mg KCl, 15 mg $CaCl_2$, 60 mg $CaSO_4{\cdot}2H_2O$, and 75 mg $MgSO_4$ per liter of deionized or distilled water; final pH 7.6 to 7.9). Eggs should be removed from the dejellying solution and rinsed with clean water as soon as dejellying is completed or survival will be reduced. Only normally cleaving embryos should be selected for use in the test. The *Atlas of Abnormalities* [22] should be consulted to determine abnormalities during the assay. Each test must use embryos derived from a single mated pair.

FETAX is a 96-hour test. Each test consists of at least five concentrations arranged in a geometric progression, three of which should fall within the 16 to 84% effect range on the mortality and malformation dose–response curves. A range-finding test should be conducted first, consisting of at least seven concentrations that differ by factors of ten. Each test concentration and the positive control should have two dishes, each containing 25 embryos and 10 mL of test solution. The negative control groups must have four dishes of 25 embryos each. The positive control (6-aminonicotinamide) consists of two dishes exposed to 2500 mg 6-aminonicotinamide per liter and two dishes exposed to 5.5 mg/L. Dishes are placed in an incubator to maintain the temperature at 24±2°C. The pH of the test solutions should be 7.7 (range, 6.5 to 9.0). The test material is renewed every 24 hours during the test. Renewal should be done by removing the old test solution with a Pasteur pipette with the orifice enlarged and fire-polished to accommodate embryos without damage if they are picked up accidentally. Fresh solution is added quickly to minimize embryo desiccation.

Dead embryos are removed at the end of each 24-hour period during the 96-hour test (at the time the solutions are changed). Death at 24 hours is ascertained by the embryo's skin pigmentation, structural integrity, and irritability. At 48, 72, and 96 hours, a lack of heartbeat indicates death. Total number dead and numbers and types of malformations occurring at the end of 96 hours are reported. The ability of the test material to inhibit growth is determined at 96 hours by recording head-to-tail length. If the embryo is curved or kinked, the measurement should be made as if the embryo were straight (i.e., following the contour of the embryo). The minimum concentration of test material that significantly inhibits growth (MCIG) should be determined from these data using the *t*-test to determine significance at $p = 0.05$. The $LC_{50}$ and $EC_{50}$ (malformation) should be determined using probit analysis.

## Plant Tests

A standard guide for conducting terrestrial plant toxicity tests has been completed and published by the ASTM [15]. Included with the guide are standard protocols for seedling emergence, root elongation, woody plant assays, and a *Brassica* life-cycle test. Additionally, a seed-germination test is available [61,74,120,186], as is a test for vegetative vigor [75,120]. These bioassays initially were developed to determine hazards due to soil contamination at hazardous waste sites or in sludge disposal areas [195]; however, they also can be applied to *a priori* safety testing requirements. A comprehensive review of plant toxicity tests is available in Kapustka [81].

The seed germination test [61,74] is a 120-hour static test and measures seed survival, germination, and seedling emergence. Butter crunch lettuce (*Lactuca sativa*) is the most commonly used test species, although over 30 species have been accepted by regulatory agencies. Butter crunch lettuce seeds and many other domestic plants can be purchased from a commercial vendor. Only one lot should be used for each test, and information on the germination percentage should be provided by the seed source. Alternatively, native seeds may be field collected. Regardless of source, only untreated seeds (no applications of fungicide, repellents, etc.) should be used. The seed lot should be examined to discard trash, empty hulls, and damaged seed. Seeds are sized by stacking four wire-mesh sizing screens on top of each other with a collection pan underneath; the largest mesh screen should be on top. The seeds are poured in and the nest of screens is shaken until no more seeds fall through the screens. The size class containing the most seed is selected and used for all of the tests. Seeds can be stored in a dessicator at 4°C in airtight, waterproof containers until used.

Test soils are first homogenized using a blender and then mixed with artificial soil (commercially available, 20-mesh, washed silica sand) to prepare 400 g each of a geometric series of test soil concentrations (e.g., 100, 50, 25, 12.5, 6.25, 3.13% w/w) plus a 100% artificial soil control. The ASTM [15] requires the use of boric acid as a positive control. The total amount for each concentration is blended to ensure even mixing prior to dividing into aliquots for the test. The soils are air dried, and 100 g of each concentration is placed in each of three replicate 150-mm plastic Petri dishes labeled with the appropriate soil dilution. The Petri dishes are then randomized, and 40 seeds are placed in each dish at least 1.25 cm from the edge. The seeds are pressed into the test soil with a glass rod or beaker bottom. Soils are hydrated with deionized water to 85% of water-holding capacity. Cover sand (90 g commercially available, 16-mesh sand passed through a 20-mesh screen to remove fines) is poured over the top of the soil and leveled with a ruler. The dishes are then placed in an incubator for the duration of the test. They should be incubated at 24±2°C in the dark for 48 hours, followed by 16 hours of light and 8 hours of dark until termination of the test at 120 hours. Light intensity should be 4300±430 lux. Soil pH should be measured at the beginning and end of the test (and should be between 4 and 10), and soil temperature should be measured every 24 hours in one replicate of each concentration and the control. At 120 hours, the number of germinated seeds in each dish is determined by counting each seedling that protrudes above the surface. The $LC_{50}$ and its 95% confidence limits are determined by probit analysis.

The seedling emergence test [15] is very similar to the seed germination test. Seeds are placed in pots, rather than Petri dishes, and need not be stored in the dark for the first 48 hours. Soil is hydrated to water-holding capacity throughout the test. Test duration is twice the amount of time required to achieve acceptable germination for whatever test species is being used. In addition to counting number of emergent seedlings at the end of the test, optional measures include shoot and root growth. Shoot measurements are made from the transition point between hypocotyl and root to the tallest point on the shoot. Roots are measured from the same transition point to the tip of the root. If sufficient growth is present, dry weight measurements may be obtained. Material is harvested and placed in a preweighed drying vessel and dried at 70°C until constant weight is achieved (approximately 24 hours). Weights are taken to the nearest 0.001 g.

Root elongation is a key component of the early stages of plant growth and development. The root elongation bioassay tests the toxicity of water or soil eluates (i.e., the water-soluble constituents) to seed germination [15,145]. It generally is used as a bioassay of hazards due to soil or water contamination at hazardous waste sites or in sludge disposal areas; however, as with the seed germination test, it also can be applied to *a priori* safety testing requirements. As a general rule, root elongation is more sensitive than seed germination. The test is a 120-hour static test that measures seedling root growth. Butter crunch lettuce is the most commonly used and is the most sensitive species, but other plants have been used as well (e.g., cucumber, wheat, alfalfa, radish, red clover, rape). Seeds are purchased, sorted, and sized as described for the seed germination assay. Soil eluates are prepared by mixing 4 mL of deionized water per gram (dry weight) of soil. The slurry is mixed in total darkness for 48 hours at 20±2°C. After mixing, the eluate is centrifuged and filtered through a 0.45-μm cellulose acetate or glass fiber filter. Water samples (or soil eluates) are diluted in deionized water to provide a geometrically spaced range of concentrations, generally using a 0.3 or 0.5 dilution factor and five concentrations. As with the seed germination study, boric acid should be used as a positive control. A sheet of Whatman No. 3 filter paper is placed in each of three replicate 100-mm plastic Petri dishes and 4 mL of the test solution is poured over the paper. Five seeds are placed in a circle on the filter paper, equidistant from the edge to the center and equally spaced from each other. The lid is placed on each Petri dish that are then set in layers in a black 33-gallon plastic garbage bag within a cardboard box. Moist paper towels are placed between layers of dishes to keep the humidity level elevated. The bag and box are sealed and placed in a controlled environmental chamber at 24±2°C for 120 hours. The pH and hardness of the test solutions are measured at the beginning of the test, and temperature is recorded every 24 hours. At the end of the 120-hour period, root lengths are measured to the nearest millimeter by placing the seedling on a glass work surface and measuring the distance from the transition point between the hypocotyl and root to the tip of the root. Swellings or deformities at the transition area should be noted. Roots may be harvested, dried to a constant weight, and weighed to the nearest 0.001 g as an optional endpoint. The percent inhibition of the treated seeds as compared to the controls is calculated, and an $EC_{50}$ (the concentration that inhibits root growth to 50% of the control length) is calculated.

Whole-plant toxicity is evaluated using a 5-day screening test (TOXSCREEN) conducted in hydroponic solutions [142] or a longer full life-cycle test using vascular plants in soils, such as *Brassica* [15]. TOXSCREEN uses whole plants such as soybean (*Glycine max*), barley (*Hordeum vulgare*), and woody perennials that have been grown in hydroponic culture for at least 28 days. Alternatively, known-age plants may be purchased commercially and adapted for hydroponic exposures [106]. Regardless, exposures occur hydroponically when young plants are transferred from nursery containers to containers filled with exposure medium. Geometrically spaced dilutions of the exposure medium or chemical concentrations within the exposure medium are used to develop a dose–response curve. Soil eluates may be used to fill exposure containers. Tests are conducted under a 16:8 L:D photoperiod (light intensity of 350 μmol/m²/s at the top of the canopy) at 25/21±2°C (light/dark) and relative humidity of 50 to 70% for 5 days. Following the exposure period, survival, root and shoot growth, and total biomass of the plants are measured. Probit analysis is used to calculate $LC_{50}$ values.

An alternative procedure to TOXSCREEN is available in the woody plants test protocol [15]. This is similar to the TOXCREEN test, but woody plants are gown in silica sand, a formulated soil, or a contaminated soil rather than in a hydroponic solution. Plants are obtained from the field or a horticulture source, and the roots washed gently by dipping the root mass into deionized water. Plants are kept moist and sorted according to size and stage of development to begin the study with uniform sized plants in similar stages of root and shoot growth. The starting conditions for all endpoints are measured and recorded. Plants are sorted randomly into treatment groups and placed in pots of sufficient dimension to contain test medium to a depth of approximately two times the root length of plants. Each replicate container should be planted with one test replicate; five replicates of each soil sample, sample dilution, and positive and negative control are required. Medium should be added to cover roots completely and to bring the level within 1 to 2 cm of the top of the container. Medium may be gently packed by hand but should not be compacted. Deionized water is used to bring the posts to water-holding capacity. A regular water application schedule should be followed thereafter for the duration of the test. Test duration should be approximately twice the length of time required to achieve amounts of shoot and root growth acceptable for statistical characterization. Tests are conducted under a 16:8 L:D photoperiod (light intensity of 100 to 200 μmol/m$^2$/s) at 20 to 30°C and relative humidity of >50%. Measured endpoints include wet and dry shoot mass, number of new shoots or leaves, number of root initiation points, and changes in total plant weight. Additional observations on general plant condition and leaf or root malformations are also noted. Plants may be harvested, dried at 85°C until constant weight (approximately 24 hours), and weighed to the nearest 0.01 g.

The full life-cycle plant test evaluates the effect of test materials on germination, growth of shoots and roots, photosynthetic systems, flower development, and reproductive capabilities of plants. It is conducted using either *Arabidopsis thaliana* [146] or *Brassica rapa* [15,151] exposed to the toxic substance for 25 to 36 or 36 to 44 days, respectively. Hydroponic exposures [146,151] occur in double-pot, static-replacement systems where a vermiculite-filled growth container (for *Arabidopsis*) or greenhouse potting soil (for *Brassica*) is nested above a second, larger pot that serves as a nutrient solution reservoir. Nutrients, toxics, and water move from the reservoir to the vermiculite or potting soil via polyester wicks that are draped between the two pots. Alternatively [15], *Brassica rapa* seeds may be planted directly into test soil as described above for root elongation or woody plant tests. Dilutions and eluates of test substances are made as described above. Seeds are uniformly planted on the surface of the vermiculite or potting soil and incubated in the conditions described for TOXSCREEN and the woody plant test. Plants will germinate, grow, mature, and set seeds. Exposure to the test substance occurs from the time seeds are planted and continues until the mature plant drops its own seeds. Observations include leaf and flower structure, total biomass, vegetative and reproductive biomass (e.g., stems and leaves vs. seeds and reproductive structures), foliar height, initial flowering date, time to flowering, stunting, chlorosis, and survival. Probit analysis is used to calculate $EC_{50}$ values for each endpoint. Analysis of variance techniques can be used to determine which test concentrations produce effects significantly different from control values.

Soil-core microcosms [9] can be used to develop site-specific or regional information on the probable chemical fate and ecological effects in a soil system resulting from release or spillage of chemicals into the environment. This test is most useful in the assessment process after preliminary knowledge of the chemical properties and biological activity of the compound of interest has been obtained. The test is designed to determine impacts in agricultural or natural field soil ecosystems and may not be applicable to forest soils. Specifically, the test will determine the effect of a chemical on growth and reproduction of either natural grassland vegetation or crops, nutrient uptake and cycling, potential for bioaccumulation in the plant tissues, and potential for and rate of transport of the chemical through soil to groundwater.

Soil cores are extracted from the site or region of interest with a specially designed, steel extraction tube. The steel tube surrounds an ultra-high-molecular-weight, high-density, nonplasticized polyethylene pipe to prevent the tube from warping or splitting during extraction. The tube is 60 cm deep and 20 cm in diameter. In agricultural systems, the plowed topsoil is moved aside prior to coring and backfilled into the upper 20 cm of the tube after collection. For the natural grassland system, the vegetation is clipped before the core is extracted. The polyethylene tube containing the soil core is removed from the steel coring device and placed within a specially designed wheeled cart that holds six to eight tubes packed within insulating Styrofoam® beads. The tube sits on a Buchner funnel that is covered by a thin layer of glass wool and connected by polyvinylchloride (PVC) tubing to a flask for collection of leachates. For the natural ecosystem, the natural plant cover collected with the soil core is suitable for the test. For agricultural systems, a mixture of grasses and broad leaves (e.g., legumes) that are typically grown together in the region of interest should be planted in the soil cores. The seed application rate should duplicate standard farming practices. Microcosms should be watered with purified water on a predetermined regimen, usually established on the basis of site history. Microcosms are leached once before dosing and once every 2 or 3 weeks after dosing. Microcosms that take longer than 2 days to produce 100 mL of leachate are not used in the test. The microcosms in their insulated carts are kept in a greenhouse or environmental chamber where temperatures and photoperiods are set to simulate outdoor conditions during a typical growing season in the region of interest.

If information is available, only one concentration of the test chemical above that known to cause at least 50% change in plant growth or 50% change in bacterial growth or respiration needs to be tested. Generally, three concentrations are used, choosing one that produced a 20 to 25% change in productivity. In any case, the lowest treatment level should not be less than 10 times greater than the analytical limits of detectability of the parent compound. A range-finding test utilizing concentrations of 0.1, 1.0, 10, 100, and 1000 μg/g in the upper 20 cm of topsoil can be used to determine the appropriate dose for the definitive test. The range-finder should last a minimum of 4 weeks, while the definitive test lasts for 12 weeks. For aqueous compounds, the test solution is mixed with water and applied as single or multiple (daily, weekly) exposures in sufficient volume to bring the microcosm to field capacity. In no case should the exposure volume be sufficient to cause leaching in any of the microcosms. Frequency of application should reflect the situation encountered in the region of interest. The use of carriers should be avoided. If the test substance does not mix with water, it should be applied evenly to the top of the unplanted microcosm and mixed into the topsoil prior to planting. If the test substance is normally sprayed on growing plants, then it should be sprayed on the plants with a nebulizer when at the seedling stage. The number of replicate microcosms per dose is determined by the desired power of the statistical test and the variability between microcosms (determined in the range-finding test). Individual microcosms should be assigned to treatment groups in a randomized block design, with the cart as a blocking variable.

The parent compound should be radiolabeled with $^{14}C$, either in an appropriate aromatic, cyclic carbon group or in a linear chain, to follow uptake and degradation rates. The following parameters are measured:

- Primary productivity, the total yield or yield of harvestable portion (e.g., grain) reported as oven-dried weight
- Physical appearance and abnormalities of plants
- Amounts of nutrients in leachate (e.g., calcium, potassium, nitrate–nitrogen, ammonium–nitrogen, *ortho*-phosphate, dissolved organic carbon)
- Amounts of parent compound in all plant parts, soil horizons, and leachate (by measuring radioactivity of parent compound)
- Amounts of chemical degradation products in plant material, soil, and leachate
- Soil properties, including pedologic identity (according to the USDA Seventh Approximation Soil Classification System), percent organic matter, hydraulic characteristics, cation exchange capacity, bulk density, macro- and micronutrient content, organic matter content, inorganic mineral content, exchange capacity, particle-size distribution, and hydraulic characteristics

Data analysis should follow standard statistical techniques as appropriate, such as analysis of variance for comparing biomass data or regression analysis on sequential measures of productivity. The 5% level should be considered as the level of significance with a power of 0.90 or 0.95.

## AIR POLLUTION

Air pollution has the potential to seriously injure plants and animals. A good example of air pollution effects is the change in the forest vegetation of the San Bernadino Mountains in southern California [23]. Here, prolonged exposure to photochemical oxidants (smog) resulted in a shift from ozone-sensitive pine trees to more ozone-tolerant oaks and shrubs. Ozone injury in pines results in decreased photosynthesis due to foliar injury and premature needle fall, reduced nutrient retention in needles, and decreased growth. In general, gaseous pollutants have the potential to disrupt plant-leaf biochemical processes through absorption by stomata or cuticle, while trace metals and organochlorine compounds tend to accumulate in humus and organic matter. Animals are affected either by uptake through the food chain or direct exposure via inhalation [114]; however, there has been little direct measurement of toxicity to wildlife from air pollutants other than studies of effects of smelter emissions and a documented reduction in small mammal diversity in the ozone-impacted San Bernadino Mountains. Many species of freshwater animals have been affected by wet deposition of sulfur dioxides, also known as "acid rain." In general, fish are most sensitive to acidification, followed by invertebrates, algae, and microbes [114].

Certainly, the potential impact to biodiversity from global climate change is significant. The amount of production or the type of plant that will grow efficiently may change in response to increasing $CO_2$, to changes in the amount and timing of precipitation, or to the amount of available solar radiation. $CO_2$ enrichment alone would increase the rate of growth of plants, while higher temperatures would increase the rate of microbial decomposition of organic matter, adversely affecting soil fertility [135]. On the other hand, higher temperatures generally hasten plant maturity in annual species, shortening the growth stages of some plants. At the same time, mid-latitude summer dryness is likely to reduce yields by 10 to 30% [148].

Air pollution levels generally are measured mechanically, through the use of biosensors or air sampling devices *in situ*; however, as with chemical analysis of soils and other media, this type of monitoring does not provide information about the toxicity of air pollutants to terrestrial organisms. Lichens are particularly sensitive to low levels of some forms of air pollution. Some are killed by pollutant levels that are too low to cause visible injury to other plants; consequently, lichens are now used as qualitative indicators of air quality in parts of the United States and Europe, in particular for sulfur dioxide, hydrogen fluoride, acidic precipitation, ozone, nitrogen dioxide, heavy metals, and radioactive compounds [110]. Lichen species in the area of study are identified, and species diversity, density, and frequency are recorded, as is percent cover. Plants also are observed for obvious signs of damage. These parameters are compared with historical accounts of lichen communities to determine if airborne pollutants have affected the plants. Additionally, some compounds (e.g., heavy metals, organochlorines) are bioaccumulated by lichens without visible signs of damage to the plants. Residue analysis of lichens has been used to map the location and distribution of these pollutants.

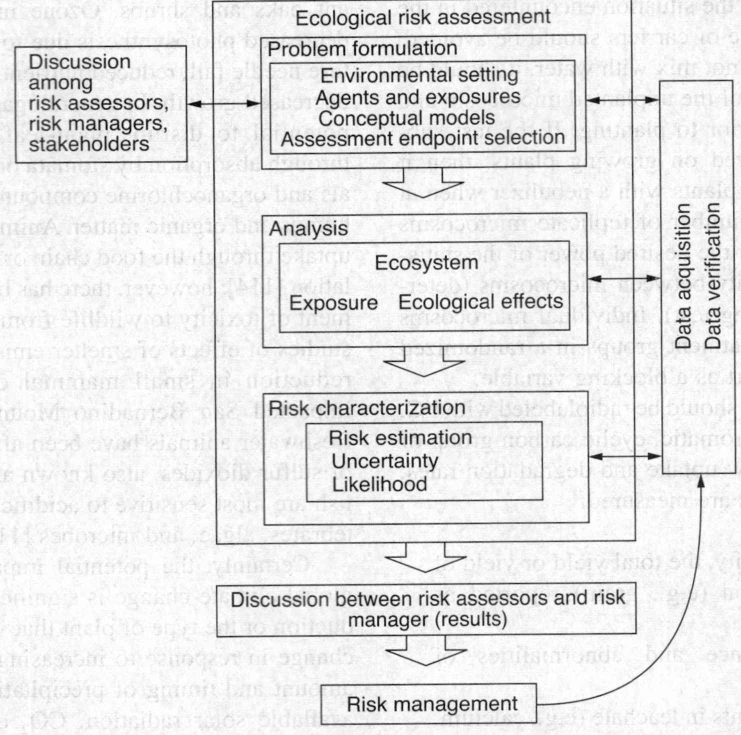


**FIGURE 42.7** The ecological risk assessment process. (From UNEP International Environmental Technology Centre, *Environmental Risk Assessment for Sustainable Cities*, Tech. Publ. Series, Issue 3, United Nations Environment Programme, Nairobi, Kenya, 1996.)

## ECOLOGICAL RISK ASSESSMENT

The toxicity tests described in this chapter ultimately become a component of the process of establishing the likelihood that adverse ecological effects are occurring or may occur as a result of exposure to one or more stressors. The process of ecological risk assessment has evolved rapidly since the USEPA issued a framework for ecological risk assessment in 1992 [168]. Since that time, many workshops have been held by governmental entities and professional societies to refine the process, culminating in the publication of *Guidelines for Ecological Risk Assessment* in 1998 [175]. The guidelines should be used for improving the quality, consistency, and applicability of assessments of the impact of environmental stressors on components of an ecosystem. A further elaboration of approaches to ecological risk assessment is the development of generic ecological assessment endpoints (GEAEs) by the USEPA [155,182]. The Agency has proposed a set of criteria and specific proposals for GEAEs to provide improved consistency in ecological risk assessment and management.

Ecological risk assessment involves three stages in a continuous process: (1) problem formulation, (2) analysis of exposure and effects, and (3) risk characterization. Because ecological risk assessment must consider effects at the population, community, and ecosystem levels, as well as to the individual species, and the relevant assessment endpoints are not universally accepted, the process is generally more complex and protracted than are most human health risk assessments. Furthermore, ecological risk assessment frequently must consider the effect of mixtures of chemicals that interact in a complex chemical and physical environment. Many examples of comprehensive ecological risk assessments can be found in the scientific literature. A good example is a series of five papers that present a probabilistic risk assessment of pyrethroid insecticides used on cotton [56,68,103,152,159].

The ecological risk assessment process is illustrated in Figure 42.7. The first, and most critical, phase of the process is problem formulation. It is during this phase that the affected environmental resource and the stressors of concern must be considered within the context of the overall situation that is to be evaluated. It is frequently useful at this stage to involve risk assessors, risk managers, and stakeholders in discussions about how to proceed so the results of the assessment will be most useful to those who will have to use them. It is also worthwhile to spend some time modeling the problem, at least in a conceptual sense, so assessment endpoints will be reasonable and achievable and a sampling and analysis plan with appropriate measurement endpoints can be established. A quality assurance project plan should also be established at this stage.

During the analysis phase of the risk assessment, the exposure level or concentration of the stressor of concern on the environmental resource must be established and the effects assessed. The effects assessment should include both potential toxicological effects, as described in this chapter, and ecological effects. Whereas toxicological effects are usually measured on individual organisms using a variety of standardized tests and are relatively easily quantified, ecological effects involve predictions about changes in populations and communities of organisms in longer term studies, which are more difficult to conduct and interpret. Studies in microcosms, mesocosms, small ponds, and streams are frequently required, as described earlier. Food web relationships may have to be determined. Finally, indirect effects of contaminants on the organisms of interest must be considered, such as changes in food availability, habitat structure, or predator abundance. The analysis of exposure and effects also must include consideration of fate and transport of the stressors. As described earlier, exposure is dependent on the bioavailability of the stressor, which depends on its chemical, physical, and environmental characteristics, and effects are entirely dependent on exposure; thus, knowledge of the physical and chemical behavior of stressors in the ecosystem is critical.

The final step in the risk assessment process is to integrate the exposure of the resource to the stressors with the observed or predicted effects within the context of the problem formulation to estimate the degree of risk and the probability of adverse environmental changes actually being observed. Very simple systems where statistical analyses can establish the confidence of the predicted result are rare in ecological risk assessment. The more usual cases involve a high degree of uncertainty that in turn require the risk manager to establish the parameters within which regulation must be effected, frequently using a worst-case scenario. Obviously, the more quantitative and precise the assessment results presented to the risk manager, the more effective the management of the ecosystem will be. In reality, as indicated on the diagram, ecological risk assessment is a continuous process in which risk managers and risk assessors must interact in concert with stakeholders to strive for continuous improvement in understanding the ecosystem and providing for optimum functionality. More detailed analysis of the ecological risk assessment process can be found in the USEPA *Risk Assessment Guidelines* [175] and in the several applications of the methodology that have been published [48,160].

## CONCLUSION

Clearly, toxicity testing under realistic environmental conditions is a much more complex enterprise than determining the effects of single chemicals on single species under controlled laboratory conditions; however, the information obtained in a well-designed, well-executed environmental study provides a valuable supplement to the toxicity data obtained in more traditional laboratory studies. Understanding the impacts of a xenobiotic chemical upon the environment, including the organisms that live in it, requires these more complex studies. Interactions of organisms with each other and with their environment are altered by the introduction of foreign chemicals. The impacts of such chemicals must be assessed in terms of these interactions as well as the effect on individuals if the overall health of an ecosystem is to be properly evaluated and understood. This is the essential difference between ecotoxicology and the discipline of toxicology as it relates to human health. Toxicologists have developed rigorous methods for assessing the effects of chemicals on single species under controlled conditions; however, effects on ecosystems in which many species are involved are much more difficult to identify and measure quantitatively. Great strides have been made in the development of controlled ecosystems, in the form of microcosms and mesocosms, for this purpose. There is still much work to be done, however, to perfect the methodology to the extent that effects can be truly assessed at the ecosystem level. The truism that "the whole is greater than the sum of its parts" definitely applies.

Much progress has been made by the various environmental regulatory authorities in codifying procedures and requirements for the use of field-derived toxicology assessments for populations and communities to evaluate the impacts of chemicals on the environment [175]. The guidelines include not only community-level toxicologic data but also the requirement to understand the spatial and temporal scales of stress and the natural variability in biotic community dynamics. Ultimately, the extra costs in terms of time and money for conducting environmental toxicology evaluations and ecological risk assessments will be returned many-fold in our wiser use of natural resources and the protection of the environment in which we live.

## QUESTIONS

1. What are the differences in experimental considerations and methodologies used to assess the toxicities of contaminants in surface water and sediment?
2. What predictive methods are available to compensate for the limitations in the chemical toxicity databases for aquatic flora and fauna? What are the advantages and disadvantages in the use of these methods?
3. What environmental factors affect the toxicity of anthropogenic chemicals to aquatic life? How do these physical and chemical factors differ between freshwater and saltwater ecosystems?
4. What parts of the terrestrial system have available standardized ecotoxicity tests?

5. What methods are available for extrapolating toxicity thresholds from tested to nontested species?
6. What are the differences between acute, subchronic, and chronic avian toxicity studies?
7. What are the advantages or disadvantages in conducting field studies as compared to laboratory toxicity bioassays?
8. Why are there no standardized rodent studies described for ecotoxicity testing?
9. What is the importance of honeybee studies?
10. What ecological role do soil invertebrates play and what are standard test species?
11. How do soil microbial assays differ from single-species toxicity tests?
12. Amphibians have both an aquatic and a terrestrial life phase. What tests are available to determine effects of soil contaminants on these species? Why are aquatic bioassays described under this section, and not under the section about aquatic toxicity tests?
13. How do terrestrial plant bioassays differ from aquatic plant tests?
14. What are soil core microcosms and what is the advantage in using these types of studies?
15. How are terrestrial ecosystem effects of air pollution (including climate-induced changes from elevated $CO_2$) measured?

## REFERENCES


1. Adams, W.J., Kimerle, R.A., Barnett, J.W., Sediment quality and aquatic life assessment, *Environ. Sci. Technol.*, 26, 1865, 1992.
2. Aldenberg, T. and Slob, W., Confidence limits for hazardous concentrations based on logistically distributed NOEC toxicity data, *Ecotoxicol. Environ. Safety*, 25, 48, 1993.
3. American Petroleum Institute, *Fathead Minnow 7-Day Test: Round Robin Study*, No. 4468, Health and Environmental Sciences Department, Washington, D.C., 1985.
4. APHA, AWWA, and WPCF, *Standard Methods for Examination of Water and Wastewater*, 20th ed. (Suppl.), American Public Health Association, American Water Works Association, and Water Pollution Control Federation, Washington, D.C., 1998.
5. ASTM, E593: standard test method for efficacy of a multiple-dose rodenticide under laboratory conditions, in *Annual Book of ASTM Standards*. Vol. 11.04. *Pesticides, Resource Recovery, Hazardous Substances and Oil Spill Responses, Waste Disposal, Biological Effects*, American Society for Testing and Materials, Philadelphia, PA, 1991, pp. 238–243.
6. ASTM, E857: standard practice for conducting subacute dietary toxicity tests with avian species, in *Annual Book of ASTM Standards*. Vol. 11.04. *Pesticides, Resource Recovery, Hazardous Substances and Oil Spill Responses, Waste Disposal, Biological Effects*, American Society for Testing and Materials, Philadelphia, PA, 1991, pp. 456–460.
7. ASTM, E1062: standard practice for conducting reproductive studies with avian species, in *Annual Book of ASTM Standards*. Vol. 11.04. *Pesticides, Resource Recovery, Hazardous Substances and Oil Spill Responses, Waste Disposal, Biological Effects*, American Society for Testing and Materials, Philadelphia, PA, 1991, pp. 678–688.
8. ASTM, E1163: standard test method for estimating acute oral toxicity in rats, in *Annual Book of ASTM Standards*. Vol. 11.04. *Pesticides, Resource Recovery, Hazardous Substances and Oil Spill Responses, Waste Disposal, Biological Effects*, American Society for Testing and Materials, Philadelphia, PA, 1991, pp. 741–746.
9. ASTM, E1197: standard guide for conducting a terrestrial soil-core microcosm test, in *Annual Book of ASTM Standards*. Vol. 11.04. *Pesticides, Resource Recovery, Hazardous Substances and Oil Spill Responses, Waste Disposal, Biological Effects*, American Society for Testing and Materials, Philadelphia, PA, 1991, pp. 819–831.
10. ASTM, E1439-91: *Standard Guide for Conducting the Frog Embryo Teratogenesis Assay–Xenopus (FETAX)*, American Society for Testing and Materials. Philadelphia, PA, 1991, 11 pp.
11. ASTM, E1676-95: standard guide for conducting a laboratory soil toxicity test with lumbricid earthworm *Eisenia foetida*, in *Annual Book of ASTM Standards*. Vol. 11.05. *Pesticides, Environmental Assessment, Hazardous Substances and Oil Spill Responses*, American Society for Testing and Materials, West Conshohocken, PA, 1996. pp. 1093–1109.
12. ASTM, E1218-90: guide for conducting static 96-h toxicity tests with microalgae, in *Annual Book of ASTM Standards*, Vol. 11.04, *Pesticides, Resource Recovery, Hazardous Substances and Oil Spill Responses, Waste Disposal, Biological Effects*, American Society for Testing and Materials, West Conshohocken, PA, 1998, pp. 168–179.
13. ASTM, E1415-91: new standard guide for conducting static toxicity tests with *Lemna gibba* G3, in *Annual Book of ASTM Standards*, Vol. 11.04, *Pesticides, Resource Recovery, Hazardous Substances and Oil Spill Responses, Waste Disposal, Biological Effects*, American Society for Testing and Materials, West Conshohocken, PA, 1998, pp. 211–220.
14. ASTM, *Biological Effects and Environmental Fate*, 2nd ed., American Society for Testing and Materials, Philadelphia, PA, 1999.
15. ASTM, E1963-02: standard guide for conducting terrestrial plant toxicity tests, in *Annual Book of ASTM Standards*, Vol. 11.05. *Pesticides, Environmental Assessment, Hazardous Substances and Oil Spill Responses*, American Society for Testing and Materials, West Conshohocken, PA, 1999.
16. American Veterinary Medical Association, Panel on Euthanasia, 1986 Report of the AVMA Panel on Euthanasia, *J. Am. Vet. Med. Assoc.*, 188, 52, 1986.
17. Anderson, S.L. and Norberg-King, T.J., Precision of short-term chronic toxicity tests in the real world, *Environ. Toxicol. Chem.*, 10, 143, 1991.
18. Asfaw, A., Ellerseick, M.R., and Mayer, F.L., Interspecies correlation estimations (ICE) for acute toxicity to aquatic organisms and wildlife. EPA/600/R-03/106, Office of Research and Development, Washington, D.C., 2003.

19. Atkins, E.L., Anderson, L.D., and Tuft, T.O., Equipment and technique used in laboratory evaluation of pesticide dusts in toxicological studies with honeybees, *J. Econ. Entomol.*, 47, 965, 1954.
20. Aulerich, R.J. and Ringer, R.K., Current status of PCB toxicity to mink, and effect on their reproduction, *Arch. Environ. Contam. Toxicol.*, 6, 279, 1977.
21. Aulerich, R.J. and Ringer, R.K., Toxic effects of dietary polybrominated biphenyls on mink, *Arch. Environ. Contam. Toxicol.*, 8, 487, 1979.
22. Bantle, J.A., Dumont, J.N., Finch, R.A., and Linder, G., Atlas of abnormalities: a guide for the performance of FETAX, Oklahoma State University Press, Stillwater, 1991, 68 pp.
23. Barker, J.R. and Tingey, D.T., *Air Pollution Effects on Biodiversity.* Van Nostrand Reinhold, New York, 1992, 322 pp.
24. Bascietto, J., *Hazard Evaluation Division Standard Evaluation Procedure: Avian Dietary LC50 Test*, USEPA 540/9–85/008, U.S. Environmental Protection Agency, Washington, D.C., 1985, 11 pp.
25. British Columbia Environment, *Criteria for Managing Contaminated Sites in British Columbia*, Environmental Protection Department, Ministry of Environment, Lands and Parks, Victoria, B.C., 1995.
26. Blaise, C.B., Microbiotests in aquatic ecotoxicology: characteristics, utility and prospects, *Environ. Toxicol. Water Qual.*, 6, 145, 1991.
27. Bleavins, M.R., Aulerich, R.J., and Ringer, R.K., Effects of chronic dietary hexachlorobenzene on the reproductive performance and survivability of mink and European ferrets, *Arch. Environ. Contam. Toxicol.*, 13, 357, 1984.
28. Boethling, R.E. and Mackay, D., Eds., *Handbook of Property Estimation Methods for Chemicals: Environmental and Health Sciences*, Lewis Publishers, Boca Raton, FL, 2000, 481 pp.
29. Bonna, R.J., Aulerich, R.J., Bursian, S.J., Poppenga, R.H., Braselton, W.E., and Watson, G.L., Efficacy of hydrated sodium calcium aluminosilicate and activated charcoal in reducing the toxicity of dietary aflatoxin to mink, *Arch. Environ. Contam. Toxicol.*, 20, 441, 1991.
30. Bookhout, T.A., Ed., *Research and Management Techniques for Wildlife and Habitats*, 5th ed., The Wildlife Society, Bethesda, MD, 1994, 740 pp.
31. Boyle, T.P., Ed., *Validation and Predictability of Laboratory Methods for Assessing the Fate and Effects of Contaminants in Aquatic Ecosystems*, ASTM Special Publ. STP 865, American Society for Testing and Materials, Philadelphia, PA, 1985, pp. 134–151.
32. Brewer, L.W. and Fagerstone, K.A., *Radiotelemetry Applications for Wildlife Toxicology Field Studies*, SETAC Press, Pensacola, FL, 1998, 201 pp.
33. Burton, G.A., Ed., *Sediment Toxicity Assessment*, Lewis Publishers, Boca Raton, FL, 1992, 457 pp.
34. Cairns, J., Ed., *Multispecies Toxicity Testing*, Society of Environmental Toxicology and Chemistry/Pergamon Press, New York, 1985, 261 pp.
35. Cairns, J., Ed., *Community Toxicity Testing*, ASTM-STP 920, American Society for Testing and Materials, Philadelphia, PA, 1986, 350 pp.
36. Canadian Council of Ministers of the Environment, *A Protocol for the Derivation of Environmental and Human Health Soil Quality Guidelines*, CCME-EPC-101E, The National Contaminated Sites Remediation Program, 1996.
37. Canton, J.H. and Adema, D.M.M., Reproducibility of short-term and reproduction toxicity experiments with *Daphnia magna* and comparison of the sensitivity of *Daphnia magna* with *Daphnia pulex* and *Daphnia cucullata* in short-term experiments, *Hydrobiologia*, 59, 135, 1978.
38. Carr, R.S. and Nipper, M., *Porewater Toxicity Testing: Biological, Chemical and Ecological Considerations*, Society of Environmental Toxicology and Chemistry, Pensacola, FL, 2003.
39. Chapman, P.M., Current approaches to developing sediment quality criteria, *Environ. Toxicol. Chem.*, 8, 589, 1989.
40. Chen, G. and White, P.A., The mutagenic hazards of aquatic sediments: a review, *Mut. Res.*, 567, 151, 2004.
41. Crawford, J.K. and Luoma, S.N., *Guidelines for Studies of Contaminants in Biological Tissues for the National Water-Quality Assessment Program*, Open-File Report 92-494, U.S. Geological Survey, Lemoyne, PA, 1994.
42. De Graeve, G.M., Cooney, J.D., Marsh,B.H., Pollock, T.L., and Reichenbach, N.G., Variability in the performance of the 7-d *Ceriodaphnia dubia* survival and reproduction tests: an intra- and interlaboratory study, *Environ. Toxicol. Chem.*, 11, 851, 1992.
43. Dourson, M.L. and Stara, J.F., Regulatory history and experimental support of uncertainty (safety) factors. *Reg. Toxicol. Pharmacol.*, 3, 224,1983.
44. Dunnett, C.W., New tables for multiple comparisons with a control. *Biometrics*, 20, 482, 1964.
45. Edwards, C.A., *Report of the Second Stage in Development of a Standardized Laboratory Method for Assessing the Toxicity of Chemical Substances to Earthworms*, Commission of the European Communities, 1984, 98 pp.
46. Ellersieck, M.R., Asfaw, A., Mayer, F.L., Krause, G.F., Sun, K., and Lee, G., *Acute to Chronic Estimation (ACE v2.0) with Time-Concentration Effect Models*, EPA/600/R-03/107, Office of Research and Development, Washington, D.C., 2003.
47. Fairbrother, A., Putting the impacts of environmental contamination in perspective, in *Ecotoxicology of Wild Mammals*, Shore, R.E. and Rattner, B.A., Eds., John Wiley & Sons, Chichester, U.K., 2001, pp. 671–689.
48. Fairbrother, A., Lines of evidence in ecological risk assessment, *Human Ecol. Risk Assess.*, 9, 1475, 2003.
49. Fairbrother, A., Kapustka, L.A., Williams, B.A., and Bennett, R.S., Effects-initiated assessments are not risk assessments, *Human Ecol. Risk Assess.*, 3, 119, 1997.
50. Fender, W.M., Earthworms of the western United States. Part I. Lumbricidae, *Megadrilogica*, 4, 93, 1985.
51. Finney, D.J., *Probit Analysis*, 3rd ed., Cambridge University Press, Cambridge, U.K., 1971, 333 pp.
52. Fite, E.C., L.W. Turner, Cook, N.J., and Stunkard, C., *Guidance Document for Conducting Terrestrial Field Studies*, EPA/540/09–88/109, U.S. Environmental Protection Agency, Washington, D.C., 1988, 67 pp.
53. Forney, D.R. and Davis, D.E., Effects of low concentration of herbicides on submersed aquatic plants, *Weed Sci.*, 29, 677, 1991.

54. Fur Farm Animal Welfare Coalition, *Standard Guidelines for Operation of Mink Farms in the United States*, 2nd ed., Fur Farm Animal Welfare Coalition, St. Paul, MN, 1988.
55. Gano, K.A., Carline, D.W., and Roger, L.E., *A Harvester Ant Bioassay for Assessing Hazardous Chemical Waste Sites*, PNL-5434, UC-11, Pacific Northwest Laboratory, Richland, WA, 1985, 12 pp.
56. Giddings, J.M., Solomon, K.R., and Maund, S.J., Probabilistic risk assessment of cotton pyrethroids. II. Aquatic mesocosm and field studies, *Environ. Toxicol. Chem.*, 20, 660, 2001.
57. Gersich, F.M., Blanchard, F.A., Applegath, S.L., and Park, C.N., The precision of daphnid (*Daphnia magna* Straus, 1820) static acute toxicity tests, *Arch. Environ. Contam. Toxicol.*, 15, 741, 1986.
58. Giesy, J. and Allred, P.M., Replicability of aquatic multispecies test systems, in *Multispecies Toxicity Testing*, Cairns, J., Ed., Society of Environmental Toxicology and Chemistry/Pergamon Press, New York, 1985, pp. 87–247.
59. Giesy, J.P. and Hoke, R.A., Freshwater sediment quality criteria: toxicity bioassessment, in *Sediments: Chemistry and Toxicity of In-Place Pollutants*, Baudo, R. et al., Eds., Lewis Publishers, Chelsea, MI, 1990, pp. 265–348.
60. Gore, R.H., *The Gulf of Mexico*, Pineapple Press, Sarasota, FL, 1992.
61. Greene, J.C., Bartels, C.L., Warren-Hicks, W.J., Parkhurst, B.R., Linder, G.L., Peterson, S.A., and Miller, W.E., *Protocols for Short-Term Toxicity Screening of Hazardous Waste Sites*, EPA/600/3–88/029, U.S. Environmental Protection Agency, Washington, D.C., 1989, 102 pp.
62. Grothe, D.E., Dickson, K.L., and Reed-Judkins, D. K., *Whole Effluent Toxicity Testing: An Evaluation of Methods and Prediction of Receiving System Impacts*, SETAC Press, Pensacola, FL, 1996.
63. Grothe, D.E. and Kimerle, D.R., Inter- and intralaboratory variability in *Daphnia magna* toxicity test results, *Environ. Toxicol. Chem.*, 4, 189, 1985.
64. Hamelink, J.L., Current bioconcentration test methods and theory, in *Aquatic Toxicology and Hazard Evaluation*, STP 634, Mayer, F.L. and J.L. Hamelink, J.L., Eds., American Society for Testing and Materials, Philadelphia, PA, 1977, pp. 149–161.
65. Hansen, P.D., Bioassays on sediment toxicity, in *Sediments and Toxic Substances*, Calmano, W. and Forstner, U., Eds., Springer-Verlag, Berlin, 1996, pp. 179–194.
66. Hansen, S.R. and Garton, R.R., Ability of standard toxicity tests to predict the effects of the insecticide diflubenzuron on laboratory stream communities, *Can. J. Fish. Aquat. Sci.*, 39, 127, 1982.
67. Harrass, M.C. and Sayre, P.G., Use of microcosm data for regulatory decisions, in *Aquatic Toxicology and Hazard Assessment*, Vol. 12, ASTM-STP 1027, Cowgill, U.M. and Williams, L.R., Eds., American Society for Testing and Materials, Philadelphia, PA, 1989, pp. 204–223.
68. Hendley, P., Holmes, C., Kay, S., Maund, S.J., Travis, K.Z., and Zhang. M., Probabilistic risk assessment of cotton pyrethroids. III. A spatial analysis of the Mississippi, USA, cotton landscape, *Environ. Toxicol. Chem.*, 20, 669, 2001.
69. Hendrix, P.E., Ed., *Earthworm Ecology and Biogeography*, CRC Press, Boca Raton, FL, 1995, 244 pp.
70. Hill, E.F., Heath, R.G., Spann, J.W., and Williams, J.D., *Lethal Dietary Toxicities of Environmental Pollutants to Birds*, U.S. Fish and Wildlife Services Special Report, No. 191, Washington, D.C., 1975, 61 pp.
71. Hill, E.F. and Camardese, M.B., *Lethal Dietary Toxicities of Environmental Contaminants and Pesticides to Coturnix*, U.S. Fish and Wildlife Services Technical Report 2, Washington, D.C., 1986, 147 pp.
72. Hill, I.R., Matthiessen, P., and Heinbach, F., Guidance document on sediment toxicity tests and bioassays for freshwater and marine environments, in *Proc. of Workshop on Sediment Toxicity Assessment*, Renesee, The Netherlands, Society of Environmental Toxicology and Chemistry, Europe, 1993.
73. Hochstein, J.R., Aulerich, R.J., and Bursian, S.J., Acute toxicity of 2,3,7,8-tetrachlorodibenzo-*p*-dioxin to mink, *Arch. Environ. Contam. Toxicol.*, 17, 33, 1988.
74. Holst, R.W., *Hazard Evaluation Division Standard Procedure Non-Target Plants: Seed Germination/Seedling Emergence, Tier 1 and 2*, EPA 5430/9-86-132, Office of Pesticides and Toxic Substances, U.S. Environmental Protection Agency, Washington, D.C., 1986.
75. Holst, R.W., *Hazard Evaluation Division Standard Procedure Non-Target Plants: Vegetative Vigor, Tiers 1 and 2*, EPA 5430/9-86-133, Office of Pesticides and Toxic Substances, U.S. Environmental Protection Agency, Washington, D.C., 1986.
76. Houk, V.S., The genotoxicity of industrial wastes and effluents: a review, *Mut. Res.*, 277, 91, 1992.
77. Hudson, R.H. et al., *Handbook of Toxicity of Pesticides to Wildlife*, 2nd ed., Fish and Wildlife Service Resource Publ. 153, U.S. Department of the Interior, Washington, D.C., 1984, 90 pp.
78. Hulbert, S.H., Pseudoreplication and the design of ecological field experiments, *Ecol. Monogr.*, 54, 187, 1984.
79. Hunn, J.B., *History of Acute Toxicity Tests with Fish, 1863–1987*, U.S. Fish and Wildlife Service, Washington, D.C., 1989, 10 pp.
80. Jarvinen, A.W. and Ankley, G.T., *Linkage of Effects to Tissue Residues: Development of a Comprehensive Database for Aquatic Organisms Exposed to Inorganic and Organic Chemicals*, Society of Environmental Toxicology and Chemistry, Pensacola, FL, 1999, 364 pp.
81. Kapustka, L.A., Selection of phytotoxicity tests for use in ecological risk assessments, in *Plants for Environmental Studies*, Wang, W. Gorsuch, J.W., and Hughes, J.S., Eds., Lewis Press, Boca Raton, FL, 1997, pp. 517–550.
82. Kenaga, R. and Moolenar, R., Fish and *Daphnia* toxicity as surrogates for aquatic and vascular plants and algae, *Environ. Sci. Technol.*, 13, 1479, 1979.
83. Kendall, R.J., Brewer, L.W., Lacher, T.E., Whitten, M.L., and Marden, B.T., *The Use of Starling Nest Boxes for Field Reproductive Studies: Provisional Guidance Document and Technical Support Document*, EPA/600/8-89/056, U.S. Environmental Protection Agency, Washington, D.C., 1989, 82 pp.
84. Kosinski, R., Artificial streams in ecotoxicological research, in *Aquatic Toxicology: Fundamental Concepts and Methodologies*, Vol. II, Boudou, A. and Ribeyre, F. Eds., CRC Press, Boca Raton, FL, 1989, pp. 297–316.

85. Krebs, C.J., *Ecology: The Experimental Analysis of Distribution and Abundance*, 4th ed., Harper Collens College Publications, New York, 1994, 801 pp.
86. Landis, W.G. and Yu, M.H., *Introduction to Environmental Toxicology*, Lewis Publishers, Boca Raton, FL, 1995, 328 pp.
87. LaPoint, T.W., Fairchild, J.F., Little, E.E., and Finger, S.E., Laboratory and field techniques in ecotoxicological research: strengths and limitations, in *Aquatic Ecotoxicology: Fundamental Concepts and Methodologies*, Vol. II, Boudou, A. and Ribeyre, F., Eds., CRC Press, Boca Raton, FL, 1989, pp. 240–255.
88. LeBlanc, G.A., Interspecies relationships in acute toxicity of chemicals to aquatic organisms, *Environ. Toxicol. Chem.*, 3, 47, 1984.
89. Lewis, M.A., Chronic toxicities of surfactants and detergent builders to algae: a review and risk assessment, *J. Ecotox. Environ. Safety*, 20, 123, 1990.
90. Lewis, M.A., Chronic toxicities of surfactants to aquatic animals: a review and risk assessment, *Water Res.*, 25, 101, 1991.
91. Lewis, M.A., Powell, R.L., Nelson, M.K., Henry, M.G., Klaine, S.J., Dickson, G.W., and Mayer, F.L., Eds., *Ecotoxicology and Risk Assessment for Wetlands*, SETAC Press, Pensacola, FL, 1999.
92. Lewis, P.A. and Weber, C.I., A study of the reliability of *Daphnia* acute toxicity tests, in *Aquatic Toxicology and Hazard Assessment: Seventh Symposium*, ASTM-STP 854, Cardwell, R.D. et al., Eds., American Society for Testing and Materials, Philadelphia, PA, 1985, pp. 73–86.
93. Linder, G., Ingham, E., Brandt, C.J., and Henderson, G., *Evaluation of Terrestrial Indicators for Use in Ecological Assessments at Hazardous Waste Sites*, EPA/600/R-92/183, U.S. Environmental Protection Agency, Washington, D.C., 1991, 63 pp.
94. Linder, G., Wyant, J., Meganck, R., and Williams, B., Evaluating amphibian responses in wetlands impacted by mining activities in the western United States, in *Issues and Technology in the Management of Impacted Wildlife*, Comer, R.D., Davis, P.R., Foster, S.Q, Grant, C.V., Rush, S., Thorne, O., and Todd, J., Eds., Thorne Ecological Institute, Boulder, CO, 1991, pp. 17–25.
95. Linz, D.G. and Nakles, D.V., *Environmentally Acceptable Endpoints in Soil: Risk-Based Approach to Contaminated Site Management Based on Availability of Chemicals in Soil*, American Academy of Environmental Engineers, Annapolis, MD, 1997, 632 pp.
96. Lokke, H., Ed., *Handbook of Soil Invertebrate Toxicity Tests*, John Wiley & Sons, New York, 1998, 304 pp.
97. Long, E.R. and MacDonald, D.D., Perspective: recommended uses of empirically derived sediment quality guidelines for marine and estuarine ecosystems, *Human Ecol. Risk Assess.*, 5, 1019, 1998.
98. Luttik, R., Assessing repellency in a modified avian $LC_{50}$ procedure removes the need for additional tests, *Ecotoxicol. Environ. Safety*, 40, 201, 1998.
99. Luttik, R. and Aldenberg, T., Extrapolation factors for small samples of pesticide toxicity data: special focus on $LD_{50}$ values for birds and mammals, *Environ. Toxicol. Chem.*, 16, 1785, 1997.
100. Lyman, W.J., Reehl, W.F., and Rosenblatt, D.H., *Handbook of Chemical Property Estimation Methods*, McGraw-Hill, New York, 1982.
101. Macek, K., Birge, W., Mayer, F., Buikema, A., and Maki, A., Discussion session synopsis, in *Estimating the Hazard of Chemical Substances to Aquatic Life*, STP 657, Cairns, J. et al., Eds., American Society for Testing and Materials, Philadelphia, PA, 1978, pp. 27–32.
102. Maki, A.S., Correlations between *Daphnia magna* and fathead minnow (*Pimephales promelas*) chronic toxicity values for several classes of test substances, *J. Fish. Res. Board Can.*, 36, 411, 1979.
103. Maund, S.J., Travis, K.Z., Hendley, P., Giddings, J.M., and Solomon, K.R., Probabilistic risk assessment of cotton pyrethroids. V. Combining landscape-level exposures and ecotoxicological effects data to characterize risks, *Environ. Toxicol. Chem.*, 20, 687, 2001.
104. Mayer, F. and Ellersieck, M.R., *Manual of Acute Toxicity: Interpretation and Data Base for 410 Chemicals and 66 Species of Freshwater Animals*, Resource Publ. 160, U.S. Department of Interior, Washington, D.C., 1986.
105. Mayfield, H.F., Suggestions for calculating nest success, *Wilson Bull.*, 87, 456, 1975.
106. McFarlane, J.C., Pfleeger, T.G., and Fletcher, J.S., Effect, uptake, and disposition of nitrobenzene in several terrestrial plants, *Environ. Toxicol. Chem.*, 9, 513, 1990.
107. McGrath, S.P. et al., Recommendations for testing toxicity to microbes in soil, in *Test Methods To Determine Hazards of Sparingly Soluble Metal Compounds in Soils*, Fairbrother, A., Glazebrook, P.W., Tarazonna, J.V., and van Straalan, N.M., Eds., SETAC Press, Pensacola, FL, 2002, pp. 17–36.
108. McLane, D.J., *Hazard Evaluation Division Standard Evaluation Procedure: Avian Reproduction Test*, EPA/540/9-86/139, U.S. Environmental Protection Agency, Washington, D.C., 1986.
109. Menzer, R.E., Selection of animal models for data interpretation, in *Toxic Substances and Human Risk*, Tardiff, R.G. and Rodricks, J.V., Eds., Plenum Press, New York, 1987, pp. 133–152.
109a. Morrison, G., Torello, E., Comeleo, R., Walsh, R., Kuhn, A., Burgess, R., Tagliabue, M., and Greene, W., Interlaboratory precision of saltwater short-term chronic toxicity tests, *Res. J. Water Pollut. Contr. Fed.*, 61, 1708, 1989.
110. Nash, T.H. and Wirth, V., Lichens, bryophytes and air quality, *Bibliotheca Lichenologica*, 30, 231, 1988.
111. National Institutes of Health, *Guide for the Care and Use of Laboratory Animals*, (NIH) 86-23, U.S. Department of Health, Education, and Welfare, Washington, D.C., 1985.
112. National Oceanic and Atmospheric Administration, Natural Resource Damage Assessments: Proposed Rule, 15 CFR Part 990, *Fed. Reg.*, 60(149), 39804–39836, 1995.
113. National Research Council, *Nutrient Requirements of Mink and Foxes*, National Academy of Sciences. Washington, D.C., 1982.
114. Newman, J.R., Schreiber, R.K., and Novakova, E., Air pollution effects on terrestrial and aquatic animals, in *Air Pollution Effects on Biodiversity*, Barker, J.R. and Tingey, D.T., Eds., Van Nostrand Reinhold, New York, 1992, pp. 177–233.

115. Norton, S., McVey, M., Colt, J., Durda, J., and Hegner, R., *Review of Ecological Risk Assessment Methods*, EPA-230/-10-88-041, U.S. Environmental Protection Agency, Washington, D.C., 1988.
116. Nyholm, N. and Källqvist, T., Methods for growth inhibition toxicity tests with freshwater algae, *Environ. Toxicol. Chem.*, 8, 689, 1989.
117. Okkerman, P.C., Plassche, E.J., Sloof, W., Van Leeuwen, C.J., and Canton, J.H., Ecotoxicological effects assessment: a comparison of several extrapolation procedures, *Ecotox. Environ. Safety*, 21, 182, 1991.
118. OECD, Test no. 205: avian dietary toxicity test, in *OECD Guidelines for the Testing of Chemicals*. Section 2. *Effects on Biotic Systems*, Environmental Health and Safety Publications Series on Testing and Assessment, Environment Directorate, Organization for Economic Cooperation and Development, Paris, 1984 (http://www.oecd.org/document/62/0,2340,en_2649_34377_2348862_1_1_1_1,00.html).
119. OECD, Test no. 206: avian reproduction test, in *OECD Guidelines for the Testing of Chemicals*. Section 2. *Effects on Biotic Systems*, Environmental Health and Safety Publications Series on Testing and Assessment, Environment Directorate, Organization for Economic Cooperation and Development, Paris, 1984 (http://www.oecd.org/document/62/0,2340,en_2649_34377_2348862_1_1_1_1,00.html).
120. OECD, Test no. 208: terrestrial plants growth test, in *OECD Guidelines for the Testing of Chemicals*. Section 2. *Effects on Biotic Systems*, Environmental Health and Safety Publications Series on Testing and Assessment, Environment Directorate, OECD, Paris, 1984 (http://www.oecd.org/document/62/0,2340,en_2649_34377_ 2348862_1_1_1_1,00.html).
121. OECD, *OECD Guidelines for the Testing of Chemicals*, Organization for Economic Cooperation and Development, Paris, France, 1993.
122. OECD, *Report of the SETAC/OECD Workshop on Avian Toxicity Testing*, OCDE/GD(96)166, Environmental Directorate, Organization for Economic Cooperation and Development, Paris, 1996.
123. OECD, Test no. 207: earthworm, acute toxicity tests, Test No., 207, in *OECD Guidelines for the Testing of Chemicals*. Section 2. *Effects on Biotic Systems*, Environmental Health and Safety Publications Series on Testing and Assessment, Environment Directorate, OECD, Paris, 1998 (http://www.oecd.org/document/62/0,2340,en_2649_34377_2348862_1_1_1_1,00.html).
124. OECD, Test no. 213: honeybees, acute oral toxicity test, in *OECD Guidelines for the Testing of Chemicals*. Section 2. *Effects on Biotic Systems*, Environmental Health and Safety Publications Series on Testing and Assessment, Environment Directorate, OECD, Paris, 1998 (http://www.oecd.org/document/62/0,2340,en_2649_34377_2348862_1_1_1_1,00.html).
125. OECD, Test no., 214: honeybees, acute contact toxicity test, in *OECD Guidelines for the Testing of Chemicals*. Section 2. *Effects on Biotic Systems*, Environmental Health and Safety Publications Series on Testing and Assessment, Environment Directorate, OECD, Paris, 1998 (http://www.oecd.org/document/62/0,2340,en_2649_34377_2348862_1_1_1_1,00.html).
126. OECD, Test no. 217: soil microorganisms: carbon transformation test, in *OECD Guidelines for the Testing of Chemicals*. Section 2. *Effects on Biotic Systems*, Environmental Health and Safety Publications Series on Testing and Assessment, Environment Directorate, OECD, Paris, 2000 (http://www.oecd.org/document/62/0,2340,en_2649_34377_2348862_1_1_1_1,00.html).
127. OECD, Test no. 216: soil microorganisms: nitrogen transformation test, in *OECD Guidelines for the Testing of Chemicals*. Section 2. *Effects on Biotic Systems*, Environment Directorate, OECD, Paris, 2000 (http://www.oecd.org/document/62/0,2340,en_2649_34377_2348862_1_1_1_1,00.html).
128. OECD, *Draft Proposal for a New Guideline 223: Avian Acute Oral Toxicity Test*, Environment Directorate, Organization for Economic Cooperation and Development, Paris, 2002 (http://www.oecd.org/dataoecd/16/41/1836204.pdf).
129. OECD, *Draft Guidance Document for on the Statistical Analysis of Ecotoxicity Data*, Environment Directorate, Organization for Economic Cooperation and Development, Paris, 2003 (http://www.oecd.org/dataoecd/25/13/2956192.pdf).
130. OECD, *Guidelines for the Testing of Chemicals: Proposal for a New Guideline; Draft Guidance Document on Testing Avian Avoidance Behaviour (Pen Test)*, Environment Directorate, Organization for Economic Cooperation and Development, Paris, 2003 (http://www.oecd.org/dataoecd/26/59/2495091.pdf).
131. OECD, *Revised Draft: Detailed Review Paper for Avian Two-Generation Toxicity Test*, Environment Directorate, Organization for Economic Cooperation and Development, Paris, 2003 (http://www.oecd.org/dataoecd/25/30/2956228.pdf).
132. OECD, Test no. 222: earthworm reproduction test (*Eisenia fetida/Eisenia andrei*), in *OECD Guidelines for the Testing of Chemicals*. Section 2. *Effects on Biotic Systems*, Environmental Health and Safety Publications Series on Testing and Assessment, Environment Directorate, OECD, Paris 2004 (http://www.oecd.org/document/62/0,2340,en_2649_34377_2348862_1_1_1_1,00.html).
133. OECD, Test no. 220: Enchytraeid reproduction test, in *OECD Guidelines for the Testing of Chemicals*. Section 2. *Effects on Biotic Systems*, Environmental Health and Safety Publications Series on Testing and Assessment, Environment Directorate, OECD, Paris, 2004 (http://www. oecd.org/document/62/0,2340,en_2649_34377_2348862_1_1_1_1,00.html).
134. Parkhurst, B.R., Forte, J.L., and Wright, G.P., Reproducibility of a life cycle toxicity test with *Daphnia magna*, *Bull. Environ. Contam. Toxicol.*, 26, 1, 1981.
135. Parry, M.L., *Climate Change and World Agriculture*, Earthscan Publications, London, 1990.
136. Pastorok, R.A., Bartell, S.M., Ferson, S., and Ginzburg, L.R., *Ecological Modeling in Risk Assessment: Chemical Effects on Populations, Ecosystems, and Landscapes*, Lewis Publishers, Boca Raton, FL, 2002, 302 pp.
137. Peakall, D.B. and Fairbrother, A., Biomarkers for monitoring and measuring effects, in *Pollution Risk Assessment and Management*, Douben, P.E., Ed., John Wiley & Sons, New York, 1998, pp. 351–376.

138. Peltier, W. and Weber, C.I., *Methods for Measuring the Acute Toxicity of Effluents to Freshwater and Marine Organisms*, EPA/600/4-85-013, U.S. Environmental Protection Agency, Cincinnati, OH, 1985.
139. Penny, C. and Adams, C., *Fourth Report*. Vol. 2. *Evidence*, Royal Commission on Pollution of Rivers in Scotland, 1863.
140. Peters, E.C., Gassman, N.J., Firman, J.C., Richmond, R.H., and Power, E.A., Ecotoxicology of tropical marine systems, *Environ. Toxicol. Chem.*, 16, 12, 1997.
141. Petrocelli, S.R., Chronic toxicity tests, in *Fundamentals of Aquatic Toxicology*, Rand, G. and Petrocelli, S., Eds., McGraw-Hill, New York, 1985, pp. 96–110.
142. Pfleeger, T., McFarlane, J.C., Sherman, R., and Volk, G., A short-term bioassay for whole plant toxicity, in *Plants for Toxicity Assessment*, Vol. 2, ASTM Publ. No. 04-011150-16, Gorsuch, J.W., Lower, W.R., Lewis, M.A., and Wang, W., Eds., American Society for Testing and Materials, Philadelphia, PA, 1991, pp. 355–364.
143. Posthuma, L., Suter, G.W., and Traas, T.P., Eds., *Species Sensitivity Distributions in Ecotoxicology*, Lewis Publishers, Boca Raton, FL, 2002, 587 pp.
144. Powers, E.B., The goldfish (*Carassisus carassius*) as a test animal in the study of toxicity, *Illinois Biol. Monogr.*, 4, 7, 1917.
145. Ratsch, H., *Interlaboratory Root Elongation Testing of Toxic Substances on Selected Plant Species*, NTIS, PB 83-226, U.S. Environmental Protection Agency, Washington, D.C., 1983.
146. Ratsch, H.C., Johndro, D.J., and McFarlane, J.C., Growth inhibition and morphological effects of several chemicals in *Arabidopsis thaliana* (L.) Heynh, *Environ. Contam. Toxicol.*, 5, 55, 1986.
147. Ringer, R.K., Hornshaw, T.C., and Aulerich, R.J., *Mammalian Wildlife (Mink and Ferret) Toxicity Test Protocols ($LC_{50}$, Reproduction, and Secondary Toxicity)*, EPA 600/3-91/043, U.S. Environmental Protection Agency, Washington, D.C., 1991, 77 pp.
148. Rosenzweig, C. and Liverman, D., Predicted effects of climate change on agriculture: a comparison of temperate and tropical regions, in *Global Climate Change: Implications, Challenges, and Mitigation Measures*, The Pennsylvania Academy of Sciences, Philadelphia, PA, 1992, pp. 342–361.
149. Rue, W.J., Fava, J.A., and Grothe, D.R., A review of inter- and intralaboratory effluent toxicity test method variability, in *Aquatic Toxicology and Hazard Assessment*, Vol. 10, STP 971, Adams, W.J., Chapman, G.A., and Landis, W.G., Eds., American Society for Testing and Materials, Philadelphia, PA, 1988, pp. 190–203.
150. Schafer, E.W., The acute oral toxicity of 369 pesticidal, pharmaceutical, and other chemicals to wild birds, *Toxicol. Appl. Pharmacol.*, 21, 315, 1972.
151. Shimabuku, R.A., Ratsch, H.C., Wise, C.M., Nwosu, J.U., and Kapustka, L.A., A new plant life-cycle bioassay for assessment of the effects of toxic chemicals using rapid cycling *Brassica*, in *Plants for Toxicity Assessment*, Vol. 2, Gorsuch, J.W., Lower, W.R., Lewis, M.A., and Wang, W., Ed., ASTM Publ. 04-011150-16, American Society for Testing and Materials, Philadelphia, PA, 1991, pp. 3365–3375.
152. Solomon, K.R., Giddings, J. M., and Maund, S.J., Probabilistic risk assessment of cotton pyrethroids. I. Distributional analyses of laboratory aquatic toxicity data, *Environ. Toxicol. Chem.*, 20, 652, 2001.
153. Sortkjaer, O., Macrophytes and macrophyte communities as test systems in ecotoxicological studies of aquatic systems, *Ecol. Bull. (Stockholm)*, 36, 75, 1984.
154. Stephan, C.E., Methods for calculating an $LC_{50}$, in *Aquatic Toxicology and Hazard Evaluation*, ASTM-STP 634, Mayer, F.L. and Hamelink, J.L., Eds., American Society for Testing and Materials, Philadelphia, PA, 1977, pp. 65–84.
155. Suter II, G. W., Generic assessment endpoints are needed for ecological risk assessment, *Risk Anal.*, 20, 173, 2000.
156. Swann, R.L., Laskowski, D.A., McCall, P.J., Vander Kuy, K., and Dishburger, H.J., A rapid method for the estimation of the environmental parameters octanol/water partition coefficient, soil sorption constant, water to air ratio, and water solubility, *Residue Rev.*, 85, 17, 1983.
157. Thursby, G.B., Anderson, B.S., Walsh, G.E., and Steele, R.L., *A Review of the Current Status of Marine Algal Toxicity Testing in the United States*, American Society for Testing and Materials, Philadelphia, PA, 1993.
158. Traunspurger, W. and Drews, C., Toxicity analysis of freshwater and marine sediments with meio- and macrobenthic organisms: a review, *Hydrobiologia*, 328, 215, 1996.
159. Travis, K.Z. and Hendley, P., Probabilistic risk assessment of cotton pyrethroids. IV. Landscape-level exposure characterization, *Environ. Toxicol. Chem.*, 20, 679, 2001.
160. UNEP International Environmental Technology Centre, *Environmental Risk Assessment for Sustainable Cities*, Tech. Publ. Series, Issue 3, United Nations Environment Programme, Nairobi, Kenya, 1996, 57 pp.
161. Urban, D.J. and Cook, N.J., *Hazard Evaluation Division Standard Evaluation Procedure: Ecological Risk Assessment*, EPA/540/9-85/001, U.S. Environmental Protection Agency, Washington, D.C., 1986, 96 pp.
162. USEPA, *Pesticide Assessment Guidelines Subdivision E, Hazard Evaluation: Wildlife and Aquatic Organisms*, EPA/540/9-82/024, U.S. Environmental Protection Agency, Washington, D.C., 1978, 91 pp.
163. USEPA, Toxic Substances Control Act: Premanufacture Testing of New Chemical Substances, *Fed. Reg.*, 44, 16240–16292, 1979.
164. USEPA, *Pesticide Assessment Guidelines Subdivision L, Hazard Evaluation: Nontarget Insects*, EPA/540/9-82/019, U.S. Environmental Protection Agency, Washington, D.C., 1982, 34 pp.
165. USEPA, *Estimating 'Concern Levels' for Concentrations of Chemical Substances in the Environment*, Environmental Effects Branch, Health and Environmental Review Division, U.S. Environmental Protection Agency, Washington, D.C., 1984.
166. USEPA, *Aquatic Mesocosm Tests To Support Pesticide Registrations*, EPA/EEB/HED/OPP, U.S. Environmental Protection Agency, Washington, D.C., 1987, 35 pp.
167. USEPA/U.S. Army Corps of Engineers, *Evaluation of Dredged Materials Proposed for Ocean Disposal: Testing Manual*, EPA/503/8-91-001, U.S. Environmental Protection Agency, Washington, D.C., 1991.

168. USEPA, *Framework for Ecological Risk Assessment*, EPA/630/R-92/001, U.S. Environmental Protection Agency, Washington, D.C., 1992.
169. USEPA, *Toxicity Identification Evaluation: Characterization of Chronically Toxic Effluents, Phase 1*, EPA/600/6-91/005F, National Effluent Toxicity Assessment Center, U.S. Environmental Protection Agency, Duluth, MN, 1992.
170. USEPA, *Methods for Measuring the Acute Toxicity of Effluents and Receiving Waters to Freshwater and Marine Organisms*, EPA/600/4-90/027F, Office of Research and Development, U.S. Environmental Protection Agency, Washington, D.C., 1993.
171. USEPA, *Short-Term Methods for Estimating the Chronic Toxicity of Effluents and Receiving Water to Freshwater Organisms*, EPA/600/4-91/002, Office of Research and Development, U.S. Environmental Protection Agency, Washington, D.C., 1994.
172. USEPA, *Great Lakes Water Quality Initiative Technical Support Document for the Procedure To Determine Bioaccumulation Factors*, EPA/820/B-95/005, Office of Water, U.S. Environmental Protection Agency, Washington, D.C., 1995.
173. USEPA, *Series 850 Ecological Effects Test Guidelines*, Office of Prevention, Pesticides, and Toxic Substances, U.S. Environmental Protection Agency, Washington, D.C., 1996.
174. USEPA, *Region 9 Preliminary Remediation Goals*, U.S. Environmental Protection Agency, San Francisco, CA, 1997 (http://www.epa.gov/region09/water/sfund/prg/index.html).
175. USEPA, *Guidelines for Ecological Risk Assessment*, EPA/630/R-95/002F, U.S. Environmental Protection Agency, Washington, D.C., 1998.
176. USEPA, *National Sediment Bioaccumulation Conference Proceedings*, EPA/823/R-98/002, Office of Water, U.S. Environmental Protection Agency, Washington, D.C., 1998.
177. USEPA, *Methods for Measuring the Toxicity and Bioaccumulation of Sediment-Associated Contaminants with Freshwater Invertebrates*, EPA/600/R-99/064, Office of Water, U.S. Environmental Protection Agency, Washington, D.C., 2000.
178. USEPA, *Methods for Assessing the Chronic Toxicity of Marine and Estuarine Sediment-Associated Contaminants with the Amphipod Leptocheirus plumulosus*, EPA/600/R-01/020, Office of Water, U.S. Environmental Protection Agency, Washington, D.C., 2001.
179. USEPA, *Methods for Measuring the Acute Toxicity of Effluents and Receiving Waters to Freshwater and Marine Organisms*, EPA/821/R-02/012, Office of Water, U.S. Environmental Protection Agency, Washington, D.C., 2002.
180. USEPA, *Short-Term Methods for Estimating the Chronic Toxicity of Effluents and Receiving Waters to Marine and Estuarine Organisms*, EPA/821/R-02/014, Office of Water, U.S. Environmental Protection Agency, Washington, D.C., 2002.
181. USEPA, *A Framework for a Computational Toxicology Research Program in ORD (Draft)*, EPA/600/R-03/065, Office of Research and Development, U.S. Environmental Protection Agency, Washington, D.C., 2003.
182. USEPA, *Generic Ecological Assessment Endpoints (GEAEs) for Ecological Risk Assessment*, EPA/630/P-02/004F, Risk Assessment Forum, U.S. Environmental Protection Agency, Washington, D.C., 2003.
183. USEPA, *Guidance for Developing Ecological Soil Screening Levels (Eco-SSLs)*, OSWER Directive 92857-55, Office of Solid Waste and Emergency Response, U.S. Environmental Protection Agency, Washington, D.C., 2003 (http://www.epa.gov/ecotox/ecossl/).
184. USEPA, Review of existing soil screening benchmarks, in *Guidance for Developing Ecological Soil Screening Levels (Eco-SSLs)*, OSWER Directive 92857-55, Office of Solid Waste and Emergency Response, U.S. Environmental Protection Agency, Washington, D.C., 2003 (http://www.epa.gov/ecotox/ecossl/pdf/ecossl_attachment_1–1.pdf).
185. U.S. Fish and Wildlife Service, Migratory bird hunting: revised test protocol for nontoxic approval procedures for shot and shot coating; final rule, *Fed. Reg.*, 62(230), 63608–63615, 1997.
186. USFDA, Seed germination and root elongation, in *Environmental Assessment Technical Handbook*, 4.06, Center for Food Safety and Applied Nutrition, Center for Veterinary Medicine, U.S. Food and Drug Administration, Washington, D.C., 1987.
187. USFDA, *Guidance for Industry: Environmental Impact Assessments for Veterinary Medical Products, Phase II*, Center for Veterinary Medicine, U.S. Food and Drug Administration, U.S. Dept. of Health and Human Services, Washington, D.C., 2003.
188. Van der Zandt, P.T.J. and Van Leeuwen, C.J., *A Proposal for Priority Setting of Existing Chemical Substances*, proposal prepared for the Directorate General for Environment, Nuclear Safety and Civil Protection of the Commission of the European Communities, Directorate General for Environmental Protection, The Hague, The Netherlands, 1992.
189. Vittozzi, L. and De Angelis, G., A critical review of comparative acute toxicity data on freshwater fish, *Aquat. Toxicol.*, 19, 167, 1991.
190. Wagner, C. and Lokke, H., Estimation of ecotoxicological protection levels from NOEC toxicity data, *Water Res.*, 10, 1237, 1991.
191. Walker, J.D., *QSARs for Pollution Prevention, Toxicity Screening, Risk Assessment and Web Applications*, Society for Environmental Toxicology and Chemistry, Pensacola, FL, 2003, 259 pp.
192. Walsh, G.E., Principles of toxicity testing with marine unicellular algae, *Environ. Toxicol. Chem.*, 7, 979, 1988.
193. Walsh, G.E., Weber, D.E., Simon, T.L., Brashers, L.K. and Moore, J.C., Use of marsh plants for toxicity testing of water and sediments, in *Plants for Toxicity Assessment*, Vol. 2, Gorsuch, J.W. et al., Eds., STP 115, American Society for Testing and Materials, Philadelphia, PA, 1991, pp. 341–354.
194. Walton, B.T., Differential life-stage susceptibility of *Acheta domesticus* to acridine, *Environ. Entomol.*, 9, 18, 1980.
195. Warren-Hicks, W., Parkhurst, B.R., and Baker, S.S., *Ecological Assessments of Hazardous Waste Sites: A Field and Laboratory Reference Document*, EPA/600/3-89/01, U.S. Environmental Protection Agency, Washington, D.C., 1989, 350 pp.
196. Wells, P., Biomonitoring the health of coastal marine ecosystems-the roles and challenges of microscale toxicity tests, *Mar. Pollut. Bull.*, 39, 39, 1999.

# Glossary

*50th Percentile*: The number in a distribution such that half the values in the distribution are greater than the number and half the values are less. The 50th percentile is equivalent to the median.

*95th Percentile*: The number in a distribution such that 95% of the values in the distribution are less than or equal to the number and 5% are greater.

*95% Upper Confidence Limit for Mean (95% UCL)*: The 95% upper confidence limit (95% UCL) for a mean is defined as a value that, when repeatedly calculated for randomly drawn subsets of size *n*, equals or exceeds the true population mean 95% of the time. Although the 95% UCL provides a conservative estimate of the mean, it should not be confused with a 95th percentile. As the sample size increases, the difference between the UCL for the mean and the true mean decreases while the 95th percentile of the distribution remains relatively unchanged at the upper end of the distribution. The EPA's Superfund program has traditionally used the 95% UCL for the mean as the concentration term in point estimates of the reasonable maximum exposure (RME) for human health risk assessment.

*Absolute White Blood Cell and Reticulocyte Counts*: Counts expressed as cell concentration (i.e., cells/volume of blood). Absolute counts, as opposed to relative counts (%), are preferred for interpretation and reporting.

*Absorbed Dose*: Energy imparted to matter when radiation passes through; measured in grays or rads.

*Absorption*: Uptake of the chemical from the site of administration into the general circulation. Absorption may involve a number of stages (e.g., dissolution) and diffusion through membranes. Chemicals may be changed during absorption due to metabolism or degradation, such that it is possible to have complete absorption and low bioavailability.

*Absorption Barrier*: Any of the exchange barriers of the body that allow differential diffusion of various substances across a boundary. Examples of absorption barriers are the skin, lung tissue, and gastrointestinal tract wall.

*ACB*: Accelerated cancer bioassay.

*Acceptable Daily Intake (ADI)*: Daily intake of a chemical (e.g., food additive, pesticide) that, during the entire lifetime, appears to be without appreciable risk (affects 1 in 1 million people or less) on the basis of all known facts at the time.

*Accessory Cells*: Cells that support T or B cells in the induction of an immune response. These cells usually express MHC class II molecules.

*Accuracy*: A measure of the extent to which the mean estimate of a quantity approaches its true value: (1) the closeness of agreement between a test method result and an accepted reference value; (2) the proportion of correct outcomes of a test method. It is a measure of test method performance and one aspect of "relevance." The term is often used interchangeably with "concordance" (*see Two-by-Two Table*). Accuracy is highly dependent on the prevalence of positives in the population being examined in the validation study.

*Acid–Base Balance*: Maintenance of pH; kidneys function in acid–base balance by the regulation of $H^+$, $HCO_3^-$, and $NH_4^+$ ions.

*ACTH*: Adrenocortical tropic hormone.

*Action Level*: Level of unavoidable contaminants in foods and feeds considered as the upper limit of safety but which are not subjected to regulatory control.

*Acute*: Characterized by a time period of short duration; commonly used to describe single-dose exposure in toxicity studies.

*Acute Toxicity*: The adverse effects occurring within a short time of administration of a single dose of a substance or multiple doses given within 24 hours.

*Acute Toxicity Study*: Usually, a single-dose study in which animals are observed for a 2-week period post-dose to determine overt signs of toxicity, normally including some form of assessment of the lethal dose.

*Acute-to-Chronic Ratio (ACR)*: A ratio determined experimentally or mathematically for a chemical that is used to predict chronic toxicity when only acute toxicity data are available.

*Ad libitum*: Available with unrestricted access; this term is commonly used in toxicity studies to describe free access by animals to feed or water.

*Adenohypophysis*: The anterior lobe of the pituitary gland.

*ADH*: Antidiuretic hormone or vasopressin; an octapeptide.

*Adjunct Test*: A test that provides information that adds to or helps interpret the results of other tests and provides information useful for the risk assessment process.

*Adjuvant*: A material that enhances an immune response, it traditionally refers to a mixture of oil and mycobacterial cell fragments.

*ADME*: Absorption, distribution, metabolism, and excretion—the processes that determine the disposition and fate of an administered molecule.

*Administered Dose*: The amount of a substance given to a test subject (human or animal) in determining dose–response relationships, especially through ingestion or inhalation. In exposure assessment, because exposure to chemicals is usually inadvertent, this quantity is called *potential dose*.

*Advance Access*: The online publication of papers in manuscript form soon after they have been peer reviewed and deemed final; they appear considerably sooner than as print publications.

*Adverse effect*: A test-compound-related effect (e.g., morphological, biochemical, developmental) that alters the function of an organ or system or alters the ability to respond to additional environmental challenges.

*AEGL*: Acute exposure guideline levels are values intended to provide estimates of concentrations and exposure durations (minutes to hours) above which one could reasonably anticipate observing adverse health effects. Compare to emergency response planning guidelines (ERPGs).

*Aerodynamic Equivalent Diameter*: The diameter of a spherical particle of unit density (1 g/mL) that has the same terminal settling velocity as the particle in question.

*Aerosol*: A suspension of either microscopic liquid or solid particles dispensed in a gas, the particles of which have a negligible falling velocity; also used to characterize a product or chemical form that contains particles that can enter the respiratory tract.

*Aglycone*: The xenobiotic substrate that is conjugated by glucuronosyltransferases.

*Air Elutriation*: A process in which particles are separated on the basis of size by pitting their settling velocity against the velocity of a current of air with which they move.

*Air Shower*: A device that uses high-velocity, ultrafiltered air to remove particulates from the surfaces of the clothing worn by personnel.

*AL*: *Ad libitum*.

*Alkaloids*: Nitrogenous heterocyclic compounds that protect plants from attack by microorganisms, pests, and herbivores. Any of a large, heterogeneous group of alkaline, bitter tasting, biologically active, usually water-insoluble, nitrogenous organic compounds produced by plants.

*Allergic Contact Dermatitis*: Chemically induced immunologic (delayed hypersensitivity) dermatitis.

*Allergy*: A state of altered immunity in which contact with an antigen (allergen) results in a hypersensitivity response.

*Allogeneic*: From a different genetic background. In the context of immunotoxicology, it generally refers to the use of genetically dissimilar cells *in vitro* assays to elicit a cell-mediated immune response.

*Allometry*: The study of the relationship between body size and various biological and physiological parameters, such as organ sizes, blood flow rates, and metabolic rates.

*Alloxan*: A chemical that can cause destruction to the pancreatic beta cell.

*Alpha Particle*: Nucleus of a helium atom emitted by certain radioisotopes upon disintegration. Contains two protons and two neutrons.

*Alveolar Macrophage*: A motile, phagocytic cell of the pulmonary region of the lung, essential to removal of particulate matter from the alveoli and sterility of alveolar surfaces.

*Alveolus*: The smallest functional gas exchange unit of the pulmonary region of the lung. The alveolus contains a very thin epithelial and endothelial surface that allows the rapid exchange of oxygen and carbon dioxide, as well as a portal for the absorption and elimination of volatile substances from the vasculature.

*Amatoxin*: One of two groups of thermostable toxins isolated from poisonous species of *Amanita*; they are extremely toxic, bicyclic octapeptides that act on the RNA polymerase II system of eukartyotic cells.

*Ambient Measurement*: A measurement (usually of the concentration of a chemical or pollutant) taken in an ambient medium, normally with the intent of relating the measured value to the exposure of an organism that contacts that medium.

*Amine Precursor Uptake and Decarboxylation (APUD) Cells*: A group of apparently unrelated endocrine cells found throughout the body which have a number of similar characteristics and which make a number of hormones with similar structures (including serotonin, epinephrine, dopamine, neurotensin, and norepinephrine). Over 60 types of endocrine cells have been identified in the APUD system which can be found in the numerous organs throughout the body.

*Ammonium Molybdate*: A metallic negative stain that is used to negative stain the Formvar[a] backing on coated electron microscopic grids for transmission electron microscopy.

*Amphibia*: A class of anamniote tetrapods comprised of 2600 known species of toads, frogs, salamanders, and newts, many of which are poisonous but not necessary dangerous to humans; any life form adapted to or able to live in both aquatic and terrestrial environments.

*Amyotrophic Lateral Sclerosis*: Fatal neurologic disease characterized by progressive degeneration of upper and lower motor neurons in the brain and spinal cord.

*Analytic Study*: A study designed to examine *a priori* hypothesized causal associations.

*Anaphylaxis*: An extreme, immediate immunologic reaction characterized by contraction of smooth muscle and dilation of capillaries due to release of pharmacologically active substances (e.g., histamine) in response to administration of a foreign material; a local or systematic immediate hypersensitivity reaction resulting from the release of mediators following exposure to antigen. A life-threatening, often fatal response.

*Androgens*: Male sex steroids (e.g., testosterone).

*Anemia*: Reduction below normal of hemoglobin concentration; usually accompanied by a similar reduction of red blood cell count and hematocrit. Functionally, anemia is characterized by a decrease in red cell mass sufficient to cause reduced oxygen delivery to peripheral tissues.

*Anemone*: Any herb of the genus *Anemone*.

*Anergy (Tolerance)*: Unresponsiveness to antigenic stimulation; also referred to as *tolerance*.

*Annexin V-FITC Staining*: Method used to assay apoptosis or necrosis. Annexin V labeled with fluorescein isothiocyanate (FITC) binds to phosphatidylserine residues of cellular membranes; cell populations with differential binding and, hence, fluorescence are detected by flow cytometry.

*Anogenital Distance (AGD)*: A primary landmark of sexual development, typically measured at birth, reflecting the linear distance between the genital tubercle and the anus. AGD is sexually dimorphic, being greater for males than females, and androgen dependent. The AGD in male rodents is decreased by development exposure to antiandrogens.

*ANOVA*: Analysis of variance.

*Antibody*: Complex molecules produced by plasma cells that recognize specific antigens. Antibodies, also called *immunoglobulins* (Ig), consist of two basic units. The antigen-binding section (Fab) contains variable regions with coding for antigen recognition. In mammals, the constant region of the molecule (Fc) may be grouped into several classes, designated IgA, IgD, IgE, IgG, and IgM, depending on the function of the molecule. Cross-linking of antibody molecules on the surface of a target leads to activation of complement, usually resulting in the destruction of the target.

*Antibody-Forming Cell (AFC)/Plaque-Forming Cell (PFC) Assay*: An assay that measures the ability of animals to produce specific antibodies against a T-dependent or T-independent antigen following *in vivo* sensitization. Due to the involvement of multiple cell populations in mounting an antibody response, the AFC assay actually evaluates several immune parameters simultaneously. It is considered to be one of the most sensitive indicator systems for immunotoxicology studies.

*Antigen*: A molecule that is the subject of a specific immune reaction. Antigens are recognized in a cognate fashion by the T-cell antigen receptor, the B-cell antigen receptor, or immunoglobulins (antibodies). Antigens generally are proteinaceous in nature.

*Antigen-Presenting Cell (APC)*: Cells that are responsible for making antigens accessible to immune effector and regulatory cells. Following internalization and degradation of the antigen (generally by phagocytosis), a fragment of the antigen molecules is presented on the APC cell surface in association with a major histocompatibility complex (MHC) molecule. This complex is recognized by either B cells via surface-bound immunoglobulin molecule or by T cells via the T-cell antigen receptor. Induction of a specific immune response then proceeds. APCs include macrophages, dendritic cells, and certain B cells.

*Antigenicity (Immunogenicity)*: The property of eliciting an immune response, characterized by an interaction of a foreign material with endogenous antibodies or immune cells, in a subject that has been previously exposed (sensitized) to that foreign material.

*Anti-Mullerian Hormone* (also called *Mullerian inhibiting substance*): A protein produced by fetal Sertoli cells that prevents formation of a female reproductive tract in male fetuses.

*Aortic Perfusion*: Aorta is cannulated for inflow of the perfusate through the heart and the perfusate exists through the left atrium.

*Aplastic Anemia*: Failure of blood cell production resulting from direct injury to pluripotent hematopoietic stem cells or their stromal microenvironment and characterized by varying degrees of pancytopenia (i.e., decreased erythrocytes, leukocytes [primarily neutrophils], and platelets) and hypocellular bone marrow.

*Apoptosis*: A series of biochemical events characterized by activation of a series of caspase enzymes that lead to the digestion of cellular DNA with the appearance of DNA laddering on agarose gels and morphological formation of intranuclear clumps in affected cells. This process is also known as *programmed cell death*, whereby cells die in a controlled, progressive manner that is regulated in part by the release of mitochondrial-initiating factors and cytochrome c. Programmed cell death is a normal cellular process for removal of unneeded cells during organogenesis (e.g., cell death and replacement) which is accelerated by a number of toxic agents, including chemotherapeutic drugs. Thus, apoptosis is a single-cell phenomenon that is energy dependent and tightly regulated and generally occurs at lower doses of many toxicants and is important in morphogenesis and development. It is a genetically programmed form of cell death, distinct from necrosis, which is accidental cell death. Apoptosis helps regulate animal cell populations by eliminating cells that have been overproduced or mutated. The process is complementary to mitosis but opposite in function.

*Apoptosis-Inducing Factor (AIF)*: A pro-apoptotic protein which when released from the mitochondrion stimulates activation of caspase 3.

*Apparent Volume of Distribution*: The volume of plasma (or blood) into which the body load appears to have been dissolved or distributed; equivalent to the body load, at any time, divided by the corresponding plasma (or blood) concentration. It is not a physiological volume but is important, as it indicates the volume of plasma that has to be cleared of chemical; independent of concentration and dose under first-order conditions.

*Applied Dose*: The amount of a substance in contact with the primary absorption boundaries of an organism (e.g., skin, lung, gastrointestinal tract) and available for absorption.

*Aquatic Toxicology*: The study of adverse effects on freshwater and saltwater biota and on the ecosystems that contain them.

*Arachnida*: A large class of invertebrates comprised chiefly of predaceous terrestrial forms such as scorpions, spiders, harvestmen, mites, ticks, and related forms; they are characterized by a cephalothorax that bears four pairs of walking appendages.

*Arithmetic Mean*: The sum of all the measurements in a data set divided by the number of measurements in the data set.

*Aspect Ratio*: Usually applied to fibers; the ratio of length to width.

*Atherosclerosis*: Nodular sclerosis characterized by irregularly distributed lipid deposits in the intima of the large and medium-sized arteries; such deposits are associated with fibrosis and calcification.

*Atopy*: General systemic or local hypersensitivity (i.e., allergy), often related to genetic predisposition; may be though of us "unwanted reactivity."

*Atrial Perfusion*: Right atrium is cannulated for inflow of the perfusate, which exits through the right ventricle via pulmonary arterial cannula or an open slit.

*AUC*: Area under the plasma concentration time course.

*Autoimmunity*: Immune reactivity toward self.

*Autologous Blood Perfusate*: Blood used as perfusate comes from the same animal from which the organ was removed for perfusion.

*Autolysis*: Enzymatic self-digestion of cells or tissues that occurs after death. Autolysis complicates detection of pathologic changes in tissues or organs during necropsy and subsequent microscopic examination and can make valid observations during these activities impossible.

*Autoradiography*: The production of an image by the emission of radioactive decay energy from a radionuclide. It provides a qualitative visual image of the tissue distribution of a xenobiotic and can provide quantitative distribution data as well. Quantitative whole-body autoradiography (QWBA) is becoming a standard tool for absorption, distribution, metabolism, and excretion (ADME) studies.

*Autosome*: A chromosome that is not a sex-determining chromosome.

*Azotemia*: Accumulation of nitrogenous wastes such as urea or creatinine in the blood.

*Background Exposure*: Exposures that are not related to the site—for example, exposure to chemicals at a different time or from locations other than the exposure unit of concern. Background sources may be either naturally occurring or anthropogenic (man-made).

*Background Level (Environmental)*: The concentration of substance in a defined control area during a fixed period of time before, during, or after a data-gathering operation.

*Basepair Substitution*: A gene mutation characterized by the replacement of one nucleotide pair for another in a codon.

*B-Cell Antigen Receptor*: A membrane-bound molecular complex responsible for antigen recognition by B cells. It comprises membrane immunoglobulin (mIg) and several accessory molecules. Functionally analogous, but structurally dissimilar, to the T-cell antigen receptor.

*B Cell/B Lymphocyte*: Lymphocytes that recognize antigen via surface-bound immunoglobulins. B cells that have been exposed to specific antigen differentiate into plasma cells that are responsible for producing specific antibodies. B cells differentiate in the bone marrow in mammals and in an organ known as the bursa in birds.

*Becquerel (bq)*: SI unit of radioactivity equaling one disintegration/second, approximately $2.7 \times 10^{-11}$ curies (Ci).

*Beer–Lambert Law*: This law relates solute concentration and cell path length to ultraviolet absorbance. The law can be expressed by the following equation:

$$A = \log I_o/I_t = abc$$

where $A$ = absorbance, $I_o$ is the original intensity incident on the cell, $I_t$ is the reduced intensity transmitted from the cell, $a$ is a proportionality constant (the absorptivity), $b$ is the path length of the solution, and $c$ is the concentration of the analyte. When $c$ is in moles per liter, the constant is the molar absorptivity or molar extinction coefficient. When $c$ is a 1% solution and $b$ is a 1-cm path length, the term *specific absorptivity* is used (a 1%/1 cm is the most common form used).

*Behavioral Teratology*: The functional deficits arising from exposure to neurotoxic agents during early development.

*Benchmark Dose (BD)*: The lower confidence limit on a dose associated with a specified level of response; a dose corresponding to a specified level of risk, generally in the range of 1 to 10%.

*Beta Cells*: Insulin-secreting cells of the endocrine pancreas.

*Bias*: Systemic error as opposed to a sampling error; for example, selection bias may occur when each member of the population does not have an equal chance of being selected for the sample.

*Bifurcation*: Usually related to airway anatomy, describing a branching of the parent airway into two or more smaller airways, often at acute angles.

*Bioaccumulation*: The net uptake of chemicals from the environment from all sources.

*Bioactivation*: The enzymatic conversion of a chemical to a more toxic form (in the body or *in vitro* by an enzyme as a model of a process in the body).

*Bioassay*: A functional assay that depends on the use of living cells or cell components as an indicator system.

*Bioavailability*: The fraction (or sometimes percentage) of the administered dose that enters the general circulation as the parent compound. A low bioavailability may be due to poor absorption or first-order conditions; the state of being capable of being absorbed and available to interact with the metabolic processes of an organism. Bioavailability is typically a function of chemical properties, physical state of the material to which an organism is exposed, and the ability of the individual organism to physiologically take up the chemical.

*Bioconcentration*: The uptake of chemicals from water alone.

*Bioconcentration Factor (BCF)*: The tendency of a chemical to be more concentrated in an aquatic organism than the concentration in its environment, calculated by dividing the concentration of the chemical in the organism (wet weight) by the concentration of the chemical in the water.

*Biocontainment*: The process and equipment used for the purpose of preventing the unwanted release of hazardous material or organisms.

*Biodegradation*: The break down of chemicals in organisms or the environment, primarily by microorganisms.

*Biodiversity*: The variety of organisms considered at all levels, from genetic variants belonging to the same species through arrays of species to arrays of genera, families, and higher taxa; includes the variety of ecosystems that comprise both the communities of organisms within particular habitats and the physical conditions under which they live.

*Bioexclusion*: The process of preventing the introduction of unwanted microorganisms into animals or their immediate environment.

*Biologically Based Dose–Response Model*: A mathematical expression of the relationship between the incidence of severity of a biological effect and a dose that is based on the biological mechanism or mode of action.

*Biologically Effective Dose*: The amount of a deposited or absorbed chemical that reaches the cells or target site where an adverse effect occurs or where that chemical interacts with a membrane surface.

*Biomagnification*: The increase in tissue contaminant concentration in higher trophic levels as a result of dietary accumulation.

*Biomarker*: Observable change (not necessarily pathological) in the function of an organism, related to a specific exposure or event; a biochemical, genetic, or molecular indicator that can be used to screen disease or toxicity; parameters that can be used as an indictor of exposure, effect, or susceptibility and may be a metabolite, enzyme, or cell surface marker, among others.

*Biotransformation*: The biochemical modification of a xenobiotic once it enters an organism. Chemical modification can be enzymatic or nonezymatic and may result in either reduced or increased toxicity. This process generally gives rise to compounds that are more readily excreted in the urine and feces and thus serves as a detoxification process; however, some xenobiotics are activated to more toxic metabolites by these enzymatic conversions.

*Birth Defect/Congenital Malformation*: An abnormality identified *in utero* or within the first 2 years postnatal (i.e., death, growth or functional retardation, or alteration or dysmorphogenesis).

*Blackfoot Disease*: A condition caused by long-term exposure to arsenic. The condition was first noted in Taiwan in regions containing high levels of arsenic in drinking water. The condition is characterized by poor circulation, leading to distal gangrene of the foot and other extremities.

*Blood*: A complex tissue composed of plasma and cellular elements with many different functions; the circulating tissue of the body; the fluid and its suspended formed elements that are circulated through the heart, arteries, capillaries, and veins; the means by which oxygen and nutrient materials are transported to the tissues and carbon dioxide and various metabolic products are removed for excretion.

*Body Burden*: The amount of a particular chemical stored in a body at a particular time, especially a potentially toxic chemical in the body as a result of exposure. Body burdens can be the result of long-term or short-term storage—for example, the amount of a metal in bone, the amount of a lipophilic substance such as polychlorinated biphenyl (PCB) in adipose tissue, or the amount of carbon monoxide (as carboxyhemoglobin) in the blood; amount of radioactive material present in a human or animal.

*Bolus Dose*: A quantity of test material administered all at once. This term is commonly applied to the single daily administration of a test material by oral gavage in toxicity studies.

*Bond Stretching Frequency*: This frequency is related to the masses of the two atoms that form the bond ($M_a$ and $M_b$, in grams), the velocity of light ($c$), and the force constant of the bond ($k$, in dynes/cm). The frequency can be expressed approximately as:

$$v\ (\text{in cm}^{-1}) = \frac{1}{2\pi c}\sqrt{\frac{k}{M_a M_b/(M_a + M_b)}}$$

The value of $k$ is unique for a specific bond type (e.g., $sp^3$, $sp^2$, and $sp$ bond have values of 5, 10, and $15 \times 10^5$ dynes/cm, respectively).

*Bootstrap*: A method of sampling actual data at random, with replacement, to derive an estimate of a population parameter such as the arithmetic mean or the standard error of the mean. The sample size of each bootstrap sample is equal to the sample size of the original data set. Both parametric and nonparametric bootstrap methods have been developed.

*Botulism*: Fatal paralytic disease resulting from the consumption of food containing preformed toxin from *Clostridium botulinum*.

*Bounding Estimate*: An estimate of exposure, dose, or risk that is higher than that incurred by the person in the population with the highest exposure, dose, or risk. Bounding estimates are useful in developing statements that exposures, doses, or risks are "not greater than" the estimated value.

*Bowman's Membrane*: An acellular layer of collagen and ground substance that provides a functional interface between the stroma and epithelium of the cornea.

*Bufotenin*: A very toxic, water-insoluble genin and serotonin derivative that is synthetically produced and is found in certain toads and in very small amounts in the usual edible mushroom.

*Bursa of Fabricius*: A structure located in the cloaca of avians, where bone-marrow-derived lymphocytes mature into immunocompetent B cells prior to moving to the peripheral lymphoid organs.

*Capture Velocity*: Air velocity at any point in front of the hood or at the hood opening necessary to overcome opposing air currents and to capture the contaminated air at that point by causing it to flow into the hood.

*Cascade Impactor*: Instrument used to collect and sort aerosols onto discreet stages by separating the particles according to their aerodynamic size.

*Case Control Study*: A study in which the past histories of those with a specific disease (the case) are compared with those who do not have the disease (the controls). The measure of association is the odds ratio (i.e., the odds of the cases having had some type of exposure compared to the odds of the controls having had that same exposure). In the context of exploratory data analysis, the case control approach has sometimes been called "a disease in search of an exposure."

*Case Reports and Case Series*: A description of a single individual or group of individuals with the same or similar disease. This type of work lacks controls; therefore, any conclusions derived from such anecdotal information must be viewed with caution. Nonetheless, case reports and case series sometimes are useful for generating hypotheses.

*Cause of Death*: The disease or injury that initiated the train of events leading directly to death, or the circumstances of an accident or violence that produced the fatal injury.

*CBG*: Corticosteriod-binding globulin.

*CBI*: Chemical-binding index.

*CD (Cluster of Differentiation)*: The CD series is used to denote cell surface markers (e.g., CD4, CD8). These markers, used experimentally as a means of identifying cell types, serve various physiological roles.

*cDNA*: Complementary DNA enzymatically synthesized as a copy of mRNA.

*C. elegans*: *Caenorhabditis elegans*, a nematode or roundworm, was the first animal to have its genome completely sequenced and all genes fully characterized.

*Central Tendency Exposure (CTE)*: A risk representing the average or typical individual in the population, usually considered to be the arithmetic mean or median of the risk distribution.

*Centrilobular Cells*: Collection of hepatic cells situated around the terminal hepatic venule (or central vein).

*Centromere*: Constriction along the length of a chromosome.

*Chelating Agent*: An organic compound that forms multiple coordinate covalent bonds with metal ions yielding stable compounds that can be excreted. Chelating agents are used in the treatment of metal toxicity. An example is dimercaptopropanol (British Anti-Lewisite) to treat arsenic poisoning.

*Chemical Shift*: The frequency at which a given nucleus absorbs in a nuclear magnetic resonance (NMR) spectrum. The precession frequency ($v$) for a given nucleus depends on its chemical environment and the shift in $v$ from the standard value is given by the fundamental NMR equation:

$$v = \frac{\gamma_0^{bm}}{2\pi}$$

is called the *chemical shift* for that nucleus. Chemical shifts are rarely (if ever) expressed in absolute frequency units; instead, they are measured relative to a standard reference compound. For $^1$H and $^{13}$C

NMR, the universally accepted reference is tetramethylsilane (TMS). The protons and carbons of TMS absorb at a lower frequency (are more shielded) than those of almost all other organic compounds, so their chemical shifts are arbitrarily set to 0 Hz, and most other chemical shifts measured relative to them are positive. Chemical shifts are expressed in dimensionless units designated δ. The chemical shift in δ units is defined according to the following relationship:

$$\delta = \frac{v_s v_{std}}{\text{Operating frequency}} \times 10^6 \text{ ppm}$$

*Chemokine*: Small peptide molecules related to cytokines and associated with a variety of physiological states, such as inflammation and immunoregulation.

*Chemotaxis*: Directed movement of cells through a concentration gradient of an attractant molecule, such as a chemokine.

*Chilopoda*: A class of about 2000 species of nocturnal, predatory, terrestrial arthropods that includes centipedes; they superficially resemble millipedes.

*Cholestasis*: Diminution or cessation of bile flow, accompanied by decreased excretion and enhanced retention of normal constituents found in bile.

*Chromatography*: A process for separation of molecules on the basis of their affinities for a stationary phase and a mobile phase.

*Chromophore*: A structural moiety in an organic molecule that absorbs light in the useful part of the ultraviolet spectrum. Common chromophores are aromatic moieties and conjugated double-bond moieties.

*Chromophoric Resonance Raman Label*: These labels provide detailed vibrational and electronic spectral data when they are incorporated into the vicinity of a biologically important site. They usually are designed to mimic natural biochemical components and are themselves biologically active compounds. These labels are useful for obtaining information on protein–ligand interactions and have been utilized for studying enzyme–substrate complexes, where vibrational spectra of the substrate during enzyme catalysis can be obtained.

*Chromosome*: Microscopically visible organelle composed of DNA and proteins.

*Chrondrichthyes*: The vertebrae class comprised of cartilaginous fishes, including sharks, rays, skates, and chimaeras; there are about 300 species, some of which are venomous.

*Chronic*: Characterized by a time period of long duration; commonly used to describe long-term (6 to 12 months) exposure in toxicity studies.

*Chronic Toxicity Study*: A multiple-dose study in which animals are treated for ≥6 months to comprehensively assess the potential toxicological effects of a compound following long-term exposure; normally, these studies are required prior to phase II testing in humans.

*Chylomicrons*: A class of large lipoprotein structures that are created by the cells of the lumen of the small intestine and transport dietary fat by exocytosis from the small intestine to the lymphatic system, where they are eventually removed by lipoprotein lipase.

*Ciguatoxins*: A group of colorless and heat-stable lipophilic polyether neurotoxins produced by 300 to 400 tropical reef and semipelagic marine animals.

*Cilia*: Flexible, microscopic projections from cell surfaces that participate in the active movement of overlying mucous; effective rhythmic ciliary motion is essential for normal particle clearance of deposited particles from airway surfaces.

*Clastogen*: An agent that causes chromosomal breakage.

*Clastogenicity*: Chromosome breakage and/or rearrangements.

*Clearance*: The volume of plasma (or blood) that is cleared of chemical per unit time; equivalent to the rate of elimination, at any time, divided by the corresponding plasma (or blood) concentration. Clearance may be dependent on the blood flow and/or the metabolic activity or extraction ratio (at steady state) of the organ(s) elimination; independent of concentration and dose under first-order conditions. The volume of plasma from which a compound is completely removed by the kidneys per unit time.

*Cleft Phallus*: *See Hypospadia*.

*Clinical Assays of Renal Function*: Assays that can be performed in humans; typically involve measurement of parameters in blood (serum, plasma) or urine; usually noninvasive.

*$C_{max}$*: Maximum achieved concentration.

*CMI (Cell-Mediated Immunity)*: Antigen-specific immune reactivity mediated primarily by T lymphocytes. Cell-mediated immunity may be expressed as immune regulatory activity (primarily mediated by CD4+ T-helper cells) or immune effector activity (mediated largely by CD8+ T-cytotoxic cells). Other forms of direct cellular activity (e.g., NK cells, macrophages) are generally not antigen specific (i.e., nonimmune) and are more accurately described as natural immunity.

*Cnidaria*: A large family of venomous marine invertebrates of the Indo-Pacific region often found in coral reefs.

*COD*: Caloric optimization diet.

*Coded Chemicals*: Chemicals labeled by code rather than names so they can be tested and evaluated without knowledge of their identity or anticipation of test results. Coded chemicals are used to avoid intentional or unintentional bias when evaluating laboratory or test method performance.

*Coded Microscopic Evaluation*: The practice of conducting the initial histopathological evaluation with the pathologist having no knowledge of treatment status of individual animals.

*Coding Regions*: Those parts of the DNA that contain the information required to form proteins. Other parts of the DNA may have noncoding functions (e.g., start–stop, pointing, or timer functions) or as yet unresolved functions or perhaps even "noise."

*Codon*: A DNA basepair triplet coding for an amino acid or stop signal.

*Coefficient of Inbreeding*: Refers to a mathematical relationship used to express the relatedness and is expressed in mathematical values ranging between 0 and 1.

*Cohort Study*: This type of epidemiology study is conceptually quite similar to the approach used in most toxicology experiments. The health experience (incidence of disease or mortality) of those exposed to some agent is compared to that of a group not so exposed. In epidemiology, however, the results usually are presented in terms of a relative risk or standardized morbidity (or mortality) ratio. In the context of exploratory data analysis, the cohort approach also has been called "an exposure in search of a disease."

*Collision-Induced Dissociation*: A method where ions are collided with neutral molecules to produce fragmentation. This process is also referred to as *MS/MS*, as a mass spectrum is produced on an ion selected from a mass spectrum.

*Colubridae*: The largest and most cosmopolitan family of snakes, comprised of more than 1700 of the known species of snakes, most of which are nonvenomous; both jaws hold solid or grooved teeth, but no enlarged or hollow fangs, and in most cases, the head is wider than the neck.

*Commensal*: An organism commonly found in association with animals or their environment that under normal circumstances does not produce disease.

*Complement*: A group of approximately 20 proteinase precursors that interact in a cascading fashion. Following activation, the various precursors interact to form a complex that eventually leads to osmotic lysis of a target cell.

*Computational Toxicology*: The application of mathematical and computer models to predict adverse effects and to better understand the mechanisms through which a given chemical induces harm (U.S. Environmental Protection Agency definition).

*Conditioned Response*: Response to an originally neutral stimulus, such as a sound or light, that has acquired the ability to evoke the response because it was paired with an eliciting stimulus.

*Confidence Interval*: A range of values (above, below, or above and below) about the midpoint of the sample (e.g., the mean, median, mode) that contains (with a specified level of probability, such as 95%, or a standard deviation or error, such as 67%) the true value of the population midpoint. The 95% confidence interval (also called the *fiducial limit*) is equivalent to the $p = 0.05$ region boundary.

*Confidence Limit*: A statistical estimate that considers the influence of experimental variation of a parameter.

*Confocal Microscope*: An instrument capable of producing high-resolution microscopic images that can be used to study ocular tissues. A confocal microscope with scanning capability can be used to determine area and depth of corneal injury.

*Confounding Variable*, *Confounder*: A confounder is an alternative cause for the disease in question that is unequally distributed among those exposed and unexposed to the putative agent of interest. As a consequence, it can confound or confuse the measure of association and any resulting interpretations of cause and effect.

*Coniine*: A highly toxic liquid alkaloid and derivative of pyridine; it is the chief toxic agent of poison hemlock.

*Conjugation*: A common mechanism during phase II metabolism where an endogenous compound is added to specific functional groups of a xenobiotic; generally, this increases the excretion of the xenobiotic and decreases its potential to interact at critical sites to produce toxicity.

*Conjunctiva*: The delicate membrane that lines the eyelid and covers the exposed surface of the eyeball. Histologically, the conjunctiva is an aqueous nonkeratinized epithelium with numerous mucous-secreting cells. In the Draize eye test, effects to the conjunctiva represent a maximum of 20 out of 110 total points.

*Coprophagy*: The process of eating one's own feces.

*Cornea*: The transparent outermost covering of the anterior portion of the eye consisting of the epithelium, the stroma, and the endothelium. In the Draize eye test, effects to the cornea represent a maximum of 80 out of 110 total points.

*Correlation*: The relationship or interdependence between measurable varieties or ranks—that is, the extent to which, as one set of values changes, another set also changes in the same (positive correlation) or an opposite (negative correlation) direction.

*Cortical Collecting Duct*: Originates at convergence of two initial collecting tubules and extends to the corticomedullary border; nephron segment responsible for regulating the final composition of urine.

*Cosmic Rays*: Radiation of many sorts, mostly nuclei (protons) with very high energies, originating outside the Earth's atmosphere.

*Coupling Constant*: The separation between nuclear magnetic resonance (NMR) spectral lines due to coupling is the *coupling constant* ($J$). The magnitude of coupling constants depends on the number of intervening bonds and the bond geometry. Coupling constants are expressed in units of hertz (Hz).

*Critical Temperature*: Maximum temperature at which a gas may be liquefied by application of pressure alone. Above this temperature, the substance may only exist as a gas. The critical temperature for $CO_2$ is 31°C.

*Cross-Sectional Study*: A prevalence study (i.e., an epidemiology study) that examines the association between health status and other variables of interest as they exist in a defined population at one particular time. This type of research can also be useful for generating etiologic hypotheses, but these hypotheses then need to be tested in analytic studies. Conceptually, it is also the first step of the more rigorous cohort method.

*Cumulative Distribution Function (CDF)*: A representation, generally a function or graph, of the cumulative probability of occurrence for a random independent variable. The CDF is obtained from the probability density function (PDF) by integration in the case of a continuous random variable and by summation for discreet random variable. Each value *c* of the function is the probability that a random observation *x* will be less than or equal to *c*.

*Curie*: Standard measure of rate of radioactive decay; based on the disintegration of 1 g of radium, or 3.7 $\times 10^{10}$ disintegrations/second.

*Cyclone Separator*: A process in which particle-laden air is introduced radially into the upper portion of a cylinder so it makes several revolutions inside the cylinder. The particles in the air are accelerated outward to the cylinder walls, where they either stick and are retained (low particle loading) or are swirled down to a collection port at the bottom of the cylinder (high particle loading).

*Cyclopeptides*: Group of toxins, produced by the *Amanita* and *Galerina* species of mushrooms, that inhibit mammalian nuclear RNA polymerase.

*Cytochrome C*: A major component of the mitochondrial electron transport chain which is localized in the inter-mitochondrial membrane space. Increased release of this protein into the cytosol is one of the major factors activating the caspase system resulting in apoptosis.

*Cytochrome P450*: Heme thiolate proteins associated with the endoplasmic reticulum originally named for their absorption maximum at 450 nm. These proteins, which serve as catalysts in a variety of oxidative reactions involving both endogenous and xenobiotic lipophilic compounds, are unique in their multiplicity of isozymes, substrates, reactions, and regulatory mechanisms. This family of heme-containing enzymes is involved in so-called phase I metabolism of xenobiotics and endogenous compounds and catalyzes myriad reactions, including oxidation, reduction, dealkylation, and hydrolysis.

*Cytokine*: Small peptide molecules that subserve a wide range of regulatory and effector mechanisms. These include interleukins, tumor necrosis factors, interferons, colony-stimulating factors, and other growth and regulatory factors. Often referred to as *lymphokines* in the older literature.

*Cytotoxic T Lymphocyte (CTL)*: A subset of T lymphocytes bearing the CD3/CD8 surface markers; CTLs are able to kill target cells following induction of a specific immune response. The mechanism of this lysis appears to be a combination of direct lysis resulting from the extrusion of lytic granules by CTLs, as well as the induction of apoptosis in the target cell. The target cells most frequently used for assessment of CTL activity are virally infected cells and tumor cells. Measurement of CTL activity provides an indication of cell-mediated immunity.

*Datum*: A single point of measurement; more than one point is referred to as *data*.

*Default Value*: A conservative value used for a model parameter or uncertainty (safety) factor when data are inadequate to justify a different value.

*Definitive Test*: A toxicity determination designed to provide an accurate quantitative result with low variability; a second-level test conducted following pretests or range-finding tests.

*Delayed-Type Hypersensitivity (DTH)*: A form of cell-mediated immunity in which recalled exposure to an antigen results in an inflammatory reaction mediated by T lymphocytes; usually referred to as *contact hypersensitivity*.

*Deposition*: The process whereby the amount and location of matter (vapor, gas, or solid) is absorbed into or onto a surface (usually described as the amount per unit area during a specified time).

*Descemet's Membrane*: An acellular layer that lies beneath the stroma and forms the basement membrane of the corneal endothelium.

*Descriptive Study*: A study designed to describe the distribution of certain variables (e.g., a health survey of a community that gathers data on disease status and presents the resulting information in the form of what disease was present when, where, and among whom). Although not useful for etiologic research, it can be used to generate hypotheses.

*Detoxication*: The enzymatic conversion of a chemical to a less toxic form or, for these purposes, the conversion of a chemical to a form that can no longer be bioactivated (distinguished from detoxification, which is the process of removing the chemical from the body by physical means).

*Developmental Toxicology*: The study of the causes, mechanisms, and sequelae of perturbed developmental events in species of animals that undergo ontogenesis; affected endpoints include death, delayed or retarded development, dysmorphology, and functional impairment.

*Diagnostic Drift*: The gradual changes in nomenclature or application of severity grading scales that may occur in a single study group, across several groups in a single study, or when several studies are compared.

*Diastole*: The dilation of the heart cavities during which they fill with blood.

*Diffuse Neuroendocrine System (DNES)*: Neurones and amine precursor uptake and decarboxylation (APUD) cells produce some identical biogenic amines and peptides in cells that in some ways are similar to nerve cells and in other ways are similar to endocrine cells. These cells are located in different organs and have a common regulatory mission. Although they do not form a specific organ, they do function to control many aspects of homeostasis and are considered to be a regulatory entity.

*Diffuse Neuroimmunoendocrine System (DNIES)*: Recently, it has been shown that there are common chemical regulatory systems present among the nervous, endocrine, and immune systems. The close relationship among the amine precursor uptake and decarboxylation (APUD) cells that produce peptidergic/aminergic neurons, the DNIES system that unties many different locations, and the peptide-producing immunocompetent cells suggests that the entire system should be combined into a single regulatory homeostatic functional system and termed the *diffuse neuroimmunoendocrine system*.

*Diffusion-Limited Uptake*: Tissue uptake is limited by the diffusion through the membranes rather than by the blood flow to the tissue.

*Dinoflagellata*: An order of predominately marine, free-swimming protists that have two flagella; they are a staple food of shellfish and some crabs and are thus in the human food chain. Certain species may become extremely abundant in warm coastal waters, causing certain of the toxic "red tides," although few are actually toxic.

*Diploid*: Two set of chromosomes, one maternal and one paternal.

*Dirty bomb*: A bomb that uses conventional explosives such as dynamite to spread radioactive material. The best known of the radiological dispersal devices, the dirty bomb does not produce a nuclear explosion but is sometimes referred to as a "weapon of mass disruption."

*Discrimination Behavior*: Behavior based on the ability of subjects to discriminate stimulus qualities. In the typical situation, because different stimulus properties require different behaviors, discrimination abilities can be measured by the type or location of responses.

*Distal Tubule*: Generic term describing epithelial cell types in the cortex comprising distal convoluted tubule, connecting segment, and initial collecting tubule.

*Distribution*: The reversible transfer of chemical from the general circulation into the body tissues.

*DNA*: Deoxyribonucleic acid, the chemical substance containing the genetic code; *see also Nucleotide*.

*DNA Adduct*: A molecules that is covalently linked to a portion of the DNA helix.

*DOCA*: Deoxycorticosterone.

*Donor Animal*: Animal from which the organ or the blood was removed.

*Dosage*: A general term encompassing the dose, its frequency, and the duration of dosing.

*Dose*: The amount of a substance available for interaction with metabolic processes or biologically significant receptors after crossing the outer boundary of an organism. The *potential dose* is the amount ingested, inhaled, or applied to the skin. The *applied dose* is the amount of a substance presented to an absorption barrier and available for absorption (although not necessarily having yet crossed the outer boundary of the organism). The *absorbed dose* is the amount crossing a specific absorption barrier (e.g., the exchange boundaries of skin, lung, and digestive tract) through uptake processes. *Internal dose* is a more general term denoting the amount absorbed without respect to specific absorption barriers or exchange boundaries. The amount of the chemical available for interaction by any particular organ or cell is termed the *delivered dose* for that organ or cell.

*Dose Equivalent (H)*: Unit of biologically effective dose, defined as the absorbed dose in rads multiplied by the quality factor ($Q$). For all x-rays, gamma rays, beta particles, and positrons likely to be used in nuclear medicine, $Q = 1$.

*Dose Rate*: Dose per unit time (e.g., in mg/day), sometimes also called *dosage*. Dose rates are often expressed on a per-unit-body-weight basis, yielding units such as mg/kg/day (mg/kg-day). They are also often expressed as averages over some time period (e.g., a lifetime).

*Dose Reconstruction*: An approach to quantifying exposure from internal dose, which is in turn reconstructed after exposure has occurred, from evidence within an organism, such as chemical levels in tissues or fluids or from evidence of other biomarkers of exposure.

*Dose–Response*: "What is there that is not poison? All things are poison and nothing [is] without poison. Solely, the dose determines that which is not a poison" (Paracelsus). The biological response to an agent is a function of the condition of exposure, including dose, duration, and route.

*Dose–Response Assessment*: That part of risk assessment associated with evaluating the relationship between the dose of an agent administered or received and the incidence or severity of an adverse health or ecological effect.

*Dose–Response Model*: A mathematical expression that relates the incidence or magnitude of a biological effect to the dose of a chemical.

*Dosimetry*: Estimating or measuring the quantity of material (mainly refers to particulate) at specific target sites at a particular point in time. The quantity can be measured in terms of mass number, surface area, or volume; process of measuring or estimating dose.

*DPA*: Decision point approach.

*DR*: Diet restricted.

*Draculin*: Anticoagulant factor present in the salvia of vampire bats.

*Duct Velocity*: Air velocity through the duct cross-section. When solid material is present in the air stream, the duct velocity must be equal to or greater than the minimum air velocity required to move the particles in the air stream.

*Dust*: Solid particles that are capable of temporary suspension in air or other gases. Usually produced from larger particles or masses through grinding, crushing, or other handling. Particles may be up to 300 to 400 μm, but those above 20 to 30 μm usually do not remain airborne.

*Dysplasia*: Abnormal tissue development.

*Early-Phase Regeneration (EPR)*: Tissue repair response, where arrested $G_2$ hepatocytes are activated to proceed through mitosis (*see SPR*).

*Early Transient Incapacitation (ETI)*: Transient performances deficits observed in animals and humans after a large, rapidly delivered dose of ionizing radiation. Five to 10 minutes after radiation exposure, behavioral performance rapidly falls to near zero, followed by partial or total recovery 10 to 15 minutes later.

*$EC_{50}$ (Effective Concentration–50%)*: A statistically or graphically determined concentration of a chemical that reduces a sublethal response parameter of interest by 50%.

*Echinodermata*: Phylum of marine invertebrate coelomate animals that includes starfish and sea urchins. Approximately 85 of the nearly 6000 species are known to be venomous or poisonous.

*Ecological Risk Assessment*: The process used to establish the likelihood that adverse ecological effects are occurring or may occur as a result of exposure to one or more stressors.

*Ectopic or Cryptorchid Testis*: Malposition or displacement of the testis outside the scrotum. Typically, such testes are found within the peritoneum or inguinal canal; however, abdominal sites outside the peritoneum have been observed.

*$ED_{50}$ (Effective Dose–50%)*: The dose of radiation or a chemical agent that would result in a given response in 50% of the population; *see also Median Effective Dose*.

*Elapidae*: A family of front-fanged, highly venomous, aquatic, burrowing, terrestrial, and arboreal snakes that includes cobras, kraits, mambas, coral snakes, death adder, the Australian copperhead, and the African garter snake; they are characterized by a pair of comparatively short, stout, permanently erect, deeply grooved fangs at the front of the mouth, and the tail is cylindrical and tapered.

*Electrocardiogram*: A bioelectric potential originating in the myocardium and recorded on the surface of the body; represents the sum of the electrical depolarizations of the myocardium syncytium as the wave of depolarization sweeps across the heart.

*Electron Paramagnetic Resonance (EPR)*: An analytical technique that is used to study the structure and properties of free radicals. EPR spectroscopy is similar in principle and practice to nuclear magnetic resonance (NMR) spectroscopy. Whereas NMR uses energy in the radiofrequency region of the electromagnetic spectrum, EPR uses energy in the microwave region.

*Elimination*: The irreversible transfer of the chemical from the circulation to the organs of elimination and its subsequent removal from the body by metabolism or excretion.

*ELISA (Enzyme-Linked Immunosorbent Assay)*: A type of immunoassay in which specific antibodies are used to both capture and detect antigens of interest. The most popular type is the "sandwich" ELISA, in which antibodies are bound be a substrate such as a plastic culture plates. These antibodies bind antigenic determinates on molecules (or alternatively on whole cells). Unrelated material is washed away and the plates are exposed to an antibody of a different specificity; this antibody is coupled to a detector molecule.

*EMIT*: Enzyme-multiplied immunoassay technique.

*Empirical Distribution*: A distribution obtained from actual data possibly smoothed with interpolation technique. Data are not fit to a particular parametric distribution (e.g., normal, lognormal) but are described by the percentile values.

*Empirical Dose–Response Model*: A mathematical model that is selected from plausible models based on agreement with experimental data.

*Enantiomer*: A type of stereoisomer, an enantiomer is one of a pair of isomers that are nonsuperimposable mirror images of each other. Although chemically identical (except for optical rotation), enantiomers may demonstrate widely different pharmacological and toxicological properties.

*Endobiotics*: Chemicals that are normally present in the biochemistry of a cell; these include chemicals that are normally found with the biochemistry of anabolic and catabolic metabolism.

*Endocrine Disruptor*: Chemicals or naturally occurring substances that can alter the endocrine system.

*Endoscopy*: *In vivo* inspection of the internal surface of a hollow organ with an instrument.

*Endpoint*: The biological or chemical process, response, or effect assessed by a test method.

*Engineered Nanoparticle*: A form of nanoparticles that are designed and fabricated by material scientists for specific physical, chemical, optical, magnetic, catalytic, or morphological properties not evident in the macroscale.

*Engineering Standards*: Specifications that do not provide for interpretation and modification of prescribed methods or procedures, even if acceptable alternative methods are available or unusual circumstances occur. This term is most commonly used in conjunction with prescribing conditions for the care and use of laboratory animals.

*Enteric Endocrine System (EES)*: An endocrine system that helps control digestive function. Over 25 hormones have been identified as affecting the gastrointestinal tract, with single-hormone-secreting cells located throughout the lumen of the stomach and small intestine.

*Enteric Nervous System (ENS)*: A nervous system that helps control digestive function; it is located primarily in the wall of the gastrointestinal tract and is an interdependent part of the autonomic nervous system that can be affected by sympathetic and parasympathetic nervous impulses. It is also refereed to as the "second brain" because it can operate the gastrointestinal tract in the absence of external stimuli.

*Enterohepatic Circulation*: Bile salts are necessary for the absorption of lipids when ingested and present in the small intestine. The bile salt forms micelles that are integral in the absorption of longer chained lipids. When the micelles are spent, they are eventually reabsorbed in the ileum of the small intestine to be returned throughout the circulatory system to the liver where the bile salts are resecreted through the bile ducts into the small intestine to again form micelles.

*Environmental Fate*: The destiny of a chemical or biological pollutant after release into the environment. Environment fate involves temporal and spatial considerations of transport, transfer, storage, and transformation.

*Enzyme*: A biological catalyst (for these purposes, proteins).

*Epididymis*: The duct that conveys sperm from the testis to the vas deferens. It consists of caput (head), corpus (body), and cauda (tail) regions that support sperm transit, maturation, and storage prior to ejaculation.

*Epitope*: The portion of an antigen that is recognized by an antibody or T-cell antigen receptor; also known as the *antigenic determinant*.

*Epizootic*: Refers to the initial period following introduction of a microorganism into a naïve population, during which time it undergoes rapid spread within the population and is often associated with a high incidence of clinical signs or disease.

*ERPGs*: Emergency response planning guidelines are values intended to provide estimates of concentration ranges for 1-hour exposures above which one could reasonably anticipate observing adverse health effects. Compare to acute exposure guideline levels (AEGLs).

*Essential Test Method Components*: Structural, functional, and procedural elements of validated test methods that should be included in the protocol of a proposed mechanistically and functionally similar test method. These components include unique characteristics of the test method, critical procedural details, and quality control measures. Adherence to essential test method components is necessary when the acceptability of a proposed test method is being evaluated based on performance standards derived from a mechanistically and functionally similar validated test method. (*Note*: Essential test method components were previously referred to as *minimum procedural standards*.)

*Estrogen*: Female sex steroid (e.g., estradiol).

*Exon*: Actively transcribed DNA in a eukaryotic gene.

*Exposure*: Contact of a chemical, physical, or biological substance with the outer boundary of an organism. Exposure is quantified as the concentration of the agent in the medium in contact integrated over the time duration of contact.

*Exposure Assessment*: The determination or estimation (qualitative or quantitative) of the magnitude, frequency, duration, and route of exposure; the qualitative or quantitative estimate (or measurement) of the magnitude, frequency, duration, and rote of exposure. A process that integrates information on chemical fate and transport environmental measurements, human behavior, and human physiology to estimate the average doses of chemicals received by individual receptors. For simplicity in this guidance, exposure encompasses concepts of absorbed dose (i.e., uptake and bioavailability).

*Exposure Concentration*: The concentration of a chemical in its transport or carrier medium at the point of contact.

*Exposure Pathway*: The physical course a chemical or pollutant takes from the source to the organism exposed.

*Exposure Point Concentration (EPC)*: The contaminant concentration within an exposure unit to which receptors are exposed. Estimates of the EPC represent the concentration term used in exposure assessment.

*Exposure Route*: The way in which a chemical enters an organism after contact (i.e., by inhalation, ingestion, or dermal absorption).

*Exposure Scenario*: A set of facts, assumptions, and inferences about how exposure takes place; this information aids the exposure assessor in evaluating, estimating, or quantifying exposures.

*Extra Risk*: $(P - P_o)(1 - P_o)$, where $P$ is the total risk and $P_o$ is the background (spontaneous) risk in control animals or unexposed humans.

*Extrapolation*: Using data or results from one set of conditions (e.g., animal experimental results) to predict results for a different set of conditions (e.g., humans).

*Extravascular Hemolysis*: Destruction of red blood cells by cells of the mononuclear phagocyte system (e.g., splenic macrophages); may be a normal process for removal of senescent red blood cells or part of a pathological process for removal of abnormal red blood cells or red blood cells coated by immunoglobulin (i.e., immune-mediated hemolysis).

*Eye Corrosion*: Irreversible ocular tissue damage following exposure to a material. Eye corrosion represents gross tissue destruction, which generally occurs rapidly after exposure.

*Eye Irritation*: Reversible inflammatory changes in the eye and its surrounding mucous membranes following direct exposure to a material on the surface of the anterior portion of the eye.

*Face Velocity*: Air velocity at the hood opening.

*FACS Analysis*: Flow-assisted cell sorting; analytical method applied to data from flow cytometry experiments.

*False Negative*: A substance incorrectly identified as negative by a test method.

*False Negative Rate*: The proportion of all positive substances falsely identified by a test method as negative (*see Two-by-Two Table*). It is one indicator of test method accuracy.

*False Positive*: A substance incorrectly identified as positive by a test method.

*False Positive Rate*: The proportion of all negative substance that are falsely identified by a test method as positive (*see Two-by-Two Table*). It is one indicator of test method accuracy.

*Favism*: A hemolytic disease in individuals deficient in glucose-6-phosphate dehydrogenase resulting from consumption of fava beans.

*Fertility*: The ability to conceive and produce a live offspring. The fertility index should be calculated separately for male and female animals and is generally calculated as the ratio of the number of animals pregnant divided by the number of animals inseminated.

*Fetal Alterations (Malformations, Variations, and Developmental Delays)*: Any morphological change identified in a term conceptus, including frank malformation, minor deviations from normal development, and reversible delays or accelerations in development, regardless of the potential effect on subsequent viability or quality of life.

*Fiber*: A particle that has a length-to-width ratio of 3:1 and a length greater than 5 µm. These can be naturally occurring, manmade mineral, or synthetic organic fibers.

*Fibrosis*: An abnormal accumulation or proliferation of fibrous connective tissue, usually in the lung.

*First-Order Process*: A process for which the rate of reaction is proportional to the available concentration—for example, diffusion, metabolism (at low concentrations), and filtration.

*Fixation*: Preparation of a histologic or pathologic specimen for the purpose of maintaining the existing form and structure of its constituent elements. Maintenance of the normal form and structure of tissues and organs is critical to the evaluation of pathologic effects during microscopic examination. Common fixatives, such as formaldehyde, result in the precipitation of proteins. This fixes them in place and preserves the morphology of the specimen.

*Flash-Point*: The lowest temperature at which vapor is given off in sufficient quantity so the air–vapor mixture above the surface of the solvent will ignite momentarily in a flame.

*Flavin-Containing Monooxygenases (FMOs)*: Enzymes containing flavin adenine dinucleotide that belong to a family of proteins that are important in the NADPH-dependent metabolism of exogenous compounds. FMOs catalyze the addition of a single oxygen atom to nucleophilic nitrogen, sulfur, and phosphorus centers of a variety of xenobiotics.

*Flavonoids*: Phenolic plant pigments belonging to flavone, flavanone, isoflavone, anthocyanidin, chalcone, or aurone groups.

*Flow cytometry*: Method to separate populations of fluorescent-labeled cells according to differences in a specific property.

*Fluid Mosaic Membrane Model*: The accepted models for the description of the biological membrane. This model is a noncontinuous phospholipid bilayer that forms planar bimolecular films that separate two aqueous compartments with hydrophobic cores. Proteins are an integral part of the membrane and span the entire thickness of the lipid layer.

*Food Poisoning*: A predominantly gastrointestinal disease resulting from the consumption of food containing any of the ever-increasing number of toxigenic microorganisms or the preformed toxins produced by them.

*Formication*: The unpleasant sensation of tiny insects (*L. formica* ant) crawling on the skin.

*Fourier Transformation (FT)*: A mathematical operation that converts information from the time domain to the frequency domain. FT can be applied to a variety of spectral techniques, including mass spectrometry, nuclear magnetic resonance, electron paramagnetic resonance, and infrared spectroscopy.

*Fractional Excretion*: The excretion of a substance in urine relative to the rate it is filtered into the urine; an index of reabsorptive function.

*Frameshift*: A gene mutation characterized by the addition or deletion of one or more basepairs in a gene.

*Free Radical*: A molecule that is inherently unstable and highly reactive with other components of living systems and produced by sufficient exposure to ionizing radiation.

*Frequency Distribution*: A graph or plot that shows the number of observations that occur within a given interval. Usually presented as a histogram showing the relative probabilities for each value, it conveys the range of values and the count (or proportion of the sample) that was observed across that range.

*FSH*: Follicle-stimulating hormone.

*Fugu*: Puffer fish that can lead to sever tetrodotoxin poisoning if improperly prepared.

*Fume*: Minute solid particles arising from high-temperature or combustion processes with subsequent condensation of vapors and often accompanied by a chemical reaction, such as oxidation. Metal vapors commonly condense to create fume; exposure to fresh metal fumes can cause metal fume fever. Fumes exist initially as very small (<100 nanometers [nm]; $10^{-9}$ m) particles but form larger (~1 μm) particle clusters through agglomerative growth.

*Functional Observation Battery*: A set of standardized observations, typically performed with rodents, designed to assess neurobehavioral functions such as reflexes.

*Functionalization*: The addition to or uncovering of functional groups that are required for subsequent phase II metabolism; for example, hydroxylation of a hydrocarbon provides a functional group from which a conjugate can be formed during phase II metabolism.

*Gametogenesis*: Production of sperm or ova.

*Gamma Rays*: High-energy, short-wavelength electromagnetic radiation emitted from the nucleus of an atom.

*Gas*: A substance that exists in the gaseous state at standard temperature and pressure.

*Gas Chromatography–Mass Spectrometry*: A method for introducing analytes volatized into the gas phase into the mass spectrometer through a chromatographic system where a carrier gas passes through a column packed with a solid-phase material.

*Gastrointestinal Transit*: The rate of passage of the luminal contents of the gastrointestinal tract in the oral to anal direction, ordinarily measured with a nonabsorbable marker.

*Gavage*: Method of oral administration of a solution or suspension using a suitable stomach tube or feeding needle attached to a syringe. In toxicity studies, gavage is a common method of test material administration by the oral route.

*Gene*: Generally described as the smallest functional unit of an organism's genome.

*Gene Pool*: The total genetic information contained in the reproductive cells of a species.

*Genome*: The total number of genes contained in the hereditary material of an organism; the chromosomal DNA information.

*Genomics*: The techniques available to identify the DNA sequence of the genome; field of study involving analysis of genetic material (i.e., DNA, RNA). Genomics can be studied at the cellular level in a population; automated methods, such as DNA microarray, have been developed for high-throughout analysis.

*Genotoxic*: Property of an agent making it capable of damaging or altering an organism's genetic composition.

*Genotoxicity*: Alteration of nucleic acids and associated components at subtoxic exposure levels, resulting in modified hereditary characteristics or DNA inactivation.

*Genotype*: The nucleotide composition of an organism's hereditary material; the full set of genes carried by an individual organism. Note that this term is more limited than the genome, as the genome also contains noncoding DNA and genes.

*Geometric Mean*: The *n*th root of the product of *n* values.

*Gestation/Pregnancy*: The interval in a pregnant animal from conception (fertilization) to the beginning of parturition.

*GH*: Growth hormone or somatotropin.

*Glomerular Filtration Rate*: The volume of plasma separated from the vascular space across the renal glomerulus, expressed per unit time. The filtrate is further modified by the nephron tubule to become the final urine that is eliminated from the body. The rate at which plasma is filtered through the glomeruler capillary bed into the tubular lumen is defined as the difference between the hydraulic and oncotic pressure across the glomerular capillary wall multiplied by an ultrafiltration coefficient and the surface area available for filtration.

*Glycone*: The activated form of glucose used in glucuronidation. The term is sometimes used for other activated forms of the endogenous compound used in xenobiotic conjugation reactions.

*Glycosides*: Chemicals of varied structure linked to a mono- or disaccharide by a β-glycosidic linkage; an acetal that yields sugar and a non-sugar on hydrolysis and is found more commonly in plants than in alkaloids. Toxicity varies from nontoxic to extremely toxic.

*GnRH (Gonadotropin-Releasing Hormone)*: A neuropeptide that regulates the release of the gonadotropins follicle-stimulating hormone (FSH) and luteinizing hormone (LH).

*Gonadotropins*: Usually considered to be follicle-stimulating hormone and luteinizing hormone; collectively includes follicle-stimulating hormone (FSH) and luteinizing hormone (LH).

*Good Laboratory Practices (GLPs)*: Regulations promulgated by the U.S. Food and Drug Administration (FDA) and the U.S. Environmental Protection Agency (EPA), as well as principles and procedures adopted by the Organization for Economic Cooperation and Development (OECD) and Japanese authorities, that describe record keeping and quality assurance procedures for laboratory records that will be the basis for data submission to national regulatory agencies.

*Graded Response*: A response to a stimulus or a treatment that can be determined quantitatively on a continuous scale. In acute toxicity studies, body weight and feed consumption are examples of measurements of graded responses.

*Gray (Gy)*: SI standard unit of absorbed dose; 1 Gy equals 1 joule of energy per kilogram of absorber or 100 rads.

*GTT*: Glucose tolerance test.

*Haber's Rule*: The Haber relationship expresses the constancy of the product of exposure concentration and duration ($CT = k$). This relationship does not hold over more than small differences in exposure time. The hypothesis states that equal values of ($C \times T$) produce equal biological effects, where $C$ is the concentration of a chemical and $T$ is the duration of exposure.

*Half-Life*: The time taken for the plasma (or tissue) concentration to change by 50%; half-life is a characteristic of first-order reactions and is independent of concentration and dose.

*Half-Life, Biologic*: Time it takes an organism to eliminate half of the radionuclide by biologic processes.

*Half-Life, Effective*: Time required for the activity of a radionuclide in a biologic system to be reduced to half its initial value as a consequence of both radioactive decay and biologic elimination.

*Half-Life, Physical*: The time necessary for a radionuclide to decay to half of its initial activity.

*Haploid*: A single set of eukaryotic chromosomes.

*Hapten*: Low-molecular-weight molecules that are not antigenic by themselves but are recognized as antigens when bound to larger molecules such as proteins.

*Hazard*: The inherent property of a single chemical or mixture to cause adverse effects if an organism is exposed to it; the potential for an adverse health or ecological effect. A hazard potential results only if an exposure occurs that leads to the possibility of an adverse effect being manifested.

*Hazard Analysis and Critical Control Point (HACCP)*: A system designed to assess various stages of food processing critical in controlling microbial contamination and to propose and implement procedures aimed at minimizing the microbial burden in foods.

*Hazard Identification*: A qualitative assessment of the types of adverse effects caused by a particular chemical, including (but not limited to) an evaluation of quality of the studies, identification of susceptible subpopulations, and assessing the relevance of humans of the effects observed in animals.

*Hazardous Substance*: Any substance or mixture of substances that is toxic, corrosive to an irritant; has the potential to cause substantial personal injury as a result of handling or use.

*Heinz Body*: Irreversibly denatured hemoglobin attached to the inner cell membrane of the red blood cell; caused by oxidizing agents.

*Heloderma*: A genus of lizards comprised of two species of heavy-bodied lizards with short stout legs; the body is covered by tubercular scales.

*Henry's Law Constant*: The relationship between the solubility of a gas in a liquid to the pressure of the gas above the liquid at equilibrium, which describes the tendency of a chemical to escape from solution.

*Hepatic First-Pass Effect*: Virtually all of the substances ingested into the circulatory system are transported directly to the liver where the substances can be metabolized via phase I and phase II reactions prior to their entry into the general circulatory system.

*Hepatocytes*: Liver cells.

*Hepatotoxicity*: Toxicity to the liver.

*Heterologous Blood Perfusate*: Blood used as perfusate that comes from another animal or a different species than the source of the organ being perfused.

*Heterologous Expression*: Production of a recombinant protein in an artificial host system.

*Heterologous Expression Systems*: Systems that allow expression of a gene in a different organism.

*Heterozygous*: Two different alleles on a chromosome pair.

*Hexagonal Lobule*: Classical morphological configuration of the functional unit of the liver as described by Kiernan in 1833.

*High-End Exposure (Dose) Estimate*: A plausible estimate of individual exposure or dose for those persons at the upper end of an exposure or dose distribution, conceptually above the 90th percentile but not higher than the individual in the population who has the highest exposure or dose.

*High-End Risk Descriptor*: A plausible estimate of individual risk for those persons at the upper end of the risk distribution, conceptually above the 90th percentile but no higher than the individual in the population with the highest risks. Note that persons in the high end of the risk distribution have high risk due to high exposure, high susceptibility, or other reasons; therefore, persons in the high end of the exposure or dose distribution are not necessarily the same individuals as those in the high end of the risk distribution.

*Hippocampus*: Radiosensitive area of the brain involved in learning and memory functions.

*Histopathology*: The study of morphological changes in tissues at the light-microscopic levels.

*Holozoic Nutrition*: Obtaining nutrients or nourishment through the ingestion of large complex organic materials or whole organisms.

*Homozygous*: Two similar alleles on a chromosome pair.

*Hormesis*: A U-shaped dose–response relationship in which adverse effects do not increase monotonically with dose but decrease initially as dose increases and then rise with higher doses. (An inverted U-shaped dose–response is observed with beneficial effects.)

*Host Defense*: The ability of an animal to protect itself against disease associated with exposure to infectious organisms, foreign tissues and chemicals, and neoplasia. Host defense may be either nonspecific or specific (immune) in nature.

*Host Resistance*: The ability of an organism to defend immunologically against infection. Host resistance may be decreased in response to an immunosuppressive insult.

*HTE*: Human time equivalents.

*Human Health Risk Assessment*: Process by which data on the metabolism and toxicity of a chemical are used to estimate the potential hazard to humans of exposure; often involves the extrapolation of data from laboratory animals (e.g., rats or mice) to humans; used by regulatory agencies to define safe levels of exposure of human populations to potentially toxic chemicals.

*Humoral-Mediated Immunity (HMI)*: Specific immune responses that are mediated primarily by humoral factors (i.e., antibodies and complement). The induction of humoral immune responses generally requires the cooperation of cellular immune mechanisms.

*Hybridization*: The formation of a double strand from two different, more or less complementary, single nucleic acid strands.

*Hydroponic*: Referring to a nutrient solution capable of supporting plant growth without the support of inert material.

*Hydrozoa*: One of three classes of the phylum Cnidaria (referred to as Coelenterata in the past) in which the stomodaeum is absent and the mesoglea has few or no cellular elements and is usually metagenetic. Most species are marine.

*Hyperplasia*: A histopathologic finding characterized by an abnormal increase in the number of cells in a tissue or organ. Hyperplasia generally results from increased cell division but can result from decreased cell death.

*Hypertrophy*: A histopathologic finding characterized by an abnormal increase in the size of cells in a tissue or organ; for example, accumulation of fat vacuoles within a cell can increase its size.

*Hypospadia*: In males, a congenital malformation in which the urethra remains open on the undersurface of the penis. An extreme expression of this malformation results in cleft phallus, with a cleft running the entire length of the penis.

*Ibotenic Acid*: A neurotoxic isoxazole that, when paired with mescimol, is largely responsible for the toxicities of *Amanita* species and certain other poisonous mushrooms.

*ICH*: International Conference on Harmonization.

*IDDM*: Insulin-dependent diabetes mellitus, type 1.

*Idiosyncrasy*: A specific (and usually unexplained) reaction of an individual to, for example, a chemical exposure to which most other individuals do not react at all (e.g., some people react to their very first aspirin with a potentially fatal shock). General allergic reactions do not fall into this category.

*Immune Reserve*: The concept that the immune response exhibits multiple redundancies capable of modulating acute reductions in certain immune functions. This reserve would theoretically prevent a severe reduction in host resistance following temporary immunosuppression of selected parameters (e.g., NK cell function).

*Immunoassay*: An assay that utilizes specific antibodies as reagents. Examples include enzyme-linked immunosorbent assays (ELISAs) and radioimmunoassays (RIAs).

*Immunochemistry*: The use of antibodies in analytical or preparative procedures.

*Immunologic Contact Urticaria (ICU)*: Contact urticaria is an immediate-type reaction, on an immunologic basis, such as in latex contact.

*Immunostimulation*: Enhancement of immune function above an established baseline (control) response.

Immunostimulation may be beneficial (e.g., therapeutics designed to restore a depressed immune response) or detrimental (e.g., the induction of allergy/hypersensitivity or autoimmunity).

*Immunosuppression*: Depression of immune function below an established baseline (control) response. Immunosuppression may result from inadvertent exposure to immunosuppressive agents, deliberate (therapeutic) immunosuppression, or exposure to certain infectious agents. An important consideration in immunotoxicology is determining the degree or nature of immunosuppression necessary to alter host defense. Immunosuppression can be said to result in a state of immunodeficiency.

*Immunotoxicity*: The condition in which a drug, chemical, or physical agent alters the structure or function of the immune system.

*Immunotoxicology*: The discipline of synergistically applying cardinal principles of both immunology and toxicology to study the ability of certain materials to alter the normal immune response.

*Impaction*: A particle deposition process related to particle inertia whereby particles contact airway surfaces by their inability to follow the air stream around directional changes from airway bends or bifurcations.

*Imposex*: A condition caused by low concentrations of organotin compounds (e.g., tributyltins and triphenyltins) characterized by the superimposing of male sex organs on otherwise normal dioecious female gastropods. The resulting pseudohermaphrodites are incapable of reproduction.

*In Situ Perfusion*: Vascularly isolated and perfused organs are left in the animal body with the neighboring tissues.

*In Vitro Hemolysis*: Rupture or lysis of red blood cells during blood collection or handling causing the release of intracellular substances (e.g., hemoglobin, enzymes, electrolytes) into serum or plasma. If severe enough, it may be responsible for artifactual test results.

*In Vivo Nuclear Magnetic Resonance (NMR) Spectroscopy*: A technique of value for measuring alterations in specific P-31 species within cells in real time using a surface coil and large-bore, high-resolution nuclear magnetic resonance (NMR) magnets.

*Incidence*: The number of new cases of a particular disease in a defined population during a specified period of time. The term *incidence* is sometimes used synonymously with *incidence rate*. Note that, in the medical literature, *prevalence* and *incidence* are often used incorrectly as equivalent terms.

*Induction*: As used with respect to xenobiotic metabolisms, induction refers to the process where exposure of an organism to a xenobiotic results in increased activity of specific biotransformation enzymes. Generally, but not always, induction results in more rapid metabolism of the inducing agent. Induction, as used in this context, does not require *de novo* protein synthesis but may result from other mechanisms.

*Inflammation*: A nonspecific host defense mechanism. It is characterized primarily by the infiltrating of leukocytes into the peripheral tissue, followed by release of various molecules that elicit nonspecific physiological defense mechanisms.

*Inhalable*: A particle size characteristic denoting that the substance can enter the respiratory tract; however, particles may not be sufficiently small to enter beyond the nose and into the airways and pulmonary regions. Contrast with *respirable*.

*Inhalation*: The breathing in of a substance in the air (gas, vapor, particulate, dust, fume, or mist).

*Inhibin*: A glycoprotein produced by Sertoli cells and by the pituitary that acts as a negative regulator of follicle-stimulating hormone (FSH) secretion.

*Initiated Cell*: A normal body (stem?) cell that has undergone the first transformation step in the process of cancer development to an intermediate state but that is not malignant.

*Innate immunity*: Host defense mechanism that does not require prior exposure to an antigen; various effector mechanisms have been described, including cellular cytotoxicity mediated by macrophages or NK cells, complement, and activity gamma delta T cells.

*Insecta*: A very large class of invertebrate animals representing more than 75% of known animal species. They have three main body parts: head, thorax, and abdomen.

*Instillation*: Slow introduction of a fluid (for a fluid-containing particulate) directly into the trachea of an experimental animal; this is a surrogate method for the introduction of fluid or particulate into the lungs following inhalation.

*Insufflation*: Introduction of particulate (without fluid) directly into the trachea (or bronchioles) of an experimental animal. This is a surrogate method for the introduction of material into the lungs following inhalation.

*Intake*: The process by which a substance crosses the outer boundary of an organism without passing an absorption barrier (e.g., through ingestion or inhalation) (*see Potential Dose*).

*Interception*: A particle deposition process related to a particle trajectory and the physical proximity to airway surfaces, resulting in particle contact and deposition. Interception is a particularly important deposition process for inhaled fibers.

*Internal Dose*: The amount of a substance penetrating across the absorption barriers (the exchange boundaries of an organism) via either physical or biological processes. This term is synonymous with *absorbed dose*.

*International Conference on Harmonization of Technical Requirements for Registration of Pharmaceuticals for Human Use*: An initiative intended to establish similar criteria to support the worldwide registration of drugs. Representatives from regulatory agencies and pharmaceutical companies from the United States, Europe, and Japan have been the primary participants.

*International Union of Toxicology (IUTOX)*: An international organization representing toxicology societies worldwide.

*Intestinal First-Pass Effect*: Evidence suggests that not only does the small intestines provide for absorption of various substances, but it can also play a significant role in phase I and phase II metabolism reactions of certain substances prior to their absorption into the circulatory system.

*Intralaboratory Repeatability*: The closeness of agreement between test results obtained within a single laboratory when the procedure is performed on the same substance under identical conditions within a given time period.

*Intralaboratory Reproducibility*: The first stage of validation; a determination of whether qualified people within the same laboratory can successfully replicate results using a specific test protocol at different times.

*Intravascular Hemolysis*: Rupture or lysis of red blood cells within the vascular system causing release of intracellular substances (e.g., hemoglobin) into the plasma. If severe enough, it may be recognized by hemoglobinema and/or hemoglobinuria.

*Intron*: Noncoding spacer DNA in a eukaryotic gene.

*Intubation*: Placement of a tube into a hollow organ. Intubation is used to place various materials into animals for toxicology studies. If the organ is in the stomach, the process is generally referred to as *gavage*.

*Inulin*: A polysaccharide (molecular weight, ~5000) that is not metabolized by mammalian cells and is too large to enter cells. The clearance of inulin is used to measure glomerular filtration rate. Inulin is used *in vitro* as an indicator of extracellular space.

*IOCA*: *In ovo* carcinogenicity assay.

*Ionization*: The formation of ions from neutral molecules. This is accomplished in mass spectrometry by a variety of methods, including bombardment with electrons, gas-phase chemical reactions, and ion desorption methods.

*IR Fingerprint Region*: This frequency region of the infrared spectrum ranges from 910 to 1430 $cm^{-1}$ and contains bending and stretching absorptions of characteristic groupings in an organic molecule. The region is abundant in absorption bands, and their frequencies and intensities are unique for a specific compound; hence, the absorption pattern in this region of the infrared spectrum is utilized as a fingerprint for the recognition of an unknown molecule by comparison with the fingerprint frequencies of an authentic standard. It is important to note that polymorphic forms of the same compound may show significant differences in the fingerprint region; therefore, comparisons should be made in the solution in the same solvent. If solid-state spectra are utilized, the unknown and authentic standard should be crystallized from the same solvent under similar conditions.

*Iris*: The structure of the eye that is anatomically located posterior to the cornea. The iris forms the pupil of the eye and functions in regulating the amount of light that reaches the retina. In the Draize eye test, effects on the iris represent a maximum of 10 out of 110 total points.

*IRMA*: Immunoradiometric assay.

*Irradiation*: Exposure to radiation.

*Irritant Dermatitis*: Chemically induced, nonimmunologic contact dermatitis; a complex syndrome of many types.

*Ischemia*: Local anemia due to mechanical obstruction (mainly arterial narrowing) of the blood supply.

*Islet of Langerhans*: Specialized cells (beta cells) in the pancreas that secrete insulin.

*Isoenzymes*: Enzymatically active proteins that catalyze the same reactions and occur in the same species but differ in their physicochemical properties (also, *isozymes* or *isoforms*).

*Isolated Epithelial Cells*: *In vitro* preparation of single tubular epithelial cells, typically isolated by enzymatic (e.g., collagenase) digestion. Cells from specific nephron regions can be enriched by various separation methods; useful for short-term (up to 4 hours) metabolic or cellular function studies.

*Isolated Perfused Kidney*: *In vitro* preparation involving the whole kidney. A cannula is inserted into the renal artery, and fluid is pumped into the kidney. The technique can be used to assess renal physiology and determine drug clearance. A chief advantage is that the model maintains an intact organ structure but does not include extrarenal tissues or factors; viability is maintained for up to 2 to 4 hours.

*Isolated Perfused Organ*: Vascularly perfused organ is physically removed from the donor animal and maintained in an artificial chamber usually kept at physiological temperature.

*Isolated Perfused Tubules*: Intact tubules from specific nephron cell types are usually isolated by mechanical means and are perfused with micropipettes; useful for short-term (up to 2 hours) metabolic, transport, and acute toxicity studies.

*Isolated Perfused Ventilated Lung*: Vascularly perfused isolated lungs are also ventilated using mechanical devices.

*Isolated Tubular Fragments*: *In vitro* preparation of fragments of nephron segments, typically isolated by enzymatic (e.g., collagenase) digestion. The segments from specific cell types can be enriched by various separation methods; useful for short-term (up to 4 hours) metabolic or cellular function studies.

*ITO Cells*: Fat-storing cells found in the liver; also called *lipocytes* or *stellate cells*.

*Juxtaglomerular Apparatus*: A specialized area of the glomerulus where the distal tubule from the nephron has contact with the arterioles entering and leaving the glomerulus; one factor in the control of renal blood flow.

*Knockout Animals*: Genetically engineered animals in which one or more genes, usually present and active in the normal animal, are absent or inactive.

*Kriging*: A statistical interpolation method that selects the best linear unbiased estimate of the parameter in question. Often used as geostatistical method of spatial statistics for predicting values at unobserved locations based on data from the surrounding area. Information on the fate and transport of chemicals within the area lacking data can be incorporated into Kriged estimates.

*Lactate Dehydrogenase Leakage*: Assay for cellular necrosis that measures the fraction of cytosolic enzyme lactate dehydrogenase (LDH) released from cells due to membrane premeabilization; assays are performed by measuring NADH oxidation spectrophotometrically.

*Large Intestine*: The region of the gastrointestinal tract from the ileum to the anus which consists of the cecum, the colon, and the rectum.

*Laser Capture Microdissection*: A potentially useful technique for excision of tissue area of interest (e.g., a tumor) via laser dissection from surrounding normal tissue so differences in gene or protein expression patterns may be compared by two-dimensional gel electrophoresis patterns.

*Latin Hypercube Sampling (LHS)*: A variant of the Monte Carlo sampling method that ensures selection of equal numbers of values from all segments of the distribution. LHS divides the distribution into regions of equal sampling coverage; hence, the values obtained will be forced to cover the entire distribution. It is more efficient than simple random sampling, as it requires fewer iterations to generate the distribution sufficiently.

*LCB*: Limited carcinogenicity bioassay.

*$LD_{50}$*: Lethal dose of radiation or chemical agent that has been determined to cause death in 50% of a defined population; expressed in terms of weight of test substance per unit of test animal (mg/kg). *See Median Lethal Dose*.

*$LD_{50/30}$*: Median lethal dose (MLD or $LD_{50}$) required to kill 50% of the population of organisms within 30 days.

*Leukopenia*: Any situation in which the total number of leukocytes in the circulation is less than normal.

*Leydig Cells*: Cells located in the interstitial compartment of the testes between the seminiferous tubules that synthesize testosterone; androgen-secreting cells present in the interstitium of the testes.

*LH*: Luteinizing hormone.

*Limit of Quantification (LOQ)*: The concentration of analyte in a specific matrix for which the probability of producing analytical values above the method detection limit is 99%.

*Limit Study*: A study in which a single, maximal dose level of test material is administered to the test animals. Limit studies are conducted when administration of test material at higher dose levels is not required either because exposure at higher levels is not physically possible or because the test material has been shown to be of extremely low toxicity.

*Linear Energy Transfer (LET)*: The amount of energy transferred by a unit dose of radiation per unit pathway traveled through matter (keV/micron of path); varies with the type of radiation. Alpha particles are high LET radiation with 10 to 100 s of keV/micron of path, whereas x-rays and gamma rays are low LET radiations (tenths to 10 keV/micron).

*Linear Extrapolation*: The process of estimating a value at conditions not directly measurable using a linear relationship between the biological effect and the dose or duration of exposure. (Technically, when a measure of the background effect is available, an *interpolation* is being performed.)

*Linear Kinetics*: With linear kinetics there is a linear relationship between dose and plasma concentrations and body loads. Linear kinetics are characteristic of first-order reactions because the rates of reactions increase as the concentration or body load increases; therefore, parameters such as bioavailability, clearance, apparent volume of distribution, and half-life are independent of dose (see definitions in text).

*Linear Model*: A model in which the change in a biological effect is proportional to the change in dose or duration of exposure.

*Lipid Peroxidation*: The chain reaction formation of the mediators of lipid degradation by ionizing radiation.

*Lipophilicity*: The tendency of a chemical to partition into a lipid medium or a fat solvent, such as hexane, from an aqueous medium.

*Liquid Chromatography–Mass Spectrometry*: A method for introducing analytes dissolved in solution into the mass spectrometer through a chromatographic system where a liquid mobile phase passes through a column packed with a solid-phase material.

*Liver Acinus*: Functional unit of hepatocytes as defined by Rappaport; consists of a small parenchymal mass arranged around an axis consisting of a terminal portal venule, a hepatic arteriole, a bile ductile, lymph vessels, and nerves.

*LOAEL*: Lowest observed-adverse-effect level.

*Low-Dose Extrapolation Model*: A model that uses information on observed dose–response relationships combined with mechanistic understanding to predict the dose–response relationship below the range of observable data.

*Lymph*: A clear, transparent, sometimes faintly yellow and slightly opalescent fluid that is collected from the tissues throughout the body. It flows into the lymphatic vessels and is eventually added to the venous blood circulation.

*Lymphoproliferation*: Proliferation of lymphocytes in response to stimulation with cellular activators, including antigens or mitogens. Because the proliferation of lymphocytes is one of the initial consequences of cellular activation, lymphoproliferation is often used as a nonspecific *in vitro* measure of immunoresponsiveness. This assay is sometimes referred to as the *blastogenesis assay*.

*Mab*: Monoclonal antibody.

*Macrocirculation*: Comprised of the heart, great vessels (both arterial and venous), and larger arteries and veins.

*Magic Angle Spinning*: A nuclear magnetic resonance (NMR)-based technique enabling the analysis of intact tissue. The term *magic angle* is derived from the fact that, when samples are spun rapidly at 54.7° relative to the applied magnetic field (the so-called magic angle), line-broadening effects that would ordinarily obfuscate a proton spectra of a solid sample can be reduced.

*Major Histocompatibility Complex (MHC)*: A complex of genes coding for tissue compatibility markers. Two major classes are recognized: class I (present on all nucleated cells) and class II (present on B cell, T cells, and macrophages). MHC molecules appear to direct the course of immune reactivity and are presented in association with antigen by antigen-presenting cells. The human equivalent is termed *human leukocyte antigen* (HLA).

*Malabsorption*: Impairment in the uptake of ingested substances from the gastrointestinal lumen into the bloodstream because of defects in luminal digestion, mucosal transport, or gastrointestinal transit.

*MALT (Mucosa-Associated Lymphoid Tissue)*: Lymphoid tissue associated with the mucosal layer in various tissues, believed to act as a primary defense at secretory surfaces. It acts, to a limited extent, independently of the systemic response. Various tissues comprise this system, including gut-associated (GALT), nasal-associated (NALT), and bronchus-associated (BALT) lymphoid tissues.

*MAP Kinase Pathway*: The mitogen-activated protein kinase pathway is a key cellular signaling pathway resulting in transcriptional activation and mitogenic proliferation. It consists of several enzymatic activation steps and can be stimulated by a number of agents, including cytokines and radiation exposure. It is believed to account for the accelerated repopulation of cells occurring after radiation exposure, which can affect the success of radiation therapy.

*Margin of Safety (MOS)*: The difference, normally expressed as fold-difference, between the dose (or exposure level) that results in toxicity and the dose that results in the desired pharmacological activity.

*Mass Cell*: A polymorphonuclear, granule-containing cell with a major role in hypersensitivity reactions.

*Mass Median Aerodynamic Diameter (MMAD)*: Standard method of characterizing the size distribution of a particulate atmosphere. It represents the size when 50% of the particles are larger (or smaller) than the stated size; aerodynamic diameter that divides the particles of a sample in half, based on their weight.

*Mass Spectrometry*: An analytical technique that determines the mass of ionized molecules and their fragments and adducts.

*Mass Spectroscopy (MS)*: An analytical technique where charged particles (ions) are created from sample molecules that are then analyzed to provide information about the molecular weight of ion and its chemical structure.

*Mating Performance*: The ratio of mated (inseminated) animals to the number cohabited, generally expressed mathematically as the ratio of the number of female animals inseminated divided by the number cohabited with a cohort male.

*Maximally Exposed Individual (MEI)*: The single individual with the highest exposure in a given population (also, most exposed individual). This term has historically been defined various ways, including as defined here and also synonymously with worst-case or bounding estimate. Assessors are cautioned to look for contextual definitions when encountering this term in the literature.

*Maximum Tolerated Dose (MTD)*: The dose that causes no more than a 10% reduction in body weight and does not produce mortality, clinical signs of toxicity, or pathologic lesions that would be predicted to shorten the natural life span of an experimental animal for any reason other than the induction of neoplasms.

*MCL*: Mononuclear cell leukemia.

*Mean*: The average value. In a normally distributed sample, this has the same value as the median (the middle value) and the mode (the most frequent value).

*Measure of Association*: A term that represents the strength of association between variables. The relative risk, odds ratio, and standardized mortality ratio (SMR) are measures of association commonly used in epidemiology.

*Median Effective Dose*: A statistically derived single dose of a substance that can be expected to produce a particular effect in 50% of the study population ($ED_{50}$). The $ED_{50}$ is expressed in terms of weight of test substance per unit weight of test animal (mg/kg).

*Median Lethal Dose*: A statistically derived single dose of a substance that can be expected to cause death in 50% of the study population ($LD_{50}$; $LC_{50}$). The $LD_{50}$ is expressed in terms of weight of test substance per unit weight of test animal (mg/kg). The $LC_{50}$ is expressed in terms of weight of test substance per unit volume (mg/L).

*Median Value*: The value in a measurement data set such that half of the measured values are greater and half are less.

*Mee's Lines*: Horizontal white lines on the fingernails that occur after exposure to arsenic. The lines appear after the exposed nail bed grows to the exterior.

*Metabolic Activation*: The process by which relatively stable substrates are converted to highly reactive, generally electrophilic products with the capability of producing damage to critical cellular macromolecules. The term is occasionally used to refer to the metabolism of therapeutically inactive prodrugs to the active form of the drug.

*Metabolome*: The set of all low-molecular-weight compounds synthesized by an organism; the total quantitative collection of small-molecular-weight biochemical metabolites present in a cell, tissue, or organism that are involved in growth, energetics, maintenance, pathology, and other biochemical functions.

*Metabolomics*: The study of the dynamic expression, production, relative concentration, and variation of small-molecular-weight, endogenous biochemicals and metabolites in living systems. Characteristic changes in these discrete biomarkers before and after exposure to xenobiotics and in pathological processes may be useful for detecting and understanding toxicological mechanisms, histopathology, and environmental stressors.

*Metabonome*: The constituent metabolites in a biological sample.

*Metabonomics*: The techniques available to identify the presence (and concentrations) of metabolites in a biological sample.

*Metal Fume Fever*: An acute condition of short duration caused by exposure to fresh fumes of zinc and other metals. The condition is characterized by fever and chills, occurring 4 to 12 hours after exposure. Recovery usually is complete within 1 day.

*Metalloid*: Any element with both metal and nonmetal characteristics (e.g., arsenic, boron, tellurium).

*Metallothionein*: Any inducible low-molecular-weight, cytosolic protein with a high cysteine content (approximately 30%) and characteristically deficient in aromatic amino acids and histidine. Presence of numerous cysteinyl thiol groups permits high-affinity binding of several metals (e.g., cadmium, lead, mercury, or zinc).

*Metals*: A grouping of elements generally characterized by opacity, ductility, luster, being electropositive with a tendency to lose electrons, and having the property of conducting heat and electricity. Heavy metals may be further defined as any element having a density greater than 5 g/cm$^3$, and those of toxicological significance preferentially bind to ligands containing sulfur or nitrogen (arsenic, cadmium, chromium, lead, and mercury).

*Methemoglobin*: A form of hemoglobin in which the ferrous ion ($Fe^{2+}$) of hemoglobin has been oxidized to the ferric state ($Fe^{3+}$); methemoglobin is unable to carry oxygen.

*Micelles*: These are discreet aggregated of 20 to 50 molecules of bile salts that mediate the absorption of fatty acids longer than 12 carbon molecules across the lumen of the small intestine. The water-soluble part of the molecules faces outward while the nonpolar nuclei face inward. Micelles are able to dissolve the aqueous component of the chyme allowing fatty acids to accumulate in the center of the micelle. The micelles are subsequently transported to the lumen, where the fatty acids are deposited and absorbed, and the micelles are then recirculated in the chyme to repeat the process.

*Microarray*: A series of molecular probes, immobilized on a solid substrate, such as nylon or glass, which can be used to detect the presence of specific biomolecules, such as RNA, DNA, or protein, in a biological sample.

*Microcirculation*: The business end of the circulation, where delivery of oxygen and nutrients and the removal of carbon dioxide and metabolites take place.

*Microcosm*: A small model or piece of an ecosystem contained in a test chamber and used to study the effect of contaminants on a biotic community in a controlled environment.

*Microenvironment Method*: A method used in predictive exposure assessments to estimate exposures by sequentially assessing exposure for a series of areas (microenvironments) that can be approximated by constant or well-characterized concentrations of a chemical or other agent.

*Microenvironments*: Well-defined surroundings, such as a home, office, automobile, kitchen, or store, that can be treated as homogeneous (or well characterized) in the concentrations of a chemical or other agent.

*Microexposure Event (MEE) Analysis*: A method accessing risk based on the aggregate sum of a receptor's contact with a contaminated medium. MEE analysis simulates lifetime exposure as the sum of many short-term or microexposures. MEE approaches can be used to explore uncertainty associated with the model time step in probabilistic risk assessment (PRA)—for example, the use of a single value to represent a long-term average phenomenon, seasonal patterns in exposures, or intraindividual variability.

*Microisolation Cage*: An animal cage usually constructed of plastic that completely surrounds the animals contained therein such that air entering and exiting the cage must pass through an integrated filter designed to stop the passage of unwanted microorganisms or fomites. Most commonly, such caging is maintained using sterile techniques to prevent the introduction of unwanted microorganisms.

*Microperfusion*: Technique to determine the function of a specific nephron segment. The segment is isolated from the glomerular filtrate between a wax and oil block, and the segment is then perfused *in situ* with a solution of controlled composition through a micropipette; the solution is then quantitatively collected from the same nephron segment.

*Micropuncture*: Specialized technique used to analyze the effects of drugs or chemicals on single nephron function in the intact kidney. It involves the insertion of a micropipette into the lumen of a specific nephron segment of interest; experiments are performed under microscopic control.

*Microsomes*: The operational definition of microsomes is the 105,000-*g* pellet produced from a tissue homogenate after removal of nuclei, mitochrondria, and cell debris by centrifugation at 9000 *g*. Microsomes are the remnants of the endoplasmic reticuluum after cellular disruption. The outer surface of microsomal vesicles represents the cytosolic side of the endoplasmic reticuluum.

*Midzonal Cells*: Collection of hepatic cells situated between the periportal area and the centrilobular area of the lobule.

*Milk Sickness*: Potentially fatal neurologic disease resulting from the consumption of unpasteurized milk containing the neurotoxin tremetol, derived from animals grazing on toxic plants, white snakeroot, or rayless goldenrod.

*Misclassification*: A type of bias in which an individual or attribute is assigned a value that is incorrect. The misclassification may be *nondifferential* (the same in all study groups) or *differential* (not equivalent across groups). Each type of misclassification may impact the results, the former usually underestimating the true measure of association and the latter often overestimating it.

*Mists*: Similar to fogs, a suspension of liquid particles in gas. These can be formed from condensation of vapors or from atomization of liquids (e.g., sprayers). Mists are characterized by particles larger than about 40 μm; fogs are categorized by their smaller size.

*Mitochondrial Membrane Permeability Transition (MPT)*: An increase in the permeability of the mitochondrial membrane due to a loss of membrane potential resulting in the release of a number of proteins such as cytochrome c and apoptosis-inducing factor (AIF).

*Mitogen*: Molecules capable of inducing cellular activation; may include sugars or peptides. The ability of a cell to respond to stimulation with mitogen (generally assessed by cellular proliferation) is believed to give an indication of the cell's immune responsiveness.

*Mitogenesis*: The induction of mitosis or cell transformation. The stimulation of cell proliferation, a natural recovery process in response to severe toxicologic insult that does not normally occur at reasonable multiples of human exposure levels, can account for the carcinogenic response toward nongenotoxic compounds that may not be meaningful in the clinical setting.

*Mixed Lymphocyte Response/Reaction (MLR)*: An *in vitro* assay that measures the ability of lymphocytes to proliferate in response to exposure to allogeneic cells. The proliferation represents the initial stage in acquisition of CTL function; the assay thus serves as a measure of cell-mediated immune function. Sometimes referred to as *mixed lymphocyte culture* (MLC).

*MOE*: Margin of exposure.

*Mole Viper*: A viper of the genus *Atractaspis*.

*Mollusk*: Any member of the phylum Mollusca that includes cones, snails, mussels, clams, and oysters, as well as octopi and squid.

*Monocyte/Macrophage*: Bone-marrow-derived mononuclear cells that serve a wide variety of host defense needs, acting as both nonspecific phagocytic cells and as regulators of other immune and nonimmune host resistance mechanisms. A variety of forms exist, including monocytes (found in the blood), macrophages (found in peripheral tissue), Kupffer cells (liver), Langerhans cells (skin), microglia (brain), veiled cells (lymph), and others.

*Monte Carlo Technique*: A repeated random sampling from the distribution of values for each of the parameters in a generic (exposure or dose) equation to derive an estimate of the distribution of (exposures or doses in) the population.

*Moribund Status*: The condition of an animal as a result of the toxic properties of a test substance where death is anticipated. For toxicity determinations, animals killed for humane reasons are considered in the same way as animals that die.

*Morphology*: Pertaining to the form or structure of a cell, organ, or whole animal.

*Motor Activity*: Spontaneous locomotion in a specified enclosure designed to provide quantitative indices of movement.

*MOU*: Memorandum of understanding.

*MRA*: Mutual recognition agreement.

*mRNA (Messenger Ribonucleic Acid)*: The substance carrying genetic information from the DNA to the protein production site.

*MSH*: Melanocyte-stimulating hormone.

*MTD*: Maximum tolerated dose.

*Mucociliary Transport*: Mucous lining of the nasal passage extending from the respiratory epithelium to the pharynx. The purpose of mucociliary transport is to move waste solids up from the deeper lung to the pharyngeal area to be swallowed. Particle matter entrained in a mucous layer over ciliated airways is moved along to the nasopharynx for removal from the respiratory tract by expectoration or swallowing.

*Mullerian Duct*: Involves a protein secreted by the fetal testes that causes the potential female Mullerian tract to regress.

*Multistage Model of Carcinogenesis*: A model that describes the carcinogenesis process as a series of mutations or mutation-like events over time that result in a malignant growth.

*Murine*: Of the mouse.

*Muscarine*: Extremely toxic alkaloid that causes muscarinism following ingestion; an extremely toxic parasynthetic poison.

*Muscimol*: A extremely toxic central nervous system (CNS) depressant and GABA agonist that is isolated from the mushroom *Amanita muscaria* and causes visual damage, mental confusion, spatiotemporal dislocation, and memory loss in humans.

*Mutagenesis*: The production of genetic alterations by exposure to chemicals or radiation. This can result in irreparable DNA damage and subsequent tumor development.

*Mutation*: A stable change in the nucleotide sequence of a gene.

*Mycotoxins*: A group of more than a hundred toxins produced by various fungal organisms in food and feed commodities.

*Myocardium*: The middle layer of the heart, consisting of cardiac muscle.

*NADPH-P450 Reductase*: A flavoprotein responsible for mediating the reduction of cytochrome P450.

*Nanoparticles*: Particles, either of manmade or biological origin, with at least one dimension less than 100 nm; also referred to as *ultrafine particles*, which have all dimensions in the nanometer range.

*Nasopharyngeal*: Region of the respiratory tract serving as the entry port for inspired air (includes the turbinates, epiglottis, glottis, pharynx, and larynx).

*Natural Immunity*: Host defense mechanisms that do not require prior exposure to an antigen. Various effector mechanisms have been described, including cellular cytotoxicity mediated by macrophages or NK cells, complement, and activity of gamma delta T cells.

*Natural Killer* (*NK*) *Cells*: A population of lymphocytes distinct from T and B lymphocytes; also referred to as *large granular lymphocytes* (LGL). NK cells exhibit cytotoxicity against virally infected cells and certain tumor cells. They are notable in that they do not require prior exposure to antigen to express cytotoxicity toward their targets. Assessment of NK activity provides a measurement of nonspecific host resistance.

*Necropsy*: Postmortem examination of a nonhuman body. A similar term, *autopsy*, generally is used to refer to postmortem examination of a human body. Necropsies are conducted in toxicology studies following spontaneous death or euthanasia of animals to detect potential pathologic effects of test material administration.

*Necrosis*: Form of cell death involving numerous cells; generally occurs at higher doses of toxicants. Necrosis *in vivo* often involves infiltration of macrophages and other inflammatory cells.

*Neoantigen*: A new antigen that appears on cells during malignant transformation, viral infection, or other process by which a naturally occurring antigen is modified.

*Neoplasia*: The pathologic process that results in the formation and growth of abnormal tissue (neoplasm). Neoplasms usually form a distinct mass of tissue (tumor) that may be benign or malignant; the growth of the tumor exceeds and is uncoordinated with that of normal tissue and persists in the same excessive manner after cessation of the stimulus that evoked the change.

*Nephrotoxicity*: Toxicity to the kidney observed *in vivo*.

*Neurohypophysis*: The posterior lobe of the pituitary.

*Neuropsychology*: A clinical discipline that specializes in the application of psychological tests to ascertain aberrations of neurobehavioral function.

*Nictitating Membrane*: An important ocular structure in many species of animals (but not humans or primates) that aids in protecting the cornea and conjunctiva when the eyeball is retracted; also called the *third eyelid*.

*NIDDM*: Non-insulin-dependent diabetes mellitus, type 2.

*NME*: New molecular (chemical) entity.

*NOAEL (No-Observed-Adverse-Effect Level)*: A number applied to the highest dose that did not elicit an adverse effect in a properly designed and executed toxicological study.

*Nongenotoxic Carcinogen*: A substance that causes cancer, not by primarily damaging the genetic material but by mechanisms that stimulate cell proliferation, thus increasing the chances for natural mutations to be reproduced, or by the selection of specific cell populations that may derange in a later stage.

*Nonimmunologic Contact Urticaria (NICU)*: Contact urticaria is an immediate-type reaction, on a nonimmunologic basis.

*Nonlinear Kinetics*: With nonlinear kinetics, a linear relationship does not exist between dose and plasma concentrations or body loads. Nonlinear kinetics is typically due to saturation of absorption, distribution, or elimination.

*Nonparametric Statistical Methods*: Methods that do not assume functional form with identifiable parameters for the statistical distribution of interest (distribution-free methods).

*Nonregenerative Anemia*: Anemia characterized by decreased production and delivery into circulation of newly formed red blood cells (reticulocytes).

*Normal Equivalent Deviate (NED)*: Units of deviation from the mean; probability of response on a transformed scale (e.g., 50% response equals a NED of 0).

*Nose-Only Exposure*: Experimental mode of inhalation exposure in which only the nose of the animal is placed in contact with the test atmosphere (variant is head-only).

*No-Threshold Dose–Response Relationship*: A dose–response relationship that assumes that any dose carries some probability of an effect.

*NTEL*: Nontumorigenic effect level.

*Nuclear Magnetic Resonance (NMR)*: A technique to identify atoms in a sample by measuring the signal given off by relaxation of, for example, protons previously aligned in a strong magnetic field.

*Nuclear Magnetic Resonance Spectroscopy*: An analytical technique that is used to study the structure of molecules. Nuclear magnetic resonance (NMR) spectroscopy takes advantage of the behavior of atomic nuclei in the presence of a strong magnetic field. When samples are placed in a magnetic field and irradiated with energy in the radiofrequency (RF) range, they absorb RF energy. The exact frequency absorbed by an atomic nucleus in the sample molecule differs depending on the environment of the nucleus (its position within the molecule). A plot of intensity of RF absorption vs. frequency is called an *NMR spectrum* and is characteristic of the structure of the sample molecule.

*Nucleotide*: In this case, the basic building block of DNA and RNA: a base–sugar–phosphate complex. Three nucleotides form a codon that codes for one amino acid.

*Null Allele*: An inactive form of a gene.

*Octanol–Water Partition Coefficient ($K_{ow}$, P)*: The ratio of the concentrations of a chemical in the water phase and the *n*-octanol phase after the chemical is equilibrated between equal volumes of the two solvents.

*Odds Ratio*: The ratio of two odds. In case-control studies, it is an *exposure odds ratio* (i.e., the odds of exposure among the cases as compared to the odds of that same exposure among the controls). Although less seldom used, an odds ratio is sometimes calculated for a cohort study, but it is a *disease odds ratio* (i.e., the odds of disease among the exposed vs. the odds of that same disease among the unexposed).

*Oligonucleotide*: A chain (usually of less than 100 units) of individual nucleotide residues used as a molecular probe to detect or quantify the presence of specific DNA or RNA molecules.

*Open Access Publishing*: Making publication available at no charge in public repositories, typically web-based.

*Open Microscopic Evaluation*: The practice of conducting the initial histopathological evaluation with the pathologists having access to all available information about the animals from which the tissues were derived.

*Opportunistic Organism*: Refers to an organism, most commonly a microorganism, that under a particular set of biological conditions can cause disease when normally it coexists with its host without producing disease.

*Organelle*: A subcellular structure with a specialized function within the cell.

*Organic Anion Transport*: Movement of negatively charged compounds across renal cellular plasma membrane; involves both reabsorption (lumen to plasma) or secretion (plasma to lumen) and is mediated by specific carrier proteins.

*Organic Cation Transport*: Movement of positively charged compounds across renal cellular membrane; involves both reabsorption (lumen to plasma) and secretion (plasma to lumen) mediated by specific carrier proteins.

*Orthologs*: Genes that are believed to have evolved from a common ancestral gene. Orthologs may have a high degree of sequence homology, but their protein products do not necessarily have a high degree of functional homology.

*Osmium Tetroxide ($OsO_4$)*: A toxic metallic cellular stain used in transmission electron microscopy.

*Osteichthyes*: The class of vertebrate animals that contains venomous fishy bones, characterized by the presence of an endoskeleton chiefly of bone. They are the dominant fishes; they invade all types of waters.

*Osteomalacia*: Inadequate or delayed mineralization of bone, resulting in an increased softness of the bone. It is the adult equivalent of rickets.

*Osteoporosis*: A loss in both the mineral and matrix phase of bone; associated with an increased tendency to fracture.

*Outlier*: A value that is far separated from other members of a sample. Outliers may be due to faulty sampling techniques (the values actually belong to a different population) or an error in measurement, or they may be real and very meaningful.

*Overload*: A term used in inhalation toxicology to describe a condition where the normal particle clearance mechanisms of the test system have been overwhelmed by the high inhaled particle concentrations, leading to particle accumulation in the lung and inflammation.

*Oxalate*: Any salt acid that contains the $(COO)_2^{2-}$ radical found in numerous plants and some fungi.

*Oxidation*: The net loss of electrons, which may also involve the addition of oxygen to a molecule.

*Oxidative Stress*: Imbalance between prooxidants and antioxidants.

*Oxygen Enhancement Ratio (OER)*: The ratio of the dose of radiation required to produce a given biological effect in the absence of oxygen (anoxia or hypoxia) compared to the same dose of radiation exposure required in the presence of oxygen. The OERs for low linear energy transfer (LET) radiation, such as x-rays or gamma rays, usually range from 2.8 to 3.0. Living systems are more radiosensitive when irradiated in the presence of oxygen.

*Oxytocin*: A peptide secreted by the neurohypophysis.

*$^{31}P$ NMR Spectroscopy*: A technique involving the use of a nuclear magnetic resonance spectrometer tuned to following phosphorous-31 resonance in a variety of phosphorylated molecules such as ATP.

*p53*: A 53,000-molecular-weight tumor suppressor/transcriptional regulating protein that is commonly mutated in cancers. It participates in cell cycle regulation, transcription, and apoptosis. Activation of p53 after radiation exposure leads to radiation-induced cell cycle delay.

*Pachymetry*: A means of obtaining quantitative measurements of corneal thickness using either optical or ultrasonic methods. Derived from the Greek words *pachys* ("thick") and *metry* ("process of measuring").

*Pancytopenia*: Reduction of all three formed elements of blood: red blood cells, white blood cells, and platelets.

*Paracelsus*: The founder of modern toxicology.

*Paracrine*: A type of cellular regulation in which a substance exerts a regulatory influence primarily on cells in close proximity as opposed to the endocrine system, which exerts regulatory action at locations distant from the release.

*Parenteral*: Introduction into the body by a route other than the alimentary canal (e.g., subcutaneous intravenous, intramuscular injection).

*Particle*: A small, discrete object.

*Particulate*: An adjective for particle-related properties.

*Partition Coefficient*: Ratio of concentration or relative distribution of a chemical in two matrices at equilibrium.

*Partitioning Factors*: Factors that determine which individuals and results are used to construct a reference interval (e.g., species, strain, sex, age, instrument, method).

*Parturition*: The interval in a pregnant animal during which delivery of an offspring occurs; it is initiated at the first signs of labor and completed at the birth of the last offspring in the litter, in multiparous animals.

*Pathogenesis*: The cellular events and reactions and other pathologic mechanism occurring in the development of disease.

*Pathology*: The medical science, and specialty practice, concerned with all aspects of disease but with special reference to the essential nature, causes, and development of abnormal conditions, as well as the structural and functional changes that result from the disease processes.

*Pathology Data Review*: A quality assurance review of pathology data to ensure the quality of the materials and procedures used to generate histopathological data.

*Pathology Peer Review*: The procedure whereby a second pathologist reviews a subset of tissues and other data from the initial pathology evaluation to verify the accuracy of toxicologically significant microscopic findings.

*Pathology Working Group (PWG)*: A panel of expert pathologists assembled to review a specific question concerning pathology study results.

*PCNA*: Proliferating cell nuclear antigen.

*PCR*: Polymerase chain reaction.

*Performance*: The accuracy and reliability characteristics of a test method (*see Accuracy, Reliability*).

*Performance Standard*: A set of specifications that define an outcome in detail and provide criteria for assessing that outcome. The term is most commonly used in conjunction with prescribing conditions for the care and use of laboratory animals. These standards are based on validated test methods, which provide a basis for evaluating the comparability of a proposed test method that is mechanistically and functionally similar. Included are (1) essential test method components; (2) a minimum list of reference chemicals selected from among the chemicals used to demonstrate the acceptable performance of the validated test method; and (3) the comparable levels of accuracy and reliability, based on what was obtained for the validated test method, that the proposed test method should demonstrate when evaluated using the minimum list of reference chemicals.

*Perfusion-Limited Uptake*: Tissue uptake limited by the rate of blood flow to the tissue.

*Peripheral Blood Leukocyte (PBL)*: Leukocytes derived from the peripheral circulatory system in humans. Due to the accessibility, these cells are often used in *ex vivo* assays of the human immune function.

*Periportal Cells*: Collection of hepatic cells situated around the portal triad (branch of the portal vein, a hepatic arteriole, and a bile duct).

*Peroxisome*: Membrane-bound cell organelles present in cells of animals, plants, fungi, and protozoa that contain oxidative enzymes responsible for the formation and degradation of hydrogen peroxide. These organelles are involved in lipid, sterol, and purine metabolism, as well as peroxidative detoxification. Nongenotoxic chemicals that cause hepatic peroxisome proliferation can cause liver tumor development in rodents. This phenomenon has not been shown to be relevant to humans.

*Persistent Light Eruption*: Photoallergic contact dermatitis that flares with ultraviolet exposure continuing past cessation of photoallergen exposure.

*Personal Measurement*: A measurement collected from an individual's immediate environment using active or passive devices to collect the samples.

*Pfiesteria*: An unclassified marine organism associated with human illness from a toxin not yet identifiable.

*Phagocytosis*: Process of ingestion by phagocytes whereby the cell membrane of the phagocyte engulfs bacteria and delivers them into the cell, where enzymatic digestion occurs; the active process of particle indigestion by cells (generally macrophages). Once within phagocytic vesicles (phagosomes), particles are subjected to hydrolytic enzymes secreted by macrophages.

*Pharmacodynamics*: The study of the biological and physiological effects of chemicals and their mechanisms of actions.

*Pharmacogenetics*: The study of inherited differences in xenobiotic metabolism. This includes genetic mechanisms of species differences and interindividual and interpopulation differences in the genotype and phenotype of cytochrome P450s and other xenobiotic metabolizing enzymes.

*Pharmacokinetics*: The study of the movement of a chemical (drug) in the body (absorption, distribution, metabolism, excretion). This study involves the time course of absorption, disruption, metabolism, and excretion of a foreign substance (e.g., a drug or pollutant) in an organism's body.

*Phase 1 Clinical Trials*: The first stage of clinical testing. These single- and multiple-dose studies are normally conducted in healthy male volunteers to assess the safety and systemic exposure of new drug candidates.

*Phase 2 Clinical Trials*: The second stage of clinical testing. These trials are designed to assess safety and efficacy in the target patient population; the length of these trials is determined by the time required to demonstrate clinical endpoints suggestive of efficacy.

*Phase 3 Clinical Trials*: Expanded controlled and uncontrolled trials. These trials are performed after preliminary evidence suggesting effectiveness of the drug has been obtained in phase 2 and are intended to gather additional information about the effectiveness and safety that is needed to evaluate the overall benefit–risk relationship of the drug. Phase 3 studies also provide an adequate basis for extrapolating the results to the general population and transmitting that information to physicians. Phase 3 studies usually include several hundred to several thousand people.

*Phase 4 Clinical Trials*: These are postmarketing trials that may be requested by regulatory agencies upon their review of the New Drug Application (NDA) or in response to effects that become evident as more patients become exposed to the drug. Sponsors may also choose to conduct these studies to support a line extension strategy (e.g., new formulations, expansion of the patient population).

*Phase I Metabolism*: The first step in drug metabolism. Its purpose is to aid the elimination of foreign compounds from the body. The main reactions involved are oxidation, reduction, and hydrolysis but could also be hydration or isomerisation. The aim of phase I is add or unmask a reactive functional group to which phase II metabolic enzymes can add a highly water soluble molecule. This is desired because the more water soluble a compound, the more readily it is excreted by the kidneys. Although some compounds can be eliminated solely as a result of phase I drug metabolism, most go on to be involved in the conjugation reactions of phase II metabolism. Also, phase I is not a necessary precursor to phase II metabolism for some foreign molecules.

*Phase II Metabolism*: Conjugation reactions by which covalent bonds are formed between chemicals (both xenobiotics and endogenous) and small, polar, endogenous compounds (e.g., glucuronic acid, sulfate, glutathione, glycine); these reactions often, but does not always, follow phase I reactions.

*Phenotype*: The total of observable features of an organism, as the result of interaction between the genetic material (genotype) and the environment.

*Photoallergic Contact Dermatitis*: Allergic contact dermatitis dependent on ultraviolet (typically UVA) exposure.

*Photoirritant (Phototoxic)*: Acute irritant dermatitis dependent upon ultraviolet (typically UVA) exposure.

*Physiological Leukocytosis*: Increased white blood cell count associated with endogenous catecholamine release, usually resulting from excitement or fear (fight-or-flight phenomenon) or pain.

*Physiologically Based Pharmacokinetic* (*PBPK*) *Model*: A mechanistic model that describes quantitatively the uptake, distribution, metabolism, and excretion of a chemical; the model can also be used to quantify the dose of an active metabolite received by the target tissue.

*Phytotoxin*: A plant toxin composed of complex proteins that are produced by a relatively small number of plants and are structurally similar to bacterial toxins.

*Pica*: Compulsive ingestion of nonnutritive items, such as dirt, flaking paint, plaster, ashes, or laundry starch. Individuals with pica often have greater exposure to toxicants (e.g., ingestion of lead in paint chips).

*Pinocytosis*: Cellular process of actively engulfing a liquid.

*Pit Viper*: Any snake of the family Crotalidae that had a depression or pit between the nostril and the eye.

*Plasma*: The liquid portion of blood *in vivo*. Plasma is obtained from blood collected with an anticoagulant and centrifuged to separate it from the cells.

*Plenum Velocity*: Air velocity in the plenum. For good air distribution with slot-type hoods, the maximum plenum velocity should be one half of the slot velocity or less.

*Point Estimate*: A quantity calculated from values in a sample to represent an unknown population parameter. Point estimates typically represent a descriptive statistic (e.g., arithmetic mean, 95th percentile).

*Point Estimate Risk Assessment*: Familiar risk assessment methodology in which a single estimate of risk is calculated from set point estimates. The results provide point estimates of risk for the central tendency exposure (CTE) and reasonable maximum exposure (RME) exposed individuals. Variability and uncertainty are discussed in a qualitative manner.

*Point-of-Control Measurement of Exposure*: An approach to quantifying exposure by taking measurements of concentration over time at or near the point of contact between the chemical and an organism while the exposure is taking place.

*Poison*: Any substance (chemical, physical, biological) that is harmful or destructive to a biological (living) system.

*Polymerase Chain Reaction*: A method in which a region of a nucleic acid is selectively amplified by cycles of nucleotide polymerization; a technique enabling a rapid multiplication of selected parts of a DNA or RNA strand.

*Polymorphisms*: Multiple phenotypes of an organism determined by different alleles; in this context, the existence of an interindividual difference in DNA sequence coding for one specific gene. The effects of such a difference may vary dramatically, ranging from no effect at all to the building of inactive proteins or not even building the protein. Two structurally distinct genes for the same protein; polymorphic genes can be produced by mutations that result in nucleotide sequence differences that can lead to the production of proteins that differ functionally or that may not have altered functionality.

*Polynomial Dose–Response Model*: A dose–response model in which the response is expressed mathematically as the sum of quantities containing increasing powers of dose.

*Polytocous*: The production of litters of offspring.

*Porifera*: An animal phylum of some 5000 species of the simplest, multicellular life forms, most of which are marine.

*Positron-Emitting Isotopes*: These isotopes have relatively short half-lives (10 to 100 minutes) and decay exclusively by emission of a positron. They produce readily detectable, tissue-penetrating, $\gamma$-ray photons and are thus commonly used in tomographic techniques (position emission tomography, or PET). Examples of such isotopes are $^{11}C$, $^{13}N$, $^{15}O$, and $^{18}F$. Because of their brief existence, positron-emitting isotopes usually have to be manufactured close to the site where they are to be used.

*Potential Dose*: The amount of a chemical contained in the material ingested, air breathed, or bulk material applied to the skin.

*Power*: The effect of the experimental conditions on the dependent variable relative to sampling fluctuation. When the effect is maximized, the experiment is more powerful. Power can also be defined as the probability that there will not be a type II error (1-beta). Conventionally, power should be at least 0.07.

*Preanalytical Variables*: Factors that occur prior to analysis of samples and influence clinical pathology results.

*Precision*: A measure of the agreement among replicate measurements of an analyte; a measure of the reproducibility of a measured value under a given set of conditions.

*Prediction Model*: In alternative method development, the tool that is used to predict the endpoint of interest; the prediction model is an algorithm that defines how to convert results from the alternative method into a prediction of the *in vivo* toxicity. A formula or algorithm is used to convert the results obtained using a test method into a prediction of the toxic effect of interest. A prediction model contains four elements: (1) a definition of the specific purpose for which the test method is to be used, (2) specifications of all possible results that may be obtained, (3) an algorithm that converts each study result into a prediction of the toxic effect of interest, and (4) specifications as to the accuracy of the prediction.

*Predictivity (Negative)*: The proportion of correct negative responses among substances testing negative by a test method (*see Two-by-Two Table*). It is one indicator of test method accuracy. Negative predictivity is a function of the sensitivity of the test method and the prevalence of negatives among the substances tested.

*Predictivity (Positive)*: The proportion of correct positive responses among materials testing positive by a test method (*see Two-by-Two Table*). It is one indicator of test method accuracy. Positive predictivity is a function of the sensitivity of the test method and the prevalence of positives among the substances tested.

*Prevalence*: The number of cases of a designated disease that exist at a particular point in time is the *point prevalence*. Various types of *period prevalence* represent the number of existing and new cases of a disease that are extant at the beginning of or at any time during a defined period of time. In the medical literature, both types of prevalence, particularly the latter, may erroneously be called *incidence*, especially if the data are presented as a *prevalence rate*. Prevalence is also defined as the proportion of positive or negative substances in the population of substances tested (*see Two-by-Two Table*).

*Prevalidation*: In alternative methods development, the preliminary phase to validation in which the purpose of a test and its capabilities, limitations, interlaboratory transferability, and predictability are determined; the process during which a standardized test method protocol is developed and evaluated for use in validation studies. Based on the outcome of those studies, the test method protocol may be modified or optimized to increase intra- or interlaboratory reproductivity for use in further validation studies.

*Primary Cell Culture*: Isolated epithelial cells, tubules, or tubular fragments are used as seed material to grow renal cells under defined conditions on a solid matrix. It is a primary cell culture when seed material is directly derived from fresh tissue. It typically grows to confluence in 5 to 7 days and has limited ability to be passaged, depending on the species.

*Primary Enclosure*: The cage, pen, or stall that forms the immediate limit of an animal's environment in a research facility.

*Primary Response*: The immune response following initial contact with an antigen, resulting in the establishment of immunologic memory; synonymous with *immunization*.

*PRL*: Prolactin.

*Probabilistic Risk Assessment (PRA)*: A risk assessment that uses probabilistic methods to derive a distribution of risk or hazard based on multiple sets of values sampled for random variables.

*Probability Density Function (PDF)*: A representation, generally a function, graph, or histogram, of the probability of occurrence of an unknown or variable quantity. The sum of the probabilities for discreet random variables and the integral for continuous random variables (i.e., the area under the curve) is equal to 1.0. PDFs can be used to display distributions used as input to a probabilistic assessment or the distribution of risks that forms the output of that assessment.

*Probit*: Inverse cumulative distribution function of the normal distribution; it is the normal equivalent deviate (NED) plus 5 (e.g., 50% response equals a probit of 5).

*Proliferation*: Process of cell growth. Uncontrolled proliferation can result in cancer.

*Proprietary Test Method*: A test method for which manufacture and distribution are restricted by patents, copyrights, trademarks, etc.

*Proteome*: The entire protein complement of a biological sample.

*Proteomics*: The techniques available to identify the proteins in biological samples; field of study involving analysis of protein levels in cells or populations; automated methods developed for high-throughput analysis.

*Protocol*: The precise step-by-step description of a test method, including listing of all necessary reagents and all criteria and procedures for generating and evaluating test data.

*Proton ($^1H$) Nuclear Magnetic Resonance (NMR) Spectroscopy*: Nuclear magnetic resonance is a phenomenon that occurs when the nuclei of certain atoms are immersed in a static magnetic field and exposed to a second oscillating magnetic field. Proton NMR spectroscopy is the use of NMR phenomenon or protons (there are other magnetic nuclei) to study physical, chemical, and biological properties of samples.

*Proximal Tubules*: Initial cell types of the tubular epithelium connected to glomerulus. They are subdivided into two overall segments—proximal convoluted (S1) and proximal straight (S2, S3) tubules. S1 cells are found in the cortex; S2 cells are found in the cortex and outer stripe of the outer medulla; and S3 cells are found in the outer stripe of the outer medulla. These are major sites in the nephron for drug metabolism, transport, reabsorption, and secretion of amino acids, glucose, organic anions, and organic cations.

*Psilocybin*: A hallucinogenic that is much less potent than LSD but much more potent than mescaline. It is moderately toxic to humans by ingestion and intraperitoneal routes and is probably toxic by all routes of exposure.

*Psychophysics*: A group of methods formulated to guide the testing of sensory capacities.

*PTH*: Parathyroid hormone.

*Pulmonary*: Region of the respiratory tract serving primarily as the air exchange region composed of respiratory bronchioles, alveolar ducts, and alveoli.

*Quality Assurance*: A management process by which adherence to laboratory testing standards, requirements, and record keeping procedures is assessed independently by individuals other than those performing the testing.

*Quality Control*: Procedures for establishing, monitoring and evaluating the quality of the analytical testing process of each method to assure accuracy and reliability of test results for test subjects or patients.

*Quantal Response*: A response to a stimulus or a treatment that can be referred to as "all or none" (it either happens or it does not happen). In acute toxicity studies, mortality is an example of a quantal response.

*Racemate*: Compound containing a 50:50 proportion of enantiomers.

*Rad*: Radiation-absorbed dose of ionizing radiation. One rad = 100 erg/g.

*Radiation*: Energy propagated through space or matter as waves (gamma rays, ultraviolet light) or as particles (alpha or beta rays).

*Radiological Dispersal Device (RDD)*: Mechanism designed to scatter radioactive material into the environment. The amount and type of radioactive material used will determine the number of causalities, the extent of evacuations, and the area of contamination.

*Radionuclide*: An element, either in the environment or internal, which emits ionizing radiation.

*Radioprotective Agents*: Chemical compounds that, when administered before irradiation, protect the organism against radiation damage.

*Radiosensitizer*: Substance that enhances the radiation response of biologic systems.

*Radiotolerance*: The eventual lack of sensitivity to radiation.

*Random*: Each individual member of the population has the same chance of being selected for the sample.

*Randomization*: Process used in toxicity studies to ensure a homogeneous population and minimize errors due to sampling bias. Randomization can be done by assigning animals to treatment groups using computer-generated random numbers or through the use of tables of random numbers.

*Rate*: An expression of the frequency with which an event occurs in a defined population; a measure of the *absolute risk* for the disease in that population. In a rate, the numerator is a subset of the denominator; therefore, all rates are ratios but not all ratios are rates. A rate usually has a time dimension. For a prevalence rate, it is the implied or explicit time at which the data were collected. For an incidence rate, it is the time over which the new events took place. For convenience, the denominator of either type of rate usually is presented as some power of 10 (i.e., the number of cases per 100 or per 1000 or, for rare events, per 100,000 or even per 1,000,000).

*Rate Difference*: The difference between two rates. In the situation where etiology has been established, the rate difference between the disease incidence in the exposed and unexposed groups is sometimes called the *excess rate* or *attributable risk*.

*Raw Data*: Any laboratory worksheets, records, memoranda, notes, or exact copies thereof that are the result of original observations and activities from a study. This may include manually recorded information, printouts from automated instruments, computer printouts, photographs, microfilm or microfiche copies, and magnetic media (including dictated observations).

*RCB*: Rodent cancer bioassay.

*$RD_{50}$*: The concentration of airborne substance that produces 50% decrease in the respiratory rate in rodents, usually mice. This is the numerical output of a sensory irritation study.

*Reactive or Secondary Thrombocytosis*: Increased platelet counts observed in conjunction with generalized bone marrow stimulation as may occur with hemolysis, blood loss, and many types of acute and chronic inflammation.

*Reasonable Maximum Exposure (RME)*: The highest exposure that is reasonably expected to occur at a site. The intent of the RME is to estimate a conservative exposure case (i.e., well above the average case) that is still within the range of possible exposures.

*Reasonable Worst Case*: A semiquantitative term referring to the lower portion of the high end of the exposure, dose, or risk distribution. The reasonable worst case has historically been loosely defined, including synonymously with *maximum exposure* or *worst case*, and assessors are cautioned to look for contextual definitions when encountering this term in the literature. As a semiquantitative term, it is sometimes useful to refer to individual exposures, doses, or risks that, while in the high end of the distribution, are not in the extreme tail. For consistency, it should refer to a range that can conceptually be described as above the 90th percentile in the distribution but below about the 98th percentile.

*Receptor*: The sensitive site for chemical–biological interaction.

*Recirculating Perfusion*: Perfusate flows from a reservoir through the organ being perfused and returns to the same reservoir.

*Red Cell Generative Response*: A response to reduced red cell mass characterized by an appropriate increase in erythropoiesis and correlative changes in related parameters.

*Red Cell Mass Parameters*: Parameters that provide an estimation of whole body red cell mass and include RBC count, hemoglobin concentration and hematocrit.

*Rederivation*: Refers to a process that utilizes removal of term fetuses with subsequent cross-fostering onto mothers of the right microbiological status or the use of embryo transfer procedures to change the microbiological status of animals.

*Reduction Alternative*: A new or modified test method that reduces the number of animals required.

*Reference Chemicals*: Chemicals selected for use during the research, development, prevalidation, and validation of a proposed test method because their response in the *in vivo* reference test method or the species of interest is known. Reference chemicals should represent the classes of chemicals for which the proposed test method is expected to be used and cover the range of expected response (negative, weak to strong positive). Different sets of reference chemicals are likely to be required for the various stages of validation. After a proposed test method has been recommended or accepted as valid for its intended purposes (i.e., has been recommended as a validated test method to federal agencies), a representative subset of chemicals used during the validation process may be selected to validate a mechanistically and functionally similar test method. To the extent possible, this subset of reference chemicals should: (1) be representative of the range of responses that the validated test method is capable of measuring or predicting; (2) have produced consistent results in the validated test method and in the reference test method or the species of interest; (3) reflect the accuracy of the validated test method; (4) have well-defined chemical structures; (5) be readily available; and (6) not be associated with excessive hazard or prohibitive disposals costs. This list of reference chemicals would represent the minimum number of chemicals that should be used to evaluate the performance of proposed mechanistically and functionally similar test methods with established performance standards. If any of the recommended chemicals are unavailable, other chemicals for which adequate reference data are available could be substituted. If desired, additional chemicals representing other chemical or product classes and for which adequate reference data are available can be used to more comprehensively evaluate the accuracy of the proposed test method.

*Reference Concentration (RfC)*: Air concentration of a chemical exposure (expressed in $mg/m^3$) that is associated with minimal or no risk of adverse effects, even in susceptible subpopulations.

*Reference Dose (RfD)*: An estimate of a daily exposure to a human population, including sensitive groups, that is likely to be without appreciable risk of deleterious health effects during a lifetime.

*Reference Interval*: The central interval of test values (usually the central 95th percentile) obtained from a defined group of apparently healthy individuals using defined methods (see partitioning factors). In contrast, a reference range is the entire range of values from those individuals, including minimum and maximum values.

*Reference Toxicant*: Chemicals of known toxicity that are used to determine the health of a population of a test species.

*Refinement Alternative*: A new or modified test method that refines procedures to lessen or eliminate pain or distress in animals or enhances the well-being of animals.

*Regenerative Anemia*: Anemia characterized by increased production and delivery of newly formed red blood cells (reticulocytes) into circulation; typically associated with hemorrhage or hemolysis.

*Regional Particle Deposition*: Particles ≤ 1 μm deposit by diffusion in the alveolar region of the respiratory tract; particles 1 to 5 μm deposit by sedimentation in the tracheal/bronchial/bronchiolar/alveolar regions; particles between 5 and 30 μm are deposited primarily by inertial impaction in the nasopharyngeal region.

*Reinforcement Schedule*: A designated relationship between a specific behavior, such as pressing a lever in an experimental chamber, and the delivery of a reinforcer, such as a feed pellet.

*Relative Biological Effectiveness (RBE)*: The ratio of a dose of a test radiation (e.g., neutron, gamma, x-ray) required to produce the same reference biological endpoint as the dose of a standard radiation of 250 kVp x-rays.

*Relative Risk*: The ratio of the absolute risk of disease or death (the incidence rate) among the exposed population to the risk of disease among the unexposed; a measure of association that sometimes is called the *risk ratio*, *rate ratio*, or *RR*.

*Relevance*: The extent to which a test method correctly predicts or measures the biological effect of interest in humans or another species of interest. Relevance incorporates consideration of the accuracy or concordance of a test method.

*Reliability*: A measure of the degrees to which a test method can be performed reproducibly within and among laboratories over time It is assessed by calculating intra- and interlaboratory reducibility and intralaboratory repeatability.

*Remediation Action Level (RAL)*: A concentration such that remediation of all concentrations above this level in an exposures unit will result in the 95% upper confidence limit being reduced to a level that does not pose an unacceptable risk to an individual experiencing random exposures. The RAL will depend on the mean, variance, and sample size of the concentrations within an exposure unit as well as considerations of short-term effects of the chemicals of concern.

*Renal Cell Lines*: Immortalized cultures of renal cells, derived from specific nephron cell types but often expressing properties of multiple cell types. They can be maintained indefinitely.

*Renal Cellular Repair and Regeneration*: Process by which the renal epithelium synthesizes new cells to replace those damaged by chemical toxicants or in various pathological states.

*Renal Slices*: *In vitro* model involving slicing renal tissue. Typically obtained with a razor and a glass microscope slide, the slices can be obtained from the cortex, outer or inner stripe of outer medulla, or the inner medulla. They are useful over the short term (up to 60 minutes) in metabolic and transport studies.

*Replacement Alternative*: A new or modified test method that replaces animals with nonanimal systems or one animal species with a phylogenetically lower one (e.g., a mammal with an invertebrate).

*Reproducibility*: The consistency of individual test results obtained in a single laboratory (*intralaboratory reproducibility; see Intralaboratory Reproducibility*) or in different laboratories (*interlaboratory reproducibility*) using the same protocol and test samples.

*Reproductive Toxicology*: The study of causes, mechanisms, and sequelae of adverse effects on the reproductive system, including alterations to the reproductive organs, related endocrine system, or pregnancy outcomes, and manifested as adverse effects on sexual maturation, gamete production and transport, cycle normality, sexual behavior, fertility, gestation, parturition, lactation, pregnancy outcomes, premature reproductive senescence, or modifications in other functions dependent on the integrity of the reproductive system.

*Residual Analysis*: Analysis of the difference between experimental and simulated data as a function of time or other controllable variables. The residuals should be random if the model is adequate.

*Residual Body*: A lobe of cytoplasm containing residual organelles (e.g., mitochondria, Golgi apparatus, endoplasmic reticulum, ribosomes, lipid droplets) that detaches from the elongated spermatid at spermiation. Residual bodies are eliminated from the seminiferous epithelium by Sertoli cell phagocytosis.

*Resinoid*: Any substance that contains or resembles a resin.

*Respirable*: Inhalable materials that are capable of getting to and being deposited in the gas-exchange region of the lung; a particle size characteristic whereby the substance can enter the pulmonary region of the respiratory tract. These are generally low-micrometer- to submicrometer-sized particles. Contrast with *inhalable*.

*Respiratory System*: Complex arrangement of organs designed primarily for the intake of oxygen and the elimination of carbon dioxide; divided into the proximal or upper airway (nose, pharynx, larynx, trachea, nonalveolarized bronchioles) and the distal or lower airway (alveolar bronchioles, alveolar ducts, alveolar air sacs).

*Respiratory Tract*: The entire breathing system, including mouth, nose, larynx, trachea, lungs, and associated nerves and blood supply.

*Retention (of Particulate Material)*: Quantity of particles present at specific respiratory tract sites that is the net difference between deposition and clearance processes; refers to particle matter and represents the quantity of particles present at specific respiratory tract sites. Retention is the net difference between deposition and clearance processes.

*RIA*: Radioimmunoassay.

*Ring Tail*: A condition seen in rodents in which annular constructions of the tail occur which may lead to necrosis of all or part of the tail. This condition is presumed to be associated with abnormal environmental conditions.

*Risk*: Proportion or probability expressed on a scale of 0 to 1, or 0 to 100%, that individuals process or develop a specified biological effect by a given time for a defined set of exposure conditions; the probability or degree of concern that exposure to an agent will cause an adverse effect in the species of interest.

*Risk Characterization*: A component of risk assessment that describes the nature and magnitude of risk, including uncertainty. In assessments of Superfund sites, it includes the summary and interpretation of information gathered from previous steps in the site risk assessment (e.g., data evaluation, exposure assessment, toxicity assessment), including the results of a probabilistic analysis.

*Robust*: Having inferences or conclusions little affected by departure from assumptions.

*Rodent Lifetime*: Generally taken to be 2 years for risk estimation.

*Roentgen (R)*: Quantity of x- or gamma radiation per cubic centimeter of air that produces one electrostatic unit of charge.

*Roentgen Equivalent Man (REM)*: Unit of dose equivalent; the absorbed dose in rads multiplied by the $Q$ of the type of radiation.

*Rough Endoplasmic Reticulum (RER)*: The endoplasmic reticulum component that contains bound ribosomes and is largely responsible for protein synthesis.

*Route of Exposure*: Portal of entry of a chemical into the body: oral, inhalation, dermal, or injection.

*S9 Mix*: A 9000 *g* supernatant fraction from a tissue homogenate (e.g., liver).

*Safety Factor*: A number or factor that considers inter- and intraspecies variability, sensitivity, and extrapolatability. It is applied to a NOAEL to establish an acceptable daily intake (ADI). *See Uncertainty Factor*.

*Safety Pharmacology*: The study of the pharmacologic effects of a drug candidate that are unrelated to the desired pharmacologic effect. These studies are generally conducted at doses similar to the anticipated therapeutic dose.

*SAM (Standardized Aquatic Microcosm)*: A defined aquatic microcosm (3-liter vessel) containing 15 species in a well-defined medium of trace metals and vitamins and 200 g of sand to which chitin and cellulose are added.

*Sample*: In statistics, the collected set of data points which (if properly collected) are representative of the population (all of the values there are).

*Saponins*: Steroidal or terpenoid glycosides from plants and animals capable of reducing surface tension and disruption of cell membranes.

*Saturation Dose*: Dose that overwhelms a toxification or detoxification mechanism such that no additional changes in the effect are incurred above that dose.

*Sawfly*: Insects, the larva of which are hepatotoxic when ingested by ruminant livestock.

*Schedule-Controlled Operant Behavior*: An approach to the study of behavioral function that relies on manipulation of reinforcement schedules.

*Scombroid Poisoning*: An allergic fish poisoning resulting from the consumption of inadequately processed fish containing histamine and saurine formed from bacterial action.

*Screen/Screening Test*: A rapid, simple test conducted for the purposes of a general classification of substances according to general categories of hazard. The results of a screen generally are used for preliminary decision-making and to set priorities for more definitive tests. A screening test may have a truncated response range (it may be able to reliably identify active chemicals but not inactive chemicals).

*Screening Tests*: In acute toxicity testing, tests designed to define the range of toxicity using fewer animals per dose level or fewer dose levels.

*Sea Snakes*: Members of the Elapidae family that spend most of their lives in oceans.

*Secondary Enclosure*: The room or space in which a primary enclosure is located.

*Secondary Phase Regeneration (SPR)*: Tissue repair response where hepatocytes are mobilized from $G_0/G_1$ to proceed through mitosis (*see Electron Paramagnetic Resonance*).

*Secondary Response*: The immune response that occurs after initial contact with an antigen (the primary response). The secondary response is quicker, of higher affinity, and more pronounced. For humoral-mediated immune reactions, the secondary response is associated with antibody class switching.

*Secondary Toxicity*: The toxicity of a chemical determined by feeding the test animals (especially ferrets or mink) with prey that have been fed the test chemical.

*SEL*: Safe exposure level.

*Selection and Selection Bias*: The process by which subjects are included in a study. If systematic differences exist between those who are selected for a study and those who are not, selection bias may occur. The bias may be introduced by the subjects themselves through self-selection (into or out of the study), as a result of the sources of the subjects, or by the study investigators.

*Seminiferous Epithelium*: The Sertoli cells and developing male germ cells within the seminiferous tubules in the testes.

*Sensitivity*: The number of subjects experiencing each experimental condition divided by the variance of scores in the sample; the proportion of all positive chemicals that are classified correctly as positive in a test method. It is a measure of test method accuracy (*see Two-by-Two Table*).

*Sensitivity Analysis*: Evaluation of the effect of changes in the value of a particular parameter on the estimates of a state variable provided by a mathematical model. Sensitivity is expressed as the magnitude of change in the endpoint of interest (e.g., tissue dose) as a function of change in the value of a particular model parameter.

*Sertoli Cell*: A somatic cell present in the seminiferous epithelium of mammals. Solidly attached to the basement membrane of seminiferous tubules, it produces important regulatory glycoproteins, including anti-Mullerian hormone, inhibin, and androgen-binding protein and is essential for male germ cell development. Junctions between Sertoli cells form the blood–testis barrier.

*Serum*: The liquid portion of blood that remains after a clot has formed. Serum is obtained from blood collected without anticoagulant and centrifuged to separate it from the cells.

*Serum Responsiveness*: Cell proliferative reaction to the addition of serum to tissue culture medium after prior deprivation.

*Severity Grading of Lesion*: The semiquantitative application of a defined severity score to specific microscopic lesions, usually to denote the extent of tissue involvement or degree of tissue damage.

*Shuttle Vector*: A DNA transfer agent capable of moving genes into or out of a cell.

*SI Units*: International system of units; SI refers to *Système International d'Unites*. Radiation units include Joule/kilogram, Gray, Sievert, and Becquerel.

*Sievert (Sv)*: SI unit of dose equivalence; the absorbed dose in grays multiplied by the *Q* of the type of radiation. One Sv = 100 rems.

*Signature Sequencing*: Sequencing of a short stretch of cDNA close to the end of the complementary mRNA. Sequence stretches of some 20 nucleotides are sufficiently discriminative to identify the transcript of an individual gene in a mammalian tissue.

*Significance Level*: The probability that a difference has been erroneously declared to be significant, typically 0.05 and 0.01, corresponding to 5% and 1% chance of error.

*Simulation*: System behavior predicted for specific exposure conditions by solving the set of differential and algebraic equations of a model.

*Single Nucleotide Polymorphism*: Interindividual variation in the genetic code at the level of one single building block (see *Nucleotide*, *Polymorphisms*).

*Single-Pass Perfusion*: Perfusate flows from a reservoir through the vasculature of a perfused organ; it is collected and not reused so the perfusate goes through only once.

*Single-Wall Carbon Nanotube (SWNT)*: An engineered nanoparticle consisting of carbon atoms arranged in a single tubular structure with a very high aspect ratio. The diameter of the nanotubes is less than 100 nm but the tube lengths can be several micrometers.

*Slit-Lamp Biomicroscope*: An instrument used to study ocular tissues that consists of a microscope and high-intensity light source. It allows the eye to be illuminated and observed from different angles and can detect lesions not observable by gross examination.

*Slope*: The difference in the incidence or magnitude of an effect divided by the difference in dose that created the effect.

*Slot Velocity*: Air velocity through the openings in a slot-type hood. It is used primarily as a means of obtaining uniform air distribution across the face of the hood.

*Small Intestine*: The region of the gastrointestinal tract from the pyloric sphincter of the stomach to the cecum which consists of the duodenum, the jejunum, and the ileum; the primary site of absorption of ingested substances.

*Smokes*: A complex mixture of solid or liquid particles, such as soot, liquid droplets, or mineral ash particles from incomplete combustion of organic materials. Smoke particles are generally ~0.5 μm.

*Smooth Endoplasmic Reticulum (SER)*: The endoplasmic component of the cell that does not contain bound ribosomes and is strongly associated with drug-metabolizing enzyme systems such as those mediated by cytochrome P450.

*Society of Toxicology (SOT)*: The preeminent professional society for toxicologists in the United States.

*Soil Aging*: The changes that occur in the interaction of chemicals with soil during which the chemicals are bound to soil particles and become less bioavailable.

*Soil Sorption Constant ($K_{oc}$)*: The extent of partitioning of an organic chemical between the solid and solution phases of a water-saturated soil or sediment.

*Somatotropin*: Growth hormone.

*Specific Activity*: The radioactivity per unit mass of material. Specific activity is used to quantify radionuclides. Specific activity can be given in Curies per mole (Ci/mol) or Becquerels per mole (Bq/mol).

*Specific Gravity*: The ratio of the density of a substance to the density of a reference material at a specified temperature. Water is the reference standard for liquids and solids (density 1 g/mL at 4°C).

*Specificity*: The proportion of all negative chemicals that are classified correctly as negative in a test method. It is a measure of test method accuracy (*see Two-by-Two Table*).

*Spermatid*: The haploid germ cell, arising from meiotic divisions of spermatocytes, that differentiates within the seminiferous epithelium into a spermatozoon.

*Spermatocytes*: Germ cells in the seminiferous epithelium derived from spermatogonia that subsequently undergo two meiotic divisions to form round spermatids.

*Spermatogenesis*: The process of germ-cell division and differentiation that begins with the multiplication of spermatogonia and ends with the release of elongated spermatids into the lumen of the seminiferous tubules (spermiation).

*Spermatogenic Cycle*: A complete sequential progression of the cellular associations (or stages) of spermatogenesis. The stages follow one another though an entire cycle, returning to the original stage and repeating the cycle approximately 4.5 times until spermatogonia become elongated spermatids and undergo spermiation.

*Spermatogonia*: The diploid germ cells in adult males that divide by mitosis to produce additional stem cell spermatogonia and spermatocytes.

*Spermiation*: The process by which elongated spermatids are released from the germinal epithelium into the seminiferous tubule lumen.

*Spermiogenesis*: Last phase of spermatogenesis during which elongated spermatids are formed from round spermatids.

*Spin Trapping*: The most commonly used indirect method for detecting free radicals. The technique involves adding a spin trap (usually a nitrone or nitroso compound) to the sample prior to radical generation. When the radical is generated, it reacts rapidly with the spin trap, producing a secondary radical or *spin adduct*, which is more stable than the parent free radical and can be detected by electron paramagnetic resonance (EPR).

*Standard Deviation*: The most widely used measure of dispersion of the points in a frequency distribution about the midpoint (usually the mean). It is equal to the square root of the variance. For a normally distributed population, the region within one standard deviation of the mean contains 67% of the distribution.

*Statistical Heterospectroscopy*: A statistical technique for combining nuclear magnetic resonance (NMR)- and mass spectroscopy (MS)-based metabonomic data. It provides a focus on metabolites that change consistently across the platforms on a sample-to-sample basis.

*Statistical Significance*: An inference that the probability is low that the observed difference in quantities being measured could be due to variability in the data rather than an actual difference in the quantities themselves. The inference that an observed difference is statistically significant is typically based on a test to reject one hypothesis and accept another.

*Steady State*: A situation in which the rate of change is equal to zero.

*Steatosis*: Accumulation of lipid within hepatocytes (fatty liver).

*Stem Cell*: A multipotential self-renewing cell in the bone marrow that serves as the precursor for all hematopoietic cell lineages, including those of the immune system.

*Stereoisomers*: A general term for isomers that differ only in the orientation of the atoms in space. Enantiomers, isomers that differ in their optical rotation, are a subclass of stereoisomers.

*Stomach*: Sac-like region of the gastrointestinal tract from the esophagus to the duodenum consisting of the cardia, fundus, corpus, antrum, and pylorus; the site of hydrogen ion secretion from parietal cells of the gastric glands which activates pepsin-mediated proteolysis.

*Stress-Response Leukogram*: Typically, an increase in neutrophil count and decreases in lymphocyte and eosinophil counts following exogenous corticosteroid administration or when stressful conditions result in increased production of endogenous corticosteroids.

*Subchronic*: Characterized by a time period of intermediate duration; commonly used to describe exposure between acute and chronic duration in toxicity studies (usually 3 months).

*Subchronic Toxicity Study*: Multiple-dose study in which animals are treated for less than 6 months. They are intended to elucidate the target organs for toxicity and demonstrate dose–response relationships. They are normally required prior to any clinical testing.

*Substitute Method*: A new or modified test method proposed for use in lieu of a currently used test method, regardless of whether that test method is for a definitive screening or adjunct purpose.

*Substrate Probe*: Individual isozymes may show significant differences in substrate specificity. A substrate that is metabolized by a specific isozyme and does not show significant overlaps with other isozymes can be used to probe for the presence and activity of the isozyme. Substrate probes can be used *in vitro* and *in vivo*. Use of substrate probes may not be as accurate as certain other techniques for isozyme identification but are sometimes more practical.

*Syncytium*: A multinucleated protoplasmic mass formed by the secondary union of originally separate cells.

*Systole*: Contraction of the heart, especially of the ventricles, by which blood is driven through the aorta and pulmonary artery to transverse the systemic and pulmonary circulation, respectively.

*$T_3$*: Triiodothyronine.

*$T_4$*: Thyroxine.

*$T_{max}$*: Time to maximum achieved concentration.

*T-Cell Receptor (TCR)*: The heterodimeric surface molecule on T cells that serves to recognize antigen. It always occurs in conjunction with the CD3 surface antigen, which is responsible for transmembrane signaling following antigen recognition.

*T Cell/T Lymphocyte*: Lymphocytes primarily responsible for the induction and maintenance of cell-mediated immunity, as well as regulating humoral-mediated immunity and certain nonimmune effector mechanisms. A variety of T-cell subtypes have been described, including T-helper cells, T-cytotoxic cells, T-suppressor cells, and T-inducer cells.

*Tannins*: Heterogeneous polyphenols of plant origin.

*Target Tissue Dose*: Concentration of a chemical in the tissue or organ where the biological effect occurs.

*TBG*: Thyroid-binding globulin.

*TeBG*: Testosterone-binding globulin.

*TEF*: Toxic equivalency factor.

*Telomere*: The terminal portion (end) of a chromosome.

*Teratology*: The study of the causes, mechanisms, and sequelae of perturbed developmental events in species of animals that undergo ontogenesis; in the past, the definition was limited to malformation, but the term is now generally accepted to be synonymous with developmental toxicology.

*Test Article*: Any food additive, color additive, drug, biological product, electronic product, medical device, pesticide, or other chemical substance to be subjected to studies.

*Test Material*: Any chemical substance to be subjected to studies. This does not include electronic products or medical devices.

*Test Method*: A process or procedure used to obtain information on the characteristics or a substance or agent. Toxicological test methods generate information regarding the ability of a substance or agent to produce a specified biological effect under specified condition. Used interchangeably with *test* and *assay* (*see also Validated Test Method*).

*Test System*: Any animal, plant, microorganism, and subparts thereof (e.g., *in vitro* organ systems) or physical matrix (e.g., soil or water) to which a test or control article is administered or added for study.

*Testosterone*: The primary male sex hormone produced by Leydig cells in the testis that, along with the metabolite dihydrotestosterone, is responsible for male reproductive tract development, maintenance of spermatogenesis, secondary sex characteristics, and sexual behavior.

*Tetrodotoxin*: An extremely toxic, highly lethal, crystalline neurotoxin that acts on both the central and peripheral nervous systems, causing nerve and skeletal muscle paralysis by selectively blocking the regenerative sodium conductance channel.

*TFM*: Test facility manager.

*Therapeutic Index (TI)*: The ratio of $LD_{50}/ED_{50}$. The therapeutic index is used to establish the safety margin of biologically active materials such as drugs. The higher the index, the greater the margin of safety.

*Thermal Neutral Zone*: The temperature or range of ambient temperatures at which an animal does not expend energy to heat or cool itself.

*Thin Ascending Limb*: Cell type of epithelium found in the inner medulla only in long-looped nephrons; begins after bend of Henle's loop.

*Thin Descending Limb*: Cell type of epithelium found in the inner medulla in long-looped nephrons and in the inner stripe of the outer medulla in short-looped nephrons. The thin descending limb follows proximal straight tubules as one follows nephrons from glomerular to urinary poles.

*Threshold Dose*: Dose below which a specified biological effect does not occur for specified exposure conditions.

*Threshold Dose–Response Relationship*: A dose–response relationship that assumes that adverse effects occur only when a threshold dose is exceeded.

*Threshold Limit Values (TLV)*: Airborne concentration of a substance that would be anticipated to produce no adverse health effects in nearly all workers exposed 8 hours per day, 5 days per week, for a working lifetime. TLVs are established by the American Conference of Governmental Industrial Hygienists.

*Thrombocytopenia*: A condition in which there is an abnormally small number of circulating platelets.

*Thymus*: Central lymphoid organ located in the thorax. Its function is to generate immunocompetent T cells from lymphocytes originating in the bone marrow.

*Tissue Time Constant*: The product of partition coefficient and volume divided by blood flow rate.

*Tolerance Level*: The maximum legally permissible concentration of residues of a pesticide in food.

*Toxic*: Substance that has the capacity to produce personal injury via ingestion, inhalation, or absorption through any body surface.

*Toxic Neutrophils*: Neutrophils with morphologic changes indicative of greatly accelerated neutrophil production, regardless of cause.

*Toxicant*: An agent that can result in the occurrence of a structural or functional adverse effect in a biological system.

*Toxicodynamics*: The process of interaction of chemical substances with target sites and the subsequent reactions leading to adverse effects.

*Toxicokinetics*: The process of the uptake of potentially toxic substances by the body, the biotransformations they undergo, the distribution of the substances and their metabolites in the tissues, and the elimination of the substances and their metabolites from the body. Both the amounts and the concentration of the substances and their metabolites are studied. The term has essentially the same meaning as *pharmacokinetics*, but the latter term is usually restricted to the study of pharmaceutical substances.

*Toxicological Pathology*: The science that integrates the disciplines of pathology and toxicology and is concerned with the effects of potentially noxious substances.

*Toxicological Profiles*: Prepared by the Agency for Toxic Substances and Disease Registry, these profiles include extensive toxicological information for substances found at the National Priorities List (NPL) and other sites.

*Toxicology*: The study of the adverse effects of biological, chemical, or physical agents on living organisms and the ecosystems, including the prevention and amelioration of such adverse effects.

*Toxicology Data Network (TOXNET)*: A major group of publicly available databases from the National Library of Medicine.

*Toxin*: A poison derived from a biological source.

*Toxinology*: The study of toxins.

*Tracheobronchial*: A region of the respiratory tract serving to deliver inspired air to deeper portions of the lung, comprised of a series of branching ducts beginning at the trachea and ending at the terminal bronchioles.

*Transcript Profiling*: *See Transcriptomics*.

*Transcription*: The formation of messenger RNA (mRNA), complementary to a string of DNA.

*Transcriptome*: The messenger RNA (mRNA) from actively transcribed genes.

*Transcriptomics*: The techniques available to identify the messenger RNA (mRNA) from actively transcribed genes.

*Transgenic*: Referring to an organism in which new DNA is introduced into the genome. Some transgenic animal models have been suggested to complement the assessment of the potential for compounds to cause tumor development.

*Transgenic Animals*: Genetically engineered animals carrying genes from a different species.

*Translocation*: Transfer of a portion of one chromosome to another chromosome.

*TRH*: Thyroid-releasing hormone; a tripeptide secreted by the hypothalamus.

*Trypan blue exclusion*: Assay for cellular necrosis; the vital dye trypan blue is only taken up by cells whose membranes have been permeabilized. The cell viability is estimated by counting the fraction of stained (i.e., blue) cells on a light microscope.

*TS*: Test substance.

*TSH*: Thyroid-stimulating hormone.

*Two-by-Two Table*: The two-by-two table can be used to calculate accuracy (concordance) ($[a + d]/[a + b + c + d]$), negative predictivity ($d/[c + d]$), positive predictivity ($a/[a + b]$), prevalence ($[a + c]/[a + b + c + d]$), sensitivity ($a/[a + c]$), specificity ($d/[b + d]$), false-positive rate ($b/[b + d]$), and false-negative rate ($c/[a + c]$).

| | | New Test Outcome | | |
|---|---|---|---|---|
| | | Positive | Negative | Total |
| Reference Test Classification | Positive | *a* | *c* | *a* + *c* |
| | Negative | *b* | *d* | *b* + *d* |
| | Total | *a* + *b* | *c* + *d* | *a* + *b* + *c* + *d* |

*Two-Dimensional Gel Electrophoresis (2-D Gel Electrophoresis)*: A series of electrophoretic techniques which separates proteins (gene products) on the basis of isoelectric points in one dimension followed by separation on the basis of molecular mass in the second dimension. This technique is useful for obtaining an overall assessment of both up- and downregulation of specific gene products as the result of cellular responses to chemical exposures. This technique separates proteins (gene products) in two dimensions on the basis of isoelectric point in the first dimension and on the basis of molecular mass in the second dimension. It is a powerful technique for proteomic research, particularly when coupled with computerized image analysis systems for quantification.

*Tyndall Phenomena*: The abnormal cloudy appearance of the anterior chamber of the eye when light passes through the pupil; also called *aqueous flair*. It is the result of protein leakage from the iris into the aqueous humor causing the scattering of light and producing cloudiness.

*Type I Error (False Positive)*: Concluding that there is an effect when there really is not an effect. Its probability is the alpha level.

*Type II Error ( False Negative)*: Concluding that there is no effect when there really is an effect. Its probability is the beta level.

*Ultrafine Particles*: Particles that have all dimensions in the nanometer range; contrast with *nanoparticles*.

*Ultrastructural Cytochemistry*: A series of *in situ* techniques for localizing organelle-specific enzymes, such as acid phosphatase for lysosomes and peroxidase for peroxisomes, at the ultrastructural level.

*Ultrastructural Morphometry*: A series of techniques for quantifying changes in organelle systems from electron micrographs *in situ*; these techniques provide useful correlative information when used in combination with biochemical measurements of chemical-induced alterations in organelle system functionality. The technique is based on the systematic evaluation of large numbers of transmission electron micrographs followed by the statistical analysis of changes in various organelle compartments induced by exposure to toxic agents. Data generated by this technique may be expressed as volume density (Vv) of an organelle compartment, surface density (Sv), or numerical density (Nv).

*Uncertainty*: Lack of knowledge about specific variables, parameters, models, or other factors. Examples include limited data regarding the concentration of a contaminant in an environmental medium and lack of information on local fish consumption practices. Uncertainty may be reduced through further study.

*Uncertainty Factor*: A factor applied to a no-observed-adverse-effect level (NOAEL) or a lowest-observed-adverse-effect level (LOAEL) which is used to derive a reference dose or reference concentration. The aim of the uncertainty factor is to account for a lack of information on inter- and intraspecies variability, study deficiencies, or an incomplete database. The value, generally from 1 to 10, based on scientific data or judgment, allows for the unknown potential differences in doses or durations of exposure that produce effects in different individuals, different species, or different routes of exposure or are due to an inadequate database on which to base a decision.

*Upper Respiratory Tract*: The mouth, nose, sinuses, and pharynx.

*Urinalysis*: Assays on chemical and biophysical parameters of urine; includes measurement of parameters such as levels of metabolites, protein, glucose and creatinine excretion, urinary enzymes, and urinary specific gravity.

*Urinary Enzymes*: Enzymes secreted into urine used to measure renal function; can often be used as biomarkers of cell-type specific toxicity and function.

*Urticaria*: Hives; an eruption of itching wheals (welts).

*UVR*: Ultraviolet radiation.

*Validated Test Method*: An accepted test method for which validation studies have been completed to determine the accuracy and reliability of this method for a specific purpose.

*Validation*: The process by which the reliability and accuracy of a procedure are established for a specific purpose.

*Vapor*: Gaseous forms of substances that normally are in the liquid or solid state; the gas-phase components of a substance that is a solid or liquid at standard temperature and pressure.

*Vapor Pressure*: The amount of pressure exerted by a saturated vapor above its own liquid in a closed container.

*Vaporization*: The transfer of a chemical from a solid surface or a solution to a gas, dependent upon the chemical's diffusivity, water solubility, and vapor pressure.

*Variability*: The range of values expected among individuals of a given population; a term used to describe the natural variation in human (or test animal) responses to chemical exposures. It is also used to describe variation in exposures to chemicals in the environment. Variability represents true heterogeneity, diversity, or a range that characterizes an exposure variable or response (e.g., differences in body weight). Further study (e.g., increasing sample size *n)* will not reduce variability, but it can provide greater confidence in quantitative characterizations of variability.

*Vasopressin (ADH, Antidiuretic Hormone)*: An octapeptide secreted by the neurohypophysis.

*Vehicle*: A substance to which a test material is added to provide a consistency or form suitable for its intended use. In toxicity studies, test materials are added to vehicles to prepare solutions (e.g., water), suspensions (e.g., methyl cellulose), ointments (e.g., petrolatum), triturates (e.g., milk, sugar), and other forms to facilitate administration to the animals.

*Venom*: An animal toxin.

*Viperidae*: True vipers; a diverse genus of venomous snakes, many of which are large and extremely dangerous, even deadly.

*Vipers*: Snakes of the families Viperidae and Crotalidae.

*Water Solubility*: The maximum amount of a chemical that may be dissolved in a given quantity of pure water at a given temperature.

*Weapon of Mass Disruption*: A terrorist weapon designed to frighten and disrupt the population.

*Whole Animal Studies*: Studies in which physiological function and toxicity are determined *in vivo*.

*Whole-Body Exposure*: Experimental mode of inhalation exposure in which the entire animal is placed within the test atmosphere.

*Wolffian Duct*: Embryonic duct from which the male reproductive duct system, accessory sex glands, and external genitalia are derived.

*Xenobiotic*: A substance that is foreign to a biological system; a substance (usually) not present in the reference organism; a chemical species, synthetic in origin, that is toxic or damaging to a biological system.

*X-Ray Microanalysis*: The *in situ* localization of trace elements within organelles at the ultrastructural level using the electron beam of either a transmission or scanning electron microscope to displace electrons from specific energy shells (K, L, M) with the resulting release of characteristic x-rays that may be monitored by energy-dispersive or wavelength-dispersive spectrometers.

*Zero-Order Process*: A process for which the rate is constant and independent of dose or concentration; characteristic of enzymatic processes under saturating concentrations.

*Zonation, Hepatic Lobule*: Quantitative (or qualitative) distribution of enzymes in liver lobules based on morphological configuration.

*Zonation, Metabolic*: Differences observed between enzyme activities in periportal and perivenous regions of the liver.

*Zoonotic*: Refers to the ability of an organism to be transmitted between species of animals; most commonly used in reference to the ability of an organism to be exchanged between animals and humans.

# Notes

# Index

## C

## L

## M

## N

## P

## S

## U

## W

## X



## Y



## Z

# Notes

# Notes